ATLAS
OF THE
HELLENIC FLORA

I

Arne Strid

Atlas of the Hellenic Flora

I

Published with support from the
Aage V. Jensen Charity Foundation

Broken Hill Publishers Ltd

ISBN: 978-9925-35-025-4

16 Zenas Kanther Princess De Tyras Str.
1065 Nicosia, CYPRUS
e-mail: info@brokenhill.com.cy
website: http://www.brokenhill.com.cy

To Niki Goulandris (1925-2019)

and

Elli Stamatiadou (1933-2015)

for their outstanding contributions to Greek botany

Malcolmia orsiniana

κυρία Νίκη

Paeonia parnassica

κυρία Έλλη

and to my wife Barbro

for her never-failing support during 58 years

Ranunculus cacuminis

Ferrum splendescat arando

INTRODUCTION

Botany like other branches of western science has its roots in ancient Greece, with Theophrastos of Eressos (372-287 B.C.) generally considered to have been the first botanist. Modern botanical studies in the area started with the travels of John Sibthorp and Ferdinand Bauer in 1786-1787, the results of which were subsequently published in the great *Flora Graeca Sibthorpiana* appearing in ten heavy folio volumes in the period 1806-1840. An annotated re-issue (five volumes in A4 format) was prepared by Strid & Strid (2009-2013). For a brief history of the botanical exploration of Greece from the time of Sibthorp to the present, see Strid (2020a).

The "Atlas of the Aegean Flora" (Strid 2016a&b) provided distribution maps and short descriptions for all species of vascular plants on the Aegean islands. The present Atlas similarly covers the whole of Greece. Maps have been generated from the Flora Hellenica Database which was started in 1989 and now comprises 1,052,338 geo-referenced records, 436,353 of which have been registered by the present author, 139,276 by Britt Snogerup (mainly specimens at Herbarium LD) and the rest by c. 40 operators. Major contributors, each with more than 5,000 registered records to their credit, are Kyriaki Athanasiou, Charilaos Chouliaras, Eva Ekeblad, Ylva Heed, Ralf Jahn, Bent Johannsen, Thomas Landström, Clara Lundsgaard, Mette Mikkelsen, Henry Nielsen, Dimitrios Perdetzoglou, Phung-Binh Quan, Barbro Jende Strid, Karen Thingsgaard and Kit Tan.

Records in the database comprise 502,882 herbarium collections and 321,759 literature reports, the rest being field notes, photographs, etc. Registrations are almost complete for specimens in some major herbaria such as ATH, C and LD; substantial parts of the holdings at B, G, UPA, W and WU have also been registered as well smaller sets at other herbaria. It can be estimated that perhaps 75 per cent of all Greek herbarium specimens are now in the database, and those that are lacking will to a large extent be redundant, representing re-collection of known species at already known localities. The Flora Hellenica Bibliography (see vol. 3 of this Atlas) comprises 16,078 titles relevant to the taxonomy and distribution of vascular plants of Greece. About 98 per cent of these have been examined by the present author. Floristic information (i.e., records of named species at reasonably well defined localities) was found in 3,259 publications. These records have been carefully vetted and registered for the database, although registration has sometimes been selective when many records for the same taxon were available from a small area. Registration of herbarium specimens and literature records is far from a trivial pursuit, but one that requires extensive knowledge of the flora and geography of the territory. Great care has been taken in data validation, and the percentage of errors is believed to be comparatively low. The present dataset is sufficient to produce reasonably complete and correct distribution pictures for almost all species. Twice as many records would have increased the density of dots but very rarely changed the general pattern for a taxon. This Atlas and the underlying database will provide a platform for more detailed and sophisticated analyses of Greek biodiversity than have hitherto been possible. Some such studies have already appeared and others will follow in separate publications.

As registered in the database, major botanical collectors in Greece are:

	Number of records	Number of species
Greuter	9,359	2,718
Phitos	5,495	1,360
Raus	23,585	2,265
Rechinger	17,380	3,041
Runemark	51,363	2,151
Snogerup	47,522	3,109
Stamatiadou	17,380	2,802
Strid	57,411	4,374
Kit Tan	8,738	2,429
Willing	226,882	3,205

Taxonomic principles and categories

Principles for taxonomy, nomenclature, geographical names, abbreviations, etc., are those outlined in Strid (2016a: 9-11).

The species is the basic category of biological taxonomy, and the only one that can be scientifically defined with a reasonable degree of accuracy. Populations are referred to different species when there is simultaneously, under natural conditions, morphological differences and a barrier to gene exchange. Species are functional biological units existing in nature regardless of those who classify them. The validity of species can be tested by observation or experiment.

Within a species, major geographical (or sometimes ecological) races have been recognized as subspecies. Varieties, often characterized by a single, sometimes conspicuous, morphological trait but without a clear pattern of distribution, are mentioned only occasionally in notes on variation, etc.

Although the principles for species-level taxonomy are relatively simple, the practical application is not necessary easy, and decisions often have to be taken on incomplete or even contradictory evidence. The Atlas follows the best available monograph, taxonomic revision or Flora account (not necessarily the most recent) as well as the author's personal experience in the field, herbarium and experimental garden. In genera such as *Allium*, *Galium*, *Limonium* and *Ophrys* there has been a tendency among some recent authors to describe poorly defined local populations as separate species (often based on a single collection). This is a practice to be discouraged.

Names of genera, species and subspecies are in *italics*. Abbreviations of authors' names (in the Index to Scientific Names as well as the chapter "Excluded Species and Taxonomic Comments") are according to the International Plant Names Index (IPNI). For geographical names, Greek forms have been preferred for Greek territory (e.g., Kriti and Kiklades rather than Crete and Cyclades) and anglicized forms for non-Greek territory (e.g., Serbia and Montenegro rather than Srbija and Crna Gora).

Contrary to species, the higher taxonomic categories (sections, subgenera, genera, tribes, subfamilies, families, orders, etc.) are not functional biological units but variously sized boxes created by taxonomists for the practical purpose of classification. Within the three major groups (vascular cryptogams, gymnosperms and angiosperms), only families, genera, species and subspecies are recognized in this Atlas.

Recent decades have seen much gerrymandering of borders between families and genera, often by followers of the cladistic faith. The main trend has been one of inflation. Clarity is rarely served by raising sections and subgenera to the rank of genus. Citing only the author of the basionym would remove 90 per cent of the incentive for nuisance recombinations.

How to read the maps

Within the three major taxonomic groups (vascular cryptogams, gymnosperms and angiosperms), maps are arranged alphabetically according to family, genus and species.

In addition to the Greek coastline and northern borders the map shows the 1000 m contour with higher-lying areas marked in light green. Degrees of longitude and latitude are shown on the upper and left-hand margins, respectively. The outlying Greek island group of Kastellorizo at c. 36°08'N and 29°37'E off the coast of S Anatolia appears as an inset in the lower right-hand corner of the map.

In most maps there is only one symbol, a solid red dot. The diameter of the dot is c. 4 km or 2.2' latitude or longitude. Neighbouring dots therefore sometimes overlap. To even out local and regional differences in the degree of botanical exploration and produce more balanced maps, a thinning procedure has usually been applied, so that only one dot appears within a 6' × 6' square; this is placed in its correct position (not in the middle of the square), but other potential dots in that square are automatically suppressed. In some cases there are up to three (exceptionally more) symbols of different shapes and colours showing, e.g., different subspecies. Dubious but plausible occurrences are sometimes indicated by a question mark.

A sequential number appears in the upper right-hand corner of the map. For species illustrated in the photo appendix in volume 3 there is also a cross reference (e.g., 216:8) just below the sequential number.

The label in the lower left-hand corner of the map gives the name of the taxon (without nomenclatural authority) followed by brief information on habitat and general distribution. A second paragraph provides an ultra-short description with emphasis on diagnostic characters, and occasionally other notes.

The vertical bar inside the right-hand margin indicates altitude for registered occurrences with the normal range in reddish-brown and exceptional values in yellow. Similarly the horizontal bar inside the upper margin indicates flowering time, also with the normal range in reddish-brown and exceptional values in yellow.

Just below the bar indicating flowering time is another horizontal bar with coded information for "Stat" (status, i.e., native or alien), "Ch" (chorology or general distribution), "Lf" (life form) and "Hab" (habitat), according to the principles adopted in the Greek checklist (Dimopoulos & al. 2013: 23-27). This information should only be taken as indicative since the categories are not necessarily sharply defined nor mutually exclusive. When two or more alternatives are given, the main one is sometimes underlined. For convenience the codes are repeated here:

Stat (status)

N = native.

X = xenophyte or alien.

r = range-restricted (native taxon with less than 500 km between the most distant occurrences).

Ch (chorology or general distribution)

• = Endemic to Greece.

Bk = Balkan (restricted to the Balkan countries or extending marginally beyond).

BI = Balkan Peninsula and Italy (amphi-Adriatic).

BC = Balkan and Central European (occurring in the Balkans, Alps, Carpathians and adjacent areas; usually mountain plants).

BA = Balkan-Anatolian (occasionally extending to Crimea and Caucasia).

EM = East Mediterranean.

Me = Mediterranean (widespread in the Mediterranean area).

MA = Mediterranean-Atlantic (occurring in the Mediterranean area and W Europe).

ME = Mediterranean-European (widespread in the Mediterranean area and Europe, sometimes extending to Caucasia, etc.).

MS = Mediterranean - SW Asian (occurring in the Mediterranean area and SW to C Asia).

Eu = European (usually widespread).

EA = European - SW Asian (occurring in Europe and SW to C Asia).

ES = Euro-Siberian (widespread in temperate Eurasia).

Pt = Paleotemperate (similar to Euro-Siberian but more southerly).

Ct = Circumtemperate (in temperate Eurasia and North America).

IT = Irano-Turanian (with main distribution in semi-arid SW to C Asia).

SS = Saharo-Sindian (with main distribution in arid N Africa and SW Asia).

ST = Subtropical-Tropical (widespread in warmer areas of both hemispheres).

Bo = Boreal (widespread in northern Eurasia and North America).

AA = Arctic-Alpine (main distribution in subarctic Eurasia and North America, extending to mountains further south).

Co = Cosmopolitan (widespread in both hemispheres).

For alien taxa, the origin is given in square brackets, e.g., [S-Afr.] or [N-Am.].

Lf (life form)

P = Phanerophyte (woody plant more than 50 cm tall).

C = Chamaephyte (woody or semi-woody plant less than 50 cm tall).

H = Hemicryptophyte (perennial plant with overwintering buds at ground level or with slender rhizomes).

G = Geophyte (perennial plant with swollen underground bulbs, corms or tubers).

T = Therophytes or annuals, completing their life cycle within one growing period.

A = Aquatic plants.

Hab (habitat)

A = Freshwater habitats.

C = Cliffs, screes, boulders.

G = Dry to mesic grassland.

H = High mountains (above timberline).

M = Maritime habitats.

P = Phrygana and garigue.

R = Agricultural and ruderal habitats.

W = Woodland and tall scrub.

Acknowledgements

Preparatory work for the Atlas has extended over many years, and a large number of persons have kindly contributed unpublished records, critical identifications, notes on taxonomy/nomenclature and other relevant data (cf. also introductions to Strid 2016a and Dimopoulos & al. 2013). Particularly important help on taxonomic matters has been provided by the following persons who should be regarded as co-authors for families/genera as indicated:

Günter Gottschlich and Franz G. Dunkel: *Hieracium* and *Pilosella*.

Richard Lansdown: *Callitriche*.

Thomas Raus: Amaranthaceae and Chenopodiaceae.

Jean-Marc Tison: *Gagea*.

Holger Uhlich: Orobanchaceae.

Aris Zografidis: *Verbascum*.

In addition, several accounts in Rosaceae and Fabaceae are partly based on unpublished draft typescripts for Flora Hellenica by a number of authors as specified under the respective genera in the Aegean Atlas (Strid 2016a).

Scandinavian interest in the Greek (particularly Aegean) flora goes back to a project started at the University of Lund in the late nineteen fifties by Hans Runemark and Sven & Britt Snogerup. In the sixties and seventies a team working on evolution and biogeography in the Aegean flora included Bengt Bentzer, Roland von Bothmer, Lennart Engstrand and myself. Although subsequently shifting to applied botany the three last-mentioned have remained my long-term friends, participating in several joint, more leisurely trips to Greece in recent years.

The most prolific contemporary collector of Greek flora is Eckhard Willing. After having worked and published extensively on Greek and European orchids he has carried out annual collecting expeditions throughout the Greek mainland from 1988 to the present. Lists of localities and specimens (kept at Herbarium B) have been provided as photocopies or computer printouts, and are now available at www.willing-botanik.de. Records have initially been registered from these lists in which identifications are sometimes preliminary. Whenever possible, herbarium specimens representing outlying or dubious records have been revised by the present author on a series of working sessions at Herbarium B. Willing's collections will remain an important source of information for Greek botany in the foreseeable future.

Thomas Raus has kindly shared his extensive knowledge of the Greek flora and has provided valuable help on scientific and practical matters on my research visits to Herbarium B. In addition to accounts of Amaranthaceae and Chenopodiaceae he has critically examined many specimens at my request, and has assisted with rare literature, interpretation of geographical names and in many other ways.

The late Elli Stamatiadou of the Goulandris Natural History Museum was the most important Greek botanical collector in the period 1967-2005, with many discoveries to her credit. Her numerous, carefully prepared and annotated specimens (at Herbarium ATH) will remain an important data source for future studies of the Greek flora. Our joint excursions are fondly remembered.

Ina Dinter has made important contributions to the botany of Samos, Lesvos, Chios, Kos, Rodos, Karpathos, Kerkira, Zakinthos, Thasos, Chalkidiki and Peloponnisos. Carefully prepared and beautifully illustrated excursion reports have been distributed privately and have been a valuable source of information for this Atlas.

Burkhard Biel, a careful and meticulous collector, has contributed extensively to the botanical exploration of the Aegean islands, notably Amorgos, Milos, Thasos and Samothaki, and has published a long series of floristic reports, often together with Kit Tan.

Kit Tan, formerly of Flora of Turkey fame, has contributed extensively to the botanical exploration of Greece over a period of 30 years, not least as co-editor and frequent contributor to the series "New floristic records in the Balkans" in the periodical Phytologia Balcanica. For this Atlas she has kindly undertaken the demanding and time-consuming job of critically examining the almost completed set of maps and text, making many valuable observations and suggestions.

Panayotis Dimopoulos, coordinator of the most successful contemporary botanical research group in Greece, has consistently encouraged and facilitated my work. His wife, Constantia Patelodimou, has helped design the maps and has given valuable advice on other IT matters. Special thanks are due to both of them as well as the team at the University of Patras.

In addition to those mentioned above, a large number of persons have helped throughout the years with the progress and completion of this work. The following should be listed: Björn Aldén, Spyros Apostolopoulos, Ulf Arup, Evangelos Axiotis, Makis Axiotis, Evangelos Baliousis, Zoltán Barina, Ioannis Bazos, Erwin Bergmeier, Jean Bienvenu, Niels Böhling, Thomas Borsch, Christian Burri, Cristina Cattaneo, Stylianos Charalambidis, Steen N. Christensen, Anastasia Christopolou, Theofanis Constantinidis, Nils Cronberg, Kiki Dima, Panayiotis Dimitrakopoulos, Eleni Eleftheriadou, Michael Foley, Georgios Fotiadis, Christina Fournaraki, Ori Fragman, Bozo Frajman, Patrik Frödén, Maria Georgiopoulou, Christian Gilli, Katerina Goula, Walter Gutermann, Ralf Hand, Per Hartvig, Gregory Iatroú, Eleni Iliadou, Vasilis Ioannidis, Armin Jagel, Jan Jordan, Joachim Kadereit, Heinz Kalheber, Zdeněk Kaplan, Elpida Karadimou, Yannis Kazoglou, Jan Kirschner, Andreas Kleinsteuber, Georgios Korakis, Konstantinos Kougioumoutzis, Konstantina Koutroumpa, Georgi Kunev, Ingvar Kärnefelt, Hans Walter Lack, Dmitar Lakušić, Per Lassen, Steve Lenton, Annita Logotheti, Myrsini Malakou, Roger Marchant, Karel Marhold, Brian Mathew, Penelope Matsoukas, Dionysis Mermygas, Stefan Meyer, Chris Mounsey, Ladislav Mucina, Alfred Mullay, Enio Nardi, Marjan Niketić, Gisela Nikolopoulou, Maria Panitsa, Jimmy & Karin Persson, Nikos Petrou, Ana Petrova, Giorgos Poulis, Eckhard Raab-Straube, Uwe Raabe, Stefan Raetzel, Michael Ristow, Dominik Roepert, Fanikos Sakellarakis, Astrid Scharlau, Andreas Schuler, Pierre Seba, Federico Selvi, Alexander Sennikov, Pascale Servais, Georgios Sfikas, Stanislav Španiel, Karl Günter Stadler, Fred Stauffer, Anastasia Stefanaki, Vladimir Stevanovic, Åke Svensson, Ricos Thanopoulos, Costas Thanos, Arne Thell, Flemming Thorning-Lund, Chris Thorogood, Gordana Tomovic, Panayotis Trigas, Spyros Tsiftsis, Ioannis Tsiripidis, Nicholas Turland, Torbjörn Tyler, Dimitrios Tzanoudakis, Dionysios Vasiliades, Ernst Vitek, Vladimir Vladimirov, Robert Vogt, Michael S. Vrahnakis, Håkan Witzell, Fotis Xystrakis, George Zarkos, Henrik Zetterlund, Jerzy Zieliński, Andreas Zikos and Lana Zoric.

Travel grants for field work and work at foreign herbaria have been received from several sources, in recent years notably from the Velux Foundation (Denmark) and the Royal Physiographic Society of Lund (Sweden). Publication of this Atlas has been generously financed by the Aage V. Jensen Charity Foundation, a Danish fund supporting nature conservation and natural history projects in Denmark and internationally.

It has been my pleasure and privilege to study the Greek flora for nearly 60 years. My field days amount to 1,619 or about 4½ years. I have ventured into the real world trying to find out how things *are* rather than preparing mathematical models of how things *ought to be*. Hopefully this Atlas will provide a baseline for future studies of Greek biodiversity, thus repaying some of my debt to this great country.

Ørbæk, Denmark, in May 2022
Arne Strid
Professor emeritus
Former director, Göteborg Botanic Garden

Foothills of Olimbos, 12.5. 1974.
Jankaea heldreichii in background.

In the Olympic fog (Mitikas, 2918 m).
24.7. 2019.

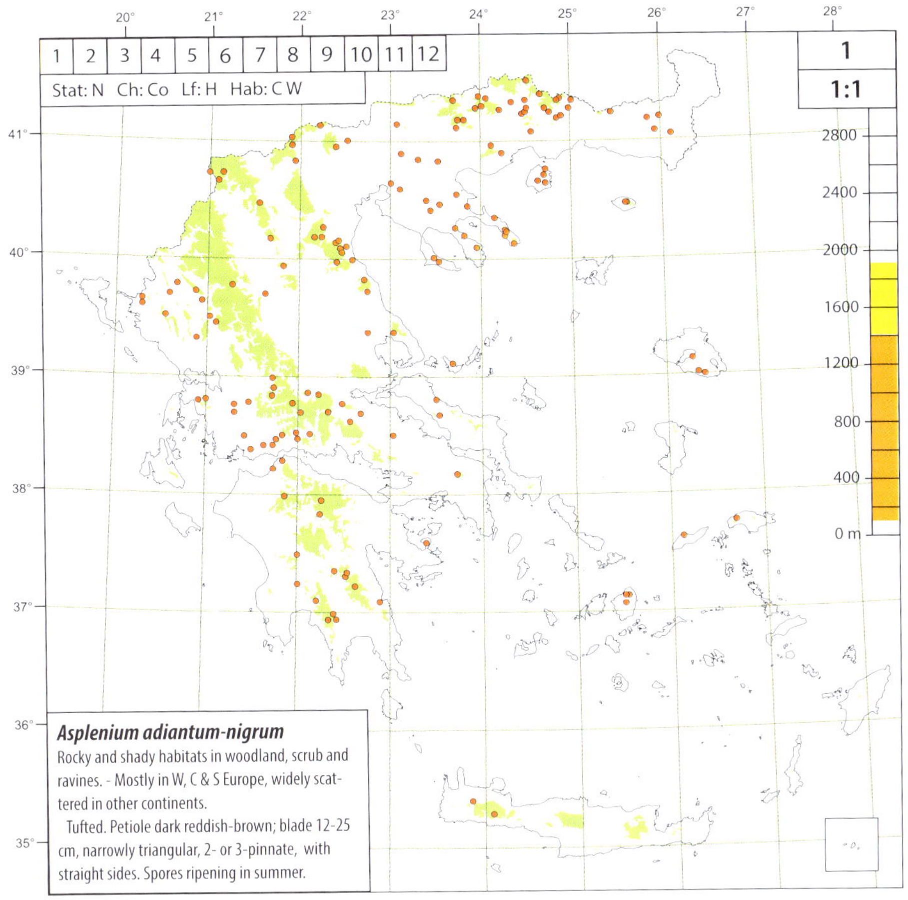

Asplenium adiantum-nigrum

Rocky and shady habitats in woodland, scrub and ravines. - Mostly in W, C & S Europe, widely scattered in other continents.

Tufted. Petiole dark reddish-brown; blade 12-25 cm, narrowly triangular, 2- or 3-pinnate, with straight sides. Spores ripening in summer.

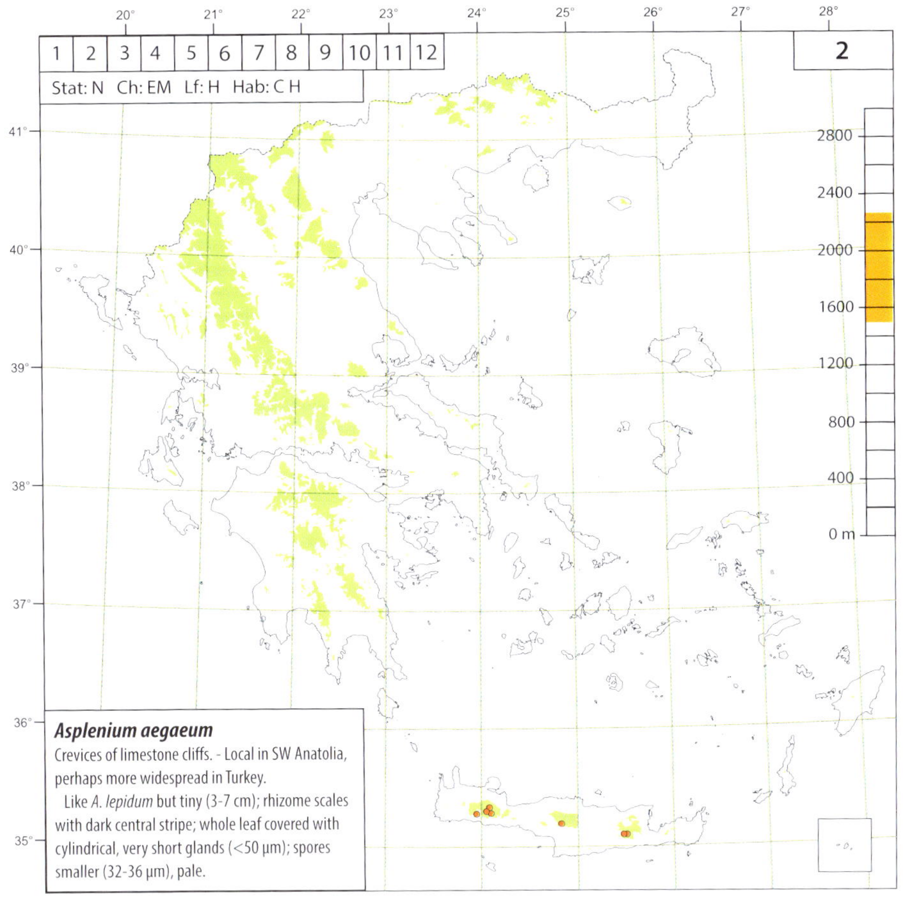

Asplenium aegaeum

Crevices of limestone cliffs. - Local in SW Anatolia, perhaps more widespread in Turkey.

Like *A. lepidum* but tiny (3-7 cm); rhizome scales with dark central stripe; whole leaf covered with cylindrical, very short glands (<50 µm); spores smaller (32-36 µm), pale.

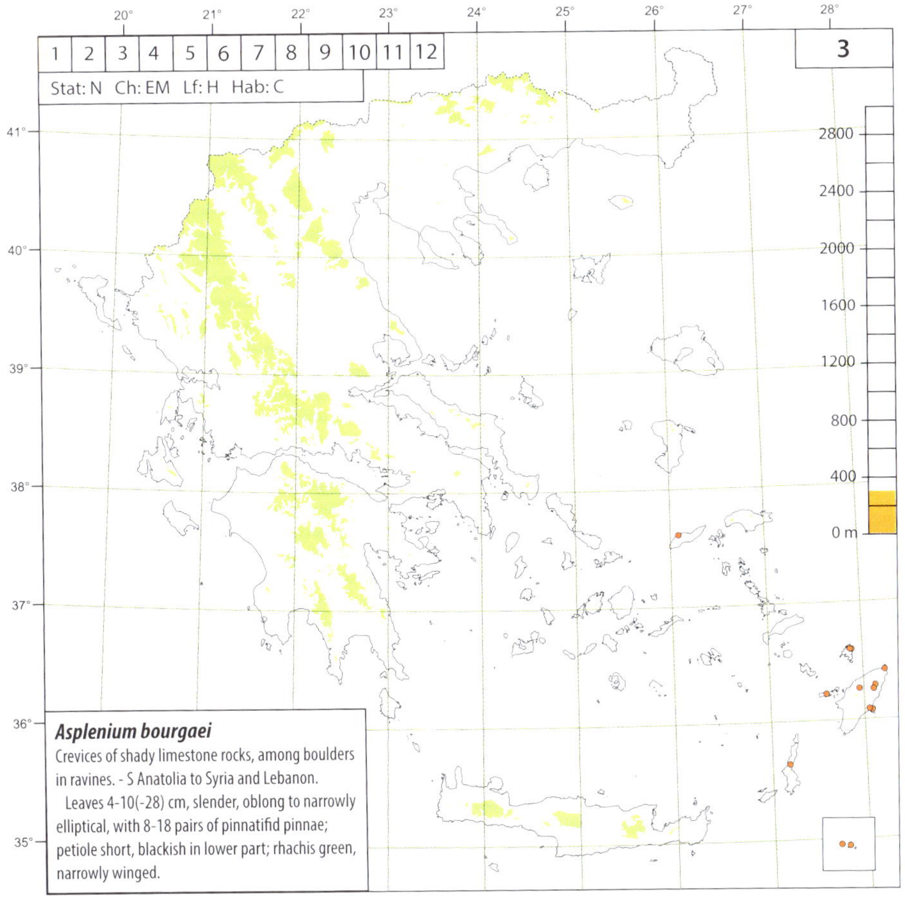

Asplenium bourgaei

Crevices of shady limestone rocks, among boulders in ravines. - S Anatolia to Syria and Lebanon.

Leaves 4-10(-28) cm, slender, oblong to narrowly elliptical, with 8-18 pairs of pinnatifid pinnae; petiole short, blackish in lower part; rhachis green, narrowly winged.

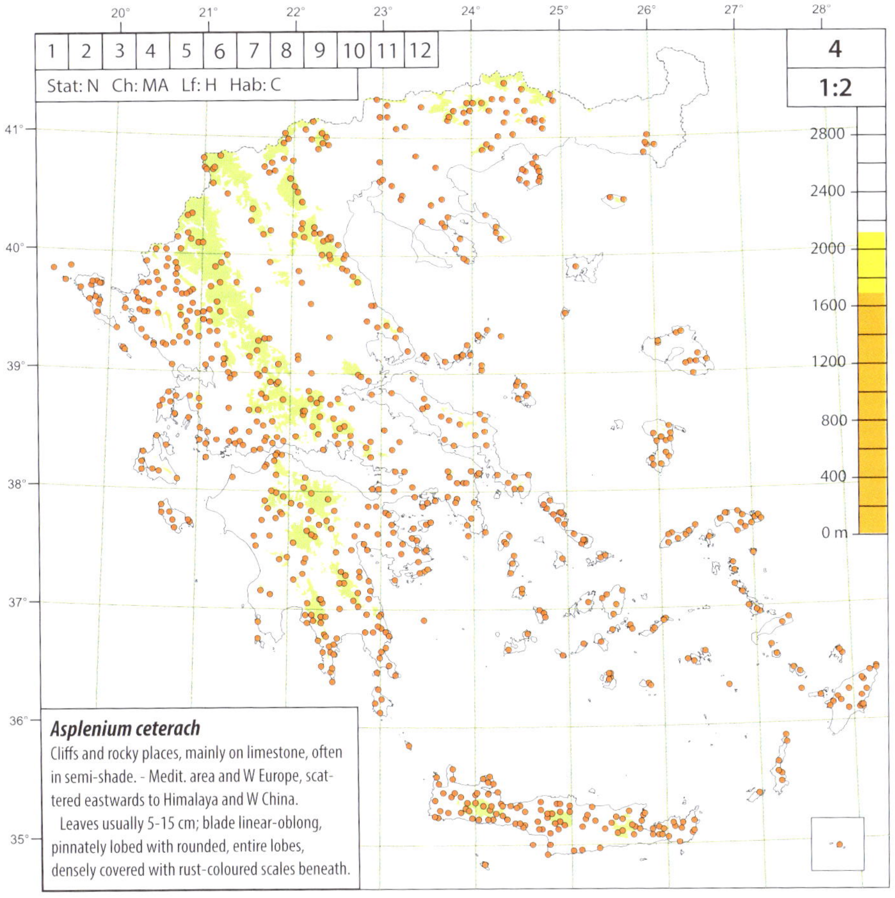

Asplenium ceterach

Cliffs and rocky places, mainly on limestone, often in semi-shade. - Medit. area and W Europe, scattered eastwards to Himalaya and W China.

Leaves usually 5-15 cm; blade linear-oblong, pinnately lobed with rounded, entire lobes, densely covered with rust-coloured scales beneath.

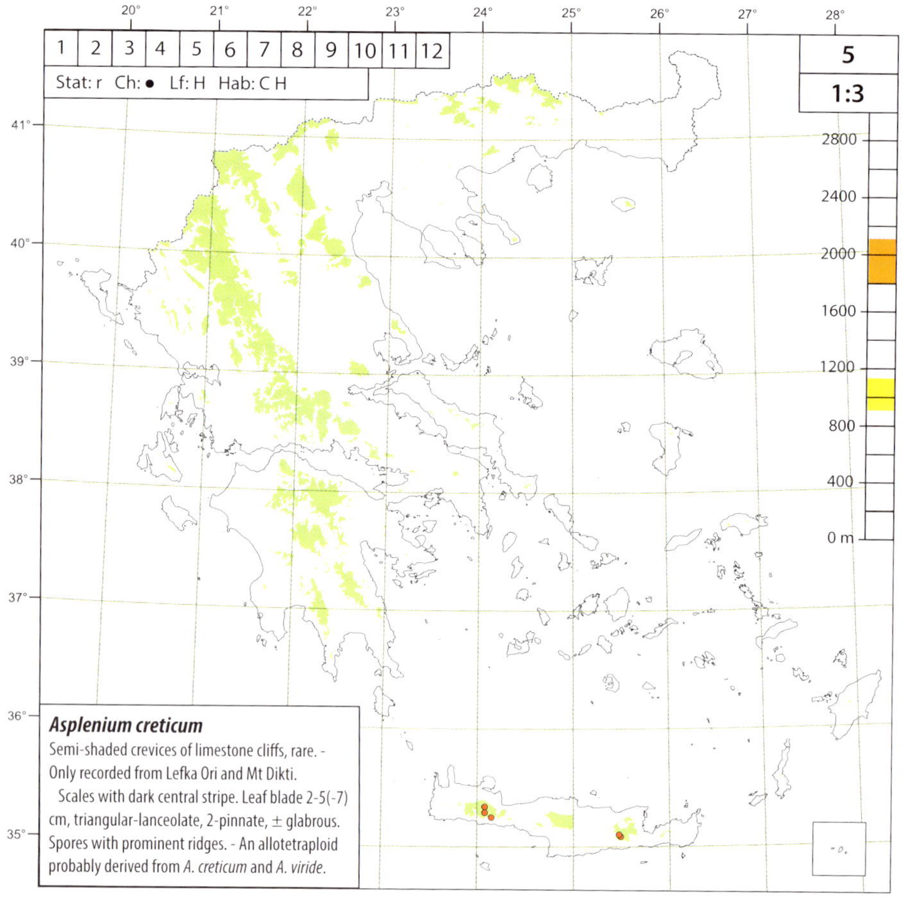

Asplenium creticum

Semi-shaded crevices of limestone cliffs, rare. - Only recorded from Lefka Ori and Mt Dikti.

Scales with dark central stripe. Leaf blade 2-5(-7) cm, triangular-lanceolate, 2-pinnate, ± glabrous. Spores with prominent ridges. - An allotetraploid probably derived from *A. creticum* and *A. viride*.

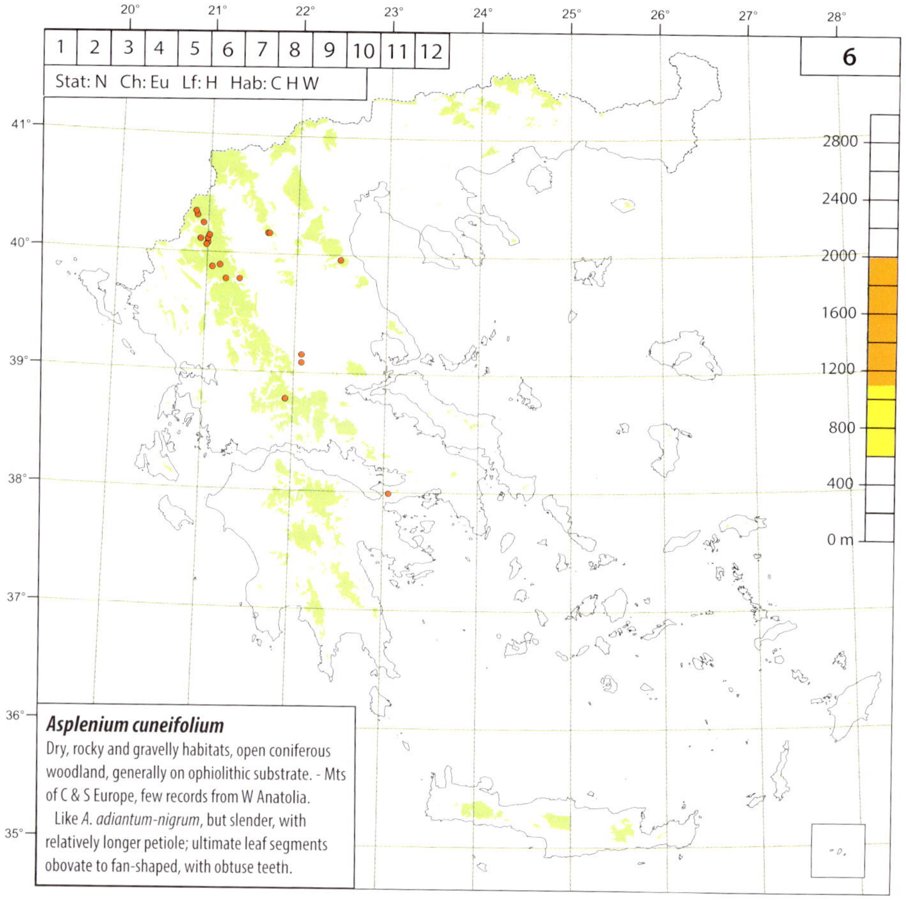

Asplenium cuneifolium

Dry, rocky and gravelly habitats, open coniferous woodland, generally on ophiolithic substrate. - Mts of C & S Europe, few records from W Anatolia.

Like *A. adiantum-nigrum*, but slender, with relatively longer petiole; ultimate leaf segments obovate to fan-shaped, with obtuse teeth.

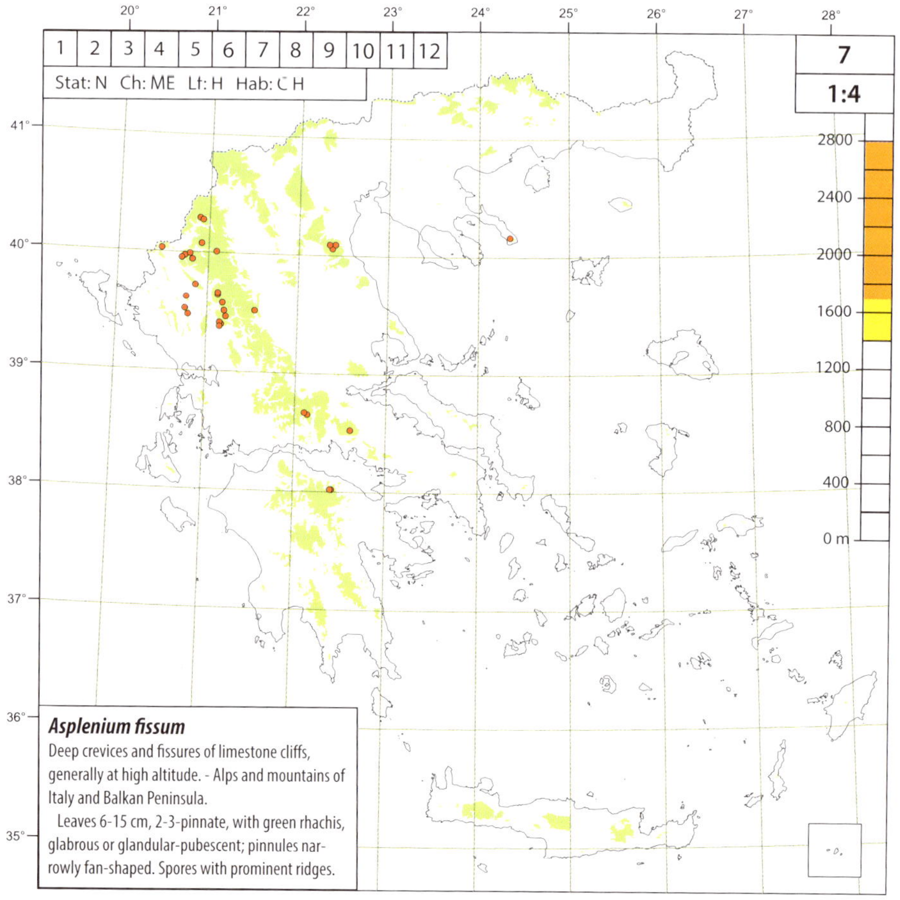

Asplenium fissum

Deep crevices and fissures of limestone cliffs, generally at high altitude. - Alps and mountains of Italy and Balkan Peninsula.

Leaves 6-15 cm, 2-3-pinnate, with green rhachis, glabrous or glandular-pubescent; pinnules narrowly fan-shaped. Spores with prominent ridges.

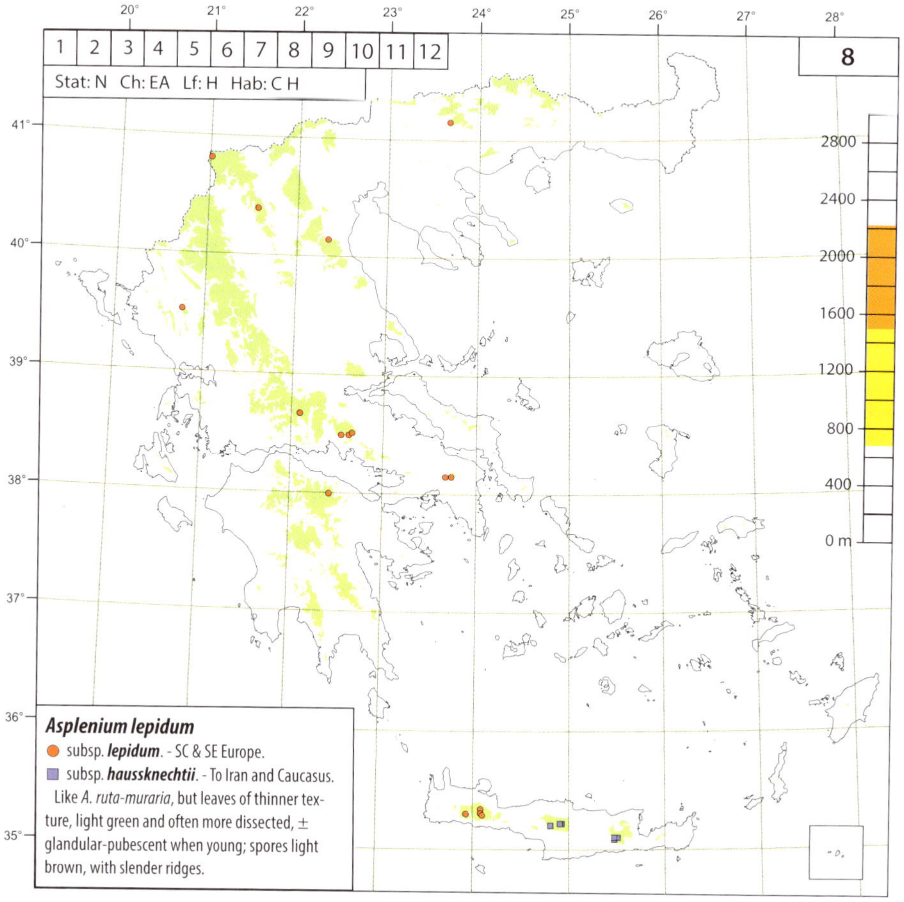

Asplenium lepidum

● subsp. ***lepidum***. - SC & SE Europe.

■ subsp. ***haussknechtii***. - To Iran and Caucasus.

Like *A. ruta-muraria*, but leaves of thinner texture, light green and often more dissected, ± glandular-pubescent when young; spores light brown, with slender ridges.

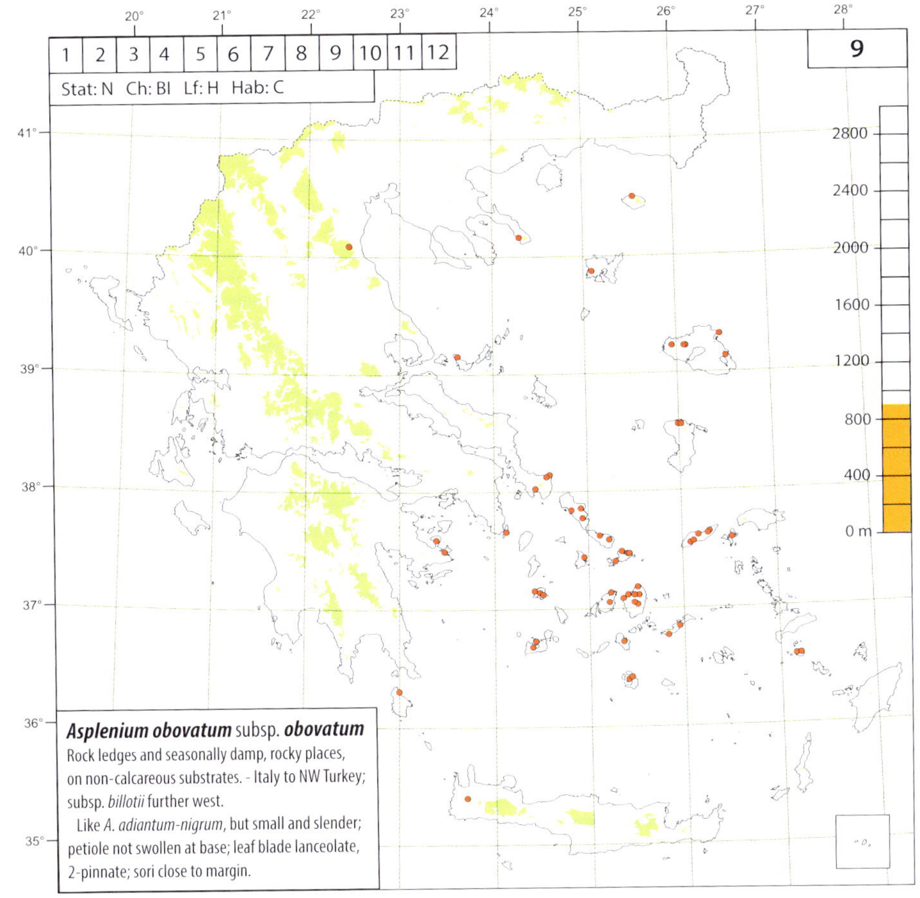

Asplenium obovatum subsp. _obovatum_
Rock ledges and seasonally damp, rocky places, on non-calcareous substrates. - Italy to NW Turkey; subsp. *billotii* further west.
Like *A. adiantum-nigrum*, but small and slender; petiole not swollen at base; leaf blade lanceolate, 2-pinnate; sori close to margin.

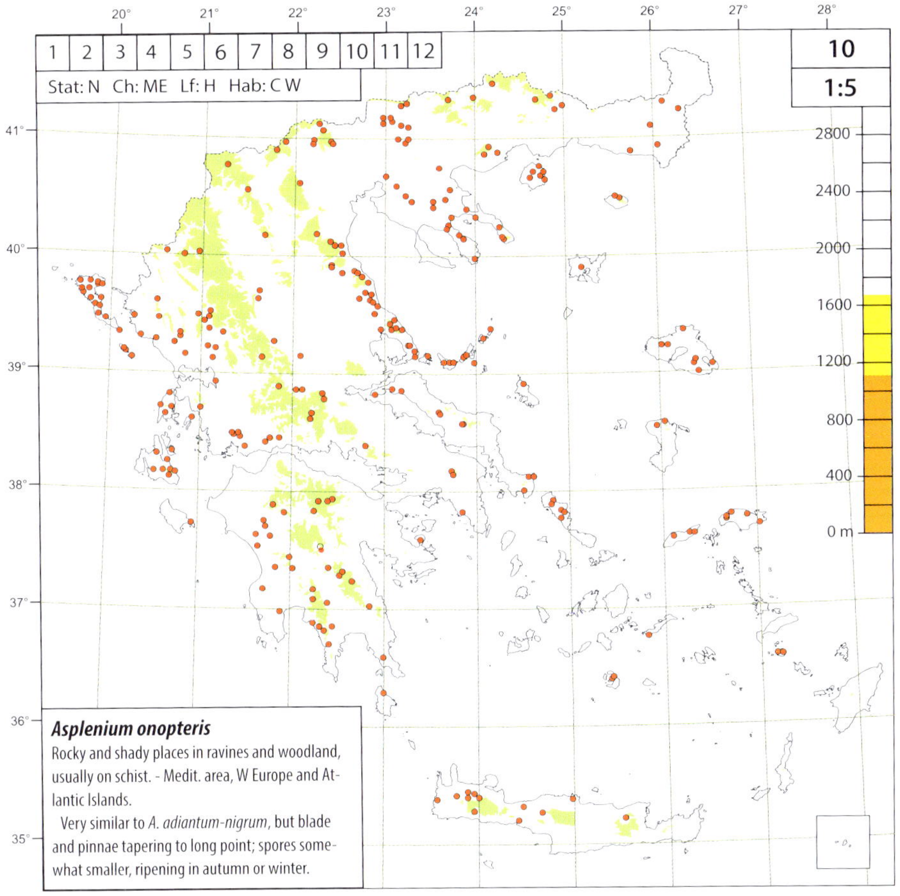

Asplenium onopteris
Rocky and shady places in ravines and woodland, usually on schist. - Medit. area, W Europe and Atlantic Islands.
Very similar to *A. adiantum-nigrum*, but blade and pinnae tapering to long point; spores somewhat smaller, ripening in autumn or winter.

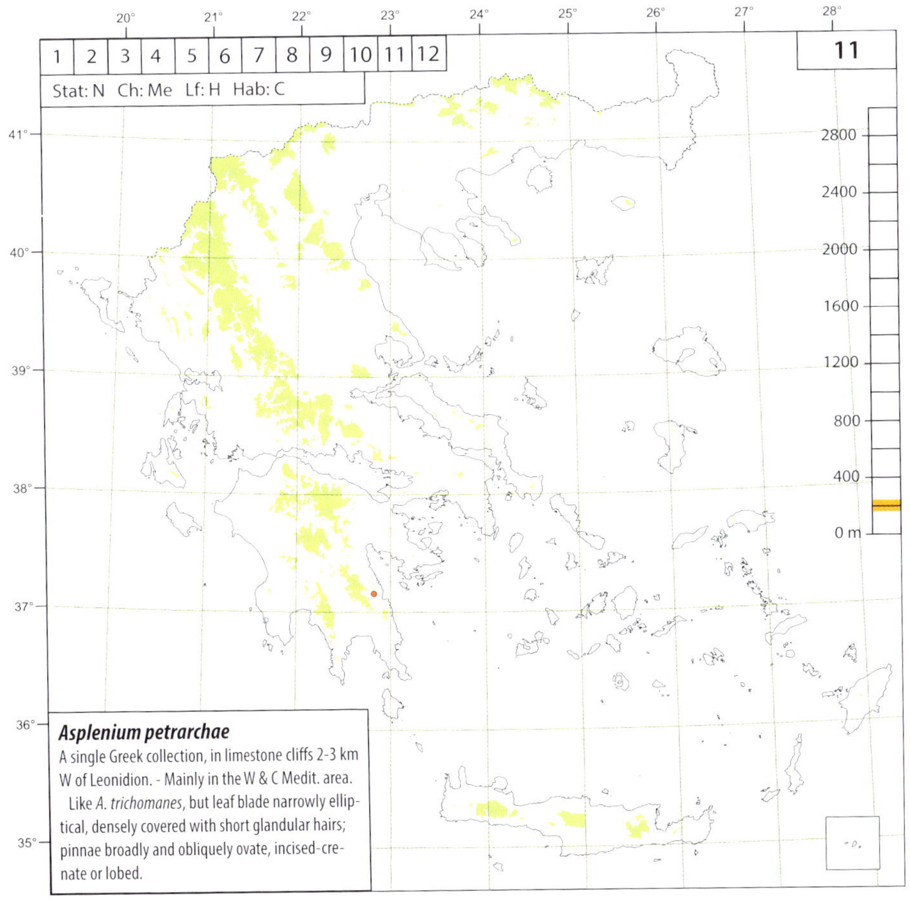

Asplenium petrarchae
A single Greek collection, in limestone cliffs 2-3 km W of Leonidion. - Mainly in the W & C Medit. area.
Like *A. trichomanes*, but leaf blade narrowly elliptical, densely covered with short glandular hairs; pinnae broadly and obliquely ovate, incised-crenate or lobed.

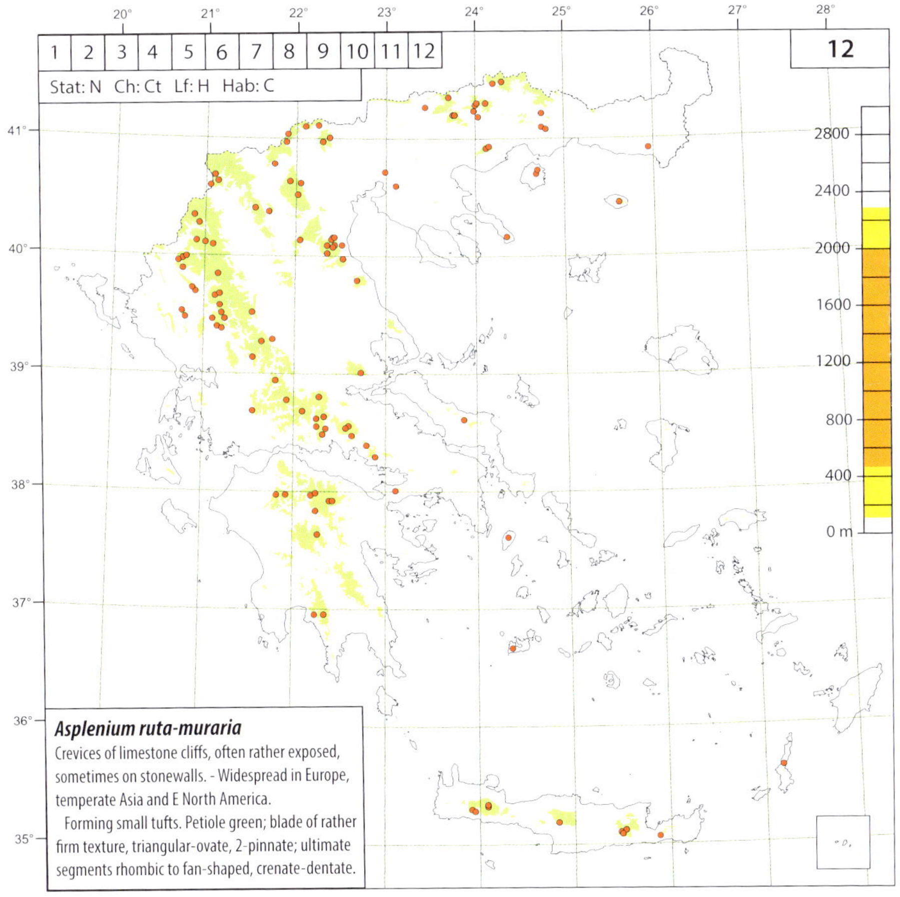

Asplenium ruta-muraria
Crevices of limestone cliffs, often rather exposed, sometimes on stonewalls. - Widespread in Europe, temperate Asia and E North America.
Forming small tufts. Petiole green; blade of rather firm texture, triangular-ovate, 2-pinnate; ultimate segments rhombic to fan-shaped, crenate-dentate.

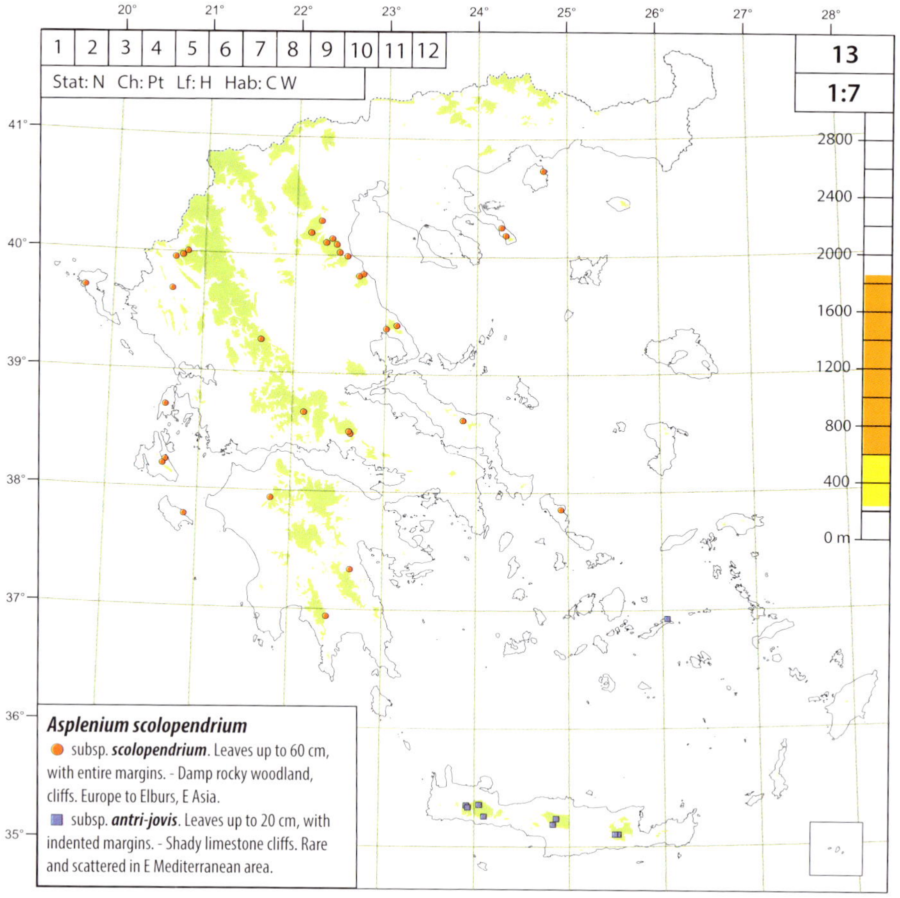

Asplenium scolopendrium
● subsp. ***scolopendrium***. Leaves up to 60 cm, with entire margins. - Damp rocky woodland, cliffs. Europe to Elburs, E Asia.
■ subsp. ***antri-jovis***. Leaves up to 20 cm, with indented margins. - Shady limestone cliffs. Rare and scattered in E Mediterranean area.

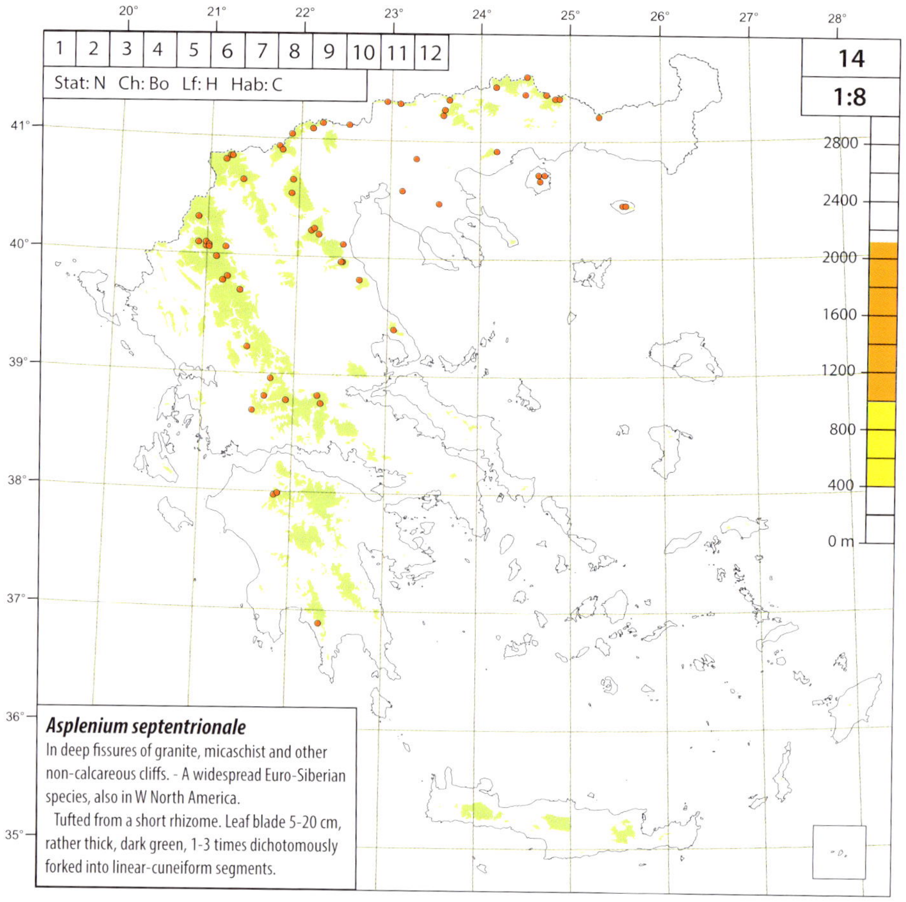

Asplenium septentrionale
In deep fissures of granite, micaschist and other non-calcareous cliffs. - A widespread Euro-Siberian species, also in W North America.
Tufted from a short rhizome. Leaf blade 5-20 cm, rather thick, dark green, 1-3 times dichotomously forked into linear-cuneiform segments.

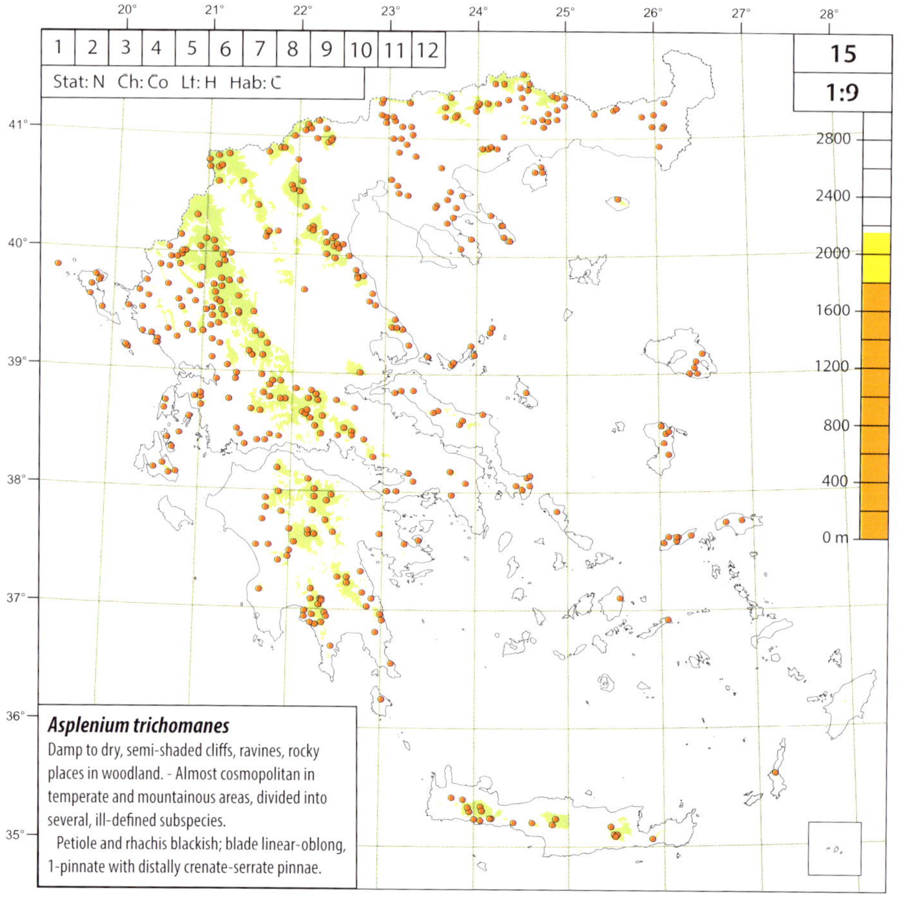

Asplenium trichomanes
Damp to dry, semi-shaded cliffs, ravines, rocky places in woodland. - Almost cosmopolitan in temperate and mountainous areas, divided into several, ill-defined subspecies.
Petiole and rhachis blackish; blade linear-oblong, 1-pinnate with distally crenate-serrate pinnae.

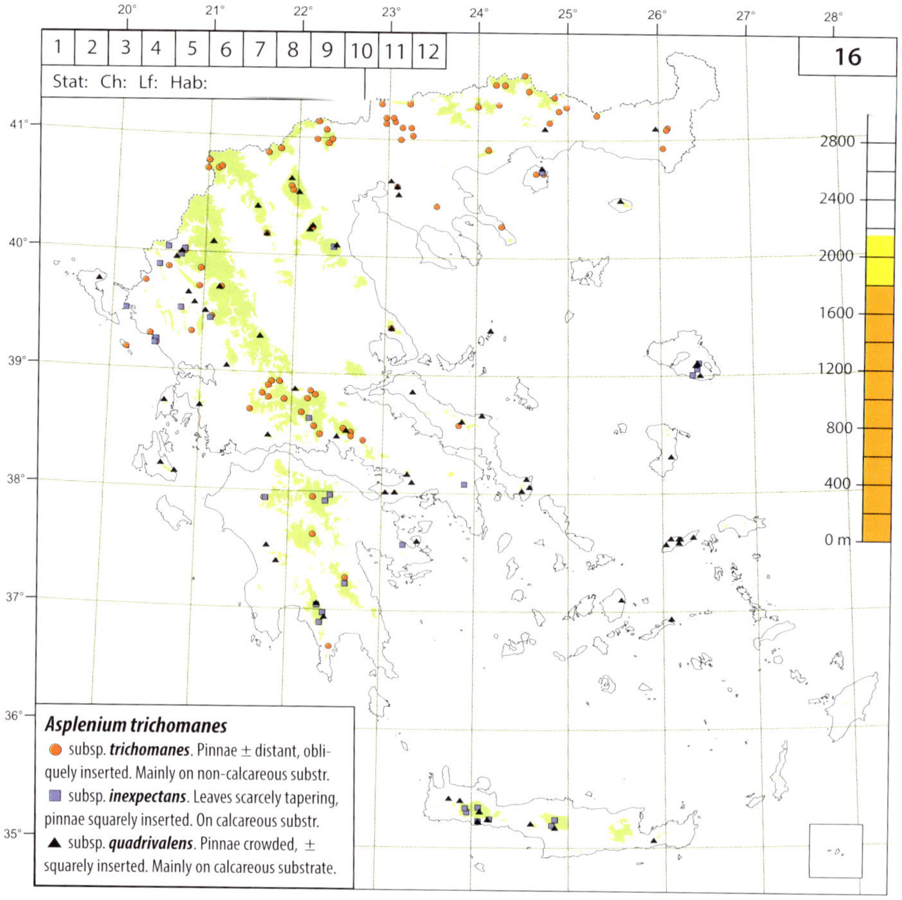

Asplenium trichomanes
● subsp. ***trichomanes***. Pinnae ± distant, obliquely inserted. Mainly on non-calcareous substr.
■ subsp. ***inexpectans***. Leaves scarcely tapering, pinnae squarely inserted. On calcareous substr.
▲ subsp. ***quadrivalens***. Pinnae crowded, ± squarely inserted. Mainly on calcareous substrate.

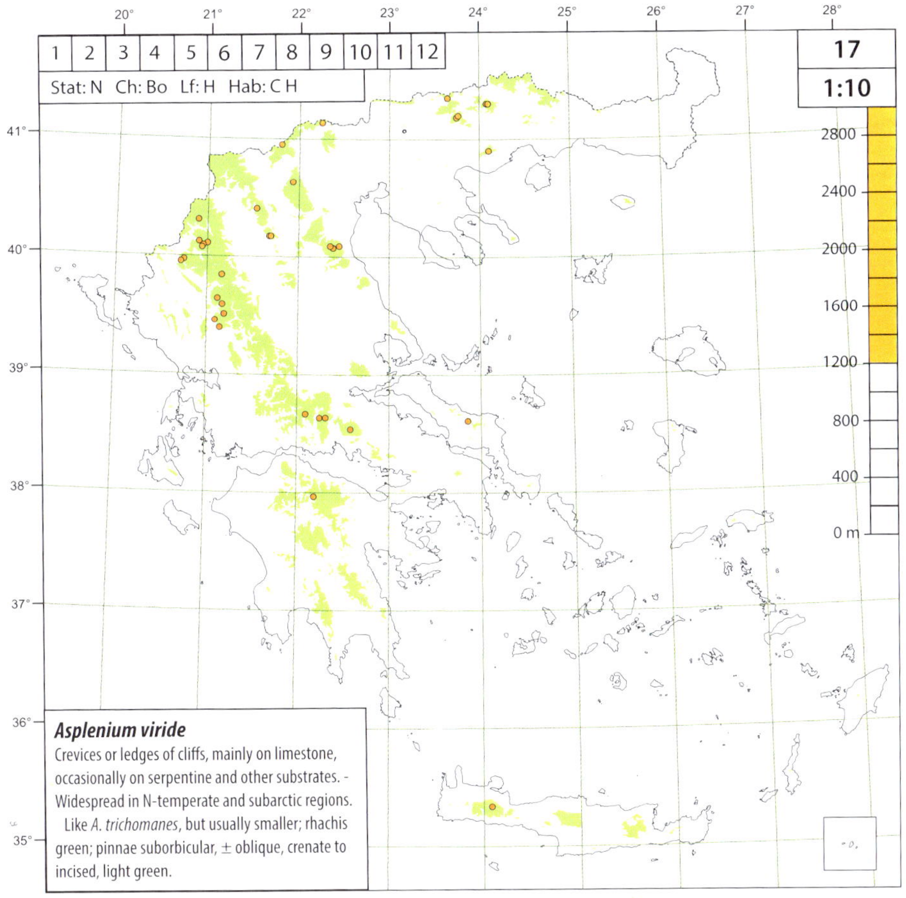

Asplenium viride

Crevices or ledges of cliffs, mainly on limestone, occasionally on serpentine and other substrates. - Widespread in N-temperate and subarctic regions.

Like *A. trichomanes*, but usually smaller; rhachis green; pinnae suborbicular, ± oblique, crenate to incised, light green.

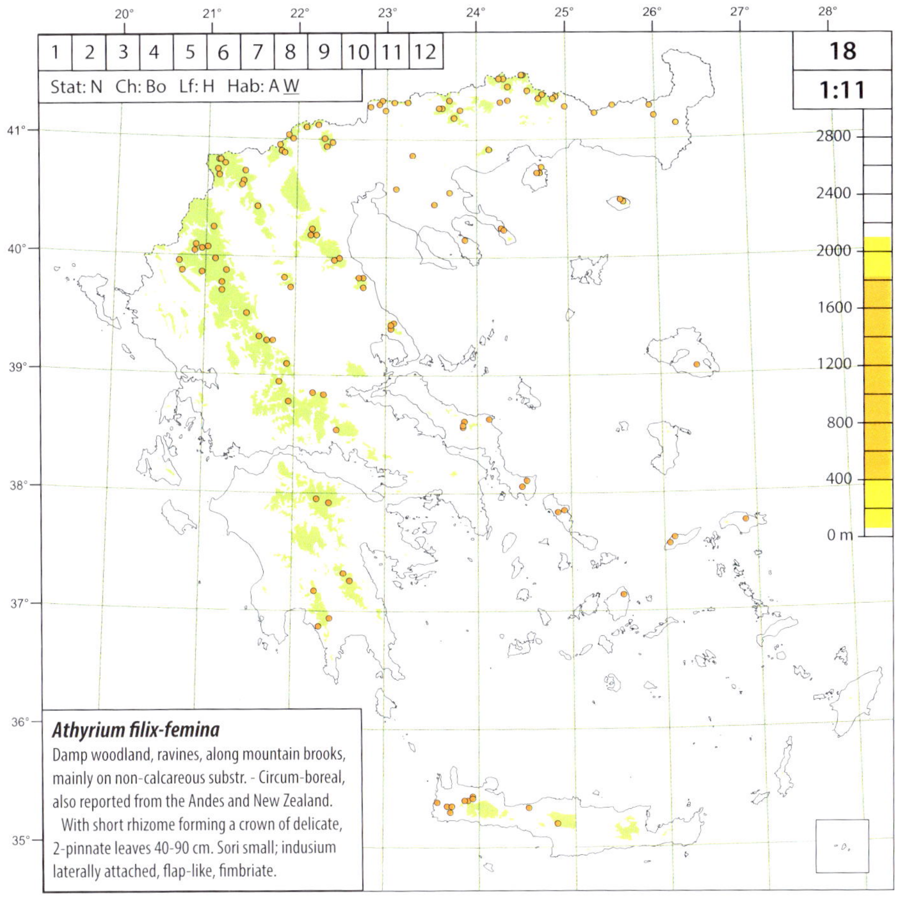

Athyrium filix-femina

Damp woodland, ravines, along mountain brooks, mainly on non-calcareous substr. - Circum-boreal, also reported from the Andes and New Zealand.

With short rhizome forming a crown of delicate, 2-pinnate leaves 40-90 cm. Sori small; indusium laterally attached, flap-like, fimbriate.

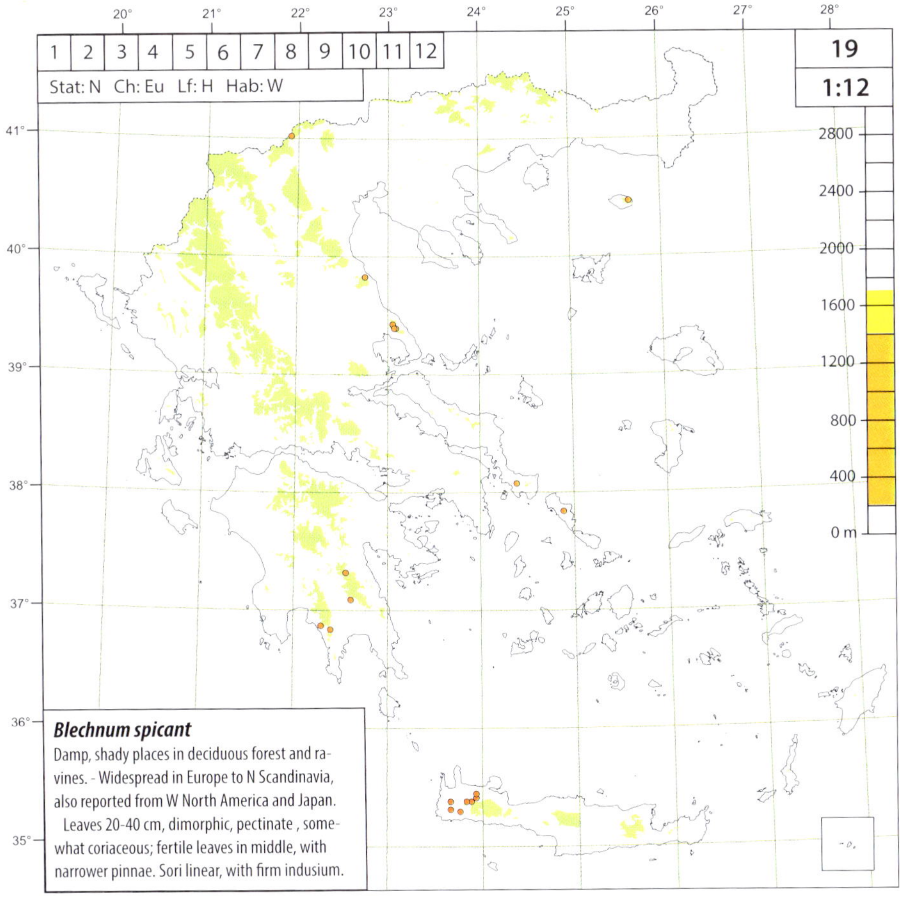

Blechnum spicant

Damp, shady places in deciduous forest and ravines. - Widespread in Europe to N Scandinavia, also reported from W North America and Japan.

Leaves 20-40 cm, dimorphic, pectinate , somewhat coriaceous; fertile leaves in middle, with narrower pinnae. Sori linear, with firm indusium.

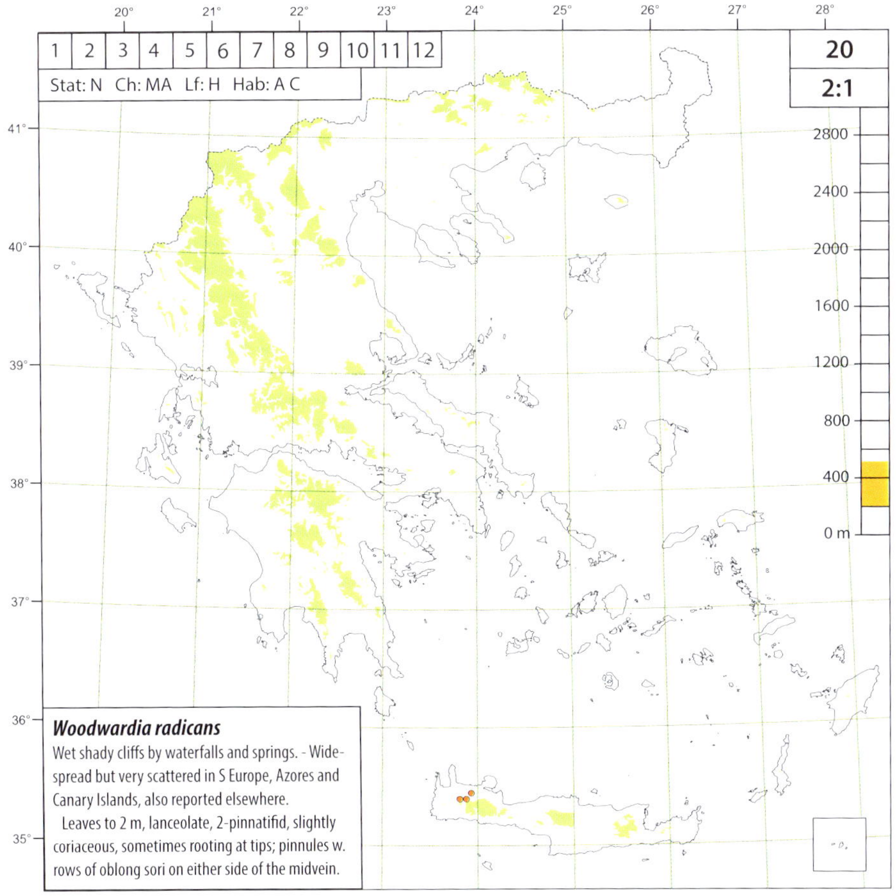

Woodwardia radicans

Wet shady cliffs by waterfalls and springs. - Widespread but very scattered in S Europe, Azores and Canary Islands, also reported elsewhere.

Leaves to 2 m, lanceolate, 2-pinnatifid, slightly coriaceous, sometimes rooting at tips; pinnules w. rows of oblong sori on either side of the midvein.

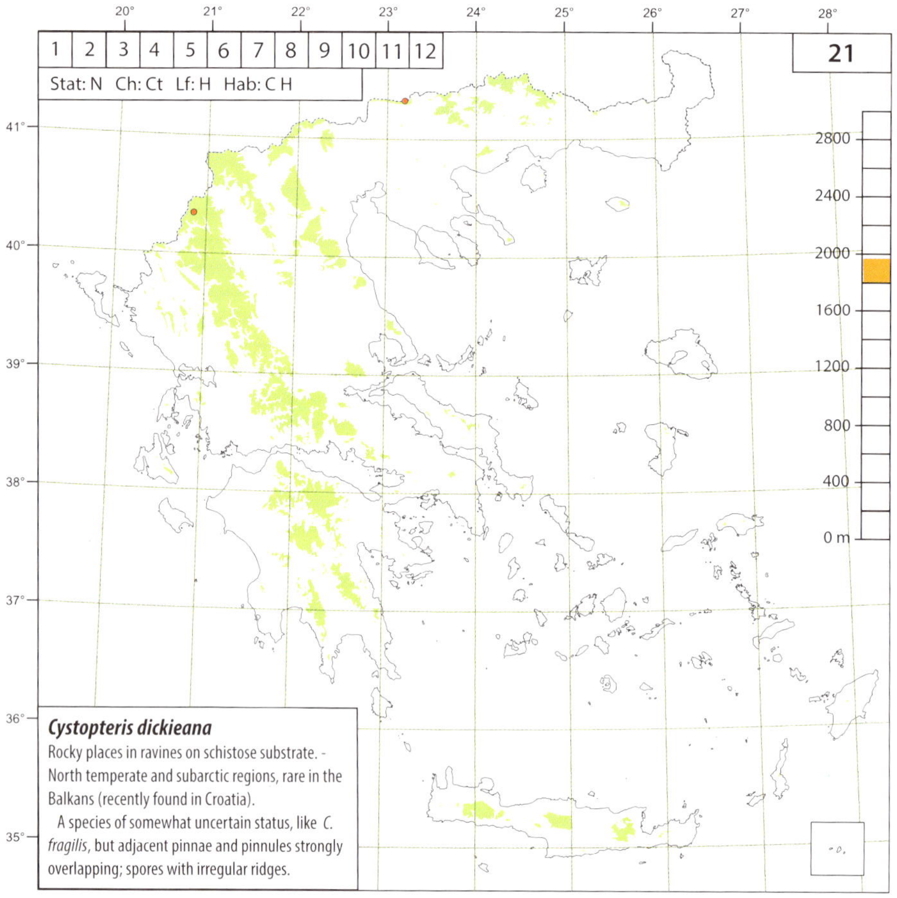

Cystopteris dickieana

Rocky places in ravines on schistose substrate. - North temperate and subarctic regions, rare in the Balkans (recently found in Croatia).

A species of somewhat uncertain status, like *C. fragilis*, but adjacent pinnae and pinnules strongly overlapping; spores with irregular ridges.

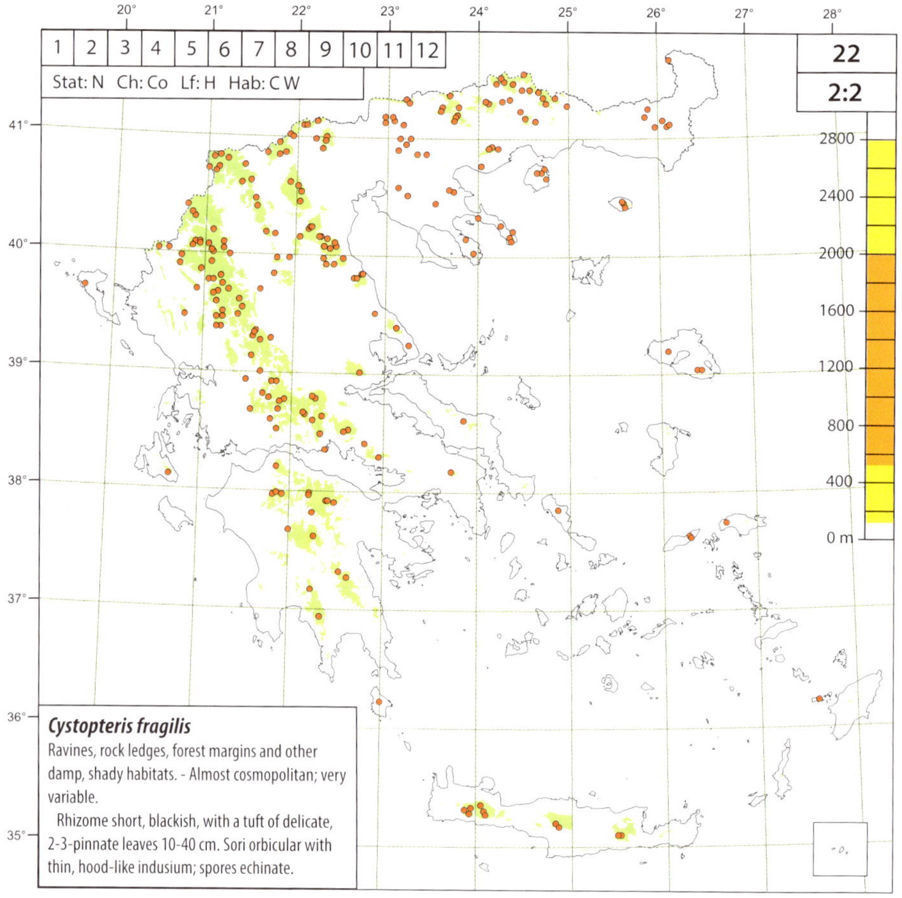

Cystopteris fragilis

Ravines, rock ledges, forest margins and other damp, shady habitats. - Almost cosmopolitan; very variable.

Rhizome short, blackish, with a tuft of delicate, 2-3-pinnate leaves 10-40 cm. Sori orbicular with thin, hood-like indusium; spores echinate.

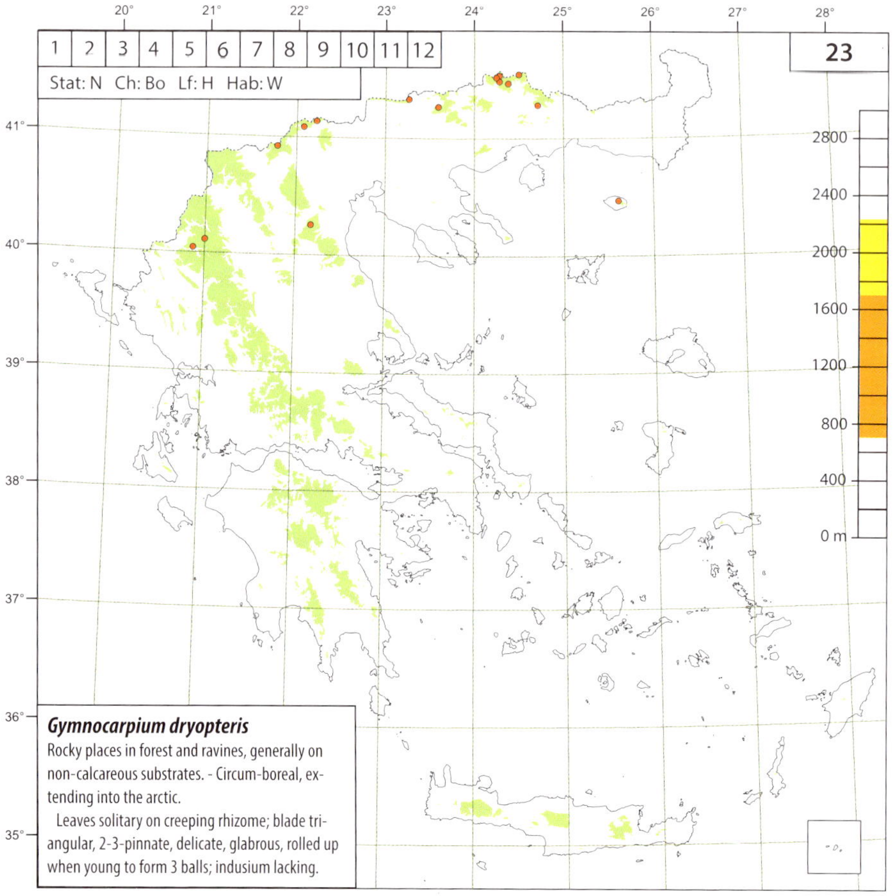

Gymnocarpium dryopteris

Rocky places in forest and ravines, generally on non-calcareous substrates. - Circum-boreal, extending into the arctic.

Leaves solitary on creeping rhizome; blade triangular, 2-3-pinnate, delicate, glabrous, rolled up when young to form 3 balls; indusium lacking.

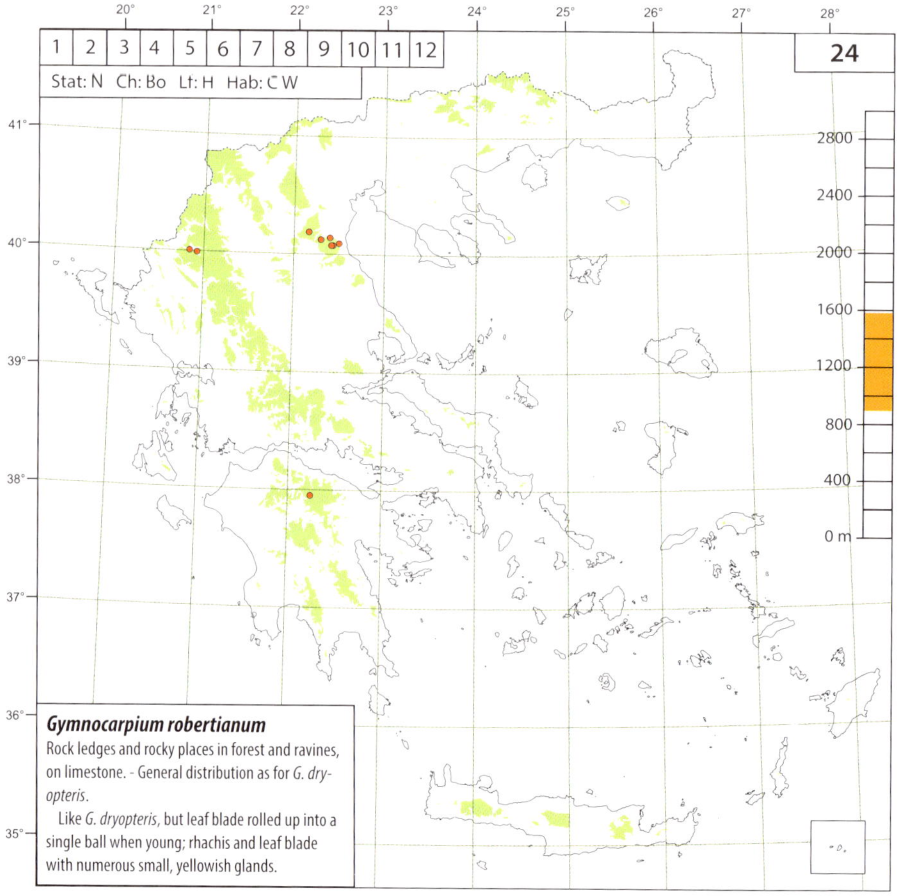

Gymnocarpium robertianum

Rock ledges and rocky places in forest and ravines, on limestone. - General distribution as for *G. dryopteris*.

Like *G. dryopteris*, but leaf blade rolled up into a single ball when young; rhachis and leaf blade with numerous small, yellowish glands.

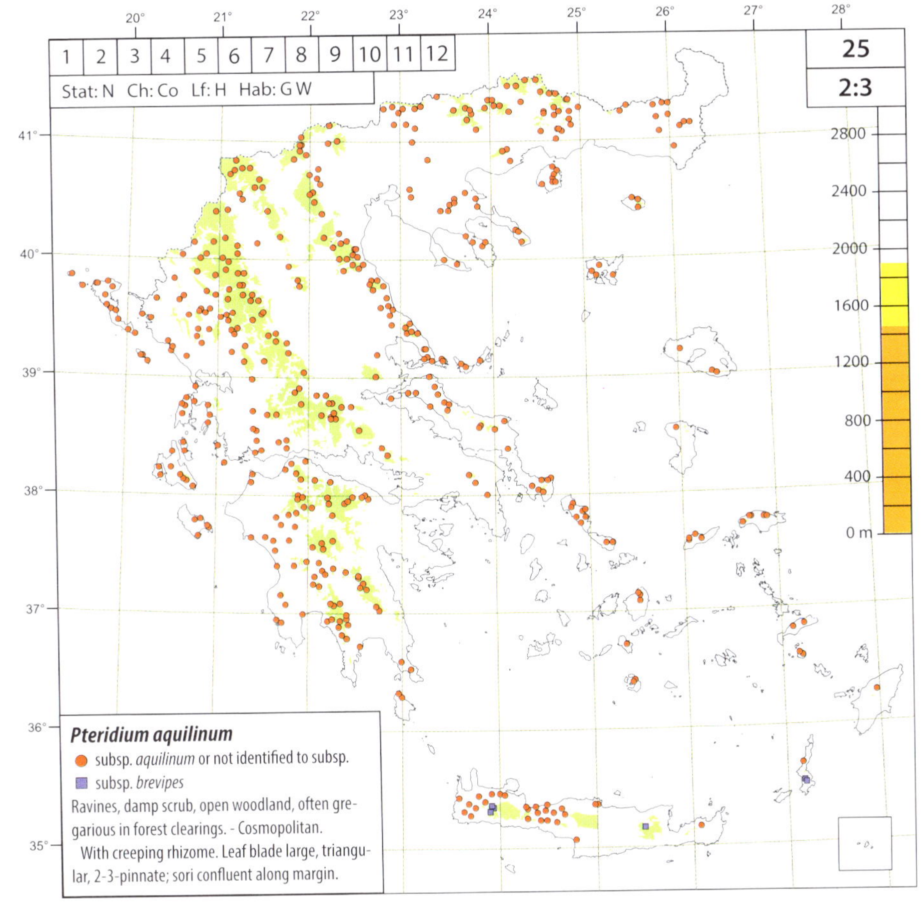

Pteridium aquilinum

- subsp. *aquilinum* or not identified to subsp.
- subsp. *brevipes*

Ravines, damp scrub, open woodland, often gregarious in forest clearings. - Cosmopolitan.

With creeping rhizome. Leaf blade large, triangular, 2-3-pinnate; sori confluent along margin.

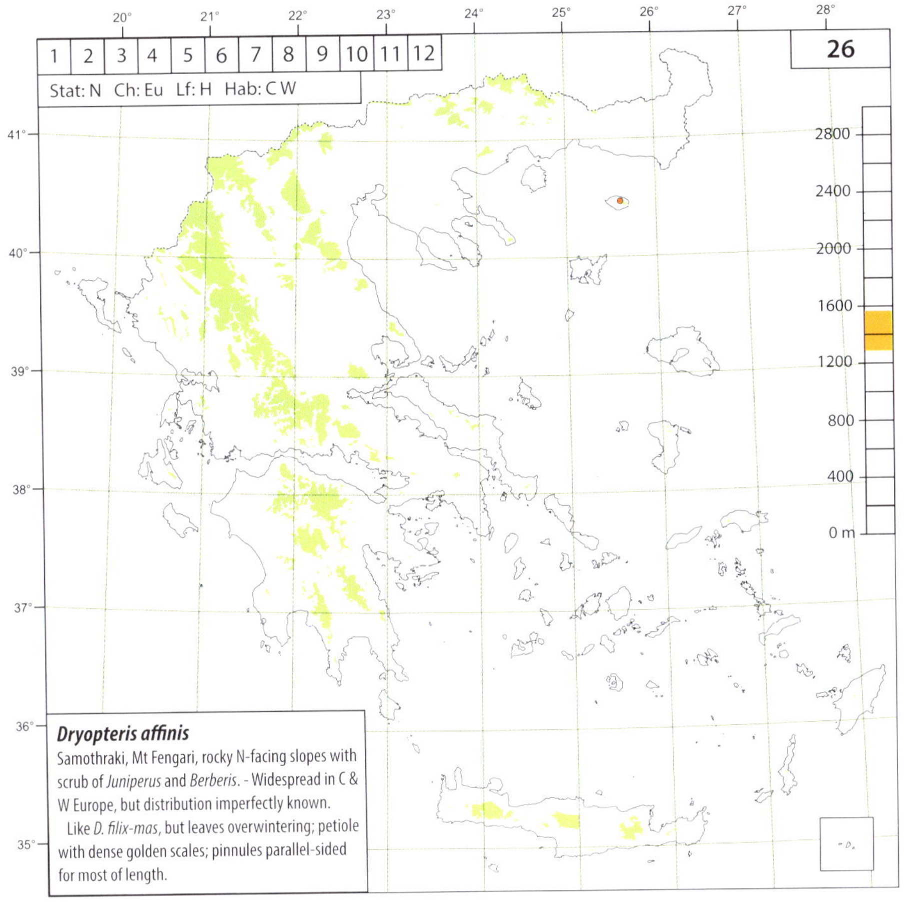

Dryopteris affinis

Samothraki, Mt Fengari, rocky N-facing slopes with scrub of *Juniperus* and *Berberis*. - Widespread in C & W Europe, but distribution imperfectly known.

Like *D. filix-mas*, but leaves overwintering; petiole with dense golden scales; pinnules parallel-sided for most of length.

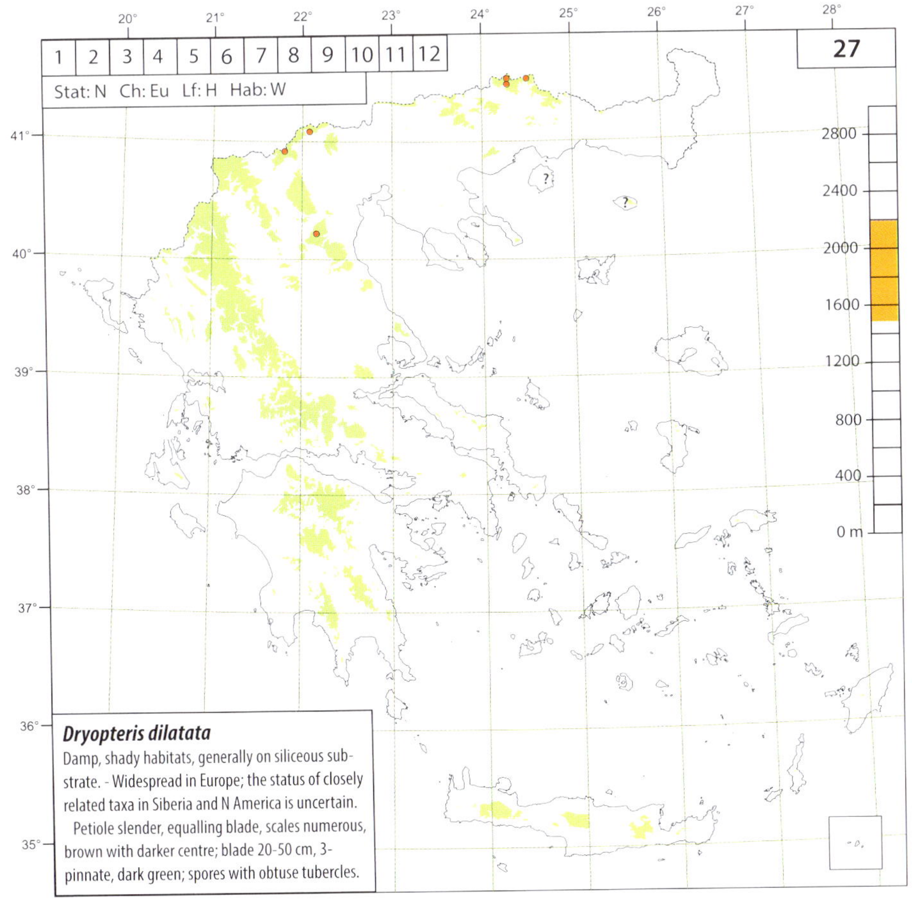

Dryopteris dilatata

Damp, shady habitats, generally on siliceous substrate. - Widespread in Europe; the status of closely related taxa in Siberia and N America is uncertain.

Petiole slender, equalling blade, scales numerous, brown with darker centre; blade 20-50 cm, 3-pinnate, dark green; spores with obtuse tubercles.

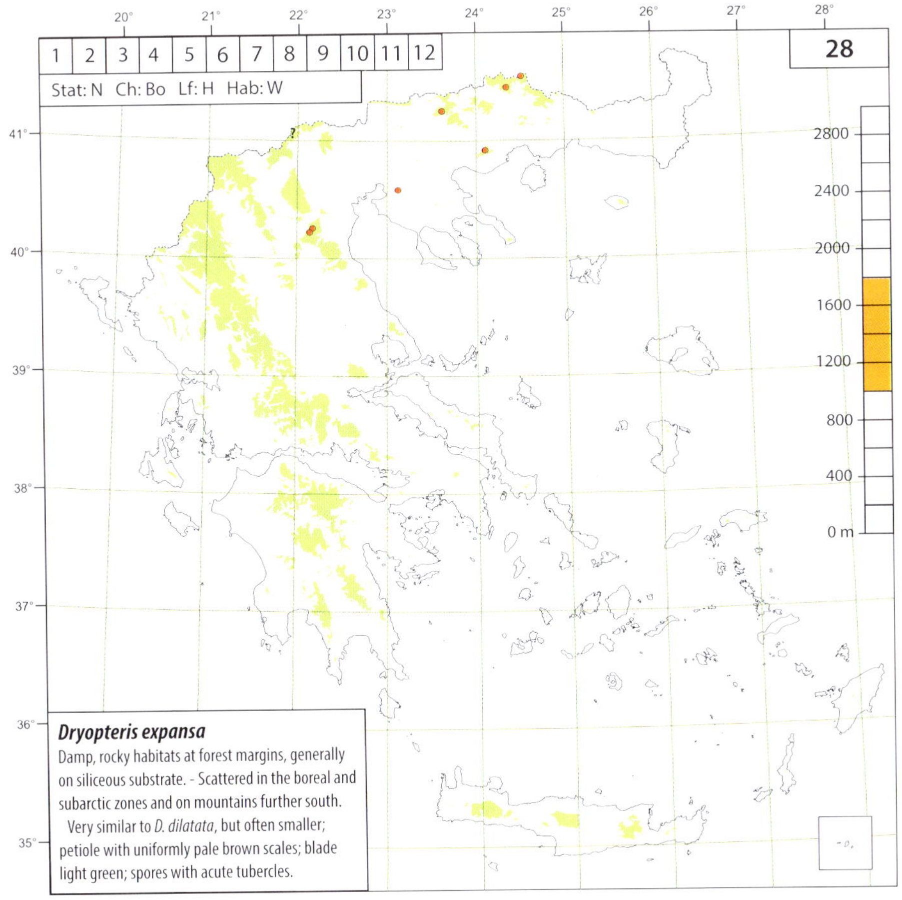

Dryopteris expansa

Damp, rocky habitats at forest margins, generally on siliceous substrate. - Scattered in the boreal and subarctic zones and on mountains further south.

Very similar to *D. dilatata*, but often smaller; petiole with uniformly pale brown scales; blade light green; spores with acute tubercles.

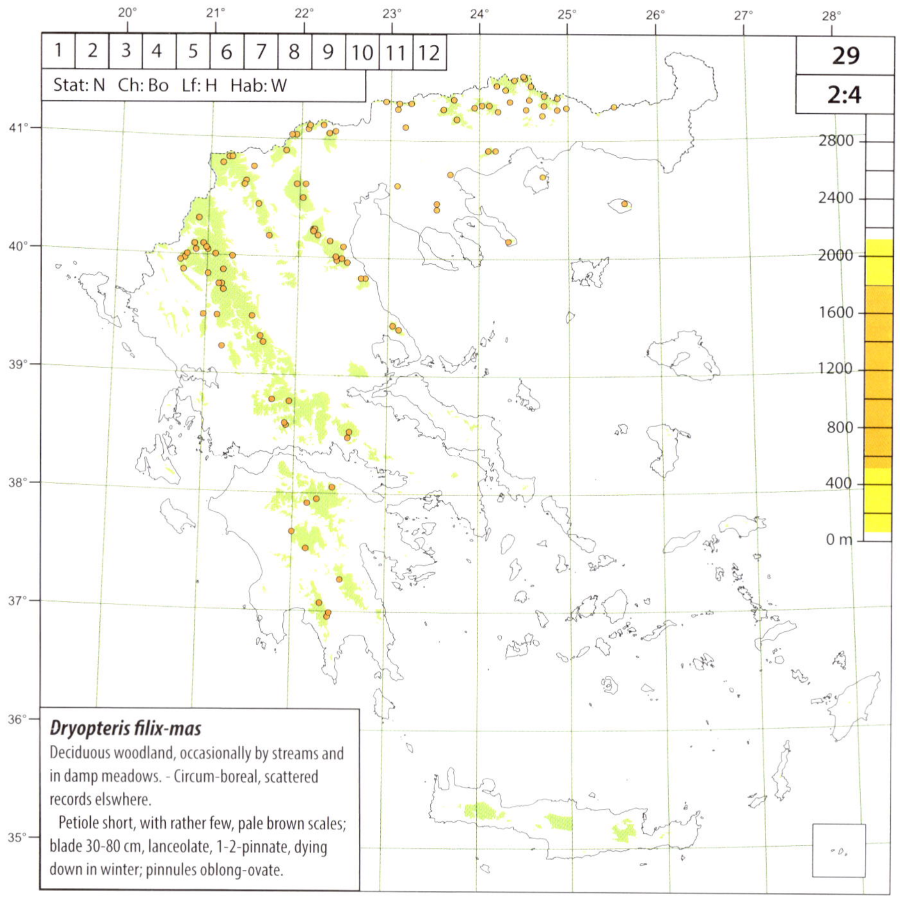

Dryopteris filix-mas
Deciduous woodland, occasionally by streams and in damp meadows. - Circum-boreal, scattered records elswhere.
Petiole short, with rather few, pale brown scales; blade 30-80 cm, lanceolate, 1-2-pinnate, dying down in winter; pinnules oblong-ovate.

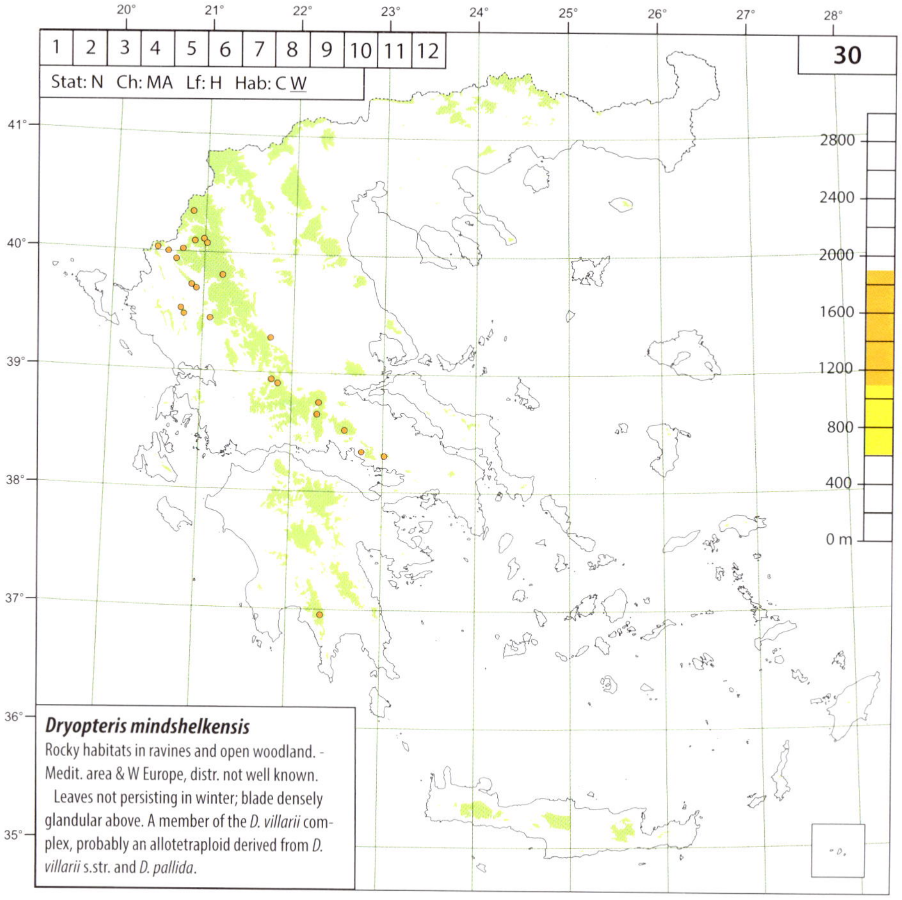

Dryopteris mindshelkensis
Rocky habitats in ravines and open woodland. - Medit. area & W Europe, distr. not well known.
Leaves not persisting in winter; blade densely glandular above. A member of the *D. villarii* complex, probably an allotetraploid derived from *D. villarii* s.str. and *D. pallida*.

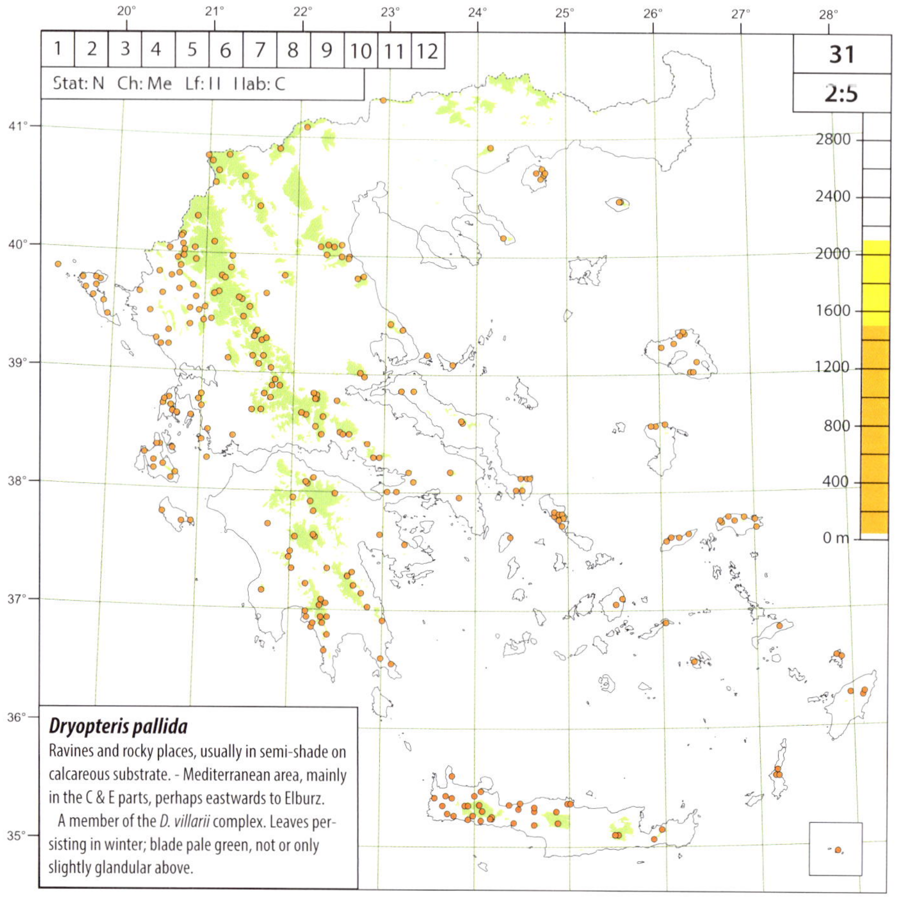

Dryopteris pallida
Ravines and rocky places, usually in semi-shade on calcareous substrate. - Mediterranean area, mainly in the C & E parts, perhaps eastwards to Elburz.
A member of the *D. villarii* complex. Leaves persisting in winter; blade pale green, not or only slightly glandular above.

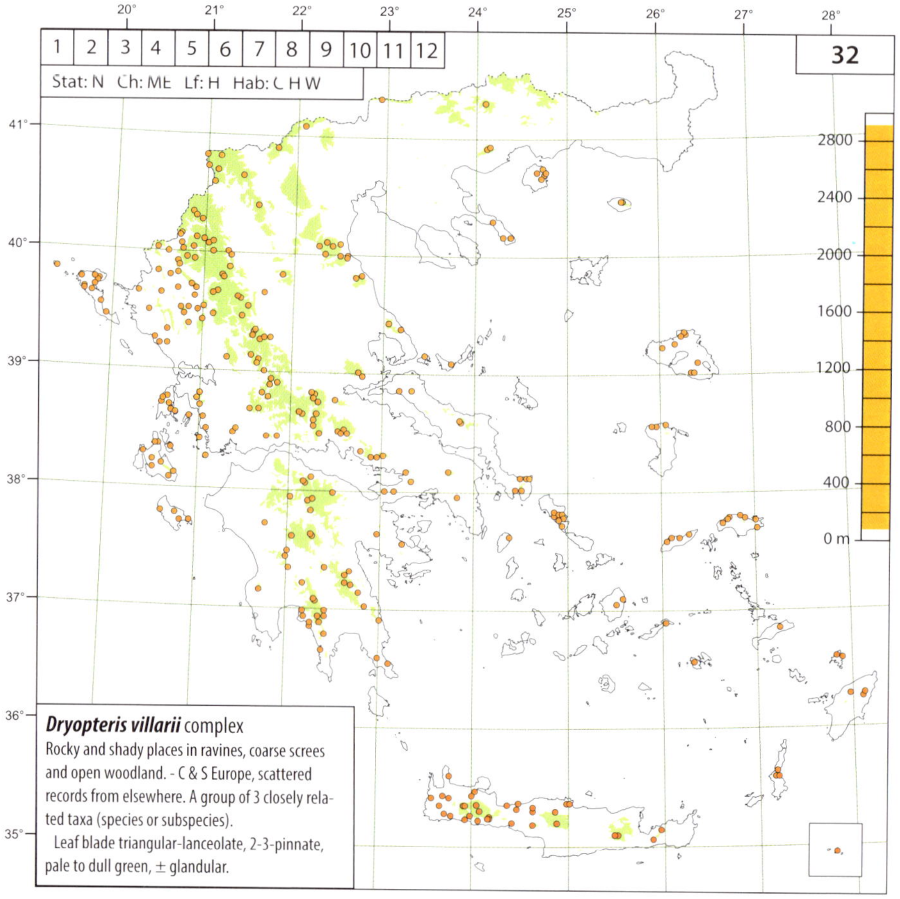

Dryopteris villarii complex
Rocky and shady places in ravines, coarse screes and open woodland. - C & S Europe, scattered records from elsewhere. A group of 3 closely related taxa (species or subspecies).
Leaf blade triangular-lanceolate, 2-3-pinnate, pale to dull green, ± glandular.

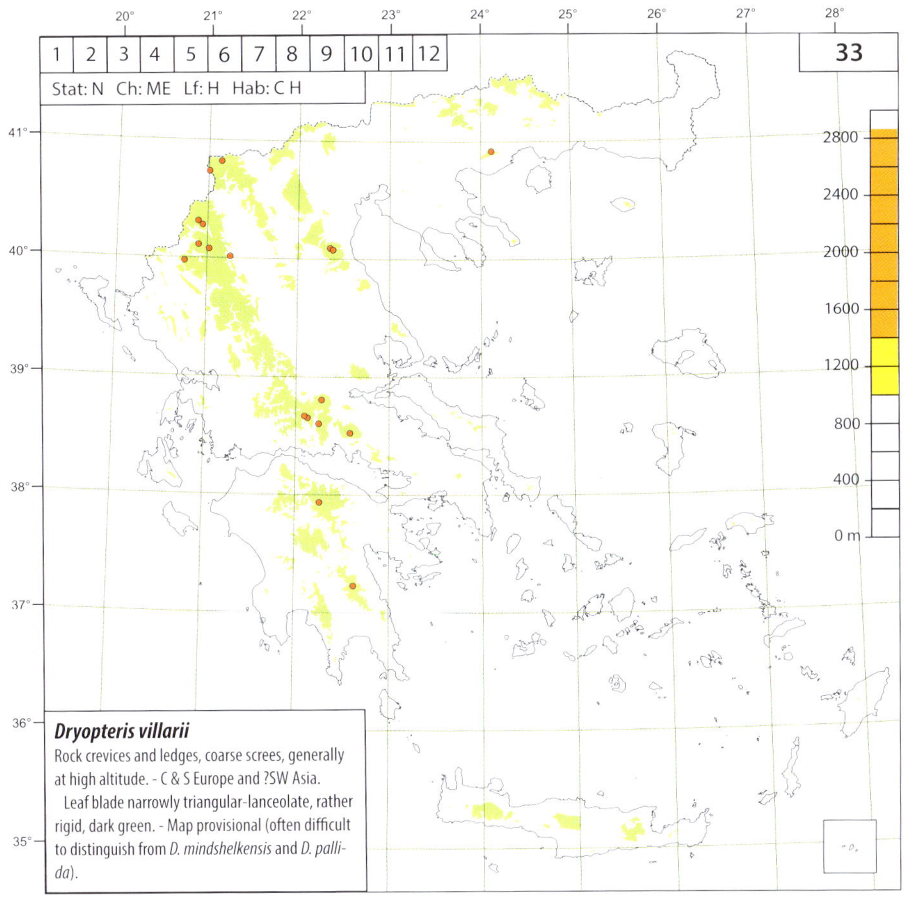

Dryopteris villarii

Rock crevices and ledges, coarse screes, generally at high altitude. - C & S Europe and ?SW Asia.

Leaf blade narrowly triangular-lanceolate, rather rigid, dark green. - Map provisional (often difficult to distinguish from *D. mindshelkensis* and *D. pallida*).

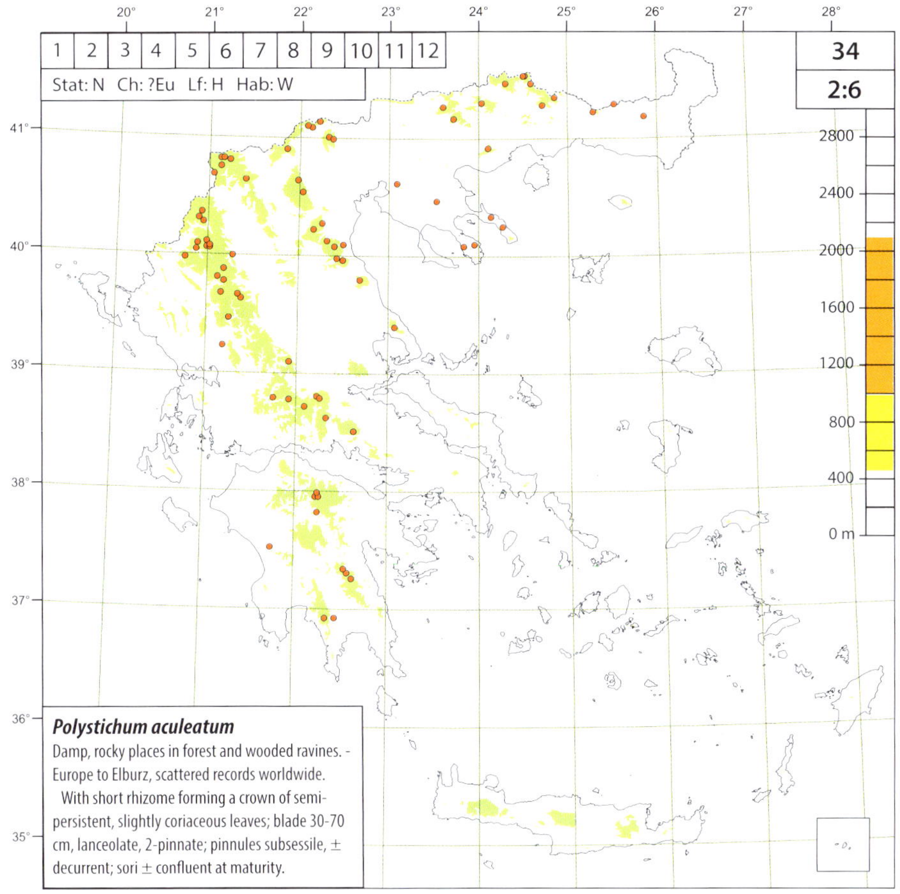

Polystichum aculeatum

Damp, rocky places in forest and wooded ravines. - Europe to Elburz, scattered records worldwide.

With short rhizome forming a crown of semi-persistent, slightly coriaceous leaves; blade 30-70 cm, lanceolate, 2-pinnate; pinnules subsessile, ± decurrent; sori ± confluent at maturity.

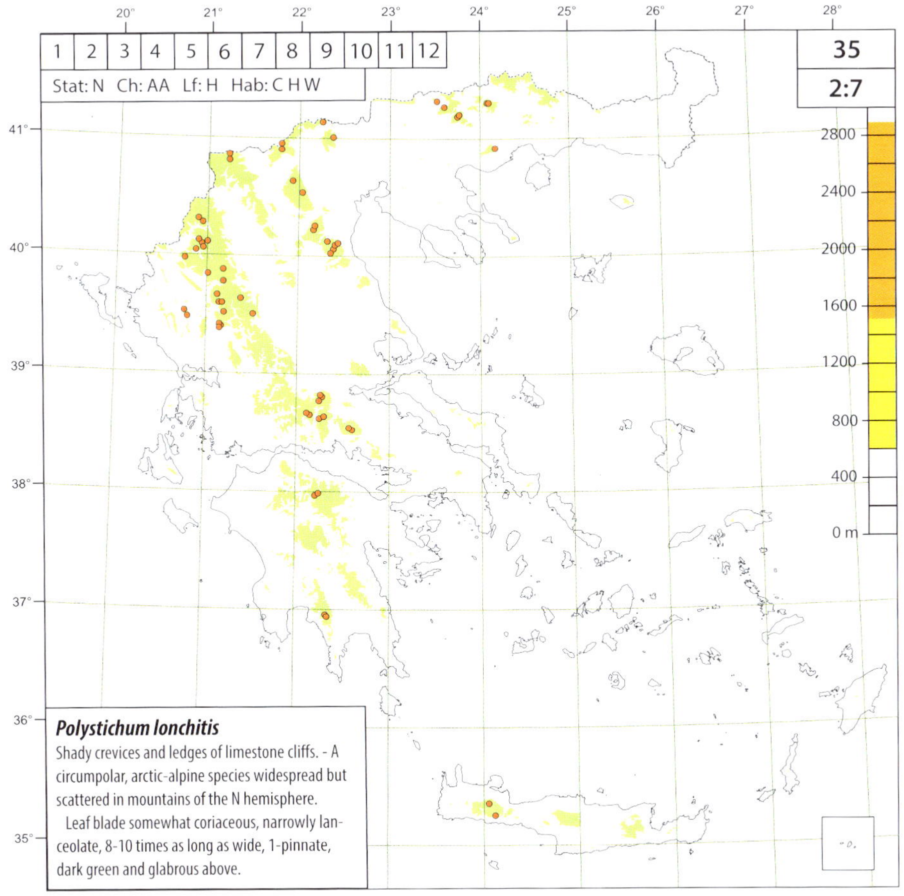

Polystichum lonchitis

Shady crevices and ledges of limestone cliffs. - A circumpolar, arctic-alpine species widespread but scattered in mountains of the N hemisphere.

Leaf blade somewhat coriaceous, narrowly lanceolate, 8-10 times as long as wide, 1-pinnate, dark green and glabrous above.

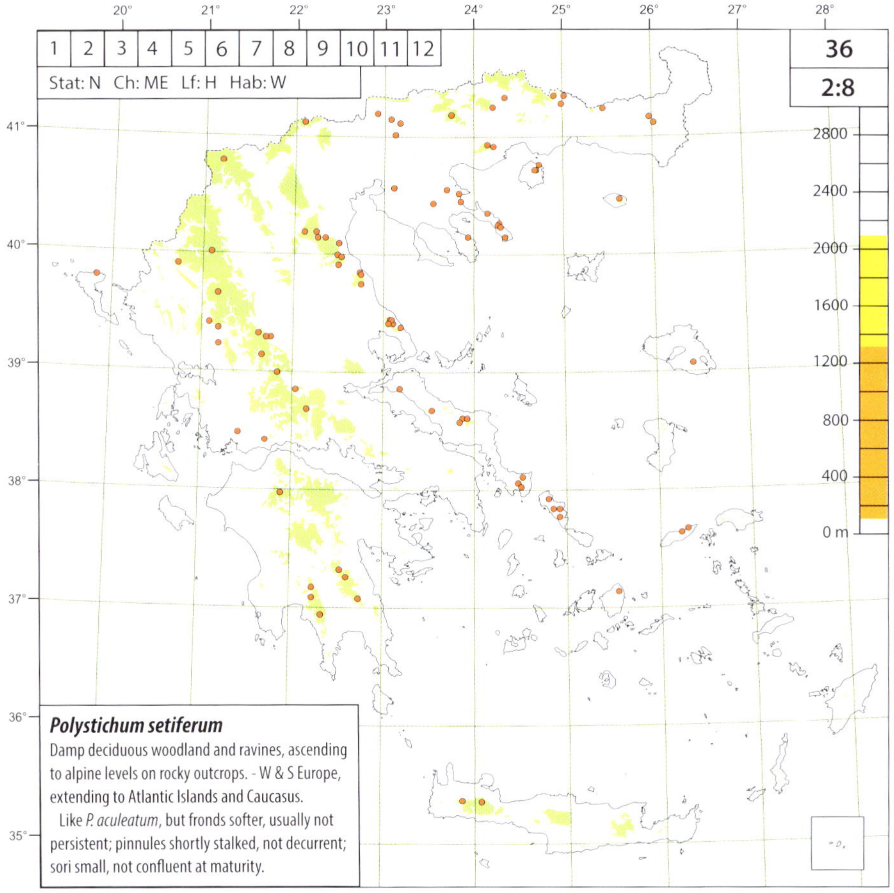

Polystichum setiferum

Damp deciduous woodland and ravines, ascending to alpine levels on rocky outcrops. - W & S Europe, extending to Atlantic Islands and Caucasus.

Like *P. aculeatum*, but fronds softer, usually not persistent; pinnules shortly stalked, not decurrent; sori small, not confluent at maturity.

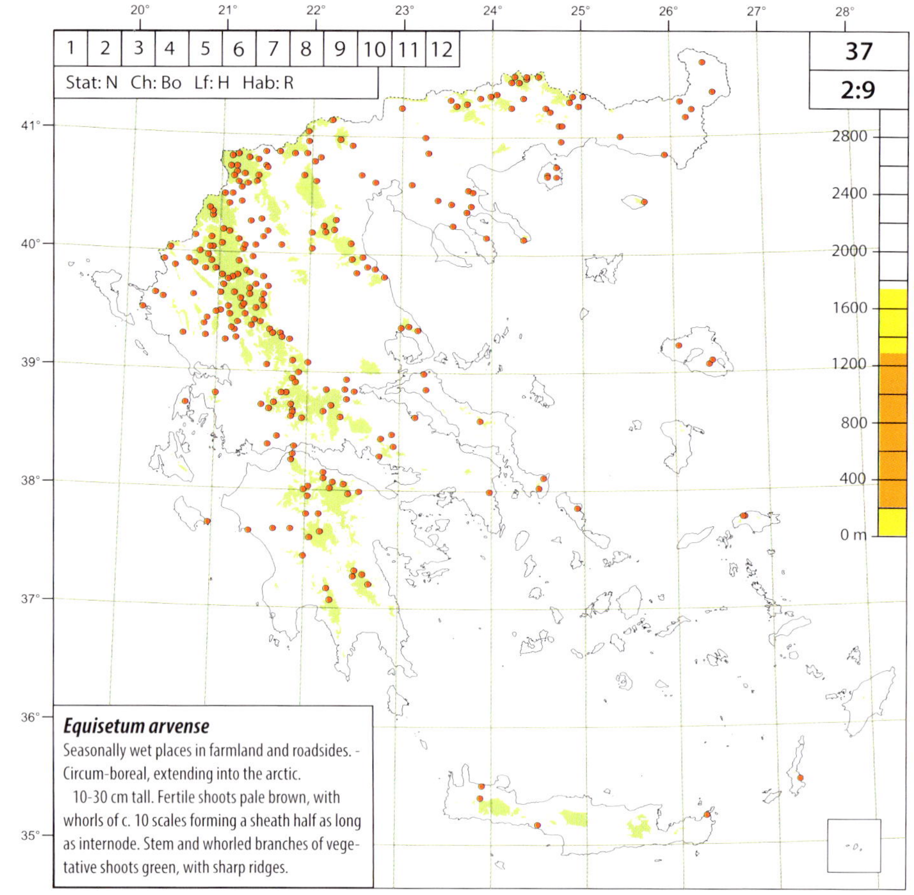

Equisetum arvense
Seasonally wet places in farmland and roadsides. - Circum-boreal, extending into the arctic.
10-30 cm tall. Fertile shoots pale brown, with whorls of c. 10 scales forming a sheath half as long as internode. Stem and whorled branches of vegetative shoots green, with sharp ridges.

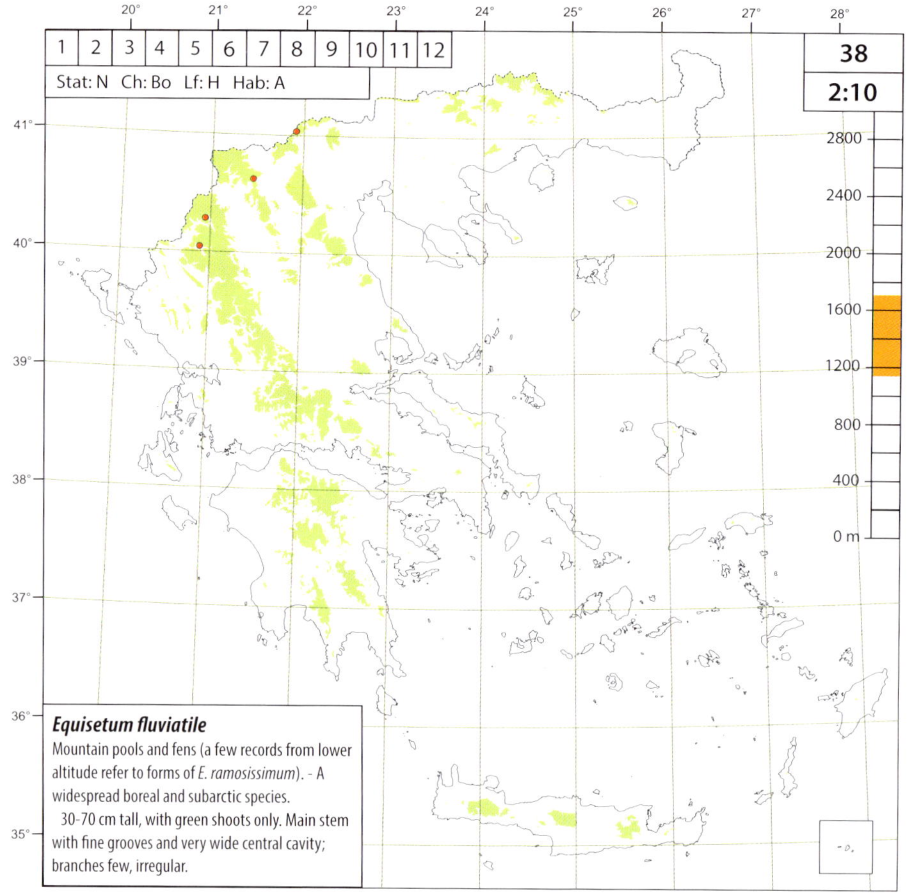

Equisetum fluviatile
Mountain pools and fens (a few records from lower altitude refer to forms of *E. ramosissimum*). - A widespread boreal and subarctic species.
30-70 cm tall, with green shoots only. Main stem with fine grooves and very wide central cavity; branches few, irregular.

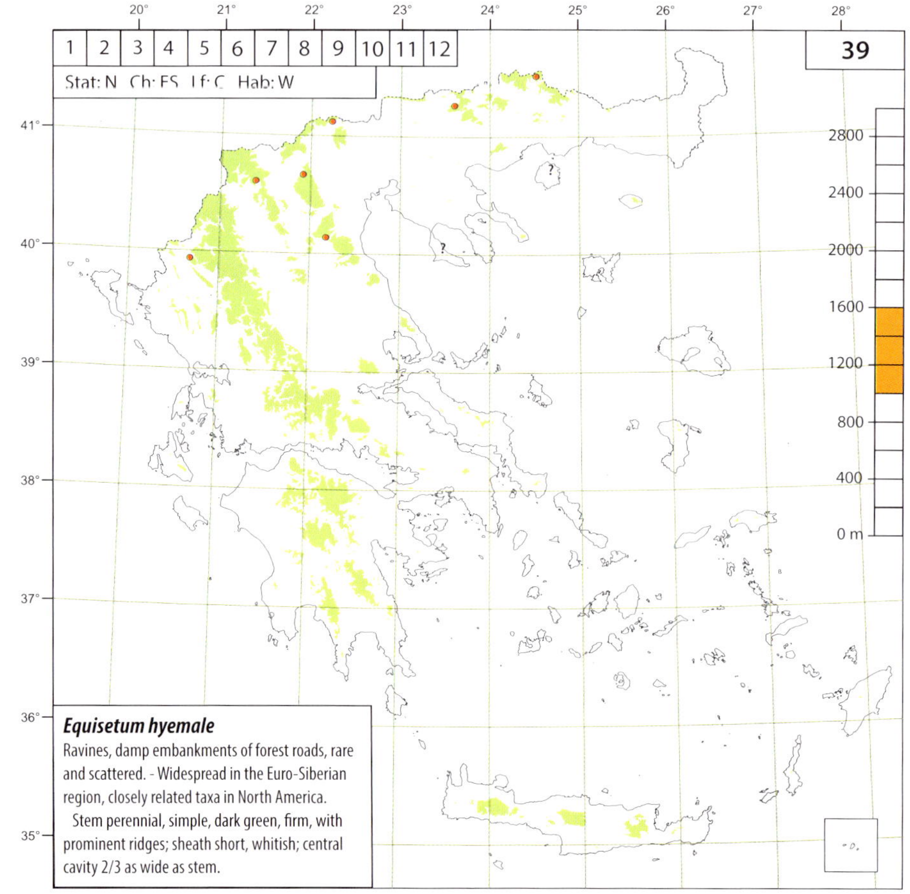

Equisetum hyemale
Ravines, damp embankments of forest roads, rare and scattered. - Widespread in the Euro-Siberian region, closely related taxa in North America.
Stem perennial, simple, dark green, firm, with prominent ridges; sheath short, whitish; central cavity 2/3 as wide as stem.

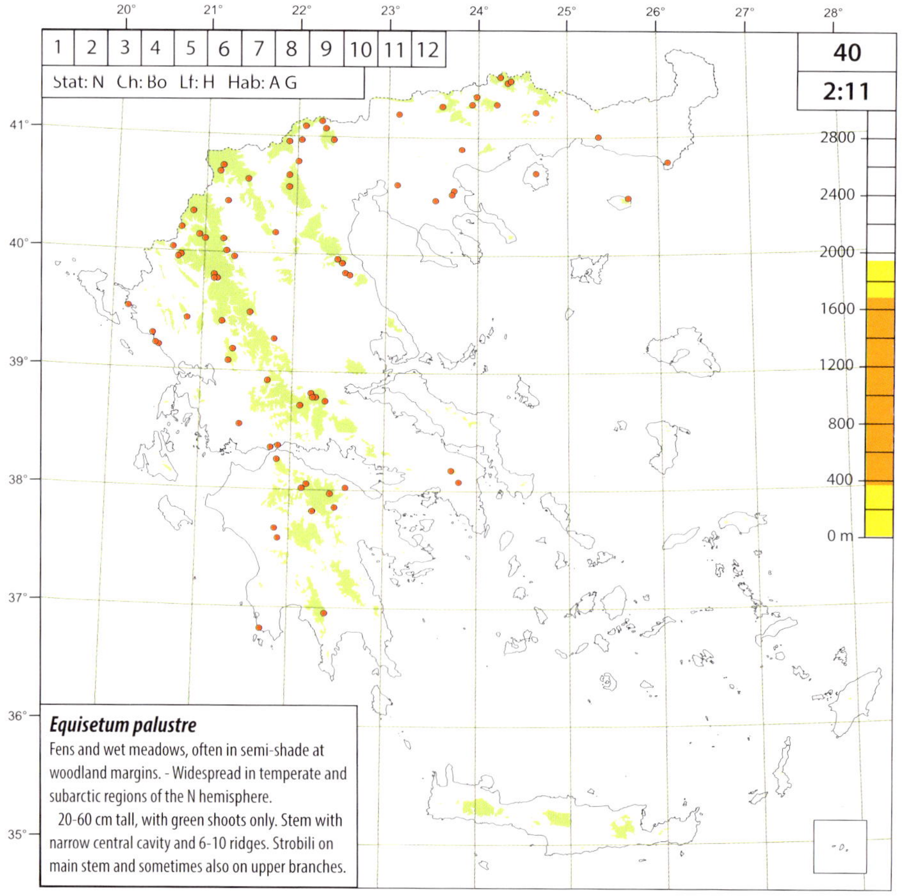

Equisetum palustre
Fens and wet meadows, often in semi-shade at woodland margins. - Widespread in temperate and subarctic regions of the N hemisphere.
20-60 cm tall, with green shoots only. Stem with narrow central cavity and 6-10 ridges. Strobili on main stem and sometimes also on upper branches.

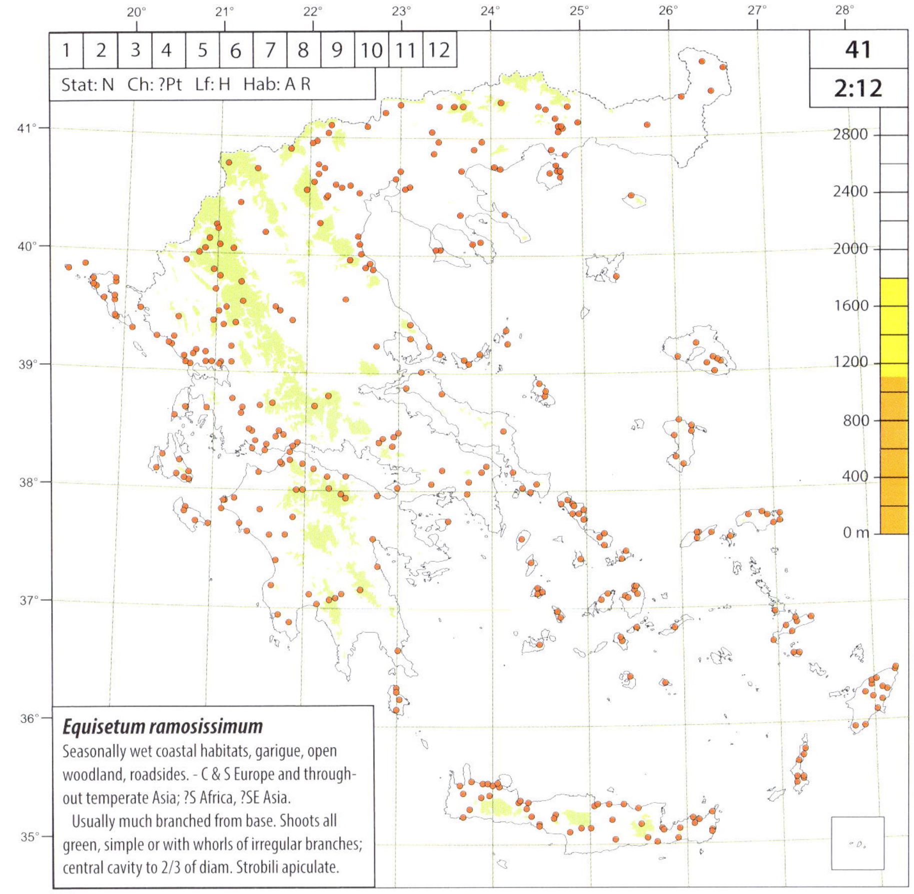

Equisetum ramosissimum

Seasonally wet coastal habitats, garigue, open woodland, roadsides. - C & S Europe and throughout temperate Asia; ?S Africa, ?SE Asia.

Usually much branched from base. Shoots all green, simple or with whorls of irregular branches; central cavity to 2/3 of diam. Strobili apiculate.

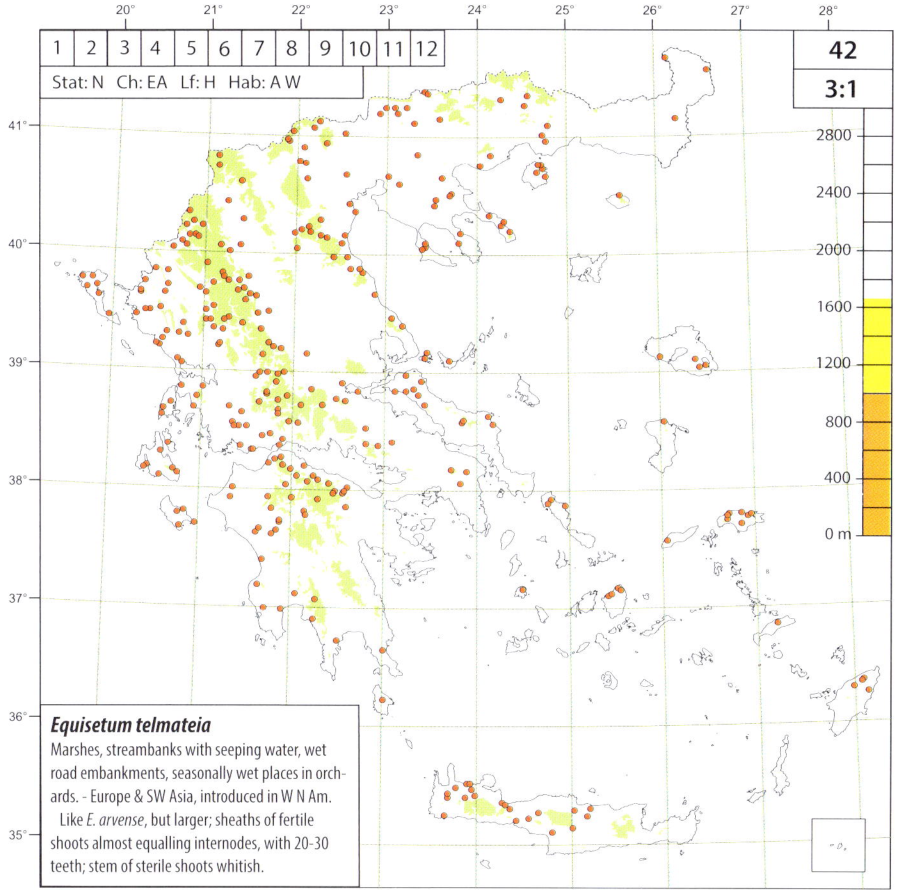

Equisetum telmateia

Marshes, streambanks with seeping water, wet road embankments, seasonally wet places in orchards. - Europe & SW Asia, introduced in W N Am.

Like *E. arvense*, but larger; sheaths of fertile shoots almost equalling internodes, with 20-30 teeth; stem of sterile shoots whitish.

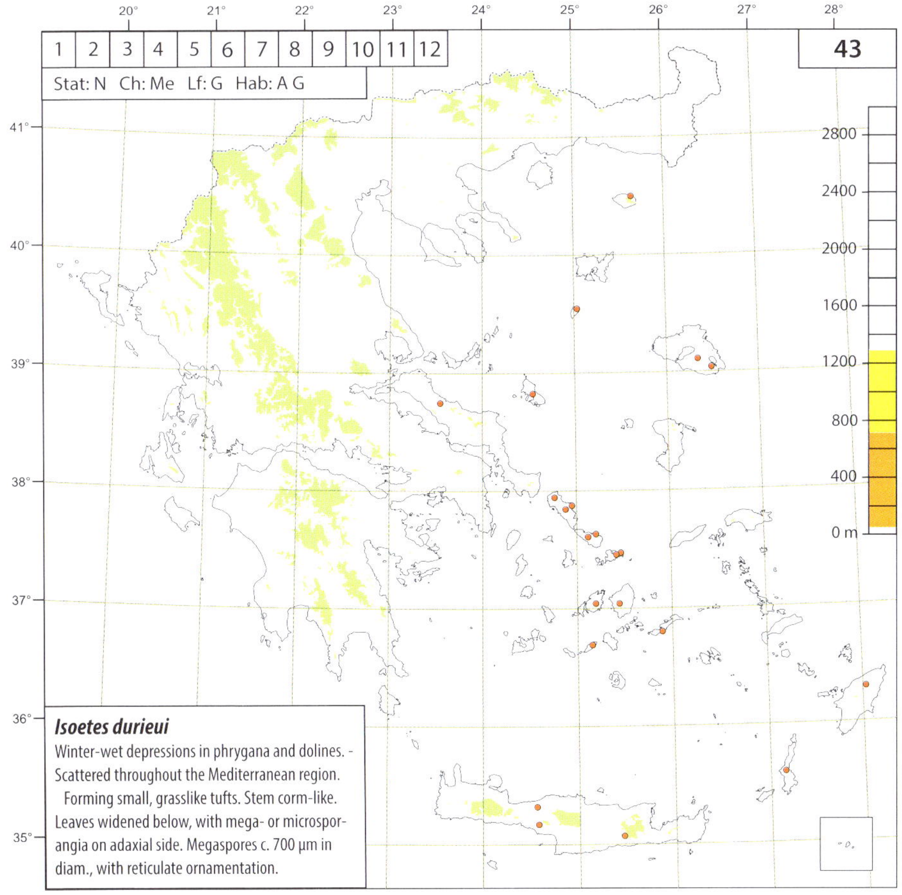

Isoetes durieui

Winter-wet depressions in phrygana and dolines. - Scattered throughout the Mediterranean region.

Forming small, grasslike tufts. Stem corm-like. Leaves widened below, with mega- or microsporangia on adaxial side. Megaspores c. 700 µm in diam., with reticulate ornamentation.

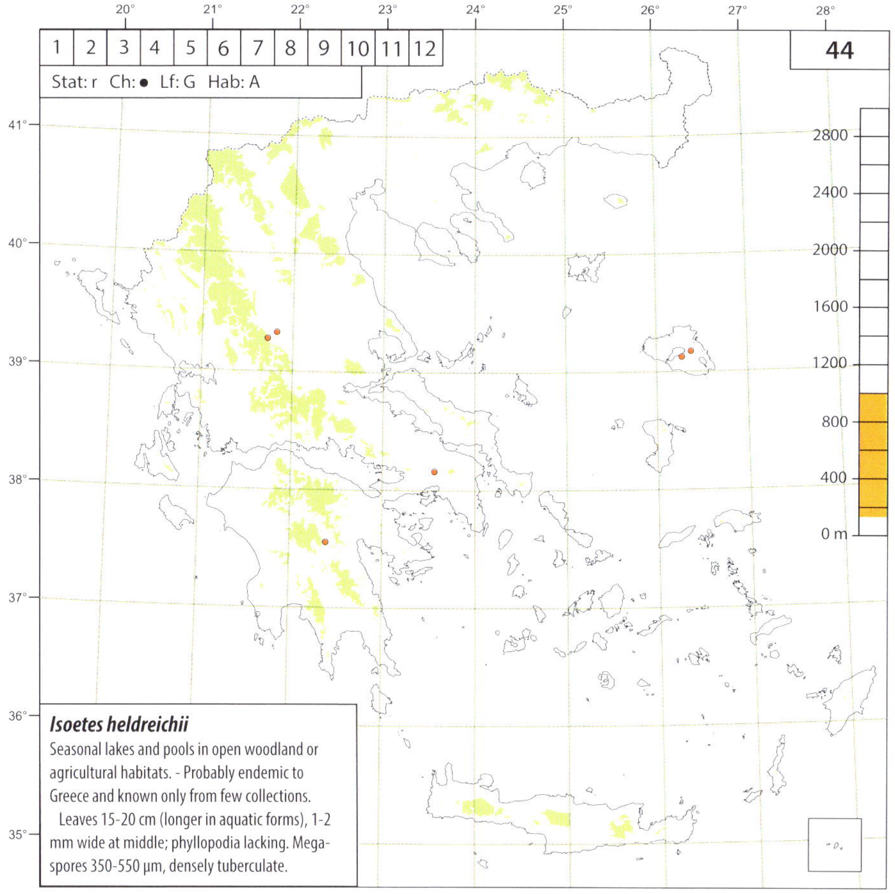

Isoetes heldreichii

Seasonal lakes and pools in open woodland or agricultural habitats. - Probably endemic to Greece and known only from few collections.

Leaves 15-20 cm (longer in aquatic forms), 1-2 mm wide at middle; phyllopodia lacking. Megaspores 350-550 µm, densely tuberculate.

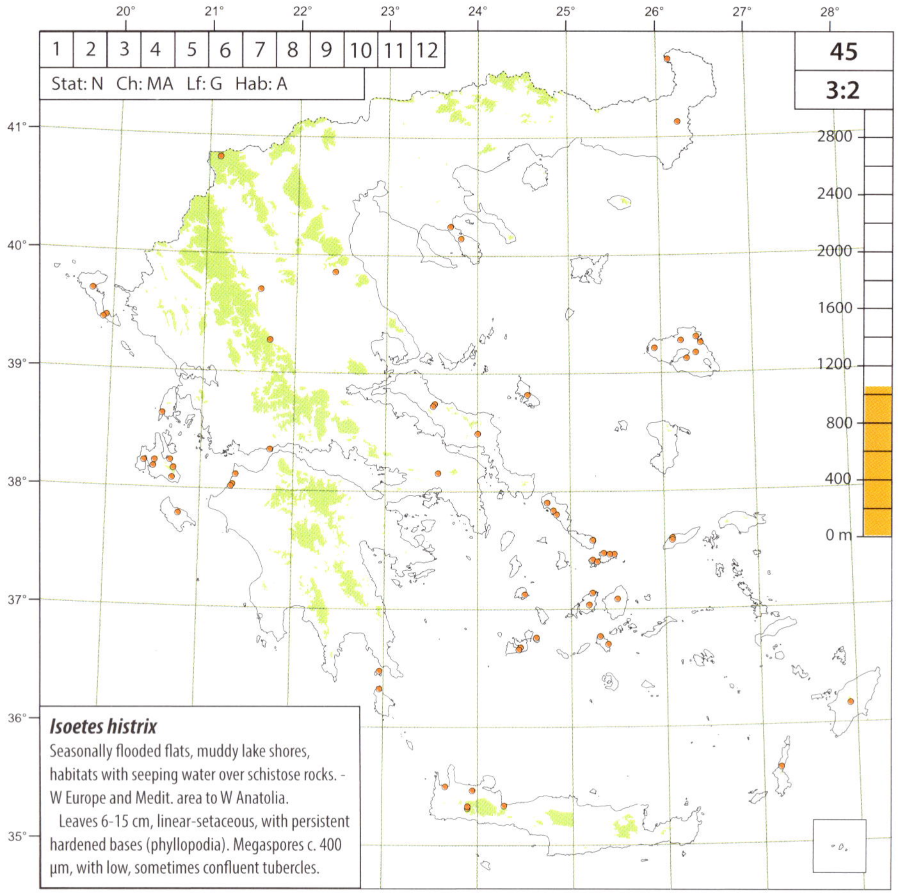

Isoetes histrix

Seasonally flooded flats, muddy lake shores, habitats with seeping water over schistose rocks. - W Europe and Medit. area to W Anatolia.

Leaves 6-15 cm, linear-setaceous, with persistent hardened bases (phyllopodia). Megaspores c. 400 µm, with low, sometimes confluent tubercles.

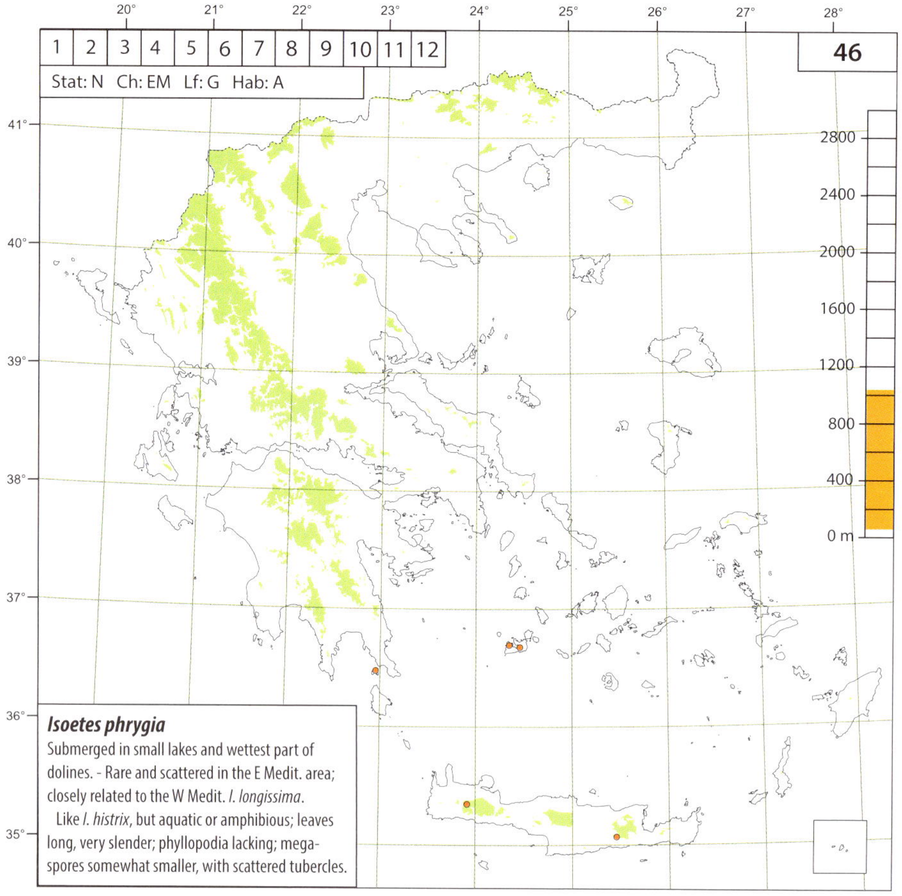

Isoetes phrygia

Submerged in small lakes and wettest part of dolines. - Rare and scattered in the E Medit. area; closely related to the W Medit. *I. longissima*.

Like *I. histrix*, but aquatic or amphibious; leaves long, very slender; phyllopodia lacking; megaspores somewhat smaller, with scattered tubercles.

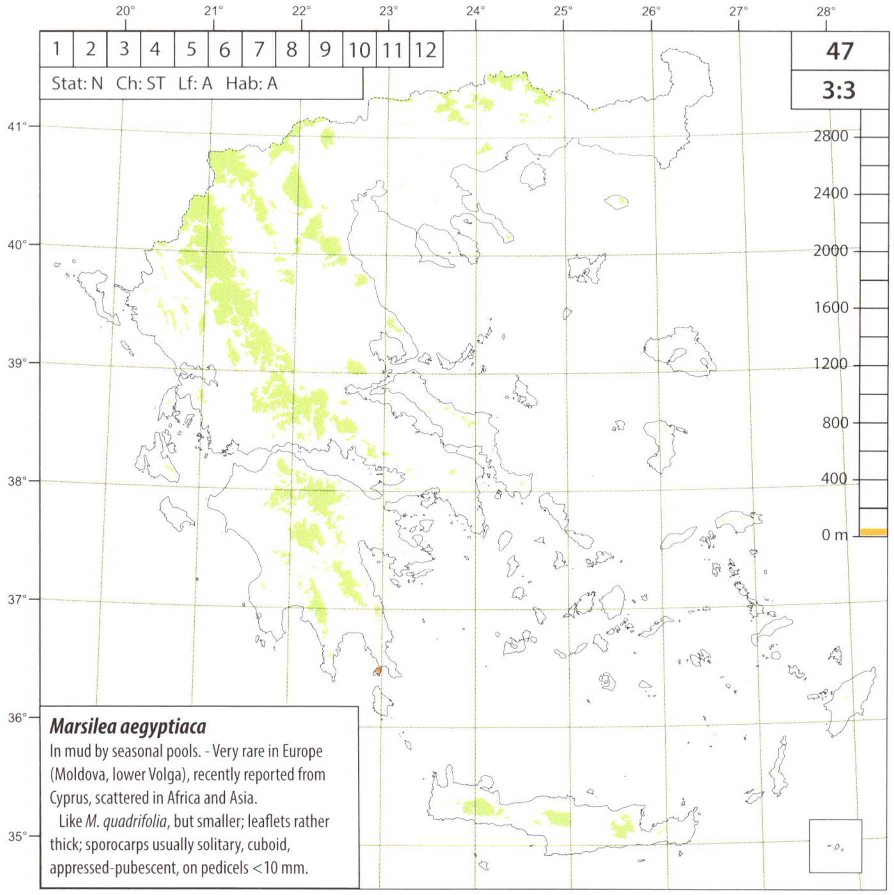

Marsilea aegyptiaca

In mud by seasonal pools. - Very rare in Europe (Moldova, lower Volga), recently reported from Cyprus, scattered in Africa and Asia.

Like *M. quadrifolia*, but smaller; leaflets rather thick; sporocarps usually solitary, cuboid, appressed-pubescent, on pedicels <10 mm.

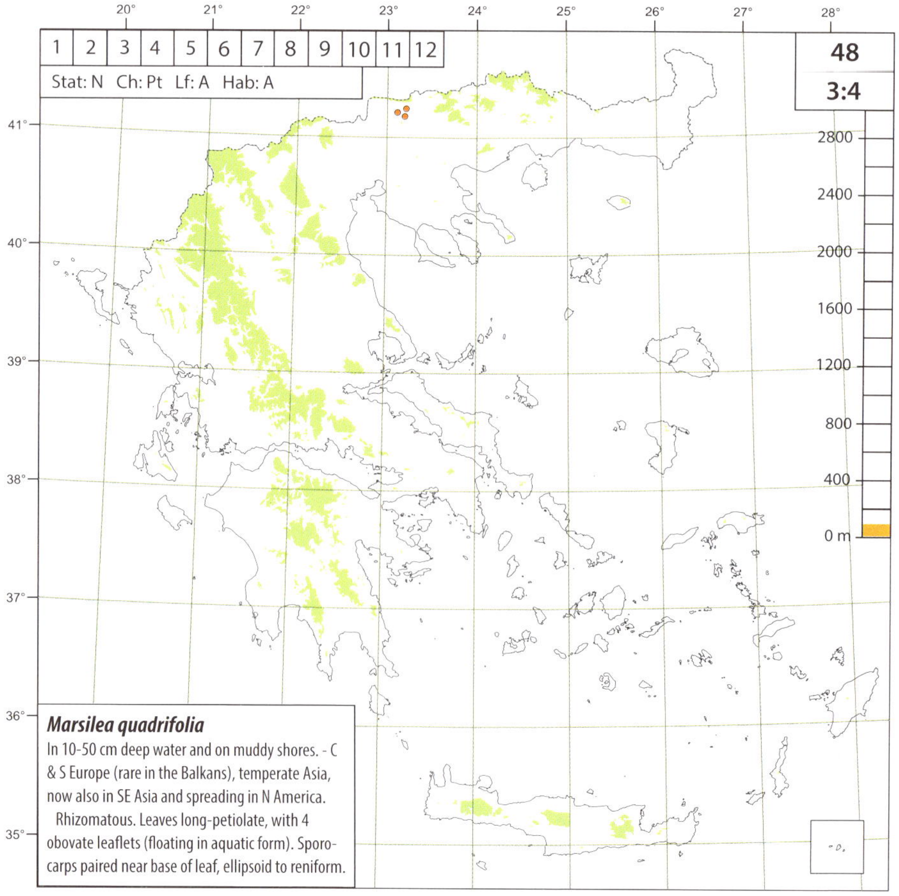

Marsilea quadrifolia

In 10-50 cm deep water and on muddy shores. - C & S Europe (rare in the Balkans), temperate Asia, now also in SE Asia and spreading in N America.

Rhizomatous. Leaves long-petiolate, with 4 obovate leaflets (floating in aquatic form). Sporocarps paired near base of leaf, ellipsoid to reniform.

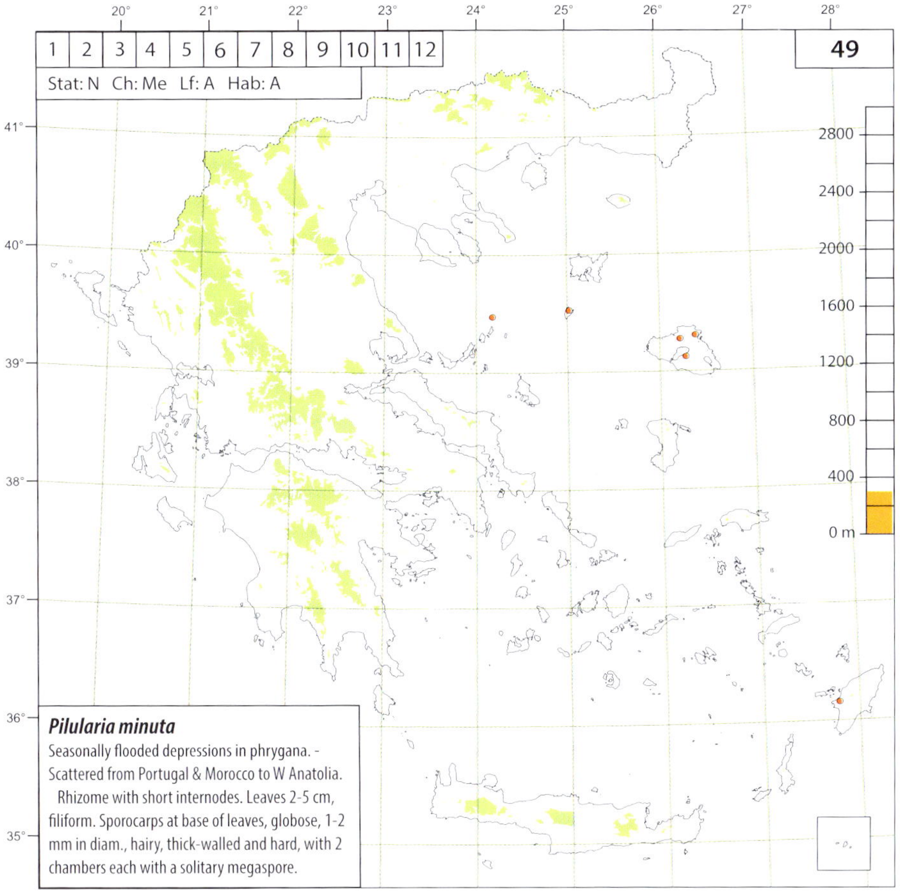

Pilularia minuta

Seasonally flooded depressions in phrygana. - Scattered from Portugal & Morocco to W Anatolia.

Rhizome with short internodes. Leaves 2-5 cm, filiform. Sporocarps at base of leaves, globose, 1-2 mm in diam., hairy, thick-walled and hard, with 2 chambers each with a solitary megaspore.

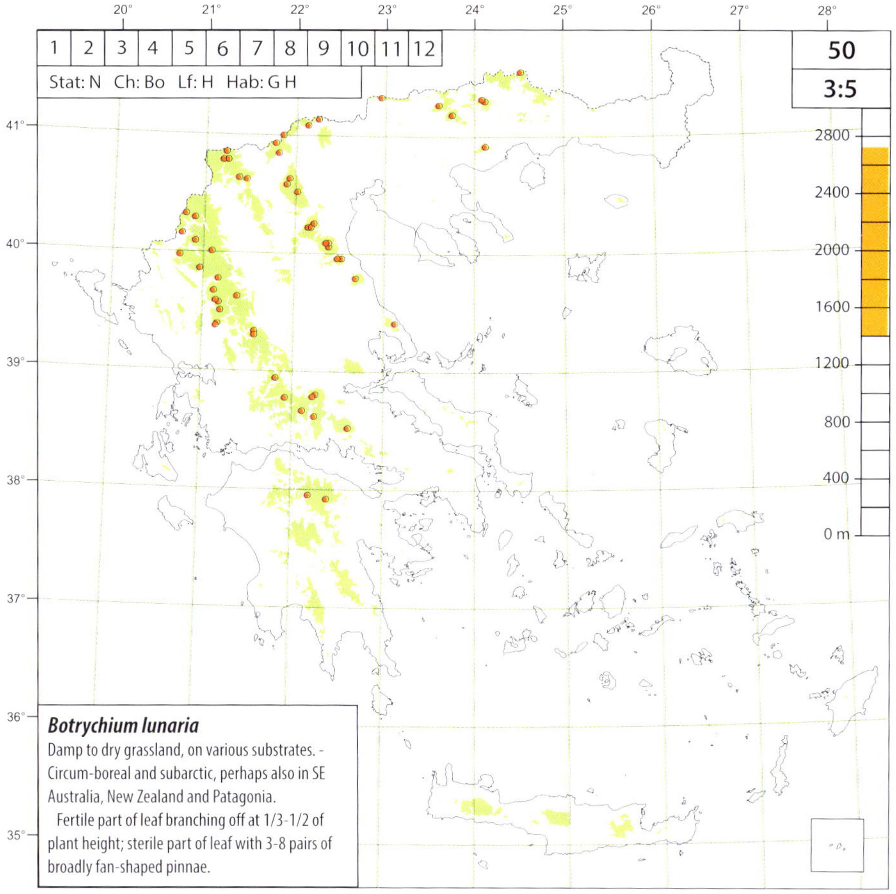

Botrychium lunaria

Damp to dry grassland, on various substrates. - Circum-boreal and subarctic, perhaps also in SE Australia, New Zealand and Patagonia.

Fertile part of leaf branching off at 1/3-1/2 of plant height; sterile part of leaf with 3-8 pairs of broadly fan-shaped pinnae.

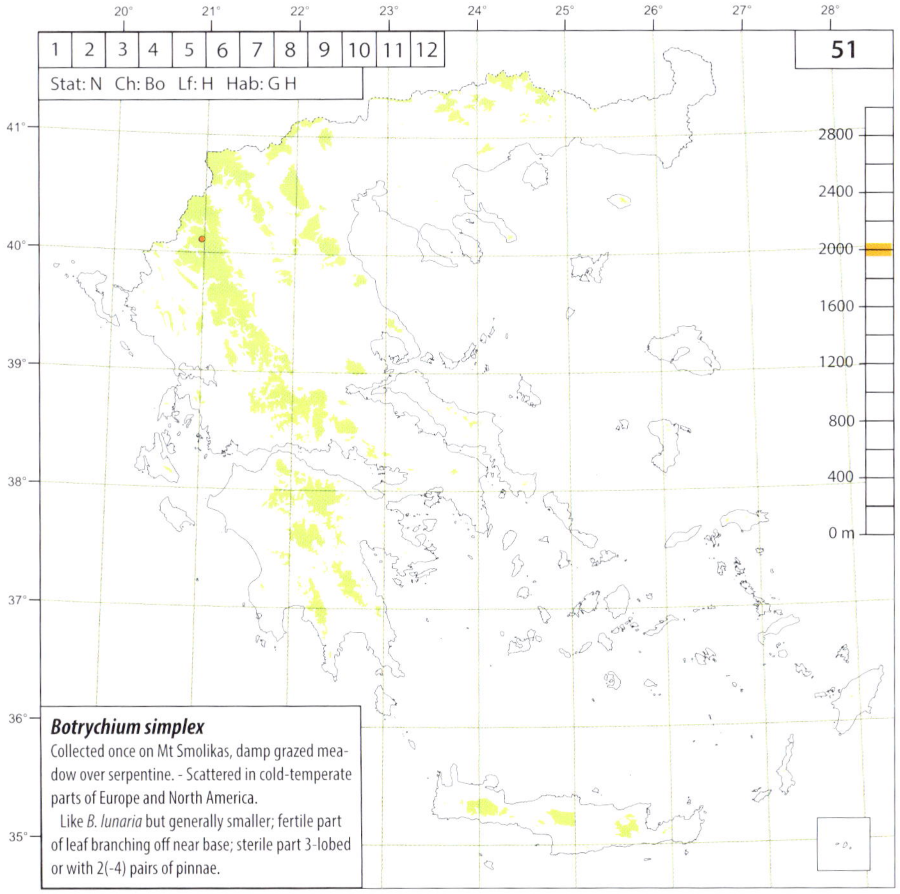

Botrychium simplex

Collected once on Mt Smolikas, damp grazed meadow over serpentine. - Scattered in cold-temperate parts of Europe and North America.

Like *B. lunaria* but generally smaller; fertile part of leaf branching off near base; sterile part 3-lobed or with 2(-4) pairs of pinnae.

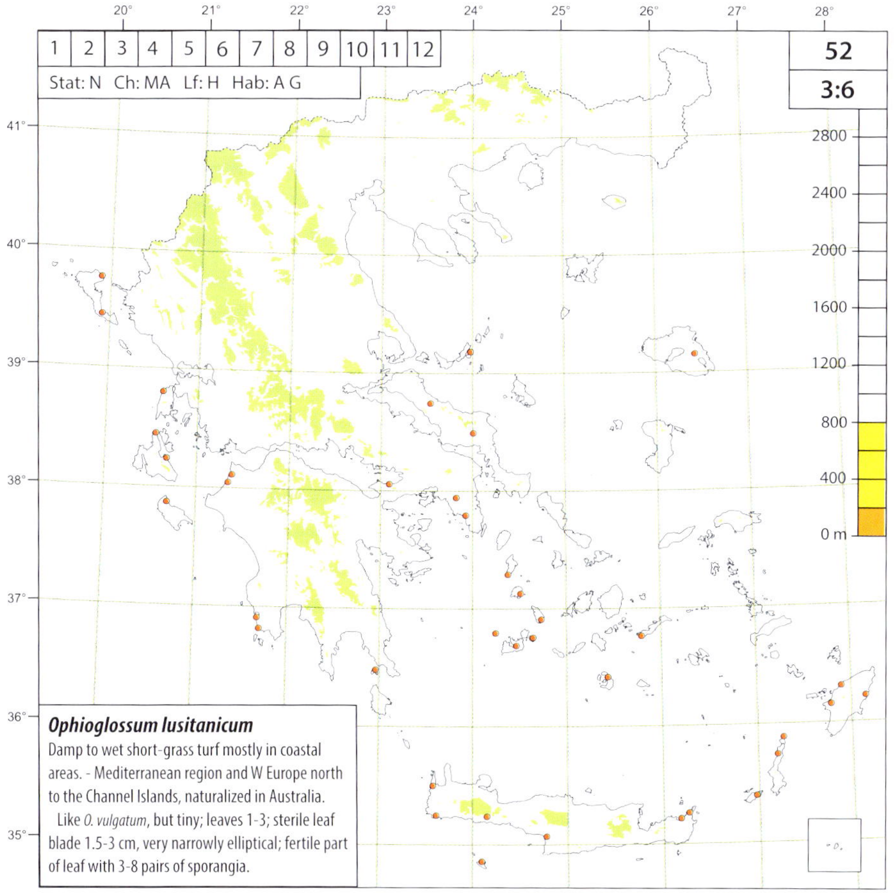

Ophioglossum lusitanicum

Damp to wet short-grass turf mostly in coastal areas. - Mediterranean region and W Europe north to the Channel Islands, naturalized in Australia.

Like *O. vulgatum*, but tiny; leaves 1-3; sterile leaf blade 1.5-3 cm, very narrowly elliptical; fertile part of leaf with 3-8 pairs of sporangia.

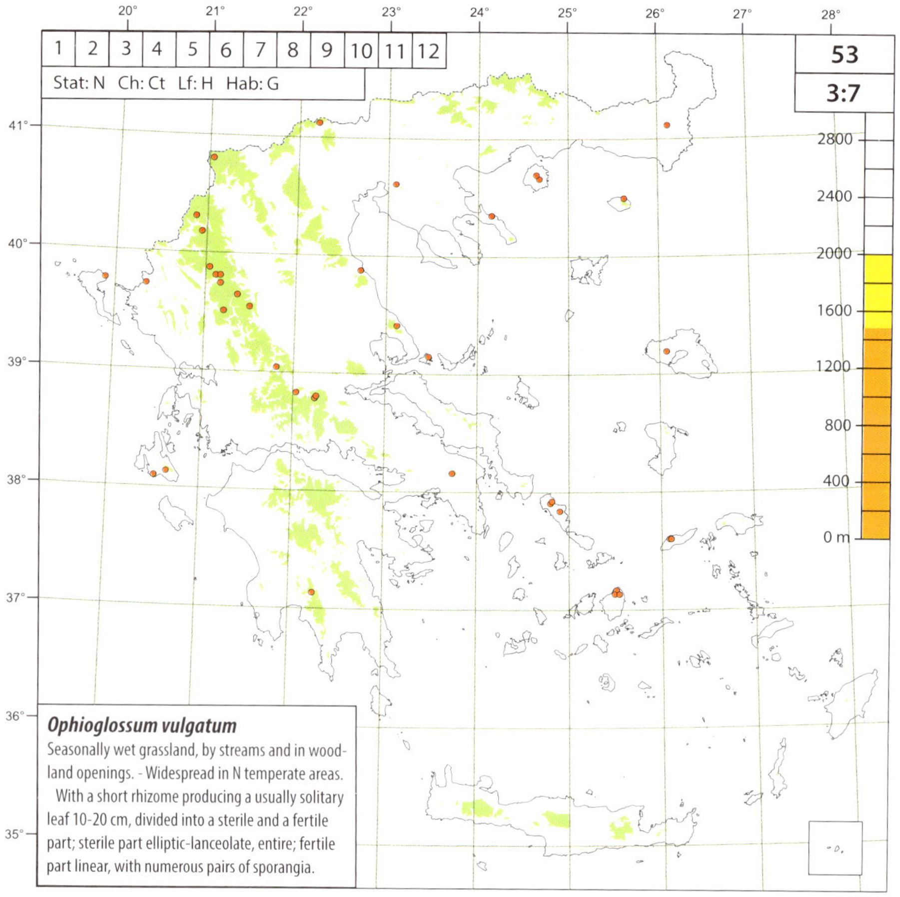

Ophioglossum vulgatum

Seasonally wet grassland, by streams and in woodland openings. - Widespread in N temperate areas.

With a short rhizome producing a usually solitary leaf 10-20 cm, divided into a sterile and a fertile part; sterile part elliptic-lanceolate, entire; fertile part linear, with numerous pairs of sporangia.

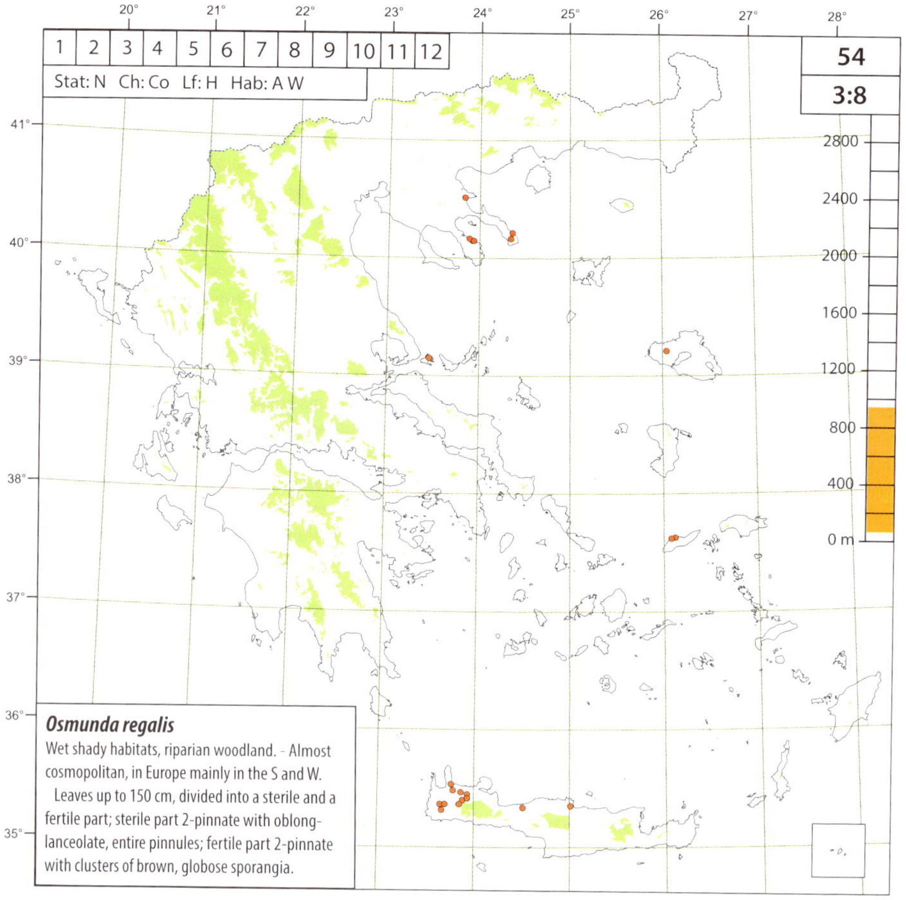

Osmunda regalis

Wet shady habitats, riparian woodland. - Almost cosmopolitan, in Europe mainly in the S and W.

Leaves up to 150 cm, divided into a sterile and a fertile part; sterile part 2-pinnate with oblong-lanceolate, entire pinnules; fertile part 2-pinnate with clusters of brown, globose sporangia.

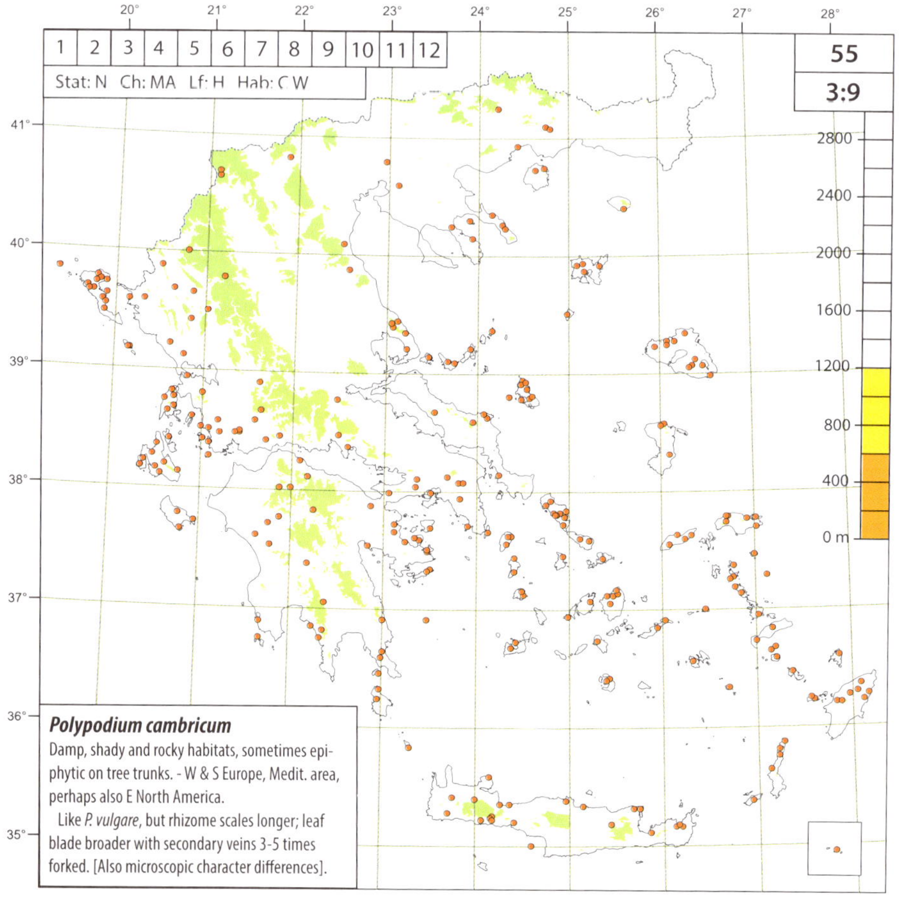

Polypodium cambricum

Damp, shady and rocky habitats, sometimes epiphytic on tree trunks. - W & S Europe, Medit. area, perhaps also E North America.

Like *P. vulgare*, but rhizome scales longer; leaf blade broader with secondary veins 3-5 times forked. [Also microscopic character differences].

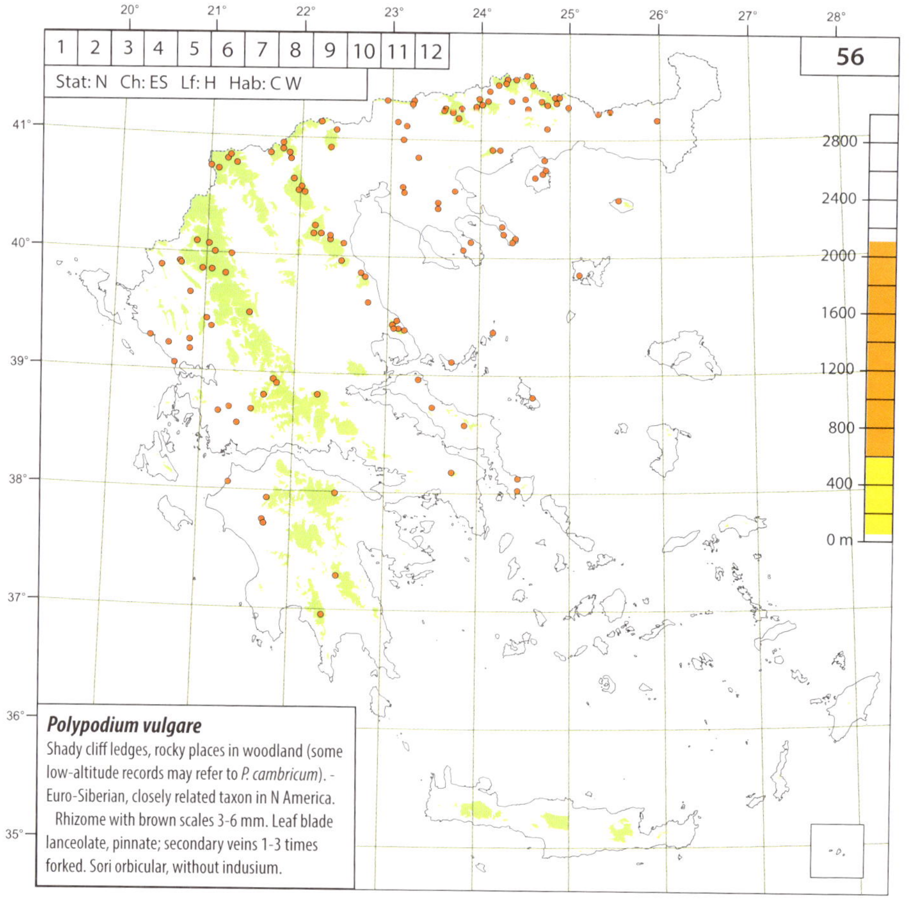

Polypodium vulgare

Shady cliff ledges, rocky places in woodland (some low-altitude records may refer to *P. cambricum*). - Euro-Siberian, closely related taxon in N America.

Rhizome with brown scales 3-6 mm. Leaf blade lanceolate, pinnate; secondary veins 1-3 times forked. Sori orbicular, without indusium.

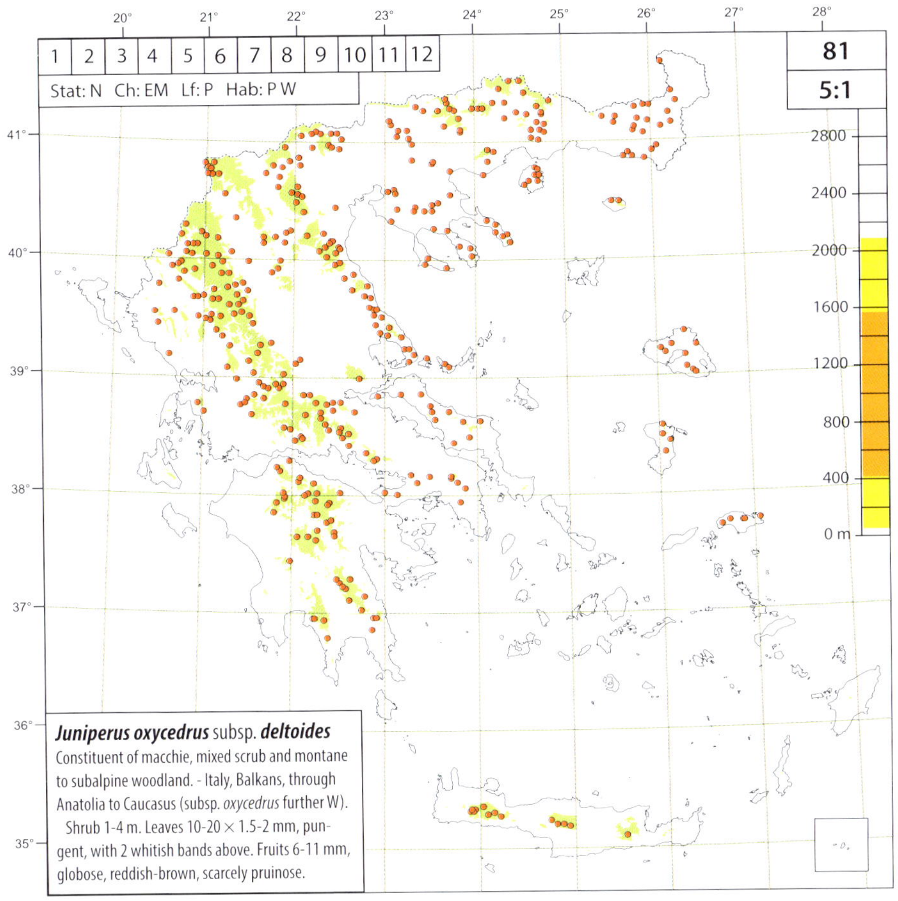

Juniperus oxycedrus **subsp. *deltoides***
Constituent of macchie, mixed scrub and montane to subalpine woodland. - Italy, Balkans, through Anatolia to Caucasus (subsp. *oxycedrus* further W).
Shrub 1-4 m. Leaves 10-20 × 1.5-2 mm, pungent, with 2 whitish bands above. Fruits 6-11 mm, globose, reddish-brown, scarcely pruinose.

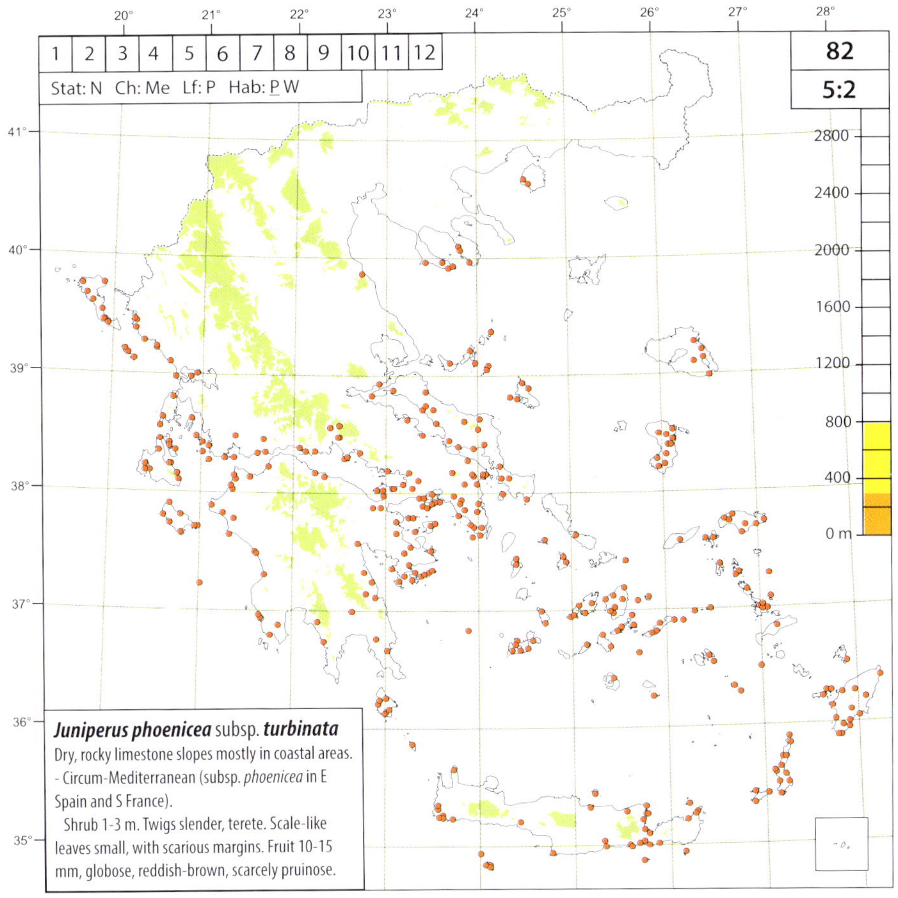

Juniperus phoenicea **subsp. *turbinata***
Dry, rocky limestone slopes mostly in coastal areas. - Circum-Mediterranean (subsp. *phoenicea* in E Spain and S France).
Shrub 1-3 m. Twigs slender, terete. Scale-like leaves small, with scarious margins. Fruit 10-15 mm, globose, reddish-brown, scarcely pruinose.

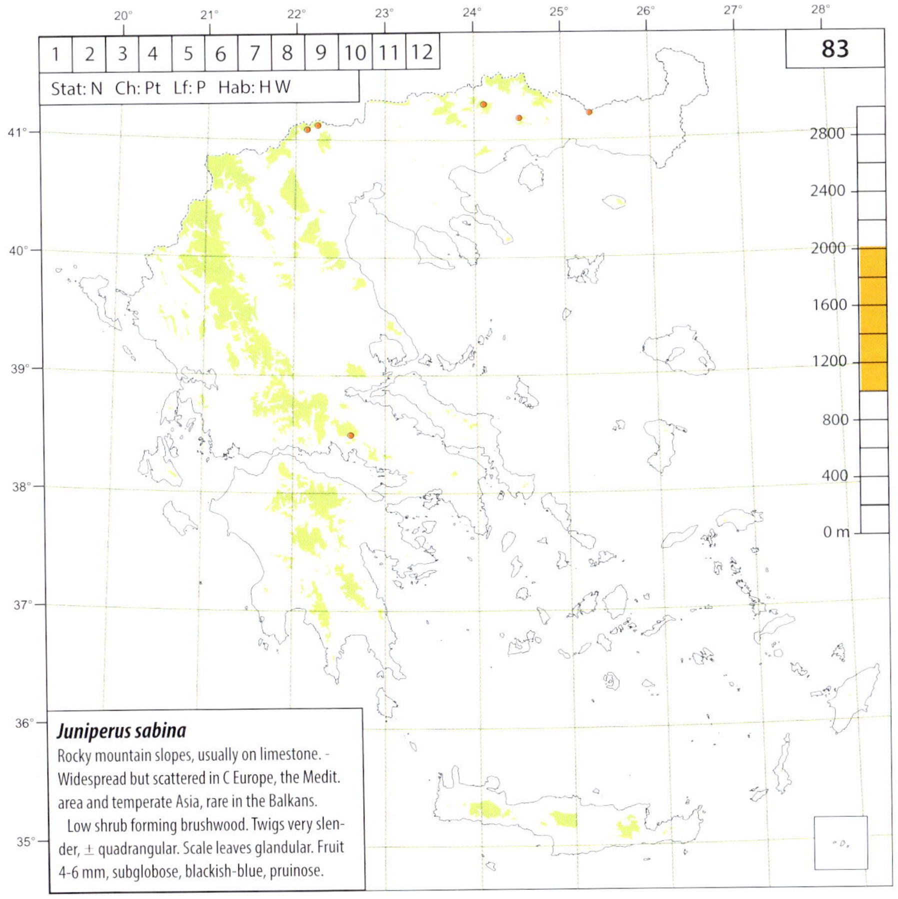

Juniperus sabina
Rocky mountain slopes, usually on limestone. - Widespread but scattered in C Europe, the Medit. area and temperate Asia, rare in the Balkans.
Low shrub forming brushwood. Twigs very slender, ± quadrangular. Scale leaves glandular. Fruit 4-6 mm, subglobose, blackish-blue, pruinose.

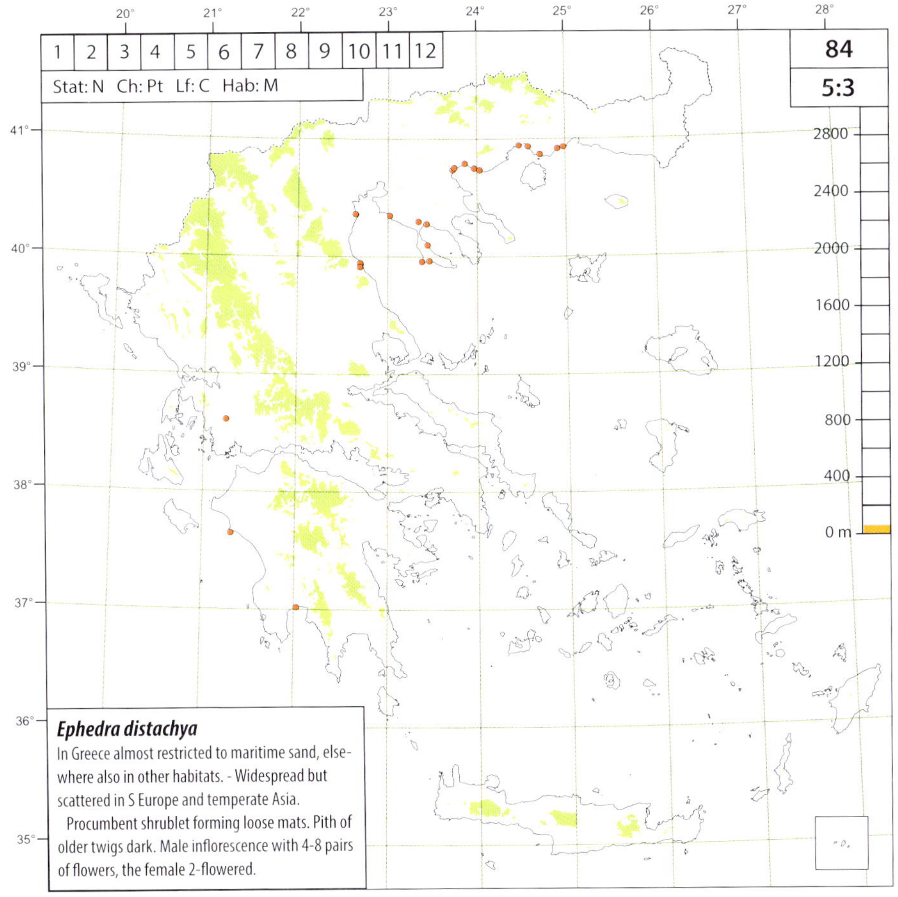

Ephedra distachya
In Greece almost restricted to maritime sand, elsewhere also in other habitats. - Widespread but scattered in S Europe and temperate Asia.
Procumbent shrublet forming loose mats. Pith of older twigs dark. Male inflorescence with 4-8 pairs of flowers, the female 2-flowered.

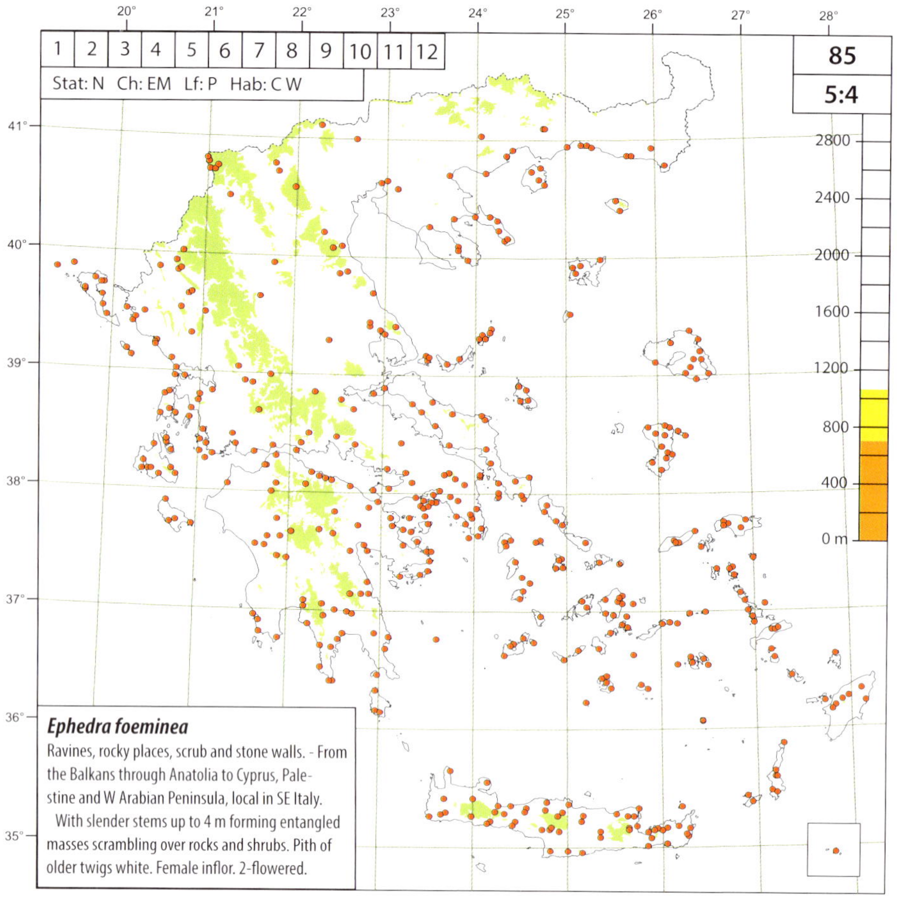

Ephedra foeminea

Ravines, rocky places, scrub and stone walls. - From the Balkans through Anatolia to Cyprus, Palestine and W Arabian Peninsula, local in SE Italy.

With slender stems up to 4 m forming entangled masses scrambling over rocks and shrubs. Pith of older twigs white. Female inflor. 2-flowered.

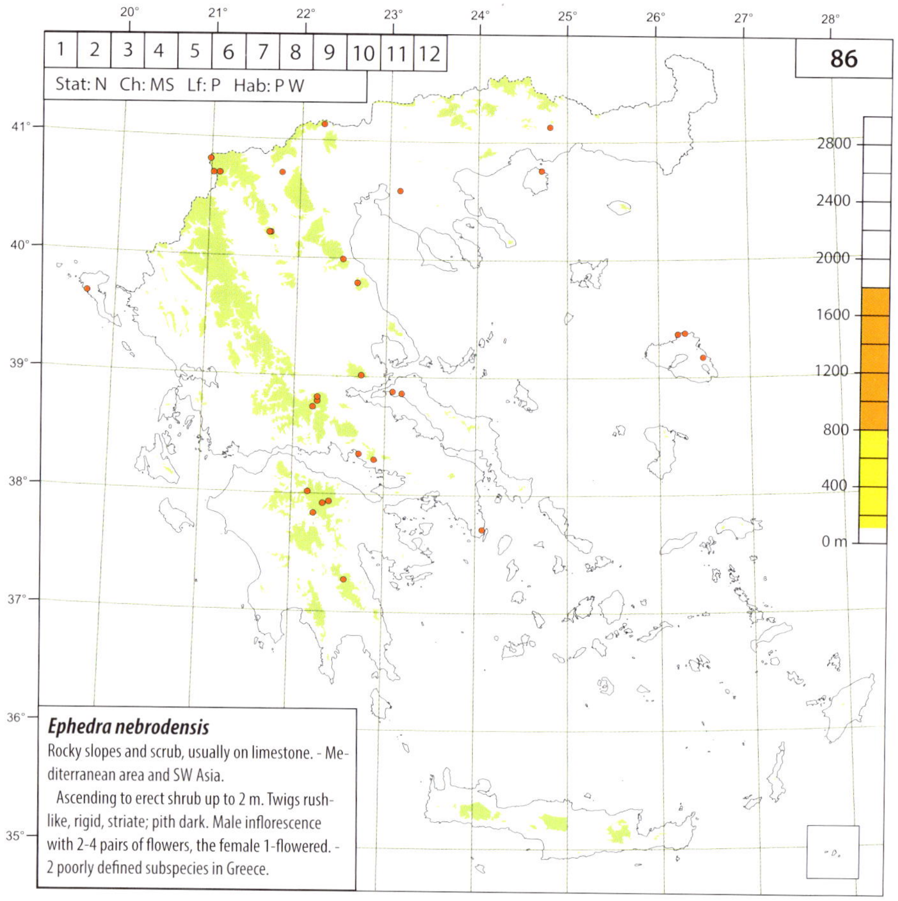

Ephedra nebrodensis

Rocky slopes and scrub, usually on limestone. - Mediterranean area and SW Asia.

Ascending to erect shrub up to 2 m. Twigs rush-like, rigid, striate; pith dark. Male inflorescence with 2-4 pairs of flowers, the female 1-flowered. - 2 poorly defined subspecies in Greece.

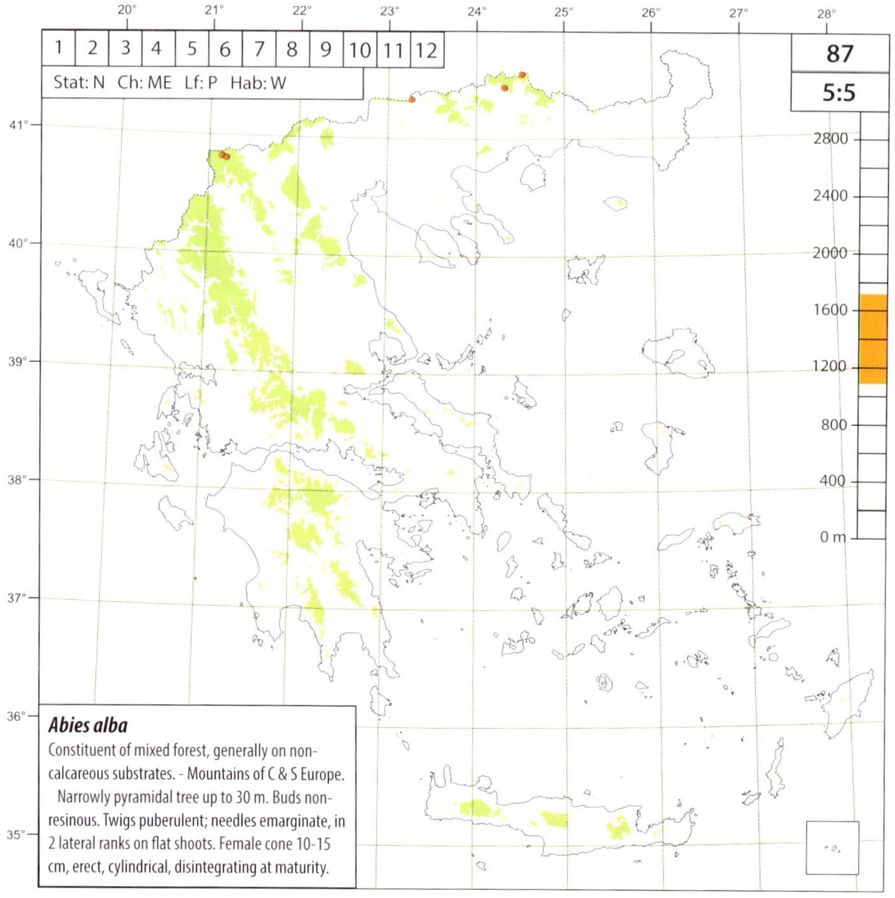

Abies alba

Constituent of mixed forest, generally on non-calcareous substrates. - Mountains of C & S Europe.

Narrowly pyramidal tree up to 30 m. Buds non-resinous. Twigs puberulent; needles emarginate, in 2 lateral ranks on flat shoots. Female cone 10-15 cm, erect, cylindrical, disintegrating at maturity.

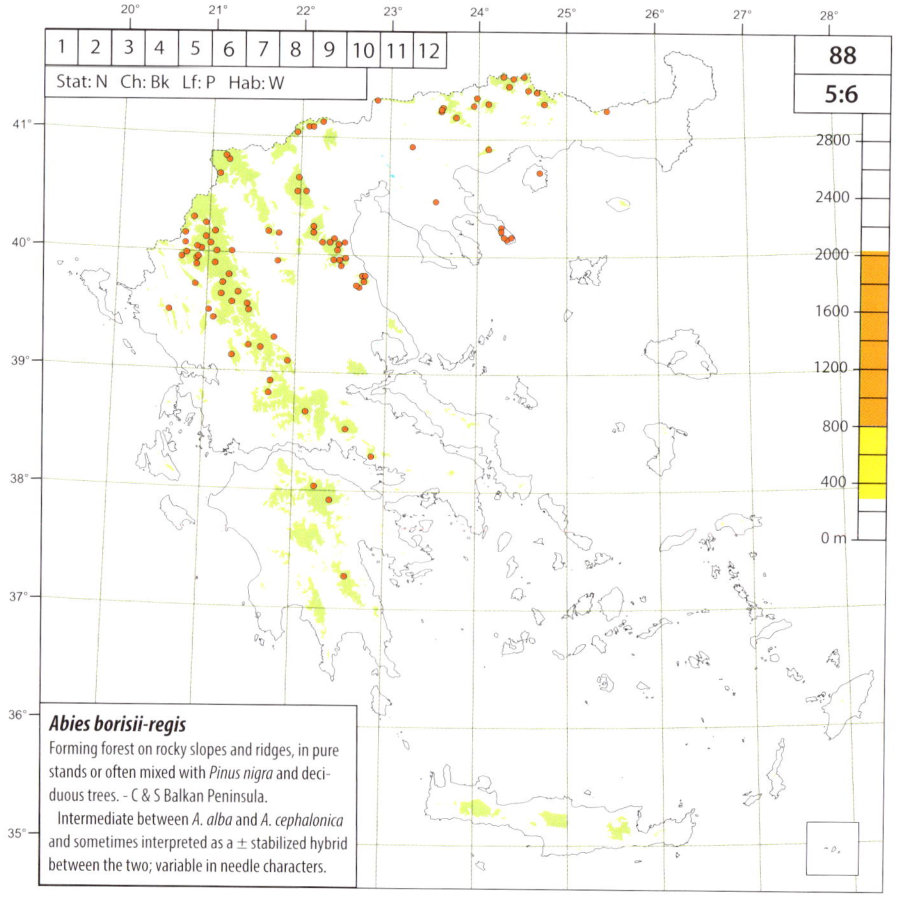

Abies borisii-regis

Forming forest on rocky slopes and ridges, in pure stands or often mixed with *Pinus nigra* and deciduous trees. - C & S Balkan Peninsula.

Intermediate between *A. alba* and *A. cephalonica* and sometimes interpreted as a ± stabilized hybrid between the two; variable in needle characters.

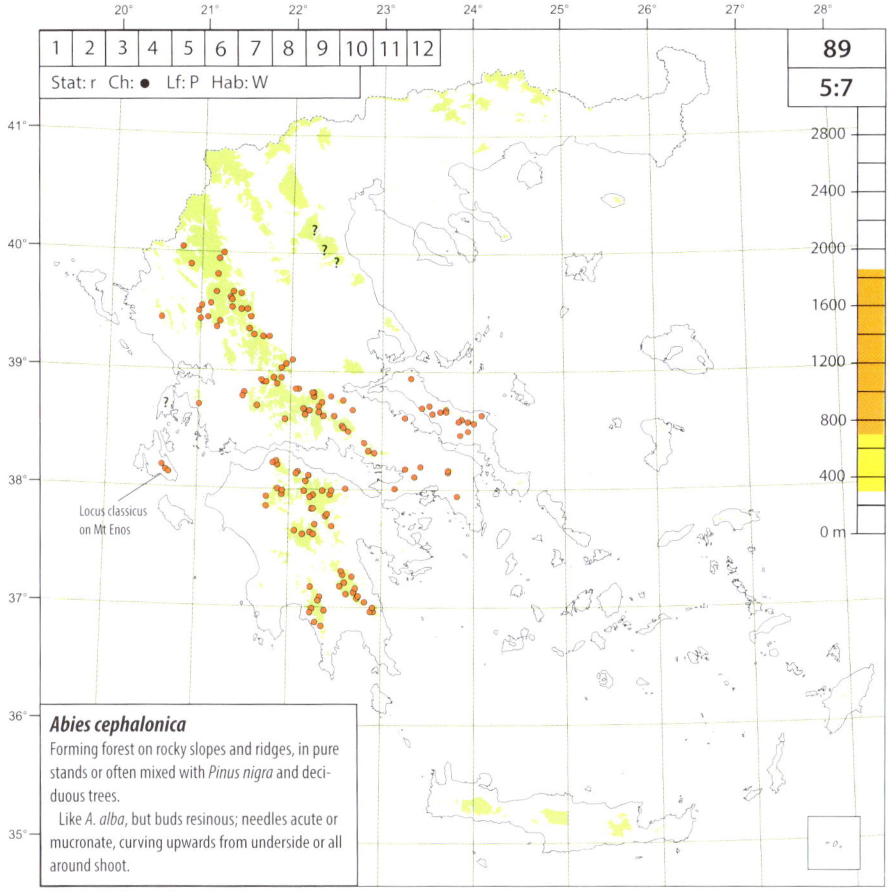

Abies cephalonica

Forming forest on rocky slopes and ridges, in pure stands or often mixed with *Pinus nigra* and deciduous trees.

Like *A. alba*, but buds resinous; needles acute or mucronate, curving upwards from underside or all around shoot.

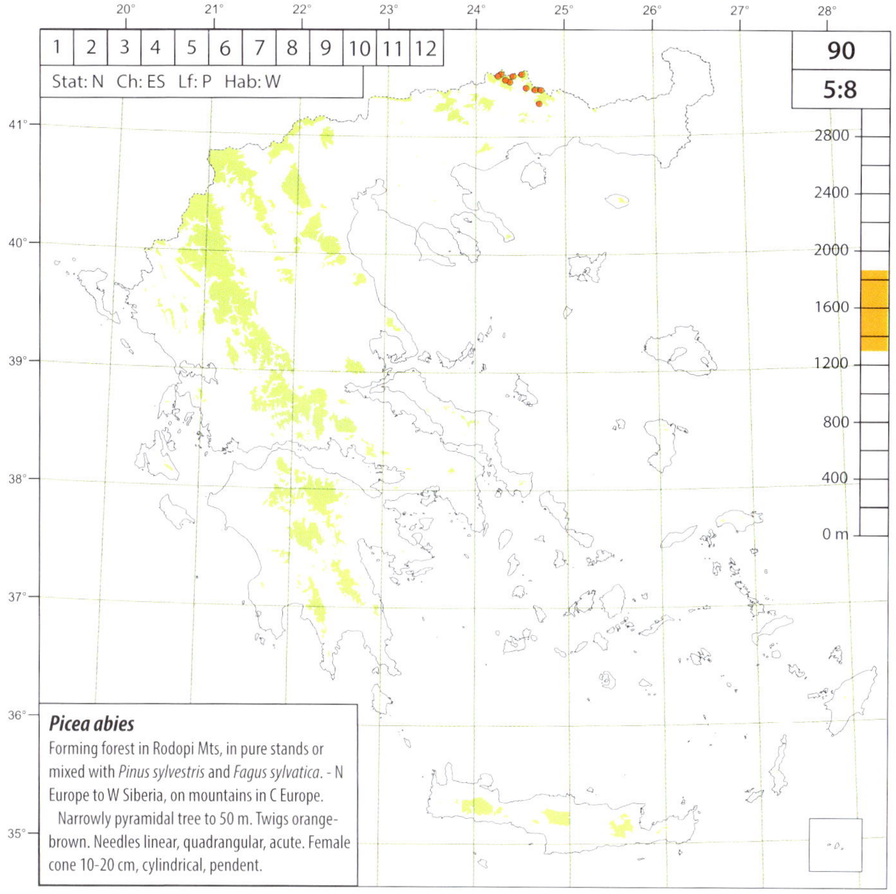

Picea abies

Forming forest in Rodopi Mts, in pure stands or mixed with *Pinus sylvestris* and *Fagus sylvatica*. - N Europe to W Siberia, on mountains in C Europe.

Narrowly pyramidal tree to 50 m. Twigs orange-brown. Needles linear, quadrangular, acute. Female cone 10-20 cm, cylindrical, pendent.

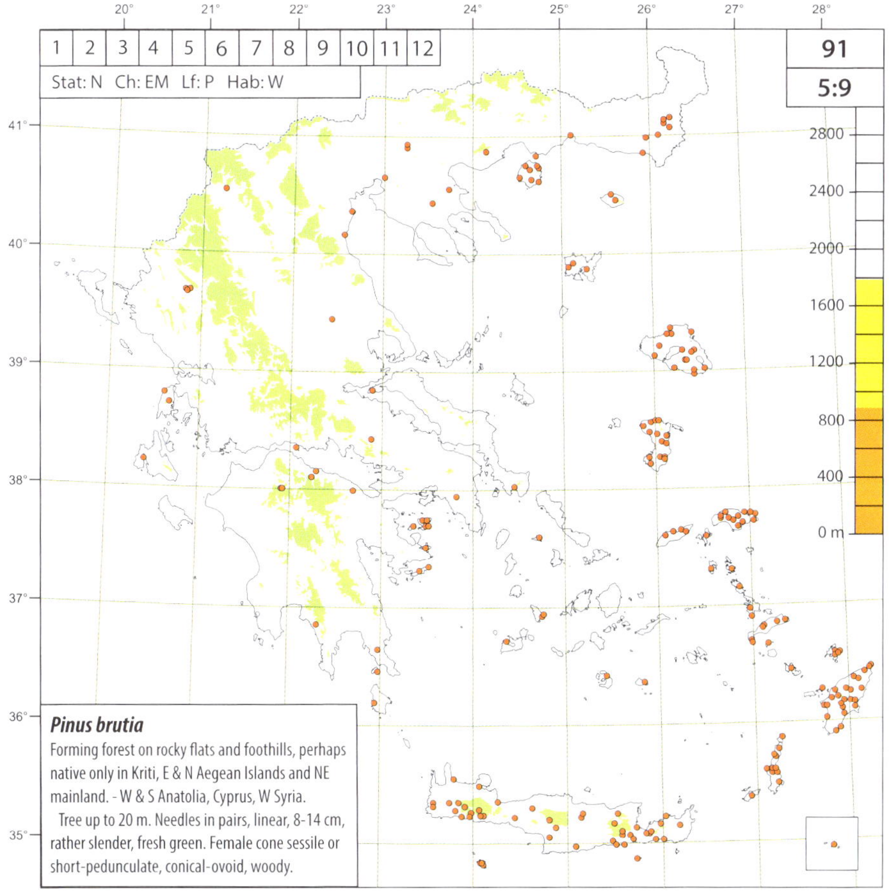

Pinus brutia

Forming forest on rocky flats and foothills, perhaps native only in Kriti, E & N Aegean Islands and NE mainland. - W & S Anatolia, Cyprus, W Syria.

Tree up to 20 m. Needles in pairs, linear, 8-14 cm, rather slender, fresh green. Female cone sessile or short-pedunculate, conical-ovoid, woody.

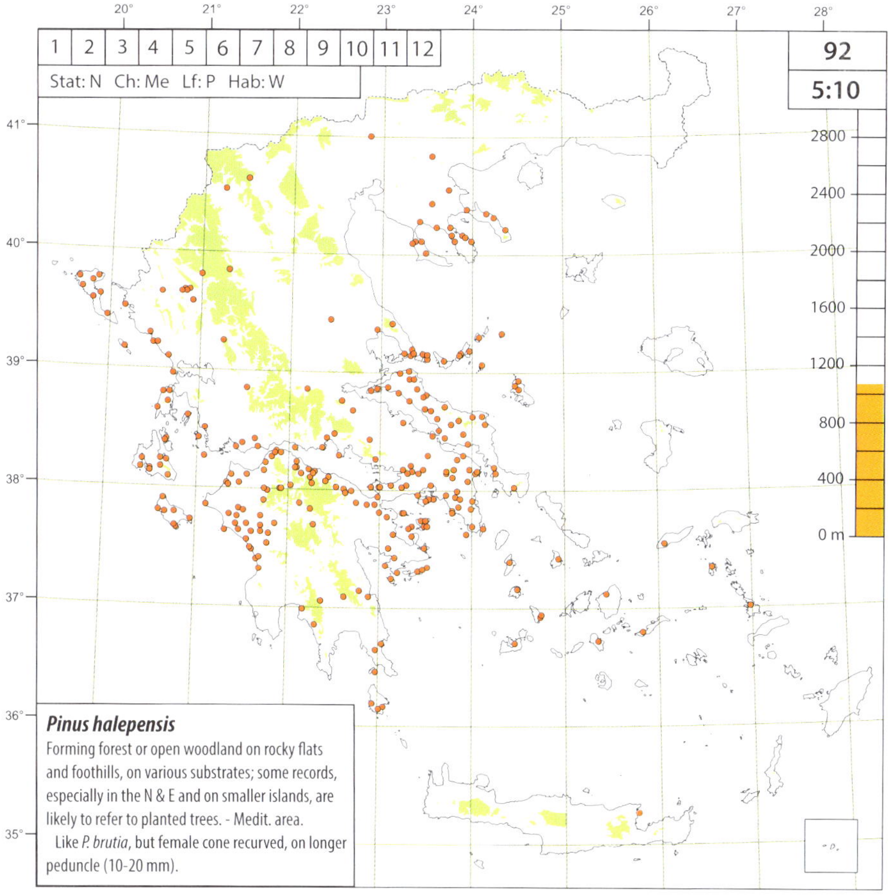

Pinus halepensis

Forming forest or open woodland on rocky flats and foothills, on various substrates; some records, especially in the N & E and on smaller islands, are likely to refer to planted trees. - Medit. area.

Like *P. brutia*, but female cone recurved, on longer peduncle (10-20 mm).

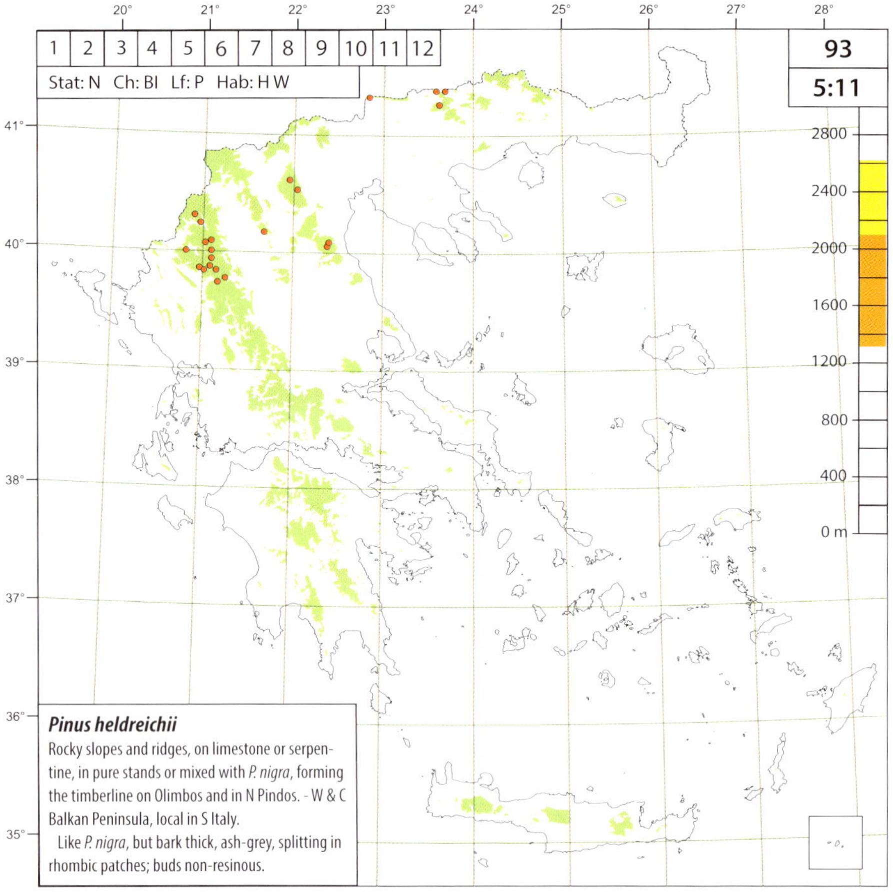

Pinus heldreichii
Rocky slopes and ridges, on limestone or serpentine, in pure stands or mixed with *P. nigra*, forming the timberline on Olimbos and in N Pindos. - W & C Balkan Peninsula, local in S Italy.
Like *P. nigra*, but bark thick, ash-grey, splitting in rhombic patches; buds non-resinous.

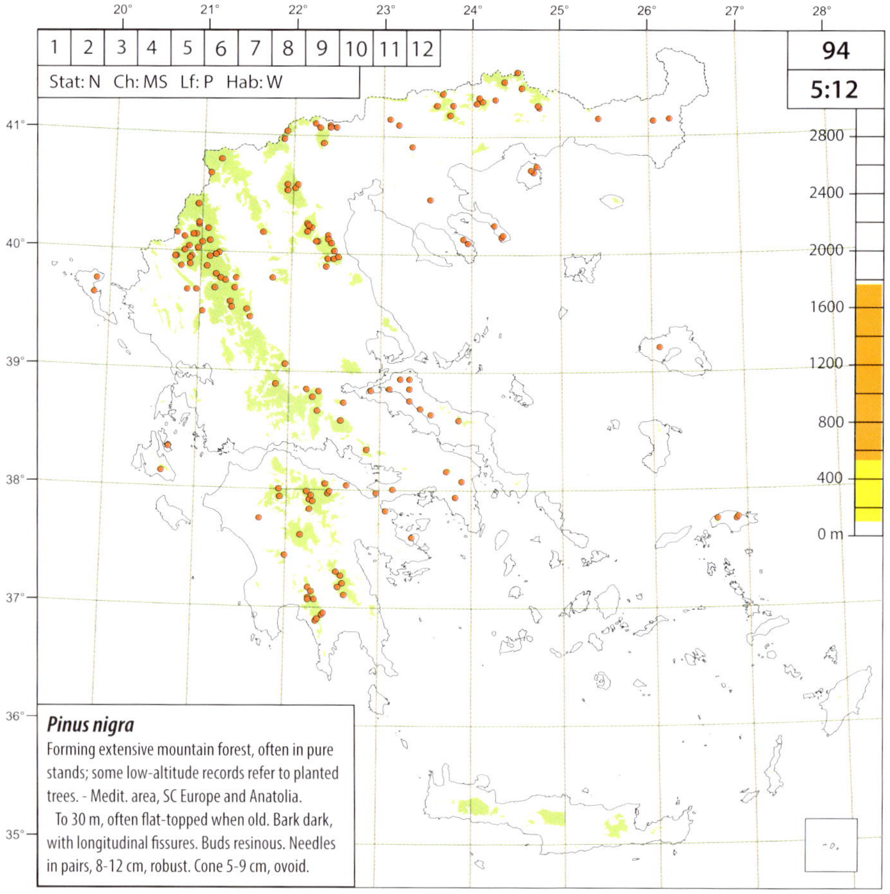

Pinus nigra
Forming extensive mountain forest, often in pure stands; some low-altitude records refer to planted trees. - Medit. area, SC Europe and Anatolia.
To 30 m, often flat-topped when old. Bark dark, with longitudinal fissures. Buds resinous. Needles in pairs, 8-12 cm, robust. Cone 5-9 cm, ovoid.

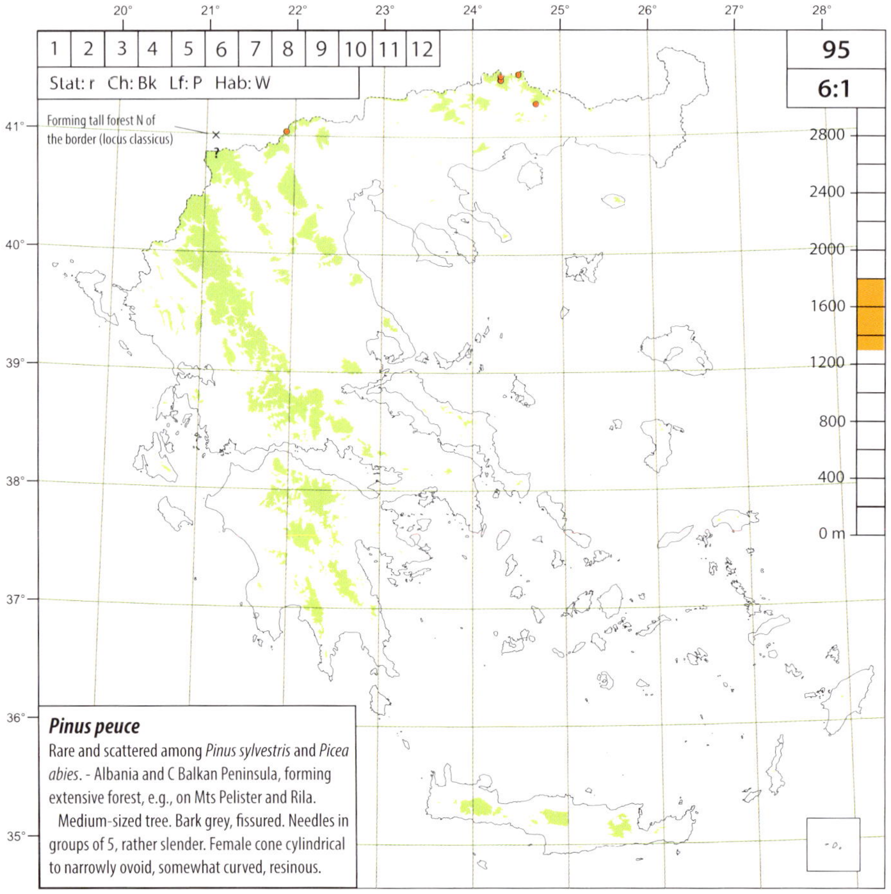

Pinus peuce
Rare and scattered among *Pinus sylvestris* and *Picea abies*. - Albania and C Balkan Peninsula, forming extensive forest, e.g., on Mts Pelister and Rila.
Medium-sized tree. Bark grey, fissured. Needles in groups of 5, rather slender. Female cone cylindrical to narrowly ovoid, somewhat curved, resinous.

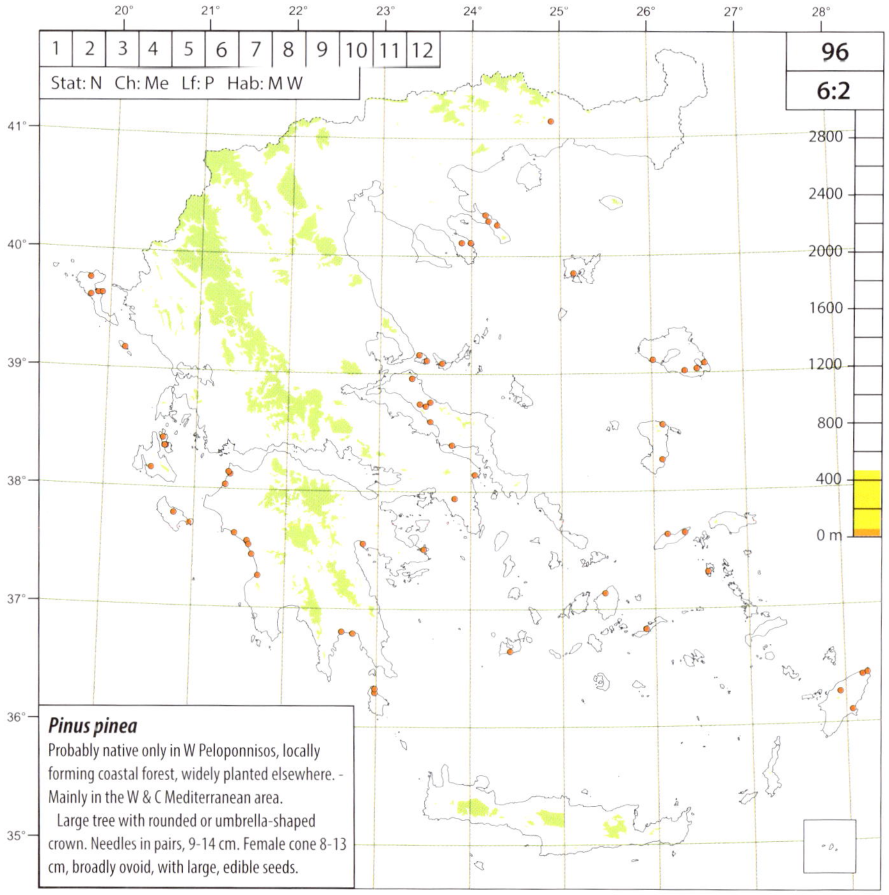

Pinus pinea
Probably native only in W Peloponnisos, locally forming coastal forest, widely planted elsewhere. - Mainly in the W & C Mediterranean area.
Large tree with rounded or umbrella-shaped crown. Needles in pairs, 9-14 cm. Female cone 8-13 cm, broadly ovoid, with large, edible seeds.

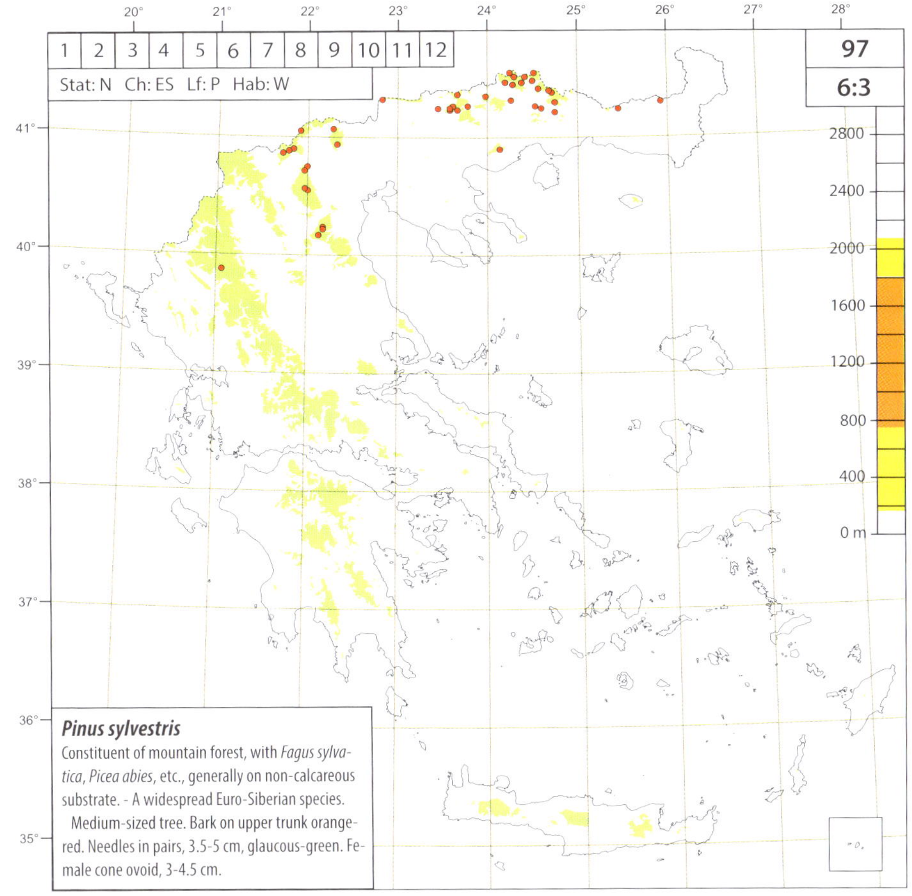

Pinus sylvestris

Constituent of mountain forest, with *Fagus sylvatica*, *Picea abies*, etc., generally on non-calcareous substrate. - A widespread Euro-Siberian species.

Medium-sized tree. Bark on upper trunk orange-red. Needles in pairs, 3.5-5 cm, glaucous-green. Female cone ovoid, 3-4.5 cm.

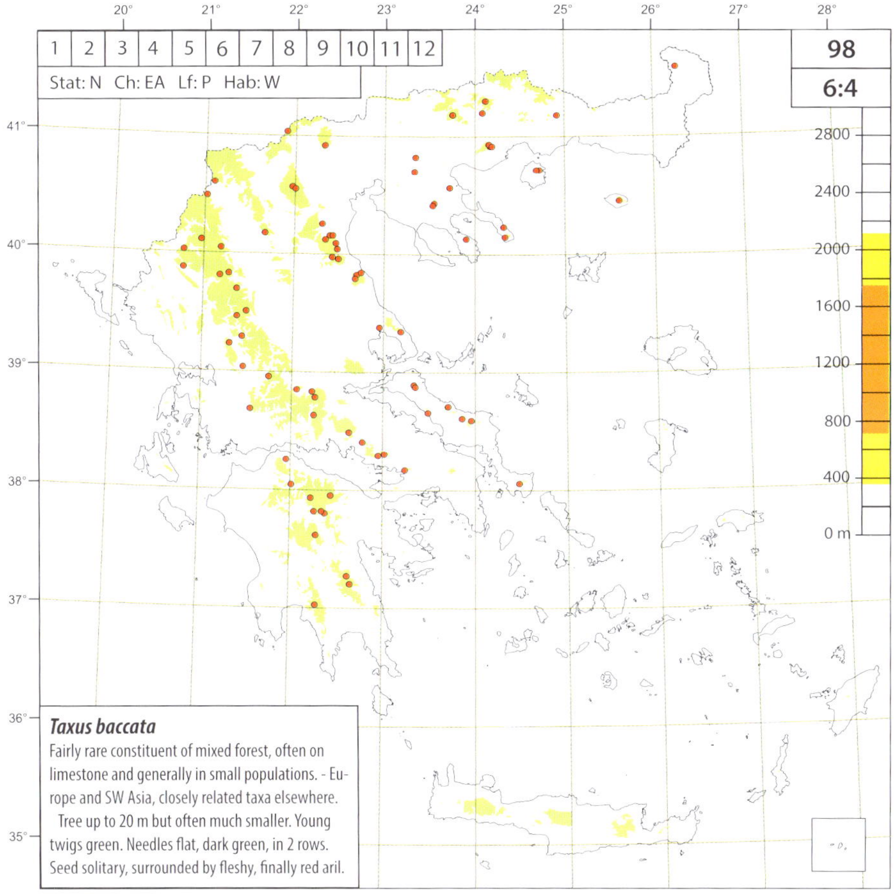

Taxus baccata

Fairly rare constituent of mixed forest, often on limestone and generally in small populations. - Europe and SW Asia, closely related taxa elsewhere.

Tree up to 20 m but often much smaller. Young twigs green. Needles flat, dark green, in 2 rows. Seed solitary, surrounded by fleshy, finally red aril.

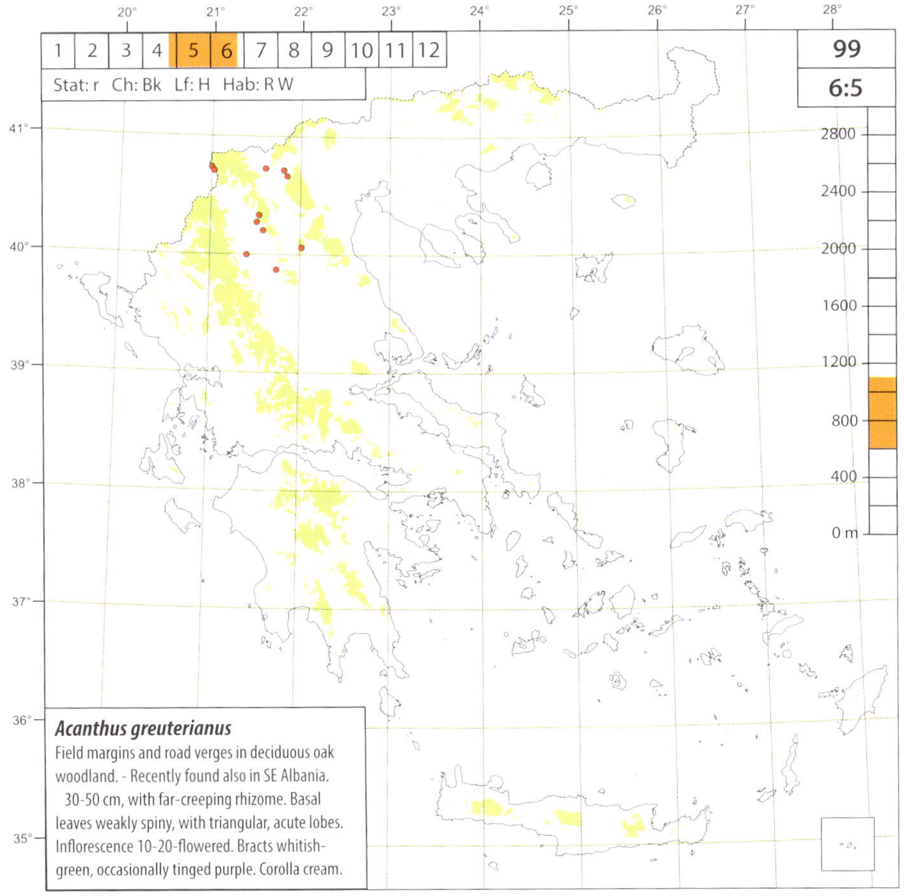

Acanthus greuterianus

Field margins and road verges in deciduous oak woodland. - Recently found also in SE Albania.

30-50 cm, with far-creeping rhizome. Basal leaves weakly spiny, with triangular, acute lobes. Inflorescence 10-20-flowered. Bracts whitish-green, occasionally tinged purple. Corolla cream.

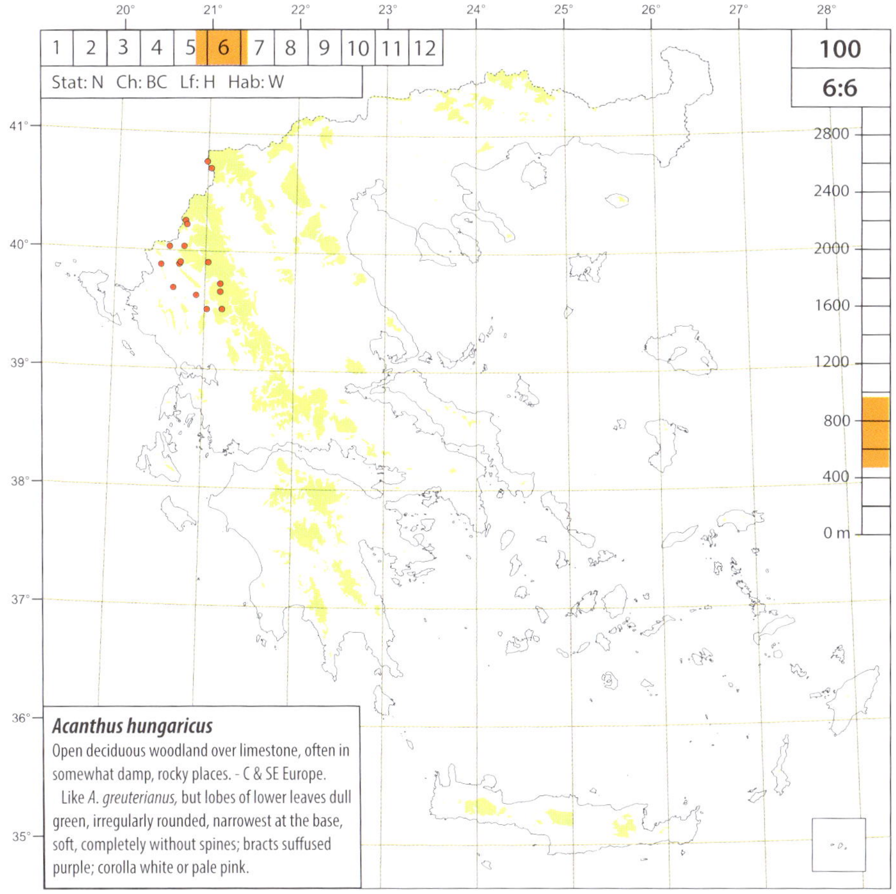

Acanthus hungaricus

Open deciduous woodland over limestone, often in somewhat damp, rocky places. - C & SE Europe.

Like *A. greuterianus*, but lobes of lower leaves dull green, irregularly rounded, narrowest at the base, soft, completely without spines; bracts suffused purple; corolla white or pale pink.

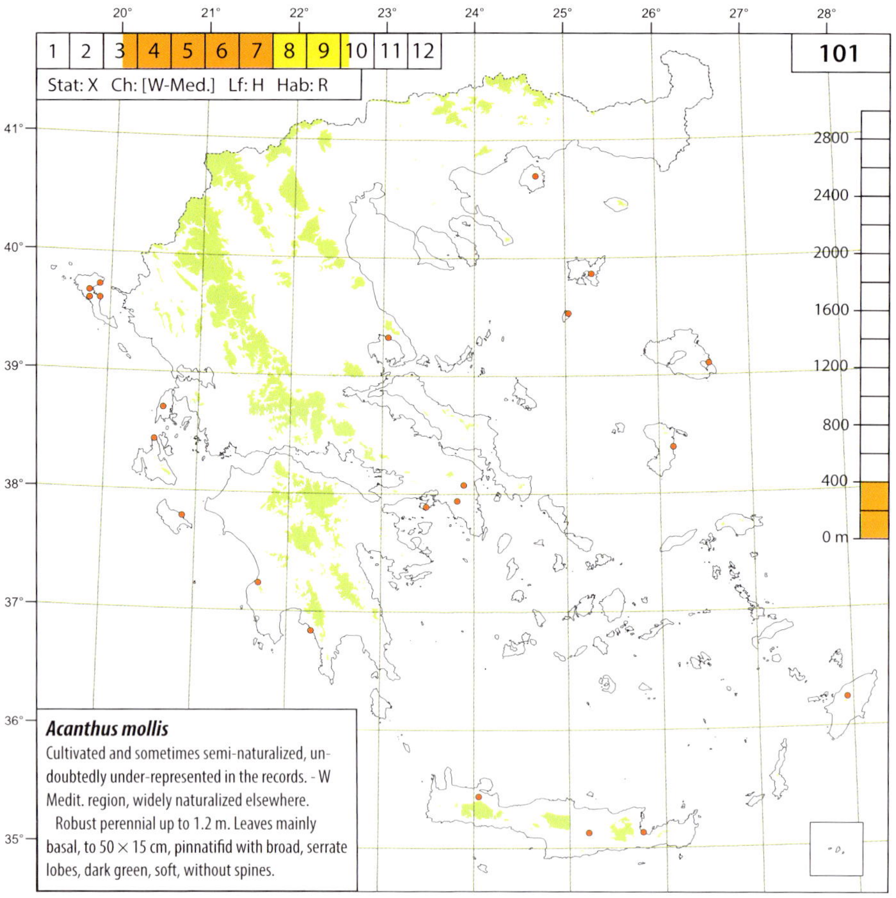

Acanthus mollis

Cultivated and sometimes semi-naturalized, undoubtedly under-represented in the records. - W Medit. region, widely naturalized elsewhere.

Robust perennial up to 1.2 m. Leaves mainly basal, to 50 × 15 cm, pinnatifid with broad, serrate lobes, dark green, soft, without spines.

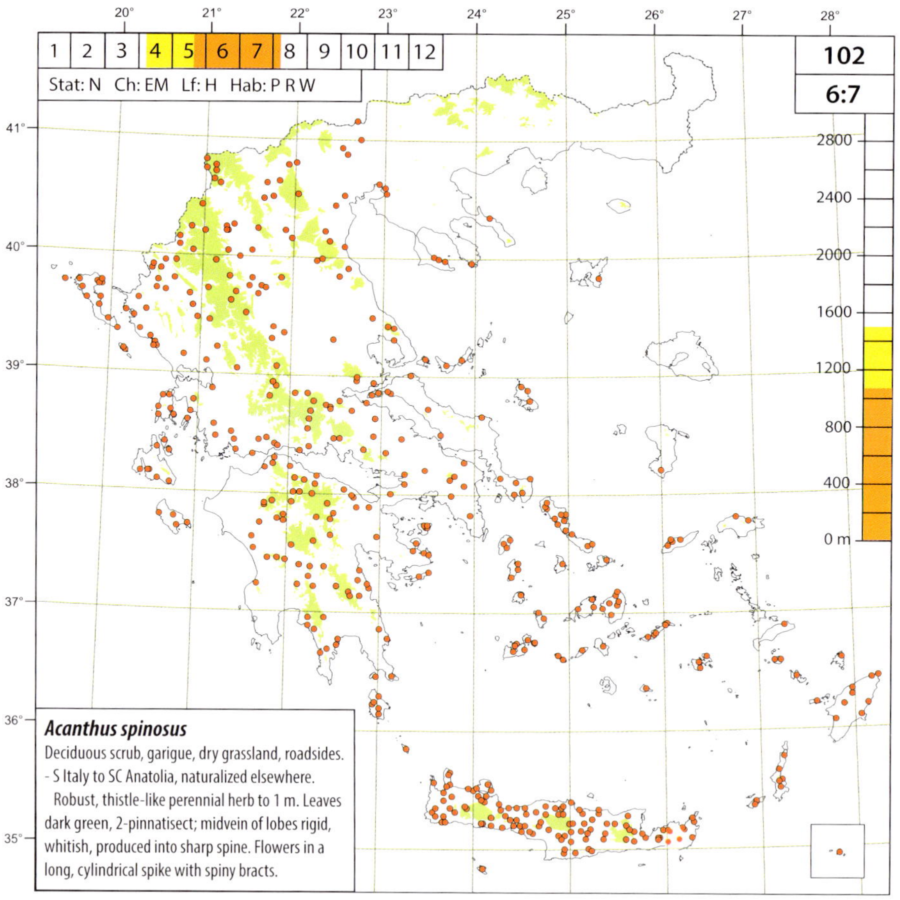

Acanthus spinosus

Deciduous scrub, garigue, dry grassland, roadsides. - S Italy to SC Anatolia, naturalized elsewhere.

Robust, thistle-like perennial herb to 1 m. Leaves dark green, 2-pinnatisect; midvein of lobes rigid, whitish, produced into sharp spine. Flowers in a long, cylindrical spike with spiny bracts.

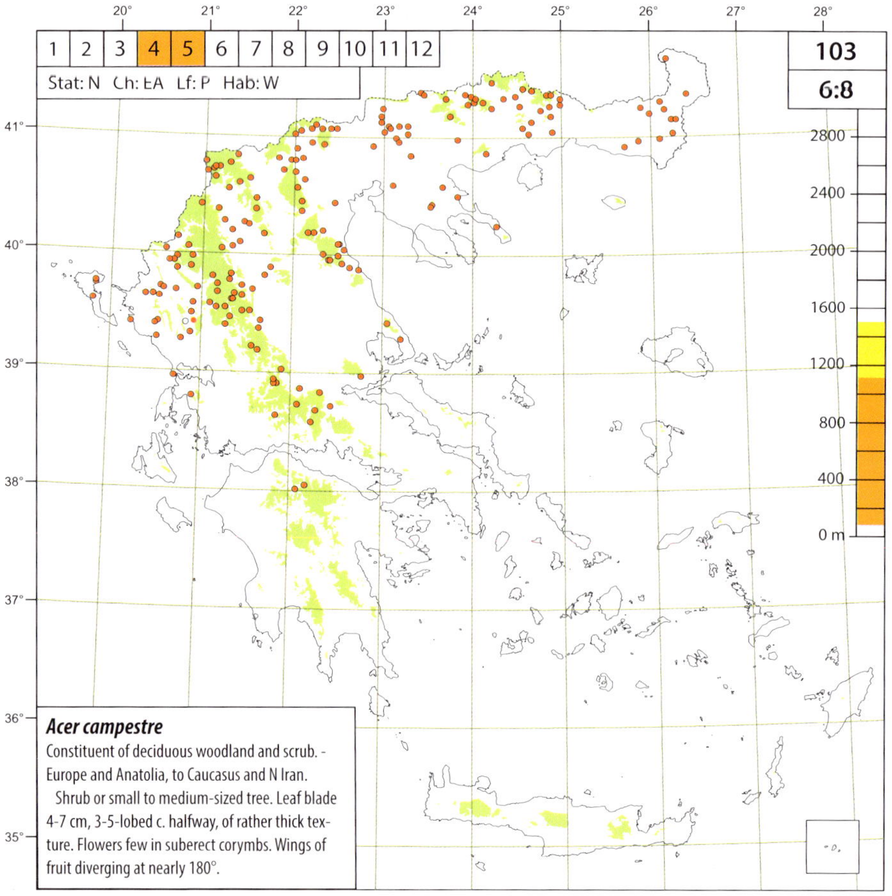

Acer campestre

Constituent of deciduous woodland and scrub. - Europe and Anatolia, to Caucasus and N Iran.

Shrub or small to medium-sized tree. Leaf blade 4-7 cm, 3-5-lobed c. halfway, of rather thick texture. Flowers few in suberect corymbs. Wings of fruit diverging at nearly 180°.

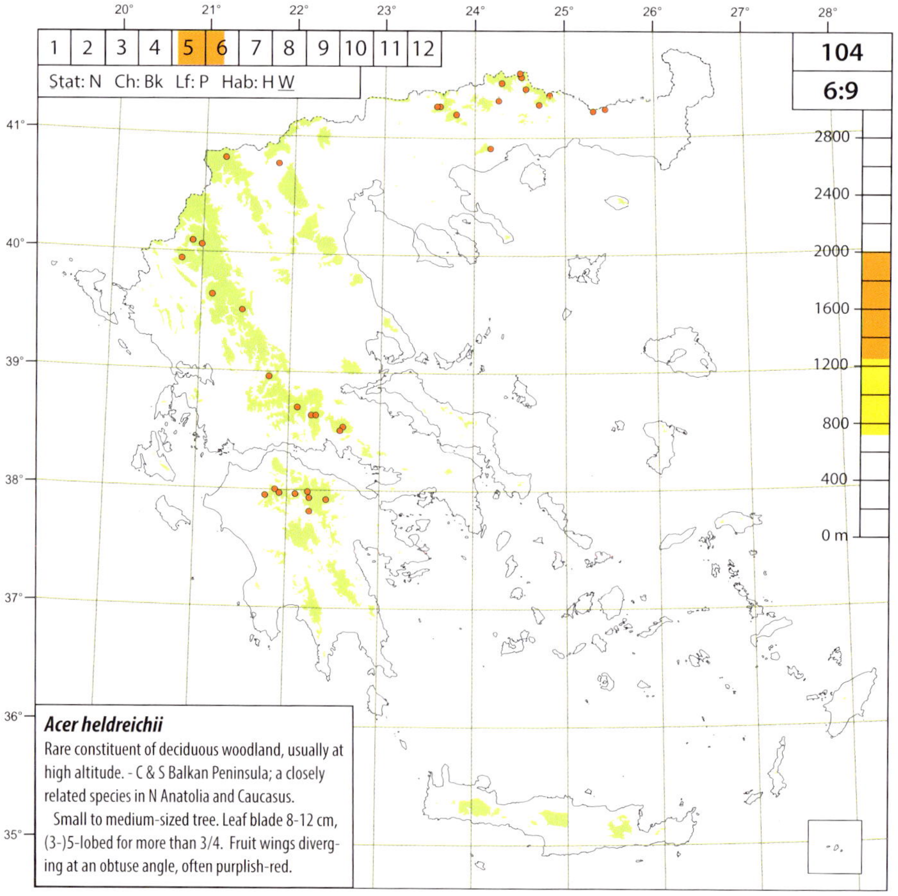

Acer heldreichii

Rare constituent of deciduous woodland, usually at high altitude. - C & S Balkan Peninsula; a closely related species in N Anatolia and Caucasus.

Small to medium-sized tree. Leaf blade 8-12 cm, (3-)5-lobed for more than 3/4. Fruit wings diverging at an obtuse angle, often purplish-red.

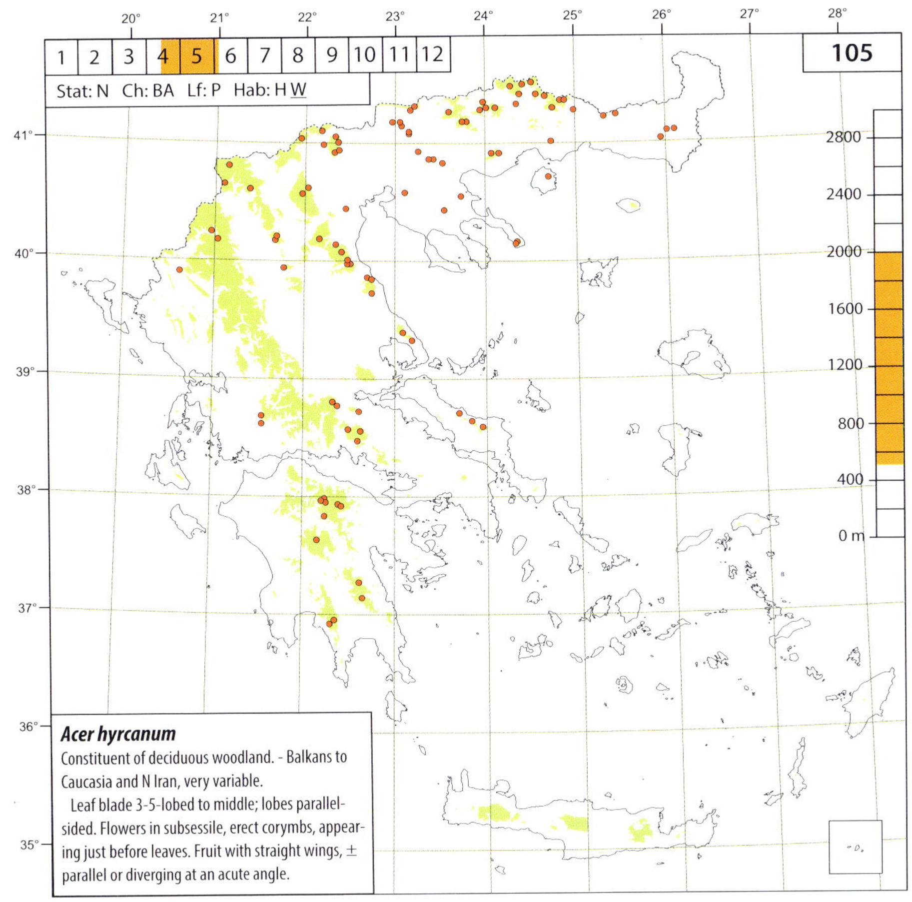

Acer hyrcanum
Constituent of deciduous woodland. - Balkans to Caucasia and N Iran, very variable.
Leaf blade 3-5-lobed to middle; lobes parallel-sided. Flowers in subsessile, erect corymbs, appearing just before leaves. Fruit with straight wings, ± parallel or diverging at an acute angle.

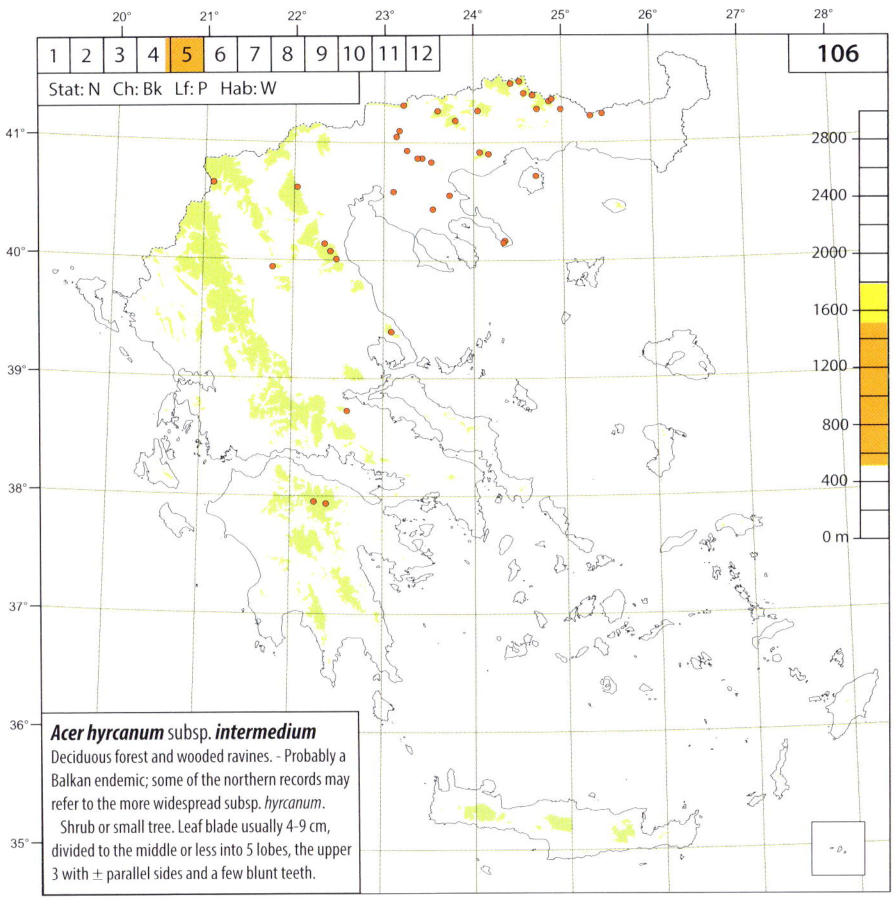

Acer hyrcanum **subsp. *intermedium***
Deciduous forest and wooded ravines. - Probably a Balkan endemic; some of the northern records may refer to the more widespread subsp. *hyrcanum*.
Shrub or small tree. Leaf blade usually 4-9 cm, divided to the middle or less into 5 lobes, the upper 3 with ± parallel sides and a few blunt teeth.

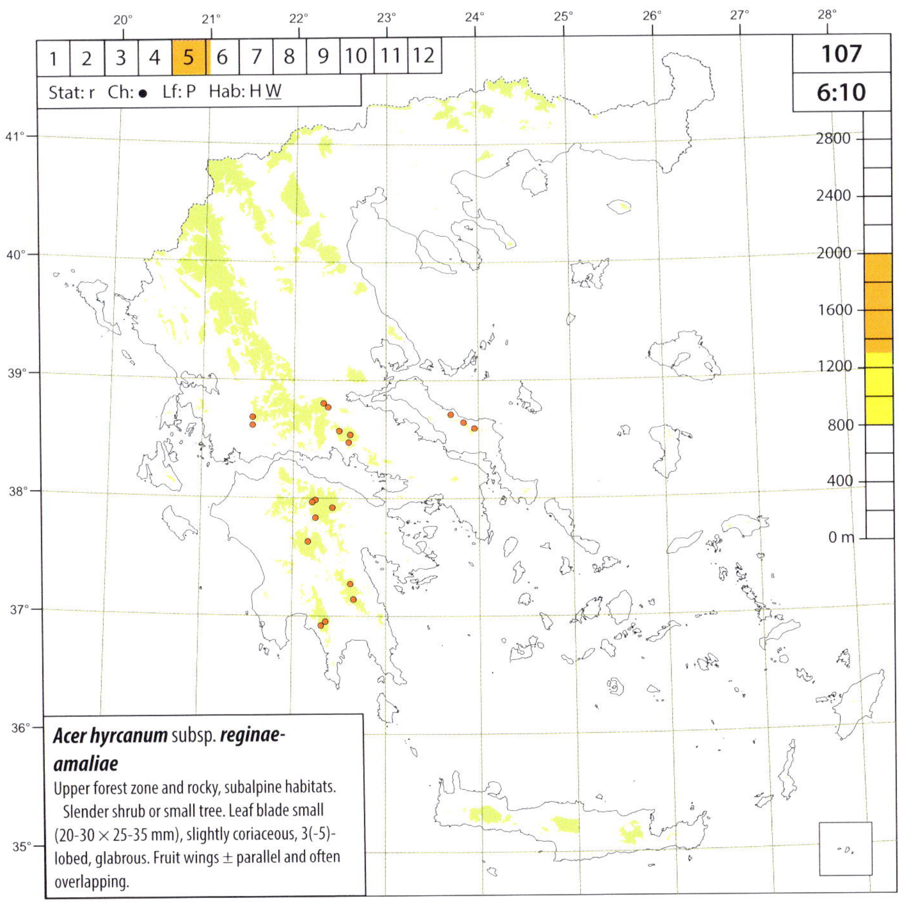

Acer hyrcanum **subsp. *reginae-amaliae***
Upper forest zone and rocky, subalpine habitats.
Slender shrub or small tree. Leaf blade small (20-30 × 25-35 mm), slightly coriaceous, 3(-5)-lobed, glabrous. Fruit wings ± parallel and often overlapping.

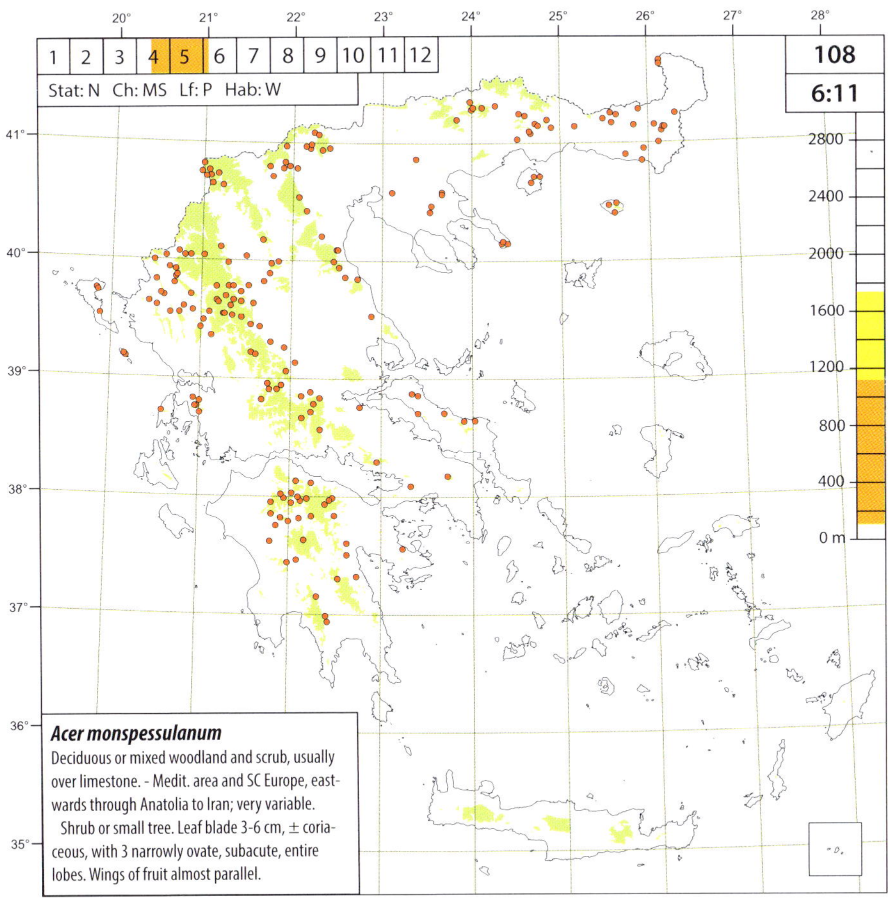

Acer monspessulanum
Deciduous or mixed woodland and scrub, usually over limestone. - Medit. area and SC Europe, eastwards through Anatolia to Iran; very variable.
Shrub or small tree. Leaf blade 3-6 cm, ± coriaceous, with 3 narrowly ovate, subacute, entire lobes. Wings of fruit almost parallel.

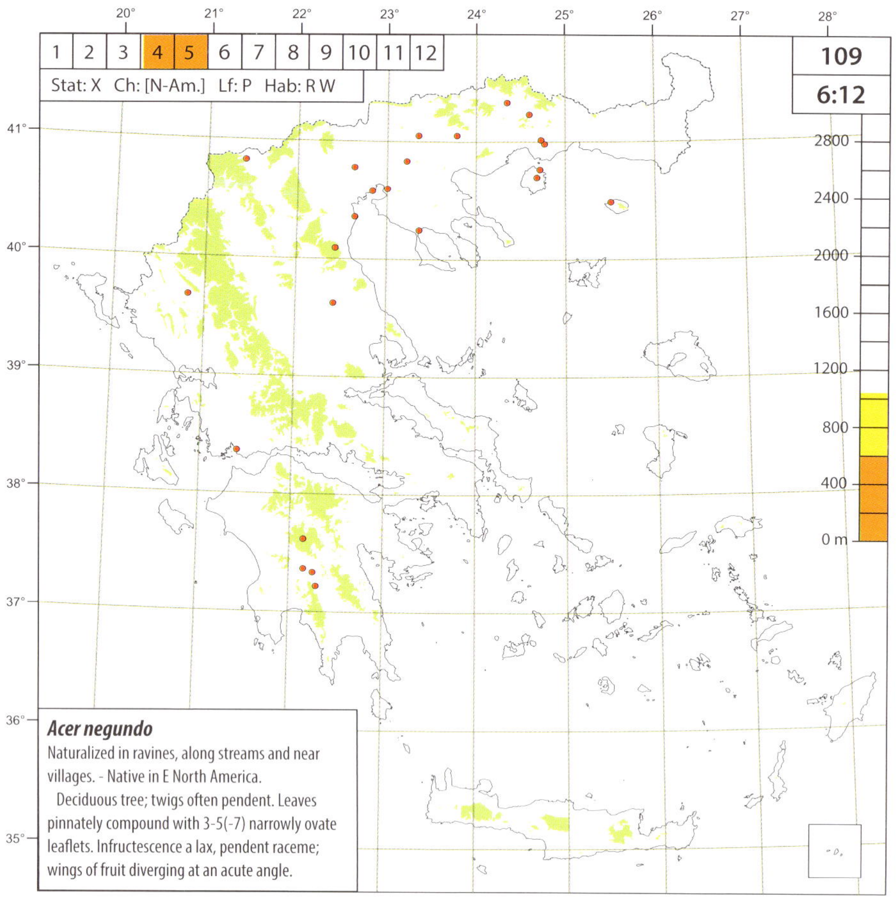

Acer negundo

Naturalized in ravines, along streams and near villages. - Native in E North America.

Deciduous tree; twigs often pendent. Leaves pinnately compound with 3-5(-7) narrowly ovate leaflets. Infructescence a lax, pendent raceme; wings of fruit diverging at an acute angle.

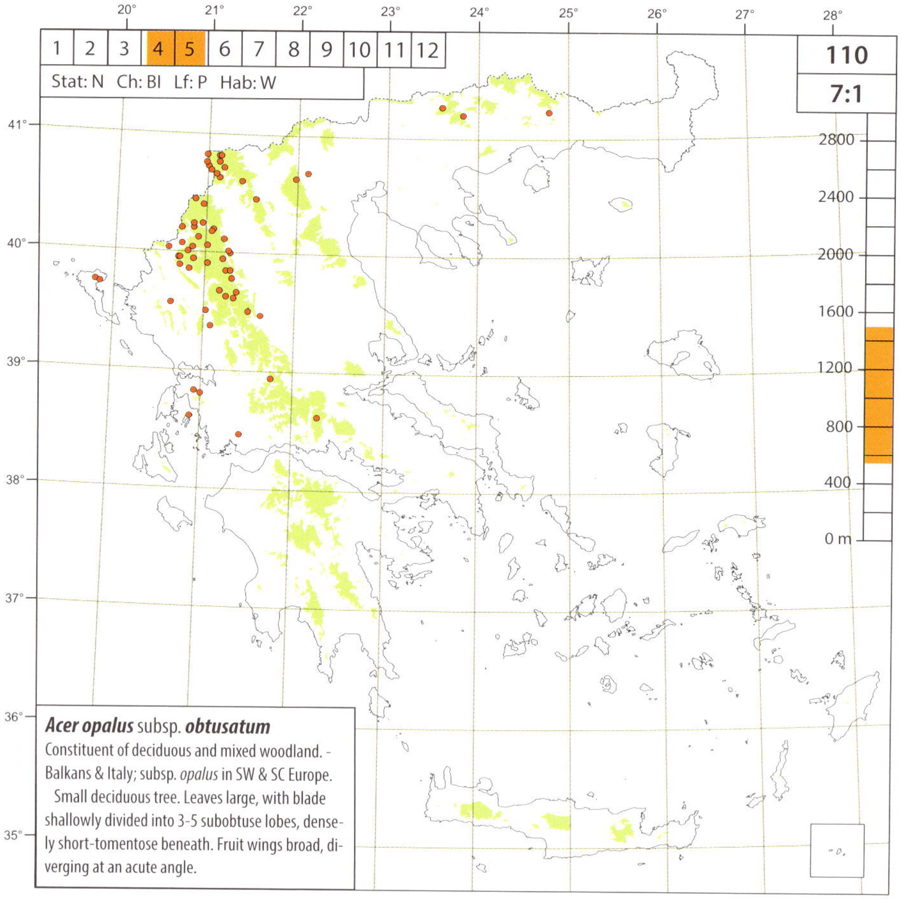

Acer opalus subsp. ***obtusatum***

Constituent of deciduous and mixed woodland. - Balkans & Italy; subsp. *opalus* in SW & SC Europe.

Small deciduous tree. Leaves large, with blade shallowly divided into 3-5 subobtuse lobes, densely short-tomentose beneath. Fruit wings broad, diverging at an acute angle.

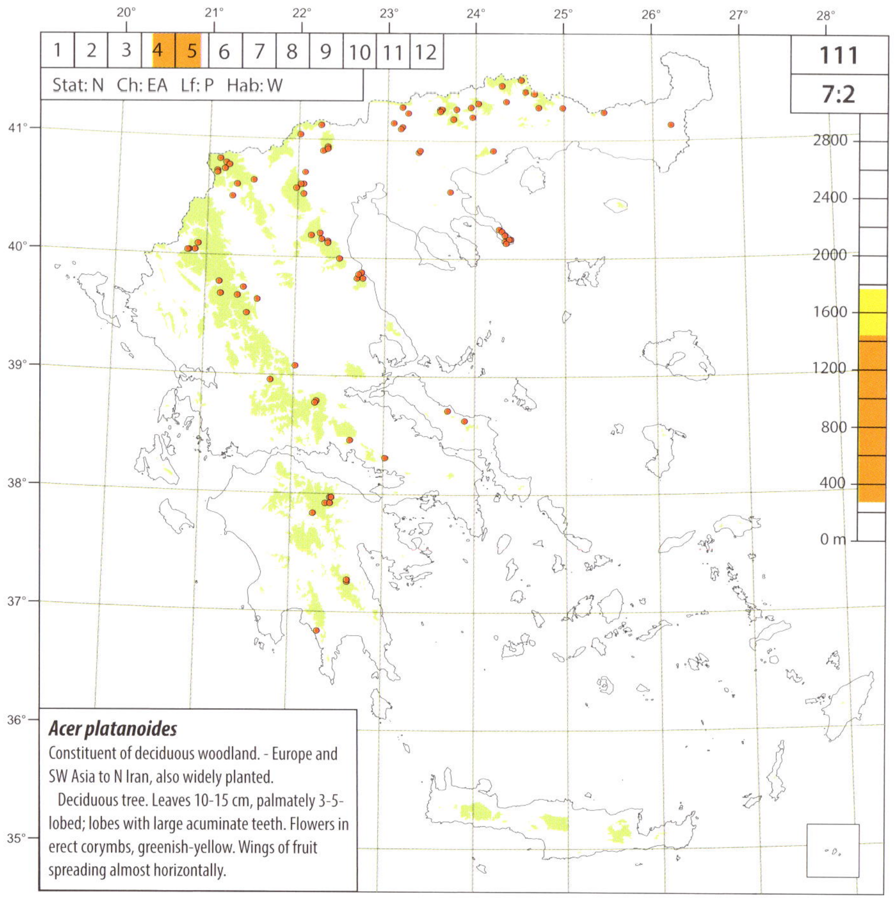

Acer platanoides

Constituent of deciduous woodland. - Europe and SW Asia to N Iran, also widely planted.

Deciduous tree. Leaves 10-15 cm, palmately 3-5-lobed; lobes with large acuminate teeth. Flowers in erect corymbs, greenish-yellow. Wings of fruit spreading almost horizontally.

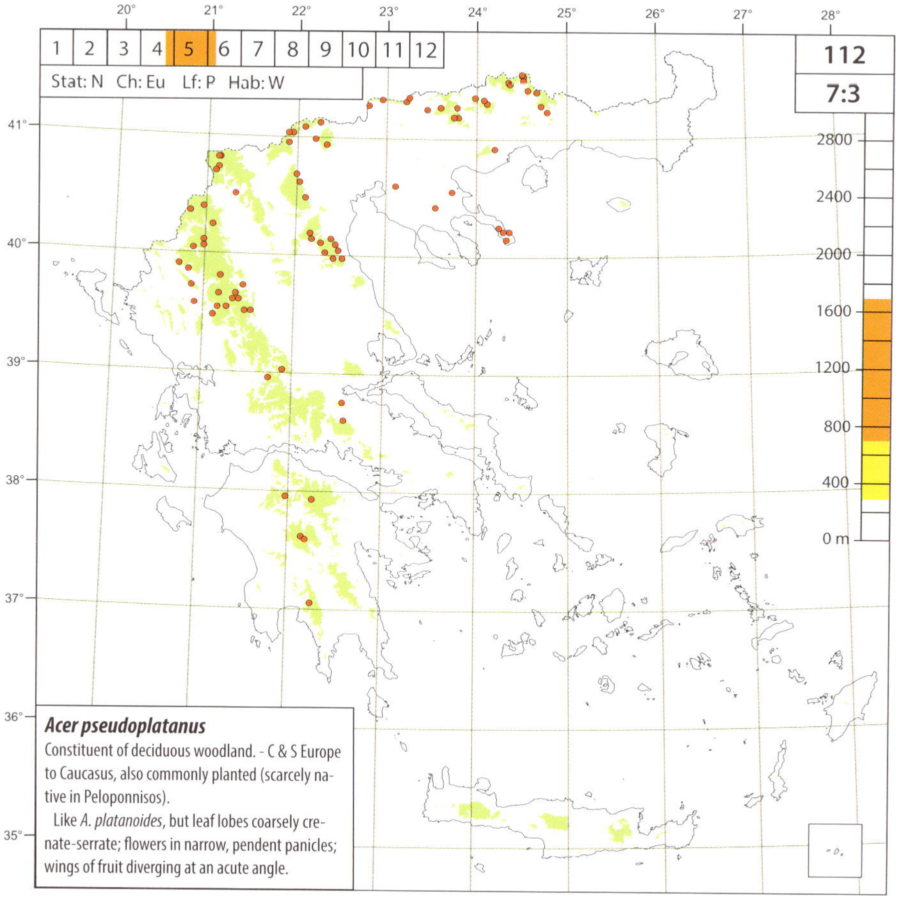

Acer pseudoplatanus

Constituent of deciduous woodland. - C & S Europe to Caucasus, also commonly planted (scarcely native in Peloponnisos).

Like *A. platanoides*, but leaf lobes coarsely crenate-serrate; flowers in narrow, pendent panicles; wings of fruit diverging at an acute angle.

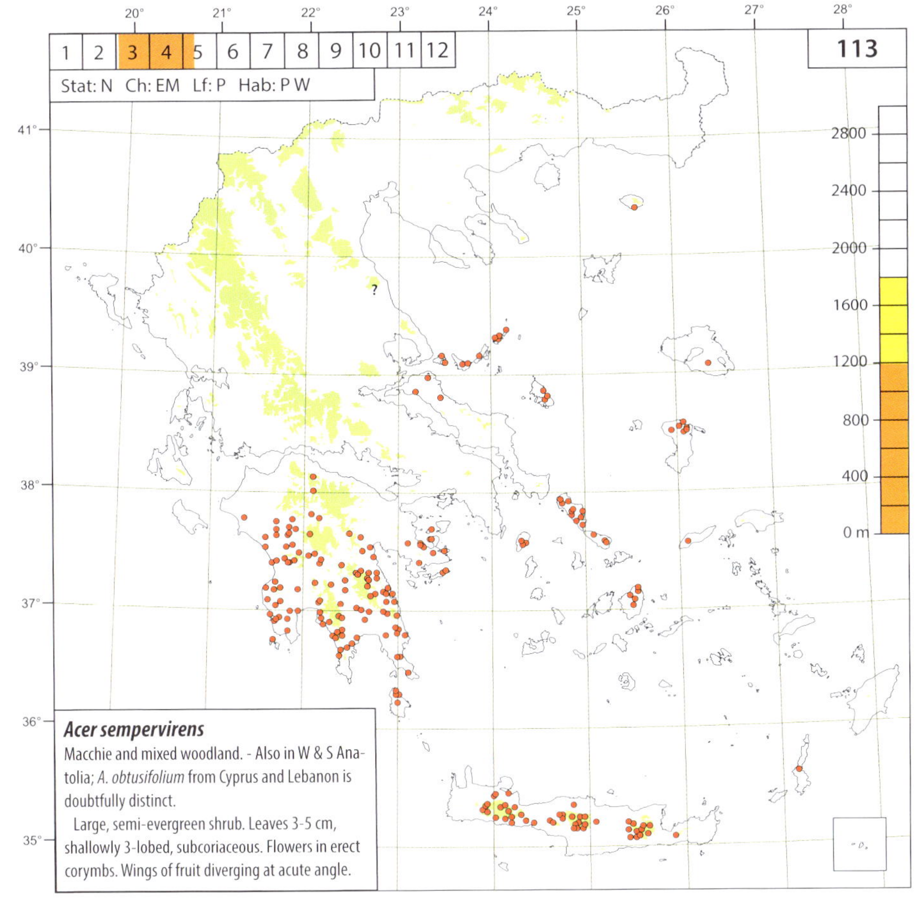

113

Stat: N Ch: EM Lf: P Hab: P W

Acer sempervirens

Macchie and mixed woodland. - Also in W & S Anatolia; *A. obtusifolium* from Cyprus and Lebanon is doubtfully distinct.

Large, semi-evergreen shrub. Leaves 3-5 cm, shallowly 3-lobed, subcoriaceous. Flowers in erect corymbs. Wings of fruit diverging at acute angle.

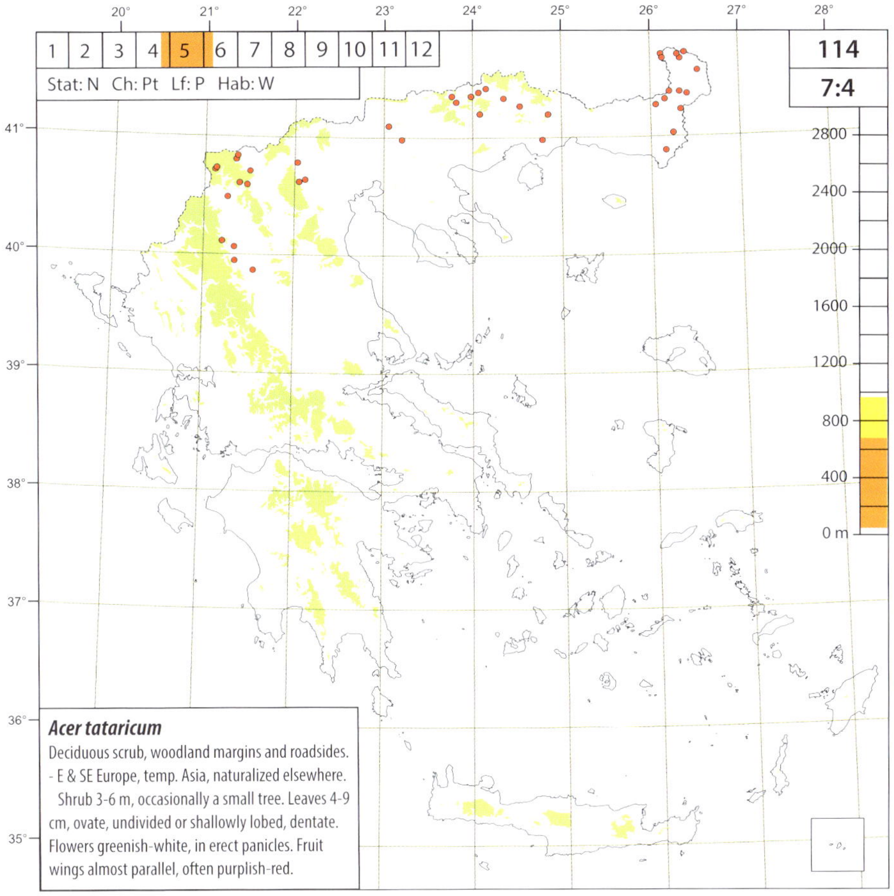

114 7:4

Stat: N Ch: Pt Lf: P Hab: W

Acer tataricum

Deciduous scrub, woodland margins and roadsides. - E & SE Europe, temp. Asia, naturalized elsewhere.

Shrub 3-6 m, occasionally a small tree. Leaves 4-9 cm, ovate, undivided or shallowly lobed, dentate. Flowers greenish-white, in erect panicles. Fruit wings almost parallel, often purplish-red.

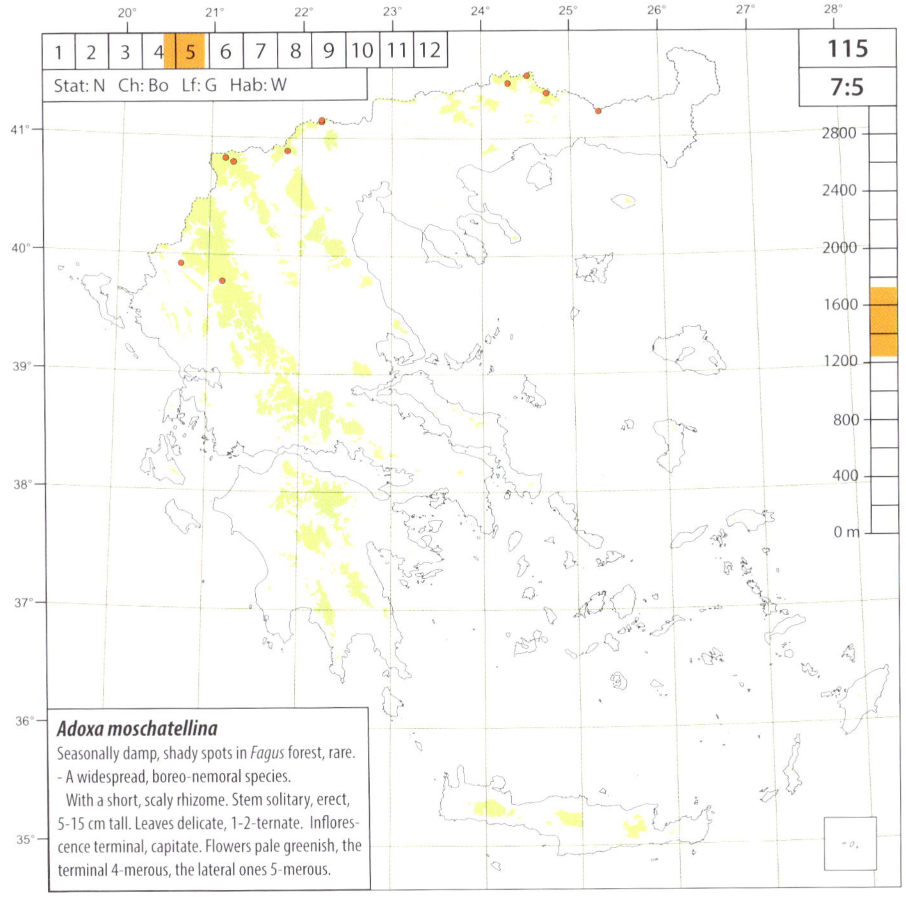

115 7:5

Stat: N Ch: Bo Lf: G Hab: W

Adoxa moschatellina

Seasonally damp, shady spots in *Fagus* forest, rare. - A widespread, boreo-nemoral species.

With a short, scaly rhizome. Stem solitary, erect, 5-15 cm tall. Leaves delicate, 1-2-ternate. Inflorescence terminal, capitate. Flowers pale greenish, the terminal 4-merous, the lateral ones 5-merous.

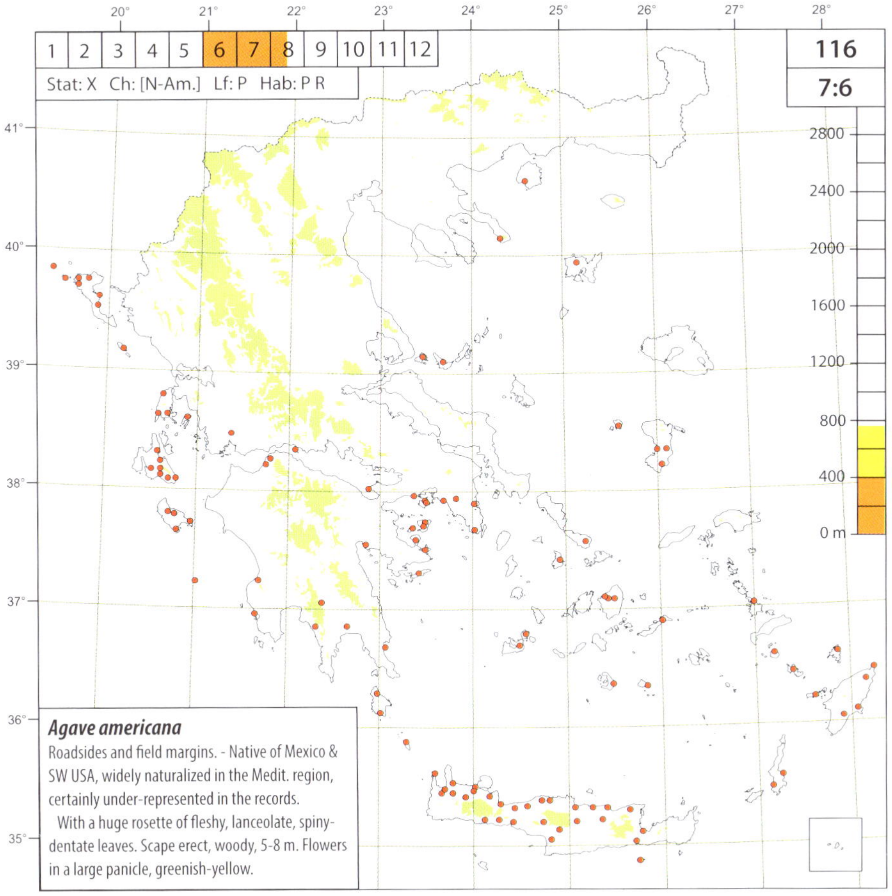

116 7:6

Stat: X Ch: [N-Am.] Lf: P Hab: P R

Agave americana

Roadsides and field margins. - Native of Mexico & SW USA, widely naturalized in the Medit. region, certainly under-represented in the records.

With a huge rosette of fleshy, lanceolate, spiny-dentate leaves. Scape erect, woody, 5-8 m. Flowers in a large panicle, greenish-yellow.

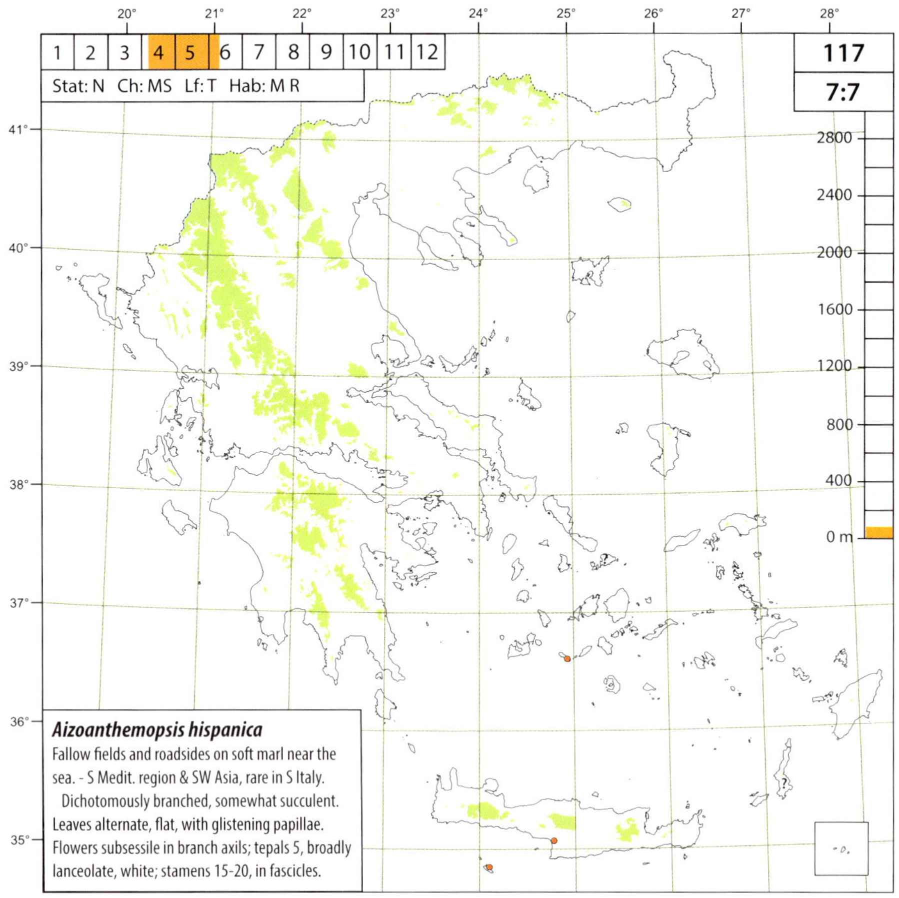

Aizoanthemopsis hispanica

Fallow fields and roadsides on soft marl near the sea. - S Medit. region & SW Asia, rare in S Italy.

Dichotomously branched, somewhat succulent. Leaves alternate, flat, with glistening papillae. Flowers subsessile in branch axils; tepals 5, broadly lanceolate, white; stamens 15-20, in fascicles.

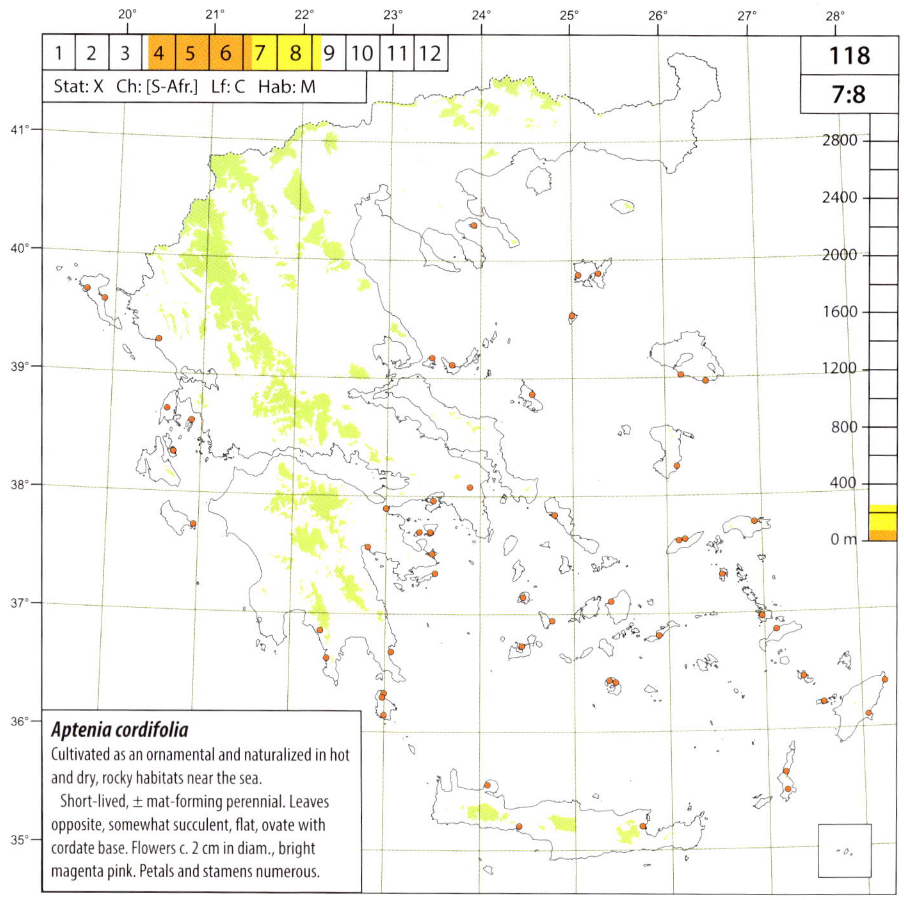

Aptenia cordifolia

Cultivated as an ornamental and naturalized in hot and dry, rocky habitats near the sea.

Short-lived, ± mat-forming perennial. Leaves opposite, somewhat succulent, flat, ovate with cordate base. Flowers c. 2 cm in diam., bright magenta pink. Petals and stamens numerous.

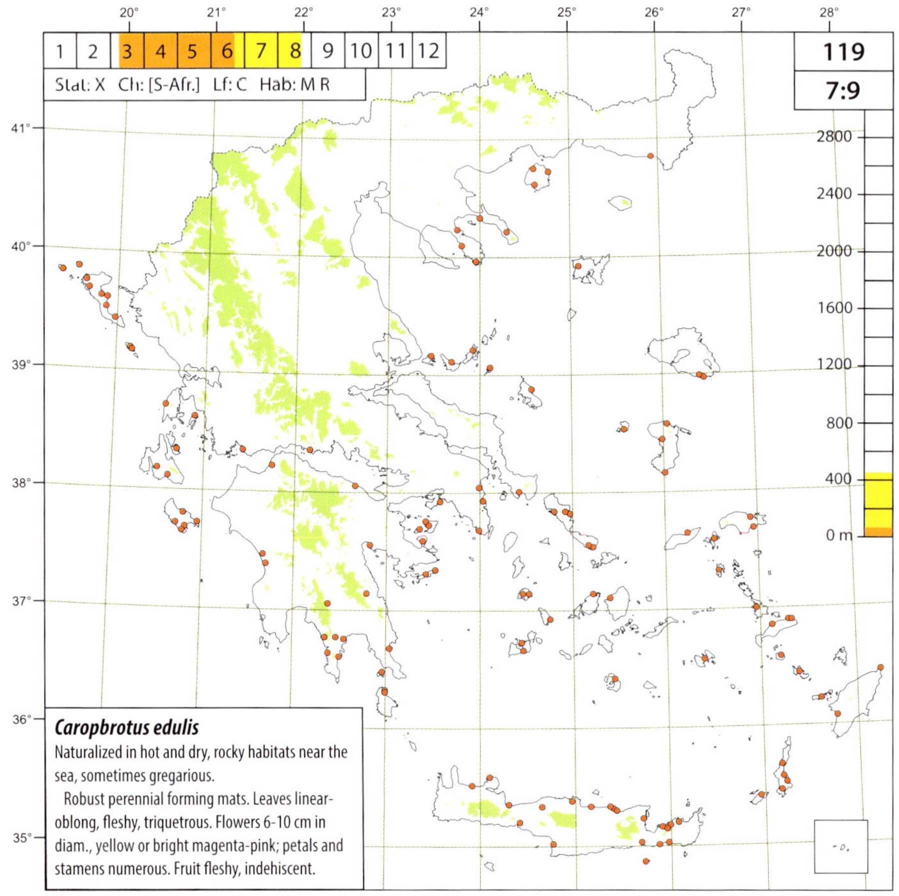

Carpobrotus edulis

Naturalized in hot and dry, rocky habitats near the sea, sometimes gregarious.

Robust perennial forming mats. Leaves linear-oblong, fleshy, triquetrous. Flowers 6-10 cm in diam., yellow or bright magenta-pink; petals and stamens numerous. Fruit fleshy, indehiscent.

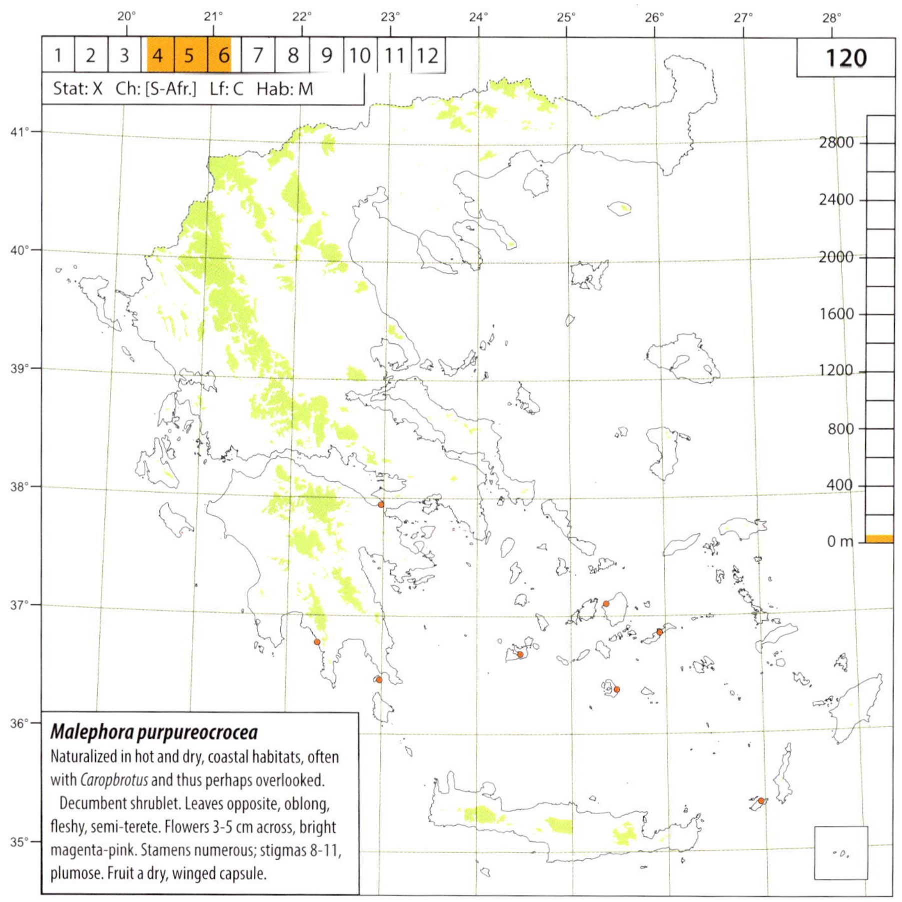

Malephora purpureocrocea

Naturalized in hot and dry, coastal habitats, often with *Carpobrotus* and thus perhaps overlooked.

Decumbent shrublet. Leaves opposite, oblong, fleshy, semi-terete. Flowers 3-5 cm across, bright magenta-pink. Stamens numerous; stigmas 8-11, plumose. Fruit a dry, winged capsule.

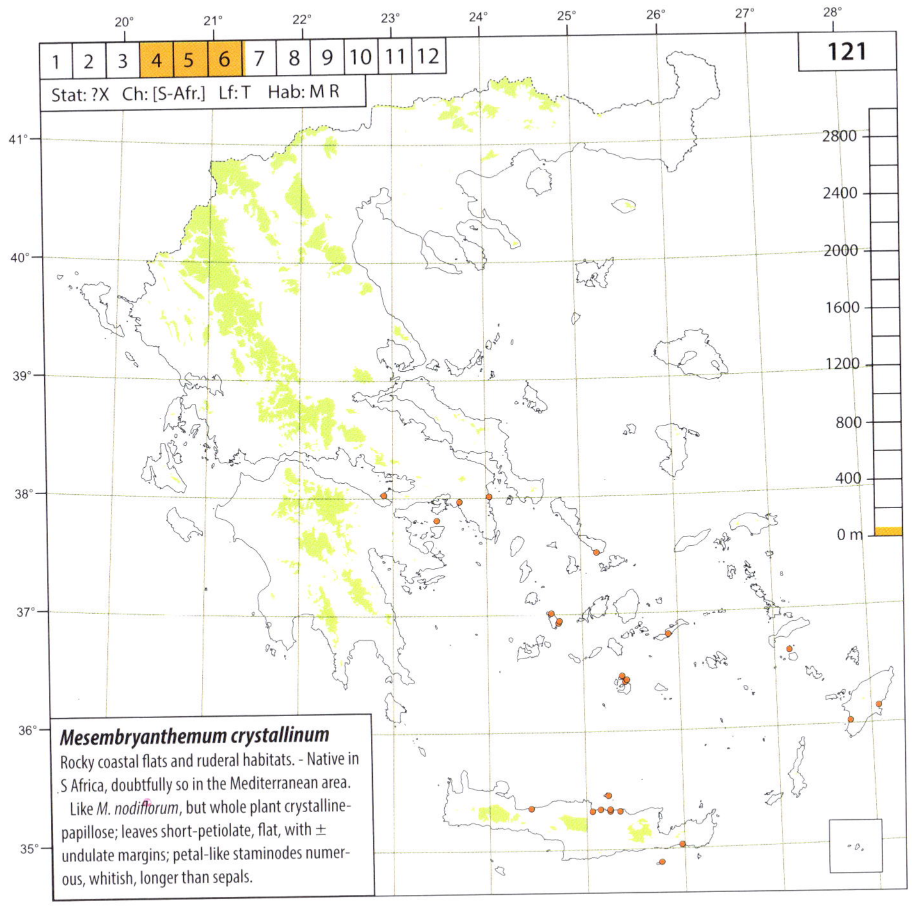

Mesembryanthemum crystallinum

Rocky coastal flats and ruderal habitats. - Native in S Africa, doubtfully so in the Mediterranean area.

Like *M. nodiflorum*, but whole plant crystalline-papillose; leaves short-petiolate, flat, with ± undulate margins; petal-like staminodes numerous, whitish, longer than sepals.

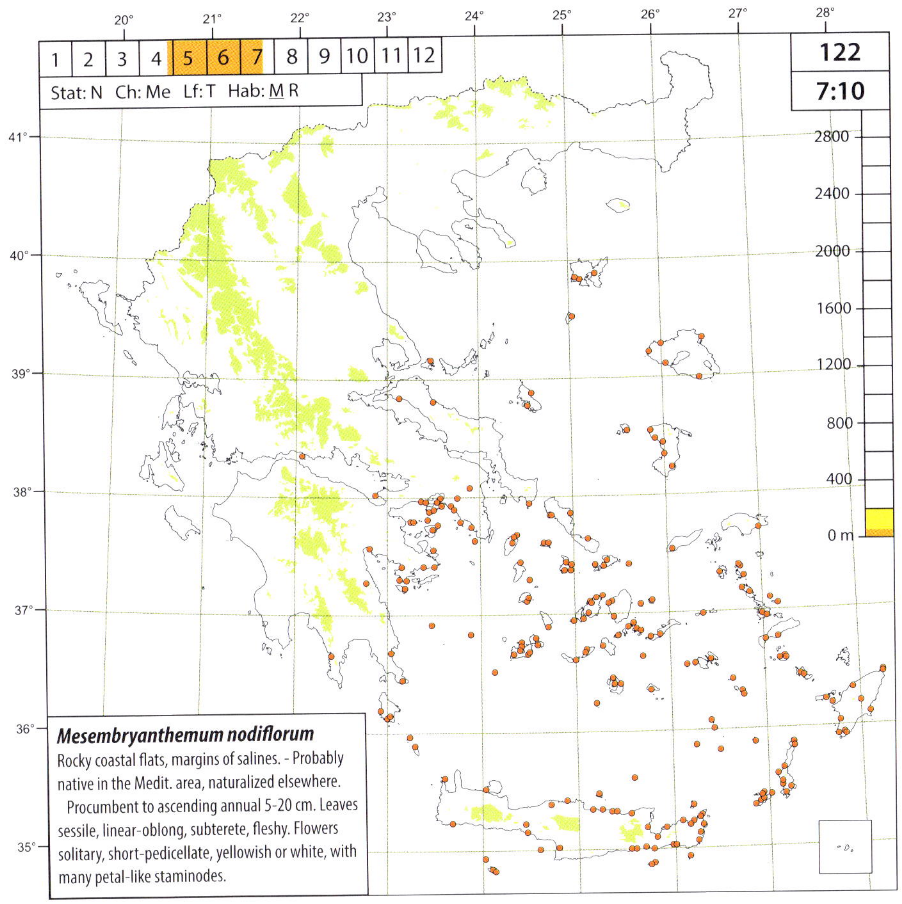

Mesembryanthemum nodiflorum

Rocky coastal flats, margins of salines. - Probably native in the Medit. area, naturalized elsewhere.

Procumbent to ascending annual 5-20 cm. Leaves sessile, linear-oblong, subterete, fleshy. Flowers solitary, short-pedicellate, yellowish or white, with many petal-like staminodes.

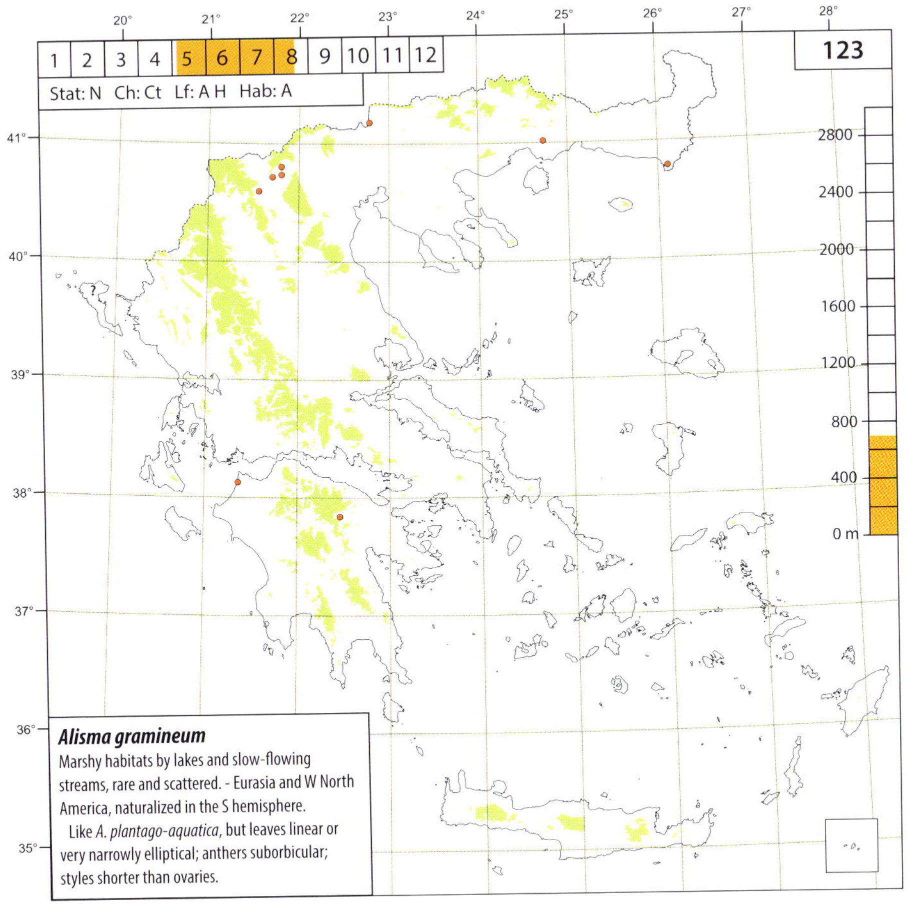

Alisma gramineum

Marshy habitats by lakes and slow-flowing streams, rare and scattered. - Eurasia and W North America, naturalized in the S hemisphere.

Like *A. plantago-aquatica*, but leaves linear or very narrowly elliptical; anthers suborbicular; styles shorter than ovaries.

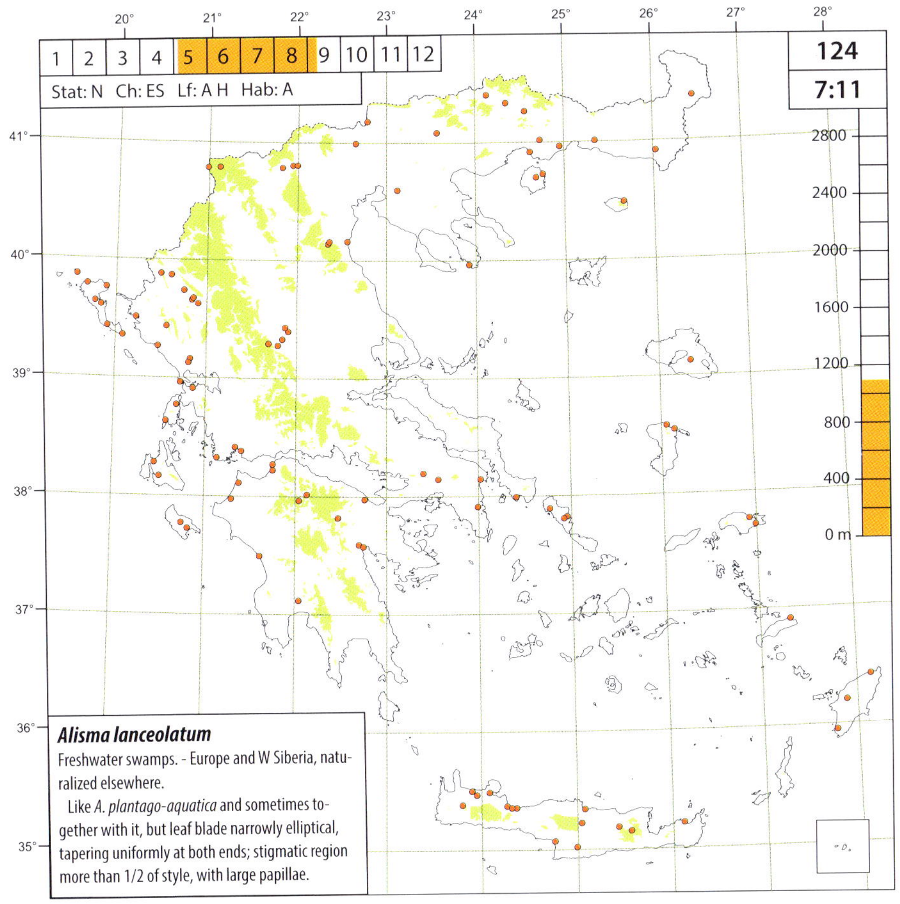

Alisma lanceolatum

Freshwater swamps. - Europe and W Siberia, naturalized elsewhere.

Like *A. plantago-aquatica* and sometimes together with it, but leaf blade narrowly elliptical, tapering uniformly at both ends; stigmatic region more than 1/2 of style, with large papillae.

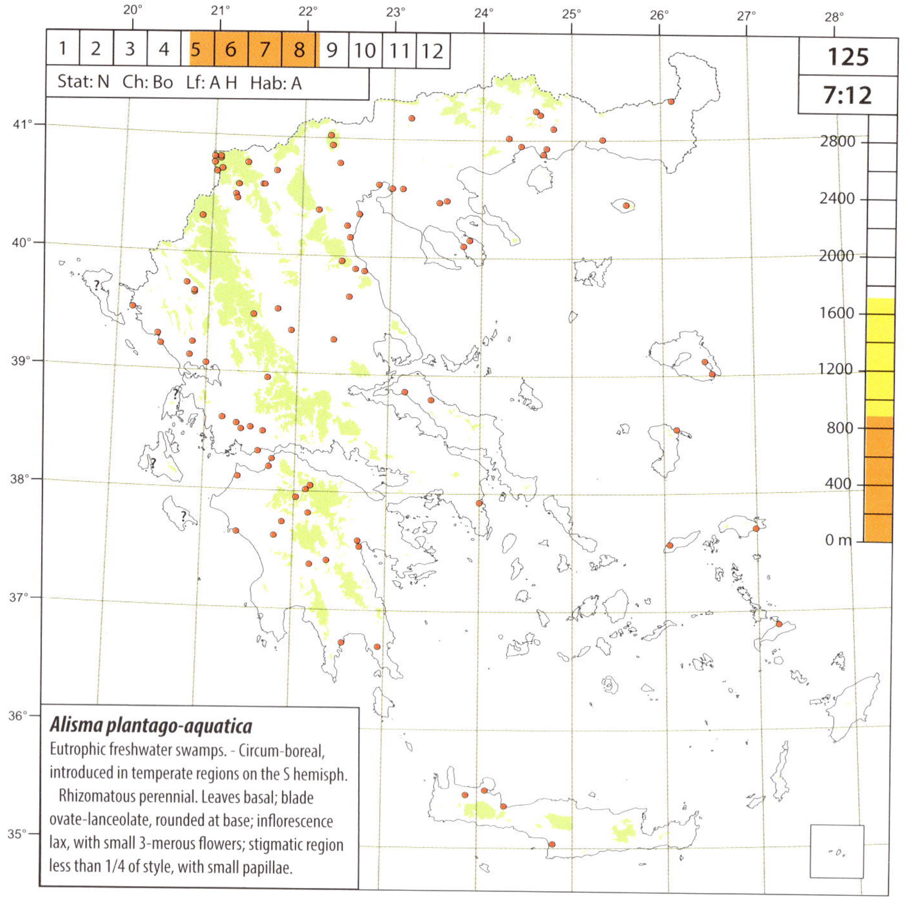

Alisma plantago-aquatica
Eutrophic freshwater swamps. - Circum-boreal, introduced in temperate regions on the S hemisph.
Rhizomatous perennial. Leaves basal; blade ovate-lanceolate, rounded at base; inflorescence lax, with small 3-merous flowers; stigmatic region less than 1/4 of style, with small papillae.

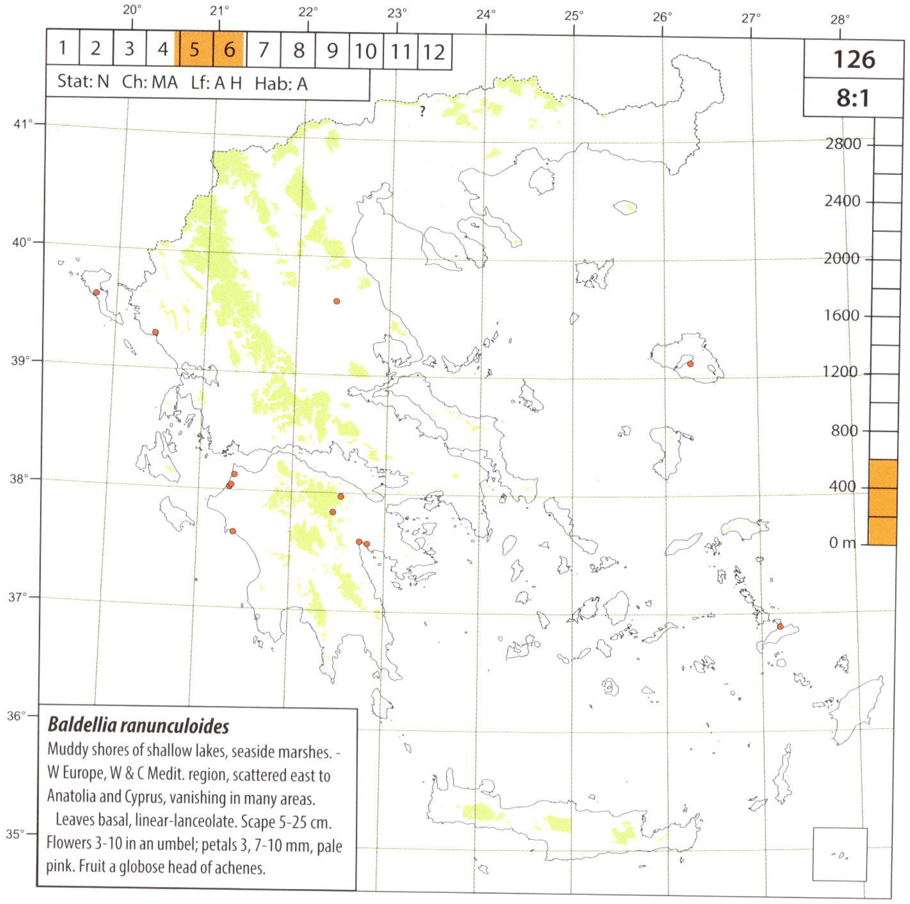

Baldellia ranunculoides
Muddy shores of shallow lakes, seaside marshes. - W Europe, W & C Medit. region, scattered east to Anatolia and Cyprus, vanishing in many areas.
Leaves basal, linear-lanceolate. Scape 5-25 cm. Flowers 3-10 in an umbel; petals 3, 7-10 mm, pale pink. Fruit a globose head of achenes.

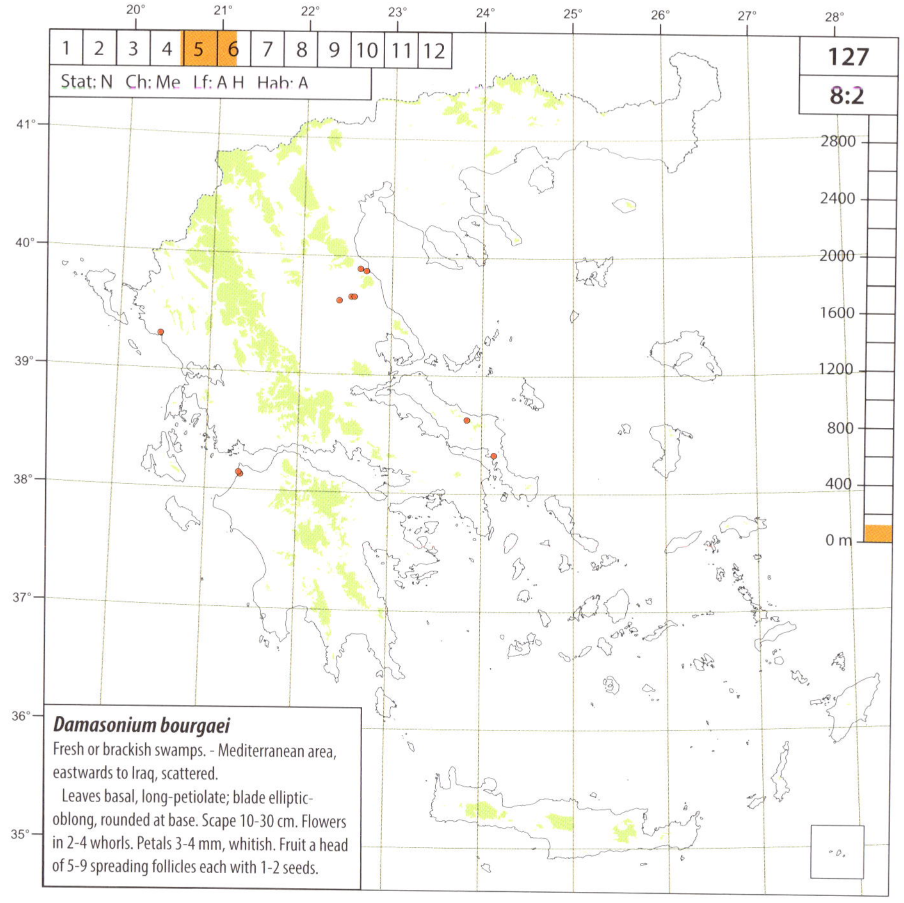

Damasonium bourgaei
Fresh or brackish swamps. - Mediterranean area, eastwards to Iraq, scattered.
Leaves basal, long-petiolate; blade elliptic-oblong, rounded at base. Scape 10-30 cm. Flowers in 2-4 whorls. Petals 3-4 mm, whitish. Fruit a head of 5-9 spreading follicles each with 1-2 seeds.

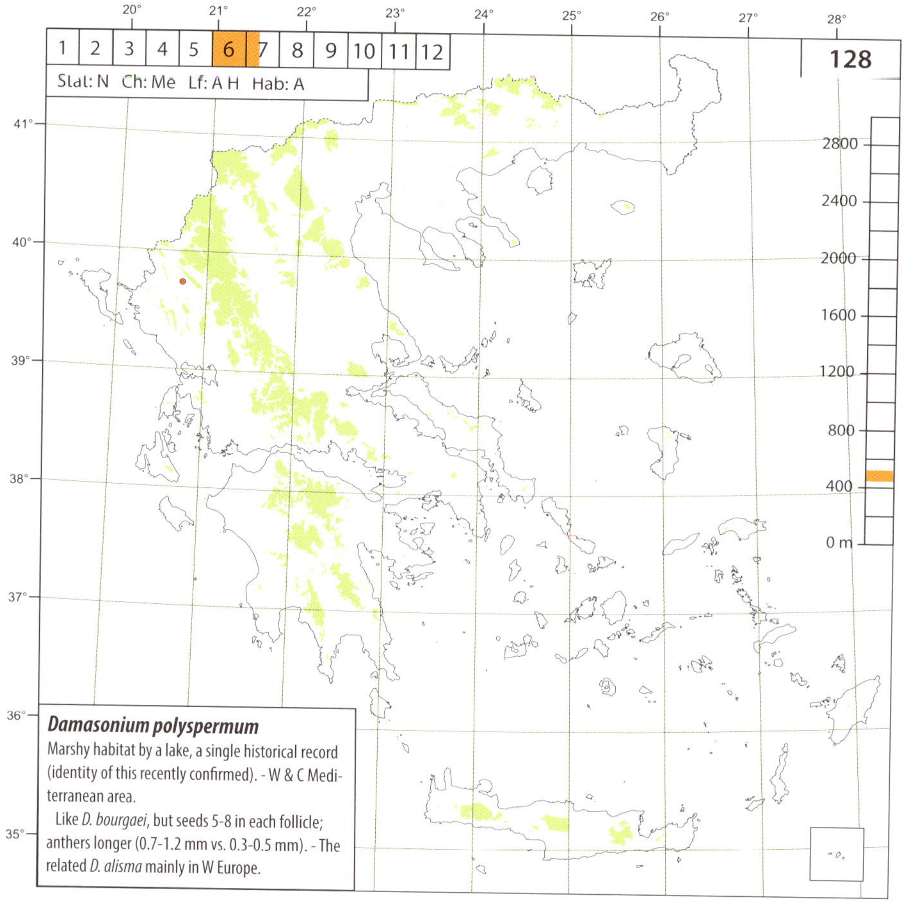

Damasonium polyspermum
Marshy habitat by a lake, a single historical record (identity of this recently confirmed). - W & C Mediterranean area.
Like *D. bourgaei*, but seeds 5-8 in each follicle; anthers longer (0.7-1.2 mm vs. 0.3-0.5 mm). - The related *D. alisma* mainly in W Europe.

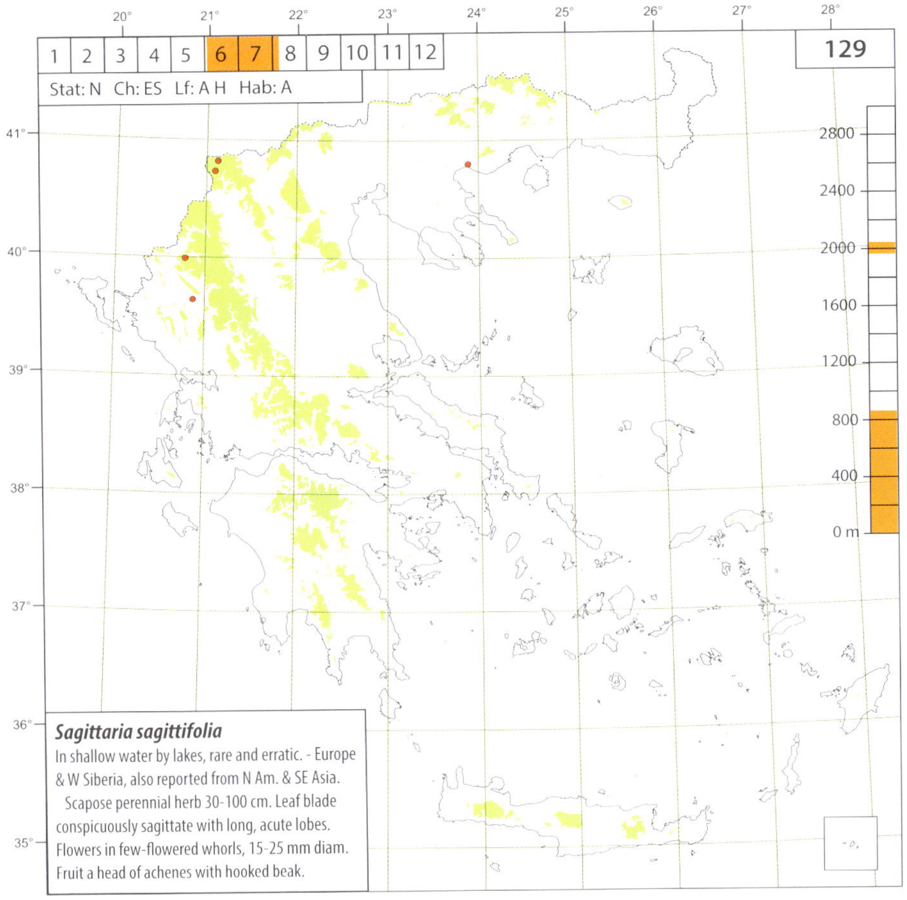

Sagittaria sagittifolia

In shallow water by lakes, rare and erratic. - Europe & W Siberia, also reported from N Am. & SE Asia.

Scapose perennial herb 30-100 cm. Leaf blade conspicuously sagittate with long, acute lobes. Flowers in few-flowered whorls, 15-25 mm diam. Fruit a head of achenes with hooked beak.

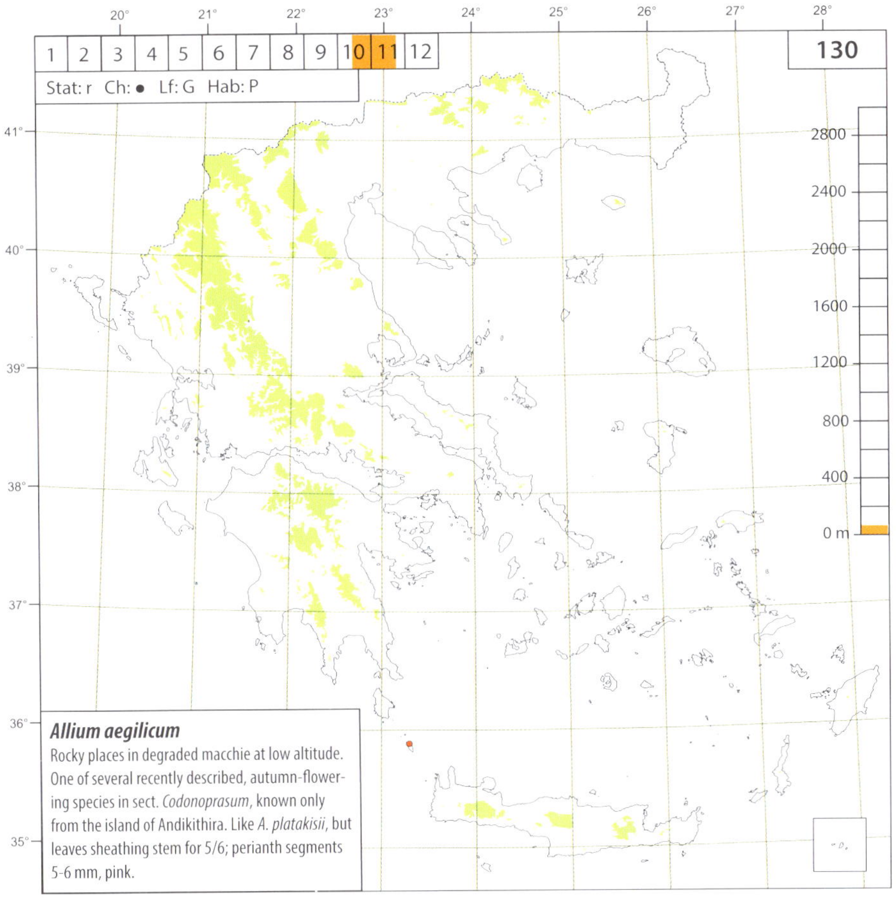

Allium aegilicum

Rocky places in degraded macchie at low altitude. One of several recently described, autumn-flowering species in sect. *Codonoprasum*, known only from the island of Andikithira. Like *A. platakisii*, but leaves sheathing stem for 5/6; perianth segments 5-6 mm, pink.

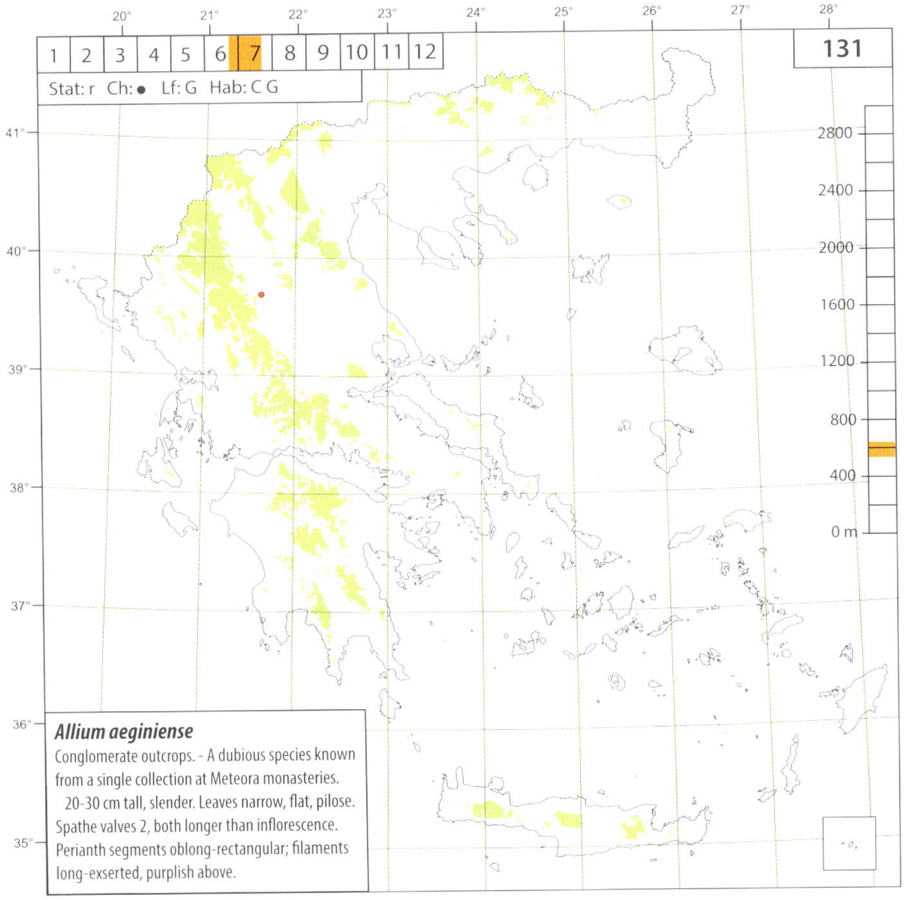

Allium aeginiense

Conglomerate outcrops. - A dubious species known from a single collection at Meteora monasteries.

20-30 cm tall, slender. Leaves narrow, flat, pilose. Spathe valves 2, both longer than inflorescence. Perianth segments oblong-rectangular; filaments long-exserted, purplish above.

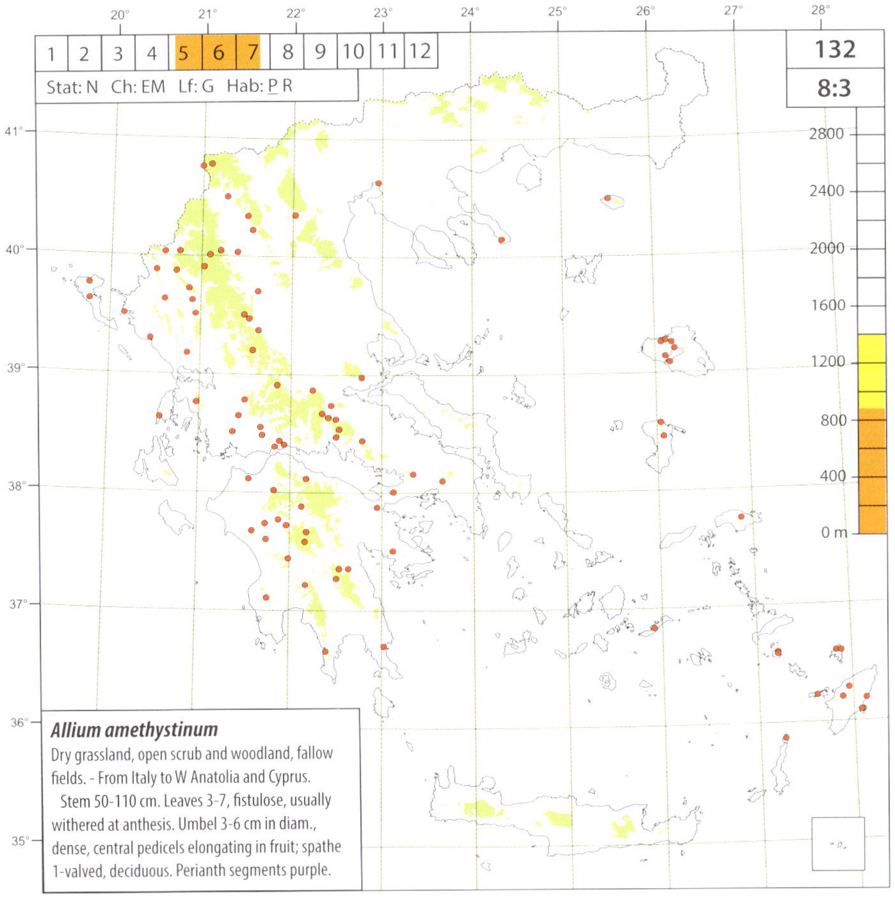

Allium amethystinum

Dry grassland, open scrub and woodland, fallow fields. - From Italy to W Anatolia and Cyprus.

Stem 50-110 cm. Leaves 3-7, fistulose, usually withered at anthesis. Umbel 3-6 cm in diam., dense, central pedicels elongating in fruit; spathe 1-valved, deciduous. Perianth segments purple.

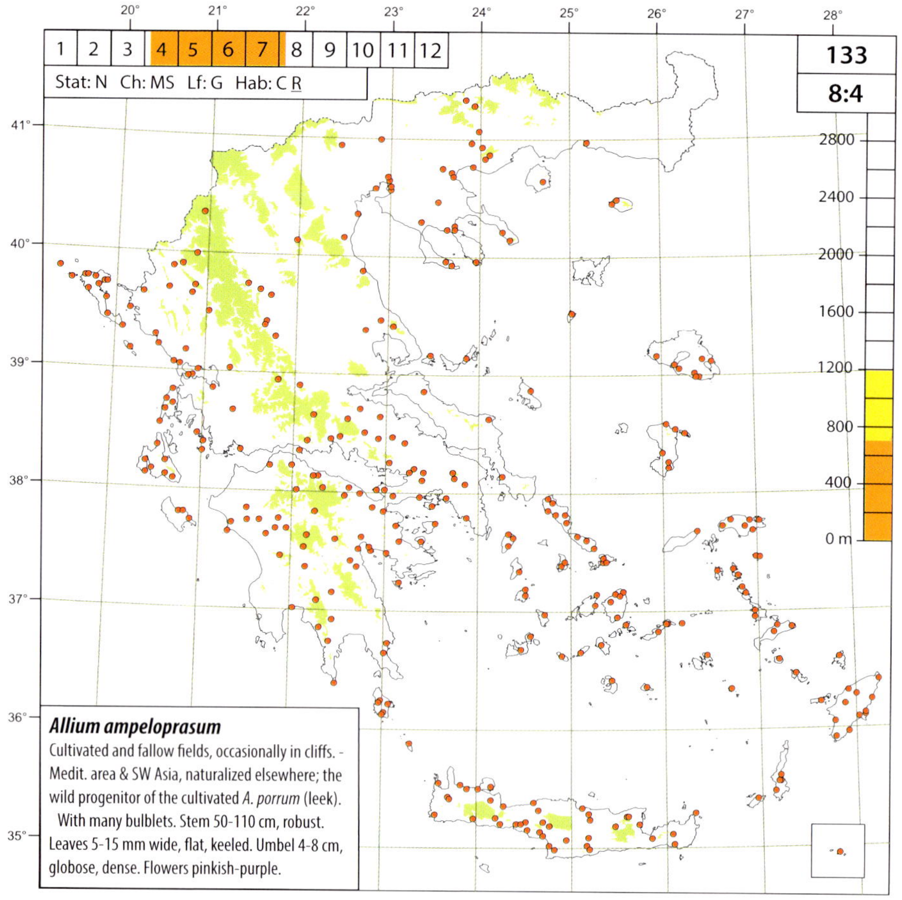

Allium ampeloprasum

Cultivated and fallow fields, occasionally in cliffs. - Medit. area & SW Asia, naturalized elsewhere; the wild progenitor of the cultivated *A. porrum* (leek).

With many bulblets. Stem 50-110 cm, robust. Leaves 5-15 mm wide, flat, keeled. Umbel 4-8 cm, globose, dense. Flowers pinkish-purple.

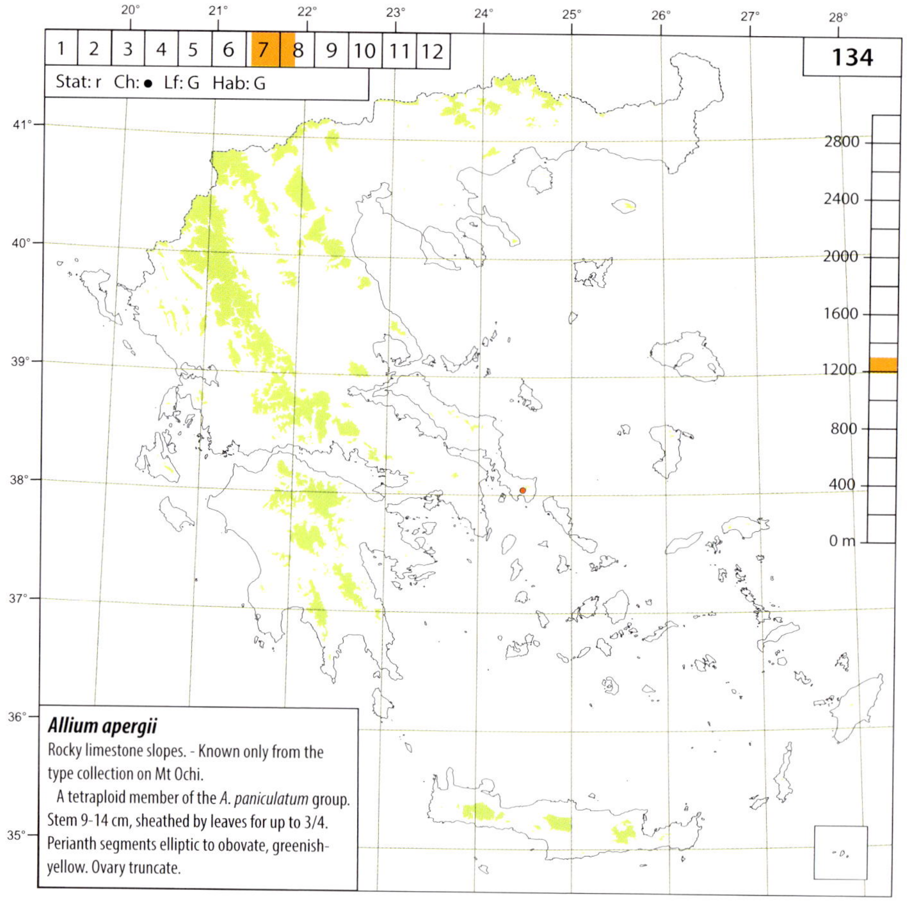

Allium apergii

Rocky limestone slopes. - Known only from the type collection on Mt Ochi.

A tetraploid member of the *A. paniculatum* group. Stem 9-14 cm, sheathed by leaves for up to 3/4. Perianth segments elliptic to obovate, greenish-yellow. Ovary truncate.

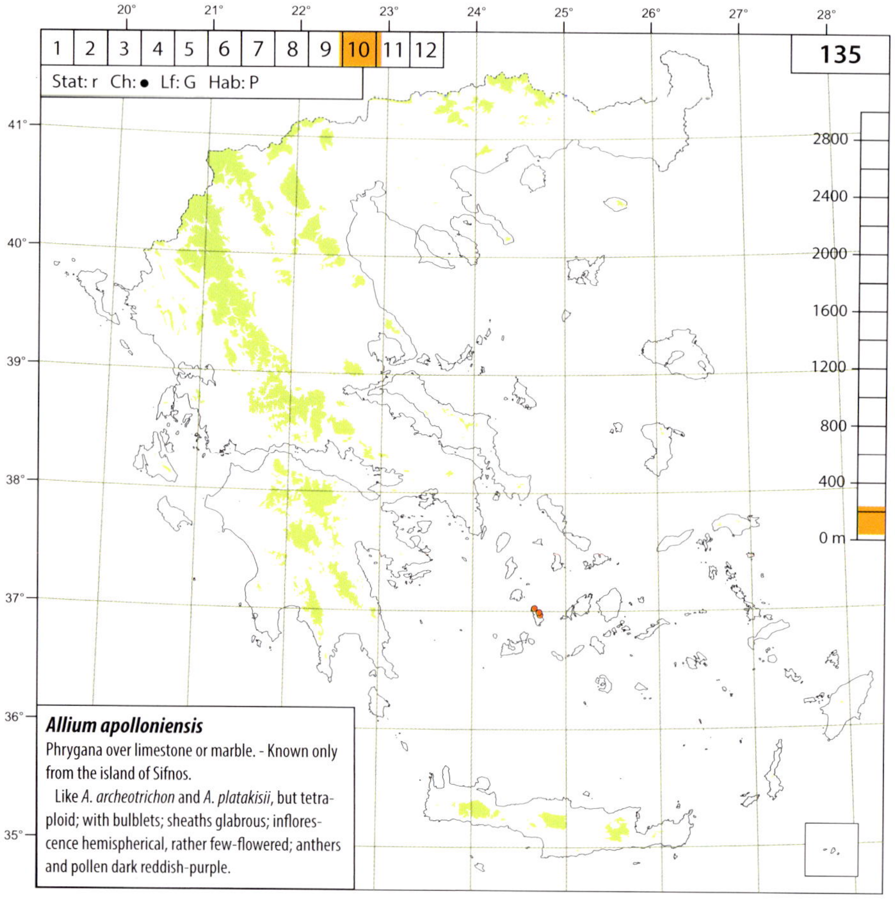

Allium apolloniensis

Phrygana over limestone or marble. - Known only from the island of Sifnos.

Like *A. archeotrichon* and *A. platakisii*, but tetraploid; with bulblets; sheaths glabrous; inflorescence hemispherical, rather few-flowered; anthers and pollen dark reddish-purple.

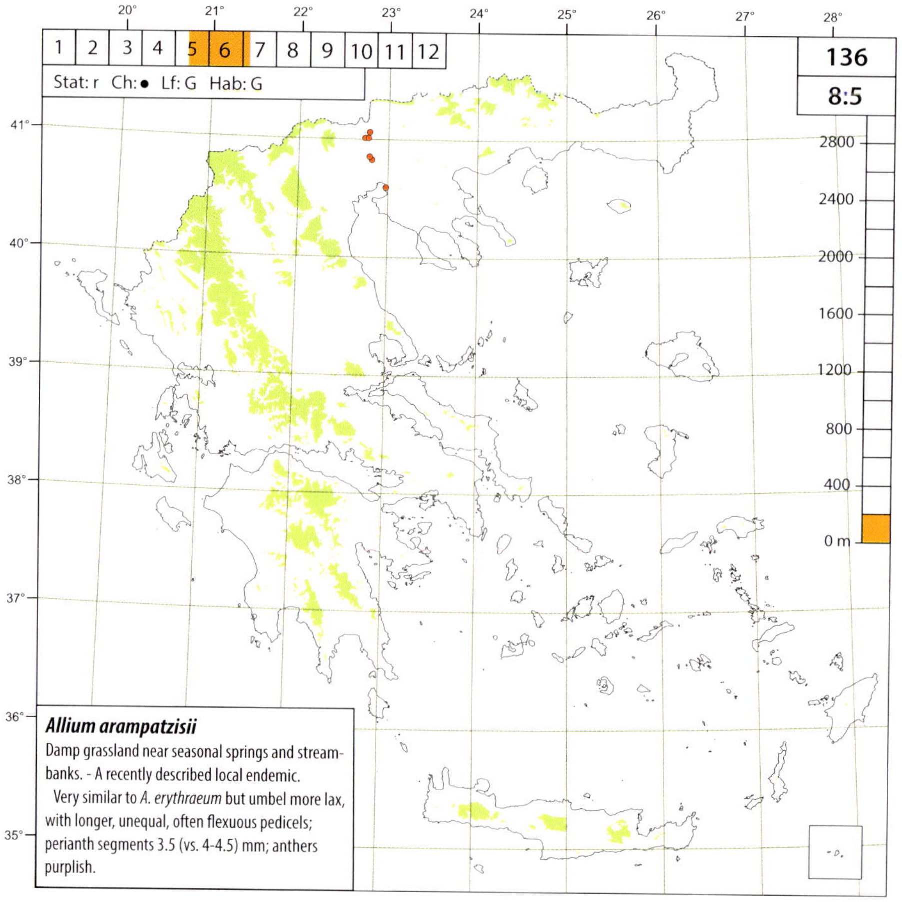

Allium arampatzisii

Damp grassland near seasonal springs and stream-banks. - A recently described local endemic.

Very similar to *A. erythraeum* but umbel more lax, with longer, unequal, often flexuous pedicels; perianth segments 3.5 (vs. 4-4.5) mm; anthers purplish.

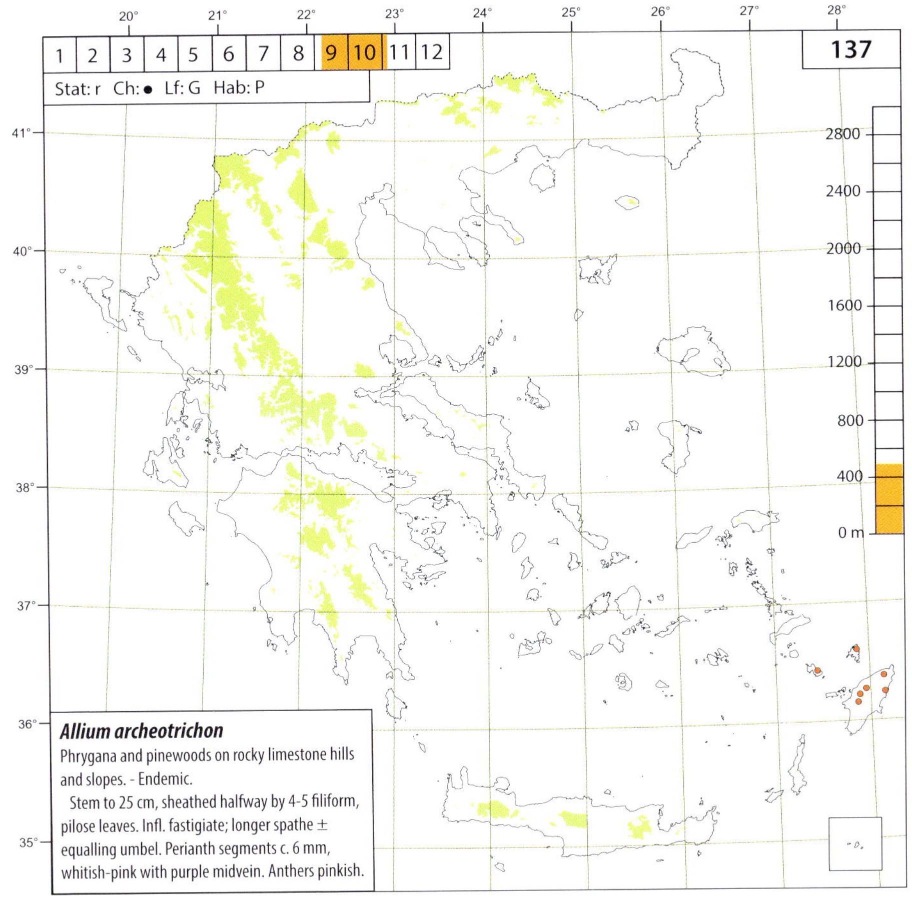

Allium archeotrichon

Phrygana and pinewoods on rocky limestone hills and slopes. - Endemic.

Stem to 25 cm, sheathed halfway by 4-5 filiform, pilose leaves. Infl. fastigiate; longer spathe ± equalling umbel. Perianth segments c. 6 mm, whitish-pink with purple midvein. Anthers pinkish.

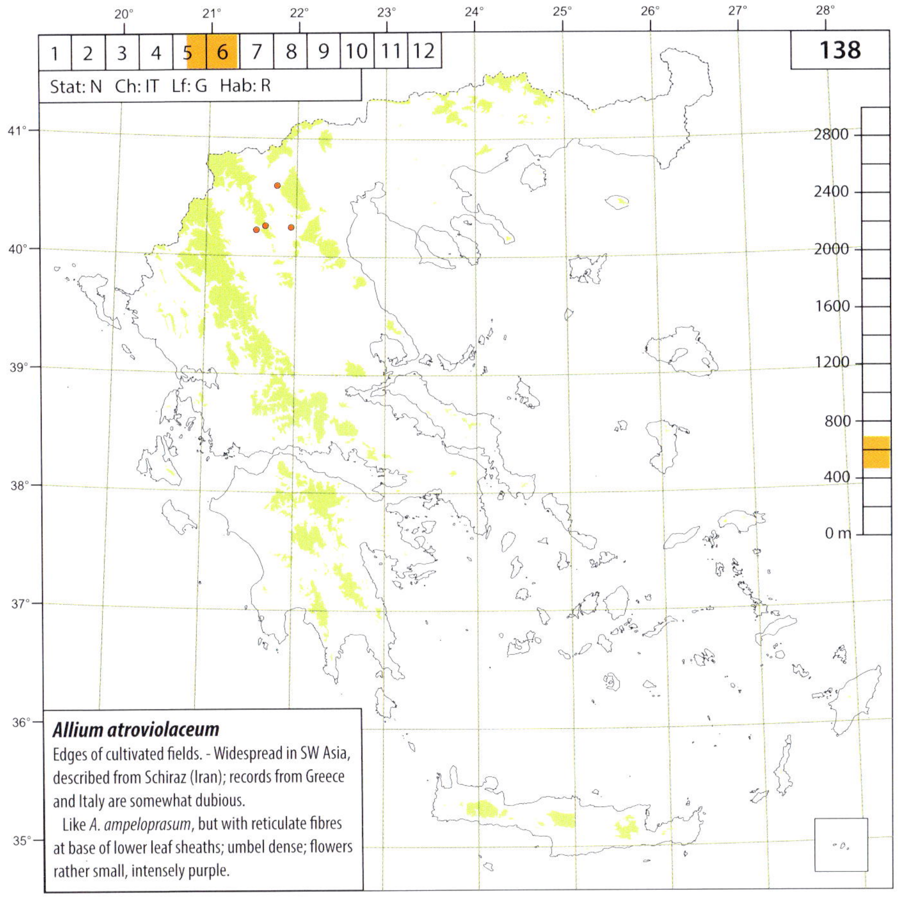

Allium atroviolaceum

Edges of cultivated fields. - Widespread in SW Asia, described from Schiraz (Iran); records from Greece and Italy are somewhat dubious.

Like *A. ampeloprasum*, but with reticulate fibres at base of lower leaf sheaths; umbel dense; flowers rather small, intensely purple.

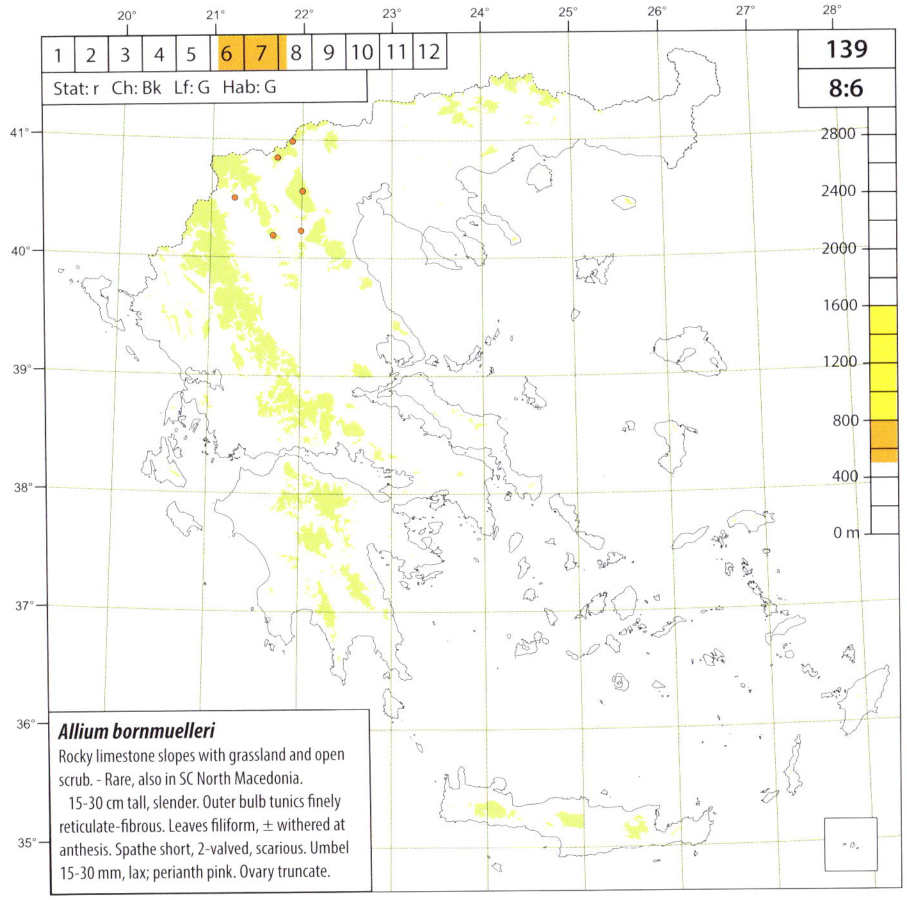

Allium bornmuelleri

Rocky limestone slopes with grassland and open scrub. - Rare, also in SC North Macedonia.

15-30 cm tall, slender. Outer bulb tunics finely reticulate-fibrous. Leaves filiform, ± withered at anthesis. Spathe short, 2-valved, scarious. Umbel 15-30 mm, lax; perianth pink. Ovary truncate.

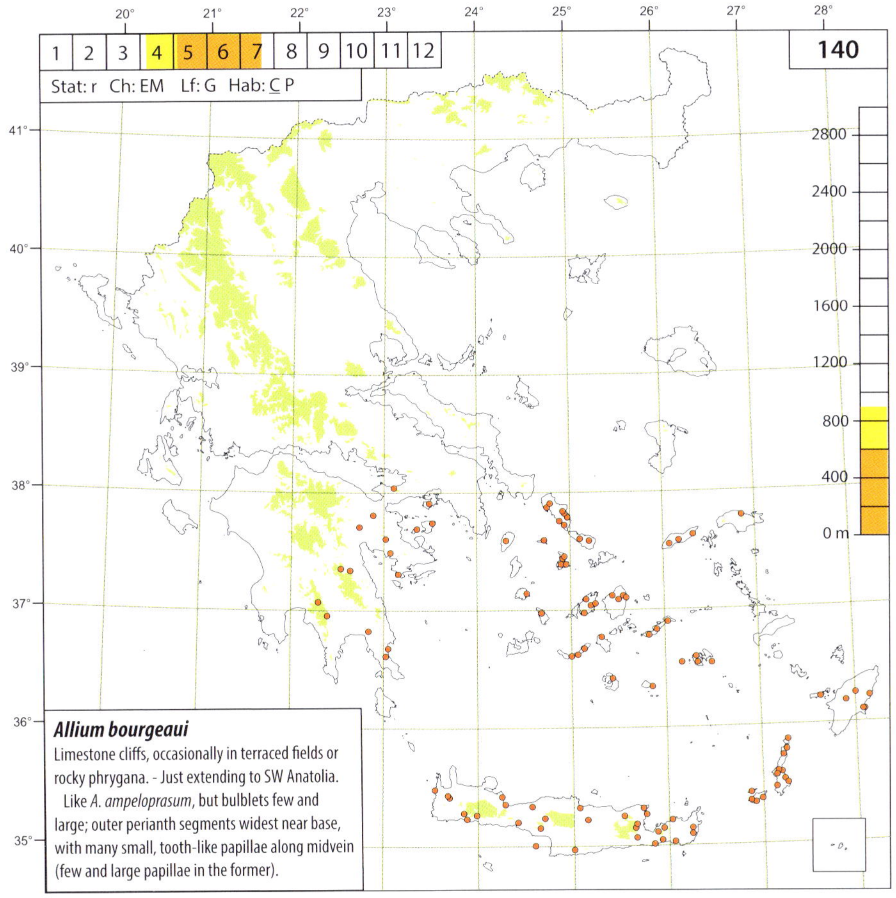

Allium bourgeaui

Limestone cliffs, occasionally in terraced fields or rocky phrygana. - Just extending to SW Anatolia.

Like *A. ampeloprasum*, but bulblets few and large; outer perianth segments widest near base, with many small, tooth-like papillae along midvein (few and large papillae in the former).

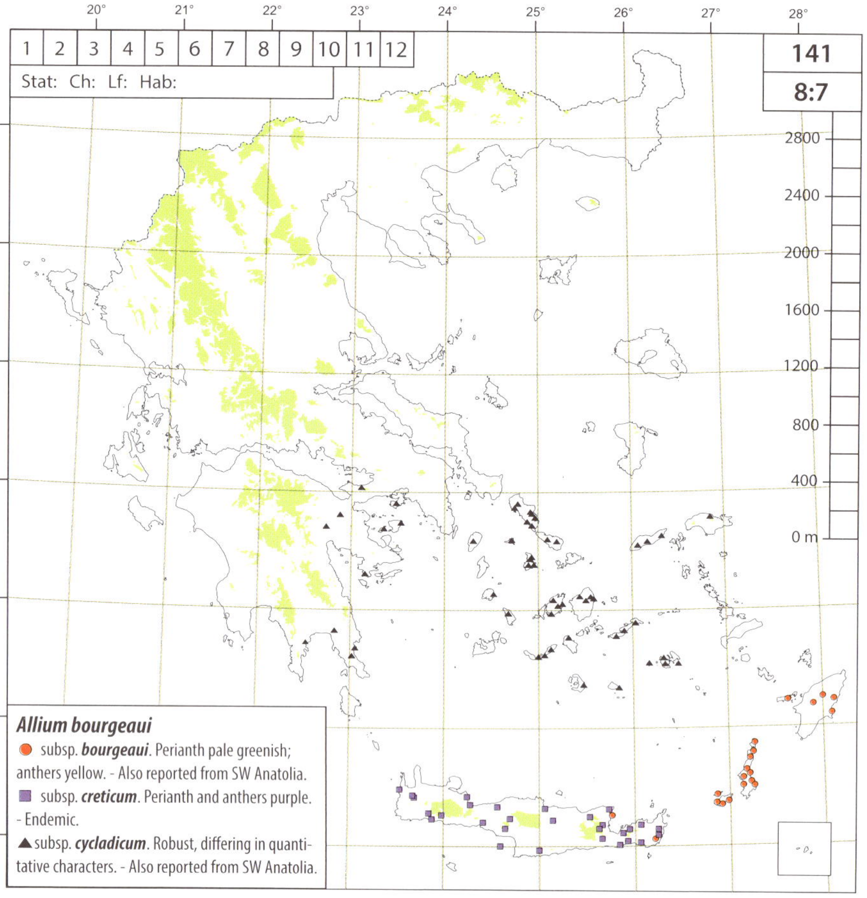

Allium bourgeaui

● subsp. ***bourgeaui***. Perianth pale greenish; anthers yellow. - Also reported from SW Anatolia.
■ subsp. ***creticum***. Perianth and anthers purple. - Endemic.
▲ subsp. ***cycladicum***. Robust, differing in quantitative characters. - Also reported from SW Anatolia.

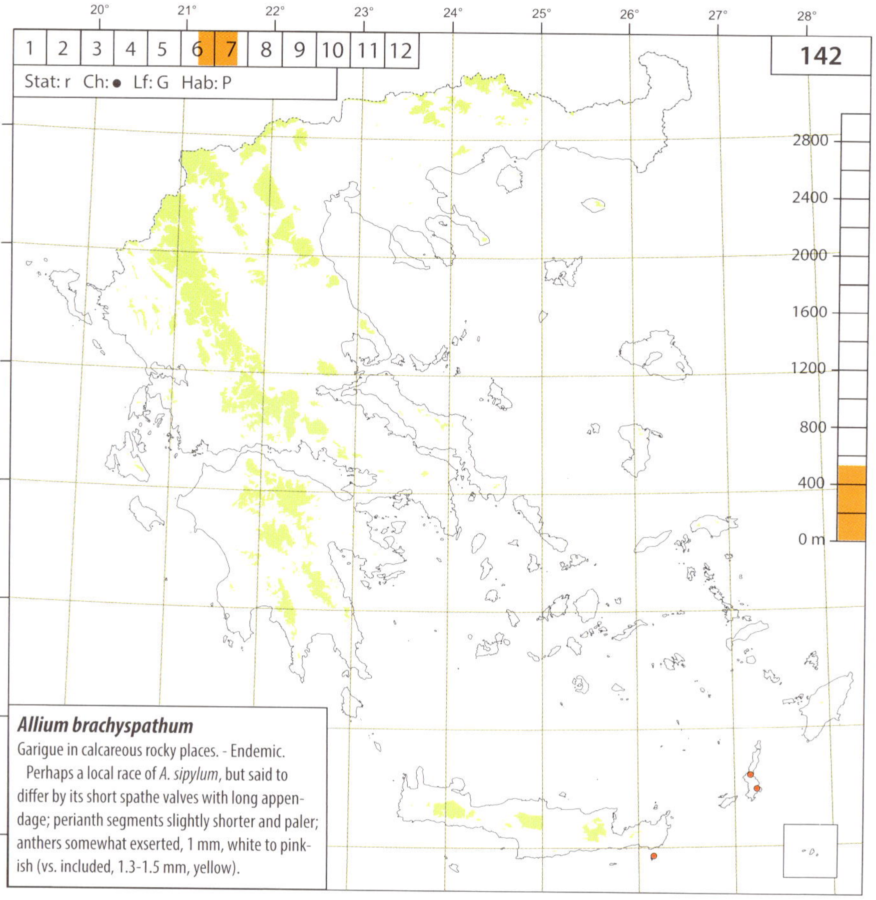

Allium brachyspathum

Garigue in calcareous rocky places. - Endemic.

Perhaps a local race of *A. sipylum*, but said to differ by its short spathe valves with long appendage; perianth segments slightly shorter and paler; anthers somewhat exserted, 1 mm, white to pinkish (vs. included, 1.3-1.5 mm, yellow).

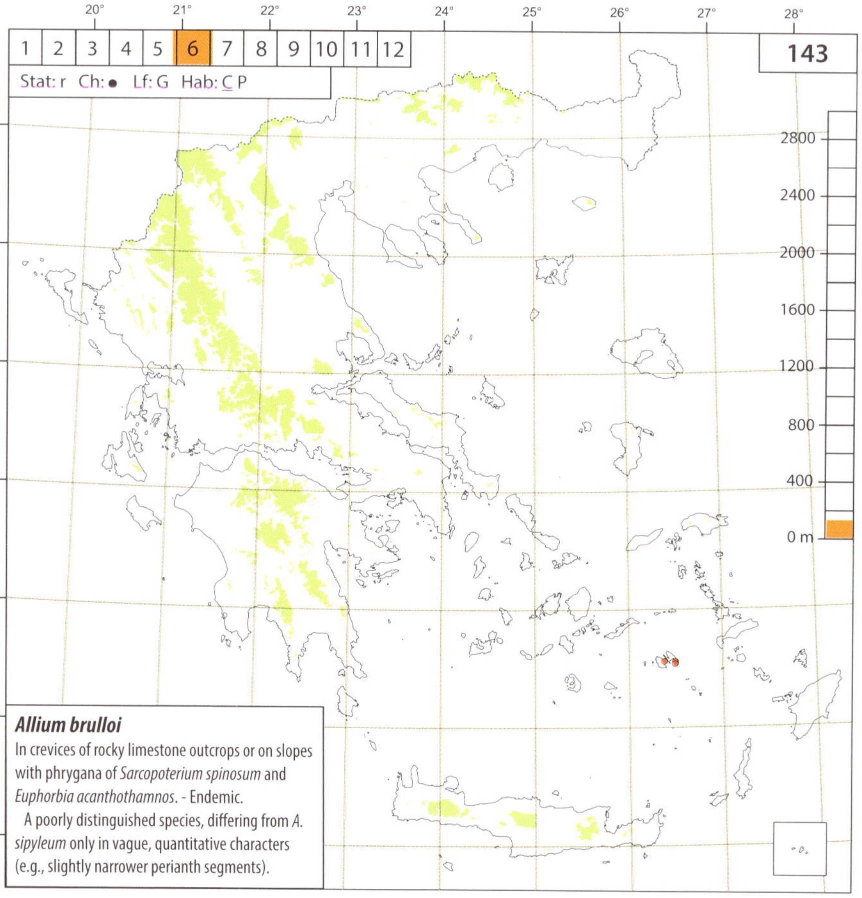

Allium brulloi

In crevices of rocky limestone outcrops or on slopes with phrygana of *Sarcopoterium spinosum* and *Euphorbia acanthothamnos*. - Endemic.

A poorly distinguished species, differing from *A. sipyleum* only in vague, quantitative characters (e.g., slightly narrower perianth segments).

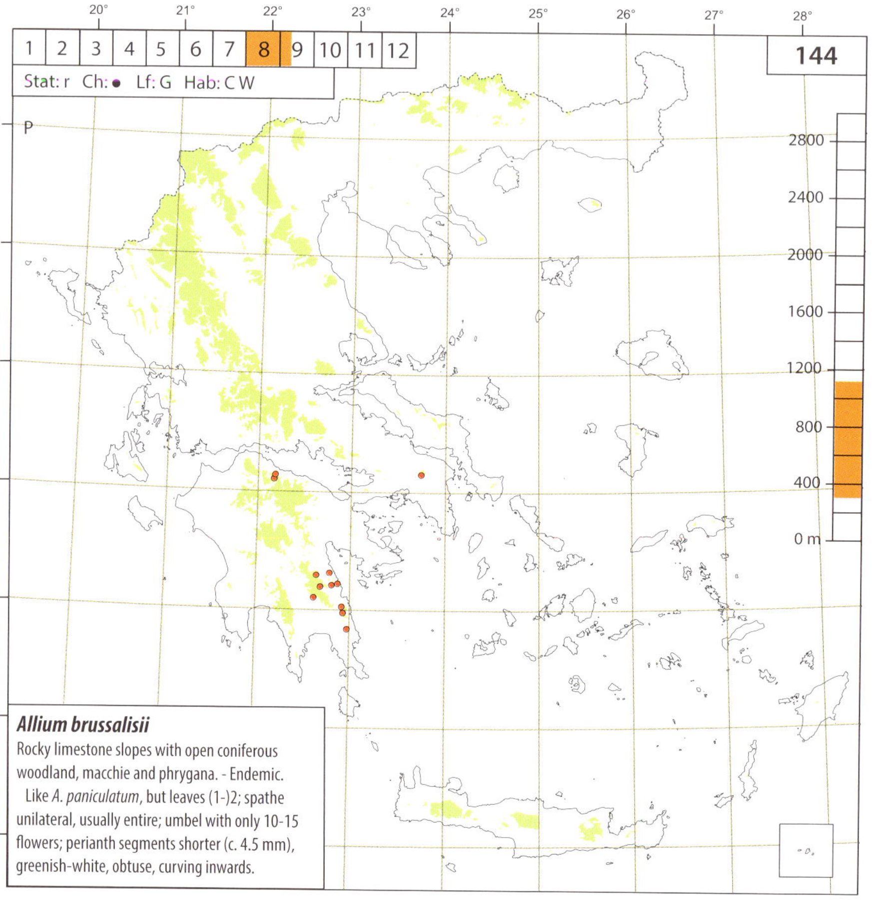

Allium brussalisii

Rocky limestone slopes with open coniferous woodland, macchie and phrygana. - Endemic.

Like *A. paniculatum*, but leaves (1-)2; spathe unilateral, usually entire; umbel with only 10-15 flowers; perianth segments shorter (c. 4.5 mm), greenish-white, obtuse, curving inwards.

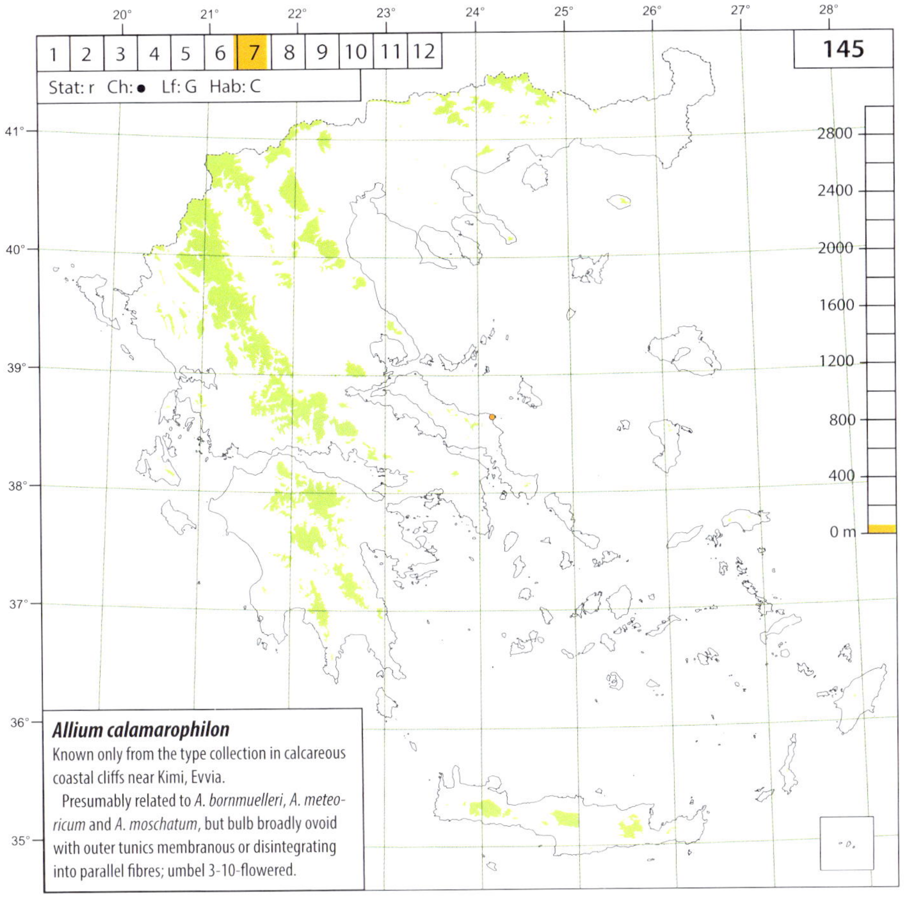

Allium calamarophilon
Known only from the type collection in calcareous coastal cliffs near Kimi, Evvia.
Presumably related to *A. bornmuelleri*, *A. meteoricum* and *A. moschatum*, but bulb broadly ovoid with outer tunics membranous or disintegrating into parallel fibres; umbel 3-10-flowered.

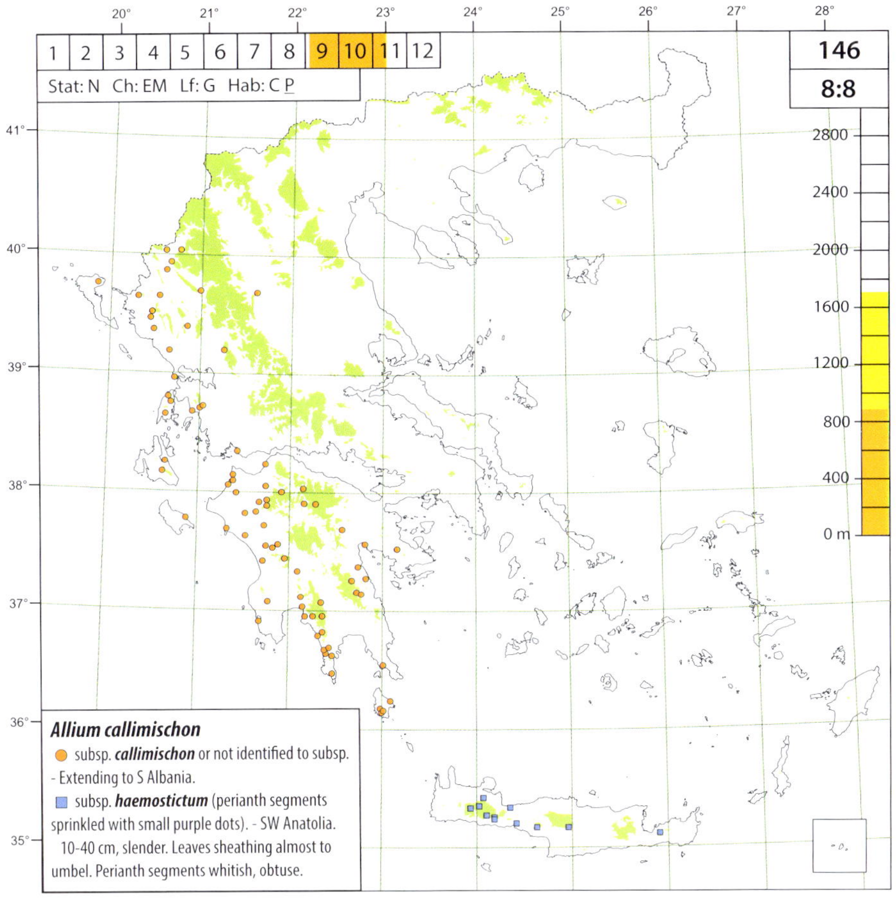

Allium callimischon
● subsp. ***callimischon*** or not identified to subsp. - Extending to S Albania.
■ subsp. ***haemostictum*** (perianth segments sprinkled with small purple dots). - SW Anatolia.
10-40 cm, slender. Leaves sheathing almost to umbel. Perianth segments whitish, obtuse.

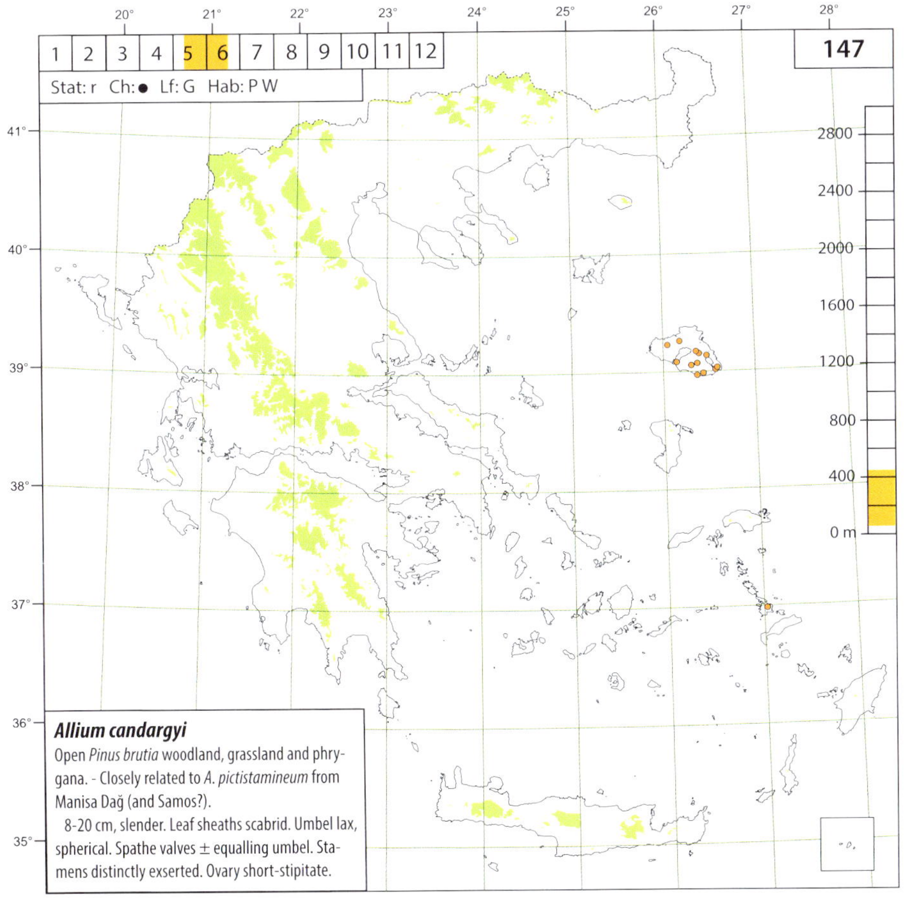

Allium candargyi
Open *Pinus brutia* woodland, grassland and phrygana. - Closely related to *A. pictistamineum* from Manisa Dağ (and Samos?).
8-20 cm, slender. Leaf sheaths scabrid. Umbel lax, spherical. Spathe valves ± equalling umbel. Stamens distinctly exserted. Ovary short-stipitate.

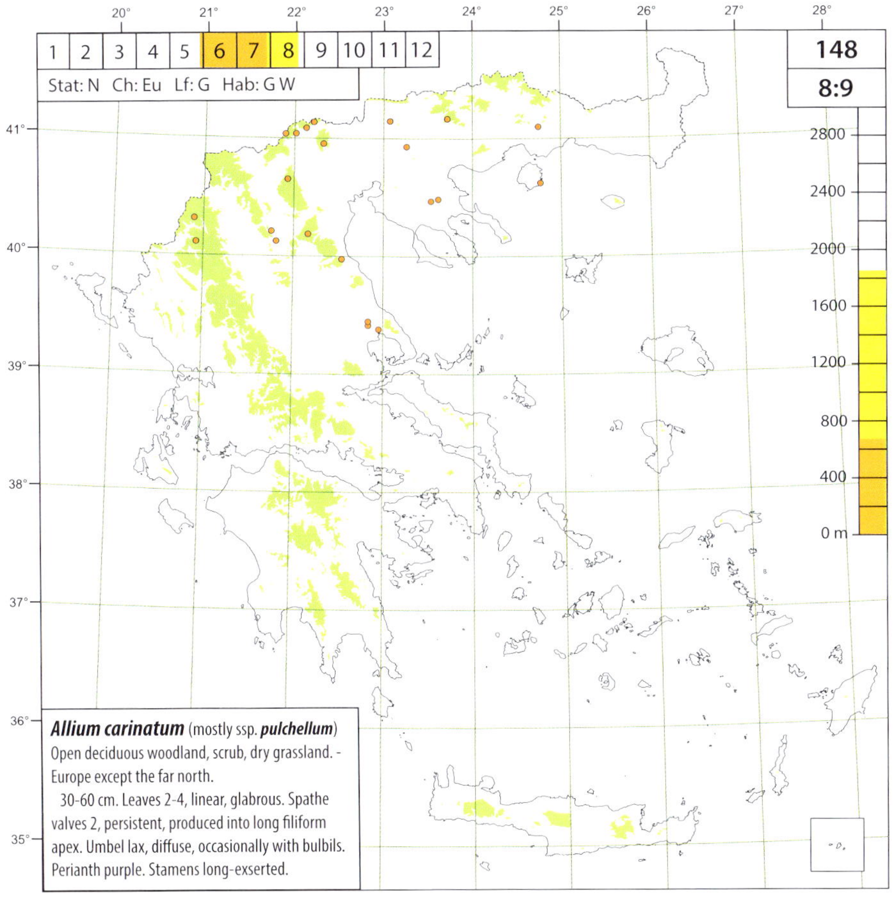

Allium carinatum (mostly ssp. ***pulchellum***)
Open deciduous woodland, scrub, dry grassland. - Europe except the far north.
30-60 cm. Leaves 2-4, linear, glabrous. Spathe valves 2, persistent, produced into long filiform apex. Umbel lax, diffuse, occasionally with bulbils. Perianth purple. Stamens long-exserted.

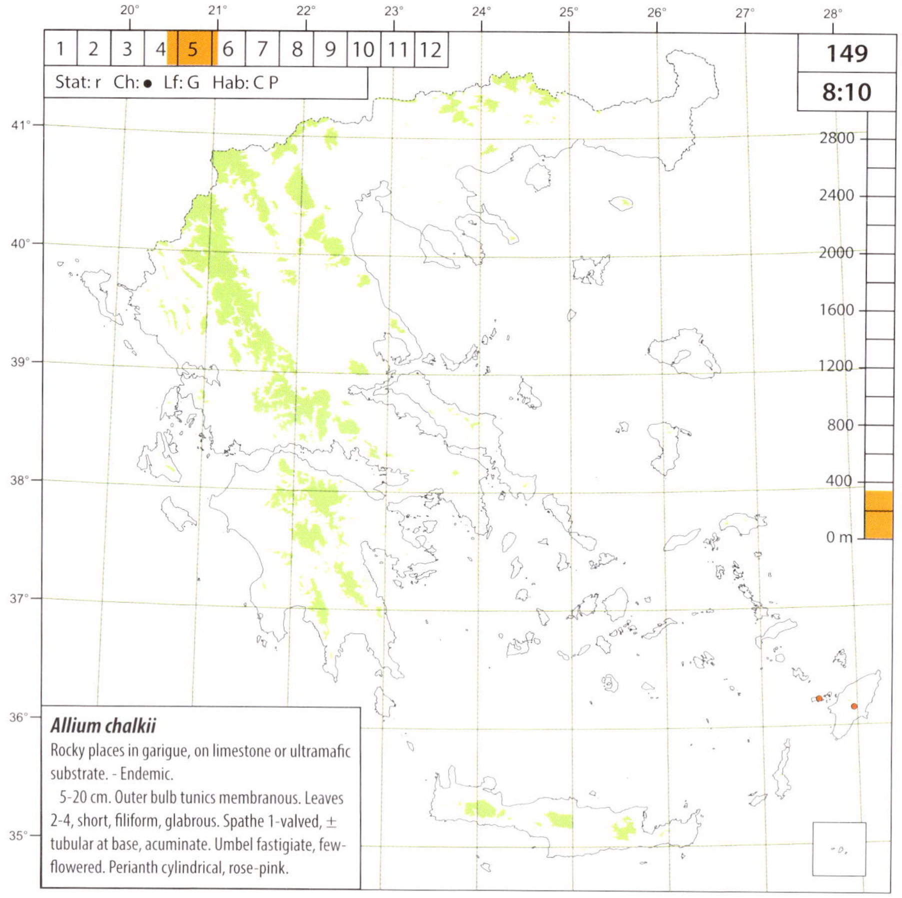

Allium chalkii

Rocky places in garigue, on limestone or ultramafic substrate. - Endemic.

5-20 cm. Outer bulb tunics membranous. Leaves 2-4, short, filiform, glabrous. Spathe 1-valved, ± tubular at base, acuminate. Umbel fastigiate, few-flowered. Perianth cylindrical, rose-pink.

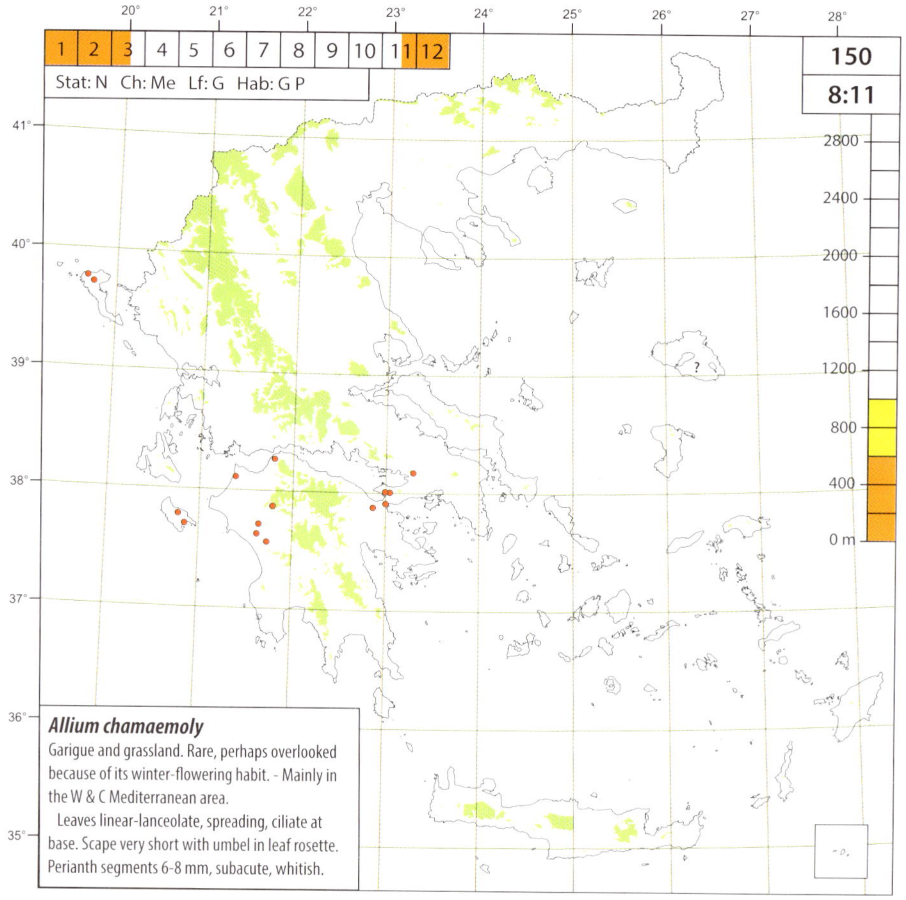

Allium chamaemoly

Garigue and grassland. Rare, perhaps overlooked because of its winter-flowering habit. - Mainly in the W & C Mediterranean area.

Leaves linear-lanceolate, spreading, ciliate at base. Scape very short with umbel in leaf rosette. Perianth segments 6-8 mm, subacute, whitish.

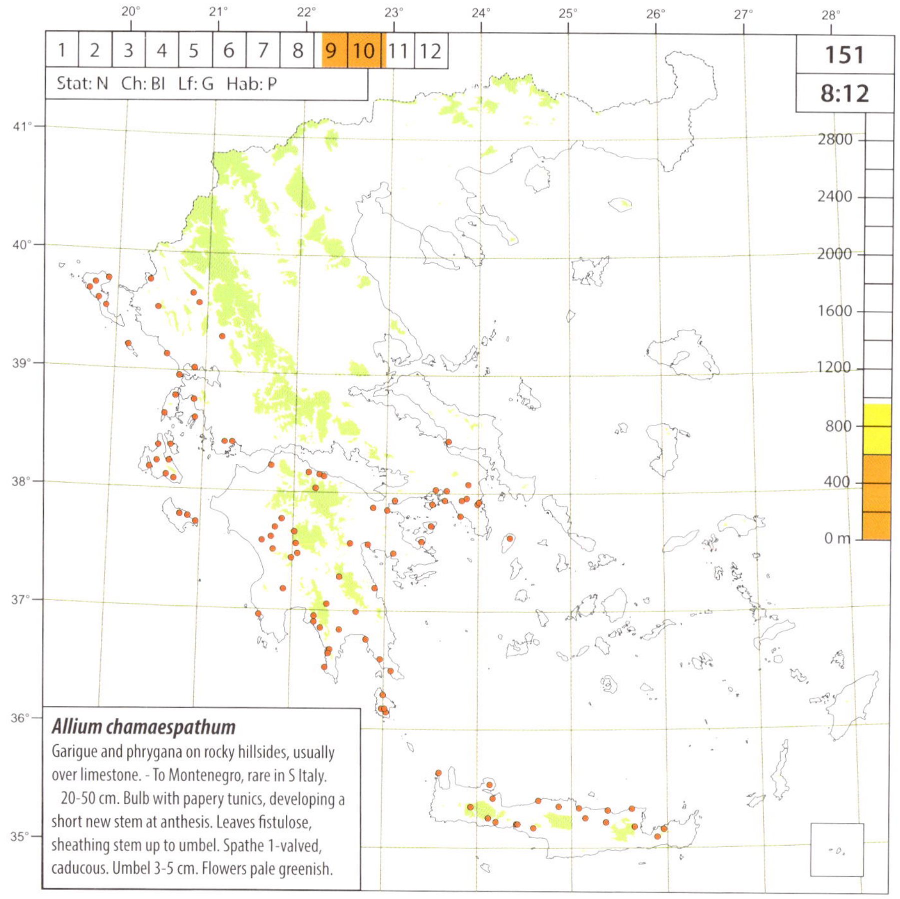

Allium chamaespathum

Garigue and phrygana on rocky hillsides, usually over limestone. - To Montenegro, rare in S Italy.

20-50 cm. Bulb with papery tunics, developing a short new stem at anthesis. Leaves fistulose, sheathing stem up to umbel. Spathe 1-valved, caducous. Umbel 3-5 cm. Flowers pale greenish.

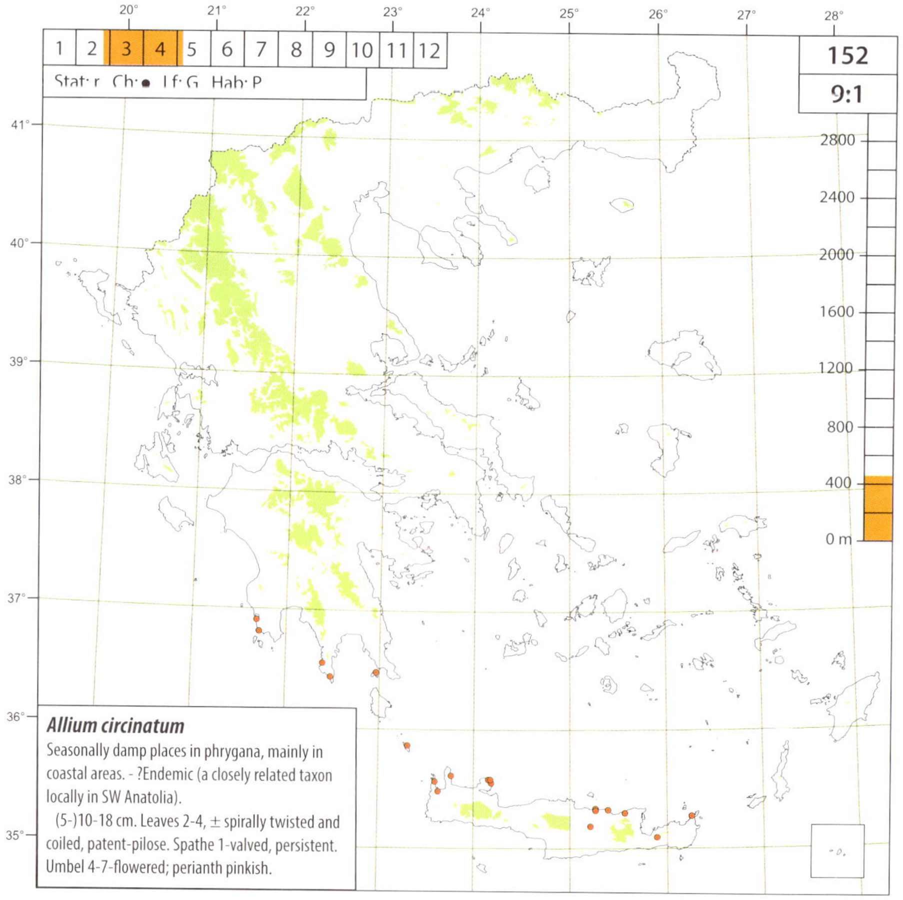

Allium circinatum

Seasonally damp places in phrygana, mainly in coastal areas. - ?Endemic (a closely related taxon locally in SW Anatolia).

(5-)10-18 cm. Leaves 2-4, ± spirally twisted and coiled, patent-pilose. Spathe 1-valved, persistent. Umbel 4-7-flowered; perianth pinkish.

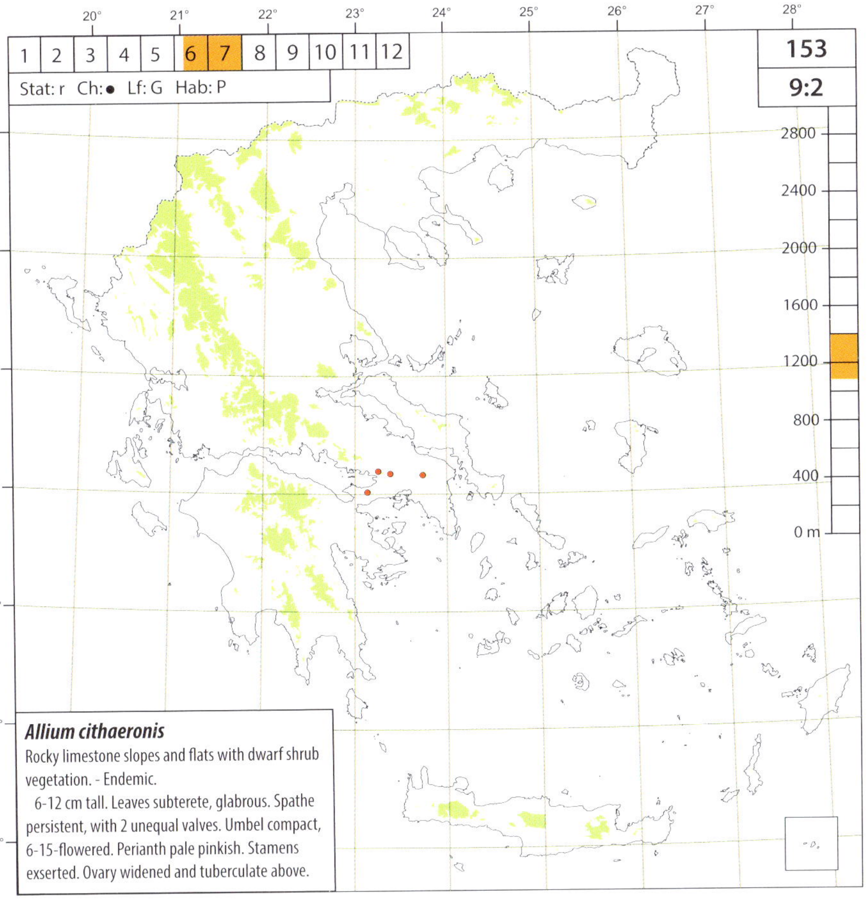

Allium cithaeronis

Rocky limestone slopes and flats with dwarf shrub vegetation. - Endemic.

6-12 cm tall. Leaves subterete, glabrous. Spathe persistent, with 2 unequal valves. Umbel compact, 6-15-flowered. Perianth pale pinkish. Stamens exserted. Ovary widened and tuberculate above.

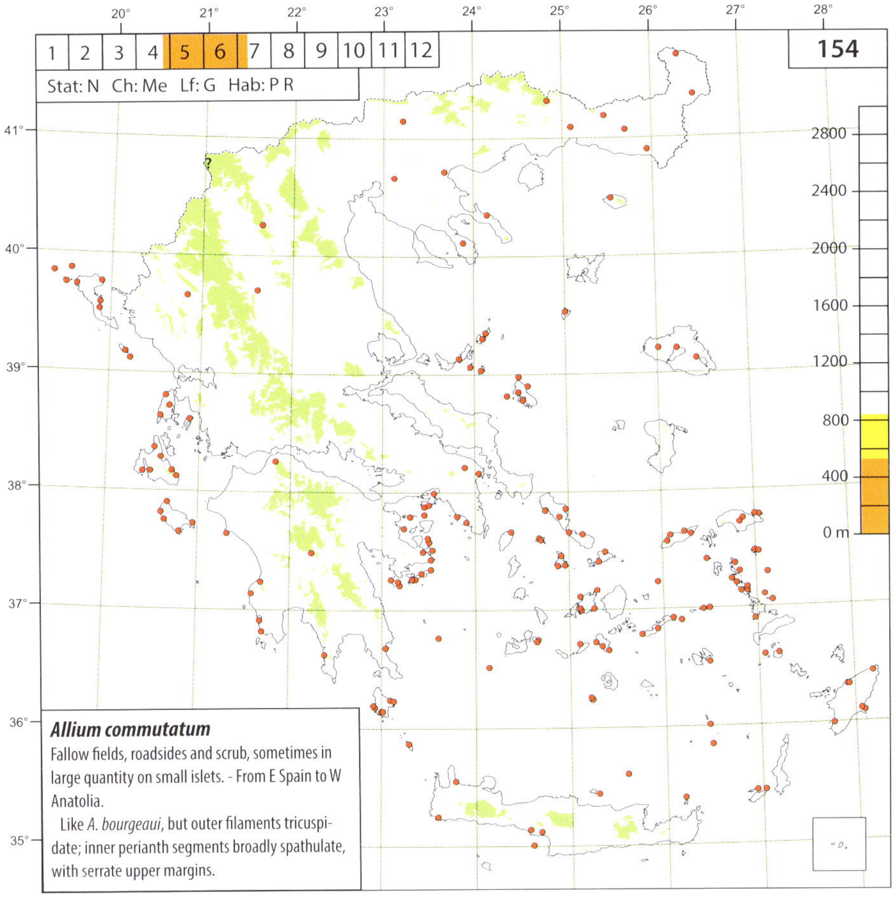

Allium commutatum

Fallow fields, roadsides and scrub, sometimes in large quantity on small islets. - From E Spain to W Anatolia.

Like *A. bourgeaui*, but outer filaments tricuspidate; inner perianth segments broadly spathulate, with serrate upper margins.

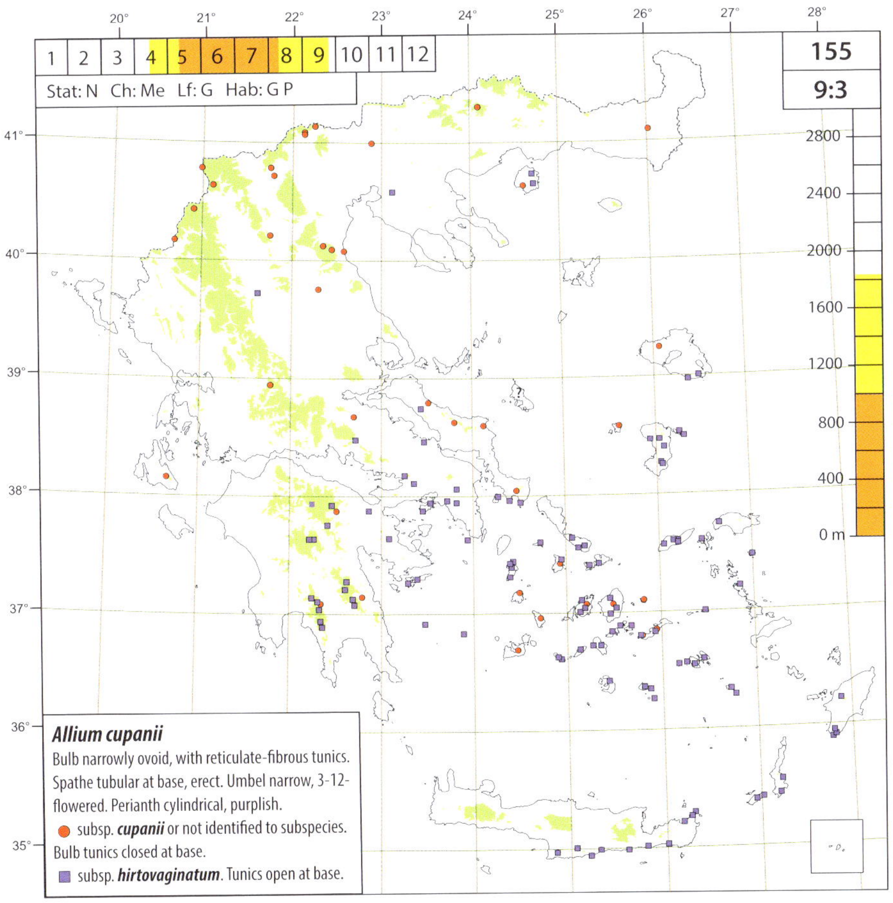

Allium cupanii

Bulb narrowly ovoid, with reticulate-fibrous tunics. Spathe tubular at base, erect. Umbel narrow, 3-12-flowered. Perianth cylindrical, purplish.

● subsp. ***cupanii*** or not identified to subspecies. Bulb tunics closed at base.

■ subsp. ***hirtovaginatum***. Tunics open at base.

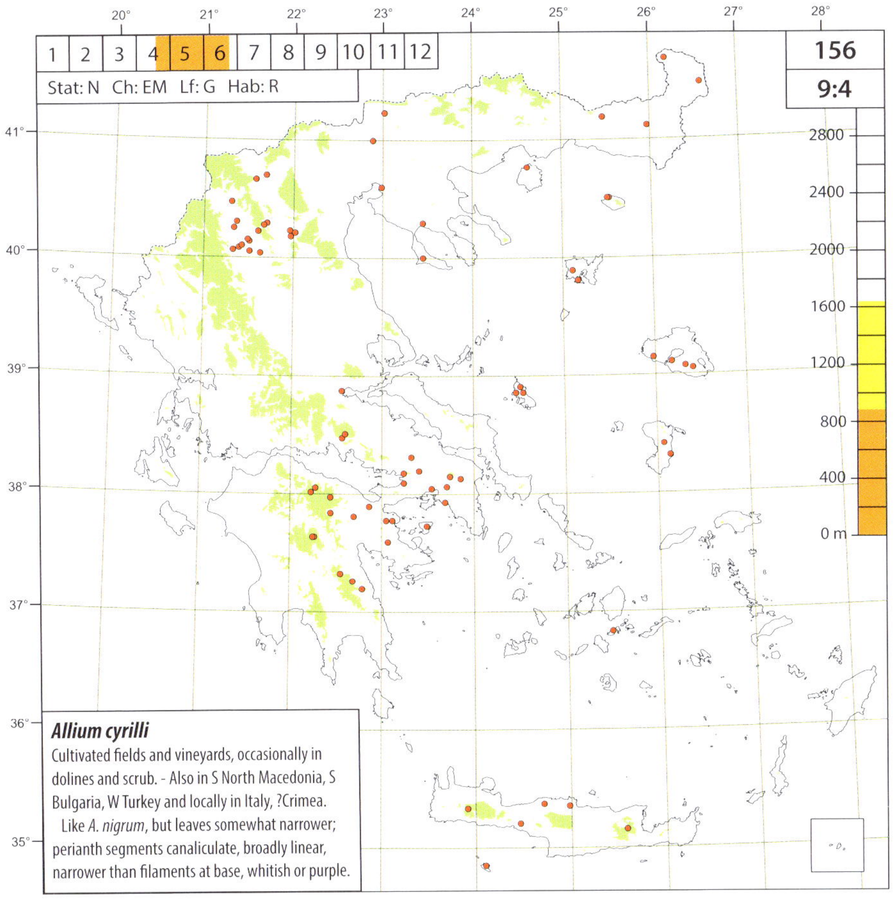

Allium cyrilli

Cultivated fields and vineyards, occasionally in dolines and scrub. - Also in S North Macedonia, S Bulgaria, W Turkey and locally in Italy, ?Crimea.

Like *A. nigrum*, but leaves somewhat narrower; perianth segments canaliculate, broadly linear, narrower than filaments at base, whitish or purple.

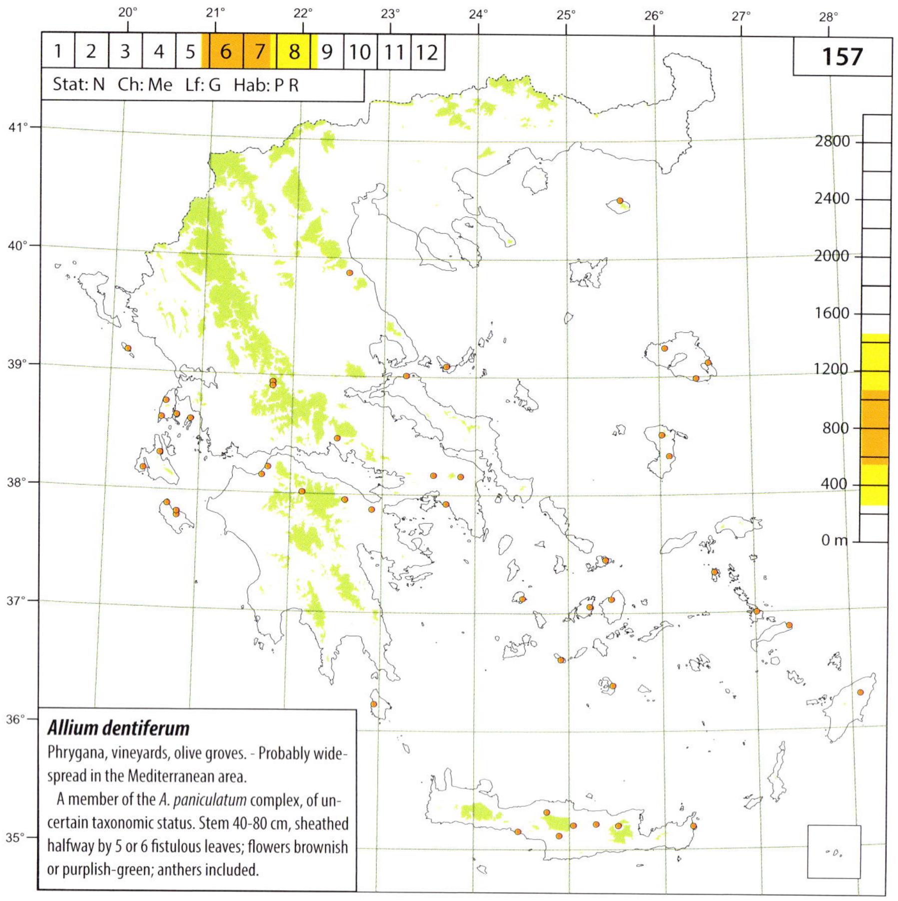

Allium dentiferum

Phrygana, vineyards, olive groves. - Probably widespread in the Mediterranean area.

A member of the *A. paniculatum* complex, of uncertain taxonomic status. Stem 40-80 cm, sheathed halfway by 5 or 6 fistulous leaves; flowers brownish or purplish-green; anthers included.

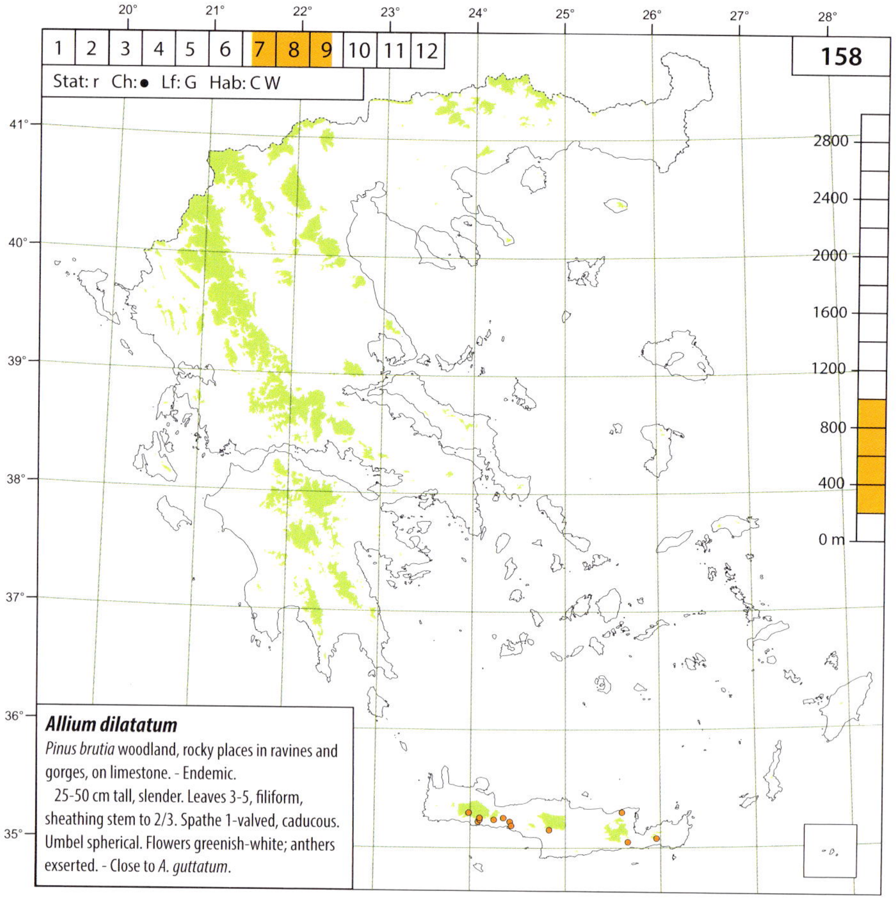

Allium dilatatum

Pinus brutia woodland, rocky places in ravines and gorges, on limestone. - Endemic.

25-50 cm tall, slender. Leaves 3-5, filiform, sheathing stem to 2/3. Spathe 1-valved, caducous. Umbel spherical. Flowers greenish-white; anthers exserted. - Close to *A. guttatum*.

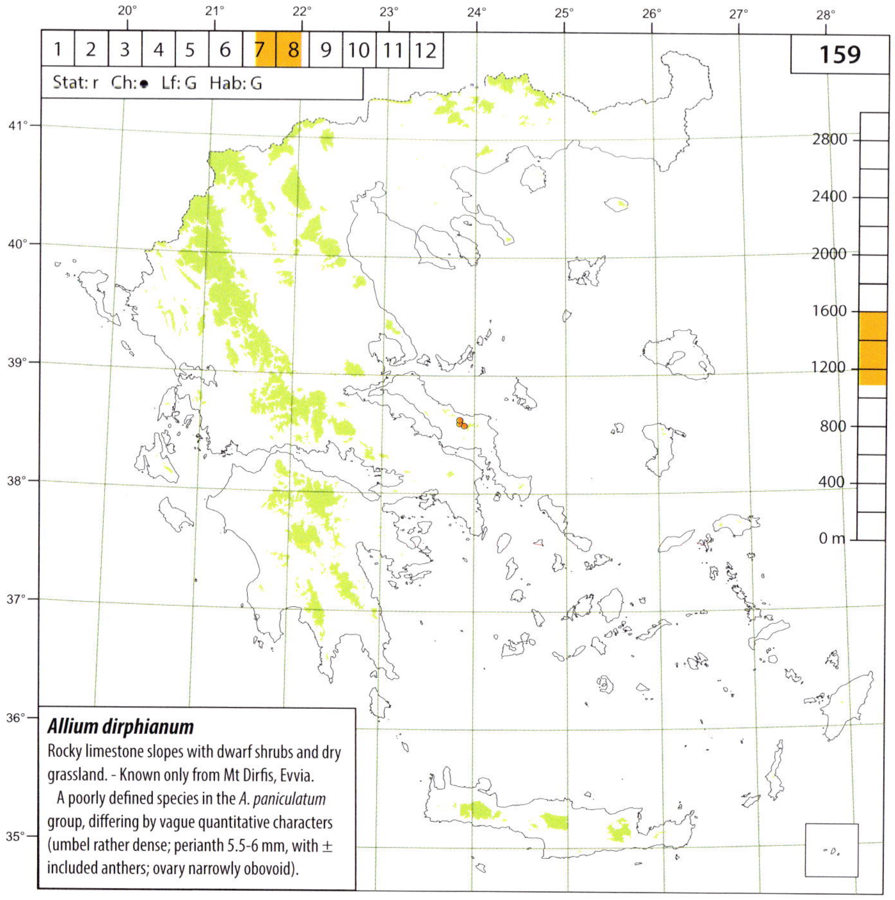

Allium dirphianum

Rocky limestone slopes with dwarf shrubs and dry grassland. - Known only from Mt Dirfis, Evvia.

A poorly defined species in the *A. paniculatum* group, differing by vague quantitative characters (umbel rather dense; perianth 5.5-6 mm, with ± included anthers; ovary narrowly obovoid).

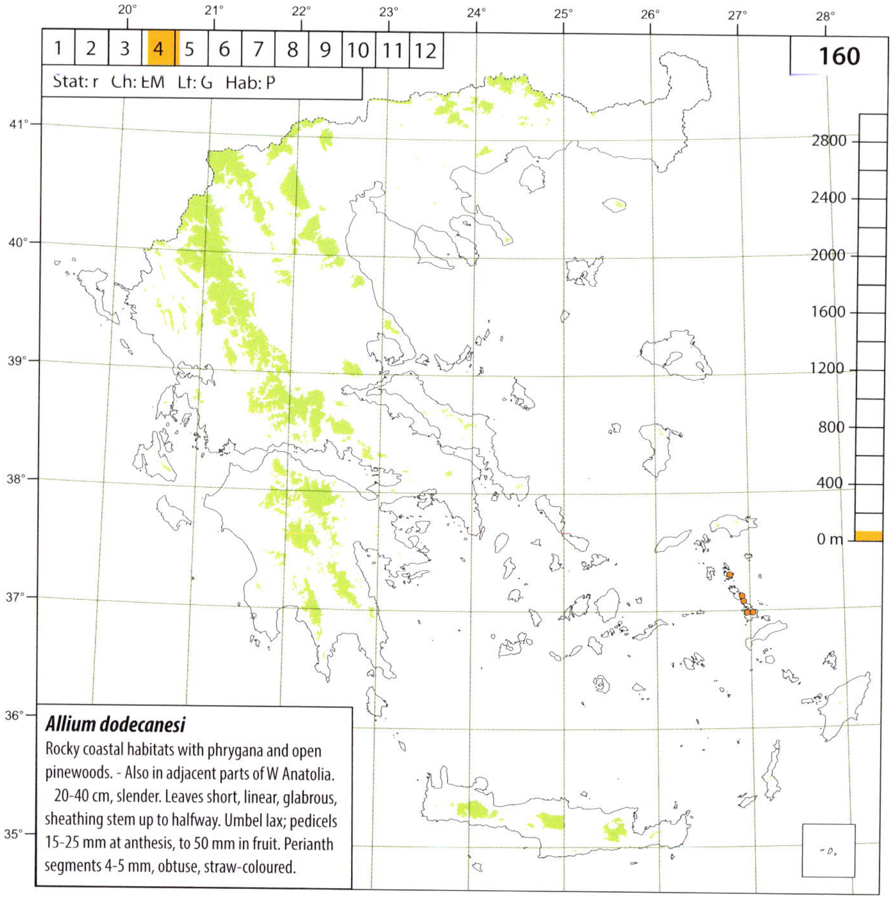

Allium dodecanesi

Rocky coastal habitats with phrygana and open pinewoods. - Also in adjacent parts of W Anatolia.

20-40 cm, slender. Leaves short, linear, glabrous, sheathing stem up to halfway. Umbel lax; pedicels 15-25 mm at anthesis, to 50 mm in fruit. Perianth segments 4-5 mm, obtuse, straw-coloured.

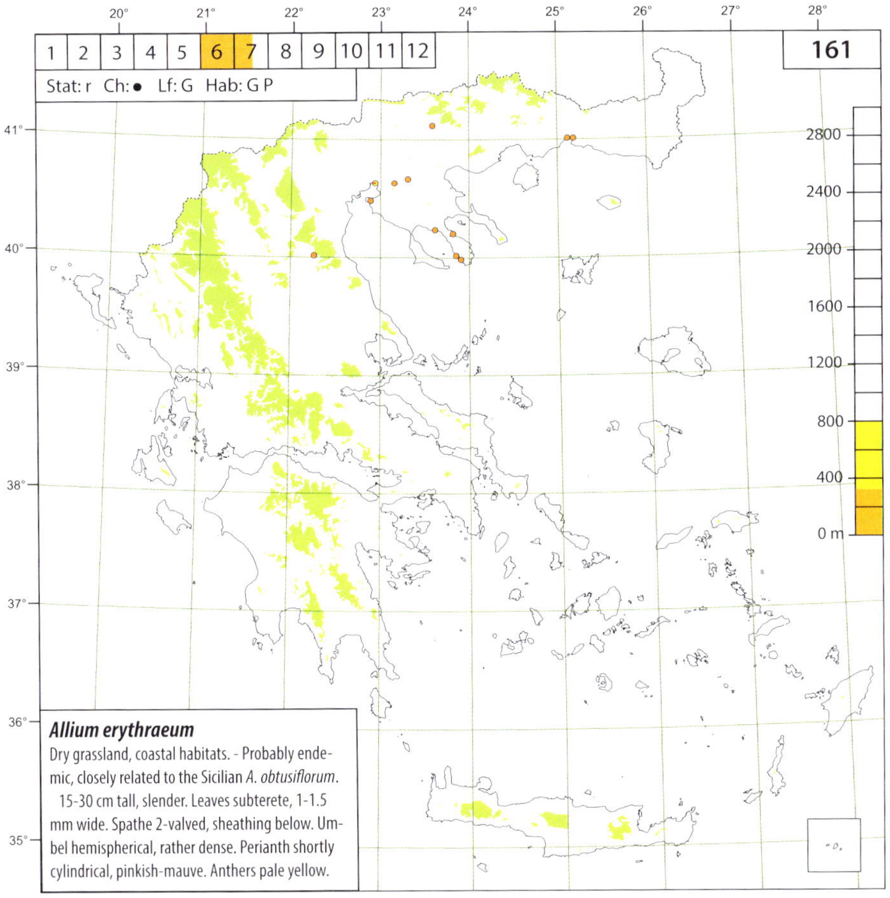

Allium erythraeum

Dry grassland, coastal habitats. - Probably endemic, closely related to the Sicilian *A. obtusiflorum*.

15-30 cm tall, slender. Leaves subterete, 1-1.5 mm wide. Spathe 2-valved, sheathing below. Umbel hemispherical, rather dense. Perianth shortly cylindrical, pinkish-mauve. Anthers pale yellow.

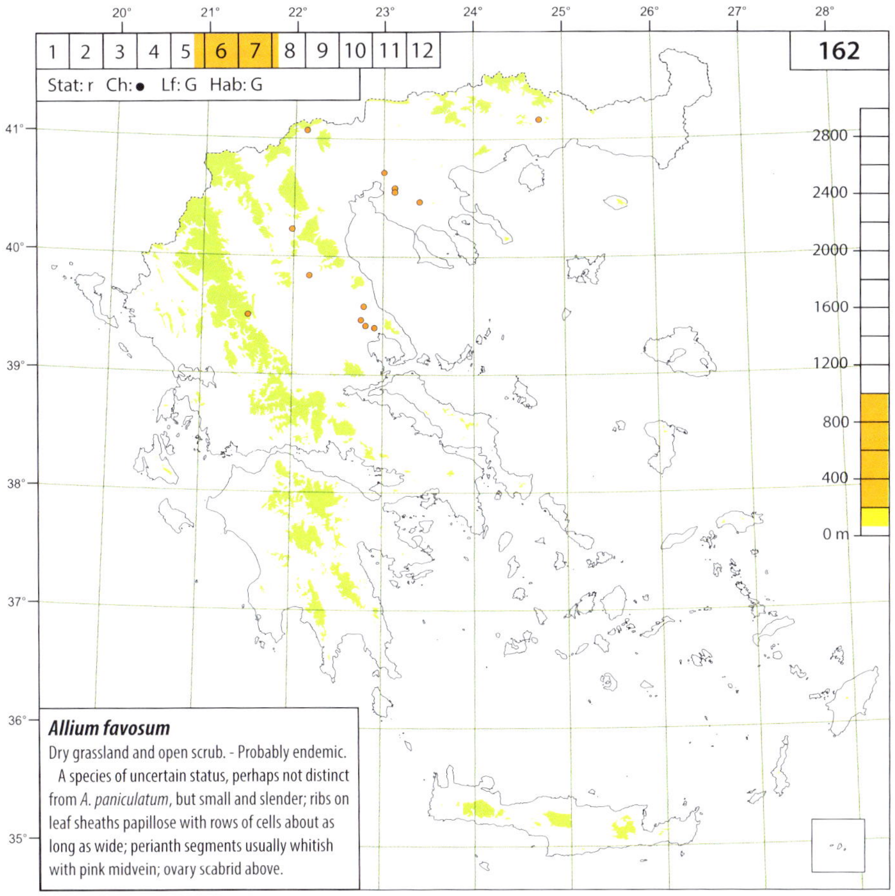

Allium favosum

Dry grassland and open scrub. - Probably endemic.

A species of uncertain status, perhaps not distinct from *A. paniculatum*, but small and slender; ribs on leaf sheaths papillose with rows of cells about as long as wide; perianth segments usually whitish with pink midvein; ovary scabrid above.

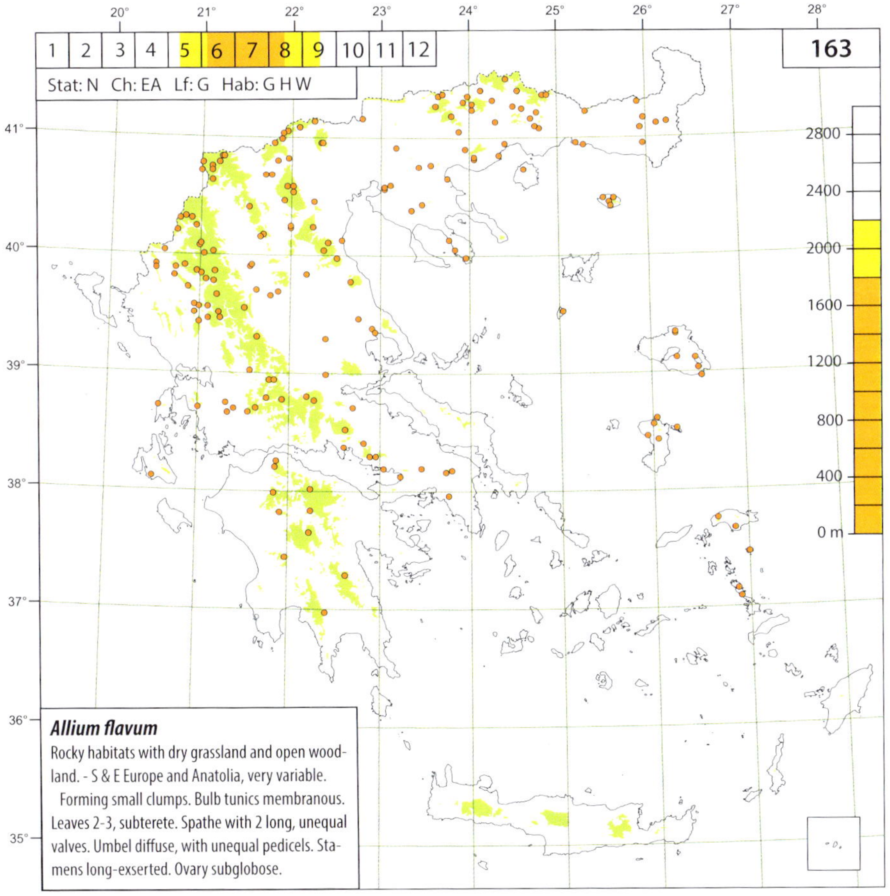

Allium flavum

Rocky habitats with dry grassland and open woodland. - S & E Europe and Anatolia, very variable.

Forming small clumps. Bulb tunics membranous. Leaves 2-3, subterete. Spathe with 2 long, unequal valves. Umbel diffuse, with unequal pedicels. Stamens long-exserted. Ovary subglobose.

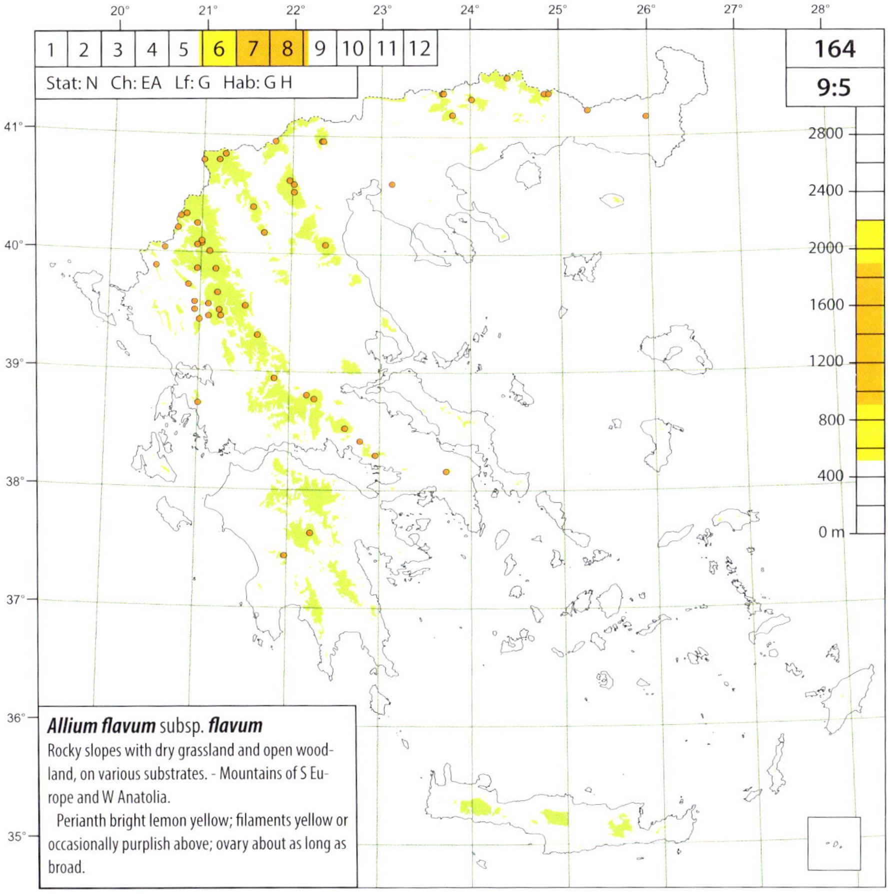

Allium flavum* subsp. *flavum

Rocky slopes with dry grassland and open woodland, on various substrates. - Mountains of S Europe and W Anatolia.

Perianth bright lemon yellow; filaments yellow or occasionally purplish above; ovary about as long as broad.

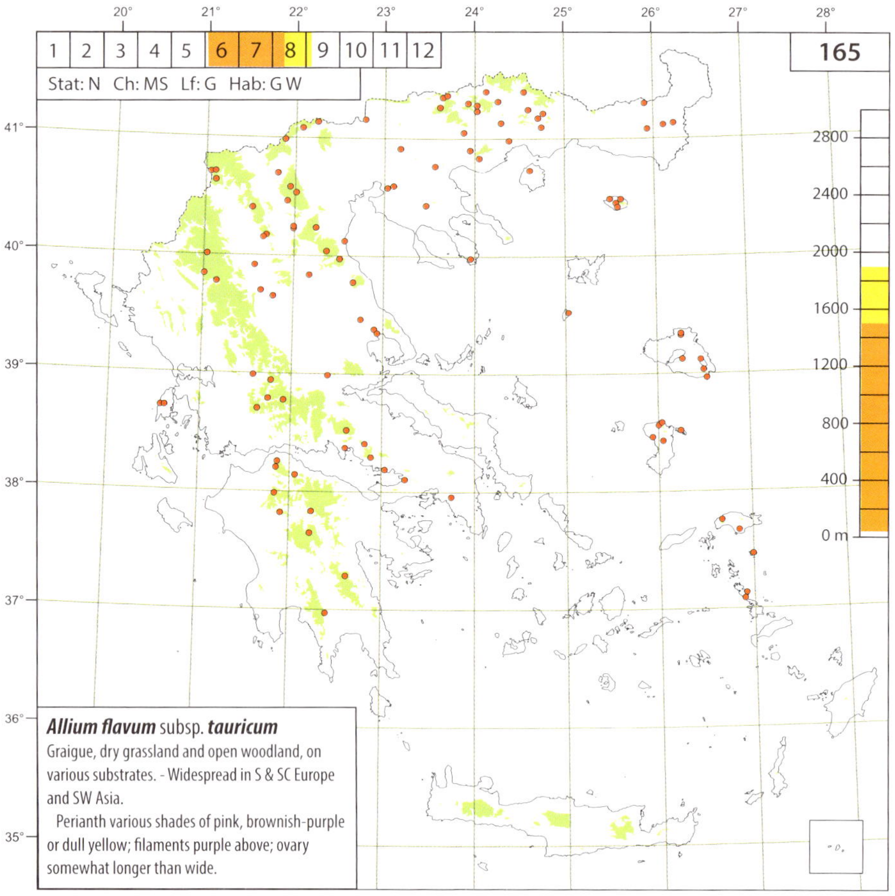

Allium flavum **subsp.** ***tauricum***
Graigue, dry grassland and open woodland, on various substrates. - Widespread in S & SC Europe and SW Asia.
Perianth various shades of pink, brownish-purple or dull yellow; filaments purple above; ovary somewhat longer than wide.

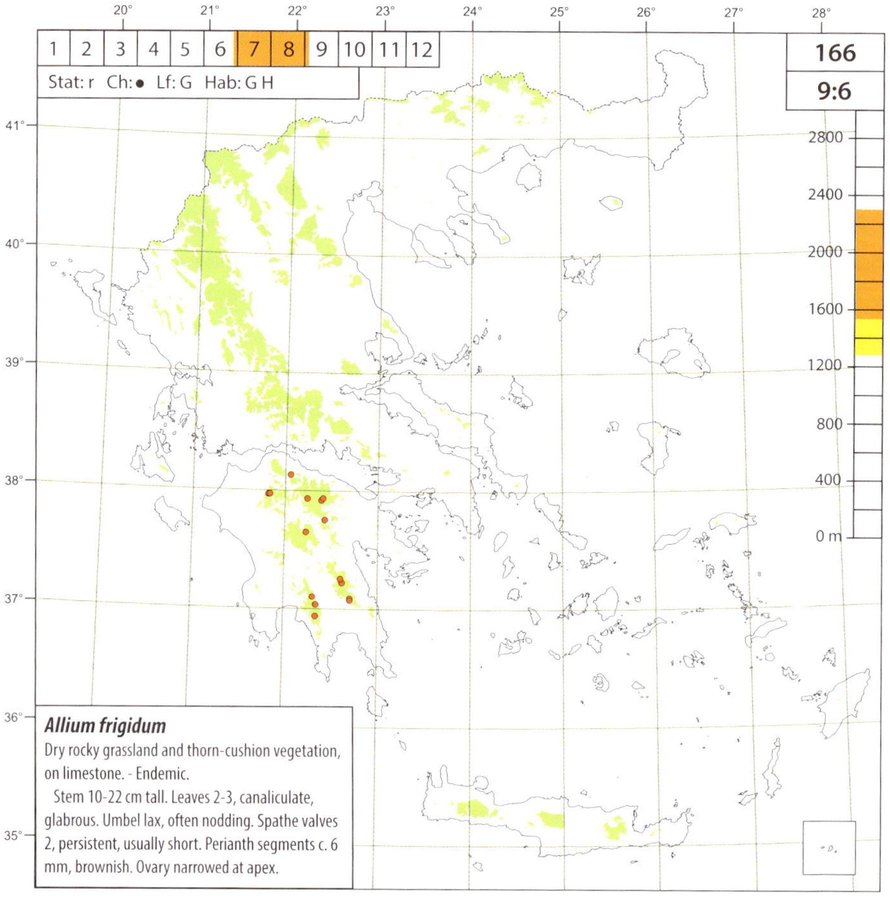

Allium frigidum
Dry rocky grassland and thorn-cushion vegetation, on limestone. - Endemic.
Stem 10-22 cm tall. Leaves 2-3, canaliculate, glabrous. Umbel lax, often nodding. Spathe valves 2, persistent, usually short. Perianth segments c. 6 mm, brownish. Ovary narrowed at apex.

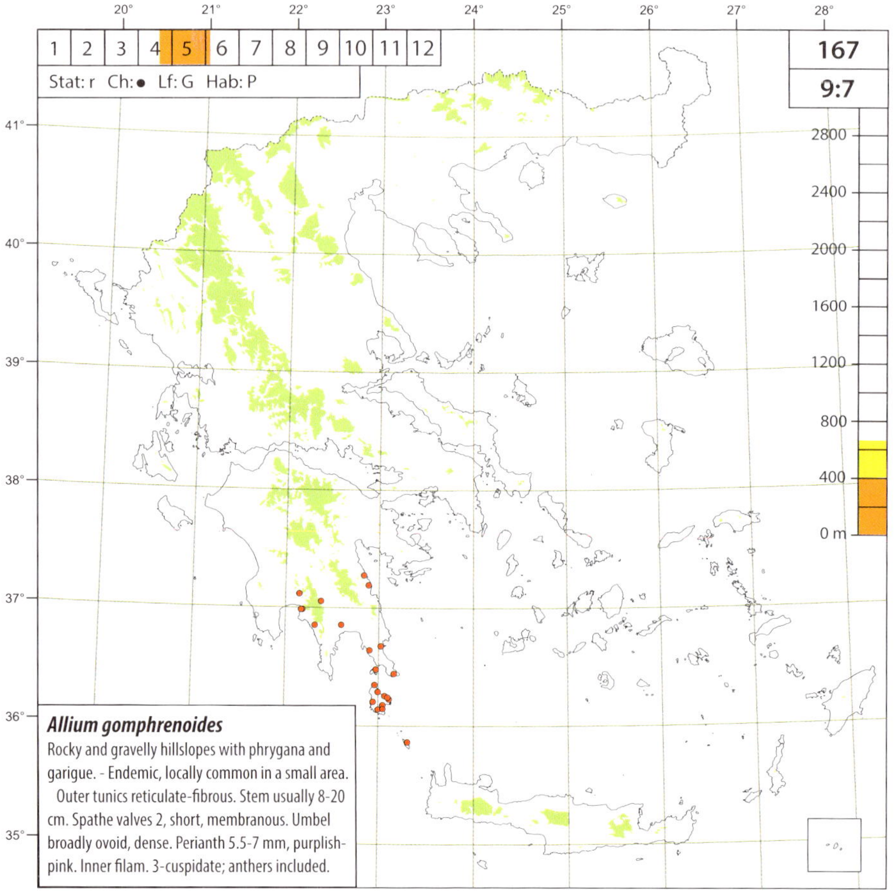

Allium gomphrenoides
Rocky and gravelly hillslopes with phrygana and garigue. - Endemic, locally common in a small area.
Outer tunics reticulate-fibrous. Stem usually 8-20 cm. Spathe valves 2, short, membranous. Umbel broadly ovoid, dense. Perianth 5.5-7 mm, purplish-pink. Inner filam. 3-cuspidate; anthers included.

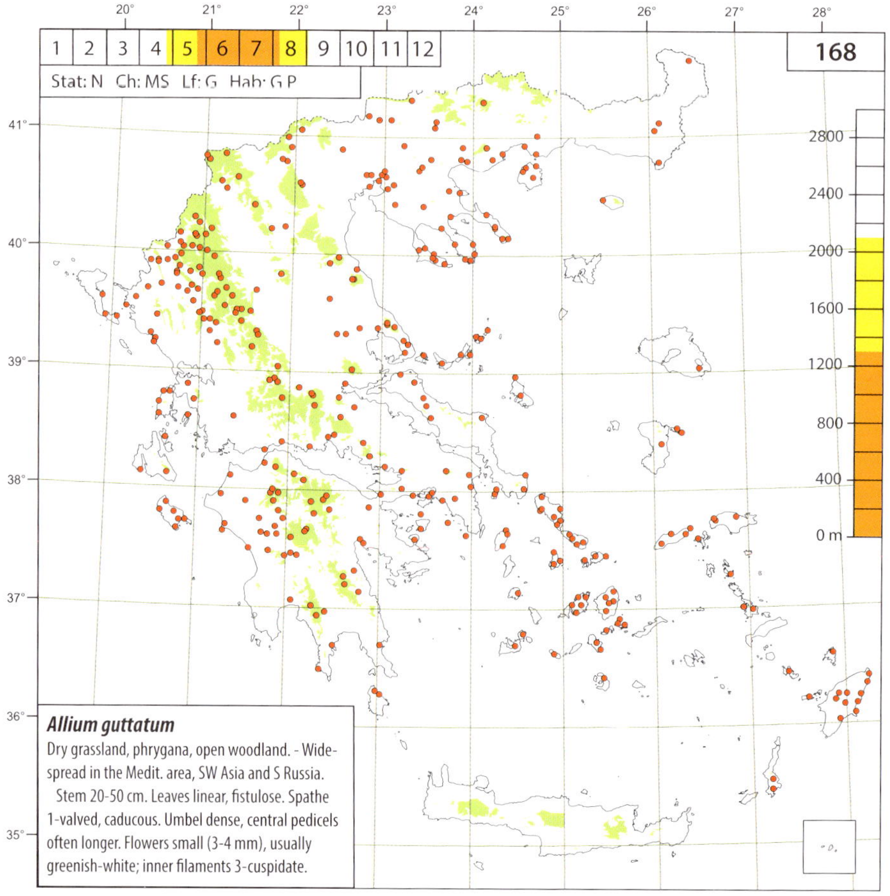

Allium guttatum
Dry grassland, phrygana, open woodland. - Widespread in the Medit. area, SW Asia and S Russia.
Stem 20-50 cm. Leaves linear, fistulose. Spathe 1-valved, caducous. Umbel dense, central pedicels often longer. Flowers small (3-4 mm), usually greenish-white; inner filaments 3-cuspidate.

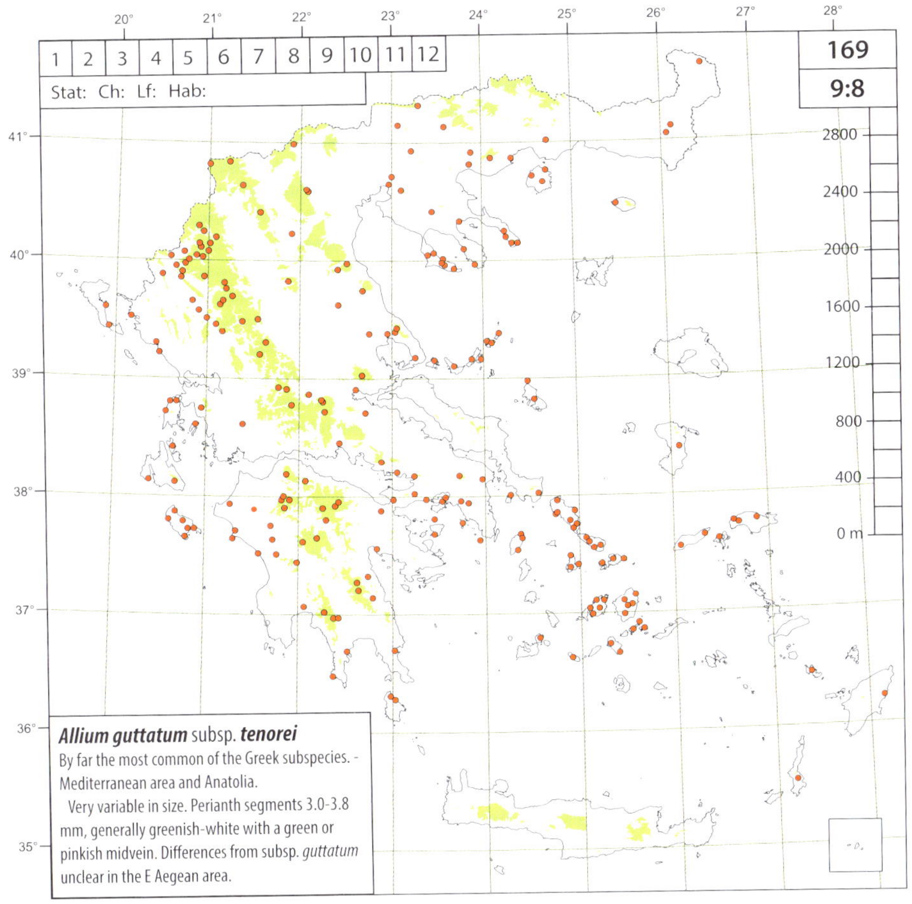

Allium guttatum **subsp.** ***tenorei***

By far the most common of the Greek subspecies. - Mediterranean area and Anatolia.

Very variable in size. Perianth segments 3.0-3.8 mm, generally greenish-white with a green or pinkish midvein. Differences from subsp. *guttatum* unclear in the E Aegean area.

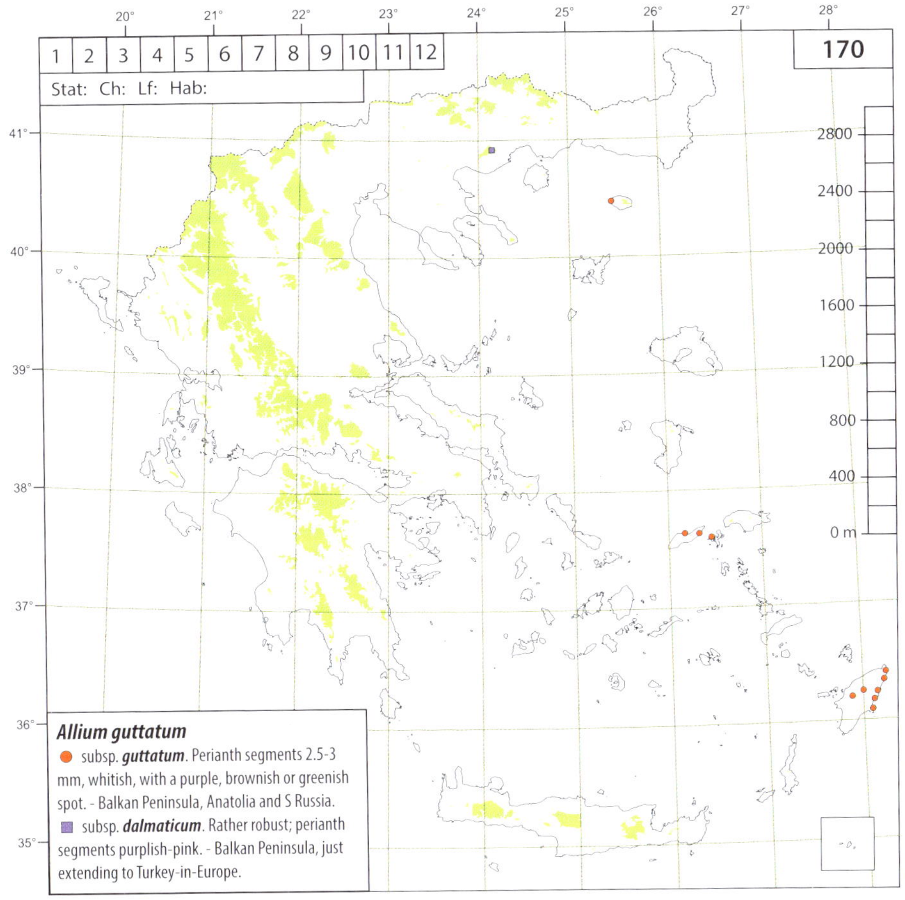

Allium guttatum

● subsp. ***guttatum***. Perianth segments 2.5-3 mm, whitish, with a purple, brownish or greenish spot. - Balkan Peninsula, Anatolia and S Russia.

■ subsp. ***dalmaticum***. Rather robust; perianth segments purplish-pink. - Balkan Peninsula, just extending to Turkey-in-Europe.

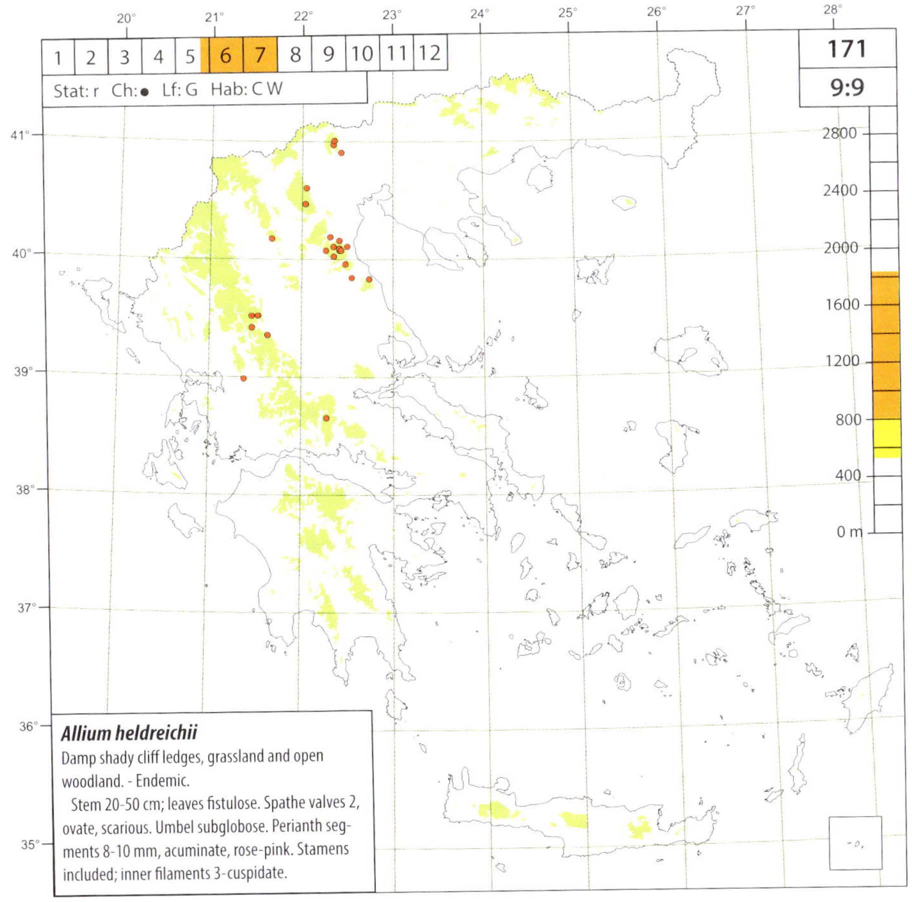

Allium heldreichii

Damp shady cliff ledges, grassland and open woodland. - Endemic.

Stem 20-50 cm; leaves fistulose. Spathe valves 2, ovate, scarious. Umbel subglobose. Perianth segments 8-10 mm, acuminate, rose-pink. Stamens included; inner filaments 3-cuspidate.

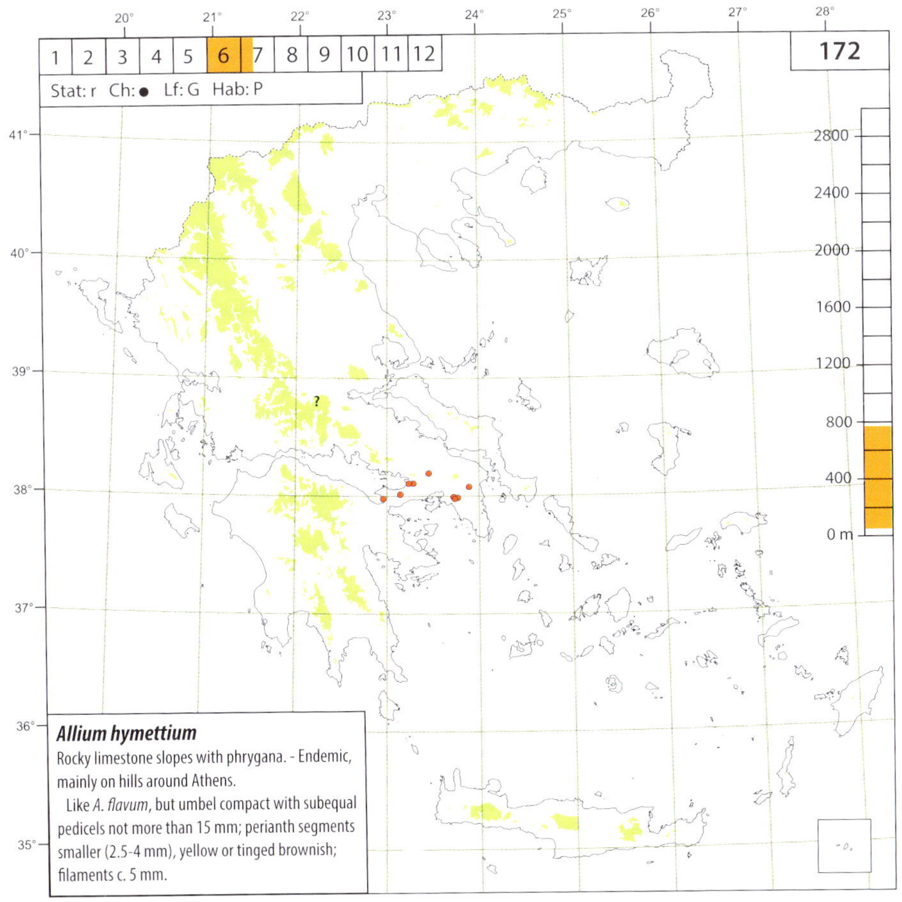

Allium hymettium

Rocky limestone slopes with phrygana. - Endemic, mainly on hills around Athens.

Like *A. flavum*, but umbel compact with subequal pedicels not more than 15 mm; perianth segments smaller (2.5-4 mm), yellow or tinged brownish; filaments c. 5 mm.

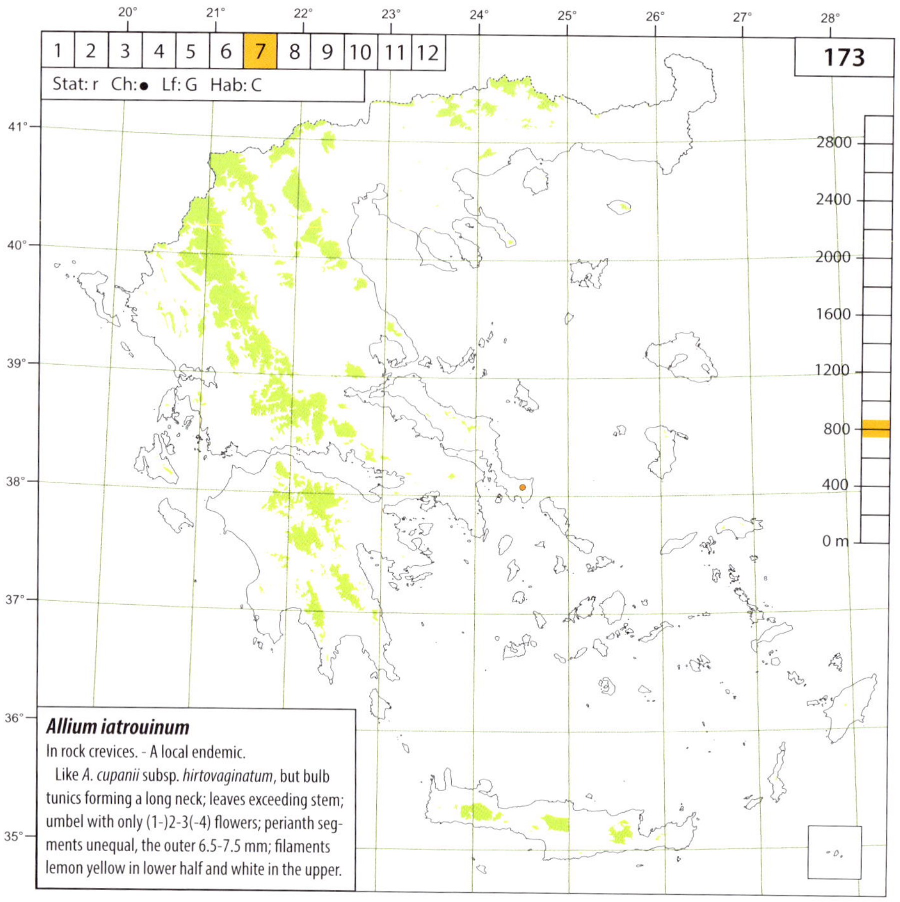

Allium iatrouinum

In rock crevices. - A local endemic.

Like *A. cupanii* subsp. *hirtovaginatum*, but bulb tunics forming a long neck; leaves exceeding stem; umbel with only (1-)2-3(-4) flowers; perianth segments unequal, the outer 6.5-7.5 mm; filaments lemon yellow in lower half and white in the upper.

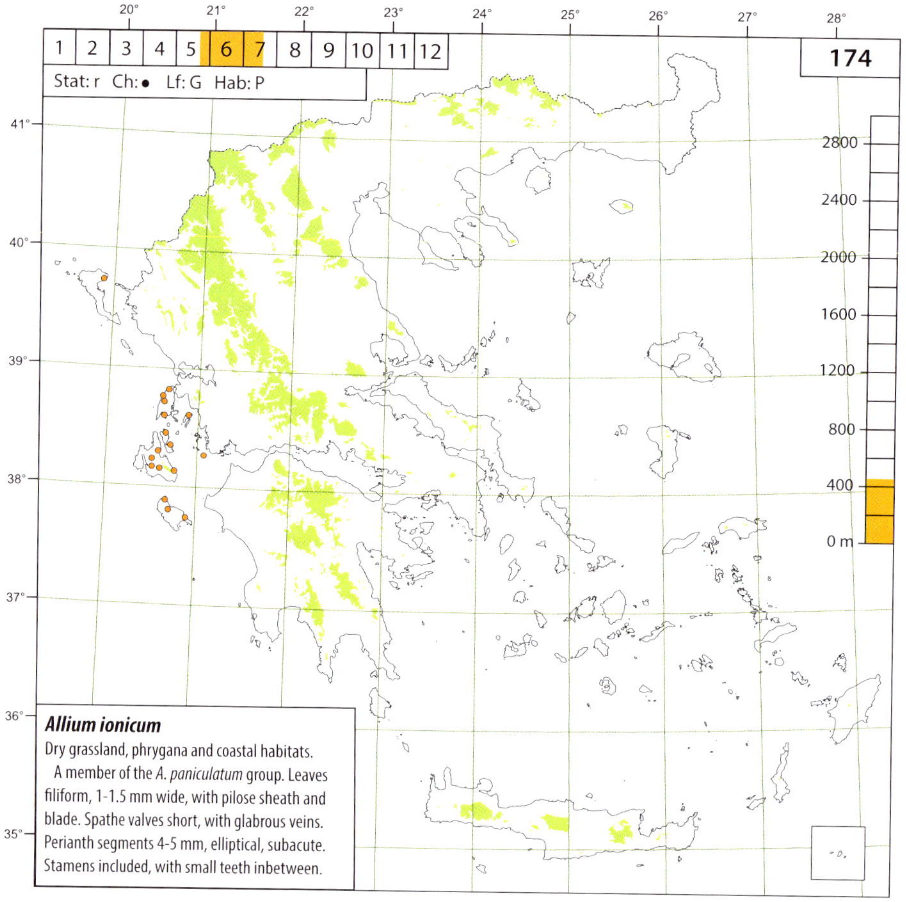

Allium ionicum

Dry grassland, phrygana and coastal habitats.

A member of the *A. paniculatum* group. Leaves filiform, 1-1.5 mm wide, with pilose sheath and blade. Spathe valves short, with glabrous veins. Perianth segments 4-5 mm, elliptical, subacute. Stamens included, with small teeth inbetween.

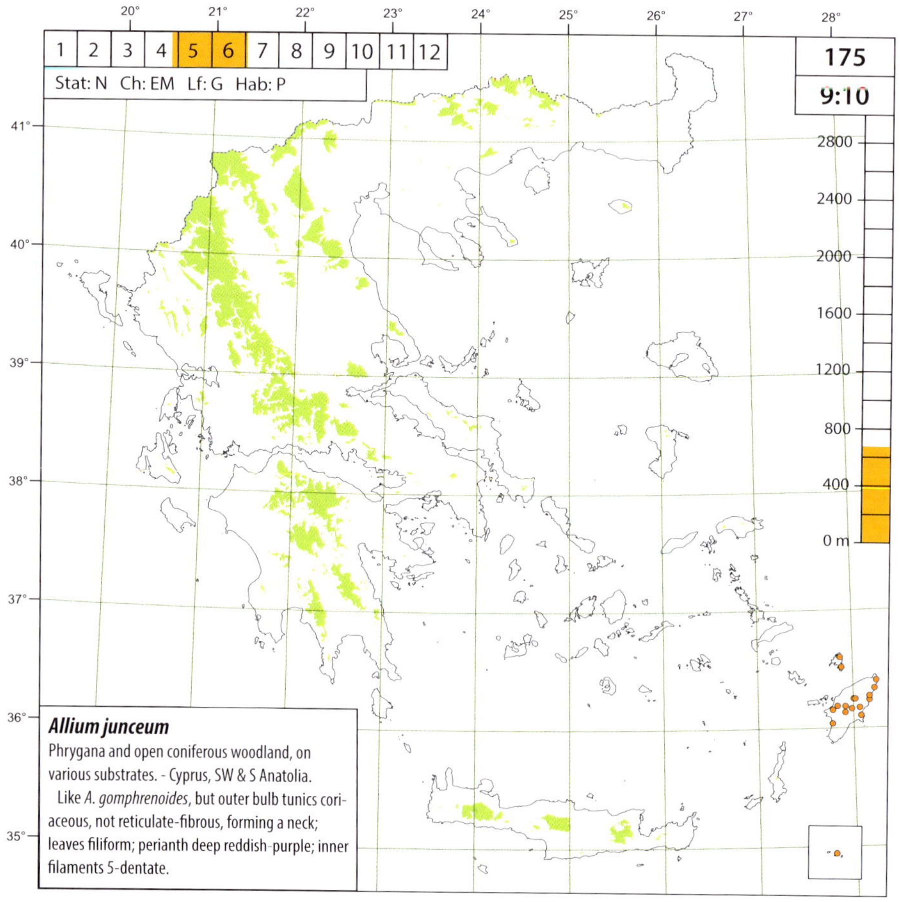

Allium junceum

Phrygana and open coniferous woodland, on various substrates. - Cyprus, SW & S Anatolia.

Like *A. gomphrenoides*, but outer bulb tunics coriaceous, not reticulate-fibrous, forming a neck; leaves filiform; perianth deep reddish-purple; inner filaments 5-dentate.

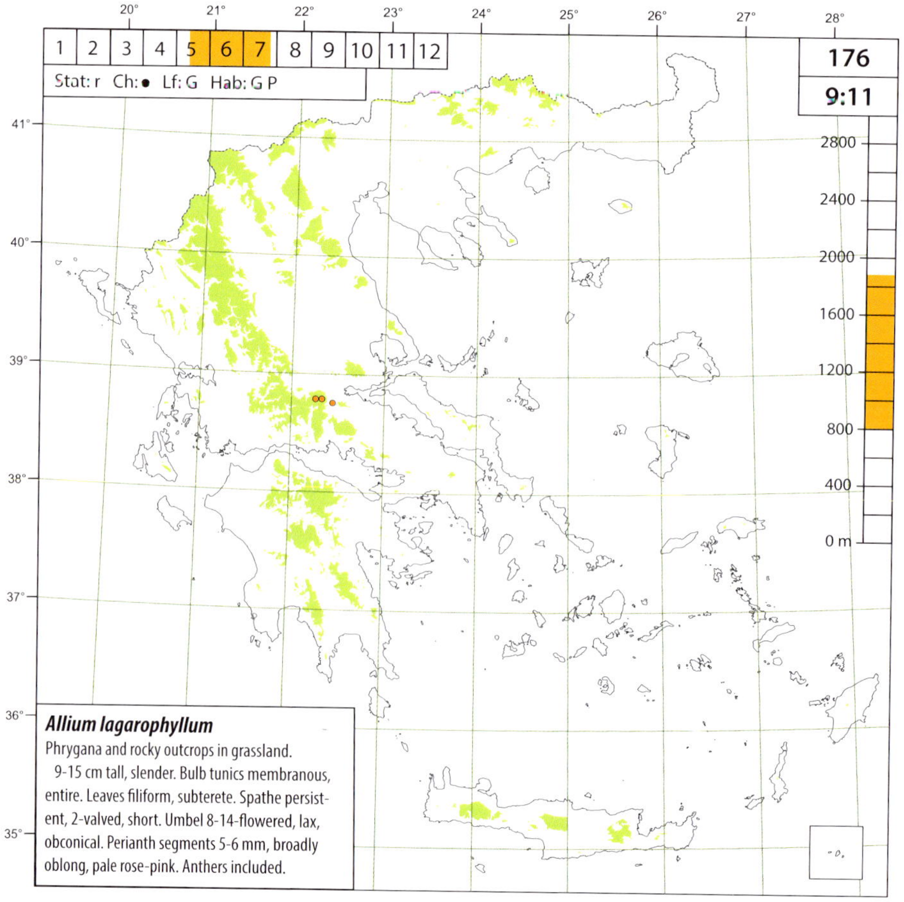

Allium lagarophyllum

Phrygana and rocky outcrops in grassland.

9-15 cm tall, slender. Bulb tunics membranous, entire. Leaves filiform, subterete. Spathe persistent, 2-valved, short. Umbel 8-14-flowered, lax, obconical. Perianth segments 5-6 mm, broadly oblong, pale rose-pink. Anthers included.

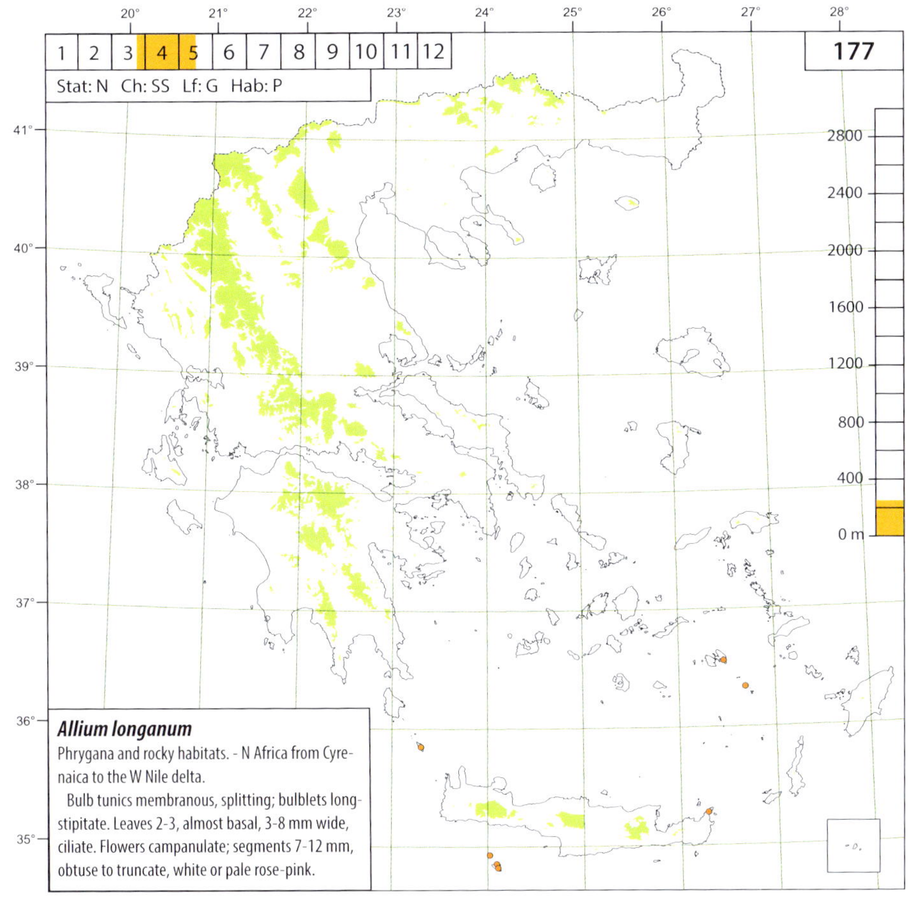

Allium longanum
Phrygana and rocky habitats. - N Africa from Cyrenaica to the W Nile delta.
Bulb tunics membranous, splitting; bulblets long-stipitate. Leaves 2-3, almost basal, 3-8 mm wide, ciliate. Flowers campanulate; segments 7-12 mm, obtuse to truncate, white or pale rose-pink.

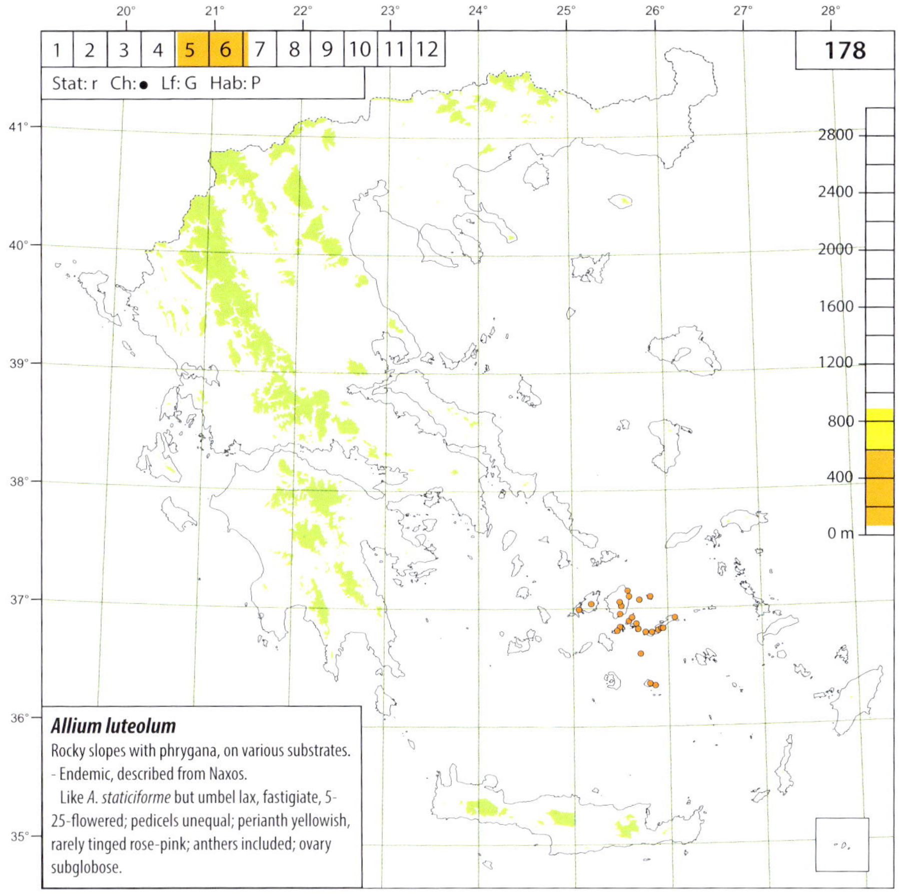

Allium luteolum
Rocky slopes with phrygana, on various substrates. - Endemic, described from Naxos.
Like *A. staticiforme* but umbel lax, fastigiate, 5-25-flowered; pedicels unequal; perianth yellowish, rarely tinged rose-pink; anthers included; ovary subglobose.

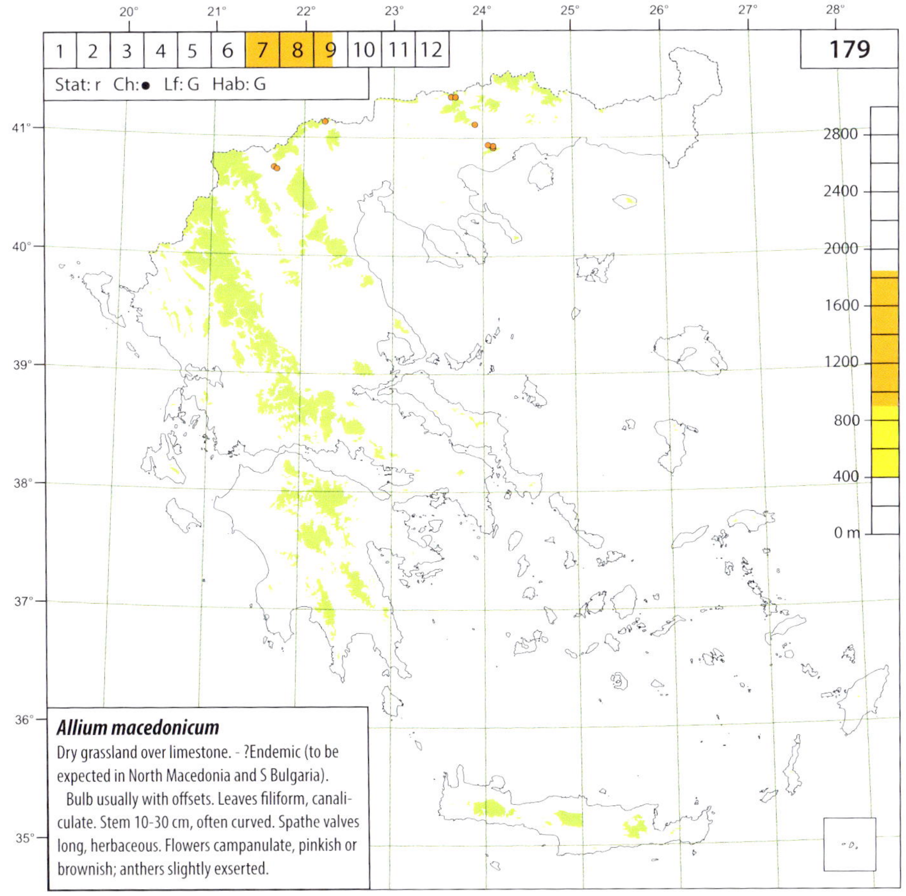

Allium macedonicum
Dry grassland over limestone. - ?Endemic (to be expected in North Macedonia and S Bulgaria).
Bulb usually with offsets. Leaves filiform, canaliculate. Stem 10-30 cm, often curved. Spathe valves long, herbaceous. Flowers campanulate, pinkish or brownish; anthers slightly exserted.

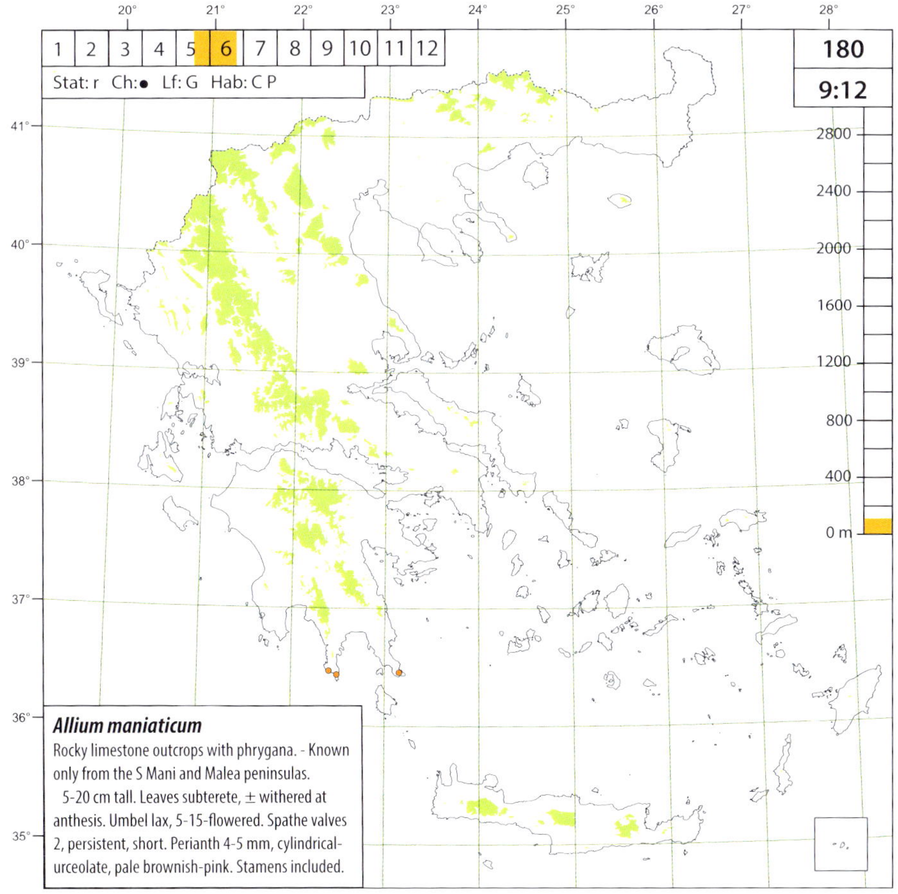

Allium maniaticum
Rocky limestone outcrops with phrygana. - Known only from the S Mani and Malea peninsulas.
5-20 cm tall. Leaves subterete, ± withered at anthesis. Umbel lax, 5-15-flowered. Spathe valves 2, persistent, short. Perianth 4-5 mm, cylindrical-urceolate, pale brownish-pink. Stamens included.

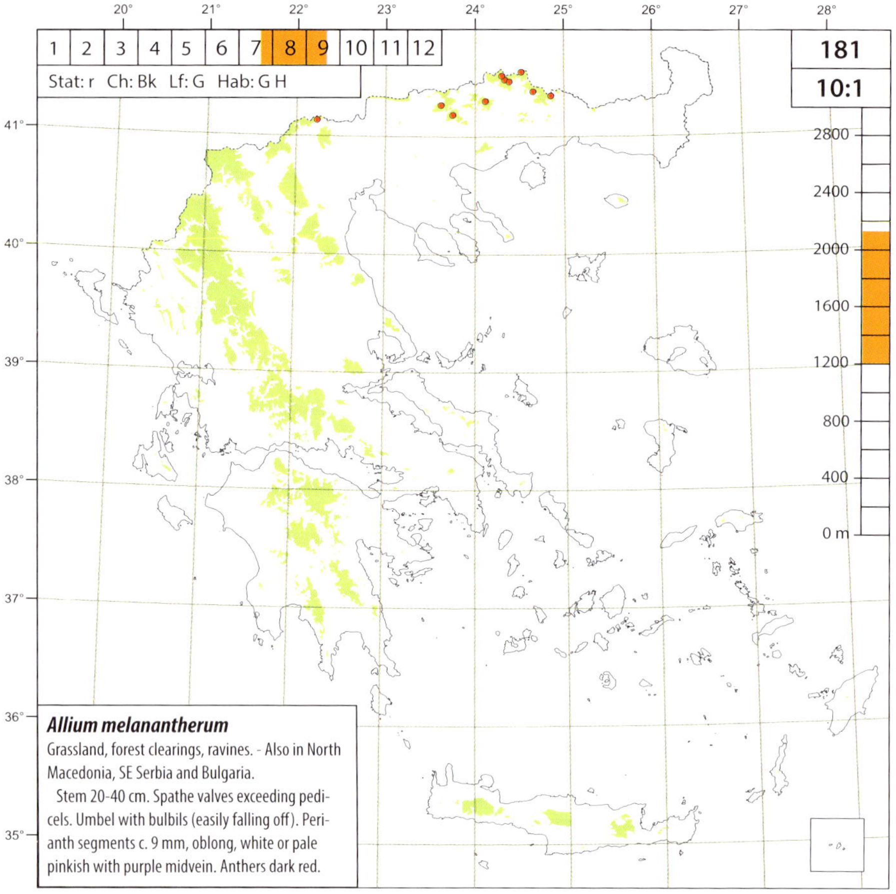

Allium melanantherum

Grassland, forest clearings, ravines. - Also in North Macedonia, SE Serbia and Bulgaria.

Stem 20-40 cm. Spathe valves exceeding pedicels. Umbel with bulbils (easily falling off). Perianth segments c. 9 mm, oblong, white or pale pinkish with purple midvein. Anthers dark red.

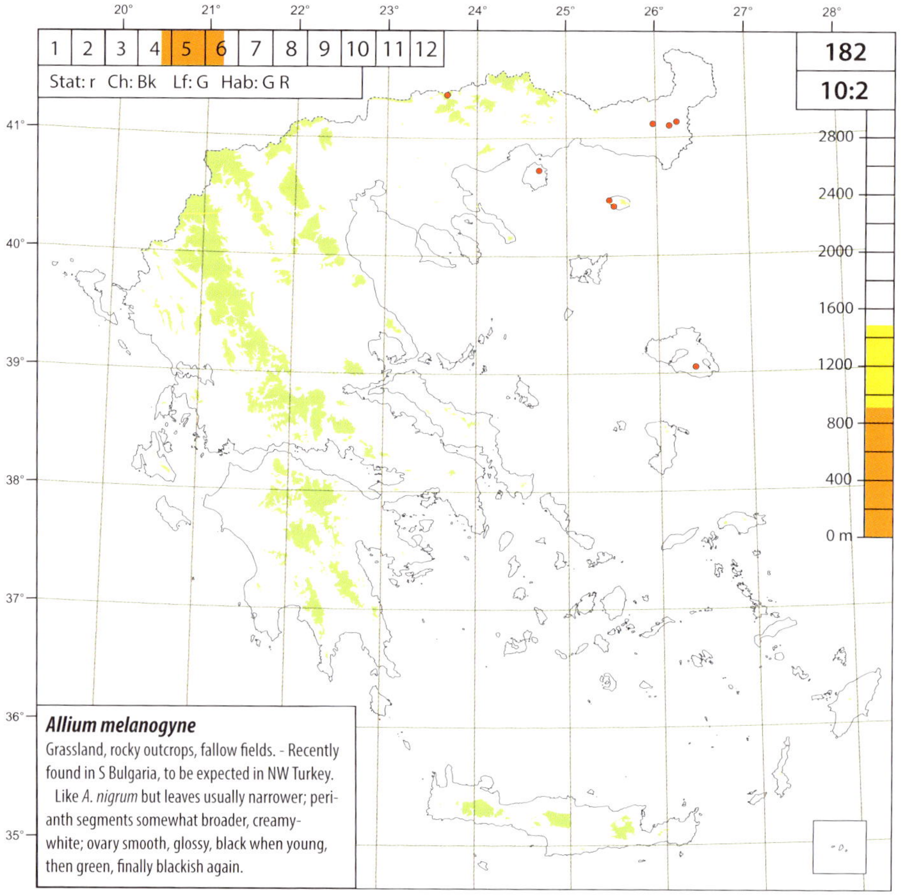

Allium melanogyne

Grassland, rocky outcrops, fallow fields. - Recently found in S Bulgaria, to be expected in NW Turkey.

Like *A. nigrum* but leaves usually narrower; perianth segments somewhat broader, creamy-white; ovary smooth, glossy, black when young, then green, finally blackish again.

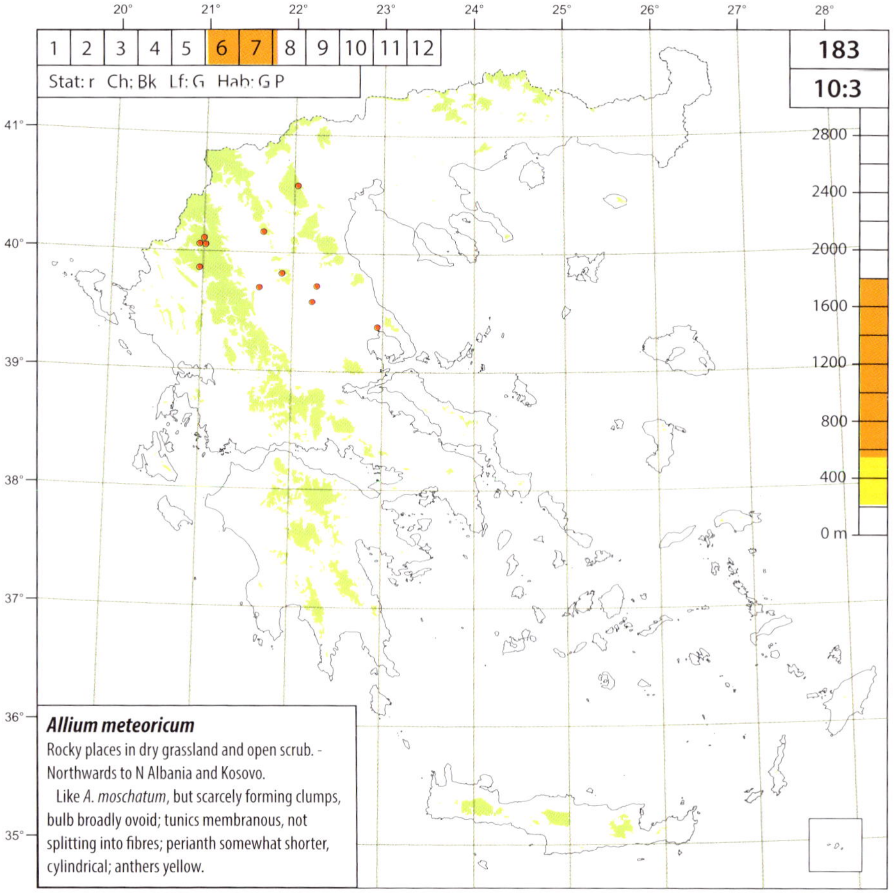

Allium meteoricum

Rocky places in dry grassland and open scrub. - Northwards to N Albania and Kosovo.

Like *A. moschatum*, but scarcely forming clumps, bulb broadly ovoid; tunics membranous, not splitting into fibres; perianth somewhat shorter, cylindrical; anthers yellow.

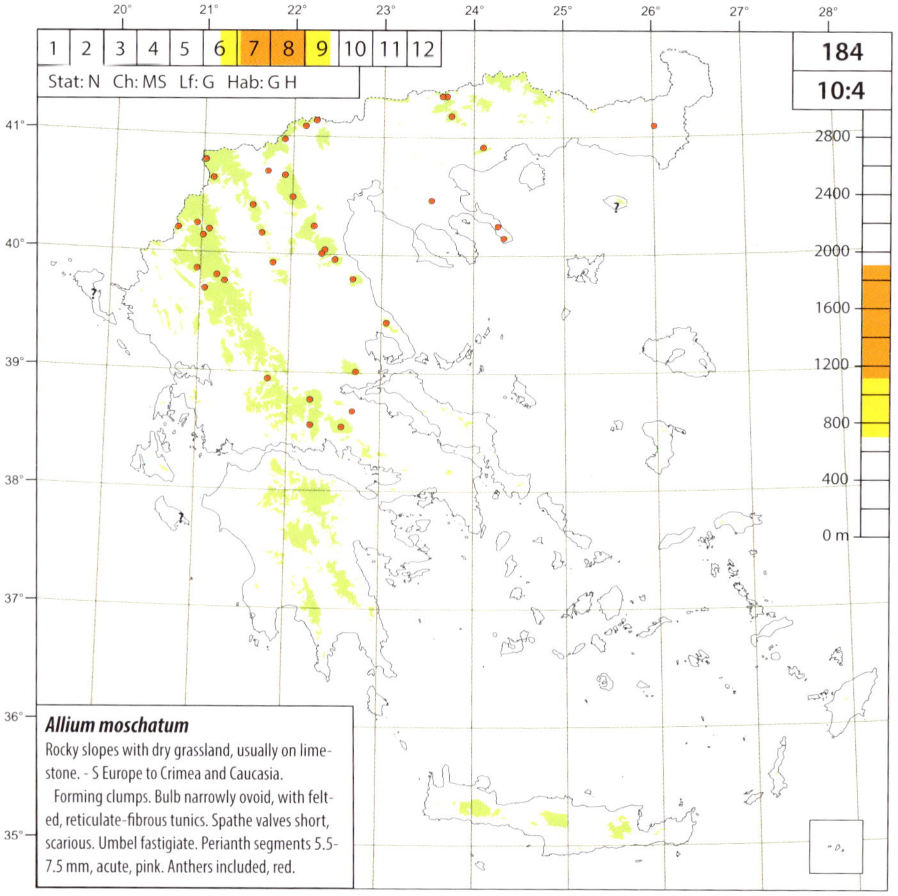

Allium moschatum

Rocky slopes with dry grassland, usually on limestone. - S Europe to Crimea and Caucasia.

Forming clumps. Bulb narrowly ovoid, with felted, reticulate-fibrous tunics. Spathe valves short, scarious. Umbel fastigiate. Perianth segments 5.5-7.5 mm, acute, pink. Anthers included, red.

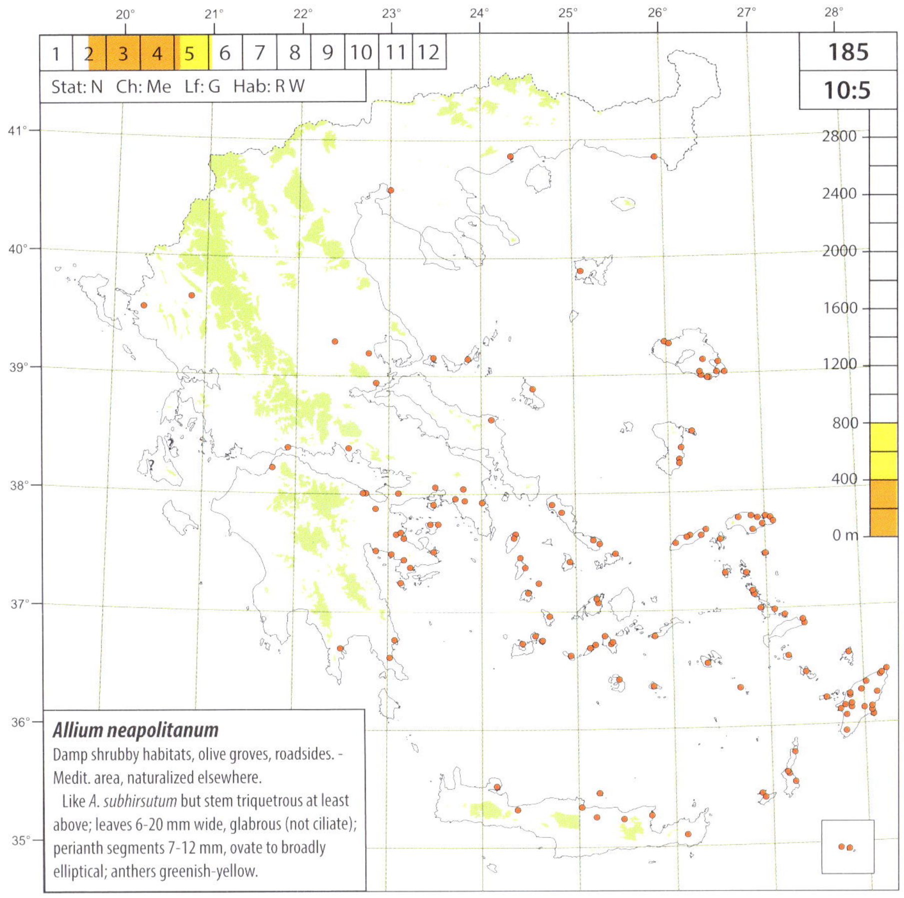

Allium neapolitanum

Damp shrubby habitats, olive groves, roadsides. - Medit. area, naturalized elsewhere.

Like *A. subhirsutum* but stem triquetrous at least above; leaves 6-20 mm wide, glabrous (not ciliate); perianth segments 7-12 mm, ovate to broadly elliptical; anthers greenish-yellow.

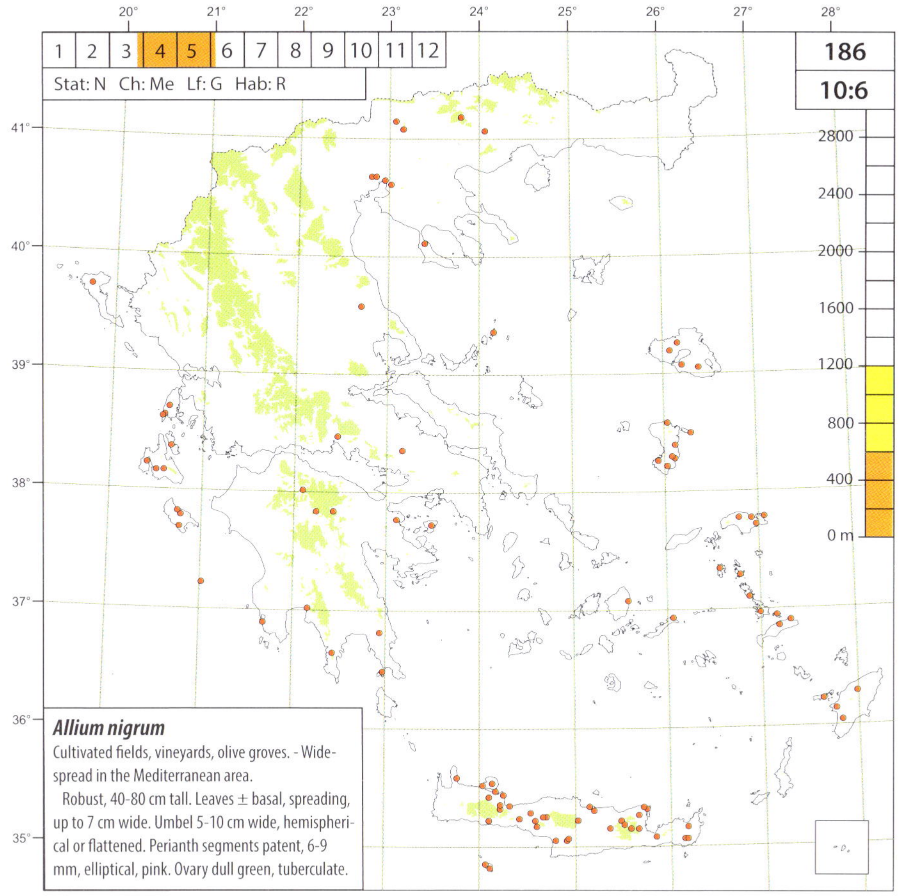

Allium nigrum

Cultivated fields, vineyards, olive groves. - Widespread in the Mediterranean area.

Robust, 40-80 cm tall. Leaves ± basal, spreading, up to 7 cm wide. Umbel 5-10 cm wide, hemispherical or flattened. Perianth segments patent, 6-9 mm, elliptical, pink. Ovary dull green, tuberculate.

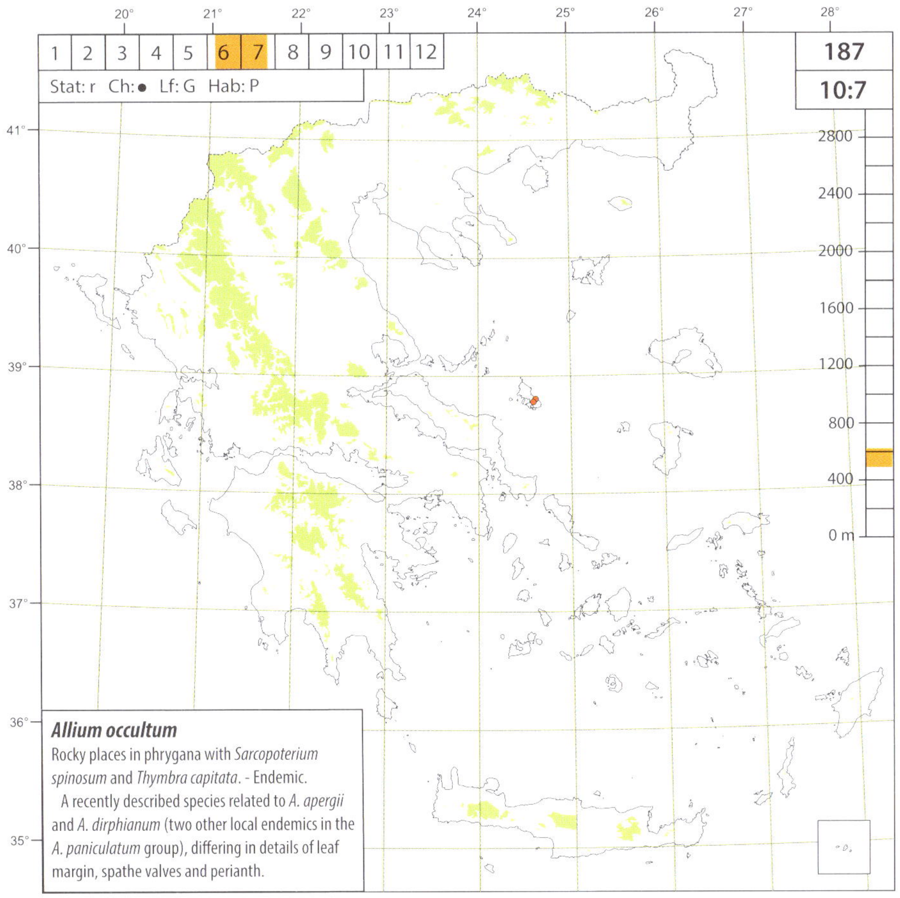

Allium occultum

Rocky places in phrygana with *Sarcopoterium spinosum* and *Thymbra capitata*. - Endemic.

A recently described species related to *A. apergii* and *A. dirphianum* (two other local endemics in the *A. paniculatum* group), differing in details of leaf margin, spathe valves and perianth.

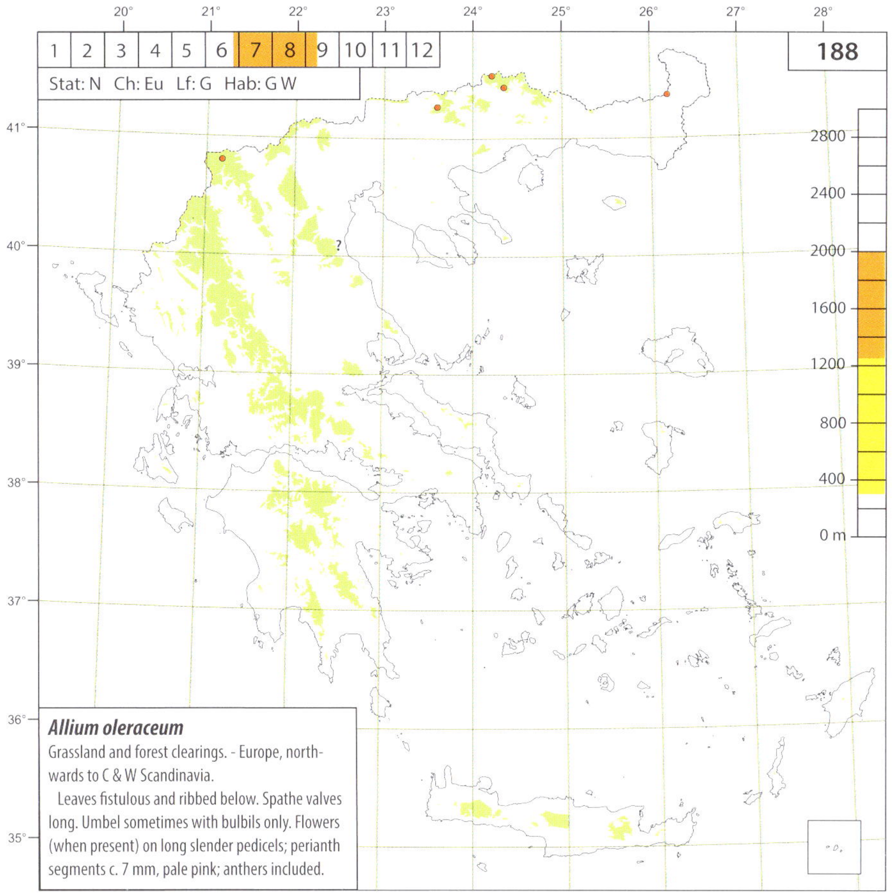

Allium oleraceum

Grassland and forest clearings. - Europe, northwards to C & W Scandinavia.

Leaves fistulous and ribbed below. Spathe valves long. Umbel sometimes with bulbils only. Flowers (when present) on long slender pedicels; perianth segments c. 7 mm, pale pink; anthers included.

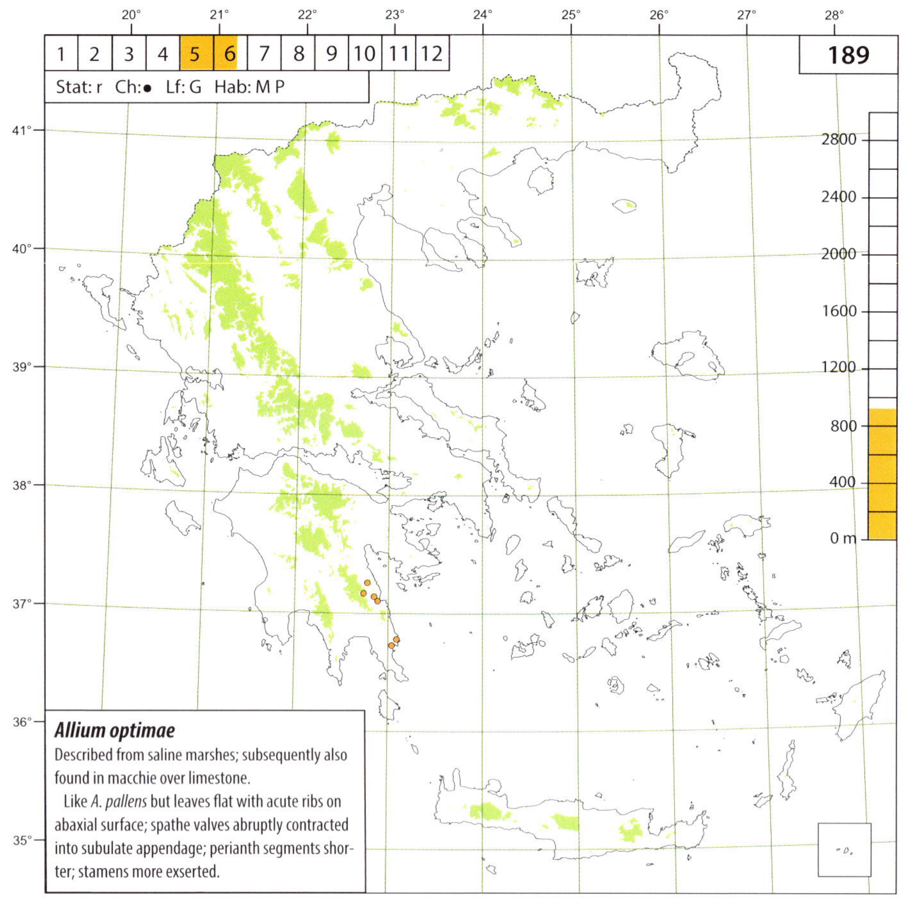

Allium optimae

Described from saline marshes; subsequently also found in macchie over limestone.

Like *A. pallens* but leaves flat with acute ribs on abaxial surface; spathe valves abruptly contracted into subulate appendage; perianth segments shorter; stamens more exserted.

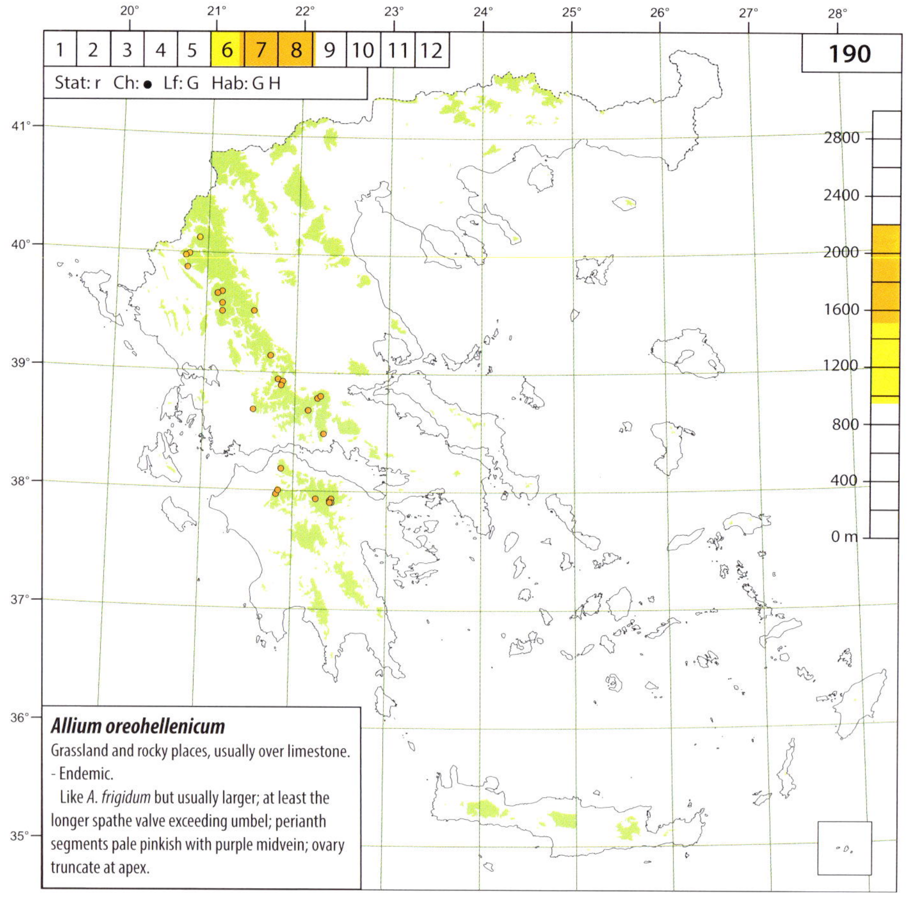

Allium oreohellenicum

Grassland and rocky places, usually over limestone. - Endemic.

Like *A. frigidum* but usually larger; at least the longer spathe valve exceeding umbel; perianth segments pale pinkish with purple midvein; ovary truncate at apex.

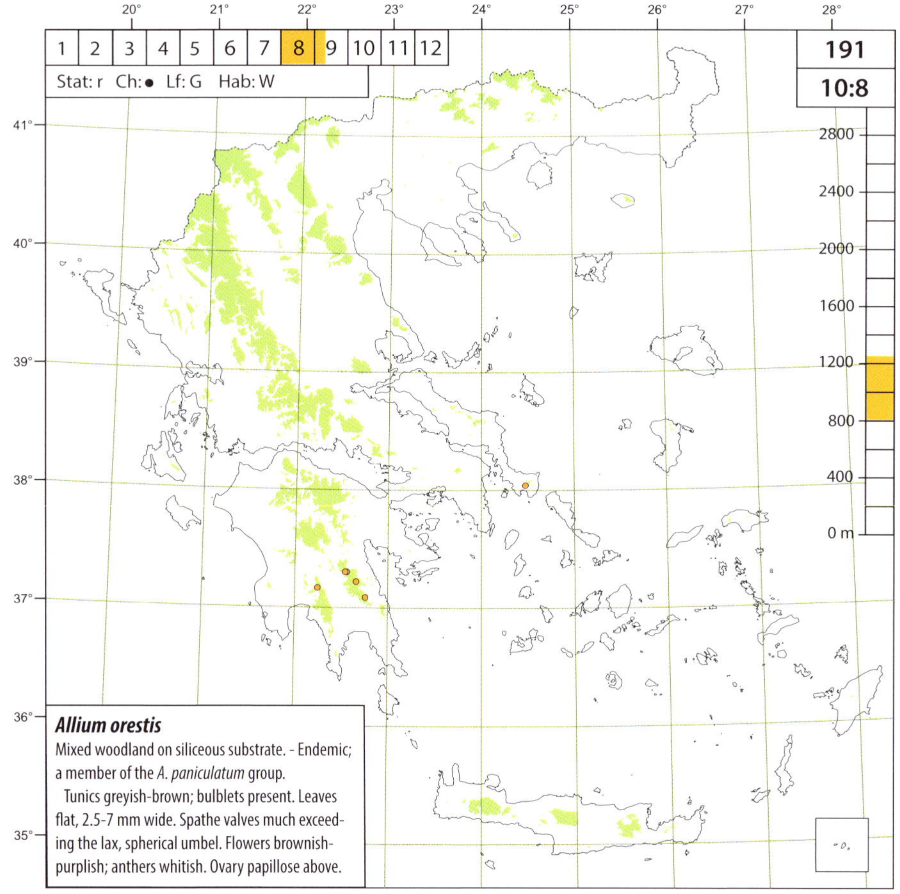

Allium orestis

Mixed woodland on siliceous substrate. - Endemic; a member of the *A. paniculatum* group.

Tunics greyish-brown; bulblets present. Leaves flat, 2.5-7 mm wide. Spathe valves much exceeding the lax, spherical umbel. Flowers brownish-purplish; anthers whitish. Ovary papillose above.

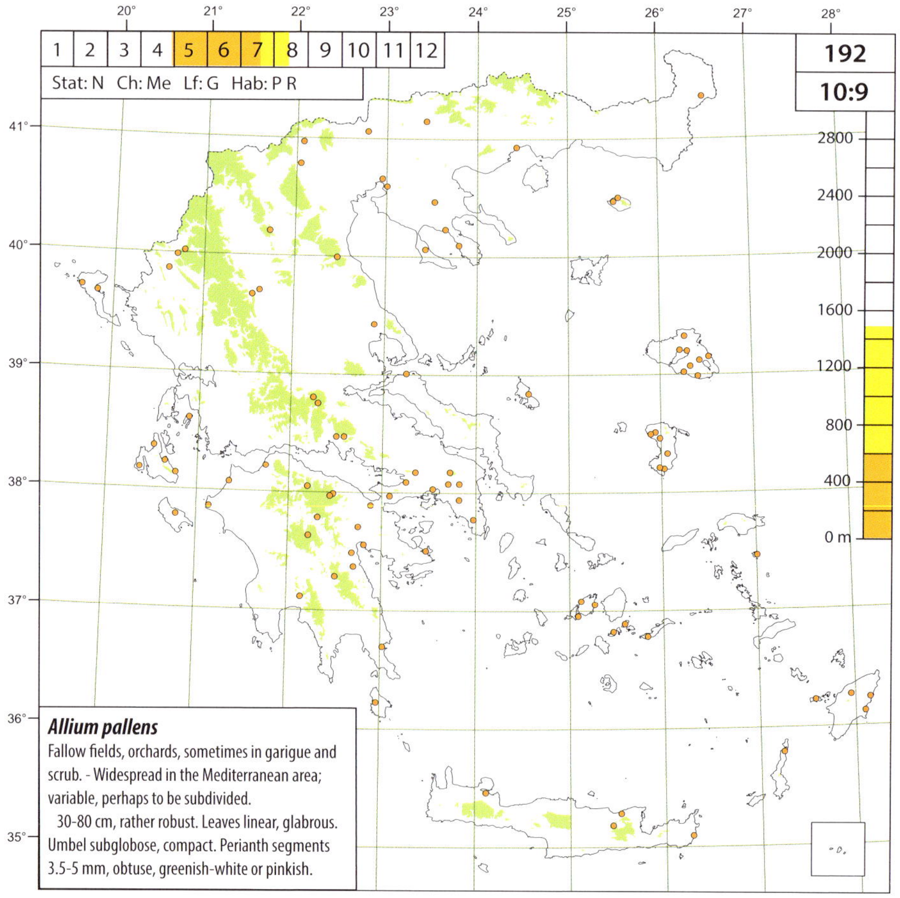

Allium pallens

Fallow fields, orchards, sometimes in garigue and scrub. - Widespread in the Mediterranean area; variable, perhaps to be subdivided.

30-80 cm, rather robust. Leaves linear, glabrous. Umbel subglobose, compact. Perianth segments 3.5-5 mm, obtuse, greenish-white or pinkish.

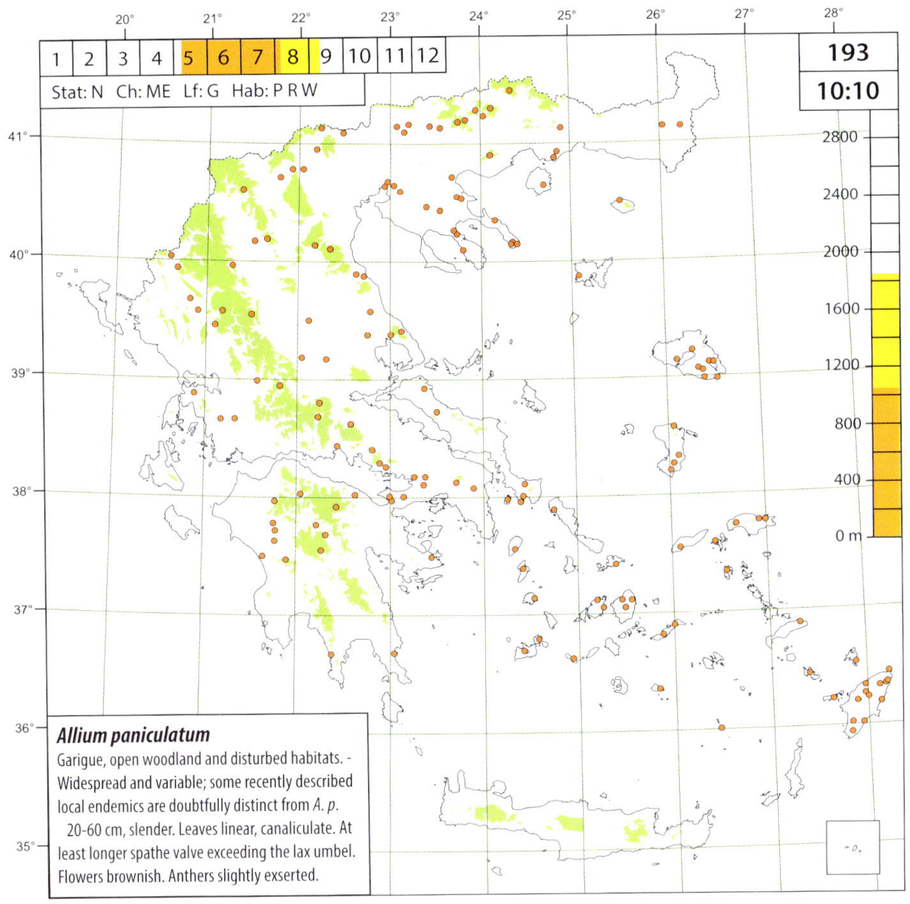

Allium paniculatum

Garigue, open woodland and disturbed habitats. - Widespread and variable; some recently described local endemics are doubtfully distinct from *A. p.*

20-60 cm, slender. Leaves linear, canaliculate. At least longer spathe valve exceeding the lax umbel. Flowers brownish. Anthers slightly exserted.

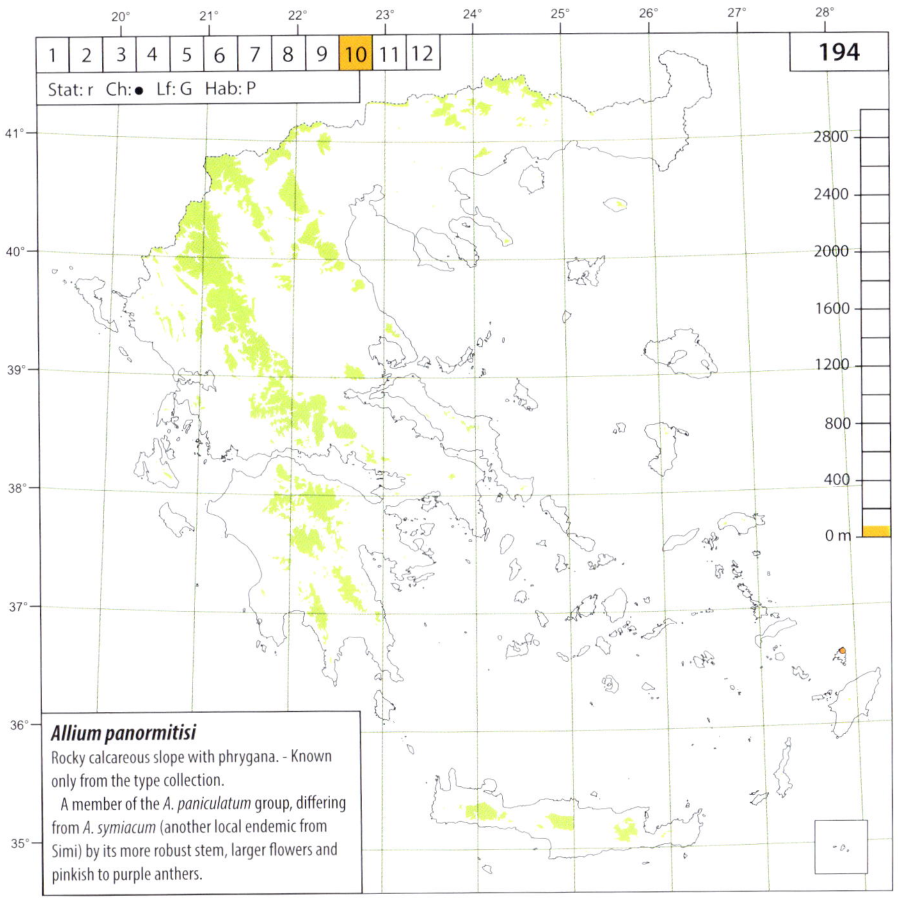

Allium panormitisi

Rocky calcareous slope with phrygana. - Known only from the type collection.

A member of the *A. paniculatum* group, differing from *A. symiacum* (another local endemic from Simi) by its more robust stem, larger flowers and pinkish to purple anthers.

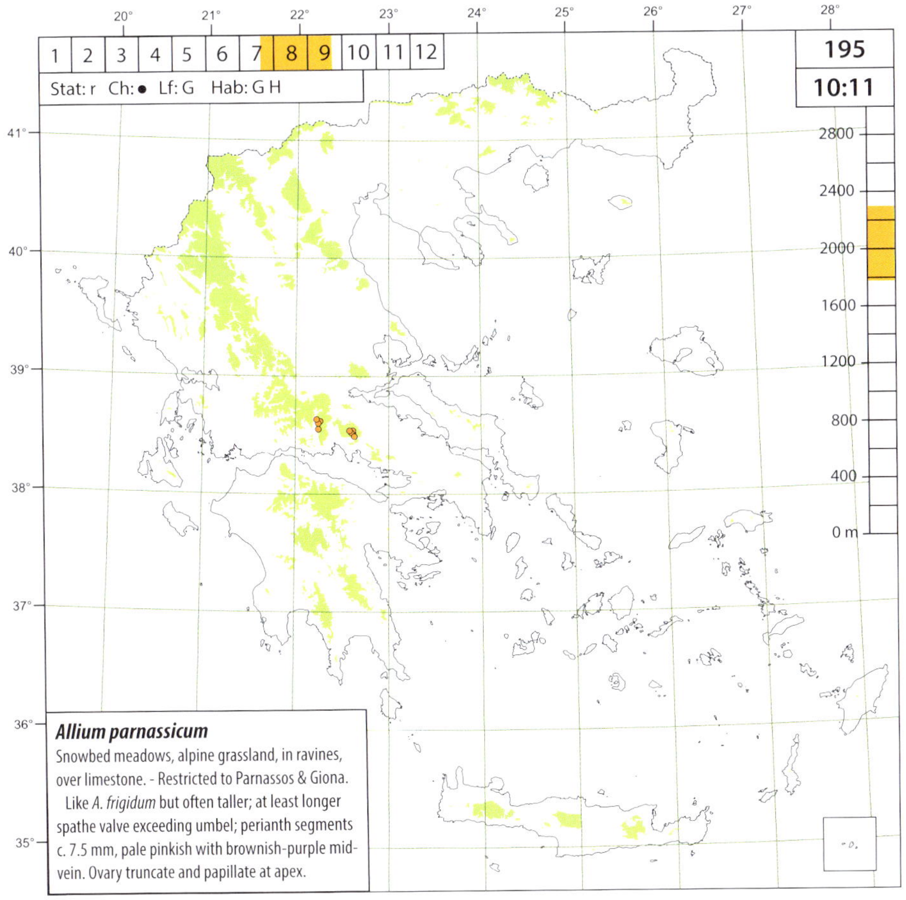

Allium parnassicum

Snowbed meadows, alpine grassland, in ravines, over limestone. - Restricted to Parnassos & Giona.

Like *A. frigidum* but often taller; at least longer spathe valve exceeding umbel; perianth segments c. 7.5 mm, pale pinkish with brownish-purple mid-vein. Ovary truncate and papillate at apex.

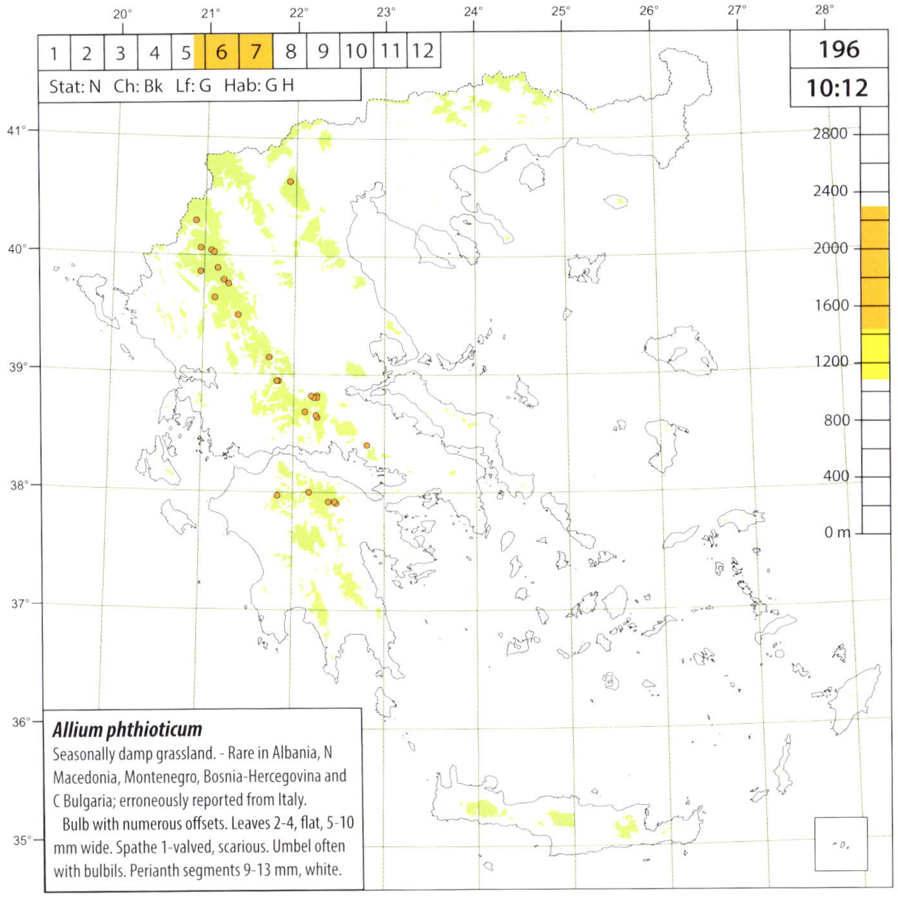

Allium phthioticum

Seasonally damp grassland. - Rare in Albania, N Macedonia, Montenegro, Bosnia-Hercegovina and C Bulgaria; erroneously reported from Italy.

Bulb with numerous offsets. Leaves 2-4, flat, 5-10 mm wide. Spathe 1-valved, scarious. Umbel often with bulbils. Perianth segments 9-13 mm, white.

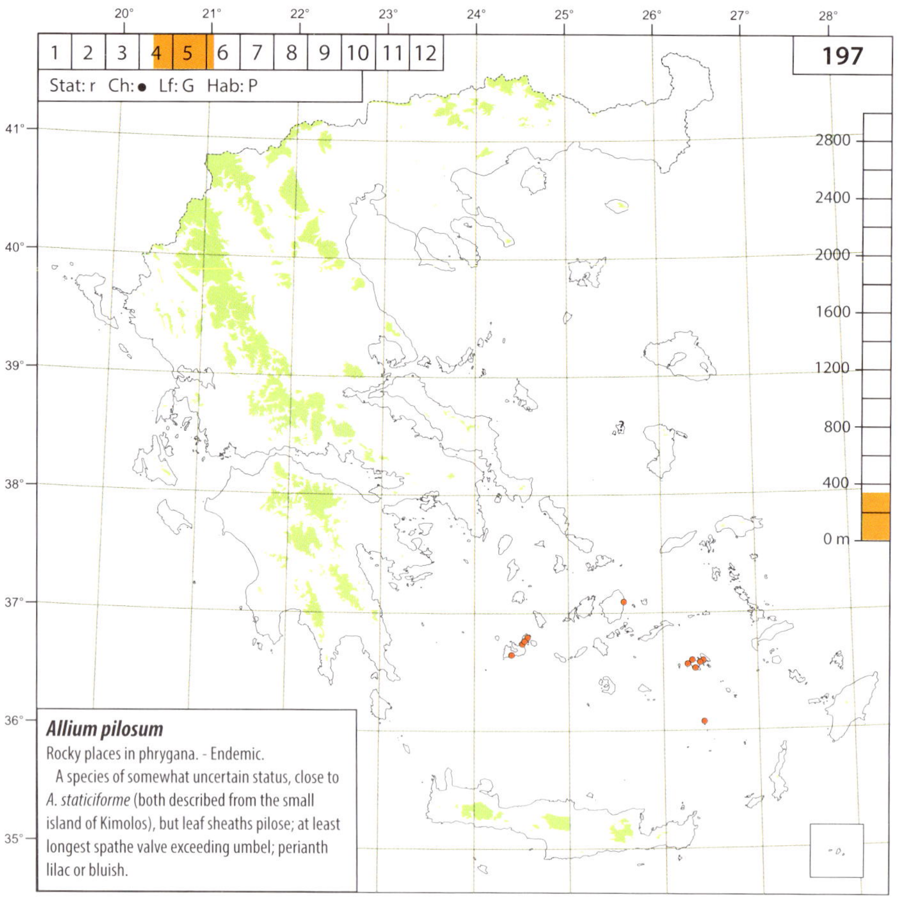

Allium pilosum

Rocky places in phrygana. - Endemic.

A species of somewhat uncertain status, close to *A. staticiforme* (both described from the small island of Kimolos), but leaf sheaths pilose; at least longest spathe valve exceeding umbel; perianth lilac or bluish.

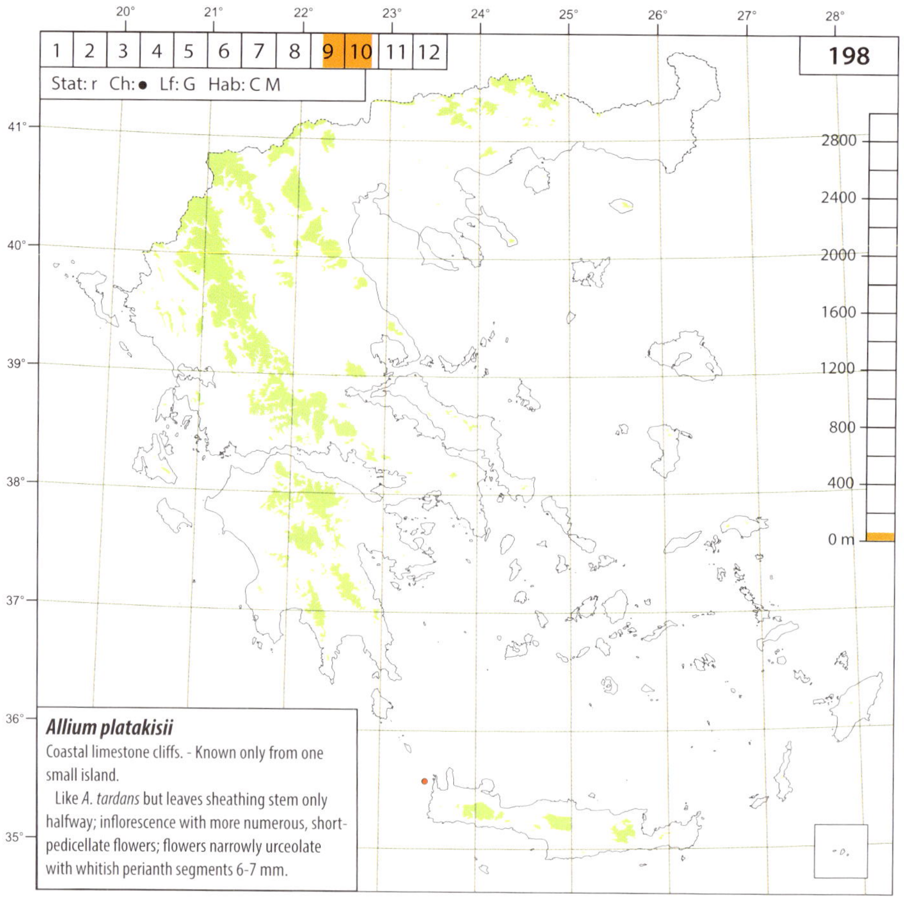

Allium platakisii

Coastal limestone cliffs. - Known only from one small island.

Like *A. tardans* but leaves sheathing stem only halfway; inflorescence with more numerous, short-pedicellate flowers; flowers narrowly urceolate with whitish perianth segments 6-7 mm.

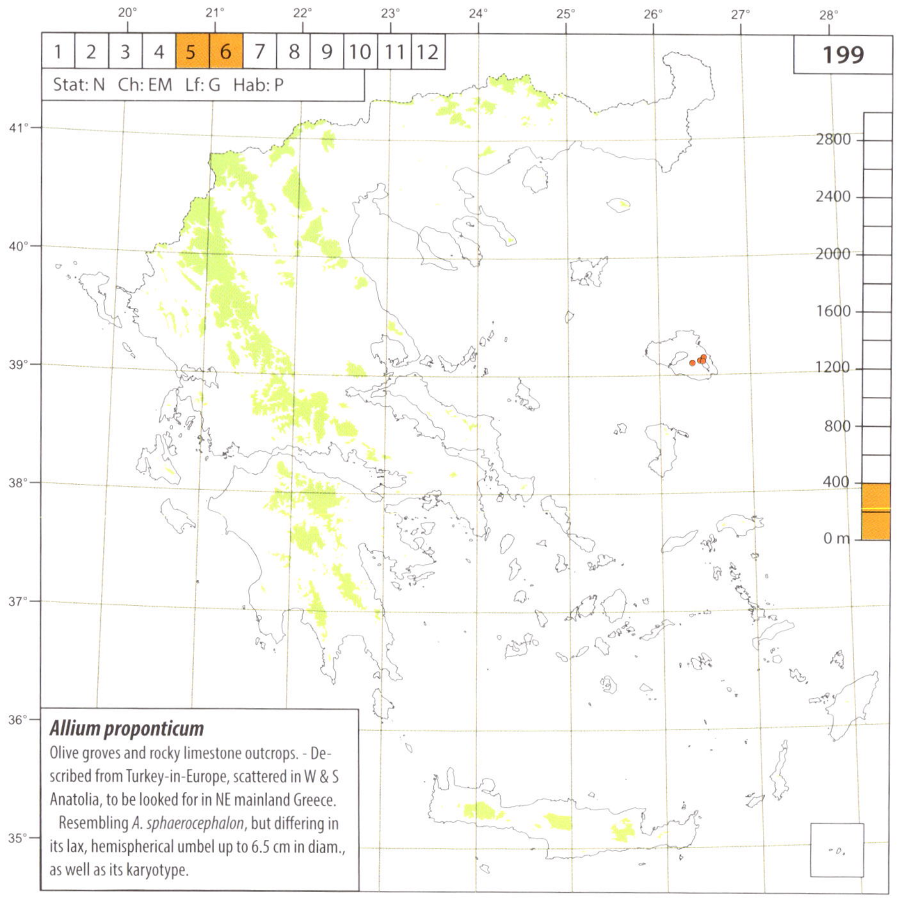

Allium proponticum

Olive groves and rocky limestone outcrops. - Described from Turkey-in-Europe, scattered in W & S Anatolia, to be looked for in NE mainland Greece.

Resembling *A. sphaerocephalon*, but differing in its lax, hemispherical umbel up to 6.5 cm in diam., as well as its karyotype.

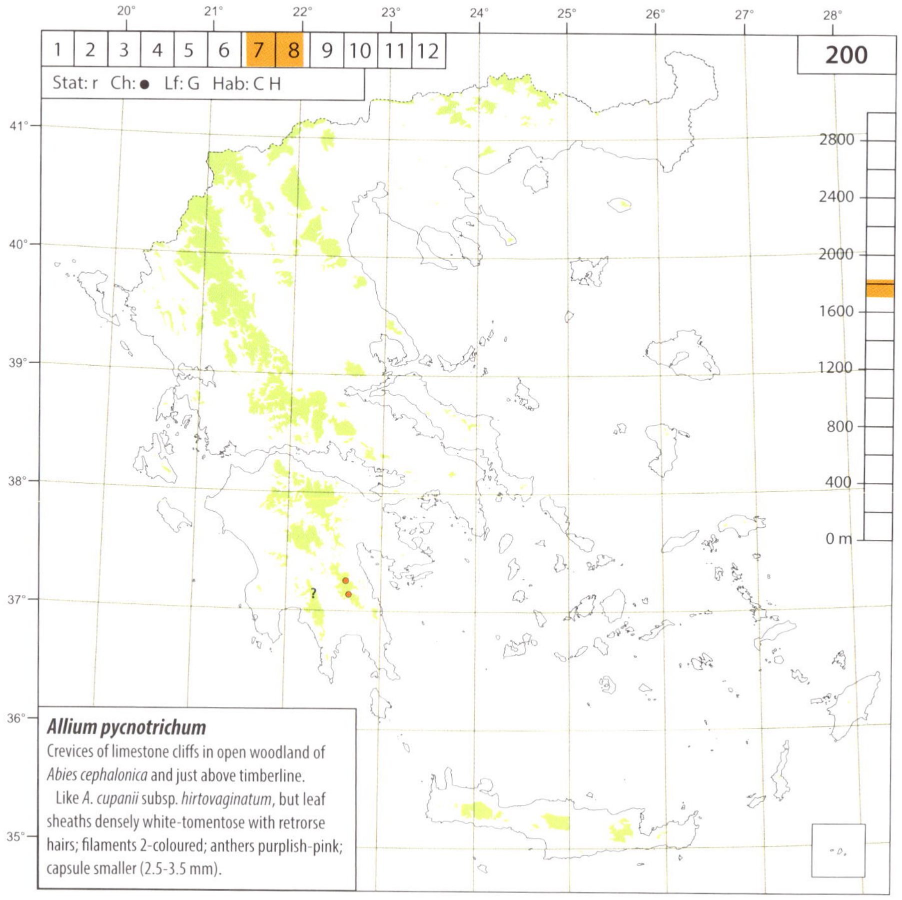

Allium pycnotrichum

Crevices of limestone cliffs in open woodland of *Abies cephalonica* and just above timberline.

Like *A. cupanii* subsp. *hirtovaginatum*, but leaf sheaths densely white-tomentose with retrorse hairs; filaments 2-coloured; anthers purplish-pink; capsule smaller (2.5-3.5 mm).

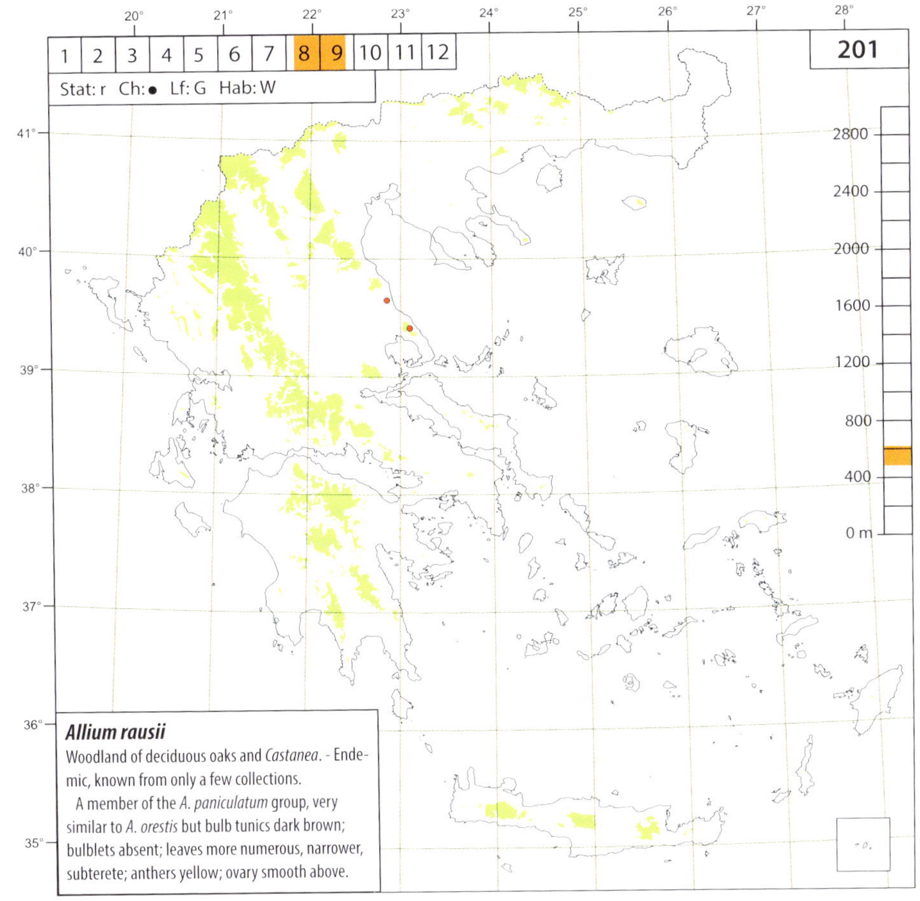

Allium rausii

Woodland of deciduous oaks and *Castanea*. - Endemic, known from only a few collections.

A member of the *A. paniculatum* group, very similar to *A. orestis* but bulb tunics dark brown; bulblets absent; leaves more numerous, narrower, subterete; anthers yellow; ovary smooth above.

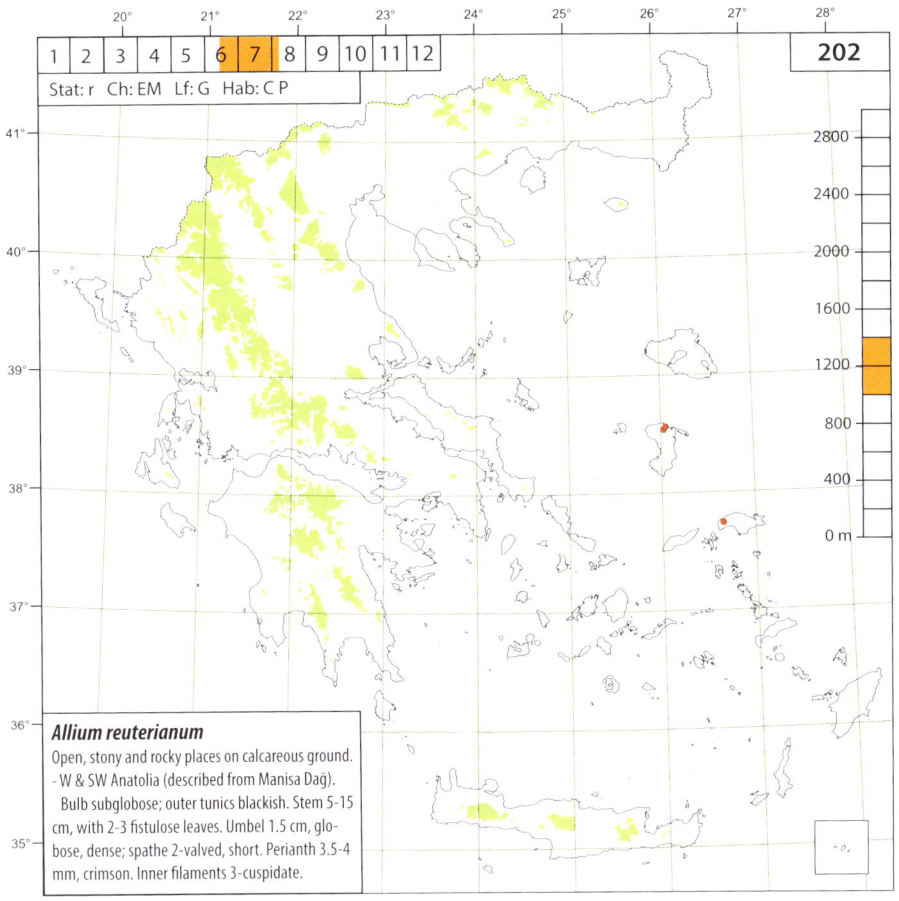

Allium reuterianum

Open, stony and rocky places on calcareous ground. - W & SW Anatolia (described from Manisa Dağ).

Bulb subglobose; outer tunics blackish. Stem 5-15 cm, with 2-3 fistulose leaves. Umbel 1.5 cm, globose, dense; spathe 2-valved, short. Perianth 3.5-4 mm, crimson. Inner filaments 3-cuspidate.

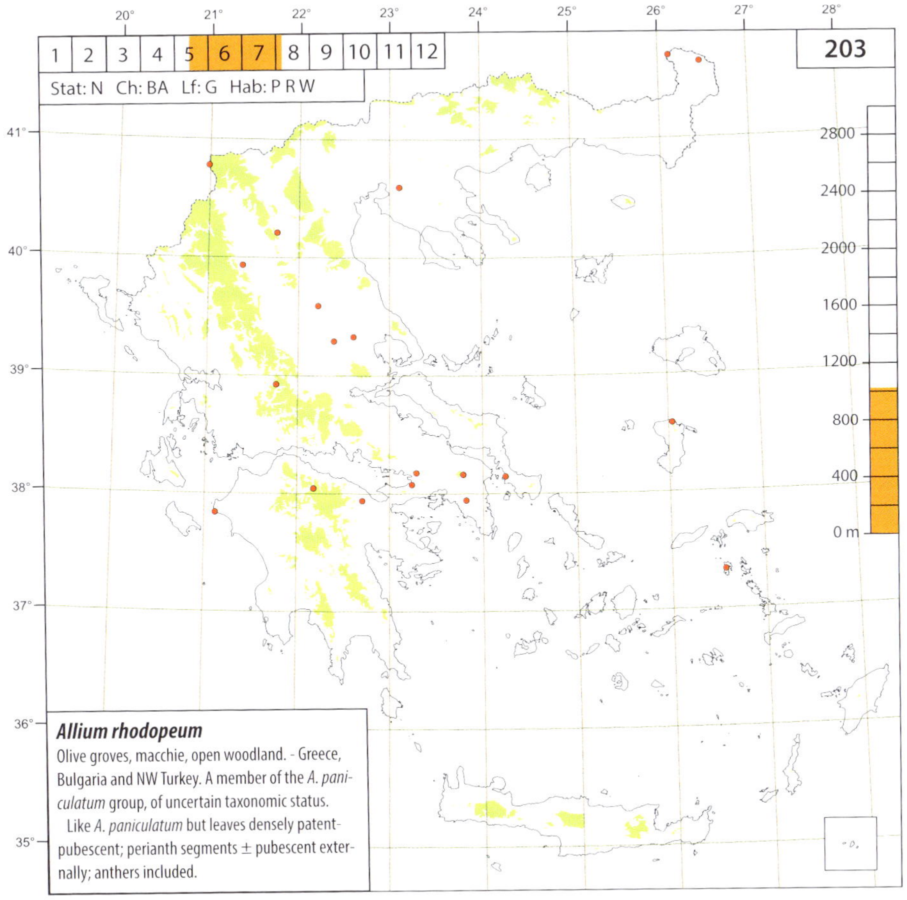

Allium rhodopeum

Olive groves, macchie, open woodland. - Greece, Bulgaria and NW Turkey. A member of the *A. paniculatum* group, of uncertain taxonomic status.

Like *A. paniculatum* but leaves densely patent-pubescent; perianth segments ± pubescent externally; anthers included.

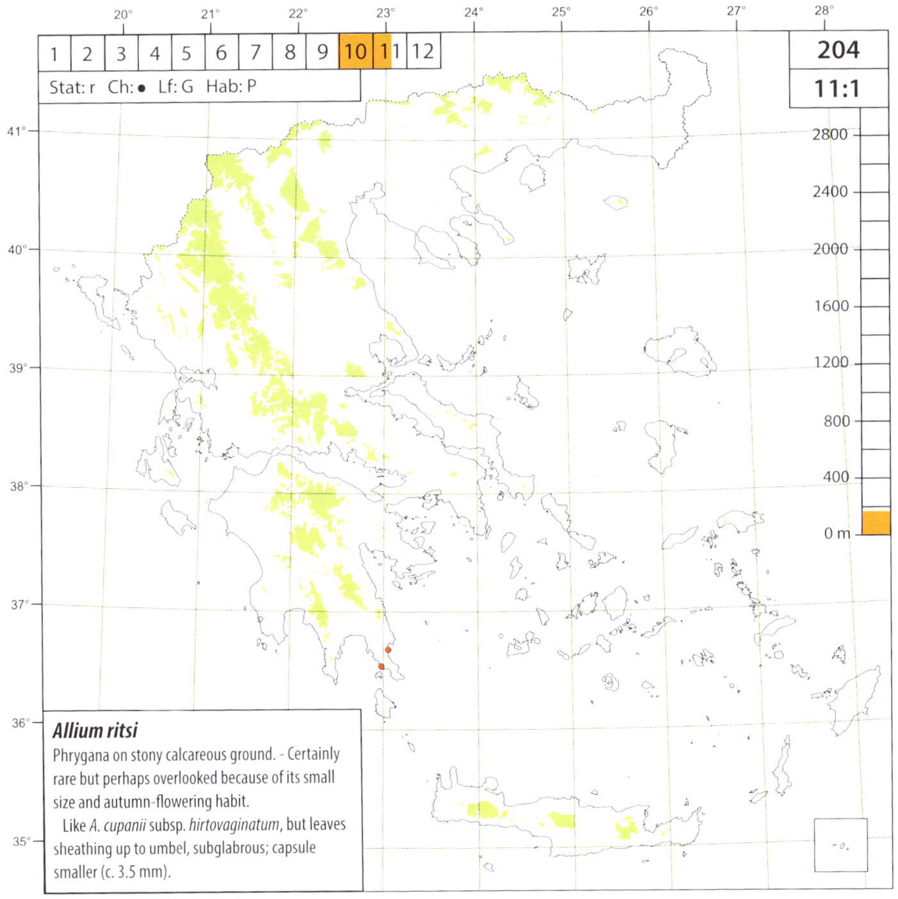

Allium ritsi

Phrygana on stony calcareous ground. - Certainly rare but perhaps overlooked because of its small size and autumn-flowering habit.

Like *A. cupanii* subsp. *hirtovaginatum*, but leaves sheathing up to umbel, subglabrous; capsule smaller (c. 3.5 mm).

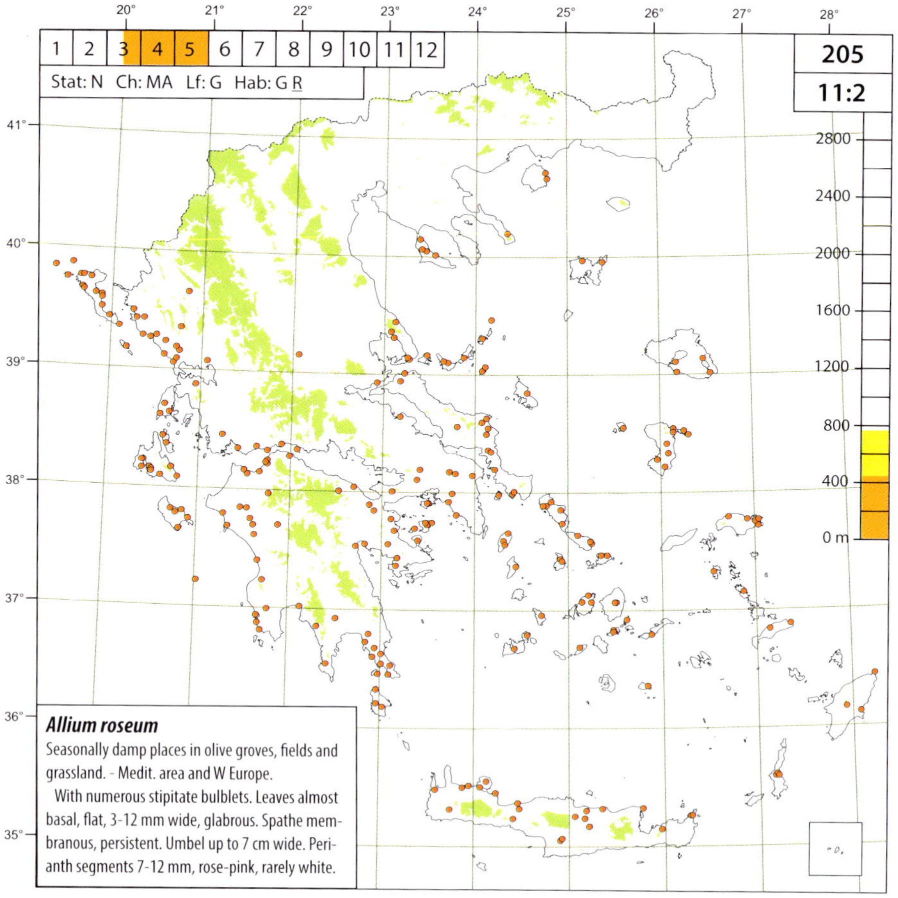

Allium roseum

Seasonally damp places in olive groves, fields and grassland. - Medit. area and W Europe.

With numerous stipitate bulblets. Leaves almost basal, flat, 3-12 mm wide, glabrous. Spathe membranous, persistent. Umbel up to 7 cm wide. Perianth segments 7-12 mm, rose-pink, rarely white.

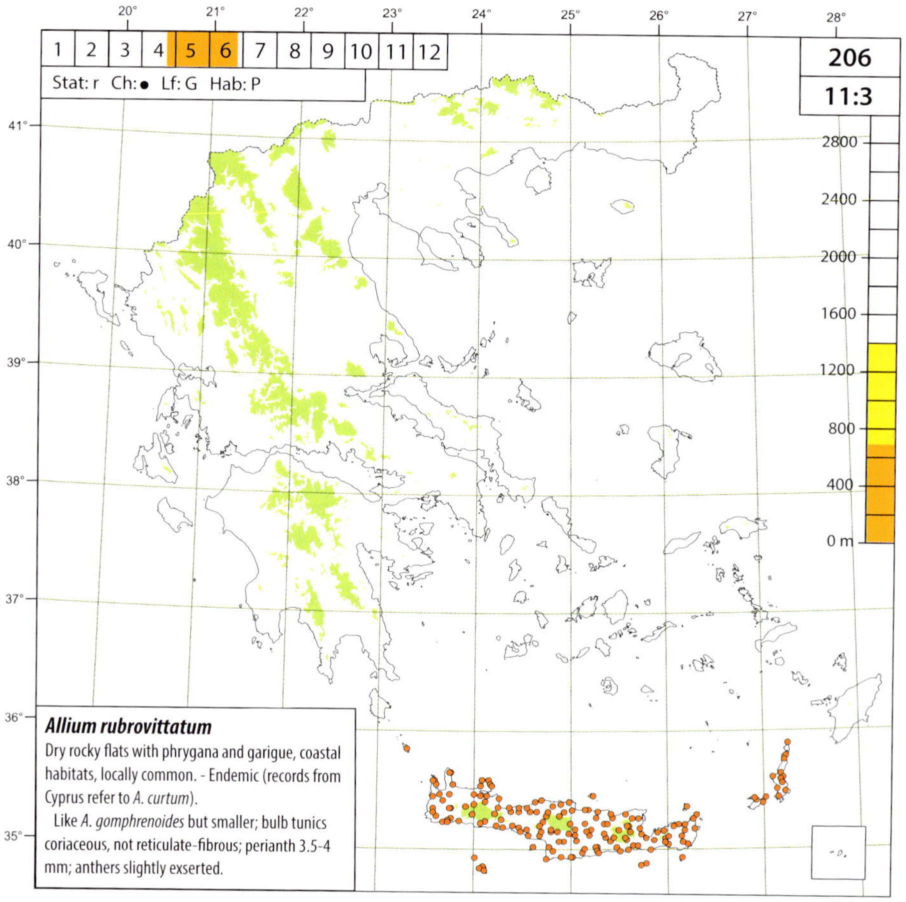

Allium rubrovittatum

Dry rocky flats with phrygana and garigue, coastal habitats, locally common. - Endemic (records from Cyprus refer to *A. curtum*).

Like *A. gomphrenoides* but smaller; bulb tunics coriaceous, not reticulate-fibrous; perianth 3.5-4 mm; anthers slightly exserted.

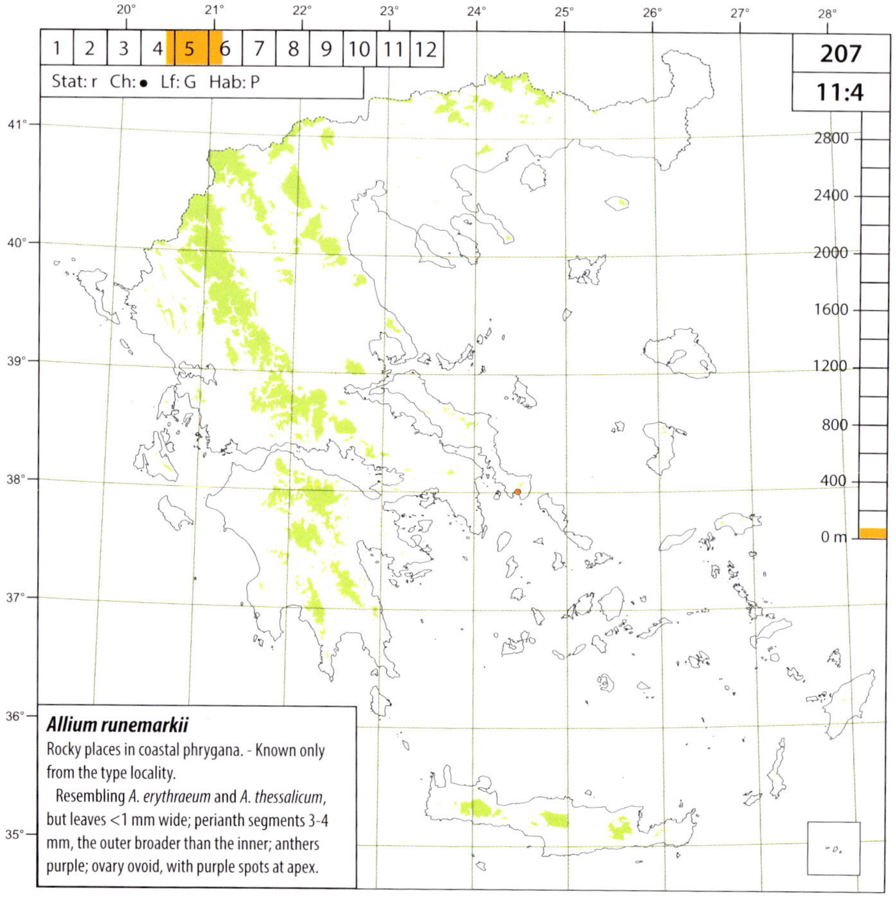

Allium runemarkii

Rocky places in coastal phrygana. - Known only from the type locality.

Resembling *A. erythraeum* and *A. thessalicum*, but leaves <1 mm wide; perianth segments 3-4 mm, the outer broader than the inner; anthers purple; ovary ovoid, with purple spots at apex.

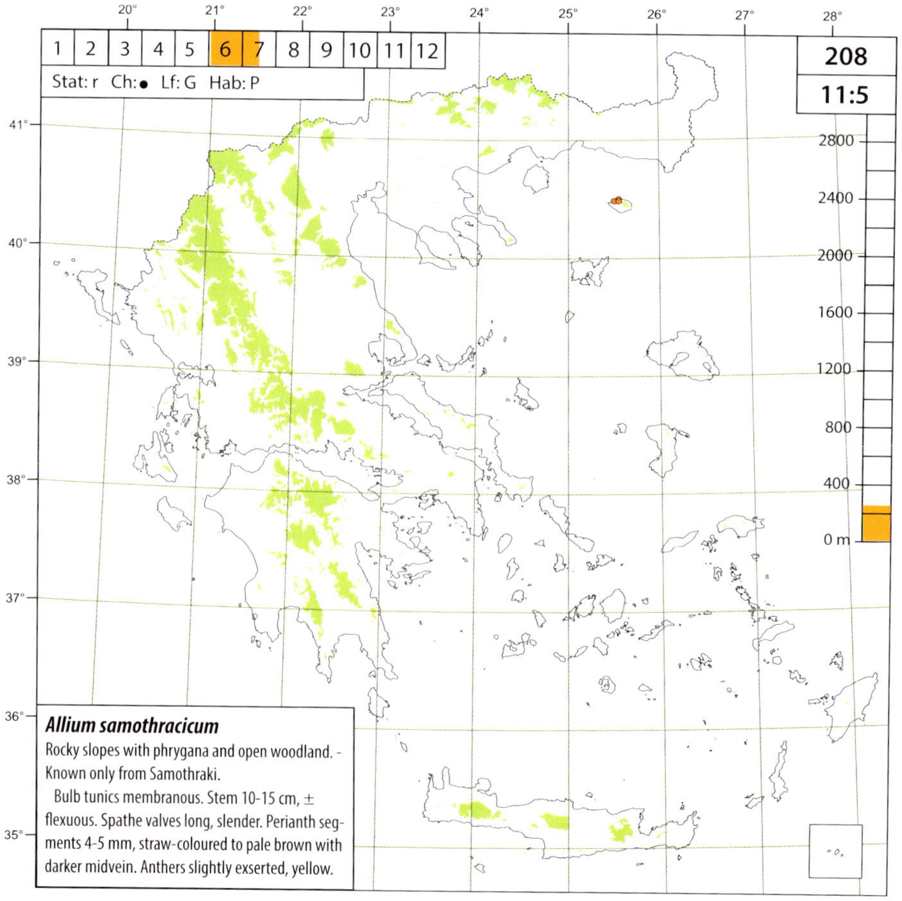

Allium samothracicum

Rocky slopes with phrygana and open woodland. - Known only from Samothraki.

Bulb tunics membranous. Stem 10-15 cm, ± flexuous. Spathe valves long, slender. Perianth segments 4-5 mm, straw-coloured to pale brown with darker midvein. Anthers slightly exserted, yellow.

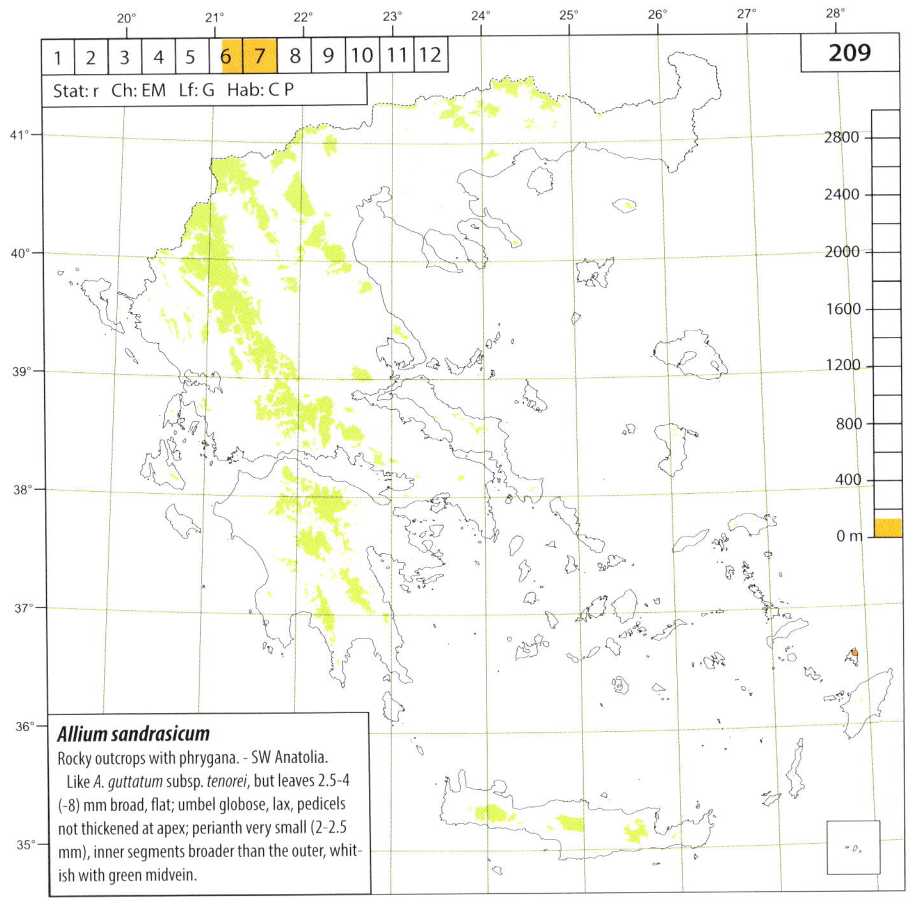

Allium sandrasicum
Rocky outcrops with phrygana. - SW Anatolia.
Like *A. guttatum* subsp. *tenorei*, but leaves 2.5-4 (-8) mm broad, flat; umbel globose, lax, pedicels not thickened at apex; perianth very small (2-2.5 mm), inner segments broader than the outer, whitish with green midvein.

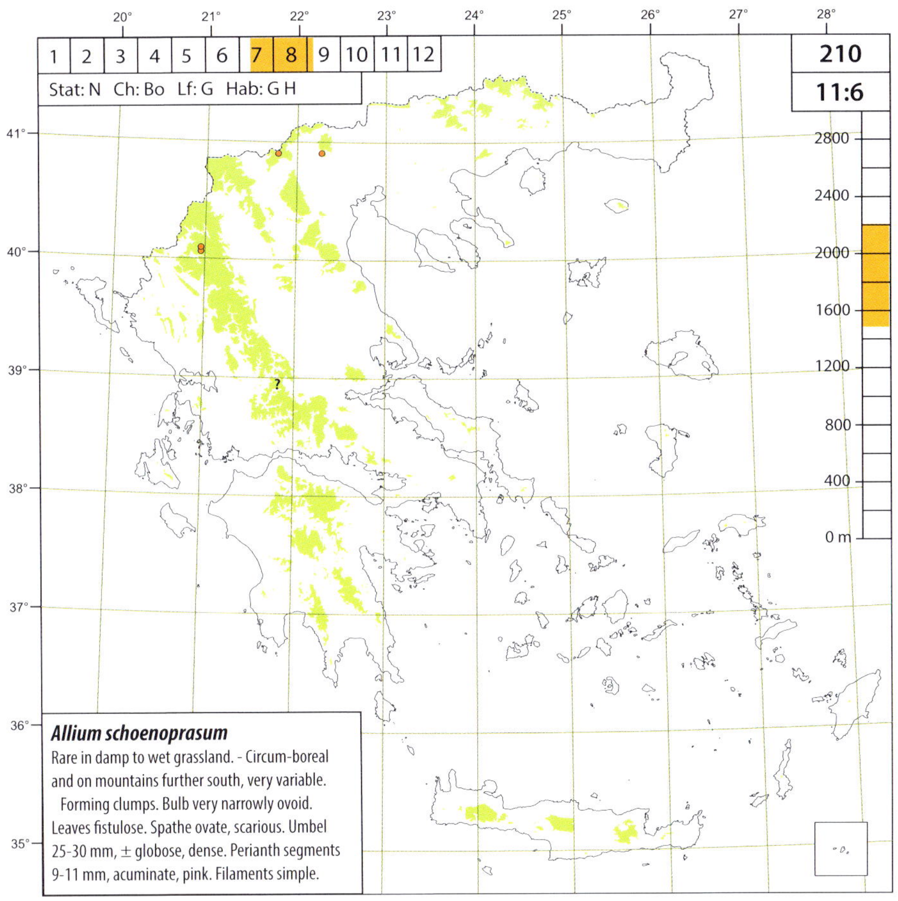

Allium schoenoprasum
Rare in damp to wet grassland. - Circum-boreal and on mountains further south, very variable.
Forming clumps. Bulb very narrowly ovoid. Leaves fistulose. Spathe ovate, scarious. Umbel 25-30 mm, ± globose, dense. Perianth segments 9-11 mm, acuminate, pink. Filaments simple.

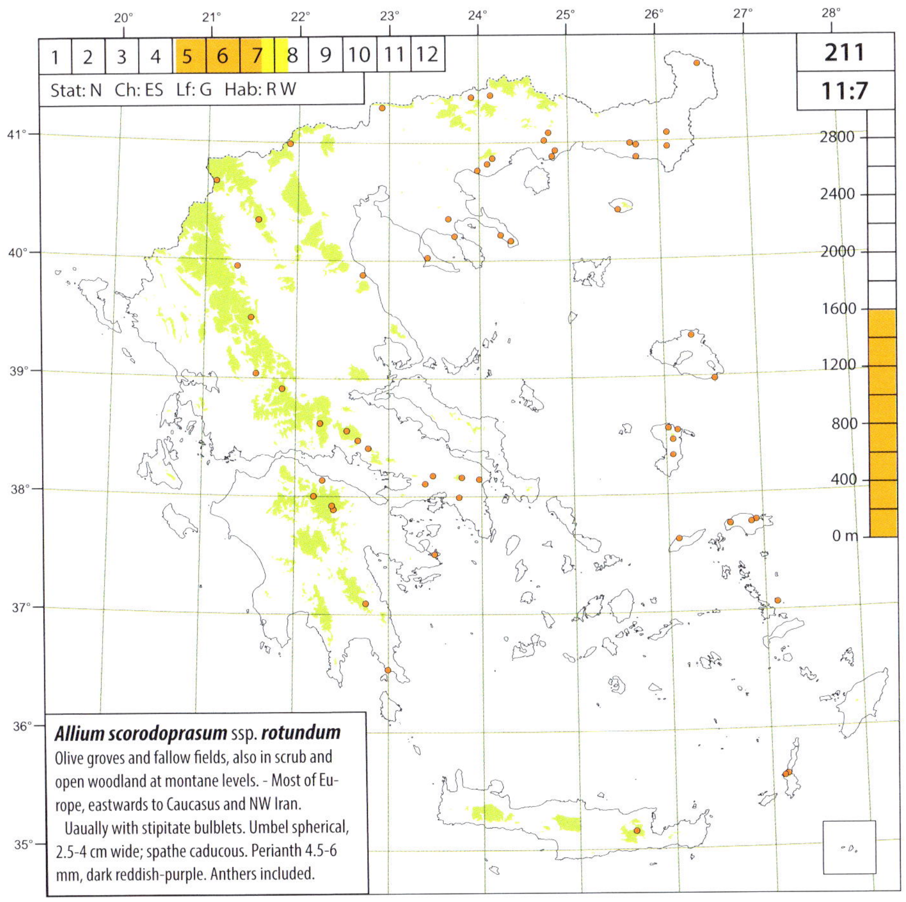

Allium scorodoprasum ssp. **_rotundum_**
Olive groves and fallow fields, also in scrub and open woodland at montane levels. - Most of Europe, eastwards to Caucasus and NW Iran.
Uaually with stipitate bulblets. Umbel spherical, 2.5-4 cm wide; spathe caducous. Perianth 4.5-6 mm, dark reddish-purple. Anthers included.

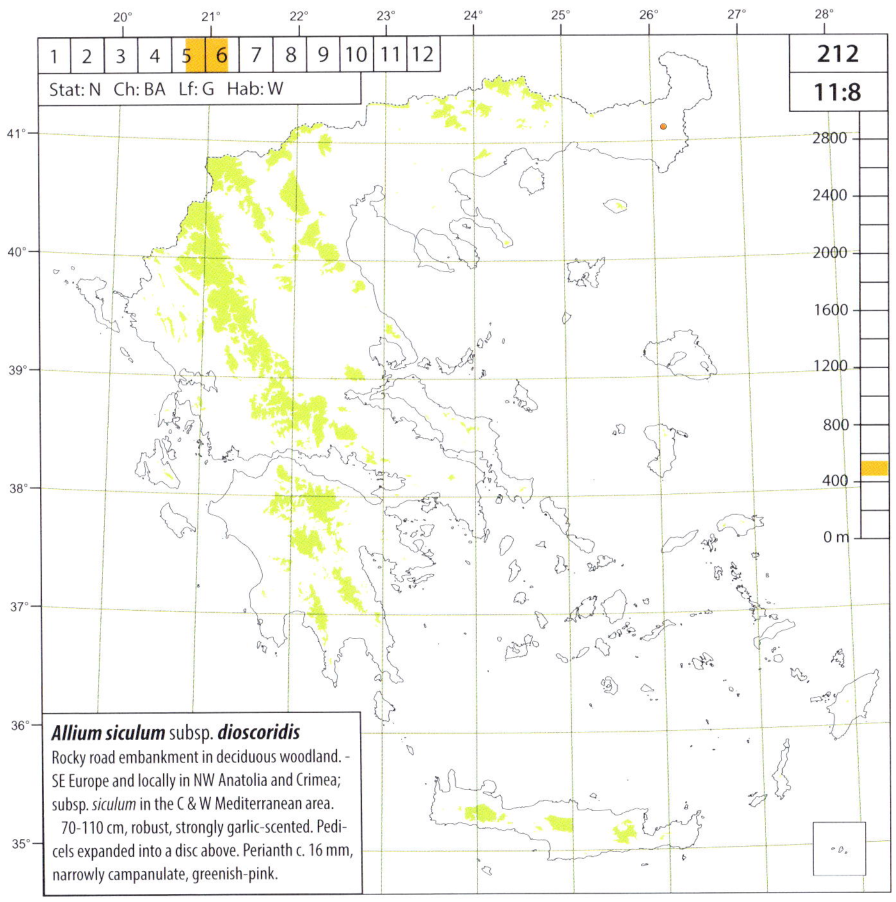

Allium siculum subsp. **_dioscoridis_**
Rocky road embankment in deciduous woodland. - SE Europe and locally in NW Anatolia and Crimea; subsp. *siculum* in the C & W Mediterranean area.
70-110 cm, robust, strongly garlic-scented. Pedicels expanded into a disc above. Perianth c. 16 mm, narrowly campanulate, greenish-pink.

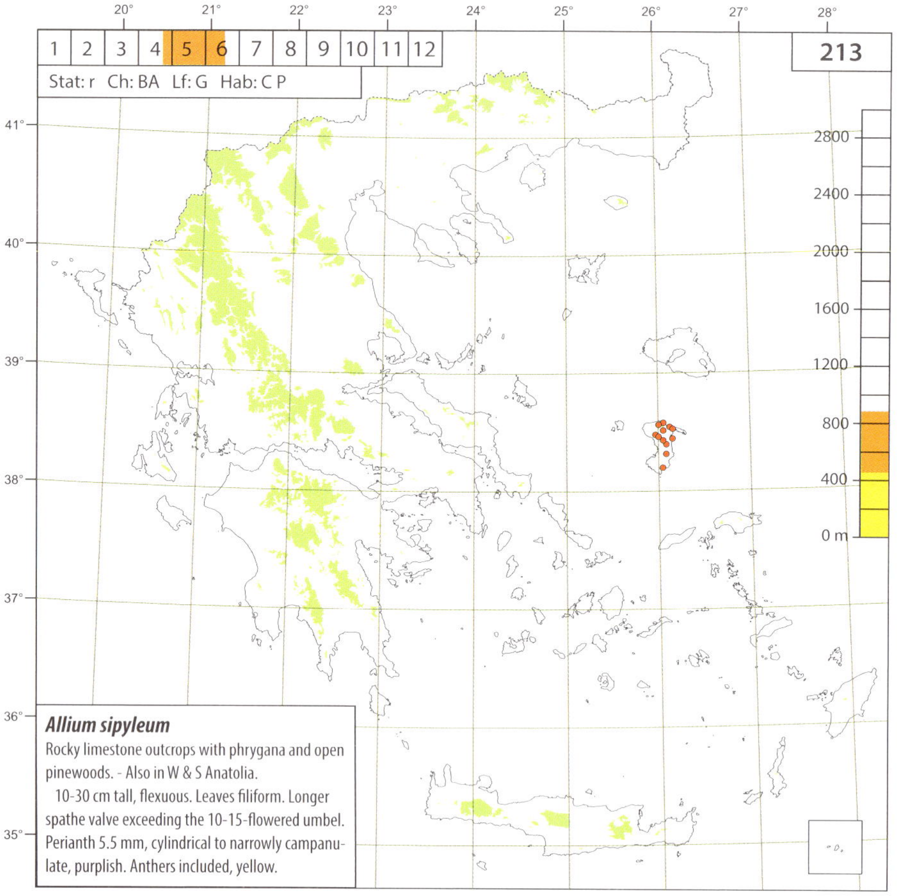

Allium sipyleum
Rocky limestone outcrops with phrygana and open pinewoods. - Also in W & S Anatolia.
10-30 cm tall, flexuous. Leaves filiform. Longer spathe valve exceeding the 10-15-flowered umbel. Perianth 5.5 mm, cylindrical to narrowly campanulate, purplish. Anthers included, yellow.

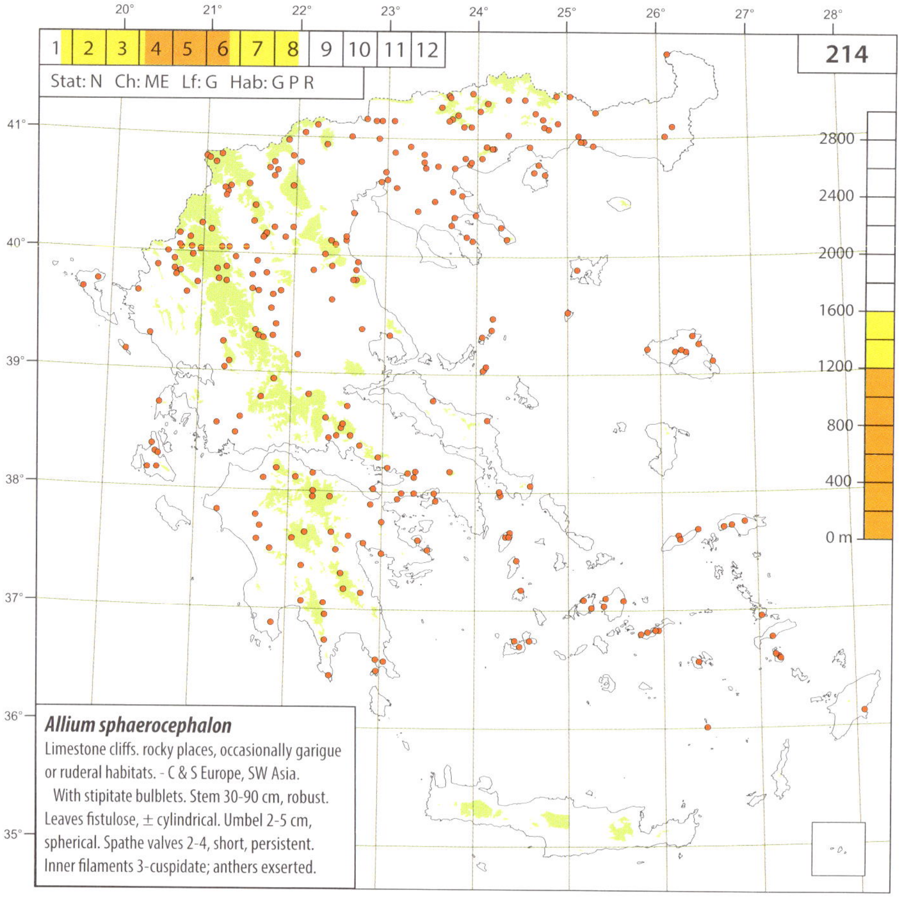

Allium sphaerocephalon
Limestone cliffs, rocky places, occasionally garigue or ruderal habitats. - C & S Europe, SW Asia.
With stipitate bulblets. Stem 30-90 cm, robust. Leaves fistulose, ± cylindrical. Umbel 2-5 cm, spherical. Spathe valves 2-4, short, persistent. Inner filaments 3-cuspidate; anthers exserted.

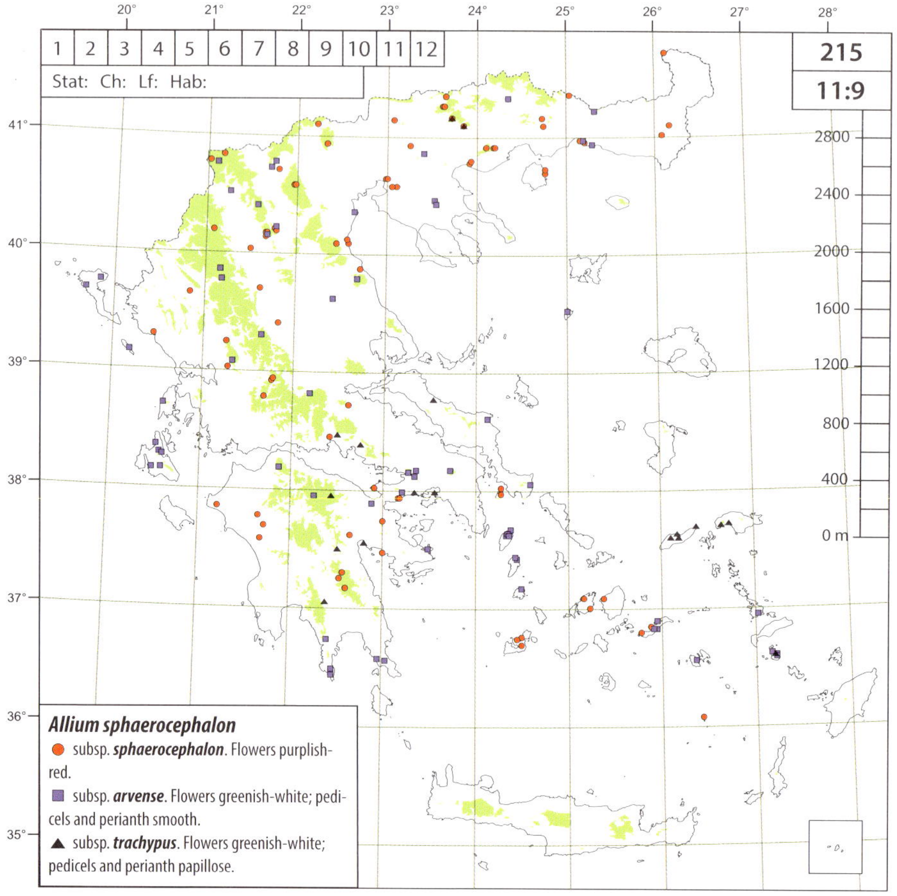

Allium sphaerocephalon
● subsp. ***sphaerocephalon***. Flowers purplish-red.
■ subsp. ***arvense***. Flowers greenish-white; pedicels and perianth smooth.
▲ subsp. ***trachypus***. Flowers greenish-white; pedicels and perianth papillose.

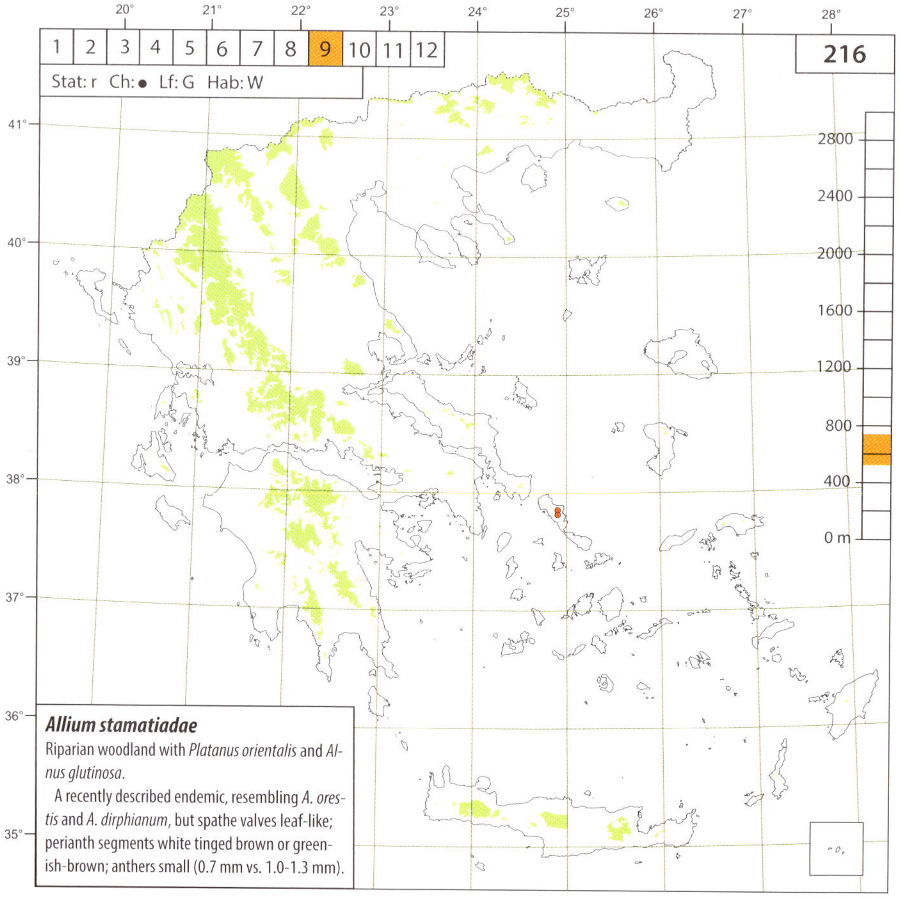

Allium stamatiadae
Riparian woodland with *Platanus orientalis* and *Alnus glutinosa*.
A recently described endemic, resembling *A. orestis* and *A. dirphianum*, but spathe valves leaf-like; perianth segments white tinged brown or greenish-brown; anthers small (0.7 mm vs. 1.0-1.3 mm).

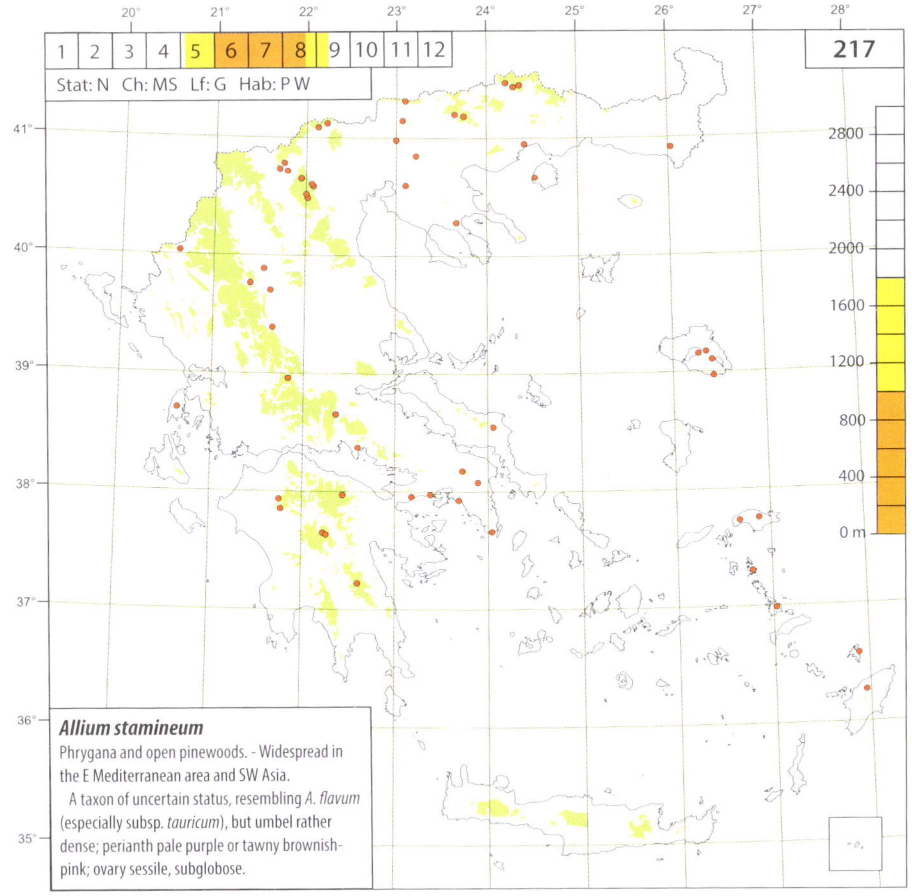

Allium stamineum

Phrygana and open pinewoods. - Widespread in the E Mediterranean area and SW Asia.

A taxon of uncertain status, resembling *A. flavum* (especially subsp. *tauricum*), but umbel rather dense; perianth pale purple or tawny brownish-pink; ovary sessile, subglobose.

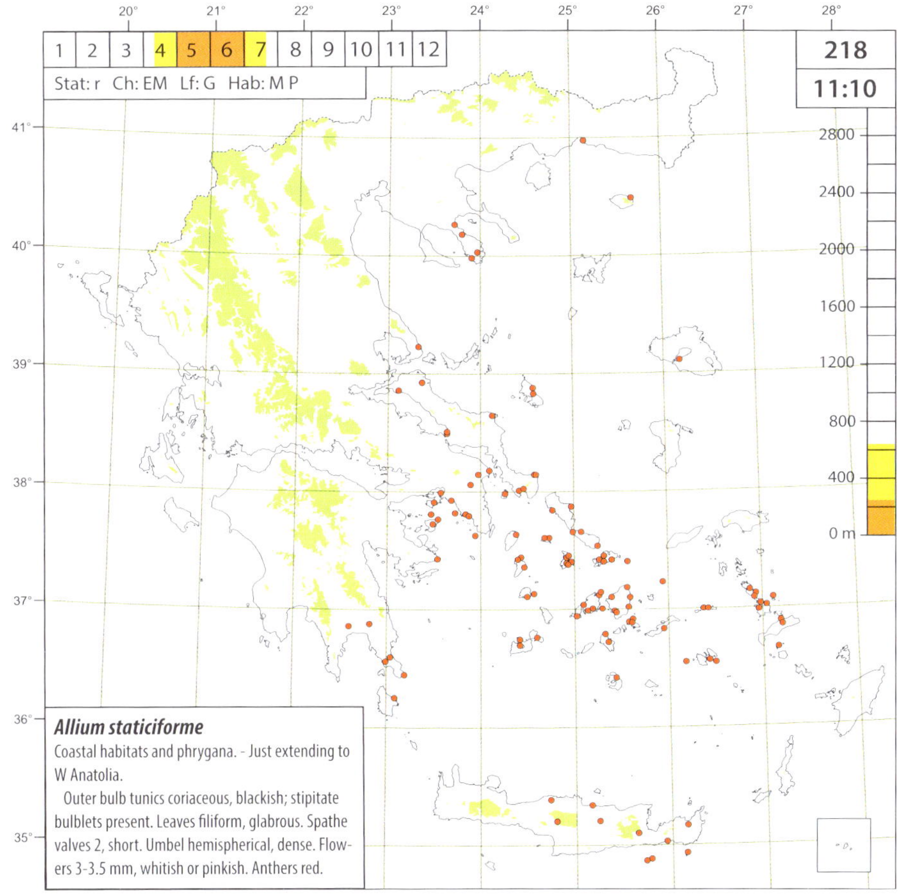

Allium staticiforme

Coastal habitats and phrygana. - Just extending to W Anatolia.

Outer bulb tunics coriaceous, blackish; stipitate bulblets present. Leaves filiform, glabrous. Spathe valves 2, short. Umbel hemispherical, dense. Flowers 3-3.5 mm, whitish or pinkish. Anthers red.

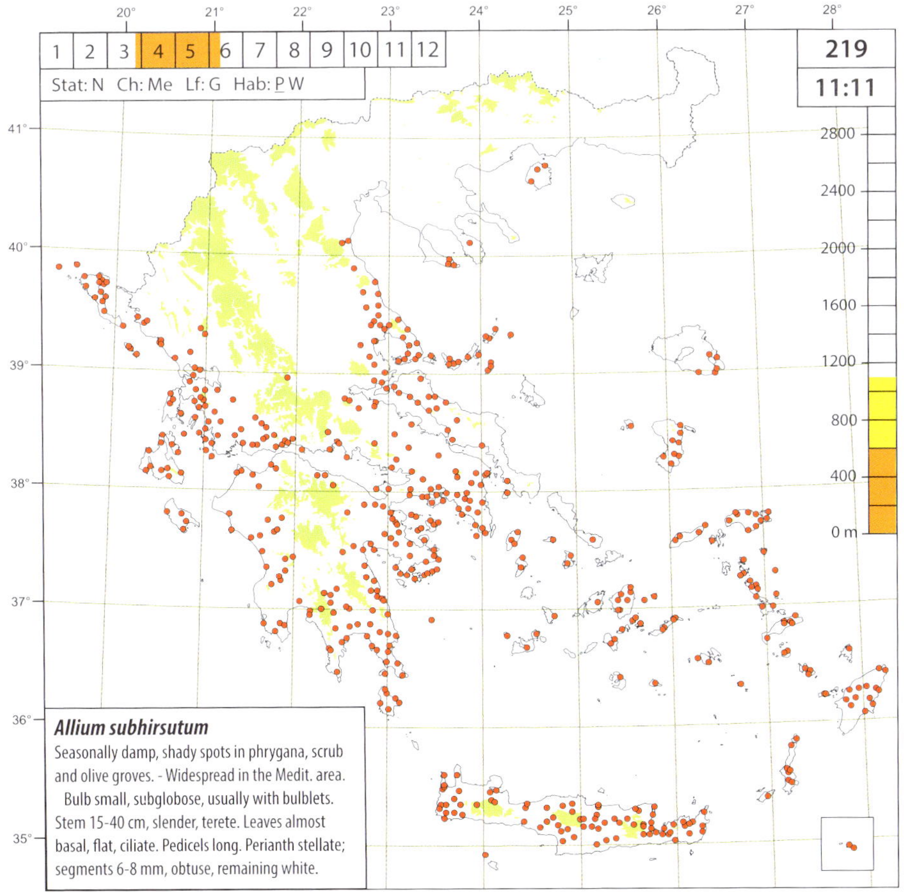

Allium subhirsutum

Seasonally damp, shady spots in phrygana, scrub and olive groves. - Widespread in the Medit. area.

Bulb small, subglobose, usually with bulblets. Stem 15-40 cm, slender, terete. Leaves almost basal, flat, ciliate. Pedicels long. Perianth stellate; segments 6-8 mm, obtuse, remaining white.

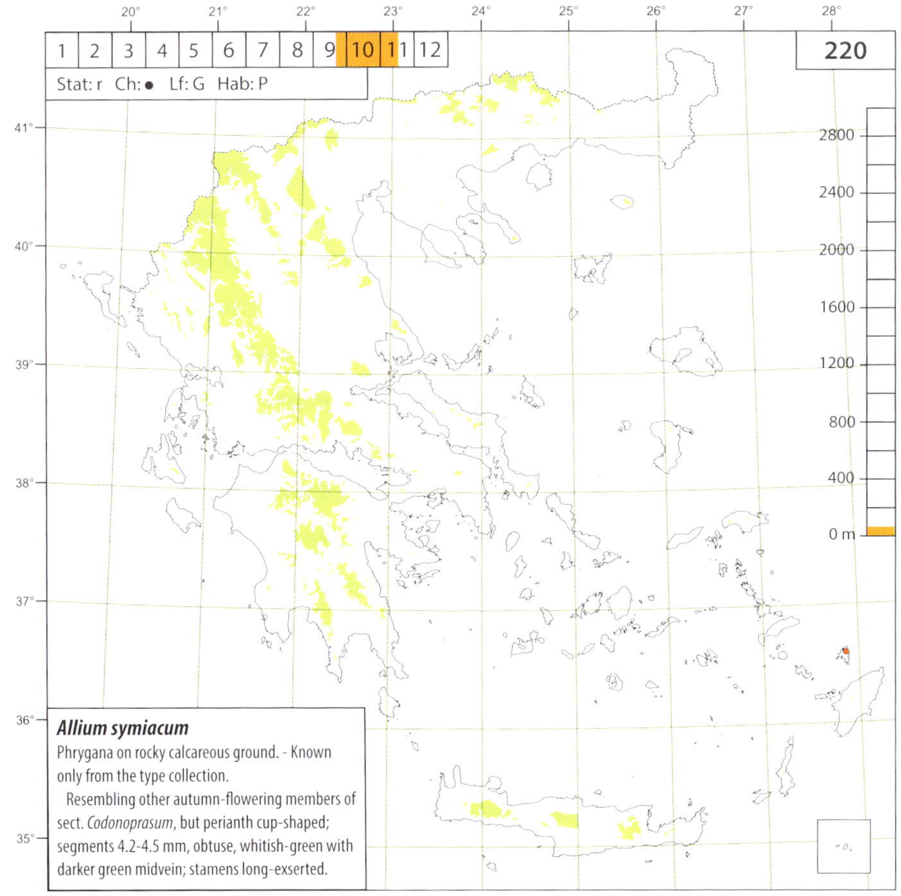

Allium symiacum

Phrygana on rocky calcareous ground. - Known only from the type collection.

Resembling other autumn-flowering members of sect. *Codonoprasum*, but perianth cup-shaped; segments 4.2-4.5 mm, obtuse, whitish-green with darker green midvein; stamens long-exserted.

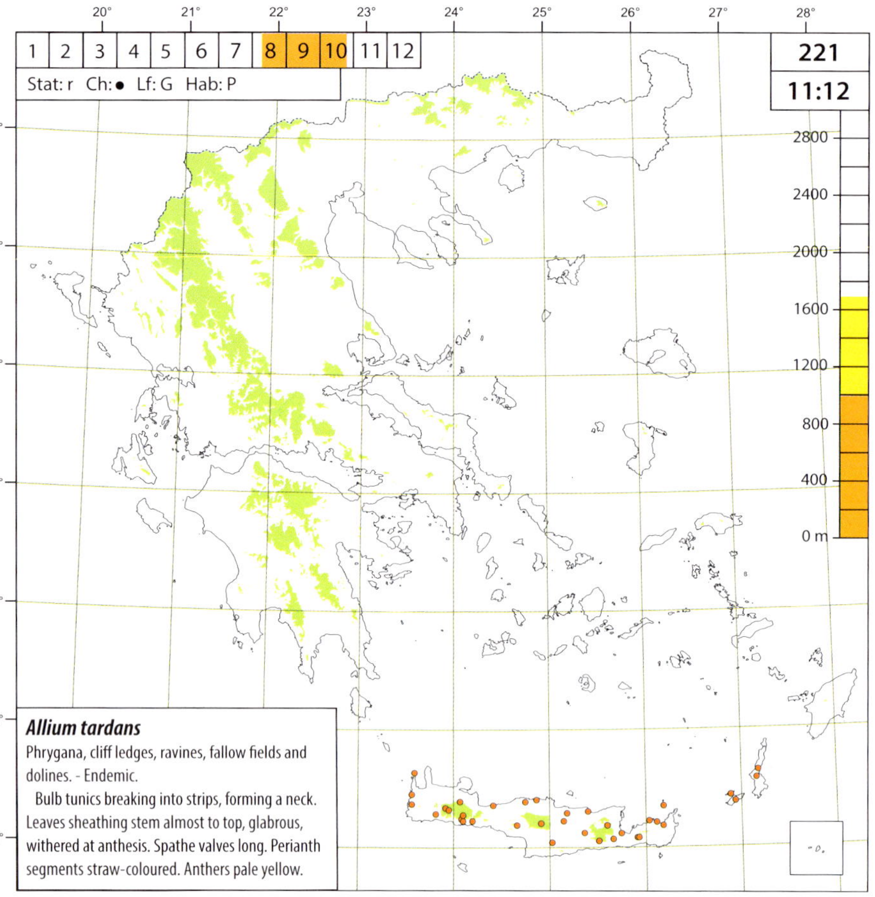

Allium tardans

Phrygana, cliff ledges, ravines, fallow fields and dolines. - Endemic.

Bulb tunics breaking into strips, forming a neck. Leaves sheathing stem almost to top, glabrous, withered at anthesis. Spathe valves long. Perianth segments straw-coloured. Anthers pale yellow.

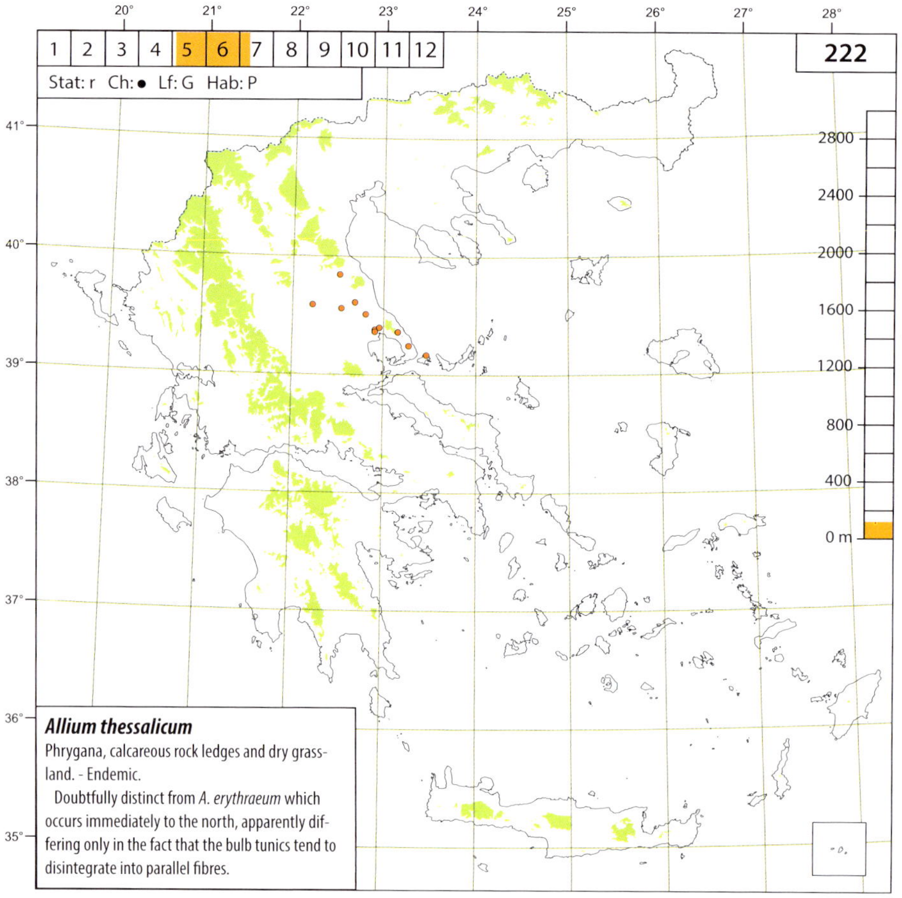

Allium thessalicum

Phrygana, calcareous rock ledges and dry grassland. - Endemic.

Doubtfully distinct from *A. erythraeum* which occurs immediately to the north, apparently differing only in the fact that the bulb tunics tend to disintegrate into parallel fibres.

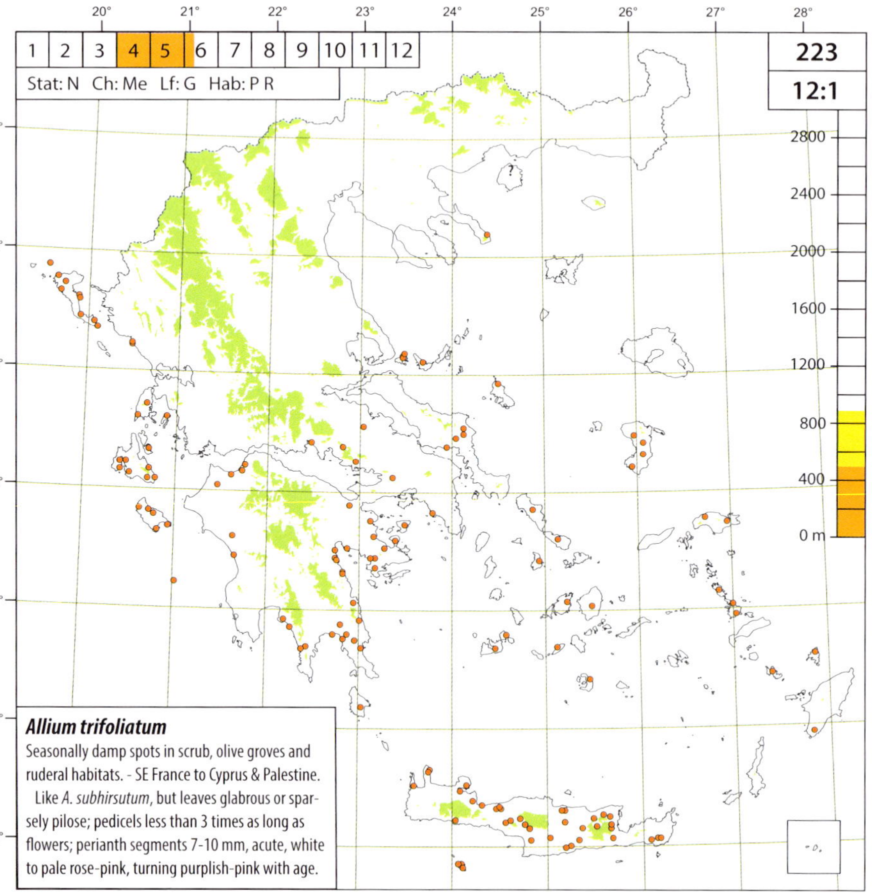

Allium trifoliatum

Seasonally damp spots in scrub, olive groves and ruderal habitats. - SE France to Cyprus & Palestine.

Like *A. subhirsutum*, but leaves glabrous or sparsely pilose; pedicels less than 3 times as long as flowers; perianth segments 7-10 mm, acute, white to pale rose-pink, turning purplish-pink with age.

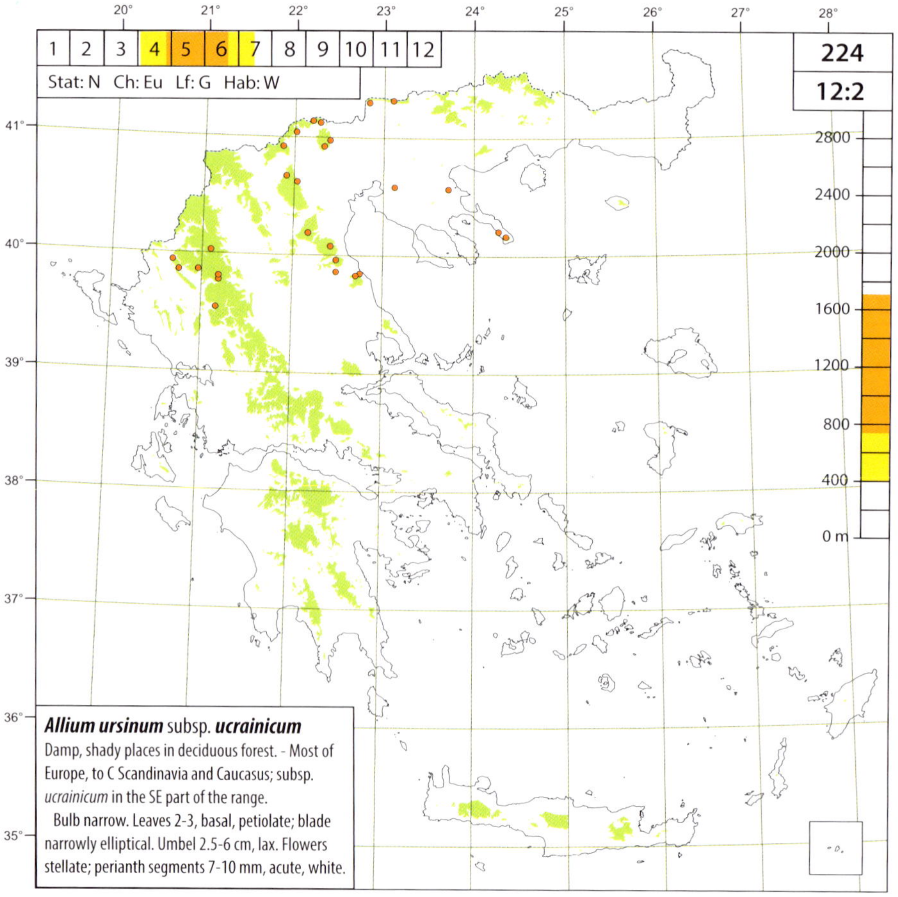

Allium ursinum subsp. ***ucrainicum***

Damp, shady places in deciduous forest. - Most of Europe, to C Scandinavia and Caucasus; subsp. *ucrainicum* in the SE part of the range.

Bulb narrow. Leaves 2-3, basal, petiolate; blade narrowly elliptical. Umbel 2.5-6 cm, lax. Flowers stellate; perianth segments 7-10 mm, acute, white.

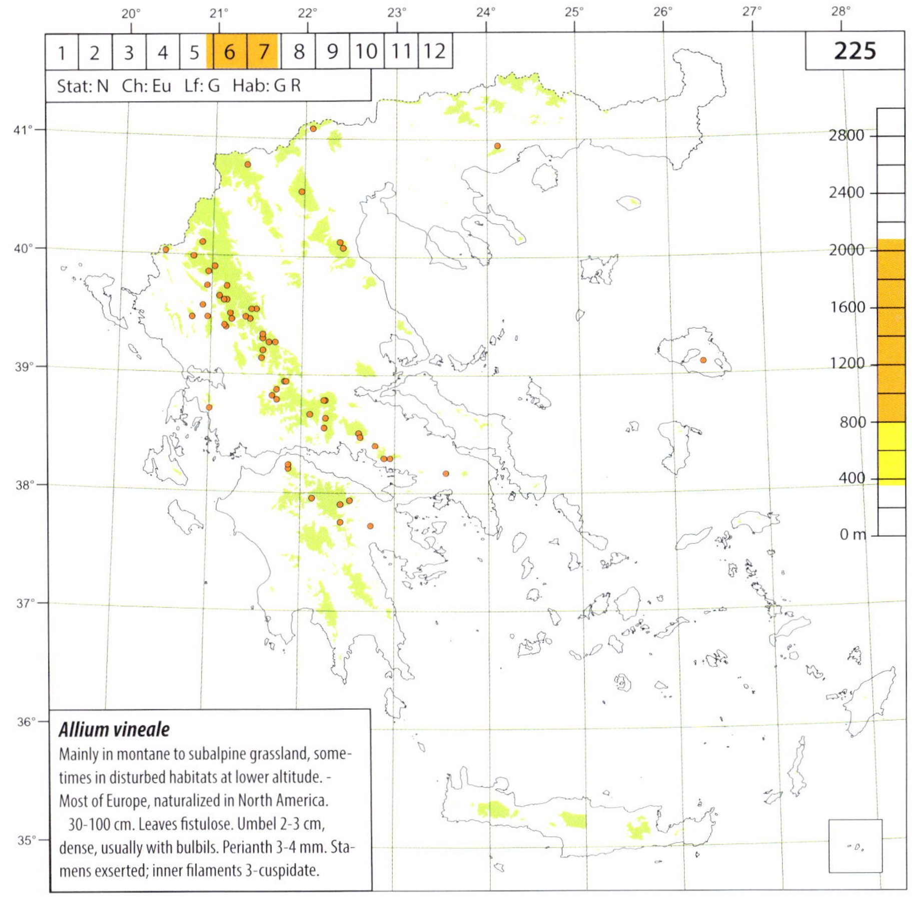

Allium vineale
Mainly in montane to subalpine grassland, sometimes in disturbed habitats at lower altitude. - Most of Europe, naturalized in North America.
30-100 cm. Leaves fistulose. Umbel 2-3 cm, dense, usually with bulbils. Perianth 3-4 mm. Stamens exserted; inner filaments 3-cuspidate.

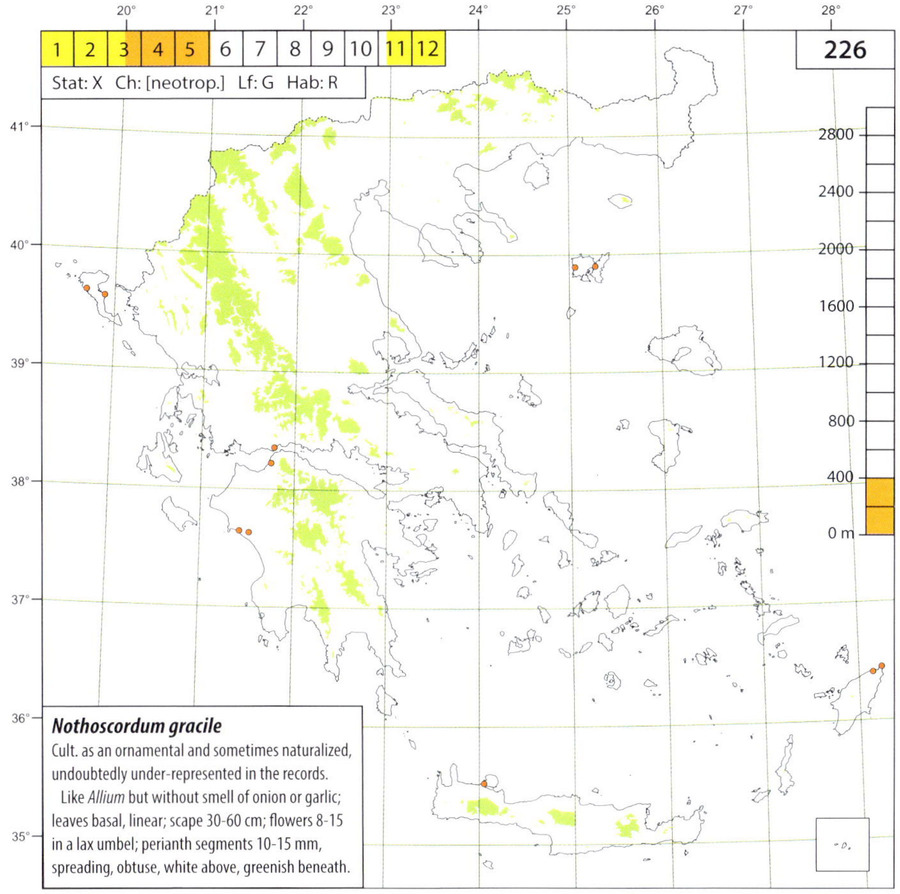

Nothoscordum gracile
Cult. as an ornamental and sometimes naturalized, undoubtedly under-represented in the records.
Like *Allium* but without smell of onion or garlic; leaves basal, linear; scape 30-60 cm; flowers 8-15 in a lax umbel; perianth segments 10-15 mm, spreading, obtuse, white above, greenish beneath.

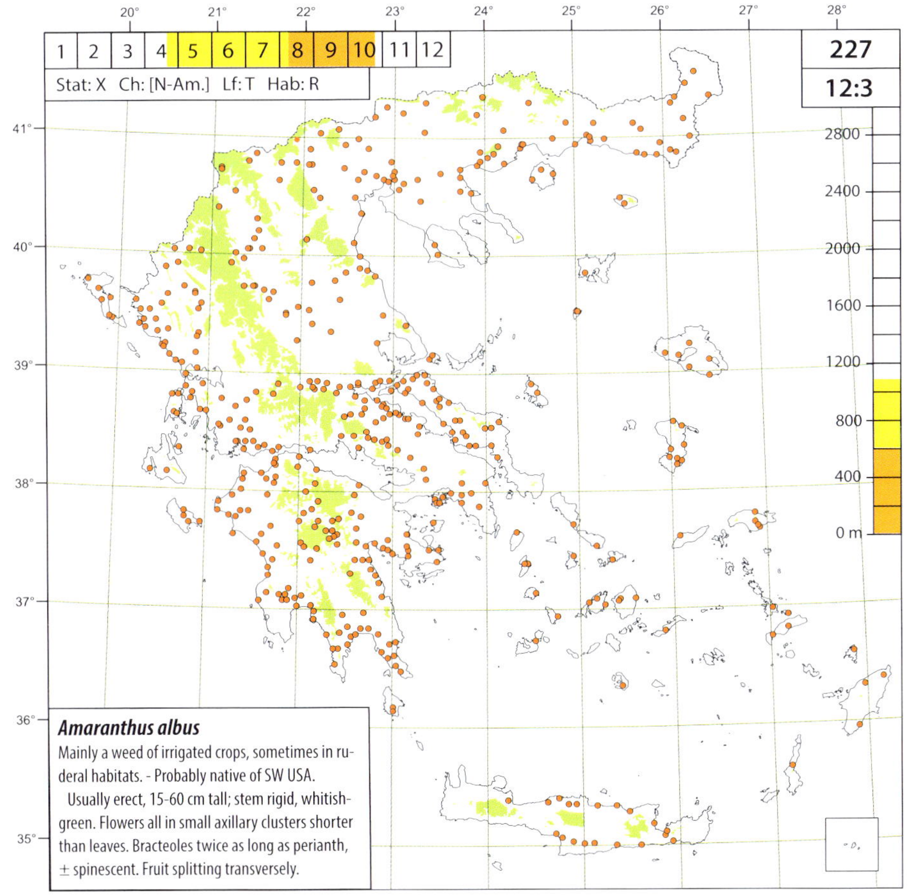

Amaranthus albus
Mainly a weed of irrigated crops, sometimes in ruderal habitats. - Probably native of SW USA.
Usually erect, 15-60 cm tall; stem rigid, whitish-green. Flowers all in small axillary clusters shorter than leaves. Bracteoles twice as long as perianth, ± spinescent. Fruit splitting transversely.

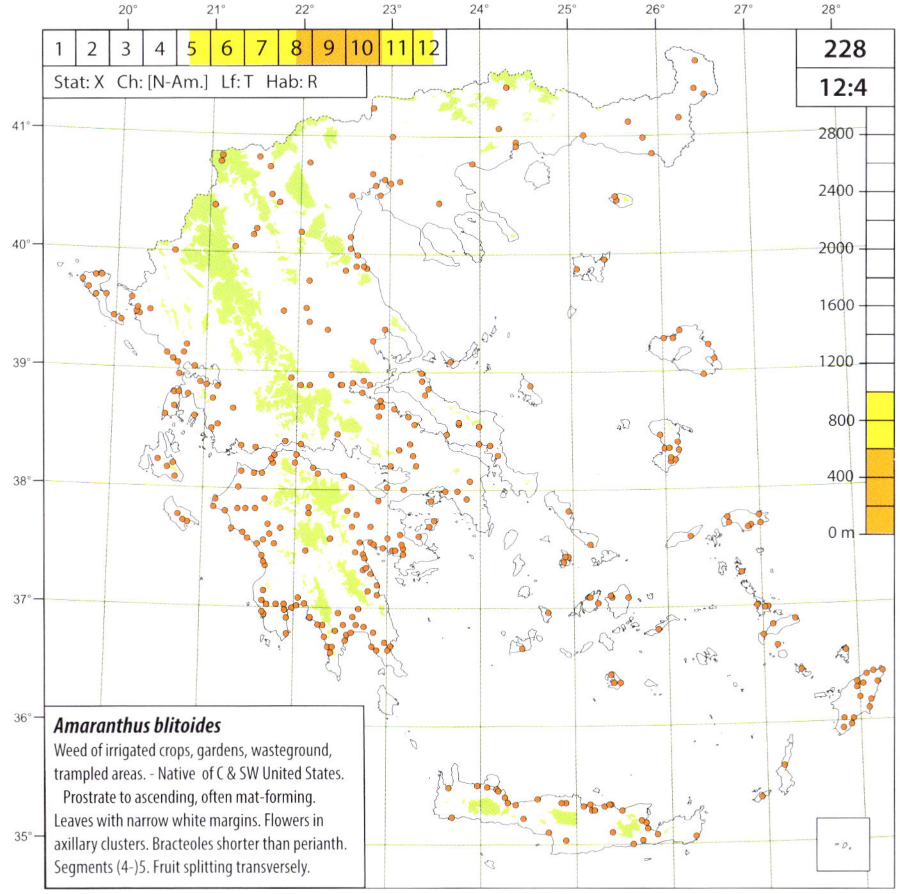

Amaranthus blitoides
Weed of irrigated crops, gardens, wasteground, trampled areas. - Native of C & SW United States.
Prostrate to ascending, often mat-forming. Leaves with narrow white margins. Flowers in axillary clusters. Bracteoles shorter than perianth. Segments (4-)5. Fruit splitting transversely.

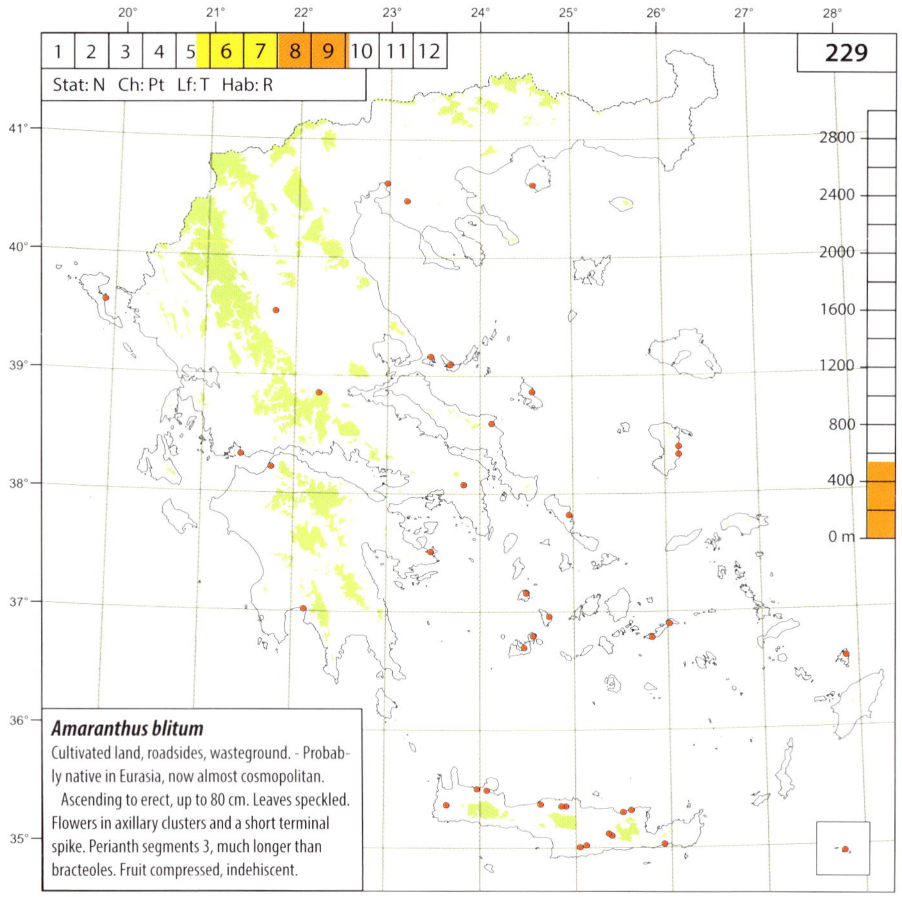

Amaranthus blitum

Cultivated land, roadsides, wasteground. - Probably native in Eurasia, now almost cosmopolitan.

Ascending to erect, up to 80 cm. Leaves speckled. Flowers in axillary clusters and a short terminal spike. Perianth segments 3, much longer than bracteoles. Fruit compressed, indehiscent.

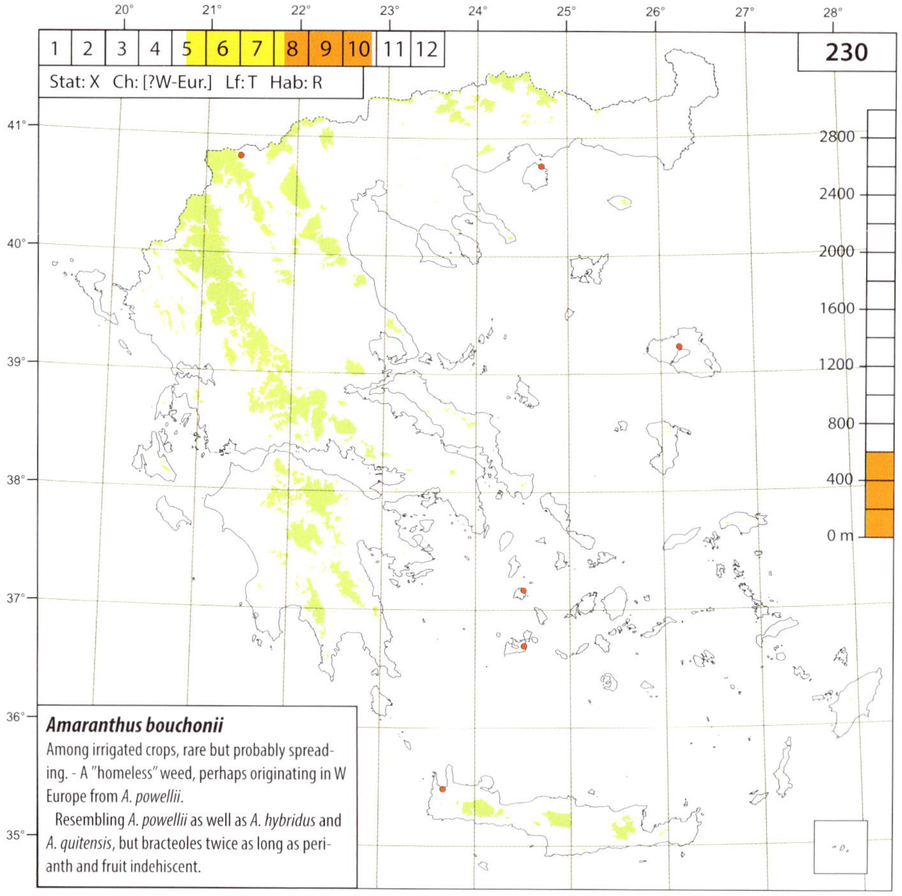

Amaranthus bouchonii

Among irrigated crops, rare but probably spreading. - A "homeless" weed, perhaps originating in W Europe from *A. powellii*.

Resembling *A. powellii* as well as *A. hybridus* and *A. quitensis*, but bracteoles twice as long as perianth and fruit indehiscent.

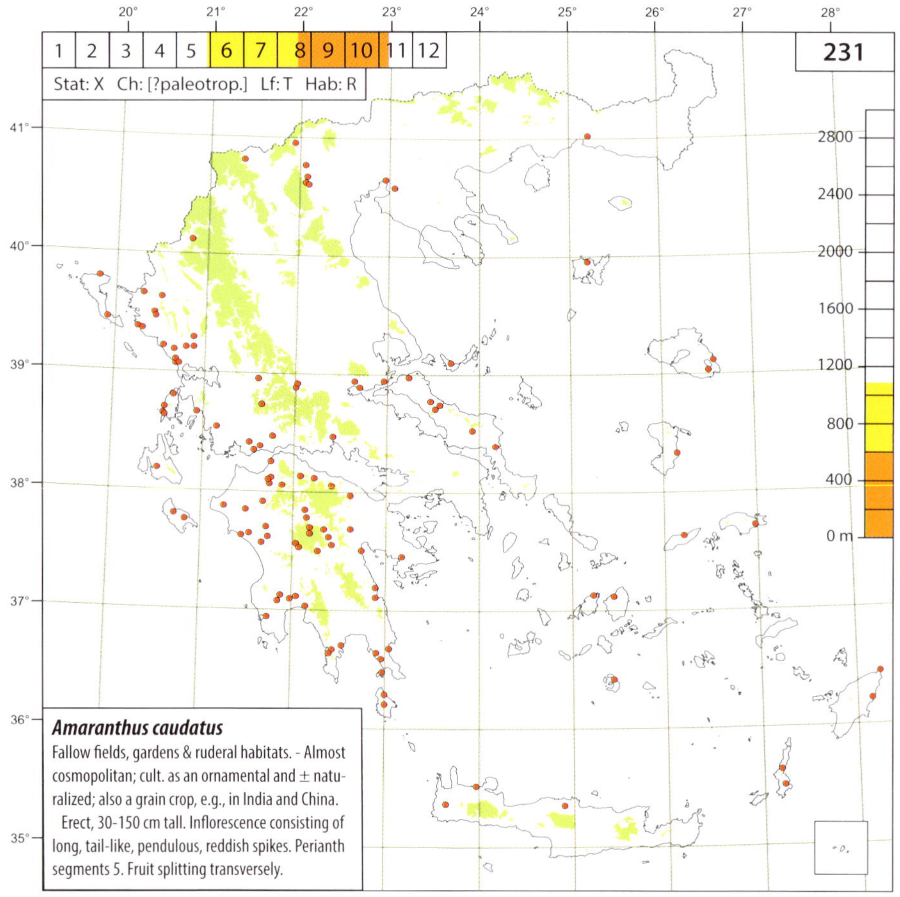

Amaranthus caudatus

Fallow fields, gardens & ruderal habitats. - Almost cosmopolitan; cult. as an ornamental and ± naturalized; also a grain crop, e.g., in India and China.

Erect, 30-150 cm tall. Inflorescence consisting of long, tail-like, pendulous, reddish spikes. Perianth segments 5. Fruit splitting transversely.

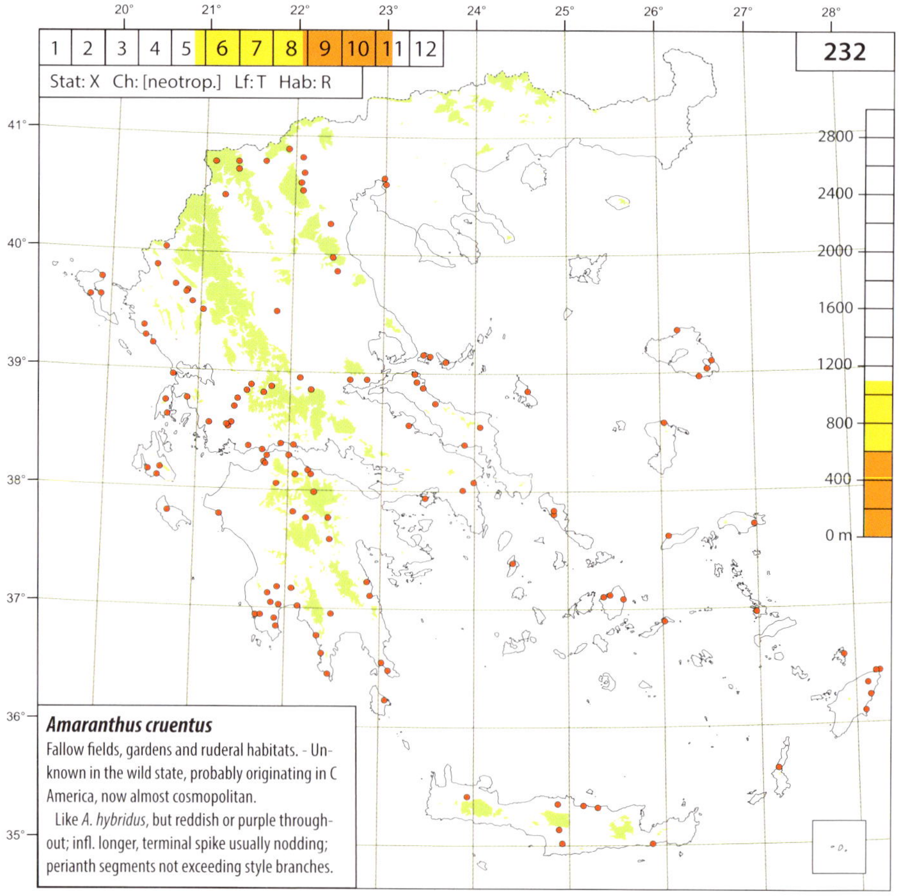

Amaranthus cruentus

Fallow fields, gardens and ruderal habitats. - Unknown in the wild state, probably originating in C America, now almost cosmopolitan.

Like *A. hybridus*, but reddish or purple throughout; infl. longer, terminal spike usually nodding; perianth segments not exceeding style branches.

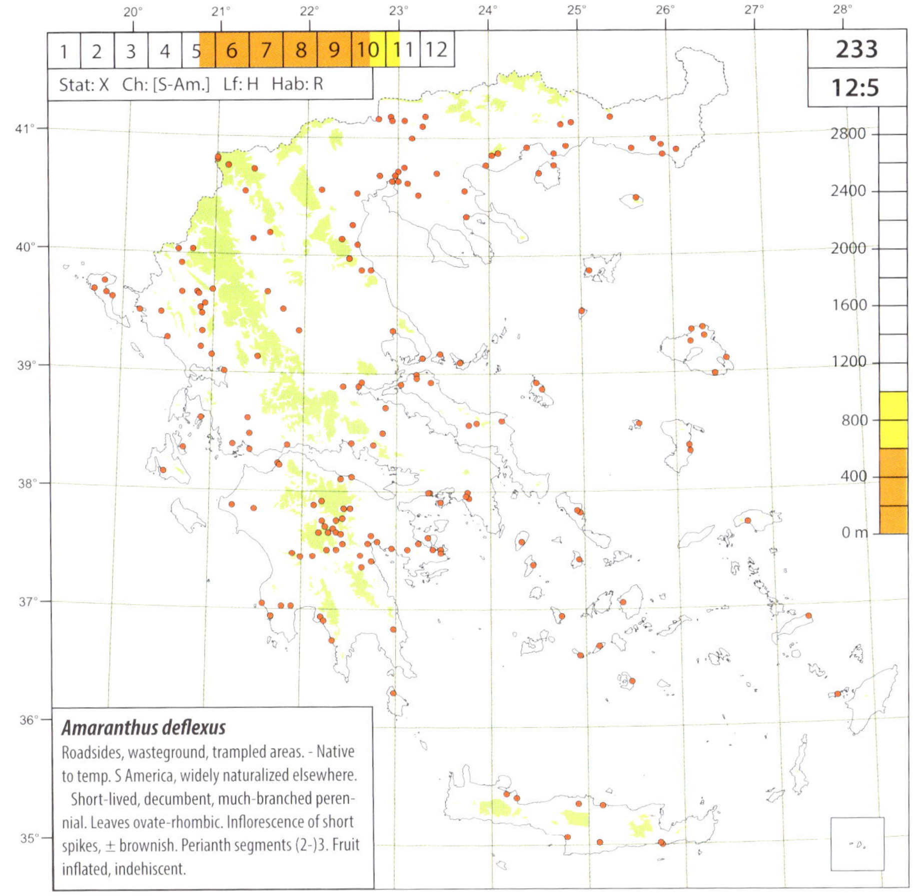

Amaranthus deflexus
Roadsides, wasteground, trampled areas. - Native to temp. S America, widely naturalized elsewhere.
Short-lived, decumbent, much-branched perennial. Leaves ovate-rhombic. Inflorescence of short spikes, ± brownish. Perianth segments (2-)3. Fruit inflated, indehiscent.

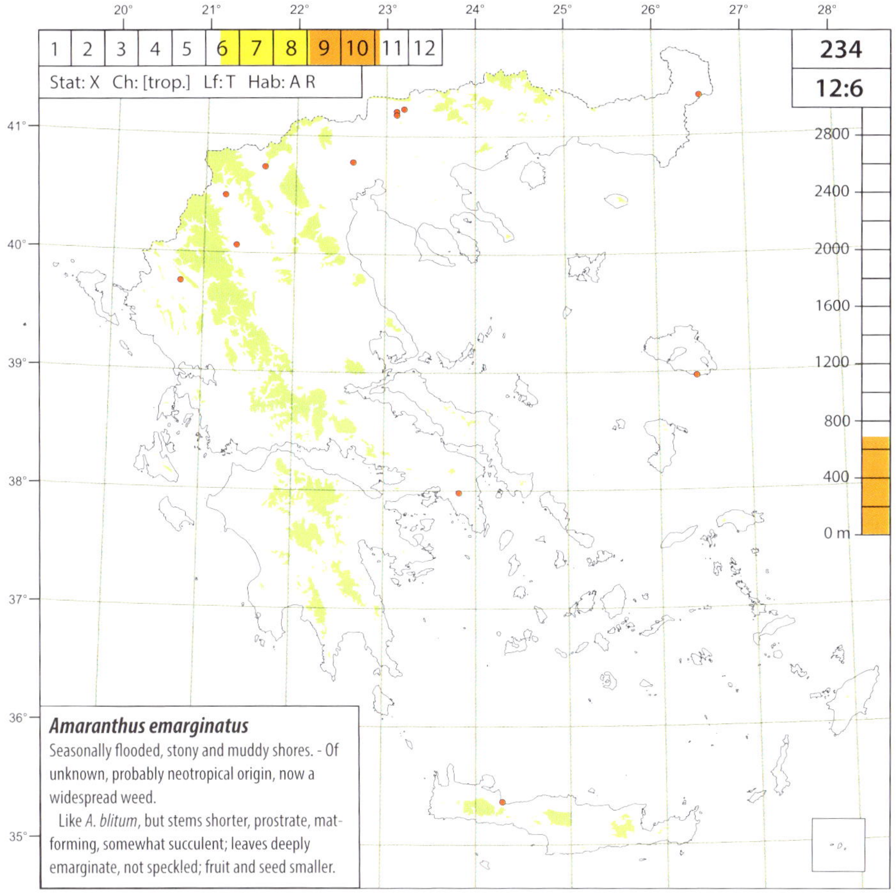

Amaranthus emarginatus
Seasonally flooded, stony and muddy shores. - Of unknown, probably neotropical origin, now a widespread weed.
Like *A. blitum*, but stems shorter, prostrate, mat-forming, somewhat succulent; leaves deeply emarginate, not speckled; fruit and seed smaller.

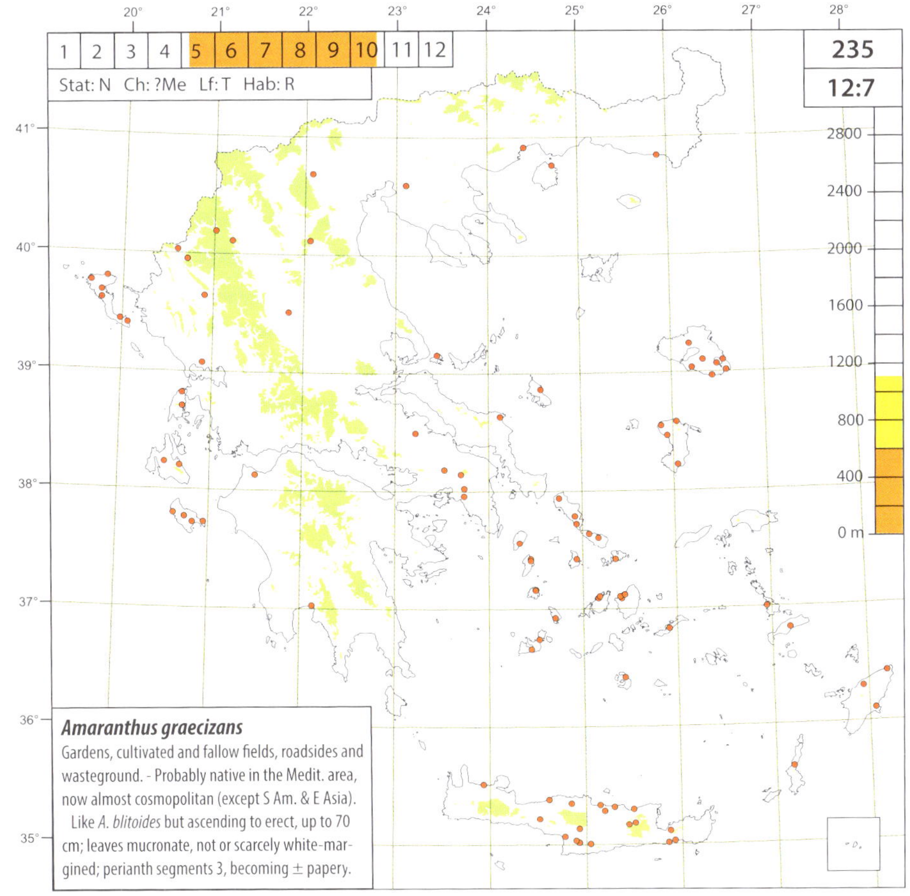

Amaranthus graecizans
Gardens, cultivated and fallow fields, roadsides and wasteground. - Probably native in the Medit. area, now almost cosmopolitan (except S Am. & E Asia).
Like *A. blitoides* but ascending to erect, up to 70 cm; leaves mucronate, not or scarcely white-margined; perianth segments 3, becoming ± papery.

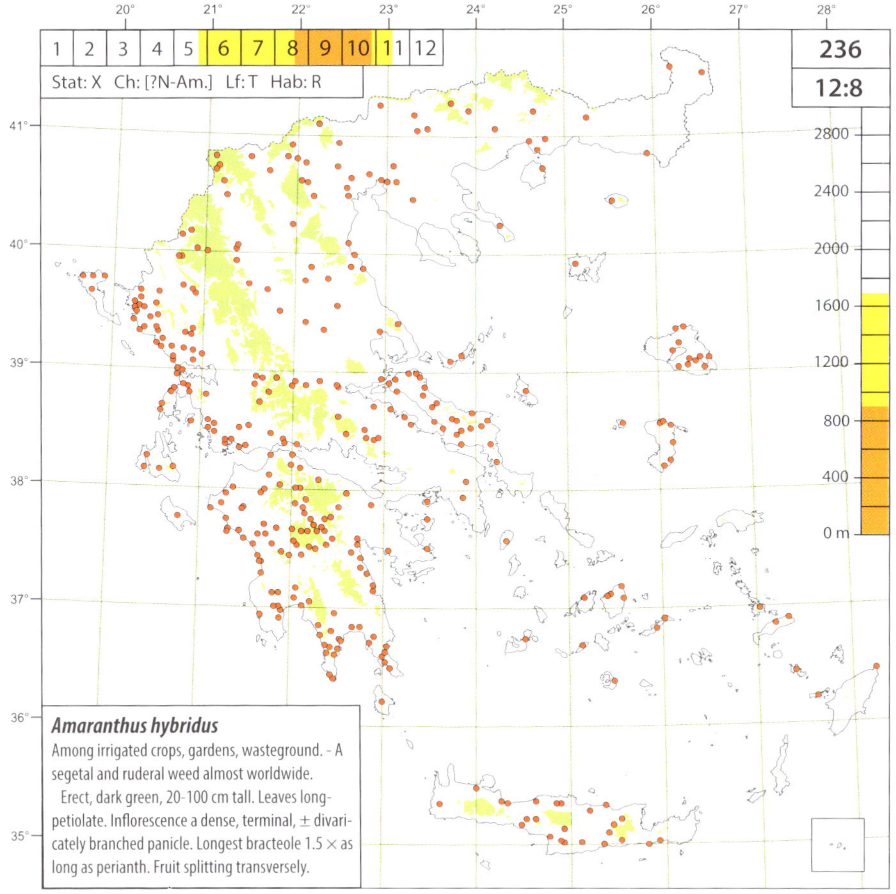

Amaranthus hybridus
Among irrigated crops, gardens, wasteground. - A segetal and ruderal weed almost worldwide.
Erect, dark green, 20-100 cm tall. Leaves long-petiolate. Inflorescence a dense, terminal, ± divaricately branched panicle. Longest bracteole 1.5 × as long as perianth. Fruit splitting transversely.

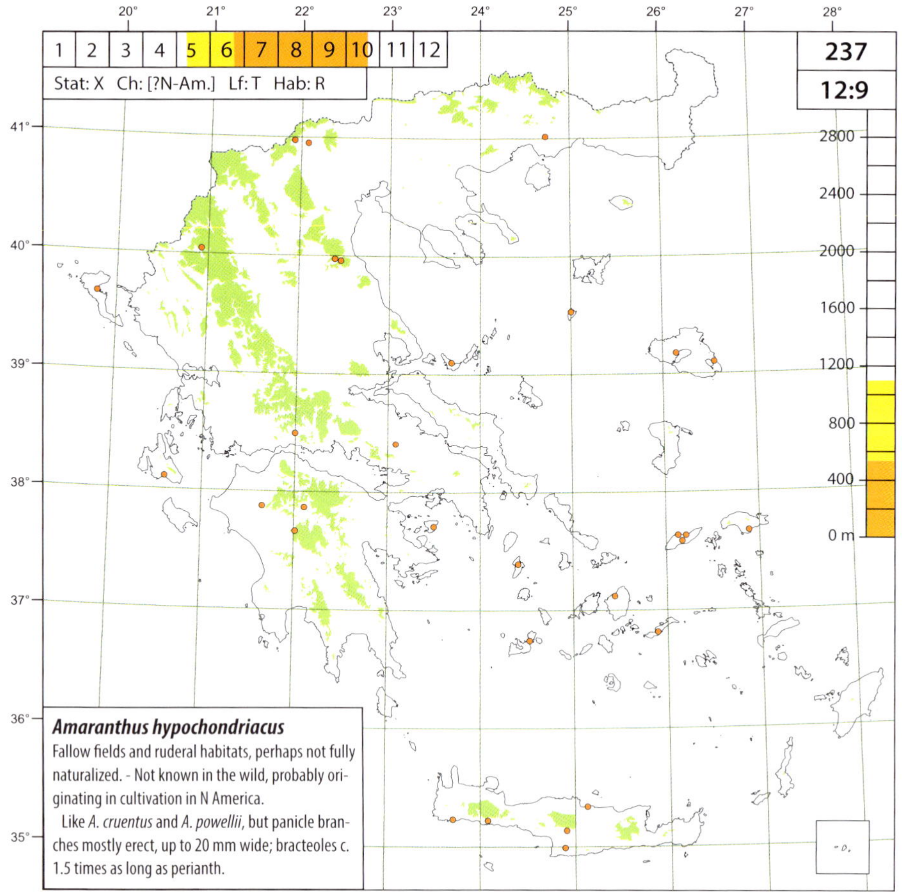

Amaranthus hypochondriacus
Fallow fields and ruderal habitats, perhaps not fully naturalized. - Not known in the wild, probably originating in cultivation in N America.
Like *A. cruentus* and *A. powellii*, but panicle branches mostly erect, up to 20 mm wide; bracteoles c. 1.5 times as long as perianth.

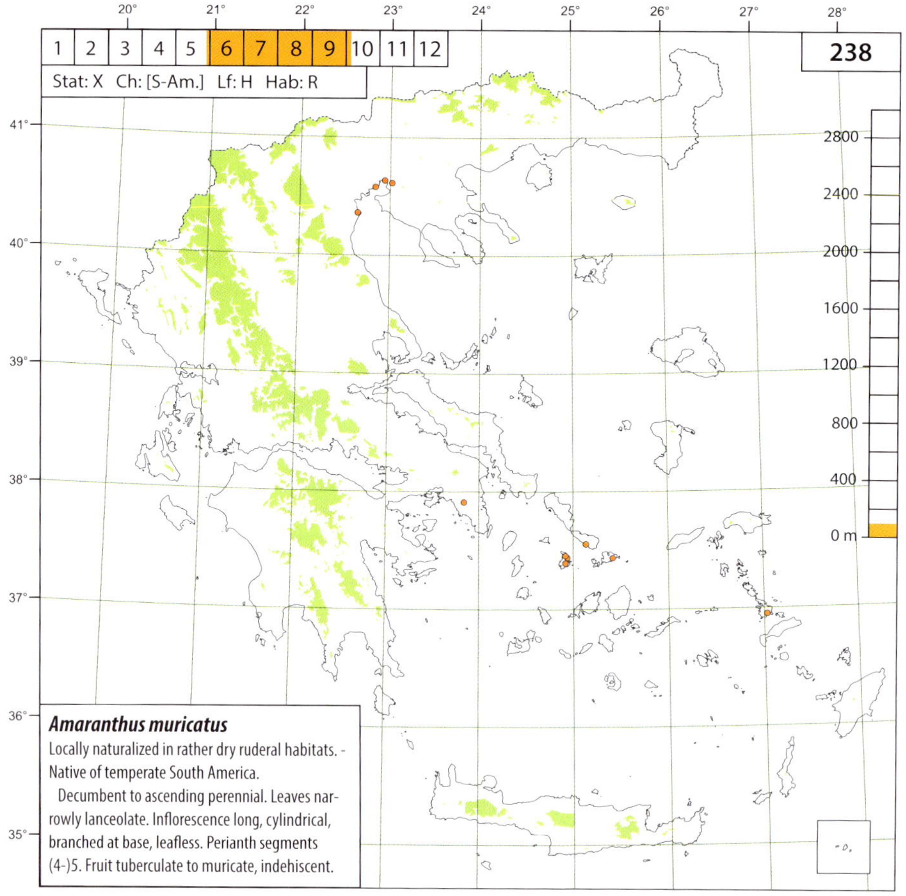

Amaranthus muricatus
Locally naturalized in rather dry ruderal habitats. - Native of temperate South America.
Decumbent to ascending perennial. Leaves narrowly lanceolate. Inflorescence long, cylindrical, branched at base, leafless. Perianth segments (4-)5. Fruit tuberculate to muricate, indehiscent.

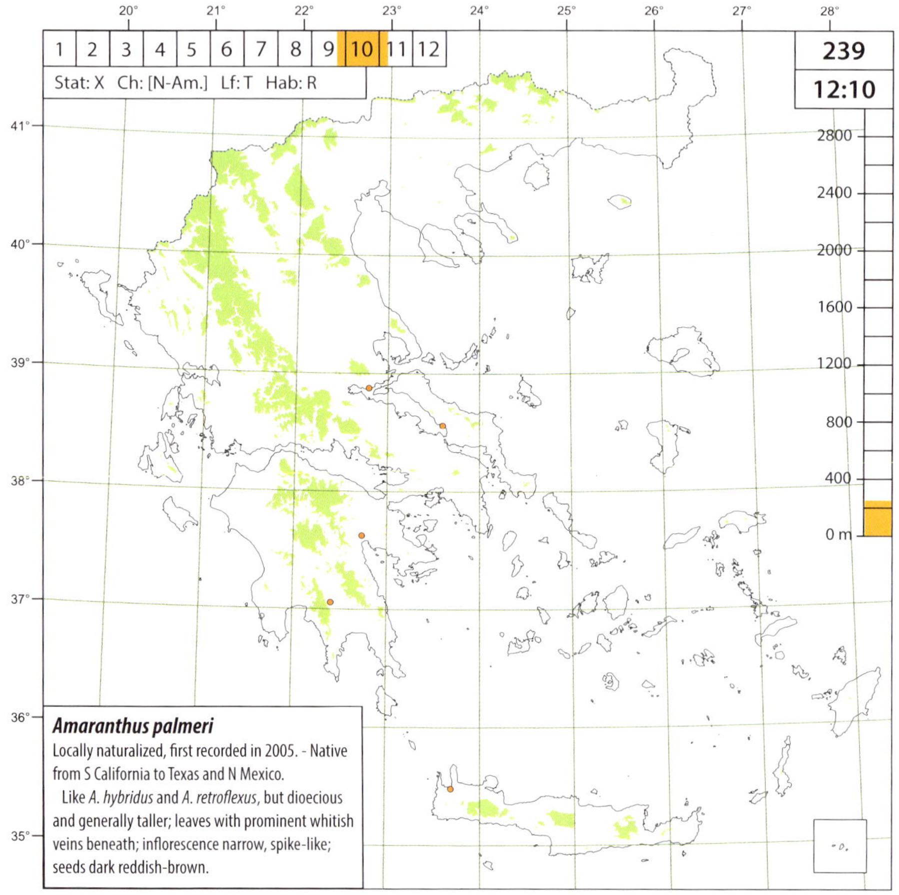

Amaranthus palmeri
Locally naturalized, first recorded in 2005. - Native from S California to Texas and N Mexico.
Like *A. hybridus* and *A. retroflexus*, but dioecious and generally taller; leaves with prominent whitish veins beneath; inflorescence narrow, spike-like; seeds dark reddish-brown.

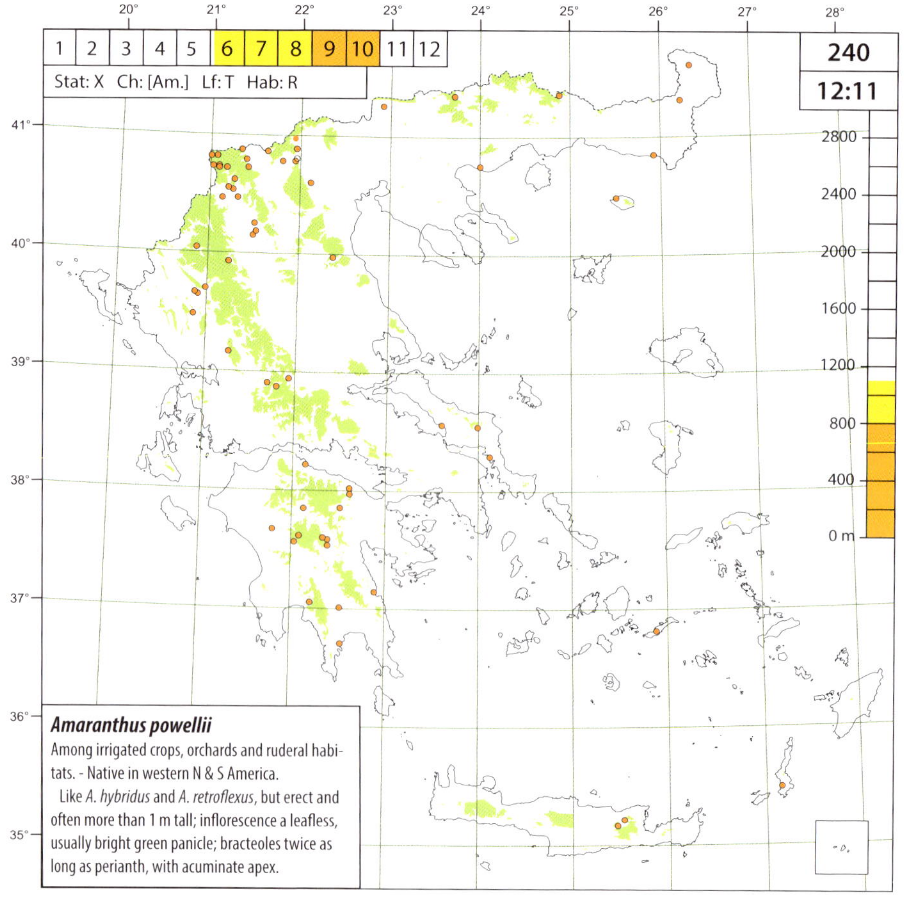

Amaranthus powellii
Among irrigated crops, orchards and ruderal habitats. - Native in western N & S America.
Like *A. hybridus* and *A. retroflexus*, but erect and often more than 1 m tall; inflorescence a leafless, usually bright green panicle; bracteoles twice as long as perianth, with acuminate apex.

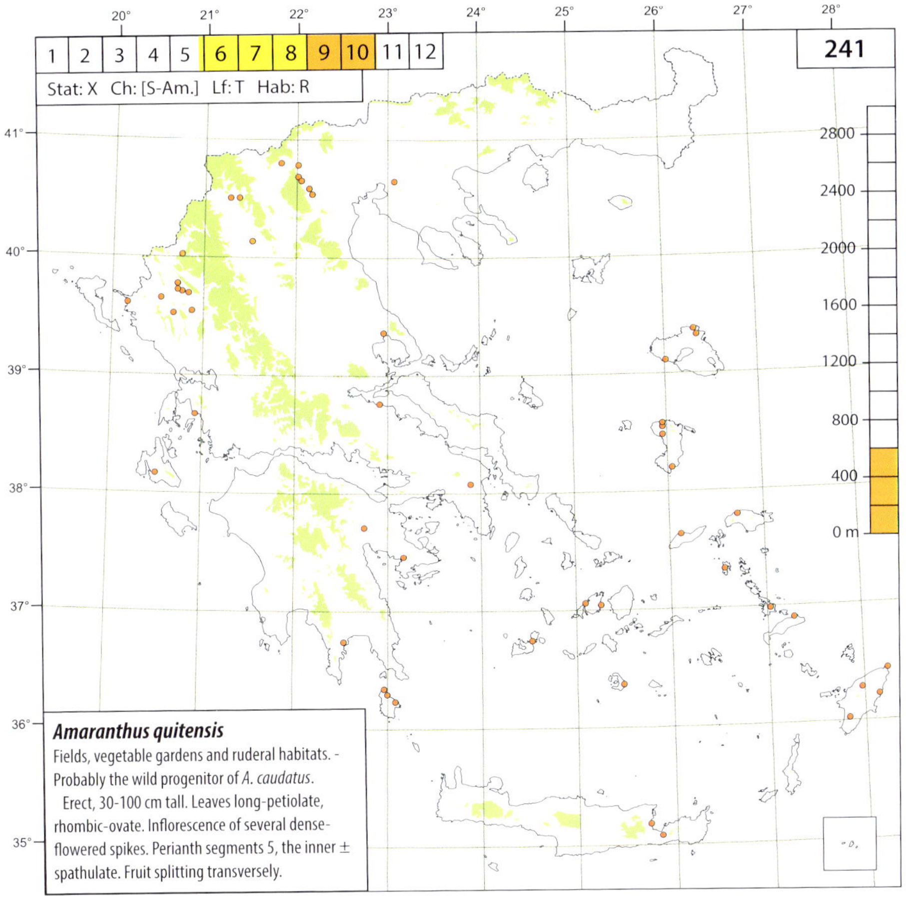

Amaranthus quitensis

Fields, vegetable gardens and ruderal habitats. - Probably the wild progenitor of *A. caudatus*.

Erect, 30-100 cm tall. Leaves long-petiolate, rhombic-ovate. Inflorescence of several dense-flowered spikes. Perianth segments 5, the inner ± spathulate. Fruit splitting transversely.

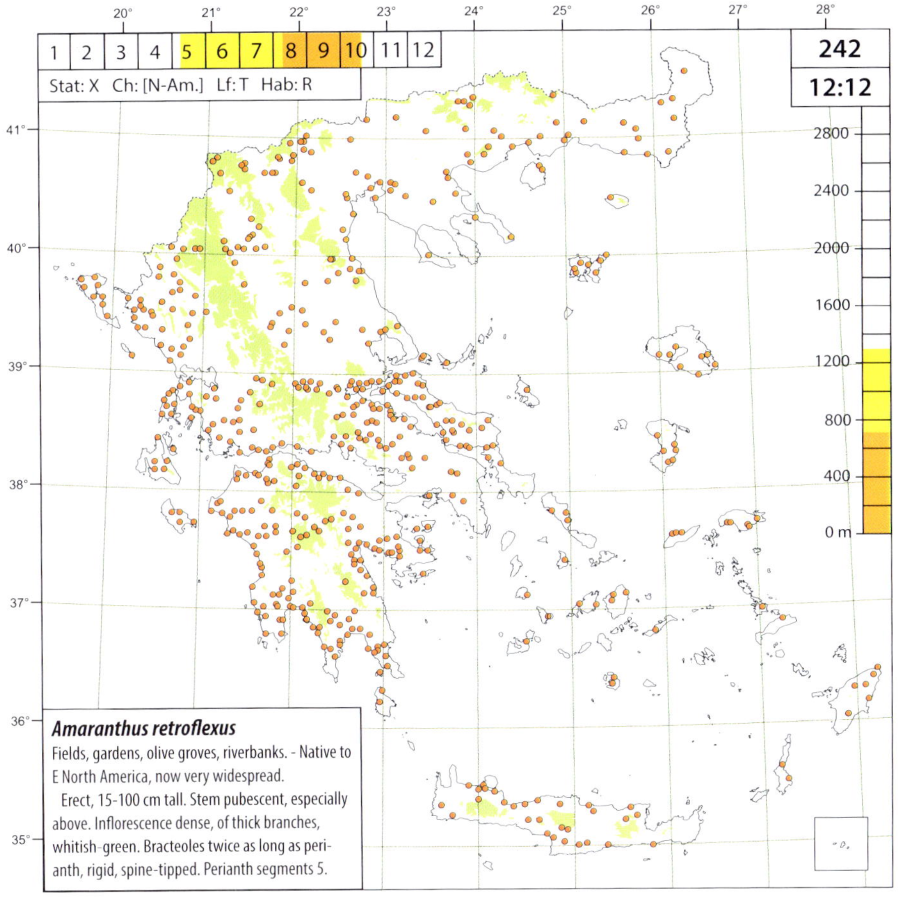

Amaranthus retroflexus

Fields, gardens, olive groves, riverbanks. - Native to E North America, now very widespread.

Erect, 15-100 cm tall. Stem pubescent, especially above. Inflorescence dense, of thick branches, whitish-green. Bracteoles twice as long as perianth, rigid, spine-tipped. Perianth segments 5.

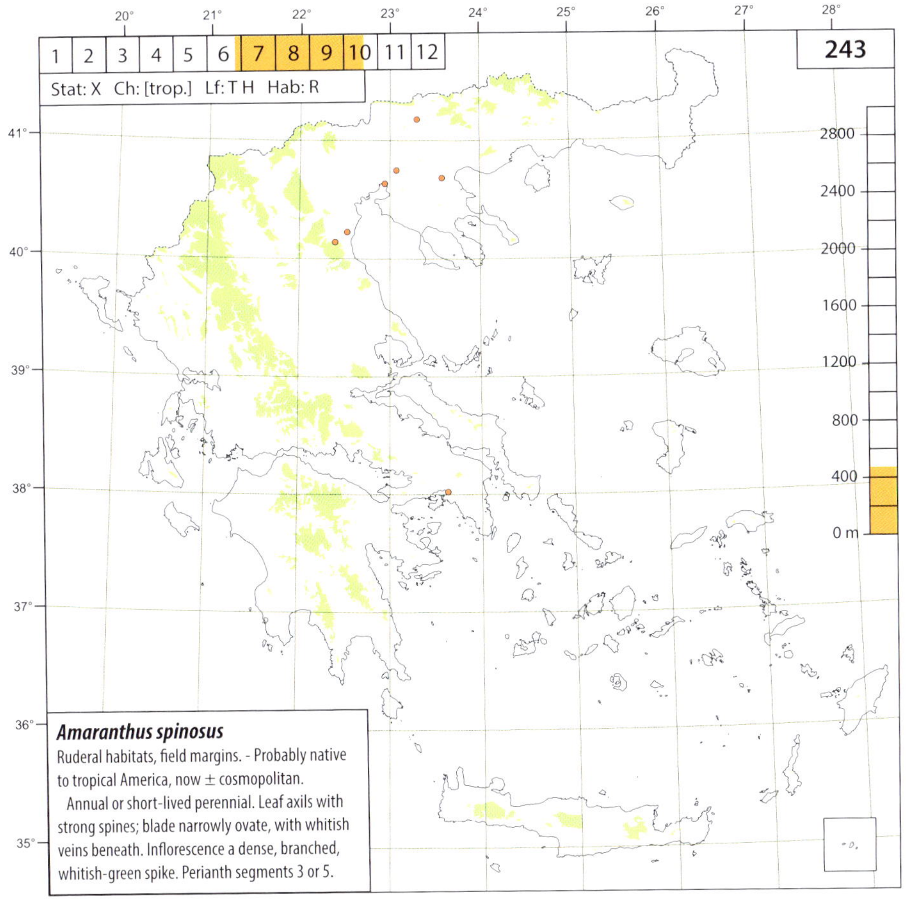

Amaranthus spinosus

Ruderal habitats, field margins. - Probably native to tropical America, now ± cosmopolitan.

Annual or short-lived perennial. Leaf axils with strong spines; blade narrowly ovate, with whitish veins beneath. Inflorescence a dense, branched, whitish-green spike. Perianth segments 3 or 5.

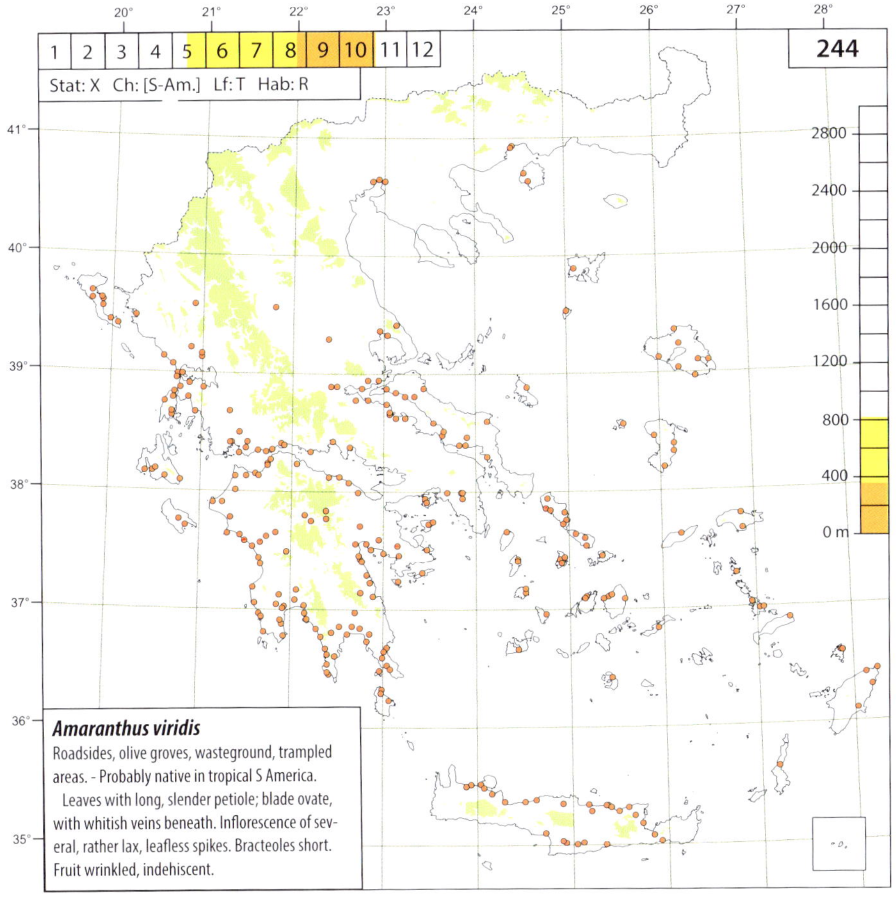

Amaranthus viridis

Roadsides, olive groves, wasteground, trampled areas. - Probably native in tropical S America.

Leaves with long, slender petiole; blade ovate, with whitish veins beneath. Inflorescence of several, rather lax, leafless spikes. Bracteoles short. Fruit wrinkled, indehiscent.

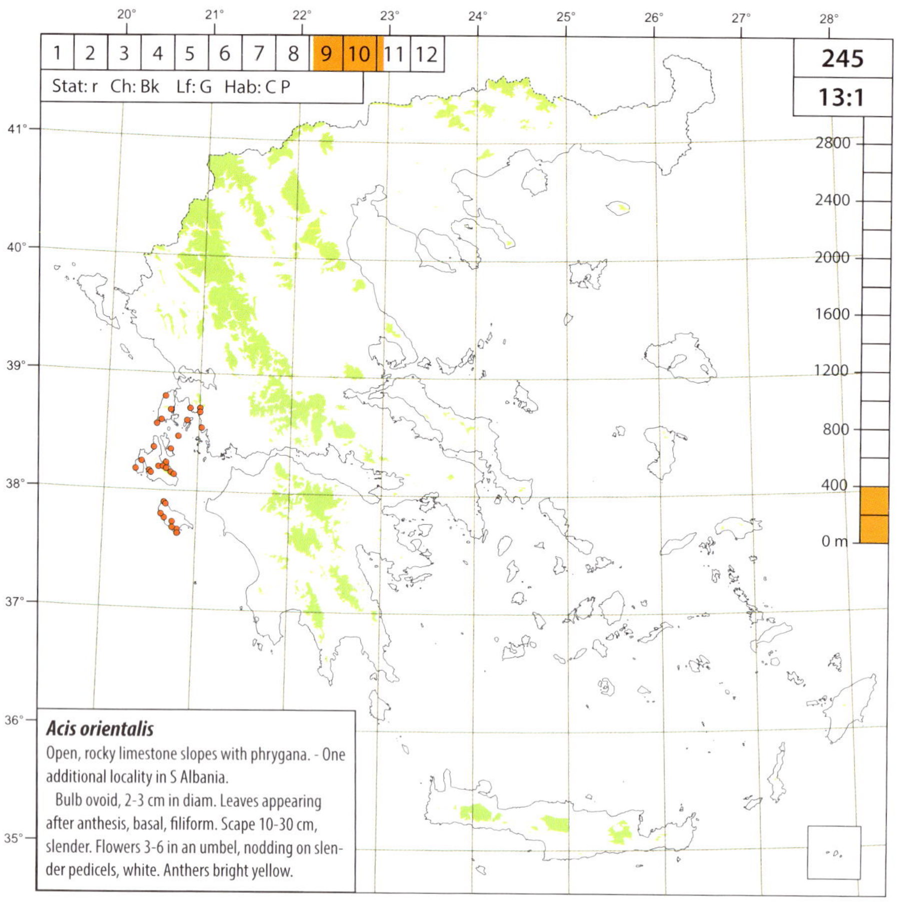

Acis orientalis

Open, rocky limestone slopes with phrygana. - One additional locality in S Albania.

Bulb ovoid, 2-3 cm in diam. Leaves appearing after anthesis, basal, filiform. Scape 10-30 cm, slender. Flowers 3-6 in an umbel, nodding on slender pedicels, white. Anthers bright yellow.

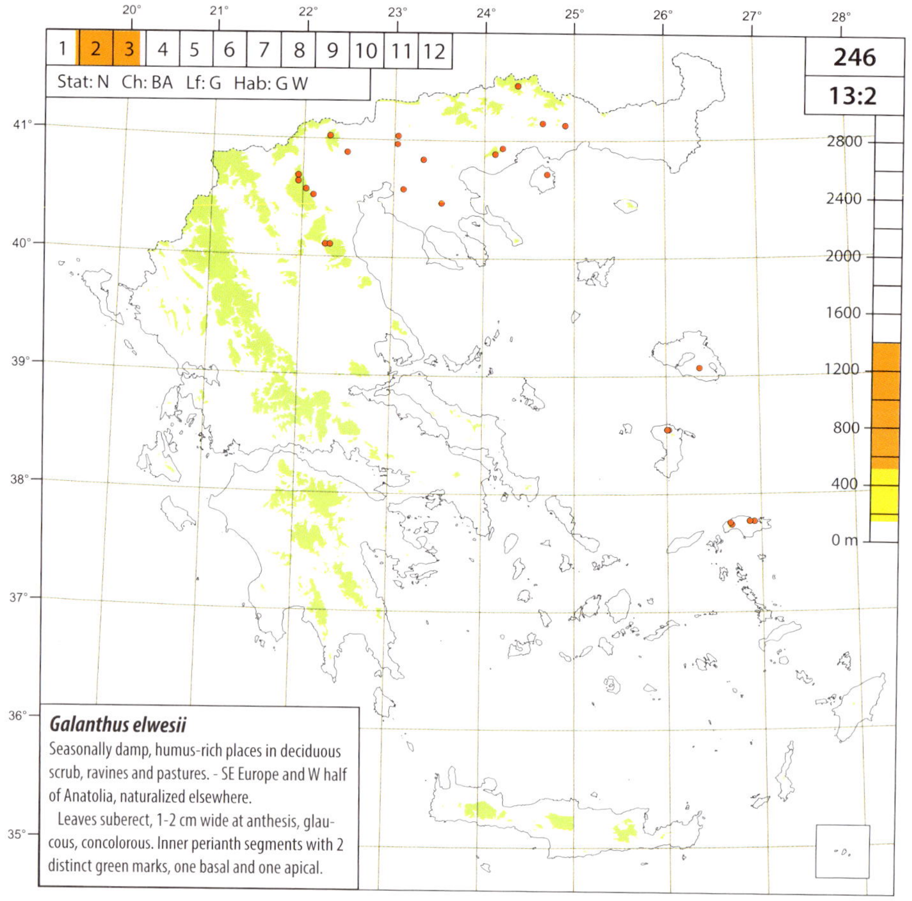

Galanthus elwesii

Seasonally damp, humus-rich places in deciduous scrub, ravines and pastures. - SE Europe and W half of Anatolia, naturalized elsewhere.

Leaves suberect, 1-2 cm wide at anthesis, glaucous, concolorous. Inner perianth segments with 2 distinct green marks, one basal and one apical.

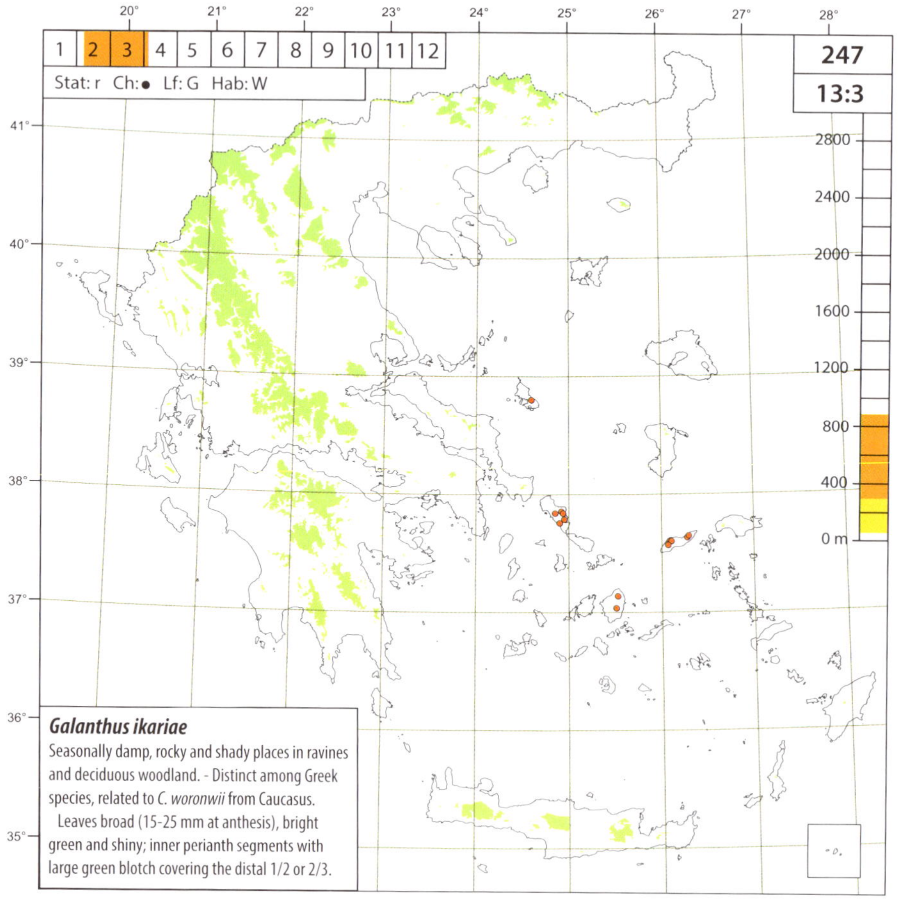

Galanthus ikariae

Seasonally damp, rocky and shady places in ravines and deciduous woodland. - Distinct among Greek species, related to *G. woronwii* from Caucasus.

Leaves broad (15-25 mm at anthesis), bright green and shiny; inner perianth segments with large green blotch covering the distal 1/2 or 2/3.

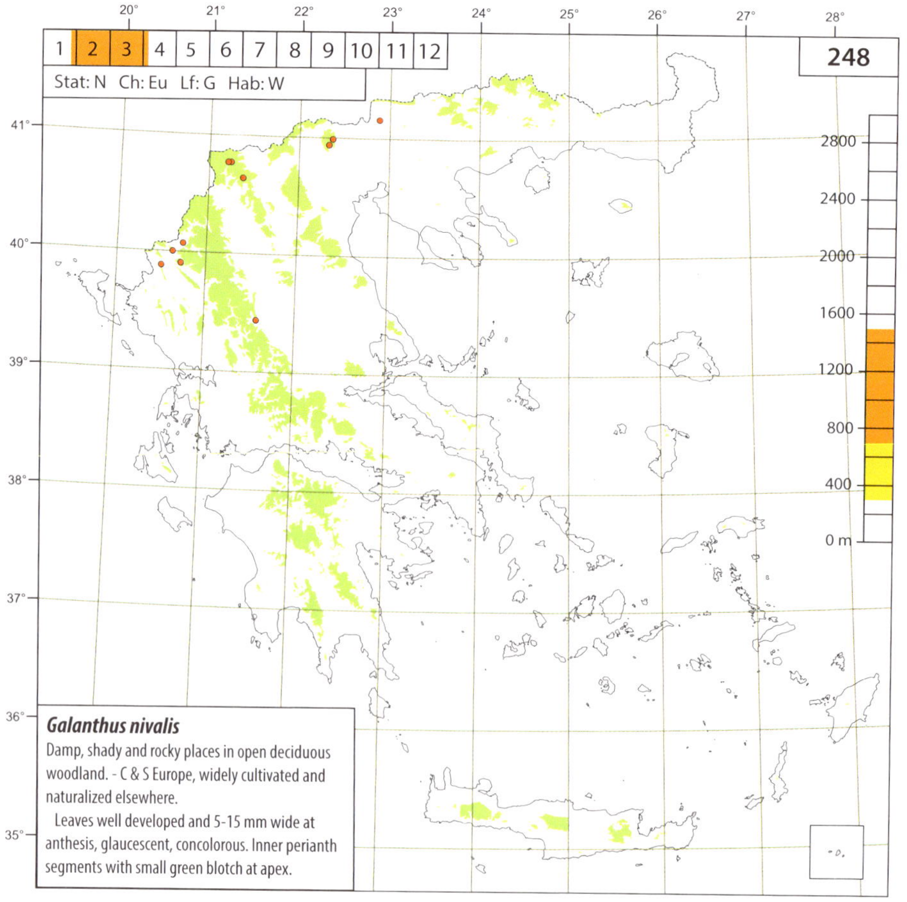

Galanthus nivalis

Damp, shady and rocky places in open deciduous woodland. - C & S Europe, widely cultivated and naturalized elsewhere.

Leaves well developed and 5-15 mm wide at anthesis, glaucescent, concolorous. Inner perianth segments with small green blotch at apex.

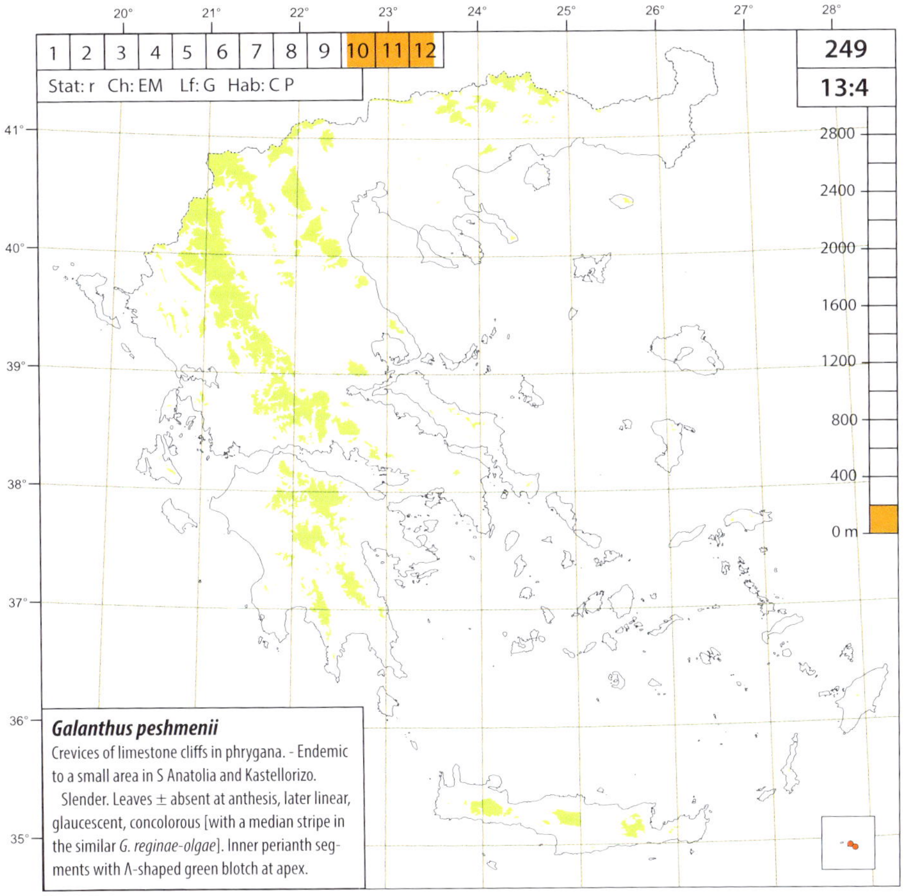

Galanthus peshmenii
Crevices of limestone cliffs in phrygana. - Endemic to a small area in S Anatolia and Kastellorizo.
Slender. Leaves ± absent at anthesis, later linear, glaucescent, concolorous [with a median stripe in the similar *G. reginae-olgae*]. Inner perianth segments with Λ-shaped green blotch at apex.

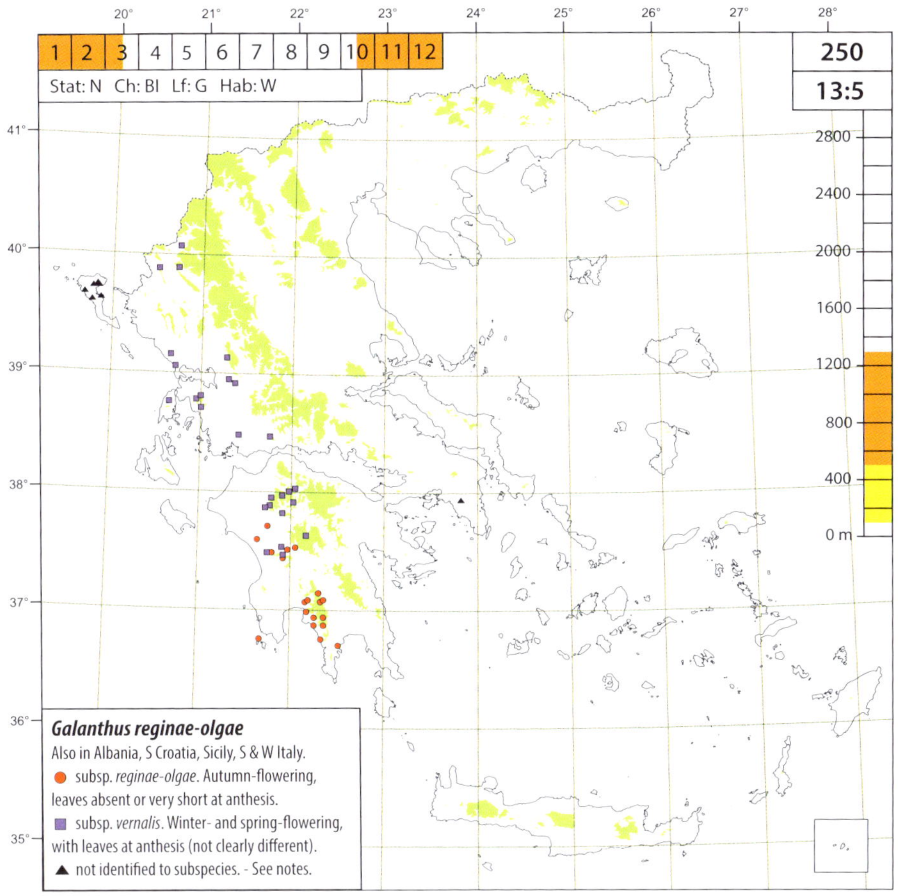

Galanthus reginae-olgae
Also in Albania, S Croatia, Sicily, S & W Italy.
● subsp. *reginae-olgae*. Autumn-flowering, leaves absent or very short at anthesis.
■ subsp. *vernalis*. Winter- and spring-flowering, with leaves at anthesis (not clearly different).
▲ not identified to subspecies. - See notes.

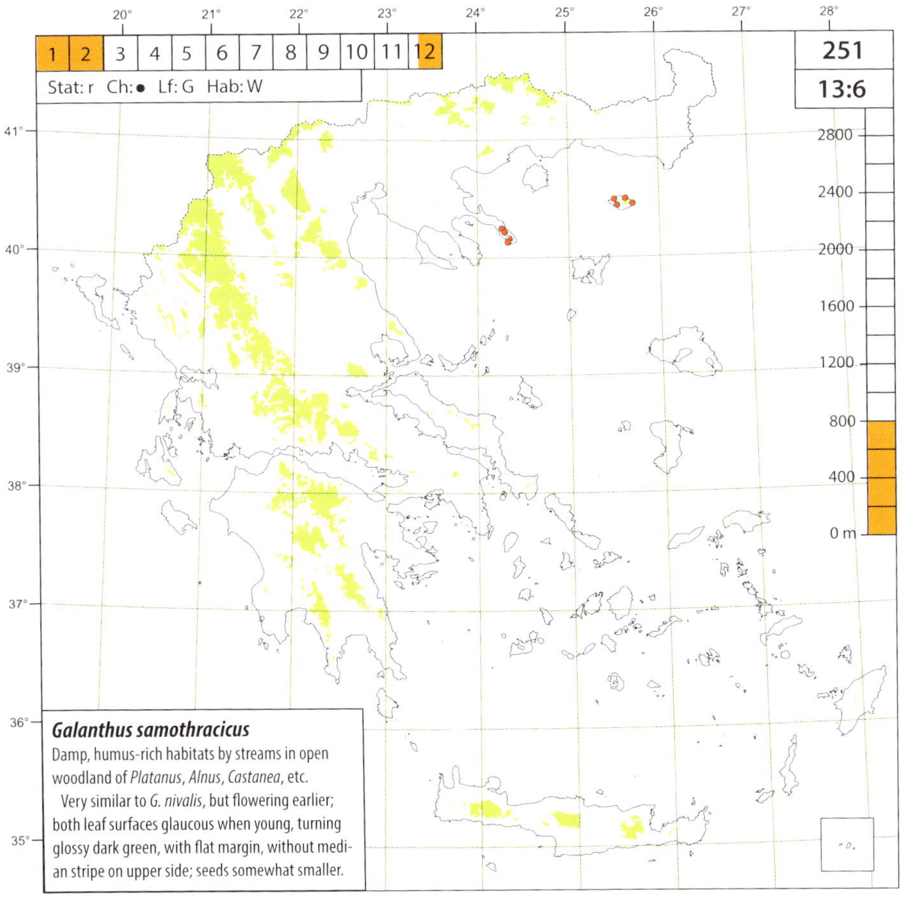

Galanthus samothracicus
Damp, humus-rich habitats by streams in open woodland of *Platanus*, *Alnus*, *Castanea*, etc.
Very similar to *G. nivalis*, but flowering earlier; both leaf surfaces glaucous when young, turning glossy dark green, with flat margin, without median stripe on upper side; seeds somewhat smaller.

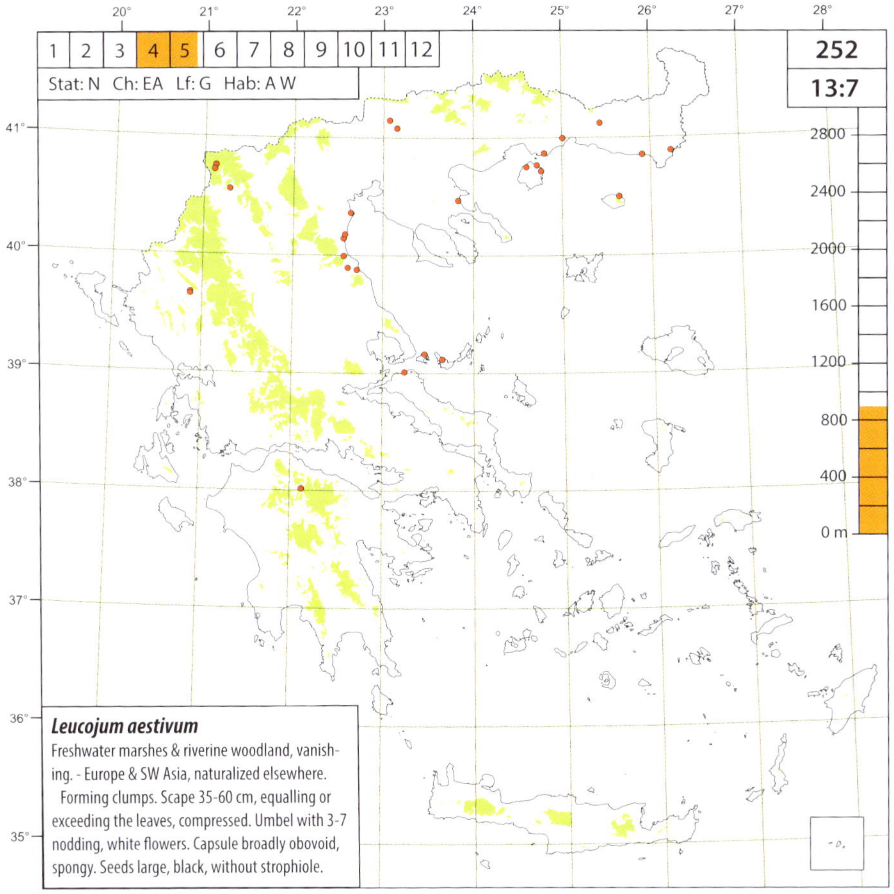

Leucojum aestivum
Freshwater marshes & riverine woodland, vanishing. - Europe & SW Asia, naturalized elsewhere.
Forming clumps. Scape 35-60 cm, equalling or exceeding the leaves, compressed. Umbel with 3-7 nodding, white flowers. Capsule broadly obovoid, spongy. Seeds large, black, without strophiole.

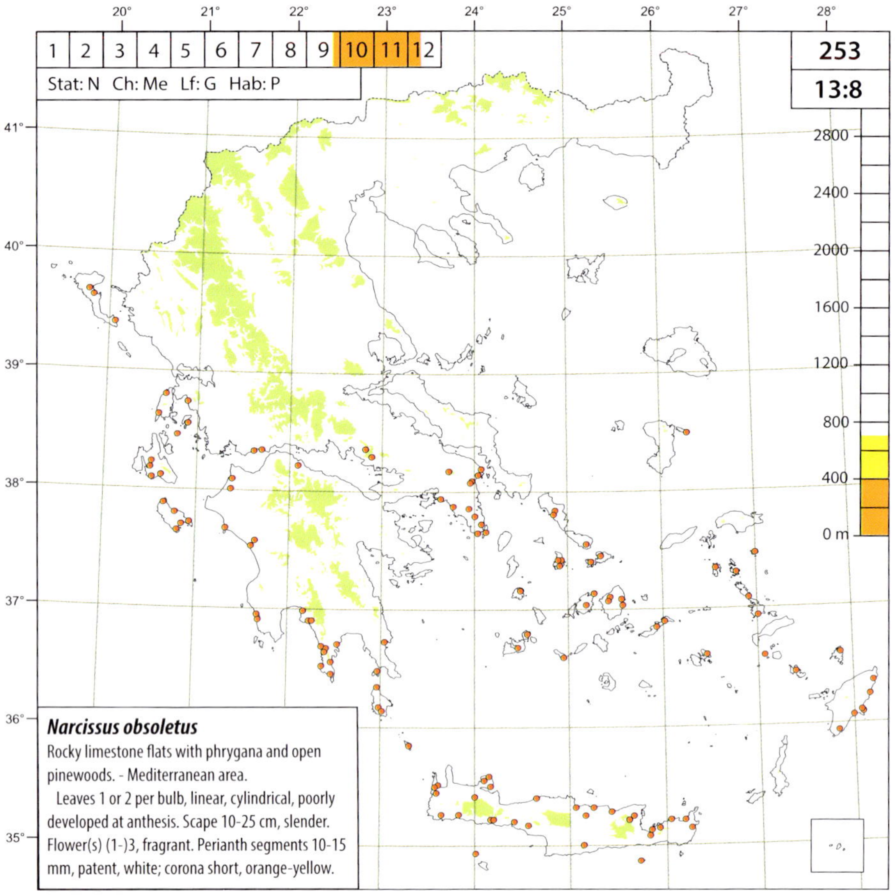

Narcissus obsoletus
Rocky limestone flats with phrygana and open pinewoods. - Mediterranean area.
Leaves 1 or 2 per bulb, linear, cylindrical, poorly developed at anthesis. Scape 10-25 cm, slender. Flower(s) (1-)3, fragrant. Perianth segments 10-15 mm, patent, white; corona short, orange-yellow.

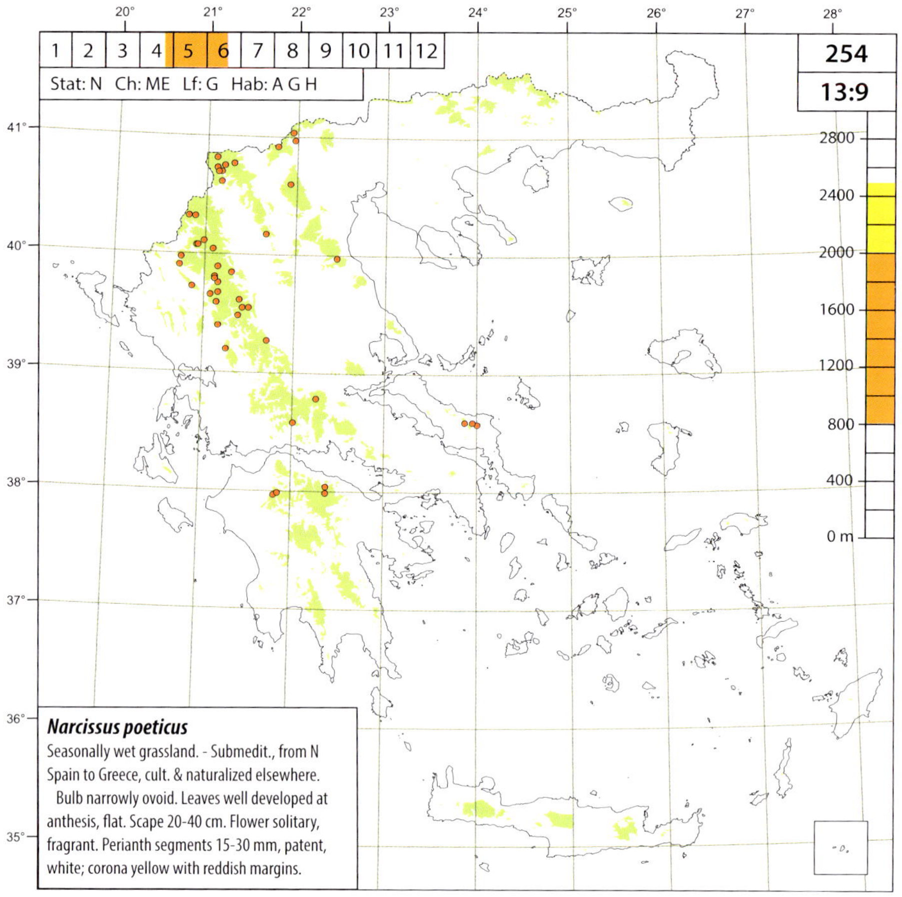

Narcissus poeticus
Seasonally wet grassland. - Submedit., from N Spain to Greece, cult. & naturalized elsewhere.
Bulb narrowly ovoid. Leaves well developed at anthesis, flat. Scape 20-40 cm. Flower solitary, fragrant. Perianth segments 15-30 mm, patent, white; corona yellow with reddish margins.

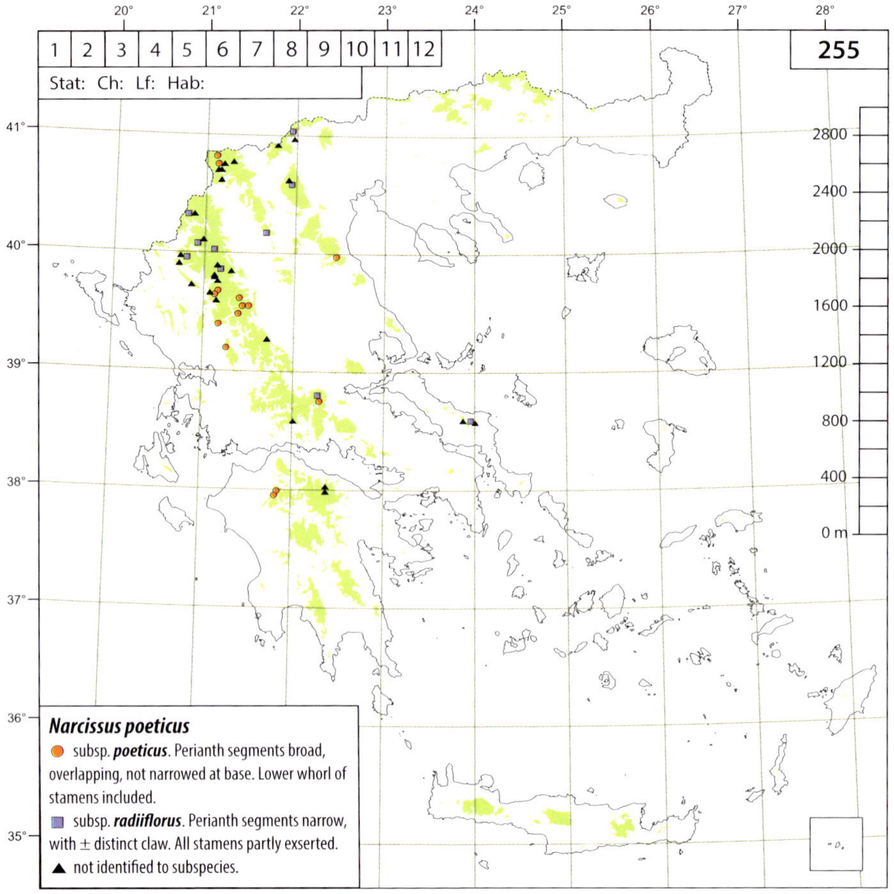

Narcissus poeticus
● subsp. ***poeticus***. Perianth segments broad, overlapping, not narrowed at base. Lower whorl of stamens included.
■ subsp. ***radiiflorus***. Perianth segments narrow, with ± distinct claw. All stamens partly exserted.
▲ not identified to subspecies.

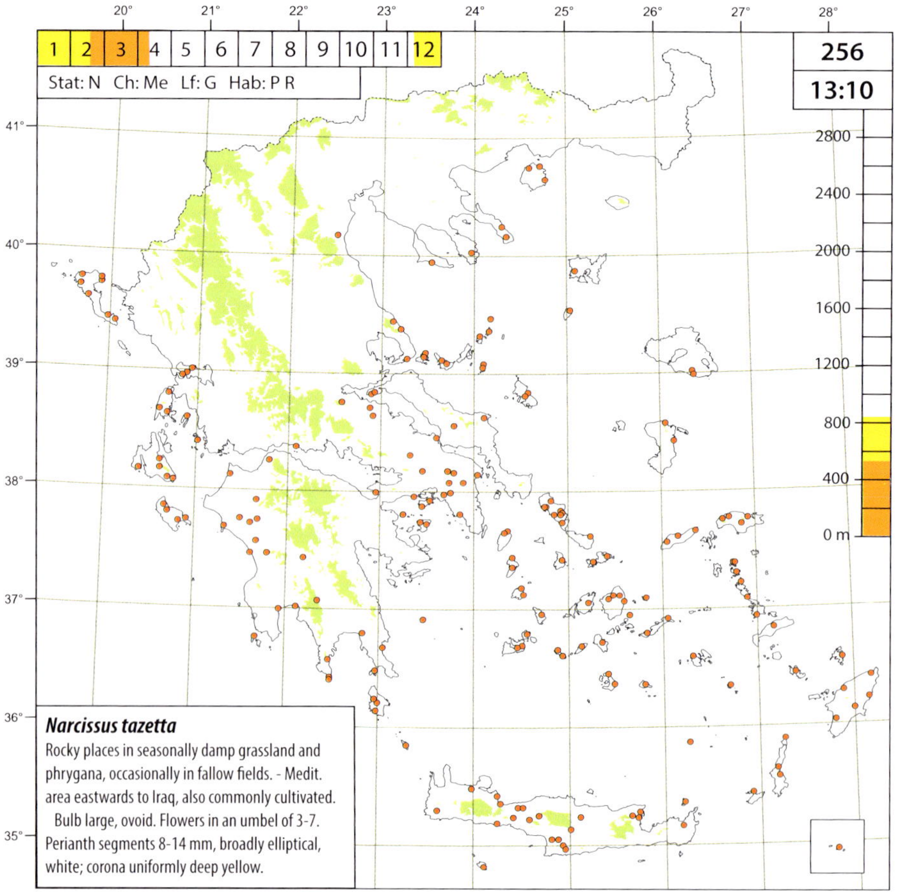

Narcissus tazetta
Rocky places in seasonally damp grassland and phrygana, occasionally in fallow fields. - Medit. area eastwards to Iraq, also commonly cultivated.
Bulb large, ovoid. Flowers in an umbel of 3-7. Perianth segments 8-14 mm, broadly elliptical, white; corona uniformly deep yellow.

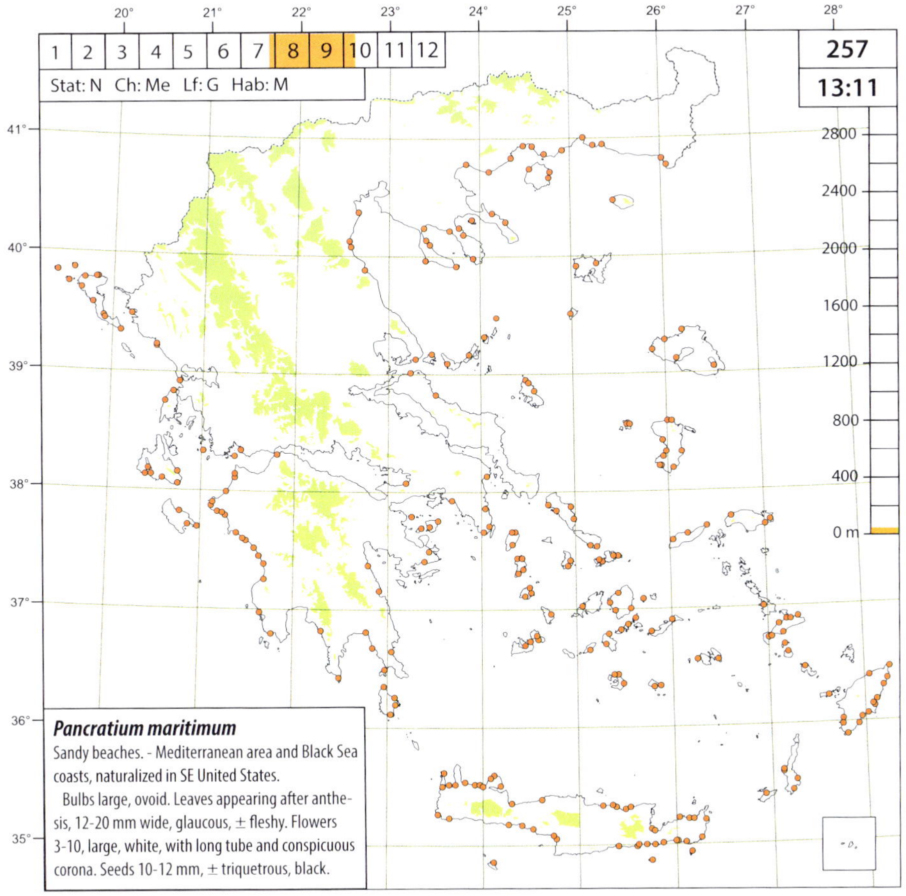

Pancratium maritimum
Sandy beaches. - Mediterranean area and Black Sea coasts, naturalized in SE United States.
Bulbs large, ovoid. Leaves appearing after anthesis, 12-20 mm wide, glaucous, ± fleshy. Flowers 3-10, large, white, with long tube and conspicuous corona. Seeds 10-12 mm, ± triquetrous, black.

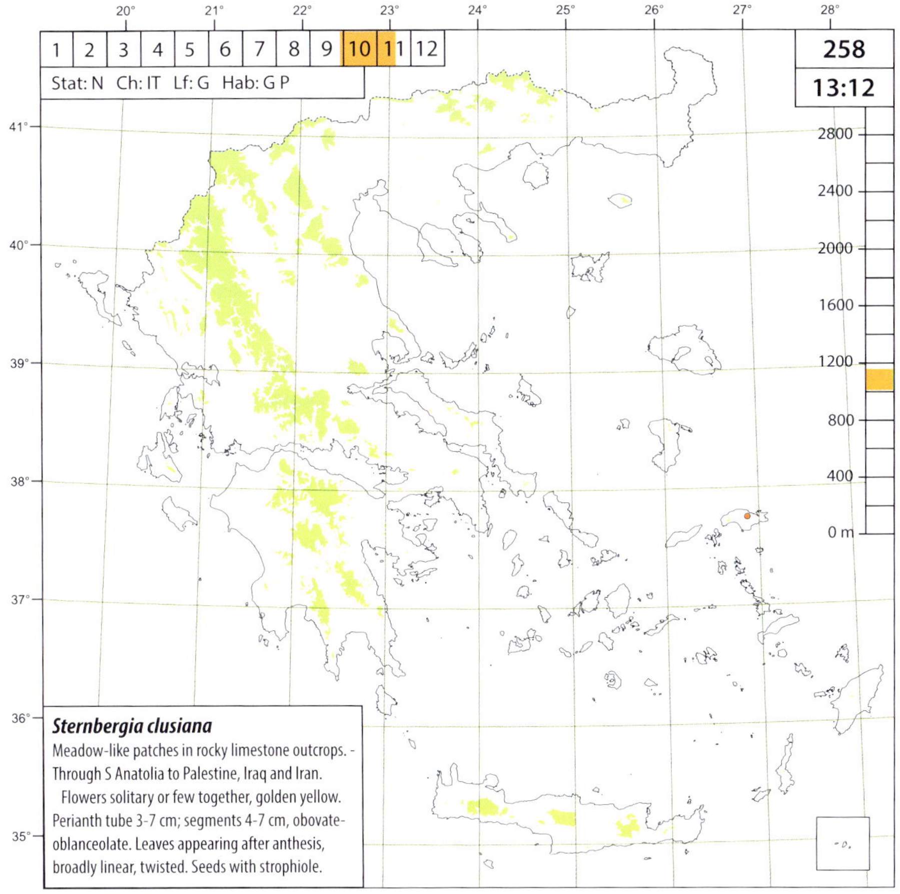

Sternbergia clusiana
Meadow-like patches in rocky limestone outcrops. - Through S Anatolia to Palestine, Iraq and Iran.
Flowers solitary or few together, golden yellow. Perianth tube 3-7 cm; segments 4-7 cm, obovate-oblanceolate. Leaves appearing after anthesis, broadly linear, twisted. Seeds with strophiole.

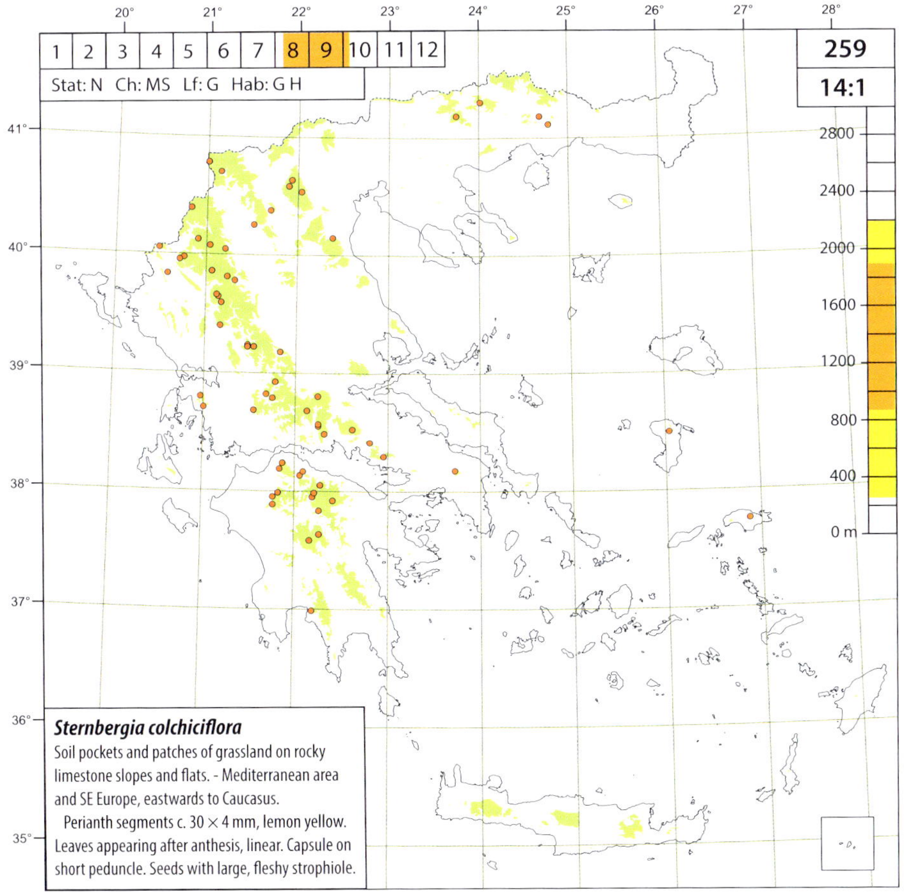

Sternbergia colchiciflora
Soil pockets and patches of grassland on rocky limestone slopes and flats. - Mediterranean area and SE Europe, eastwards to Caucasus.
Perianth segments c. 30 × 4 mm, lemon yellow. Leaves appearing after anthesis, linear. Capsule on short peduncle. Seeds with large, fleshy strophiole.

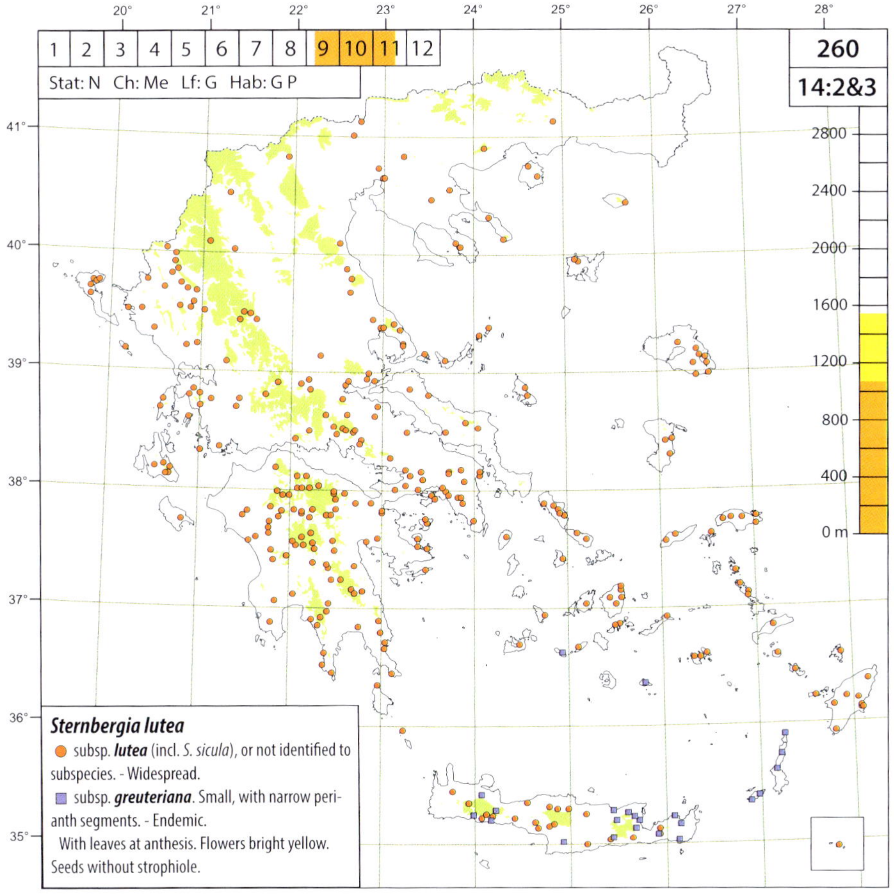

Sternbergia lutea
● subsp. ***lutea*** (incl. *S. sicula*), or not identified to subspecies. - Widespread.
■ subsp. ***greuteriana***. Small, with narrow perianth segments. - Endemic.
With leaves at anthesis. Flowers bright yellow. Seeds without strophiole.

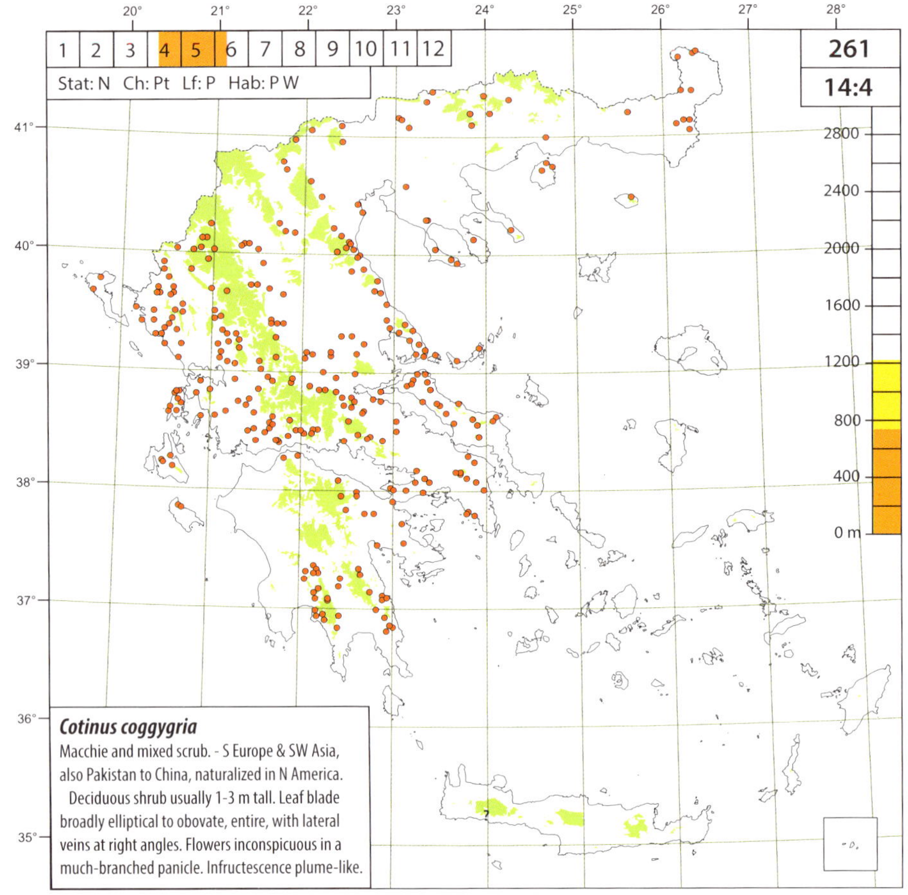

Cotinus coggygria
Macchie and mixed scrub. - S Europe & SW Asia, also Pakistan to China, naturalized in N America.
Deciduous shrub usually 1-3 m tall. Leaf blade broadly elliptical to obovate, entire, with lateral veins at right angles. Flowers inconspicuous in a much-branched panicle. Infructescence plume-like.

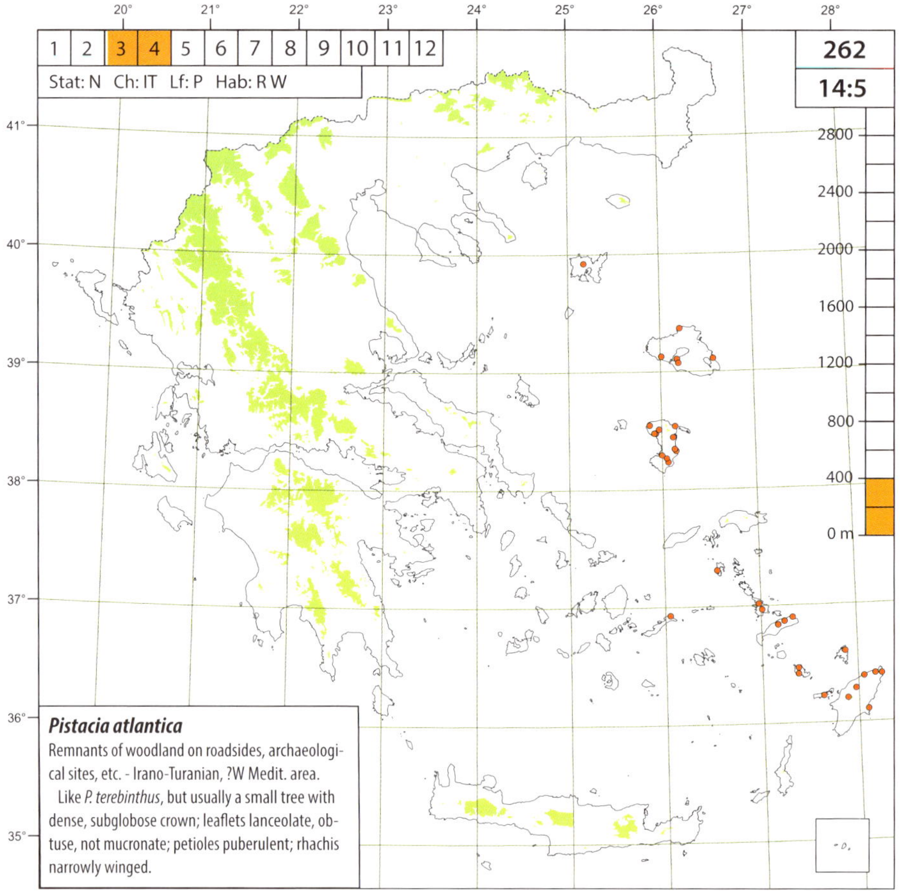

Pistacia atlantica
Remnants of woodland on roadsides, archaeological sites, etc. - Irano-Turanian, ?W Medit. area.
Like *P. terebinthus*, but usually a small tree with dense, subglobose crown; leaflets lanceolate, obtuse, not mucronate; petioles puberulent; rhachis narrowly winged.

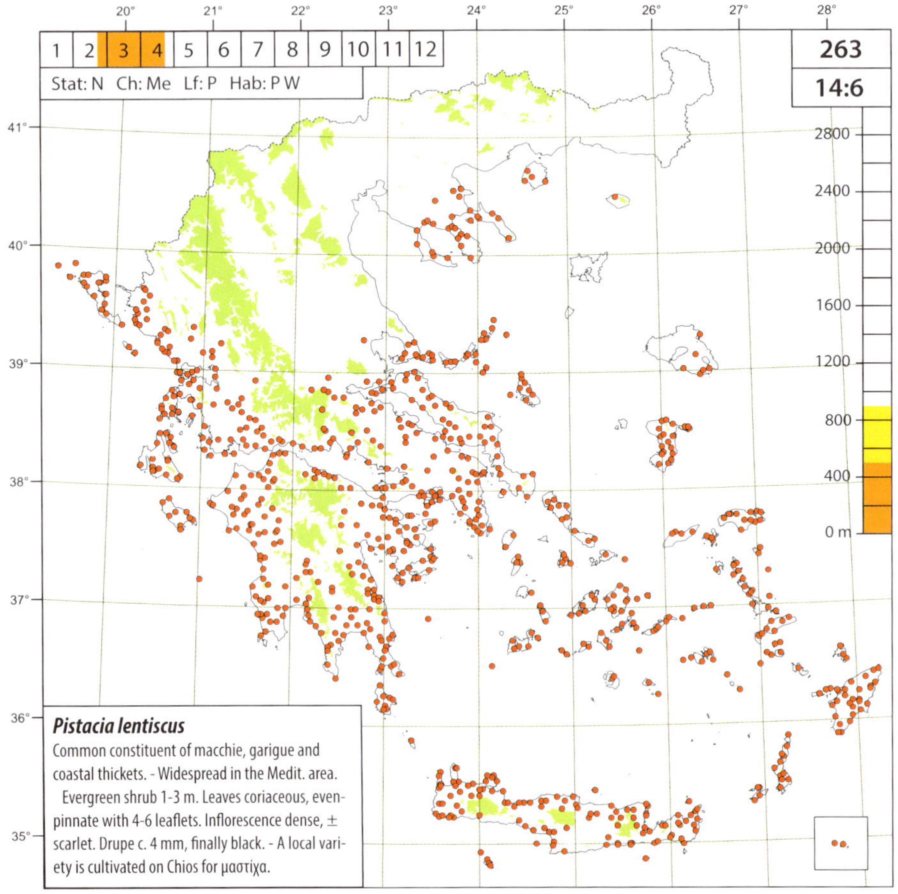

Pistacia lentiscus
Common constituent of macchie, garigue and coastal thickets. - Widespread in the Medit. area.
Evergreen shrub 1-3 m. Leaves coriaceous, evenpinnate with 4-6 leaflets. Inflorescence dense, ± scarlet. Drupe c. 4 mm, finally black. - A local variety is cultivated on Chios for μαστίχα.

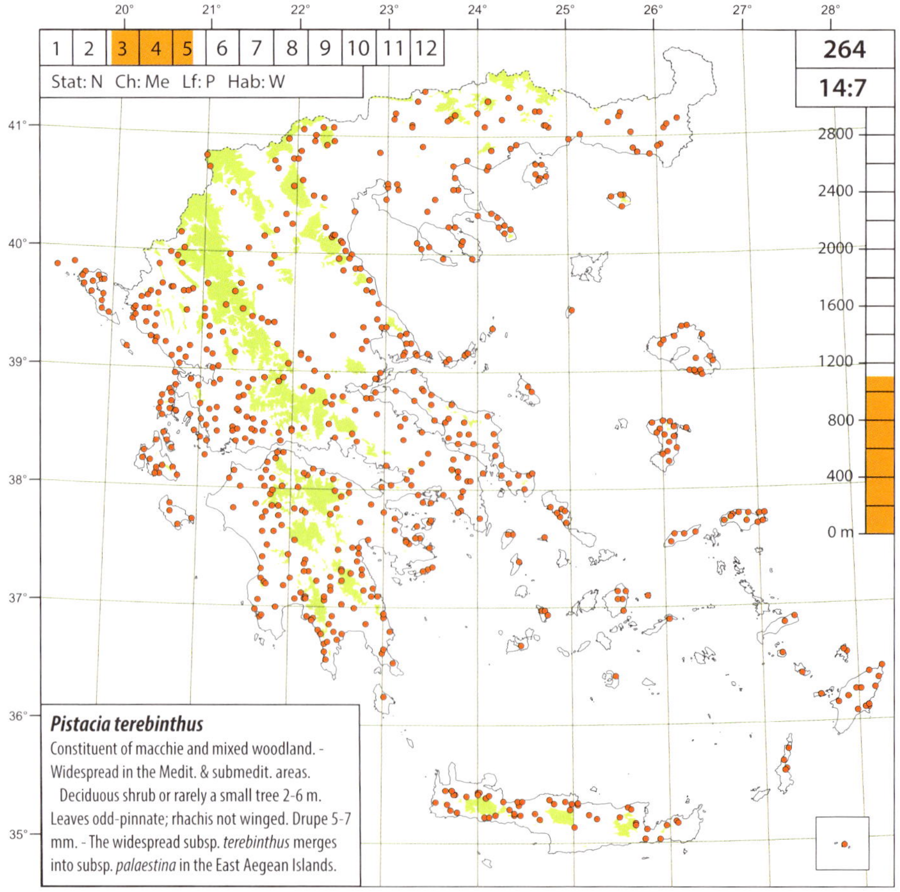

Pistacia terebinthus
Constituent of macchie and mixed woodland. - Widespread in the Medit. & submedit. areas.
Deciduous shrub or rarely a small tree 2-6 m. Leaves odd-pinnate; rhachis not winged. Drupe 5-7 mm. - The widespread subsp. *terebinthus* merges into subsp. *palaestina* in the East Aegean Islands.

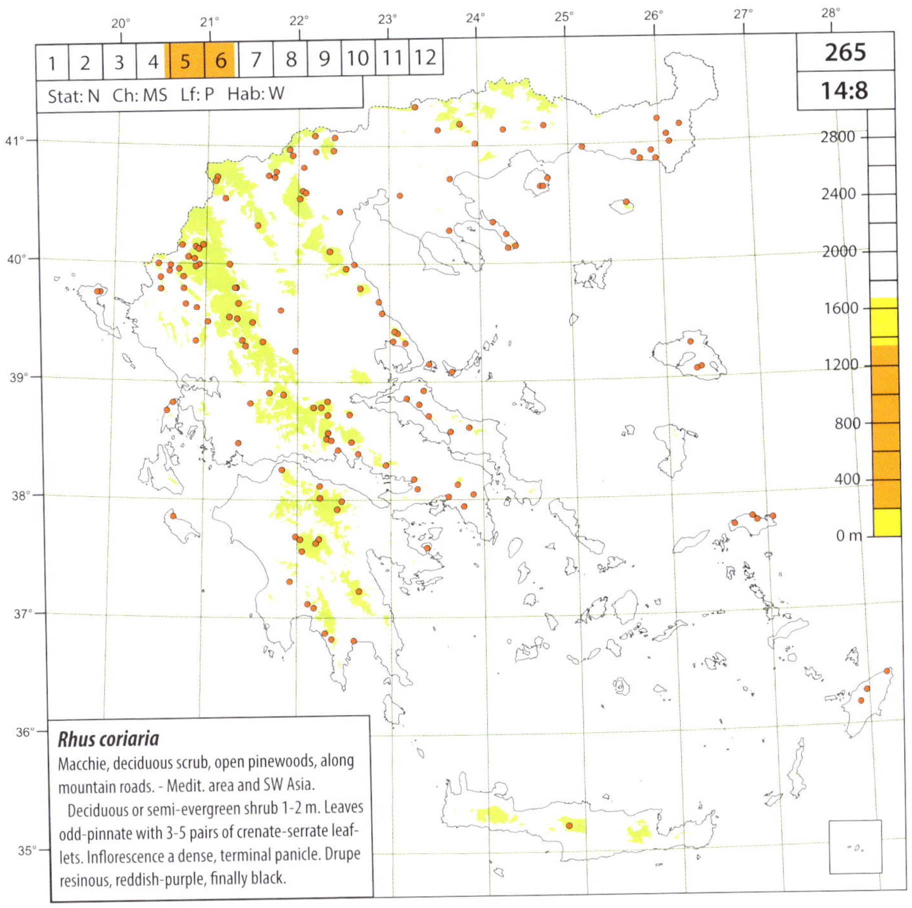

265 14:8

Stat: N Ch: MS Lf: P Hab: W

Rhus coriaria

Macchie, deciduous scrub, open pinewoods, along mountain roads. - Medit. area and SW Asia.

Deciduous or semi-evergreen shrub 1-2 m. Leaves odd-pinnate with 3-5 pairs of crenate-serrate leaflets. Inflorescence a dense, terminal panicle. Drupe resinous, reddish-purple, finally black.

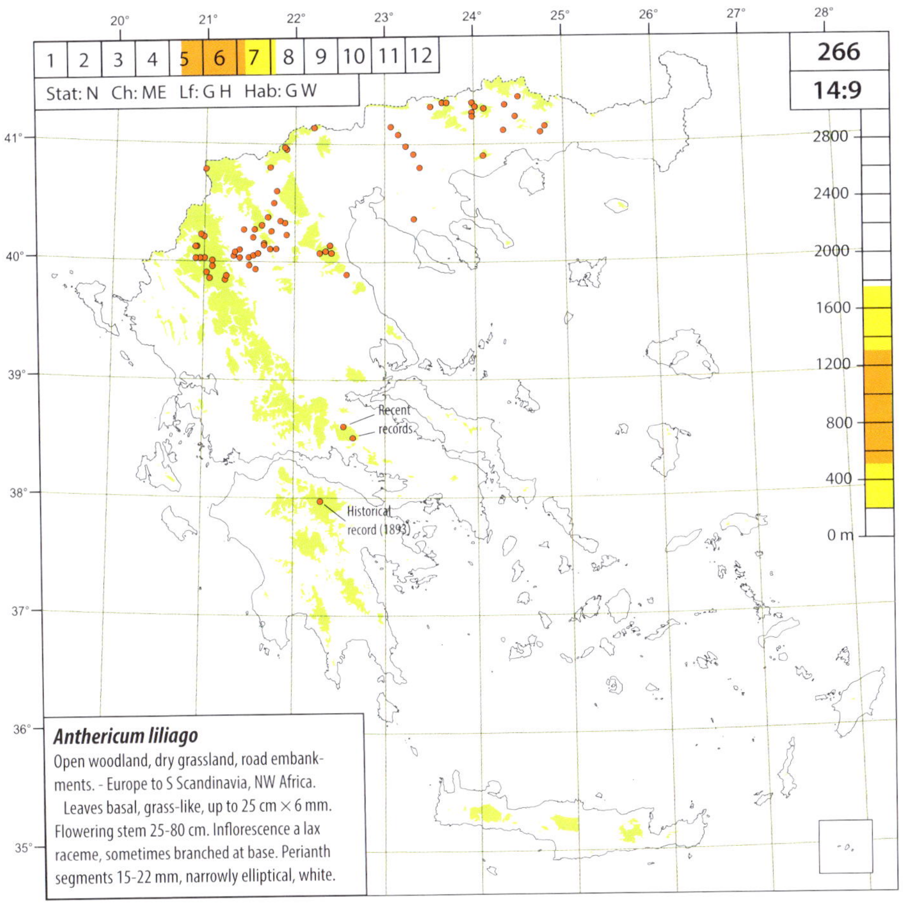

266 14:9

Stat: N Ch: ME Lf: G H Hab: G W

Anthericum liliago

Open woodland, dry grassland, road embankments. - Europe to S Scandinavia, NW Africa.

Leaves basal, grass-like, up to 25 cm × 6 mm. Flowering stem 25-80 cm. Inflorescence a lax raceme, sometimes branched at base. Perianth segments 15-22 mm, narrowly elliptical, white.

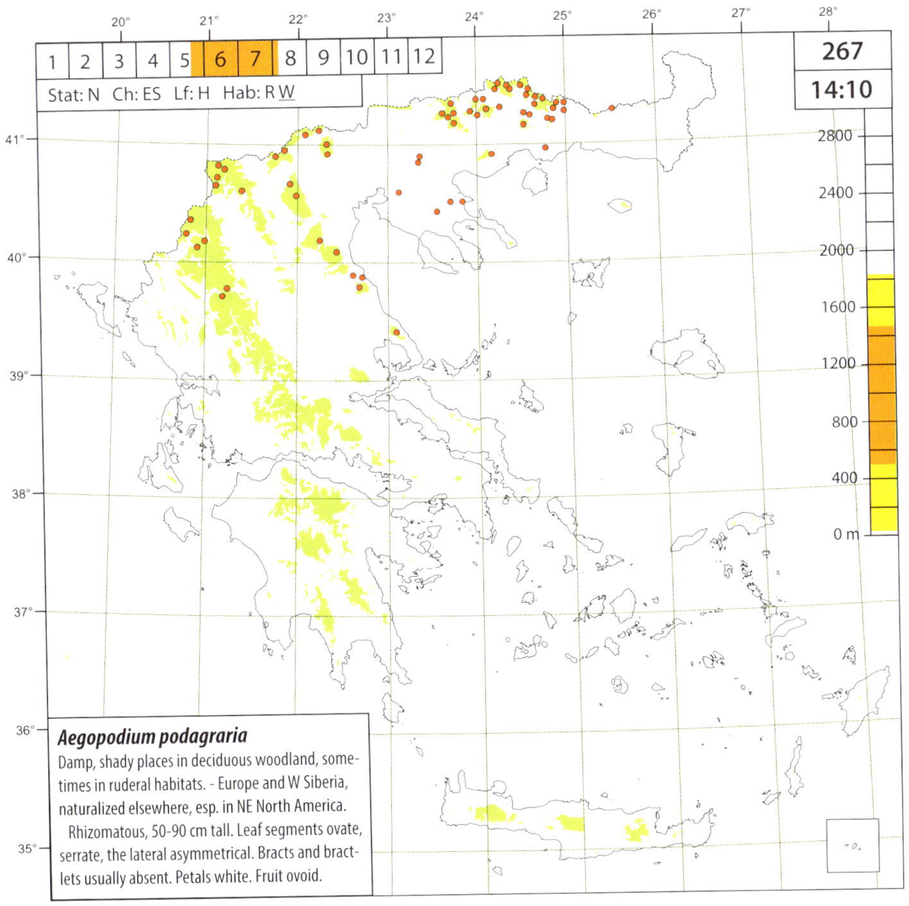

267 14:10

Stat: N Ch: ES Lf: H Hab: R W

Aegopodium podagraria

Damp, shady places in deciduous woodland, sometimes in ruderal habitats. - Europe and W Siberia, naturalized elsewhere, esp. in NE North America.

Rhizomatous, 50-90 cm tall. Leaf segments ovate, serrate, the lateral asymmetrical. Bracts and bractlets usually absent. Petals white. Fruit ovoid.

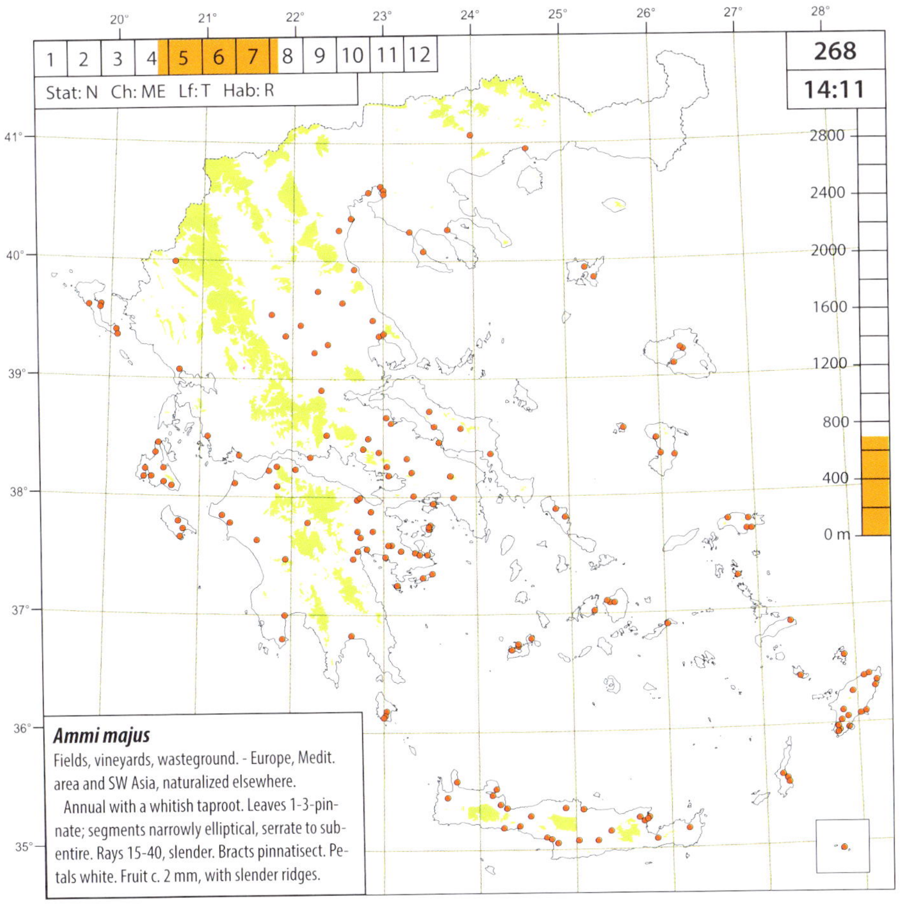

268 14:11

Stat: N Ch: ME Lf: T Hab: R

Ammi majus

Fields, vineyards, wasteground. - Europe, Medit. area and SW Asia, naturalized elsewhere.

Annual with a whitish taproot. Leaves 1-3-pinnate; segments narrowly elliptical, serrate to subentire. Rays 15-40, slender. Bracts pinnatisect. Petals white. Fruit c. 2 mm, with slender ridges.

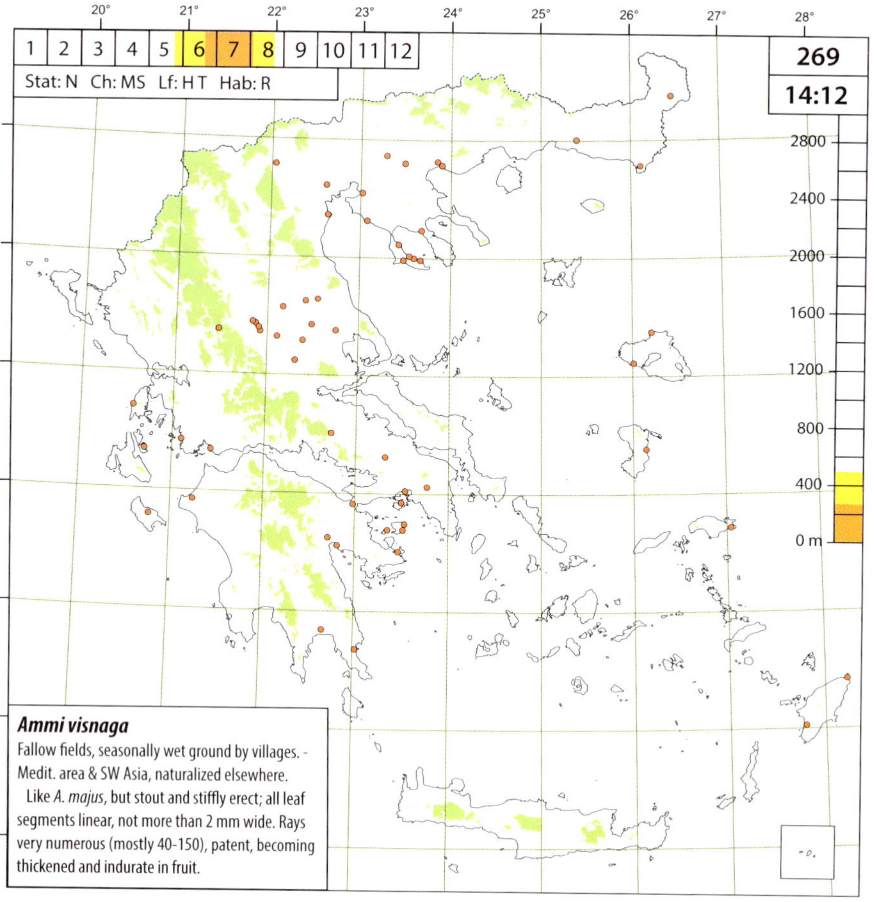

Ammi visnaga

Fallow fields, seasonally wet ground by villages. - Medit. area & SW Asia, naturalized elsewhere.

Like *A. majus*, but stout and stiffly erect; all leaf segments linear, not more than 2 mm wide. Rays very numerous (mostly 40-150), patent, becoming thickened and indurate in fruit.

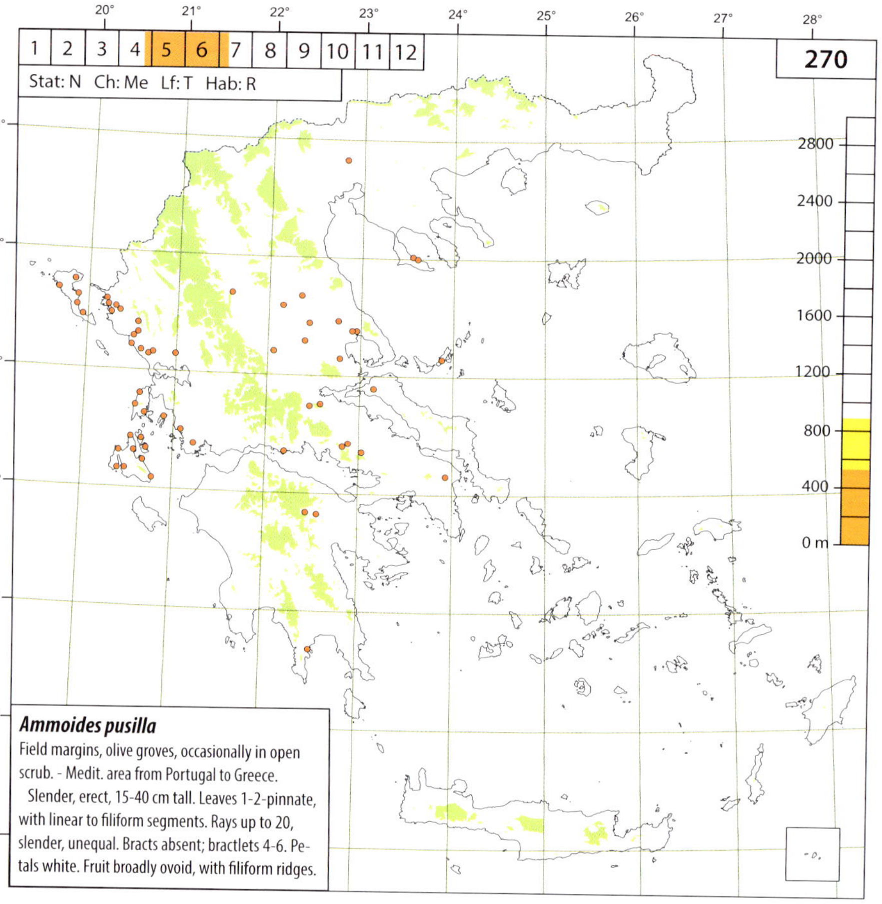

Ammoides pusilla

Field margins, olive groves, occasionally in open scrub. - Medit. area from Portugal to Greece.

Slender, erect, 15-40 cm tall. Leaves 1-2-pinnate, with linear to filiform segments. Rays up to 20, slender, unequal. Bracts absent; bractlets 4-6. Petals white. Fruit broadly ovoid, with filiform ridges.

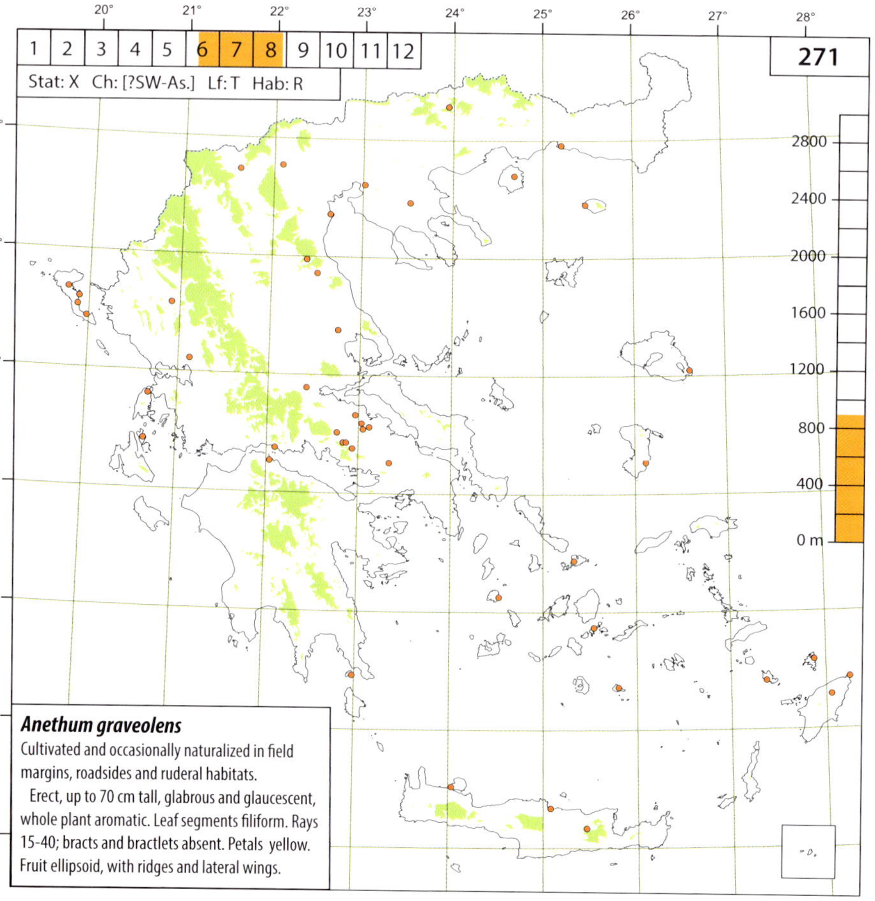

Anethum graveolens

Cultivated and occasionally naturalized in field margins, roadsides and ruderal habitats.

Erect, up to 70 cm tall, glabrous and glaucescent, whole plant aromatic. Leaf segments filiform. Rays 15-40; bracts and bractlets absent. Petals yellow. Fruit ellipsoid, with ridges and lateral wings.

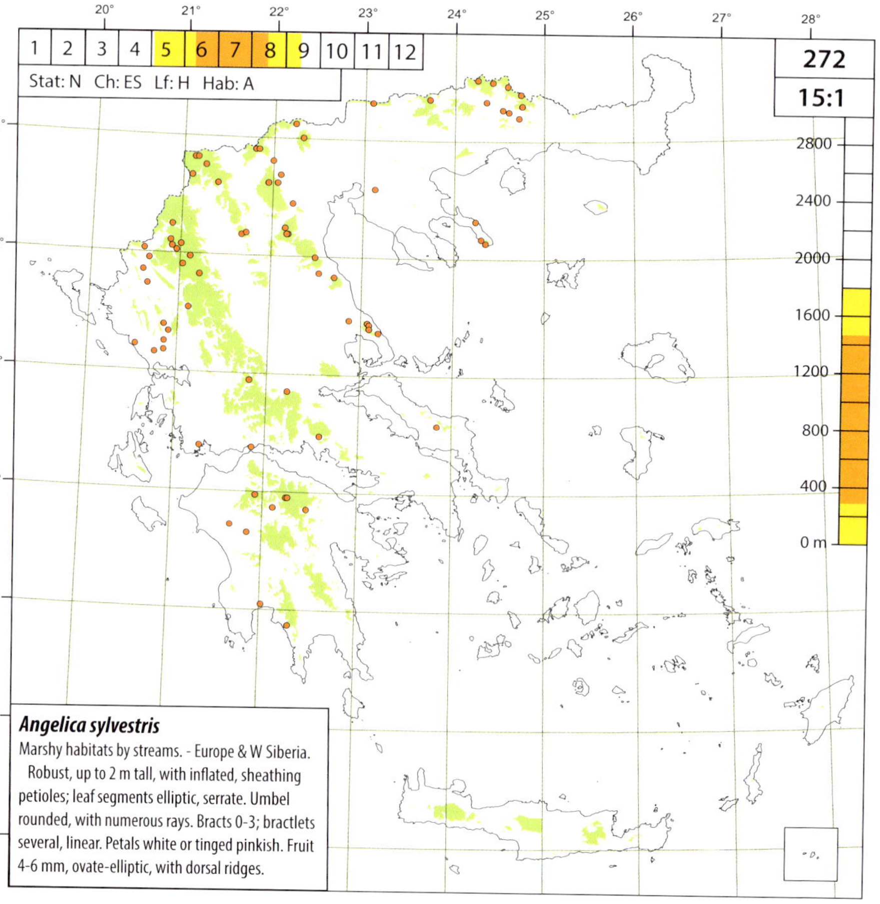

Angelica sylvestris

Marshy habitats by streams. - Europe & W Siberia.

Robust, up to 2 m tall, with inflated, sheathing petioles; leaf segments elliptic, serrate. Umbel rounded, with numerous rays. Bracts 0-3; bractlets several, linear. Petals white or tinged pinkish. Fruit 4-6 mm, ovate-elliptic, with dorsal ridges.

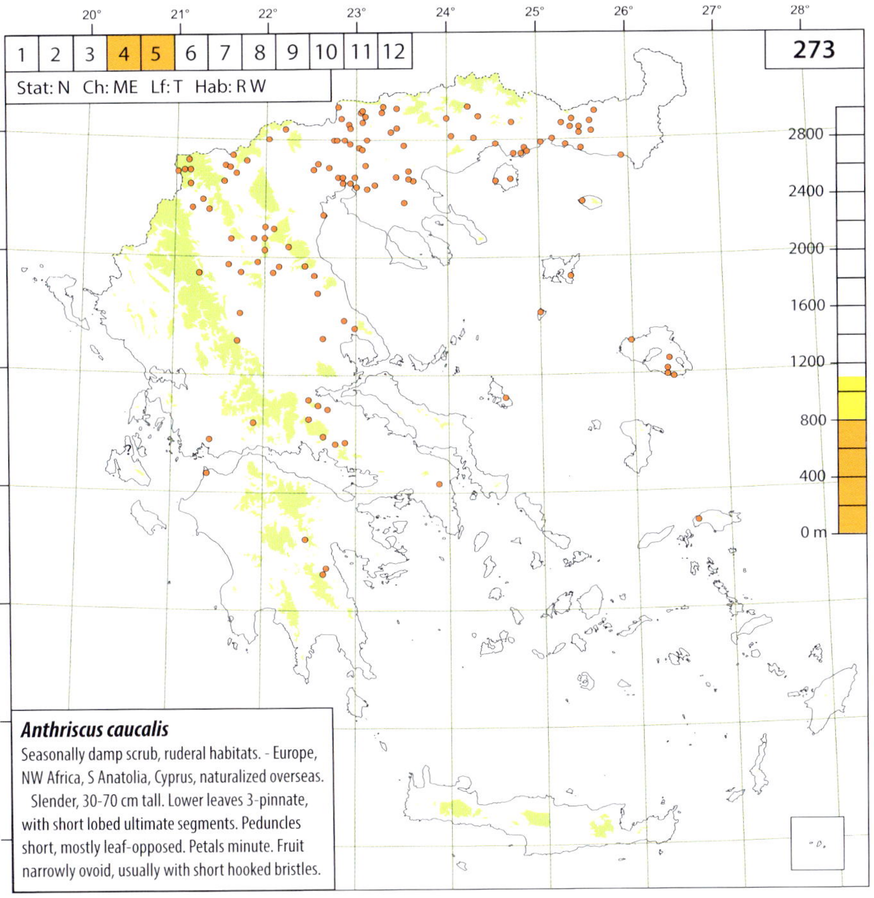

Anthriscus caucalis
Seasonally damp scrub, ruderal habitats. - Europe, NW Africa, S Anatolia, Cyprus, naturalized overseas.
Slender, 30-70 cm tall. Lower leaves 3-pinnate, with short lobed ultimate segments. Peduncles short, mostly leaf-opposed. Petals minute. Fruit narrowly ovoid, usually with short hooked bristles.

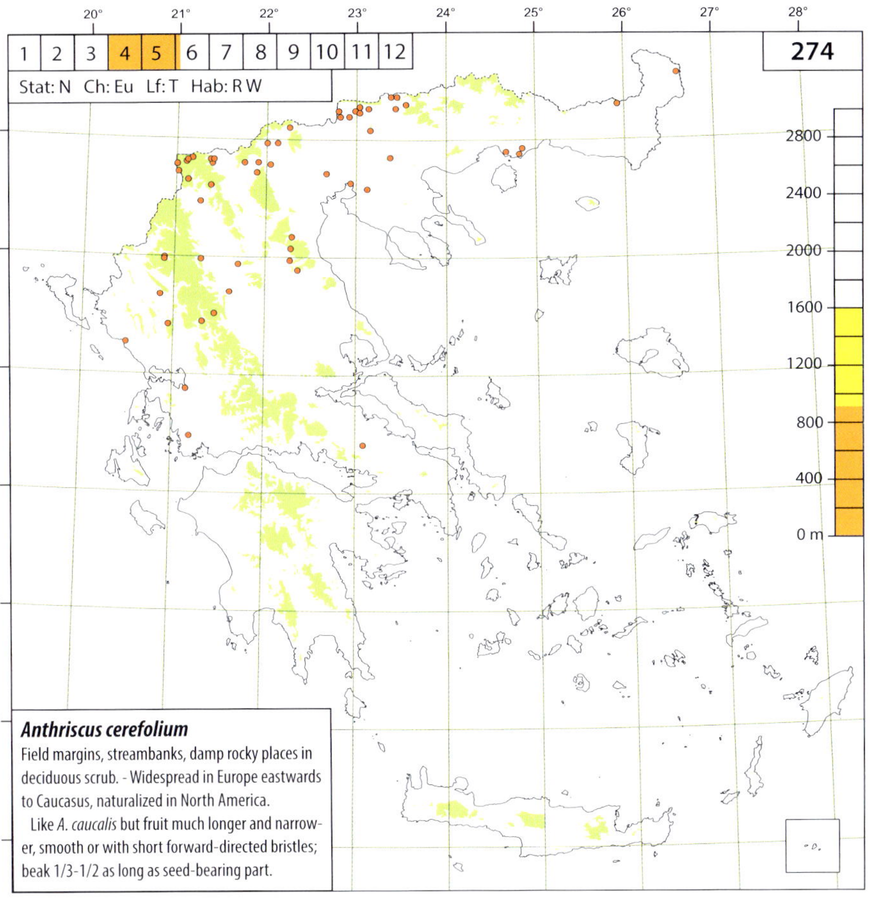

Anthriscus cerefolium
Field margins, streambanks, damp rocky places in deciduous scrub. - Widespread in Europe eastwards to Caucasus, naturalized in North America.
Like *A. caucalis* but fruit much longer and narrower, smooth or with short forward-directed bristles; beak 1/3-1/2 as long as seed-bearing part.

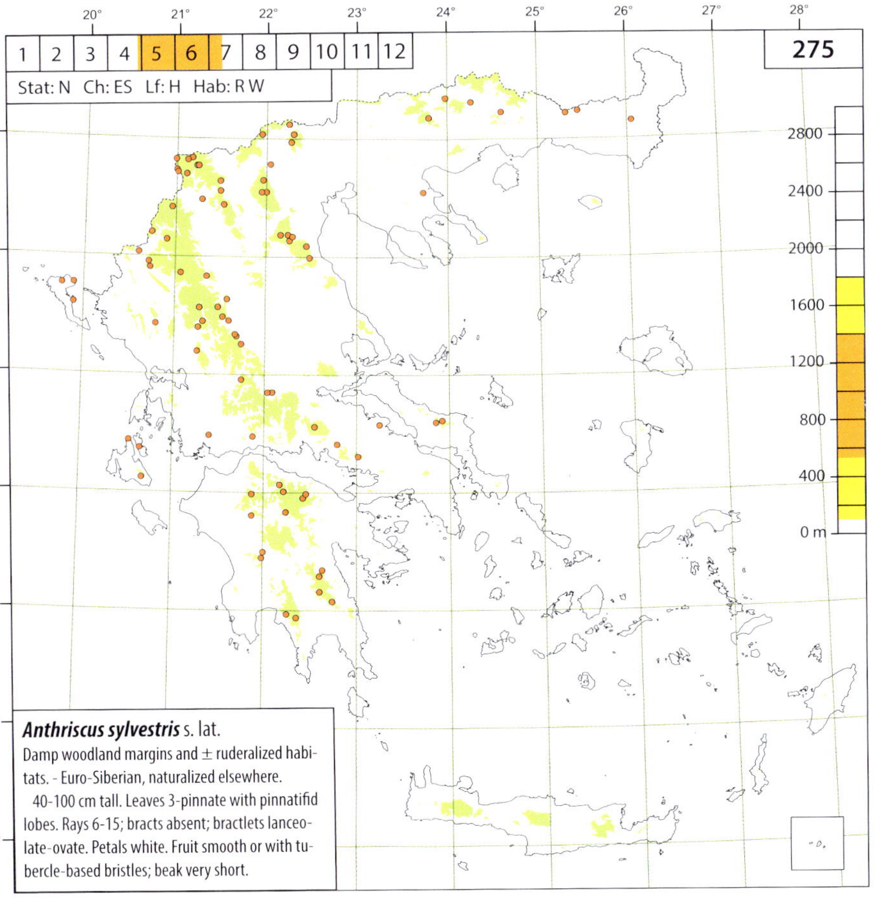

Anthriscus sylvestris s. lat.
Damp woodland margins and ± ruderalized habitats. - Euro-Siberian, naturalized elsewhere.
40-100 cm tall. Leaves 3-pinnate with pinnatifid lobes. Rays 6-15; bracts absent; bractlets lanceolate-ovate. Petals white. Fruit smooth or with tubercle-based bristles; beak very short.

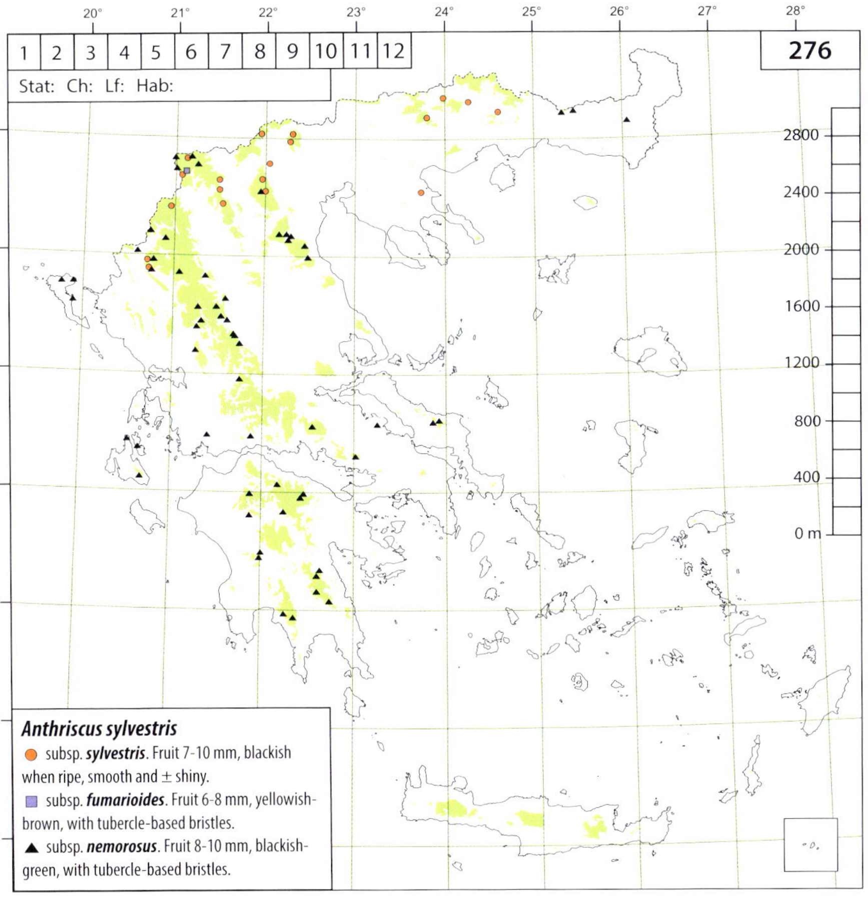

Anthriscus sylvestris
● subsp. ***sylvestris***. Fruit 7-10 mm, blackish when ripe, smooth and ± shiny.
■ subsp. ***fumarioides***. Fruit 6-8 mm, yellowish-brown, with tubercle-based bristles.
▲ subsp. ***nemorosus***. Fruit 8-10 mm, blackish-green, with tubercle-based bristles.

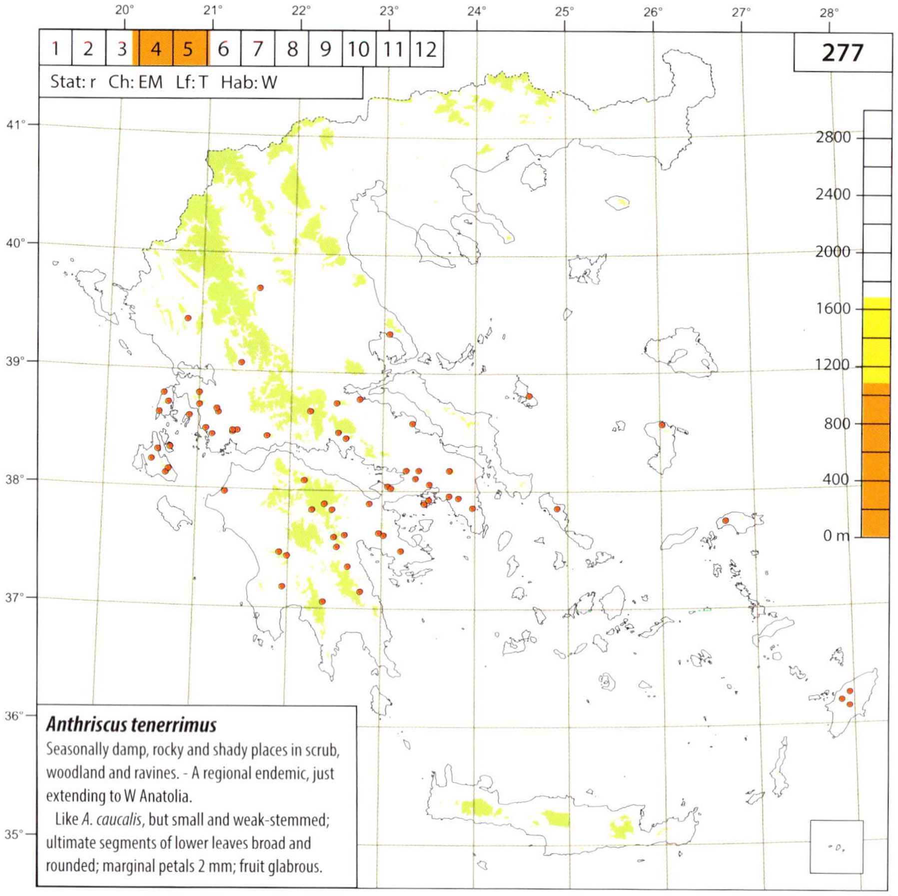

Anthriscus tenerrimus
Seasonally damp, rocky and shady places in scrub, woodland and ravines. - A regional endemic, just extending to W Anatolia.
Like *A. caucalis*, but small and weak-stemmed; ultimate segments of lower leaves broad and rounded; marginal petals 2 mm; fruit glabrous.

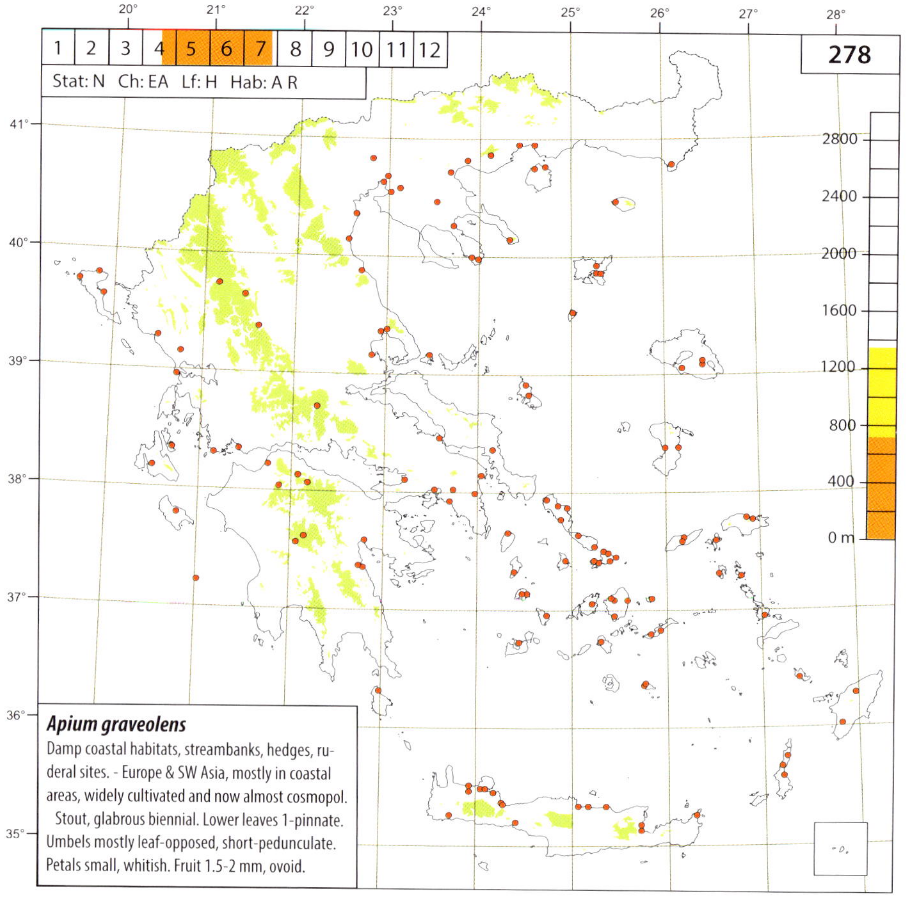

Apium graveolens
Damp coastal habitats, streambanks, hedges, ruderal sites. - Europe & SW Asia, mostly in coastal areas, widely cultivated and now almost cosmopol.
Stout, glabrous biennial. Lower leaves 1-pinnate. Umbels mostly leaf-opposed, short-pedunculate. Petals small, whitish. Fruit 1.5-2 mm, ovoid.

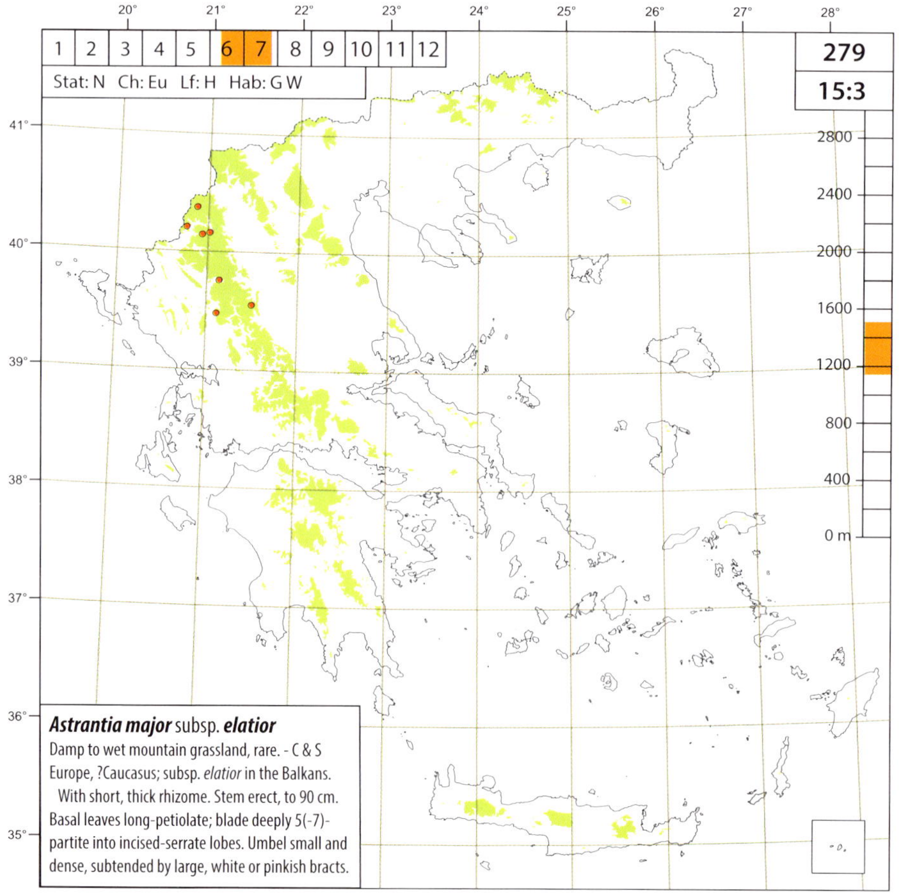

Astrantia major subsp. ***elatior***
Damp to wet mountain grassland, rare. - C & S Europe, ?Caucasus; subsp. *elatior* in the Balkans.
With short, thick rhizome. Stem erect, to 90 cm. Basal leaves long-petiolate; blade deeply 5(-7)-partite into incised-serrate lobes. Umbel small and dense, subtended by large, white or pinkish bracts.

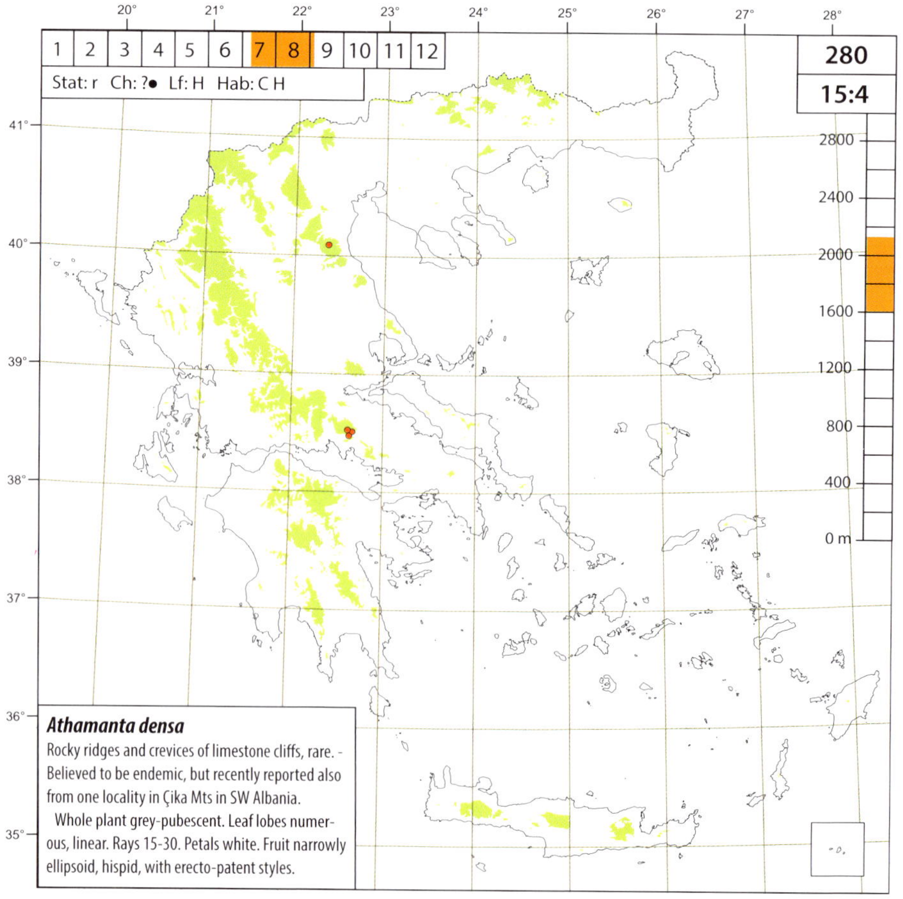

Athamanta densa
Rocky ridges and crevices of limestone cliffs, rare. - Believed to be endemic, but recently reported also from one locality in Çika Mts in SW Albania.
Whole plant grey-pubescent. Leaf lobes numerous, linear. Rays 15-30. Petals white. Fruit narrowly ellipsoid, hispid, with erecto-patent styles.

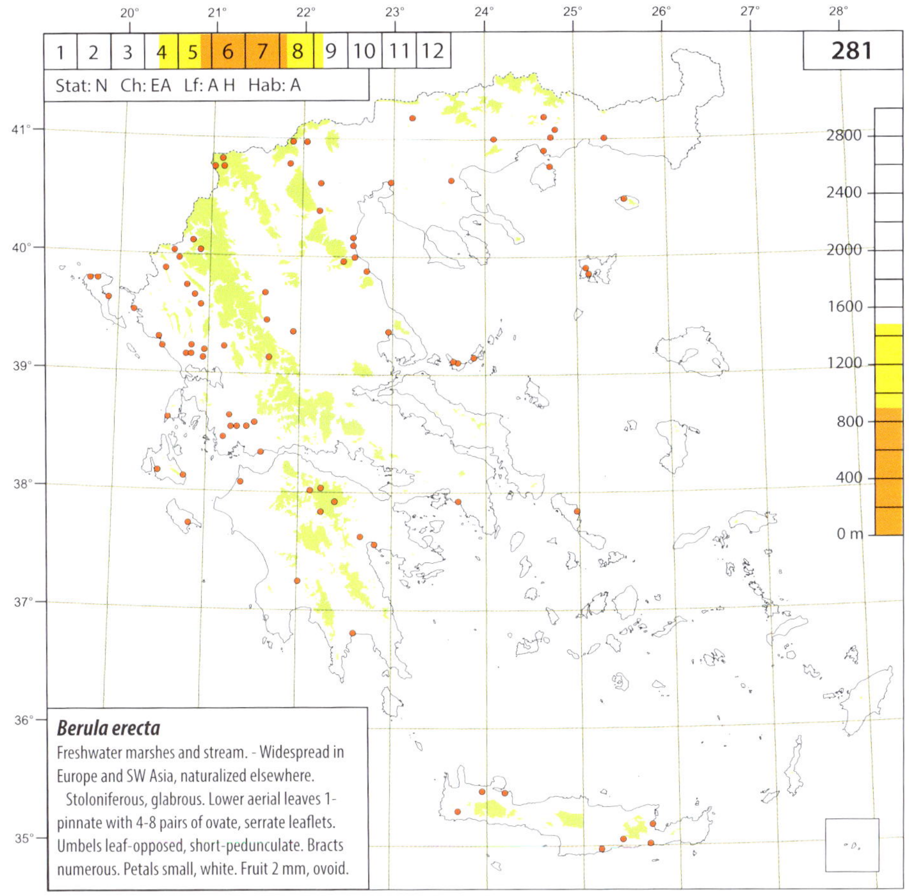

Berula erecta
Freshwater marshes and stream. - Widespread in Europe and SW Asia, naturalized elsewhere.
Stoloniferous, glabrous. Lower aerial leaves 1-pinnate with 4-8 pairs of ovate, serrate leaflets. Umbels leaf-opposed, short-pedunculate. Bracts numerous. Petals small, white. Fruit 2 mm, ovoid.

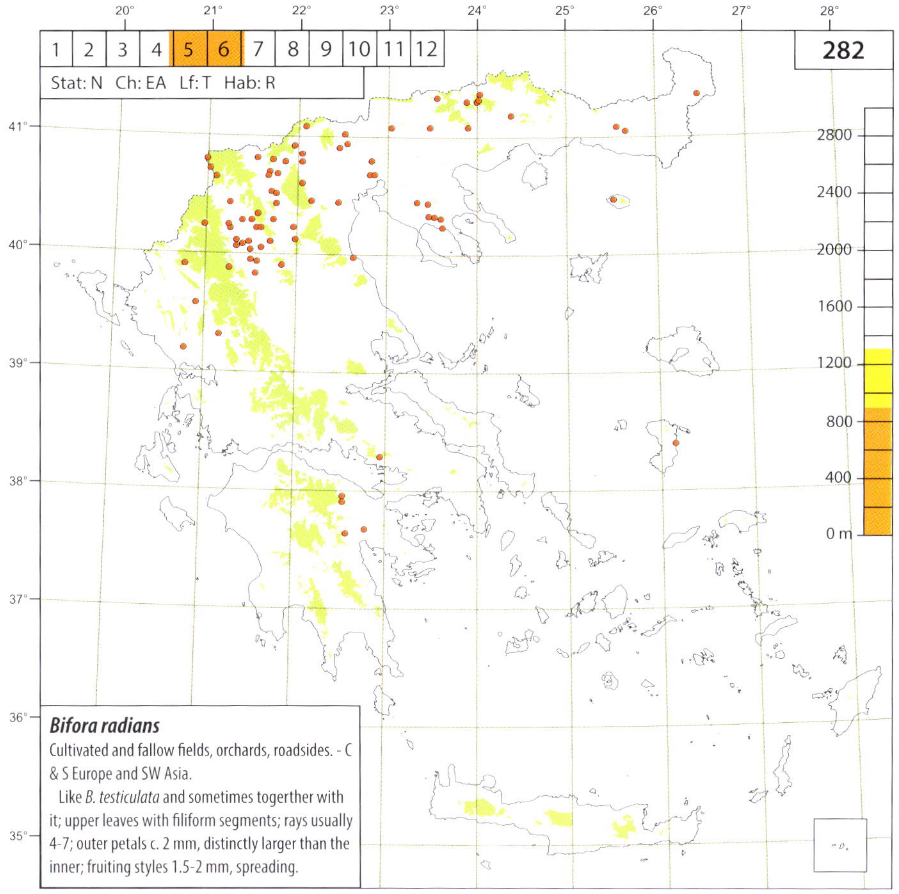

Bifora radians
Cultivated and fallow fields, orchards, roadsides. - C & S Europe and SW Asia.
Like *B. testiculata* and sometimes togerther with it; upper leaves with filiform segments; rays usually 4-7; outer petals c. 2 mm, distinctly larger than the inner; fruiting styles 1.5-2 mm, spreading.

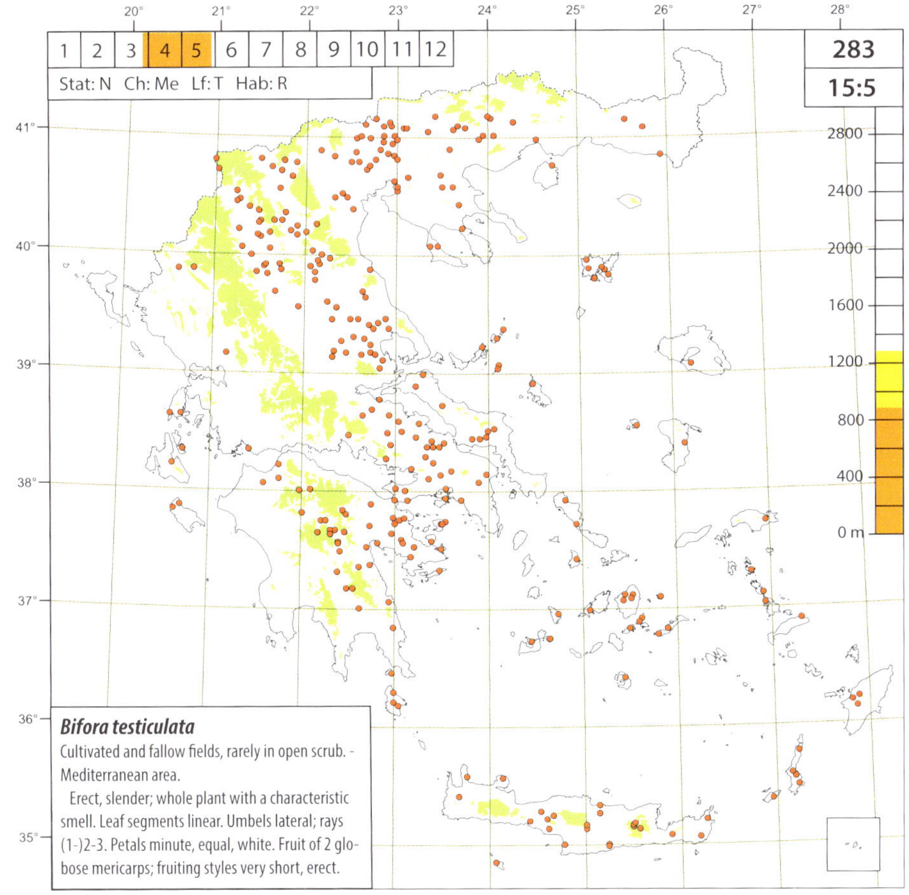

Bifora testiculata
Cultivated and fallow fields, rarely in open scrub. - Mediterranean area.
Erect, slender; whole plant with a characteristic smell. Leaf segments linear. Umbels lateral; rays (1-)2-3. Petals minute, equal, white. Fruit of 2 globose mericarps; fruiting styles very short, erect.

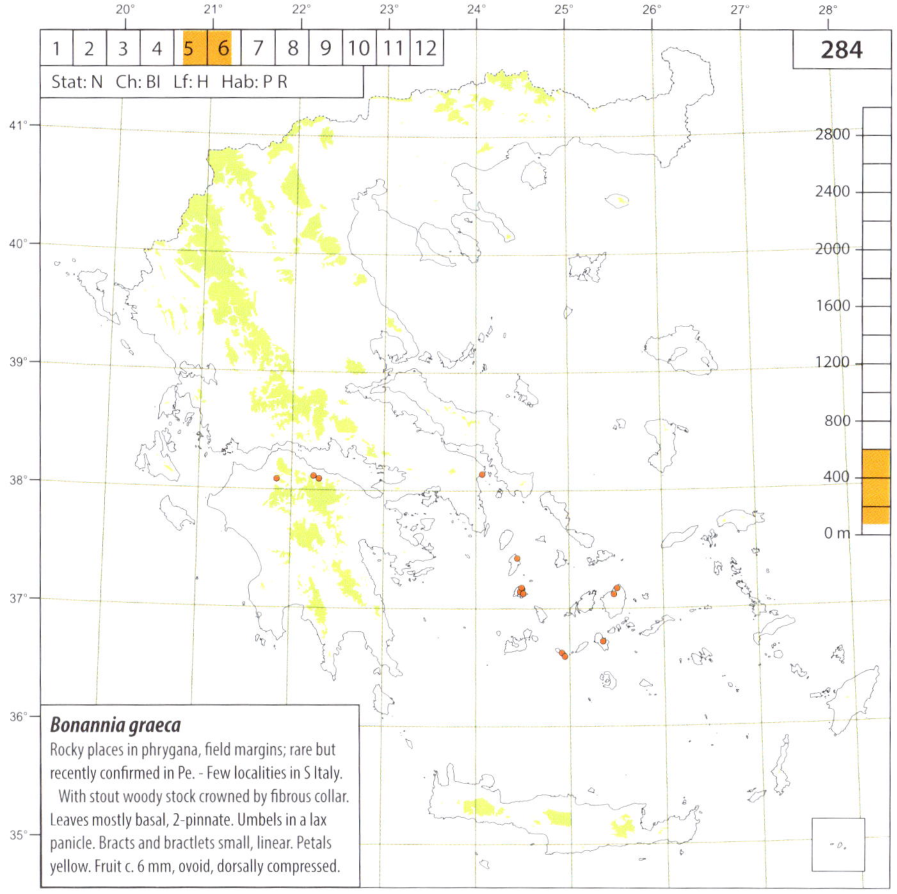

Bonannia graeca
Rocky places in phrygana, field margins; rare but recently confirmed in Pe. - Few localities in S Italy.
With stout woody stock crowned by fibrous collar. Leaves mostly basal, 2-pinnate. Umbels in a lax panicle. Bracts and bractlets small, linear. Petals yellow. Fruit c. 6 mm, ovoid, dorsally compressed.

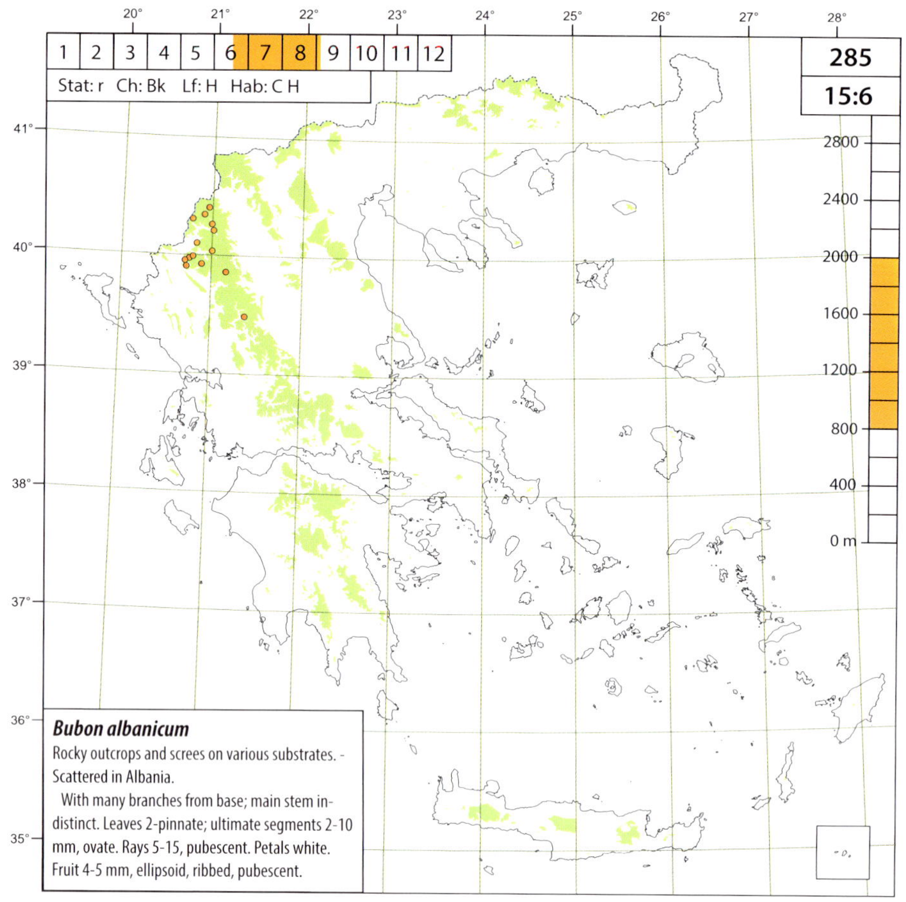

Bubon albanicum
Rocky outcrops and screes on various substrates. - Scattered in Albania.
With many branches from base; main stem indistinct. Leaves 2-pinnate; ultimate segments 2-10 mm, ovate. Rays 5-15, pubescent. Petals white. Fruit 4-5 mm, ellipsoid, ribbed, pubescent.

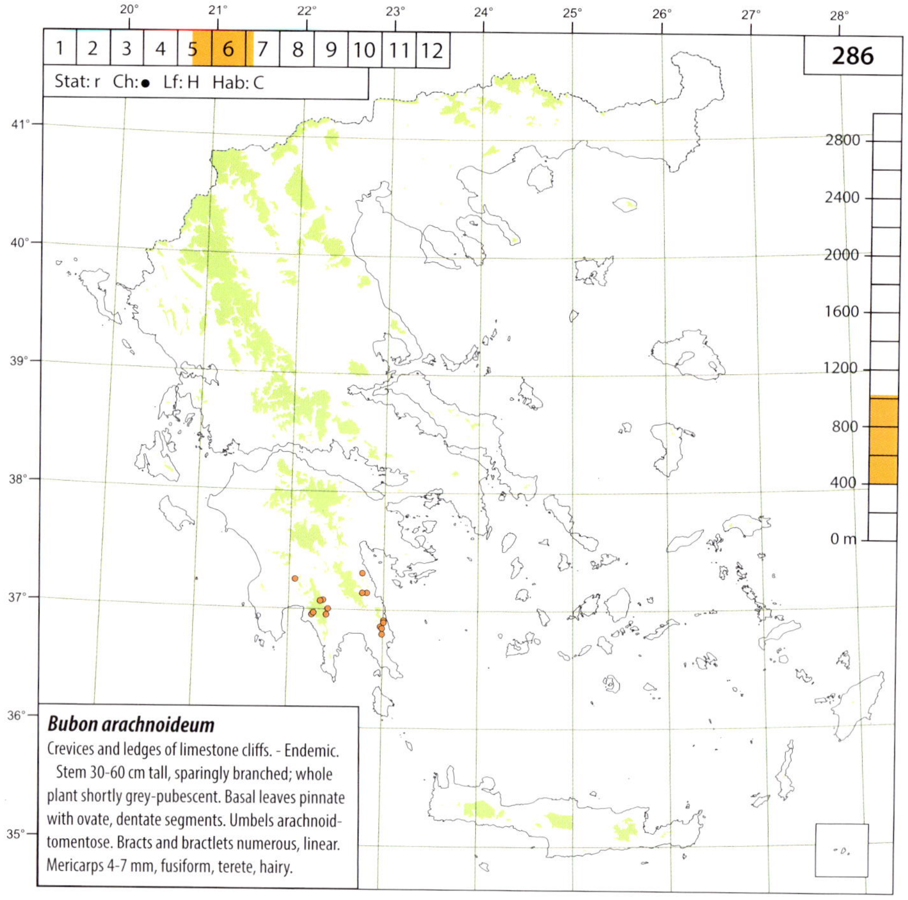

Bubon arachnoideum
Crevices and ledges of limestone cliffs. - Endemic.
Stem 30-60 cm tall, sparingly branched; whole plant shortly grey-pubescent. Basal leaves pinnate with ovate, dentate segments. Umbels arachnoid-tomentose. Bracts and bractlets numerous, linear. Mericarps 4-7 mm, fusiform, terete, hairy.

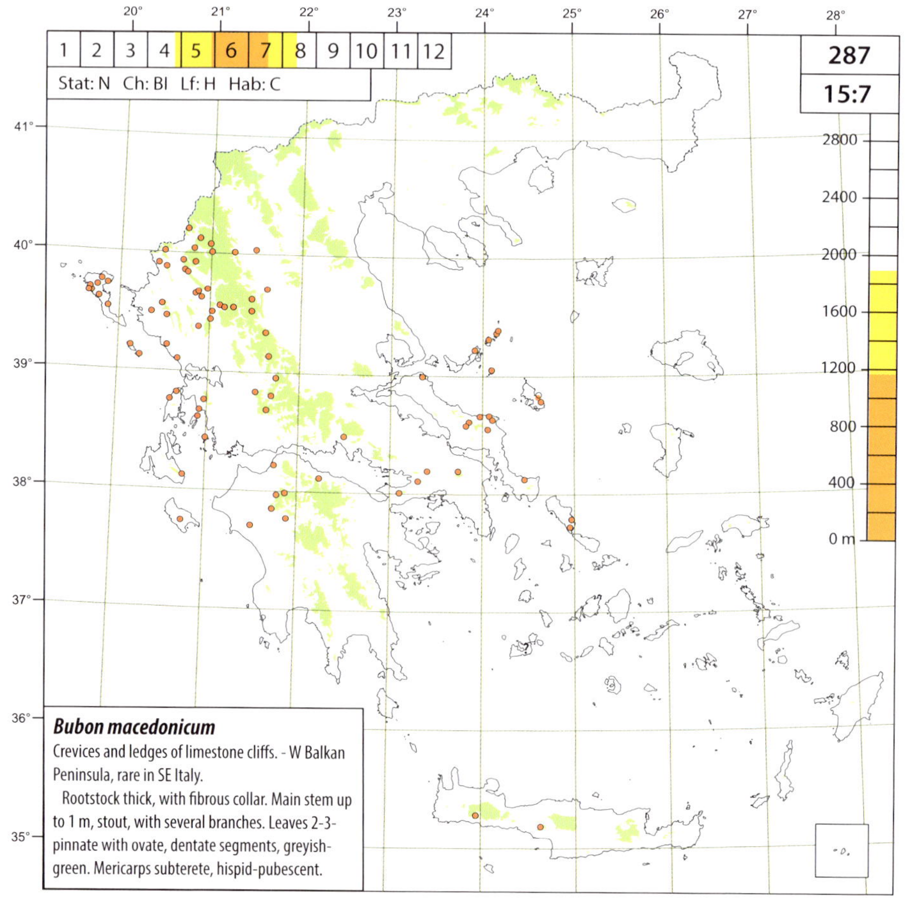

Bubon macedonicum
Crevices and ledges of limestone cliffs. - W Balkan Peninsula, rare in SE Italy.
Rootstock thick, with fibrous collar. Main stem up to 1 m, stout, with several branches. Leaves 2-3-pinnate with ovate, dentate segments, greyish-green. Mericarps subterete, hispid-pubescent.

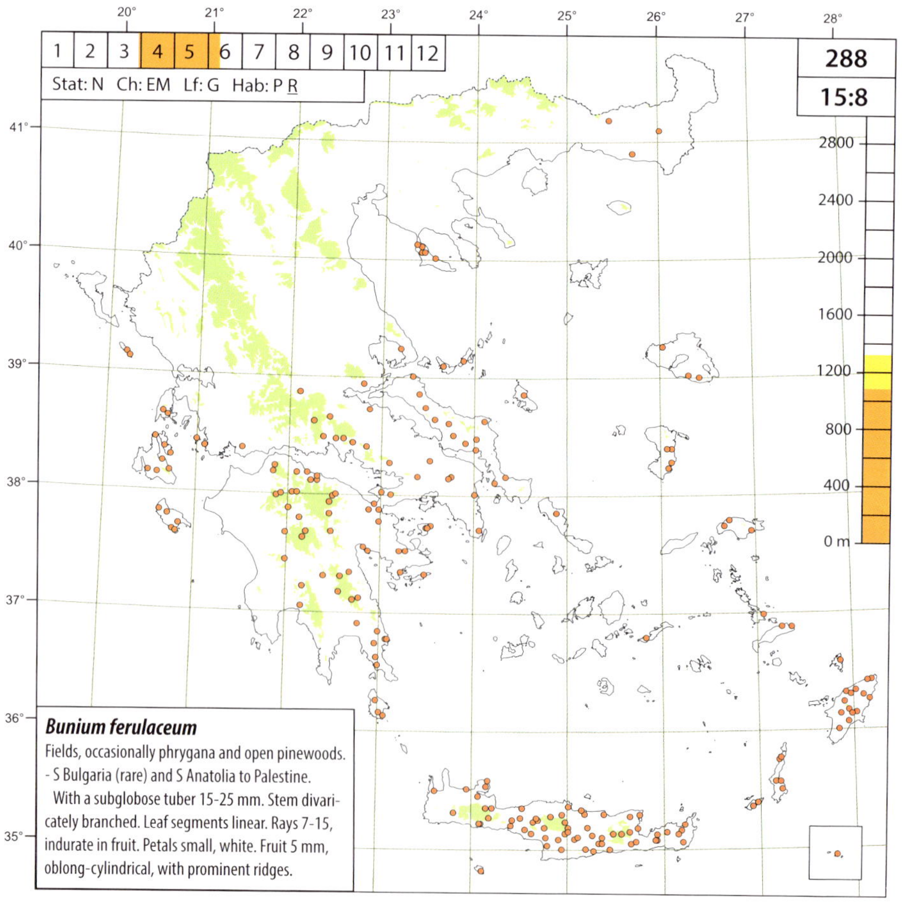

Bunium ferulaceum
Fields, occasionally phrygana and open pinewoods. - S Bulgaria (rare) and S Anatolia to Palestine.
With a subglobose tuber 15-25 mm. Stem divaricately branched. Leaf segments linear. Rays 7-15, indurate in fruit. Petals small, white. Fruit 5 mm, oblong-cylindrical, with prominent ridges.

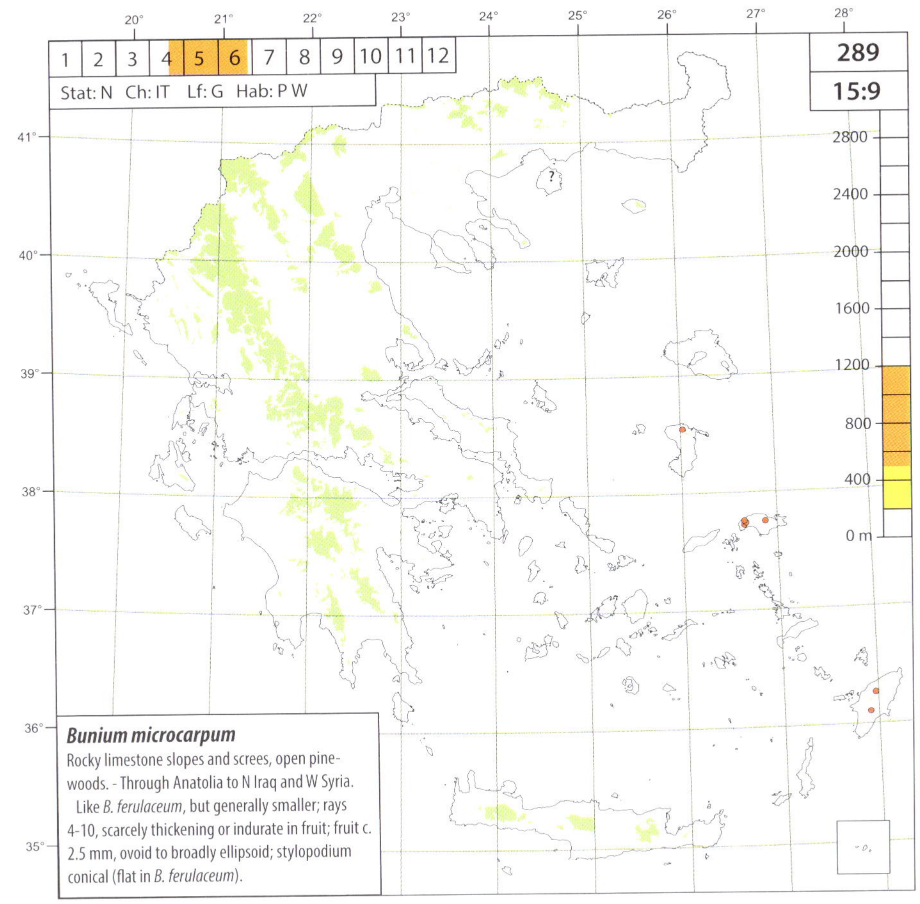

Bunium microcarpum

Rocky limestone slopes and screes, open pine-woods. - Through Anatolia to N Iraq and W Syria.

Like *B. ferulaceum*, but generally smaller; rays 4-10, scarcely thickening or indurate in fruit; fruit c. 2.5 mm, ovoid to broadly ellipsoid; stylopodium conical (flat in *B. ferulaceum*).

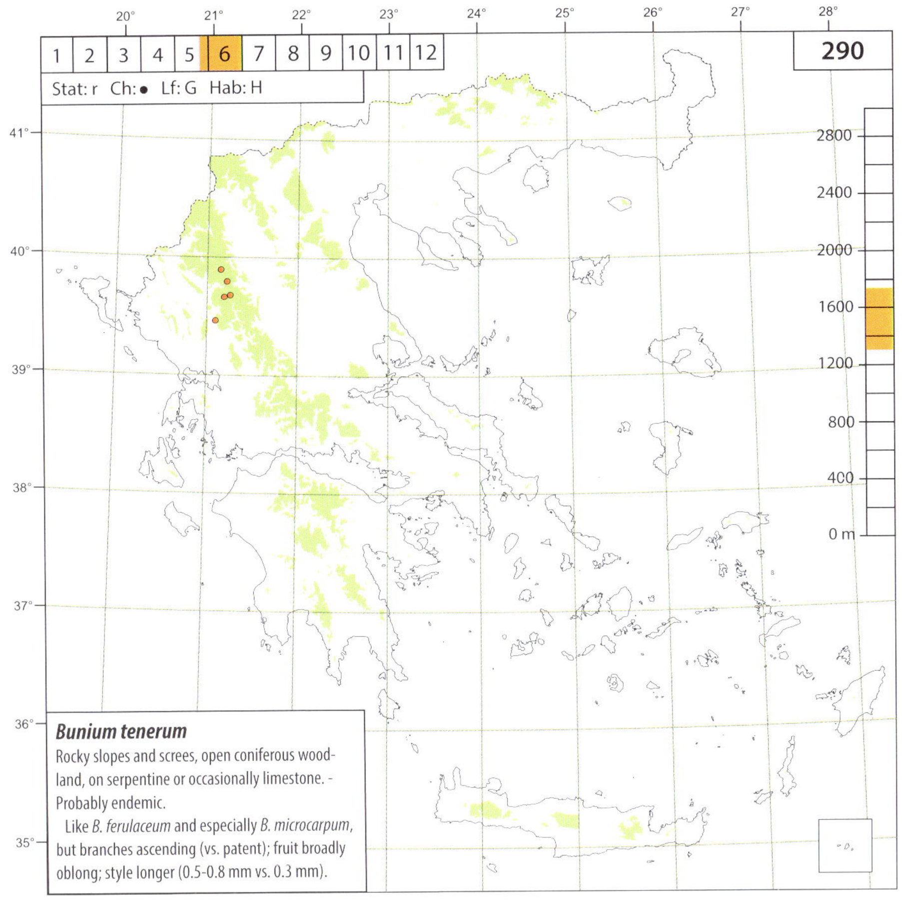

Bunium tenerum

Rocky slopes and screes, open coniferous woodland, on serpentine or occasionally limestone. - Probably endemic.

Like *B. ferulaceum* and especially *B. microcarpum*, but branches ascending (vs. patent); fruit broadly oblong; style longer (0.5-0.8 mm vs. 0.3 mm).

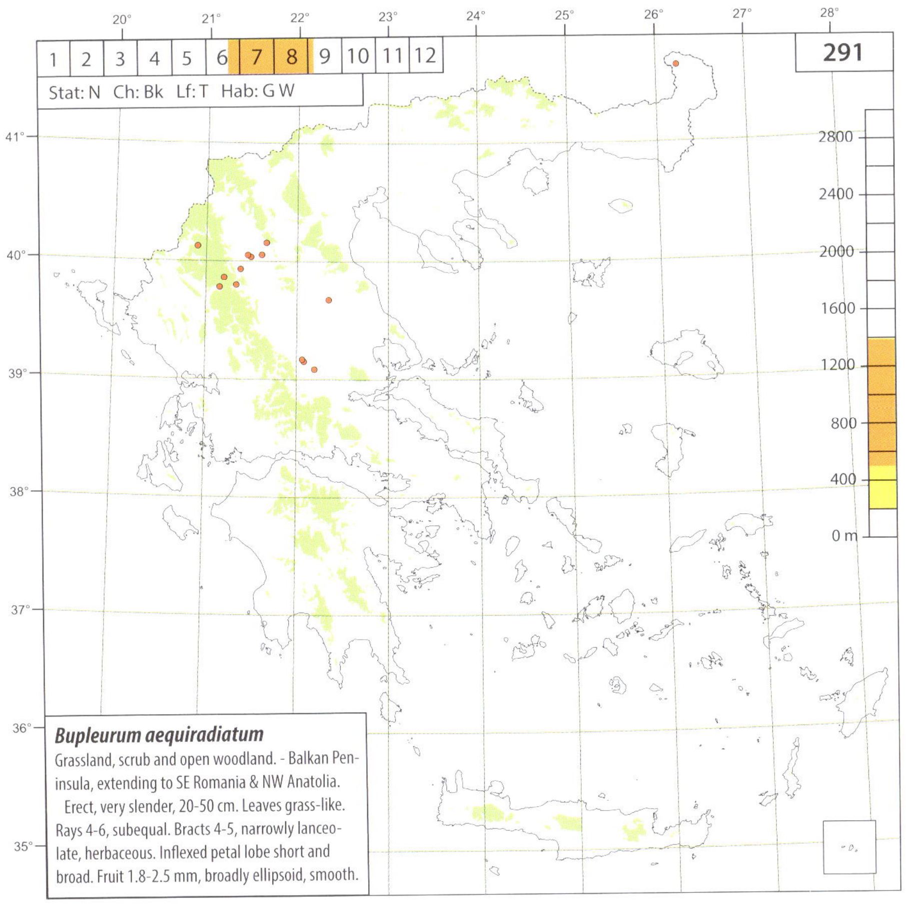

Bupleurum aequiradiatum

Grassland, scrub and open woodland. - Balkan Peninsula, extending to SE Romania & NW Anatolia.

Erect, very slender, 20-50 cm. Leaves grass-like. Rays 4-6, subequal. Bracts 4-5, narrowly lanceolate, herbaceous. Inflexed petal lobe short and broad. Fruit 1.8-2.5 mm, broadly ellipsoid, smooth.

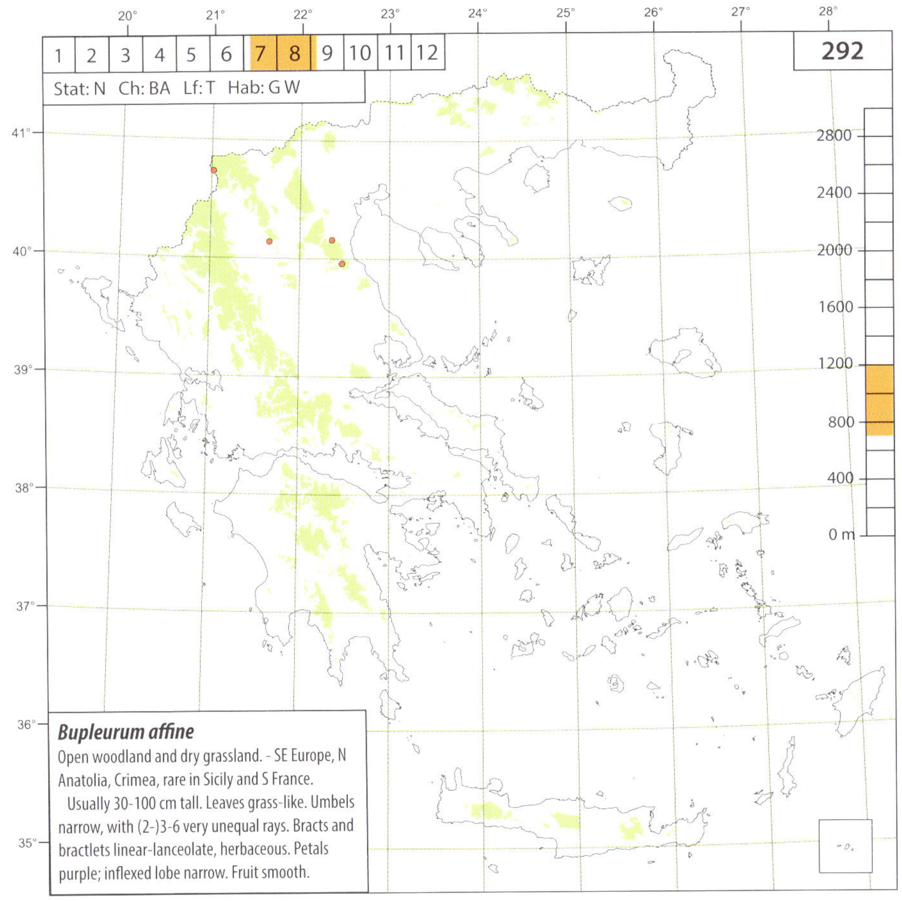

Bupleurum affine

Open woodland and dry grassland. - SE Europe, N Anatolia, Crimea, rare in Sicily and S France.

Usually 30-100 cm tall. Leaves grass-like. Umbels narrow, with (2-)3-6 very unequal rays. Bracts and bractlets linear-lanceolate, herbaceous. Petals purple; inflexed lobe narrow. Fruit smooth.

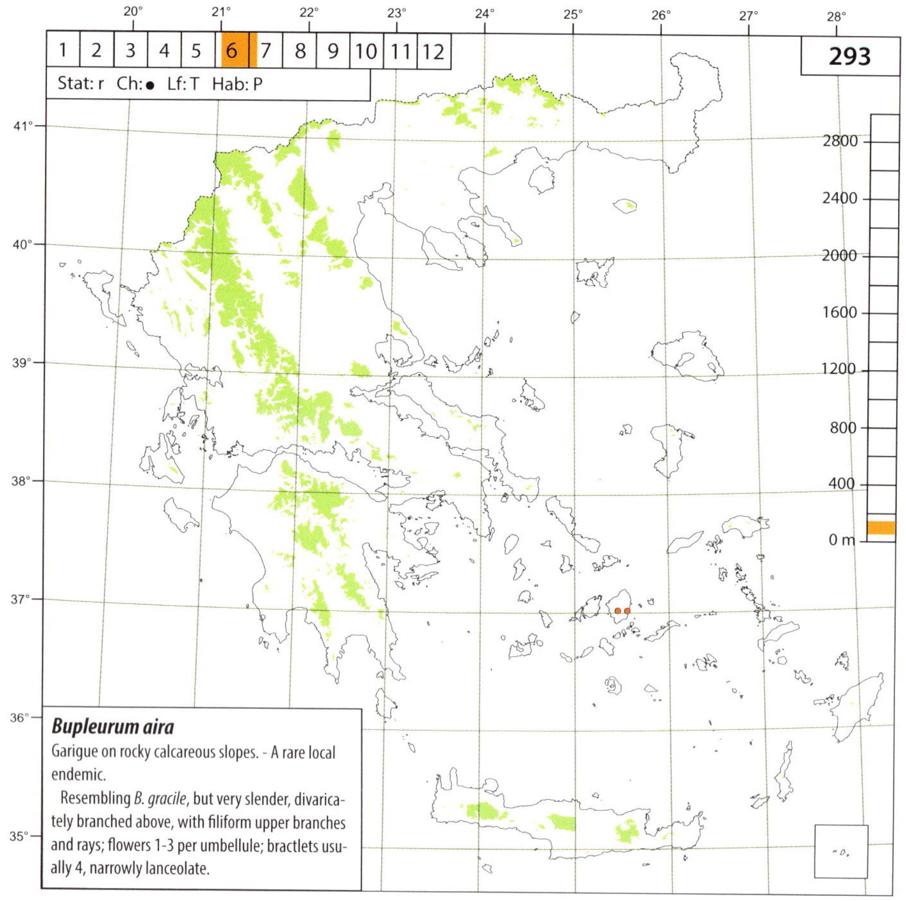

Bupleurum aira

Garigue on rocky calcareous slopes. - A rare local endemic.

Resembling *B. gracile*, but very slender, divaricately branched above, with filiform upper branches and rays; flowers 1-3 per umbellule; bractlets usually 4, narrowly lanceolate.

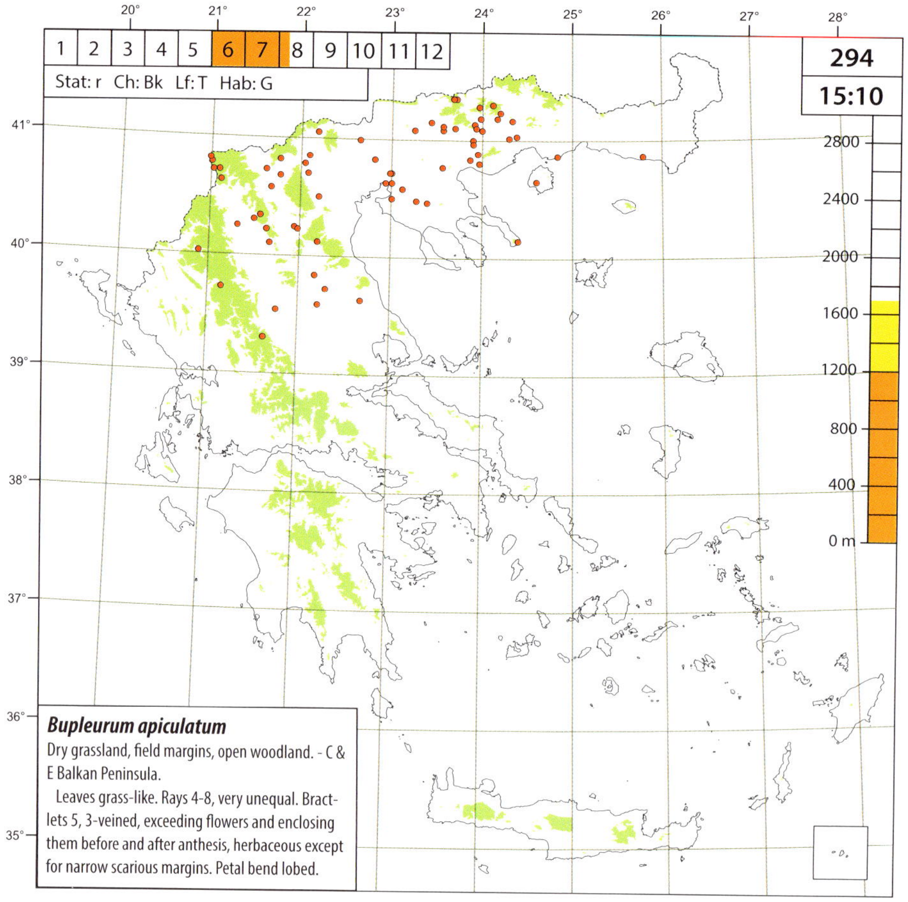

Bupleurum apiculatum

Dry grassland, field margins, open woodland. - C & E Balkan Peninsula.

Leaves grass-like. Rays 4-8, very unequal. Bractlets 5, 3-veined, exceeding flowers and enclosing them before and after anthesis, herbaceous except for narrow scarious margins. Petal bend lobed.

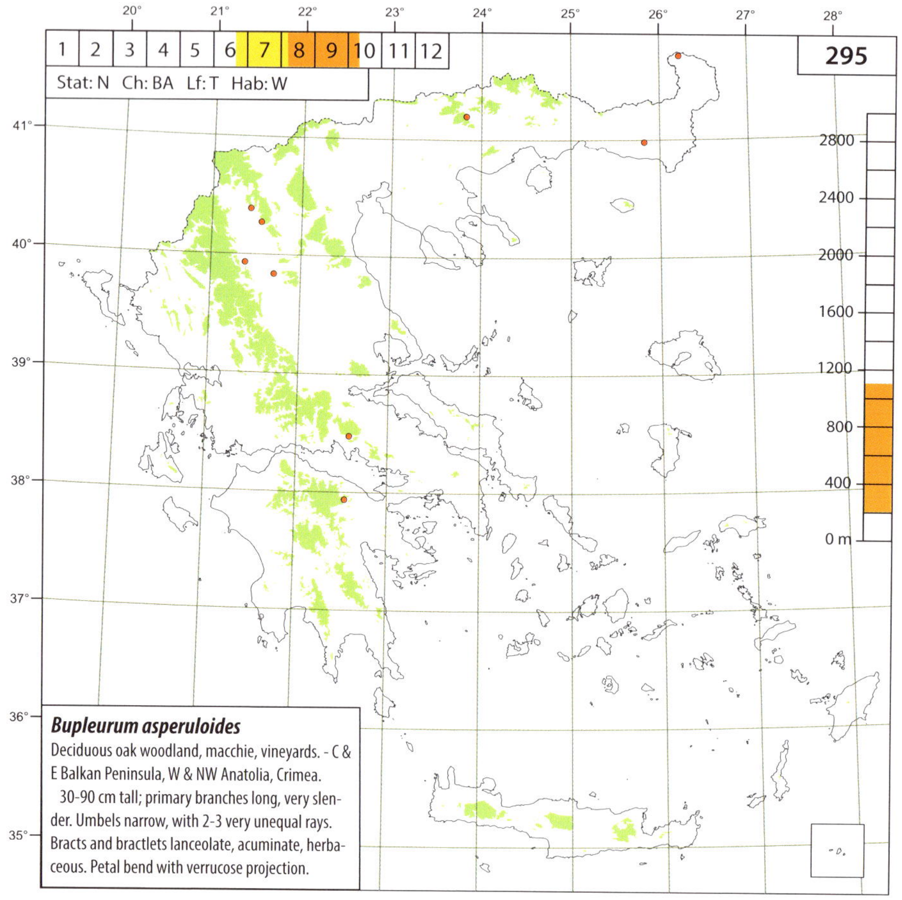

Bupleurum asperuloides

Deciduous oak woodland, macchie, vineyards. - C & E Balkan Peninsula, W & NW Anatolia, Crimea.

30-90 cm tall; primary branches long, very slender. Umbels narrow, with 2-3 very unequal rays. Bracts and bractlets lanceolate, acuminate, herbaceous. Petal bend with verrucose projection.

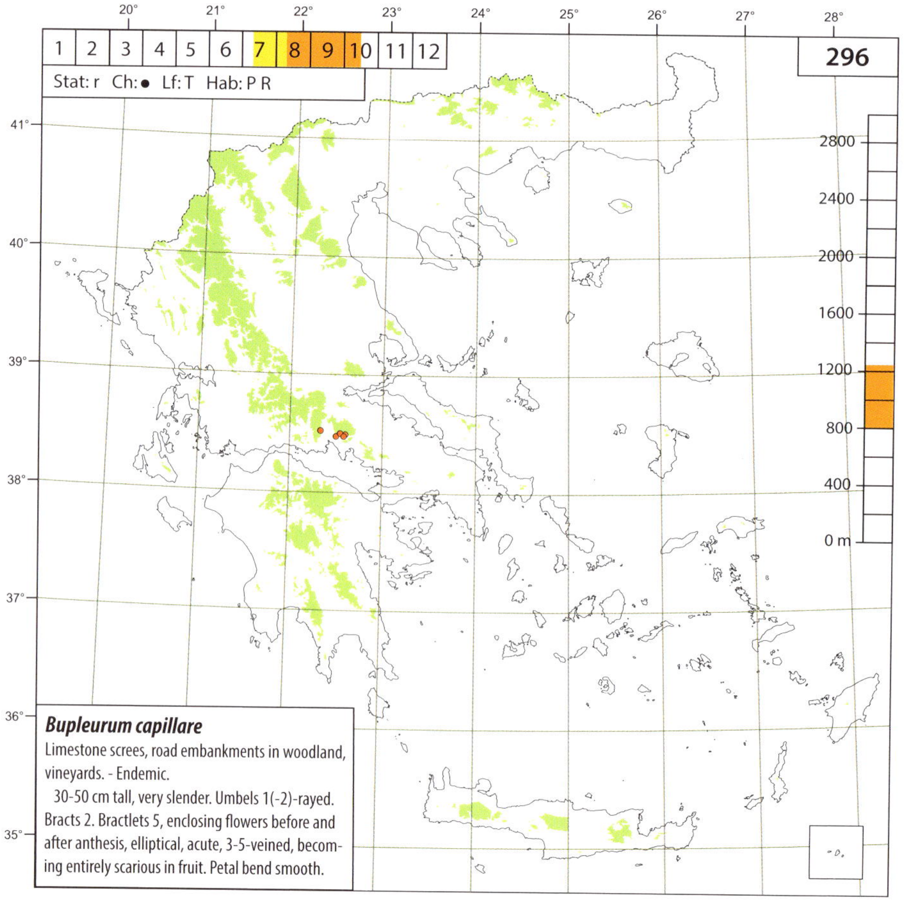

Bupleurum capillare

Limestone screes, road embankments in woodland, vineyards. - Endemic.

30-50 cm tall, very slender. Umbels 1(-2)-rayed. Bracts 2. Bractlets 5, enclosing flowers before and after anthesis, elliptical, acute, 3-5-veined, becoming entirely scarious in fruit. Petal bend smooth.

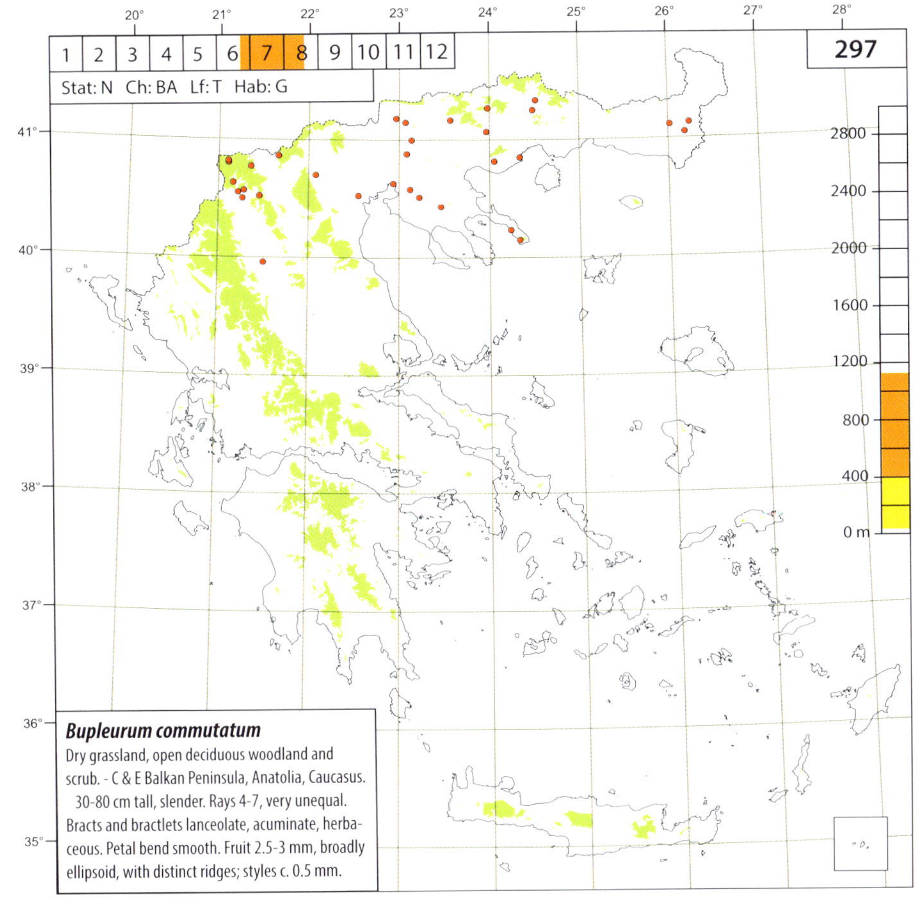

Bupleurum commutatum
Dry grassland, open deciduous woodland and scrub. - C & E Balkan Peninsula, Anatolia, Caucasus.
30-80 cm tall, slender. Rays 4-7, very unequal. Bracts and bractlets lanceolate, acuminate, herbaceous. Petal bend smooth. Fruit 2.5-3 mm, broadly ellipsoid, with distinct ridges; styles c. 0.5 mm.

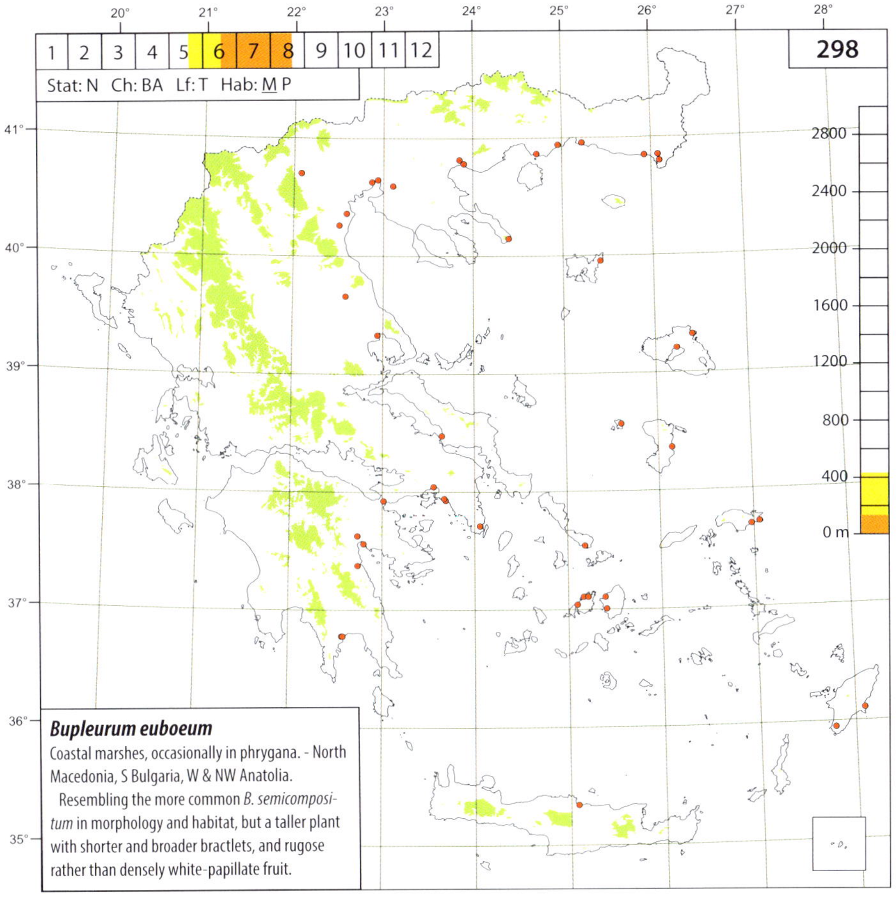

Bupleurum euboeum
Coastal marshes, occasionally in phrygana. - North Macedonia, S Bulgaria, W & NW Anatolia.
Resembling the more common *B. semicompositum* in morphology and habitat, but a taller plant with shorter and broader bractlets, and rugose rather than densely white-papillate fruit.

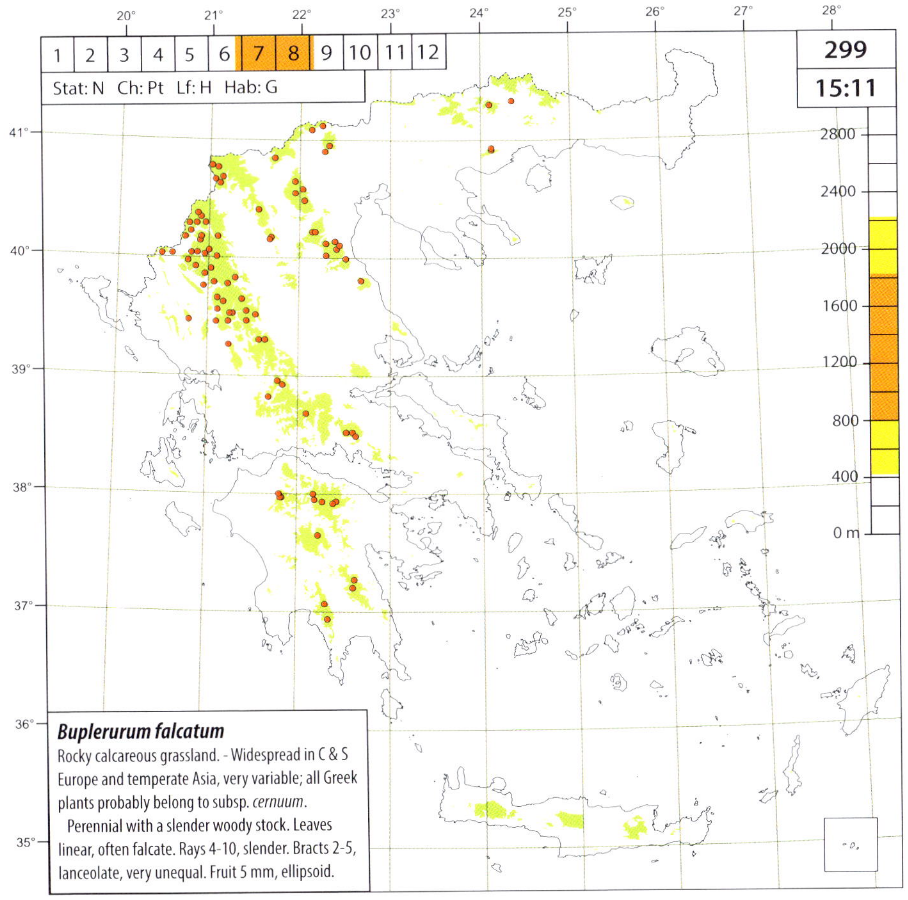

Buplerurum falcatum
Rocky calcareous grassland. - Widespread in C & S Europe and temperate Asia, very variable; all Greek plants probably belong to subsp. *cernuum*.
Perennial with a slender woody stock. Leaves linear, often falcate. Rays 4-10, slender. Bracts 2-5, lanceolate, very unequal. Fruit 5 mm, ellipsoid.

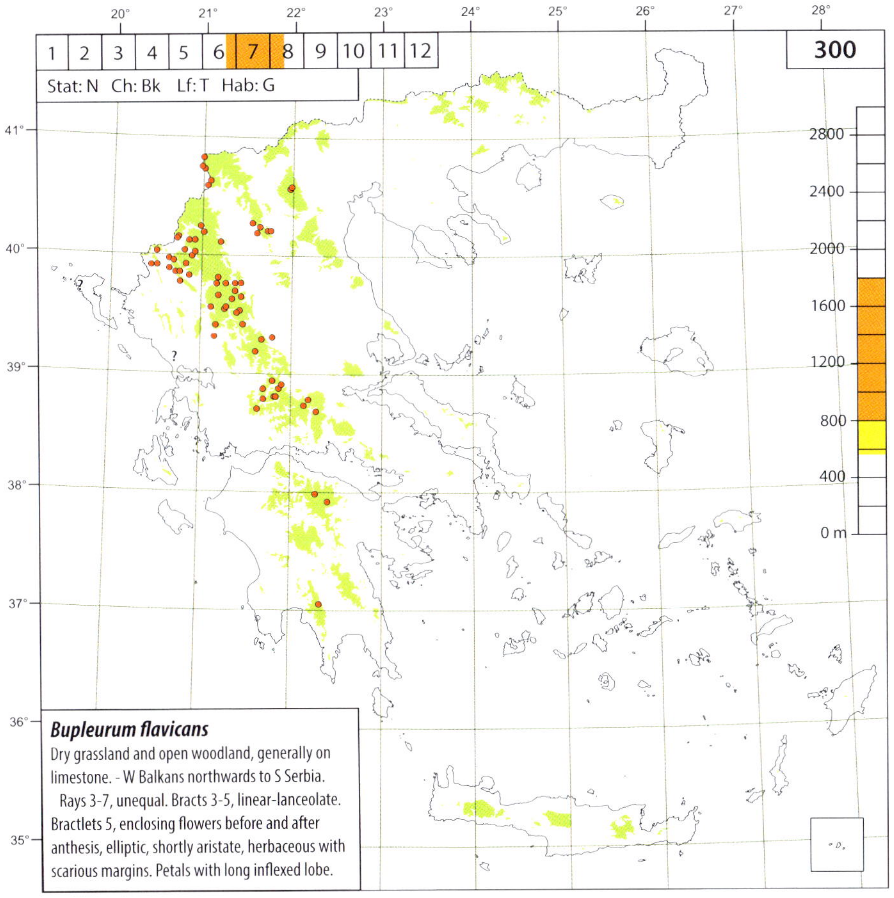

Bupleurum flavicans
Dry grassland and open woodland, generally on limestone. - W Balkans northwards to S Serbia.
Rays 3-7, unequal. Bracts 3-5, linear-lanceolate. Bractlets 5, enclosing flowers before and after anthesis, elliptic, shortly aristate, herbaceous with scarious margins. Petals with long inflexed lobe.

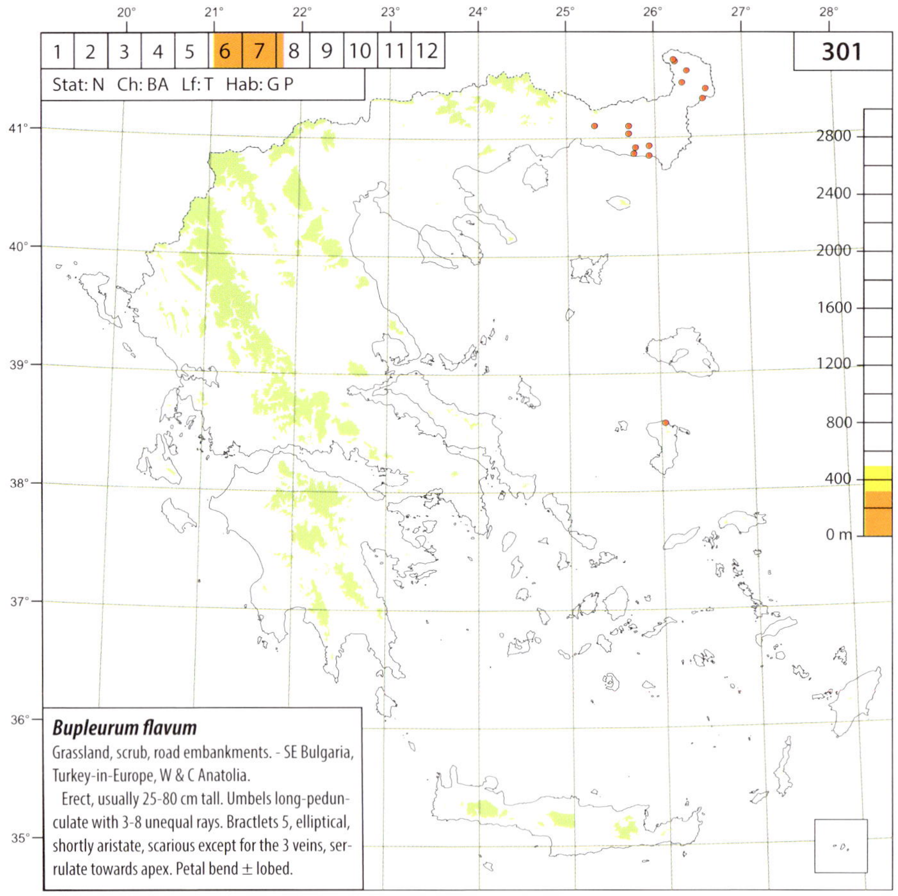

Bupleurum flavum

Grassland, scrub, road embankments. - SE Bulgaria, Turkey-in-Europe, W & C Anatolia.

Erect, usually 25-80 cm tall. Umbels long-pedunculate with 3-8 unequal rays. Bractlets 5, elliptical, shortly aristate, scarious except for the 3 veins, serrulate towards apex. Petal bend ± lobed.

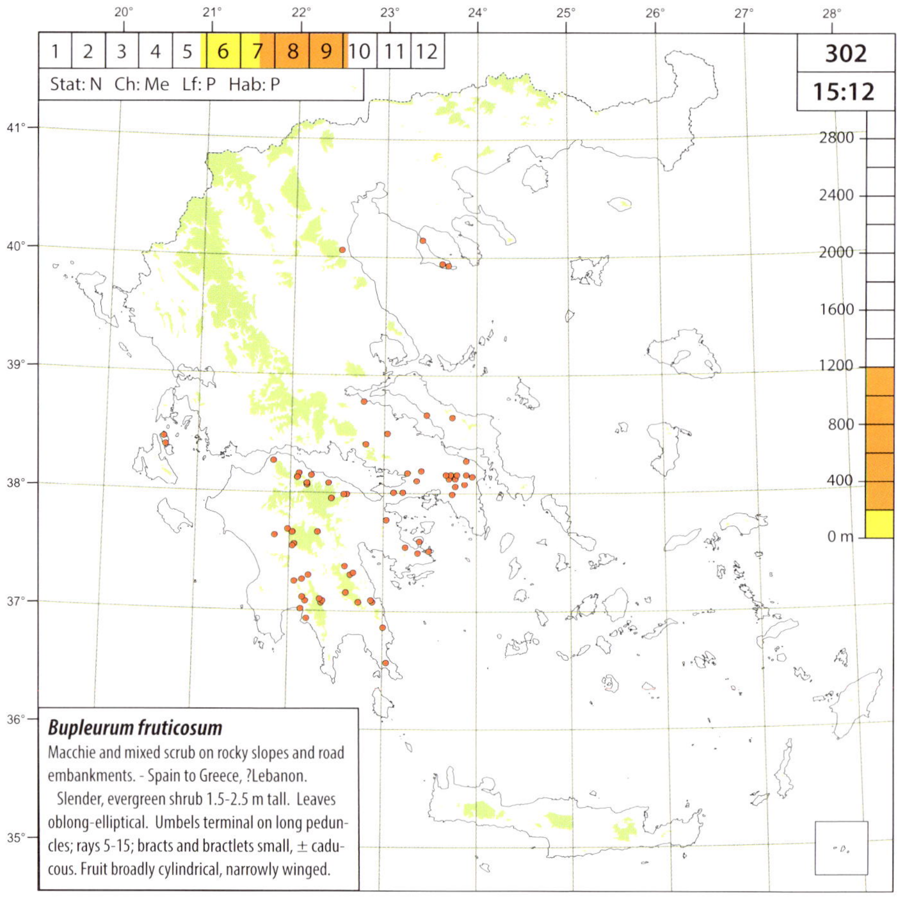

Bupleurum fruticosum

Macchie and mixed scrub on rocky slopes and road embankments. - Spain to Greece, ?Lebanon.

Slender, evergreen shrub 1.5-2.5 m tall. Leaves oblong-elliptical. Umbels terminal on long peduncles; rays 5-15; bracts and bractlets small, ± caducous. Fruit broadly cylindrical, narrowly winged.

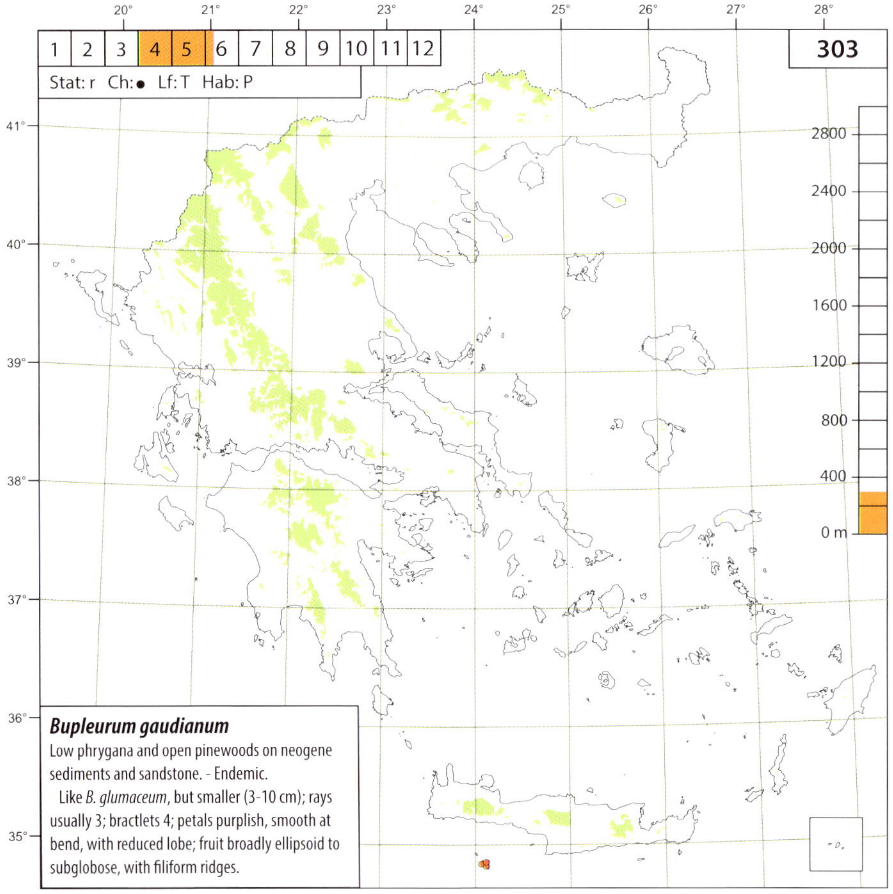

Bupleurum gaudianum

Low phrygana and open pinewoods on neogene sediments and sandstone. - Endemic.

Like *B. glumaceum*, but smaller (3-10 cm); rays usually 3; bractlets 4; petals purplish, smooth at bend, with reduced lobe; fruit broadly ellipsoid to subglobose, with filiform ridges.

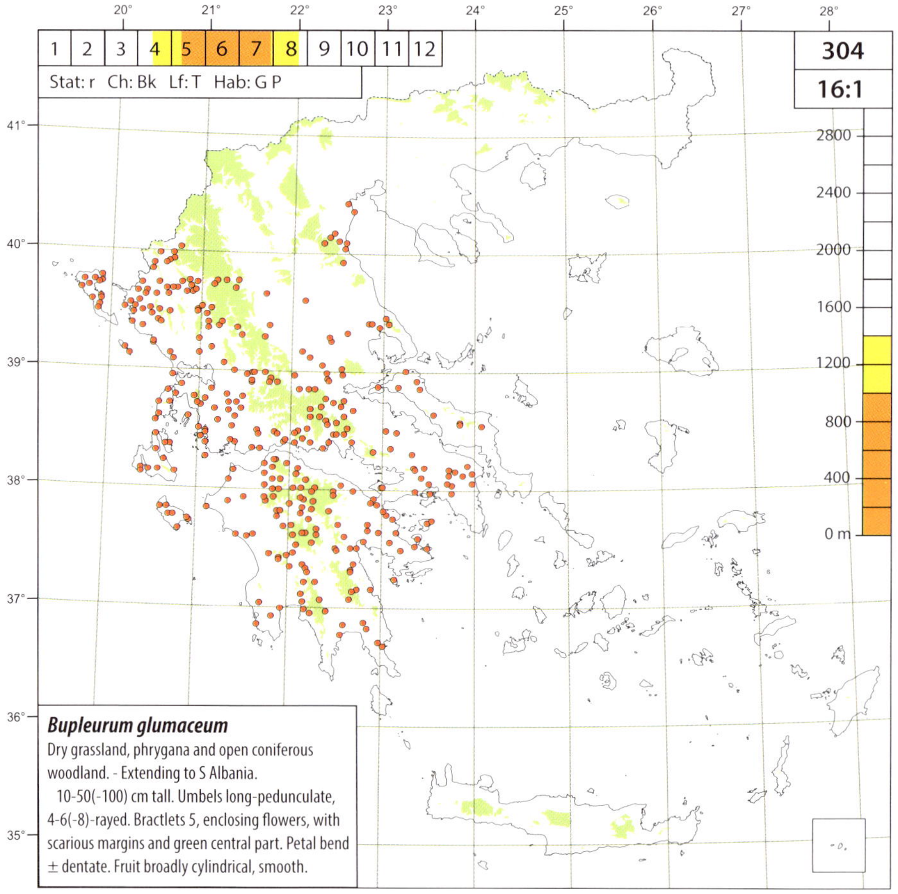

Bupleurum glumaceum

Dry grassland, phrygana and open coniferous woodland. - Extending to S Albania.

10-50(-100) cm tall. Umbels long-pedunculate, 4-6(-8)-rayed. Bractlets 5, enclosing flowers, with scarious margins and green central part. Petal bend ± dentate. Fruit broadly cylindrical, smooth.

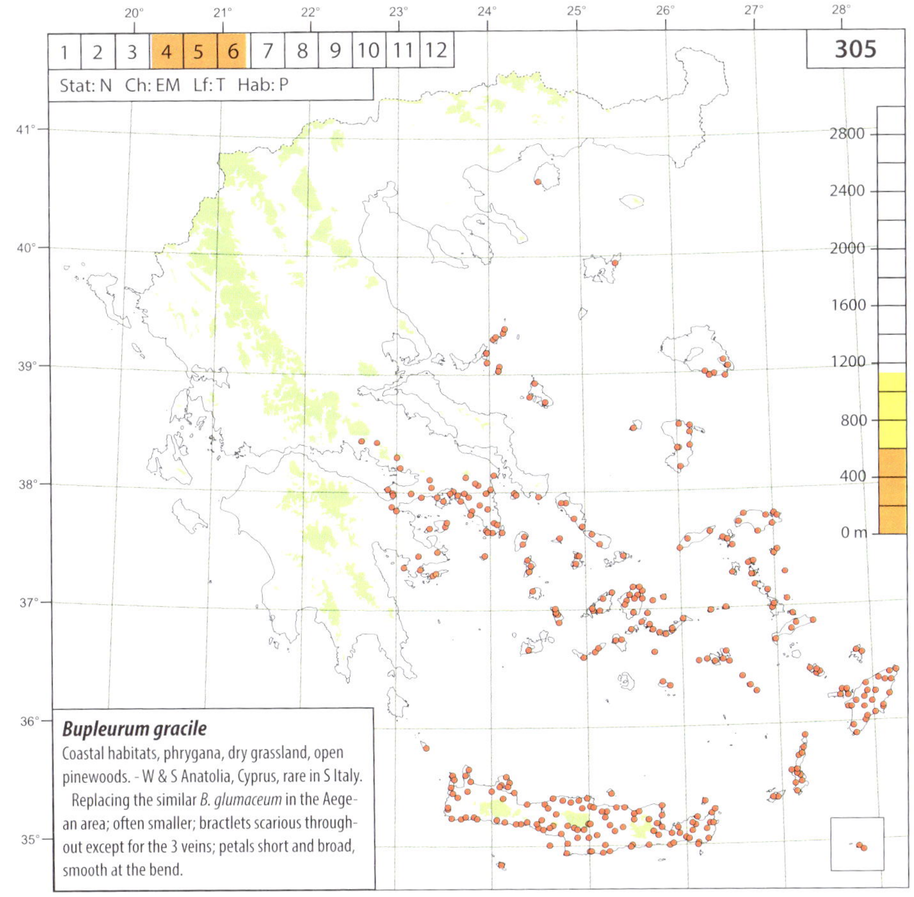

Bupleurum gracile

Coastal habitats, phrygana, dry grassland, open pinewoods. - W & S Anatolia, Cyprus, rare in S Italy.

Replacing the similar *B. glumaceum* in the Aegean area; often smaller; bractlets scarious throughout except for the 3 veins; petals short and broad, smooth at the bend.

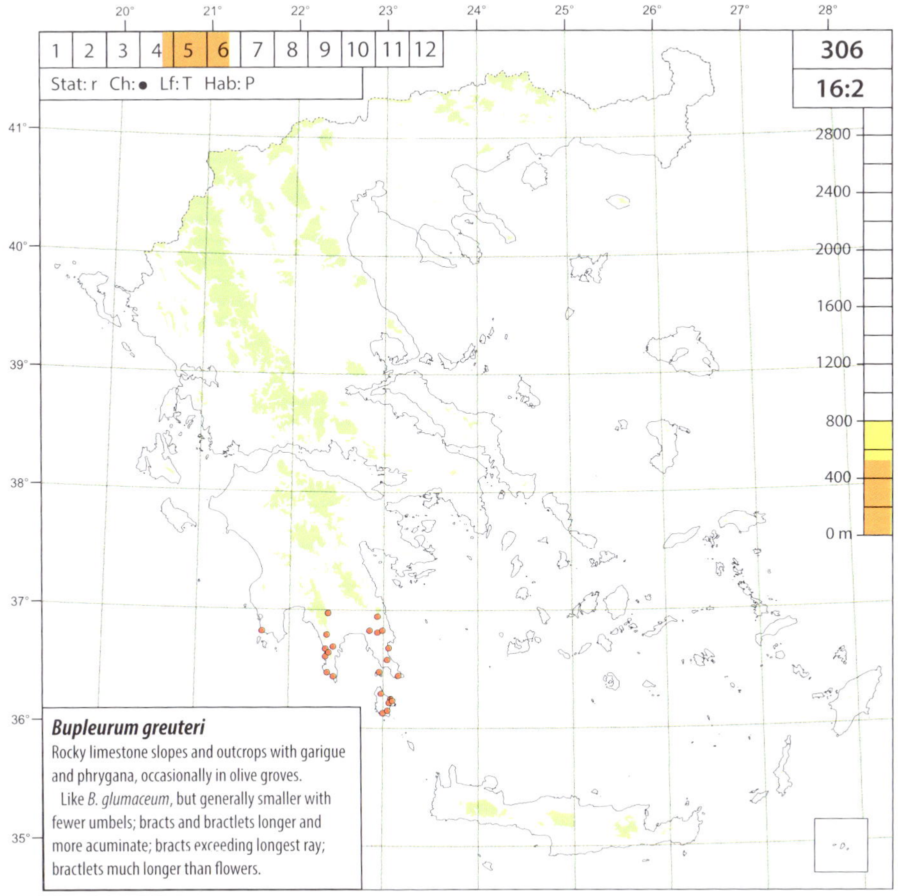

Bupleurum greuteri

Rocky limestone slopes and outcrops with garigue and phrygana, occasionally in olive groves.

Like *B. glumaceum*, but generally smaller with fewer umbels; bracts and bractlets longer and more acuminate; bracts exceeding longest ray; bractlets much longer than flowers.

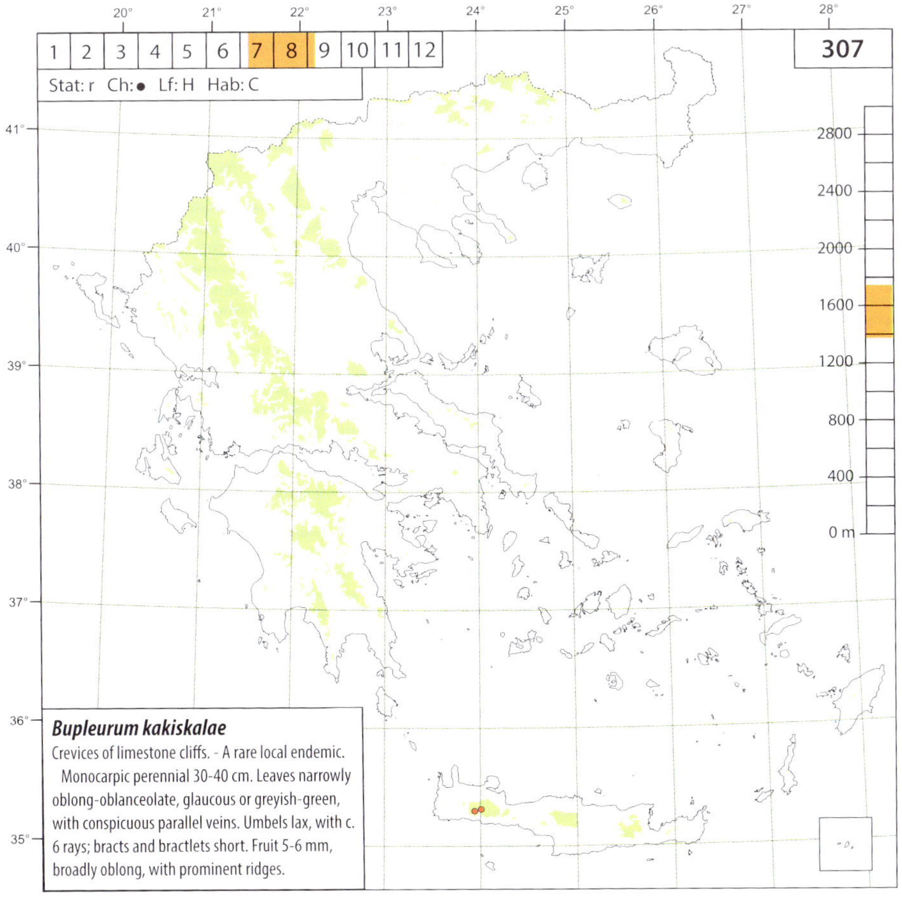

Bupleurum kakiskalae

Crevices of limestone cliffs. - A rare local endemic.

Monocarpic perennial 30-40 cm. Leaves narrowly oblong-oblanceolate, glaucous or greyish-green, with conspicuous parallel veins. Umbels lax, with c. 6 rays; bracts and bractlets short. Fruit 5-6 mm, broadly oblong, with prominent ridges.

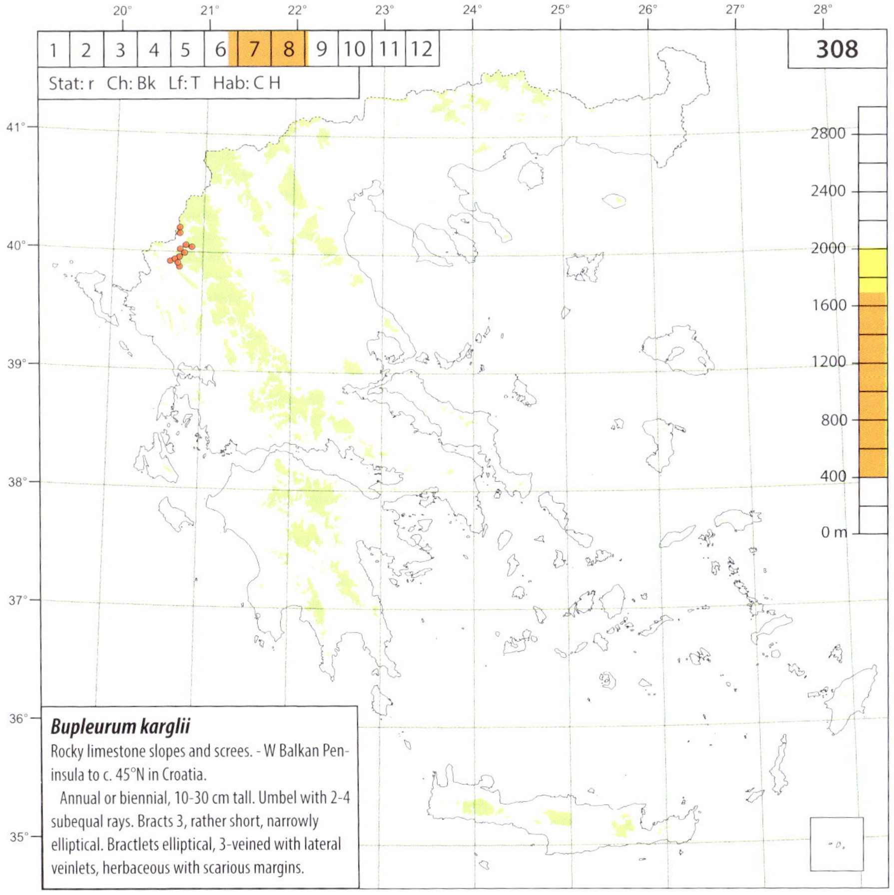

Bupleurum karglii

Rocky limestone slopes and screes. - W Balkan Peninsula to c. 45°N in Croatia.

Annual or biennial, 10-30 cm tall. Umbel with 2-4 subequal rays. Bracts 3, rather short, narrowly elliptical. Bractlets elliptical, 3-veined with lateral veinlets, herbaceous with scarious margins.

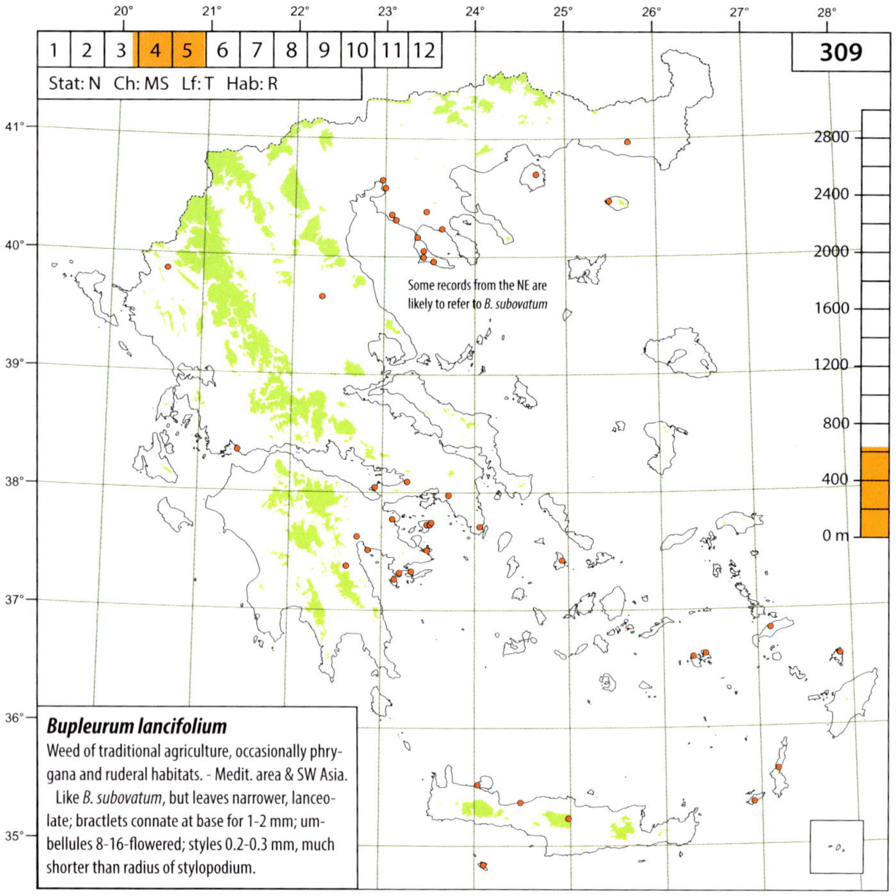

Bupleurum lancifolium
Weed of traditional agriculture, occasionally phrygana and ruderal habitats. - Medit. area & SW Asia.
Like *B. subovatum*, but leaves narrower, lanceolate; bractlets connate at base for 1-2 mm; umbellules 8-16-flowered; styles 0.2-0.3 mm, much shorter than radius of stylopodium.

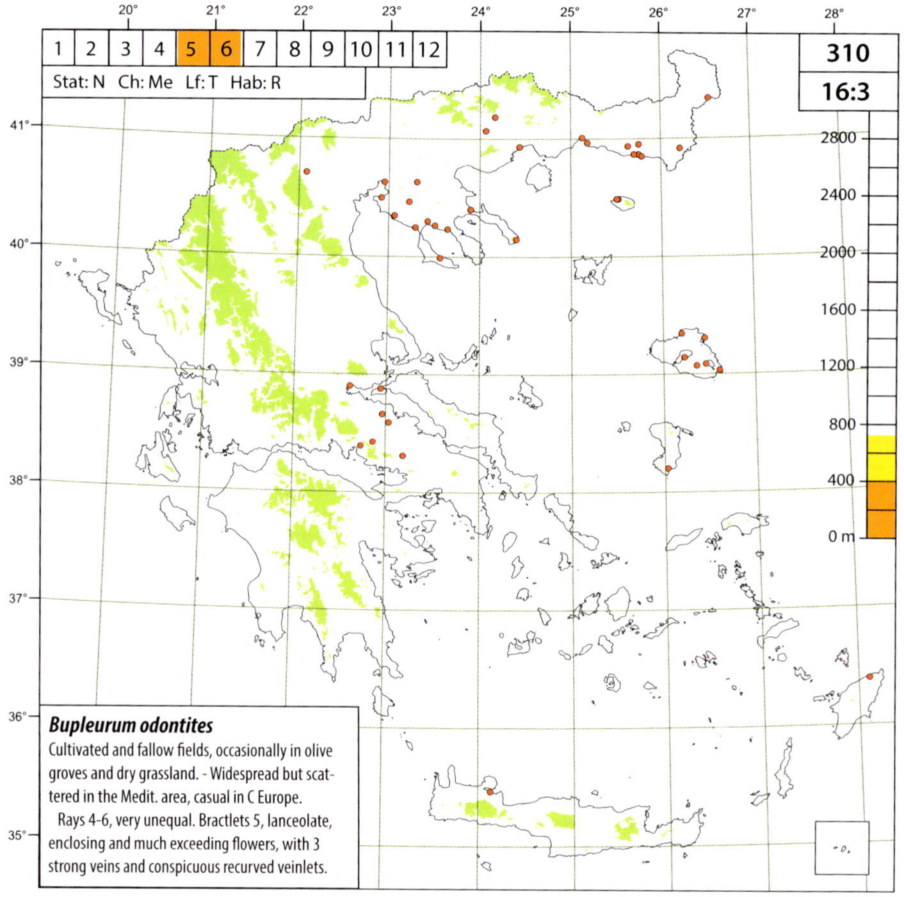

Bupleurum odontites
Cultivated and fallow fields, occasionally in olive groves and dry grassland. - Widespread but scattered in the Medit. area, casual in C Europe.
Rays 4-6, very unequal. Bractlets 5, lanceolate, enclosing and much exceeding flowers, with 3 strong veins and conspicuous recurved veinlets.

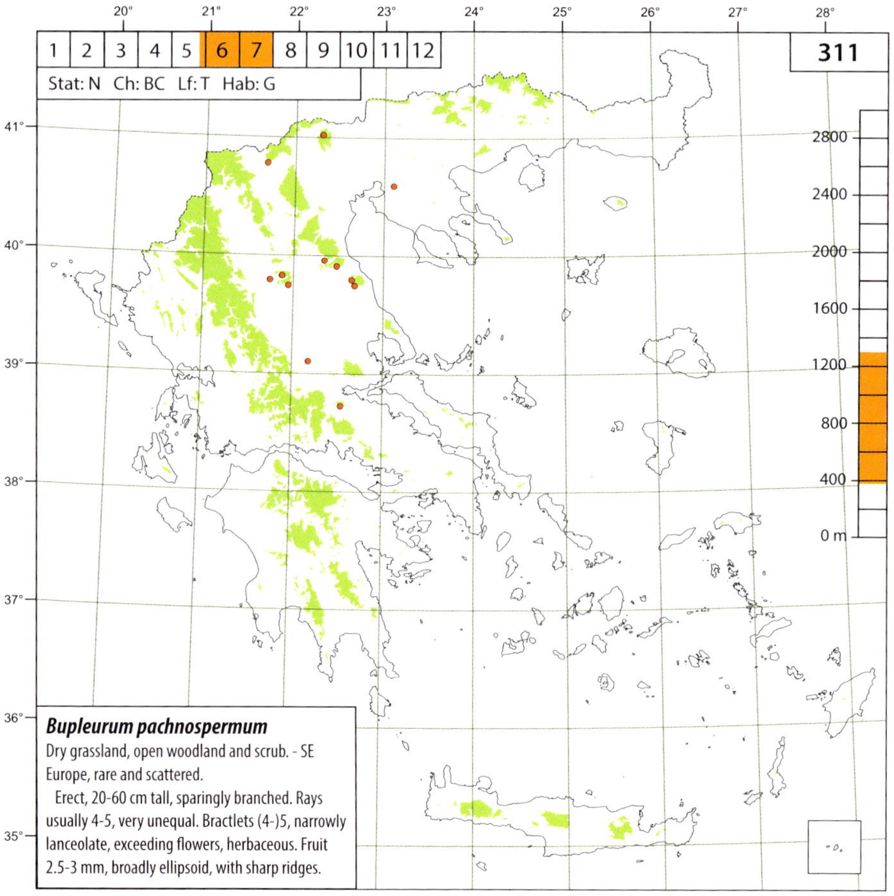

Bupleurum pachnospermum
Dry grassland, open woodland and scrub. - SE Europe, rare and scattered.
Erect, 20-60 cm tall, sparingly branched. Rays usually 4-5, very unequal. Bractlets (4-)5, narrowly lanceolate, exceeding flowers, herbaceous. Fruit 2.5-3 mm, broadly ellipsoid, with sharp ridges.

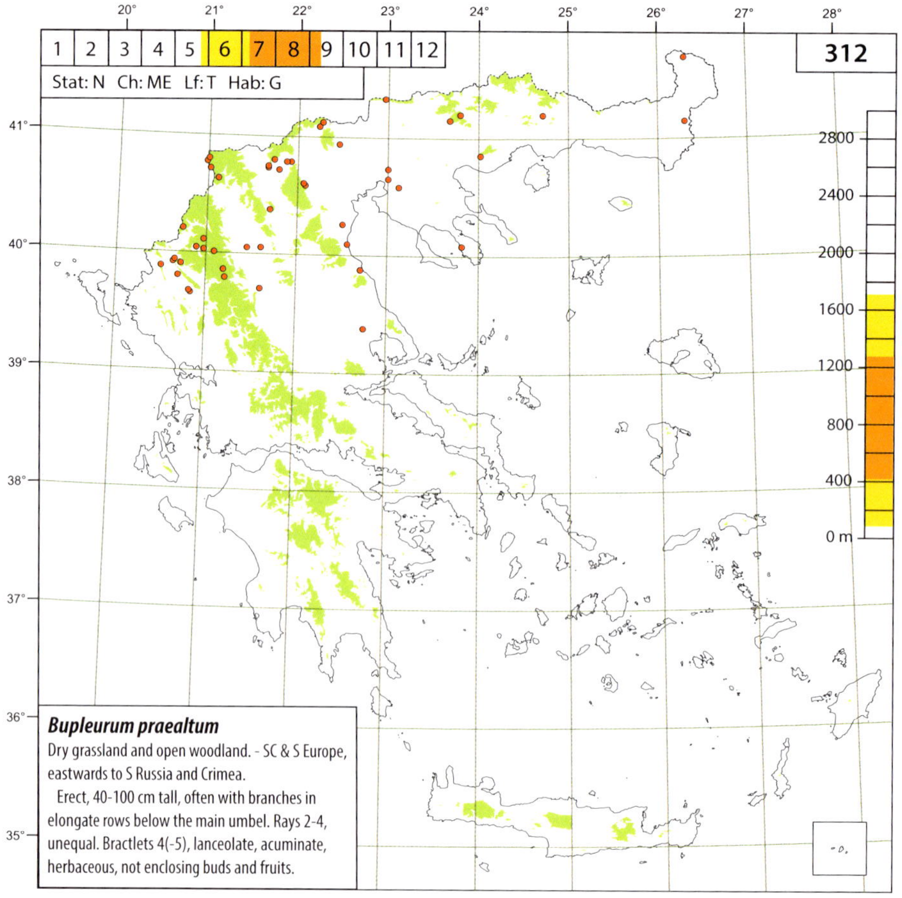

Bupleurum praealtum
Dry grassland and open woodland. - SC & S Europe, eastwards to S Russia and Crimea.
Erect, 40-100 cm tall, often with branches in elongate rows below the main umbel. Rays 2-4, unequal. Bractlets 4(-5), lanceolate, acuminate, herbaceous, not enclosing buds and fruits.

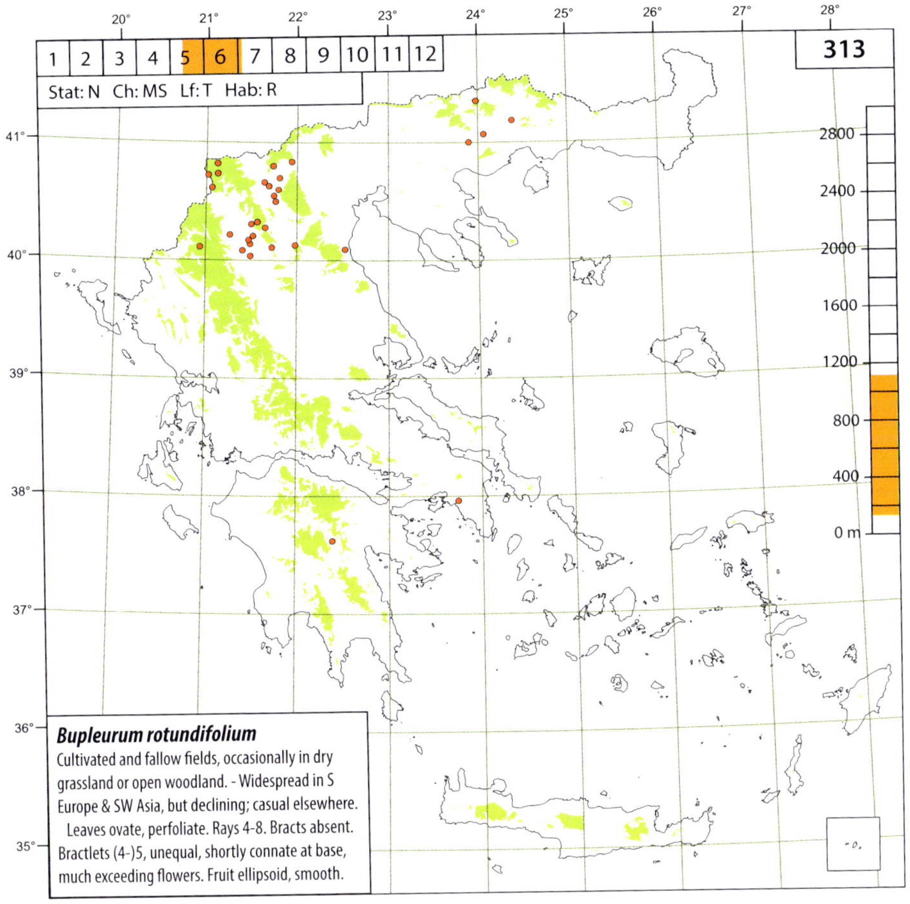

Bupleurum rotundifolium
Cultivated and fallow fields, occasionally in dry grassland or open woodland. - Widespread in S Europe & SW Asia, but declining; casual elsewhere.
Leaves ovate, perfoliate. Rays 4-8. Bracts absent. Bractlets (4-)5, unequal, shortly connate at base, much exceeding flowers. Fruit ellipsoid, smooth.

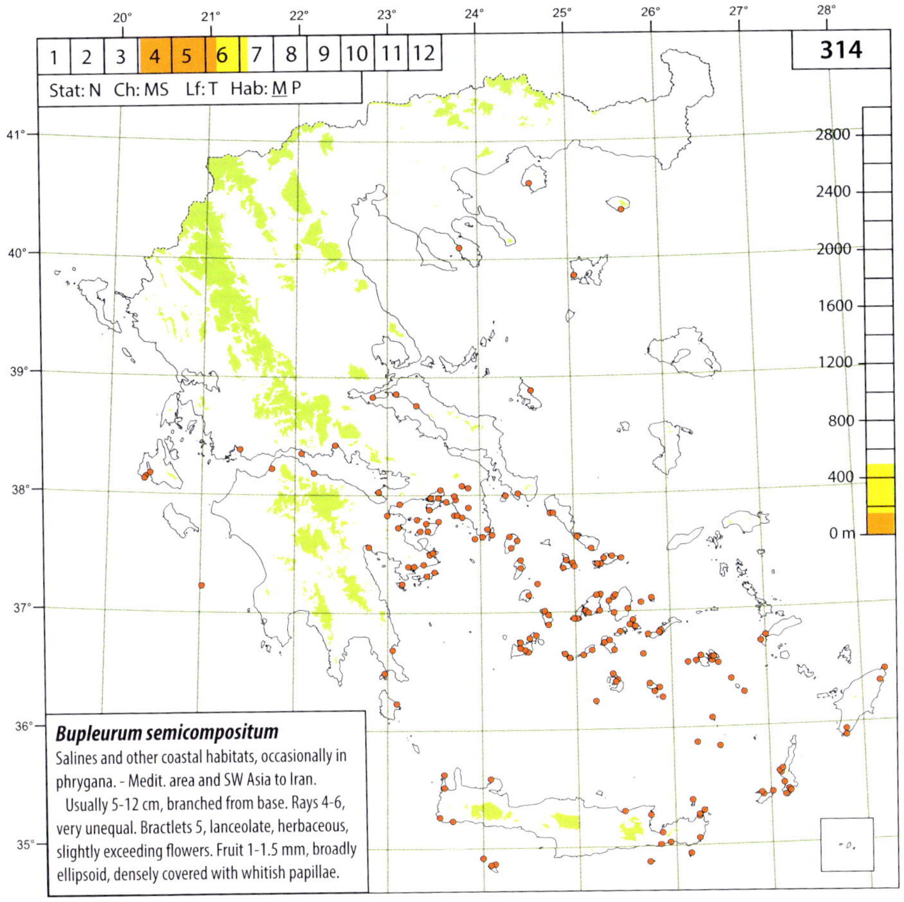

Bupleurum semicompositum
Salines and other coastal habitats, occasionally in phrygana. - Medit. area and SW Asia to Iran.
Usually 5-12 cm, branched from base. Rays 4-6, very unequal. Bractlets 5, lanceolate, herbaceous, slightly exceeding flowers. Fruit 1-1.5 mm, broadly ellipsoid, densely covered with whitish papillae.

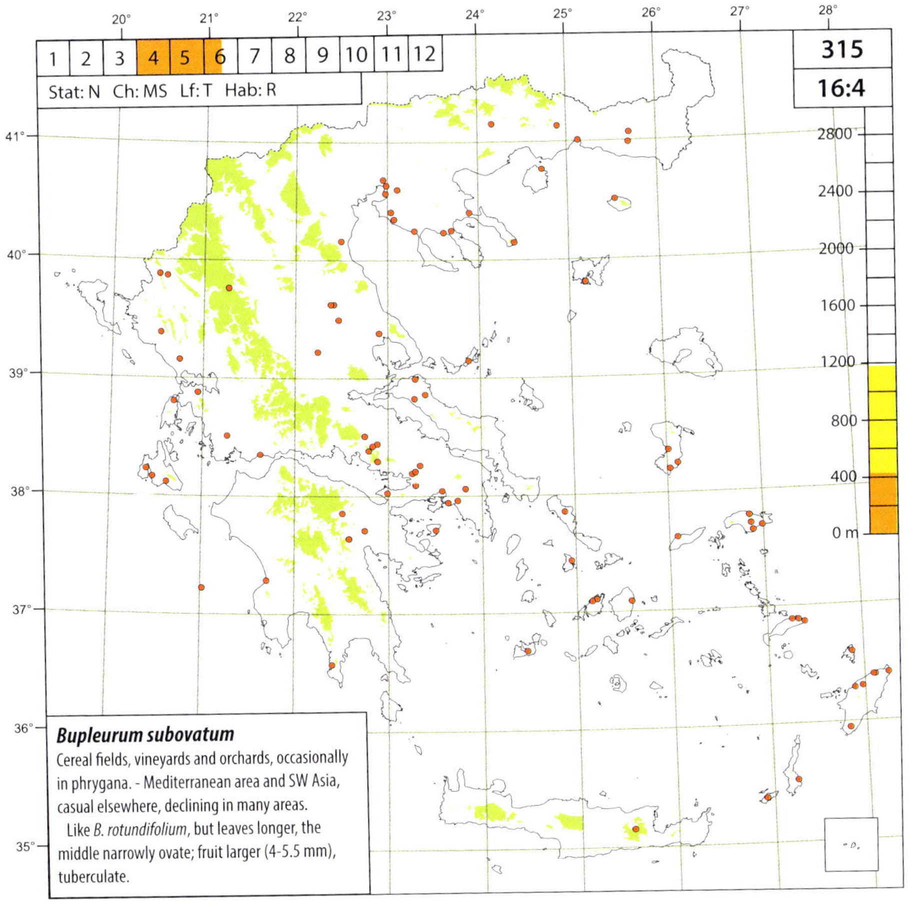

Bupleurum subovatum
Cereal fields, vineyards and orchards, occasionally in phrygana. - Mediterranean area and SW Asia, casual elsewhere, declining in many areas.
Like *B. rotundifolium*, but leaves longer, the middle narrowly ovate; fruit larger (4-5.5 mm), tuberculate.

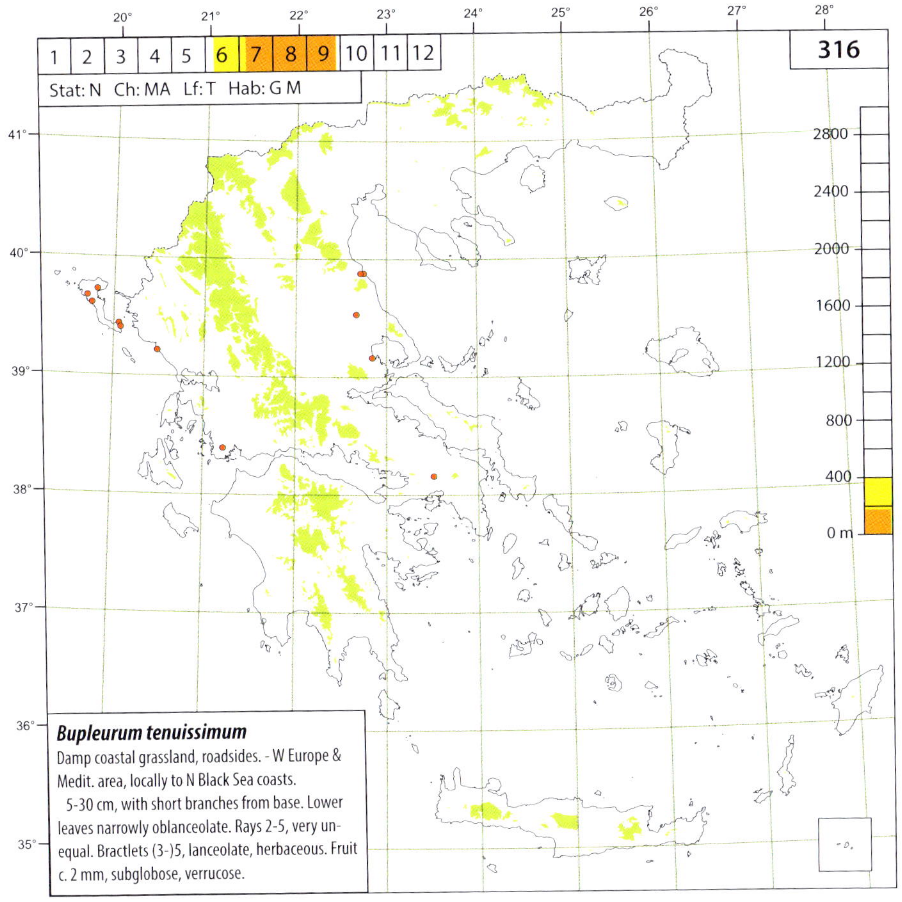

Bupleurum tenuissimum
Damp coastal grassland, roadsides. - W Europe & Medit. area, locally to N Black Sea coasts.
5-30 cm, with short branches from base. Lower leaves narrowly oblanceolate. Rays 2-5, very unequal. Bractlets (3-)5, lanceolate, herbaceous. Fruit c. 2 mm, subglobose, verrucose.

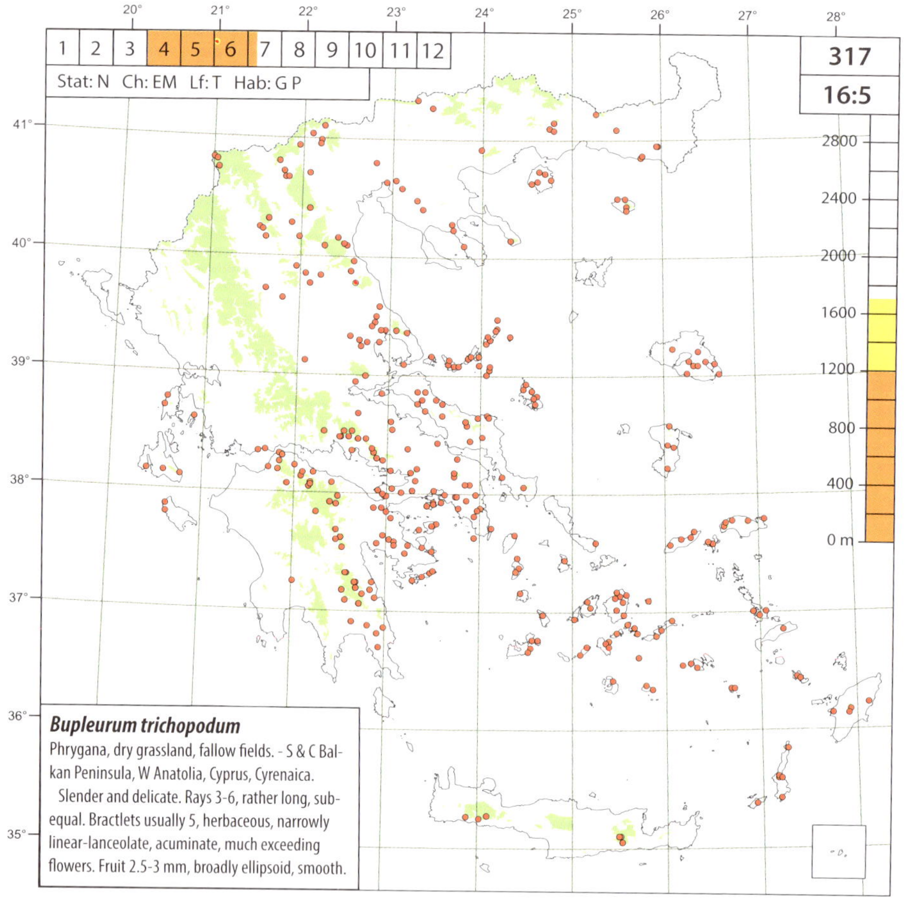

Bupleurum trichopodum
Phrygana, dry grassland, fallow fields. - S & C Balkan Peninsula, W Anatolia, Cyprus, Cyrenaica.
Slender and delicate. Rays 3-6, rather long, subequal. Bractlets usually 5, herbaceous, narrowly linear-lanceolate, acuminate, much exceeding flowers. Fruit 2.5-3 mm, broadly ellipsoid, smooth.

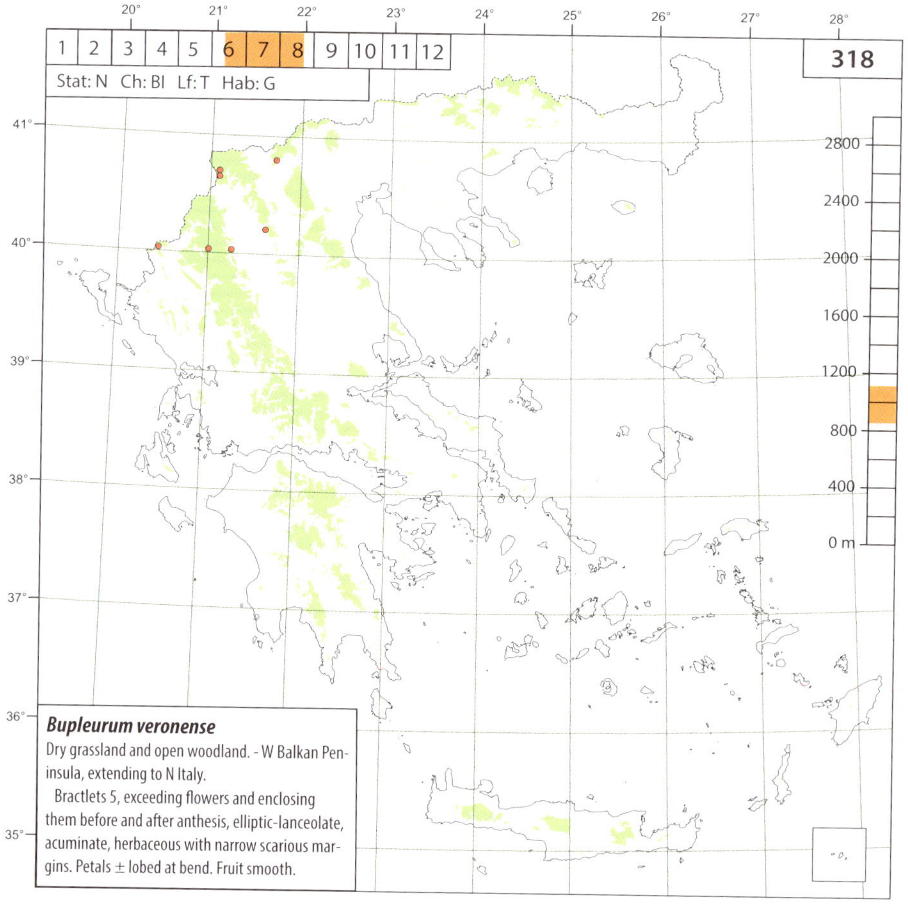

Bupleurum veronense
Dry grassland and open woodland. - W Balkan Peninsula, extending to N Italy.
Bractlets 5, exceeding flowers and enclosing them before and after anthesis, elliptic-lanceolate, acuminate, herbaceous with narrow scarious margins. Petals ± lobed at bend. Fruit smooth.

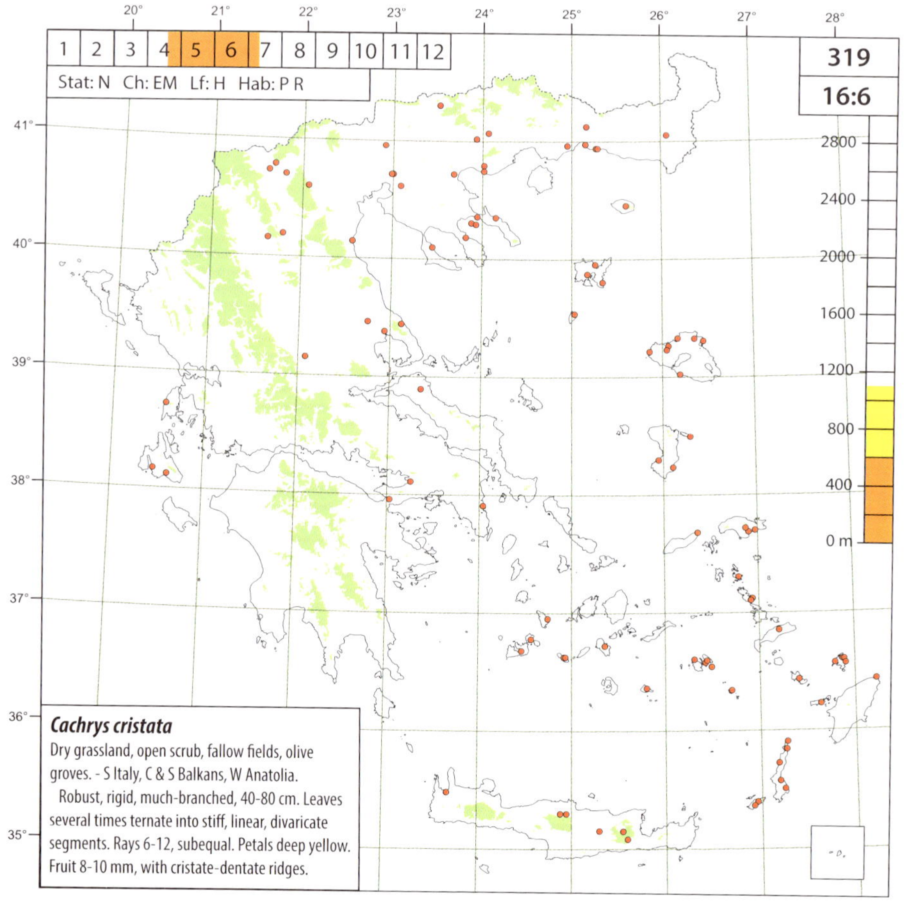

Cachrys cristata
Dry grassland, open scrub, fallow fields, olive groves. - S Italy, C & S Balkans, W Anatolia.
Robust, rigid, much-branched, 40-80 cm. Leaves several times ternate into stiff, linear, divaricate segments. Rays 6-12, subequal. Petals deep yellow. Fruit 8-10 mm, with cristate-dentate ridges.

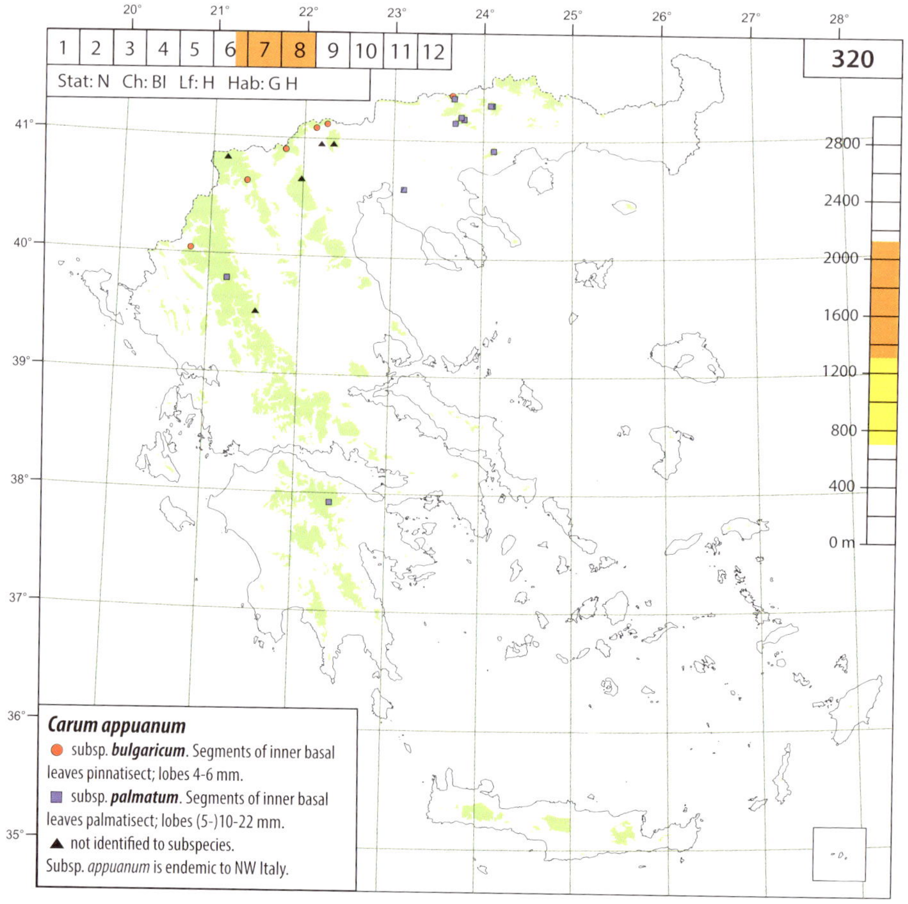

Carum appuanum
● subsp. ***bulgaricum***. Segments of inner basal leaves pinnatisect; lobes 4-6 mm.
■ subsp. ***palmatum***. Segments of inner basal leaves palmatisect; lobes (5-)10-22 mm.
▲ not identified to subspecies.
Subsp. *appuanum* is endemic to NW Italy.

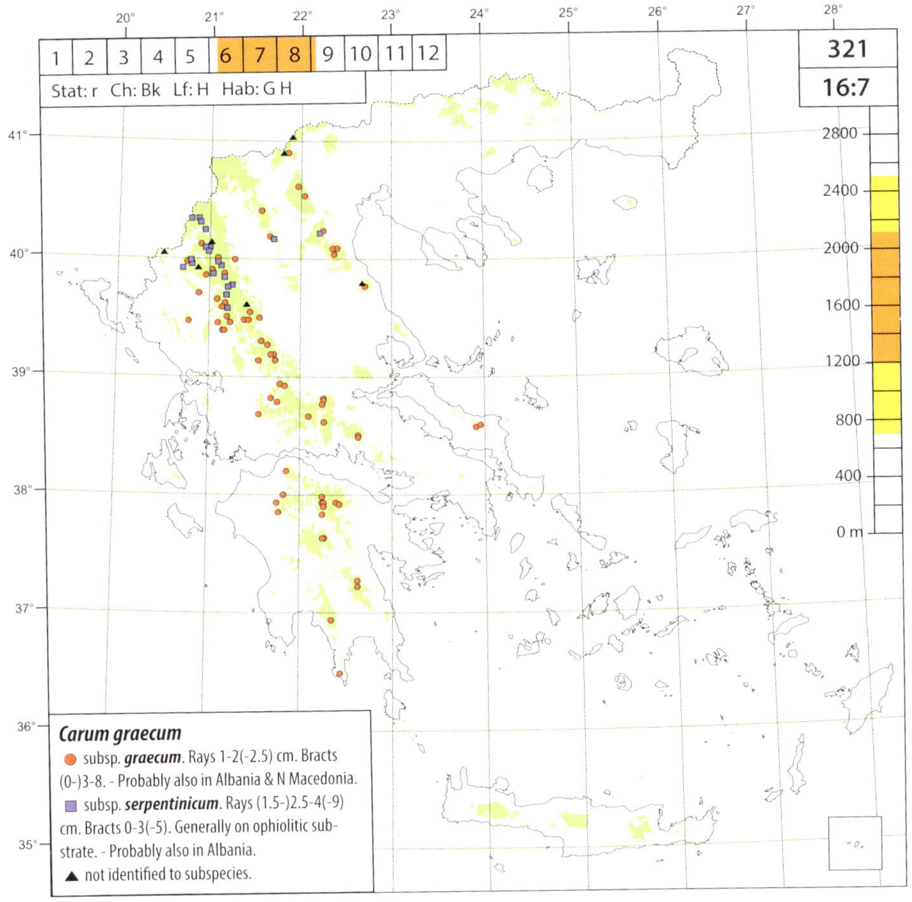

Carum graecum
● subsp. **graecum**. Rays 1-2(-2.5) cm. Bracts (0-)3-8. - Probably also in Albania & N Macedonia.
■ subsp. **serpentinicum**. Rays (1.5-)2.5-4(-9) cm. Bracts 0-3(-5). Generally on ophiolitic substrate. - Probably also in Albania.
▲ not identified to subspecies.

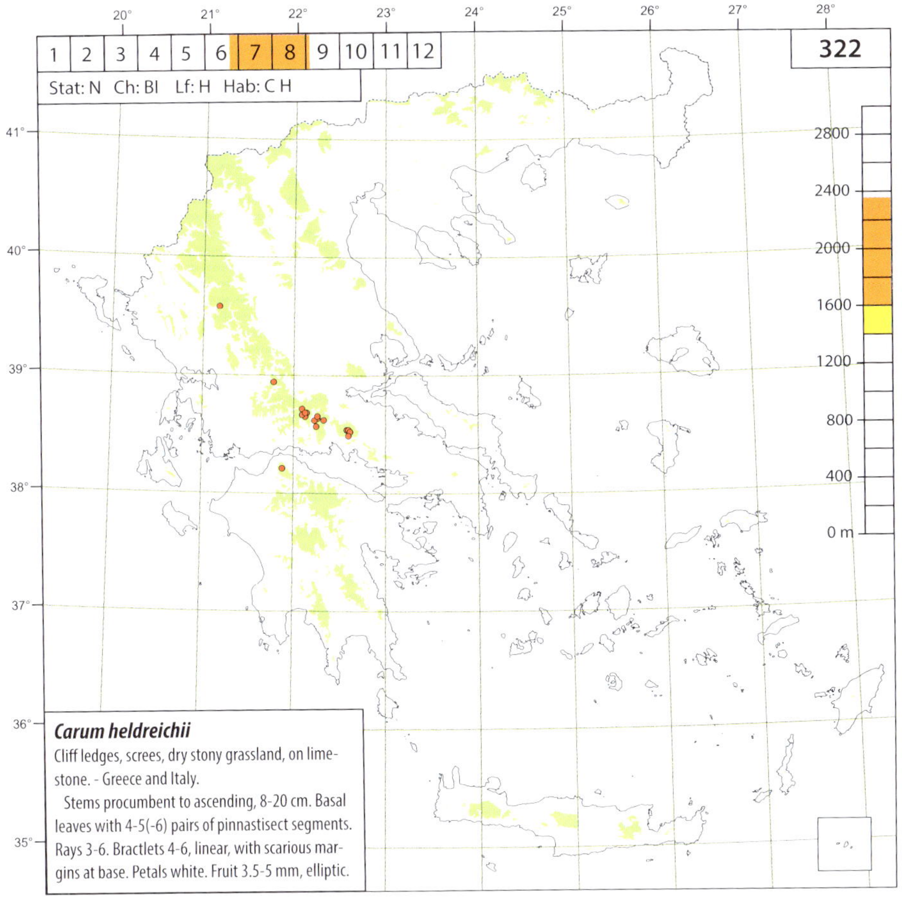

Carum heldreichii
Cliff ledges, screes, dry stony grassland, on limestone. - Greece and Italy.
Stems procumbent to ascending, 8-20 cm. Basal leaves with 4-5(-6) pairs of pinnatisect segments. Rays 3-6. Bractlets 4-6, linear, with scarious margins at base. Petals white. Fruit 3.5-5 mm, elliptic.

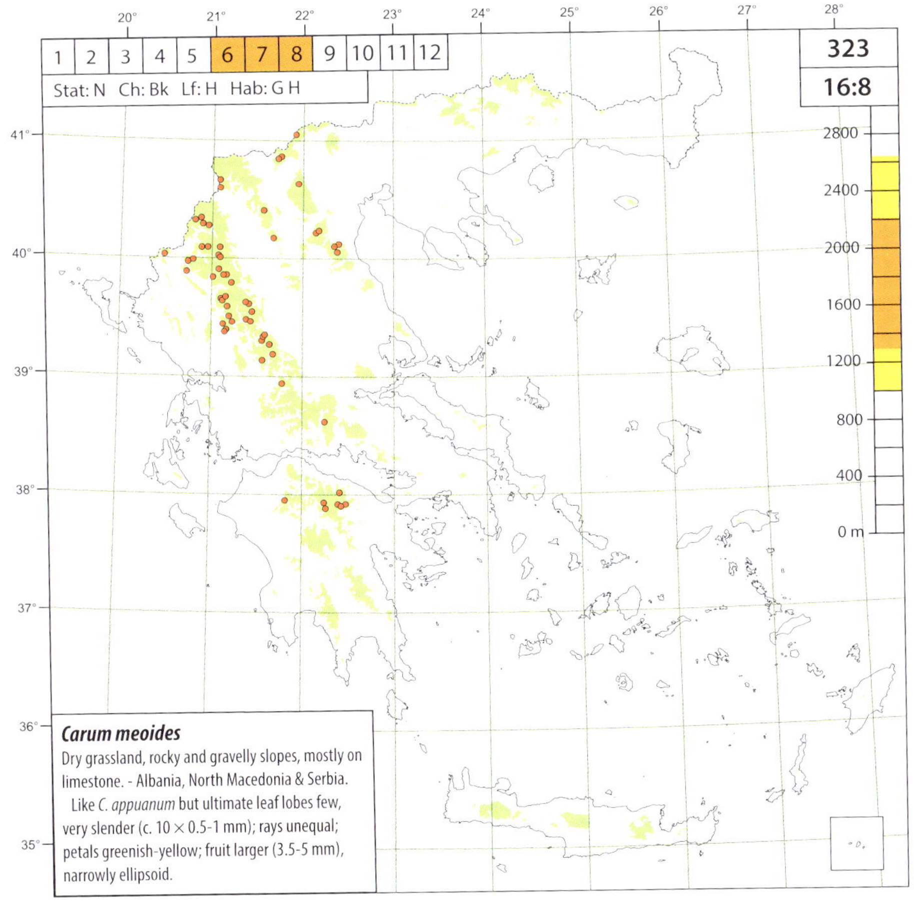

Carum meoides
Dry grassland, rocky and gravelly slopes, mostly on limestone. - Albania, North Macedonia & Serbia.
Like *C. appuanum* but ultimate leaf lobes few, very slender (c. 10 × 0.5-1 mm); rays unequal; petals greenish-yellow; fruit larger (3.5-5 mm), narrowly ellipsoid.

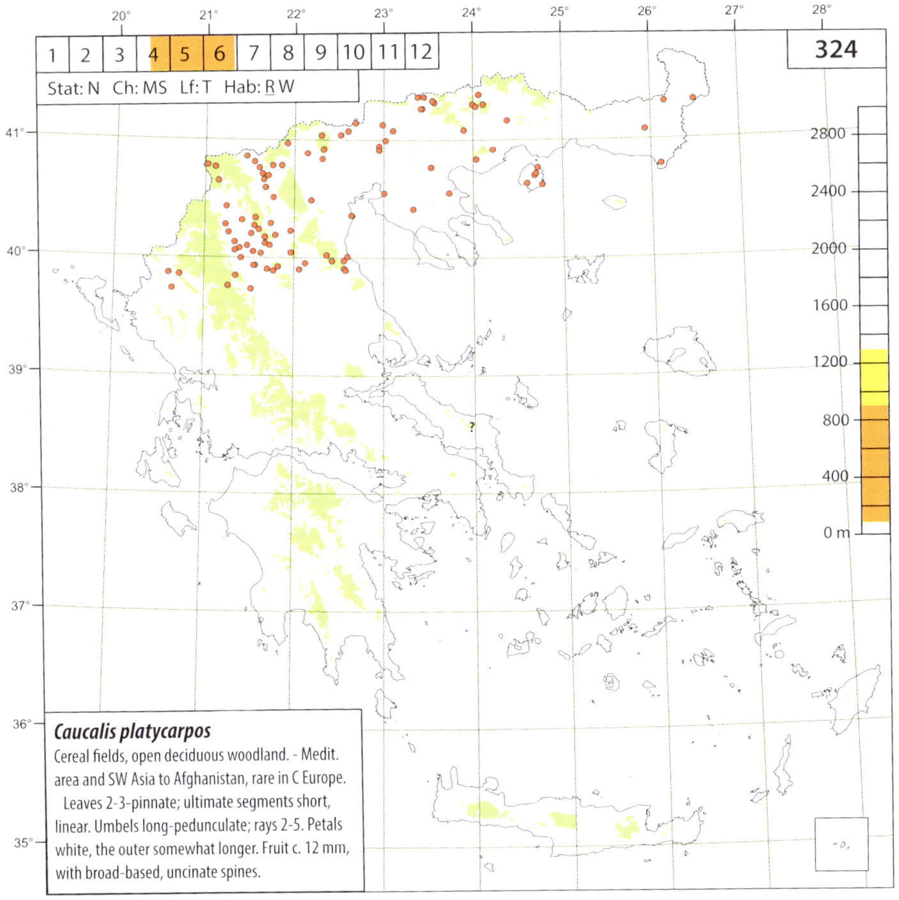

Caucalis platycarpos
Cereal fields, open deciduous woodland. - Medit. area and SW Asia to Afghanistan, rare in C Europe.
Leaves 2-3-pinnate; ultimate segments short, linear. Umbels long-pedunculate; rays 2-5. Petals white, the outer somewhat longer. Fruit c. 12 mm, with broad-based, uncinate spines.

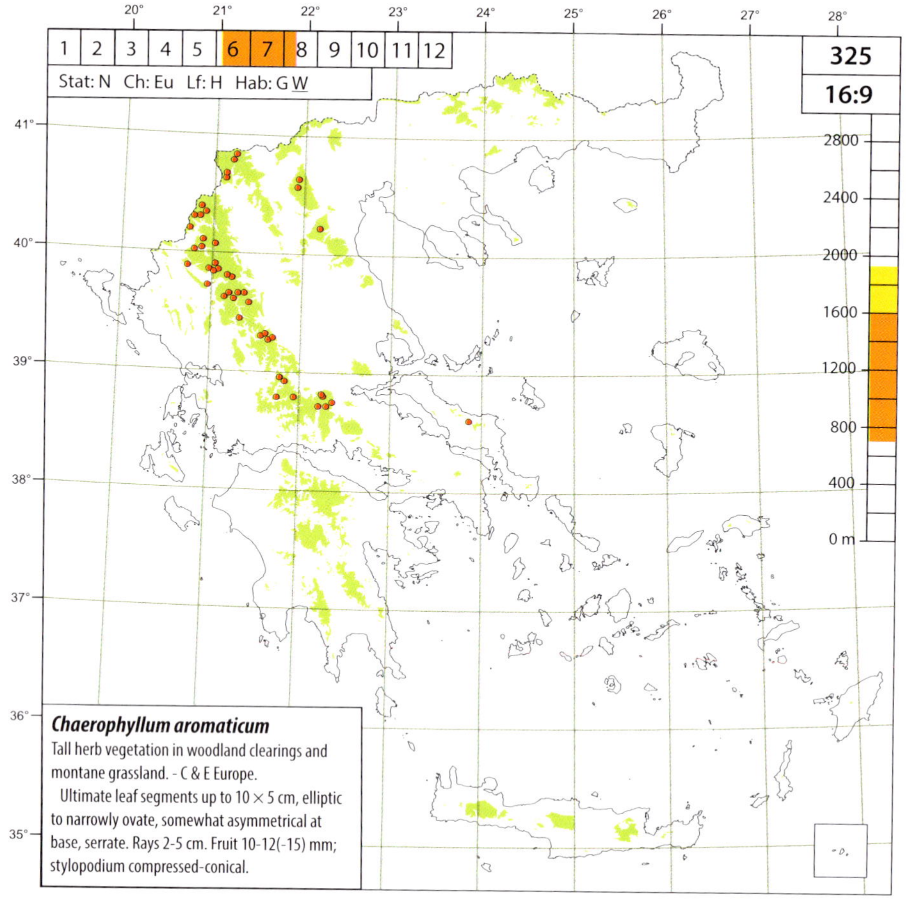

Chaerophyllum aromaticum
Tall herb vegetation in woodland clearings and montane grassland. - C & E Europe.
Ultimate leaf segments up to 10 × 5 cm, elliptic to narrowly ovate, somewhat asymmetrical at base, serrate. Rays 2-5 cm. Fruit 10-12(-15) mm; stylopodium compressed-conical.

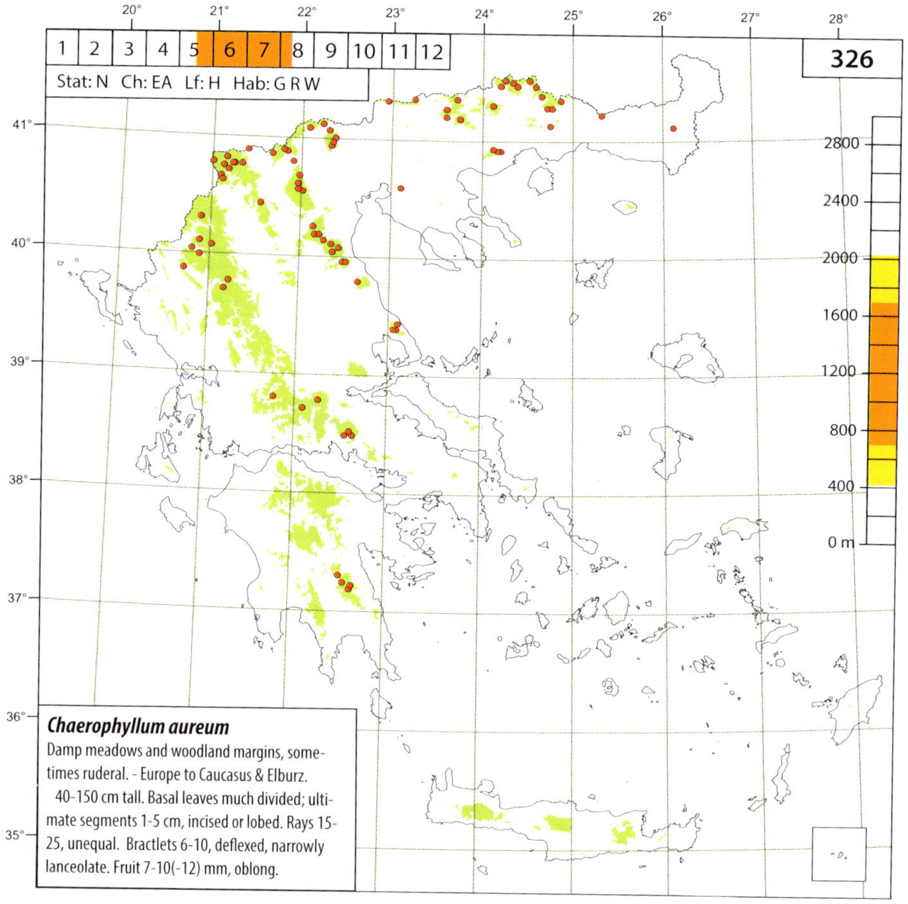

Chaerophyllum aureum
Damp meadows and woodland margins, sometimes ruderal. - Europe to Caucasus & Elburz.
40-150 cm tall. Basal leaves much divided; ultimate segments 1-5 cm, incised or lobed. Rays 15-25, unequal. Bractlets 6-10, deflexed, narrowly lanceolate. Fruit 7-10(-12) mm, oblong.

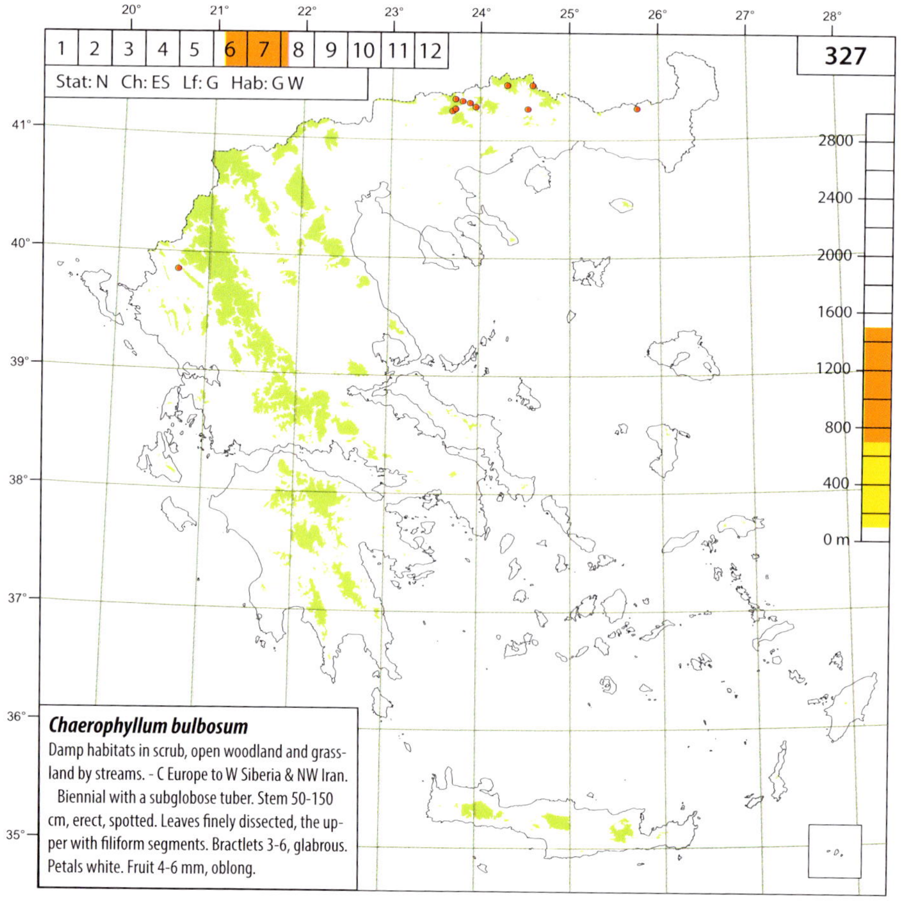

Chaerophyllum bulbosum
Damp habitats in scrub, open woodland and grassland by streams. - C Europe to W Siberia & NW Iran.
Biennial with a subglobose tuber. Stem 50-150 cm, erect, spotted. Leaves finely dissected, the upper with filiform segments. Bractlets 3-6, glabrous. Petals white. Fruit 4-6 mm, oblong.

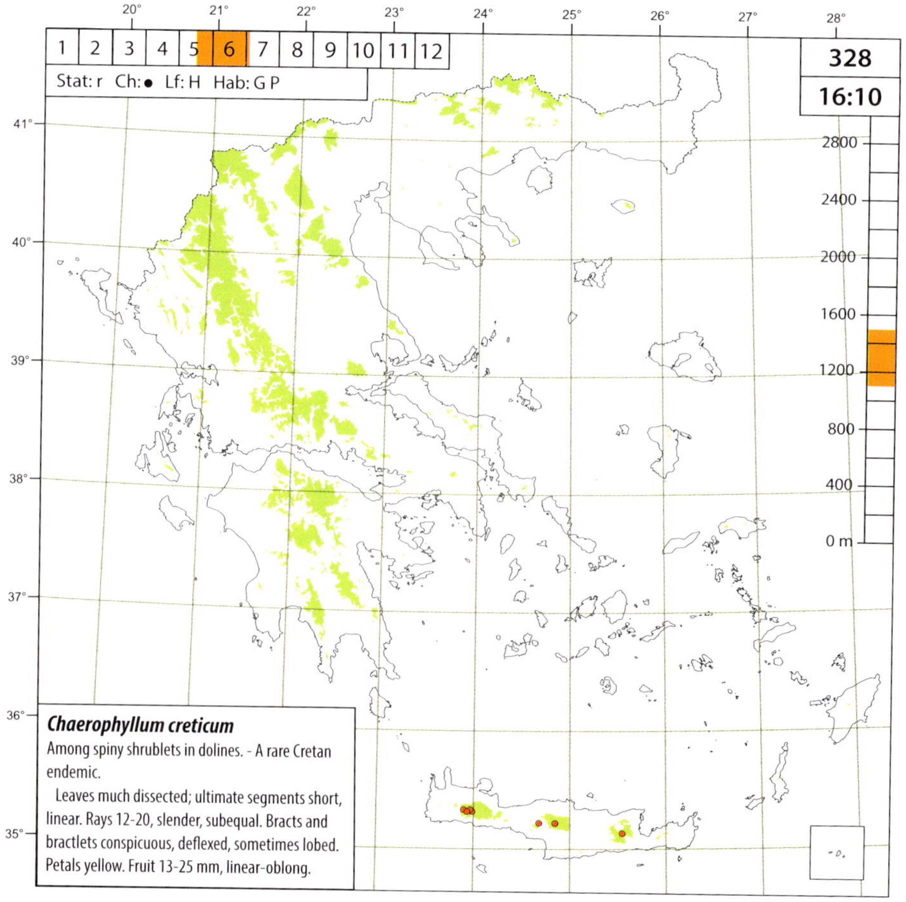

Chaerophyllum creticum
Among spiny shrublets in dolines. - A rare Cretan endemic.
Leaves much dissected; ultimate segments short, linear. Rays 12-20, slender, subequal. Bracts and bractlets conspicuous, deflexed, sometimes lobed. Petals yellow. Fruit 13-25 mm, linear-oblong.

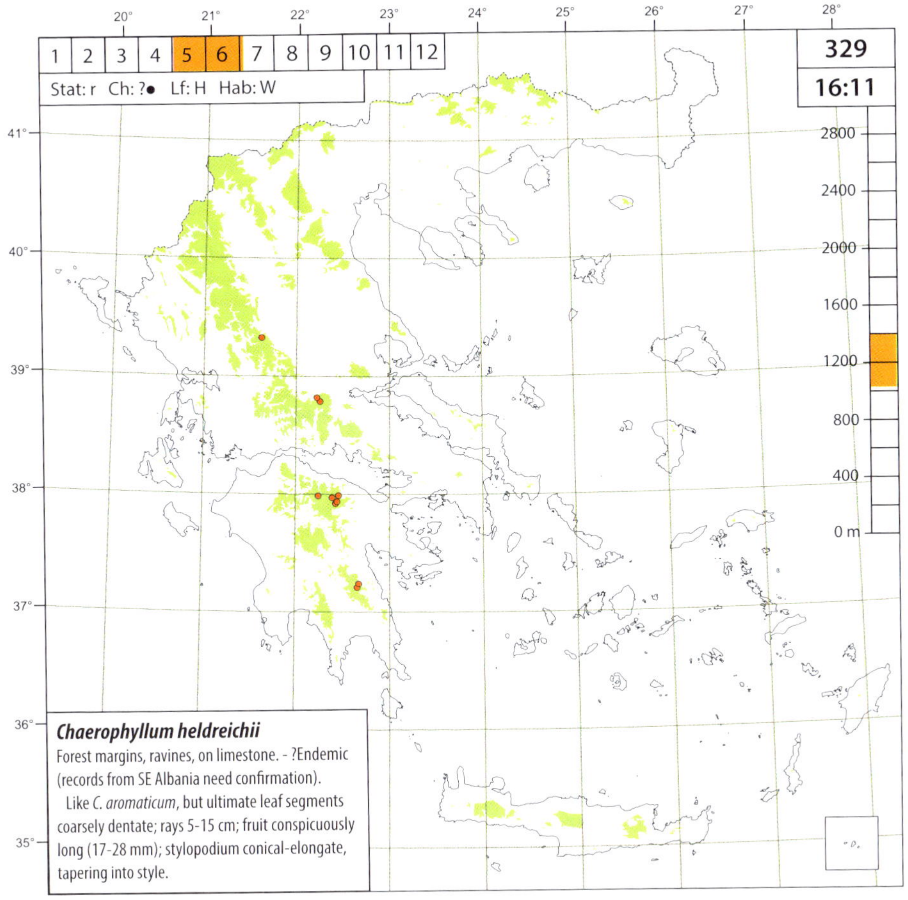

Chaerophyllum heldreichii

Forest margins, ravines, on limestone. - ?Endemic (records from SE Albania need confirmation).

Like *C. aromaticum*, but ultimate leaf segments coarsely dentate; rays 5-15 cm; fruit conspicuously long (17-28 mm); stylopodium conical-elongate, tapering into style.

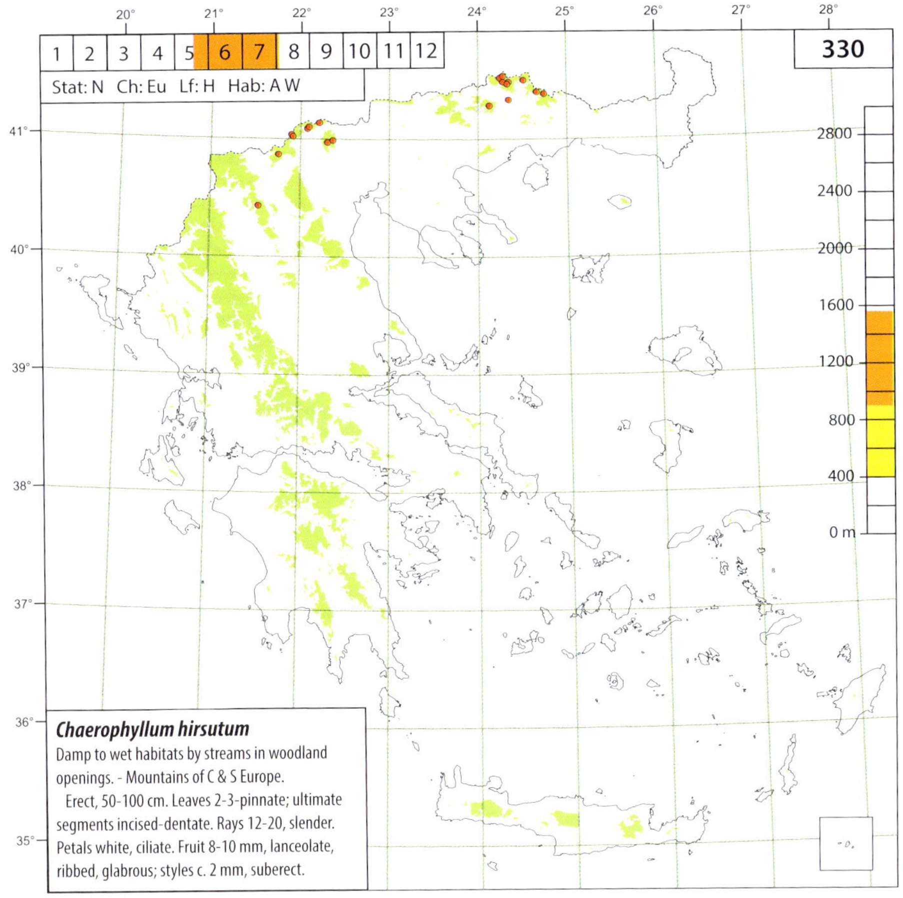

Chaerophyllum hirsutum

Damp to wet habitats by streams in woodland openings. - Mountains of C & S Europe.

Erect, 50-100 cm. Leaves 2-3-pinnate; ultimate segments incised-dentate. Rays 12-20, slender. Petals white, ciliate. Fruit 8-10 mm, lanceolate, ribbed, glabrous; styles c. 2 mm, suberect.

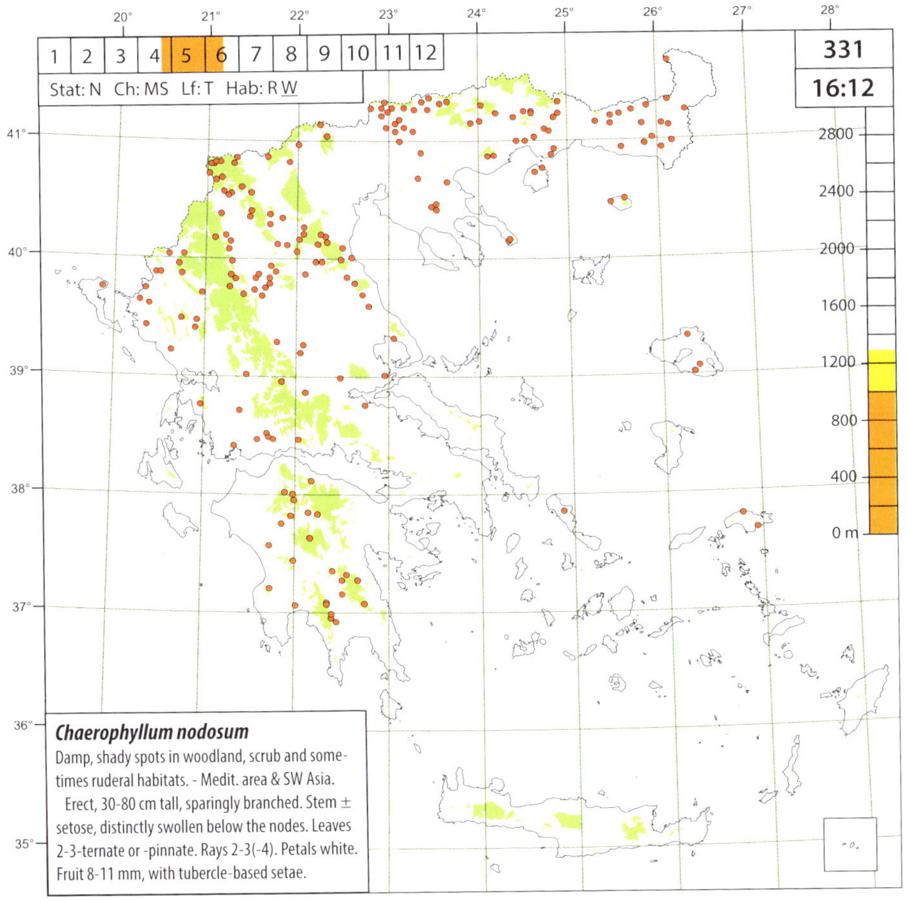

Chaerophyllum nodosum

Damp, shady spots in woodland, scrub and sometimes ruderal habitats. - Medit. area & SW Asia.

Erect, 30-80 cm tall, sparingly branched. Stem ± setose, distinctly swollen below the nodes. Leaves 2-3-ternate or -pinnate. Rays 2-3(-4). Petals white. Fruit 8-11 mm, with tubercle-based setae.

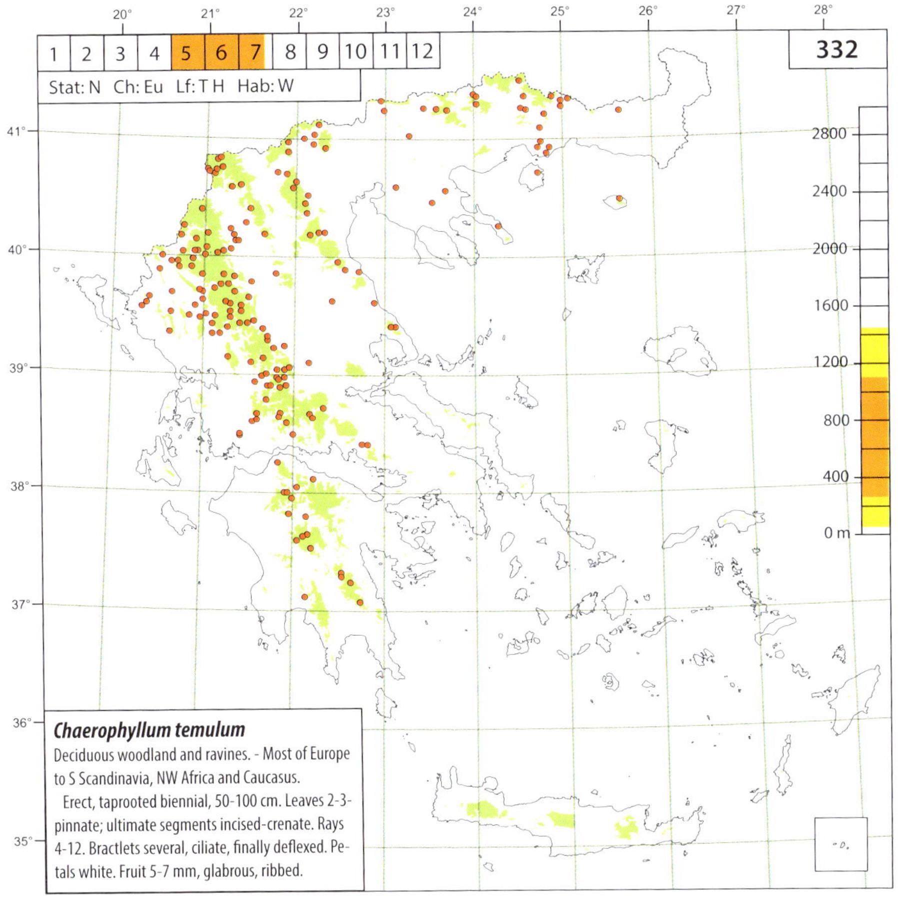

Chaerophyllum temulum

Deciduous woodland and ravines. - Most of Europe to S Scandinavia, NW Africa and Caucasus.

Erect, taprooted biennial, 50-100 cm. Leaves 2-3-pinnate; ultimate segments incised-crenate. Rays 4-12. Bractlets several, ciliate, finally deflexed. Petals white. Fruit 5-7 mm, glabrous, ribbed.

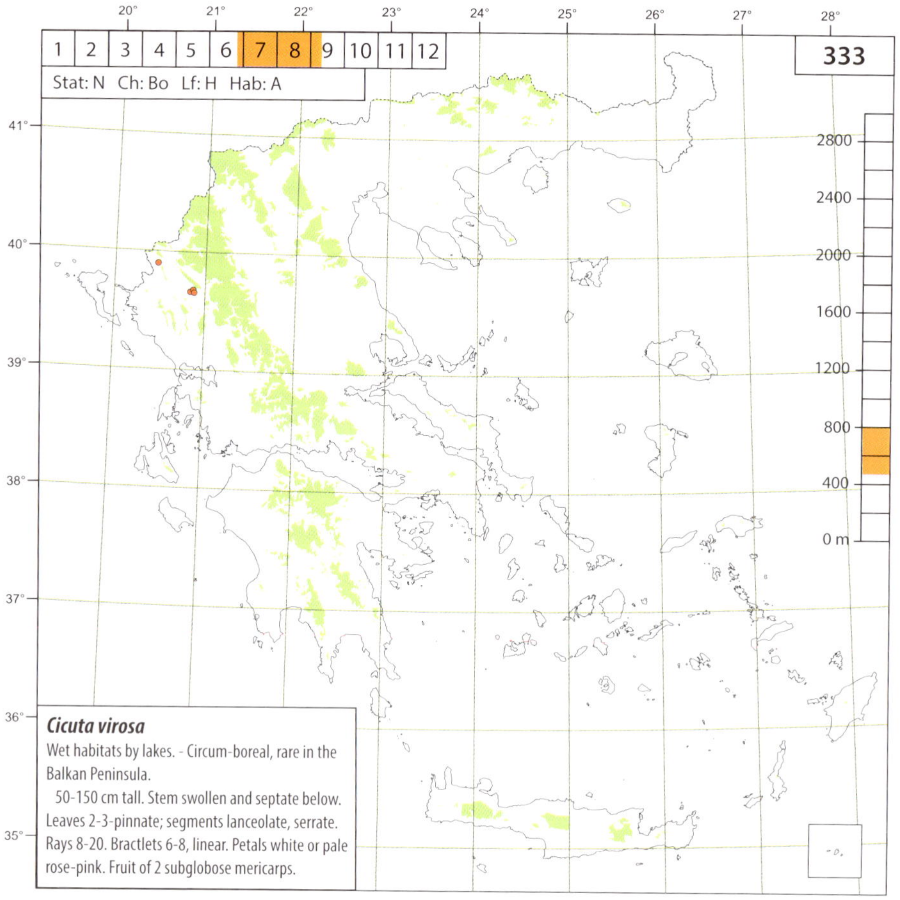

Cicuta virosa

Wet habitats by lakes. - Circum-boreal, rare in the Balkan Peninsula.

50-150 cm tall. Stem swollen and septate below. Leaves 2-3-pinnate; segments lanceolate, serrate. Rays 8-20. Bractlets 6-8, linear. Petals white or pale rose-pink. Fruit of 2 subglobose mericarps.

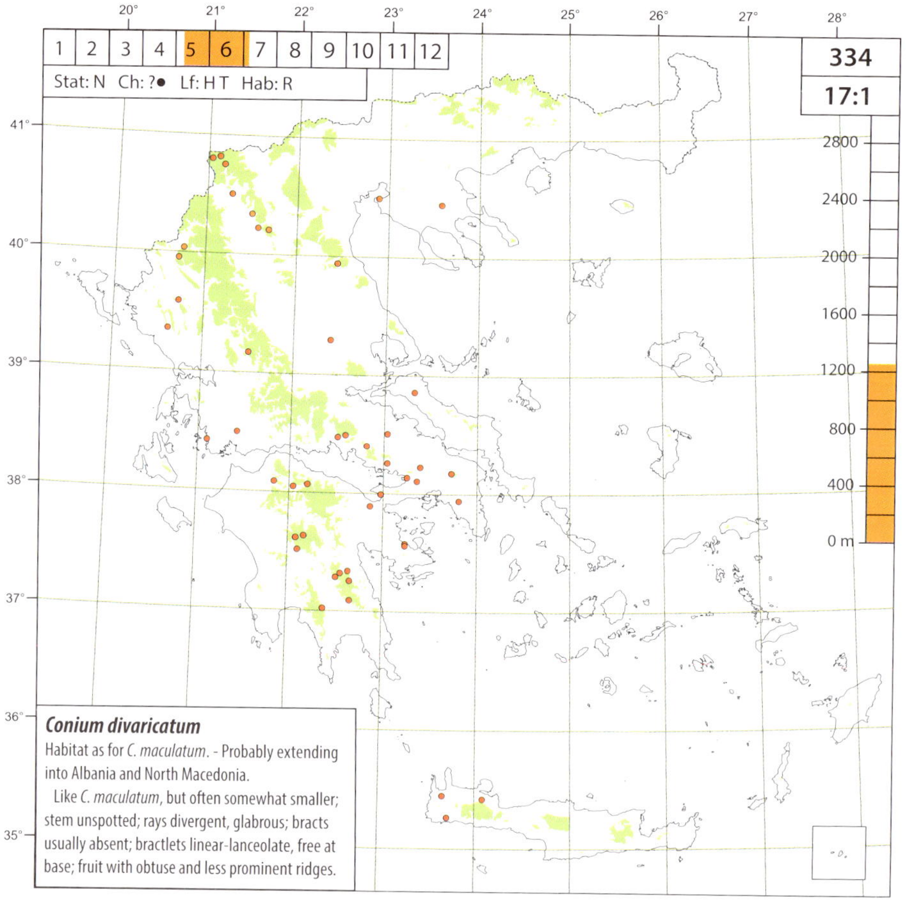

Conium divaricatum

Habitat as for *C. maculatum*. - Probably extending into Albania and North Macedonia.

Like *C. maculatum*, but often somewhat smaller; stem unspotted; rays divergent, glabrous; bracts usually absent; bractlets linear-lanceolate, free at base; fruit with obtuse and less prominent ridges.

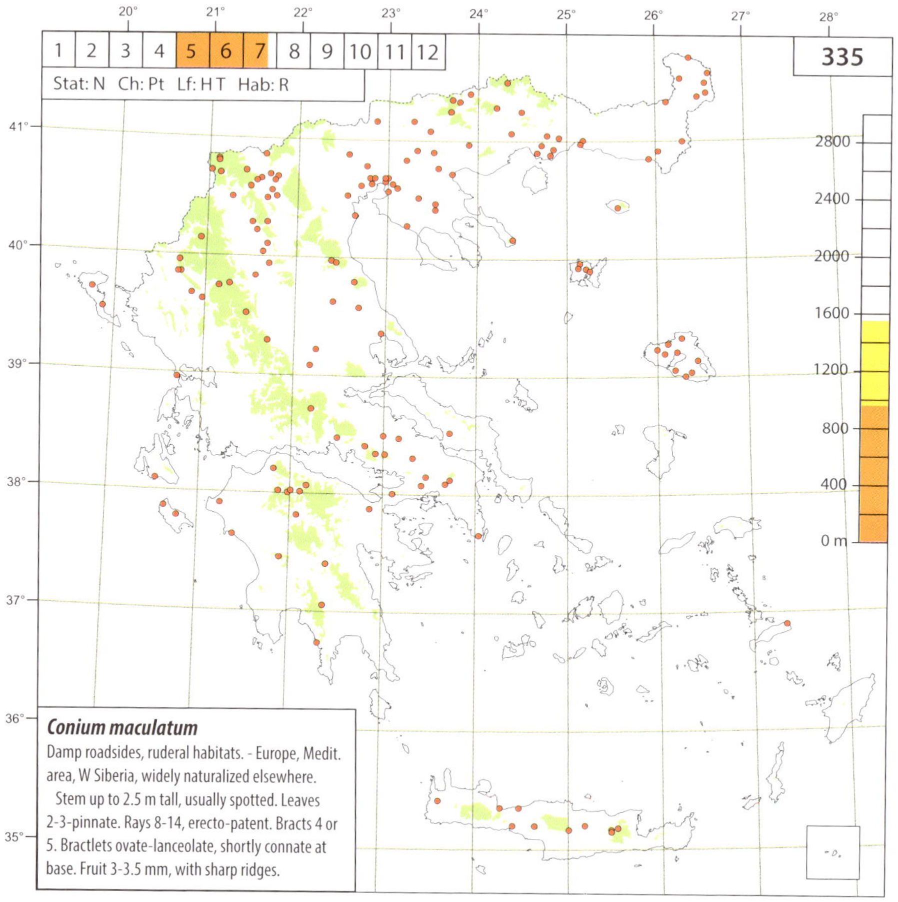

Conium maculatum

Damp roadsides, ruderal habitats. - Europe, Medit. area, W Siberia, widely naturalized elsewhere.

Stem up to 2.5 m tall, usually spotted. Leaves 2-3-pinnate. Rays 8-14, erecto-patent. Bracts 4 or 5. Bractlets ovate-lanceolate, shortly connate at base. Fruit 3-3.5 mm, with sharp ridges.

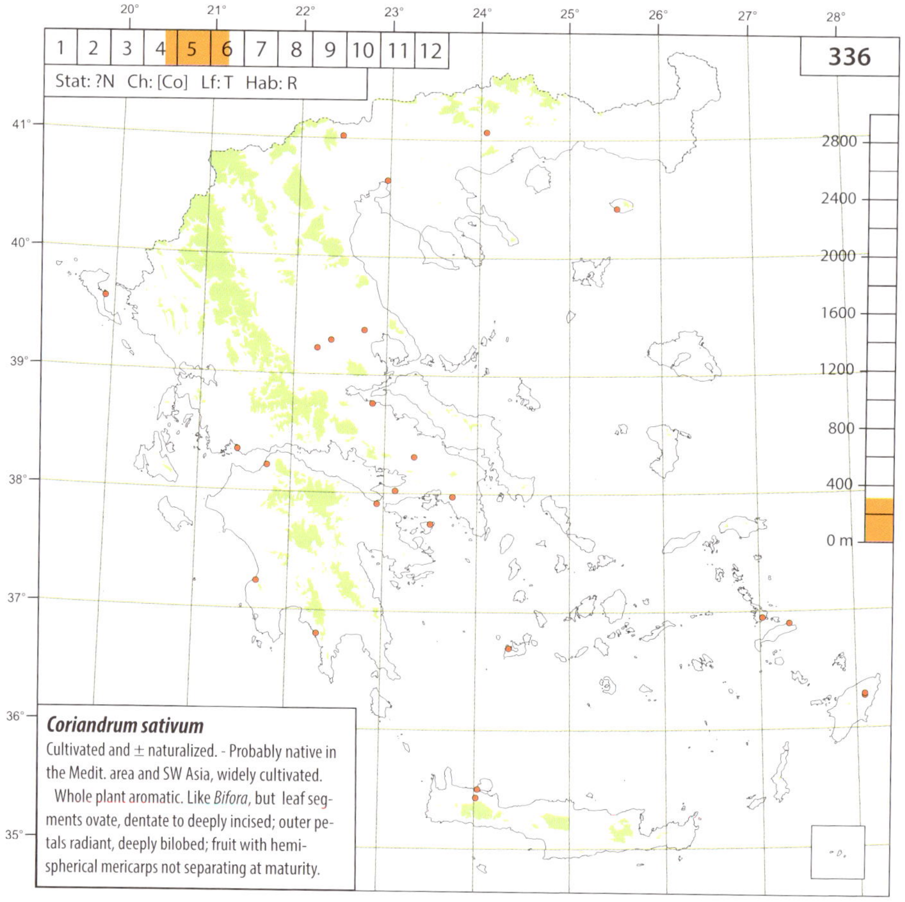

Coriandrum sativum

Cultivated and ± naturalized. - Probably native in the Medit. area and SW Asia, widely cultivated.

Whole plant aromatic. Like *Bifora*, but leaf segments ovate, dentate to deeply incised; outer petals radiant, deeply bilobed; fruit with hemispherical mericarps not separating at maturity.

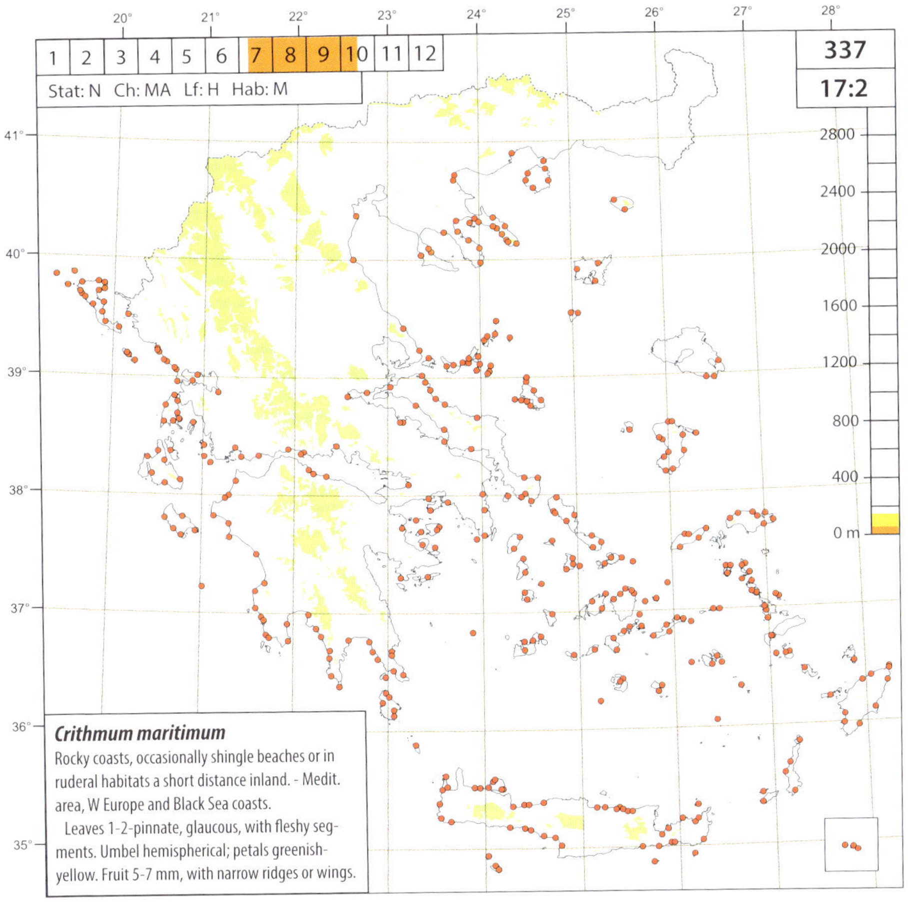

Crithmum maritimum
Rocky coasts, occasionally shingle beaches or in ruderal habitats a short distance inland. - Medit. area, W Europe and Black Sea coasts.
Leaves 1-2-pinnate, glaucous, with fleshy segments. Umbel hemispherical; petals greenish-yellow. Fruit 5-7 mm, with narrow ridges or wings.

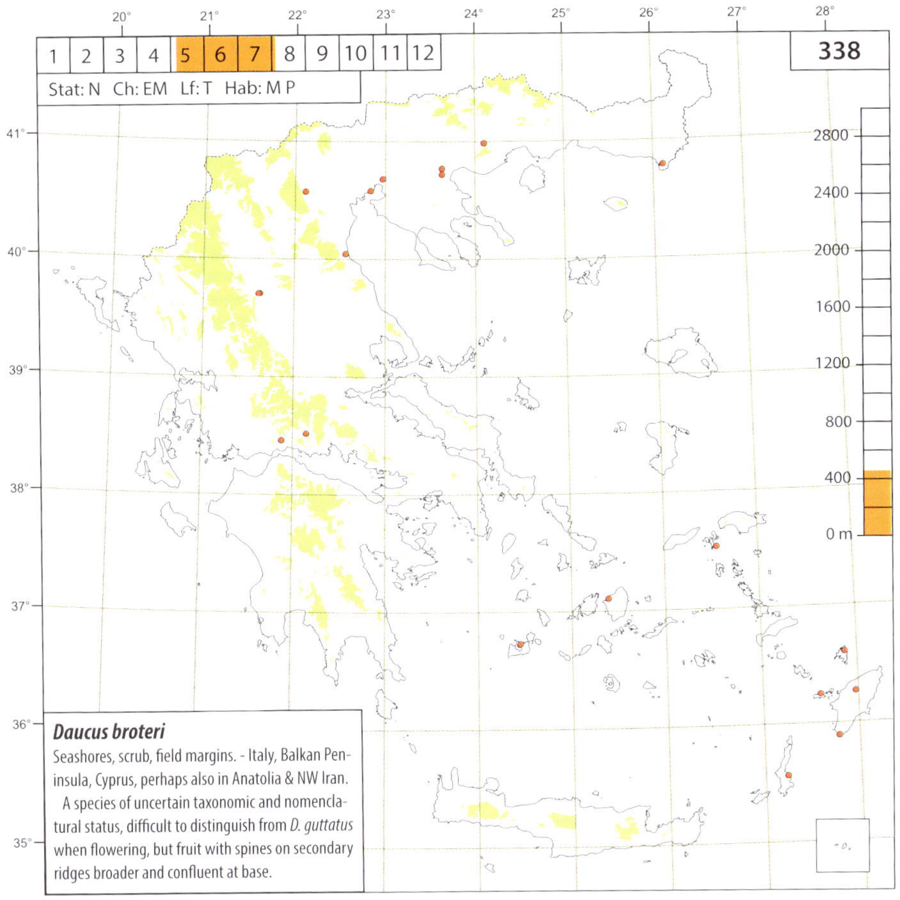

Daucus broteri
Seashores, scrub, field margins. - Italy, Balkan Peninsula, Cyprus, perhaps also in Anatolia & NW Iran.
A species of uncertain taxonomic and nomenclatural status, difficult to distinguish from *D. guttatus* when flowering, but fruit with spines on secondary ridges broader and confluent at base.

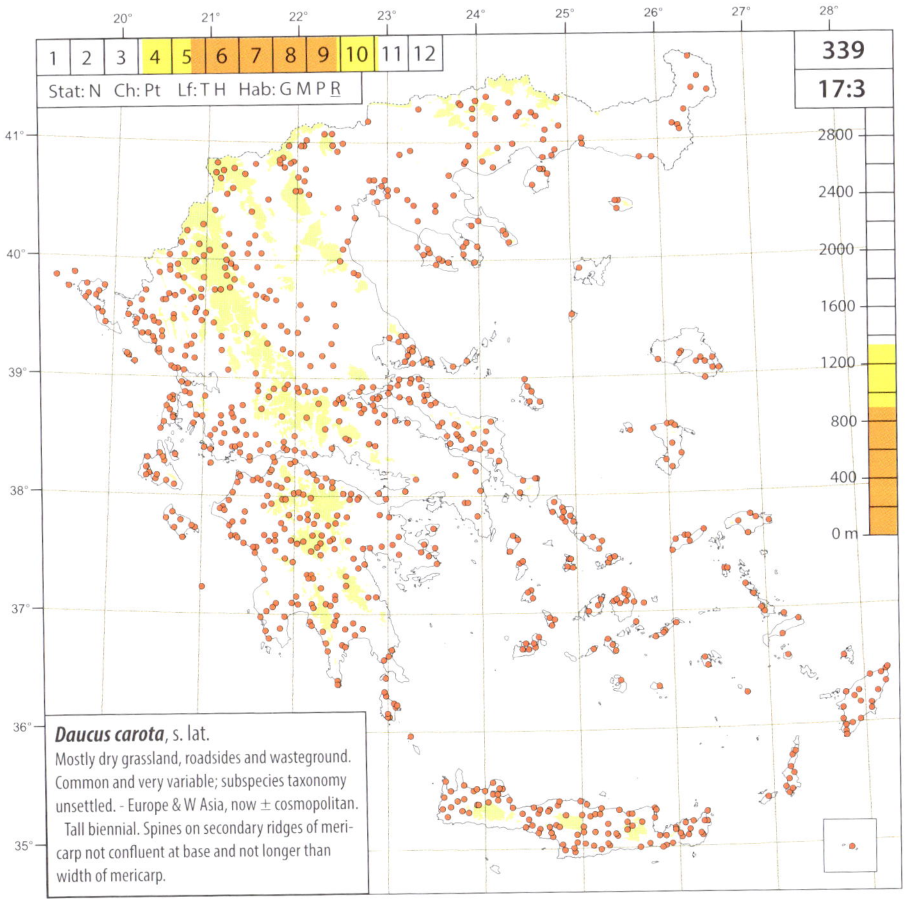

Daucus carota, s. lat.
Mostly dry grassland, roadsides and wasteground. Common and very variable; subspecies taxonomy unsettled. - Europe & W Asia, now ± cosmopolitan.
Tall biennial. Spines on secondary ridges of mericarp not confluent at base and not longer than width of mericarp.

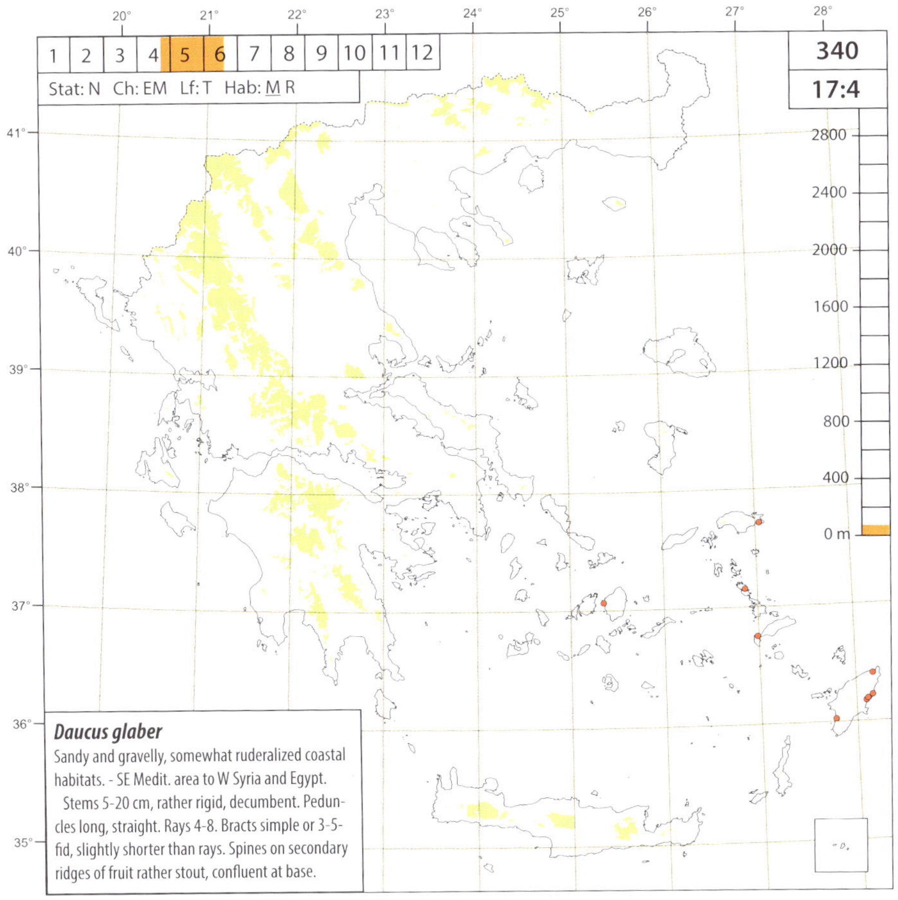

Daucus glaber
Sandy and gravelly, somewhat ruderalized coastal habitats. - SE Medit. area to W Syria and Egypt.
Stems 5-20 cm, rather rigid, decumbent. Peduncles long, straight. Rays 4-8. Bracts simple or 3-5-fid, slightly shorter than rays. Spines on secondary ridges of fruit rather stout, confluent at base.

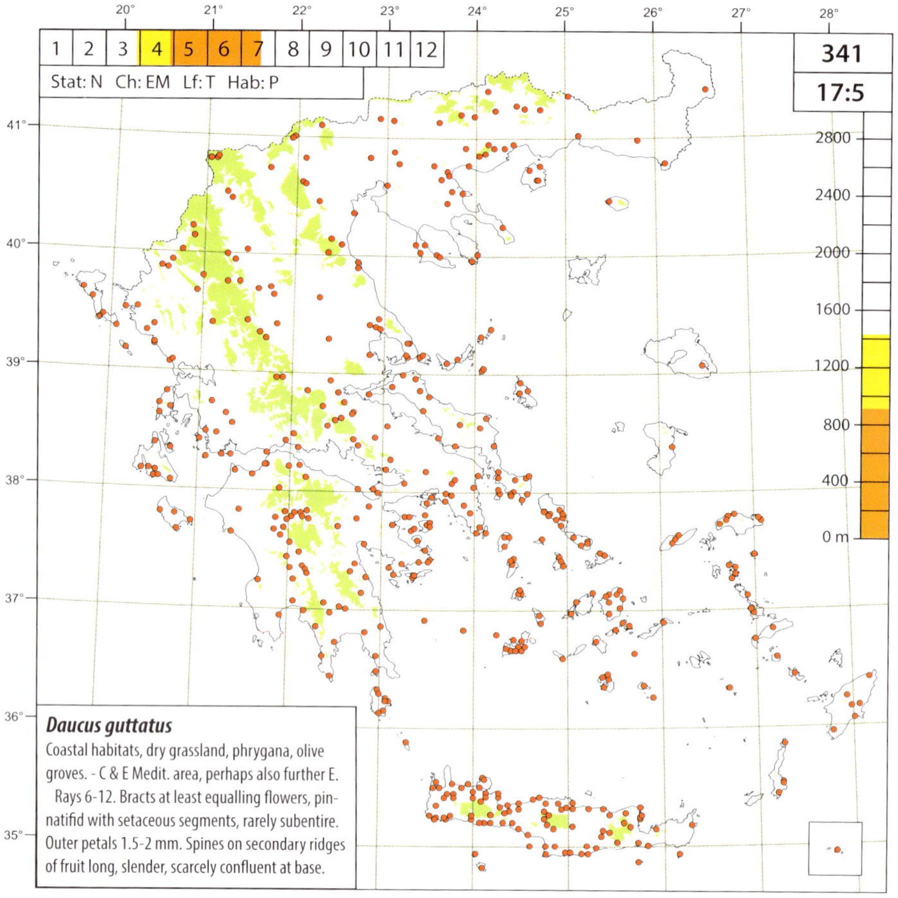

Daucus guttatus

Coastal habitats, dry grassland, phrygana, olive groves. - C & E Medit. area, perhaps also further E.

Rays 6-12. Bracts at least equalling flowers, pinnatifid with setaceous segments, rarely subentire. Outer petals 1.5-2 mm. Spines on secondary ridges of fruit long, slender, scarcely confluent at base.

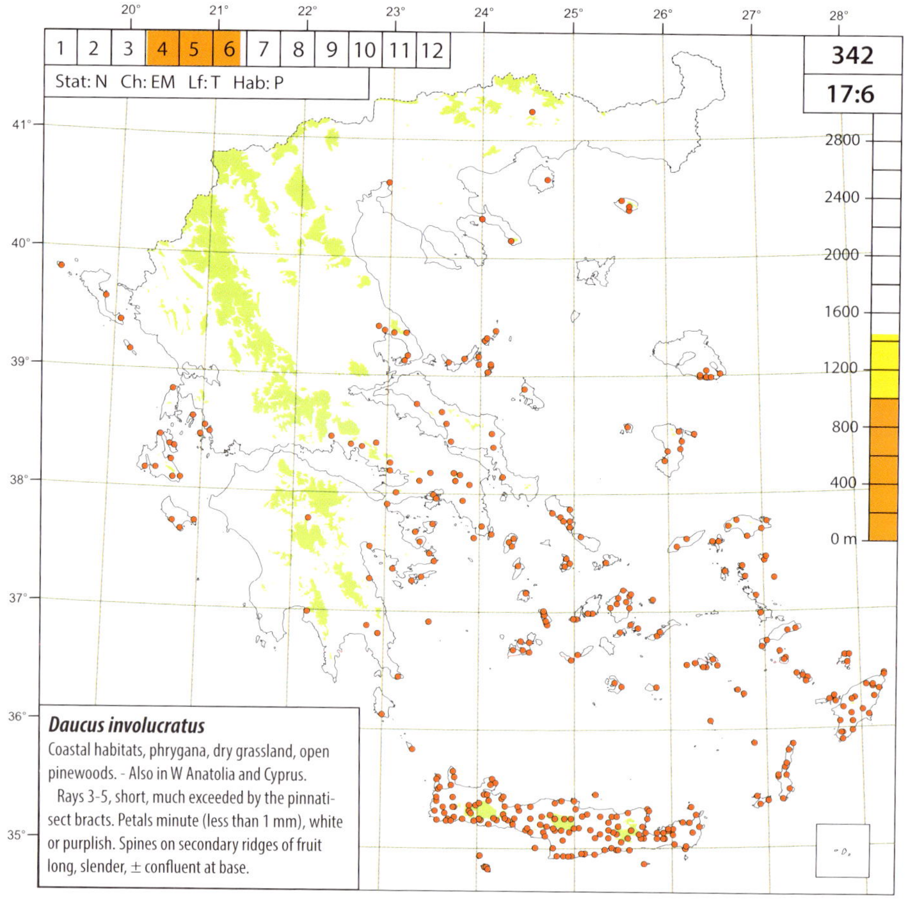

Daucus involucratus

Coastal habitats, phrygana, dry grassland, open pinewoods. - Also in W Anatolia and Cyprus.

Rays 3-5, short, much exceeded by the pinnatisect bracts. Petals minute (less than 1 mm), white or purplish. Spines on secondary ridges of fruit long, slender, ± confluent at base.

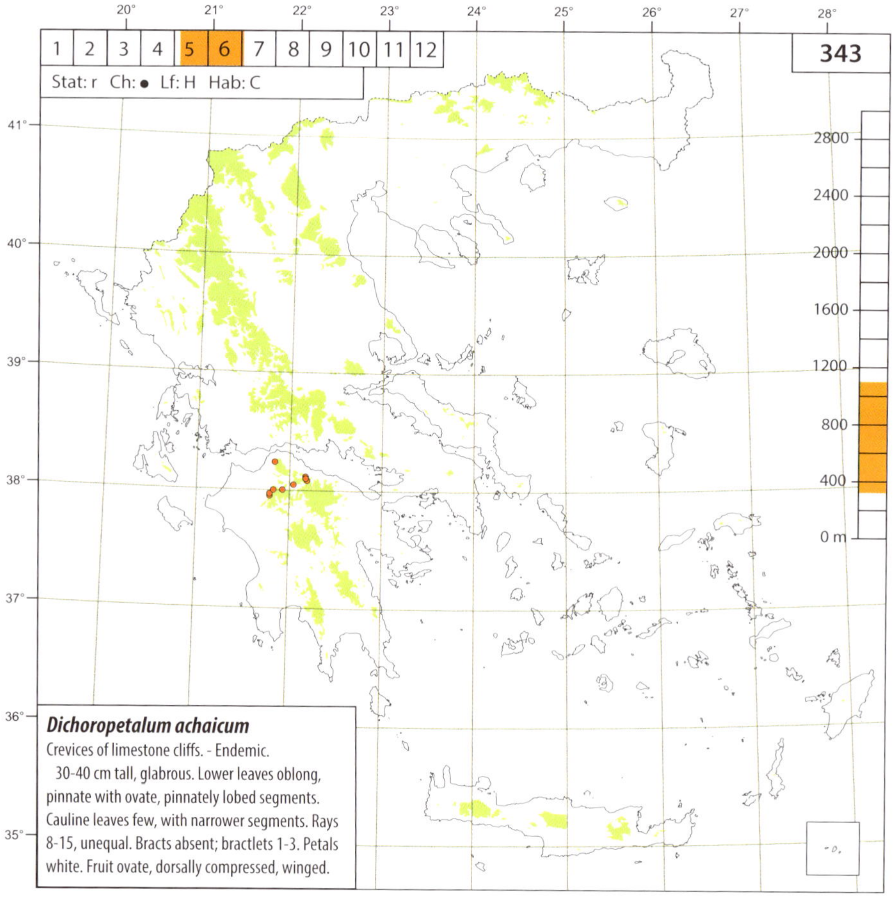

Dichoropetalum achaicum

Crevices of limestone cliffs. - Endemic.

30-40 cm tall, glabrous. Lower leaves oblong, pinnate with ovate, pinnately lobed segments. Cauline leaves few, with narrower segments. Rays 8-15, unequal. Bracts absent; bractlets 1-3. Petals white. Fruit ovate, dorsally compressed, winged.

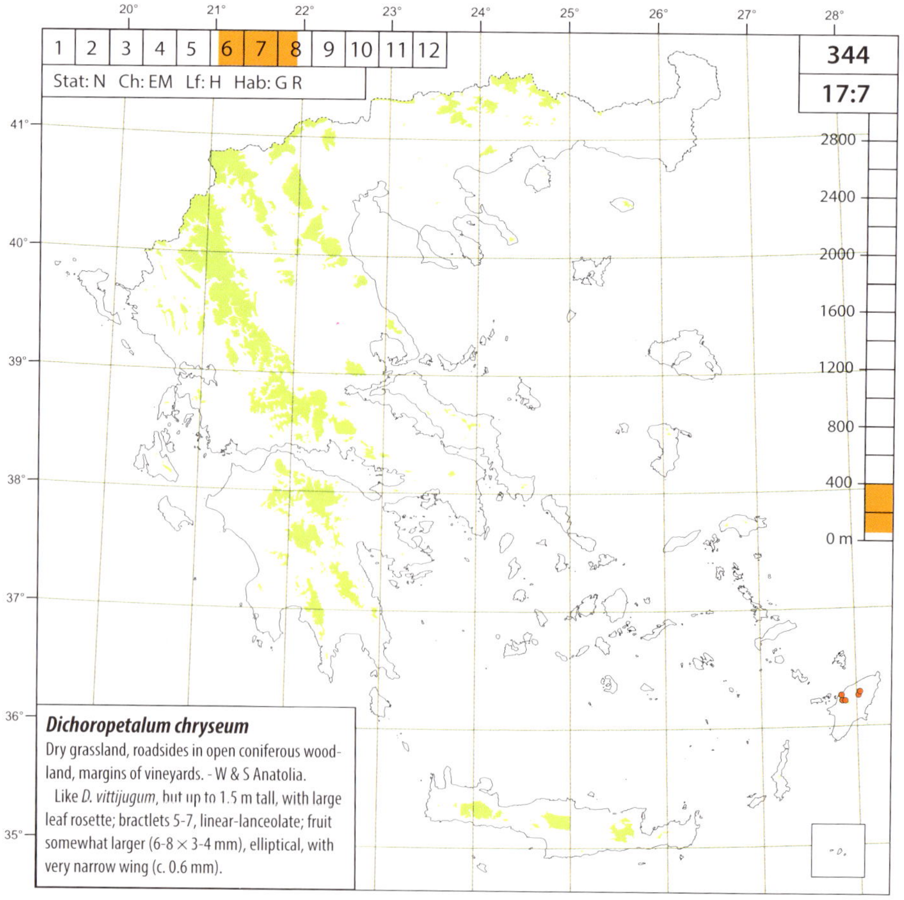

Dichoropetalum chryseum

Dry grassland, roadsides in open coniferous woodland, margins of vineyards. - W & S Anatolia.

Like *D. vittijugum*, but up to 1.5 m tall, with large leaf rosette; bractlets 5-7, linear-lanceolate; fruit somewhat larger (6-8 × 3-4 mm), elliptical, with very narrow wing (c. 0.6 mm).

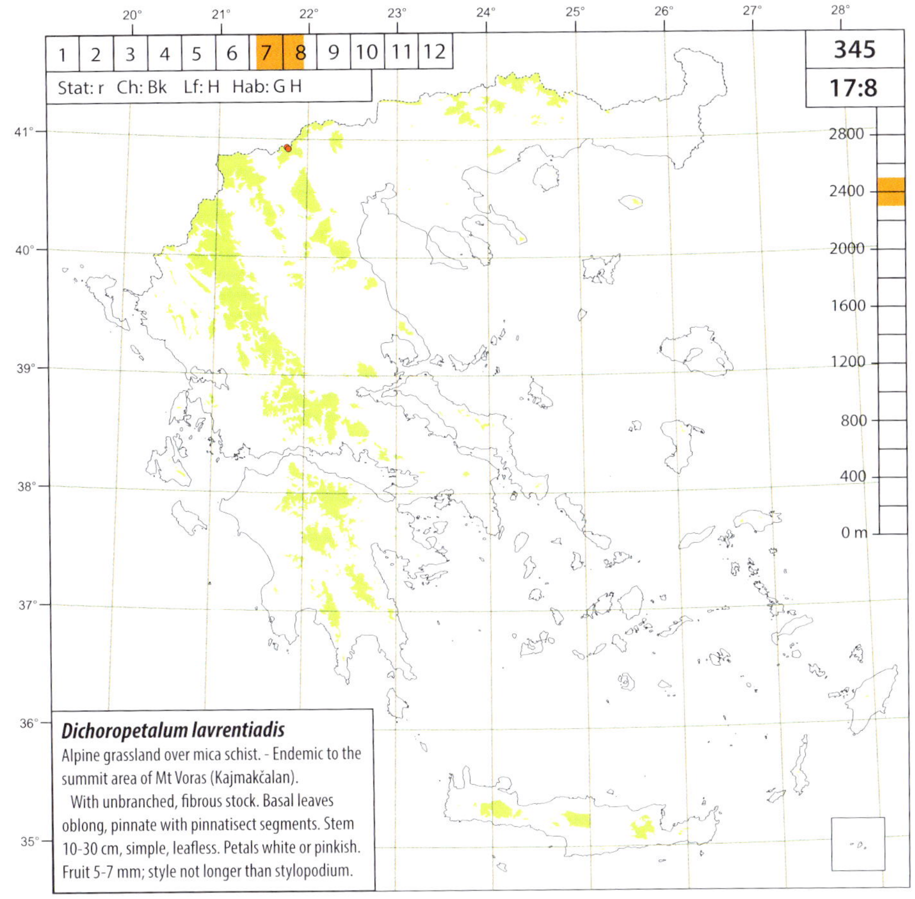

Dichoropetalum lavrentiadis

Alpine grassland over mica schist. - Endemic to the summit area of Mt Voras (Kajmakčalan).

With unbranched, fibrous stock. Basal leaves oblong, pinnate with pinnatisect segments. Stem 10-30 cm, simple, leafless. Petals white or pinkish. Fruit 5-7 mm; style not longer than stylopodium.

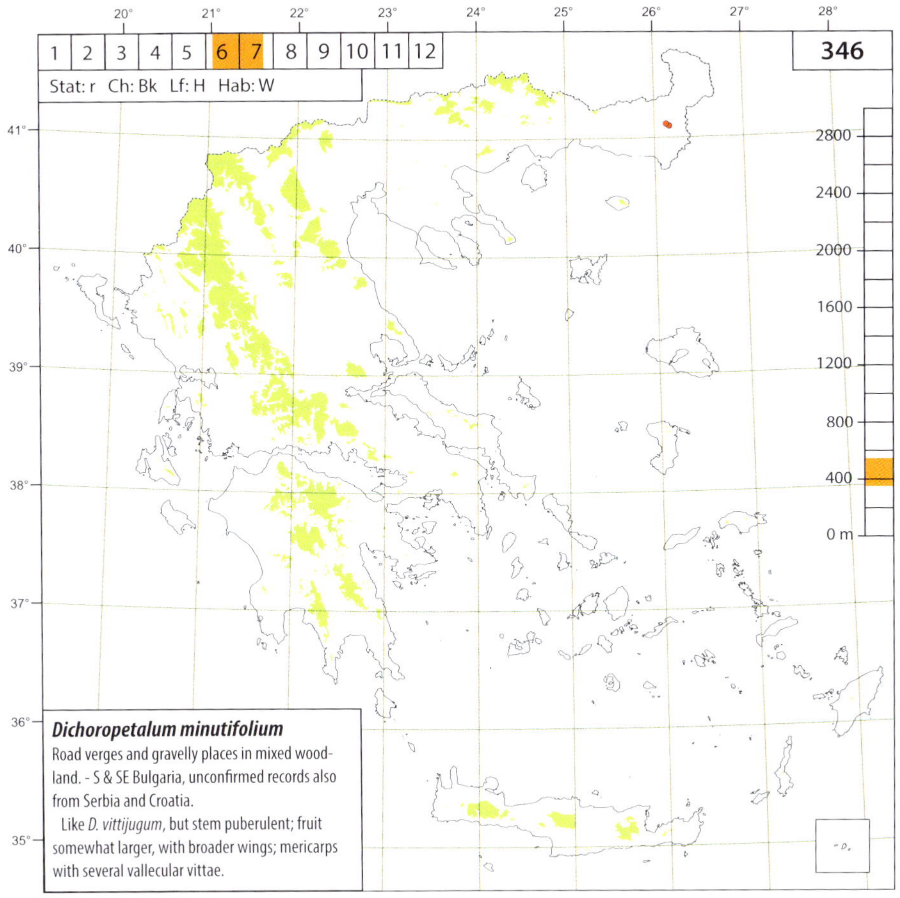

Dichoropetalum minutifolium

Road verges and gravelly places in mixed woodland. - S & SE Bulgaria, unconfirmed records also from Serbia and Croatia.

Like *D. vittijugum*, but stem puberulent; fruit somewhat larger, with broader wings; mericarps with several vallecular vittae.

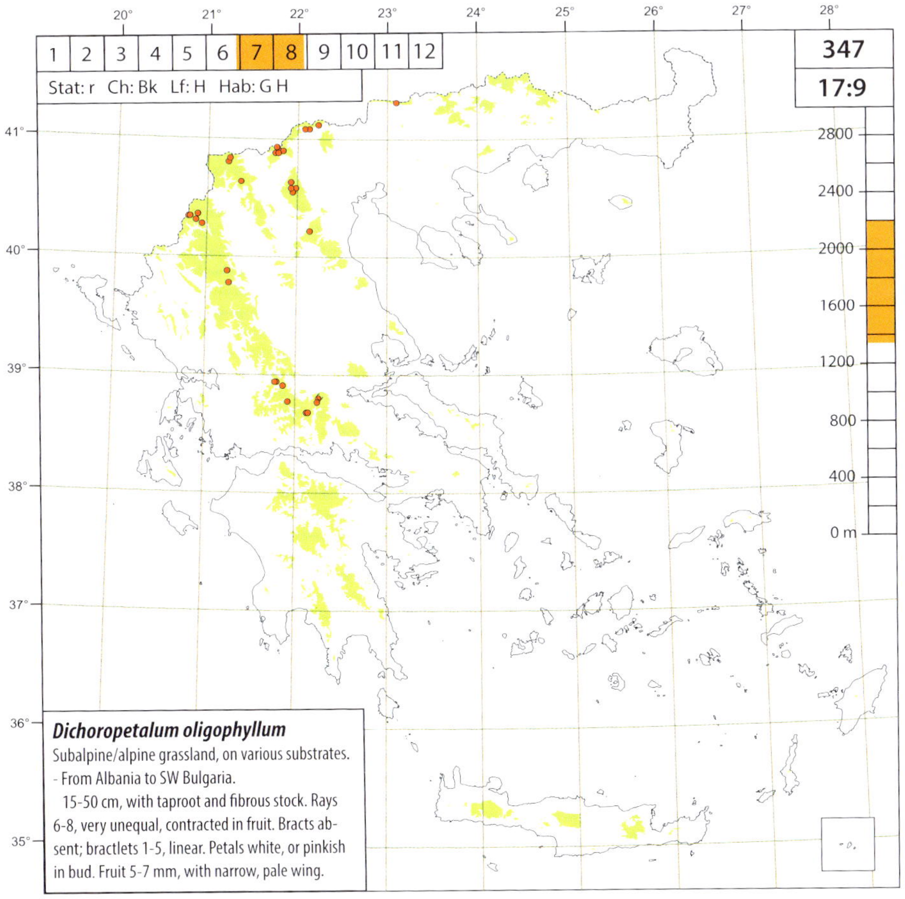

Dichoropetalum oligophyllum

Subalpine/alpine grassland, on various substrates. - From Albania to SW Bulgaria.

15-50 cm, with taproot and fibrous stock. Rays 6-8, very unequal, contracted in fruit. Bracts absent; bractlets 1-5, linear. Petals white, or pinkish in bud. Fruit 5-7 mm, with narrow, pale wing.

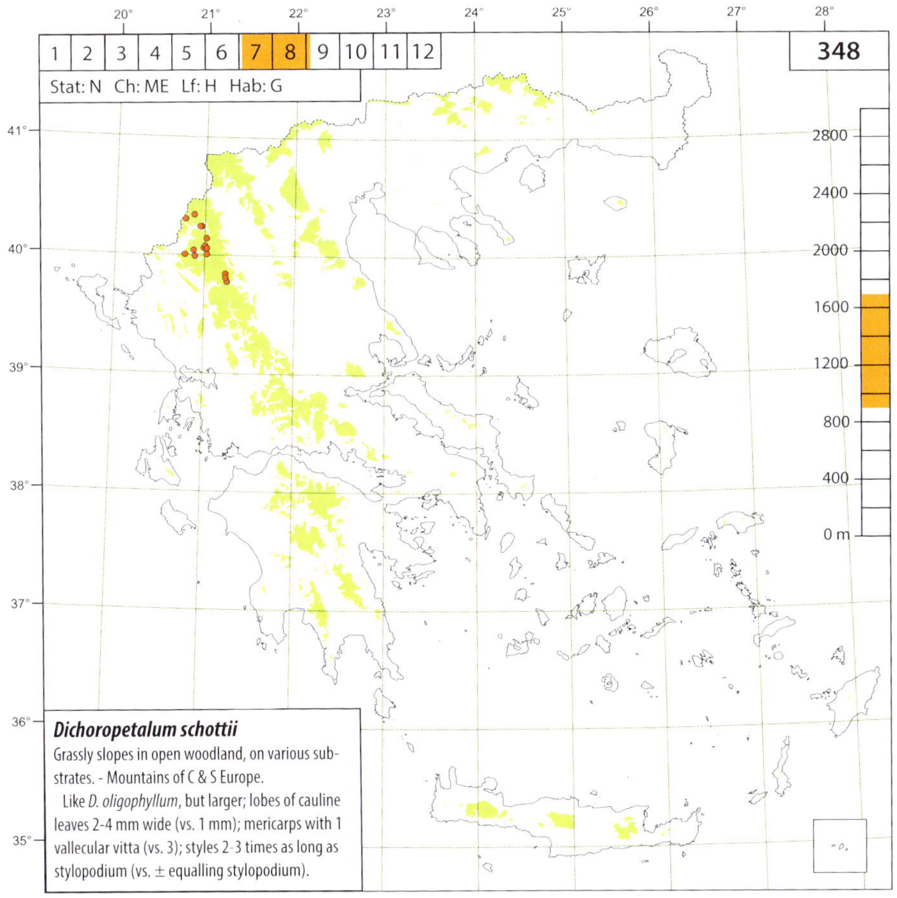

Dichoropetalum schottii

Grassy slopes in open woodland, on various substrates. - Mountains of C & S Europe.

Like *D. oligophyllum*, but larger; lobes of cauline leaves 2-4 mm wide (vs. 1 mm); mericarps with 1 vallecular vitta (vs. 3); styles 2-3 times as long as stylopodium (vs. ± equalling stylopodium).

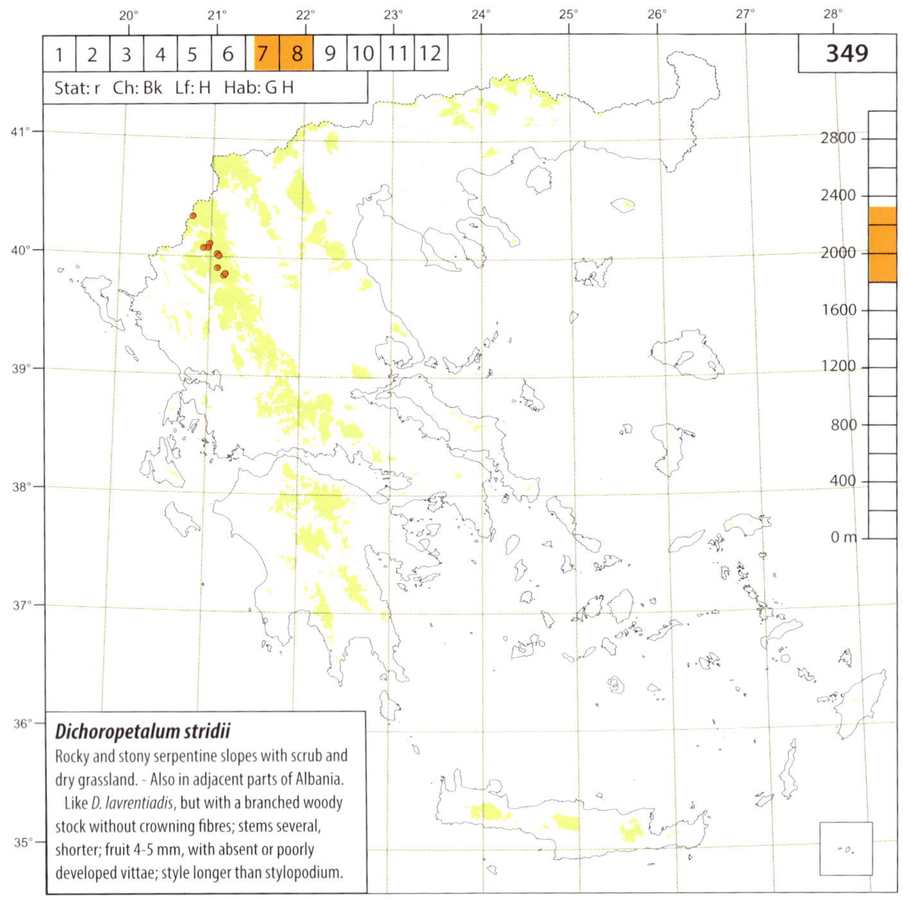

Dichoropetalum stridii
Rocky and stony serpentine slopes with scrub and dry grassland. - Also in adjacent parts of Albania.
Like *D. lavrentiadis*, but with a branched woody stock without crowning fibres; stems several, shorter; fruit 4-5 mm, with absent or poorly developed vittae; style longer than stylopodium.

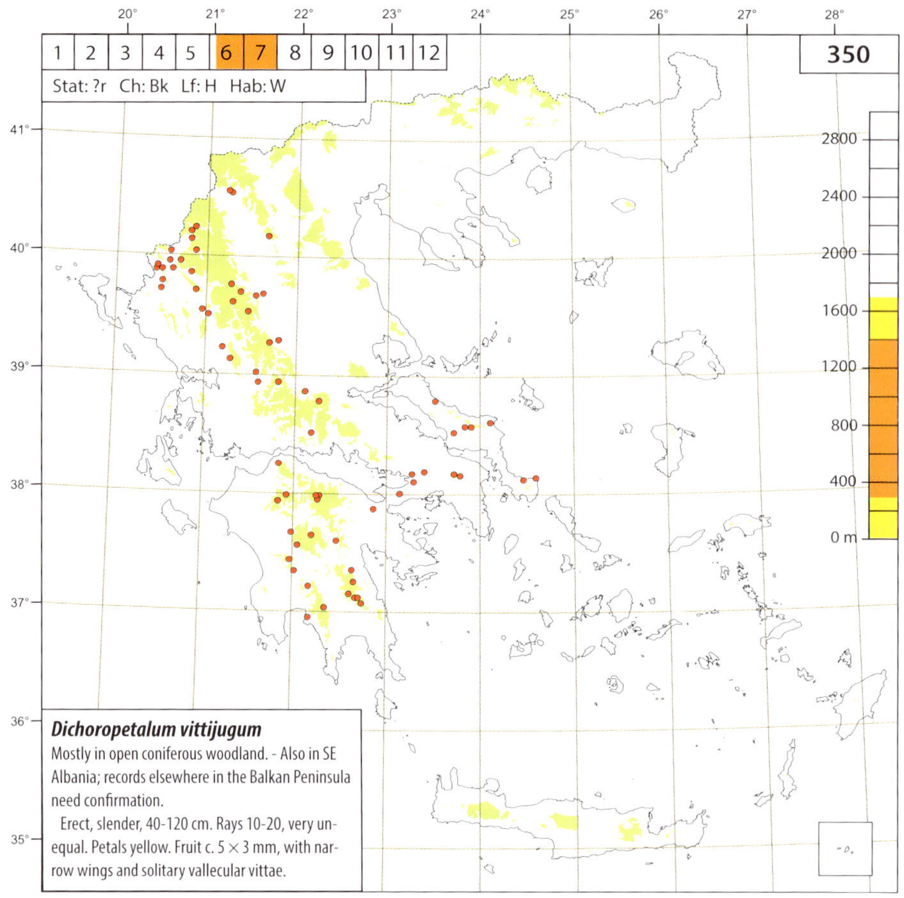

Dichoropetalum vittijugum
Mostly in open coniferous woodland. - Also in SE Albania; records elsewhere in the Balkan Peninsula need confirmation.
Erect, slender, 40-120 cm. Rays 10-20, very unequal. Petals yellow. Fruit c. 5 × 3 mm, with narrow wings and solitary vallecular vittae.

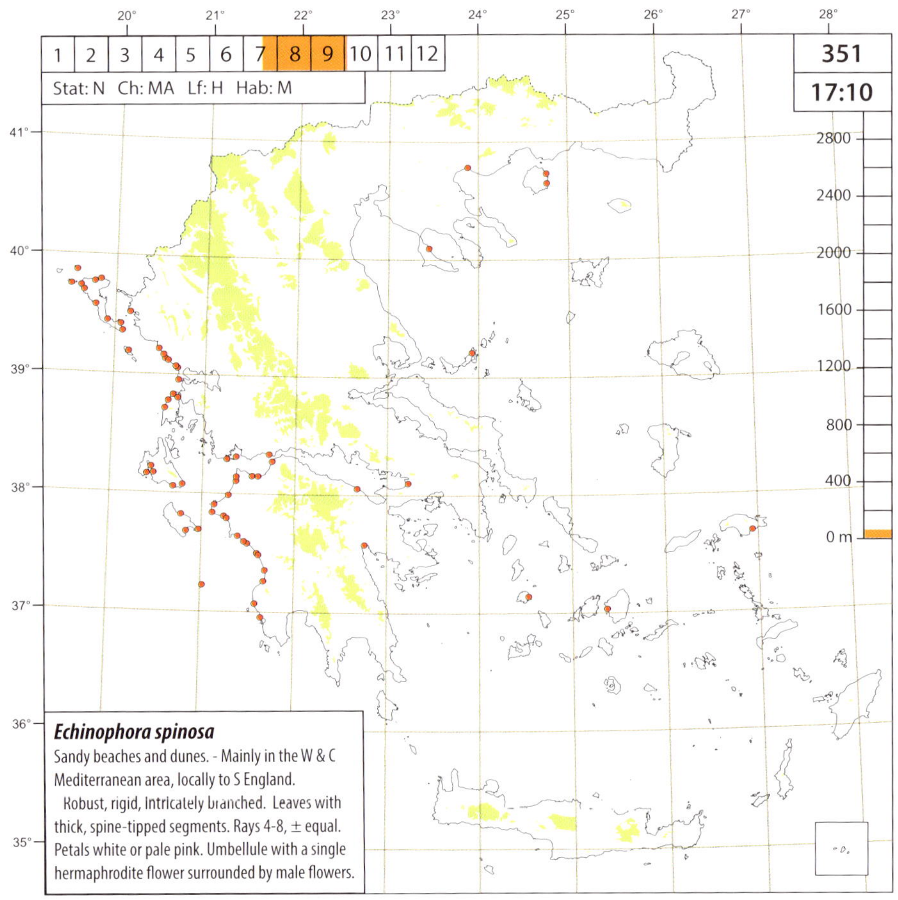

Echinophora spinosa
Sandy beaches and dunes. - Mainly in the W & C Mediterranean area, locally to S England.
Robust, rigid, intricately branched. Leaves with thick, spine-tipped segments. Rays 4-8, ± equal. Petals white or pale pink. Umbellule with a single hermaphrodite flower surrounded by male flowers.

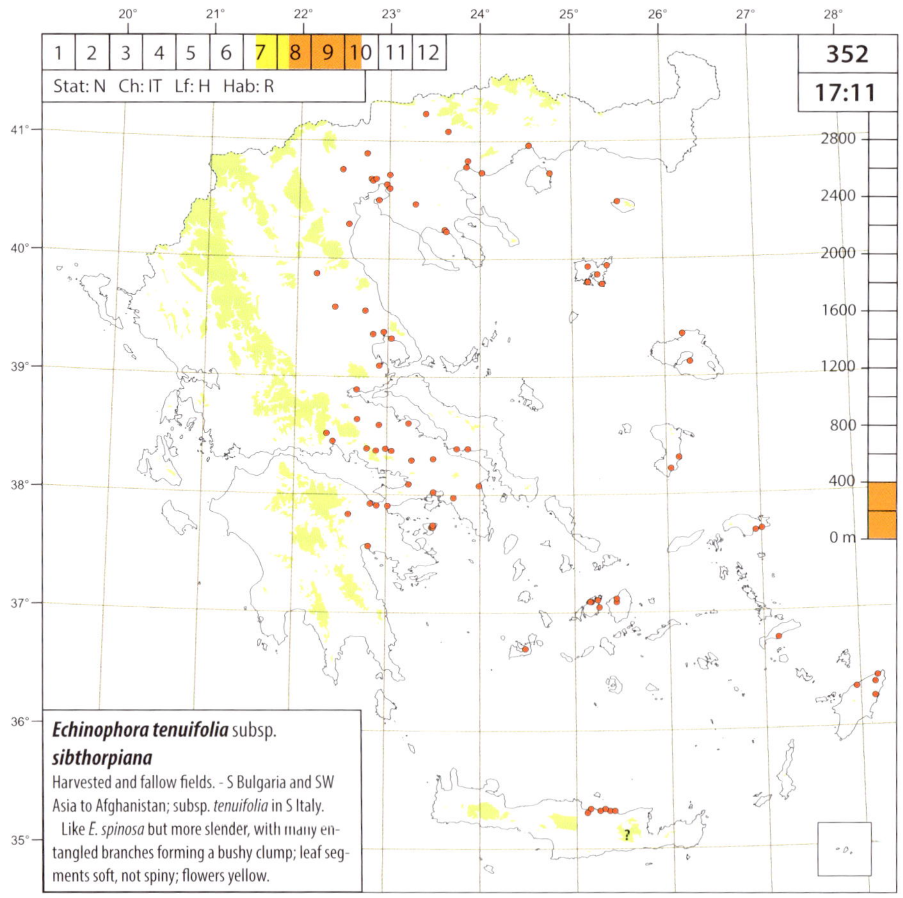

Echinophora tenuifolia subsp. ***sibthorpiana***
Harvested and fallow fields. - S Bulgaria and SW Asia to Afghanistan; subsp. *tenuifolia* in S Italy.
Like *E. spinosa* but more slender, with many entangled branches forming a bushy clump; leaf segments soft, not spiny; flowers yellow.

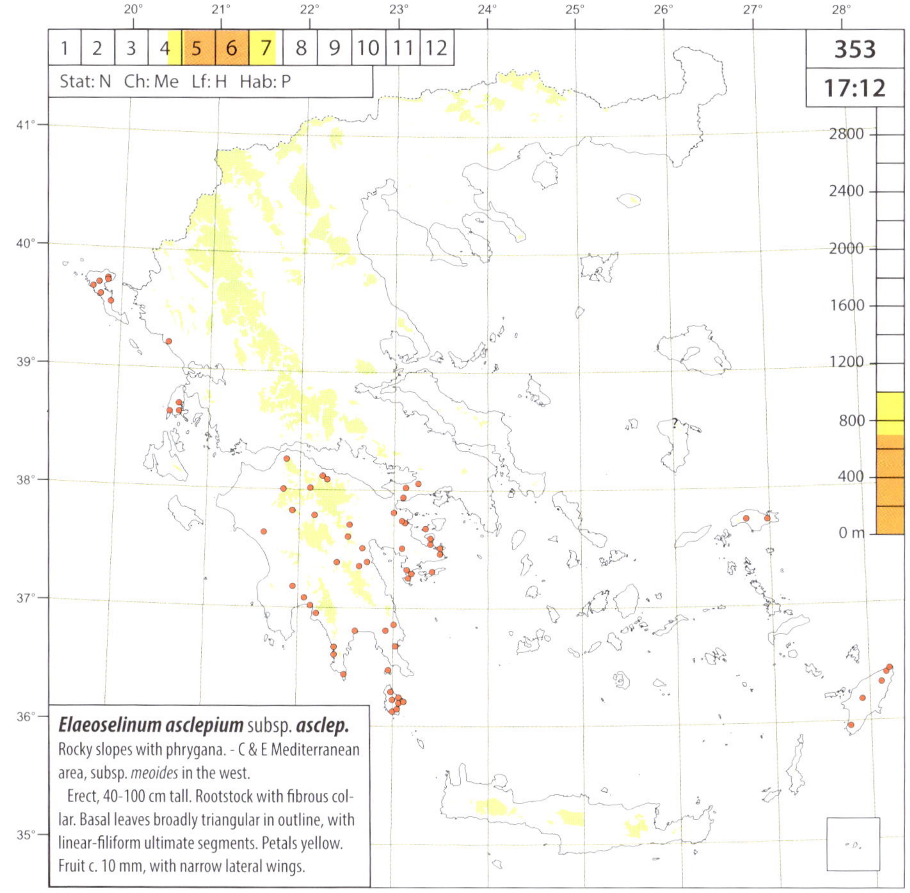

Elaeoselinum asclepium **subsp. *asclep.***
Rocky slopes with phrygana. - C & E Mediterranean area, subsp. *meoides* in the west.
Erect, 40-100 cm tall. Rootstock with fibrous collar. Basal leaves broadly triangular in outline, with linear-filiform ultimate segments. Petals yellow. Fruit c. 10 mm, with narrow lateral wings.

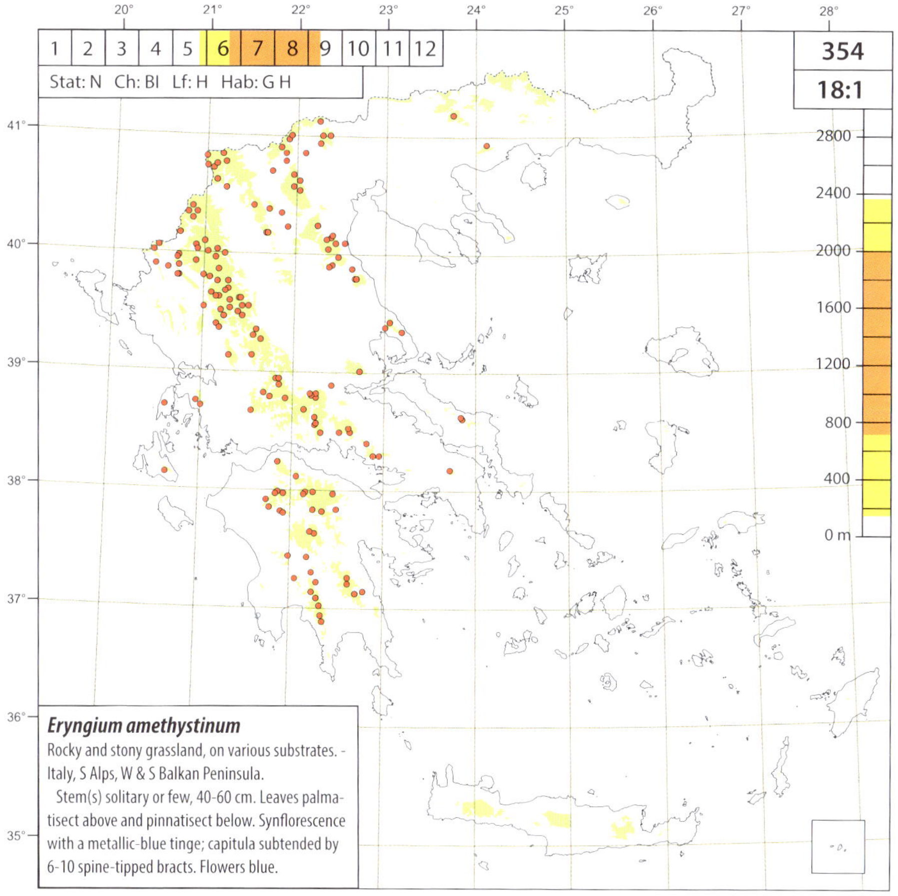

Eryngium amethystinum
Rocky and stony grassland, on various substrates. - Italy, S Alps, W & S Balkan Peninsula.
Stem(s) solitary or few, 40-60 cm. Leaves palmatisect above and pinnatisect below. Synflorescence with a metallic-blue tinge; capitula subtended by 6-10 spine-tipped bracts. Flowers blue.

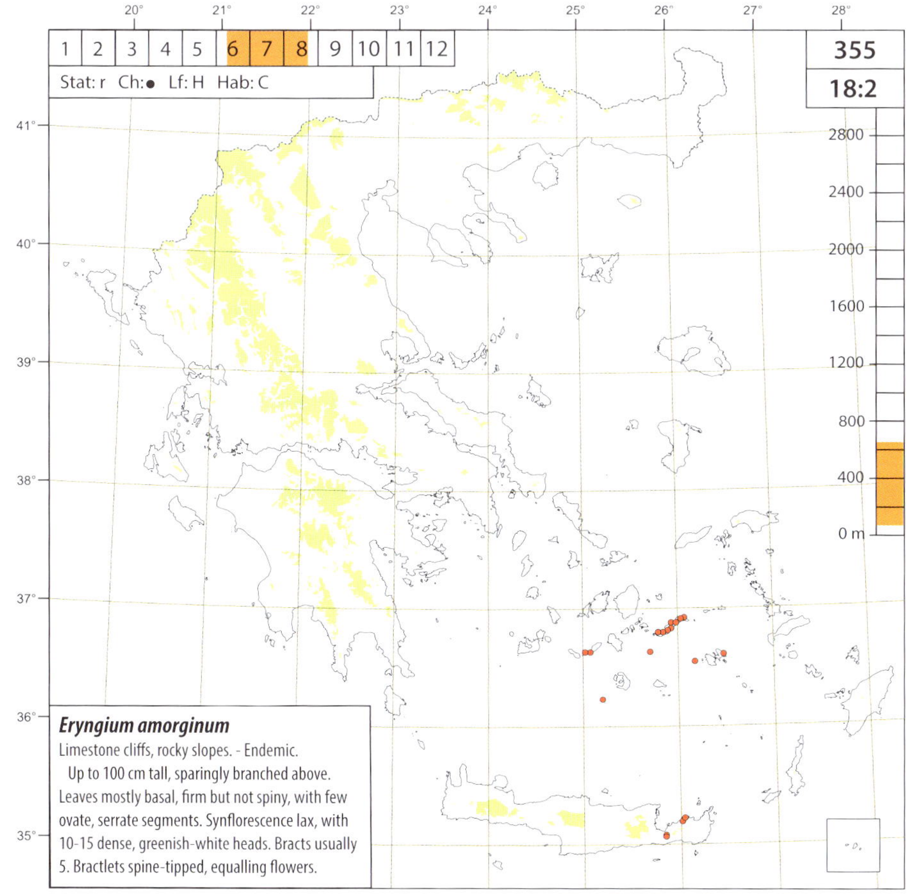

Eryngium amorginum
Limestone cliffs, rocky slopes. - Endemic.
Up to 100 cm tall, sparingly branched above. Leaves mostly basal, firm but not spiny, with few ovate, serrate segments. Synflorescence lax, with 10-15 dense, greenish-white heads. Bracts usually 5. Bractlets spine-tipped, equalling flowers.

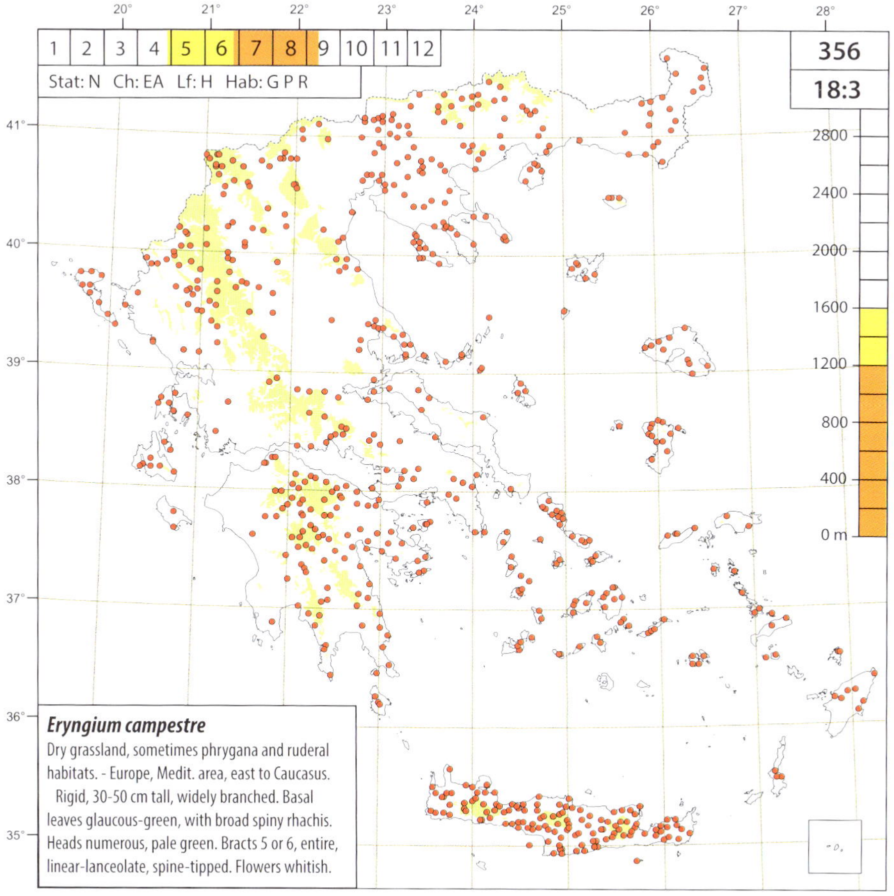

Eryngium campestre
Dry grassland, sometimes phrygana and ruderal habitats. - Europe, Medit. area, east to Caucasus.
Rigid, 30-50 cm tall, widely branched. Basal leaves glaucous-green, with broad spiny rhachis. Heads numerous, pale green. Bracts 5 or 6, entire, linear-lanceolate, spine-tipped. Flowers whitish.

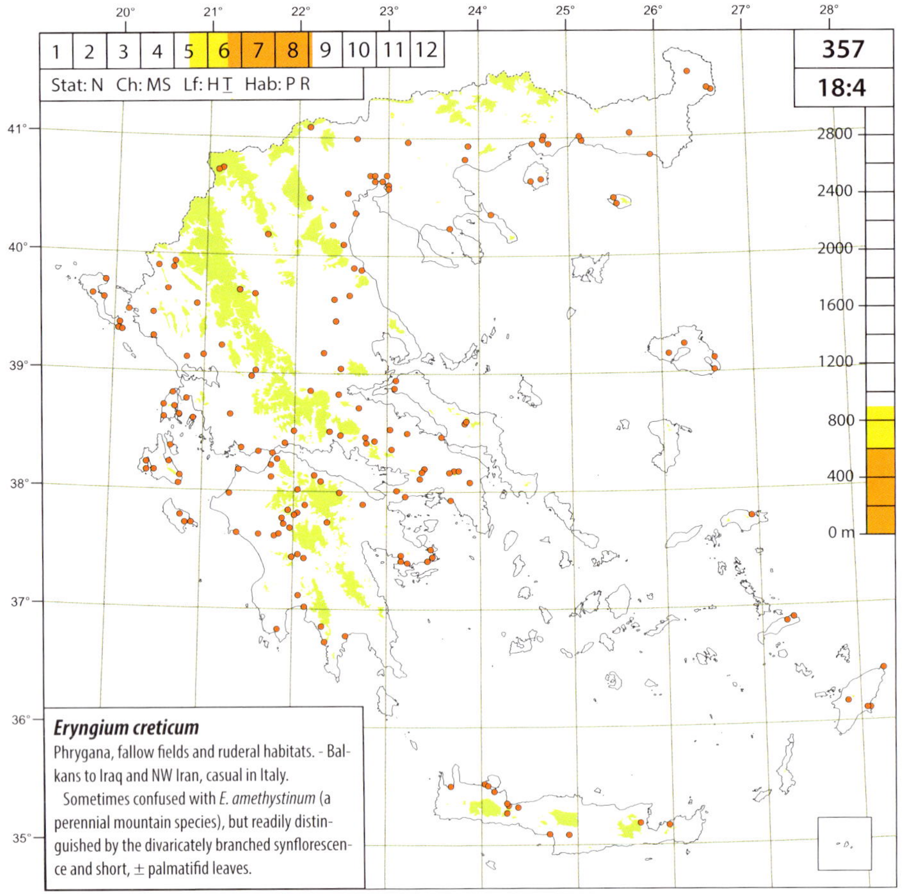

Eryngium creticum
Phrygana, fallow fields and ruderal habitats. - Balkans to Iraq and NW Iran, casual in Italy.
Sometimes confused with *E. amethystinum* (a perennial mountain species), but readily distinguished by the divaricately branched synflorescence and short, ± palmatifid leaves.

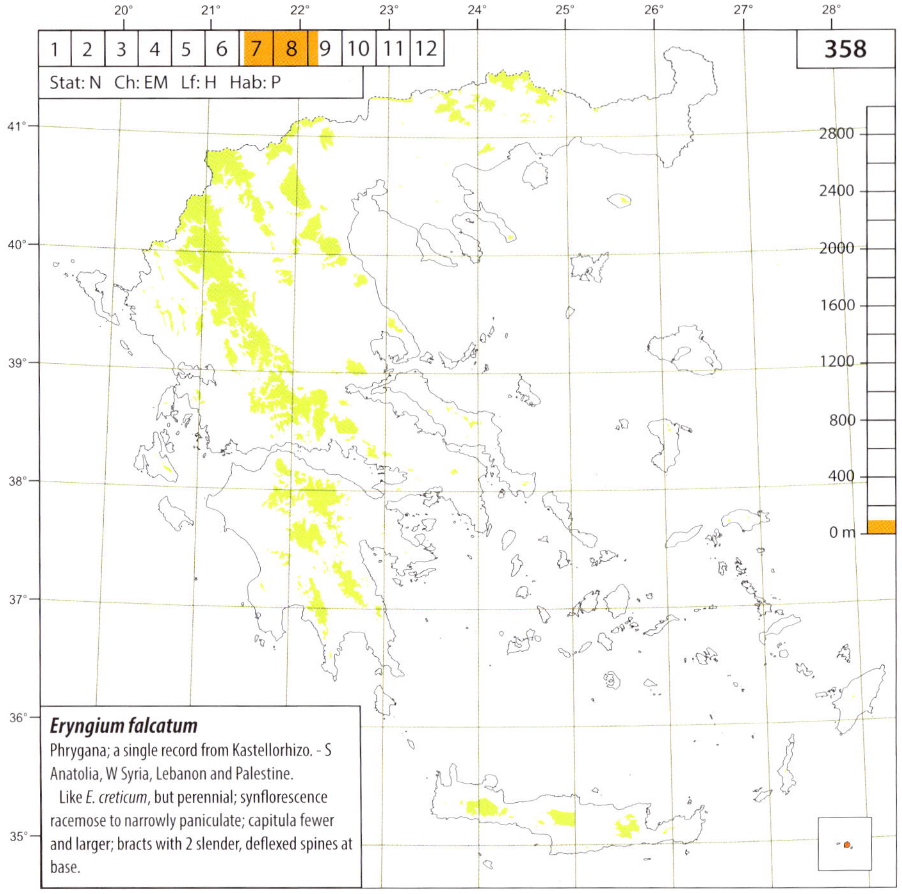

Eryngium falcatum
Phrygana; a single record from Kastellorhizo. - S Anatolia, W Syria, Lebanon and Palestine.
Like *E. creticum*, but perennial; synflorescence racemose to narrowly paniculate; capitula fewer and larger; bracts with 2 slender, deflexed spines at base.

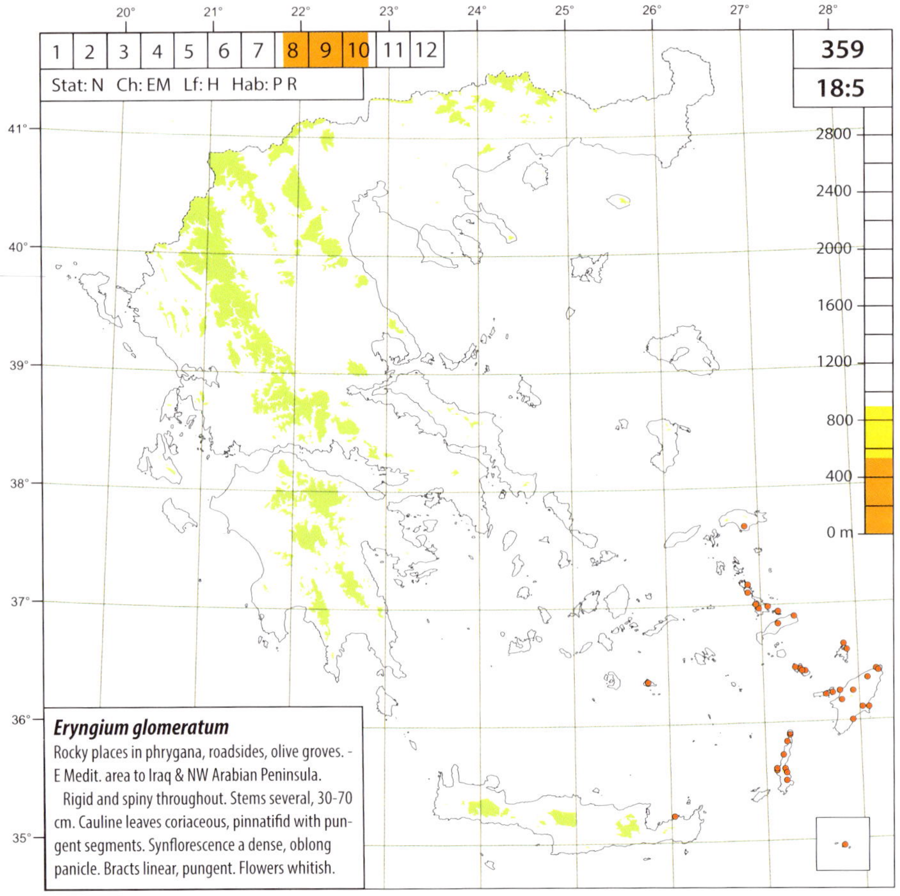

Eryngium glomeratum
Rocky places in phrygana, roadsides, olive groves. - E Medit. area to Iraq & NW Arabian Peninsula.
Rigid and spiny throughout. Stems several, 30-70 cm. Cauline leaves coriaceous, pinnatifid with pungent segments. Synflorescence a dense, oblong panicle. Bracts linear, pungent. Flowers whitish.

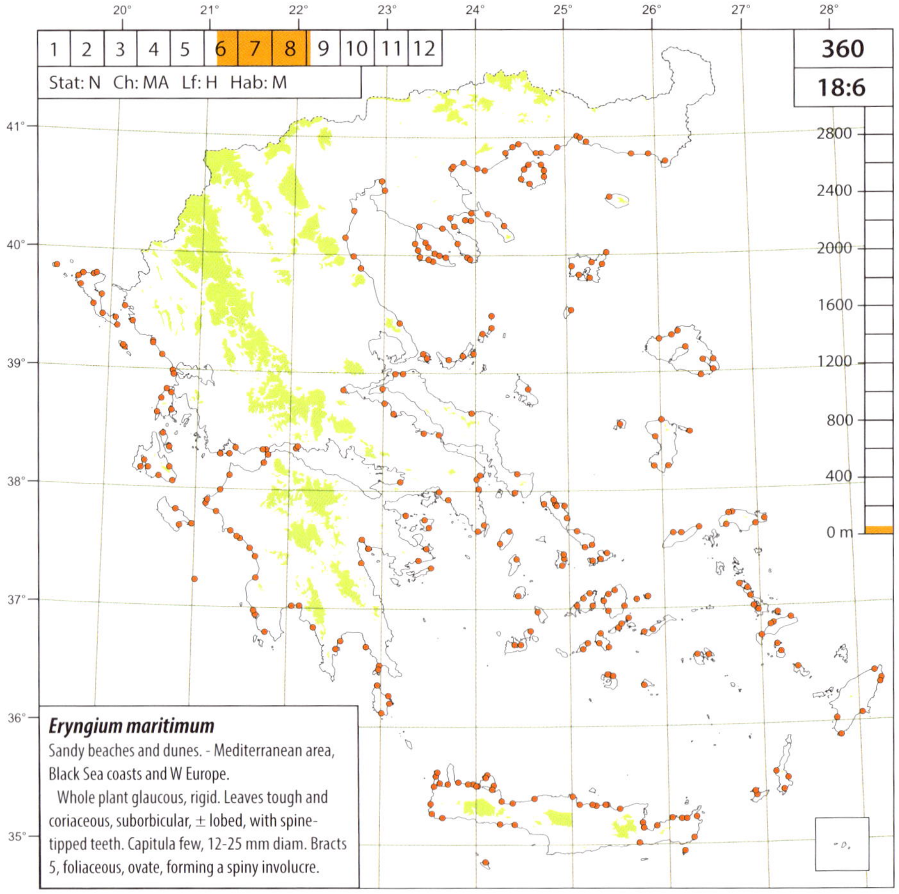

Eryngium maritimum
Sandy beaches and dunes. - Mediterranean area, Black Sea coasts and W Europe.
Whole plant glaucous, rigid. Leaves tough and coriaceous, suborbicular, ± lobed, with spine-tipped teeth. Capitula few, 12-25 mm diam. Bracts 5, foliaceous, ovate, forming a spiny involucre.

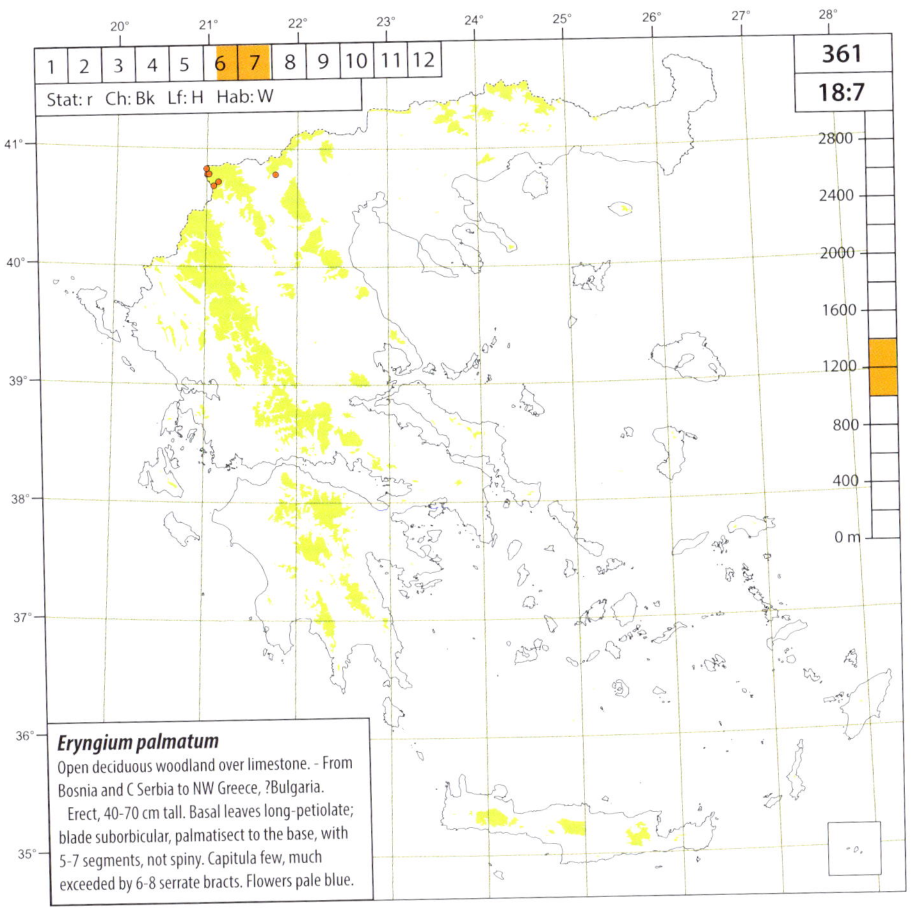

Eryngium palmatum
Open deciduous woodland over limestone. - From Bosnia and C Serbia to NW Greece, ?Bulgaria.
Erect, 40-70 cm tall. Basal leaves long-petiolate; blade suborbicular, palmatisect to the base, with 5-7 segments, not spiny. Capitula few, much exceeded by 6-8 serrate bracts. Flowers pale blue.

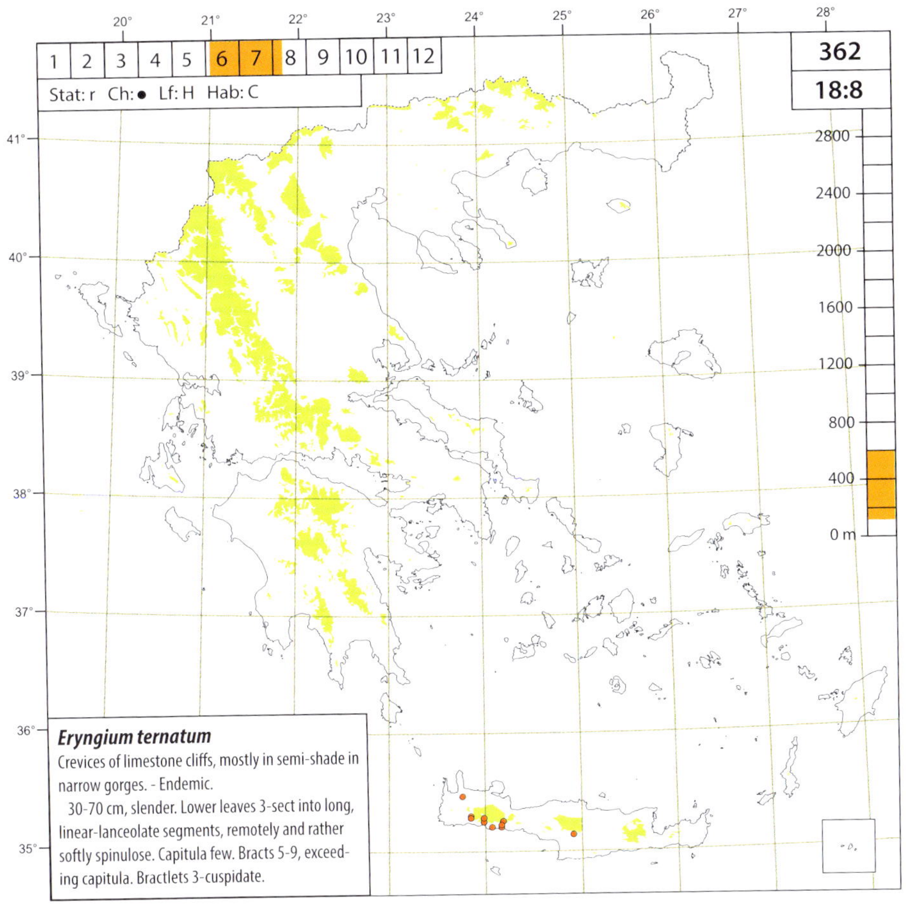

Eryngium ternatum
Crevices of limestone cliffs, mostly in semi-shade in narrow gorges. - Endemic.
30-70 cm, slender. Lower leaves 3-sect into long, linear-lanceolate segments, remotely and rather softly spinulose. Capitula few. Bracts 5-9, exceeding capitula. Bractlets 3-cuspidate.

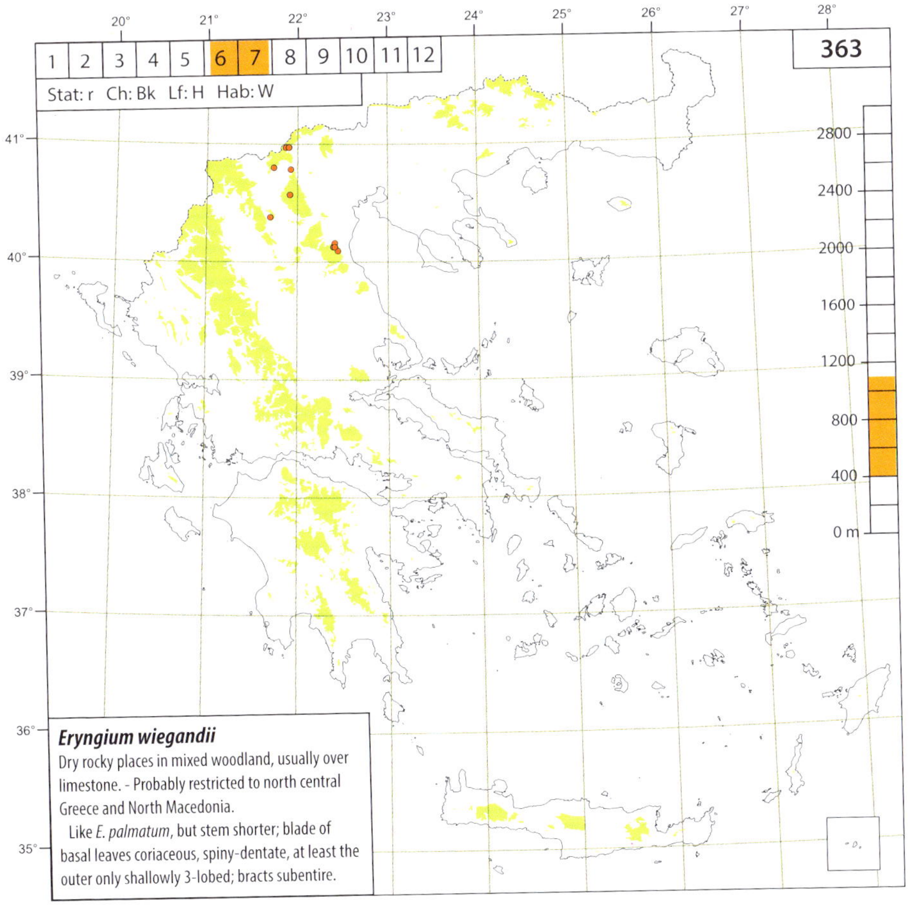

Eryngium wiegandii
Dry rocky places in mixed woodland, usually over limestone. - Probably restricted to north central Greece and North Macedonia.
Like *E. palmatum*, but stem shorter; blade of basal leaves coriaceous, spiny-dentate, at least the outer only shallowly 3-lobed; bracts subentire.

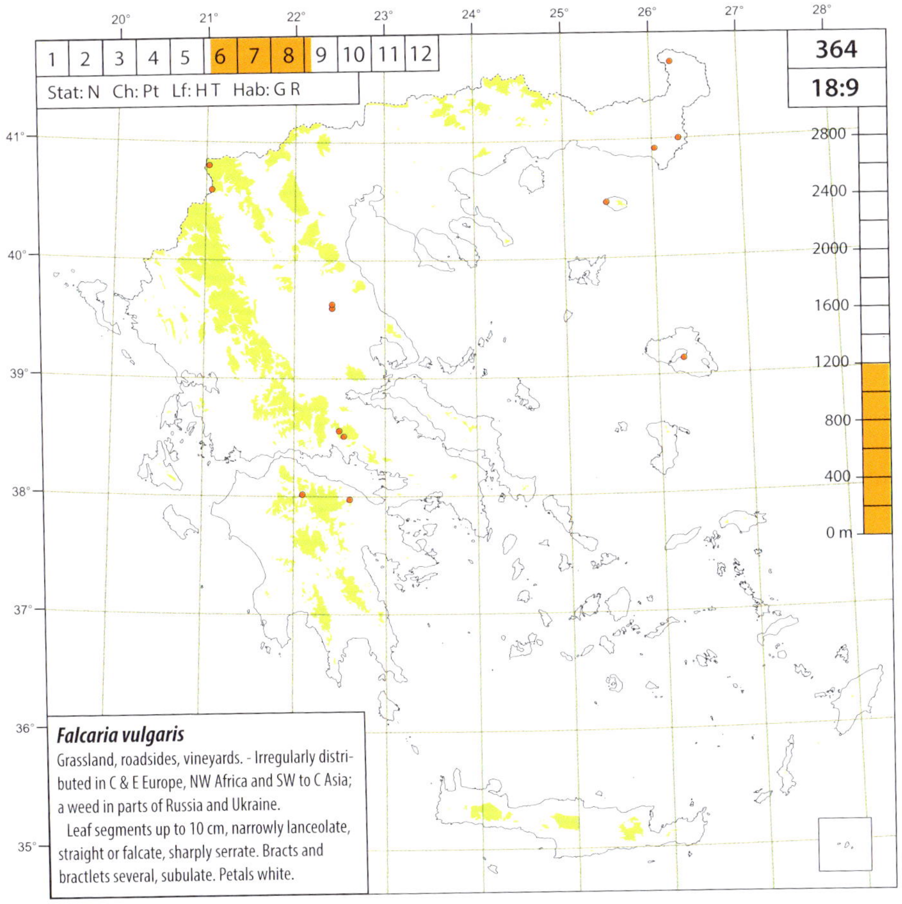

Falcaria vulgaris
Grassland, roadsides, vineyards. - Irregularly distributed in C & E Europe, NW Africa and SW to C Asia; a weed in parts of Russia and Ukraine.
Leaf segments up to 10 cm, narrowly lanceolate, straight or falcate, sharply serrate. Bracts and bractlets several, subulate. Petals white.

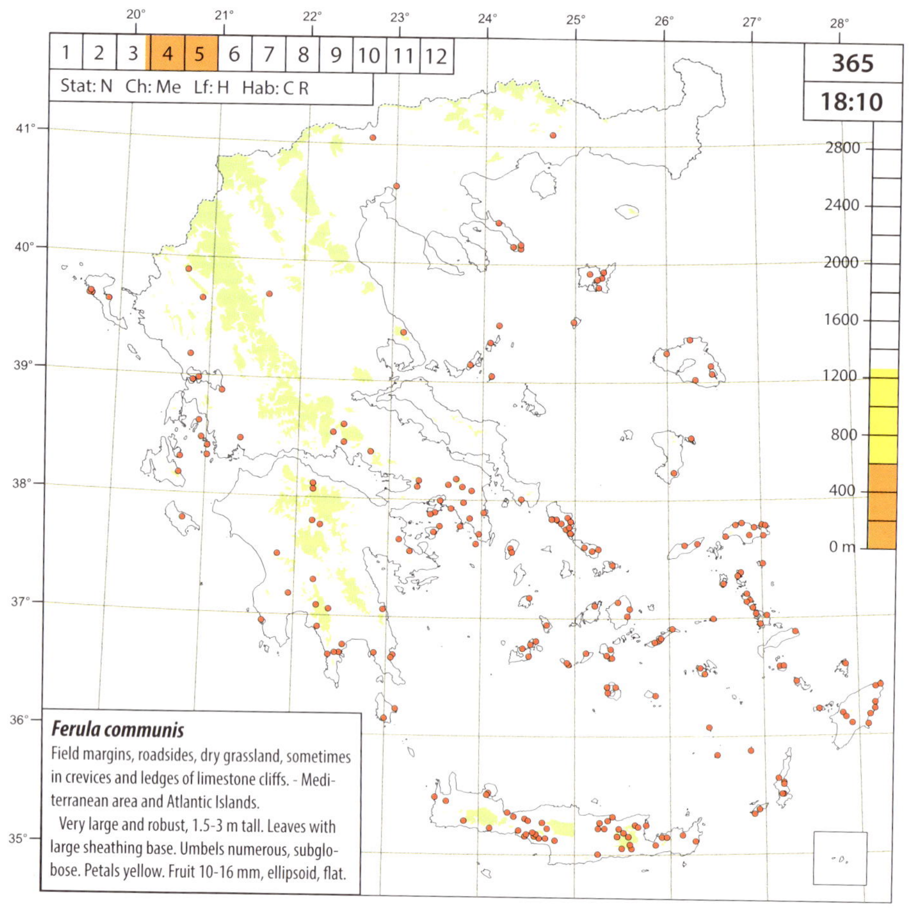

Ferula communis

Field margins, roadsides, dry grassland, sometimes in crevices and ledges of limestone cliffs. - Mediterranean area and Atlantic Islands.

Very large and robust, 1.5-3 m tall. Leaves with large sheathing base. Umbels numerous, subglobose. Petals yellow. Fruit 10-16 mm, ellipsoid, flat.

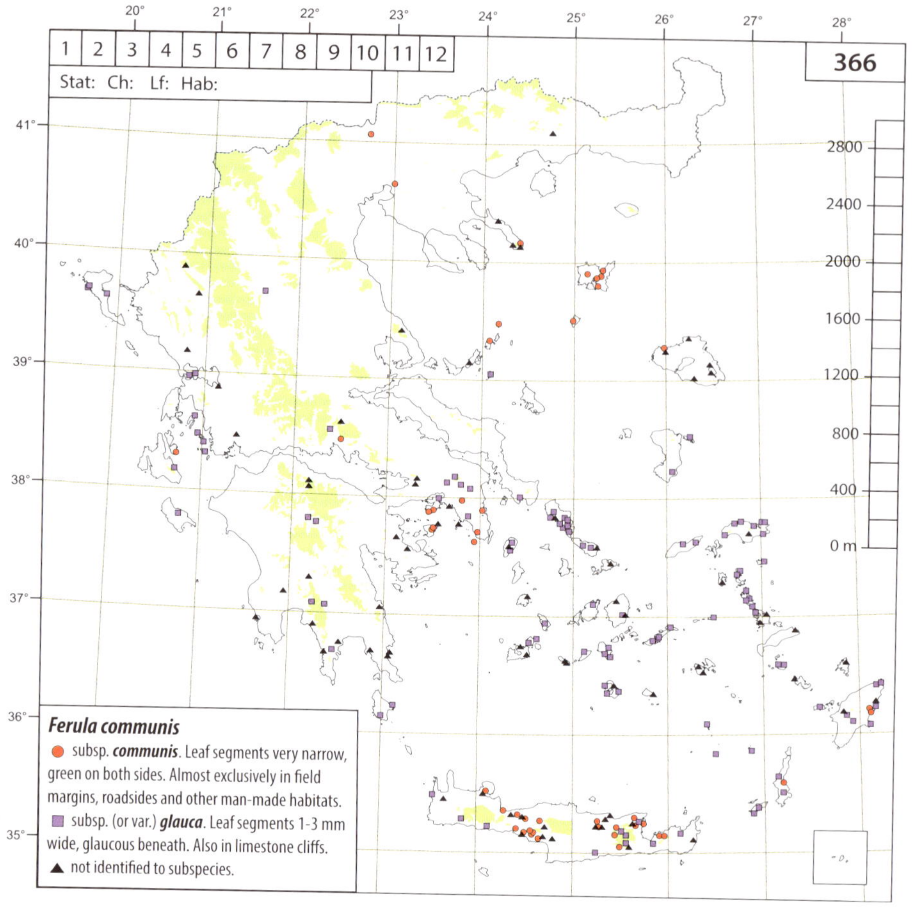

Ferula communis

● subsp. ***communis***. Leaf segments very narrow, green on both sides. Almost exclusively in field margins, roadsides and other man-made habitats.

■ subsp. (or var.) ***glauca***. Leaf segments 1-3 mm wide, glaucous beneath. Also in limestone cliffs.

▲ not identified to subspecies.

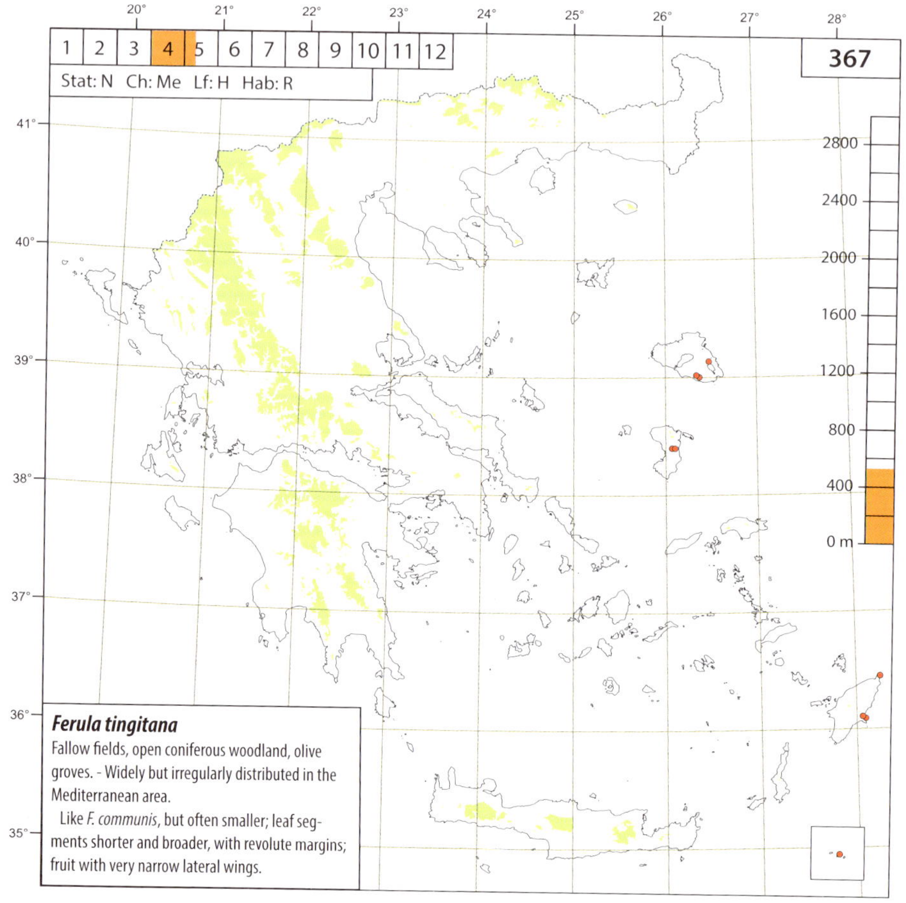

Ferula tingitana

Fallow fields, open coniferous woodland, olive groves. - Widely but irregularly distributed in the Mediterranean area.

Like *F. communis*, but often smaller; leaf segments shorter and broader, with revolute margins; fruit with very narrow lateral wings.

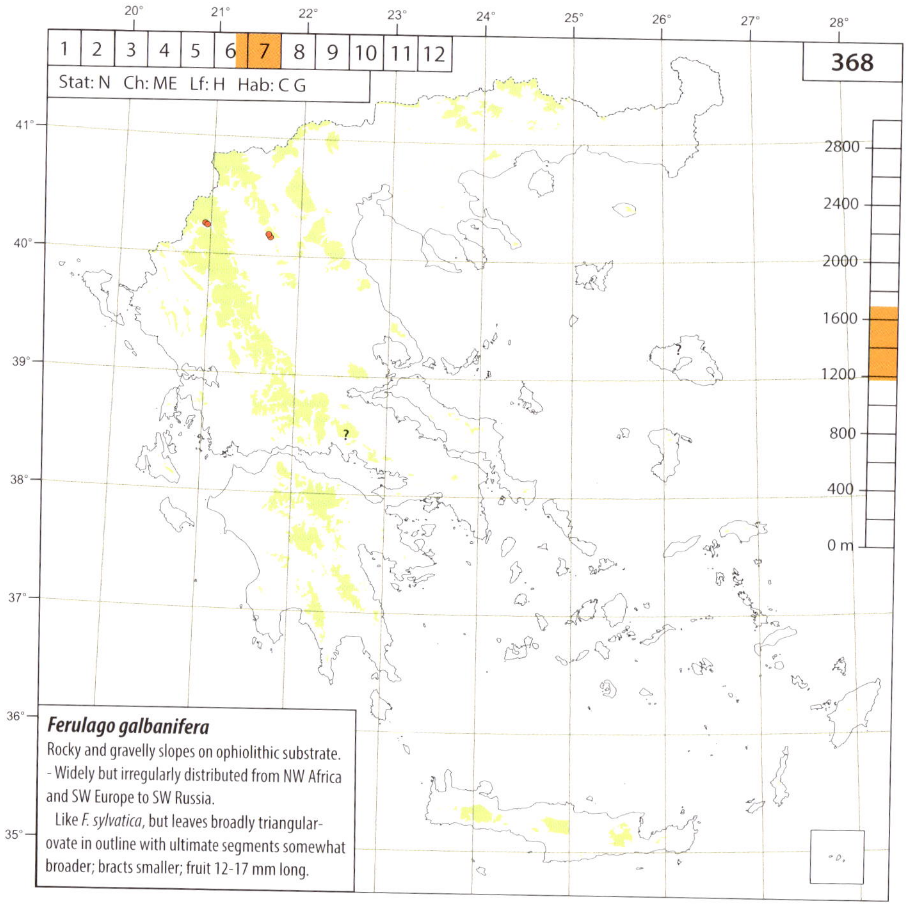

Ferulago galbanifera

Rocky and gravelly slopes on ophiolithic substrate. - Widely but irregularly distributed from NW Africa and SW Europe to SW Russia.

Like *F. sylvatica*, but leaves broadly triangular-ovate in outline with ultimate segments somewhat broader; bracts smaller; fruit 12-17 mm long.

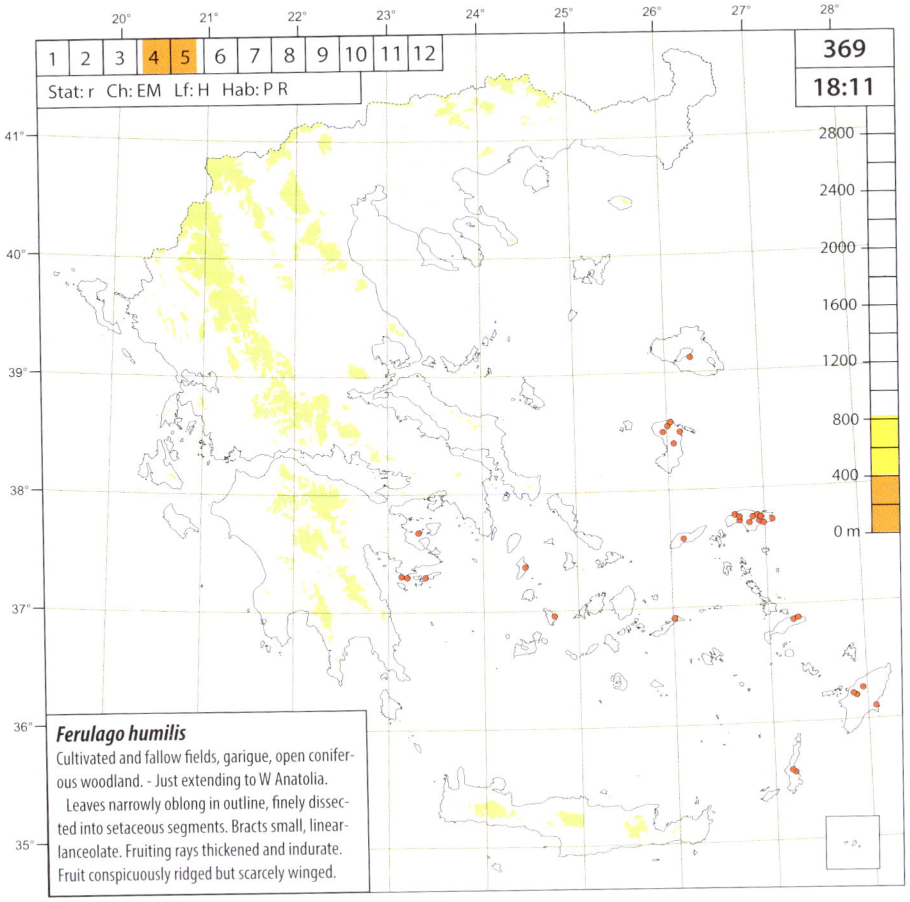

Ferulago humilis
Cultivated and fallow fields, garigue, open coniferous woodland. - Just extending to W Anatolia.
Leaves narrowly oblong in outline, finely dissected into setaceous segments. Bracts small, linear-lanceolate. Fruiting rays thickened and indurate. Fruit conspicuously ridged but scarcely winged.

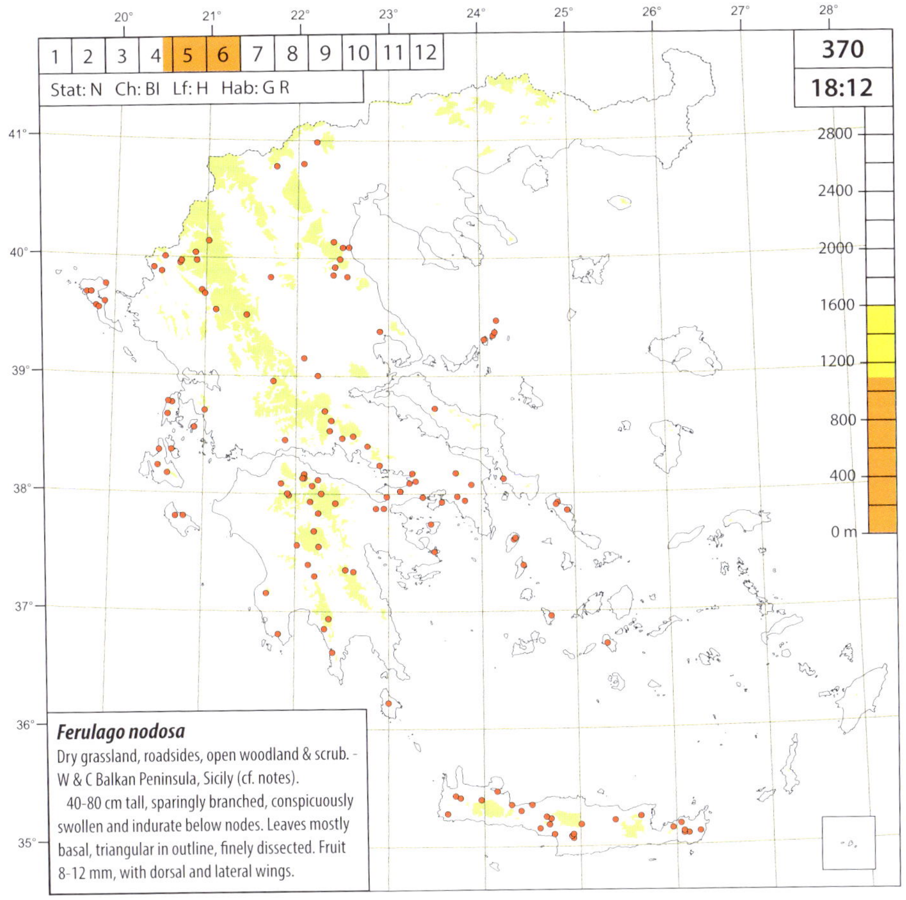

Ferulago nodosa
Dry grassland, roadsides, open woodland & scrub. - W & C Balkan Peninsula, Sicily (cf. notes).
40-80 cm tall, sparingly branched, conspicuously swollen and indurate below nodes. Leaves mostly basal, triangular in outline, finely dissected. Fruit 8-12 mm, with dorsal and lateral wings.

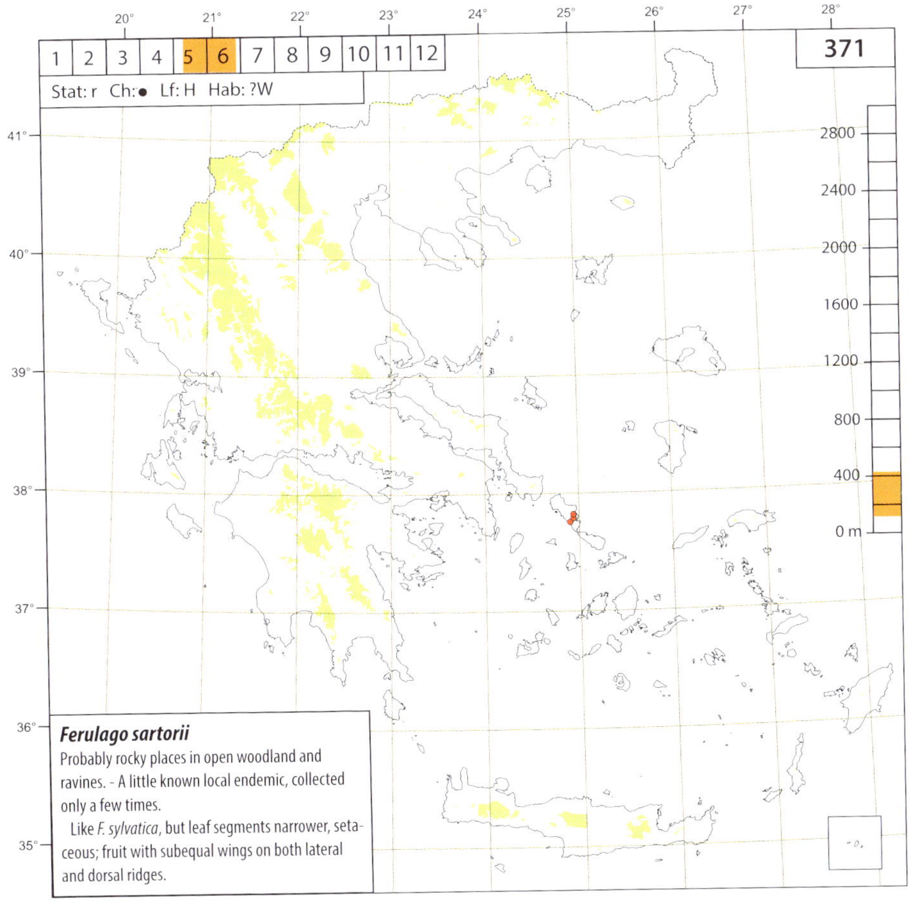

Ferulago sartorii
Probably rocky places in open woodland and ravines. - A little known local endemic, collected only a few times.
Like *F. sylvatica*, but leaf segments narrower, setaceous; fruit with subequal wings on both lateral and dorsal ridges.

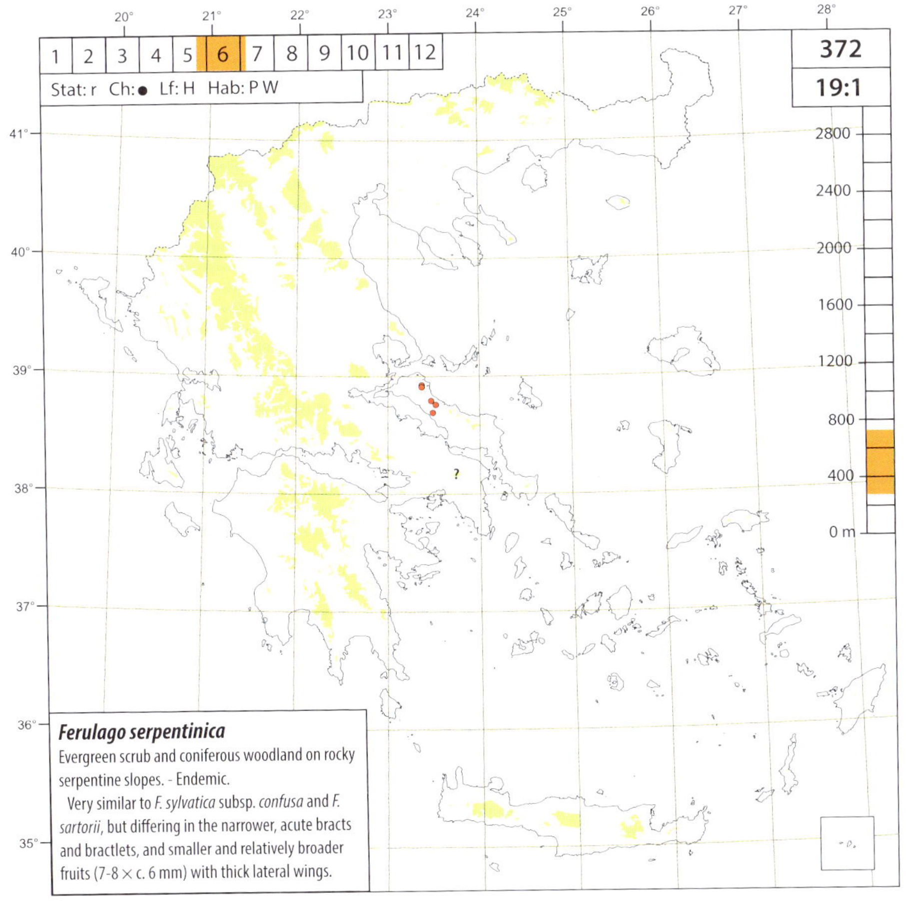

Ferulago serpentinica
Evergreen scrub and coniferous woodland on rocky serpentine slopes. - Endemic.
Very similar to *F. sylvatica* subsp. *confusa* and *F. sartorii*, but differing in the narrower, acute bracts and bractlets, and smaller and relatively broader fruits (7-8 × c. 6 mm) with thick lateral wings.

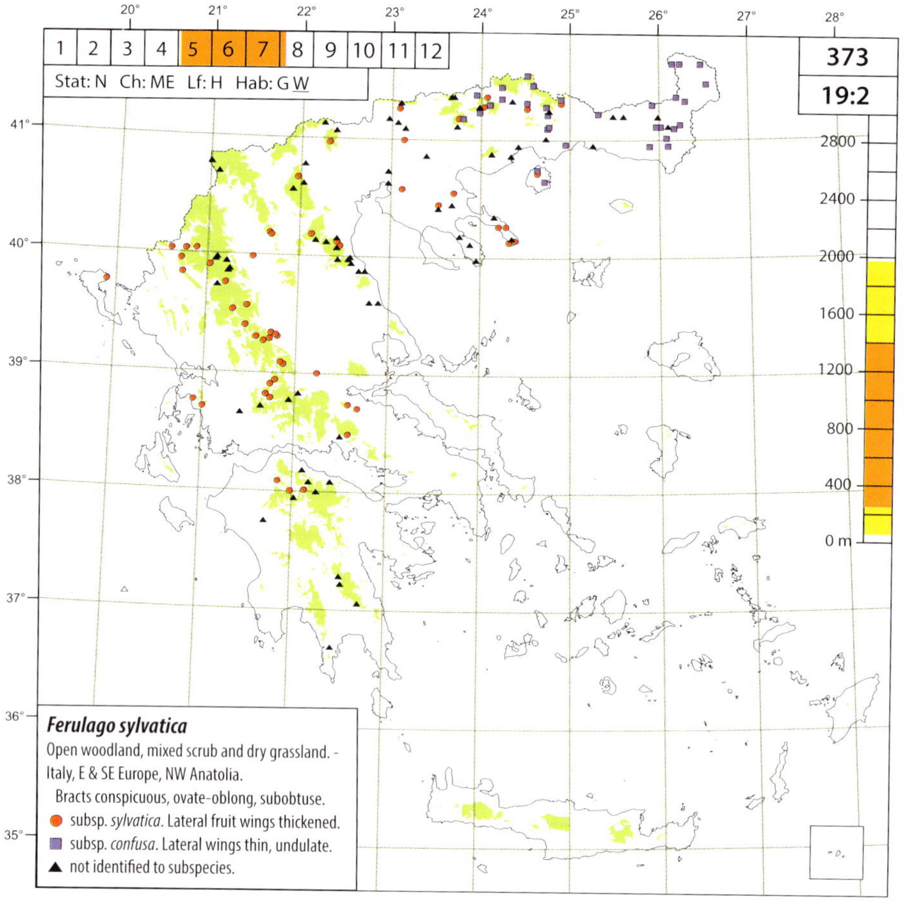

Ferulago sylvatica
Open woodland, mixed scrub and dry grassland. - Italy, E & SE Europe, NW Anatolia.
Bracts conspicuous, ovate-oblong, subobtuse.
● subsp. *sylvatica*. Lateral fruit wings thickened.
■ subsp. *confusa*. Lateral wings thin, undulate.
▲ not identified to subspecies.

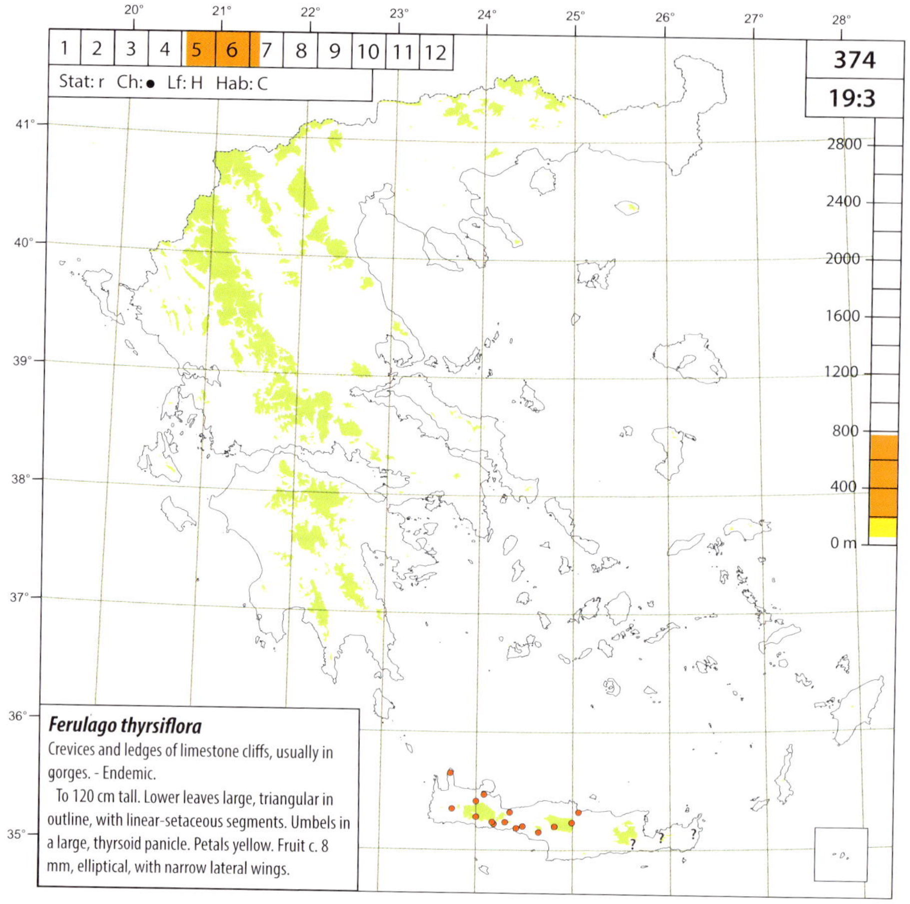

Ferulago thyrsiflora
Crevices and ledges of limestone cliffs, usually in gorges. - Endemic.
To 120 cm tall. Lower leaves large, triangular in outline, with linear-setaceous segments. Umbels in a large, thyrsoid panicle. Petals yellow. Fruit c. 8 mm, elliptical, with narrow lateral wings.

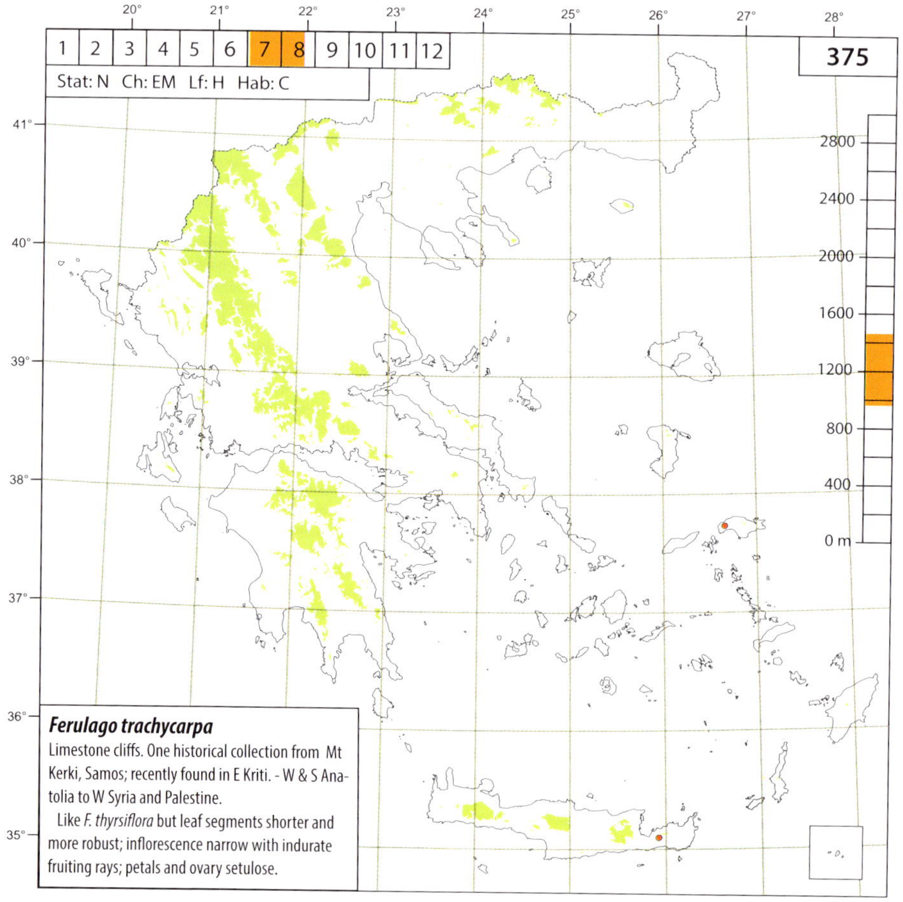

Ferulago trachycarpa
Limestone cliffs. One historical collection from Mt Kerki, Samos; recently found in E Kriti. - W & S Anatolia to W Syria and Palestine.
Like *F. thyrsiflora* but leaf segments shorter and more robust; inflorescence narrow with indurate fruiting rays; petals and ovary setulose.

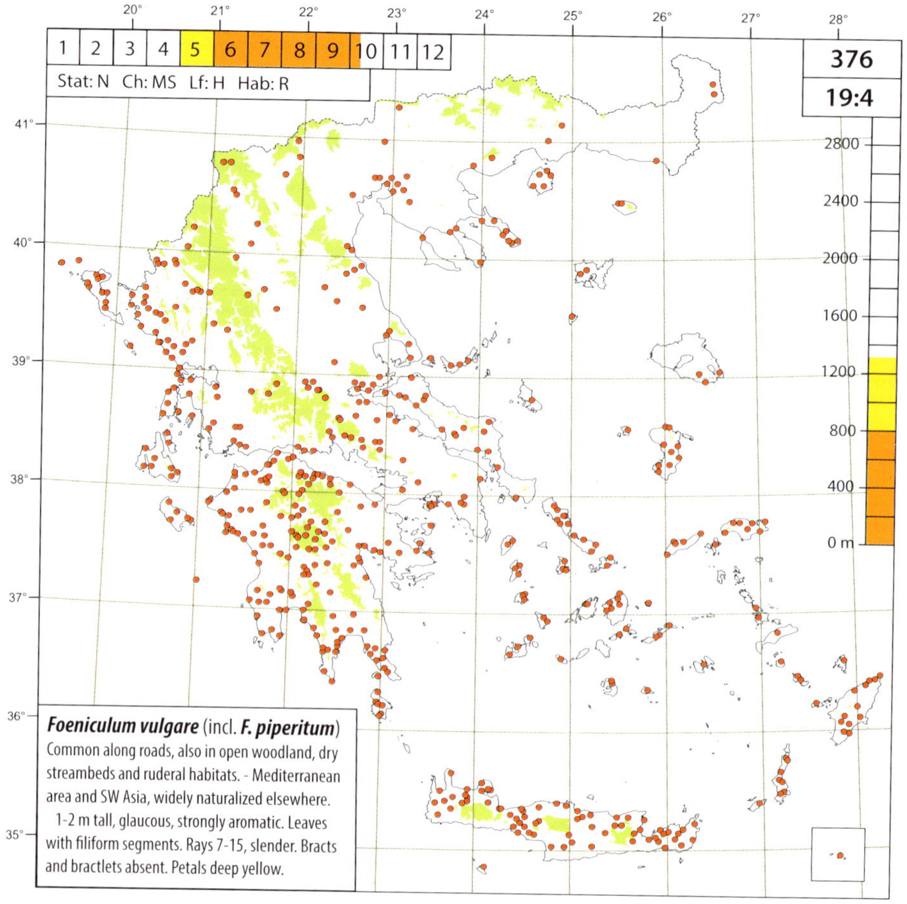

Foeniculum vulgare (incl. ***F. piperitum***)
Common along roads, also in open woodland, dry streambeds and ruderal habitats. - Mediterranean area and SW Asia, widely naturalized elsewhere.
1-2 m tall, glaucous, strongly aromatic. Leaves with filiform segments. Rays 7-15, slender. Bracts and bractlets absent. Petals deep yellow.

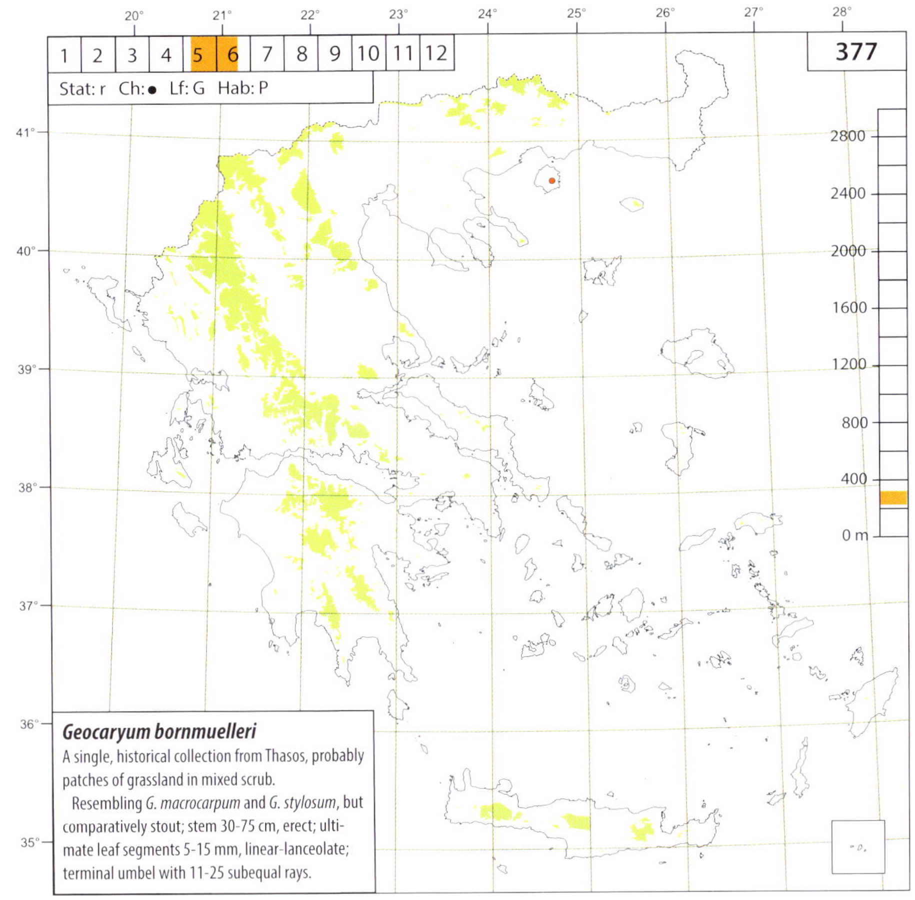

Geocaryum bornmuelleri

A single, historical collection from Thasos, probably patches of grassland in mixed scrub.

Resembling *G. macrocarpum* and *G. stylosum*, but comparatively stout; stem 30-75 cm, erect; ultimate leaf segments 5-15 mm, linear-lanceolate; terminal umbel with 11-25 subequal rays.

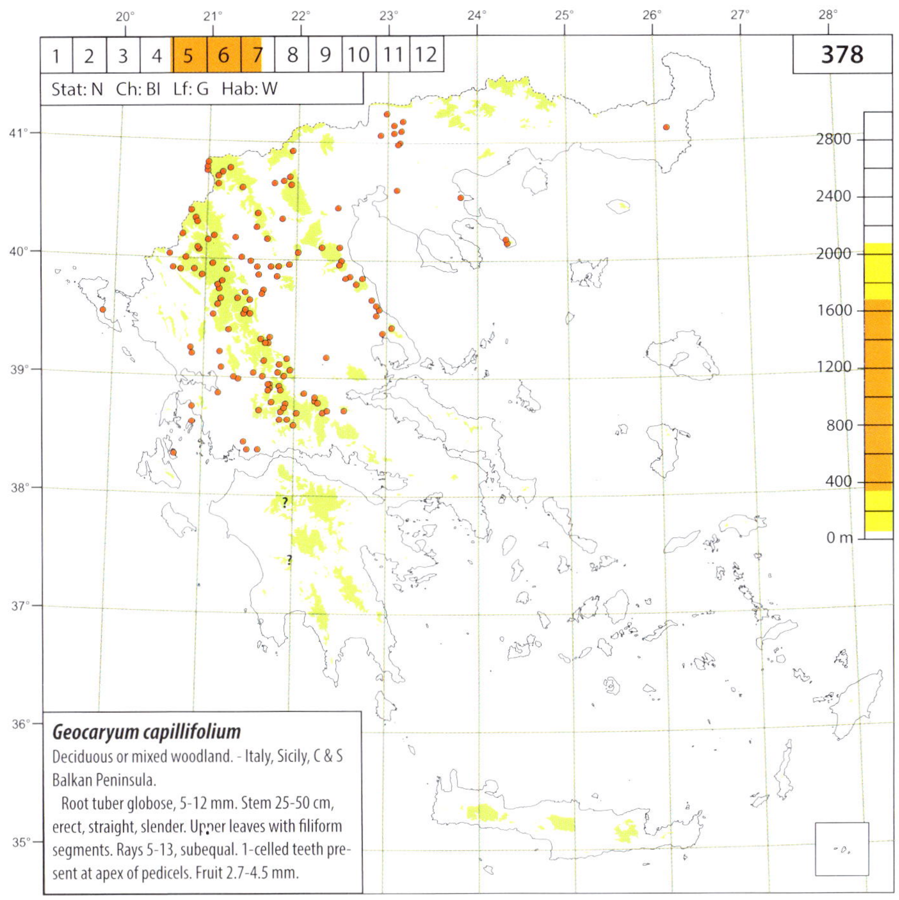

Geocaryum capillifolium

Deciduous or mixed woodland. - Italy, Sicily, C & S Balkan Peninsula.

Root tuber globose, 5-12 mm. Stem 25-50 cm, erect, straight, slender. Upper leaves with filiform segments. Rays 5-13, subequal. 1-celled teeth present at apex of pedicels. Fruit 2.7-4.5 mm.

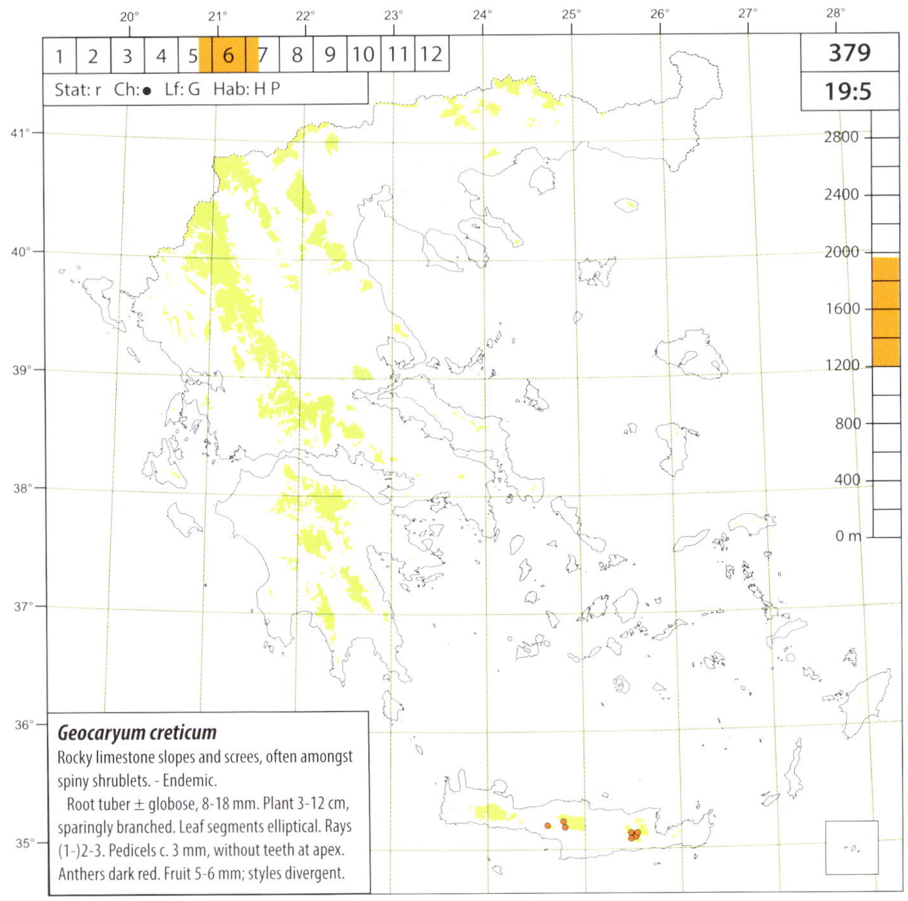

Geocaryum creticum

Rocky limestone slopes and screes, often amongst spiny shrublets. - Endemic.

Root tuber ± globose, 8-18 mm. Plant 3-12 cm, sparingly branched. Leaf segments elliptical. Rays (1-)2-3. Pedicels c. 3 mm, without teeth at apex. Anthers dark red. Fruit 5-6 mm; styles divergent.

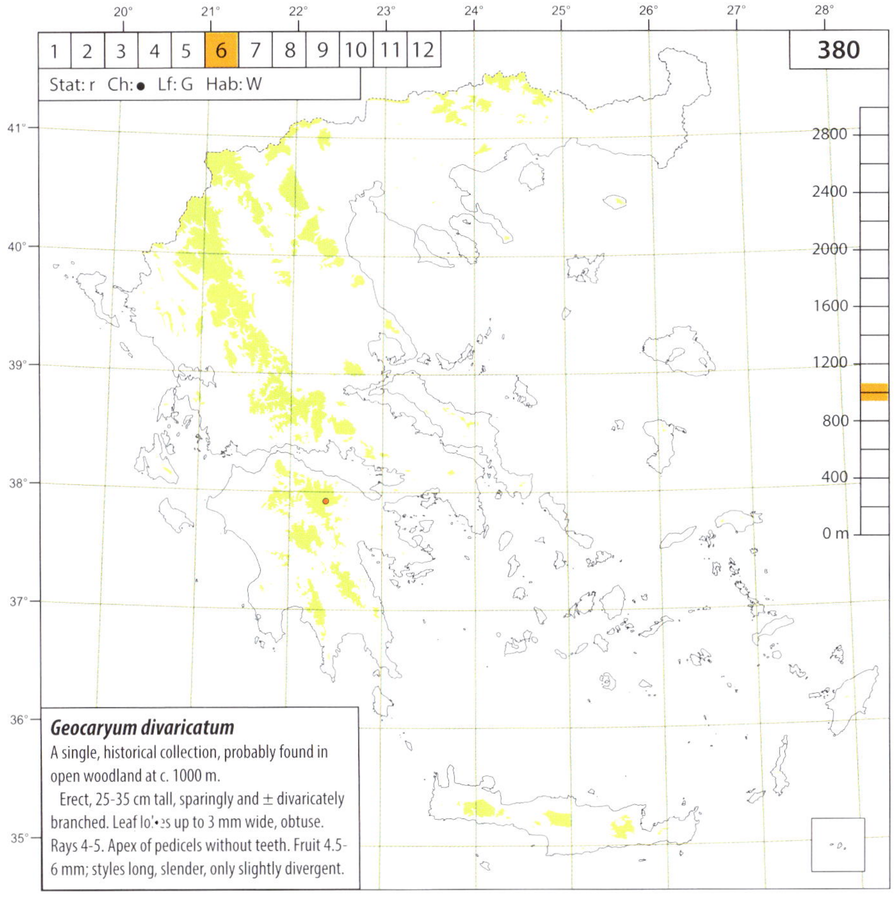

Geocaryum divaricatum

A single, historical collection, probably found in open woodland at c. 1000 m.

Erect, 25-35 cm tall, sparingly and ± divaricately branched. Leaf lobes up to 3 mm wide, obtuse. Rays 4-5. Apex of pedicels without teeth. Fruit 4.5-6 mm; styles long, slender, only slightly divergent.

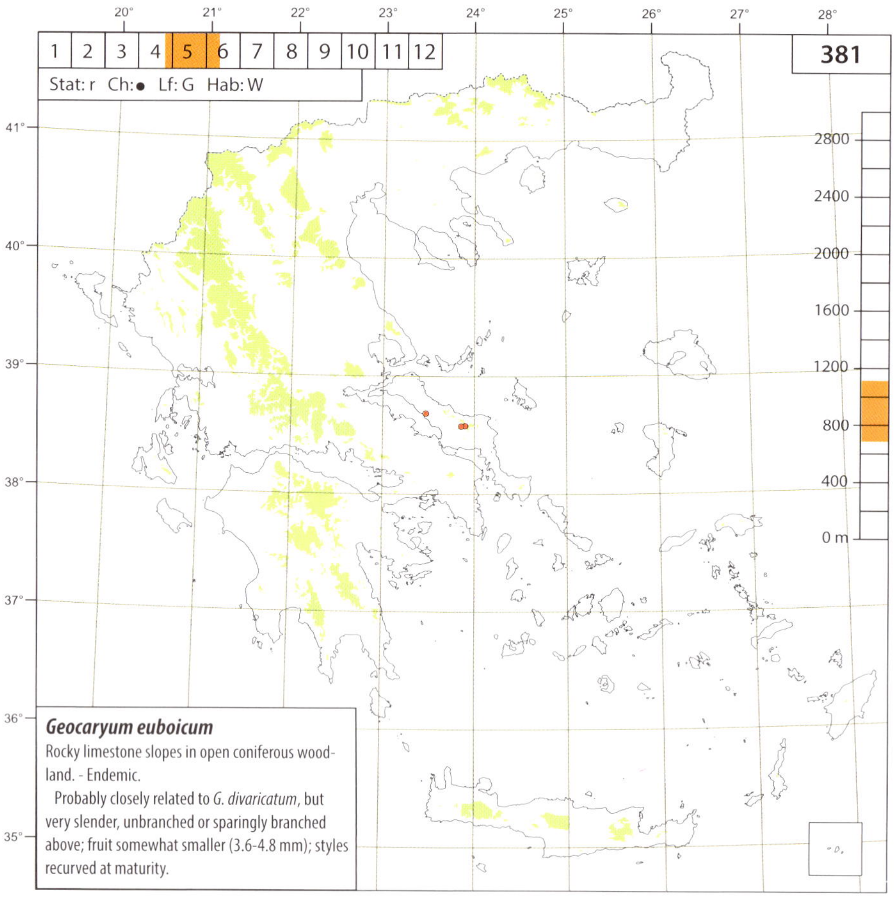

Geocaryum euboicum
Rocky limestone slopes in open coniferous woodland. - Endemic.
Probably closely related to *G. divaricatum*, but very slender, unbranched or sparingly branched above; fruit somewhat smaller (3.6-4.8 mm); styles recurved at maturity.

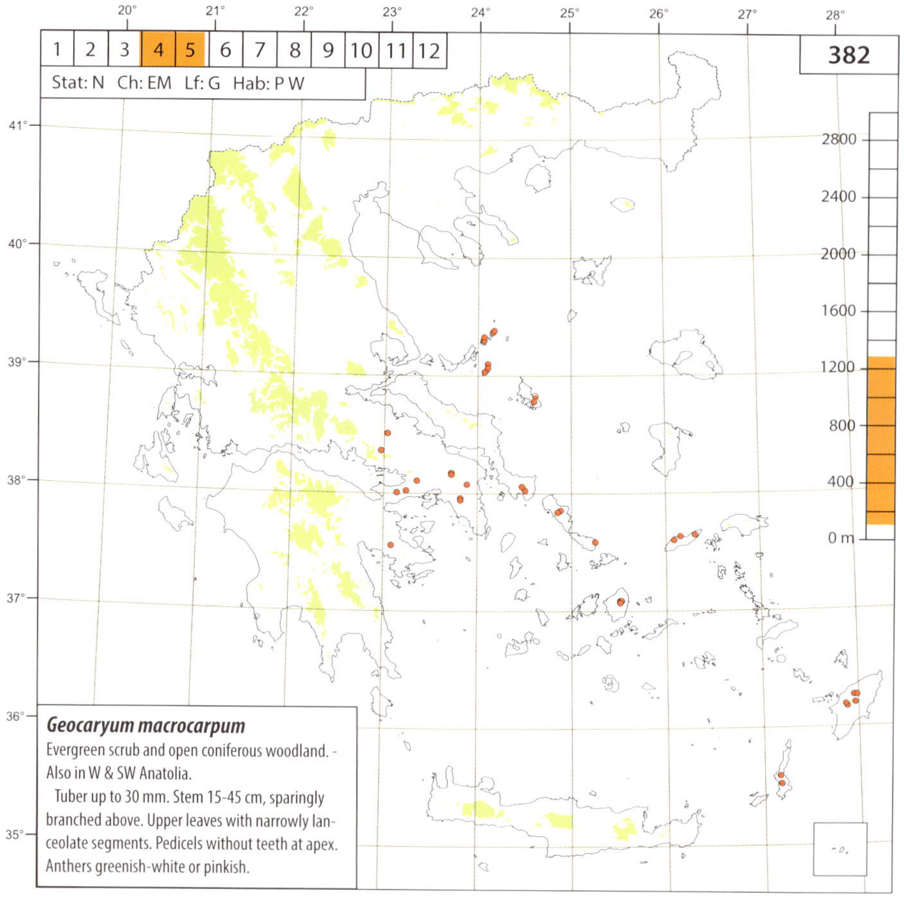

Geocaryum macrocarpum
Evergreen scrub and open coniferous woodland. - Also in W & SW Anatolia.
Tuber up to 30 mm. Stem 15-45 cm, sparingly branched above. Upper leaves with narrowly lanceolate segments. Pedicels without teeth at apex. Anthers greenish-white or pinkish.

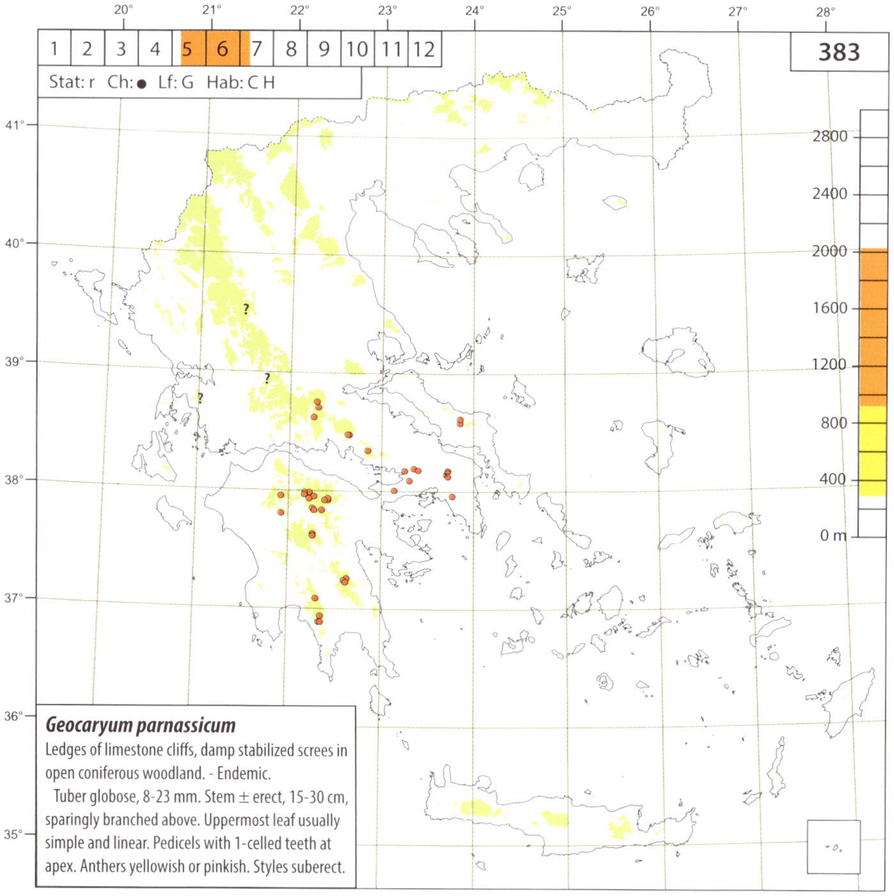

Geocaryum parnassicum
Ledges of limestone cliffs, damp stabilized screes in open coniferous woodland. - Endemic.
Tuber globose, 8-23 mm. Stem ± erect, 15-30 cm, sparingly branched above. Uppermost leaf usually simple and linear. Pedicels with 1-celled teeth at apex. Anthers yellowish or pinkish. Styles suberect.

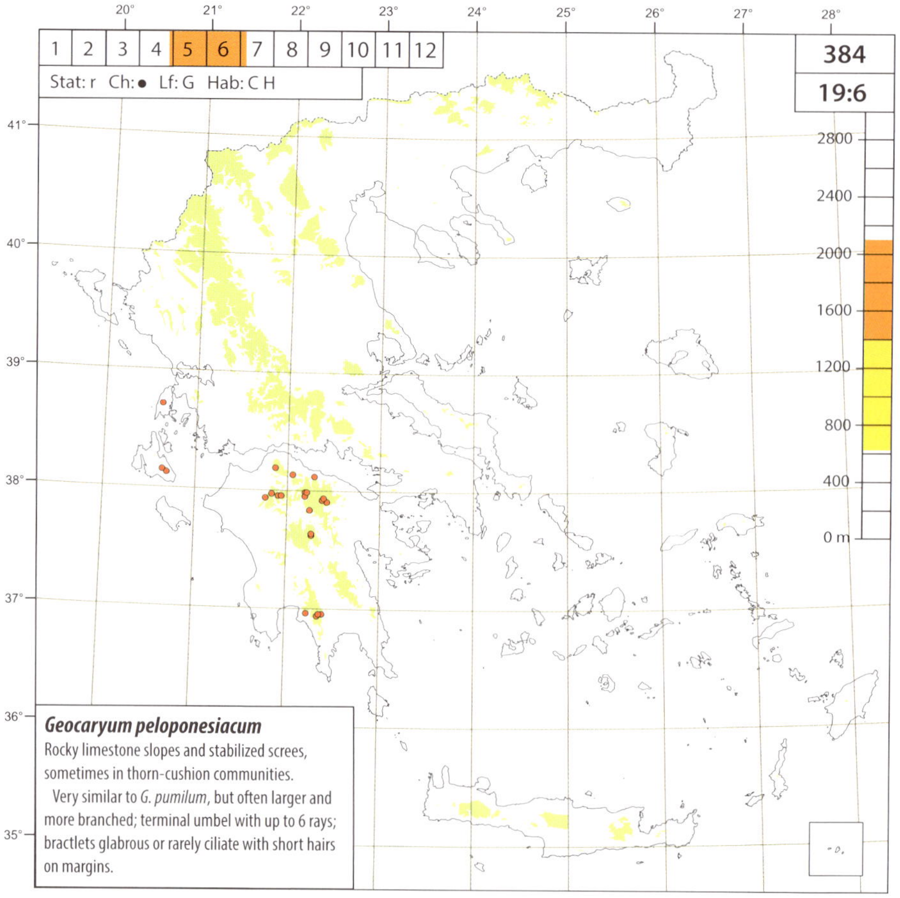

Geocaryum peloponesiacum
Rocky limestone slopes and stabilized screes, sometimes in thorn-cushion communities.
Very similar to *G. pumilum*, but often larger and more branched; terminal umbel with up to 6 rays; bractlets glabrous or rarely ciliate with short hairs on margins.

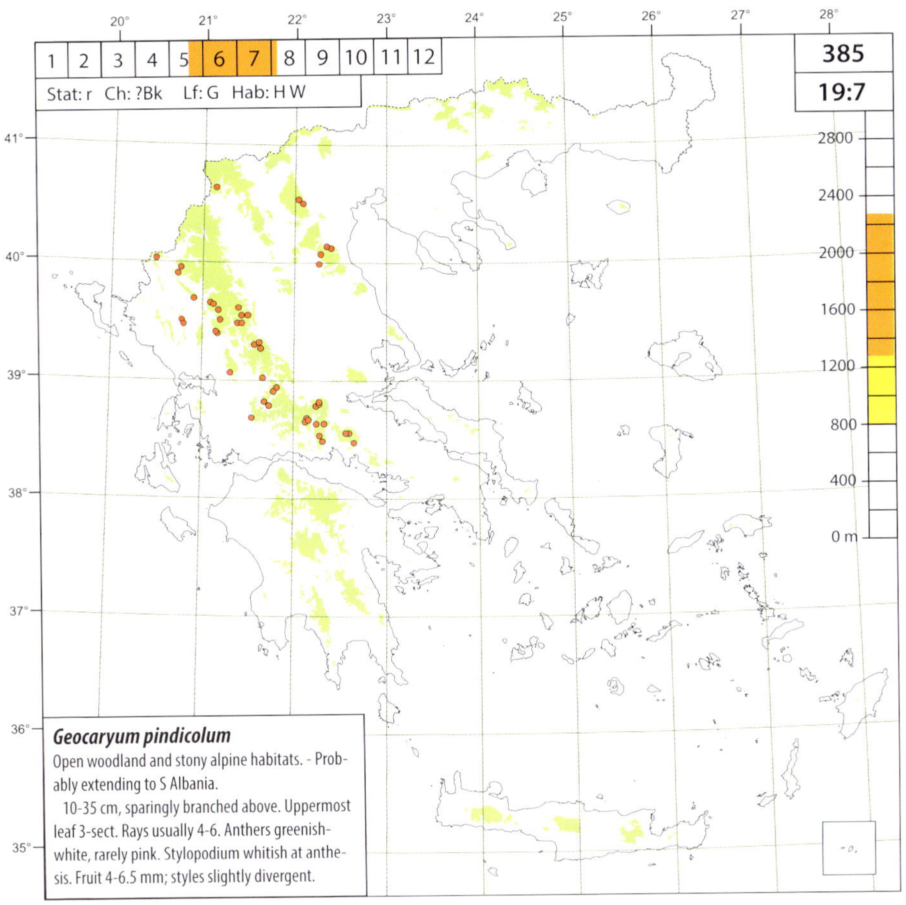

Geocaryum pindicolum

Open woodland and stony alpine habitats. - Probably extending to S Albania.

10-35 cm, sparingly branched above. Uppermost leaf 3-sect. Rays usually 4-6. Anthers greenish-white, rarely pink. Stylopodium whitish at anthesis. Fruit 4-6.5 mm; styles slightly divergent.

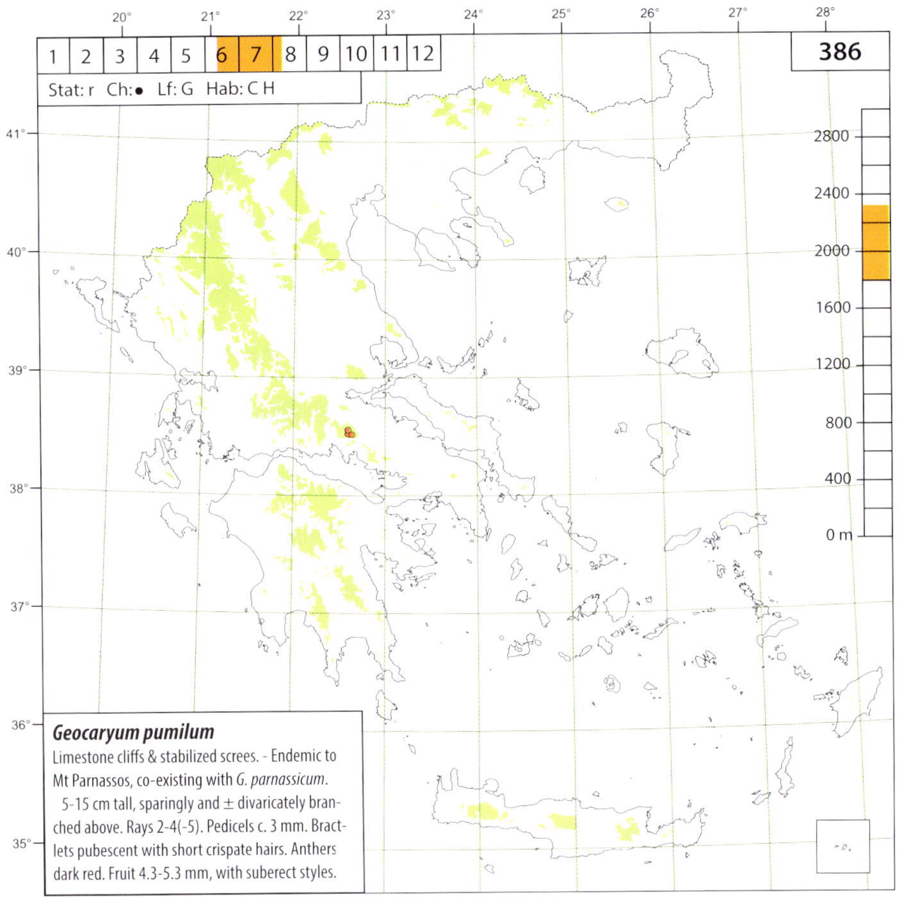

Geocaryum pumilum

Limestone cliffs & stabilized screes. - Endemic to Mt Parnassos, co-existing with *G. parnassicum*.

5-15 cm tall, sparingly and ± divaricately branched above. Rays 2-4(-5). Pedicels c. 3 mm. Bractlets pubescent with short crispate hairs. Anthers dark red. Fruit 4.3-5.3 mm, with suberect styles.

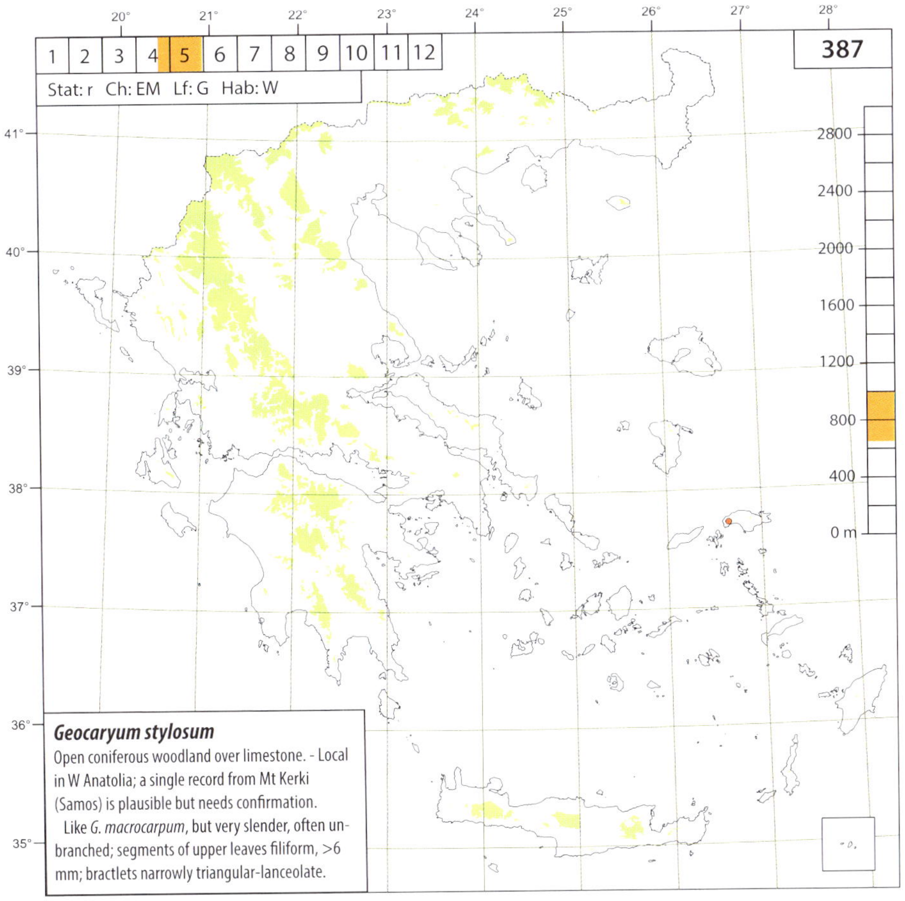

Geocaryum stylosum

Open coniferous woodland over limestone. - Local in W Anatolia; a single record from Mt Kerki (Samos) is plausible but needs confirmation.

Like *G. macrocarpum*, but very slender, often unbranched; segments of upper leaves filiform, >6 mm; bractlets narrowly triangular-lanceolate.

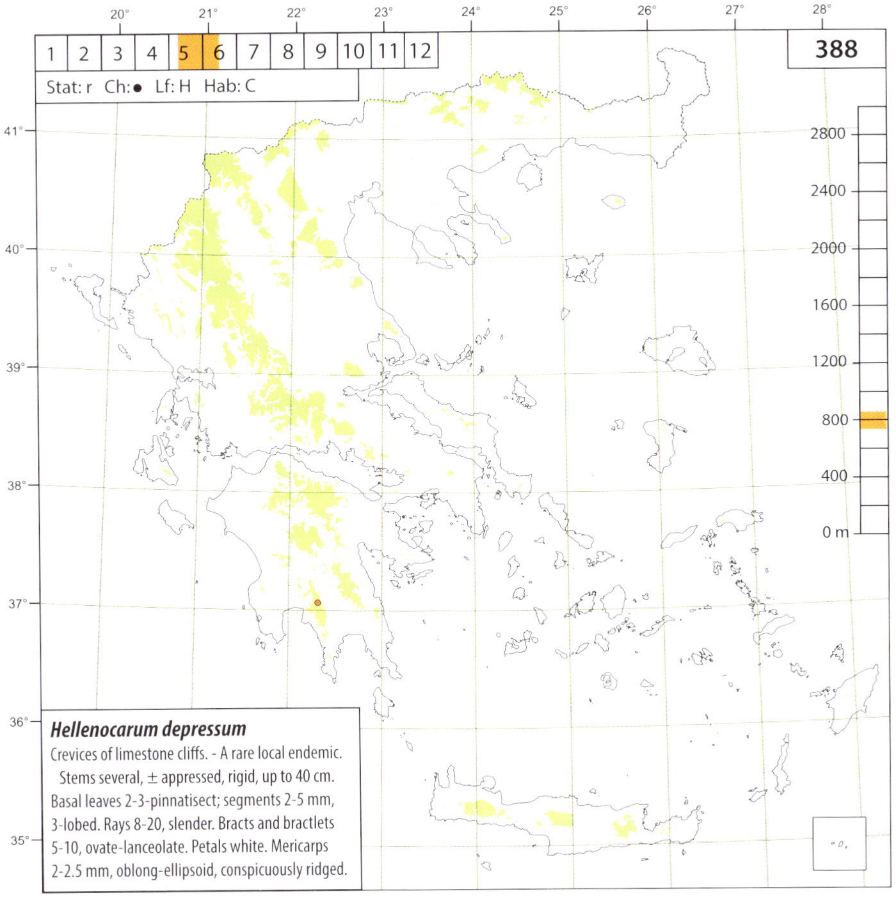

Hellenocarum depressum

Crevices of limestone cliffs. - A rare local endemic.

Stems several, ± appressed, rigid, up to 40 cm. Basal leaves 2-3-pinnatisect; segments 2-5 mm, 3-lobed. Rays 8-20, slender. Bracts and bractlets 5-10, ovate-lanceolate. Petals white. Mericarps 2-2.5 mm, oblong-ellipsoid, conspicuously ridged.

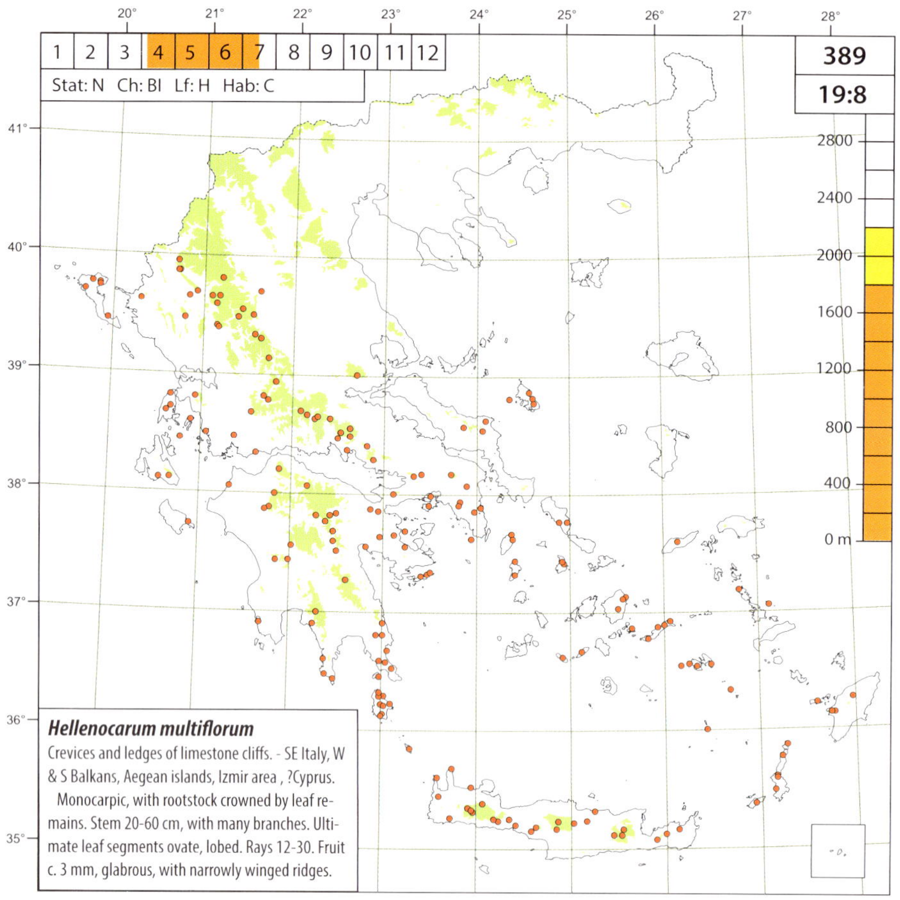

Hellenocarum multiflorum
Crevices and ledges of limestone cliffs. - SE Italy, W & S Balkans, Aegean islands, Izmir area, ?Cyprus.
Monocarpic, with rootstock crowned by leaf remains. Stem 20-60 cm, with many branches. Ultimate leaf segments ovate, lobed. Rays 12-30. Fruit c. 3 mm, glabrous, with narrowly winged ridges.

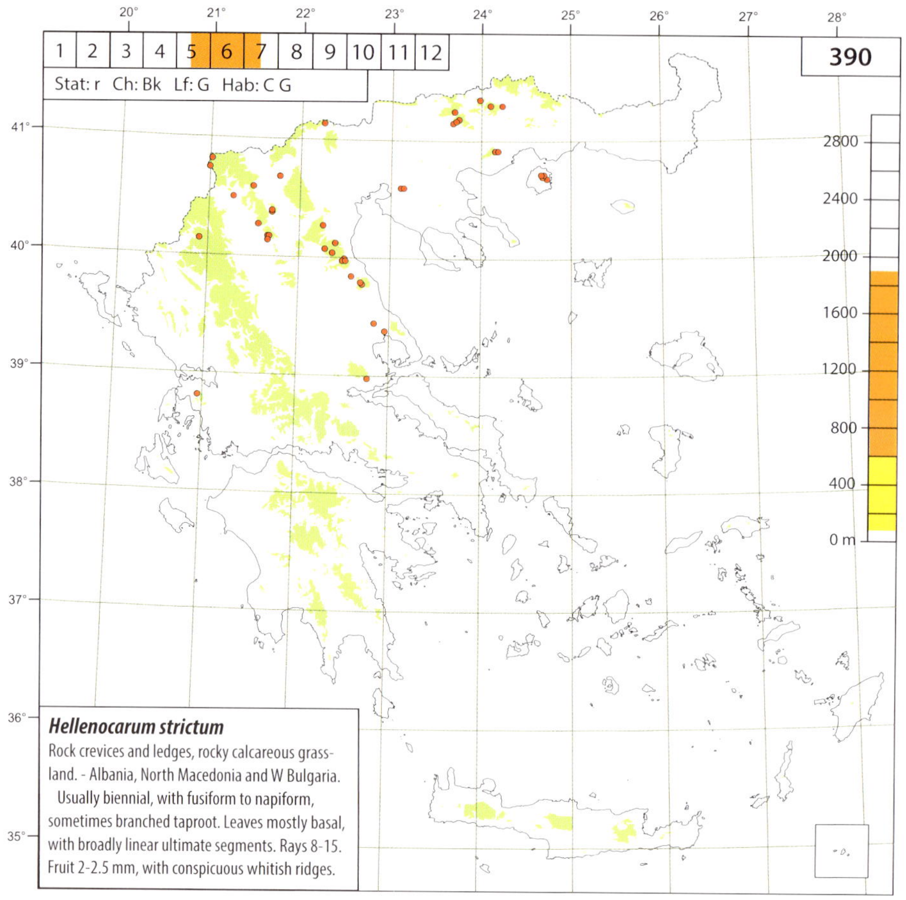

Hellenocarum strictum
Rock crevices and ledges, rocky calcareous grassland. - Albania, North Macedonia and W Bulgaria.
Usually biennial, with fusiform to napiform, sometimes branched taproot. Leaves mostly basal, with broadly linear ultimate segments. Rays 8-15. Fruit 2-2.5 mm, with conspicuous whitish ridges.

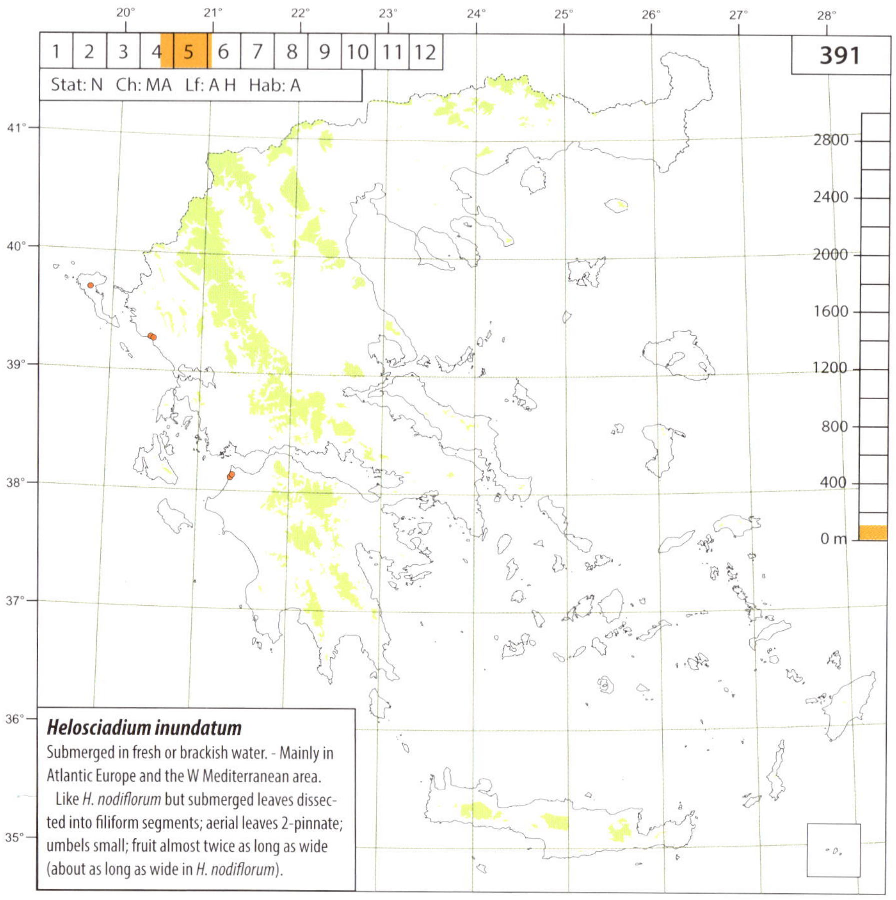

Helosciadium inundatum
Submerged in fresh or brackish water. - Mainly in Atlantic Europe and the W Mediterranean area.
Like *H. nodiflorum* but submerged leaves dissected into filiform segments; aerial leaves 2-pinnate; umbels small; fruit almost twice as long as wide (about as long as wide in *H. nodiflorum*).

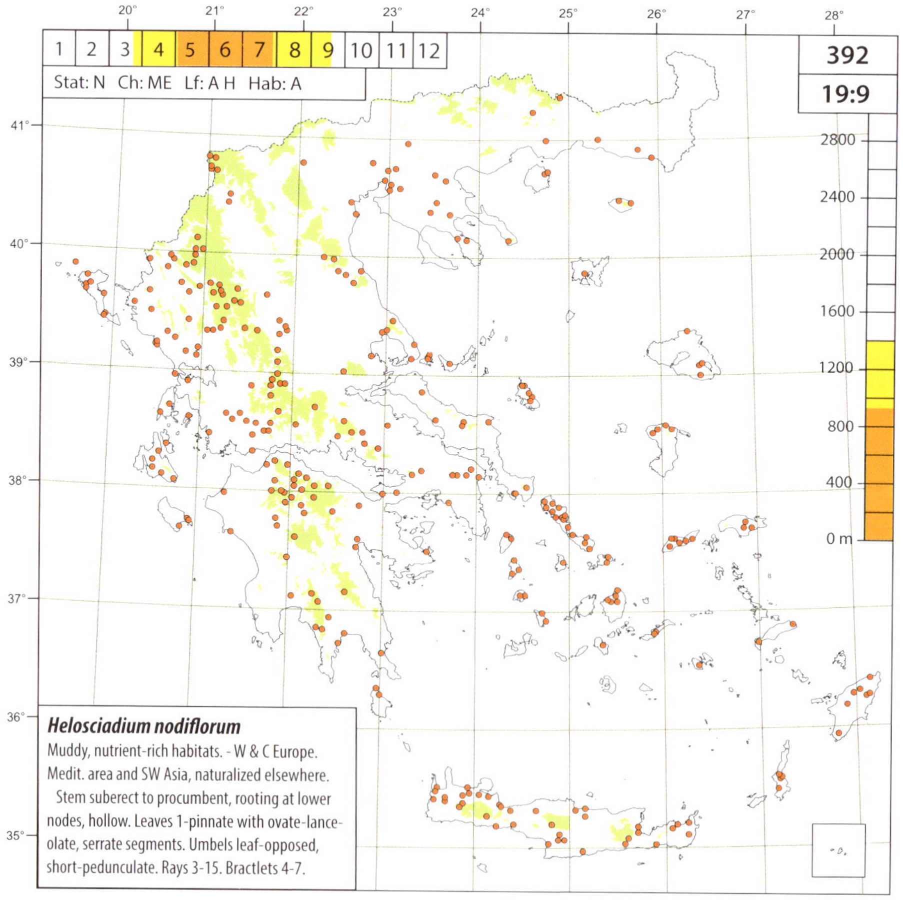

Helosciadium nodiflorum
Muddy, nutrient-rich habitats. - W & C Europe. Medit. area and SW Asia, naturalized elsewhere.
Stem suberect to procumbent, rooting at lower nodes, hollow. Leaves 1-pinnate with ovate-lanceolate, serrate segments. Umbels leaf-opposed, short-pedunculate. Rays 3-15. Bractlets 4-7.

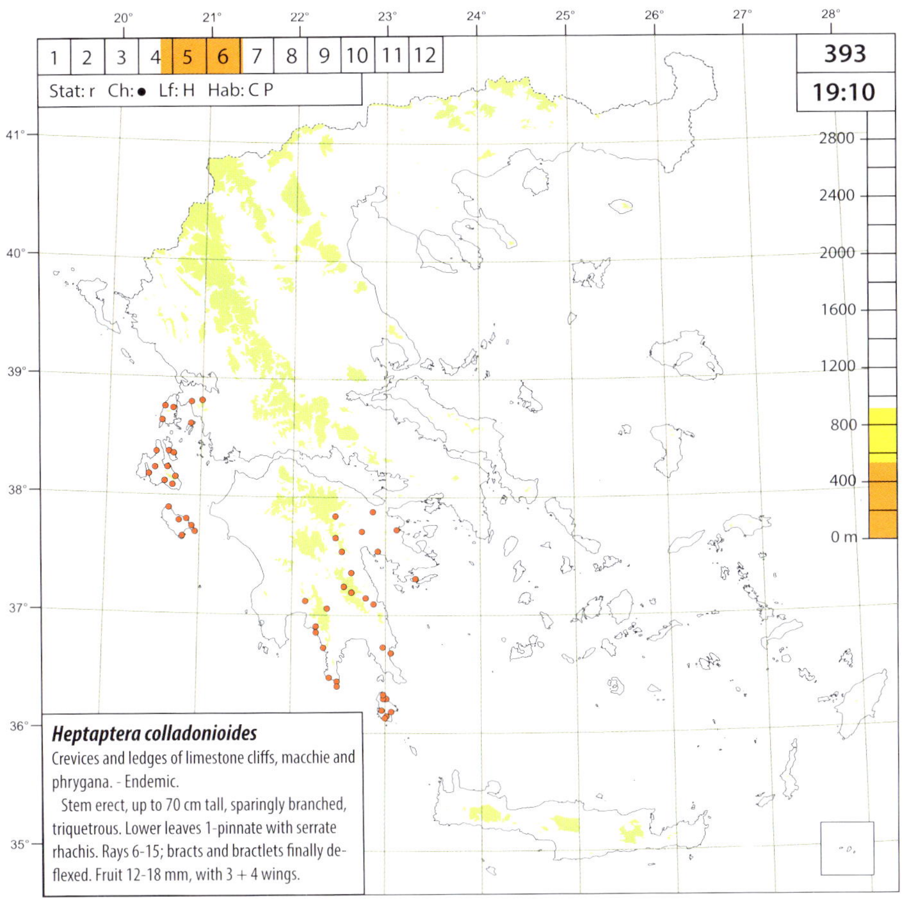

Heptaptera colladonioides
Crevices and ledges of limestone cliffs, macchie and phrygana. - Endemic.
Stem erect, up to 70 cm tall, sparingly branched, triquetrous. Lower leaves 1-pinnate with serrate rhachis. Rays 6-15; bracts and bractlets finally deflexed. Fruit 12-18 mm, with 3 + 4 wings.

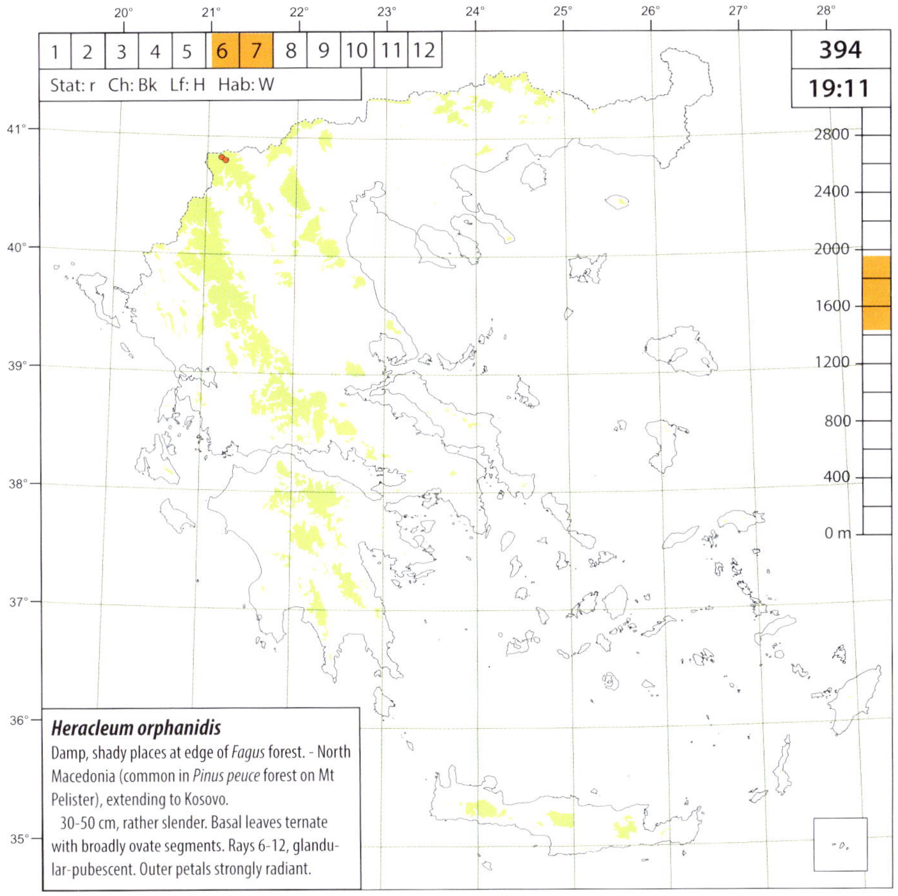

Heracleum orphanidis
Damp, shady places at edge of *Fagus* forest. - North Macedonia (common in *Pinus peuce* forest on Mt Pelister), extending to Kosovo.
30-50 cm, rather slender. Basal leaves ternate with broadly ovate segments. Rays 6-12, glandular-pubescent. Outer petals strongly radiant.

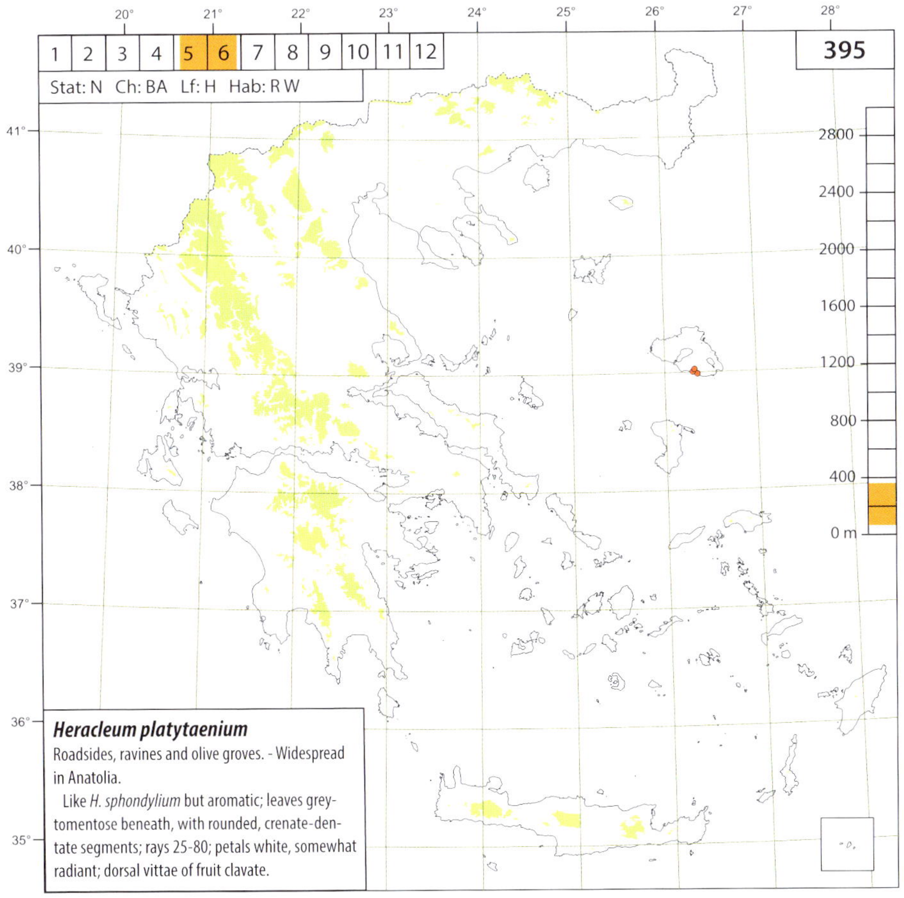

Heracleum platytaenium
Roadsides, ravines and olive groves. - Widespread in Anatolia.
Like *H. sphondylium* but aromatic; leaves grey-tomentose beneath, with rounded, crenate-dentate segments; rays 25-80; petals white, somewhat radiant; dorsal vittae of fruit clavate.

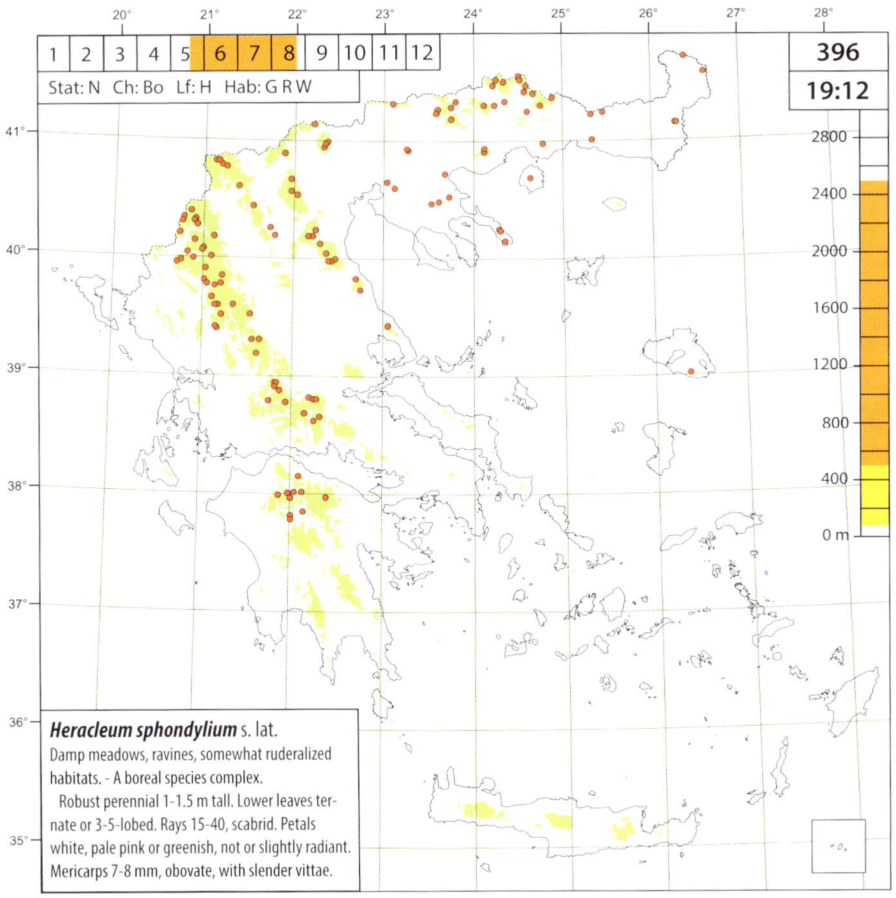

Heracleum sphondylium s. lat.
Damp meadows, ravines, somewhat ruderalized habitats. - A boreal species complex.
Robust perennial 1-1.5 m tall. Lower leaves ternate or 3-5-lobed. Rays 15-40, scabrid. Petals white, pale pink or greenish, not or slightly radiant. Mericarps 7-8 mm, obovate, with slender vittae.

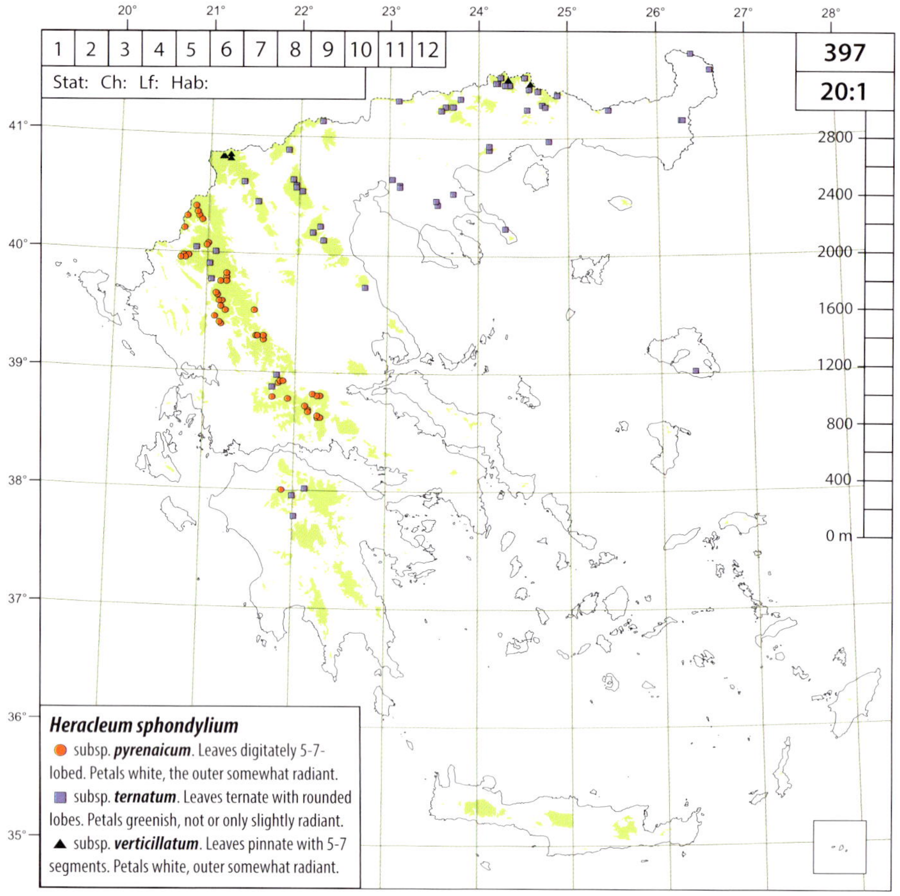

Heracleum sphondylium
● subsp. ***pyrenaicum***. Leaves digitately 5-7-lobed. Petals white, the outer somewhat radiant.
■ subsp. ***ternatum***. Leaves ternate with rounded lobes. Petals greenish, not or only slightly radiant.
▲ subsp. ***verticillatum***. Leaves pinnate with 5-7 segments. Petals white, outer somewhat radiant.

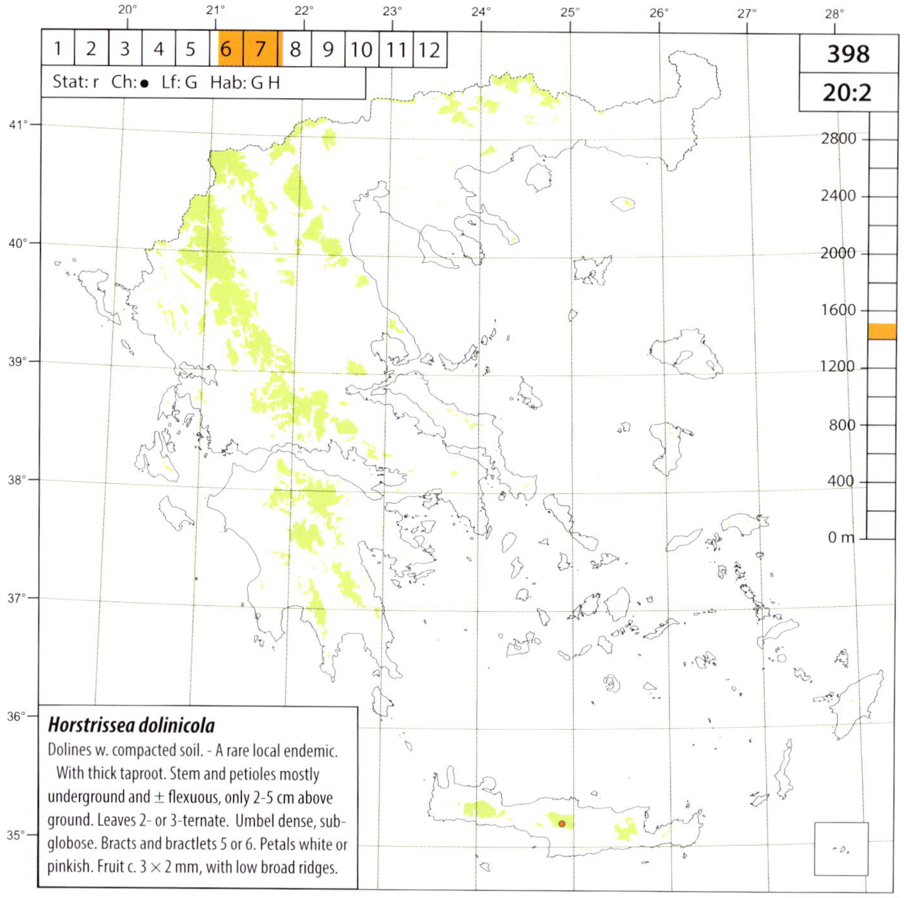

Horstrissea dolinicola
Dolines w. compacted soil. - A rare local endemic.
With thick taproot. Stem and petioles mostly underground and ± flexuous, only 2-5 cm above ground. Leaves 2- or 3-ternate. Umbel dense, subglobose. Bracts and bractlets 5 or 6. Petals white or pinkish. Fruit c. 3 × 2 mm, with low broad ridges.

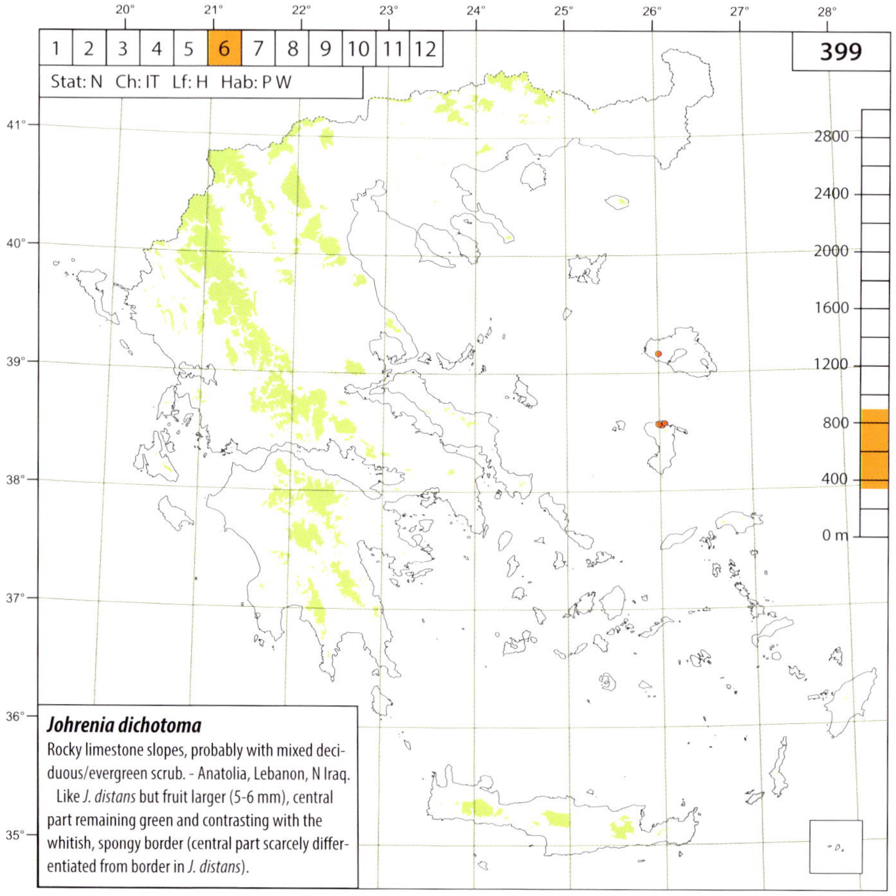

Johrenia dichotoma
Rocky limestone slopes, probably with mixed deciduous/evergreen scrub. - Anatolia, Lebanon, N Iraq.
Like *J. distans* but fruit larger (5-6 mm), central part remaining green and contrasting with the whitish, spongy border (central part scarcely differentiated from border in *J. distans*).

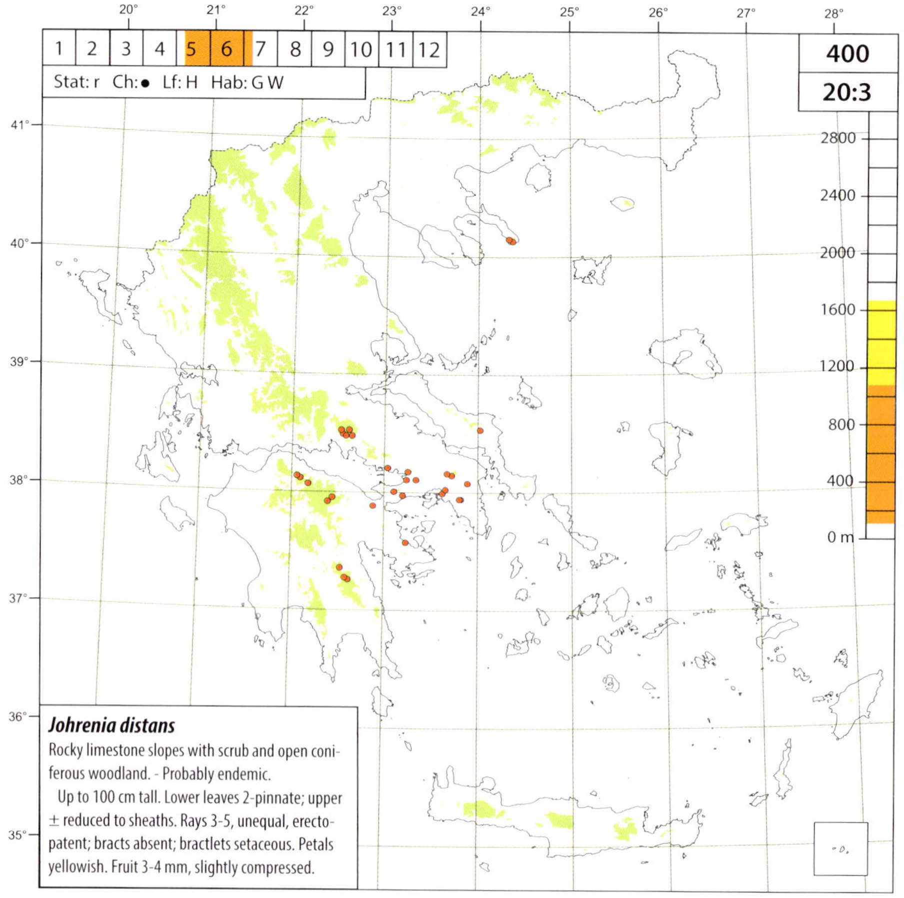

Johrenia distans
Rocky limestone slopes with scrub and open coniferous woodland. - Probably endemic.
Up to 100 cm tall. Lower leaves 2-pinnate; upper ± reduced to sheaths. Rays 3-5, unequal, erecto-patent; bracts absent; bractlets setaceous. Petals yellowish. Fruit 3-4 mm, slightly compressed.

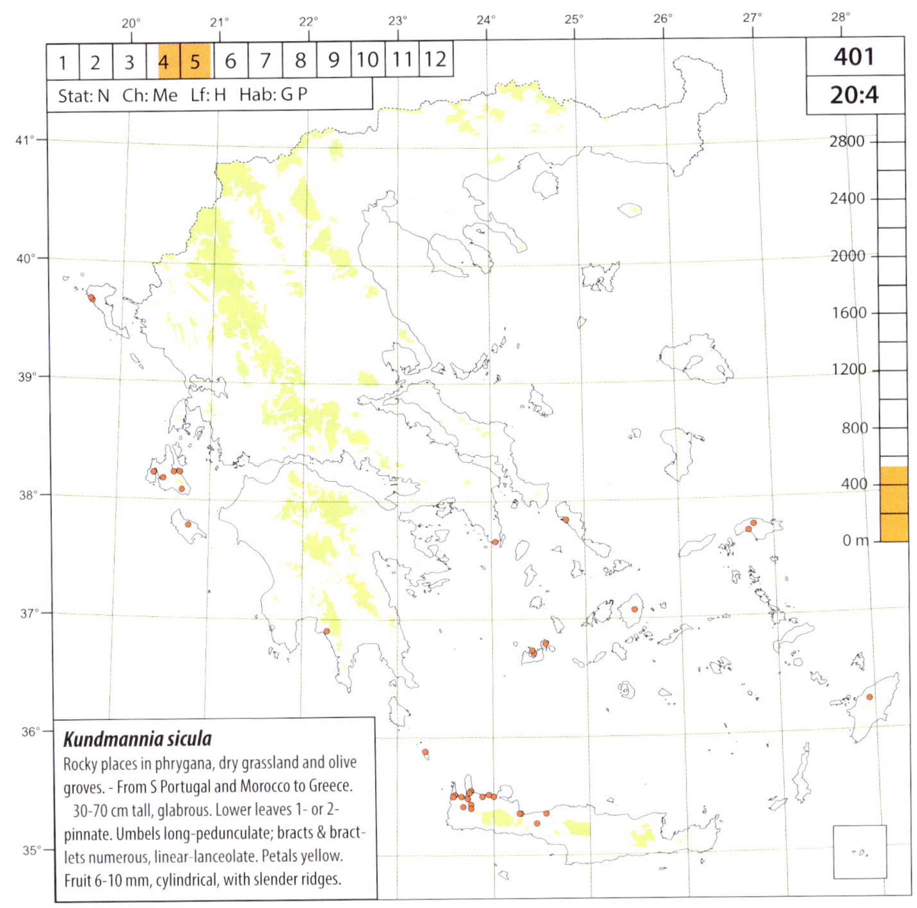

Kundmannia sicula

Rocky places in phrygana, dry grassland and olive groves. - From S Portugal and Morocco to Greece.

30-70 cm tall, glabrous. Lower leaves 1- or 2-pinnate. Umbels long-pedunculate; bracts & bractlets numerous, linear-lanceolate. Petals yellow. Fruit 6-10 mm, cylindrical, with slender ridges.

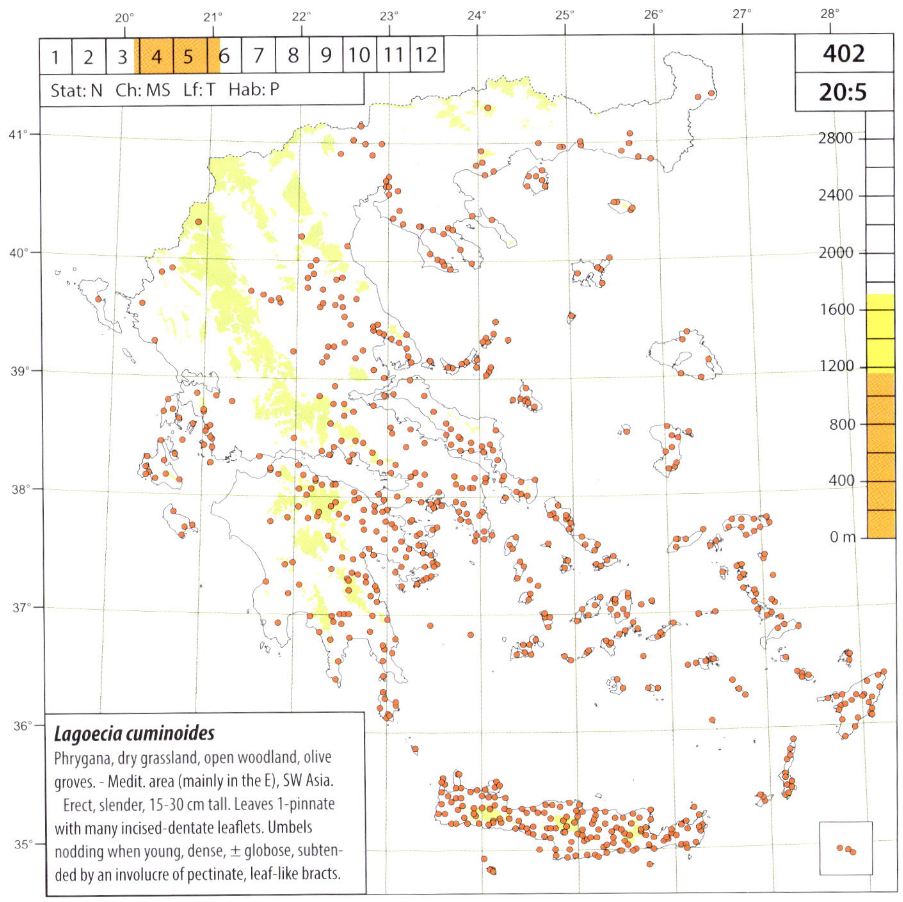

Lagoecia cuminoides

Phrygana, dry grassland, open woodland, olive groves. - Medit. area (mainly in the E), SW Asia.

Erect, slender, 15-30 cm tall. Leaves 1-pinnate with many incised-dentate leaflets. Umbels nodding when young, dense, ± globose, subtended by an involucre of pectinate, leaf-like bracts.

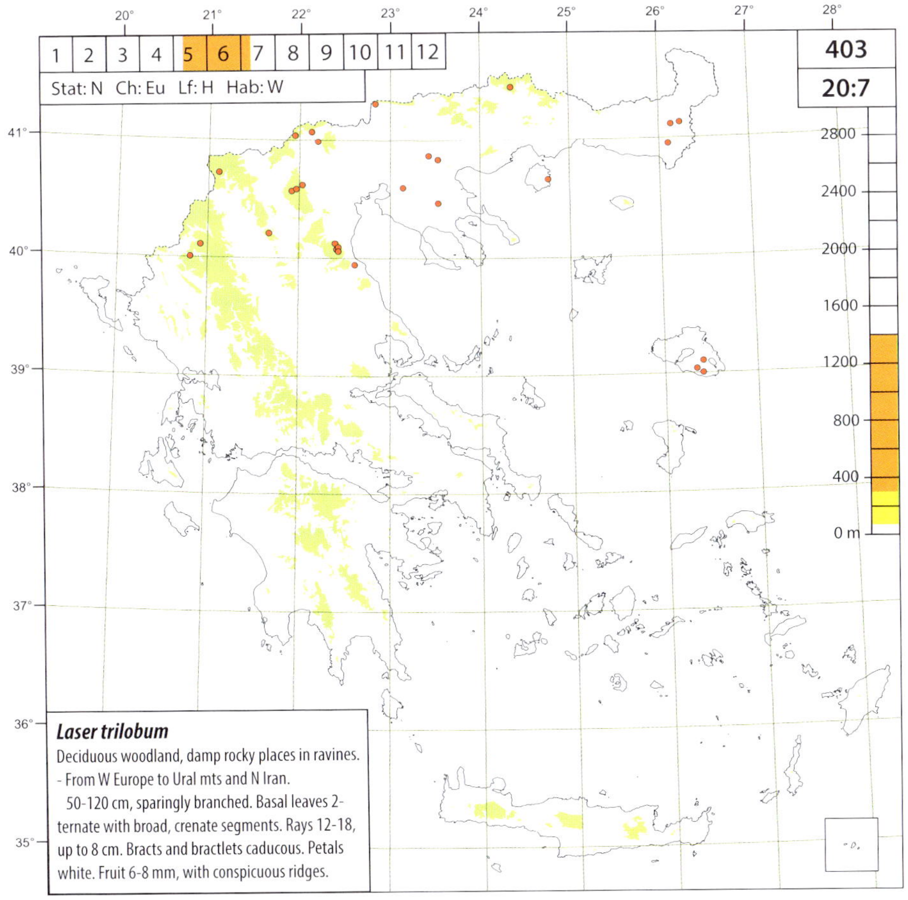

Laser trilobum

Deciduous woodland, damp rocky places in ravines. - From W Europe to Ural mts and N Iran.

50-120 cm, sparingly branched. Basal leaves 2-ternate with broad, crenate segments. Rays 12-18, up to 8 cm. Bracts and bractlets caducous. Petals white. Fruit 6-8 mm, with conspicuous ridges.

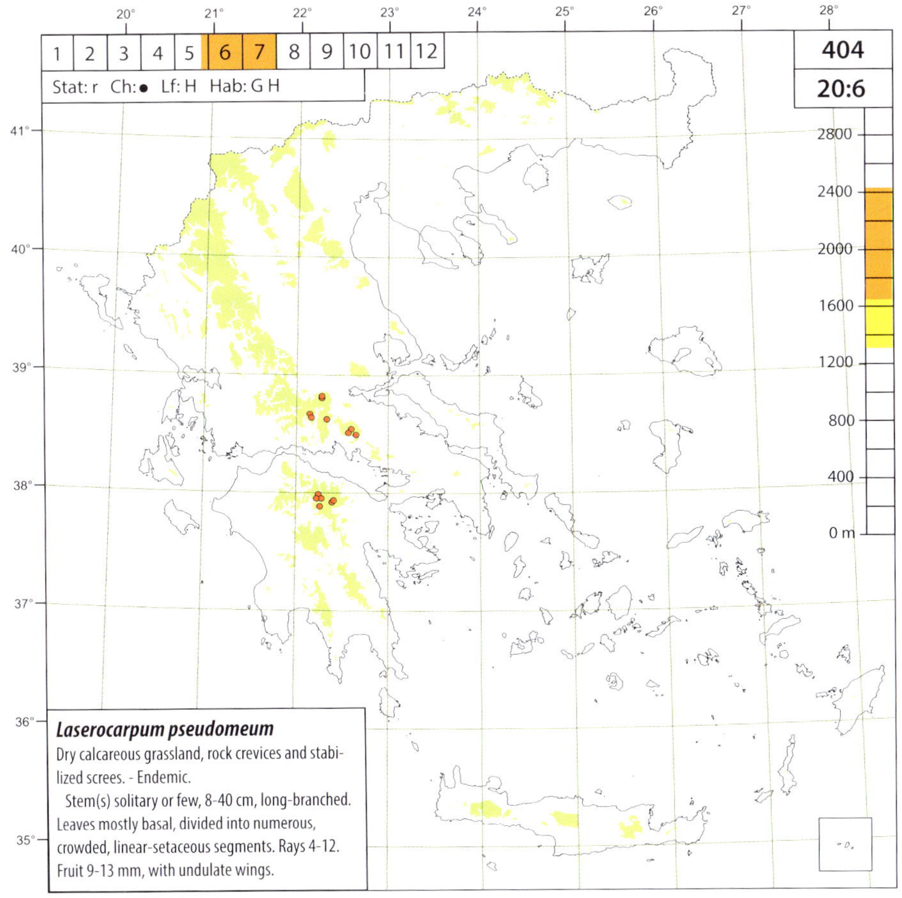

Laserocarpum pseudomeum

Dry calcareous grassland, rock crevices and stabilized screes. - Endemic.

Stem(s) solitary or few, 8-40 cm, long-branched. Leaves mostly basal, divided into numerous, crowded, linear-setaceous segments. Rays 4-12. Fruit 9-13 mm, with undulate wings.

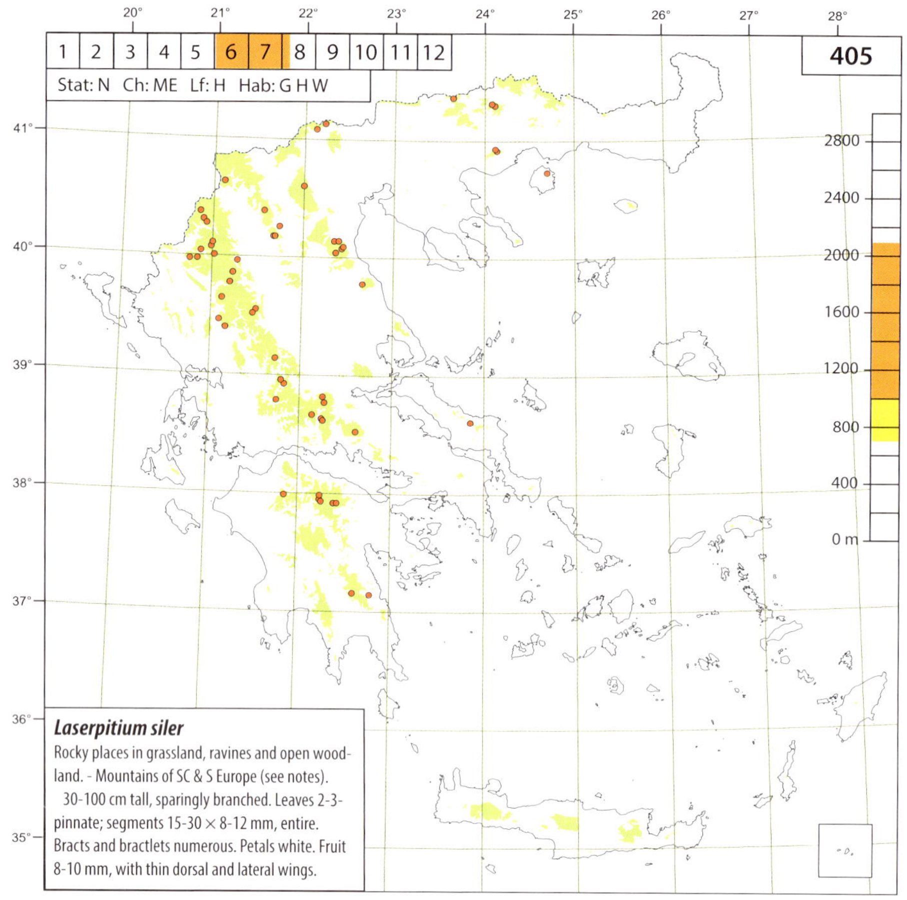

Laserpitium siler

Rocky places in grassland, ravines and open woodland. - Mountains of SC & S Europe (see notes).

30-100 cm tall, sparingly branched. Leaves 2-3-pinnate; segments 15-30 × 8-12 mm, entire. Bracts and bractlets numerous. Petals white. Fruit 8-10 mm, with thin dorsal and lateral wings.

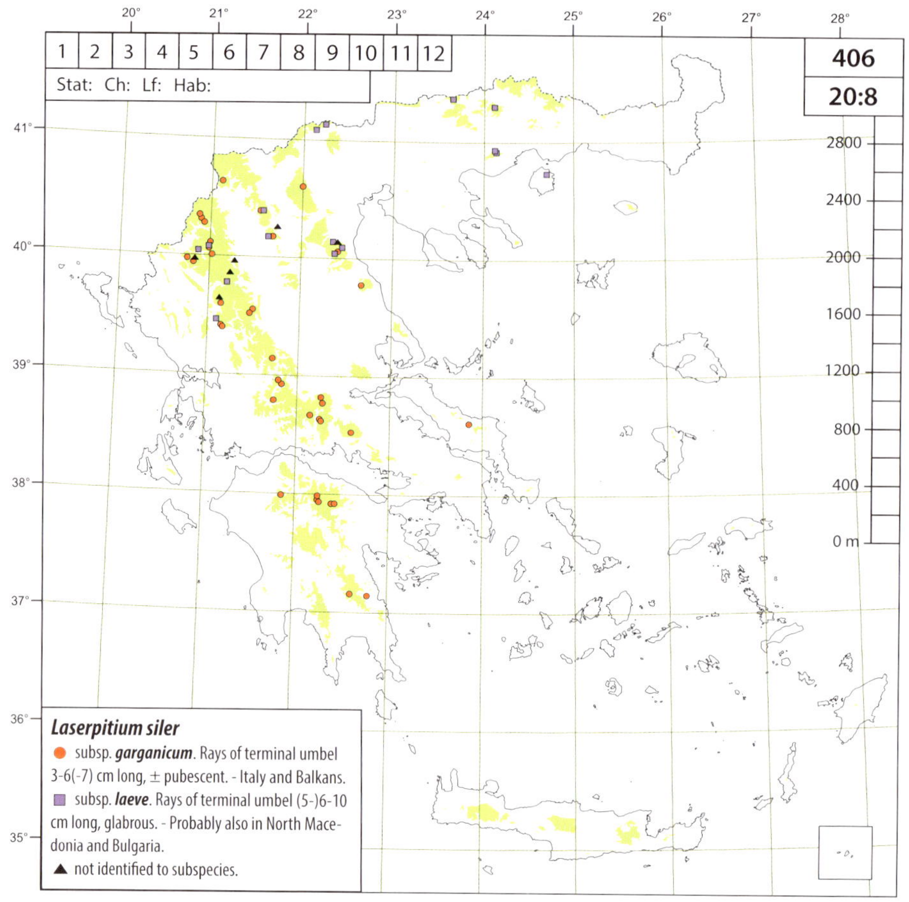

Laserpitium siler

● subsp. ***garganicum***. Rays of terminal umbel 3-6(-7) cm long, ± pubescent. - Italy and Balkans.

■ subsp. ***laeve***. Rays of terminal umbel (5-)6-10 cm long, glabrous. - Probably also in North Macedonia and Bulgaria.

▲ not identified to subspecies.

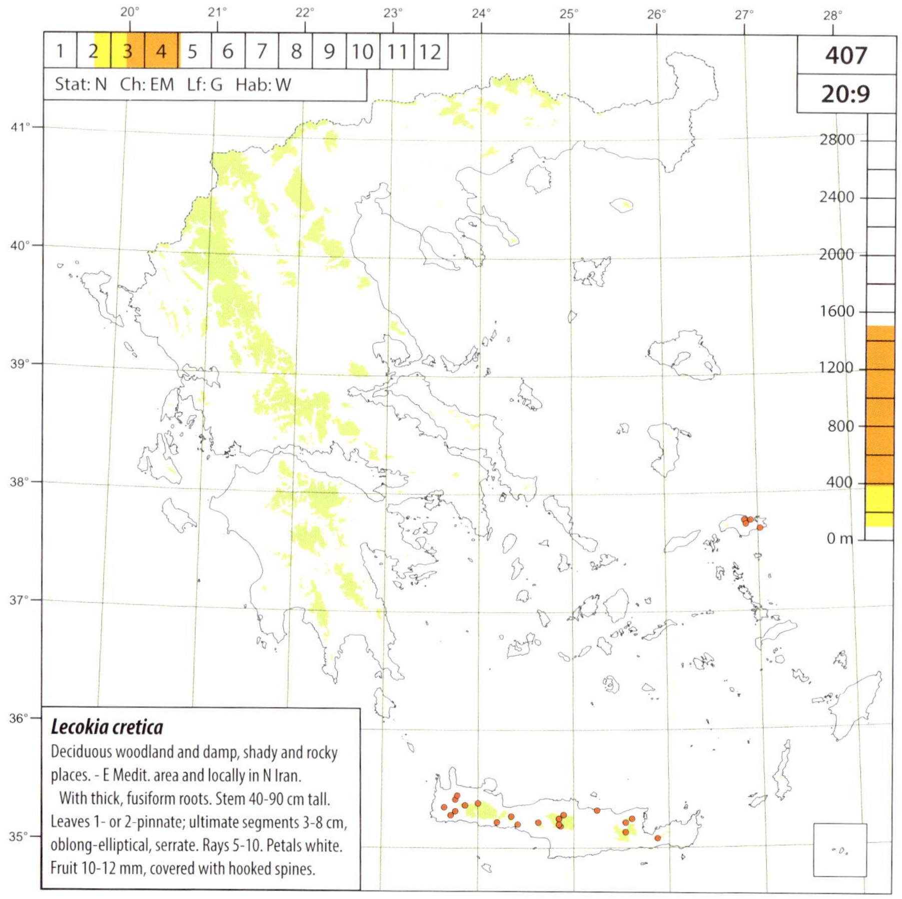

Lecokia cretica

Deciduous woodland and damp, shady and rocky places. - E Medit. area and locally in N Iran.

With thick, fusiform roots. Stem 40-90 cm tall. Leaves 1- or 2-pinnate; ultimate segments 3-8 cm, oblong-elliptical, serrate. Rays 5-10. Petals white. Fruit 10-12 mm, covered with hooked spines.

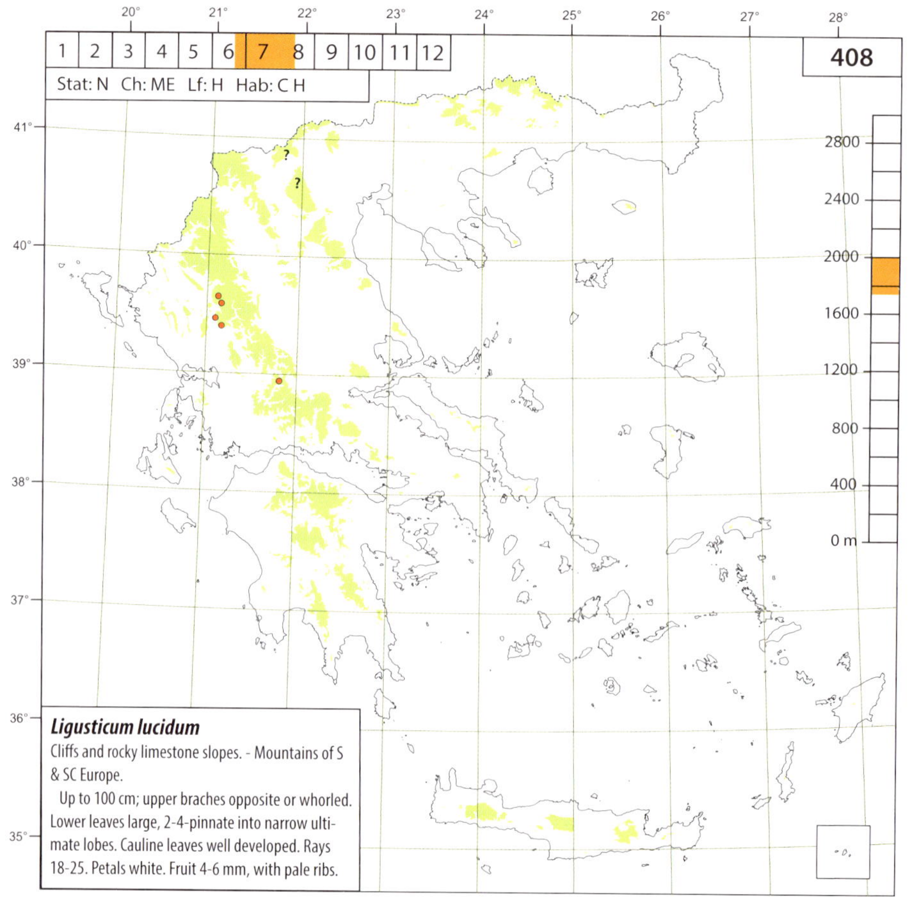

Ligusticum lucidum

Cliffs and rocky limestone slopes. - Mountains of S & SC Europe.

Up to 100 cm; upper braches opposite or whorled. Lower leaves large, 2-4-pinnate into narrow ultimate lobes. Cauline leaves well developed. Rays 18-25. Petals white. Fruit 4-6 mm, with pale ribs.

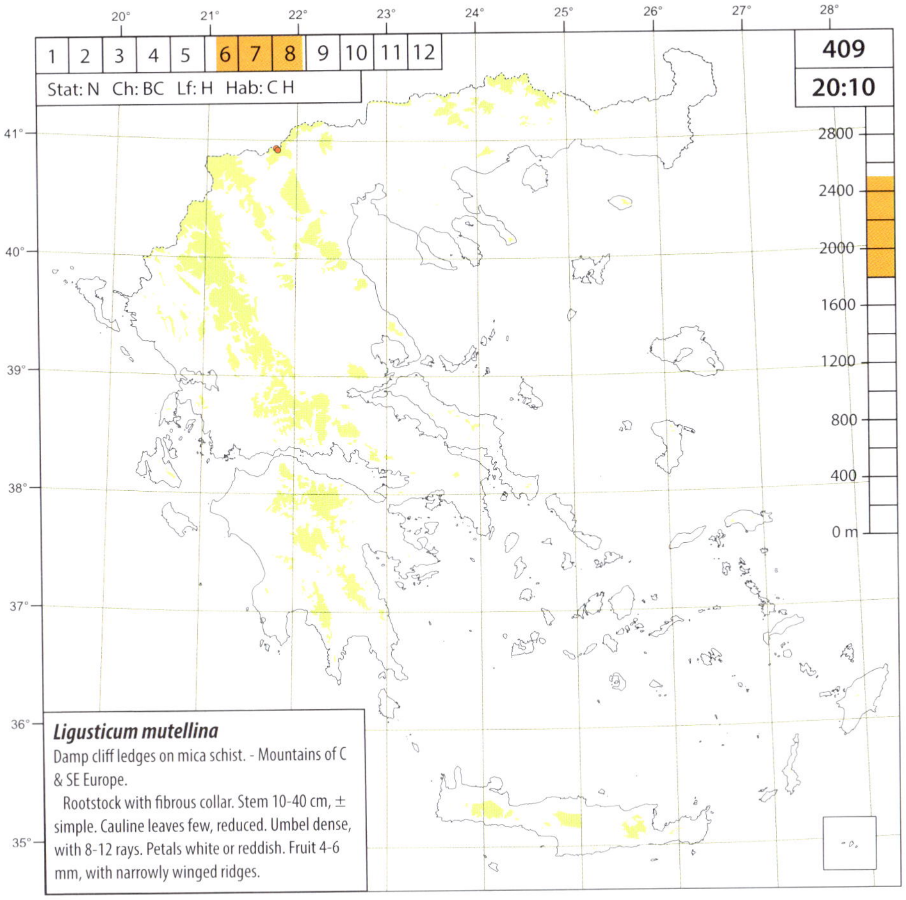

Ligusticum mutellina
Damp cliff ledges on mica schist. - Mountains of C & SE Europe.
Rootstock with fibrous collar. Stem 10-40 cm, ± simple. Cauline leaves few, reduced. Umbel dense, with 8-12 rays. Petals white or reddish. Fruit 4-6 mm, with narrowly winged ridges.

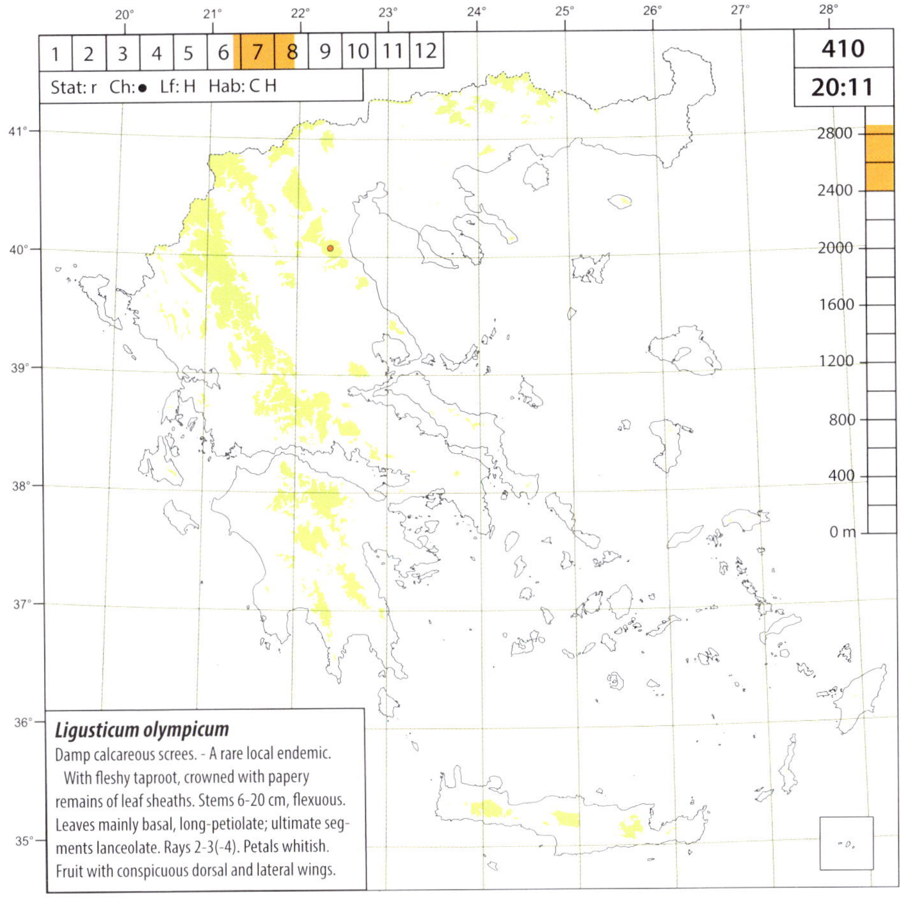

Ligusticum olympicum
Damp calcareous screes. - A rare local endemic.
With fleshy taproot, crowned with papery remains of leaf sheaths. Stems 6-20 cm, flexuous. Leaves mainly basal, long-petiolate; ultimate segments lanceolate. Rays 2-3(-4). Petals whitish. Fruit with conspicuous dorsal and lateral wings.

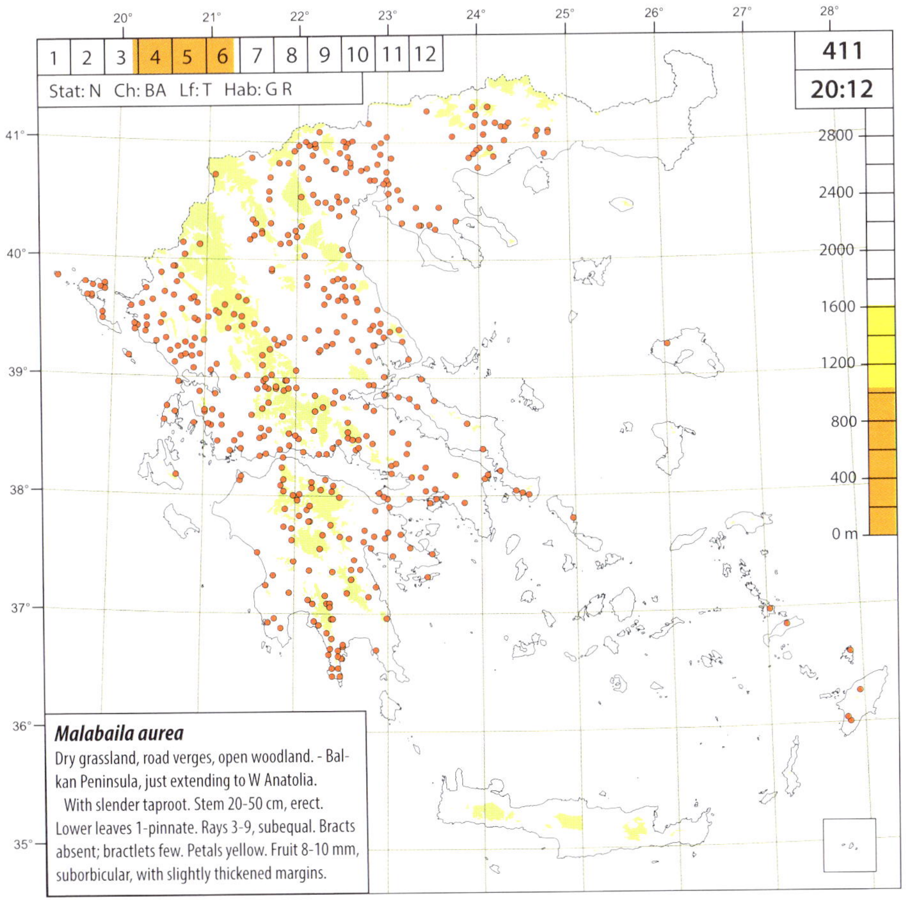

Malabaila aurea
Dry grassland, road verges, open woodland. - Balkan Peninsula, just extending to W Anatolia.
With slender taproot. Stem 20-50 cm, erect. Lower leaves 1-pinnate. Rays 3-9, subequal. Bracts absent; bractlets few. Petals yellow. Fruit 8-10 mm, suborbicular, with slightly thickened margins.

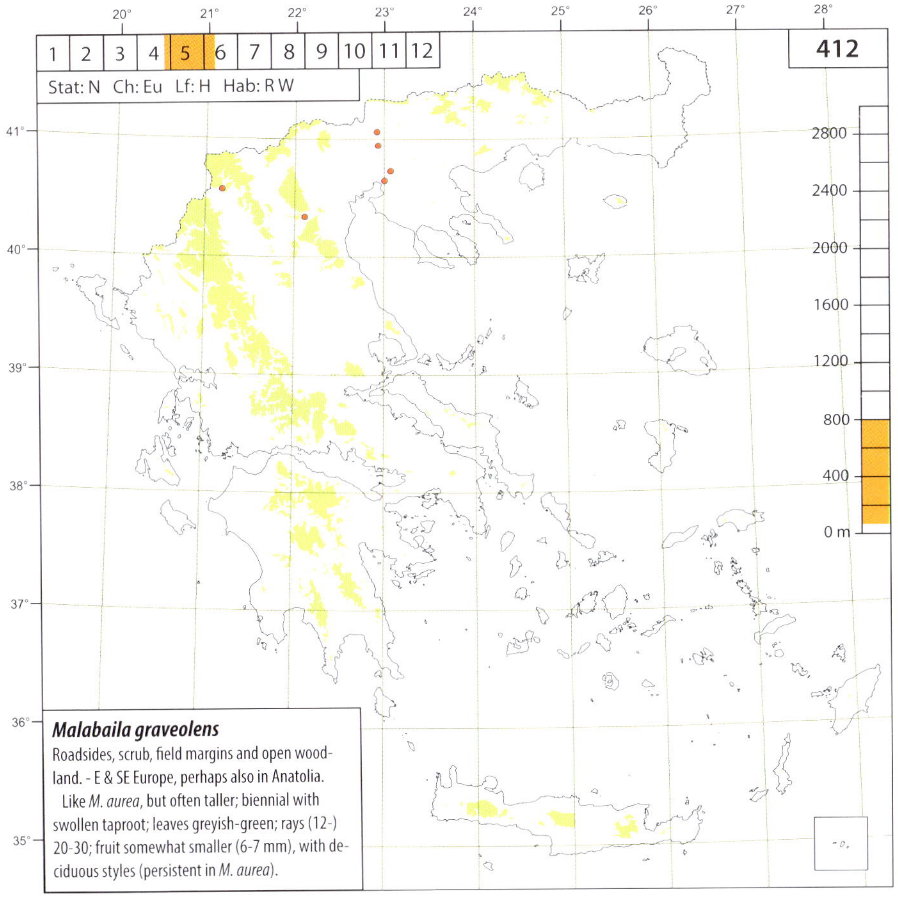

Malabaila graveolens
Roadsides, scrub, field margins and open woodland. - E & SE Europe, perhaps also in Anatolia.
Like *M. aurea*, but often taller; biennial with swollen taproot; leaves greyish-green; rays (12-) 20-30; fruit somewhat smaller (6-7 mm), with deciduous styles (persistent in *M. aurea*).

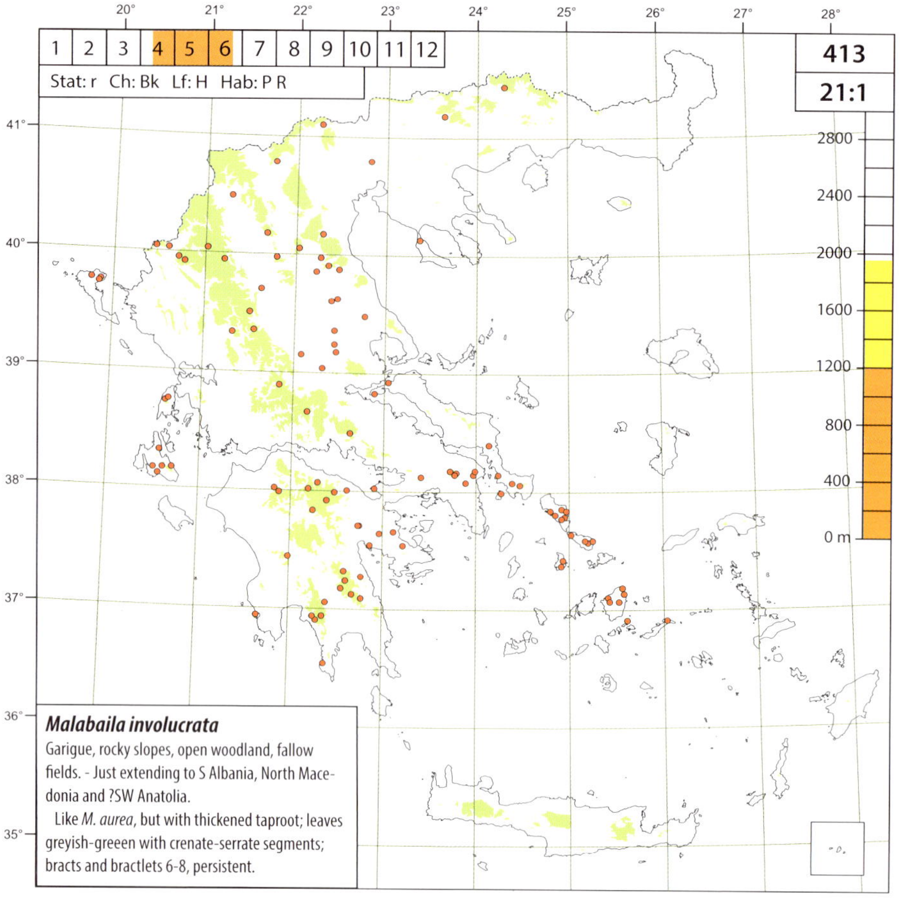

Malabaila involucrata

Garigue, rocky slopes, open woodland, fallow fields. - Just extending to S Albania, North Macedonia and ?SW Anatolia.

Like *M. aurea*, but with thickened taproot; leaves greyish-green with crenate-serrate segments; bracts and bractlets 6-8, persistent.

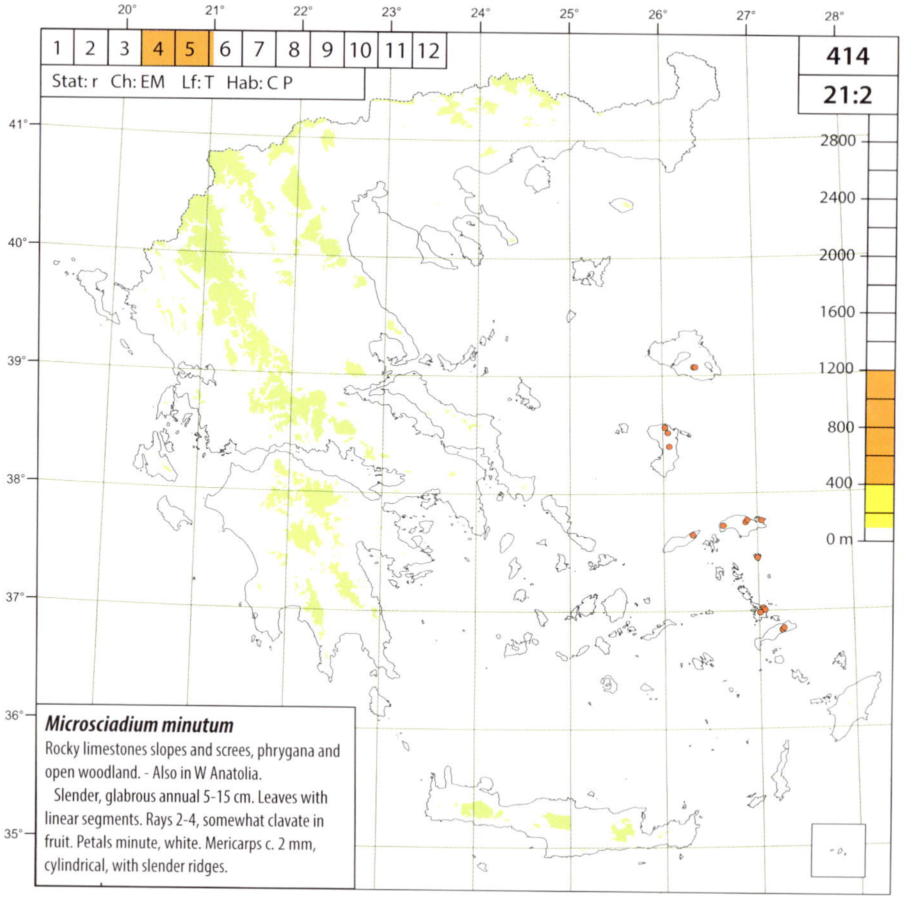

Microsciadium minutum

Rocky limestones slopes and screes, phrygana and open woodland. - Also in W Anatolia.

Slender, glabrous annual 5-15 cm. Leaves with linear segments. Rays 2-4, somewhat clavate in fruit. Petals minute, white. Mericarps c. 2 mm, cylindrical, with slender ridges.

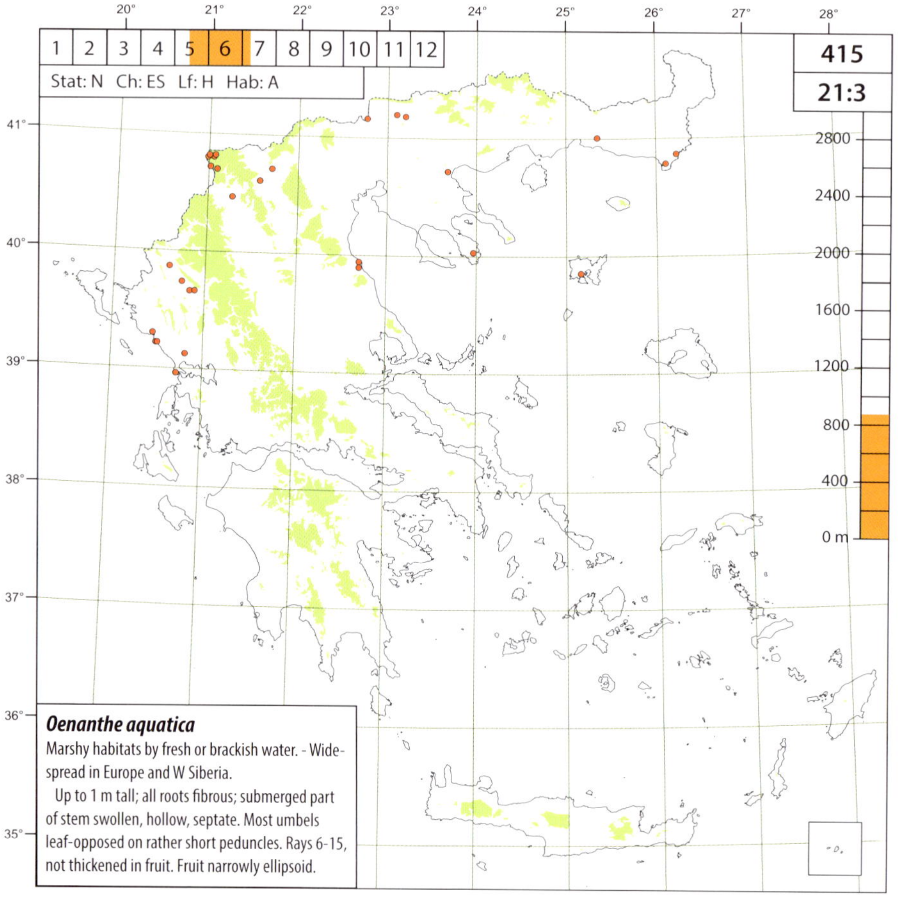

Oenanthe aquatica

Marshy habitats by fresh or brackish water. - Widespread in Europe and W Siberia.

Up to 1 m tall; all roots fibrous; submerged part of stem swollen, hollow, septate. Most umbels leaf-opposed on rather short peduncles. Rays 6-15, not thickened in fruit. Fruit narrowly ellipsoid.

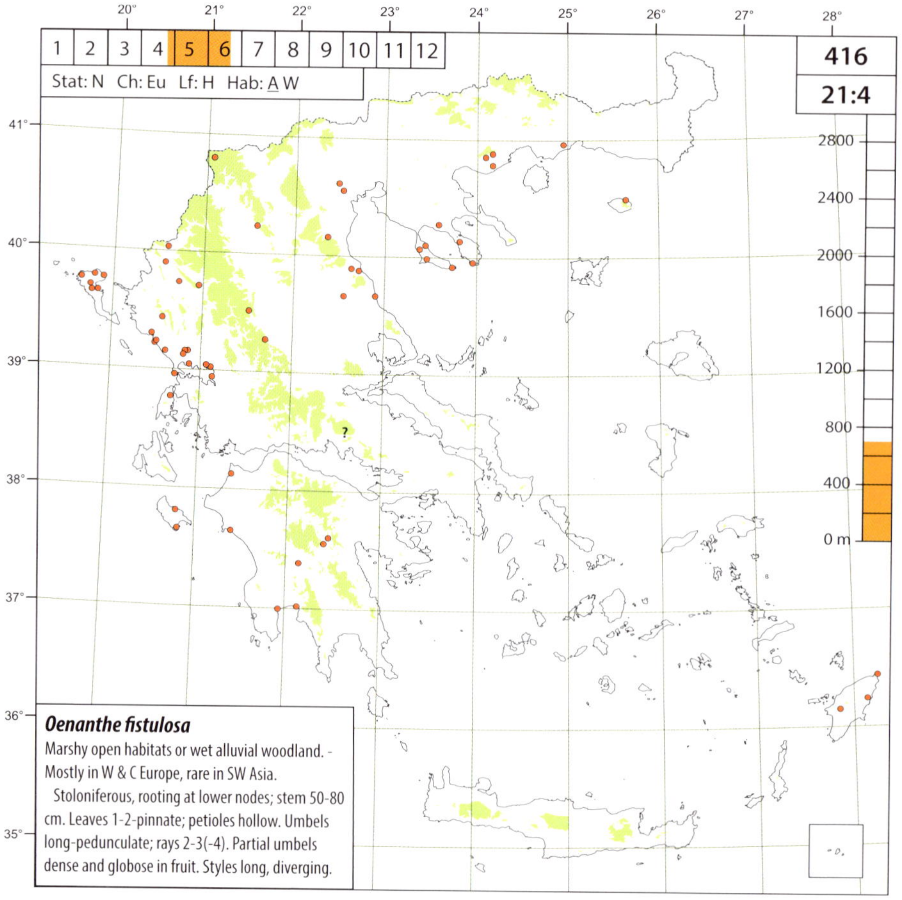

Oenanthe fistulosa

Marshy open habitats or wet alluvial woodland. - Mostly in W & C Europe, rare in SW Asia.

Stoloniferous, rooting at lower nodes; stem 50-80 cm. Leaves 1-2-pinnate; petioles hollow. Umbels long-pedunculate; rays 2-3(-4). Partial umbels dense and globose in fruit. Styles long, diverging.

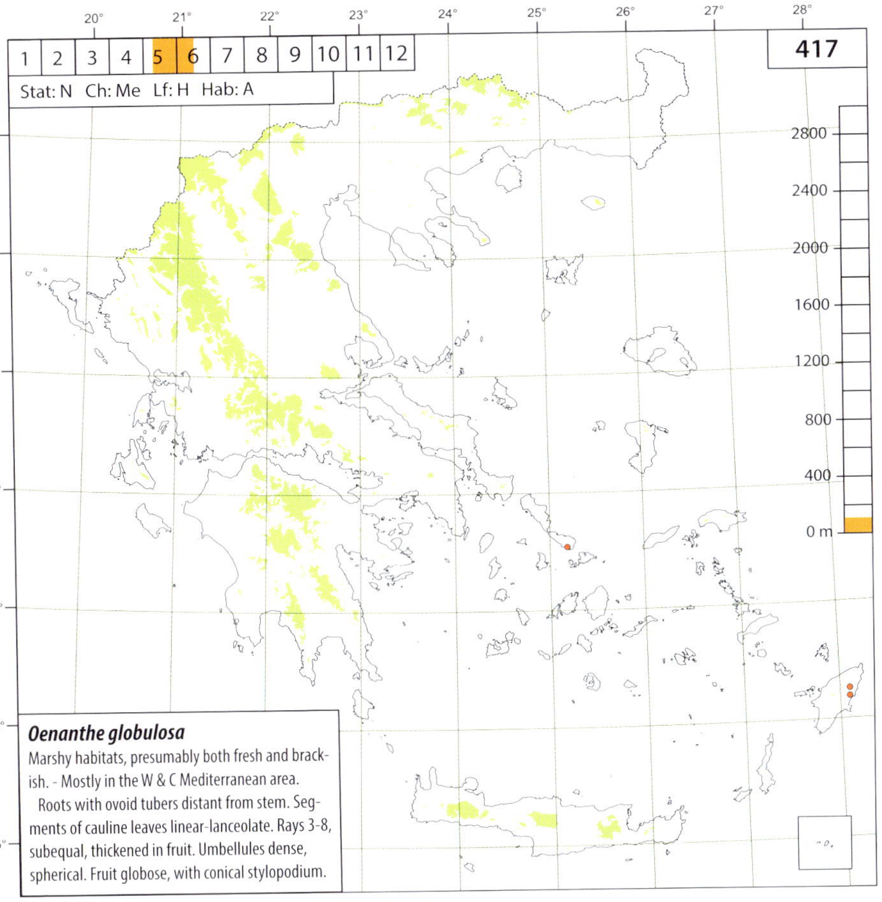

Oenanthe globulosa

Marshy habitats, presumably both fresh and brackish. - Mostly in the W & C Mediterranean area.

Roots with ovoid tubers distant from stem. Segments of cauline leaves linear-lanceolate. Rays 3-8, subequal, thickened in fruit. Umbellules dense, spherical. Fruit globose, with conical stylopodium.

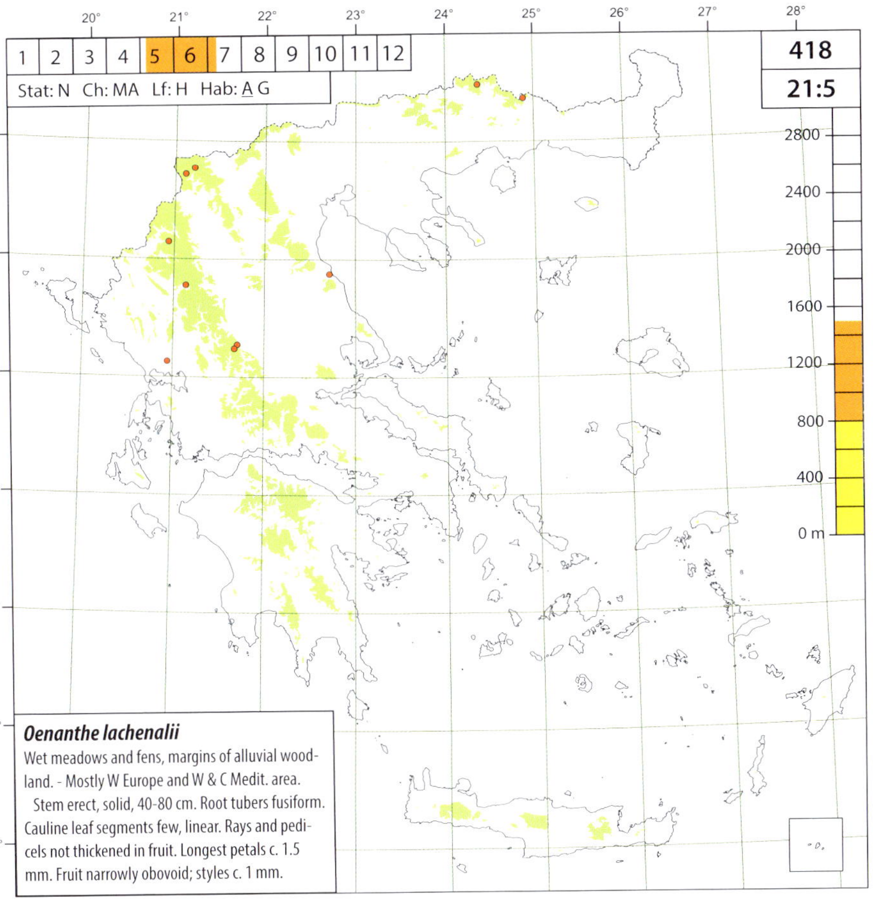

Oenanthe lachenalii

Wet meadows and fens, margins of alluvial woodland. - Mostly W Europe and W & C Medit. area.

Stem erect, solid, 40-80 cm. Root tubers fusiform. Cauline leaf segments few, linear. Rays and pedicels not thickened in fruit. Longest petals c. 1.5 mm. Fruit narrowly obovoid; styles c. 1 mm.

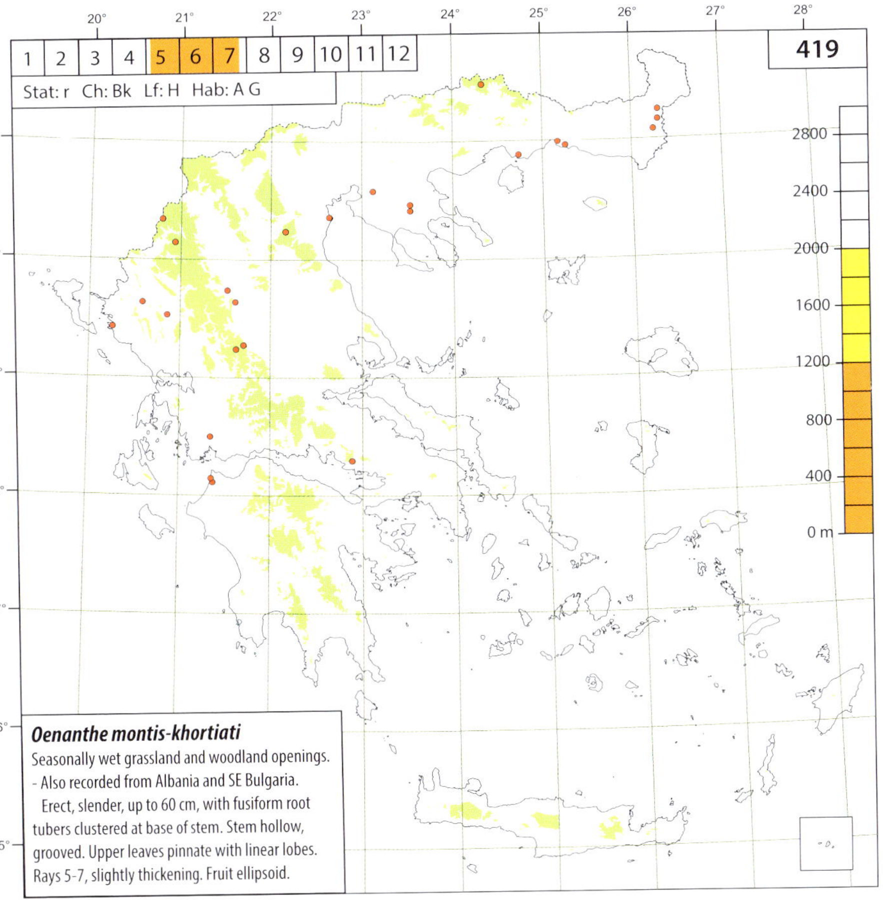

Oenanthe montis-khortiati

Seasonally wet grassland and woodland openings. - Also recorded from Albania and SE Bulgaria.

Erect, slender, up to 60 cm, with fusiform root tubers clustered at base of stem. Stem hollow, grooved. Upper leaves pinnate with linear lobes. Rays 5-7, slightly thickening. Fruit ellipsoid.

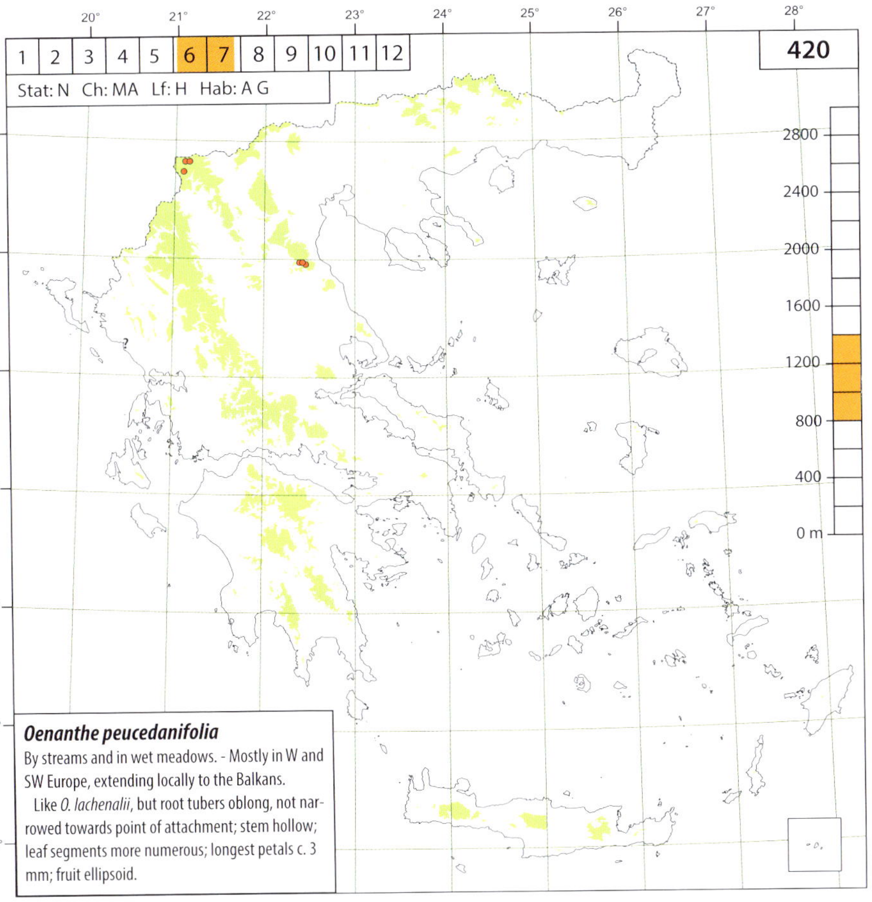

Oenanthe peucedanifolia

By streams and in wet meadows. - Mostly in W and SW Europe, extending locally to the Balkans.

Like *O. lachenalii*, but root tubers oblong, not narrowed towards point of attachment; stem hollow; leaf segments more numerous; longest petals c. 3 mm; fruit ellipsoid.

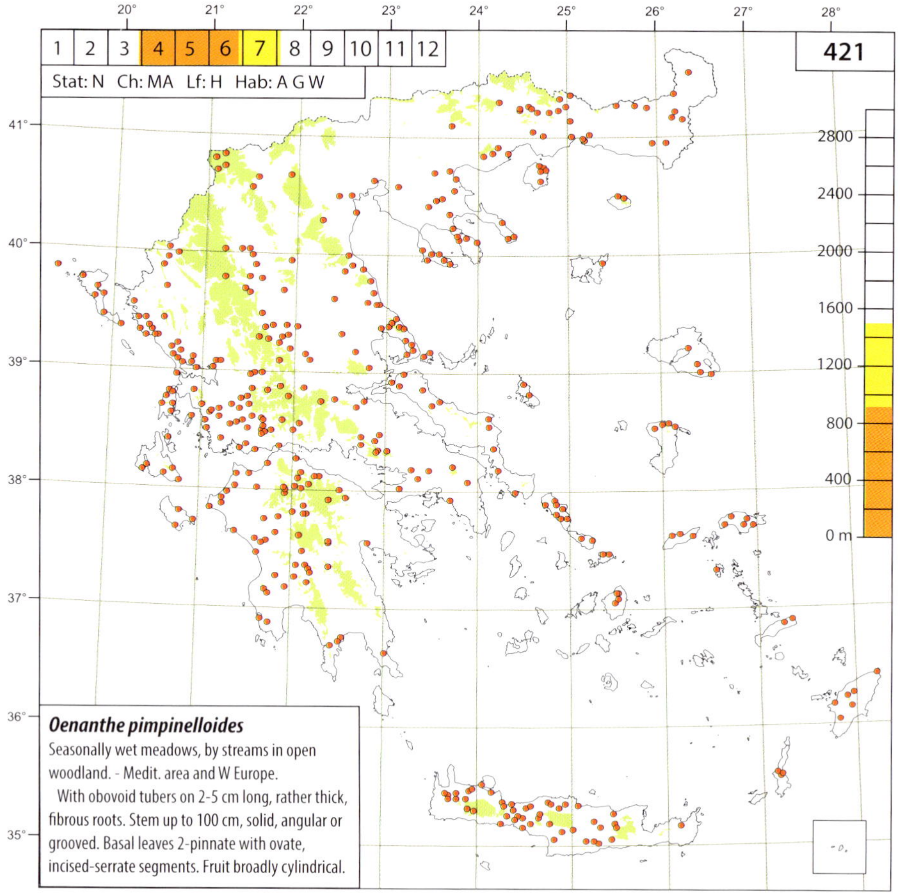

Oenanthe pimpinelloides
Seasonally wet meadows, by streams in open woodland. - Medit. area and W Europe.
With obovoid tubers on 2-5 cm long, rather thick, fibrous roots. Stem up to 100 cm, solid, angular or grooved. Basal leaves 2-pinnate with ovate, incised-serrate segments. Fruit broadly cylindrical.

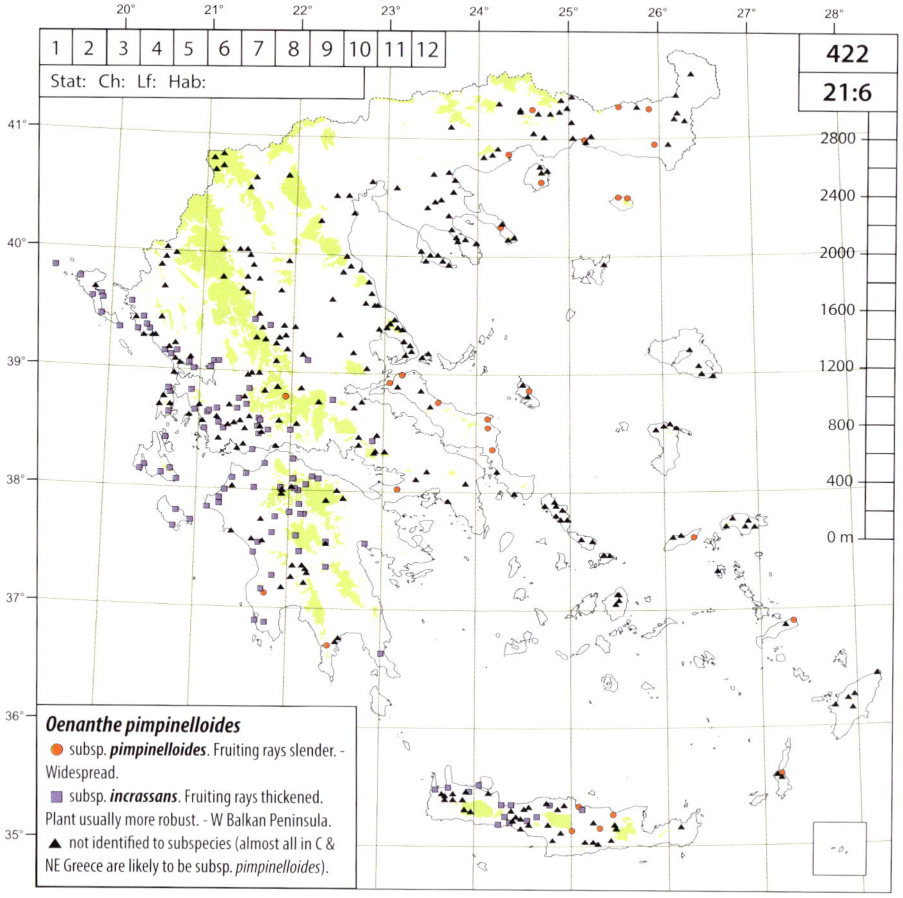

Oenanthe pimpinelloides
● subsp. ***pimpinelloides***. Fruiting rays slender. - Widespread.
■ subsp. ***incrassans***. Fruiting rays thickened. Plant usually more robust. - W Balkan Peninsula.
▲ not identified to subspecies (almost all in C & NE Greece are likely to be subsp. *pimpinelloides*).

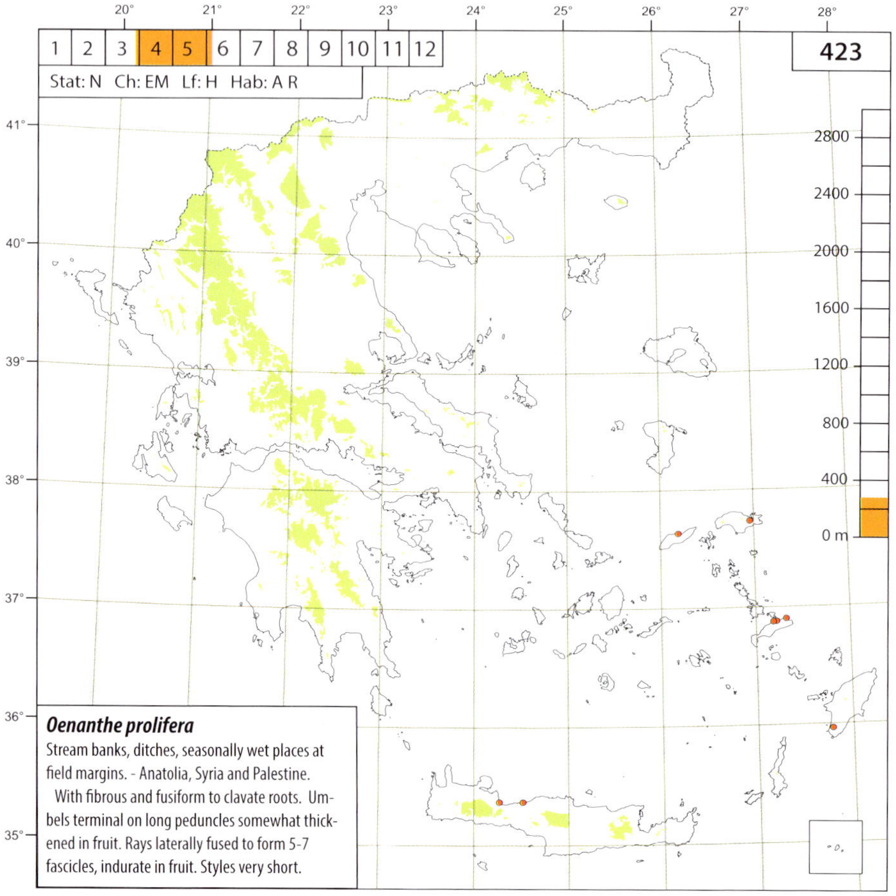

Oenanthe prolifera
Stream banks, ditches, seasonally wet places at field margins. - Anatolia, Syria and Palestine.
With fibrous and fusiform to clavate roots. Umbels terminal on long peduncles somewhat thickened in fruit. Rays laterally fused to form 5-7 fascicles, indurate in fruit. Styles very short.

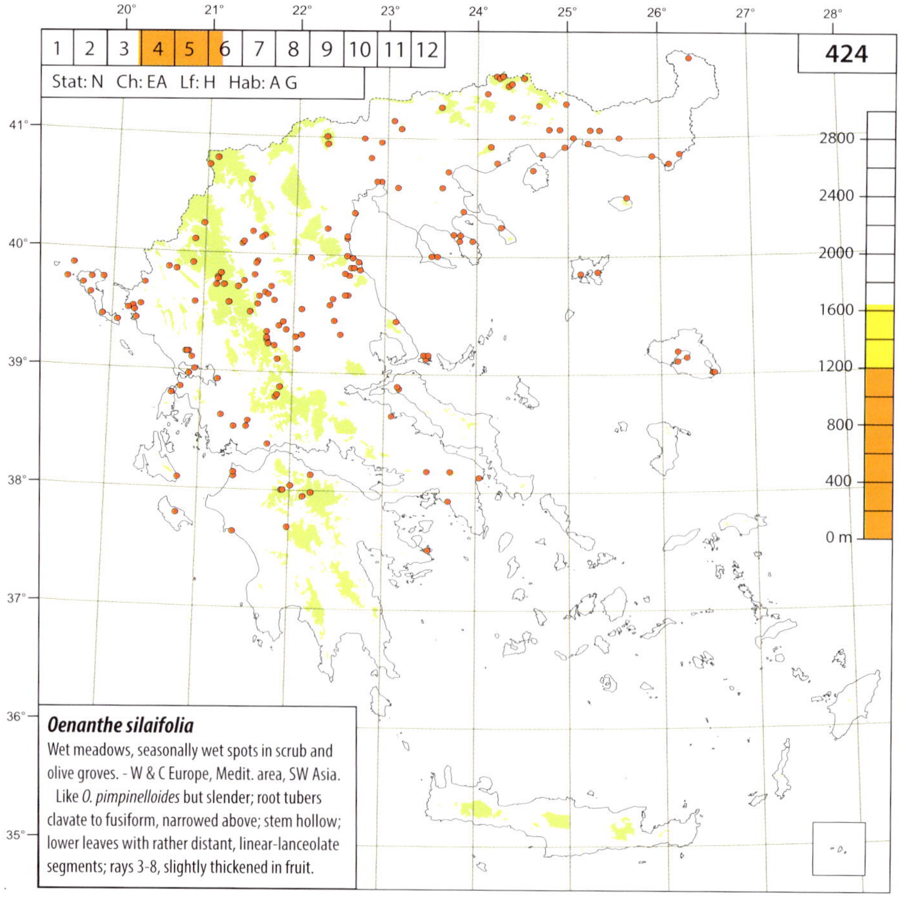

Oenanthe silaifolia
Wet meadows, seasonally wet spots in scrub and olive groves. - W & C Europe, Medit. area, SW Asia.
Like *O. pimpinelloides* but slender; root tubers clavate to fusiform, narrowed above; stem hollow; lower leaves with rather distant, linear-lanceolate segments; rays 3-8, slightly thickened in fruit.

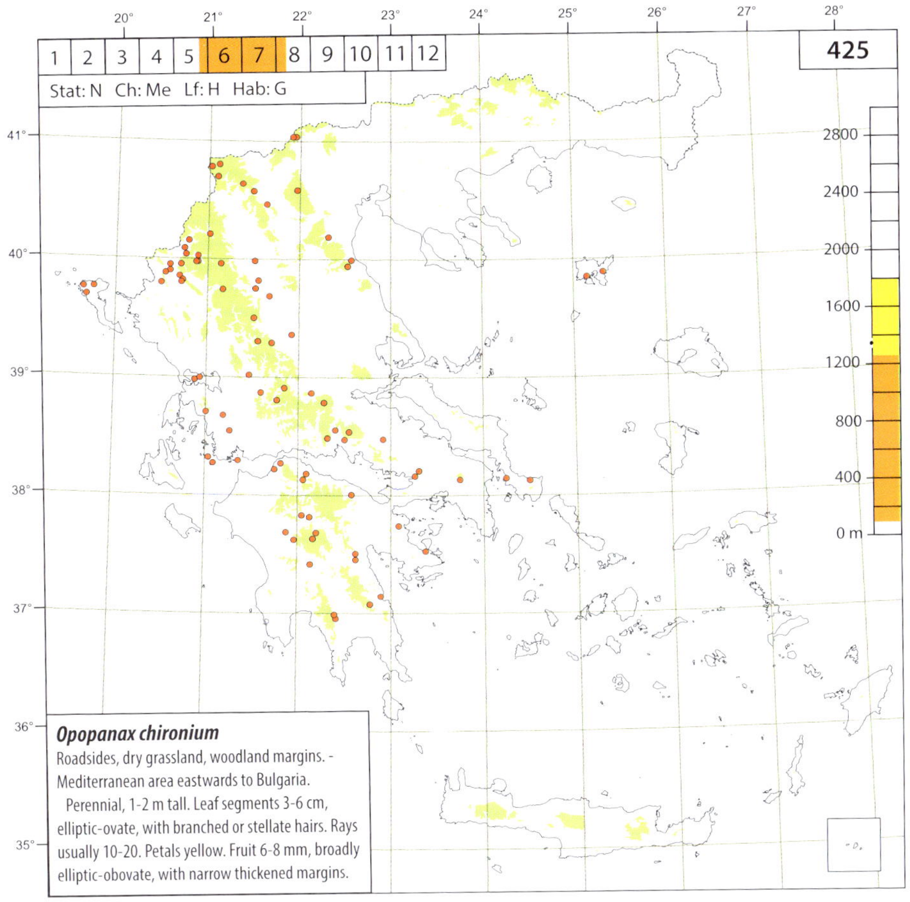

Opopanax chironium

Roadsides, dry grassland, woodland margins. - Mediterranean area eastwards to Bulgaria.

Perennial, 1-2 m tall. Leaf segments 3-6 cm, elliptic-ovate, with branched or stellate hairs. Rays usually 10-20. Petals yellow. Fruit 6-8 mm, broadly elliptic-obovate, with narrow thickened margins.

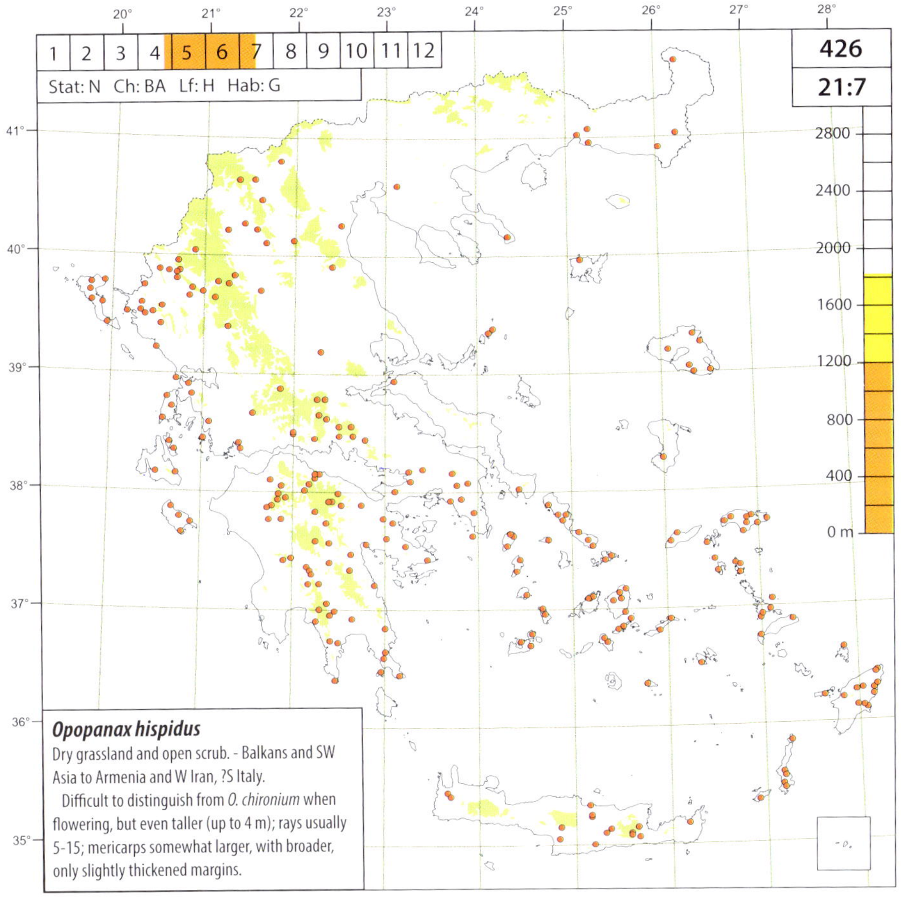

Opopanax hispidus

Dry grassland and open scrub. - Balkans and SW Asia to Armenia and W Iran, ?S Italy.

Difficult to distinguish from *O. chironium* when flowering, but even taller (up to 4 m); rays usually 5-15; mericarps somewhat larger, with broader, only slightly thickened margins.

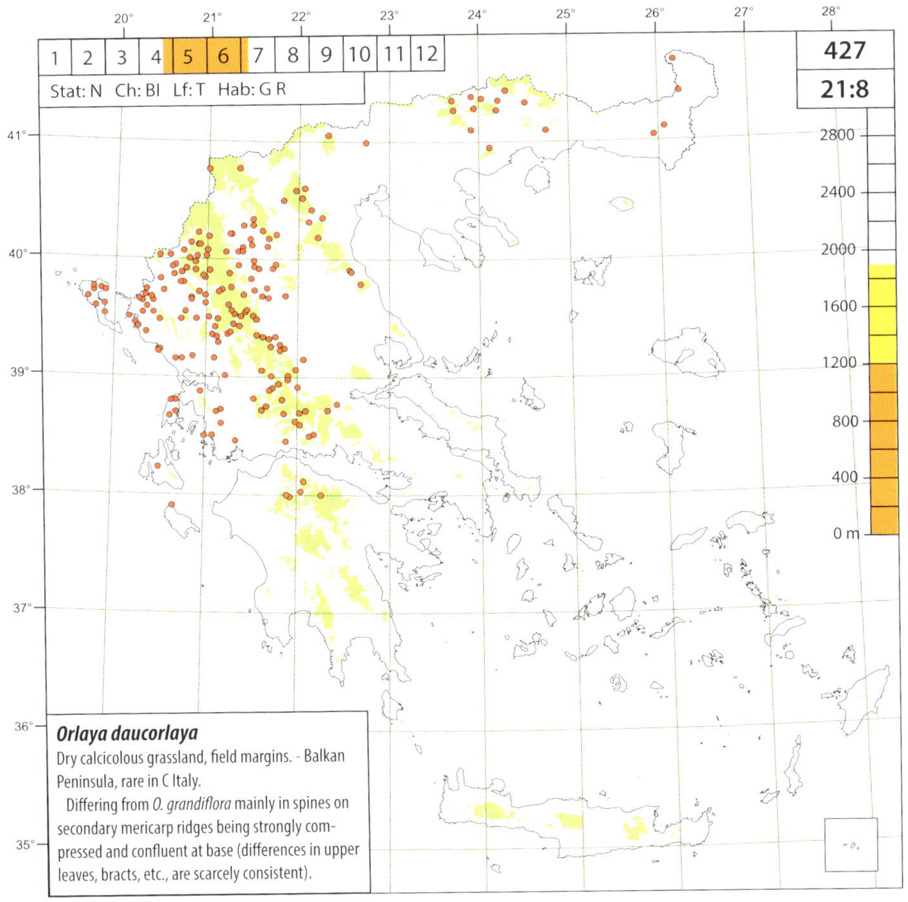

Orlaya daucorlaya

Dry calcicolous grassland, field margins. - Balkan Peninsula, rare in C Italy.

Differing from *O. grandiflora* mainly in spines on secondary mericarp ridges being strongly compressed and confluent at base (differences in upper leaves, bracts, etc., are scarcely consistent).

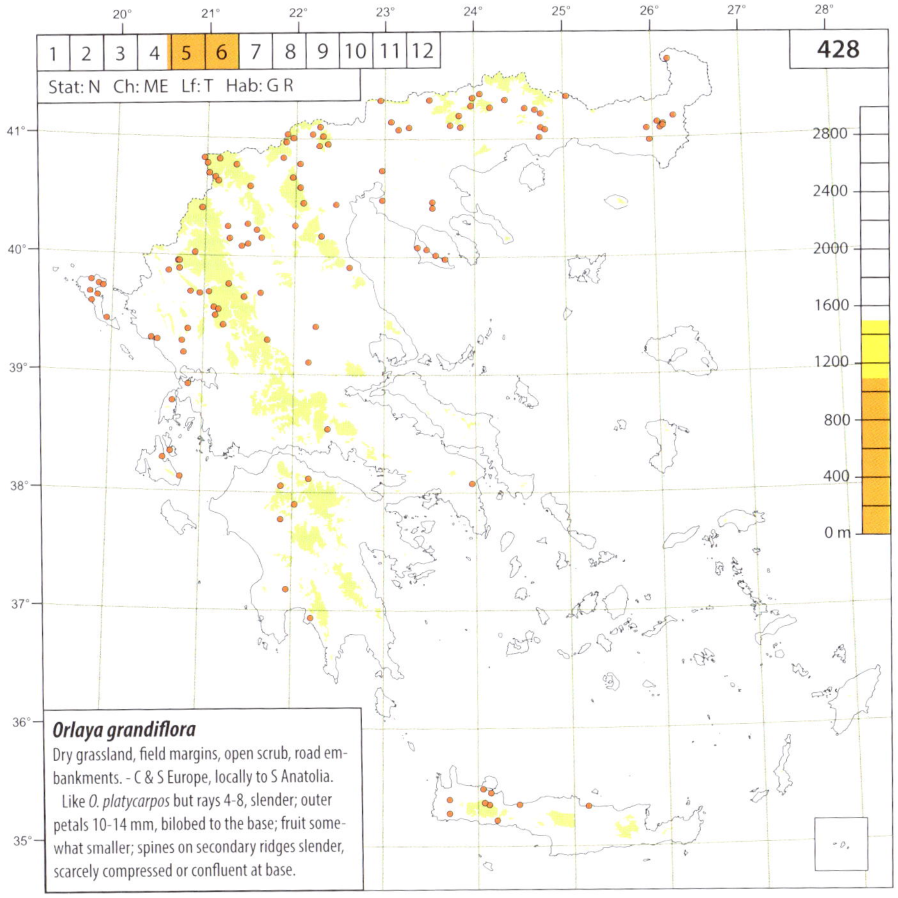

Orlaya grandiflora

Dry grassland, field margins, open scrub, road embankments. - C & S Europe, locally to S Anatolia.

Like *O. platycarpos* but rays 4-8, slender; outer petals 10-14 mm, bilobed to the base; fruit somewhat smaller; spines on secondary ridges slender, scarcely compressed or confluent at base.

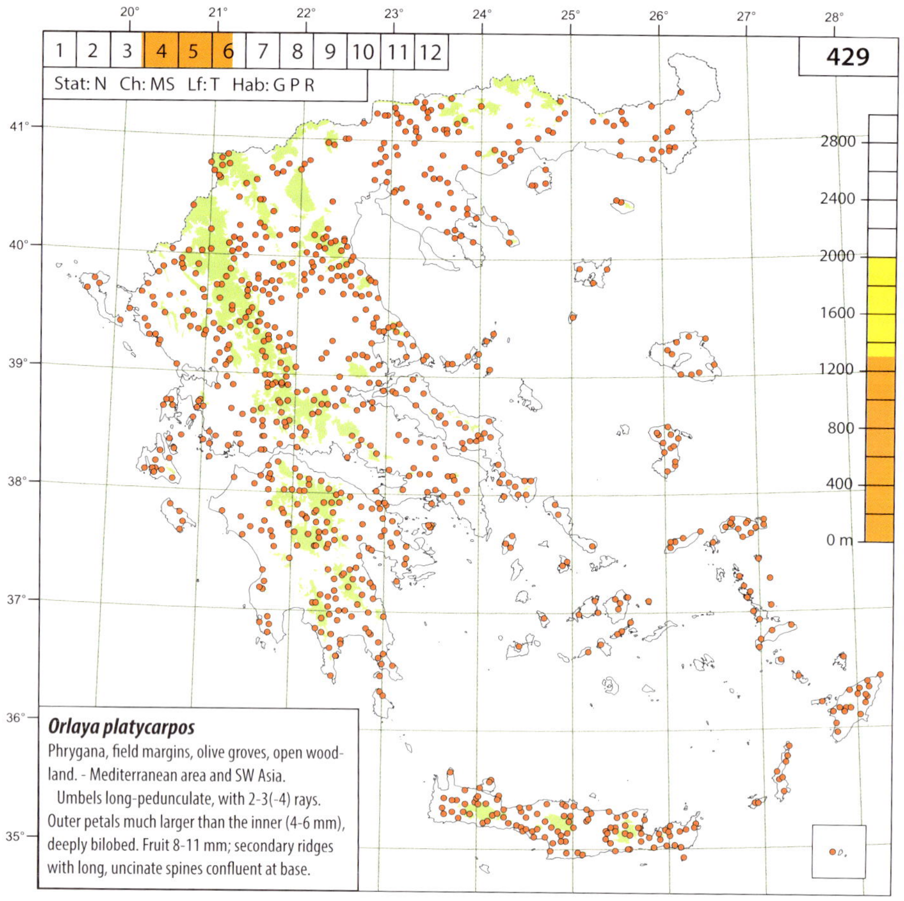

Orlaya platycarpos

Phrygana, field margins, olive groves, open woodland. - Mediterranean area and SW Asia.

Umbels long-pedunculate, with 2-3(-4) rays. Outer petals much larger than the inner (4-6 mm), deeply bilobed. Fruit 8-11 mm; secondary ridges with long, uncinate spines confluent at base.

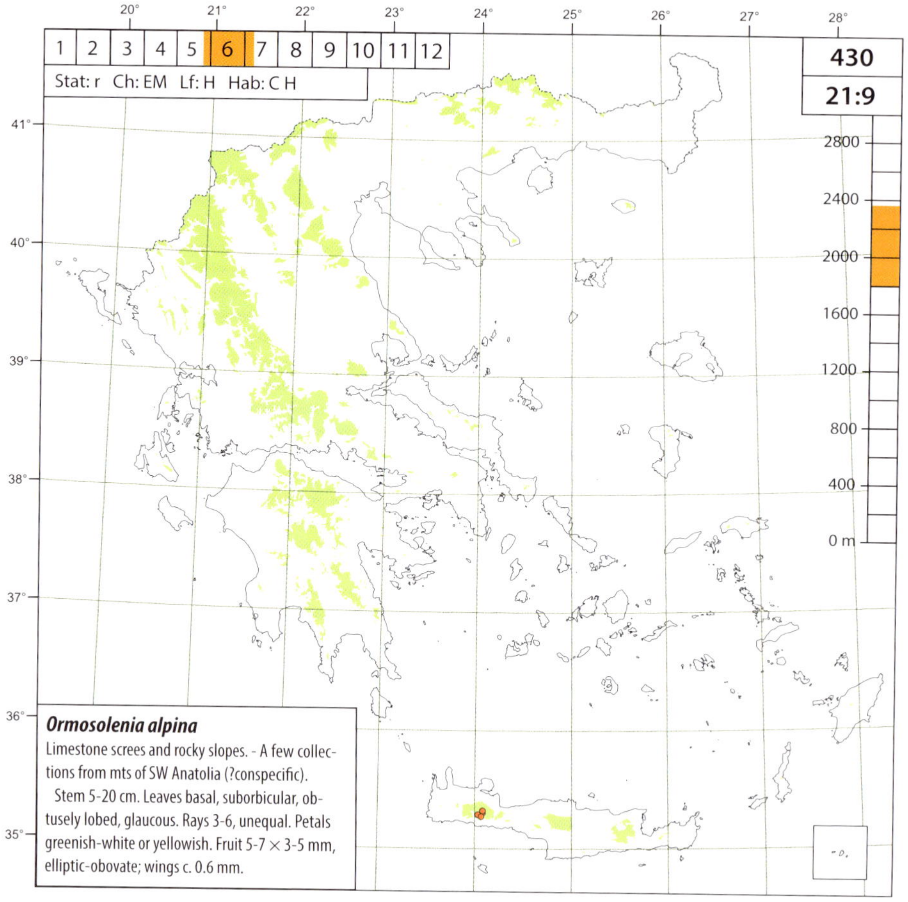

Ormosolenia alpina

Limestone screes and rocky slopes. - A few collections from mts of SW Anatolia (?conspecific).

Stem 5-20 cm. Leaves basal, suborbicular, obtusely lobed, glaucous. Rays 3-6, unequal. Petals greenish-white or yellowish. Fruit 5-7 × 3-5 mm, elliptic-obovate; wings c. 0.6 mm.

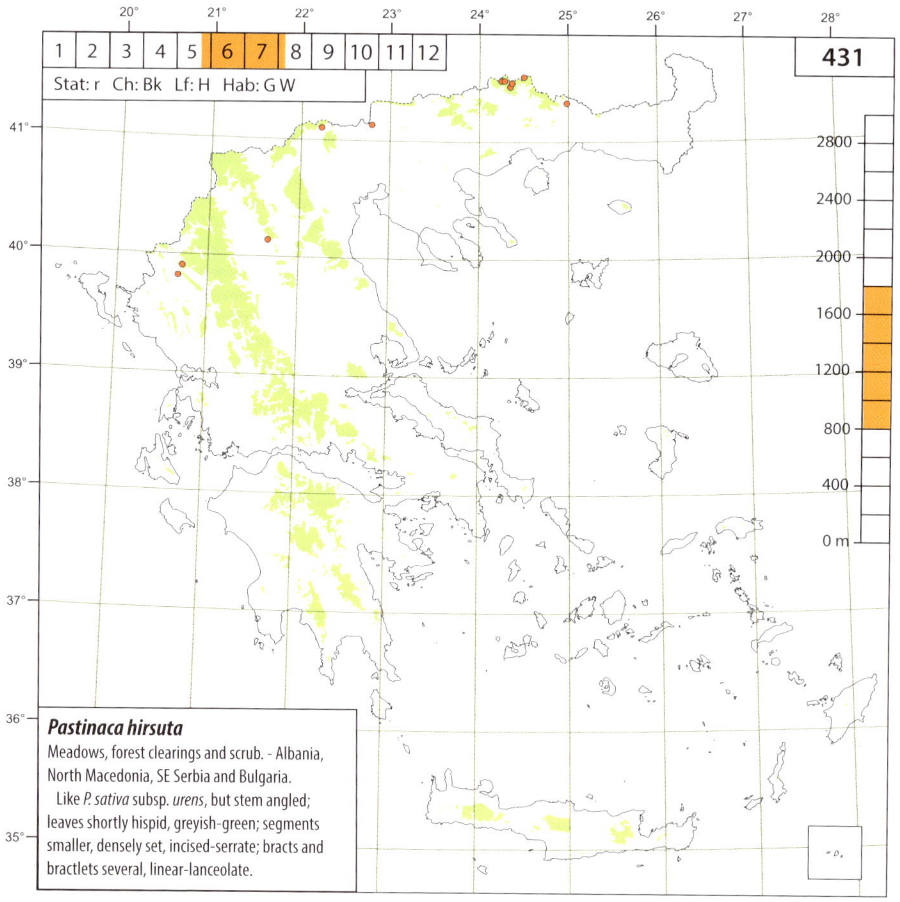

Pastinaca hirsuta

Meadows, forest clearings and scrub. - Albania, North Macedonia, SE Serbia and Bulgaria.

Like *P. sativa* subsp. *urens*, but stem angled; leaves shortly hispid, greyish-green; segments smaller, densely set, incised-serrate; bracts and bractlets several, linear-lanceolate.

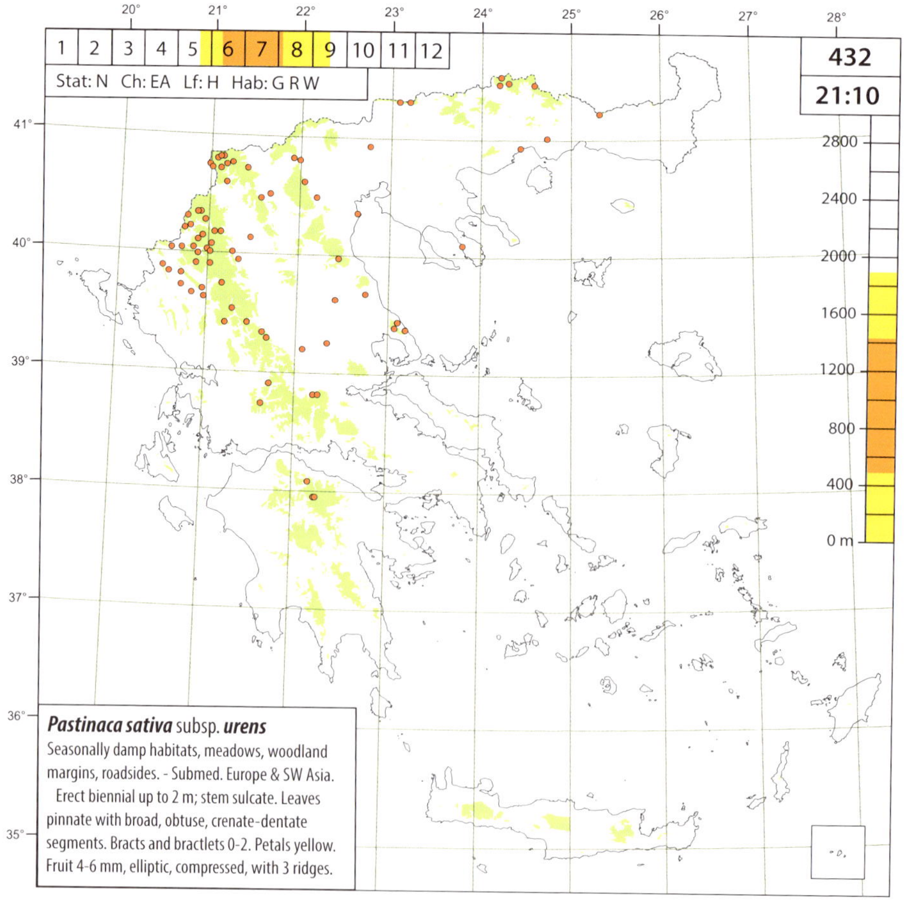

Pastinaca sativa subsp. *urens*

Seasonally damp habitats, meadows, woodland margins, roadsides. - Submed. Europe & SW Asia.

Erect biennial up to 2 m; stem sulcate. Leaves pinnate with broad, obtuse, crenate-dentate segments. Bracts and bractlets 0-2. Petals yellow. Fruit 4-6 mm, elliptic, compressed, with 3 ridges.

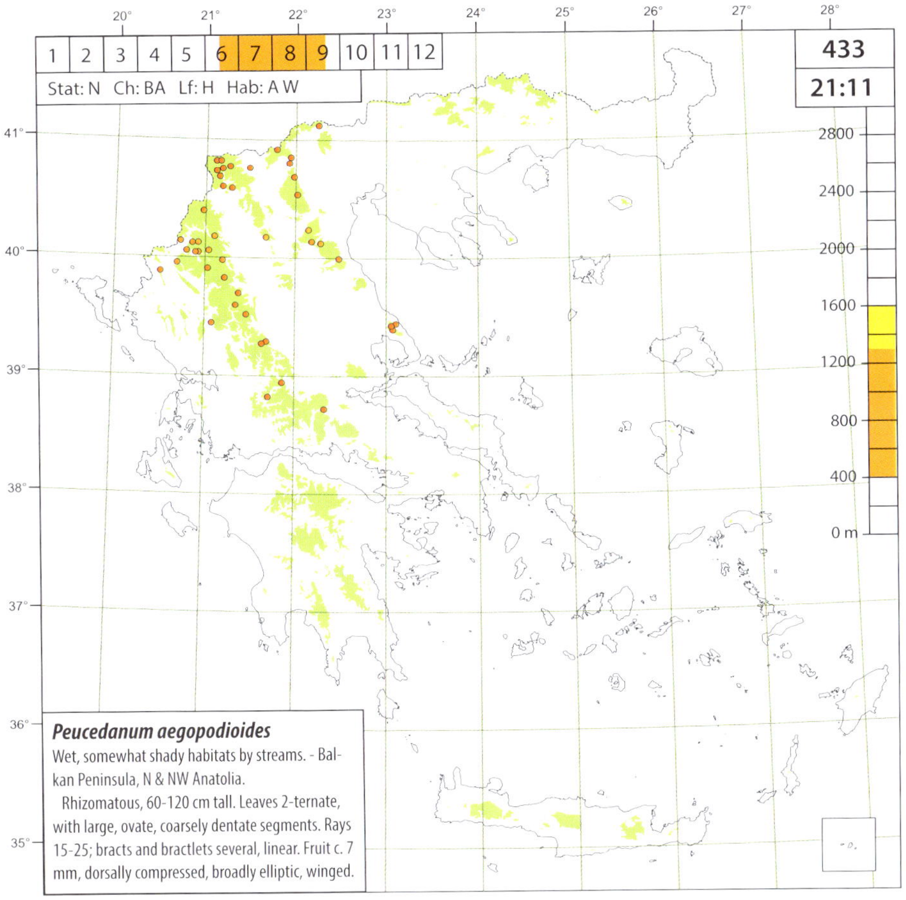

Peucedanum aegopodioides
Wet, somewhat shady habitats by streams. - Balkan Peninsula, N & NW Anatolia.
Rhizomatous, 60-120 cm tall. Leaves 2-ternate, with large, ovate, coarsely dentate segments. Rays 15-25; bracts and bractlets several, linear. Fruit c. 7 mm, dorsally compressed, broadly elliptic, winged.

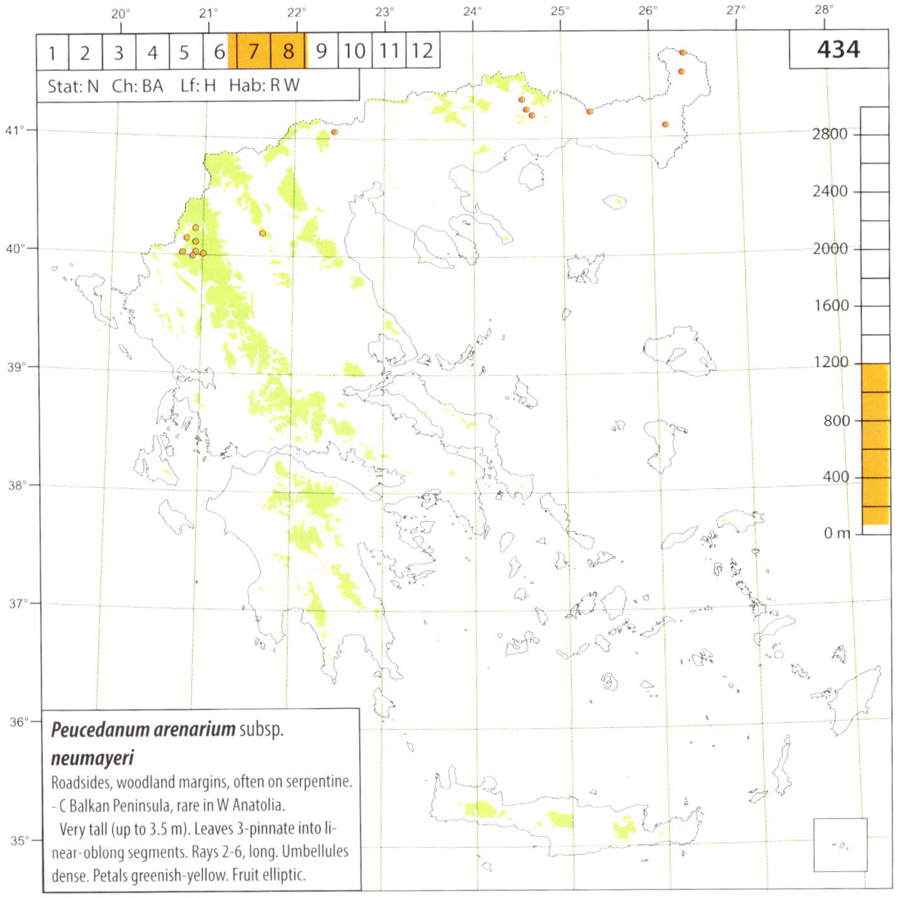

Peucedanum arenarium subsp. ***neumayeri***
Roadsides, woodland margins, often on serpentine. - C Balkan Peninsula, rare in W Anatolia.
Very tall (up to 3.5 m). Leaves 3-pinnate into linear-oblong segments. Rays 2-6, long. Umbellules dense. Petals greenish-yellow. Fruit elliptic.

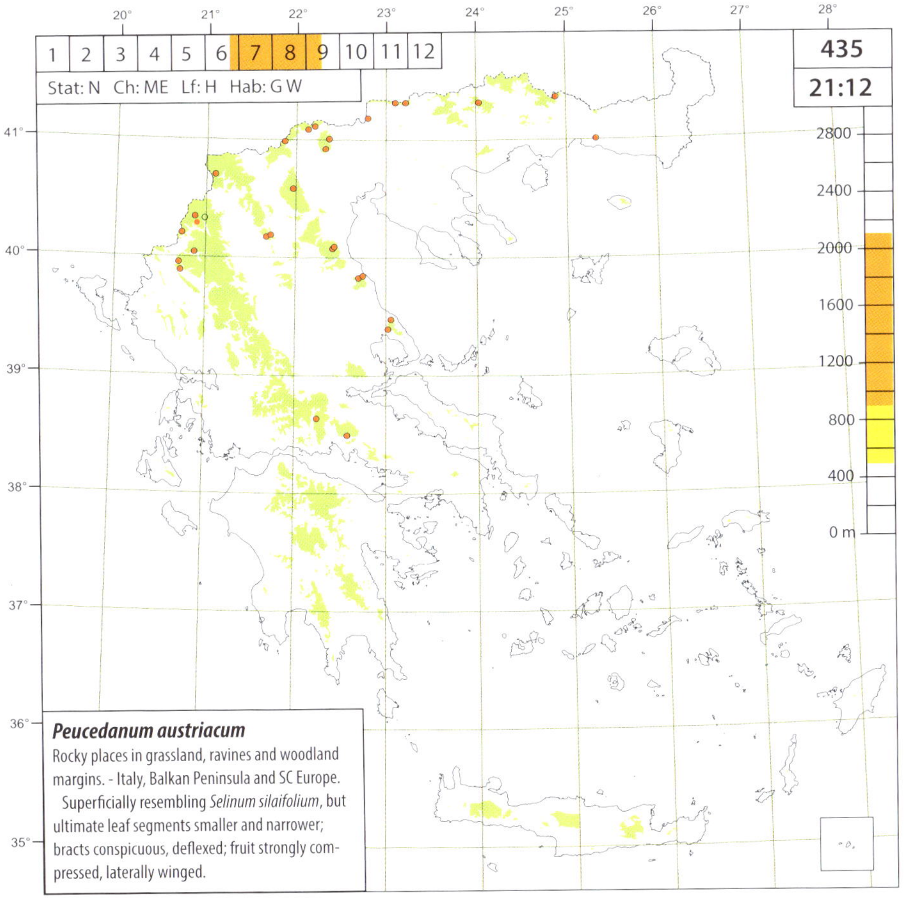

Peucedanum austriacum
Rocky places in grassland, ravines and woodland margins. - Italy, Balkan Peninsula and SC Europe.
Superficially resembling *Selinum silaifolium*, but ultimate leaf segments smaller and narrower; bracts conspicuous, deflexed; fruit strongly compressed, laterally winged.

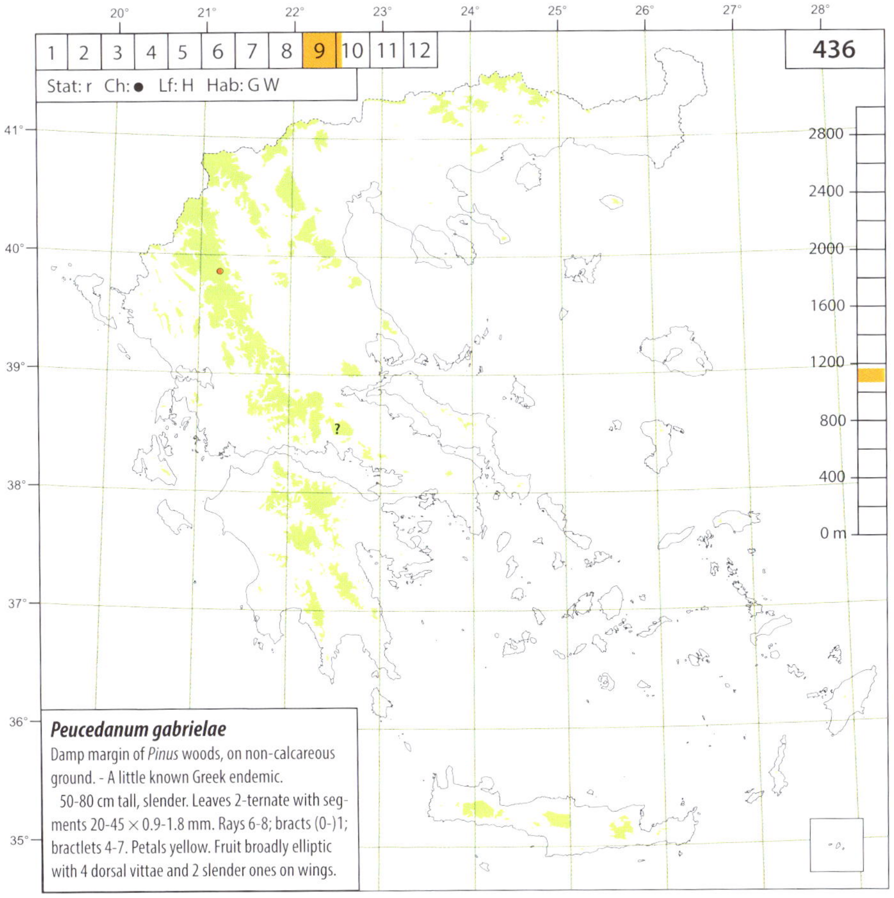

Peucedanum gabrielae
Damp margin of *Pinus* woods, on non-calcareous ground. - A little known Greek endemic.
50-80 cm tall, slender. Leaves 2-ternate with segments 20-45 × 0.9-1.8 mm. Rays 6-8; bracts (0-)1; bractlets 4-7. Petals yellow. Fruit broadly elliptic with 4 dorsal vittae and 2 slender ones on wings.

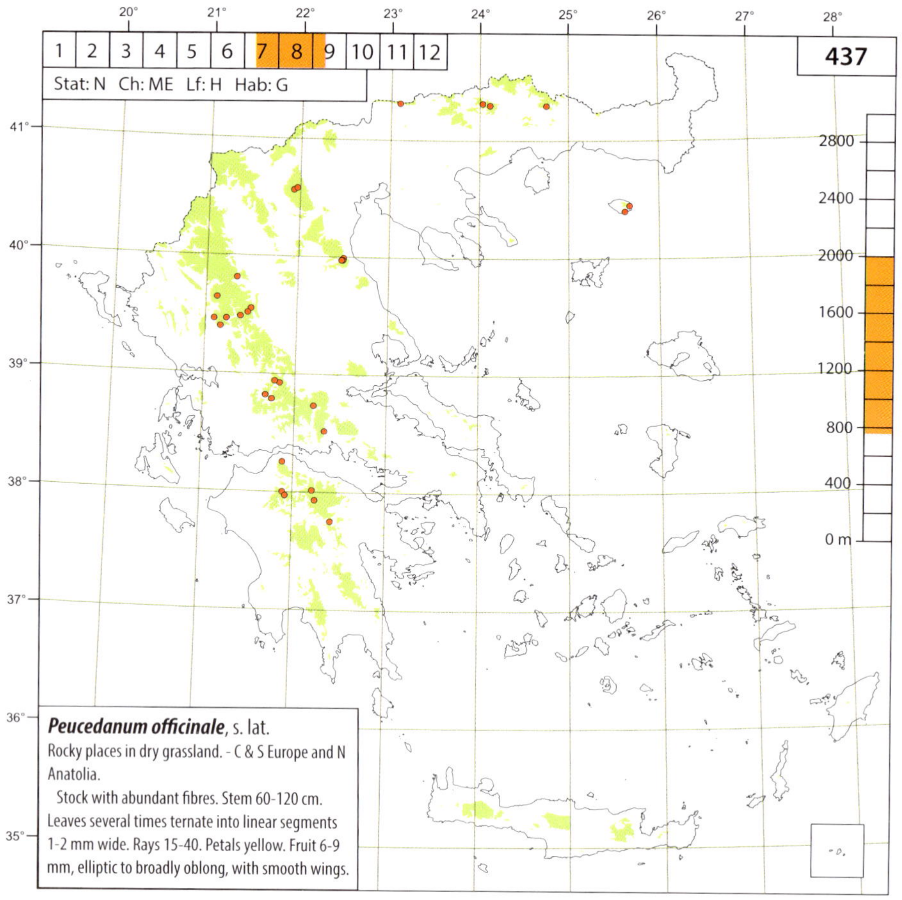

***Peucedanum officinale*, s. lat.**
Rocky places in dry grassland. - C & S Europe and N Anatolia.
Stock with abundant fibres. Stem 60-120 cm. Leaves several times ternate into linear segments 1-2 mm wide. Rays 15-40. Petals yellow. Fruit 6-9 mm, elliptic to broadly oblong, with smooth wings.

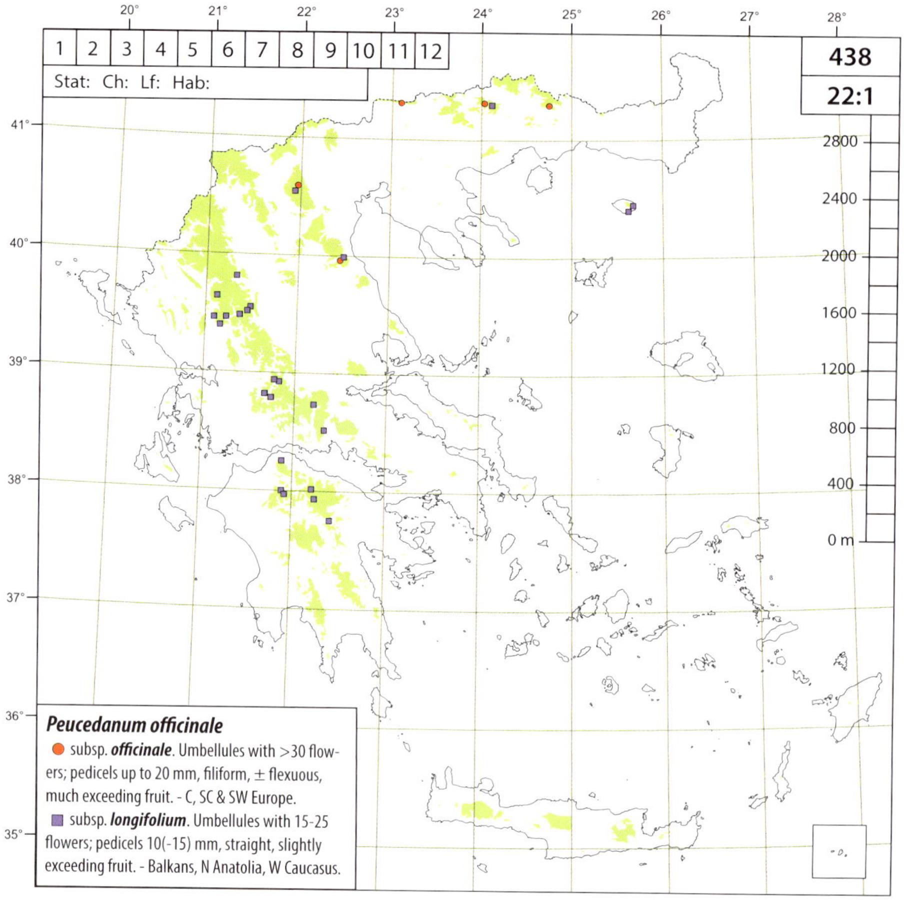

Peucedanum officinale
● subsp. ***officinale***. Umbellules with >30 flowers; pedicels up to 20 mm, filiform, ± flexuous, much exceeding fruit. - C, SC & SW Europe.
■ subsp. ***longifolium***. Umbellules with 15-25 flowers; pedicels 10(-15) mm, straight, slightly exceeding fruit. - Balkans, N Anatolia, W Caucasus.

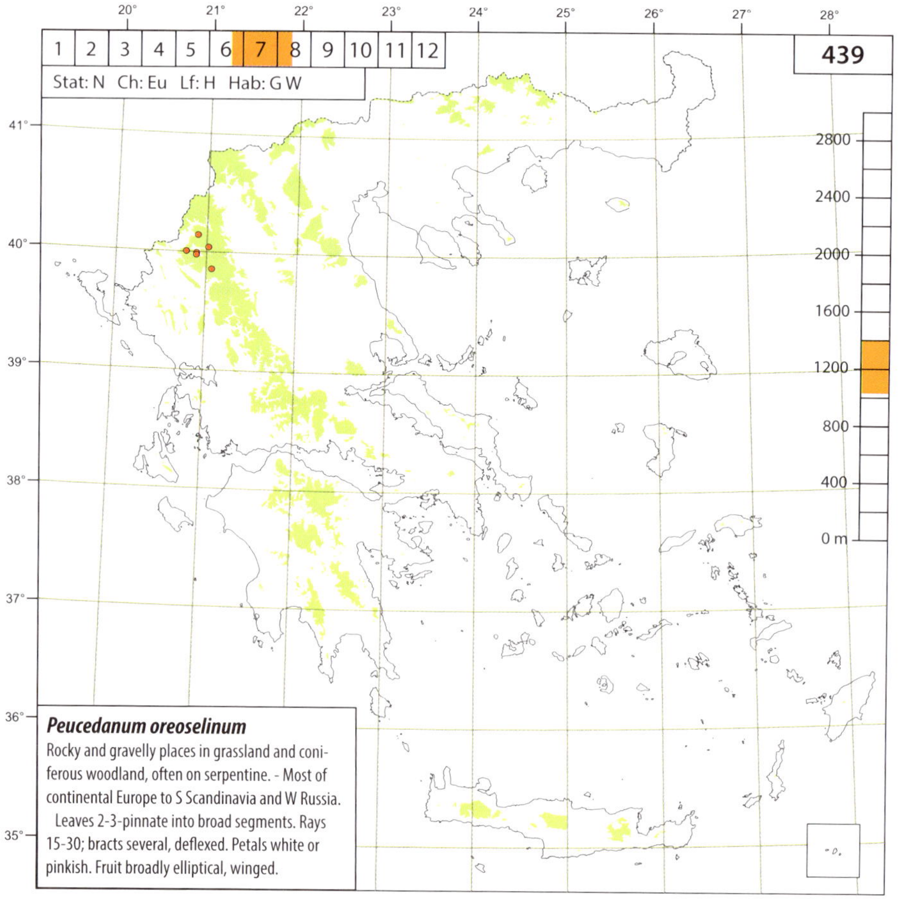

Peucedanum oreoselinum
Rocky and gravelly places in grassland and coniferous woodland, often on serpentine. - Most of continental Europe to S Scandinavia and W Russia.
Leaves 2-3-pinnate into broad segments. Rays 15-30; bracts several, deflexed. Petals white or pinkish. Fruit broadly elliptical, winged.

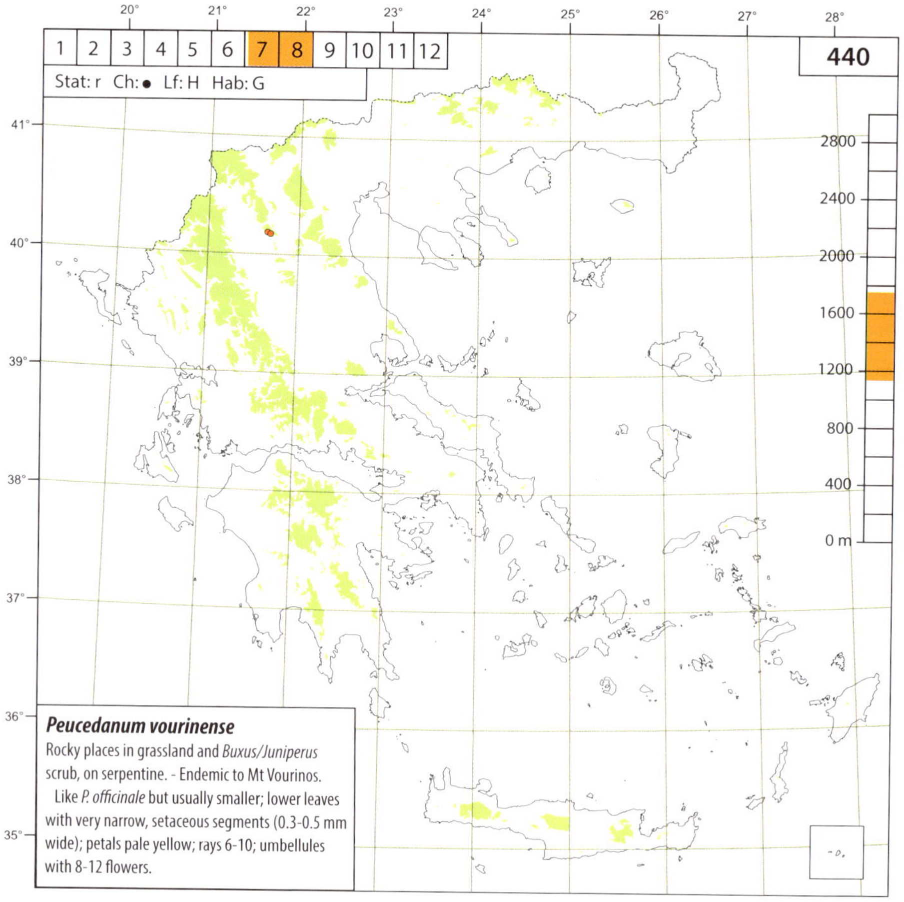

Peucedanum vourinense
Rocky places in grassland and *Buxus*/*Juniperus* scrub, on serpentine. - Endemic to Mt Vourinos.
Like *P. officinale* but usually smaller; lower leaves with very narrow, setaceous segments (0.3-0.5 mm wide); petals pale yellow; rays 6-10; umbellules with 8-12 flowers.

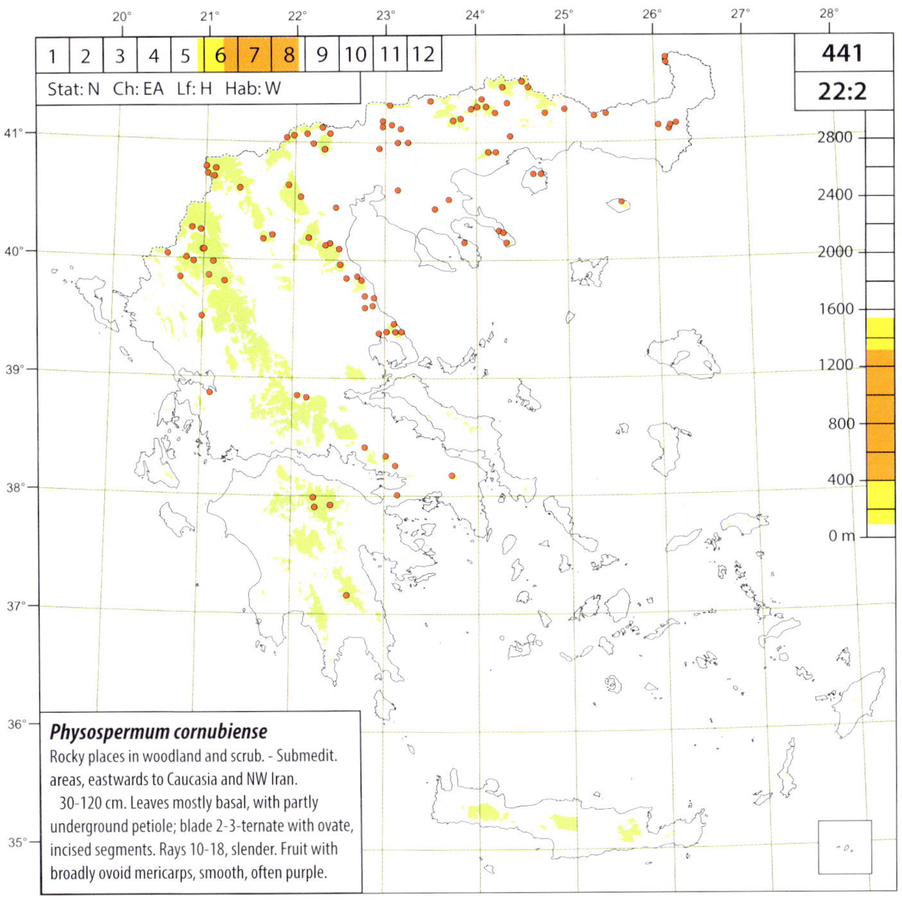

Physospermum cornubiense

Rocky places in woodland and scrub. - Submedit. areas, eastwards to Caucasia and NW Iran.

30-120 cm. Leaves mostly basal, with partly underground petiole; blade 2-3-ternate with ovate, incised segments. Rays 10-18, slender. Fruit with broadly ovoid mericarps, smooth, often purple.

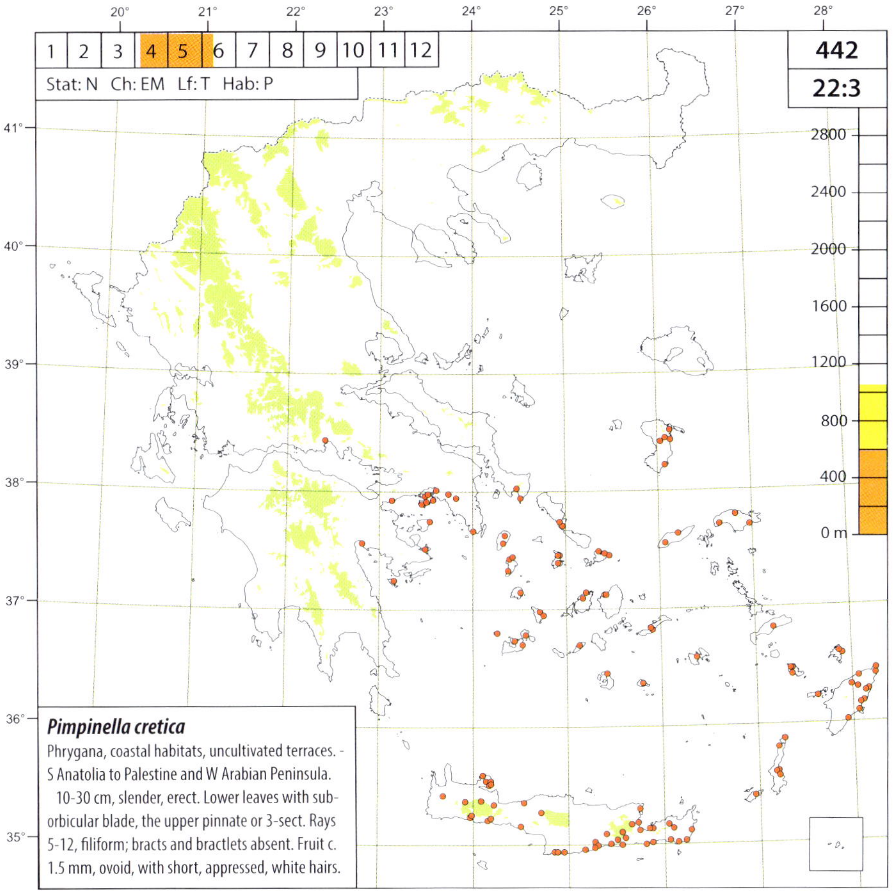

Pimpinella cretica

Phrygana, coastal habitats, uncultivated terraces. - S Anatolia to Palestine and W Arabian Peninsula.

10-30 cm, slender, erect. Lower leaves with suborbicular blade, the upper pinnate or 3-sect. Rays 5-12, filiform; bracts and bractlets absent. Fruit c. 1.5 mm, ovoid, with short, appressed, white hairs.

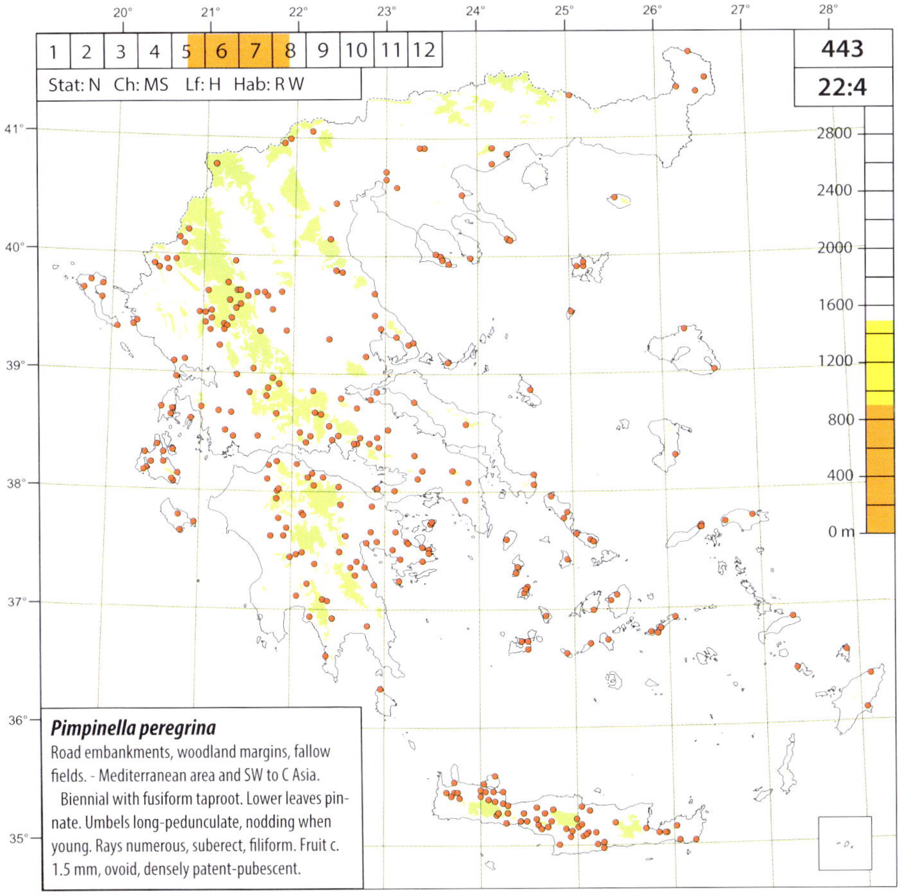

Pimpinella peregrina

Road embankments, woodland margins, fallow fields. - Mediterranean area and SW to C Asia.

Biennial with fusiform taproot. Lower leaves pinnate. Umbels long-pedunculate, nodding when young. Rays numerous, suberect, filiform. Fruit c. 1.5 mm, ovoid, densely patent-pubescent.

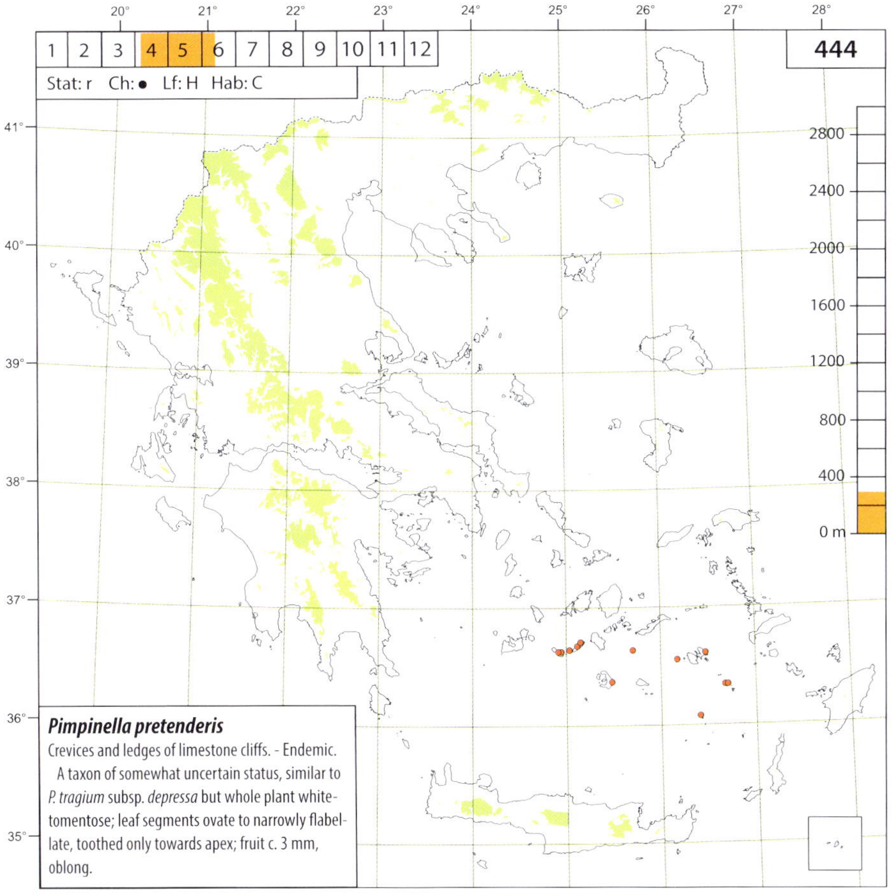

Pimpinella pretenderis

Crevices and ledges of limestone cliffs. - Endemic.

A taxon of somewhat uncertain status, similar to *P. tragium* subsp. *depressa* but whole plant white-tomentose; leaf segments ovate to narrowly flabellate, toothed only towards apex; fruit c. 3 mm, oblong.

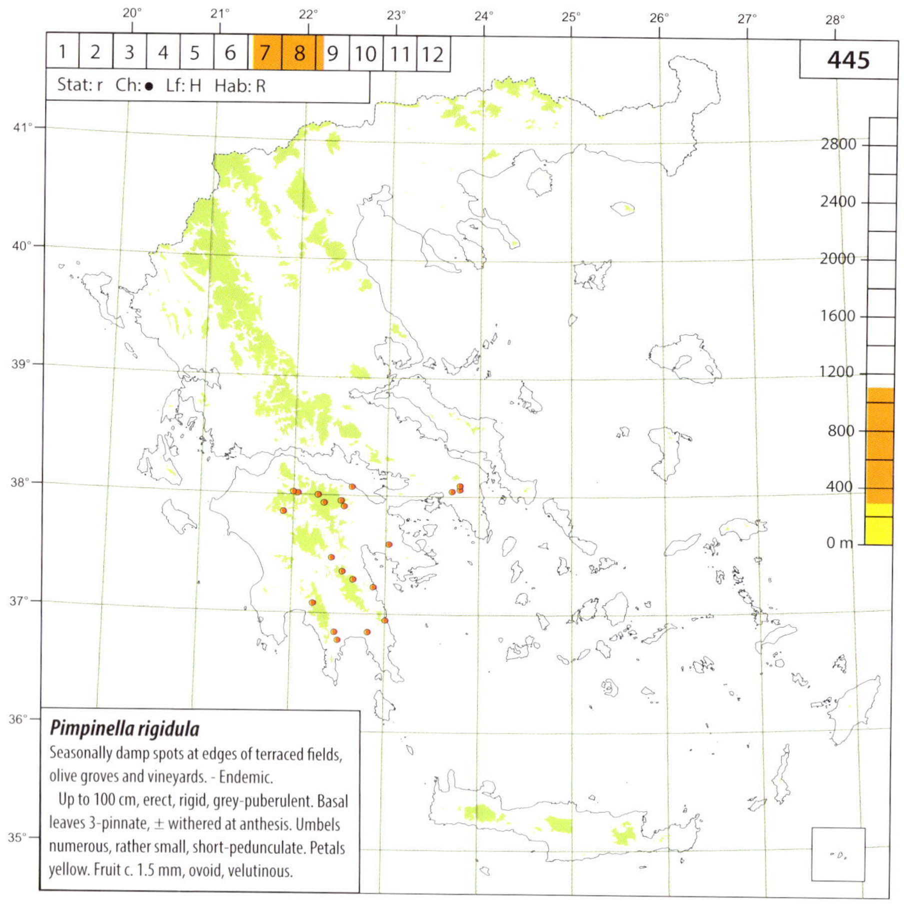

Pimpinella rigidula
Seasonally damp spots at edges of terraced fields, olive groves and vineyards. - Endemic.
Up to 100 cm, erect, rigid, grey-puberulent. Basal leaves 3-pinnate, ± withered at anthesis. Umbels numerous, rather small, short-pedunculate. Petals yellow. Fruit c. 1.5 mm, ovoid, velutinous.

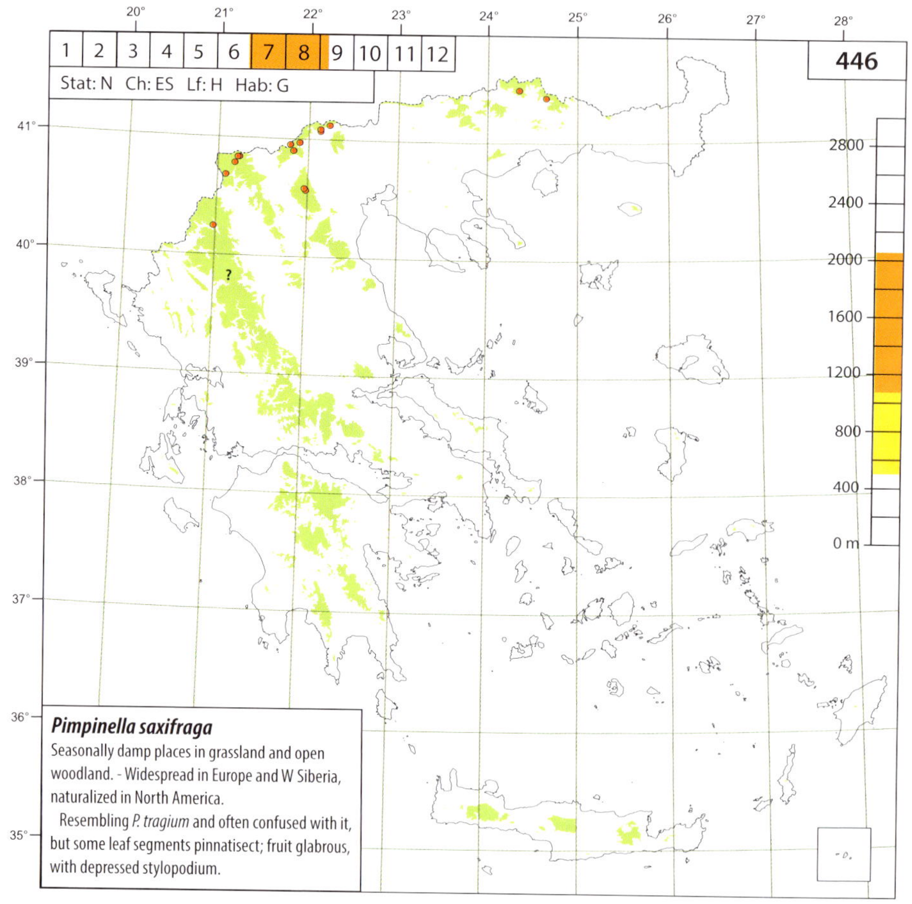

Pimpinella saxifraga
Seasonally damp places in grassland and open woodland. - Widespread in Europe and W Siberia, naturalized in North America.
Resembling *P. tragium* and often confused with it, but some leaf segments pinnatisect; fruit glabrous, with depressed stylopodium.

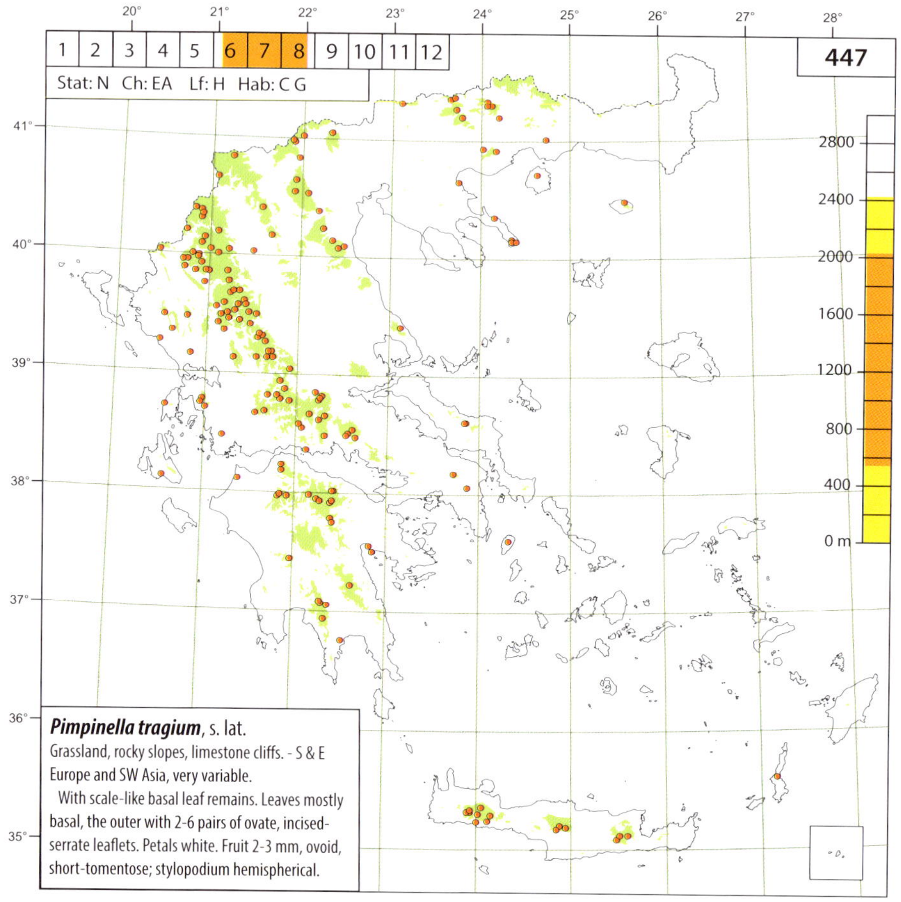

Pimpinella tragium, s. lat.
Grassland, rocky slopes, limestone cliffs. - S & E Europe and SW Asia, very variable.
With scale-like basal leaf remains. Leaves mostly basal, the outer with 2-6 pairs of ovate, incised-serrate leaflets. Petals white. Fruit 2-3 mm, ovoid, short-tomentose; stylopodium hemispherical.

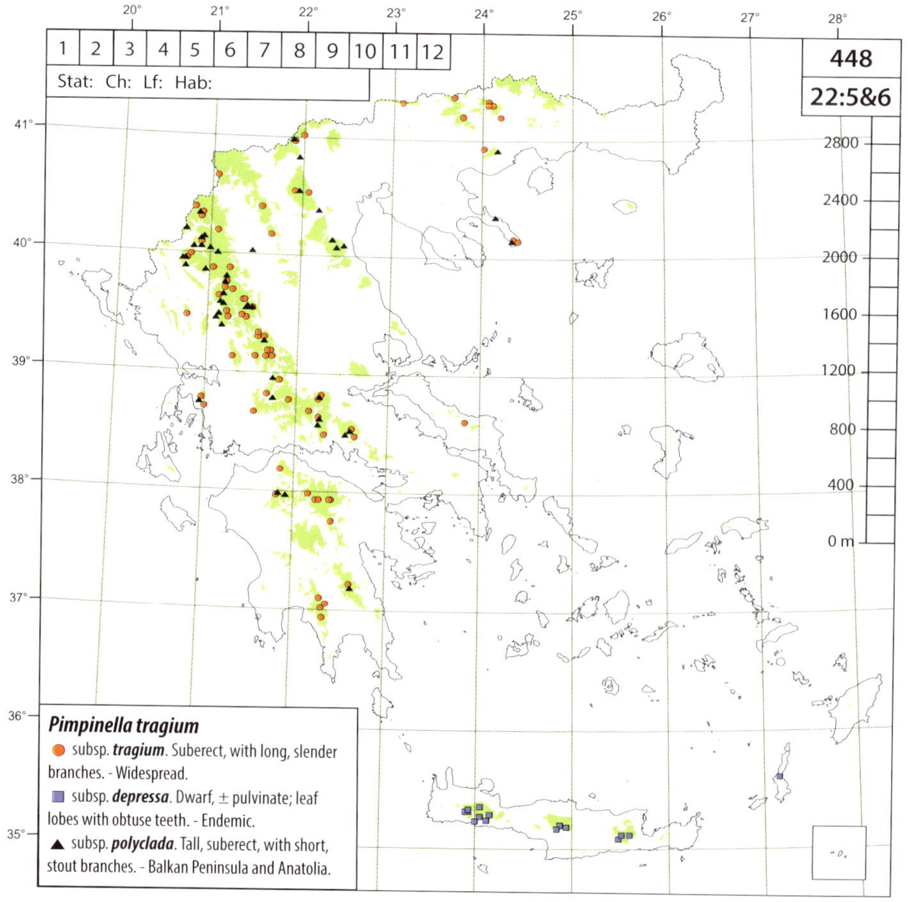

Pimpinella tragium
● subsp. ***tragium***. Suberect, with long, slender branches. - Widespread.
■ subsp. ***depressa***. Dwarf, ± pulvinate; leaf lobes with obtuse teeth. - Endemic.
▲ subsp. ***polyclada***. Tall, suberect, with short, stout branches. - Balkan Peninsula and Anatolia.

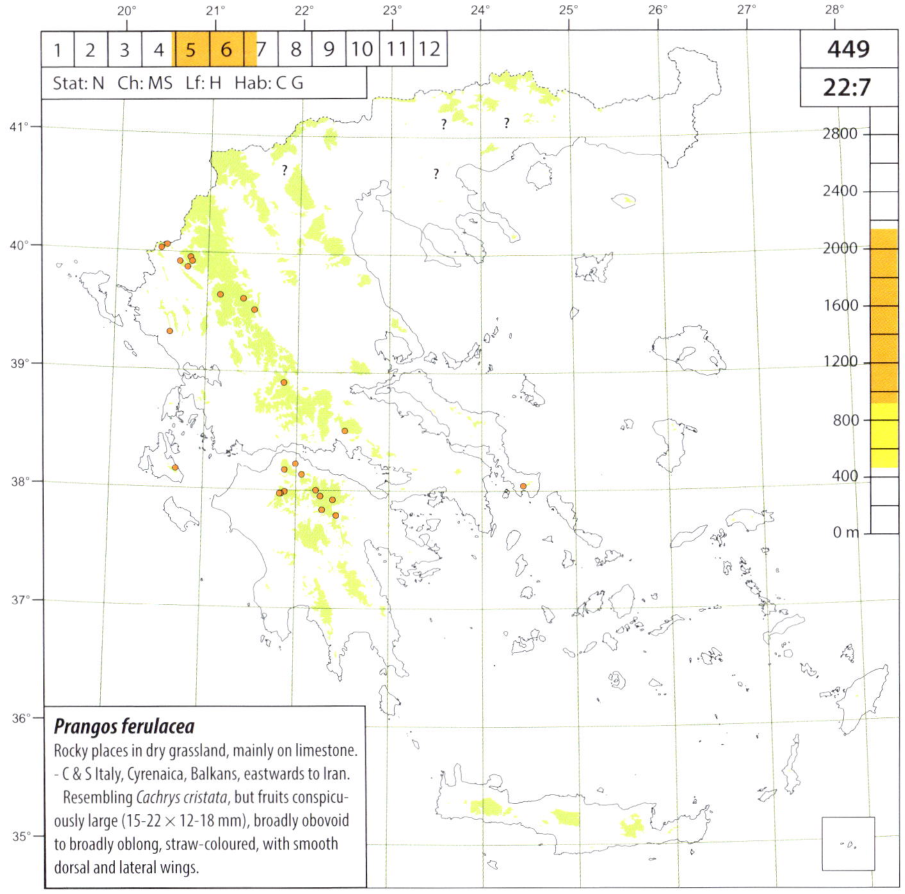

Prangos ferulacea

Rocky places in dry grassland, mainly on limestone. - C & S Italy, Cyrenaica, Balkans, eastwards to Iran.

Resembling *Cachrys cristata*, but fruits conspicuously large (15-22 × 12-18 mm), broadly obovoid to broadly oblong, straw-coloured, with smooth dorsal and lateral wings.

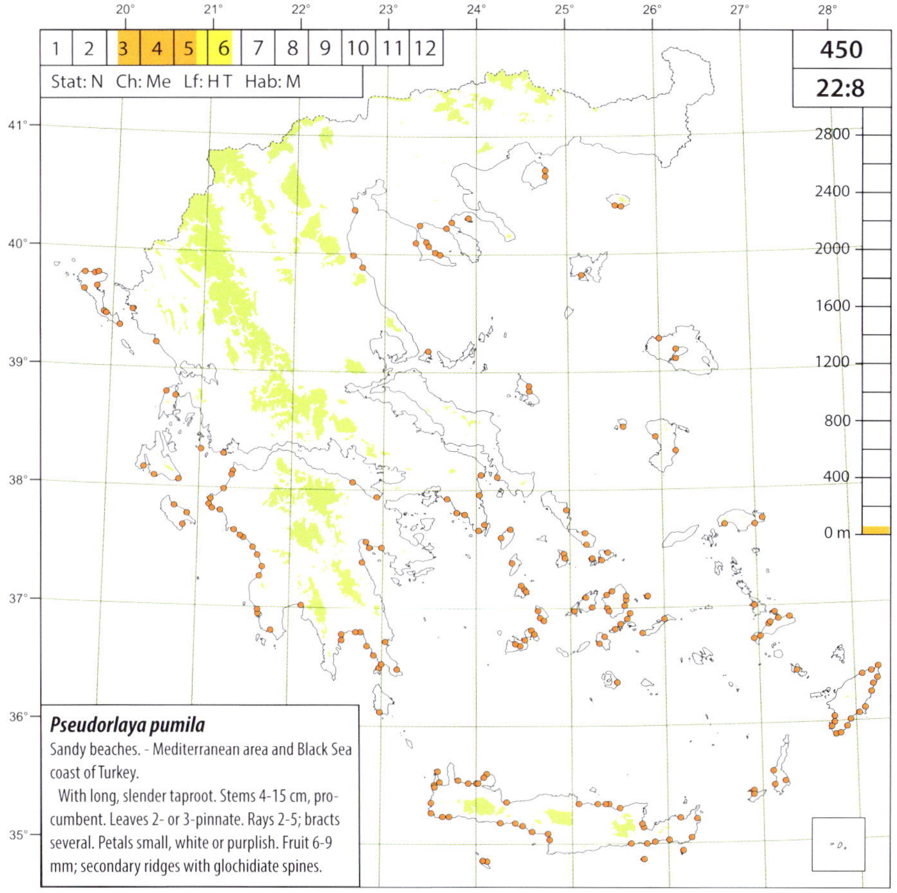

Pseudorlaya pumila

Sandy beaches. - Mediterranean area and Black Sea coast of Turkey.

With long, slender taproot. Stems 4-15 cm, procumbent. Leaves 2- or 3-pinnate. Rays 2-5; bracts several. Petals small, white or purplish. Fruit 6-9 mm; secondary ridges with glochidiate spines.

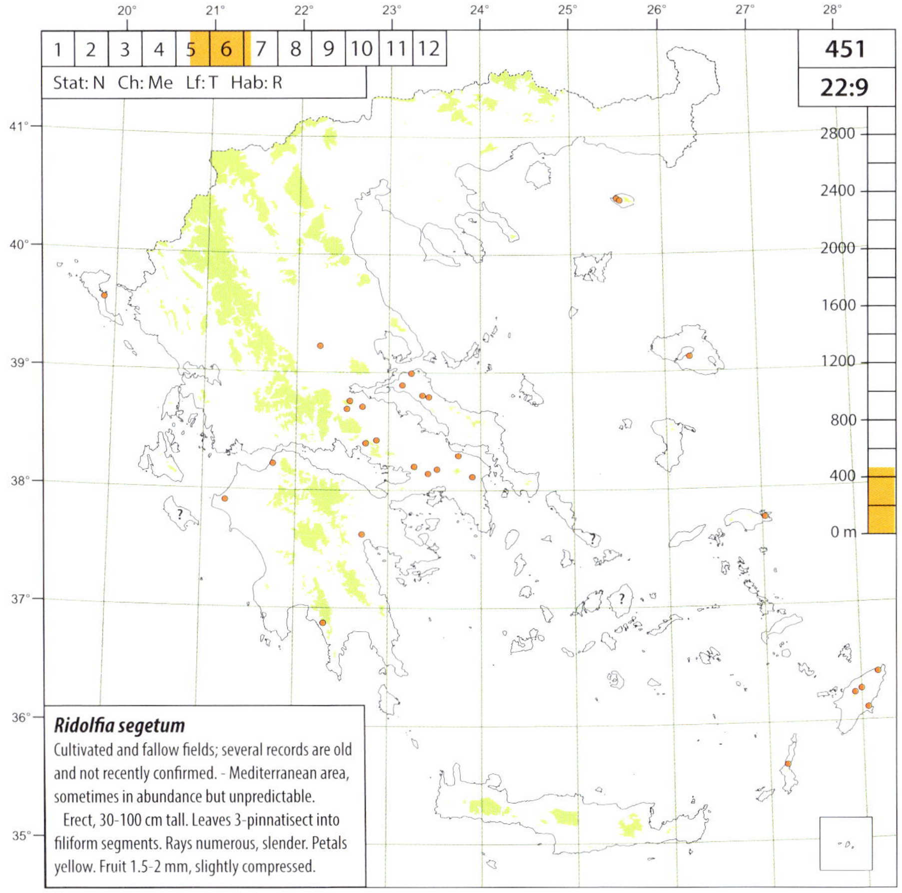

Ridolfia segetum

Cultivated and fallow fields; several records are old and not recently confirmed. - Mediterranean area, sometimes in abundance but unpredictable.

Erect, 30-100 cm tall. Leaves 3-pinnatisect into filiform segments. Rays numerous, slender. Petals yellow. Fruit 1.5-2 mm, slightly compressed.

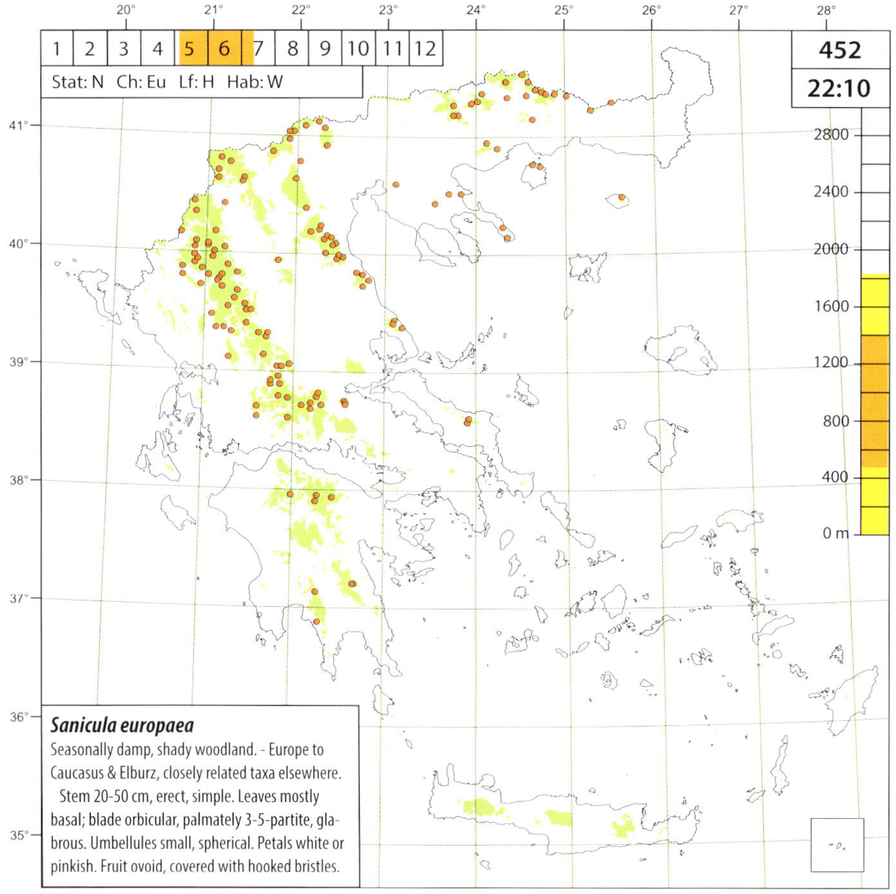

Sanicula europaea

Seasonally damp, shady woodland. - Europe to Caucasus & Elburz, closely related taxa elsewhere.

Stem 20-50 cm, erect, simple. Leaves mostly basal; blade orbicular, palmately 3-5-partite, glabrous. Umbellules small, spherical. Petals white or pinkish. Fruit ovoid, covered with hooked bristles.

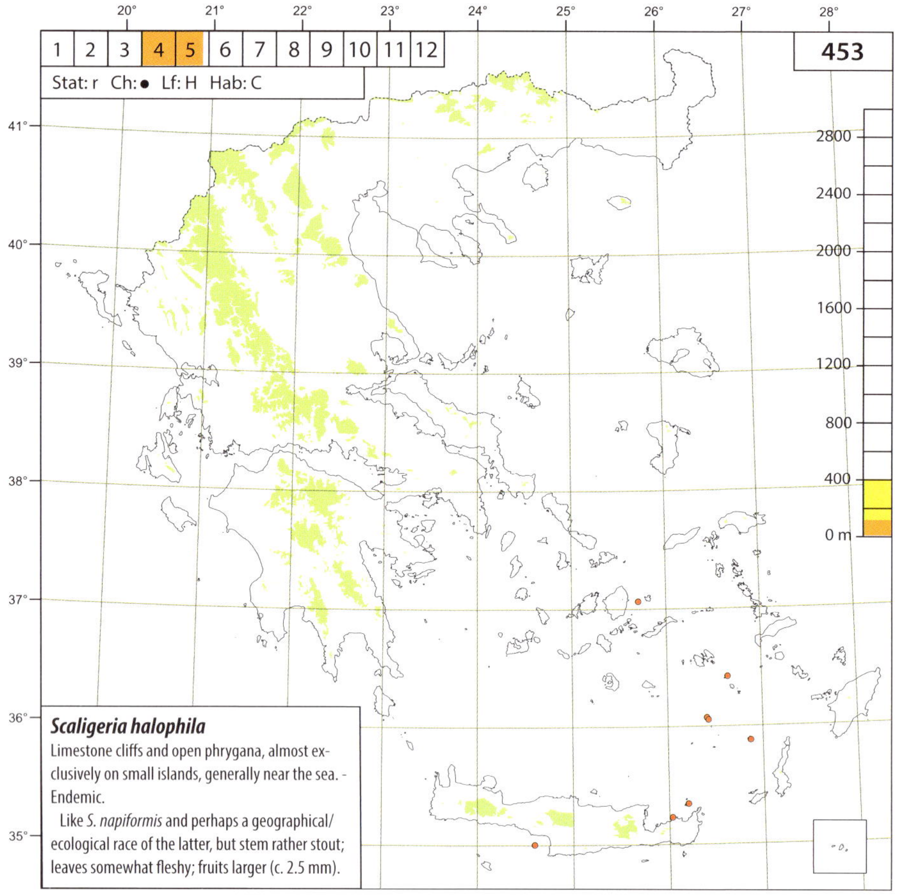

Scaligeria halophila
Limestone cliffs and open phrygana, almost exclusively on small islands, generally near the sea. - Endemic.
Like *S. napiformis* and perhaps a geographical/ecological race of the latter, but stem rather stout; leaves somewhat fleshy; fruits larger (c. 2.5 mm).

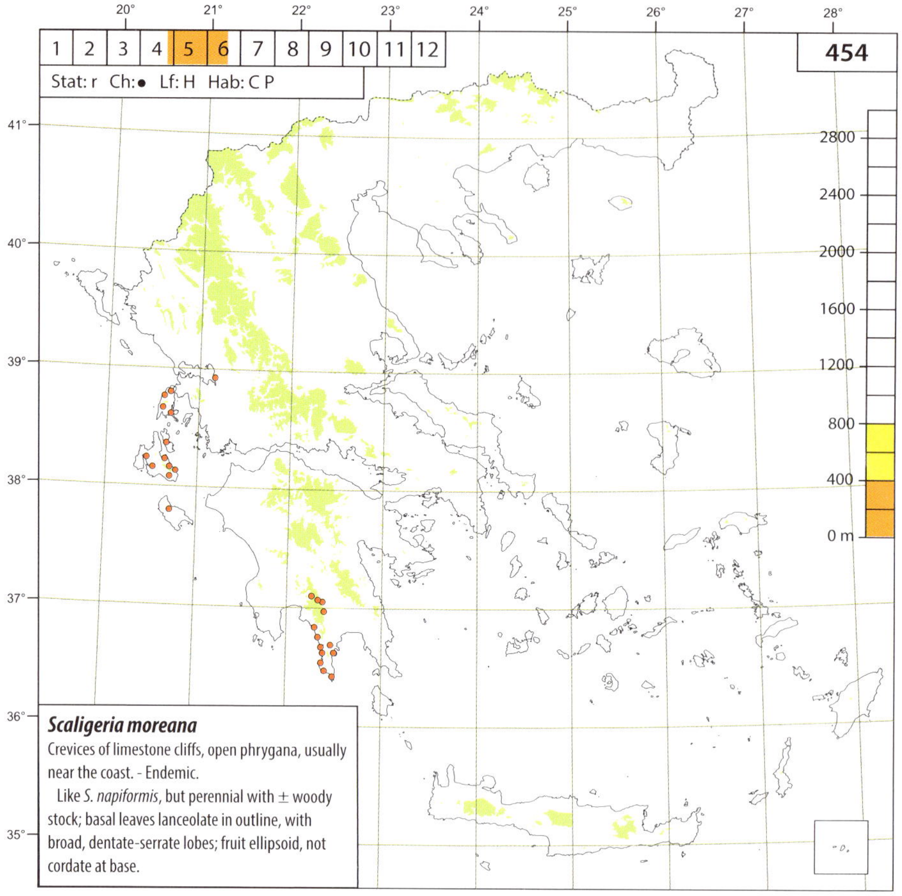

Scaligeria moreana
Crevices of limestone cliffs, open phrygana, usually near the coast. - Endemic.
Like *S. napiformis*, but perennial with ± woody stock; basal leaves lanceolate in outline, with broad, dentate-serrate lobes; fruit ellipsoid, not cordate at base.

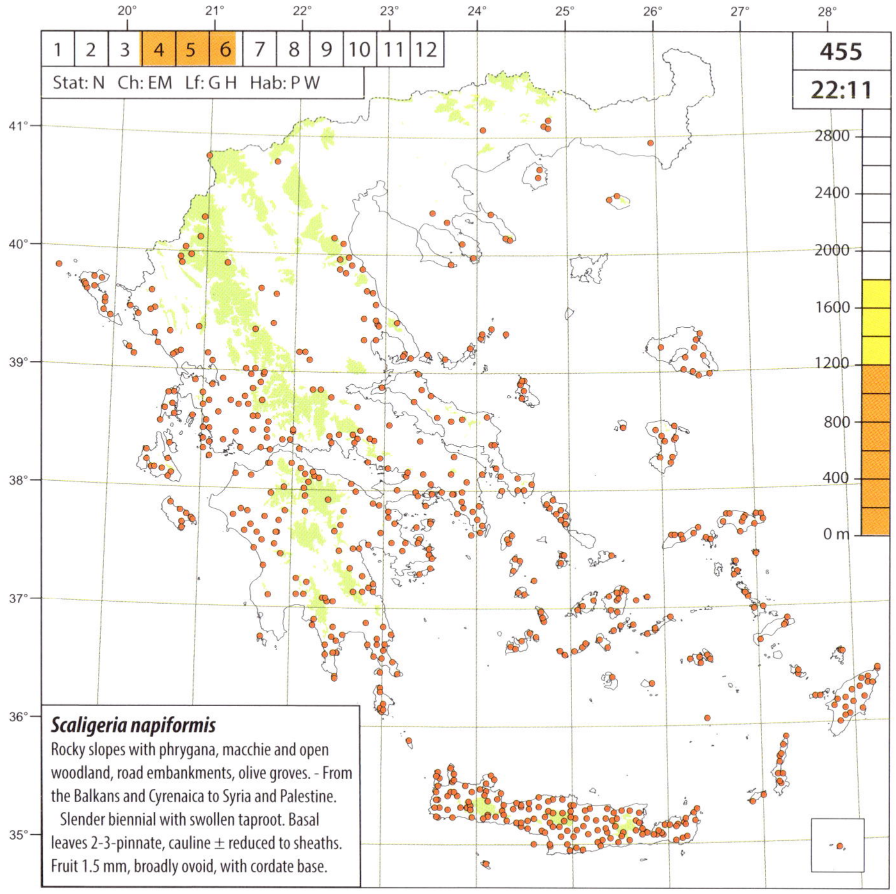

Scaligeria napiformis
Rocky slopes with phrygana, macchie and open woodland, road embankments, olive groves. - From the Balkans and Cyrenaica to Syria and Palestine.
Slender biennial with swollen taproot. Basal leaves 2-3-pinnate, cauline ± reduced to sheaths. Fruit 1.5 mm, broadly ovoid, with cordate base.

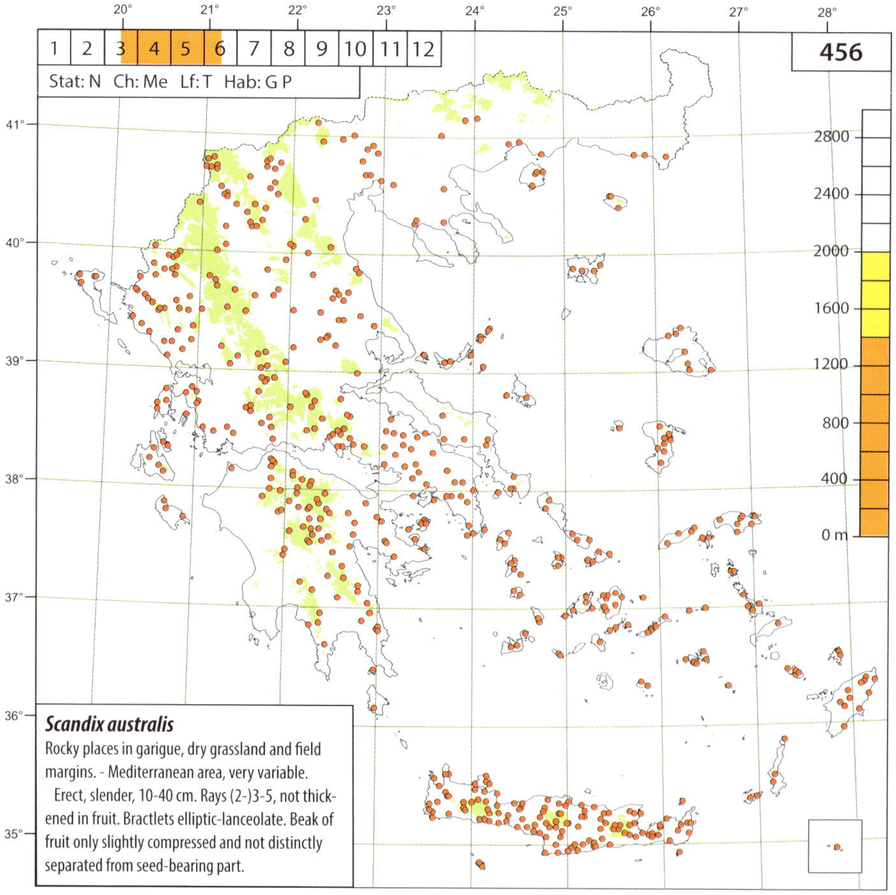

Scandix australis
Rocky places in garigue, dry grassland and field margins. - Mediterranean area, very variable.
Erect, slender, 10-40 cm. Rays (2-)3-5, not thickened in fruit. Bractlets elliptic-lanceolate. Beak of fruit only slightly compressed and not distinctly separated from seed-bearing part.

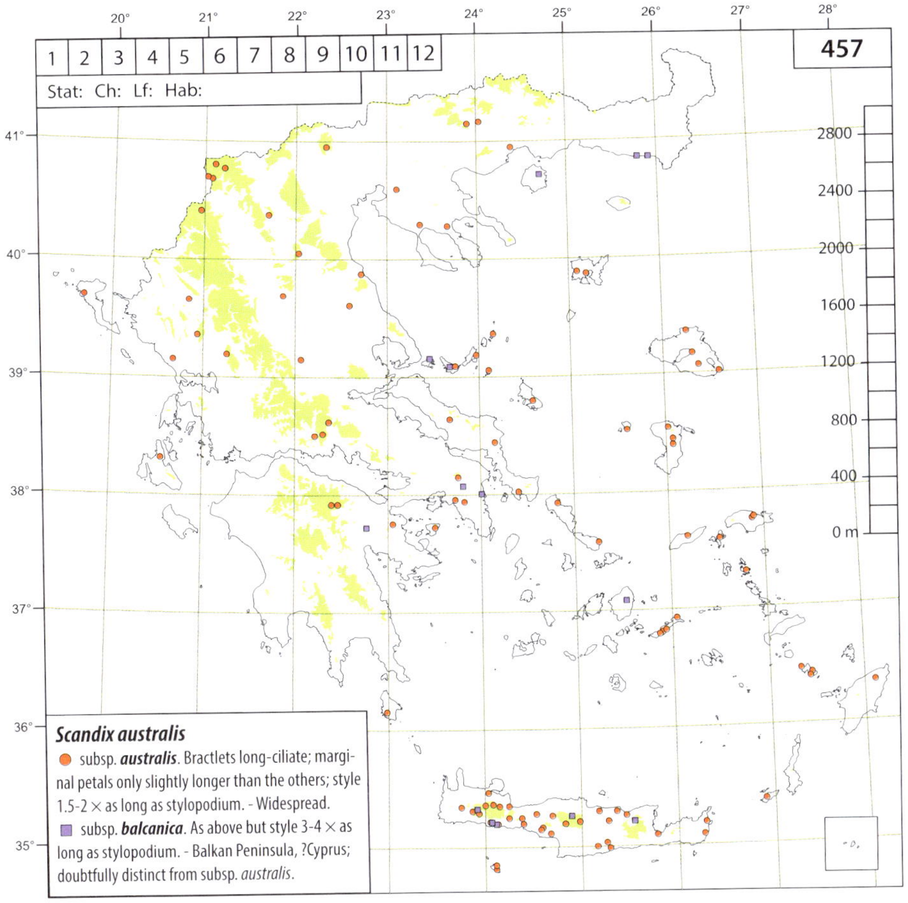

Scandix australis
● subsp. ***australis***. Bractlets long-ciliate; marginal petals only slightly longer than the others; style 1.5-2 × as long as stylopodium. - Widespread.
■ subsp. ***balcanica***. As above but style 3-4 × as long as stylopodium. - Balkan Peninsula, ?Cyprus; doubtfully distinct from subsp. *australis*.

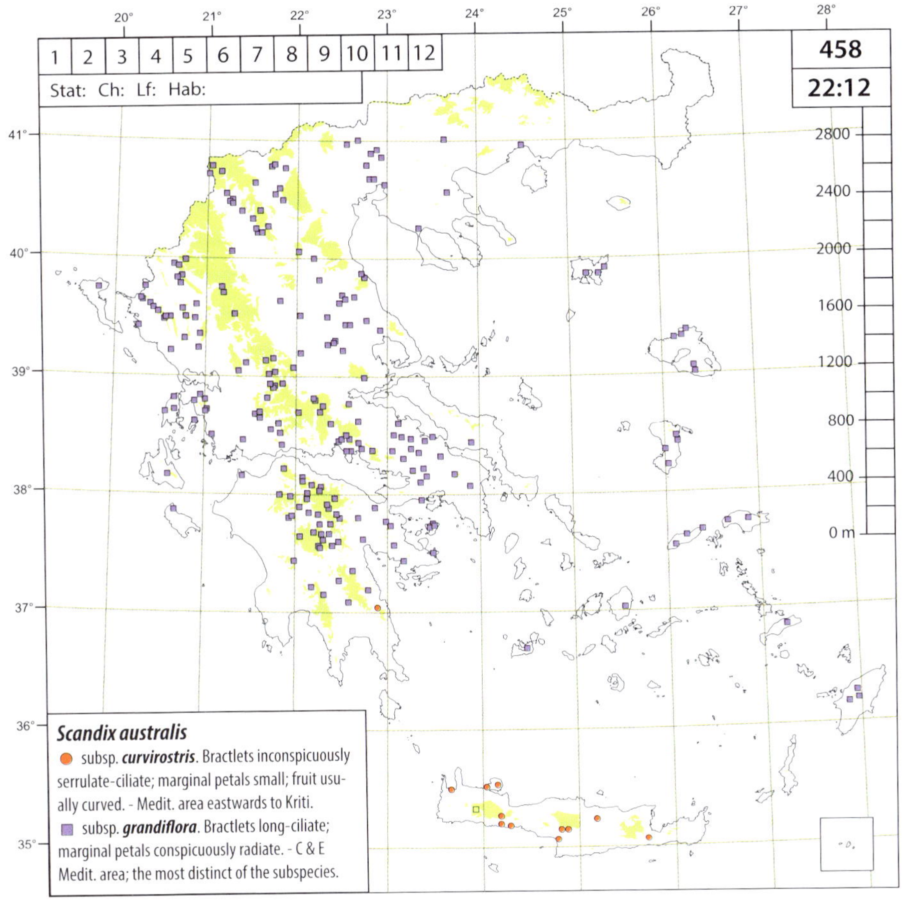

Scandix australis
● subsp. ***curvirostris***. Bractlets inconspicuously serrulate-ciliate; marginal petals small; fruit usually curved. - Medit. area eastwards to Kriti.
■ subsp. ***grandiflora***. Bractlets long-ciliate; marginal petals conspicuously radiate. - C & E Medit. area; the most distinct of the subspecies.

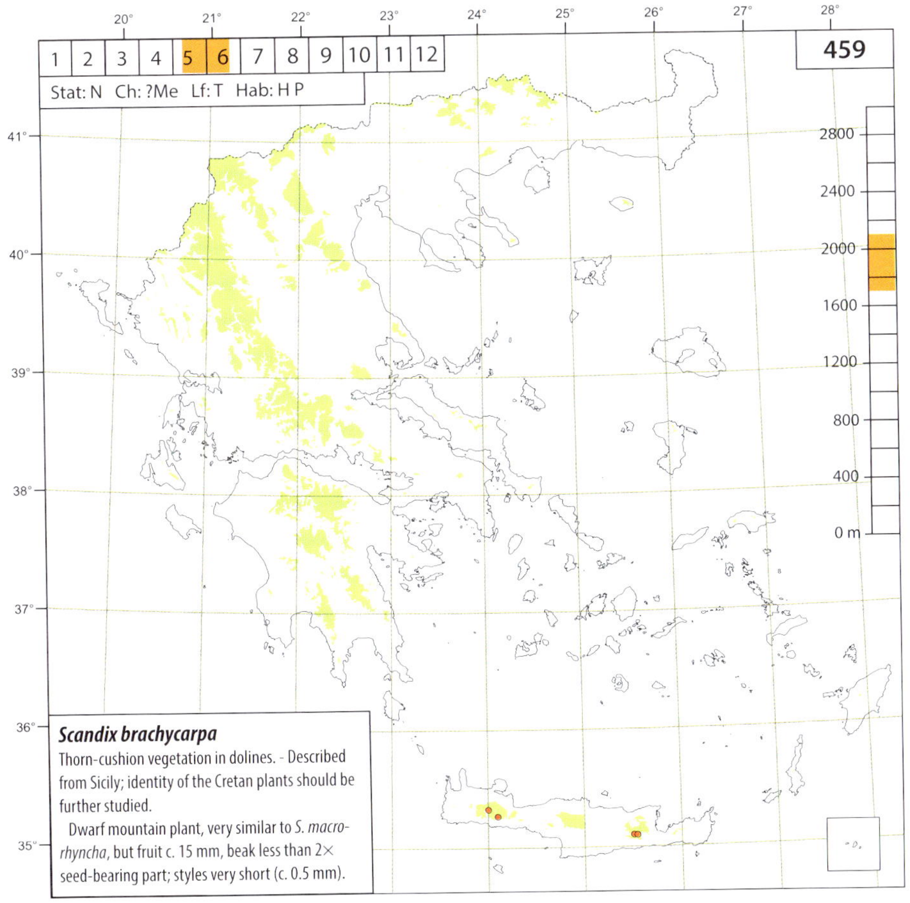

Scandix brachycarpa
Thorn-cushion vegetation in dolines. - Described from Sicily; identity of the Cretan plants should be further studied.
Dwarf mountain plant, very similar to *S. macrorhyncha*, but fruit c. 15 mm, beak less than 2× seed-bearing part; styles very short (c. 0.5 mm).

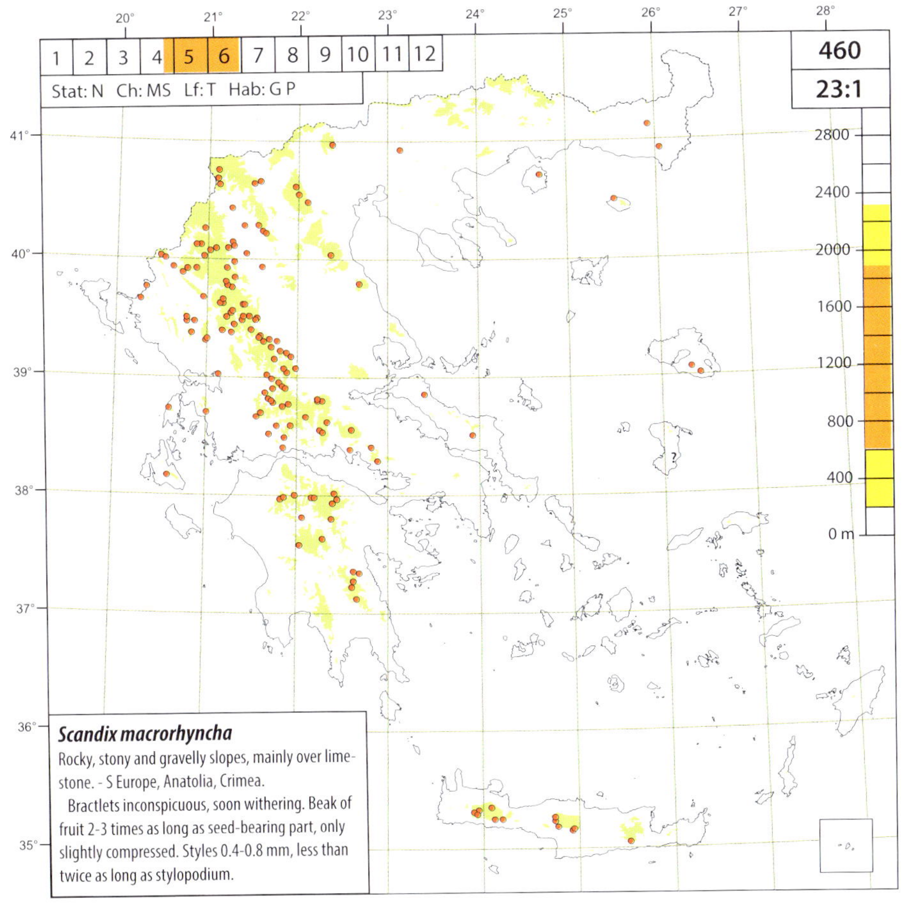

Scandix macrorhyncha
Rocky, stony and gravelly slopes, mainly over limestone. - S Europe, Anatolia, Crimea.
Bractlets inconspicuous, soon withering. Beak of fruit 2-3 times as long as seed-bearing part, only slightly compressed. Styles 0.4-0.8 mm, less than twice as long as stylopodium.

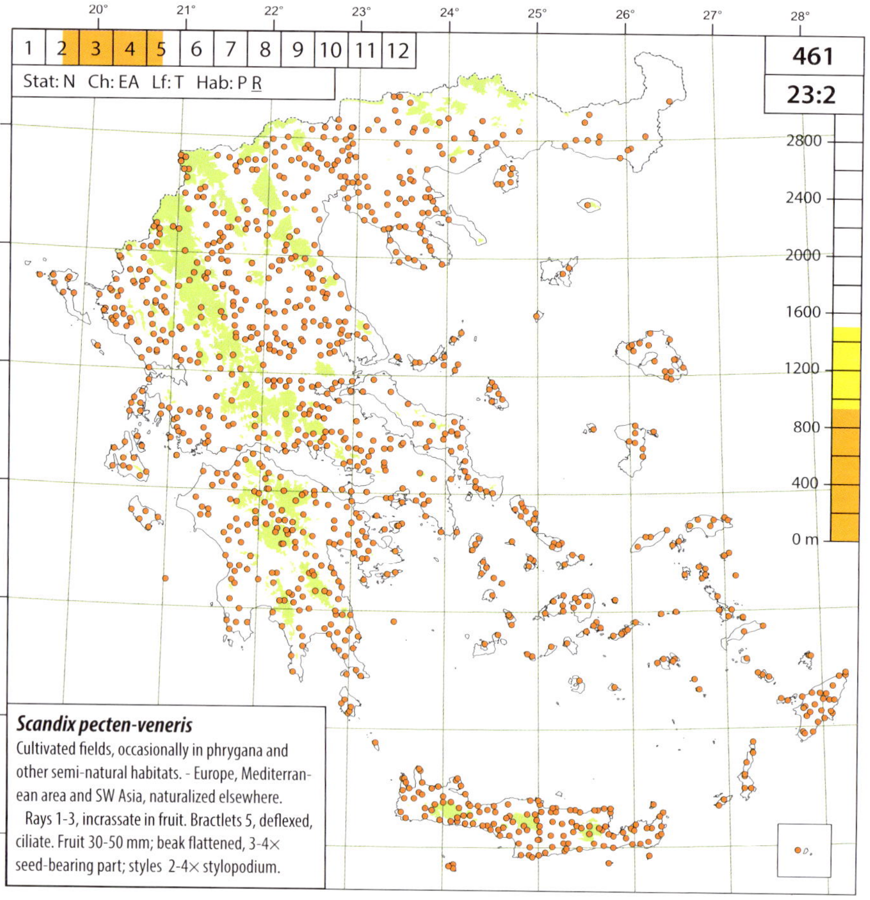

Scandix pecten-veneris
Cultivated fields, occasionally in phrygana and other semi-natural habitats. - Europe, Mediterranean area and SW Asia, naturalized elsewhere.
Rays 1-3, incrassate in fruit. Bractlets 5, deflexed, ciliate. Fruit 30-50 mm; beak flattened, 3-4× seed-bearing part; styles 2-4× stylopodium.

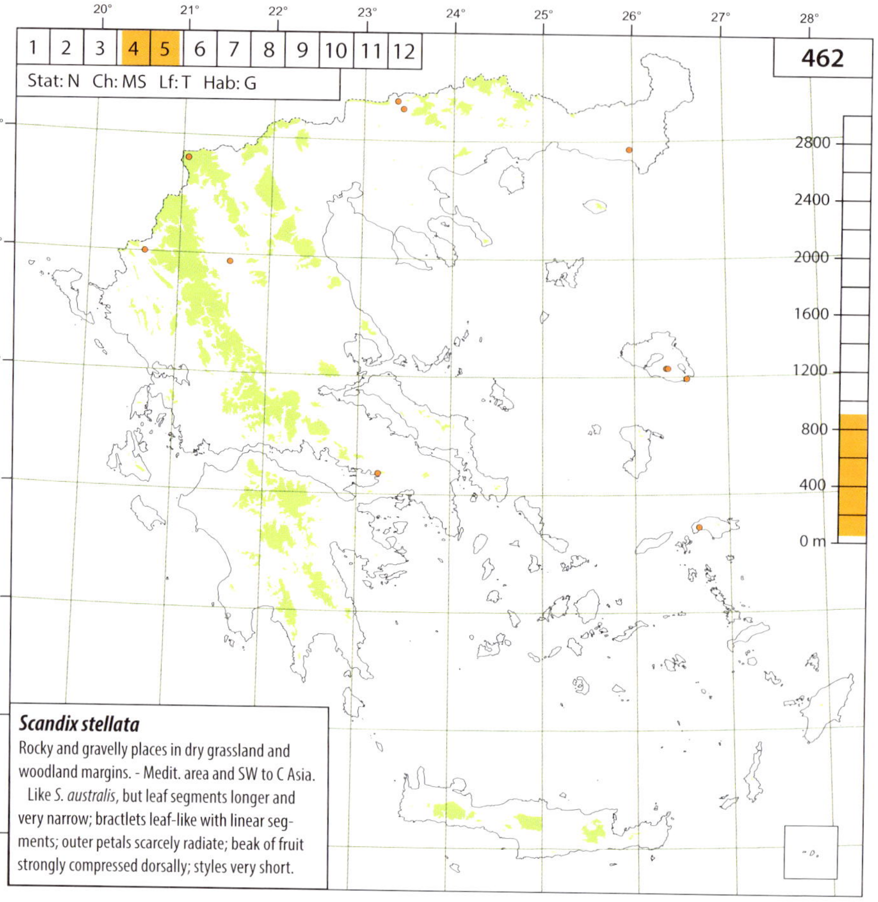

Scandix stellata
Rocky and gravelly places in dry grassland and woodland margins. - Medit. area and SW to C Asia.
Like *S. australis*, but leaf segments longer and very narrow; bractlets leaf-like with linear segments; outer petals scarcely radiate; beak of fruit strongly compressed dorsally; styles very short.

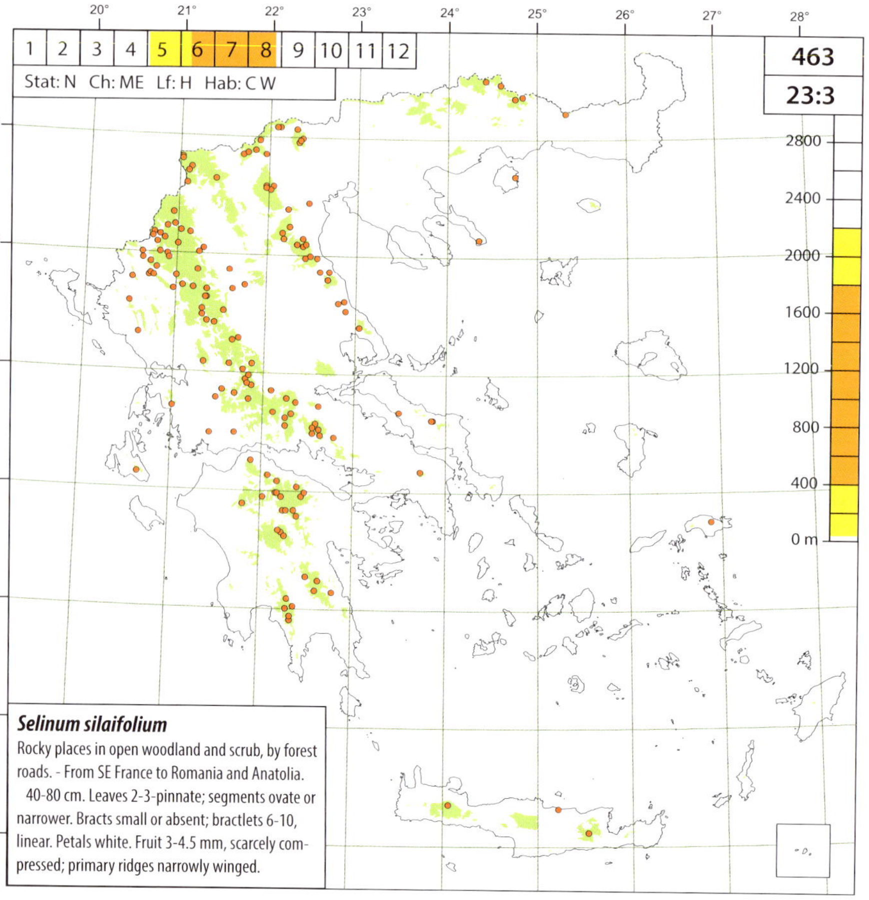

Selinum silaifolium
Rocky places in open woodland and scrub, by forest roads. - From SE France to Romania and Anatolia.
40-80 cm. Leaves 2-3-pinnate; segments ovate or narrower. Bracts small or absent; bractlets 6-10, linear. Petals white. Fruit 3-4.5 mm, scarcely compressed; primary ridges narrowly winged.

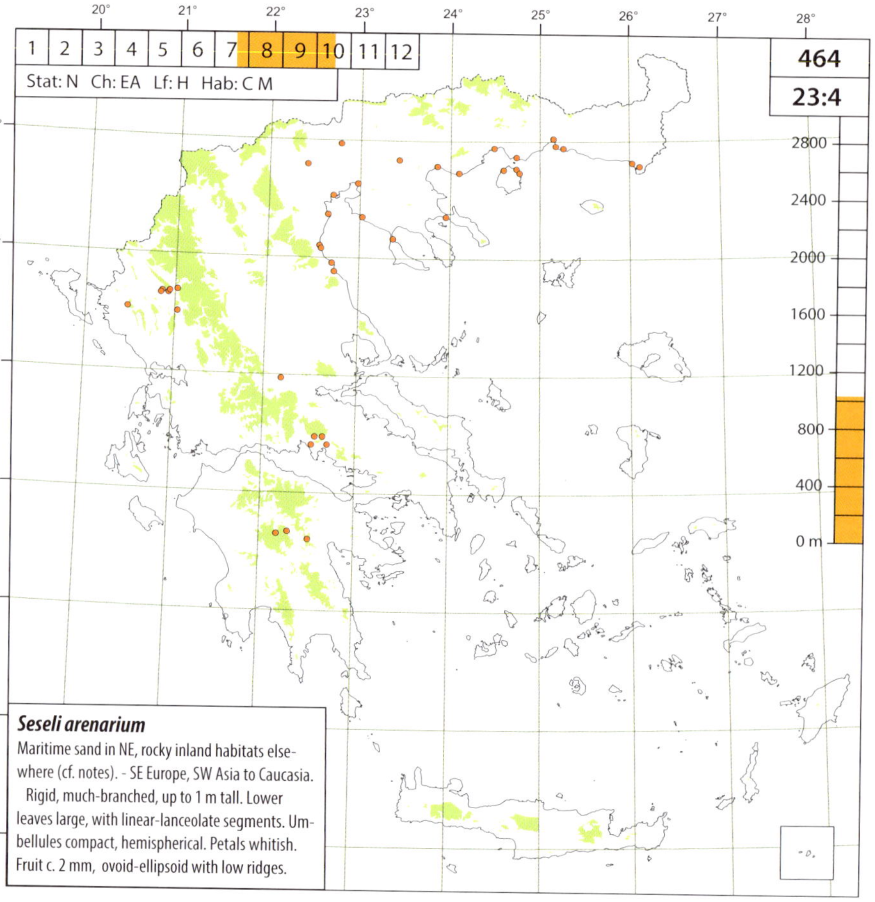

Seseli arenarium
Maritime sand in NE, rocky inland habitats elsewhere (cf. notes). - SE Europe, SW Asia to Caucasia.
Rigid, much-branched, up to 1 m tall. Lower leaves large, with linear-lanceolate segments. Umbellules compact, hemispherical. Petals whitish. Fruit c. 2 mm, ovoid-ellipsoid with low ridges.

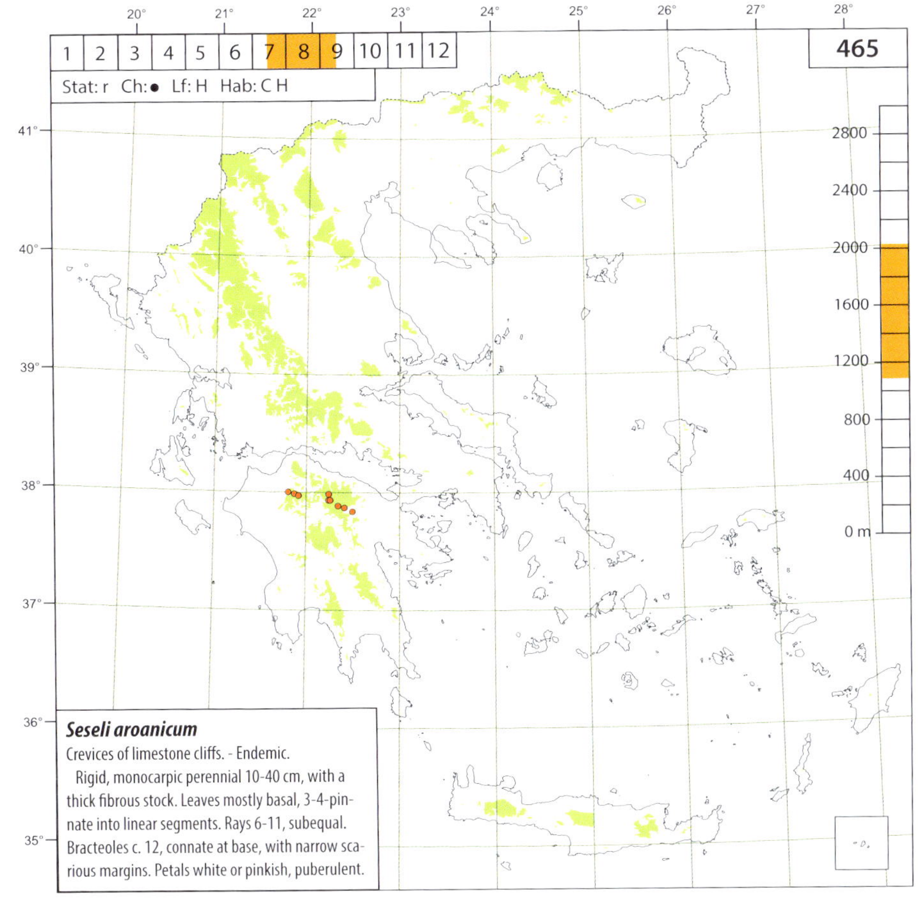

Seseli aroanicum
Crevices of limestone cliffs. - Endemic.
Rigid, monocarpic perennial 10-40 cm, with a thick fibrous stock. Leaves mostly basal, 3-4-pinnate into linear segments. Rays 6-11, subequal. Bracteoles c. 12, connate at base, with narrow scarious margins. Petals white or pinkish, puberulent.

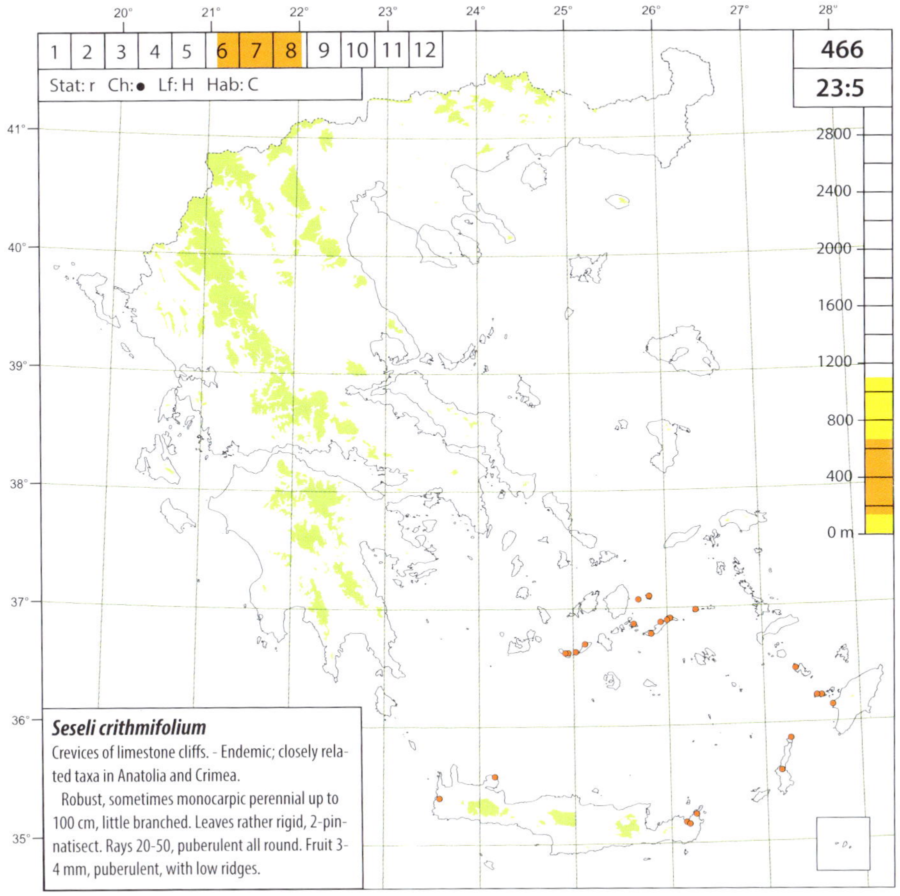

Seseli crithmifolium
Crevices of limestone cliffs. - Endemic; closely related taxa in Anatolia and Crimea.
Robust, sometimes monocarpic perennial up to 100 cm, little branched. Leaves rather rigid, 2-pinnatisect. Rays 20-50, puberulent all round. Fruit 3-4 mm, puberulent, with low ridges.

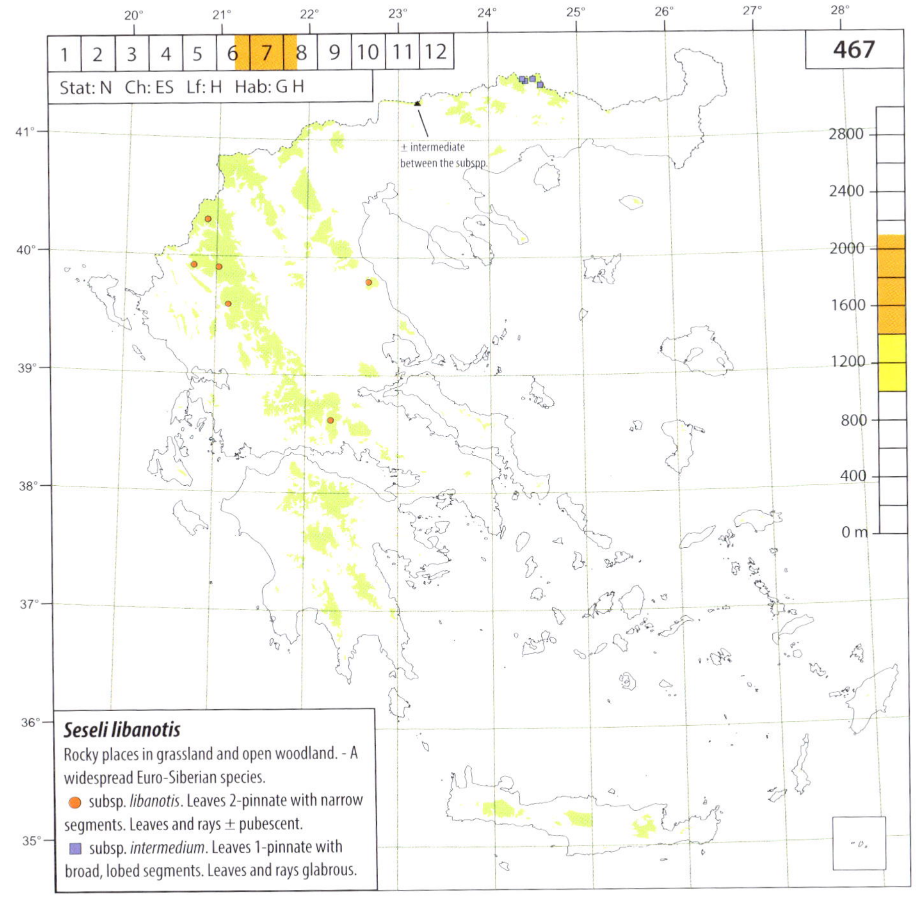

Seseli libanotis
Rocky places in grassland and open woodland. - A widespread Euro-Siberian species.
● subsp. *libanotis*. Leaves 2-pinnate with narrow segments. Leaves and rays ± pubescent.
■ subsp. *intermedium*. Leaves 1-pinnate with broad, lobed segments. Leaves and rays glabrous.

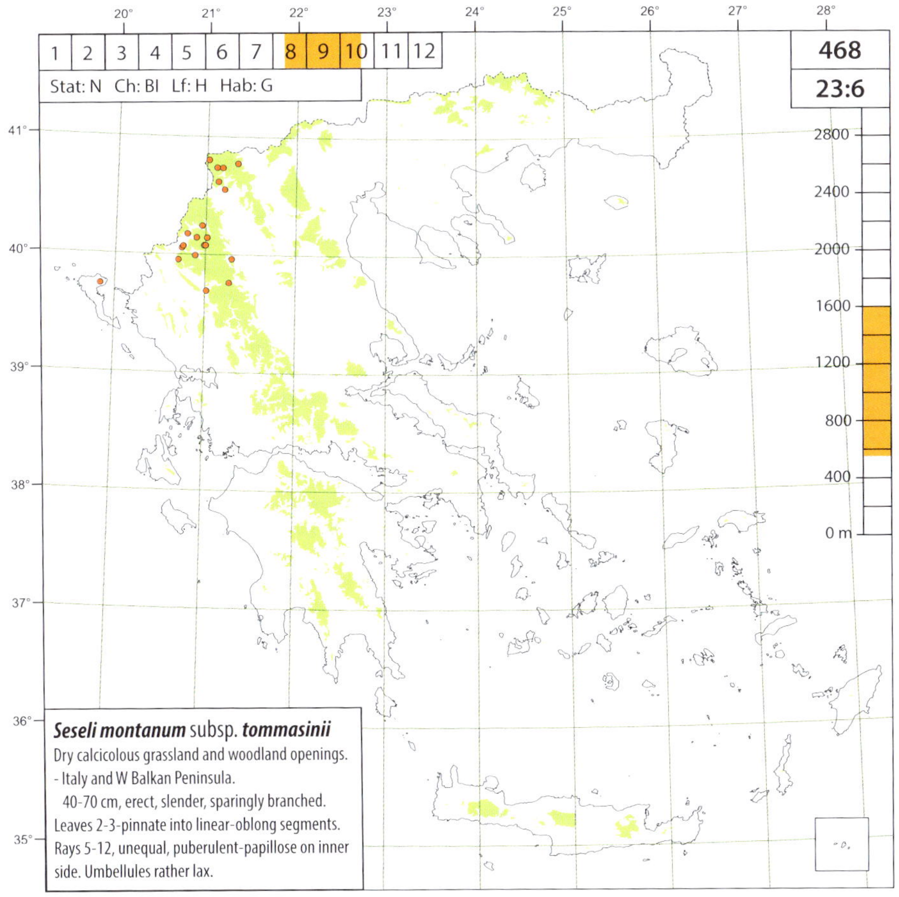

Seseli montanum* subsp. *tommasinii
Dry calcicolous grassland and woodland openings. - Italy and W Balkan Peninsula.
40-70 cm, erect, slender, sparingly branched. Leaves 2-3-pinnate into linear-oblong segments. Rays 5-12, unequal, puberulent-papillose on inner side. Umbellules rather lax.

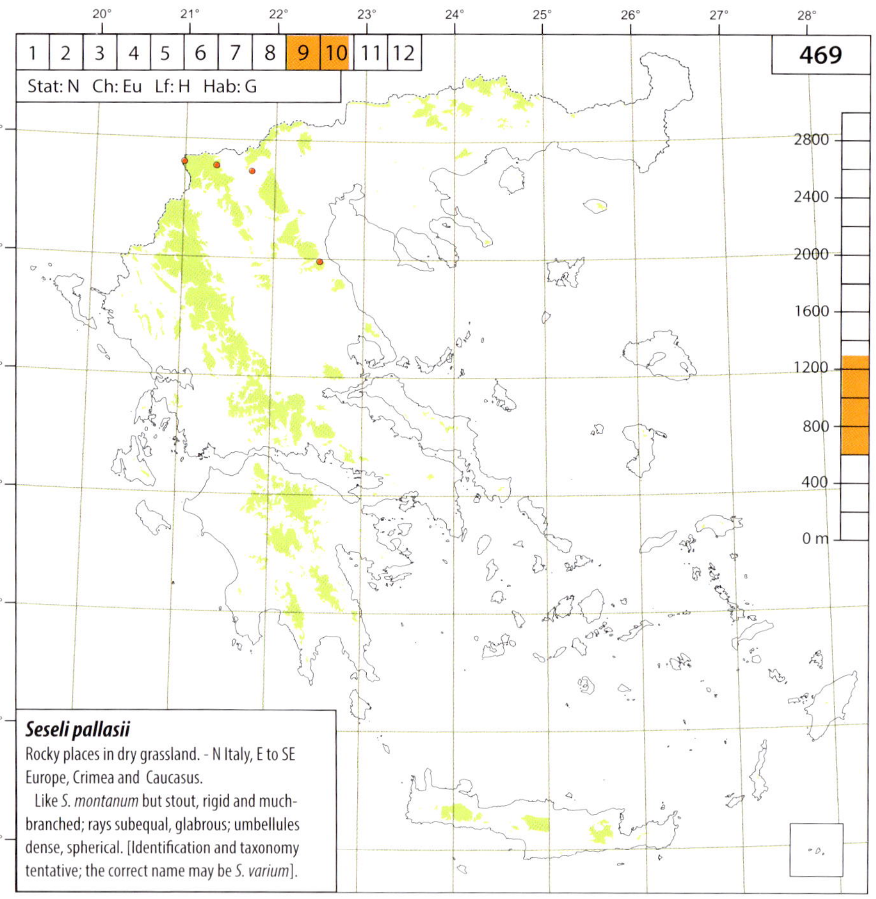

469

1 2 3 4 5 6 7 8 9 10 11 12

Stat: N Ch: Eu Lf: H Hab: G

Seseli pallasii

Rocky places in dry grassland. - N Italy, E to SE Europe, Crimea and Caucasus.

Like *S. montanum* but stout, rigid and much-branched; rays subequal, glabrous; umbellules dense, spherical. [Identification and taxonomy tentative; the correct name may be *S. varium*].

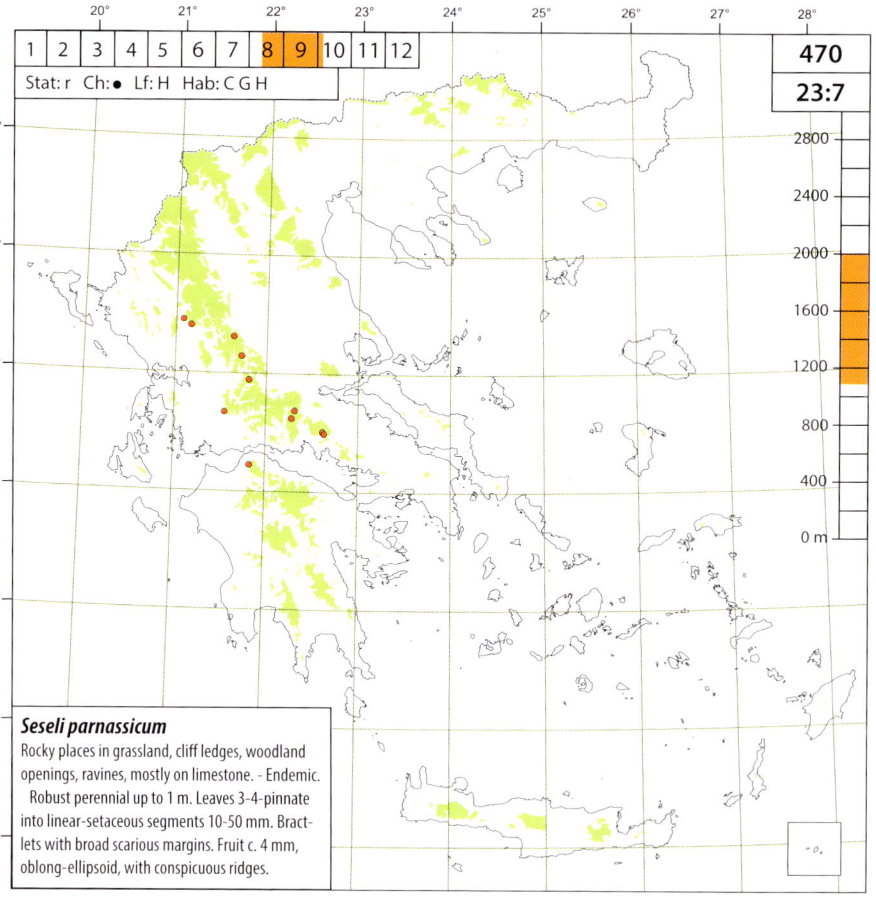

470

23:7

1 2 3 4 5 6 7 8 9 10 11 12

Stat: r Ch: ● Lf: H Hab: C G H

Seseli parnassicum

Rocky places in grassland, cliff ledges, woodland openings, ravines, mostly on limestone. - Endemic.

Robust perennial up to 1 m. Leaves 3-4-pinnate into linear-setaceous segments 10-50 mm. Bractlets with broad scarious margins. Fruit c. 4 mm, oblong-ellipsoid, with conspicuous ridges.

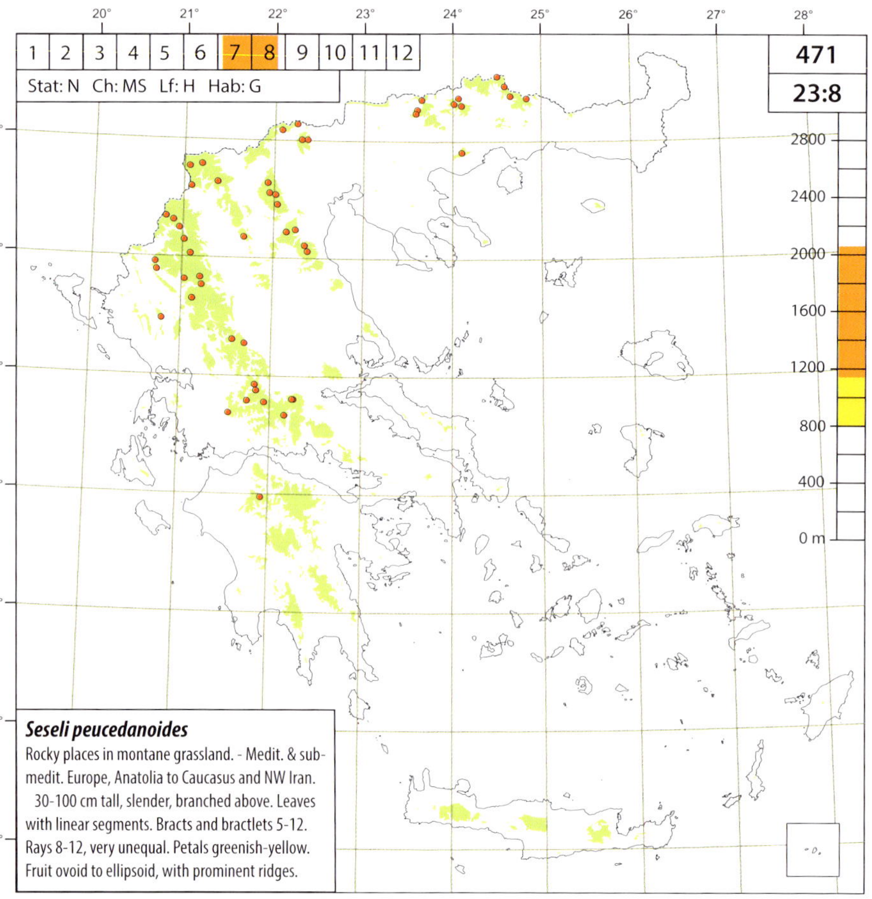

471

23:8

1 2 3 4 5 6 7 8 9 10 11 12

Stat: N Ch: MS Lf: H Hab: G

Seseli peucedanoides

Rocky places in montane grassland. - Medit. & submedit. Europe, Anatolia to Caucasus and NW Iran.

30-100 cm tall, slender, branched above. Leaves with linear segments. Bracts and bractlets 5-12. Rays 8-12, very unequal. Petals greenish-yellow. Fruit ovoid to ellipsoid, with prominent ridges.

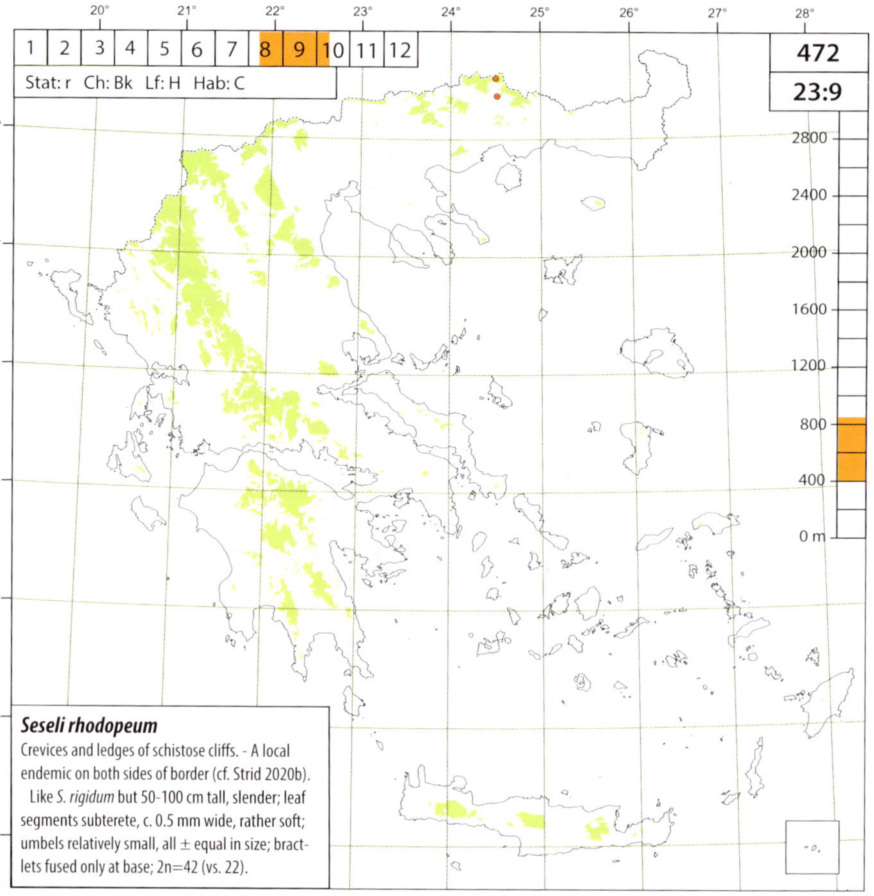

472

23:9

1 2 3 4 5 6 7 8 9 10 11 12

Stat: r Ch: Bk Lf: H Hab: C

Seseli rhodopeum

Crevices and ledges of schistose cliffs. - A local endemic on both sides of border (cf. Strid 2020b).

Like *S. rigidum* but 50-100 cm tall, slender; leaf segments subterete, c. 0.5 mm wide, rather soft; umbels relatively small, all ± equal in size; bractlets fused only at base; 2n=42 (vs. 22).

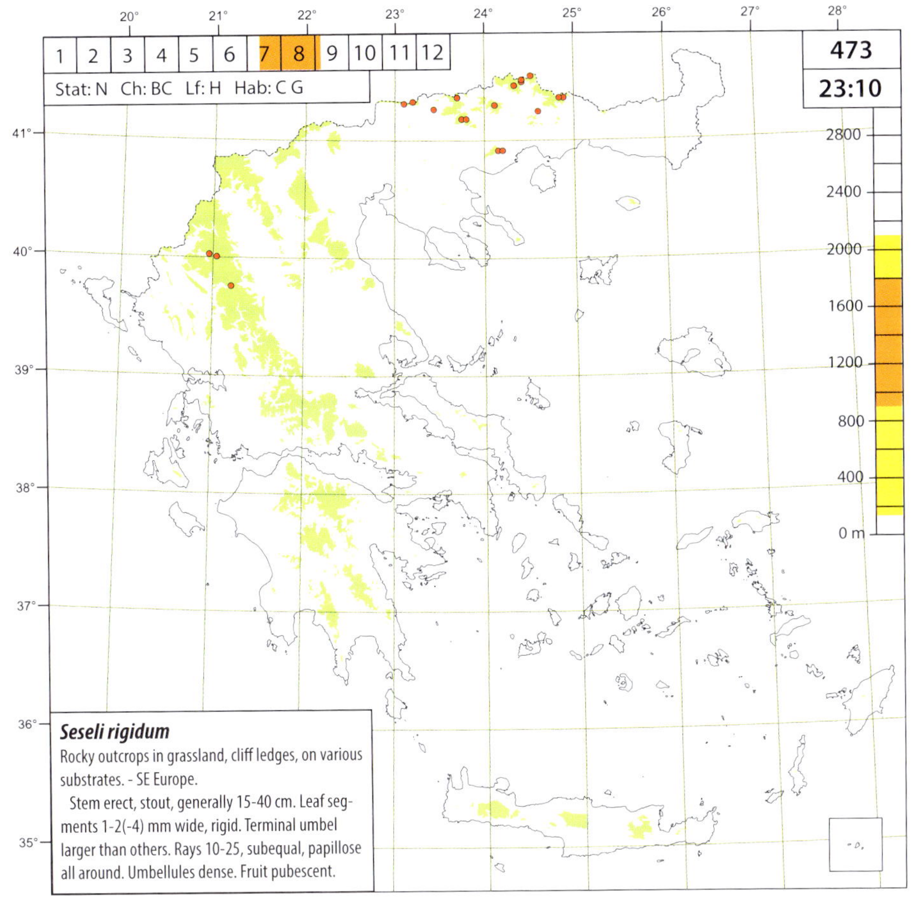

Seseli rigidum
Rocky outcrops in grassland, cliff ledges, on various substrates. - SE Europe.
Stem erect, stout, generally 15-40 cm. Leaf segments 1-2(-4) mm wide, rigid. Terminal umbel larger than others. Rays 10-25, subequal, papillose all around. Umbellules dense. Fruit pubescent.

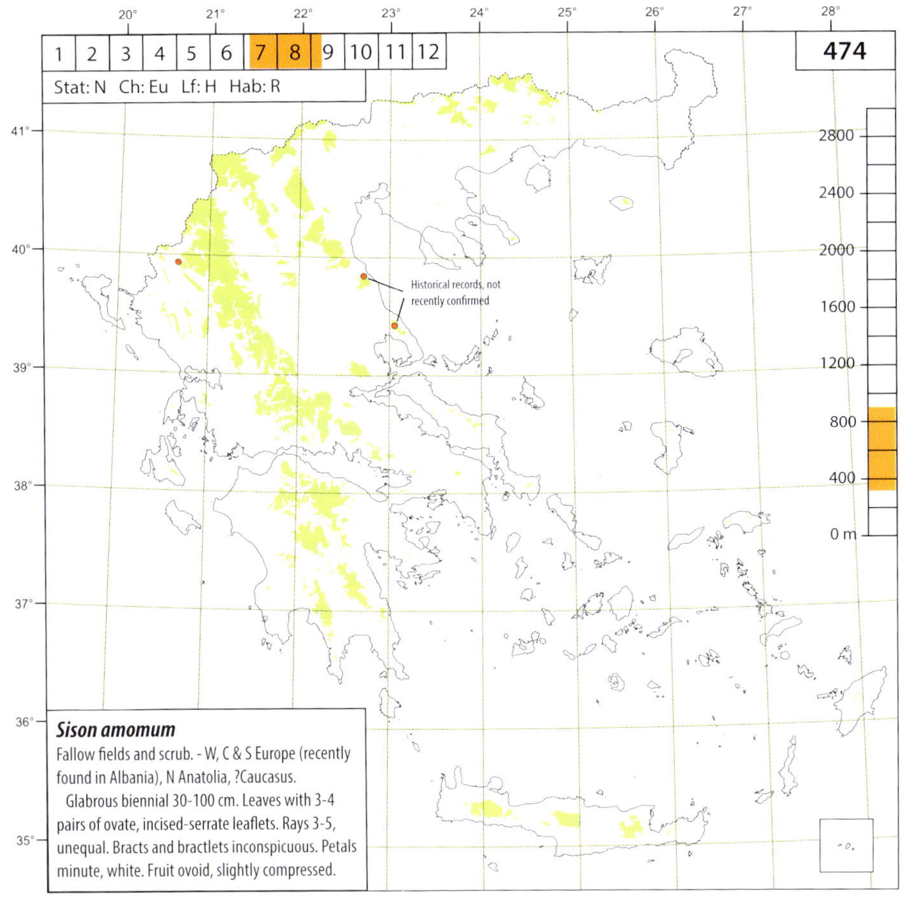

Sison amomum
Fallow fields and scrub. - W, C & S Europe (recently found in Albania), N Anatolia, ?Caucasus.
Glabrous biennial 30-100 cm. Leaves with 3-4 pairs of ovate, incised-serrate leaflets. Rays 3-5, unequal. Bracts and bractlets inconspicuous. Petals minute, white. Fruit ovoid, slightly compressed.

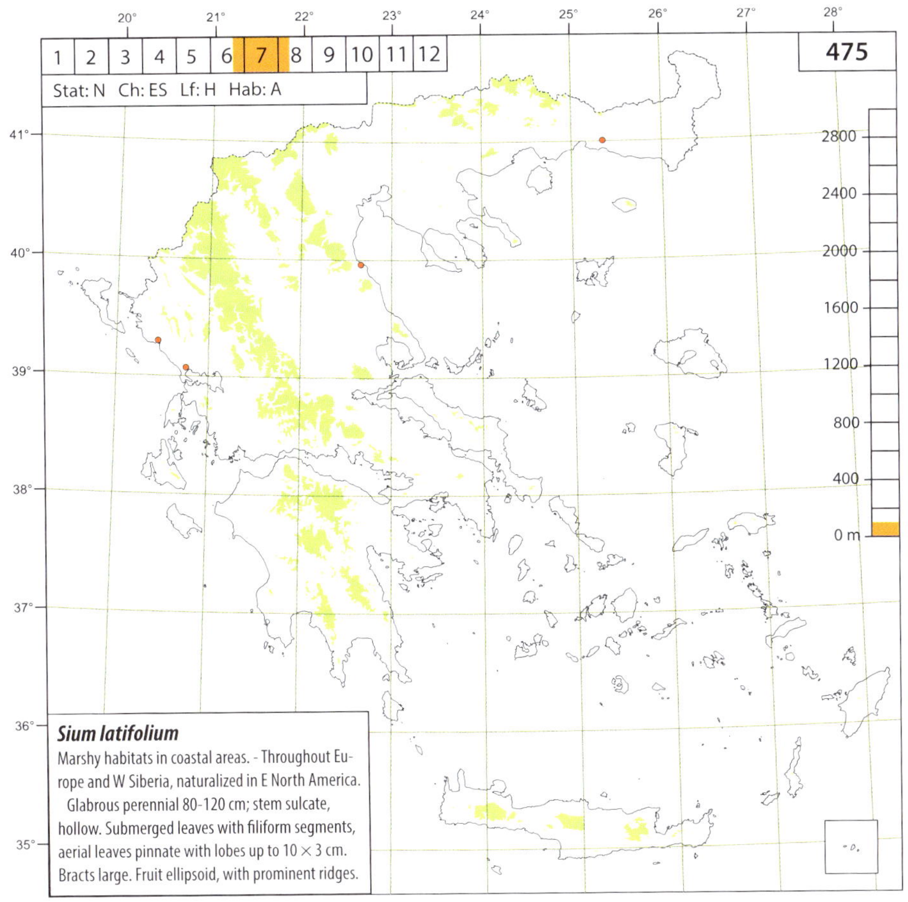

Sium latifolium
Marshy habitats in coastal areas. - Throughout Europe and W Siberia, naturalized in E North America.
Glabrous perennial 80-120 cm; stem sulcate, hollow. Submerged leaves with filiform segments, aerial leaves pinnate with lobes up to 10 × 3 cm. Bracts large. Fruit ellipsoid, with prominent ridges.

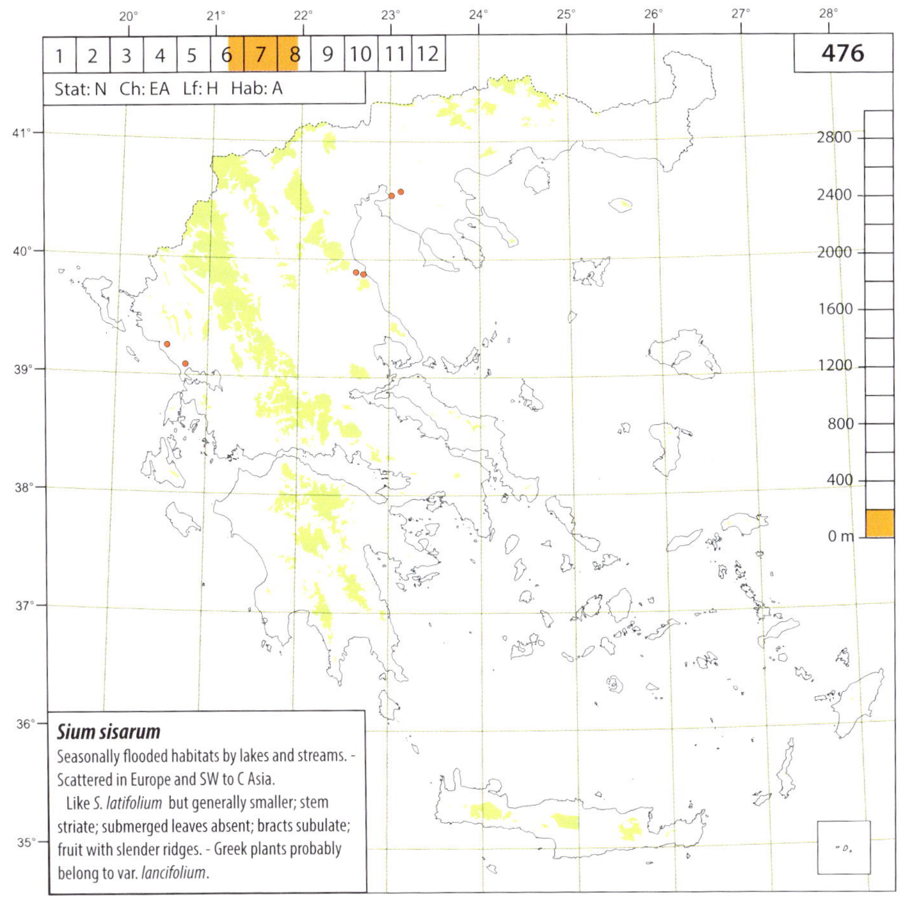

Sium sisarum
Seasonally flooded habitats by lakes and streams. - Scattered in Europe and SW to C Asia.
Like *S. latifolium* but generally smaller; stem striate; submerged leaves absent; bracts subulate; fruit with slender ridges. - Greek plants probably belong to var. *lancifolium*.

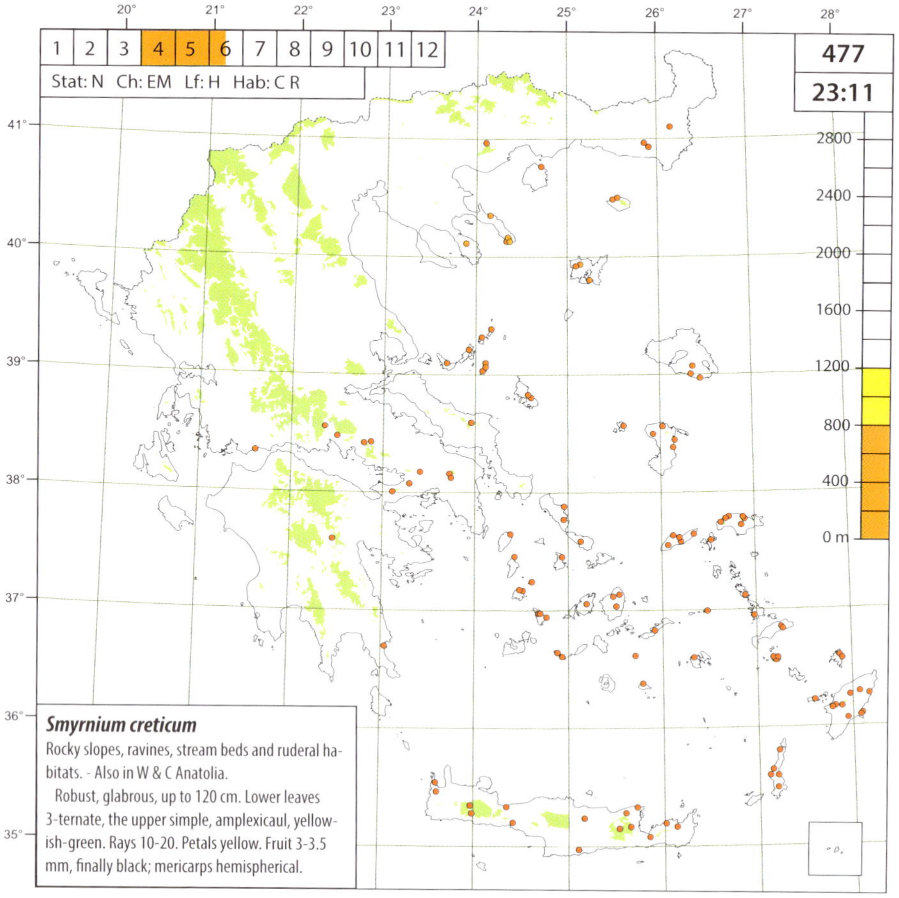

Smyrnium creticum

Rocky slopes, ravines, stream beds and ruderal habitats. - Also in W & C Anatolia.

Robust, glabrous, up to 120 cm. Lower leaves 3-ternate, the upper simple, amplexicaul, yellowish-green. Rays 10-20. Petals yellow. Fruit 3-3.5 mm, finally black; mericarps hemispherical.

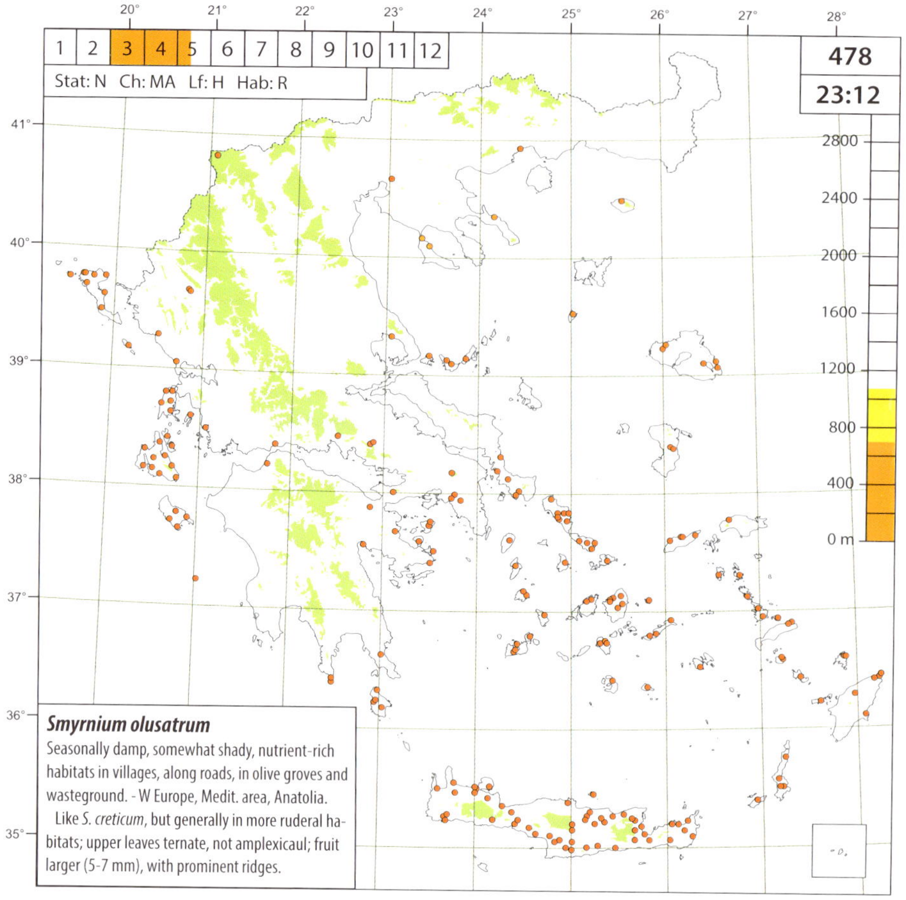

Smyrnium olusatrum

Seasonally damp, somewhat shady, nutrient-rich habitats in villages, along roads, in olive groves and wasteground. - W Europe, Medit. area, Anatolia.

Like *S. creticum*, but generally in more ruderal habitats; upper leaves ternate, not amplexicaul; fruit larger (5-7 mm), with prominent ridges.

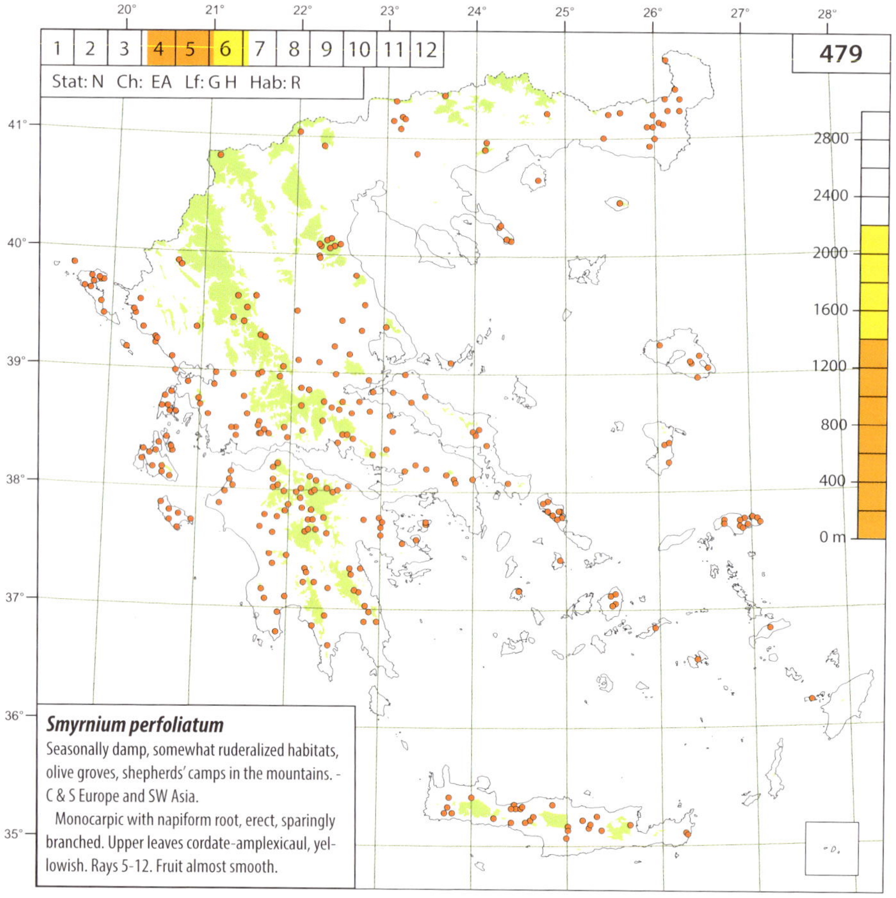

Smyrnium perfoliatum

Seasonally damp, somewhat ruderalized habitats, olive groves, shepherds' camps in the mountains. - C & S Europe and SW Asia.

Monocarpic with napiform root, erect, sparingly branched. Upper leaves cordate-amplexicaul, yellowish. Rays 5-12. Fruit almost smooth.

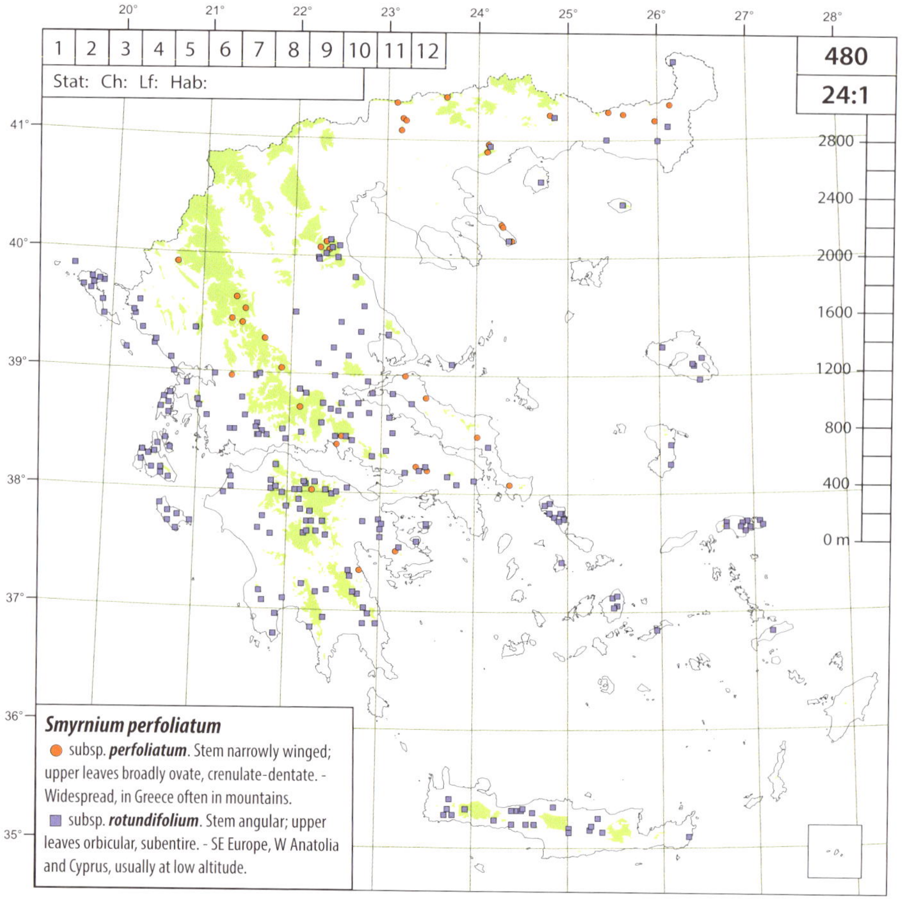

Smyrnium perfoliatum

● subsp. ***perfoliatum***. Stem narrowly winged; upper leaves broadly ovate, crenulate-dentate. - Widespread, in Greece often in mountains.

■ subsp. ***rotundifolium***. Stem angular; upper leaves orbicular, subentire. - SE Europe, W Anatolia and Cyprus, usually at low altitude.

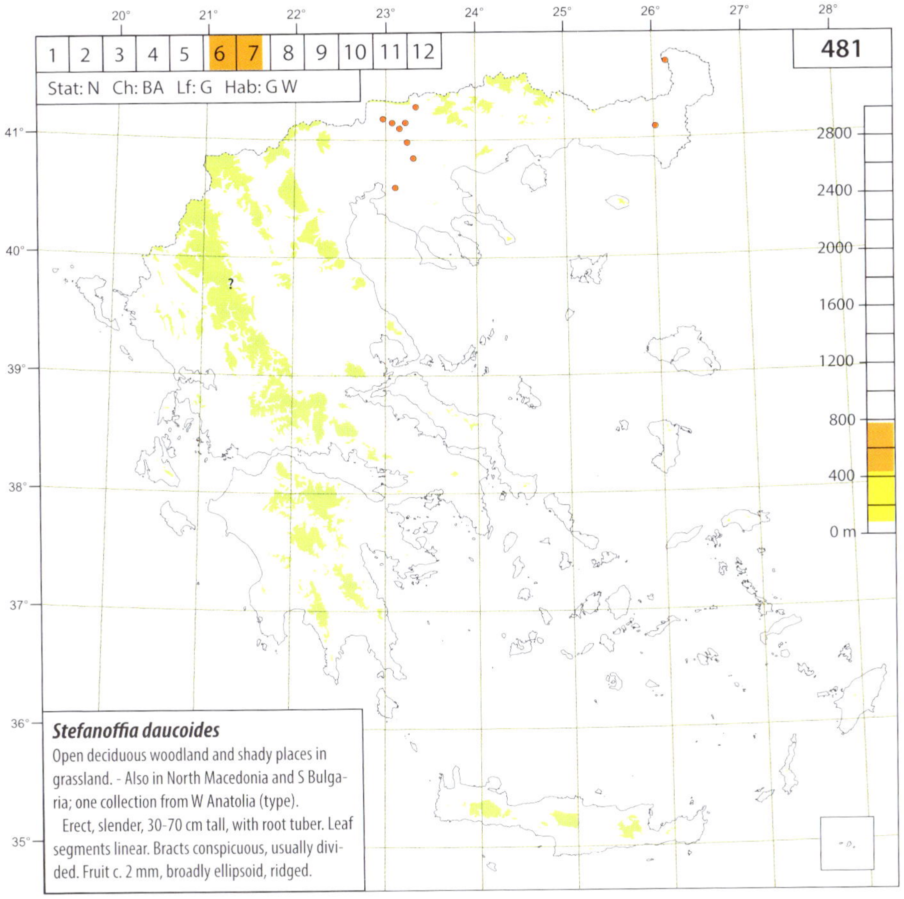

Stefanoffia daucoides

Open deciduous woodland and shady places in grassland. - Also in North Macedonia and S Bulgaria; one collection from W Anatolia (type).

Erect, slender, 30-70 cm tall, with root tuber. Leaf segments linear. Bracts conspicuous, usually divided. Fruit c. 2 mm, broadly ellipsoid, ridged.

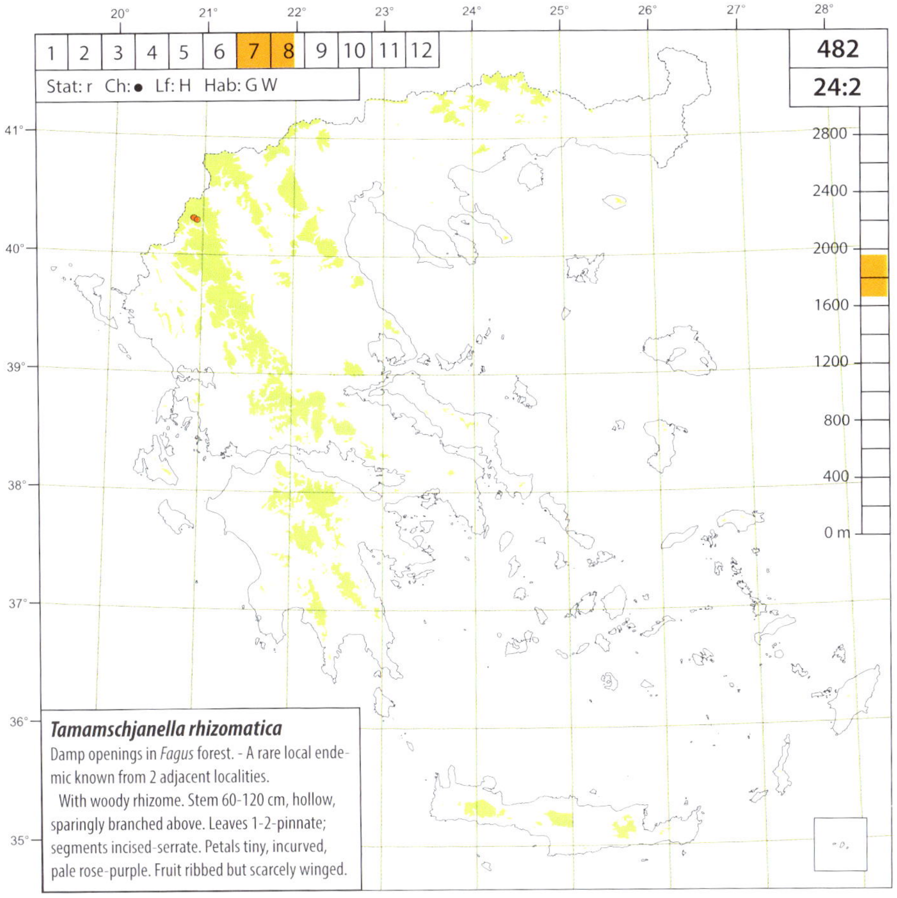

Tamamschjanella rhizomatica

Damp openings in *Fagus* forest. - A rare local endemic known from 2 adjacent localities.

With woody rhizome. Stem 60-120 cm, hollow, sparingly branched above. Leaves 1-2-pinnate; segments incised-serrate. Petals tiny, incurved, pale rose-purple. Fruit ribbed but scarcely winged.

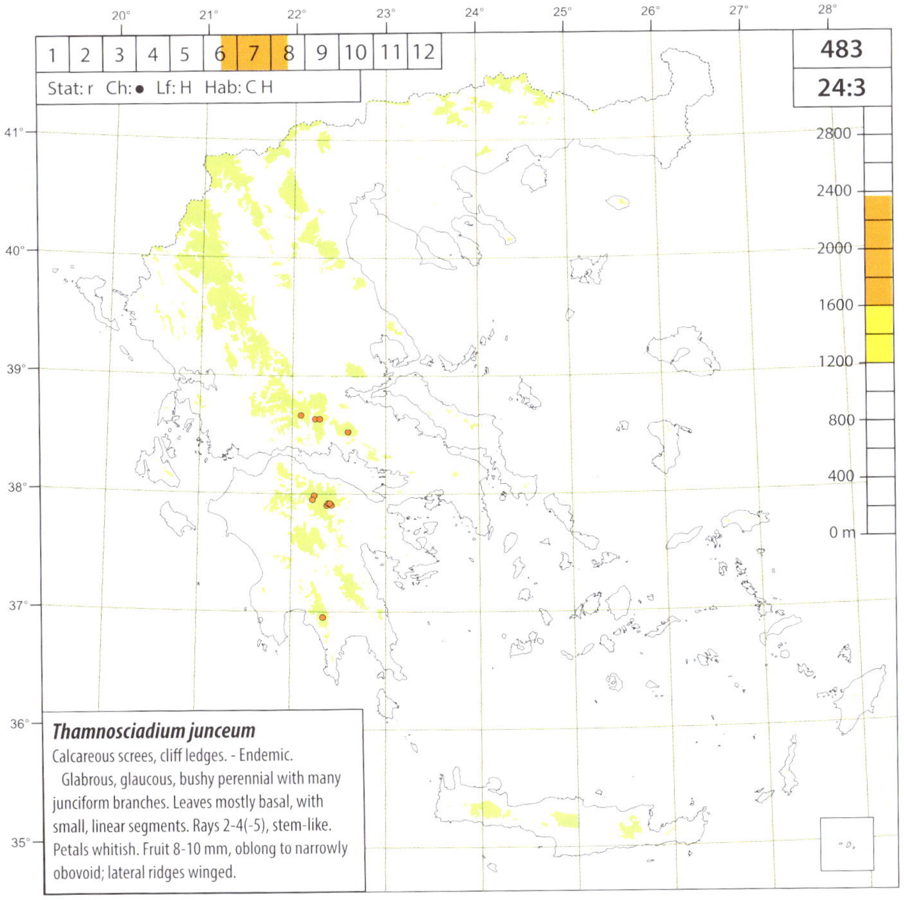

Thamnosciadium junceum

Calcareous screes, cliff ledges. - Endemic.

Glabrous, glaucous, bushy perennial with many junciform branches. Leaves mostly basal, with small, linear segments. Rays 2-4(-5), stem-like. Petals whitish. Fruit 8-10 mm, oblong to narrowly obovoid; lateral ridges winged.

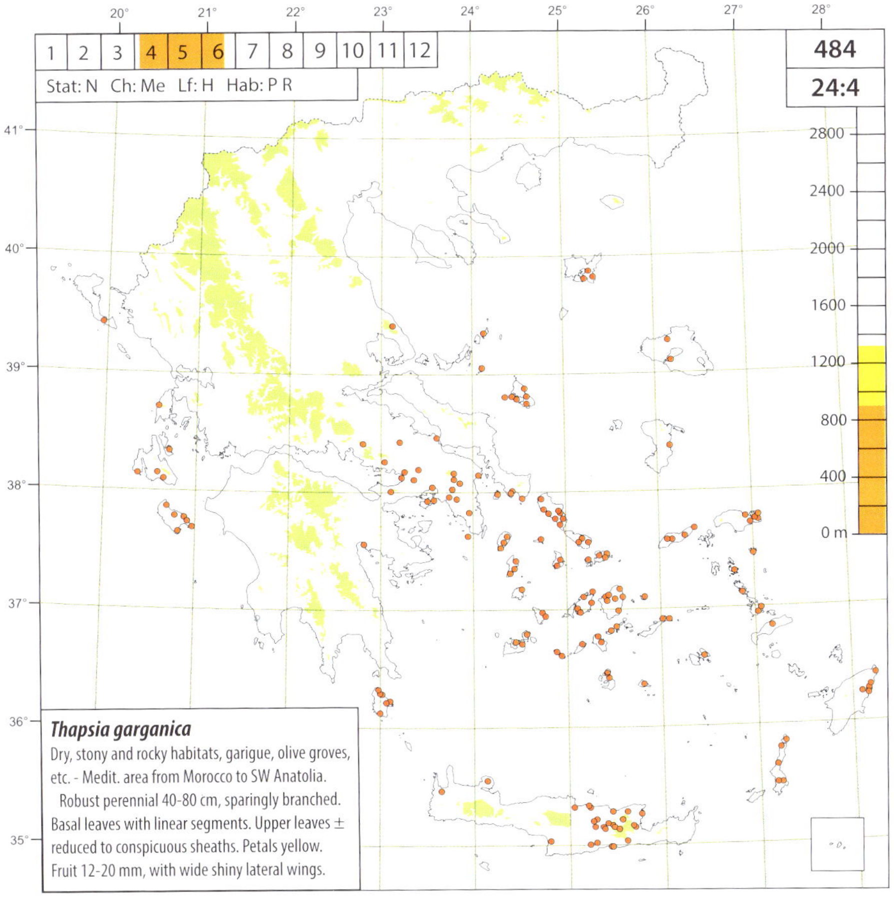

Thapsia garganica

Dry, stony and rocky habitats, garigue, olive groves, etc. - Medit. area from Morocco to SW Anatolia.

Robust perennial 40-80 cm, sparingly branched. Basal leaves with linear segments. Upper leaves ± reduced to conspicuous sheaths. Petals yellow. Fruit 12-20 mm, with wide shiny lateral wings.

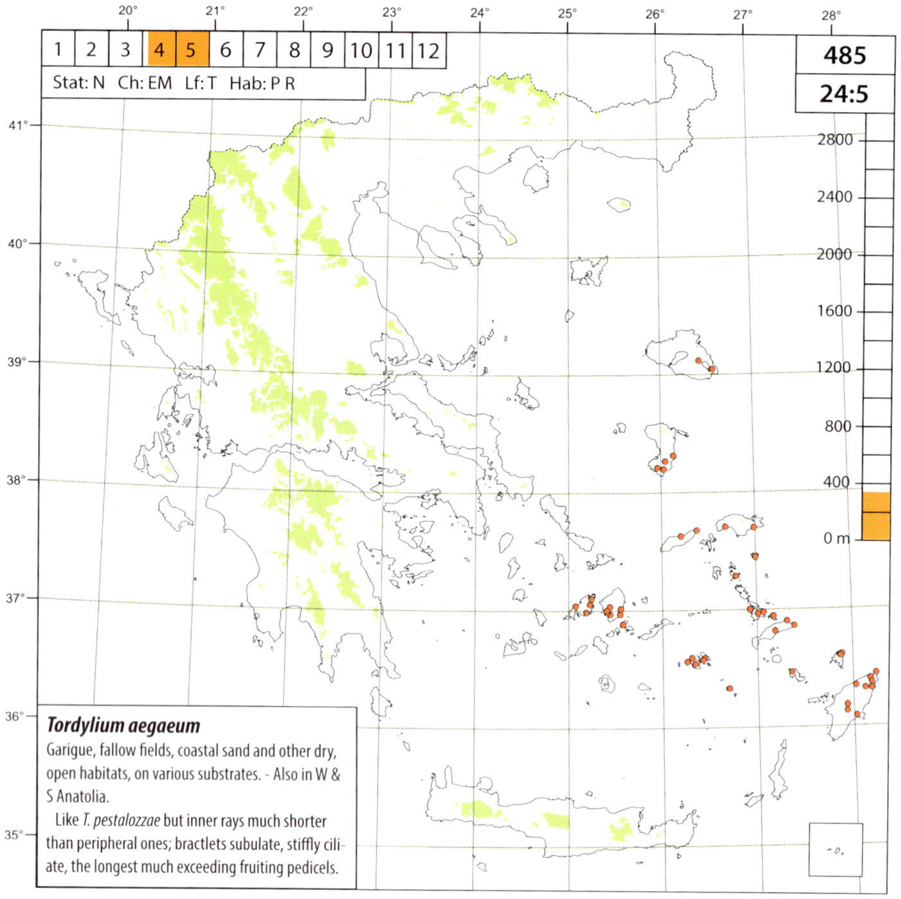

Tordylium aegaeum
Garigue, fallow fields, coastal sand and other dry, open habitats, on various substrates. - Also in W & S Anatolia.

Like *T. pestalozzae* but inner rays much shorter than peripheral ones; bractlets subulate, stiffly ciliate, the longest much exceeding fruiting pedicels.

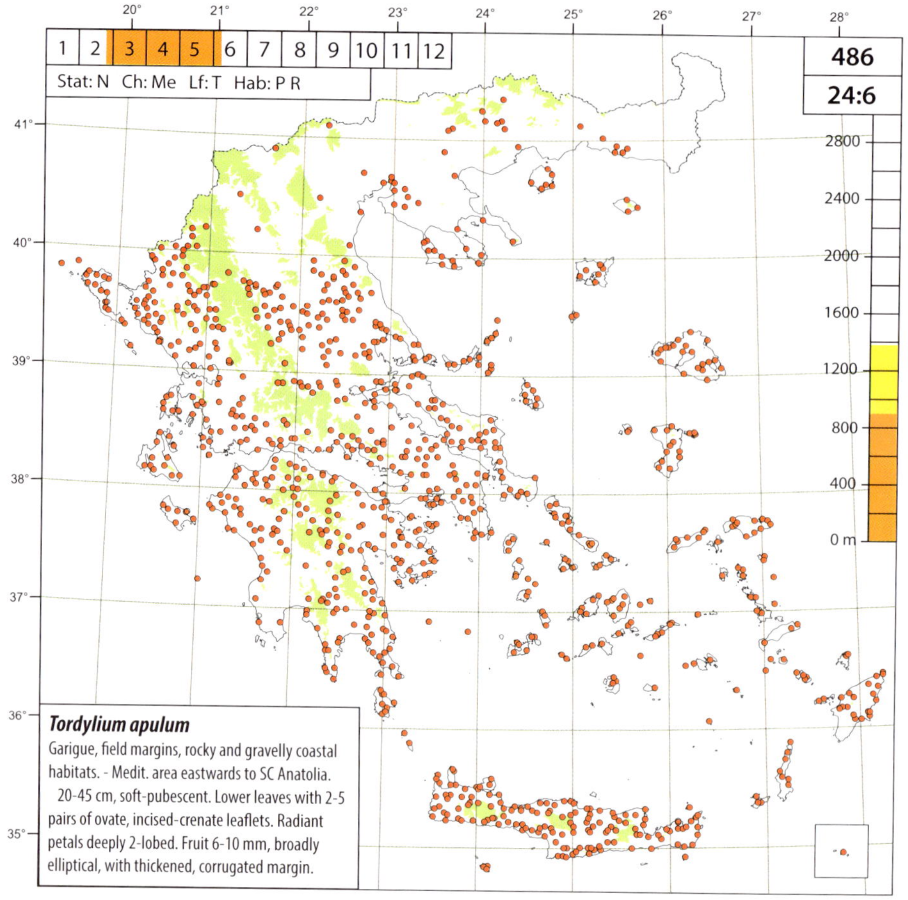

Tordylium apulum
Garigue, field margins, rocky and gravelly coastal habitats. - Medit. area eastwards to SC Anatolia.

20-45 cm, soft-pubescent. Lower leaves with 2-5 pairs of ovate, incised-crenate leaflets. Radiant petals deeply 2-lobed. Fruit 6-10 mm, broadly elliptical, with thickened, corrugated margin.

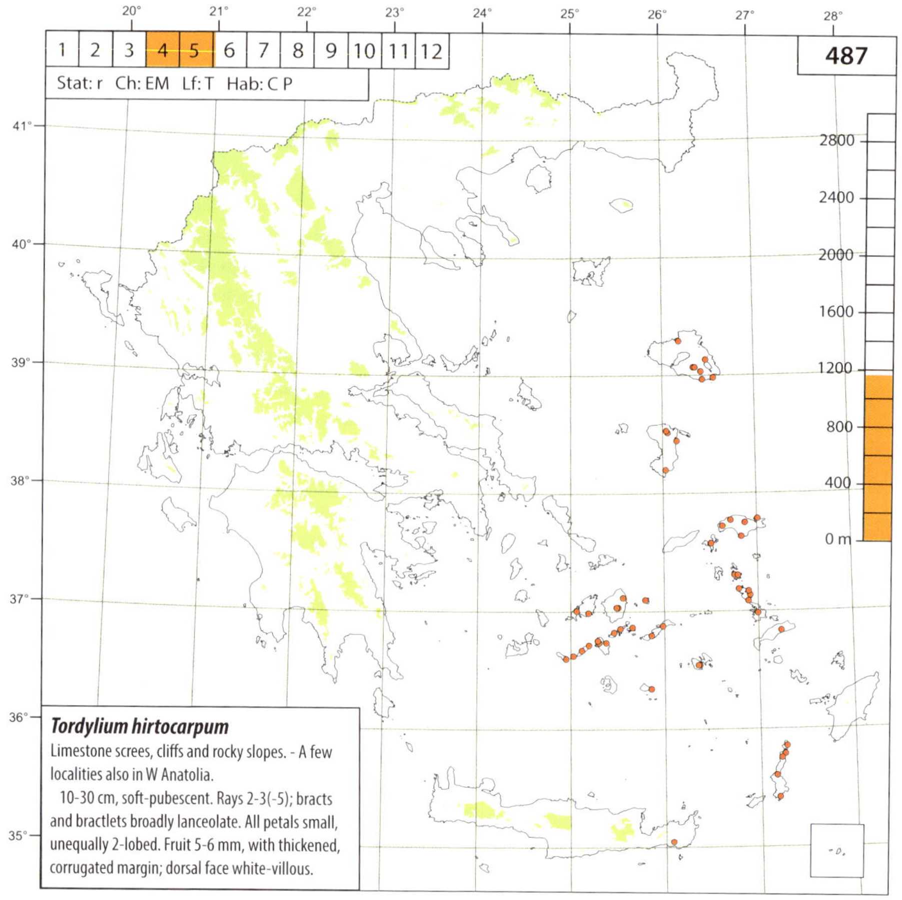

Tordylium hirtocarpum
Limestone screes, cliffs and rocky slopes. - A few localities also in W Anatolia.

10-30 cm, soft-pubescent. Rays 2-3(-5); bracts and bractlets broadly lanceolate. All petals small, unequally 2-lobed. Fruit 5-6 mm, with thickened, corrugated margin; dorsal face white-villous.

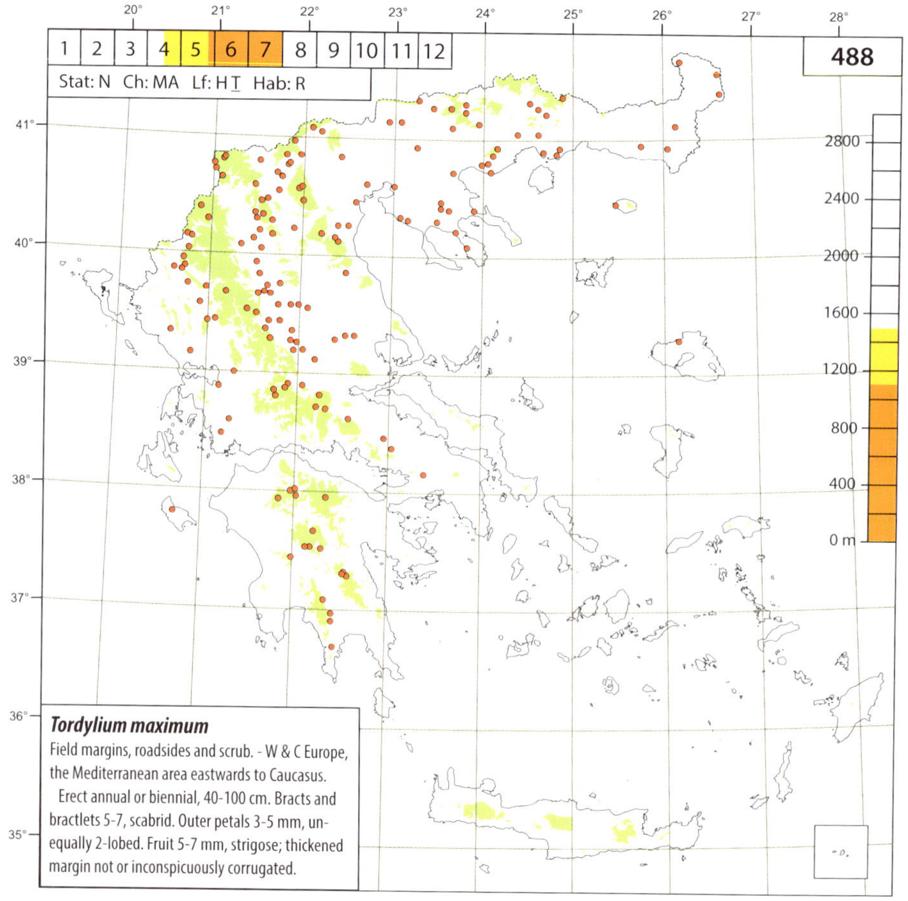

Tordylium maximum
Field margins, roadsides and scrub. - W & C Europe, the Mediterranean area eastwards to Caucasus.

Erect annual or biennial, 40-100 cm. Bracts and bractlets 5-7, scabrid. Outer petals 3-5 mm, unequally 2-lobed. Fruit 5-7 mm, strigose; thickened margin not or inconspicuously corrugated.

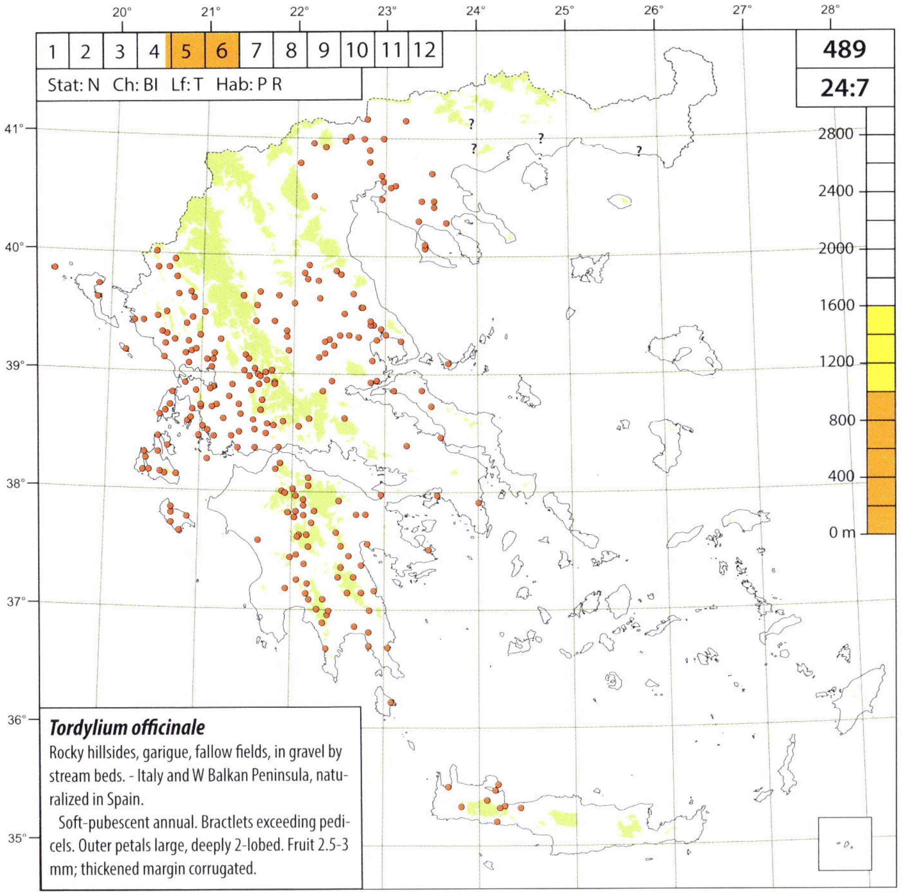

Tordylium officinale

Rocky hillsides, garigue, fallow fields, in gravel by stream beds. - Italy and W Balkan Peninsula, naturalized in Spain.

Soft-pubescent annual. Bractlets exceeding pedicels. Outer petals large, deeply 2-lobed. Fruit 2.5-3 mm; thickened margin corrugated.

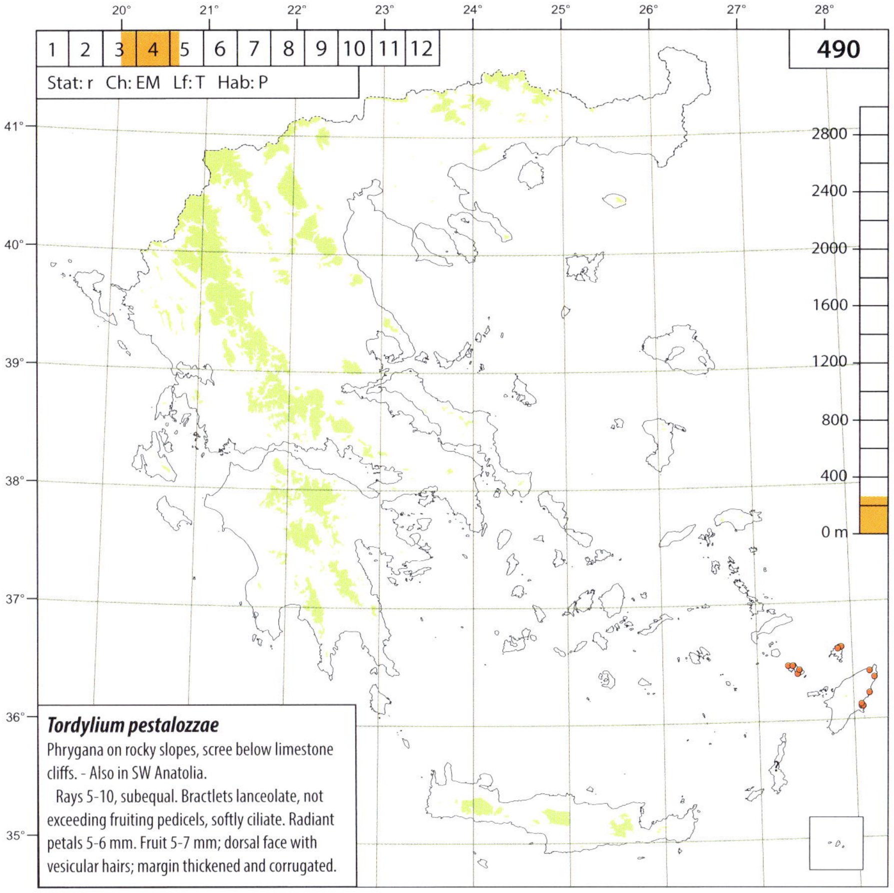

Tordylium pestalozzae

Phrygana on rocky slopes, scree below limestone cliffs. - Also in SW Anatolia.

Rays 5-10, subequal. Bractlets lanceolate, not exceeding fruiting pedicels, softly ciliate. Radiant petals 5-6 mm. Fruit 5-7 mm; dorsal face with vesicular hairs; margin thickened and corrugated.

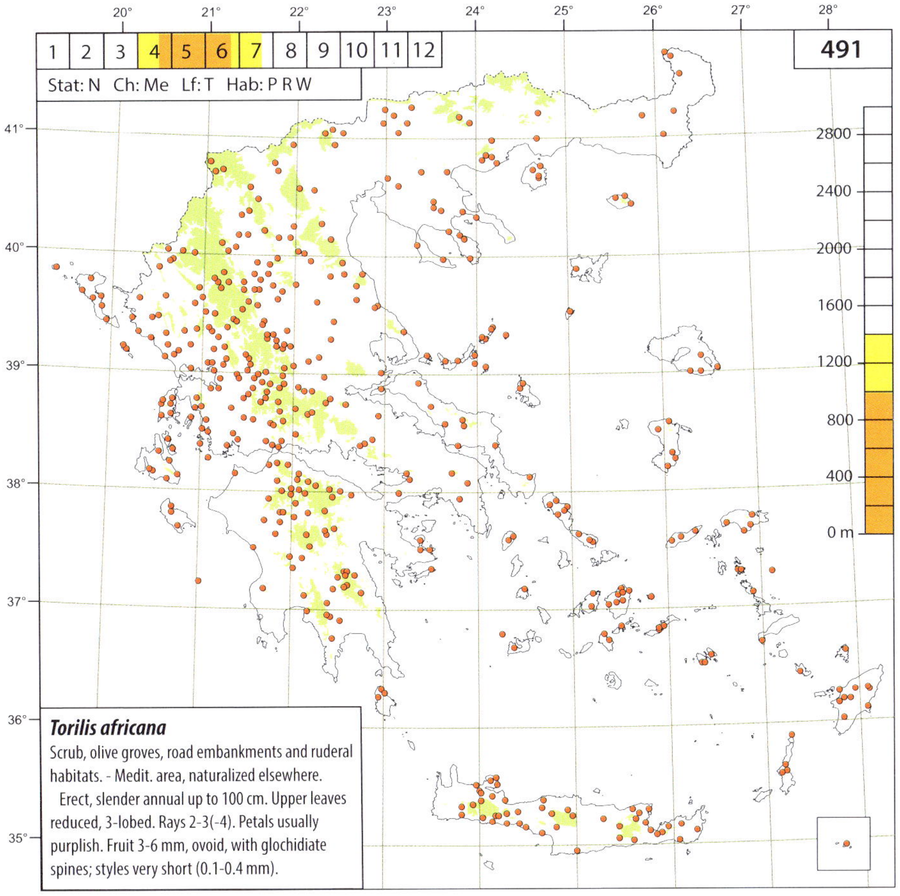

Torilis africana

Scrub, olive groves, road embankments and ruderal habitats. - Medit. area, naturalized elsewhere.

Erect, slender annual up to 100 cm. Upper leaves reduced, 3-lobed. Rays 2-3(-4). Petals usually purplish. Fruit 3-6 mm, ovoid, with glochidiate spines; styles very short (0.1-0.4 mm).

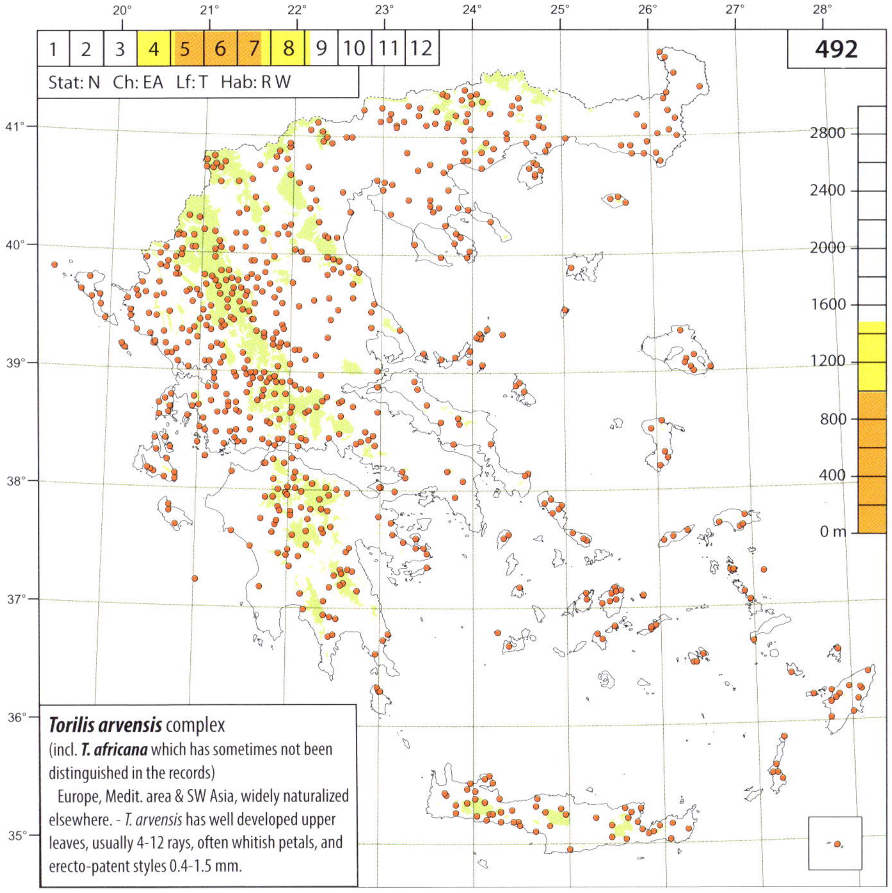

***Torilis arvensis* complex**

(incl. ***T. africana*** which has sometimes not been distinguished in the records)

Europe, Medit. area & SW Asia, widely naturalized elsewhere. - *T. arvensis* has well developed upper leaves, usually 4-12 rays, often whitish petals, and erecto-patent styles 0.4-1.5 mm.

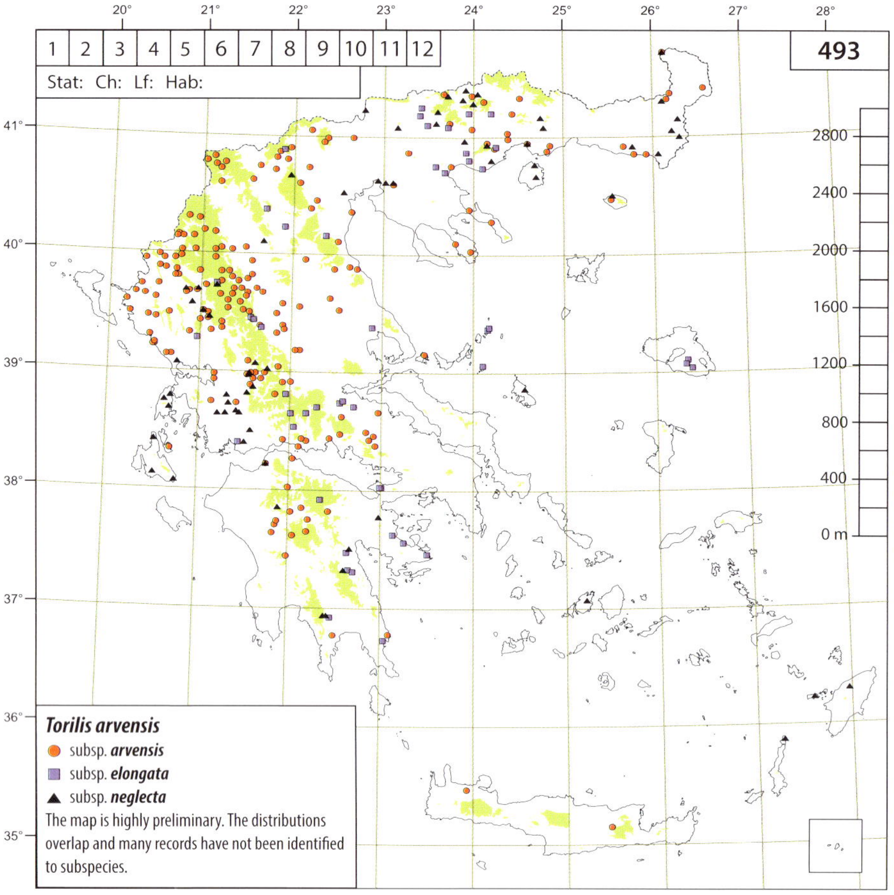

Torilis arvensis

- subsp. ***arvensis***
- subsp. ***elongata***
- subsp. ***neglecta***

The map is highly preliminary. The distributions overlap and many records have not been identified to subspecies.

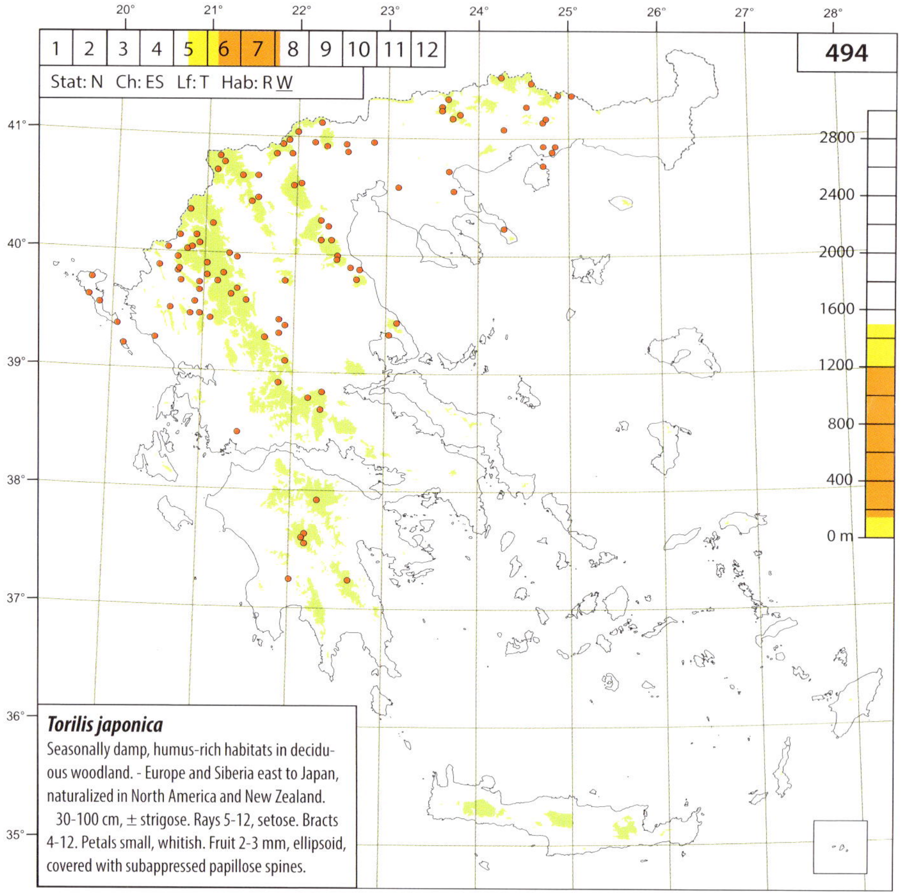

Torilis japonica

Seasonally damp, humus-rich habitats in deciduous woodland. - Europe and Siberia east to Japan, naturalized in North America and New Zealand.

30-100 cm, ± strigose. Rays 5-12, setose. Bracts 4-12. Petals small, whitish. Fruit 2-3 mm, ellipsoid, covered with subappressed papillose spines.

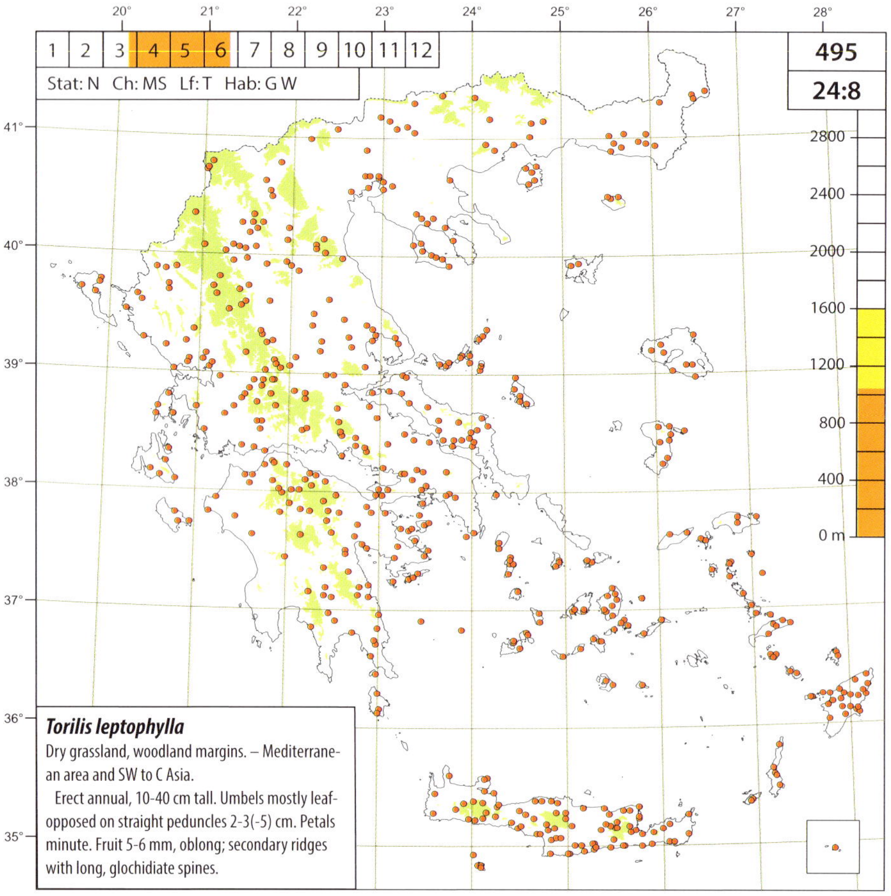

Torilis leptophylla

Dry grassland, woodland margins. – Mediterranean area and SW to C Asia.

Erect annual, 10-40 cm tall. Umbels mostly leaf-opposed on straight peduncles 2-3(-5) cm. Petals minute. Fruit 5-6 mm, oblong; secondary ridges with long, glochidiate spines.

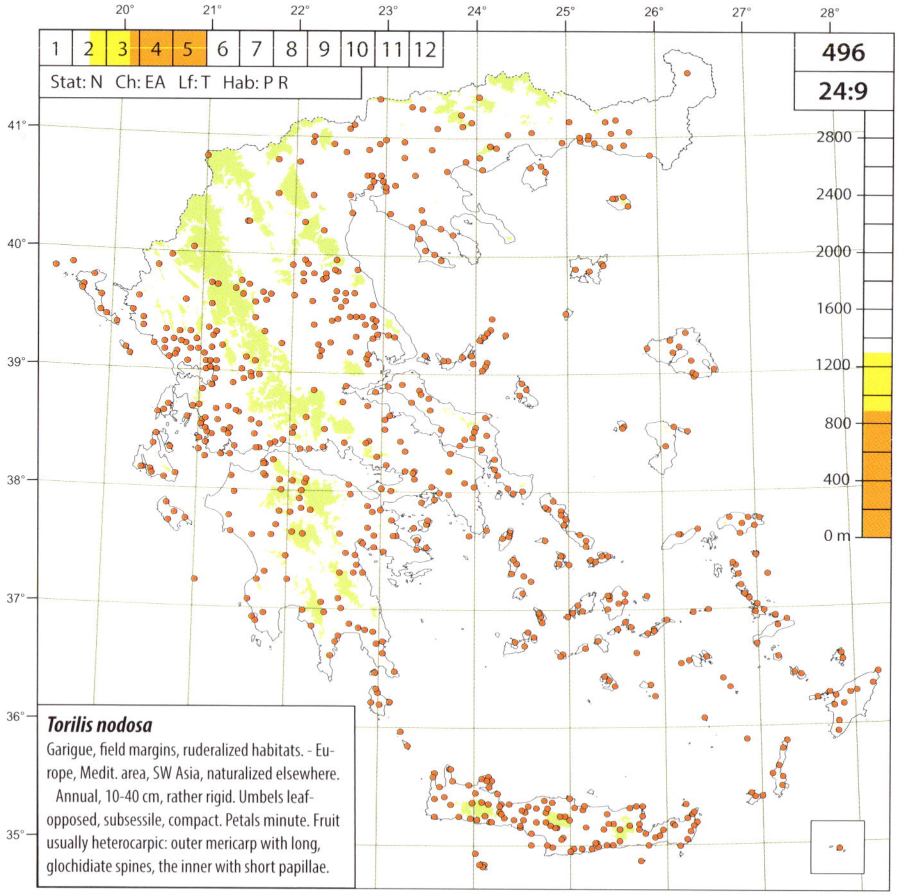

Torilis nodosa

Garigue, field margins, ruderalized habitats. - Europe, Medit. area, SW Asia, naturalized elsewhere.

Annual, 10-40 cm, rather rigid. Umbels leaf-opposed, subsessile, compact. Petals minute. Fruit usually heterocarpic: outer mericarp with long, glochidiate spines, the inner with short papillae.

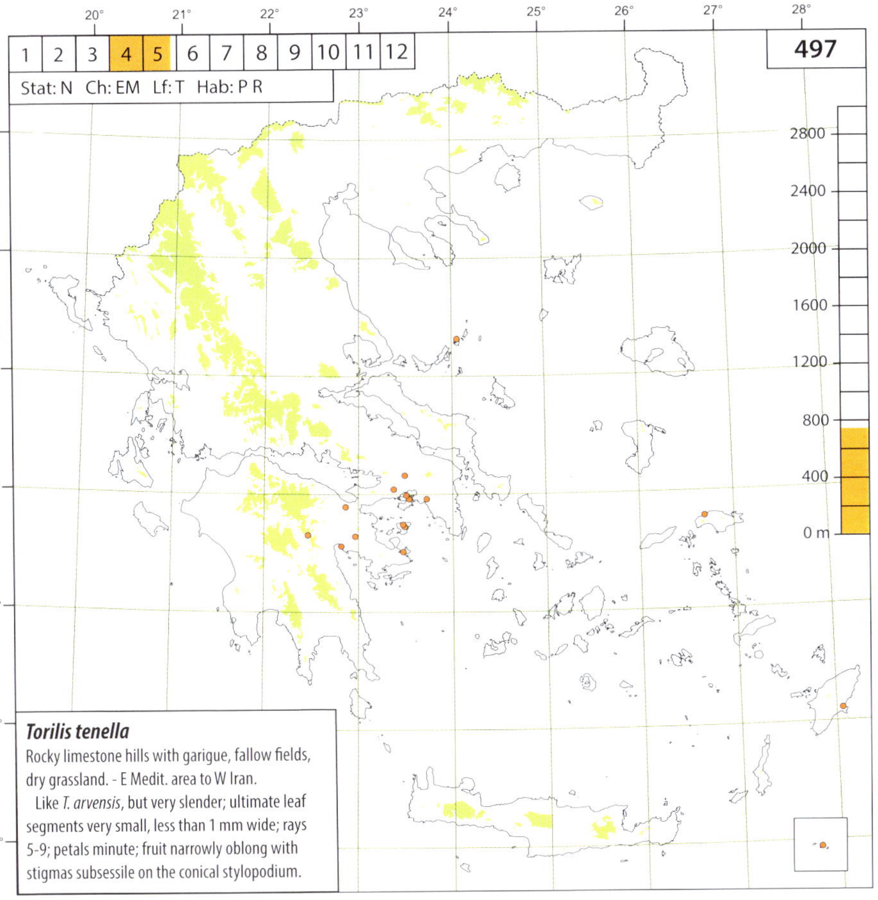

Torilis tenella

Rocky limestone hills with garigue, fallow fields, dry grassland. - E Medit. area to W Iran.

Like *T. arvensis*, but very slender; ultimate leaf segments very small, less than 1 mm wide; rays 5-9; petals minute; fruit narrowly oblong with stigmas subsessile on the conical stylopodium.

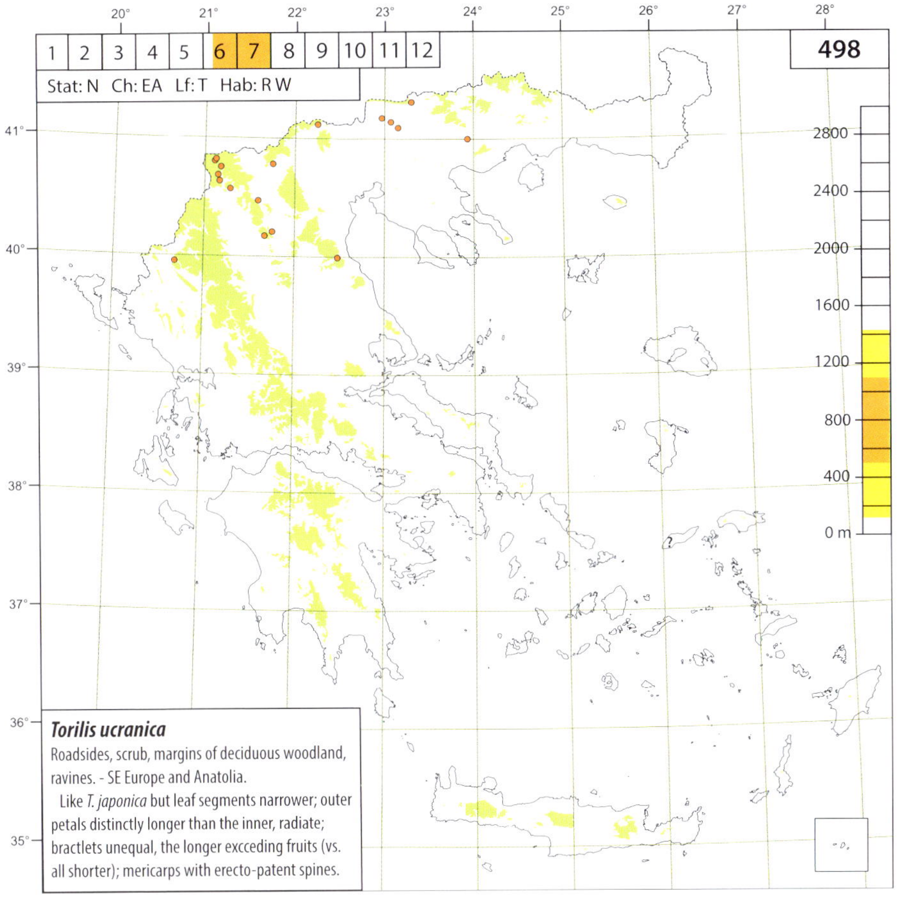

Torilis ucranica

Roadsides, scrub, margins of deciduous woodland, ravines. - SE Europe and Anatolia.

Like *T. japonica* but leaf segments narrower; outer petals distinctly longer than the inner, radiate; bractlets unequal, the longer exceeding fruits (vs. all shorter); mericarps with erecto-patent spines.

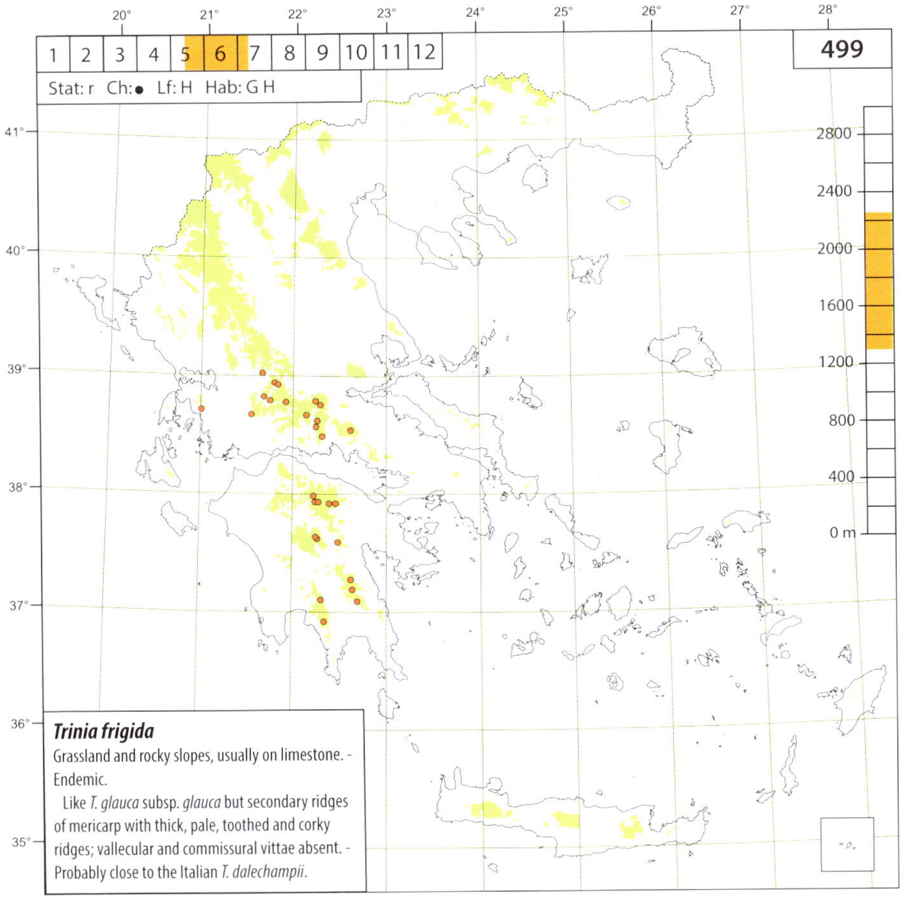

Trinia frigida

Grassland and rocky slopes, usually on limestone. - Endemic.

Like *T. glauca* subsp. *glauca* but secondary ridges of mericarp with thick, pale, toothed and corky ridges; vallecular and commissural vittae absent. - Probably close to the Italian *T. dalechampii*.

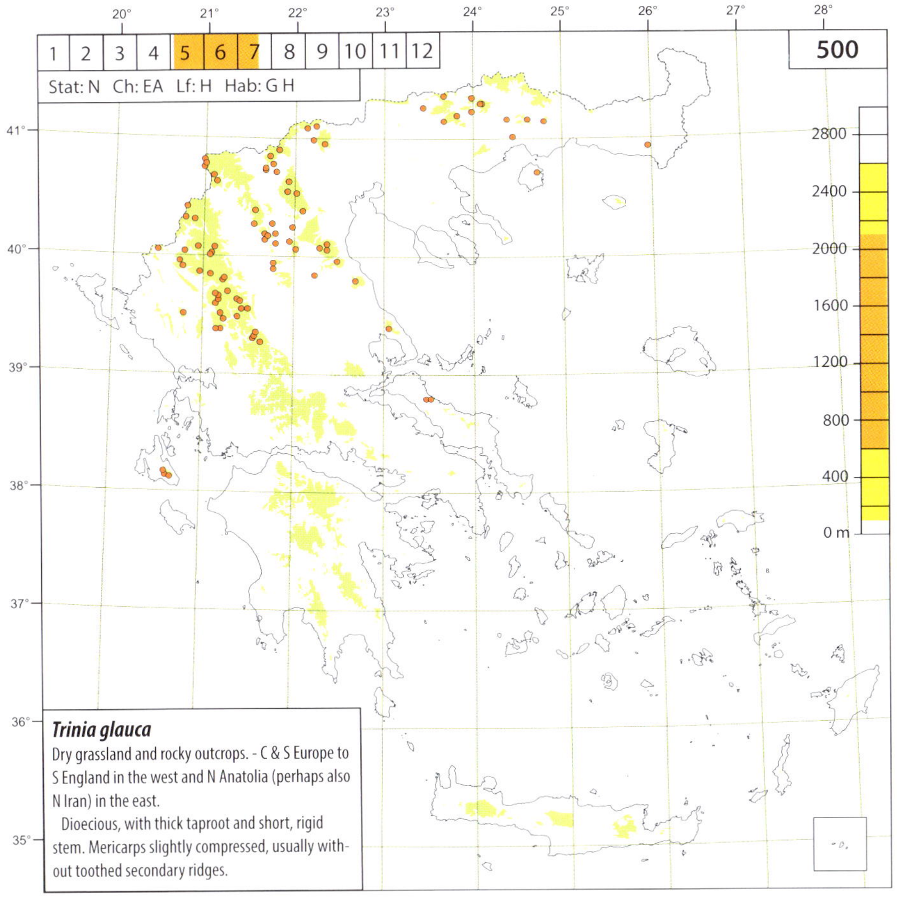

Trinia glauca

Dry grassland and rocky outcrops. - C & S Europe to S England in the west and N Anatolia (perhaps also N Iran) in the east.

Dioecious, with thick taproot and short, rigid stem. Mericarps slightly compressed, usually without toothed secondary ridges.

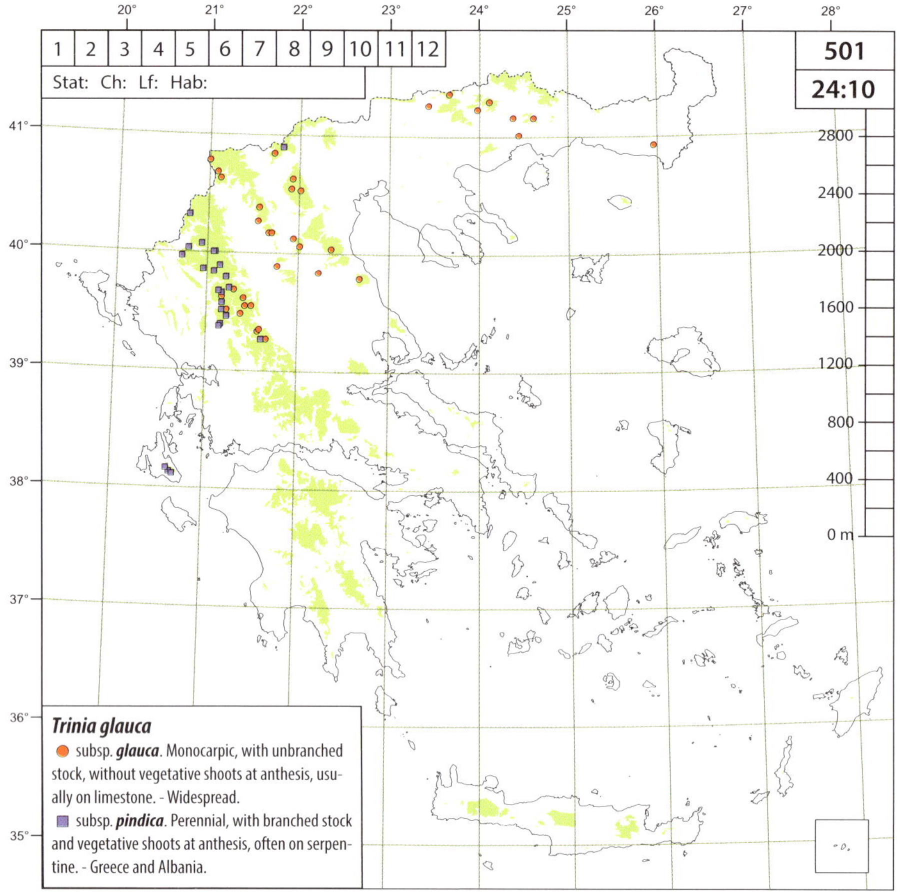

Trinia glauca
● subsp. ***glauca***. Monocarpic, with unbranched stock, without vegetative shoots at anthesis, usually on limestone. - Widespread.
■ subsp. ***pindica***. Perennial, with branched stock and vegetative shoots at anthesis, often on serpentine. - Greece and Albania.

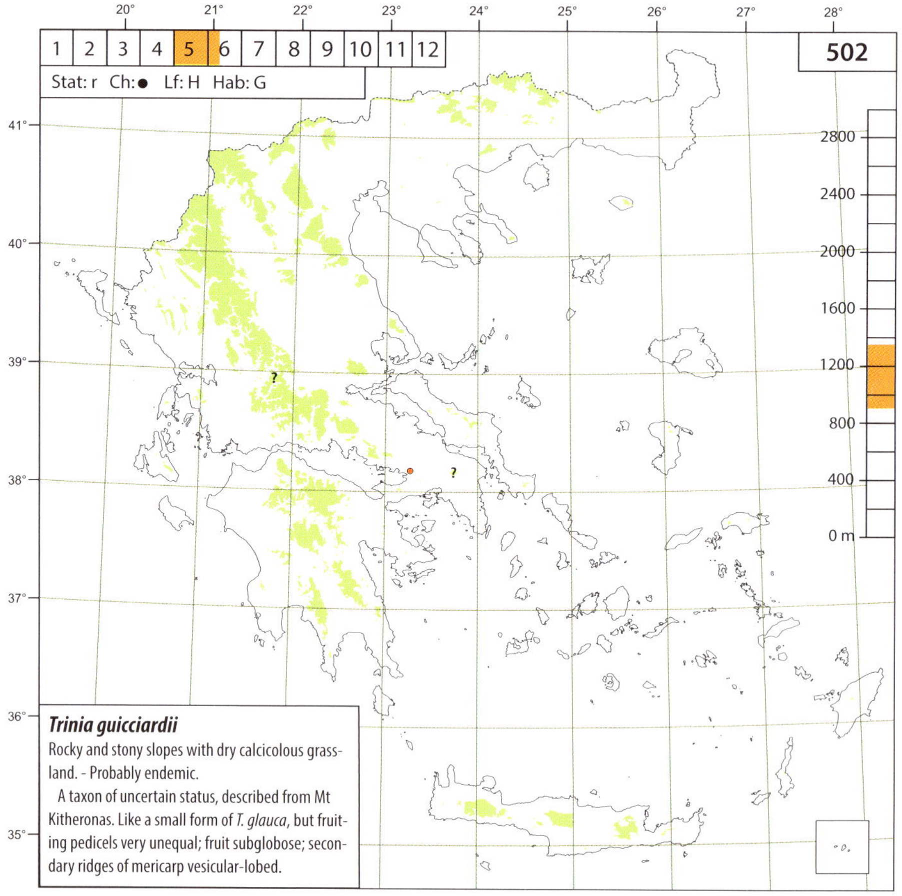

Trinia guicciardii
Rocky and stony slopes with dry calcicolous grassland. - Probably endemic.
A taxon of uncertain status, described from Mt Kitheronas. Like a small form of *T. glauca*, but fruiting pedicels very unequal; fruit subglobose; secondary ridges of mericarp vesicular-lobed.

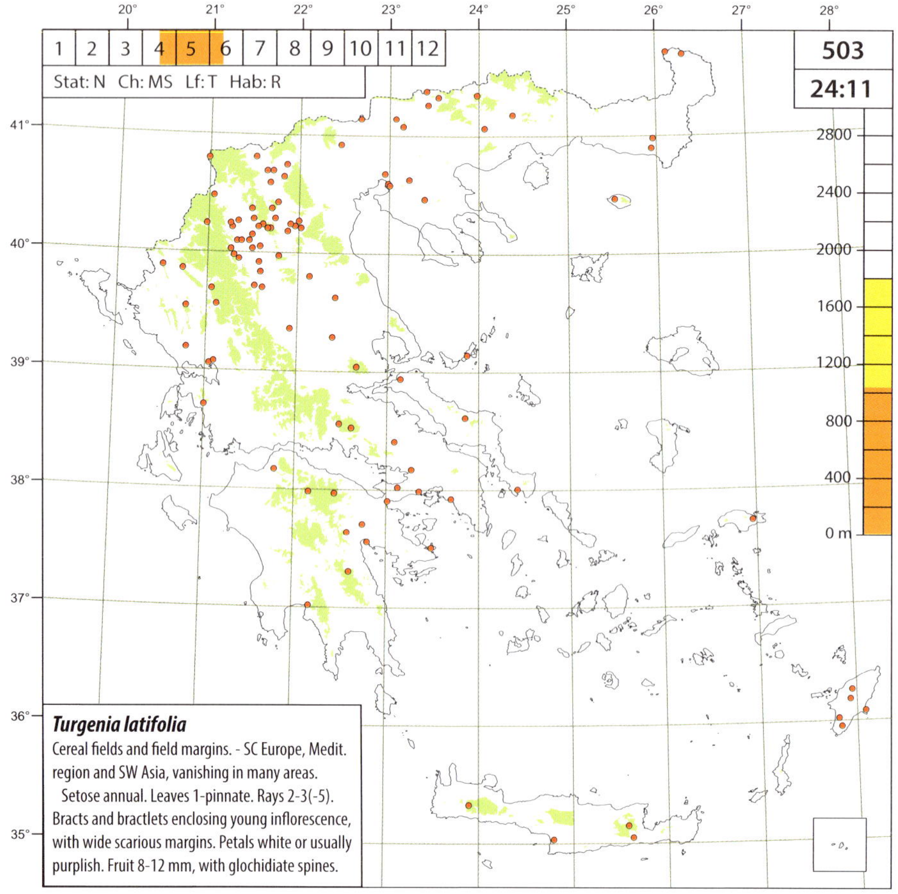

Turgenia latifolia
Cereal fields and field margins. - SC Europe, Medit. region and SW Asia, vanishing in many areas.
Setose annual. Leaves 1-pinnate. Rays 2-3(-5). Bracts and bractlets enclosing young inflorescence, with wide scarious margins. Petals white or usually purplish. Fruit 8-12 mm, with glochidiate spines.

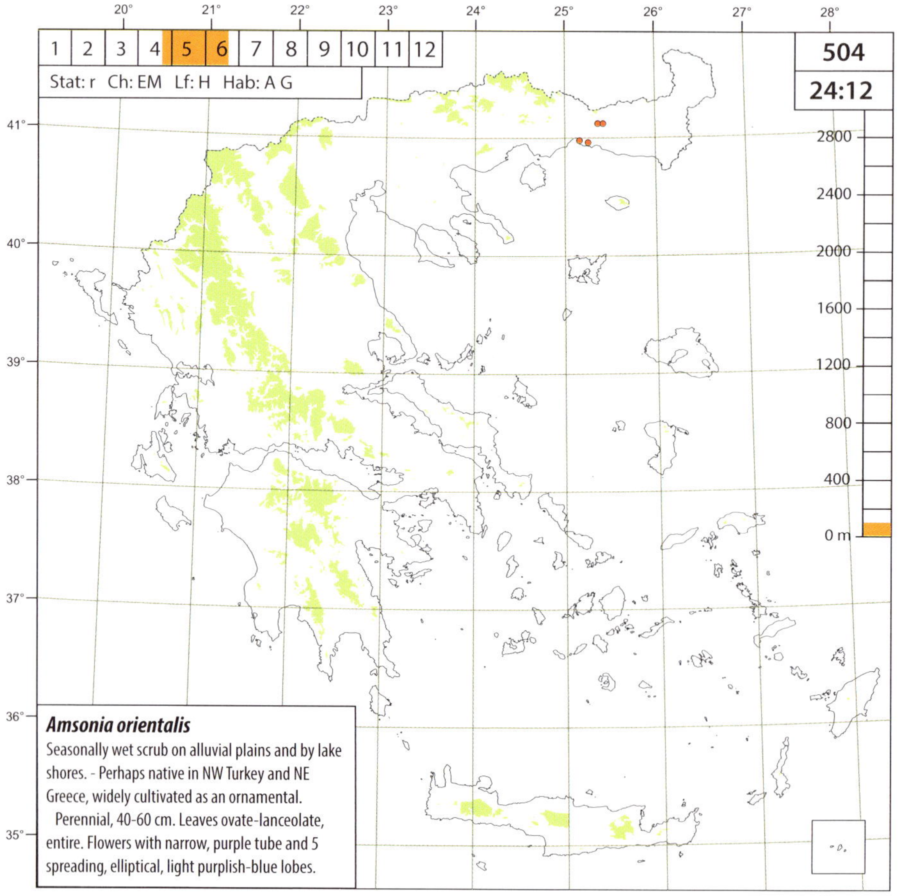

Amsonia orientalis
Seasonally wet scrub on alluvial plains and by lake shores. - Perhaps native in NW Turkey and NE Greece, widely cultivated as an ornamental.
Perennial, 40-60 cm. Leaves ovate-lanceolate, entire. Flowers with narrow, purple tube and 5 spreading, elliptical, light purplish-blue lobes.

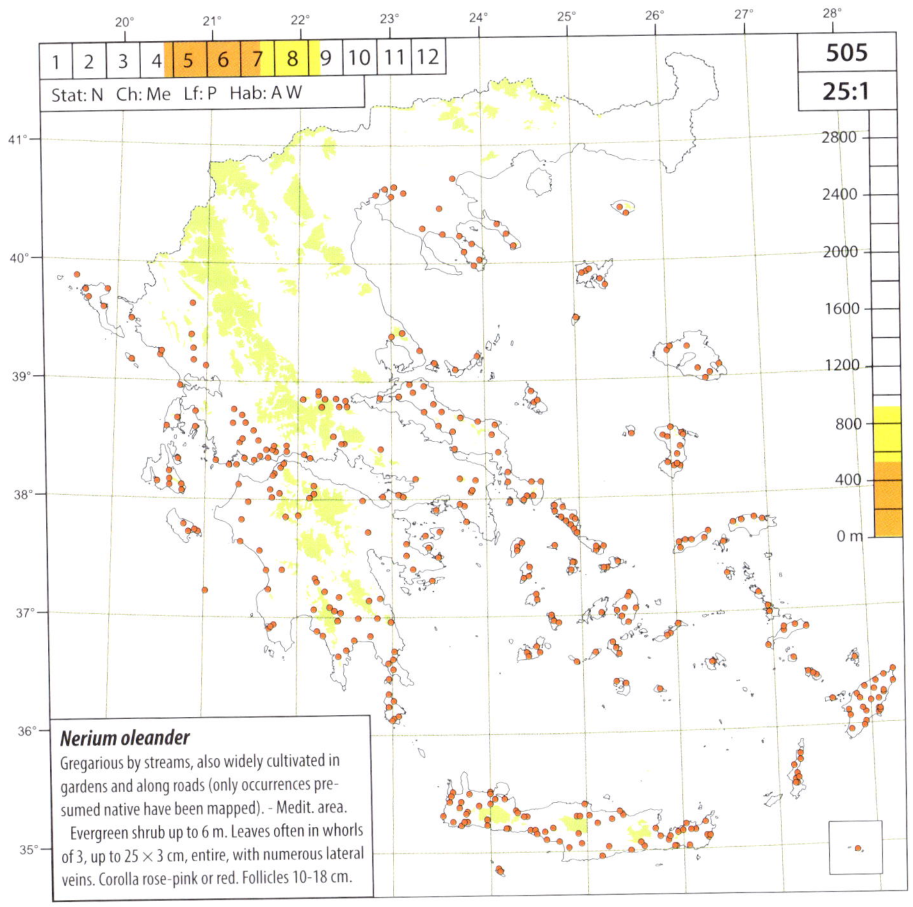

Nerium oleander

Gregarious by streams, also widely cultivated in gardens and along roads (only occurrences presumed native have been mapped). - Medit. area.

Evergreen shrub up to 6 m. Leaves often in whorls of 3, up to 25 × 3 cm, entire, with numerous lateral veins. Corolla rose-pink or red. Follicles 10-18 cm.

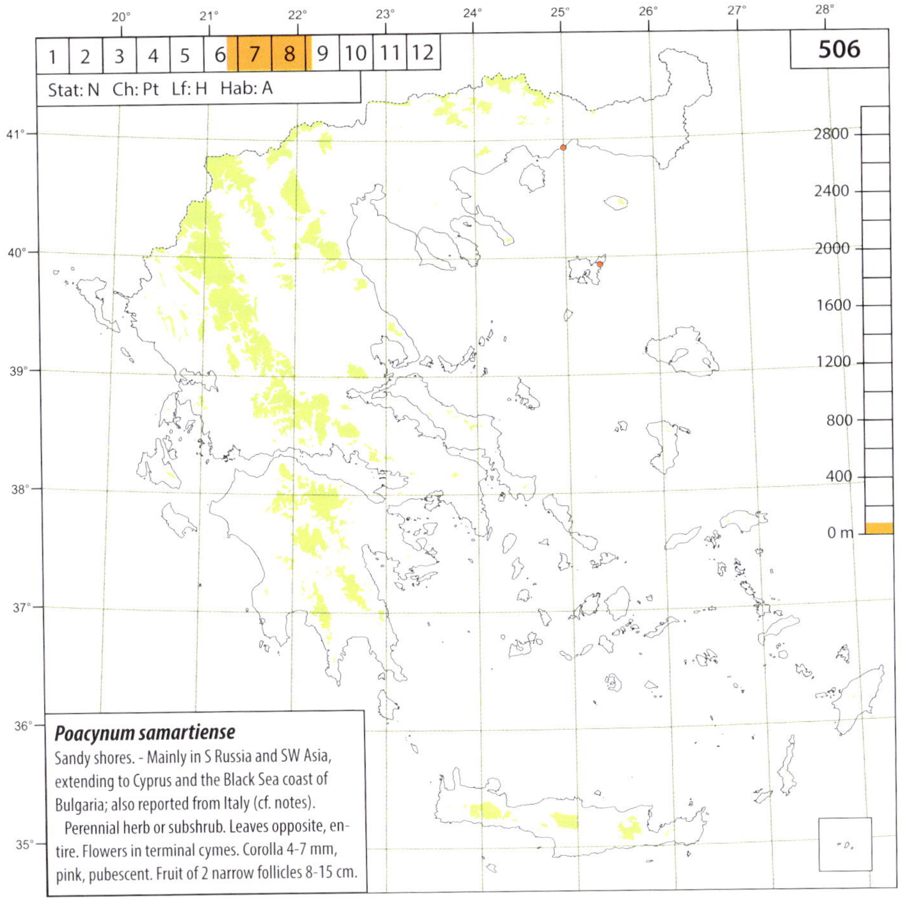

Poacynum samartiense

Sandy shores. - Mainly in S Russia and SW Asia, extending to Cyprus and the Black Sea coast of Bulgaria; also reported from Italy (cf. notes).

Perennial herb or subshrub. Leaves opposite, entire. Flowers in terminal cymes. Corolla 4-7 mm, pink, pubescent. Fruit of 2 narrow follicles 8-15 cm.

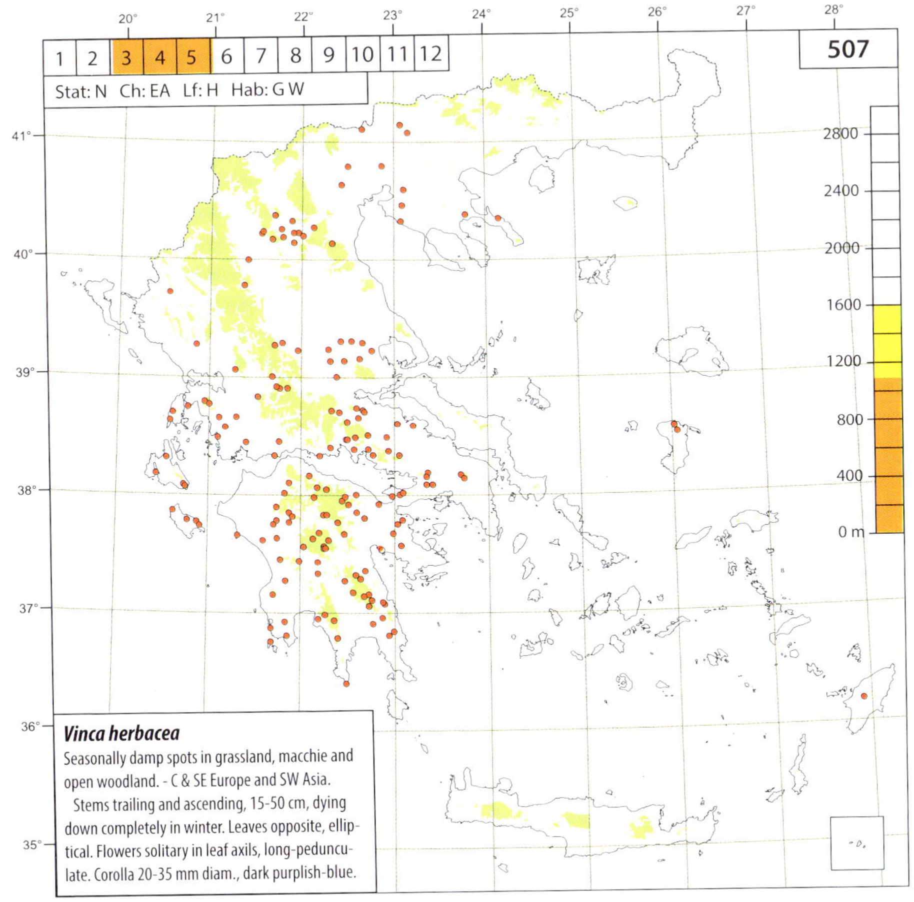

Vinca herbacea

Seasonally damp spots in grassland, macchie and open woodland. - C & SE Europe and SW Asia.

Stems trailing and ascending, 15-50 cm, dying down completely in winter. Leaves opposite, elliptical. Flowers solitary in leaf axils, long-pedunculate. Corolla 20-35 mm diam., dark purplish-blue.

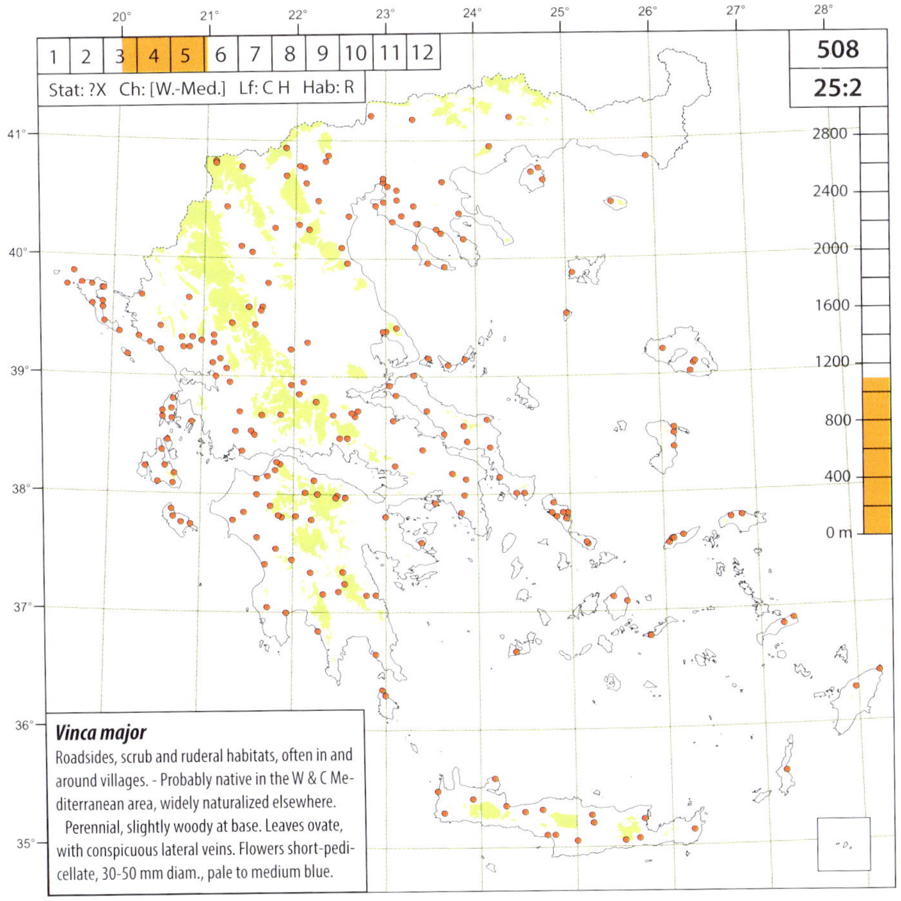

Vinca major

Roadsides, scrub and ruderal habitats, often in and around villages. - Probably native in the W & C Mediterranean area, widely naturalized elsewhere.

Perennial, slightly woody at base. Leaves ovate, with conspicuous lateral veins. Flowers short-pedicellate, 30-50 mm diam., pale to medium blue.

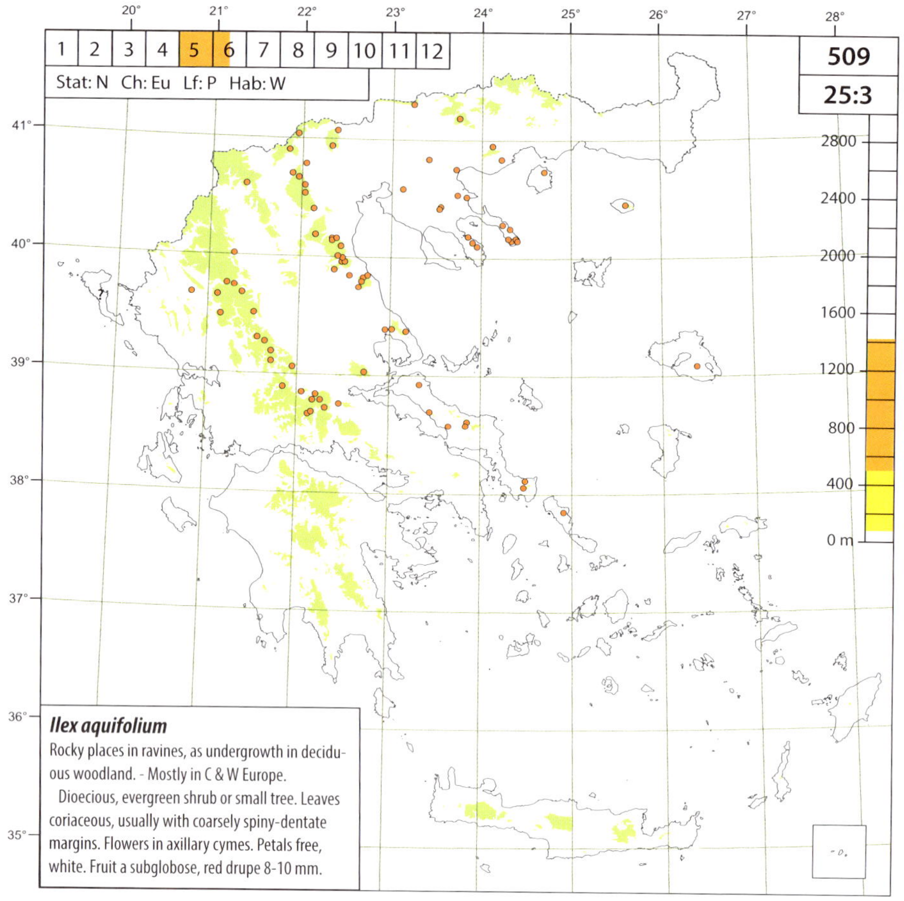

Ilex aquifolium

Rocky places in ravines, as undergrowth in deciduous woodland. - Mostly in C & W Europe.

Dioecious, evergreen shrub or small tree. Leaves coriaceous, usually with coarsely spiny-dentate margins. Flowers in axillary cymes. Petals free, white. Fruit a subglobose, red drupe 8-10 mm.

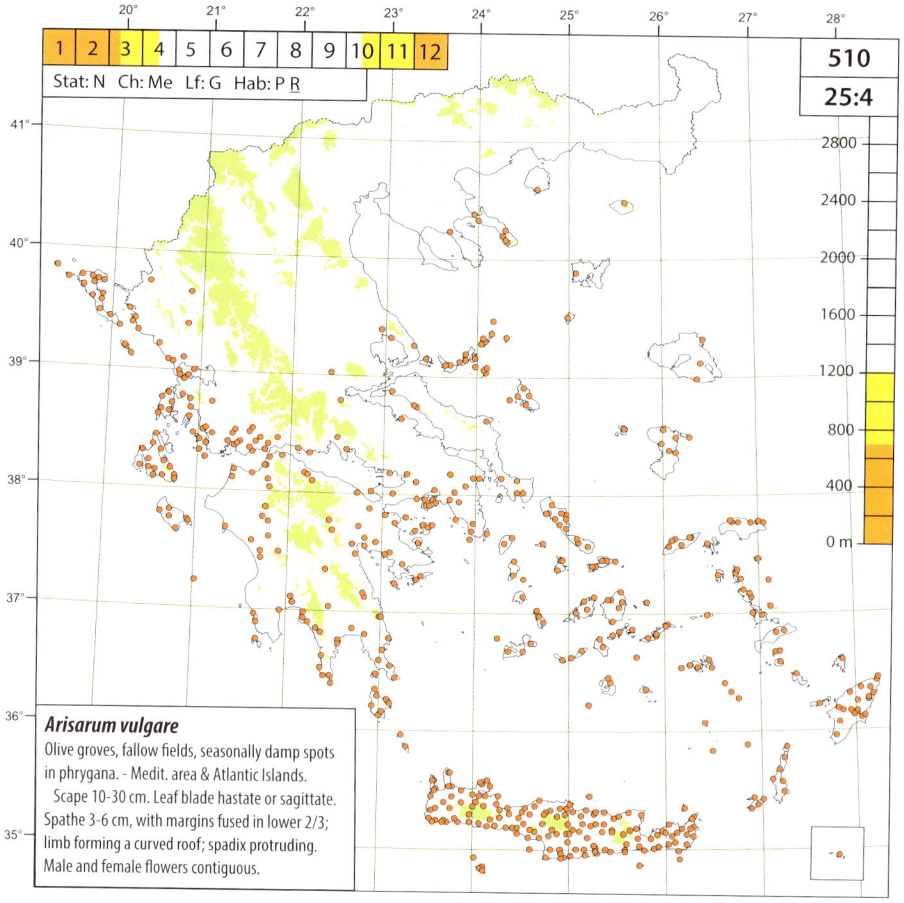

Arisarum vulgare

Olive groves, fallow fields, seasonally damp spots in phrygana. - Medit. area & Atlantic Islands.

Scape 10-30 cm. Leaf blade hastate or sagittate. Spathe 3-6 cm, with margins fused in lower 2/3; limb forming a curved roof; spadix protruding. Male and female flowers contiguous.

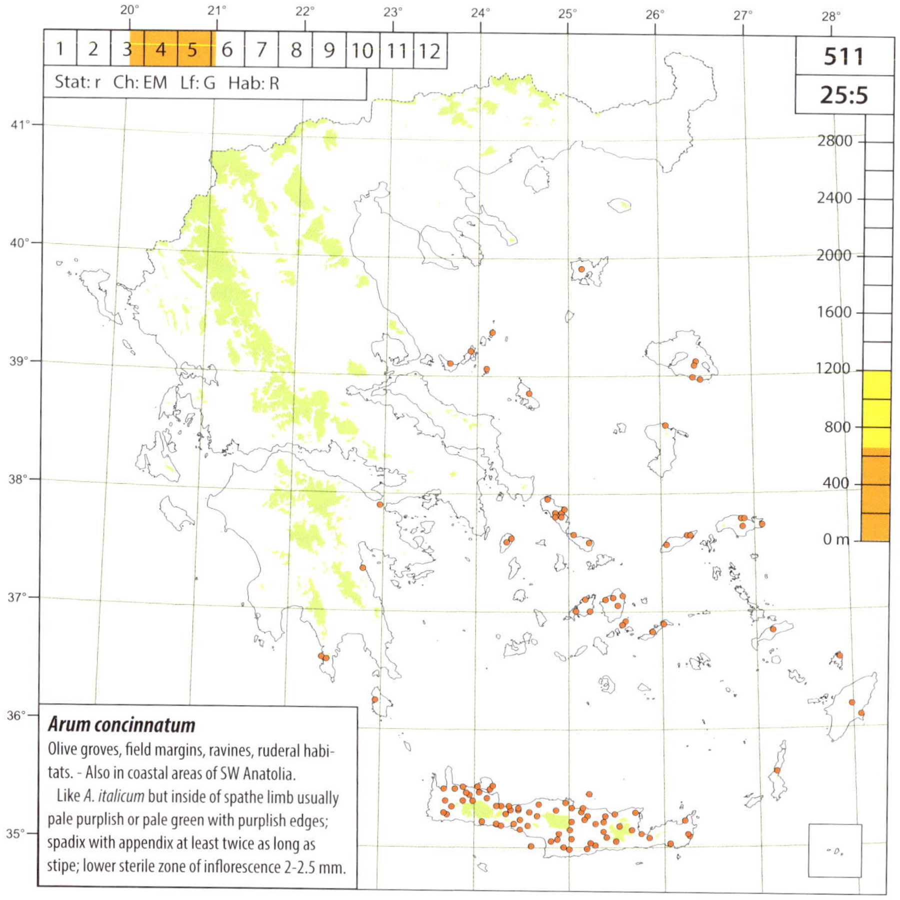

Arum concinnatum

Olive groves, field margins, ravines, ruderal habitats. - Also in coastal areas of SW Anatolia.

Like *A. italicum* but inside of spathe limb usually pale purplish or pale green with purplish edges; spadix with appendix at least twice as long as stipe; lower sterile zone of inflorescence 2-2.5 mm.

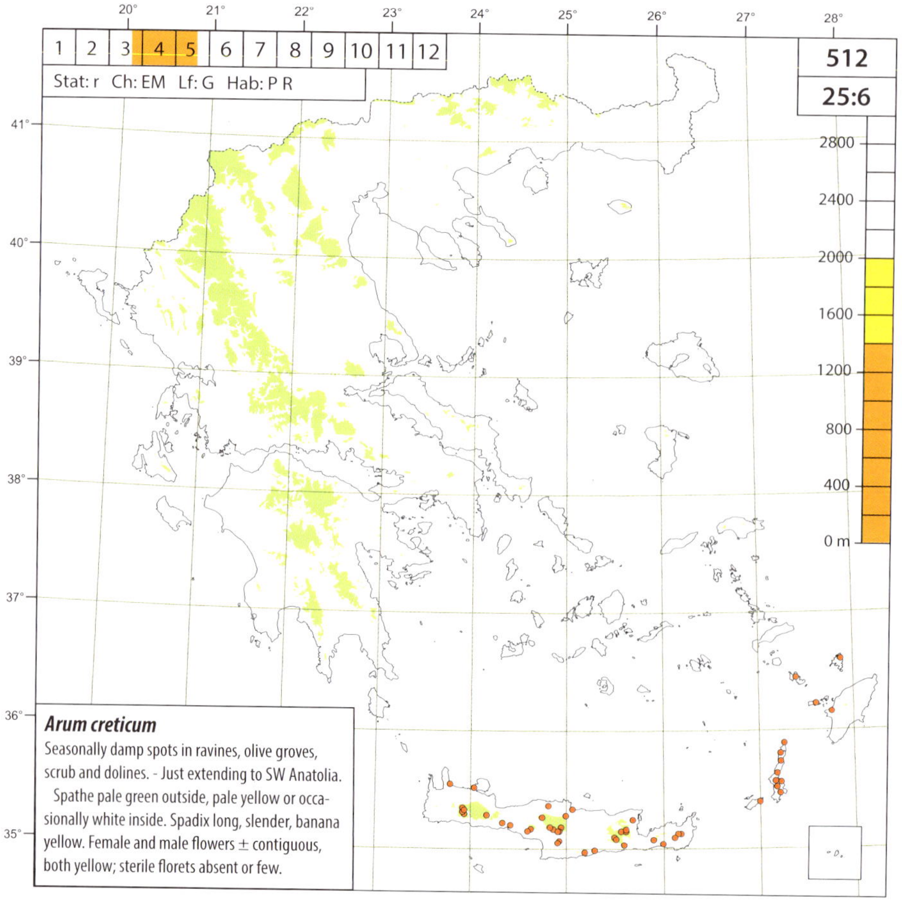

Arum creticum

Seasonally damp spots in ravines, olive groves, scrub and dolines. - Just extending to SW Anatolia.

Spathe pale green outside, pale yellow or occasionally white inside. Spadix long, slender, banana yellow. Female and male flowers ± contiguous, both yellow; sterile florets absent or few.

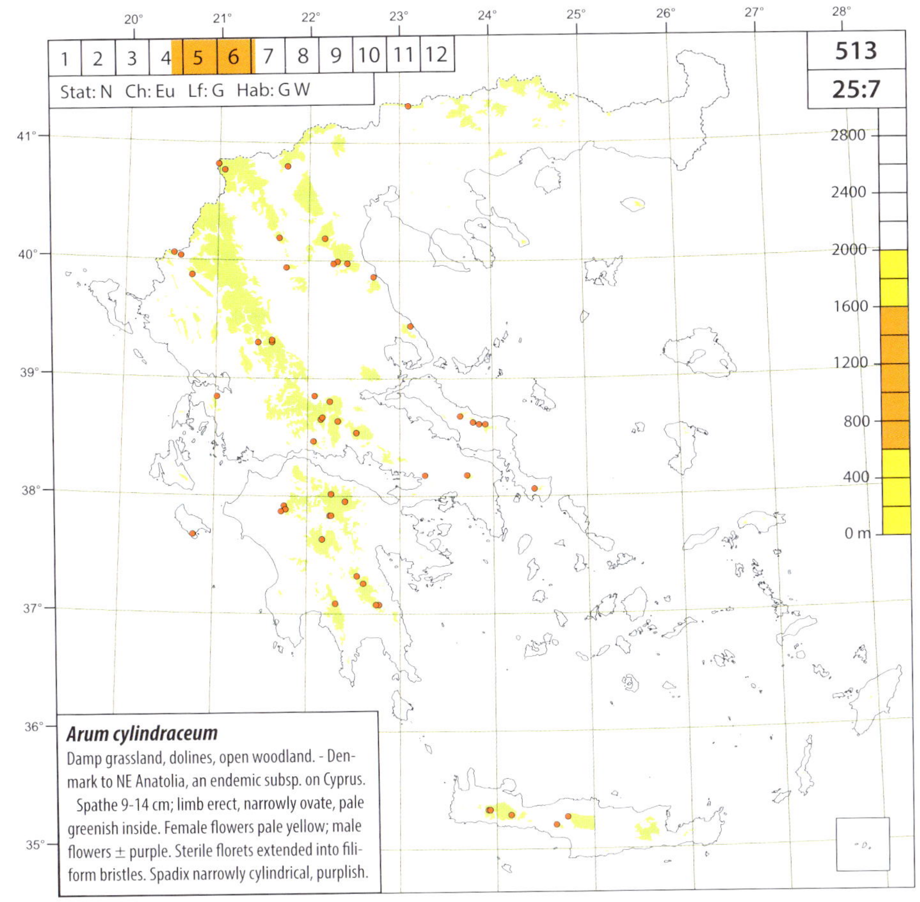

Arum cylindraceum

Damp grassland, dolines, open woodland. - Denmark to NE Anatolia, an endemic subsp. on Cyprus. Spathe 9-14 cm; limb erect, narrowly ovate, pale greenish inside. Female flowers pale yellow; male flowers ± purple. Sterile florets extended into filiform bristles. Spadix narrowly cylindrical, purplish.

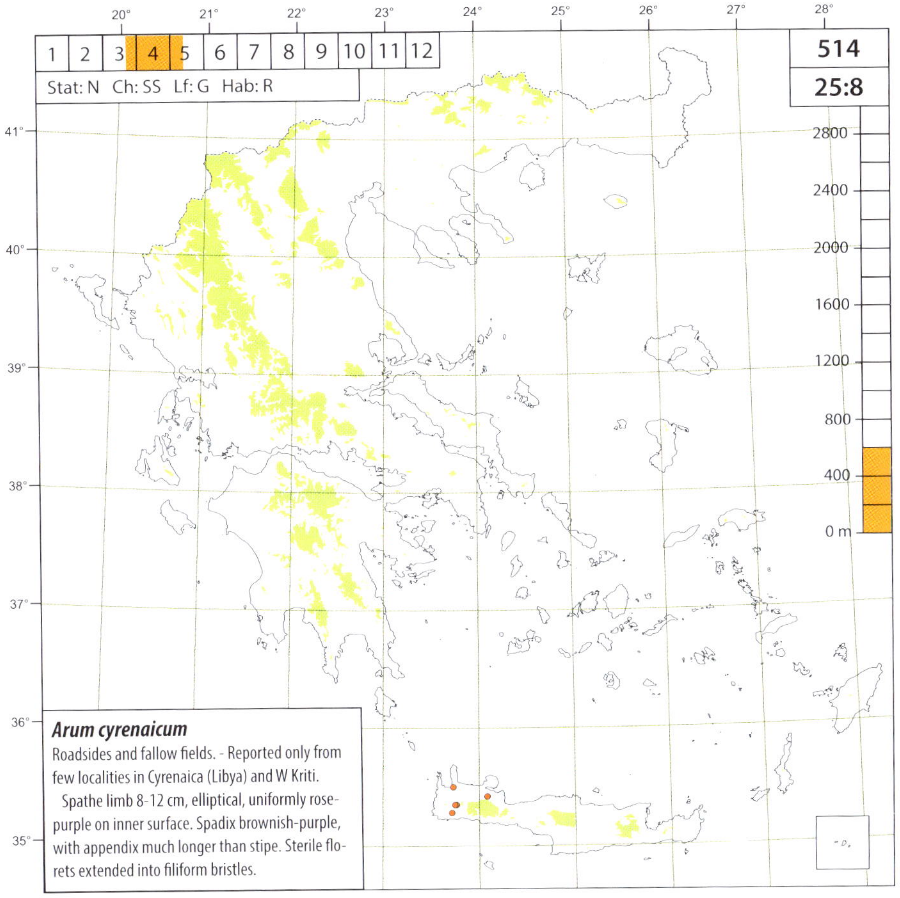

Arum cyrenaicum

Roadsides and fallow fields. - Reported only from few localities in Cyrenaica (Libya) and W Kriti. Spathe limb 8-12 cm, elliptical, uniformly rose-purple on inner surface. Spadix brownish-purple, with appendix much longer than stipe. Sterile florets extended into filiform bristles.

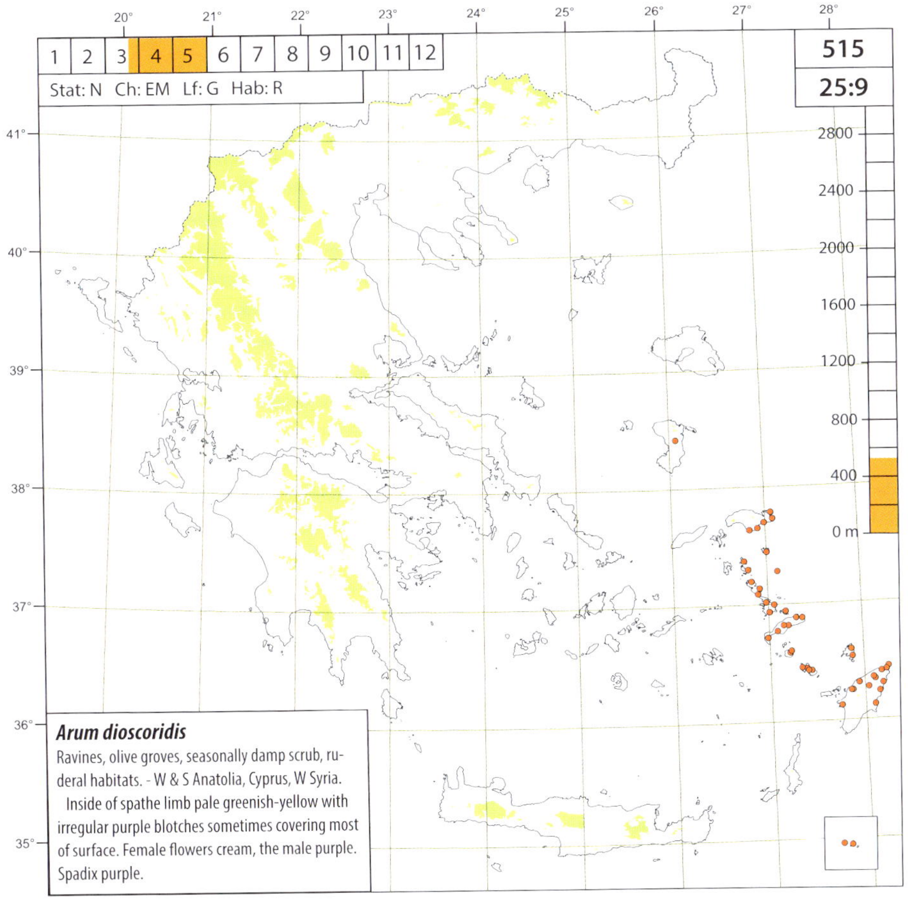

Arum dioscoridis

Ravines, olive groves, seasonally damp scrub, ruderal habitats. - W & S Anatolia, Cyprus, W Syria. Inside of spathe limb pale greenish-yellow with irregular purple blotches sometimes covering most of surface. Female flowers cream, the male purple. Spadix purple.

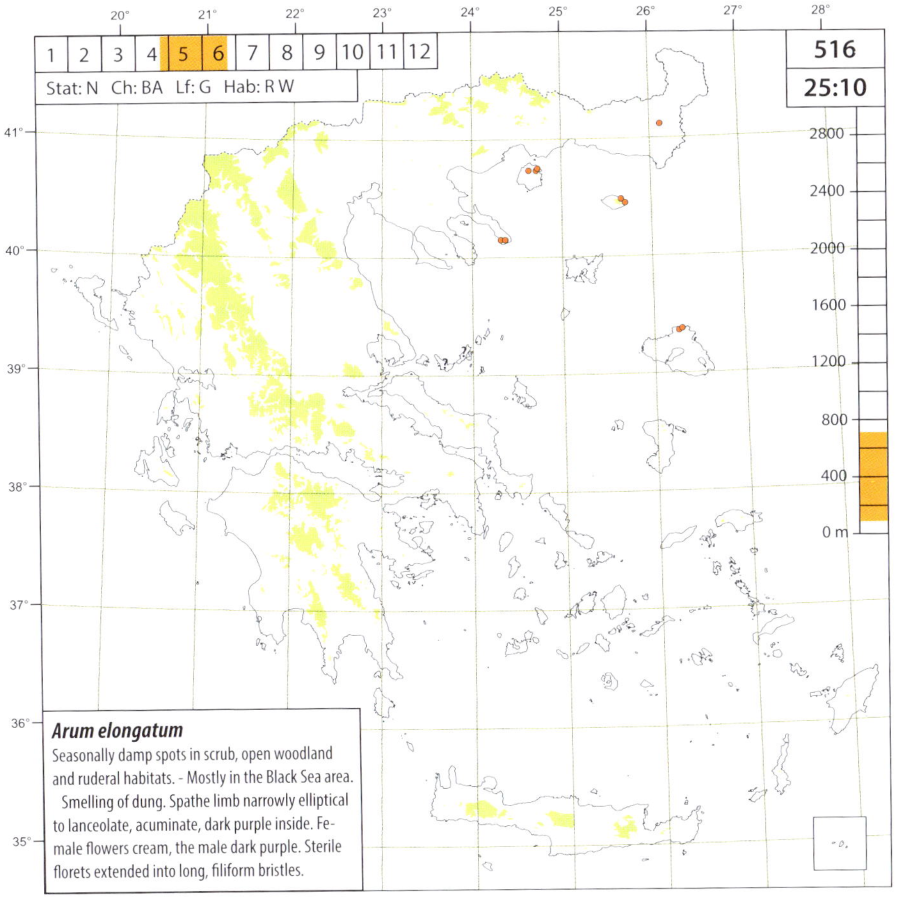

Arum elongatum

Seasonally damp spots in scrub, open woodland and ruderal habitats. - Mostly in the Black Sea area. Smelling of dung. Spathe limb narrowly elliptical to lanceolate, acuminate, dark purple inside. Female flowers cream, the male dark purple. Sterile florets extended into long, filiform bristles.

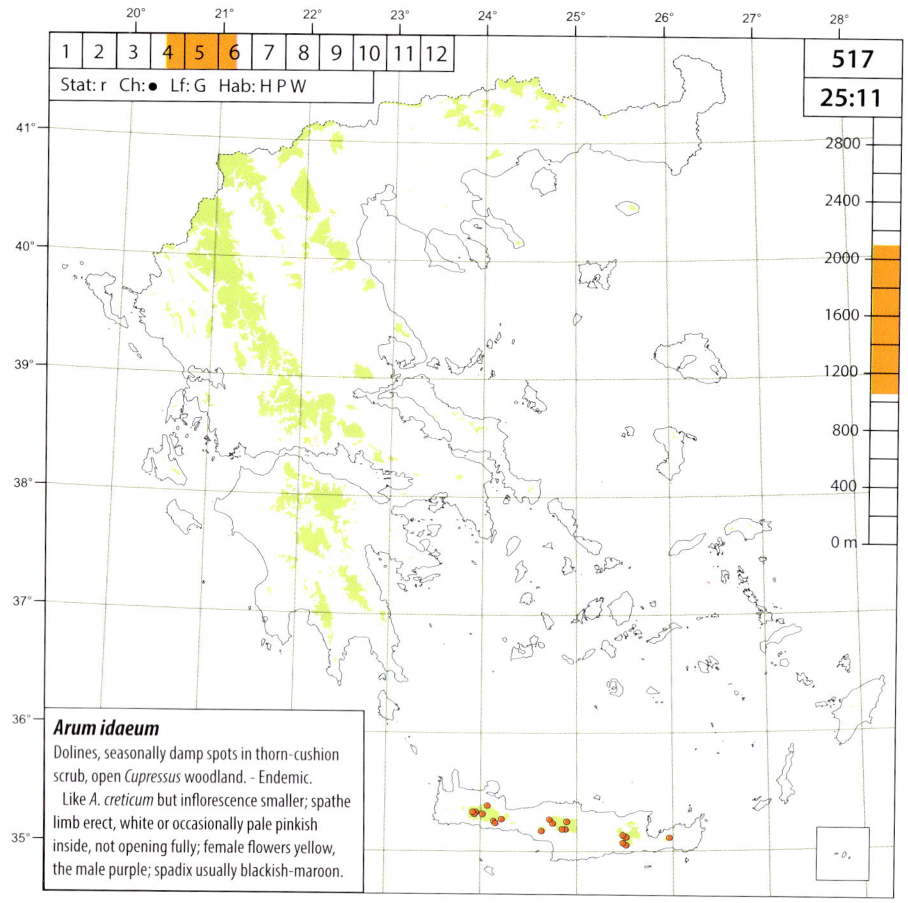

Arum idaeum
Dolines, seasonally damp spots in thorn-cushion scrub, open *Cupressus* woodland. - Endemic.
Like *A. creticum* but inflorescence smaller; spathe limb erect, white or occasionally pale pinkish inside, not opening fully; female flowers yellow, the male purple; spadix usually blackish-maroon.

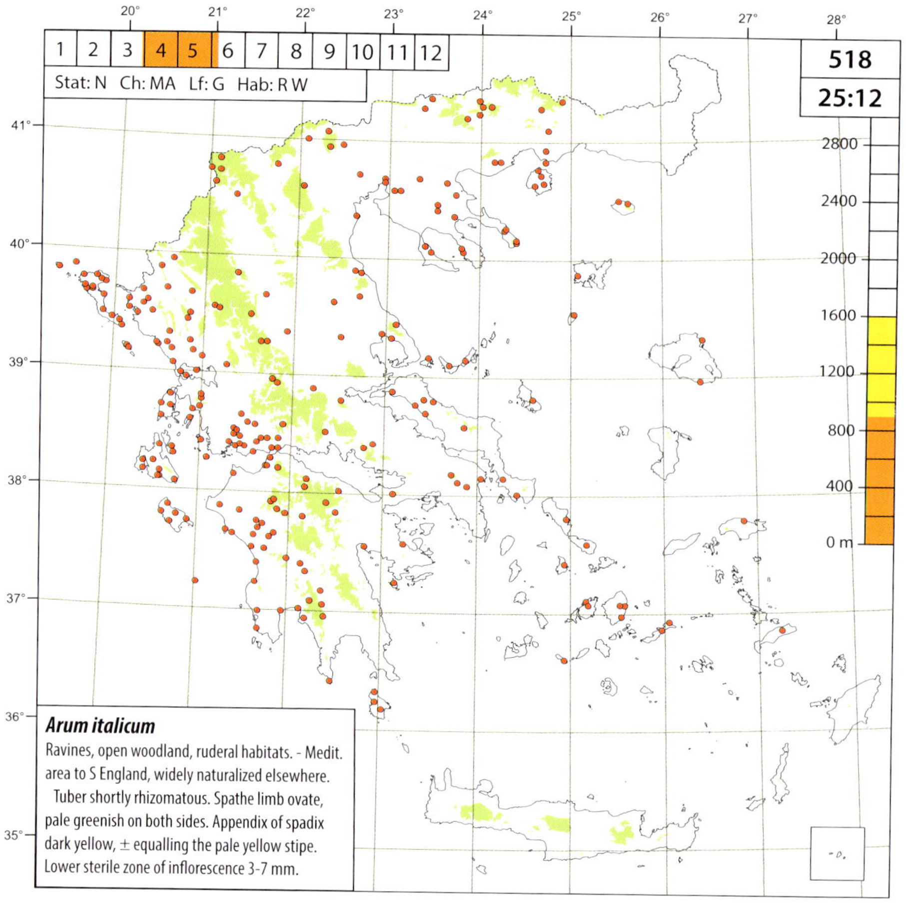

Arum italicum
Ravines, open woodland, ruderal habitats. - Medit. area to S England, widely naturalized elsewhere.
Tuber shortly rhizomatous. Spathe limb ovate, pale greenish on both sides. Appendix of spadix dark yellow, ± equalling the pale yellow stipe. Lower sterile zone of inflorescence 3-7 mm.

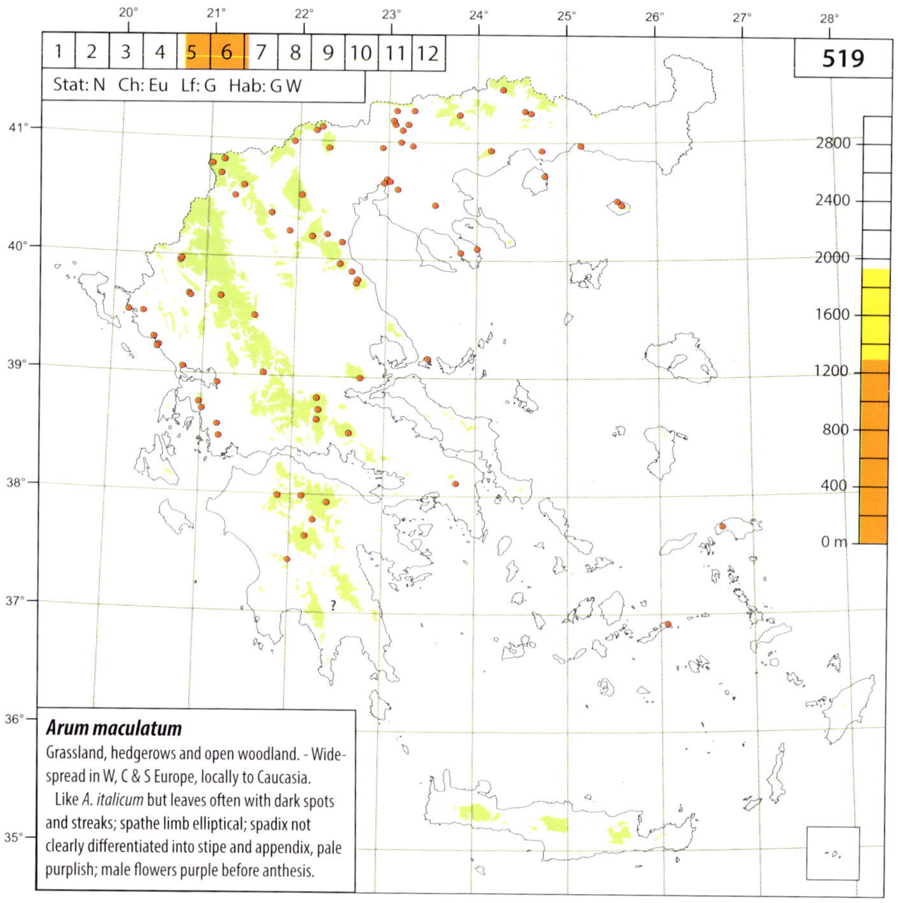

Arum maculatum
Grassland, hedgerows and open woodland. - Widespread in W, C & S Europe, locally to Caucasia.
Like *A. italicum* but leaves often with dark spots and streaks; spathe limb elliptical; spadix not clearly differentiated into stipe and appendix, pale purplish; male flowers purple before anthesis.

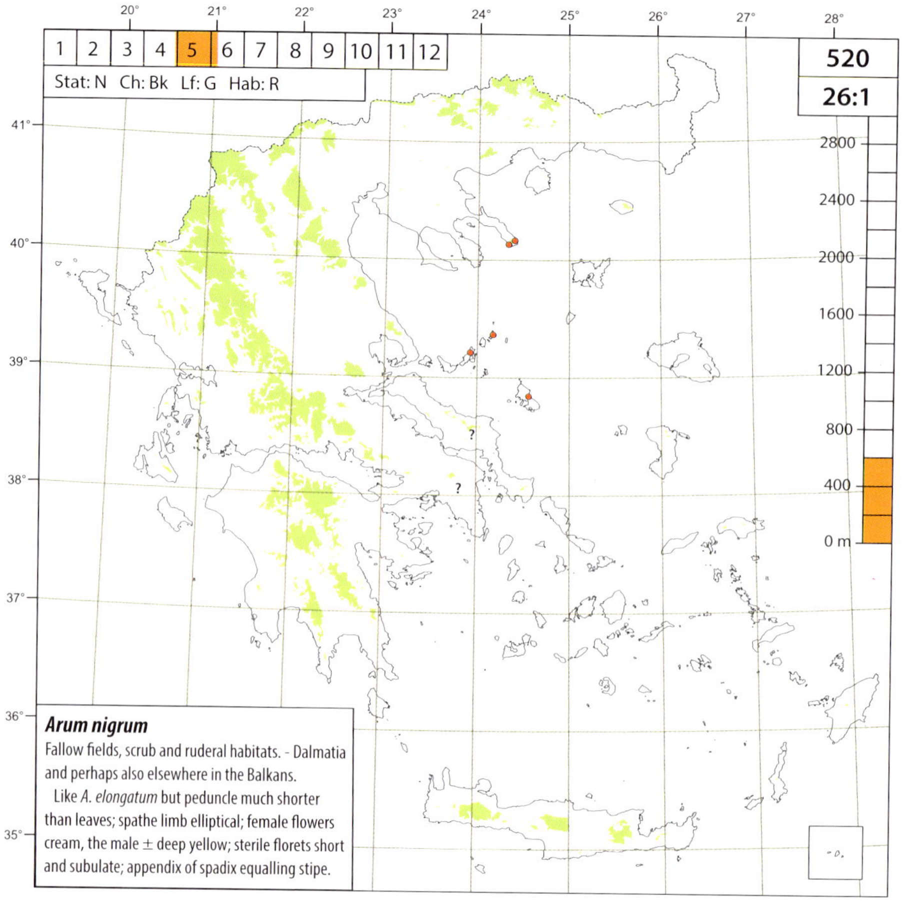

Arum nigrum
Fallow fields, scrub and ruderal habitats. - Dalmatia and perhaps also elsewhere in the Balkans.
Like *A. elongatum* but peduncle much shorter than leaves; spathe limb elliptical; female flowers cream, the male ± deep yellow; sterile florets short and subulate; appendix of spadix equalling stipe.

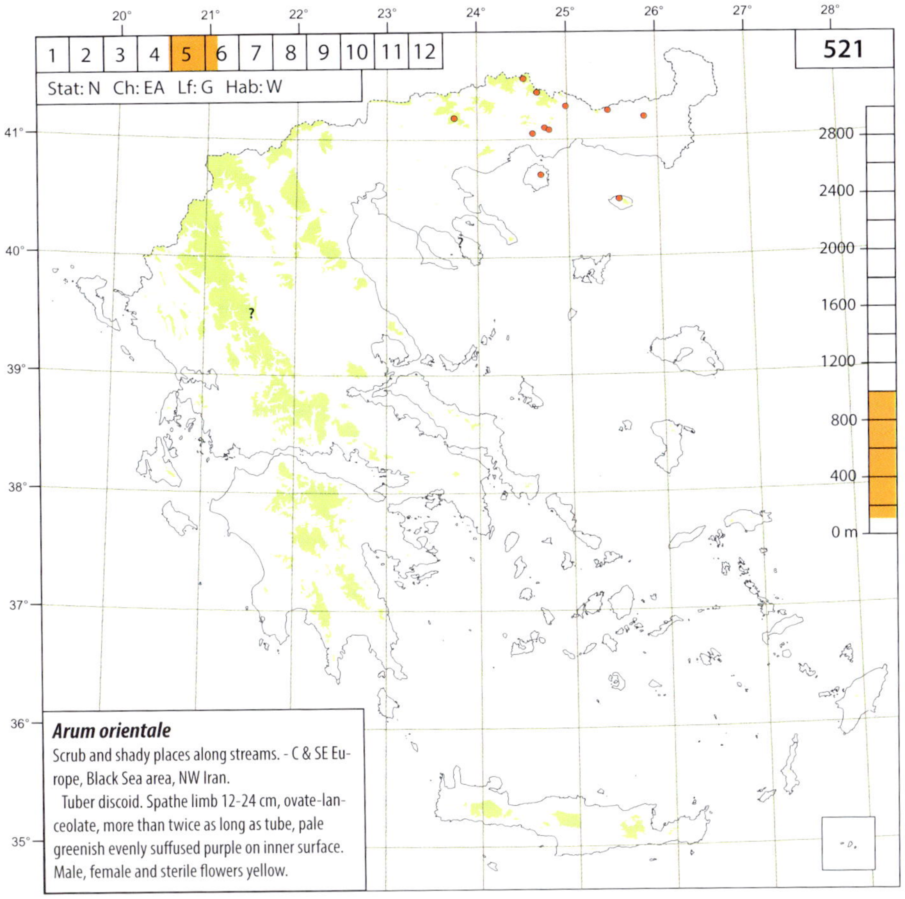

Arum orientale

Scrub and shady places along streams. - C & SE Europe, Black Sea area, NW Iran.

Tuber discoid. Spathe limb 12-24 cm, ovate-lanceolate, more than twice as long as tube, pale greenish evenly suffused purple on inner surface. Male, female and sterile flowers yellow.

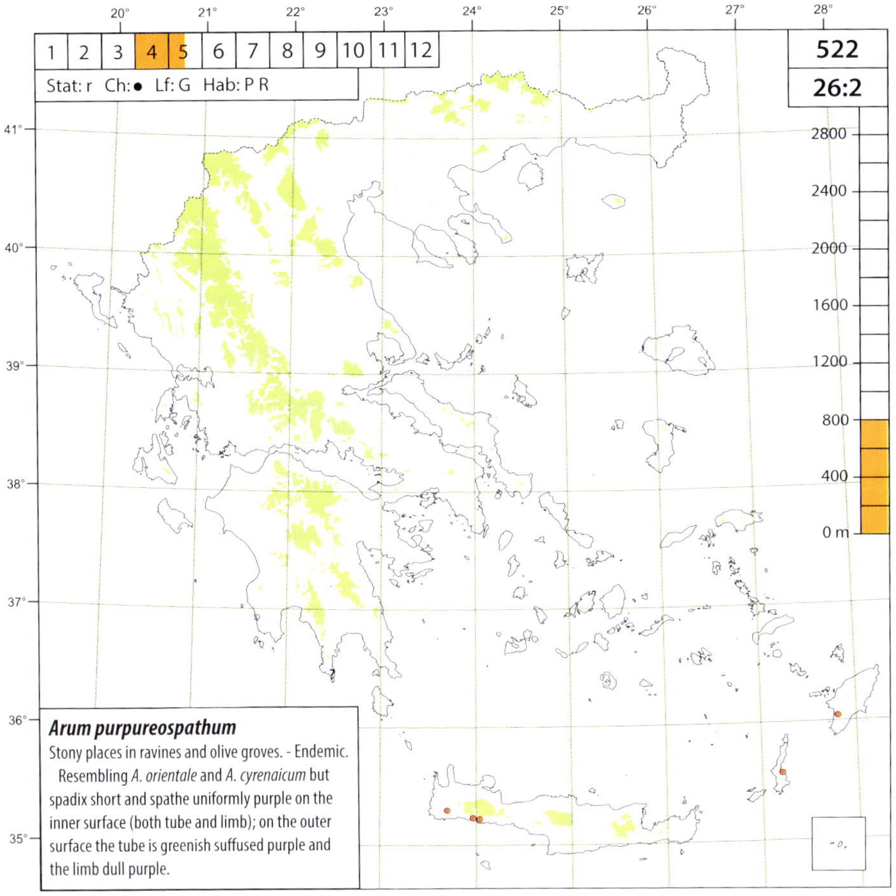

Arum purpureospathum

Stony places in ravines and olive groves. - Endemic.

Resembling *A. orientale* and *A. cyrenaicum* but spadix short and spathe uniformly purple on the inner surface (both tube and limb); on the outer surface the tube is greenish suffused purple and the limb dull purple.

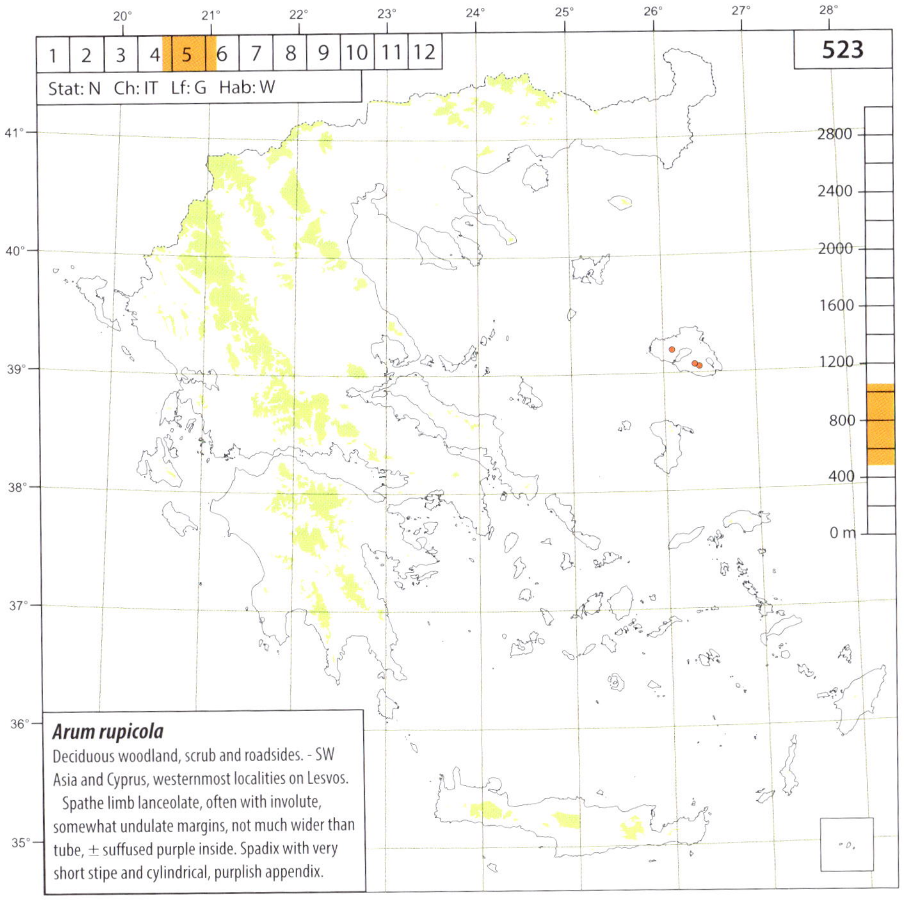

Arum rupicola

Deciduous woodland, scrub and roadsides. - SW Asia and Cyprus, westernmost localities on Lesvos.

Spathe limb lanceolate, often with involute, somewhat undulate margins, not much wider than tube, ± suffused purple inside. Spadix with very short stipe and cylindrical, purplish appendix.

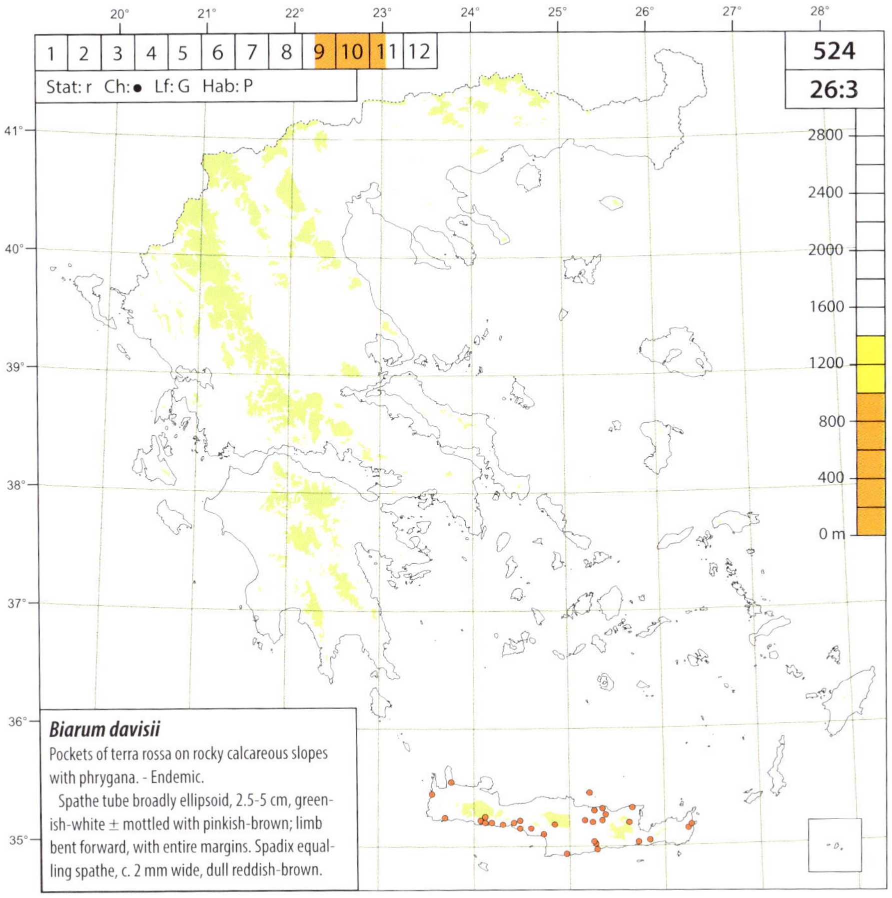

Biarum davisii

Pockets of terra rossa on rocky calcareous slopes with phrygana. - Endemic.

Spathe tube broadly ellipsoid, 2.5-5 cm, greenish-white ± mottled with pinkish-brown; limb bent forward, with entire margins. Spadix equalling spathe, c. 2 mm wide, dull reddish-brown.

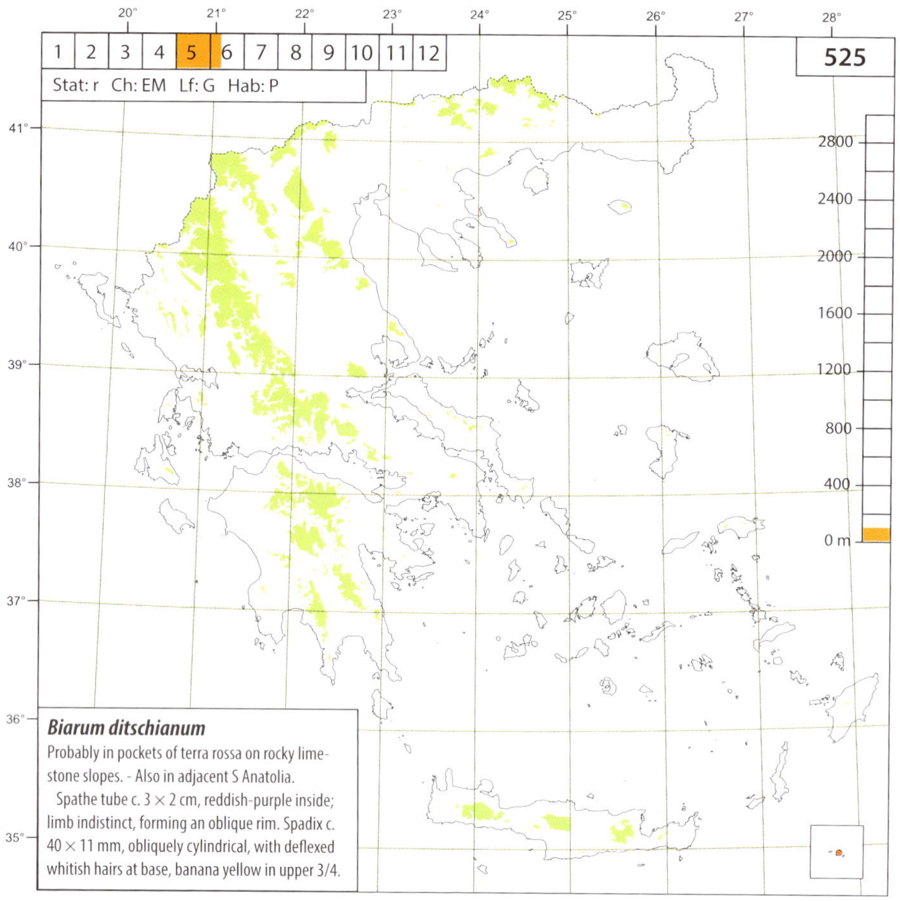

Biarum ditschianum

Probably in pockets of terra rossa on rocky limestone slopes. - Also in adjacent S Anatolia.

Spathe tube c. 3 × 2 cm, reddish-purple inside; limb indistinct, forming an oblique rim. Spadix c. 40 × 11 mm, obliquely cylindrical, with deflexed whitish hairs at base, banana yellow in upper 3/4.

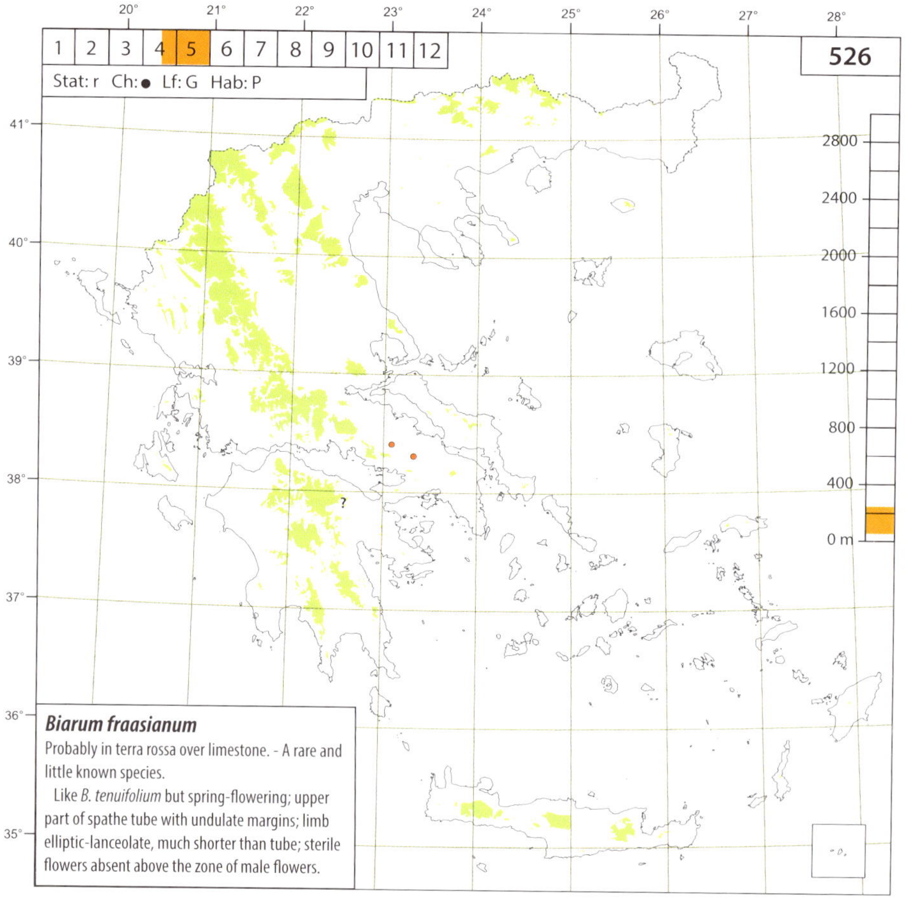

Biarum fraasianum

Probably in terra rossa over limestone. - A rare and little known species.

Like *B. tenuifolium* but spring-flowering; upper part of spathe tube with undulate margins; limb elliptic-lanceolate, much shorter than tube; sterile flowers absent above the zone of male flowers.

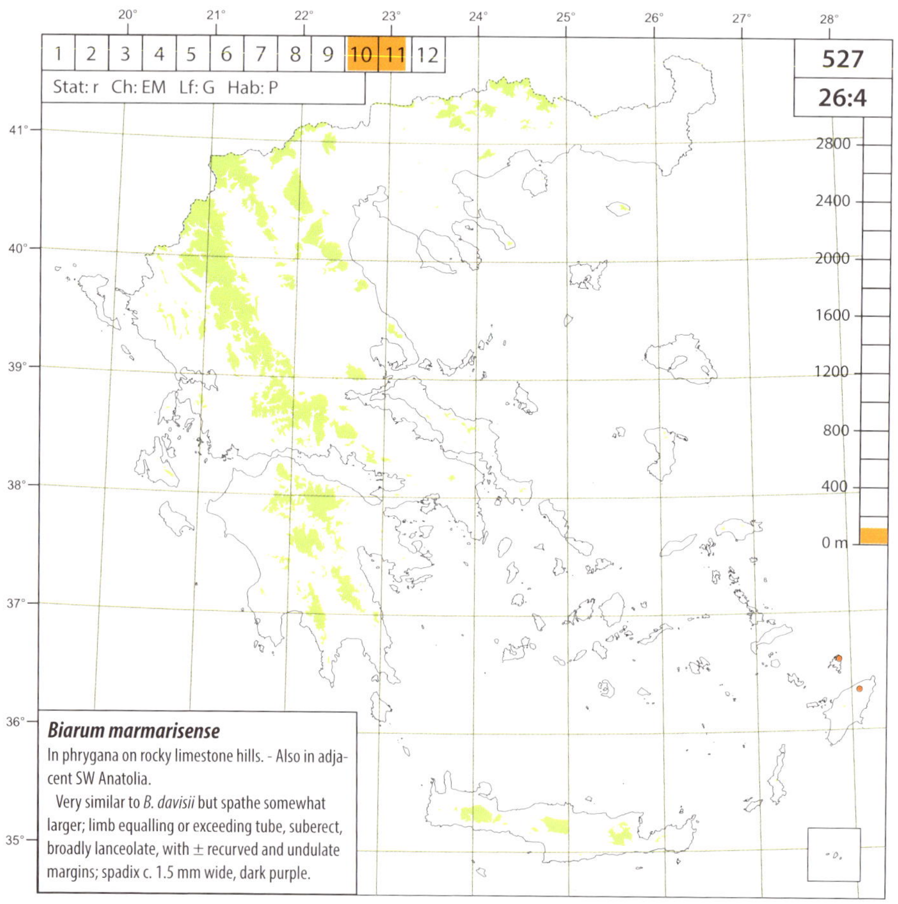

Biarum marmarisense

In phrygana on rocky limestone hills. - Also in adjacent SW Anatolia.

Very similar to *B. davisii* but spathe somewhat larger; limb equalling or exceeding tube, suberect, broadly lanceolate, with ± recurved and undulate margins; spadix c. 1.5 mm wide, dark purple.

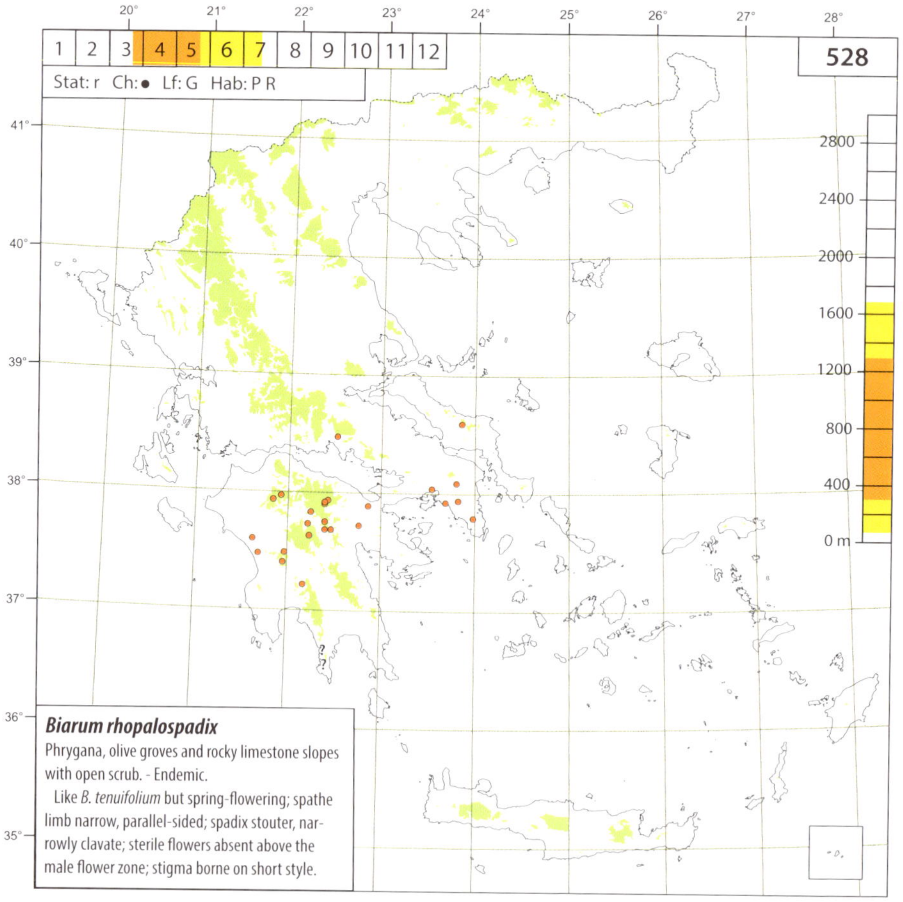

Biarum rhopalospadix

Phrygana, olive groves and rocky limestone slopes with open scrub. - Endemic.

Like *B. tenuifolium* but spring-flowering; spathe limb narrow, parallel-sided; spadix stouter, narrowly clavate; sterile flowers absent above the male flower zone; stigma borne on short style.

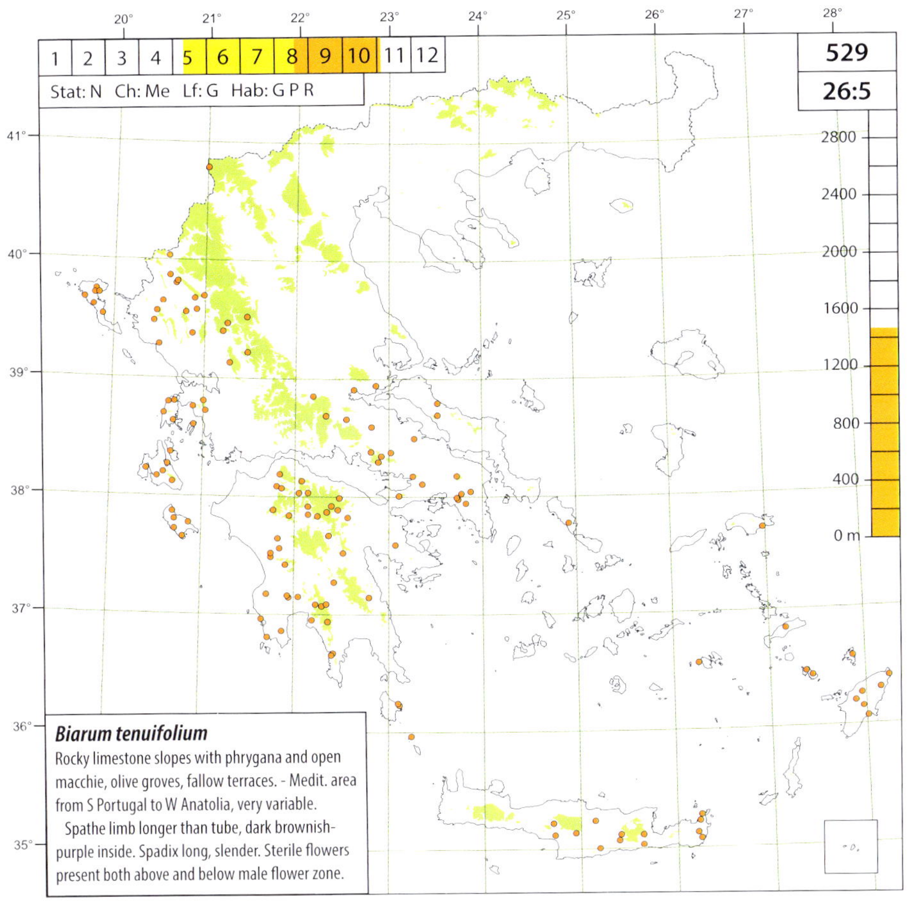

Biarum tenuifolium
Rocky limestone slopes with phrygana and open macchie, olive groves, fallow terraces. - Medit. area from S Portugal to W Anatolia, very variable.
Spathe limb longer than tube, dark brownish-purple inside. Spadix long, slender. Sterile flowers present both above and below male flower zone.

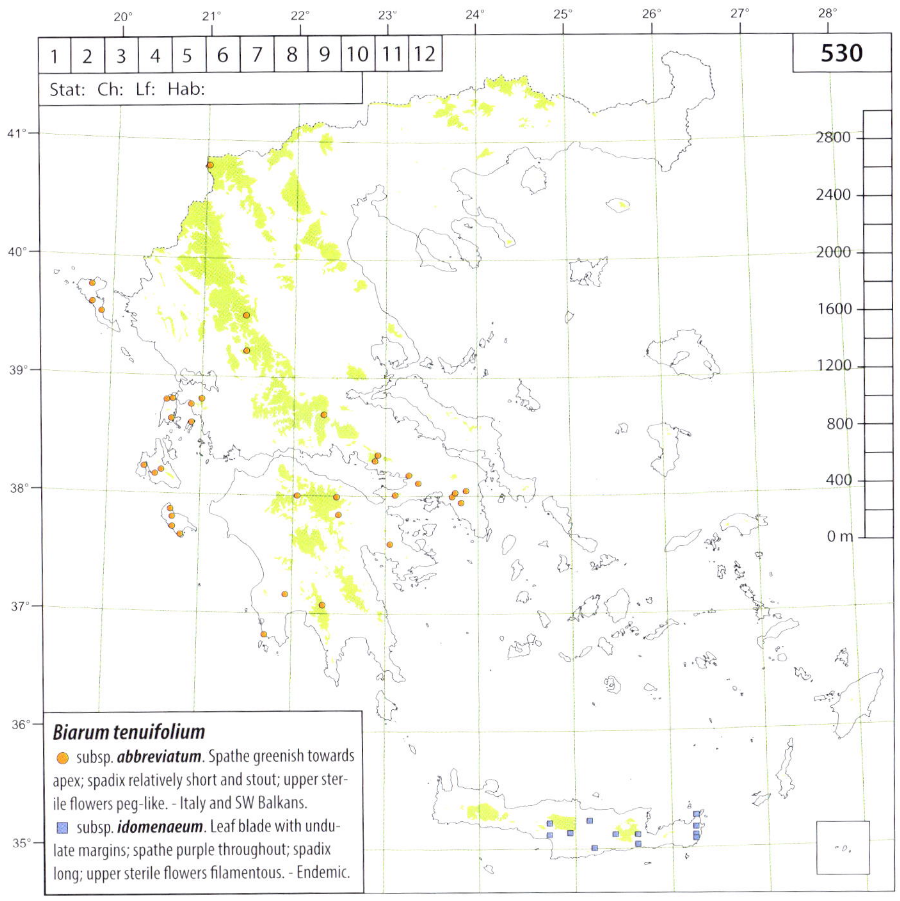

Biarum tenuifolium
● subsp. ***abbreviatum***. Spathe greenish towards apex; spadix relatively short and stout; upper sterile flowers peg-like. - Italy and SW Balkans.
■ subsp. ***idomenaeum***. Leaf blade with undulate margins; spathe purple throughout; spadix long; upper sterile flowers filamentous. - Endemic.

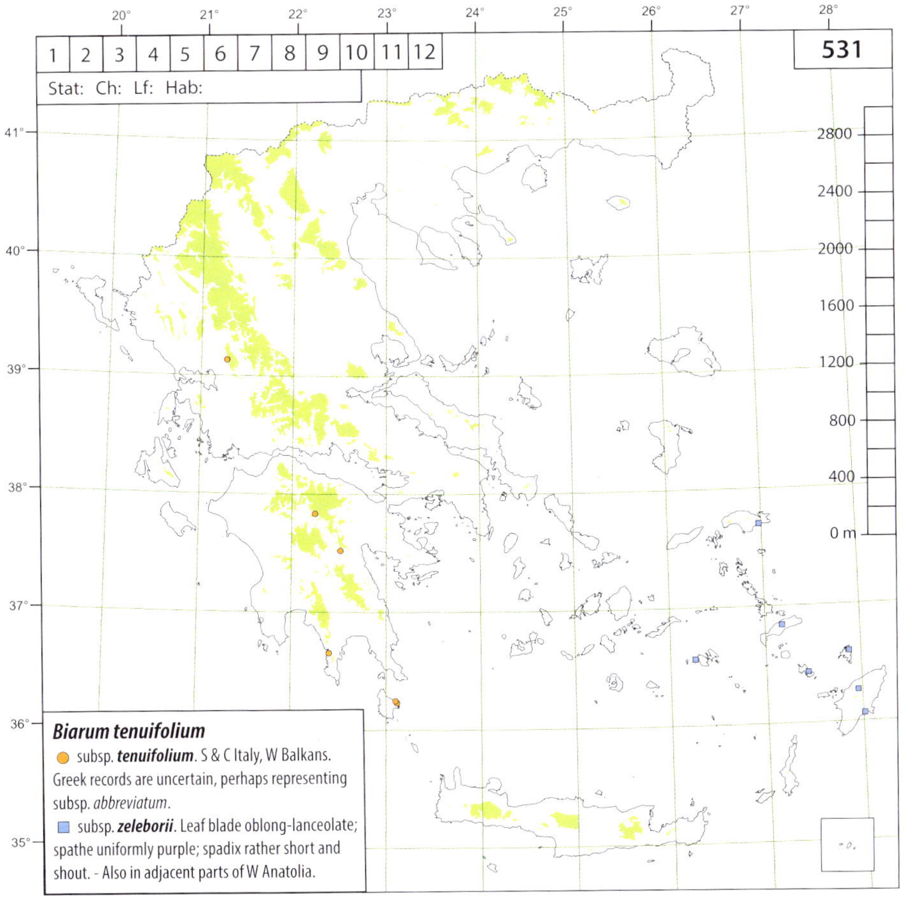

Biarum tenuifolium
● subsp. ***tenuifolium***. S & C Italy, W Balkans. Greek records are uncertain, perhaps representing subsp. *abbreviatum*.
■ subsp. ***zeleborii***. Leaf blade oblong-lanceolate; spathe uniformly purple; spadix rather short and shout. - Also in adjacent parts of W Anatolia.

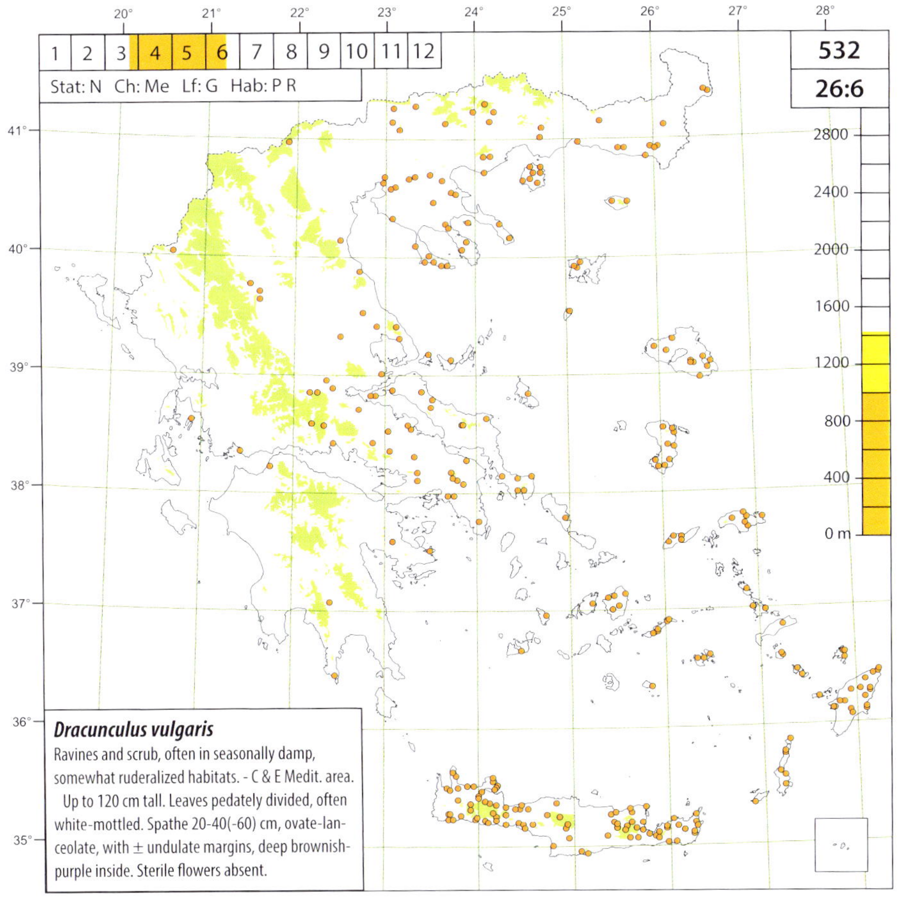

Dracunculus vulgaris
Ravines and scrub, often in seasonally damp, somewhat ruderalized habitats. - C & E Medit. area.
Up to 120 cm tall. Leaves pedately divided, often white-mottled. Spathe 20-40(-60) cm, ovate-lanceolate, with ± undulate margins, deep brownish-purple inside. Sterile flowers absent.

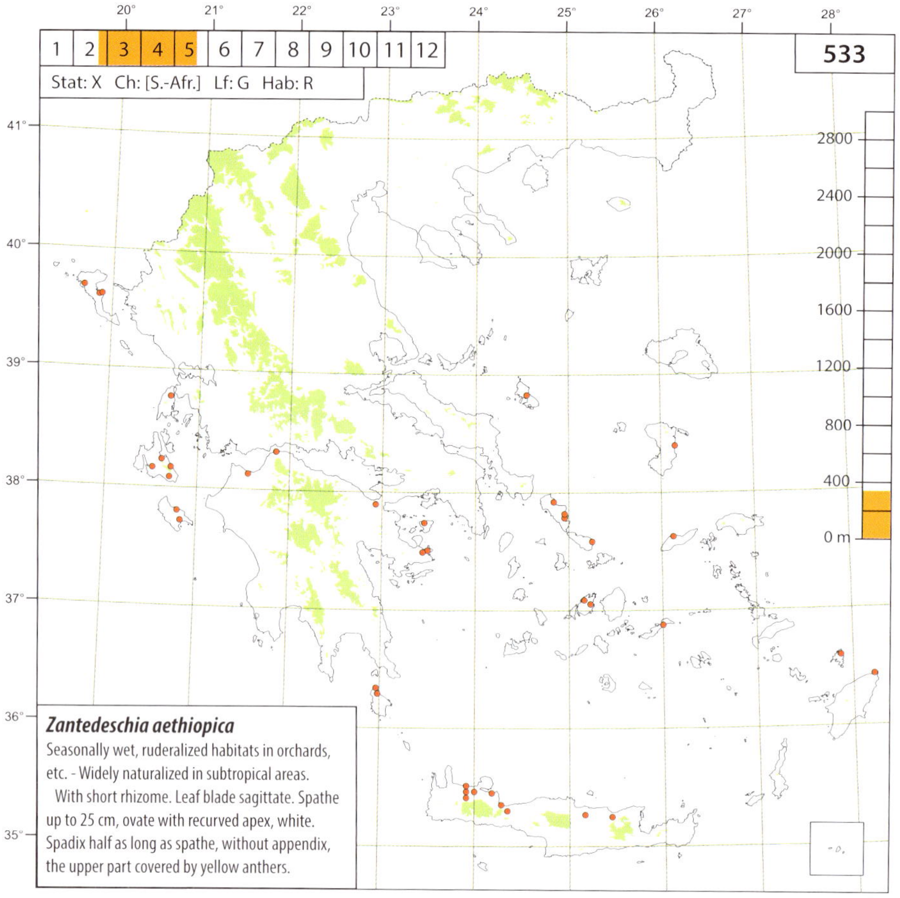

Zantedeschia aethiopica

Seasonally wet, ruderalized habitats in orchards, etc. - Widely naturalized in subtropical areas.

With short rhizome. Leaf blade sagittate. Spathe up to 25 cm, ovate with recurved apex, white. Spadix half as long as spathe, without appendix, the upper part covered by yellow anthers.

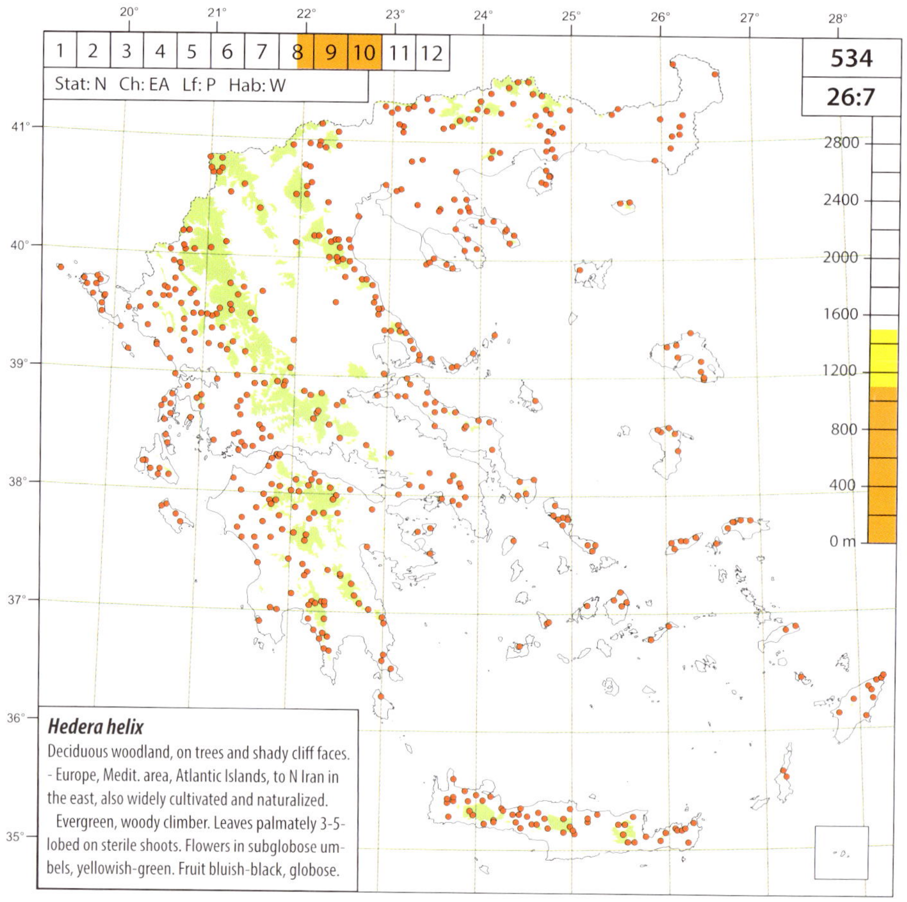

Hedera helix

Deciduous woodland, on trees and shady cliff faces. - Europe, Medit. area, Atlantic Islands, to N Iran in the east, also widely cultivated and naturalized.

Evergreen, woody climber. Leaves palmately 3-5-lobed on sterile shoots. Flowers in subglobose umbels, yellowish-green. Fruit bluish-black, globose.

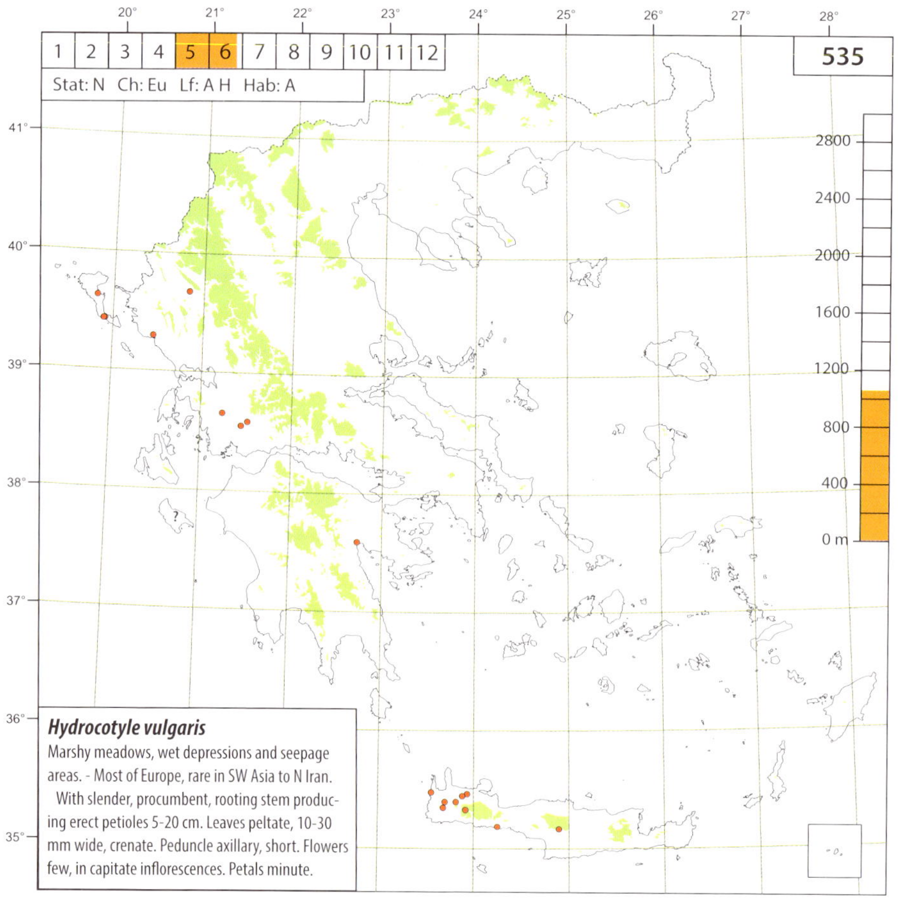

Hydrocotyle vulgaris

Marshy meadows, wet depressions and seepage areas. - Most of Europe, rare in SW Asia to N Iran.

With slender, procumbent, rooting stem producing erect petioles 5-20 cm. Leaves peltate, 10-30 mm wide, crenate. Peduncle axillary, short. Flowers few, in capitate inflorescences. Petals minute.

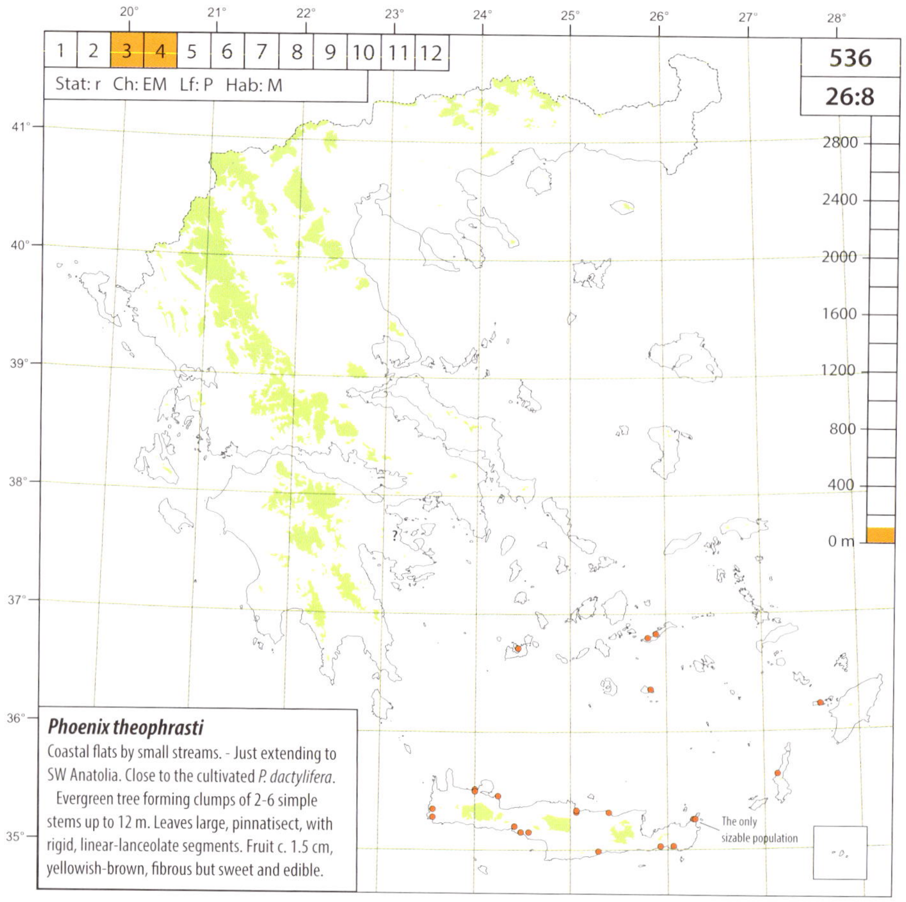

Phoenix theophrasti

Coastal flats by small streams. - Just extending to SW Anatolia. Close to the cultivated *P. dactylifera*.

Evergreen tree forming clumps of 2-6 simple stems up to 12 m. Leaves large, pinnatisect, with rigid, linear-lanceolate segments. Fruit c. 1.5 cm, yellowish-brown, fibrous but sweet and edible.

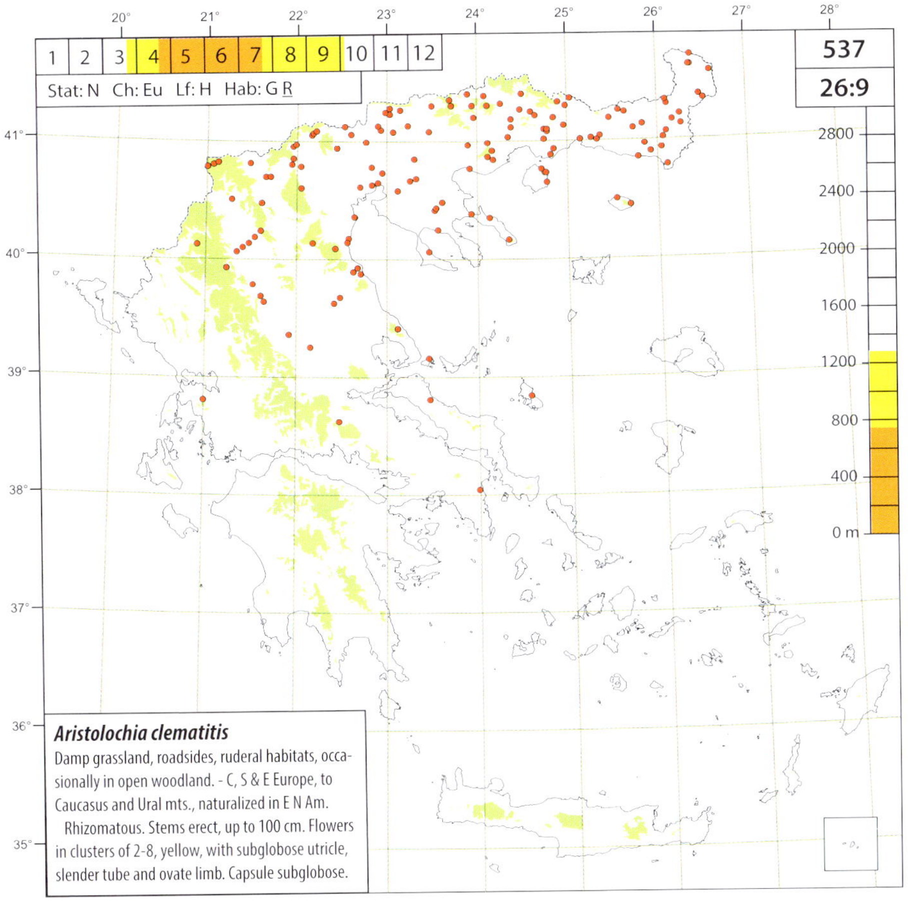

Aristolochia clematitis

Damp grassland, roadsides, ruderal habitats, occasionally in open woodland. - C, S & E Europe, to Caucasus and Ural mts., naturalized in E N Am.

Rhizomatous. Stems erect, up to 100 cm. Flowers in clusters of 2-8, yellow, with subglobose utricle, slender tube and ovate limb. Capsule subglobose.

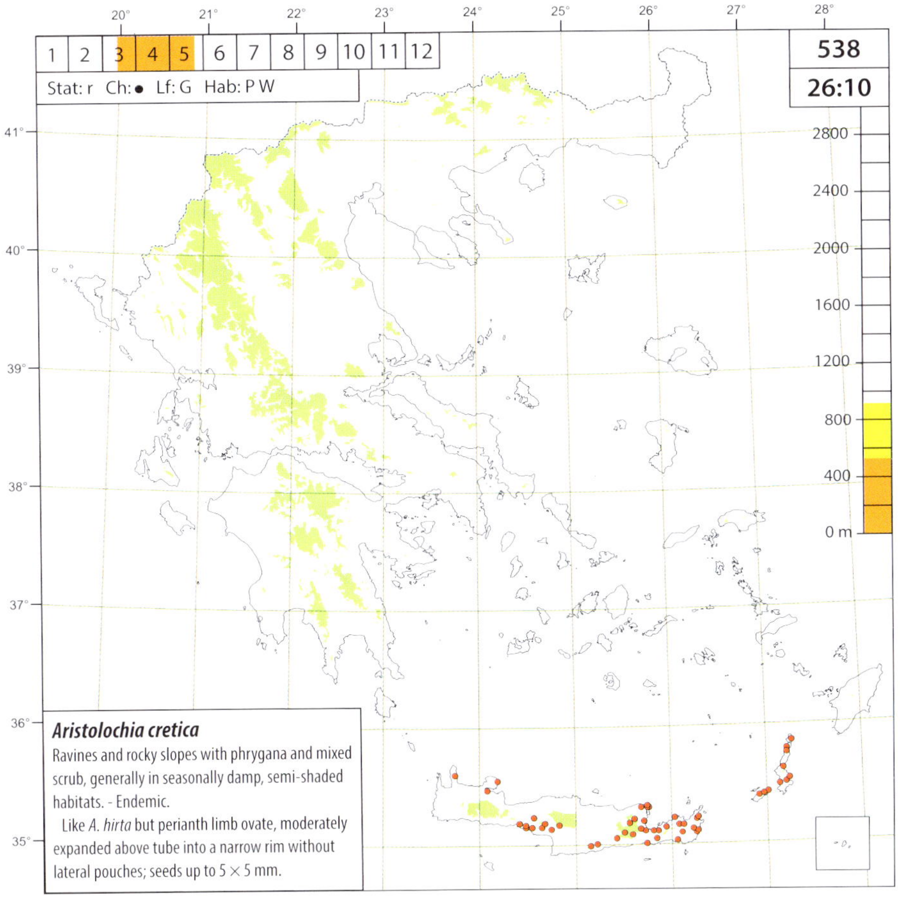

Aristolochia cretica

Ravines and rocky slopes with phrygana and mixed scrub, generally in seasonally damp, semi-shaded habitats. - Endemic.

Like *A. hirta* but perianth limb ovate, moderately expanded above tube into a narrow rim without lateral pouches; seeds up to 5 × 5 mm.

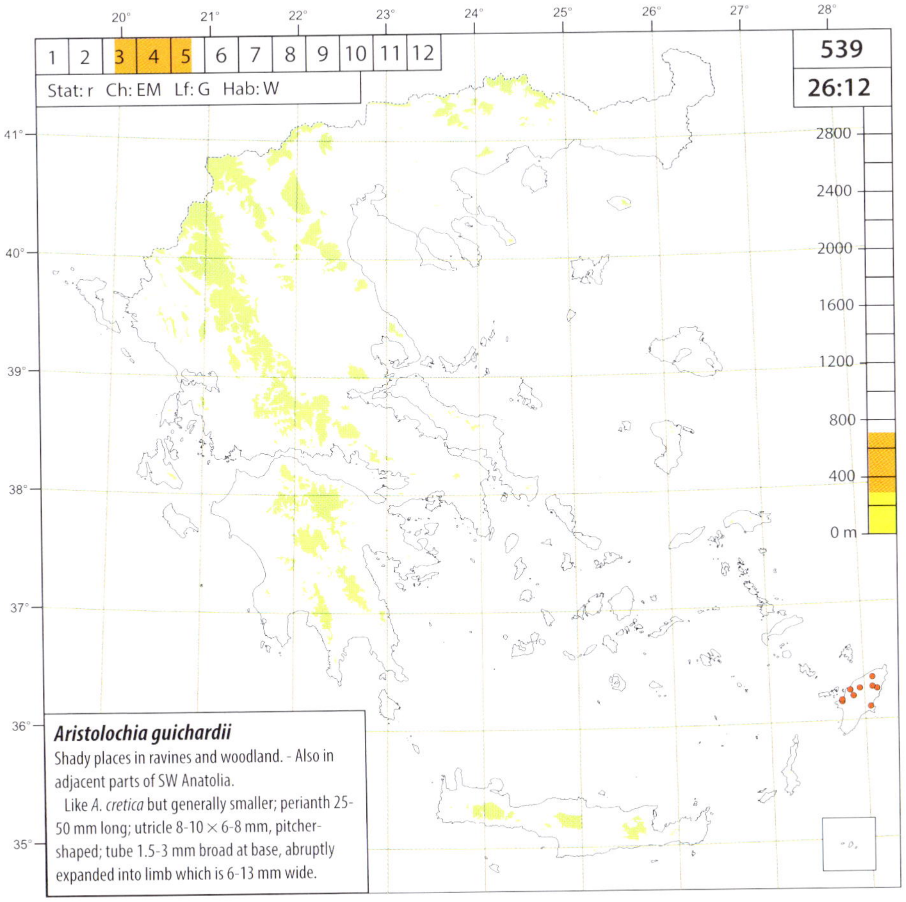

Aristolochia guichardii

Shady places in ravines and woodland. - Also in adjacent parts of SW Anatolia.

Like *A. cretica* but generally smaller; perianth 25-50 mm long; utricle 8-10 × 6-8 mm, pitcher-shaped; tube 1.5-3 mm broad at base, abruptly expanded into limb which is 6-13 mm wide.

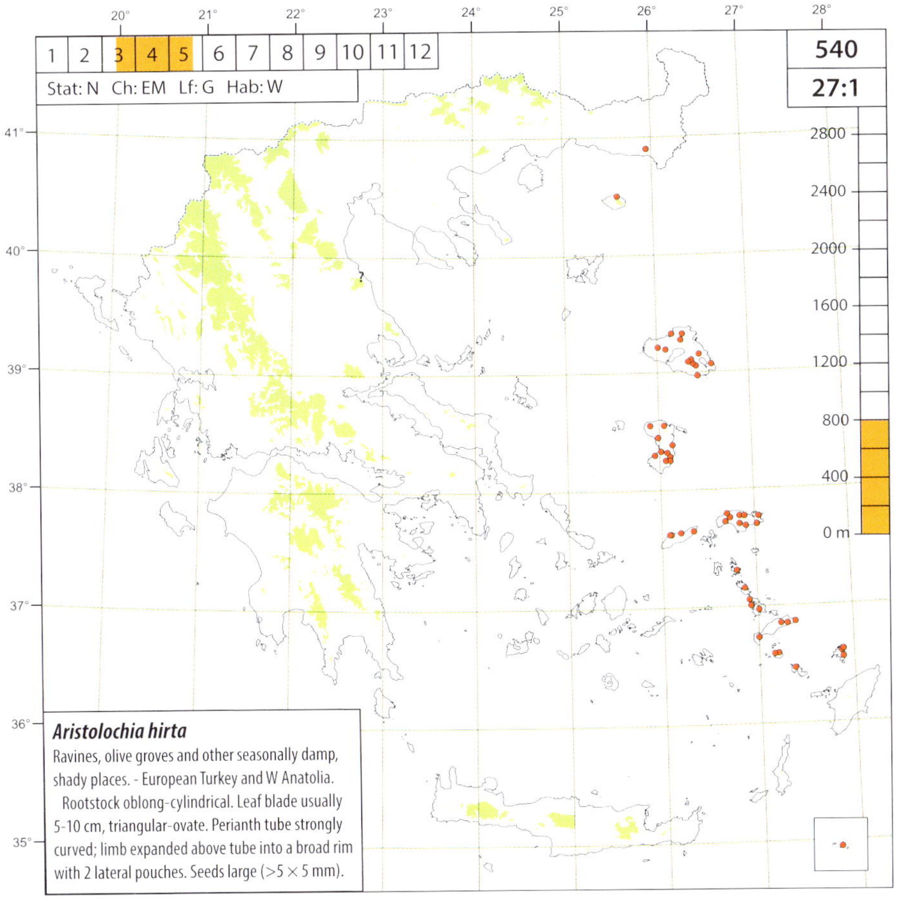

Aristolochia hirta

Ravines, olive groves and other seasonally damp, shady places. - European Turkey and W Anatolia.

Rootstock oblong-cylindrical. Leaf blade usually 5-10 cm, triangular-ovate. Perianth tube strongly curved; limb expanded above tube into a broad rim with 2 lateral pouches. Seeds large (>5 × 5 mm).

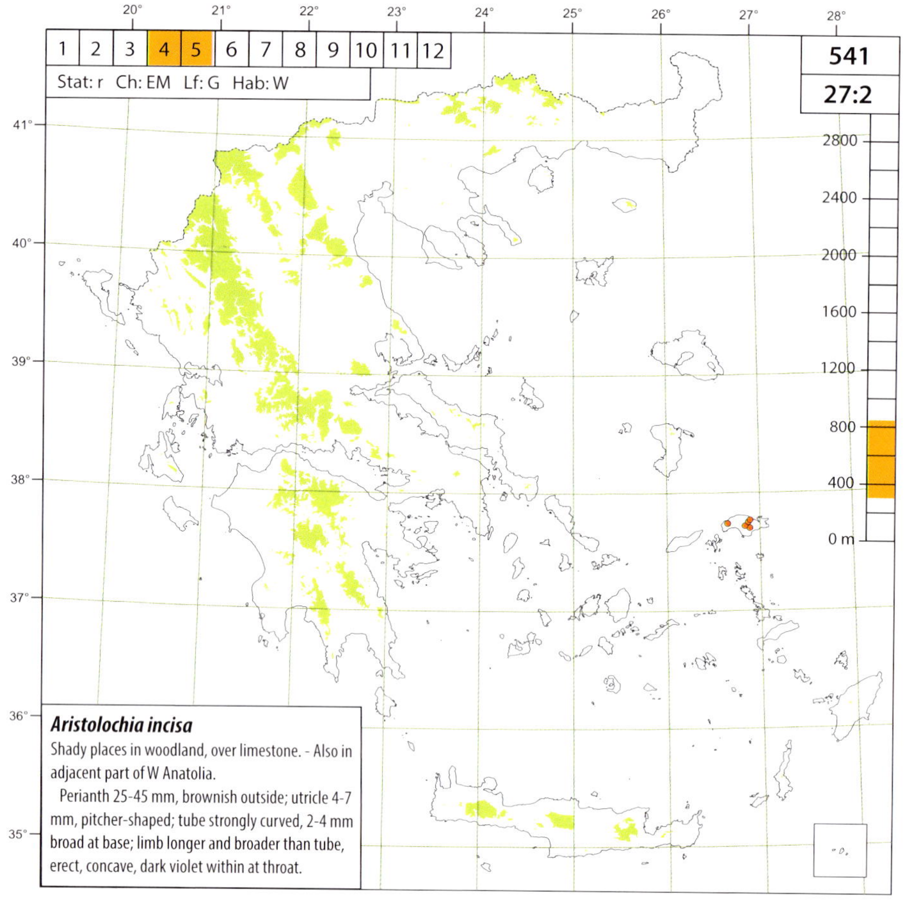

Aristolochia incisa

Shady places in woodland, over limestone. - Also in adjacent part of W Anatolia.
Perianth 25-45 mm, brownish outside; utricle 4-7 mm, pitcher-shaped; tube strongly curved, 2-4 mm broad at base; limb longer and broader than tube, erect, concave, dark violet within at throat.

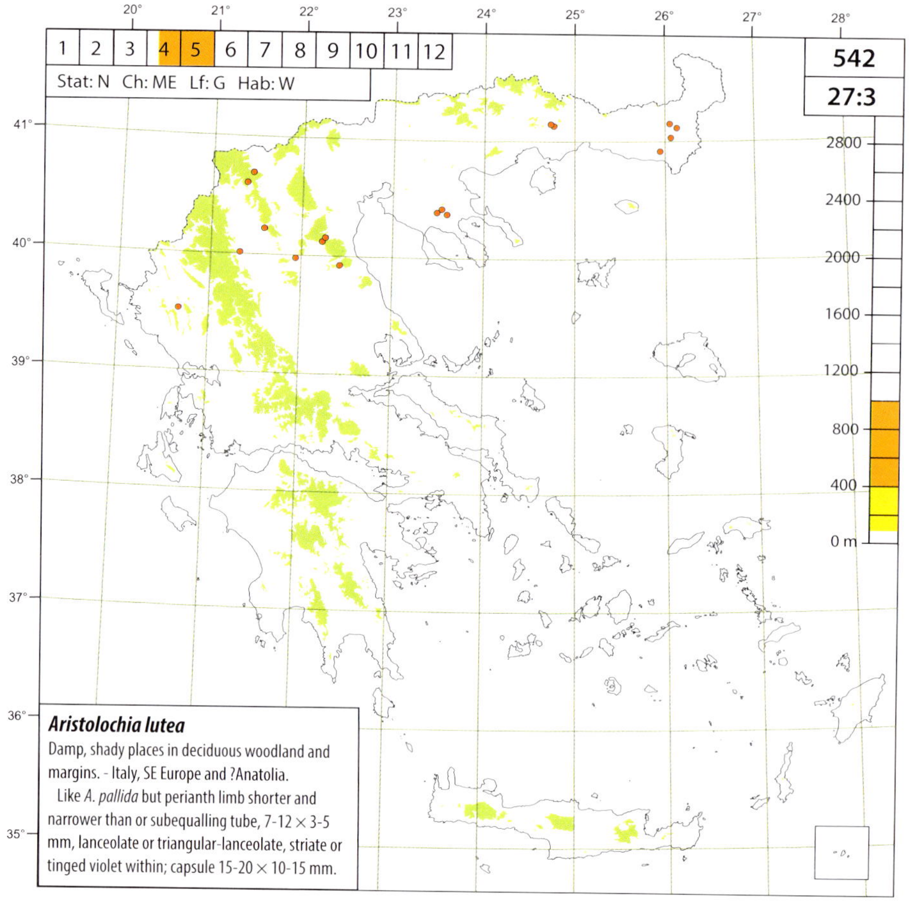

Aristolochia lutea

Damp, shady places in deciduous woodland and margins. - Italy, SE Europe and ?Anatolia.
Like *A. pallida* but perianth limb shorter and narrower than or subequalling tube, 7-12 × 3-5 mm, lanceolate or triangular-lanceolate, striate or tinged violet within; capsule 15-20 × 10-15 mm.

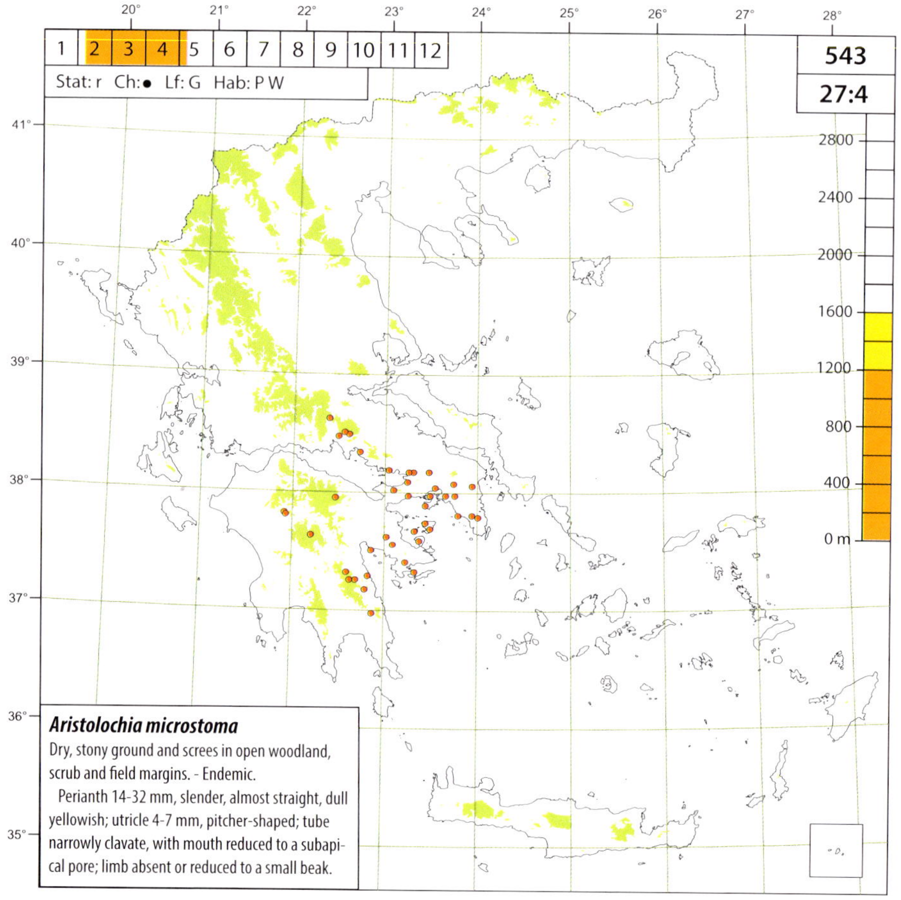

Aristolochia microstoma

Dry, stony ground and screes in open woodland, scrub and field margins. - Endemic.
Perianth 14-32 mm, slender, almost straight, dull yellowish; utricle 4-7 mm, pitcher-shaped; tube narrowly clavate, with mouth reduced to a subapical pore; limb absent or reduced to a small beak.

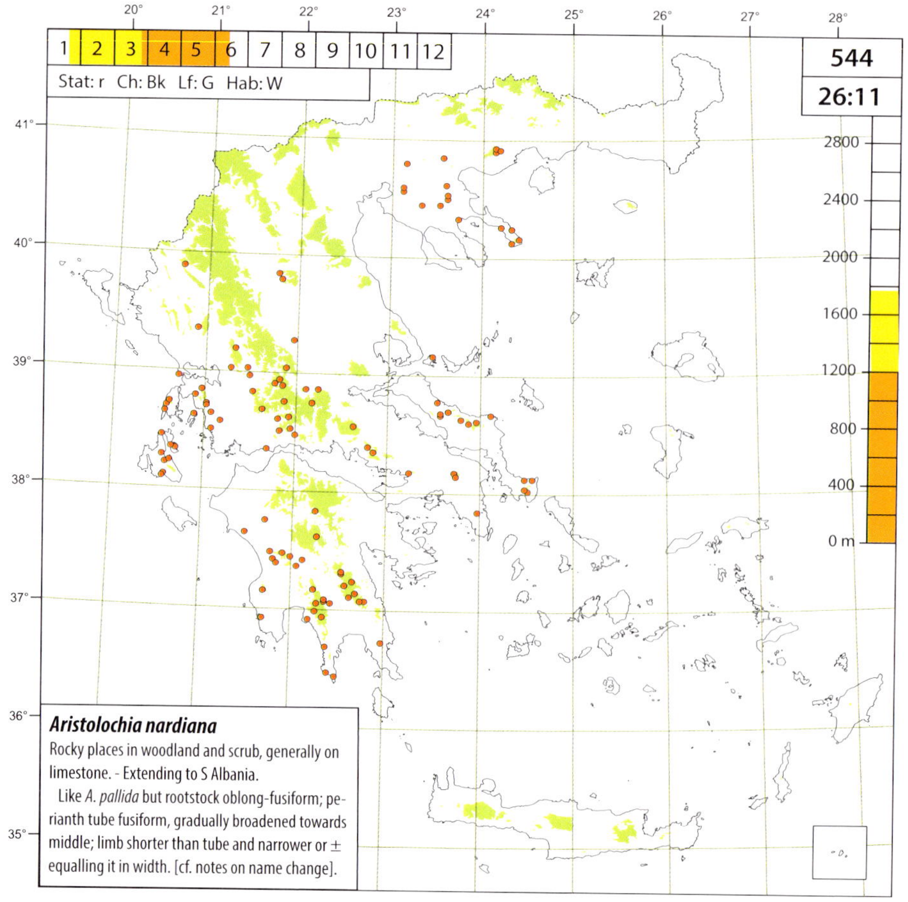

Aristolochia nardiana

Rocky places in woodland and scrub, generally on limestone. - Extending to S Albania.
Like *A. pallida* but rootstock oblong-fusiform; perianth tube fusiform, gradually broadened towards middle; limb shorter than tube and narrower or ± equalling it in width. [cf. notes on name change].

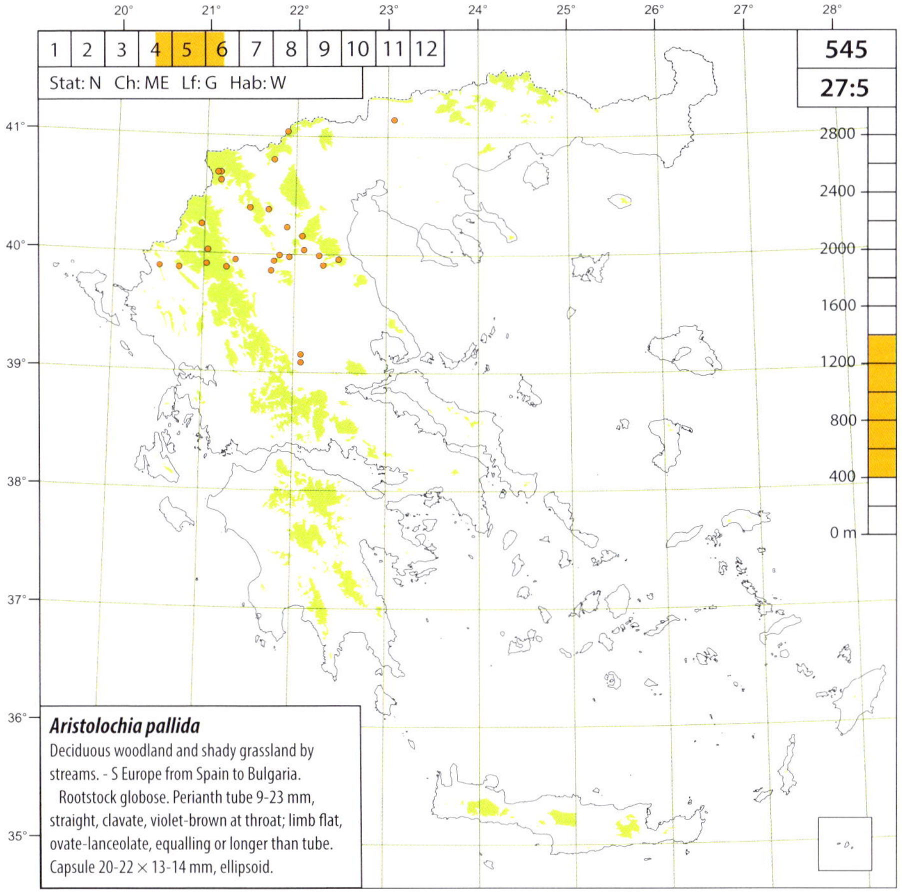

Aristolochia pallida
Deciduous woodland and shady grassland by streams. - S Europe from Spain to Bulgaria.
Rootstock globose. Perianth tube 9-23 mm, straight, clavate, violet-brown at throat; limb flat, ovate-lanceolate, equalling or longer than tube. Capsule 20-22 × 13-14 mm, ellipsoid.

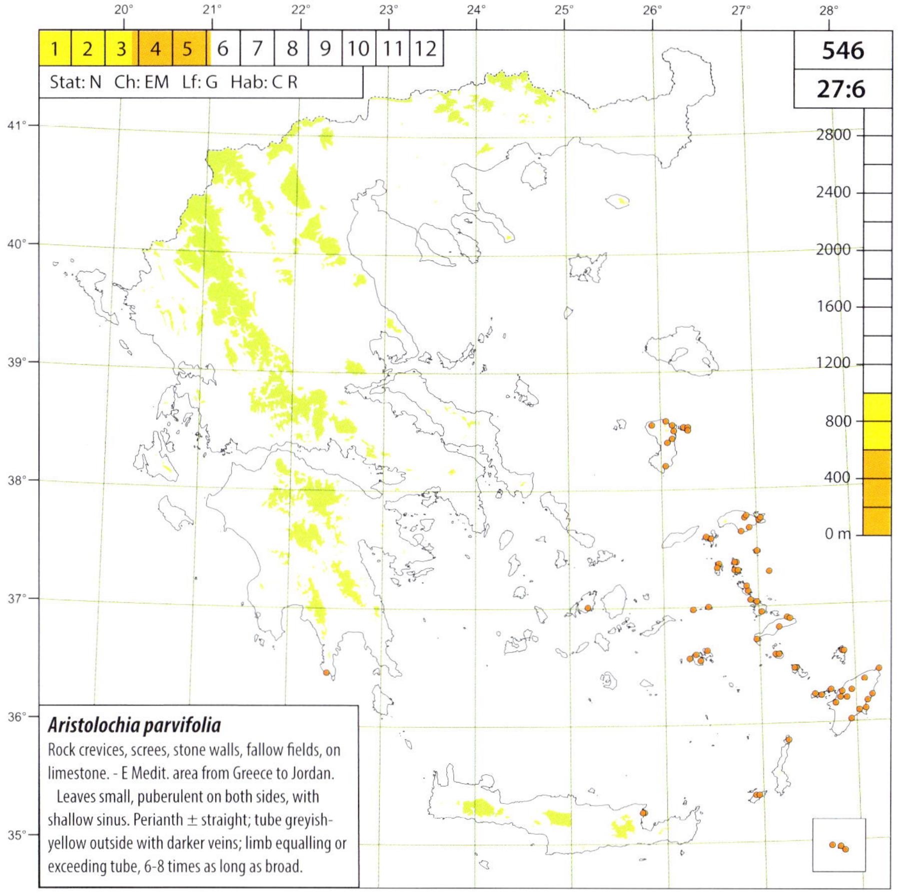

Aristolochia parvifolia
Rock crevices, screes, stone walls, fallow fields, on limestone. - E Medit. area from Greece to Jordan.
Leaves small, puberulent on both sides, with shallow sinus. Perianth ± straight; tube greyish-yellow outside with darker veins; limb equalling or exceeding tube, 6-8 times as long as broad.

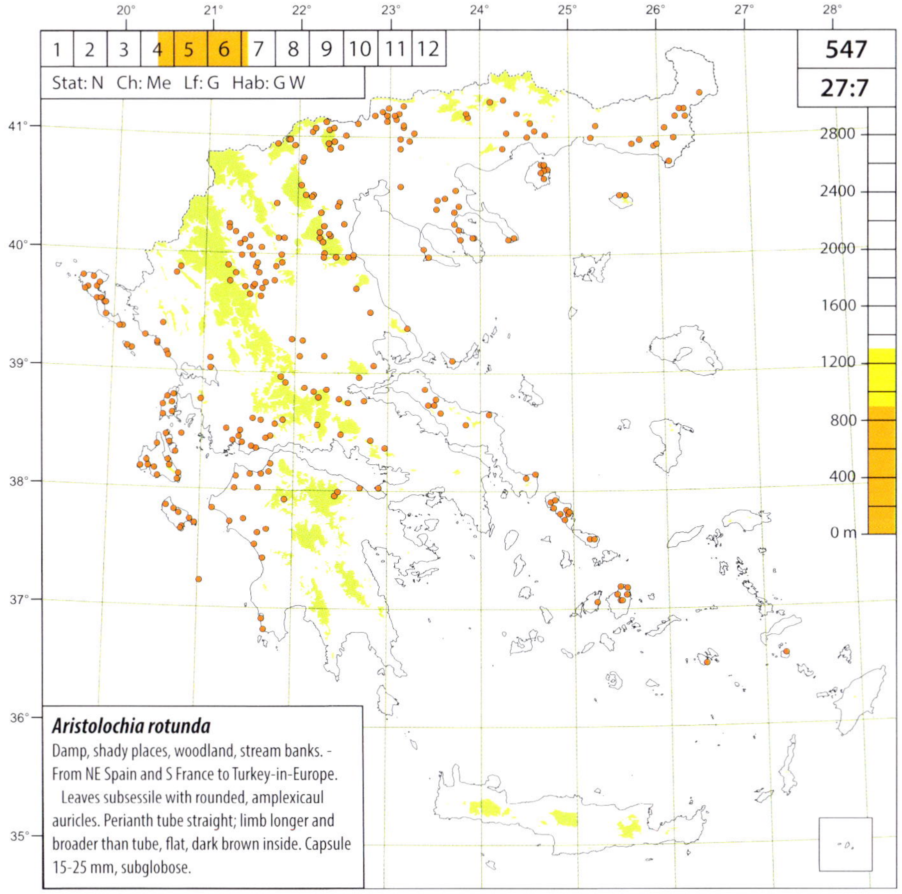

Aristolochia rotunda
Damp, shady places, woodland, stream banks. - From NE Spain and S France to Turkey-in-Europe.
Leaves subsessile with rounded, amplexicaul auricles. Perianth tube straight; limb longer and broader than tube, flat, dark brown inside. Capsule 15-25 mm, subglobose.

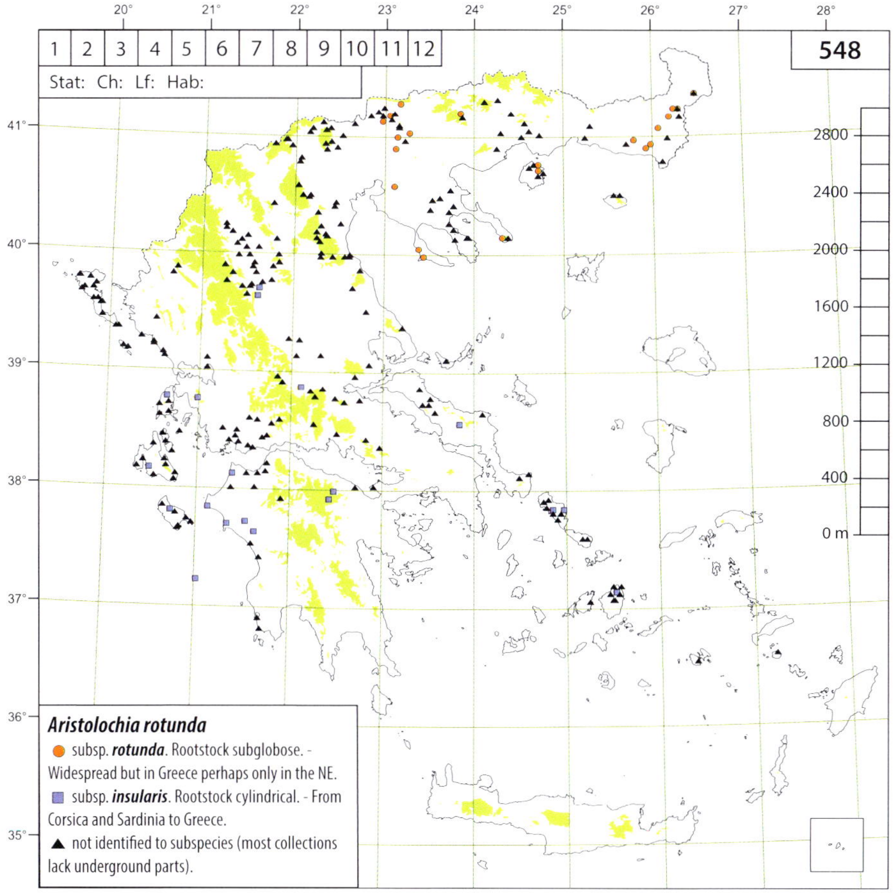

Aristolochia rotunda
● subsp. ***rotunda***. Rootstock subglobose. - Widespread but in Greece perhaps only in the NE.
■ subsp. ***insularis***. Rootstock cylindrical. - From Corsica and Sardinia to Greece.
▲ not identified to subspecies (most collections lack underground parts).

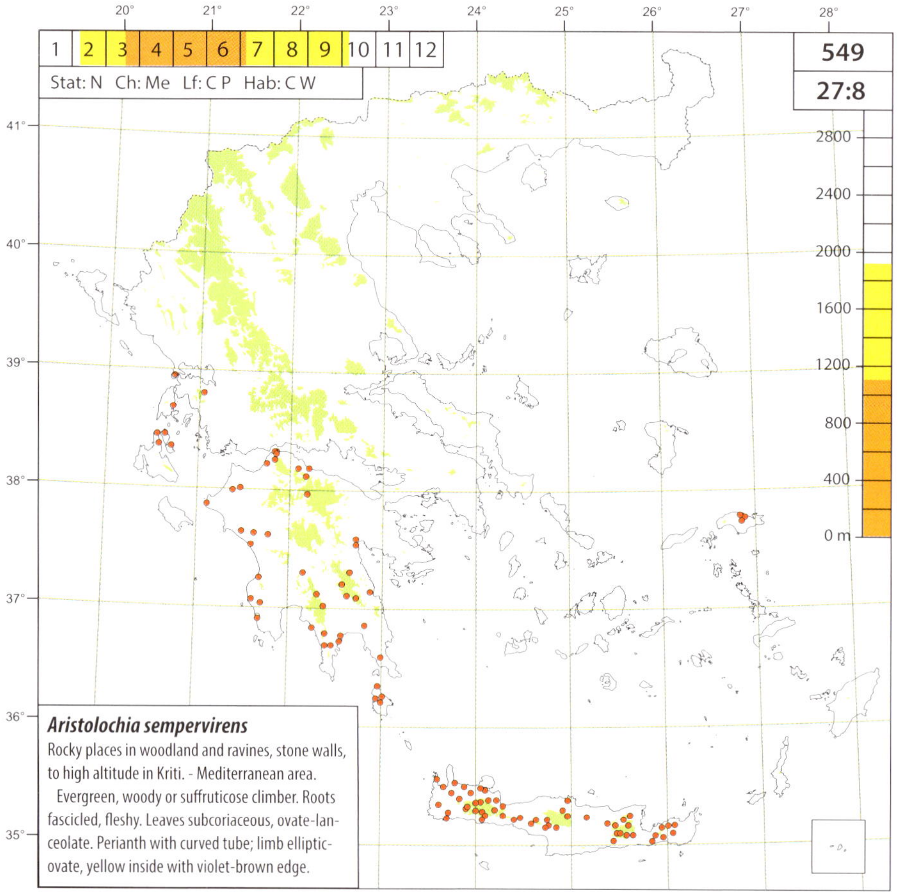

Aristolochia sempervirens
Rocky places in woodland and ravines, stone walls, to high altitude in Kriti. - Mediterranean area.
Evergreen, woody or suffruticose climber. Roots fascicled, fleshy. Leaves subcoriaceous, ovate-lanceolate. Perianth with curved tube; limb elliptic-ovate, yellow inside with violet-brown edge.

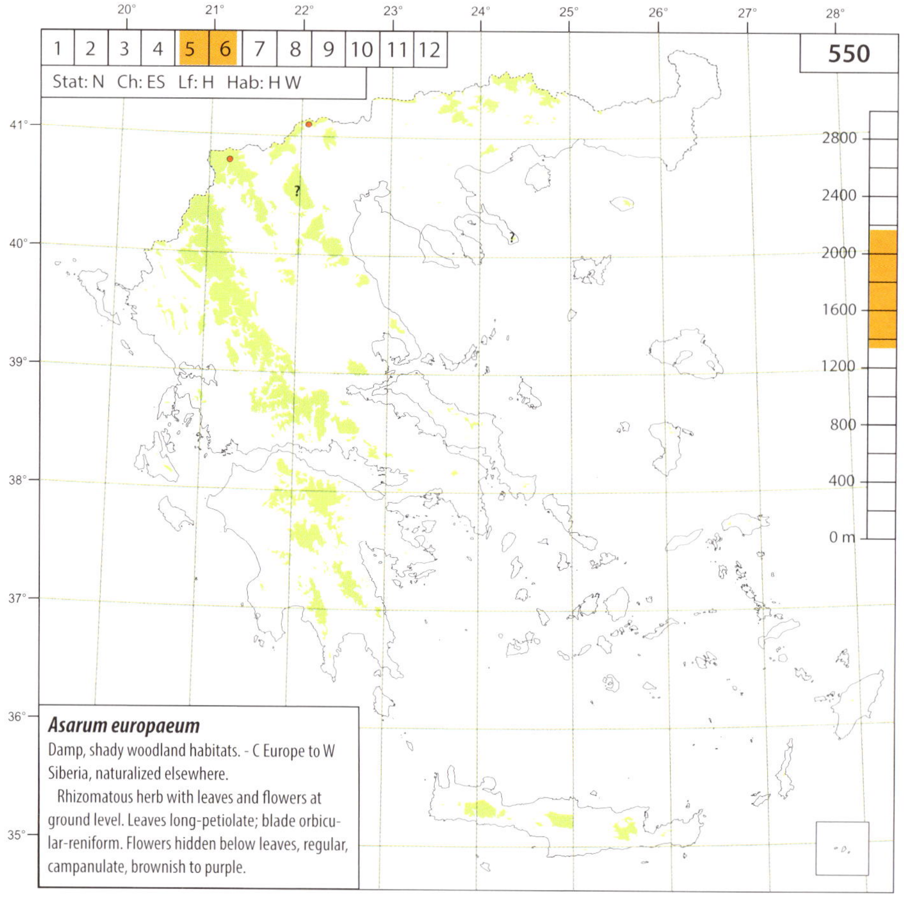

Asarum europaeum
Damp, shady woodland habitats. - C Europe to W Siberia, naturalized elsewhere.
Rhizomatous herb with leaves and flowers at ground level. Leaves long-petiolate; blade orbicular-reniform. Flowers hidden below leaves, regular, campanulate, brownish to purple.

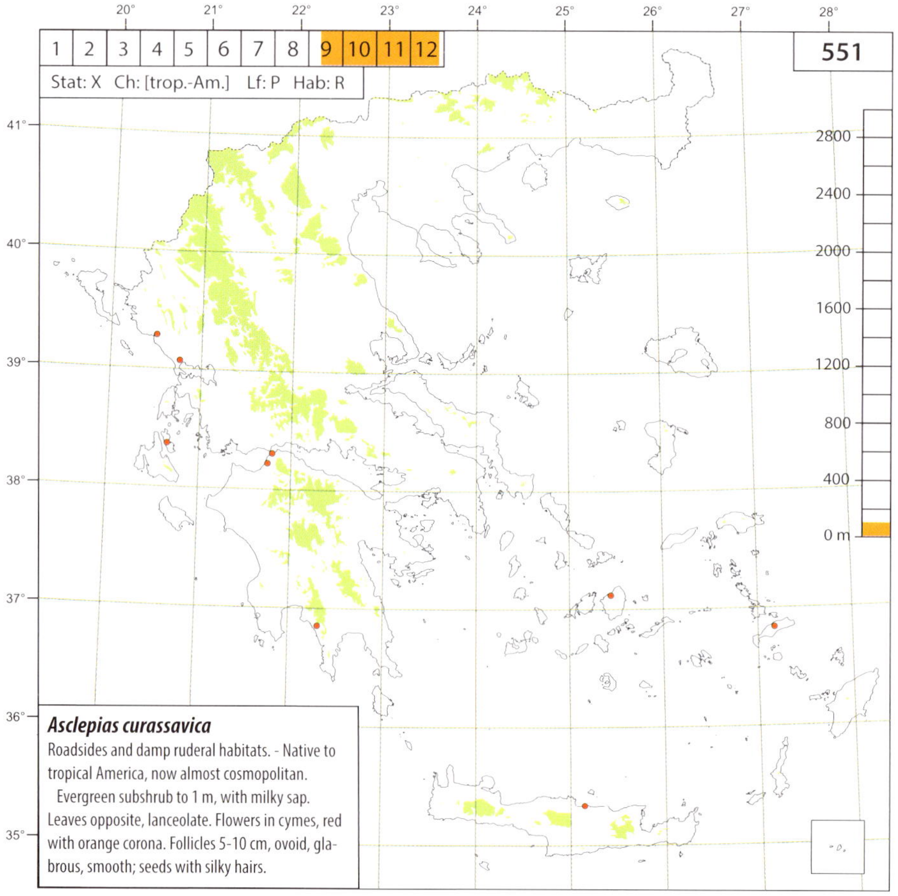

Asclepias curassavica
Roadsides and damp ruderal habitats. - Native to tropical America, now almost cosmopolitan.
Evergreen subshrub to 1 m, with milky sap. Leaves opposite, lanceolate. Flowers in cymes, red with orange corona. Follicles 5-10 cm, ovoid, glabrous, smooth; seeds with silky hairs.

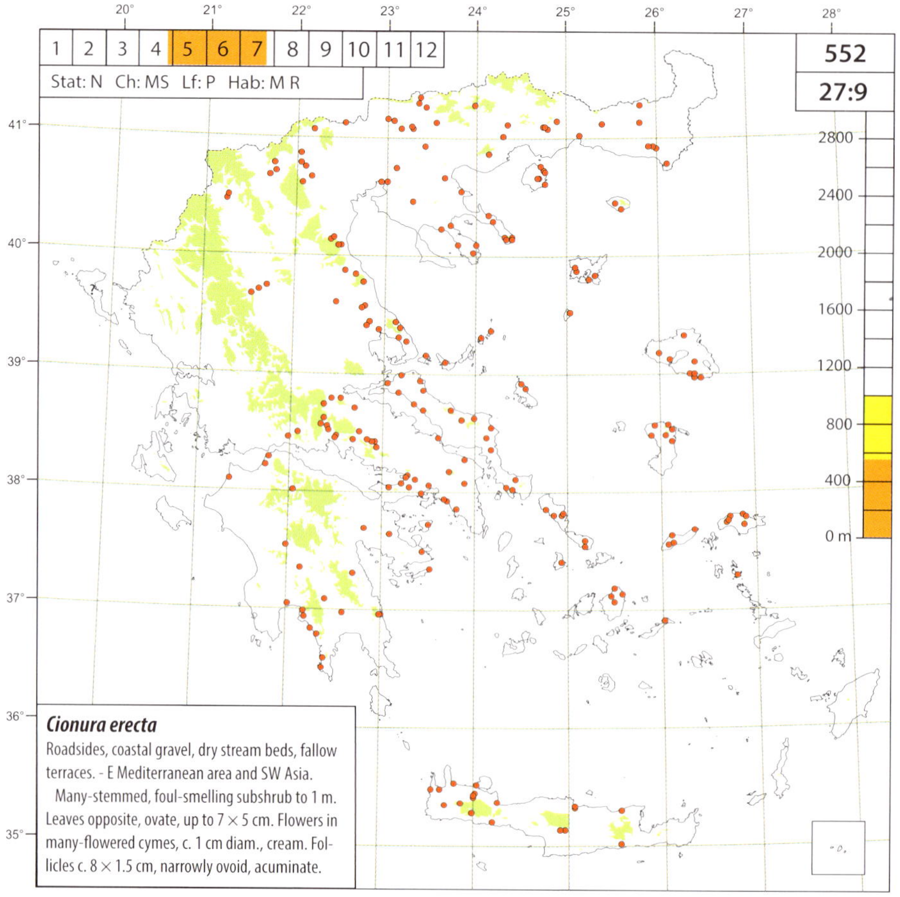

Cionura erecta
Roadsides, coastal gravel, dry stream beds, fallow terraces. - E Mediterranean area and SW Asia.
Many-stemmed, foul-smelling subshrub to 1 m. Leaves opposite, ovate, up to 7 × 5 cm. Flowers in many-flowered cymes, c. 1 cm diam., cream. Follicles c. 8 × 1.5 cm, narrowly ovoid, acuminate.

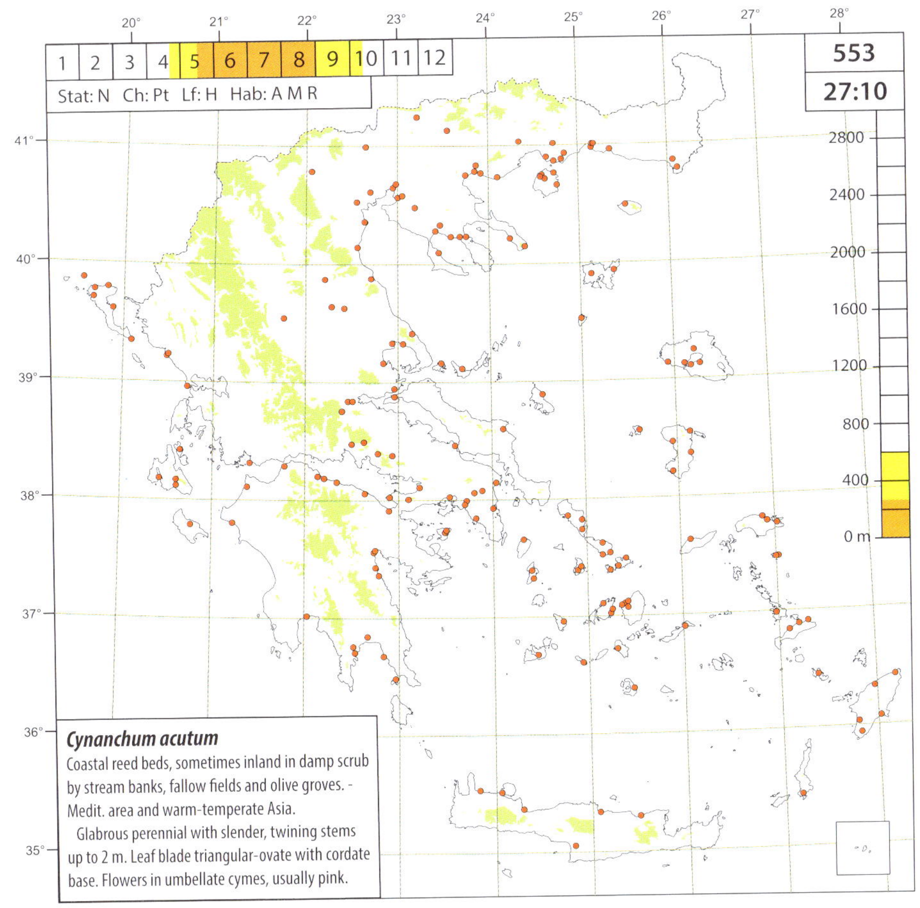

Cynanchum acutum

Coastal reed beds, sometimes inland in damp scrub by stream banks, fallow fields and olive groves. - Medit. area and warm-temperate Asia.

Glabrous perennial with slender, twining stems up to 2 m. Leaf blade triangular-ovate with cordate base. Flowers in umbellate cymes, usually pink.

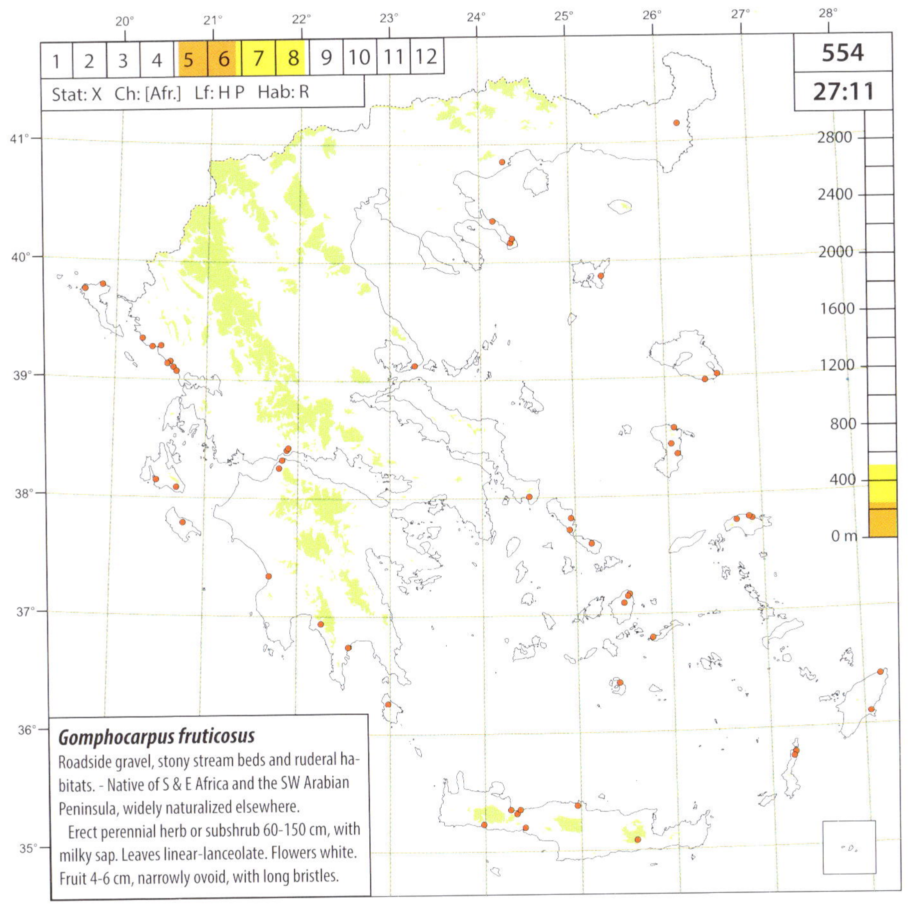

Gomphocarpus fruticosus

Roadside gravel, stony stream beds and ruderal habitats. - Native of S & E Africa and the SW Arabian Peninsula, widely naturalized elsewhere.

Erect perennial herb or subshrub 60-150 cm, with milky sap. Leaves linear-lanceolate. Flowers white. Fruit 4-6 cm, narrowly ovoid, with long bristles.

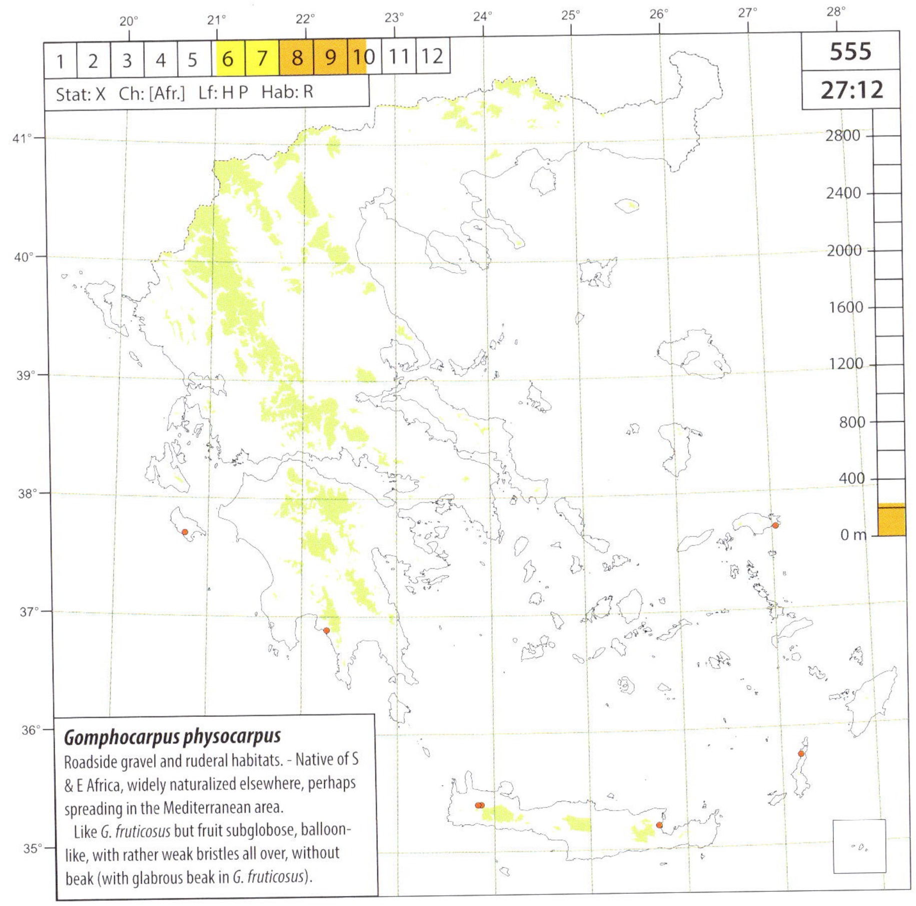

Gomphocarpus physocarpus

Roadside gravel and ruderal habitats. - Native of S & E Africa, widely naturalized elsewhere, perhaps spreading in the Mediterranean area.

Like *G. fruticosus* but fruit subglobose, balloon-like, with rather weak bristles all over, without beak (with glabrous beak in *G. fruticosus*).

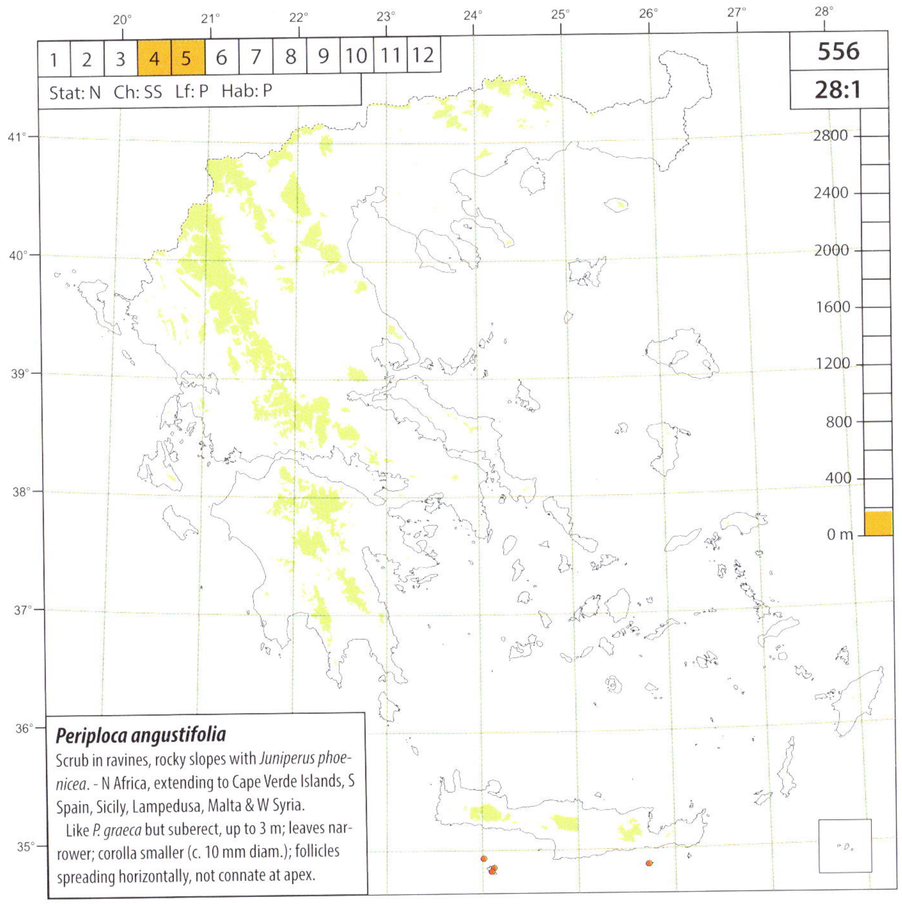

Periploca angustifolia

Scrub in ravines, rocky slopes with *Juniperus phoenicea*. - N Africa, extending to Cape Verde Islands, S Spain, Sicily, Lampedusa, Malta & W Syria.

Like *P. graeca* but suberect, up to 3 m; leaves narrower; corolla smaller (c. 10 mm diam.); follicles spreading horizontally, not connate at apex.

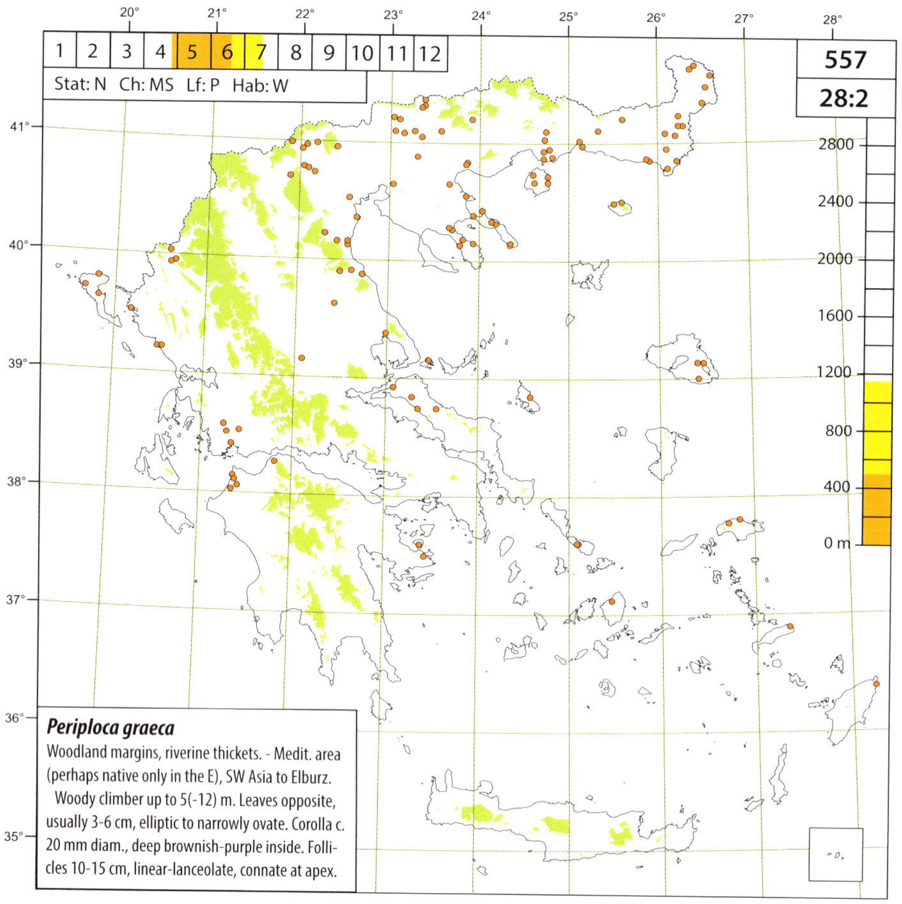

Periploca graeca

Woodland margins, riverine thickets. - Medit. area (perhaps native only in the E), SW Asia to Elburz.

Woody climber up to 5(-12) m. Leaves opposite, usually 3-6 cm, elliptic to narrowly ovate. Corolla c. 20 mm diam., deep brownish-purple inside. Follicles 10-15 cm, linear-lanceolate, connate at apex.

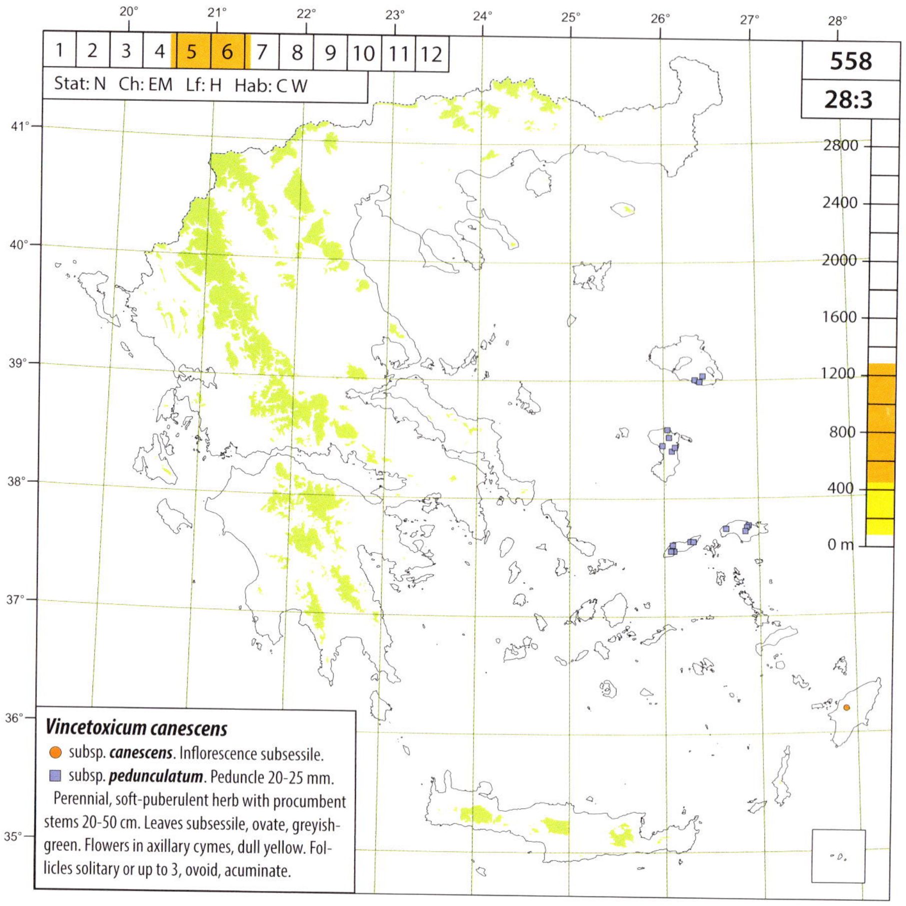

Vincetoxicum canescens

● subsp. ***canescens***. Inflorescence subsessile.

■ subsp. ***pedunculatum***. Peduncle 20-25 mm.

Perennial, soft-puberulent herb with procumbent stems 20-50 cm. Leaves subsessile, ovate, greyish-green. Flowers in axillary cymes, dull yellow. Follicles solitary or up to 3, ovoid, acuminate.

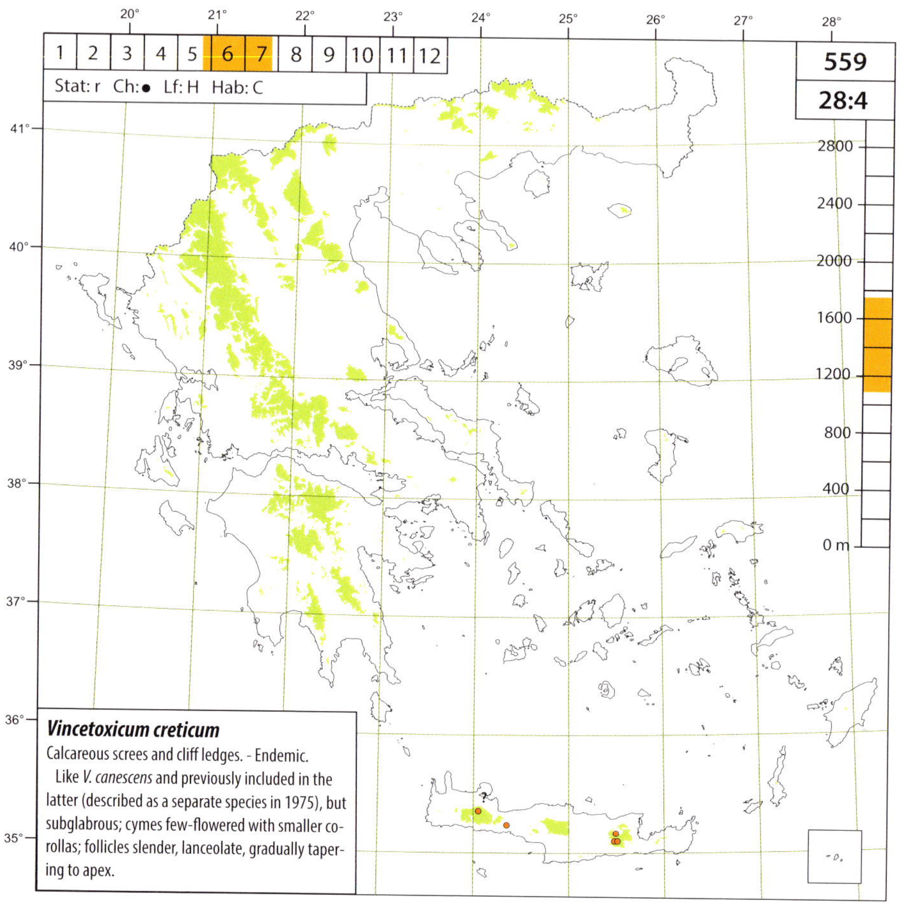

Vincetoxicum creticum

Calcareous screes and cliff ledges. - Endemic.

Like *V. canescens* and previously included in the latter (described as a separate species in 1975), but subglabrous; cymes few-flowered with smaller corollas; follicles slender, lanceolate, gradually tapering to apex.

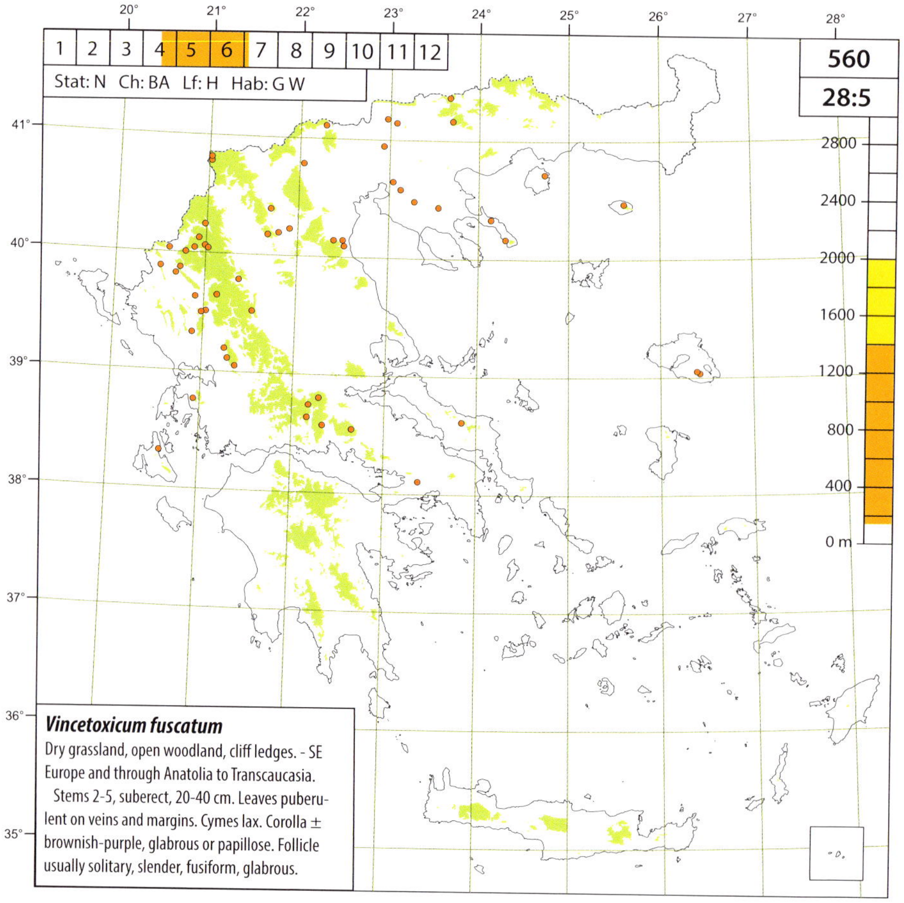

Vincetoxicum fuscatum

Dry grassland, open woodland, cliff ledges. - SE Europe and through Anatolia to Transcaucasia.

Stems 2-5, suberect, 20-40 cm. Leaves puberulent on veins and margins. Cymes lax. Corolla ± brownish-purple, glabrous or papillose. Follicle usually solitary, slender, fusiform, glabrous.

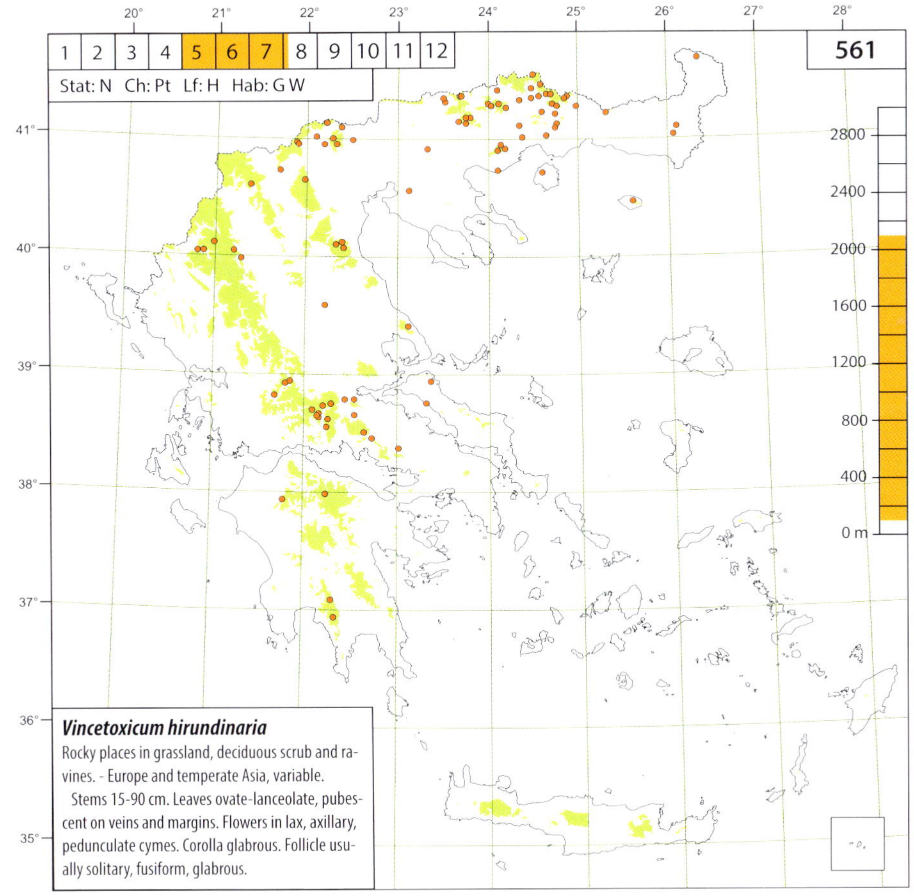

Vincetoxicum hirundinaria

Rocky places in grassland, deciduous scrub and ravines. - Europe and temperate Asia, variable.

Stems 15-90 cm. Leaves ovate-lanceolate, pubescent on veins and margins. Flowers in lax, axillary, pedunculate cymes. Corolla glabrous. Follicle usually solitary, fusiform, glabrous.

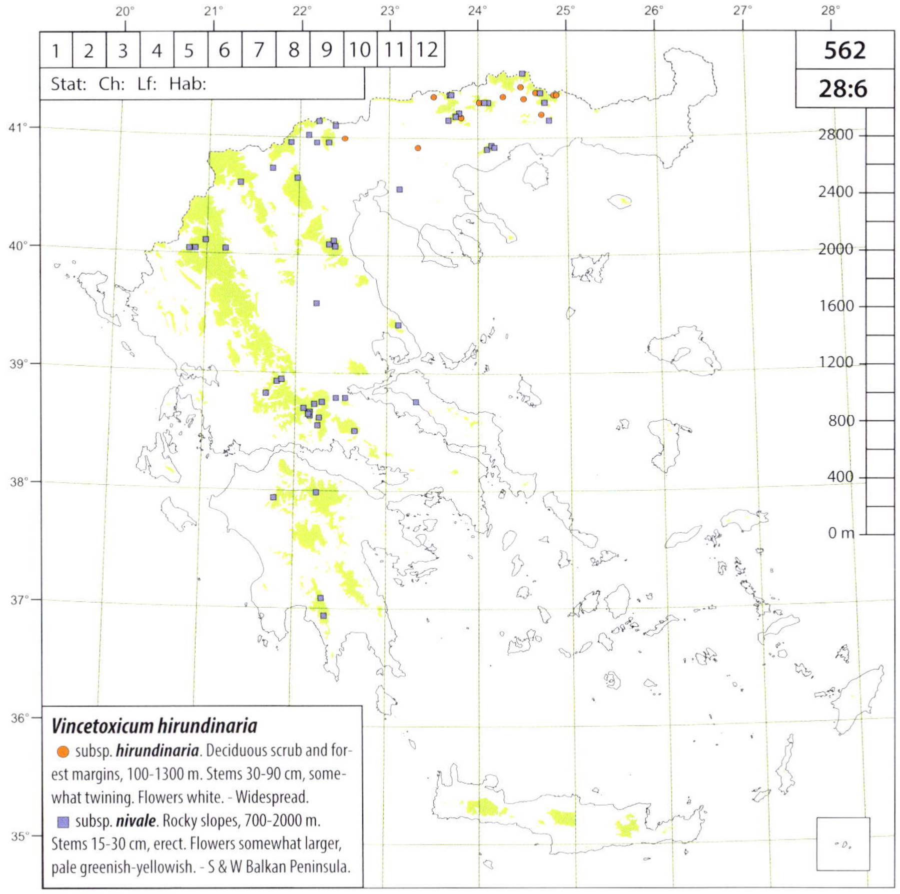

Vincetoxicum hirundinaria

● subsp. ***hirundinaria***. Deciduous scrub and forest margins, 100-1300 m. Stems 30-90 cm, somewhat twining. Flowers white. - Widespread.

■ subsp. ***nivale***. Rocky slopes, 700-2000 m. Stems 15-30 cm, erect. Flowers somewhat larger, pale greenish-yellowish. - S & W Balkan Peninsula.

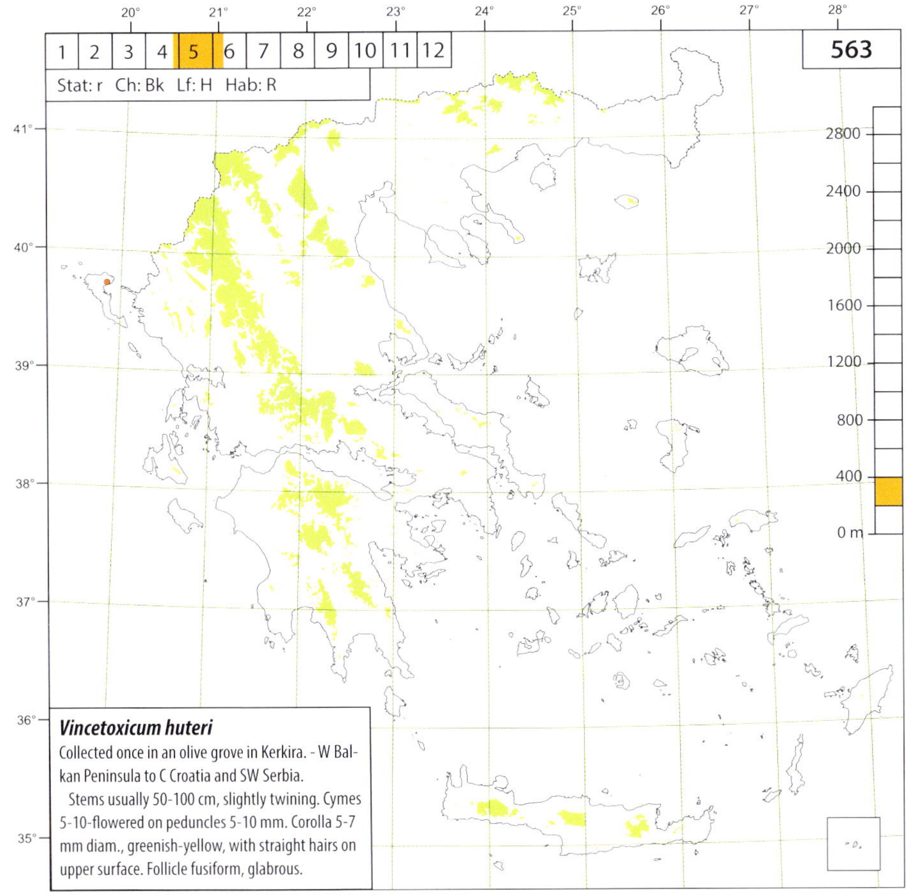

Vincetoxicum huteri

Collected once in an olive grove in Kerkira. - W Balkan Peninsula to C Croatia and SW Serbia.

Stems usually 50-100 cm, slightly twining. Cymes 5-10-flowered on peduncles 5-10 mm. Corolla 5-7 mm diam., greenish-yellow, with straight hairs on upper surface. Follicle fusiform, glabrous.

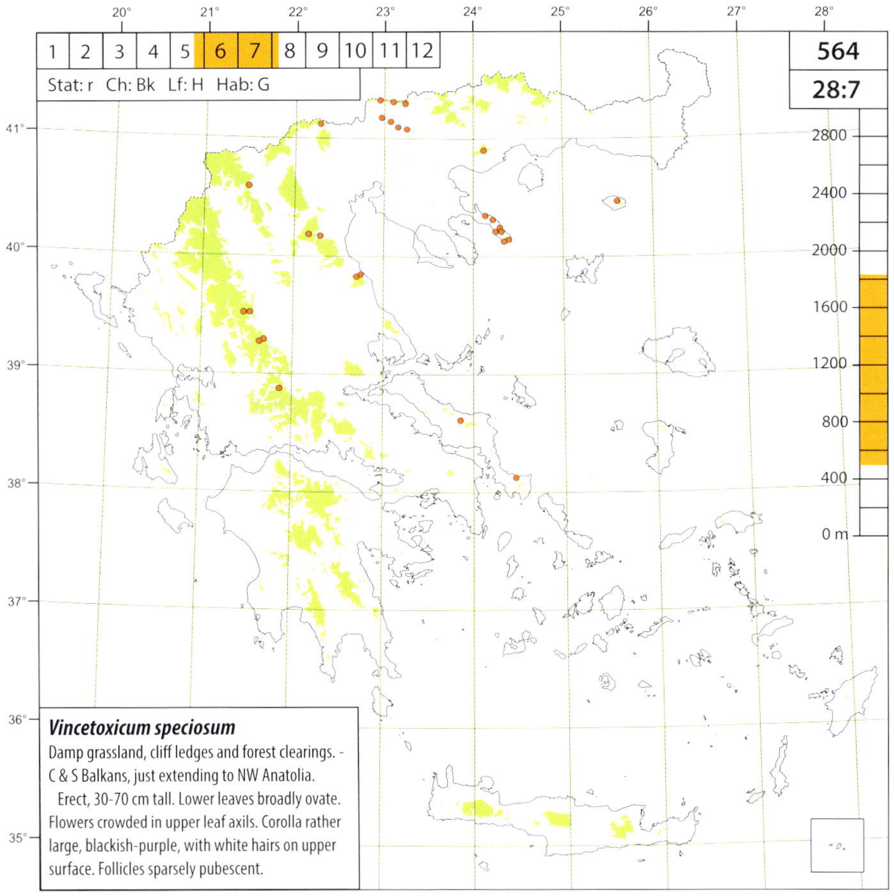

Vincetoxicum speciosum

Damp grassland, cliff ledges and forest clearings. - C & S Balkans, just extending to NW Anatolia.

Erect, 30-70 cm tall. Lower leaves broadly ovate. Flowers crowded in upper leaf axils. Corolla rather large, blackish-purple, with white hairs on upper surface. Follicles sparsely pubescent.

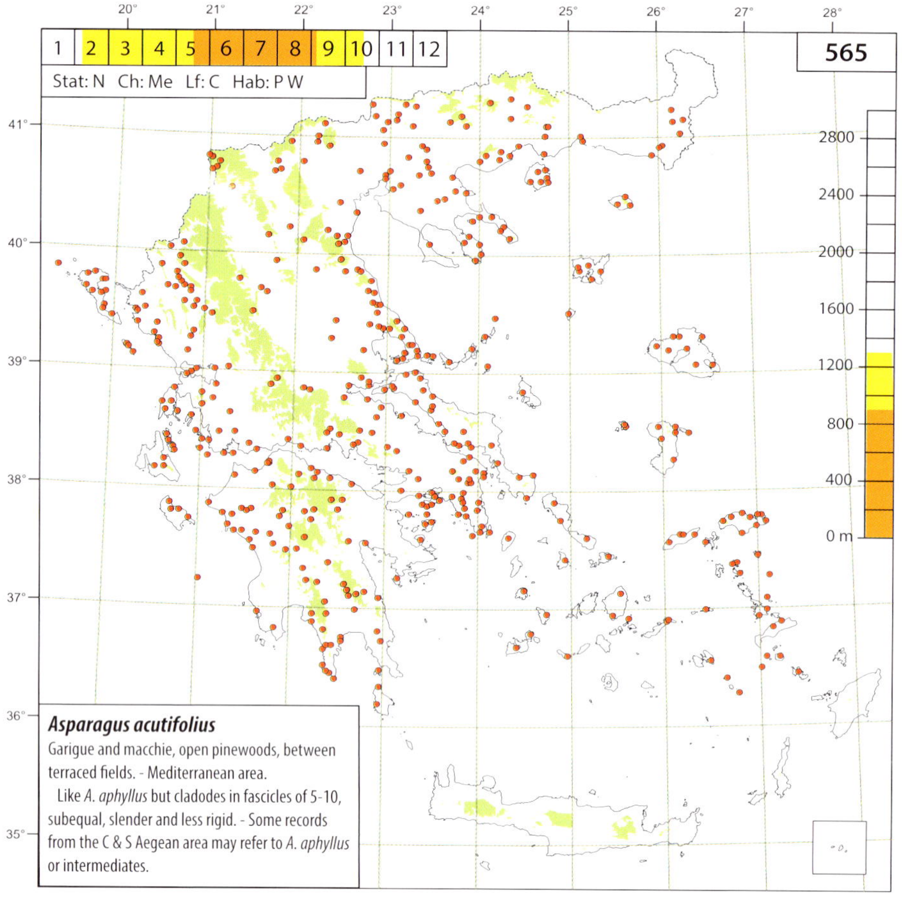

Asparagus acutifolius
Garigue and macchie, open pinewoods, between terraced fields. - Mediterranean area.
Like *A. aphyllus* but cladodes in fascicles of 5-10, subequal, slender and less rigid. - Some records from the C & S Aegean area may refer to *A. aphyllus* or intermediates.

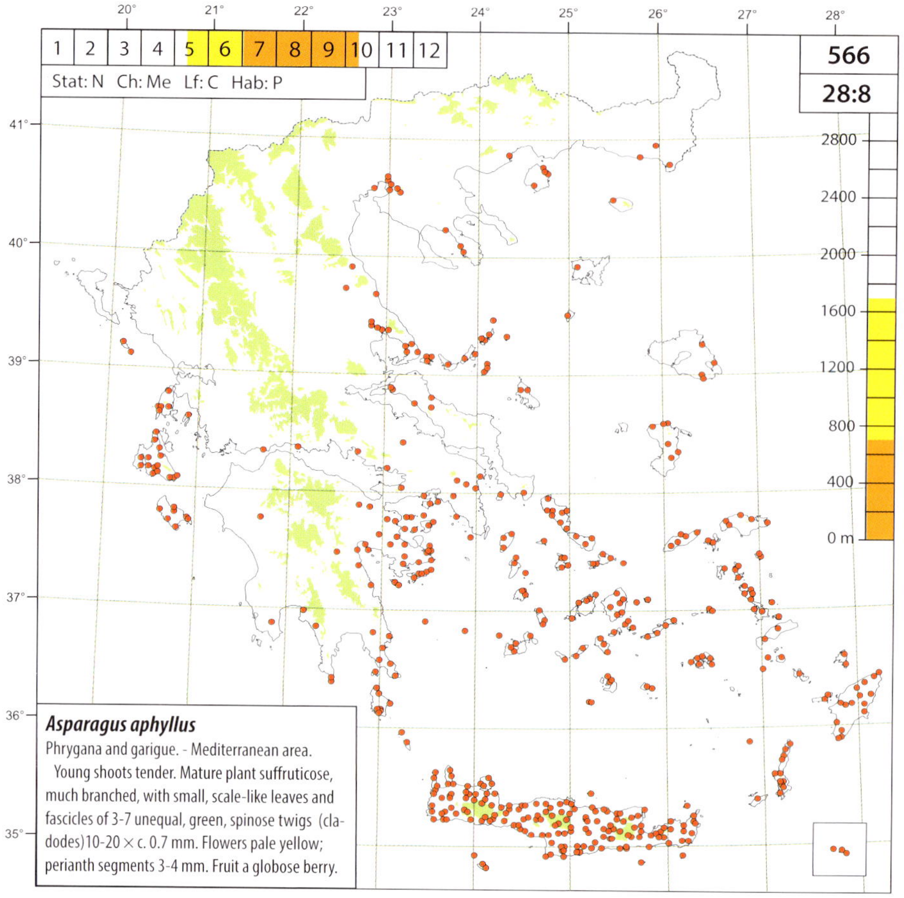

Asparagus aphyllus
Phrygana and garigue. - Mediterranean area.
Young shoots tender. Mature plant suffruticose, much branched, with small, scale-like leaves and fascicles of 3-7 unequal, green, spinose twigs (cladodes) 10-20 × c. 0.7 mm. Flowers pale yellow; perianth segments 3-4 mm. Fruit a globose berry.

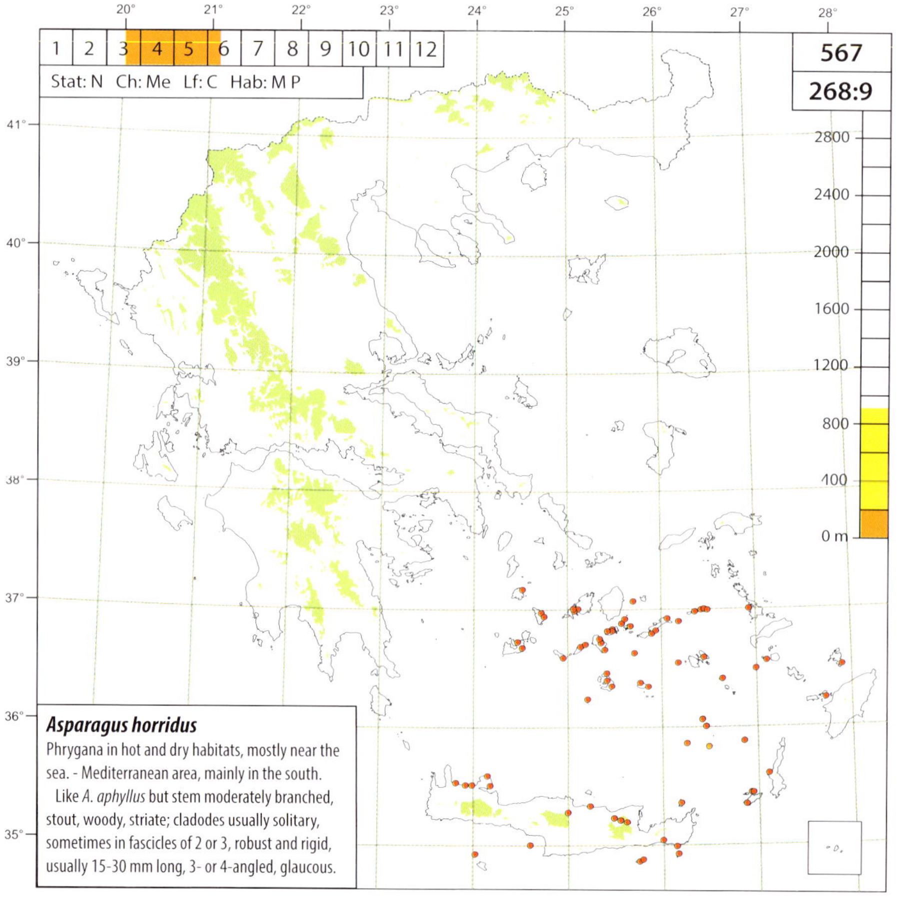

Asparagus horridus
Phrygana in hot and dry habitats, mostly near the sea. - Mediterranean area, mainly in the south.
Like *A. aphyllus* but stem moderately branched, stout, woody, striate; cladodes usually solitary, sometimes in fascicles of 2 or 3, robust and rigid, usually 15-30 mm long, 3- or 4-angled, glaucous.

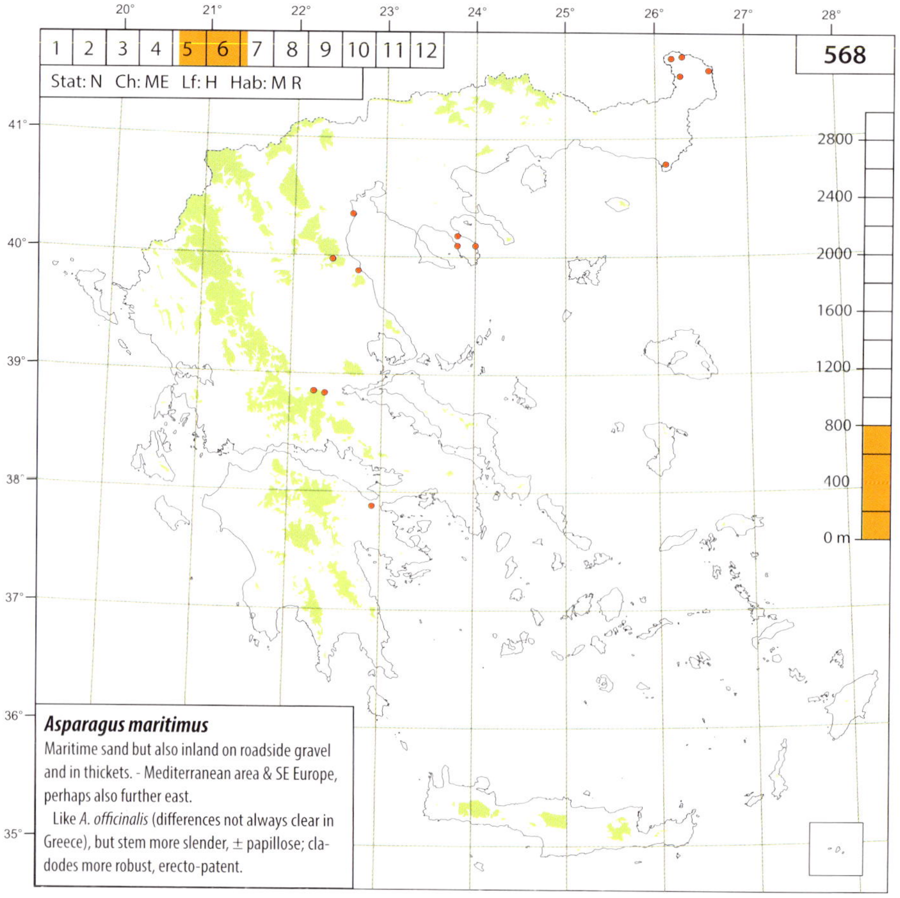

Asparagus maritimus
Maritime sand but also inland on roadside gravel and in thickets. - Mediterranean area & SE Europe, perhaps also further east.
Like *A. officinalis* (differences not always clear in Greece), but stem more slender, ± papillose; cladodes more robust, erecto-patent.

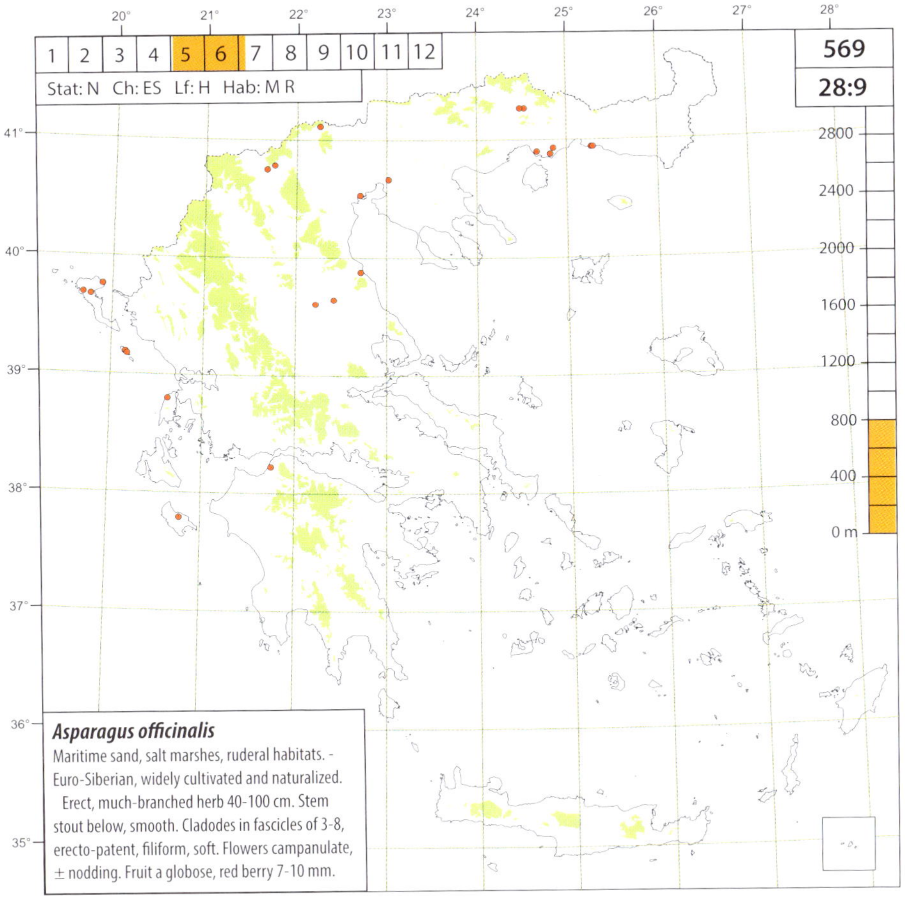

Asparagus officinalis
Maritime sand, salt marshes, ruderal habitats. - Euro-Siberian, widely cultivated and naturalized.
Erect, much-branched herb 40-100 cm. Stem stout below, smooth. Cladodes in fascicles of 3-8, erecto-patent, filiform, soft. Flowers campanulate, ± nodding. Fruit a globose, red berry 7-10 mm.

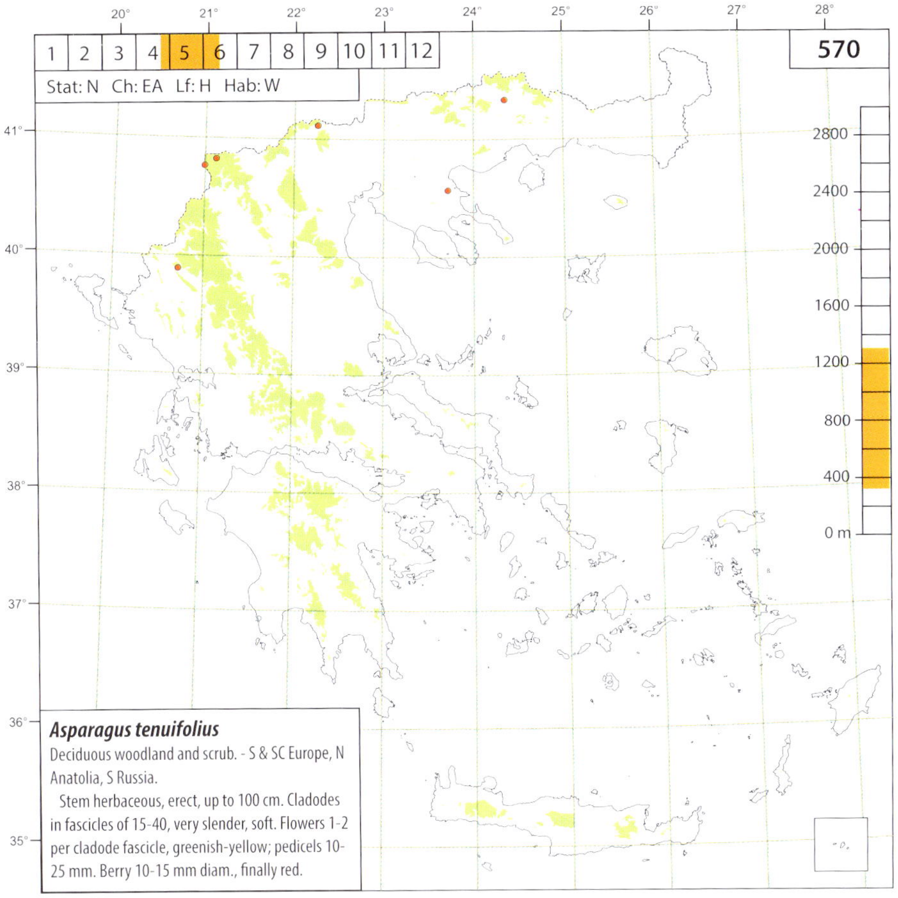

Asparagus tenuifolius
Deciduous woodland and scrub. - S & SC Europe, N Anatolia, S Russia.
Stem herbaceous, erect, up to 100 cm. Cladodes in fascicles of 15-40, very slender, soft. Flowers 1-2 per cladode fascicle, greenish-yellow; pedicels 10-25 mm. Berry 10-15 mm diam., finally red.

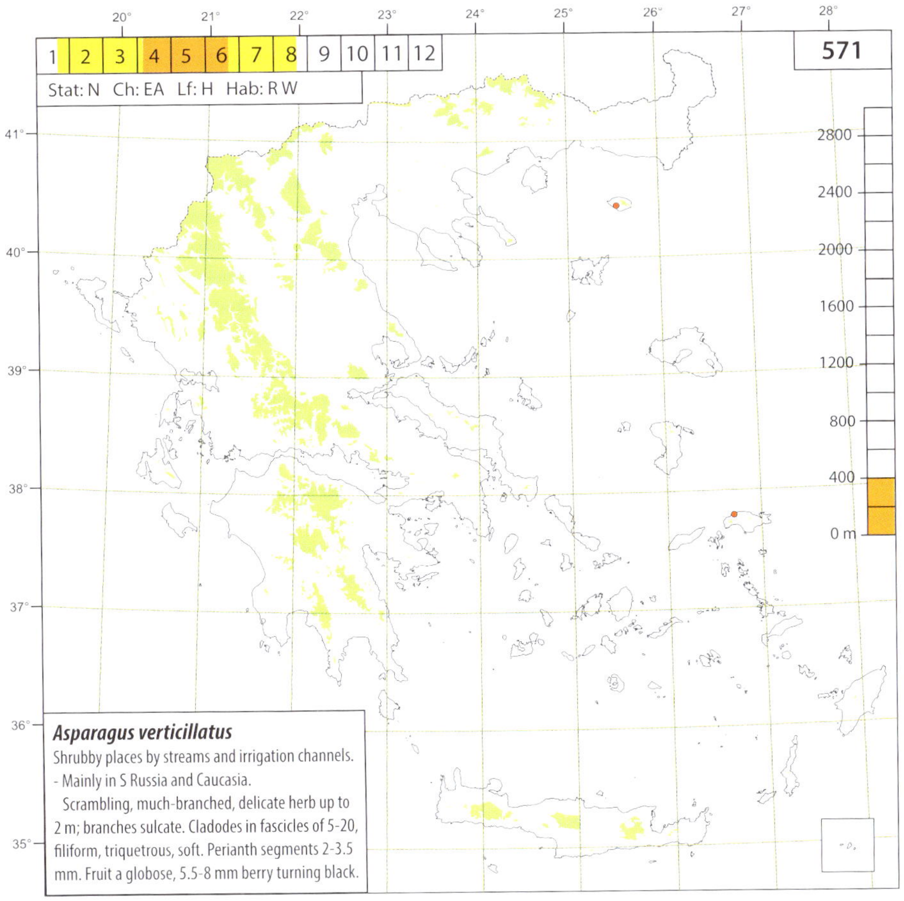

Asparagus verticillatus
Shrubby places by streams and irrigation channels. - Mainly in S Russia and Caucasia.
Scrambling, much-branched, delicate herb up to 2 m; branches sulcate. Cladodes in fascicles of 5-20, filiform, triquetrous, soft. Perianth segments 2-3.5 mm. Fruit a globose, 5.5-8 mm berry turning black.

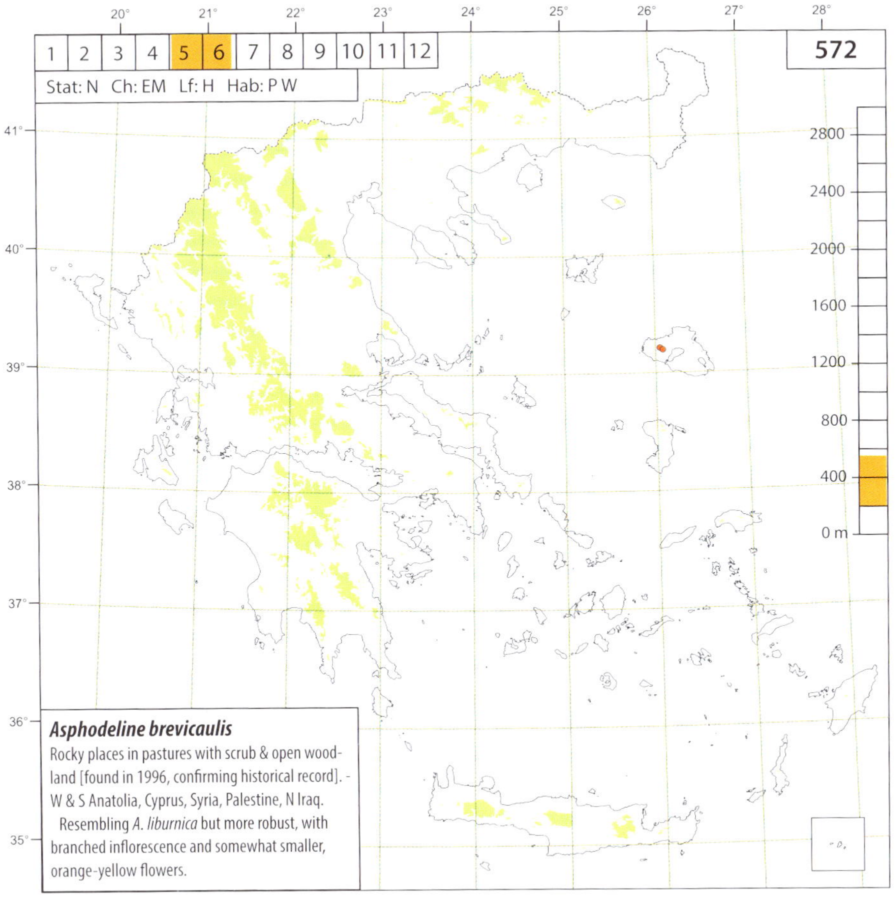

Asphodeline brevicaulis
Rocky places in pastures with scrub & open woodland [found in 1996, confirming historical record]. - W & S Anatolia, Cyprus, Syria, Palestine, N Iraq.
Resembling *A. liburnica* but more robust, with branched inflorescence and somewhat smaller, orange-yellow flowers.

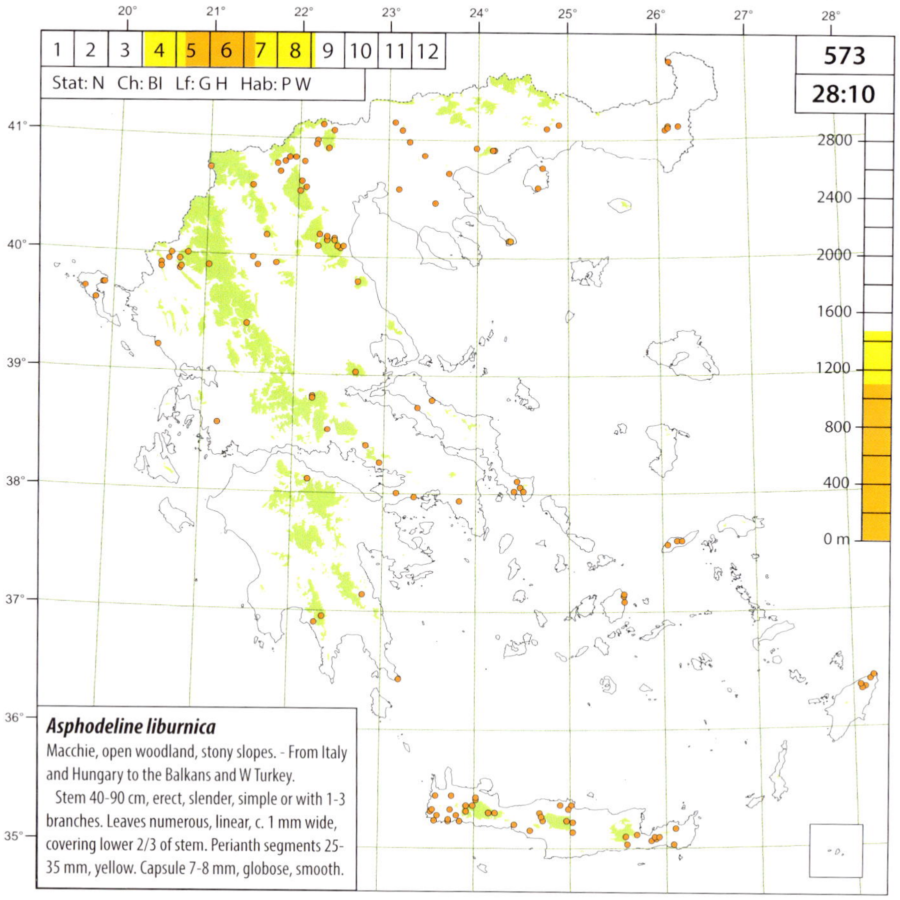

Asphodeline liburnica

Macchie, open woodland, stony slopes. - From Italy and Hungary to the Balkans and W Turkey.

Stem 40-90 cm, erect, slender, simple or with 1-3 branches. Leaves numerous, linear, c. 1 mm wide, covering lower 2/3 of stem. Perianth segments 25-35 mm, yellow. Capsule 7-8 mm, globose, smooth.

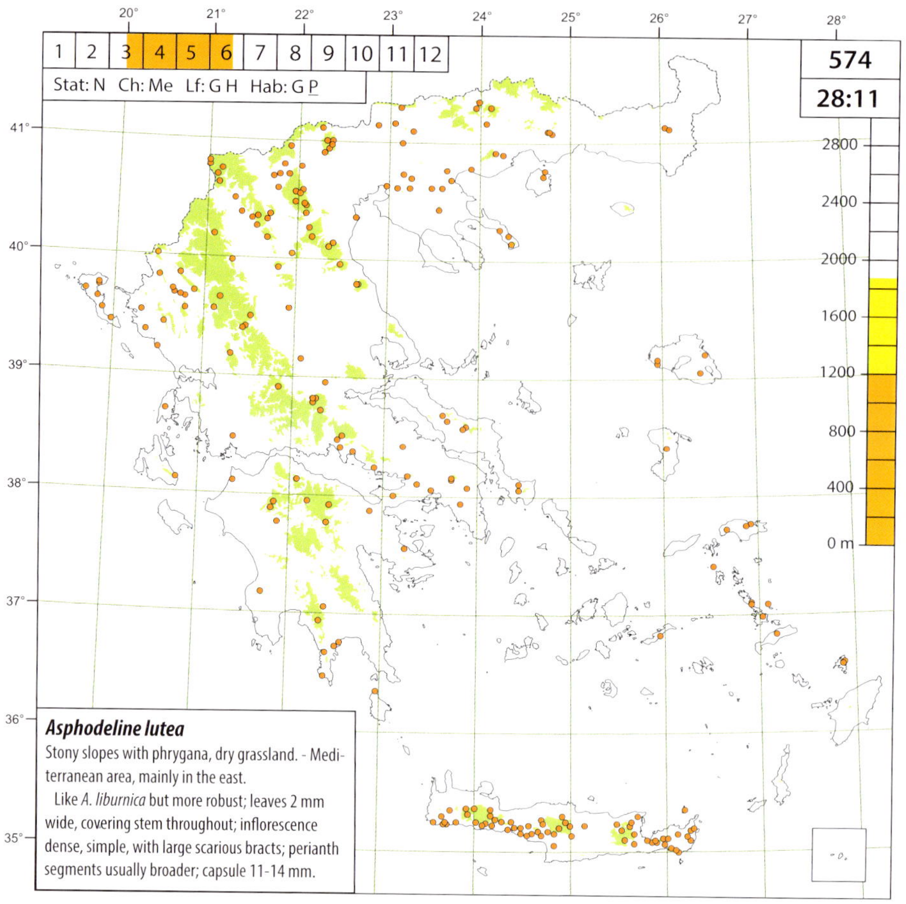

Asphodeline lutea

Stony slopes with phrygana, dry grassland. - Mediterranean area, mainly in the east.

Like *A. liburnica* but more robust; leaves 2 mm wide, covering stem throughout; inflorescence dense, simple, with large scarious bracts; perianth segments usually broader; capsule 11-14 mm.

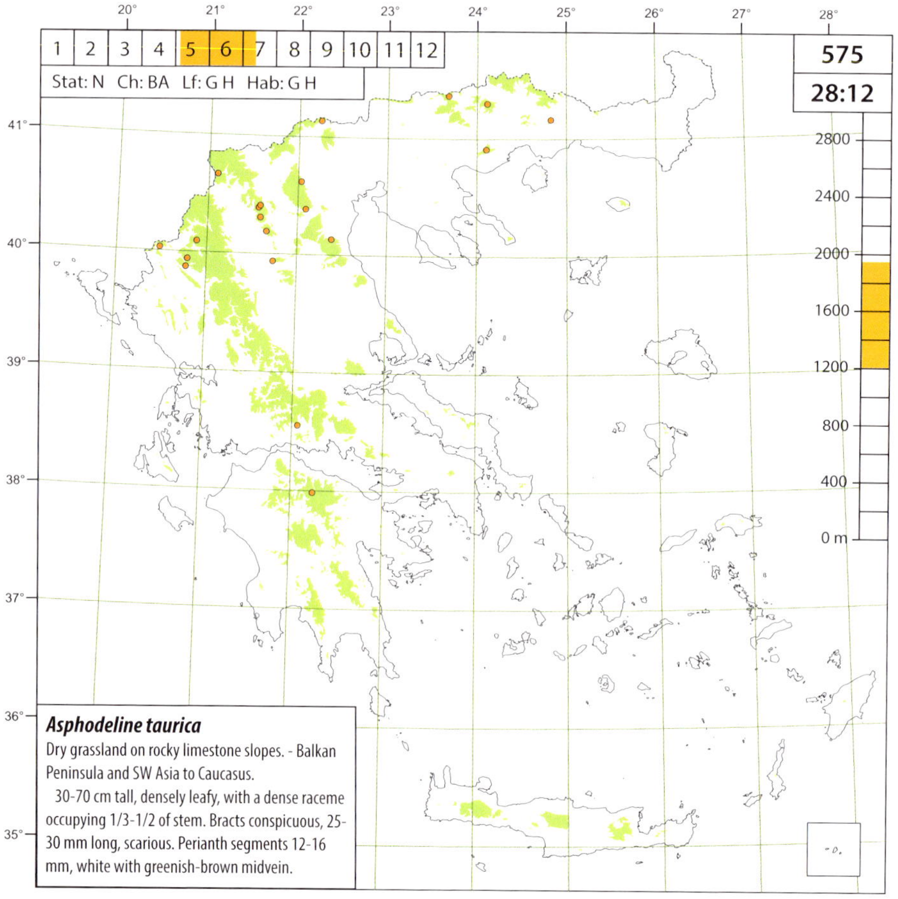

Asphodeline taurica

Dry grassland on rocky limestone slopes. - Balkan Peninsula and SW Asia to Caucasus.

30-70 cm tall, densely leafy, with a dense raceme occupying 1/3-1/2 of stem. Bracts conspicuous, 25-30 mm long, scarious. Perianth segments 12-16 mm, white with greenish-brown midvein.

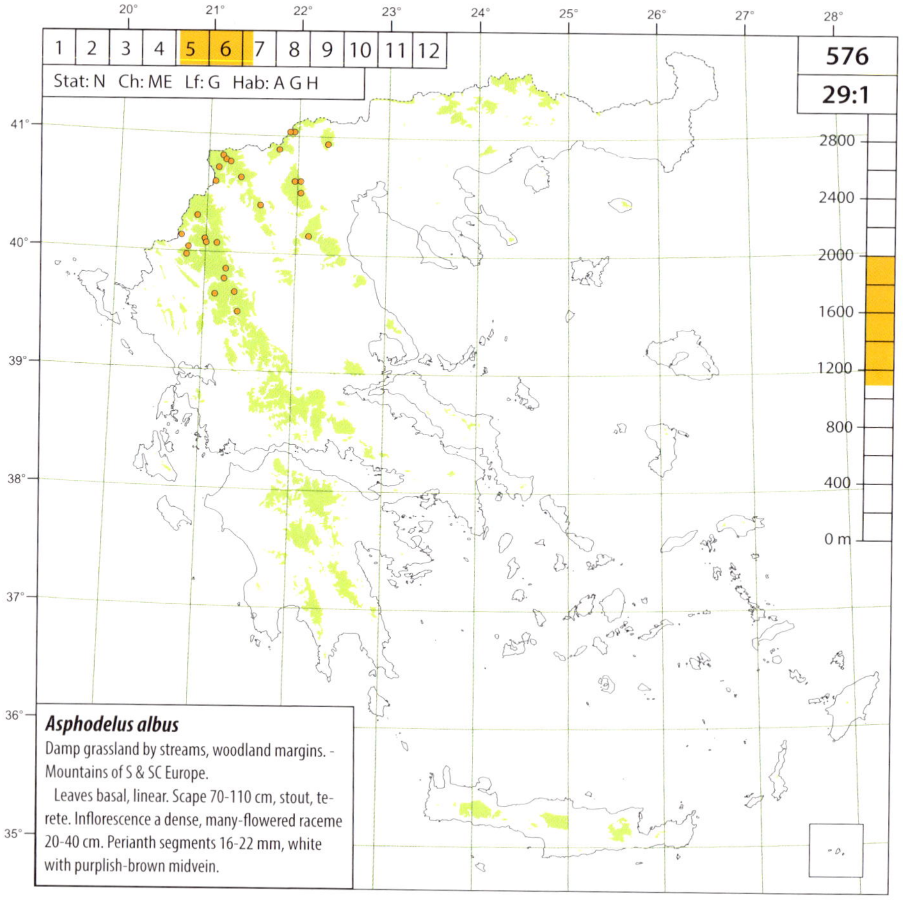

Asphodelus albus

Damp grassland by streams, woodland margins. - Mountains of S & SC Europe.

Leaves basal, linear. Scape 70-110 cm, stout, terete. Inflorescence a dense, many-flowered raceme 20-40 cm. Perianth segments 16-22 mm, white with purplish-brown midvein.

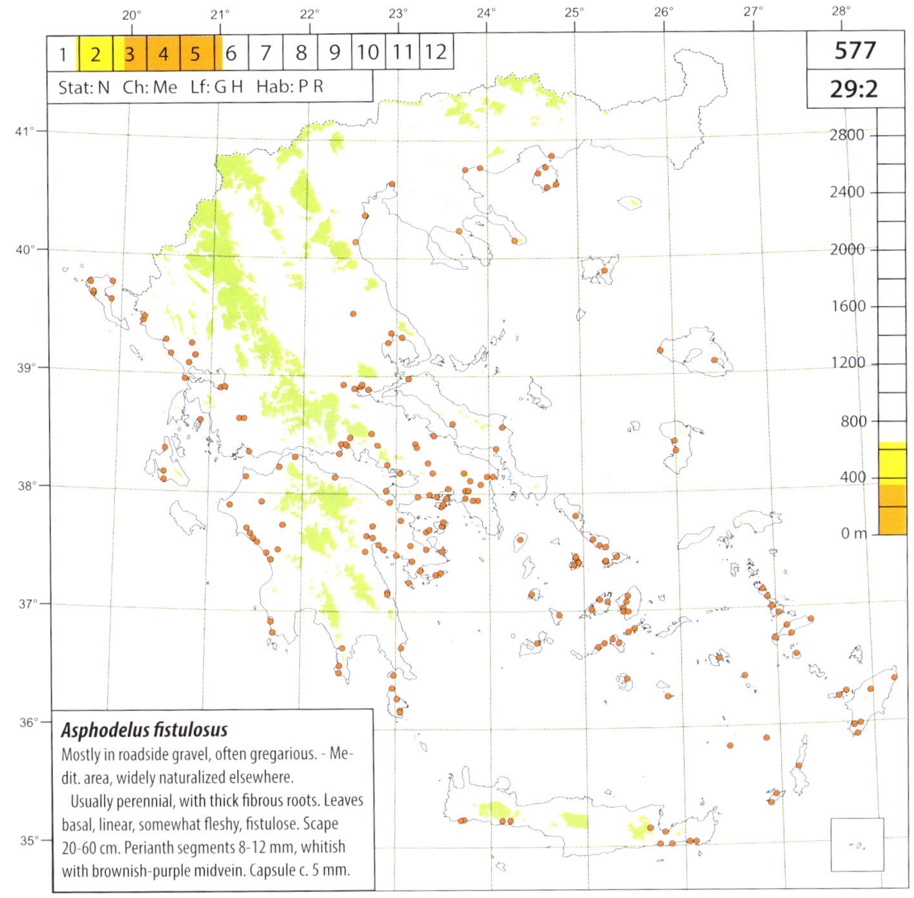

Asphodelus fistulosus

Mostly in roadside gravel, often gregarious. - Medit. area, widely naturalized elsewhere.

Usually perennial, with thick fibrous roots. Leaves basal, linear, somewhat fleshy, fistulose. Scape 20-60 cm. Perianth segments 8-12 mm, whitish with brownish-purple midvein. Capsule c. 5 mm.

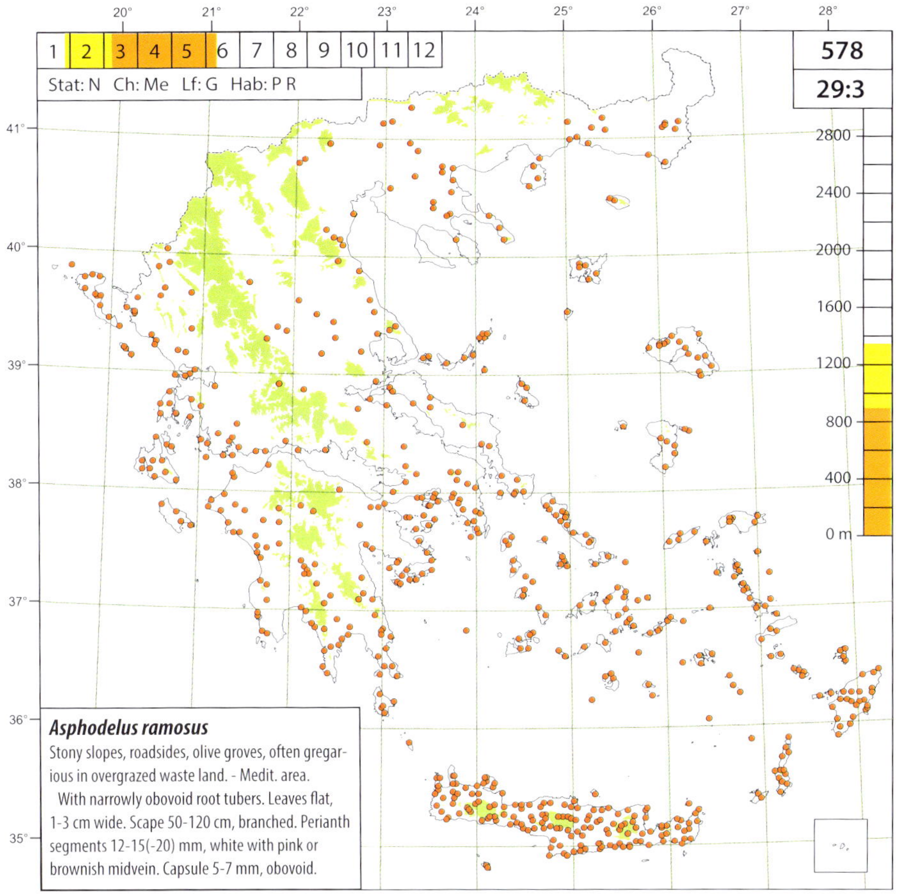

Asphodelus ramosus

Stony slopes, roadsides, olive groves, often gregarious in overgrazed waste land. - Medit. area.

With narrowly obovoid root tubers. Leaves flat, 1-3 cm wide. Scape 50-120 cm, branched. Perianth segments 12-15(-20) mm, white with pink or brownish midvein. Capsule 5-7 mm, obovoid.

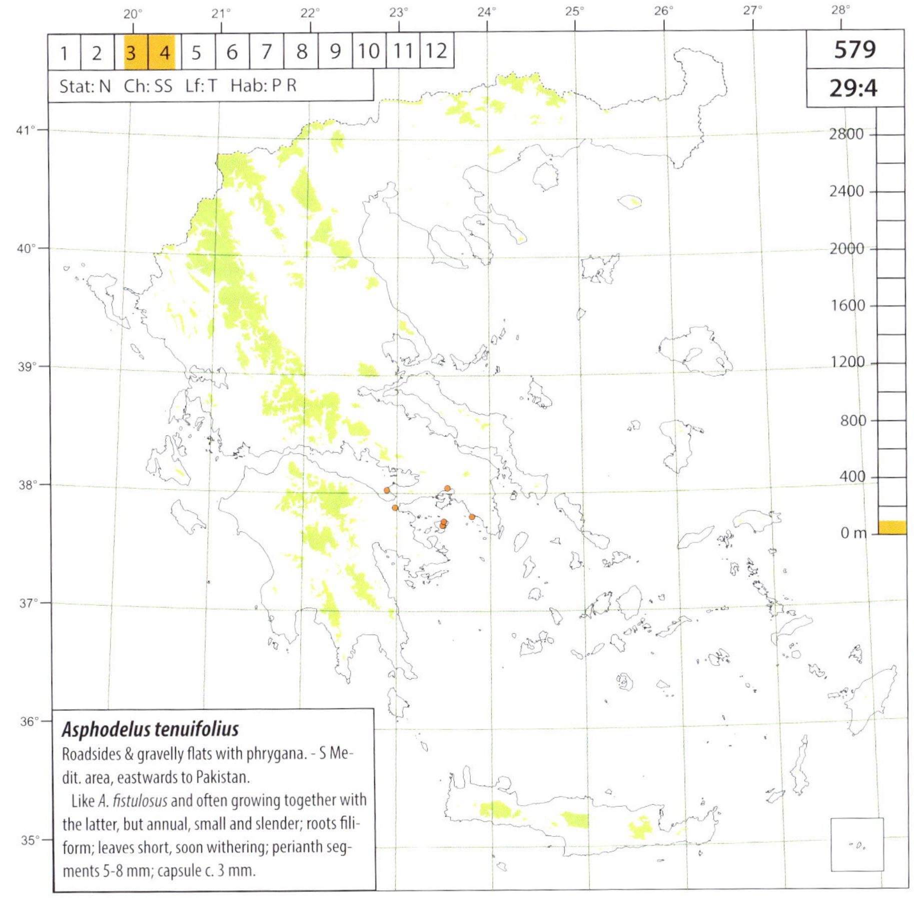

Asphodelus tenuifolius

Roadsides & gravelly flats with phrygana. - S Medit. area, eastwards to Pakistan.

Like *A. fistulosus* and often growing together with the latter, but annual, small and slender; roots filiform; leaves short, soon withering; perianth segments 5-8 mm; capsule c. 3 mm.

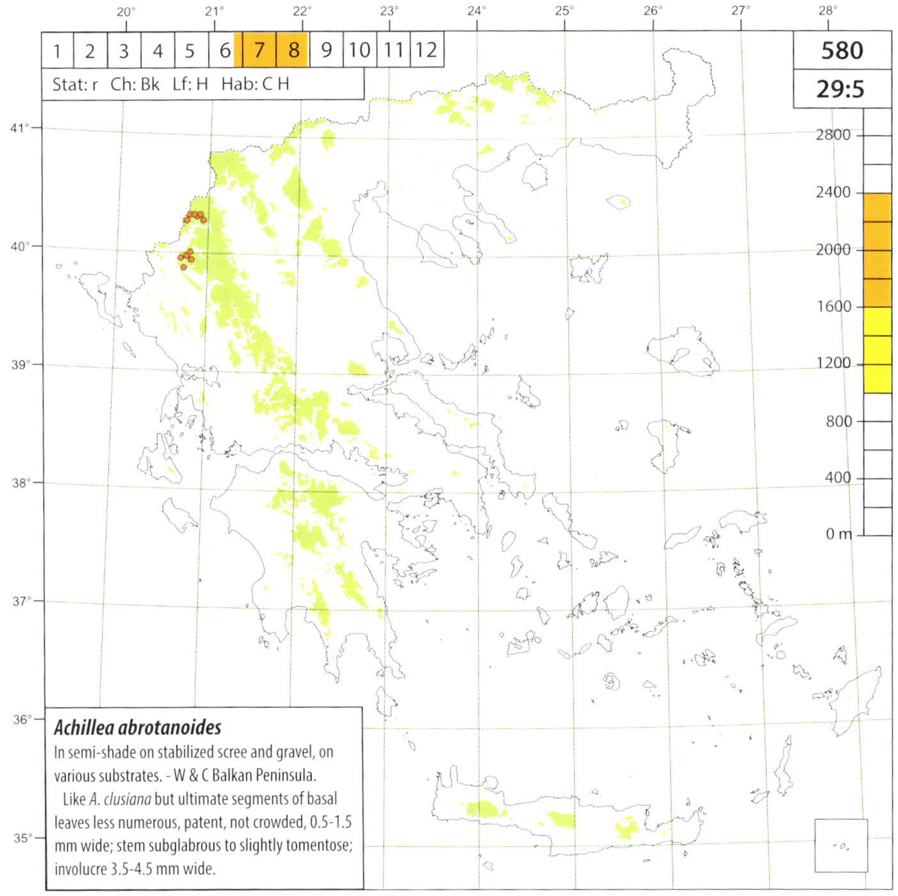

Achillea abrotanoides

In semi-shade on stabilized scree and gravel, on various substrates. - W & C Balkan Peninsula.

Like *A. clusiana* but ultimate segments of basal leaves less numerous, patent, not crowded, 0.5-1.5 mm wide; stem subglabrous to slightly tomentose; involucre 3.5-4.5 mm wide.

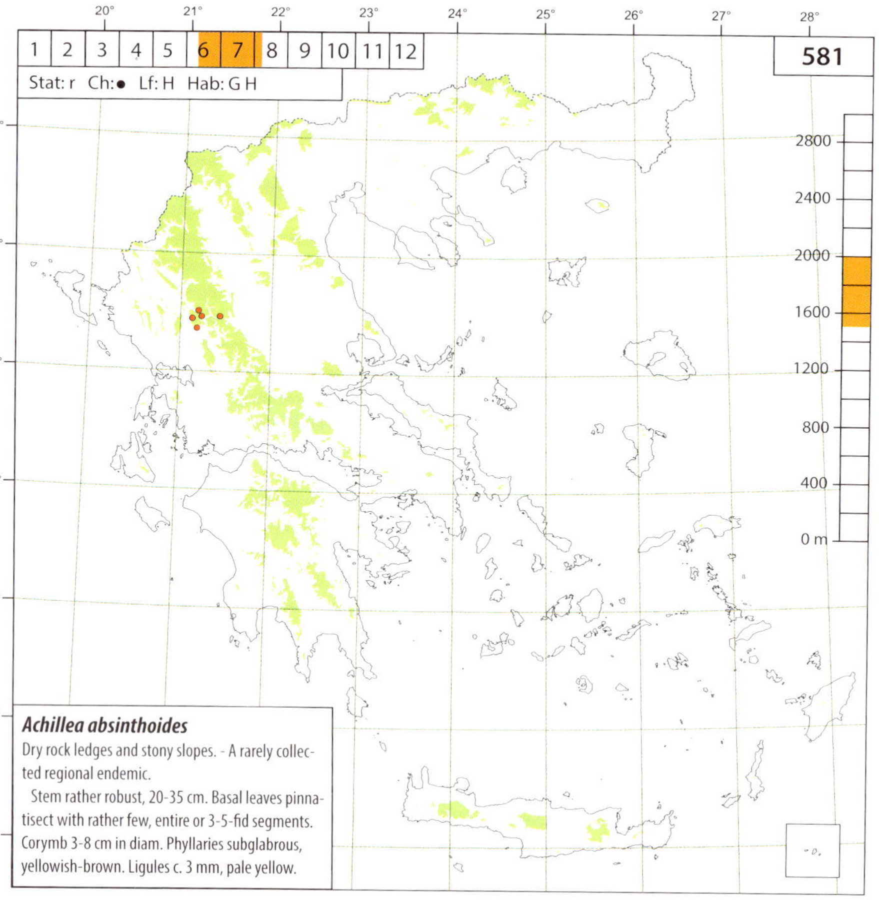

Achillea absinthoides

Dry rock ledges and stony slopes. - A rarely collected regional endemic.

Stem rather robust, 20-35 cm. Basal leaves pinnatisect with rather few, entire or 3-5-fid segments. Corymb 3-8 cm in diam. Phyllaries subglabrous, yellowish-brown. Ligules c. 3 mm, pale yellow.

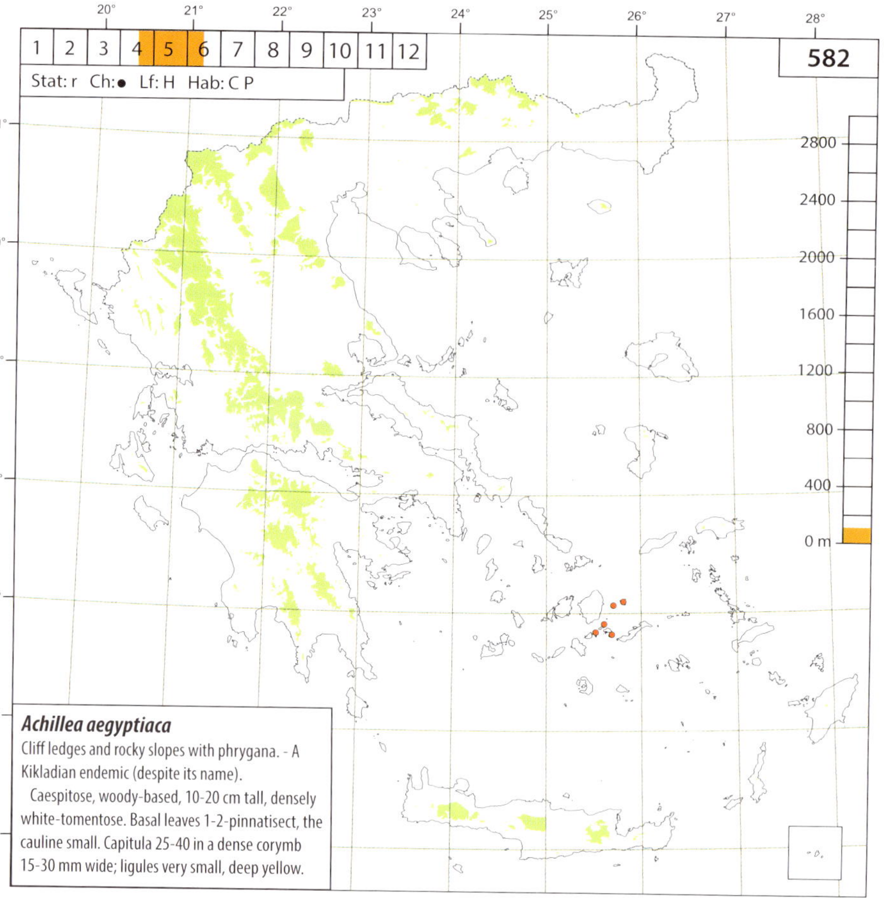

Achillea aegyptiaca

Cliff ledges and rocky slopes with phrygana. - A Kikladian endemic (despite its name).

Caespitose, woody-based, 10-20 cm tall, densely white-tomentose. Basal leaves 1-2-pinnatisect, the cauline small. Capitula 25-40 in a dense corymb 15-30 mm wide; ligules very small, deep yellow.

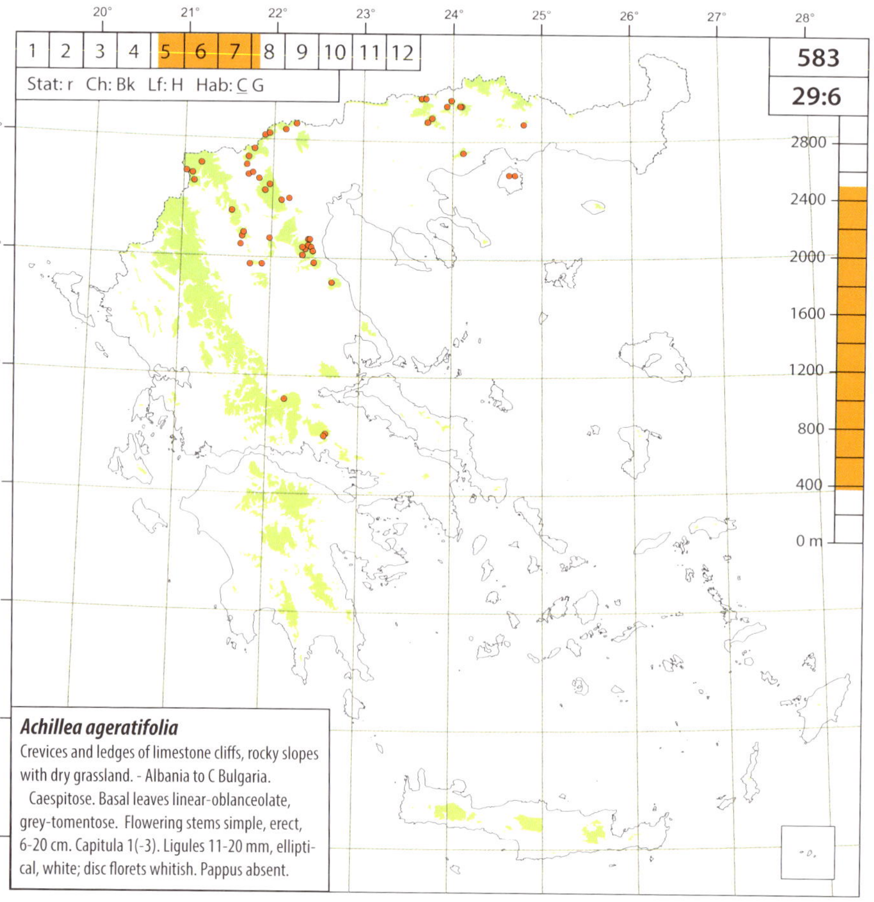

Achillea ageratifolia

Crevices and ledges of limestone cliffs, rocky slopes with dry grassland. - Albania to C Bulgaria.

Caespitose. Basal leaves linear-oblanceolate, grey-tomentose. Flowering stems simple, erect, 6-20 cm. Capitula 1(-3). Ligules 11-20 mm, elliptical, white; disc florets whitish. Pappus absent.

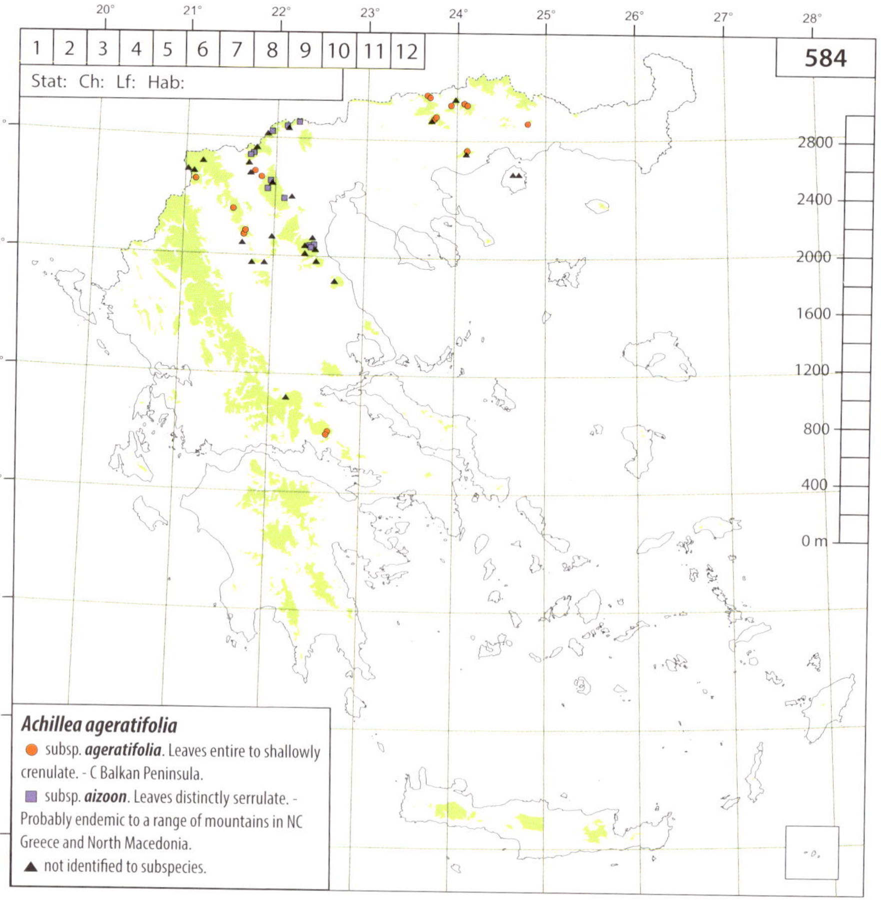

Achillea ageratifolia

● subsp. ***ageratifolia***. Leaves entire to shallowly crenulate. - C Balkan Peninsula.

■ subsp. ***aizoon***. Leaves distinctly serrulate. - Probably endemic to a range of mountains in NC Greece and North Macedonia.

▲ not identified to subspecies.

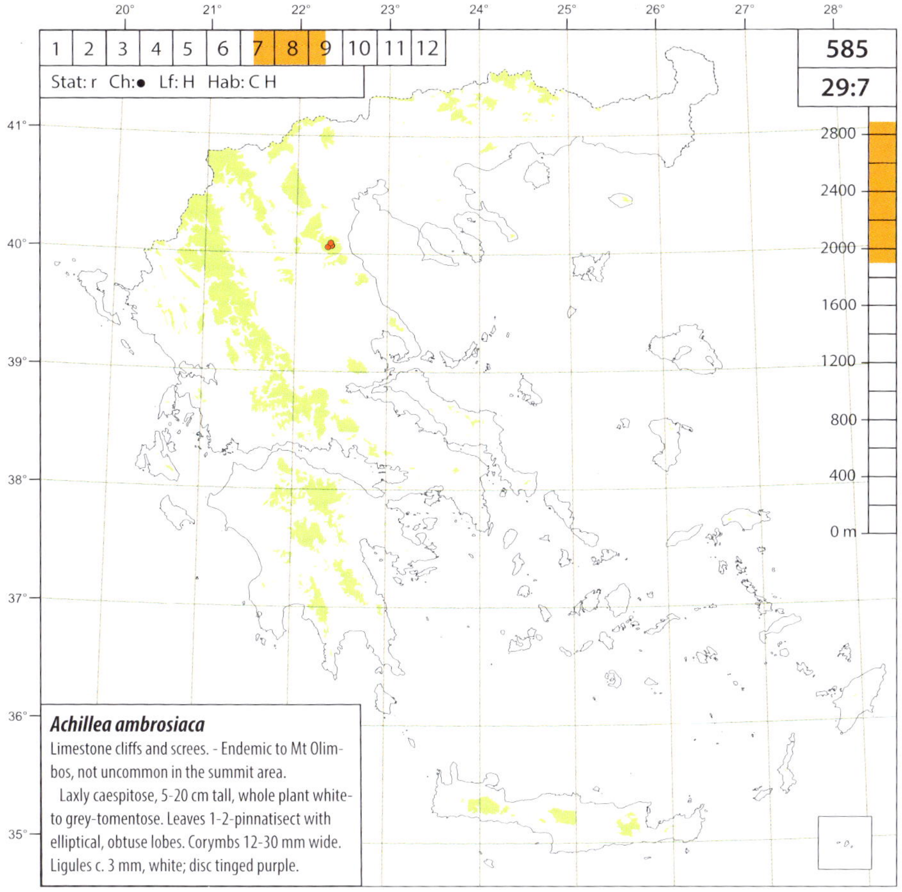

Achillea ambrosiaca

Limestone cliffs and screes. - Endemic to Mt Olimbos, not uncommon in the summit area.

Laxly caespitose, 5-20 cm tall, whole plant white- to grey-tomentose. Leaves 1-2-pinnatisect with elliptical, obtuse lobes. Corymbs 12-30 mm wide. Ligules c. 3 mm, white; disc tinged purple.

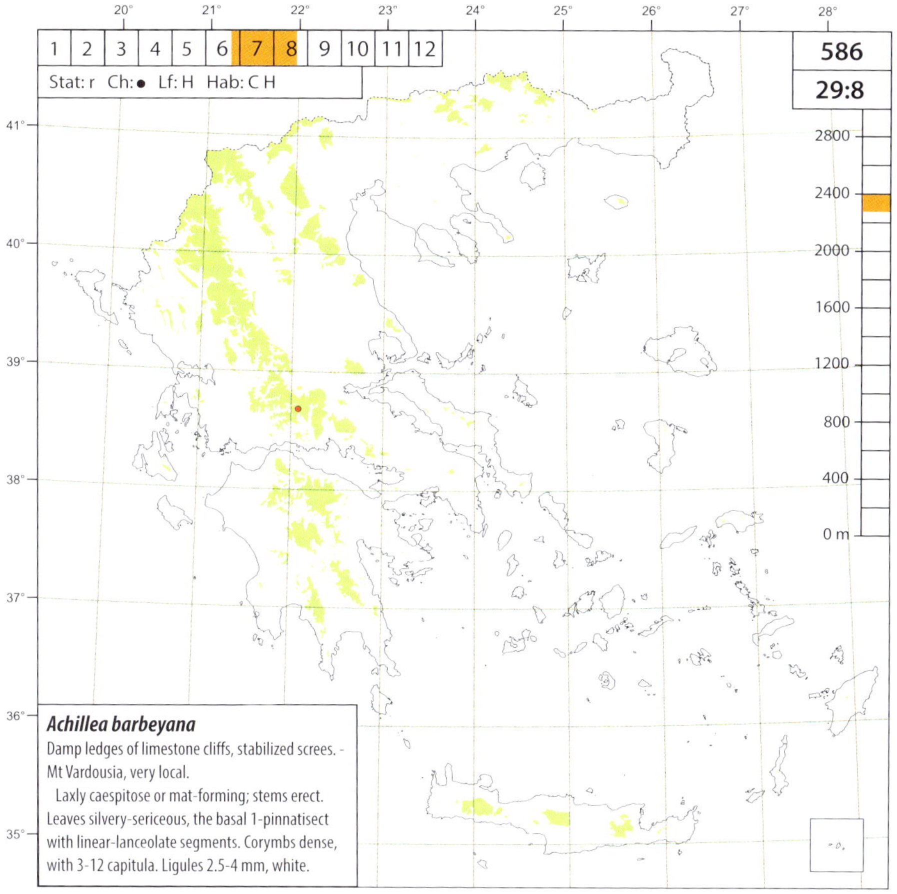

Achillea barbeyana

Damp ledges of limestone cliffs, stabilized screes. - Mt Vardousia, very local.

Laxly caespitose or mat-forming; stems erect. Leaves silvery-sericeous, the basal 1-pinnatisect with linear-lanceolate segments. Corymbs dense, with 3-12 capitula. Ligules 2.5-4 mm, white.

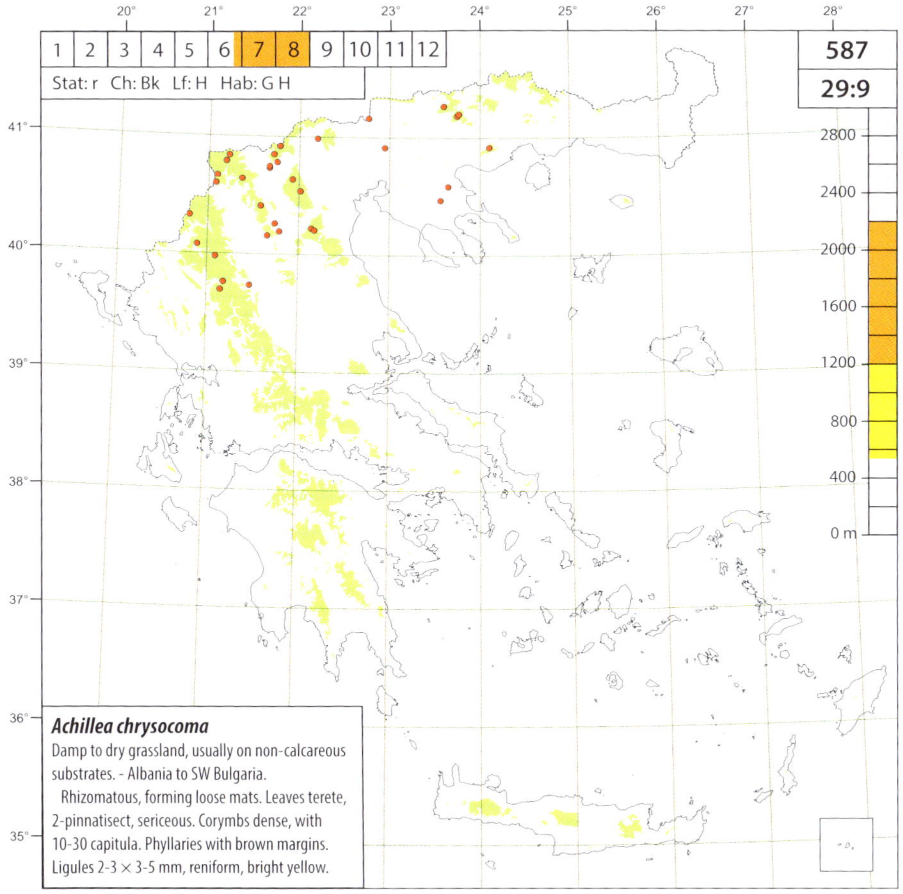

Achillea chrysocoma

Damp to dry grassland, usually on non-calcareous substrates. - Albania to SW Bulgaria.

Rhizomatous, forming loose mats. Leaves terete, 2-pinnatisect, sericeous. Corymbs dense, with 10-30 capitula. Phyllaries with brown margins. Ligules 2-3 × 3-5 mm, reniform, bright yellow.

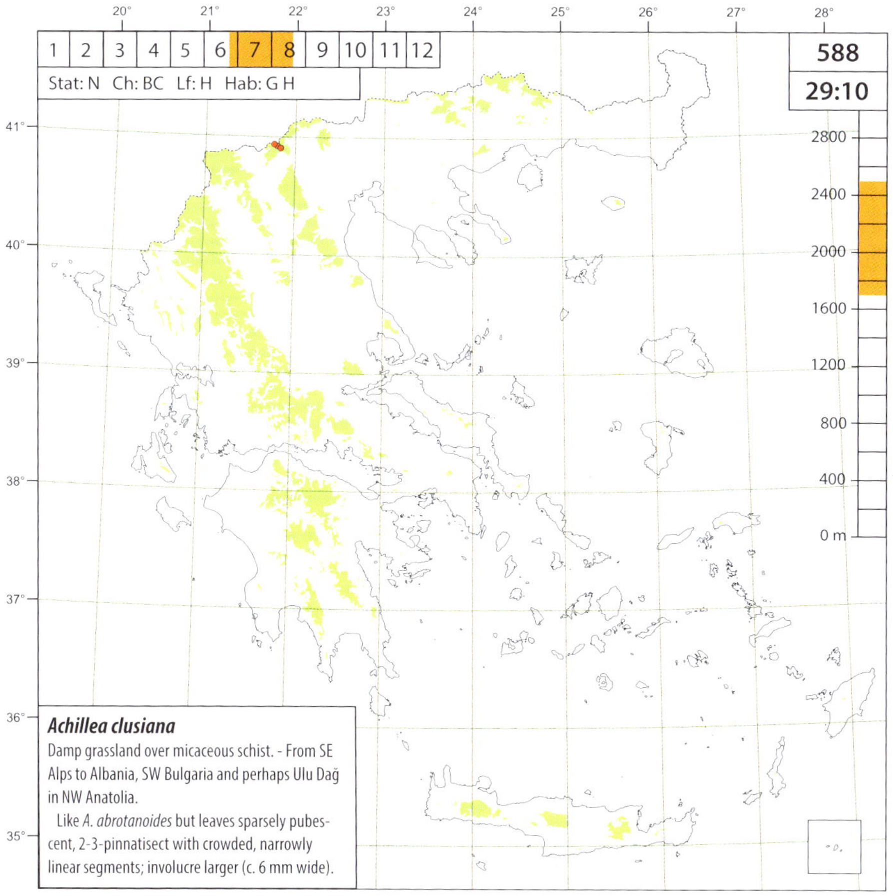

Achillea clusiana

Damp grassland over micaceous schist. - From SE Alps to Albania, SW Bulgaria and perhaps Ulu Dağ in NW Anatolia.

Like *A. abrotanoides* but leaves sparsely pubescent, 2-3-pinnatisect with crowded, narrowly linear segments; involucre larger (c. 6 mm wide).

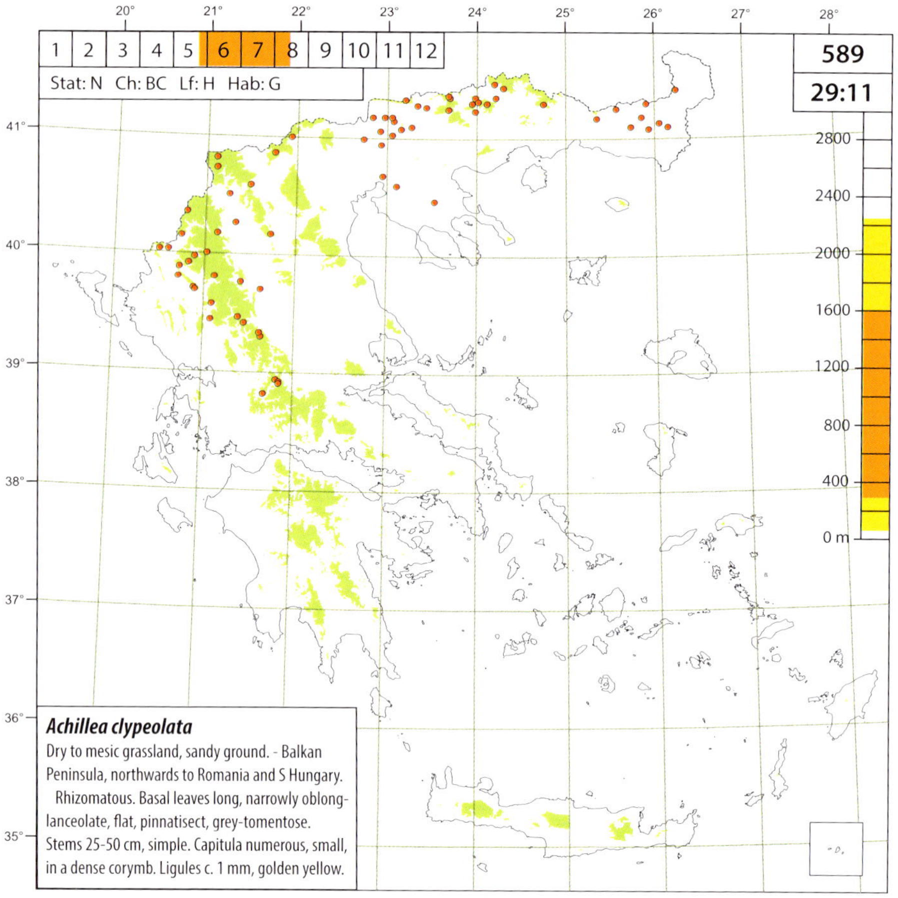

Achillea clypeolata

Dry to mesic grassland, sandy ground. - Balkan Peninsula, northwards to Romania and S Hungary.

Rhizomatous. Basal leaves long, narrowly oblong-lanceolate, flat, pinnatisect, grey-tomentose. Stems 25-50 cm, simple. Capitula numerous, small, in a dense corymb. Ligules c. 1 mm, golden yellow.

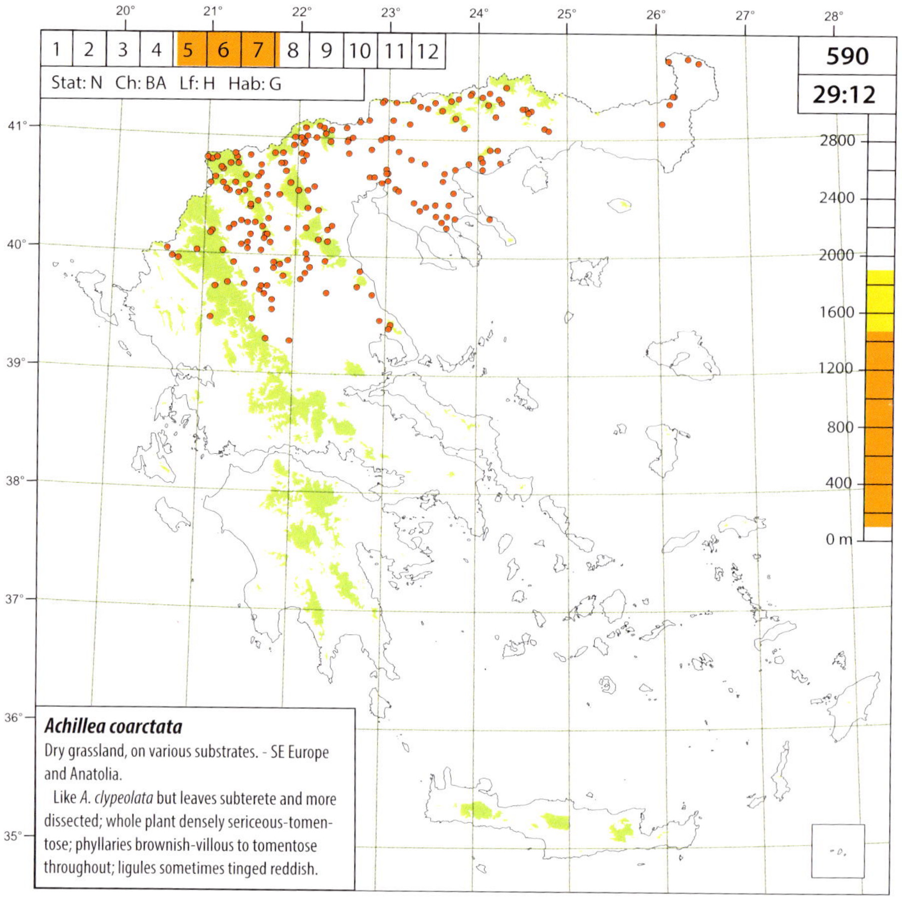

Achillea coarctata

Dry grassland, on various substrates. - SE Europe and Anatolia.

Like *A. clypeolata* but leaves subterete and more dissected; whole plant densely sericeous-tomentose; phyllaries brownish-villous to tomentose throughout; ligules sometimes tinged reddish.

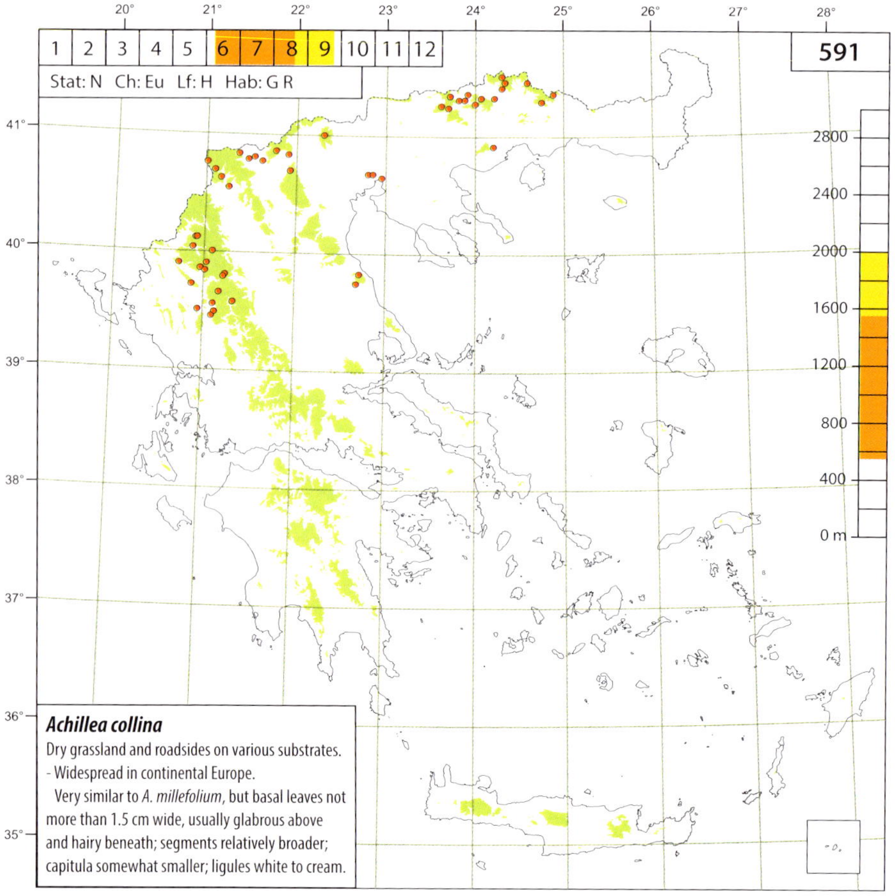

Achillea collina

Dry grassland and roadsides on various substrates. - Widespread in continental Europe.

Very similar to *A. millefolium*, but basal leaves not more than 1.5 cm wide, usually glabrous above and hairy beneath; segments relatively broader; capitula somewhat smaller; ligules white to cream.

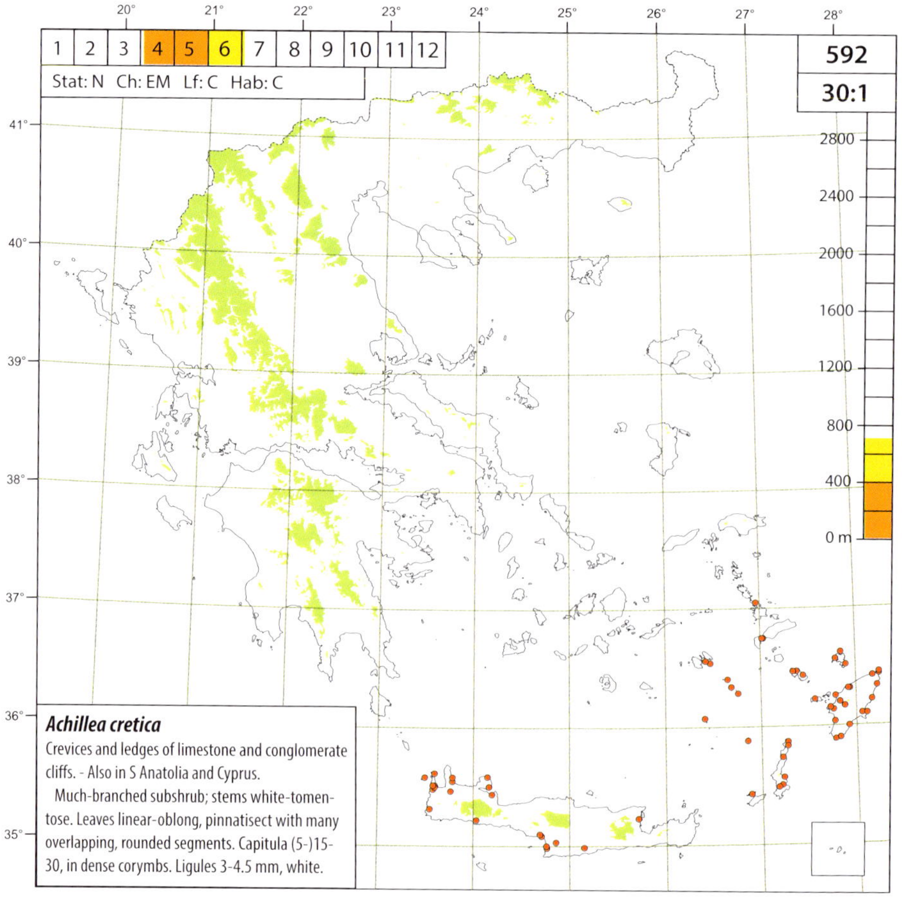

Achillea cretica

Crevices and ledges of limestone and conglomerate cliffs. - Also in S Anatolia and Cyprus.

Much-branched subshrub; stems white-tomentose. Leaves linear-oblong, pinnatisect with many overlapping, rounded segments. Capitula (5-)15-30, in dense corymbs. Ligules 3-4.5 mm, white.

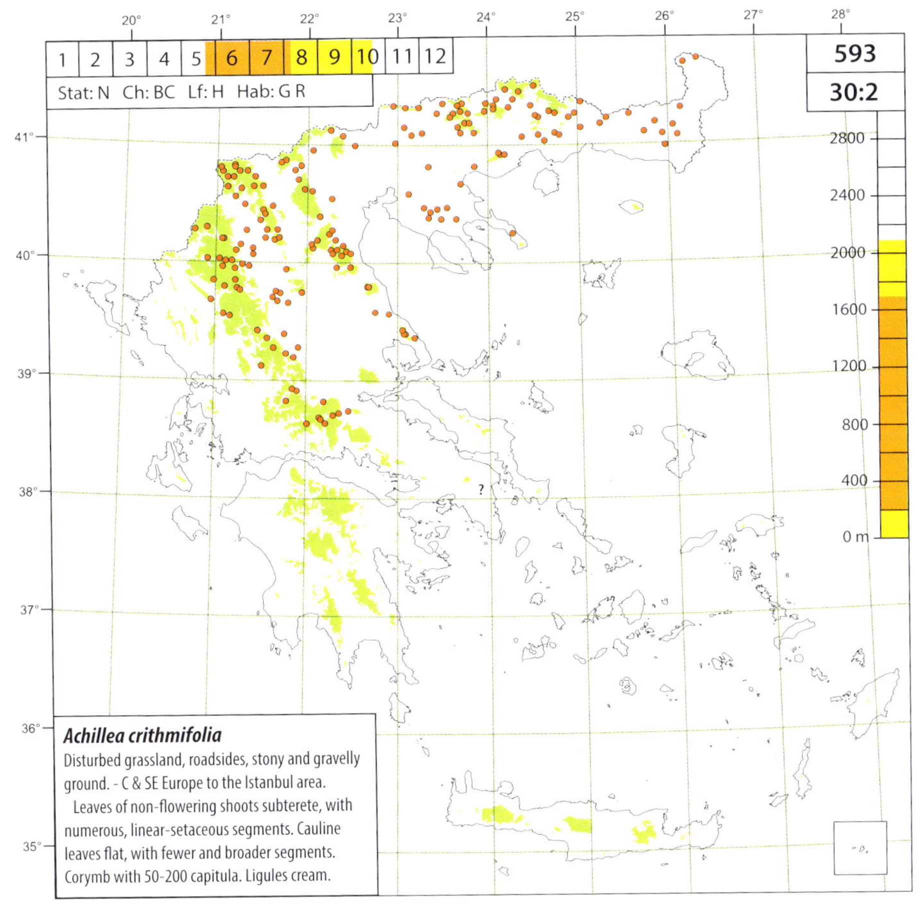

Achillea crithmifolia
Disturbed grassland, roadsides, stony and gravelly ground. - C & SE Europe to the Istanbul area.
Leaves of non-flowering shoots subterete, with numerous, linear-setaceous segments. Cauline leaves flat, with fewer and broader segments. Corymb with 50-200 capitula. Ligules cream.

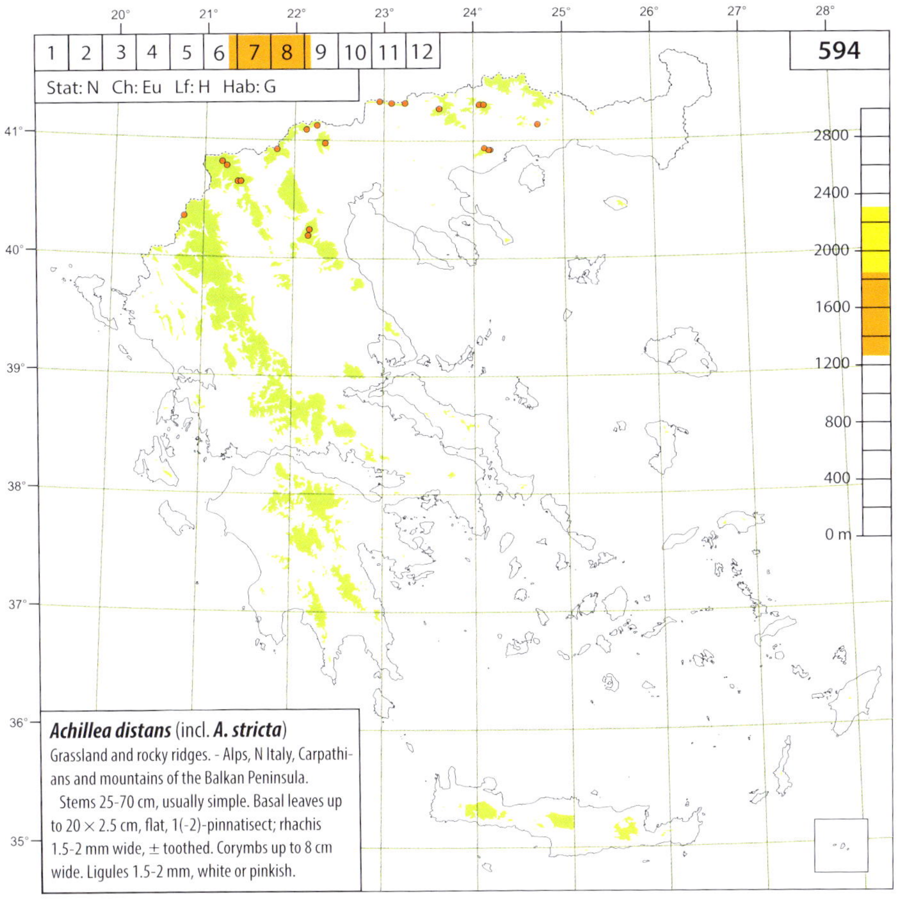

Achillea distans (incl. ***A. stricta***)
Grassland and rocky ridges. - Alps, N Italy, Carpathians and mountains of the Balkan Peninsula.
Stems 25-70 cm, usually simple. Basal leaves up to 20 × 2.5 cm, flat, 1(-2)-pinnatisect; rhachis 1.5-2 mm wide, ± toothed. Corymbs up to 8 cm wide. Ligules 1.5-2 mm, white or pinkish.

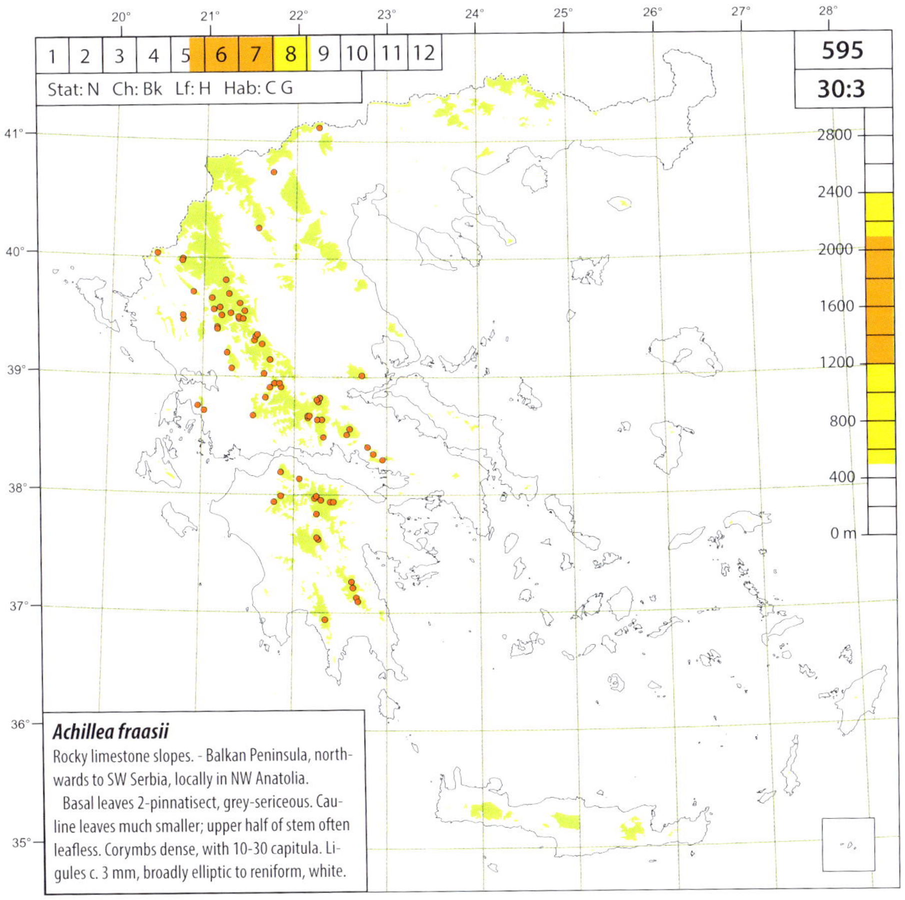

Achillea fraasii
Rocky limestone slopes. - Balkan Peninsula, northwards to SW Serbia, locally in NW Anatolia.
Basal leaves 2-pinnatisect, grey-sericeous. Cauline leaves much smaller; upper half of stem often leafless. Corymbs dense, with 10-30 capitula. Ligules c. 3 mm, broadly elliptic to reniform, white.

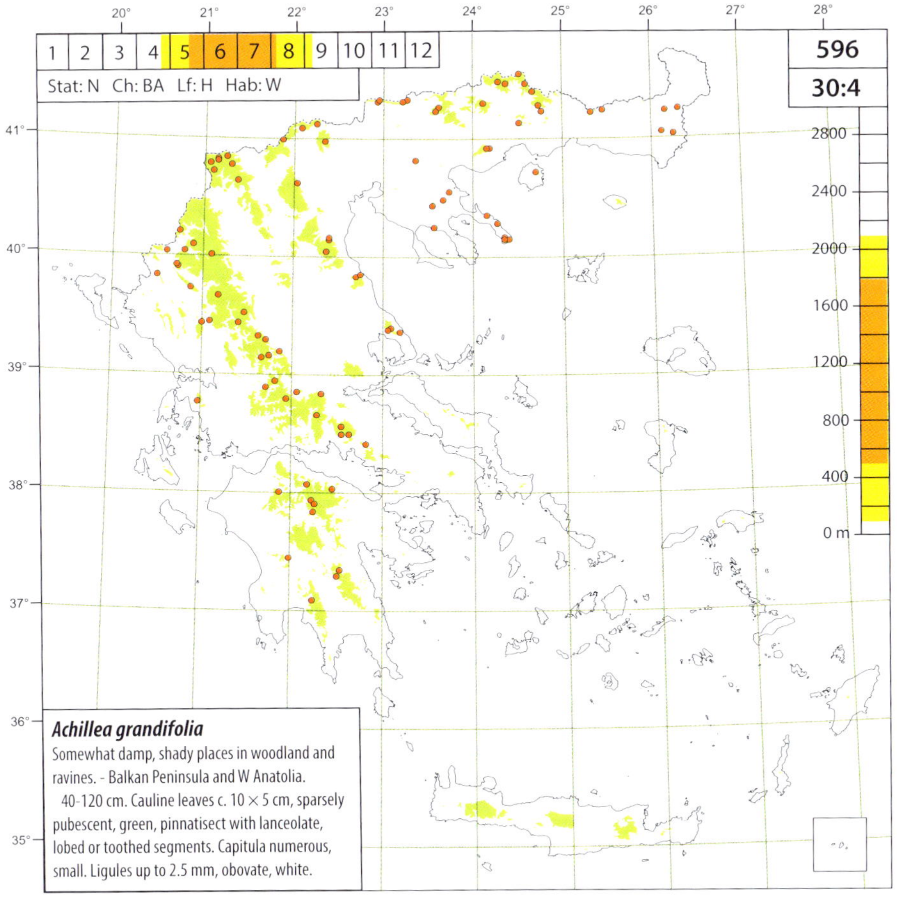

Achillea grandifolia
Somewhat damp, shady places in woodland and ravines. - Balkan Peninsula and W Anatolia.
40-120 cm. Cauline leaves c. 10 × 5 cm, sparsely pubescent, green, pinnatisect with lanceolate, lobed or toothed segments. Capitula numerous, small. Ligules up to 2.5 mm, obovate, white.

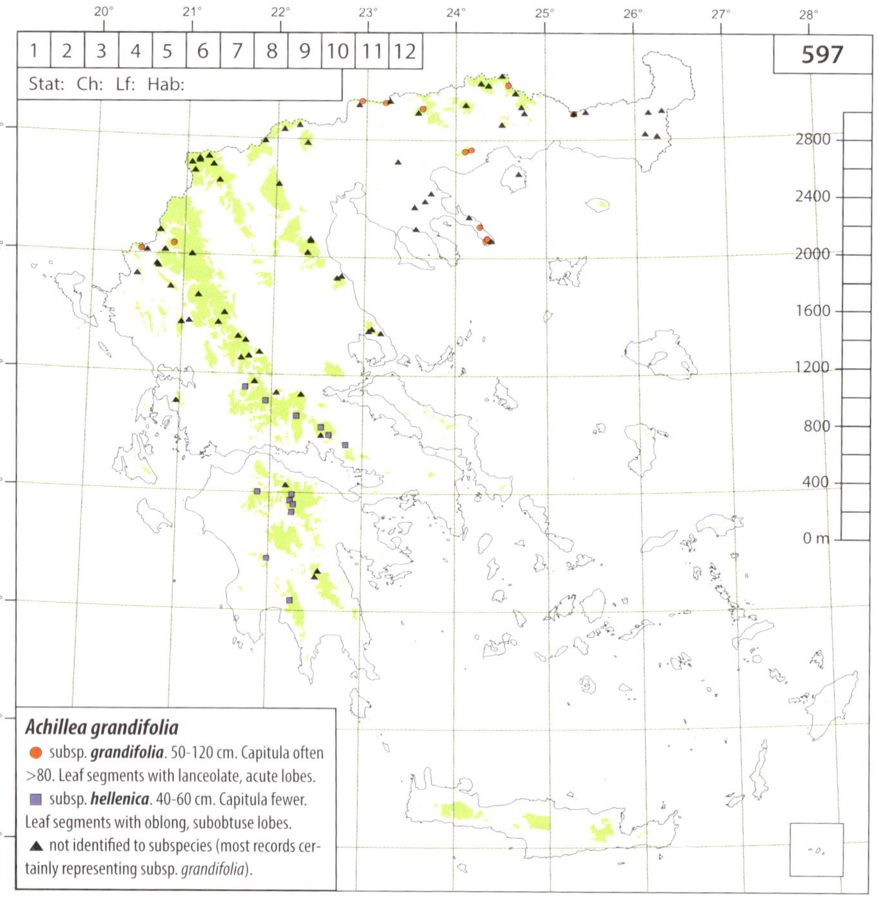

Achillea grandifolia

● subsp. ***grandifolia***. 50-120 cm. Capitula often >80. Leaf segments with lanceolate, acute lobes.
■ subsp. ***hellenica***. 40-60 cm. Capitula fewer. Leaf segments with oblong, subobtuse lobes.
▲ not identified to subspecies (most records certainly representing subsp. *grandifolia*).

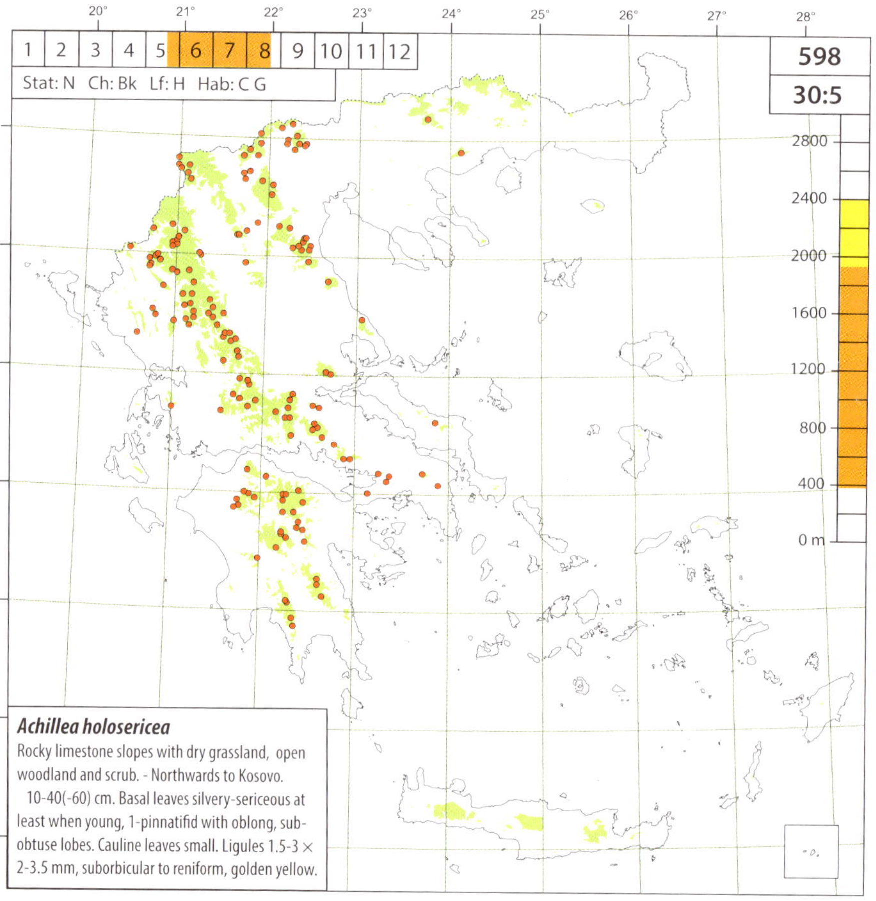

Achillea holosericea

Rocky limestone slopes with dry grassland, open woodland and scrub. - Northwards to Kosovo.
10-40(-60) cm. Basal leaves silvery-sericeous at least when young, 1-pinnatifid with oblong, subobtuse lobes. Cauline leaves small. Ligules 1.5-3 × 2-3.5 mm, suborbicular to reniform, golden yellow.

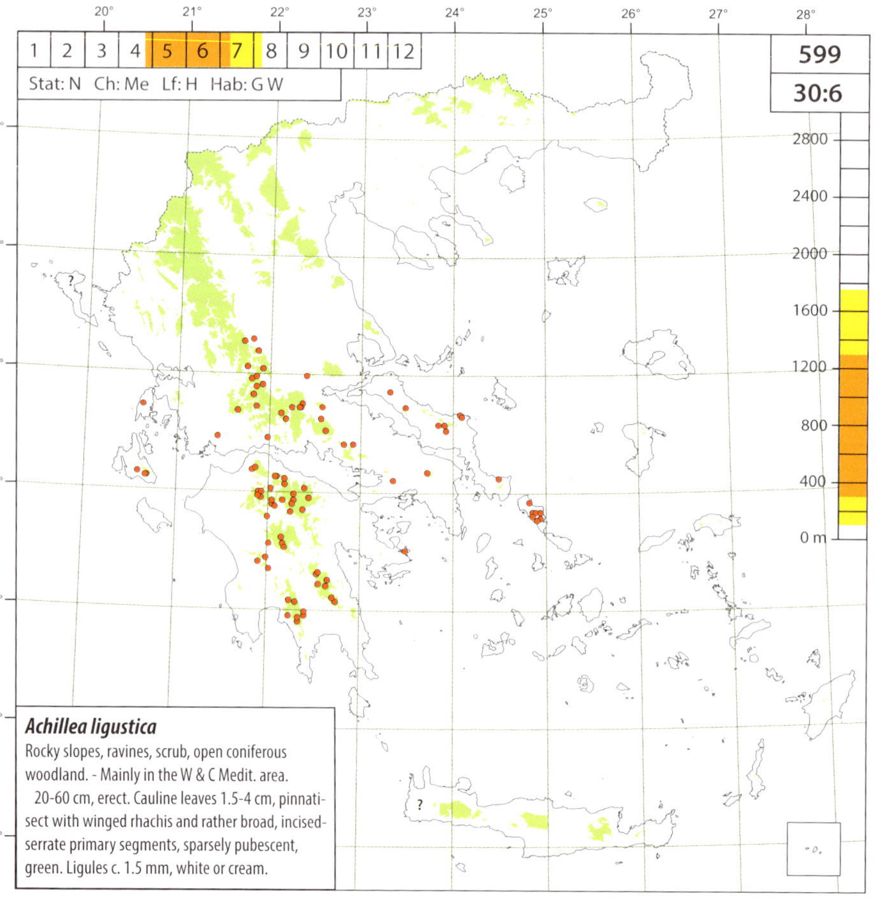

Achillea ligustica

Rocky slopes, ravines, scrub, open coniferous woodland. - Mainly in the W & C Medit. area.
20-60 cm, erect. Cauline leaves 1.5-4 cm, pinnatisect with winged rhachis and rather broad, incised-serrate primary segments, sparsely pubescent, green. Ligules c. 1.5 mm, white or cream.

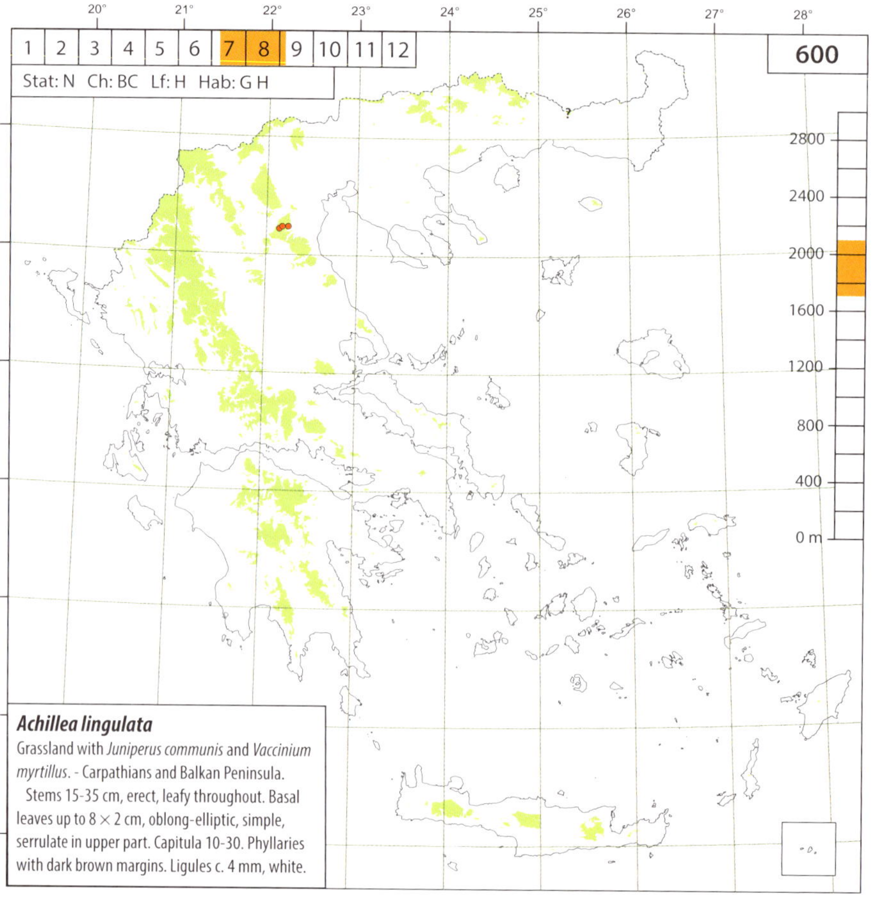

Achillea lingulata

Grassland with *Juniperus communis* and *Vaccinium myrtillus*. - Carpathians and Balkan Peninsula.
Stems 15-35 cm, erect, leafy throughout. Basal leaves up to 8 × 2 cm, oblong-elliptic, simple, serrulate in upper part. Capitula 10-30. Phyllaries with dark brown margins. Ligules c. 4 mm, white.

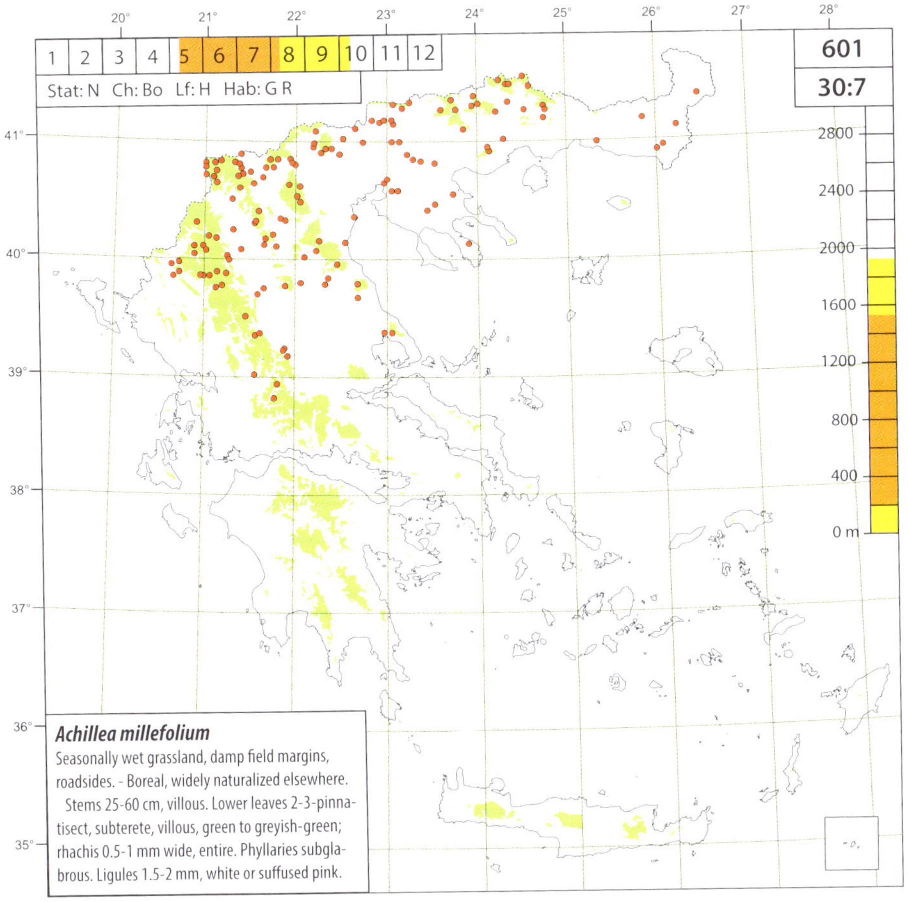

Achillea millefolium

Seasonally wet grassland, damp field margins, roadsides. - Boreal, widely naturalized elsewhere.

Stems 25-60 cm, villous. Lower leaves 2-3-pinnatisect, subterete, villous, green to greyish-green; rhachis 0.5-1 mm wide, entire. Phyllaries subglabrous. Ligules 1.5-2 mm, white or suffused pink.

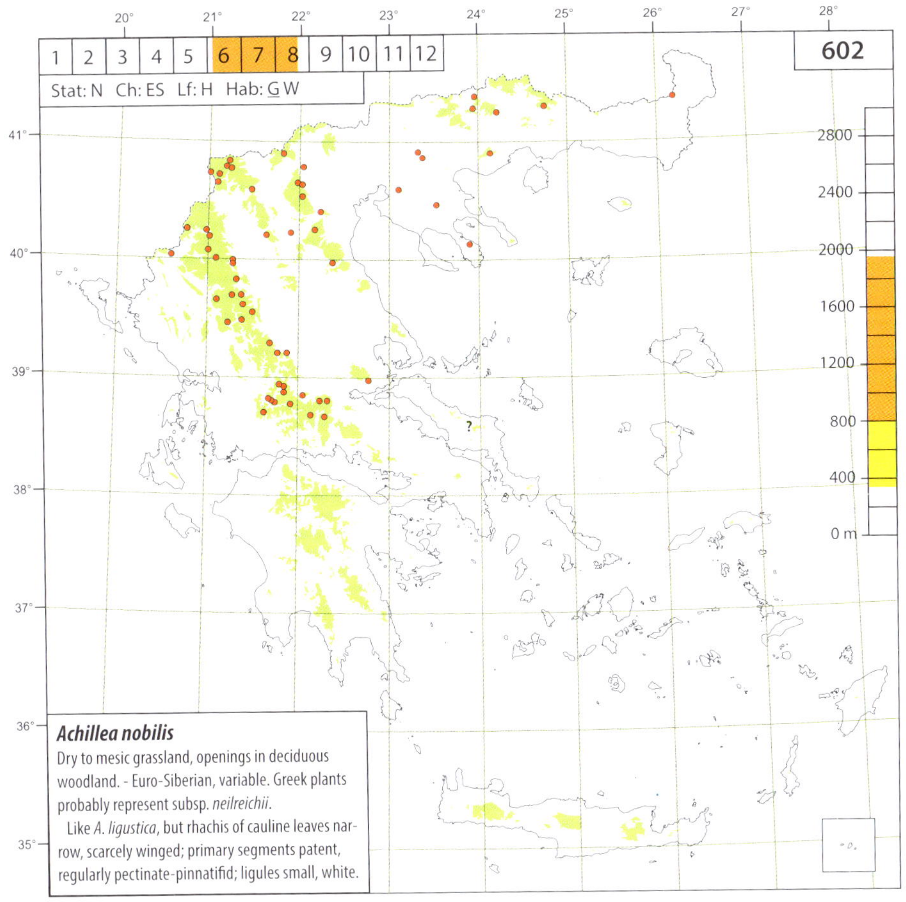

Achillea nobilis

Dry to mesic grassland, openings in deciduous woodland. - Euro-Siberian, variable. Greek plants probably represent subsp. *neilreichii*.

Like *A. ligustica*, but rhachis of cauline leaves narrow, scarcely winged; primary segments patent, regularly pectinate-pinnatifid; ligules small, white.

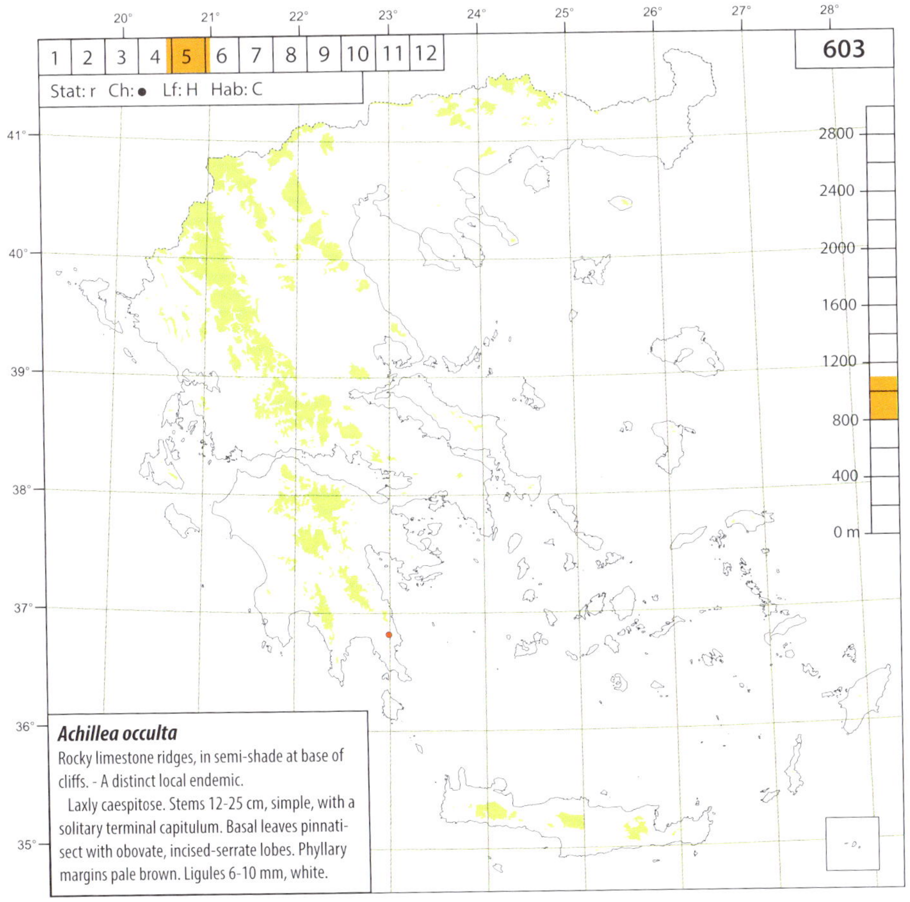

Achillea occulta

Rocky limestone ridges, in semi-shade at base of cliffs. - A distinct local endemic.

Laxly caespitose. Stems 12-25 cm, simple, with a solitary terminal capitulum. Basal leaves pinnatisect with obovate, incised-serrate lobes. Phyllary margins pale brown. Ligules 6-10 mm, white.

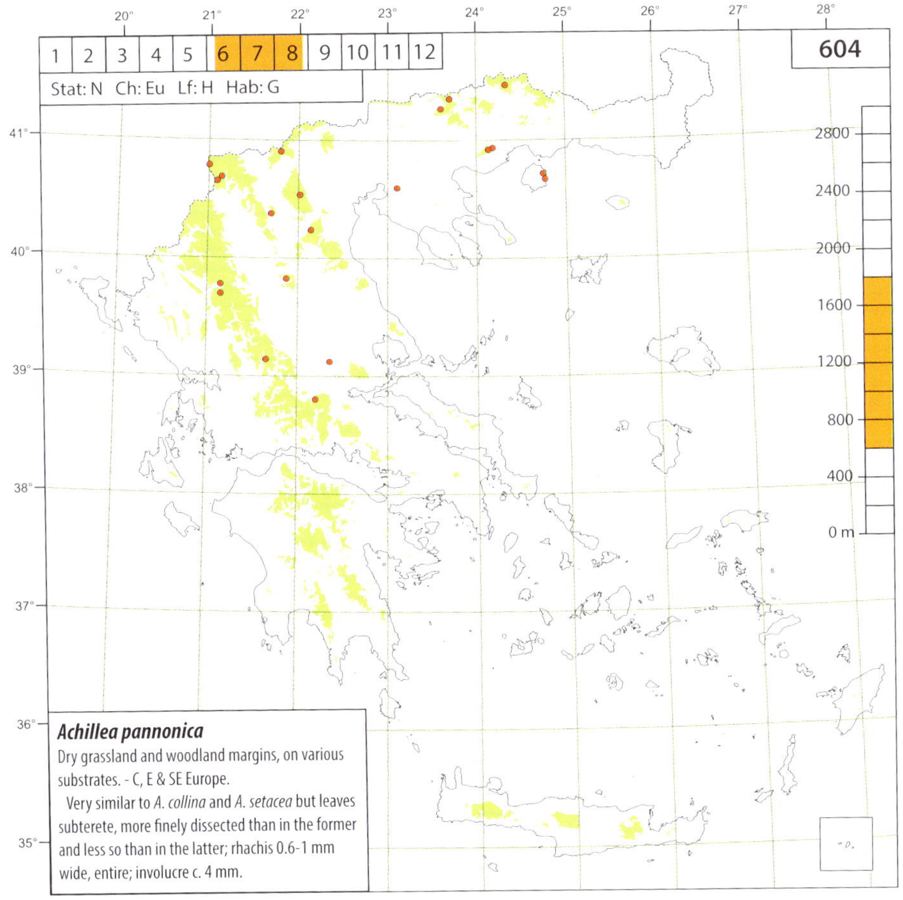

Achillea pannonica

Dry grassland and woodland margins, on various substrates. - C, E & SE Europe.

Very similar to *A. collina* and *A. setacea* but leaves subterete, more finely dissected than in the former and less so than in the latter; rhachis 0.6-1 mm wide, entire; involucre c. 4 mm.

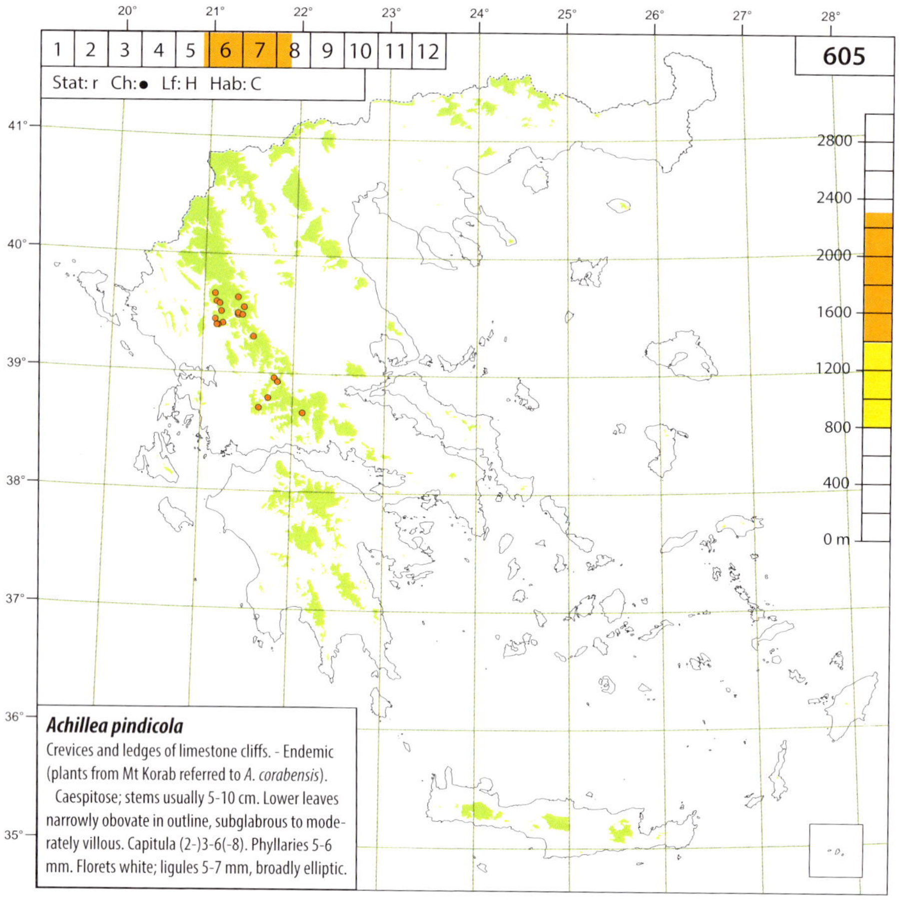

Achillea pindicola
Crevices and ledges of limestone cliffs. - Endemic (plants from Mt Korab referred to *A. corabensis*).
Caespitose; stems usually 5-10 cm. Lower leaves narrowly obovate in outline, subglabrous to moderately villous. Capitula (2-)3-6(-8). Phyllaries 5-6 mm. Florets white; ligules 5-7 mm, broadly elliptic.

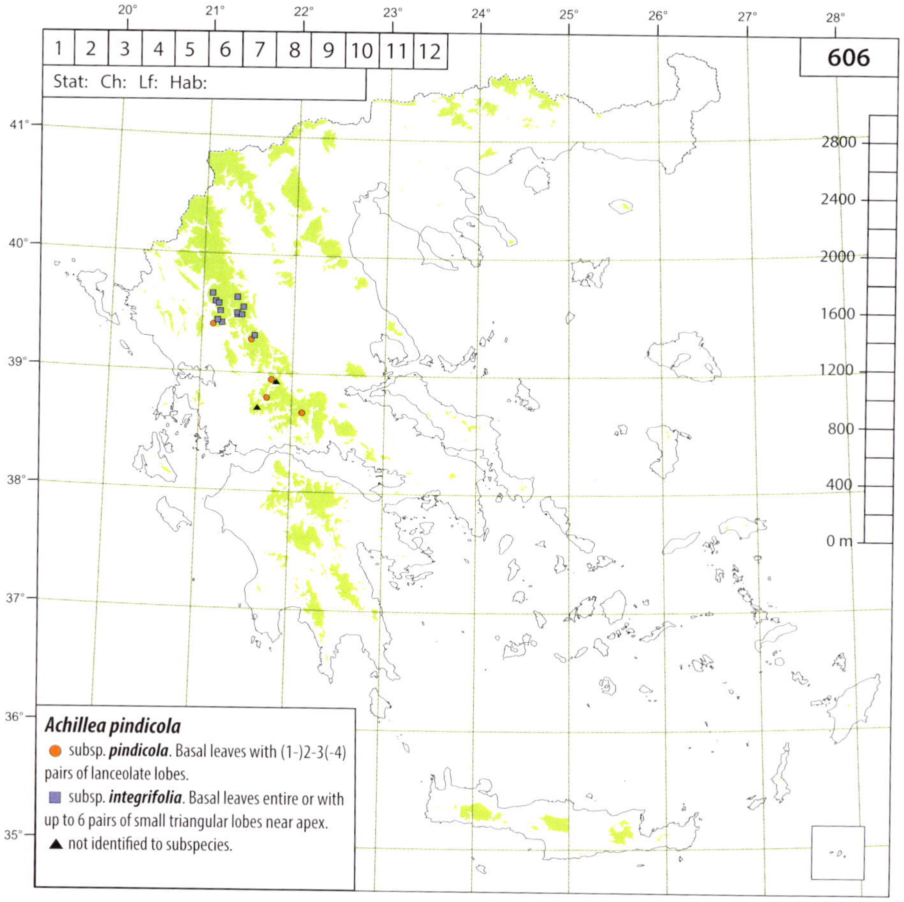

Achillea pindicola
● subsp. ***pindicola***. Basal leaves with (1-)2-3(-4) pairs of lanceolate lobes.
■ subsp. ***integrifolia***. Basal leaves entire or with up to 6 pairs of small triangular lobes near apex.
▲ not identified to subspecies.

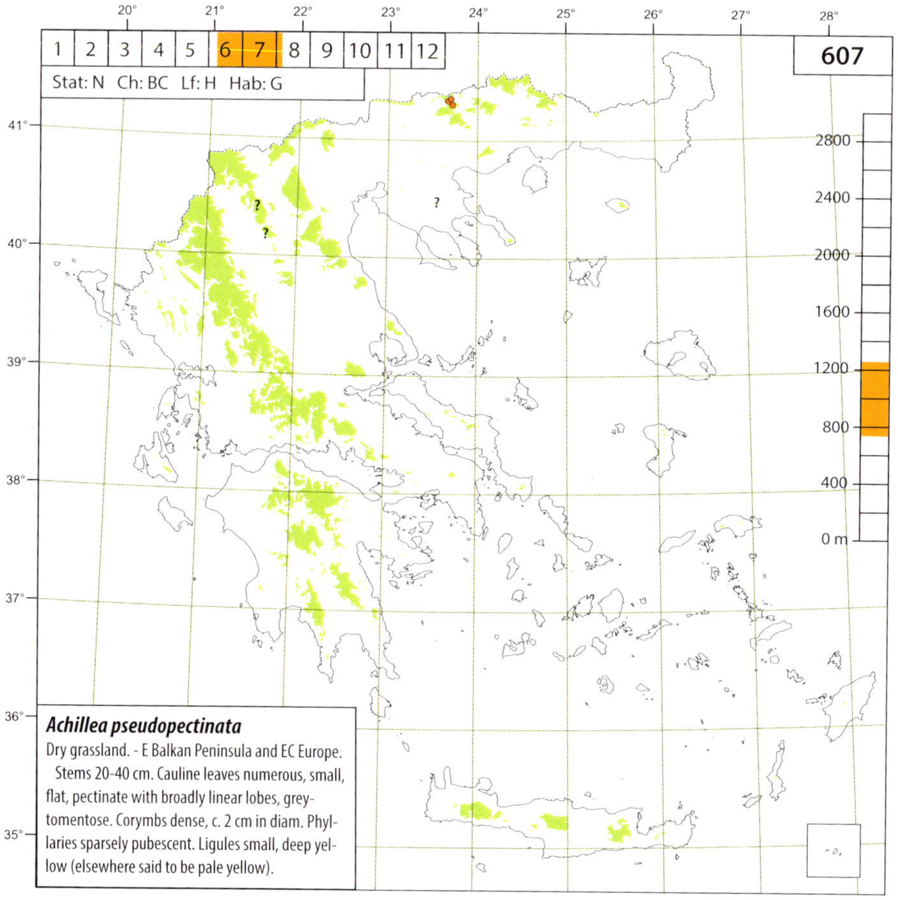

Achillea pseudopectinata
Dry grassland. - E Balkan Peninsula and EC Europe.
Stems 20-40 cm. Cauline leaves numerous, small, flat, pectinate with broadly linear lobes, grey-tomentose. Corymbs dense, c. 2 cm in diam. Phyllaries sparsely pubescent. Ligules small, deep yellow (elsewhere said to be pale yellow).

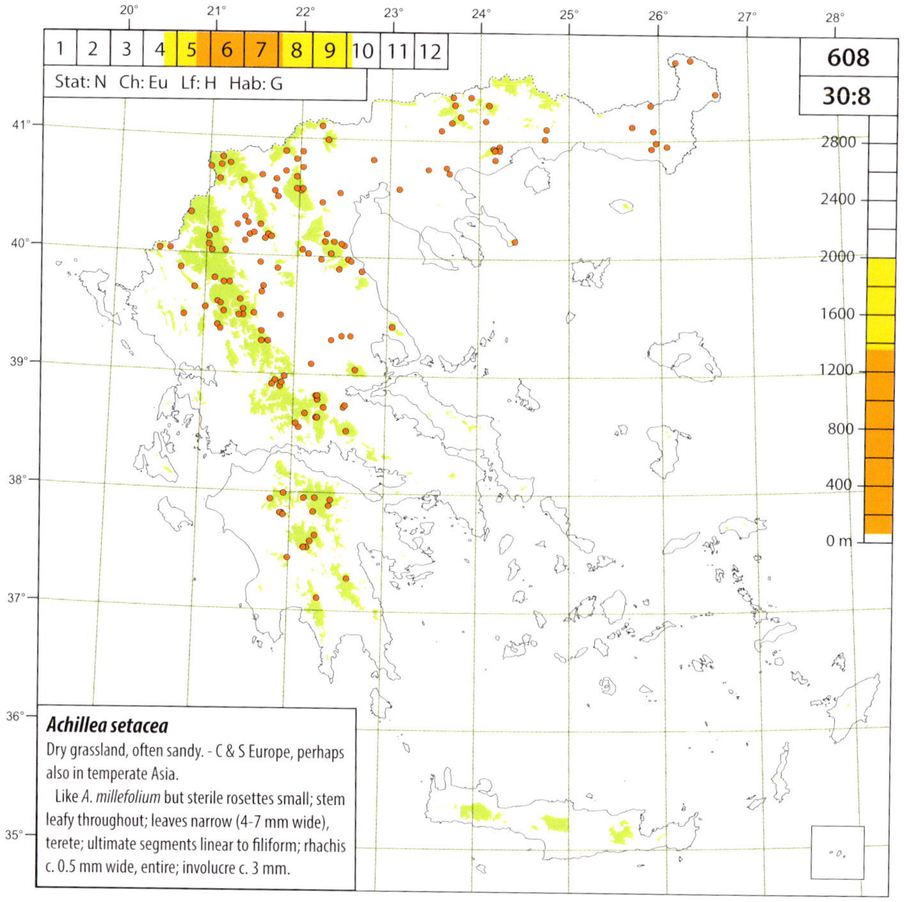

Achillea setacea
Dry grassland, often sandy. - C & S Europe, perhaps also in temperate Asia.
Like *A. millefolium* but sterile rosettes small; stem leafy throughout; leaves narrow (4-7 mm wide), terete; ultimate segments linear to filiform; rhachis c. 0.5 mm wide, entire; involucre c. 3 mm.

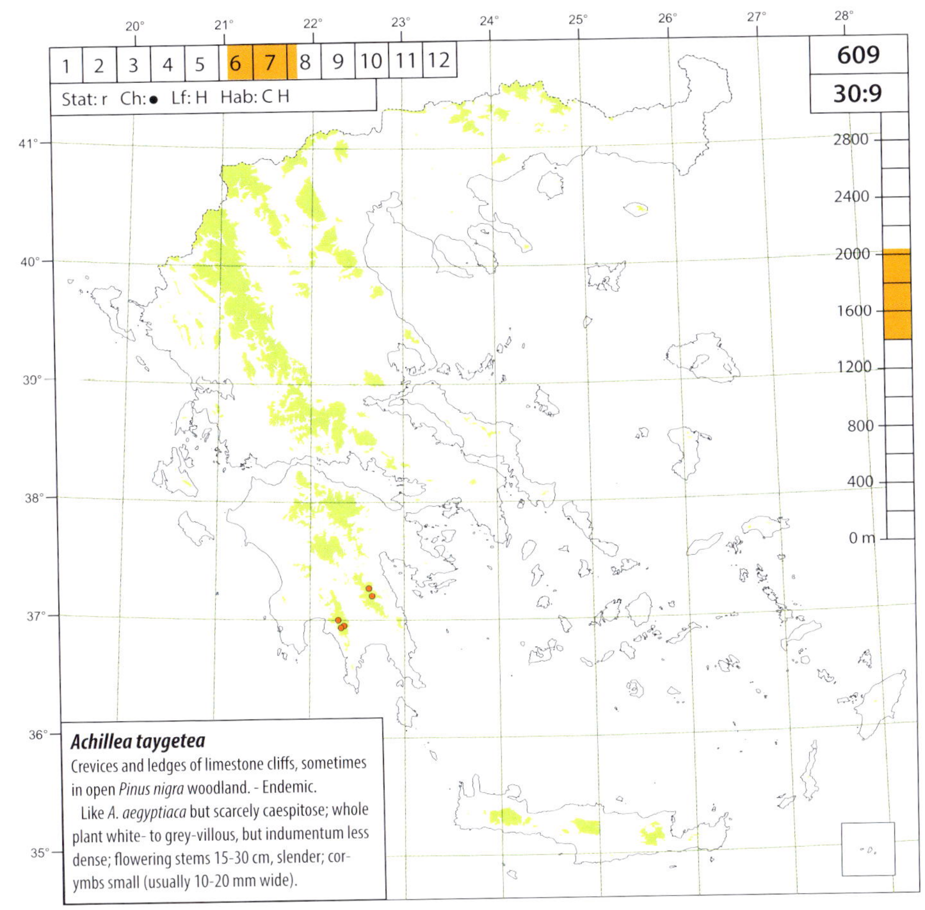

Achillea taygetea

Crevices and ledges of limestone cliffs, sometimes in open *Pinus nigra* woodland. - Endemic.

Like *A. aegyptiaca* but scarcely caespitose; whole plant white- to grey-villous, but indumentum less dense; flowering stems 15-30 cm, slender; corymbs small (usually 10-20 mm wide).

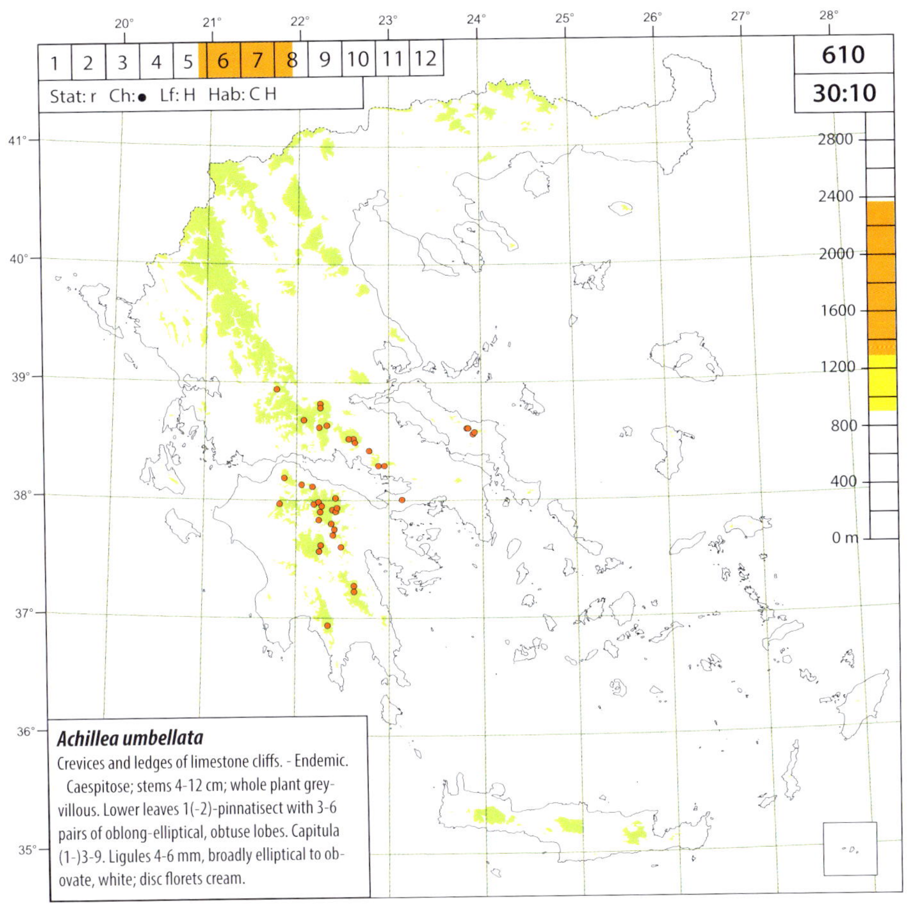

Achillea umbellata

Crevices and ledges of limestone cliffs. - Endemic.

Caespitose; stems 4-12 cm; whole plant grey-villous. Lower leaves 1(-2)-pinnatisect with 3-6 pairs of oblong-elliptical, obtuse lobes. Capitula (1-)3-9. Ligules 4-6 mm, broadly elliptical to obovate, white; disc florets cream.

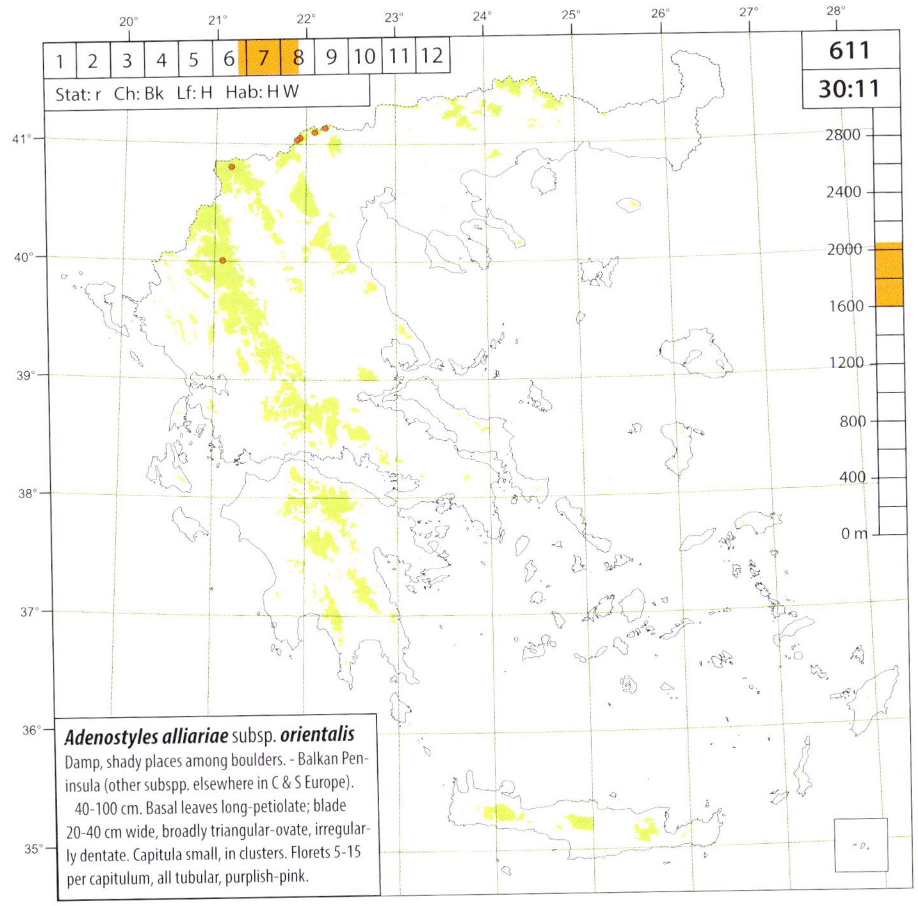

Adenostyles alliariae* subsp. *orientalis

Damp, shady places among boulders. - Balkan Peninsula (other subspp. elsewhere in C & S Europe).

40-100 cm. Basal leaves long-petiolate; blade 20-40 cm wide, broadly triangular-ovate, irregularly dentate. Capitula small, in clusters. Florets 5-15 per capitulum, all tubular, purplish-pink.

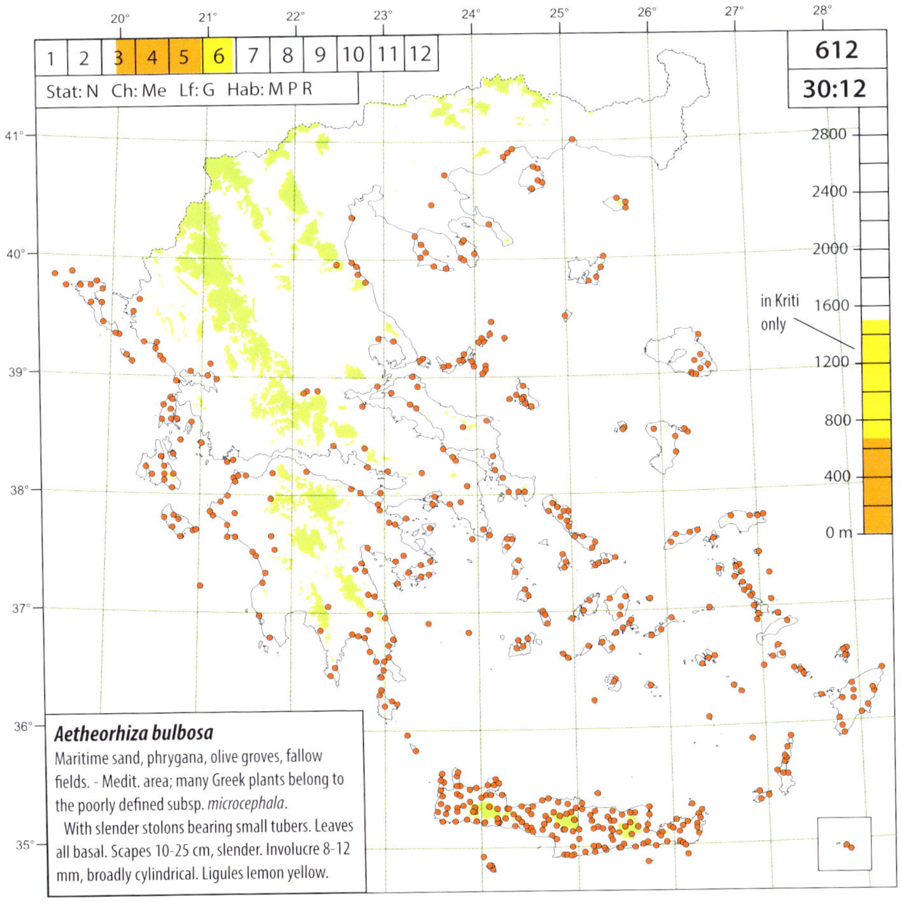

Aetheorhiza bulbosa

Maritime sand, phrygana, olive groves, fallow fields. - Medit. area; many Greek plants belong to the poorly defined subsp. *microcephala*.

With slender stolons bearing small tubers. Leaves all basal. Scapes 10-25 cm, slender. Involucre 8-12 mm, broadly cylindrical. Ligules lemon yellow.

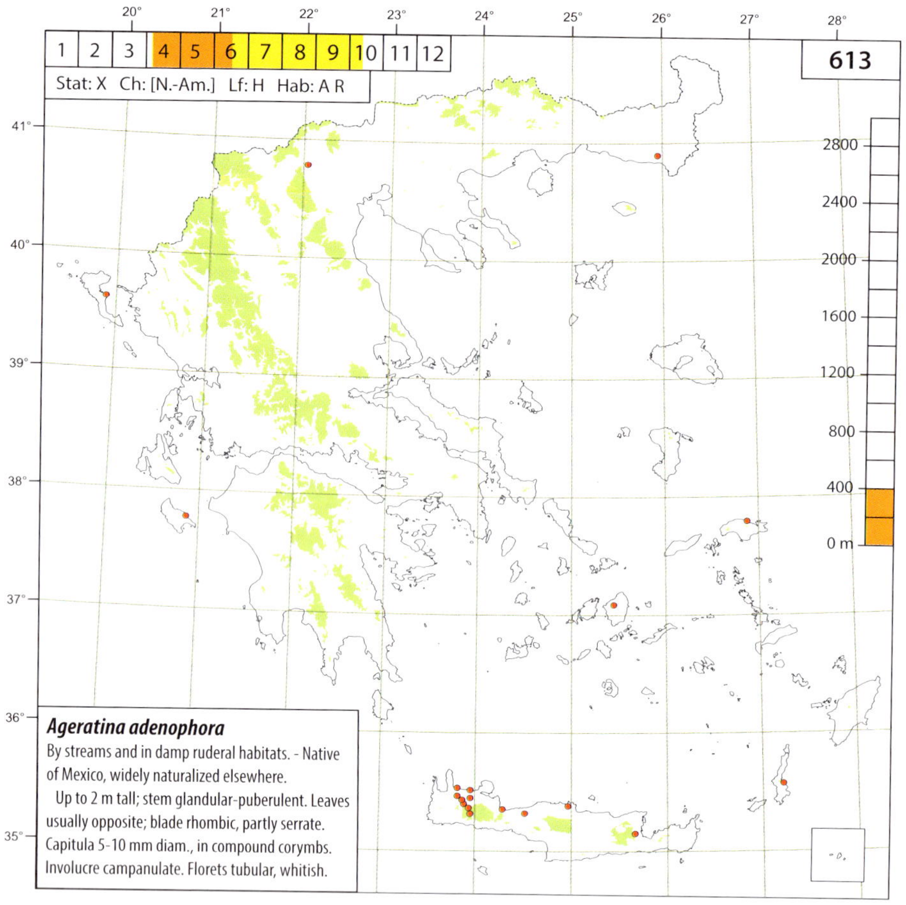

Ageratina adenophora

By streams and in damp ruderal habitats. - Native of Mexico, widely naturalized elsewhere.

Up to 2 m tall; stem glandular-puberulent. Leaves usually opposite; blade rhombic, partly serrate. Capitula 5-10 mm diam., in compound corymbs. Involucre campanulate. Florets tubular, whitish.

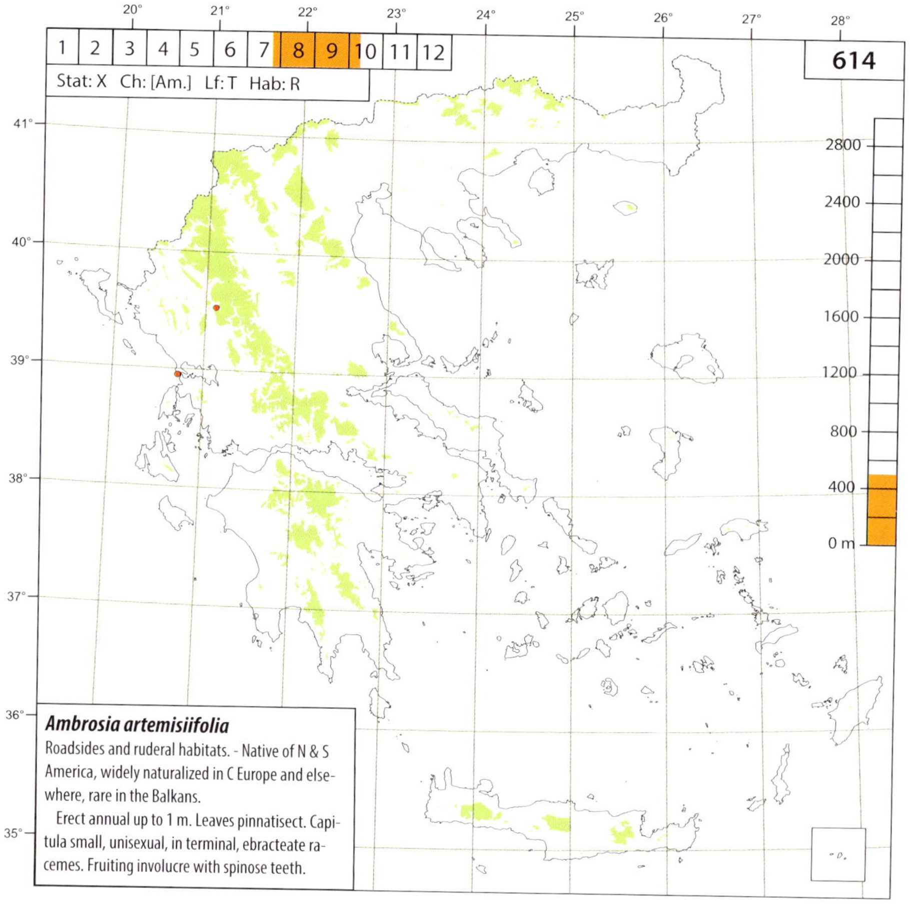

Ambrosia artemisiifolia

Roadsides and ruderal habitats. - Native of N & S America, widely naturalized in C Europe and elsewhere, rare in the Balkans.

Erect annual up to 1 m. Leaves pinnatisect. Capitula small, unisexual, in terminal, ebracteate racemes. Fruiting involucre with spinose teeth.

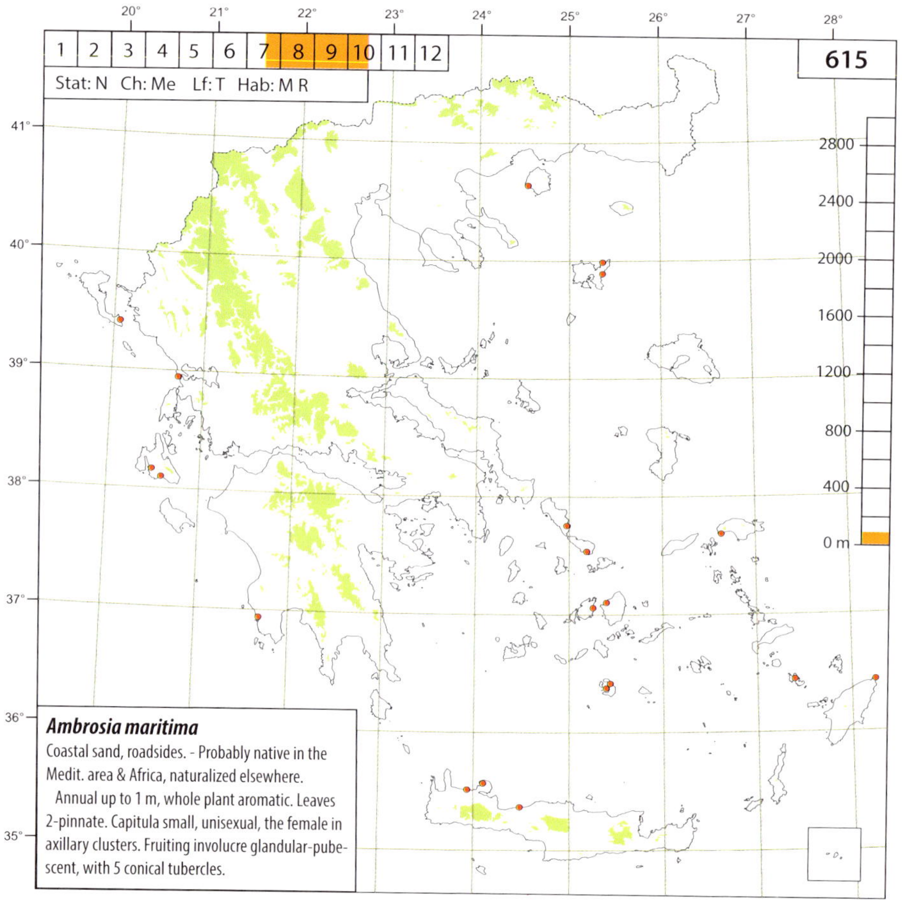

Ambrosia maritima

Coastal sand, roadsides. - Probably native in the Medit. area & Africa, naturalized elsewhere.

Annual up to 1 m, whole plant aromatic. Leaves 2-pinnate. Capitula small, unisexual, the female in axillary clusters. Fruiting involucre glandular-pubescent, with 5 conical tubercles.

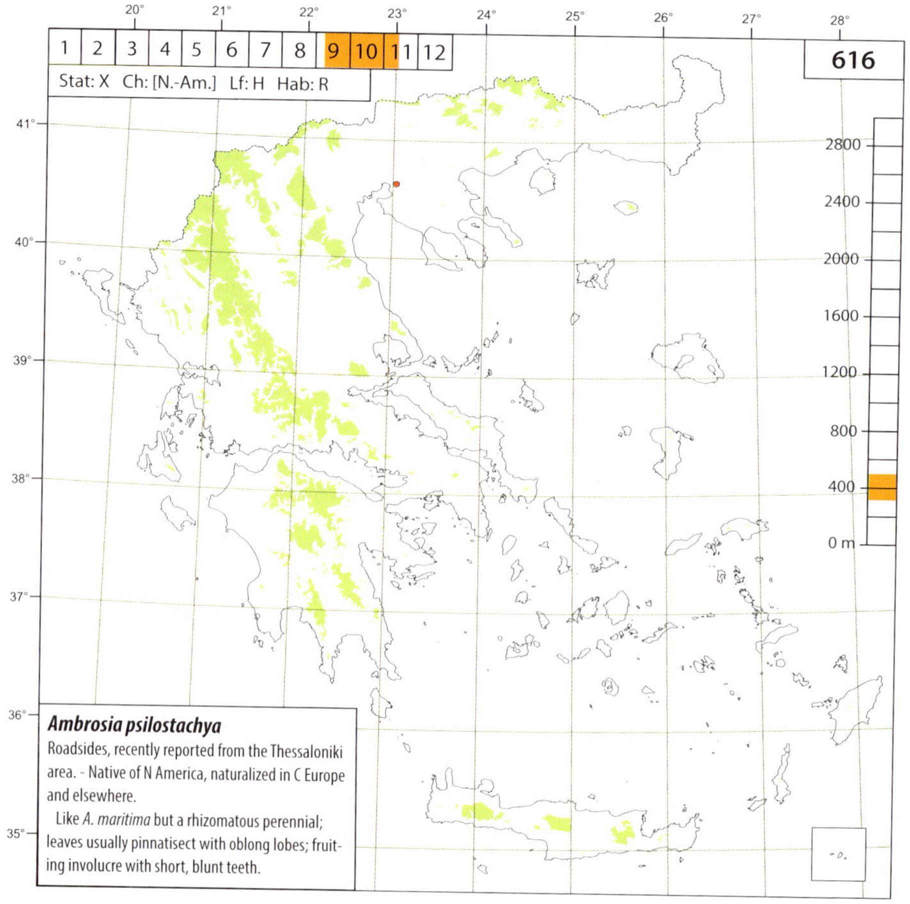

Ambrosia psilostachya

Roadsides, recently reported from the Thessaloniki area. - Native of N America, naturalized in C Europe and elsewhere.

Like *A. maritima* but a rhizomatous perennial; leaves usually pinnatisect with oblong lobes; fruiting involucre with short, blunt teeth.

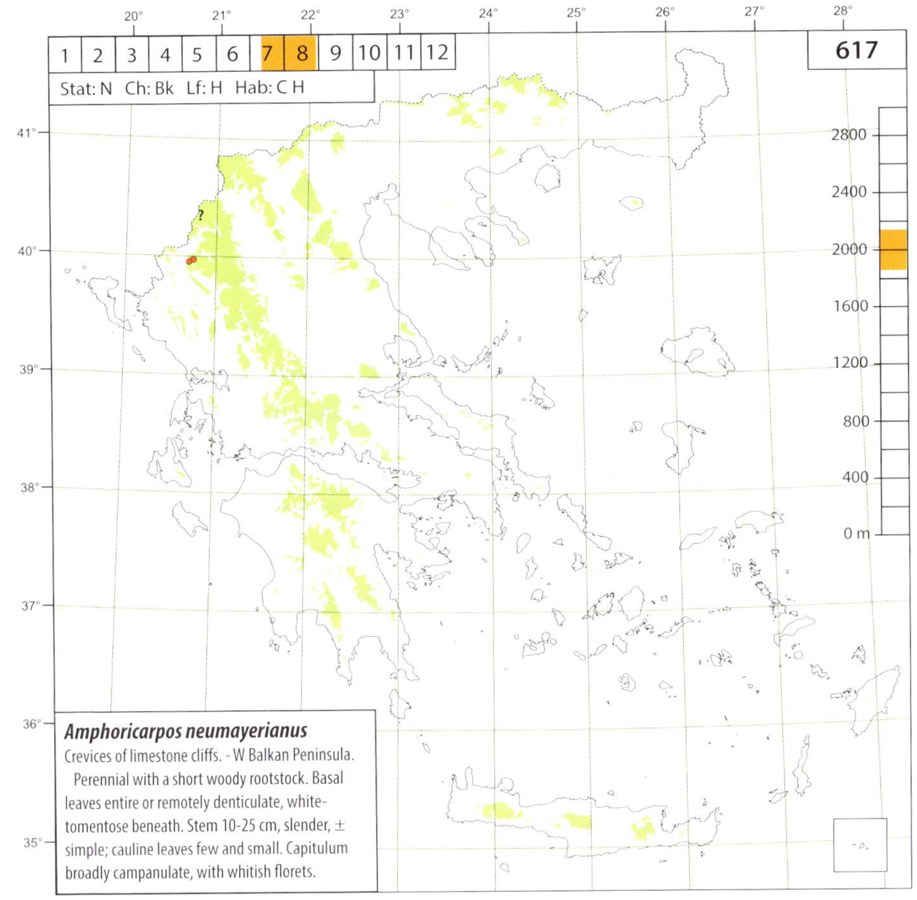

Amphoricarpos neumayerianus
Crevices of limestone cliffs. - W Balkan Peninsula.
Perennial with a short woody rootstock. Basal leaves entire or remotely denticulate, white-tomentose beneath. Stem 10-25 cm, slender, ± simple; cauline leaves few and small. Capitulum broadly campanulate, with whitish florets.

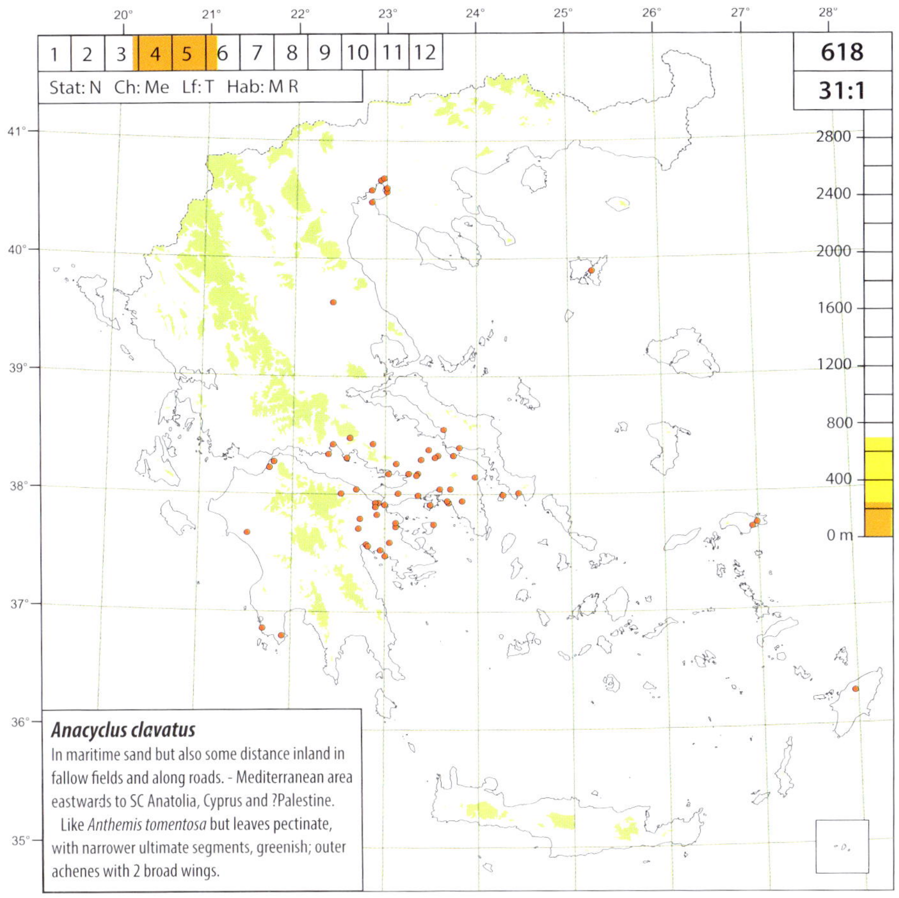

Anacyclus clavatus
In maritime sand but also some distance inland in fallow fields and along roads. - Mediterranean area eastwards to SC Anatolia, Cyprus and ?Palestine.
Like *Anthemis tomentosa* but leaves pectinate, with narrower ultimate segments, greenish; outer achenes with 2 broad wings.

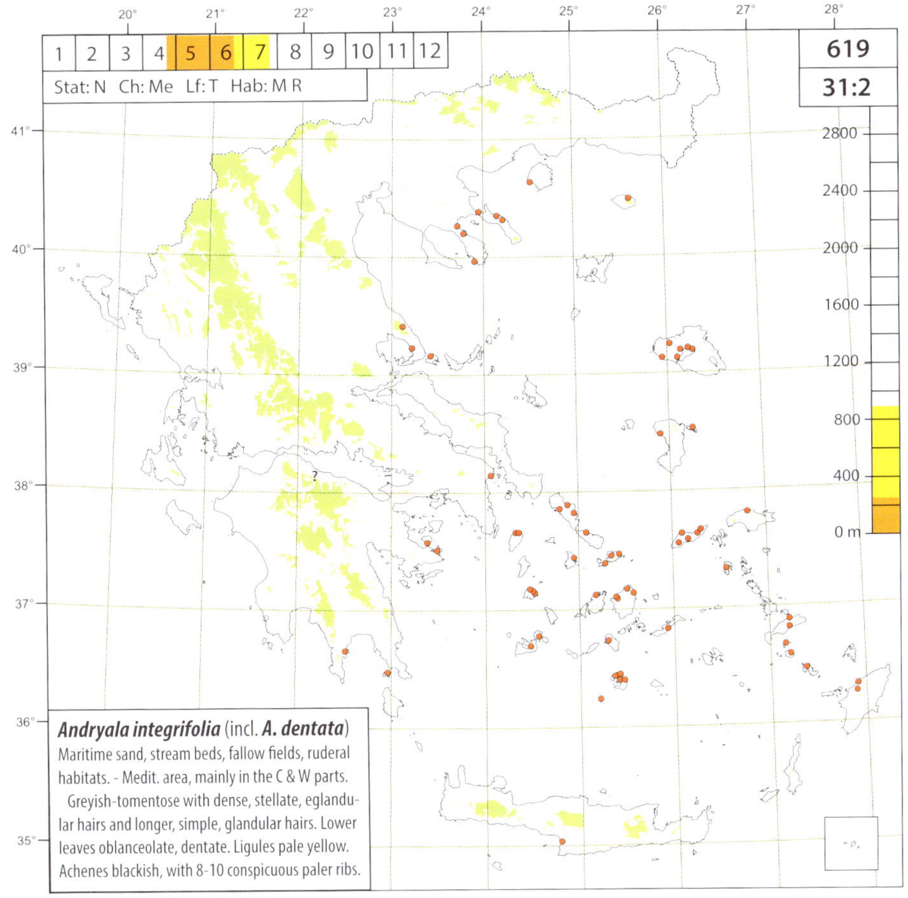

Andryala integrifolia (incl. **A. dentata**)
Maritime sand, stream beds, fallow fields, ruderal habitats. - Medit. area, mainly in the C & W parts.
Greyish-tomentose with dense, stellate, eglandular hairs and longer, simple, glandular hairs. Lower leaves oblanceolate, dentate. Ligules pale yellow. Achenes blackish, with 8-10 conspicuous paler ribs.

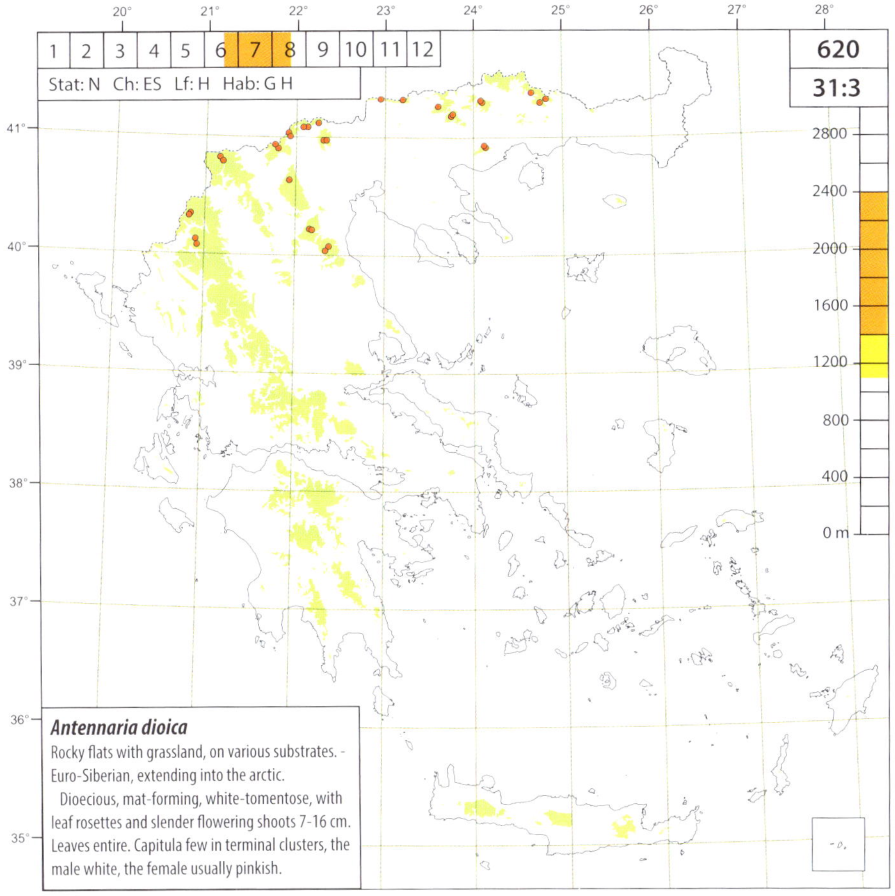

Antennaria dioica
Rocky flats with grassland, on various substrates. - Euro-Siberian, extending into the arctic.
Dioecious, mat-forming, white-tomentose, with leaf rosettes and slender flowering shoots 7-16 cm. Leaves entire. Capitula few in terminal clusters, the male white, the female usually pinkish.

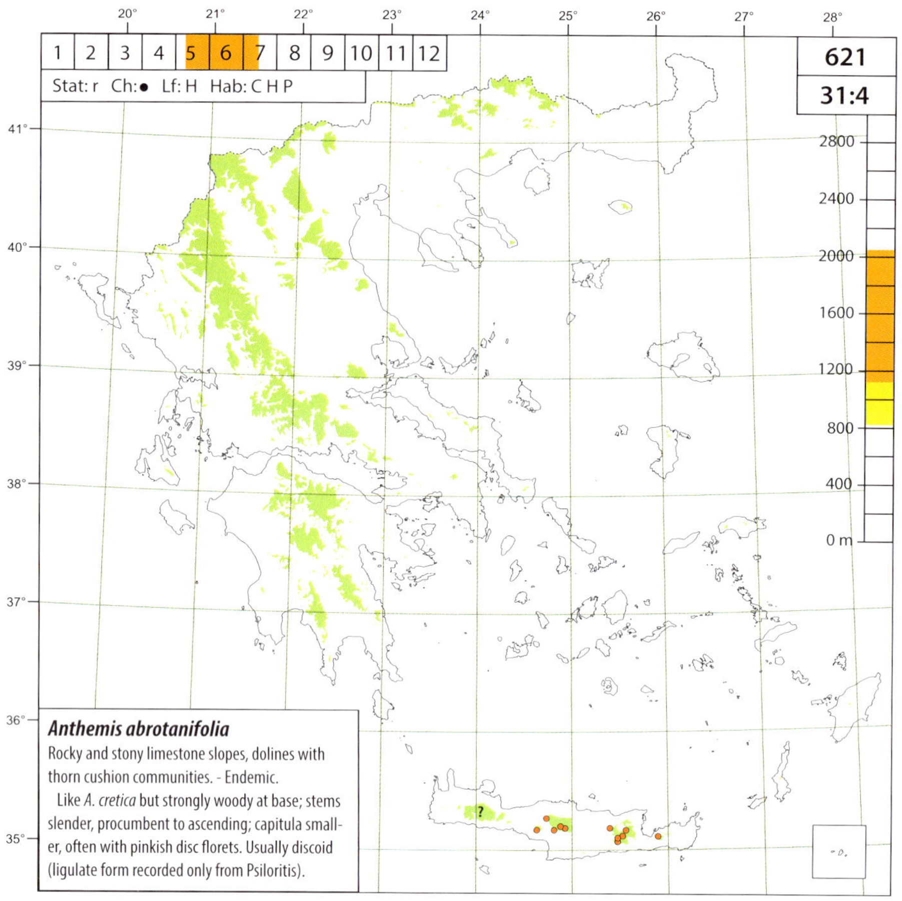

Anthemis abrotanifolia

Rocky and stony limestone slopes, dolines with thorn cushion communities. - Endemic.

Like *A. cretica* but strongly woody at base; stems slender, procumbent to ascending; capitula smaller, often with pinkish disc florets. Usually discoid (ligulate form recorded only from Psiloritis).

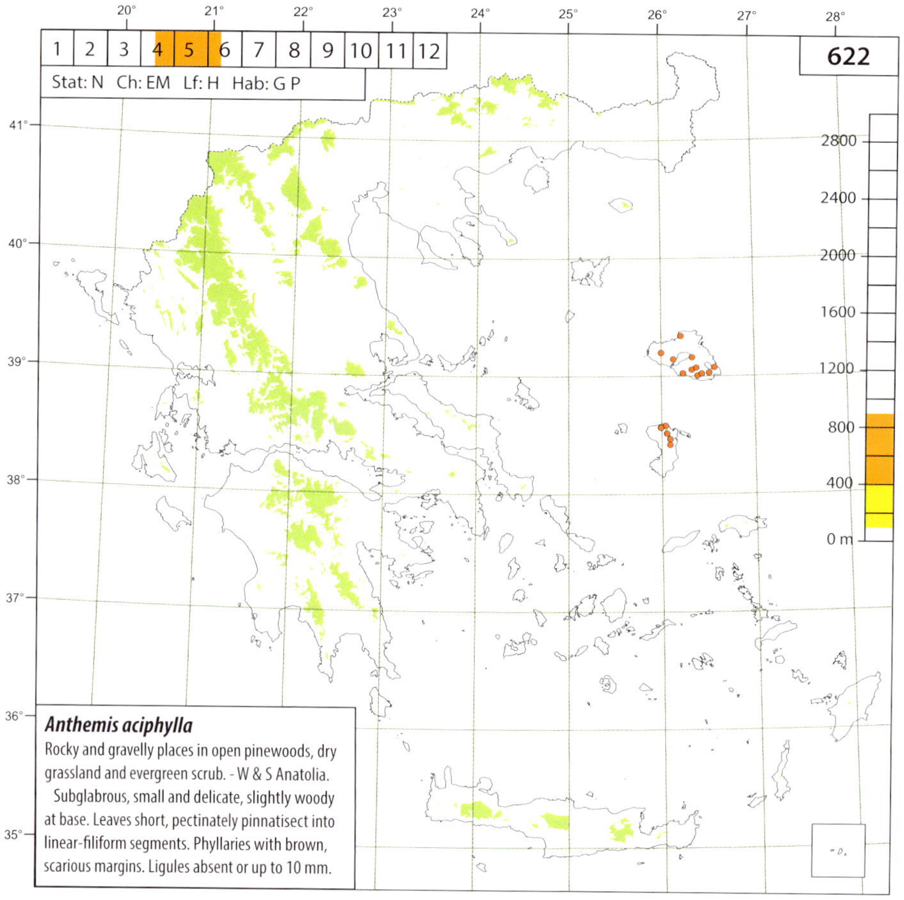

Anthemis aciphylla

Rocky and gravelly places in open pinewoods, dry grassland and evergreen scrub. - W & S Anatolia.

Subglabrous, small and delicate, slightly woody at base. Leaves short, pectinately pinnatisect into linear-filiform segments. Phyllaries with brown, scarious margins. Ligules absent or up to 10 mm.

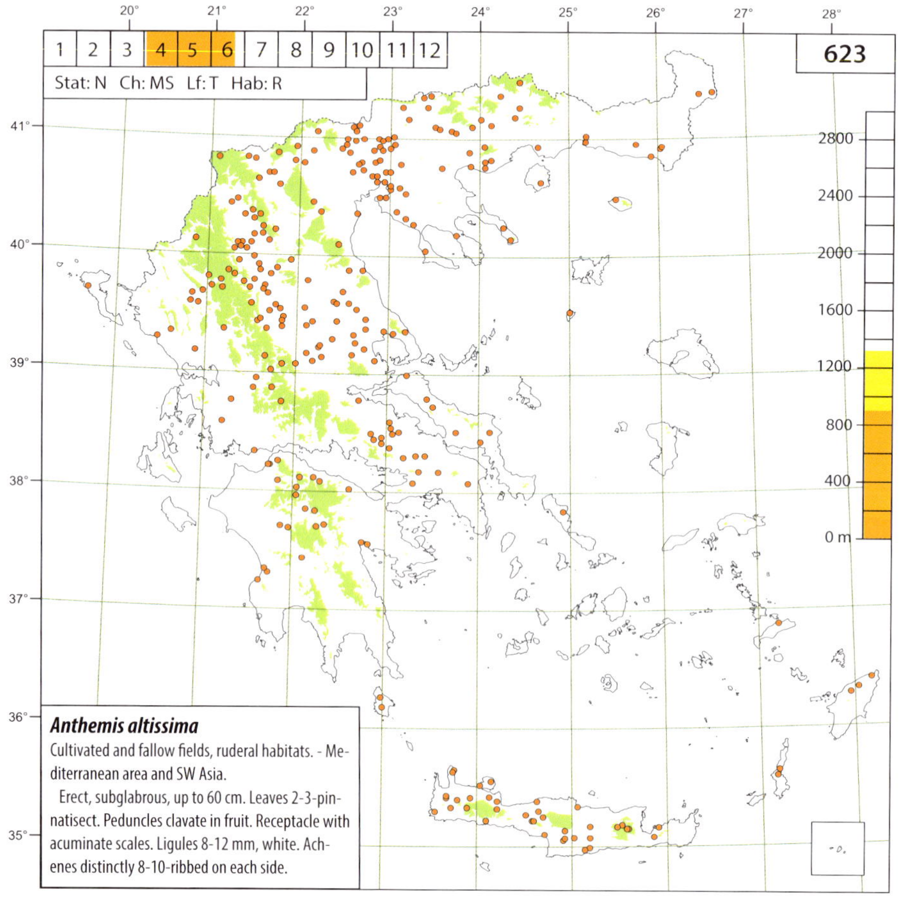

Anthemis altissima

Cultivated and fallow fields, ruderal habitats. - Mediterranean area and SW Asia.

Erect, subglabrous, up to 60 cm. Leaves 2-3-pinnatisect. Peduncles clavate in fruit. Receptacle with acuminate scales. Ligules 8-12 mm, white. Achenes distinctly 8-10-ribbed on each side.

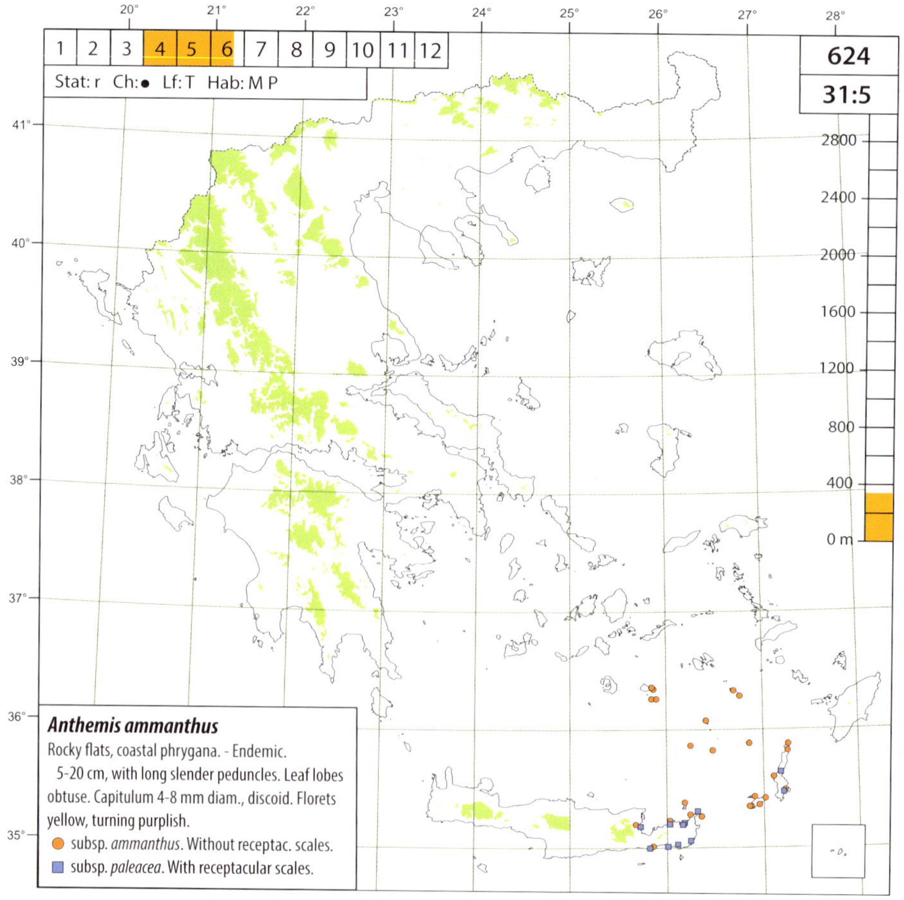

Anthemis ammanthus

Rocky flats, coastal phrygana. - Endemic.

5-20 cm, with long slender peduncles. Leaf lobes obtuse. Capitulum 4-8 mm diam., discoid. Florets yellow, turning purplish.

● subsp. *ammanthus*. Without receptac. scales.

■ subsp. *paleacea*. With receptacular scales.

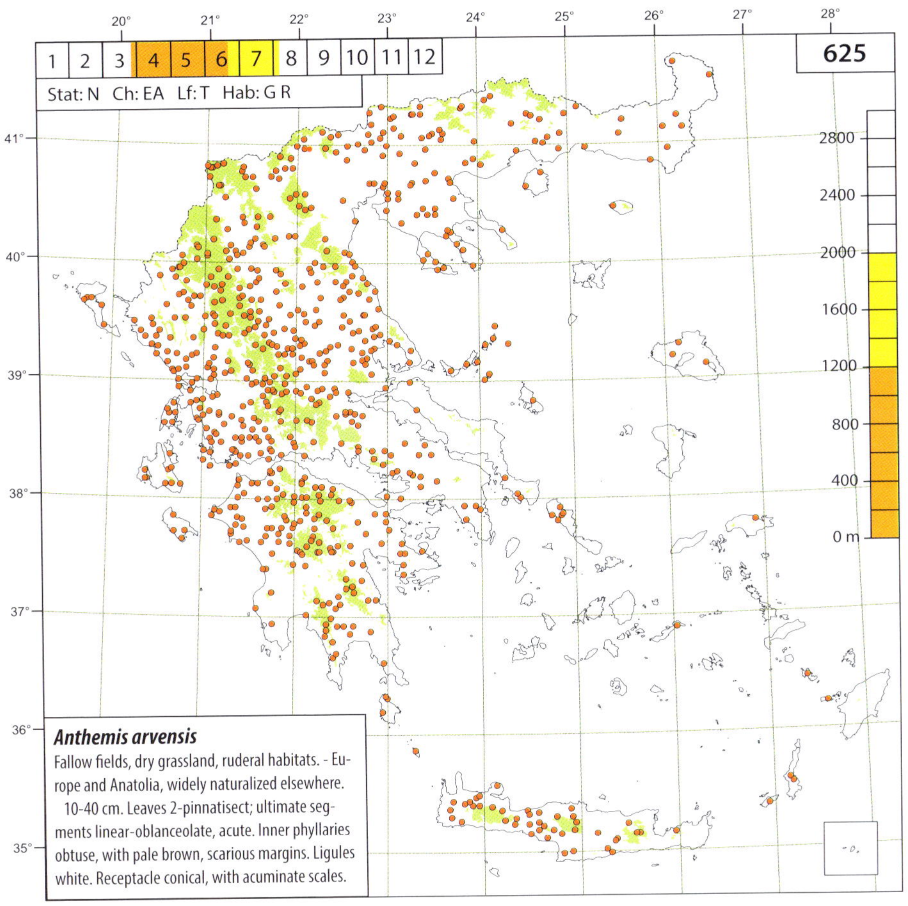

Anthemis arvensis

Fallow fields, dry grassland, ruderal habitats. - Europe and Anatolia, widely naturalized elsewhere.

10-40 cm. Leaves 2-pinnatisect; ultimate segments linear-oblanceolate, acute. Inner phyllaries obtuse, with pale brown, scarious margins. Ligules white. Receptacle conical, with acuminate scales.

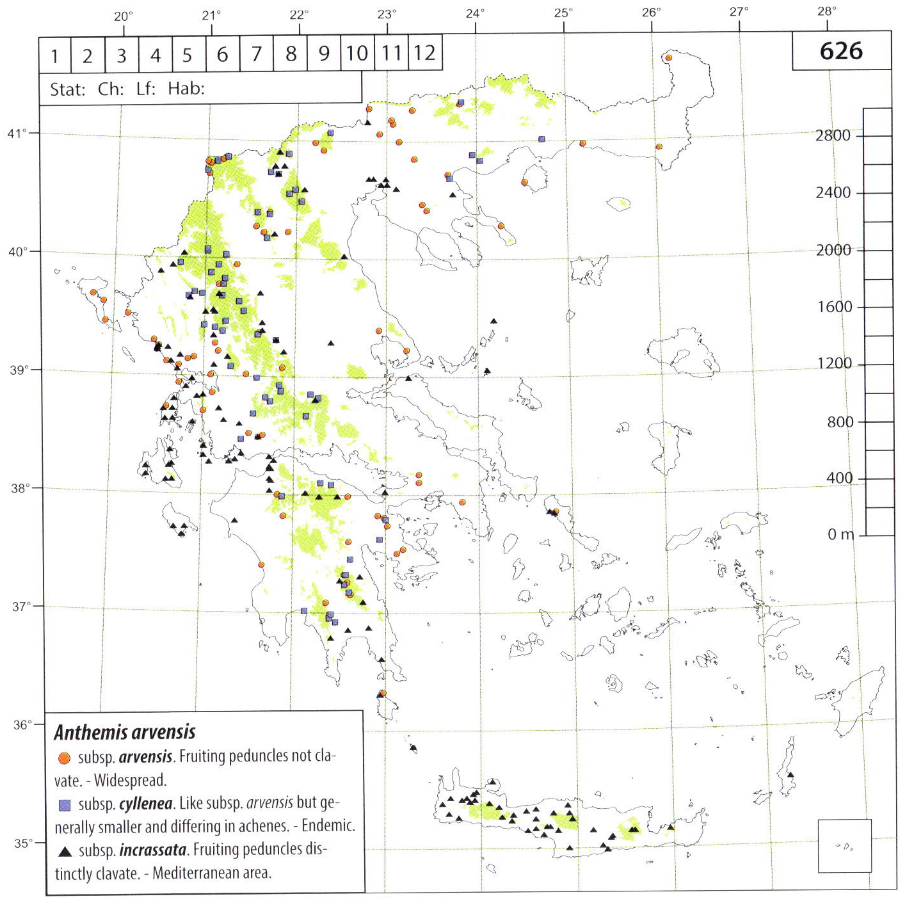

Anthemis arvensis

● subsp. ***arvensis***. Fruiting peduncles not clavate. - Widespread.

■ subsp. ***cyllenea***. Like subsp. *arvensis* but generally smaller and differing in achenes. - Endemic.

▲ subsp. ***incrassata***. Fruiting peduncles distinctly clavate. - Mediterranean area.

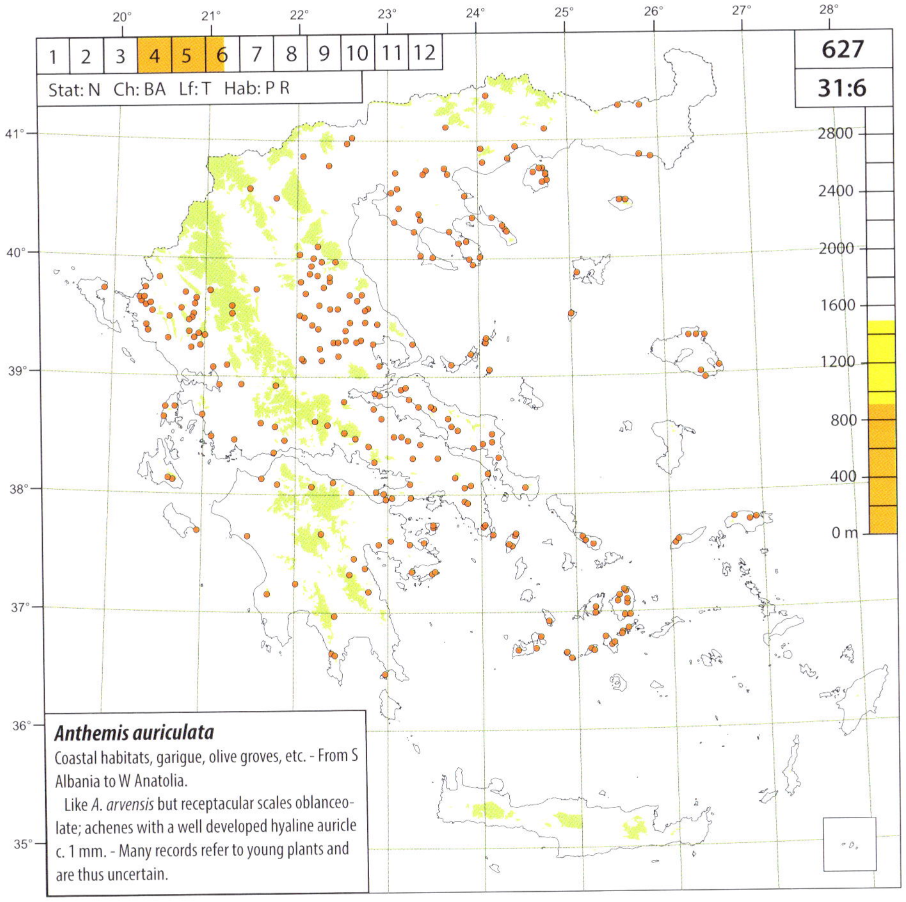

Anthemis auriculata

Coastal habitats, garigue, olive groves, etc. - From S Albania to W Anatolia.

Like *A. arvensis* but receptacular scales oblanceolate; achenes with a well developed hyaline auricle c. 1 mm. - Many records refer to young plants and are thus uncertain.

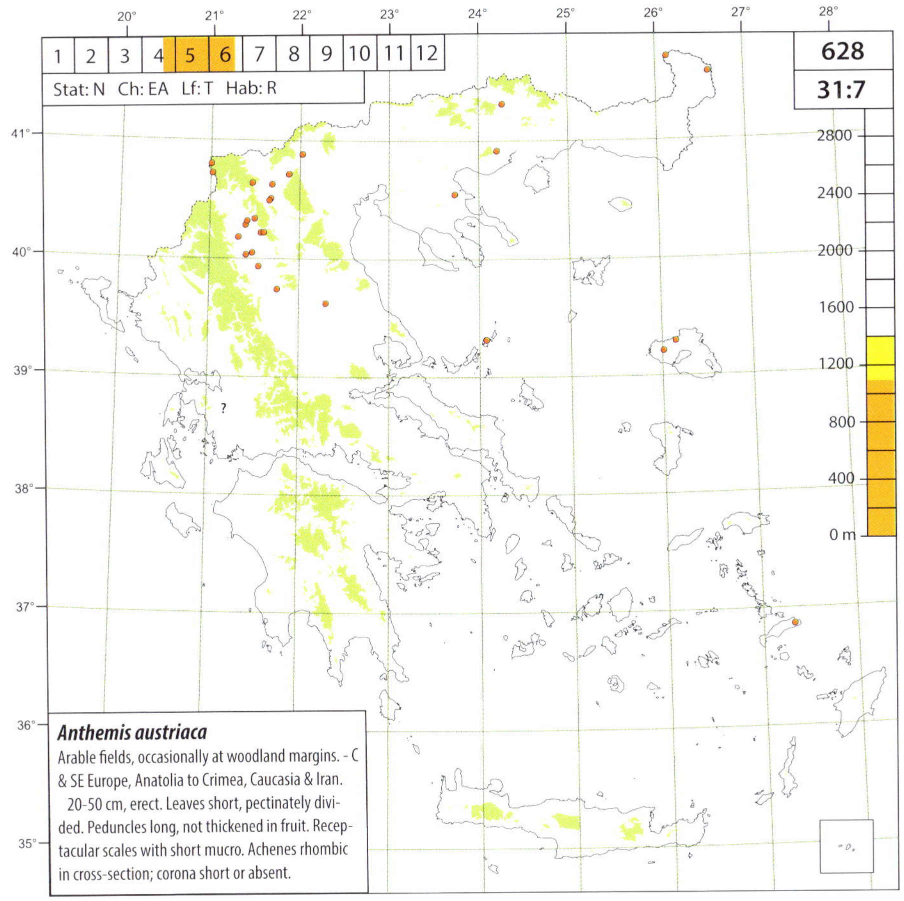

Anthemis austriaca

Arable fields, occasionally at woodland margins. - C & SE Europe, Anatolia to Crimea, Caucasia & Iran.

20-50 cm, erect. Leaves short, pectinately divided. Peduncles long, not thickened in fruit. Receptacular scales with short mucro. Achenes rhombic in cross-section; corona short or absent.

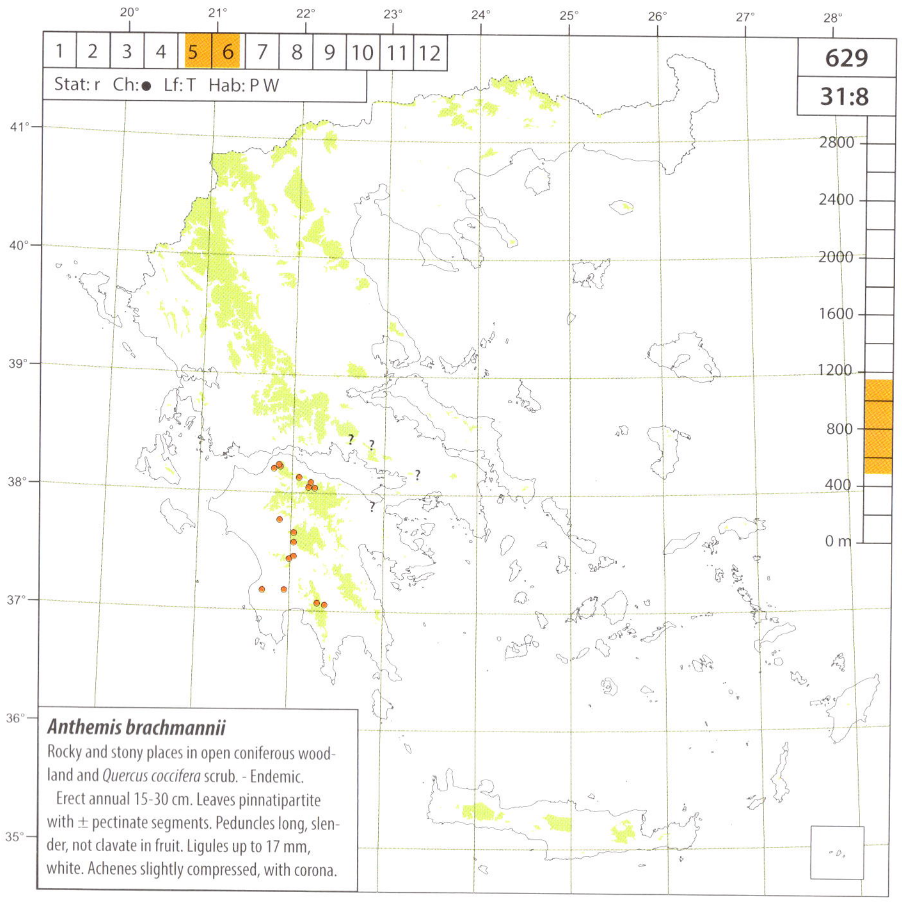

Anthemis brachmannii
Rocky and stony places in open coniferous woodland and *Quercus coccifera* scrub. - Endemic.
Erect annual 15-30 cm. Leaves pinnatipartite with ± pectinate segments. Peduncles long, slender, not clavate in fruit. Ligules up to 17 mm, white. Achenes slightly compressed, with corona.

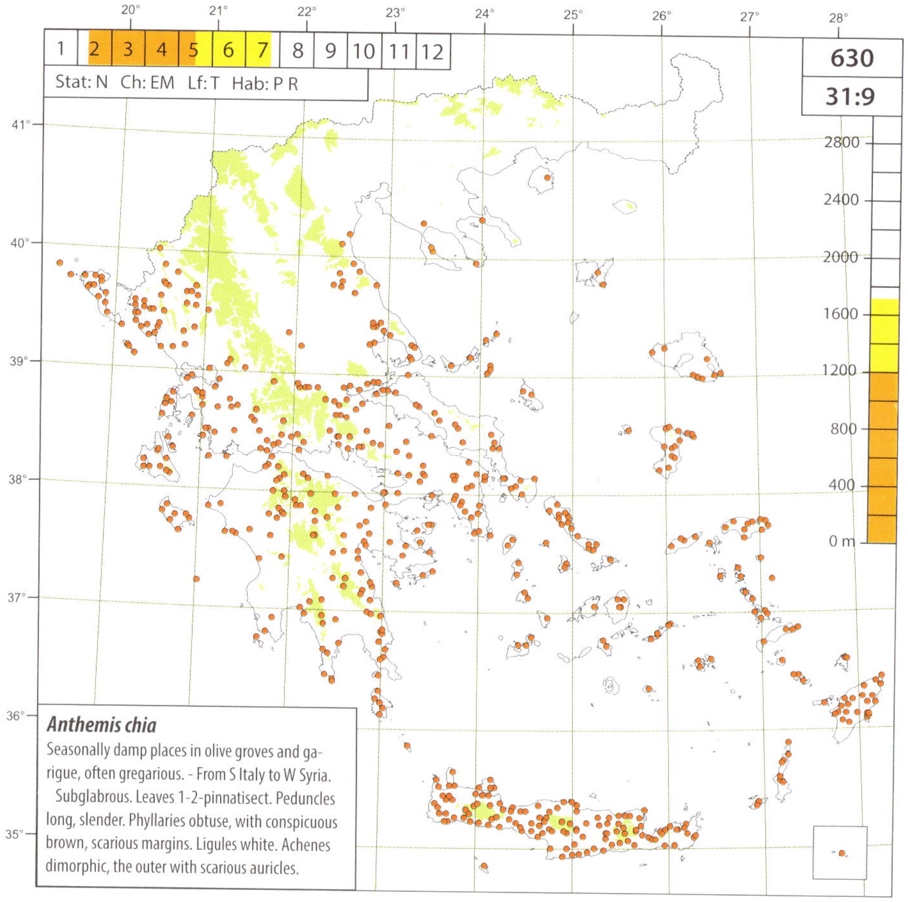

Anthemis chia
Seasonally damp places in olive groves and garigue, often gregarious. - From S Italy to W Syria.
Subglabrous. Leaves 1-2-pinnatisect. Peduncles long, slender. Phyllaries obtuse, with conspicuous brown, scarious margins. Ligules white. Achenes dimorphic, the outer with scarious auricles.

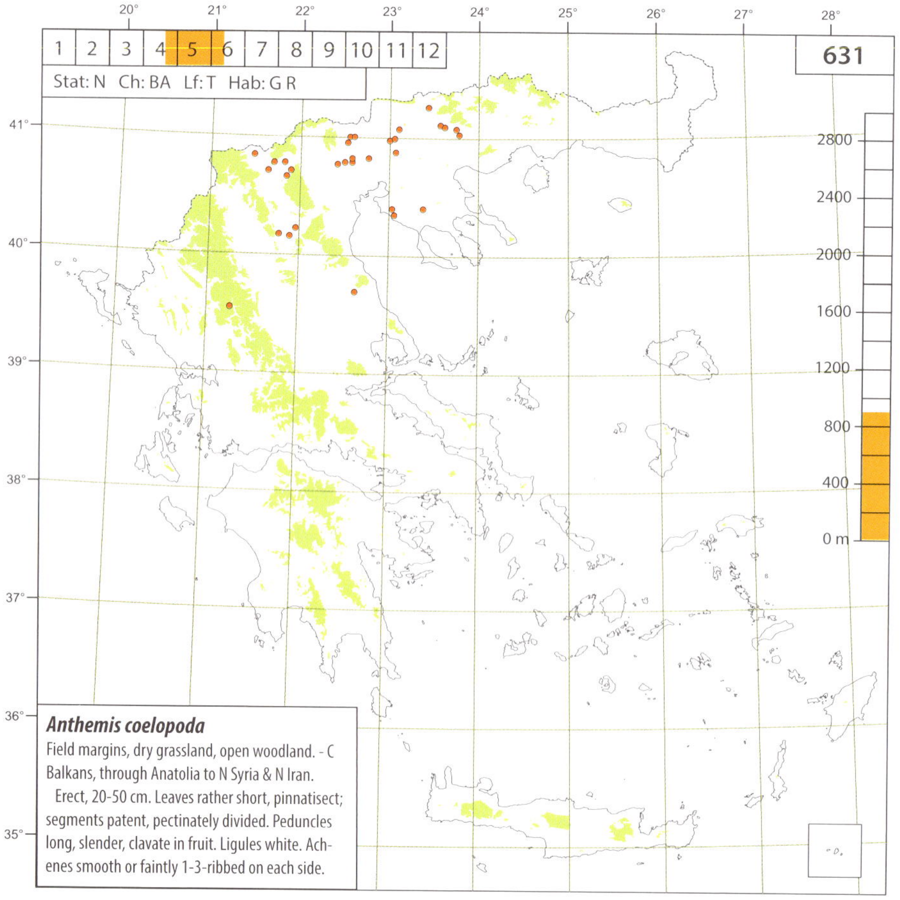

Anthemis coelopoda
Field margins, dry grassland, open woodland. - C Balkans, through Anatolia to N Syria & N Iran.
Erect, 20-50 cm. Leaves rather short, pinnatisect; segments patent, pectinately divided. Peduncles long, slender, clavate in fruit. Ligules white. Achenes smooth or faintly 1-3-ribbed on each side.

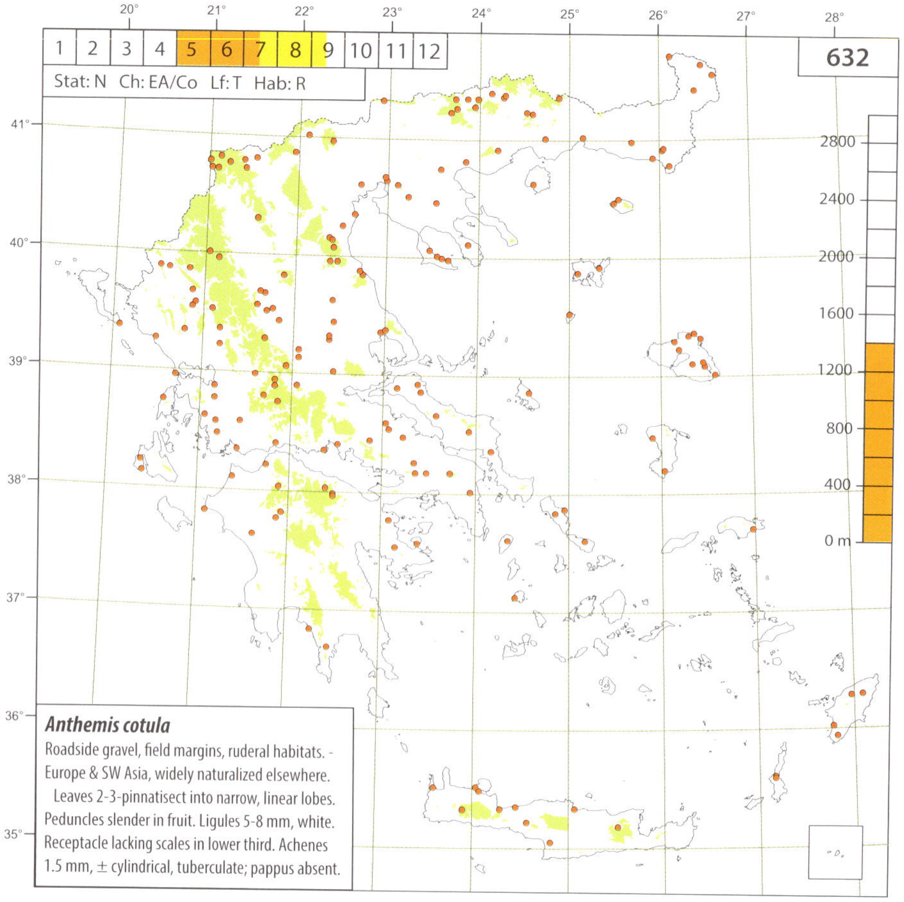

Anthemis cotula
Roadside gravel, field margins, ruderal habitats. - Europe & SW Asia, widely naturalized elsewhere.
Leaves 2-3-pinnatisect into narrow, linear lobes. Peduncles slender in fruit. Ligules 5-8 mm, white. Receptacle lacking scales in lower third. Achenes 1.5 mm, ± cylindrical, tuberculate; pappus absent.

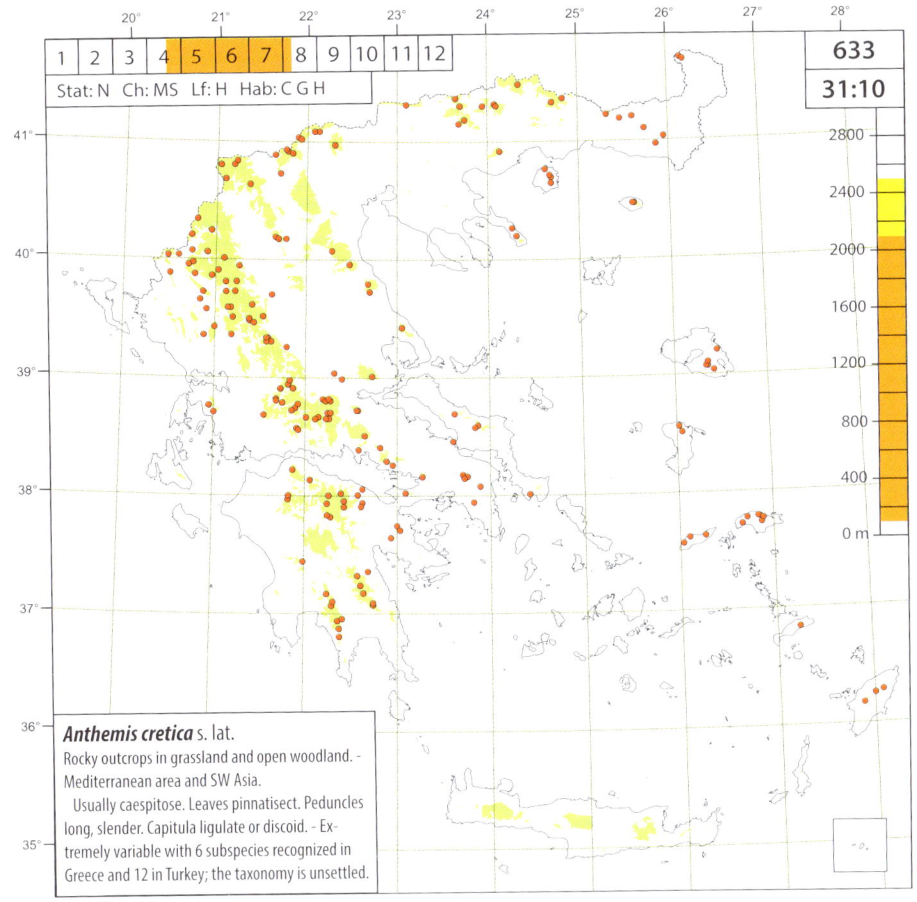

Anthemis cretica s. lat.
Rocky outcrops in grassland and open woodland. - Mediterranean area and SW Asia.
Usually caespitose. Leaves pinnatisect. Peduncles long, slender. Capitula ligulate or discoid. - Extremely variable with 6 subspecies recognized in Greece and 12 in Turkey; the taxonomy is unsettled.

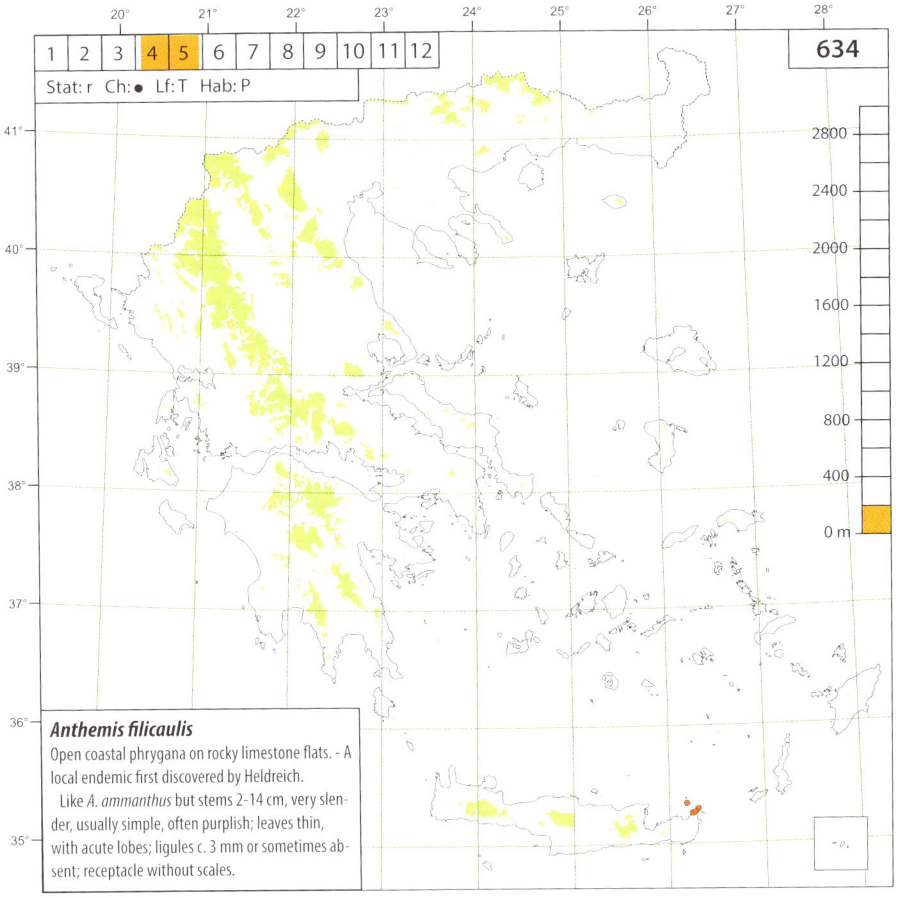

Anthemis filicaulis
Open coastal phrygana on rocky limestone flats. - A local endemic first discovered by Heldreich.
Like *A. ammanthus* but stems 2-14 cm, very slender, usually simple, often purplish; leaves thin, with acute lobes; ligules c. 3 mm or sometimes absent; receptacle without scales.

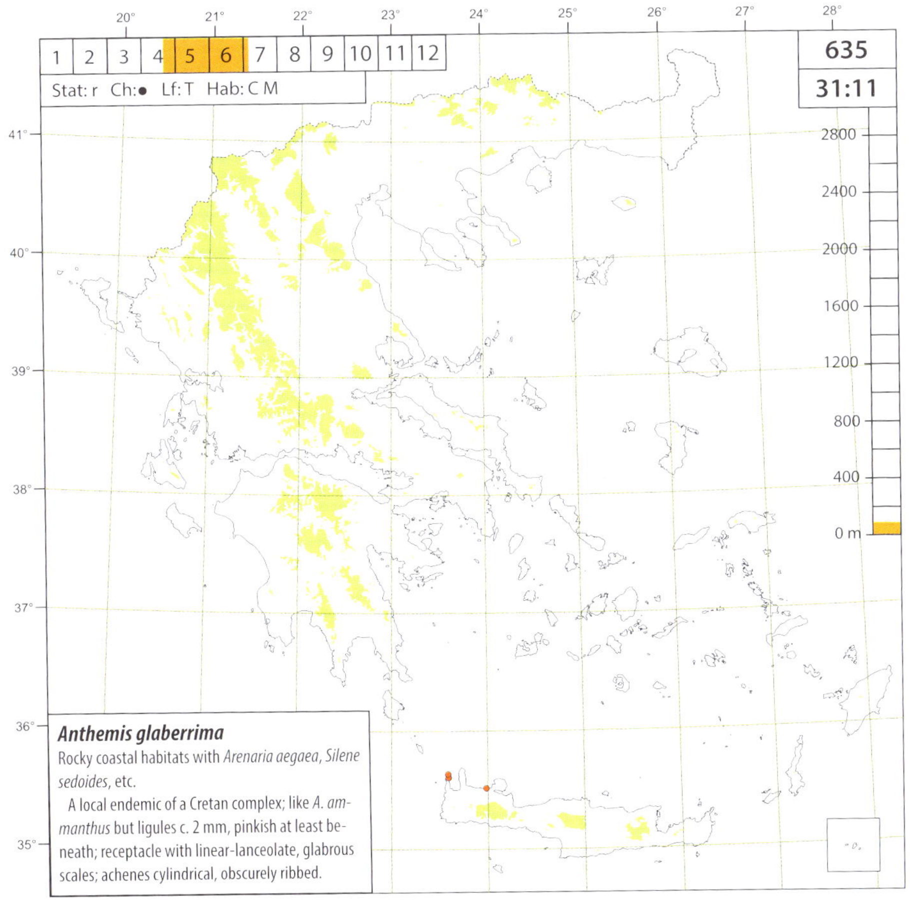

Anthemis glaberrima
Rocky coastal habitats with *Arenaria aegaea*, *Silene sedoides*, etc.
A local endemic of a Cretan complex; like *A. ammanthus* but ligules c. 2 mm, pinkish at least beneath; receptacle with linear-lanceolate, glabrous scales; achenes cylindrical, obscurely ribbed.

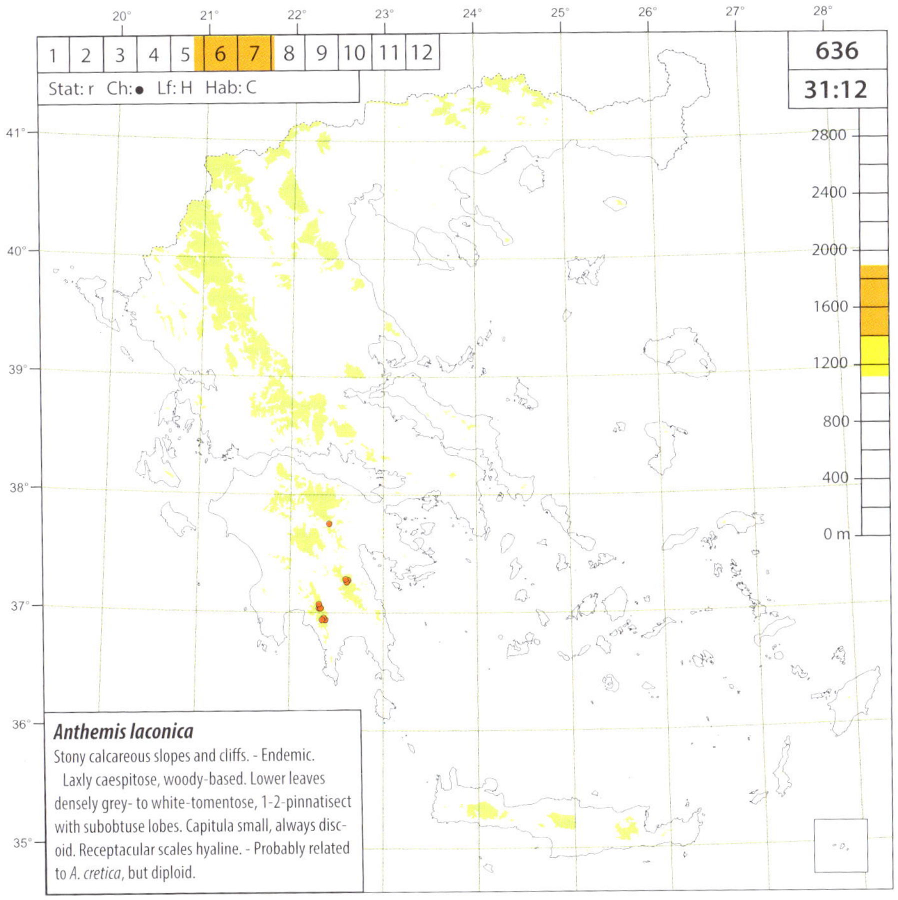

Anthemis laconica
Stony calcareous slopes and cliffs. - Endemic.
Laxly caespitose, woody-based. Lower leaves densely grey- to white-tomentose, 1-2-pinnatisect with subobtuse lobes. Capitula small, always discoid. Receptacular scales hyaline. - Probably related to *A. cretica*, but diploid.

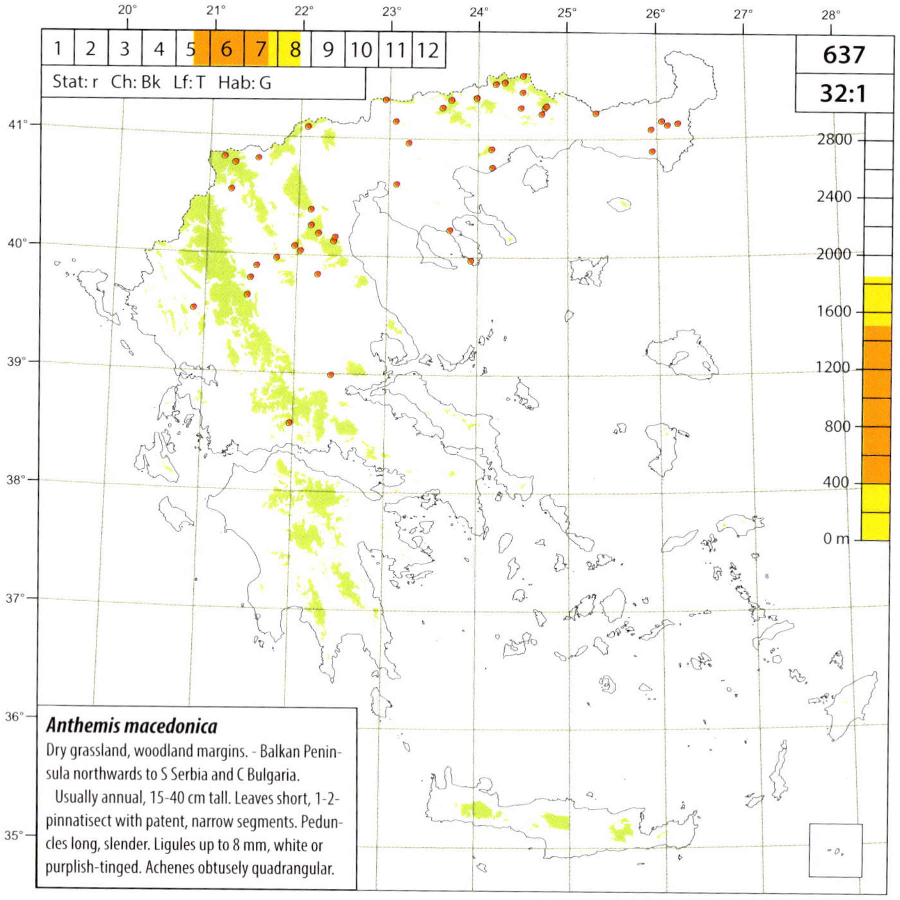

Anthemis macedonica

Dry grassland, woodland margins. - Balkan Peninsula northwards to S Serbia and C Bulgaria.

Usually annual, 15-40 cm tall. Leaves short, 1-2-pinnatisect with patent, narrow segments. Peduncles long, slender. Ligules up to 8 mm, white or purplish-tinged. Achenes obtusely quadrangular.

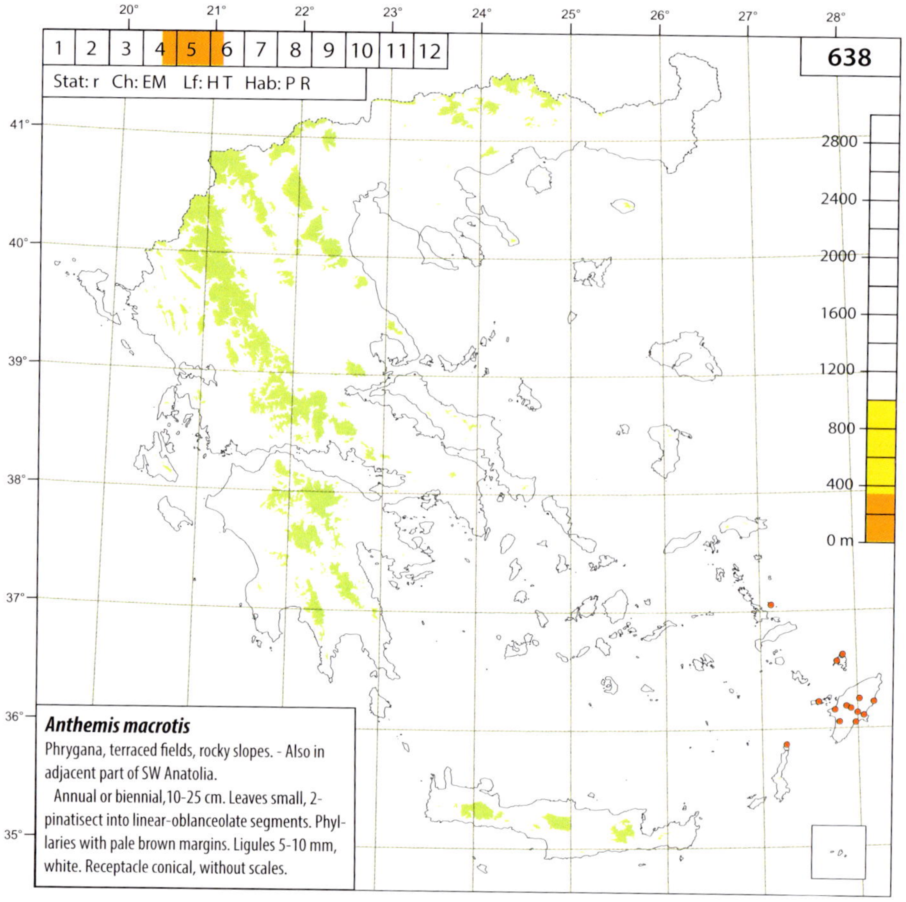

Anthemis macrotis

Phrygana, terraced fields, rocky slopes. - Also in adjacent part of SW Anatolia.

Annual or biennial, 10-25 cm. Leaves small, 2-pinatisect into linear-oblanceolate segments. Phyllaries with pale brown margins. Ligules 5-10 mm, white. Receptacle conical, without scales.

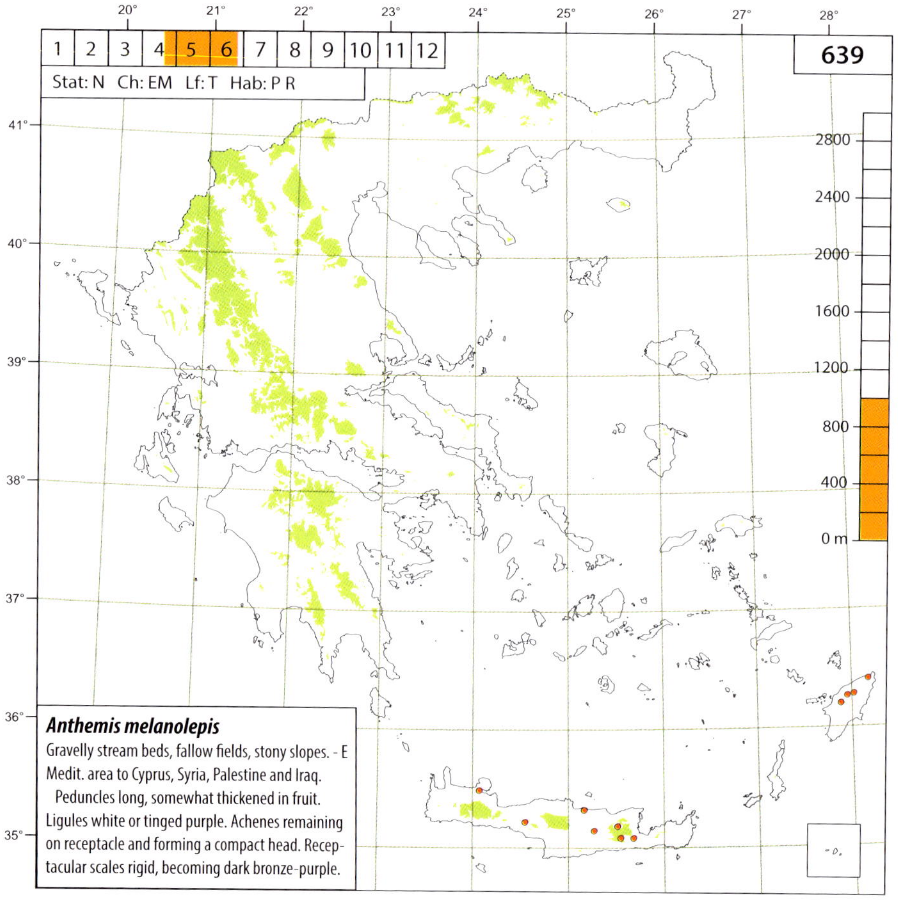

Anthemis melanolepis

Gravelly stream beds, fallow fields, stony slopes. - E Medit. area to Cyprus, Syria, Palestine and Iraq.

Peduncles long, somewhat thickened in fruit. Ligules white or tinged purple. Achenes remaining on receptacle and forming a compact head. Receptacular scales rigid, becoming dark bronze-purple.

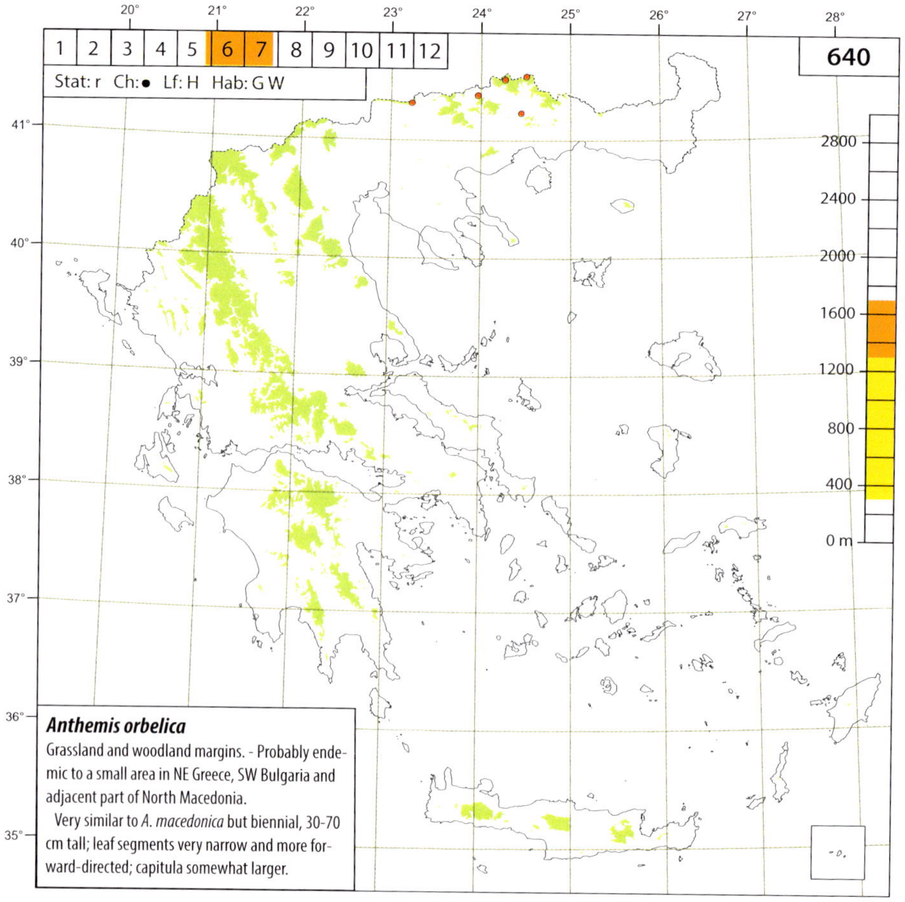

Anthemis orbelica

Grassland and woodland margins. - Probably endemic to a small area in NE Greece, SW Bulgaria and adjacent part of North Macedonia.

Very similar to *A. macedonica* but biennial, 30-70 cm tall; leaf segments very narrow and more forward-directed; capitula somewhat larger.

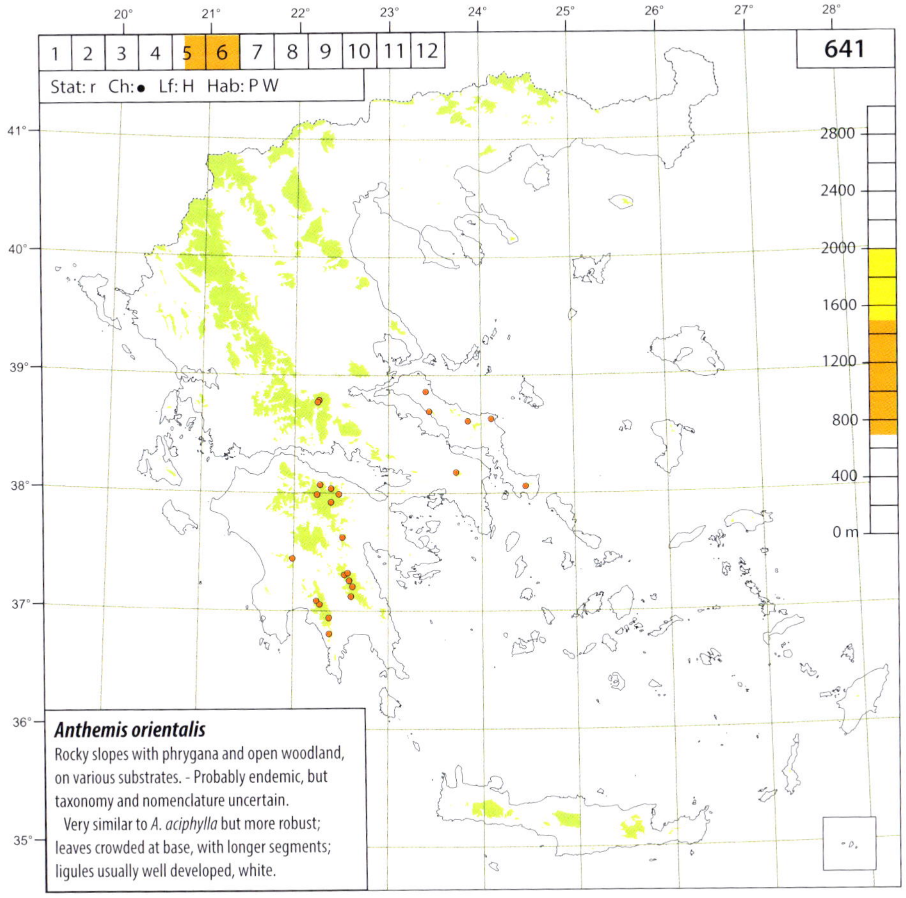

Anthemis orientalis

Rocky slopes with phrygana and open woodland, on various substrates. - Probably endemic, but taxonomy and nomenclature uncertain.

Very similar to *A. aciphylla* but more robust; leaves crowded at base, with longer segments; ligules usually well developed, white.

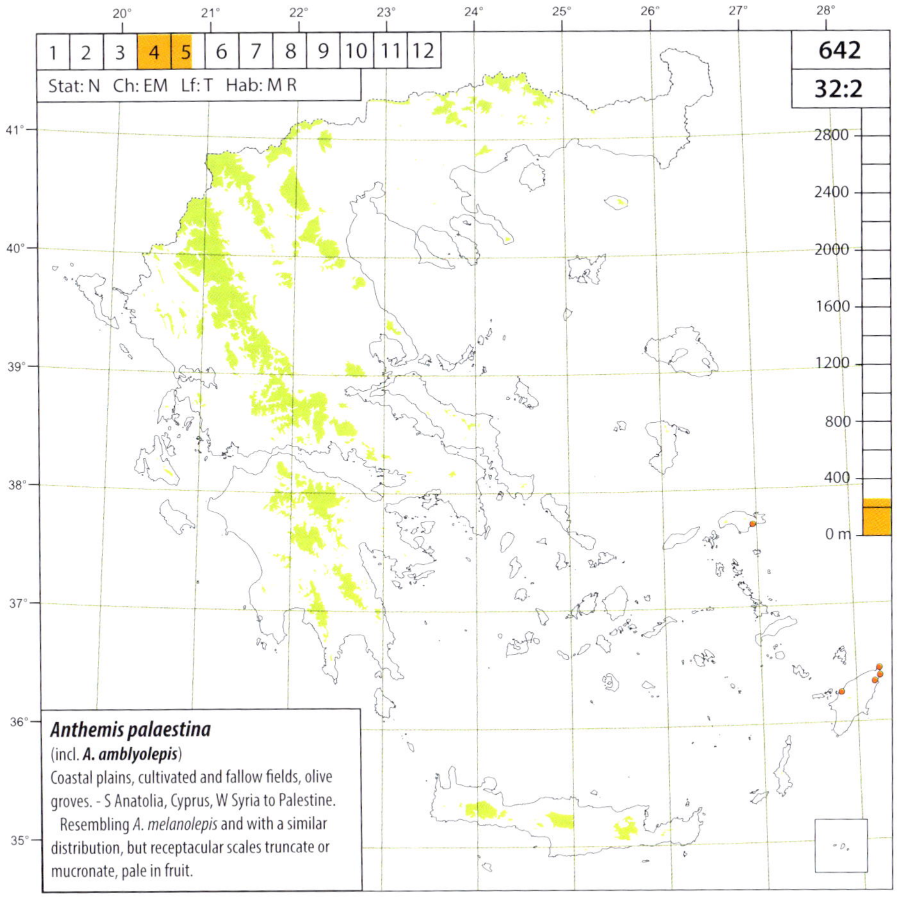

Anthemis palaestina

(incl. ***A. amblyolepis***)

Coastal plains, cultivated and fallow fields, olive groves. - S Anatolia, Cyprus, W Syria to Palestine.

Resembling *A. melanolepis* and with a similar distribution, but receptacular scales truncate or mucronate, pale in fruit.

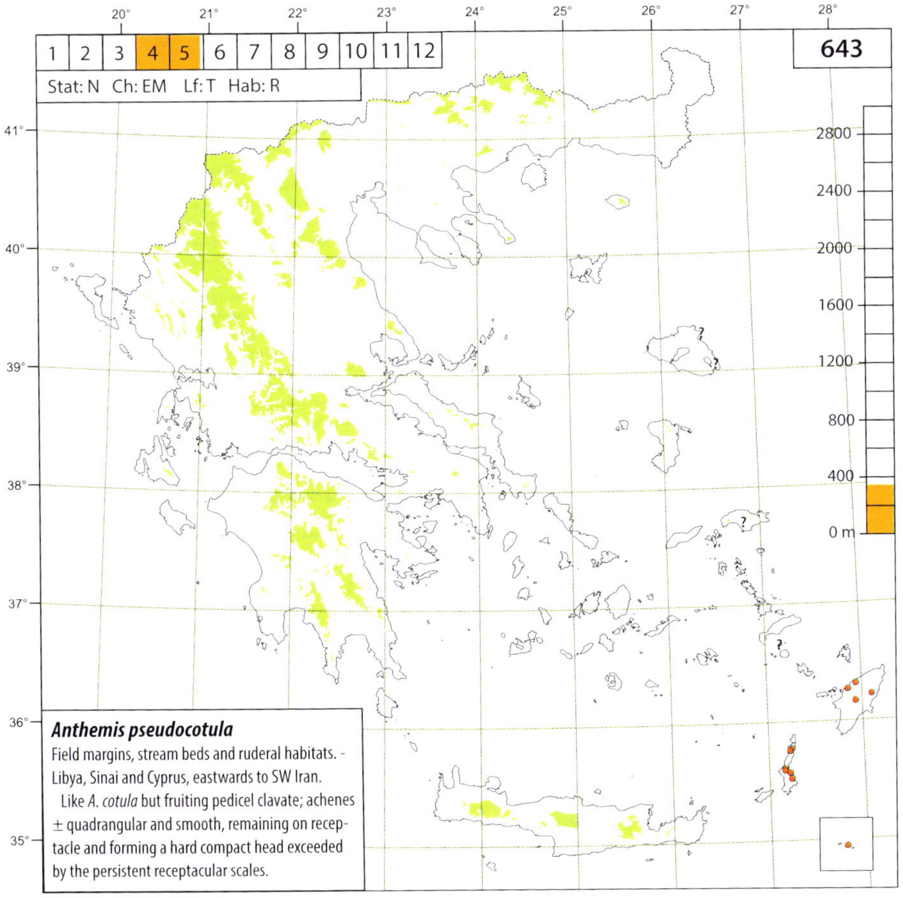

Anthemis pseudocotula

Field margins, stream beds and ruderal habitats. - Libya, Sinai and Cyprus, eastwards to SW Iran.

Like *A. cotula* but fruiting pedicel clavate; achenes ± quadrangular and smooth, remaining on receptacle and forming a hard compact head exceeded by the persistent receptacular scales.

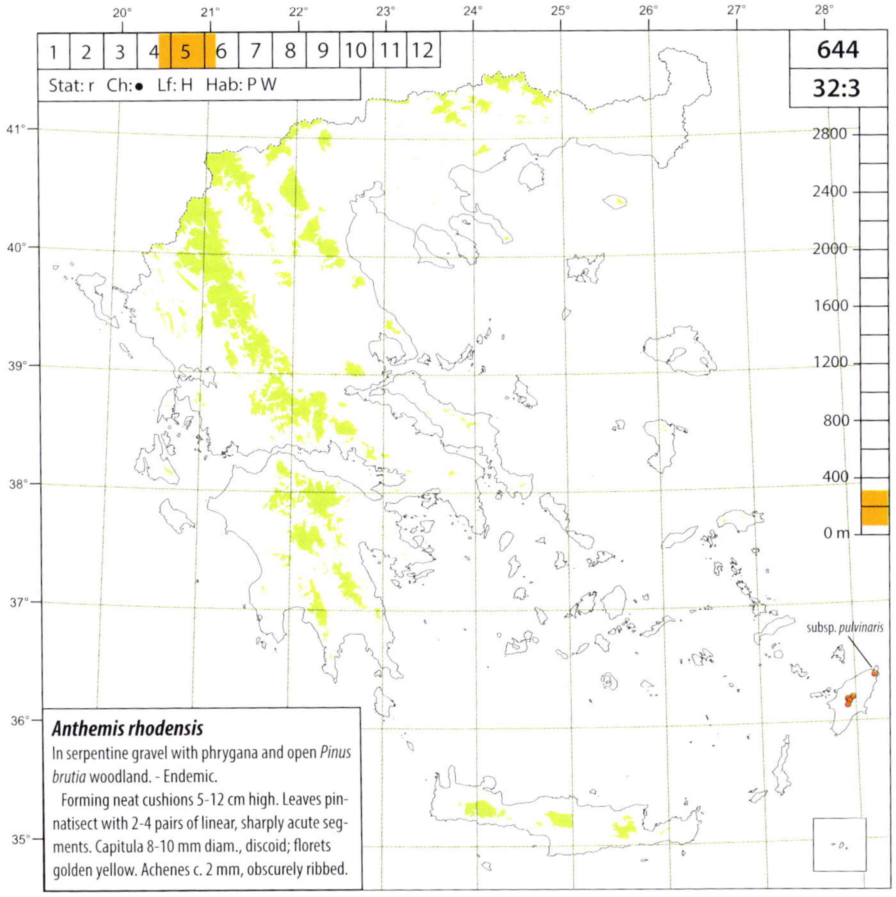

Anthemis rhodensis

In serpentine gravel with phrygana and open *Pinus brutia* woodland. - Endemic.

Forming neat cushions 5-12 cm high. Leaves pinnatisect with 2-4 pairs of linear, sharply acute segments. Capitula 8-10 mm diam., discoid; florets golden yellow. Achenes c. 2 mm, obscurely ribbed.

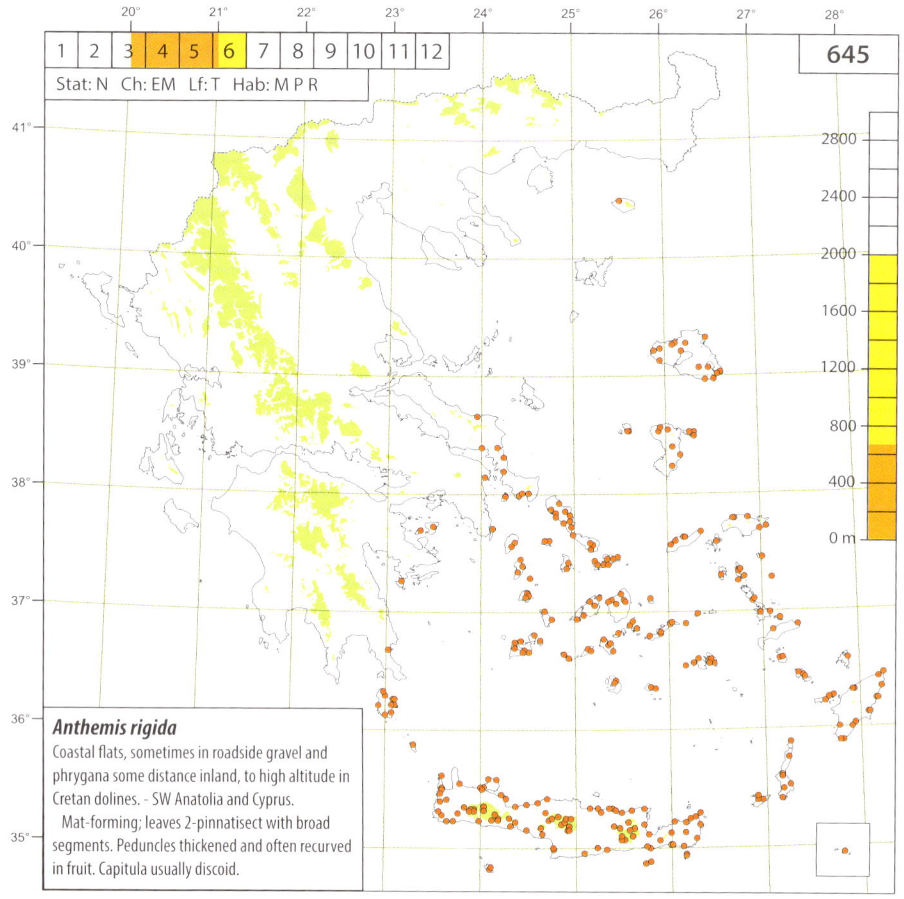

Anthemis rigida

Coastal flats, sometimes in roadside gravel and phrygana some distance inland, to high altitude in Cretan dolines. - SW Anatolia and Cyprus.

Mat-forming; leaves 2-pinnatisect with broad segments. Peduncles thickened and often recurved in fruit. Capitula usually discoid.

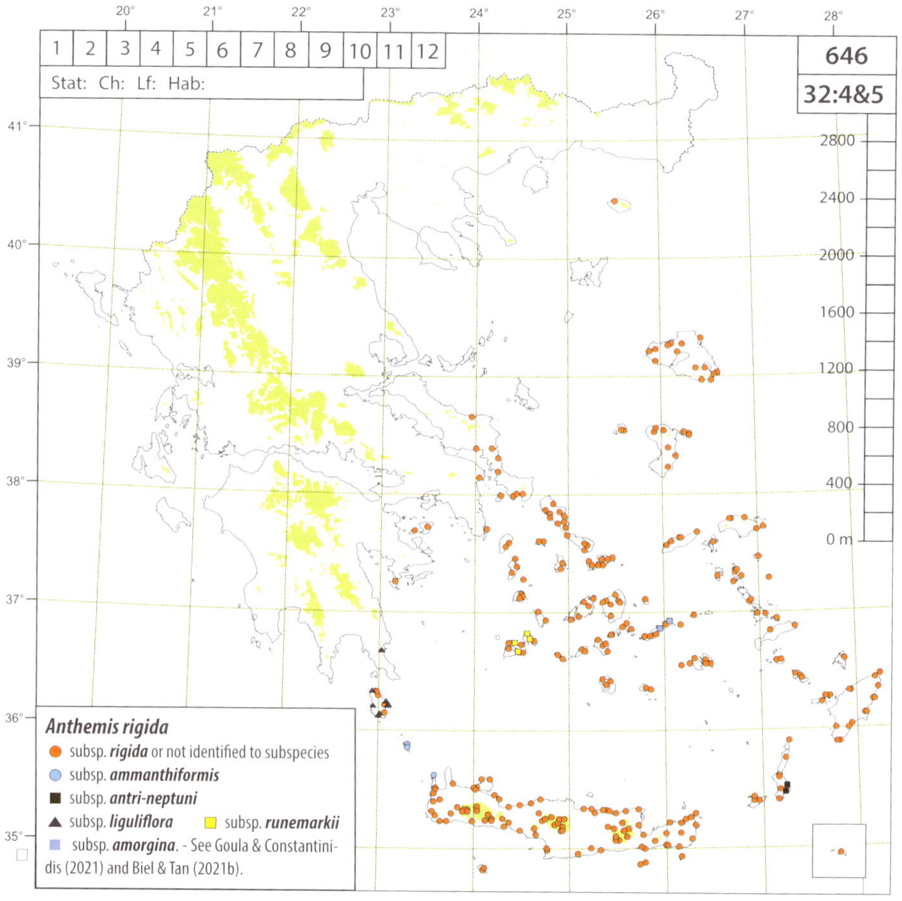

Anthemis rigida

- subsp. ***rigida*** or not identified to subspecies
- subsp. ***ammanthiformis***
- subsp. ***antri-neptuni***
- subsp. ***liguliflora***
- subsp. ***runemarkii***
- subsp. ***amorgina***. - See Goula & Constantinidis (2021) and Biel & Tan (2021b).

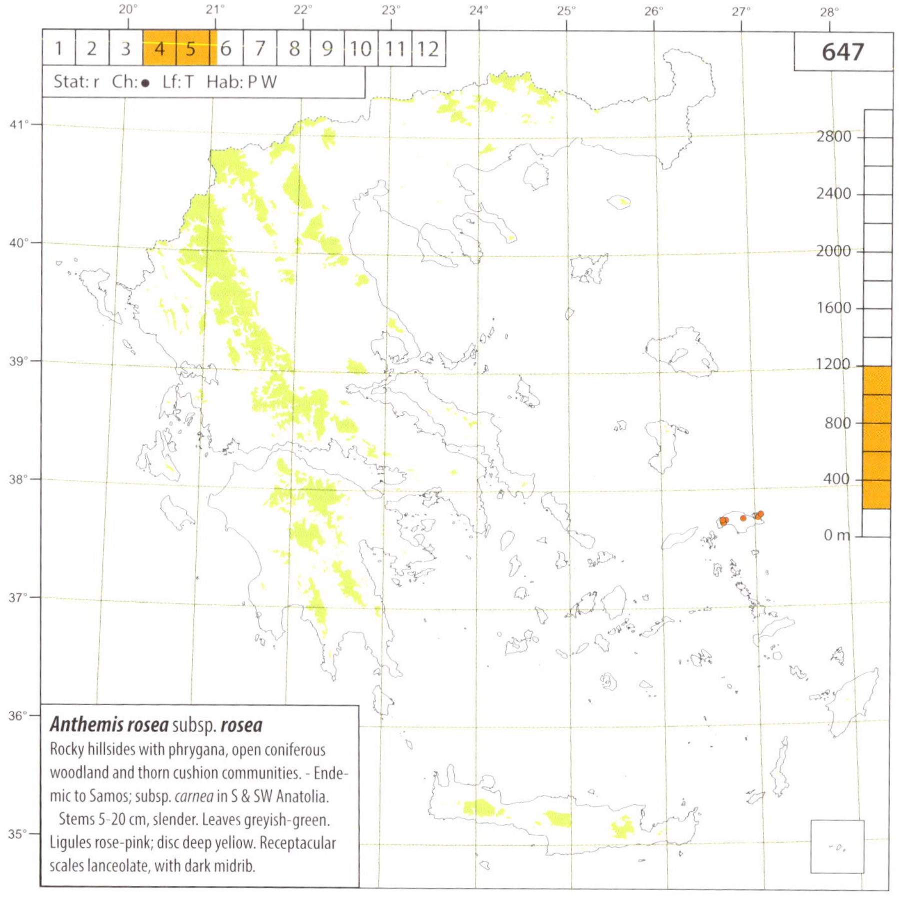

Anthemis rosea subsp. ***rosea***

Rocky hillsides with phrygana, open coniferous woodland and thorn cushion communities. - Endemic to Samos; subsp. *carnea* in S & SW Anatolia.

Stems 5-20 cm, slender. Leaves greyish-green. Ligules rose-pink; disc deep yellow. Receptacular scales lanceolate, with dark midrib.

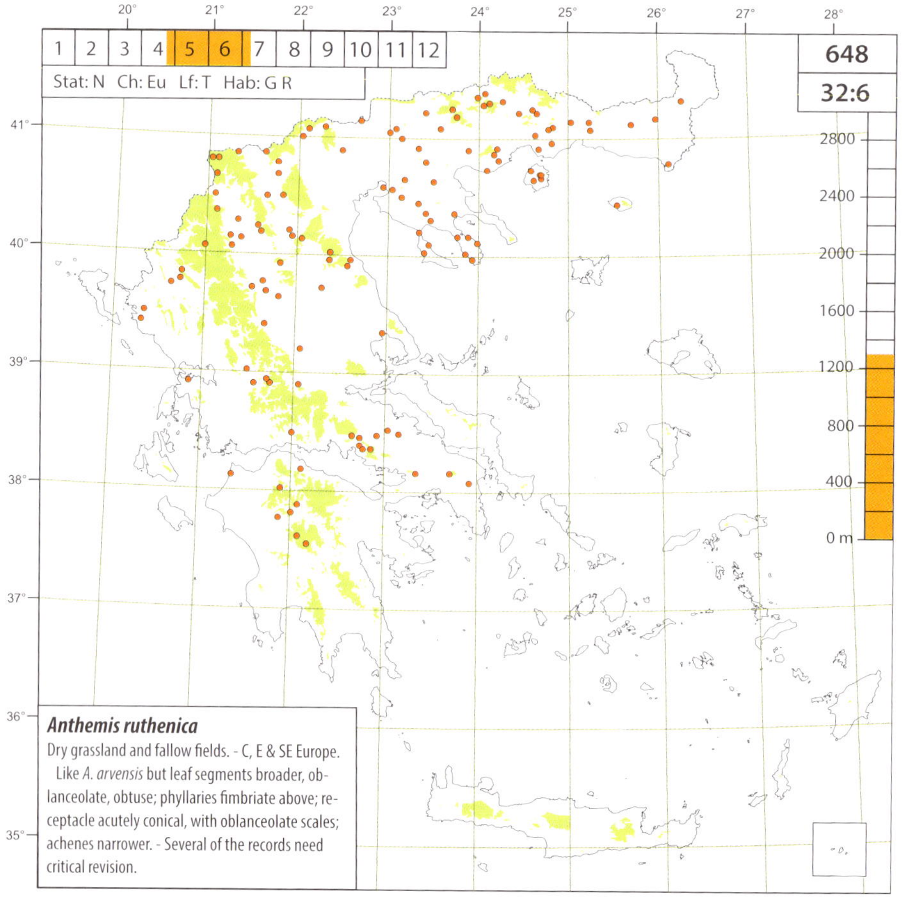

Anthemis ruthenica

Dry grassland and fallow fields. - C, E & SE Europe.

Like *A. arvensis* but leaf segments broader, oblanceolate, obtuse; phyllaries fimbriate above; receptacle acutely conical, with oblanceolate scales; achenes narrower. - Several of the records need critical revision.

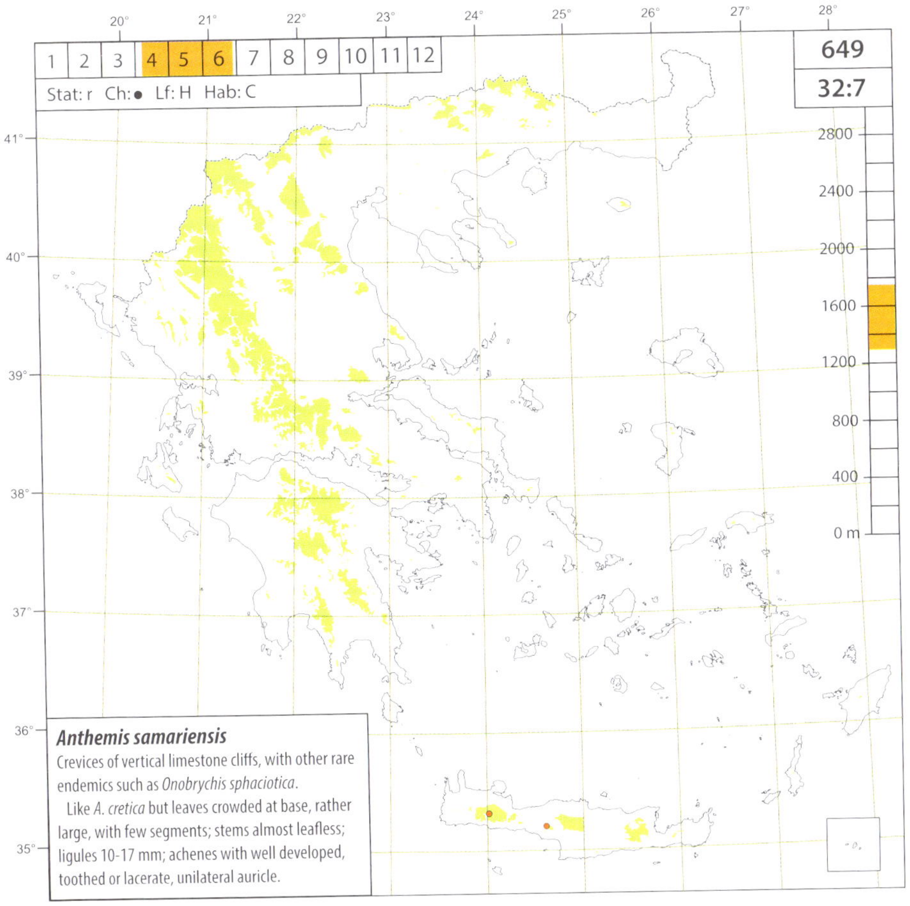

Anthemis samariensis

Crevices of vertical limestone cliffs, with other rare endemics such as *Onobrychis sphaciotica*.

Like *A. cretica* but leaves crowded at base, rather large, with few segments; stems almost leafless; ligules 10-17 mm; achenes with well developed, toothed or lacerate, unilateral auricle.

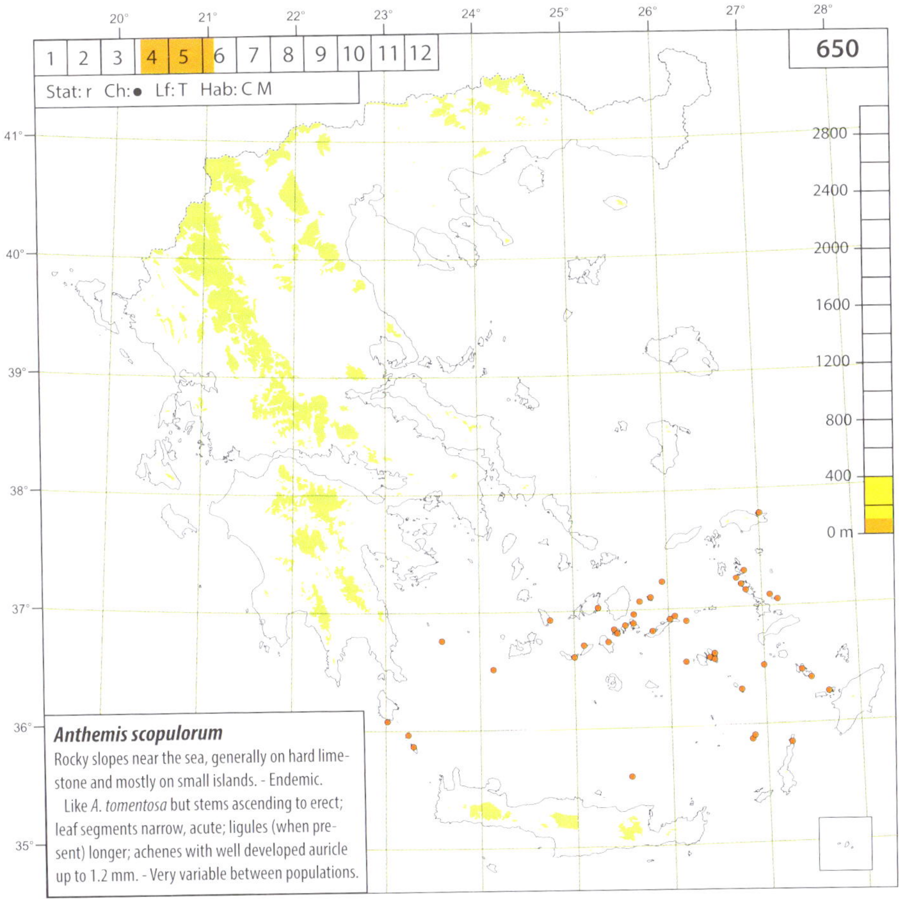

Anthemis scopulorum

Rocky slopes near the sea, generally on hard limestone and mostly on small islands. - Endemic.

Like *A. tomentosa* but stems ascending to erect; leaf segments narrow, acute; ligules (when present) longer; achenes with well developed auricle up to 1.2 mm. - Very variable between populations.

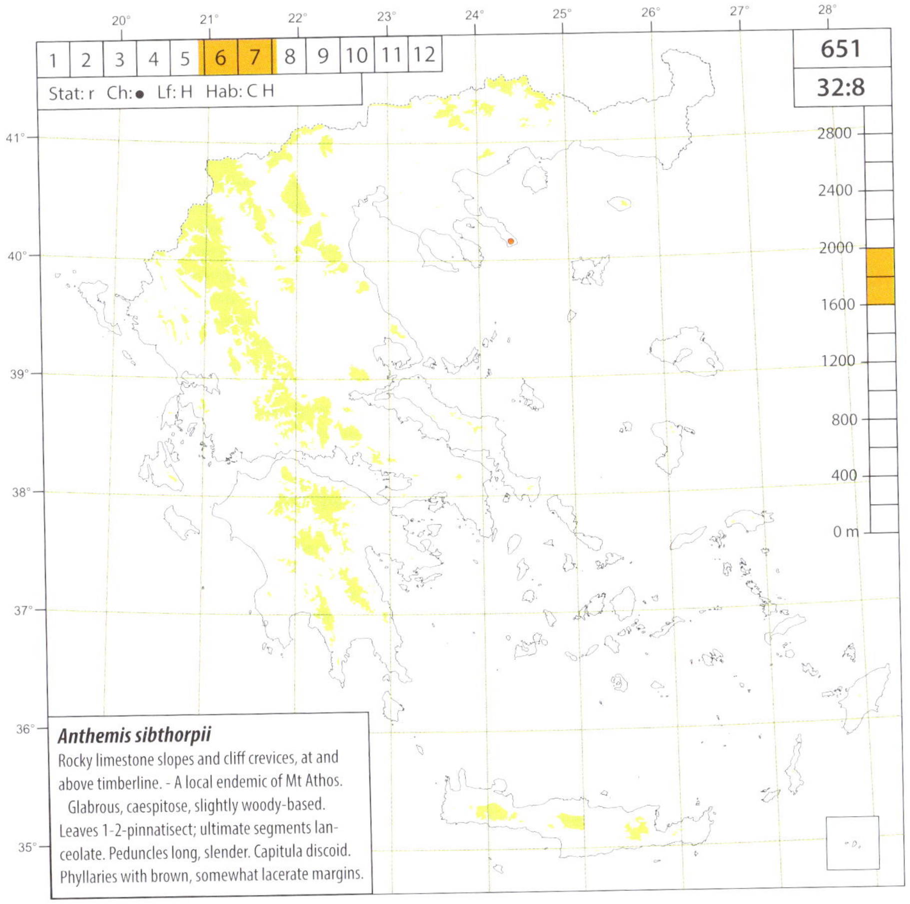

Anthemis sibthorpii

Rocky limestone slopes and cliff crevices, at and above timberline. - A local endemic of Mt Athos.

Glabrous, caespitose, slightly woody-based. Leaves 1-2-pinnatisect; ultimate segments lanceolate. Peduncles long, slender. Capitula discoid. Phyllaries with brown, somewhat lacerate margins.

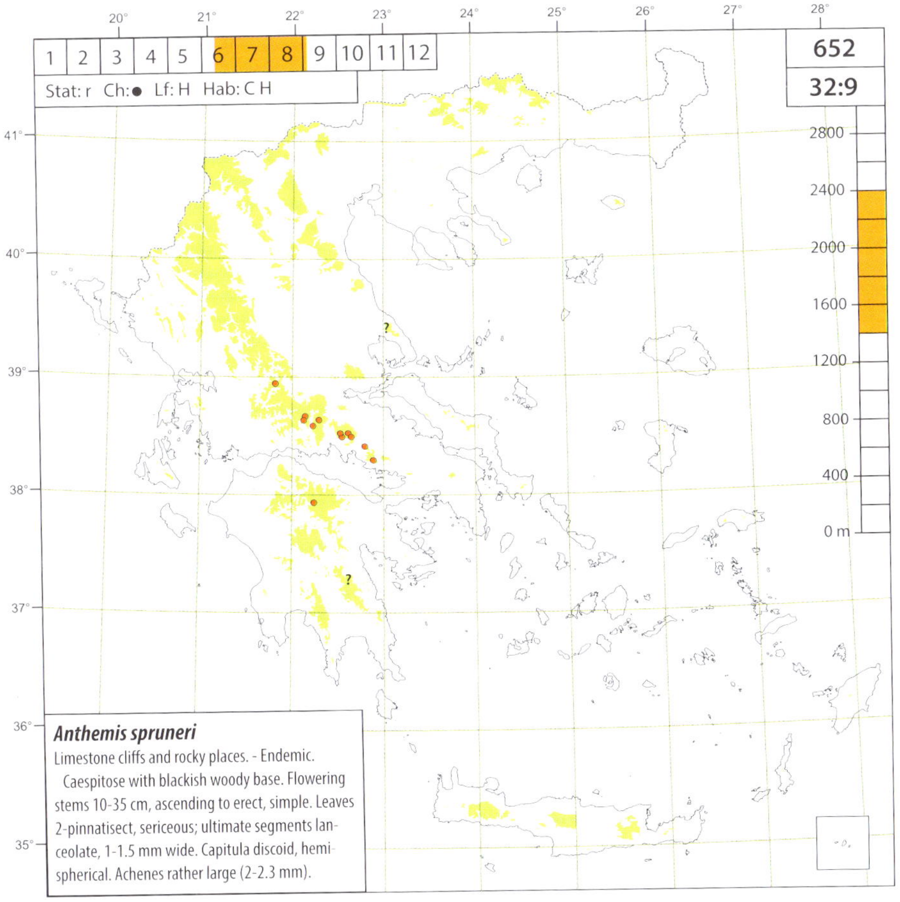

Anthemis spruneri

Limestone cliffs and rocky places. - Endemic.

Caespitose with blackish woody base. Flowering stems 10-35 cm, ascending to erect, simple. Leaves 2-pinnatisect, sericeous; ultimate segments lanceolate, 1-1.5 mm wide. Capitula discoid, hemispherical. Achenes rather large (2-2.3 mm).

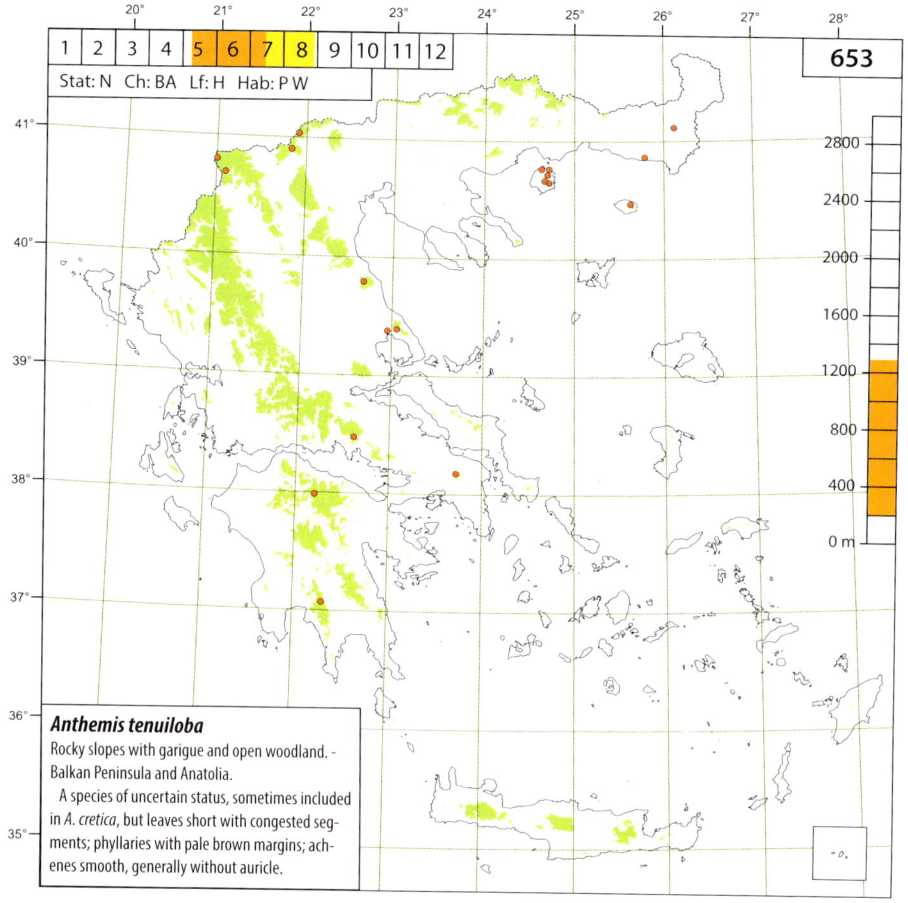

Anthemis tenuiloba
Rocky slopes with garigue and open woodland. - Balkan Peninsula and Anatolia.
A species of uncertain status, sometimes included in *A. cretica*, but leaves short with congested segments; phyllaries with pale brown margins; achenes smooth, generally without auricle.

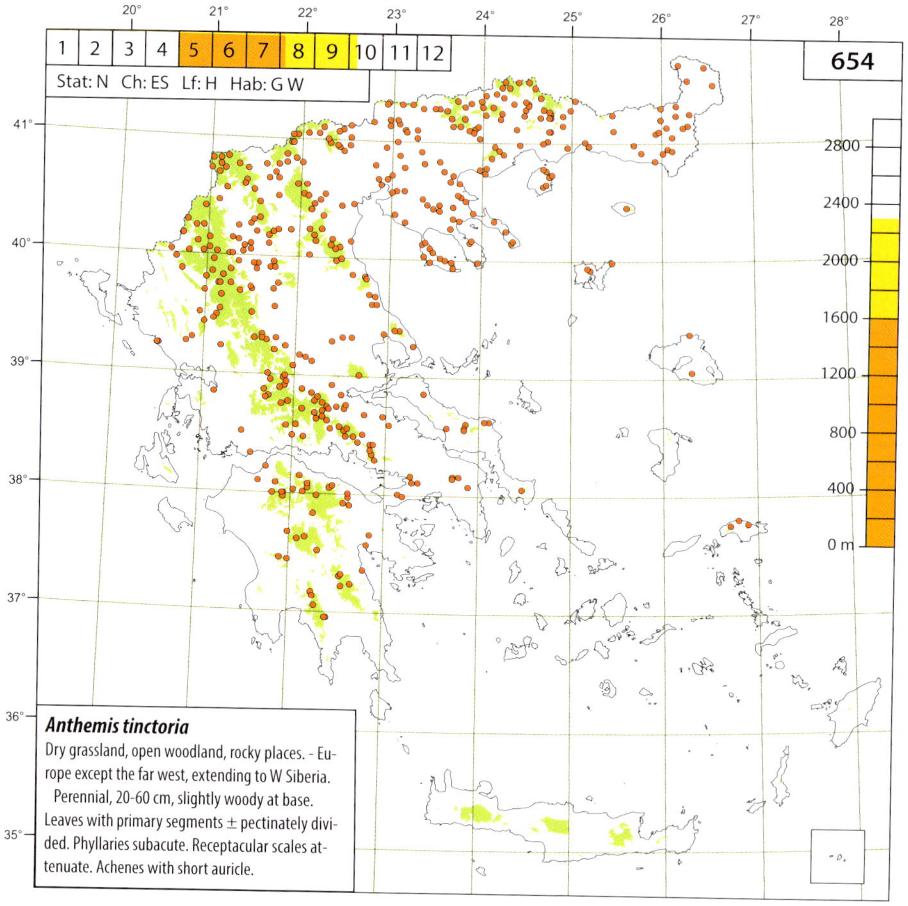

Anthemis tinctoria
Dry grassland, open woodland, rocky places. - Europe except the far west, extending to W Siberia.
Perennial, 20-60 cm, slightly woody at base. Leaves with primary segments ± pectinately divided. Phyllaries subacute. Receptacular scales attenuate. Achenes with short auricle.

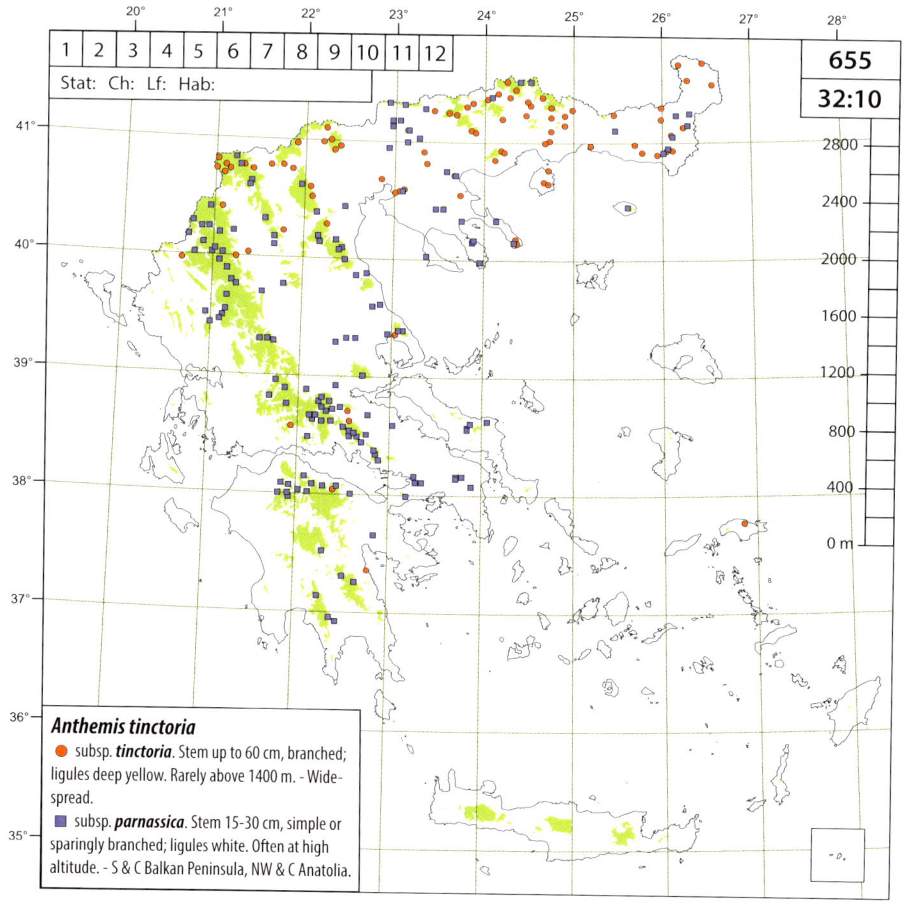

Anthemis tinctoria
● subsp. ***tinctoria***. Stem up to 60 cm, branched; ligules deep yellow. Rarely above 1400 m. - Widespread.
■ subsp. ***parnassica***. Stem 15-30 cm, simple or sparingly branched; ligules white. Often at high altitude. - S & C Balkan Peninsula, NW & C Anatolia.

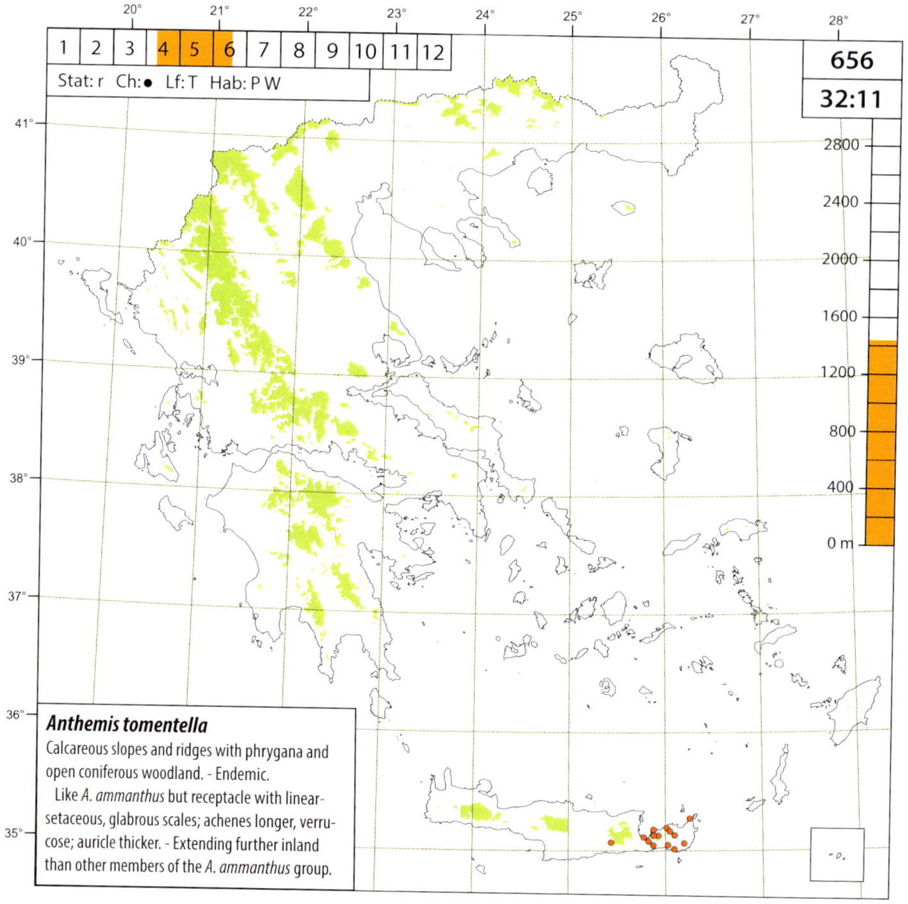

Anthemis tomentella
Calcareous slopes and ridges with phrygana and open coniferous woodland. - Endemic.
Like *A. ammanthus* but receptacle with linear-setaceous, glabrous scales; achenes longer, verrucose; auricle thicker. - Extending further inland than other members of the *A. ammanthus* group.

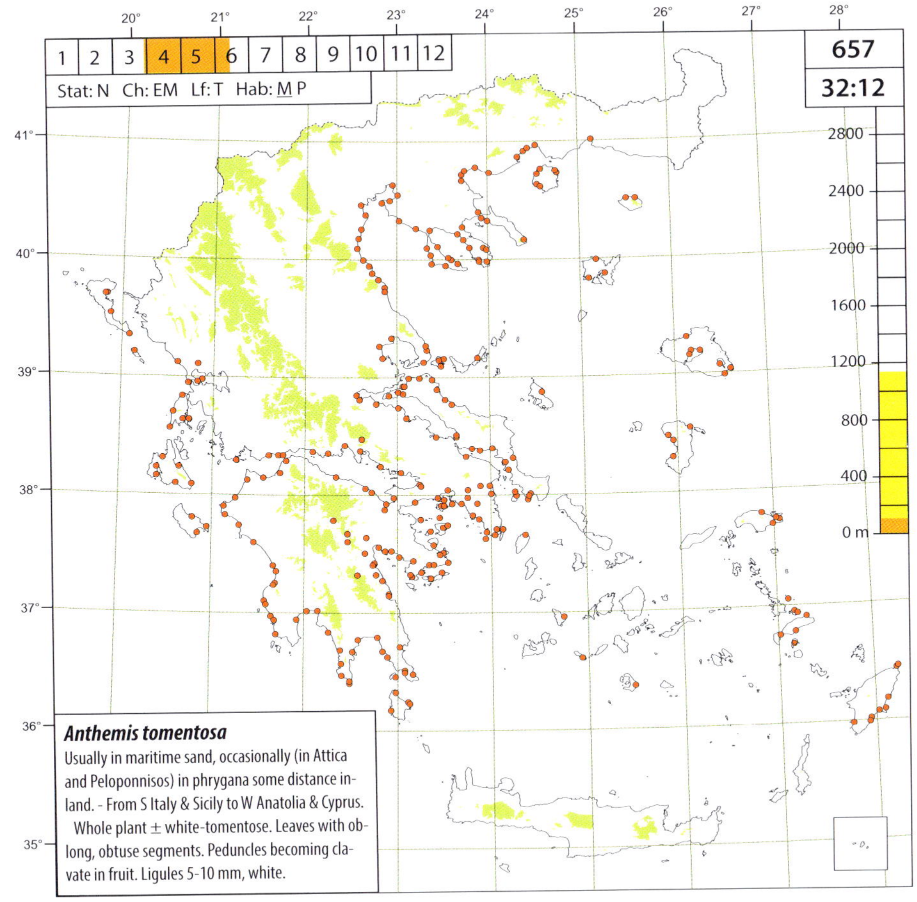

Anthemis tomentosa
Usually in maritime sand, occasionally (in Attica and Peloponnisos) in phrygana some distance inland. - From S Italy & Sicily to W Anatolia & Cyprus.
Whole plant ± white-tomentose. Leaves with oblong, obtuse segments. Peduncles becoming clavate in fruit. Ligules 5-10 mm, white.

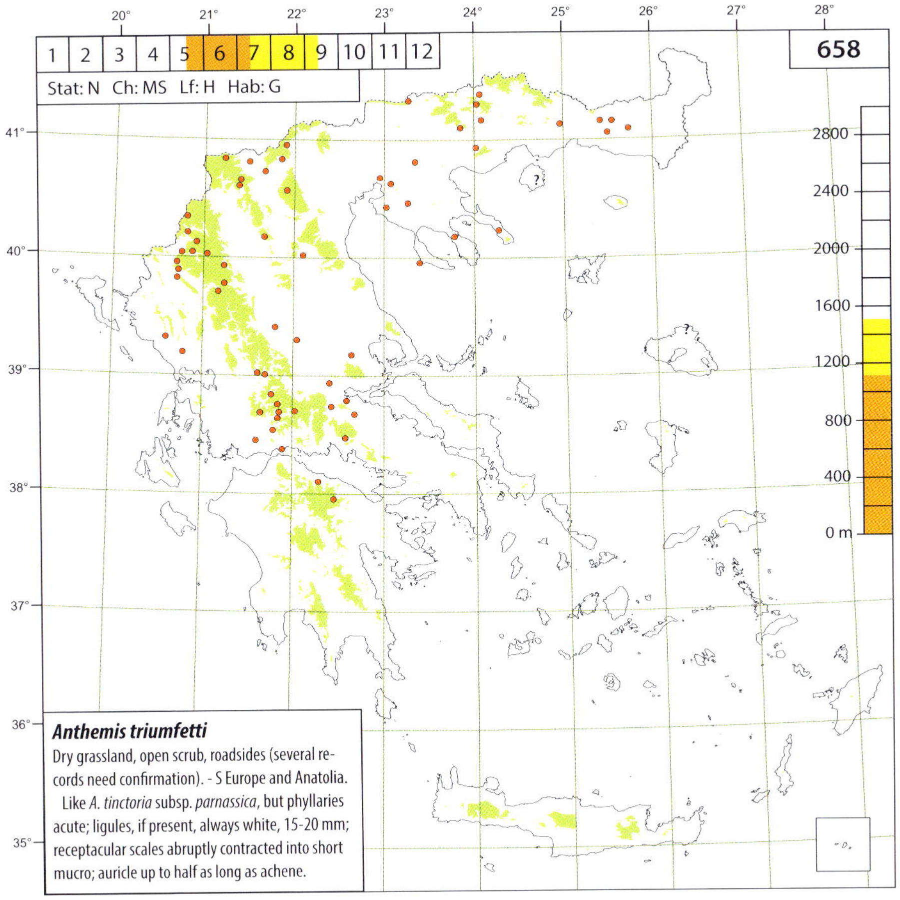

Anthemis triumfetti
Dry grassland, open scrub, roadsides (several records need confirmation). - S Europe and Anatolia.
Like *A. tinctoria* subsp. *parnassica*, but phyllaries acute; ligules, if present, always white, 15-20 mm; receptacular scales abruptly contracted into short mucro; auricle up to half as long as achene.

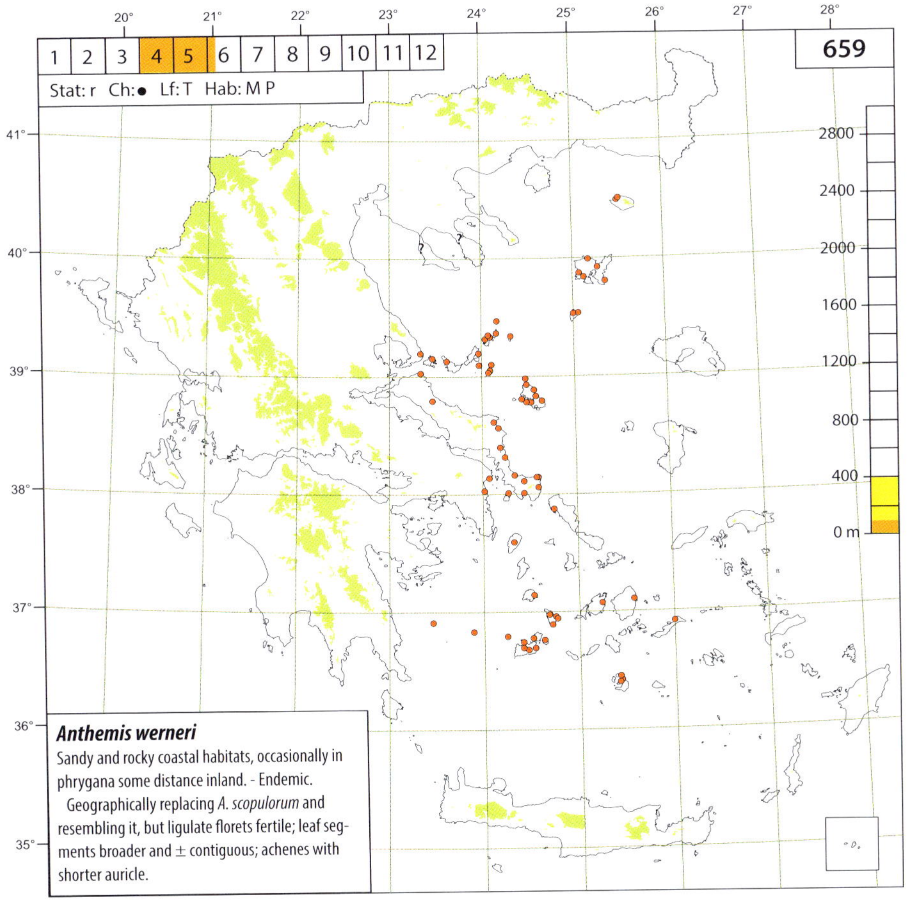

Anthemis werneri
Sandy and rocky coastal habitats, occasionally in phrygana some distance inland. - Endemic.
Geographically replacing *A. scopulorum* and resembling it, but ligulate florets fertile; leaf segments broader and ± contiguous; achenes with shorter auricle.

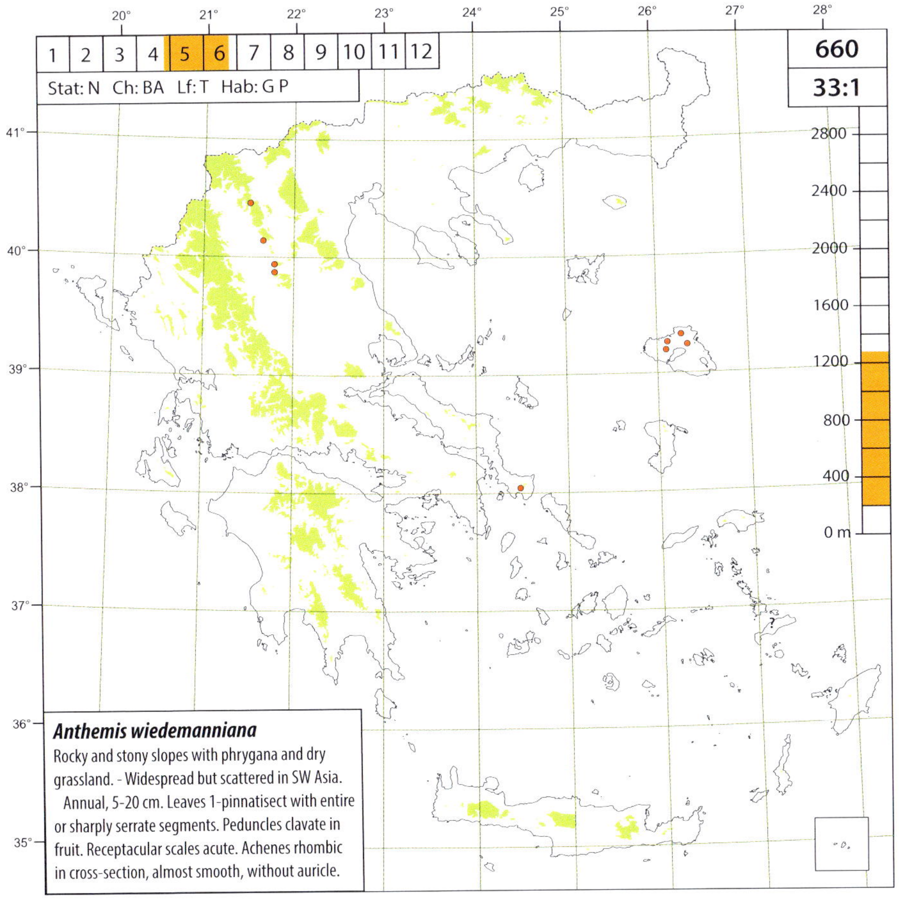

Anthemis wiedemanniana
Rocky and stony slopes with phrygana and dry grassland. - Widespread but scattered in SW Asia.
Annual, 5-20 cm. Leaves 1-pinnatisect with entire or sharply serrate segments. Peduncles clavate in fruit. Receptacular scales acute. Achenes rhombic in cross-section, almost smooth, without auricle.

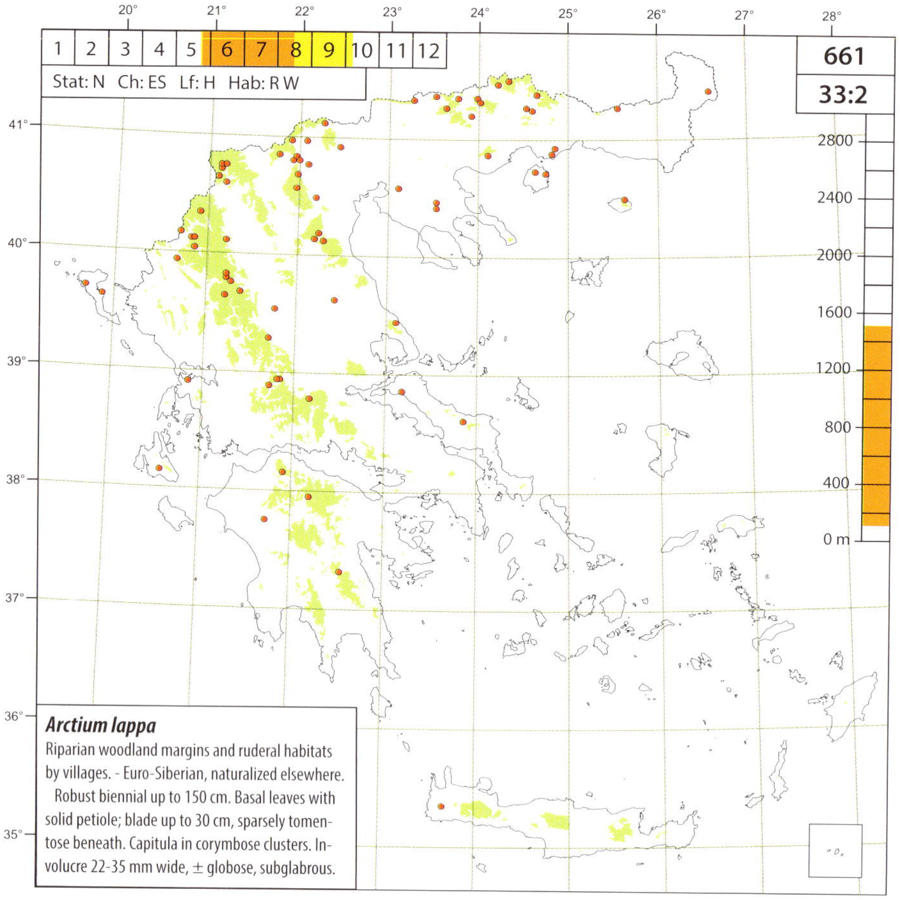

Arctium lappa
Riparian woodland margins and ruderal habitats by villages. - Euro-Siberian, naturalized elsewhere.
Robust biennial up to 150 cm. Basal leaves with solid petiole; blade up to 30 cm, sparsely tomentose beneath. Capitula in corymbose clusters. Involucre 22-35 mm wide, ± globose, subglabrous.

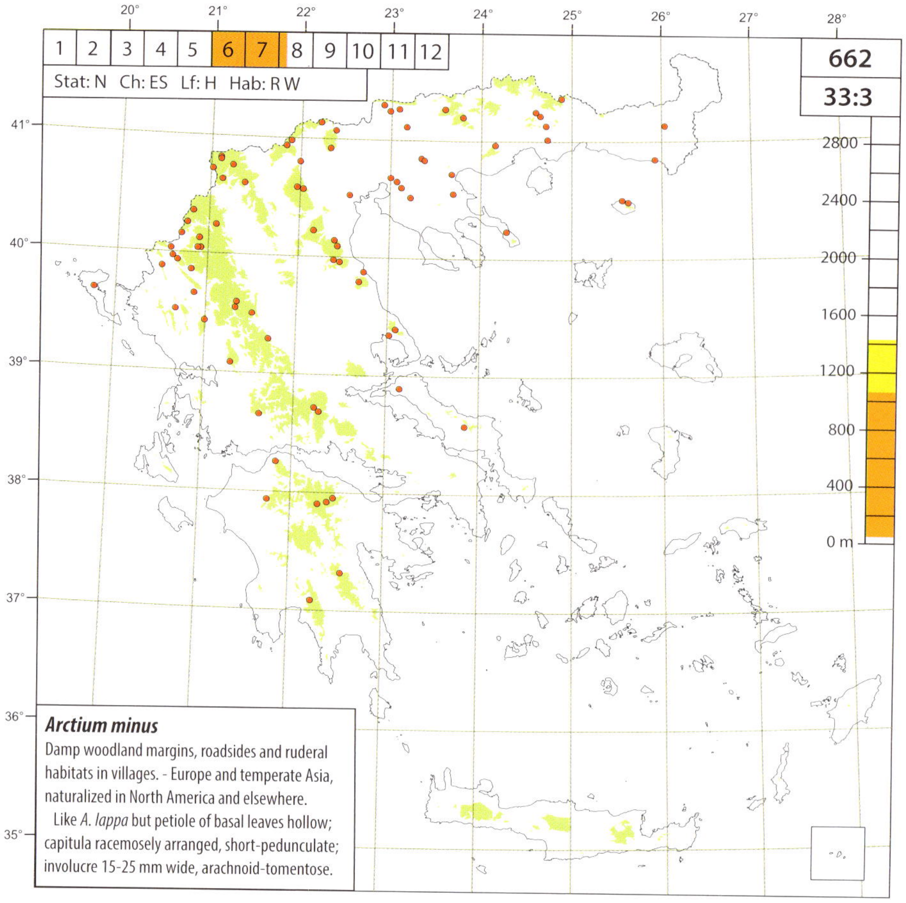

Arctium minus
Damp woodland margins, roadsides and ruderal habitats in villages. - Europe and temperate Asia, naturalized in North America and elsewhere.
Like *A. lappa* but petiole of basal leaves hollow; capitula racemosely arranged, short-pedunculate; involucre 15-25 mm wide, arachnoid-tomentose.

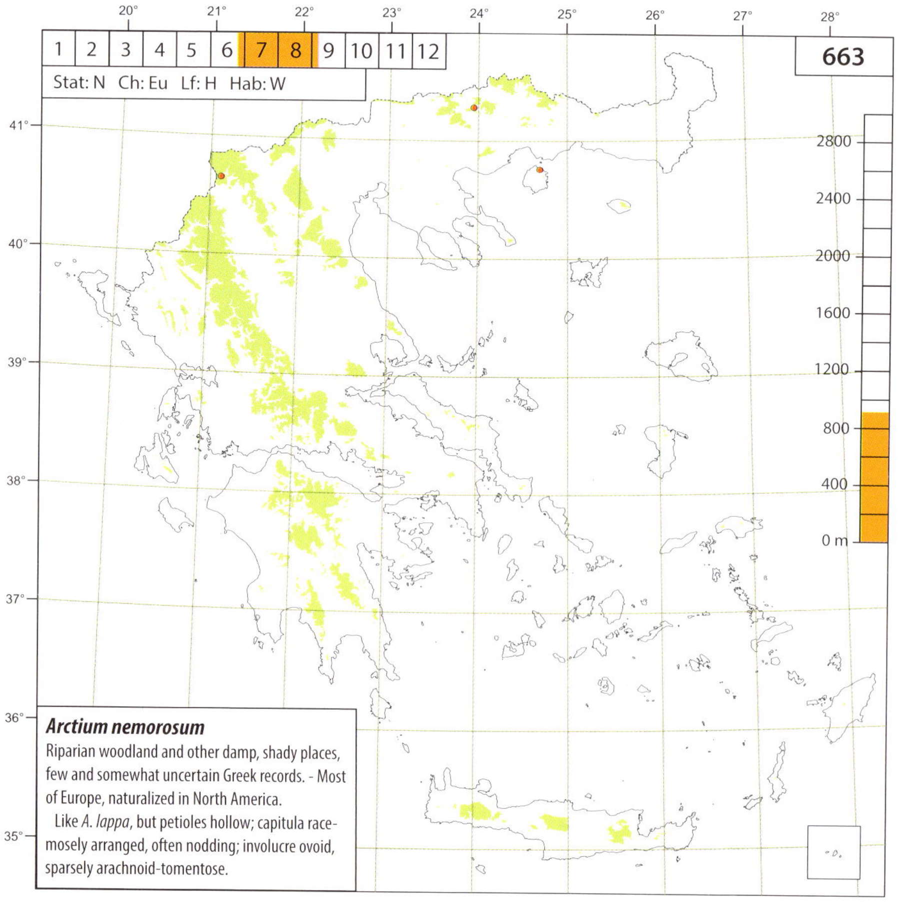

Arctium nemorosum
Riparian woodland and other damp, shady places, few and somewhat uncertain Greek records. - Most of Europe, naturalized in North America.
Like *A. lappa*, but petioles hollow; capitula racemosely arranged, often nodding; involucre ovoid, sparsely arachnoid-tomentose.

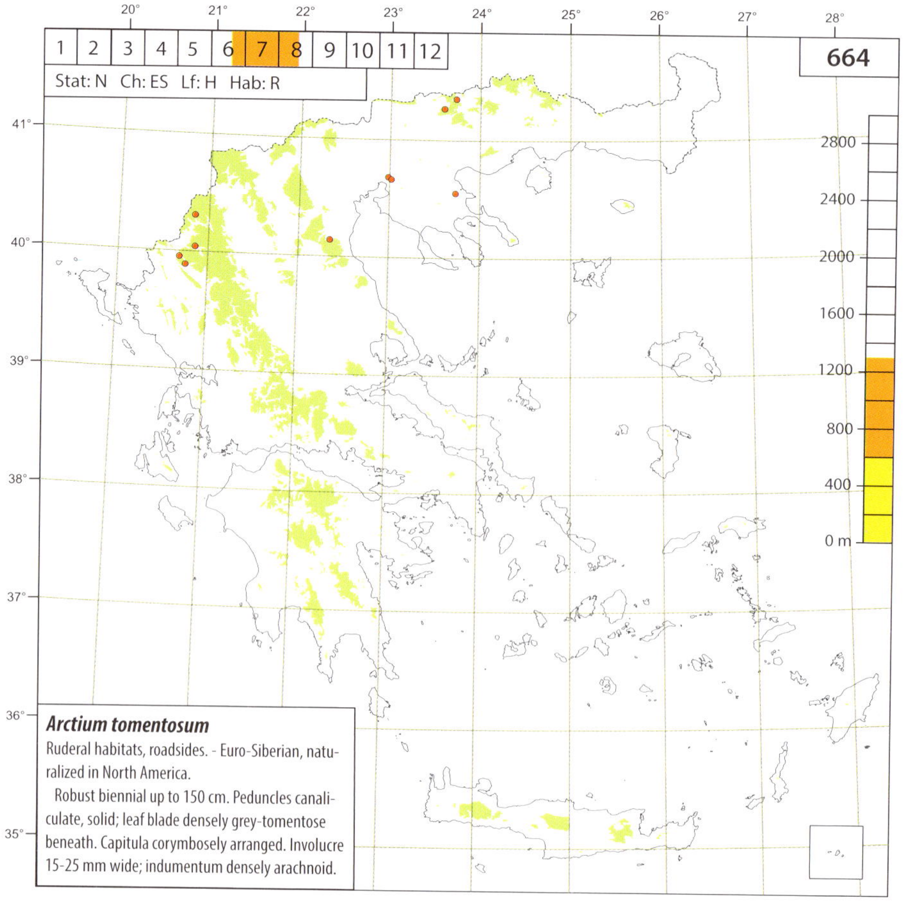

Arctium tomentosum
Ruderal habitats, roadsides. - Euro-Siberian, naturalized in North America.
Robust biennial up to 150 cm. Peduncles canaliculate, solid; leaf blade densely grey-tomentose beneath. Capitula corymbosely arranged. Involucre 15-25 mm wide; indumentum densely arachnoid.

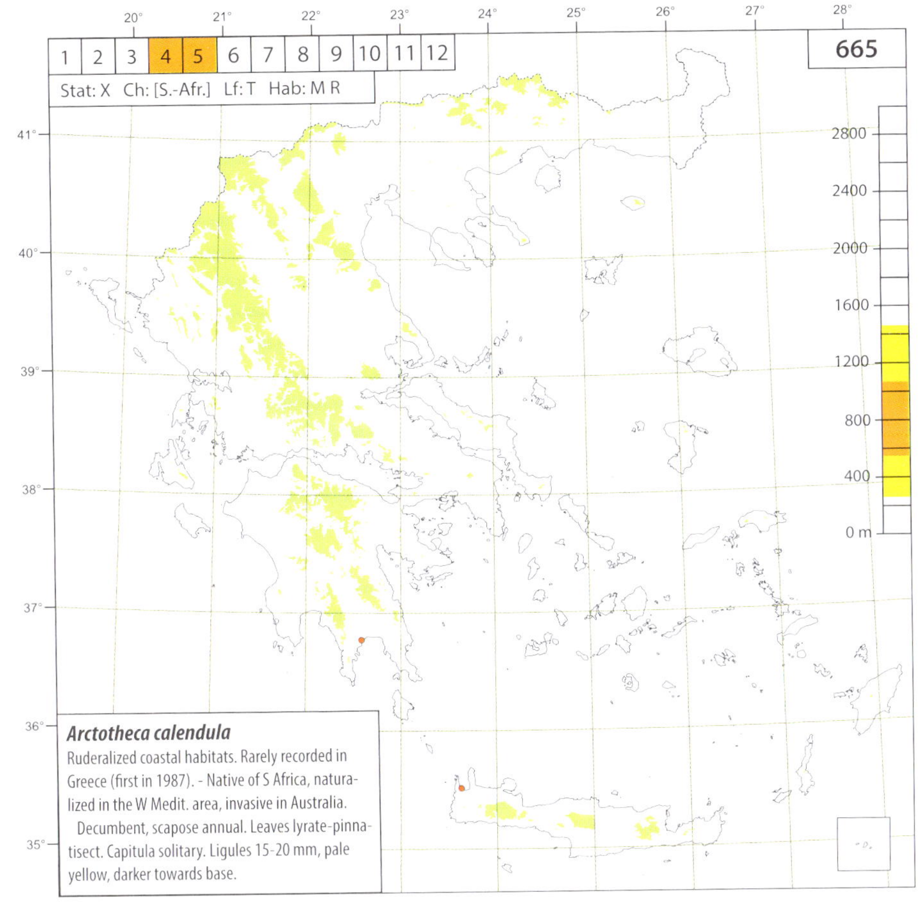

Arctotheca calendula
Ruderalized coastal habitats. Rarely recorded in Greece (first in 1987). - Native of S Africa, naturalized in the W Medit. area, invasive in Australia.
Decumbent, scapose annual. Leaves lyrate-pinnatisect. Capitula solitary. Ligules 15-20 mm, pale yellow, darker towards base.

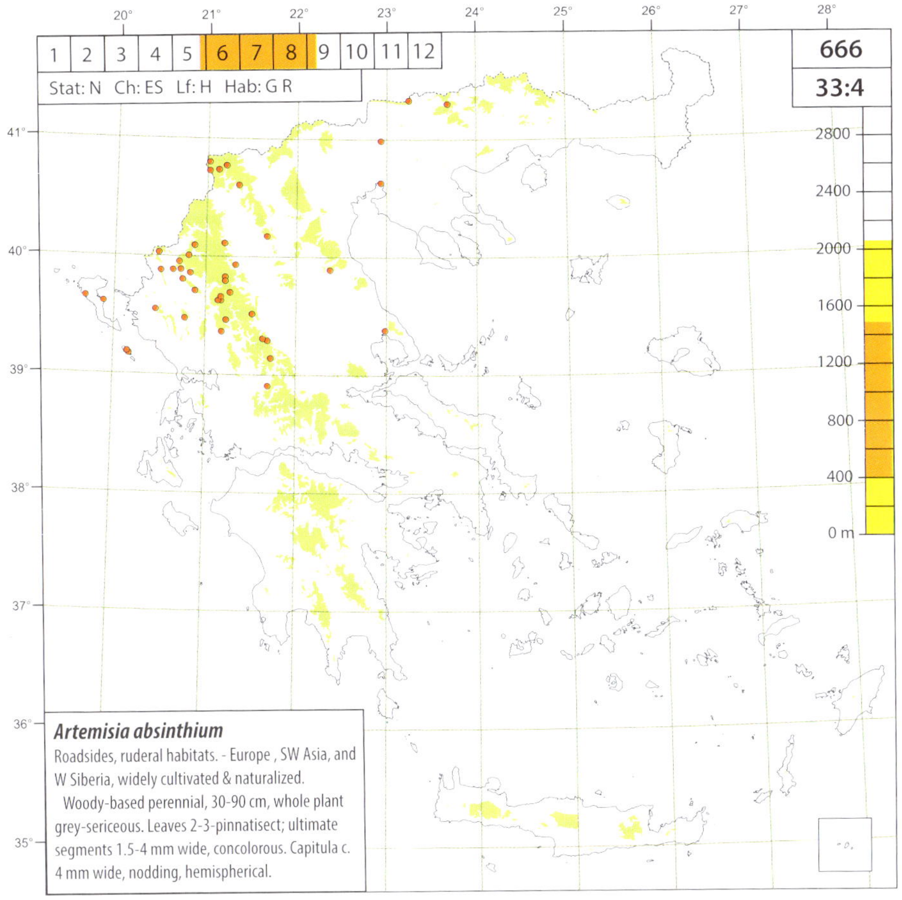

Artemisia absinthium
Roadsides, ruderal habitats. - Europe, SW Asia, and W Siberia, widely cultivated & naturalized.
Woody-based perennial, 30-90 cm, whole plant grey-sericeous. Leaves 2-3-pinnatisect; ultimate segments 1.5-4 mm wide, concolorous. Capitula c. 4 mm wide, nodding, hemispherical.

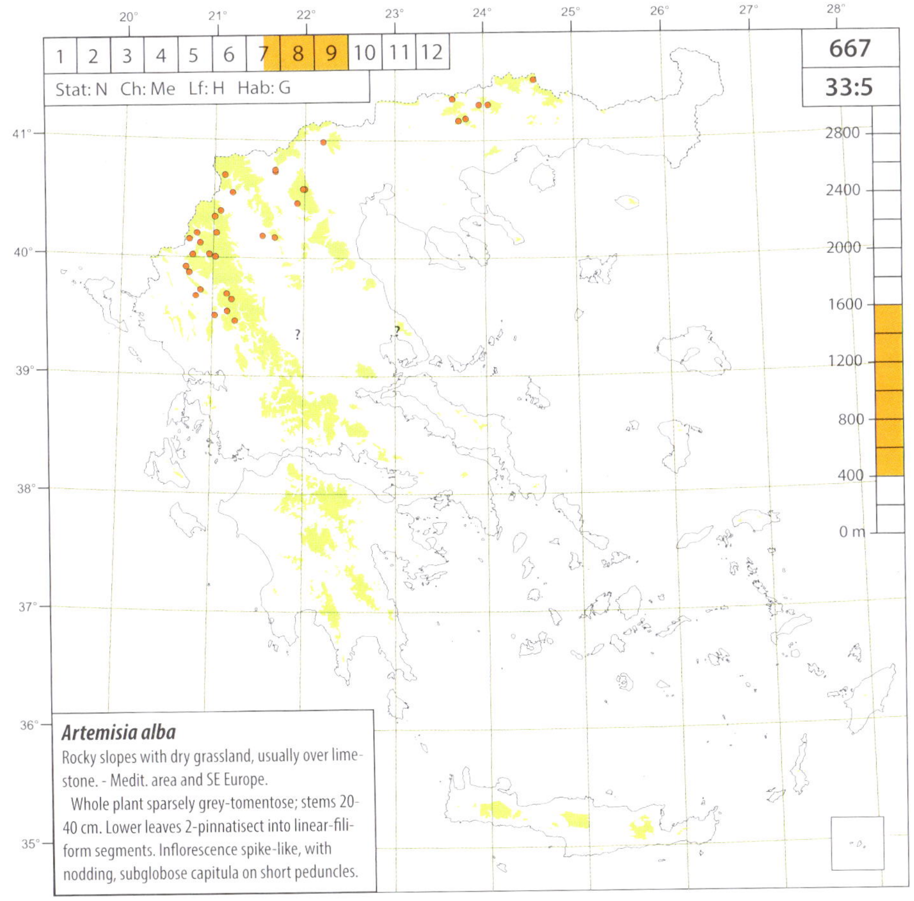

Artemisia alba
Rocky slopes with dry grassland, usually over limestone. - Medit. area and SE Europe.
Whole plant sparsely grey-tomentose; stems 20-40 cm. Lower leaves 2-pinnatisect into linear-filiform segments. Inflorescence spike-like, with nodding, subglobose capitula on short peduncles.

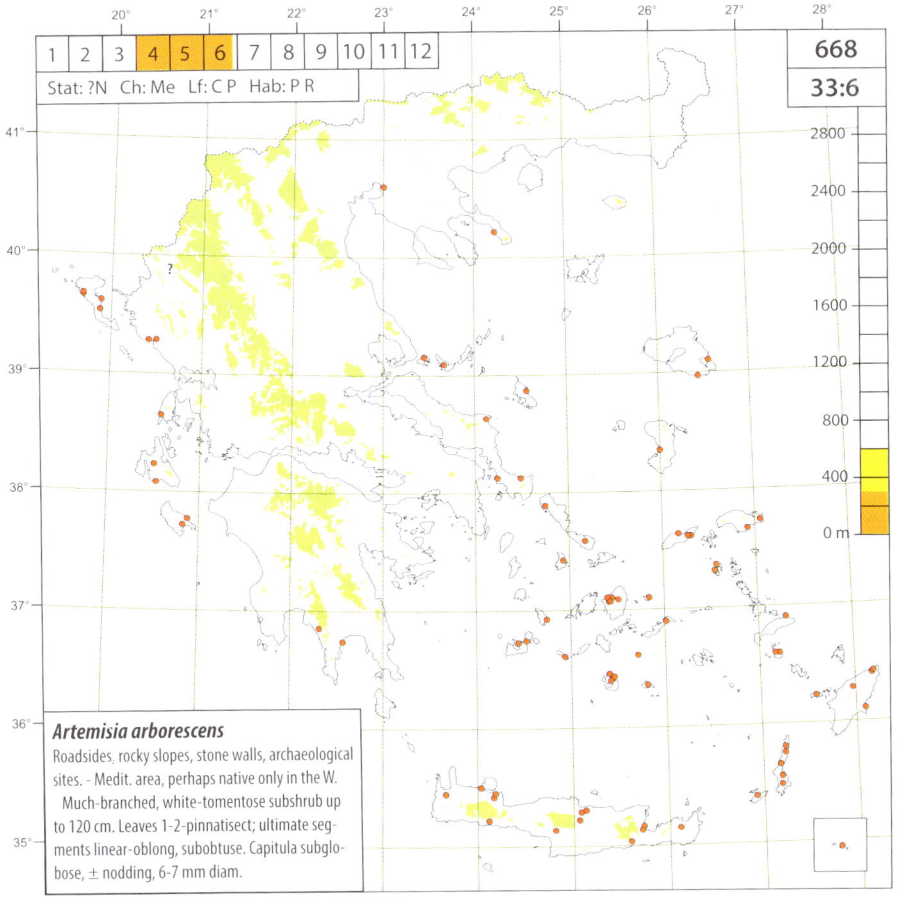

Artemisia arborescens
Roadsides, rocky slopes, stone walls, archaeological sites. - Medit. area, perhaps native only in the W.
Much-branched, white-tomentose subshrub up to 120 cm. Leaves 1-2-pinnatisect; ultimate segments linear-oblong, subobtuse. Capitula subglobose, ± nodding, 6-7 mm diam.

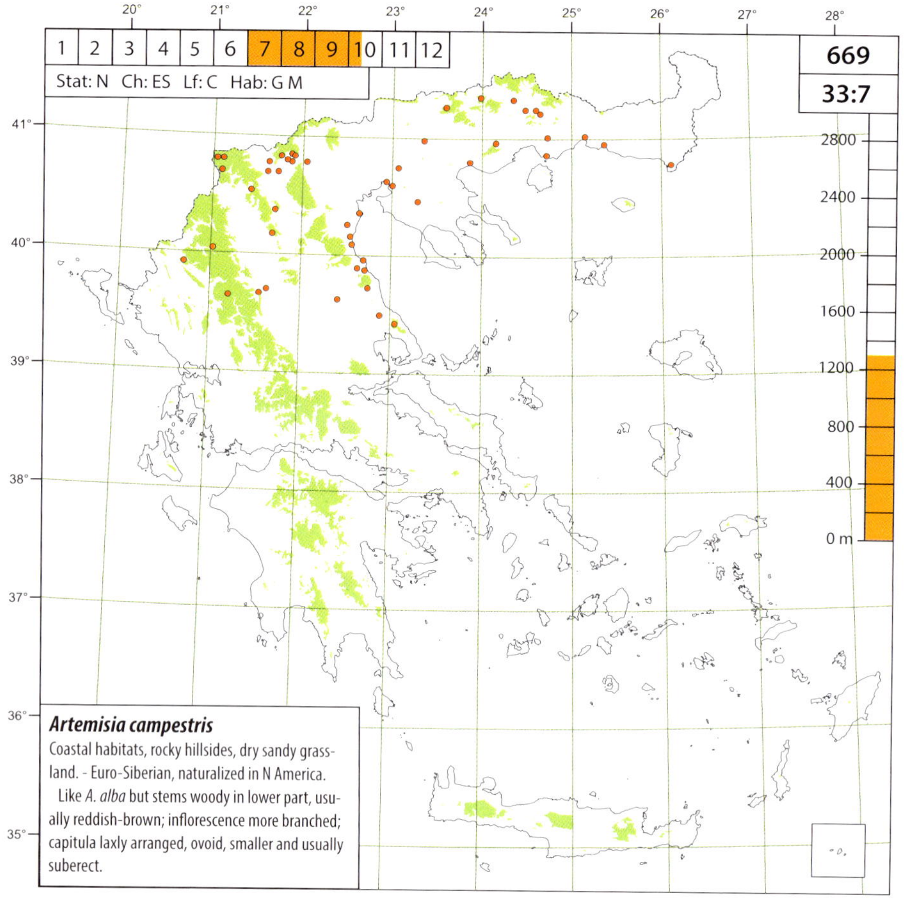

Artemisia campestris
Coastal habitats, rocky hillsides, dry sandy grassland. - Euro-Siberian, naturalized in N America.
Like *A. alba* but stems woody in lower part, usually reddish-brown; inflorescence more branched; capitula laxly arranged, ovoid, smaller and usually suberect.

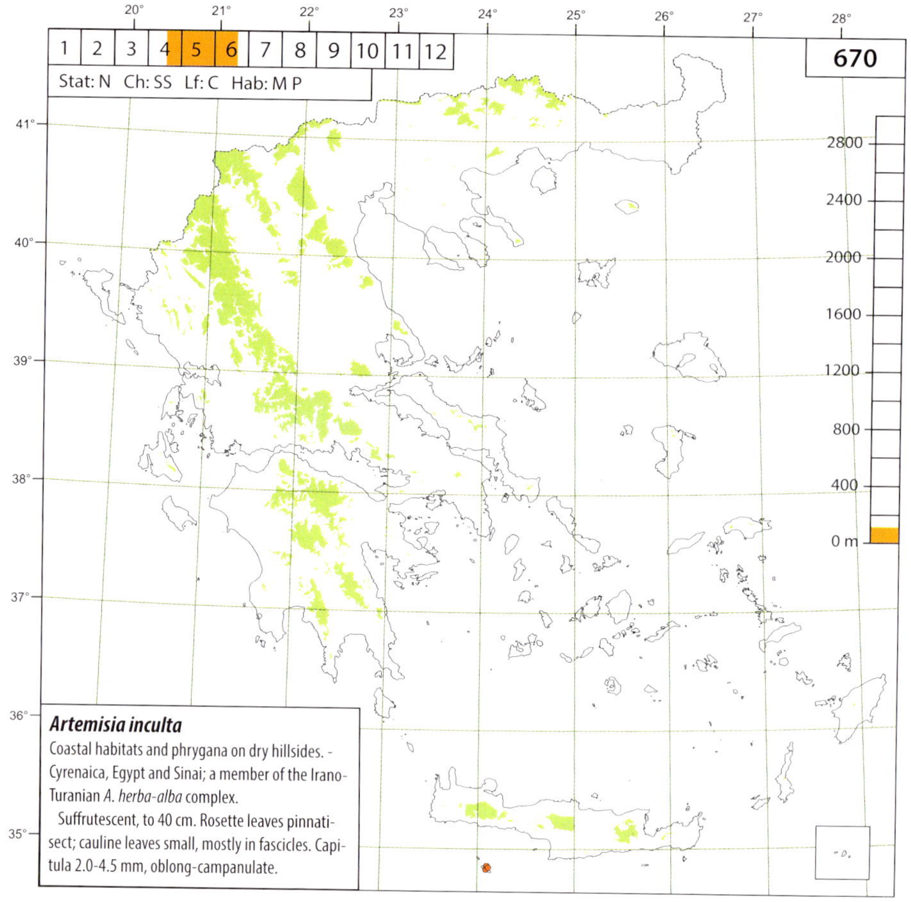

Artemisia inculta
Coastal habitats and phrygana on dry hillsides. - Cyrenaica, Egypt and Sinai; a member of the Irano-Turanian *A. herba-alba* complex.
Suffrutescent, to 40 cm. Rosette leaves pinnatisect; cauline leaves small, mostly in fascicles. Capitula 2.0-4.5 mm, oblong-campanulate.

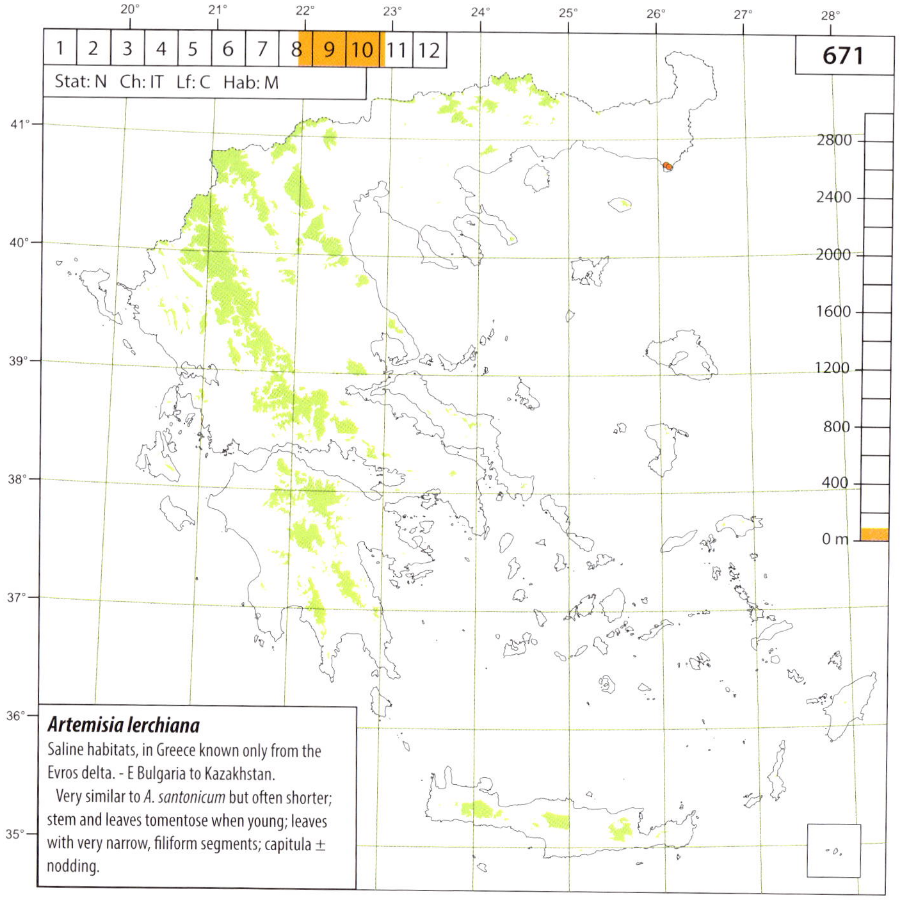

Artemisia lerchiana
Saline habitats, in Greece known only from the Evros delta. - E Bulgaria to Kazakhstan.
Very similar to *A. santonicum* but often shorter; stem and leaves tomentose when young; leaves with very narrow, filiform segments; capitula ± nodding.

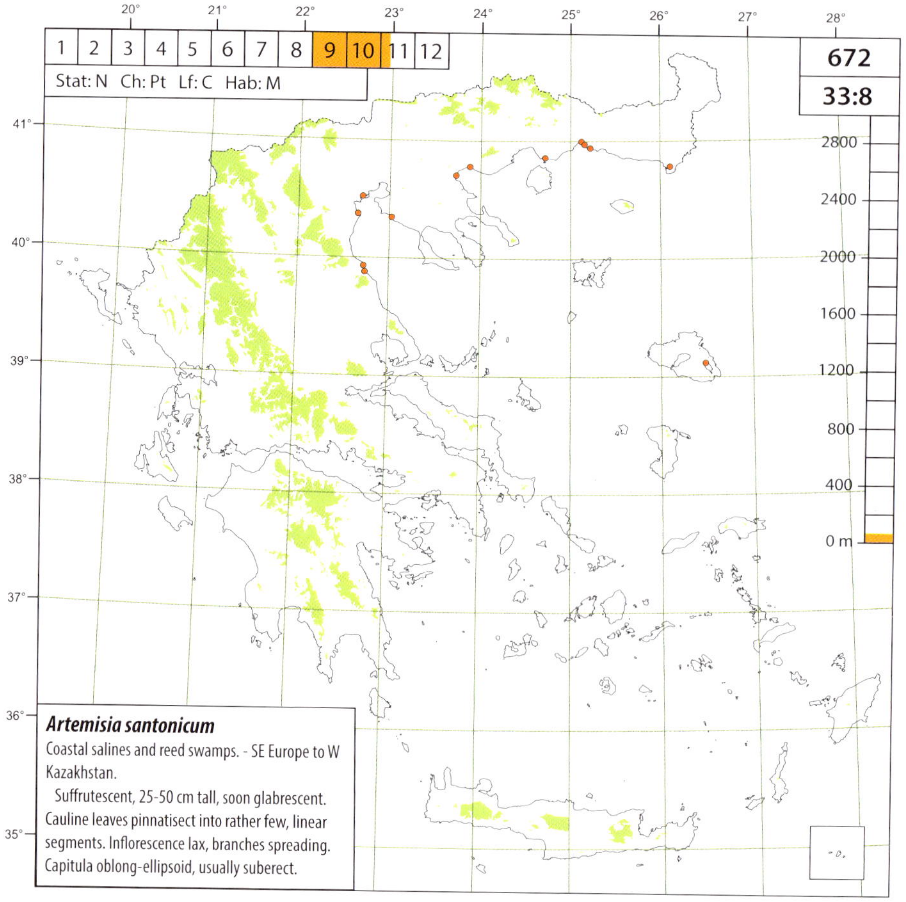

Artemisia santonicum
Coastal salines and reed swamps. - SE Europe to W Kazakhstan.
Suffrutescent, 25-50 cm tall, soon glabrescent. Cauline leaves pinnatisect into rather few, linear segments. Inflorescence lax, branches spreading. Capitula oblong-ellipsoid, usually suberect.

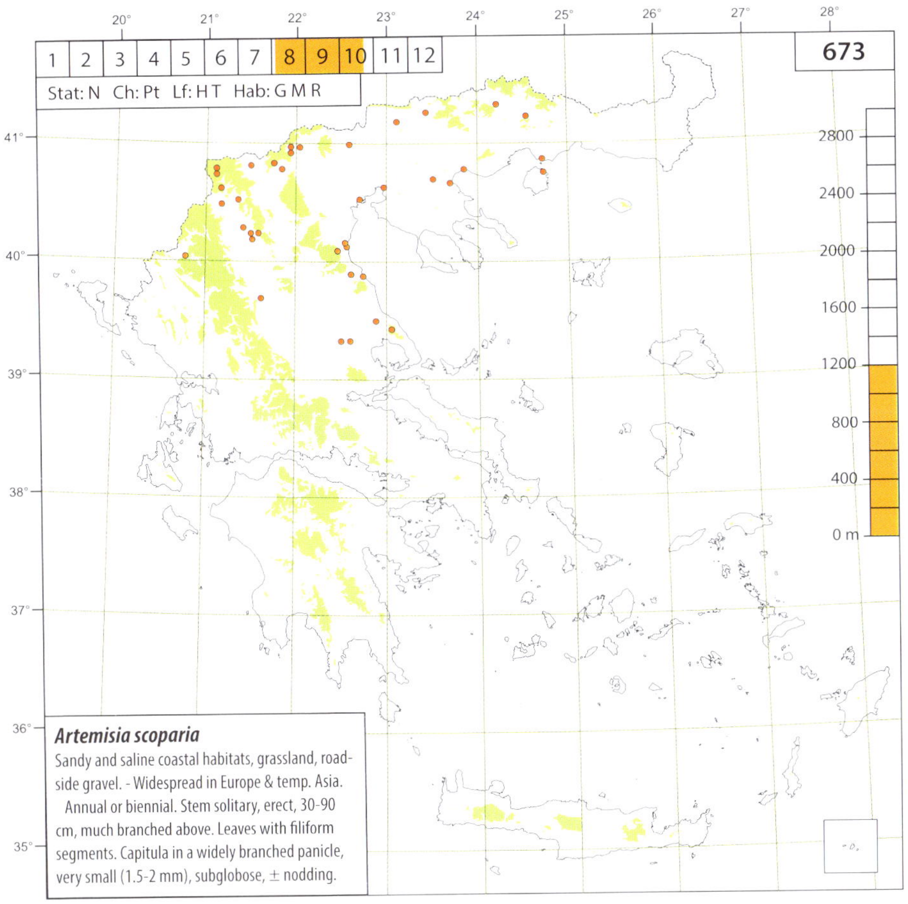

Artemisia scoparia
Sandy and saline coastal habitats, grassland, roadside gravel. - Widespread in Europe & temp. Asia.
Annual or biennial. Stem solitary, erect, 30-90 cm, much branched above. Leaves with filiform segments. Capitula in a widely branched panicle, very small (1.5-2 mm), subglobose, ± nodding.

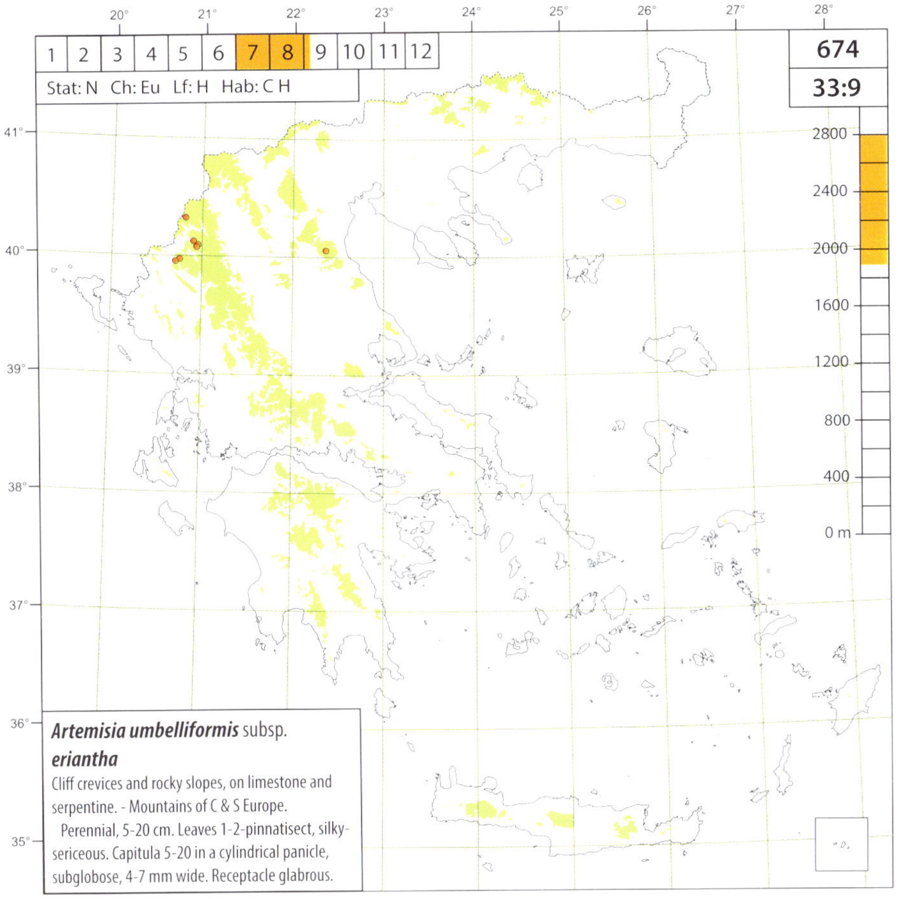

Artemisia umbelliformis subsp. ***eriantha***
Cliff crevices and rocky slopes, on limestone and serpentine. - Mountains of C & S Europe.
Perennial, 5-20 cm. Leaves 1-2-pinnatisect, silky-sericeous. Capitula 5-20 in a cylindrical panicle, subglobose, 4-7 mm wide. Receptacle glabrous.

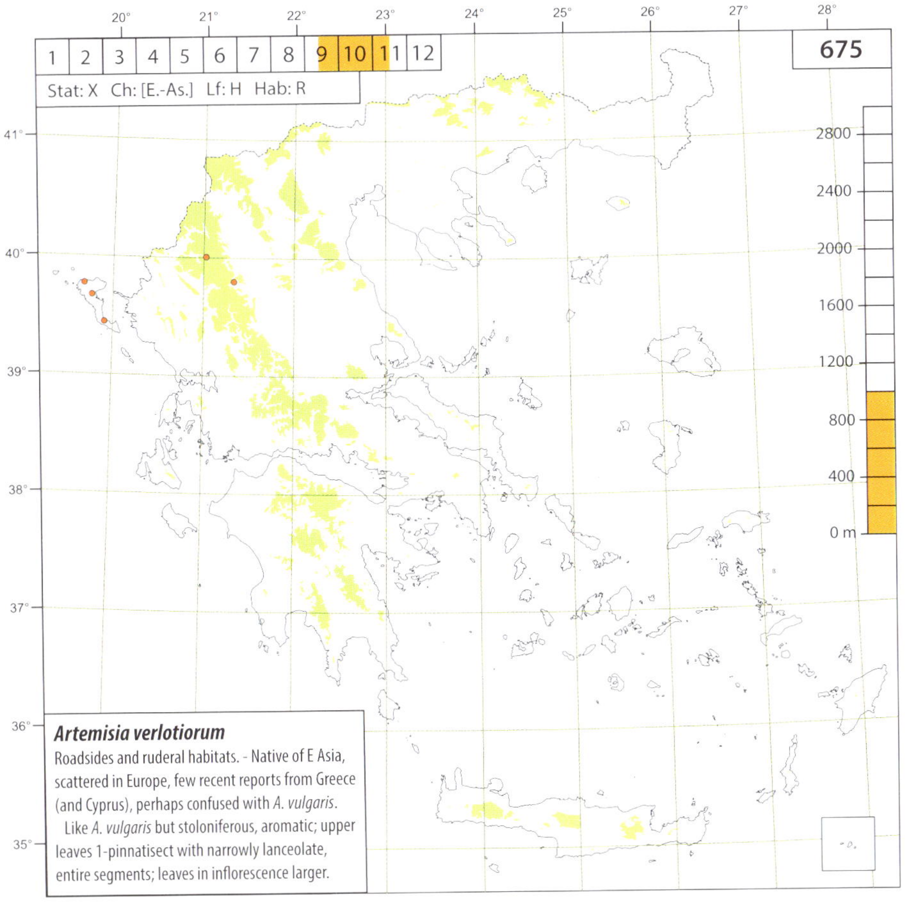

Artemisia verlotiorum
Roadsides and ruderal habitats. - Native of E Asia, scattered in Europe, few recent reports from Greece (and Cyprus), perhaps confused with *A. vulgaris*.
Like *A. vulgaris* but stoloniferous, aromatic; upper leaves 1-pinnatisect with narrowly lanceolate, entire segments; leaves in inflorescence larger.

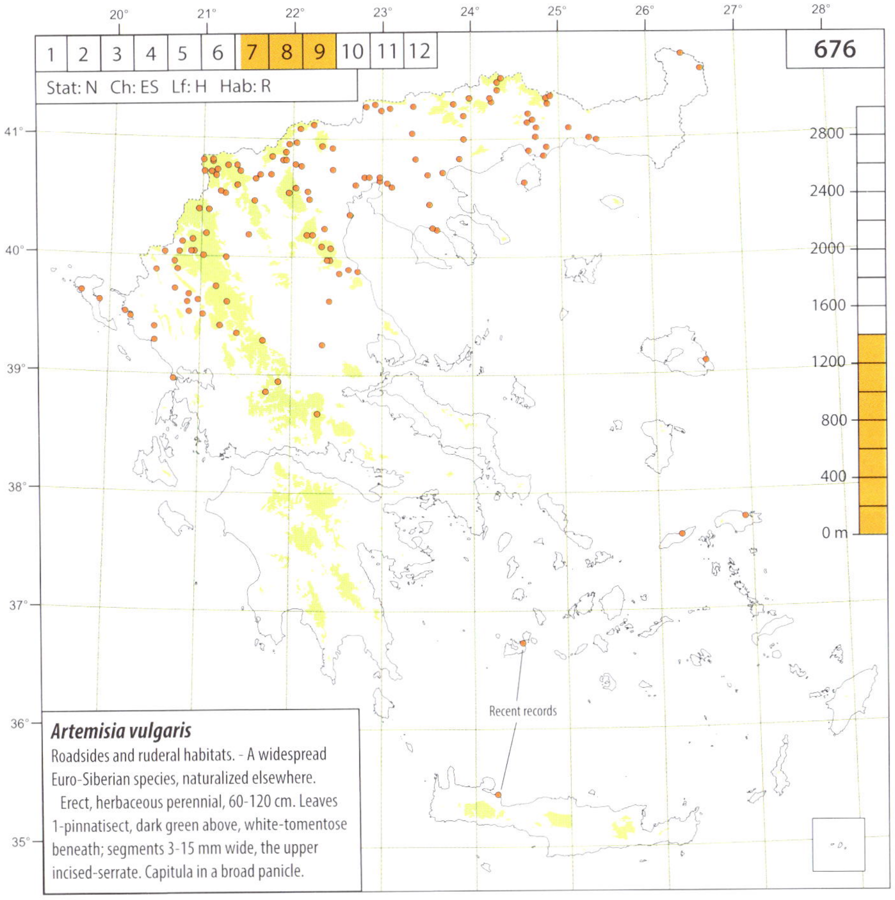

Artemisia vulgaris
Roadsides and ruderal habitats. - A widespread Euro-Siberian species, naturalized elsewhere.
Erect, herbaceous perennial, 60-120 cm. Leaves 1-pinnatisect, dark green above, white-tomentose beneath; segments 3-15 mm wide, the upper incised-serrate. Capitula in a broad panicle.

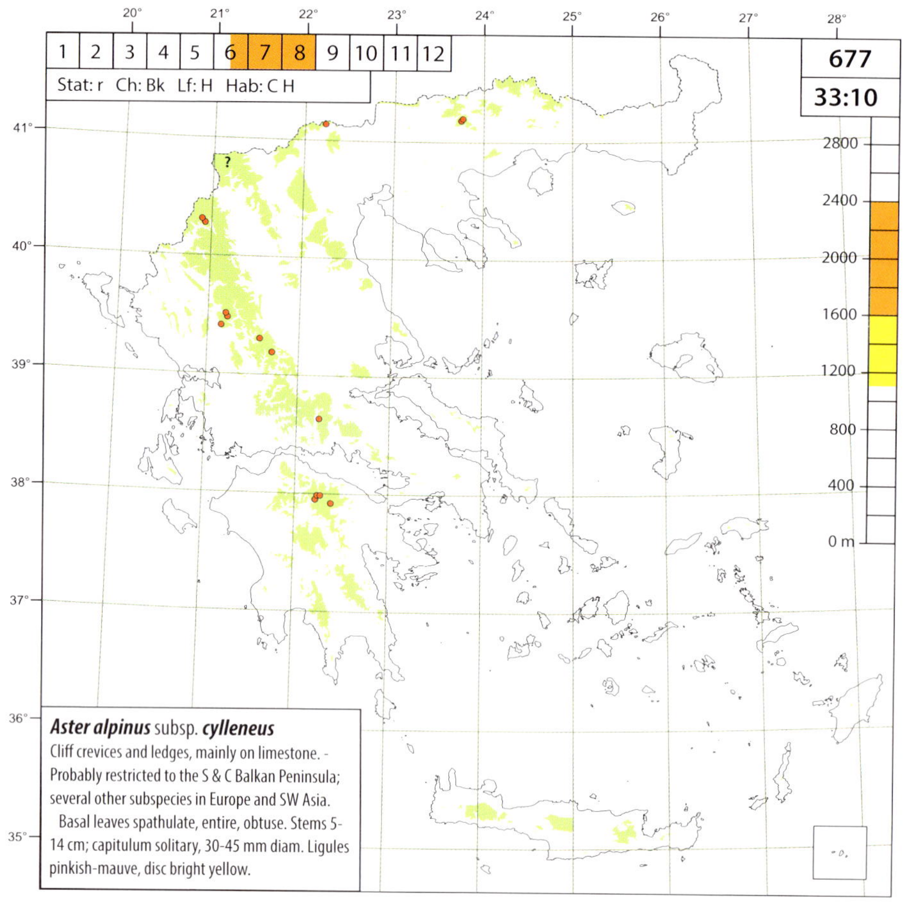

Aster alpinus* subsp. *cylleneus
Cliff crevices and ledges, mainly on limestone. - Probably restricted to the S & C Balkan Peninsula; several other subspecies in Europe and SW Asia.
Basal leaves spathulate, entire, obtuse. Stems 5-14 cm; capitulum solitary, 30-45 mm diam. Ligules pinkish-mauve, disc bright yellow.

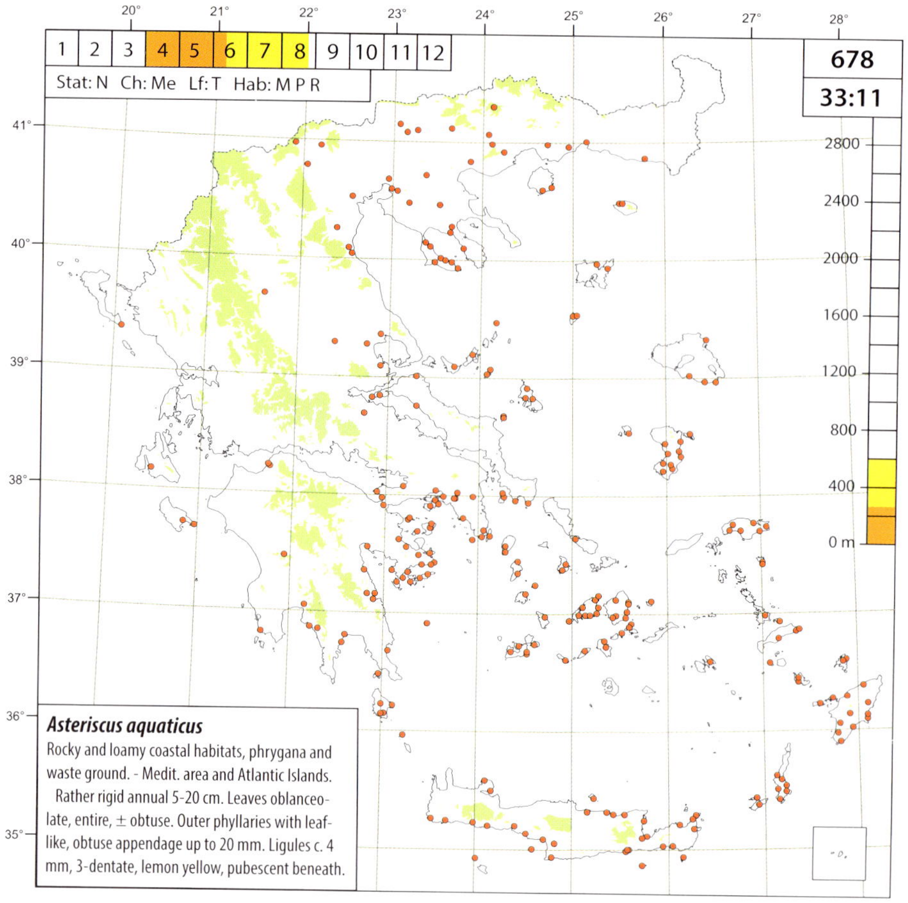

Asteriscus aquaticus
Rocky and loamy coastal habitats, phrygana and waste ground. - Medit. area and Atlantic Islands.
Rather rigid annual 5-20 cm. Leaves oblanceolate, entire, ± obtuse. Outer phyllaries with leaf-like, obtuse appendage up to 20 mm. Ligules c. 4 mm, 3-dentate, lemon yellow, pubescent beneath.

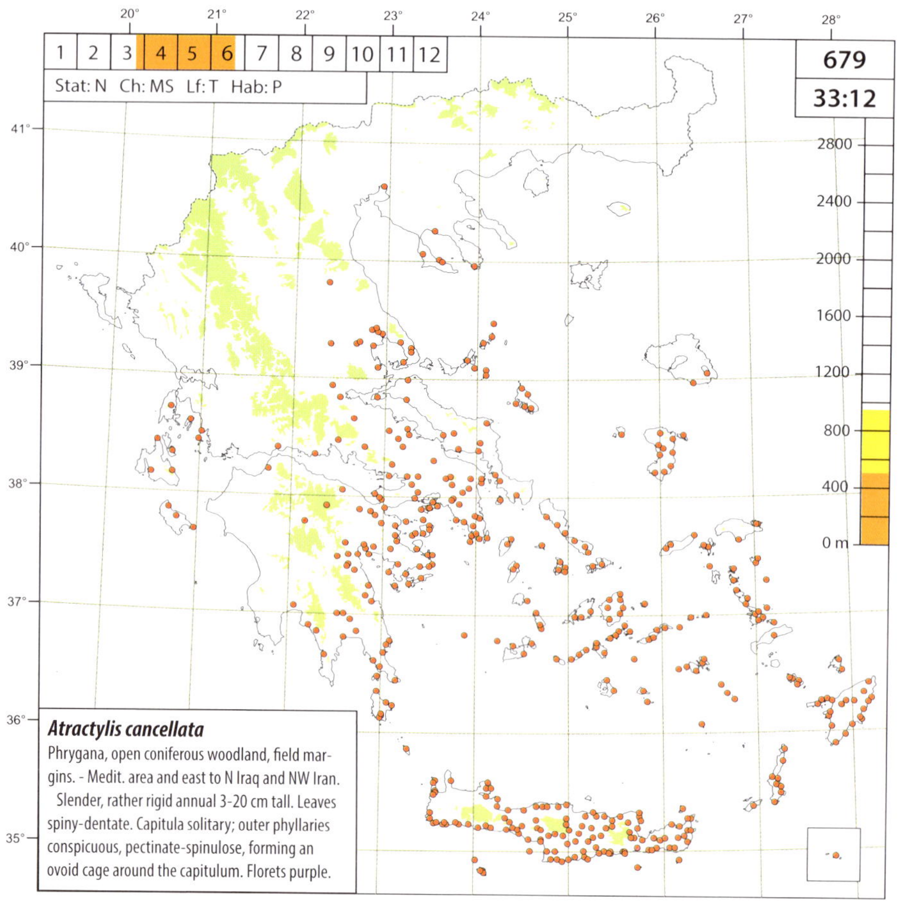

Atractylis cancellata
Phrygana, open coniferous woodland, field margins. - Medit. area and east to N Iraq and NW Iran.
Slender, rather rigid annual 3-20 cm tall. Leaves spiny-dentate. Capitula solitary; outer phyllaries conspicuous, pectinate-spinulose, forming an ovoid cage around the capitulum. Florets purple.

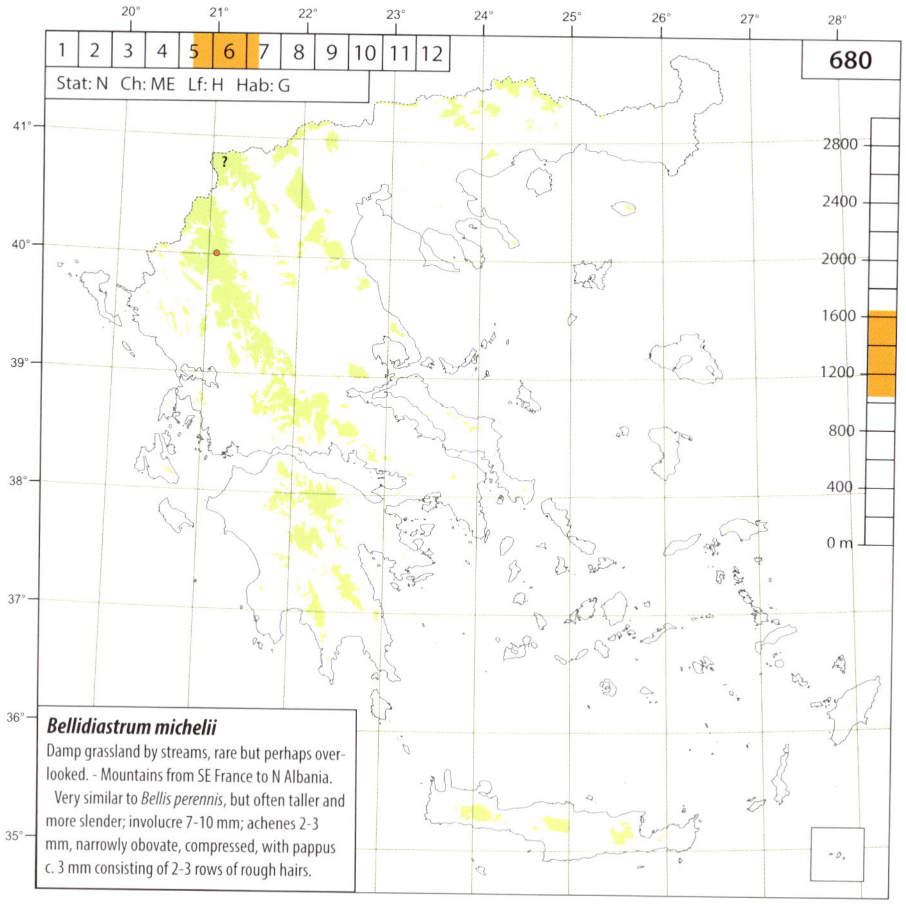

Bellidiastrum michelii
Damp grassland by streams, rare but perhaps overlooked. - Mountains from SE France to N Albania.
Very similar to *Bellis perennis*, but often taller and more slender; involucre 7-10 mm; achenes 2-3 mm, narrowly obovate, compressed, with pappus c. 3 mm consisting of 2-3 rows of rough hairs.

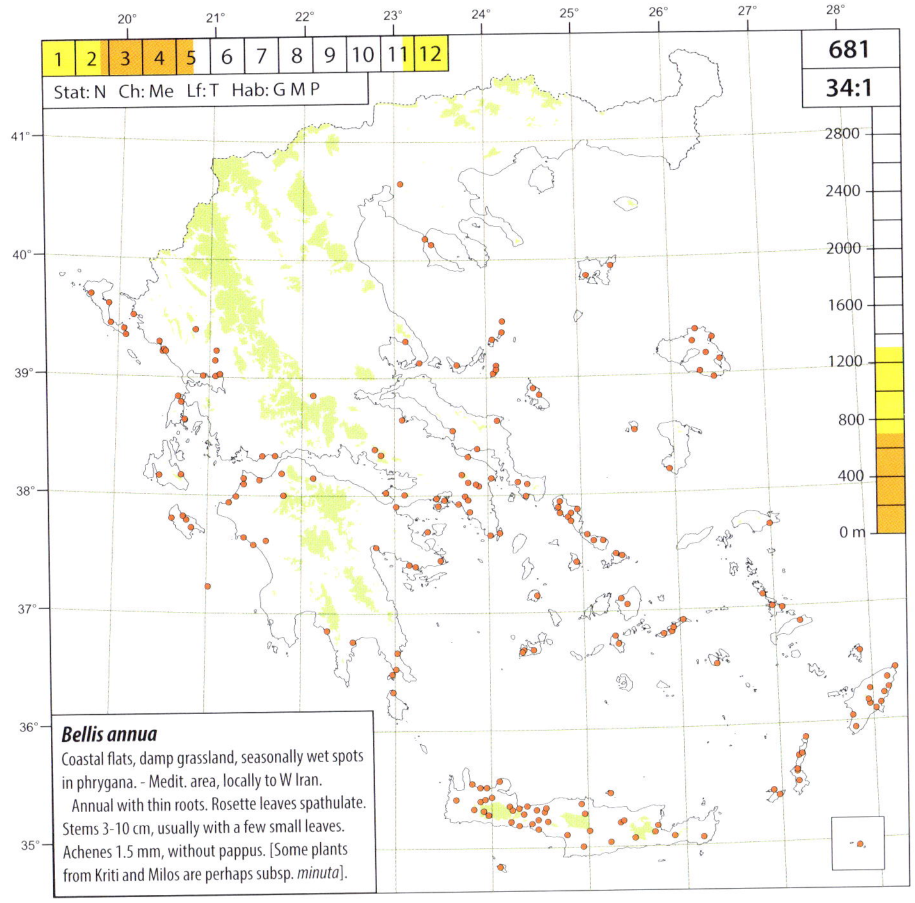

Bellis annua
Coastal flats, damp grassland, seasonally wet spots in phrygana. - Medit. area, locally to W Iran.
Annual with thin roots. Rosette leaves spathulate. Stems 3-10 cm, usually with a few small leaves. Achenes 1.5 mm, without pappus. [Some plants from Kriti and Milos are perhaps subsp. *minuta*].

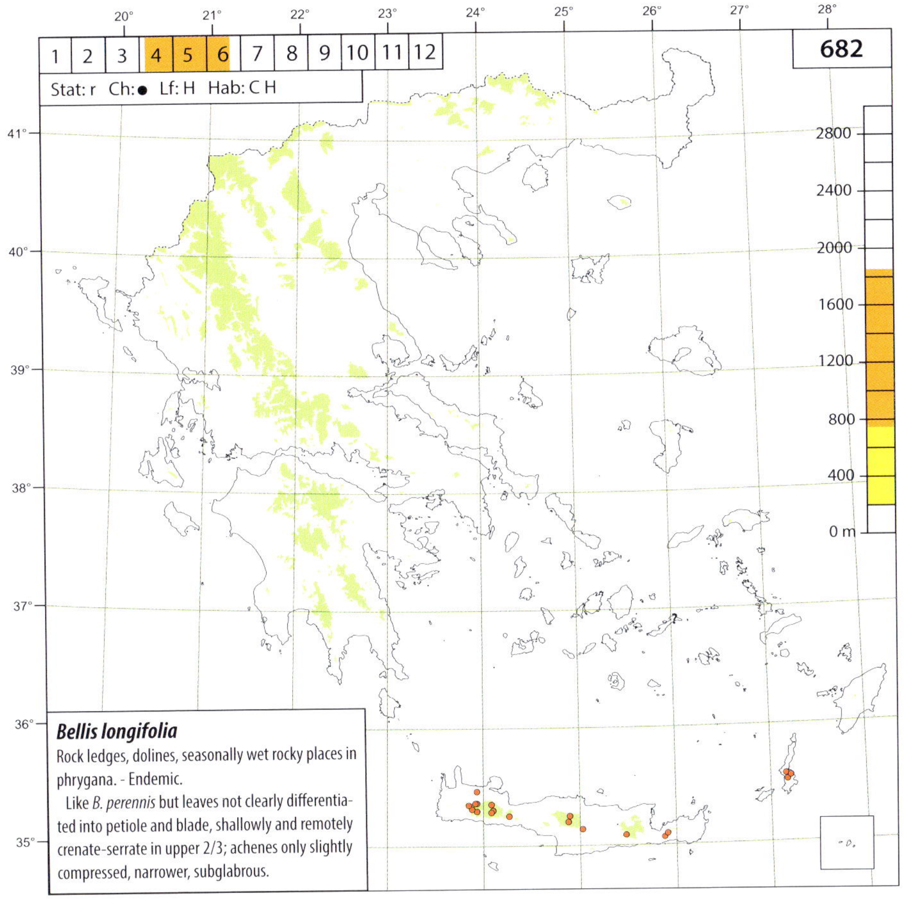

Bellis longifolia
Rock ledges, dolines, seasonally wet rocky places in phrygana. - Endemic.
Like *B. perennis* but leaves not clearly differentiated into petiole and blade, shallowly and remotely crenate-serrate in upper 2/3; achenes only slightly compressed, narrower, subglabrous.

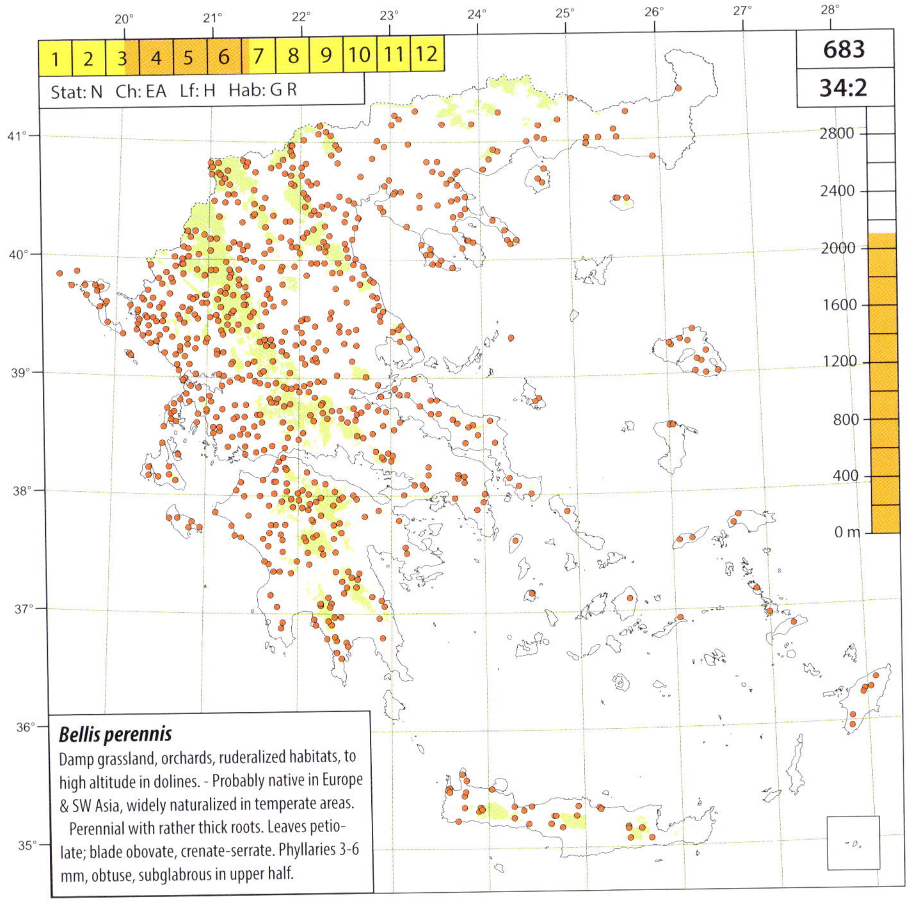

Bellis perennis
Damp grassland, orchards, ruderalized habitats, to high altitude in dolines. - Probably native in Europe & SW Asia, widely naturalized in temperate areas.
Perennial with rather thick roots. Leaves petiolate; blade obovate, crenate-serrate. Phyllaries 3-6 mm, obtuse, subglabrous in upper half.

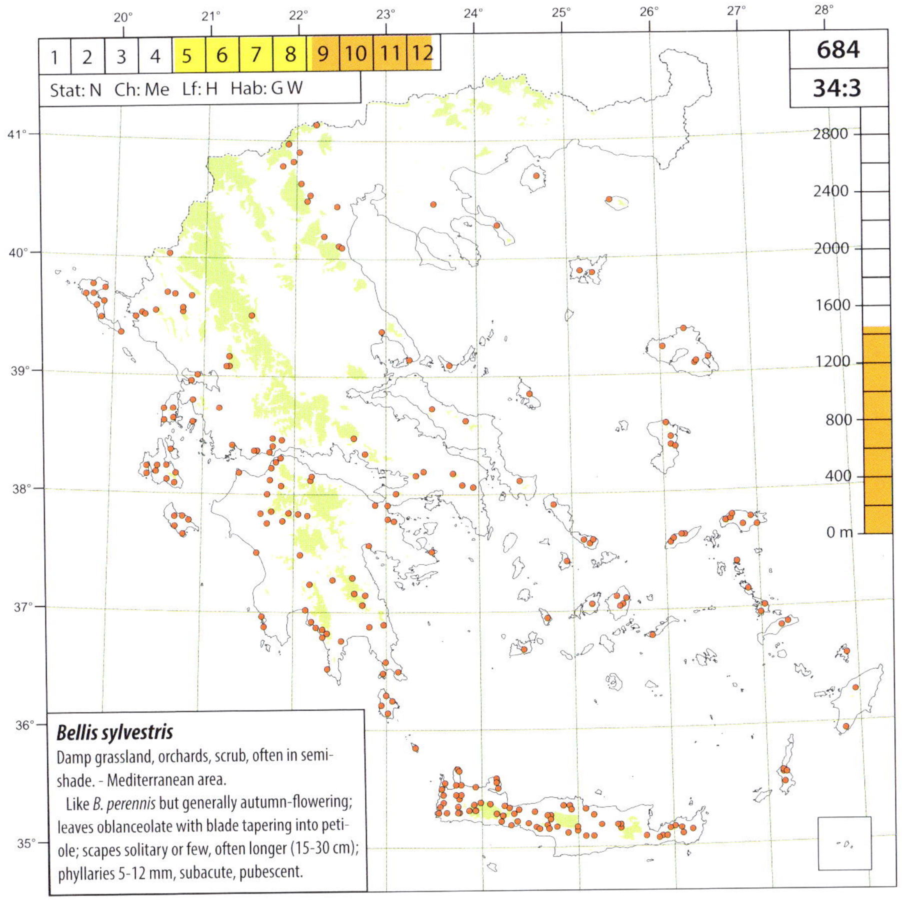

Bellis sylvestris
Damp grassland, orchards, scrub, often in semi-shade. - Mediterranean area.
Like *B. perennis* but generally autumn-flowering; leaves oblanceolate with blade tapering into petiole; scapes solitary or few, often longer (15-30 cm); phyllaries 5-12 mm, subacute, pubescent.

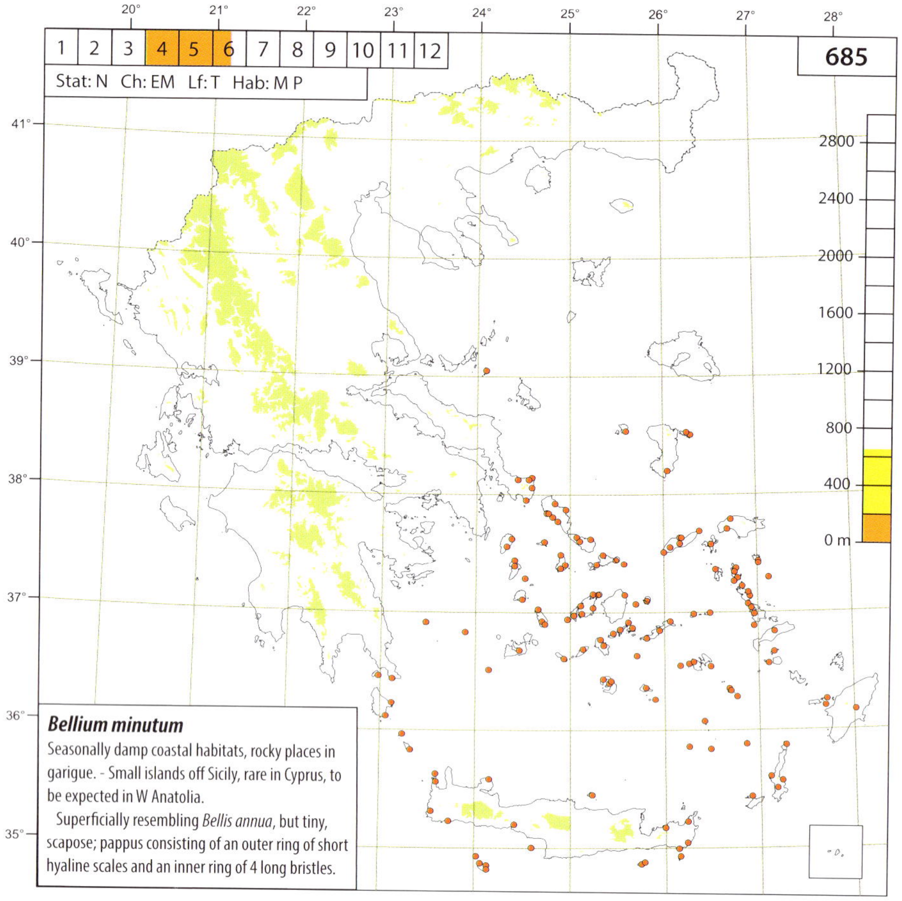

Bellium minutum

Seasonally damp coastal habitats, rocky places in garigue. - Small islands off Sicily, rare in Cyprus, to be expected in W Anatolia.

Superficially resembling *Bellis annua*, but tiny, scapose; pappus consisting of an outer ring of short hyaline scales and an inner ring of 4 long bristles.

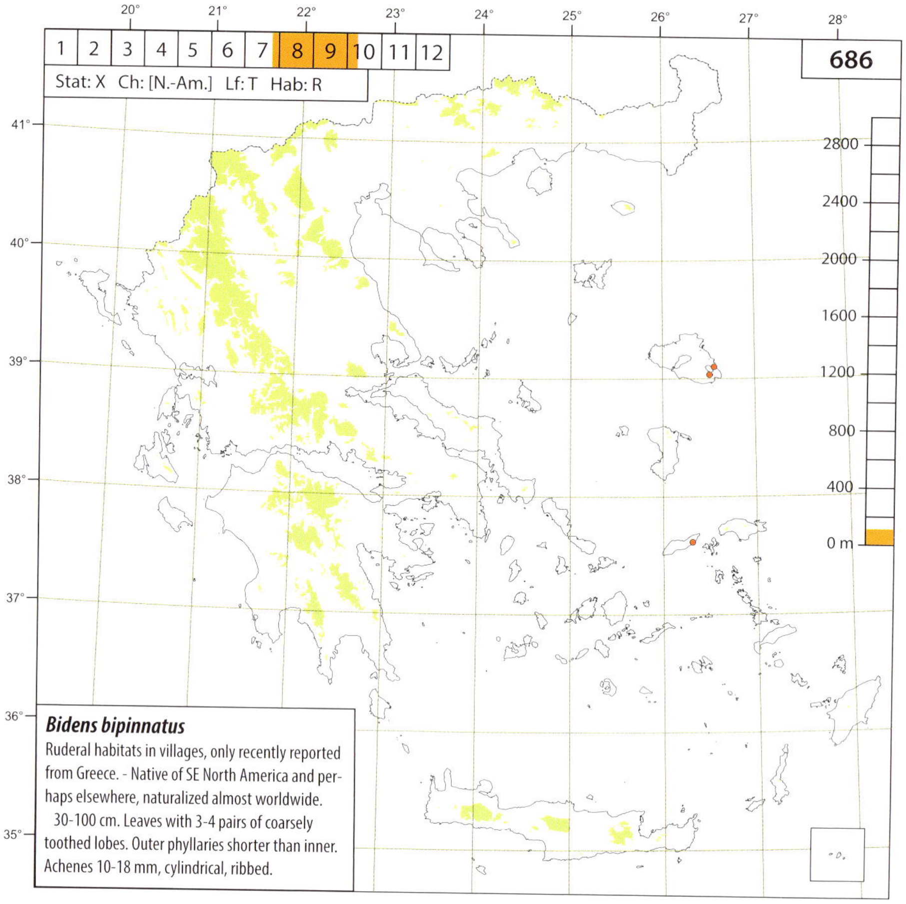

Bidens bipinnatus

Ruderal habitats in villages, only recently reported from Greece. - Native of SE North America and perhaps elsewhere, naturalized almost worldwide.

30-100 cm. Leaves with 3-4 pairs of coarsely toothed lobes. Outer phyllaries shorter than inner. Achenes 10-18 mm, cylindrical, ribbed.

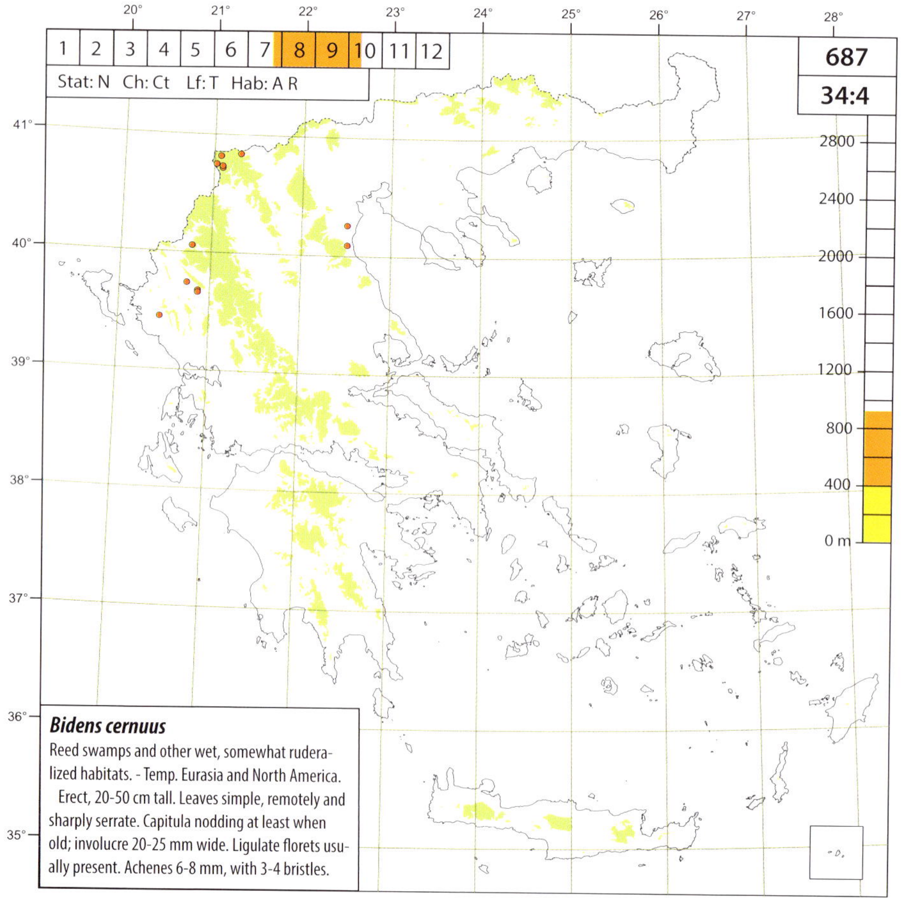

Bidens cernuus

Reed swamps and other wet, somewhat ruderalized habitats. - Temp. Eurasia and North America.

Erect, 20-50 cm tall. Leaves simple, remotely and sharply serrate. Capitula nodding at least when old; involucre 20-25 mm wide. Ligulate florets usually present. Achenes 6-8 mm, with 3-4 bristles.

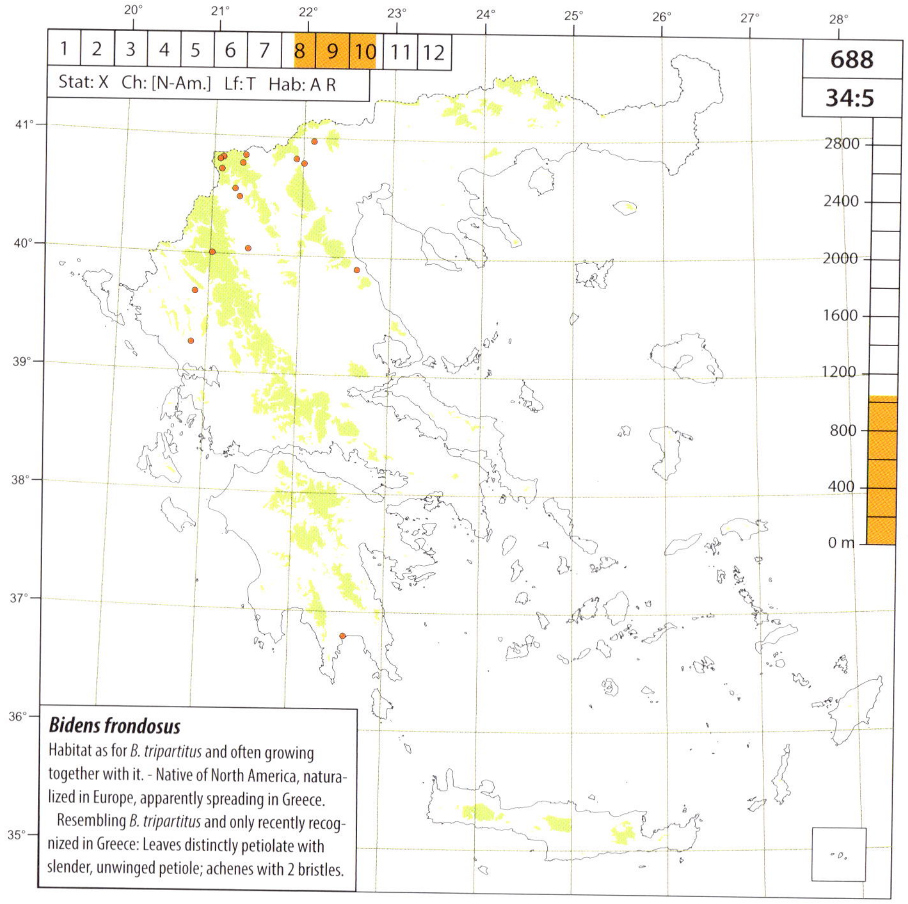

Bidens frondosus

Habitat as for *B. tripartitus* and often growing together with it. - Native of North America, naturalized in Europe, apparently spreading in Greece.

Resembling *B. tripartitus* and only recently recognized in Greece: Leaves distinctly petiolate with slender, unwinged petiole; achenes with 2 bristles.

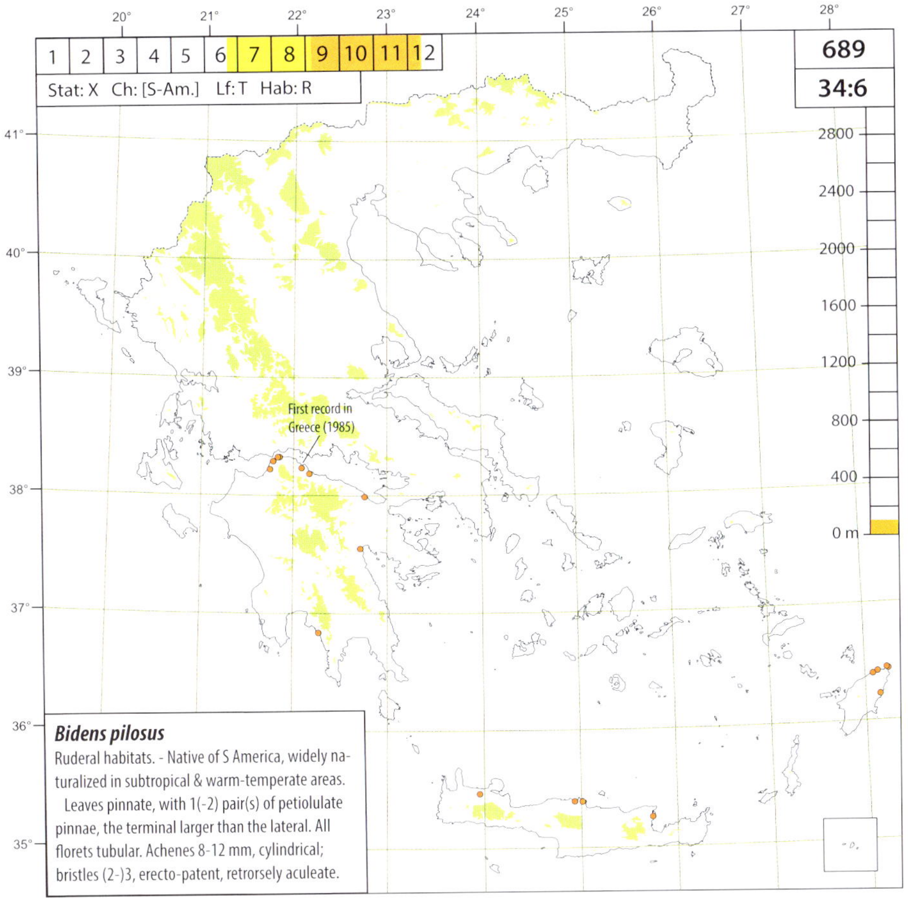

Bidens pilosus
Ruderal habitats. - Native of S America, widely naturalized in subtropical & warm-temperate areas.
Leaves pinnate, with 1(-2) pair(s) of petiolulate pinnae, the terminal larger than the lateral. All florets tubular. Achenes 8-12 mm, cylindrical; bristles (2-)3, erecto-patent, retrorsely aculeate.

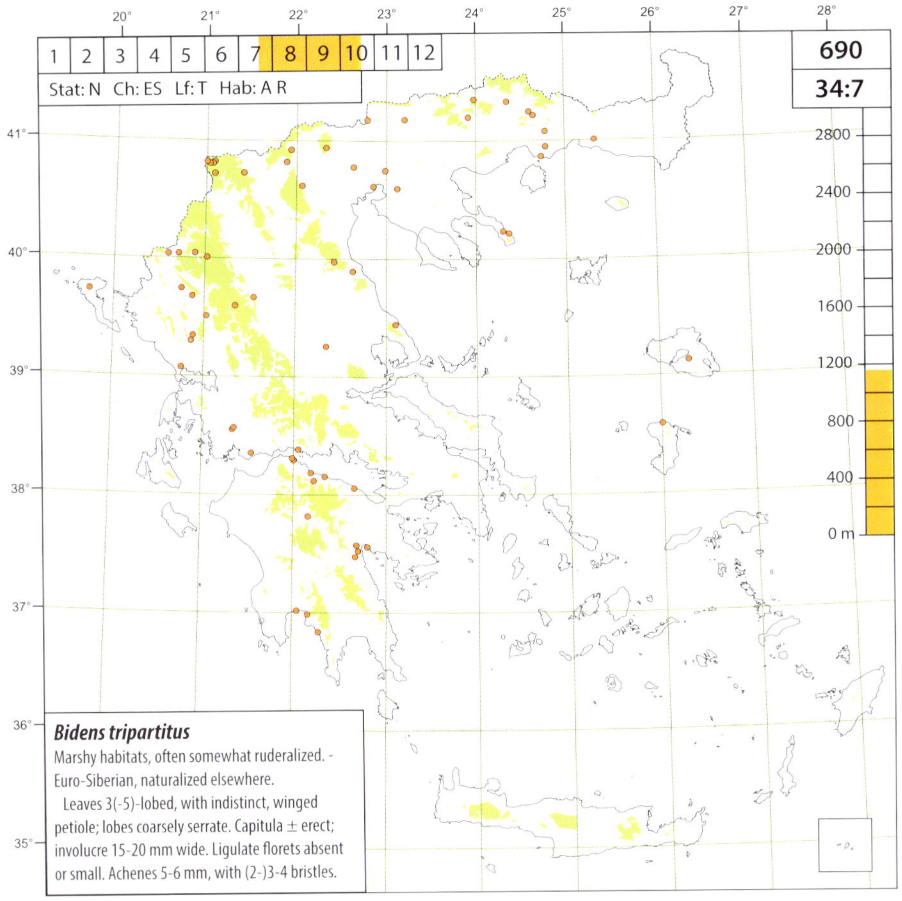

Bidens tripartitus
Marshy habitats, often somewhat ruderalized. - Euro-Siberian, naturalized elsewhere.
Leaves 3(-5)-lobed, with indistinct, winged petiole; lobes coarsely serrate. Capitula ± erect; involucre 15-20 mm wide. Ligulate florets absent or small. Achenes 5-6 mm, with (2-)3-4 bristles.

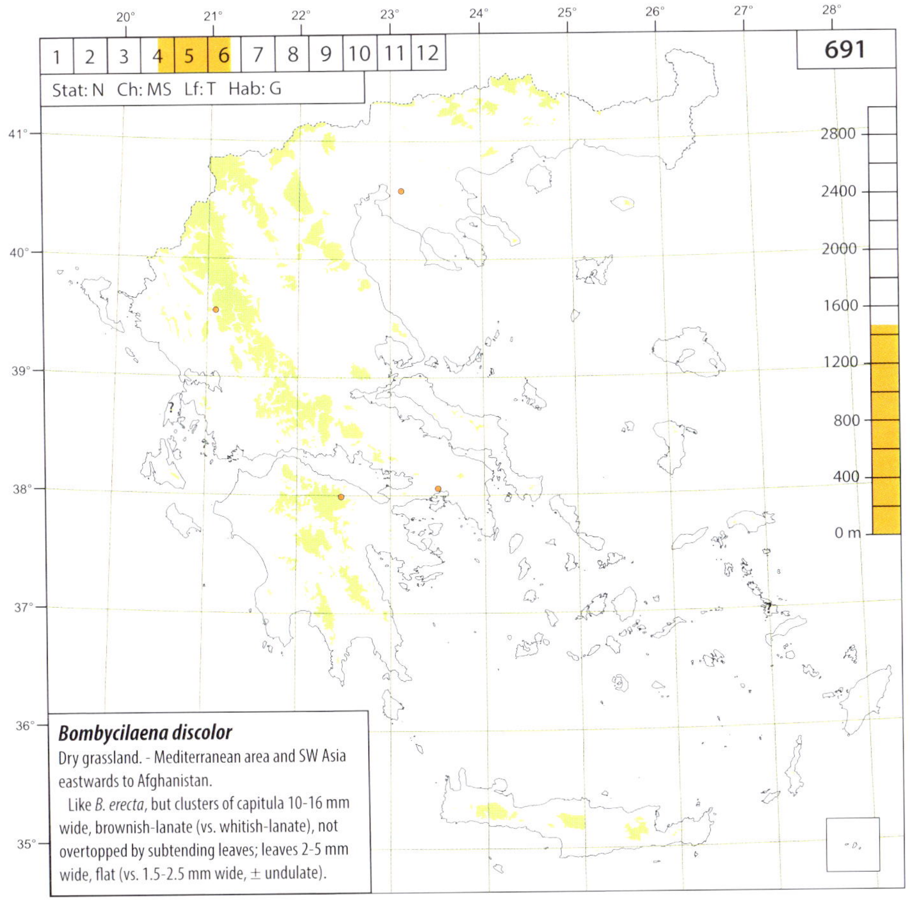

Bombycilaena discolor
Dry grassland. - Mediterranean area and SW Asia eastwards to Afghanistan.
Like *B. erecta*, but clusters of capitula 10-16 mm wide, brownish-lanate (vs. whitish-lanate), not overtopped by subtending leaves; leaves 2-5 mm wide, flat (vs. 1.5-2.5 mm wide, ± undulate).

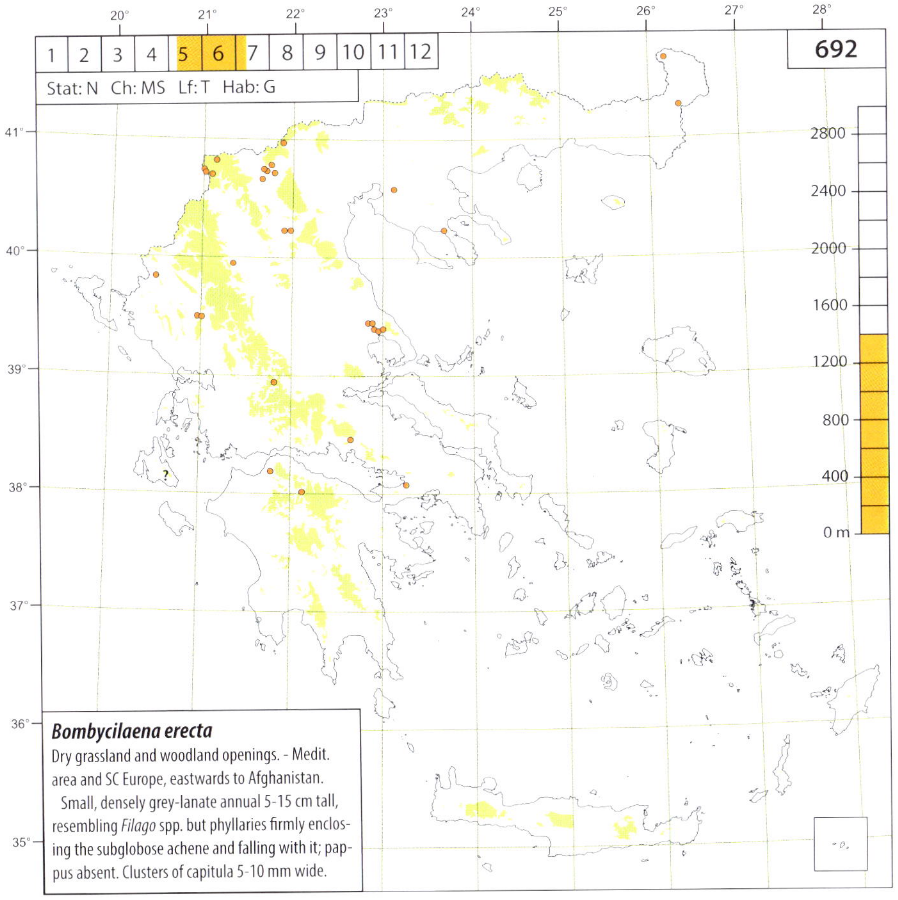

Bombycilaena erecta
Dry grassland and woodland openings. - Medit. area and SC Europe, eastwards to Afghanistan.
Small, densely grey-lanate annual 5-15 cm tall, resembling *Filago* spp. but phyllaries firmly enclosing the subglobose achene and falling with it; pappus absent. Clusters of capitula 5-10 mm wide.

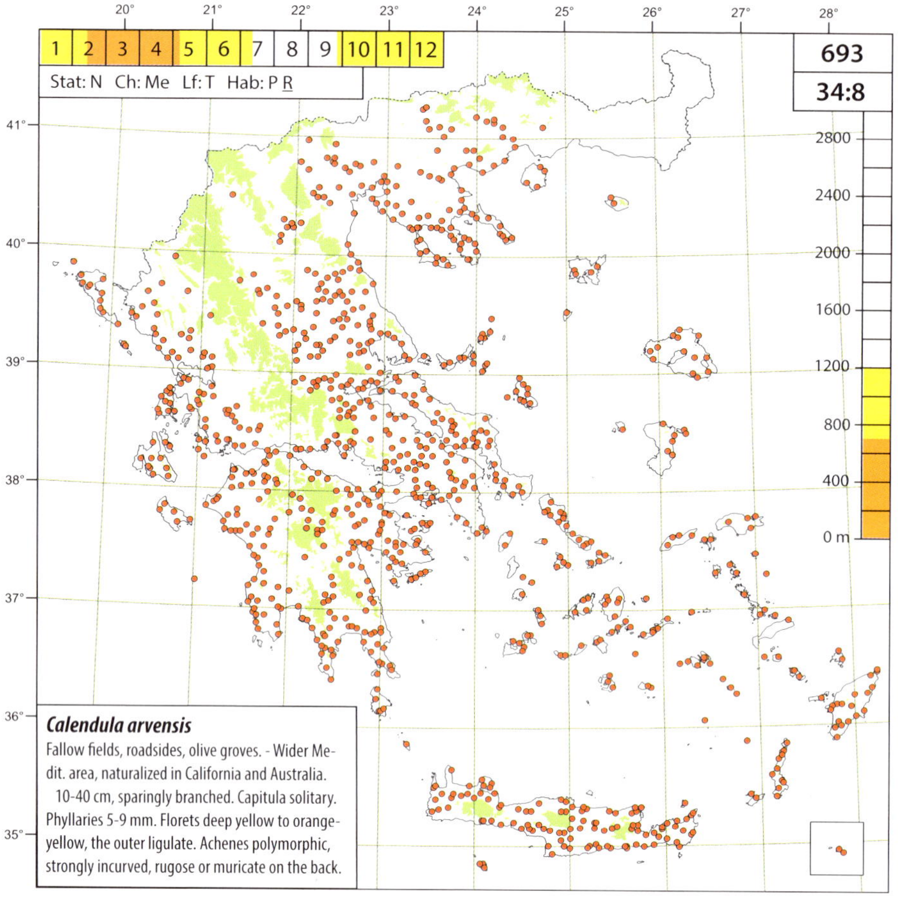

Calendula arvensis

Fallow fields, roadsides, olive groves. - Wider Medit. area, naturalized in California and Australia.

10-40 cm, sparingly branched. Capitula solitary. Phyllaries 5-9 mm. Florets deep yellow to orange-yellow, the outer ligulate. Achenes polymorphic, strongly incurved, rugose or muricate on the back.

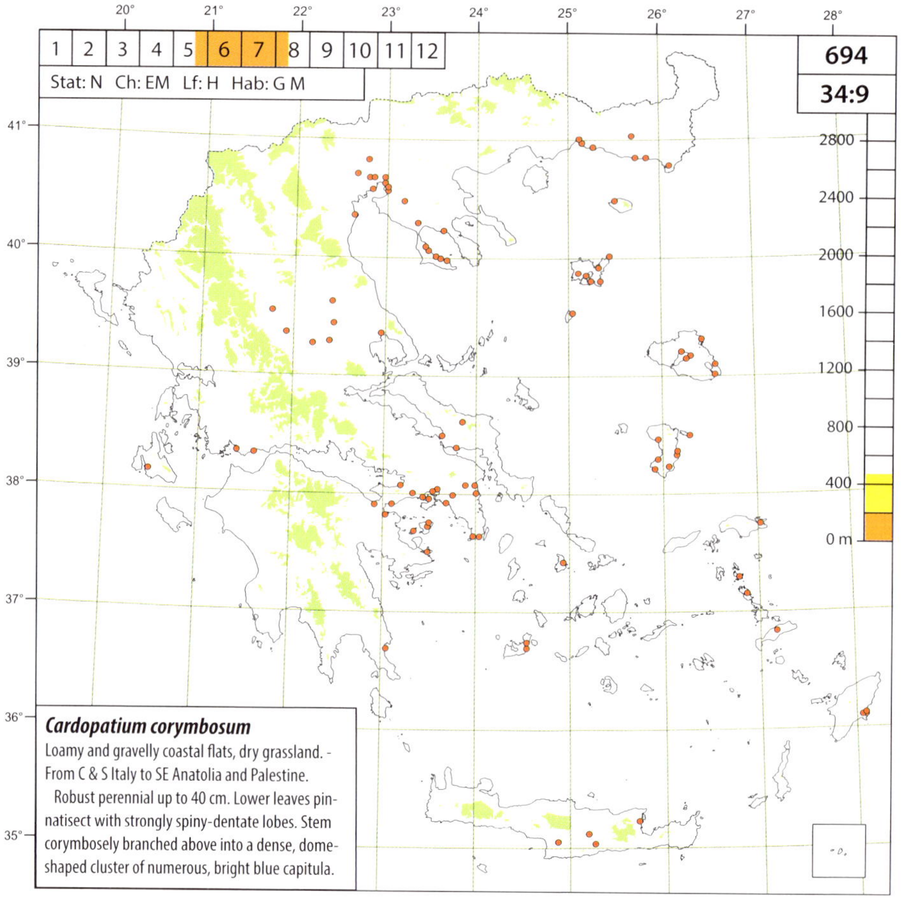

Cardopatium corymbosum

Loamy and gravelly coastal flats, dry grassland. - From C & S Italy to SE Anatolia and Palestine.

Robust perennial up to 40 cm. Lower leaves pinnatisect with strongly spiny-dentate lobes. Stem corymbosely branched above into a dense, dome-shaped cluster of numerous, bright blue capitula.

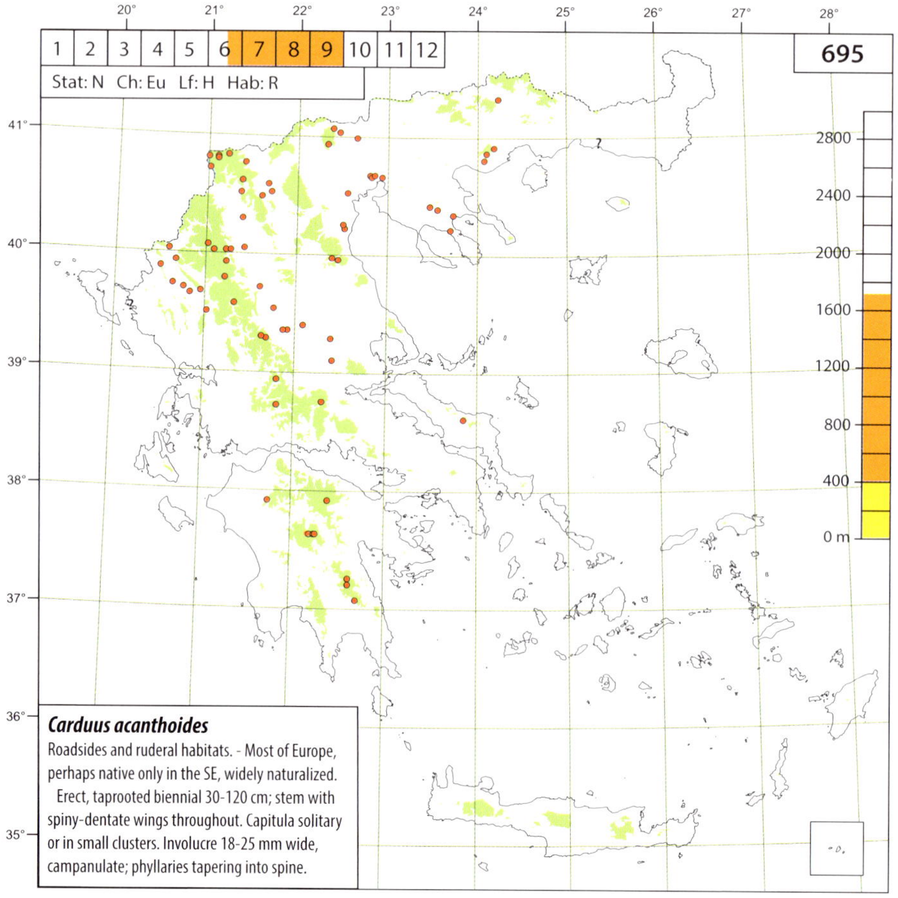

Carduus acanthoides

Roadsides and ruderal habitats. - Most of Europe, perhaps native only in the SE, widely naturalized.

Erect, taprooted biennial 30-120 cm; stem with spiny-dentate wings throughout. Capitula solitary or in small clusters. Involucre 18-25 mm wide, campanulate; phyllaries tapering into spine.

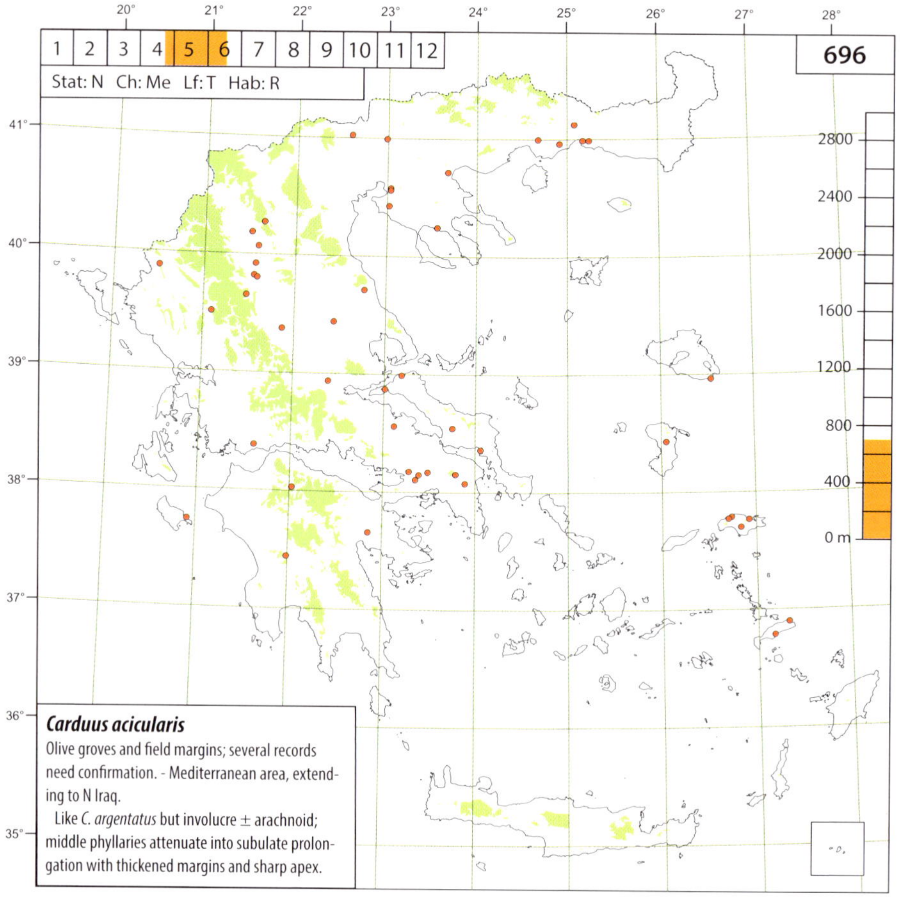

Carduus acicularis

Olive groves and field margins; several records need confirmation. - Mediterranean area, extending to N Iraq.

Like *C. argentatus* but involucre ± arachnoid; middle phyllaries attenuate into subulate prolongation with thickened margins and sharp apex.

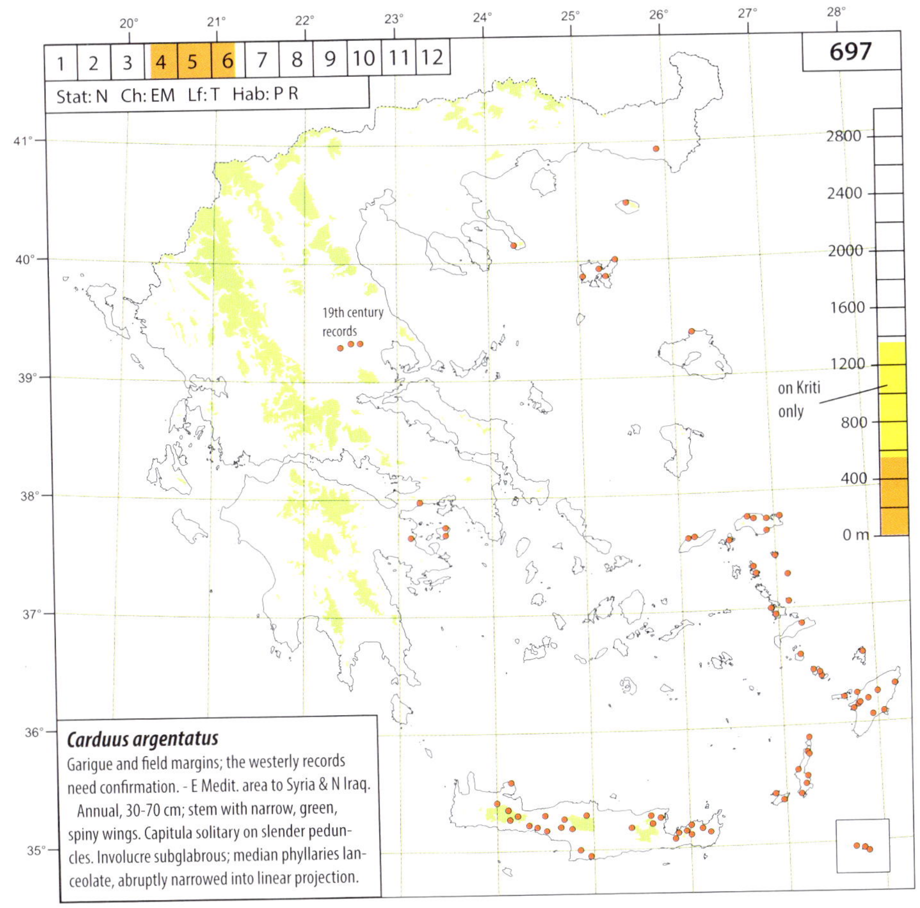

Carduus argentatus

Garigue and field margins; the westerly records need confirmation. - E Medit. area to Syria & N Iraq.

Annual, 30-70 cm; stem with narrow, green, spiny wings. Capitula solitary on slender peduncles. Involucre subglabrous; median phyllaries lanceolate, abruptly narrowed into linear projection.

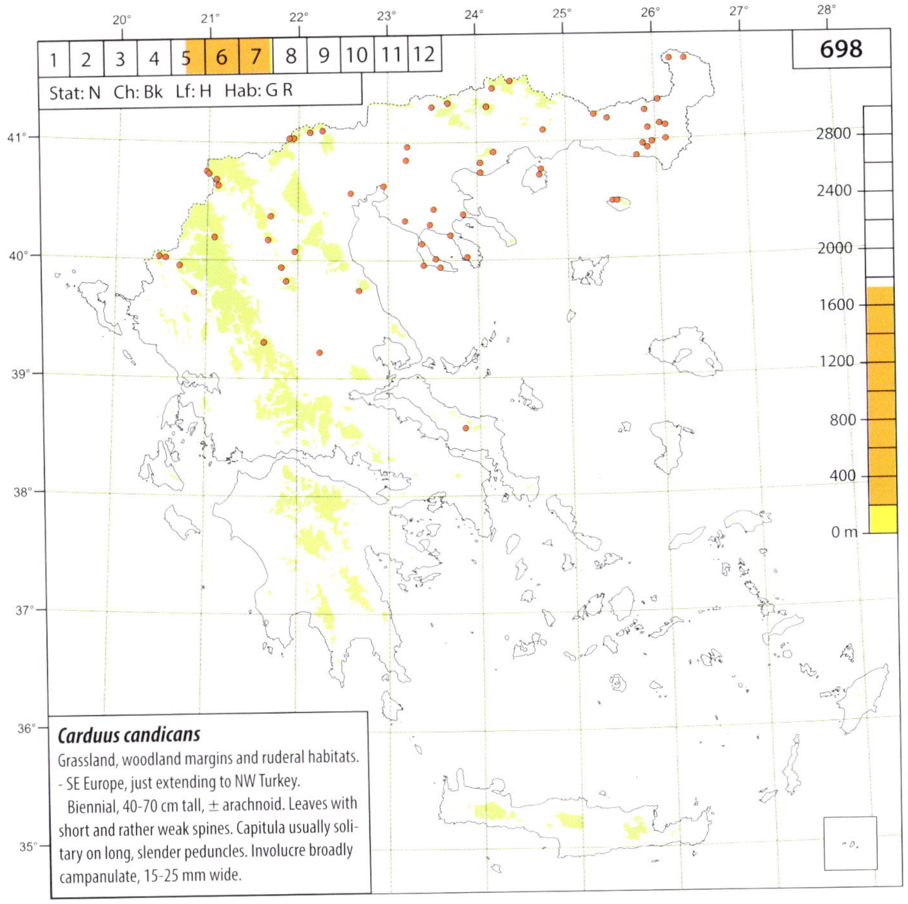

Carduus candicans

Grassland, woodland margins and ruderal habitats. - SE Europe, just extending to NW Turkey.

Biennial, 40-70 cm tall, ± arachnoid. Leaves with short and rather weak spines. Capitula usually solitary on long, slender peduncles. Involucre broadly campanulate, 15-25 mm wide.

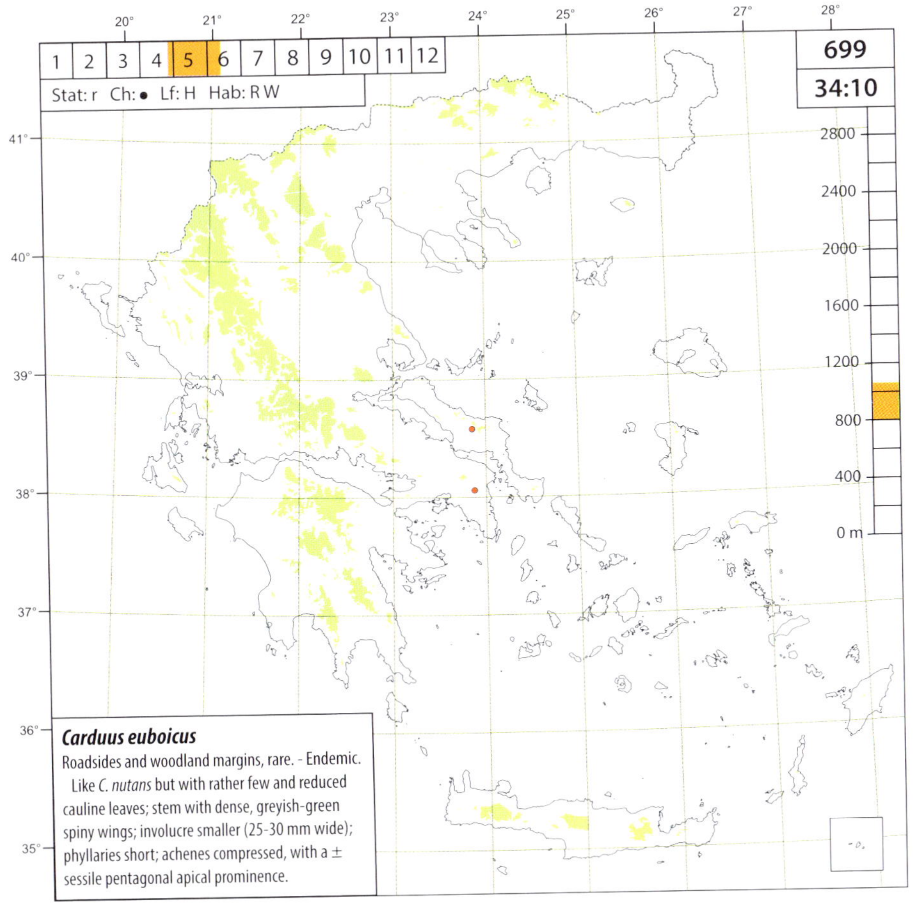

Carduus euboicus

Roadsides and woodland margins, rare. - Endemic.

Like *C. nutans* but with rather few and reduced cauline leaves; stem with dense, greyish-green spiny wings; involucre smaller (25-30 mm wide); phyllaries short; achenes compressed, with a ± sessile pentagonal apical prominence.

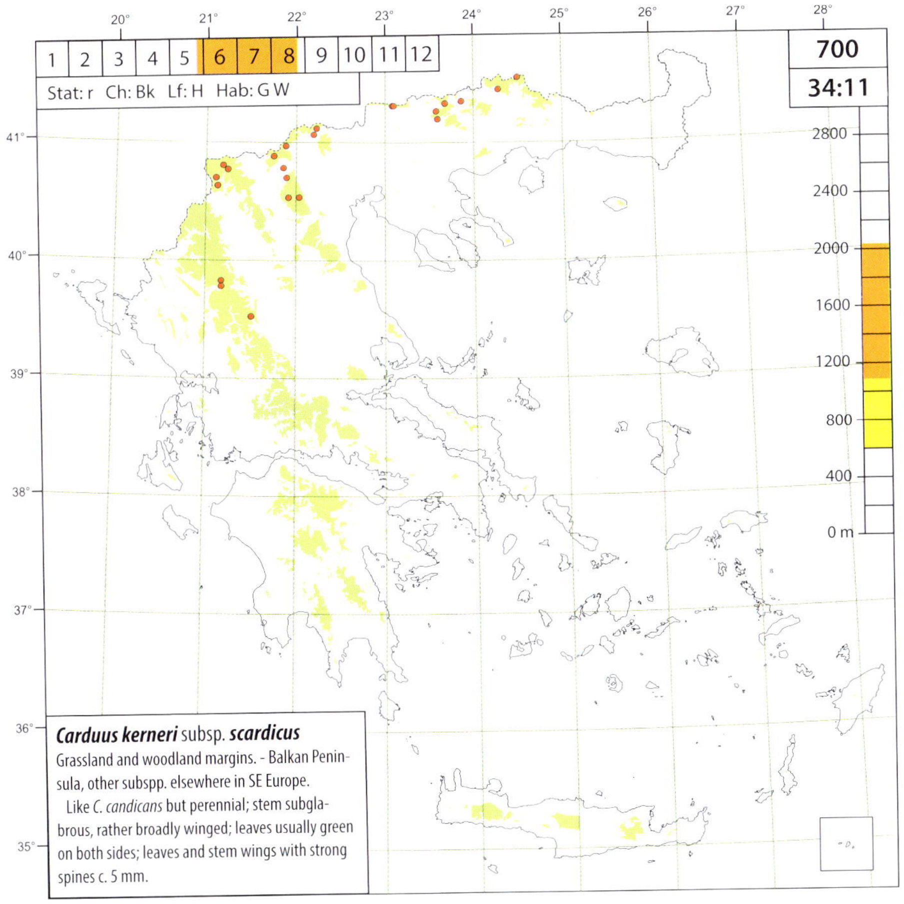

Carduus kerneri* subsp. *scardicus

Grassland and woodland margins. - Balkan Peninsula, other subspp. elsewhere in SE Europe.

Like *C. candicans* but perennial; stem subglabrous, rather broadly winged; leaves usually green on both sides; leaves and stem wings with strong spines c. 5 mm.

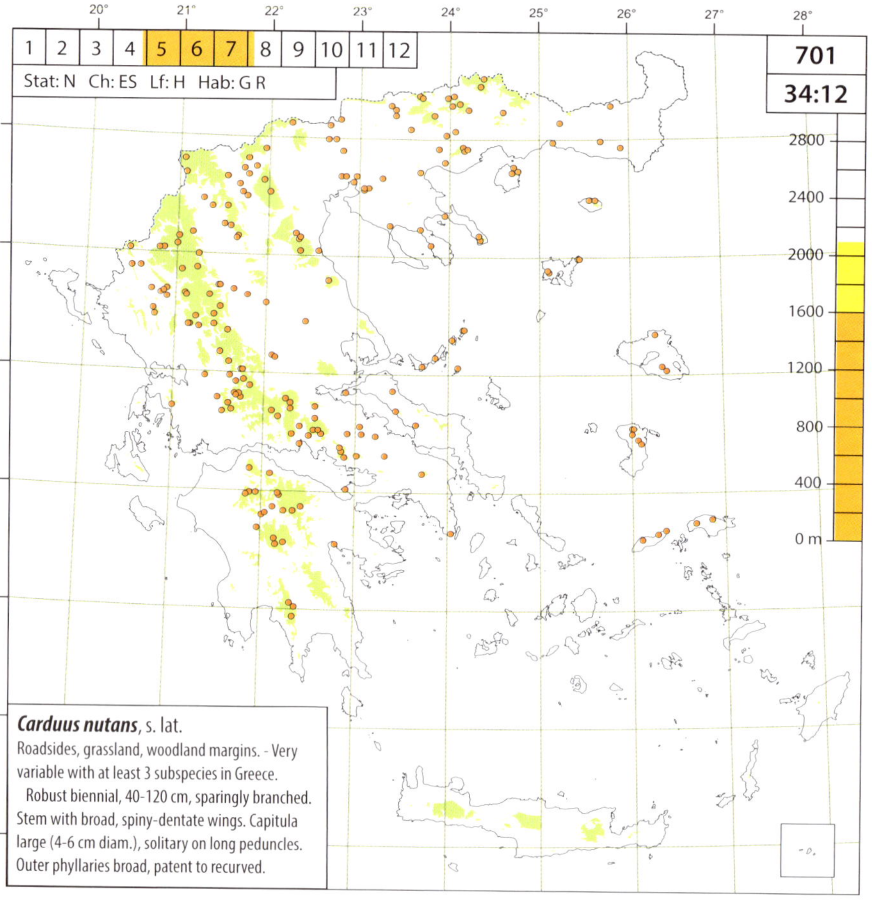

Carduus nutans, s. lat.
Roadsides, grassland, woodland margins. - Very variable with at least 3 subspecies in Greece.
Robust biennial, 40-120 cm, sparingly branched. Stem with broad, spiny-dentate wings. Capitula large (4-6 cm diam.), solitary on long peduncles. Outer phyllaries broad, patent to recurved.

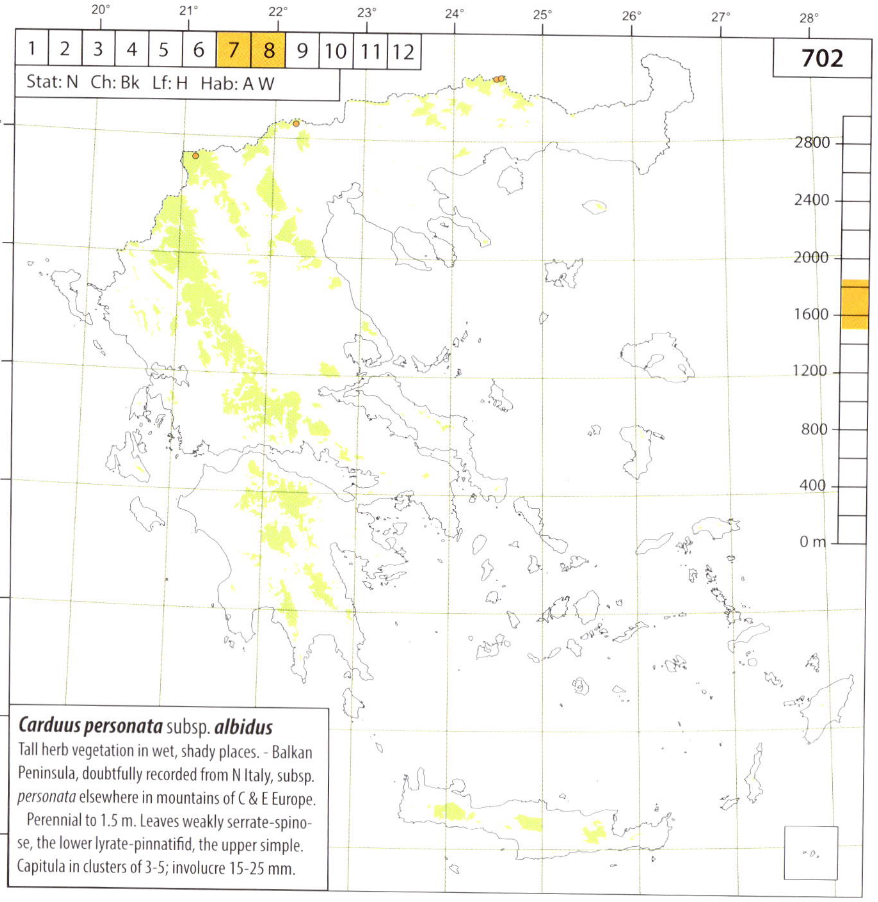

Carduus personata subsp. ***albidus***
Tall herb vegetation in wet, shady places. - Balkan Peninsula, doubtfully recorded from N Italy, subsp. *personata* elsewhere in mountains of C & E Europe.
Perennial to 1.5 m. Leaves weakly serrate-spinose, the lower lyrate-pinnatifid, the upper simple. Capitula in clusters of 3-5; involucre 15-25 mm.

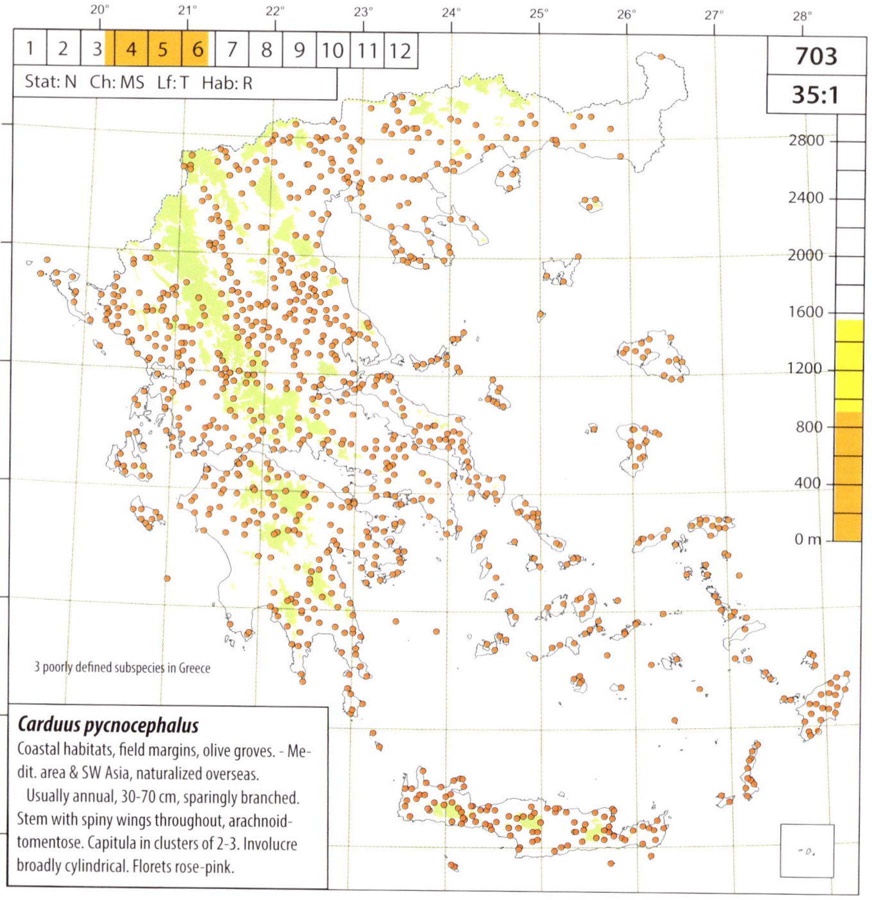

Carduus pycnocephalus
Coastal habitats, field margins, olive groves. - Medit. area & SW Asia, naturalized overseas.
Usually annual, 30-70 cm, sparingly branched. Stem with spiny wings throughout, arachnoid-tomentose. Capitula in clusters of 2-3. Involucre broadly cylindrical. Florets rose-pink.

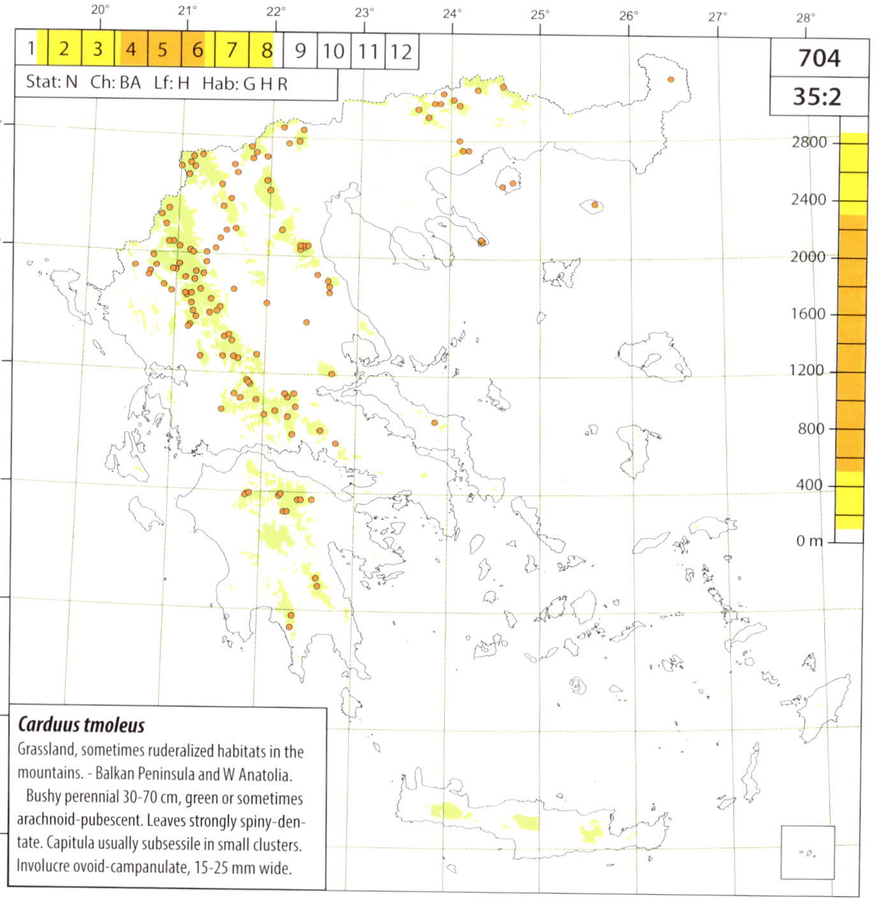

Carduus tmoleus
Grassland, sometimes ruderalized habitats in the mountains. - Balkan Peninsula and W Anatolia.
Bushy perennial 30-70 cm, green or sometimes arachnoid-pubescent. Leaves strongly spiny-dentate. Capitula usually subsessile in small clusters. Involucre ovoid-campanulate, 15-25 mm wide.

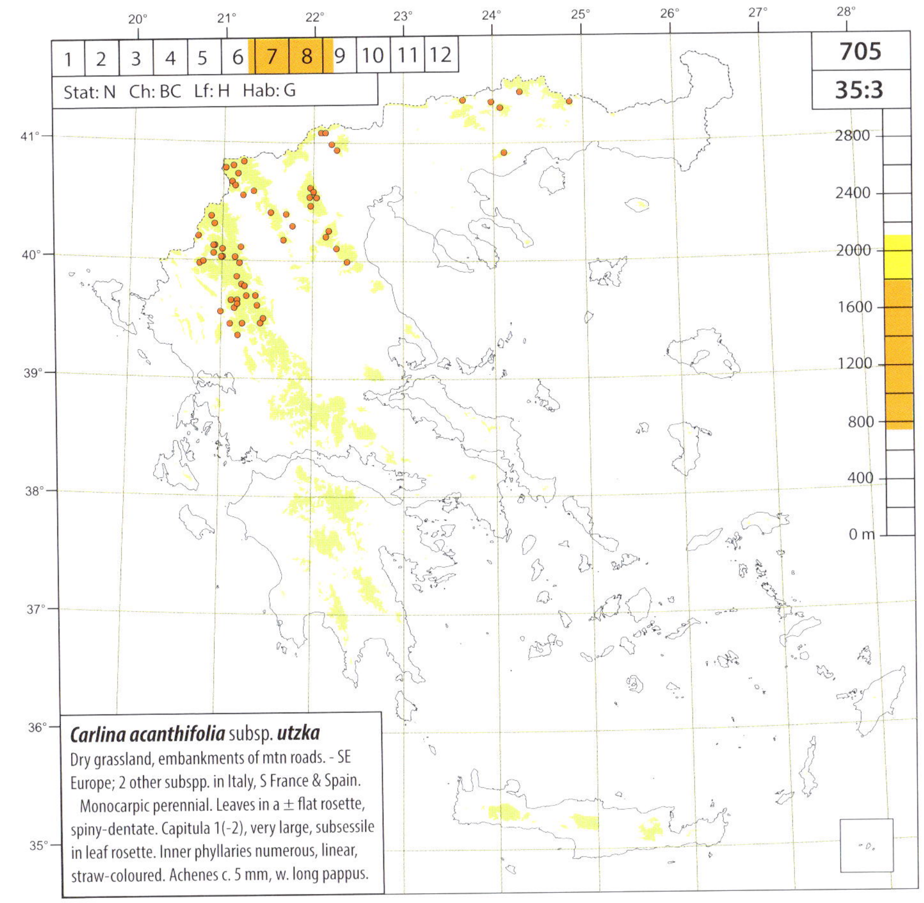

Carlina acanthifolia subsp. ***utzka***

Dry grassland, embankments of mtn roads. - SE Europe; 2 other subspp. in Italy, S France & Spain.

Monocarpic perennial. Leaves in a ± flat rosette, spiny-dentate. Capitula 1(-2), very large, subsessile in leaf rosette. Inner phyllaries numerous, linear, straw-coloured. Achenes c. 5 mm, w. long pappus.

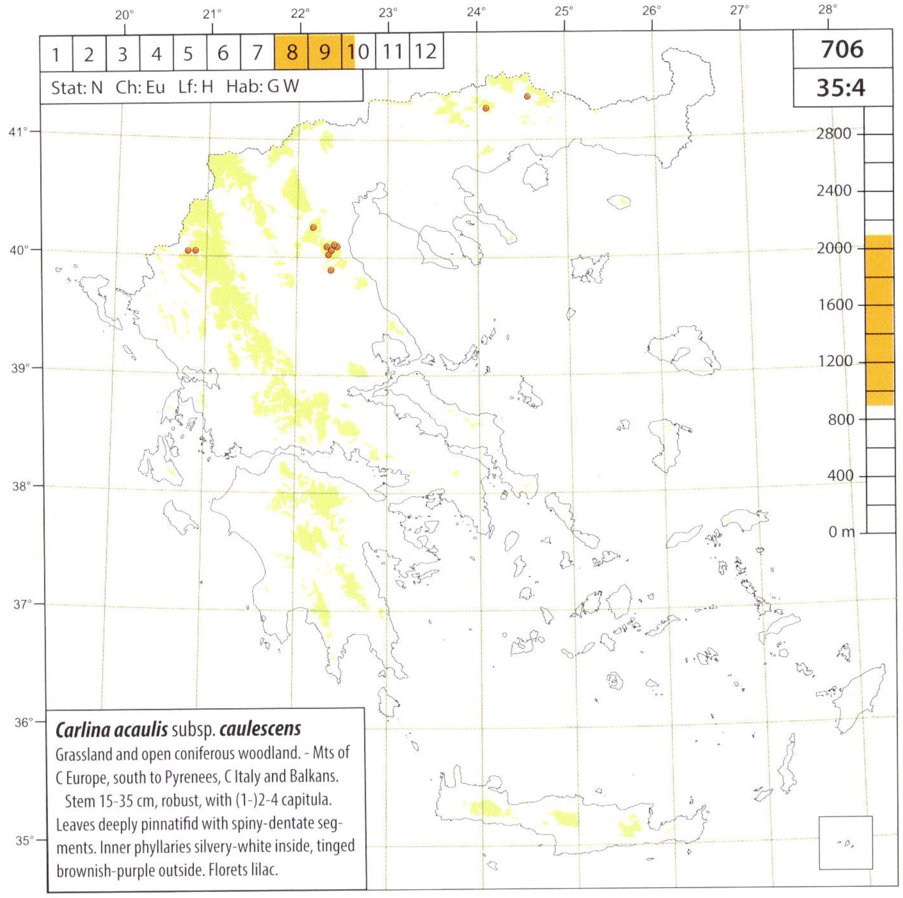

Carlina acaulis subsp. ***caulescens***

Grassland and open coniferous woodland. - Mts of C Europe, south to Pyrenees, C Italy and Balkans.

Stem 15-35 cm, robust, with (1-)2-4 capitula. Leaves deeply pinnatifid with spiny-dentate segments. Inner phyllaries silvery-white inside, tinged brownish-purple outside. Florets lilac.

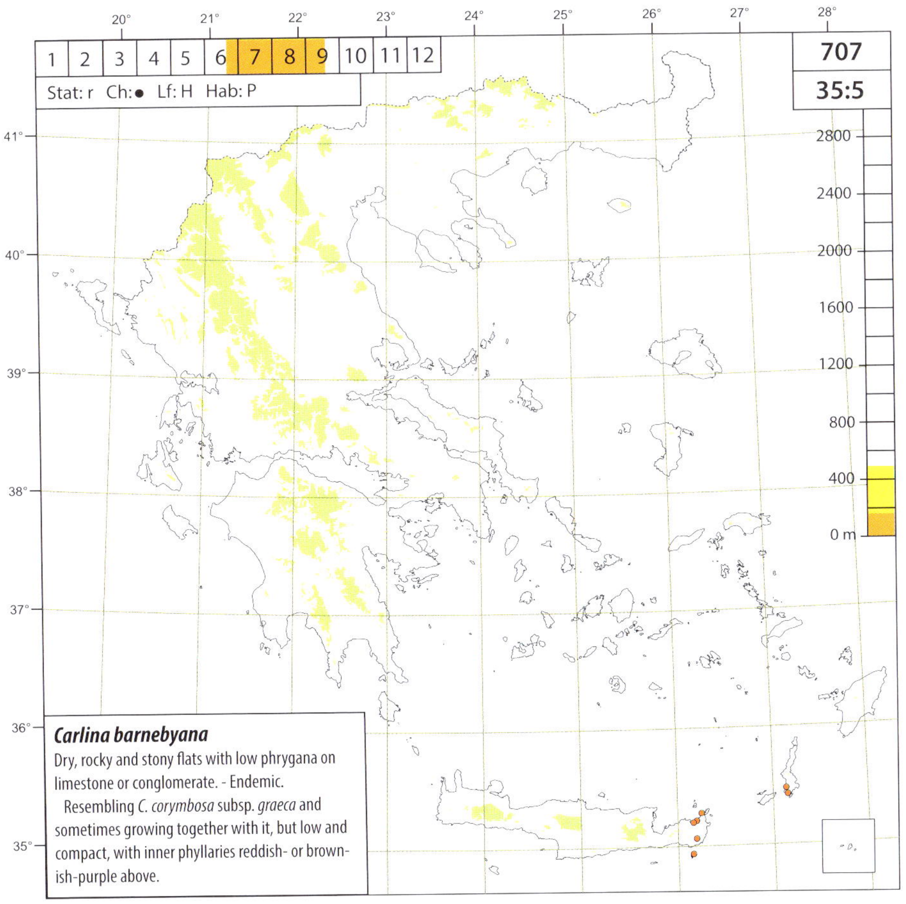

Carlina barnebyana

Dry, rocky and stony flats with low phrygana on limestone or conglomerate. - Endemic.

Resembling *C. corymbosa* subsp. *graeca* and sometimes growing together with it, but low and compact, with inner phyllaries reddish- or brownish-purple above.

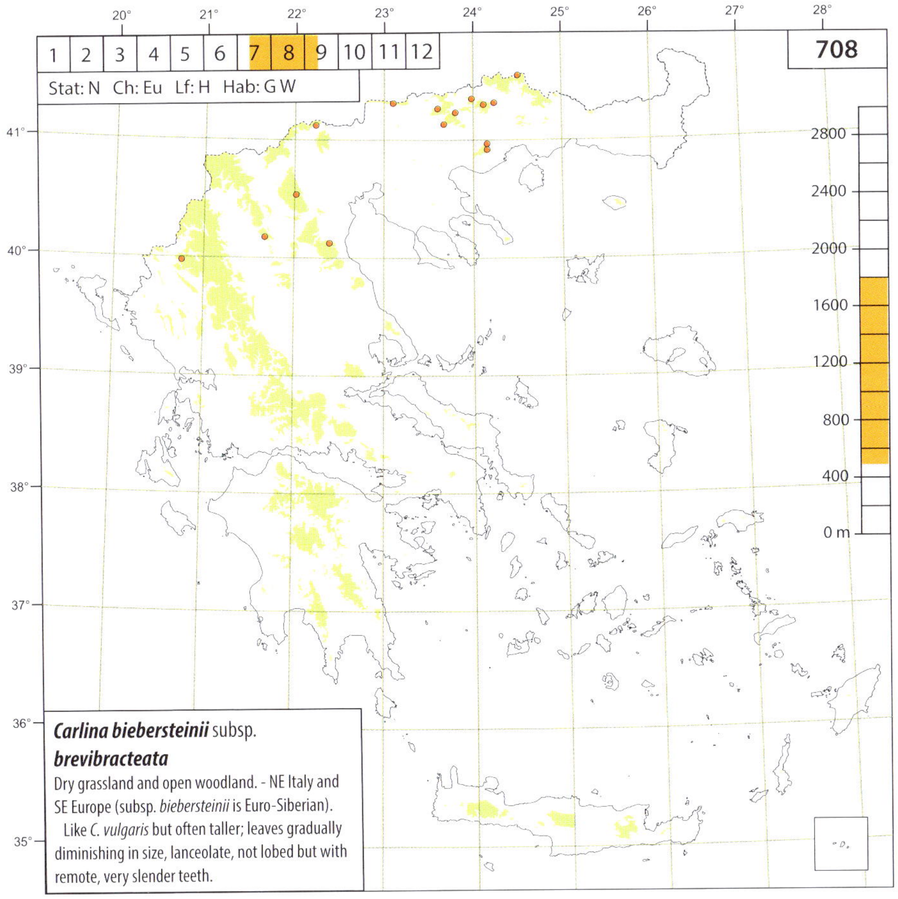

Carlina biebersteinii subsp. ***brevibracteata***

Dry grassland and open woodland. - NE Italy and SE Europe (subsp. *biebersteinii* is Euro-Siberian).

Like *C. vulgaris* but often taller; leaves gradually diminishing in size, lanceolate, not lobed but with remote, very slender teeth.

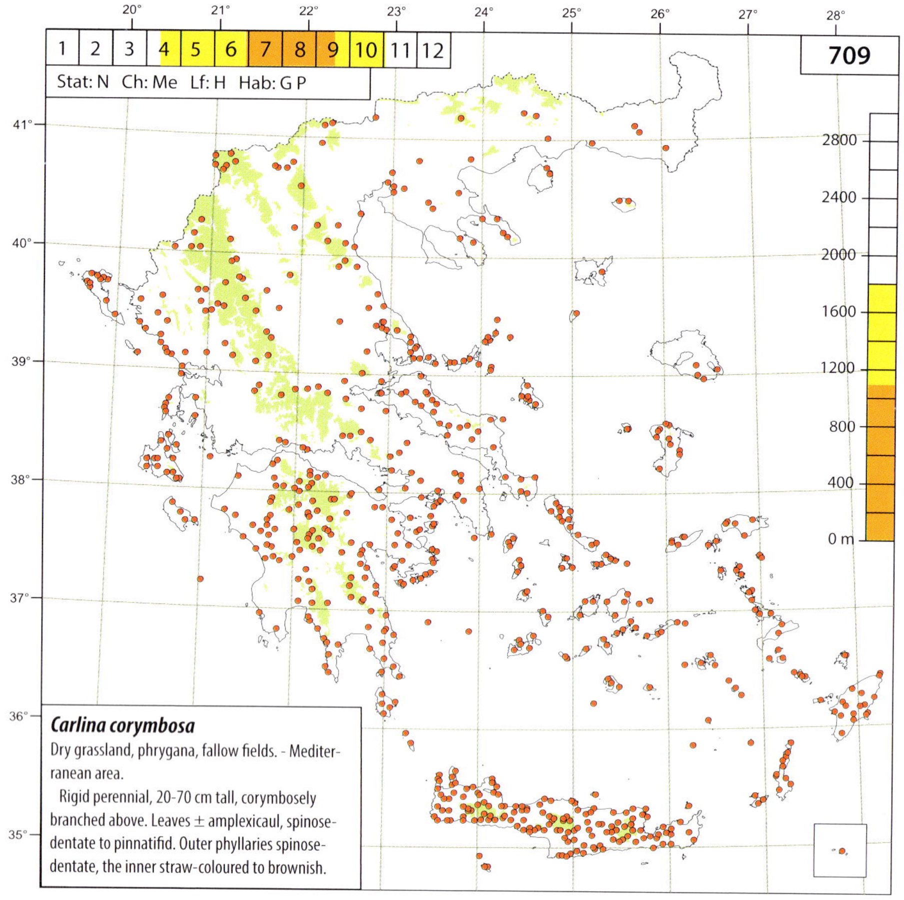

Carlina corymbosa

Dry grassland, phrygana, fallow fields. - Mediterranean area.

Rigid perennial, 20-70 cm tall, corymbosely branched above. Leaves ± amplexicaul, spinose-dentate to pinnatifid. Outer phyllaries spinose-dentate, the inner straw-coloured to brownish.

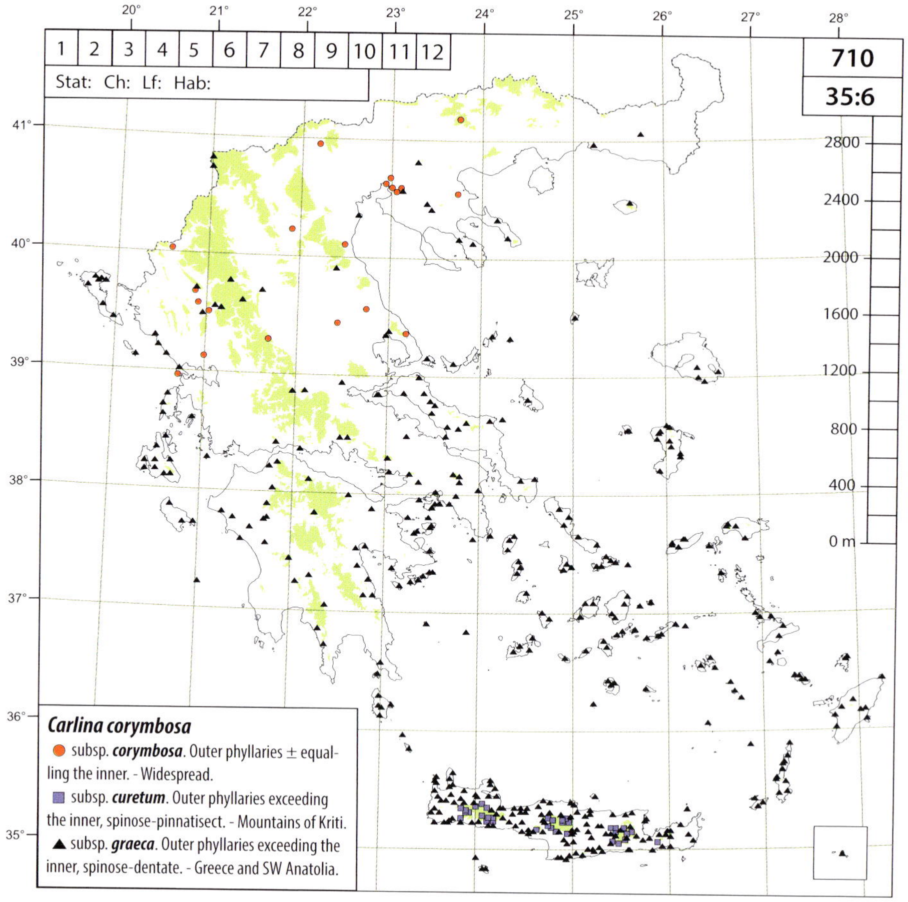

Carlina corymbosa

● subsp. ***corymbosa***. Outer phyllaries ± equalling the inner. - Widespread.

■ subsp. ***curetum***. Outer phyllaries exceeding the inner, spinose-pinnatisect. - Mountains of Kriti.

▲ subsp. ***graeca***. Outer phyllaries exceeding the inner, spinose-dentate. - Greece and SW Anatolia.

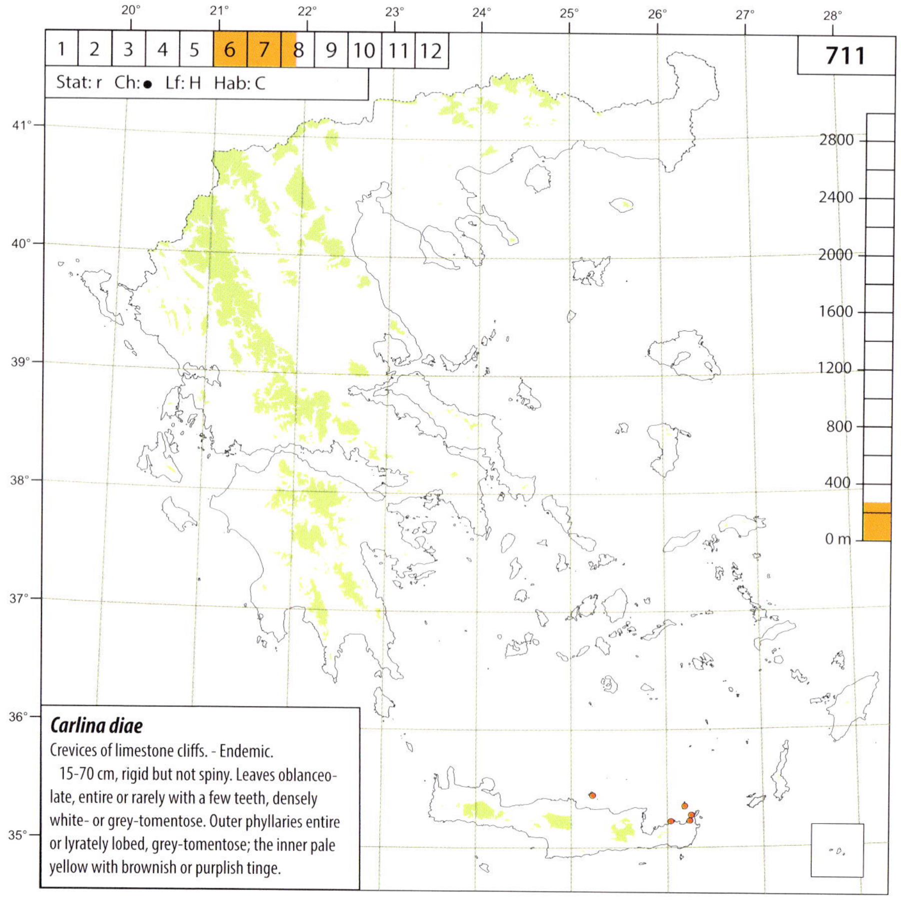

Carlina diae

Crevices of limestone cliffs. - Endemic.

15-70 cm, rigid but not spiny. Leaves oblanceolate, entire or rarely with a few teeth, densely white- or grey-tomentose. Outer phyllaries entire or lyrately lobed, grey-tomentose; the inner pale yellow with brownish or purplish tinge.

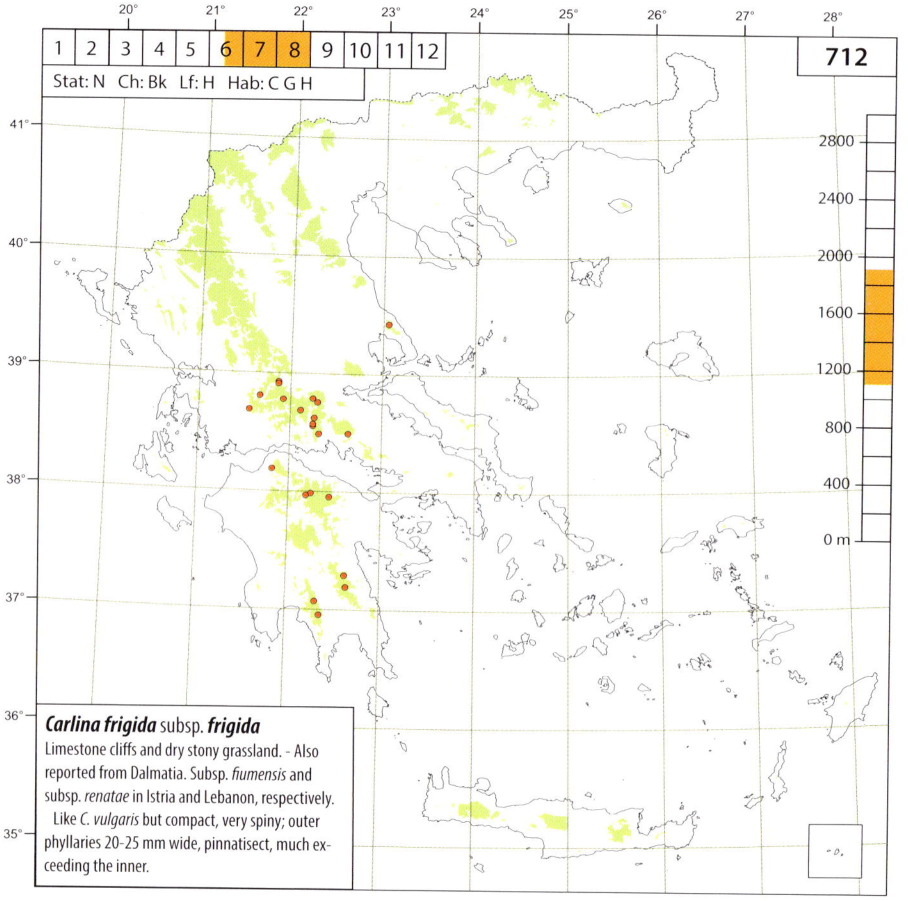

Carlina frigida **subsp. *frigida***

Limestone cliffs and dry stony grassland. - Also reported from Dalmatia. Subsp. *fiumensis* and subsp. *renatae* in Istria and Lebanon, respectively.

Like *C. vulgaris* but compact, very spiny; outer phyllaries 20-25 mm wide, pinnatisect, much exceeding the inner.

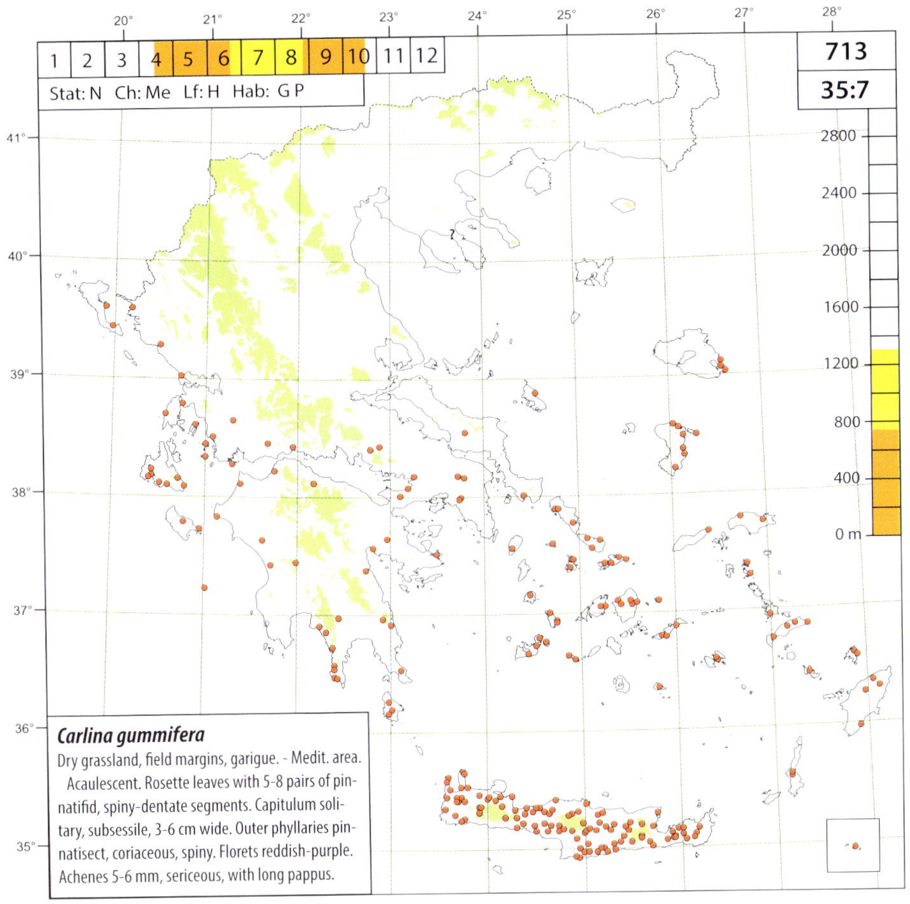

Carlina gummifera
Dry grassland, field margins, garigue. - Medit. area.
Acaulescent. Rosette leaves with 5-8 pairs of pinnatifid, spiny-dentate segments. Capitulum solitary, subsessile, 3-6 cm wide. Outer phyllaries pinnatisect, coriaceous, spiny. Florets reddish-purple. Achenes 5-6 mm, sericeous, with long pappus.

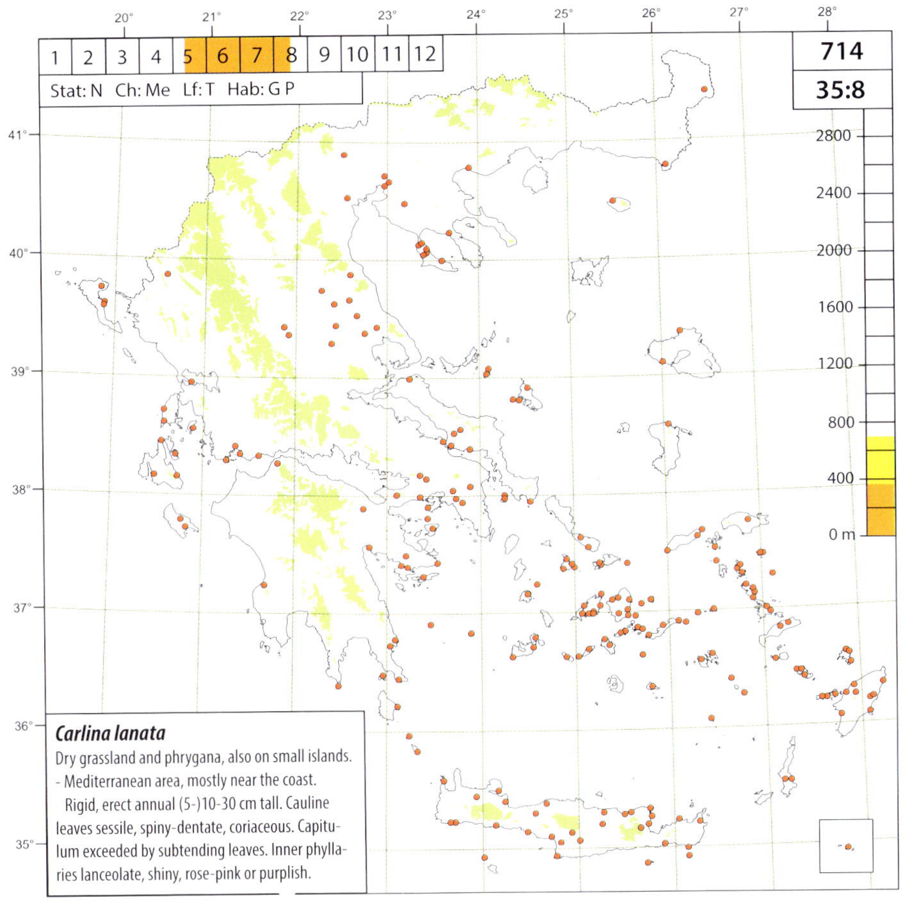

Carlina lanata
Dry grassland and phrygana, also on small islands. - Mediterranean area, mostly near the coast.
Rigid, erect annual (5-)10-30 cm tall. Cauline leaves sessile, spiny-dentate, coriaceous. Capitulum exceeded by subtending leaves. Inner phyllaries lanceolate, shiny, rose-pink or purplish.

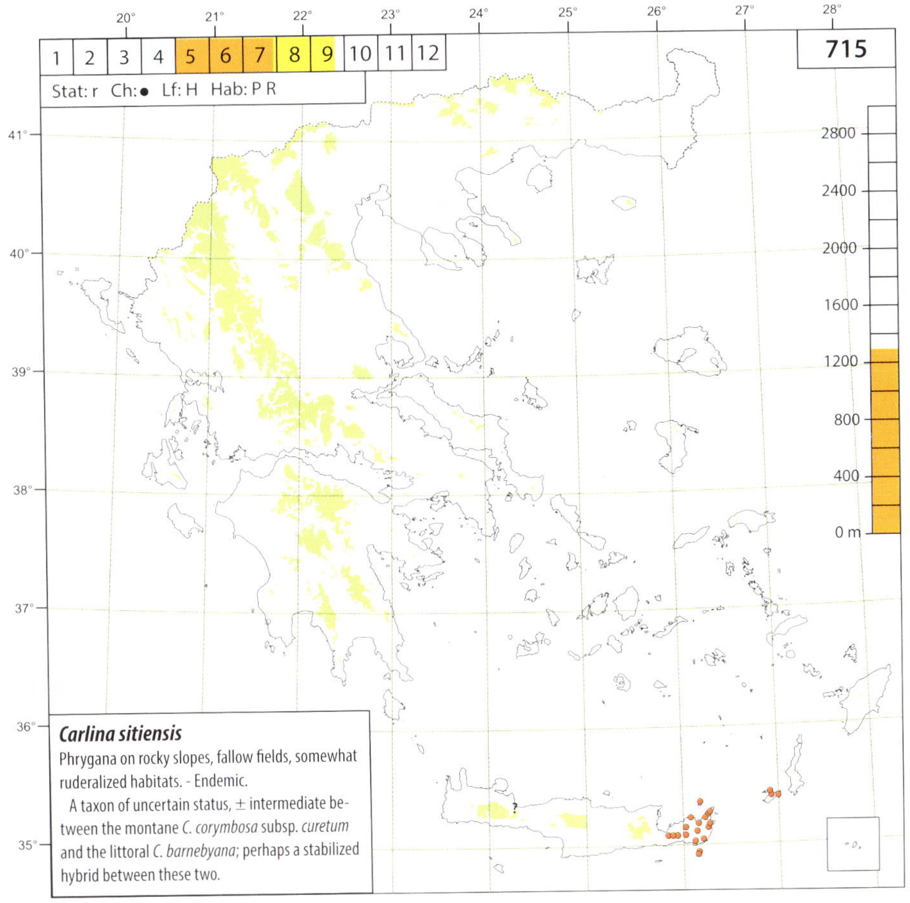

Carlina sitiensis
Phrygana on rocky slopes, fallow fields, somewhat ruderalized habitats. - Endemic.
A taxon of uncertain status, ± intermediate between the montane *C. corymbosa* subsp. *curetum* and the littoral *C. barnebyana*; perhaps a stabilized hybrid between these two.

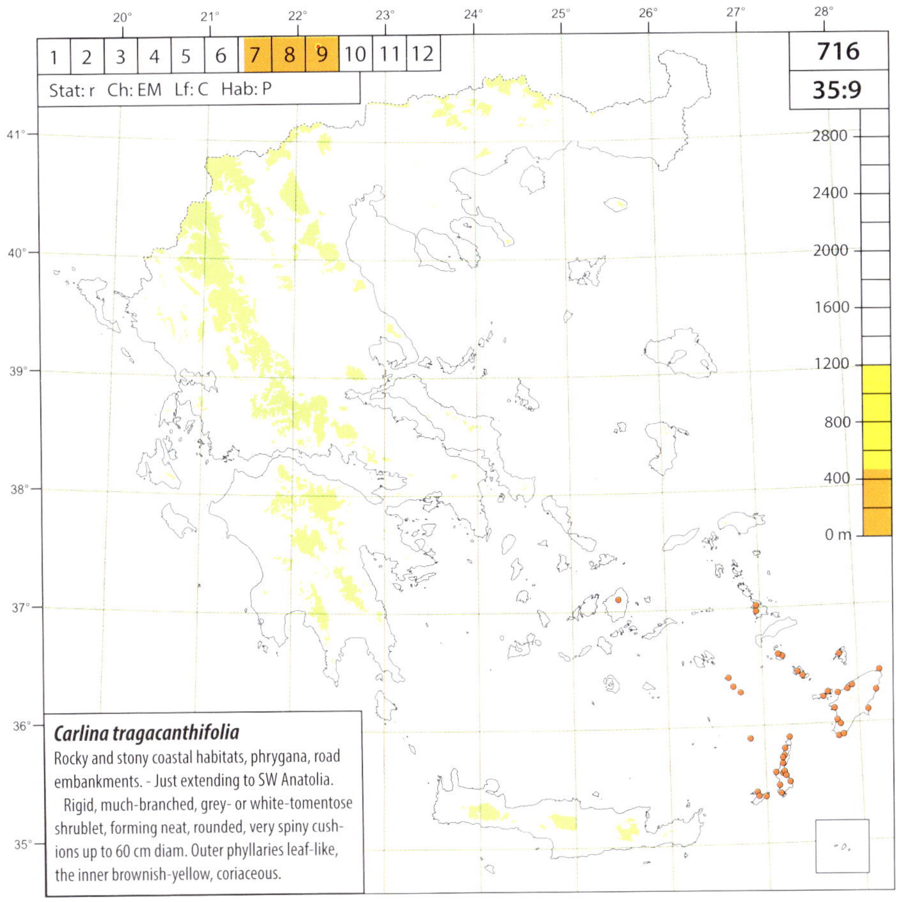

Carlina tragacanthifolia
Rocky and stony coastal habitats, phrygana, road embankments. - Just extending to SW Anatolia.
Rigid, much-branched, grey- or white-tomentose shrublet, forming neat, rounded, very spiny cushions up to 60 cm diam. Outer phyllaries leaf-like, the inner brownish-yellow, coriaceous.

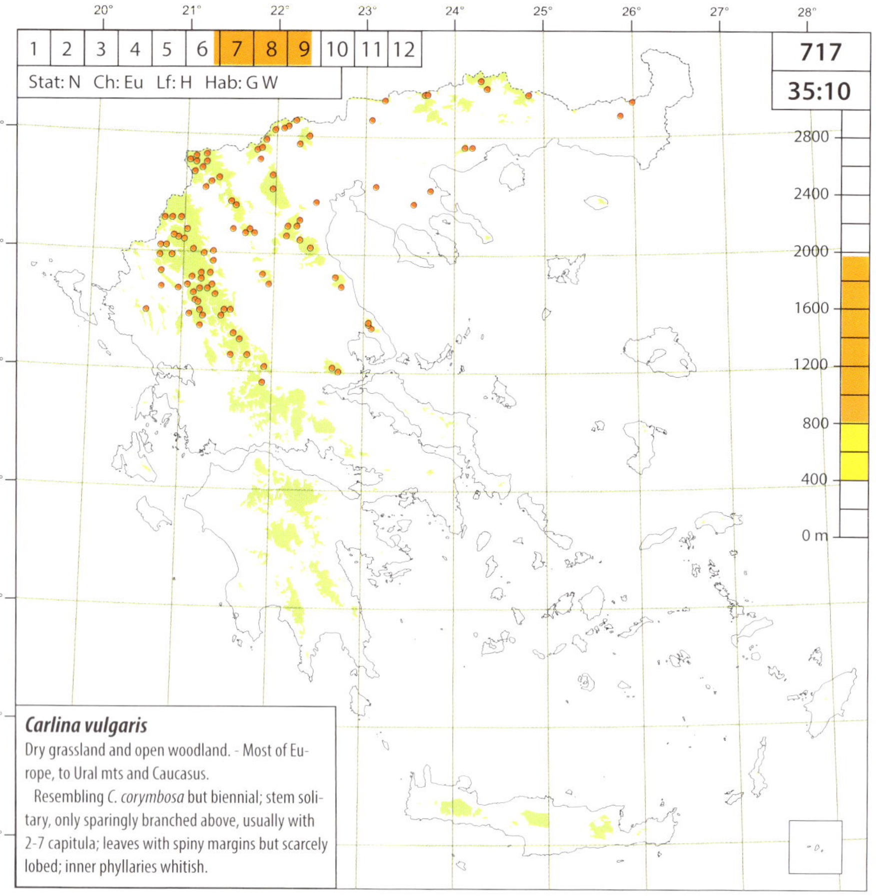

Carlina vulgaris
Dry grassland and open woodland. - Most of Europe, to Ural mts and Caucasus.
Resembling *C. corymbosa* but biennial; stem solitary, only sparingly branched above, usually with 2-7 capitula; leaves with spiny margins but scarcely lobed; inner phyllaries whitish.

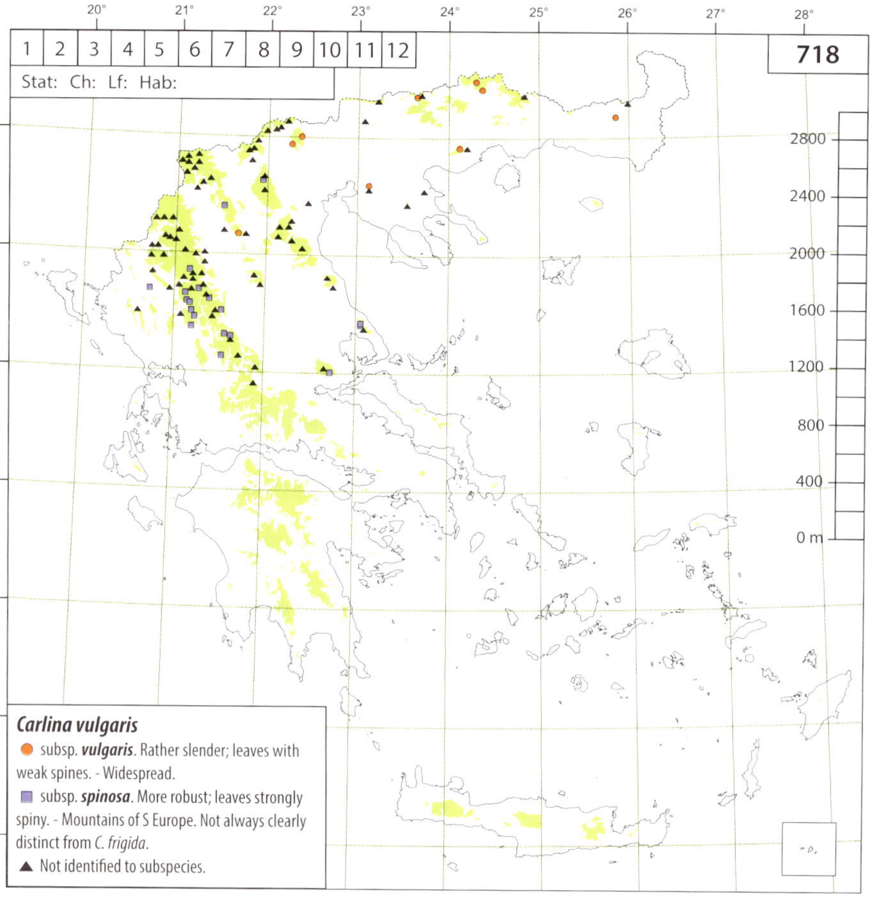

Carlina vulgaris
● subsp. ***vulgaris***. Rather slender; leaves with weak spines. - Widespread.
■ subsp. ***spinosa***. More robust; leaves strongly spiny. - Mountains of S Europe. Not always clearly distinct from *C. frigida*.
▲ Not identified to subspecies.

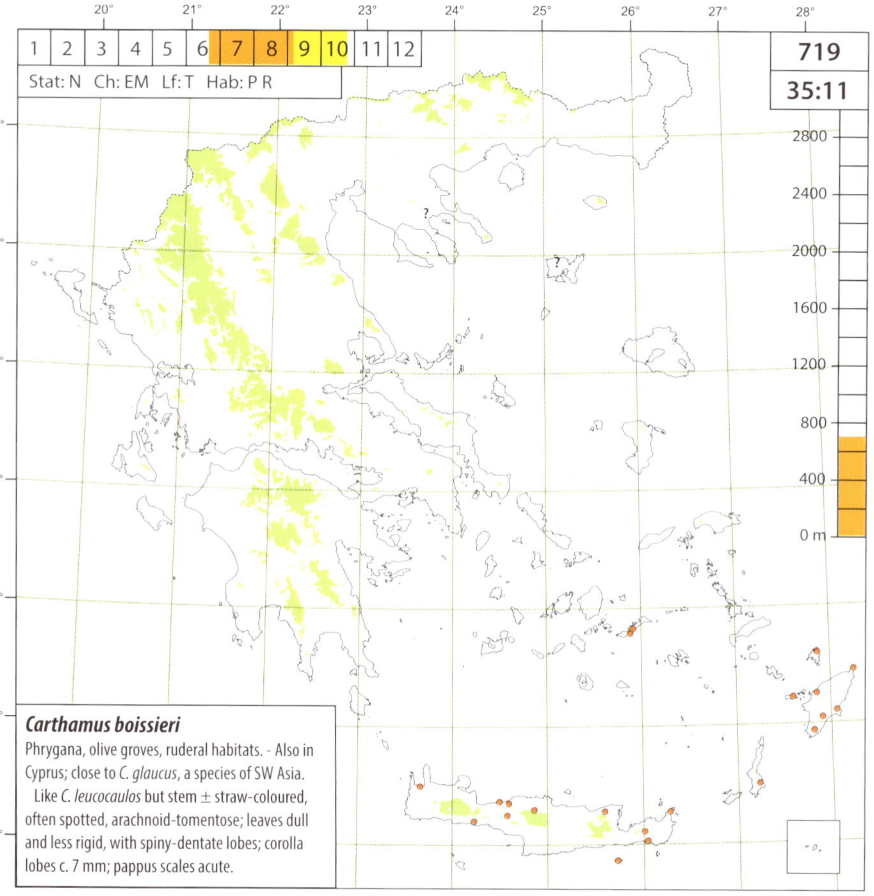

Carthamus boissieri
Phrygana, olive groves, ruderal habitats. - Also in Cyprus; close to *C. glaucus*, a species of SW Asia.
Like *C. leucocaulos* but stem ± straw-coloured, often spotted, arachnoid-tomentose; leaves dull and less rigid, with spiny-dentate lobes; corolla lobes c. 7 mm; pappus scales acute.

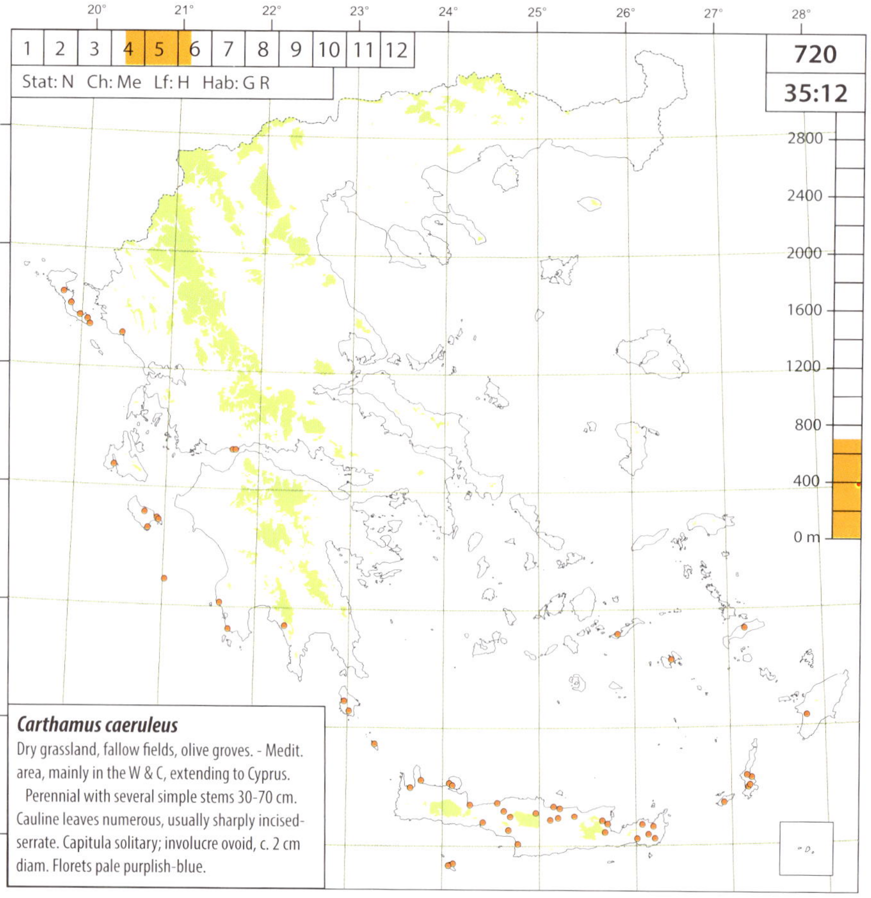

Carthamus caeruleus
Dry grassland, fallow fields, olive groves. - Medit. area, mainly in the W & C, extending to Cyprus.
Perennial with several simple stems 30-70 cm. Cauline leaves numerous, usually sharply incised-serrate. Capitula solitary; involucre ovoid, c. 2 cm diam. Florets pale purplish-blue.

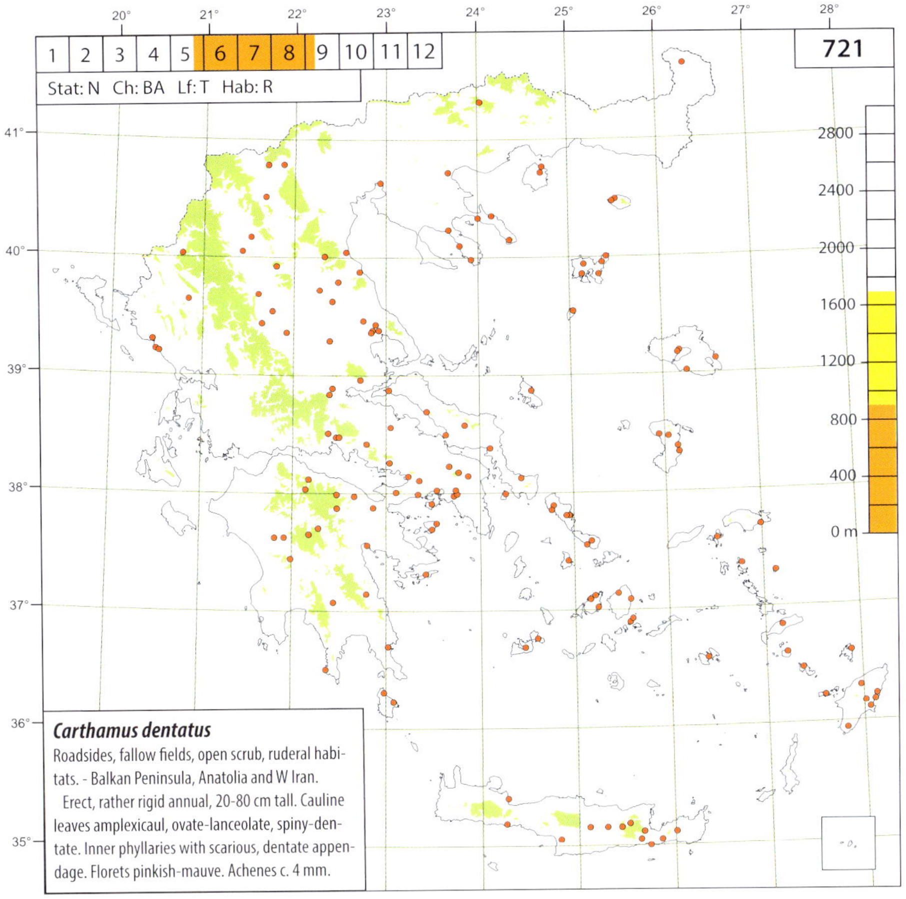

Carthamus dentatus
Roadsides, fallow fields, open scrub, ruderal habitats. - Balkan Peninsula, Anatolia and W Iran.
Erect, rather rigid annual, 20-80 cm tall. Cauline leaves amplexicaul, ovate-lanceolate, spiny-dentate. Inner phyllaries with scarious, dentate appendage. Florets pinkish-mauve. Achenes c. 4 mm.

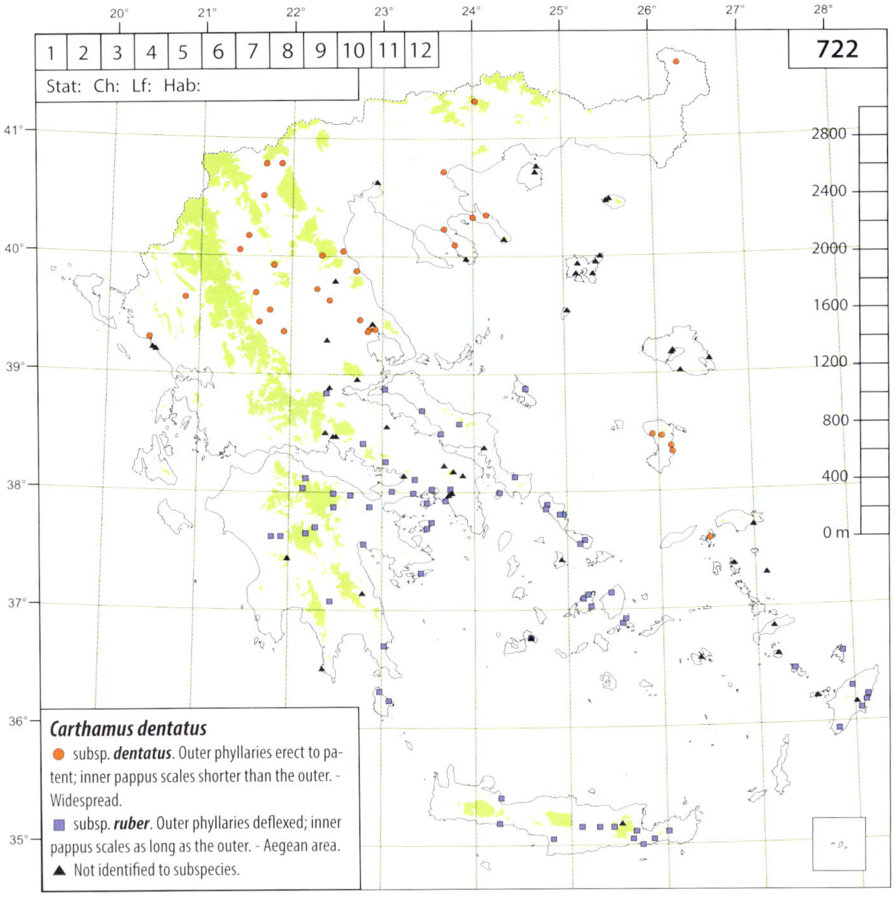

Carthamus dentatus
● subsp. **_dentatus_**. Outer phyllaries erect to patent; inner pappus scales shorter than the outer. - Widespread.
■ subsp. **_ruber_**. Outer phyllaries deflexed; inner pappus scales as long as the outer. - Aegean area.
▲ Not identified to subspecies.

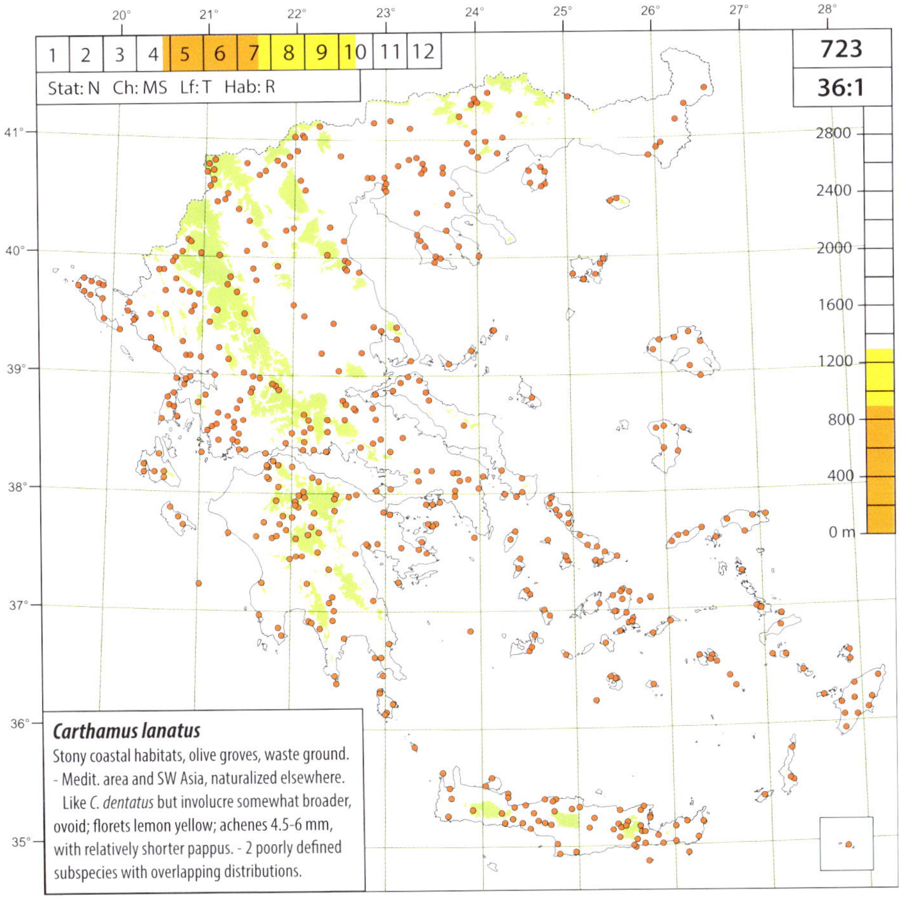

Carthamus lanatus
Stony coastal habitats, olive groves, waste ground. - Medit. area and SW Asia, naturalized elsewhere.
Like _C. dentatus_ but involucre somewhat broader, ovoid; florets lemon yellow; achenes 4.5-6 mm, with relatively shorter pappus. - 2 poorly defined subspecies with overlapping distributions.

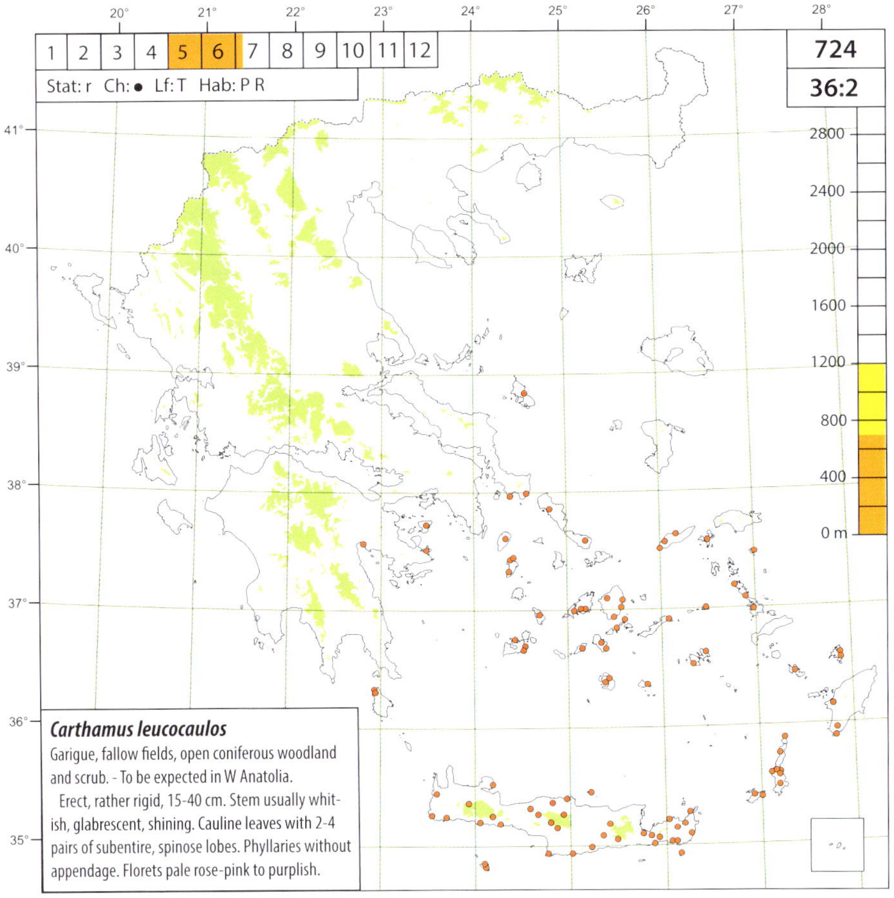

Carthamus leucocaulos
Garigue, fallow fields, open coniferous woodland and scrub. - To be expected in W Anatolia.
Erect, rather rigid, 15-40 cm. Stem usually whitish, glabrescent, shining. Cauline leaves with 2-4 pairs of subentire, spinose lobes. Phyllaries without appendage. Florets pale rose-pink to purplish.

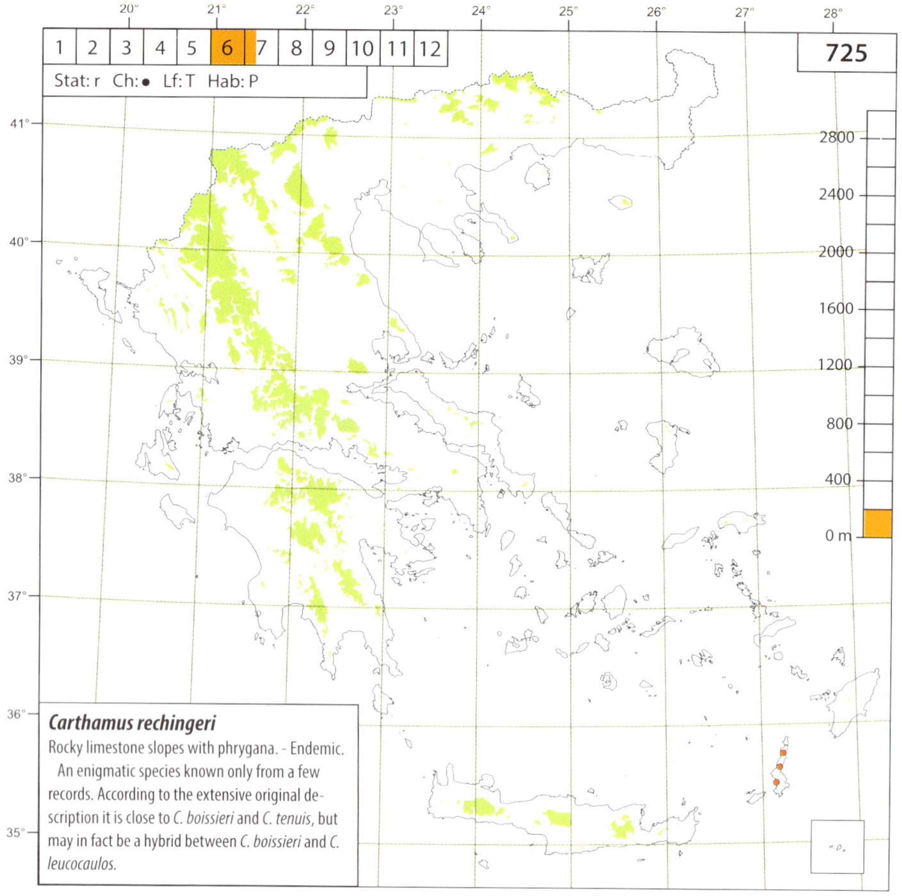

Carthamus rechingeri
Rocky limestone slopes with phrygana. - Endemic.
An enigmatic species known only from a few records. According to the extensive original description it is close to *C. boissieri* and *C. tenuis*, but may in fact be a hybrid between *C. boissieri* and *C. leucocaulos*.

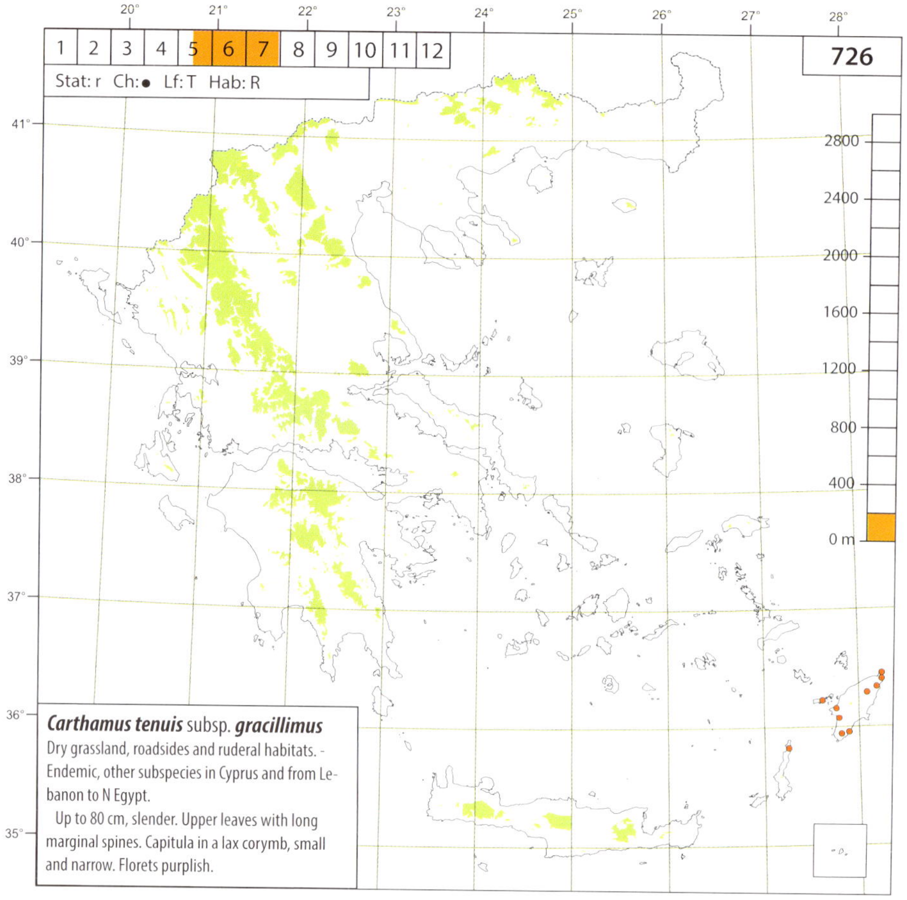

Carthamus tenuis **subsp. *gracillimus***
Dry grassland, roadsides and ruderal habitats. - Endemic, other subspecies in Cyprus and from Lebanon to N Egypt.
Up to 80 cm, slender. Upper leaves with long marginal spines. Capitula in a lax corymb, small and narrow. Florets purplish.

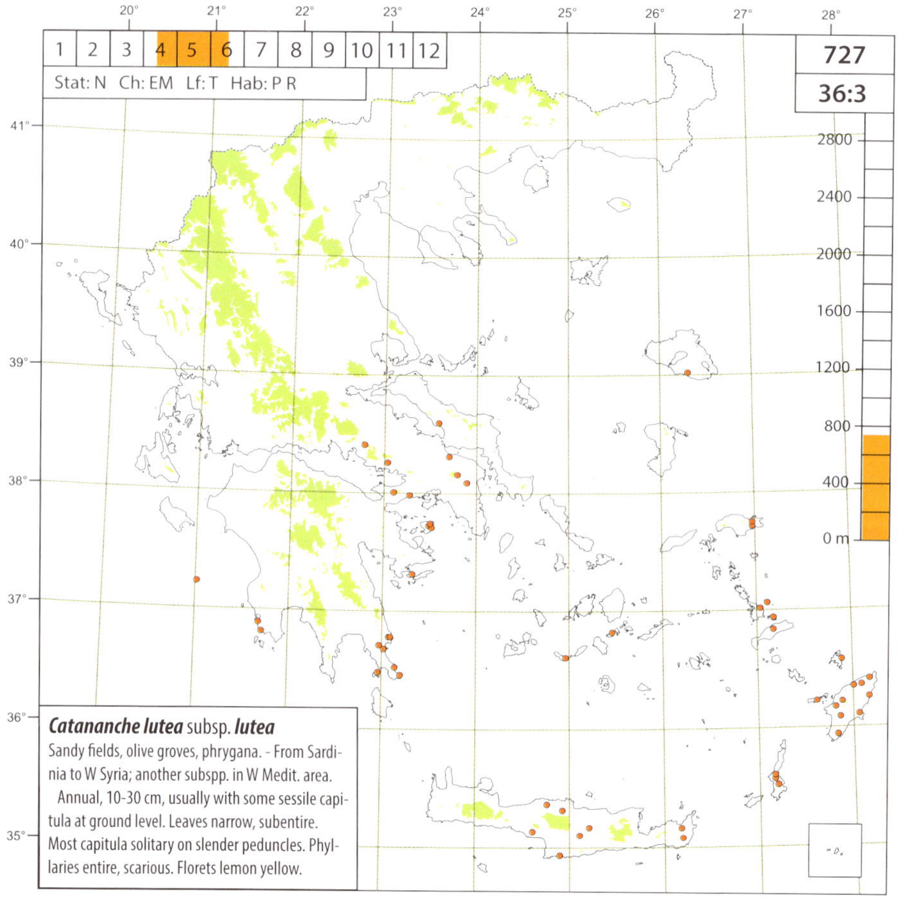

Catananche lutea **subsp. *lutea***
Sandy fields, olive groves, phrygana. - From Sardinia to W Syria; another subspp. in W Medit. area.
Annual, 10-30 cm, usually with some sessile capitula at ground level. Leaves narrow, subentire. Most capitula solitary on slender peduncles. Phyllaries entire, scarious. Florets lemon yellow.

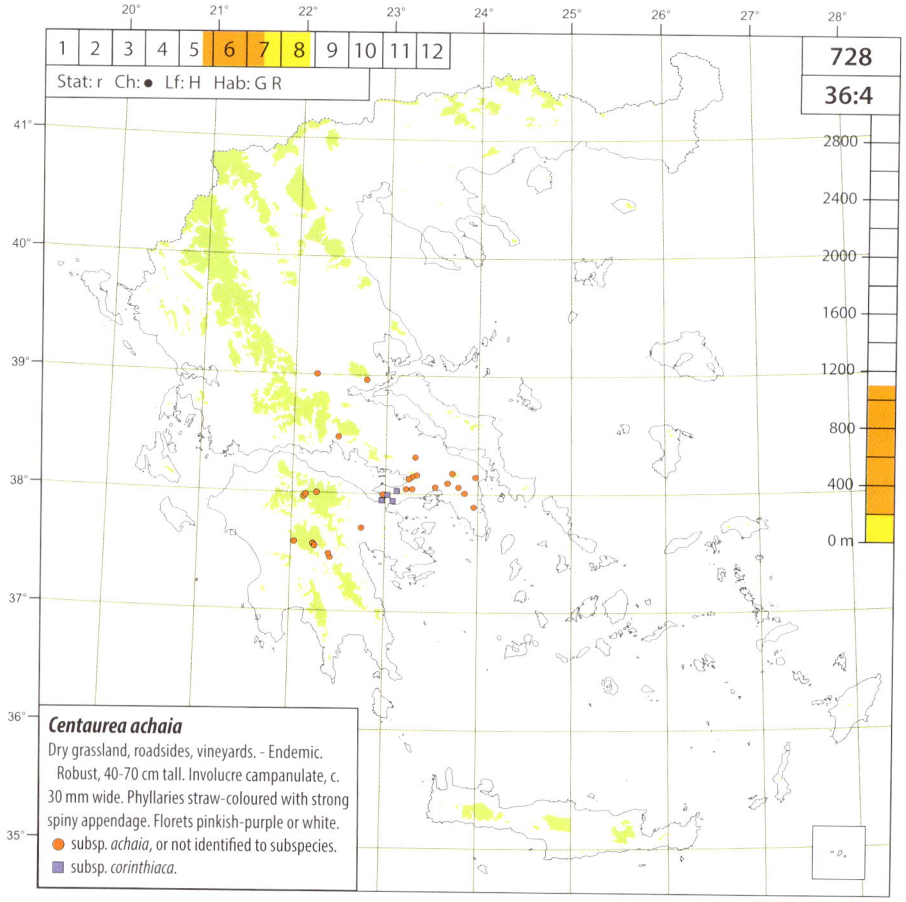

Centaurea achaia
Dry grassland, roadsides, vineyards. - Endemic.
Robust, 40-70 cm tall. Involucre campanulate, c. 30 mm wide. Phyllaries straw-coloured with strong spiny appendage. Florets pinkish-purple or white.
● subsp. *achaia*, or not identified to subspecies.
■ subsp. *corinthiaca*.

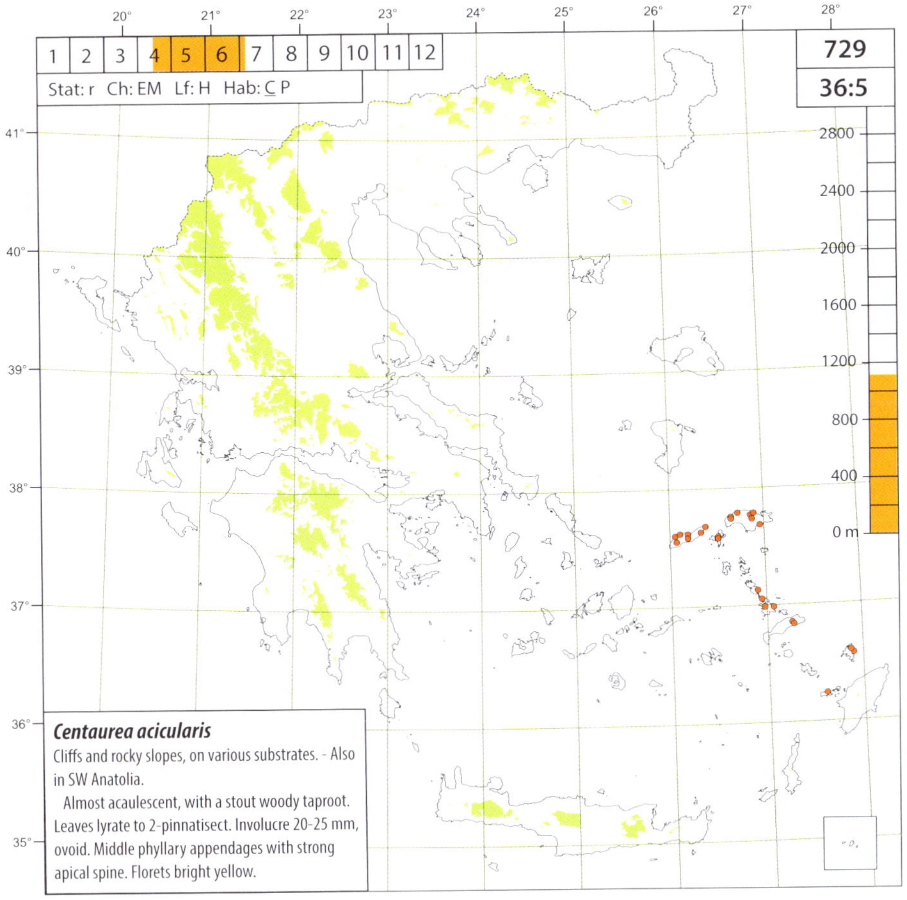

Centaurea acicularis
Cliffs and rocky slopes, on various substrates. - Also in SW Anatolia.
Almost acaulescent, with a stout woody taproot. Leaves lyrate to 2-pinnatisect. Involucre 20-25 mm, ovoid. Middle phyllary appendages with strong apical spine. Florets bright yellow.

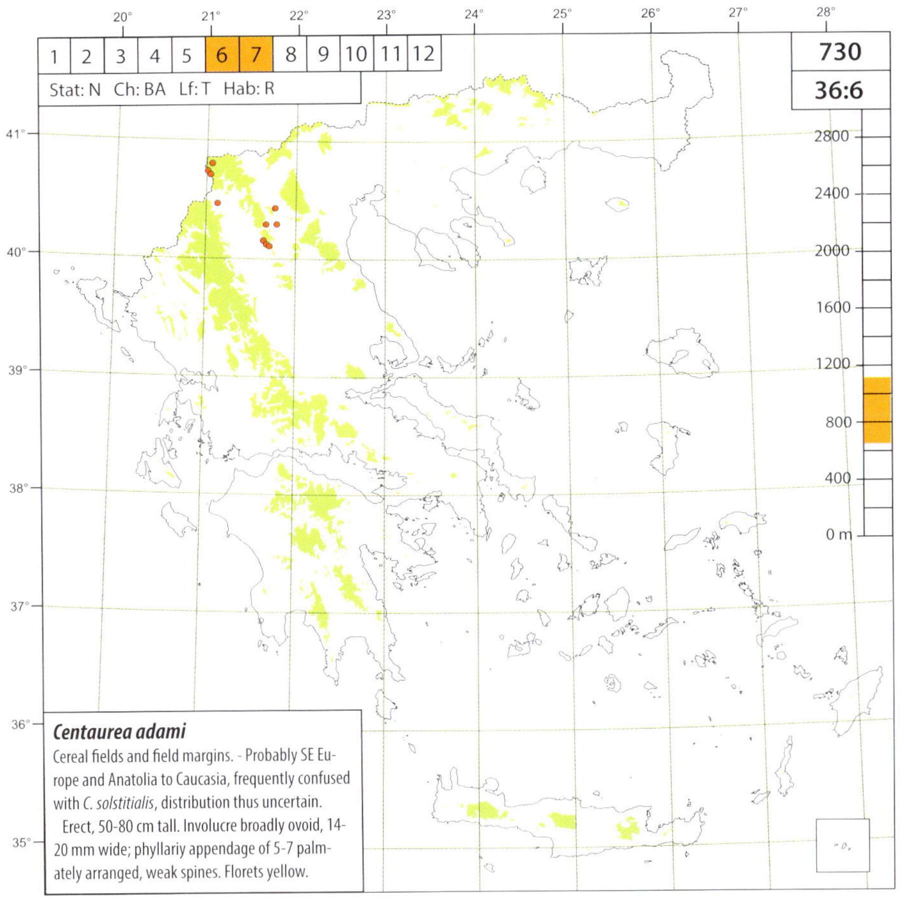

Centaurea adami
Cereal fields and field margins. - Probably SE Europe and Anatolia to Caucasia, frequently confused with *C. solstitialis*, distribution thus uncertain.
Erect, 50-80 cm tall. Involucre broadly ovoid, 14-20 mm wide; phyllariy appendage of 5-7 palmately arranged, weak spines. Florets yellow.

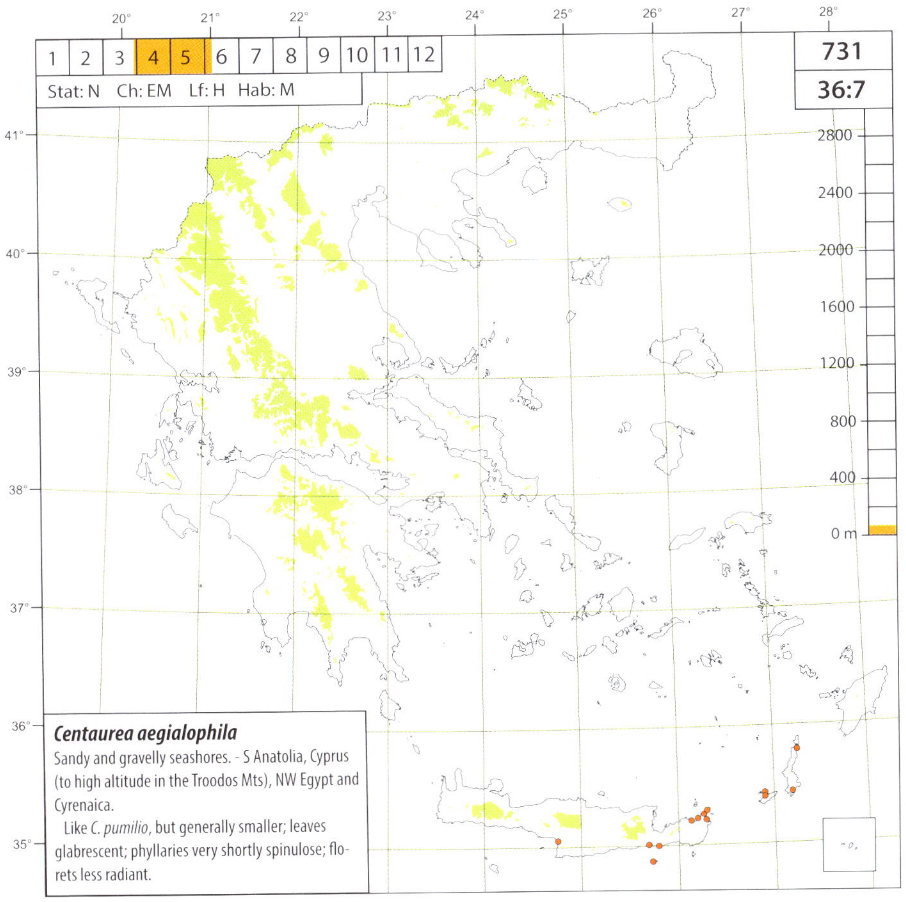

Centaurea aegialophila
Sandy and gravelly seashores. - S Anatolia, Cyprus (to high altitude in the Troodos Mts), NW Egypt and Cyrenaica.
Like *C. pumilio*, but generally smaller; leaves glabrescent; phyllaries very shortly spinulose; florets less radiant.

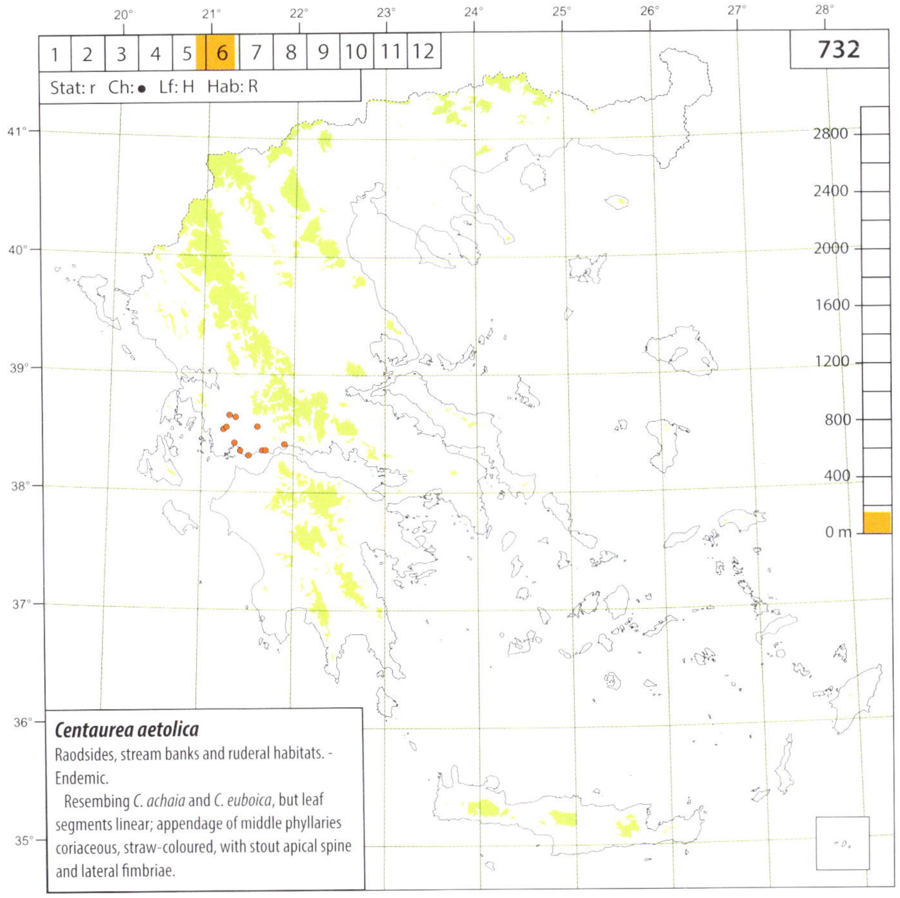

Centaurea aetolica
Raodsides, stream banks and ruderal habitats. - Endemic.
Resembling *C. achaia* and *C. euboica*, but leaf segments linear; appendage of middle phyllaries coriaceous, straw-coloured, with stout apical spine and lateral fimbriae.

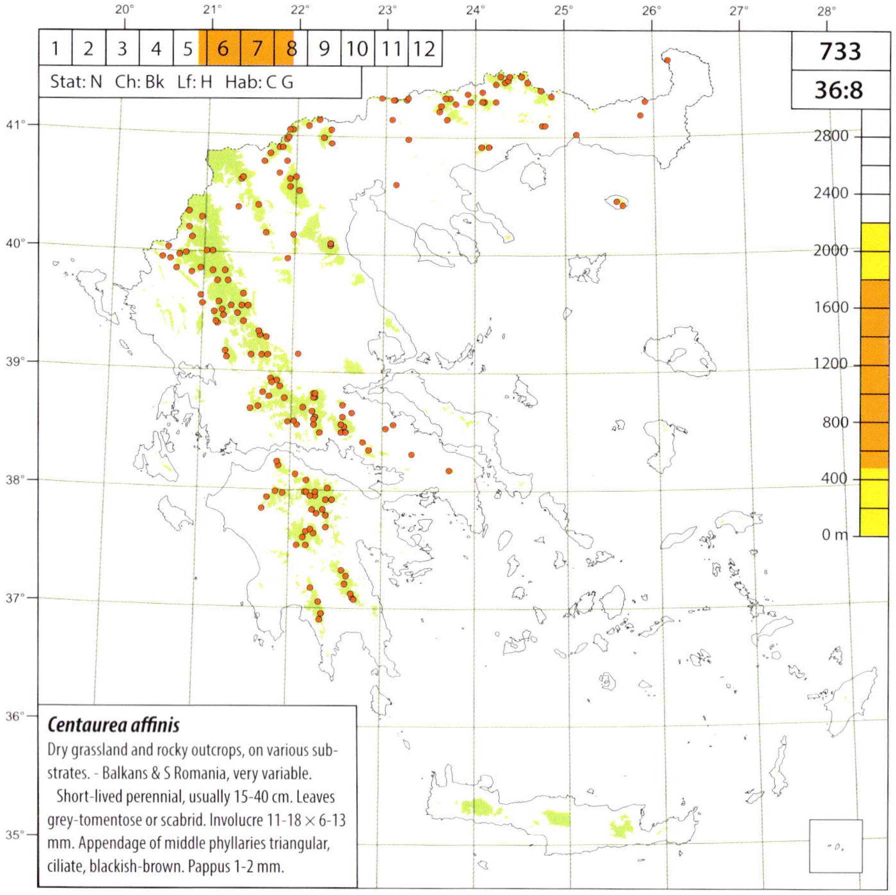
1 2 3 4 5 6 7 8 9 10 11 12
Stat: N Ch: Bk Lf: H Hab: C G
733
36:8
2800
2400
2000
1600
1200
800
400
0 m
Centaurea affinis
Dry grassland and rocky outcrops, on various substrates. - Balkans & S Romania, very variable.
Short-lived perennial, usually 15-40 cm. Leaves grey-tomentose or scabrid. Involucre 11-18 × 6-13 mm. Appendage of middle phyllaries triangular, ciliate, blackish-brown. Pappus 1-2 mm.

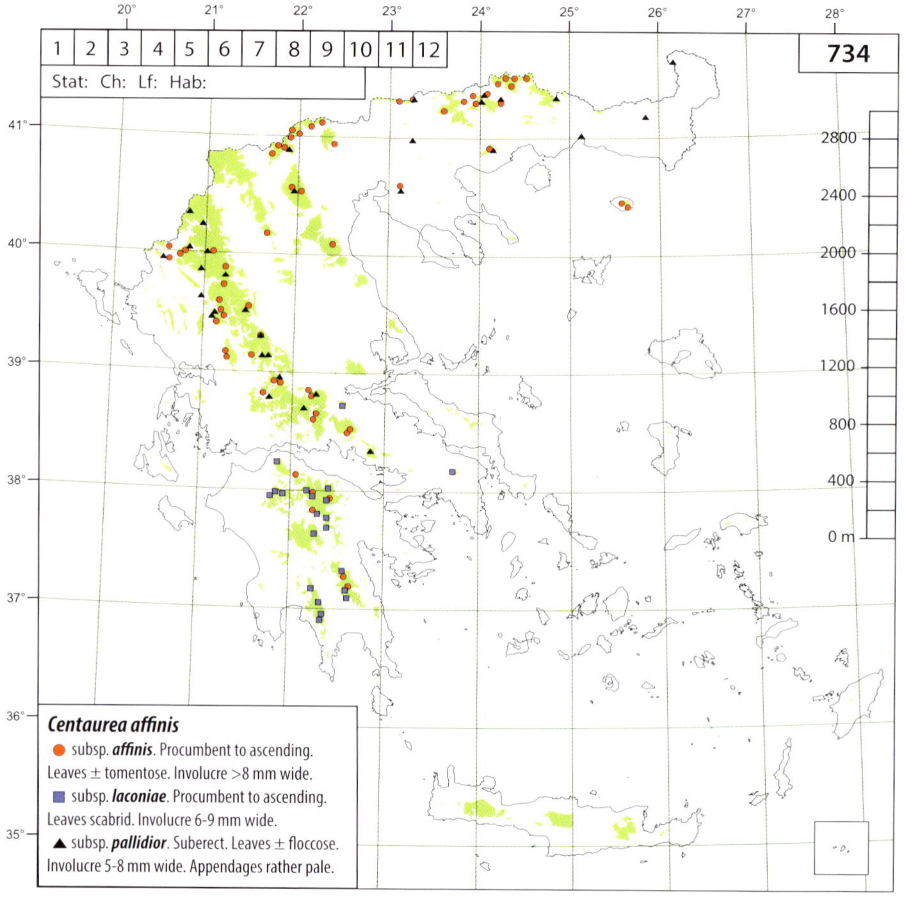
1 2 3 4 5 6 7 8 9 10 11 12
Stat: Ch: Lf: Hab:
734
2800
2400
2000
1600
1200
800
400
0 m
Centaurea affinis
● subsp. ***affinis***. Procumbent to ascending. Leaves ± tomentose. Involucre >8 mm wide.
■ subsp. ***laconiae***. Procumbent to ascending. Leaves scabrid. Involucre 6-9 mm wide.
▲ subsp. ***pallidior***. Suberect. Leaves ± floccose. Involucre 5-8 mm wide. Appendages rather pale.

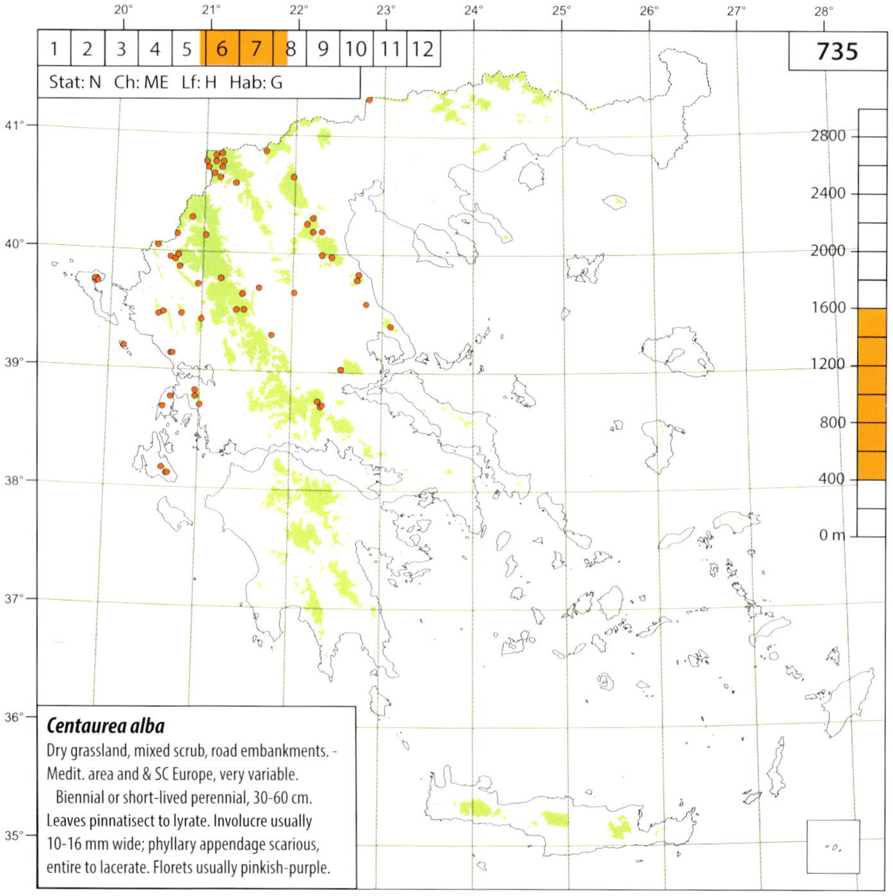
1 2 3 4 5 6 7 8 9 10 11 12
Stat: N Ch: ME Lf: H Hab: G
735
2800
2400
2000
1600
1200
800
400
0 m
Centaurea alba
Dry grassland, mixed scrub, road embankments. - Medit. area and & SC Europe, very variable.
Biennial or short-lived perennial, 30-60 cm. Leaves pinnatisect to lyrate. Involucre usually 10-16 mm wide; phyllary appendage scarious, entire to lacerate. Florets usually pinkish-purple.

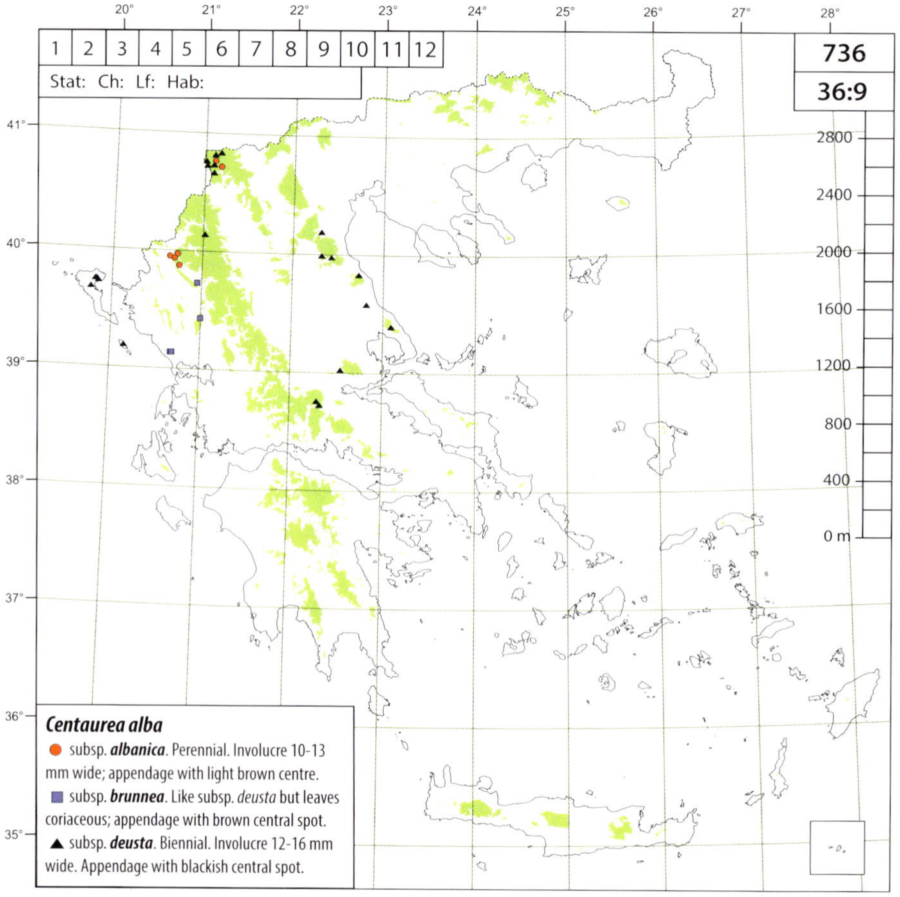
1 2 3 4 5 6 7 8 9 10 11 12
Stat: Ch: Lf: Hab:
736
36:9
2800
2400
2000
1600
1200
800
400
0 m
Centaurea alba
● subsp. ***albanica***. Perennial. Involucre 10-13 mm wide; appendage with light brown centre.
■ subsp. ***brunnea***. Like subsp. *deusta* but leaves coriaceous; appendage with brown central spot.
▲ subsp. ***deusta***. Biennial. Involucre 12-16 mm wide. Appendage with blackish central spot.

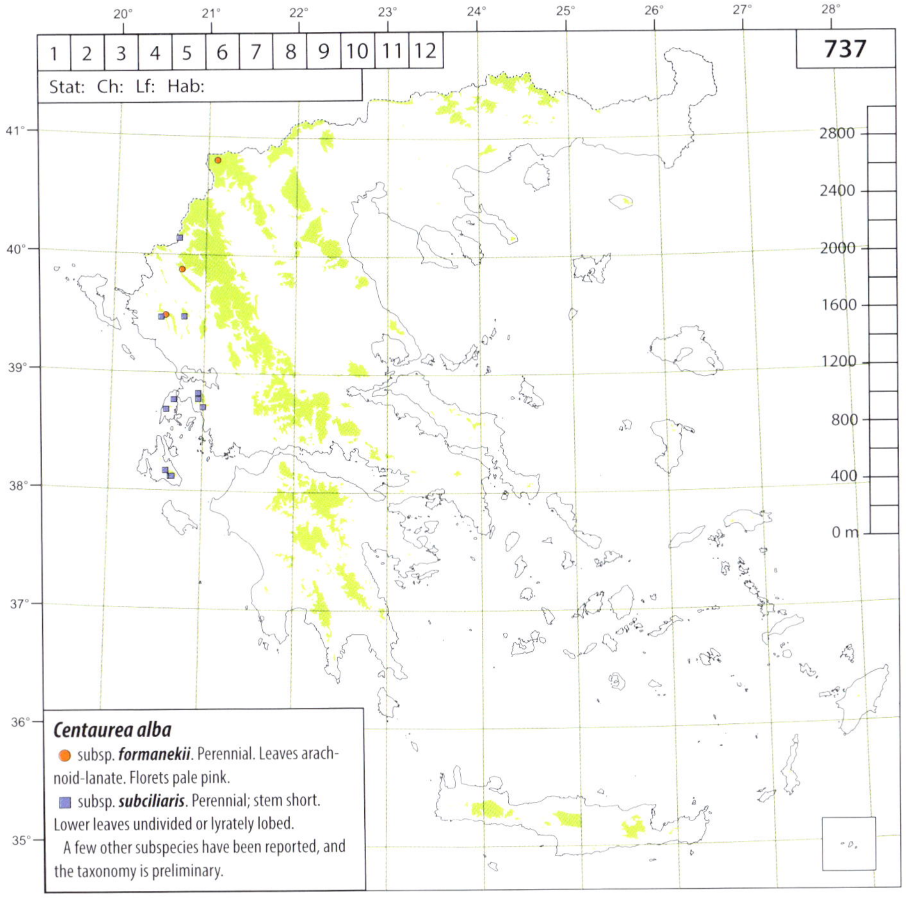

Centaurea alba

● subsp. ***formanekii***. Perennial. Leaves arachnoid-lanate. Florets pale pink.

■ subsp. ***subciliaris***. Perennial; stem short. Lower leaves undivided or lyrately lobed.

A few other subspecies have been reported, and the taxonomy is preliminary.

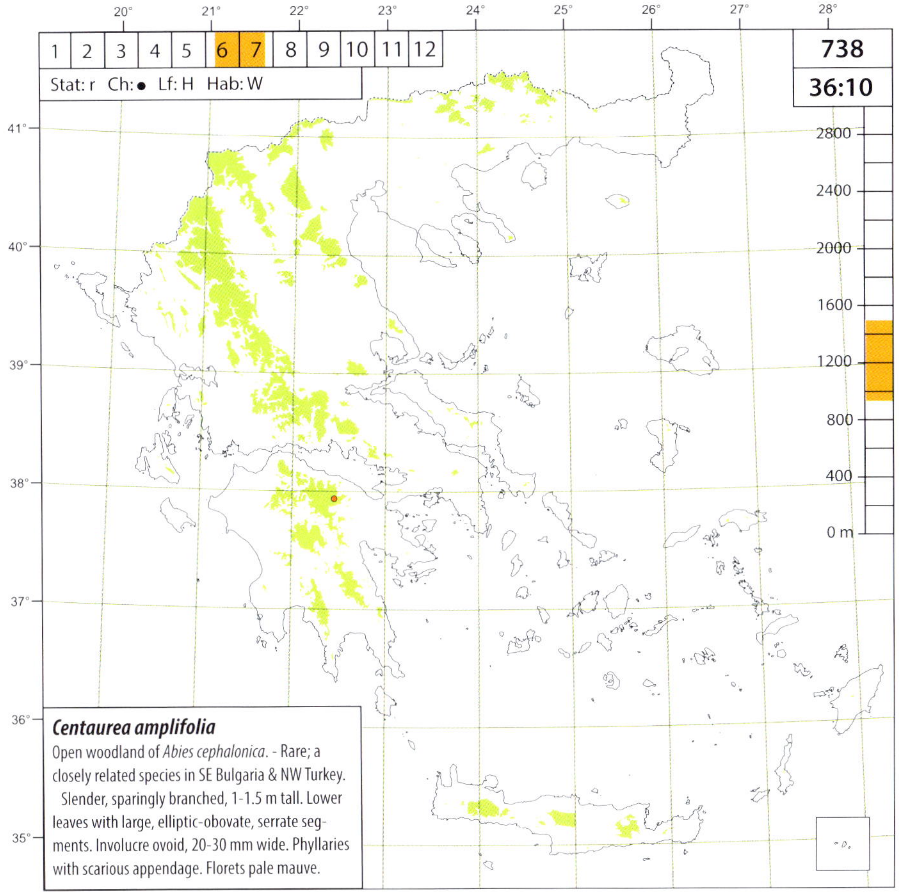

Centaurea amplifolia

Open woodland of *Abies cephalonica*. - Rare; a closely related species in SE Bulgaria & NW Turkey.

Slender, sparingly branched, 1-1.5 m tall. Lower leaves with large, elliptic-obovate, serrate segments. Involucre ovoid, 20-30 mm wide. Phyllaries with scarious appendage. Florets pale mauve.

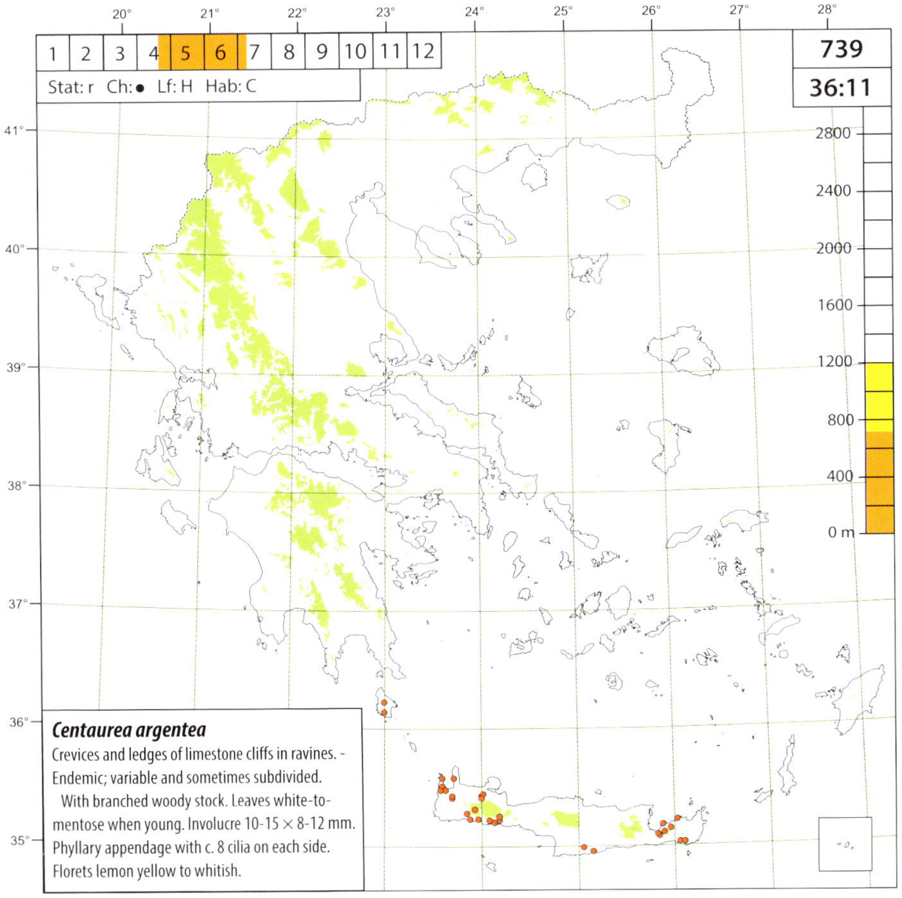

Centaurea argentea

Crevices and ledges of limestone cliffs in ravines. - Endemic; variable and sometimes subdivided.

With branched woody stock. Leaves white-to-mentose when young. Involucre 10-15 × 8-12 mm. Phyllary appendage with c. 8 cilia on each side. Florets lemon yellow to whitish.

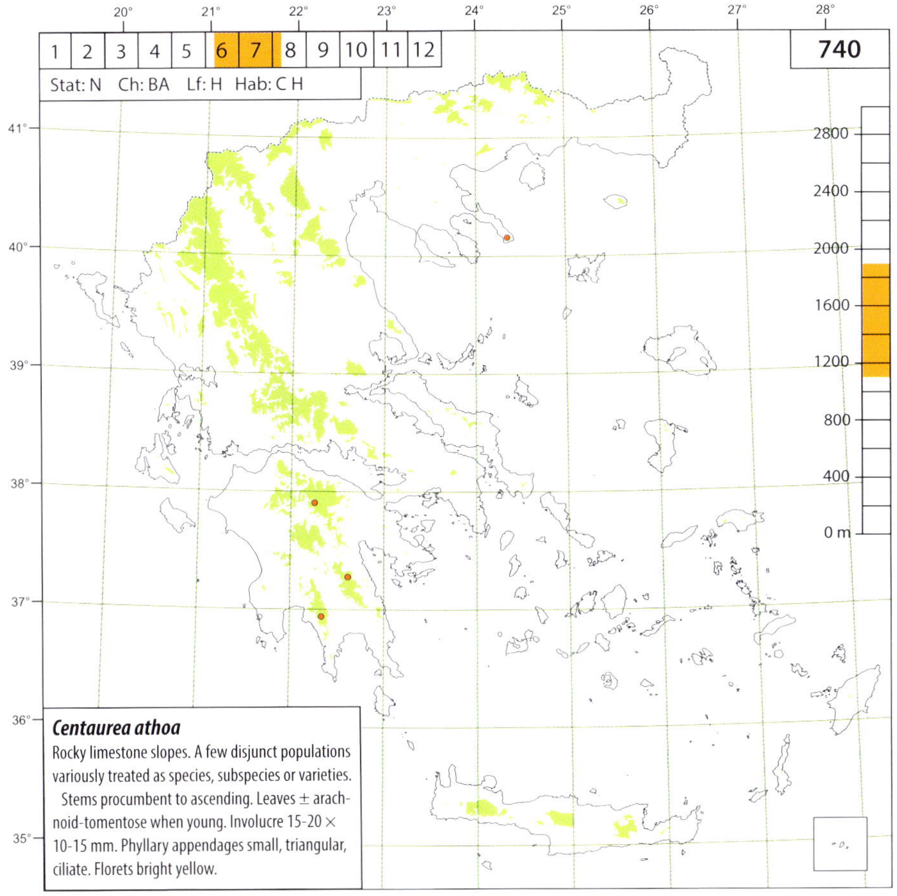

Centaurea athoa

Rocky limestone slopes. A few disjunct populations variously treated as species, subspecies or varieties.

Stems procumbent to ascending. Leaves ± arachnoid-tomentose when young. Involucre 15-20 × 10-15 mm. Phyllary appendages small, triangular, ciliate. Florets bright yellow.

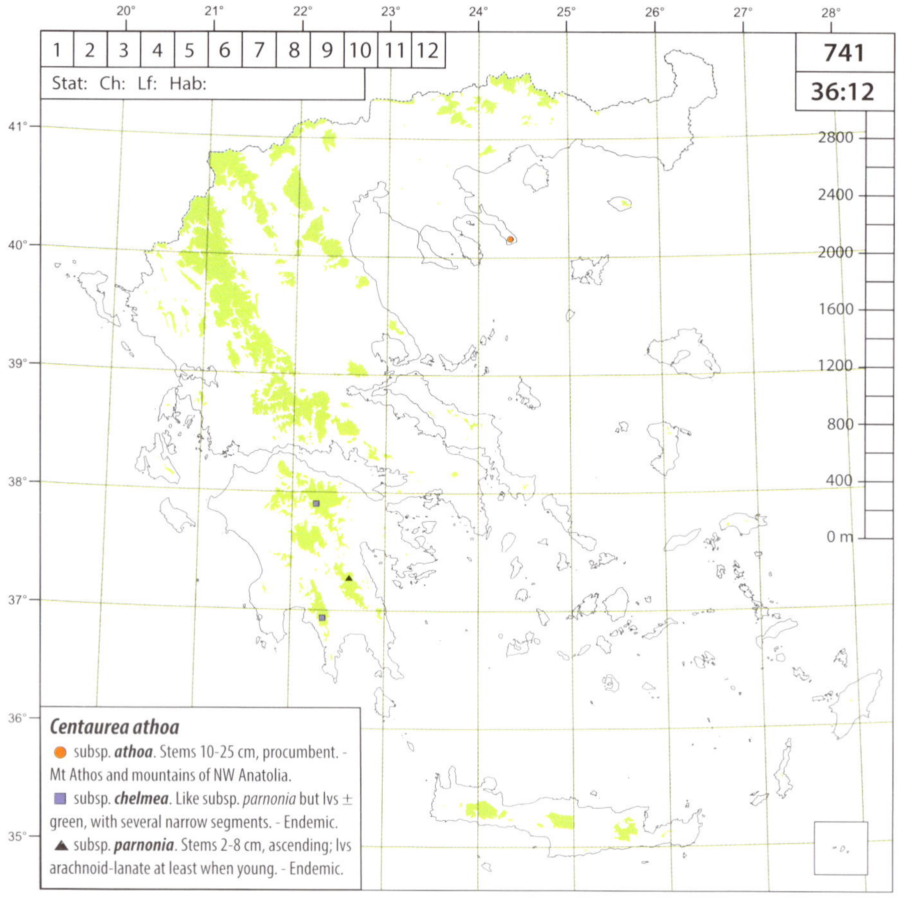

Centaurea athoa

● subsp. ***athoa***. Stems 10-25 cm, procumbent. - Mt Athos and mountains of NW Anatolia.

■ subsp. ***chelmea***. Like subsp. *parnonia* but lvs ± green, with several narrow segments. - Endemic.

▲ subsp. ***parnonia***. Stems 2-8 cm, ascending; lvs arachnoid-lanate at least when young. - Endemic.

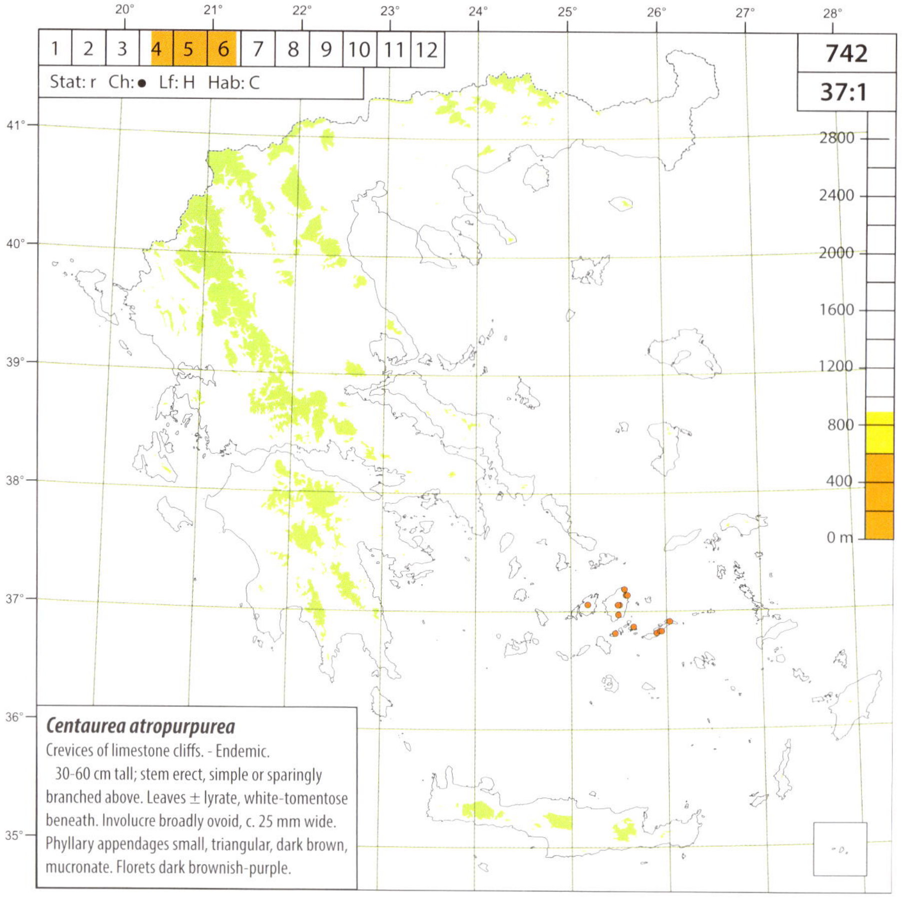

Centaurea atropurpurea

Crevices of limestone cliffs. - Endemic.

30-60 cm tall; stem erect, simple or sparingly branched above. Leaves ± lyrate, white-tomentose beneath. Involucre broadly ovoid, c. 25 mm wide. Phyllary appendages small, triangular, dark brown, mucronate. Florets dark brownish-purple.

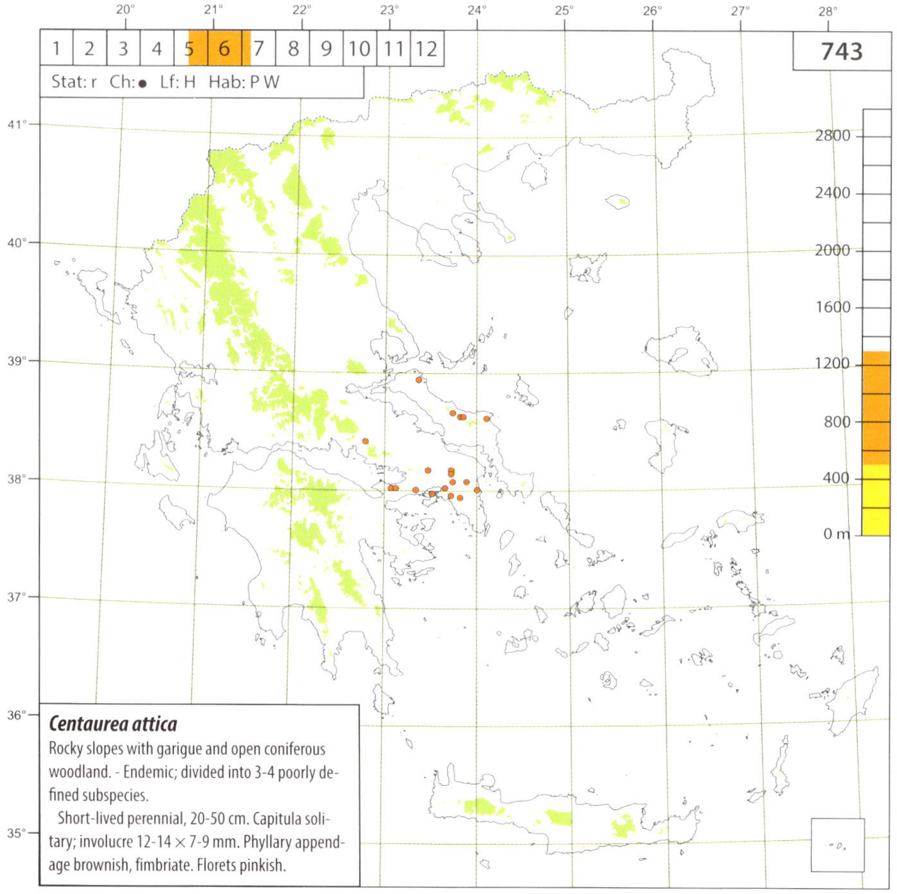

Centaurea attica

Rocky slopes with garigue and open coniferous woodland. - Endemic; divided into 3-4 poorly defined subspecies.

Short-lived perennial, 20-50 cm. Capitula solitary; involucre 12-14 × 7-9 mm. Phyllary appendage brownish, fimbriate. Florets pinkish.

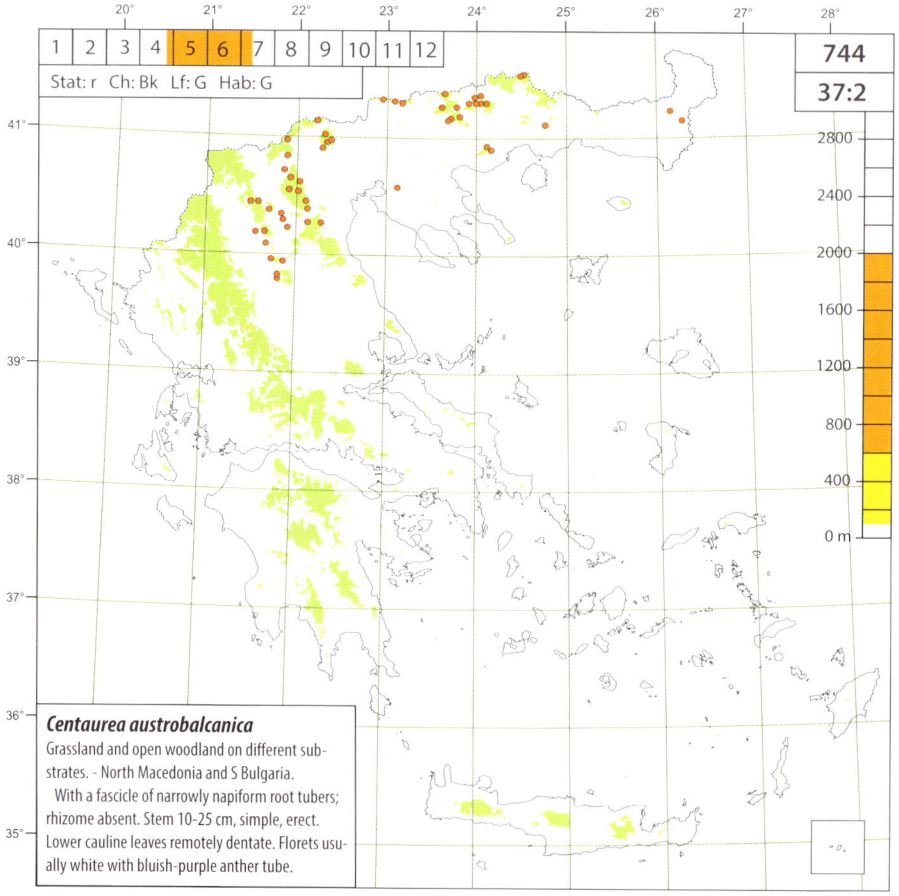

Centaurea austrobalcanica

Grassland and open woodland on different substrates. - North Macedonia and S Bulgaria.

With a fascicle of narrowly napiform root tubers; rhizome absent. Stem 10-25 cm, simple, erect. Lower cauline leaves remotely dentate. Florets usually white with bluish-purple anther tube.

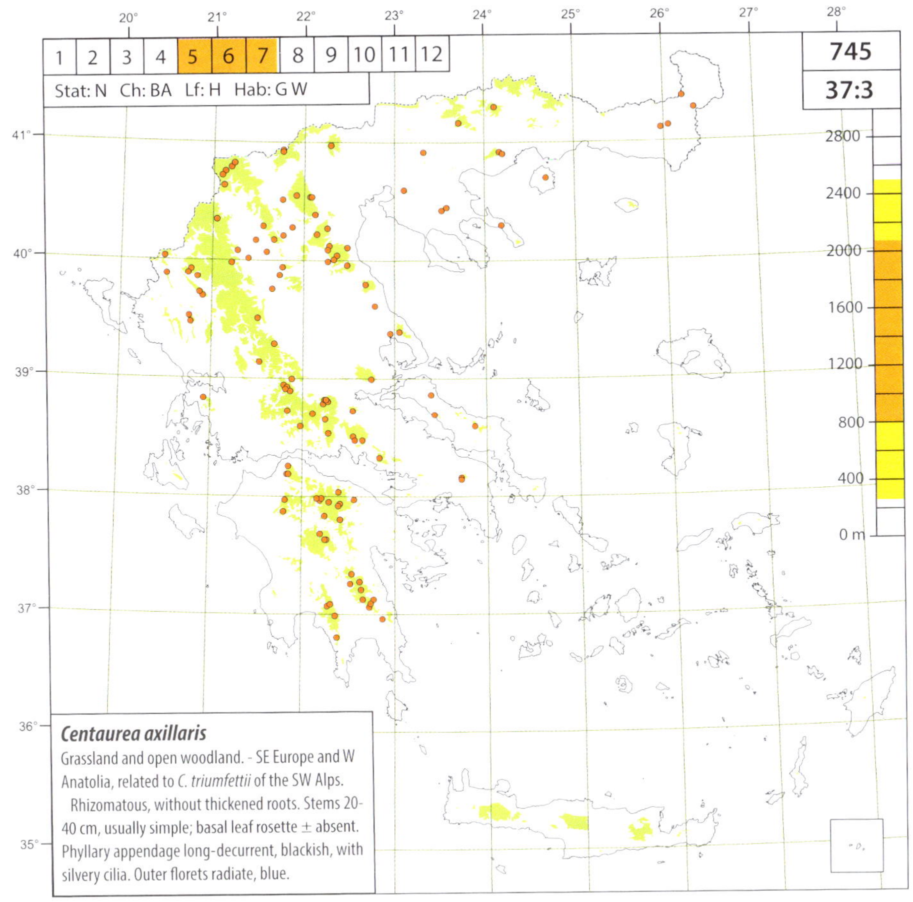

Centaurea axillaris

Grassland and open woodland. - SE Europe and W Anatolia, related to *C. triumfettii* of the SW Alps.

Rhizomatous, without thickened roots. Stems 20-40 cm, usually simple; basal leaf rosette ± absent. Phyllary appendage long-decurrent, blackish, with silvery cilia. Outer florets radiate, blue.

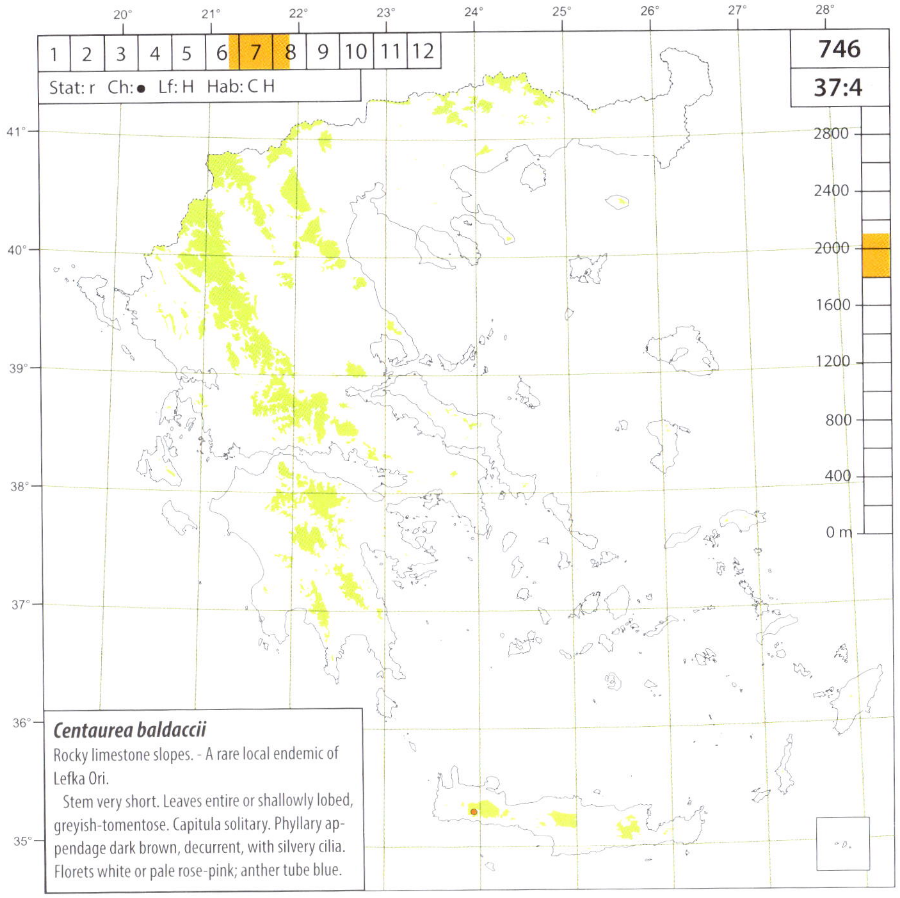

Centaurea baldaccii

Rocky limestone slopes. - A rare local endemic of Lefka Ori.

Stem very short. Leaves entire or shallowly lobed, greyish-tomentose. Capitula solitary. Phyllary appendage dark brown, decurrent, with silvery cilia. Florets white or pale rose-pink; anther tube blue.

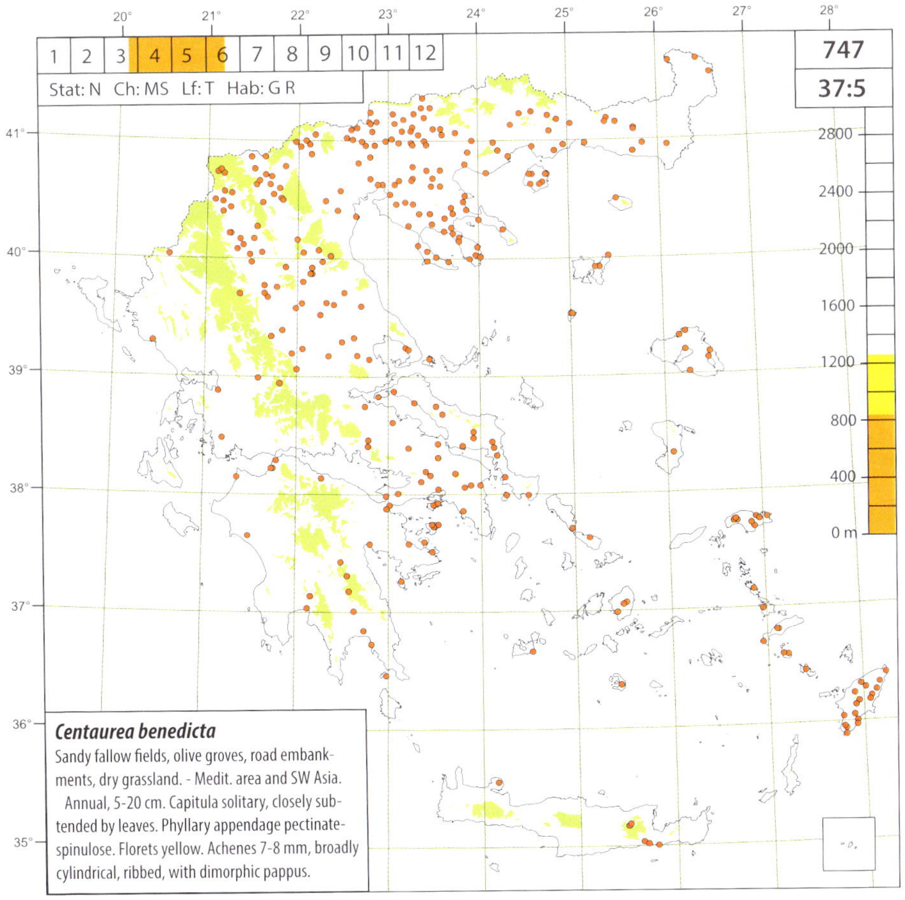

Centaurea benedicta

Sandy fallow fields, olive groves, road embankments, dry grassland. - Medit. area and SW Asia.

Annual, 5-20 cm. Capitula solitary, closely subtended by leaves. Phyllary appendage pectinate-spinulose. Florets yellow. Achenes 7-8 mm, broadly cylindrical, ribbed, with dimorphic pappus.

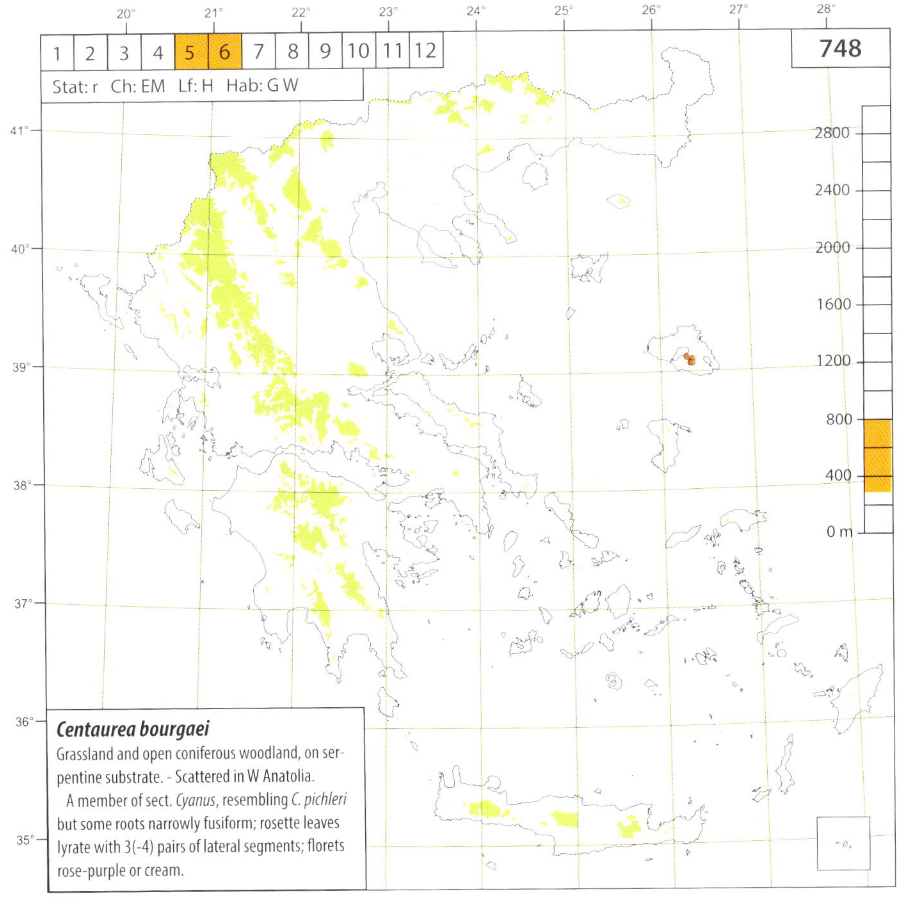

Centaurea bourgaei

Grassland and open coniferous woodland, on serpentine substrate. - Scattered in W Anatolia.

A member of sect. *Cyanus*, resembling *C. pichleri* but some roots narrowly fusiform; rosette leaves lyrate with 3(-4) pairs of lateral segments; florets rose-purple or cream.

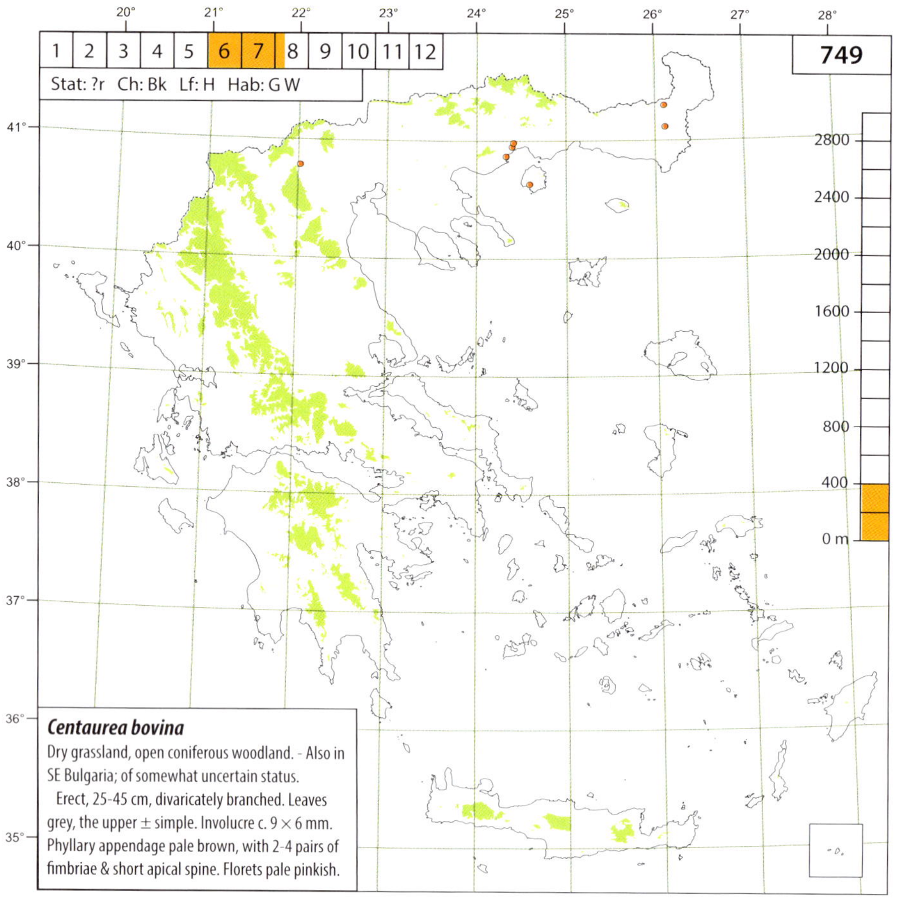

Centaurea bovina
Dry grassland, open coniferous woodland. - Also in SE Bulgaria; of somewhat uncertain status.
Erect, 25-45 cm, divaricately branched. Leaves grey, the upper ± simple. Involucre c. 9 × 6 mm. Phyllary appendage pale brown, with 2-4 pairs of fimbriae & short apical spine. Florets pale pinkish.

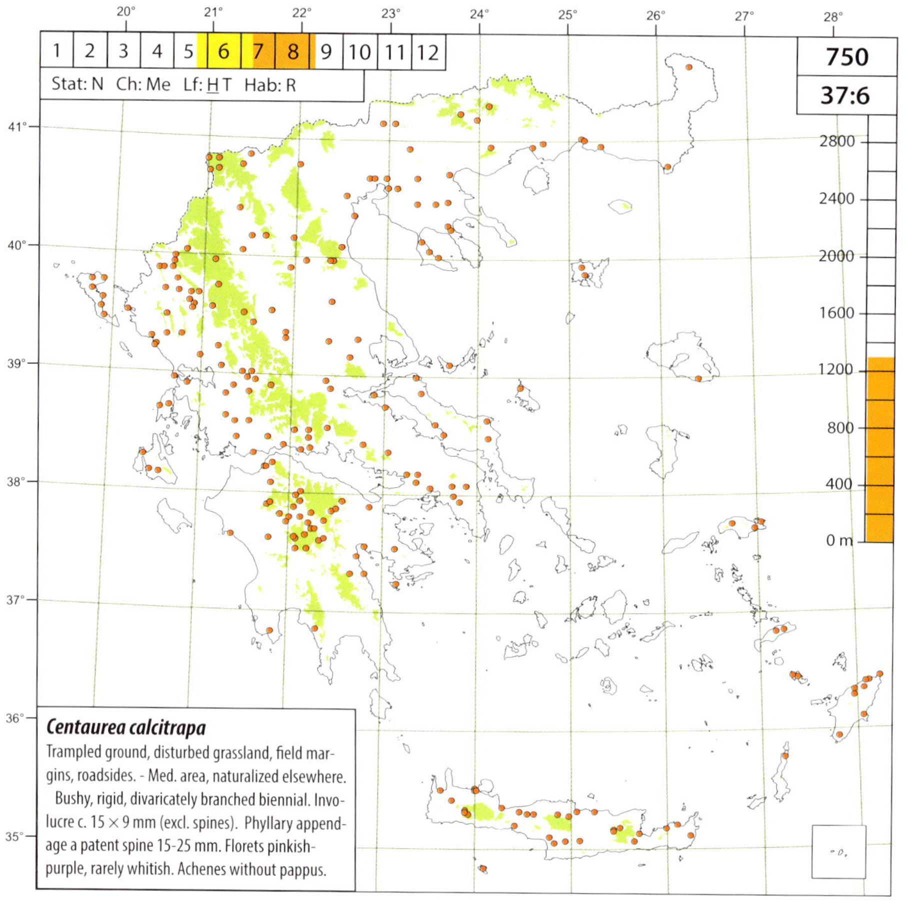

Centaurea calcitrapa
Trampled ground, disturbed grassland, field margins, roadsides. - Med. area, naturalized elsewhere.
Bushy, rigid, divaricately branched biennial. Involucre c. 15 × 9 mm (excl. spines). Phyllary appendage a patent spine 15-25 mm. Florets pinkish-purple, rarely whitish. Achenes without pappus.

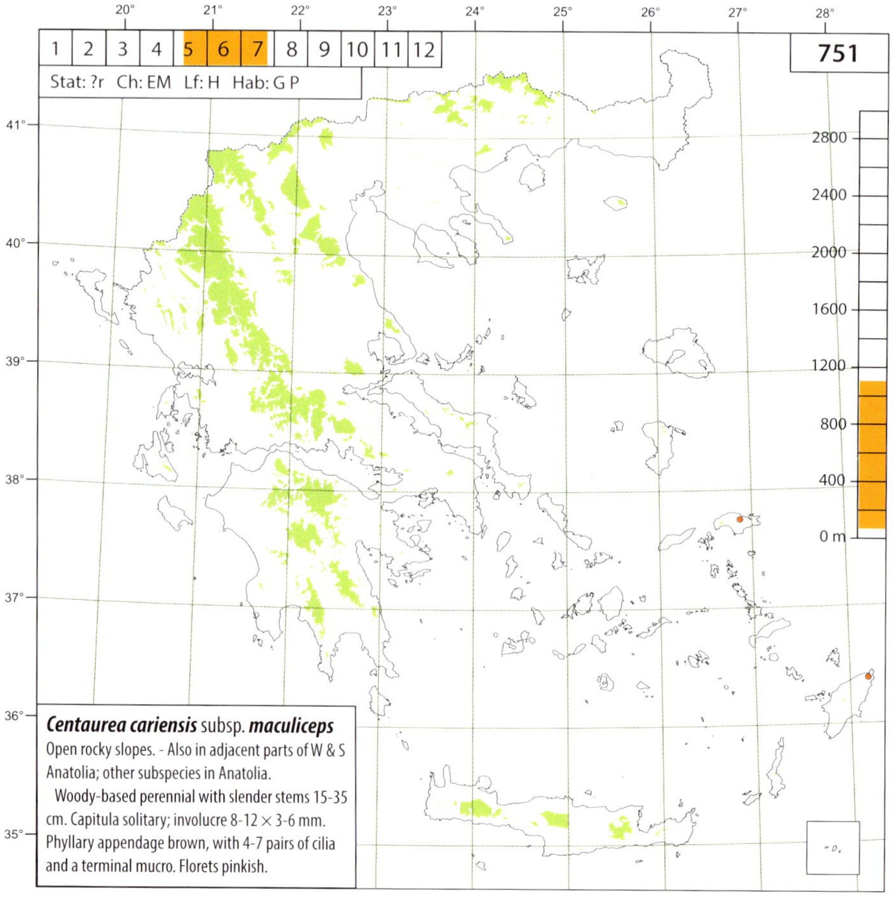

Centaurea cariensis subsp. ***maculiceps***
Open rocky slopes. - Also in adjacent parts of W & S Anatolia; other subspecies in Anatolia.
Woody-based perennial with slender stems 15-35 cm. Capitula solitary; involucre 8-12 × 3-6 mm. Phyllary appendage brown, with 4-7 pairs of cilia and a terminal mucro. Florets pinkish.

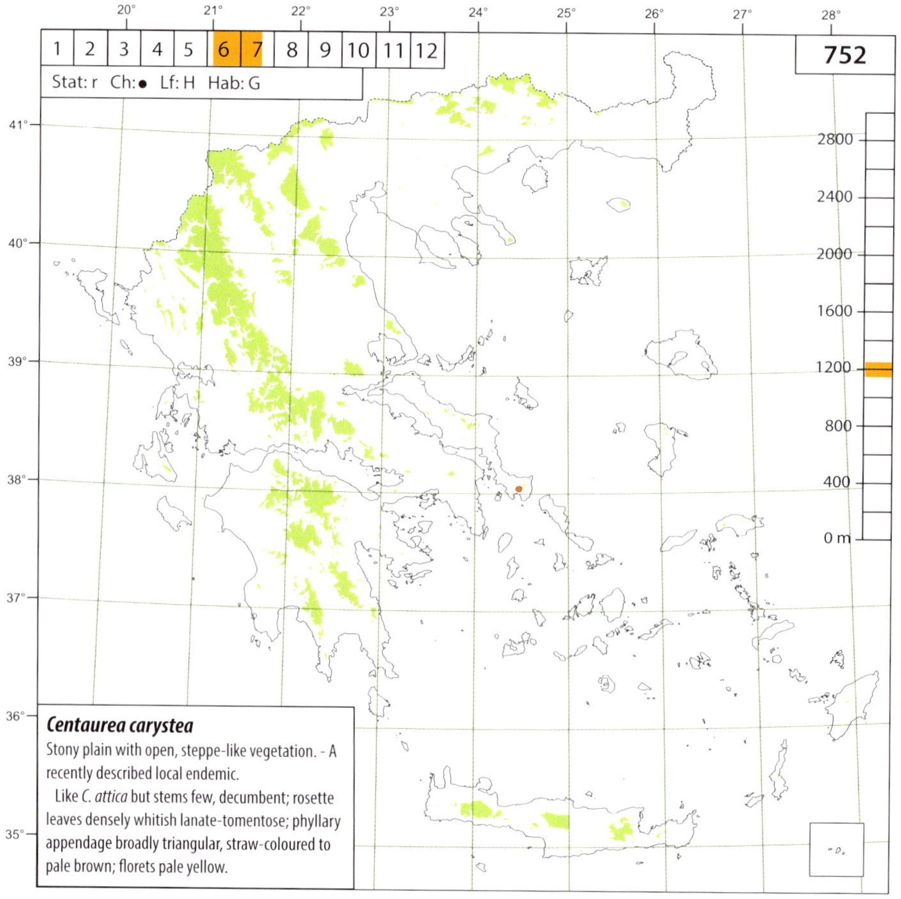

Centaurea carystea
Stony plain with open, steppe-like vegetation. - A recently described local endemic.
Like *C. attica* but stems few, decumbent; rosette leaves densely whitish lanate-tomentose; phyllary appendage broadly triangular, straw-coloured to pale brown; florets pale yellow.

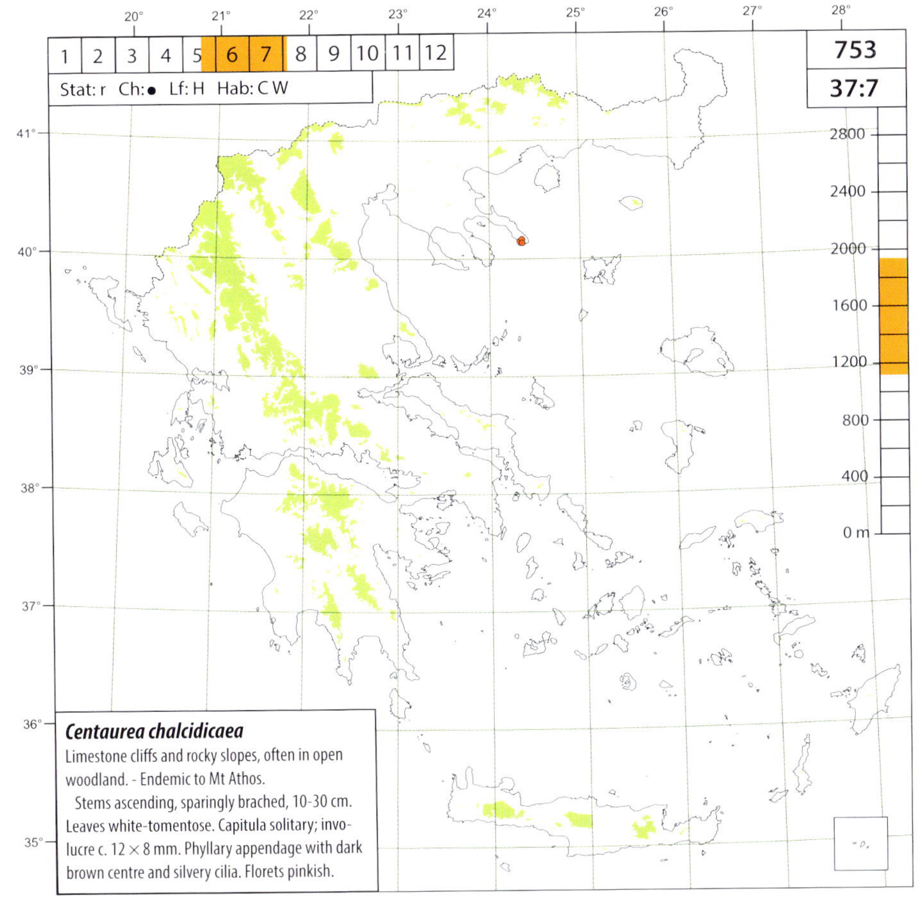

Centaurea chalcidicaea
Limestone cliffs and rocky slopes, often in open woodland. - Endemic to Mt Athos.
Stems ascending, sparingly brached, 10-30 cm. Leaves white-tomentose. Capitula solitary; involucre c. 12 × 8 mm. Phyllary appendage with dark brown centre and silvery cilia. Florets pinkish.

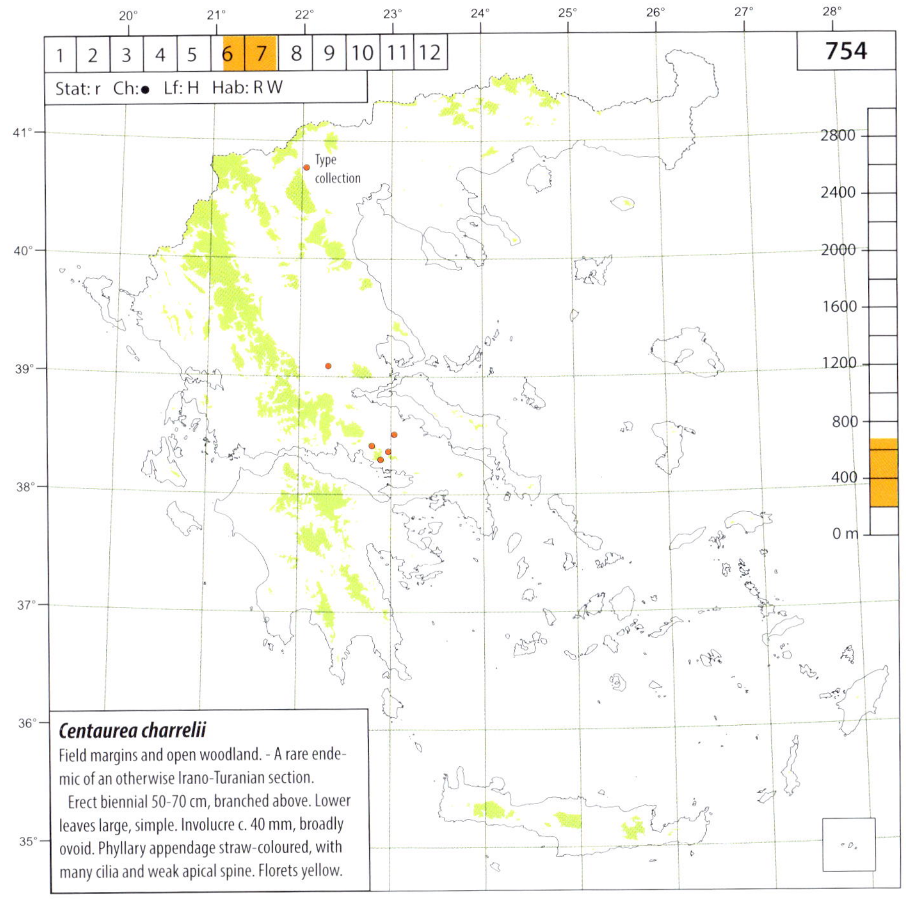

Centaurea charrelii
Field margins and open woodland. - A rare endemic of an otherwise Irano-Turanian section.
Erect biennial 50-70 cm, branched above. Lower leaves large, simple. Involucre c. 40 mm, broadly ovoid. Phyllary appendage straw-coloured, with many cilia and weak apical spine. Florets yellow.

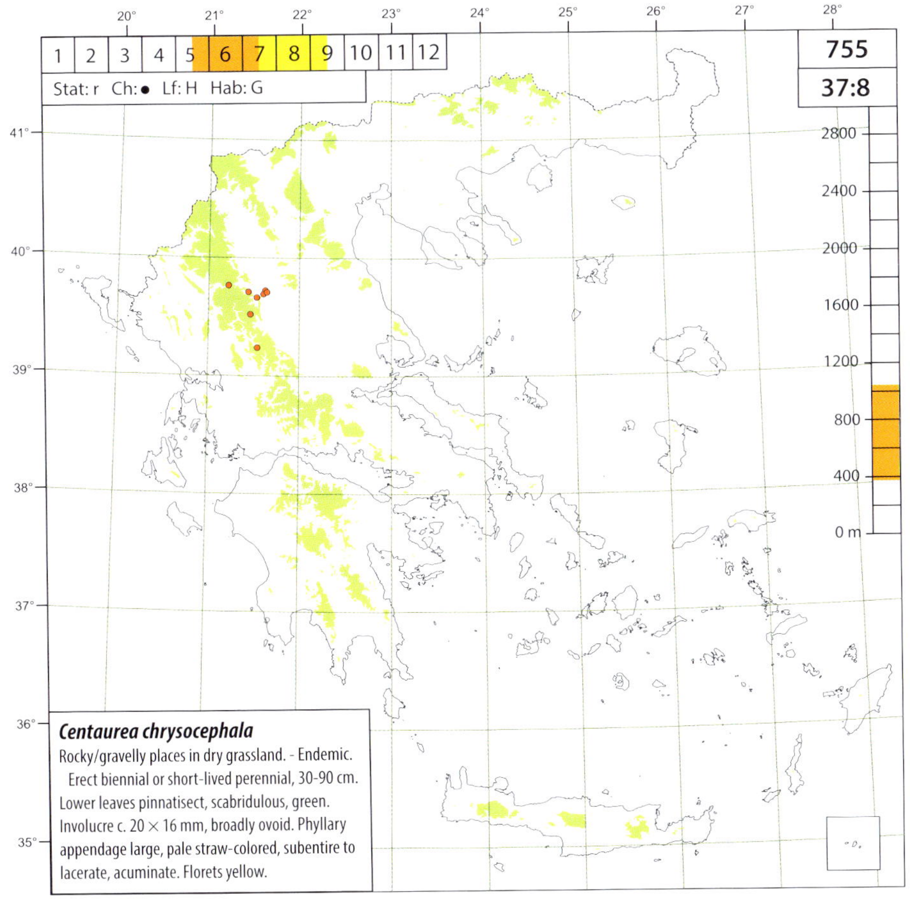

Centaurea chrysocephala
Rocky/gravelly places in dry grassland. - Endemic.
Erect biennial or short-lived perennial, 30-90 cm. Lower leaves pinnatisect, scabridulous, green. Involucre c. 20 × 16 mm, broadly ovoid. Phyllary appendage large, pale straw-colored, subentire to lacerate, acuminate. Florets yellow.

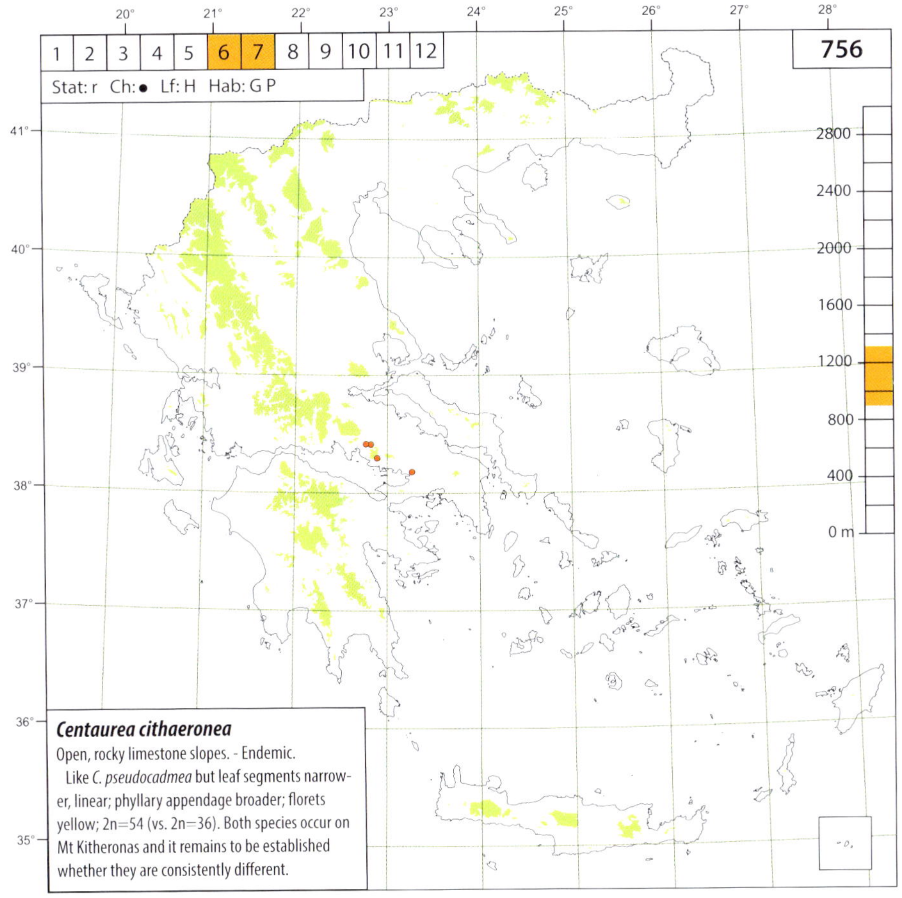

Centaurea cithaeronea
Open, rocky limestone slopes. - Endemic.
Like *C. pseudocadmea* but leaf segments narrower, linear; phyllary appendage broader; florets yellow; 2n=54 (vs. 2n=36). Both species occur on Mt Kitheronas and it remains to be established whether they are consistently different.

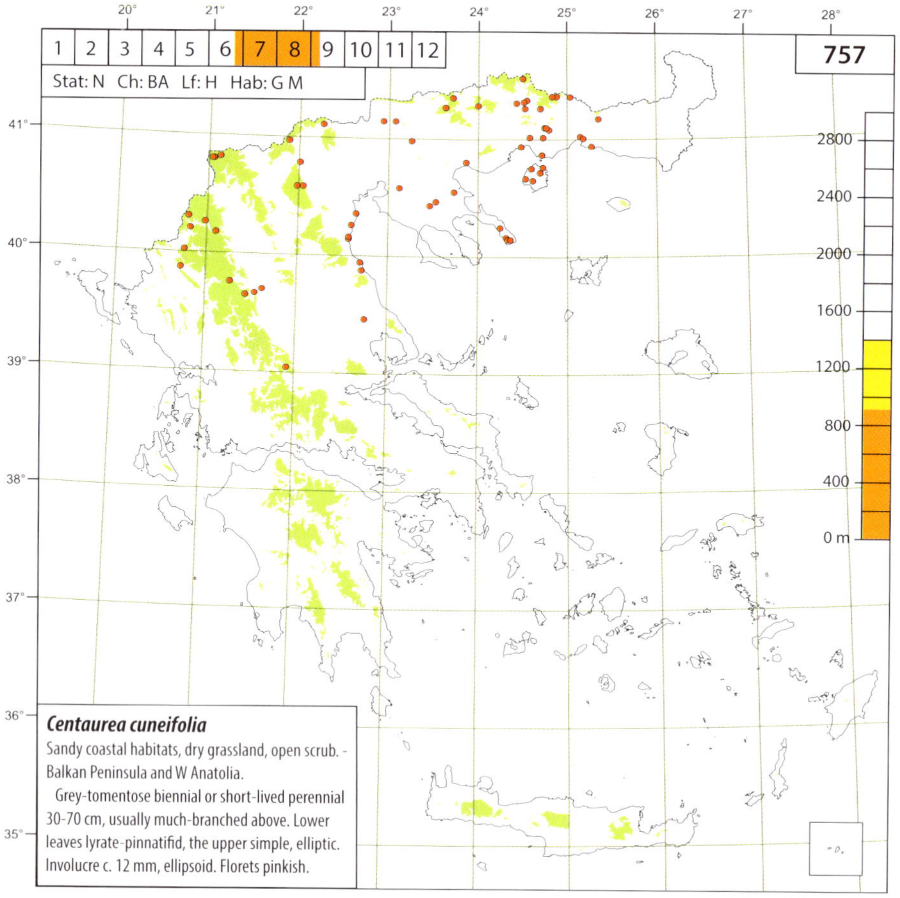

Centaurea cuneifolia

Sandy coastal habitats, dry grassland, open scrub. - Balkan Peninsula and W Anatolia.

Grey-tomentose biennial or short-lived perennial 30-70 cm, usually much-branched above. Lower leaves lyrate-pinnatifid, the upper simple, elliptic. Involucre c. 12 mm, ellipsoid. Florets pinkish.

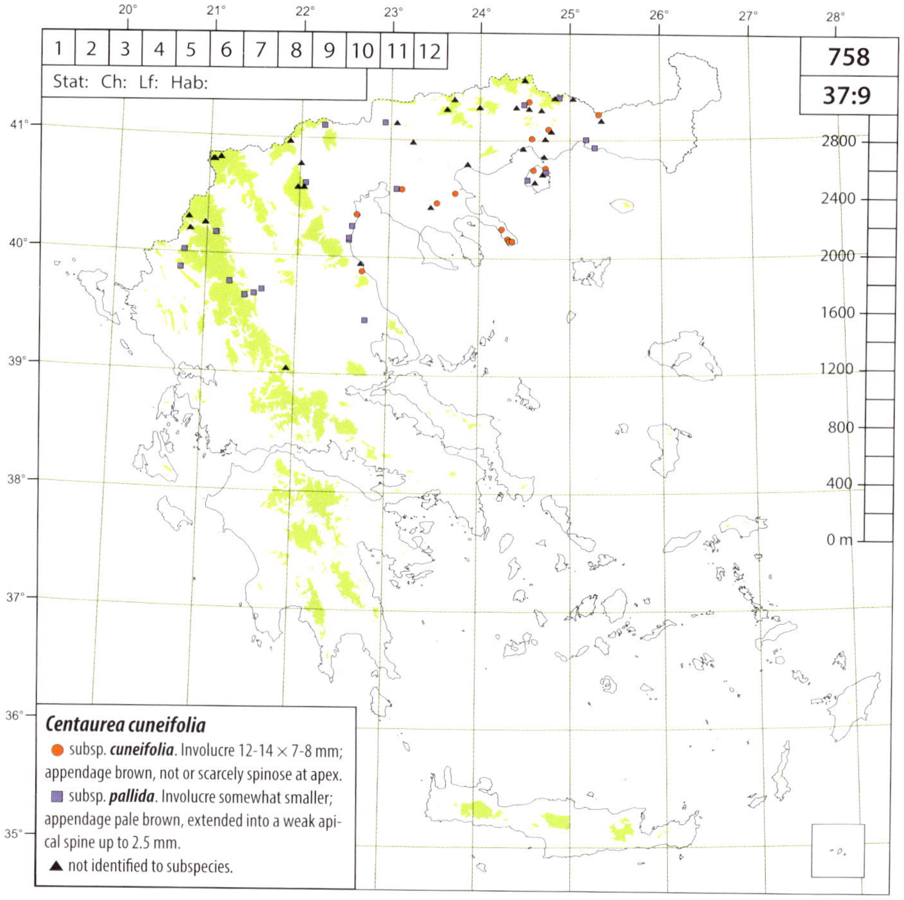

Centaurea cuneifolia

● subsp. ***cuneifolia***. Involucre 12-14 × 7-8 mm; appendage brown, not or scarcely spinose at apex.

■ subsp. ***pallida***. Involucre somewhat smaller; appendage pale brown, extended into a weak apical spine up to 2.5 mm.

▲ not identified to subspecies.

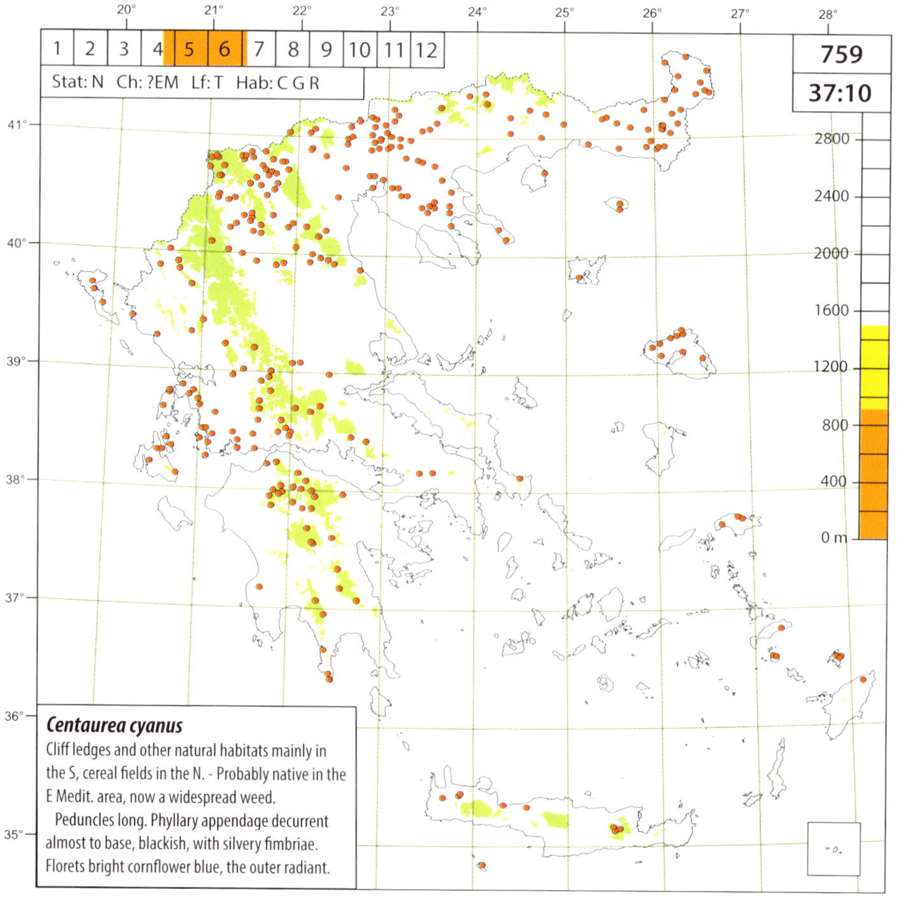

Centaurea cyanus

Cliff ledges and other natural habitats mainly in the S, cereal fields in the N. - Probably native in the E Medit. area, now a widespread weed.

Peduncles long. Phyllary appendage decurrent almost to base, blackish, with silvery fimbriae. Florets bright cornflower blue, the outer radiant.

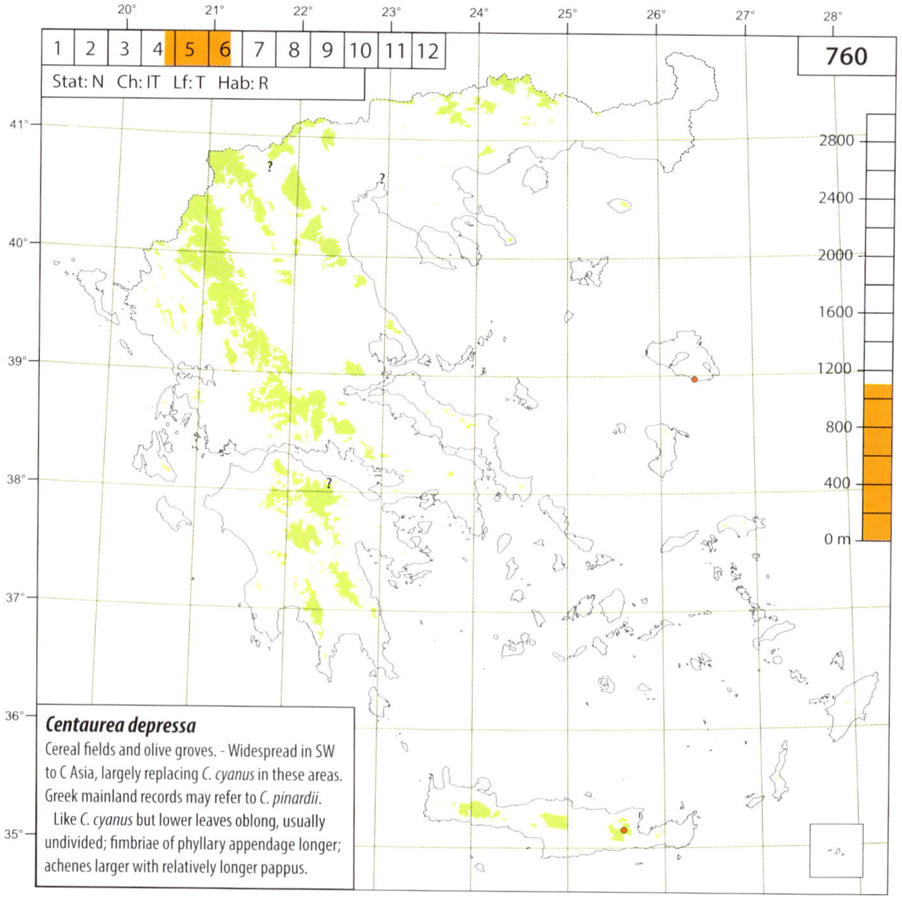

Centaurea depressa

Cereal fields and olive groves. - Widespread in SW to C Asia, largely replacing *C. cyanus* in these areas. Greek mainland records may refer to *C. pinardii*.

Like *C. cyanus* but lower leaves oblong, usually undivided; fimbriae of phyllary appendage longer; achenes larger with relatively longer pappus.

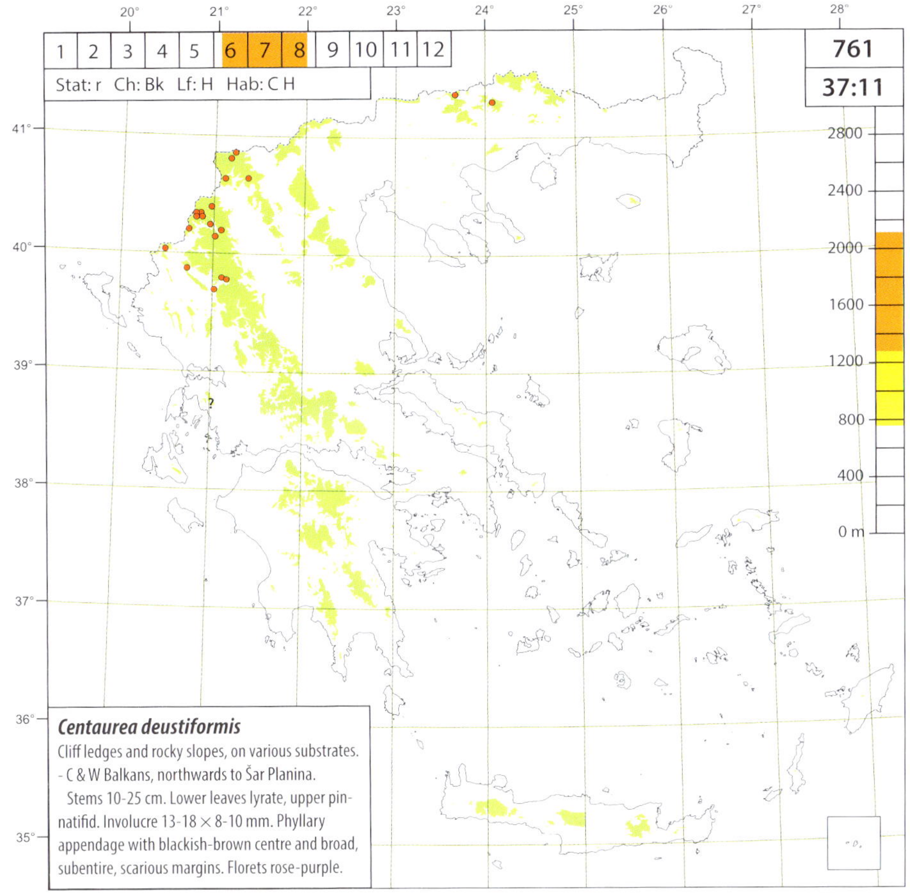

Centaurea deustiformis

Cliff ledges and rocky slopes, on various substrates. - C & W Balkans, northwards to Šar Planina.

Stems 10-25 cm. Lower leaves lyrate, upper pinnatifid. Involucre 13-18 × 8-10 mm. Phyllary appendage with blackish-brown centre and broad, subentire, scarious margins. Florets rose-purple.

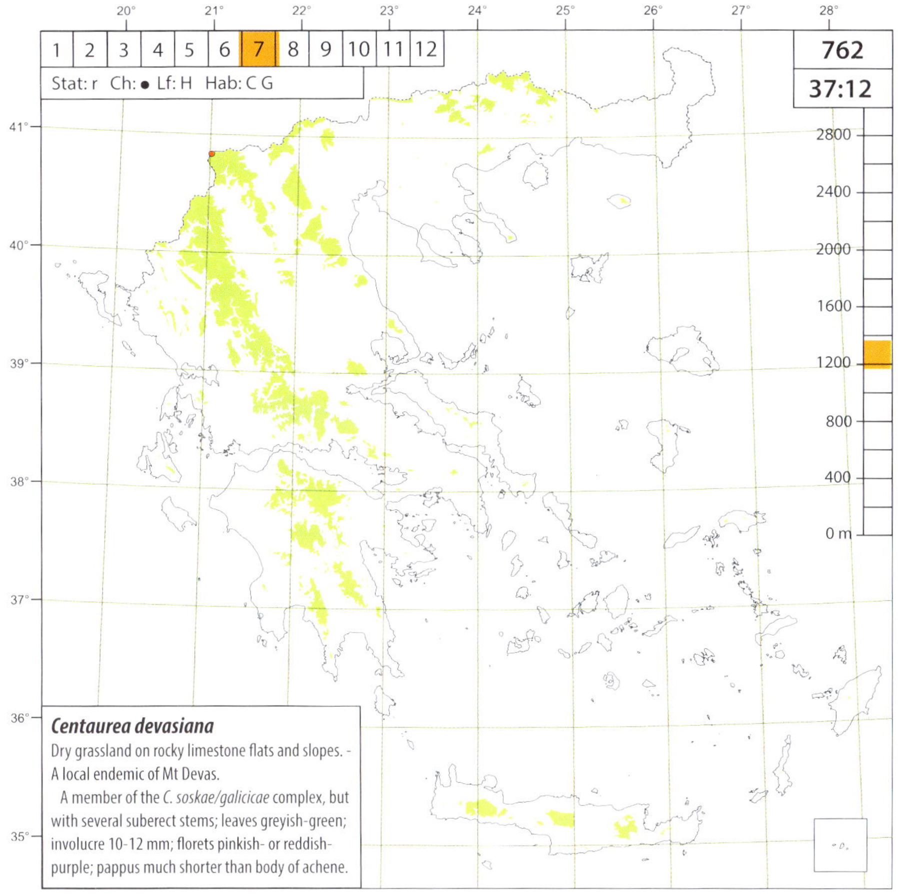

Centaurea devasiana

Dry grassland on rocky limestone flats and slopes. - A local endemic of Mt Devas.

A member of the *C. soskae/galicicae* complex, but with several suberect stems; leaves greyish-green; involucre 10-12 mm; florets pinkish- or reddish-purple; pappus much shorter than body of achene.

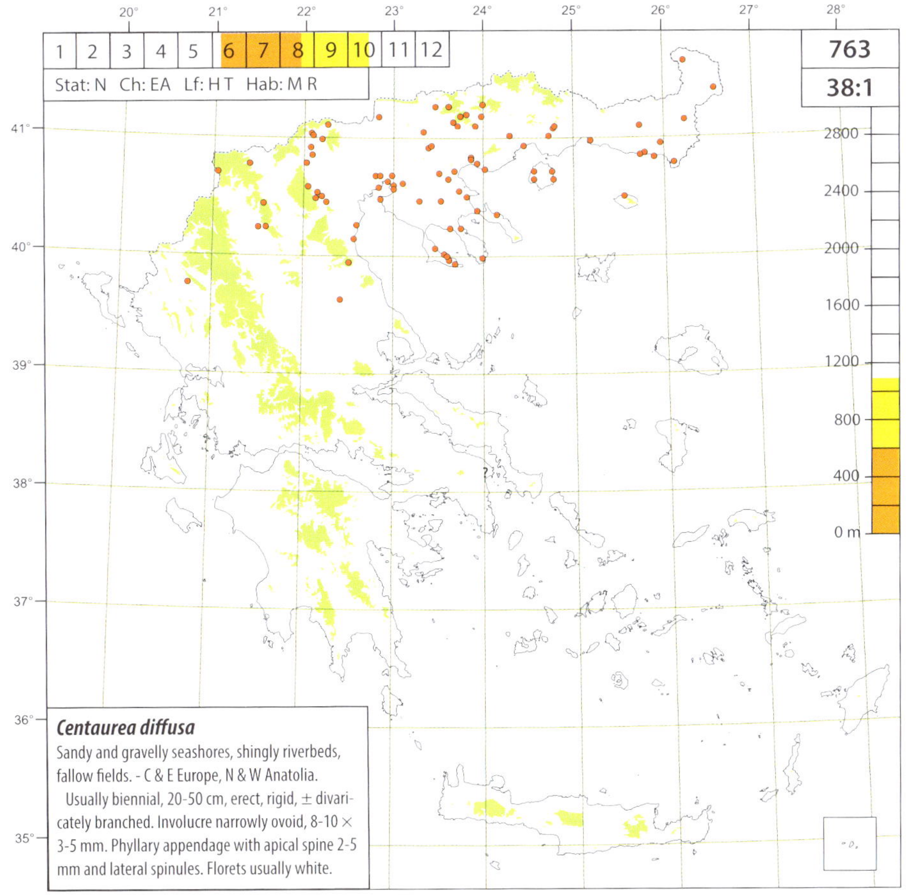

Centaurea diffusa

Sandy and gravelly seashores, shingly riverbeds, fallow fields. - C & E Europe, N & W Anatolia.

Usually biennial, 20-50 cm, erect, rigid, ± divaricately branched. Involucre narrowly ovoid, 8-10 × 3-5 mm. Phyllary appendage with apical spine 2-5 mm and lateral spinules. Florets usually white.

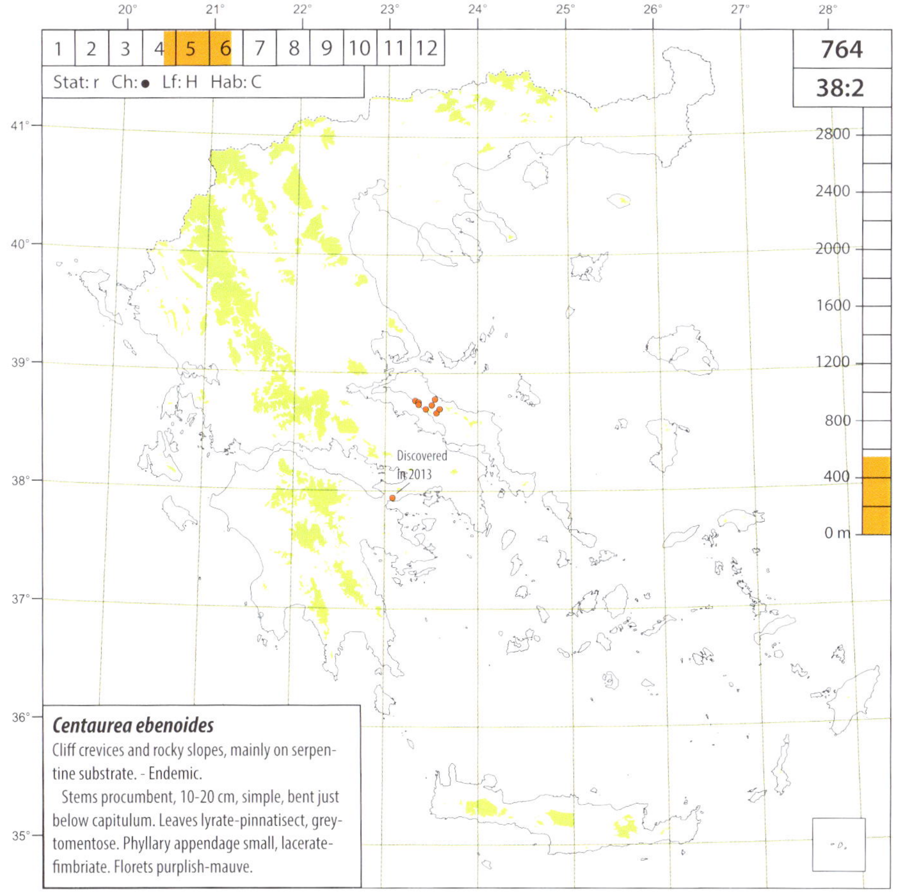

Centaurea ebenoides

Cliff crevices and rocky slopes, mainly on serpentine substrate. - Endemic.

Stems procumbent, 10-20 cm, simple, bent just below capitulum. Leaves lyrate-pinnatisect, grey-tomentose. Phyllary appendage small, lacerate-fimbriate. Florets purplish-mauve.

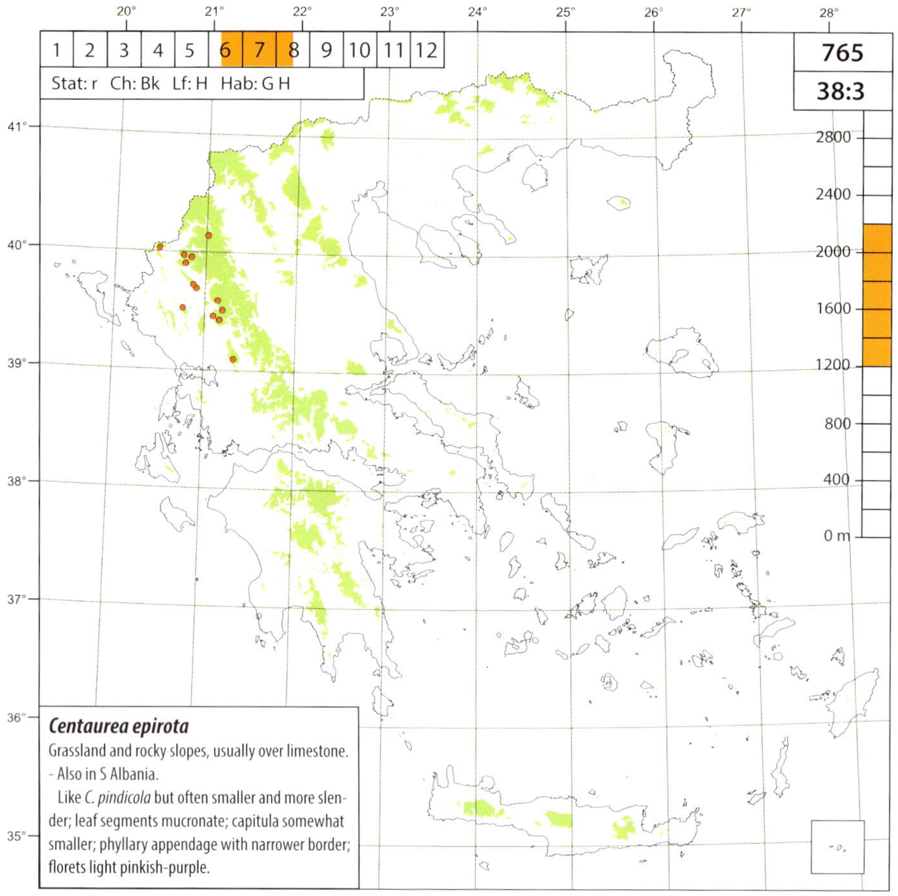

Centaurea epirota

Grassland and rocky slopes, usually over limestone. - Also in S Albania.

Like *C. pindicola* but often smaller and more slender; leaf segments mucronate; capitula somewhat smaller; phyllary appendage with narrower border; florets light pinkish-purple.

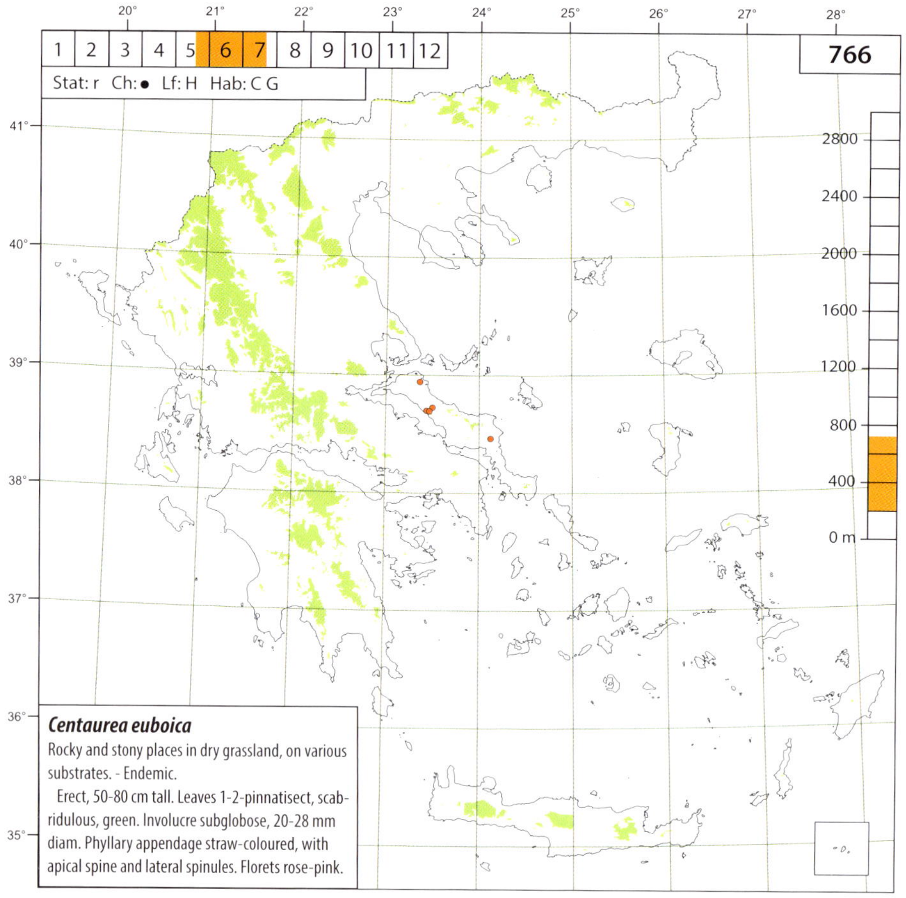

Centaurea euboica

Rocky and stony places in dry grassland, on various substrates. - Endemic.

Erect, 50-80 cm tall. Leaves 1-2-pinnatisect, scabridulous, green. Involucre subglobose, 20-28 mm diam. Phyllary appendage straw-coloured, with apical spine and lateral spinules. Florets rose-pink.

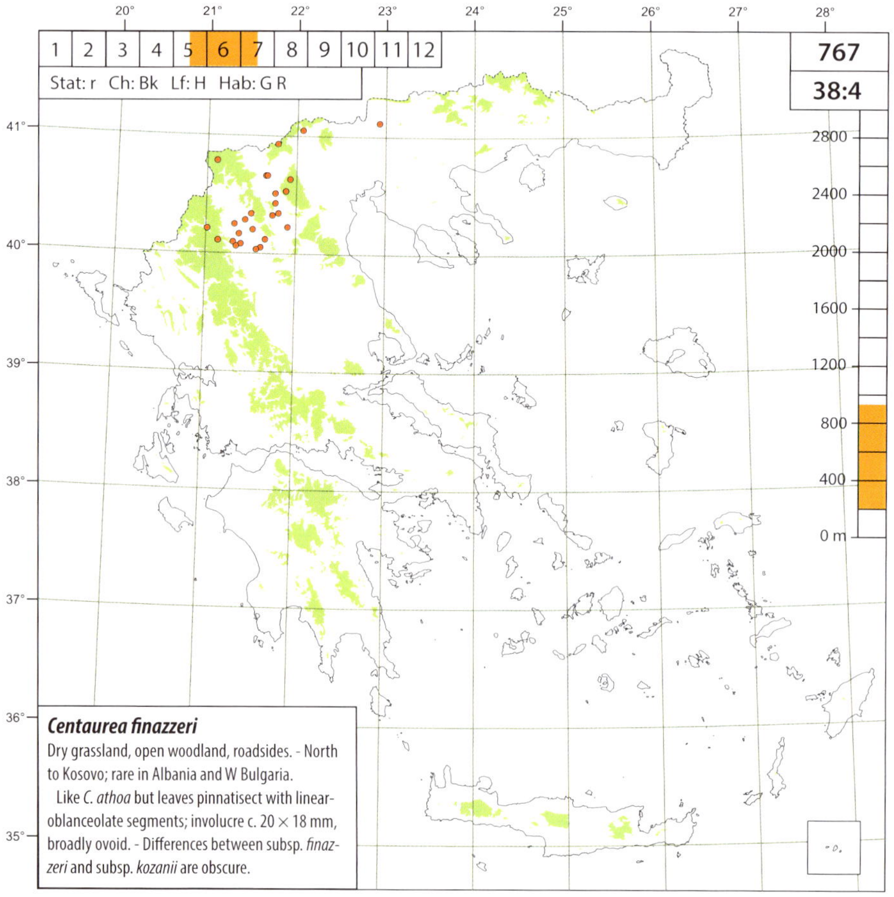

Centaurea finazzeri

Dry grassland, open woodland, roadsides. - North to Kosovo; rare in Albania and W Bulgaria.

Like *C. athoa* but leaves pinnatisect with linear-oblanceolate segments; involucre c. 20 × 18 mm, broadly ovoid. - Differences between subsp. *finazzeri* and subsp. *kozanii* are obscure.

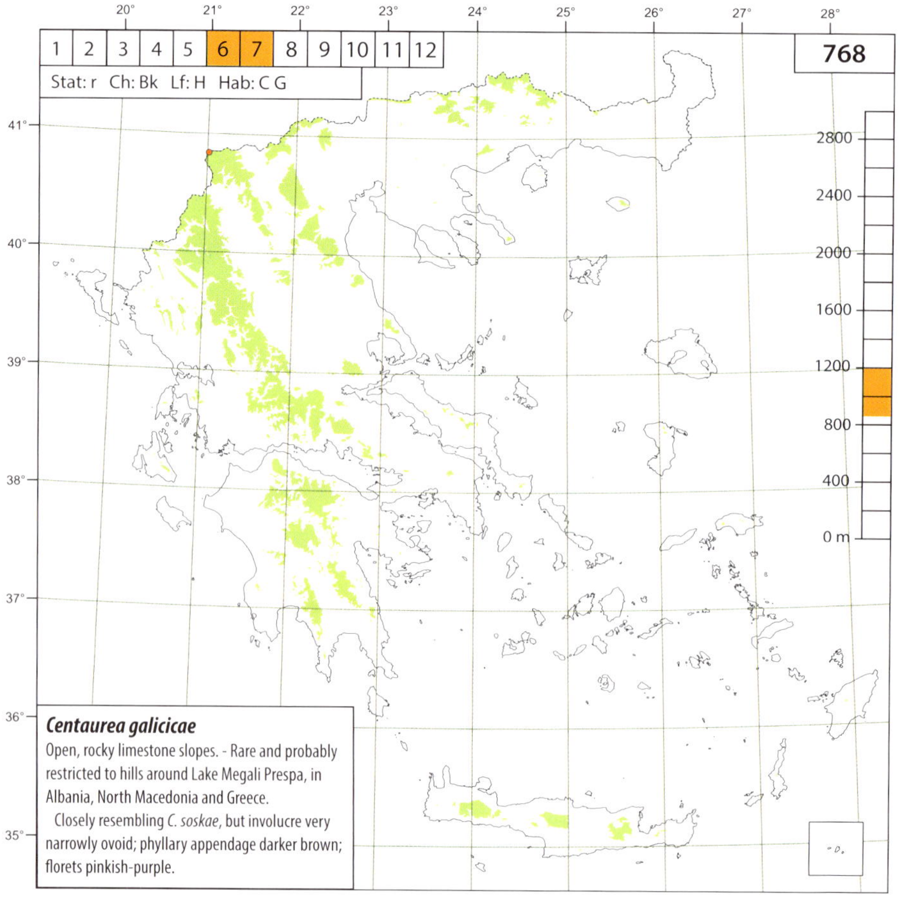

Centaurea galicicae

Open, rocky limestone slopes. - Rare and probably restricted to hills around Lake Megali Prespa, in Albania, North Macedonia and Greece.

Closely resembling *C. soskae*, but involucre very narrowly ovoid; phyllary appendage darker brown; florets pinkish-purple.

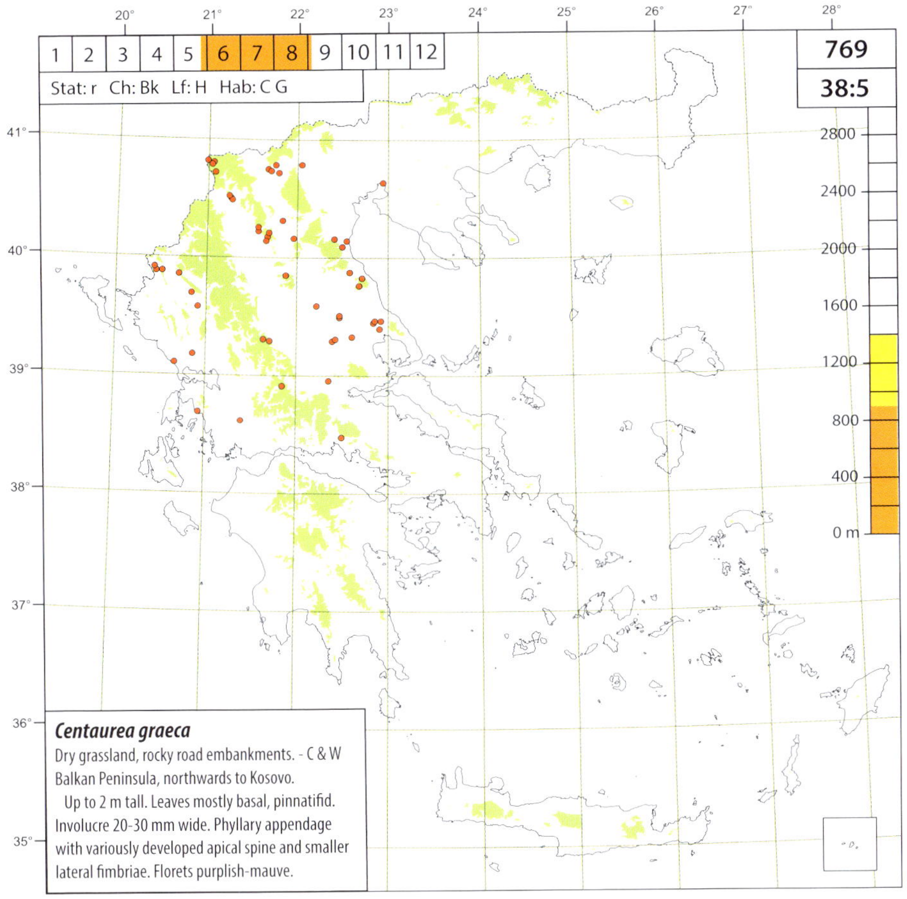

Centaurea graeca

Dry grassland, rocky road embankments. - C & W Balkan Peninsula, northwards to Kosovo.

Up to 2 m tall. Leaves mostly basal, pinnatifid. Involucre 20-30 mm wide. Phyllary appendage with variously developed apical spine and smaller lateral fimbriae. Florets purplish-mauve.

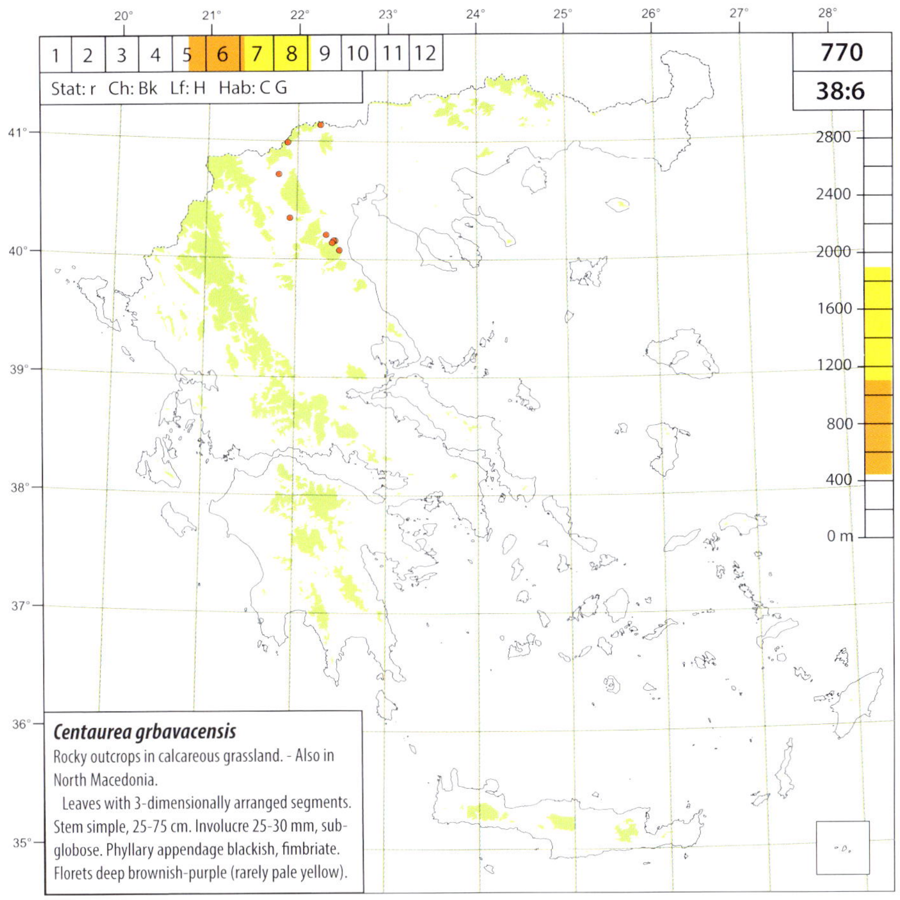

Centaurea grbavacensis

Rocky outcrops in calcareous grassland. - Also in North Macedonia.

Leaves with 3-dimensionally arranged segments. Stem simple, 25-75 cm. Involucre 25-30 mm, subglobose. Phyllary appendage blackish, fimbriate. Florets deep brownish-purple (rarely pale yellow).

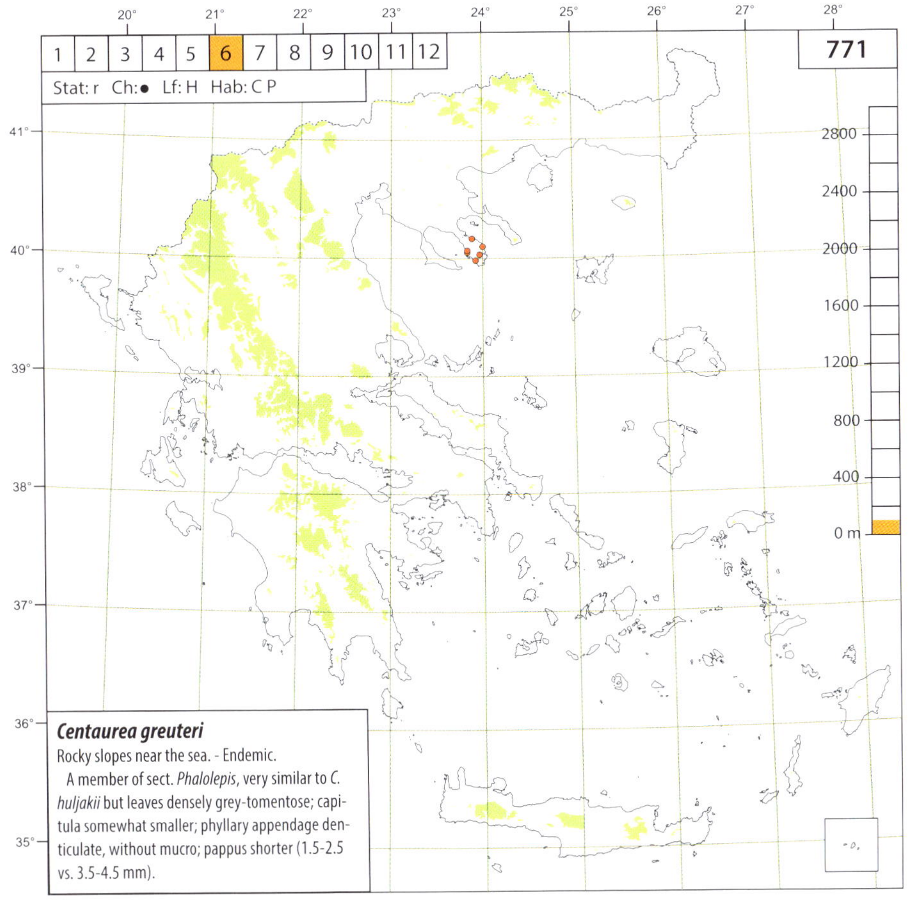

Centaurea greuteri

Rocky slopes near the sea. - Endemic.

A member of sect. *Phalolepis*, very similar to *C. huljakii* but leaves densely grey-tomentose; capitula somewhat smaller; phyllary appendage denticulate, without mucro; pappus shorter (1.5-2.5 vs. 3.5-4.5 mm).

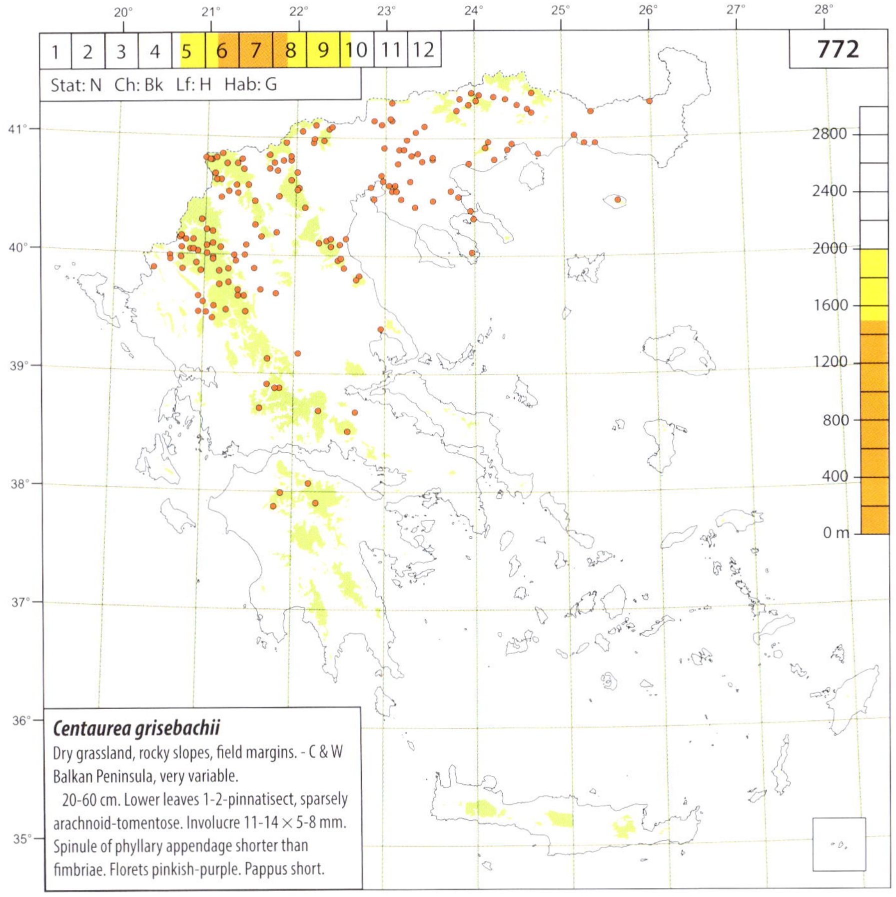

Centaurea grisebachii

Dry grassland, rocky slopes, field margins. - C & W Balkan Peninsula, very variable.

20-60 cm. Lower leaves 1-2-pinnatisect, sparsely arachnoid-tomentose. Involucre 11-14 × 5-8 mm. Spinule of phyllary appendage shorter than fimbriae. Florets pinkish-purple. Pappus short.

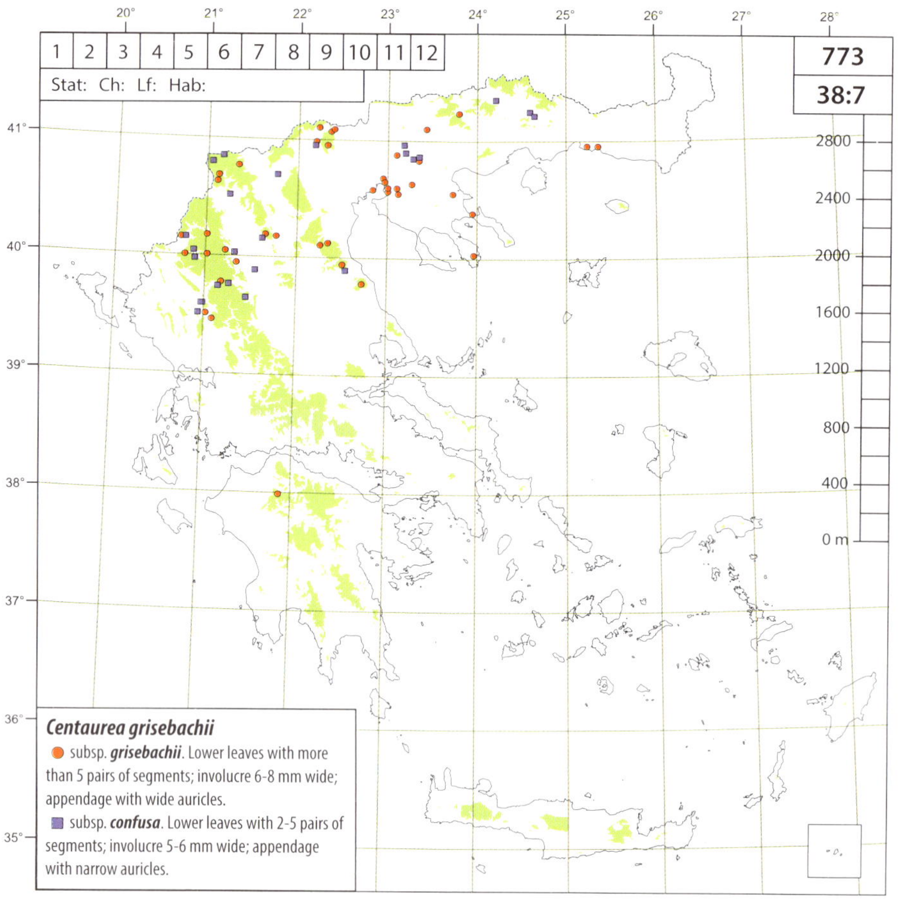

Centaurea grisebachii
● subsp. ***grisebachii***. Lower leaves with more than 5 pairs of segments; involucre 6-8 mm wide; appendage with wide auricles.
■ subsp. ***confusa***. Lower leaves with 2-5 pairs of segments; involucre 5-6 mm wide; appendage with narrow auricles.

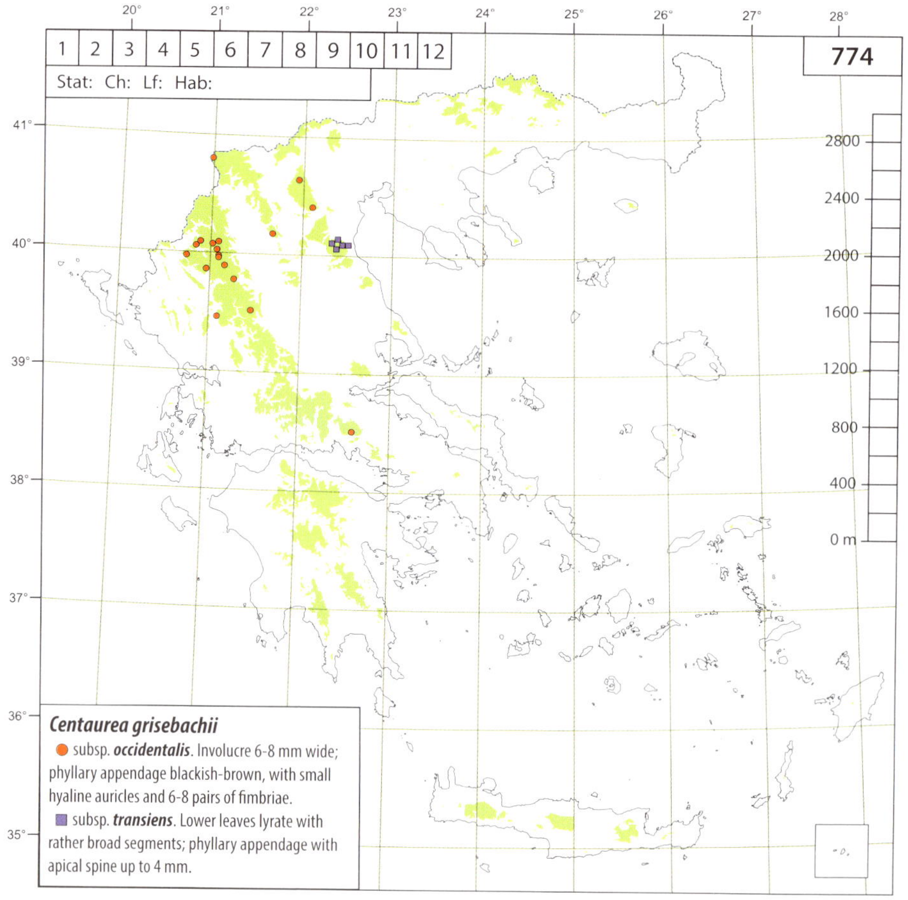

Centaurea grisebachii
● subsp. ***occidentalis***. Involucre 6-8 mm wide; phyllary appendage blackish-brown, with small hyaline auricles and 6-8 pairs of fimbriae.
■ subsp. ***transiens***. Lower leaves lyrate with rather broad segments; phyllary appendage with apical spine up to 4 mm.

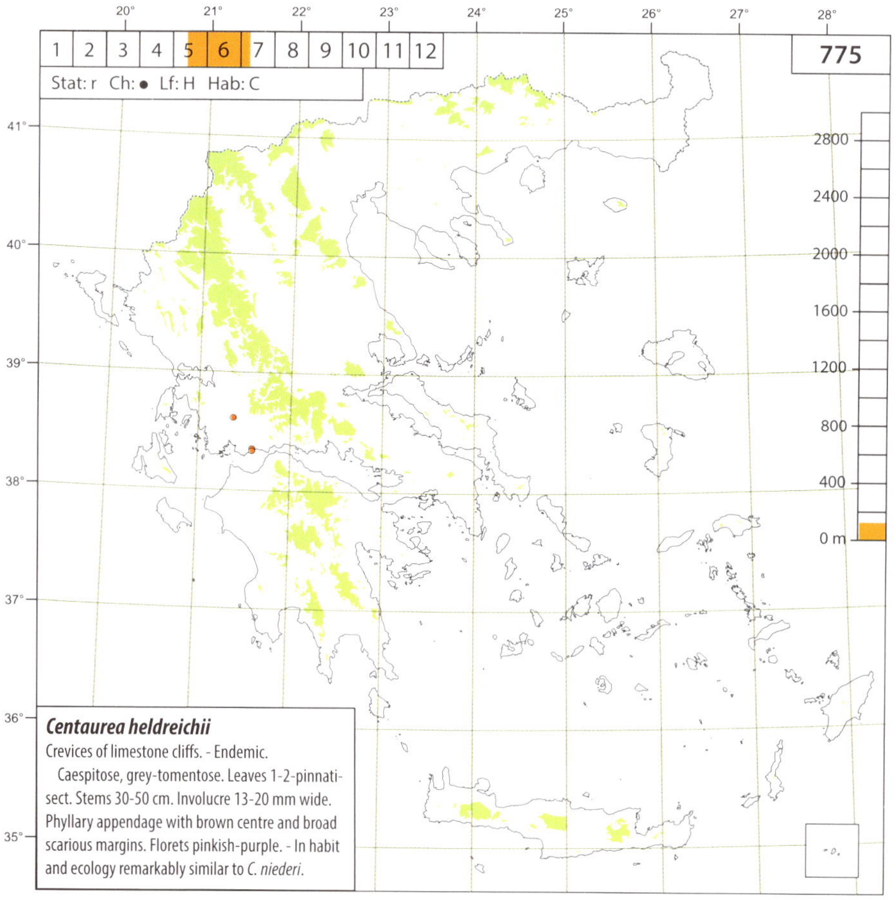

Centaurea heldreichii
Crevices of limestone cliffs. - Endemic.
Caespitose, grey-tomentose. Leaves 1-2-pinnatisect. Stems 30-50 cm. Involucre 13-20 mm wide. Phyllary appendage with brown centre and broad scarious margins. Florets pinkish-purple. - In habit and ecology remarkably similar to *C. niederi*.

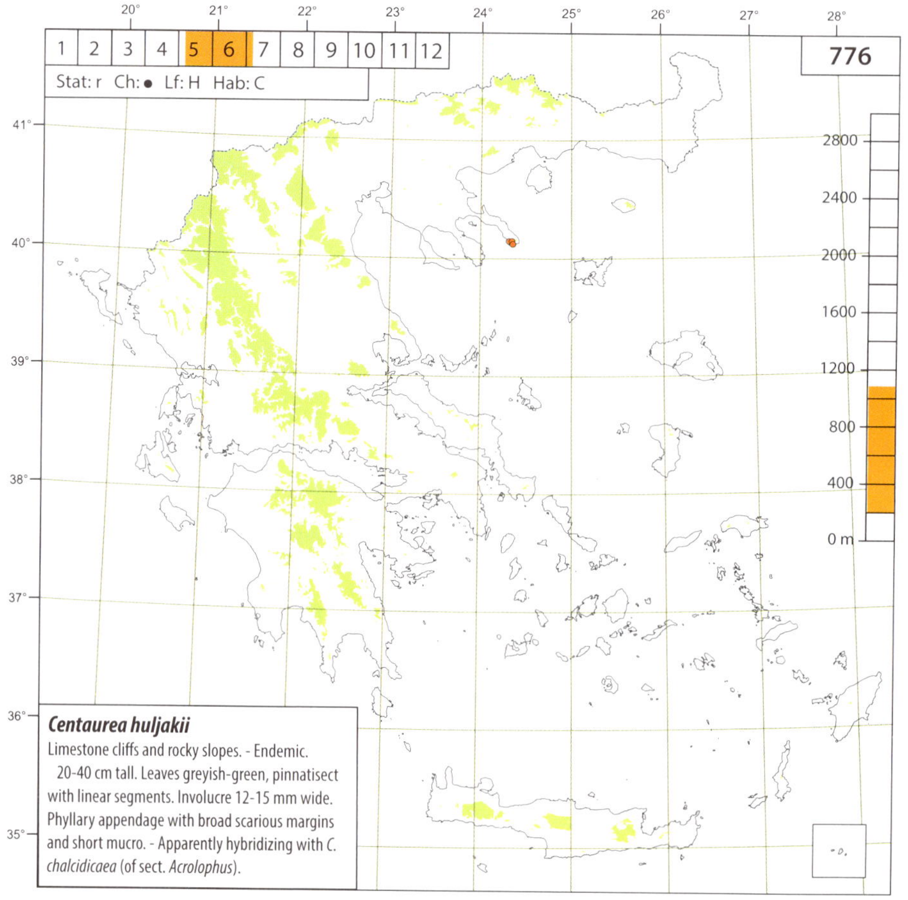

Centaurea huljakii
Limestone cliffs and rocky slopes. - Endemic.
20-40 cm tall. Leaves greyish-green, pinnatisect with linear segments. Involucre 12-15 mm wide. Phyllary appendage with broad scarious margins and short mucro. - Apparently hybridizing with *C. chalcidicaea* (of sect. *Acrolophus*).

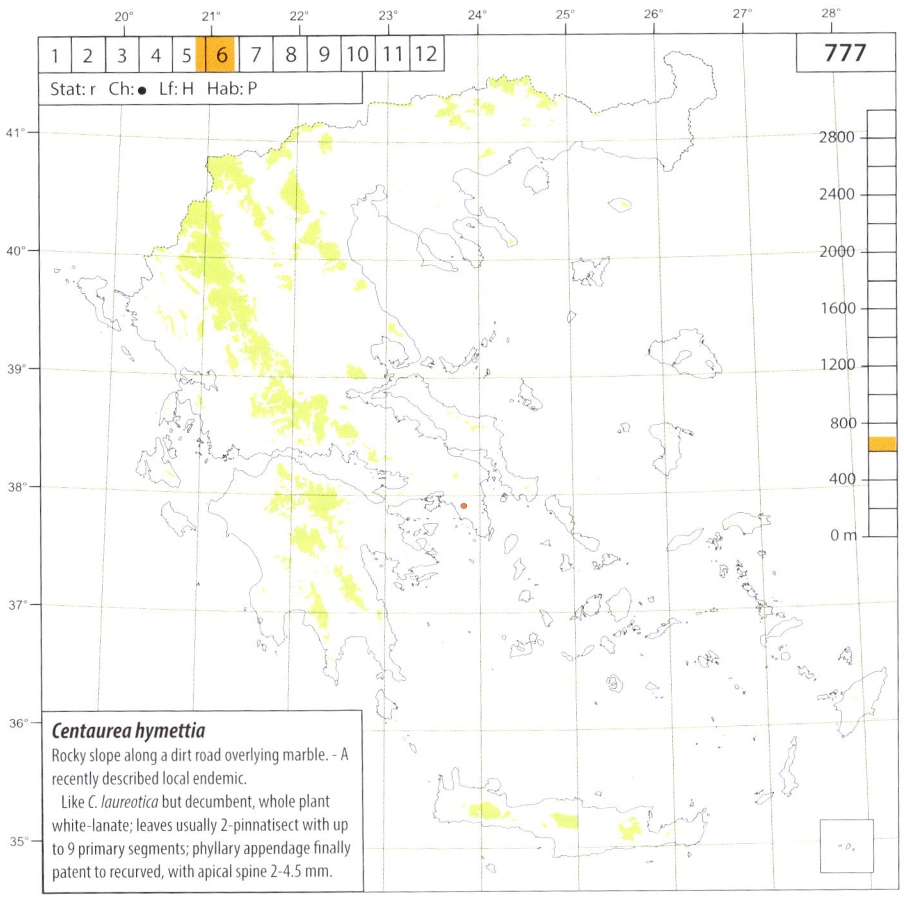

Centaurea hymettia

Rocky slope along a dirt road overlying marble. - A recently described local endemic.

Like *C. laureotica* but decumbent, whole plant white-lanate; leaves usually 2-pinnatisect with up to 9 primary segments; phyllary appendage finally patent to recurved, with apical spine 2-4.5 mm.

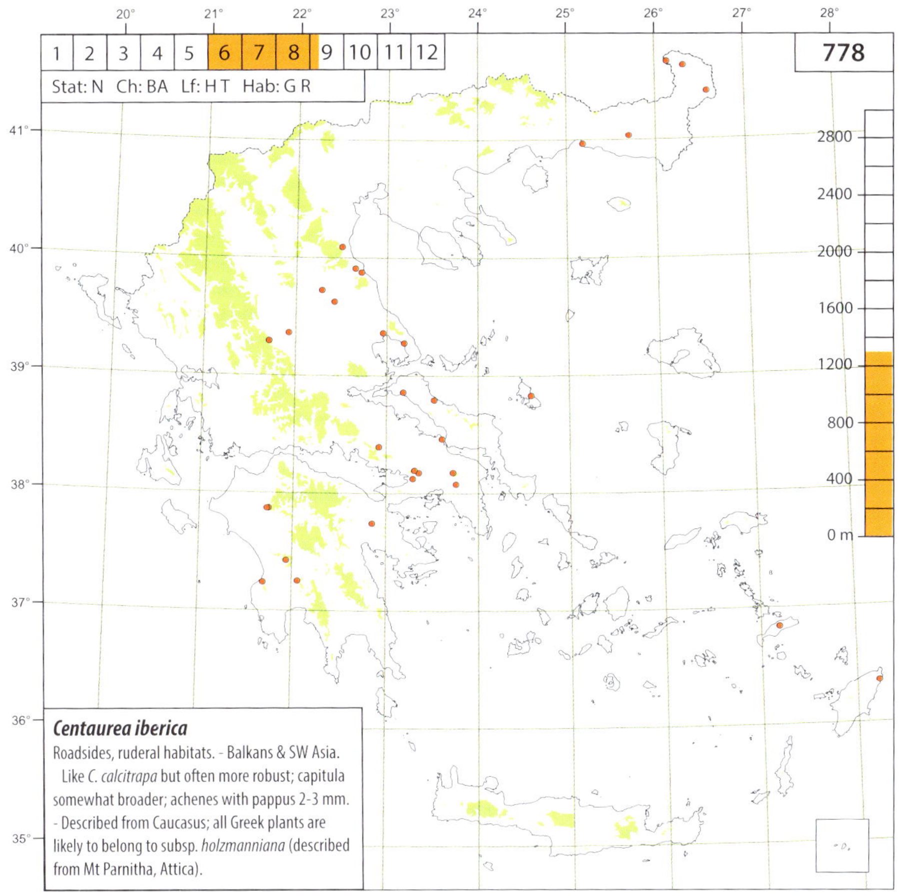

Centaurea iberica

Roadsides, ruderal habitats. - Balkans & SW Asia.

Like *C. calcitrapa* but often more robust; capitula somewhat broader; achenes with pappus 2-3 mm. - Described from Caucasus; all Greek plants are likely to belong to subsp. *holzmanniana* (described from Mt Parnitha, Attica).

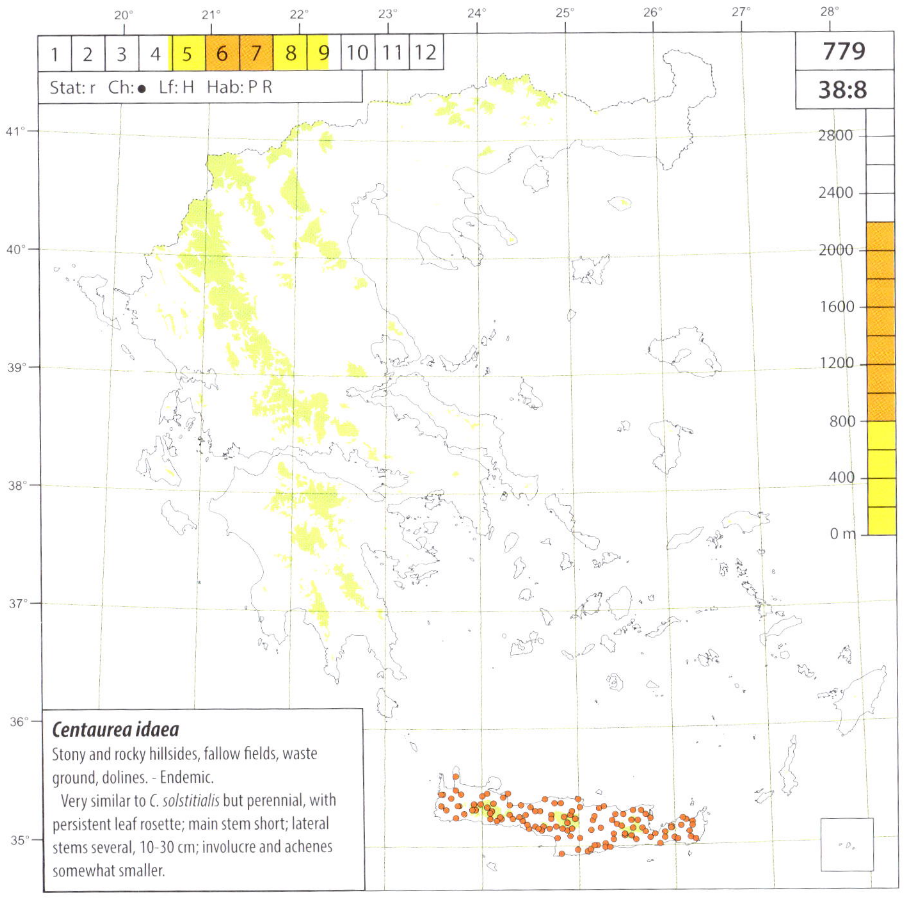

Centaurea idaea

Stony and rocky hillsides, fallow fields, waste ground, dolines. - Endemic.

Very similar to *C. solstitialis* but perennial, with persistent leaf rosette; main stem short; lateral stems several, 10-30 cm; involucre and achenes somewhat smaller.

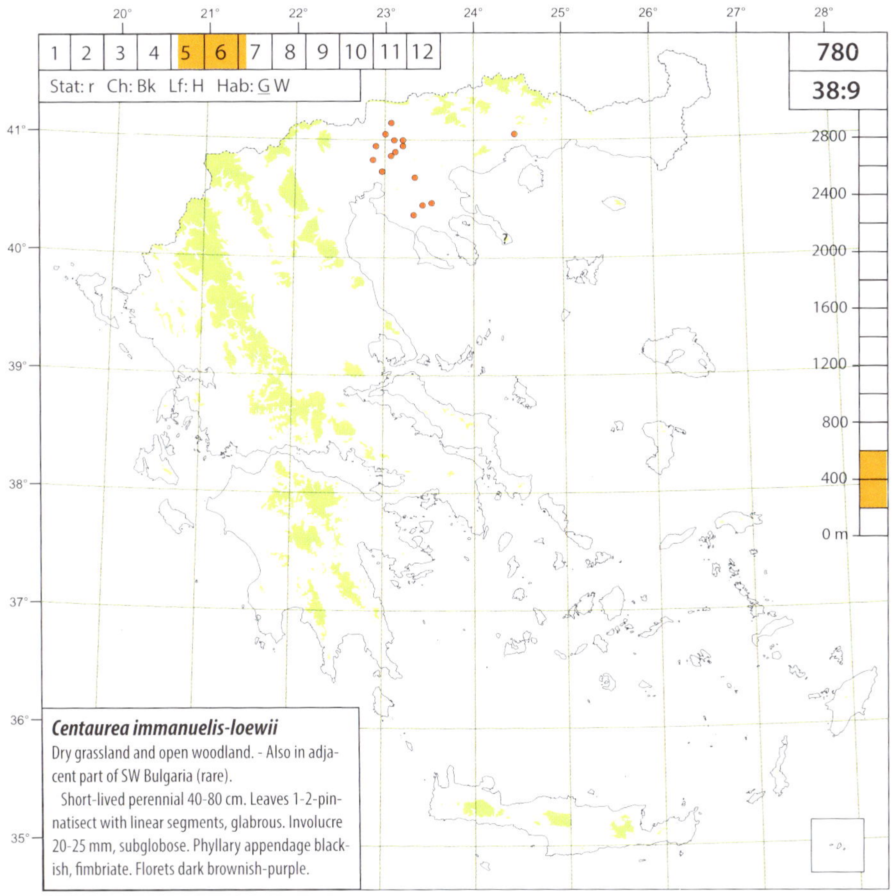

Centaurea immanuelis-loewii

Dry grassland and open woodland. - Also in adjacent part of SW Bulgaria (rare).

Short-lived perennial 40-80 cm. Leaves 1-2-pinnatisect with linear segments, glabrous. Involucre 20-25 mm, subglobose. Phyllary appendage blackish, fimbriate. Florets dark brownish-purple.

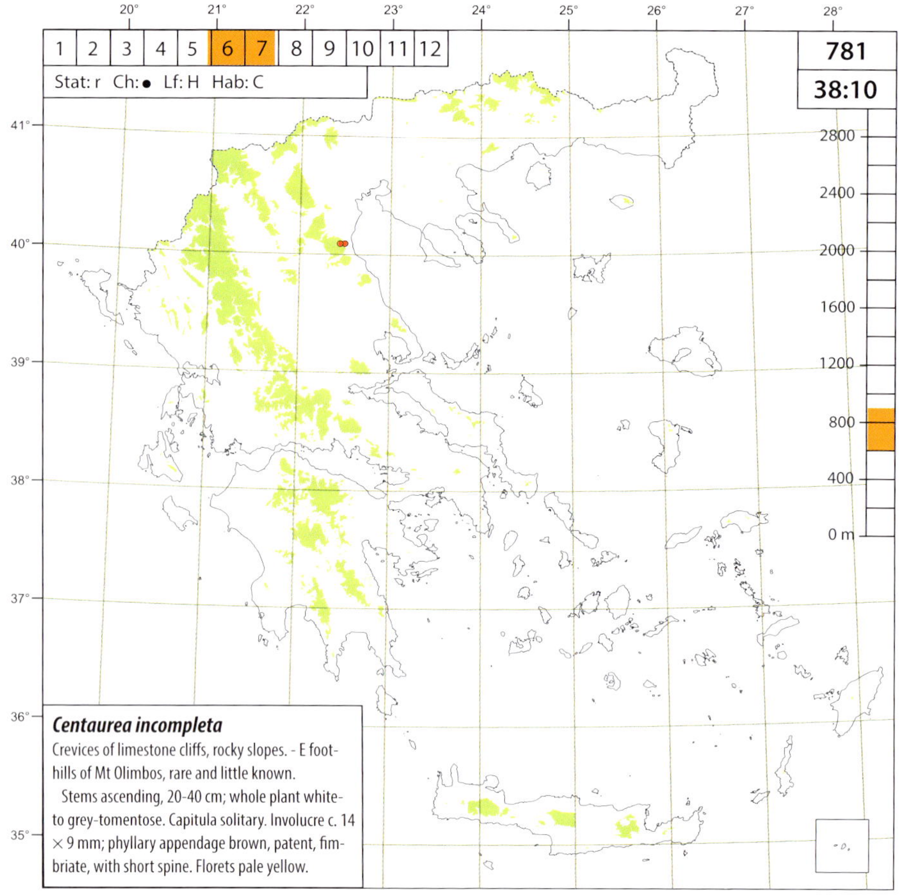

Centaurea incompleta
Crevices of limestone cliffs, rocky slopes. - E foothills of Mt Olimbos, rare and little known.
Stems ascending, 20-40 cm; whole plant white- to grey-tomentose. Capitula solitary. Involucre c. 14 × 9 mm; phyllary appendage brown, patent, fimbriate, with short spine. Florets pale yellow.

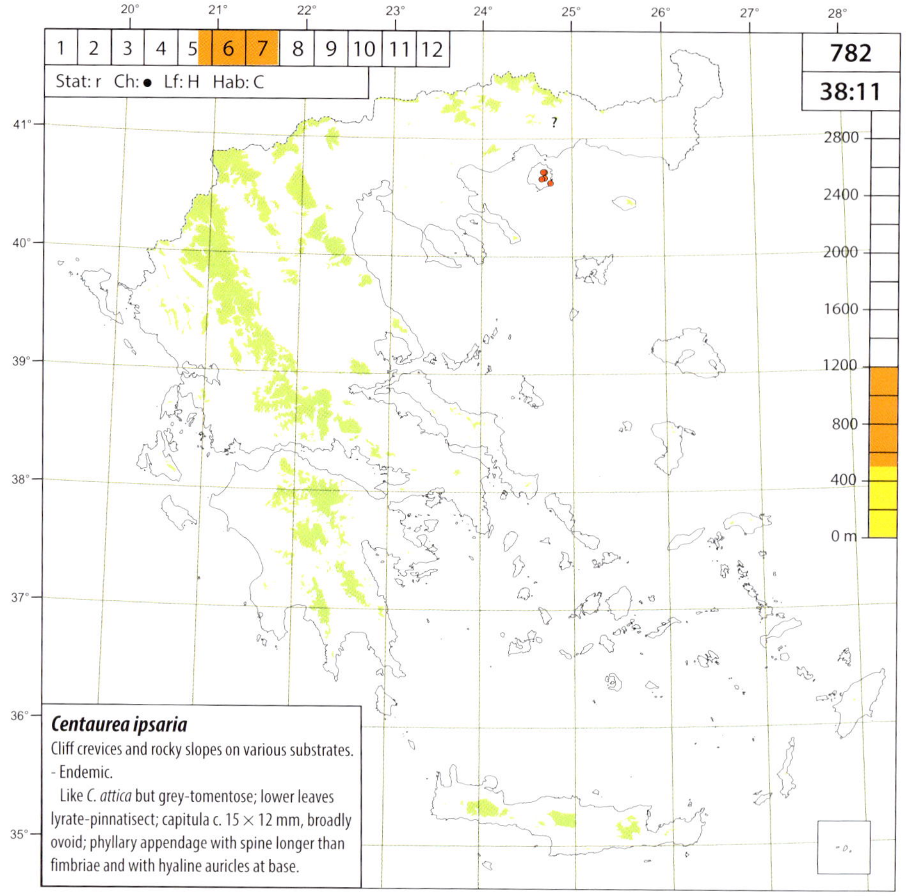

Centaurea ipsaria
Cliff crevices and rocky slopes on various substrates. - Endemic.
Like *C. attica* but grey-tomentose; lower leaves lyrate-pinnatisect; capitula c. 15 × 12 mm, broadly ovoid; phyllary appendage with spine longer than fimbriae and with hyaline auricles at base.

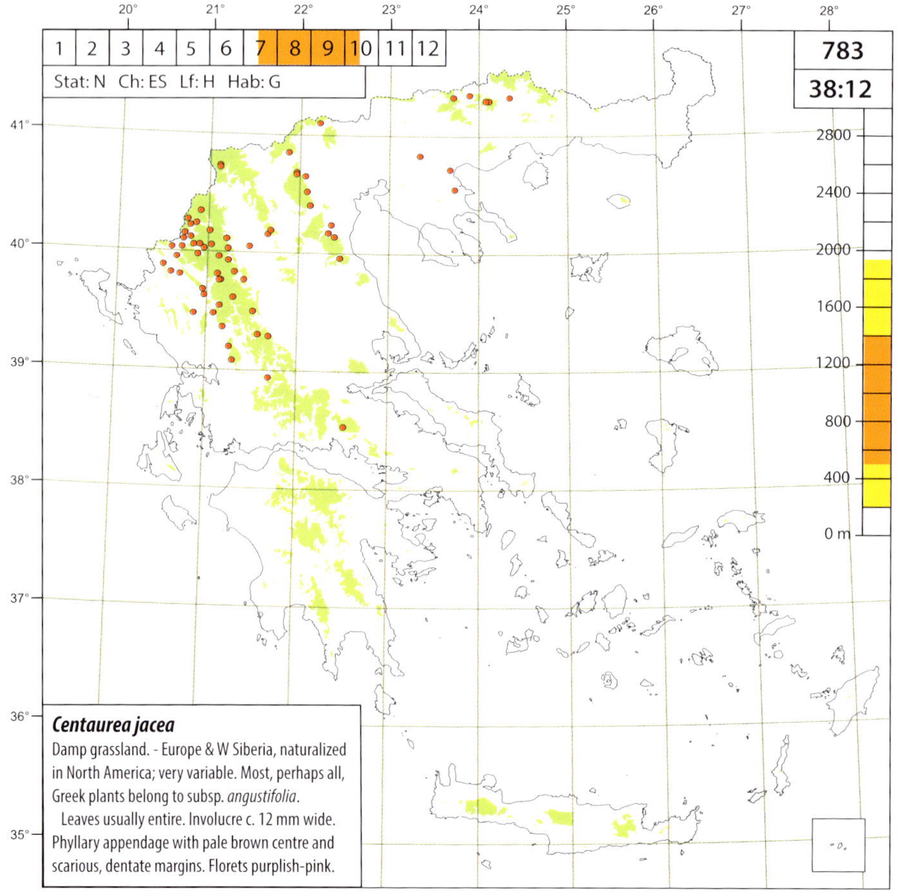

Centaurea jacea
Damp grassland. - Europe & W Siberia, naturalized in North America; very variable. Most, perhaps all, Greek plants belong to subsp. *angustifolia*.
Leaves usually entire. Involucre c. 12 mm wide. Phyllary appendage with pale brown centre and scarious, dentate margins. Florets purplish-pink.

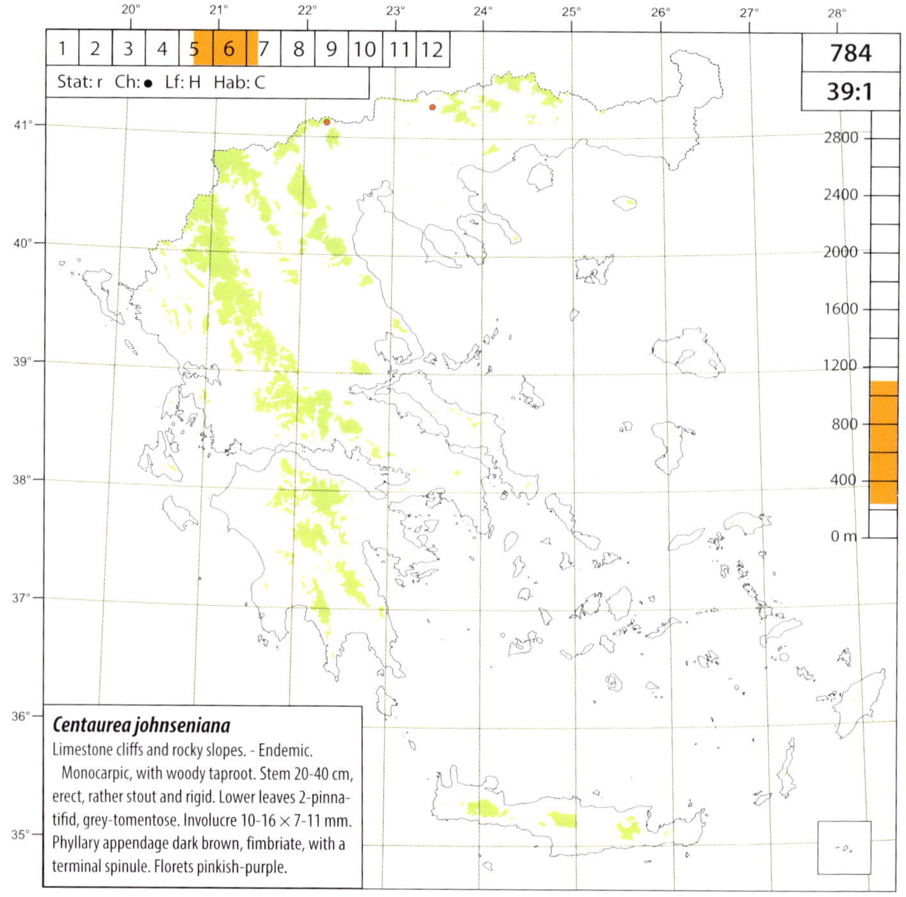

Centaurea johnseniana
Limestone cliffs and rocky slopes. - Endemic.
Monocarpic, with woody taproot. Stem 20-40 cm, erect, rather stout and rigid. Lower leaves 2-pinnatifid, grey-tomentose. Involucre 10-16 × 7-11 mm. Phyllary appendage dark brown, fimbriate, with a terminal spinule. Florets pinkish-purple.

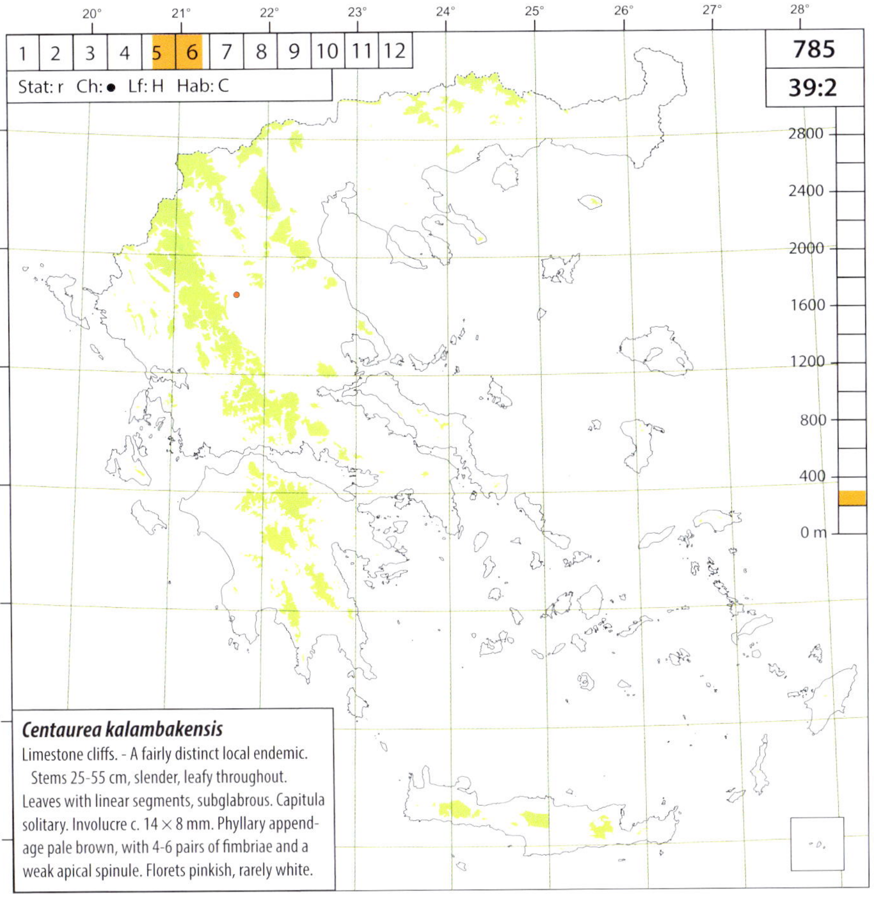

Centaurea kalambakensis

Limestone cliffs. - A fairly distinct local endemic.

Stems 25-55 cm, slender, leafy throughout. Leaves with linear segments, subglabrous. Capitula solitary. Involucre c. 14 × 8 mm. Phyllary appendage pale brown, with 4-6 pairs of fimbriae and a weak apical spinule. Florets pinkish, rarely white.

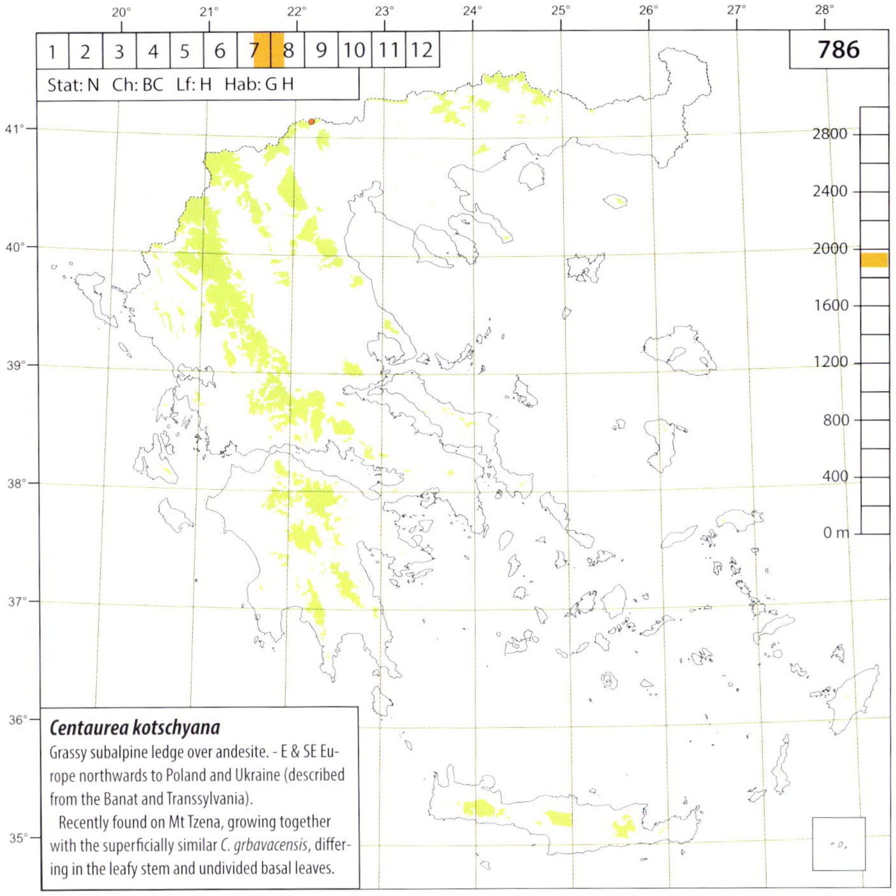

Centaurea kotschyana

Grassy subalpine ledge over andesite. - E & SE Europe northwards to Poland and Ukraine (described from the Banat and Transsylvania).

Recently found on Mt Tzena, growing together with the superficially similar *C. grbavacensis*, differing in the leafy stem and undivided basal leaves.

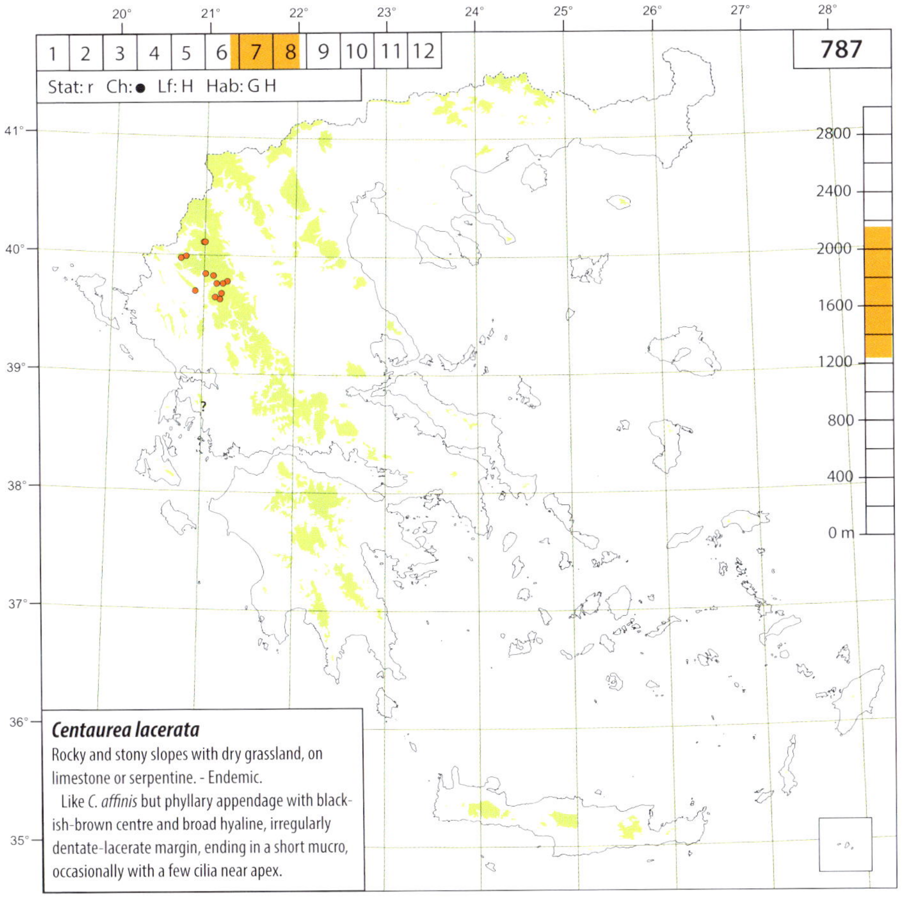

Centaurea lacerata

Rocky and stony slopes with dry grassland, on limestone or serpentine. - Endemic.

Like *C. affinis* but phyllary appendage with blackish-brown centre and broad hyaline, irregularly dentate-lacerate margin, ending in a short mucro, occasionally with a few cilia near apex.

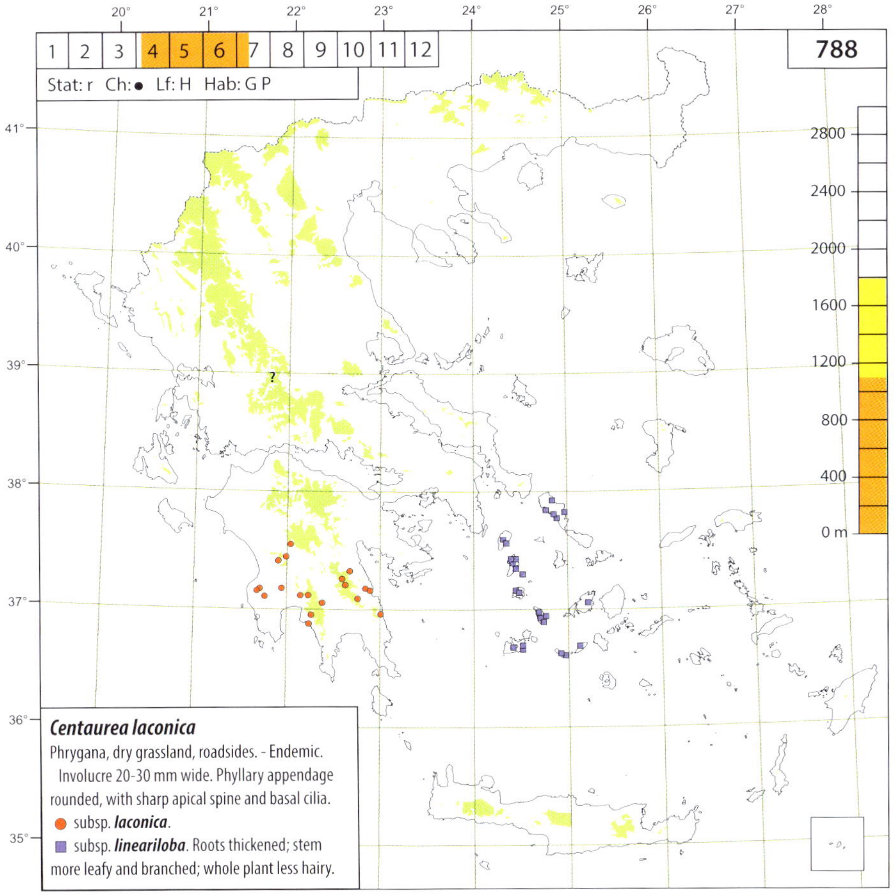

Centaurea laconica

Phrygana, dry grassland, roadsides. - Endemic.

Involucre 20-30 mm wide. Phyllary appendage rounded, with sharp apical spine and basal cilia.

● subsp. ***laconica***.

■ subsp. ***leariloba***. Roots thickened; stem more leafy and branched; whole plant less hairy.

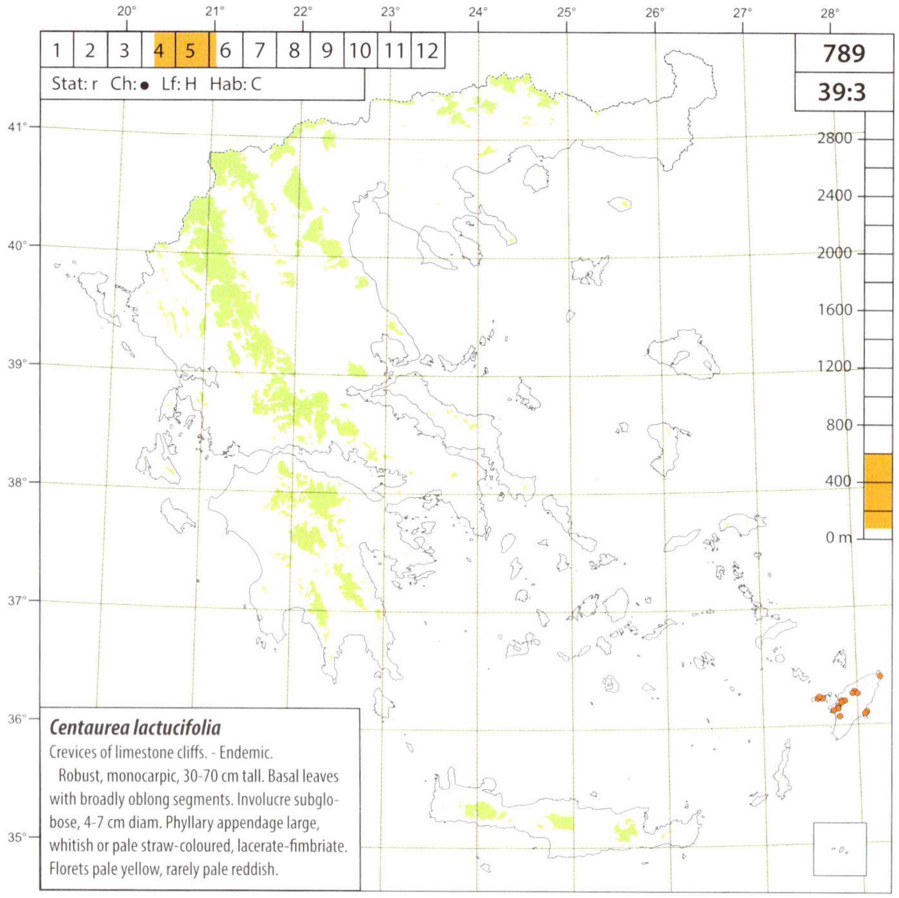

Centaurea lactucifolia

Crevices of limestone cliffs. - Endemic.

Robust, monocarpic, 30-70 cm tall. Basal leaves with broadly oblong segments. Involucre subglobose, 4-7 cm diam. Phyllary appendage large, whitish or pale straw-coloured, lacerate-fimbriate. Florets pale yellow, rarely pale reddish.

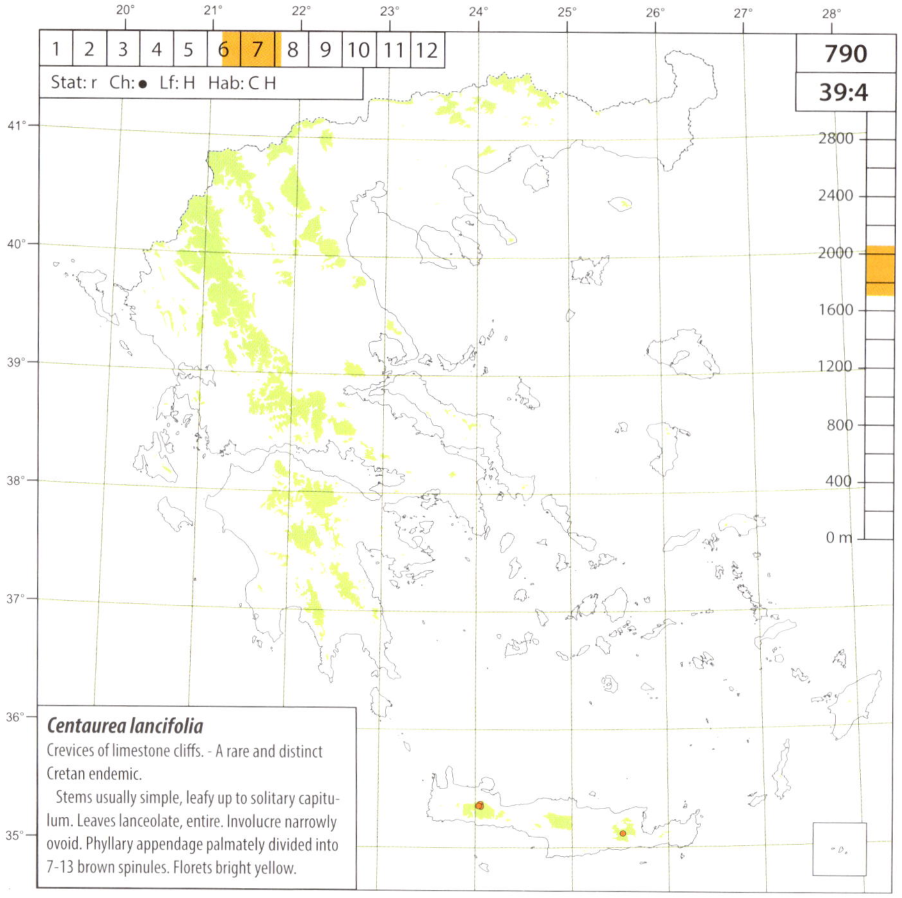

Centaurea lancifolia

Crevices of limestone cliffs. - A rare and distinct Cretan endemic.

Stems usually simple, leafy up to solitary capitulum. Leaves lanceolate, entire. Involucre narrowly ovoid. Phyllary appendage palmately divided into 7-13 brown spinules. Florets bright yellow.

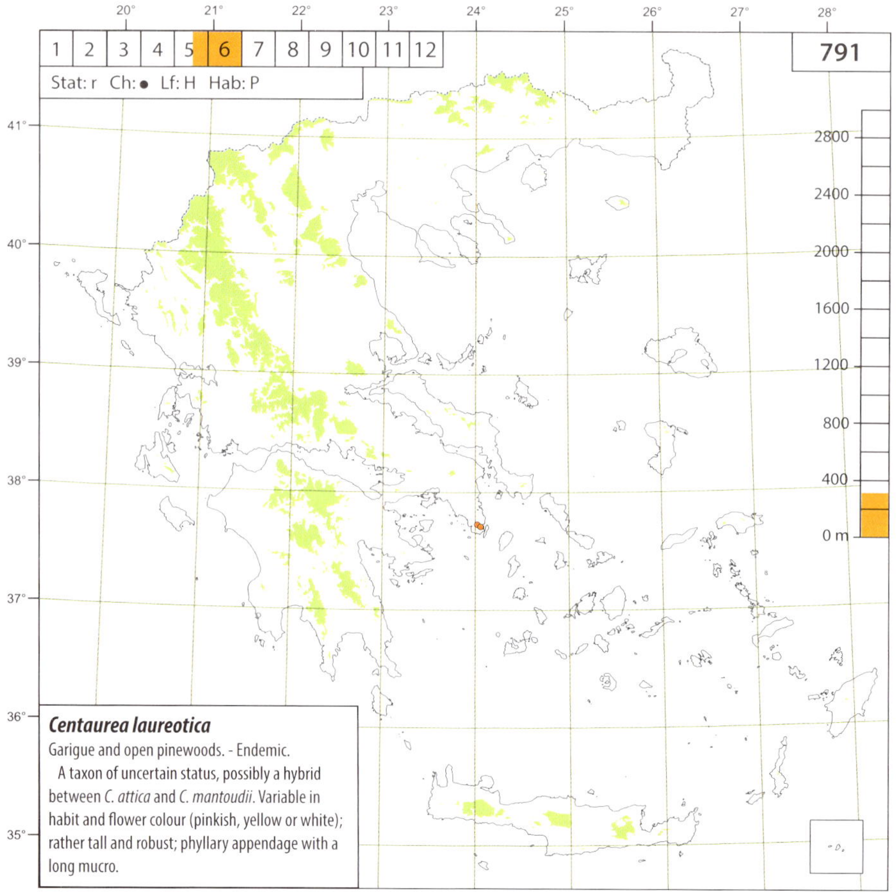

Centaurea laureotica

Garigue and open pinewoods. - Endemic.

A taxon of uncertain status, possibly a hybrid between *C. attica* and *C. mantoudii*. Variable in habit and flower colour (pinkish, yellow or white); rather tall and robust; phyllary appendage with a long mucro.

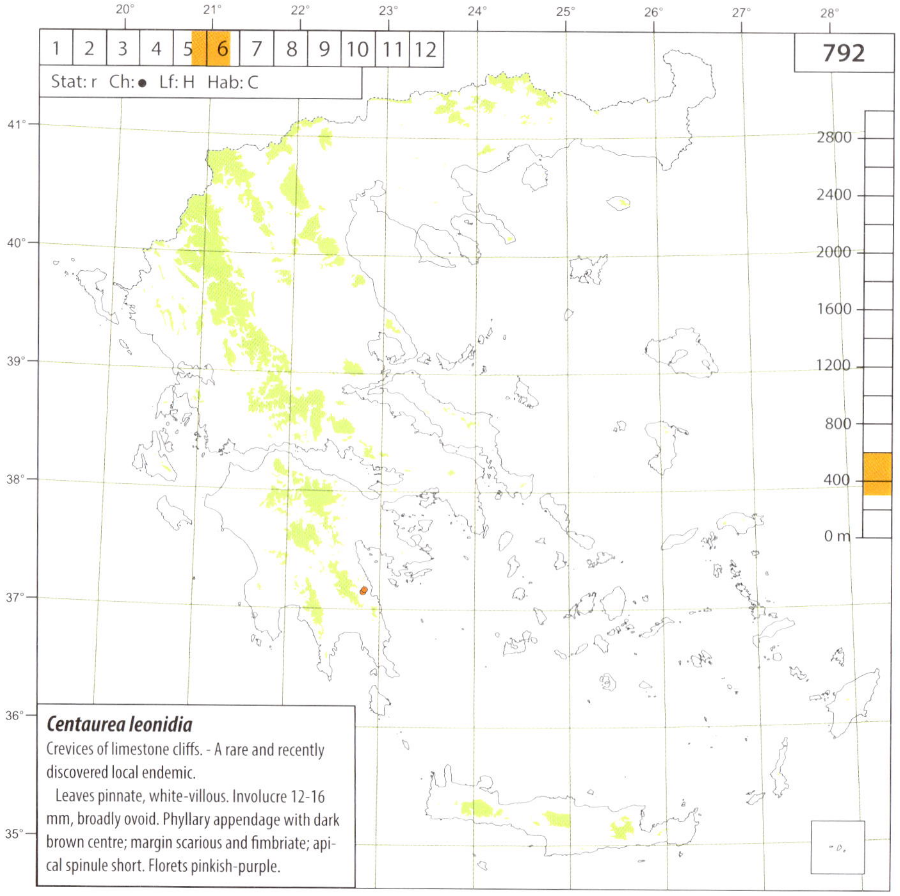

Centaurea leonidia

Crevices of limestone cliffs. - A rare and recently discovered local endemic.

Leaves pinnate, white-villous. Involucre 12-16 mm, broadly ovoid. Phyllary appendage with dark brown centre; margin scarious and fimbriate; apical spinule short. Florets pinkish-purple.

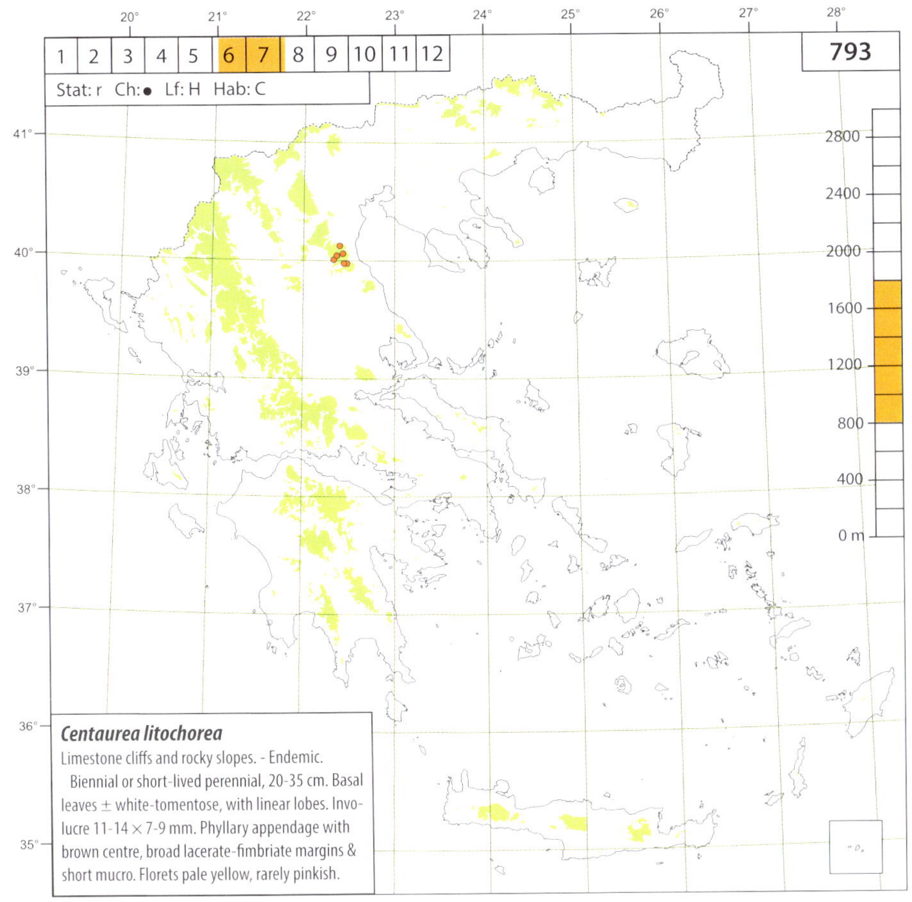

Centaurea litochorea

Limestone cliffs and rocky slopes. - Endemic.

Biennial or short-lived perennial, 20-35 cm. Basal leaves ± white-tomentose, with linear lobes. Involucre 11-14 × 7-9 mm. Phyllary appendage with brown centre, broad lacerate-fimbriate margins & short mucro. Florets pale yellow, rarely pinkish.

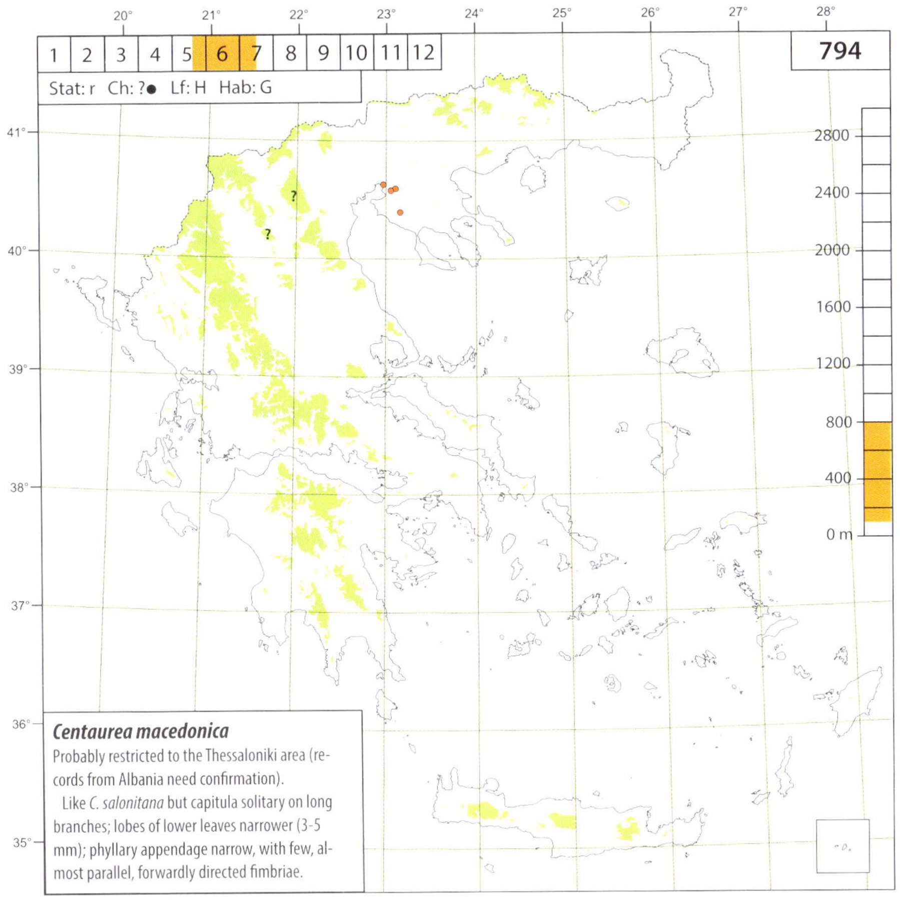

Centaurea macedonica

Probably restricted to the Thessaloniki area (records from Albania need confirmation).

Like *C. salonitana* but capitula solitary on long branches; lobes of lower leaves narrower (3-5 mm); phyllary appendage narrow, with few, almost parallel, forwardly directed fimbriae.

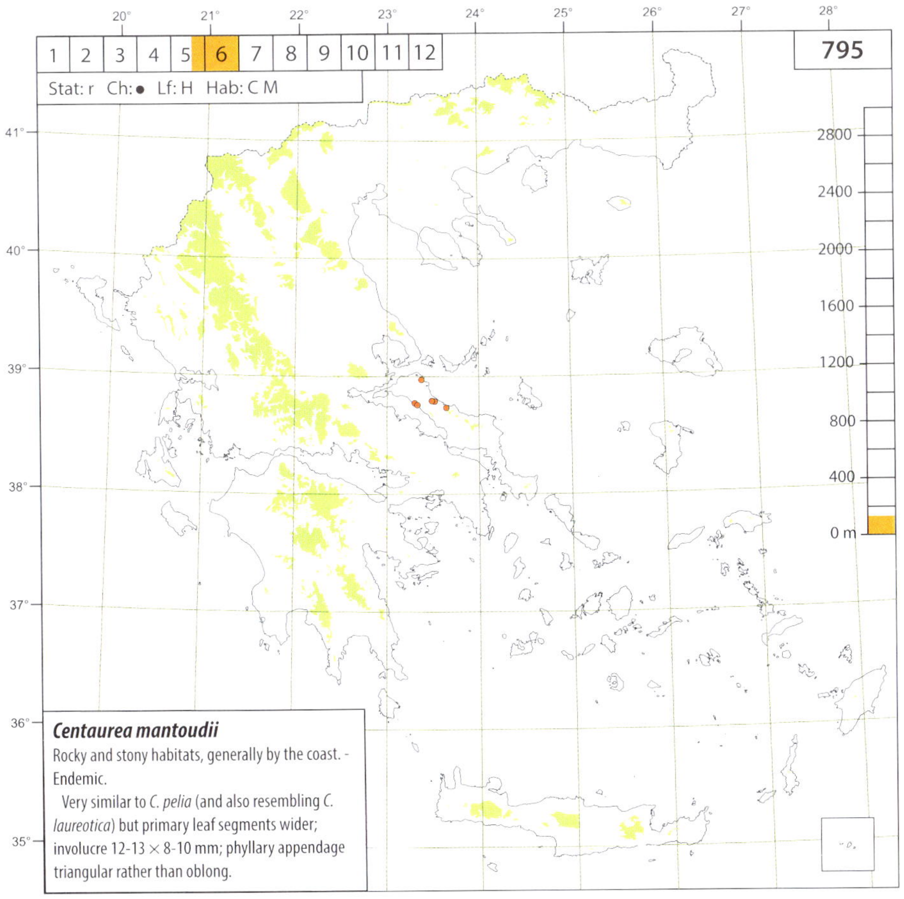

Centaurea mantoudii

Rocky and stony habitats, generally by the coast. - Endemic.

Very similar to *C. pelia* (and also resembling *C. laureotica*) but primary leaf segments wider; involucre 12-13 × 8-10 mm; phyllary appendage triangular rather than oblong.

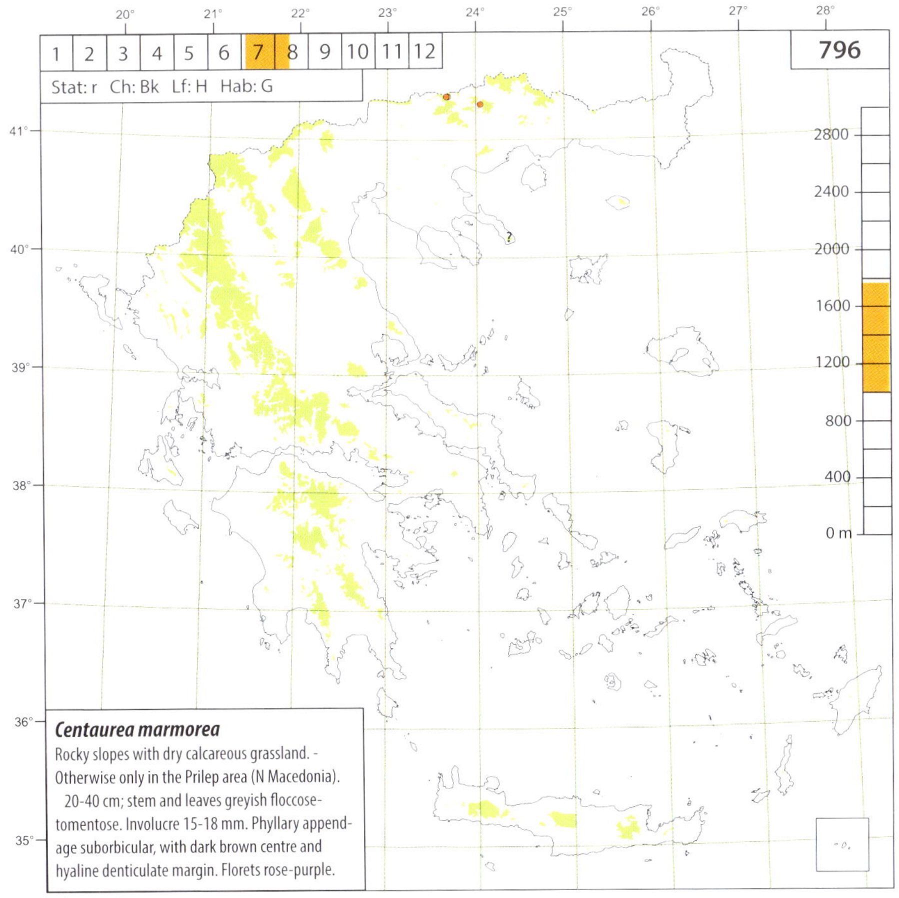

Centaurea marmorea

Rocky slopes with dry calcareous grassland. - Otherwise only in the Prilep area (N Macedonia).

20-40 cm; stem and leaves greyish floccose-tomentose. Involucre 15-18 mm. Phyllary appendage suborbicular, with dark brown centre and hyaline denticulate margin. Florets rose-purple.

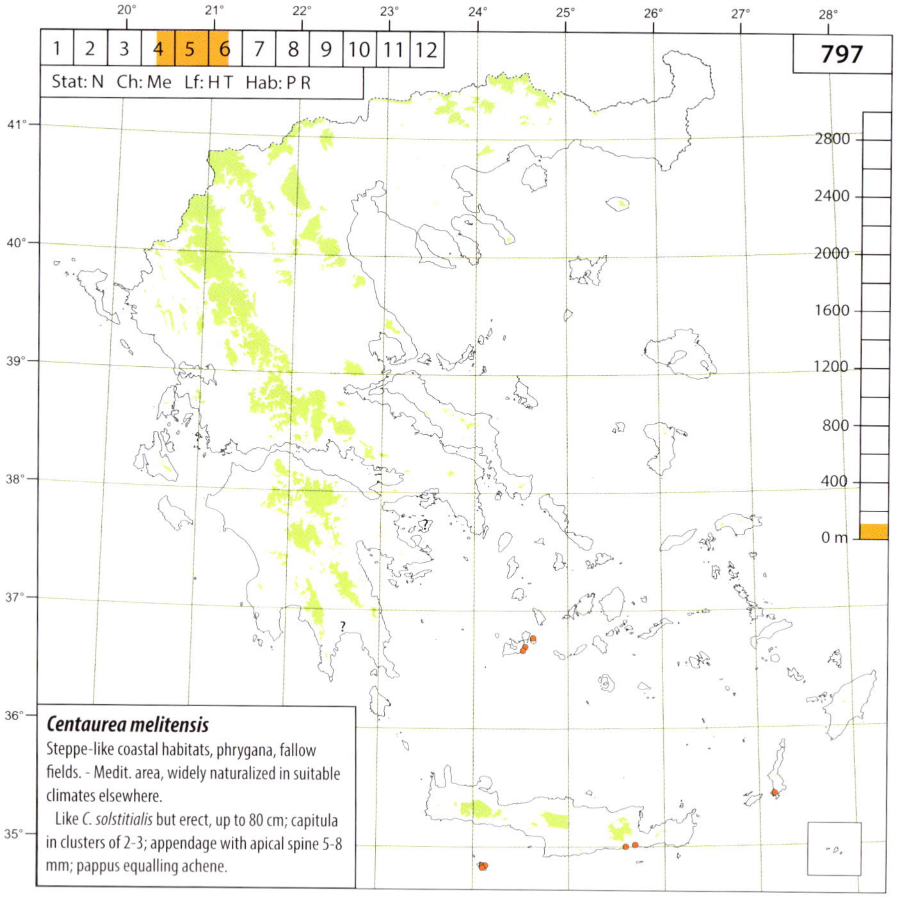

Centaurea melitensis

Steppe-like coastal habitats, phrygana, fallow fields. - Medit. area, widely naturalized in suitable climates elsewhere.

Like *C. solstitialis* but erect, up to 80 cm; capitula in clusters of 2-3; appendage with apical spine 5-8 mm; pappus equalling achene.

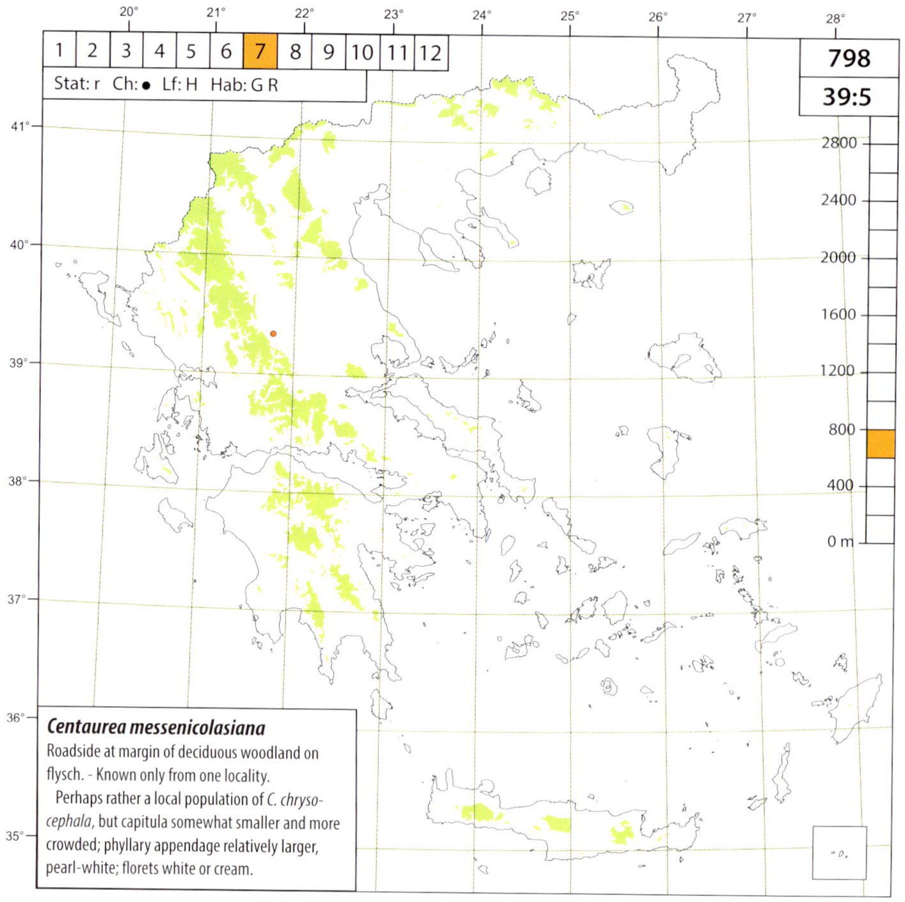

Centaurea messenicolasiana

Roadside at margin of deciduous woodland on flysch. - Known only from one locality.

Perhaps rather a local population of *C. chrysocephala*, but capitula somewhat smaller and more crowded; phyllary appendage relatively larger, pearl-white; florets white or cream.

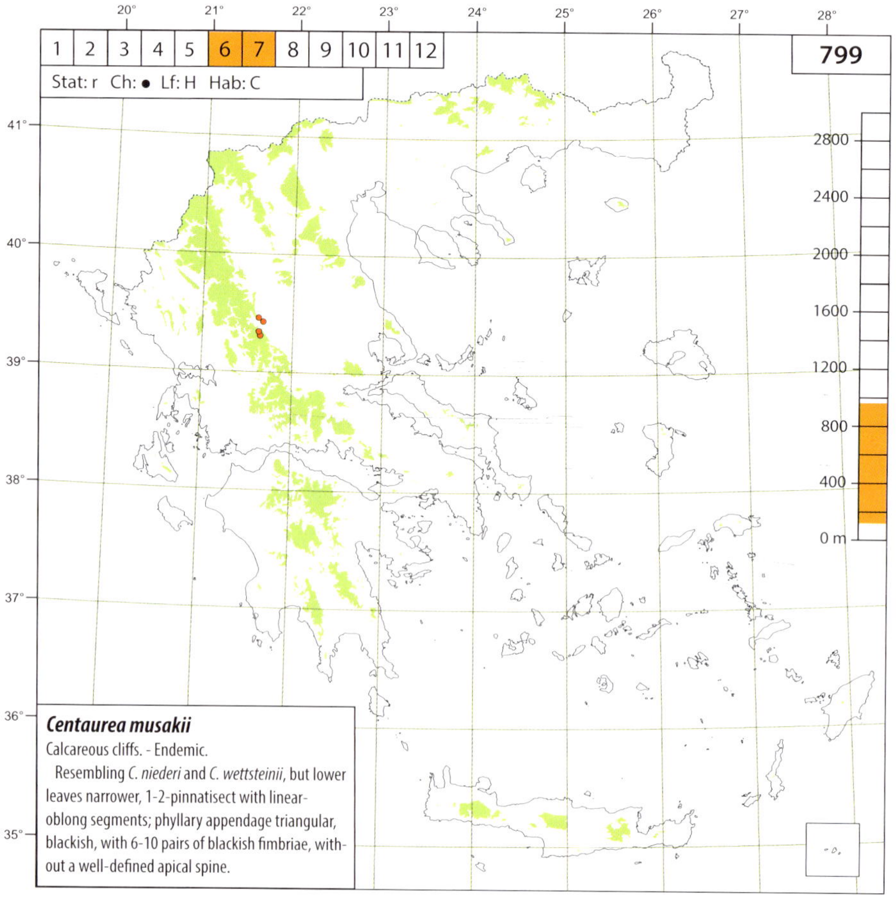

Centaurea musakii

Calcareous cliffs. - Endemic.

Resembling *C. niederi* and *C. wettsteinii*, but lower leaves narrower, 1-2-pinnatisect with linear-oblong segments; phyllary appendage triangular, blackish, with 6-10 pairs of blackish fimbriae, without a well-defined apical spine.

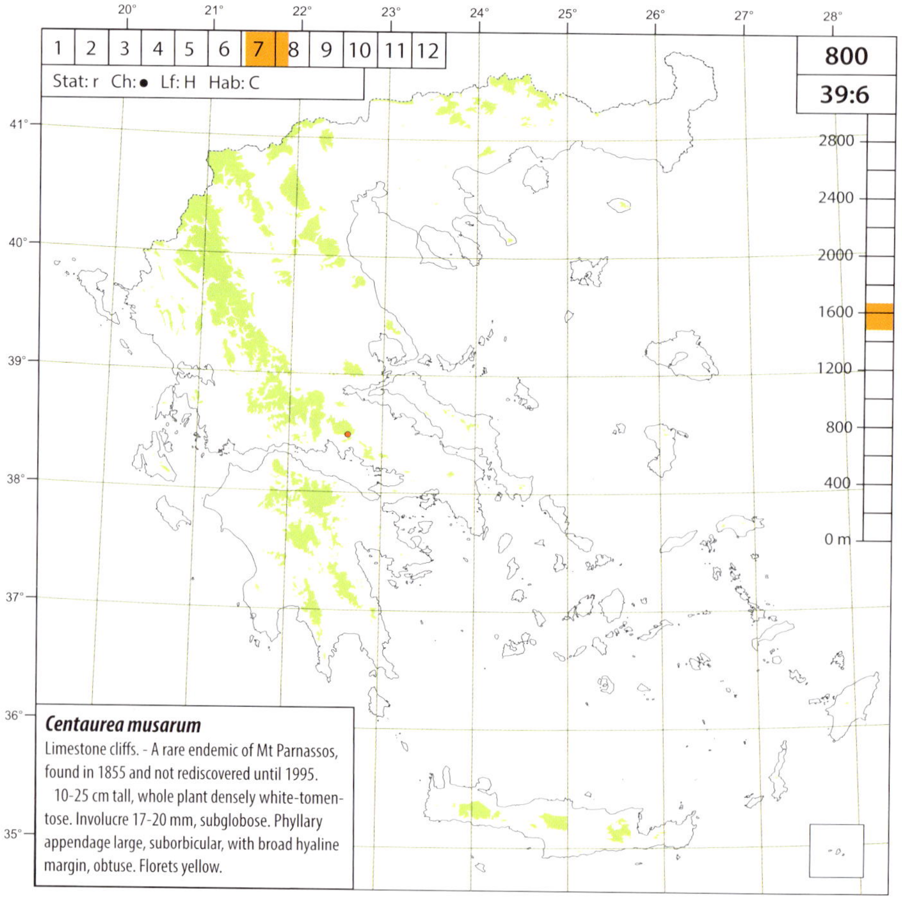

Centaurea musarum

Limestone cliffs. - A rare endemic of Mt Parnassos, found in 1855 and not rediscovered until 1995.

10-25 cm tall, whole plant densely white-tomentose. Involucre 17-20 mm, subglobose. Phyllary appendage large, suborbicular, with broad hyaline margin, obtuse. Florets yellow.

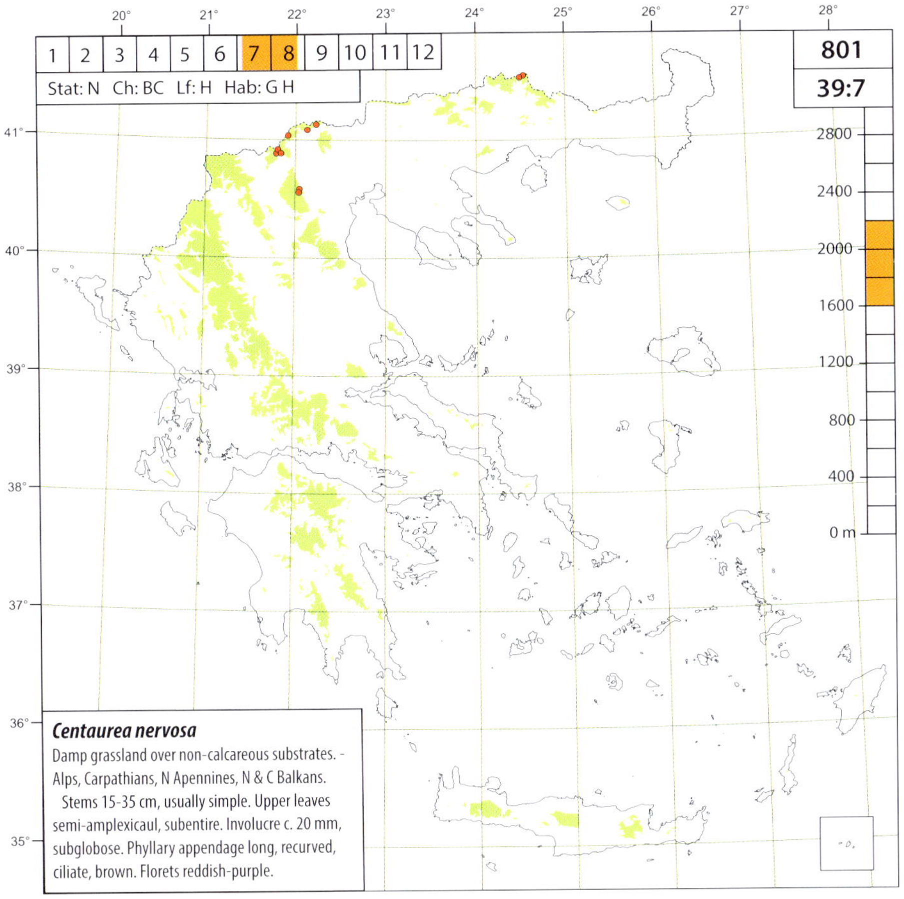

Centaurea nervosa

Damp grassland over non-calcareous substrates. - Alps, Carpathians, N Apennines, N & C Balkans.

Stems 15-35 cm, usually simple. Upper leaves semi-amplexicaul, subentire. Involucre c. 20 mm, subglobose. Phyllary appendage long, recurved, ciliate, brown. Florets reddish-purple.

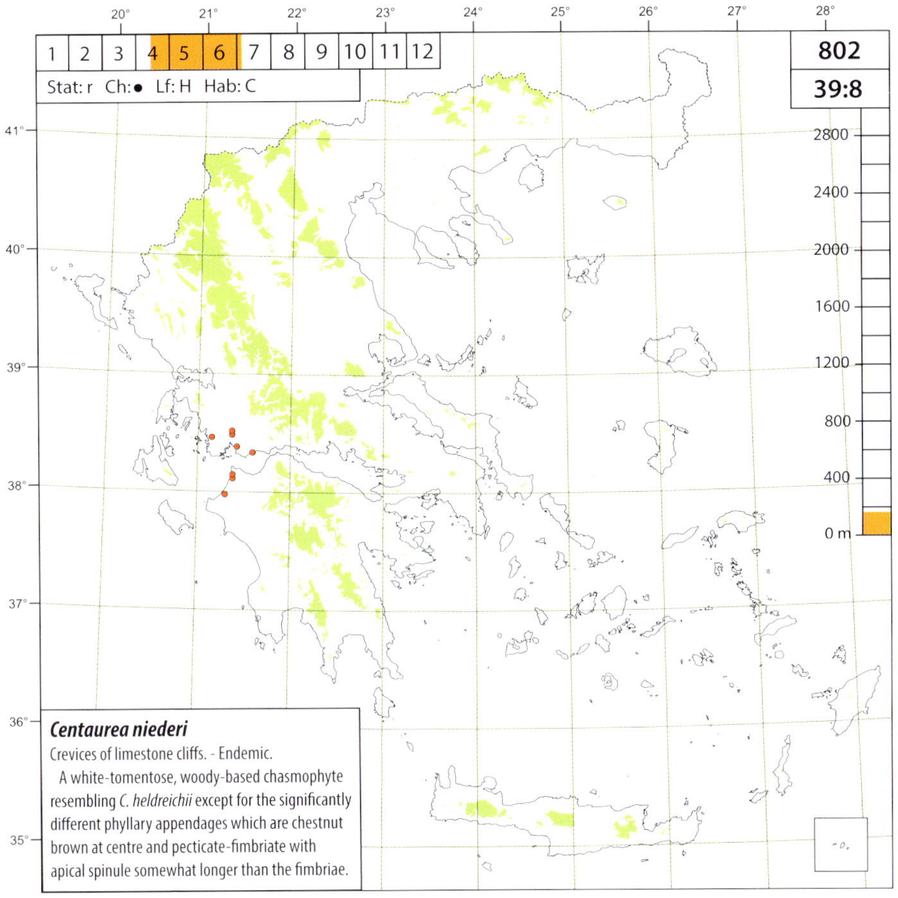

Centaurea niederi

Crevices of limestone cliffs. - Endemic.

A white-tomentose, woody-based chasmophyte resembling *C. heldreichii* except for the significantly different phyllary appendages which are chestnut brown at centre and pecticate-fimbriate with apical spinule somewhat longer than the fimbriae.

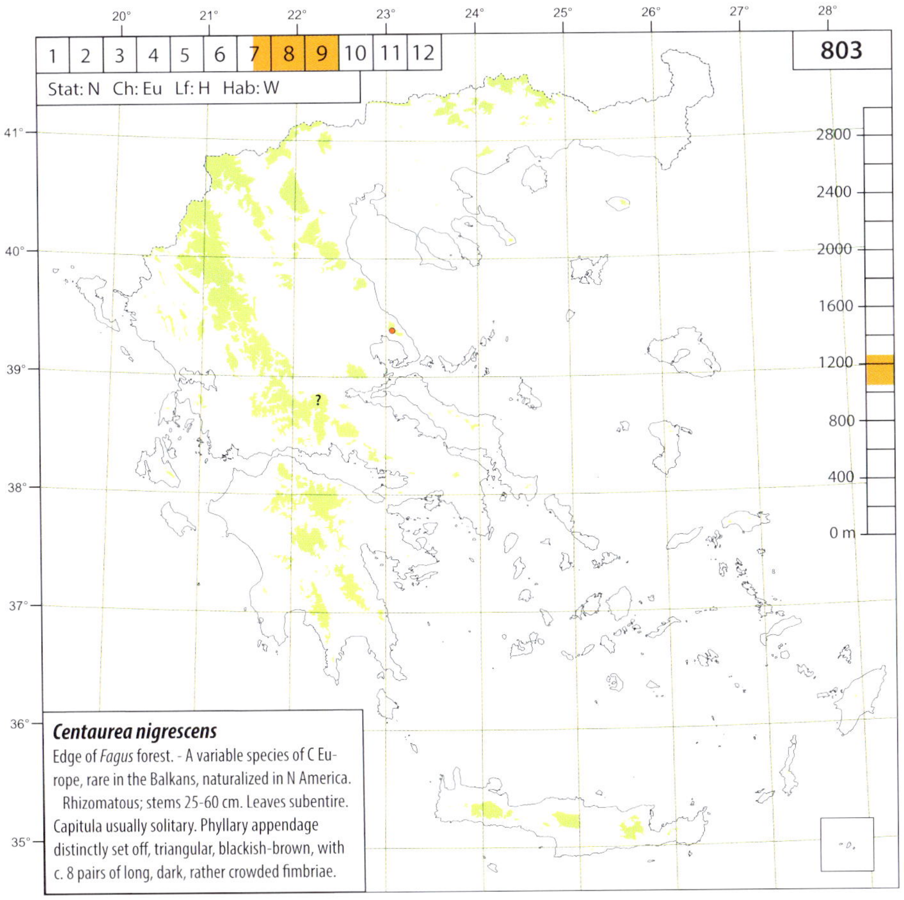

Centaurea nigrescens

Edge of *Fagus* forest. - A variable species of C Europe, rare in the Balkans, naturalized in N America.

Rhizomatous; stems 25-60 cm. Leaves subentire. Capitula usually solitary. Phyllary appendage distinctly set off, triangular, blackish-brown, with c. 8 pairs of long, dark, rather crowded fimbriae.

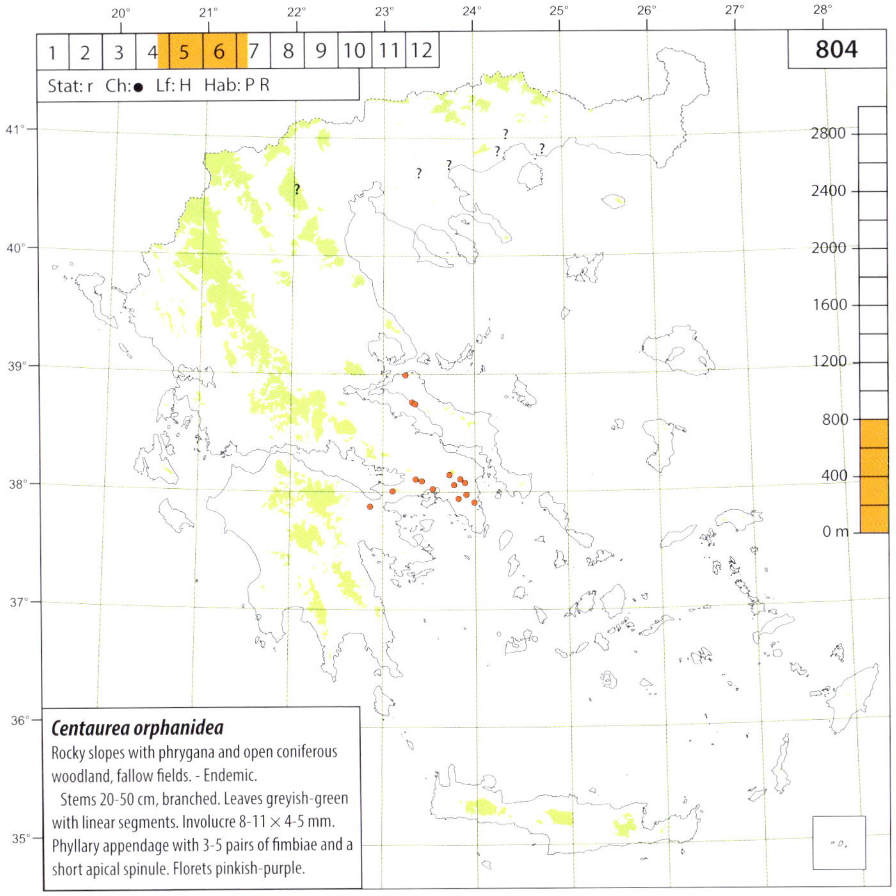

Centaurea orphanidea

Rocky slopes with phrygana and open coniferous woodland, fallow fields. - Endemic.

Stems 20-50 cm, branched. Leaves greyish-green with linear segments. Involucre 8-11 × 4-5 mm. Phyllary appendage with 3-5 pairs of fimbiae and a short apical spinule. Florets pinkish-purple.

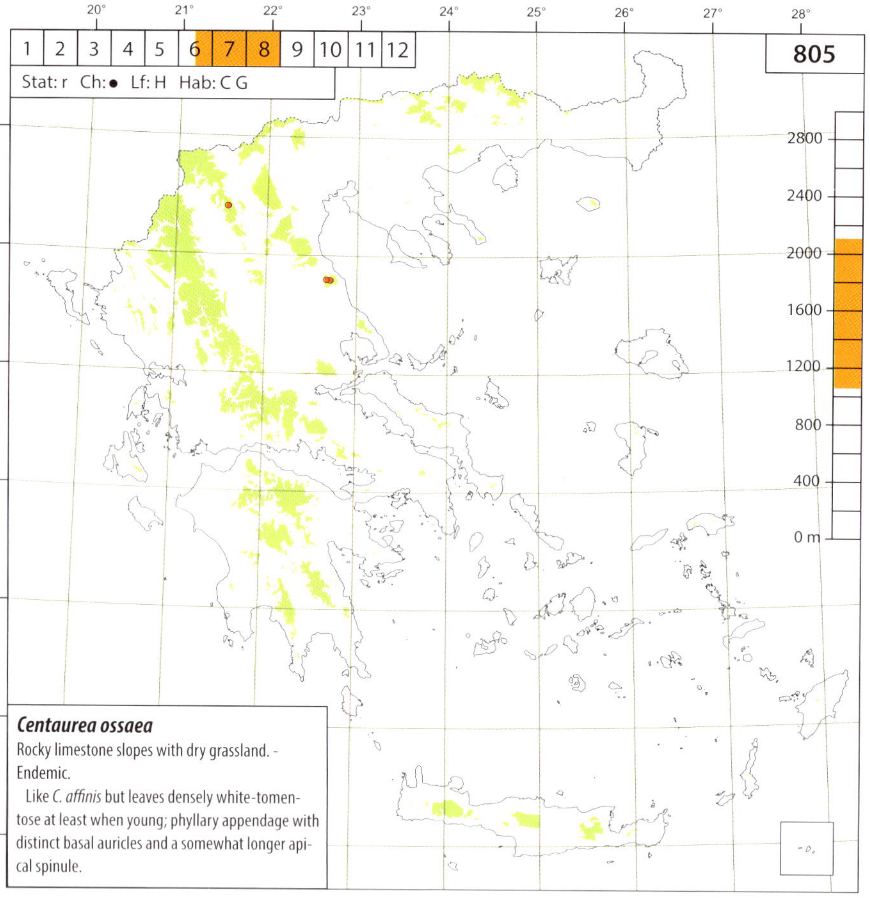

Centaurea ossaea
Rocky limestone slopes with dry grassland. - Endemic.
Like *C. affinis* but leaves densely white-tomentose at least when young; phyllary appendage with distinct basal auricles and a somewhat longer apical spinule.

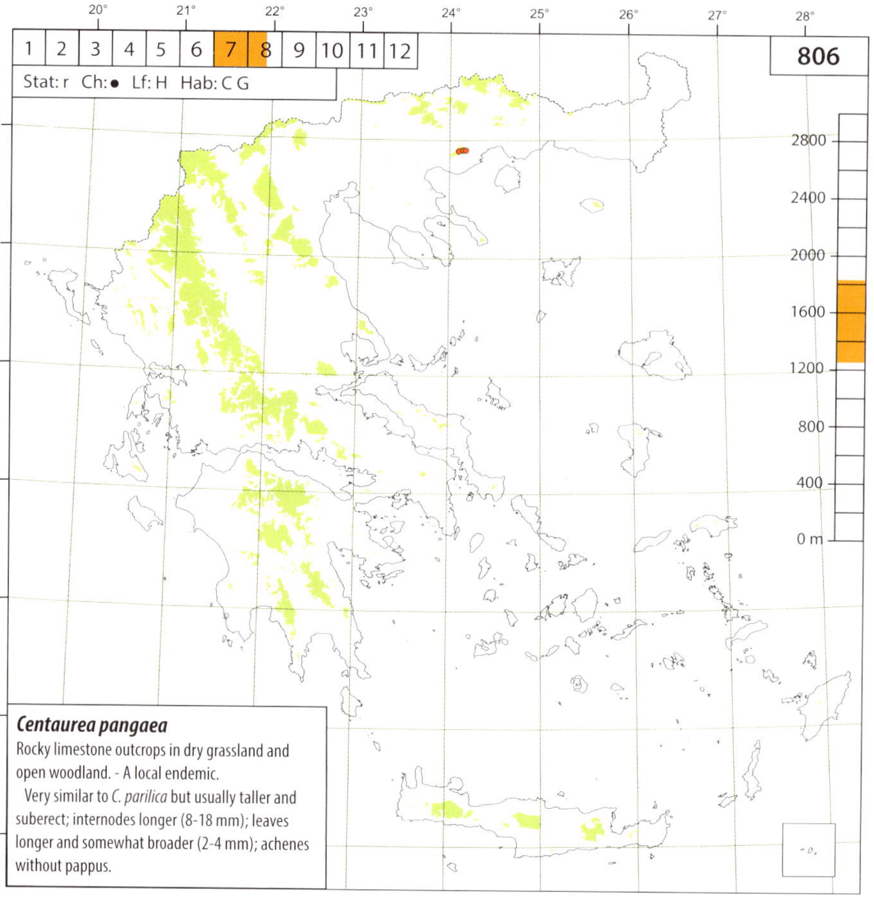

Centaurea pangaea
Rocky limestone outcrops in dry grassland and open woodland. - A local endemic.
Very similar to *C. parilica* but usually taller and suberect; internodes longer (8-18 mm); leaves longer and somewhat broader (2-4 mm); achenes without pappus.

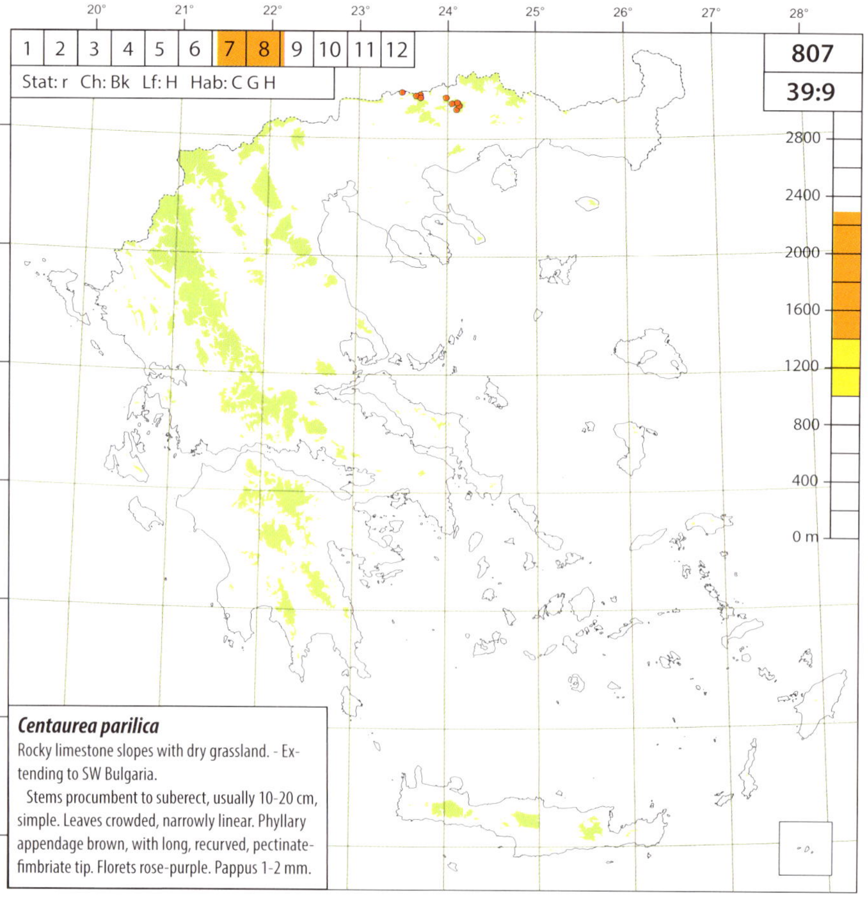

Centaurea parilica
Rocky limestone slopes with dry grassland. - Extending to SW Bulgaria.
Stems procumbent to suberect, usually 10-20 cm, simple. Leaves crowded, narrowly linear. Phyllary appendage brown, with long, recurved, pectinate-fimbriate tip. Florets rose-purple. Pappus 1-2 mm.

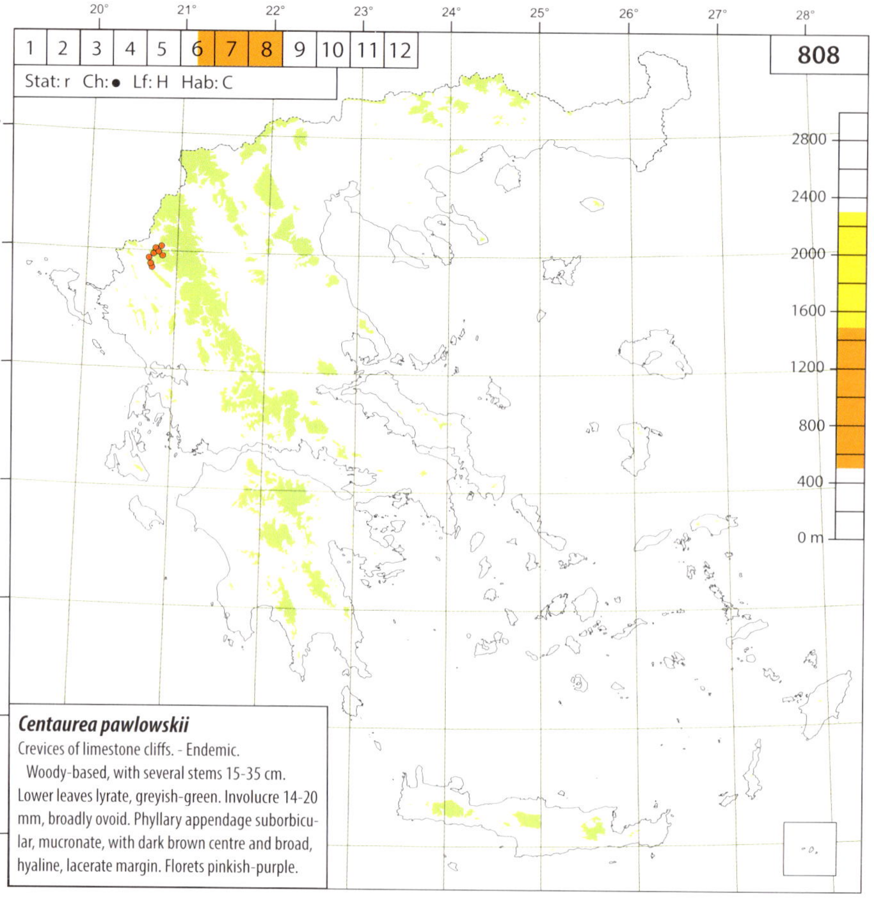

Centaurea pawlowskii
Crevices of limestone cliffs. - Endemic.
Woody-based, with several stems 15-35 cm. Lower leaves lyrate, greyish-green. Involucre 14-20 mm, broadly ovoid. Phyllary appendage suborbicular, mucronate, with dark brown centre and broad, hyaline, lacerate margin. Florets pinkish-purple.

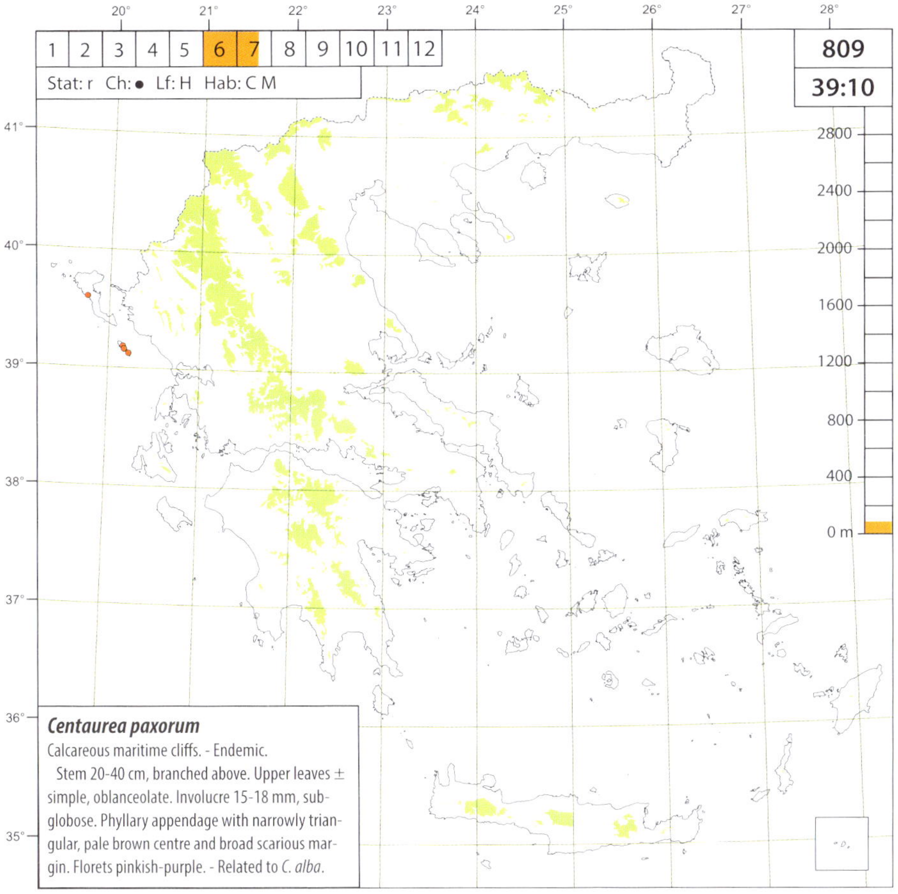

Centaurea paxorum

Calcareous maritime cliffs. - Endemic.

Stem 20-40 cm, branched above. Upper leaves ± simple, oblanceolate. Involucre 15-18 mm, subglobose. Phyllary appendage with narrowly triangular, pale brown centre and broad scarious margin. Florets pinkish-purple. - Related to *C. alba*.

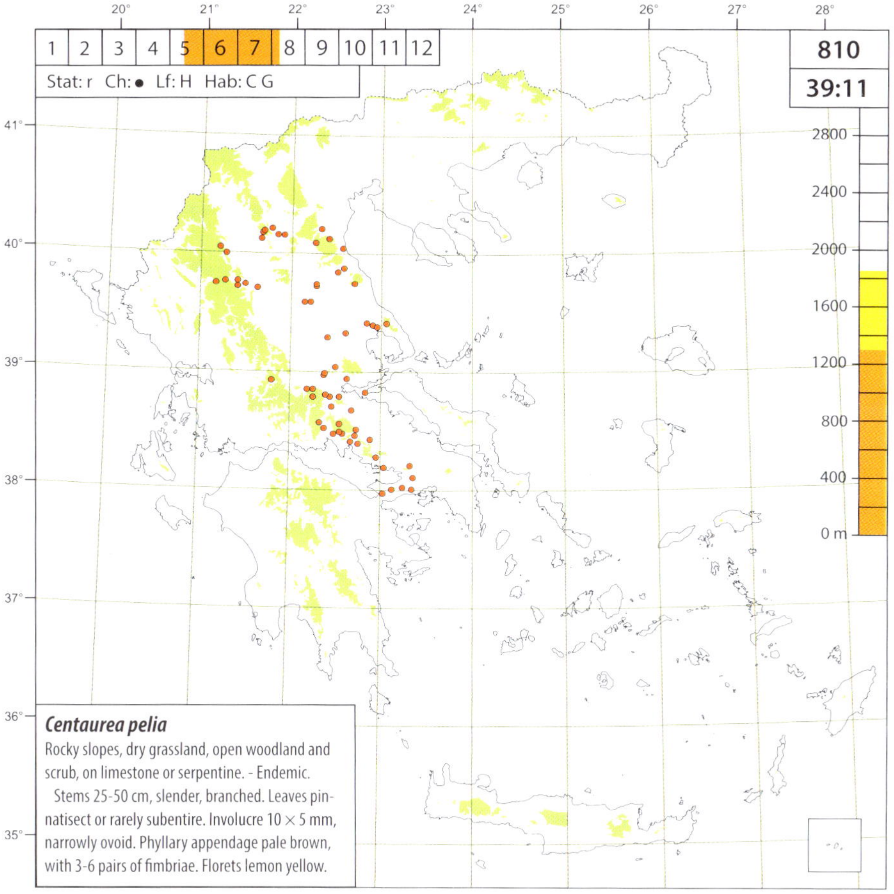

Centaurea pelia

Rocky slopes, dry grassland, open woodland and scrub, on limestone or serpentine. - Endemic.

Stems 25-50 cm, slender, branched. Leaves pinnatisect or rarely subentire. Involucre 10 × 5 mm, narrowly ovoid. Phyllary appendage pale brown, with 3-6 pairs of fimbriae. Florets lemon yellow.

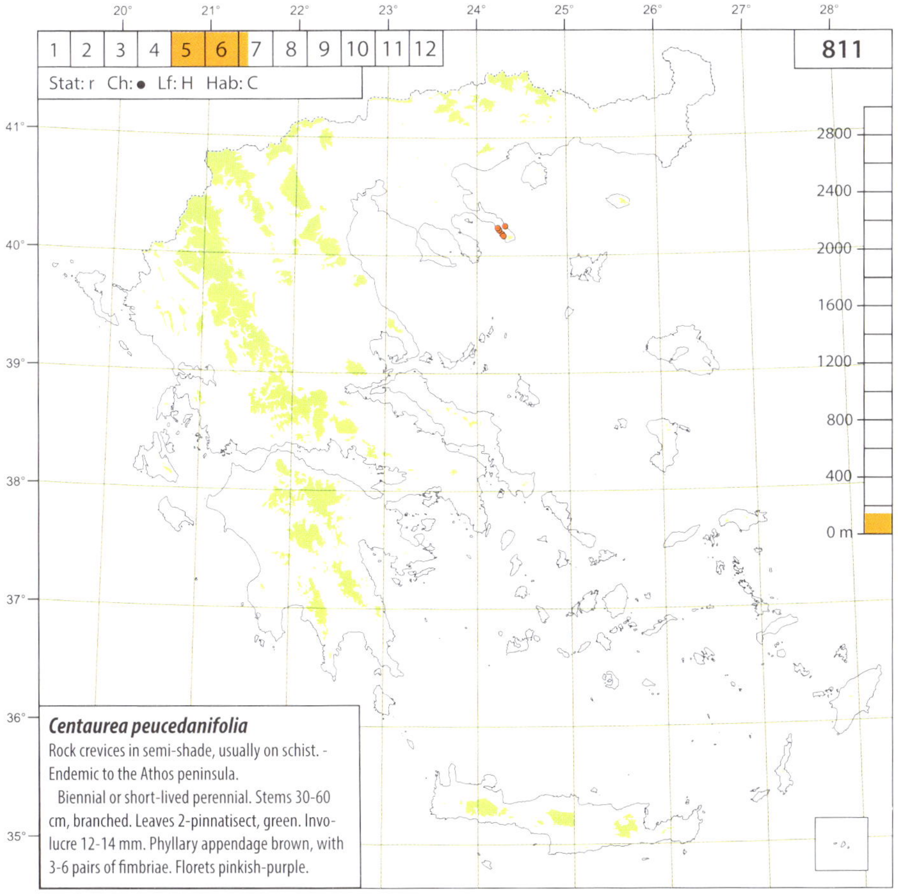

Centaurea peucedanifolia

Rock crevices in semi-shade, usually on schist. - Endemic to the Athos peninsula.

Biennial or short-lived perennial. Stems 30-60 cm, branched. Leaves 2-pinnatisect, green. Involucre 12-14 mm. Phyllary appendage brown, with 3-6 pairs of fimbriae. Florets pinkish-purple.

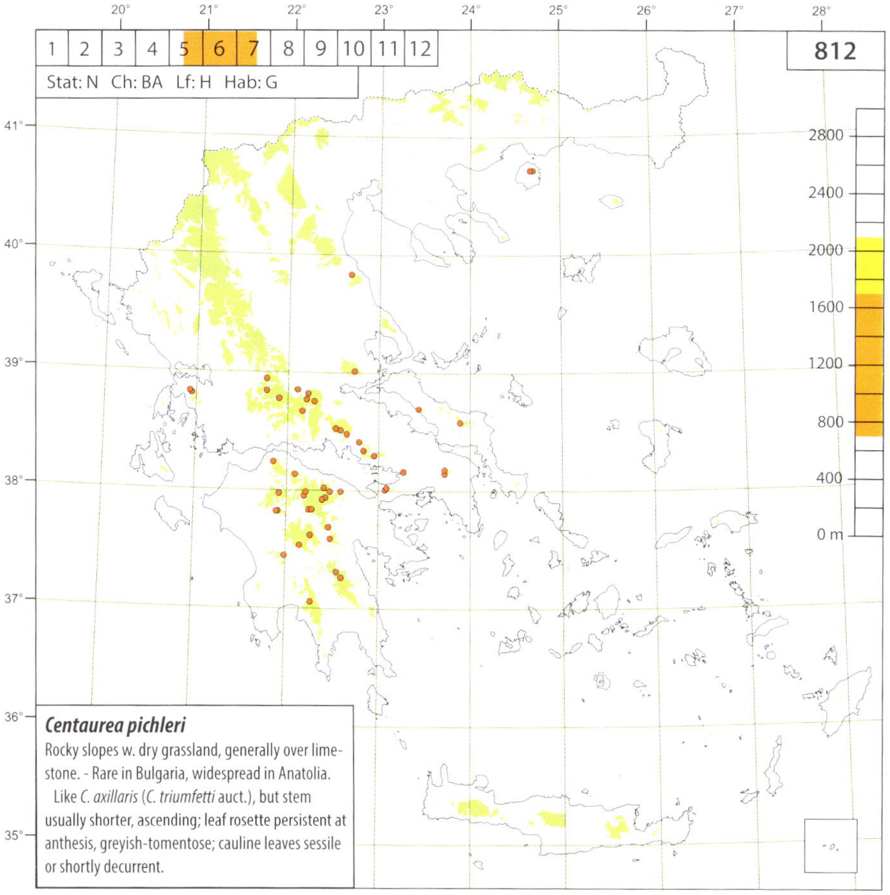

Centaurea pichleri

Rocky slopes w. dry grassland, generally over limestone. - Rare in Bulgaria, widespread in Anatolia.

Like *C. axillaris* (*C. triumfetti* auct.), but stem usually shorter, ascending; leaf rosette persistent at anthesis, greyish-tomentose; cauline leaves sessile or shortly decurrent.

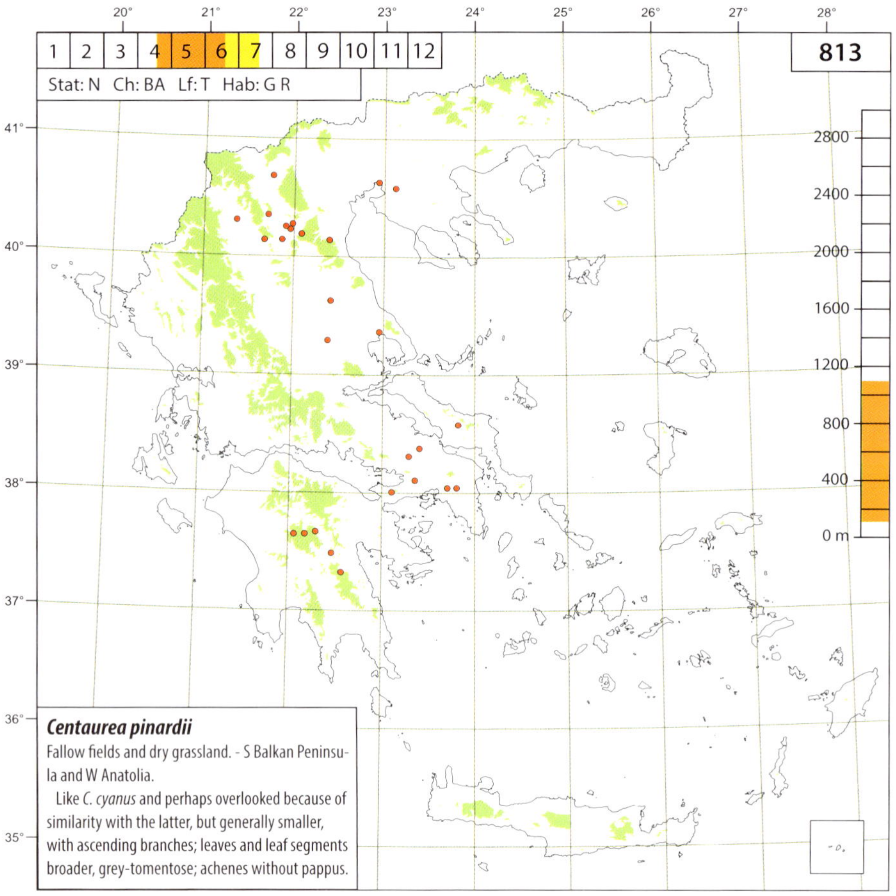

Centaurea pinardii

Fallow fields and dry grassland. - S Balkan Peninsula and W Anatolia.

Like *C. cyanus* and perhaps overlooked because of similarity with the latter, but generally smaller, with ascending branches; leaves and leaf segments broader, grey-tomentose; achenes without pappus.

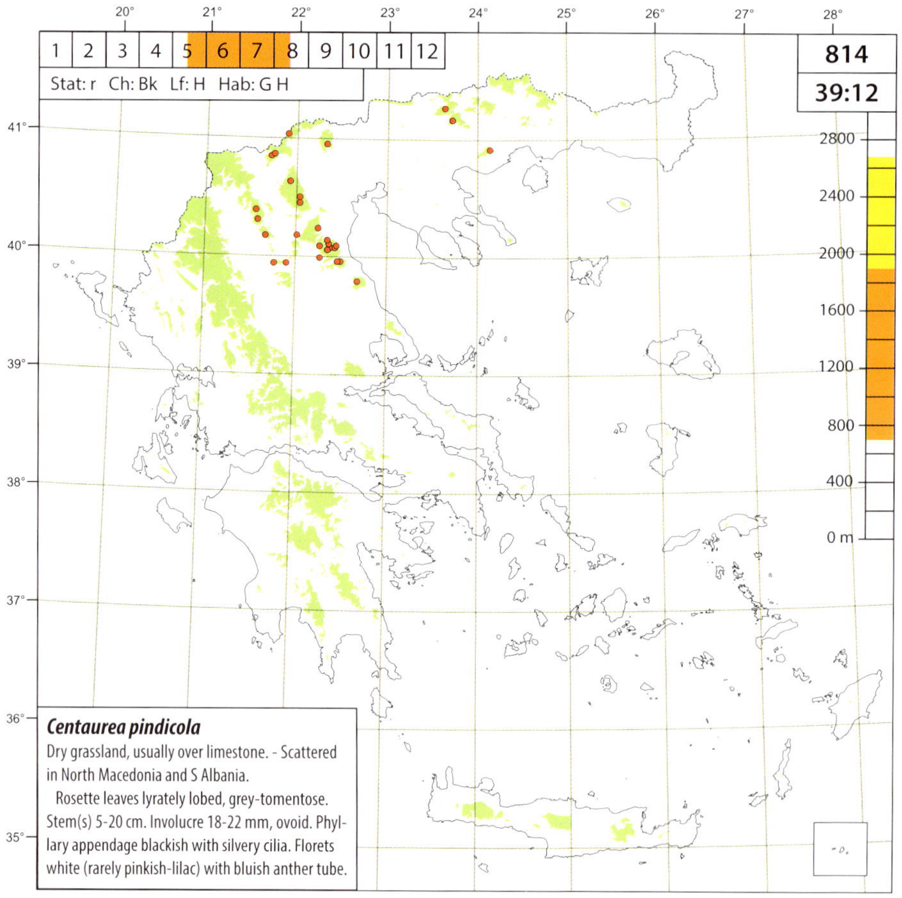

Centaurea pindicola

Dry grassland, usually over limestone. - Scattered in North Macedonia and S Albania.

Rosette leaves lyrately lobed, grey-tomentose. Stem(s) 5-20 cm. Involucre 18-22 mm, ovoid. Phyllary appendage blackish with silvery cilia. Florets white (rarely pinkish-lilac) with bluish anther tube.

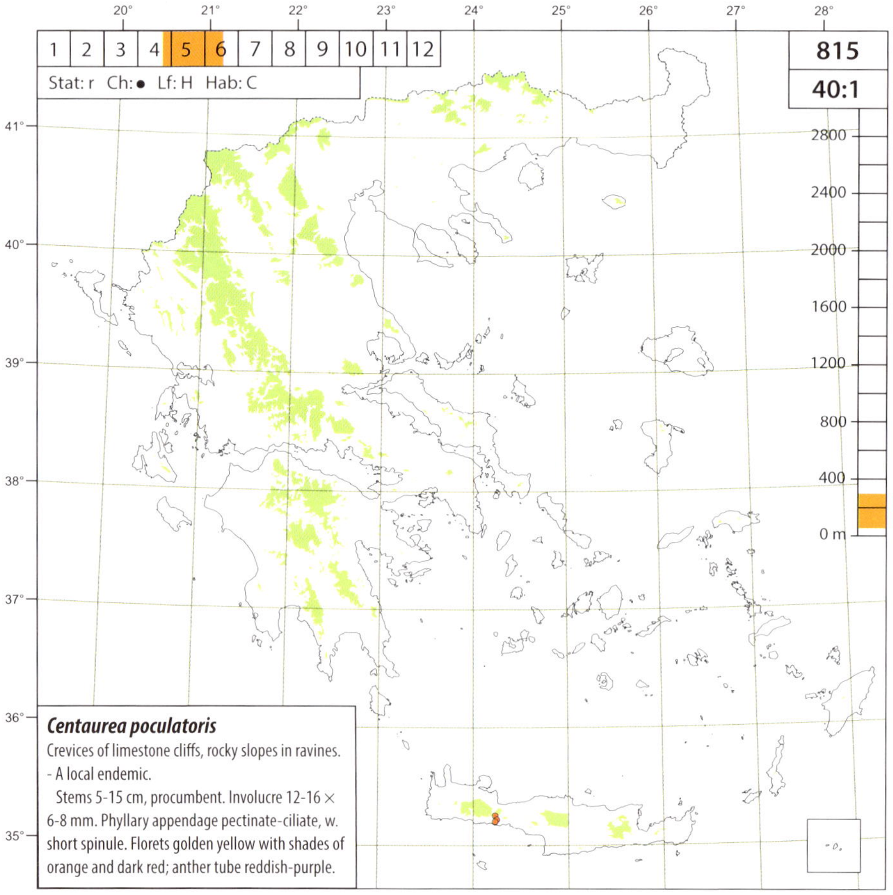

Centaurea poculatoris

Crevices of limestone cliffs, rocky slopes in ravines. - A local endemic.

Stems 5-15 cm, procumbent. Involucre 12-16 × 6-8 mm. Phyllary appendage pectinate-ciliate, w. short spinule. Florets golden yellow with shades of orange and dark red; anther tube reddish-purple.

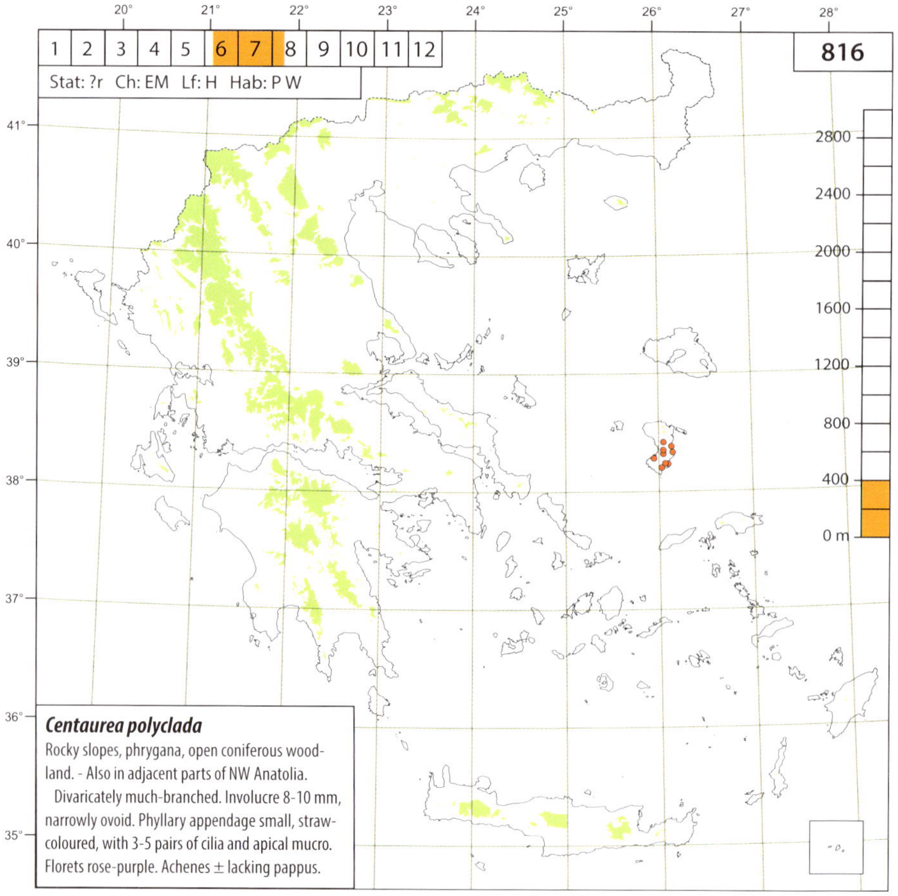

Centaurea polyclada

Rocky slopes, phrygana, open coniferous woodland. - Also in adjacent parts of NW Anatolia.

Divaricately much-branched. Involucre 8-10 mm, narrowly ovoid. Phyllary appendage small, straw-coloured, with 3-5 pairs of cilia and apical mucro. Florets rose-purple. Achenes ± lacking pappus.

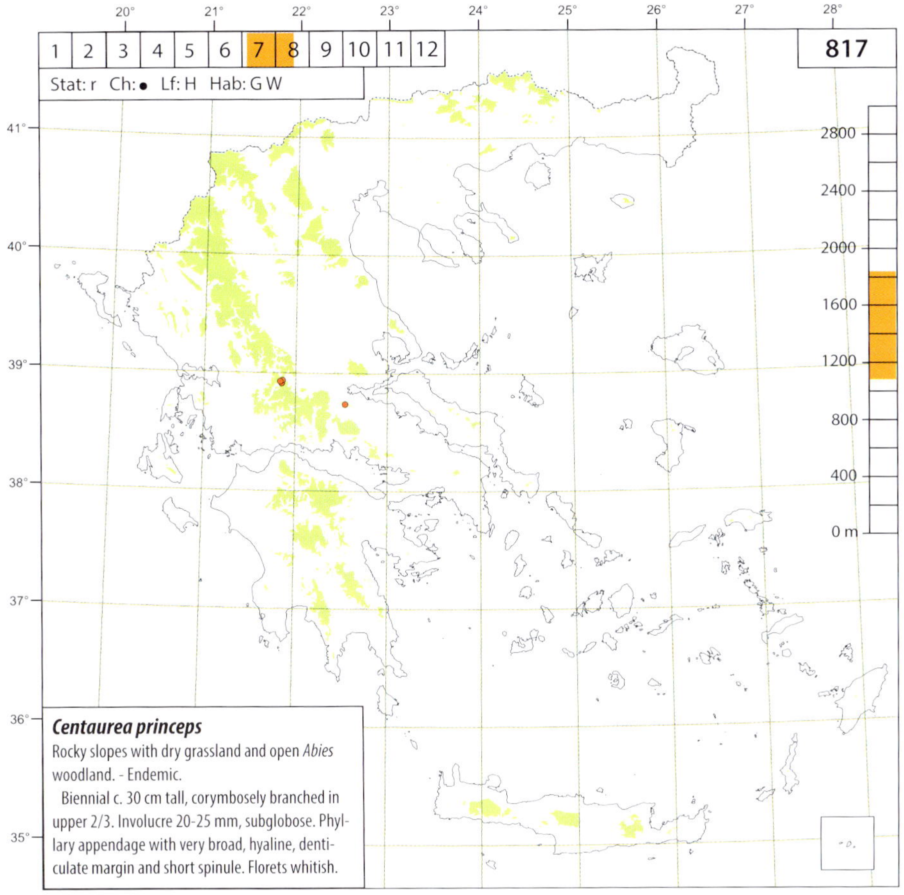

Centaurea princeps

Rocky slopes with dry grassland and open *Abies* woodland. - Endemic.

Biennial c. 30 cm tall, corymbosely branched in upper 2/3. Involucre 20-25 mm, subglobose. Phyllary appendage with very broad, hyaline, denticulate margin and short spinule. Florets whitish.

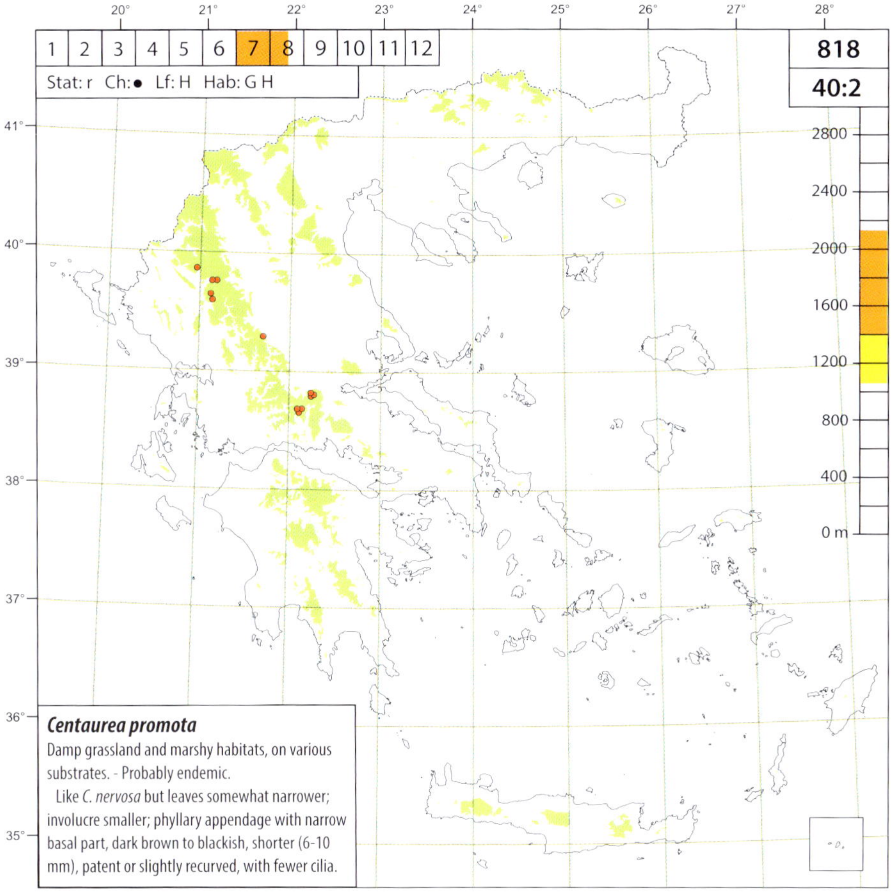

Centaurea promota

Damp grassland and marshy habitats, on various substrates. - Probably endemic.

Like *C. nervosa* but leaves somewhat narrower; involucre smaller; phyllary appendage with narrow basal part, dark brown to blackish, shorter (6-10 mm), patent or slightly recurved, with fewer cilia.

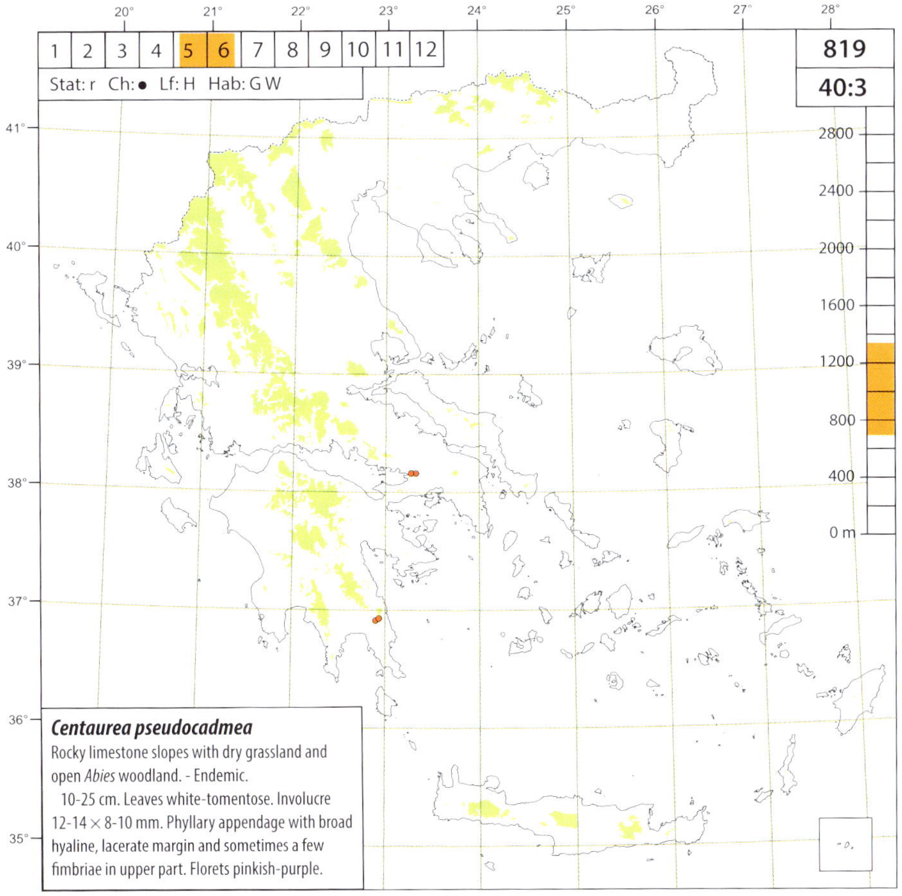

Centaurea pseudocadmea

Rocky limestone slopes with dry grassland and open *Abies* woodland. - Endemic.

10-25 cm. Leaves white-tomentose. Involucre 12-14 × 8-10 mm. Phyllary appendage with broad hyaline, lacerate margin and sometimes a few fimbriae in upper part. Florets pinkish-purple.

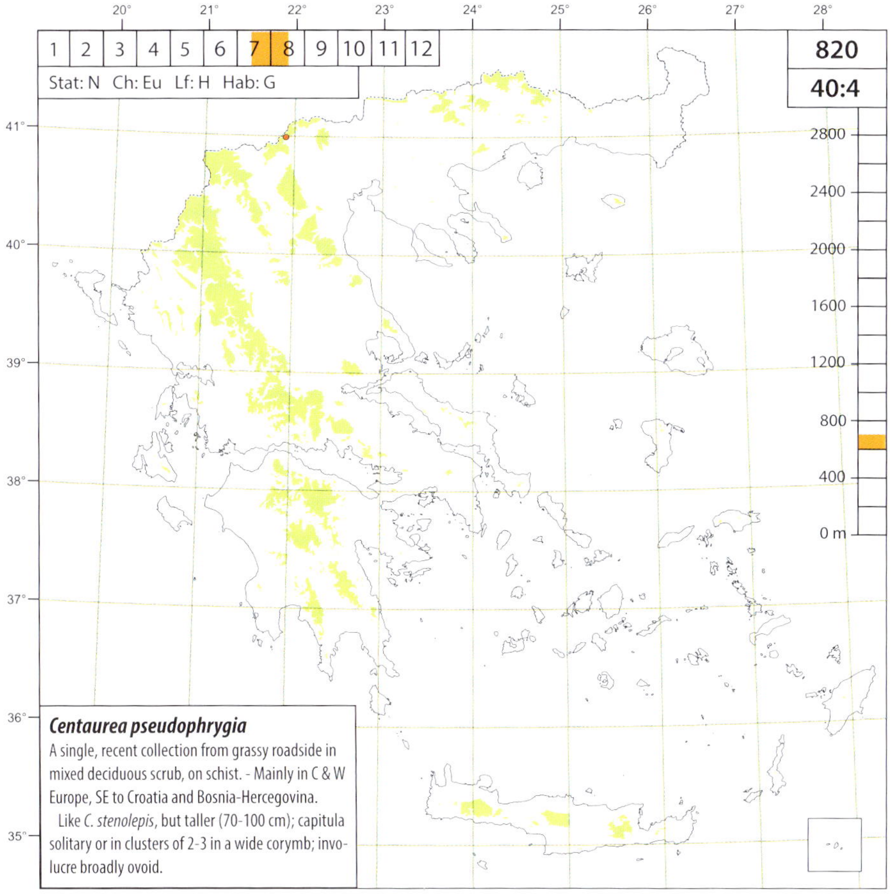

Centaurea pseudophrygia

A single, recent collection from grassy roadside in mixed deciduous scrub, on schist. - Mainly in C & W Europe, SE to Croatia and Bosnia-Hercegovina.

Like *C. stenolepis*, but taller (70-100 cm); capitula solitary or in clusters of 2-3 in a wide corymb; involucre broadly ovoid.

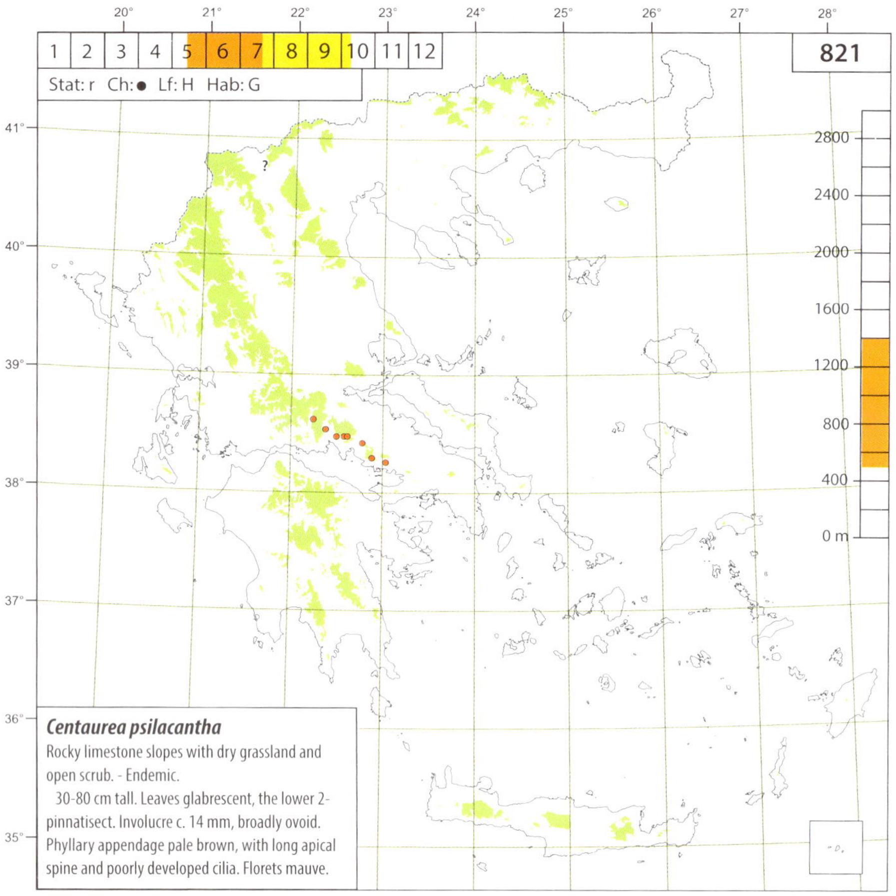

Centaurea psilacantha
Rocky limestone slopes with dry grassland and open scrub. - Endemic.
30-80 cm tall. Leaves glabrescent, the lower 2-pinnatisect. Involucre c. 14 mm, broadly ovoid. Phyllary appendage pale brown, with long apical spine and poorly developed cilia. Florets mauve.

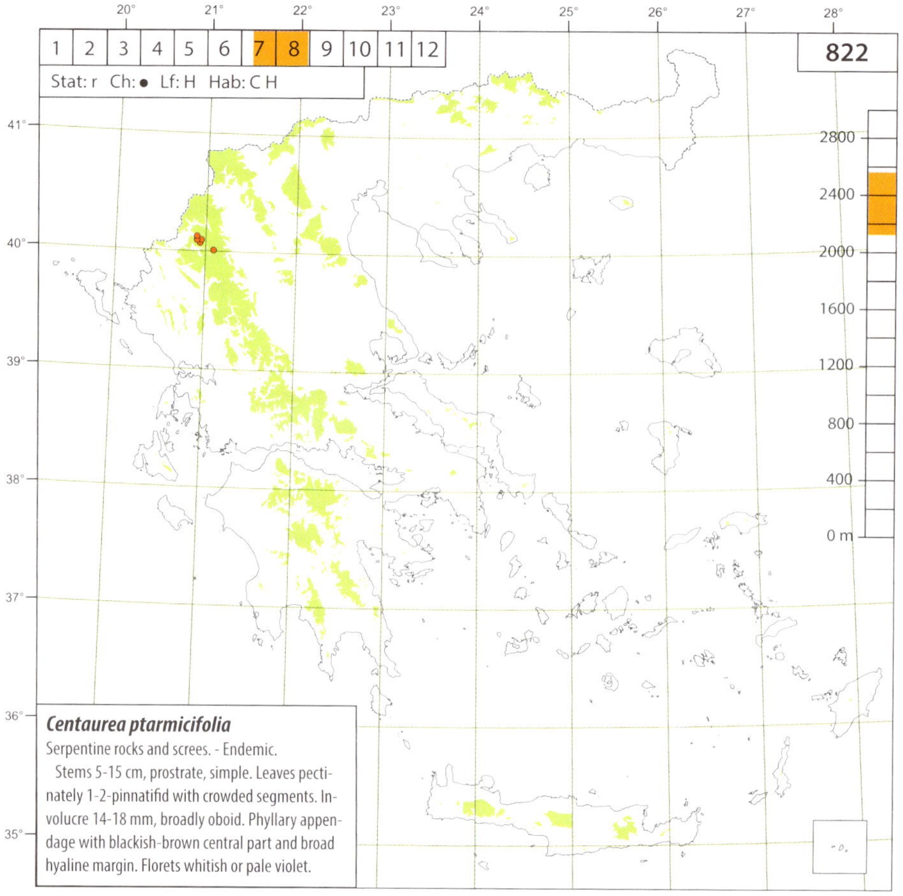

Centaurea ptarmicifolia
Serpentine rocks and screes. - Endemic.
Stems 5-15 cm, prostrate, simple. Leaves pectinately 1-2-pinnatifid with crowded segments. Involucre 14-18 mm, broadly oboid. Phyllary appendage with blackish-brown central part and broad hyaline margin. Florets whitish or pale violet.

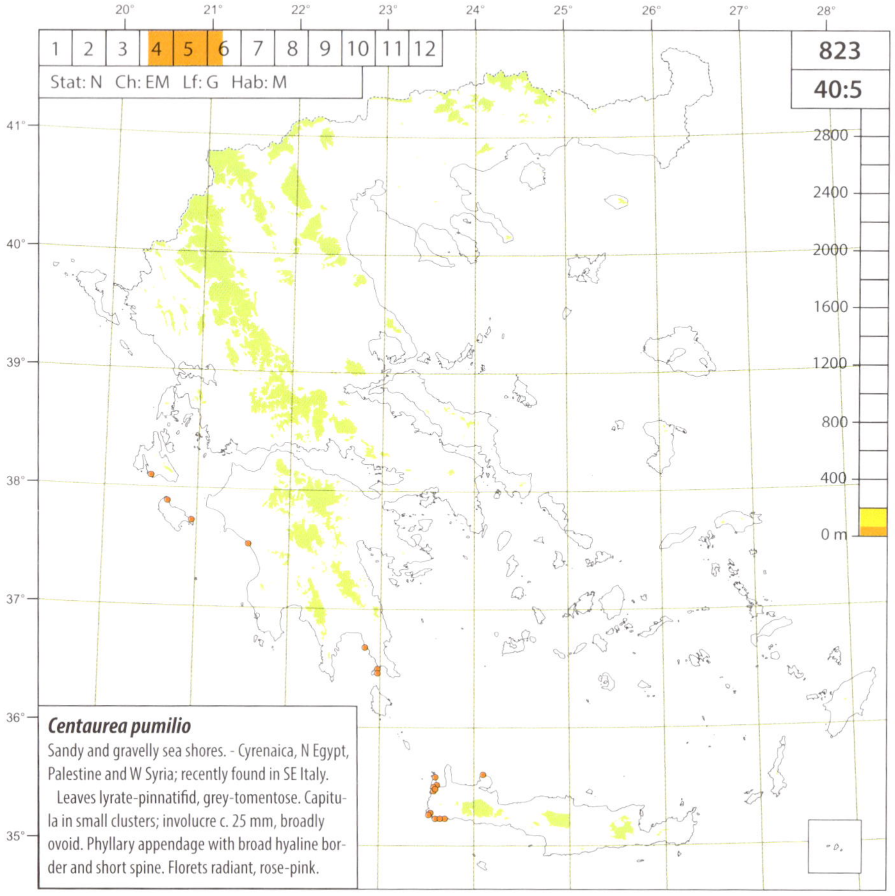

Centaurea pumilio
Sandy and gravelly sea shores. - Cyrenaica, N Egypt, Palestine and W Syria; recently found in SE Italy.
Leaves lyrate-pinnatifid, grey-tomentose. Capitula in small clusters; involucre c. 25 mm, broadly ovoid. Phyllary appendage with broad hyaline border and short spine. Florets radiant, rose-pink.

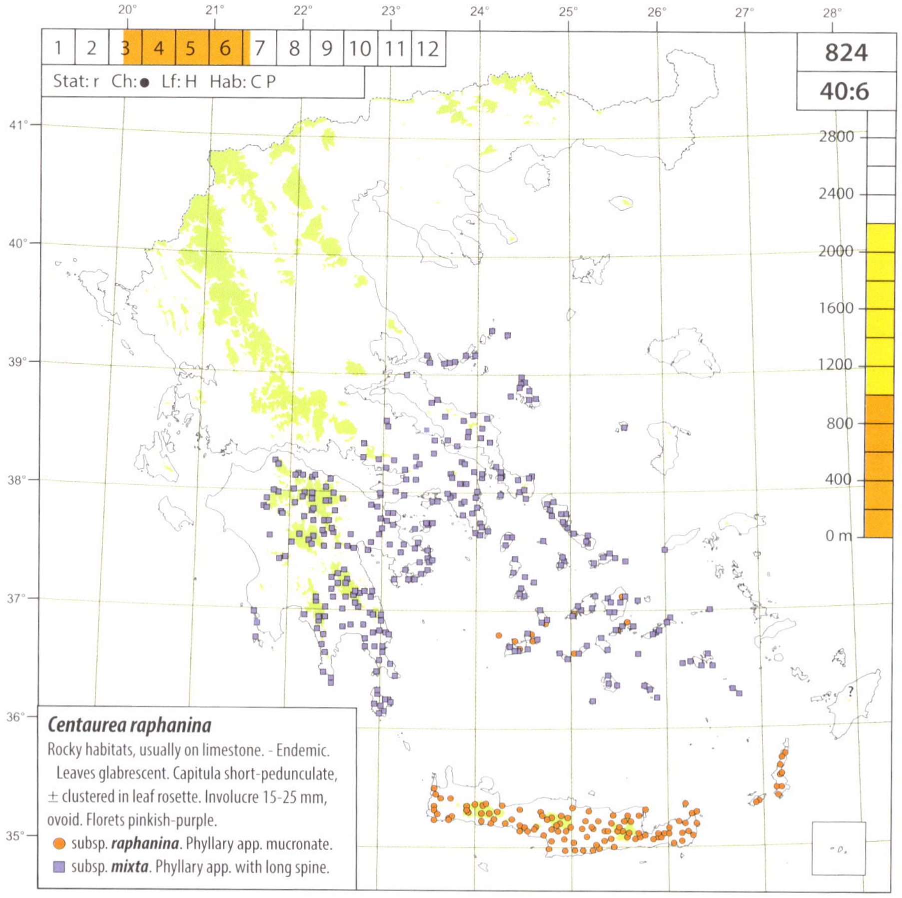

Centaurea raphanina
Rocky habitats, usually on limestone. - Endemic.
Leaves glabrescent. Capitula short-pedunculate, ± clustered in leaf rosette. Involucre 15-25 mm, ovoid. Florets pinkish-purple.
● subsp. ***raphanina***. Phyllary app. mucronate.
■ subsp. ***mixta***. Phyllary app. with long spine.

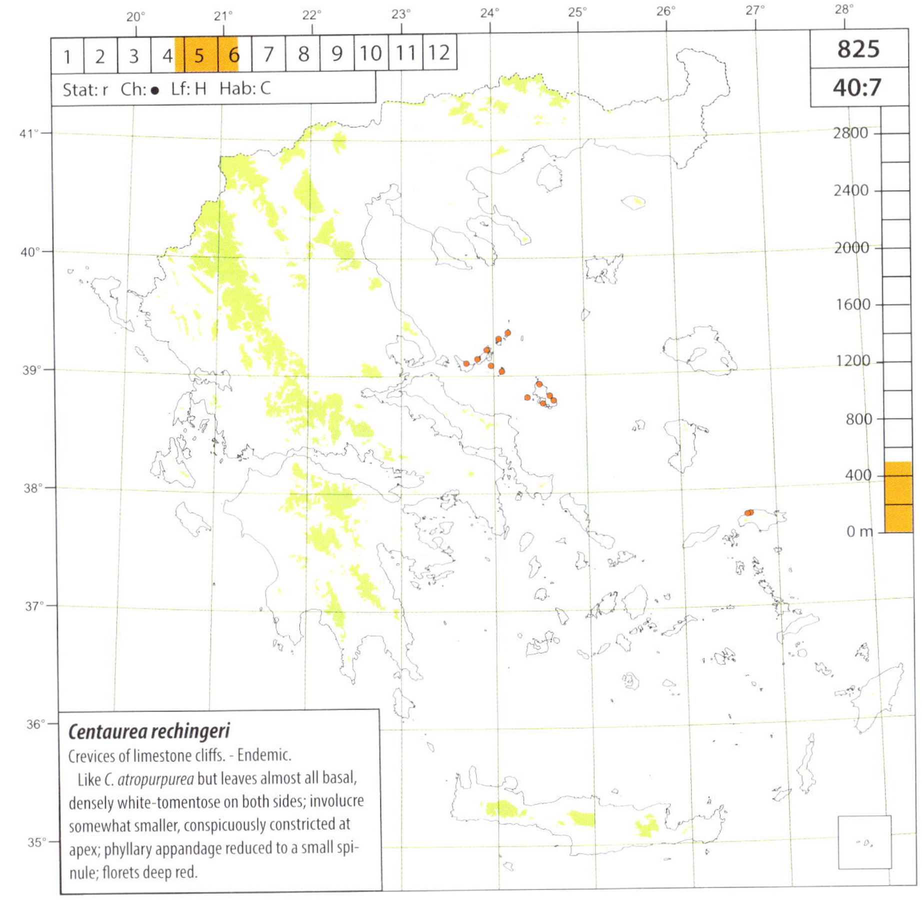

Centaurea rechingeri
Crevices of limestone cliffs. - Endemic.
Like *C. atropurpurea* but leaves almost all basal, densely white-tomentose on both sides; involucre somewhat smaller, conspicuously constricted at apex; phyllary appandage reduced to a small spinule; florets deep red.

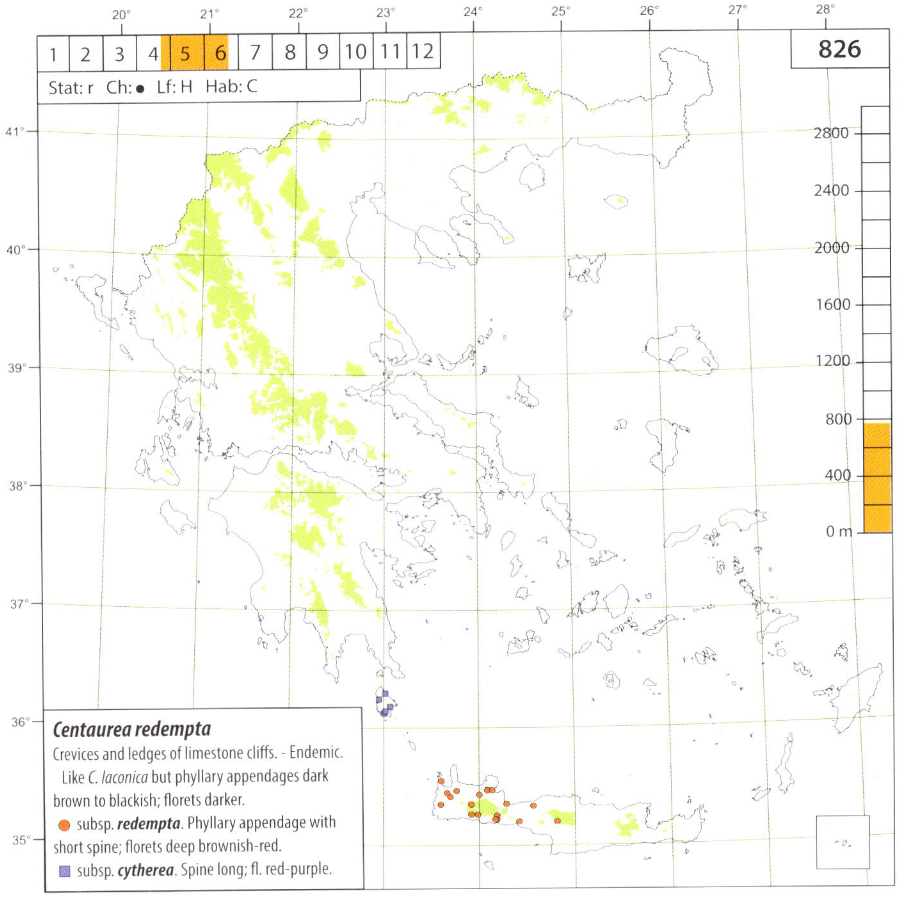

Centaurea redempta
Crevices and ledges of limestone cliffs. - Endemic.
Like *C. laconica* but phyllary appendages dark brown to blackish; florets darker.
● subsp. ***redempta***. Phyllary appendage with short spine; florets deep brownish-red.
■ subsp. ***cytherea***. Spine long; fl. red-purple.

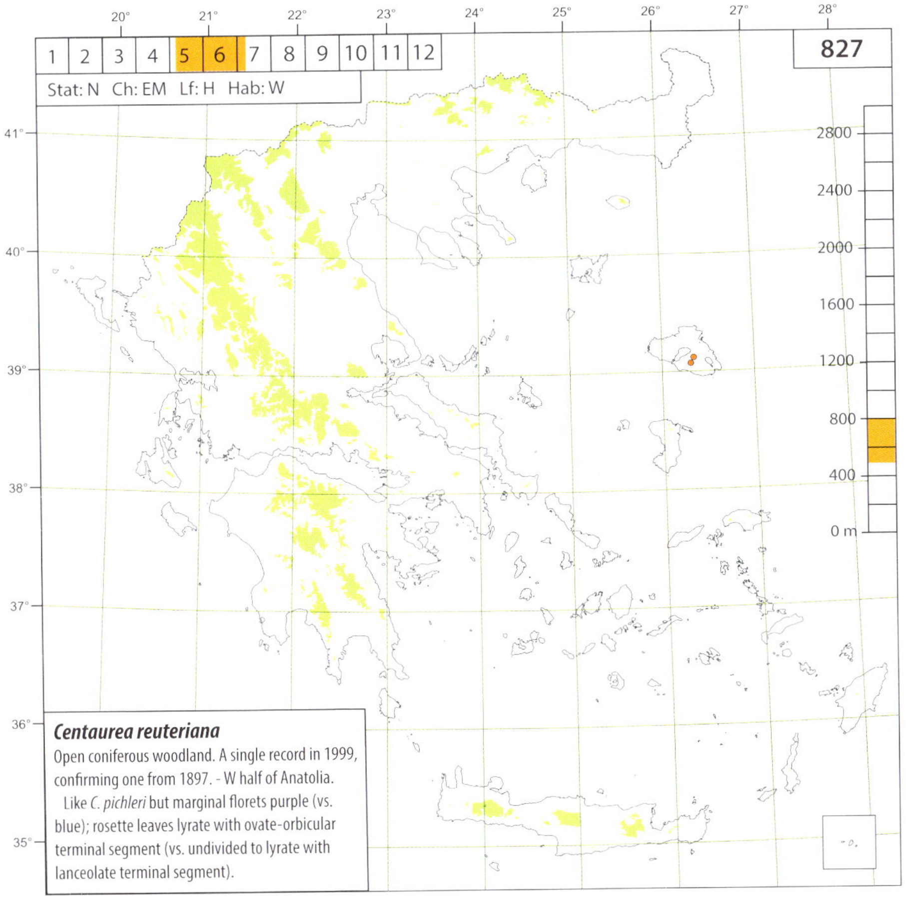

Centaurea reuteriana
Open coniferous woodland. A single record in 1999, confirming one from 1897. - W half of Anatolia.
Like *C. pichleri* but marginal florets purple (vs. blue); rosette leaves lyrate with ovate-orbicular terminal segment (vs. undivided to lyrate with lanceolate terminal segment).

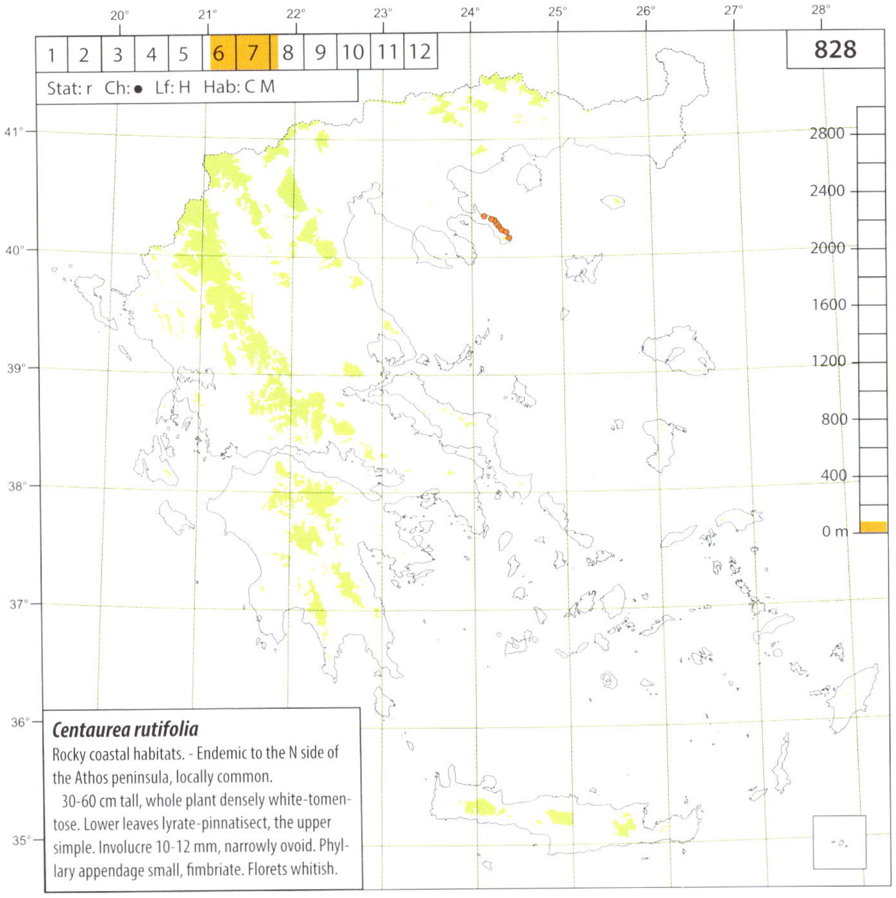

Centaurea rutifolia
Rocky coastal habitats. - Endemic to the N side of the Athos peninsula, locally common.
30-60 cm tall, whole plant densely white-tomentose. Lower leaves lyrate-pinnatisect, the upper simple. Involucre 10-12 mm, narrowly ovoid. Phyllary appendage small, fimbriate. Florets whitish.

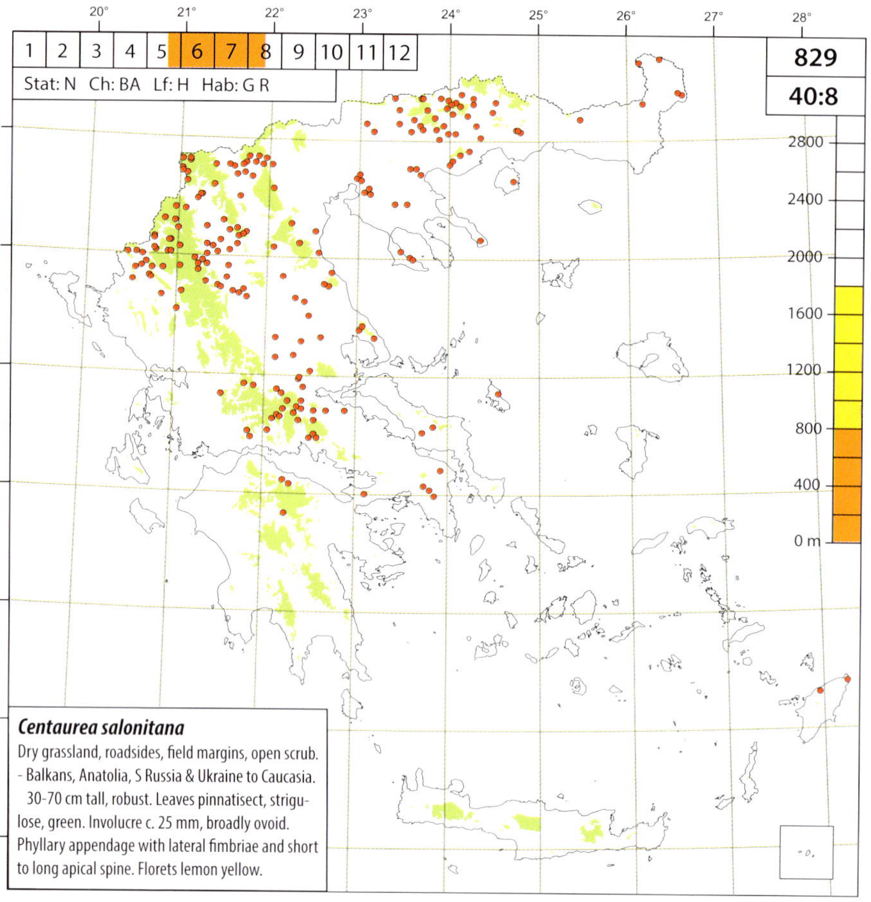

Centaurea salonitana

Dry grassland, roadsides, field margins, open scrub. - Balkans, Anatolia, S Russia & Ukraine to Caucasia.

30-70 cm tall, robust. Leaves pinnatisect, strigulose, green. Involucre c. 25 mm, broadly ovoid. Phyllary appendage with lateral fimbriae and short to long apical spine. Florets lemon yellow.

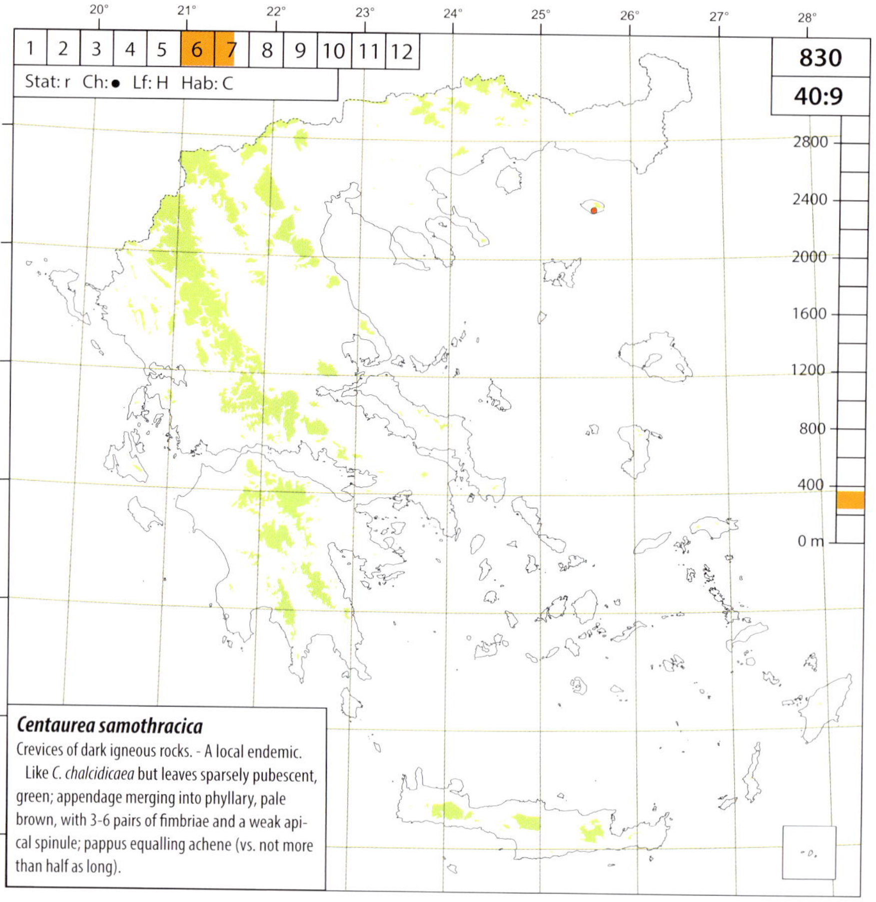

Centaurea samothracica

Crevices of dark igneous rocks. - A local endemic.

Like *C. chalcidicaea* but leaves sparsely pubescent, green; appendage merging into phyllary, pale brown, with 3-6 pairs of fimbriae and a weak apical spinule; pappus equalling achene (vs. not more than half as long).

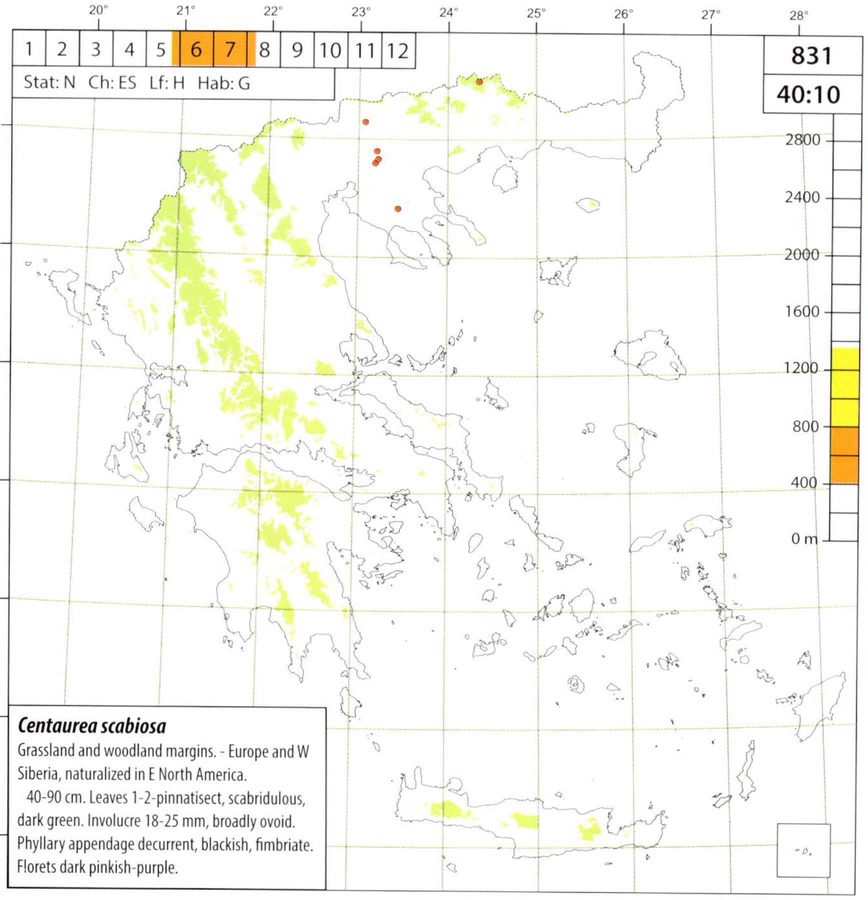

Centaurea scabiosa

Grassland and woodland margins. - Europe and W Siberia, naturalized in E North America.

40-90 cm. Leaves 1-2-pinnatisect, scabridulous, dark green. Involucre 18-25 mm, broadly ovoid. Phyllary appendage decurrent, blackish, fimbriate. Florets dark pinkish-purple.

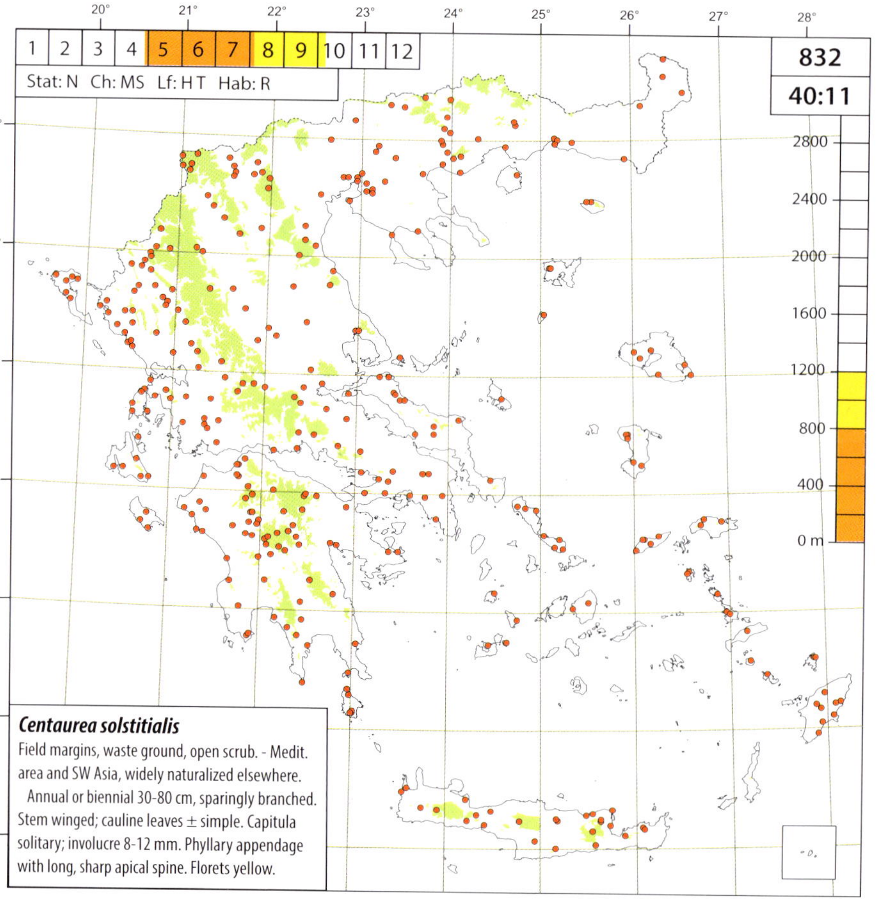

Centaurea solstitialis

Field margins, waste ground, open scrub. - Medit. area and SW Asia, widely naturalized elsewhere.

Annual or biennial 30-80 cm, sparingly branched. Stem winged; cauline leaves ± simple. Capitula solitary; involucre 8-12 mm. Phyllary appendage with long, sharp apical spine. Florets yellow.

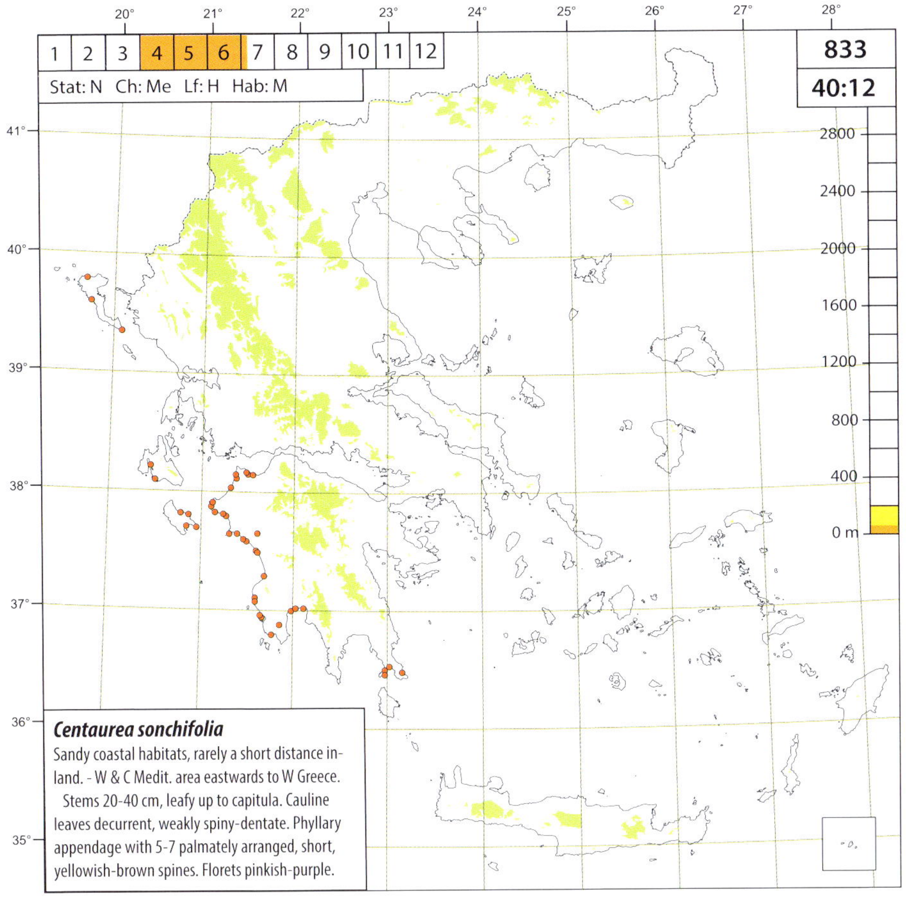

Centaurea sonchifolia
Sandy coastal habitats, rarely a short distance inland. - W & C Medit. area eastwards to W Greece.
Stems 20-40 cm, leafy up to capitula. Cauline leaves decurrent, weakly spiny-dentate. Phyllary appendage with 5-7 palmately arranged, short, yellowish-brown spines. Florets pinkish-purple.

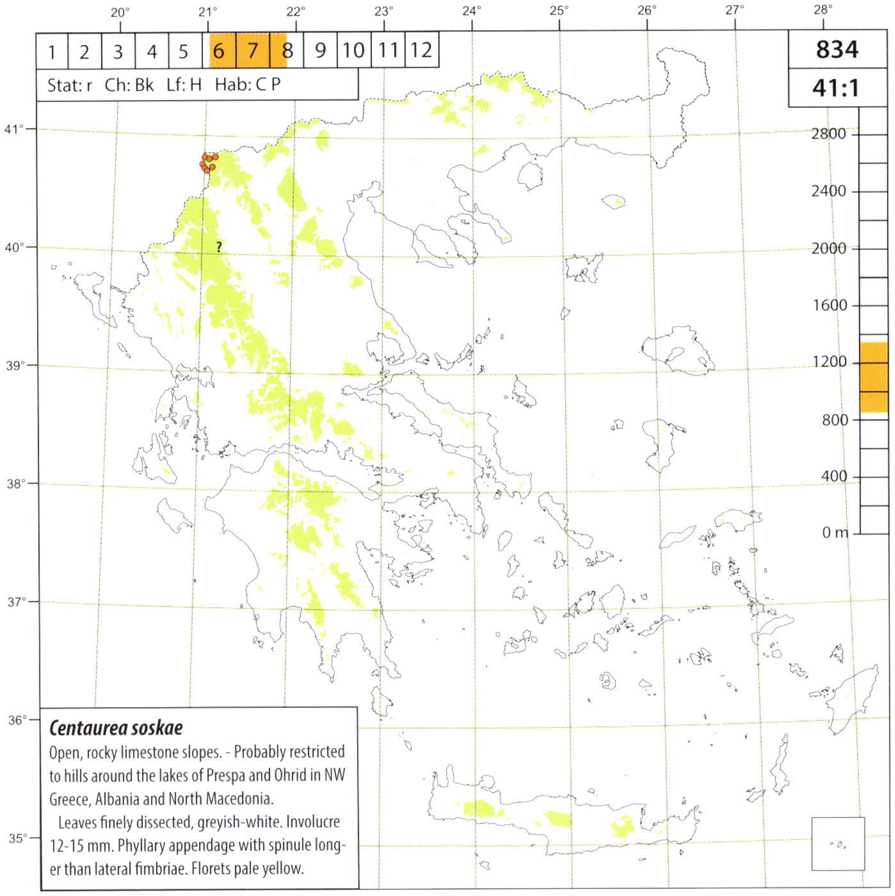

Centaurea soskae
Open, rocky limestone slopes. - Probably restricted to hills around the lakes of Prespa and Ohrid in NW Greece, Albania and North Macedonia.
Leaves finely dissected, greyish-white. Involucre 12-15 mm. Phyllary appendage with spinule longer than lateral fimbriae. Florets pale yellow.

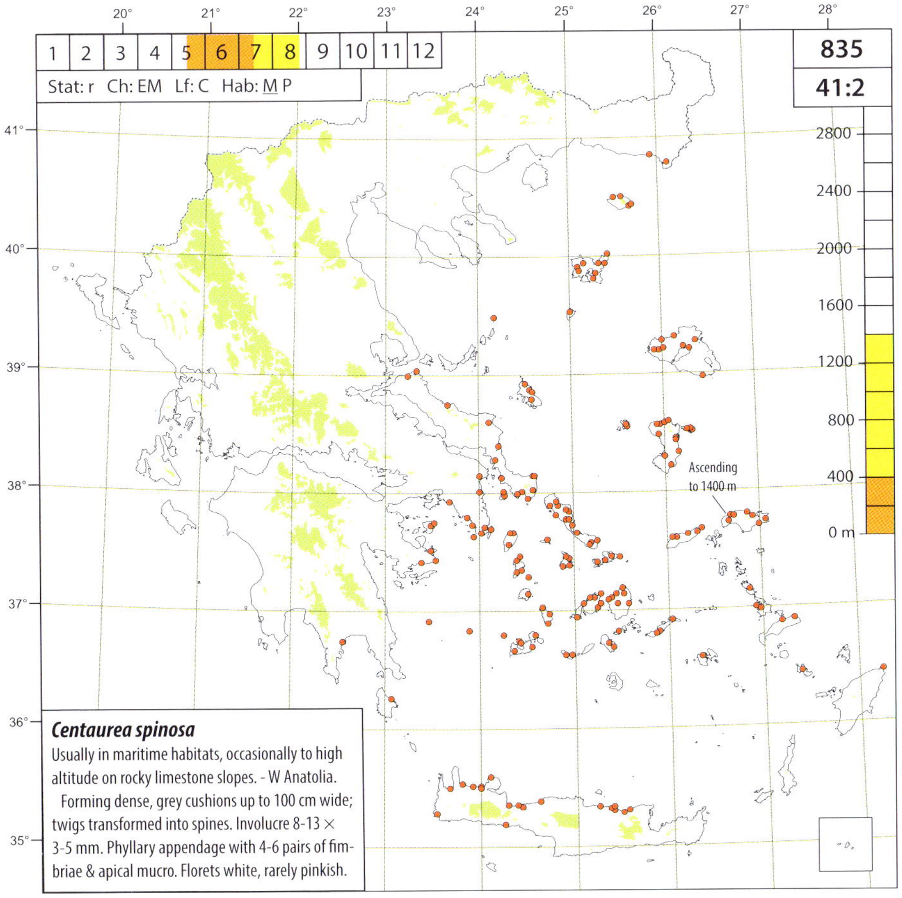

Centaurea spinosa
Usually in maritime habitats, occasionally to high altitude on rocky limestone slopes. - W Anatolia.
Forming dense, grey cushions up to 100 cm wide; twigs transformed into spines. Involucre 8-13 × 3-5 mm. Phyllary appendage with 4-6 pairs of fimbriae & apical mucro. Florets white, rarely pinkish.

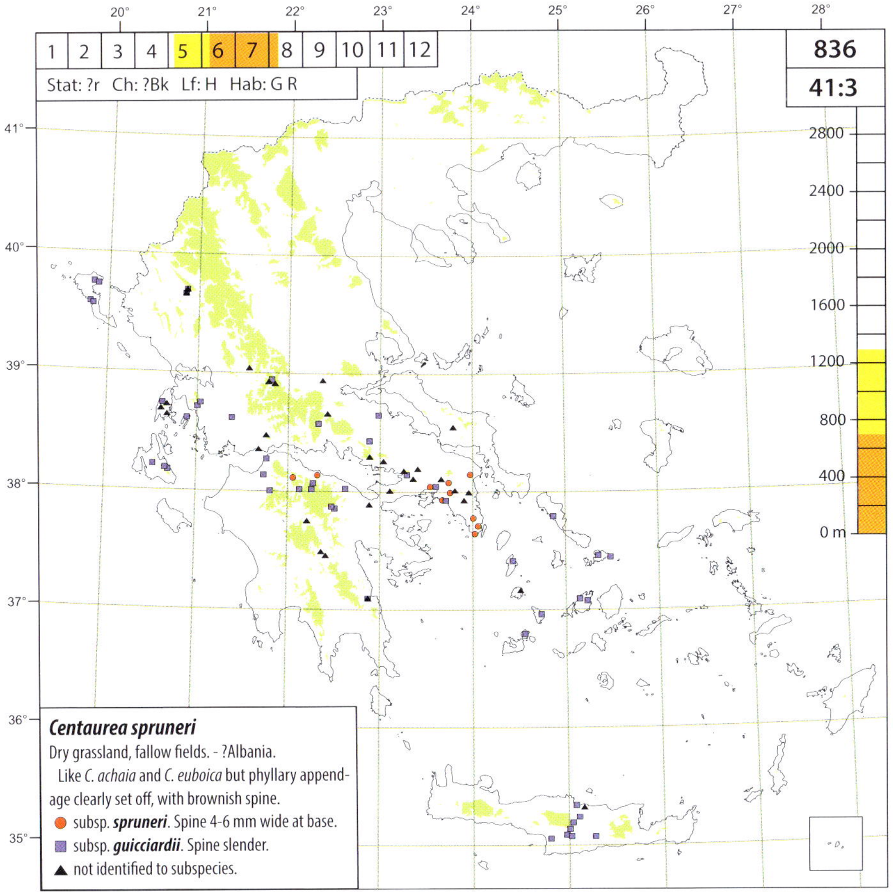

Centaurea spruneri
Dry grassland, fallow fields. - ?Albania.
Like *C. achaia* and *C. euboica* but phyllary appendage clearly set off, with brownish spine.
● subsp. ***spruneri***. Spine 4-6 mm wide at base.
■ subsp. ***guicciardii***. Spine slender.
▲ not identified to subspecies.

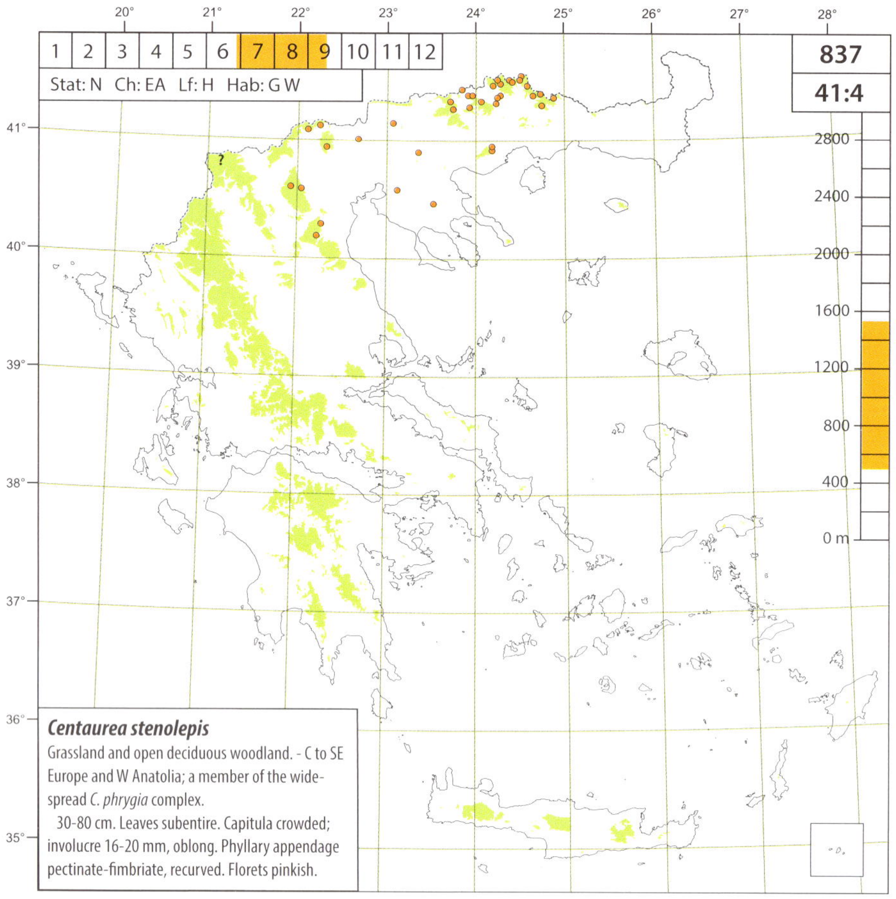

Centaurea stenolepis
Grassland and open deciduous woodland. - C to SE Europe and W Anatolia; a member of the widespread *C. phrygia* complex.
30-80 cm. Leaves subentire. Capitula crowded; involucre 16-20 mm, oblong. Phyllary appendage pectinate-fimbriate, recurved. Florets pinkish.

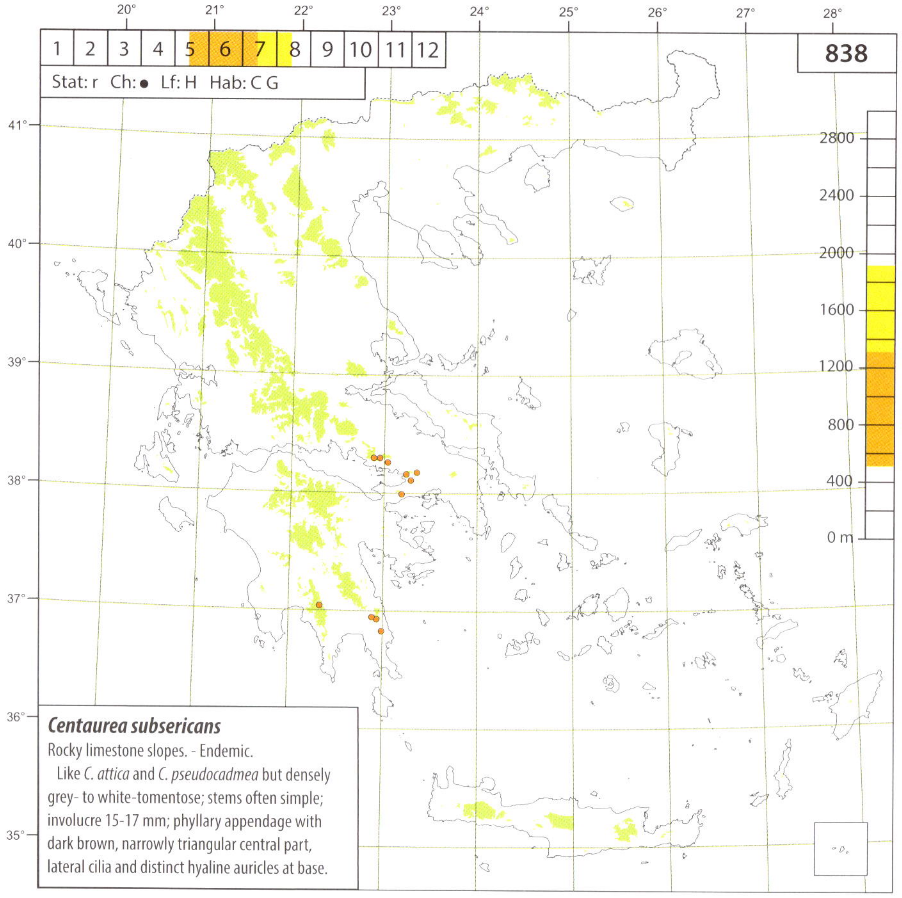

Centaurea subsericans
Rocky limestone slopes. - Endemic.
Like *C. attica* and *C. pseudocadmea* but densely grey- to white-tomentose; stems often simple; involucre 15-17 mm; phyllary appendage with dark brown, narrowly triangular central part, lateral cilia and distinct hyaline auricles at base.

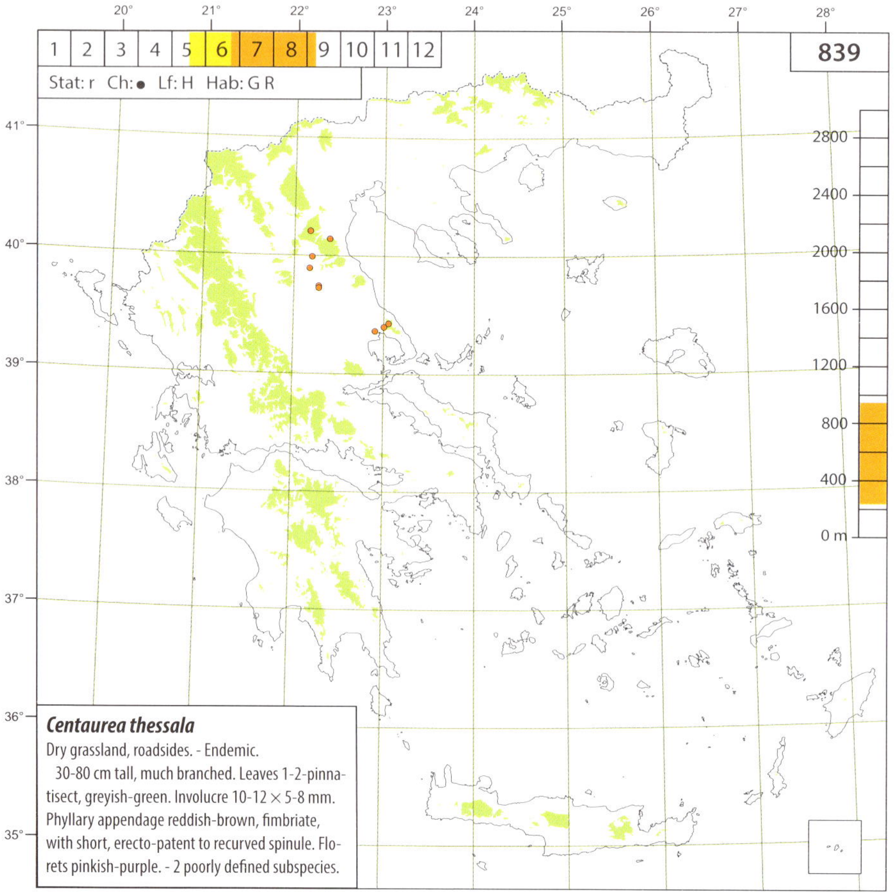

Centaurea thessala
Dry grassland, roadsides. - Endemic.
30-80 cm tall, much branched. Leaves 1-2-pinnatisect, greyish-green. Involucre 10-12 × 5-8 mm. Phyllary appendage reddish-brown, fimbriate, with short, erecto-patent to recurved spinule. Florets pinkish-purple. - 2 poorly defined subspecies.

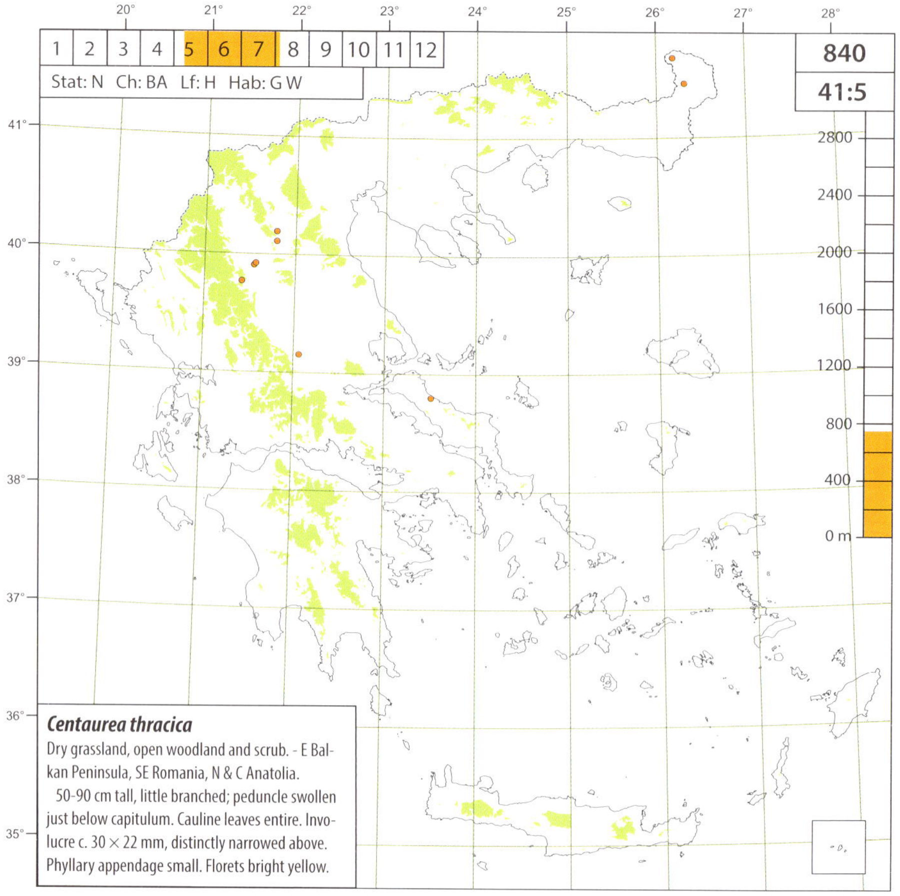

Centaurea thracica
Dry grassland, open woodland and scrub. - E Balkan Peninsula, SE Romania, N & C Anatolia.
50-90 cm tall, little branched; peduncle swollen just below capitulum. Cauline leaves entire. Involucre c. 30 × 22 mm, distinctly narrowed above. Phyllary appendage small. Florets bright yellow.

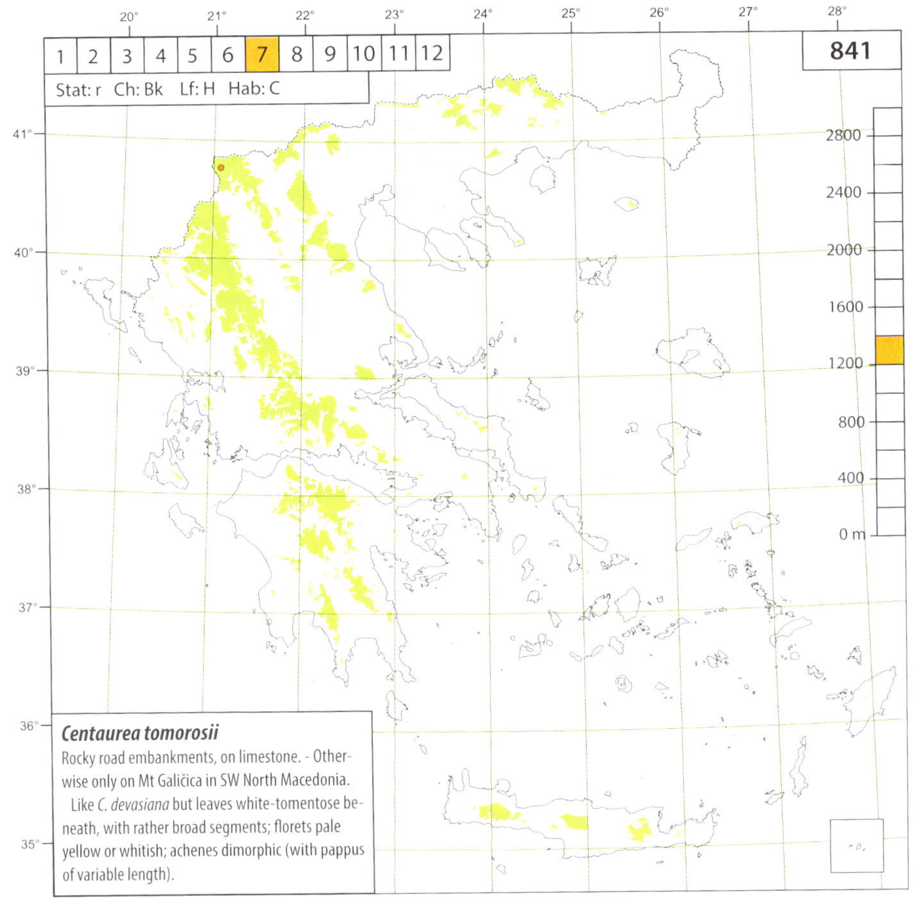

Centaurea tomorosii
Rocky road embankments, on limestone. - Otherwise only on Mt Galičica in SW North Macedonia.
Like *C. devasiana* but leaves white-tomentose beneath, with rather broad segments; florets pale yellow or whitish; achenes dimorphic (with pappus of variable length).

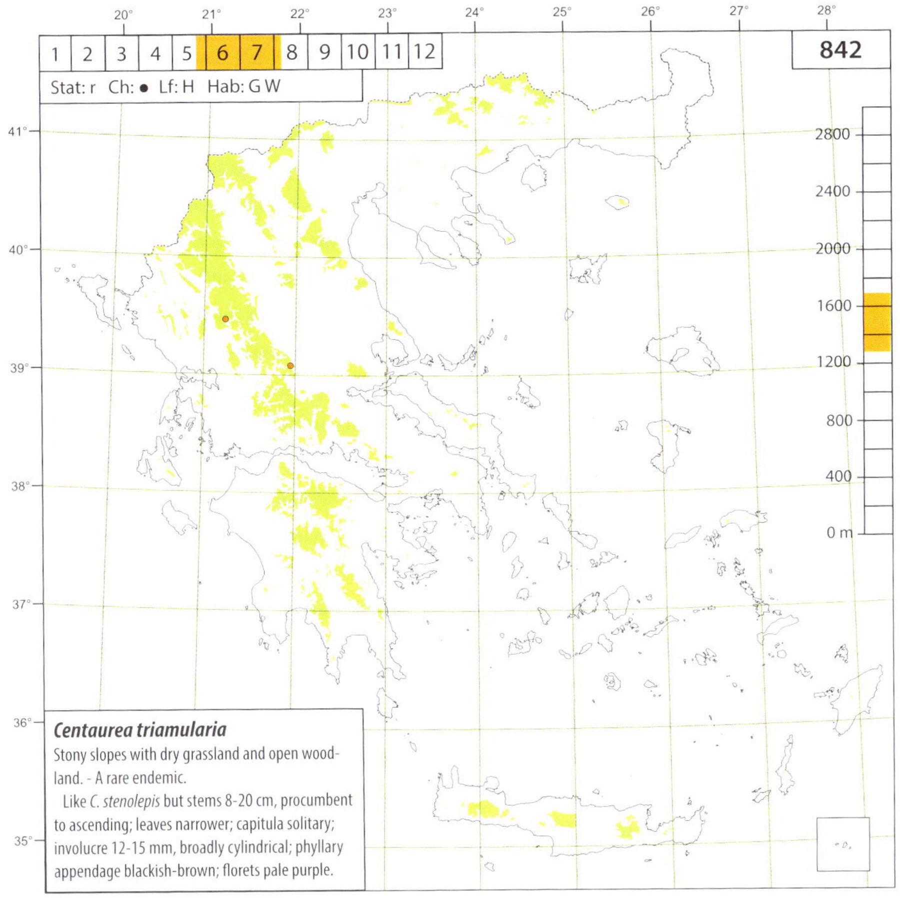

Centaurea triamularia
Stony slopes with dry grassland and open woodland. - A rare endemic.
Like *C. stenolepis* but stems 8-20 cm, procumbent to ascending; leaves narrower; capitula solitary; involucre 12-15 mm, broadly cylindrical; phyllary appendage blackish-brown; florets pale purple.

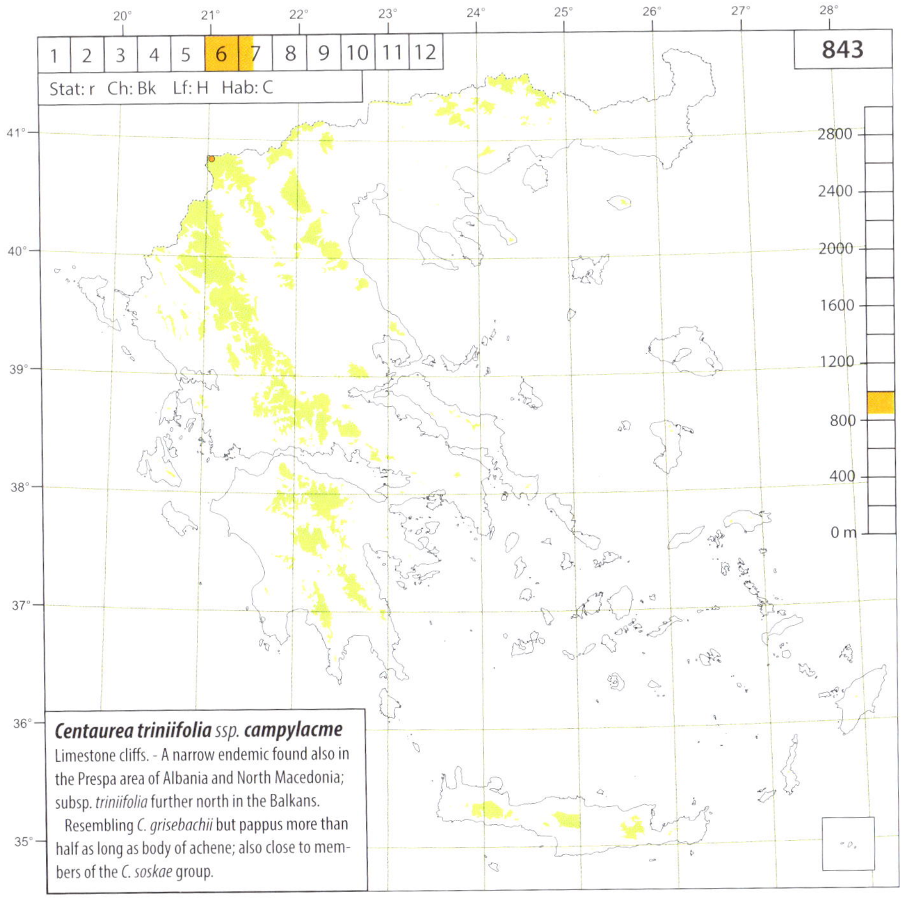

Centaurea triniifolia* ssp. *campylacme
Limestone cliffs. - A narrow endemic found also in the Prespa area of Albania and North Macedonia; subsp. *triniifolia* further north in the Balkans.
Resembling *C. grisebachii* but pappus more than half as long as body of achene; also close to members of the *C. soskae* group.

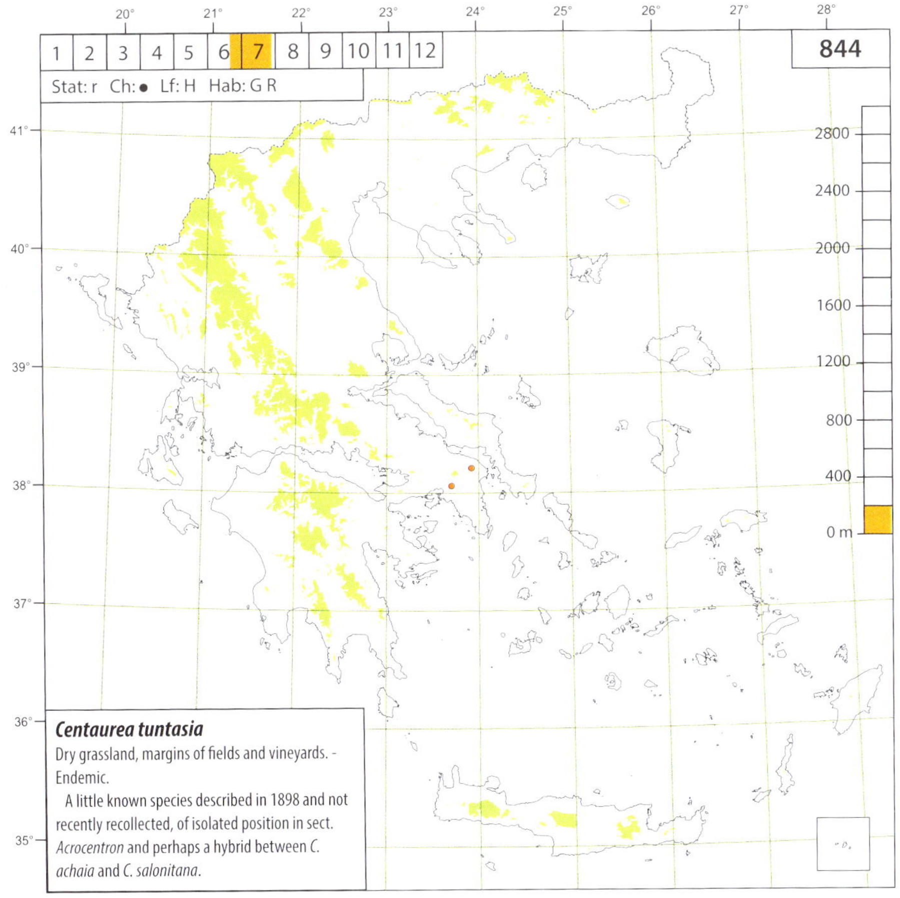

Centaurea tuntasia
Dry grassland, margins of fields and vineyards. - Endemic.
A little known species described in 1898 and not recently recollected, of isolated position in sect. *Acrocentron* and perhaps a hybrid between *C. achaia* and *C. salonitana*.

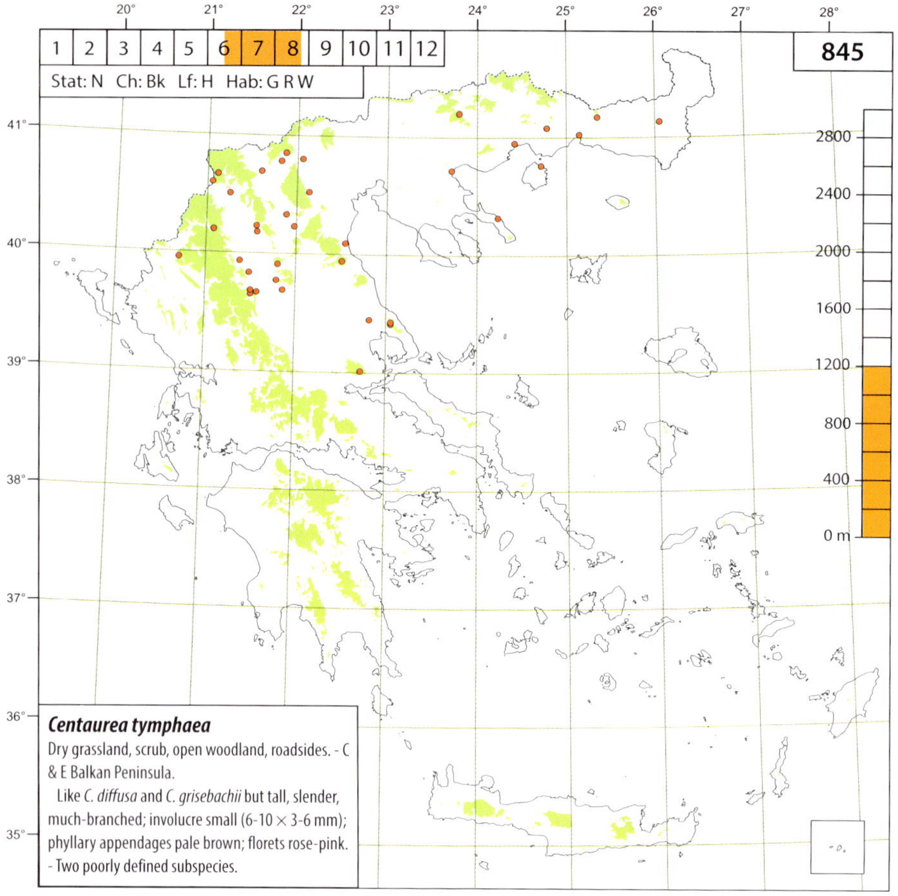

Centaurea tymphaea

Dry grassland, scrub, open woodland, roadsides. - C & E Balkan Peninsula.

Like *C. diffusa* and *C. grisebachii* but tall, slender, much-branched; involucre small (6-10 × 3-6 mm); phyllary appendages pale brown; florets rose-pink. - Two poorly defined subspecies.

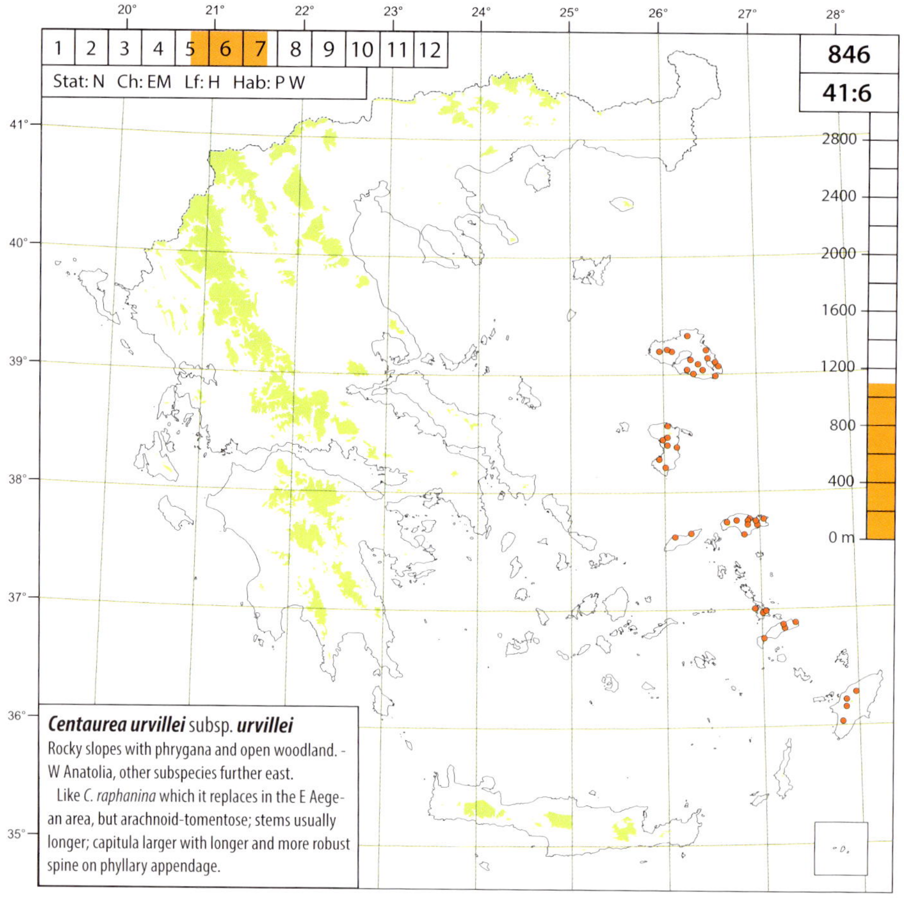

Centaurea urvillei subsp. _urvillei_

Rocky slopes with phrygana and open woodland. - W Anatolia, other subspecies further east.

Like *C. raphanina* which it replaces in the E Aegean area, but arachnoid-tomentose; stems usually longer; capitula larger with longer and more robust spine on phyllary appendage.

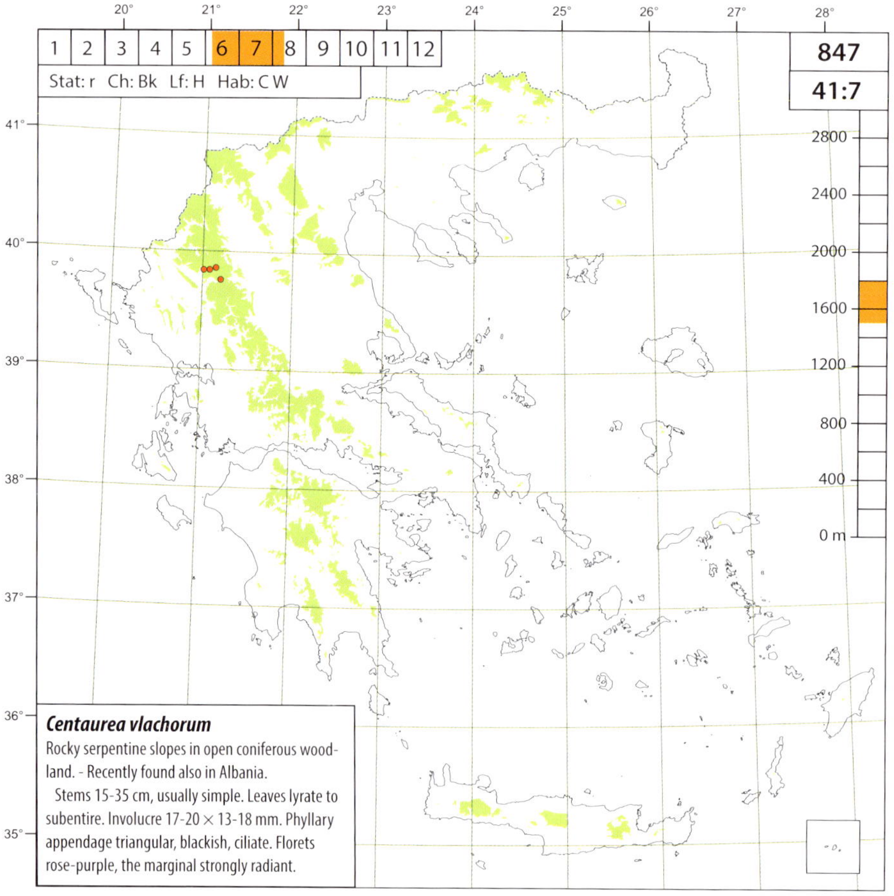

Centaurea vlachorum

Rocky serpentine slopes in open coniferous woodland. - Recently found also in Albania.

Stems 15-35 cm, usually simple. Leaves lyrate to subentire. Involucre 17-20 × 13-18 mm. Phyllary appendage triangular, blackish, ciliate. Florets rose-purple, the marginal strongly radiant.

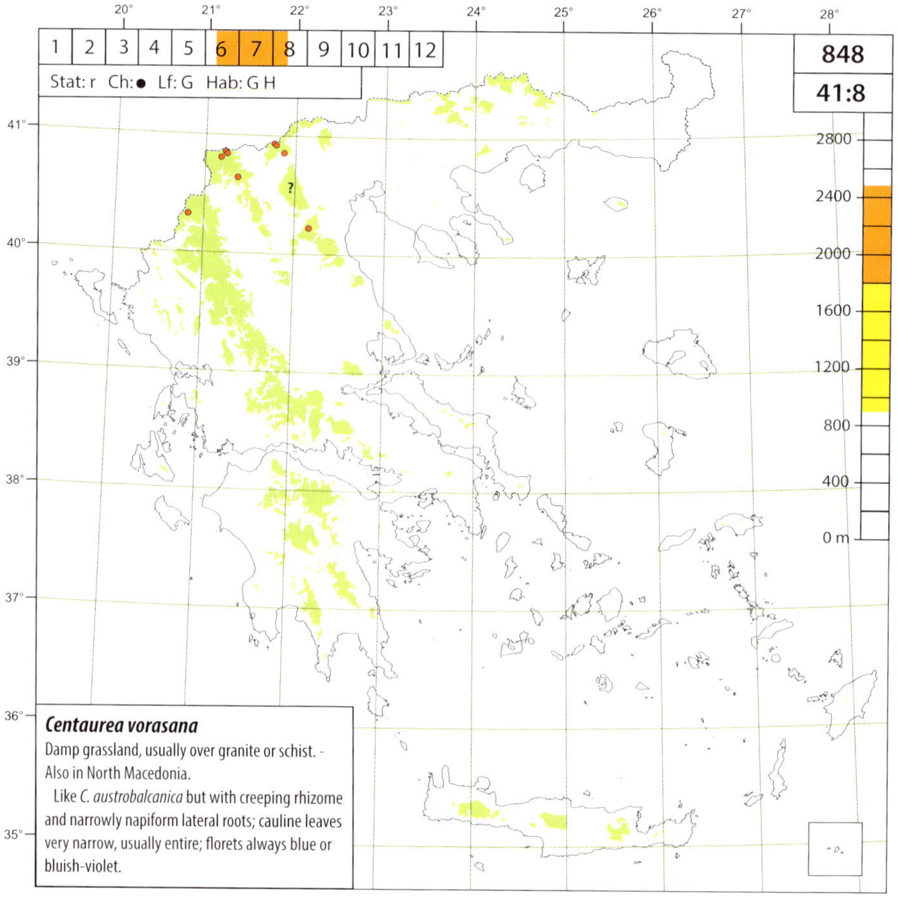

Centaurea vorasana

Damp grassland, usually over granite or schist. - Also in North Macedonia.

Like *C. austrobalcanica* but with creeping rhizome and narrowly napiform lateral roots; cauline leaves very narrow, usually entire; florets always blue or bluish-violet.

849

Stat: r Ch: ?● Lf: H T Hab: C

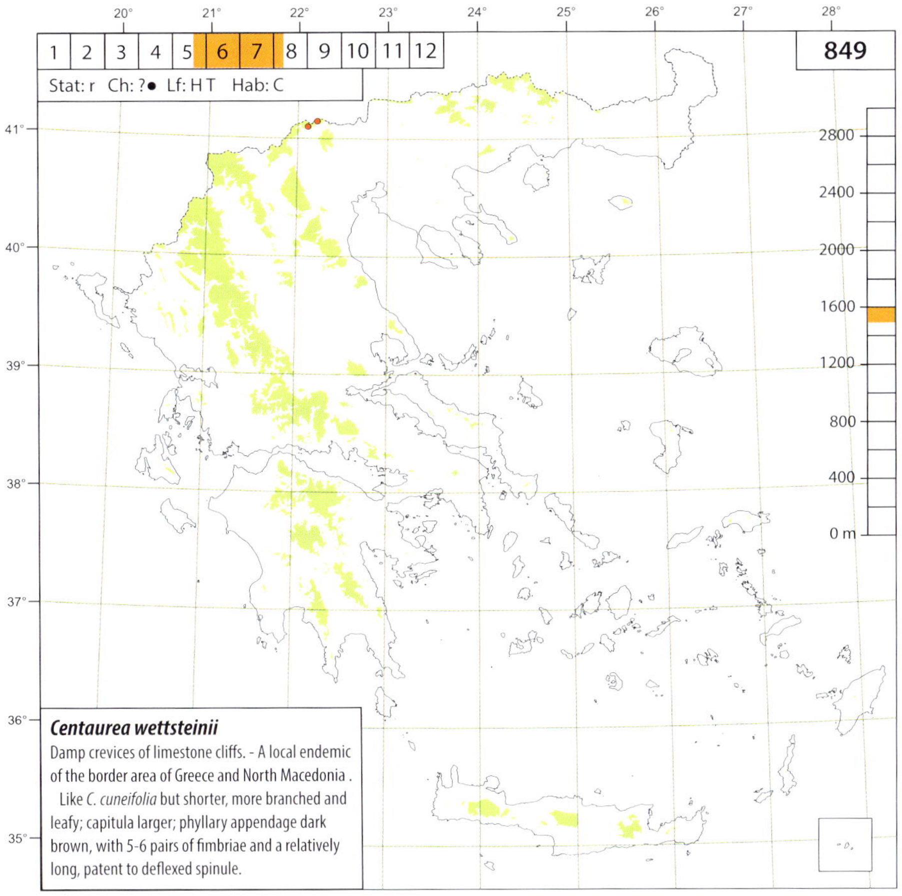

Centaurea wettsteinii

Damp crevices of limestone cliffs. - A local endemic of the border area of Greece and North Macedonia .

Like *C. cuneifolia* but shorter, more branched and leafy; capitula larger; phyllary appendage dark brown, with 5-6 pairs of fimbriae and a relatively long, patent to deflexed spinule.

850

41:9

Stat: r Ch: ● Lf: C H Hab: C

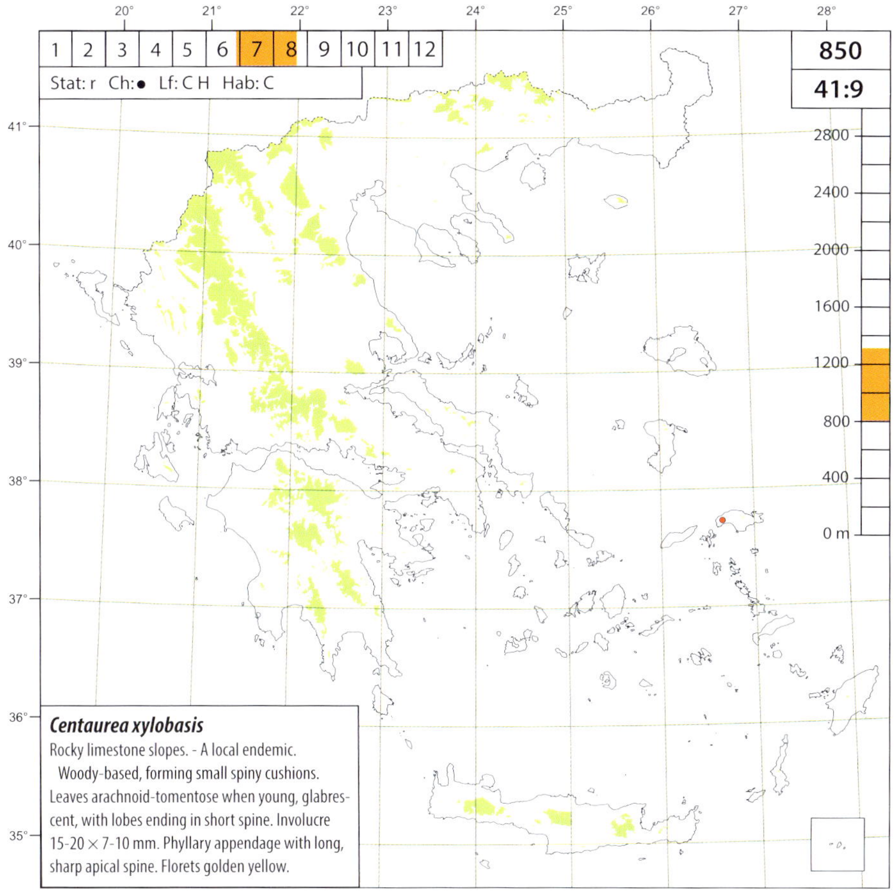

Centaurea xylobasis

Rocky limestone slopes. - A local endemic.

Woody-based, forming small spiny cushions. Leaves arachnoid-tomentose when young, glabrescent, with lobes ending in short spine. Involucre 15-20 × 7-10 mm. Phyllary appendage with long, sharp apical spine. Florets golden yellow.

851

41:10

Stat: r Ch: Bk Lf: H Hab: G W

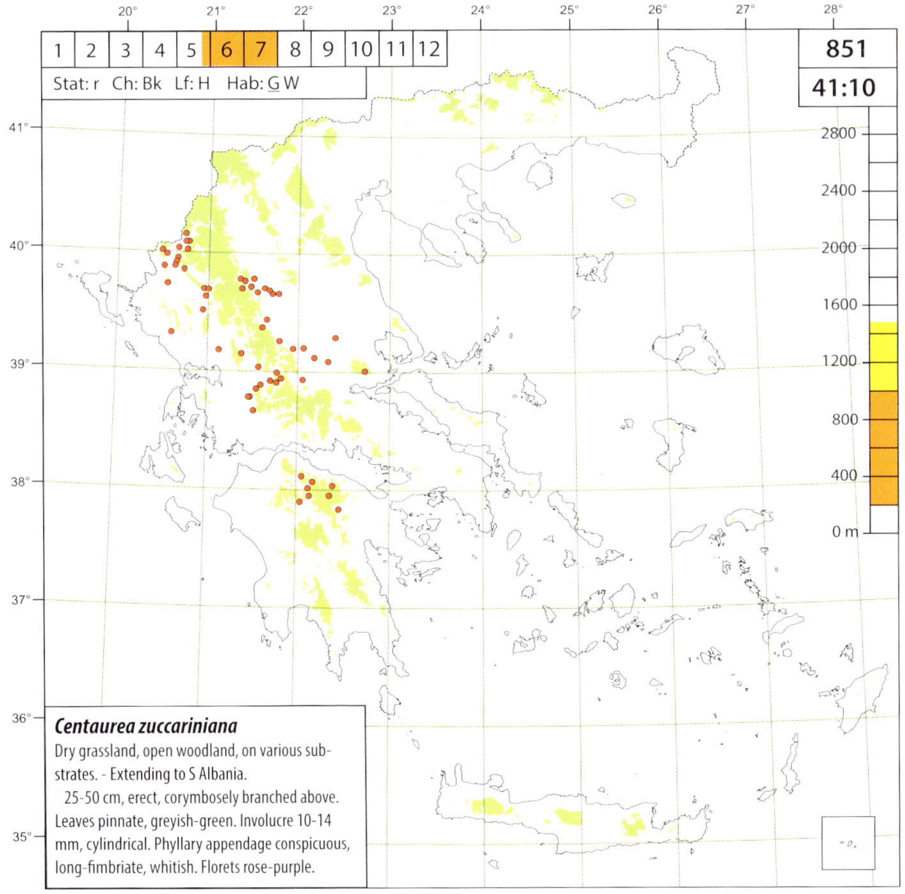

Centaurea zuccariniana

Dry grassland, open woodland, on various substrates. - Extending to S Albania.

25-50 cm, erect, corymbosely branched above. Leaves pinnate, greyish-green. Involucre 10-14 mm, cylindrical. Phyllary appendage conspicuous, long-fimbriate, whitish. Florets rose-purple.

852

268:10

Stat: N Ch: SS Lf: T Hab: P M

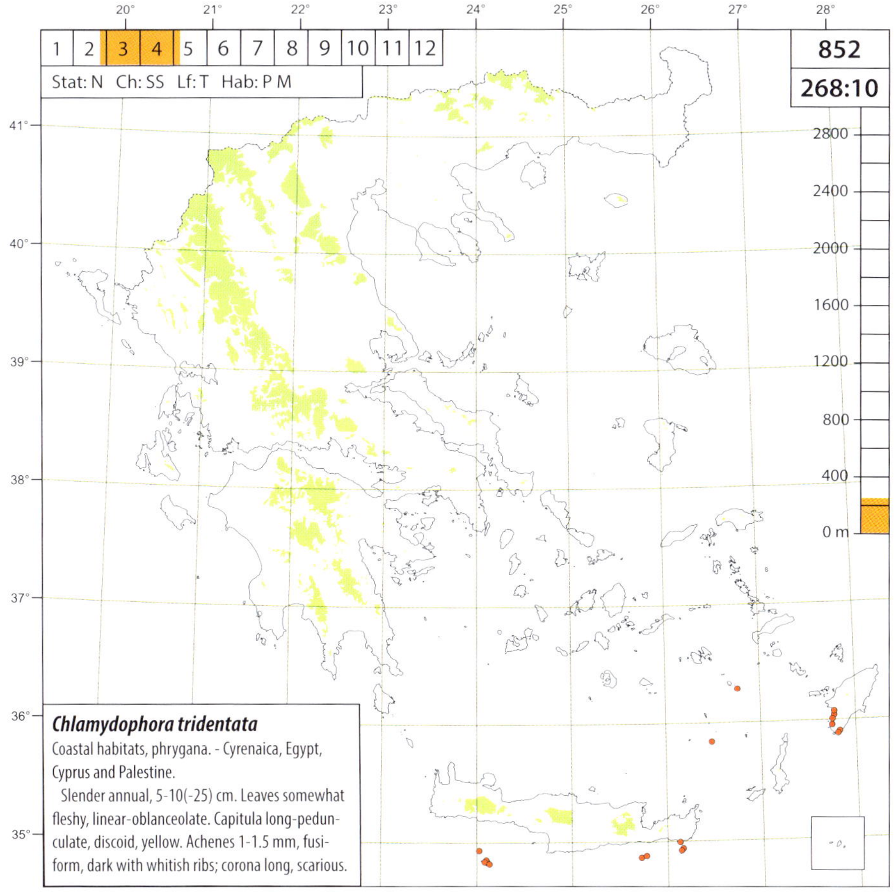

Chlamydophora tridentata

Coastal habitats, phrygana. - Cyrenaica, Egypt, Cyprus and Palestine.

Slender annual, 5-10(-25) cm. Leaves somewhat fleshy, linear-oblanceolate. Capitula long-pedunculate, discoid, yellow. Achenes 1-1.5 mm, fusiform, dark with whitish ribs; corona long, scarious.

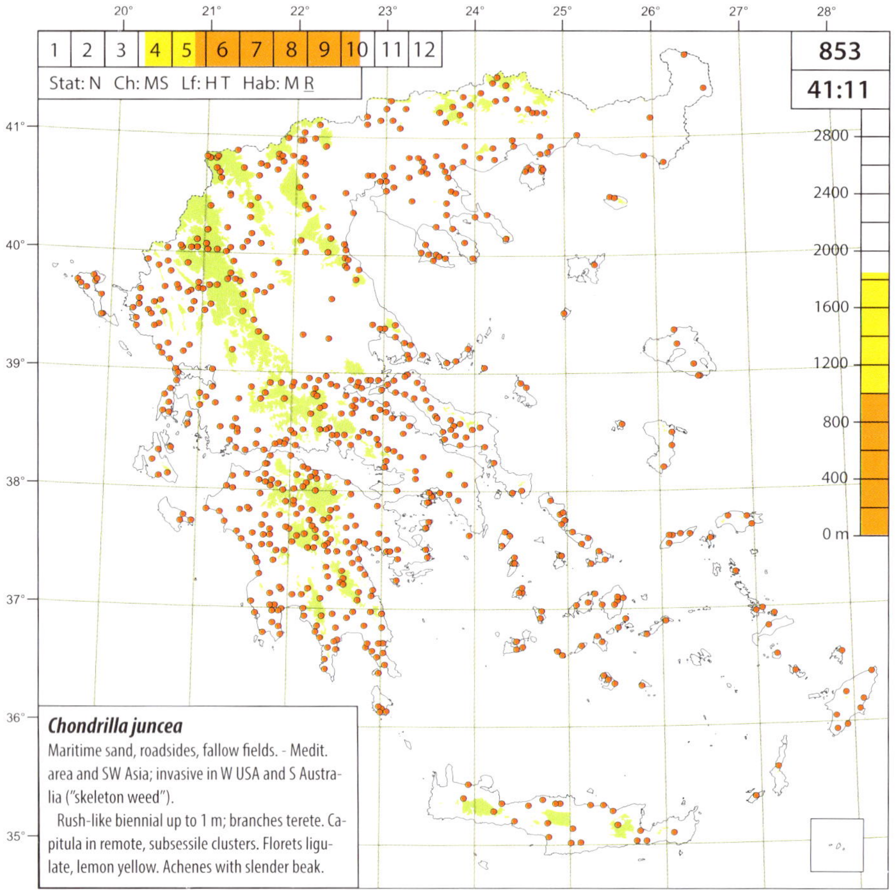

Chondrilla juncea

Maritime sand, roadsides, fallow fields. - Medit. area and SW Asia; invasive in W USA and S Australia ("skeleton weed").

Rush-like biennial up to 1 m; branches terete. Capitula in remote, subsessile clusters. Florets ligulate, lemon yellow. Achenes with slender beak.

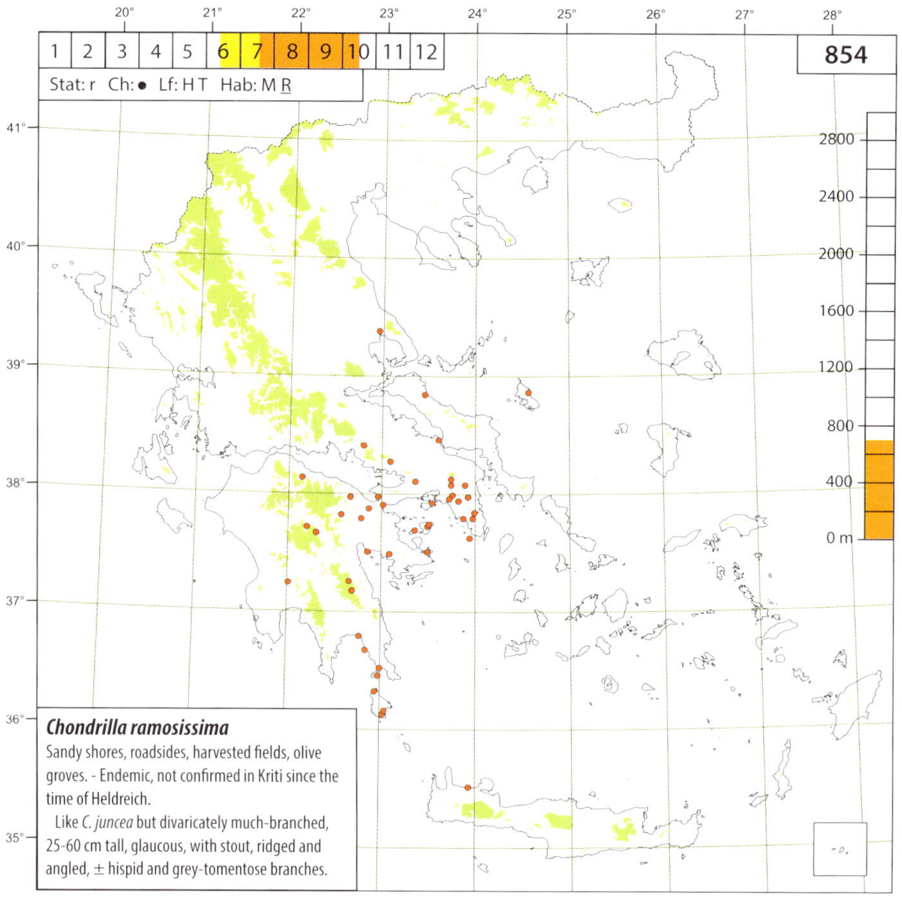

Chondrilla ramosissima

Sandy shores, roadsides, harvested fields, olive groves. - Endemic, not confirmed in Kriti since the time of Heldreich.

Like *C. juncea* but divaricately much-branched, 25-60 cm tall, glaucous, with stout, ridged and angled, ± hispid and grey-tomentose branches.

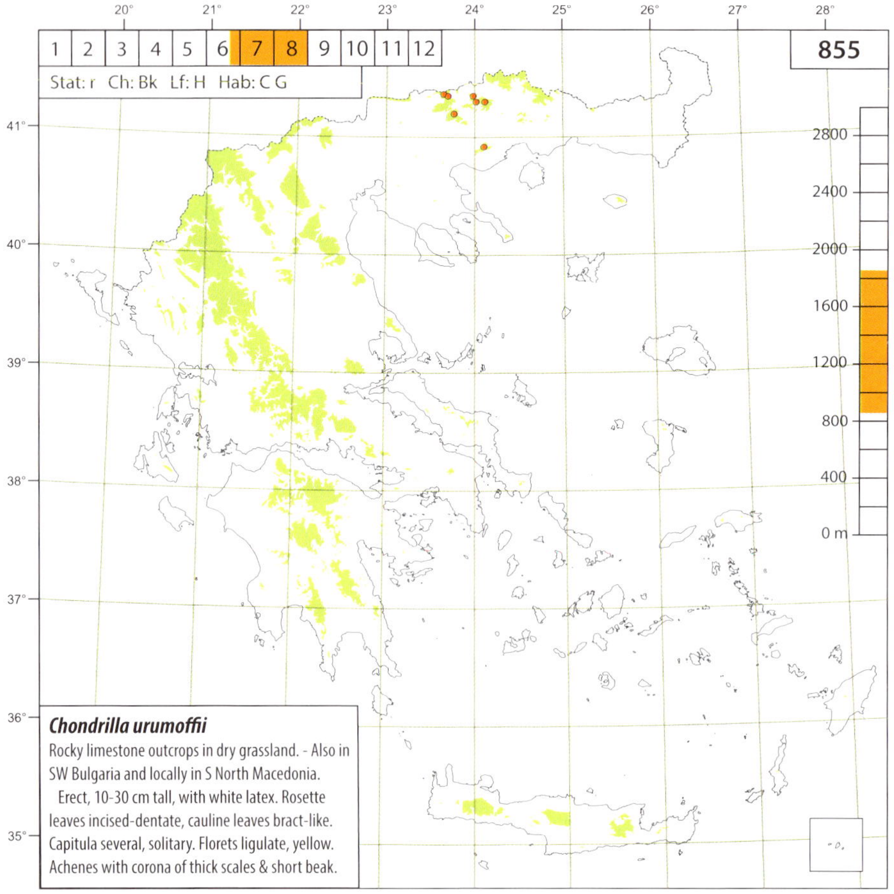

Chondrilla urumoffii

Rocky limestone outcrops in dry grassland. - Also in SW Bulgaria and locally in S North Macedonia.

Erect, 10-30 cm tall, with white latex. Rosette leaves incised-dentate, cauline leaves bract-like. Capitula several, solitary. Florets ligulate, yellow. Achenes with corona of thick scales & short beak.

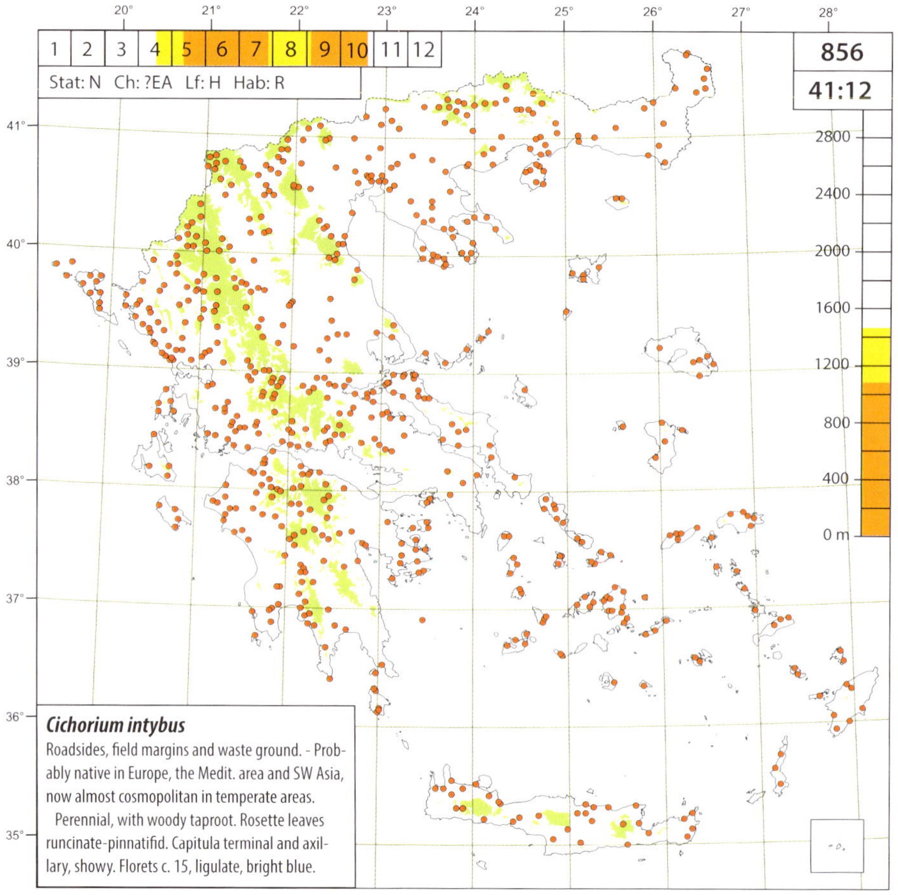

Cichorium intybus

Roadsides, field margins and waste ground. - Probably native in Europe, the Medit. area and SW Asia, now almost cosmopolitan in temperate areas.

Perennial, with woody taproot. Rosette leaves runcinate-pinnatifid. Capitula terminal and axillary, showy. Florets c. 15, ligulate, bright blue.

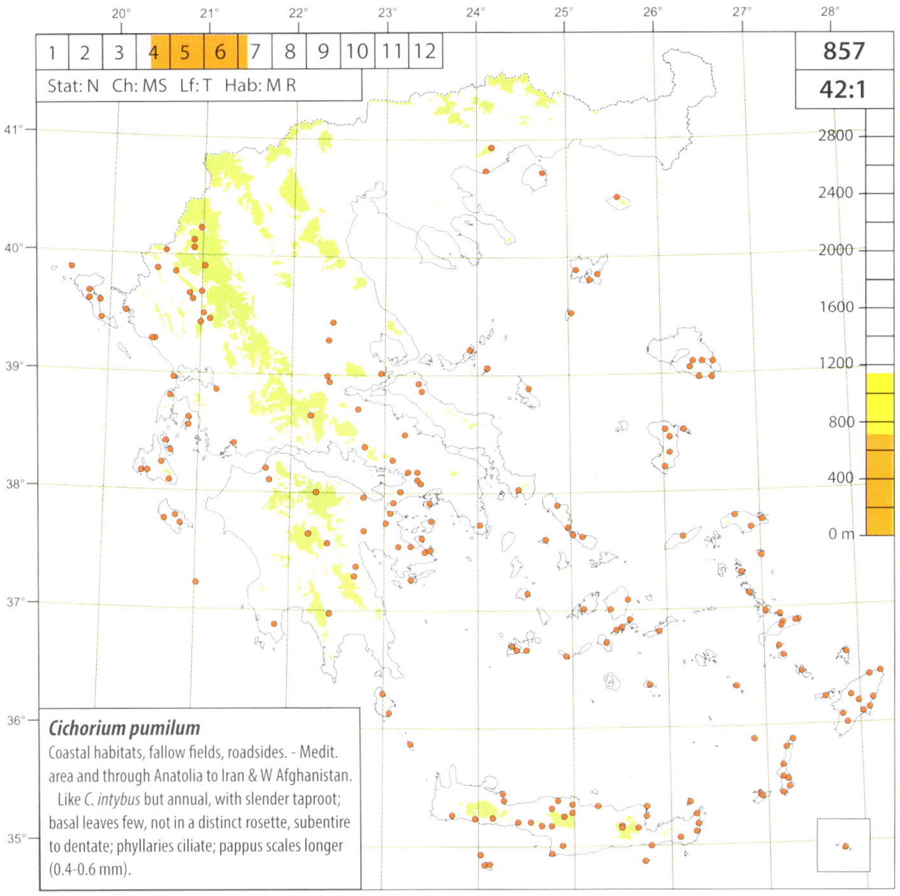

Cichorium pumilum

Coastal habitats, fallow fields, roadsides. - Medit. area and through Anatolia to Iran & W Afghanistan.

Like *C. intybus* but annual, with slender taproot; basal leaves few, not in a distinct rosette, subentire to dentate; phyllaries ciliate; pappus scales longer (0.4-0.6 mm).

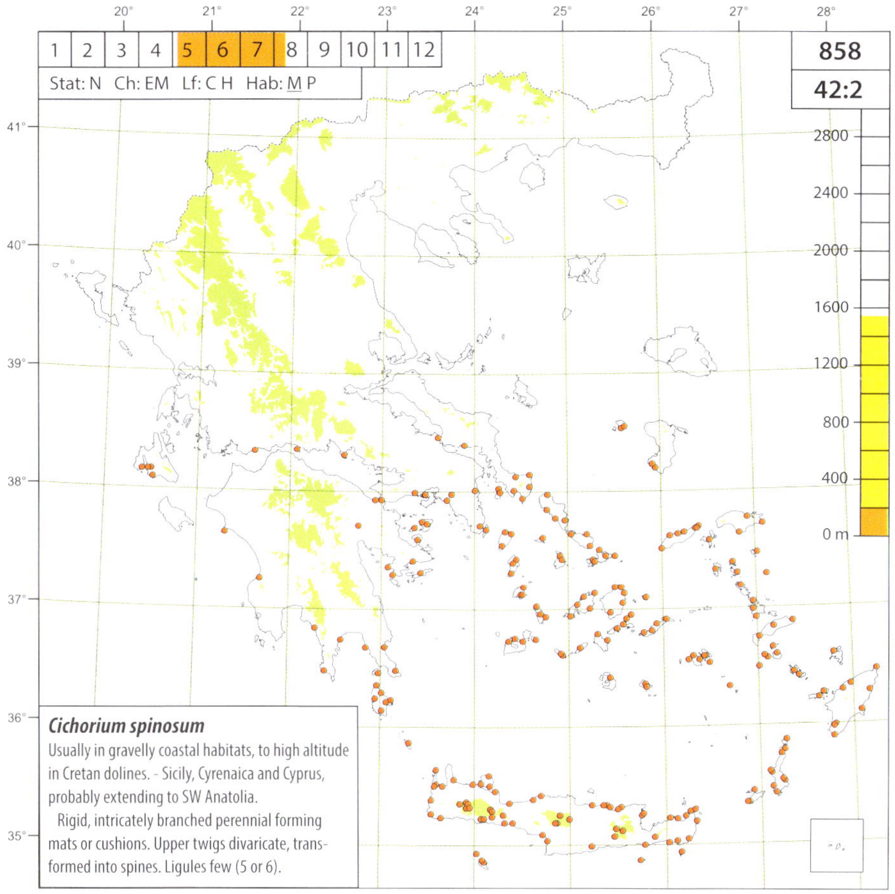

Cichorium spinosum

Usually in gravelly coastal habitats, to high altitude in Cretan dolines. - Sicily, Cyrenaica and Cyprus, probably extending to SW Anatolia.

Rigid, intricately branched perennial forming mats or cushions. Upper twigs divaricate, transformed into spines. Ligules few (5 or 6).

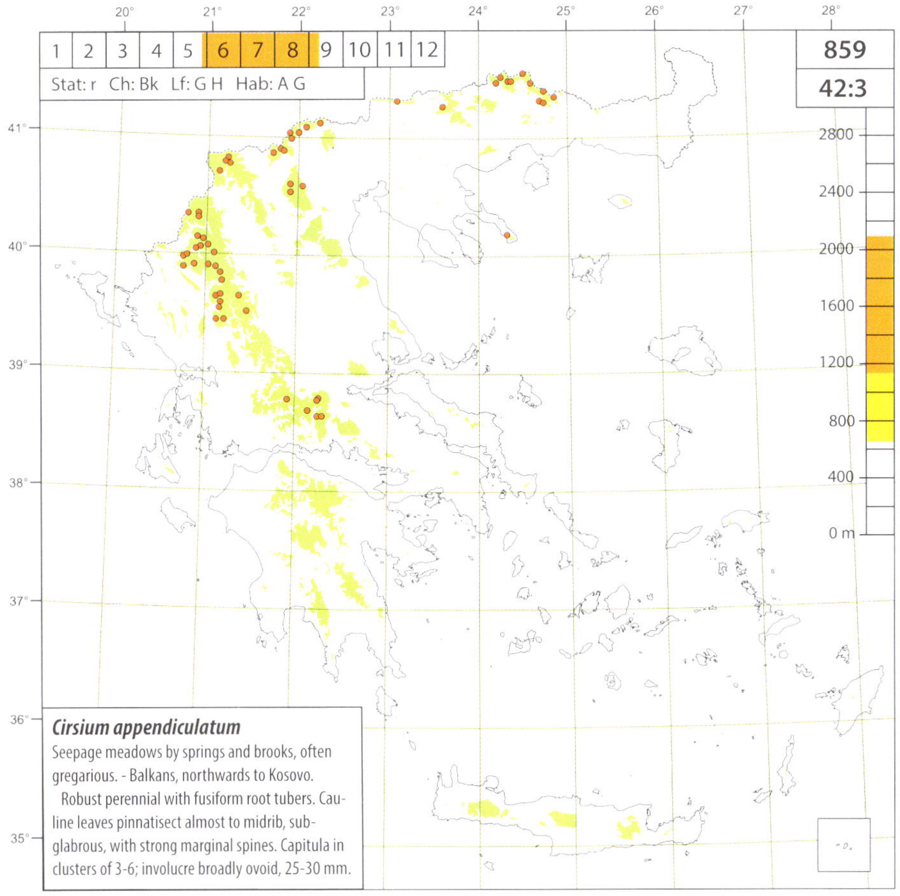

Cirsium appendiculatum

Seepage meadows by springs and brooks, often gregarious. - Balkans, northwards to Kosovo.

Robust perennial with fusiform root tubers. Cauline leaves pinnatisect almost to midrib, subglabrous, with strong marginal spines. Capitula in clusters of 3-6; involucre broadly ovoid, 25-30 mm.

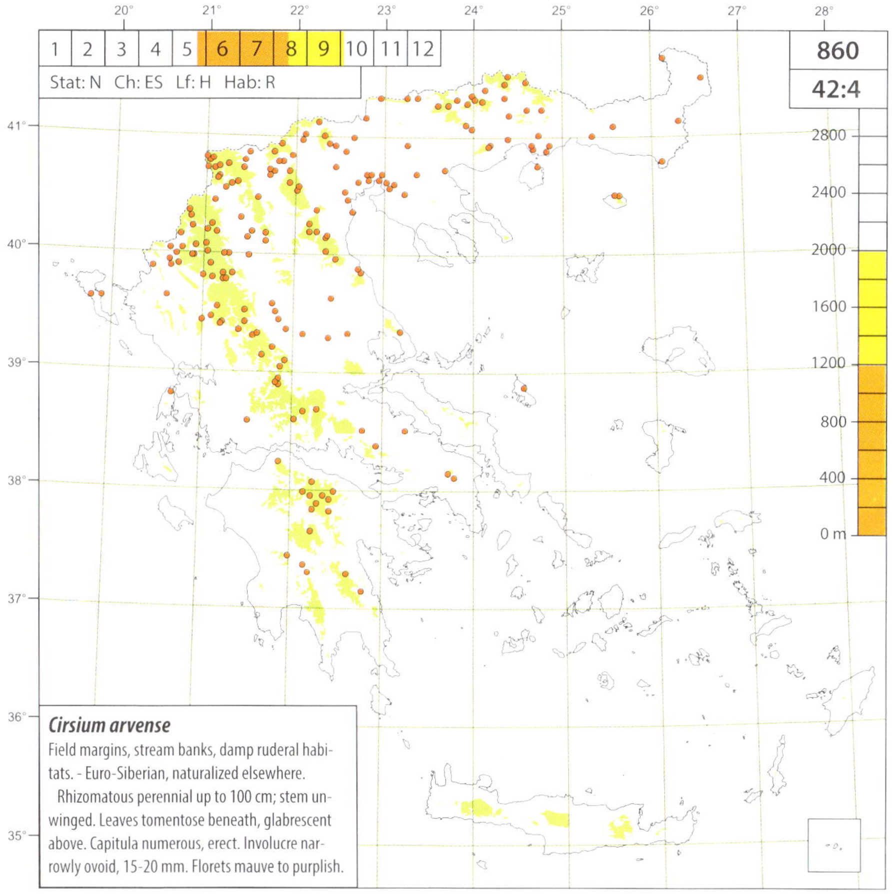

Cirsium arvense

Field margins, stream banks, damp ruderal habitats. - Euro-Siberian, naturalized elsewhere.

Rhizomatous perennial up to 100 cm; stem unwinged. Leaves tomentose beneath, glabrescent above. Capitula numerous, erect. Involucre narrowly ovoid, 15-20 mm. Florets mauve to purplish.

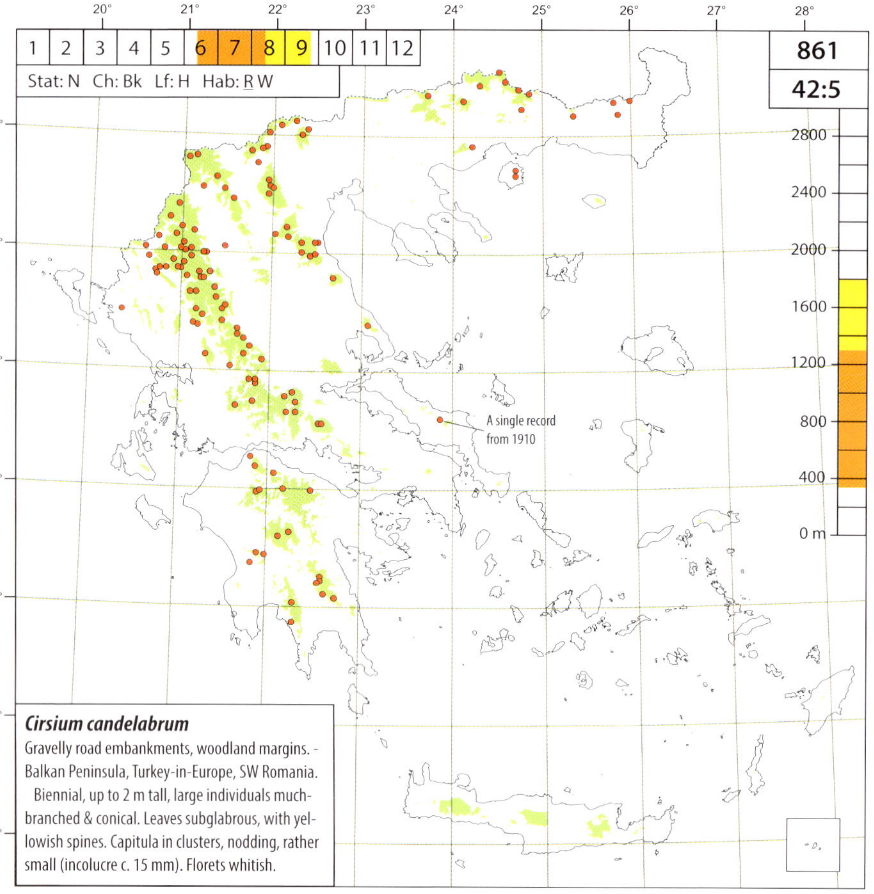

Cirsium candelabrum
Gravelly road embankments, woodland margins. - Balkan Peninsula, Turkey-in-Europe, SW Romania.
Biennial, up to 2 m tall, large individuals much-branched & conical. Leaves subglabrous, with yellowish spines. Capitula in clusters, nodding, rather small (incolucre c. 15 mm). Florets whitish.

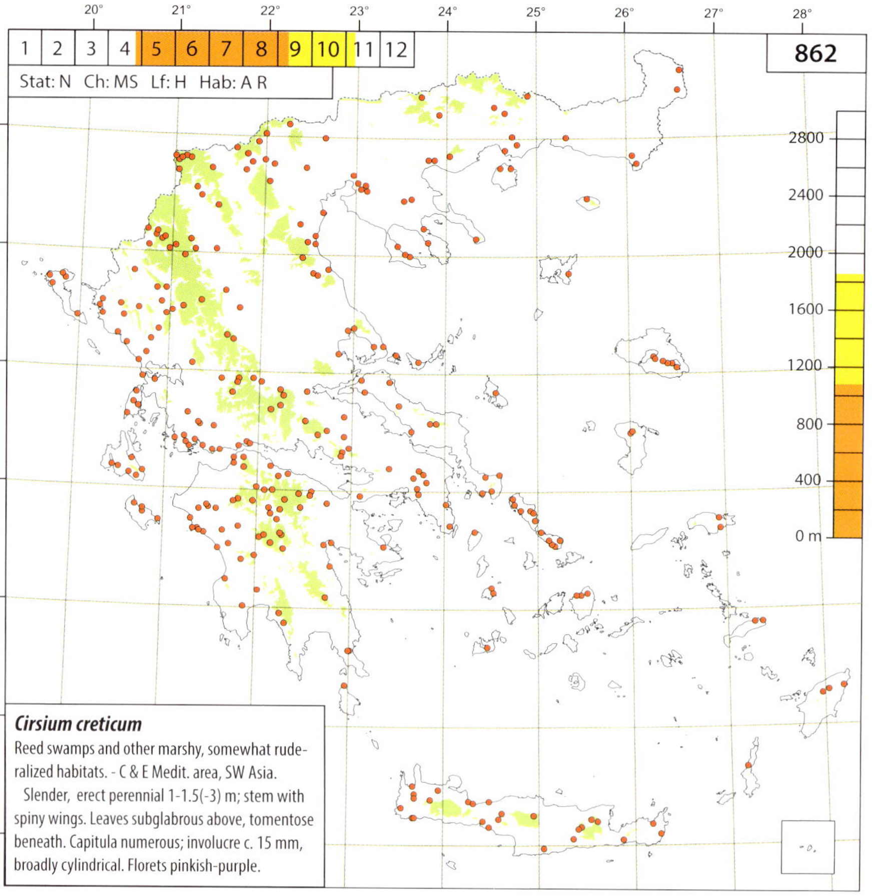

Cirsium creticum
Reed swamps and other marshy, somewhat ruderalized habitats. - C & E Medit. area, SW Asia.
Slender, erect perennial 1-1.5(-3) m; stem with spiny wings. Leaves subglabrous above, tomentose beneath. Capitula numerous; involucre c. 15 mm, broadly cylindrical. Florets pinkish-purple.

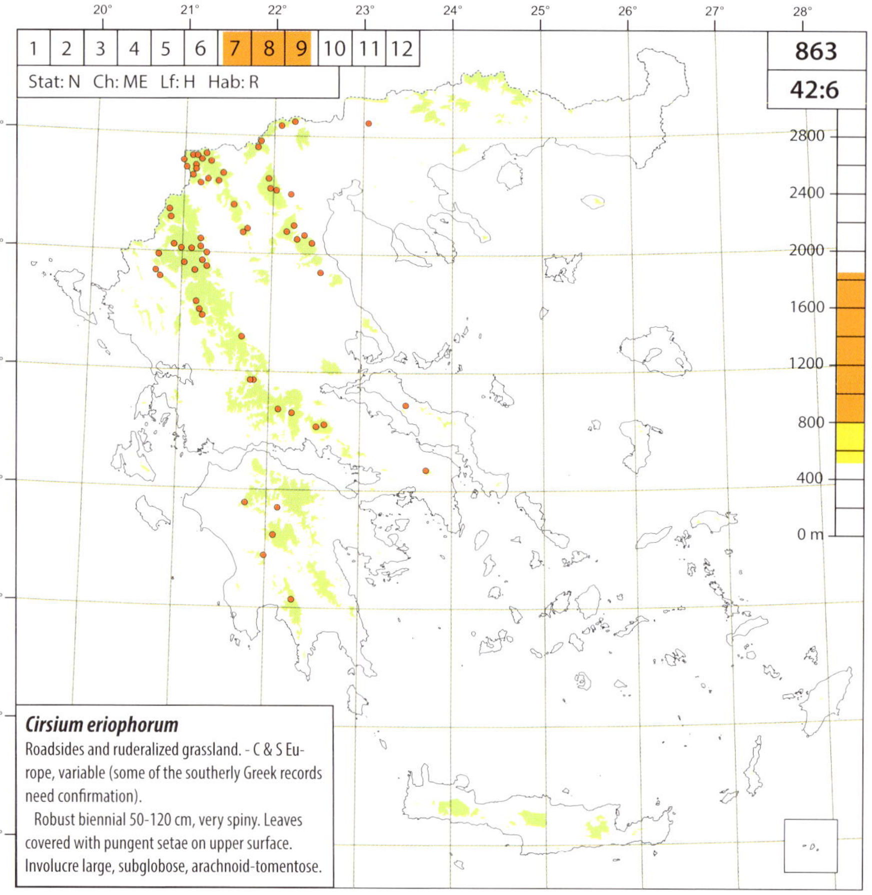

Cirsium eriophorum
Roadsides and ruderalized grassland. - C & S Europe, variable (some of the southerly Greek records need confirmation).
Robust biennial 50-120 cm, very spiny. Leaves covered with pungent setae on upper surface. Involucre large, subglobose, arachnoid-tomentose.

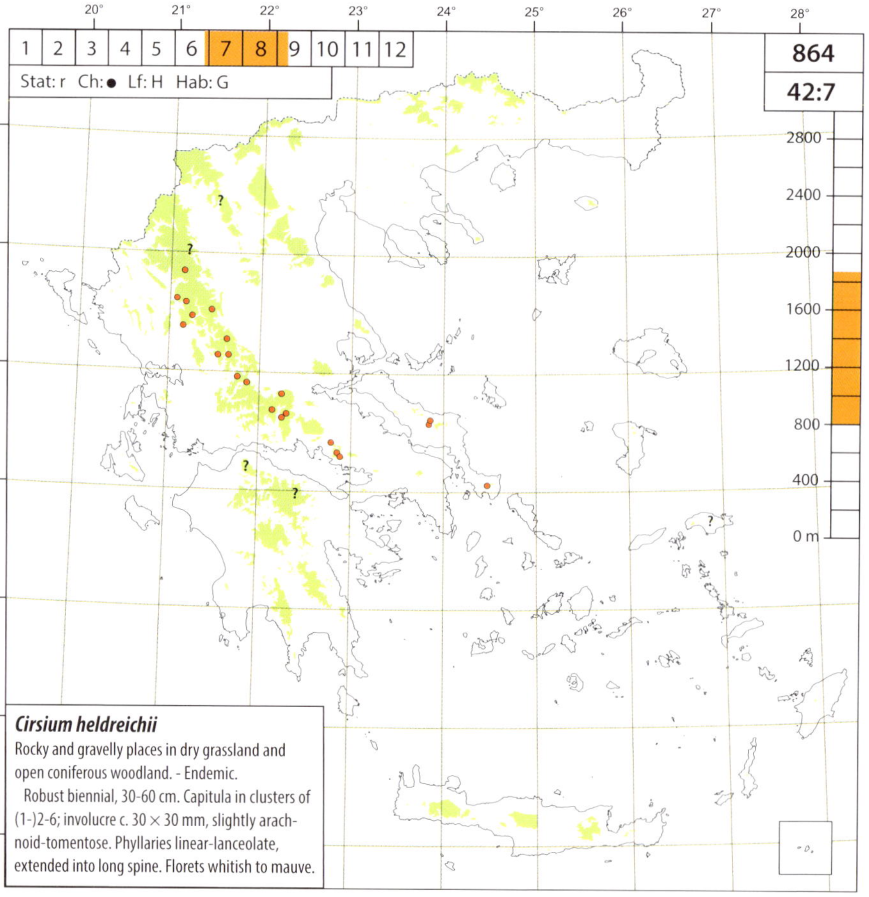

Cirsium heldreichii
Rocky and gravelly places in dry grassland and open coniferous woodland. - Endemic.
Robust biennial, 30-60 cm. Capitula in clusters of (1-)2-6; involucre c. 30 × 30 mm, slightly arachnoid-tomentose. Phyllaries linear-lanceolate, extended into long spine. Florets whitish to mauve.

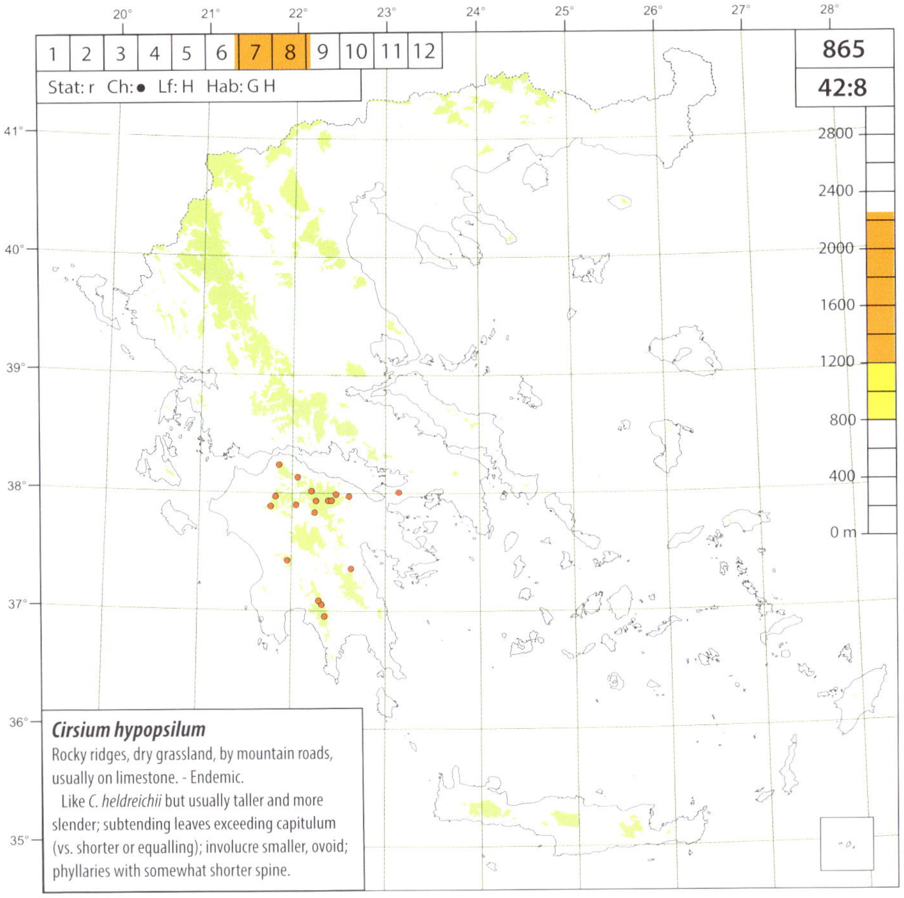

Cirsium hypopsilum

Rocky ridges, dry grassland, by mountain roads, usually on limestone. - Endemic.

Like *C. heldreichii* but usually taller and more slender; subtending leaves exceeding capitulum (vs. shorter or equalling); involucre smaller, ovoid; phyllaries with somewhat shorter spine.

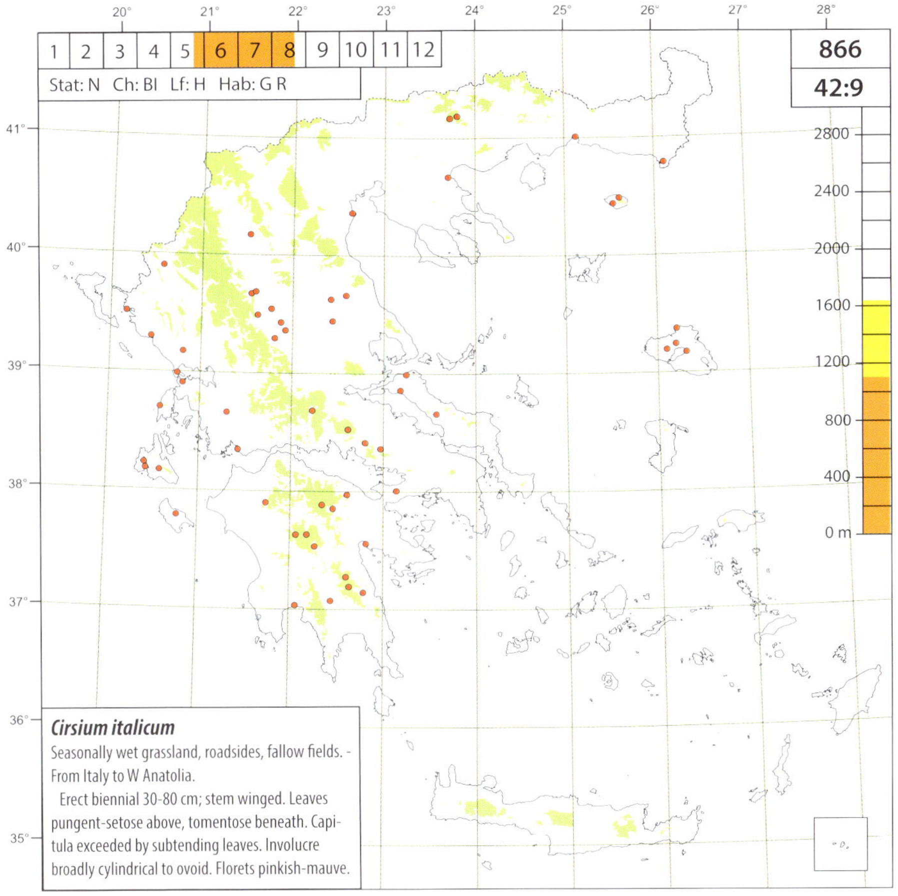

Cirsium italicum

Seasonally wet grassland, roadsides, fallow fields. - From Italy to W Anatolia.

Erect biennial 30-80 cm; stem winged. Leaves pungent-setose above, tomentose beneath. Capitula exceeded by subtending leaves. Involucre broadly cylindrical to ovoid. Florets pinkish-mauve.

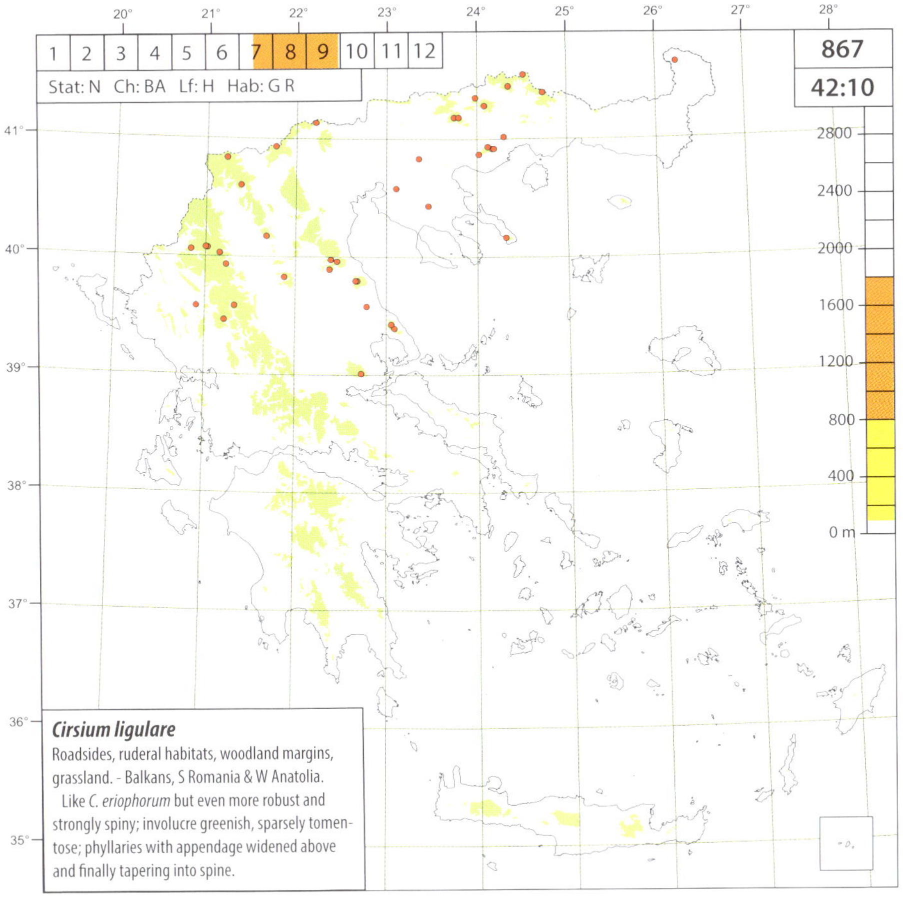

Cirsium ligulare

Roadsides, ruderal habitats, woodland margins, grassland. - Balkans, S Romania & W Anatolia.

Like *C. eriophorum* but even more robust and strongly spiny; involucre greenish, sparsely tomentose; phyllaries with appendage widened above and finally tapering into spine.

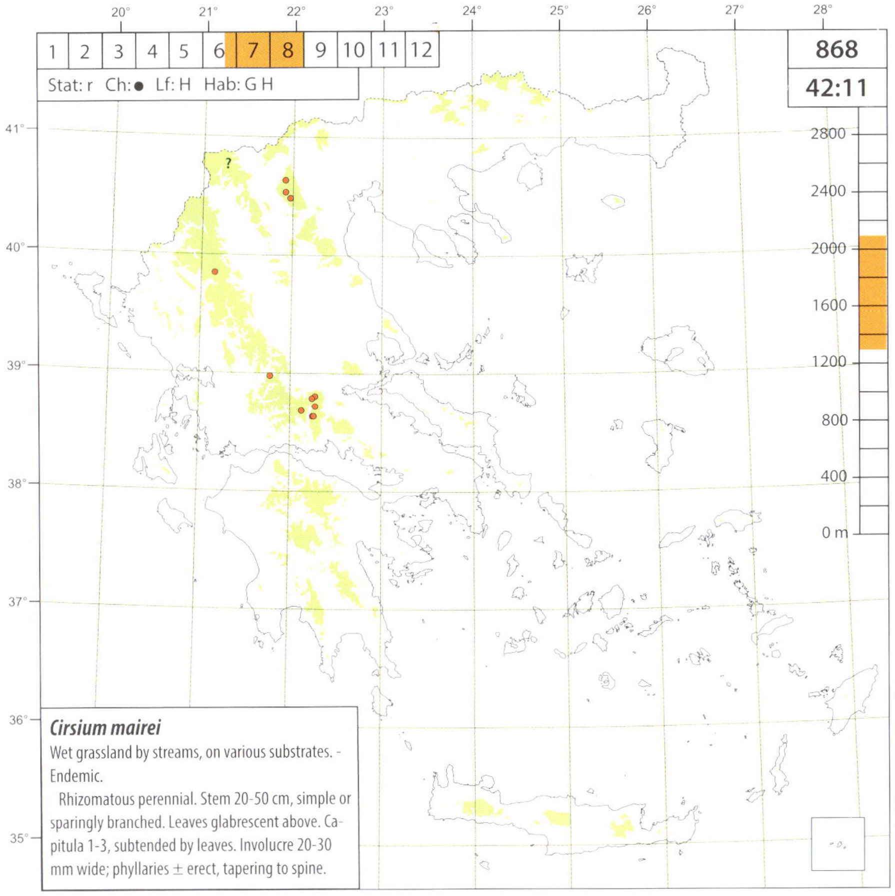

Cirsium mairei

Wet grassland by streams, on various substrates. - Endemic.

Rhizomatous perennial. Stem 20-50 cm, simple or sparingly branched. Leaves glabrescent above. Capitula 1-3, subtended by leaves. Involucre 20-30 mm wide; phyllaries ± erect, tapering to spine.

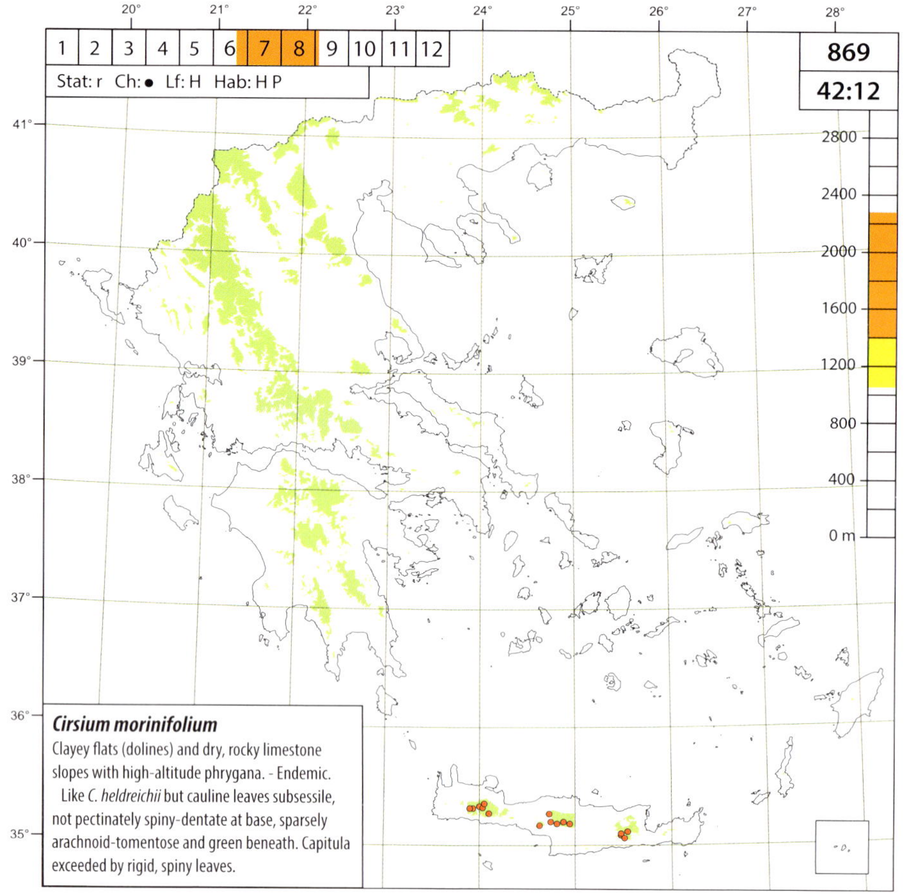

Cirsium morinifolium

Clayey flats (dolines) and dry, rocky limestone slopes with high-altitude phrygana. - Endemic.

Like *C. heldreichii* but cauline leaves subsessile, not pectinately spiny-dentate at base, sparsely arachnoid-tomentose and green beneath. Capitula exceeded by rigid, spiny leaves.

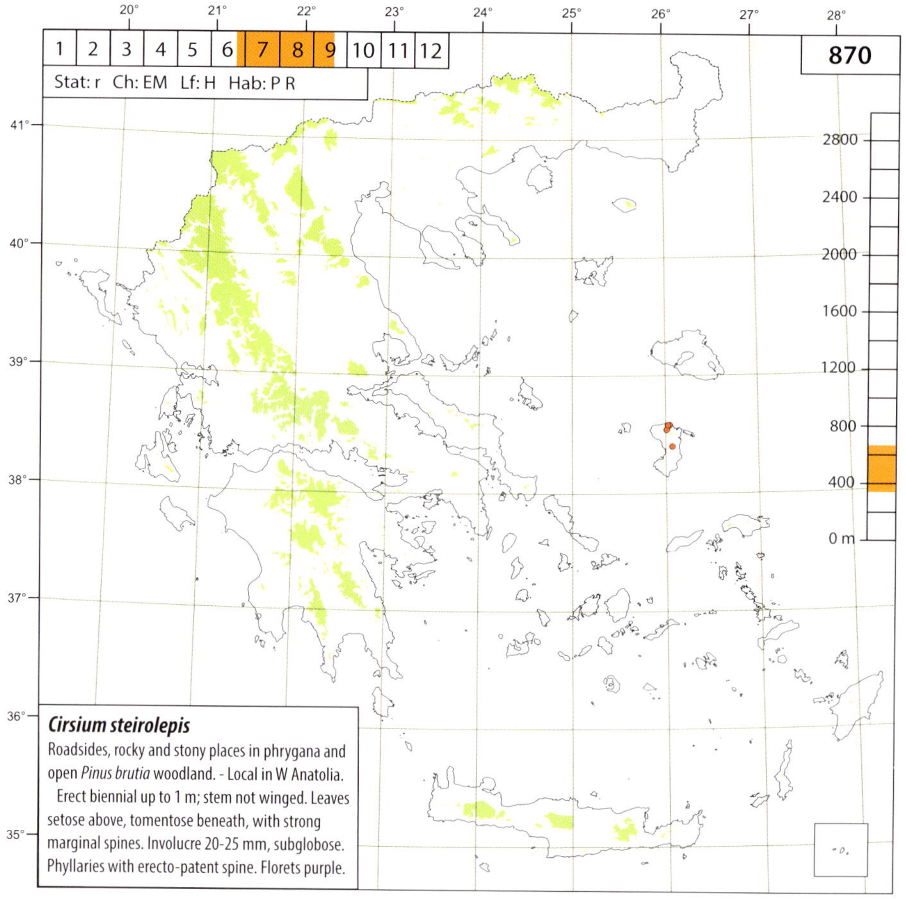

Cirsium steirolepis

Roadsides, rocky and stony places in phrygana and open *Pinus brutia* woodland. - Local in W Anatolia.

Erect biennial up to 1 m; stem not winged. Leaves setose above, tomentose beneath, with strong marginal spines. Involucre 20-25 mm, subglobose. Phyllaries with erecto-patent spine. Florets purple.

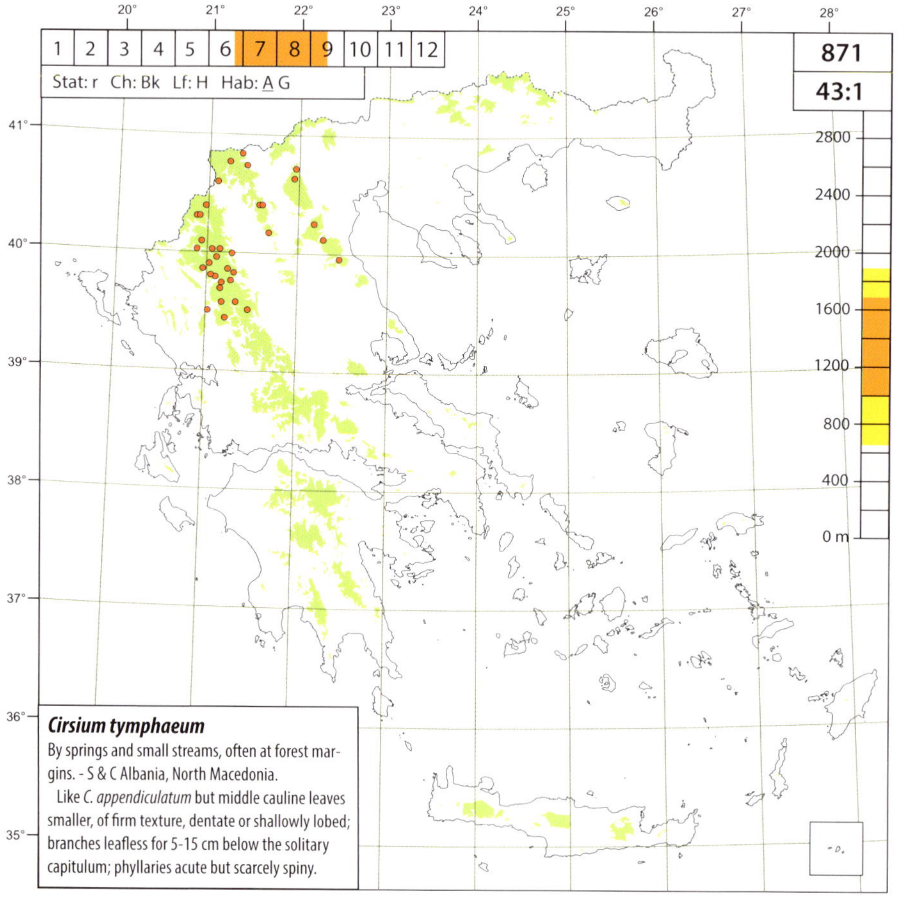

Cirsium tymphaeum

By springs and small streams, often at forest margins. - S & C Albania, North Macedonia.

Like *C. appendiculatum* but middle cauline leaves smaller, of firm texture, dentate or shallowly lobed; branches leafless for 5-15 cm below the solitary capitulum; phyllaries acute but scarcely spiny.

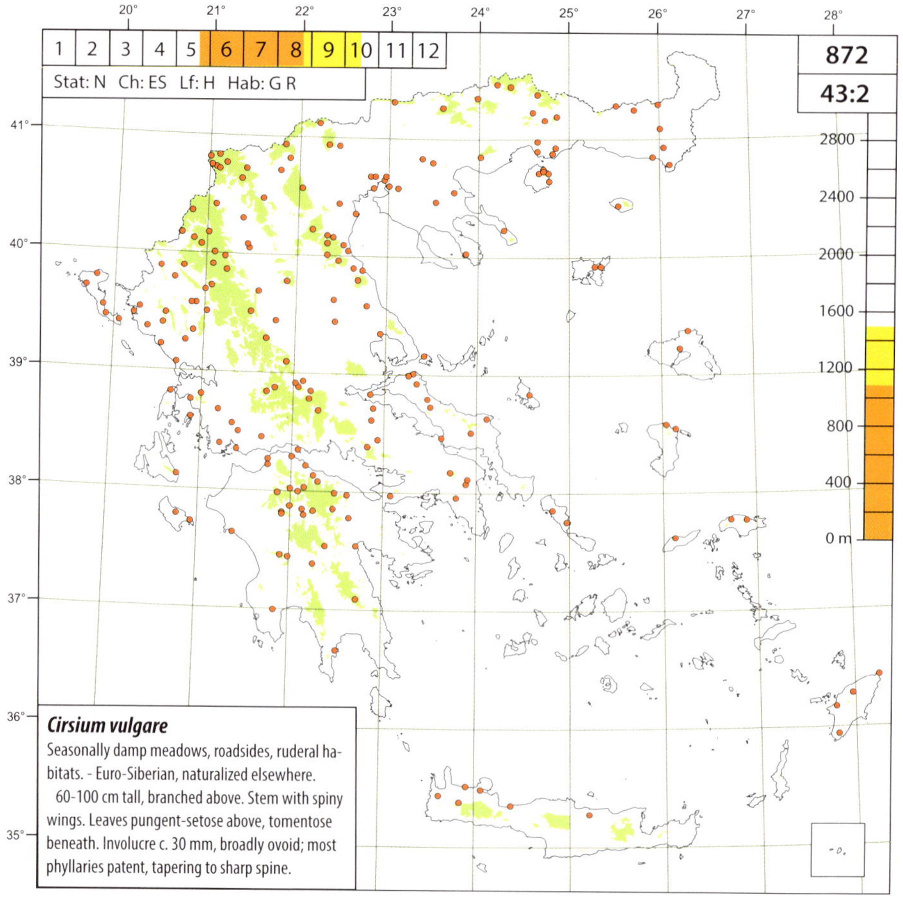

Cirsium vulgare

Seasonally damp meadows, roadsides, ruderal habitats. - Euro-Siberian, naturalized elsewhere.

60-100 cm tall, branched above. Stem with spiny wings. Leaves pungent-setose above, tomentose beneath. Involucre c. 30 mm, broadly ovoid; most phyllaries patent, tapering to sharp spine.

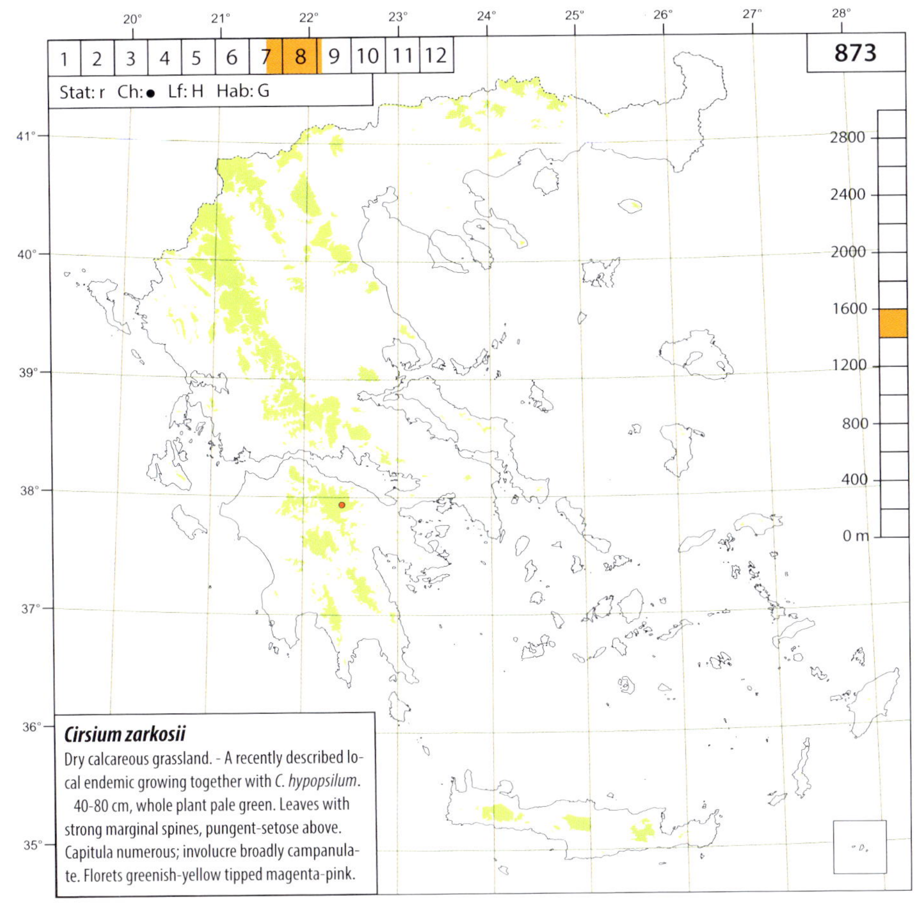

Cirsium zarkosii

Dry calcareous grassland. - A recently described local endemic growing together with *C. hypopsilum*.

40-80 cm, whole plant pale green. Leaves with strong marginal spines, pungent-setose above. Capitula numerous; involucre broadly campanulate. Florets greenish-yellow tipped magenta-pink.

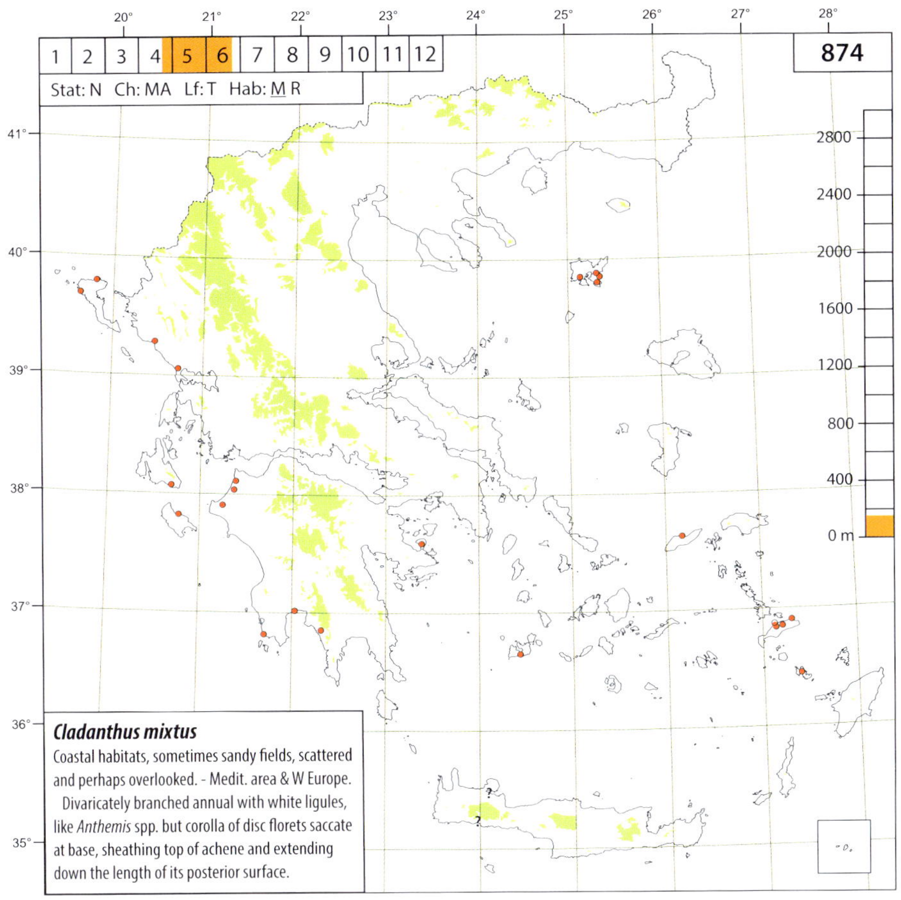

Cladanthus mixtus

Coastal habitats, sometimes sandy fields, scattered and perhaps overlooked. - Medit. area & W Europe.

Divaricately branched annual with white ligules, like *Anthemis* spp. but corolla of disc florets saccate at base, sheathing top of achene and extending down the length of its posterior surface.

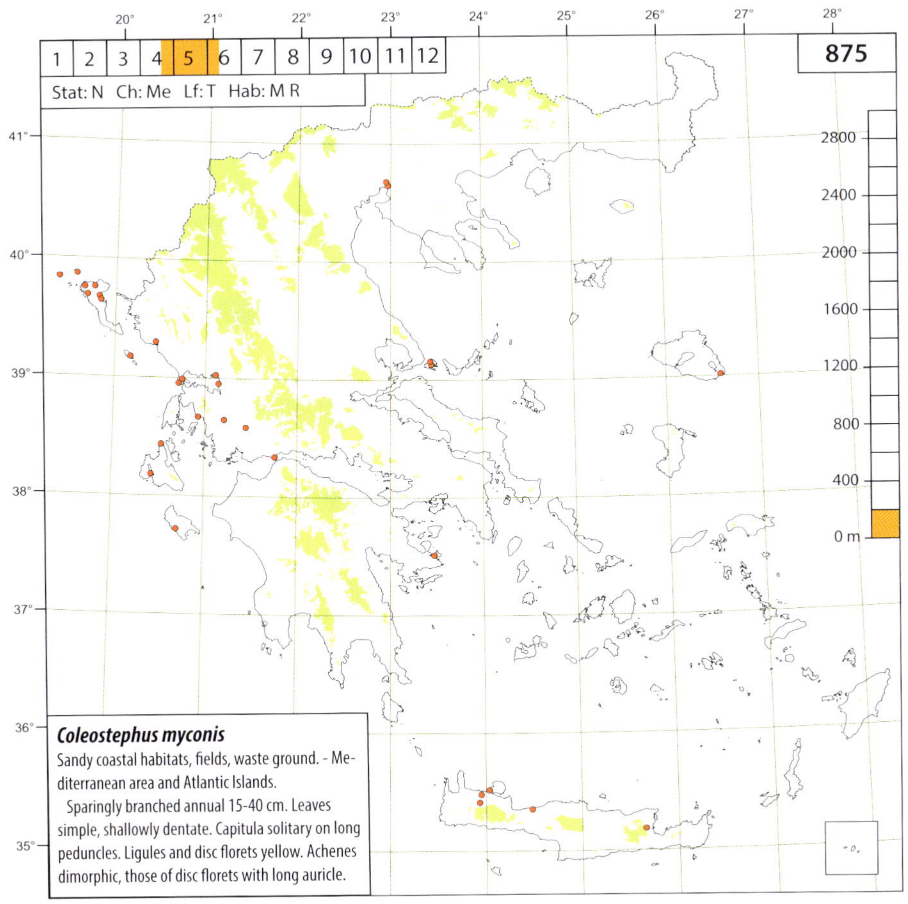

Coleostephus myconis

Sandy coastal habitats, fields, waste ground. - Mediterranean area and Atlantic Islands.

Sparingly branched annual 15-40 cm. Leaves simple, shallowly dentate. Capitula solitary on long peduncles. Ligules and disc florets yellow. Achenes dimorphic, those of disc florets with long auricle.

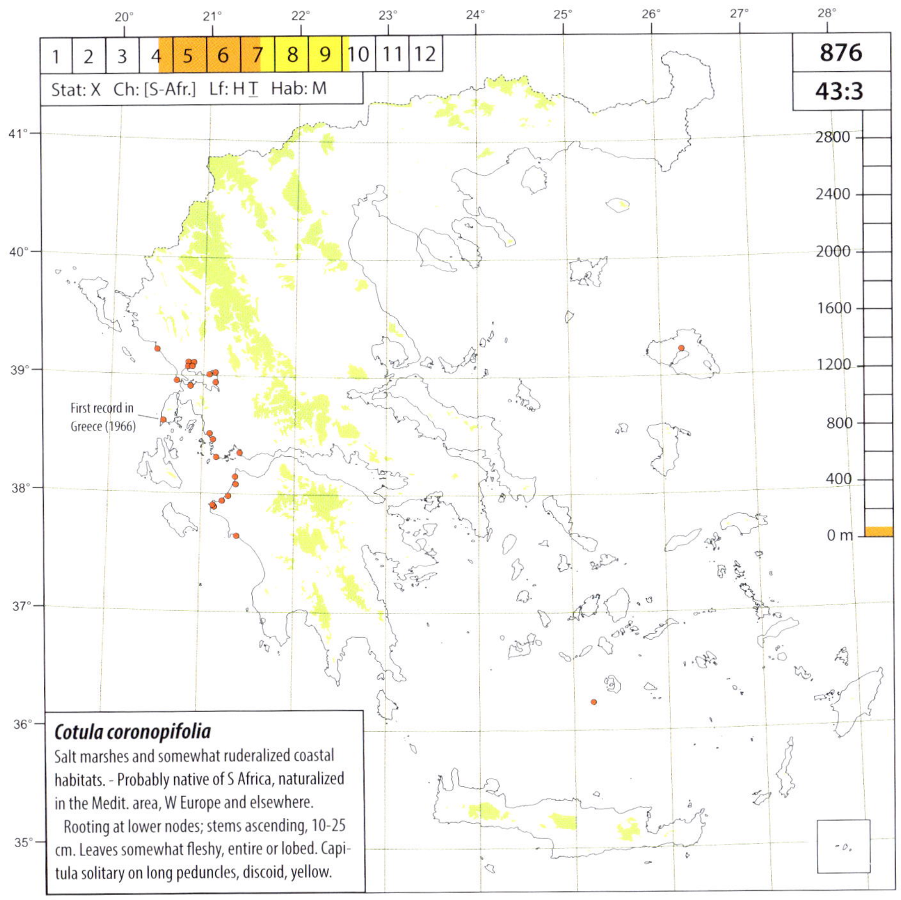

Cotula coronopifolia

Salt marshes and somewhat ruderalized coastal habitats. - Probably native of S Africa, naturalized in the Medit. area, W Europe and elsewhere.

Rooting at lower nodes; stems ascending, 10-25 cm. Leaves somewhat fleshy, entire or lobed. Capitula solitary on long peduncles, discoid, yellow.

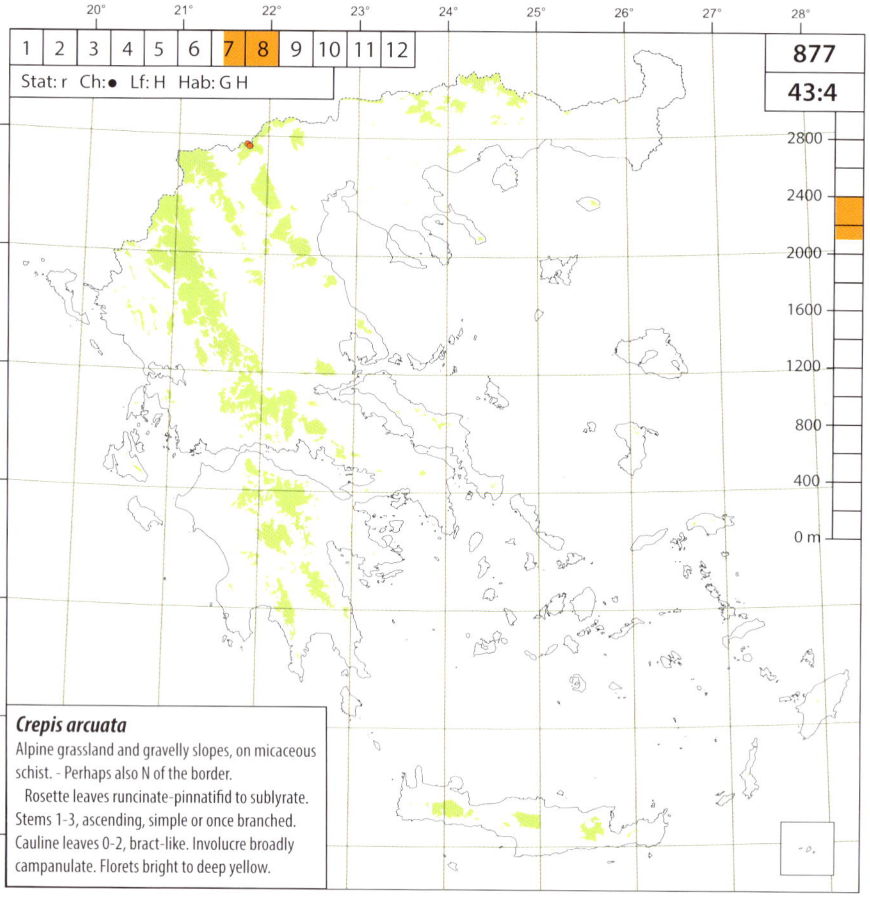

Crepis arcuata

Alpine grassland and gravelly slopes, on micaceous schist. - Perhaps also N of the border.

Rosette leaves runcinate-pinnatifid to sublyrate. Stems 1-3, ascending, simple or once branched. Cauline leaves 0-2, bract-like. Involucre broadly campanulate. Florets bright to deep yellow.

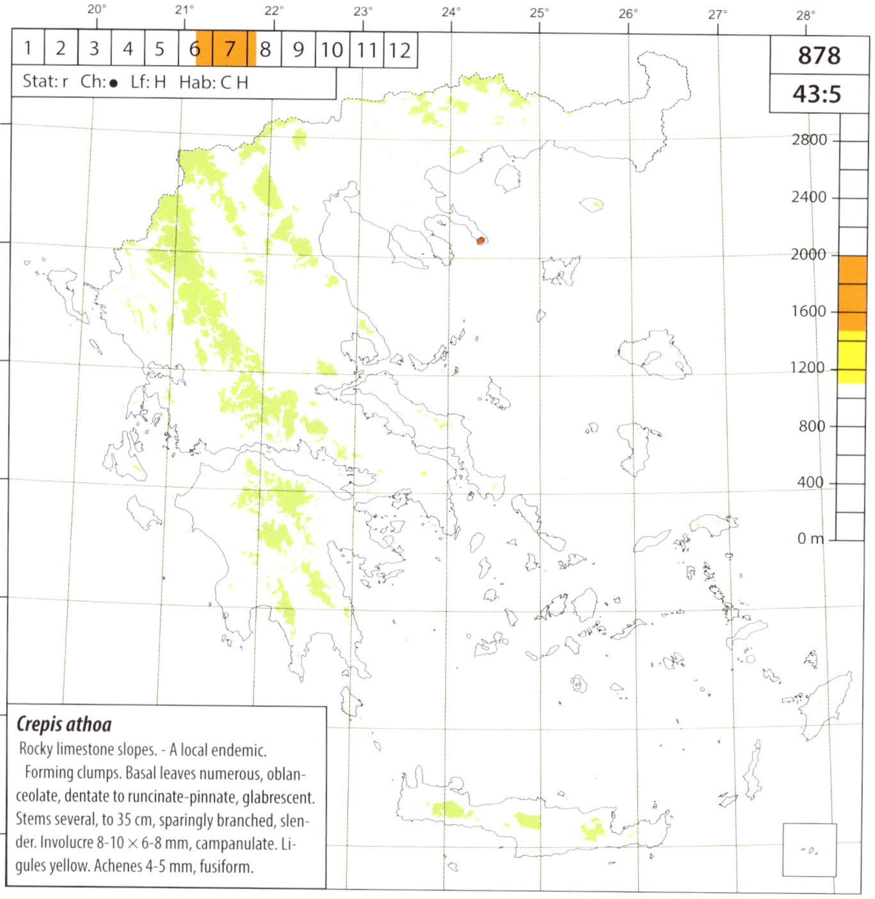

Crepis athoa

Rocky limestone slopes. - A local endemic.

Forming clumps. Basal leaves numerous, oblanceolate, dentate to runcinate-pinnate, glabrescent. Stems several, to 35 cm, sparingly branched, slender. Involucre 8-10 × 6-8 mm, campanulate. Ligules yellow. Achenes 4-5 mm, fusiform.

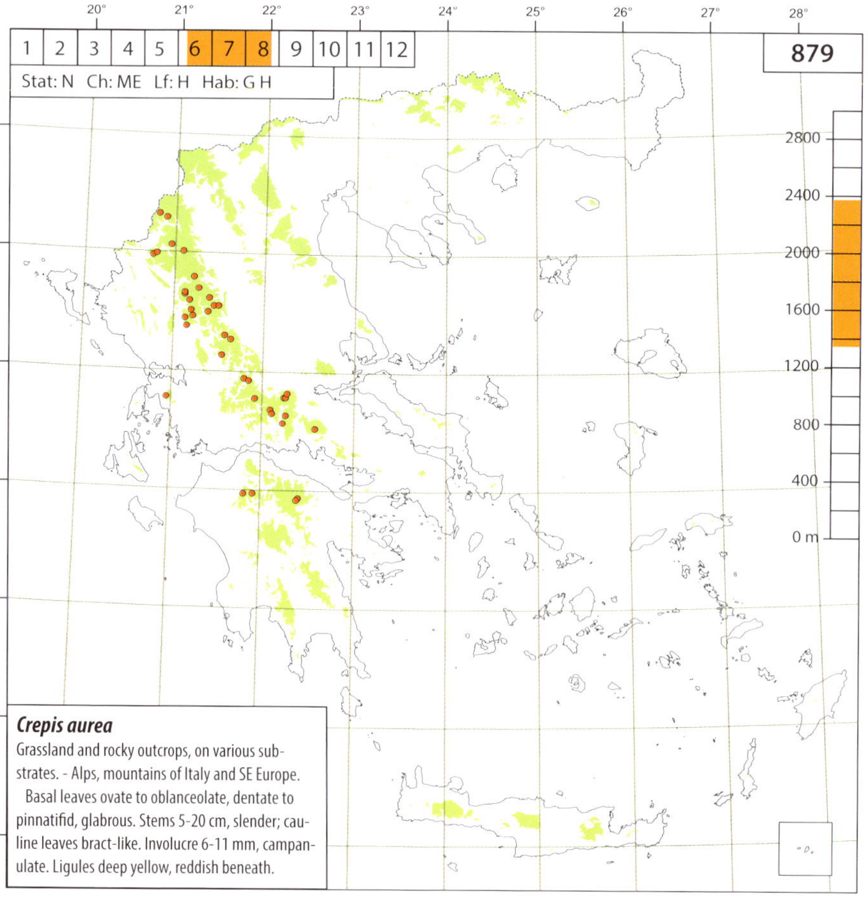

Crepis aurea

Grassland and rocky outcrops, on various substrates. - Alps, mountains of Italy and SE Europe.

Basal leaves ovate to oblanceolate, dentate to pinnatifid, glabrous. Stems 5-20 cm, slender; cauline leaves bract-like. Involucre 6-11 mm, campanulate. Ligules deep yellow, reddish beneath.

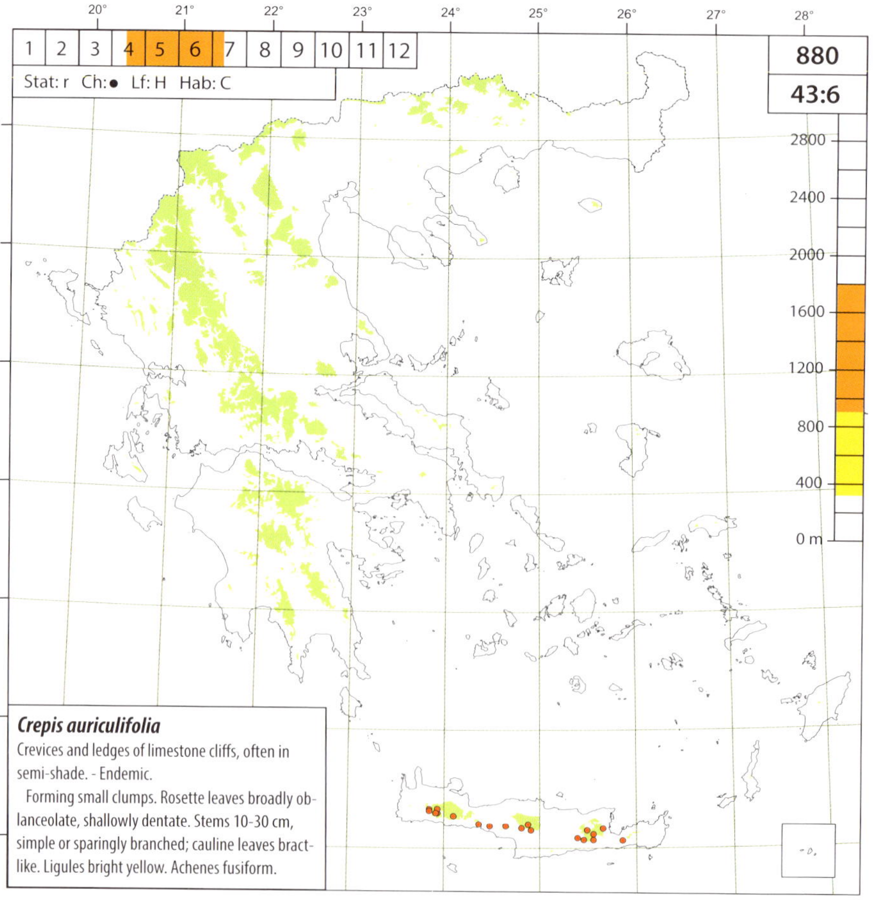

Crepis auriculifolia

Crevices and ledges of limestone cliffs, often in semi-shade. - Endemic.

Forming small clumps. Rosette leaves broadly oblanceolate, shallowly dentate. Stems 10-30 cm, simple or sparingly branched; cauline leaves bract-like. Ligules bright yellow. Achenes fusiform.

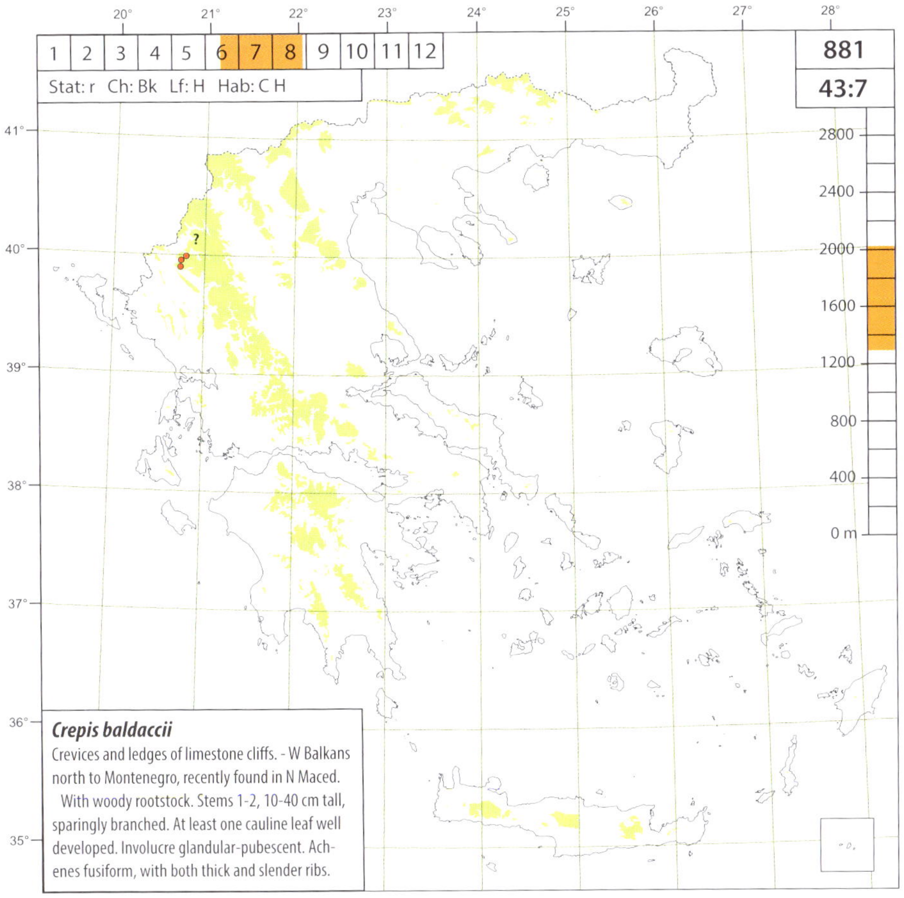

Crepis baldaccii

Crevices and ledges of limestone cliffs. - W Balkans north to Montenegro, recently found in N Maced.

With woody rootstock. Stems 1-2, 10-40 cm tall, sparingly branched. At least one cauline leaf well developed. Involucre glandular-pubescent. Achenes fusiform, with both thick and slender ribs.

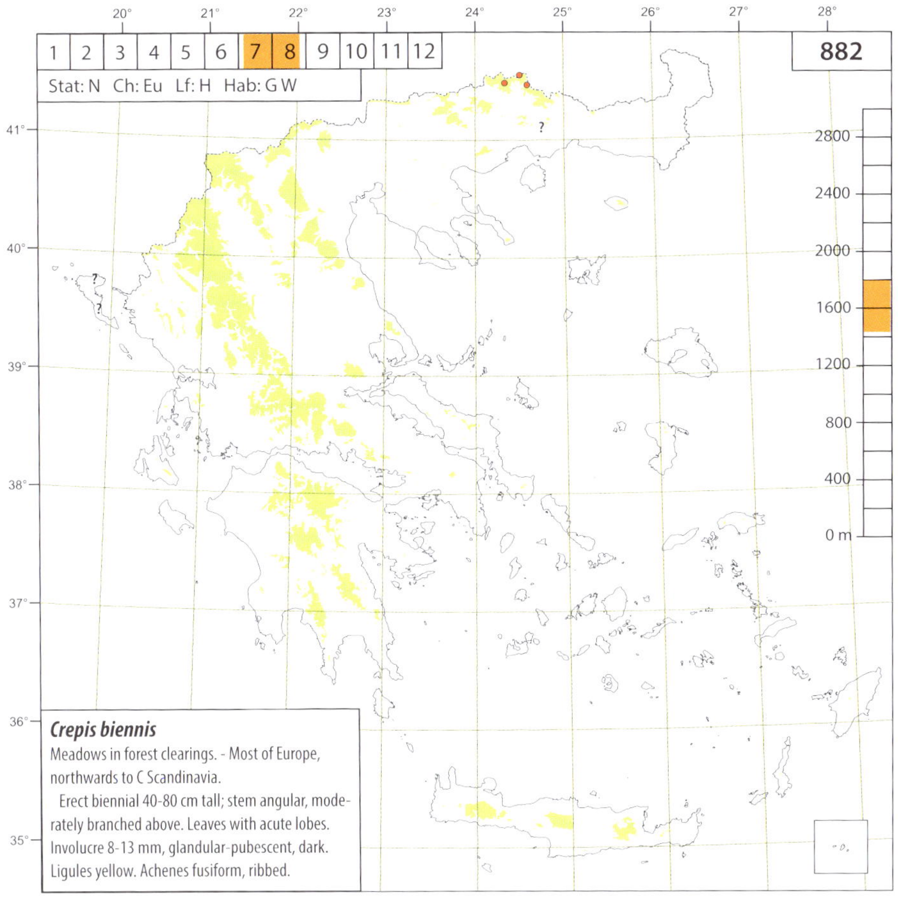

Crepis biennis

Meadows in forest clearings. - Most of Europe, northwards to C Scandinavia.

Erect biennial 40-80 cm tall; stem angular, moderately branched above. Leaves with acute lobes. Involucre 8-13 mm, glandular-pubescent, dark. Ligules yellow. Achenes fusiform, ribbed.

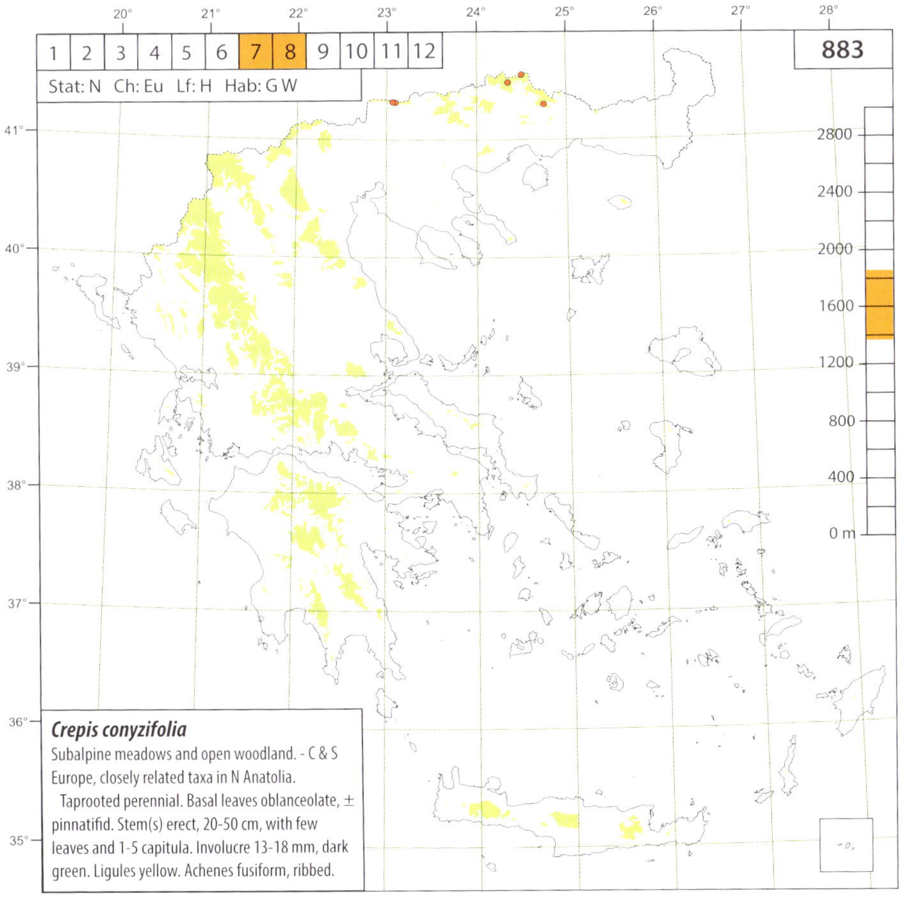

Crepis conyzifolia

Subalpine meadows and open woodland. - C & S Europe, closely related taxa in N Anatolia.

Taprooted perennial. Basal leaves oblanceolate, ± pinnatifid. Stem(s) erect, 20-50 cm, with few leaves and 1-5 capitula. Involucre 13-18 mm, dark green. Ligules yellow. Achenes fusiform, ribbed.

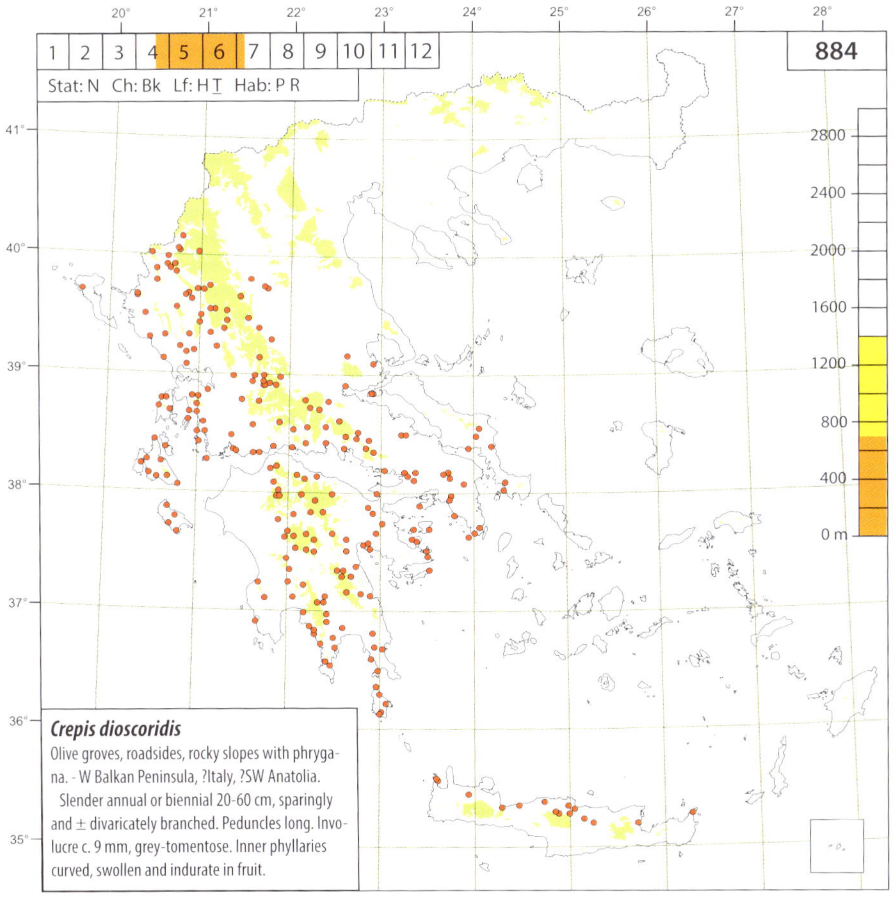

Crepis dioscoridis

Olive groves, roadsides, rocky slopes with phrygana. - W Balkan Peninsula, ?Italy, ?SW Anatolia.

Slender annual or biennial 20-60 cm, sparingly and ± divaricately branched. Peduncles long. Involucre c. 9 mm, grey-tomentose. Inner phyllaries curved, swollen and indurate in fruit.

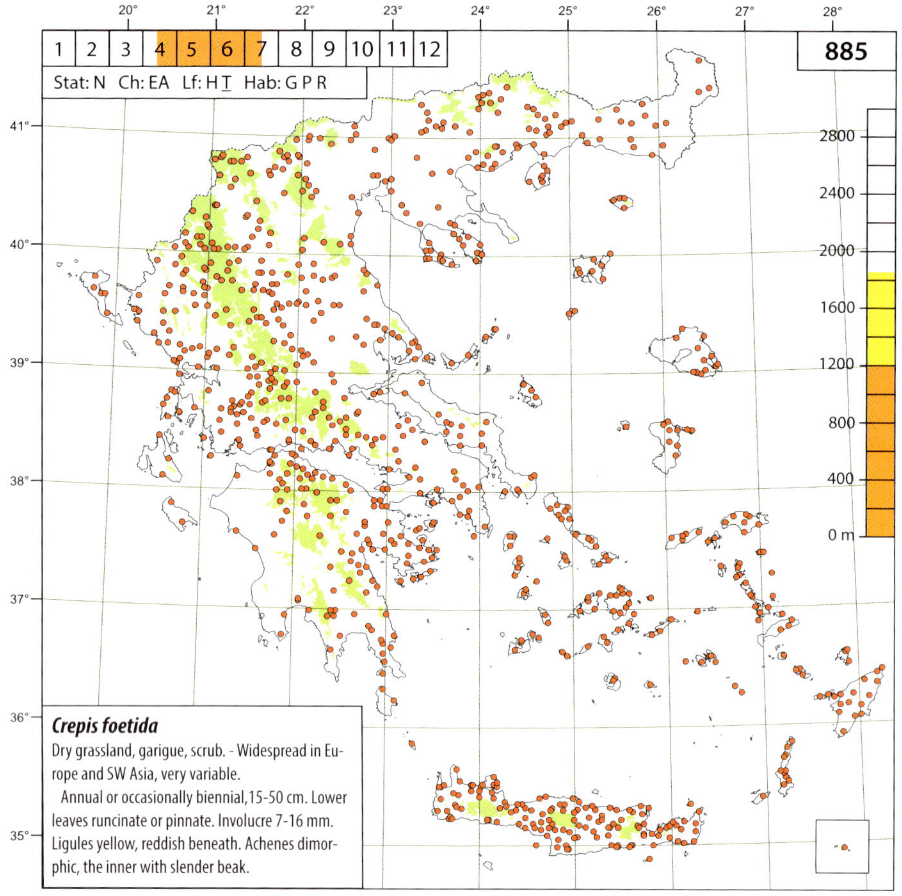

Crepis foetida

Dry grassland, garigue, scrub. - Widespread in Europe and SW Asia, very variable.

Annual or occasionally biennial, 15-50 cm. Lower leaves runcinate or pinnate. Involucre 7-16 mm. Ligules yellow, reddish beneath. Achenes dimorphic, the inner with slender beak.

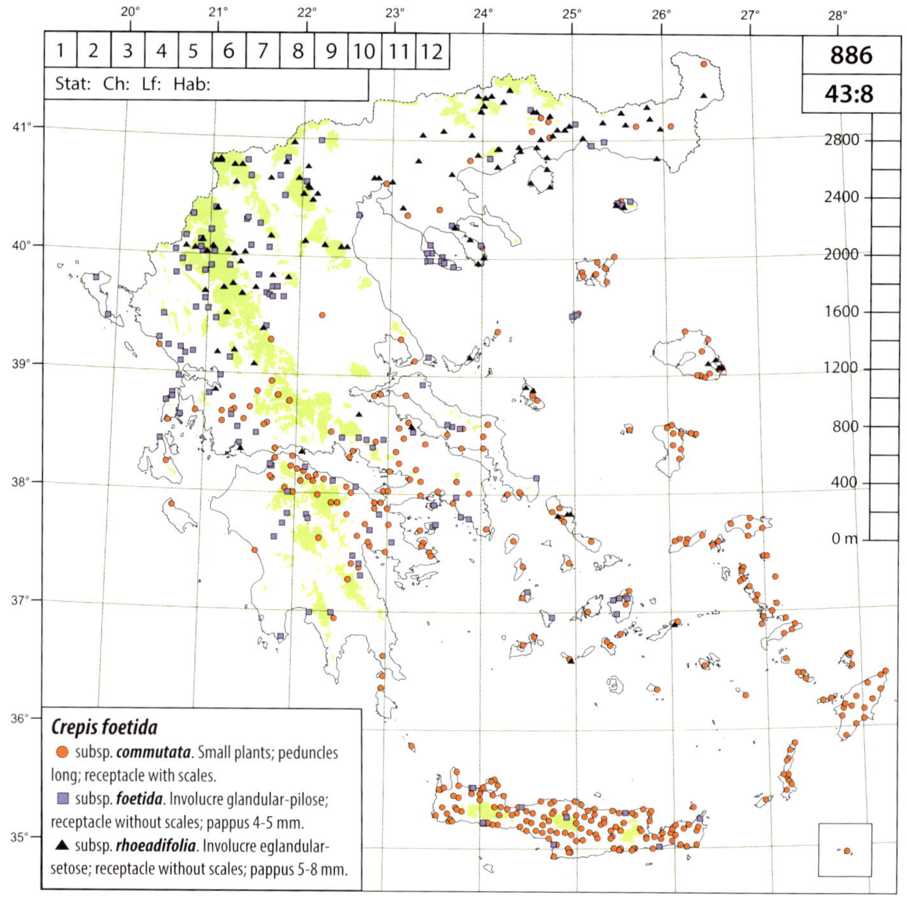

Crepis foetida

● subsp. ***commutata***. Small plants; peduncles long; receptacle with scales.

■ subsp. ***foetida***. Involucre glandular-pilose; receptacle without scales; pappus 4-5 mm.

▲ subsp. ***rhoeadifolia***. Involucre eglandular-setose; receptacle without scales; pappus 5-8 mm.

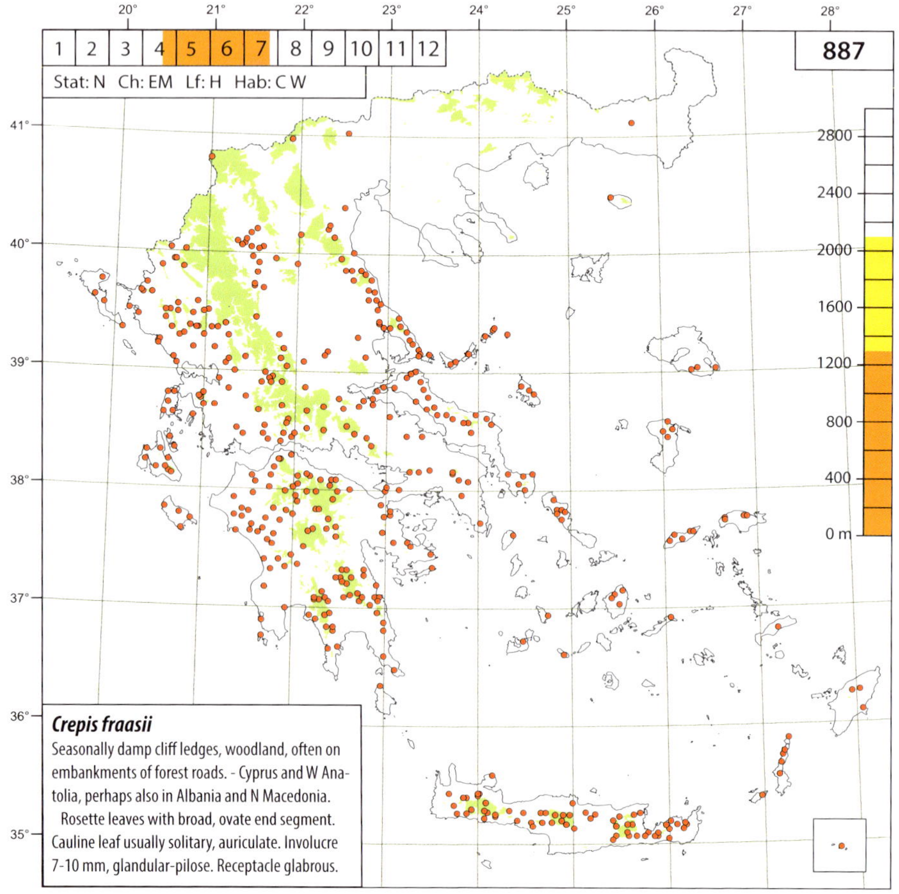

Crepis fraasii

Seasonally damp cliff ledges, woodland, often on embankments of forest roads. - Cyprus and W Anatolia, perhaps also in Albania and N Macedonia.

Rosette leaves with broad, ovate end segment. Cauline leaf usually solitary, auriculate. Involucre 7-10 mm, glandular-pilose. Receptacle glabrous.

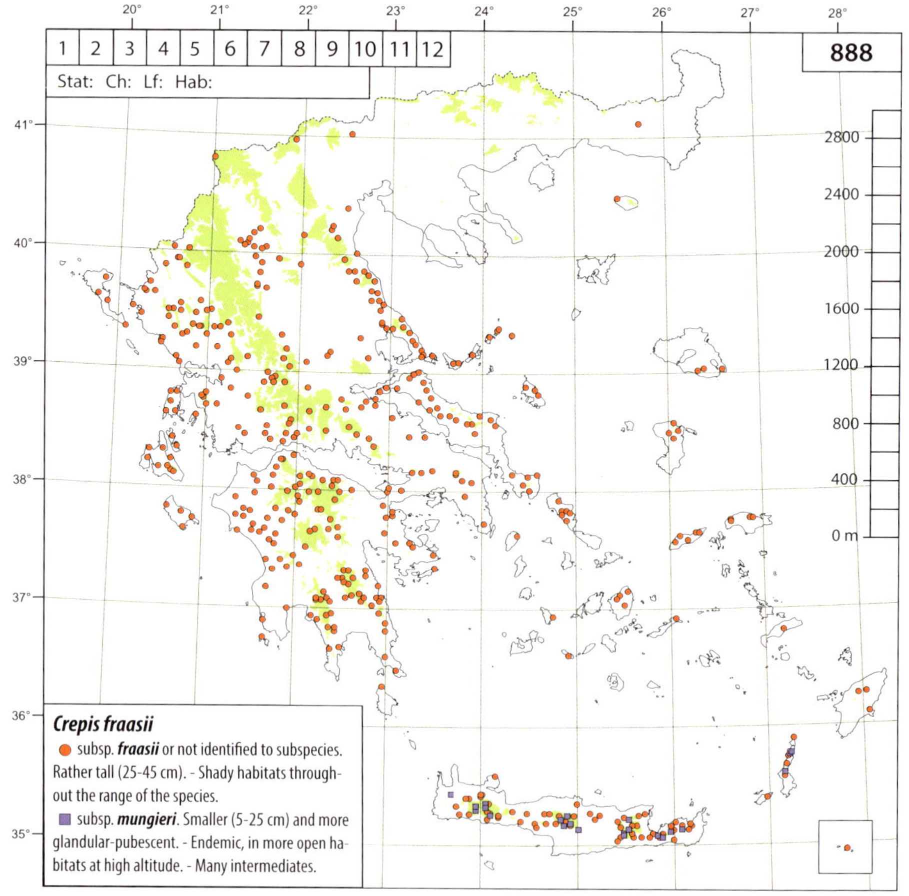

Crepis fraasii

● subsp. ***fraasii*** or not identified to subspecies. Rather tall (25-45 cm). - Shady habitats throughout the range of the species.

■ subsp. ***mungieri***. Smaller (5-25 cm) and more glandular-pubescent. - Endemic, in more open habitats at high altitude. - Many intermediates.

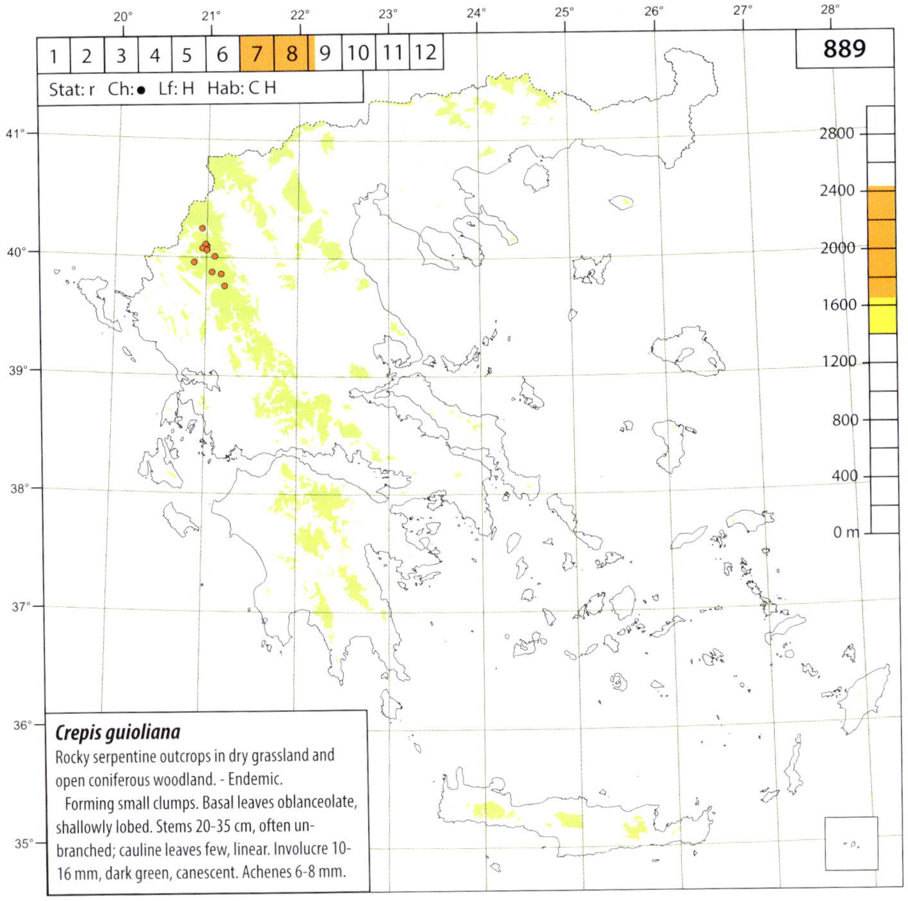

Crepis guioliana
Rocky serpentine outcrops in dry grassland and open coniferous woodland. - Endemic.
Forming small clumps. Basal leaves oblanceolate, shallowly lobed. Stems 20-35 cm, often unbranched; cauline leaves few, linear. Involucre 10-16 mm, dark green, canescent. Achenes 6-8 mm.

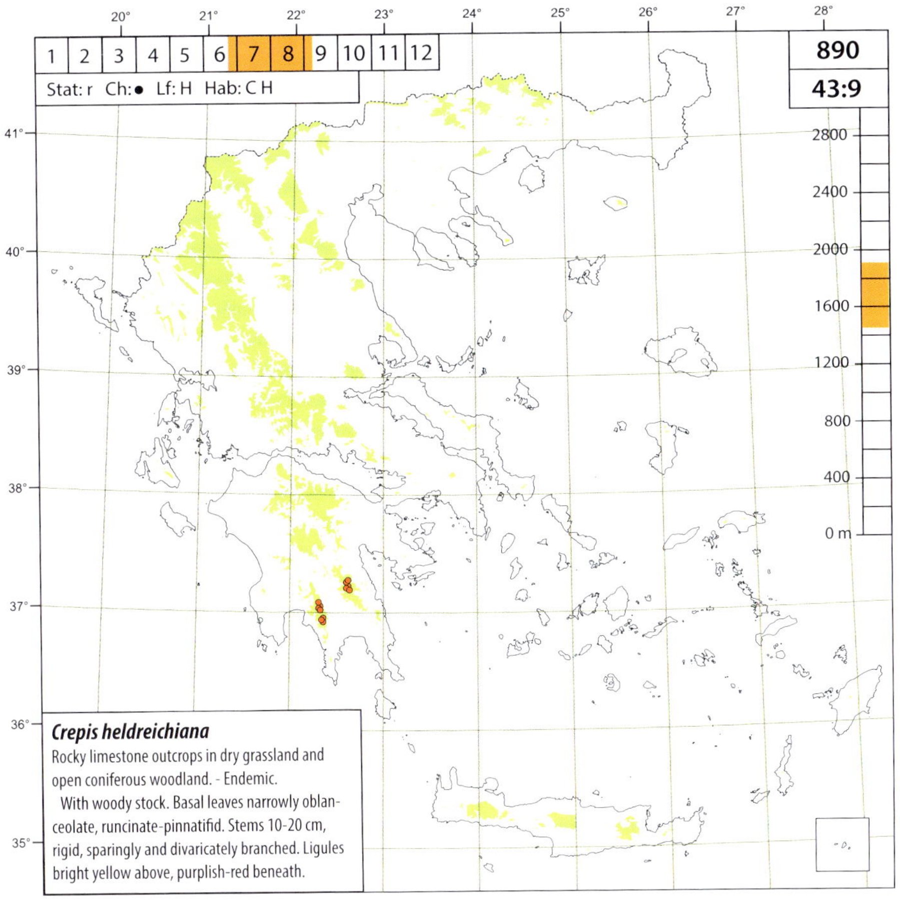

Crepis heldreichiana
Rocky limestone outcrops in dry grassland and open coniferous woodland. - Endemic.
With woody stock. Basal leaves narrowly oblanceolate, runcinate-pinnatifid. Stems 10-20 cm, rigid, sparingly and divaricately branched. Ligules bright yellow above, purplish-red beneath.

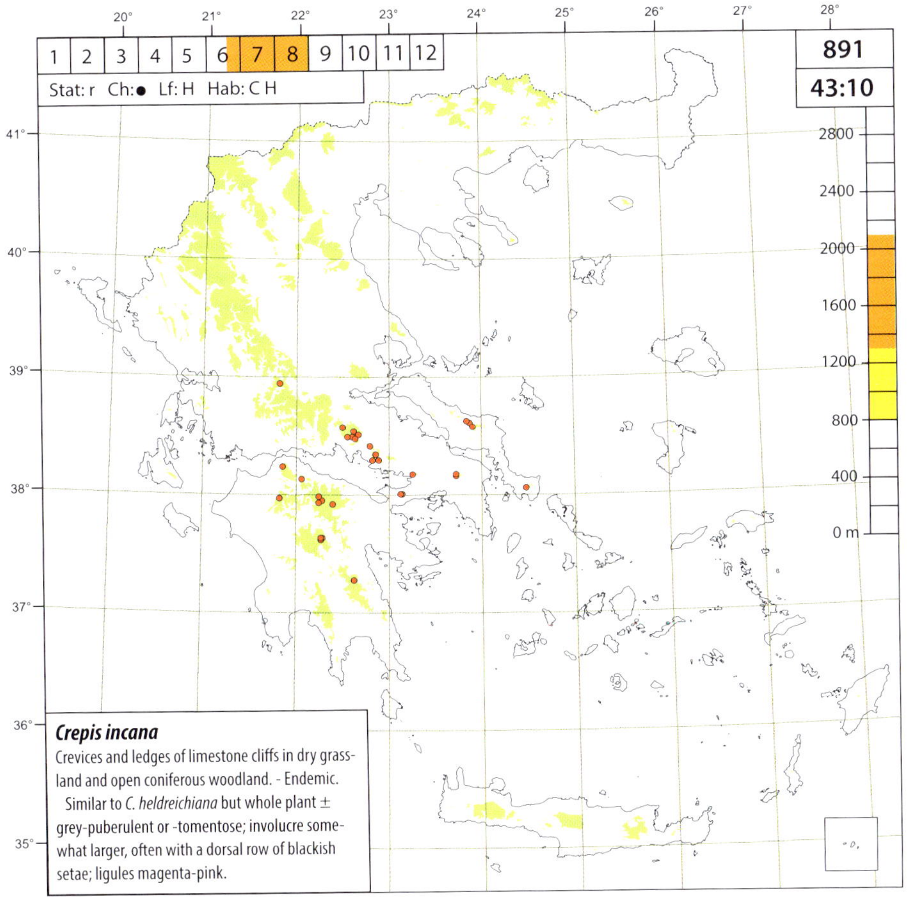

Crepis incana
Crevices and ledges of limestone cliffs in dry grassland and open coniferous woodland. - Endemic.
Similar to *C. heldreichiana* but whole plant ± grey-puberulent or -tomentose; involucre somewhat larger, often with a dorsal row of blackish setae; ligules magenta-pink.

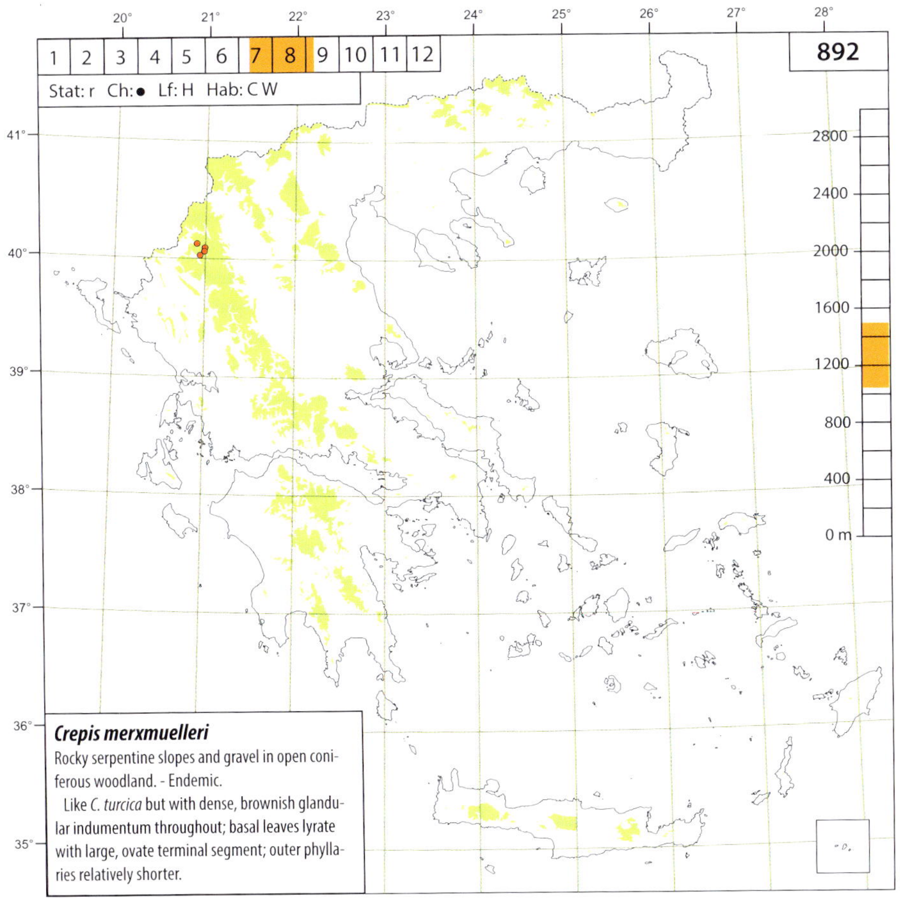

Crepis merxmuelleri
Rocky serpentine slopes and gravel in open coniferous woodland. - Endemic.
Like *C. turcica* but with dense, brownish glandular indumentum throughout; basal leaves lyrate with large, ovate terminal segment; outer phyllaries relatively shorter.

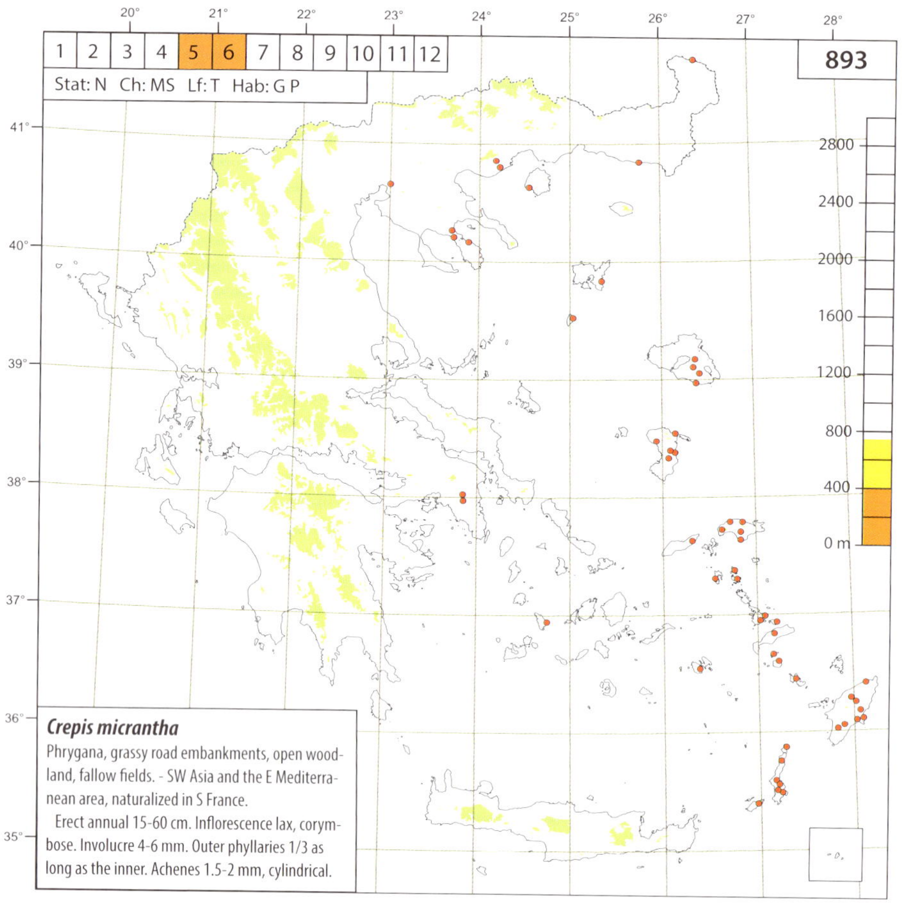

Crepis micrantha

Phrygana, grassy road embankments, open woodland, fallow fields. - SW Asia and the E Mediterranean area, naturalized in S France.

Erect annual 15-60 cm. Inflorescence lax, corymbose. Involucre 4-6 mm. Outer phyllaries 1/3 as long as the inner. Achenes 1.5-2 mm, cylindrical.

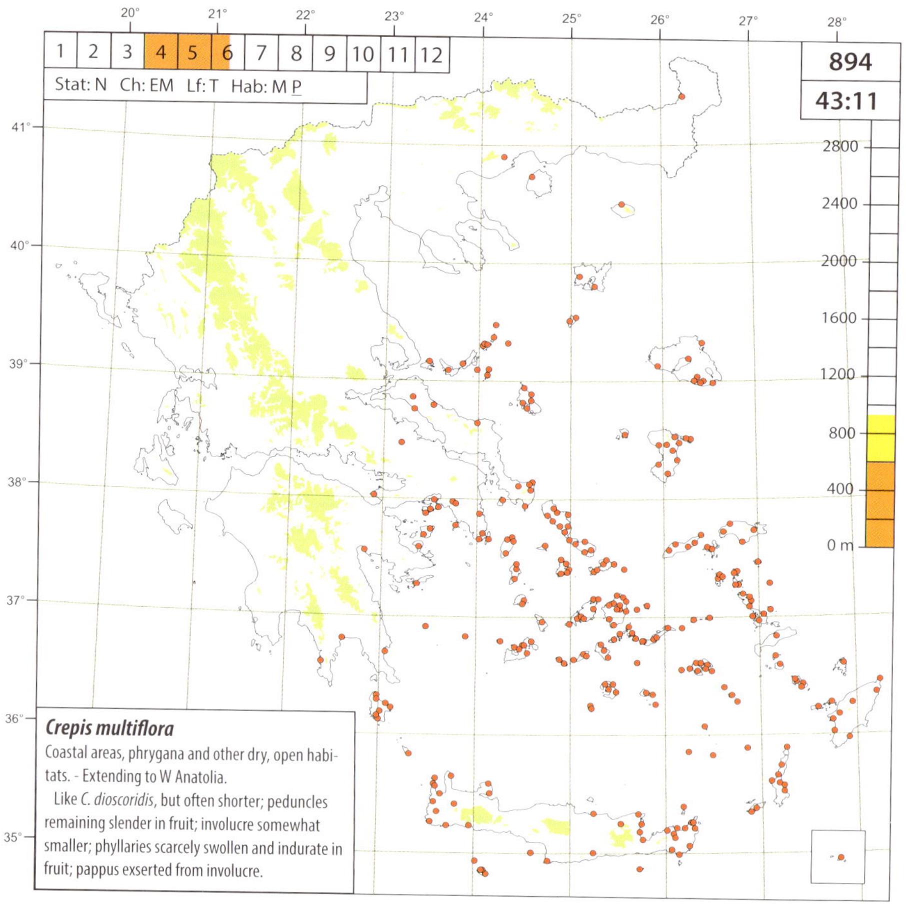

Crepis multiflora

Coastal areas, phrygana and other dry, open habitats. - Extending to W Anatolia.

Like *C. dioscoridis*, but often shorter; peduncles remaining slender in fruit; involucre somewhat smaller; phyllaries scarcely swollen and indurate in fruit; pappus exserted from involucre.

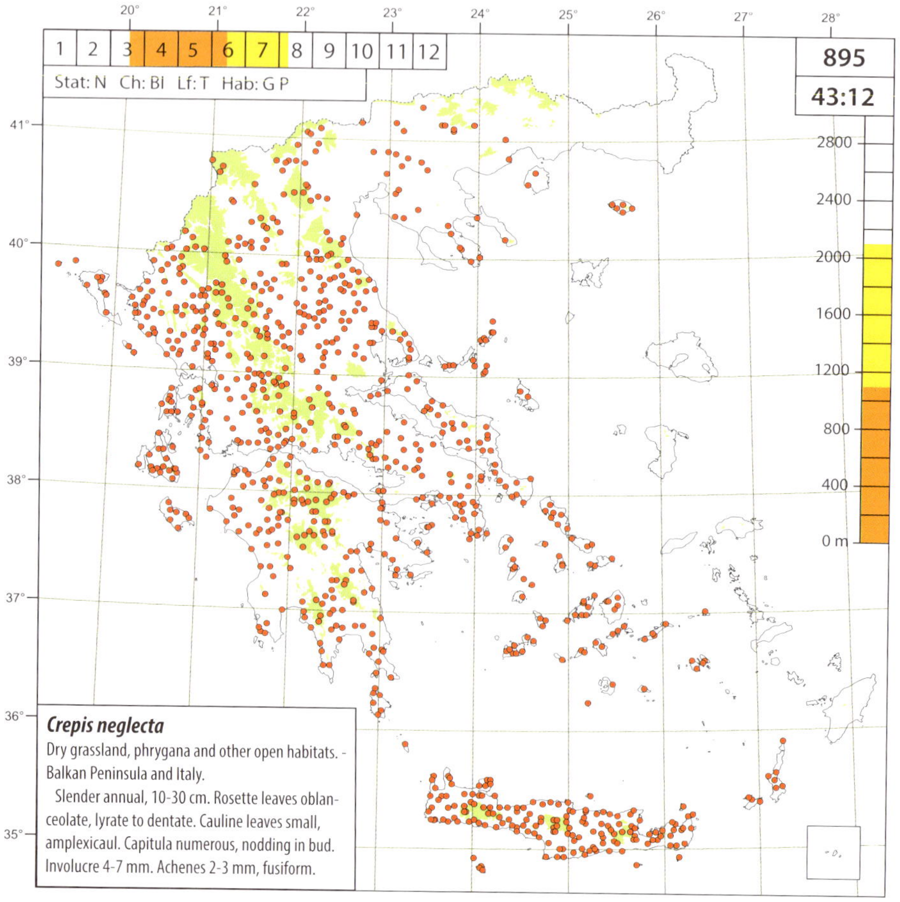

Crepis neglecta

Dry grassland, phrygana and other open habitats. - Balkan Peninsula and Italy.

Slender annual, 10-30 cm. Rosette leaves oblanceolate, lyrate to dentate. Cauline leaves small, amplexicaul. Capitula numerous, nodding in bud. Involucre 4-7 mm. Achenes 2-3 mm, fusiform.

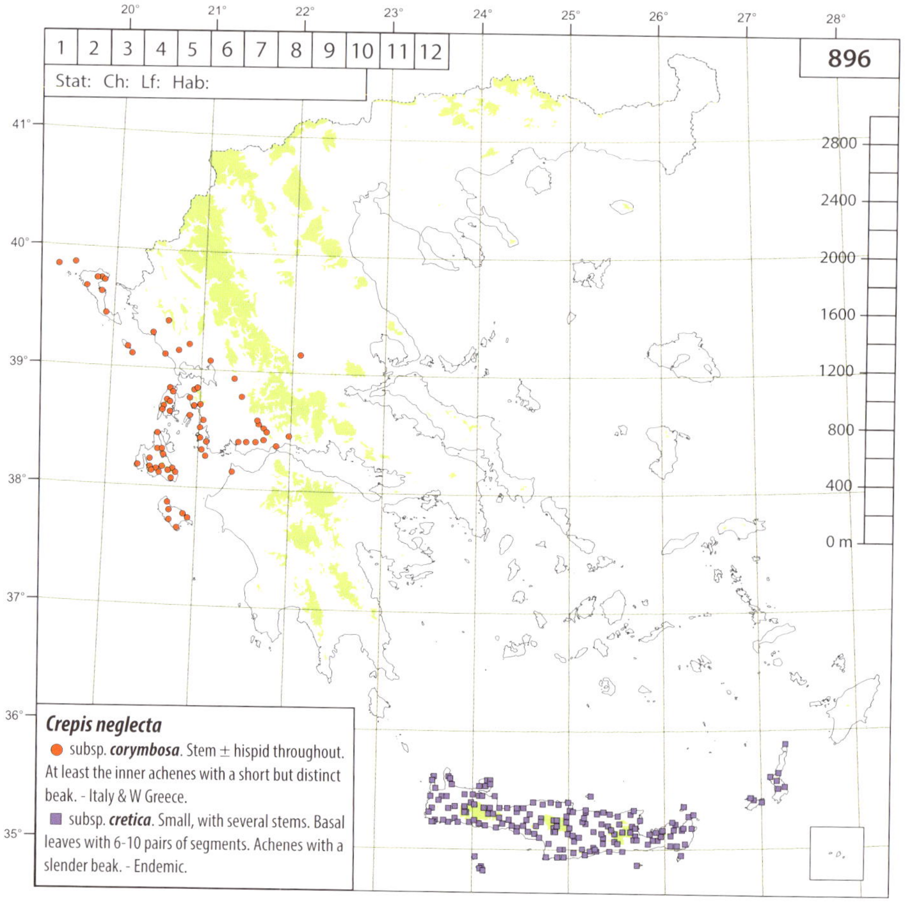

Crepis neglecta

● subsp. ***corymbosa***. Stem ± hispid throughout. At least the inner achenes with a short but distinct beak. - Italy & W Greece.

■ subsp. ***cretica***. Small, with several stems. Basal leaves with 6-10 pairs of segments. Achenes with a slender beak. - Endemic.

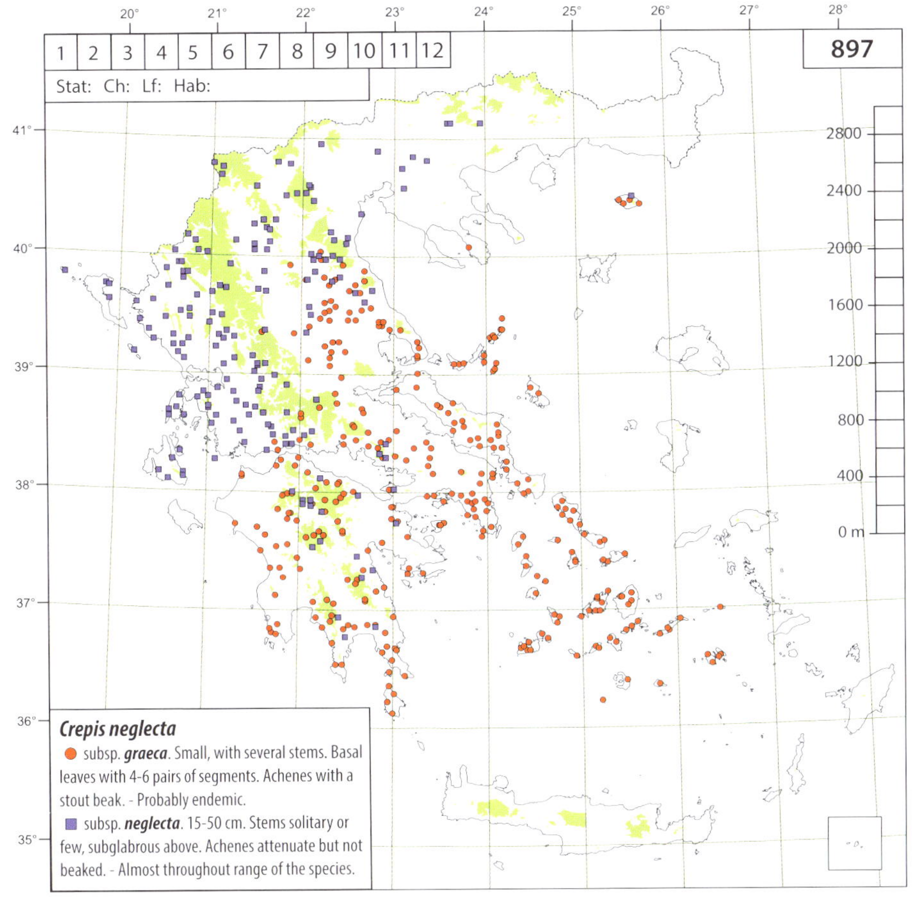

Crepis neglecta

● subsp. ***graeca***. Small, with several stems. Basal leaves with 4-6 pairs of segments. Achenes with a stout beak. - Probably endemic.

■ subsp. ***neglecta***. 15-50 cm. Stems solitary or few, subglabrous above. Achenes attenuate but not beaked. - Almost throughout range of the species.

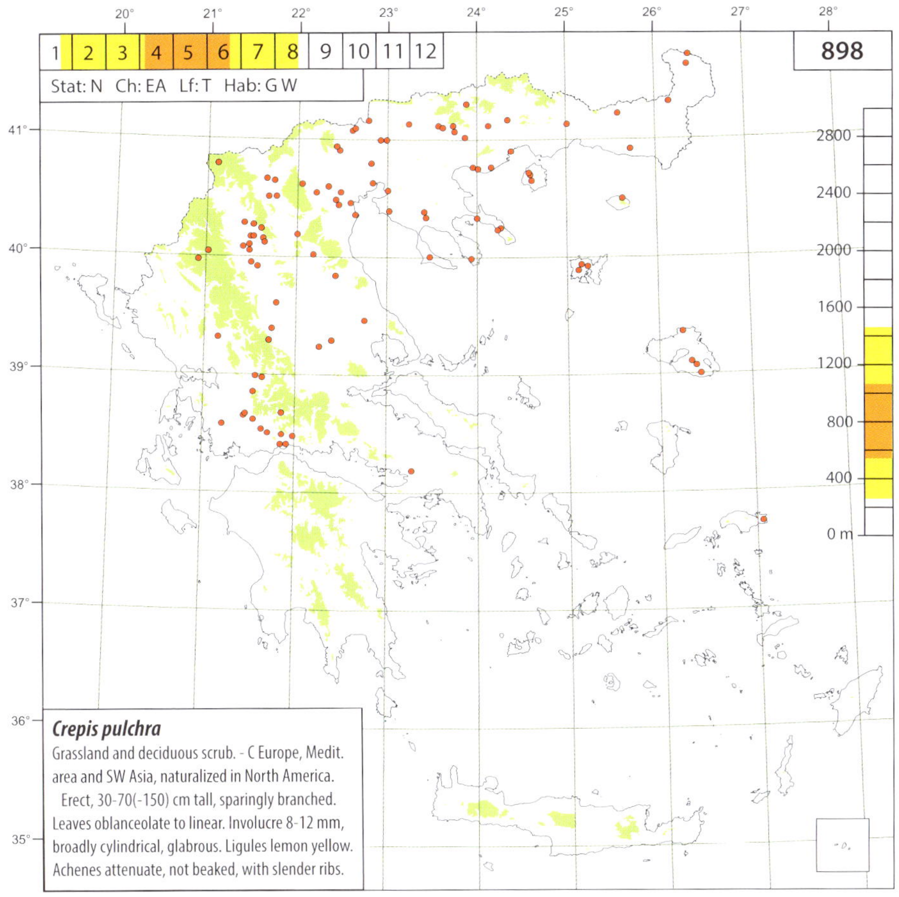

Crepis pulchra

Grassland and deciduous scrub. - C Europe, Medit. area and SW Asia, naturalized in North America.

Erect, 30-70(-150) cm tall, sparingly branched. Leaves oblanceolate to linear. Involucre 8-12 mm, broadly cylindrical, glabrous. Ligules lemon yellow. Achenes attenuate, not beaked, with slender ribs.

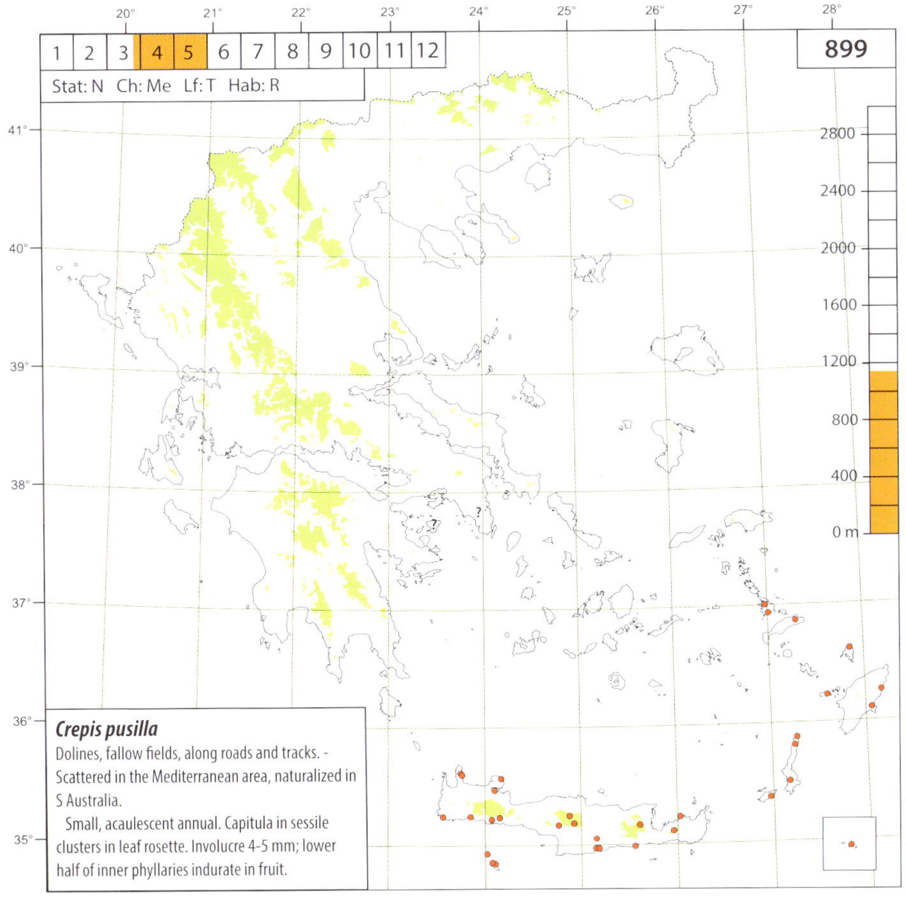

Crepis pusilla

Dolines, fallow fields, along roads and tracks. - Scattered in the Mediterranean area, naturalized in S Australia.

Small, acaulescent annual. Capitula in sessile clusters in leaf rosette. Involucre 4-5 mm; lower half of inner phyllaries indurate in fruit.

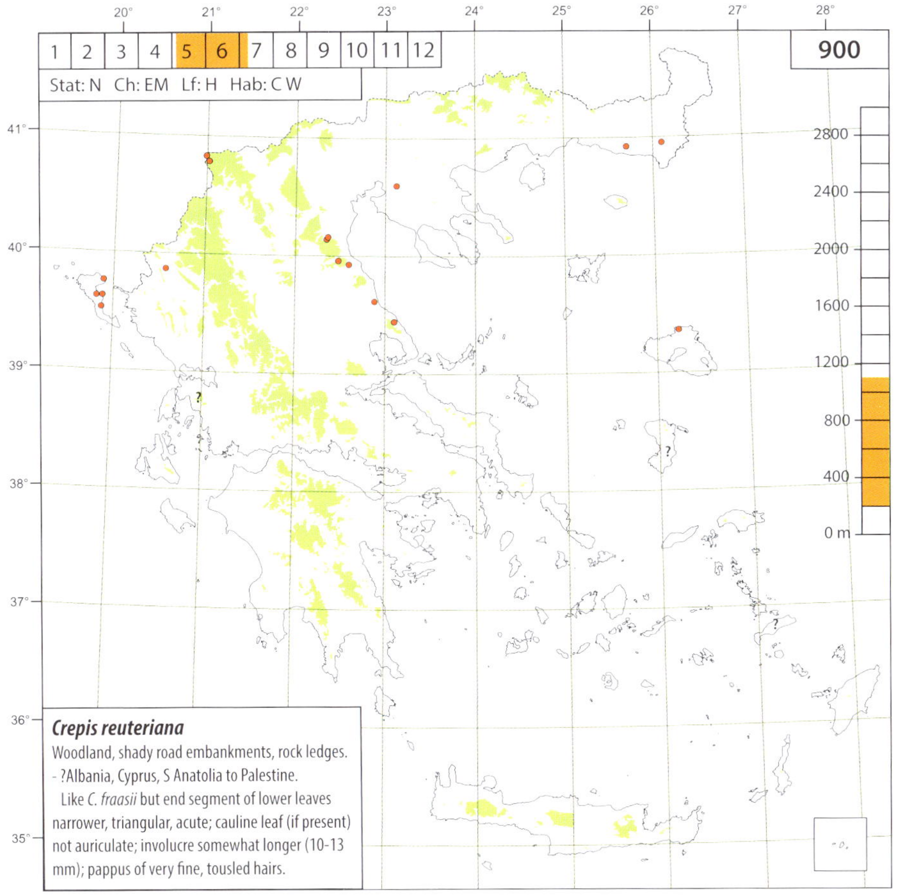

Crepis reuteriana

Woodland, shady road embankments, rock ledges. - ?Albania, Cyprus, S Anatolia to Palestine.

Like *C. fraasii* but end segment of lower leaves narrower, triangular, acute; cauline leaf (if present) not auriculate; involucre somewhat longer (10-13 mm); pappus of very fine, tousled hairs.

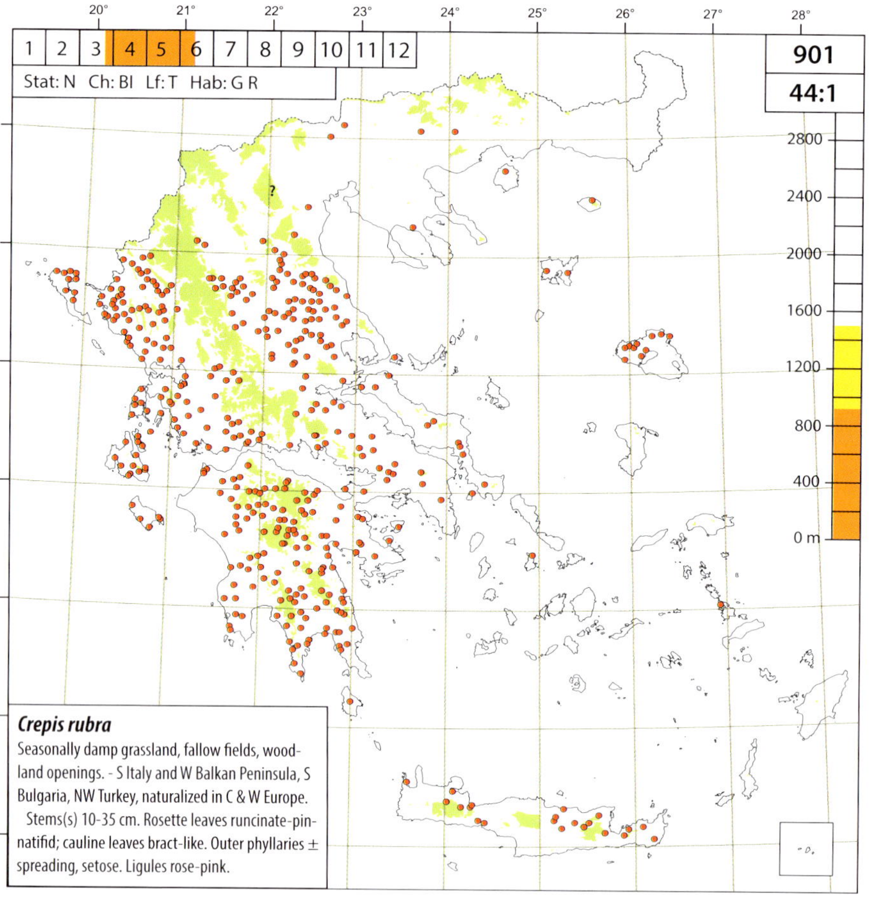

Crepis rubra
Seasonally damp grassland, fallow fields, woodland openings. - S Italy and W Balkan Peninsula, S Bulgaria, NW Turkey, naturalized in C & W Europe.
Stem(s) 10-35 cm. Rosette leaves runcinate-pinnatifid; cauline leaves bract-like. Outer phyllaries ± spreading, setose. Ligules rose-pink.

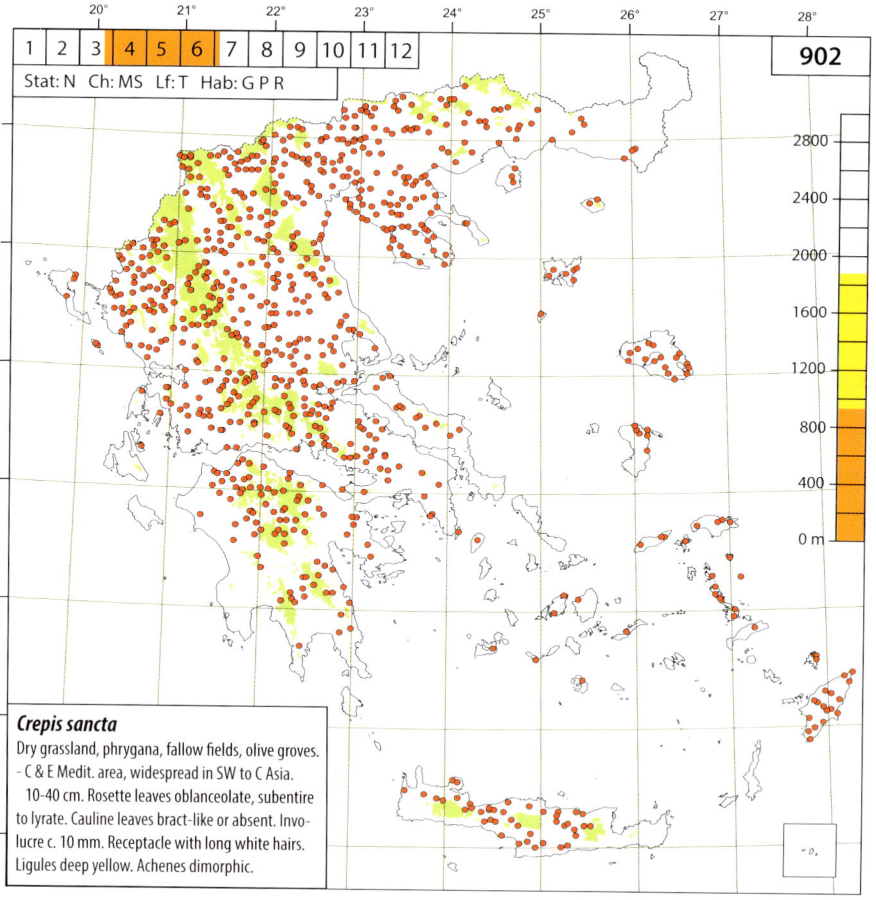

Crepis sancta
Dry grassland, phrygana, fallow fields, olive groves. - C & E Medit. area, widespread in SW to C Asia.
10-40 cm. Rosette leaves oblanceolate, subentire to lyrate. Cauline leaves bract-like or absent. Involucre c. 10 mm. Receptacle with long white hairs. Ligules deep yellow. Achenes dimorphic.

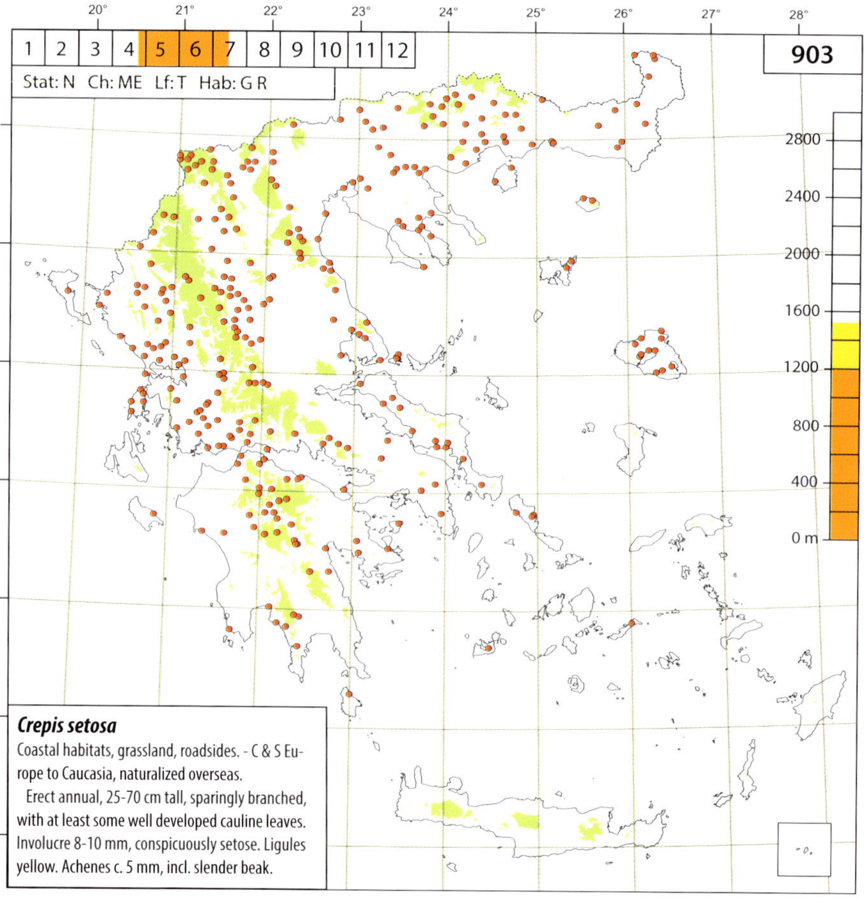

Crepis setosa
Coastal habitats, grassland, roadsides. - C & S Europe to Caucasia, naturalized overseas.
Erect annual, 25-70 cm tall, sparingly branched, with at least some well developed cauline leaves. Involucre 8-10 mm, conspicuously setose. Ligules yellow. Achenes c. 5 mm, incl. slender beak.

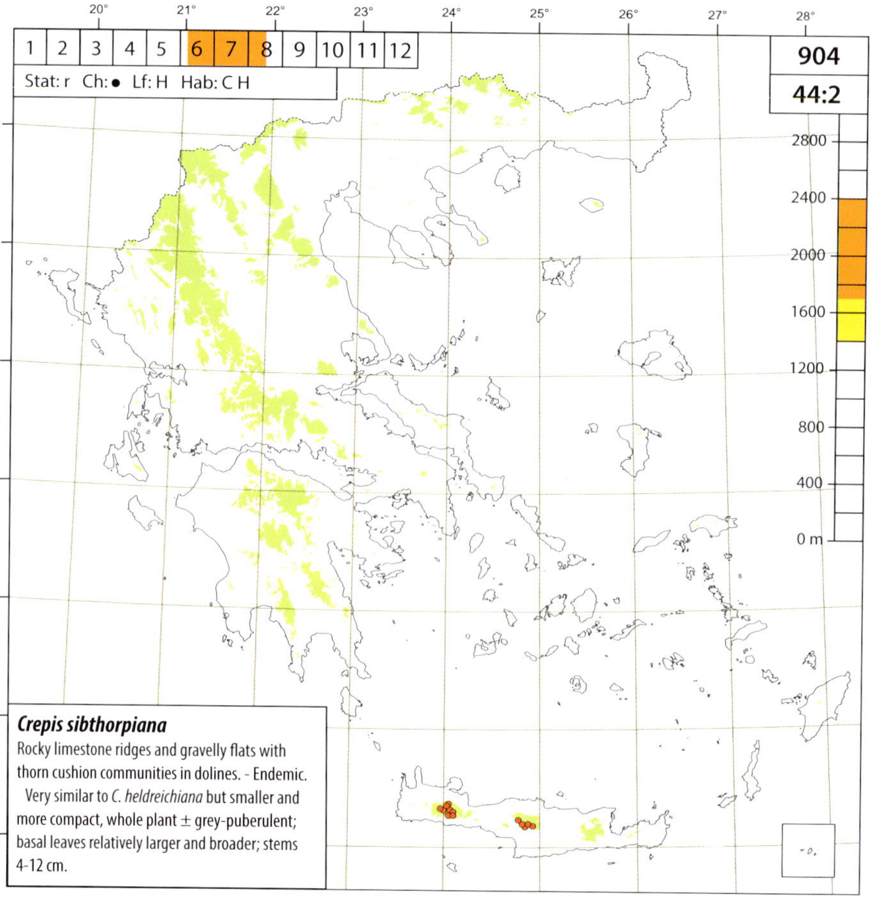

Crepis sibthorpiana
Rocky limestone ridges and gravelly flats with thorn cushion communities in dolines. - Endemic.
Very similar to *C. heldreichiana* but smaller and more compact, whole plant ± grey-puberulent; basal leaves relatively larger and broader; stems 4-12 cm.

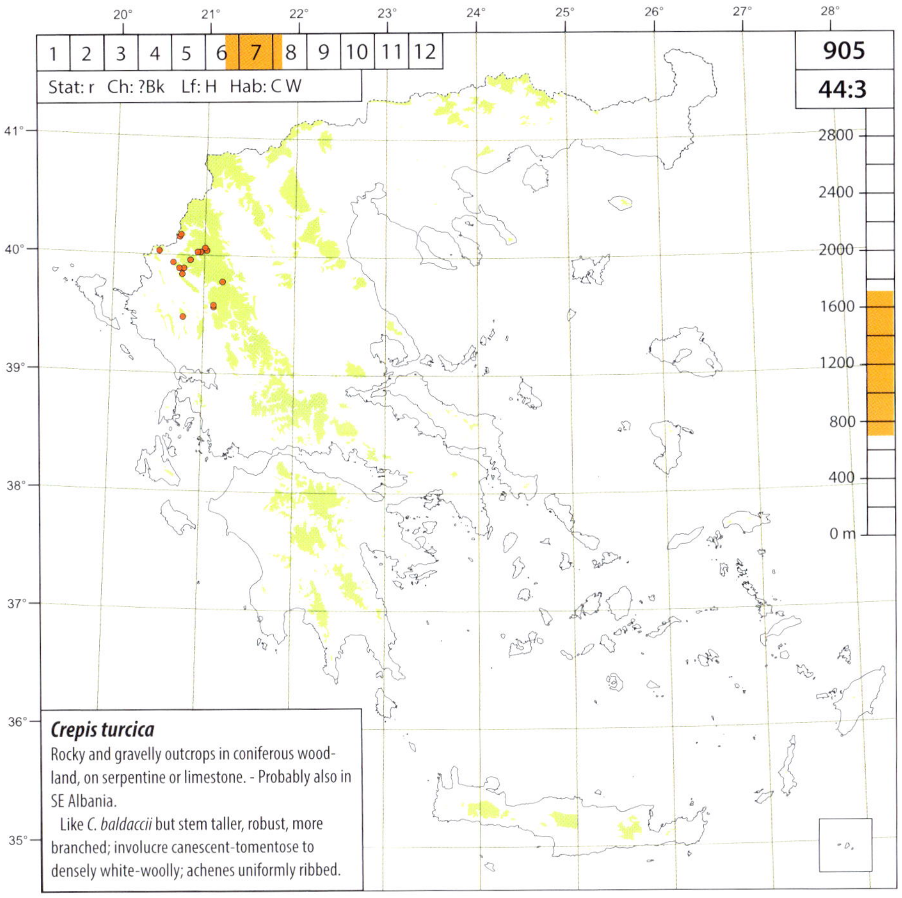

Crepis turcica
Rocky and gravelly outcrops in coniferous woodland, on serpentine or limestone. - Probably also in SE Albania.
Like *C. baldaccii* but stem taller, robust, more branched; involucre canescent-tomentose to densely white-woolly; achenes uniformly ribbed.

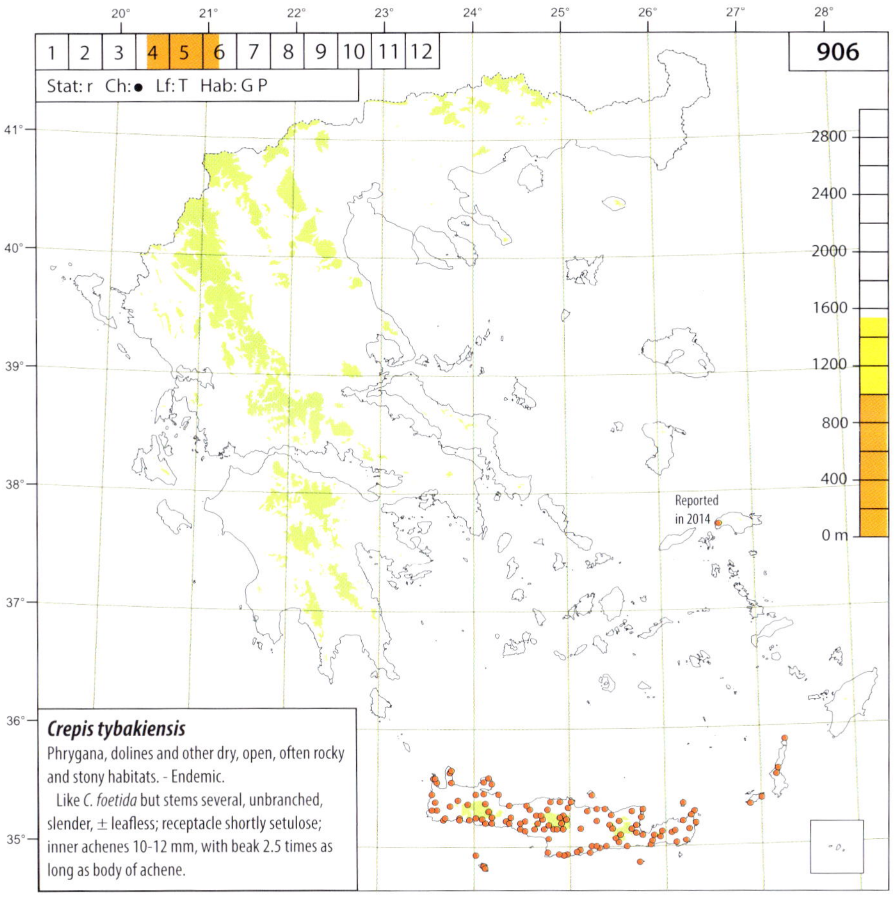

Crepis tybakiensis
Phrygana, dolines and other dry, open, often rocky and stony habitats. - Endemic.
Like *C. foetida* but stems several, unbranched, slender, ± leafless; receptacle shortly setulose; inner achenes 10-12 mm, with beak 2.5 times as long as body of achene.

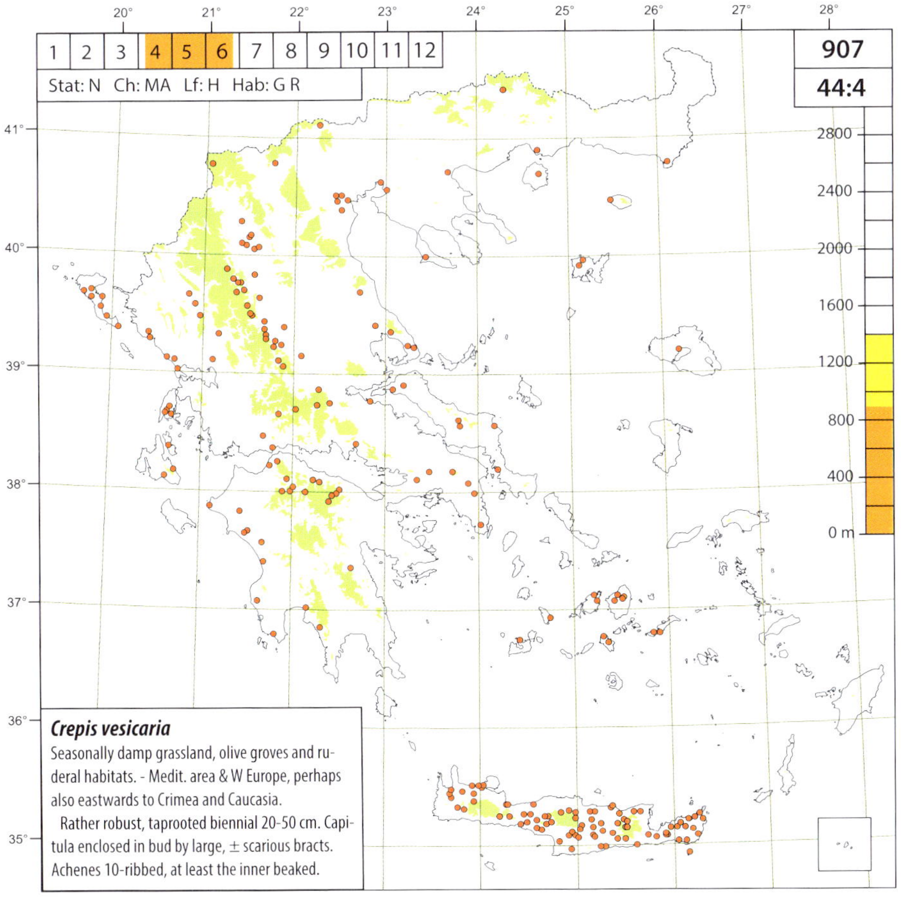

Crepis vesicaria
Seasonally damp grassland, olive groves and ruderal habitats. - Medit. area & W Europe, perhaps also eastwards to Crimea and Caucasia.
Rather robust, taprooted biennial 20-50 cm. Capitula enclosed in bud by large, ± scarious bracts. Achenes 10-ribbed, at least the inner beaked.

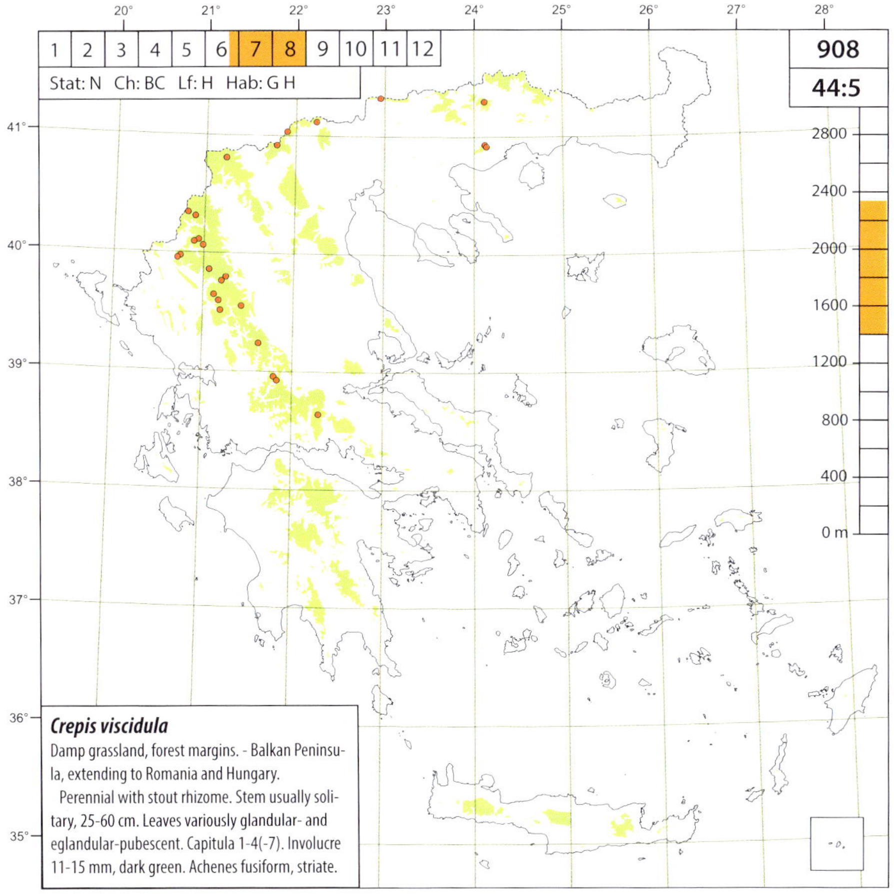

Crepis viscidula
Damp grassland, forest margins. - Balkan Peninsula, extending to Romania and Hungary.
Perennial with stout rhizome. Stem usually solitary, 25-60 cm. Leaves variously glandular- and eglandular-pubescent. Capitula 1-4(-7). Involucre 11-15 mm, dark green. Achenes fusiform, striate.

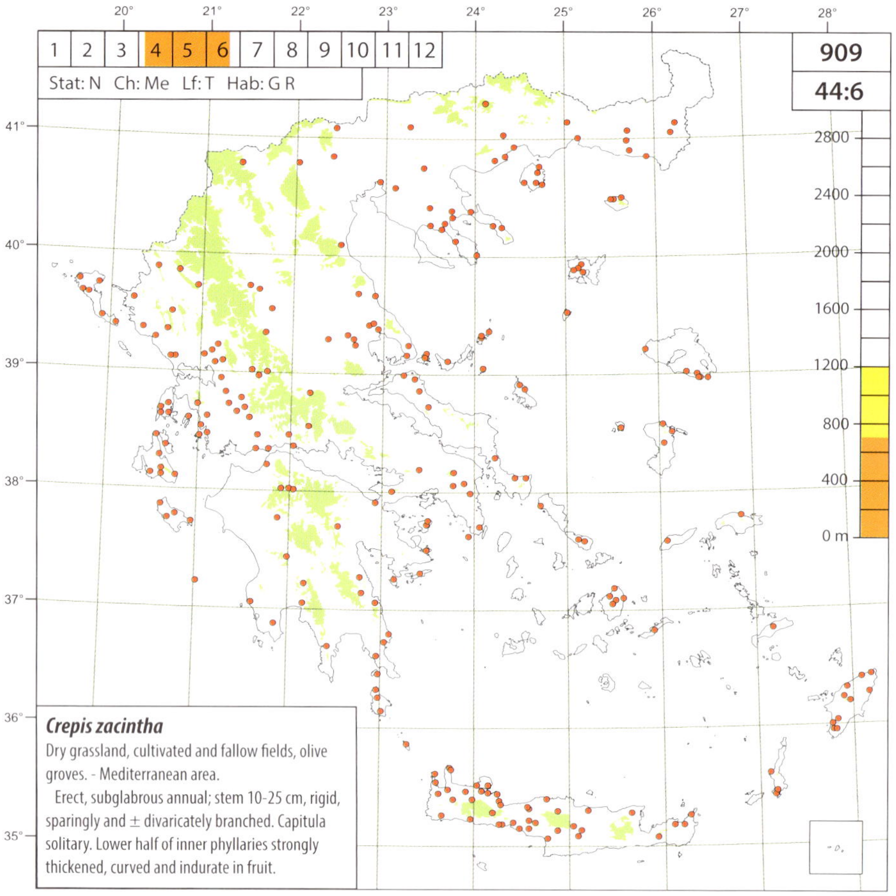

Crepis zacintha
Dry grassland, cultivated and fallow fields, olive groves. - Mediterranean area.
Erect, subglabrous annual; stem 10-25 cm, rigid, sparingly and ± divaricately branched. Capitula solitary. Lower half of inner phyllaries strongly thickened, curved and indurate in fruit.

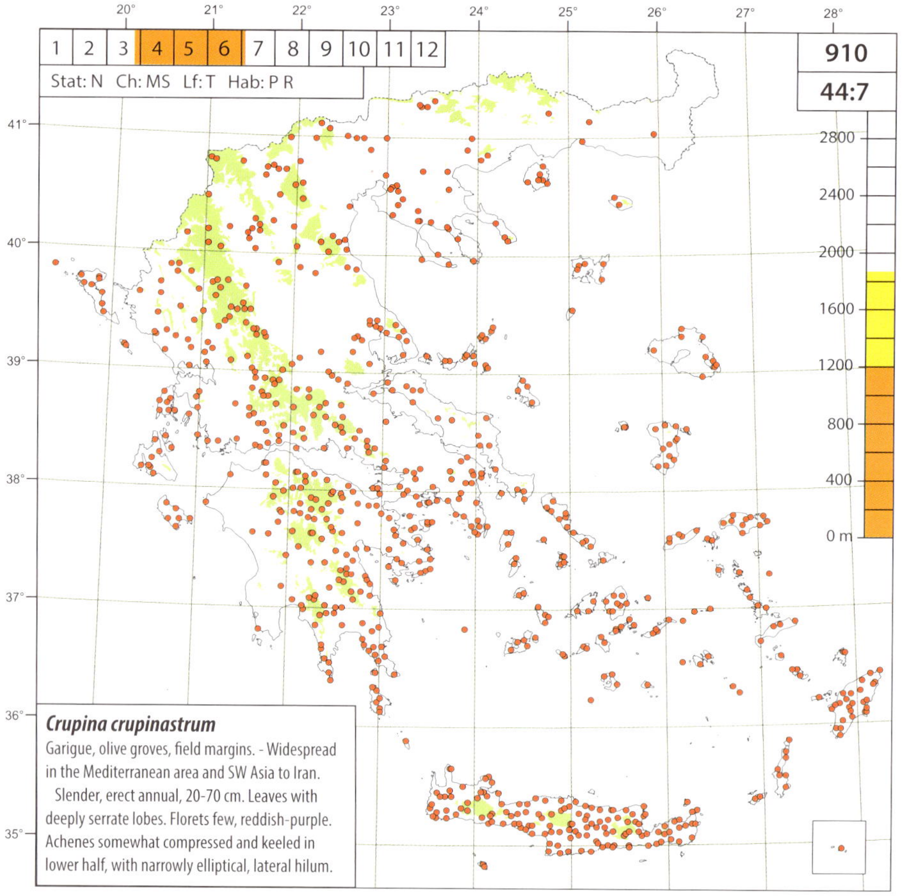

Crupina crupinastrum
Garigue, olive groves, field margins. - Widespread in the Mediterranean area and SW Asia to Iran.
Slender, erect annual, 20-70 cm. Leaves with deeply serrate lobes. Florets few, reddish-purple. Achenes somewhat compressed and keeled in lower half, with narrowly elliptical, lateral hilum.

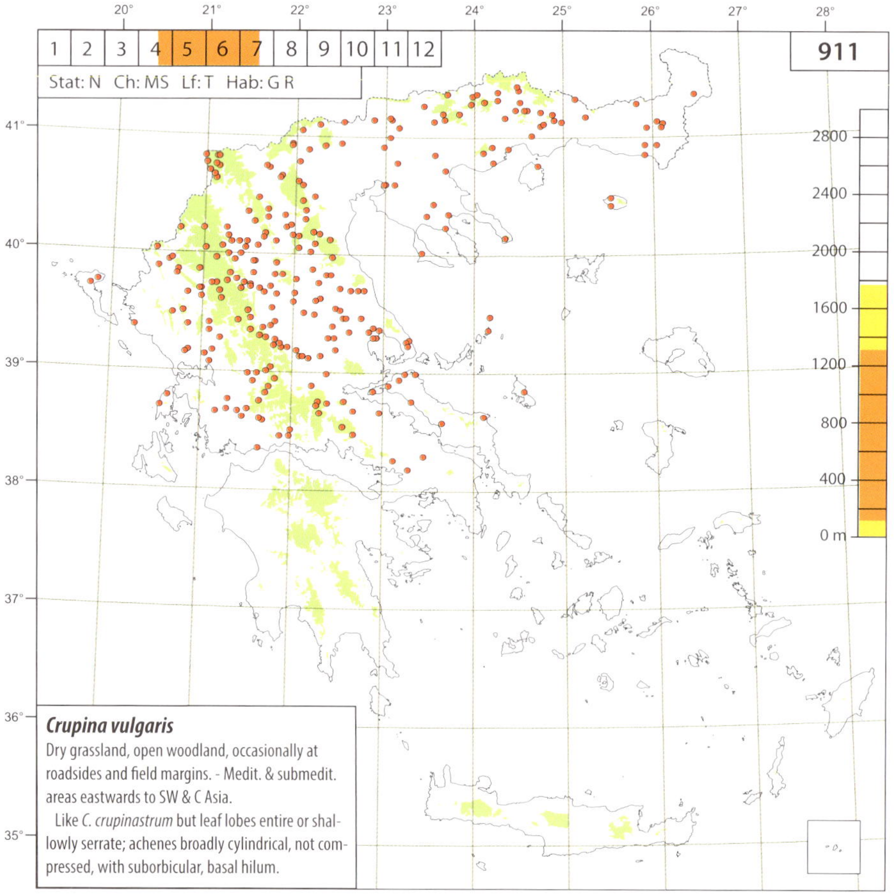

Crupina vulgaris
Dry grassland, open woodland, occasionally at roadsides and field margins. - Medit. & submedit. areas eastwards to SW & C Asia.
Like *C. crupinastrum* but leaf lobes entire or shallowly serrate; achenes broadly cylindrical, not compressed, with suborbicular, basal hilum.

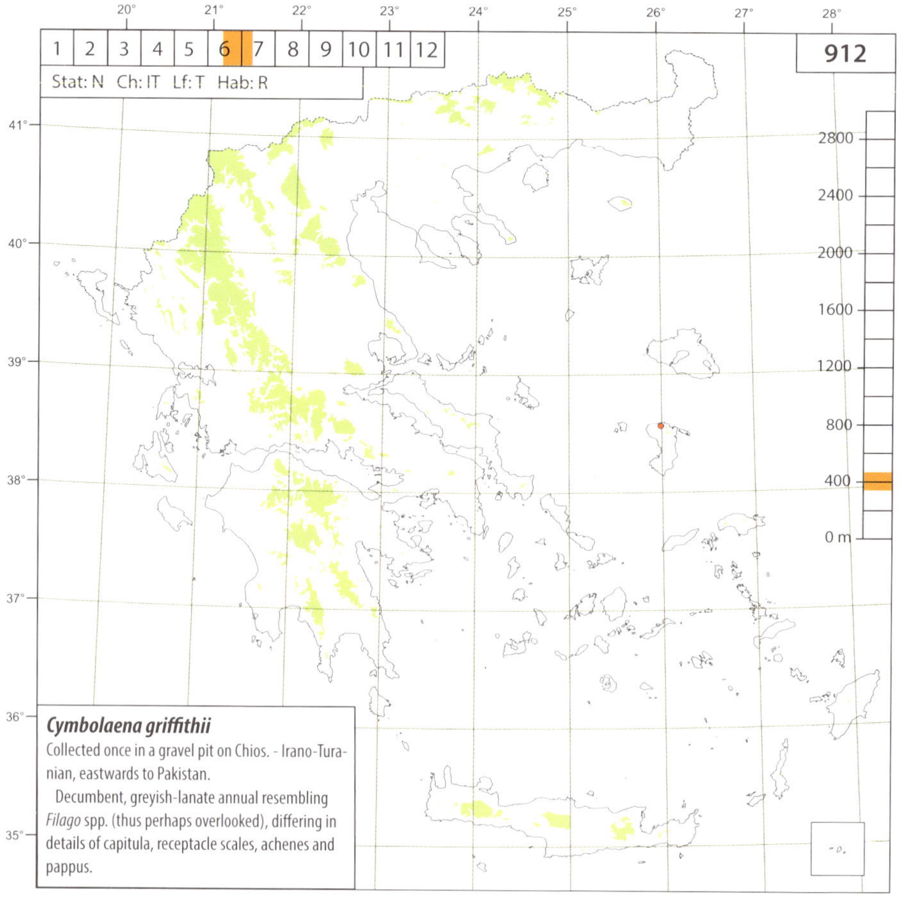

Cymbolaena griffithii
Collected once in a gravel pit on Chios. - Irano-Turanian, eastwards to Pakistan.
Decumbent, greyish-lanate annual resembling *Filago* spp. (thus perhaps overlooked), differing in details of capitula, receptacle scales, achenes and pappus.

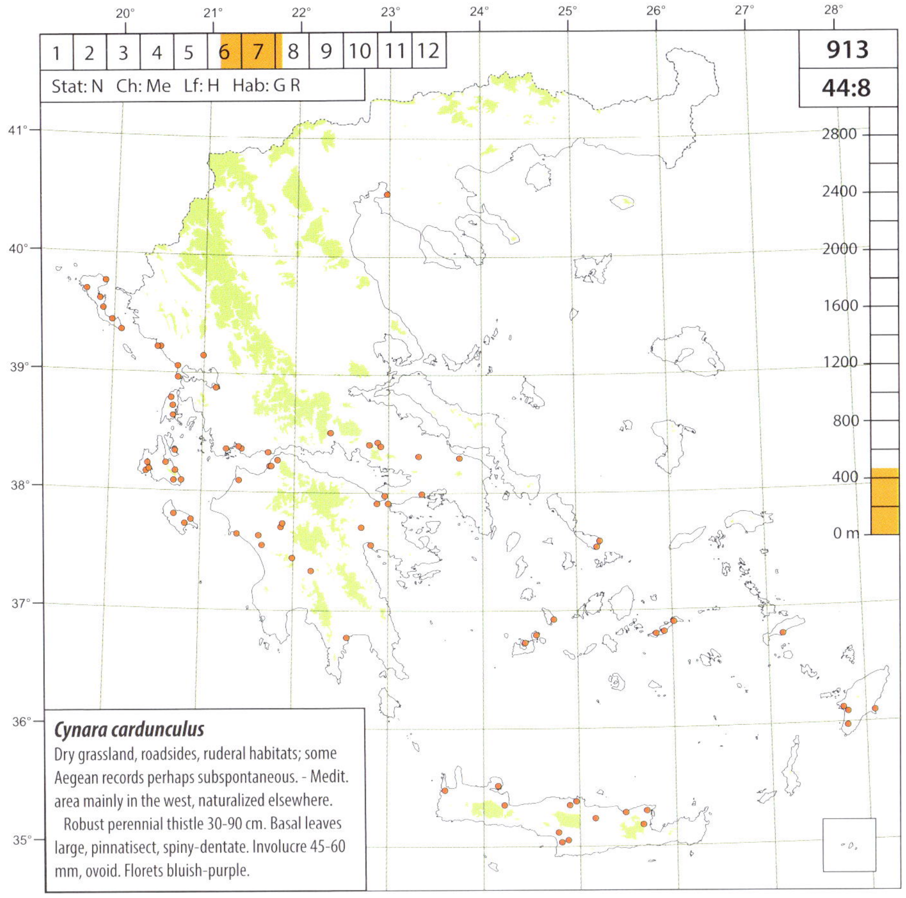

Cynara cardunculus

Dry grassland, roadsides, ruderal habitats; some Aegean records perhaps subspontaneous. - Medit. area mainly in the west, naturalized elsewhere.

Robust perennial thistle 30-90 cm. Basal leaves large, pinnatisect, spiny-dentate. Involucre 45-60 mm, ovoid. Florets bluish-purple.

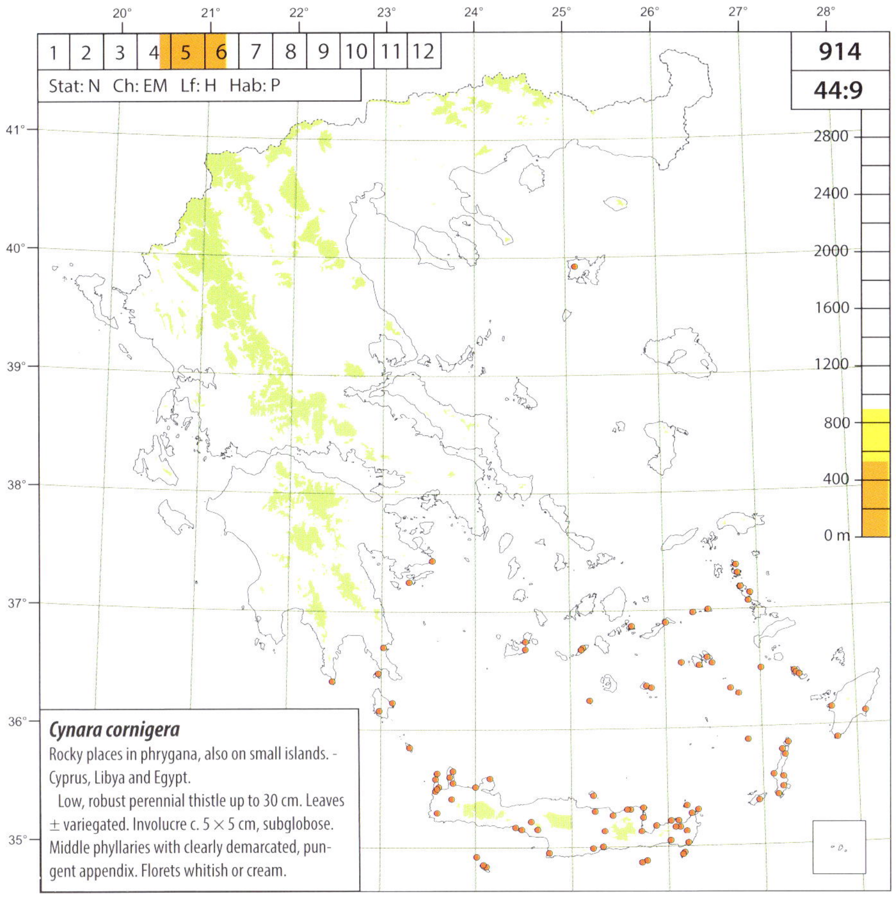

Cynara cornigera

Rocky places in phrygana, also on small islands. - Cyprus, Libya and Egypt.

Low, robust perennial thistle up to 30 cm. Leaves ± variegated. Involucre c. 5 × 5 cm, subglobose. Middle phyllaries with clearly demarcated, pungent appendix. Florets whitish or cream.

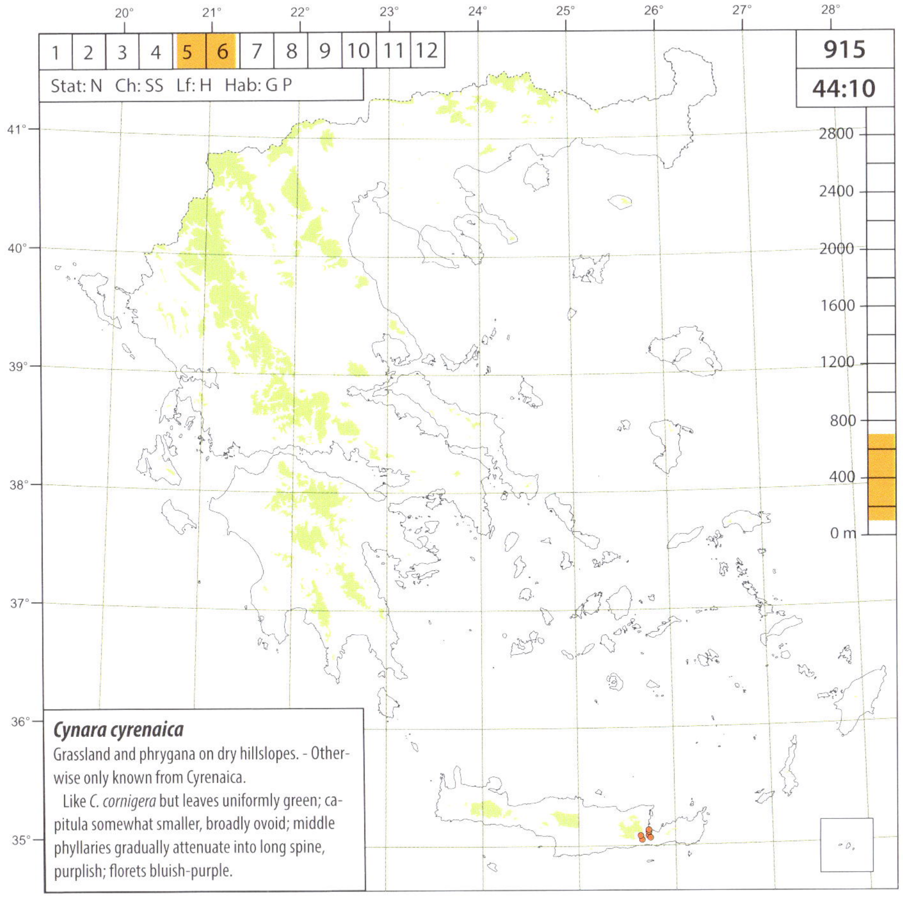

Cynara cyrenaica

Grassland and phrygana on dry hillslopes. - Otherwise only known from Cyrenaica.

Like *C. cornigera* but leaves uniformly green; capitula somewhat smaller, broadly ovoid; middle phyllaries gradually attenuate into long spine, purplish; florets bluish-purple.

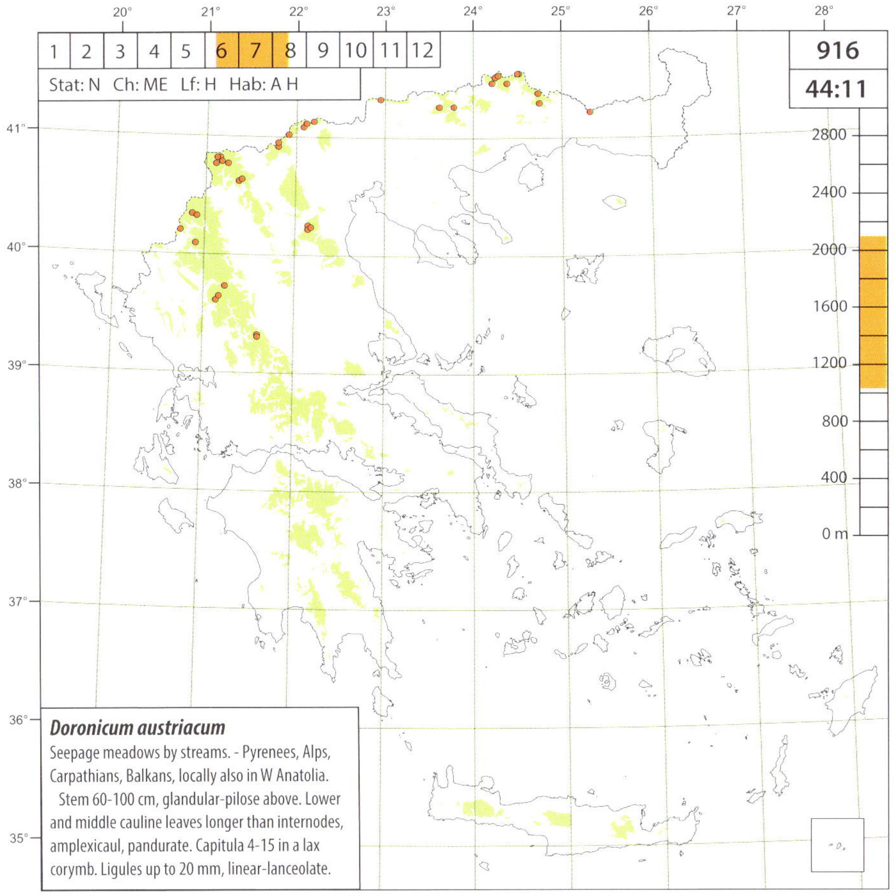

Doronicum austriacum

Seepage meadows by streams. - Pyrenees, Alps, Carpathians, Balkans, locally also in W Anatolia.

Stem 60-100 cm, glandular-pilose above. Lower and middle cauline leaves longer than internodes, amplexicaul, pandurate. Capitula 4-15 in a lax corymb. Ligules up to 20 mm, linear-lanceolate.

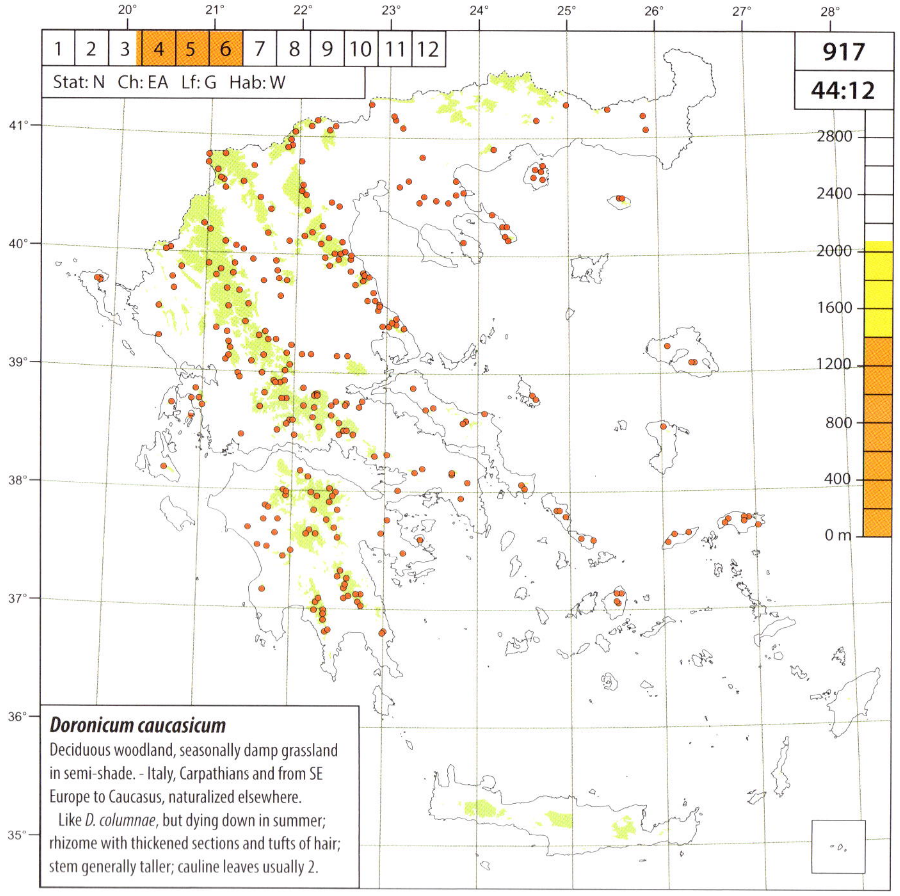

Doronicum caucasicum
Deciduous woodland, seasonally damp grassland in semi-shade. - Italy, Carpathians and from SE Europe to Caucasus, naturalized elsewhere.
Like *D. columnae*, but dying down in summer; rhizome with thickened sections and tufts of hair; stem generally taller; cauline leaves usually 2.

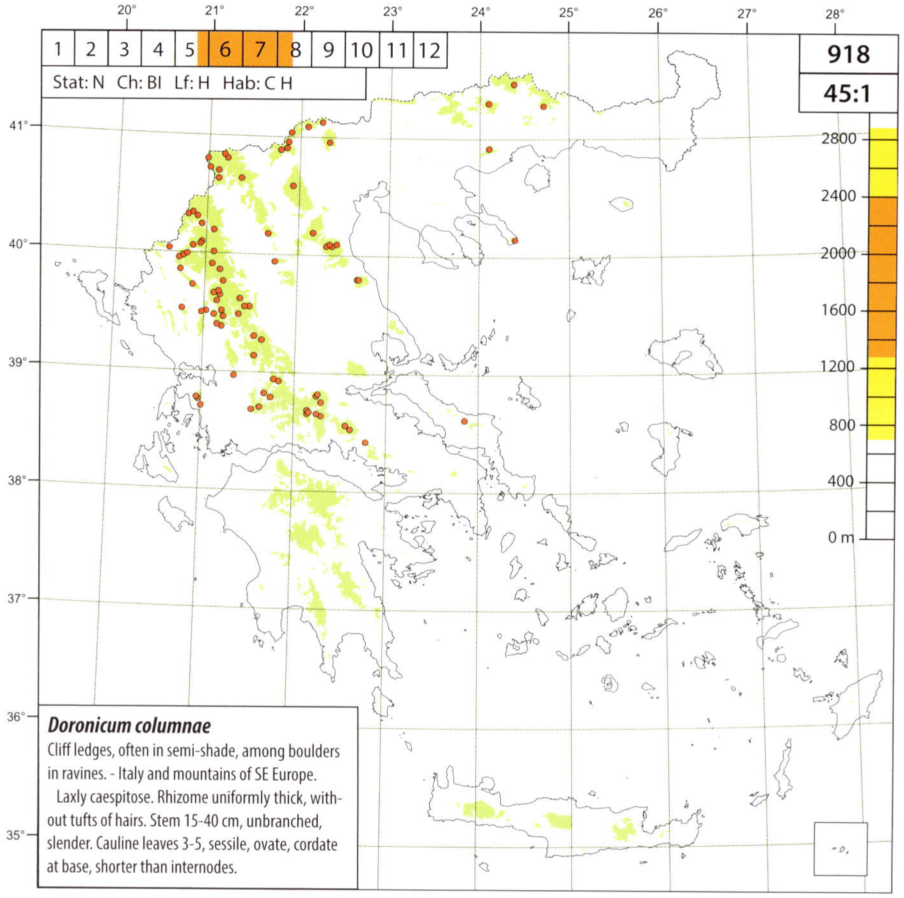

Doronicum columnae
Cliff ledges, often in semi-shade, among boulders in ravines. - Italy and mountains of SE Europe.
Laxly caespitose. Rhizome uniformly thick, without tufts of hairs. Stem 15-40 cm, unbranched, slender. Cauline leaves 3-5, sessile, ovate, cordate at base, shorter than internodes.

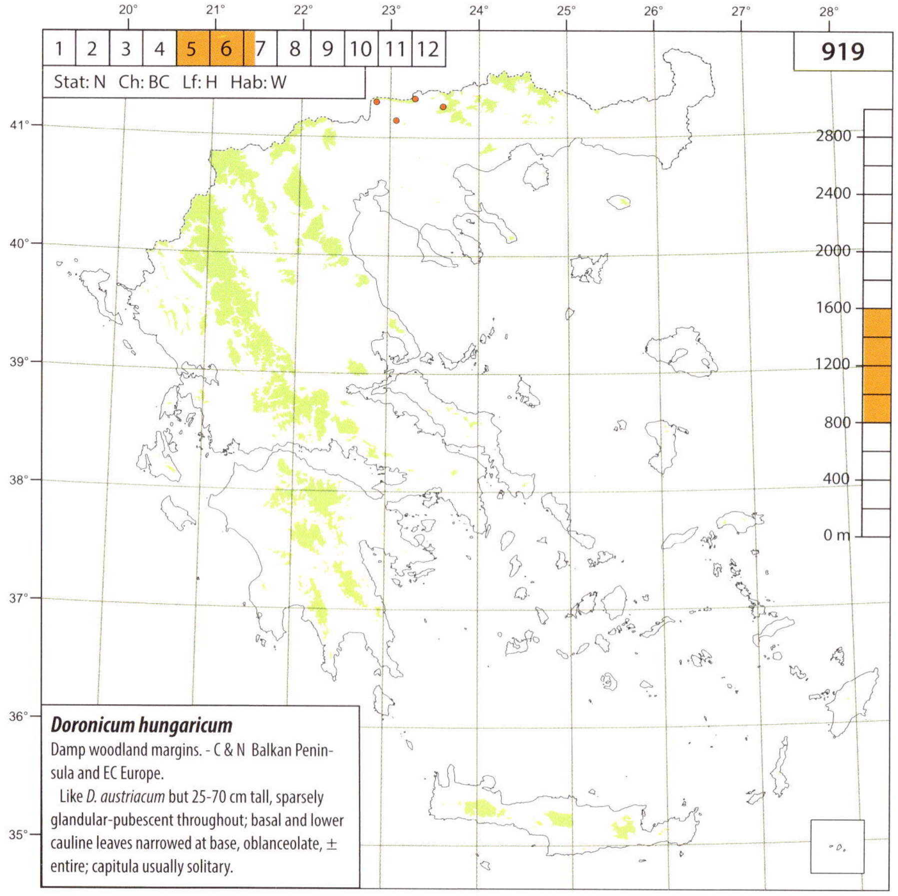

Doronicum hungaricum
Damp woodland margins. - C & N Balkan Peninsula and EC Europe.
Like *D. austriacum* but 25-70 cm tall, sparsely glandular-pubescent throughout; basal and lower cauline leaves narrowed at base, oblanceolate, ± entire; capitula usually solitary.

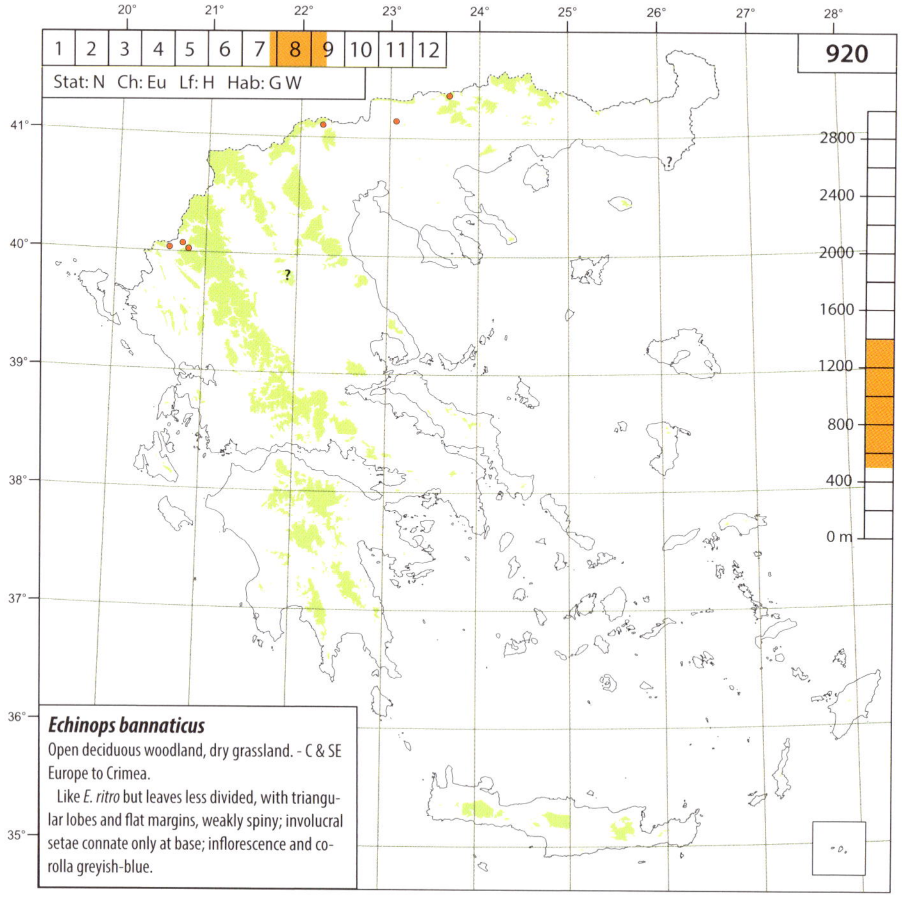

Echinops bannaticus
Open deciduous woodland, dry grassland. - C & SE Europe to Crimea.
Like *E. ritro* but leaves less divided, with triangular lobes and flat margins, weakly spiny; involucral setae connate only at base; inflorescence and corolla greyish-blue.

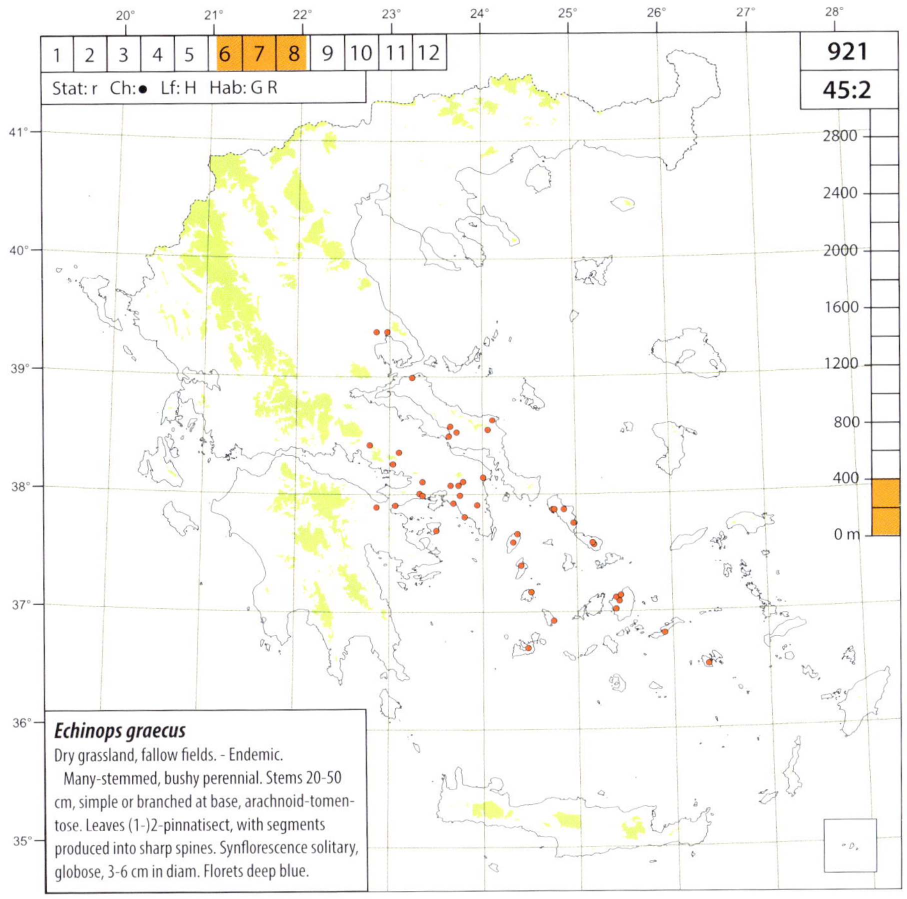

Echinops graecus

Dry grassland, fallow fields. - Endemic.

Many-stemmed, bushy perennial. Stems 20-50 cm, simple or branched at base, arachnoid-tomentose. Leaves (1-)2-pinnatisect, with segments produced into sharp spines. Synflorescence solitary, globose, 3-6 cm in diam. Florets deep blue.

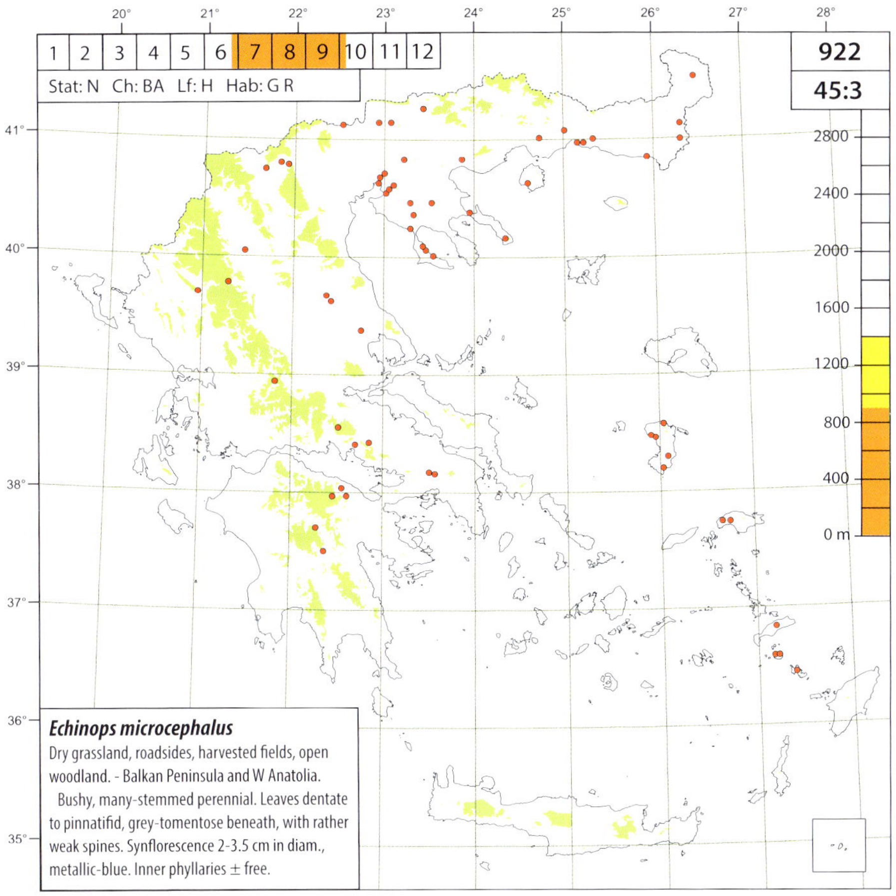

Echinops microcephalus

Dry grassland, roadsides, harvested fields, open woodland. - Balkan Peninsula and W Anatolia.

Bushy, many-stemmed perennial. Leaves dentate to pinnatifid, grey-tomentose beneath, with rather weak spines. Synflorescence 2-3.5 cm in diam., metallic-blue. Inner phyllaries ± free.

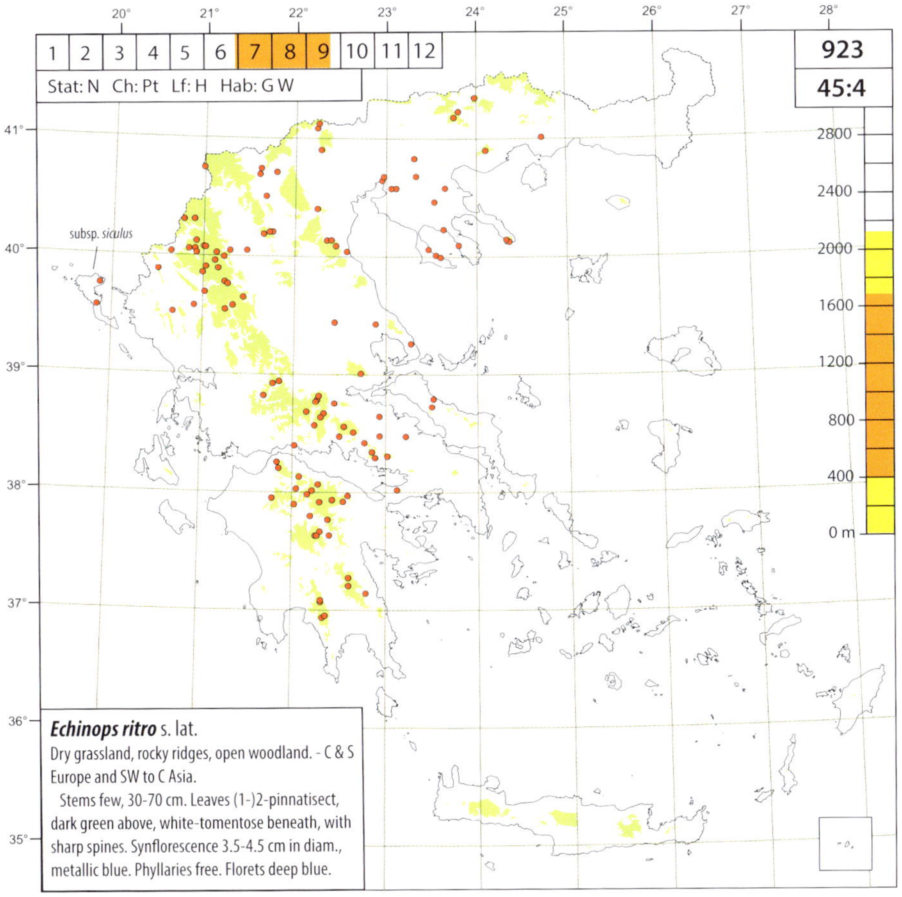

Echinops ritro s. lat.

Dry grassland, rocky ridges, open woodland. - C & S Europe and SW to C Asia.

Stems few, 30-70 cm. Leaves (1-)2-pinnatisect, dark green above, white-tomentose beneath, with sharp spines. Synflorescence 3.5-4.5 cm in diam., metallic blue. Phyllaries free. Florets deep blue.

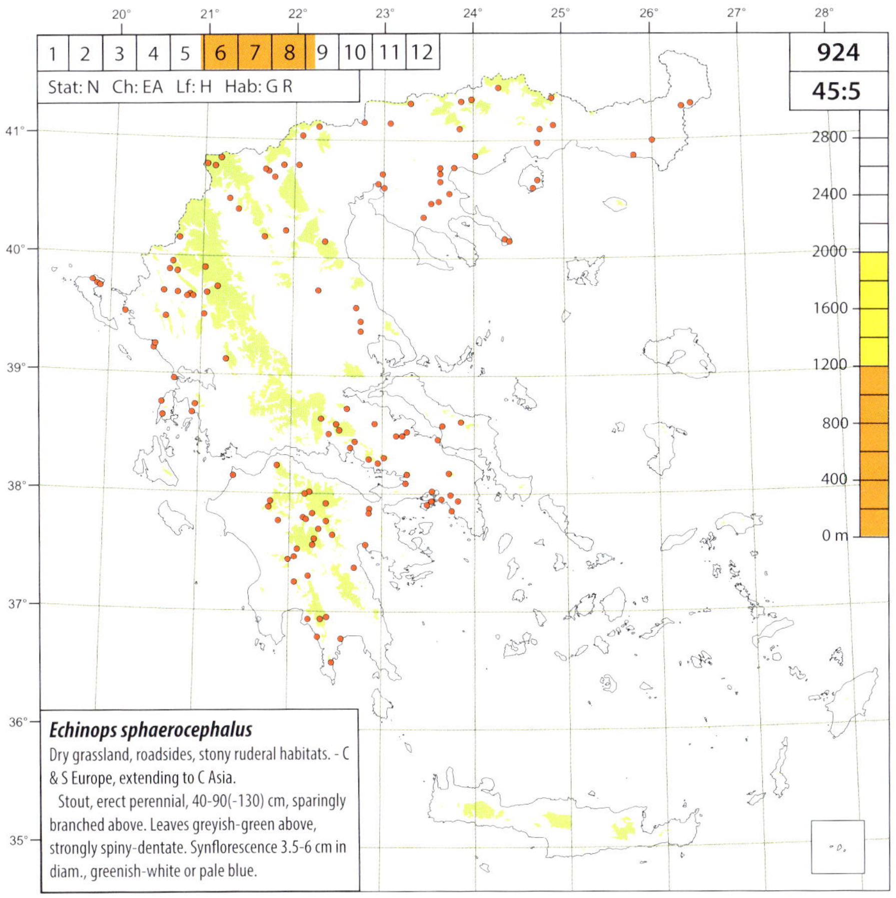

Echinops sphaerocephalus

Dry grassland, roadsides, stony ruderal habitats. - C & S Europe, extending to C Asia.

Stout, erect perennial, 40-90(-130) cm, sparingly branched above. Leaves greyish-green above, strongly spiny-dentate. Synflorescence 3.5-6 cm in diam., greenish-white or pale blue.

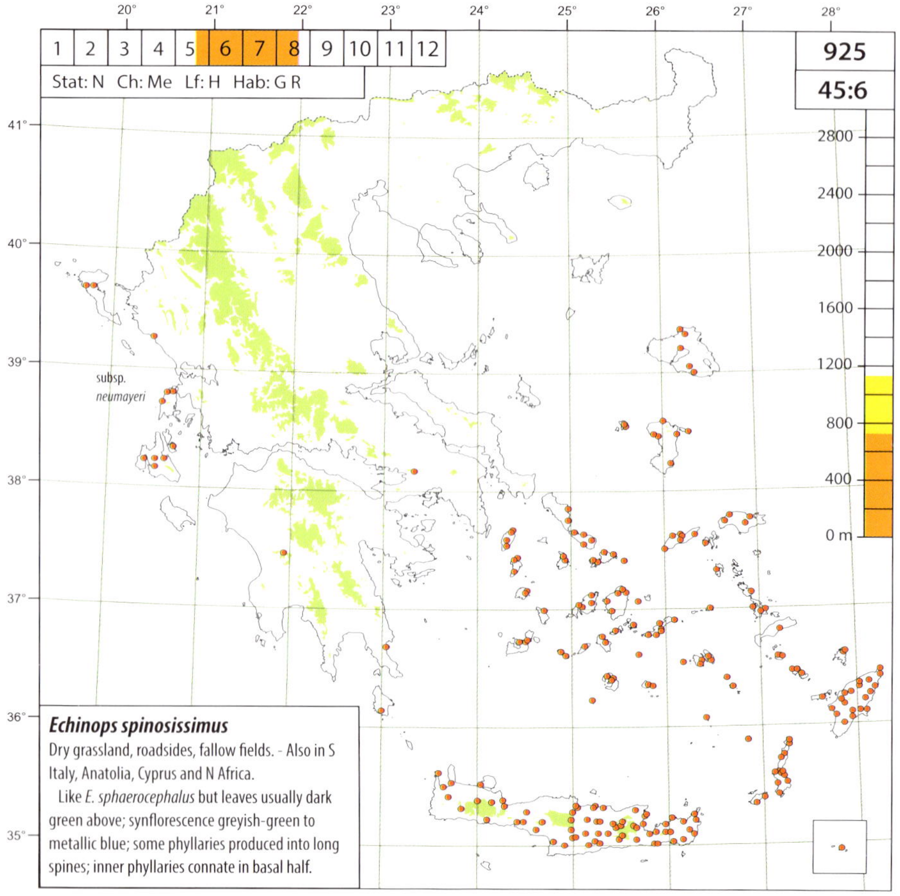

Echinops spinosissimus
Dry grassland, roadsides, fallow fields. - Also in S Italy, Anatolia, Cyprus and N Africa.
Like *E. sphaerocephalus* but leaves usually dark green above; synflorescence greyish-green to metallic blue; some phyllaries produced into long spines; inner phyllaries connate in basal half.

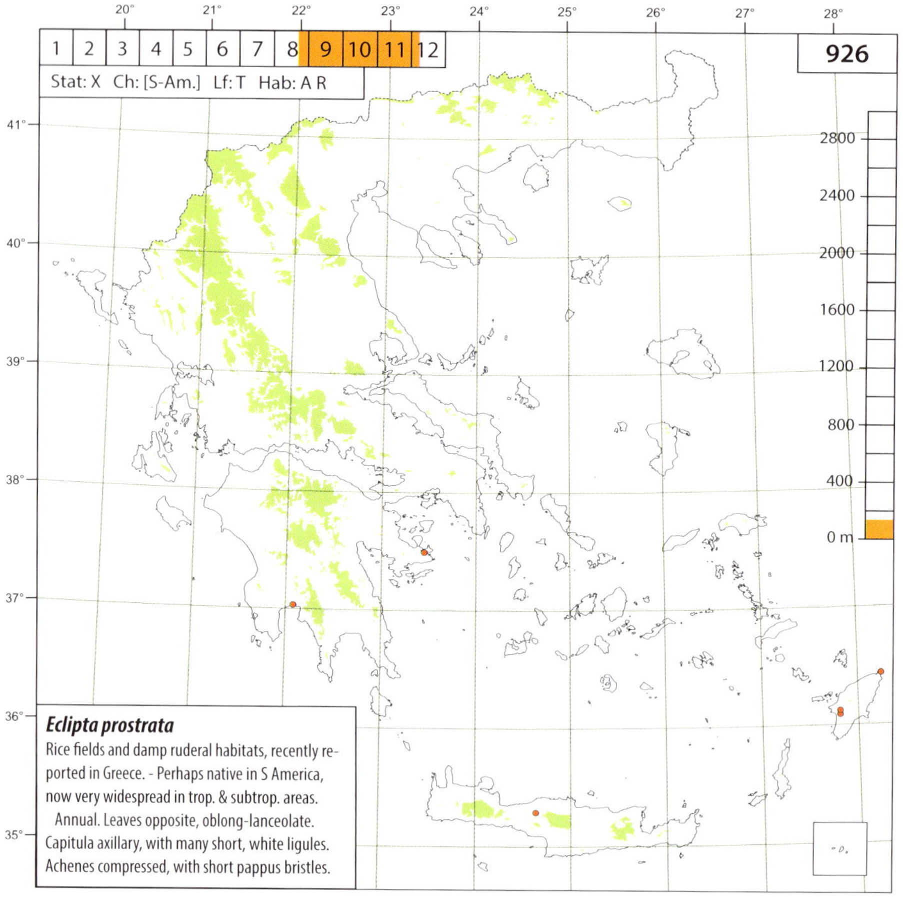

Eclipta prostrata
Rice fields and damp ruderal habitats, recently reported in Greece. - Perhaps native in S America, now very widespread in trop. & subtrop. areas.
Annual. Leaves opposite, oblong-lanceolate. Capitula axillary, with many short, white ligules. Achenes compressed, with short pappus bristles.

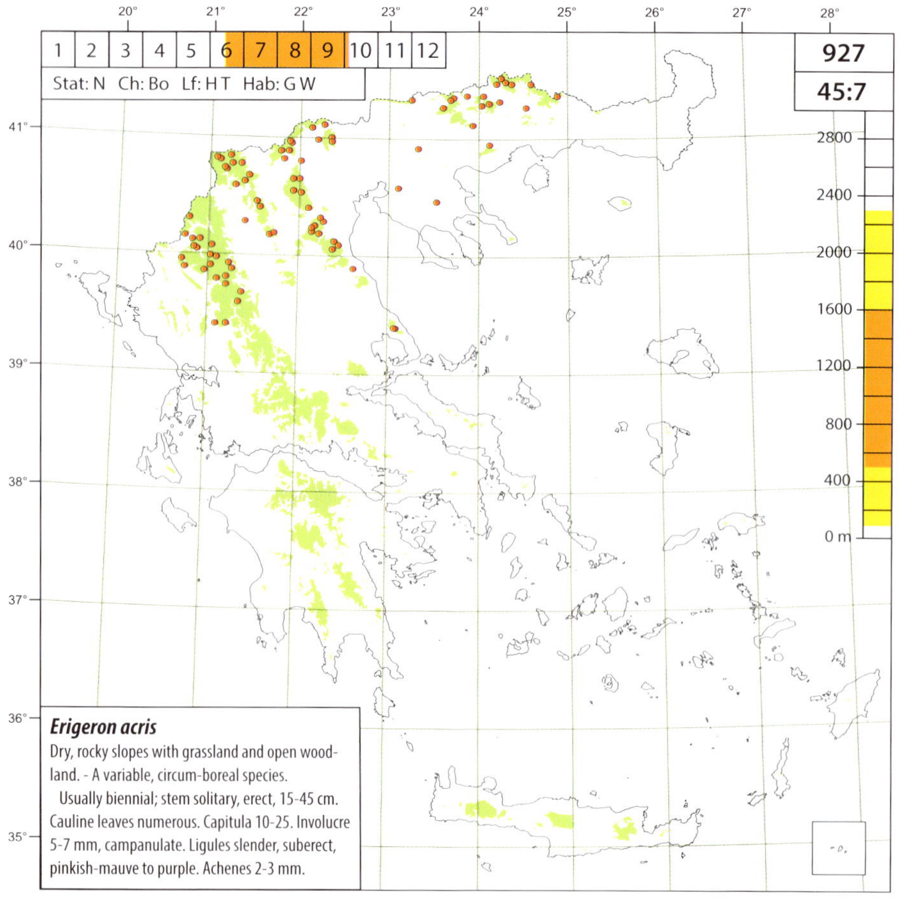

Erigeron acris
Dry, rocky slopes with grassland and open woodland. - A variable, circum-boreal species.
Usually biennial; stem solitary, erect, 15-45 cm. Cauline leaves numerous. Capitula 10-25. Involucre 5-7 mm, campanulate. Ligules slender, suberect, pinkish-mauve to purple. Achenes 2-3 mm.

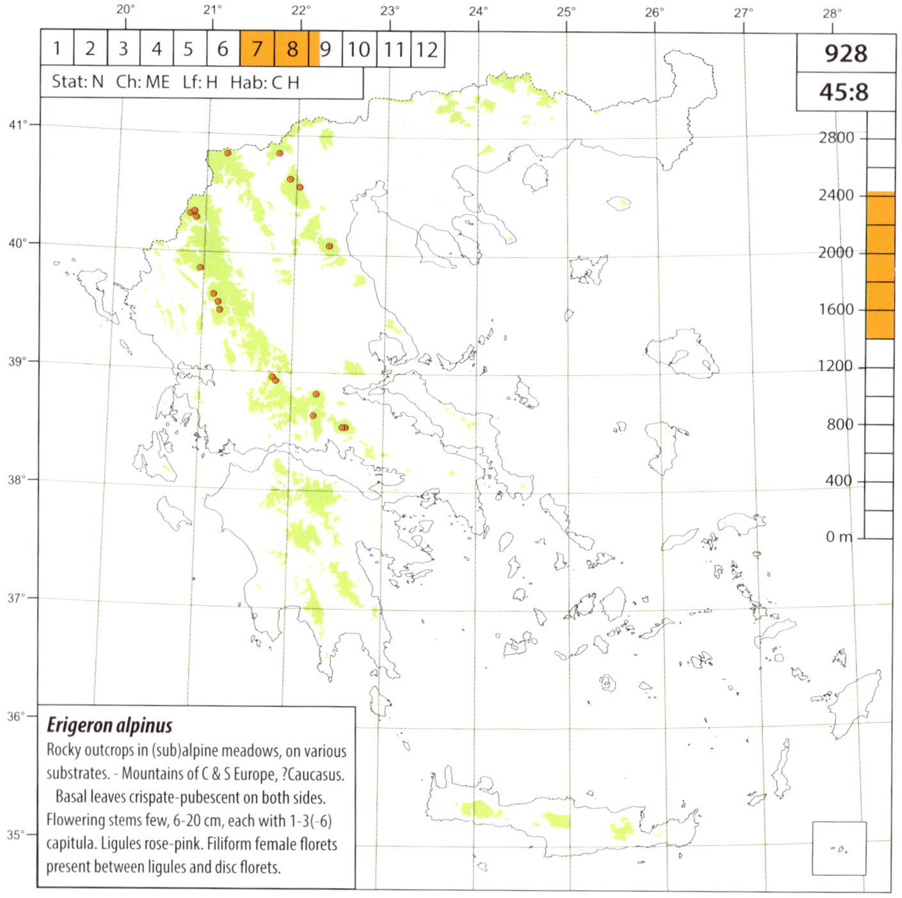

Erigeron alpinus
Rocky outcrops in (sub)alpine meadows, on various substrates. - Mountains of C & S Europe, ?Caucasus.
Basal leaves crispate-pubescent on both sides. Flowering stems few, 6-20 cm, each with 1-3(-6) capitula. Ligules rose-pink. Filiform female florets present between ligules and disc florets.

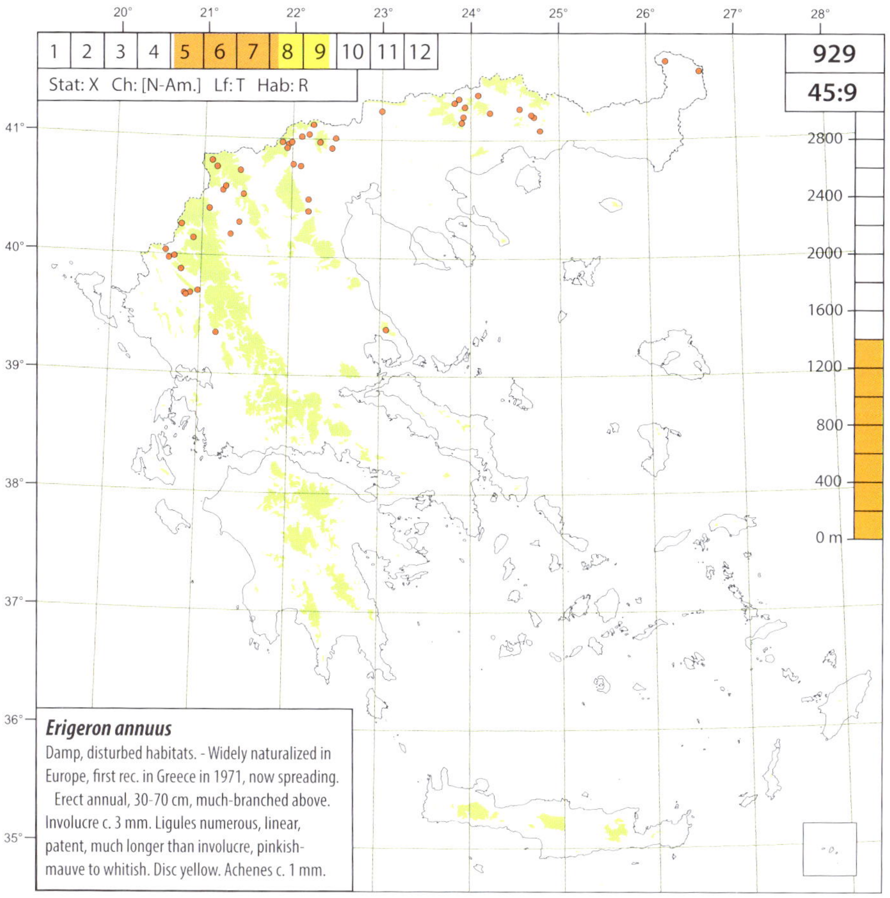

Erigeron annuus
Damp, disturbed habitats. - Widely naturalized in Europe, first rec. in Greece in 1971, now spreading.
Erect annual, 30-70 cm, much-branched above. Involucre c. 3 mm. Ligules numerous, linear, patent, much longer than involucre, pinkish-mauve to whitish. Disc yellow. Achenes c. 1 mm.

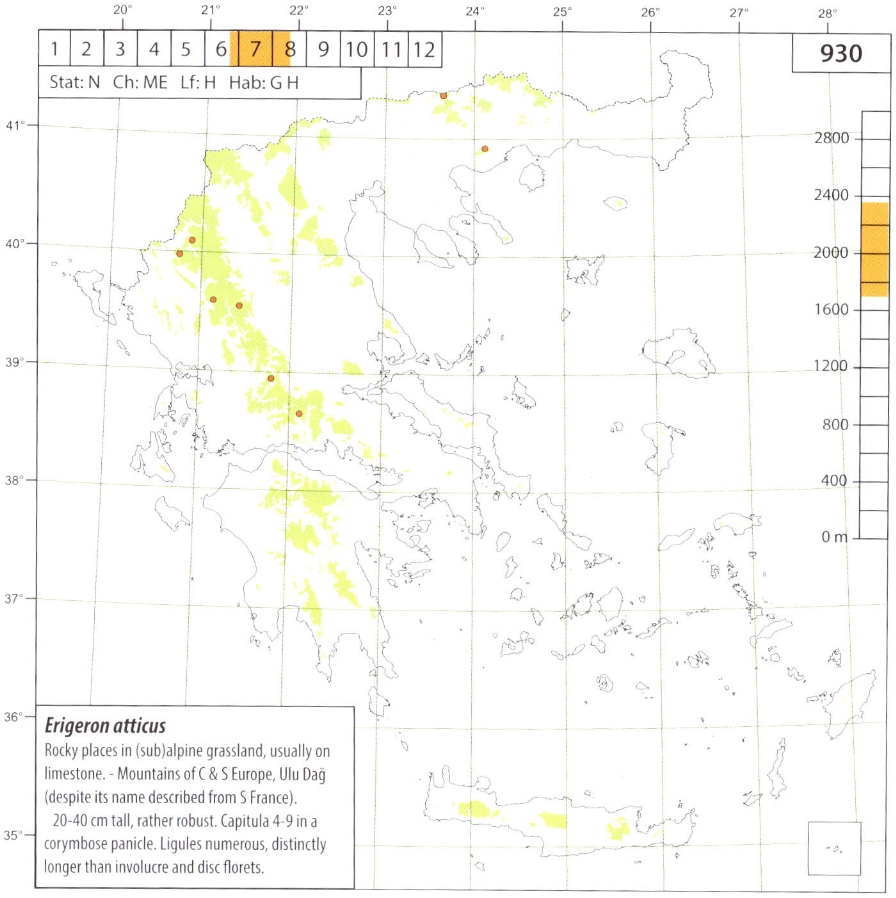

Erigeron atticus
Rocky places in (sub)alpine grassland, usually on limestone. - Mountains of C & S Europe, Ulu Dağ (despite its name described from S France).
20-40 cm tall, rather robust. Capitula 4-9 in a corymbose panicle. Ligules numerous, distinctly longer than involucre and disc florets.

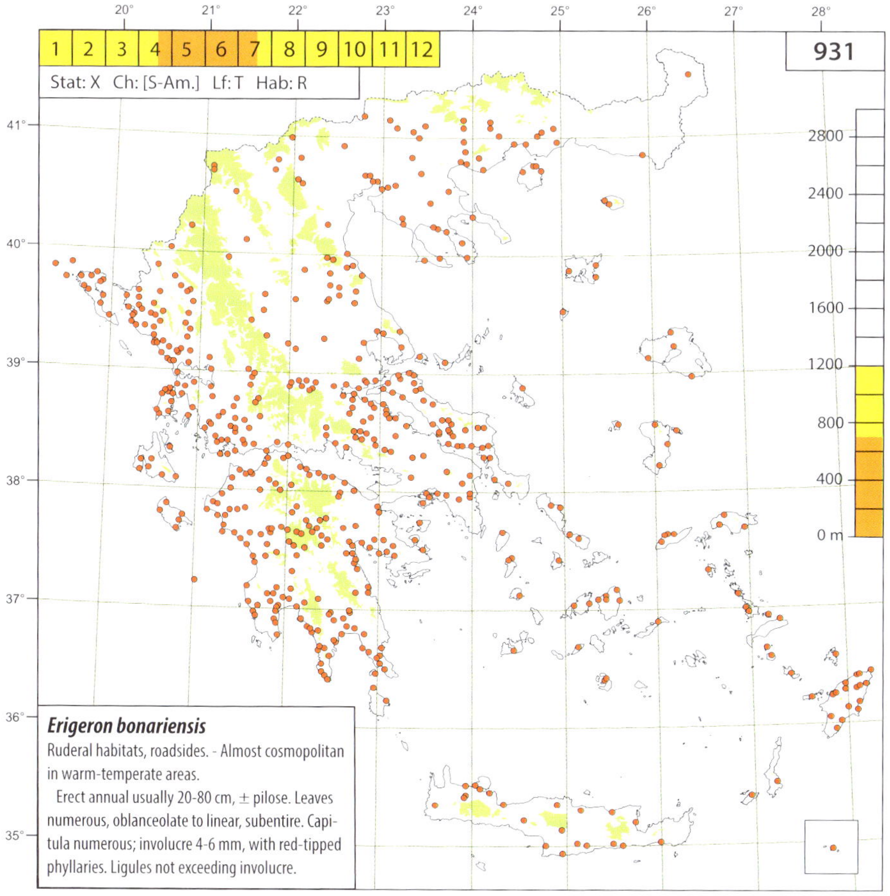

Erigeron bonariensis
Ruderal habitats, roadsides. - Almost cosmopolitan in warm-temperate areas.
Erect annual usually 20-80 cm, ± pilose. Leaves numerous, oblanceolate to linear, subentire. Capitula numerous; involucre 4-6 mm, with red-tipped phyllaries. Ligules not exceeding involucre.

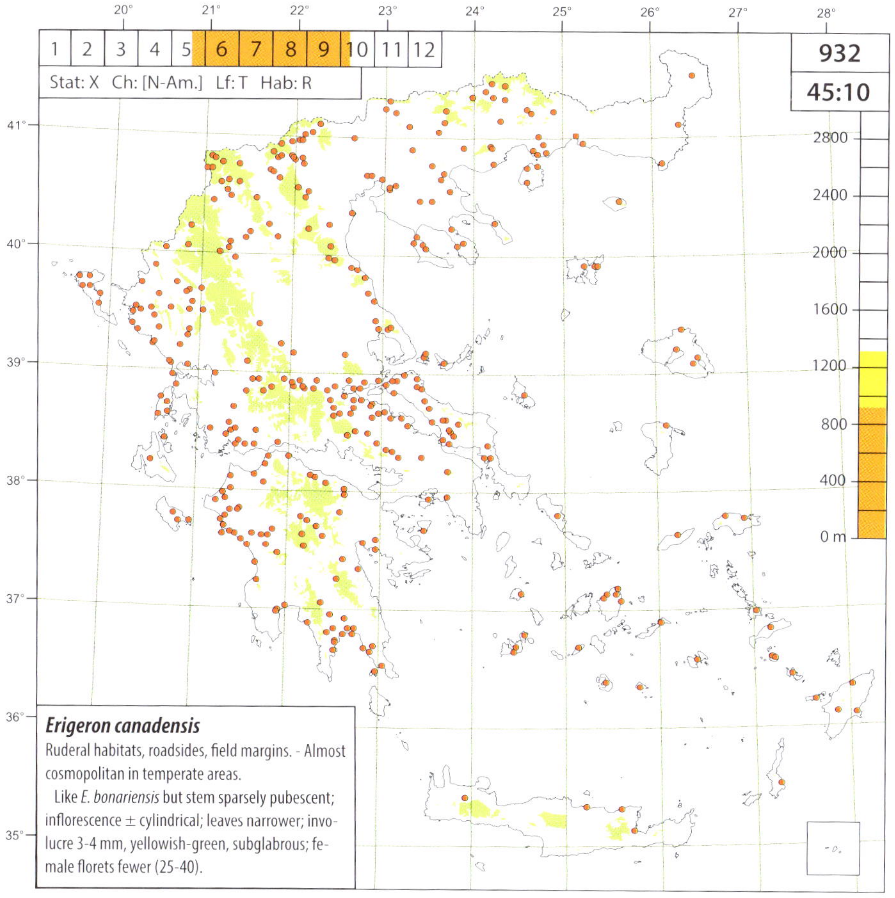

Erigeron canadensis
Ruderal habitats, roadsides, field margins. - Almost cosmopolitan in temperate areas.
Like *E. bonariensis* but stem sparsely pubescent; inflorescence ± cylindrical; leaves narrower; involucre 3-4 mm, yellowish-green, subglabrous; female florets fewer (25-40).

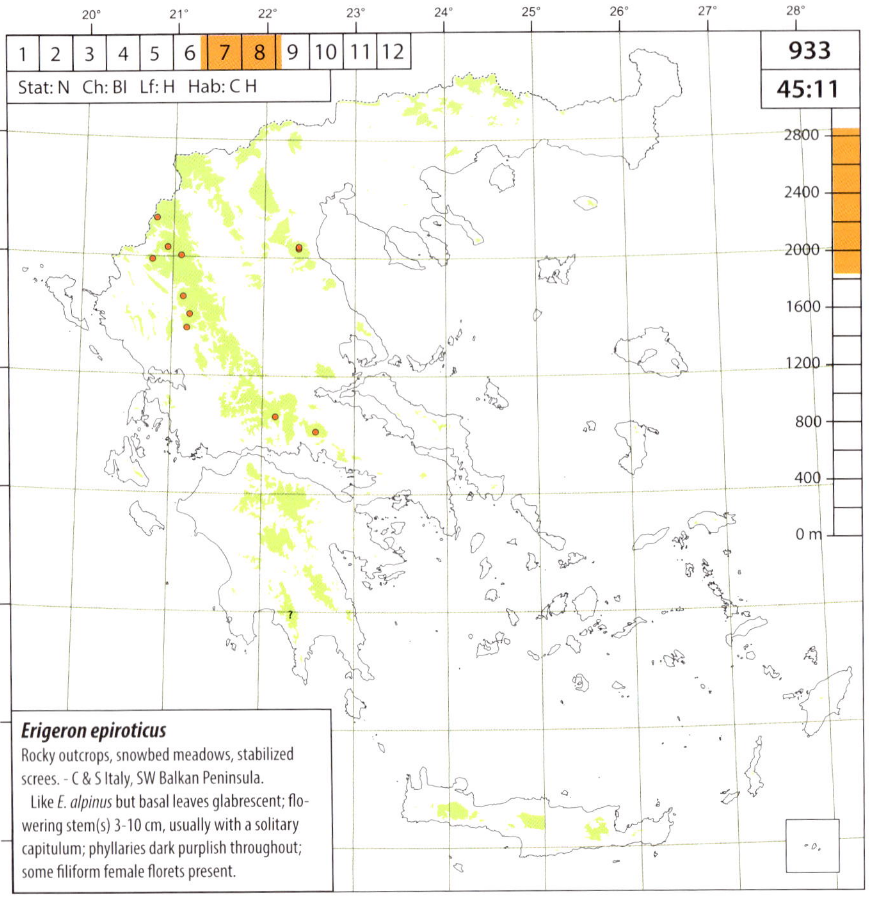

Erigeron epiroticus

Rocky outcrops, snowbed meadows, stabilized screes. - C & S Italy, SW Balkan Peninsula.

Like *E. alpinus* but basal leaves glabrescent; flowering stem(s) 3-10 cm, usually with a solitary capitulum; phyllaries dark purplish throughout; some filiform female florets present.

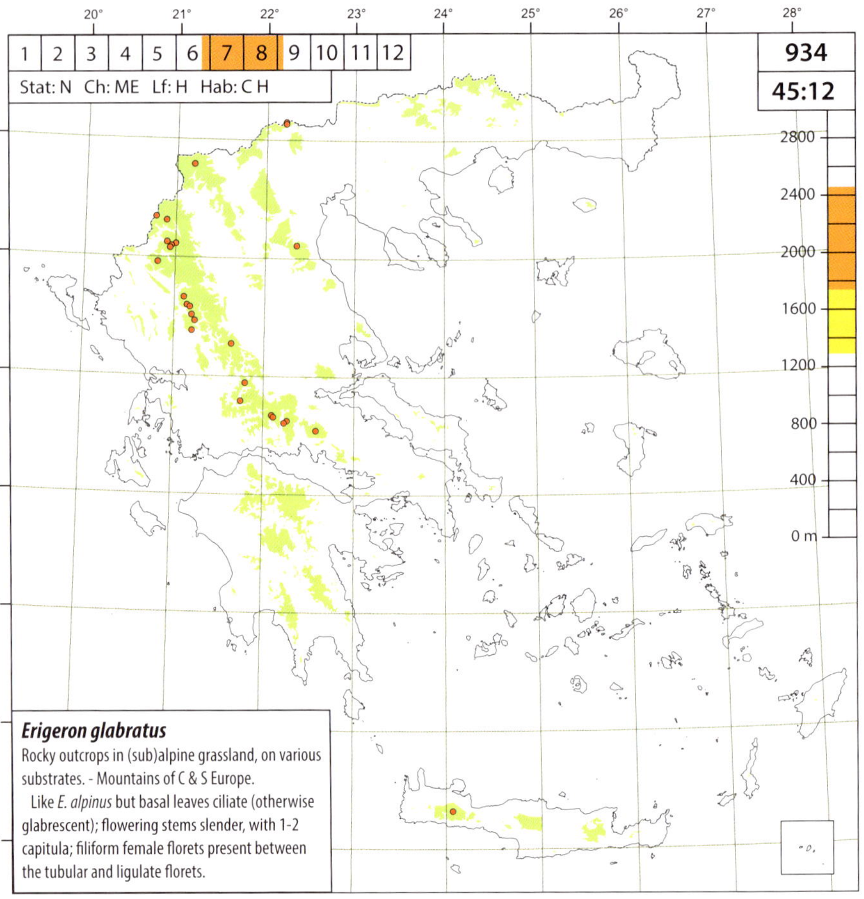

Erigeron glabratus

Rocky outcrops in (sub)alpine grassland, on various substrates. - Mountains of C & S Europe.

Like *E. alpinus* but basal leaves ciliate (otherwise glabrescent); flowering stems slender, with 1-2 capitula; filiform female florets present between the tubular and ligulate florets.

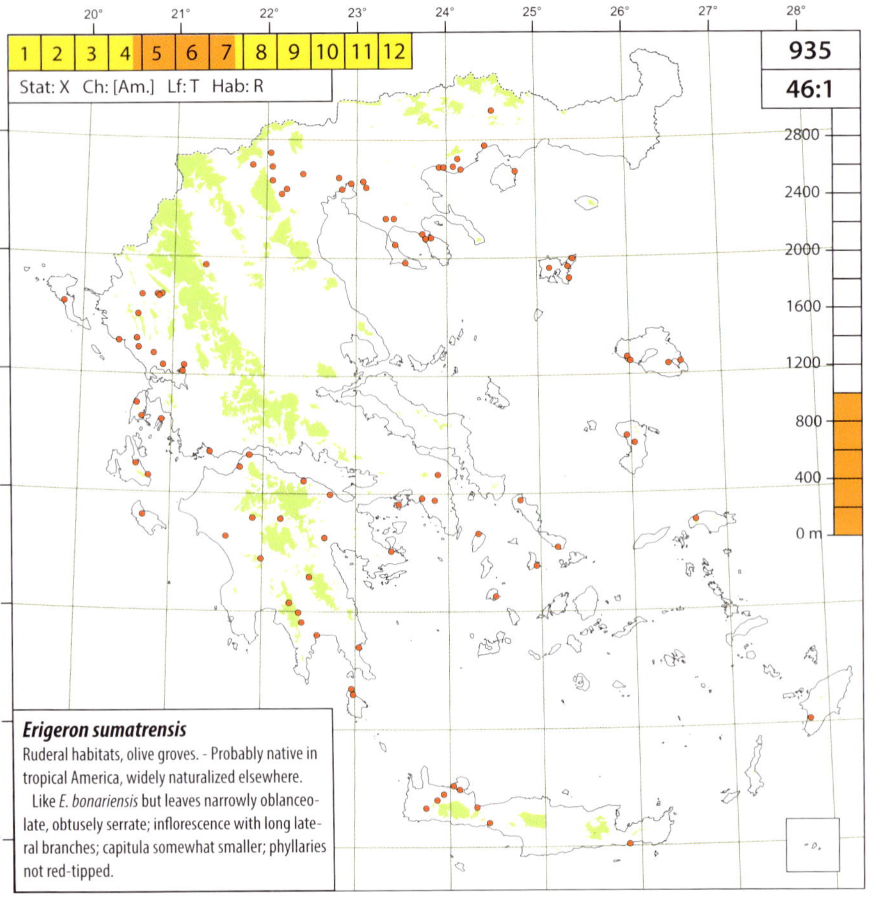

Erigeron sumatrensis

Ruderal habitats, olive groves. - Probably native in tropical America, widely naturalized elsewhere.

Like *E. bonariensis* but leaves narrowly oblanceolate, obtusely serrate; inflorescence with long lateral branches; capitula somewhat smaller; phyllaries not red-tipped.

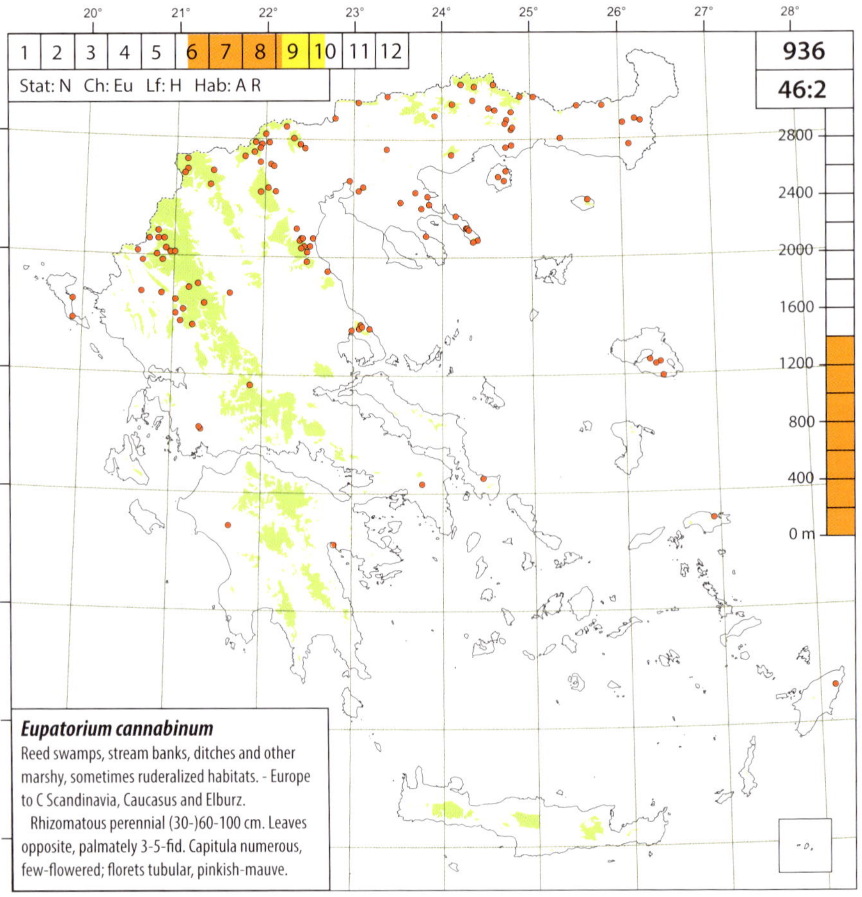

Eupatorium cannabinum

Reed swamps, stream banks, ditches and other marshy, sometimes ruderalized habitats. - Europe to C Scandinavia, Caucasus and Elburz.

Rhizomatous perennial (30-)60-100 cm. Leaves opposite, palmately 3-5-fid. Capitula numerous, few-flowered; florets tubular, pinkish-mauve.

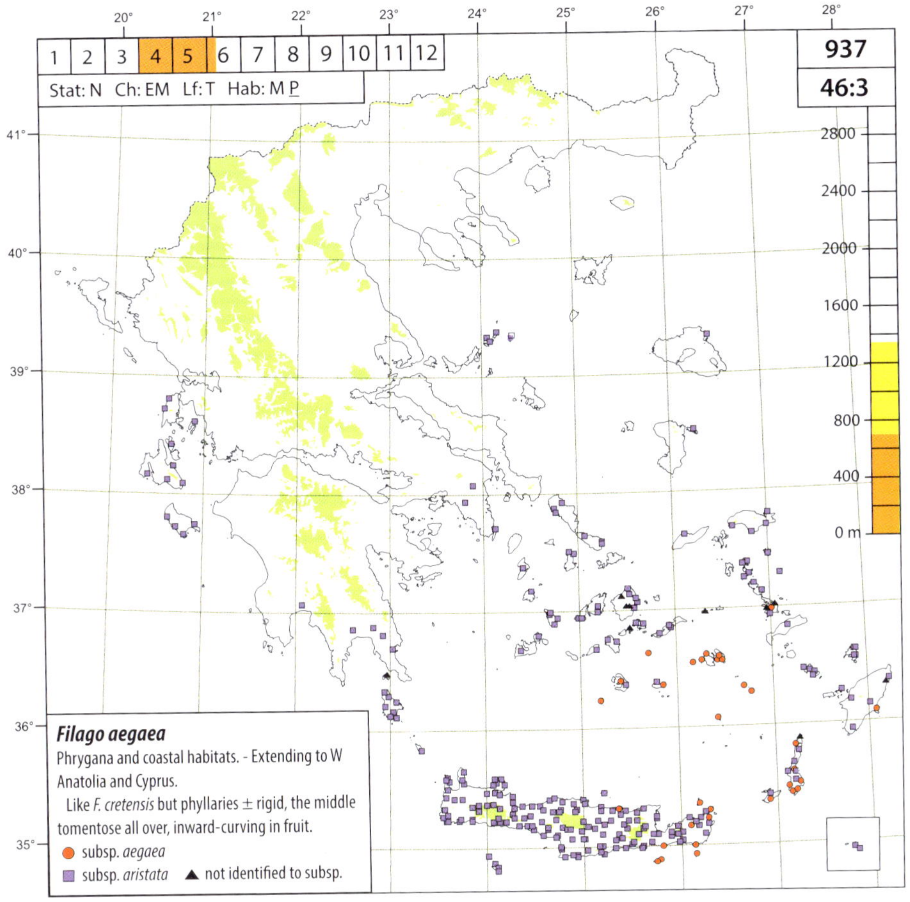

Filago aegaea

Phrygana and coastal habitats. - Extending to W Anatolia and Cyprus.

Like *F. cretensis* but phyllaries ± rigid, the middle tomentose all over, inward-curving in fruit.

● subsp. *aegaea*
■ subsp. *aristata* ▲ not identified to subsp.

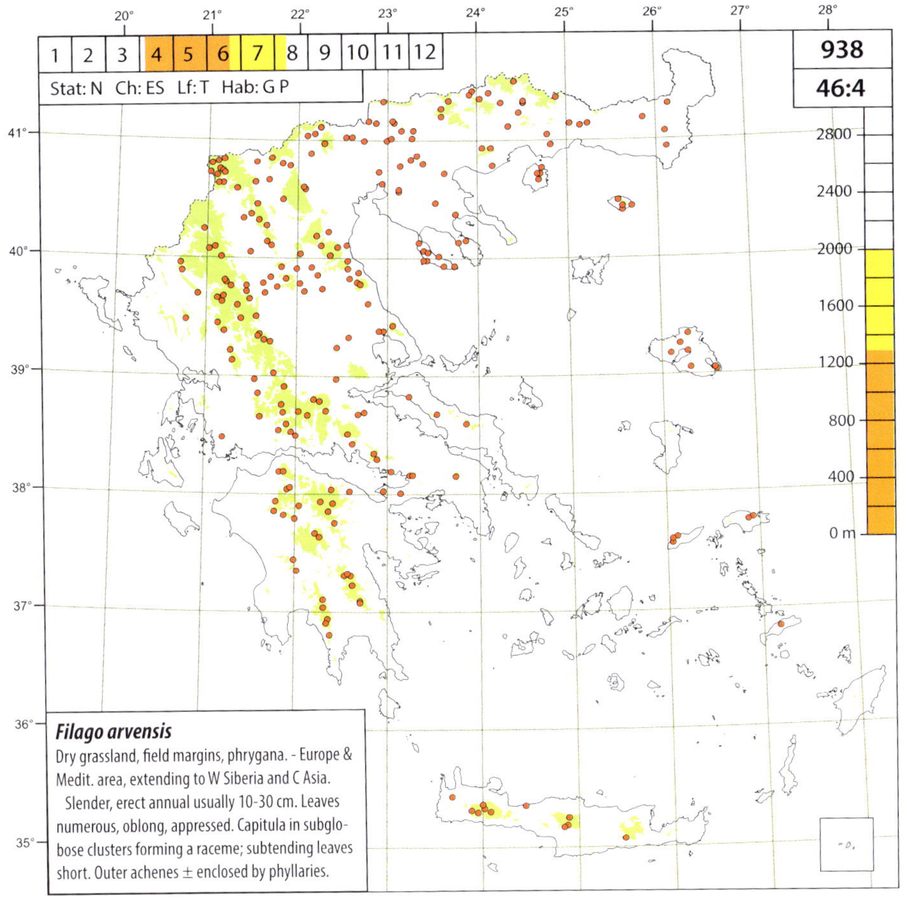

Filago arvensis

Dry grassland, field margins, phrygana. - Europe & Medit. area, extending to W Siberia and C Asia.

Slender, erect annual usually 10-30 cm. Leaves numerous, oblong, appressed. Capitula in subglobose clusters forming a raceme; subtending leaves short. Outer achenes ± enclosed by phyllaries.

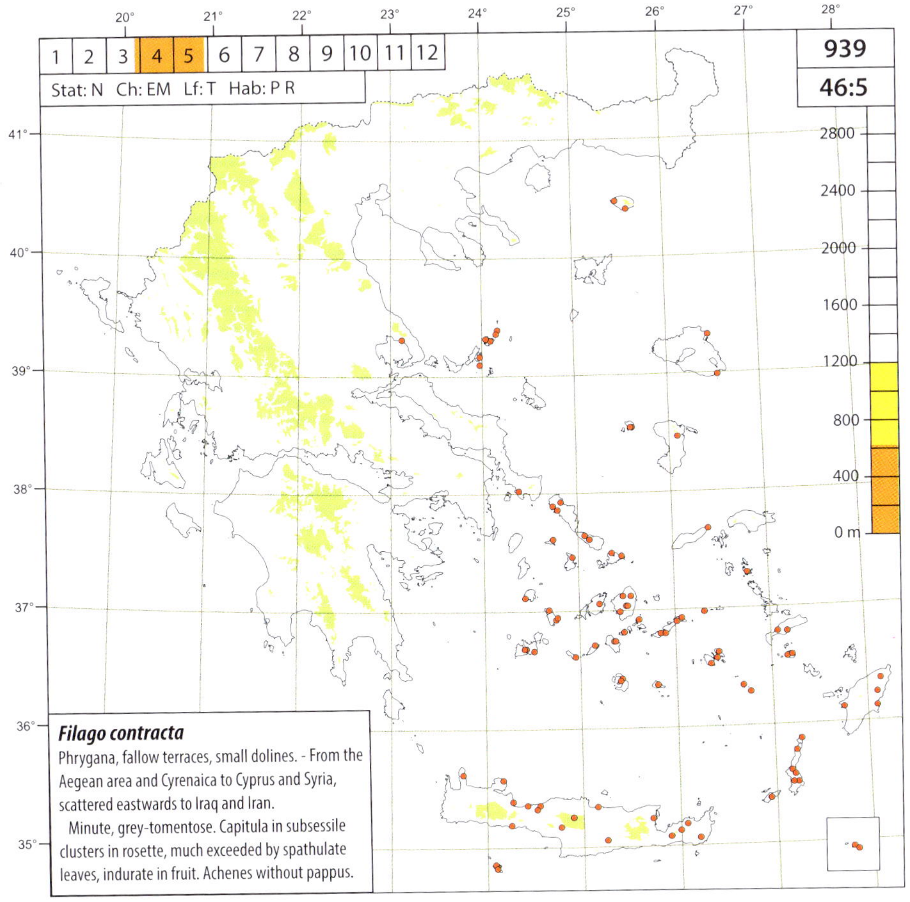

Filago contracta

Phrygana, fallow terraces, small dolines. - From the Aegean area and Cyrenaica to Cyprus and Syria, scattered eastwards to Iraq and Iran.

Minute, grey-tomentose. Capitula in subsessile clusters in rosette, much exceeded by spathulate leaves, indurate in fruit. Achenes without pappus.

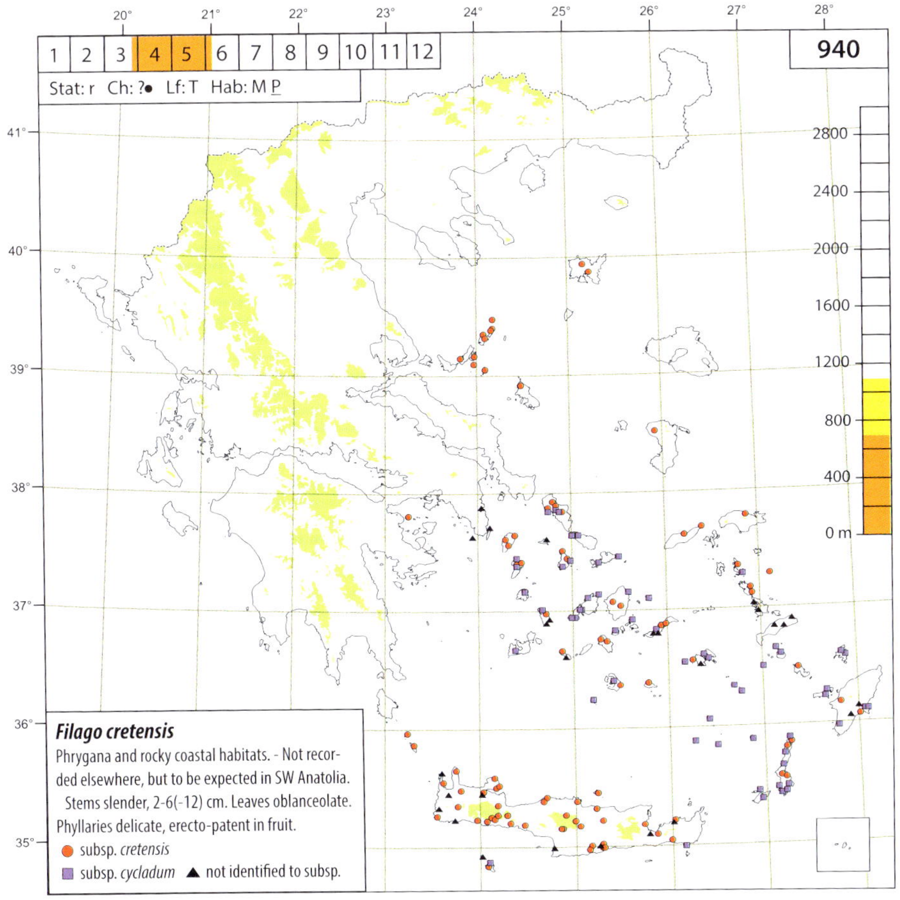

Filago cretensis

Phrygana and rocky coastal habitats. - Not recorded elsewhere, but to be expected in SW Anatolia.

Stems slender, 2-6(-12) cm. Leaves oblanceolate. Phyllaries delicate, erecto-patent in fruit.

● subsp. *cretensis*
■ subsp. *cycladum* ▲ not identified to subsp.

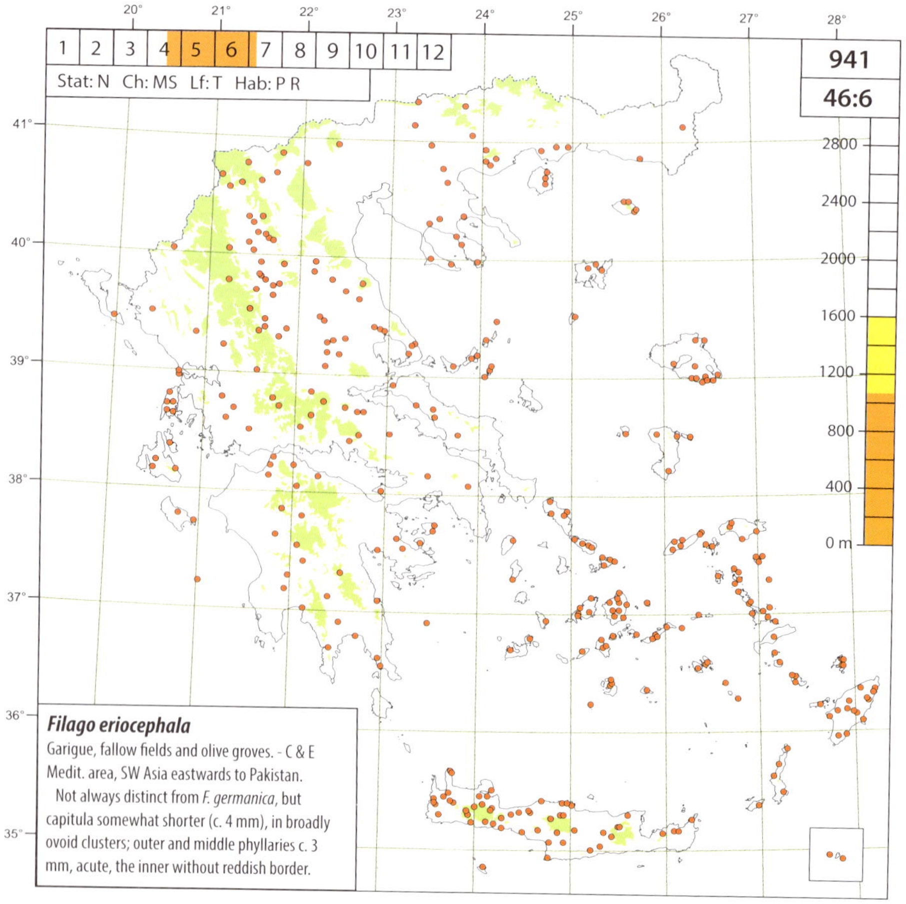

Filago eriocephala
Garigue, fallow fields and olive groves. - C & E Medit. area, SW Asia eastwards to Pakistan.
Not always distinct from *F. germanica*, but capitula somewhat shorter (c. 4 mm), in broadly ovoid clusters; outer and middle phyllaries c. 3 mm, acute, the inner without reddish border.

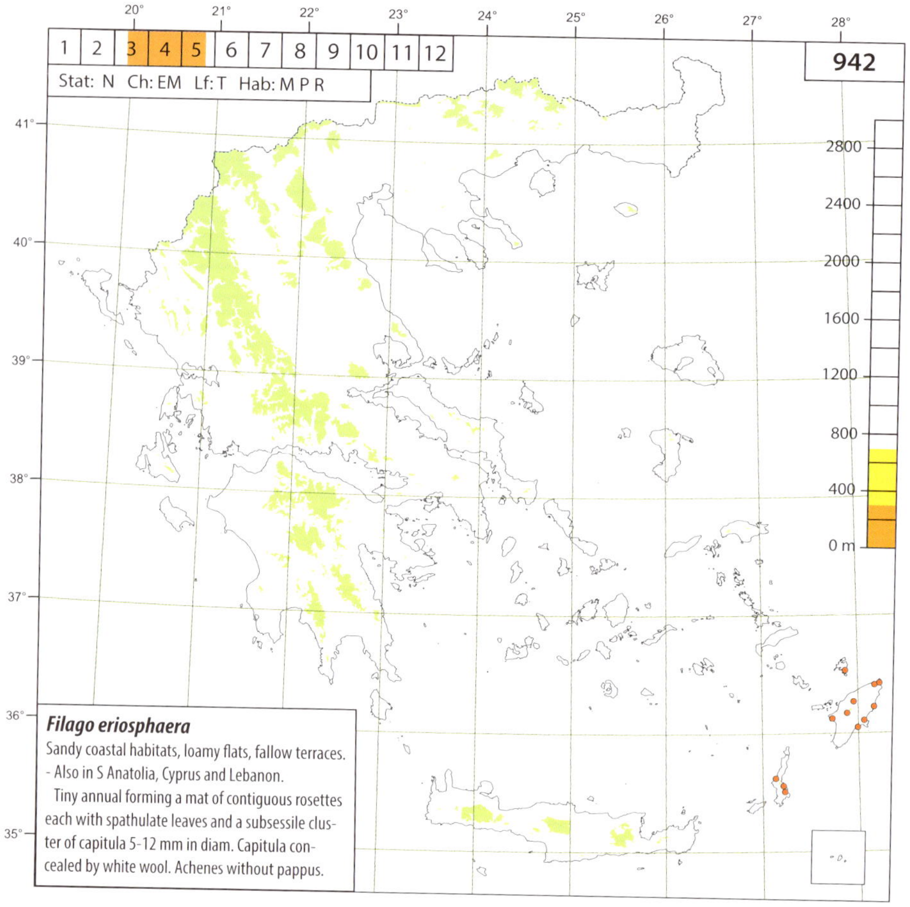

Filago eriosphaera
Sandy coastal habitats, loamy flats, fallow terraces. - Also in S Anatolia, Cyprus and Lebanon.
Tiny annual forming a mat of contiguous rosettes each with spathulate leaves and a subsessile cluster of capitula 5-12 mm in diam. Capitula concealed by white wool. Achenes without pappus.

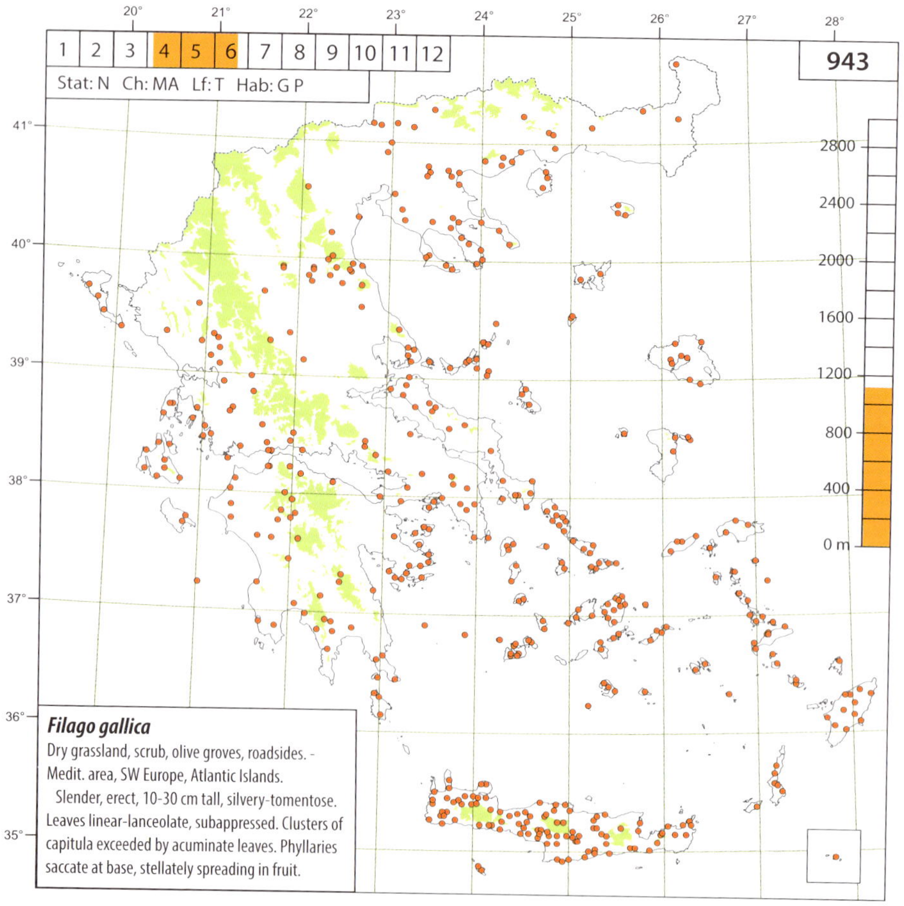

Filago gallica
Dry grassland, scrub, olive groves, roadsides. - Medit. area, SW Europe, Atlantic Islands.
Slender, erect, 10-30 cm tall, silvery-tomentose. Leaves linear-lanceolate, subappressed. Clusters of capitula exceeded by acuminate leaves. Phyllaries saccate at base, stellately spreading in fruit.

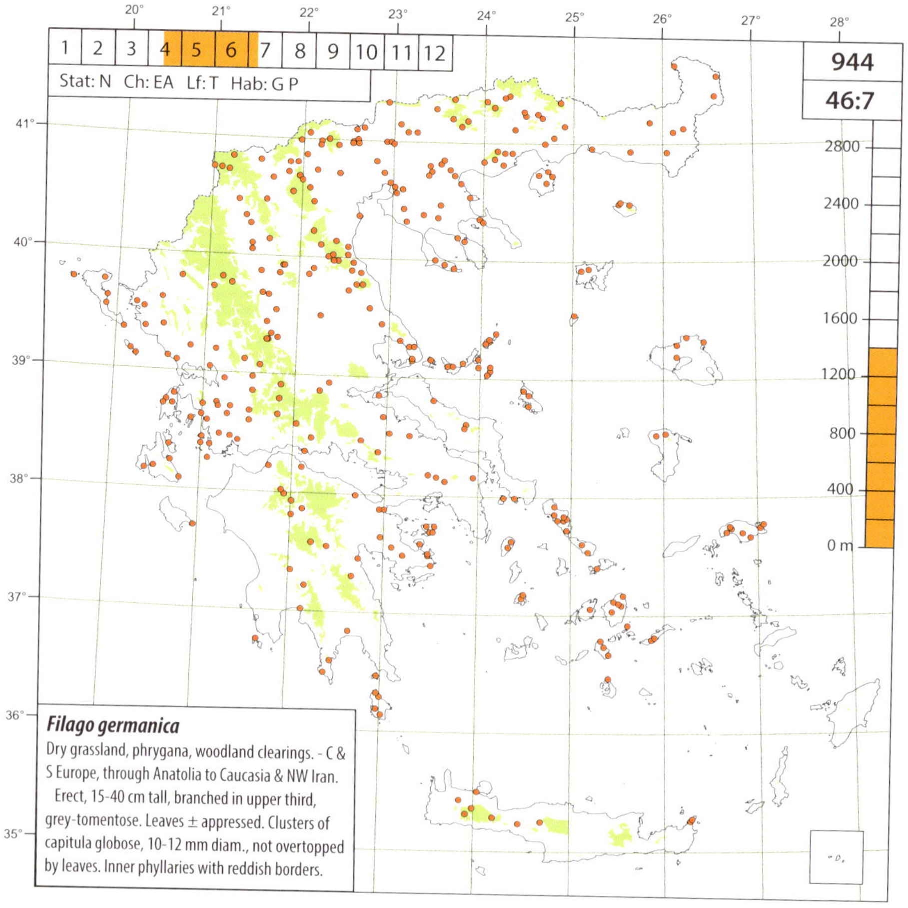

Filago germanica
Dry grassland, phrygana, woodland clearings. - C & S Europe, through Anatolia to Caucasia & NW Iran.
Erect, 15-40 cm tall, branched in upper third, grey-tomentose. Leaves ± appressed. Clusters of capitula globose, 10-12 mm diam., not overtopped by leaves. Inner phyllaries with reddish borders.

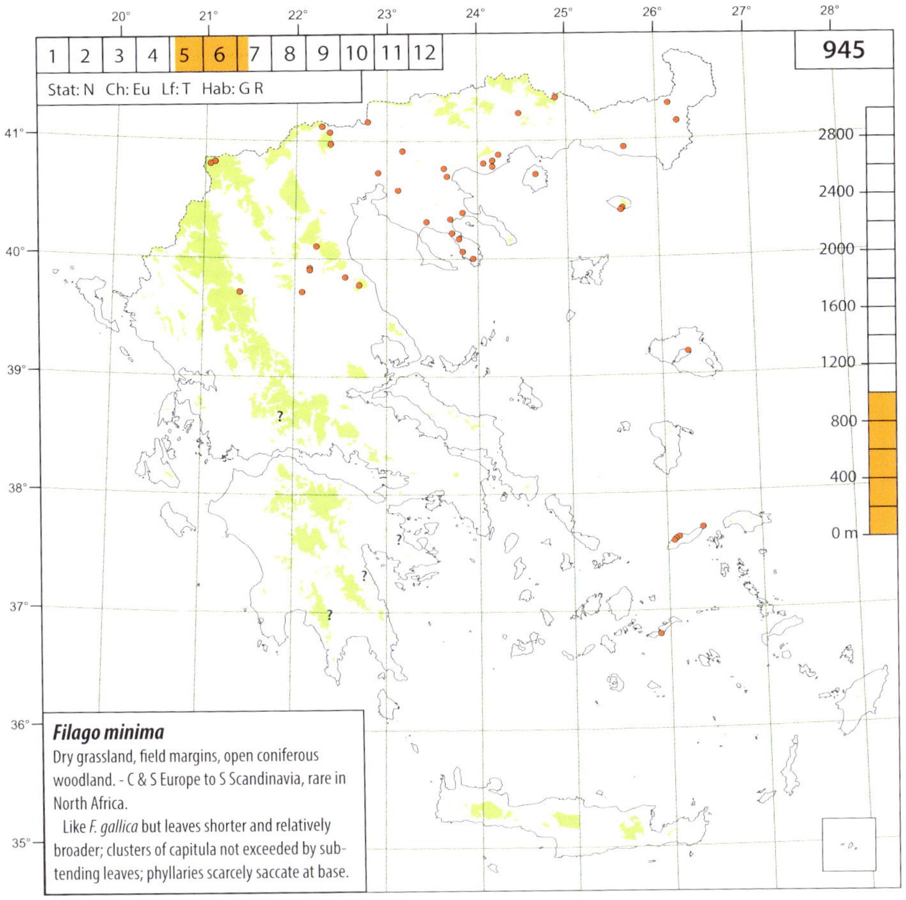

Filago minima

Dry grassland, field margins, open coniferous woodland. - C & S Europe to S Scandinavia, rare in North Africa.

Like *F. gallica* but leaves shorter and relatively broader; clusters of capitula not exceeded by subtending leaves; phyllaries scarcely saccate at base.

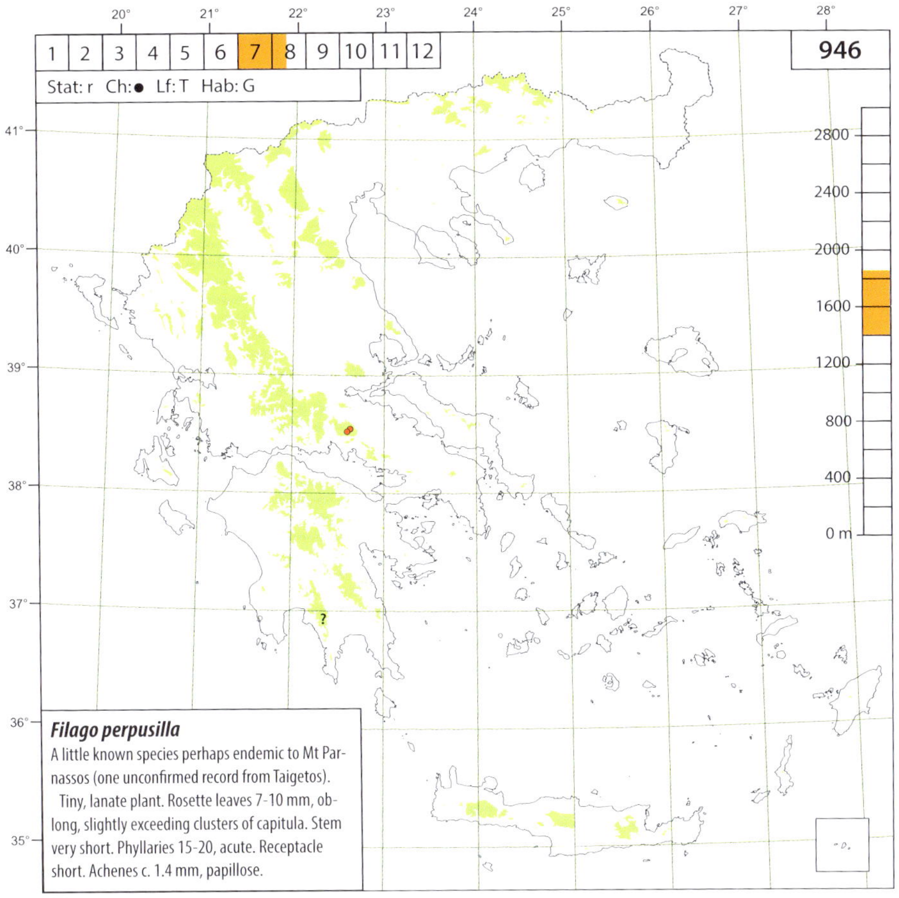

Filago perpusilla

A little known species perhaps endemic to Mt Parnassos (one unconfirmed record from Taigetos).

Tiny, lanate plant. Rosette leaves 7-10 mm, oblong, slightly exceeding clusters of capitula. Stem very short. Phyllaries 15-20, acute. Receptacle short. Achenes c. 1.4 mm, papillose.

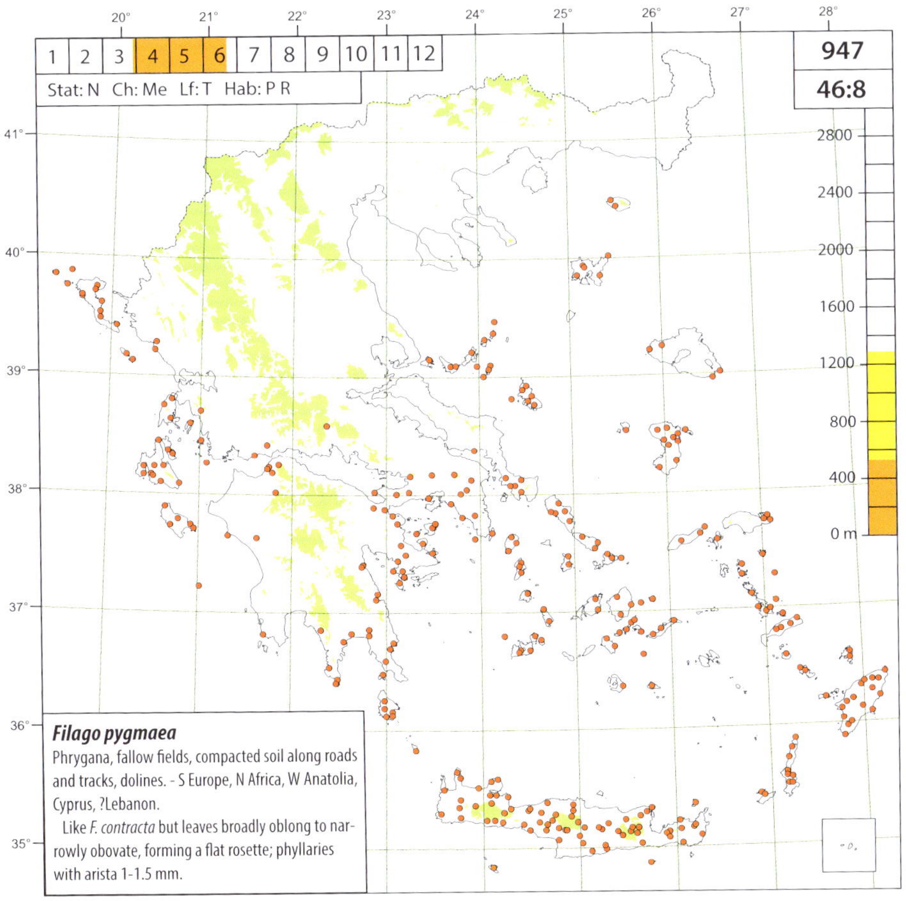

Filago pygmaea

Phrygana, fallow fields, compacted soil along roads and tracks, dolines. - S Europe, N Africa, W Anatolia, Cyprus, ?Lebanon.

Like *F. contracta* but leaves broadly oblong to narrowly obovate, forming a flat rosette; phyllaries with arista 1-1.5 mm.

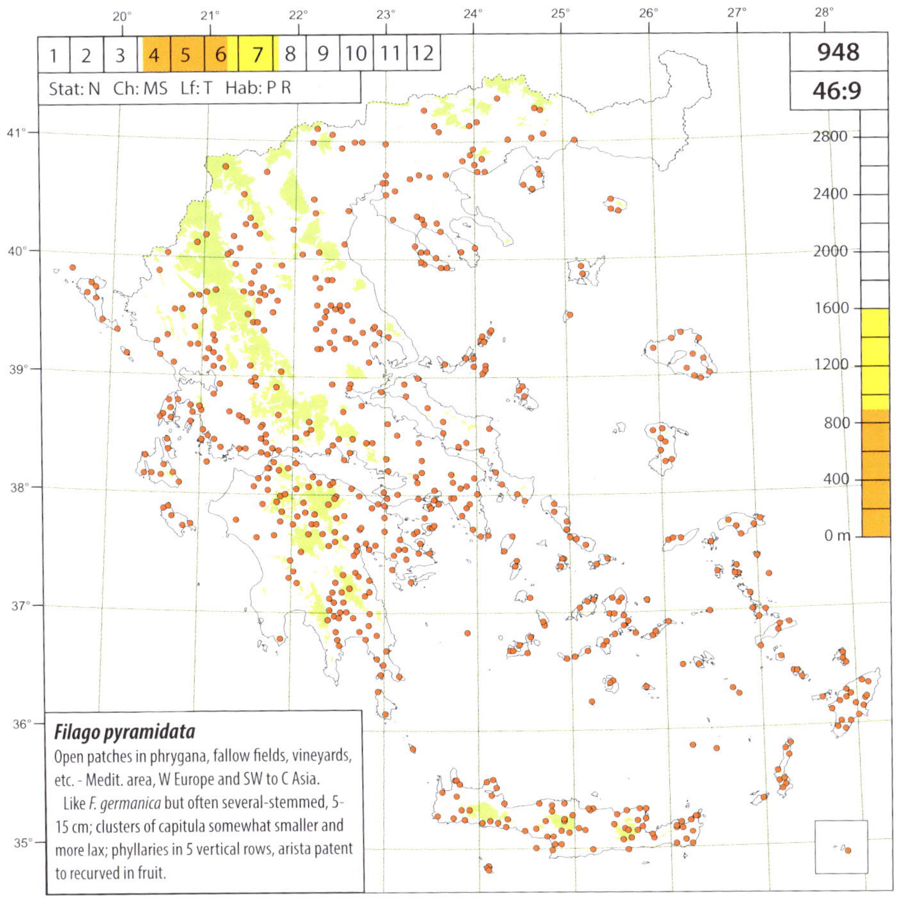

Filago pyramidata

Open patches in phrygana, fallow fields, vineyards, etc. - Medit. area, W Europe and SW to C Asia.

Like *F. germanica* but often several-stemmed, 5-15 cm; clusters of capitula somewhat smaller and more lax; phyllaries in 5 vertical rows, arista patent to recurved in fruit.

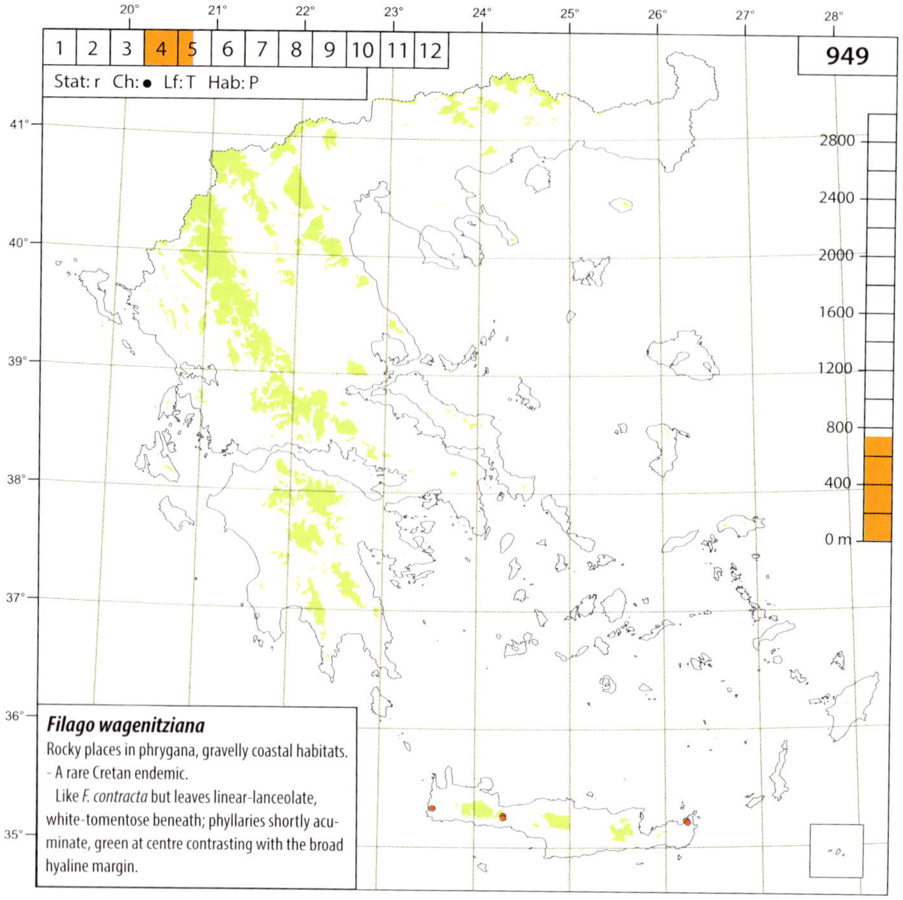

Filago wagenitziana

Rocky places in phrygana, gravelly coastal habitats. - A rare Cretan endemic.

Like *F. contracta* but leaves linear-lanceolate, white-tomentose beneath; phyllaries shortly acuminate, green at centre contrasting with the broad hyaline margin.

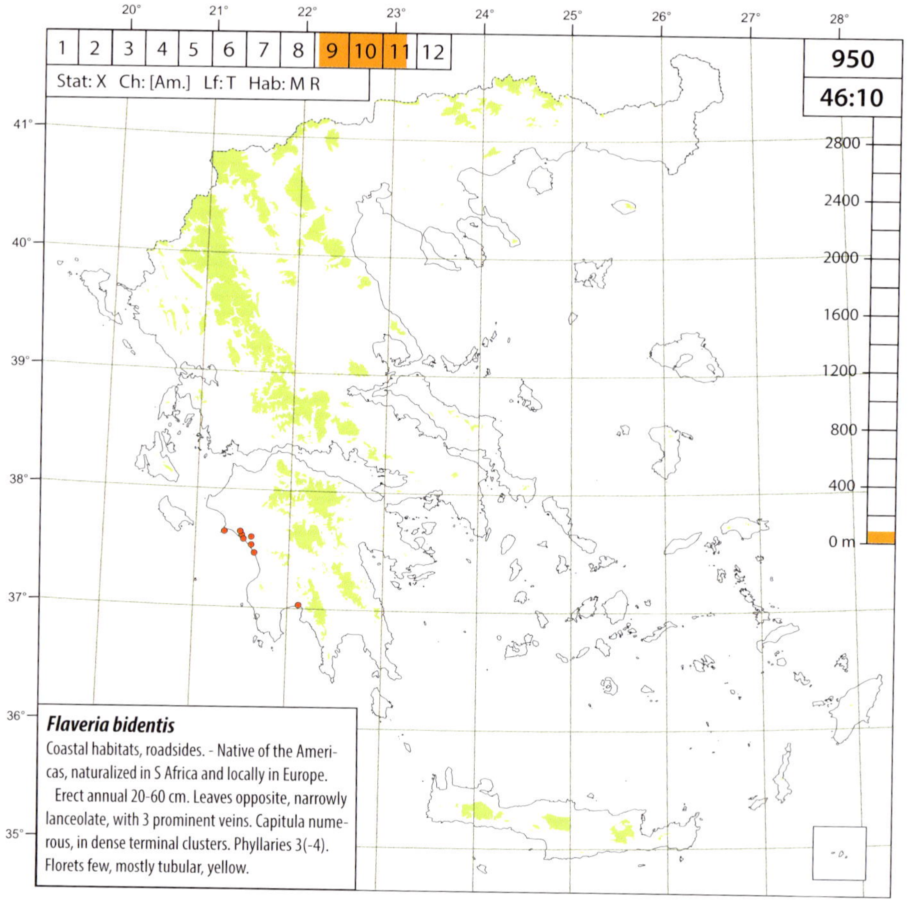

Flaveria bidentis

Coastal habitats, roadsides. - Native of the Americas, naturalized in S Africa and locally in Europe.

Erect annual 20-60 cm. Leaves opposite, narrowly lanceolate, with 3 prominent veins. Capitula numerous, in dense terminal clusters. Phyllaries 3(-4). Florets few, mostly tubular, yellow.

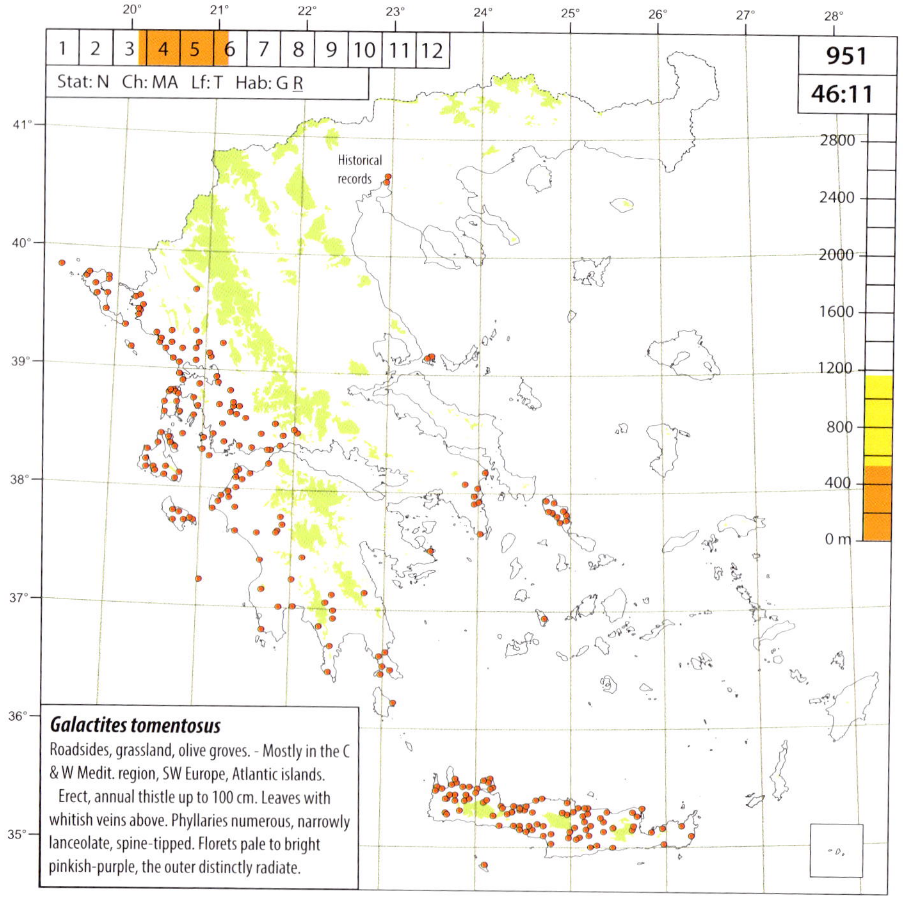

Galactites tomentosus

Roadsides, grassland, olive groves. - Mostly in the C & W Medit. region, SW Europe, Atlantic islands.

Erect, annual thistle up to 100 cm. Leaves with whitish veins above. Phyllaries numerous, narrowly lanceolate, spine-tipped. Florets pale to bright pinkish-purple, the outer distinctly radiate.

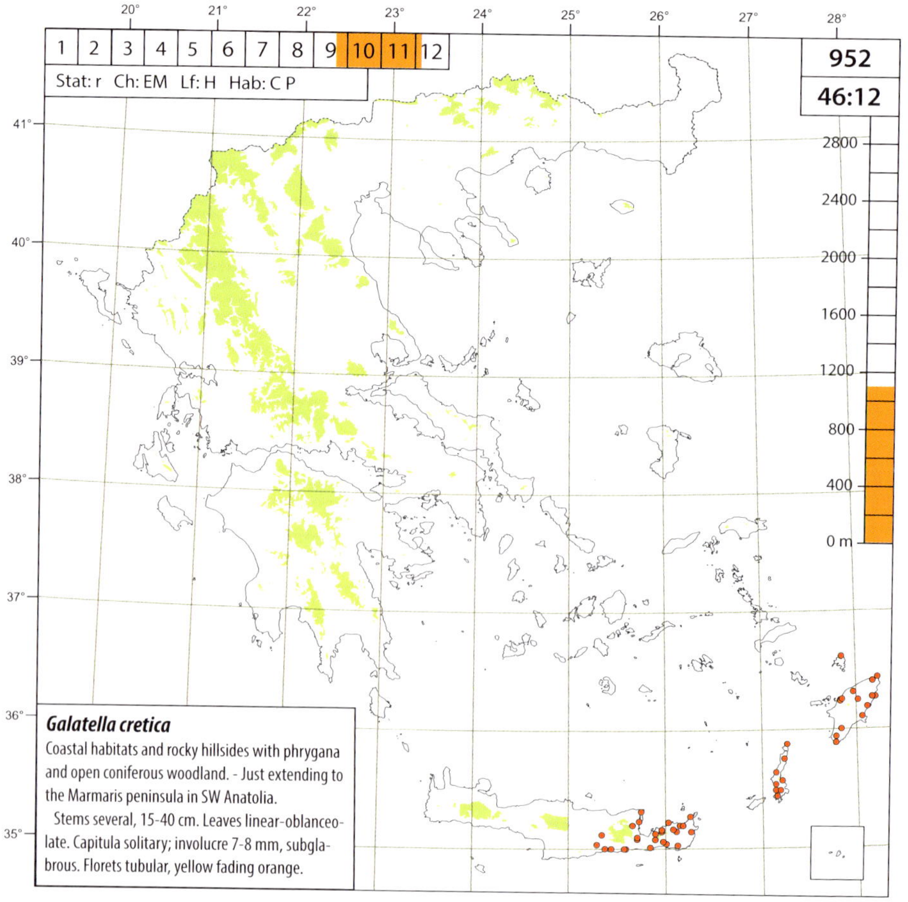

Galatella cretica

Coastal habitats and rocky hillsides with phrygana and open coniferous woodland. - Just extending to the Marmaris peninsula in SW Anatolia.

Stems several, 15-40 cm. Leaves linear-oblanceolate. Capitula solitary; involucre 7-8 mm, subglabrous. Florets tubular, yellow fading orange.

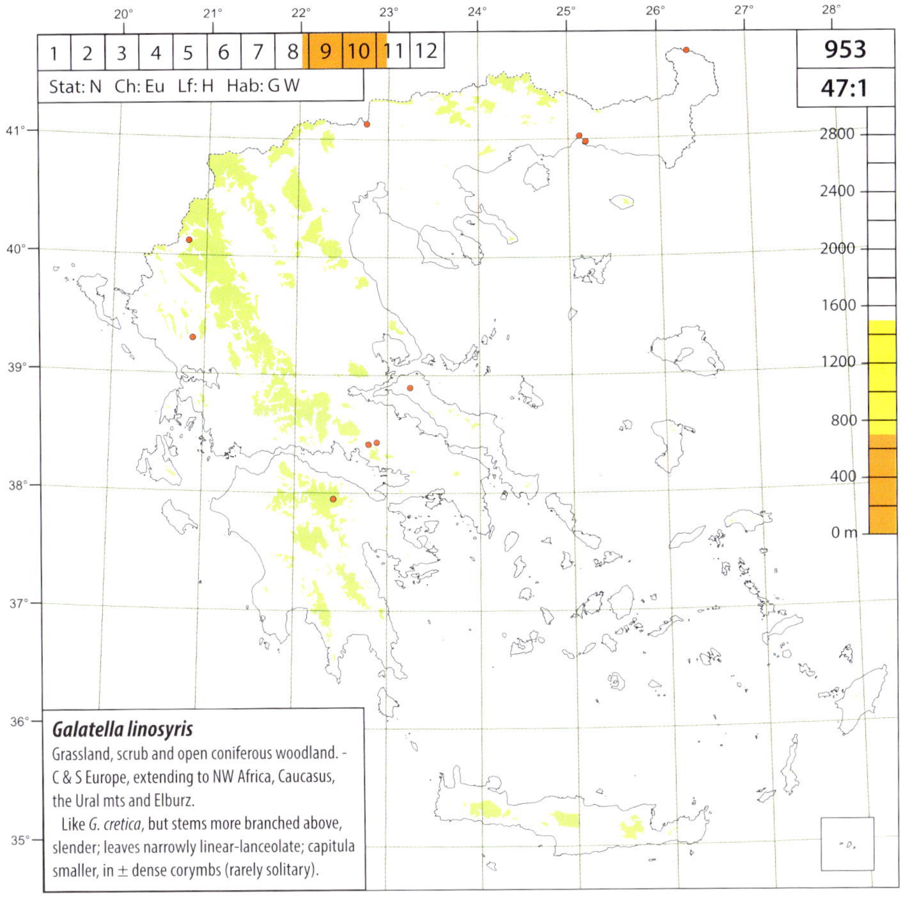

Galatella linosyris

Grassland, scrub and open coniferous woodland. - C & S Europe, extending to NW Africa, Caucasus, the Ural mts and Elburz.

Like *G. cretica*, but stems more branched above, slender; leaves narrowly linear-lanceolate; capitula smaller, in ± dense corymbs (rarely solitary).

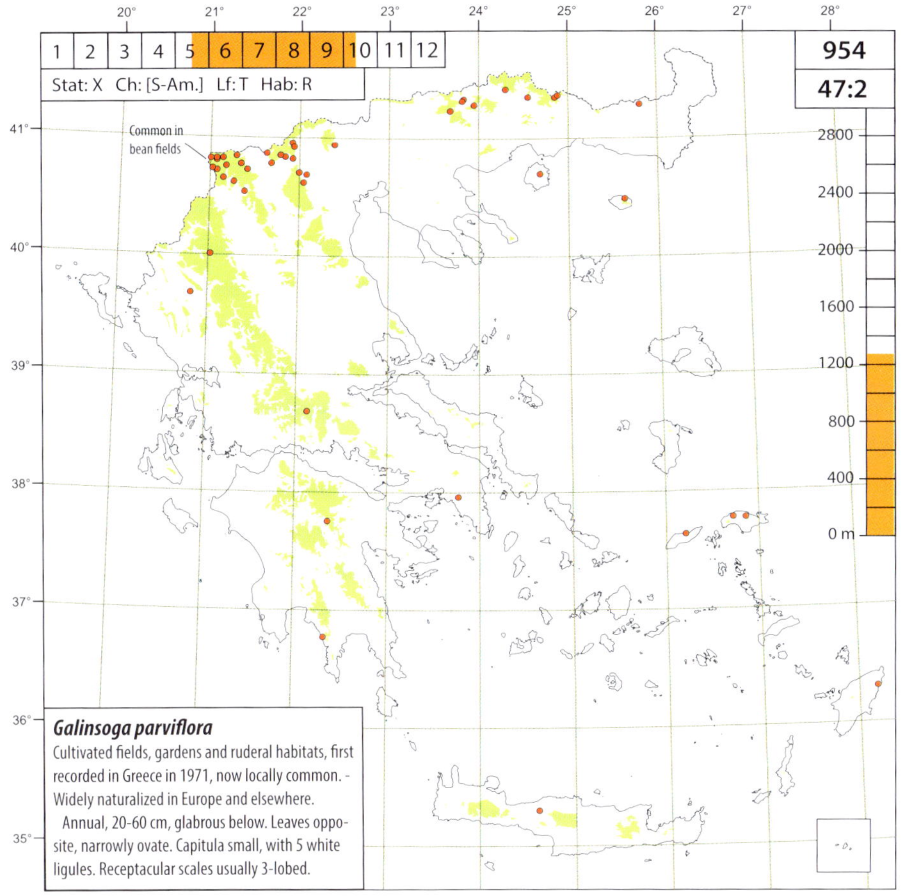

Galinsoga parviflora

Cultivated fields, gardens and ruderal habitats, first recorded in Greece in 1971, now locally common. - Widely naturalized in Europe and elsewhere.

Annual, 20-60 cm, glabrous below. Leaves opposite, narrowly ovate. Capitula small, with 5 white ligules. Receptacular scales usually 3-lobed.

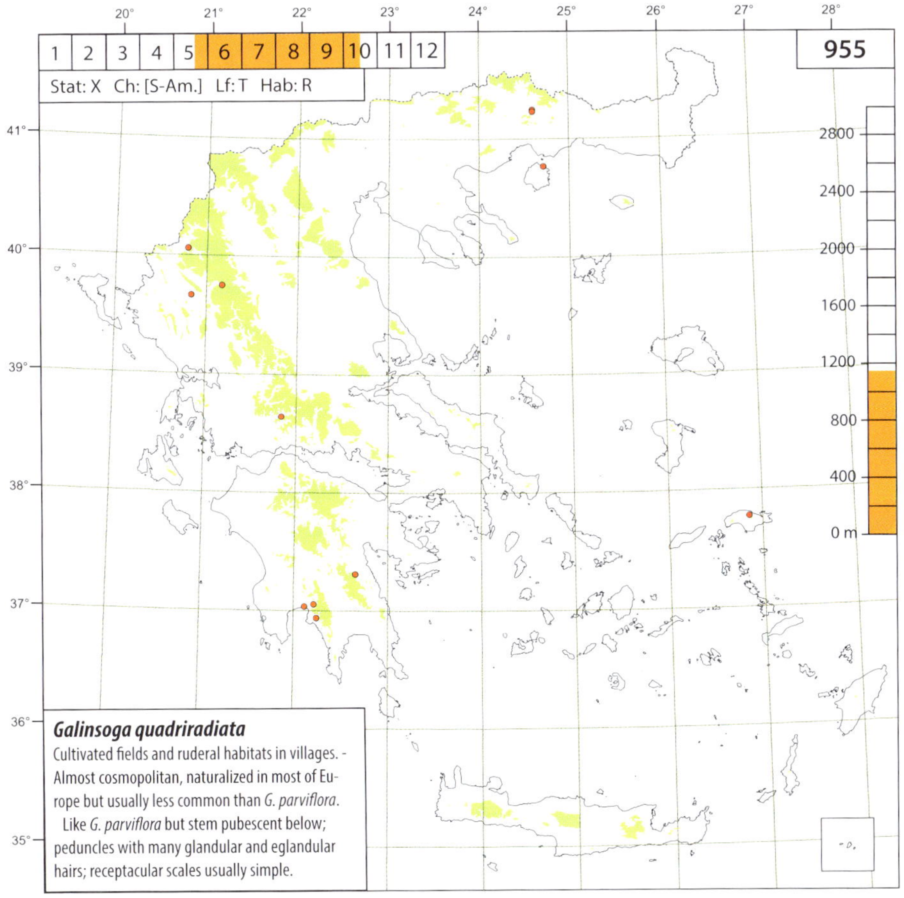

Galinsoga quadriradiata

Cultivated fields and ruderal habitats in villages. - Almost cosmopolitan, naturalized in most of Europe but usually less common than *G. parviflora*.

Like *G. parviflora* but stem pubescent below; peduncles with many glandular and eglandular hairs; receptacular scales usually simple.

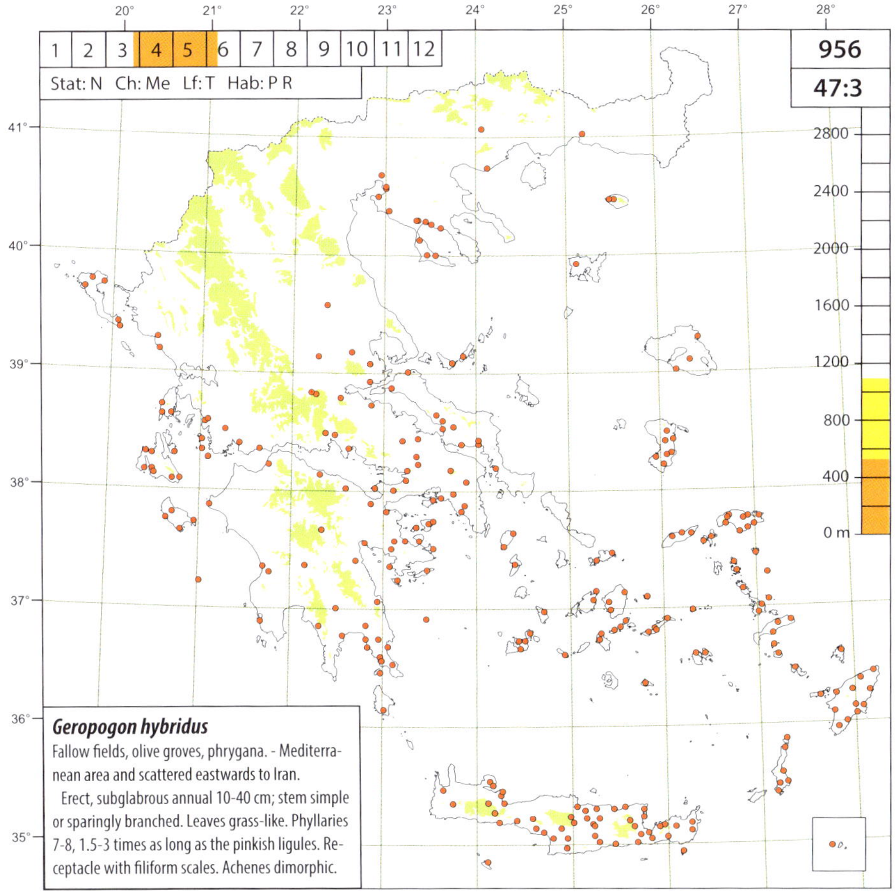

Geropogon hybridus

Fallow fields, olive groves, phrygana. - Mediterranean area and scattered eastwards to Iran.

Erect, subglabrous annual 10-40 cm; stem simple or sparingly branched. Leaves grass-like. Phyllaries 7-8, 1.5-3 times as long as the pinkish ligules. Receptacle with filiform scales. Achenes dimorphic.

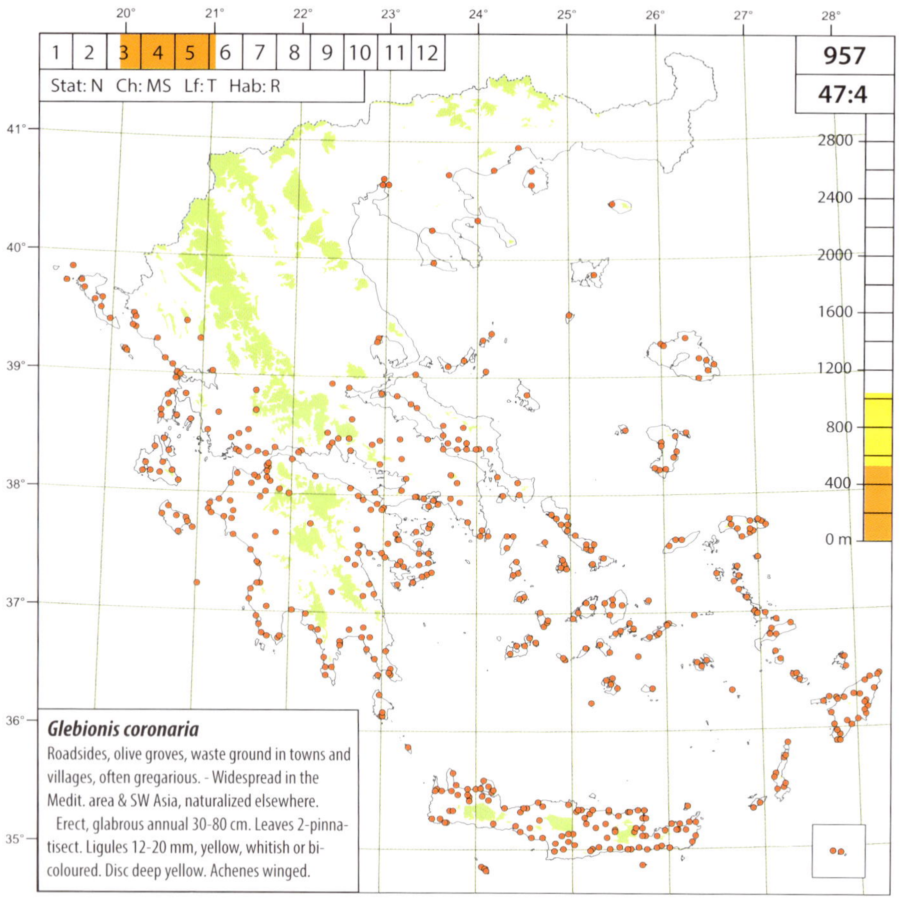

Glebionis coronaria

Roadsides, olive groves, waste ground in towns and villages, often gregarious. - Widespread in the Medit. area & SW Asia, naturalized elsewhere.

Erect, glabrous annual 30-80 cm. Leaves 2-pinnatisect. Ligules 12-20 mm, yellow, whitish or bi-coloured. Disc deep yellow. Achenes winged.

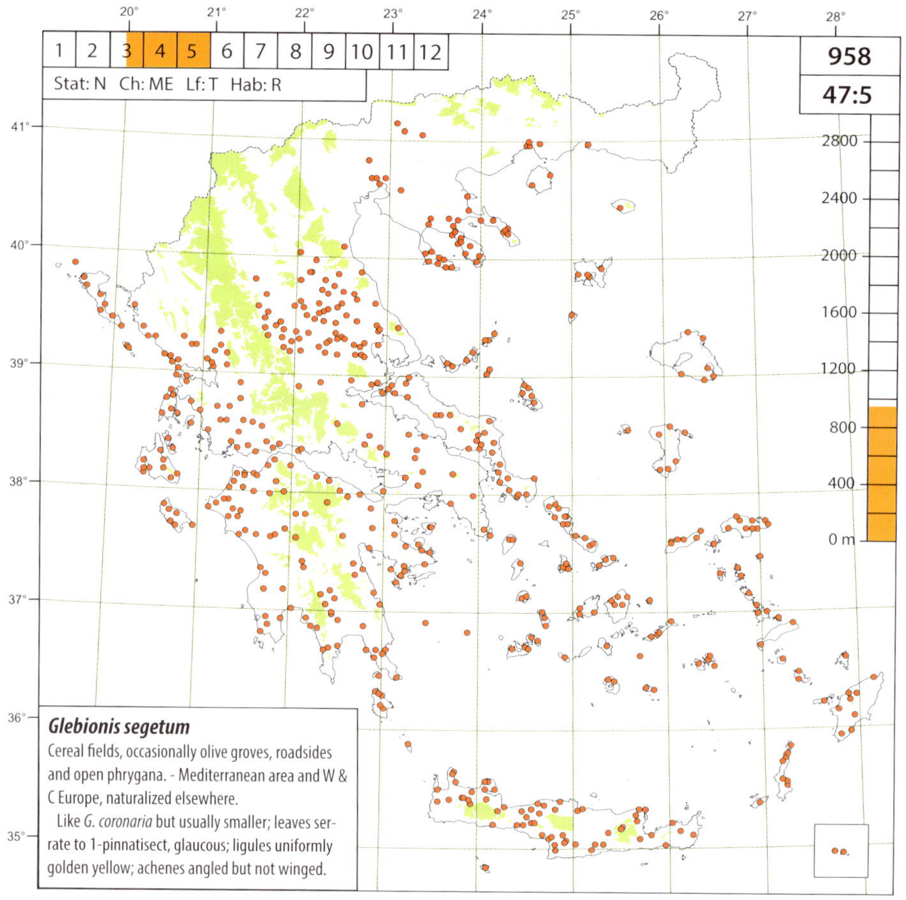

Glebionis segetum

Cereal fields, occasionally olive groves, roadsides and open phrygana. - Mediterranean area and W & C Europe, naturalized elsewhere.

Like *G. coronaria* but usually smaller; leaves serrate to 1-pinnatisect, glaucous; ligules uniformly golden yellow; achenes angled but not winged.

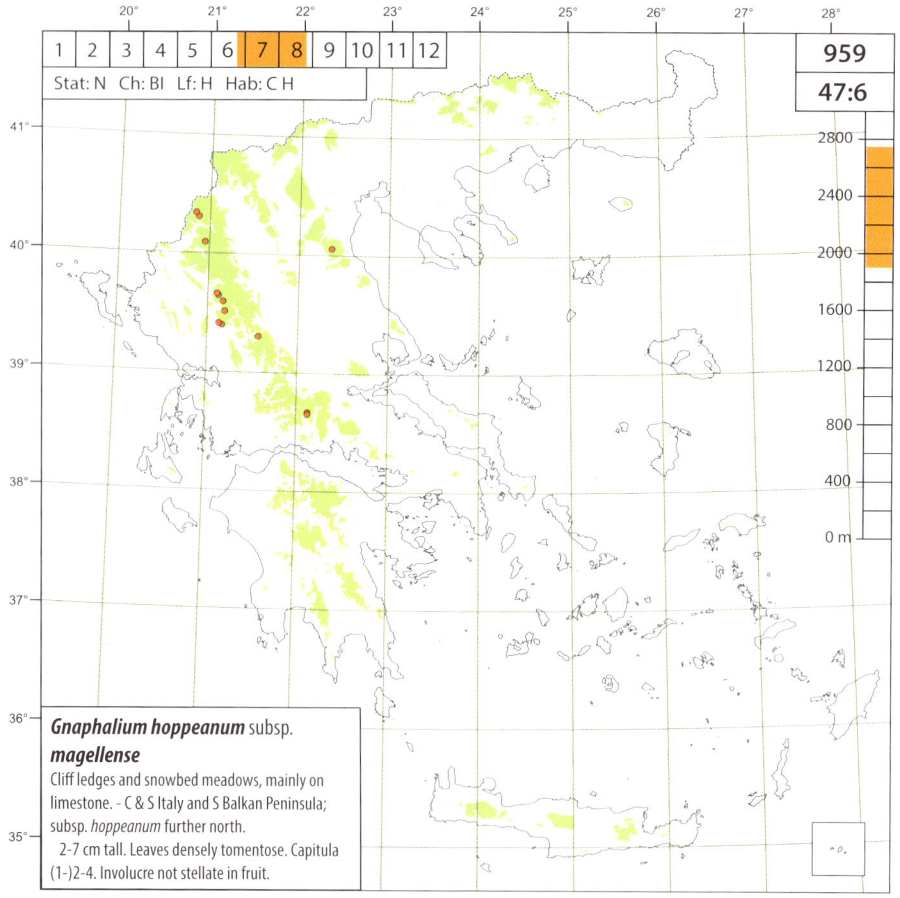

Gnaphalium hoppeanum* subsp. *magellense

Cliff ledges and snowbed meadows, mainly on limestone. - C & S Italy and S Balkan Peninsula; subsp. *hoppeanum* further north.

2-7 cm tall. Leaves densely tomentose. Capitula (1-)2-4. Involucre not stellate in fruit.

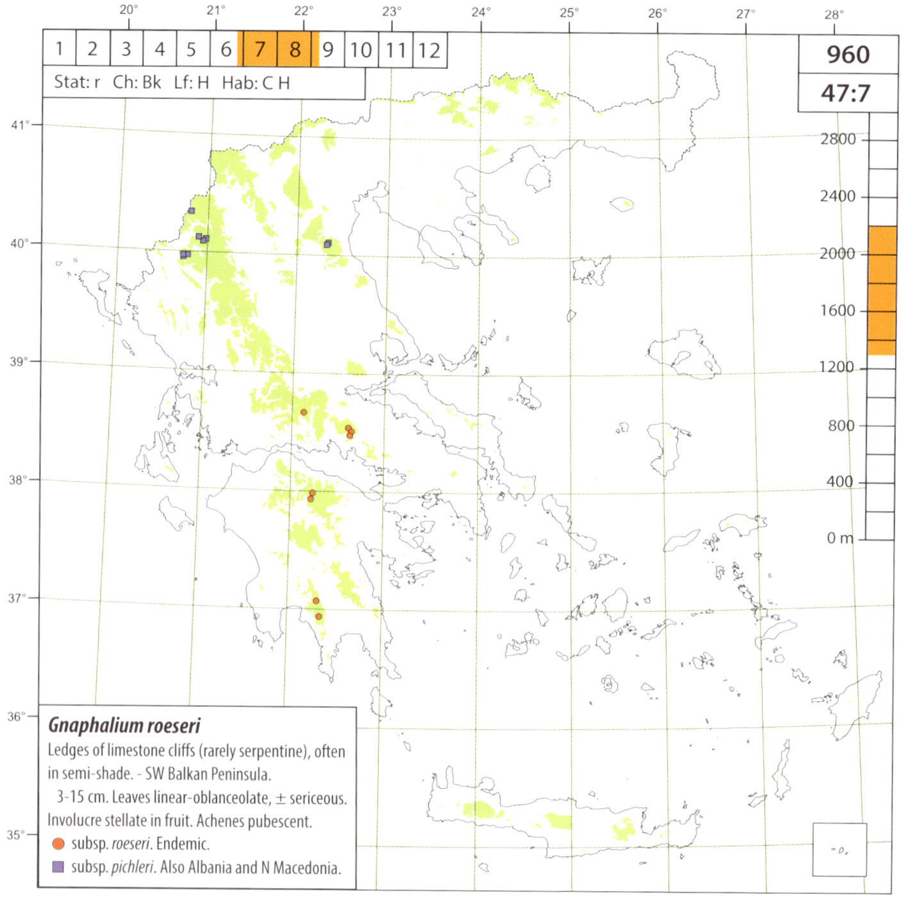

Gnaphalium roeseri

Ledges of limestone cliffs (rarely serpentine), often in semi-shade. - SW Balkan Peninsula.

3-15 cm. Leaves linear-oblanceolate, ± sericeous. Involucre stellate in fruit. Achenes pubescent.

- subsp. *roeseri*. Endemic.
- subsp. *pichleri*. Also Albania and N Macedonia.

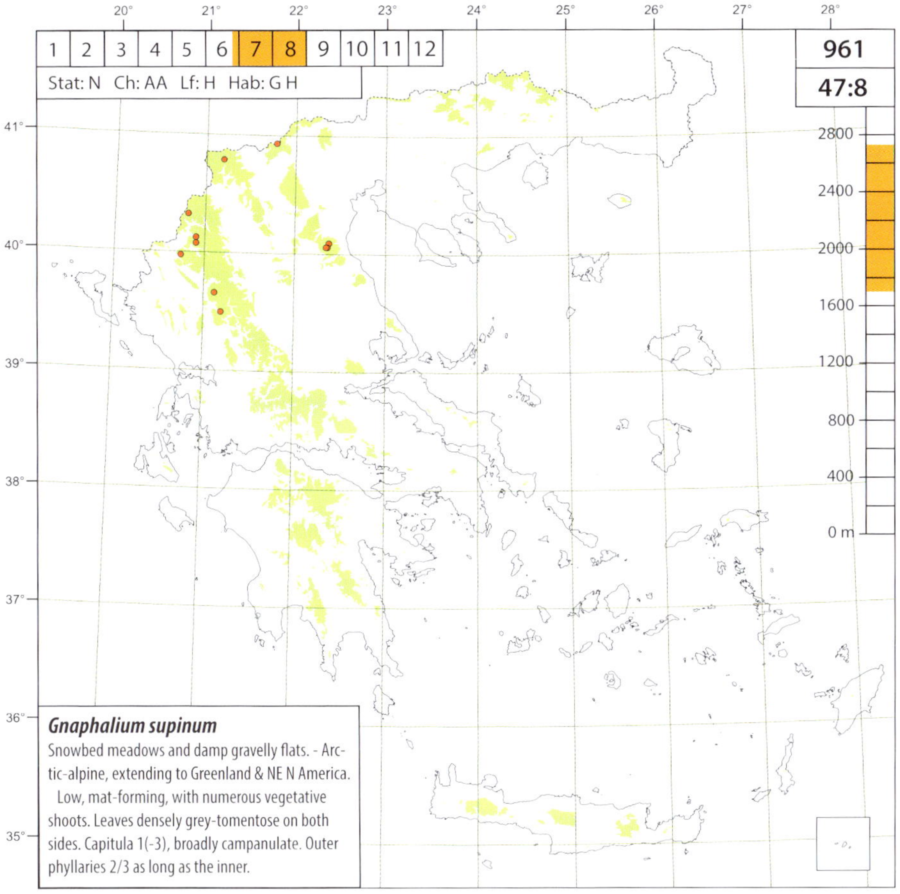

Gnaphalium supinum
Snowbed meadows and damp gravelly flats. - Arctic-alpine, extending to Greenland & NE N America.
Low, mat-forming, with numerous vegetative shoots. Leaves densely grey-tomentose on both sides. Capitula 1(-3), broadly campanulate. Outer phyllaries 2/3 as long as the inner.

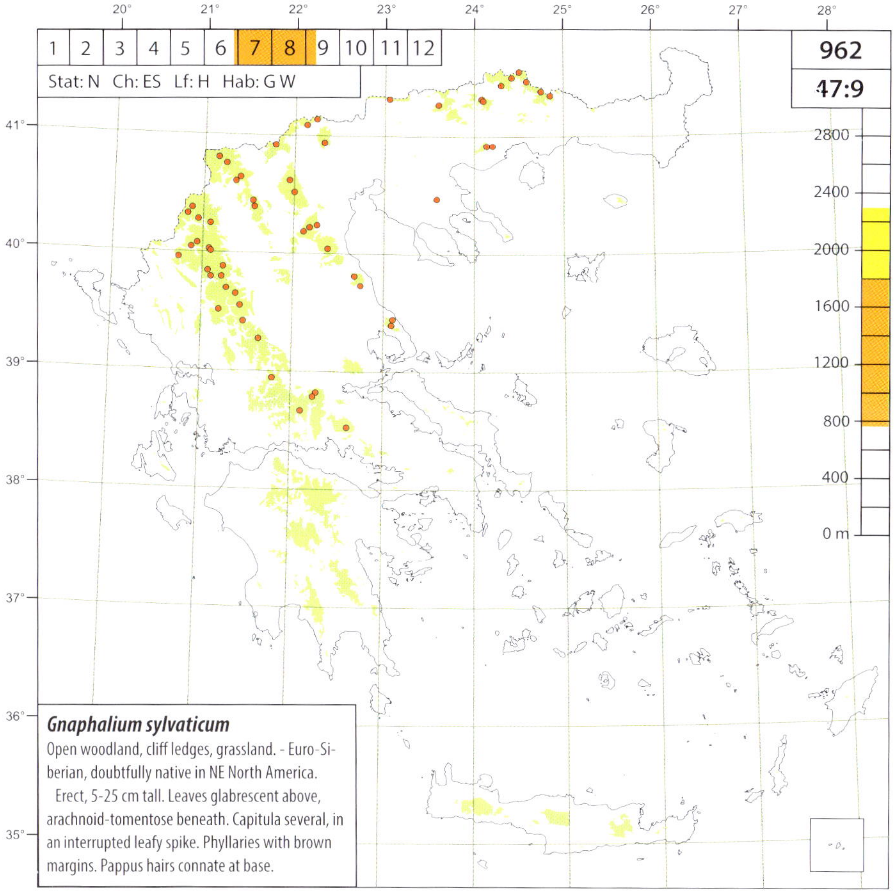

Gnaphalium sylvaticum
Open woodland, cliff ledges, grassland. - Euro-Siberian, doubtfully native in NE North America.
Erect, 5-25 cm tall. Leaves glabrescent above, arachnoid-tomentose beneath. Capitula several, in an interrupted leafy spike. Phyllaries with brown margins. Pappus hairs connate at base.

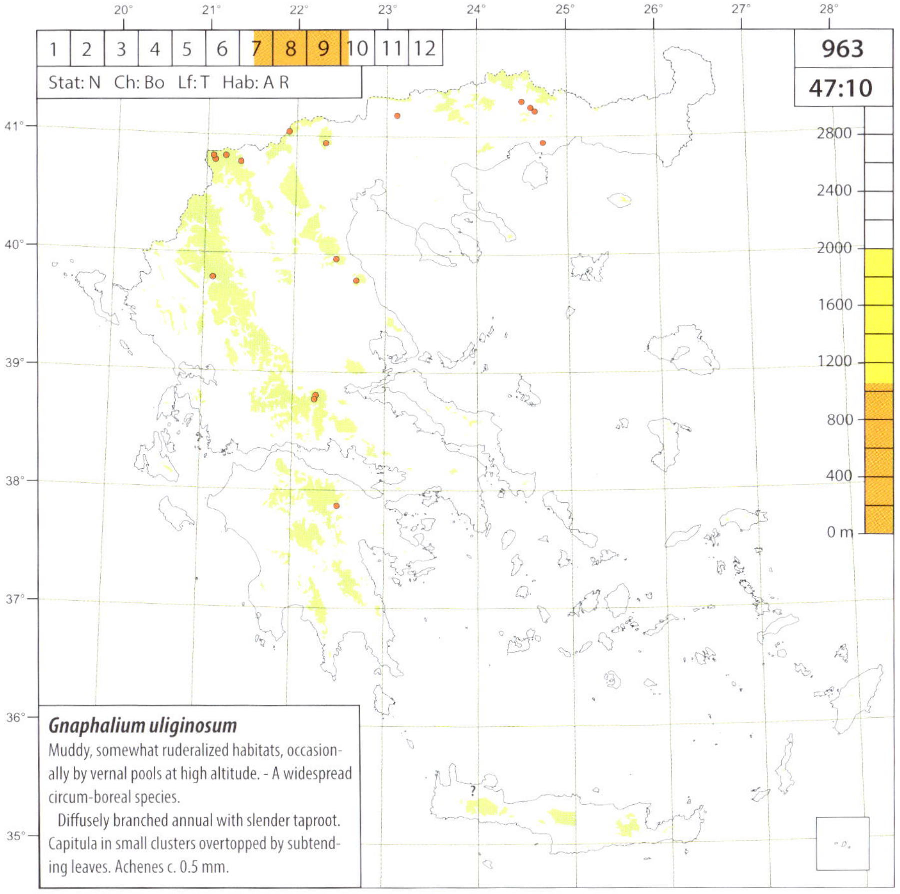

Gnaphalium uliginosum
Muddy, somewhat ruderalized habitats, occasionally by vernal pools at high altitude. - A widespread circum-boreal species.
Diffusely branched annual with slender taproot. Capitula in small clusters overtopped by subtending leaves. Achenes c. 0.5 mm.

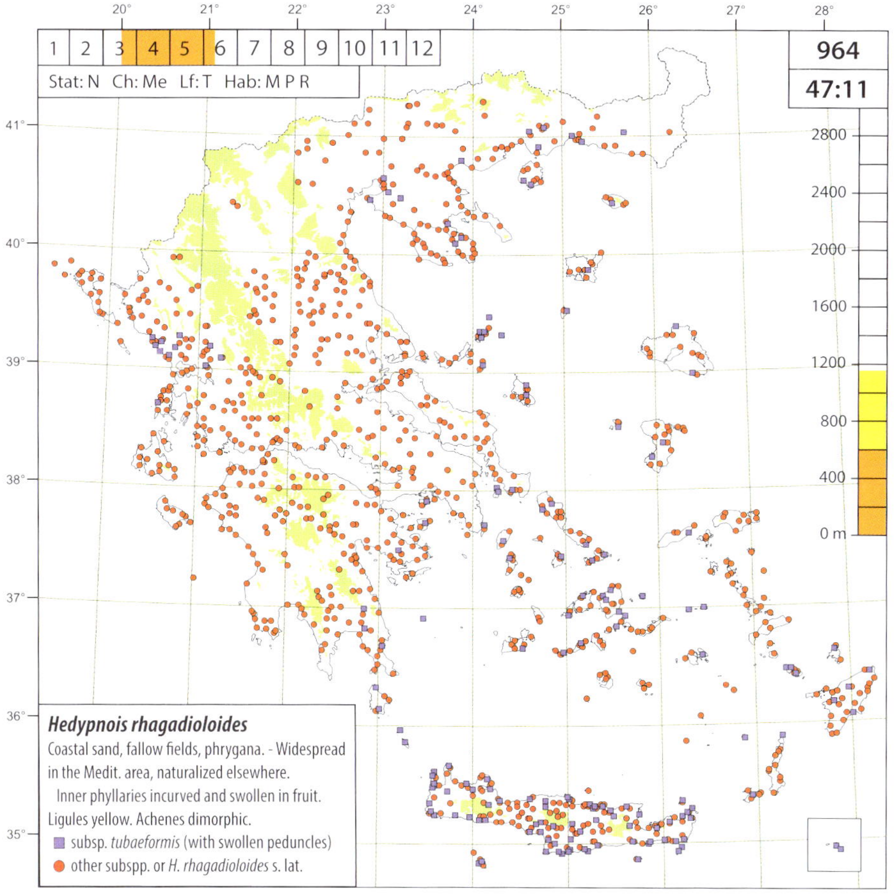

Hedypnois rhagadioloides
Coastal sand, fallow fields, phrygana. - Widespread in the Medit. area, naturalized elsewhere.
Inner phyllaries incurved and swollen in fruit. Ligules yellow. Achenes dimorphic.
■ subsp. *tubaeformis* (with swollen peduncles)
● other subspp. or *H. rhagadioloides* s. lat.

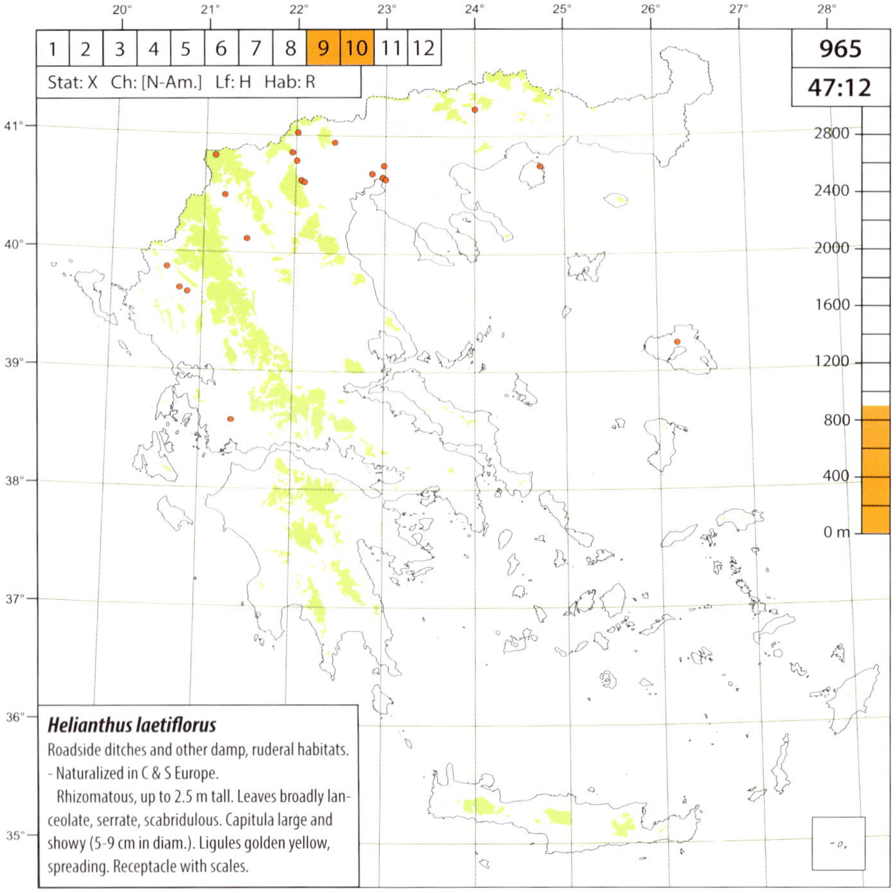

Helianthus laetiflorus

Roadside ditches and other damp, ruderal habitats. - Naturalized in C & S Europe.

Rhizomatous, up to 2.5 m tall. Leaves broadly lanceolate, serrate, scabridulous. Capitula large and showy (5-9 cm in diam.). Ligules golden yellow, spreading. Receptacle with scales.

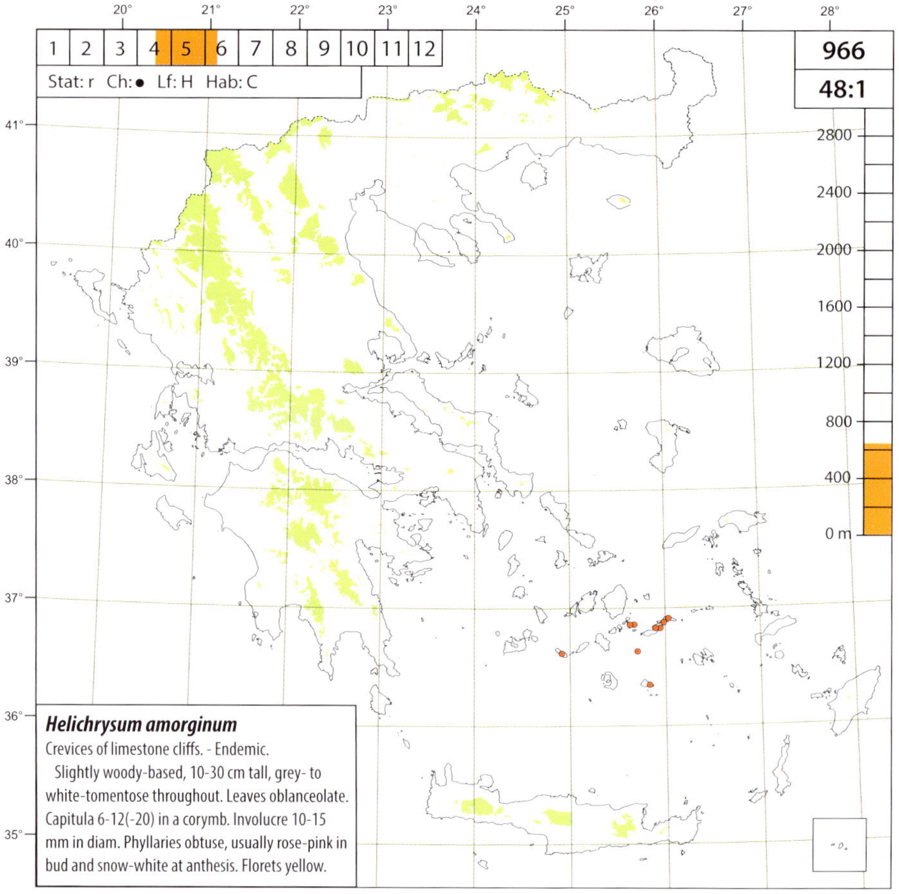

Helichrysum amorginum

Crevices of limestone cliffs. - Endemic.

Slightly woody-based, 10-30 cm tall, grey- to white-tomentose throughout. Leaves oblanceolate. Capitula 6-12(-20) in a corymb. Involucre 10-15 mm in diam. Phyllaries obtuse, usually rose-pink in bud and snow-white at anthesis. Florets yellow.

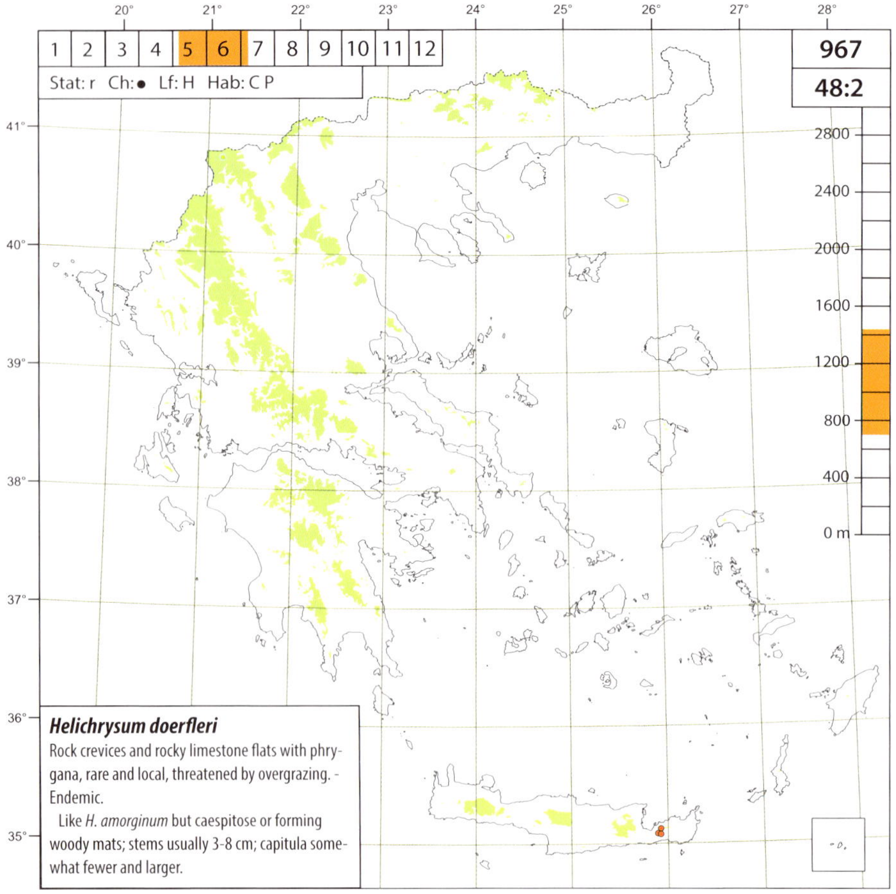

Helichrysum doerfleri

Rock crevices and rocky limestone flats with phrygana, rare and local, threatened by overgrazing. - Endemic.

Like *H. amorginum* but caespitose or forming woody mats; stems usually 3-8 cm; capitula somewhat fewer and larger.

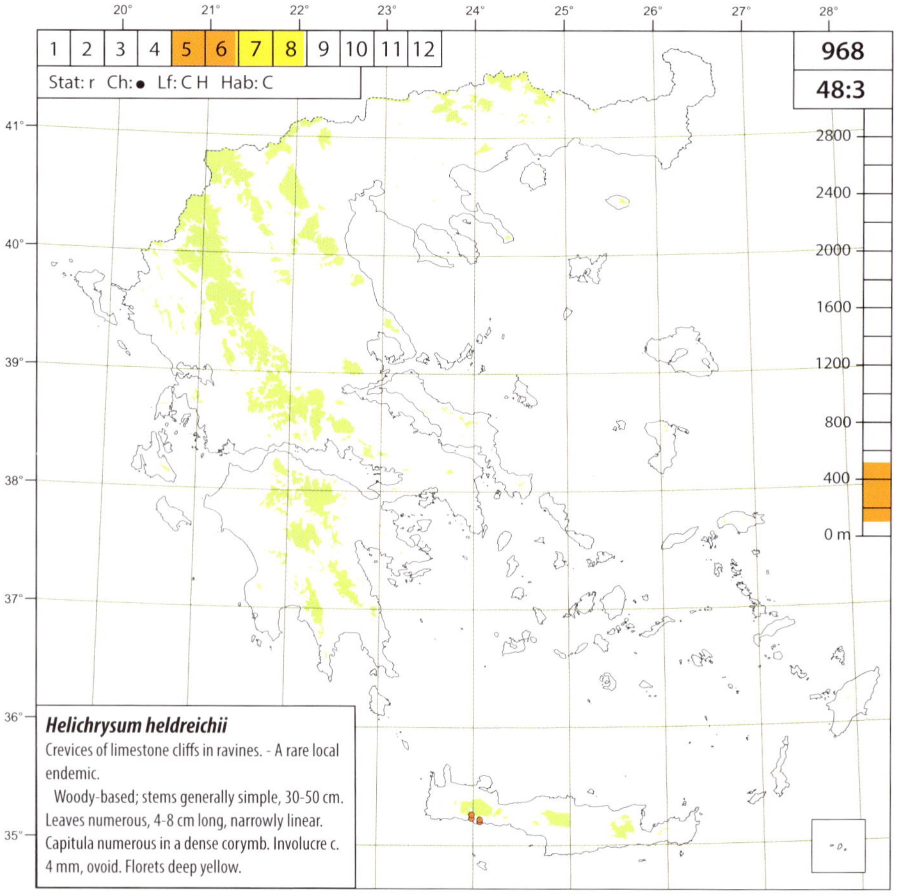

Helichrysum heldreichii

Crevices of limestone cliffs in ravines. - A rare local endemic.

Woody-based; stems generally simple, 30-50 cm. Leaves numerous, 4-8 cm long, narrowly linear. Capitula numerous in a dense corymb. Involucre c. 4 mm, ovoid. Florets deep yellow.

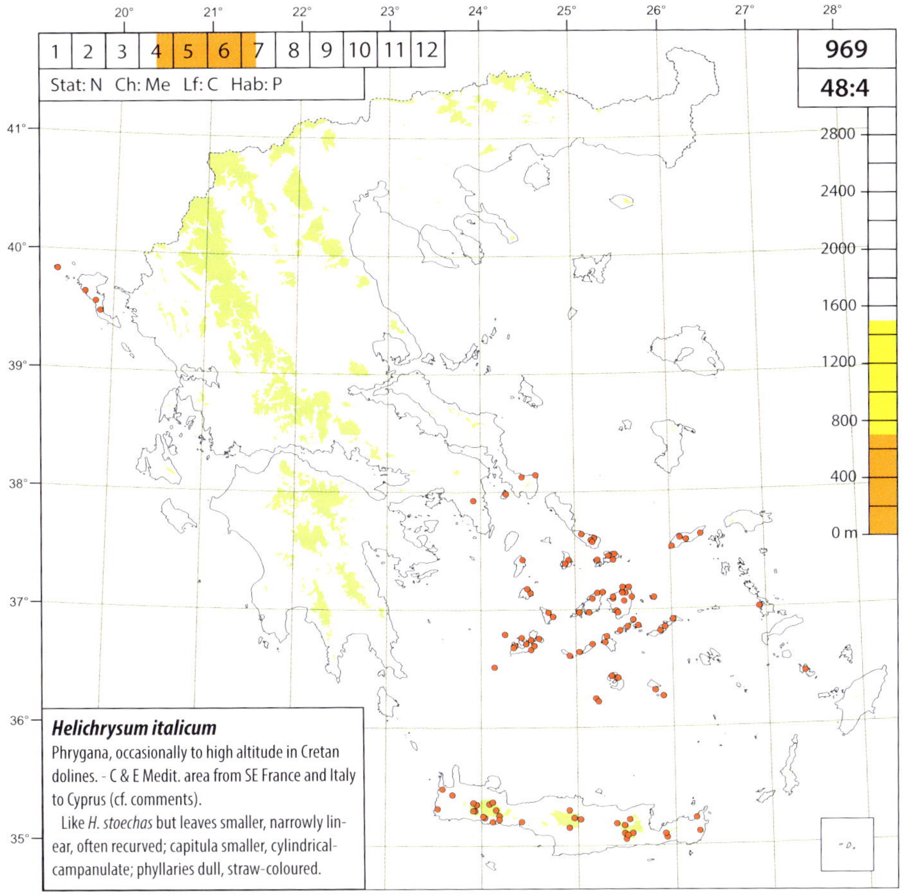

Helichrysum italicum
Phrygana, occasionally to high altitude in Cretan dolines. - C & E Medit. area from SE France and Italy to Cyprus (cf. comments).
Like *H. stoechas* but leaves smaller, narrowly linear, often recurved; capitula smaller, cylindrical-campanulate; phyllaries dull, straw-coloured.

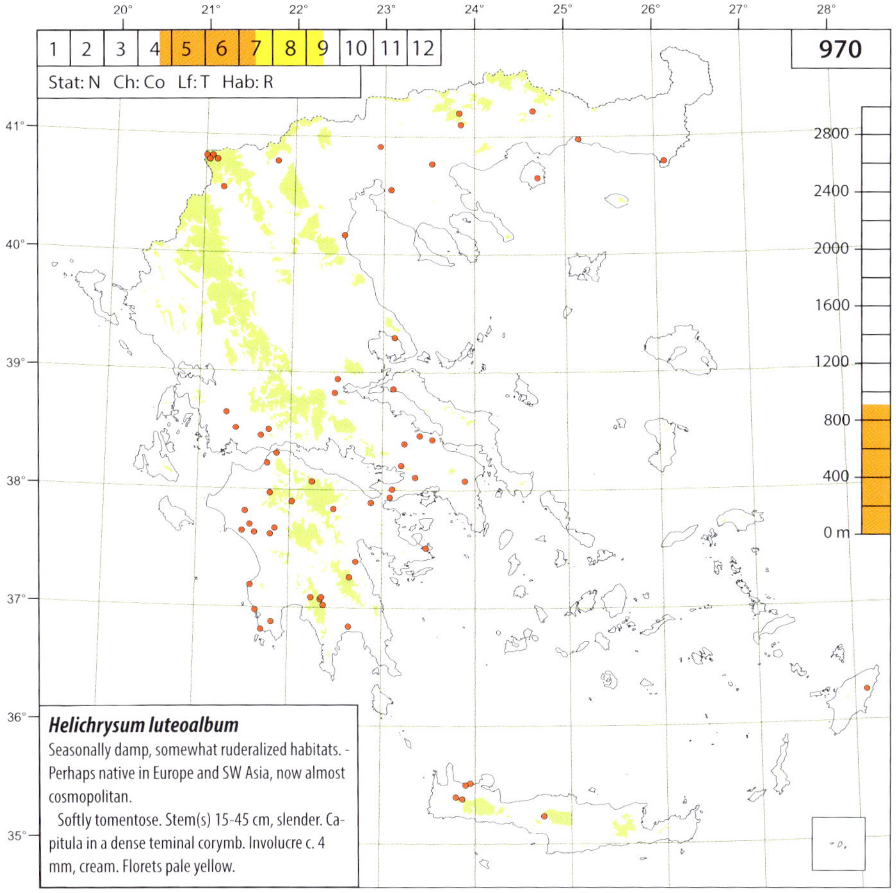

Helichrysum luteoalbum
Seasonally damp, somewhat ruderalized habitats. - Perhaps native in Europe and SW Asia, now almost cosmopolitan.
Softly tomentose. Stem(s) 15-45 cm, slender. Capitula in a dense teminal corymb. Involucre c. 4 mm, cream. Florets pale yellow.

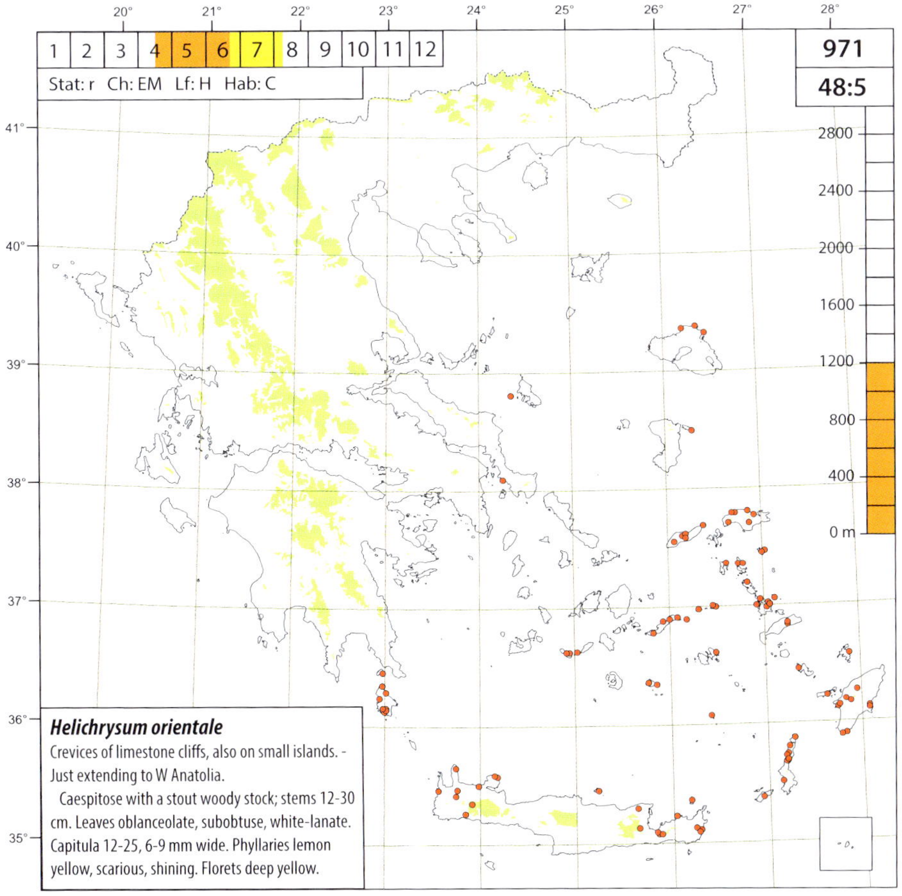

Helichrysum orientale
Crevices of limestone cliffs, also on small islands. - Just extending to W Anatolia.
Caespitose with a stout woody stock; stems 12-30 cm. Leaves oblanceolate, subobtuse, white-lanate. Capitula 12-25, 6-9 mm wide. Phyllaries lemon yellow, scarious, shining. Florets deep yellow.

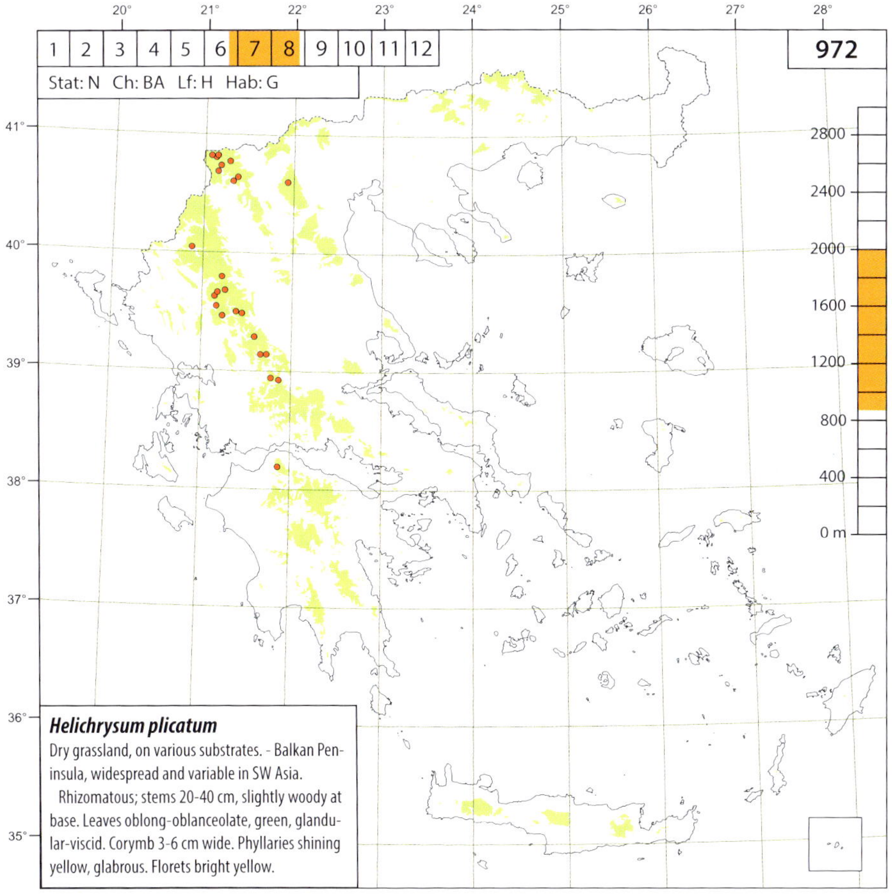

Helichrysum plicatum
Dry grassland, on various substrates. - Balkan Peninsula, widespread and variable in SW Asia.
Rhizomatous; stems 20-40 cm, slightly woody at base. Leaves oblong-oblanceolate, green, glandular-viscid. Corymb 3-6 cm wide. Phyllaries shining yellow, glabrous. Florets bright yellow.

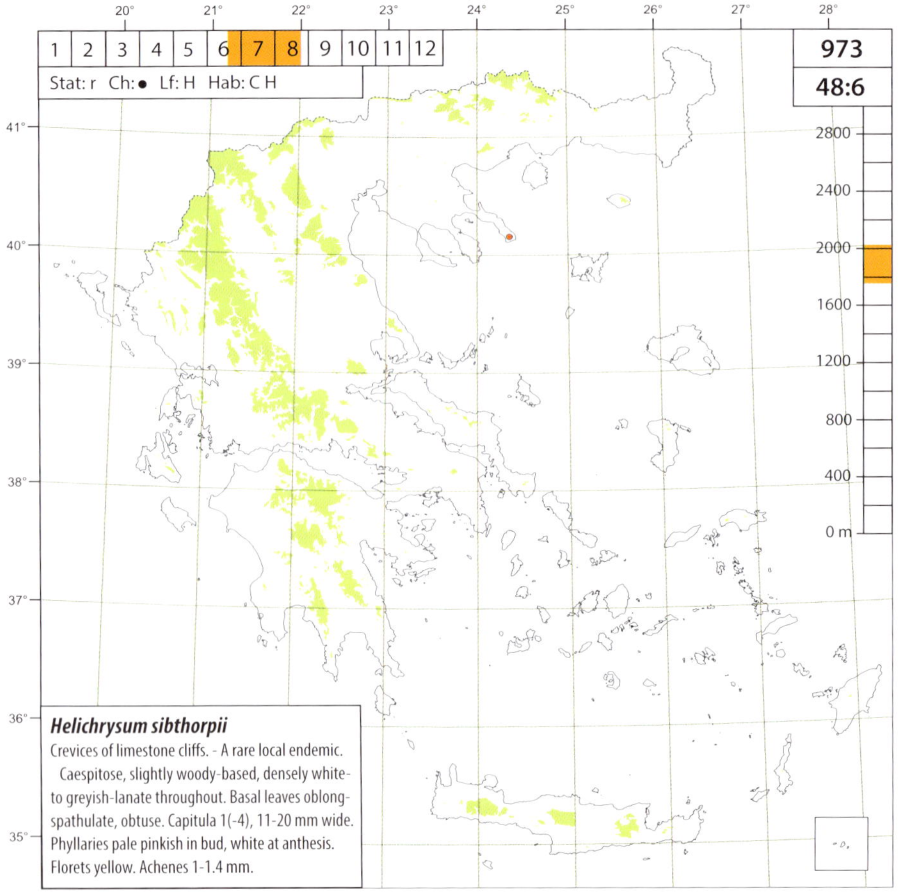

Helichrysum sibthorpii
Crevices of limestone cliffs. - A rare local endemic.
Caespitose, slightly woody-based, densely white-to greyish-lanate throughout. Basal leaves oblong-spathulate, obtuse. Capitula 1(-4), 11-20 mm wide. Phyllaries pale pinkish in bud, white at anthesis. Florets yellow. Achenes 1-1.4 mm.

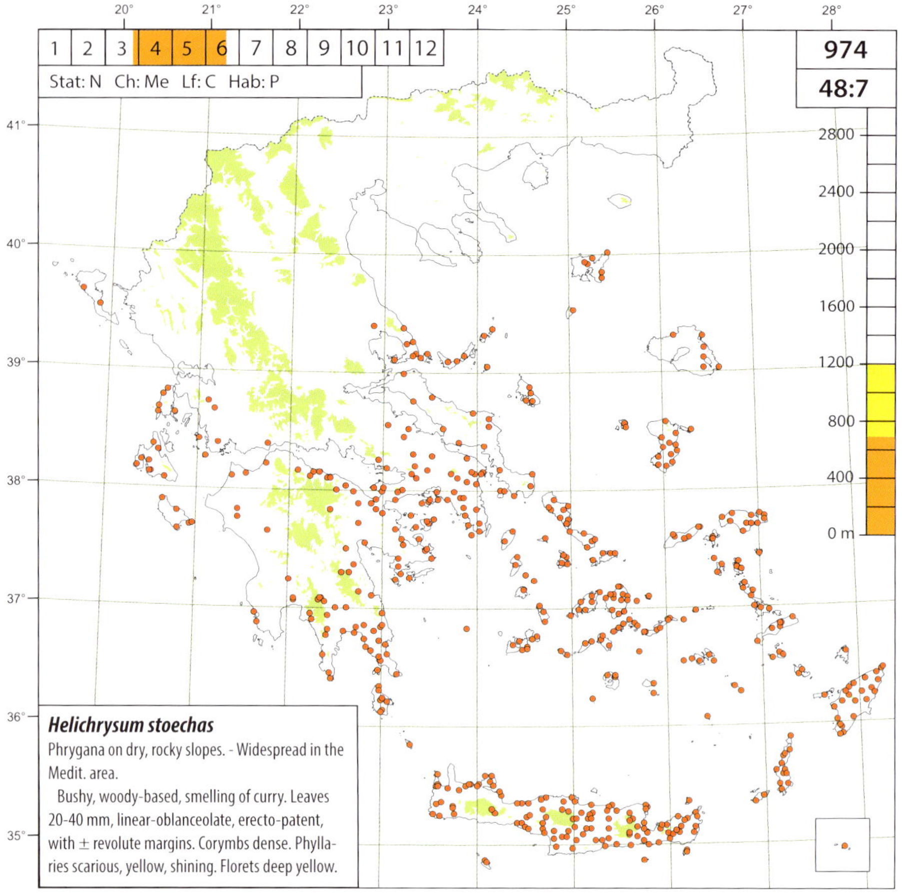

Helichrysum stoechas
Phrygana on dry, rocky slopes. - Widespread in the Medit. area.
Bushy, woody-based, smelling of curry. Leaves 20-40 mm, linear-oblanceolate, erecto-patent, with ± revolute margins. Corymbs dense. Phyllaries scarious, yellow, shining. Florets deep yellow.

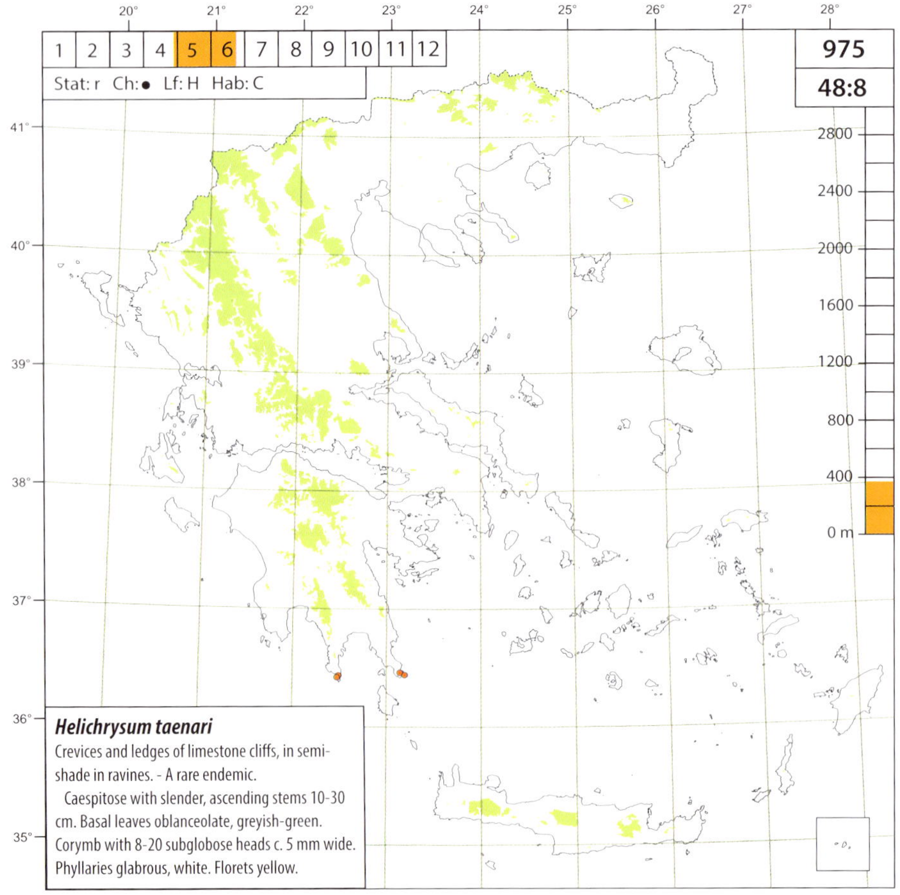

Helichrysum taenari
Crevices and ledges of limestone cliffs, in semi-shade in ravines. - A rare endemic.
Caespitose with slender, ascending stems 10-30 cm. Basal leaves oblanceolate, greyish-green. Corymb with 8-20 subglobose heads c. 5 mm wide. Phyllaries glabrous, white. Florets yellow.

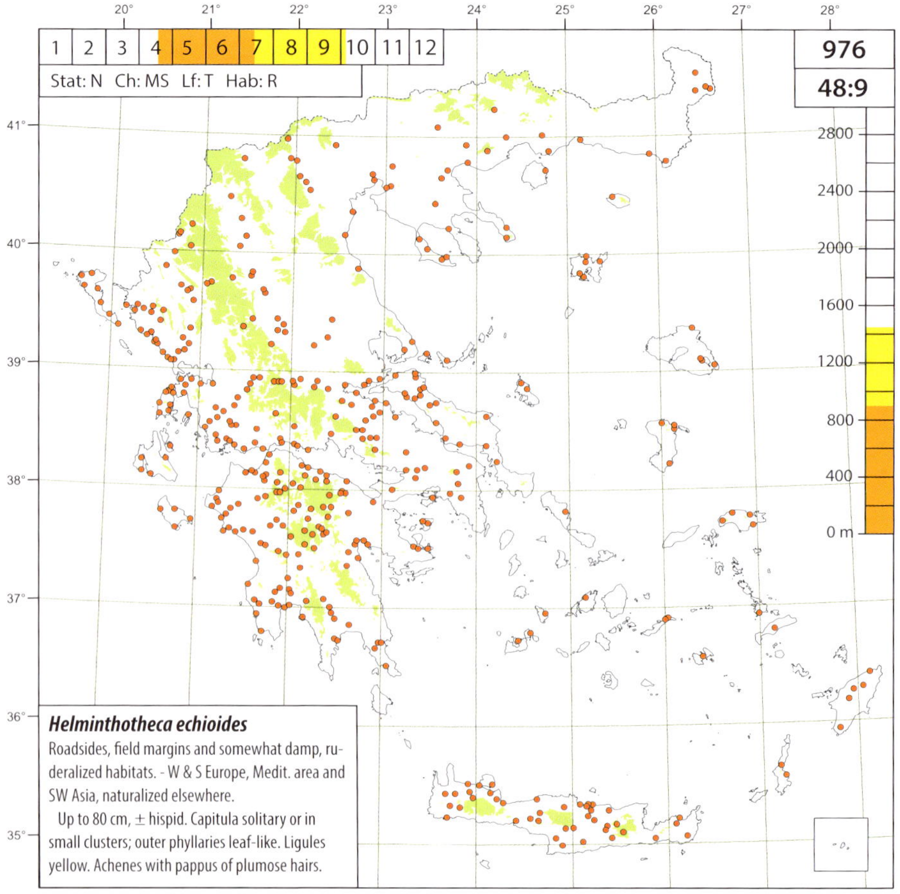

Helminthotheca echioides
Roadsides, field margins and somewhat damp, ruderalized habitats. - W & S Europe, Medit. area and SW Asia, naturalized elsewhere.
Up to 80 cm, ± hispid. Capitula solitary or in small clusters; outer phyllaries leaf-like. Ligules yellow. Achenes with pappus of plumose hairs.

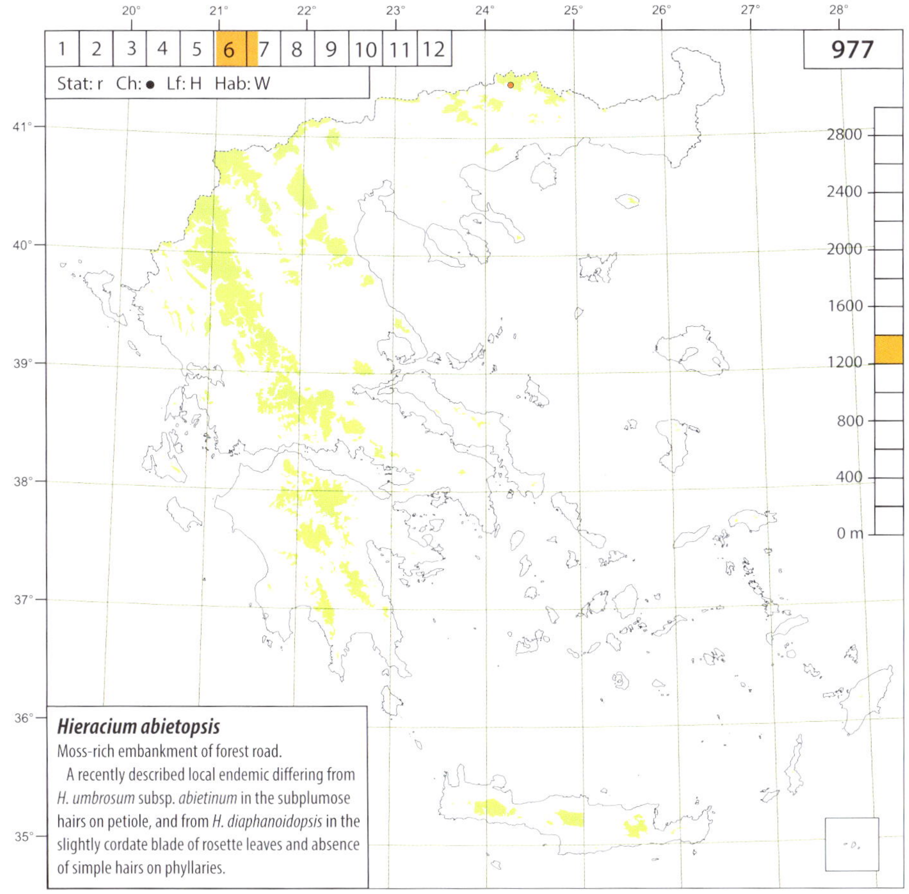

Hieracium abietopsis
Moss-rich embankment of forest road.
A recently described local endemic differing from *H. umbrosum* subsp. *abietinum* in the subplumose hairs on petiole, and from *H. diaphanoidopsis* in the slightly cordate blade of rosette leaves and absence of simple hairs on phyllaries.

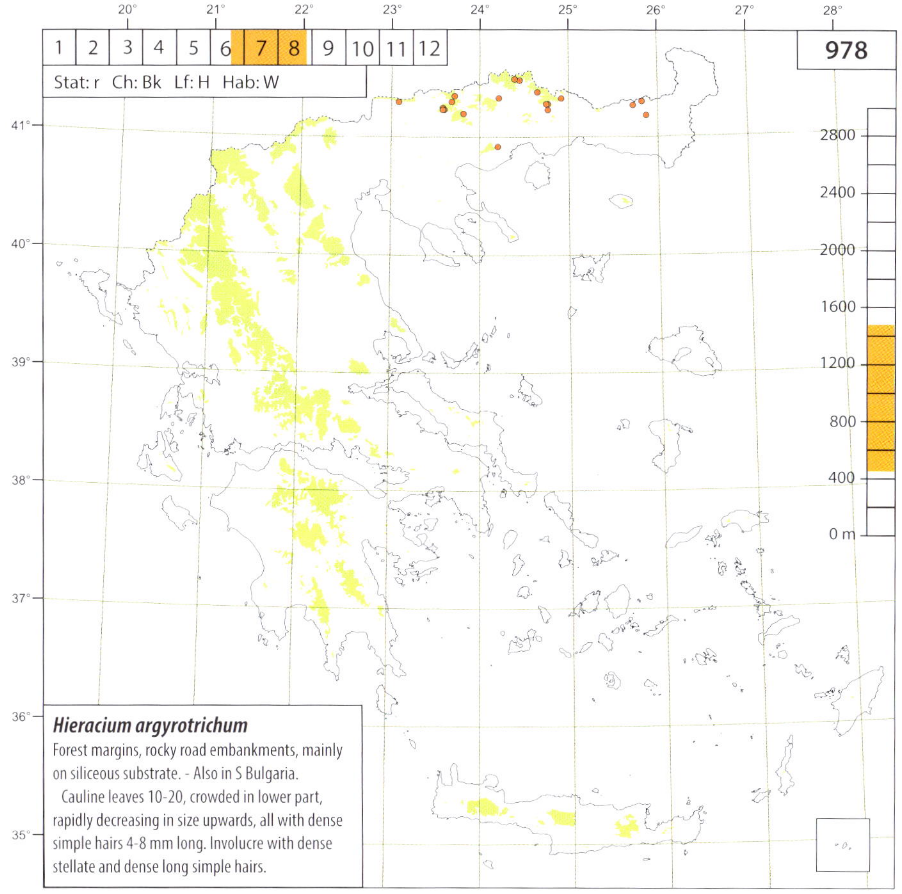

Hieracium argyrotrichum
Forest margins, rocky road embankments, mainly on siliceous substrate. - Also in S Bulgaria.
Cauline leaves 10-20, crowded in lower part, rapidly decreasing in size upwards, all with dense simple hairs 4-8 mm long. Involucre with dense stellate and dense long simple hairs.

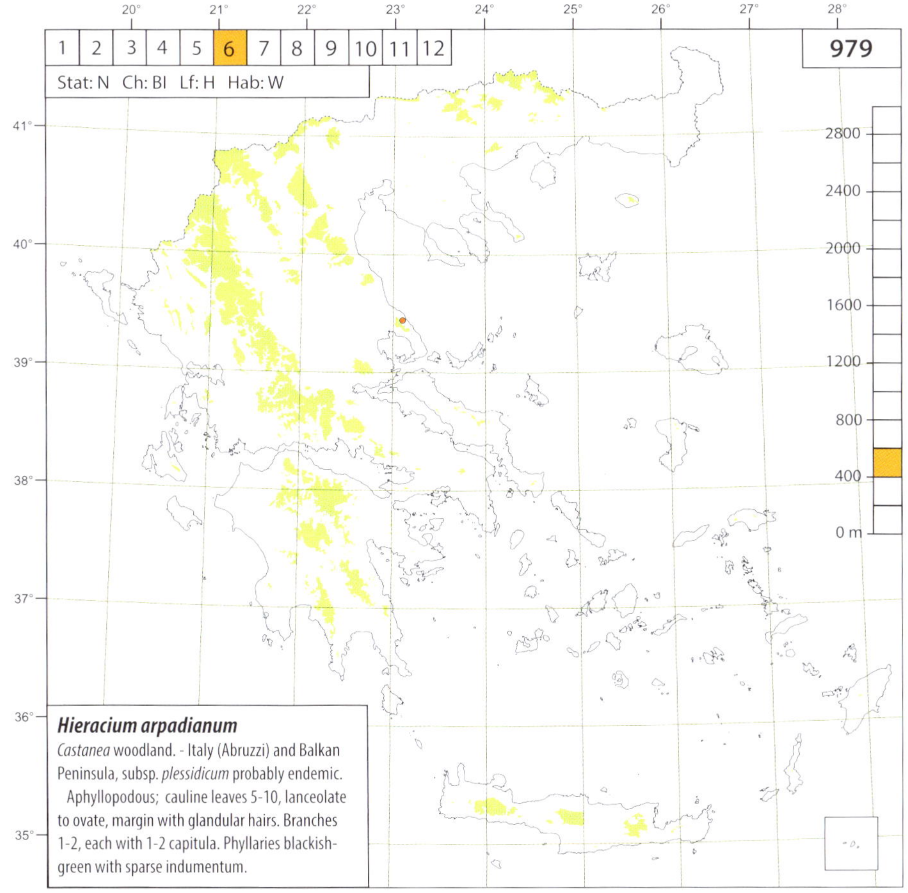

Hieracium arpadianum
Castanea woodland. - Italy (Abruzzi) and Balkan Peninsula, subsp. *plessidicum* probably endemic.
Aphyllopodous; cauline leaves 5-10, lanceolate to ovate, margin with glandular hairs. Branches 1-2, each with 1-2 capitula. Phyllaries blackish-green with sparse indumentum.

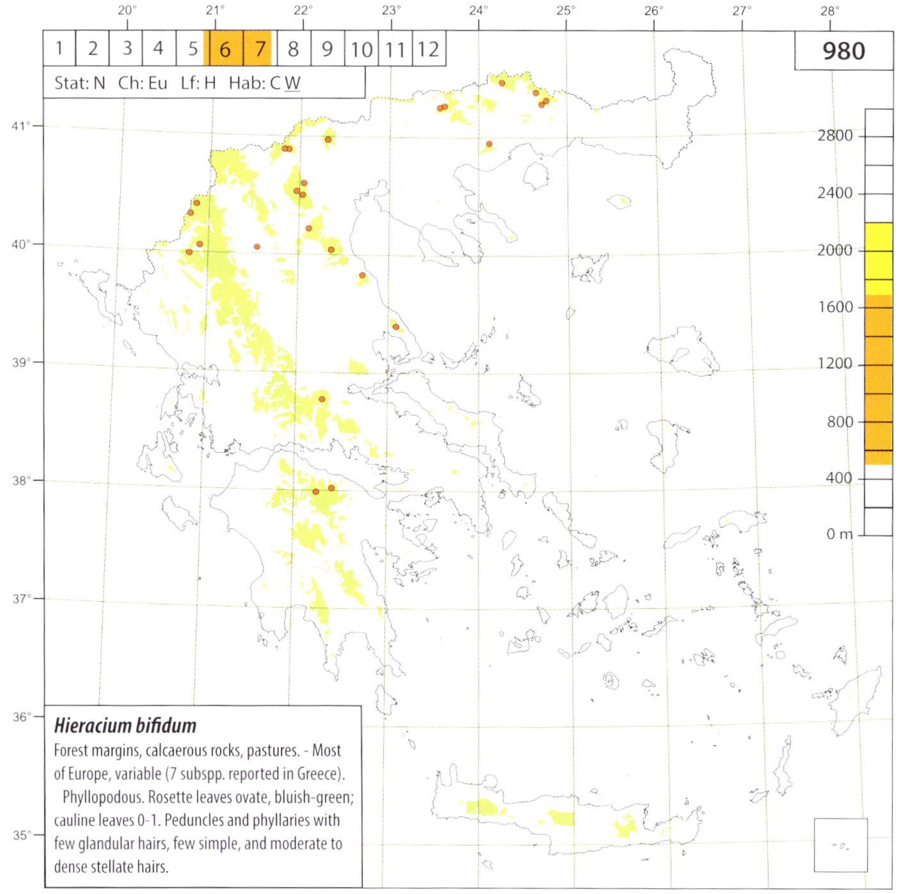

Hieracium bifidum
Forest margins, calcaerous rocks, pastures. - Most of Europe, variable (7 subspp. reported in Greece).
Phyllopodous. Rosette leaves ovate, bluish-green; cauline leaves 0-1. Peduncles and phyllaries with few glandular hairs, few simple, and moderate to dense stellate hairs.

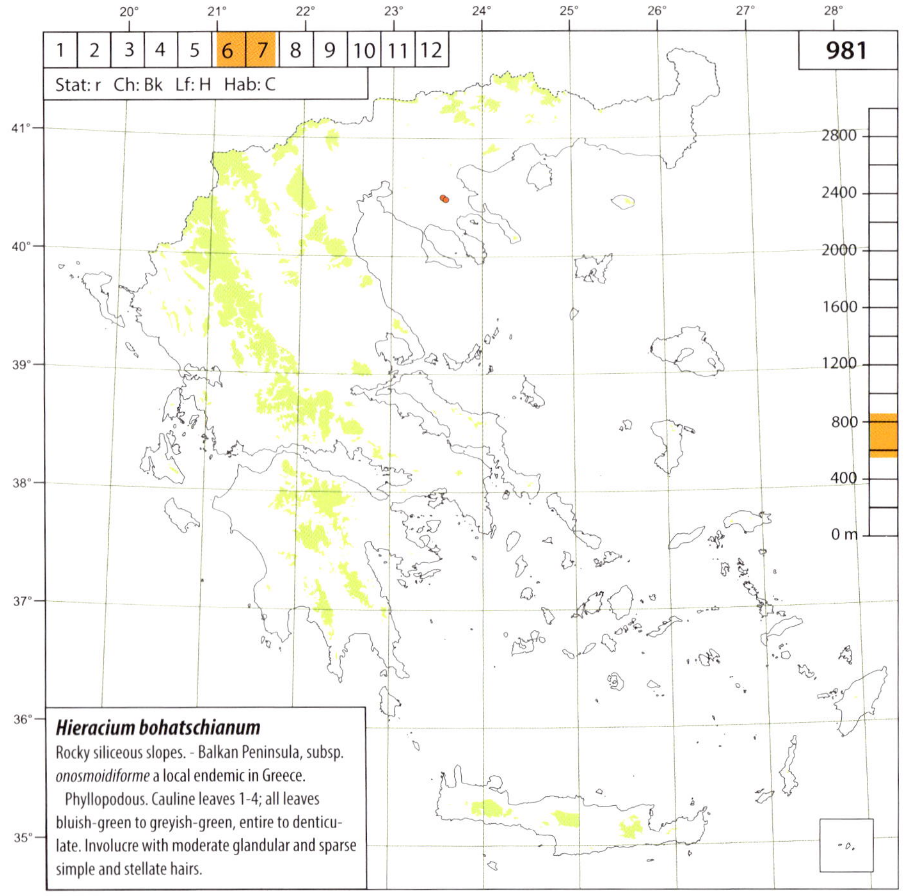

Hieracium bohatschianum

Rocky siliceous slopes. - Balkan Peninsula, subsp. *onosmoidiforme* a local endemic in Greece.

Phyllopodous. Cauline leaves 1-4; all leaves bluish-green to greyish-green, entire to denticulate. Involucre with moderate glandular and sparse simple and stellate hairs.

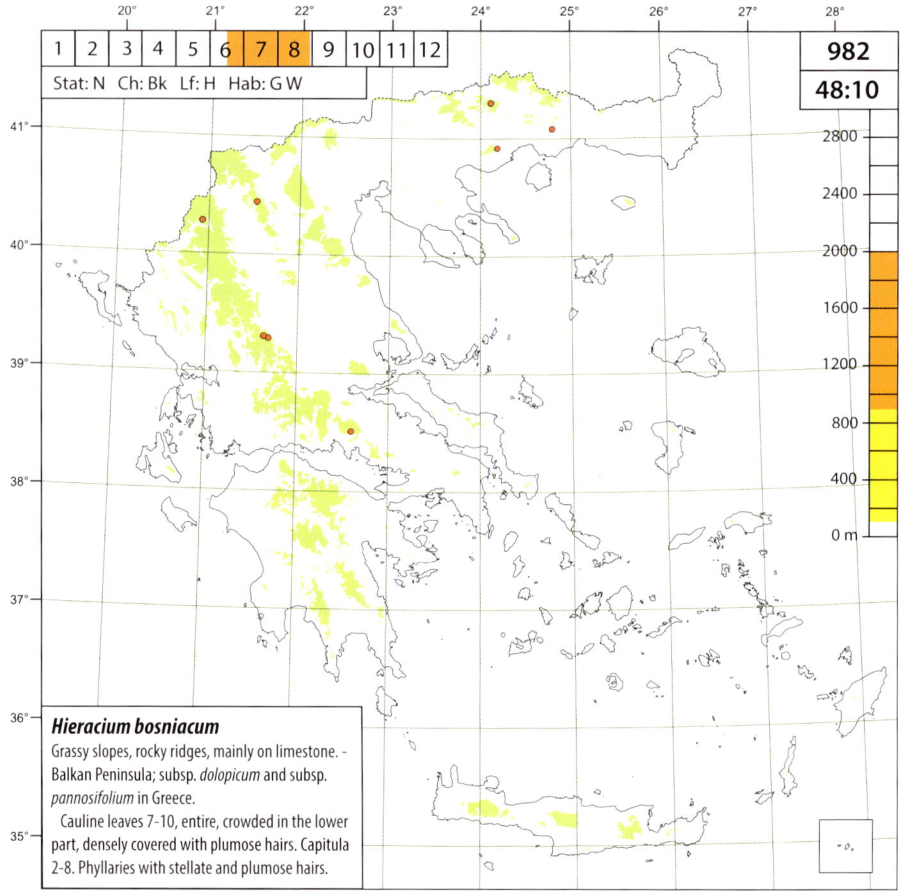

Hieracium bosniacum

Grassy slopes, rocky ridges, mainly on limestone. - Balkan Peninsula; subsp. *dolopicum* and subsp. *pannosifolium* in Greece.

Cauline leaves 7-10, entire, crowded in the lower part, densely covered with plumose hairs. Capitula 2-8. Phyllaries with stellate and plumose hairs.

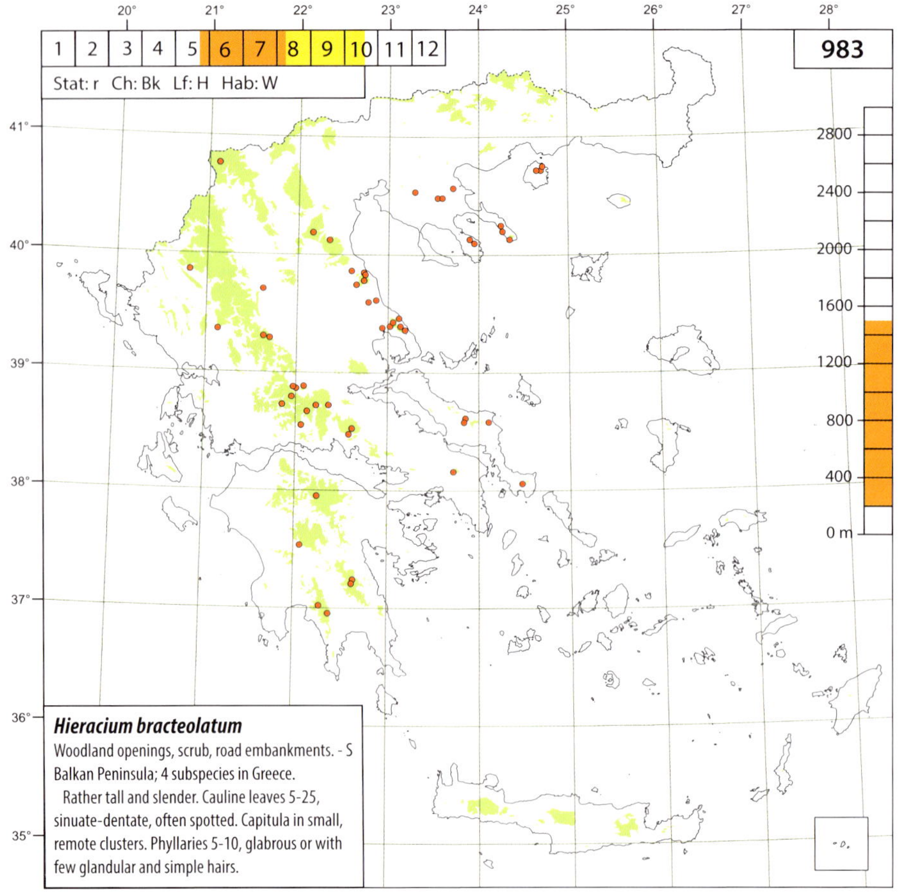

Hieracium bracteolatum

Woodland openings, scrub, road embankments. - S Balkan Peninsula; 4 subspecies in Greece.

Rather tall and slender. Cauline leaves 5-25, sinuate-dentate, often spotted. Capitula in small, remote clusters. Phyllaries 5-10, glabrous or with few glandular and simple hairs.

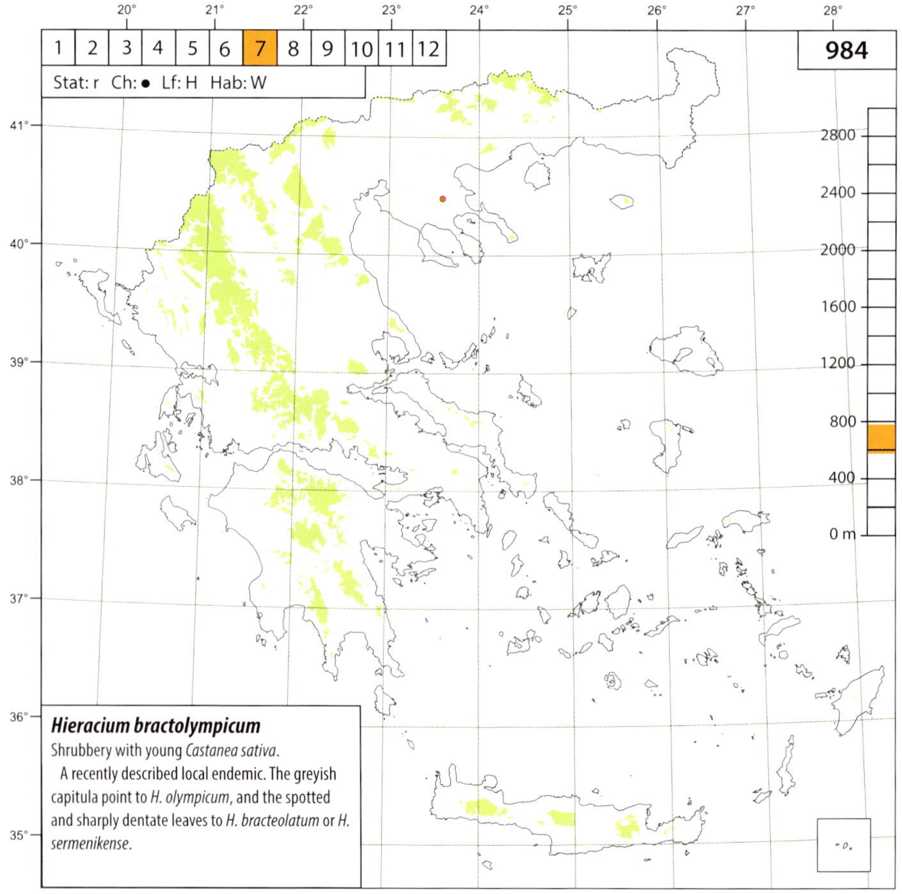

Hieracium bractolympicum

Shrubbery with young *Castanea sativa*.

A recently described local endemic. The greyish capitula point to *H. olympicum*, and the spotted and sharply dentate leaves to *H. bracteolatum* or *H. sermenikense*.

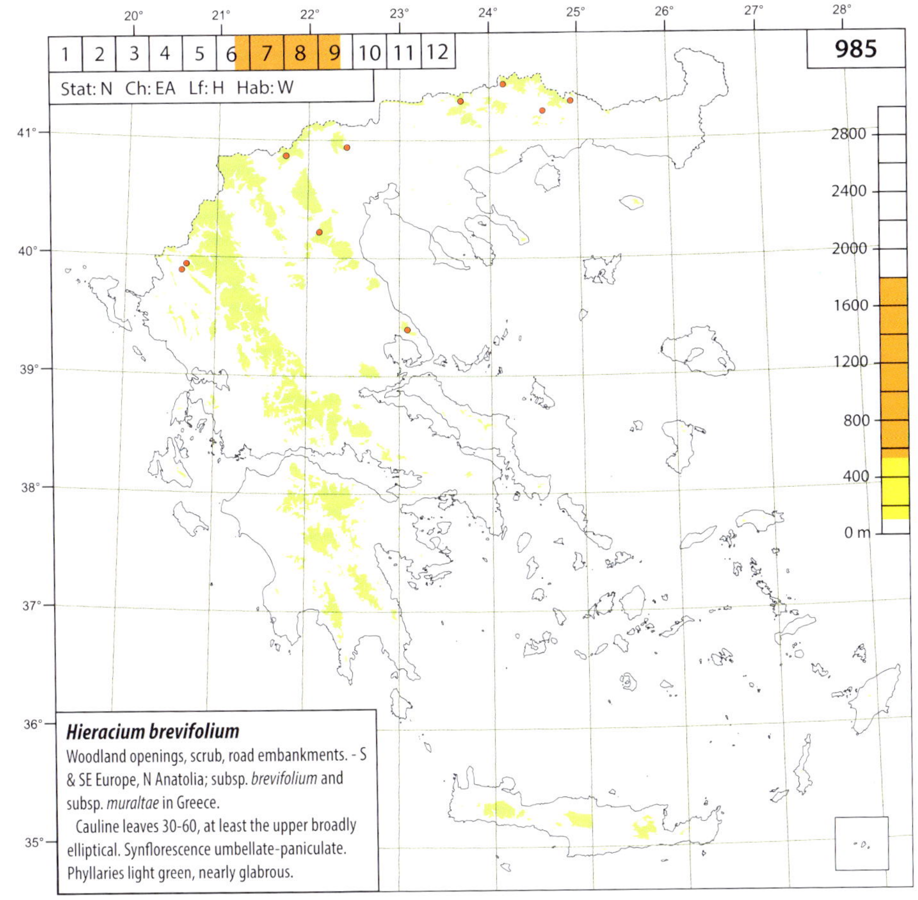

Hieracium brevifolium

Woodland openings, scrub, road embankments. - S & SE Europe, N Anatolia; subsp. *brevifolium* and subsp. *muraltae* in Greece.

Cauline leaves 30-60, at least the upper broadly elliptical. Synflorescence umbellate-paniculate. Phyllaries light green, nearly glabrous.

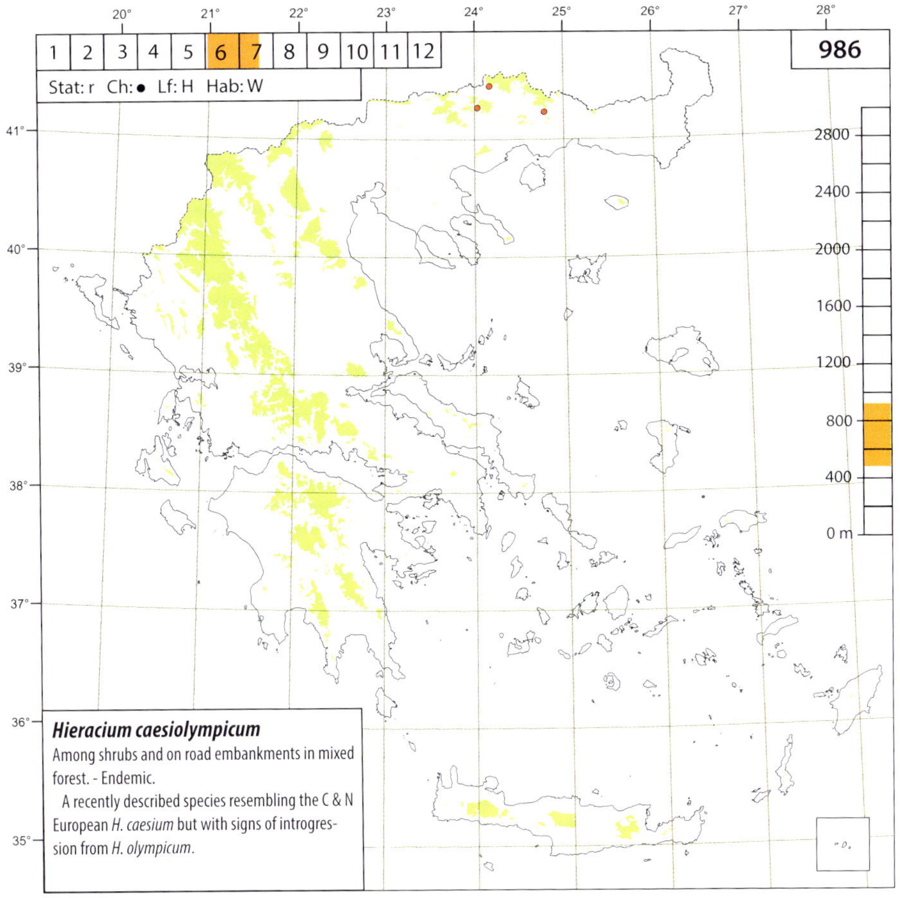

Hieracium caesiolympicum

Among shrubs and on road embankments in mixed forest. - Endemic.

A recently described species resembling the C & N European *H. caesium* but with signs of introgression from *H. olympicum*.

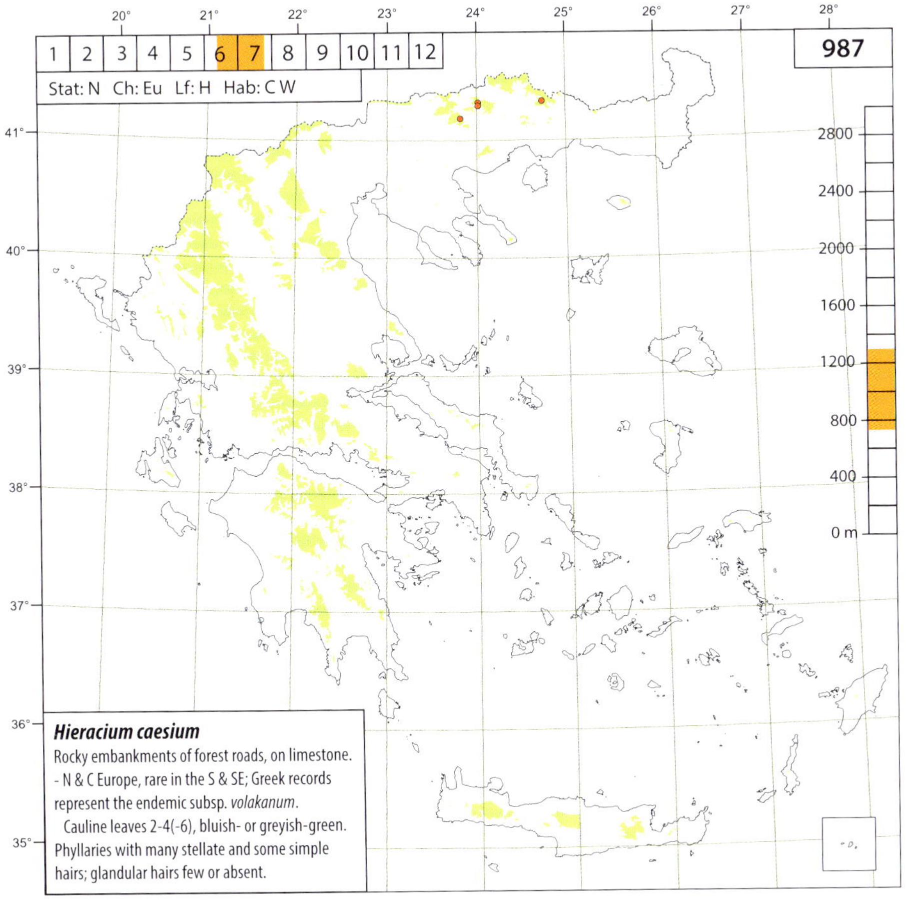

Hieracium caesium

Rocky embankments of forest roads, on limestone. - N & C Europe, rare in the S & SE; Greek records represent the endemic subsp. *volakanum*.

Cauline leaves 2-4(-6), bluish- or greyish-green. Phyllaries with many stellate and some simple hairs; glandular hairs few or absent.

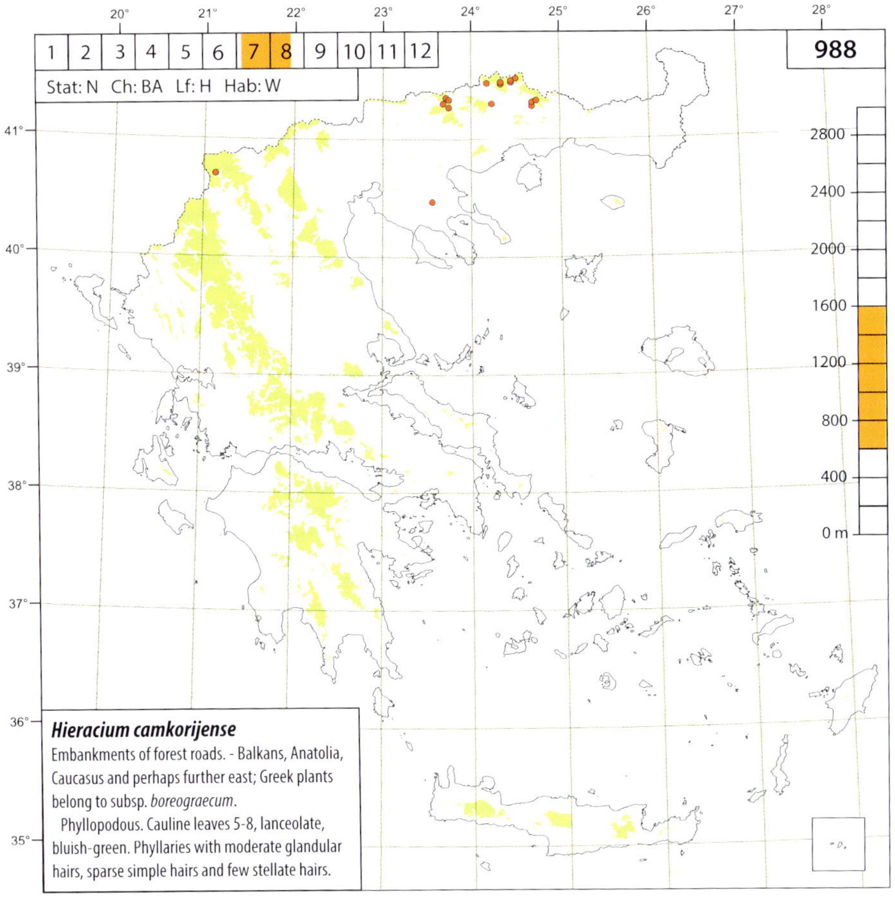

Hieracium camkorijense

Embankments of forest roads. - Balkans, Anatolia, Caucasus and perhaps further east; Greek plants belong to subsp. *boreograecum*.

Phyllopodous. Cauline leaves 5-8, lanceolate, bluish-green. Phyllaries with moderate glandular hairs, sparse simple hairs and few stellate hairs.

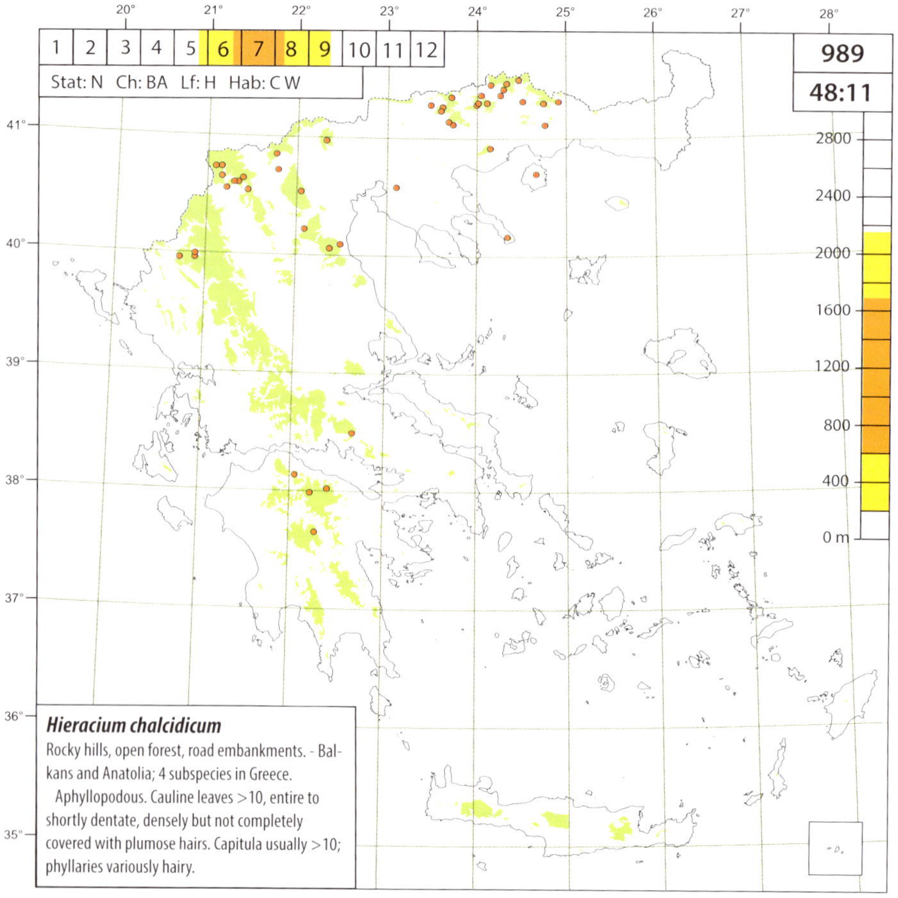

Hieracium chalcidicum
Rocky hills, open forest, road embankments. - Balkans and Anatolia; 4 subspecies in Greece.
Aphyllopodous. Cauline leaves >10, entire to shortly dentate, densely but not completely covered with plumose hairs. Capitula usually >10; phyllaries variously hairy.

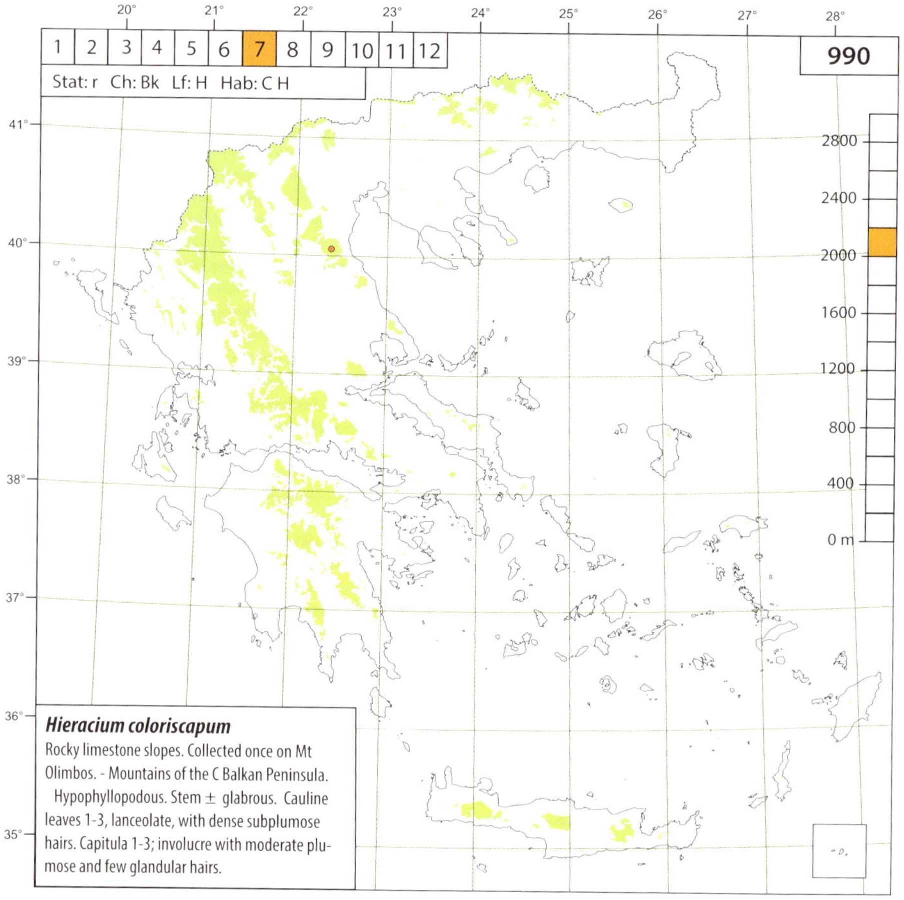

Hieracium coloriscapum
Rocky limestone slopes. Collected once on Mt Olimbos. - Mountains of the C Balkan Peninsula.
Hypophyllopodous. Stem ± glabrous. Cauline leaves 1-3, lanceolate, with dense subplumose hairs. Capitula 1-3; involucre with moderate plumose and few glandular hairs.

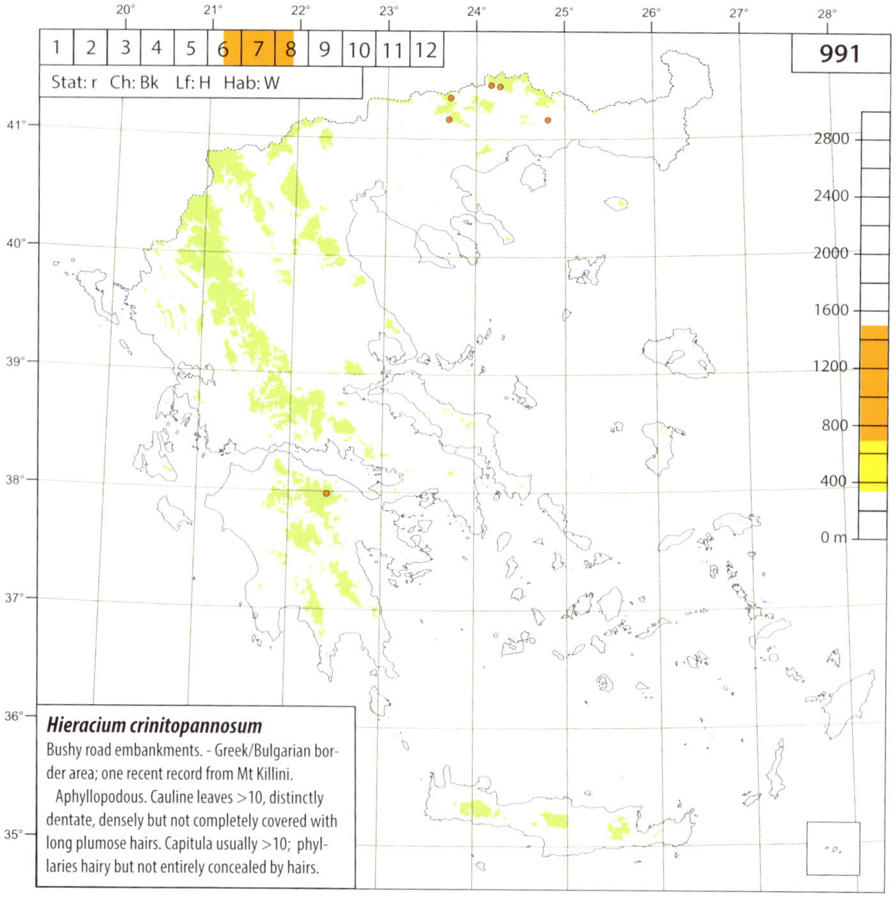

Hieracium crinitopannosum
Bushy road embankments. - Greek/Bulgarian border area; one recent record from Mt Killini.
Aphyllopodous. Cauline leaves >10, distinctly dentate, densely but not completely covered with long plumose hairs. Capitula usually >10; phyllaries hairy but not entirely concealed by hairs.

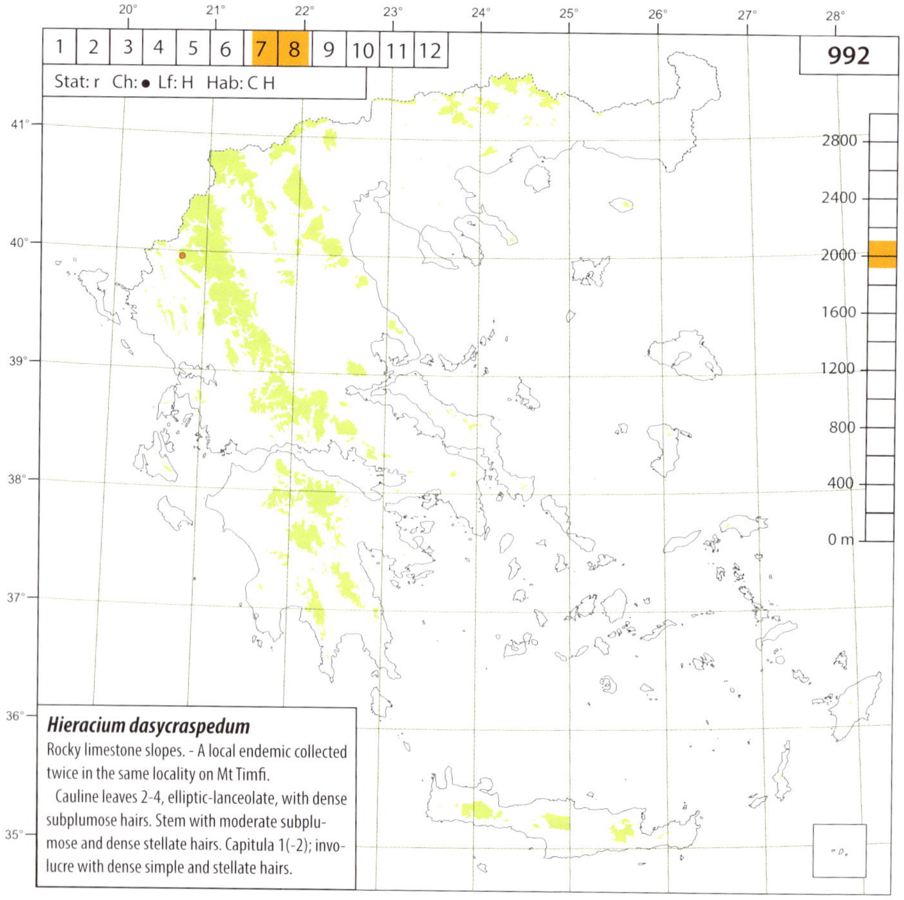

Hieracium dasycraspedum
Rocky limestone slopes. - A local endemic collected twice in the same locality on Mt Timfi.
Cauline leaves 2-4, elliptic-lanceolate, with dense subplumose hairs. Stem with moderate subplumose and dense stellate hairs. Capitula 1(-2); involucre with dense simple and stellate hairs.

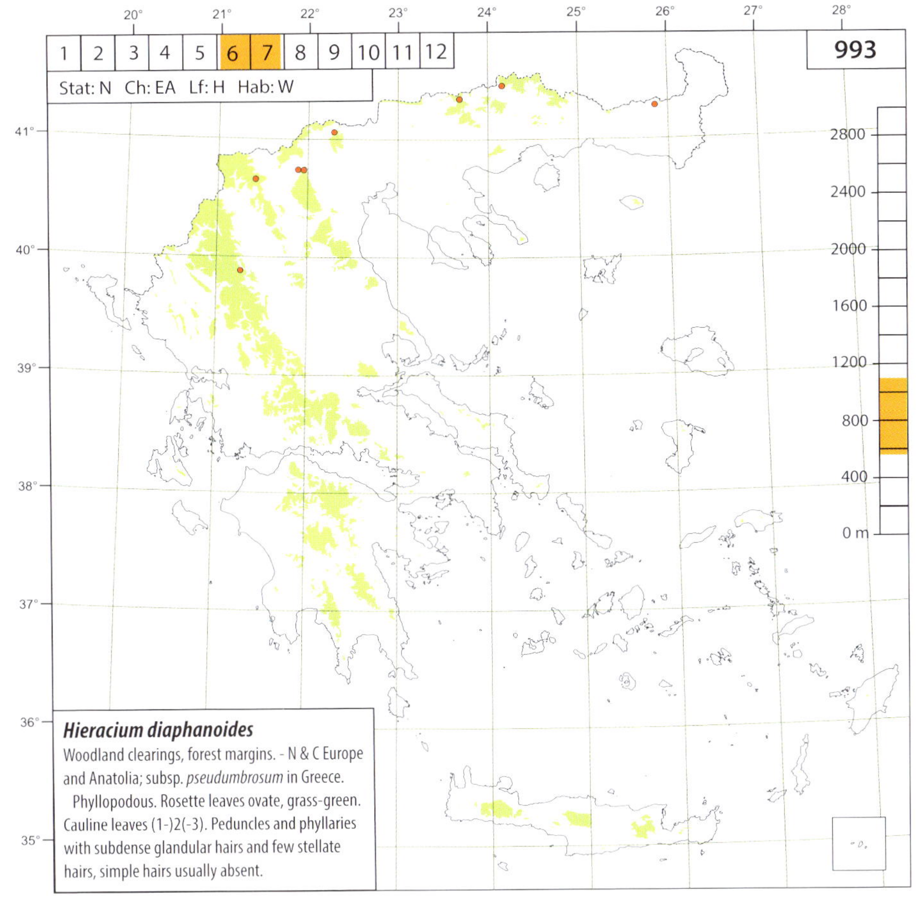

Hieracium diaphanoides

Woodland clearings, forest margins. - N & C Europe and Anatolia; subsp. *pseudumbrosum* in Greece.

Phyllopodous. Rosette leaves ovate, grass-green. Cauline leaves (1-)2(-3). Peduncles and phyllaries with subdense glandular hairs and few stellate hairs, simple hairs usually absent.

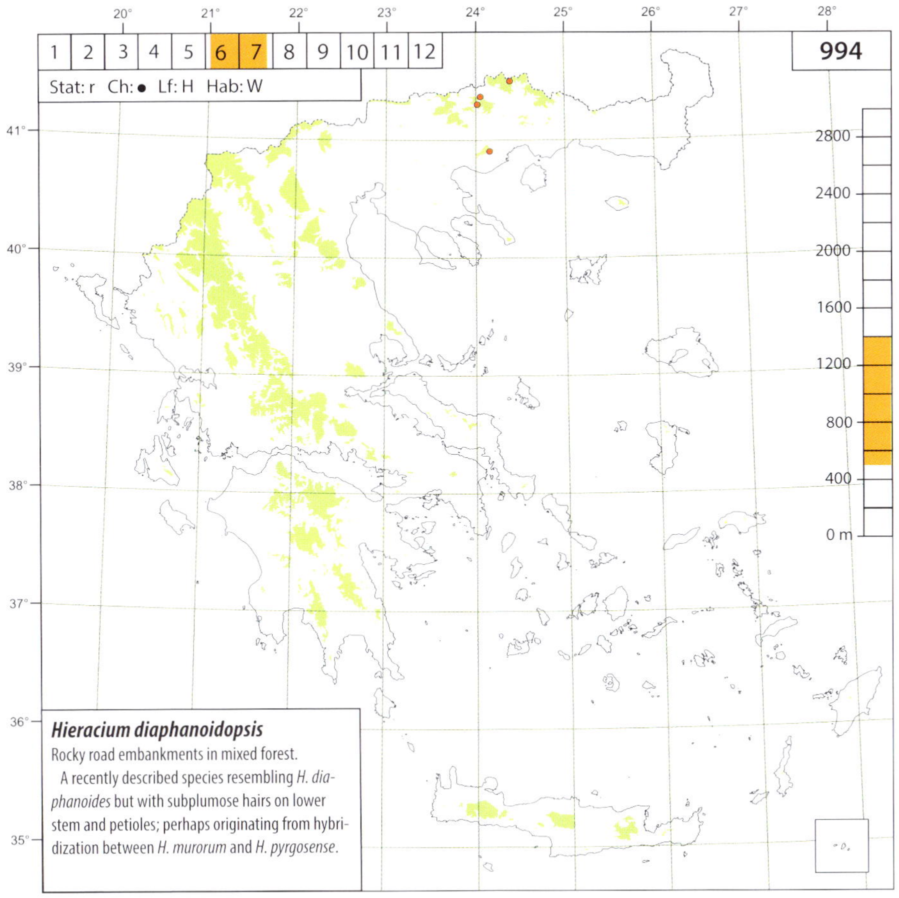

Hieracium diaphanoidopsis

Rocky road embankments in mixed forest.

A recently described species resembling *H. diaphanoides* but with subplumose hairs on lower stem and petioles; perhaps originating from hybridization between *H. murorum* and *H. pyrgosense*.

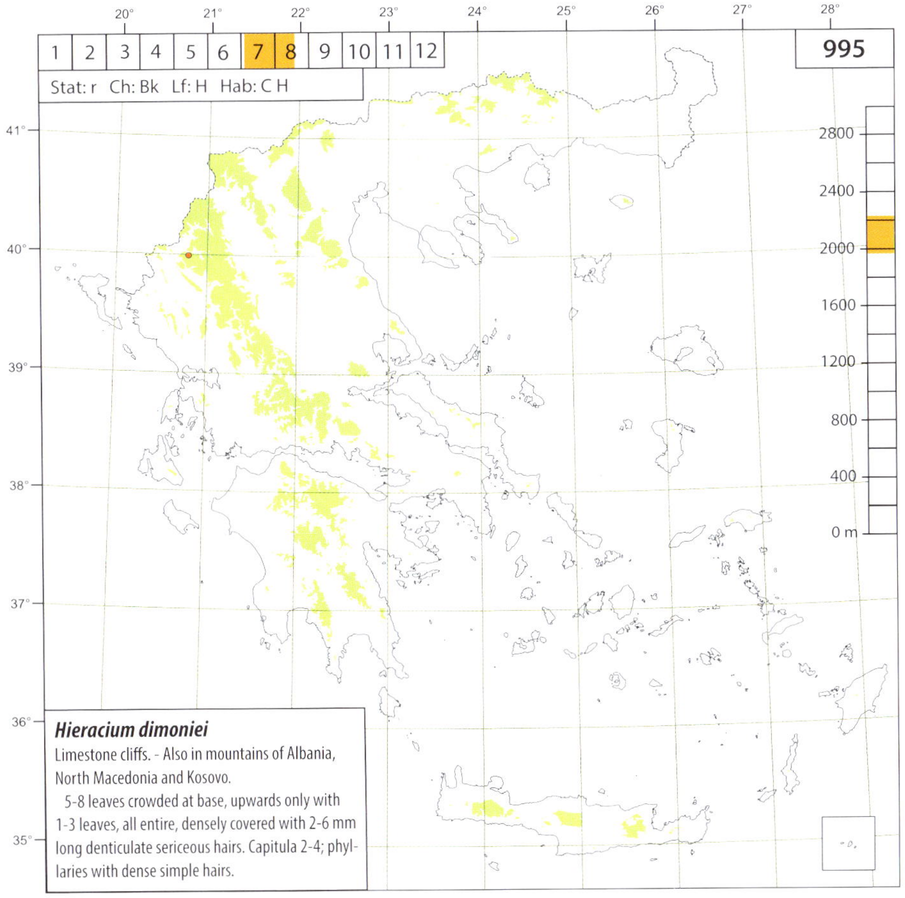

Hieracium dimoniei

Limestone cliffs. - Also in mountains of Albania, North Macedonia and Kosovo.

5-8 leaves crowded at base, upwards only with 1-3 leaves, all entire, densely covered with 2-6 mm long denticulate sericeous hairs. Capitula 2-4; phyllaries with dense simple hairs.

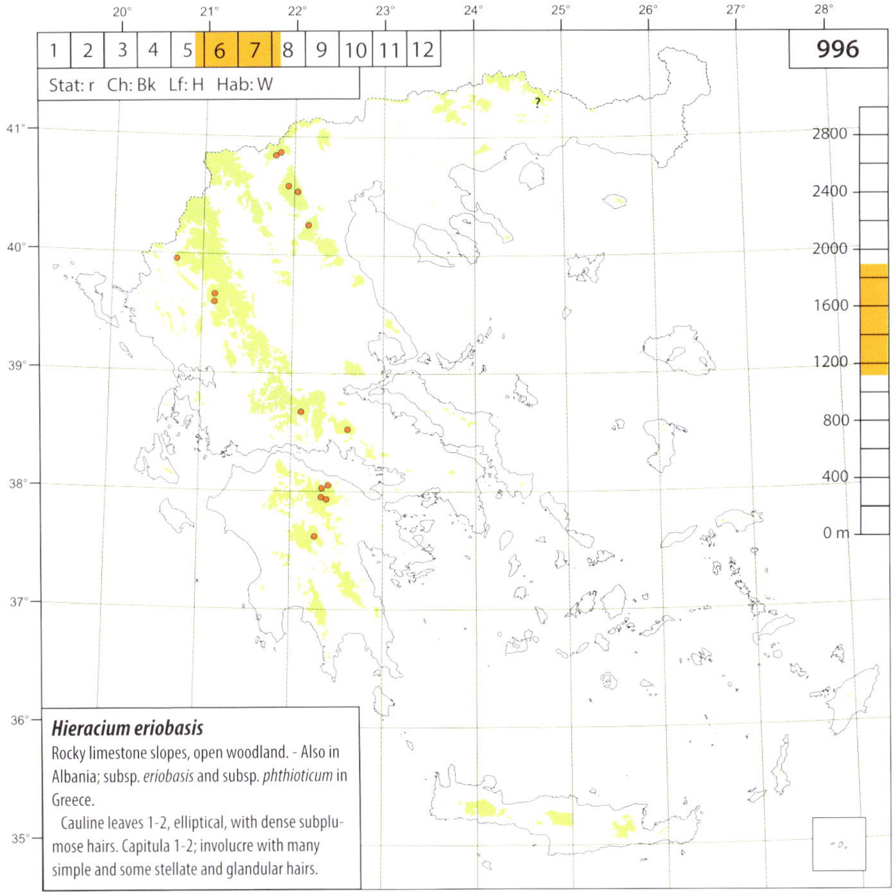

Hieracium eriobasis

Rocky limestone slopes, open woodland. - Also in Albania; subsp. *eriobasis* and subsp. *phthioticum* in Greece.

Cauline leaves 1-2, elliptical, with dense subplumose hairs. Capitula 1-2; involucre with many simple and some stellate and glandular hairs.

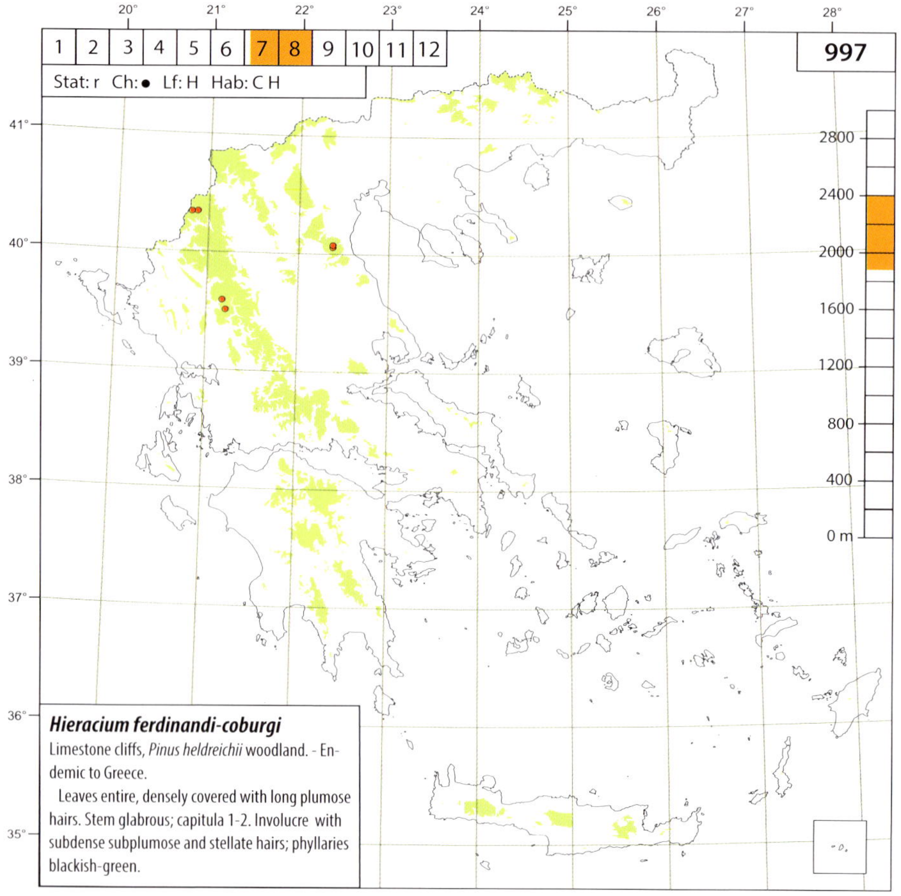

Hieracium ferdinandi-coburgi
Limestone cliffs, *Pinus heldreichii* woodland. - Endemic to Greece.
Leaves entire, densely covered with long plumose hairs. Stem glabrous; capitula 1-2. Involucre with subdense subplumose and stellate hairs; phyllaries blackish-green.

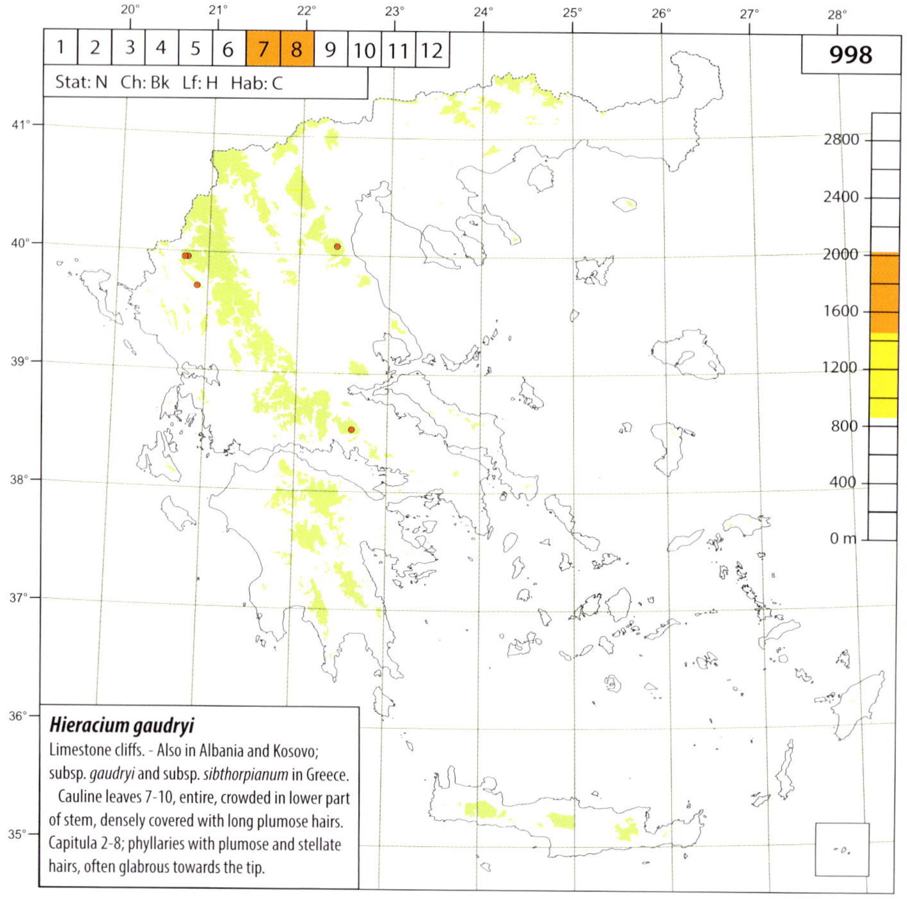

Hieracium gaudryi
Limestone cliffs. - Also in Albania and Kosovo; subsp. *gaudryi* and subsp. *sibthorpianum* in Greece.
Cauline leaves 7-10, entire, crowded in lower part of stem, densely covered with long plumose hairs. Capitula 2-8; phyllaries with plumose and stellate hairs, often glabrous towards the tip.

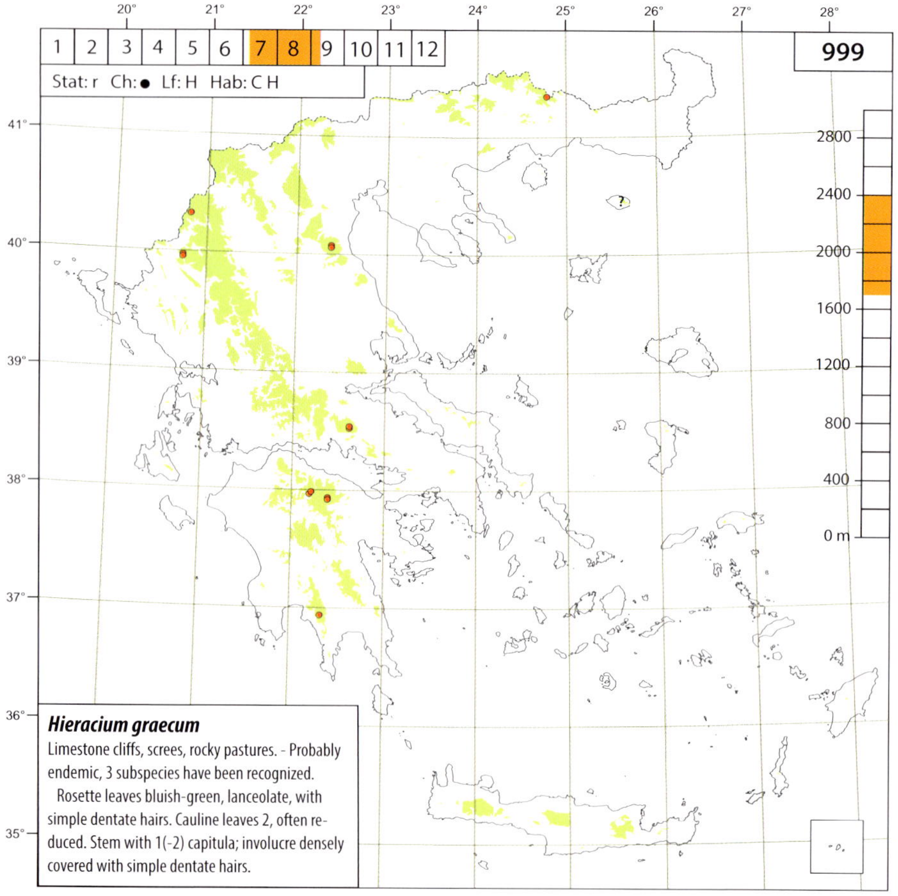

Hieracium graecum
Limestone cliffs, screes, rocky pastures. - Probably endemic, 3 subspecies have been recognized.
Rosette leaves bluish-green, lanceolate, with simple dentate hairs. Cauline leaves 2, often reduced. Stem with 1(-2) capitula; involucre densely covered with simple dentate hairs.

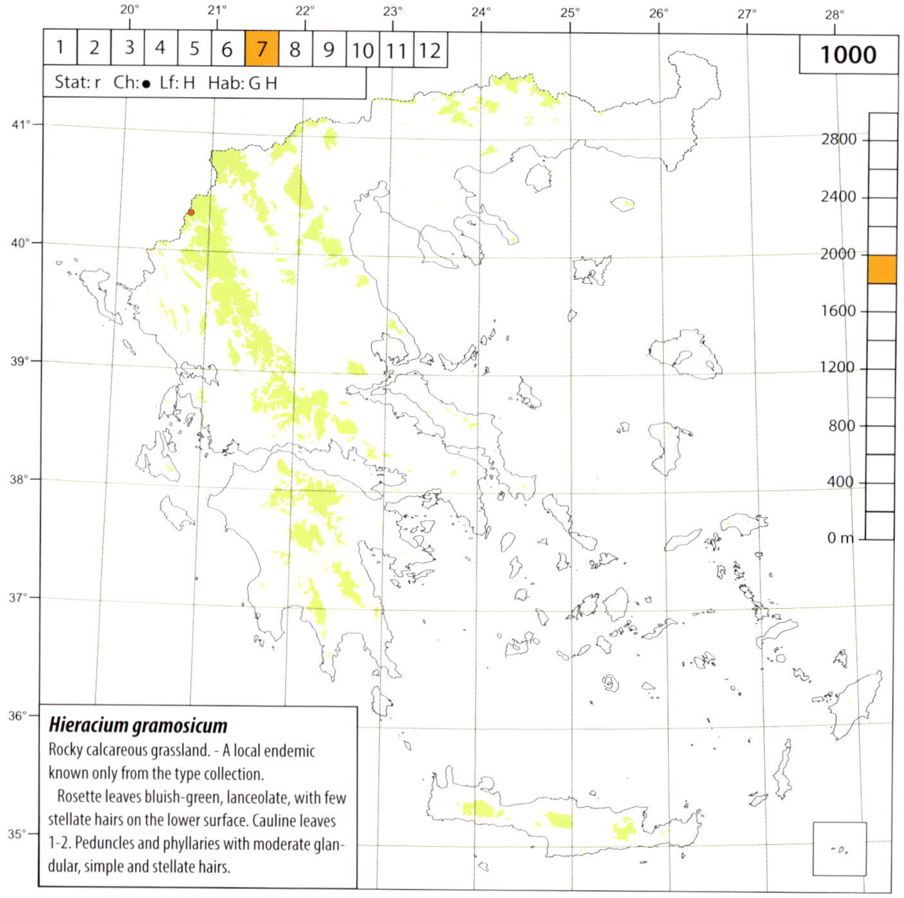

Hieracium gramosicum
Rocky calcareous grassland. - A local endemic known only from the type collection.
Rosette leaves bluish-green, lanceolate, with few stellate hairs on the lower surface. Cauline leaves 1-2. Peduncles and phyllaries with moderate glandular, simple and stellate hairs.

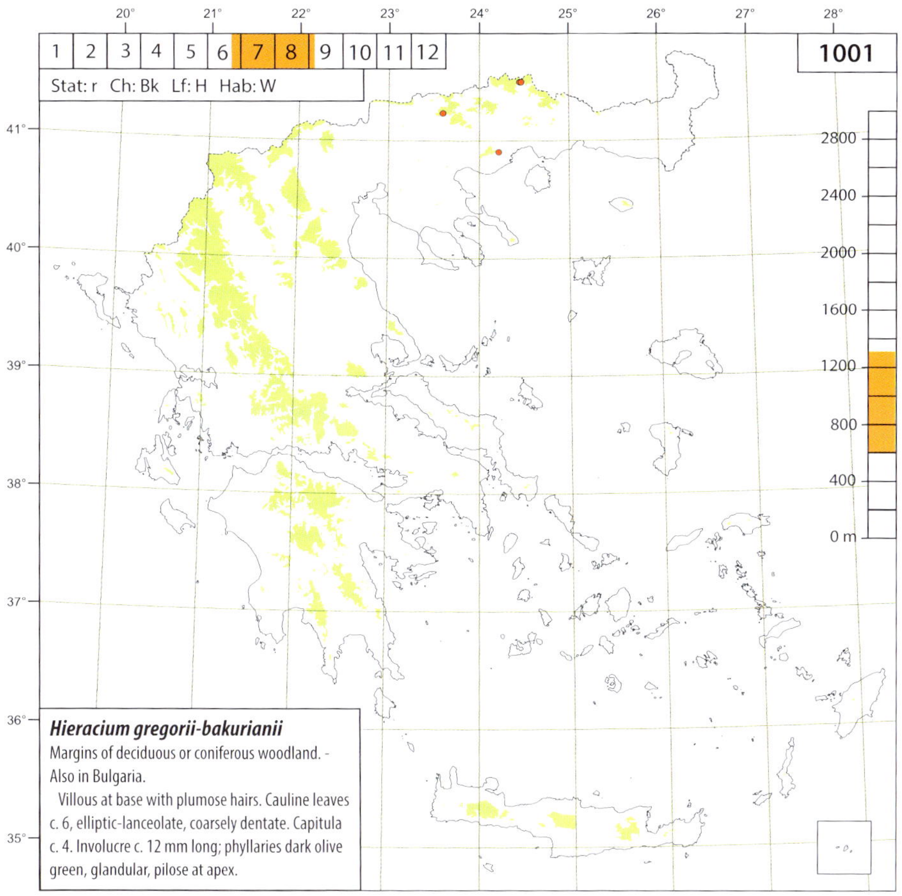

Hieracium gregorii-bakurianii
Margins of deciduous or coniferous woodland. - Also in Bulgaria.
Villous at base with plumose hairs. Cauline leaves c. 6, elliptic-lanceolate, coarsely dentate. Capitula c. 4. Involucre c. 12 mm long; phyllaries dark olive green, glandular, pilose at apex.

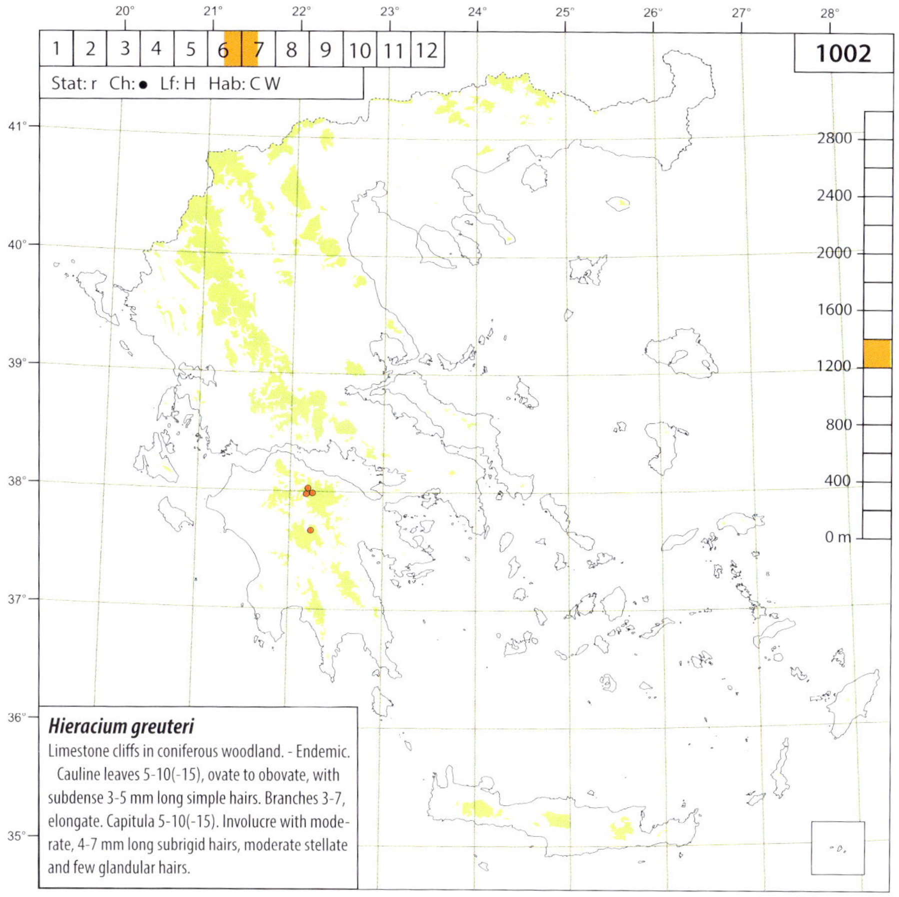

Hieracium greuteri
Limestone cliffs in coniferous woodland. - Endemic.
Cauline leaves 5-10(-15), ovate to obovate, with subdense 3-5 mm long simple hairs. Branches 3-7, elongate. Capitula 5-10(-15). Involucre with moderate, 4-7 mm long subrigid hairs, moderate stellate and few glandular hairs.

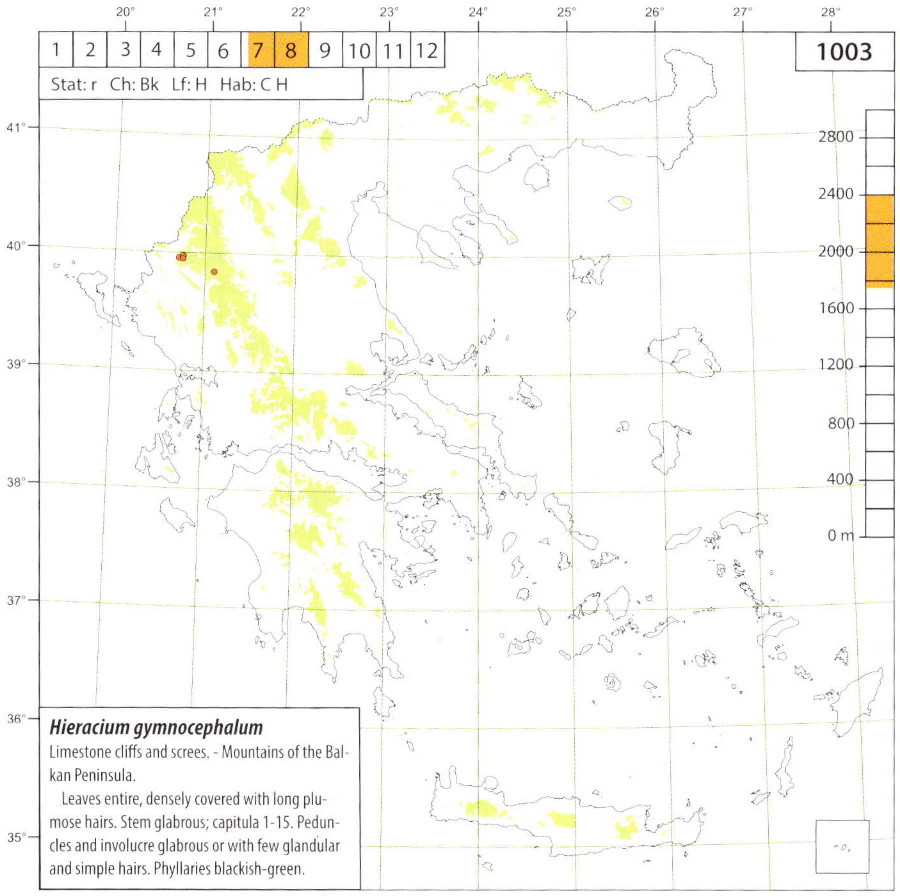

Hieracium gymnocephalum
Limestone cliffs and screes. - Mountains of the Balkan Peninsula.
Leaves entire, densely covered with long plumose hairs. Stem glabrous; capitula 1-15. Peduncles and involucre glabrous or with few glandular and simple hairs. Phyllaries blackish-green.

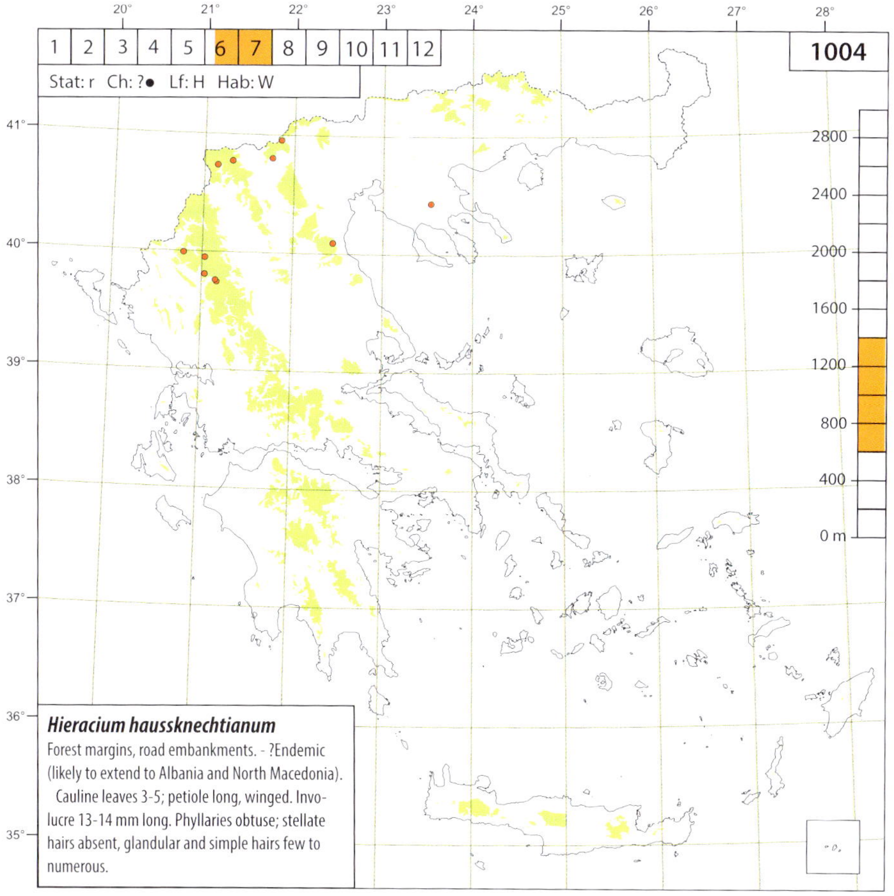

Hieracium haussknechtianum
Forest margins, road embankments. - ?Endemic (likely to extend to Albania and North Macedonia).
Cauline leaves 3-5; petiole long, winged. Involucre 13-14 mm long. Phyllaries obtuse; stellate hairs absent, glandular and simple hairs few to numerous.

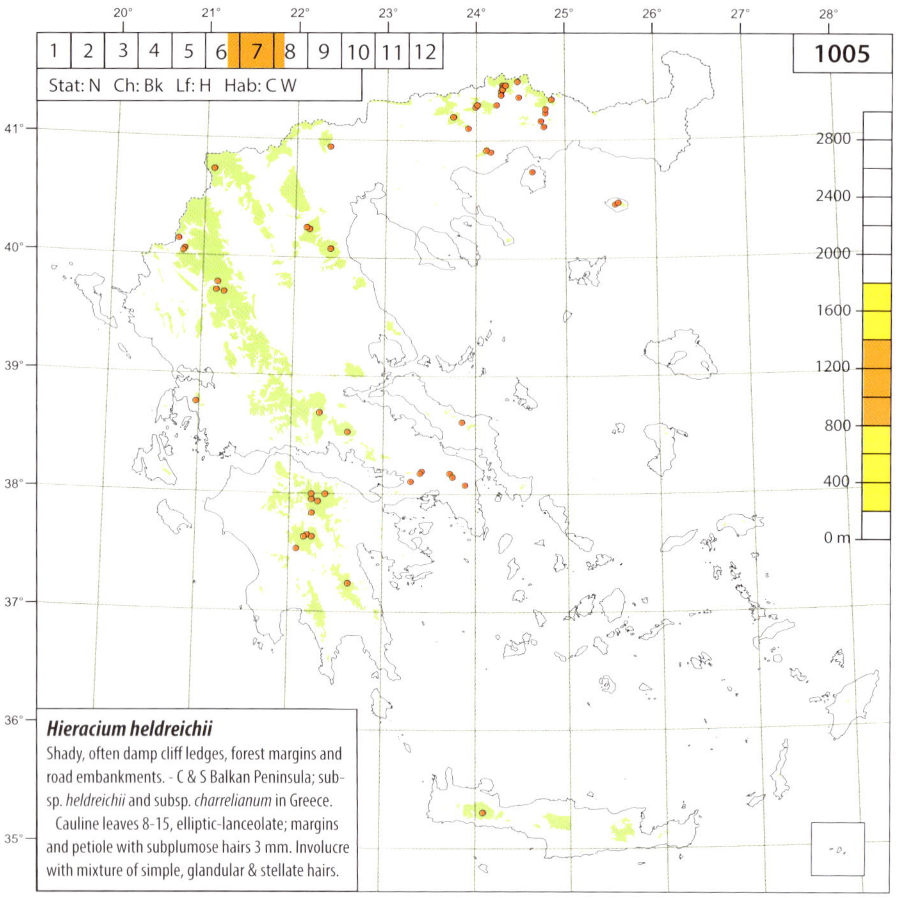

Hieracium heldreichii
Shady, often damp cliff ledges, forest margins and road embankments. - C & S Balkan Peninsula; subsp. *heldreichii* and subsp. *charrelianum* in Greece.
Cauline leaves 8-15, elliptic-lanceolate; margins and petiole with subplumose hairs 3 mm. Involucre with mixture of simple, glandular & stellate hairs.

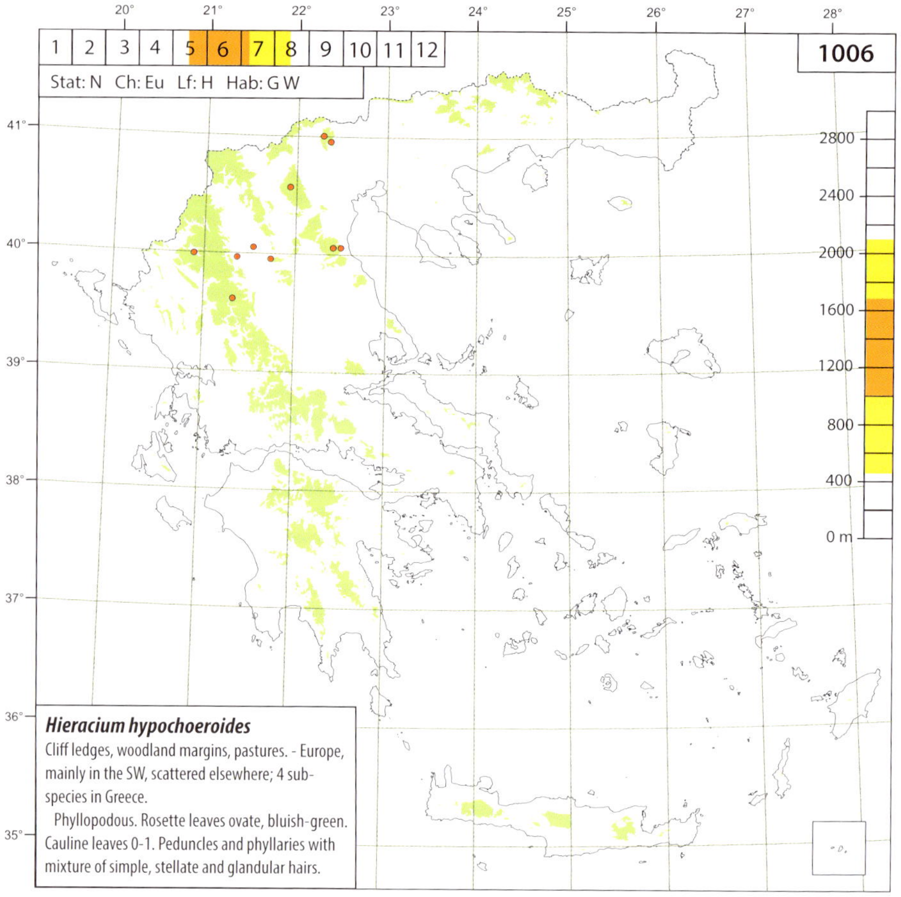

Hieracium hypochoeroides
Cliff ledges, woodland margins, pastures. - Europe, mainly in the SW, scattered elsewhere; 4 subspecies in Greece.
Phyllopodous. Rosette leaves ovate, bluish-green. Cauline leaves 0-1. Peduncles and phyllaries with mixture of simple, stellate and glandular hairs.

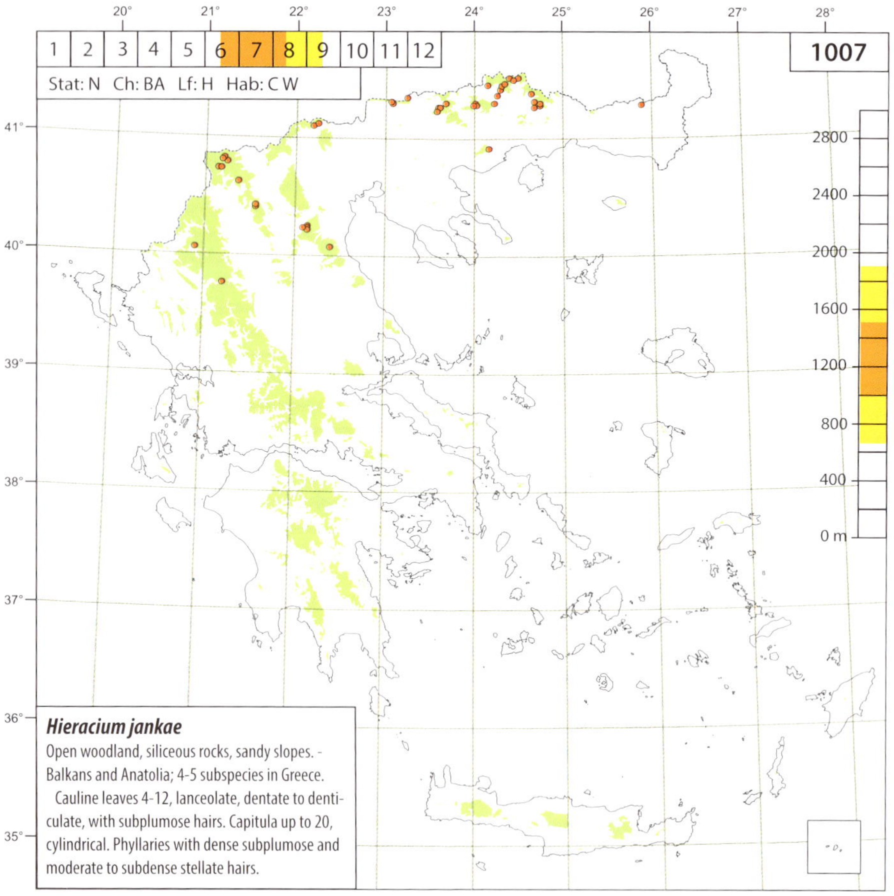

Hieracium jankae
Open woodland, siliceous rocks, sandy slopes. - Balkans and Anatolia; 4-5 subspecies in Greece.
Cauline leaves 4-12, lanceolate, dentate to denticulate, with subplumose hairs. Capitula up to 20, cylindrical. Phyllaries with dense subplumose and moderate to subdense stellate hairs.

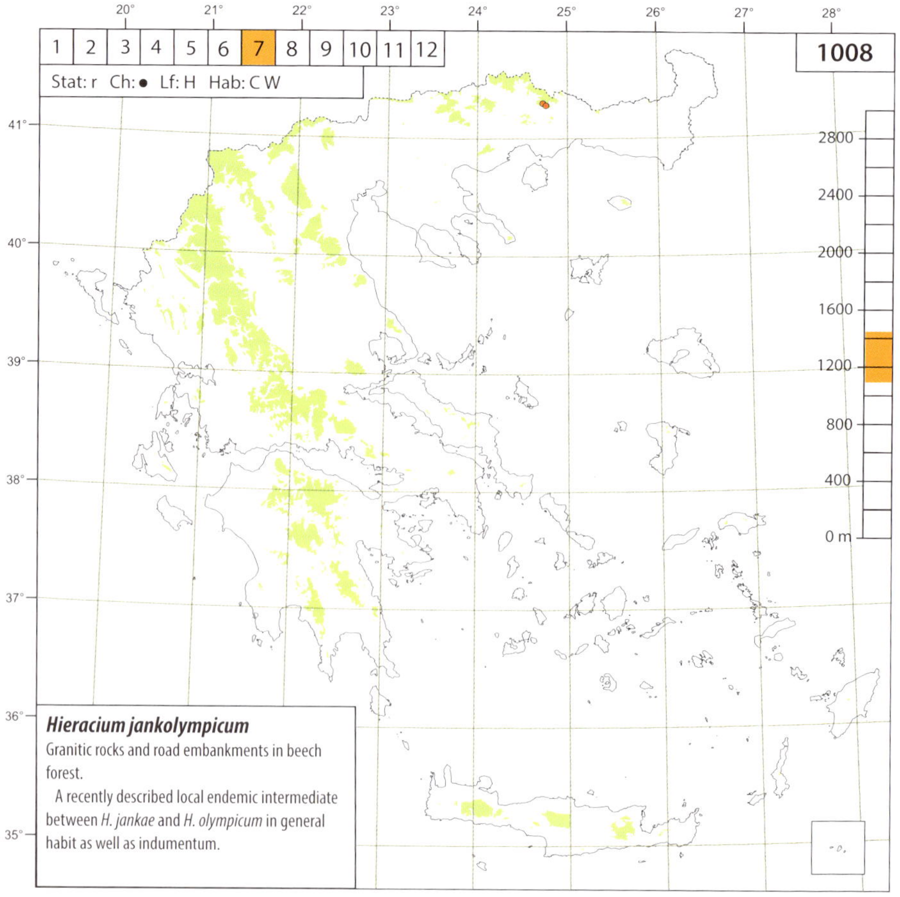

Hieracium jankolympicum
Granitic rocks and road embankments in beech forest.
A recently described local endemic intermediate between *H. jankae* and *H. olympicum* in general habit as well as indumentum.

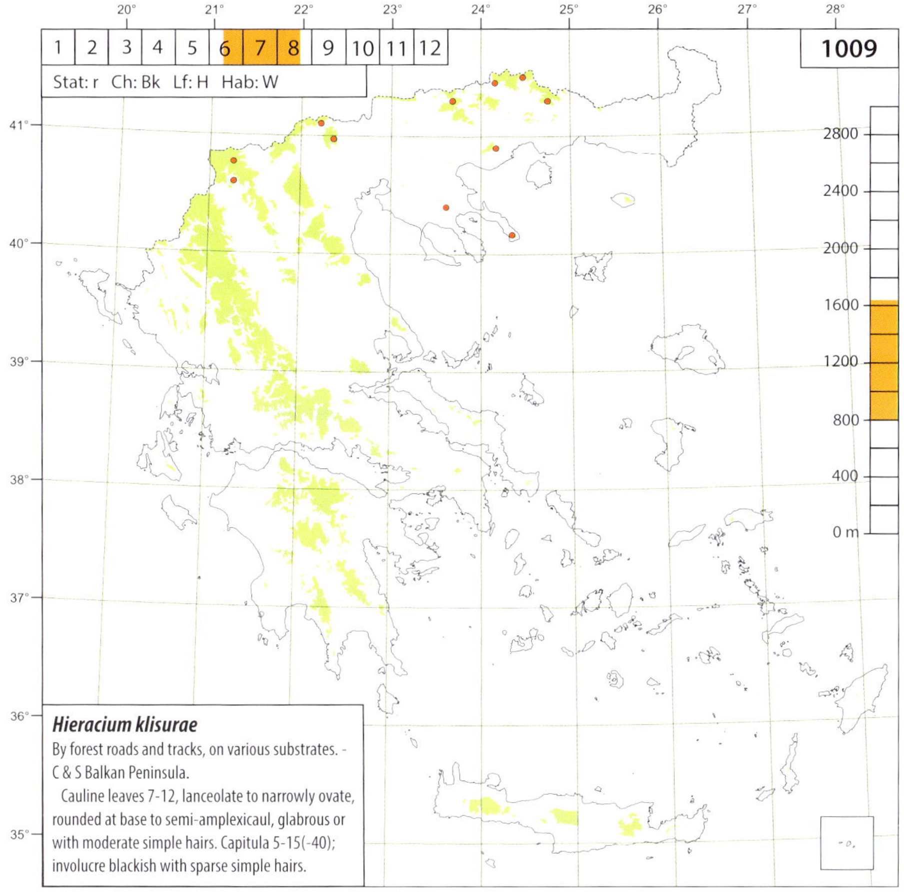

Hieracium klisurae

By forest roads and tracks, on various substrates. - C & S Balkan Peninsula.

Cauline leaves 7-12, lanceolate to narrowly ovate, rounded at base to semi-amplexicaul, glabrous or with moderate simple hairs. Capitula 5-15(-40); involucre blackish with sparse simple hairs.

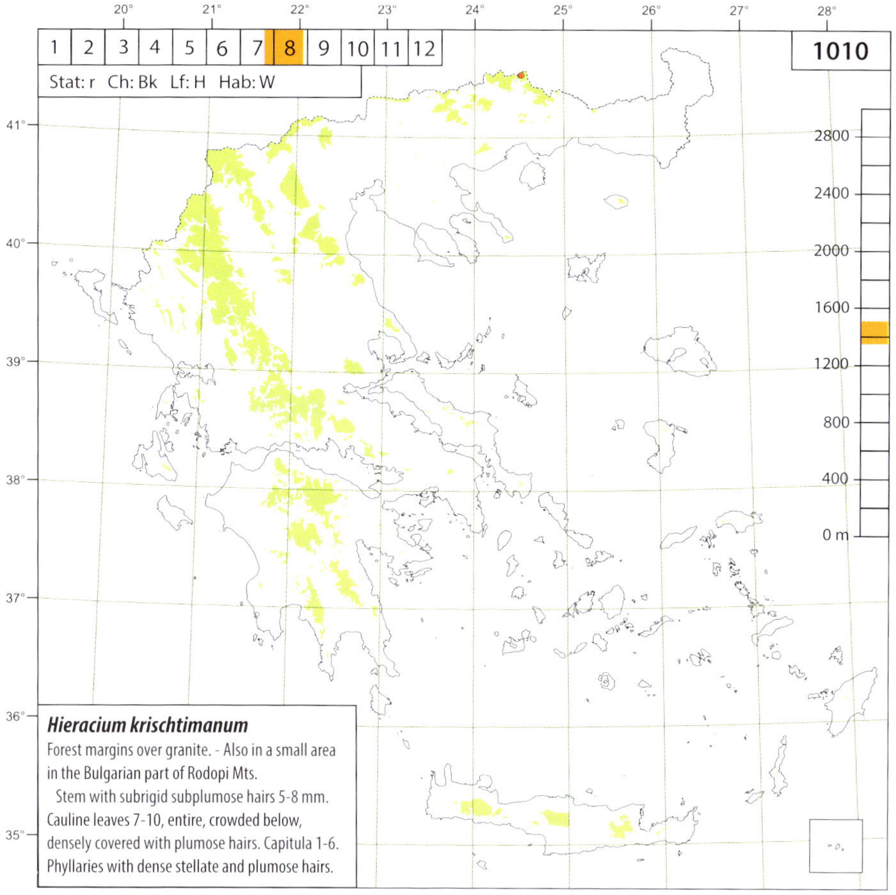

Hieracium krischtimanum

Forest margins over granite. - Also in a small area in the Bulgarian part of Rodopi Mts.

Stem with subrigid subplumose hairs 5-8 mm. Cauline leaves 7-10, entire, crowded below, densely covered with plumose hairs. Capitula 1-6. Phyllaries with dense stellate and plumose hairs.

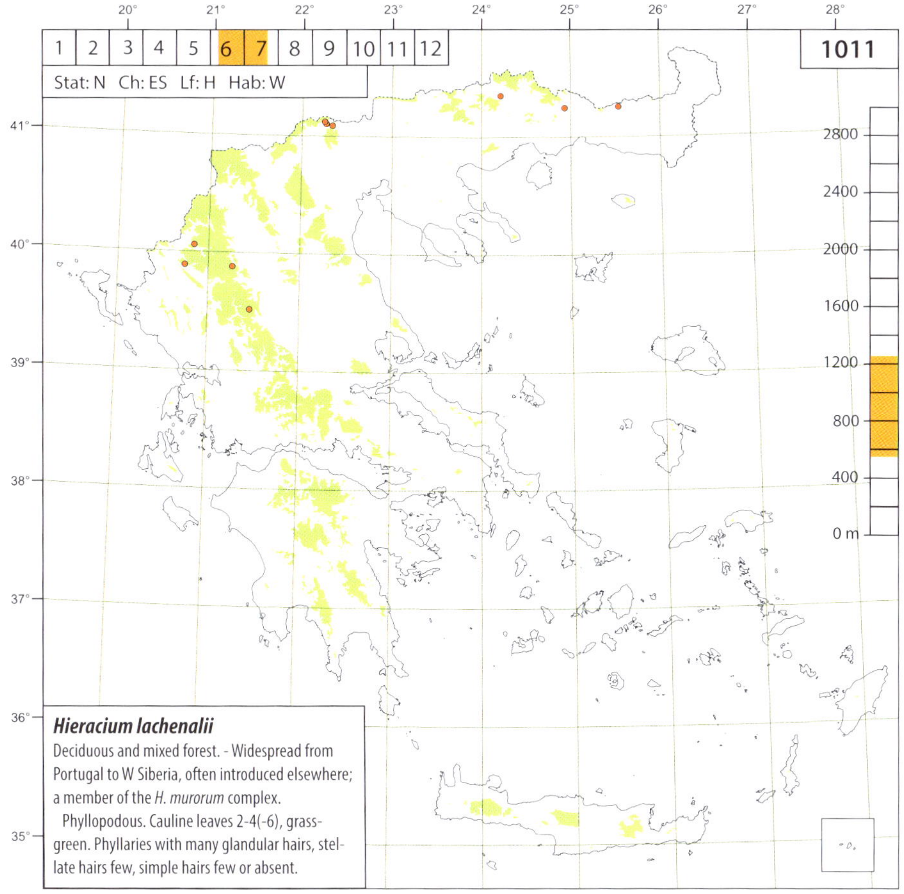

Hieracium lachenalii

Deciduous and mixed forest. - Widespread from Portugal to W Siberia, often introduced elsewhere; a member of the *H. murorum* complex.

Phyllopodous. Cauline leaves 2-4(-6), grass-green. Phyllaries with many glandular hairs, stellate hairs few, simple hairs few or absent.

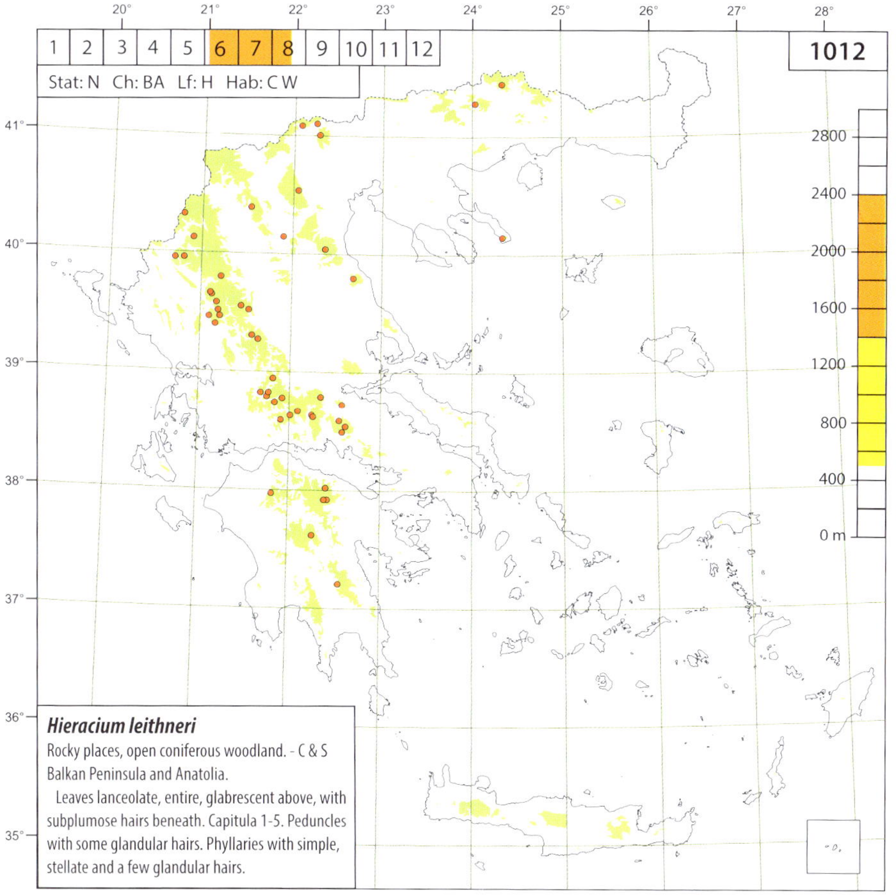

Hieracium leithneri

Rocky places, open coniferous woodland. - C & S Balkan Peninsula and Anatolia.

Leaves lanceolate, entire, glabrescent above, with subplumose hairs beneath. Capitula 1-5. Peduncles with some glandular hairs. Phyllaries with simple, stellate and a few glandular hairs.

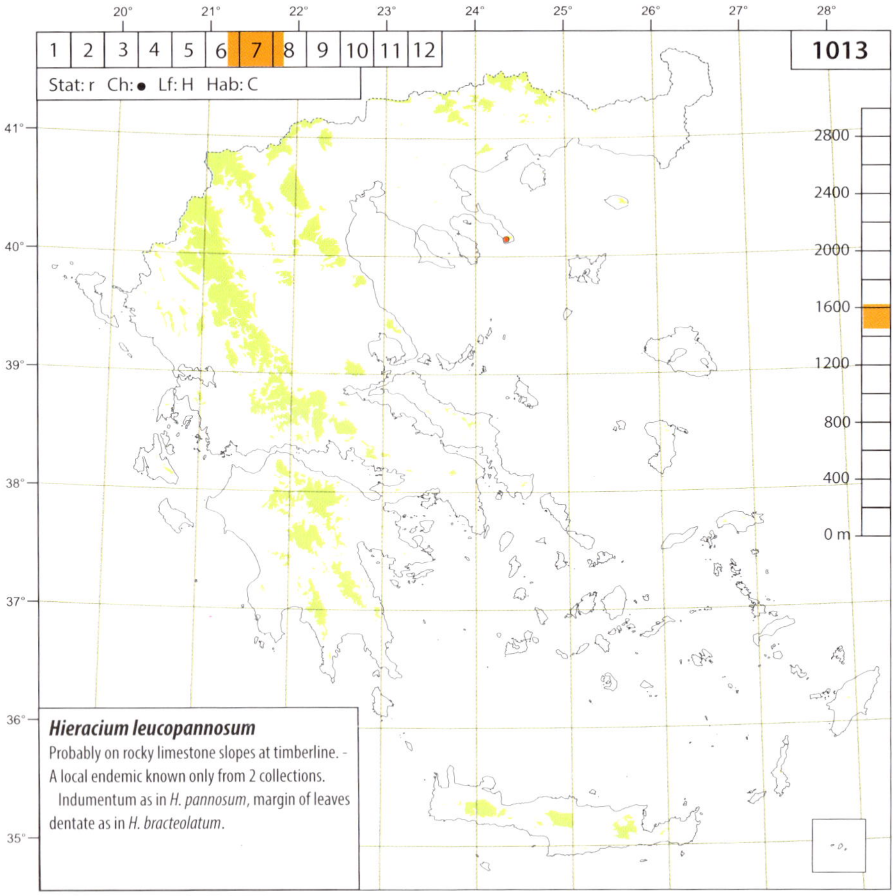

1013

Stat: r Ch: ● Lf: H Hab: C

Hieracium leucopannosum

Probably on rocky limestone slopes at timberline. - A local endemic known only from 2 collections.

Indumentum as in *H. pannosum*, margin of leaves dentate as in *H. bracteolatum*.

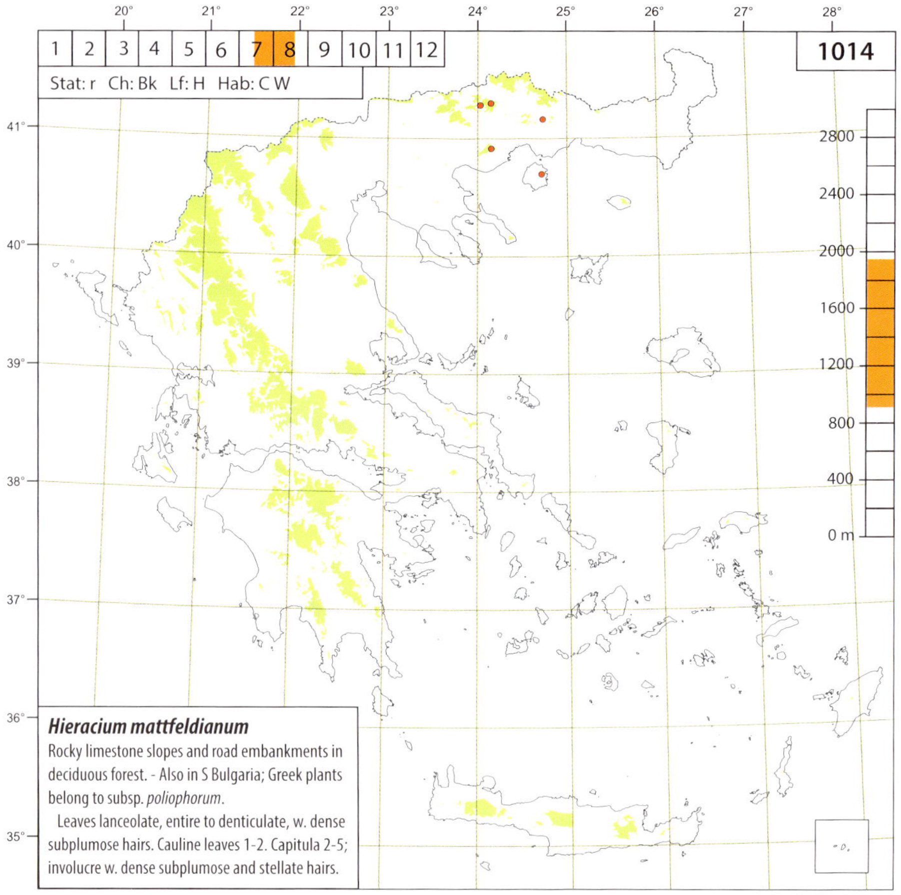

1014

Stat: r Ch: Bk Lf: H Hab: C W

Hieracium mattfeldianum

Rocky limestone slopes and road embankments in deciduous forest. - Also in S Bulgaria; Greek plants belong to subsp. *poliophorum*.

Leaves lanceolate, entire to denticulate, w. dense subplumose hairs. Cauline leaves 1-2. Capitula 2-5; involucre w. dense subplumose and stellate hairs.

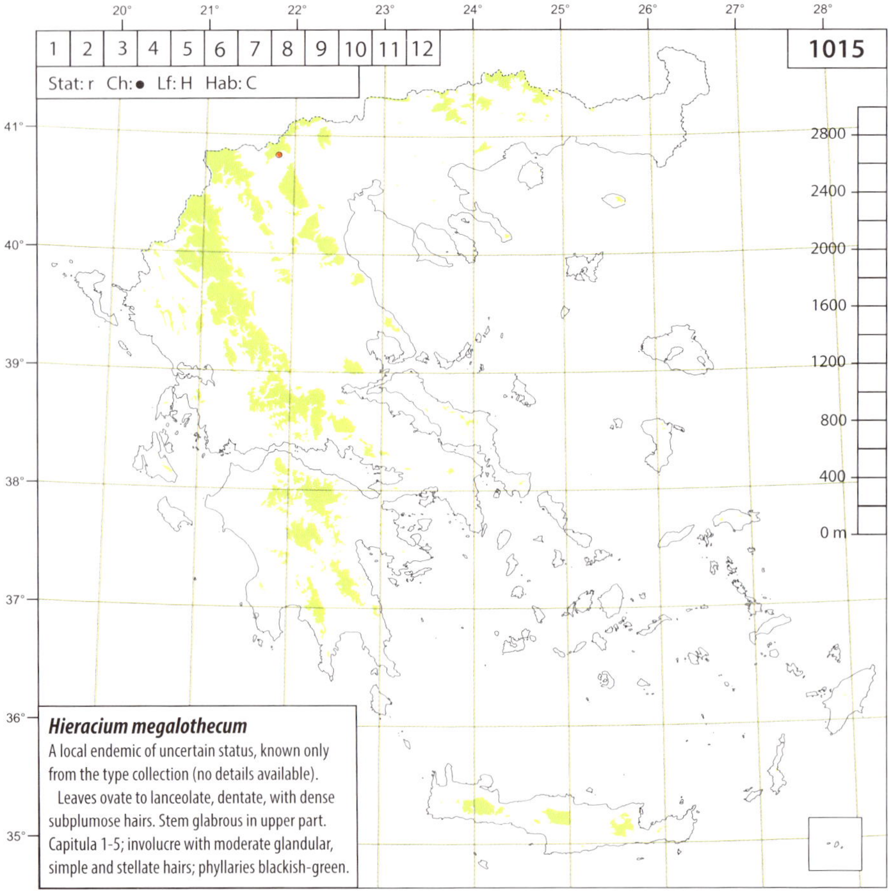

1015

Stat: r Ch: ● Lf: H Hab: C

Hieracium megalothecum

A local endemic of uncertain status, known only from the type collection (no details available).

Leaves ovate to lanceolate, dentate, with dense subplumose hairs. Stem glabrous in upper part. Capitula 1-5; involucre with moderate glandular, simple and stellate hairs; phyllaries blackish-green.

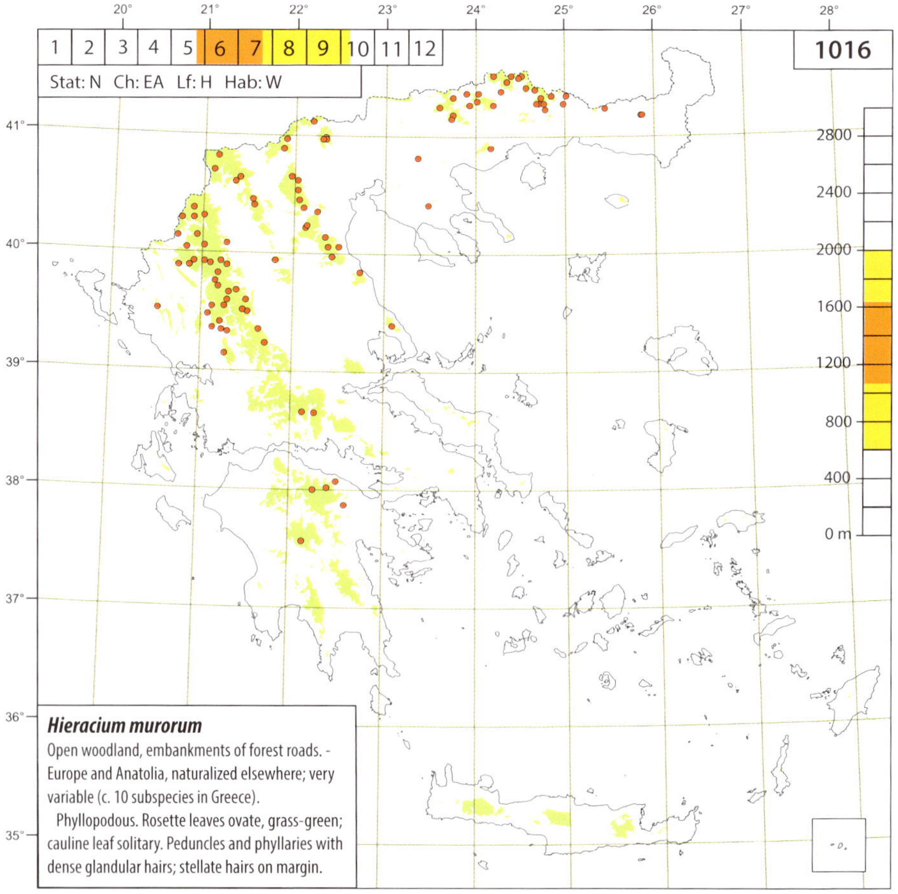

1016

Stat: N Ch: EA Lf: H Hab: W

Hieracium murorum

Open woodland, embankments of forest roads. - Europe and Anatolia, naturalized elsewhere; very variable (c. 10 subspecies in Greece).

Phyllopodous. Rosette leaves ovate, grass-green; cauline leaf solitary. Peduncles and phyllaries with dense glandular hairs; stellate hairs on margin.

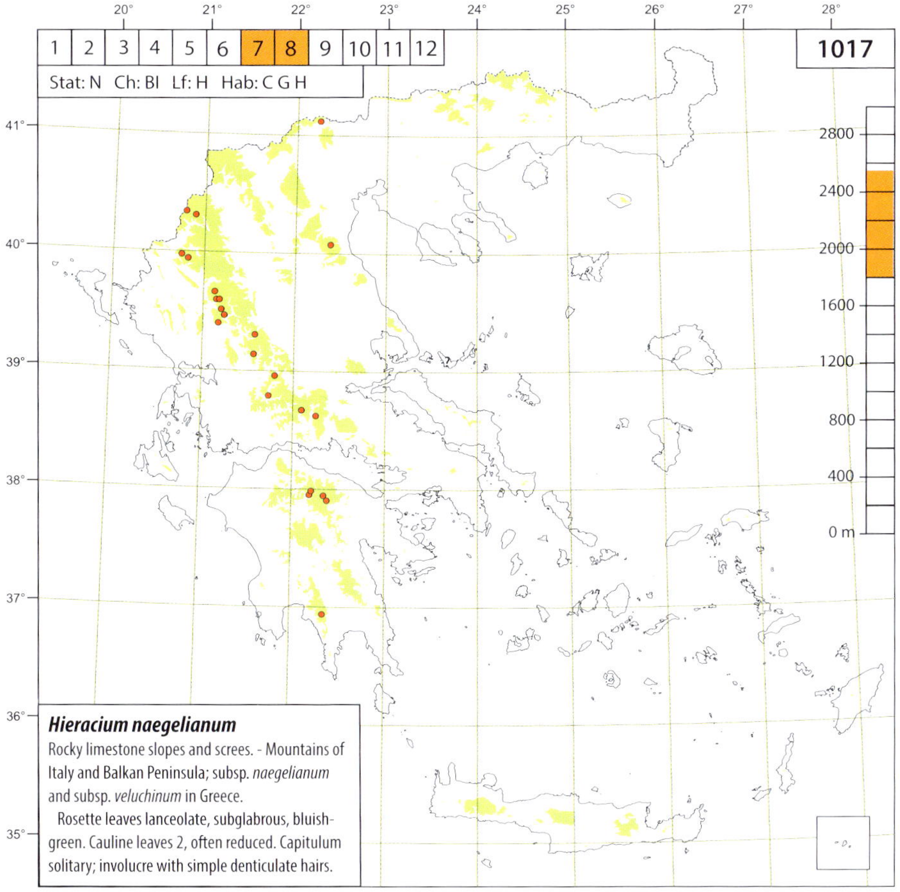

Hieracium naegelianum
Rocky limestone slopes and screes. - Mountains of Italy and Balkan Peninsula; subsp. *naegelianum* and subsp. *veluchinum* in Greece.
Rosette leaves lanceolate, subglabrous, bluish-green. Cauline leaves 2, often reduced. Capitulum solitary; involucre with simple denticulate hairs.

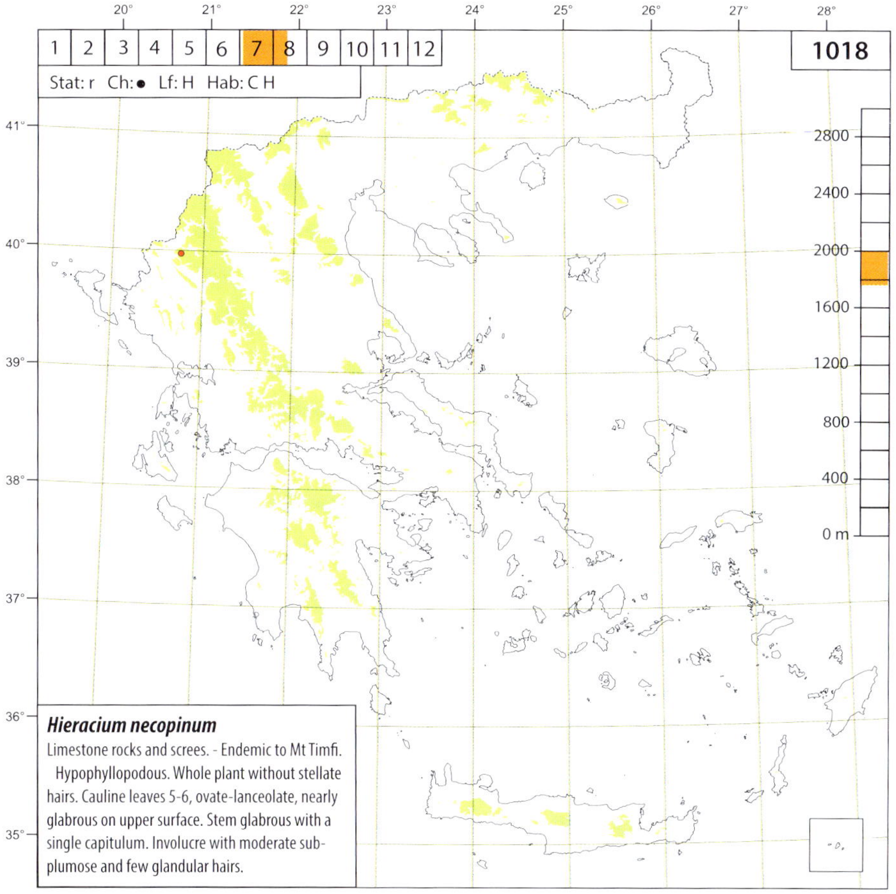

Hieracium necopinum
Limestone rocks and screes. - Endemic to Mt Timfi.
Hypophyllopodous. Whole plant without stellate hairs. Cauline leaves 5-6, ovate-lanceolate, nearly glabrous on upper surface. Stem glabrous with a single capitulum. Involucre with moderate sub-plumose and few glandular hairs.

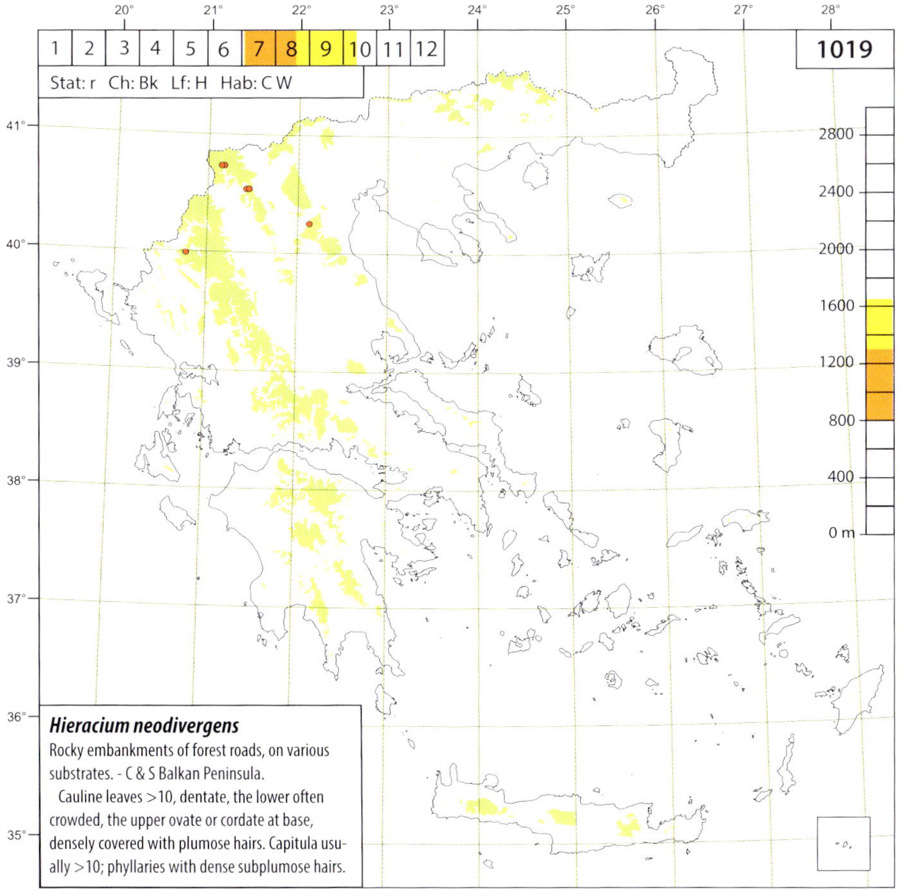

Hieracium neodivergens
Rocky embankments of forest roads, on various substrates. - C & S Balkan Peninsula.
Cauline leaves >10, dentate, the lower often crowded, the upper ovate or cordate at base, densely covered with plumose hairs. Capitula usually >10; phyllaries with dense subplumose hairs.

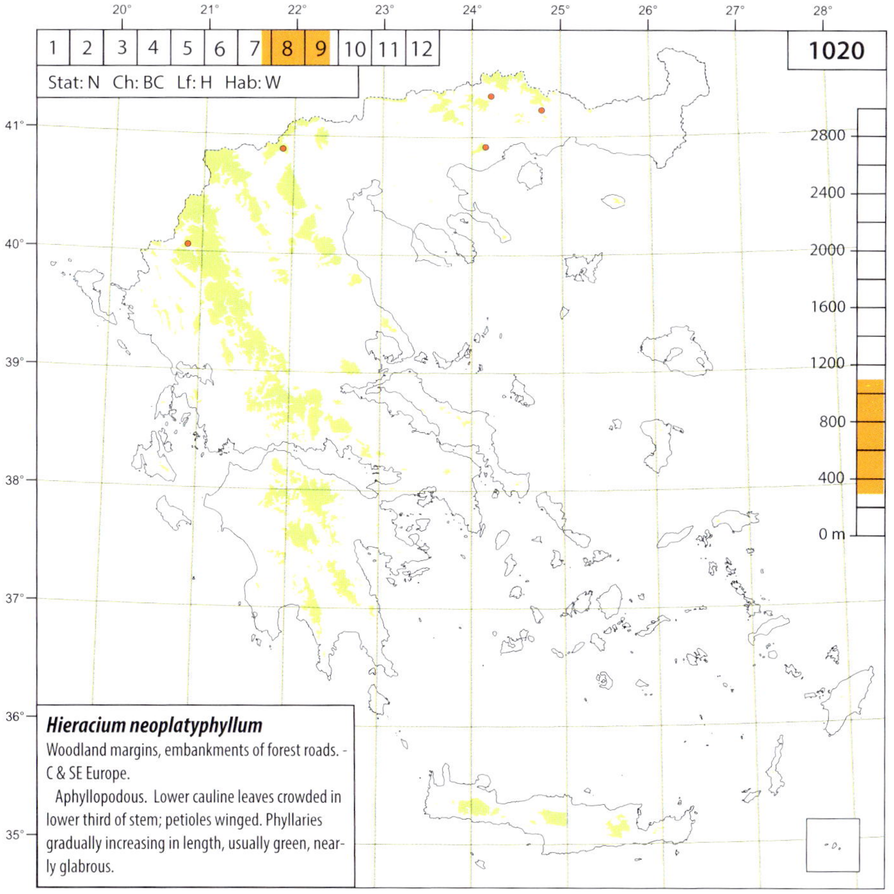

Hieracium neoplatyphyllum
Woodland margins, embankments of forest roads. - C & SE Europe.
Aphyllopodous. Lower cauline leaves crowded in lower third of stem; petioles winged. Phyllaries gradually increasing in length, usually green, nearly glabrous.

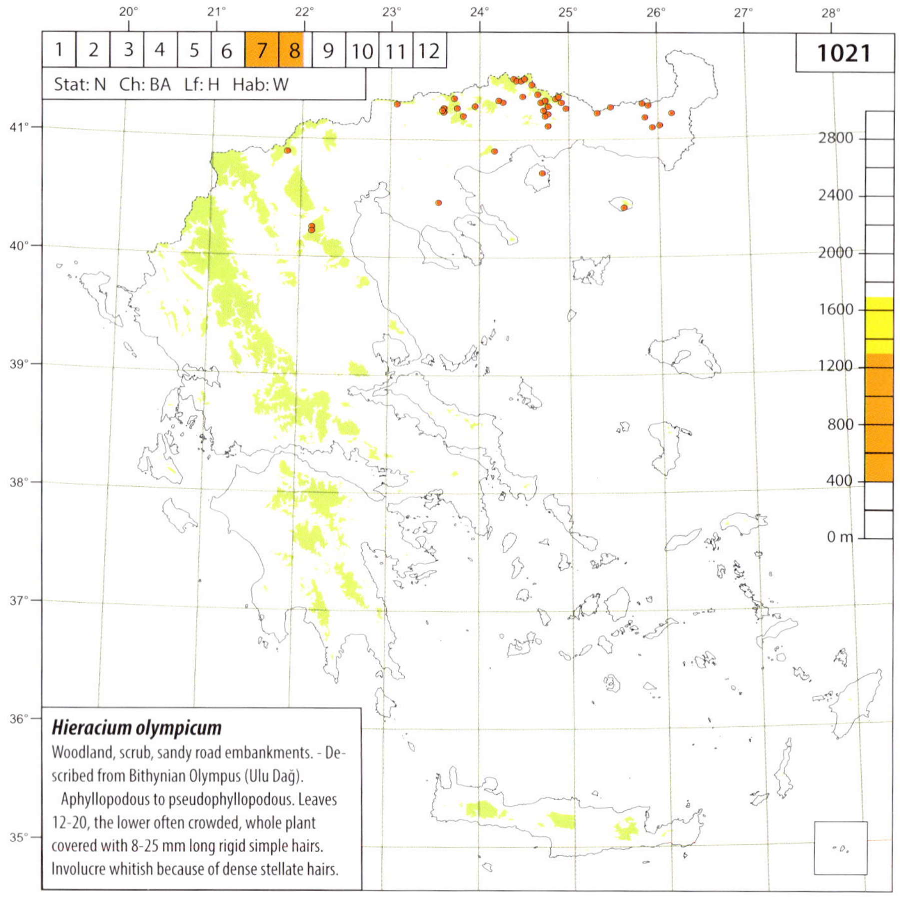

Hieracium olympicum

Woodland, scrub, sandy road embankments. - Described from Bithynian Olympus (Ulu Dağ).

Aphyllopodous to pseudophyllopodous. Leaves 12-20, the lower often crowded, whole plant covered with 8-25 mm long rigid simple hairs. Involucre whitish because of dense stellate hairs.

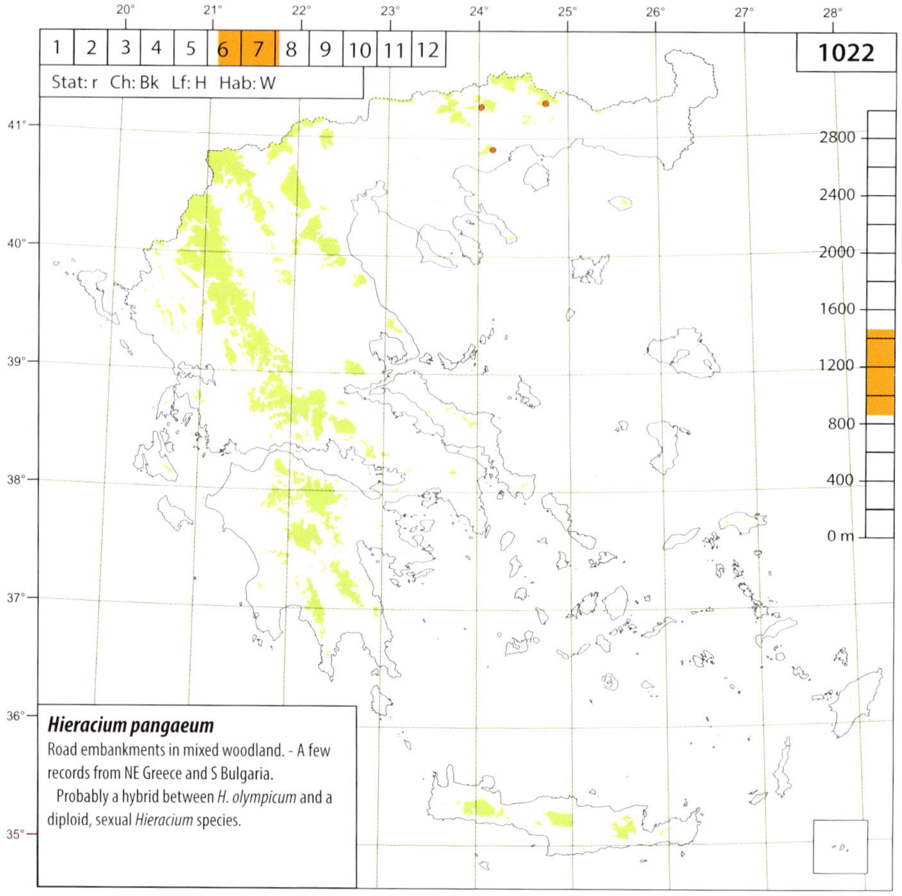

Hieracium pangaeum

Road embankments in mixed woodland. - A few records from NE Greece and S Bulgaria.

Probably a hybrid between *H. olympicum* and a diploid, sexual *Hieracium* species.

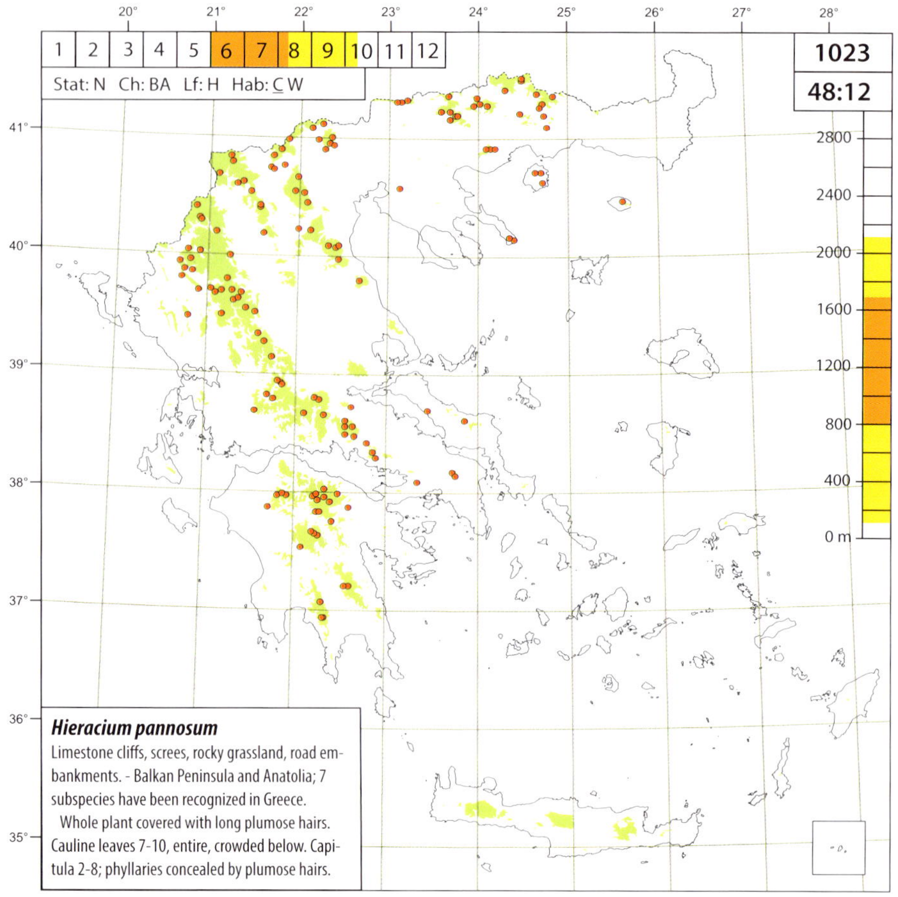

Hieracium pannosum

Limestone cliffs, screes, rocky grassland, road embankments. - Balkan Peninsula and Anatolia; 7 subspecies have been recognized in Greece.

Whole plant covered with long plumose hairs. Cauline leaves 7-10, entire, crowded below. Capitula 2-8; phyllaries concealed by plumose hairs.

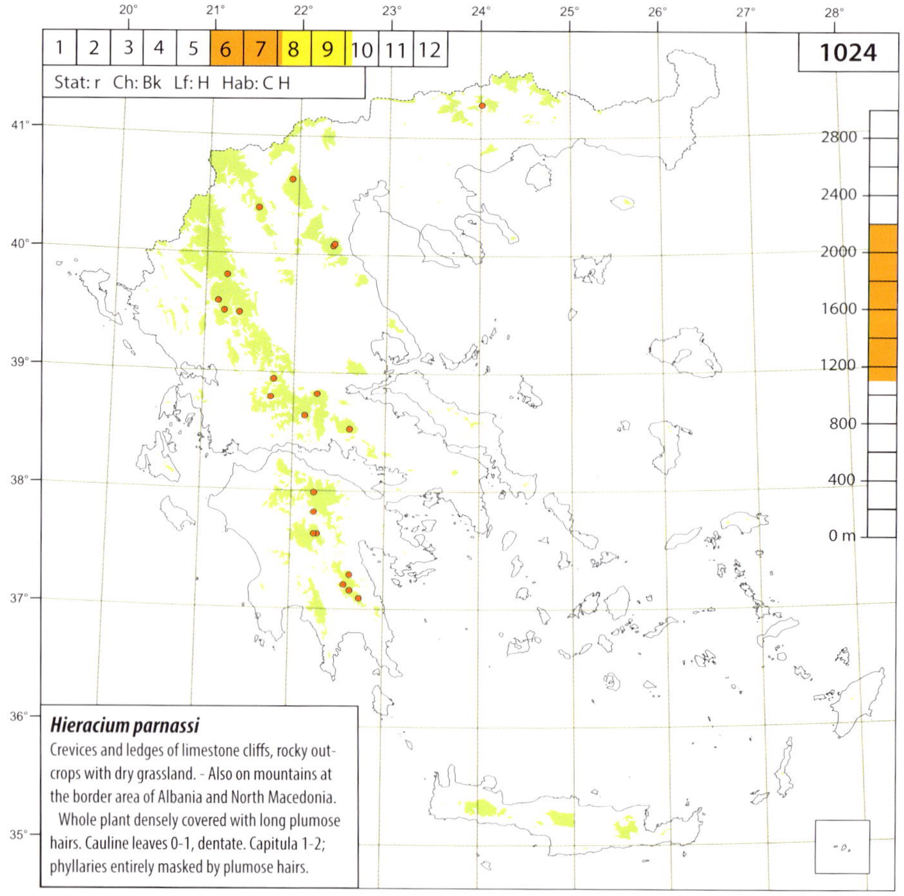

Hieracium parnassi

Crevices and ledges of limestone cliffs, rocky outcrops with dry grassland. - Also on mountains at the border area of Albania and North Macedonia.

Whole plant densely covered with long plumose hairs. Cauline leaves 0-1, dentate. Capitula 1-2; phyllaries entirely masked by plumose hairs.

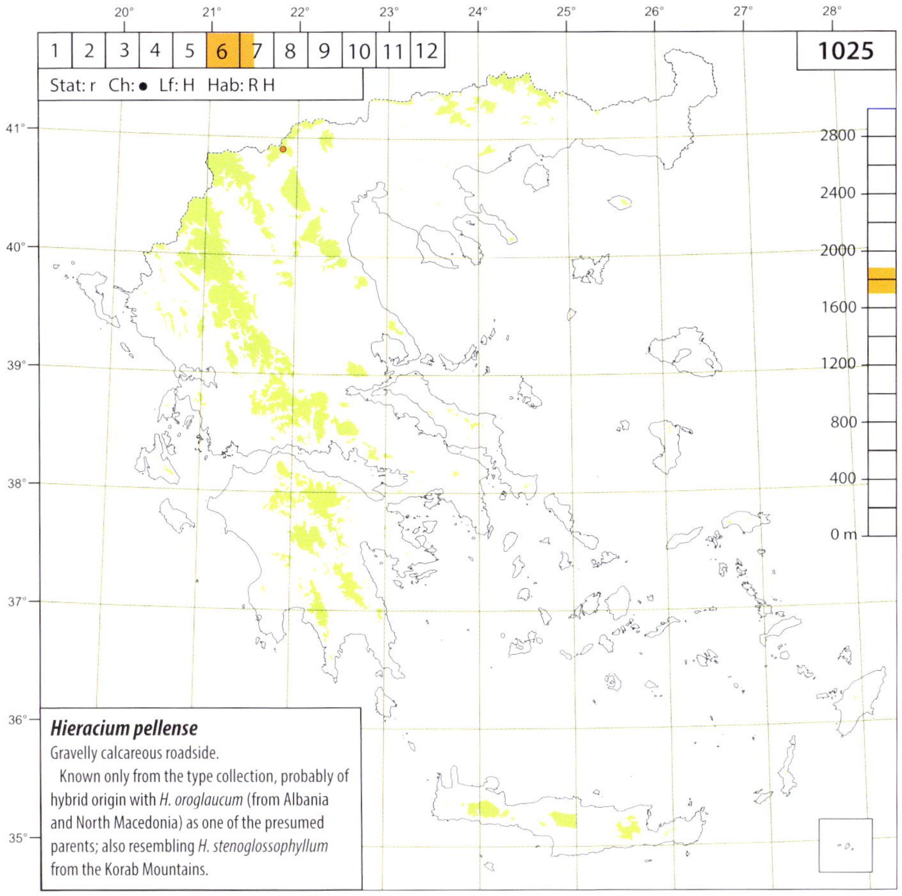

Hieracium pellense

Gravelly calcareous roadside.

Known only from the type collection, probably of hybrid origin with *H. oroglaucum* (from Albania and North Macedonia) as one of the presumed parents; also resembling *H. stenoglossophyllum* from the Korab Mountains.

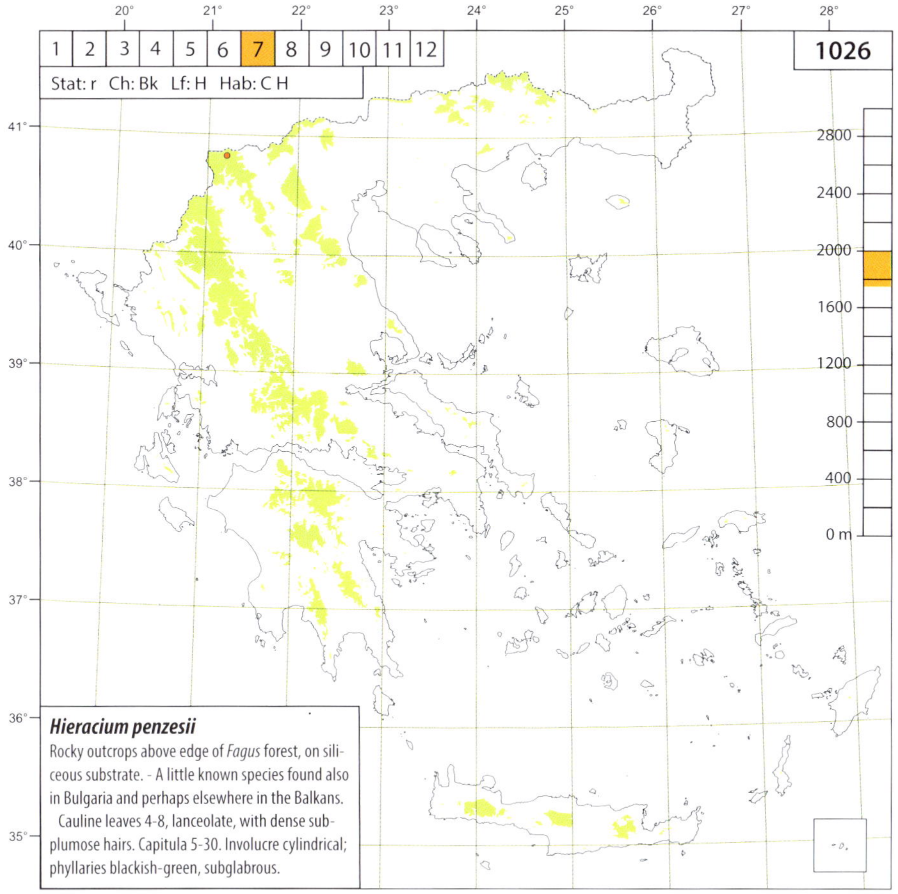

Hieracium penzesii

Rocky outcrops above edge of *Fagus* forest, on siliceous substrate. - A little known species found also in Bulgaria and perhaps elsewhere in the Balkans.

Cauline leaves 4-8, lanceolate, with dense subplumose hairs. Capitula 5-30. Involucre cylindrical; phyllaries blackish-green, subglabrous.

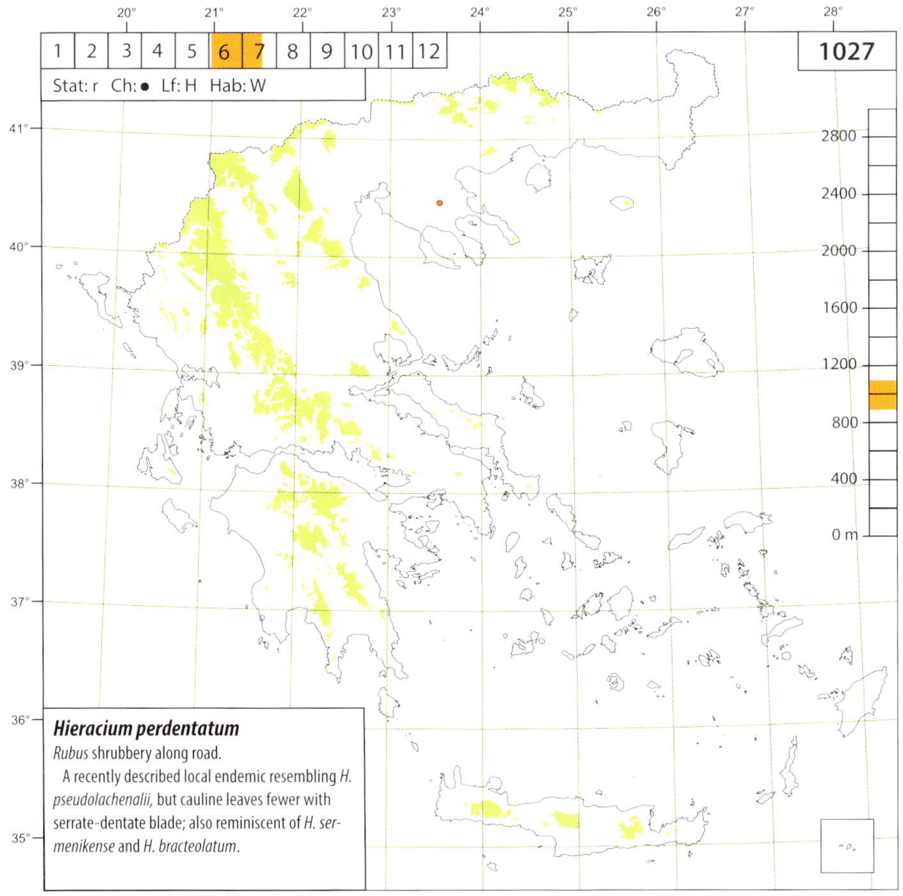

Hieracium perdentatum

Rubus shrubbery along road.

A recently described local endemic resembling *H. pseudolachenalii*, but cauline leaves fewer with serrate-dentate blade; also reminiscent of *H. sermenikense* and *H. bracteolatum*.

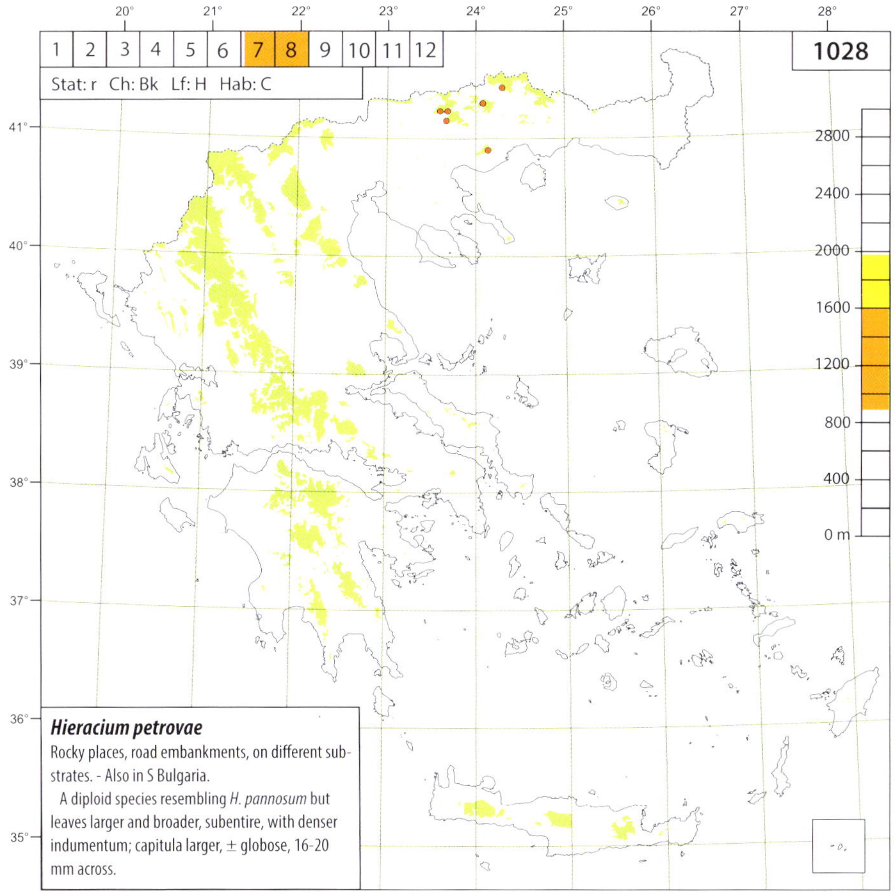

Hieracium petrovae

Rocky places, road embankments, on different substrates. - Also in S Bulgaria.

A diploid species resembling *H. pannosum* but leaves larger and broader, subentire, with denser indumentum; capitula larger, ± globose, 16-20 mm across.

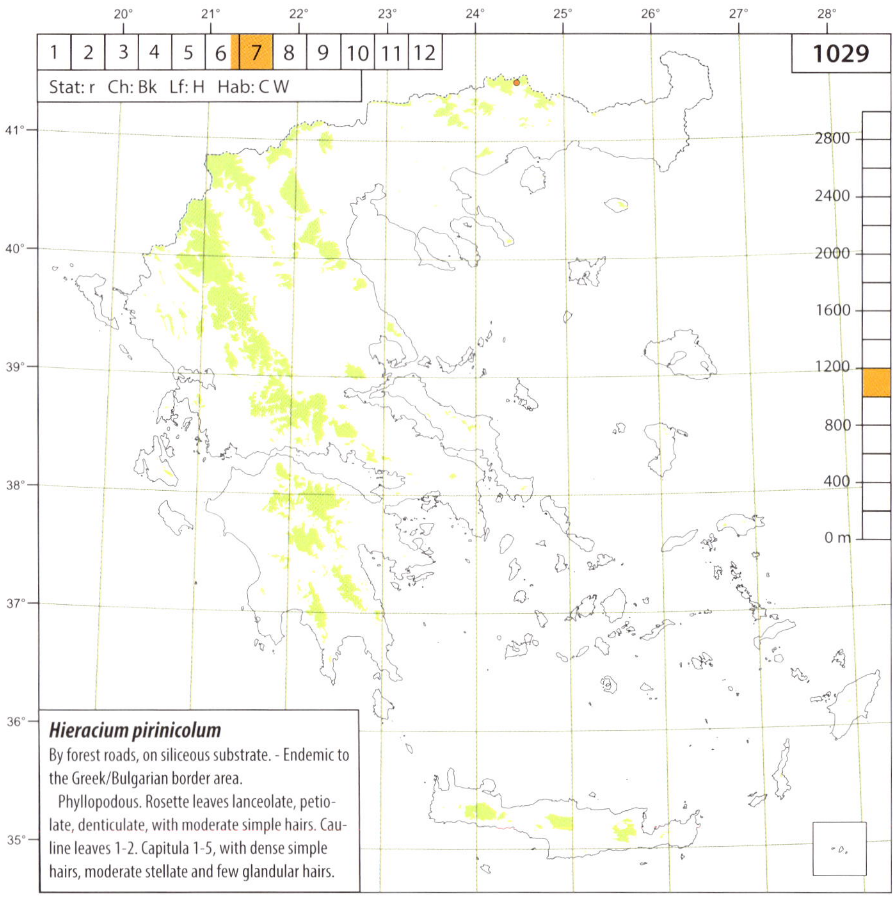

Hieracium pirinicolum

By forest roads, on siliceous substrate. - Endemic to the Greek/Bulgarian border area.

Phyllopodous. Rosette leaves lanceolate, petiolate, denticulate, with moderate simple hairs. Cauline leaves 1-2. Capitula 1-5, with dense simple hairs, moderate stellate and few glandular hairs.

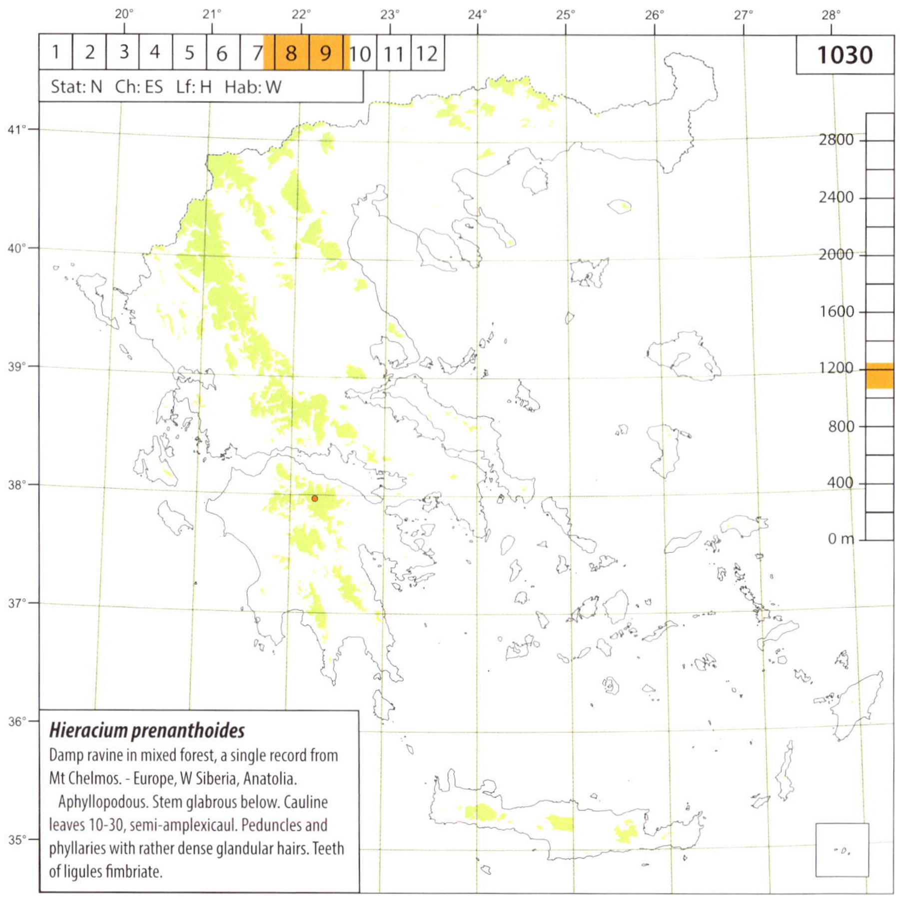

Hieracium prenanthoides

Damp ravine in mixed forest, a single record from Mt Chelmos. - Europe, W Siberia, Anatolia.

Aphyllopodous. Stem glabrous below. Cauline leaves 10-30, semi-amplexicaul. Peduncles and phyllaries with rather dense glandular hairs. Teeth of ligules fimbriate.

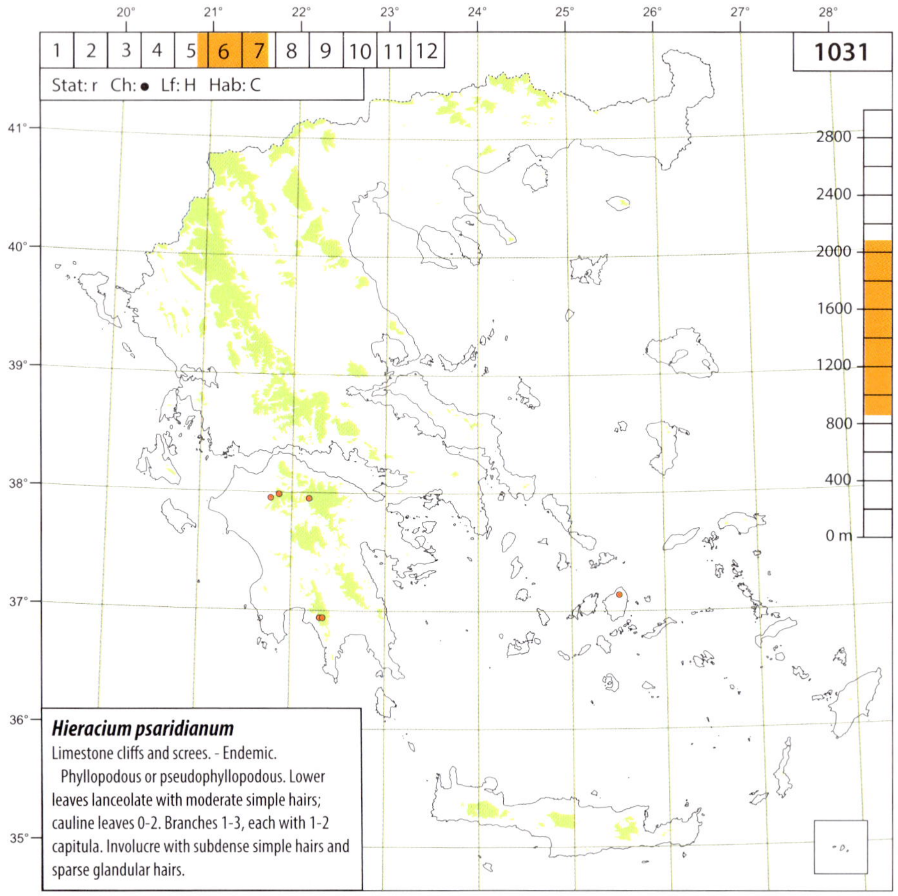

Hieracium psaridianum

Limestone cliffs and screes. - Endemic.

Phyllopodous or pseudophyllopodous. Lower leaves lanceolate with moderate simple hairs; cauline leaves 0-2. Branches 1-3, each with 1-2 capitula. Involucre with subdense simple hairs and sparse glandular hairs.

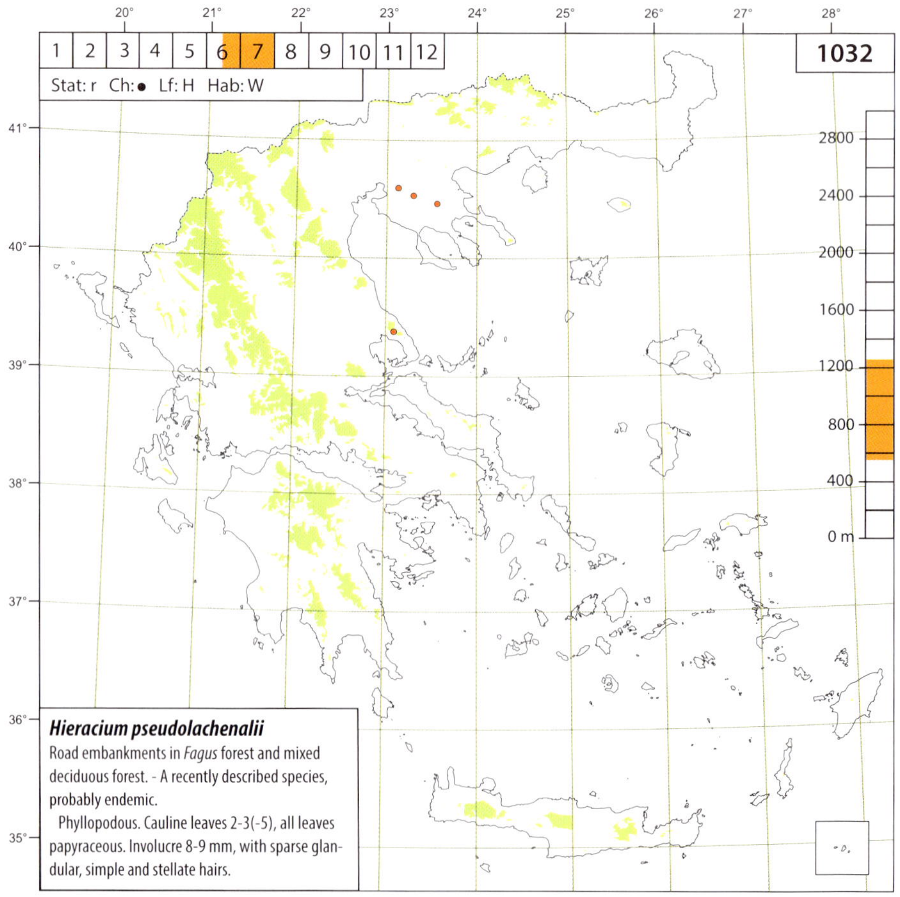

Hieracium pseudolachenalii

Road embankments in *Fagus* forest and mixed deciduous forest. - A recently described species, probably endemic.

Phyllopodous. Cauline leaves 2-3(-5), all leaves papyraceous. Involucre 8-9 mm, with sparse glandular, simple and stellate hairs.

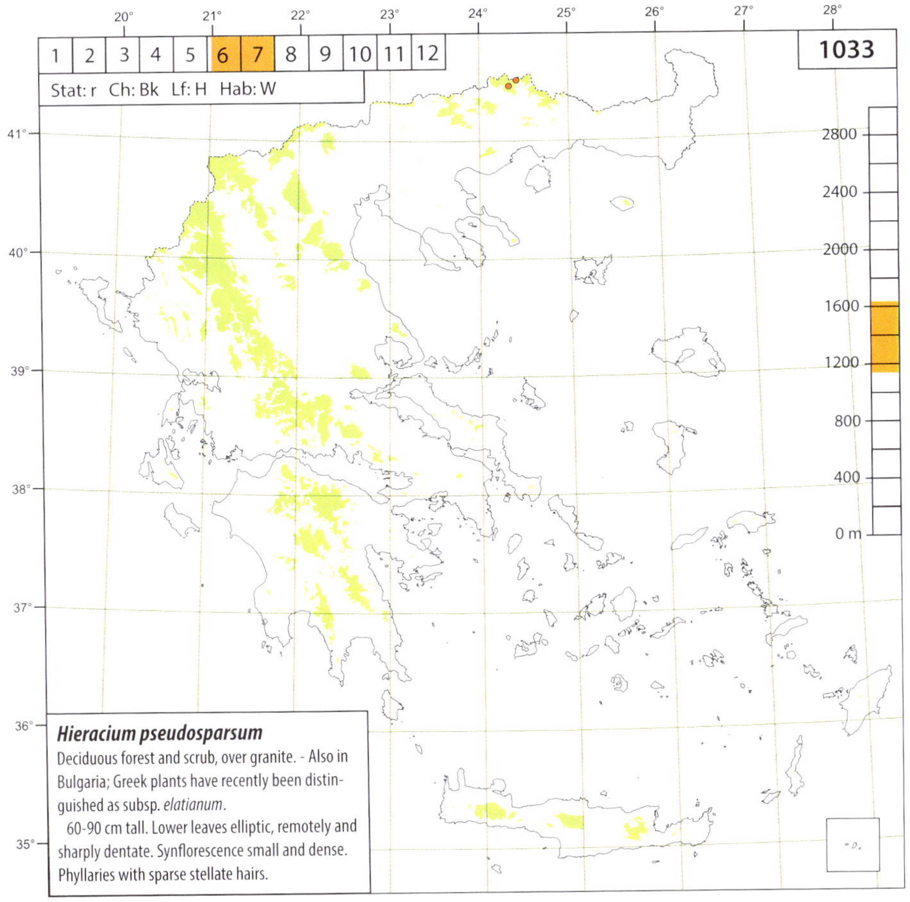

Hieracium pseudosparsum

Deciduous forest and scrub, over granite. - Also in Bulgaria; Greek plants have recently been distinguished as subsp. *elatianum*.

60-90 cm tall. Lower leaves elliptic, remotely and sharply dentate. Synflorescence small and dense. Phyllaries with sparse stellate hairs.

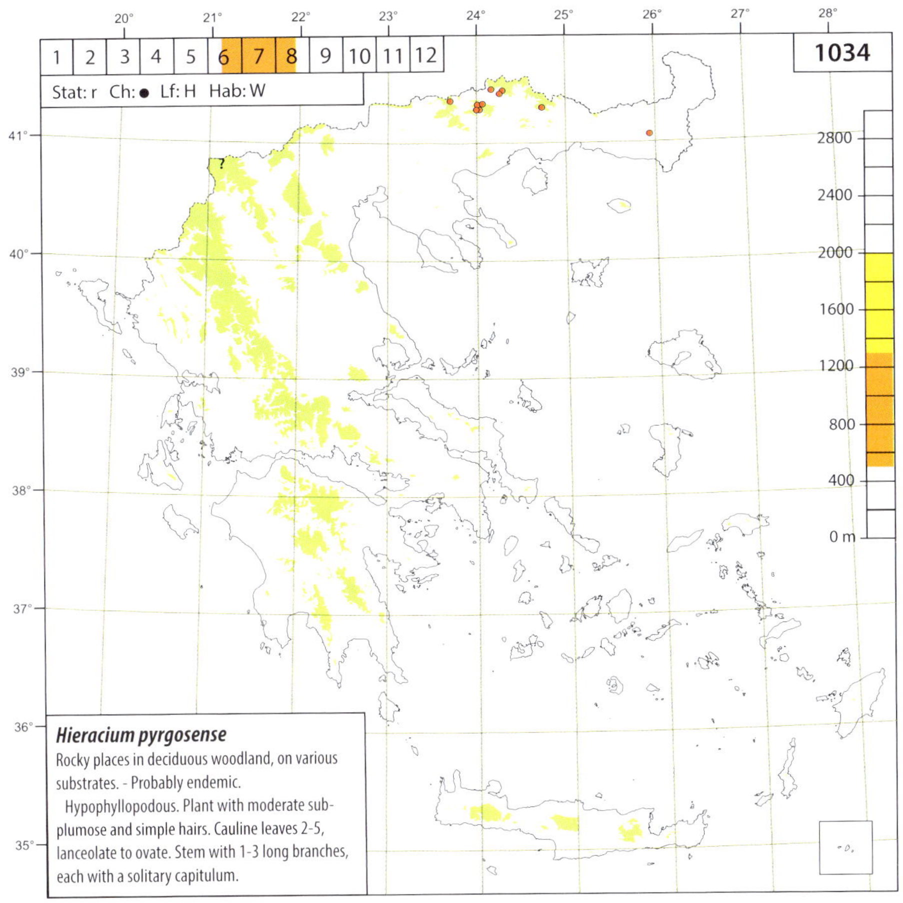

Hieracium pyrgosense

Rocky places in deciduous woodland, on various substrates. - Probably endemic.

Hypophyllopodous. Plant with moderate subplumose and simple hairs. Cauline leaves 2-5, lanceolate to ovate. Stem with 1-3 long branches, each with a solitary capitulum.

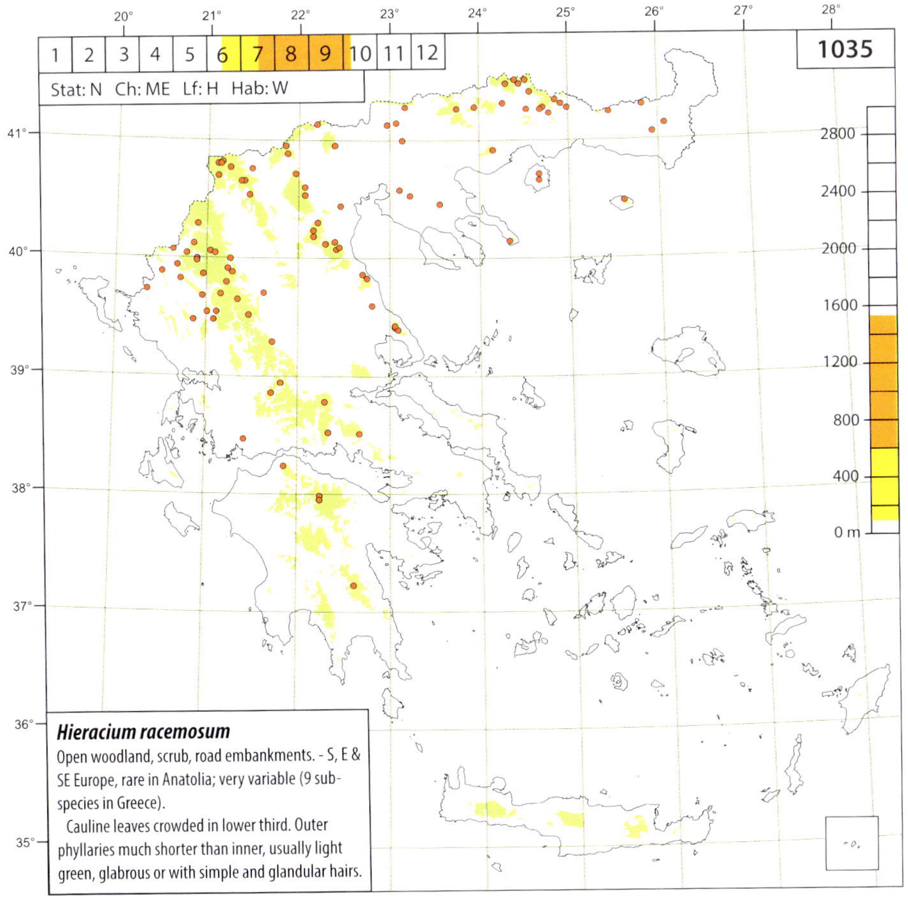

Hieracium racemosum

Open woodland, scrub, road embankments. - S, E & SE Europe, rare in Anatolia; very variable (9 subspecies in Greece).

Cauline leaves crowded in lower third. Outer phyllaries much shorter than inner, usually light green, glabrous or with simple and glandular hairs.

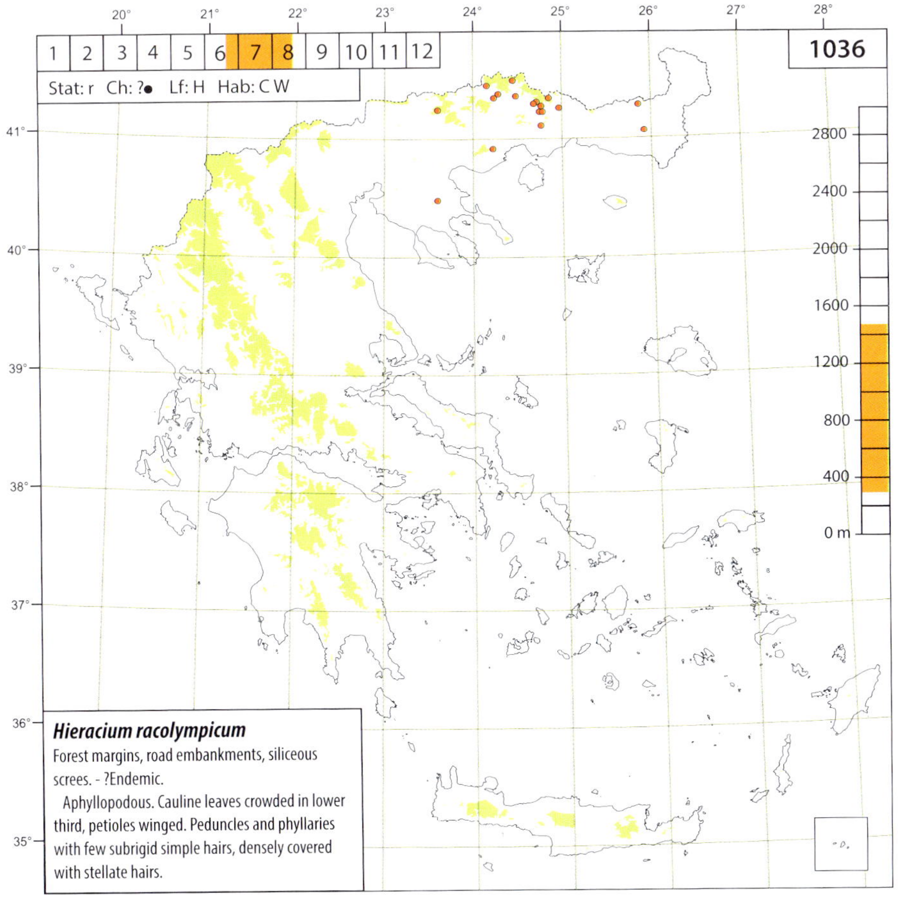

Hieracium racolympicum

Forest margins, road embankments, siliceous screes. - ?Endemic.

Aphyllopodous. Cauline leaves crowded in lower third, petioles winged. Peduncles and phyllaries with few subrigid simple hairs, densely covered with stellate hairs.

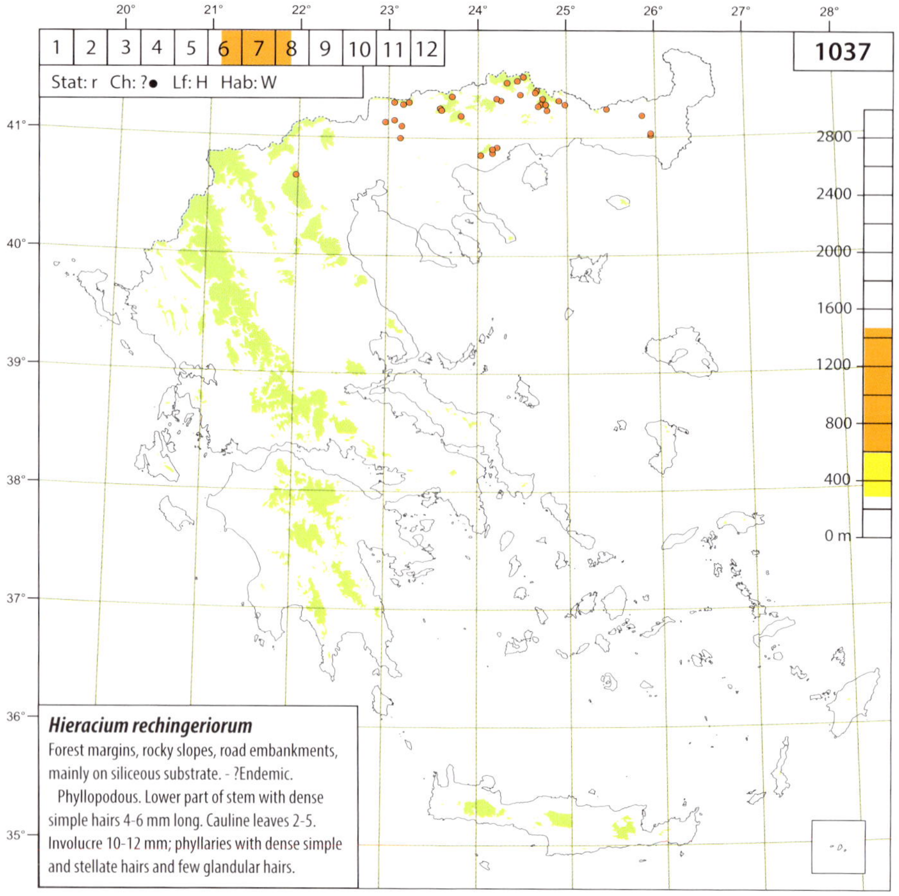

Hieracium rechingeriorum
Forest margins, rocky slopes, road embankments, mainly on siliceous substrate. - ?Endemic.
Phyllopodous. Lower part of stem with dense simple hairs 4-6 mm long. Cauline leaves 2-5. Involucre 10-12 mm; phyllaries with dense simple and stellate hairs and few glandular hairs.

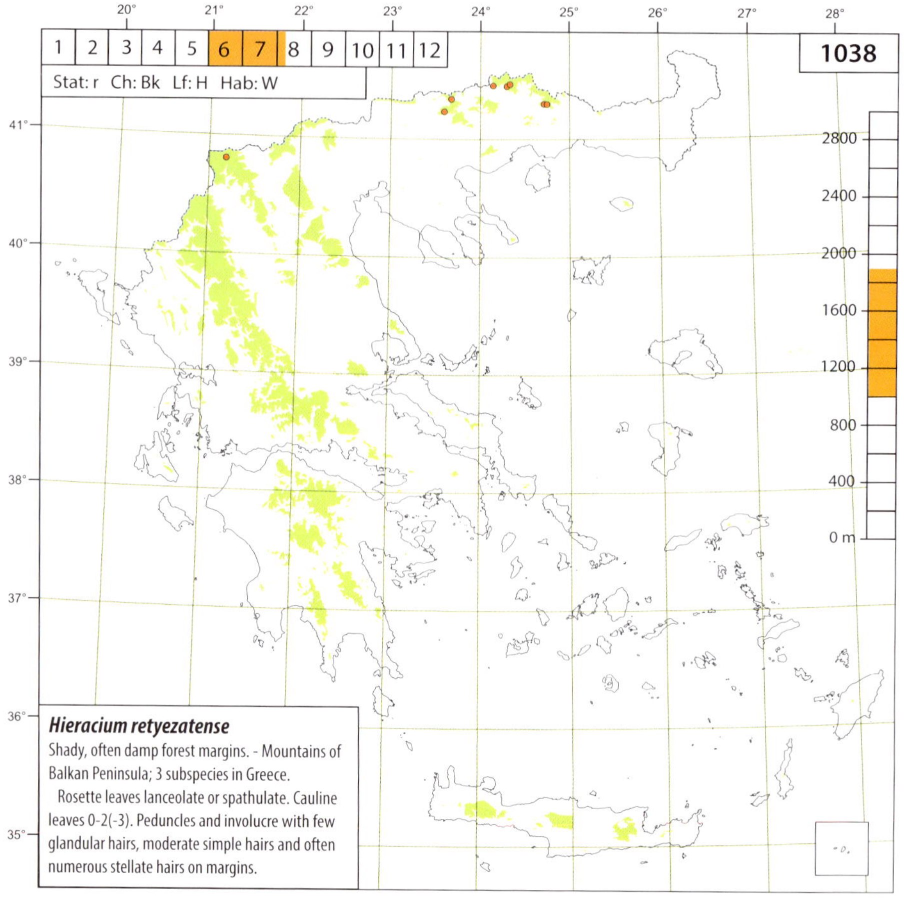

Hieracium retyezatense
Shady, often damp forest margins. - Mountains of Balkan Peninsula; 3 subspecies in Greece.
Rosette leaves lanceolate or spathulate. Cauline leaves 0-2(-3). Peduncles and involucre with few glandular hairs, moderate simple hairs and often numerous stellate hairs on margins.

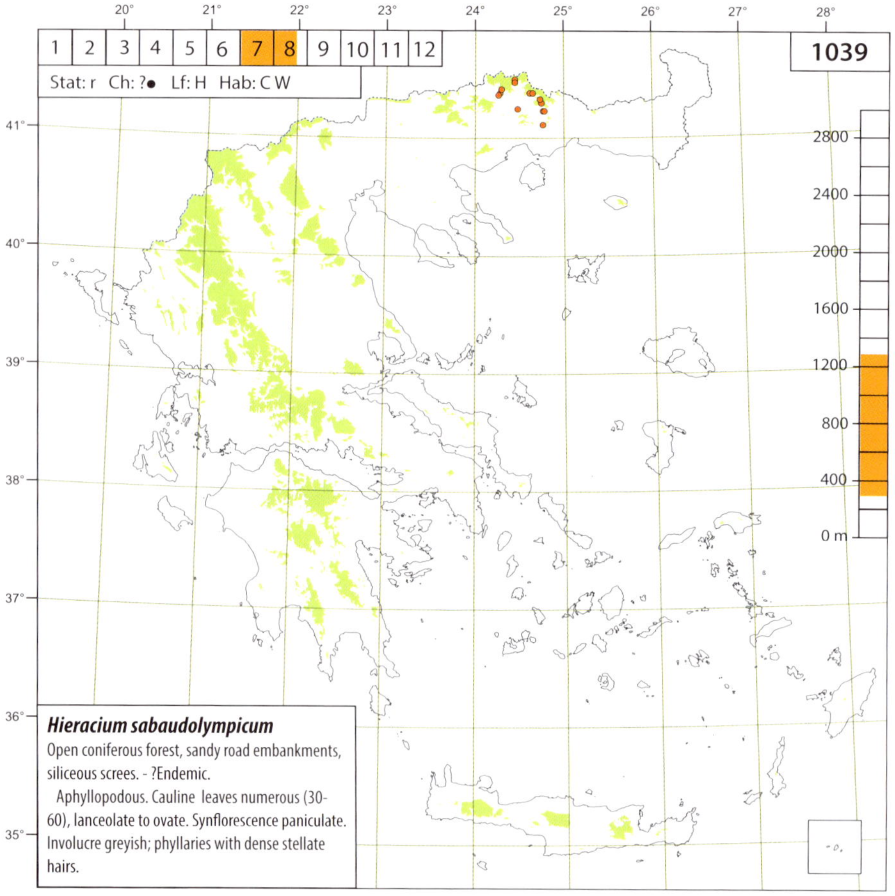

Hieracium sabaudolympicum
Open coniferous forest, sandy road embankments, siliceous screes. - ?Endemic.
Aphyllopodous. Cauline leaves numerous (30-60), lanceolate to ovate. Synflorescence paniculate. Involucre greyish; phyllaries with dense stellate hairs.

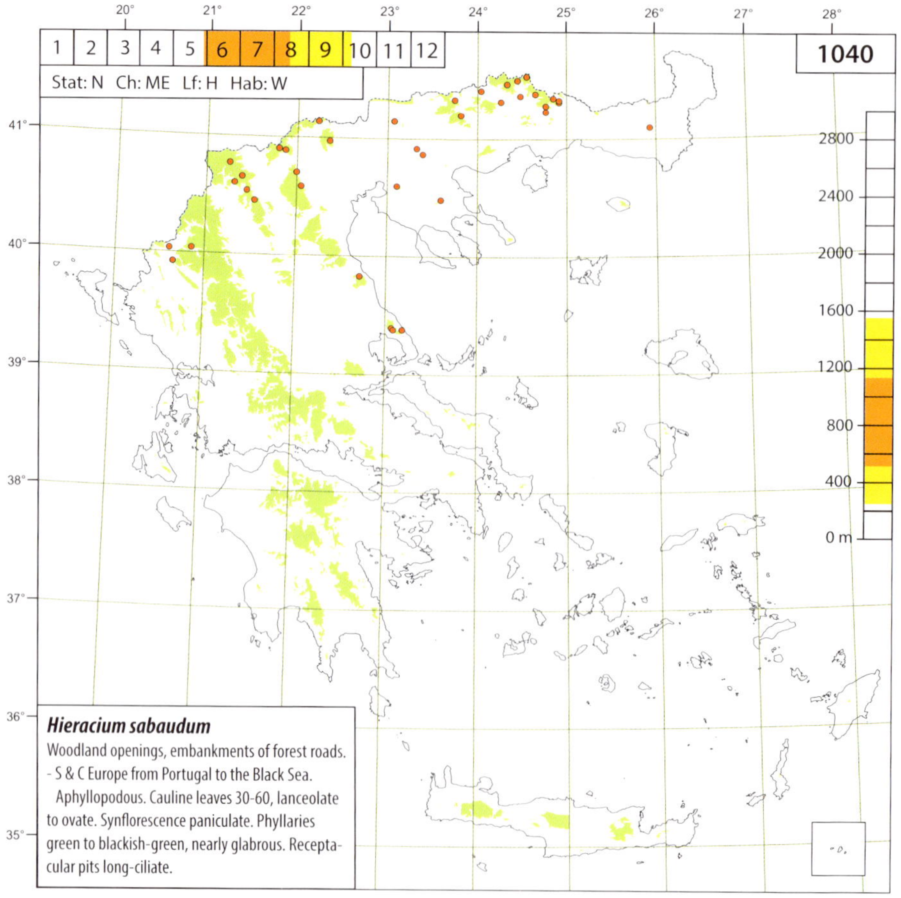

Hieracium sabaudum
Woodland openings, embankments of forest roads. - S & C Europe from Portugal to the Black Sea.
Aphyllopodous. Cauline leaves 30-60, lanceolate to ovate. Synflorescence paniculate. Phyllaries green to blackish-green, nearly glabrous. Receptacular pits long-ciliate.

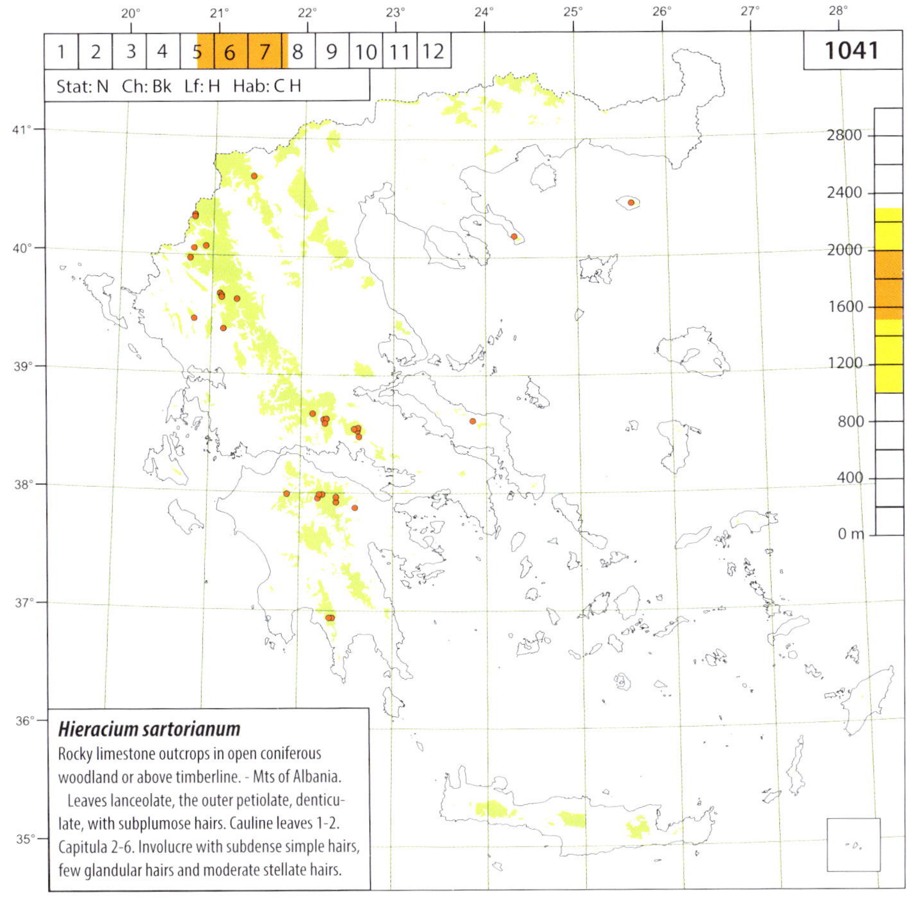

Hieracium sartorianum

Rocky limestone outcrops in open coniferous woodland or above timberline. - Mts of Albania.

Leaves lanceolate, the outer petiolate, denticulate, with subplumose hairs. Cauline leaves 1-2. Capitula 2-6. Involucre with subdense simple hairs, few glandular hairs and moderate stellate hairs.

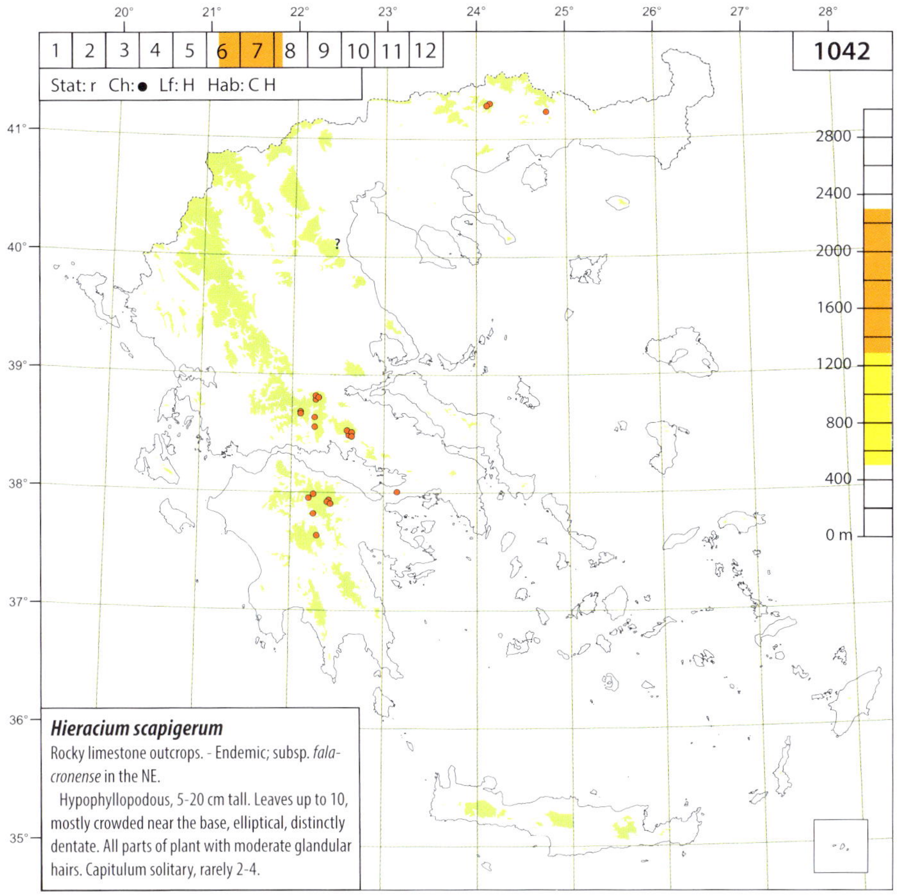

Hieracium scapigerum

Rocky limestone outcrops. - Endemic; subsp. *falacronense* in the NE.

Hypophyllopodous, 5-20 cm tall. Leaves up to 10, mostly crowded near the base, elliptical, distinctly dentate. All parts of plant with moderate glandular hairs. Capitulum solitary, rarely 2-4.

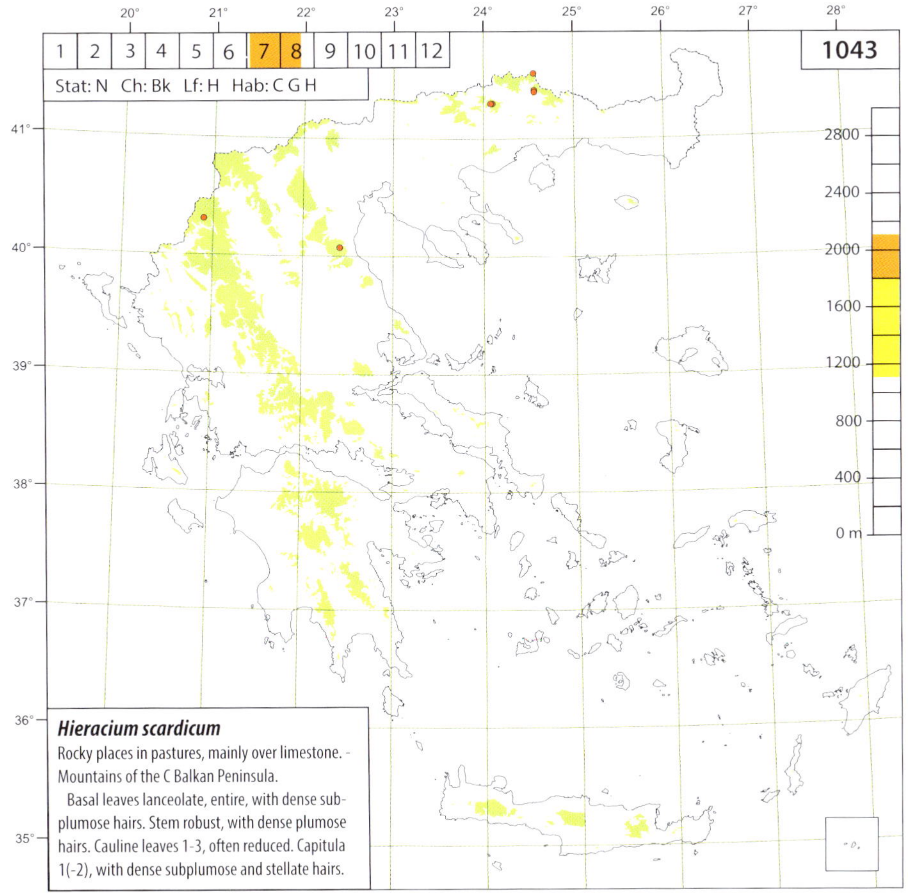

Hieracium scardicum

Rocky places in pastures, mainly over limestone. - Mountains of the C Balkan Peninsula.

Basal leaves lanceolate, entire, with dense subplumose hairs. Stem robust, with dense plumose hairs. Cauline leaves 1-3, often reduced. Capitula 1(-2), with dense subplumose and stellate hairs.

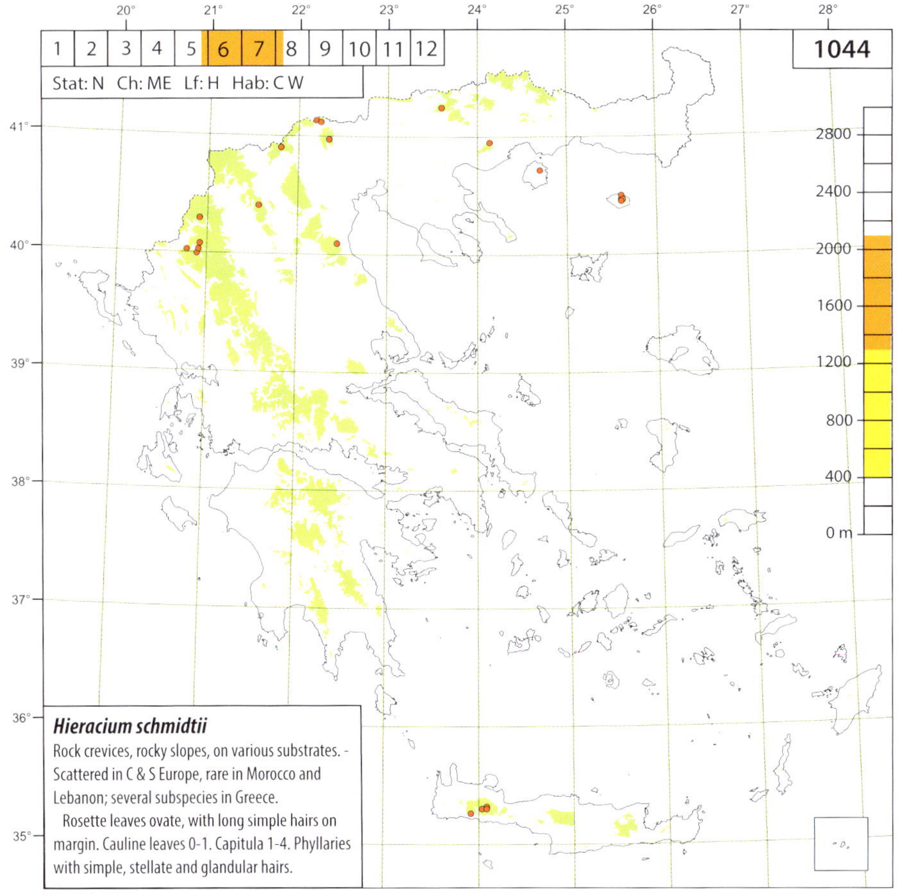

Hieracium schmidtii

Rock crevices, rocky slopes, on various substrates. - Scattered in C & S Europe, rare in Morocco and Lebanon; several subspecies in Greece.

Rosette leaves ovate, with long simple hairs on margin. Cauline leaves 0-1. Capitula 1-4. Phyllaries with simple, stellate and glandular hairs.

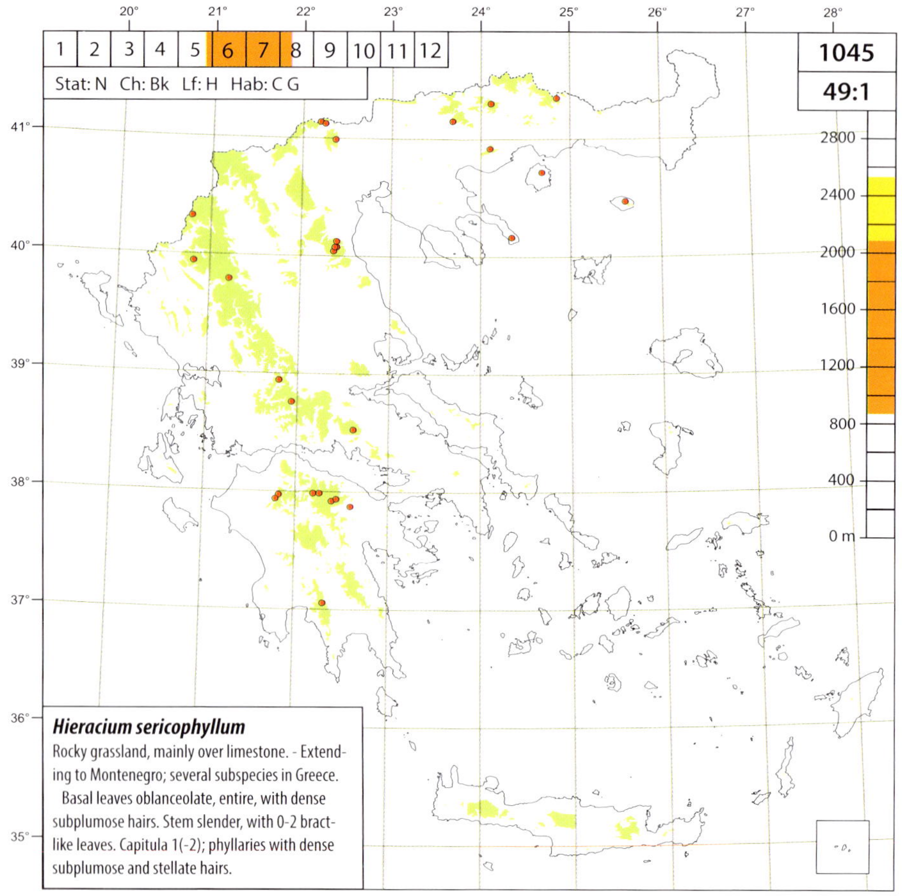

Hieracium sericophyllum

Rocky grassland, mainly over limestone. - Extending to Montenegro; several subspecies in Greece.

Basal leaves oblanceolate, entire, with dense subplumose hairs. Stem slender, with 0-2 bract-like leaves. Capitula 1(-2); phyllaries with dense subplumose and stellate hairs.

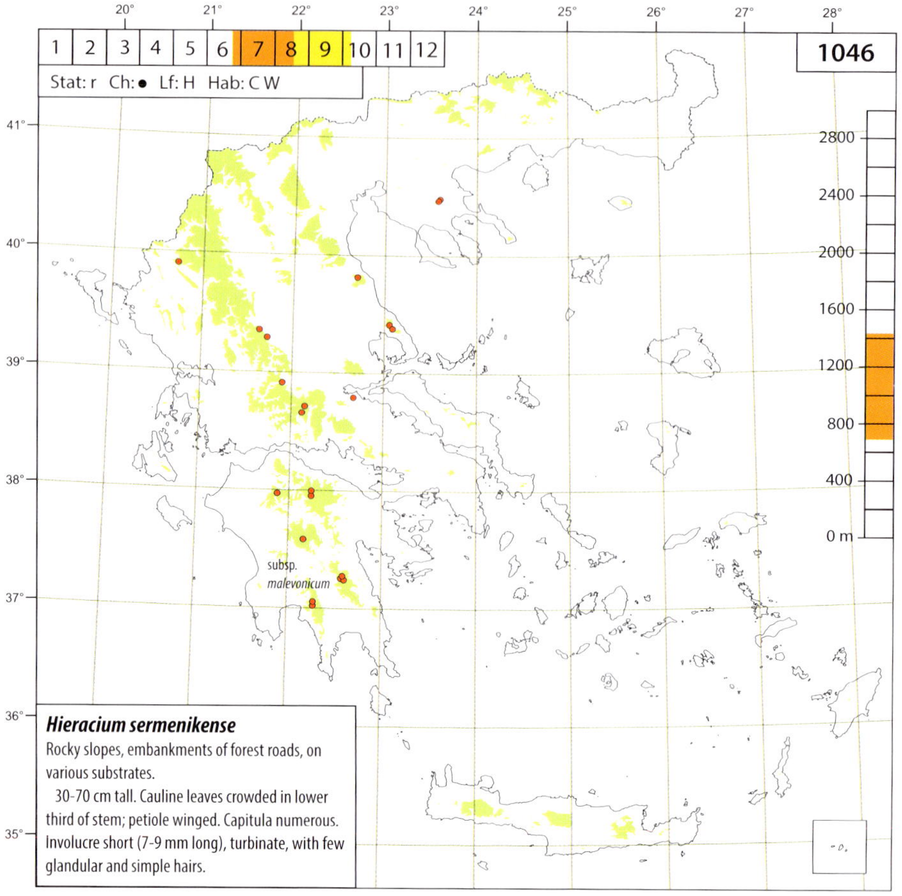

Hieracium sermenikense

Rocky slopes, embankments of forest roads, on various substrates.

30-70 cm tall. Cauline leaves crowded in lower third of stem; petiole winged. Capitula numerous. Involucre short (7-9 mm long), turbinate, with few glandular and simple hairs.

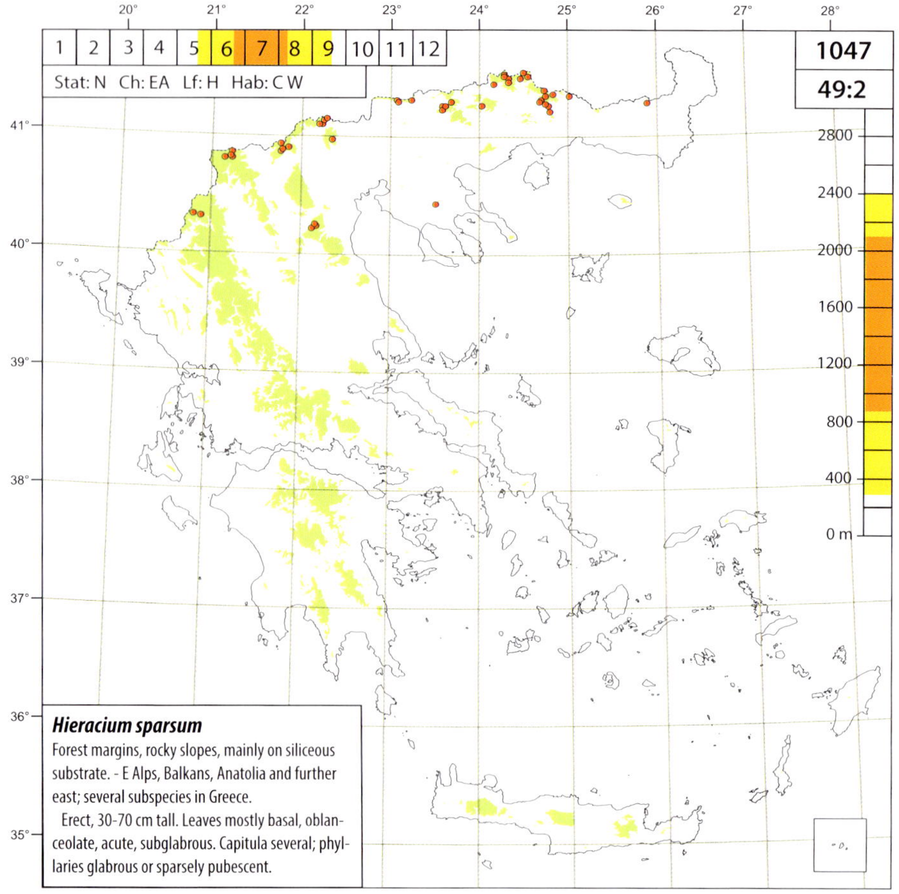

Hieracium sparsum

Forest margins, rocky slopes, mainly on siliceous substrate. - E Alps, Balkans, Anatolia and further east; several subspecies in Greece.

Erect, 30-70 cm tall. Leaves mostly basal, oblanceolate, acute, subglabrous. Capitula several; phyllaries glabrous or sparsely pubescent.

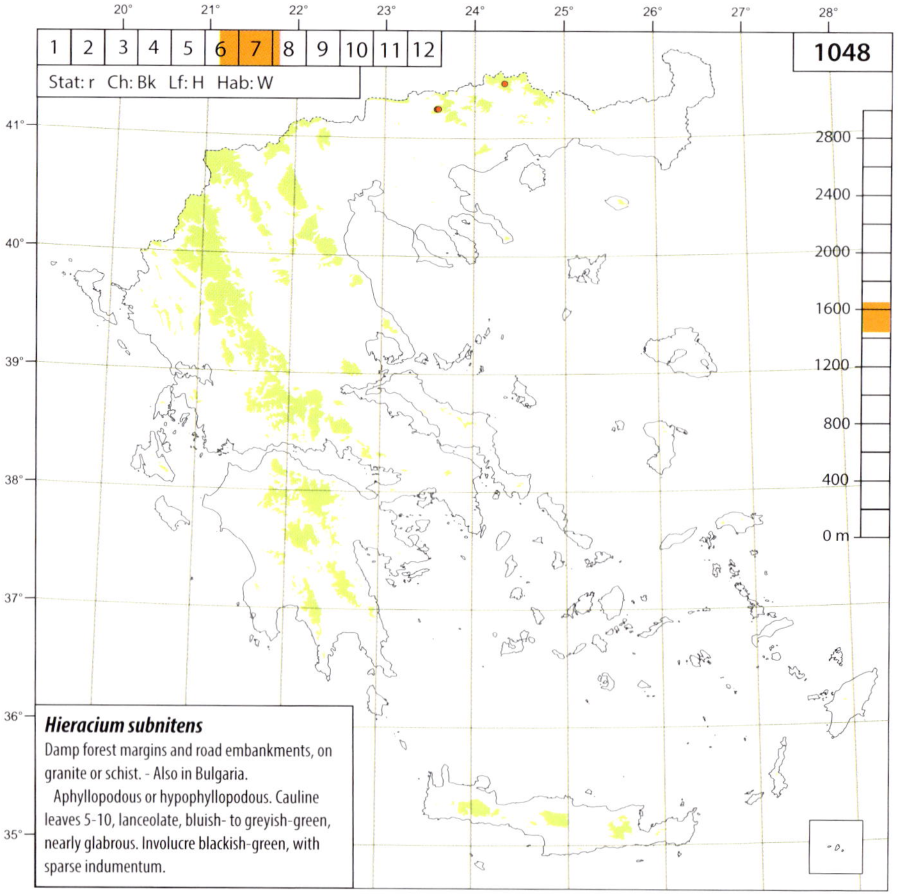

Hieracium subnitens

Damp forest margins and road embankments, on granite or schist. - Also in Bulgaria.

Aphyllopodous or hypophyllopodous. Cauline leaves 5-10, lanceolate, bluish- to greyish-green, nearly glabrous. Involucre blackish-green, with sparse indumentum.

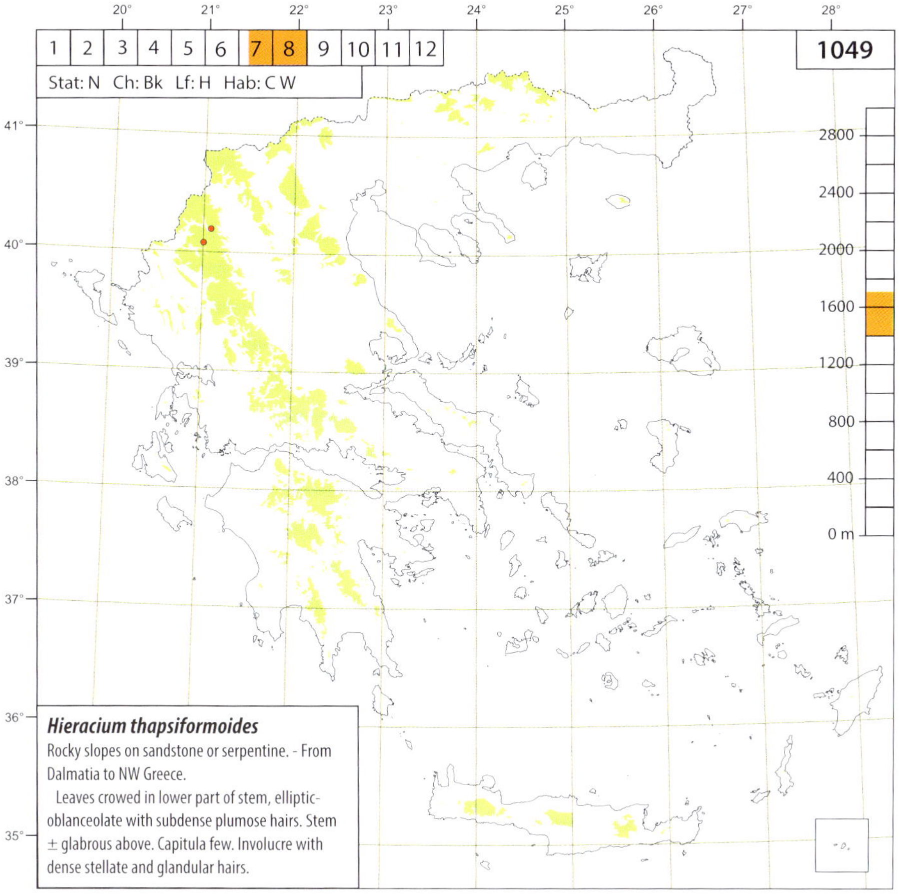

1049

Stat: N Ch: Bk Lf: H Hab: C W

Hieracium thapsiformoides

Rocky slopes on sandstone or serpentine. - From Dalmatia to NW Greece.

Leaves crowed in lower part of stem, elliptic-oblanceolate with subdense plumose hairs. Stem ± glabrous above. Capitula few. Involucre with dense stellate and glandular hairs.

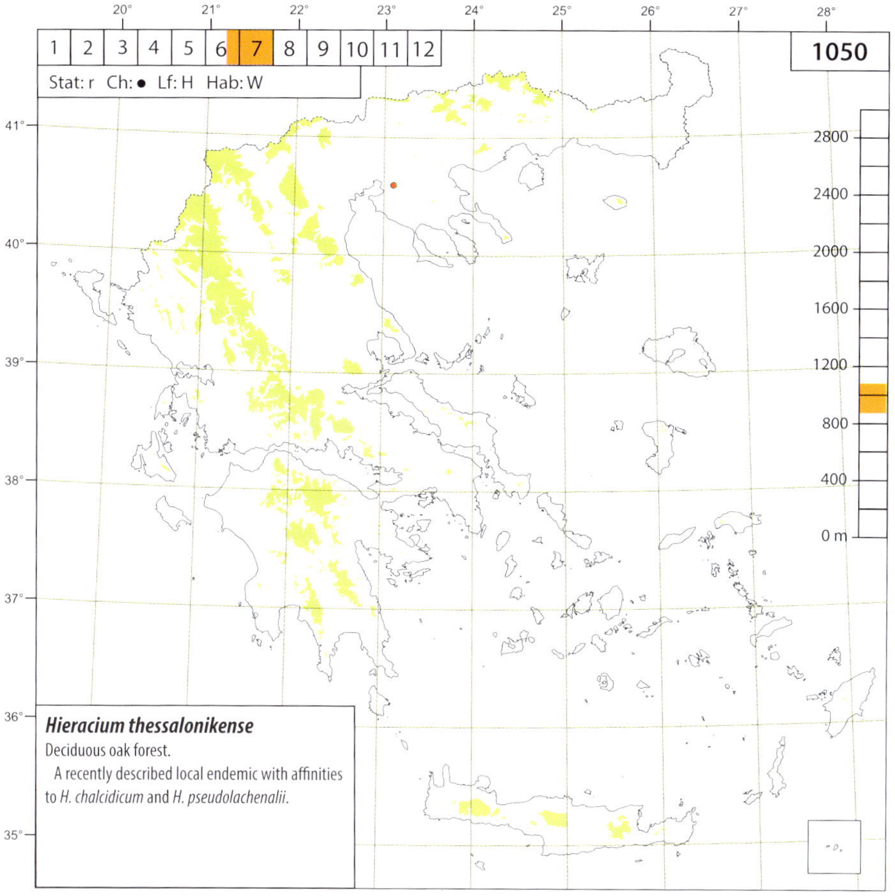

1050

Stat: r Ch: ● Lf: H Hab: W

Hieracium thessalonikense

Deciduous oak forest.

A recently described local endemic with affinities to *H. chalcidicum* and *H. pseudolachenalii*.

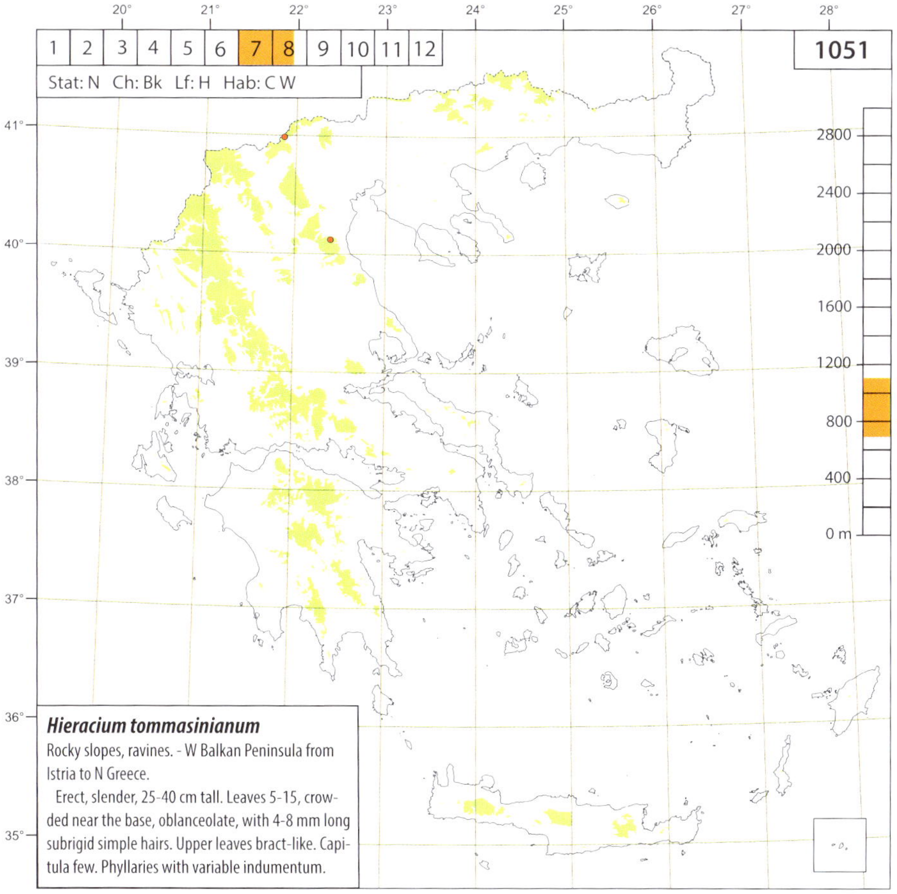

1051

Stat: N Ch: Bk Lf: H Hab: C W

Hieracium tommasinianum

Rocky slopes, ravines. - W Balkan Peninsula from Istria to N Greece.

Erect, slender, 25-40 cm tall. Leaves 5-15, crowded near the base, oblanceolate, with 4-8 mm long subrigid simple hairs. Upper leaves bract-like. Capitula few. Phyllaries with variable indumentum.

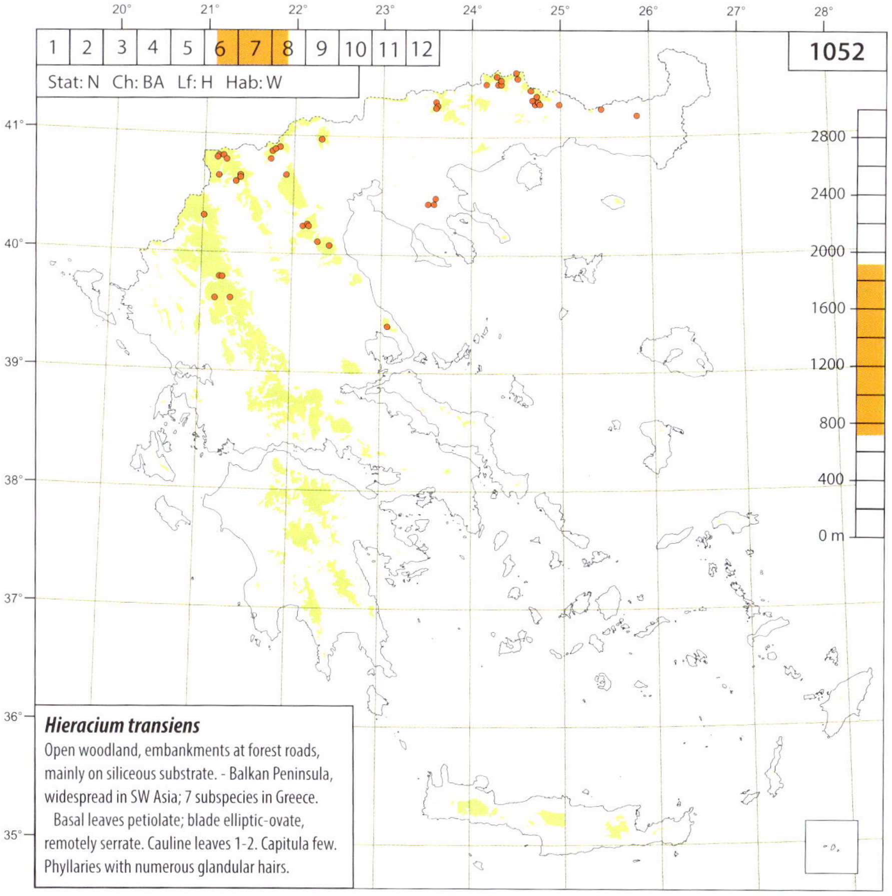

1052

Stat: N Ch: BA Lf: H Hab: W

Hieracium transiens

Open woodland, embankments at forest roads, mainly on siliceous substrate. - Balkan Peninsula, widespread in SW Asia; 7 subspecies in Greece.

Basal leaves petiolate; blade elliptic-ovate, remotely serrate. Cauline leaves 1-2. Capitula few. Phyllaries with numerous glandular hairs.

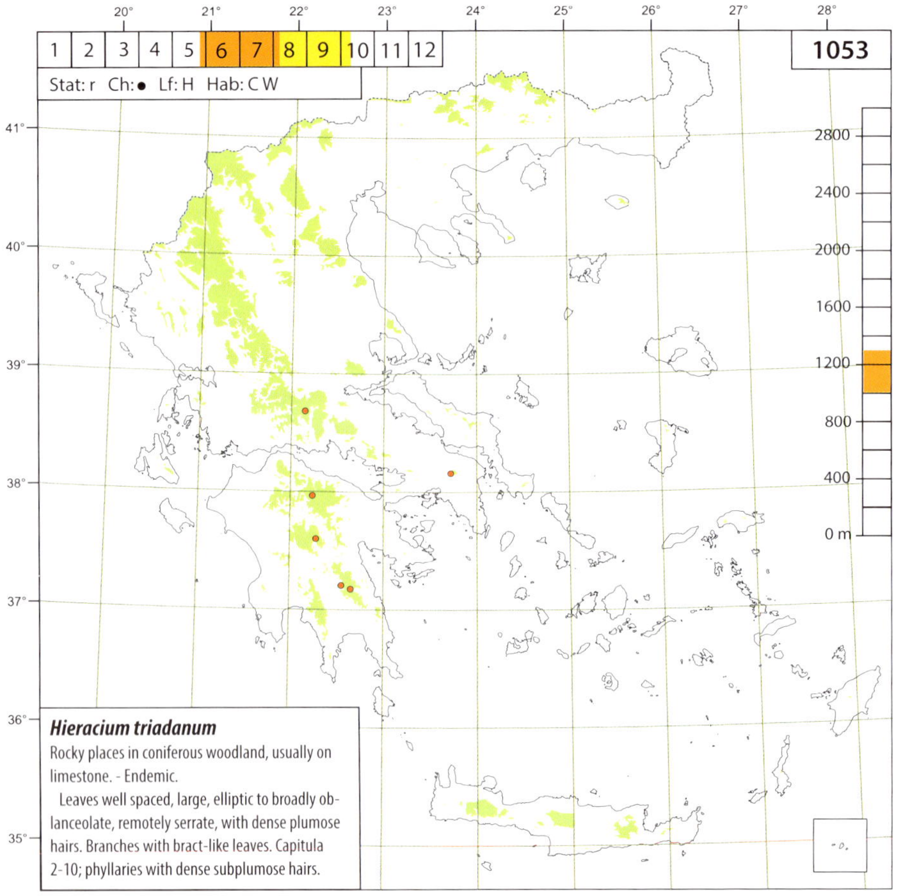

Hieracium triadanum
Rocky places in coniferous woodland, usually on limestone. - Endemic.
Leaves well spaced, large, elliptic to broadly oblanceolate, remotely serrate, with dense plumose hairs. Branches with bract-like leaves. Capitula 2-10; phyllaries with dense subplumose hairs.

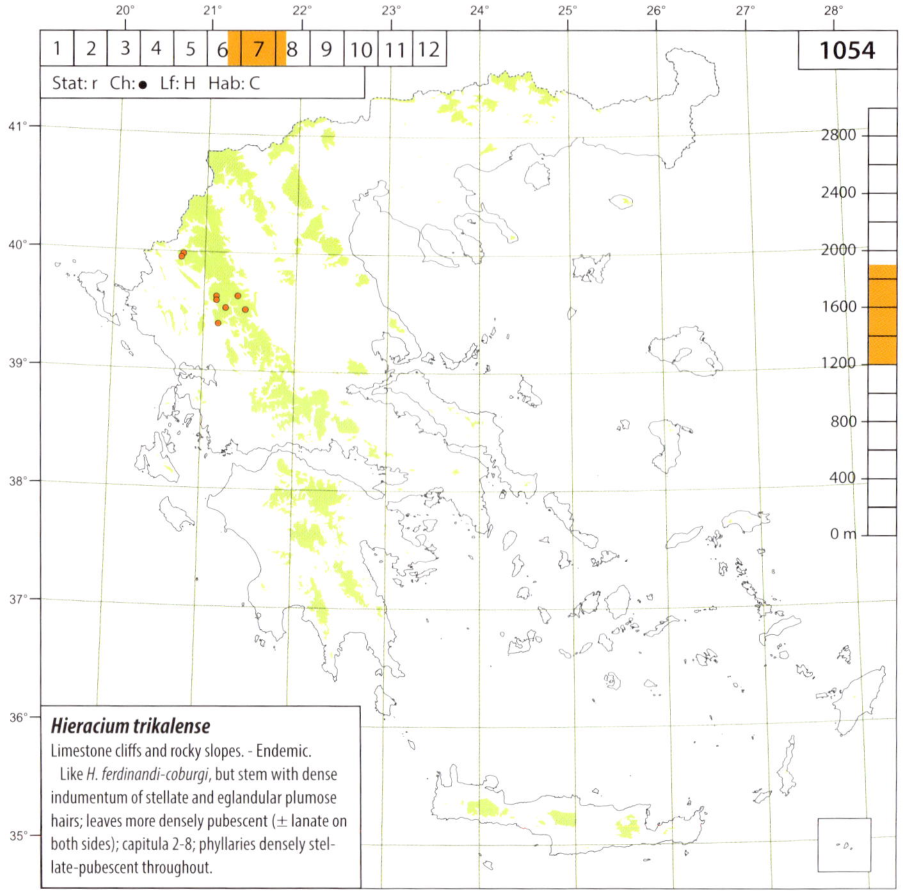

Hieracium trikalense
Limestone cliffs and rocky slopes. - Endemic.
Like *H. ferdinandi-coburgi*, but stem with dense indumentum of stellate and eglandular plumose hairs; leaves more densely pubescent (± lanate on both sides); capitula 2-8; phyllaries densely stellate-pubescent throughout.

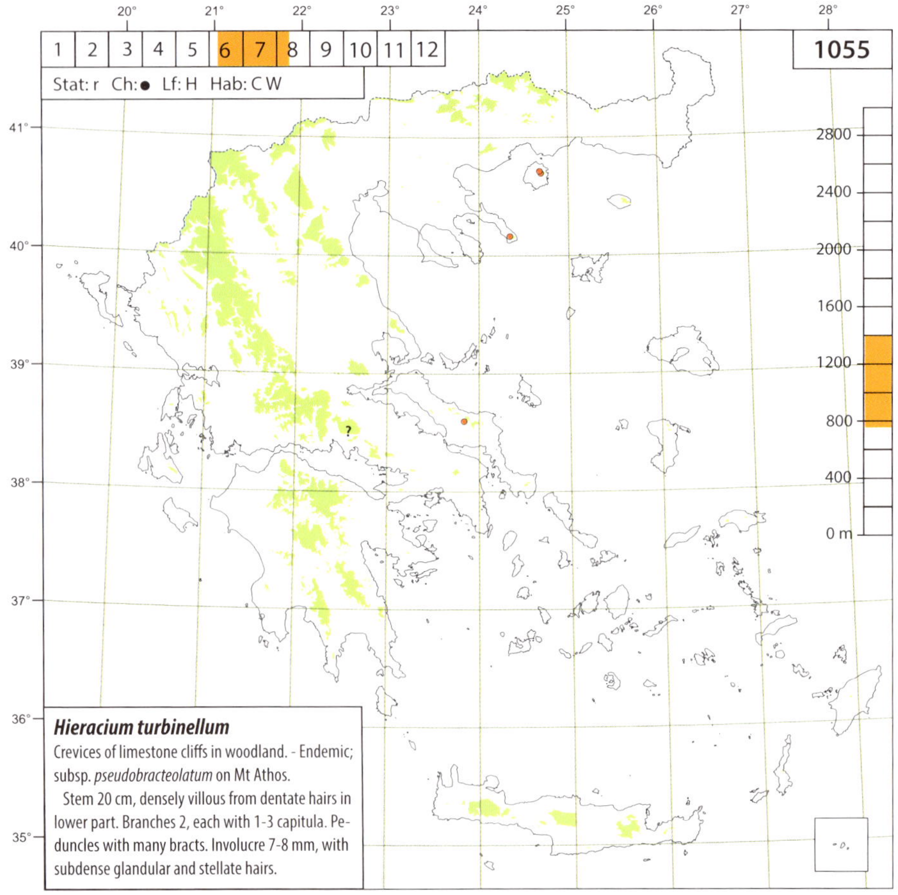

Hieracium turbinellum
Crevices of limestone cliffs in woodland. - Endemic; subsp. *pseudobracteolatum* on Mt Athos.
Stem 20 cm, densely villous from dentate hairs in lower part. Branches 2, each with 1-3 capitula. Peduncles with many bracts. Involucre 7-8 mm, with subdense glandular and stellate hairs.

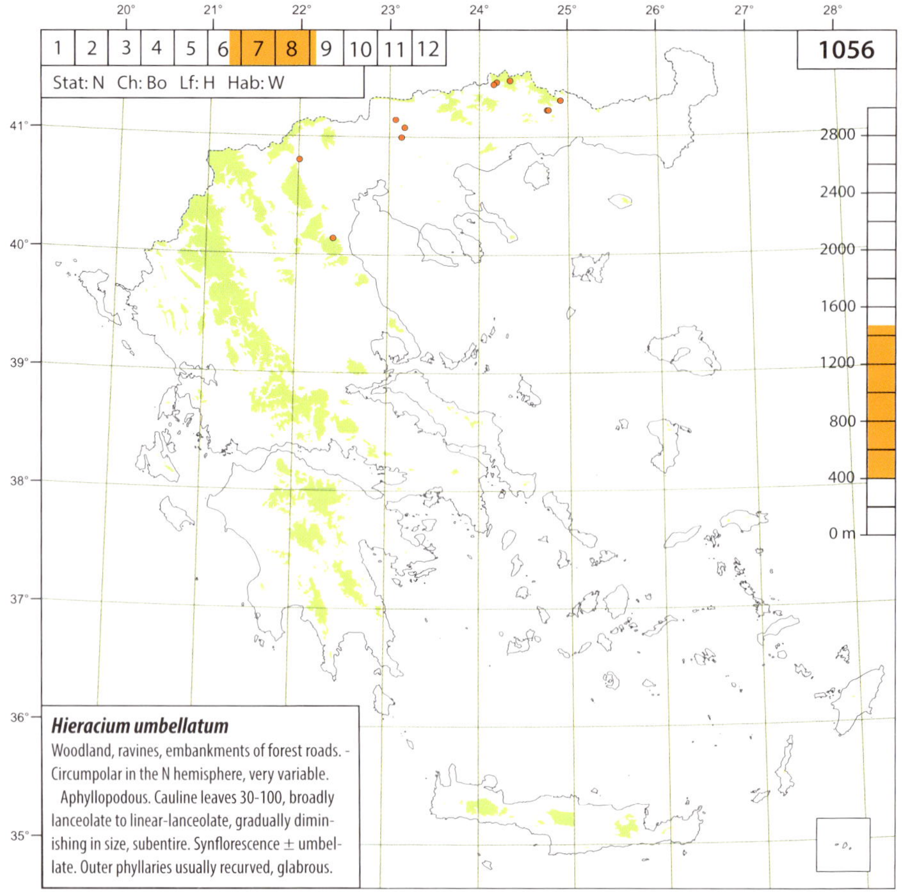

Hieracium umbellatum
Woodland, ravines, embankments of forest roads. - Circumpolar in the N hemisphere, very variable.
Aphyllopodous. Cauline leaves 30-100, broadly lanceolate to linear-lanceolate, gradually diminishing in size, subentire. Synflorescence ± umbellate. Outer phyllaries usually recurved, glabrous.

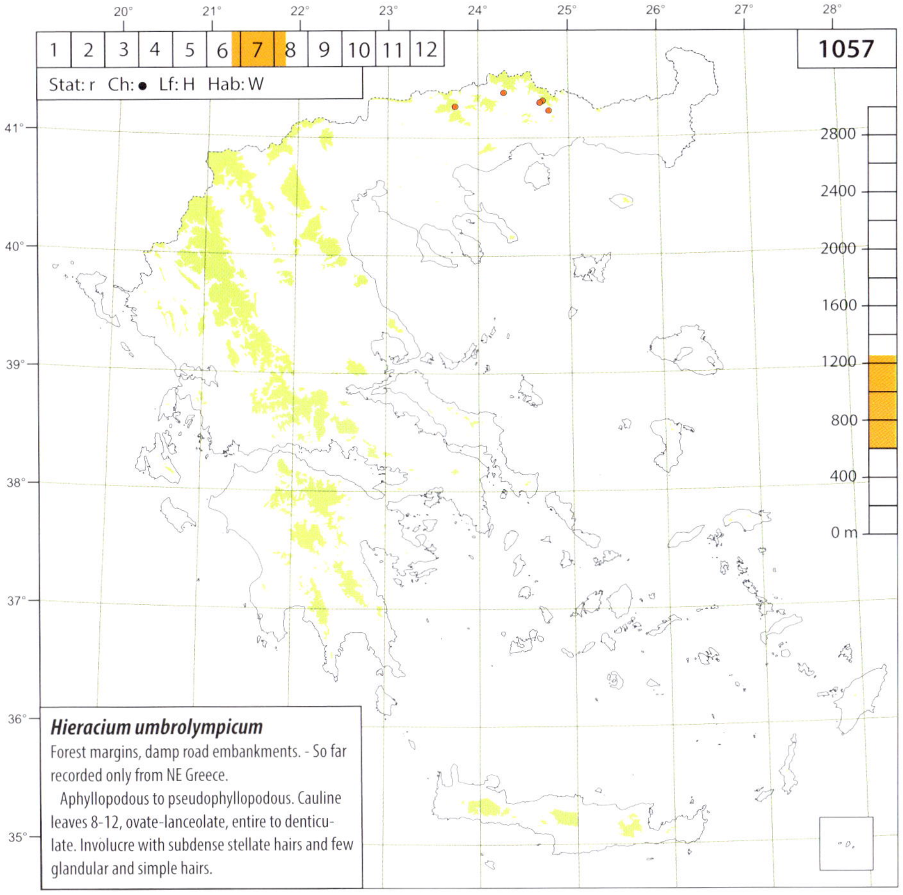

Hieracium umbrolympicum

Forest margins, damp road embankments. - So far recorded only from NE Greece.

Aphyllopodous to pseudophyllopodous. Cauline leaves 8-12, ovate-lanceolate, entire to denticulate. Involucre with subdense stellate hairs and few glandular and simple hairs.

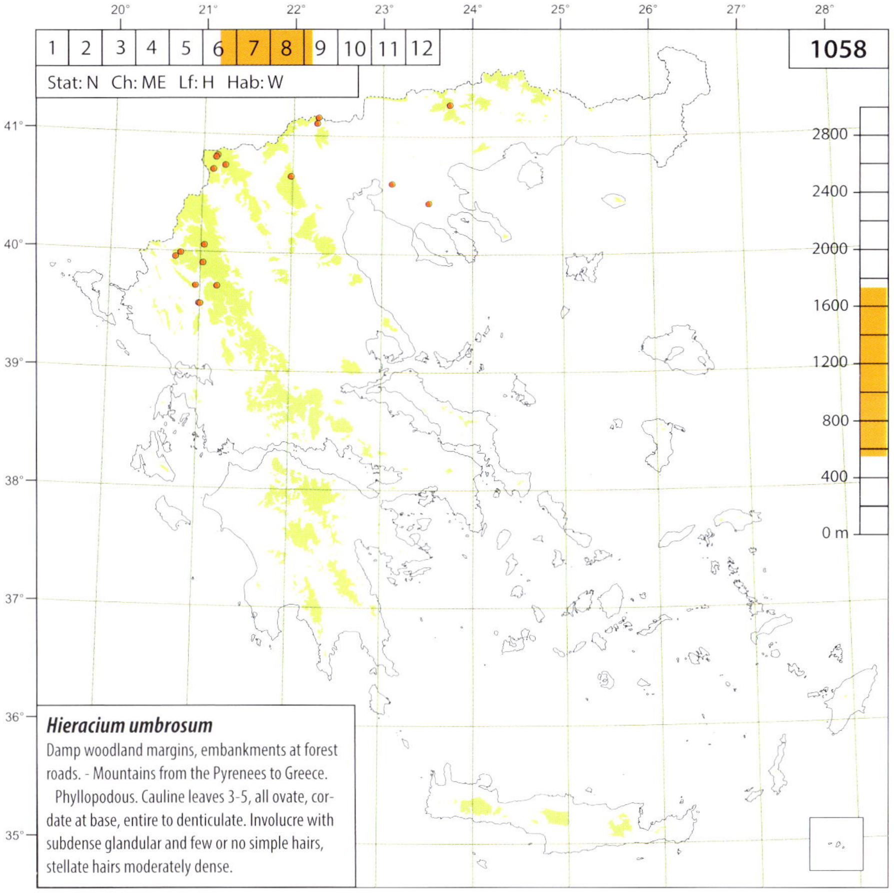

Hieracium umbrosum

Damp woodland margins, embankments at forest roads. - Mountains from the Pyrenees to Greece.

Phyllopodous. Cauline leaves 3-5, all ovate, cordate at base, entire to denticulate. Involucre with subdense glandular and few or no simple hairs, stellate hairs moderately dense.

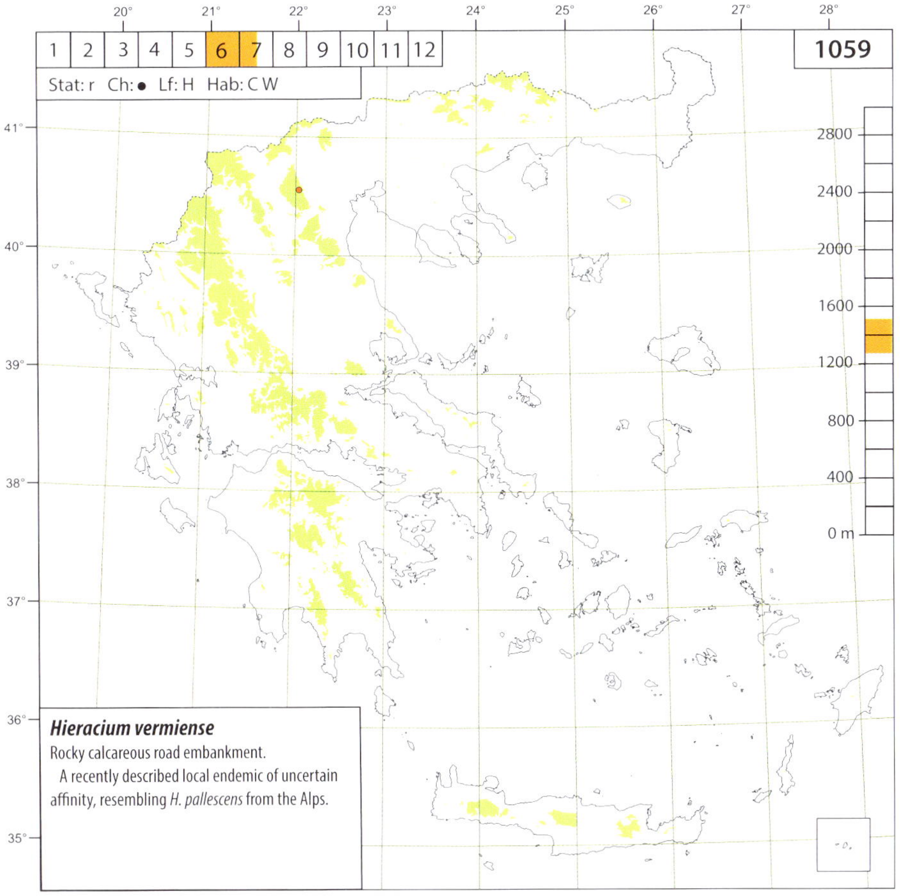

Hieracium vermiense

Rocky calcareous road embankment.

A recently described local endemic of uncertain affinity, resembling *H. pallescens* from the Alps.

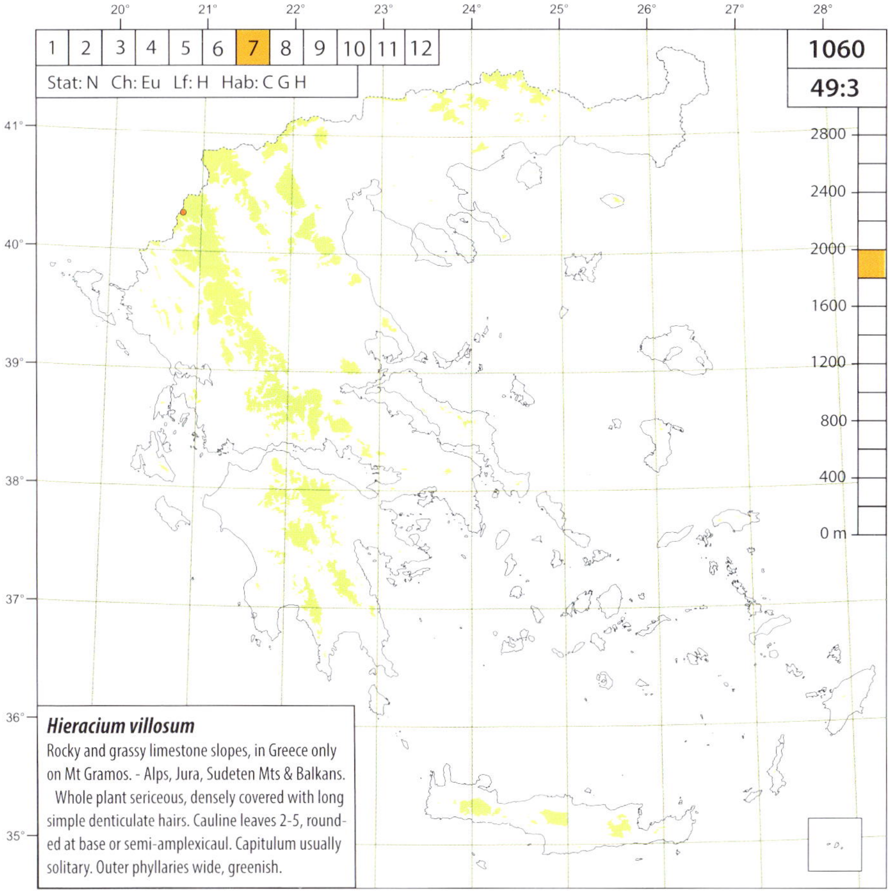

Hieracium villosum

Rocky and grassy limestone slopes, in Greece only on Mt Gramos. - Alps, Jura, Sudeten Mts & Balkans.

Whole plant sericeous, densely covered with long simple denticulate hairs. Cauline leaves 2-5, rounded at base or semi-amplexicaul. Capitulum usually solitary. Outer phyllaries wide, greenish.

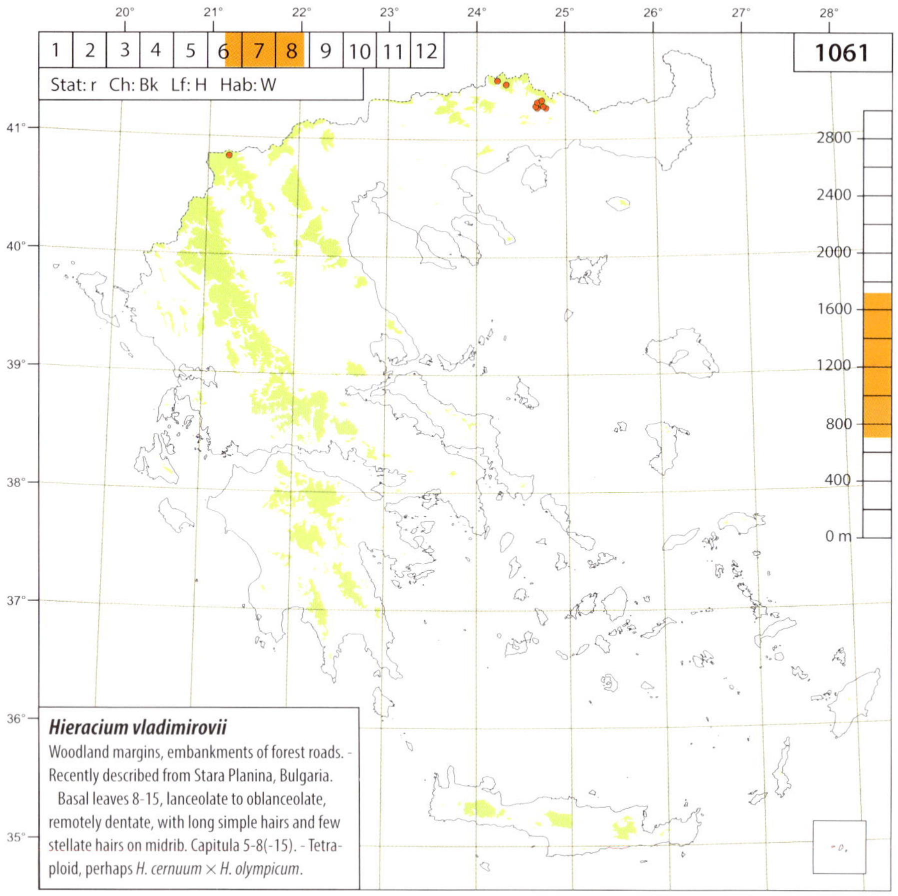

Hieracium vladimirovii
Woodland margins, embankments of forest roads. - Recently described from Stara Planina, Bulgaria.
Basal leaves 8-15, lanceolate to oblanceolate, remotely dentate, with long simple hairs and few stellate hairs on midrib. Capitula 5-8(-15). - Tetraploid, perhaps *H. cernuum* × *H. olympicum*.

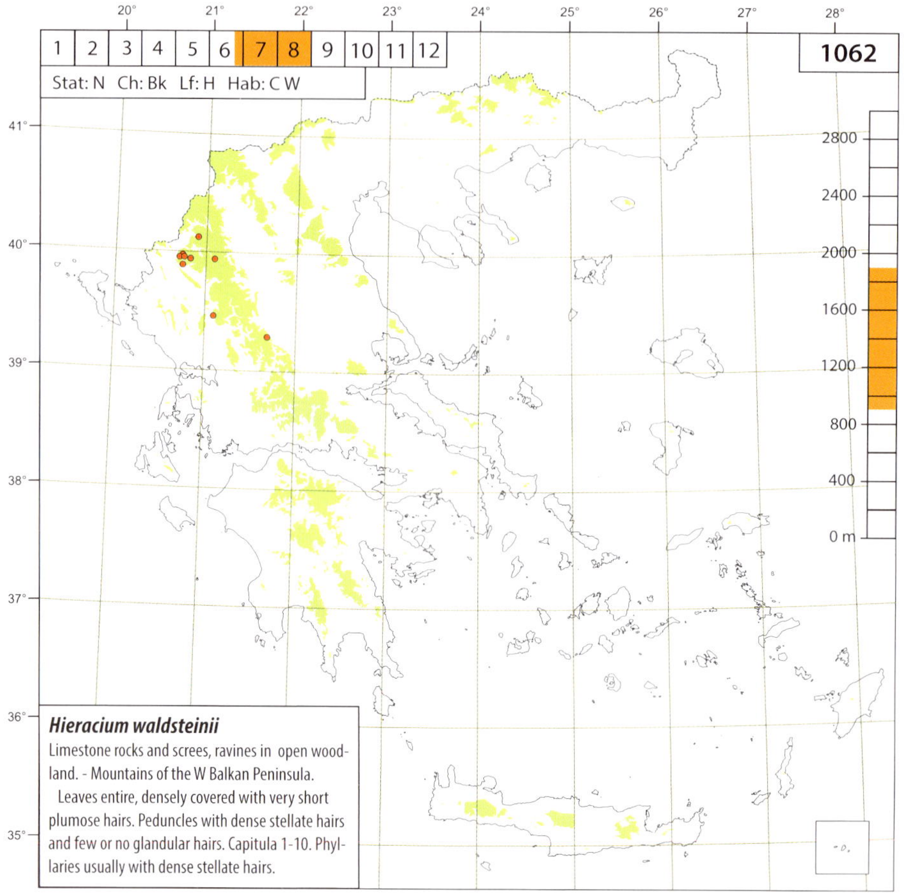

Hieracium waldsteinii
Limestone rocks and screes, ravines in open woodland. - Mountains of the W Balkan Peninsula.
Leaves entire, densely covered with very short plumose hairs. Peduncles with dense stellate hairs and few or no glandular hairs. Capitula 1-10. Phyllaries usually with dense stellate hairs.

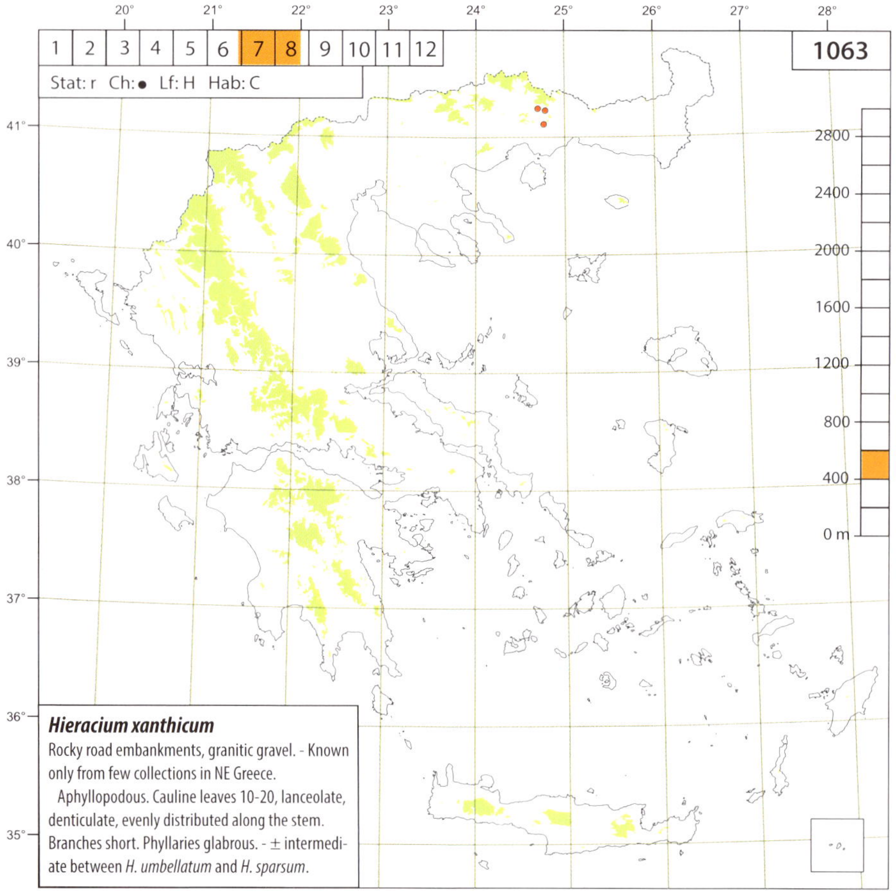

Hieracium xanthicum
Rocky road embankments, granitic gravel. - Known only from few collections in NE Greece.
Aphyllopodous. Cauline leaves 10-20, lanceolate, denticulate, evenly distributed along the stem. Branches short. Phyllaries glabrous. - ± intermediate between *H. umbellatum* and *H. sparsum*.

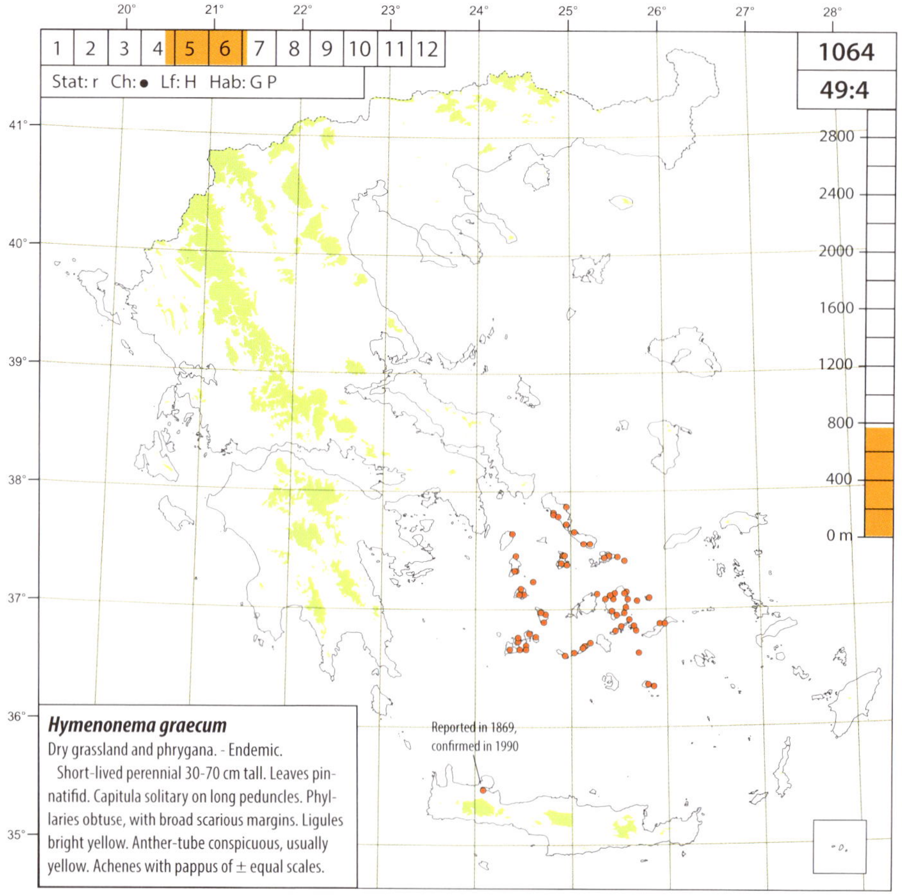

Hymenonema graecum
Dry grassland and phrygana. - Endemic.
Short-lived perennial 30-70 cm tall. Leaves pinnatifid. Capitula solitary on long peduncles. Phyllaries obtuse, with broad scarious margins. Ligules bright yellow. Anther-tube conspicuous, usually yellow. Achenes with pappus of ± equal scales.

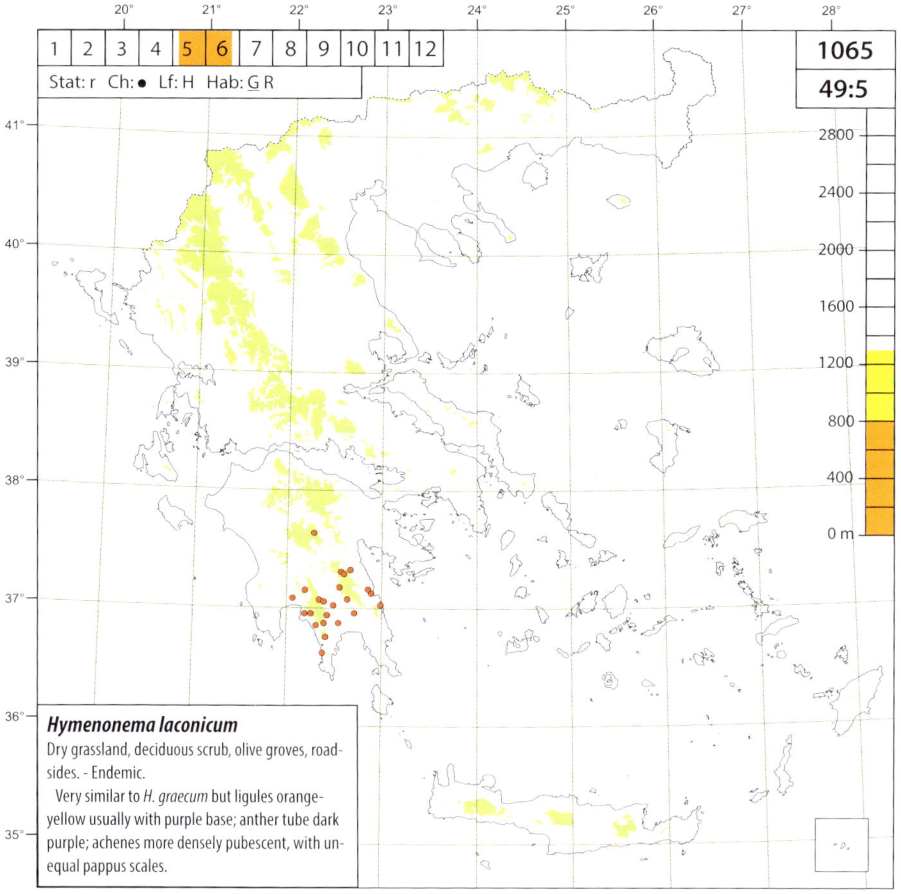

Hymenonema laconicum
Dry grassland, deciduous scrub, olive groves, roadsides. - Endemic.
Very similar to *H. graecum* but ligules orange-yellow usually with purple base; anther tube dark purple; achenes more densely pubescent, with unequal pappus scales.

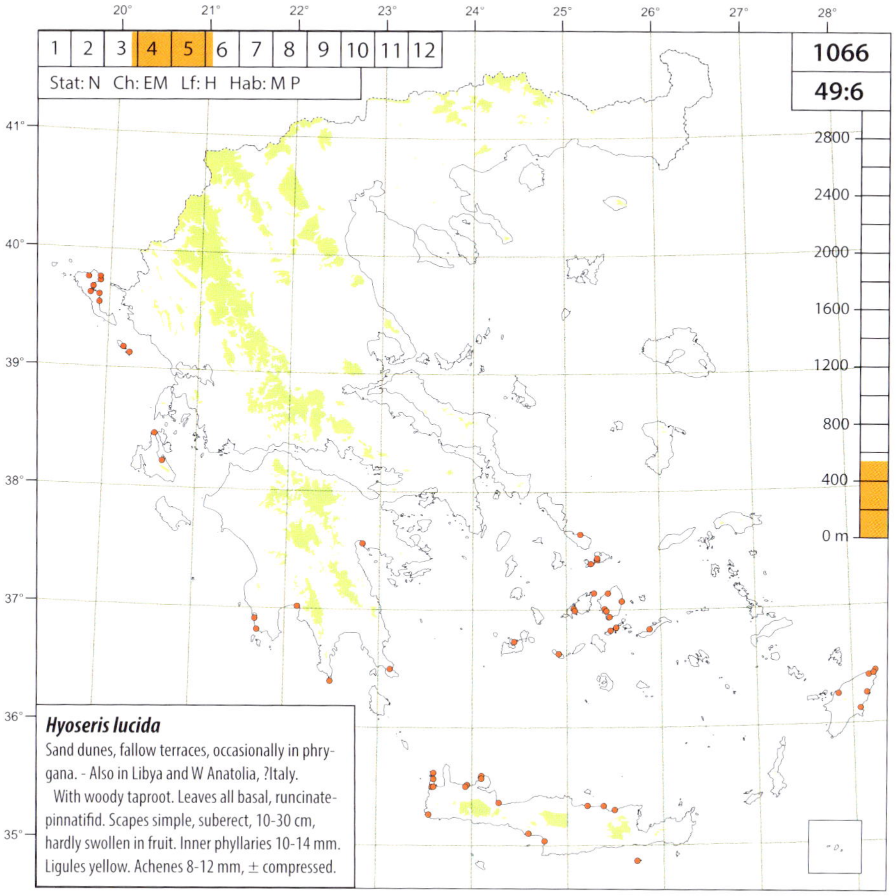

Hyoseris lucida
Sand dunes, fallow terraces, occasionally in phrygana. - Also in Libya and W Anatolia, ?Italy.
With woody taproot. Leaves all basal, runcinate-pinnatifid. Scapes simple, suberect, 10-30 cm, hardly swollen in fruit. Inner phyllaries 10-14 mm. Ligules yellow. Achenes 8-12 mm, ± compressed.

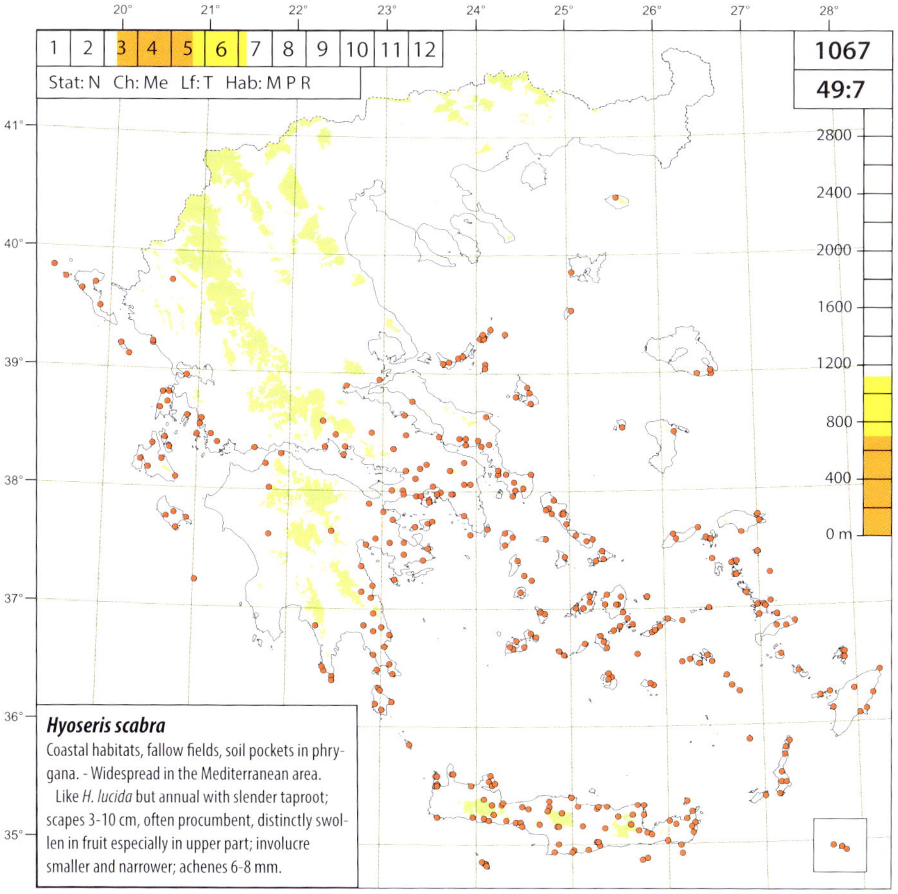

Hyoseris scabra
Coastal habitats, fallow fields, soil pockets in phrygana. - Widespread in the Mediterranean area.
Like *H. lucida* but annual with slender taproot; scapes 3-10 cm, often procumbent, distinctly swollen in fruit especially in upper part; involucre smaller and narrower; achenes 6-8 mm.

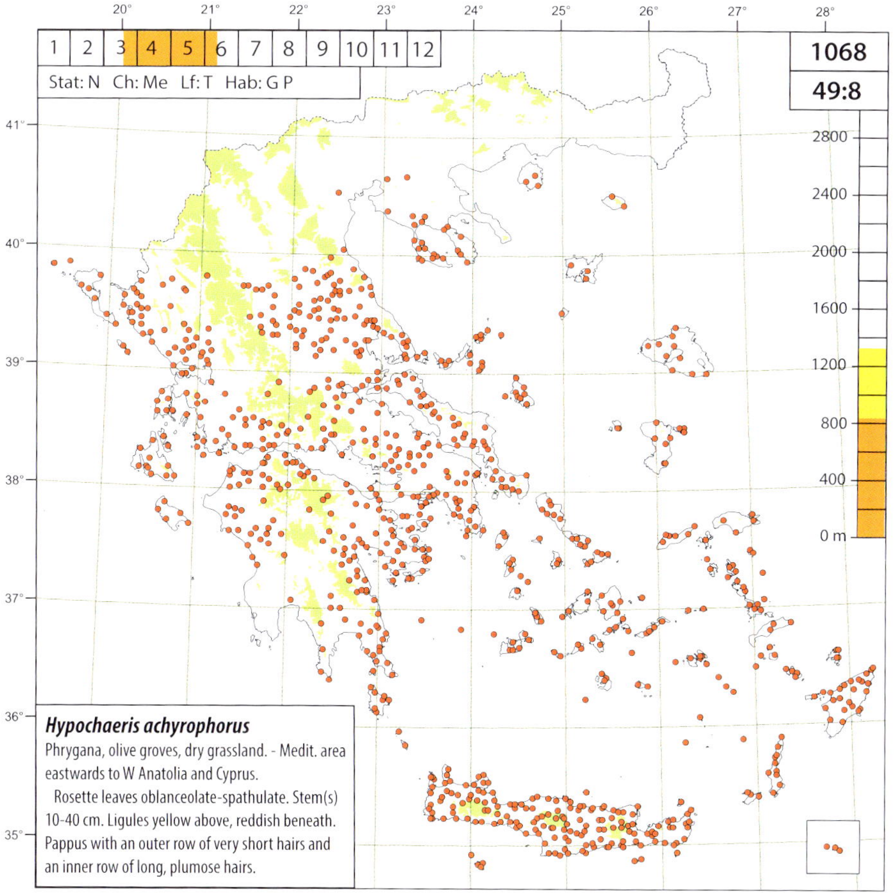

Hypochaeris achyrophorus
Phrygana, olive groves, dry grassland. - Medit. area eastwards to W Anatolia and Cyprus.
Rosette leaves oblanceolate-spathulate. Stem(s) 10-40 cm. Ligules yellow above, reddish beneath. Pappus with an outer row of very short hairs and an inner row of long, plumose hairs.

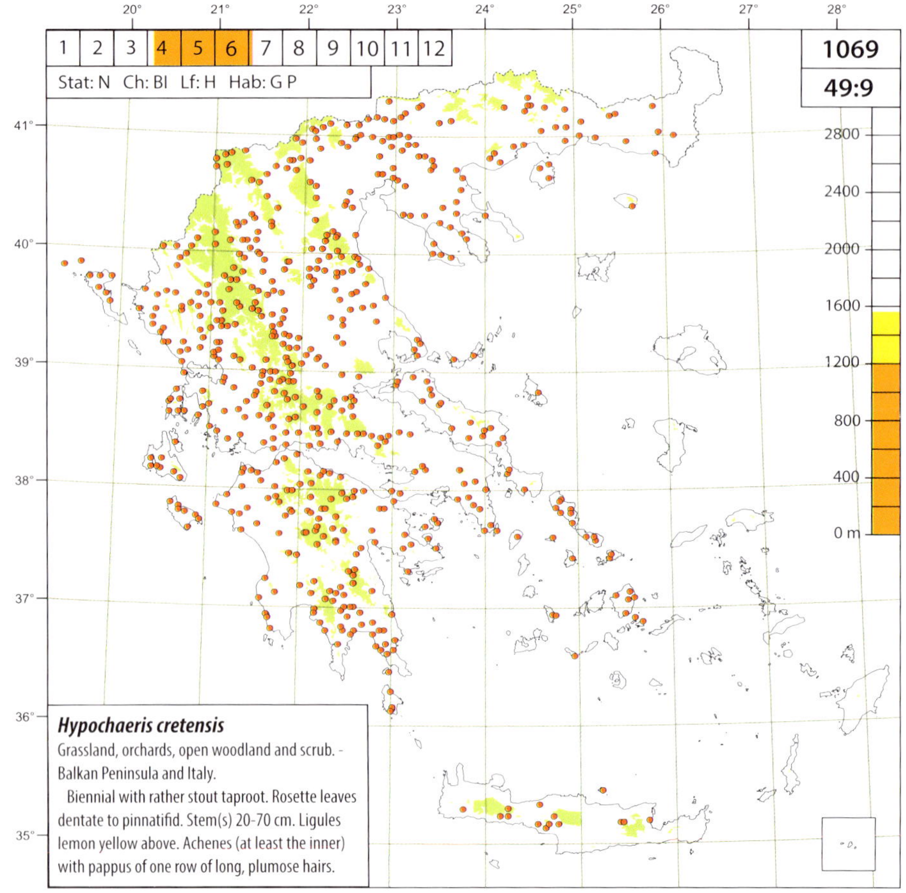

Hypochaeris cretensis
Grassland, orchards, open woodland and scrub. - Balkan Peninsula and Italy.
Biennial with rather stout taproot. Rosette leaves dentate to pinnatifid. Stem(s) 20-70 cm. Ligules lemon yellow above. Achenes (at least the inner) with pappus of one row of long, plumose hairs.

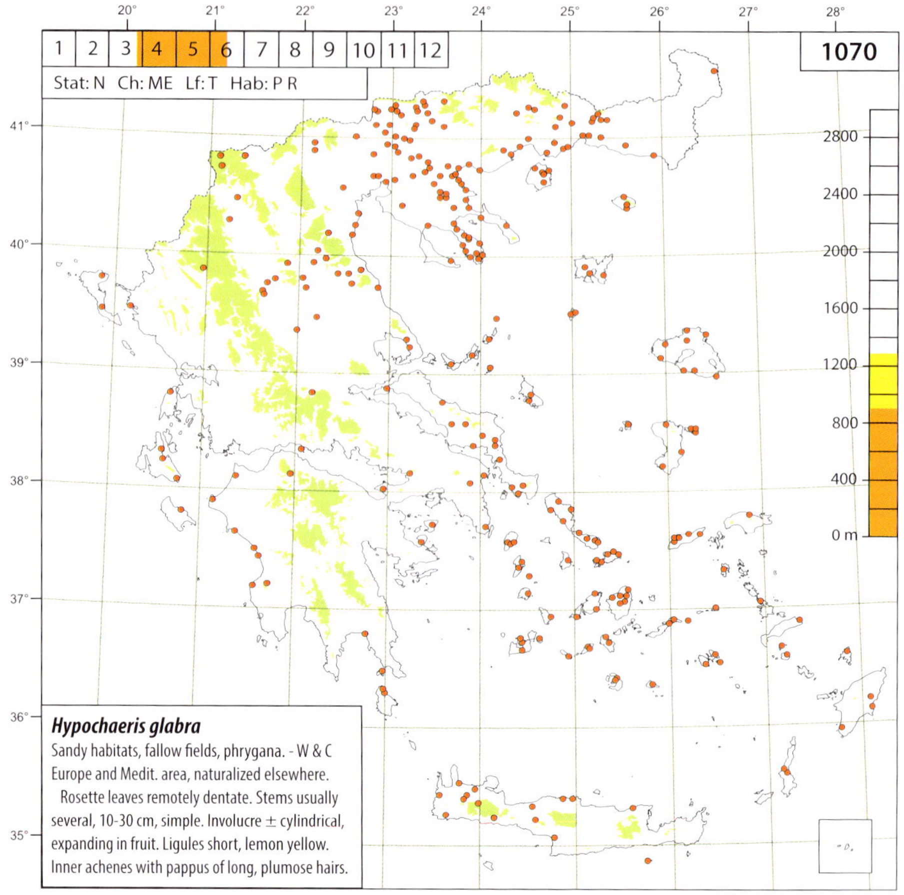

Hypochaeris glabra
Sandy habitats, fallow fields, phrygana. - W & C Europe and Medit. area, naturalized elsewhere.
Rosette leaves remotely dentate. Stems usually several, 10-30 cm, simple. Involucre ± cylindrical, expanding in fruit. Ligules short, lemon yellow. Inner achenes with pappus of long, plumose hairs.

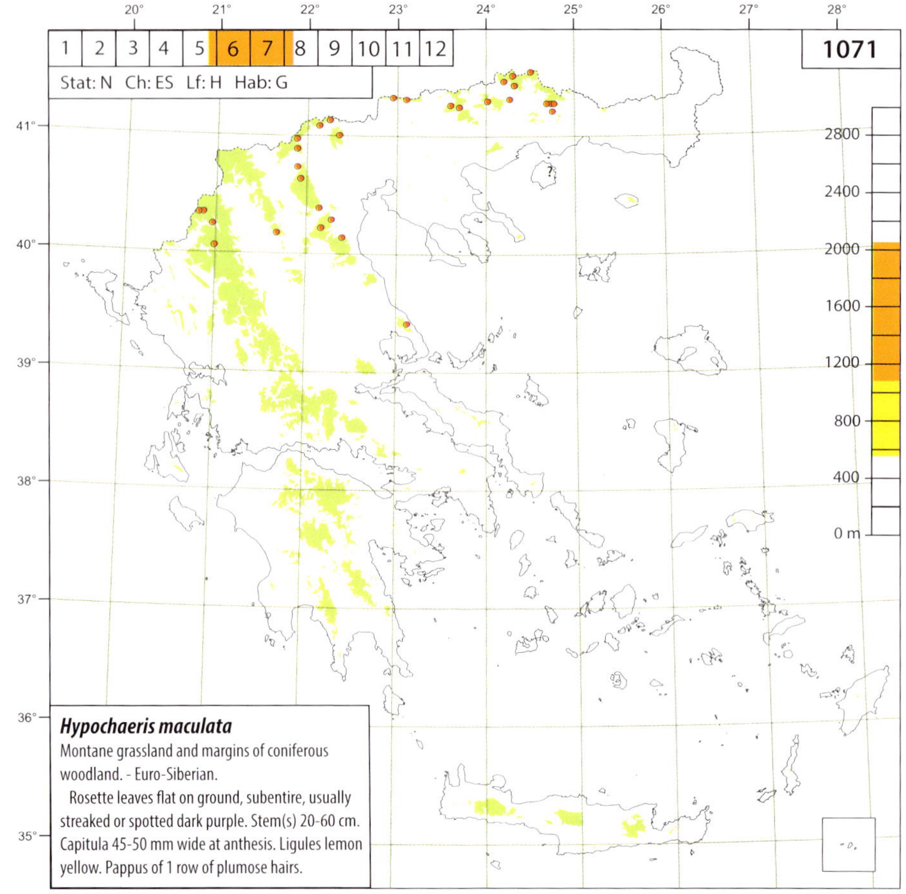

Hypochaeris maculata
Montane grassland and margins of coniferous woodland. - Euro-Siberian.
Rosette leaves flat on ground, subentire, usually streaked or spotted dark purple. Stem(s) 20-60 cm. Capitula 45-50 mm wide at anthesis. Ligules lemon yellow. Pappus of 1 row of plumose hairs.

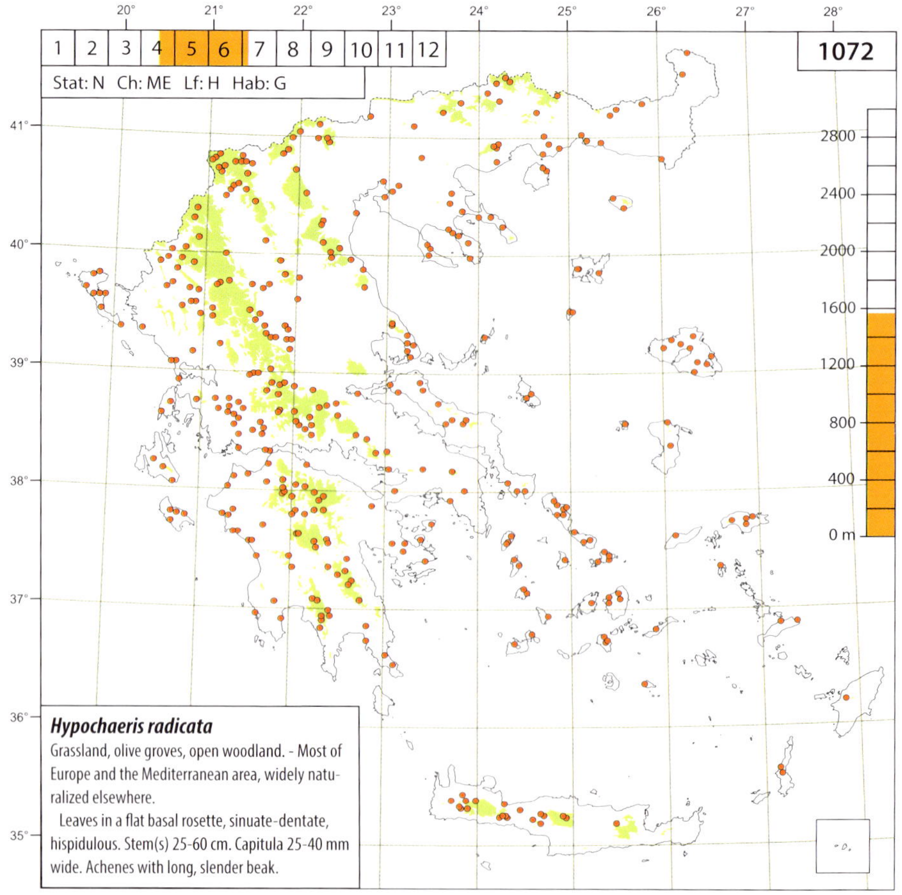

Hypochaeris radicata
Grassland, olive groves, open woodland. - Most of Europe and the Mediterranean area, widely naturalized elsewhere.
Leaves in a flat basal rosette, sinuate-dentate, hispidulous. Stem(s) 25-60 cm. Capitula 25-40 mm wide. Achenes with long, slender beak.

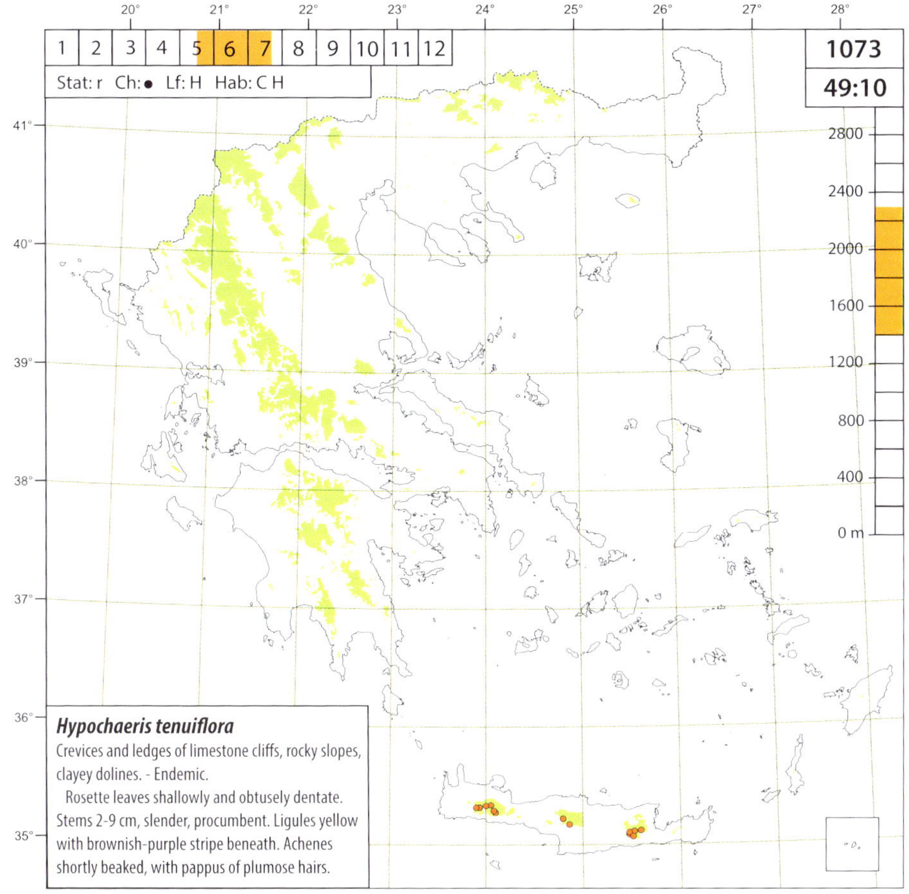

Hypochaeris tenuiflora

Crevices and ledges of limestone cliffs, rocky slopes, clayey dolines. - Endemic.

Rosette leaves shallowly and obtusely dentate. Stems 2-9 cm, slender, procumbent. Ligules yellow with brownish-purple stripe beneath. Achenes shortly beaked, with pappus of plumose hairs.

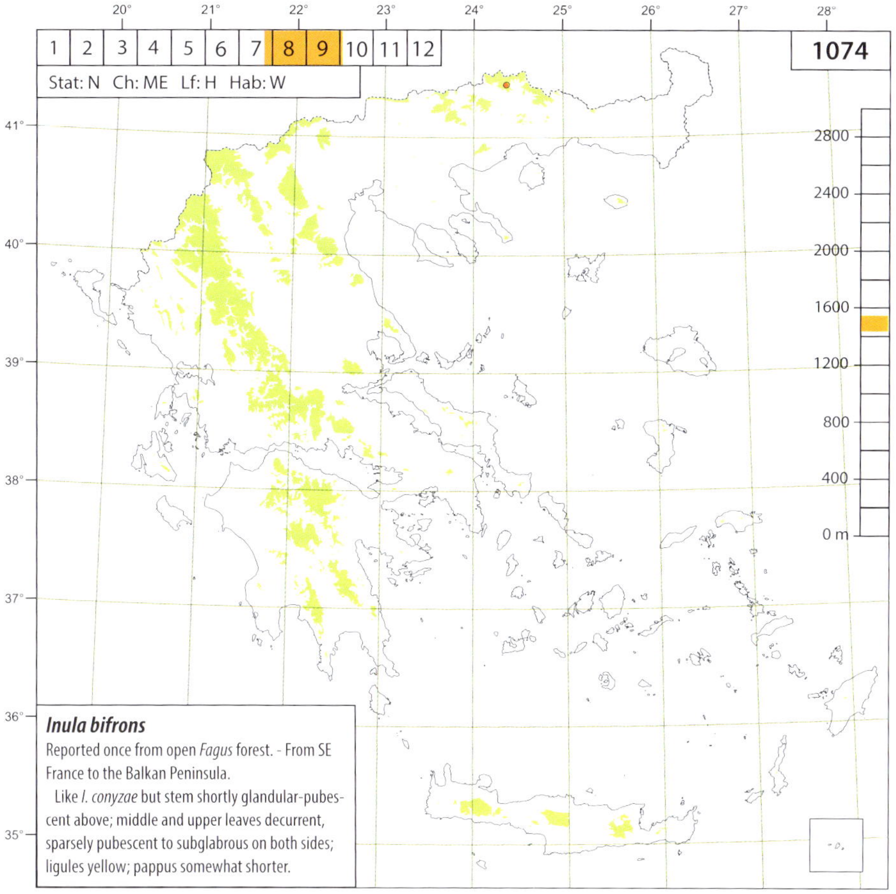

Inula bifrons

Reported once from open *Fagus* forest. - From SE France to the Balkan Peninsula.

Like *I. conyzae* but stem shortly glandular-pubescent above; middle and upper leaves decurrent, sparsely pubescent to subglabrous on both sides; ligules yellow; pappus somewhat shorter.

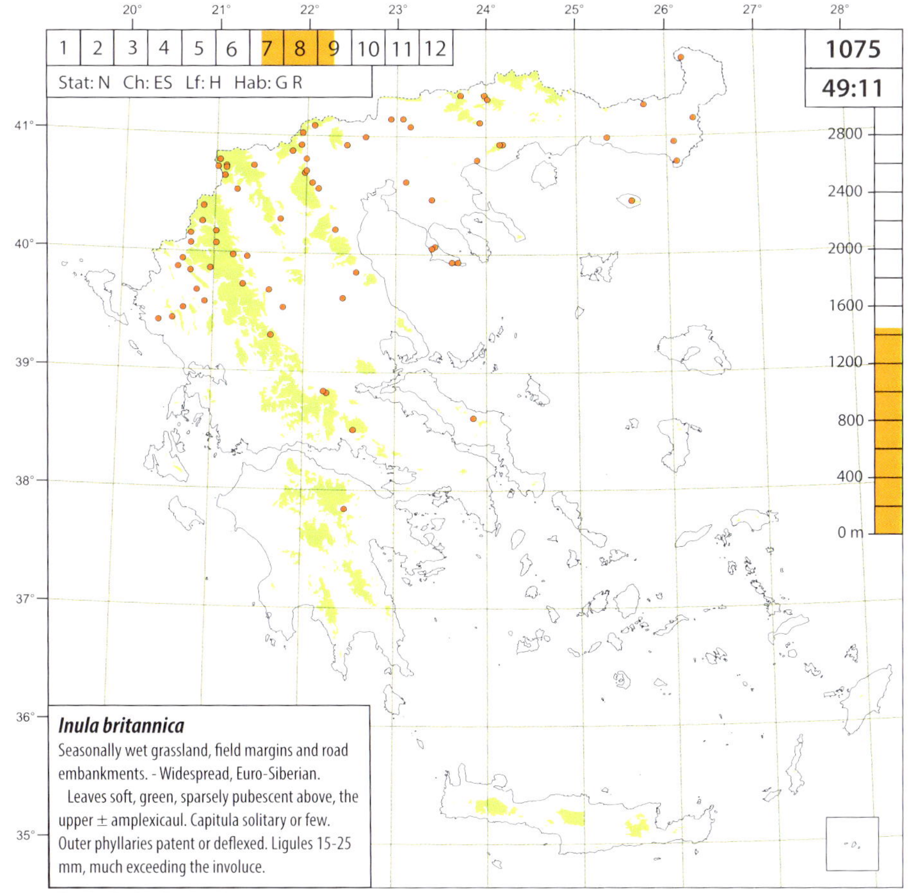

Inula britannica

Seasonally wet grassland, field margins and road embankments. - Widespread, Euro-Siberian.

Leaves soft, green, sparsely pubescent above, the upper ± amplexicaul. Capitula solitary or few. Outer phyllaries patent or deflexed. Ligules 15-25 mm, much exceeding the involuce.

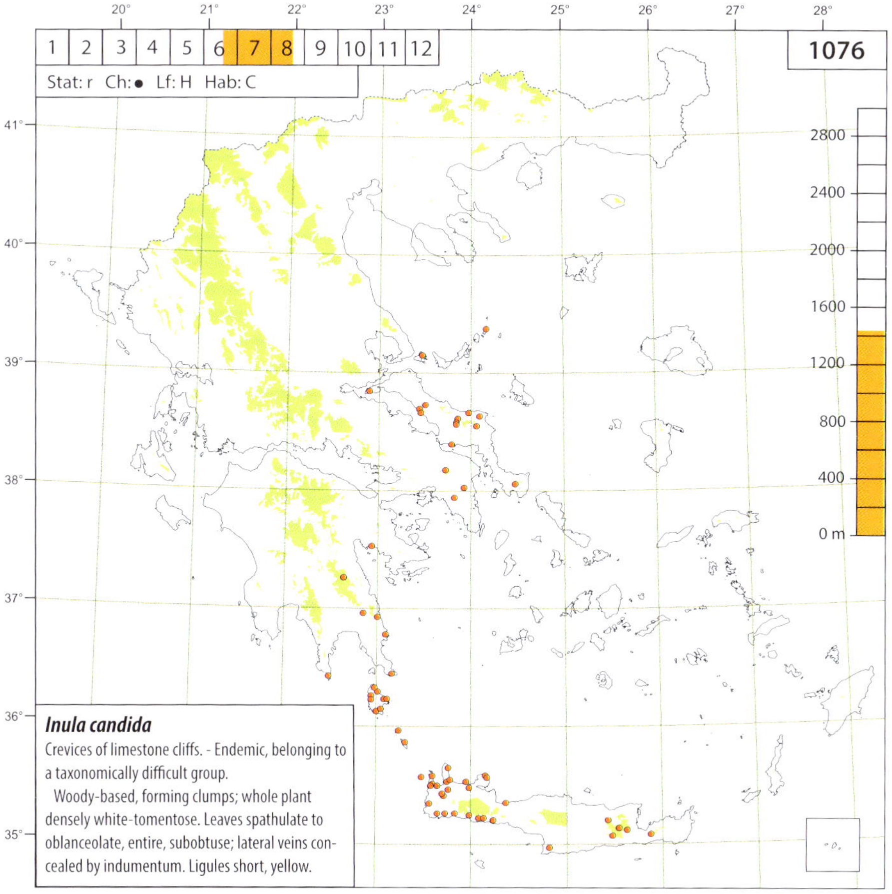

Inula candida

Crevices of limestone cliffs. - Endemic, belonging to a taxonomically difficult group.

Woody-based, forming clumps; whole plant densely white-tomentose. Leaves spathulate to oblanceolate, entire, subobtuse; lateral veins concealed by indumentum. Ligules short, yellow.

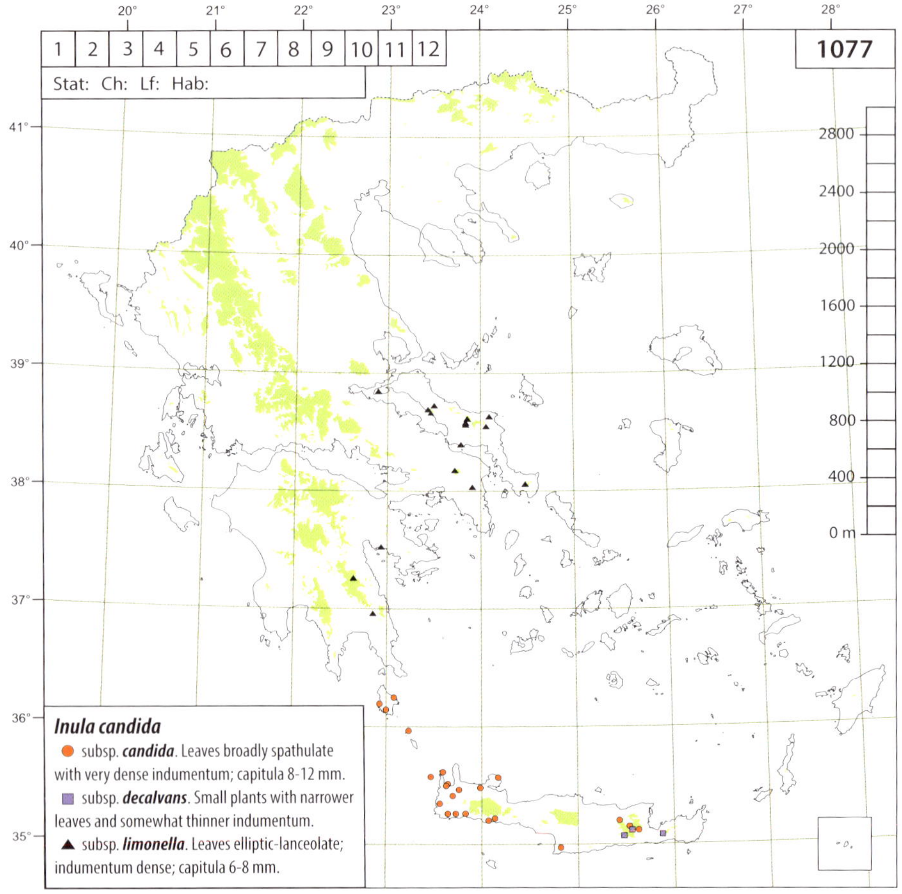

Inula candida
● subsp. ***candida***. Leaves broadly spathulate with very dense indumentum; capitula 8-12 mm.
■ subsp. ***decalvans***. Small plants with narrower leaves and somewhat thinner indumentum.
▲ subsp. ***limonella***. Leaves elliptic-lanceolate; indumentum dense; capitula 6-8 mm.

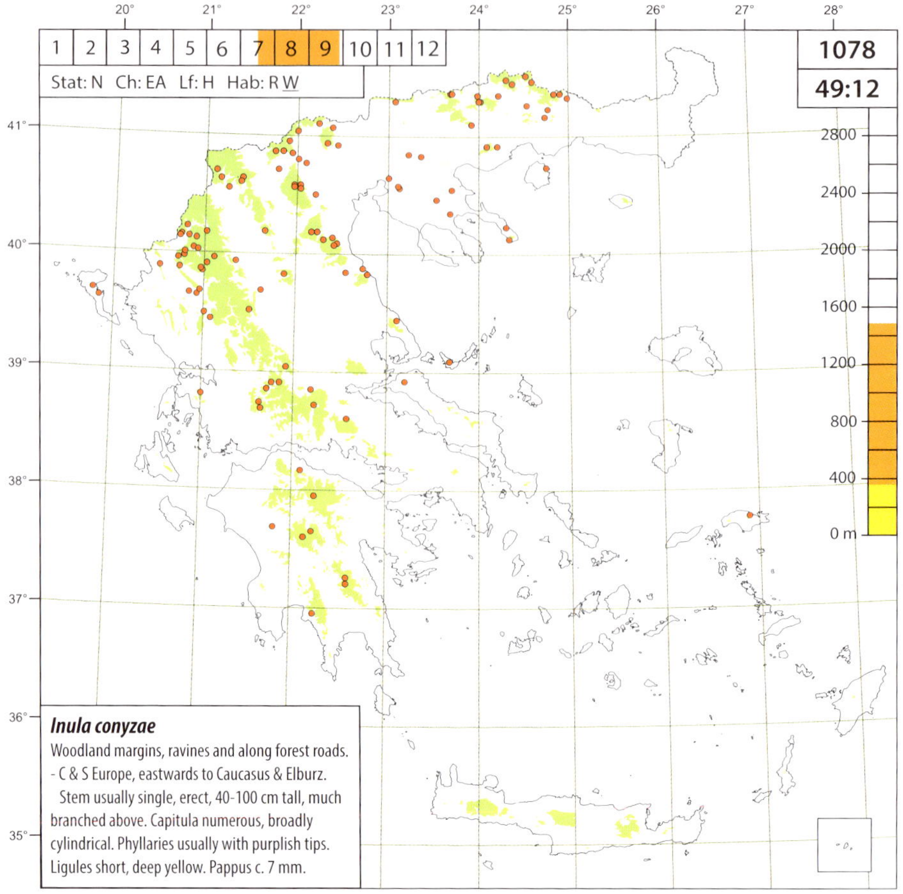

Inula conyzae
Woodland margins, ravines and along forest roads. - C & S Europe, eastwards to Caucasus & Elburz.
Stem usually single, erect, 40-100 cm tall, much branched above. Capitula numerous, broadly cylindrical. Phyllaries usually with purplish tips. Ligules short, deep yellow. Pappus c. 7 mm.

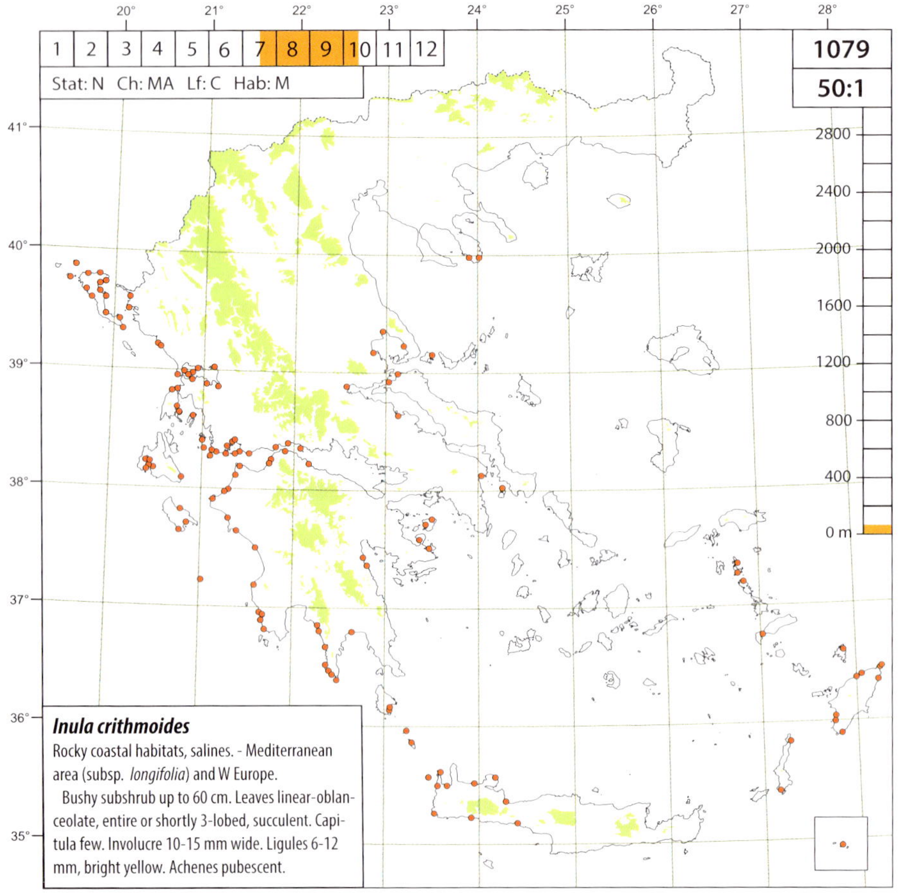

Inula crithmoides
Rocky coastal habitats, salines. - Mediterranean area (subsp. *longifolia*) and W Europe.
Bushy subshrub up to 60 cm. Leaves linear-oblanceolate, entire or shortly 3-lobed, succulent. Capitula few. Involucre 10-15 mm wide. Ligules 6-12 mm, bright yellow. Achenes pubescent.

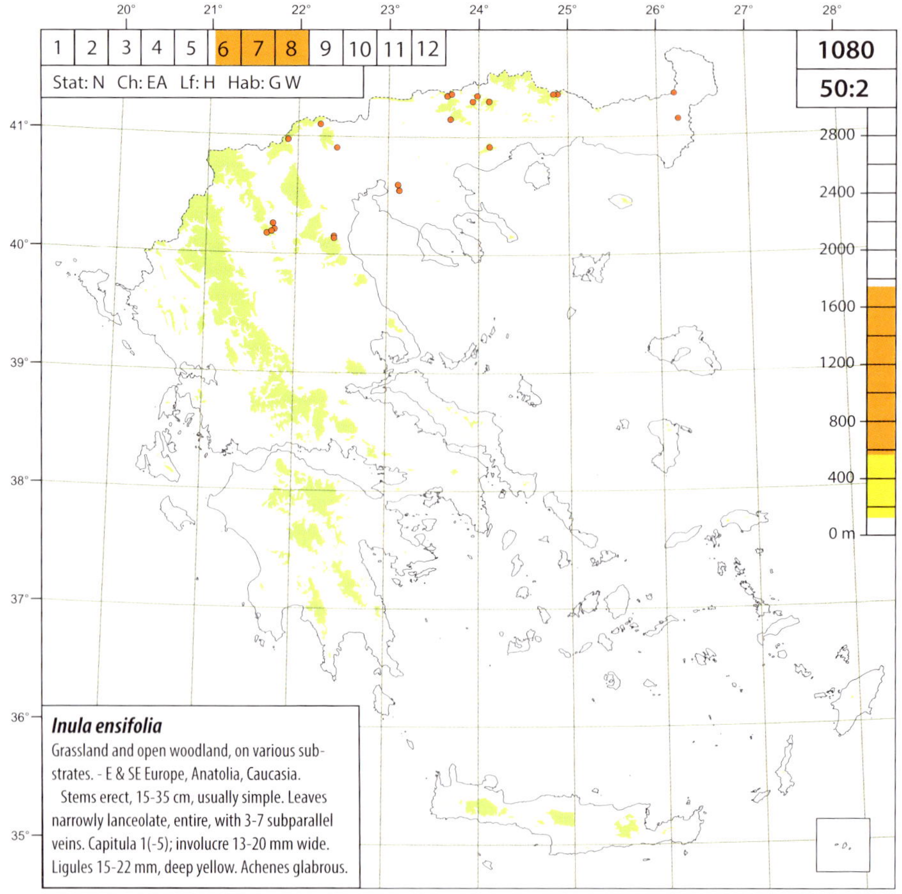

Inula ensifolia
Grassland and open woodland, on various substrates. - E & SE Europe, Anatolia, Caucasia.
Stems erect, 15-35 cm, usually simple. Leaves narrowly lanceolate, entire, with 3-7 subparallel veins. Capitula 1(-5); involucre 13-20 mm wide. Ligules 15-22 mm, deep yellow. Achenes glabrous.

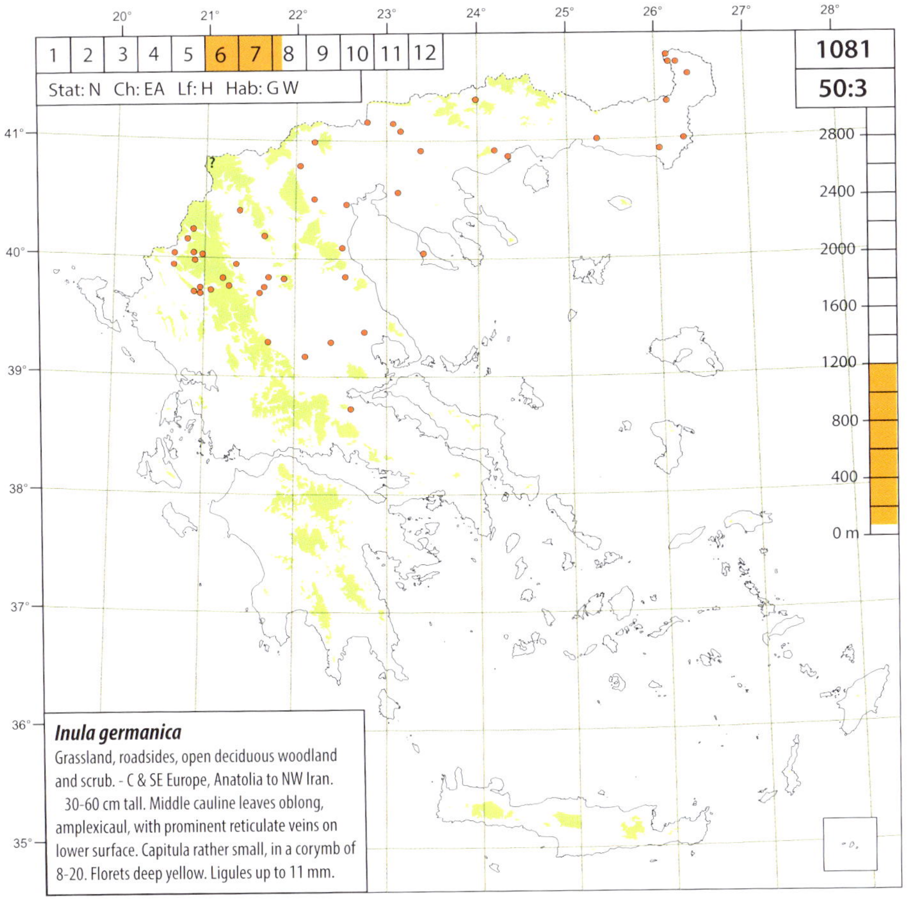

Inula germanica
Grassland, roadsides, open deciduous woodland and scrub. - C & SE Europe, Anatolia to NW Iran.
30-60 cm tall. Middle cauline leaves oblong, amplexicaul, with prominent reticulate veins on lower surface. Capitula rather small, in a corymb of 8-20. Florets deep yellow. Ligules up to 11 mm.

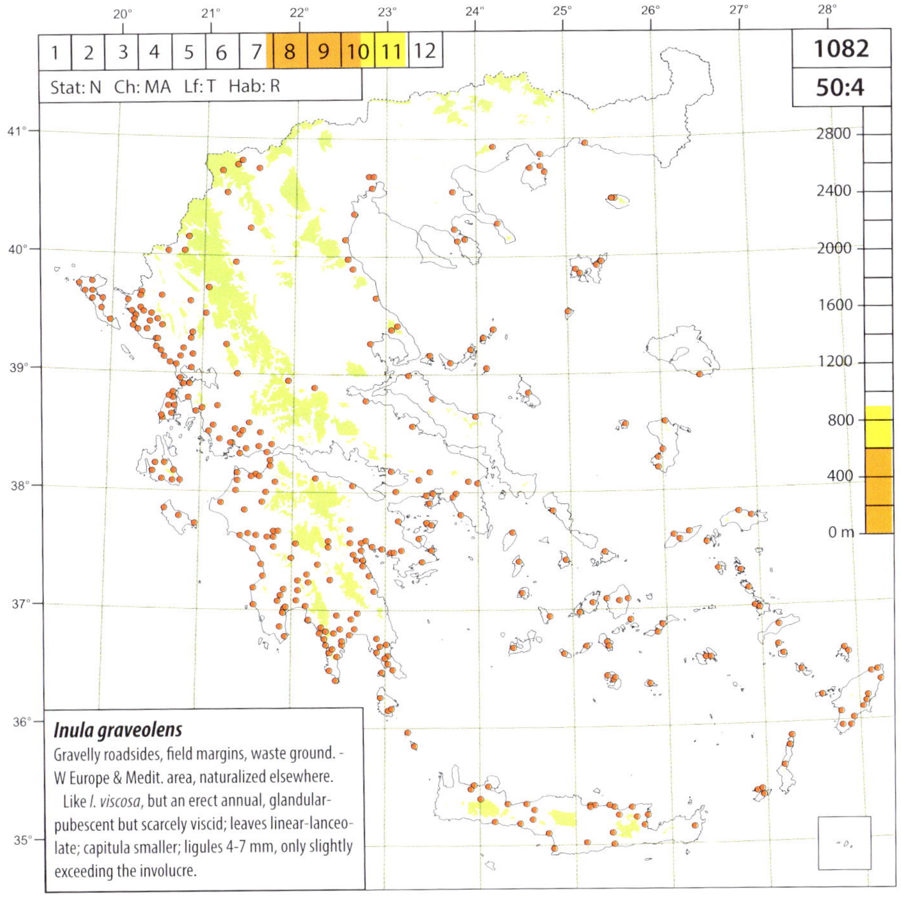

Inula graveolens
Gravelly roadsides, field margins, waste ground. - W Europe & Medit. area, naturalized elsewhere.
Like *I. viscosa*, but an erect annual, glandular-pubescent but scarcely viscid; leaves linear-lanceolate; capitula smaller; ligules 4-7 mm, only slightly exceeding the involucre.

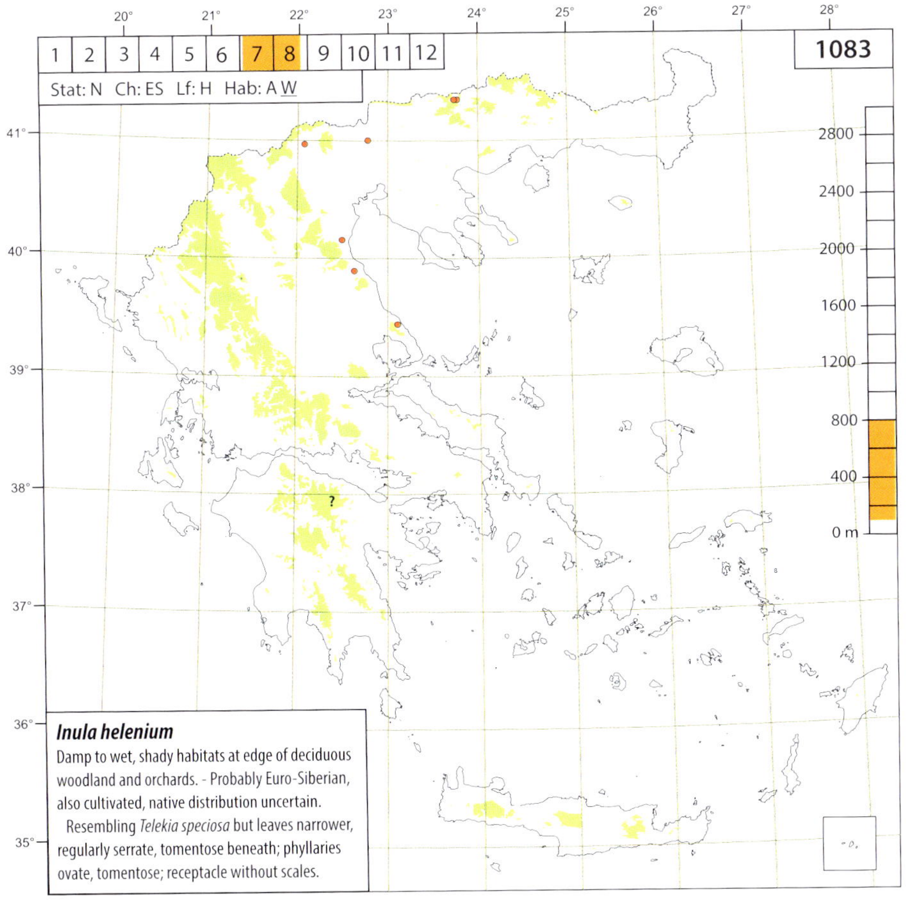

Inula helenium
Damp to wet, shady habitats at edge of deciduous woodland and orchards. - Probably Euro-Siberian, also cultivated, native distribution uncertain.
Resembling *Telekia speciosa* but leaves narrower, regularly serrate, tomentose beneath; phyllaries ovate, tomentose; receptacle without scales.

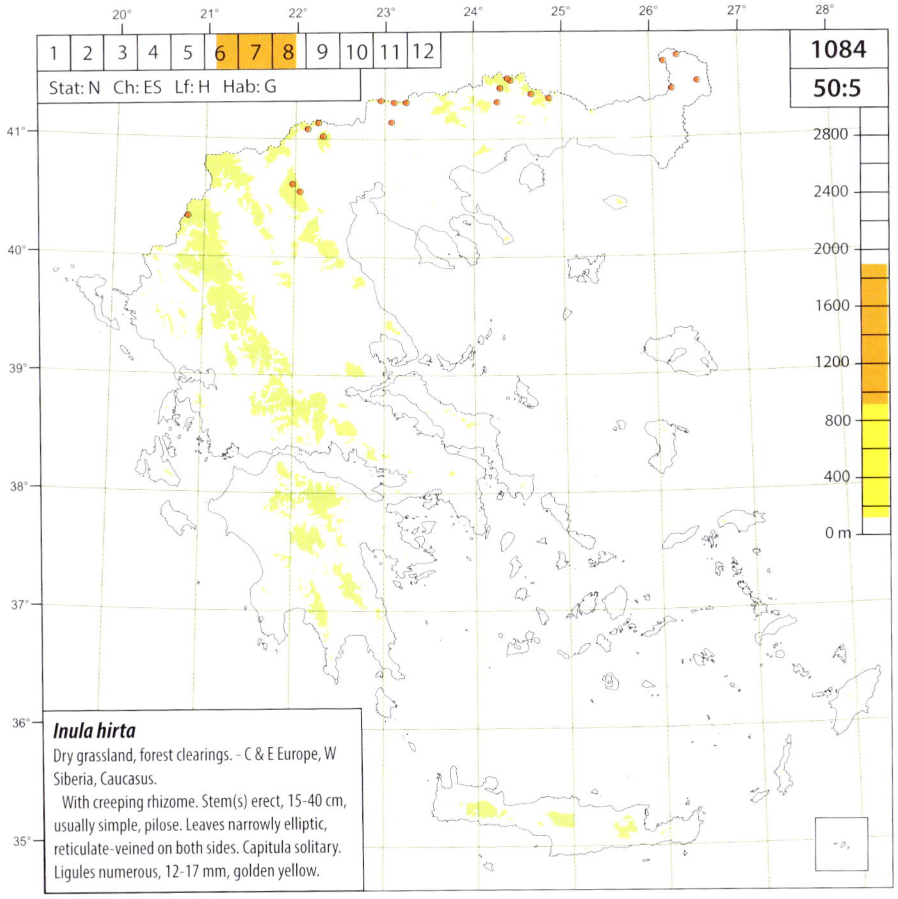

Inula hirta
Dry grassland, forest clearings. - C & E Europe, W Siberia, Caucasus.
With creeping rhizome. Stem(s) erect, 15-40 cm, usually simple, pilose. Leaves narrowly elliptic, reticulate-veined on both sides. Capitula solitary. Ligules numerous, 12-17 mm, golden yellow.

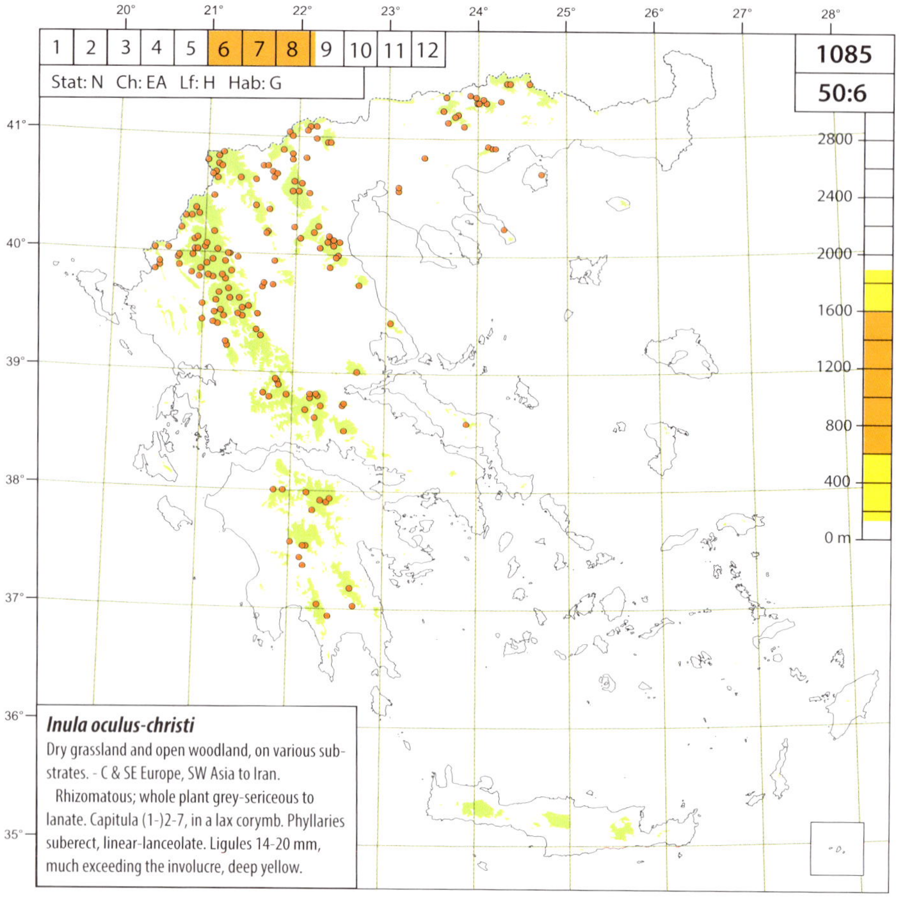

Inula oculus-christi

Dry grassland and open woodland, on various substrates. - C & SE Europe, SW Asia to Iran.

Rhizomatous; whole plant grey-sericeous to lanate. Capitula (1-)2-7, in a lax corymb. Phyllaries suberect, linear-lanceolate. Ligules 14-20 mm, much exceeding the involucre, deep yellow.

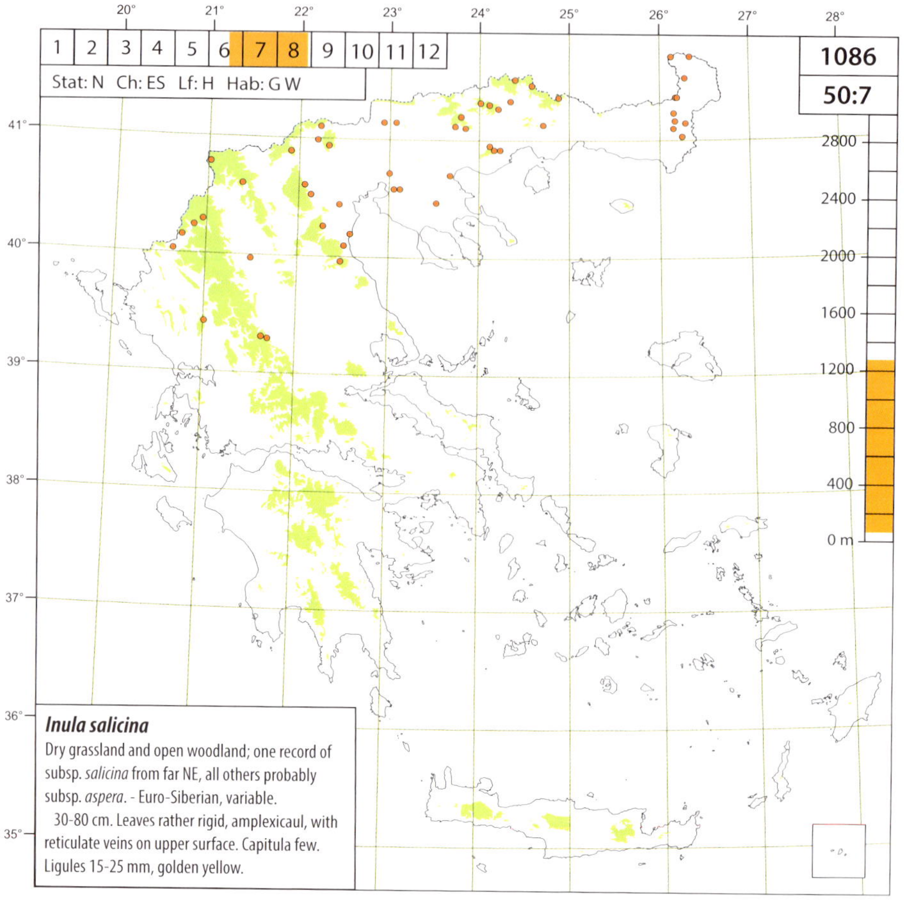

Inula salicina

Dry grassland and open woodland; one record of subsp. *salicina* from far NE, all others probably subsp. *aspera*. - Euro-Siberian, variable.

30-80 cm. Leaves rather rigid, amplexicaul, with reticulate veins on upper surface. Capitula few. Ligules 15-25 mm, golden yellow.

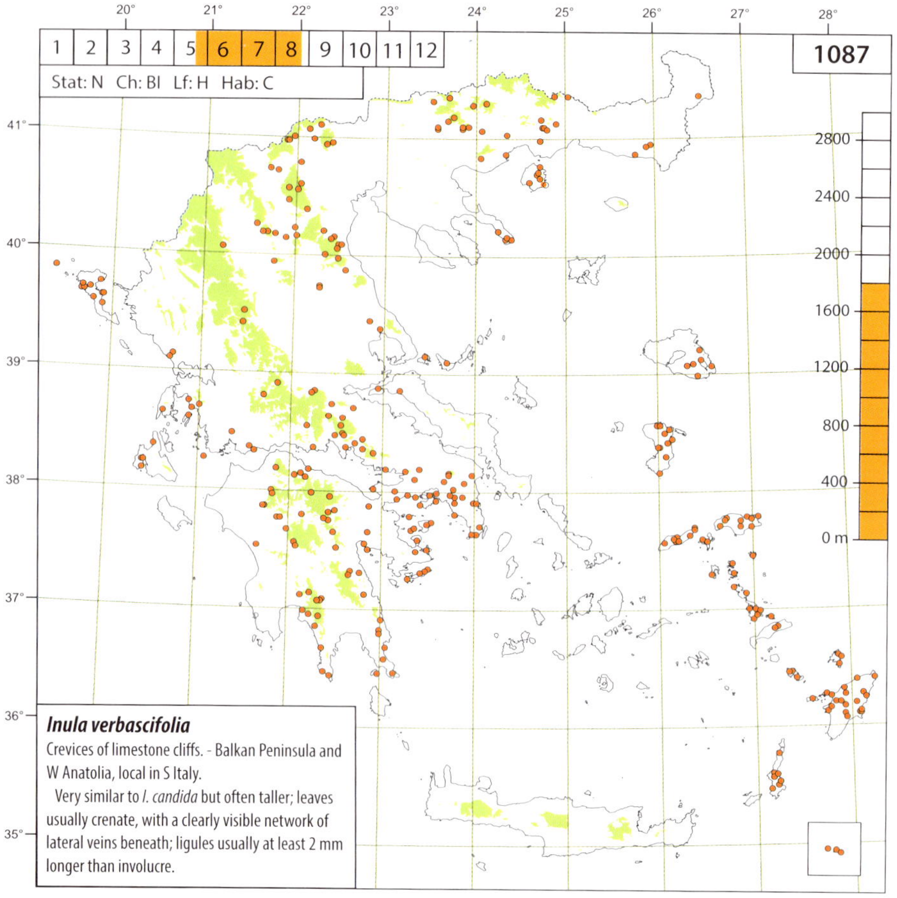

Inula verbascifolia

Crevices of limestone cliffs. - Balkan Peninsula and W Anatolia, local in S Italy.

Very similar to *I. candida* but often taller; leaves usually crenate, with a clearly visible network of lateral veins beneath; ligules usually at least 2 mm longer than involucre.

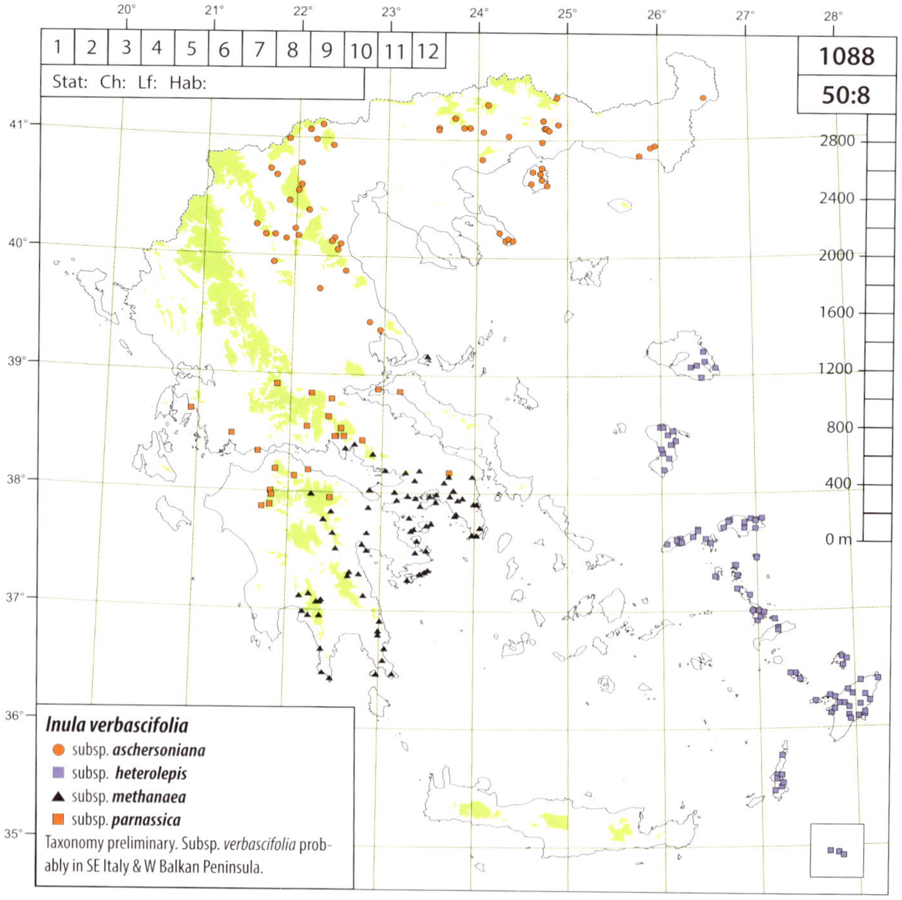

Inula verbascifolia

- subsp. ***aschersoniana***
- subsp. ***heterolepis***
- subsp. ***methanaea***
- subsp. ***parnassica***

Taxonomy preliminary. Subsp. *verbascifolia* probably in SE Italy & W Balkan Peninsula.

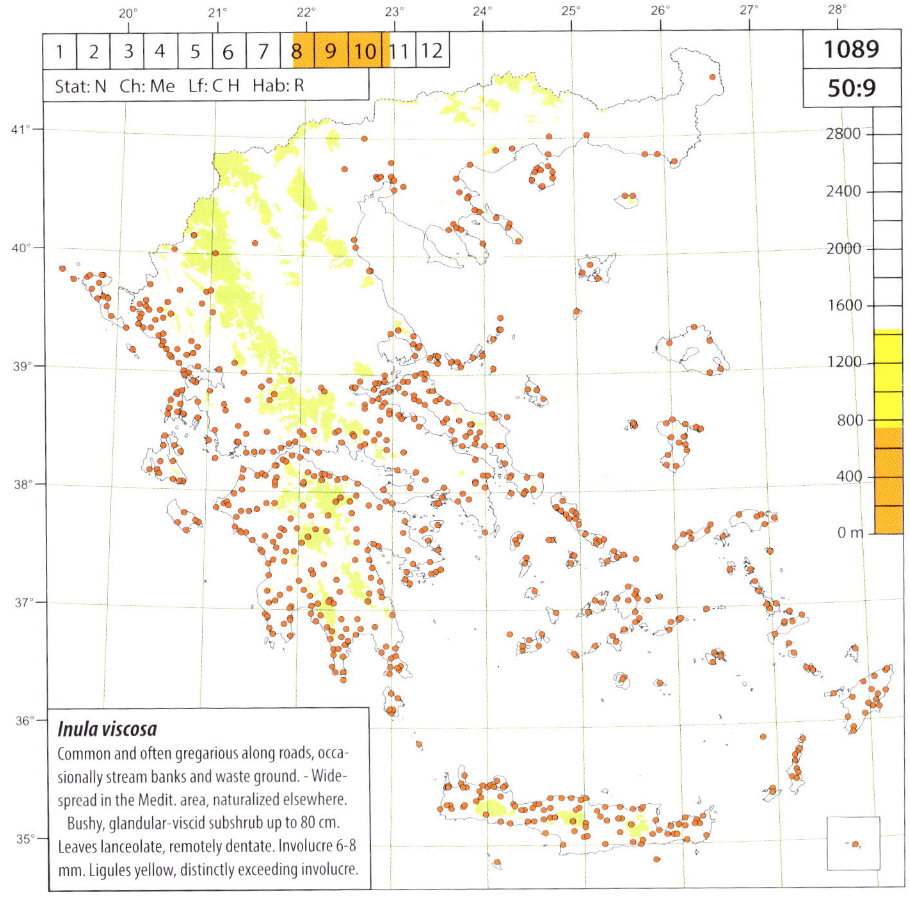

Inula viscosa
Common and often gregarious along roads, occasionally stream banks and waste ground. - Widespread in the Medit. area, naturalized elsewhere.
Bushy, glandular-viscid subshrub up to 80 cm. Leaves lanceolate, remotely dentate. Involucre 6-8 mm. Ligules yellow, distinctly exceeding involucre.

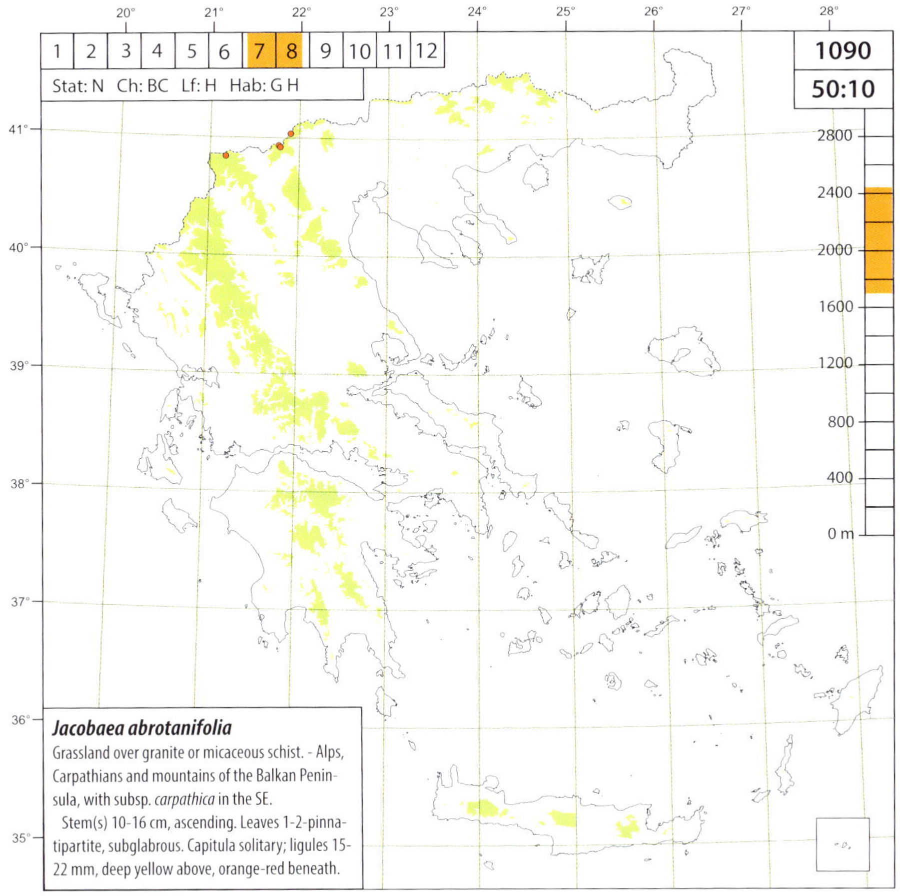

Jacobaea abrotanifolia
Grassland over granite or micaceous schist. - Alps, Carpathians and mountains of the Balkan Peninsula, with subsp. *carpathica* in the SE.
Stem(s) 10-16 cm, ascending. Leaves 1-2-pinnatipartite, subglabrous. Capitula solitary; ligules 15-22 mm, deep yellow above, orange-red beneath.

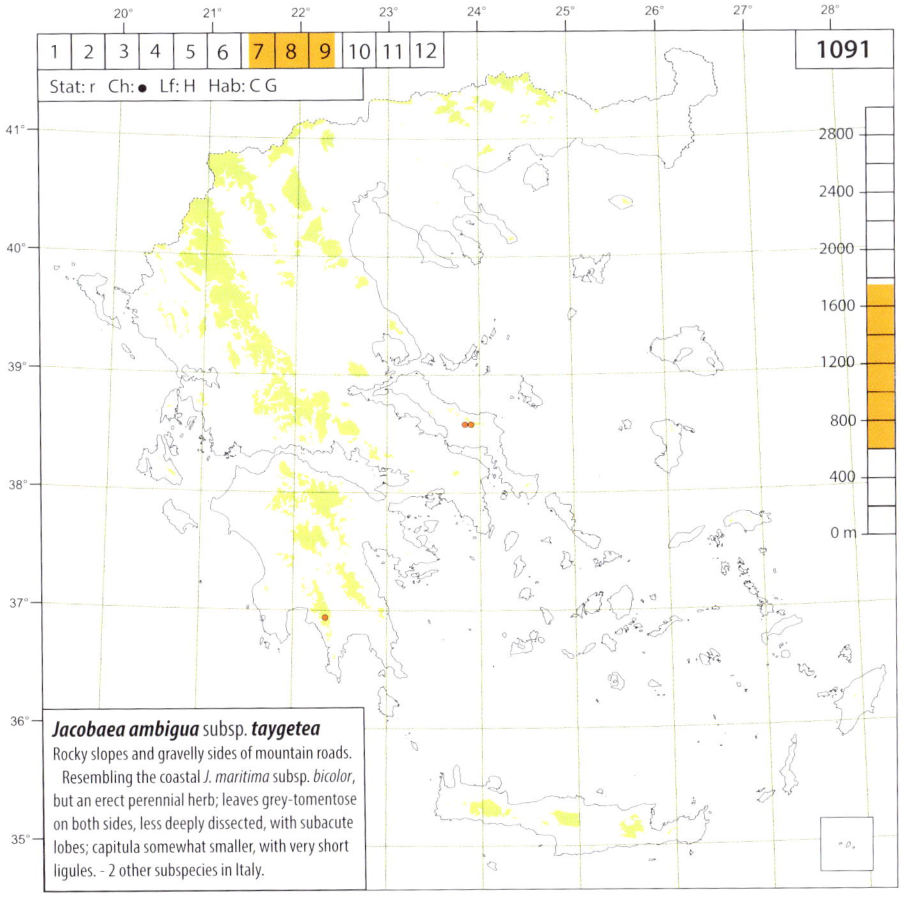

Jacobaea ambigua* subsp. *taygetea
Rocky slopes and gravelly sides of mountain roads.
Resembling the coastal *J. maritima* subsp. *bicolor*, but an erect perennial herb; leaves grey-tomentose on both sides, less deeply dissected, with subacute lobes; capitula somewhat smaller, with very short ligules. - 2 other subspecies in Italy.

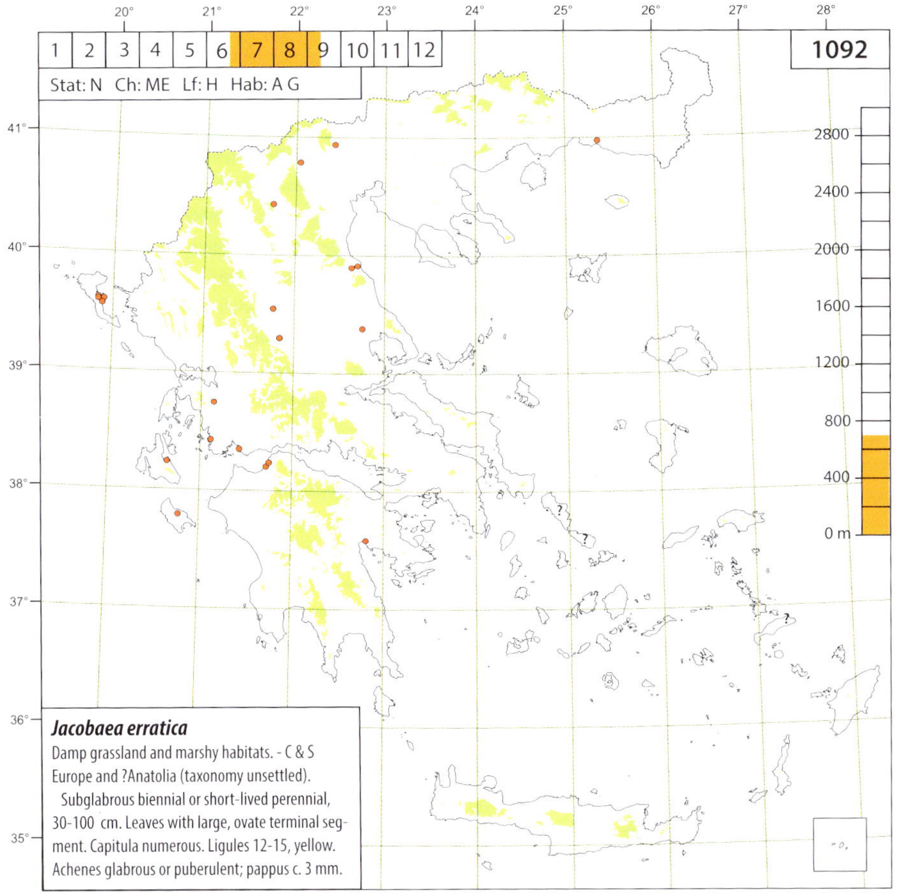

Jacobaea erratica
Damp grassland and marshy habitats. - C & S Europe and ?Anatolia (taxonomy unsettled).
Subglabrous biennial or short-lived perennial, 30-100 cm. Leaves with large, ovate terminal segment. Capitula numerous. Ligules 12-15, yellow. Achenes glabrous or puberulent; pappus c. 3 mm.

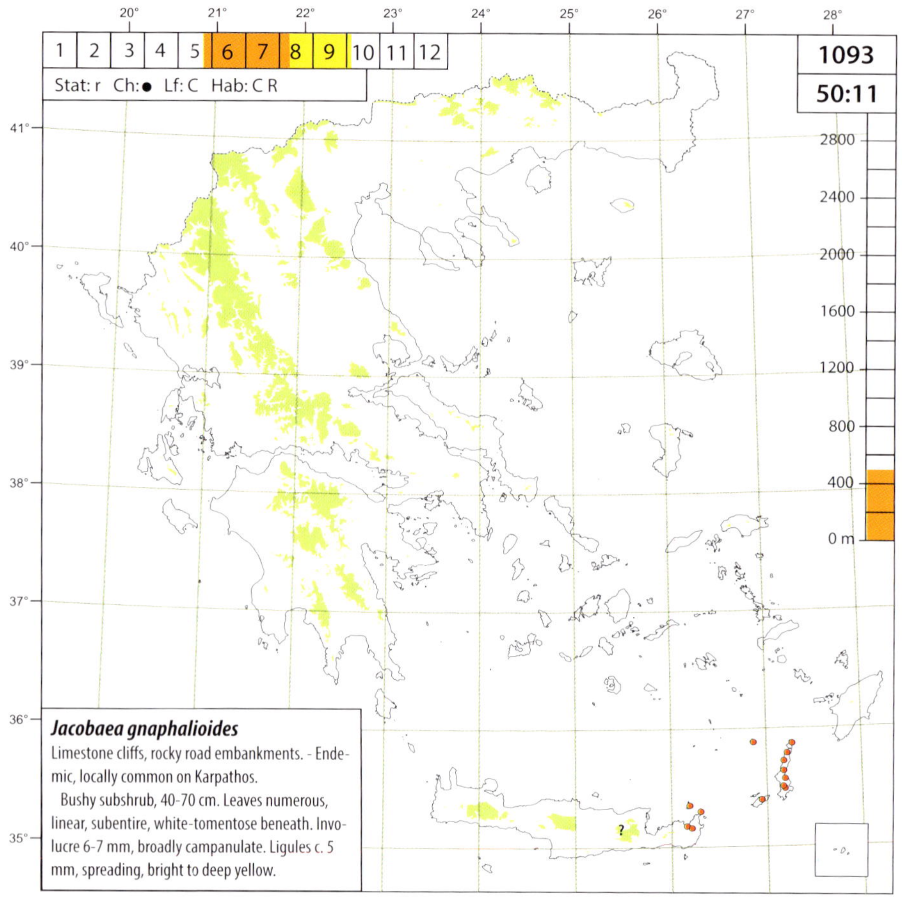

Jacobaea gnaphalioides

Limestone cliffs, rocky road embankments. - Endemic, locally common on Karpathos.

Bushy subshrub, 40-70 cm. Leaves numerous, linear, subentire, white-tomentose beneath. Involucre 6-7 mm, broadly campanulate. Ligules c. 5 mm, spreading, bright to deep yellow.

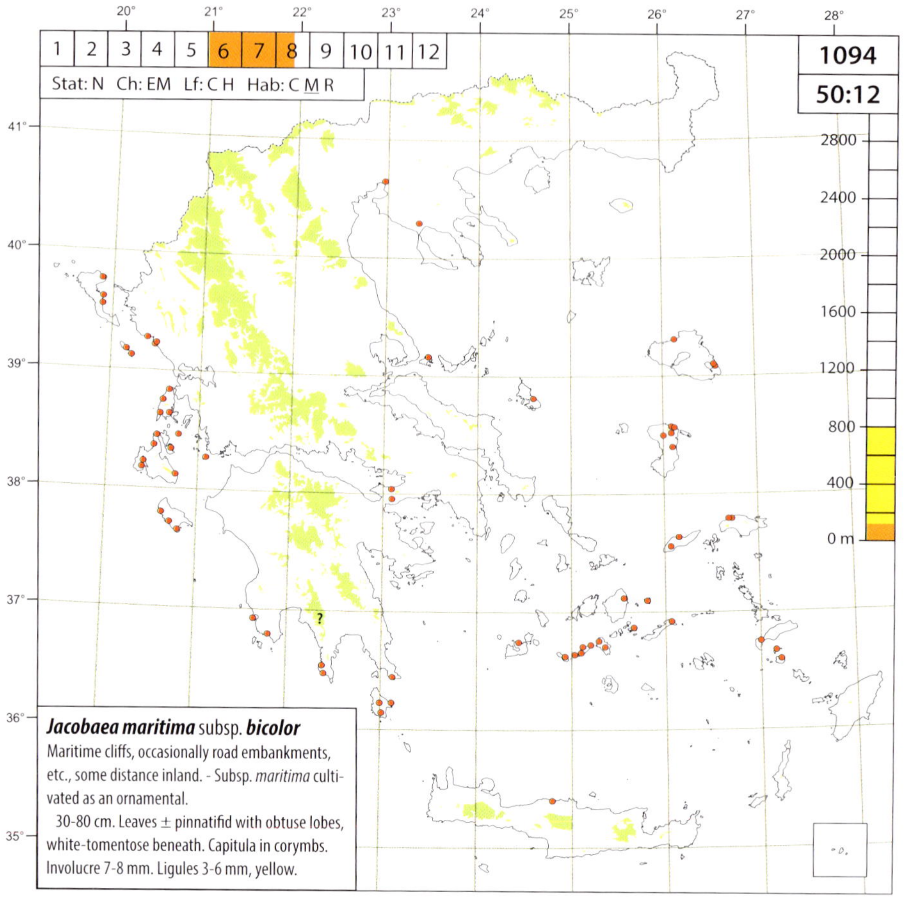

Jacobaea maritima subsp. ***bicolor***

Maritime cliffs, occasionally road embankments, etc., some distance inland. - Subsp. *maritima* cultivated as an ornamental.

30-80 cm. Leaves ± pinnatifid with obtuse lobes, white-tomentose beneath. Capitula in corymbs. Involucre 7-8 mm. Ligules 3-6 mm, yellow.

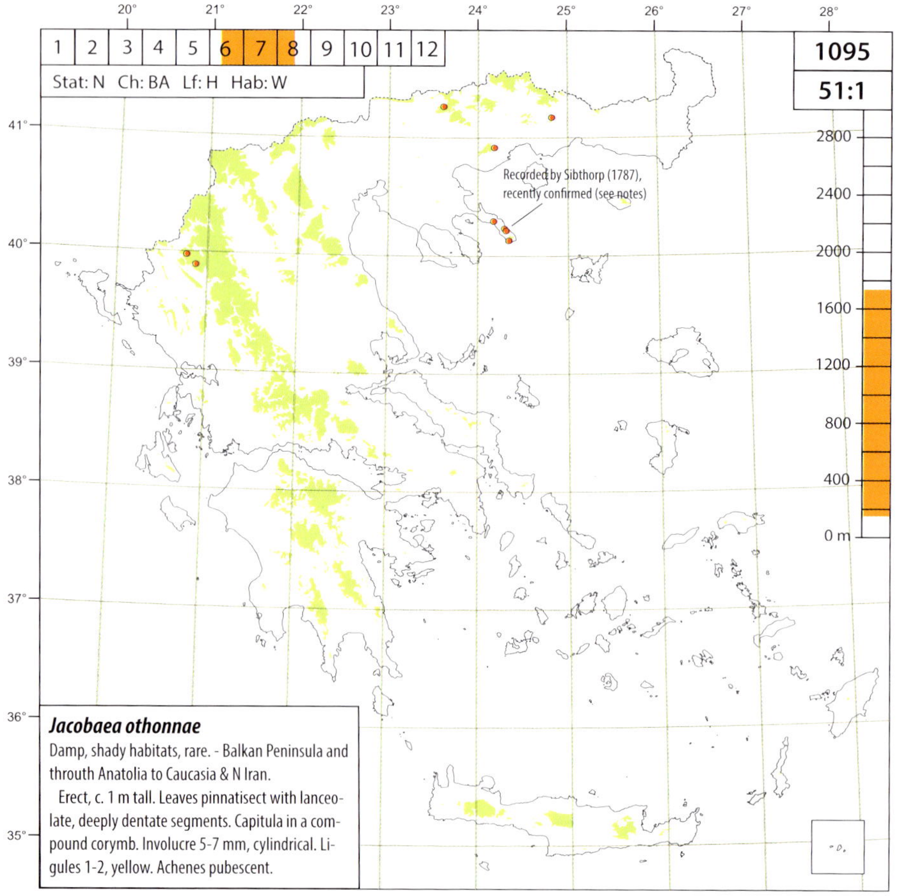

Jacobaea othonnae

Damp, shady habitats, rare. - Balkan Peninsula and throuth Anatolia to Caucasia & N Iran.

Erect, c. 1 m tall. Leaves pinnatisect with lanceolate, deeply dentate segments. Capitula in a compound corymb. Involucre 5-7 mm, cylindrical. Ligules 1-2, yellow. Achenes pubescent.

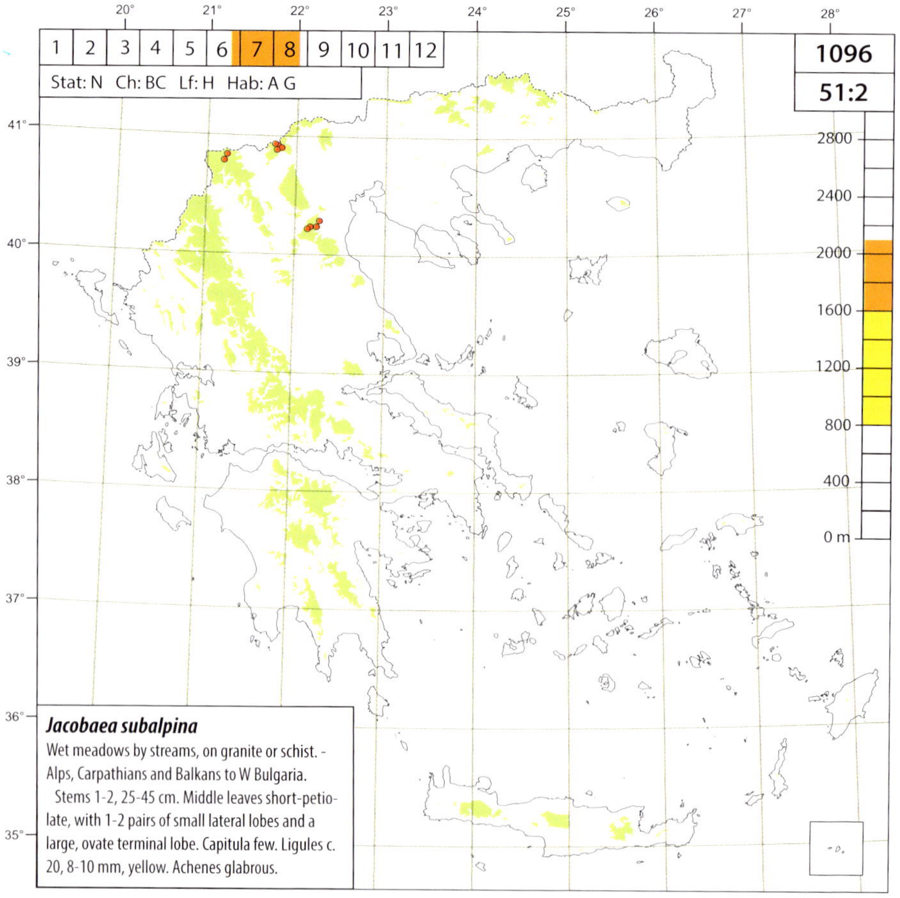

Jacobaea subalpina

Wet meadows by streams, on granite or schist. - Alps, Carpathians and Balkans to W Bulgaria.

Stems 1-2, 25-45 cm. Middle leaves short-petiolate, with 1-2 pairs of small lateral lobes and a large, ovate terminal lobe. Capitula few. Ligules c. 20, 8-10 mm, yellow. Achenes glabrous.

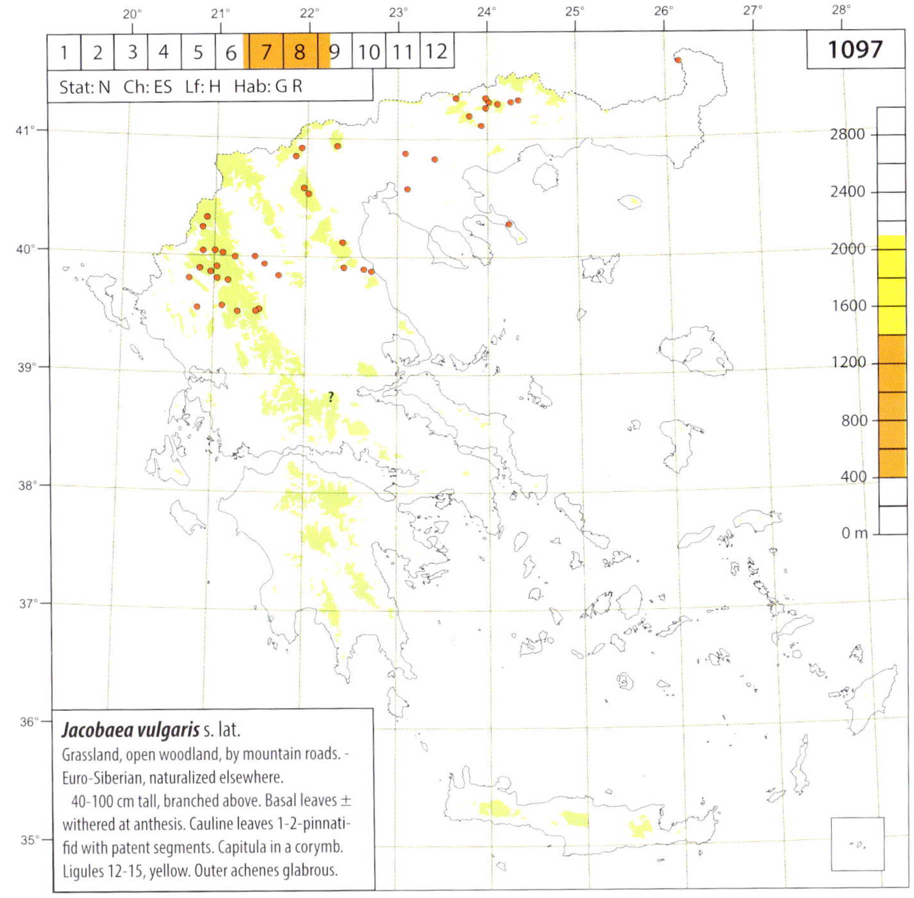

Jacobaea vulgaris s. lat.
Grassland, open woodland, by mountain roads. - Euro-Siberian, naturalized elsewhere.
40-100 cm tall, branched above. Basal leaves ± withered at anthesis. Cauline leaves 1-2-pinnatifid with patent segments. Capitula in a corymb. Ligules 12-15, yellow. Outer achenes glabrous.

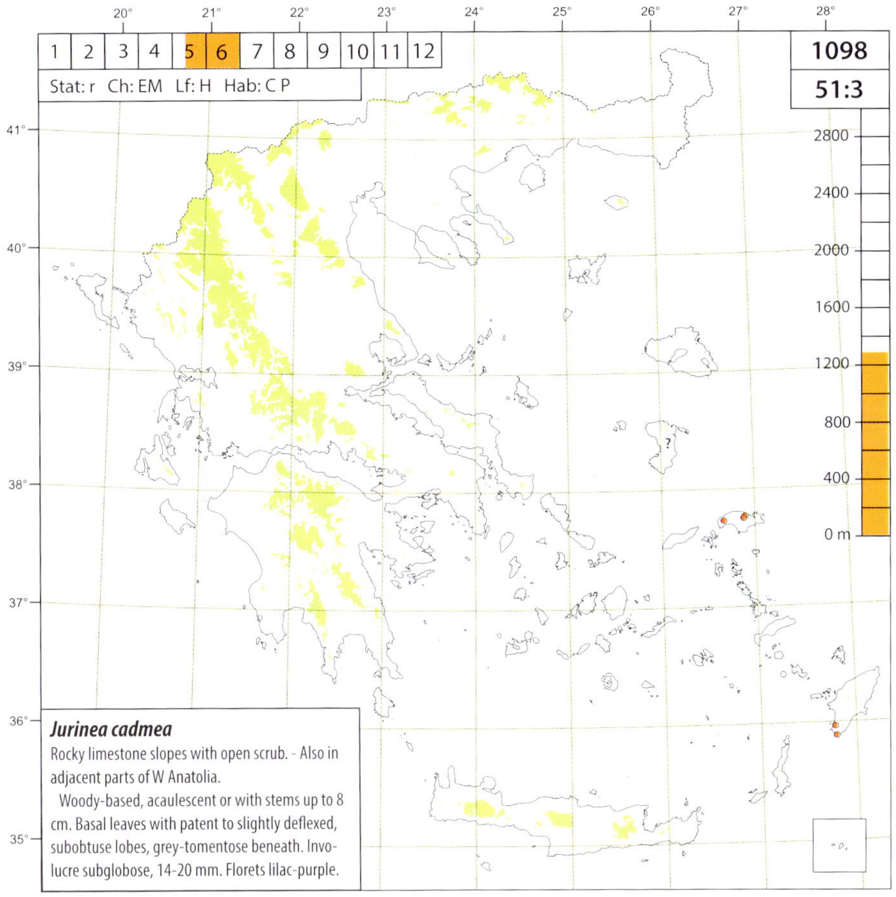

Jurinea cadmea
Rocky limestone slopes with open scrub. - Also in adjacent parts of W Anatolia.
Woody-based, acaulescent or with stems up to 8 cm. Basal leaves with patent to slightly deflexed, subobtuse lobes, grey-tomentose beneath. Involucre subglobose, 14-20 mm. Florets lilac-purple.

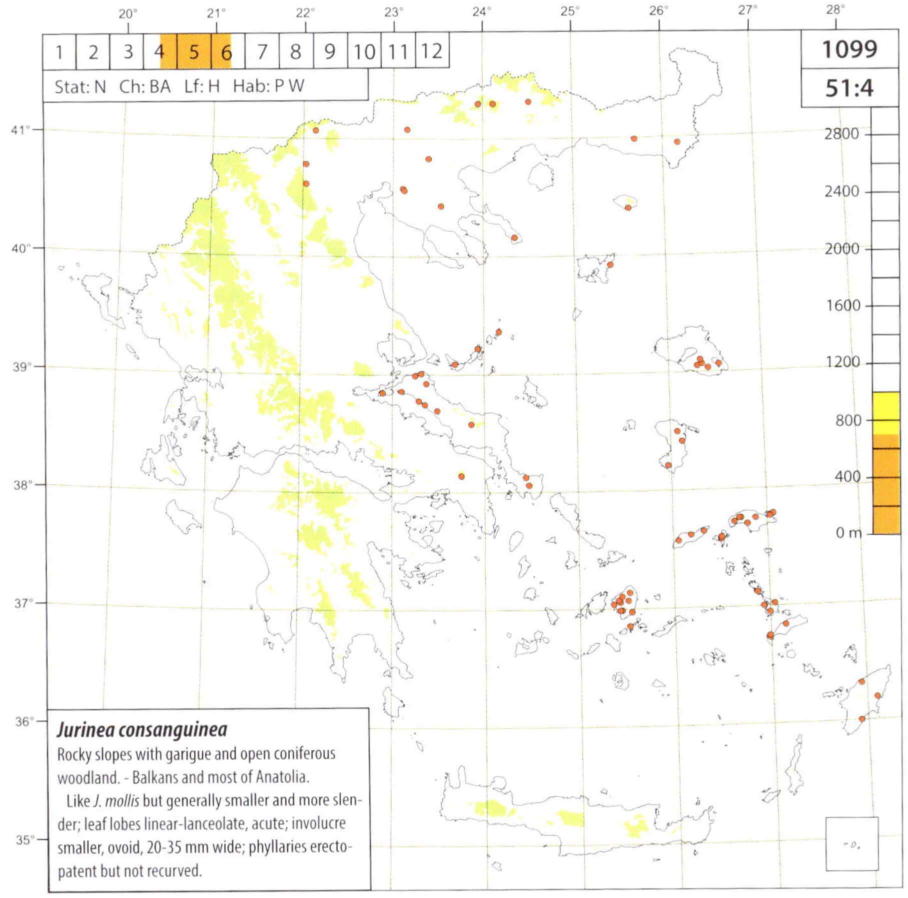

Jurinea consanguinea
Rocky slopes with garigue and open coniferous woodland. - Balkans and most of Anatolia.
Like *J. mollis* but generally smaller and more slender; leaf lobes linear-lanceolate, acute; involucre smaller, ovoid, 20-35 mm wide; phyllaries erecto-patent but not recurved.

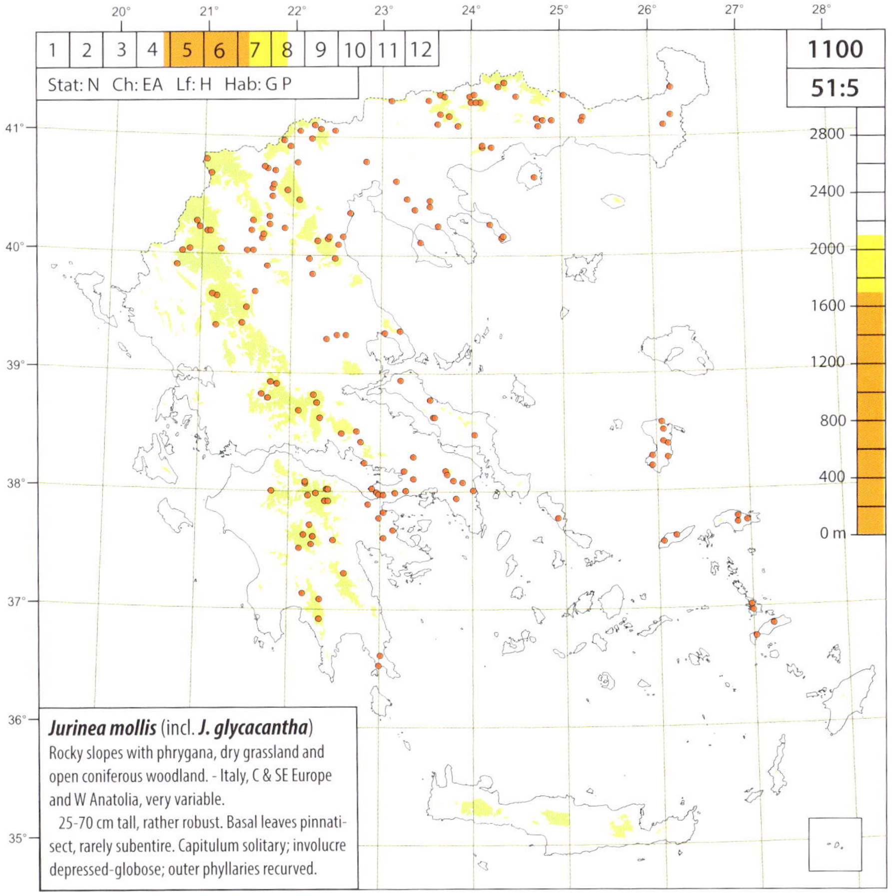

Jurinea mollis (incl. ***J. glycacantha***)
Rocky slopes with phrygana, dry grassland and open coniferous woodland. - Italy, C & SE Europe and W Anatolia, very variable.
25-70 cm tall, rather robust. Basal leaves pinnatisect, rarely subentire. Capitulum solitary; involucre depressed-globose; outer phyllaries recurved.

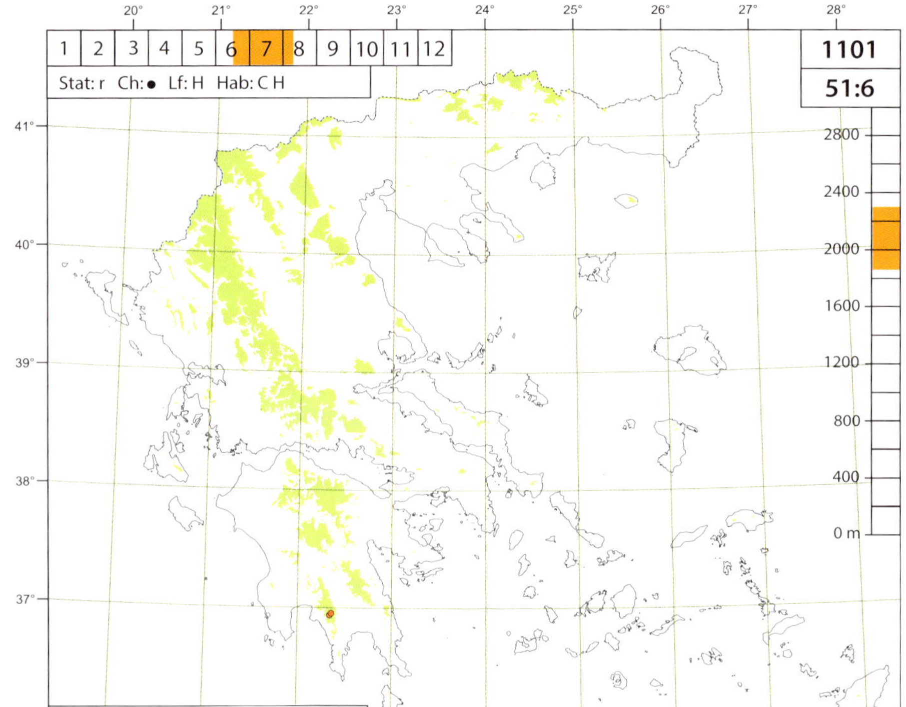

Jurinea taygetea
Cliff ledges and rocky limestone slopes. - A rare local endemic.
Acaulescent or with stem up to 6 cm. Leaves usually pinnatifid with obtuse lobes. Involucre 12-22 mm wide; outer phyllaries spreading. Florets pale pinkish. Pappus of slender hairs united at base.

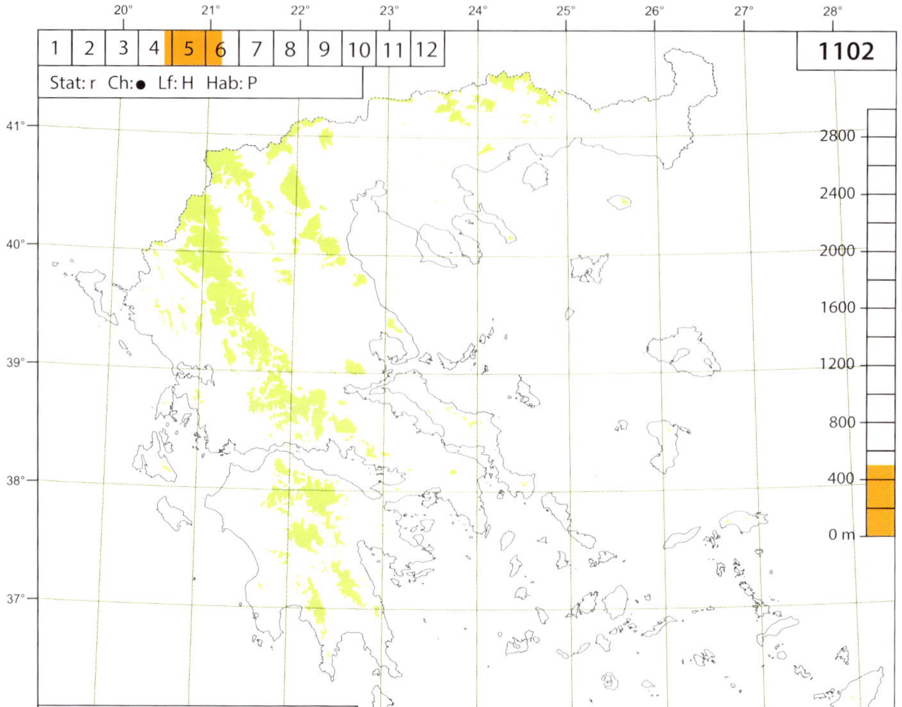

Klasea cretica
Rocky slopes with phrygana and open coniferous woodland. - Endemic.
25-40 cm tall, slender; stem narrowly winged. Cauline leaves decurrent, sinuate-dentate. Capitulum solitary; involucre 20-25 mm, ovoid. Phyllaries with mucro 4-5 mm. Florets purplish-mauve.

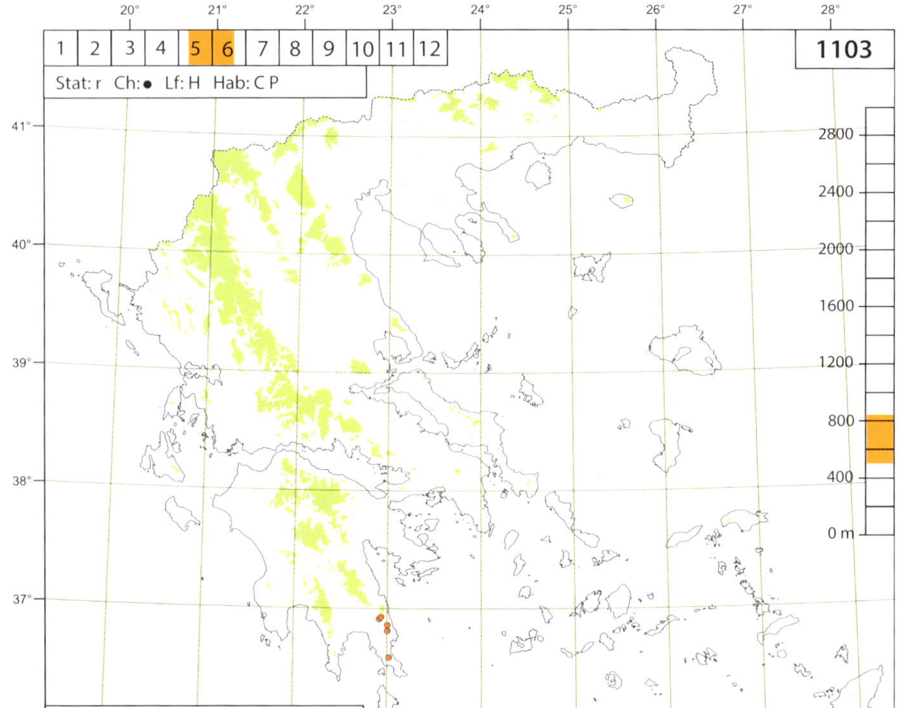

Klasea moreana
Rocky limestone slopes with phrygana. - Endemic.
Like *K. cretica* but stem unwinged; cauline leaves firm, auriculate, entire, with a network of prominent white or purplish veins; upper leaves reduced and bract-like; phyllaries somewhat narrower, with mucro 3-3.5 mm.

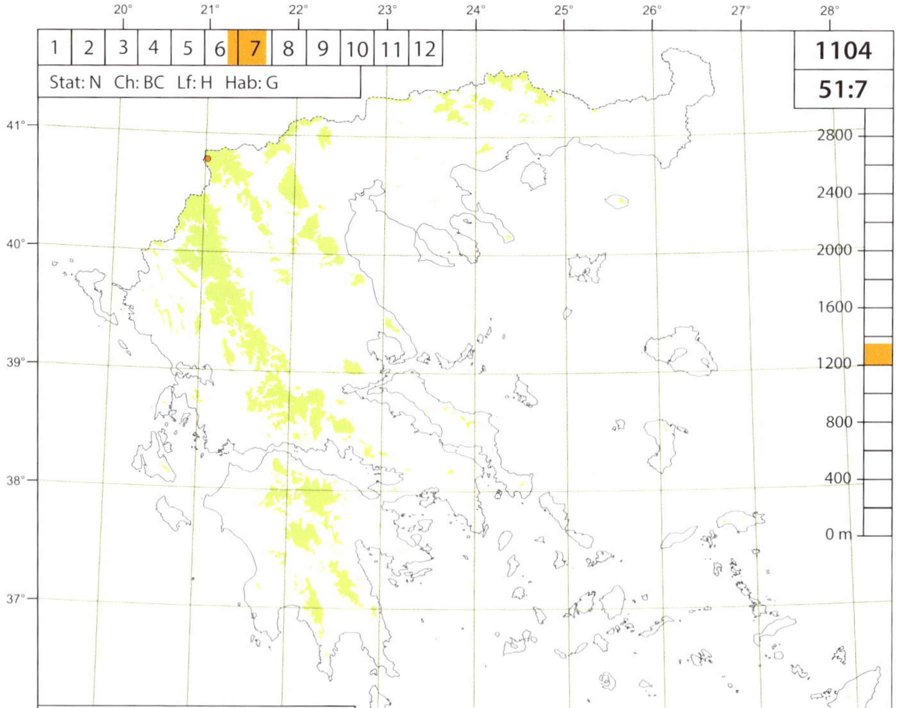

Klasea radiata
Rocky limestone flats with dry grassland and open deciduous woodland. - E & SE Europe, with subsp. *cetinjensis* in the SW part of the range.
30-50 cm. Leaves pinnatifid. Capitulum solitary; involucre c. 17 × 11 mm, ovoid. Phyllaries with weak spinule 1-1.5 mm. Florets lilac-purple.

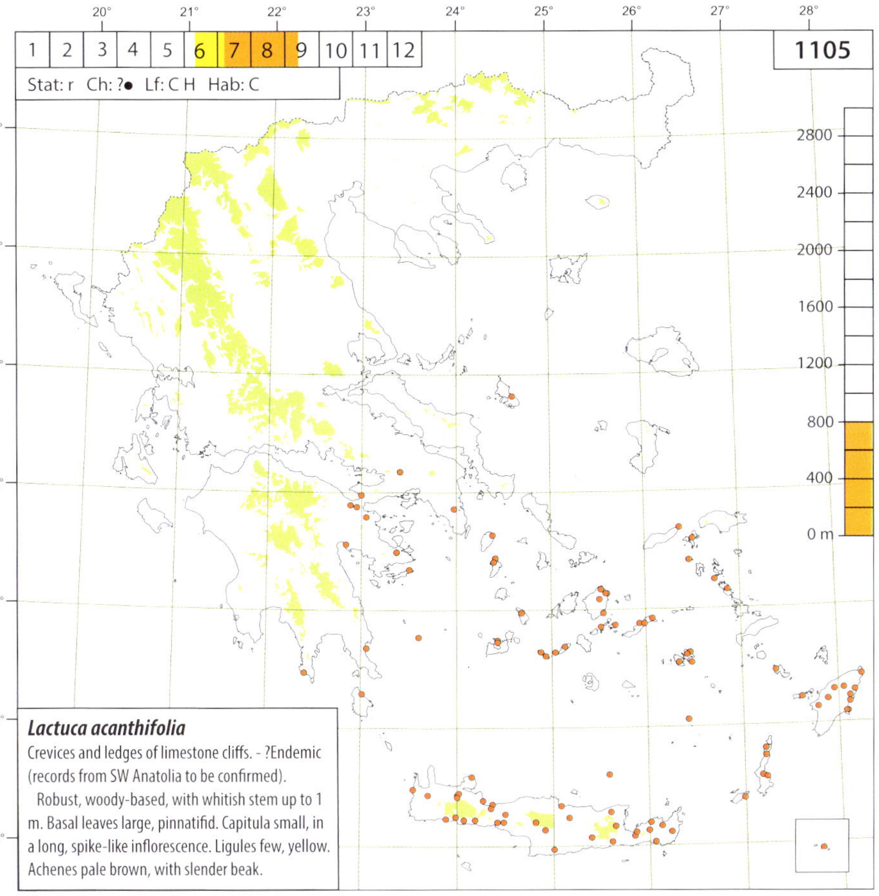

Lactuca acanthifolia

Crevices and ledges of limestone cliffs. - ?Endemic (records from SW Anatolia to be confirmed).

Robust, woody-based, with whitish stem up to 1 m. Basal leaves large, pinnatifid. Capitula small, in a long, spike-like inflorescence. Ligules few, yellow. Achenes pale brown, with slender beak.

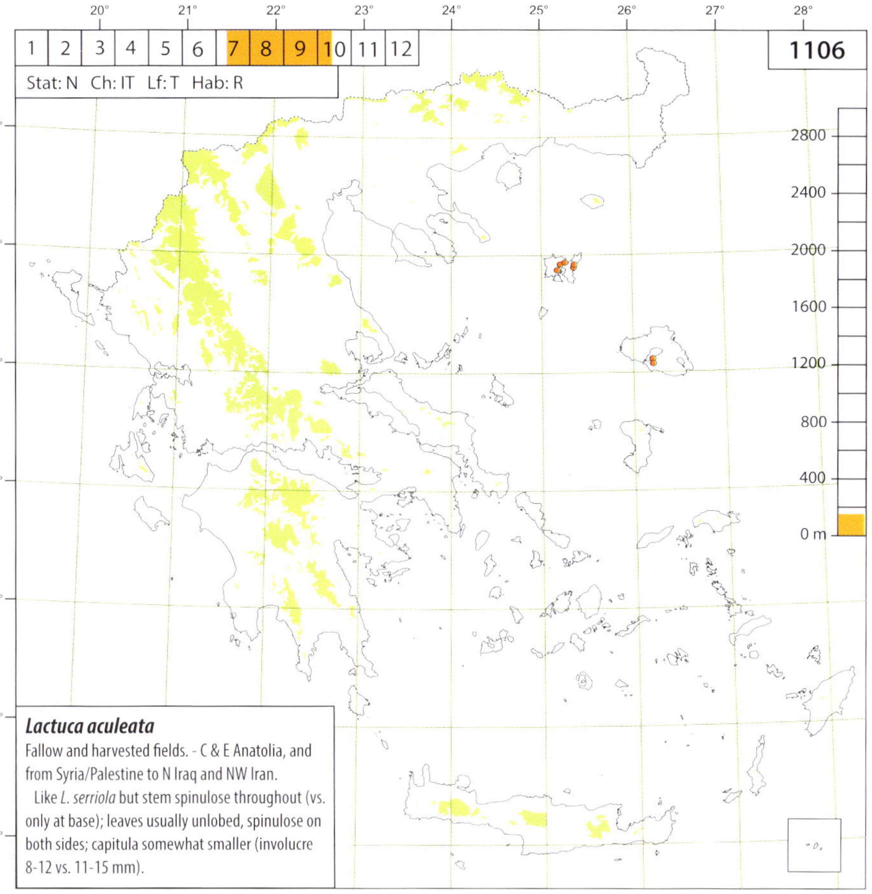

Lactuca aculeata

Fallow and harvested fields. - C & E Anatolia, and from Syria/Palestine to N Iraq and NW Iran.

Like *L. serriola* but stem spinulose throughout (vs. only at base); leaves usually unlobed, spinulose on both sides; capitula somewhat smaller (involucre 8-12 vs. 11-15 mm).

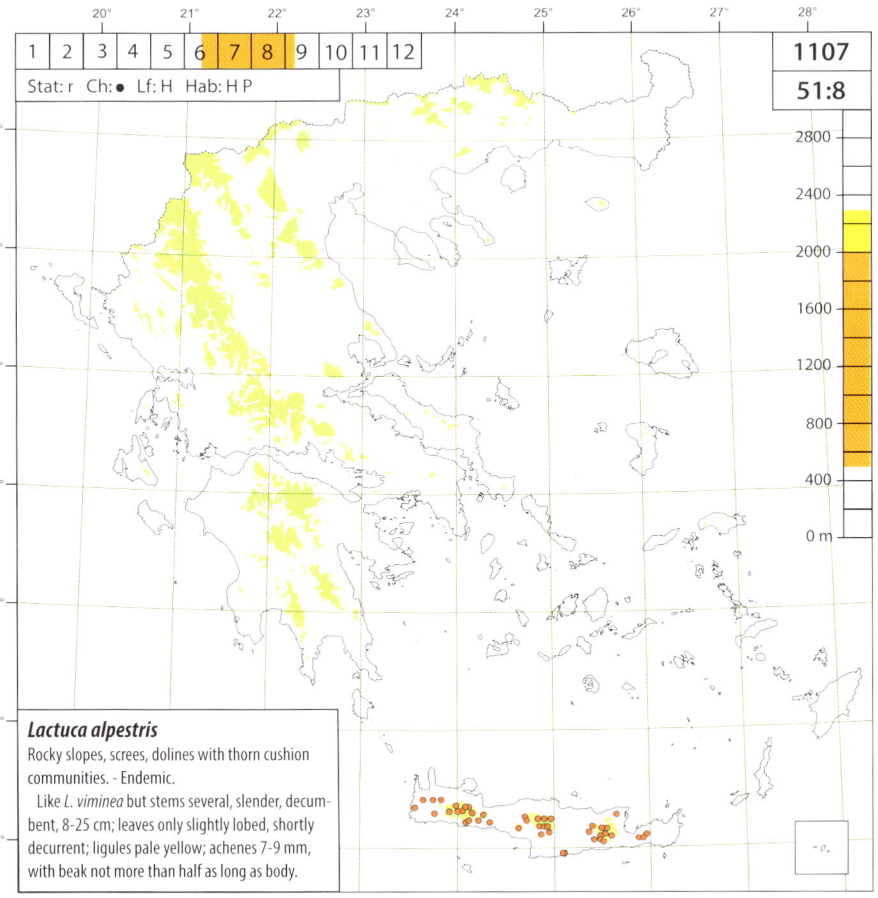

Lactuca alpestris

Rocky slopes, screes, dolines with thorn cushion communities. - Endemic.

Like *L. viminea* but stems several, slender, decumbent, 8-25 cm; leaves only slightly lobed, shortly decurrent; ligules pale yellow; achenes 7-9 mm, with beak not more than half as long as body.

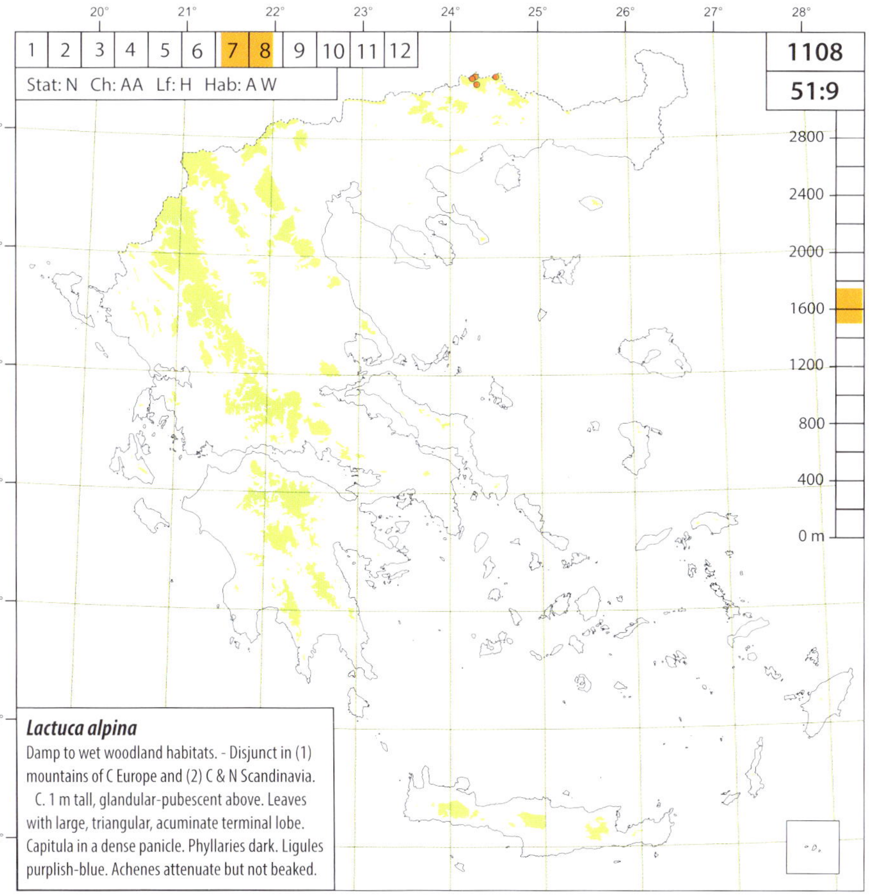

Lactuca alpina

Damp to wet woodland habitats. - Disjunct in (1) mountains of C Europe and (2) C & N Scandinavia.

C. 1 m tall, glandular-pubescent above. Leaves with large, triangular, acuminate terminal lobe. Capitula in a dense panicle. Phyllaries dark. Ligules purplish-blue. Achenes attenuate but not beaked.

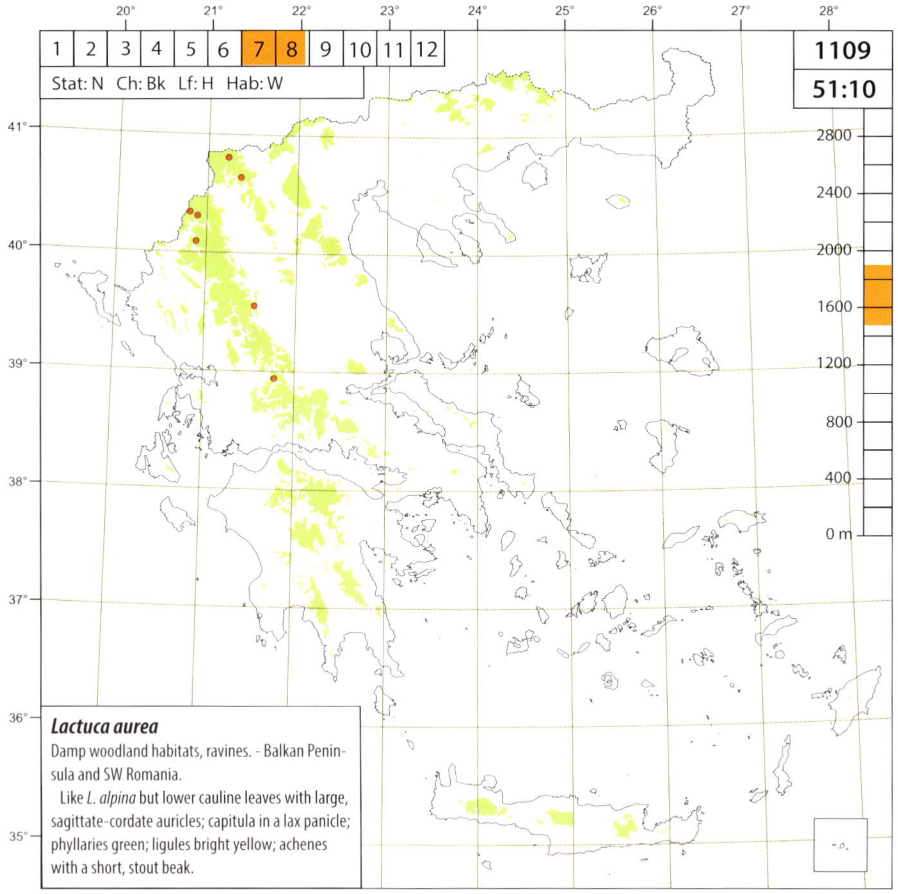

Lactuca aurea

Damp woodland habitats, ravines. - Balkan Peninsula and SW Romania.

Like *L. alpina* but lower cauline leaves with large, sagittate-cordate auricles; capitula in a lax panicle; phyllaries green; ligules bright yellow; achenes with a short, stout beak.

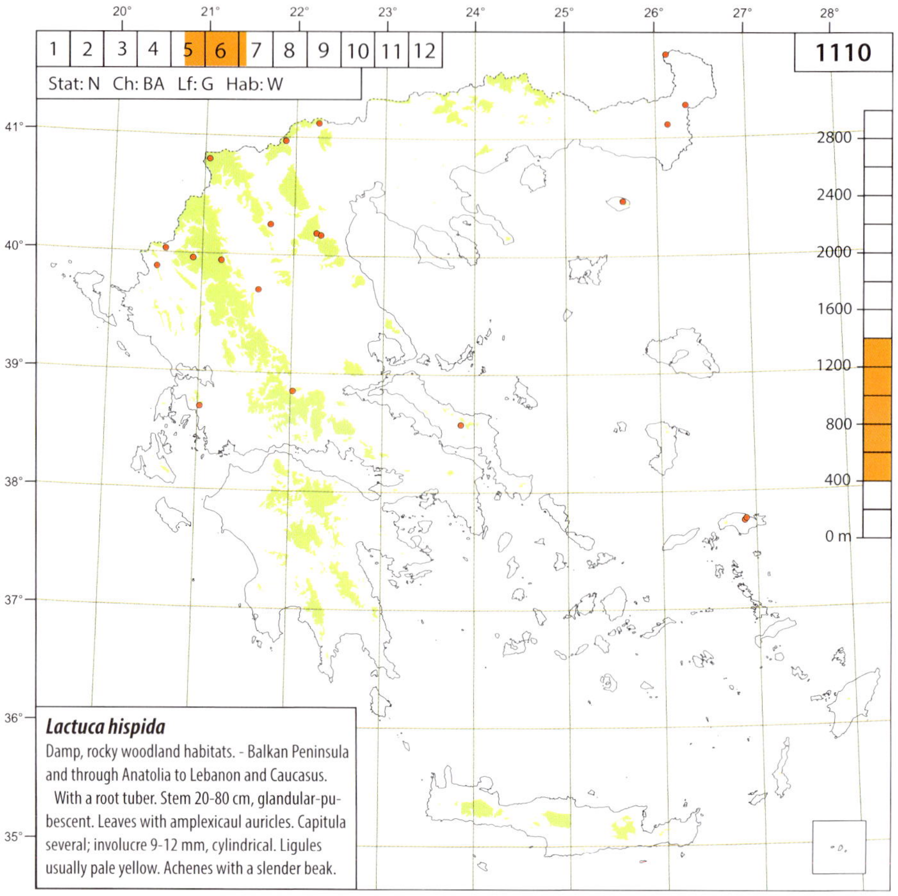

Lactuca hispida

Damp, rocky woodland habitats. - Balkan Peninsula and through Anatolia to Lebanon and Caucasus.

With a root tuber. Stem 20-80 cm, glandular-pubescent. Leaves with amplexicaul auricles. Capitula several; involucre 9-12 mm, cylindrical. Ligules usually pale yellow. Achenes with a slender beak.

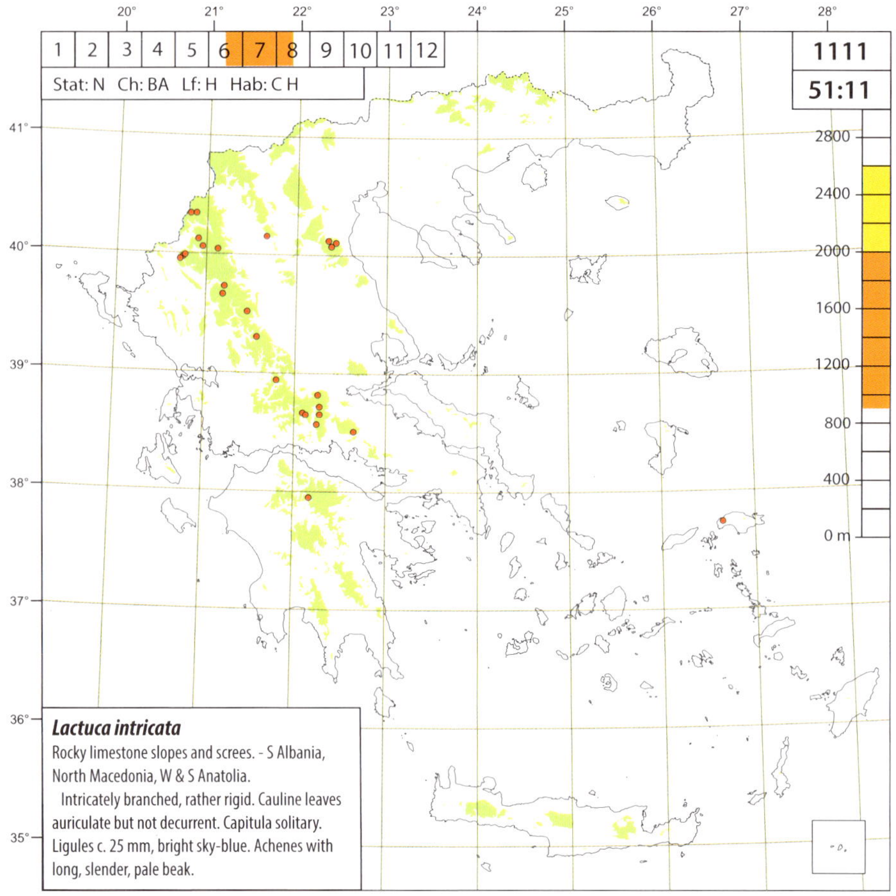

Lactuca intricata

Rocky limestone slopes and screes. - S Albania, North Macedonia, W & S Anatolia.

Intricately branched, rather rigid. Cauline leaves auriculate but not decurrent. Capitula solitary. Ligules c. 25 mm, bright sky-blue. Achenes with long, slender, pale beak.

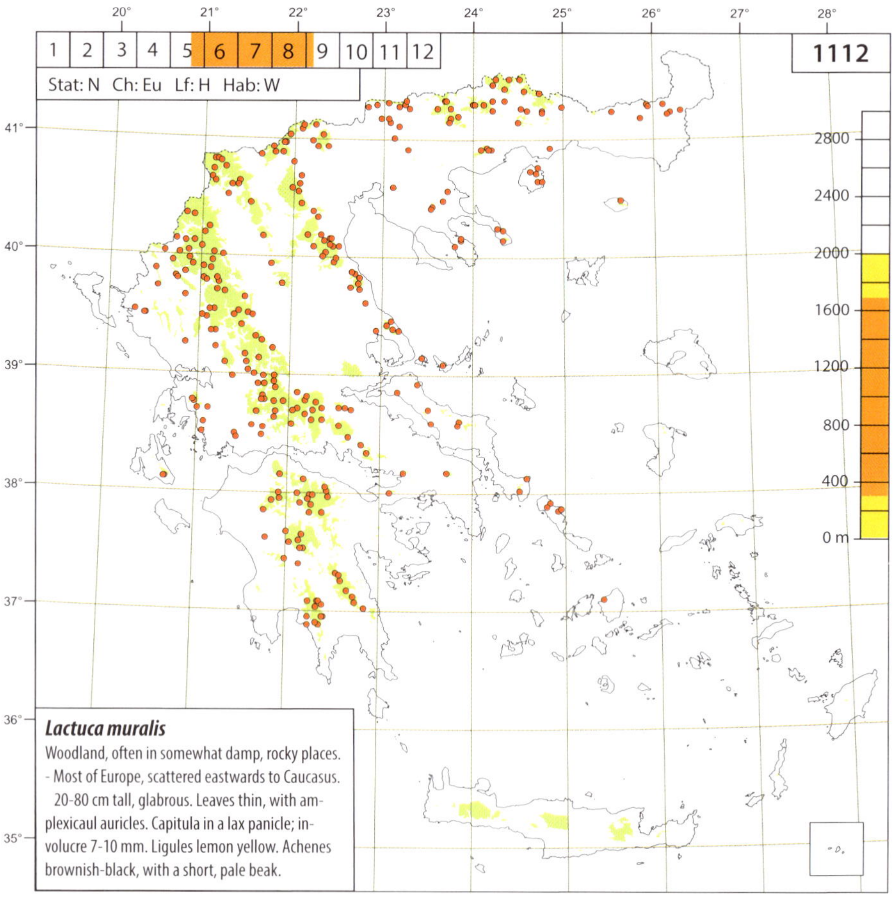

Lactuca muralis

Woodland, often in somewhat damp, rocky places. - Most of Europe, scattered eastwards to Caucasus.

20-80 cm tall, glabrous. Leaves thin, with amplexicaul auricles. Capitula in a lax panicle; involucre 7-10 mm. Ligules lemon yellow. Achenes brownish-black, with a short, pale beak.

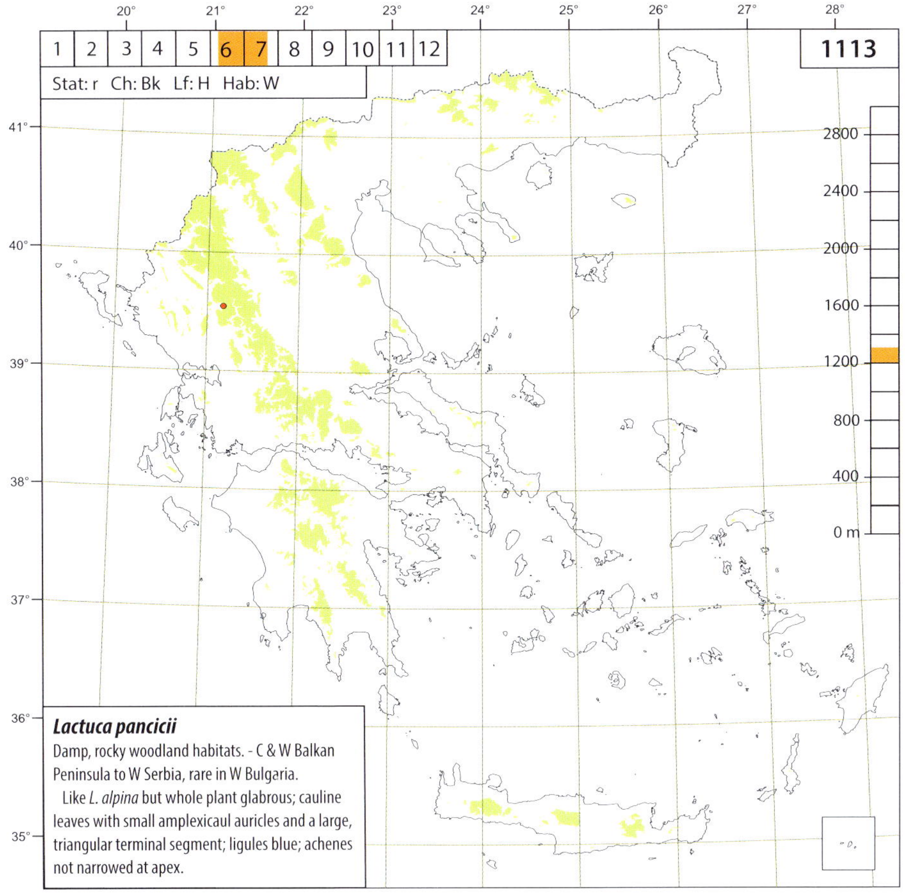

Lactuca pancicii
Damp, rocky woodland habitats. - C & W Balkan Peninsula to W Serbia, rare in W Bulgaria.
Like *L. alpina* but whole plant glabrous; cauline leaves with small amplexicaul auricles and a large, triangular terminal segment; ligules blue; achenes not narrowed at apex.

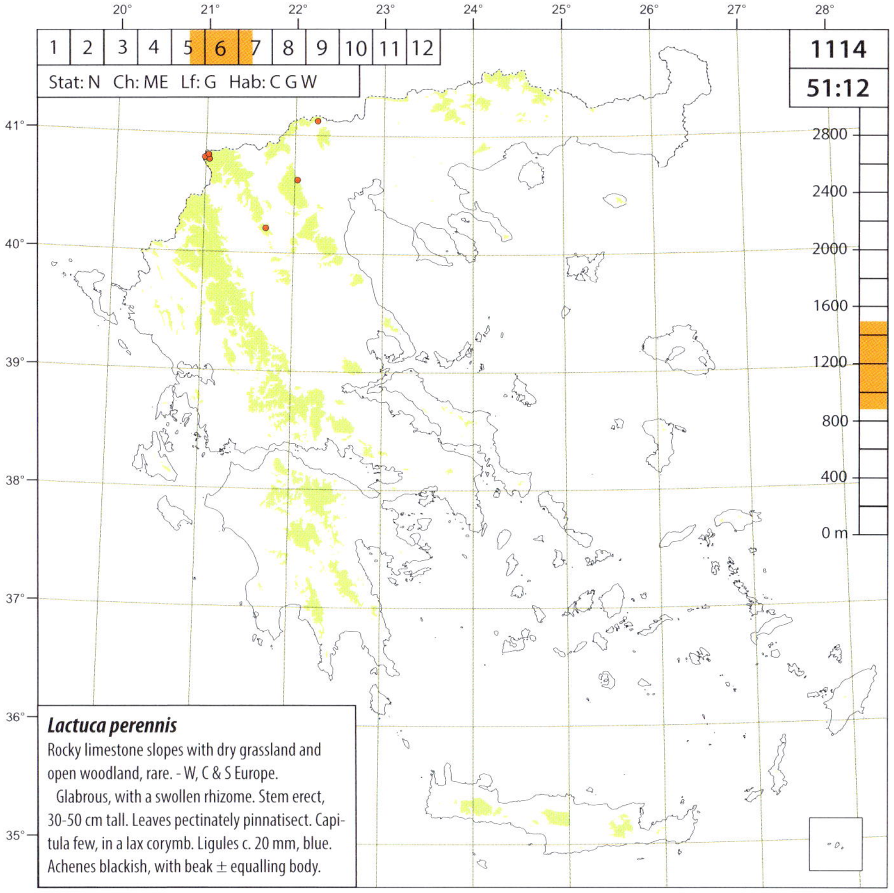

Lactuca perennis
Rocky limestone slopes with dry grassland and open woodland, rare. - W, C & S Europe.
Glabrous, with a swollen rhizome. Stem erect, 30-50 cm tall. Leaves pectinately pinnatisect. Capitula few, in a lax corymb. Ligules c. 20 mm, blue. Achenes blackish, with beak ± equalling body.

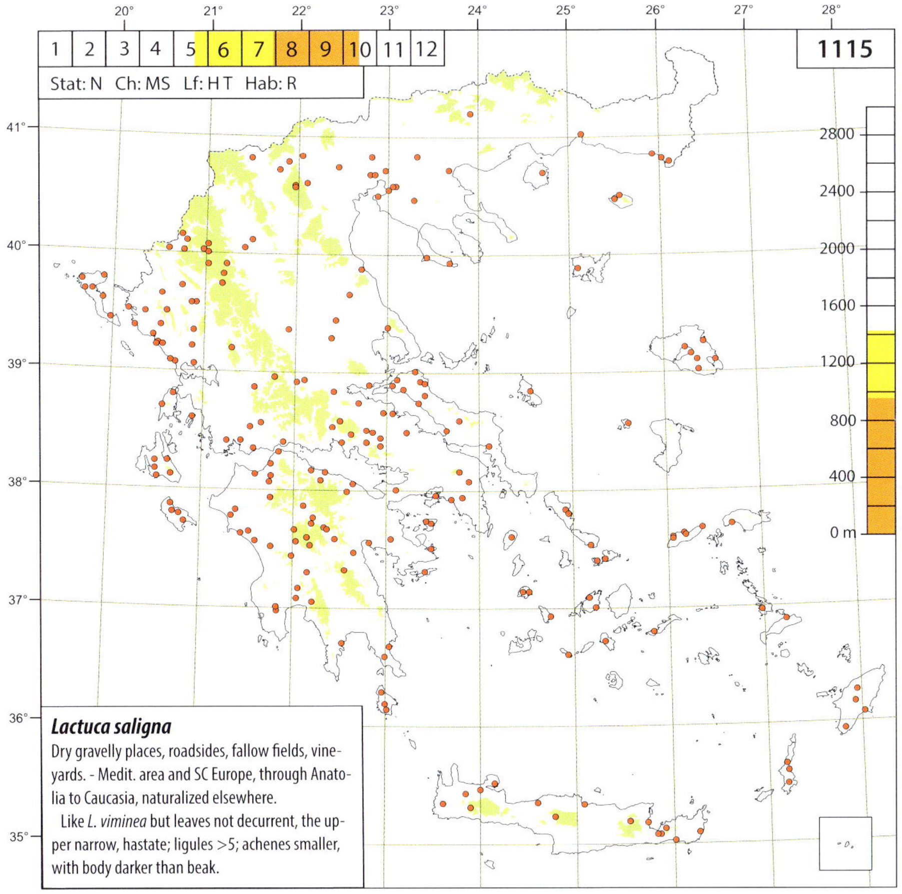

Lactuca saligna
Dry gravelly places, roadsides, fallow fields, vineyards. - Medit. area and SC Europe, through Anatolia to Caucasia, naturalized elsewhere.
Like *L. viminea* but leaves not decurrent, the upper narrow, hastate; ligules >5; achenes smaller, with body darker than beak.

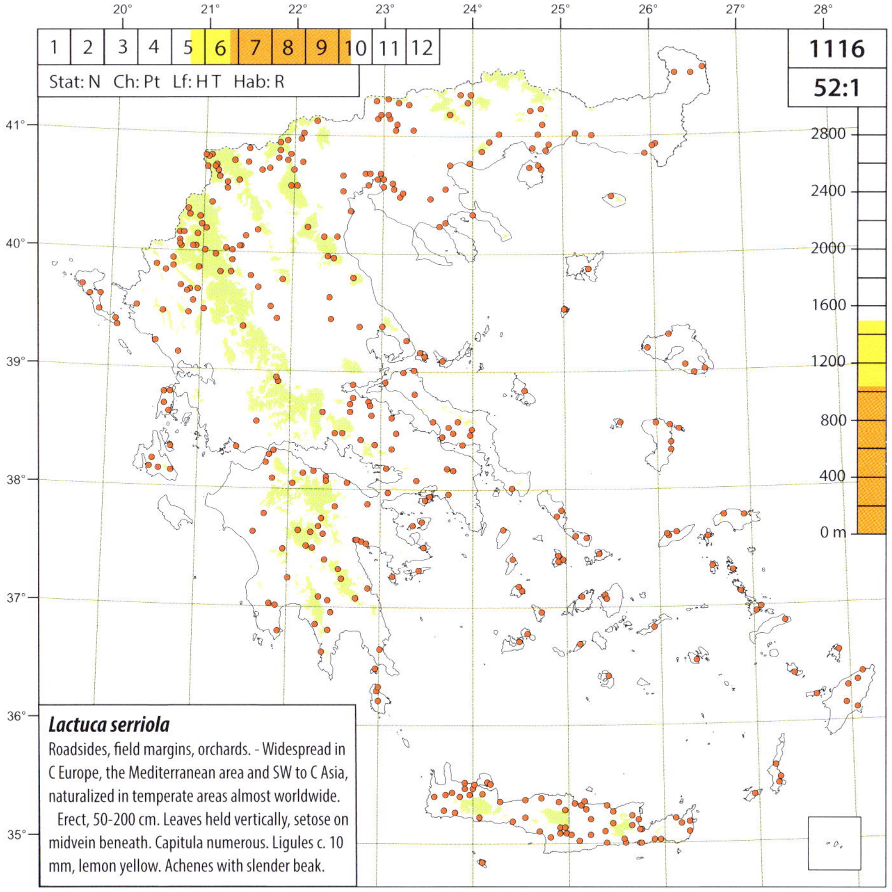

Lactuca serriola
Roadsides, field margins, orchards. - Widespread in C Europe, the Mediterranean area and SW to C Asia, naturalized in temperate areas almost worldwide.
Erect, 50-200 cm. Leaves held vertically, setose on midvein beneath. Capitula numerous. Ligules c. 10 mm, lemon yellow. Achenes with slender beak.

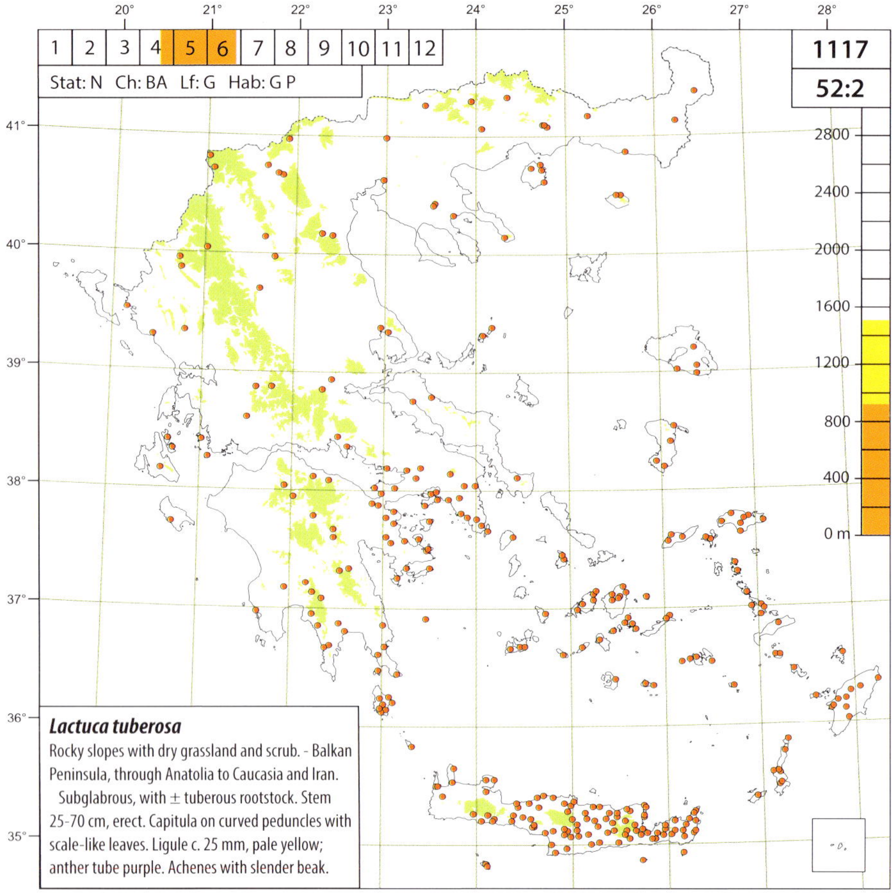

Lactuca tuberosa
Rocky slopes with dry grassland and scrub. - Balkan Peninsula, through Anatolia to Caucasia and Iran.
Subglabrous, with ± tuberous rootstock. Stem 25-70 cm, erect. Capitula on curved peduncles with scale-like leaves. Ligule c. 25 mm, pale yellow; anther tube purple. Achenes with slender beak.

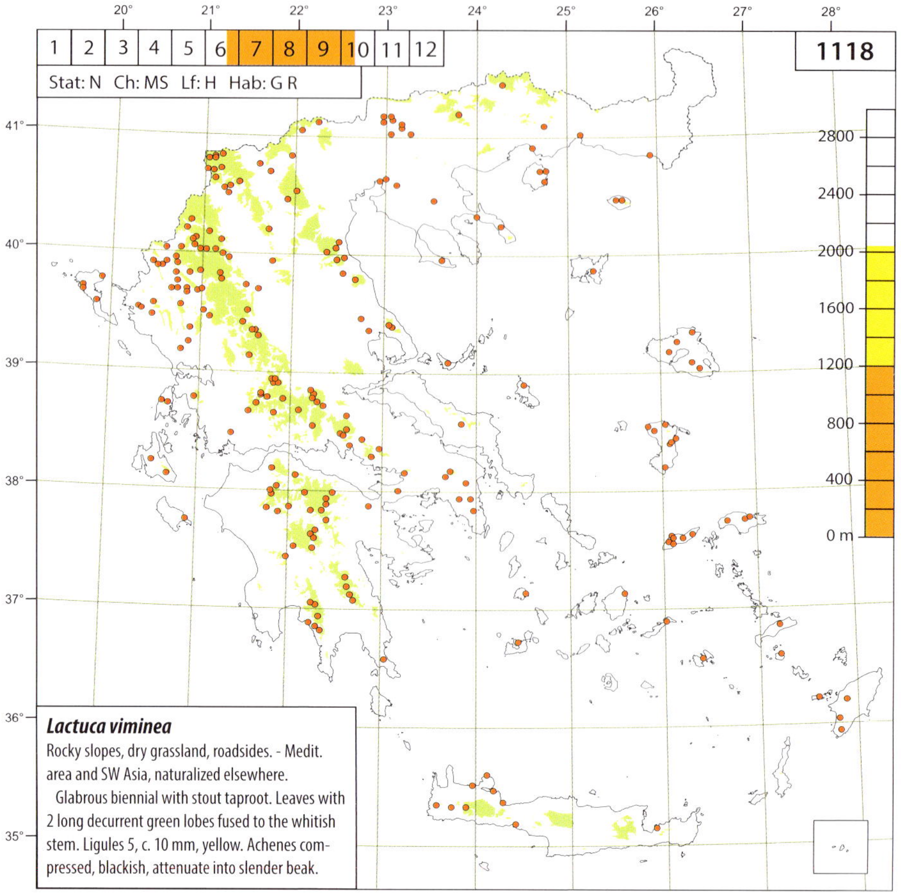

Lactuca viminea
Rocky slopes, dry grassland, roadsides. - Medit. area and SW Asia, naturalized elsewhere.
Glabrous biennial with stout taproot. Leaves with 2 long decurrent green lobes fused to the whitish stem. Ligules 5, c. 10 mm, yellow. Achenes compressed, blackish, attenuate into slender beak.

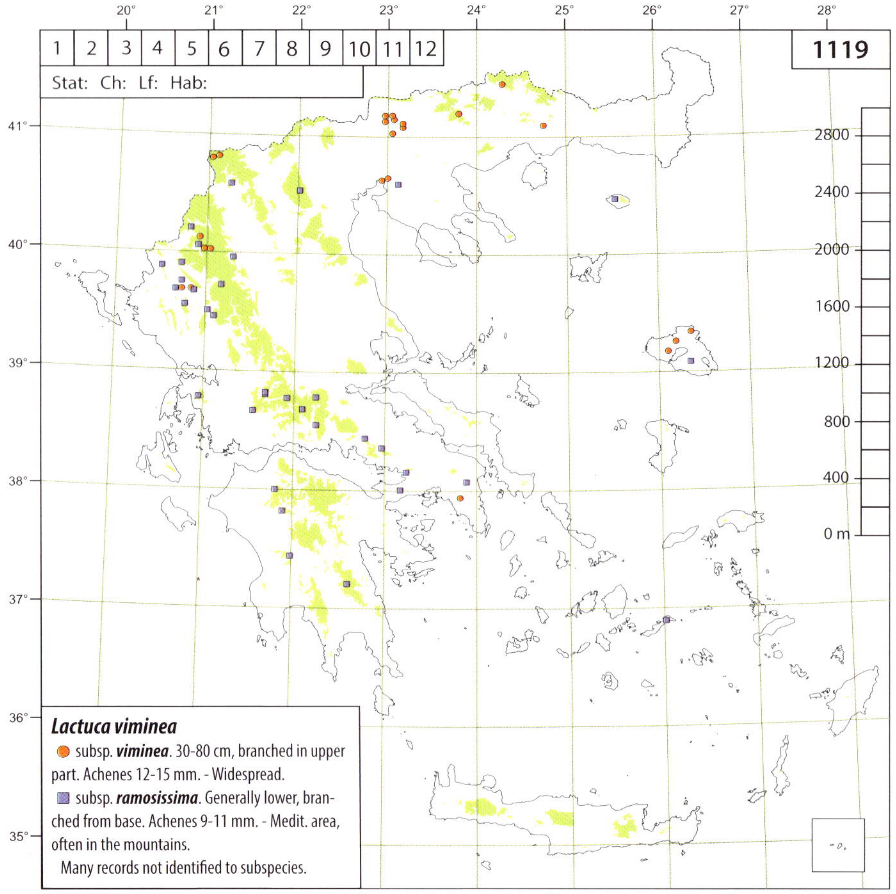

Lactuca viminea
● subsp. ***viminea***. 30-80 cm, branched in upper part. Achenes 12-15 mm. - Widespread.
■ subsp. ***ramosissima***. Generally lower, branched from base. Achenes 9-11 mm. - Medit. area, often in the mountains.
Many records not identified to subspecies.

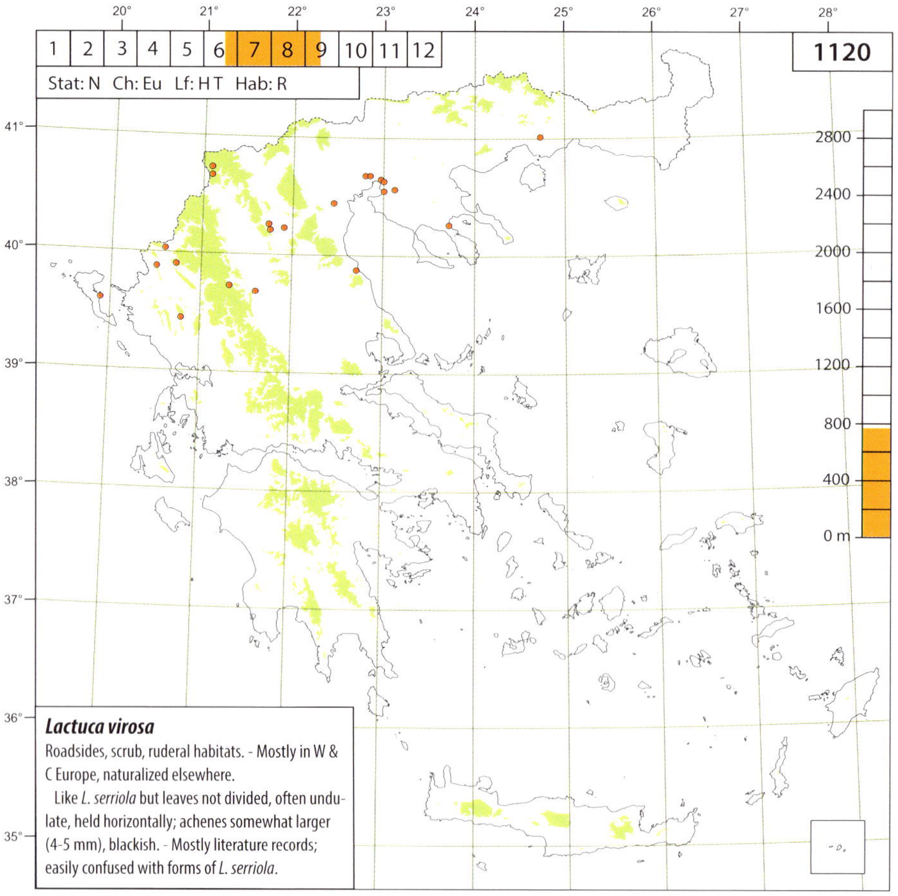

Lactuca virosa
Roadsides, scrub, ruderal habitats. - Mostly in W & C Europe, naturalized elsewhere.
Like *L. serriola* but leaves not divided, often undulate, held horizontally; achenes somewhat larger (4-5 mm), blackish. - Mostly literature records; easily confused with forms of *L. serriola*.

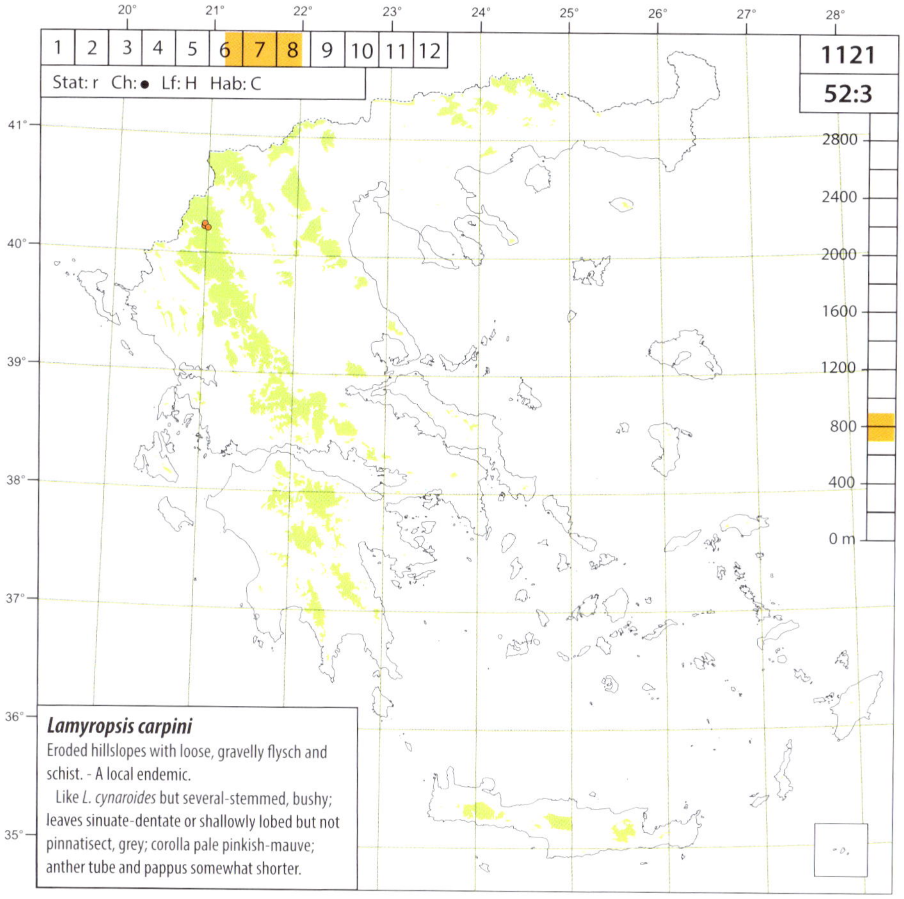

Lamyropsis carpini
Eroded hillslopes with loose, gravelly flysch and schist. - A local endemic.
Like *L. cynaroides* but several-stemmed, bushy; leaves sinuate-dentate or shallowly lobed but not pinnatisect, grey; corolla pale pinkish-mauve; anther tube and pappus somewhat shorter.

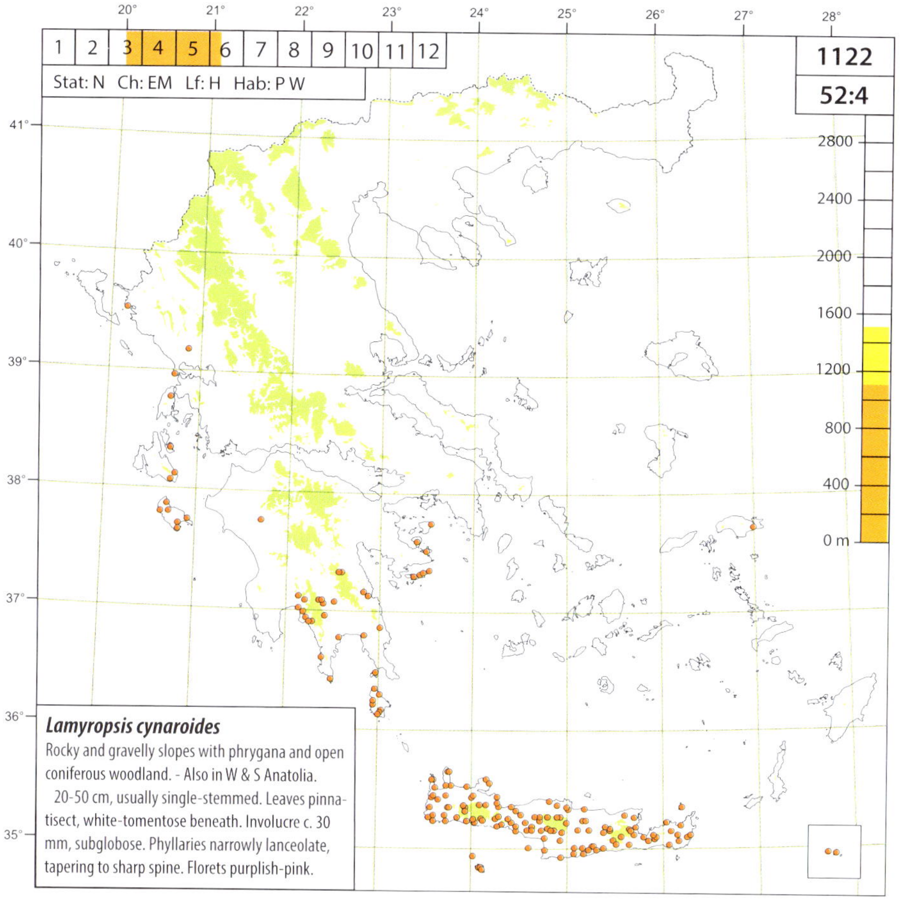

Lamyropsis cynaroides
Rocky and gravelly slopes with phrygana and open coniferous woodland. - Also in W & S Anatolia.
20-50 cm, usually single-stemmed. Leaves pinnatisect, white-tomentose beneath. Involucre c. 30 mm, subglobose. Phyllaries narrowly lanceolate, tapering to sharp spine. Florets purplish-pink.

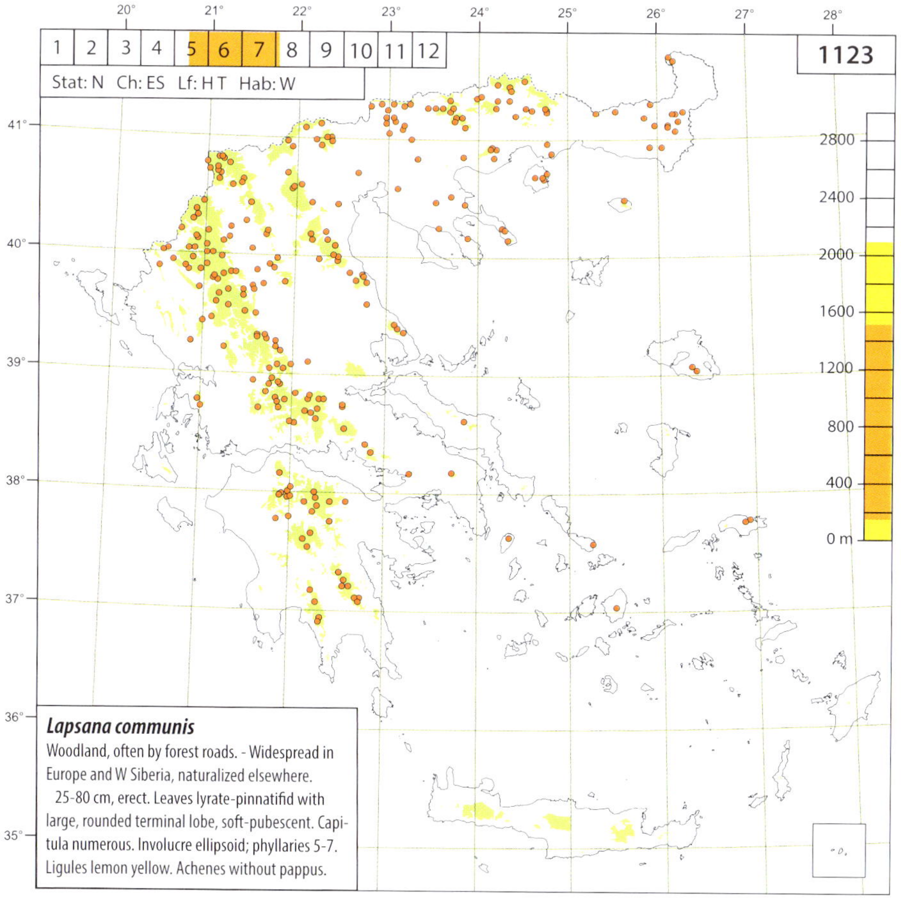

Lapsana communis
Woodland, often by forest roads. - Widespread in Europe and W Siberia, naturalized elsewhere.
25-80 cm, erect. Leaves lyrate-pinnatifid with large, rounded terminal lobe, soft-pubescent. Capitula numerous. Involucre ellipsoid; phyllaries 5-7. Ligules lemon yellow. Achenes without pappus.

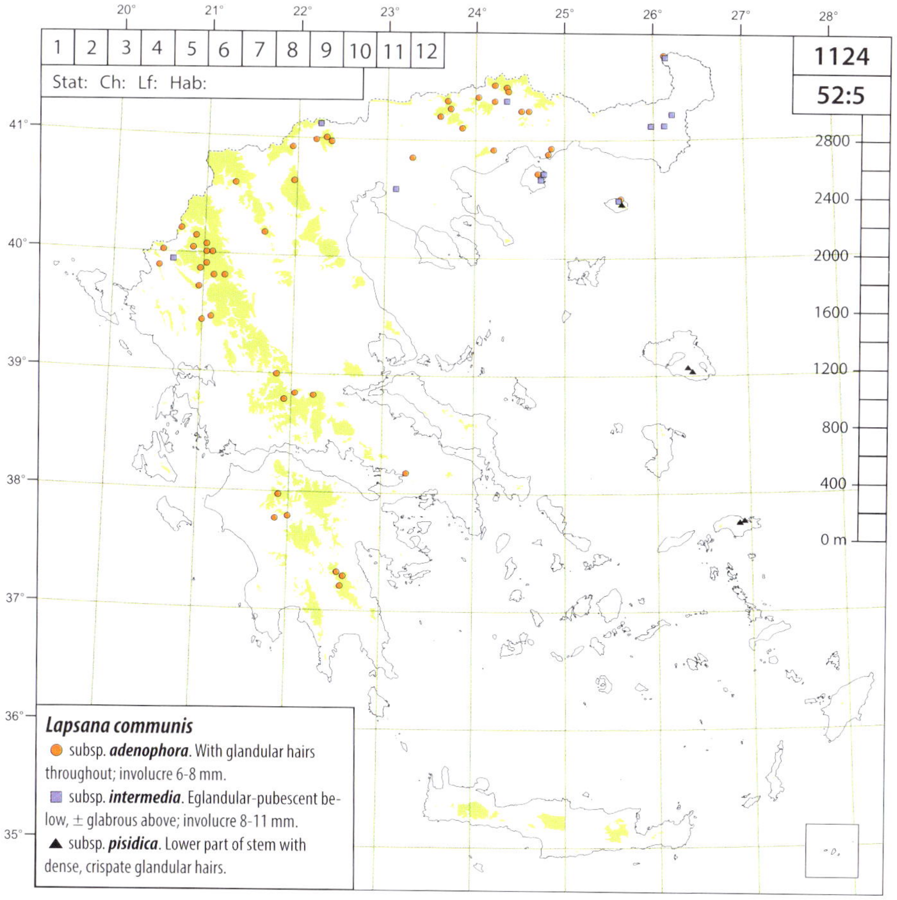

Lapsana communis
● subsp. ***adenophora***. With glandular hairs throughout; involucre 6-8 mm.
■ subsp. ***intermedia***. Eglandular-pubescent below, ± glabrous above; involucre 8-11 mm.
▲ subsp. ***pisidica***. Lower part of stem with dense, crispate glandular hairs.

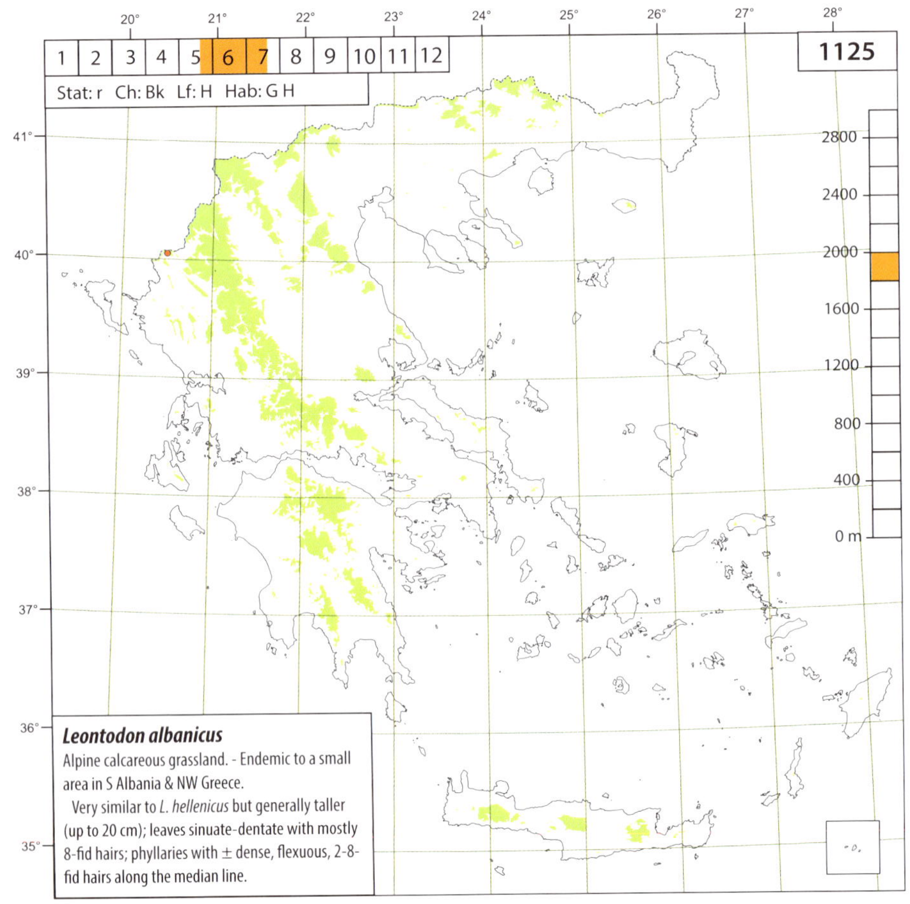

Leontodon albanicus

Alpine calcareous grassland. - Endemic to a small area in S Albania & NW Greece.

Very similar to *L. hellenicus* but generally taller (up to 20 cm); leaves sinuate-dentate with mostly 8-fid hairs; phyllaries with ± dense, flexuous, 2-8-fid hairs along the median line.

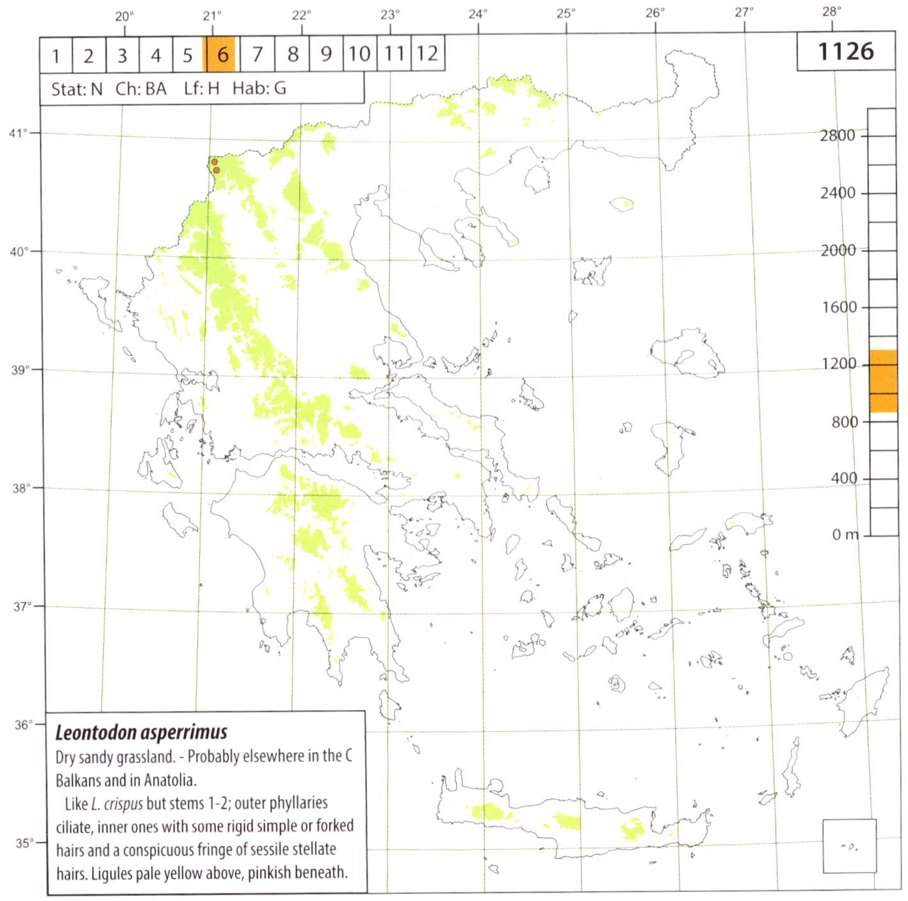

Leontodon asperrimus

Dry sandy grassland. - Probably elsewhere in the C Balkans and in Anatolia.

Like *L. crispus* but stems 1-2; outer phyllaries ciliate, inner ones with some rigid simple or forked hairs and a conspicuous fringe of sessile stellate hairs. Ligules pale yellow above, pinkish beneath.

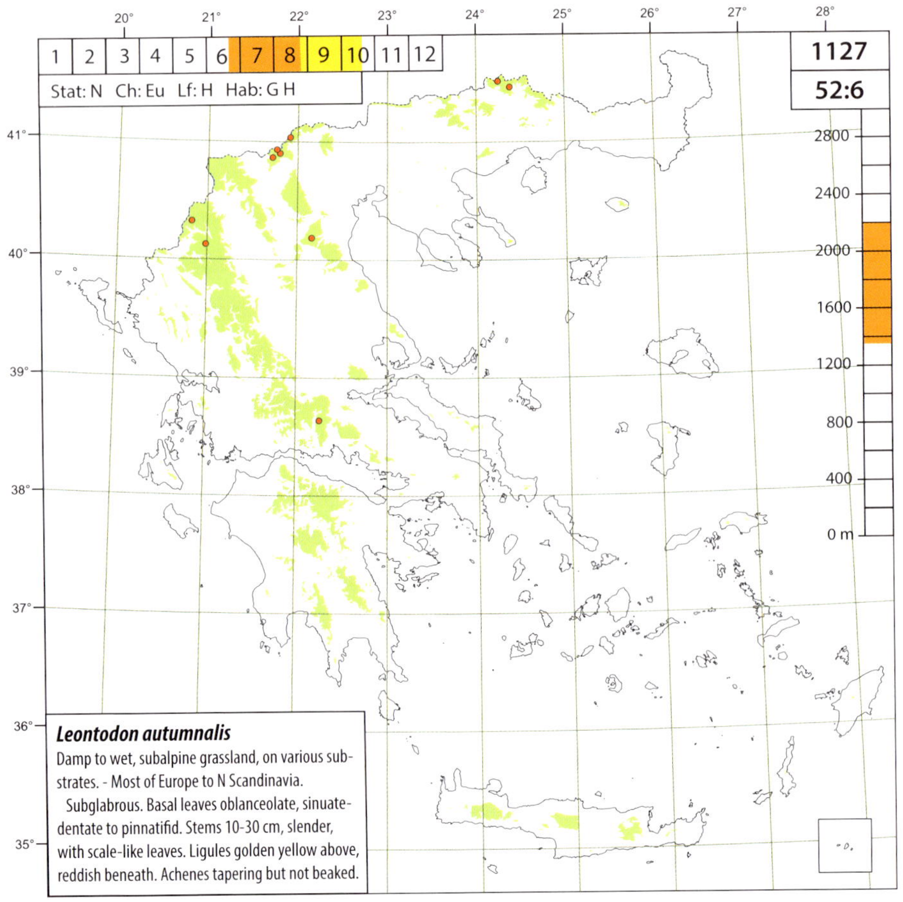

Leontodon autumnalis

Damp to wet, subalpine grassland, on various substrates. - Most of Europe to N Scandinavia.

Subglabrous. Basal leaves oblanceolate, sinuate-dentate to pinnatifid. Stems 10-30 cm, slender, with scale-like leaves. Ligules golden yellow above, reddish beneath. Achenes tapering but not beaked.

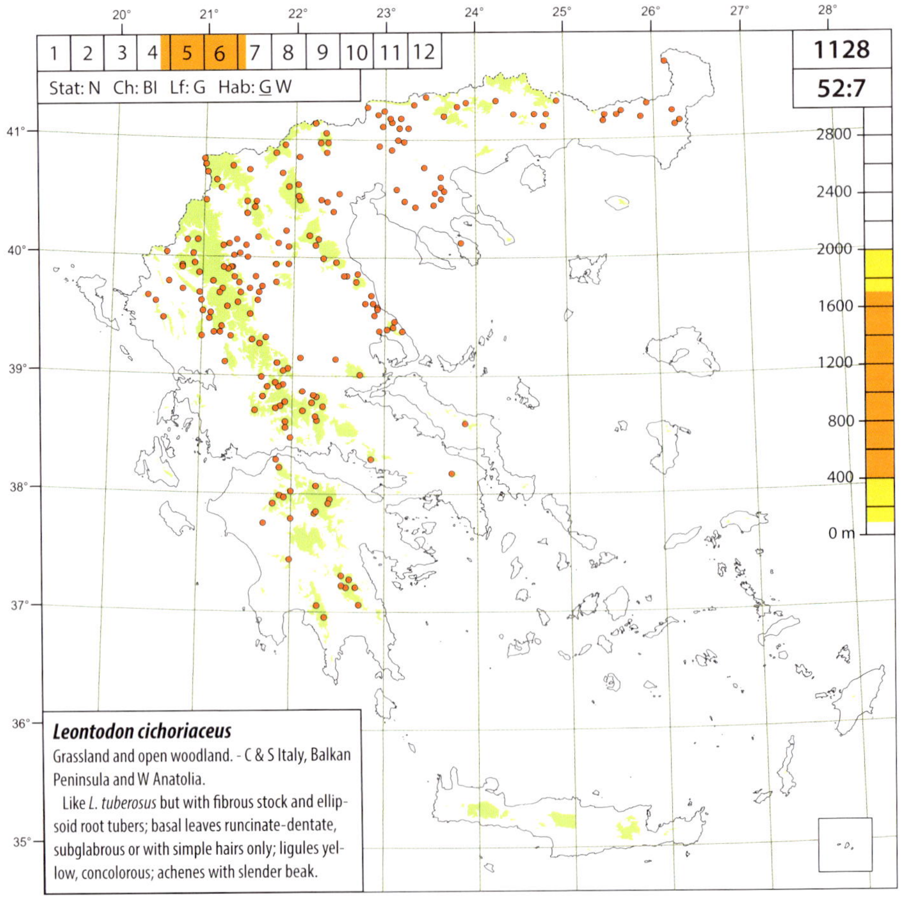

Leontodon cichoriaceus

Grassland and open woodland. - C & S Italy, Balkan Peninsula and W Anatolia.

Like *L. tuberosus* but with fibrous stock and ellipsoid root tubers; basal leaves runcinate-dentate, subglabrous or with simple hairs only; ligules yellow, concolorous; achenes with slender beak.

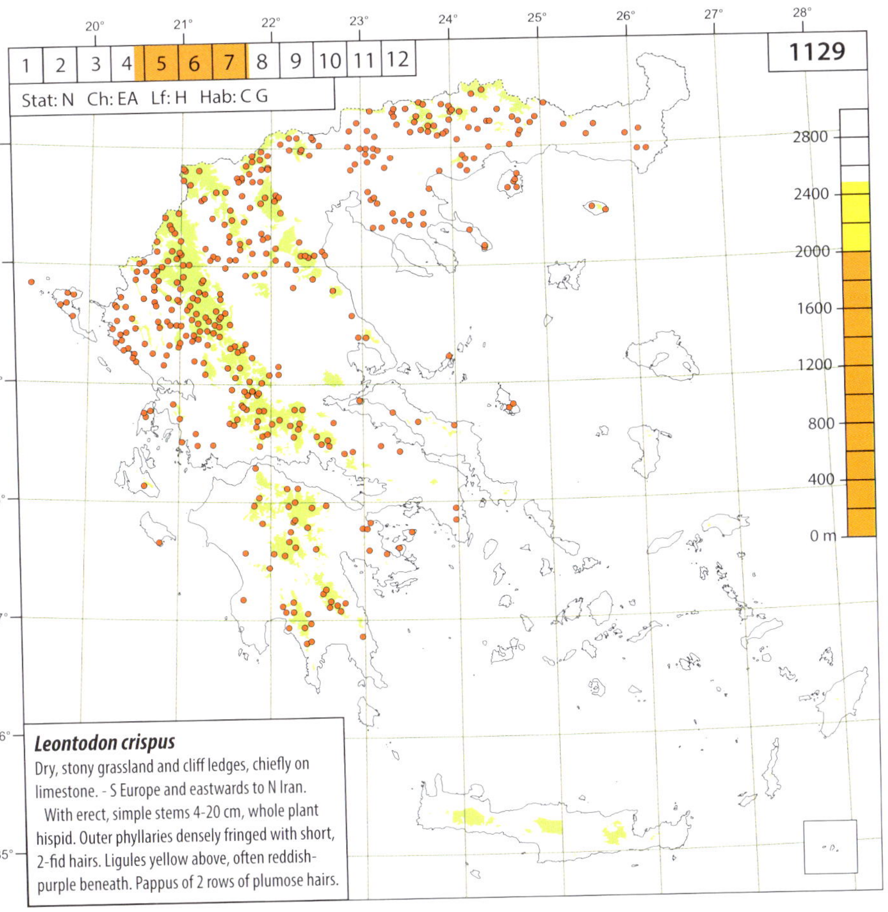

Leontodon crispus

Dry, stony grassland and cliff ledges, chiefly on limestone. - S Europe and eastwards to N Iran.

With erect, simple stems 4-20 cm, whole plant hispid. Outer phyllaries densely fringed with short, 2-fid hairs. Ligules yellow above, often reddish-purple beneath. Pappus of 2 rows of plumose hairs.

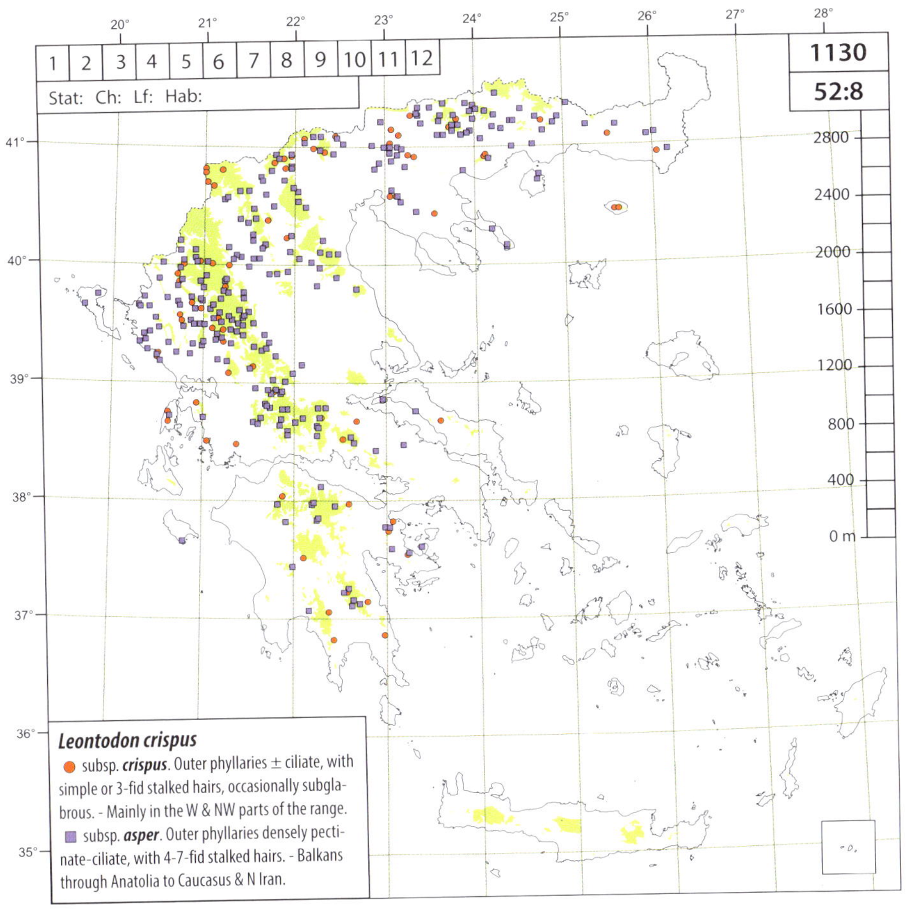

Leontodon crispus

● subsp. ***crispus***. Outer phyllaries ± ciliate, with simple or 3-fid stalked hairs, occasionally subglabrous. - Mainly in the W & NW parts of the range.

■ subsp. ***asper***. Outer phyllaries densely pectinate-ciliate, with 4-7-fid stalked hairs. - Balkans through Anatolia to Caucasus & N Iran.

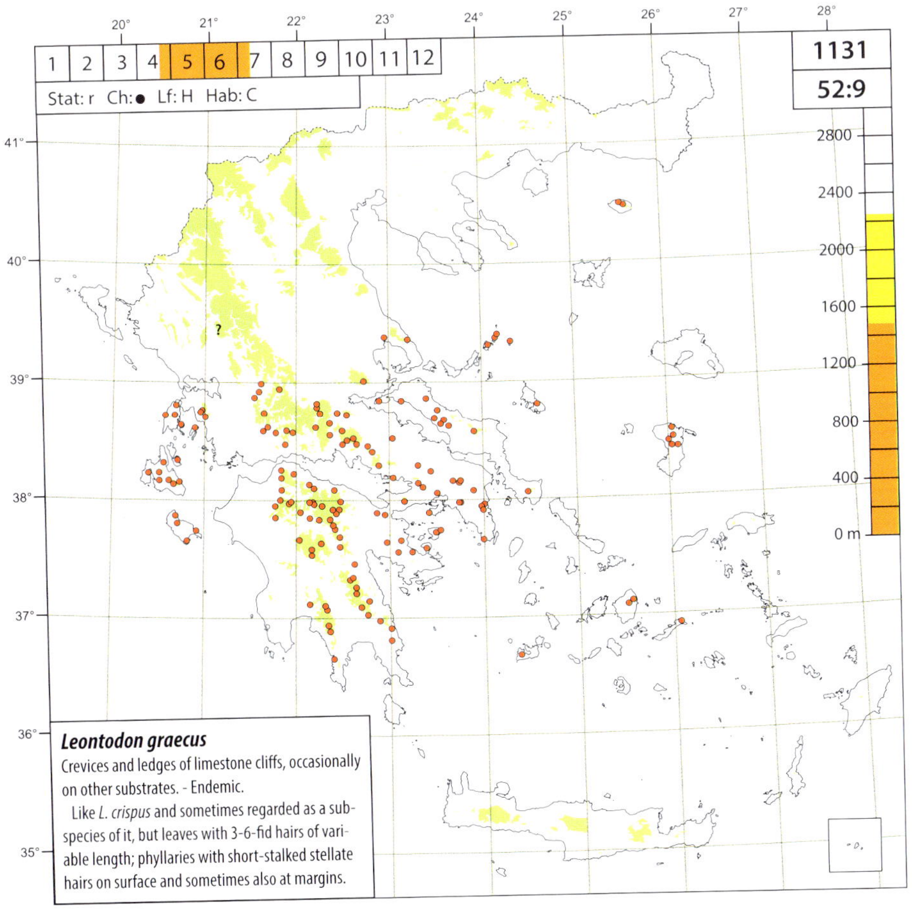

Leontodon graecus

Crevices and ledges of limestone cliffs, occasionally on other substrates. - Endemic.

Like *L. crispus* and sometimes regarded as a subspecies of it, but leaves with 3-6-fid hairs of variable length; phyllaries with short-stalked stellate hairs on surface and sometimes also at margins.

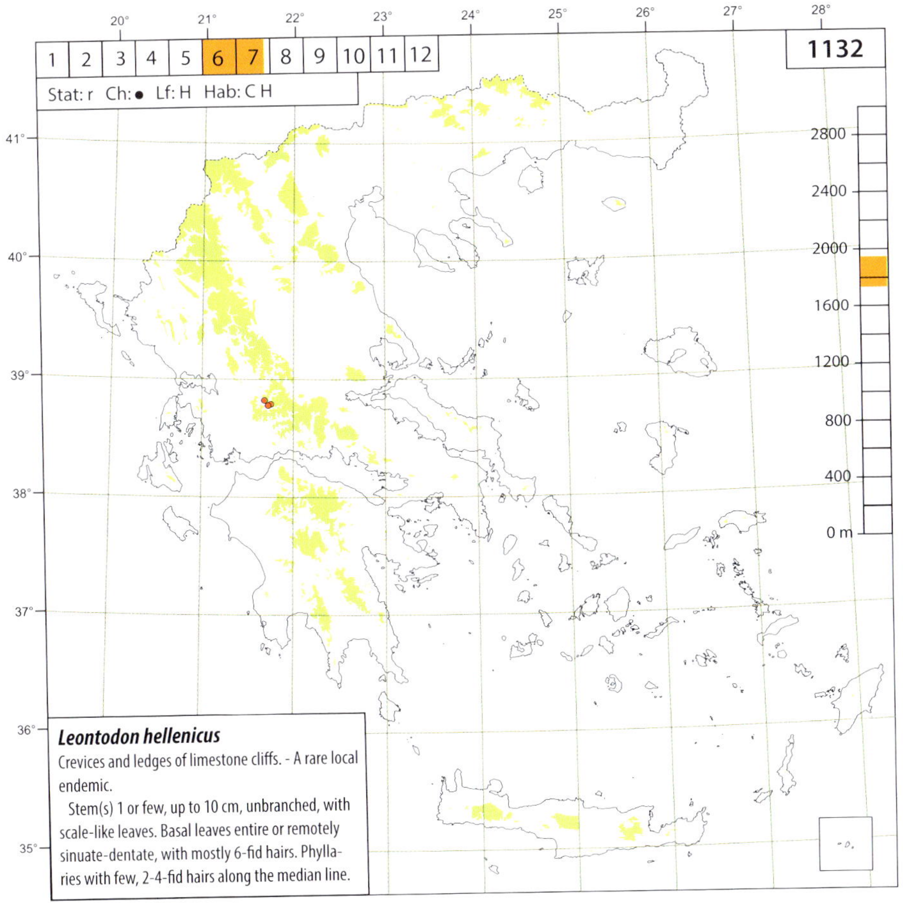

Leontodon hellenicus

Crevices and ledges of limestone cliffs. - A rare local endemic.

Stem(s) 1 or few, up to 10 cm, unbranched, with scale-like leaves. Basal leaves entire or remotely sinuate-dentate, with mostly 6-fid hairs. Phyllaries with few, 2-4-fid hairs along the median line.

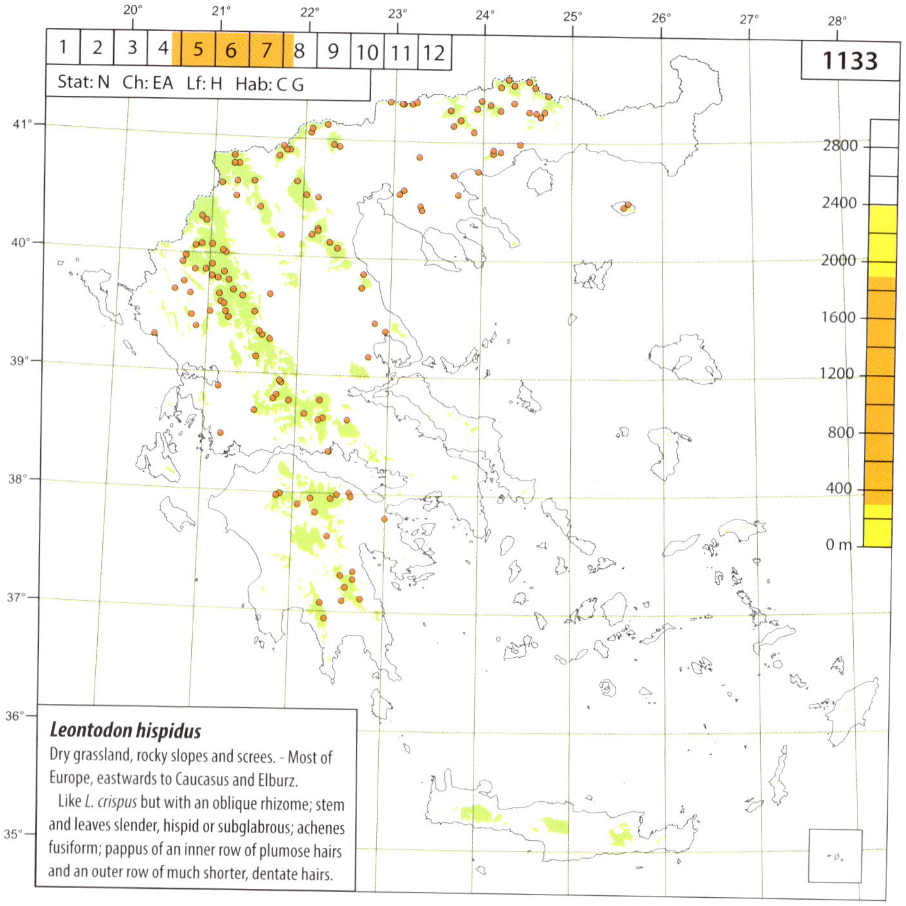

Leontodon hispidus
Dry grassland, rocky slopes and screes. - Most of Europe, eastwards to Caucasus and Elburz.
Like *L. crispus* but with an oblique rhizome; stem and leaves slender, hispid or subglabrous; achenes fusiform; pappus of an inner row of plumose hairs and an outer row of much shorter, dentate hairs.

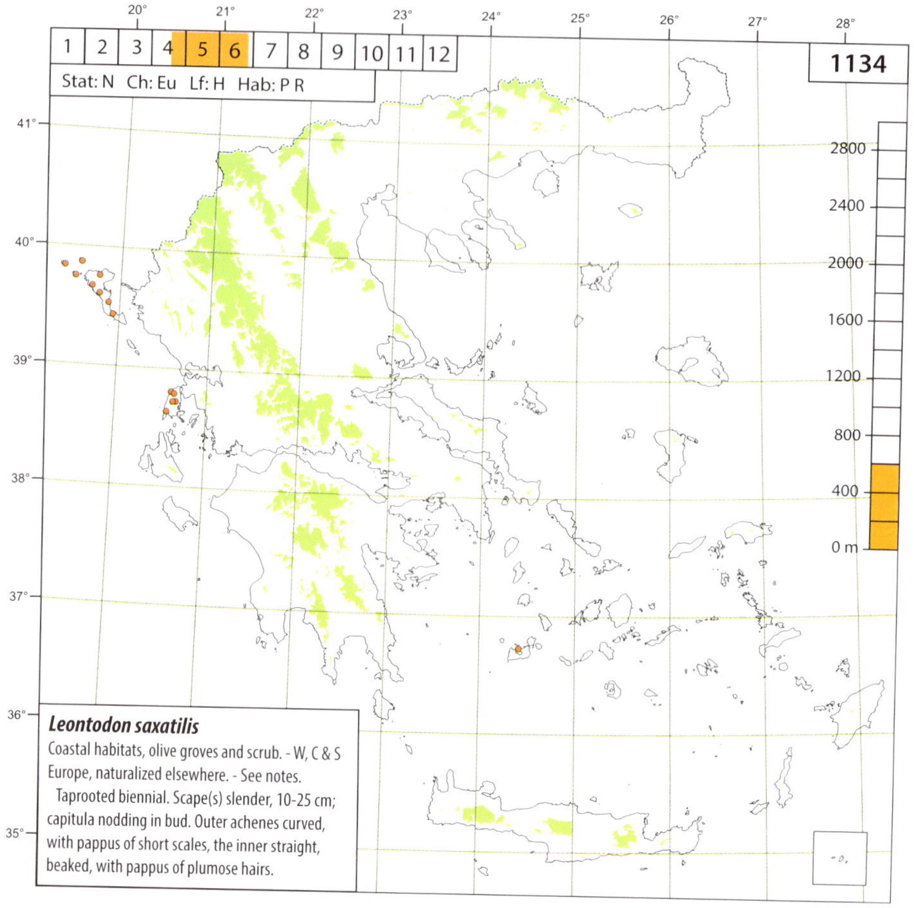

Leontodon saxatilis
Coastal habitats, olive groves and scrub. - W, C & S Europe, naturalized elsewhere. - See notes.
Taprooted biennial. Scape(s) slender, 10-25 cm; capitula nodding in bud. Outer achenes curved, with pappus of short scales, the inner straight, beaked, with pappus of plumose hairs.

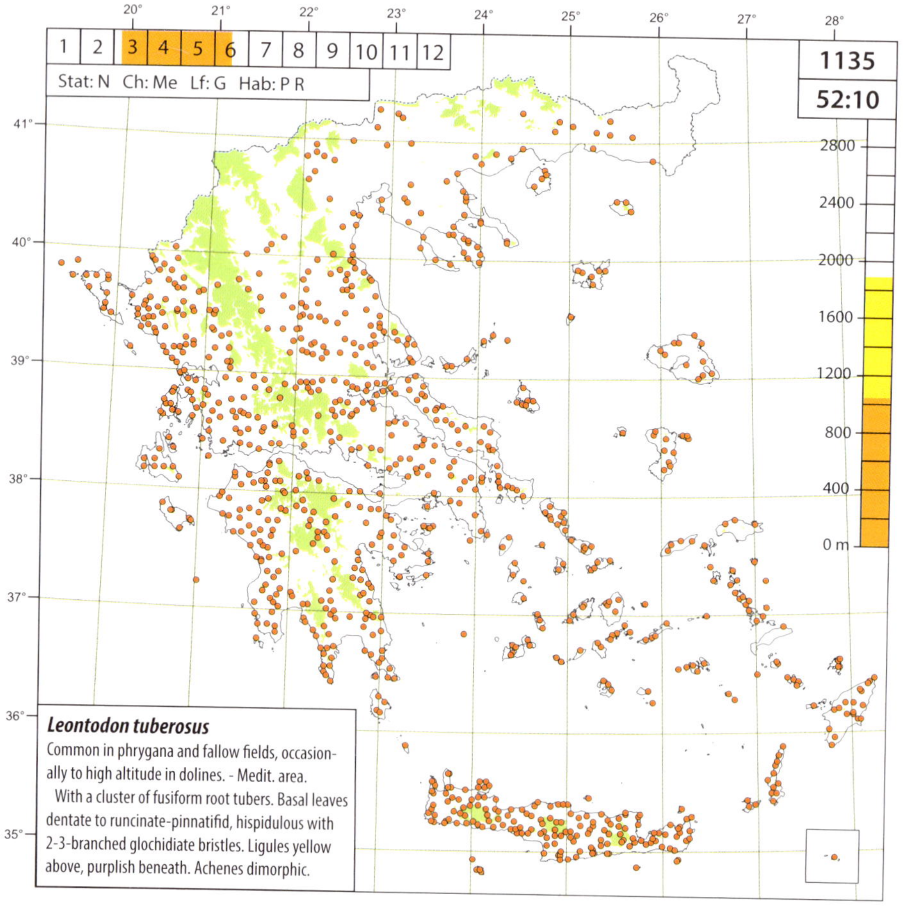

Leontodon tuberosus
Common in phrygana and fallow fields, occasionally to high altitude in dolines. - Medit. area.
With a cluster of fusiform root tubers. Basal leaves dentate to runcinate-pinnatifid, hispidulous with 2-3-branched glochidiate bristles. Ligules yellow above, purplish beneath. Achenes dimorphic.

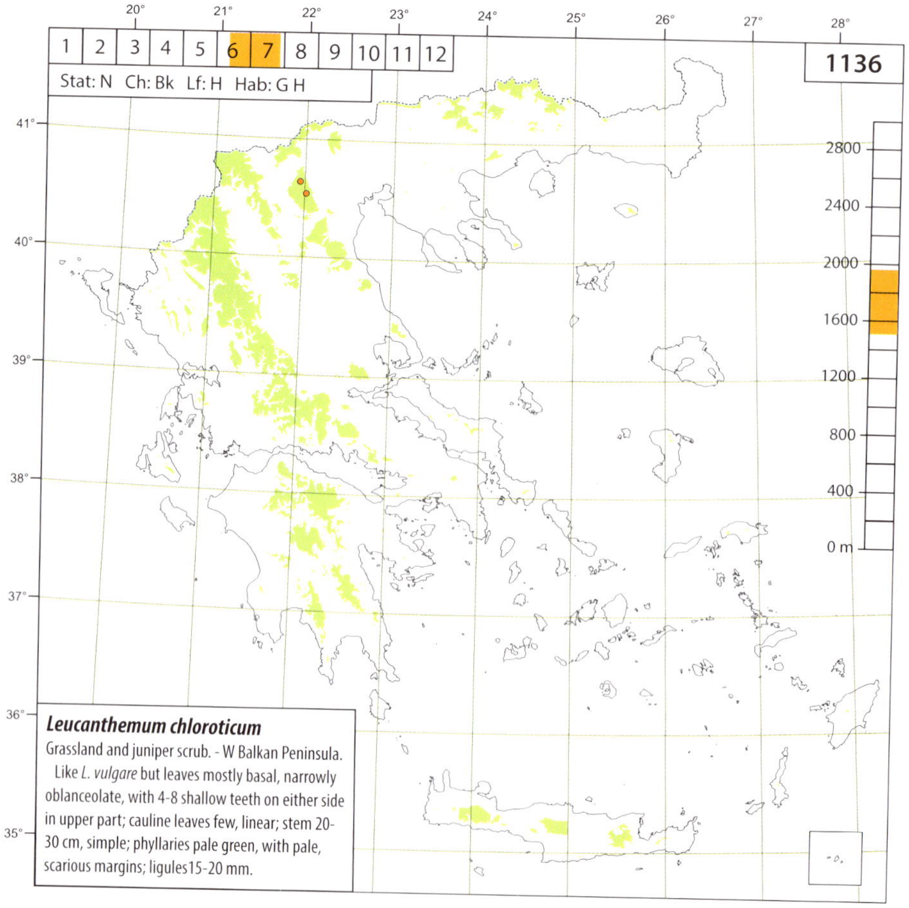

Leucanthemum chloroticum
Grassland and juniper scrub. - W Balkan Peninsula.
Like *L. vulgare* but leaves mostly basal, narrowly oblanceolate, with 4-8 shallow teeth on either side in upper part; cauline leaves few, linear; stem 20-30 cm, simple; phyllaries pale green, with pale, scarious margins; ligules15-20 mm.

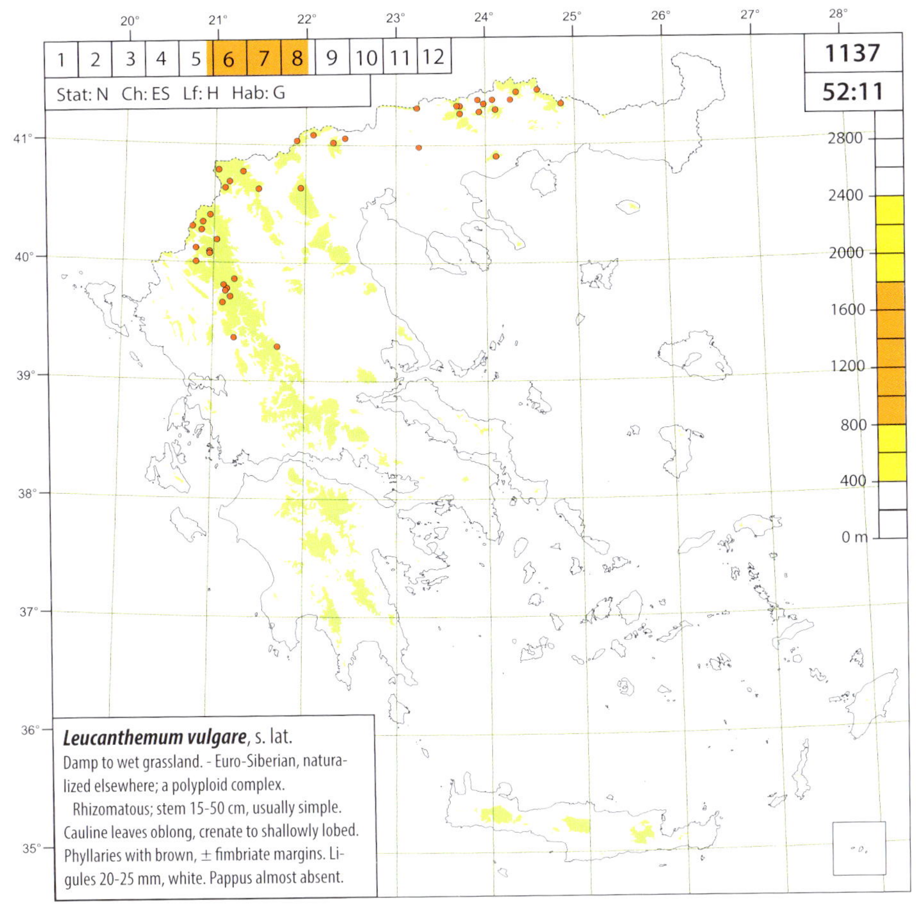

***Leucanthemum vulgare*, s. lat.**
Damp to wet grassland. - Euro-Siberian, naturalized elsewhere; a polyploid complex.
Rhizomatous; stem 15-50 cm, usually simple. Cauline leaves oblong, crenate to shallowly lobed. Phyllaries with brown, ± fimbriate margins. Ligules 20-25 mm, white. Pappus almost absent.

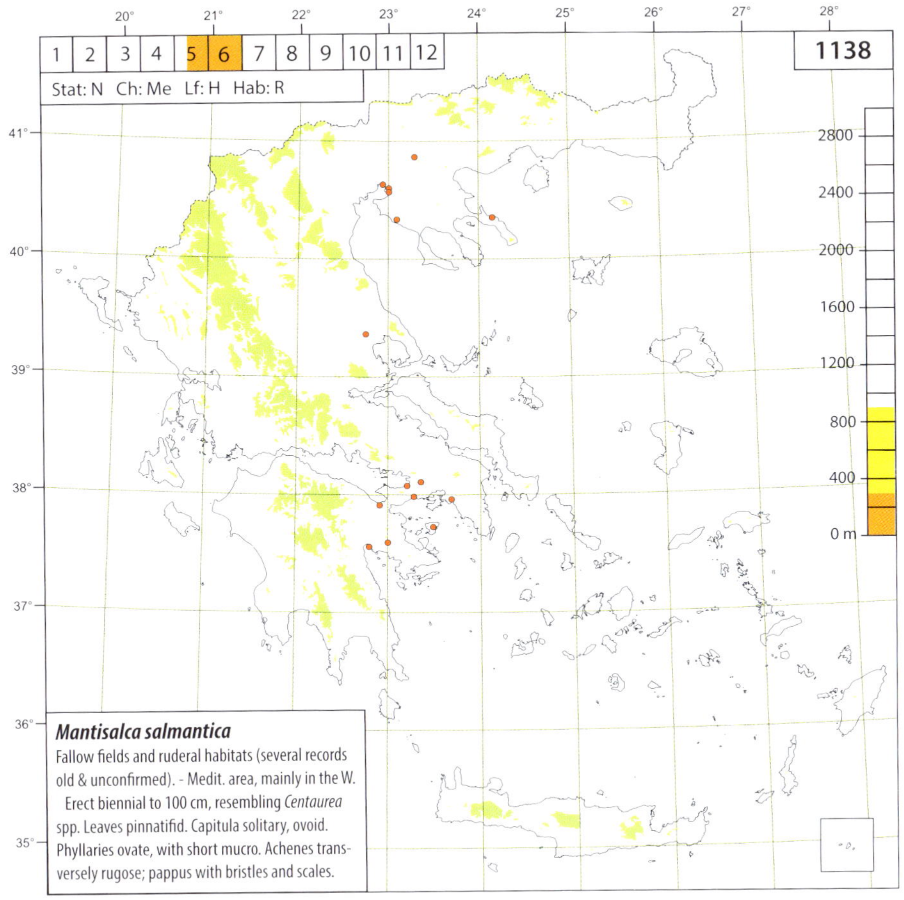

Mantisalca salmantica
Fallow fields and ruderal habitats (several records old & unconfirmed). - Medit. area, mainly in the W.
Erect biennial to 100 cm, resembling *Centaurea* spp. Leaves pinnatifid. Capitula solitary, ovoid. Phyllaries ovate, with short mucro. Achenes transversely rugose; pappus with bristles and scales.

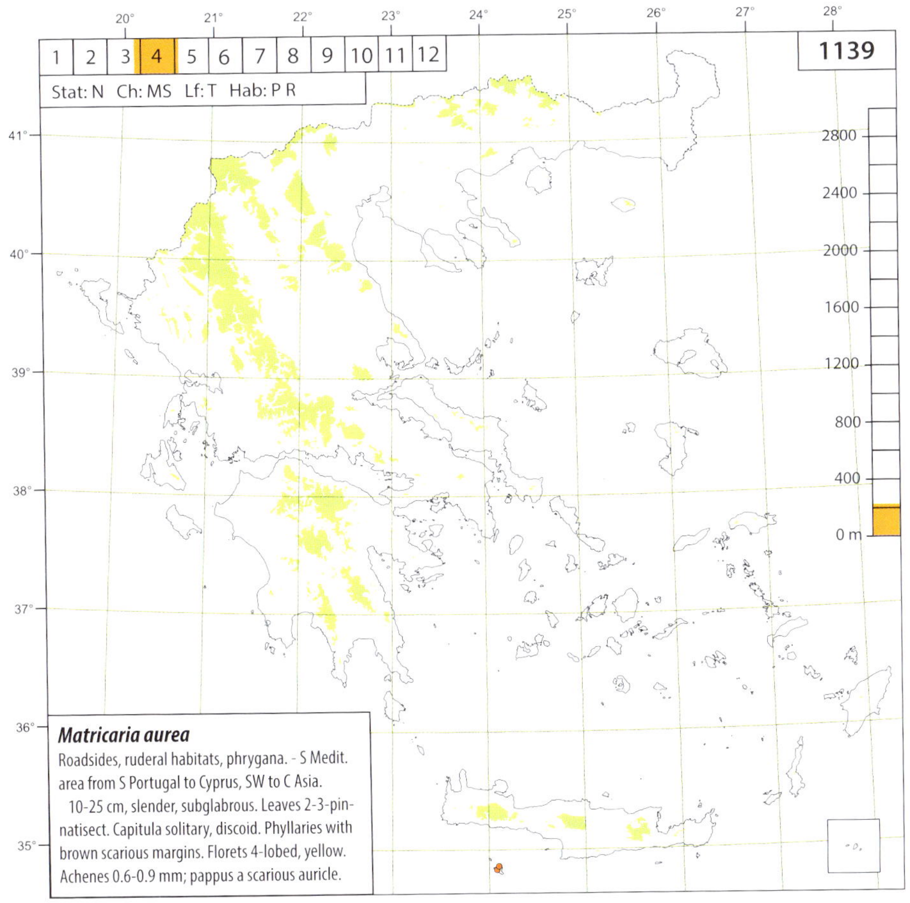

Matricaria aurea
Roadsides, ruderal habitats, phrygana. - S Medit. area from S Portugal to Cyprus, SW to C Asia.
10-25 cm, slender, subglabrous. Leaves 2-3-pinnatisect. Capitula solitary, discoid. Phyllaries with brown scarious margins. Florets 4-lobed, yellow. Achenes 0.6-0.9 mm; pappus a scarious auricle.

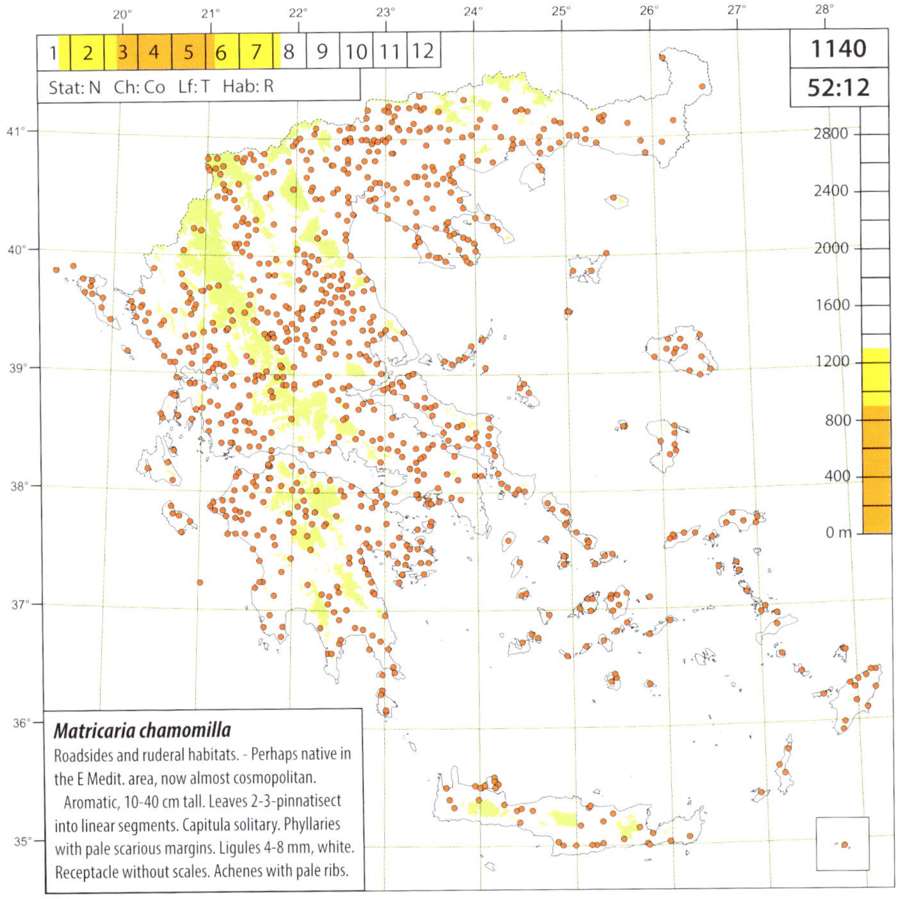

Matricaria chamomilla
Roadsides and ruderal habitats. - Perhaps native in the E Medit. area, now almost cosmopolitan.
Aromatic, 10-40 cm tall. Leaves 2-3-pinnatisect into linear segments. Capitula solitary. Phyllaries with pale scarious margins. Ligules 4-8 mm, white. Receptacle without scales. Achenes with pale ribs.

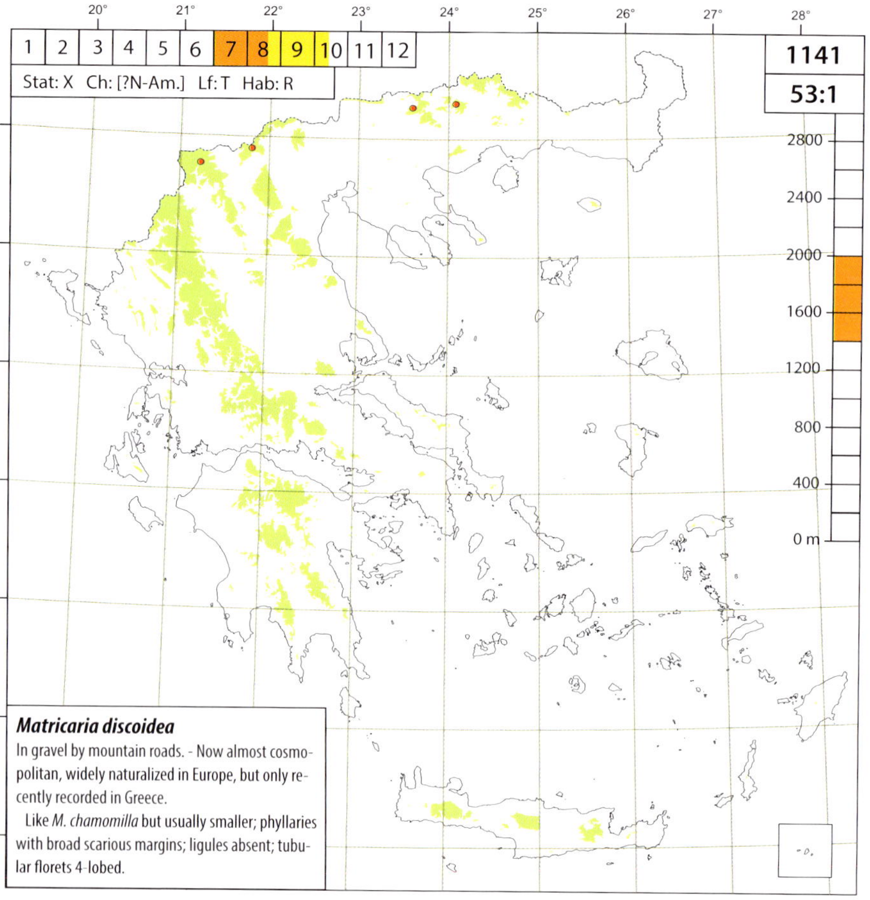

Matricaria discoidea

In gravel by mountain roads. - Now almost cosmopolitan, widely naturalized in Europe, but only recently recorded in Greece.

Like *M. chamomilla* but usually smaller; phyllaries with broad scarious margins; ligules absent; tubular florets 4-lobed.

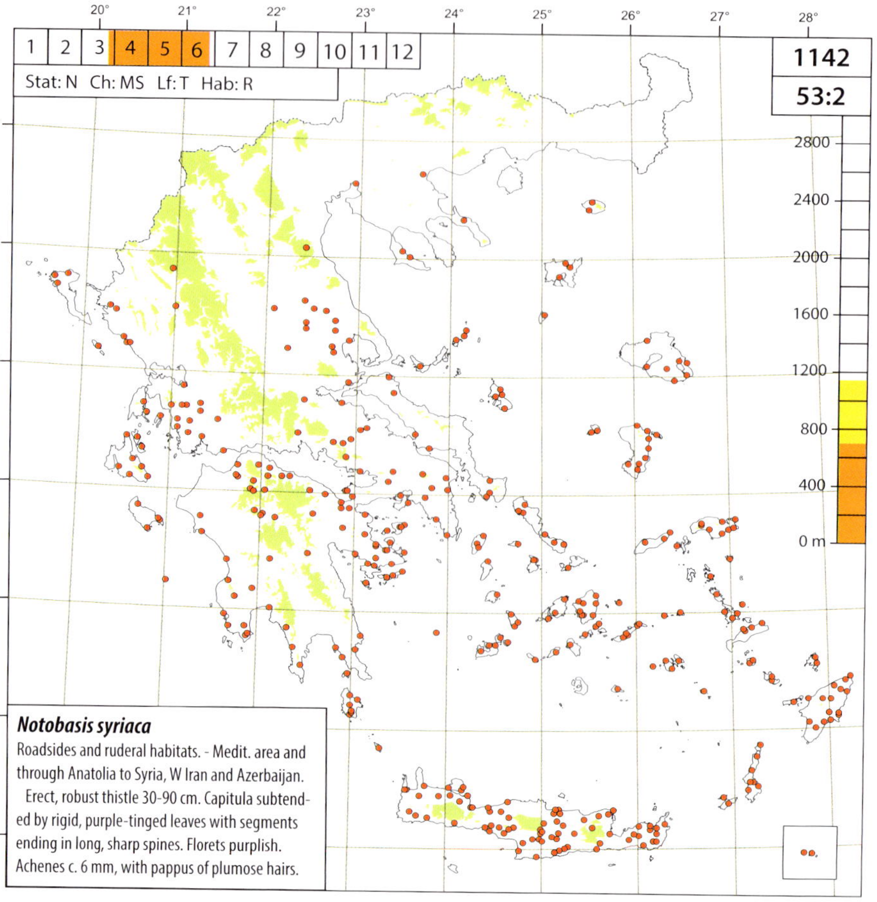

Notobasis syriaca

Roadsides and ruderal habitats. - Medit. area and through Anatolia to Syria, W Iran and Azerbaijan.

Erect, robust thistle 30-90 cm. Capitula subtended by rigid, purple-tinged leaves with segments ending in long, sharp spines. Florets purplish. Achenes c. 6 mm, with pappus of plumose hairs.

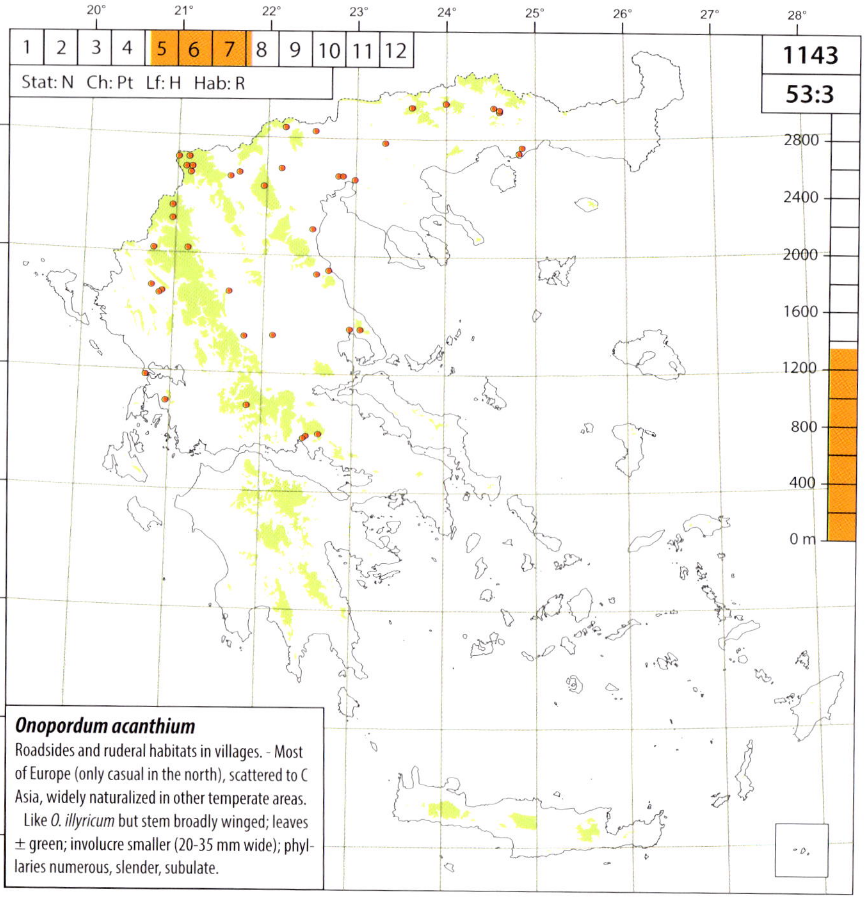

Onopordum acanthium

Roadsides and ruderal habitats in villages. - Most of Europe (only casual in the north), scattered to C Asia, widely naturalized in other temperate areas.

Like *O. illyricum* but stem broadly winged; leaves ± green; involucre smaller (20-35 mm wide); phyllaries numerous, slender, subulate.

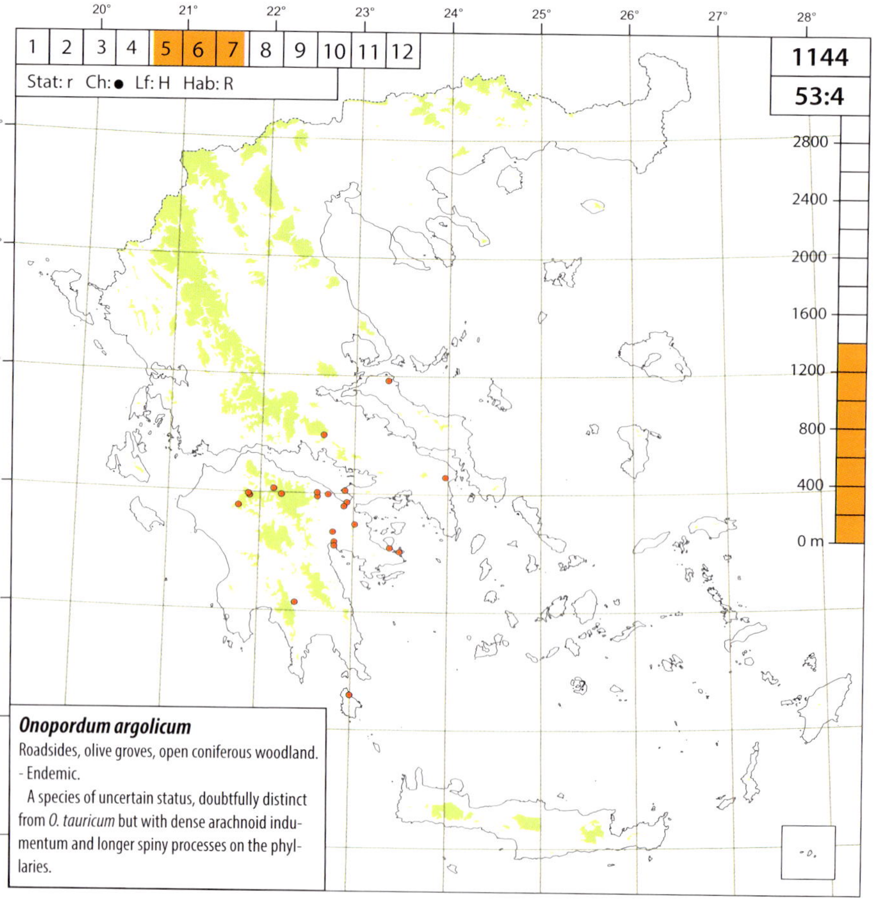

Onopordum argolicum

Roadsides, olive groves, open coniferous woodland. - Endemic.

A species of uncertain status, doubtfully distinct from *O. tauricum* but with dense arachnoid indumentum and longer spiny processes on the phyllaries.

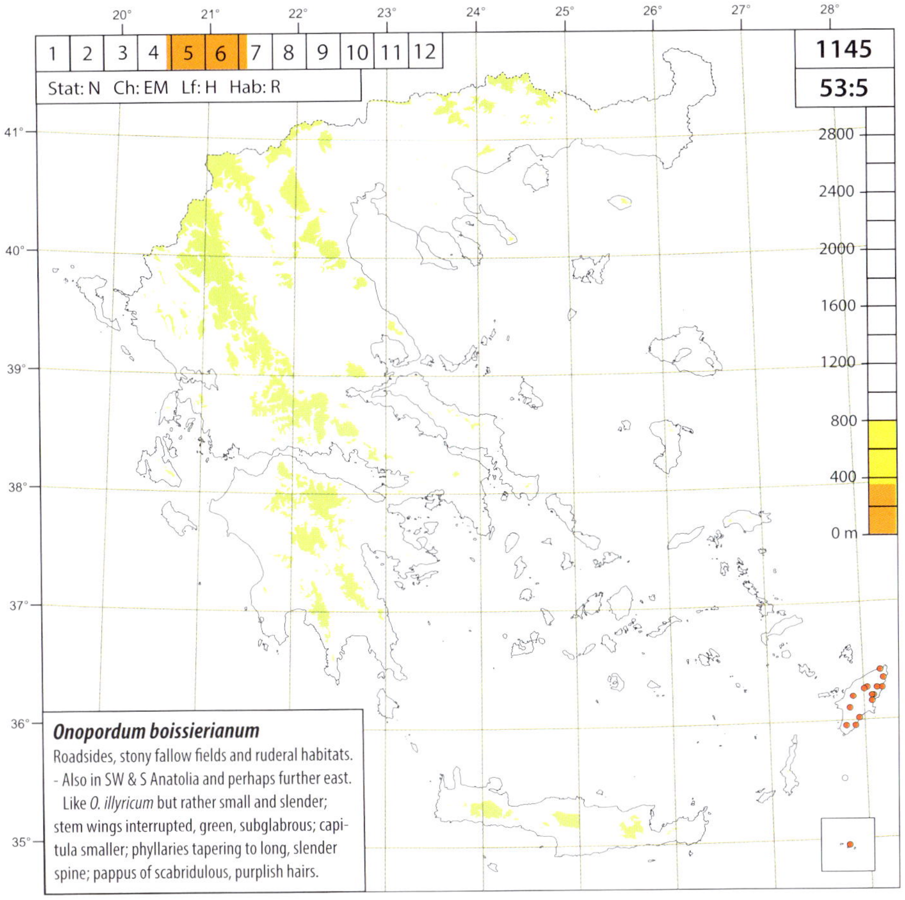

Onopordum boissierianum

Roadsides, stony fallow fields and ruderal habitats. - Also in SW & S Anatolia and perhaps further east.

Like *O. illyricum* but rather small and slender; stem wings interrupted, green, subglabrous; capitula smaller; phyllaries tapering to long, slender spine; pappus of scabridulous, purplish hairs.

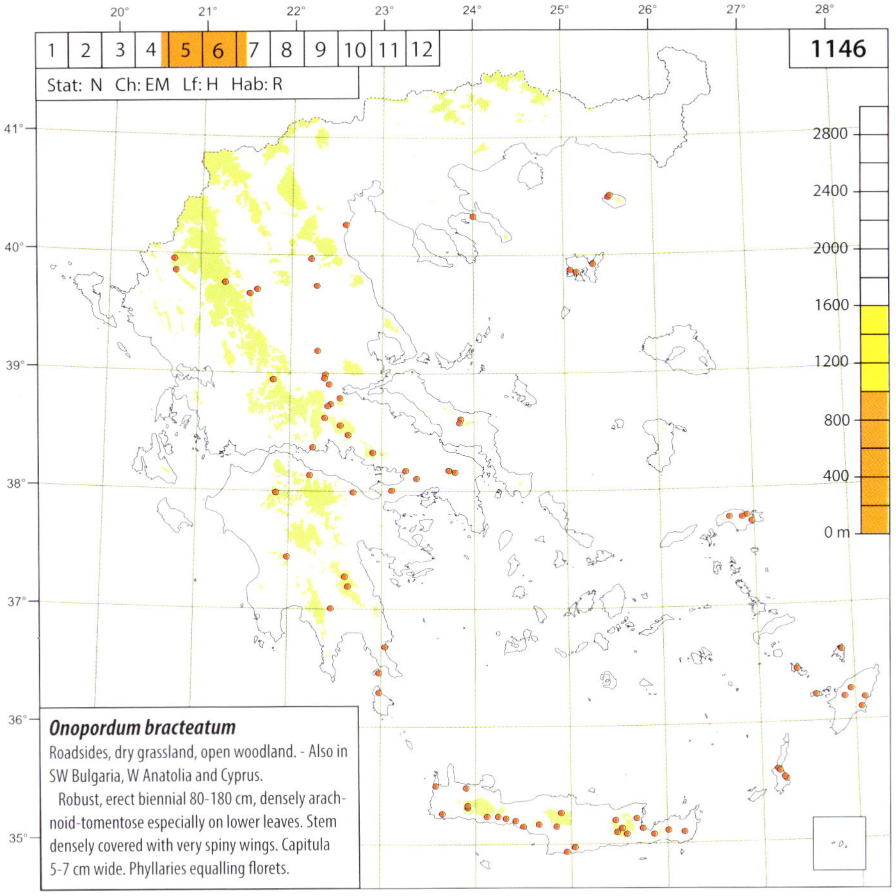

Onopordum bracteatum

Roadsides, dry grassland, open woodland. - Also in SW Bulgaria, W Anatolia and Cyprus.

Robust, erect biennial 80-180 cm, densely arachnoid-tomentose especially on lower leaves. Stem densely covered with very spiny wings. Capitula 5-7 cm wide. Phyllaries equalling florets.

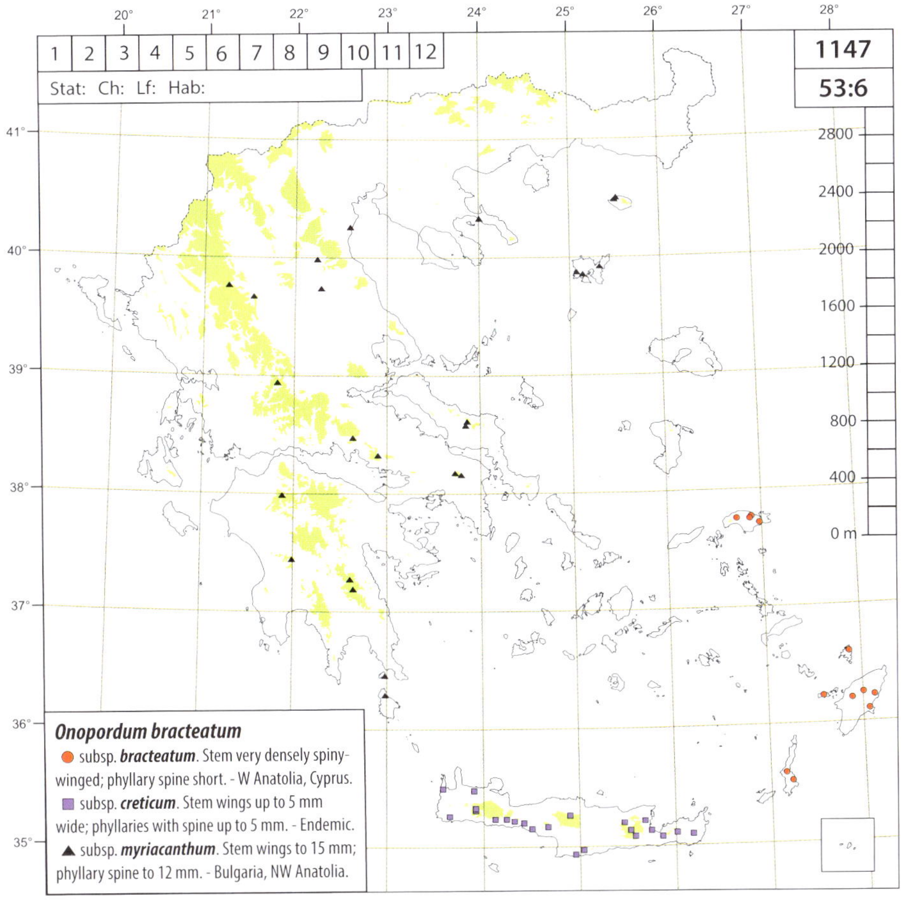

Onopordum bracteatum

● subsp. ***bracteatum***. Stem very densely spiny-winged; phyllary spine short. - W Anatolia, Cyprus.

■ subsp. ***creticum***. Stem wings up to 5 mm wide; phyllaries with spine up to 5 mm. - Endemic.

▲ subsp. ***myriacanthum***. Stem wings to 15 mm; phyllary spine to 12 mm. - Bulgaria, NW Anatolia.

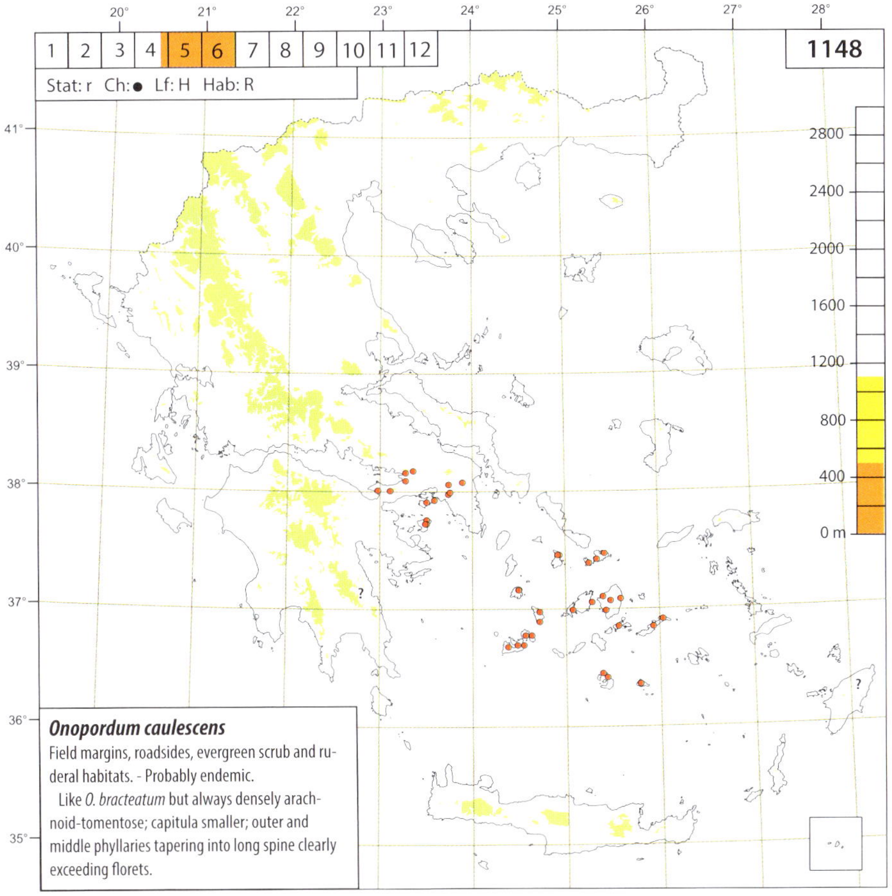

Onopordum caulescens

Field margins, roadsides, evergreen scrub and ruderal habitats. - Probably endemic.

Like *O. bracteatum* but always densely arachnoid-tomentose; capitula smaller; outer and middle phyllaries tapering into long spine clearly exceeding florets.

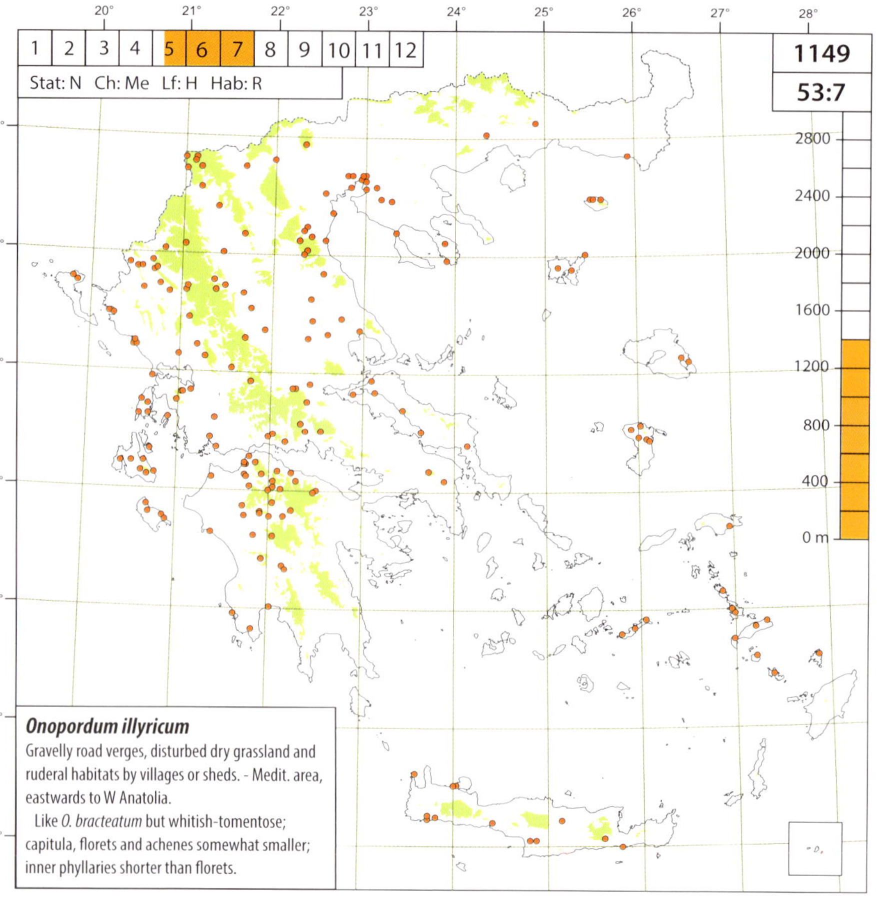

Onopordum illyricum

Gravelly road verges, disturbed dry grassland and ruderal habitats by villages or sheds. - Medit. area, eastwards to W Anatolia.

Like *O. bracteatum* but whitish-tomentose; capitula, florets and achenes somewhat smaller; inner phyllaries shorter than florets.

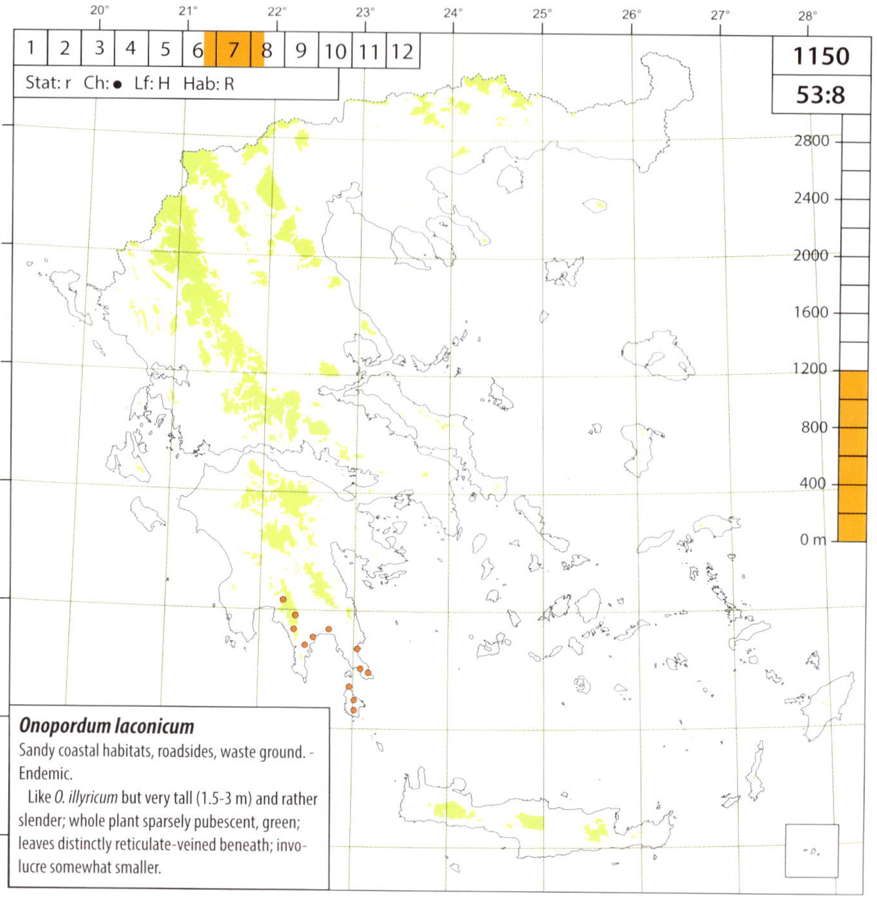

Onopordum laconicum

Sandy coastal habitats, roadsides, waste ground. - Endemic.

Like *O. illyricum* but very tall (1.5-3 m) and rather slender; whole plant sparsely pubescent, green; leaves distinctly reticulate-veined beneath; involucre somewhat smaller.

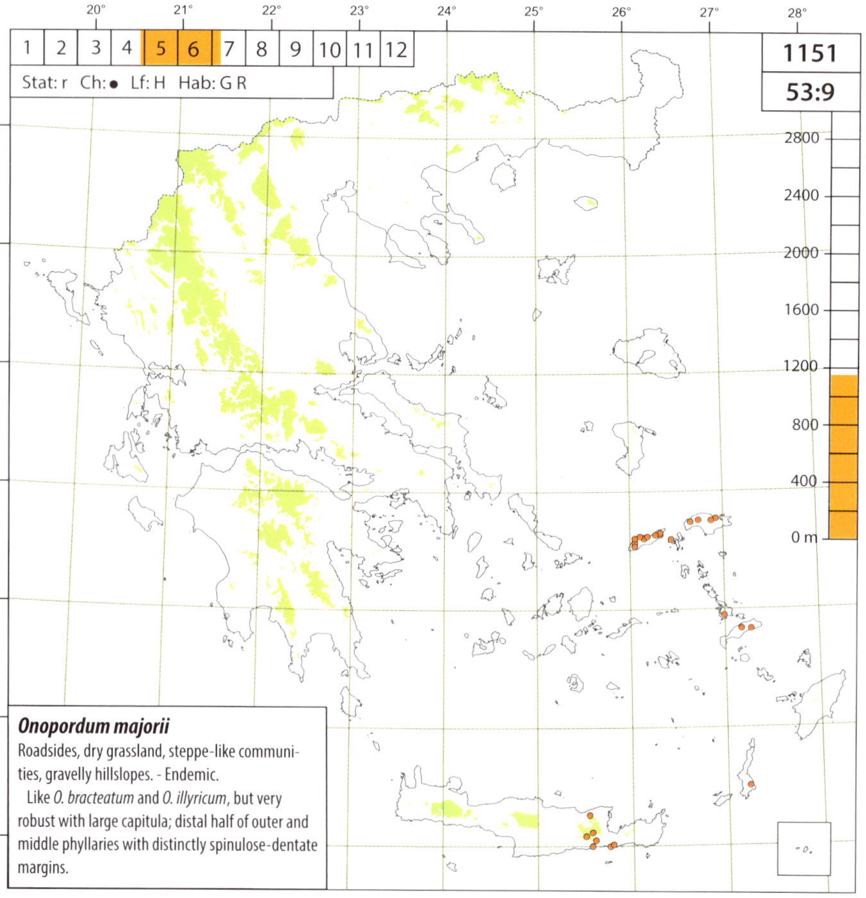

Onopordum majorii

Roadsides, dry grassland, steppe-like communities, gravelly hillslopes. - Endemic.

Like *O. bracteatum* and *O. illyricum*, but very robust with large capitula; distal half of outer and middle phyllaries with distinctly spinulose-dentate margins.

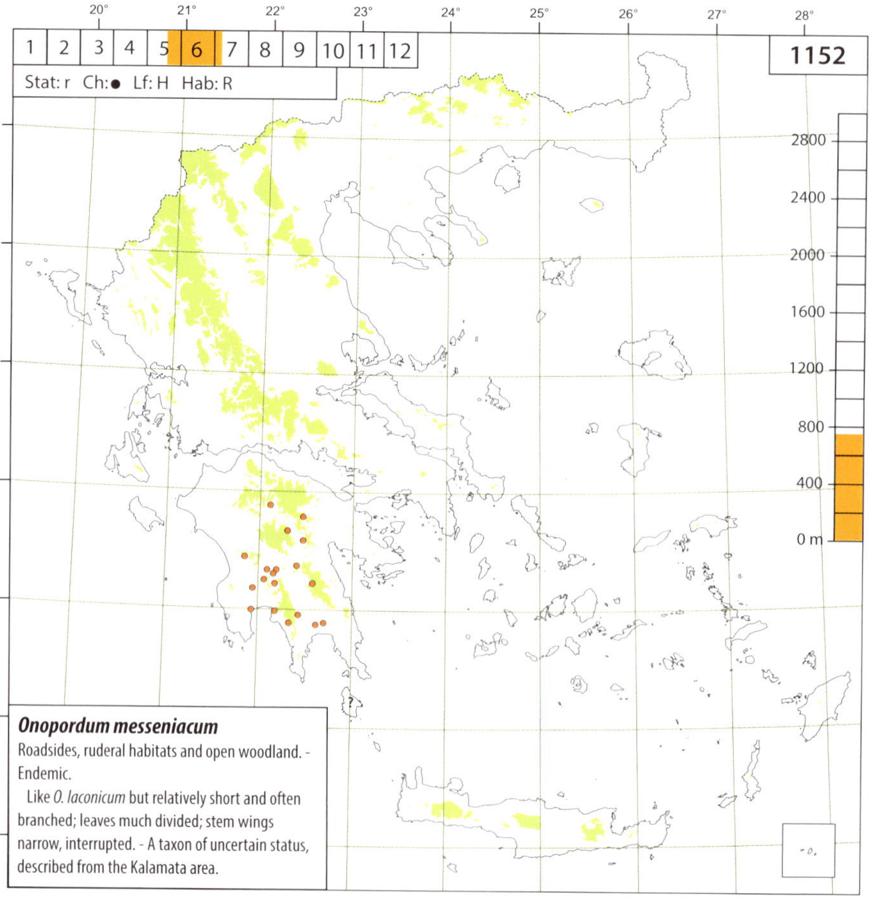

Onopordum messeniacum

Roadsides, ruderal habitats and open woodland. - Endemic.

Like *O. laconicum* but relatively short and often branched; leaves much divided; stem wings narrow, interrupted. - A taxon of uncertain status, described from the Kalamata area.

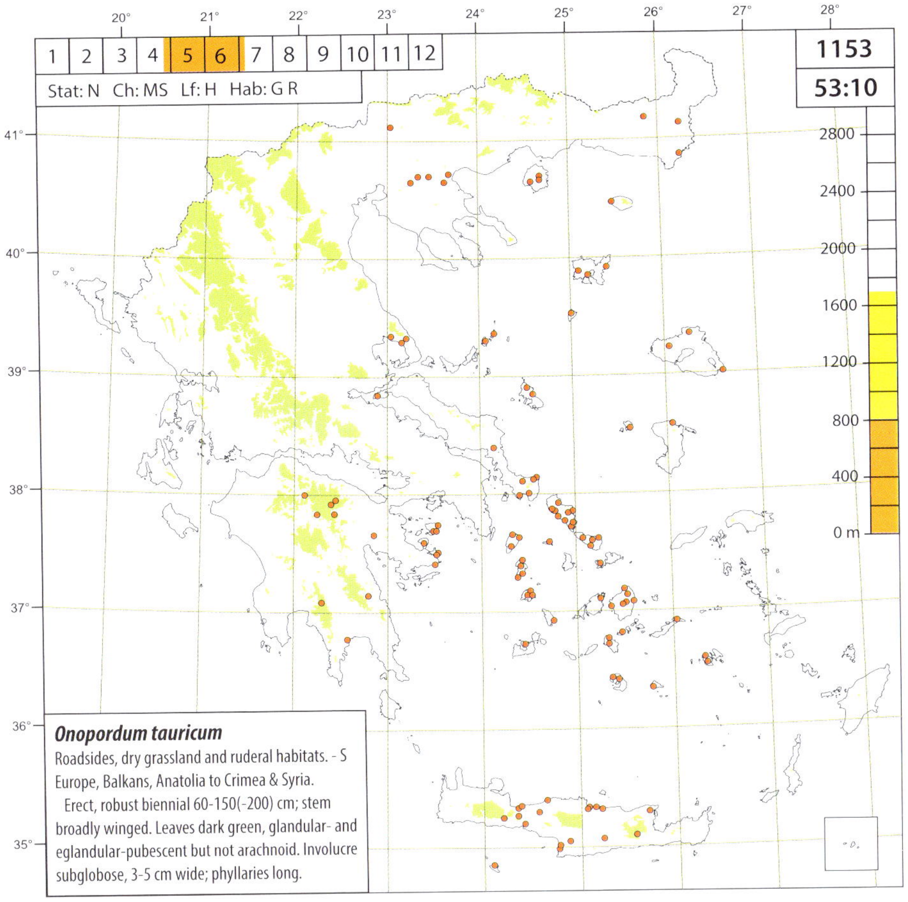

Onopordum tauricum
Roadsides, dry grassland and ruderal habitats. - S Europe, Balkans, Anatolia to Crimea & Syria.
Erect, robust biennial 60-150(-200) cm; stem broadly winged. Leaves dark green, glandular- and eglandular-pubescent but not arachnoid. Involucre subglobose, 3-5 cm wide; phyllaries long.

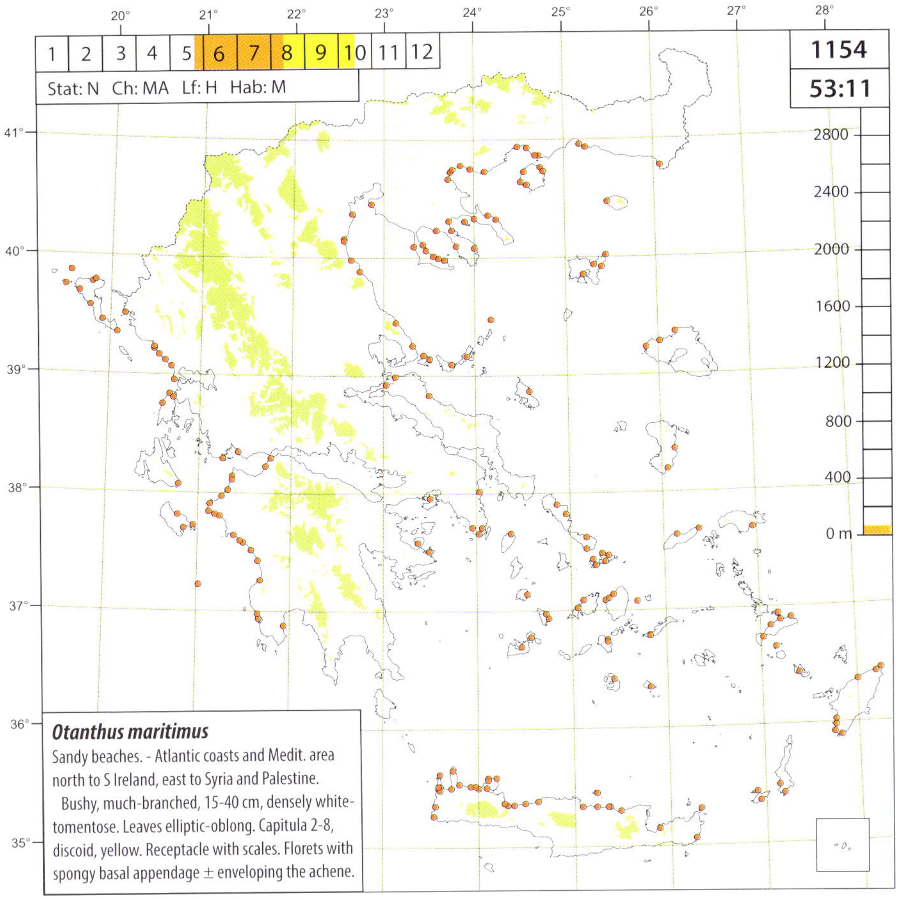

Otanthus maritimus
Sandy beaches. - Atlantic coasts and Medit. area north to S Ireland, east to Syria and Palestine.
Bushy, much-branched, 15-40 cm, densely white-tomentose. Leaves elliptic-oblong. Capitula 2-8, discoid, yellow. Receptacle with scales. Florets with spongy basal appendage ± enveloping the achene.

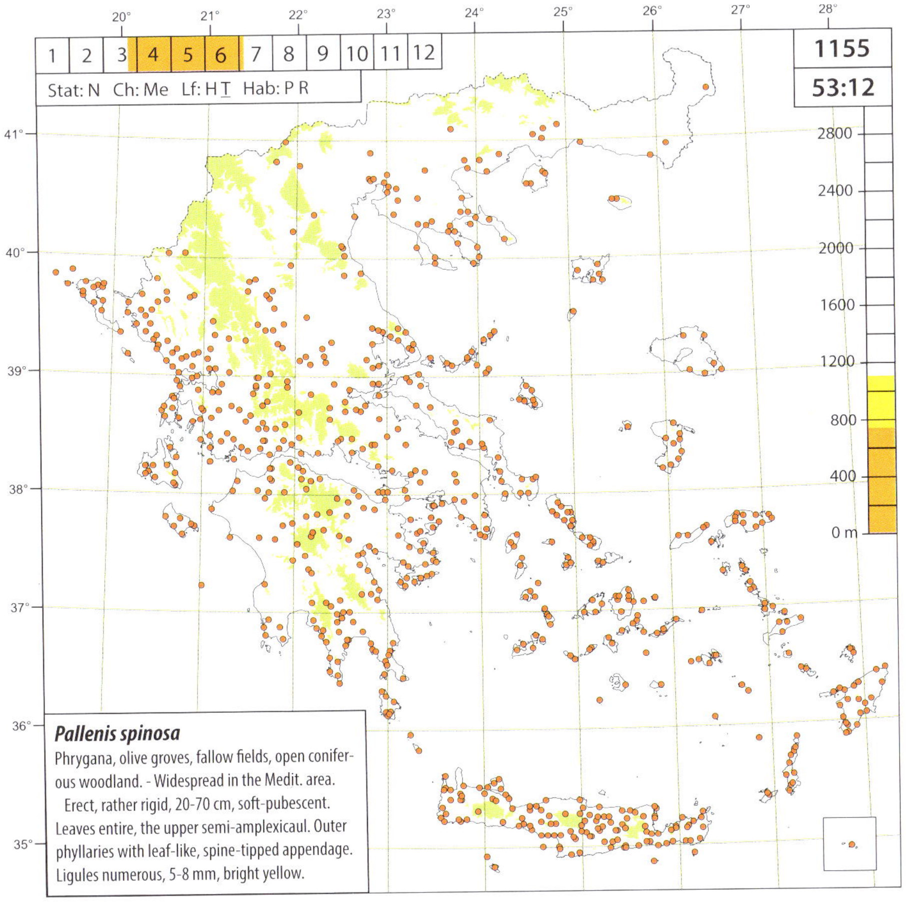

Pallenis spinosa
Phrygana, olive groves, fallow fields, open coniferous woodland. - Widespread in the Medit. area.
Erect, rather rigid, 20-70 cm, soft-pubescent. Leaves entire, the upper semi-amplexicaul. Outer phyllaries with leaf-like, spine-tipped appendage. Ligules numerous, 5-8 mm, bright yellow.

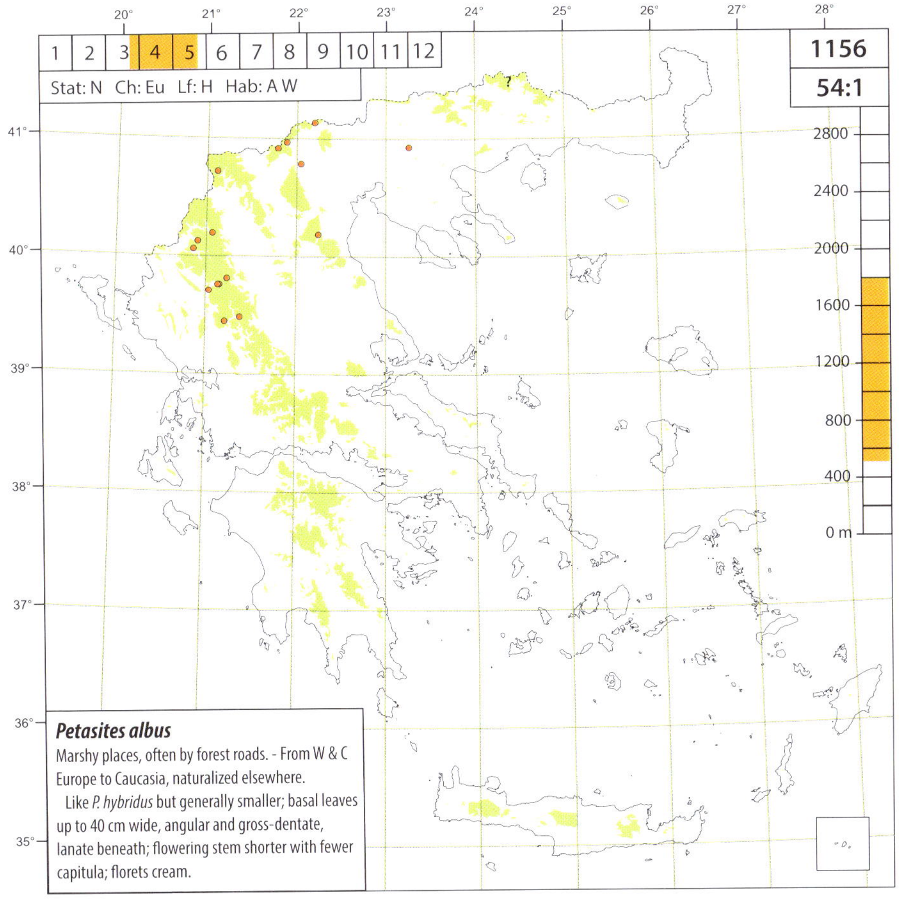

Petasites albus
Marshy places, often by forest roads. - From W & C Europe to Caucasia, naturalized elsewhere.
Like *P. hybridus* but generally smaller; basal leaves up to 40 cm wide, angular and gross-dentate, lanate beneath; flowering stem shorter with fewer capitula; florets cream.

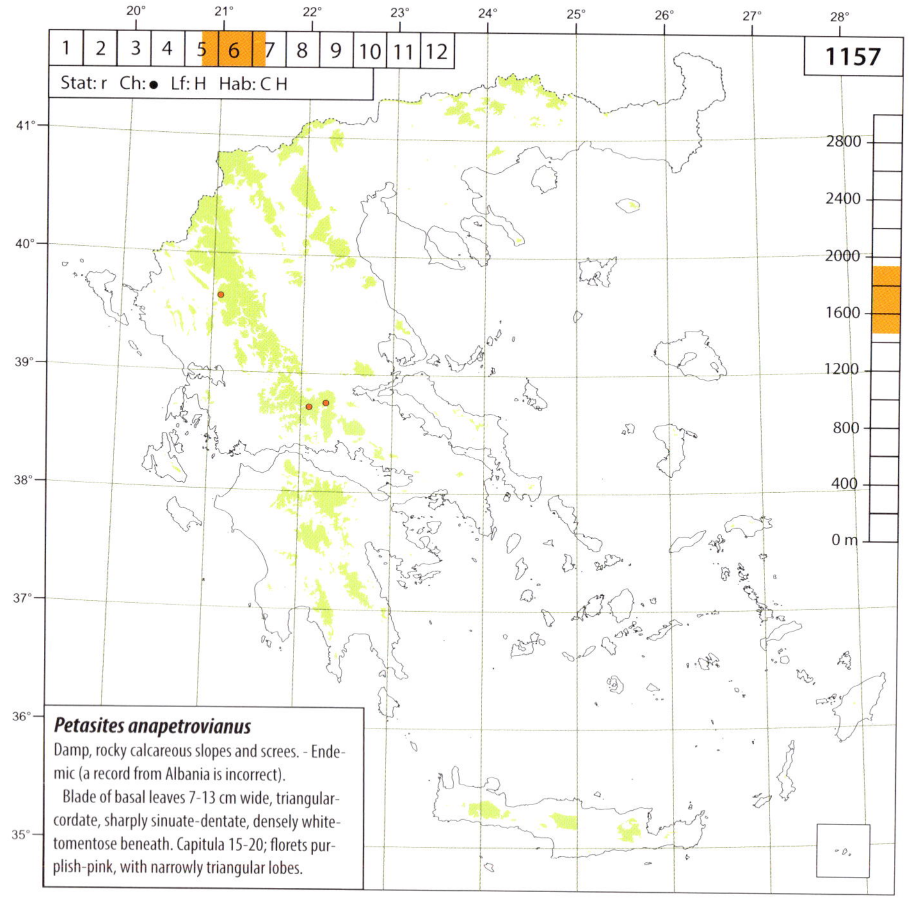

Petasites anapetrovianus
Damp, rocky calcareous slopes and screes. - Endemic (a record from Albania is incorrect).
Blade of basal leaves 7-13 cm wide, triangular-cordate, sharply sinuate-dentate, densely white-tomentose beneath. Capitula 15-20; florets purplish-pink, with narrowly triangular lobes.

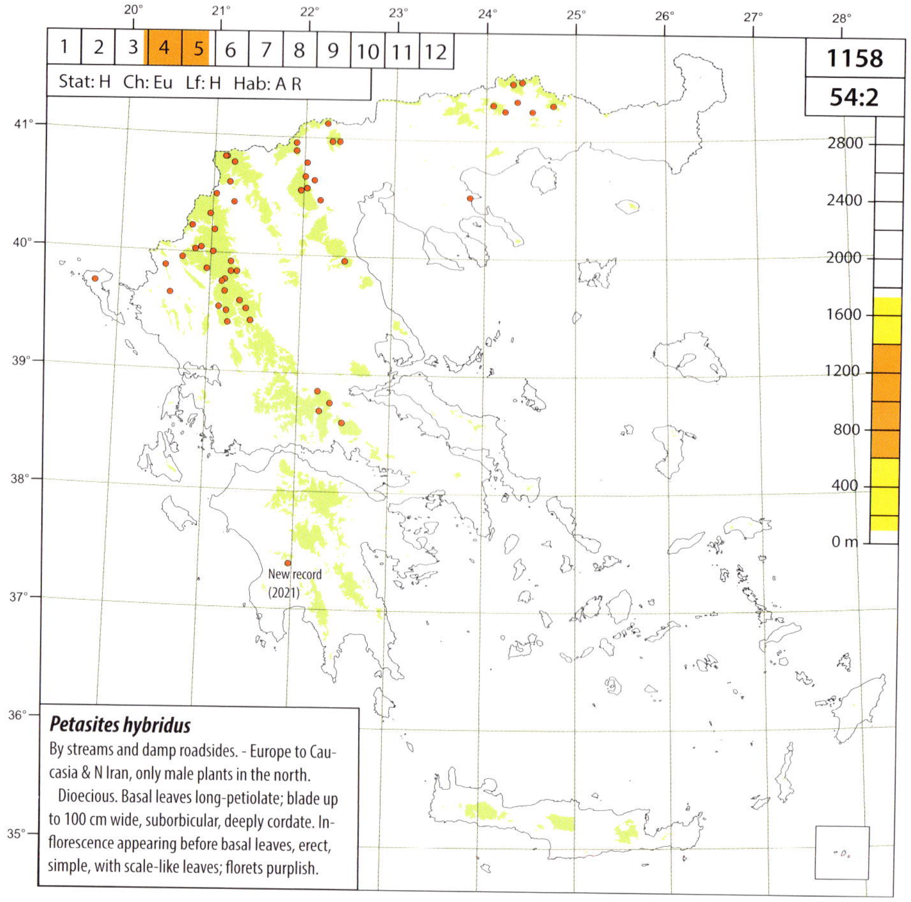

Petasites hybridus
By streams and damp roadsides. - Europe to Caucasia & N Iran, only male plants in the north.
Dioecious. Basal leaves long-petiolate; blade up to 100 cm wide, suborbicular, deeply cordate. Inflorescence appearing before basal leaves, erect, simple, with scale-like leaves; florets purplish.

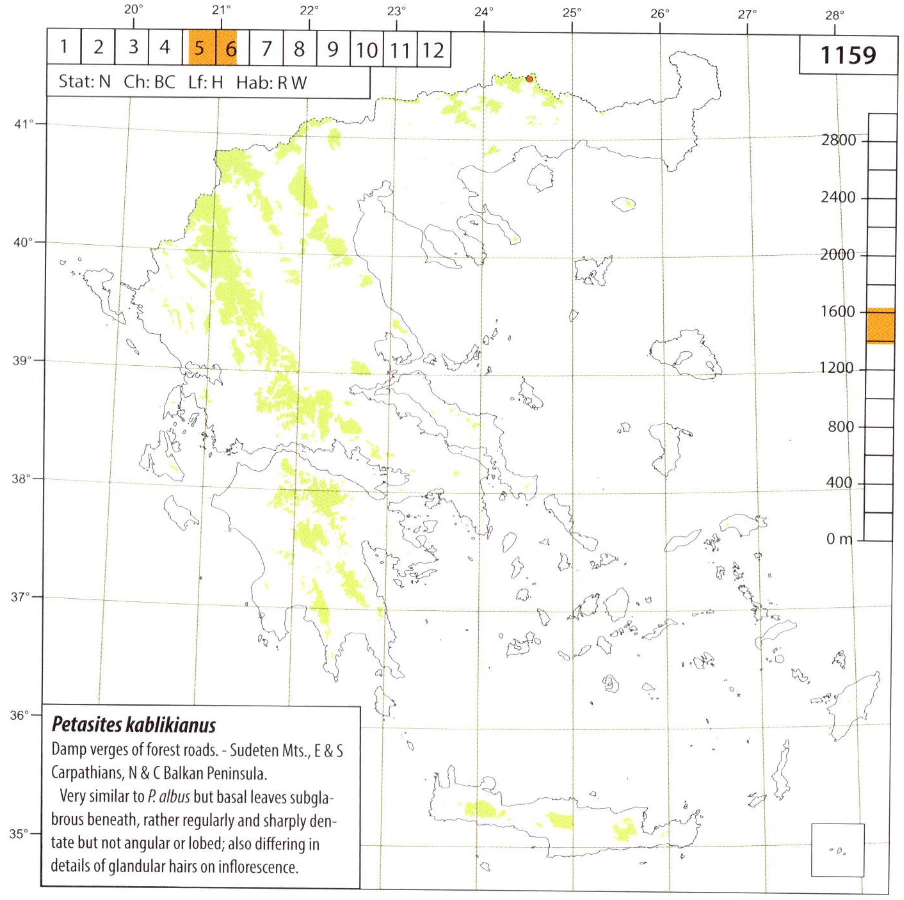

Petasites kablikianus
Damp verges of forest roads. - Sudeten Mts., E & S Carpathians, N & C Balkan Peninsula.
Very similar to *P. albus* but basal leaves subglabrous beneath, rather regularly and sharply dentate but not angular or lobed; also differing in details of glandular hairs on inflorescence.

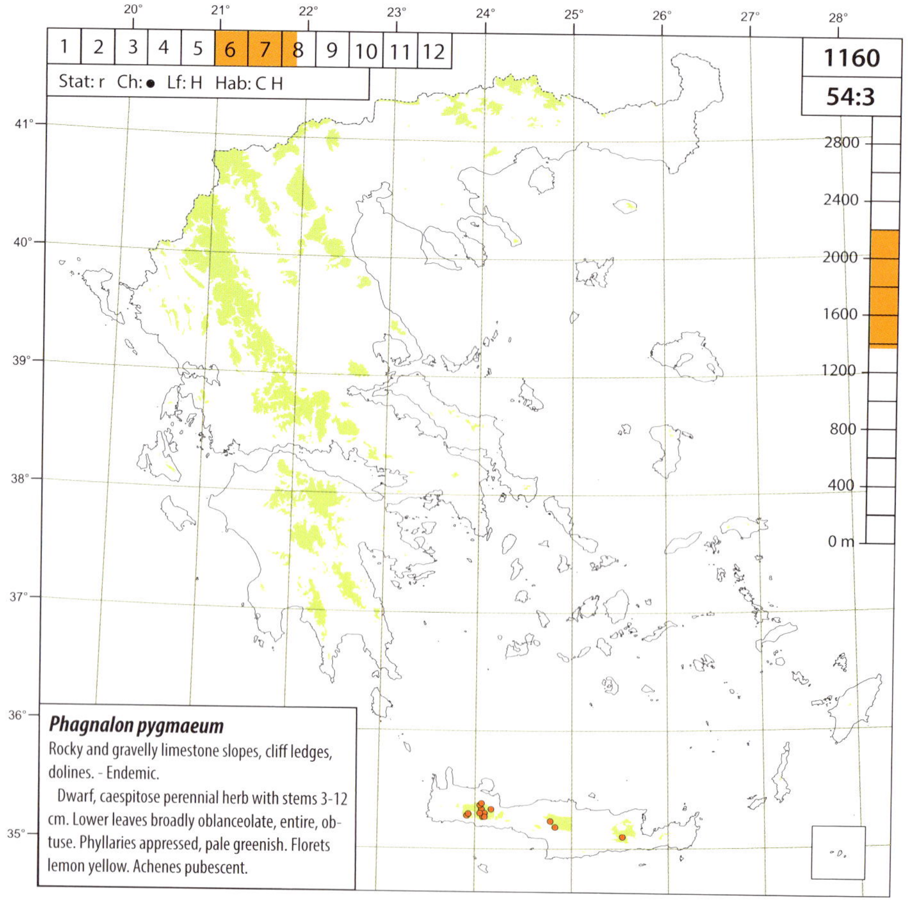

Phagnalon pygmaeum
Rocky and gravelly limestone slopes, cliff ledges, dolines. - Endemic.
Dwarf, caespitose perennial herb with stems 3-12 cm. Lower leaves broadly oblanceolate, entire, obtuse. Phyllaries appressed, pale greenish. Florets lemon yellow. Achenes pubescent.

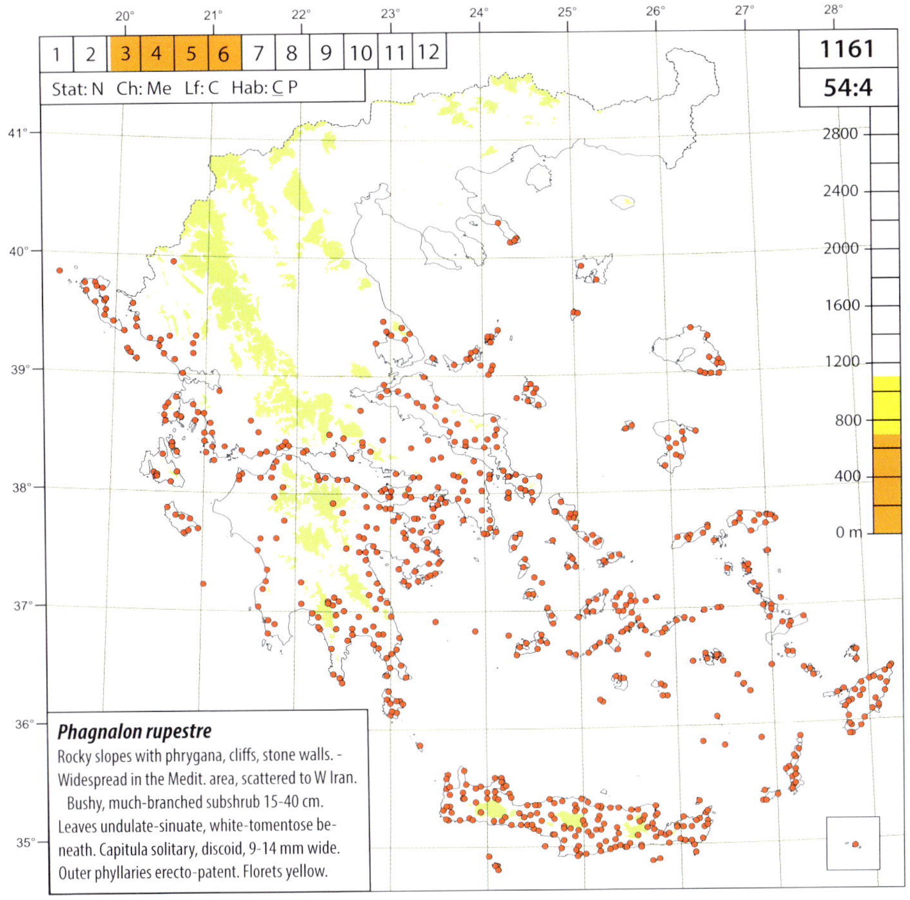

Phagnalon rupestre
Rocky slopes with phrygana, cliffs, stone walls. - Widespread in the Medit. area, scattered to W Iran.
Bushy, much-branched subshrub 15-40 cm. Leaves undulate-sinuate, white-tomentose beneath. Capitula solitary, discoid, 9-14 mm wide. Outer phyllaries erecto-patent. Florets yellow.

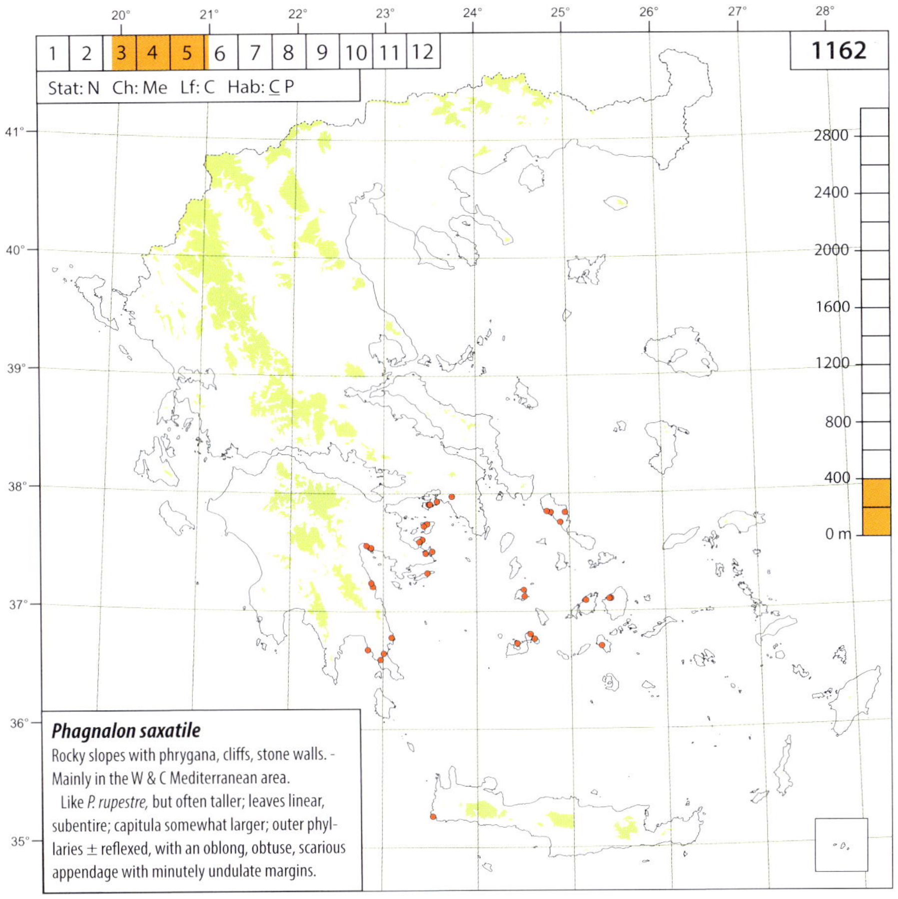

Phagnalon saxatile
Rocky slopes with phrygana, cliffs, stone walls. - Mainly in the W & C Mediterranean area.
Like *P. rupestre*, but often taller; leaves linear, subentire; capitula somewhat larger; outer phyllaries ± reflexed, with an oblong, obtuse, scarious appendage with minutely undulate margins.

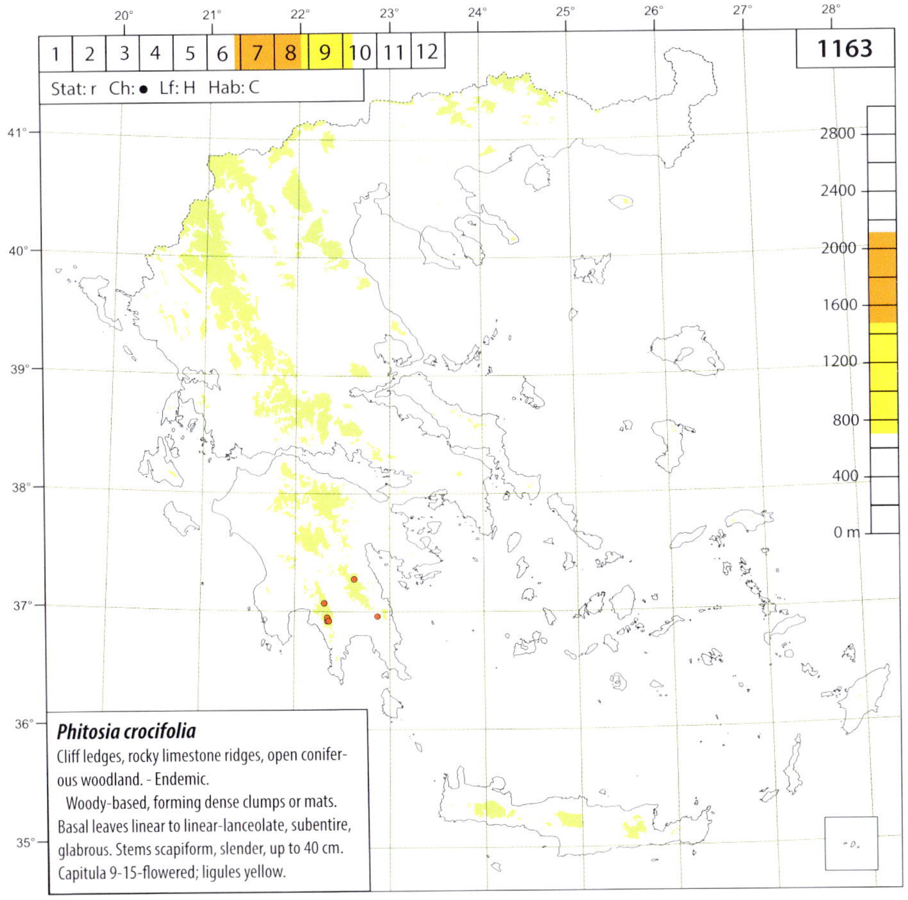

Phitosia crocifolia
Cliff ledges, rocky limestone ridges, open coniferous woodland. - Endemic.
Woody-based, forming dense clumps or mats. Basal leaves linear to linear-lanceolate, subentire, glabrous. Stems scapiform, slender, up to 40 cm. Capitula 9-15-flowered; ligules yellow.

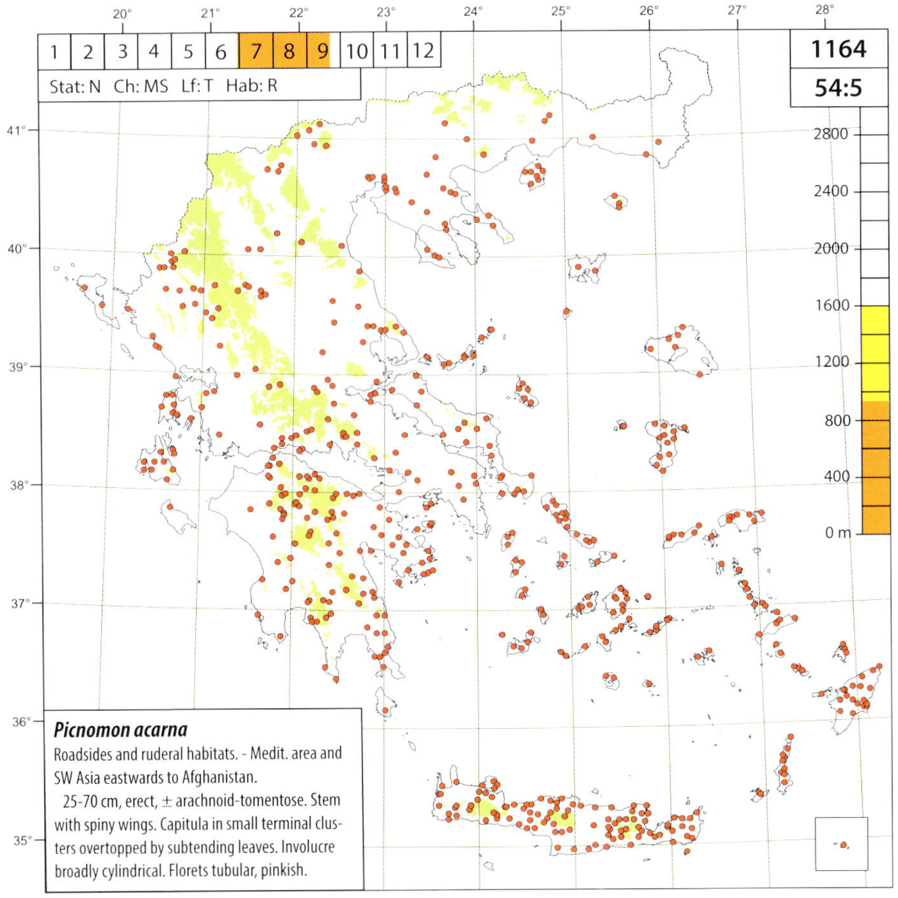

Picnomon acarna
Roadsides and ruderal habitats. - Medit. area and SW Asia eastwards to Afghanistan.
25-70 cm, erect, ± arachnoid-tomentose. Stem with spiny wings. Capitula in small terminal clusters overtopped by subtending leaves. Involucre broadly cylindrical. Florets tubular, pinkish.

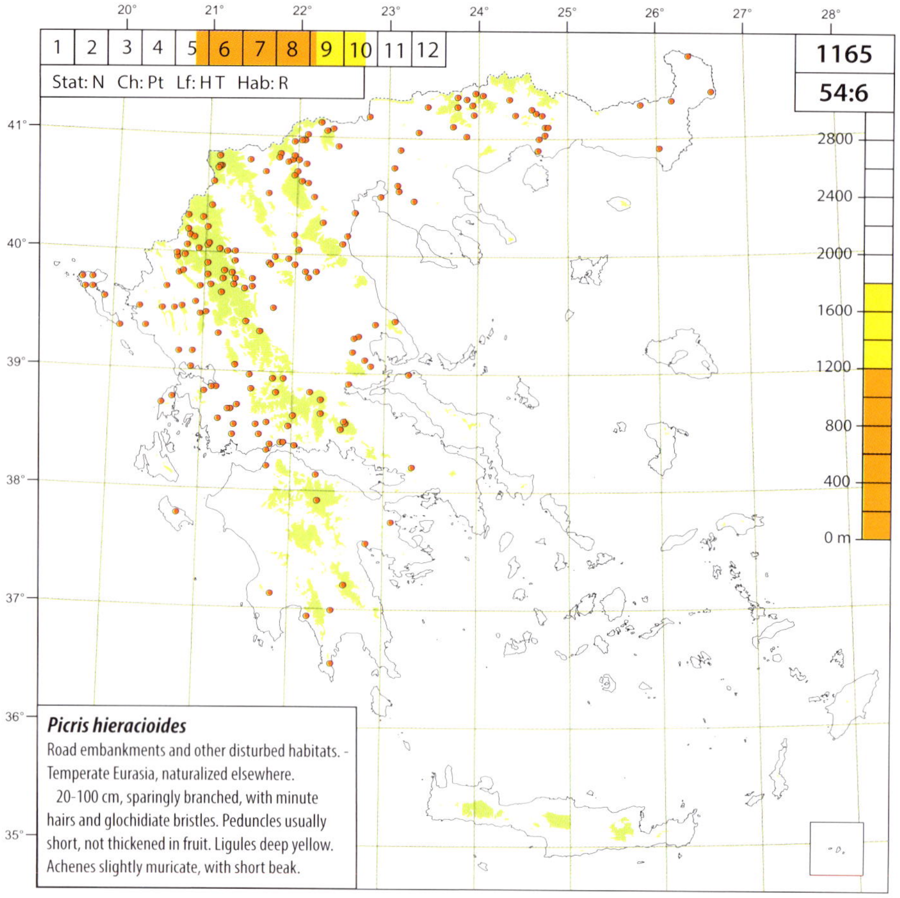

Picris hieracioides
Road embankments and other disturbed habitats. - Temperate Eurasia, naturalized elsewhere.
20-100 cm, sparingly branched, with minute hairs and glochidiate bristles. Peduncles usually short, not thickened in fruit. Ligules deep yellow. Achenes slightly muricate, with short beak.

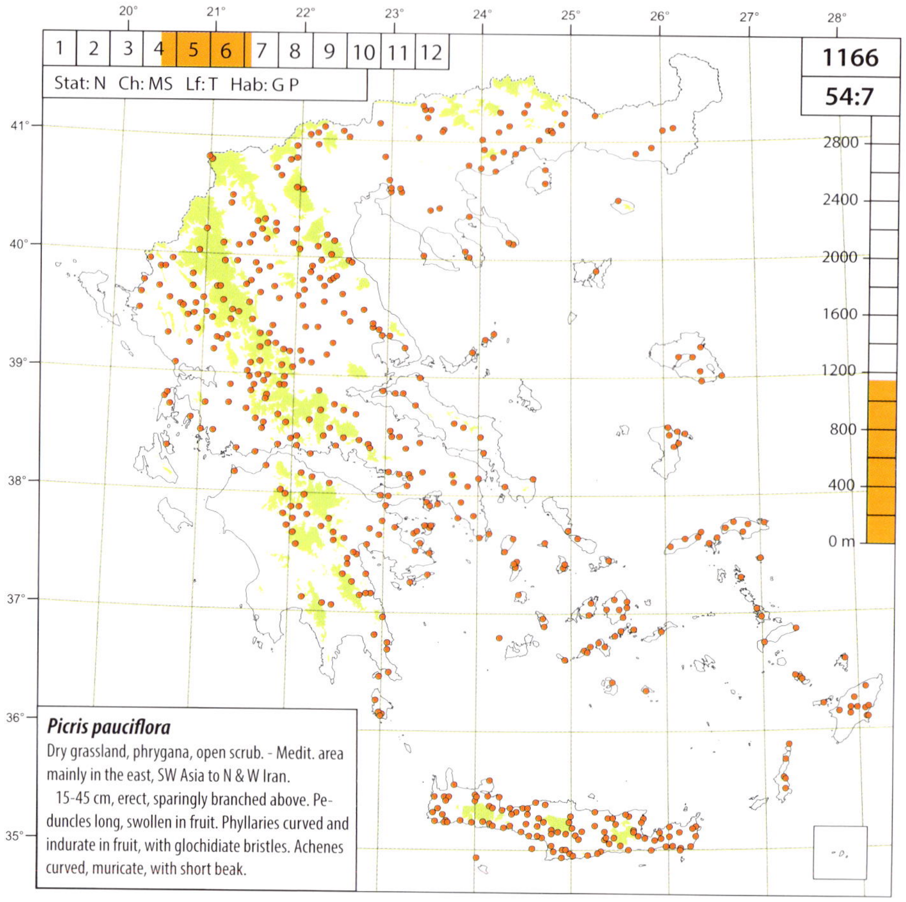

Picris pauciflora
Dry grassland, phrygana, open scrub. - Medit. area mainly in the east, SW Asia to N & W Iran.
15-45 cm, erect, sparingly branched above. Peduncles long, swollen in fruit. Phyllaries curved and indurate in fruit, with glochidiate bristles. Achenes curved, muricate, with short beak.

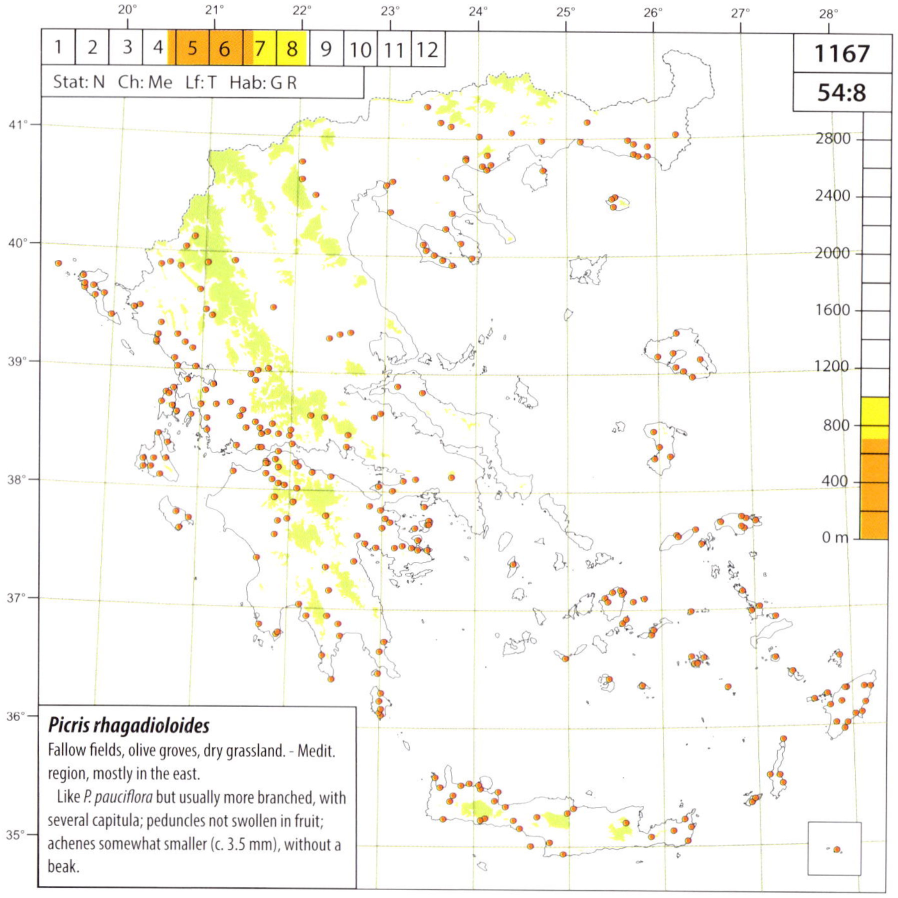

Picris rhagadioloides
Fallow fields, olive groves, dry grassland. - Medit. region, mostly in the east.
Like *P. pauciflora* but usually more branched, with several capitula; peduncles not swollen in fruit; achenes somewhat smaller (c. 3.5 mm), without a beak.

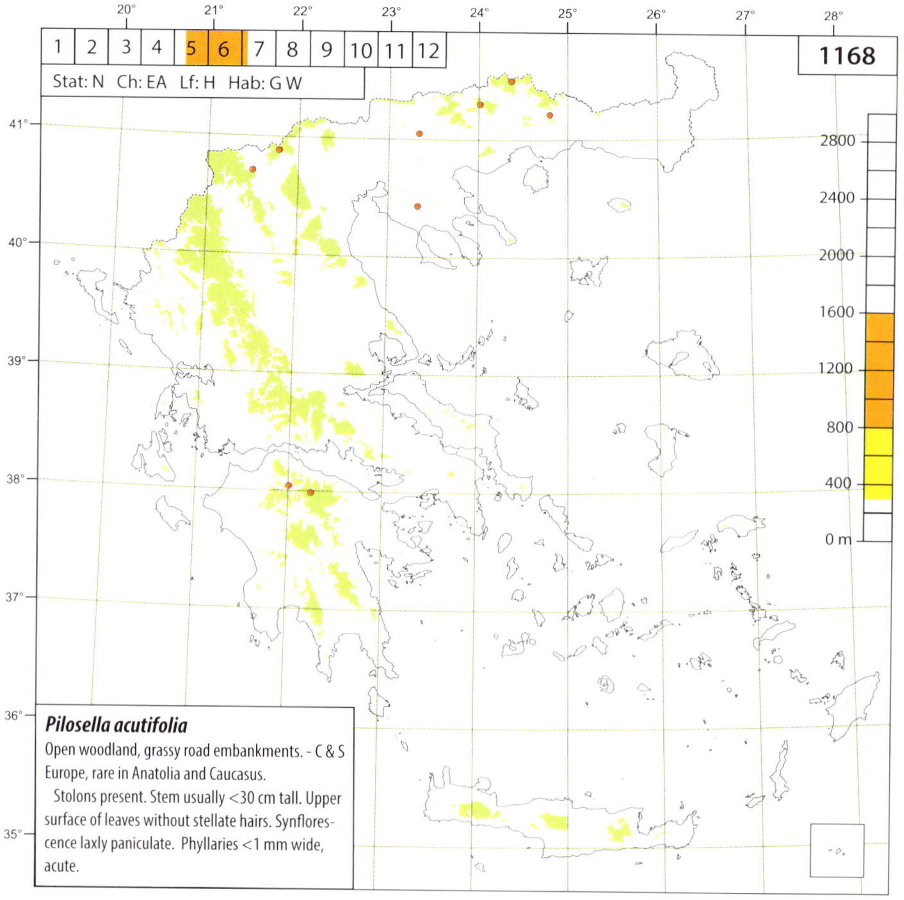

Pilosella acutifolia
Open woodland, grassy road embankments. - C & S Europe, rare in Anatolia and Caucasus.
Stolons present. Stem usually <30 cm tall. Upper surface of leaves without stellate hairs. Synflorescence laxly paniculate. Phyllaries <1 mm wide, acute.

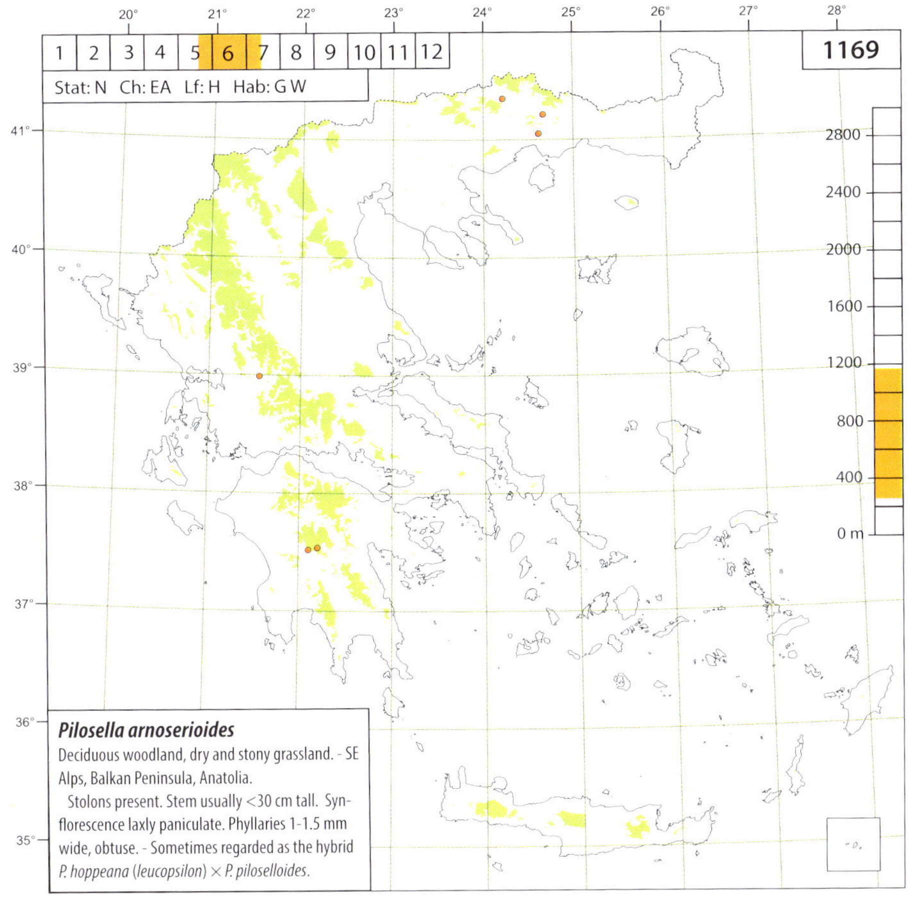

Pilosella arnoserioides
Deciduous woodland, dry and stony grassland. - SE Alps, Balkan Peninsula, Anatolia.
Stolons present. Stem usually <30 cm tall. Synflorescence laxly paniculate. Phyllaries 1-1.5 mm wide, obtuse. - Sometimes regarded as the hybrid *P. hoppeana* (*leucopsilon*) × *P. piloselloides*.

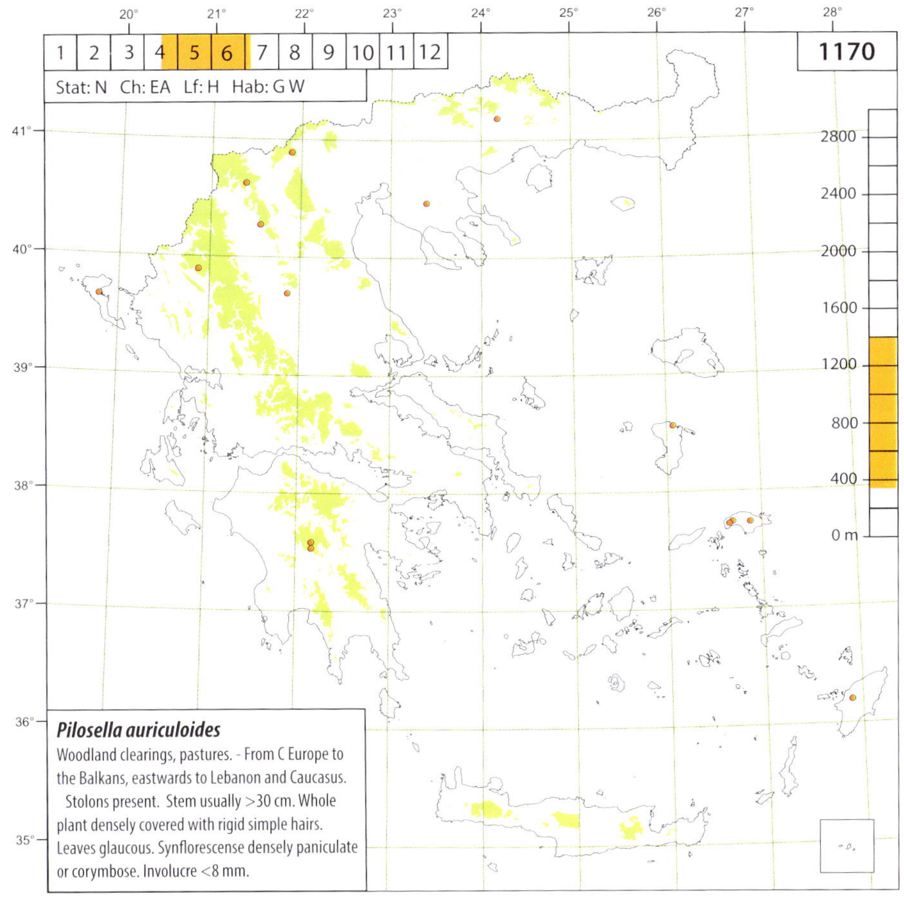

Pilosella auriculoides
Woodland clearings, pastures. - From C Europe to the Balkans, eastwards to Lebanon and Caucasus.
Stolons present. Stem usually >30 cm. Whole plant densely covered with rigid simple hairs. Leaves glaucous. Synflorescense densely paniculate or corymbose. Involucre <8 mm.

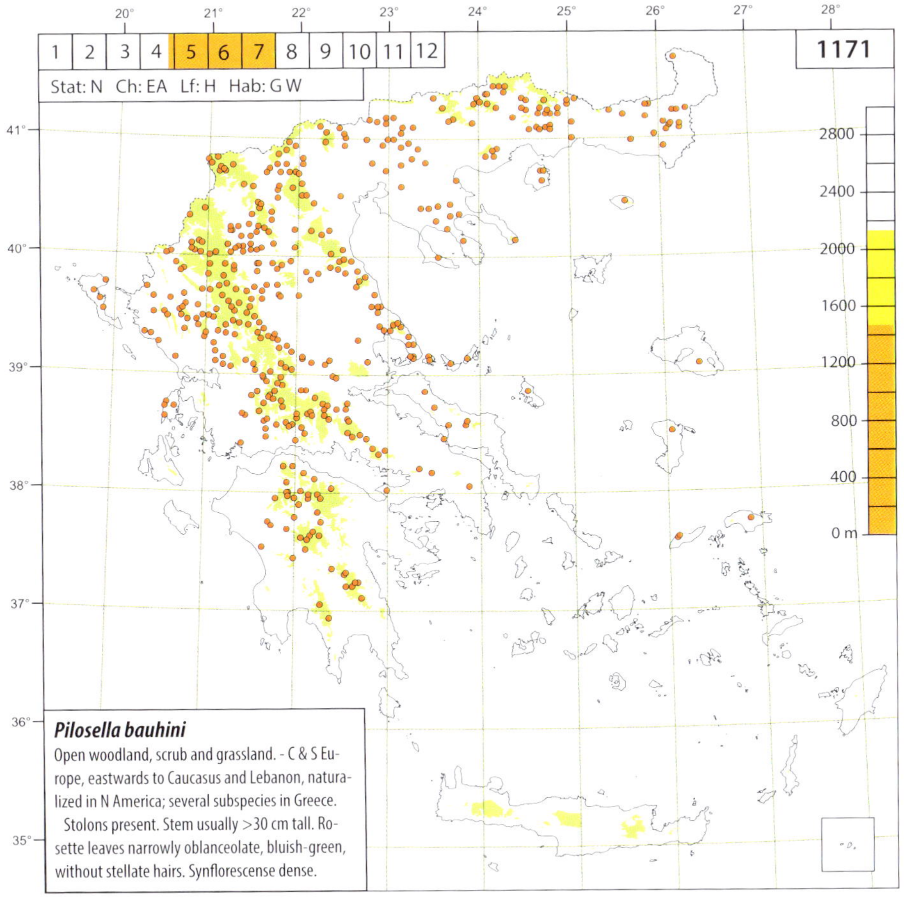

Pilosella bauhini
Open woodland, scrub and grassland. - C & S Europe, eastwards to Caucasus and Lebanon, naturalized in N America; several subspecies in Greece.
Stolons present. Stem usually >30 cm tall. Rosette leaves narrowly oblanceolate, bluish-green, without stellate hairs. Synflorescense dense.

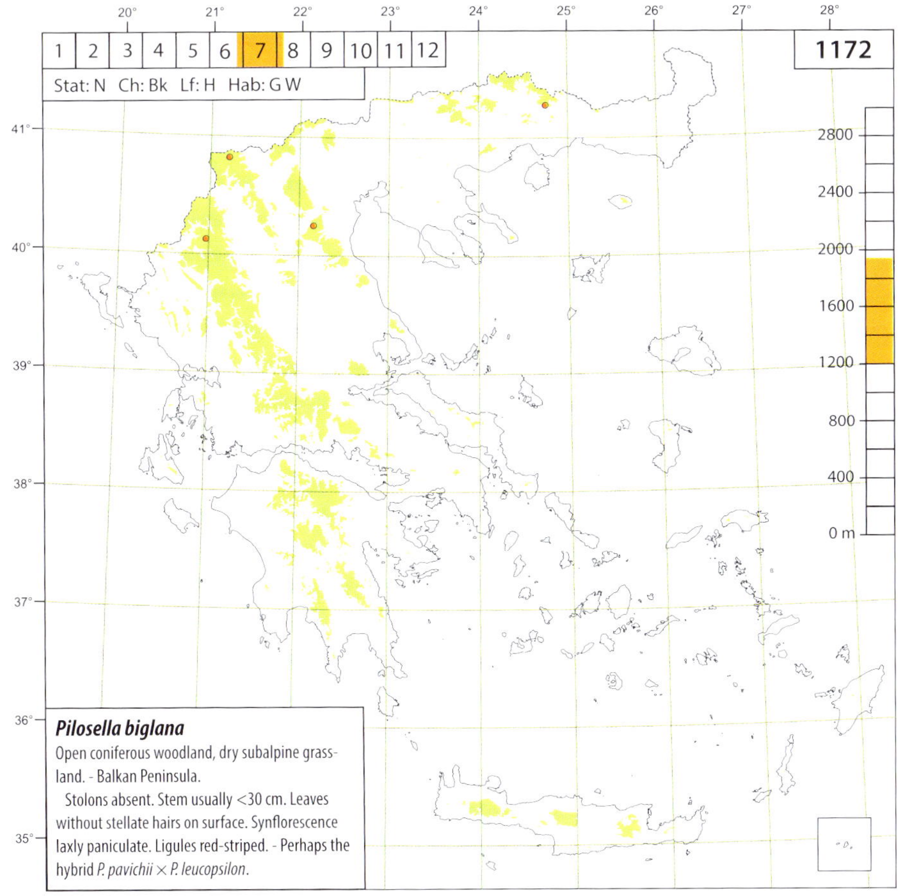

Pilosella biglana
Open coniferous woodland, dry subalpine grassland. - Balkan Peninsula.
Stolons absent. Stem usually <30 cm. Leaves without stellate hairs on surface. Synflorescence laxly paniculate. Ligules red-striped. - Perhaps the hybrid *P. pavichii* × *P. leucopsilon*.

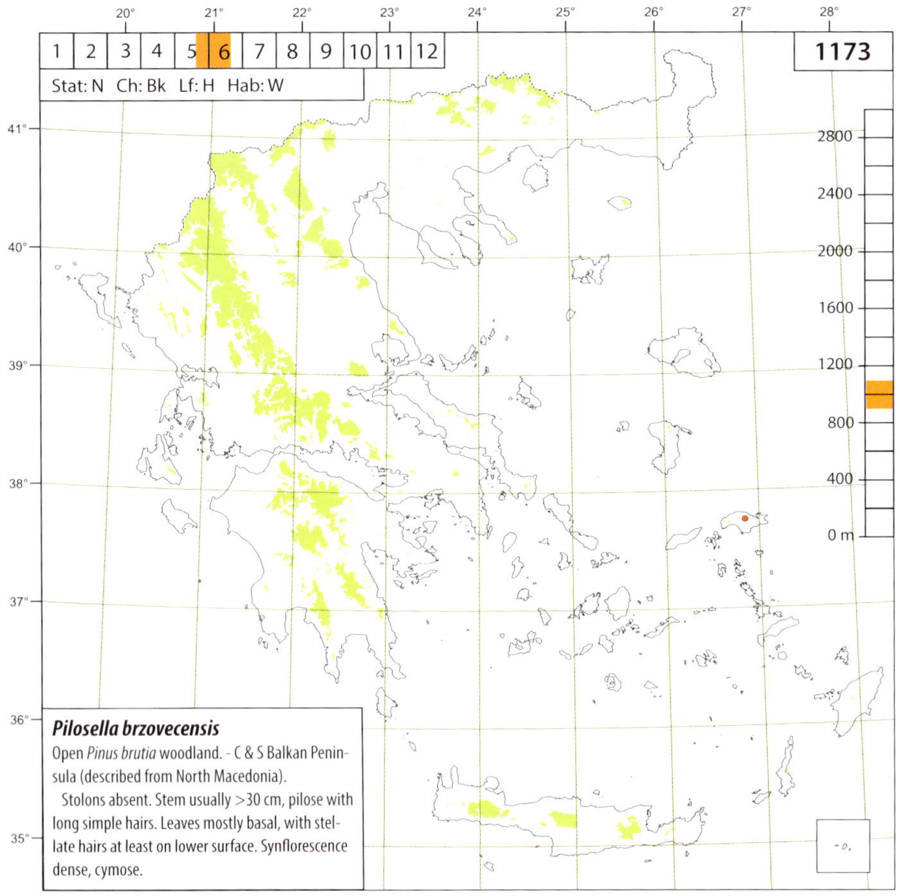

Pilosella brzovecensis
Open *Pinus brutia* woodland. - C & S Balkan Peninsula (described from North Macedonia).
Stolons absent. Stem usually >30 cm, pilose with long simple hairs. Leaves mostly basal, with stellate hairs at least on lower surface. Synflorescence dense, cymose.

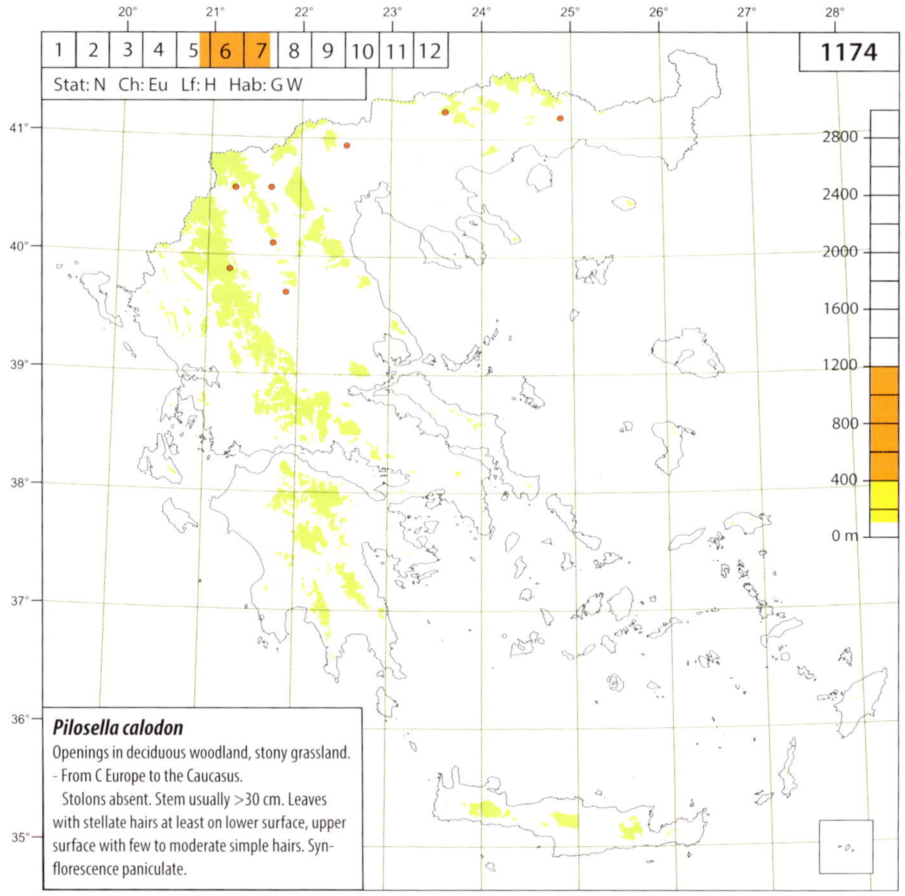

Pilosella calodon
Openings in deciduous woodland, stony grassland. - From C Europe to the Caucasus.
Stolons absent. Stem usually >30 cm. Leaves with stellate hairs at least on lower surface, upper surface with few to moderate simple hairs. Synflorescence paniculate.

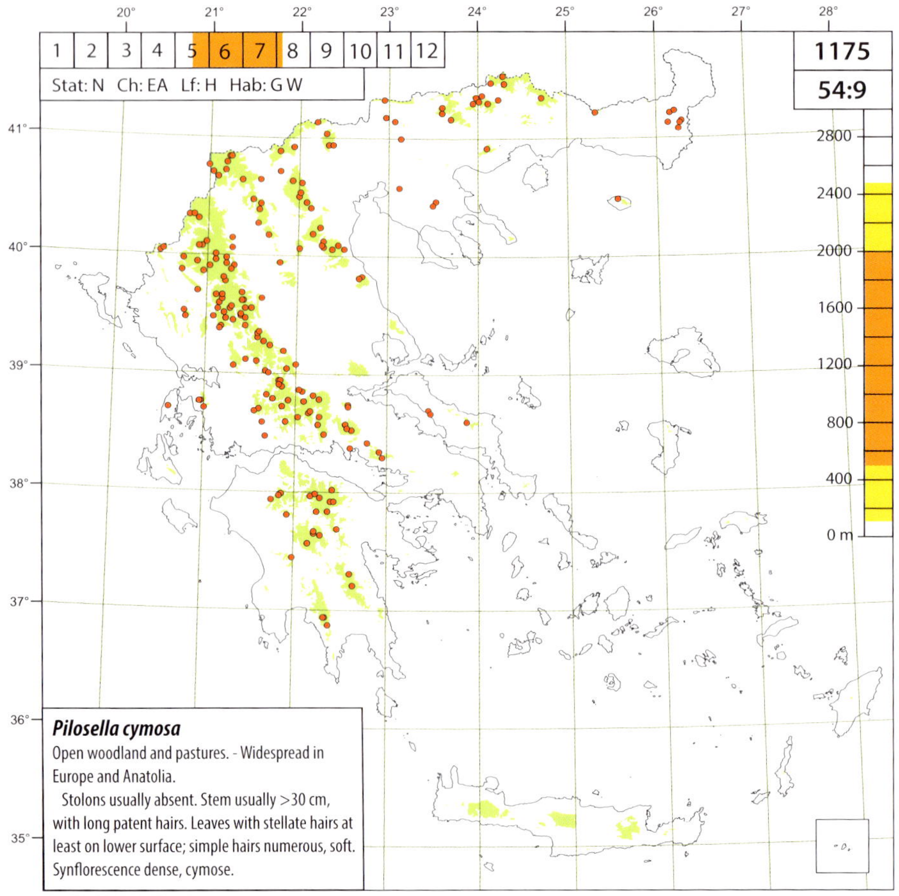

Pilosella cymosa
Open woodland and pastures. - Widespread in Europe and Anatolia.
Stolons usually absent. Stem usually >30 cm, with long patent hairs. Leaves with stellate hairs at least on lower surface; simple hairs numerous, soft. Synflorescence dense, cymose.

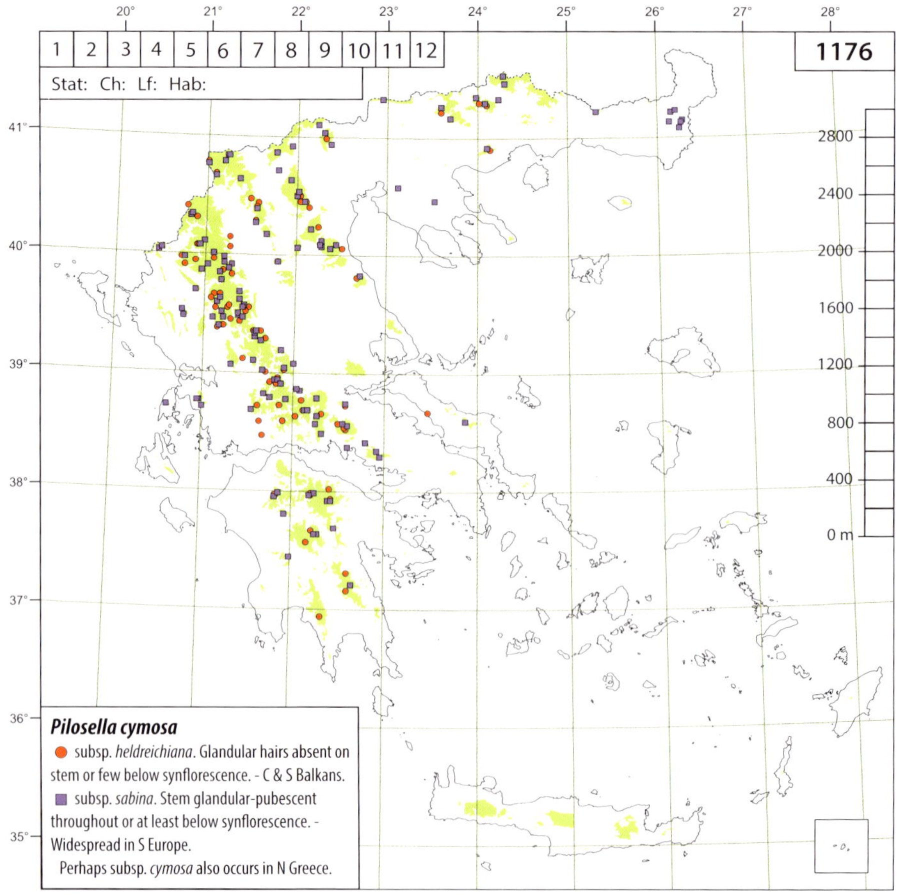

Pilosella cymosa
● subsp. *heldreichiana*. Glandular hairs absent on stem or few below synflorescence. - C & S Balkans.
■ subsp. *sabina*. Stem glandular-pubescent throughout or at least below synflorescence. - Widespread in S Europe.
Perhaps subsp. *cymosa* also occurs in N Greece.

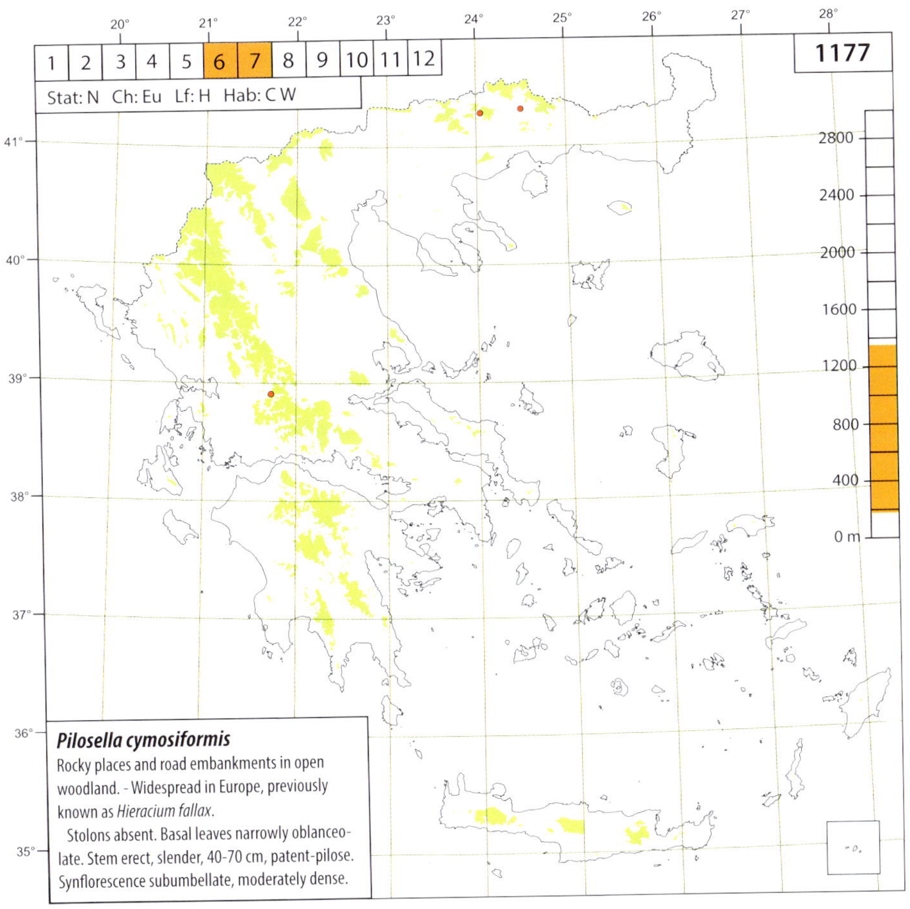

Pilosella cymosiformis

Rocky places and road embankments in open woodland. - Widespread in Europe, previously known as *Hieracium fallax*.

Stolons absent. Basal leaves narrowly oblanceolate. Stem erect, slender, 40-70 cm, patent-pilose. Synflorescence subumbellate, moderately dense.

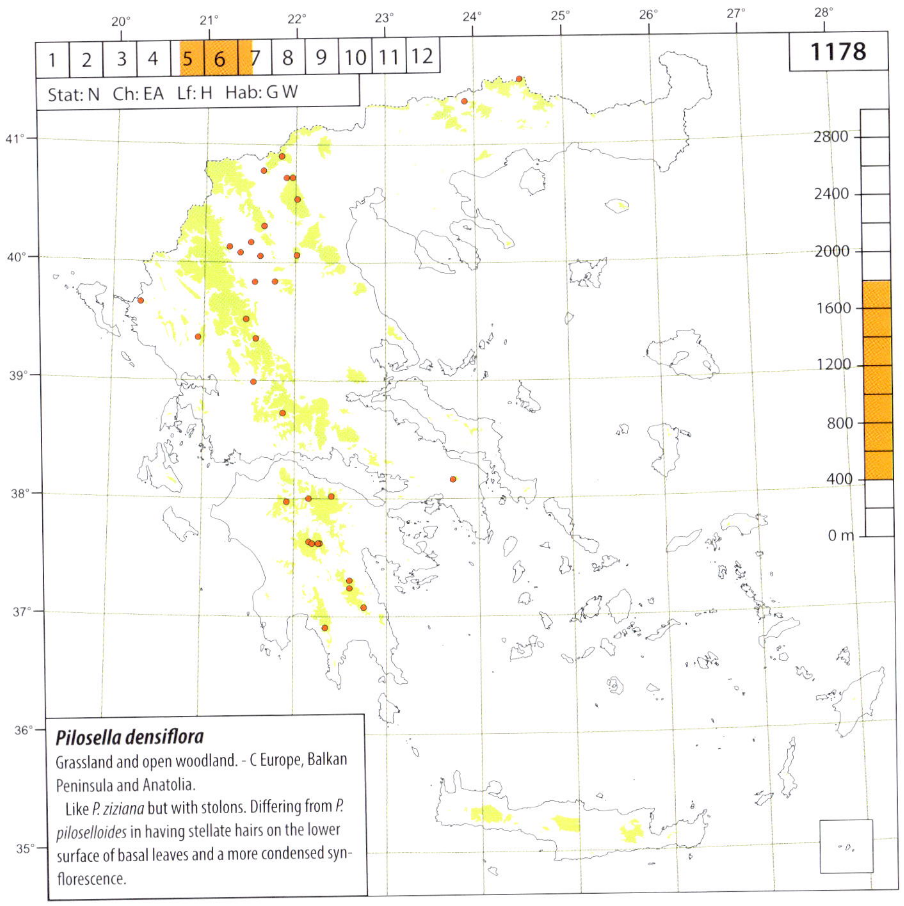

Pilosella densiflora

Grassland and open woodland. - C Europe, Balkan Peninsula and Anatolia.

Like *P. ziziana* but with stolons. Differing from *P. piloselloides* in having stellate hairs on the lower surface of basal leaves and a more condensed synflorescence.

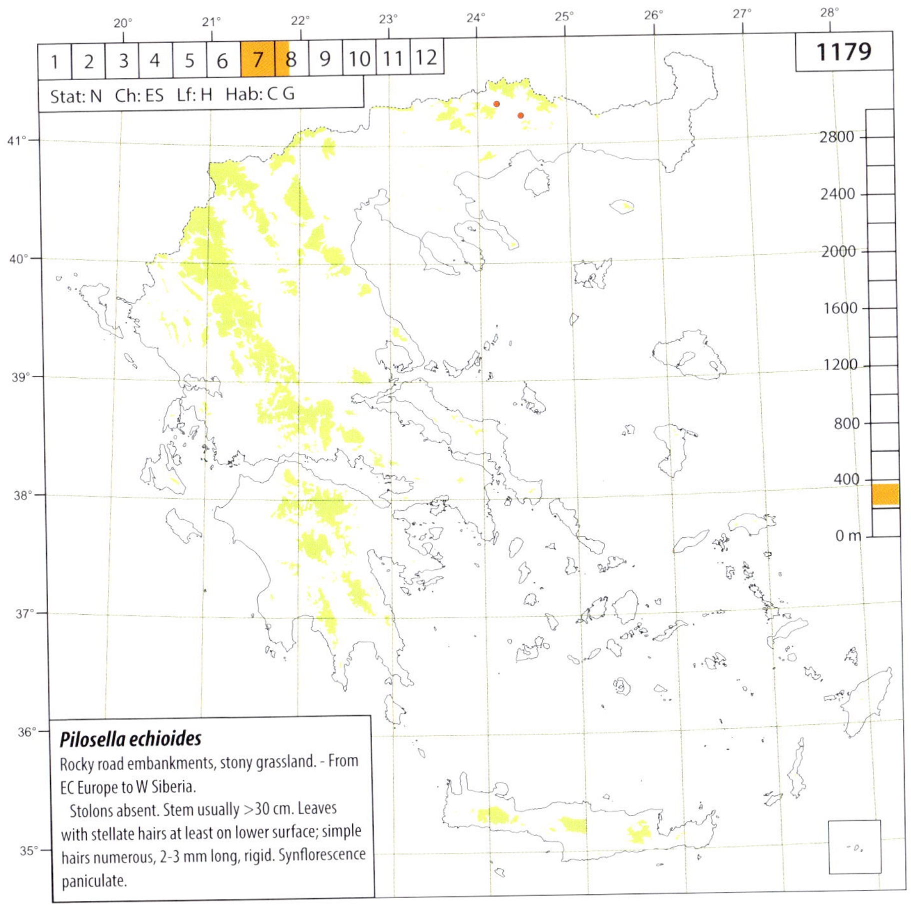

Pilosella echioides

Rocky road embankments, stony grassland. - From EC Europe to W Siberia.

Stolons absent. Stem usually >30 cm. Leaves with stellate hairs at least on lower surface; simple hairs numerous, 2-3 mm long, rigid. Synflorescence paniculate.

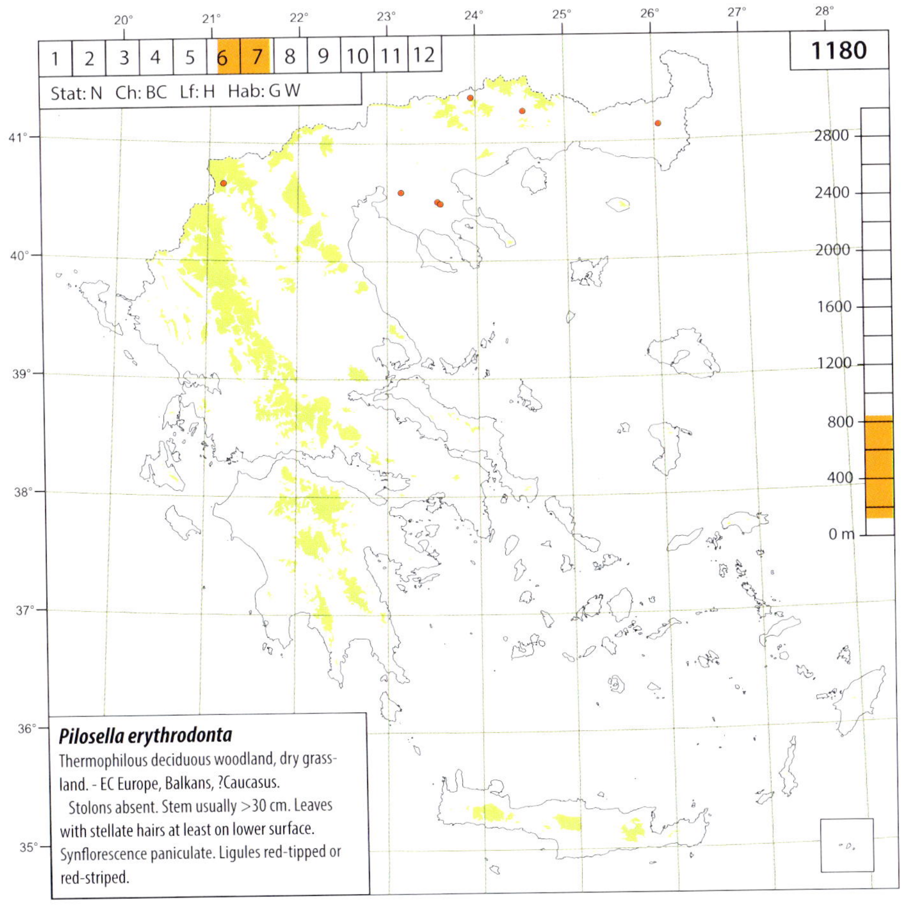

Pilosella erythrodonta

Thermophilous deciduous woodland, dry grassland. - EC Europe, Balkans, ?Caucasus.

Stolons absent. Stem usually >30 cm. Leaves with stellate hairs at least on lower surface. Synflorescence paniculate. Ligules red-tipped or red-striped.

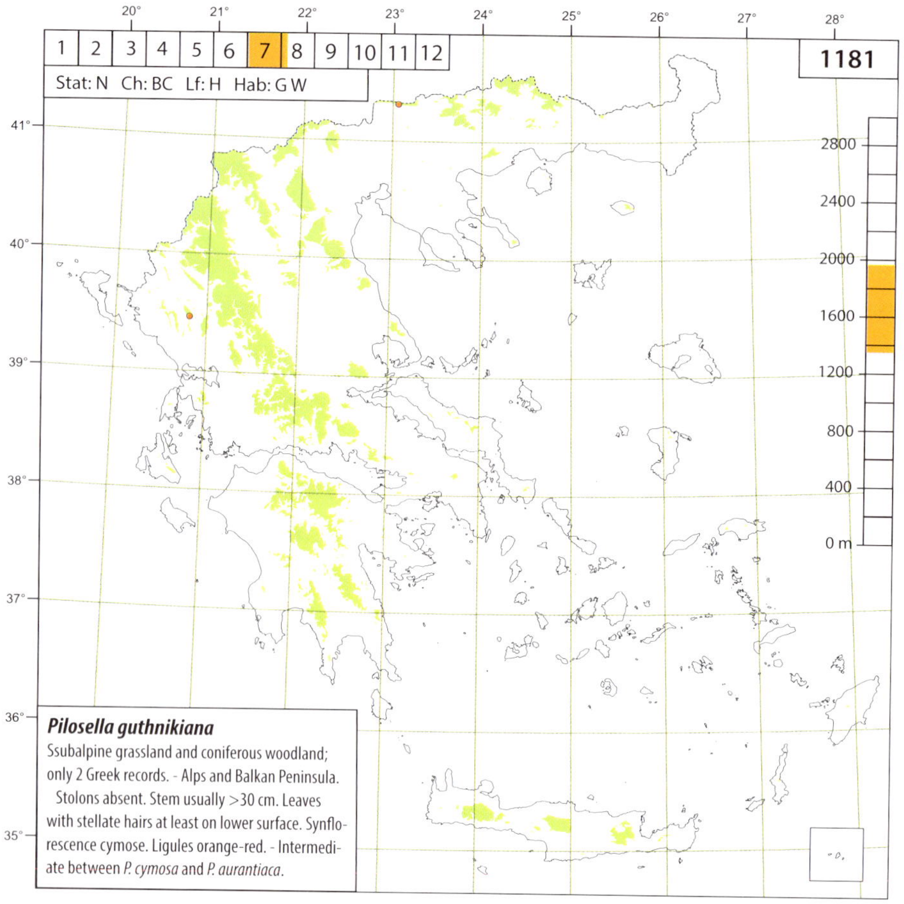

Pilosella guthnikiana
Ssubalpine grassland and coniferous woodland; only 2 Greek records. - Alps and Balkan Peninsula.
Stolons absent. Stem usually >30 cm. Leaves with stellate hairs at least on lower surface. Synflorescence cymose. Ligules orange-red. - Intermediate between *P. cymosa* and *P. aurantiaca*.

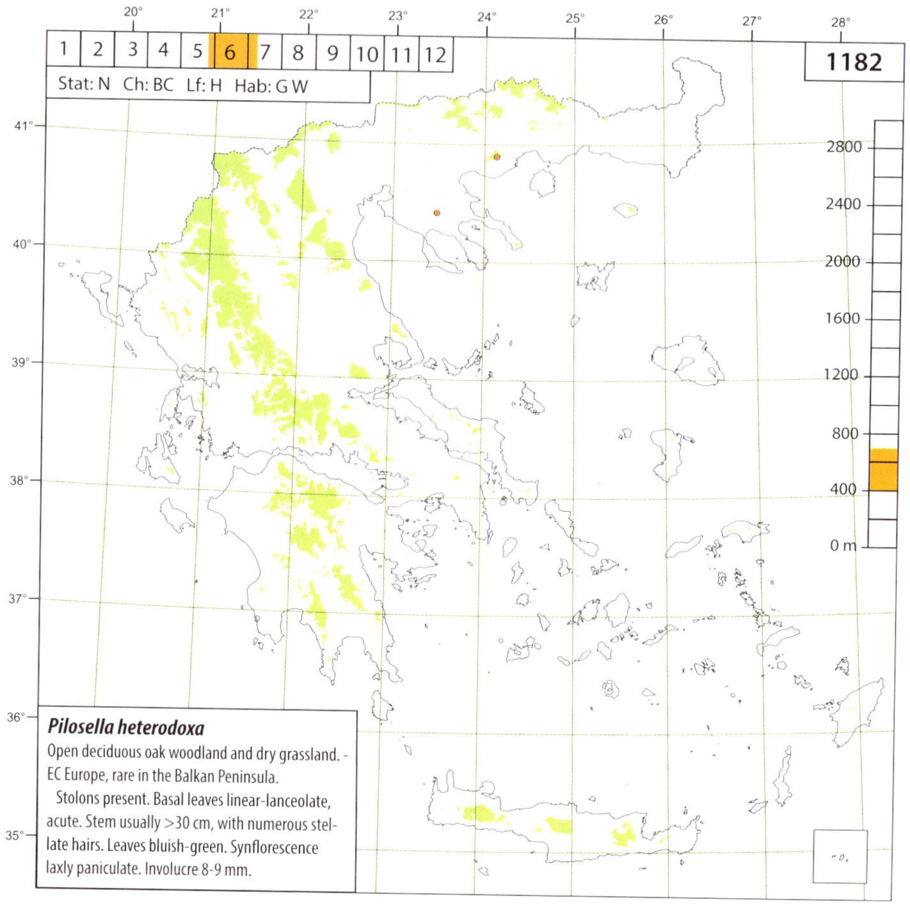

Pilosella heterodoxa
Open deciduous oak woodland and dry grassland. - EC Europe, rare in the Balkan Peninsula.
Stolons present. Basal leaves linear-lanceolate, acute. Stem usually >30 cm, with numerous stellate hairs. Leaves bluish-green. Synflorescence laxly paniculate. Involucre 8-9 mm.

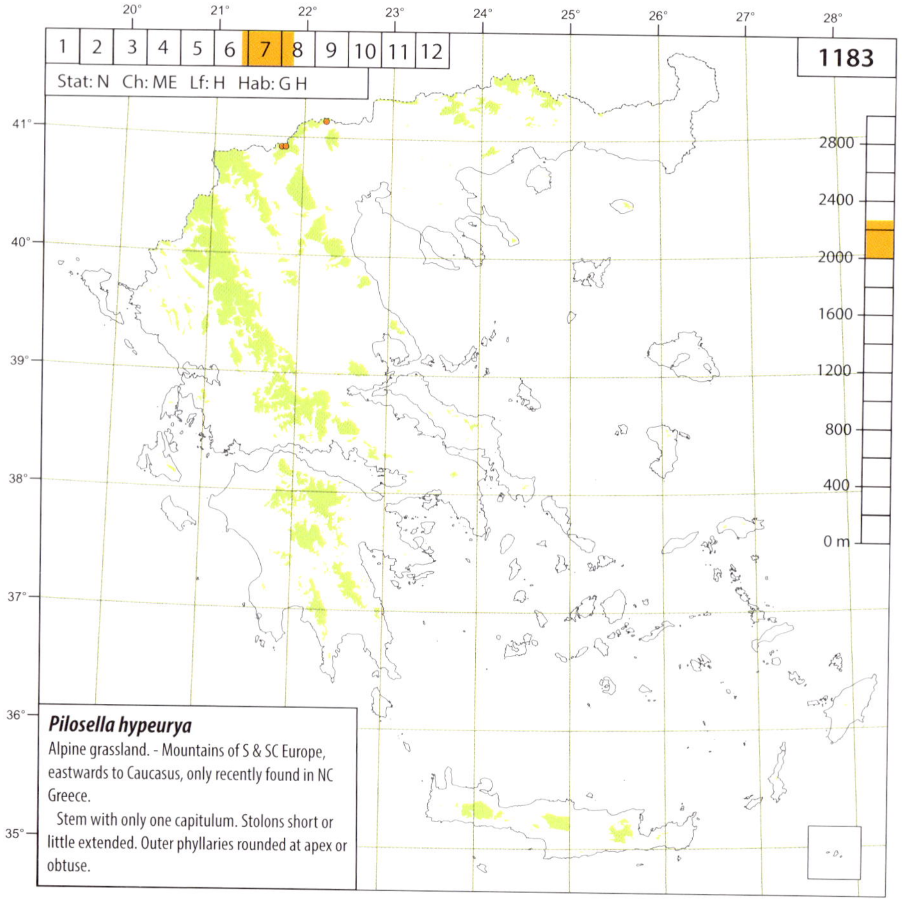

Pilosella hypeurya
Alpine grassland. - Mountains of S & SC Europe, eastwards to Caucasus, only recently found in NC Greece.
Stem with only one capitulum. Stolons short or little extended. Outer phyllaries rounded at apex or obtuse.

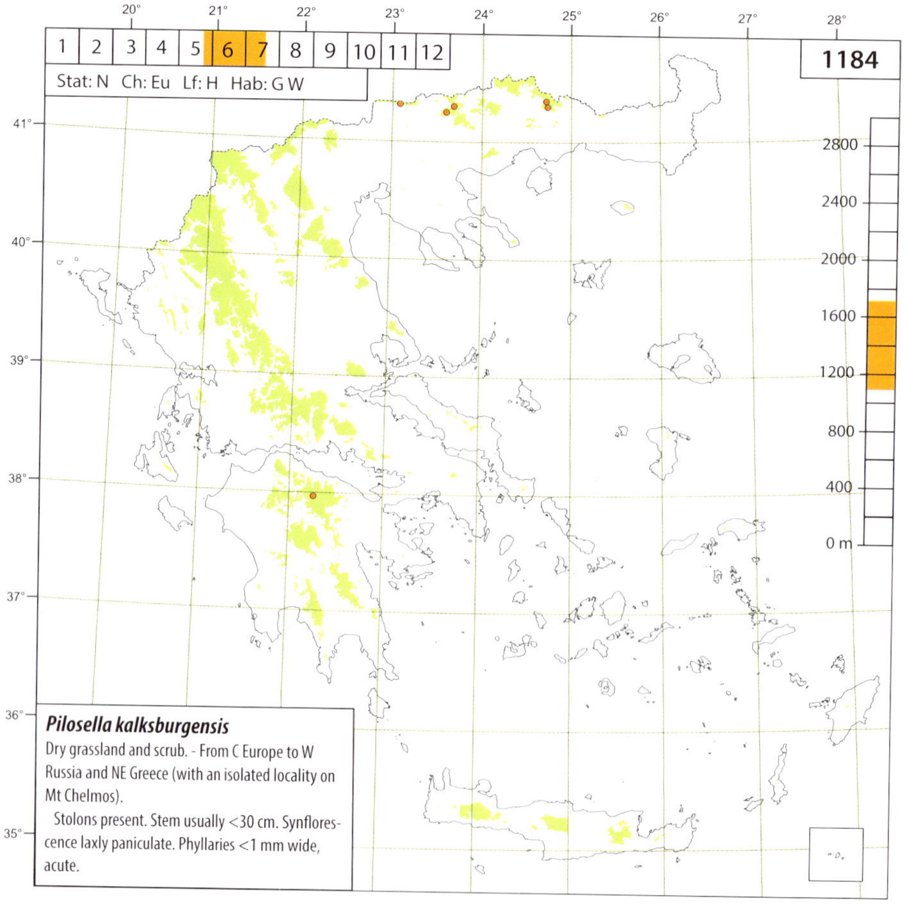

Pilosella kalksburgensis
Dry grassland and scrub. - From C Europe to W Russia and NE Greece (with an isolated locality on Mt Chelmos).
Stolons present. Stem usually <30 cm. Synflorescence laxly paniculate. Phyllaries <1 mm wide, acute.

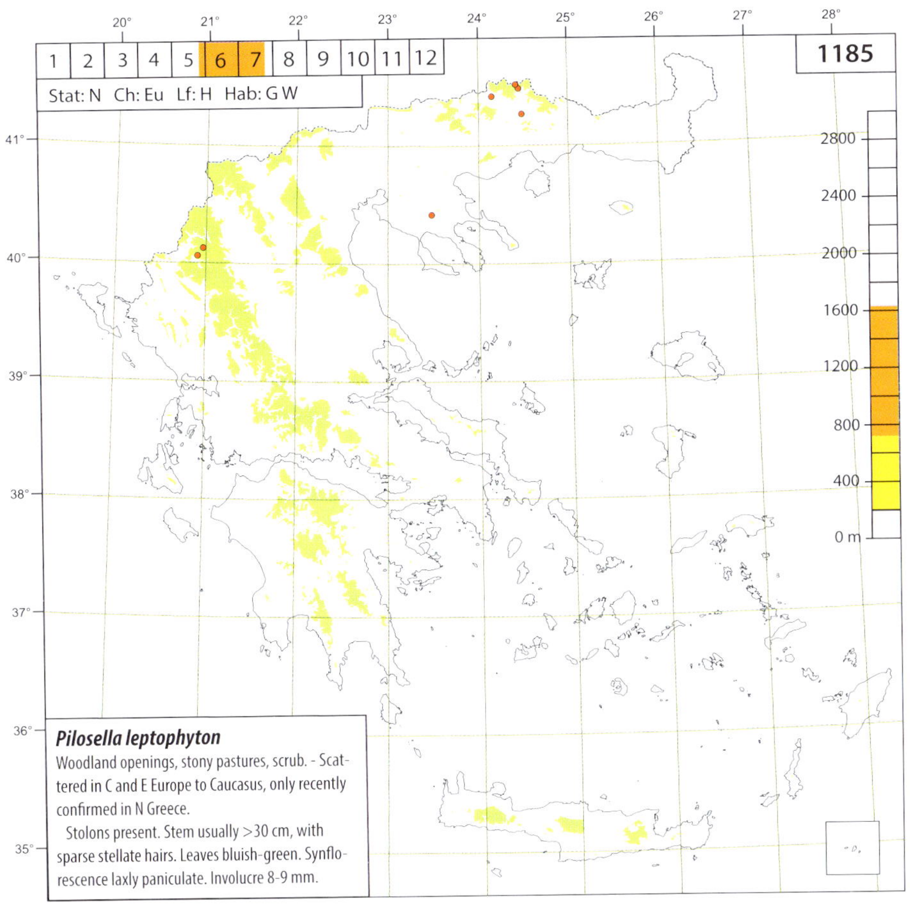

Pilosella leptophyton
Woodland openings, stony pastures, scrub. - Scattered in C and E Europe to Caucasus, only recently confirmed in N Greece.
Stolons present. Stem usually >30 cm, with sparse stellate hairs. Leaves bluish-green. Synflorescence laxly paniculate. Involucre 8-9 mm.

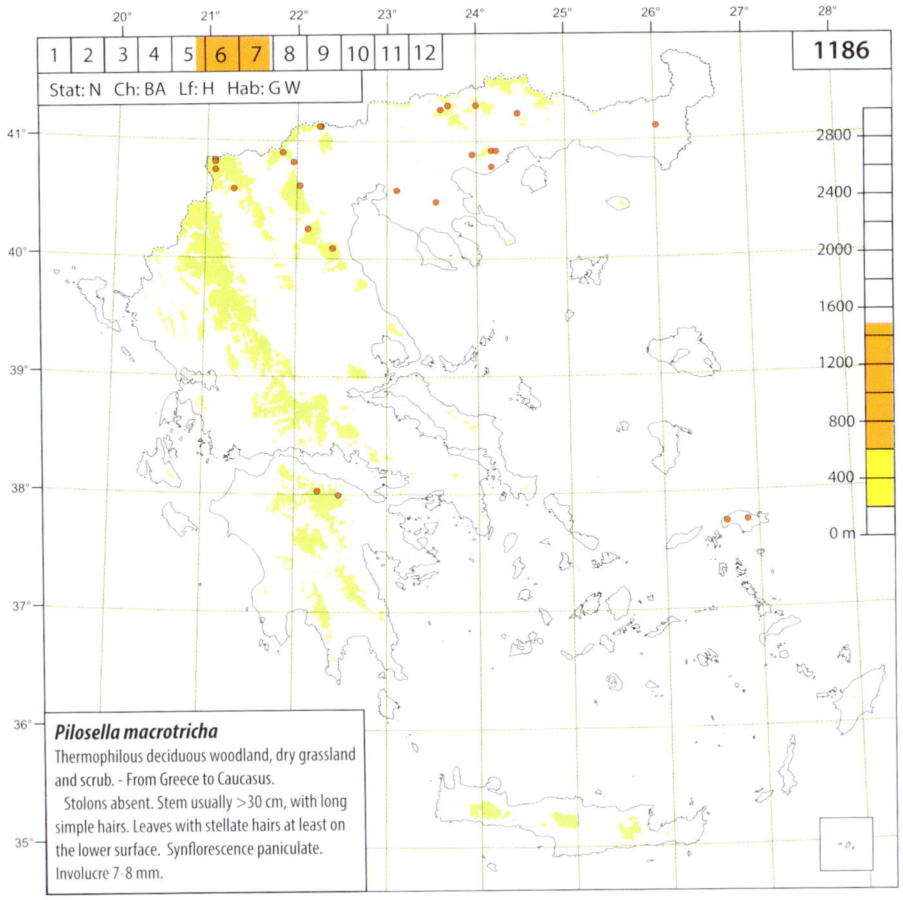

Pilosella macrotricha
Thermophilous deciduous woodland, dry grassland and scrub. - From Greece to Caucasus.
Stolons absent. Stem usually >30 cm, with long simple hairs. Leaves with stellate hairs at least on the lower surface. Synflorescence paniculate. Involucre 7-8 mm.

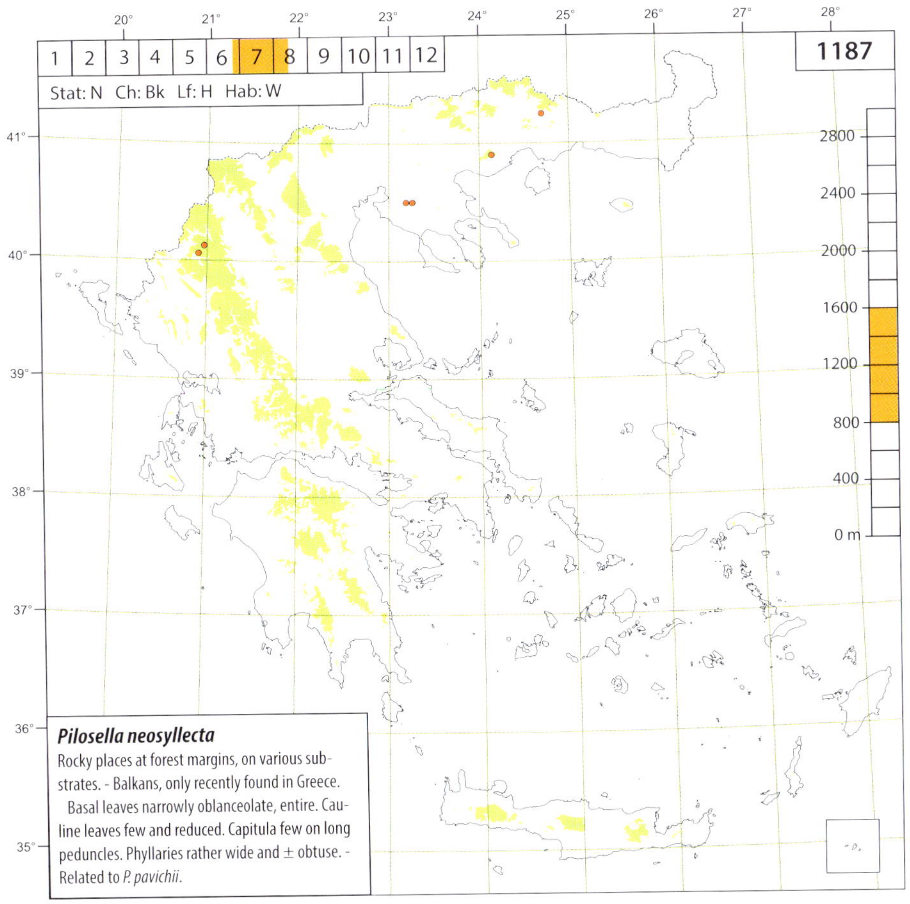

Pilosella neosyllecta
Rocky places at forest margins, on various substrates. - Balkans, only recently found in Greece.
Basal leaves narrowly oblanceolate, entire. Cauline leaves few and reduced. Capitula few on long peduncles. Phyllaries rather wide and ± obtuse. - Related to *P. pavichii*.

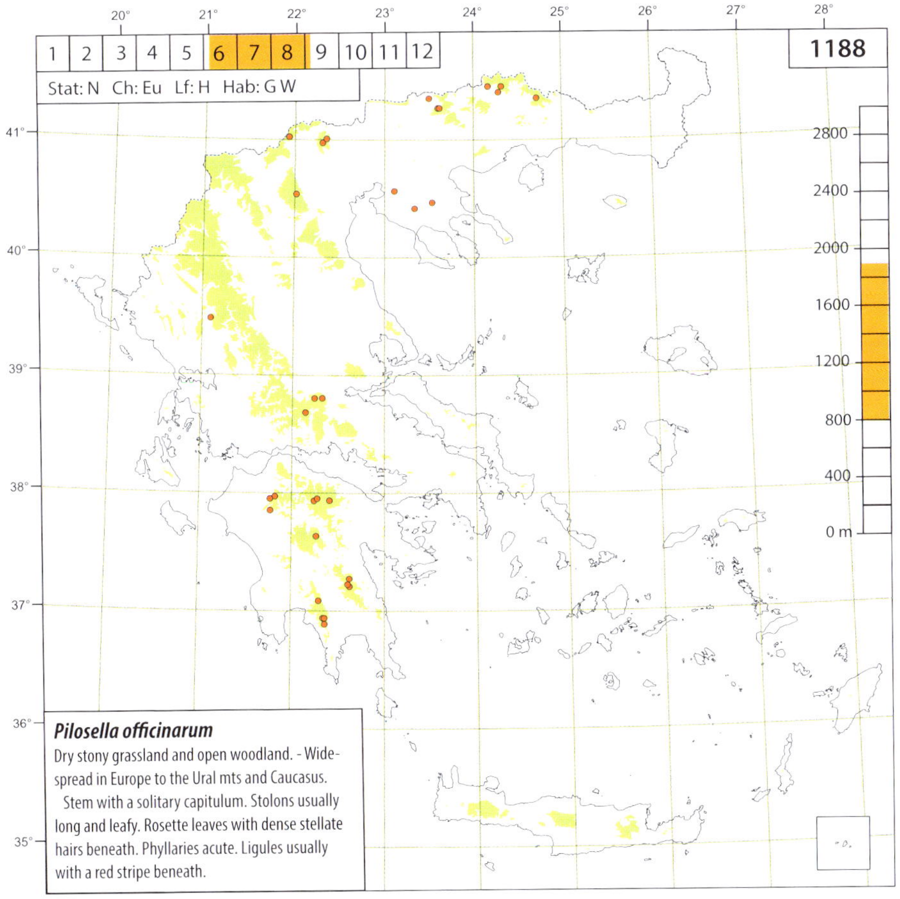

Pilosella officinarum
Dry stony grassland and open woodland. - Widespread in Europe to the Ural mts and Caucasus.
Stem with a solitary capitulum. Stolons usually long and leafy. Rosette leaves with dense stellate hairs beneath. Phyllaries acute. Ligules usually with a red stripe beneath.

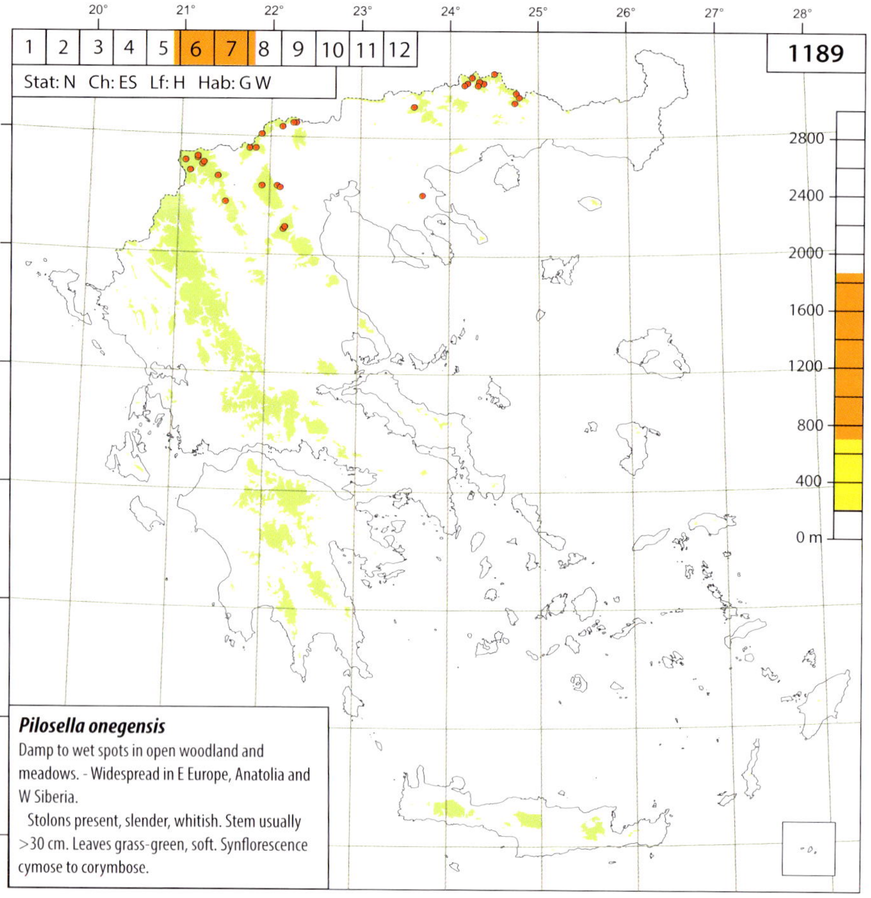

Pilosella onegensis
Damp to wet spots in open woodland and meadows. - Widespread in E Europe, Anatolia and W Siberia.
Stolons present, slender, whitish. Stem usually >30 cm. Leaves grass-green, soft. Synflorescence cymose to corymbose.

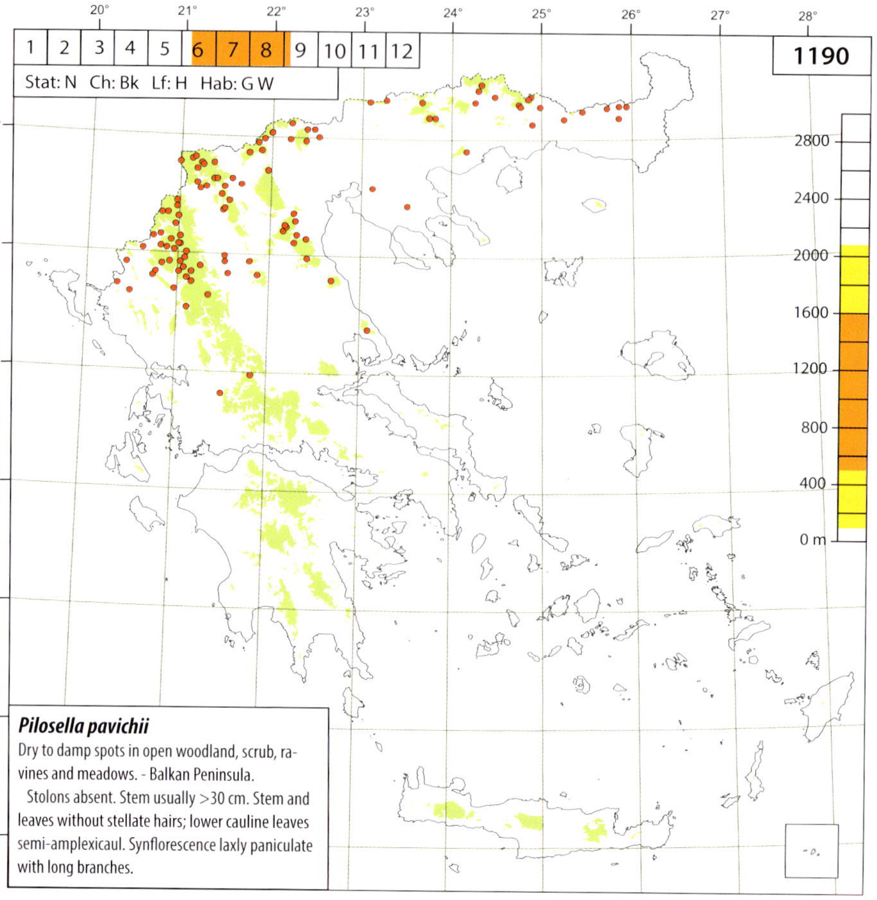

Pilosella pavichii
Dry to damp spots in open woodland, scrub, ravines and meadows. - Balkan Peninsula.
Stolons absent. Stem usually >30 cm. Stem and leaves without stellate hairs; lower cauline leaves semi-amplexicaul. Synflorescence laxly paniculate with long branches.

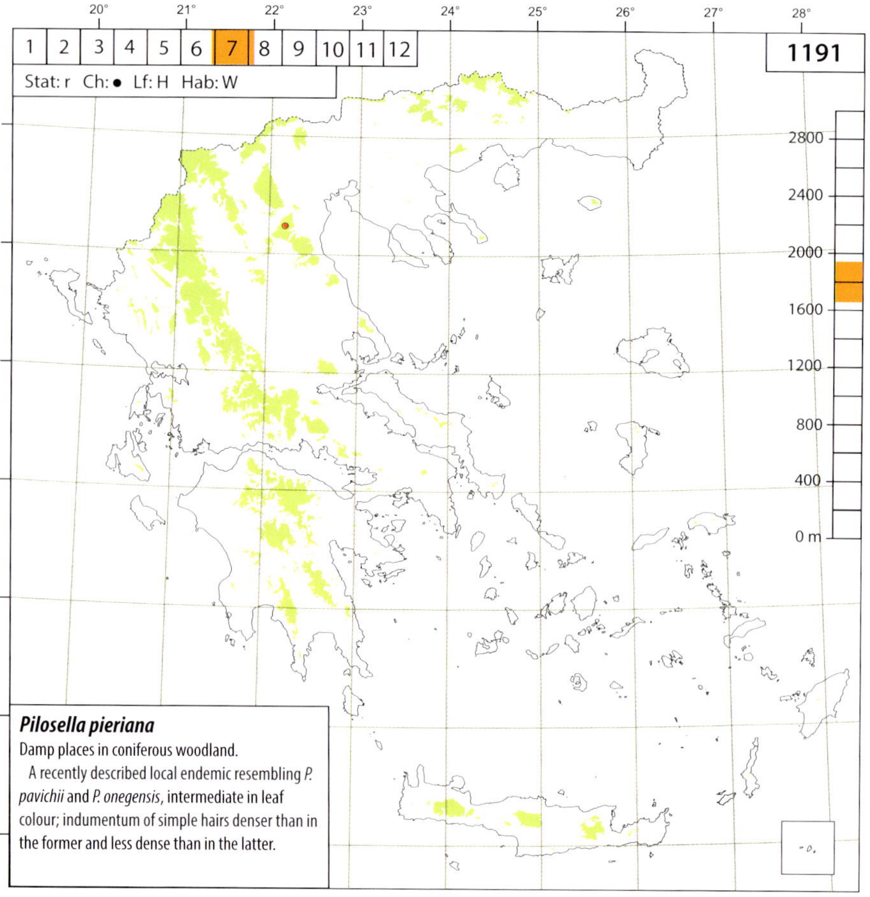

Pilosella pieriana
Damp places in coniferous woodland.
A recently described local endemic resembling *P. pavichii* and *P. onegensis*, intermediate in leaf colour; indumentum of simple hairs denser than in the former and less dense than in the latter.

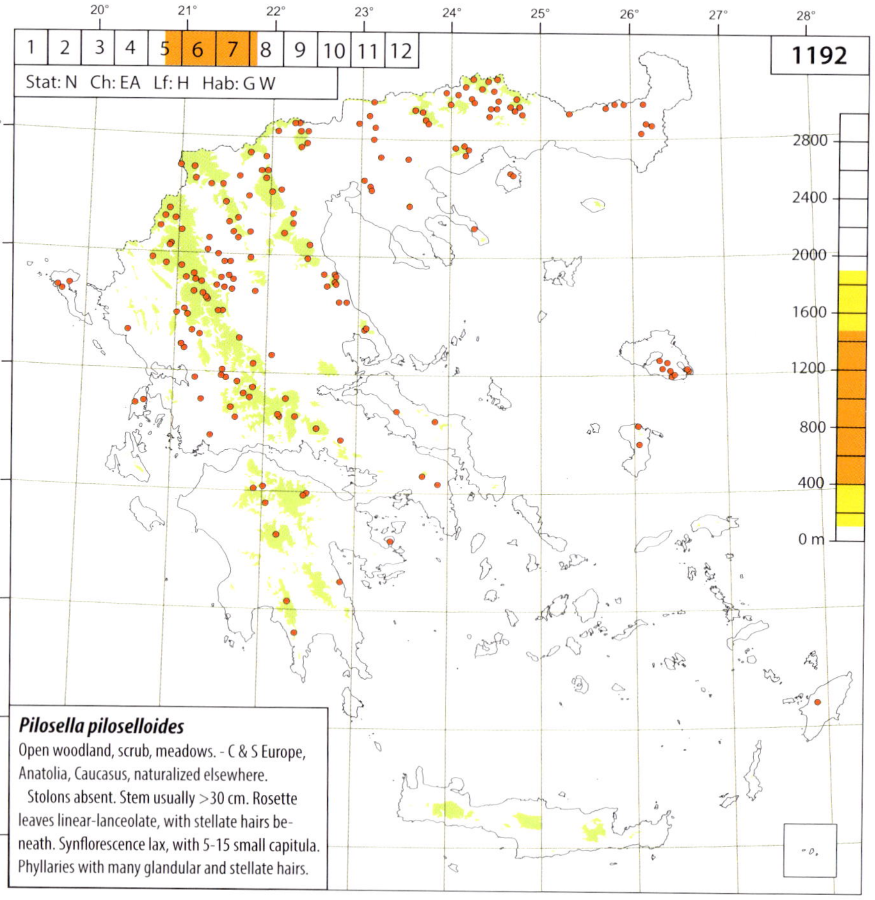

Pilosella piloselloides
Open woodland, scrub, meadows. - C & S Europe, Anatolia, Caucasus, naturalized elsewhere.
Stolons absent. Stem usually >30 cm. Rosette leaves linear-lanceolate, with stellate hairs beneath. Synflorescence lax, with 5-15 small capitula. Phyllaries with many glandular and stellate hairs.

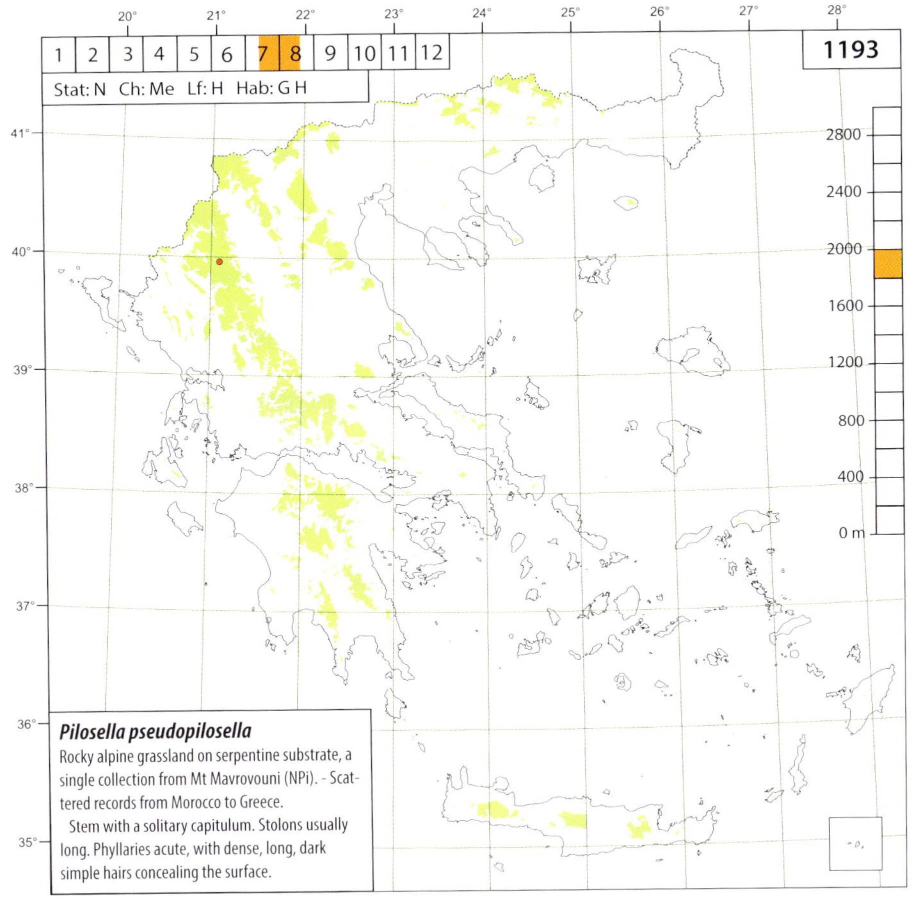

Pilosella pseudopilosella

Rocky alpine grassland on serpentine substrate, a single collection from Mt Mavrovouni (NPi). - Scattered records from Morocco to Greece.

Stem with a solitary capitulum. Stolons usually long. Phyllaries acute, with dense, long, dark simple hairs concealing the surface.

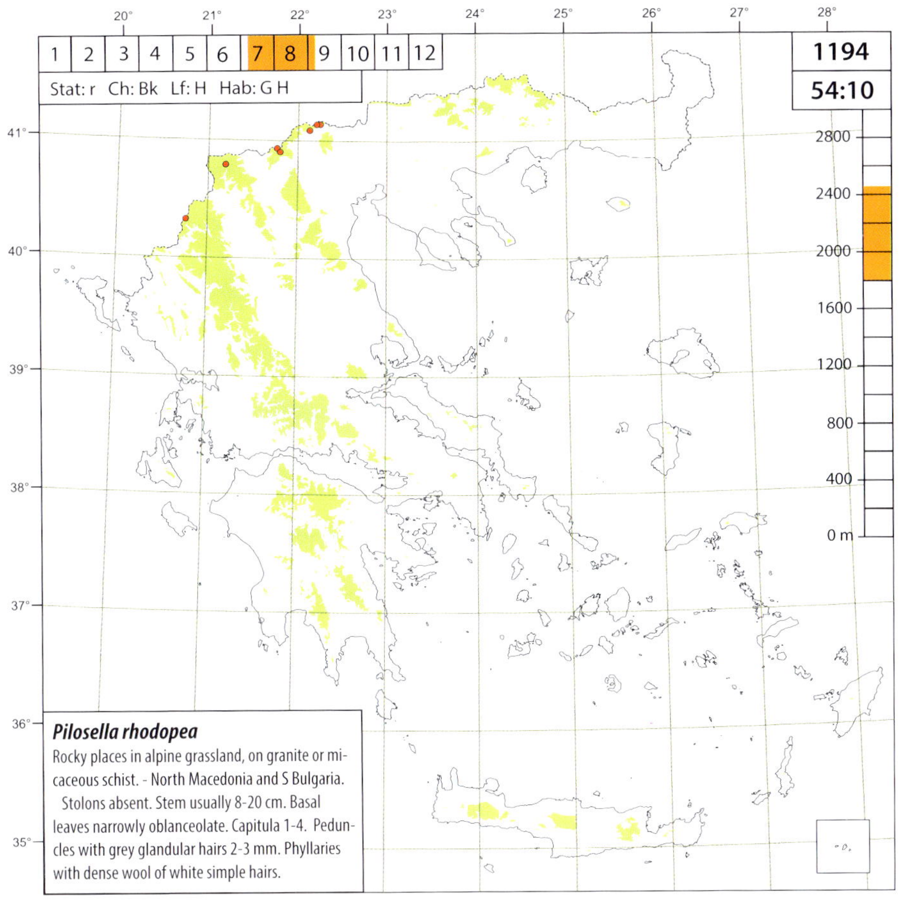

Pilosella rhodopea

Rocky places in alpine grassland, on granite or micaceous schist. - North Macedonia and S Bulgaria.

Stolons absent. Stem usually 8-20 cm. Basal leaves narrowly oblanceolate. Capitula 1-4. Peduncles with grey glandular hairs 2-3 mm. Phyllaries with dense wool of white simple hairs.

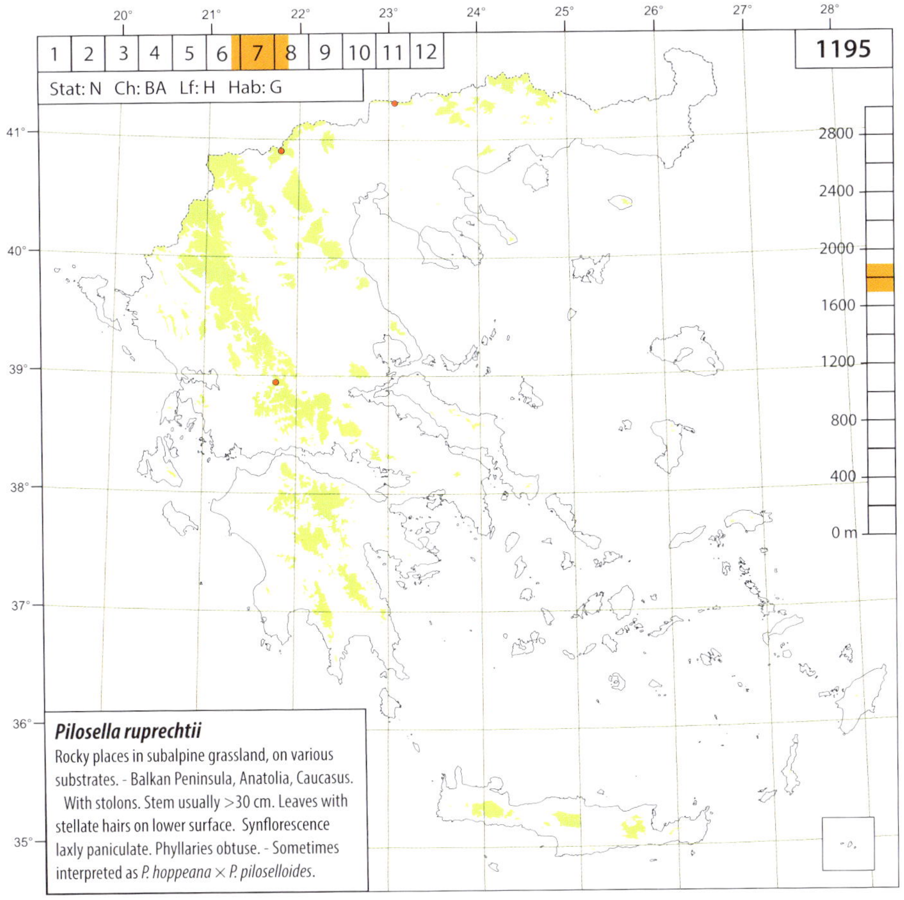

Pilosella ruprechtii

Rocky places in subalpine grassland, on various substrates. - Balkan Peninsula, Anatolia, Caucasus.

With stolons. Stem usually >30 cm. Leaves with stellate hairs on lower surface. Synflorescence laxly paniculate. Phyllaries obtuse. - Sometimes interpreted as *P. hoppeana* × *P. piloselloides*.

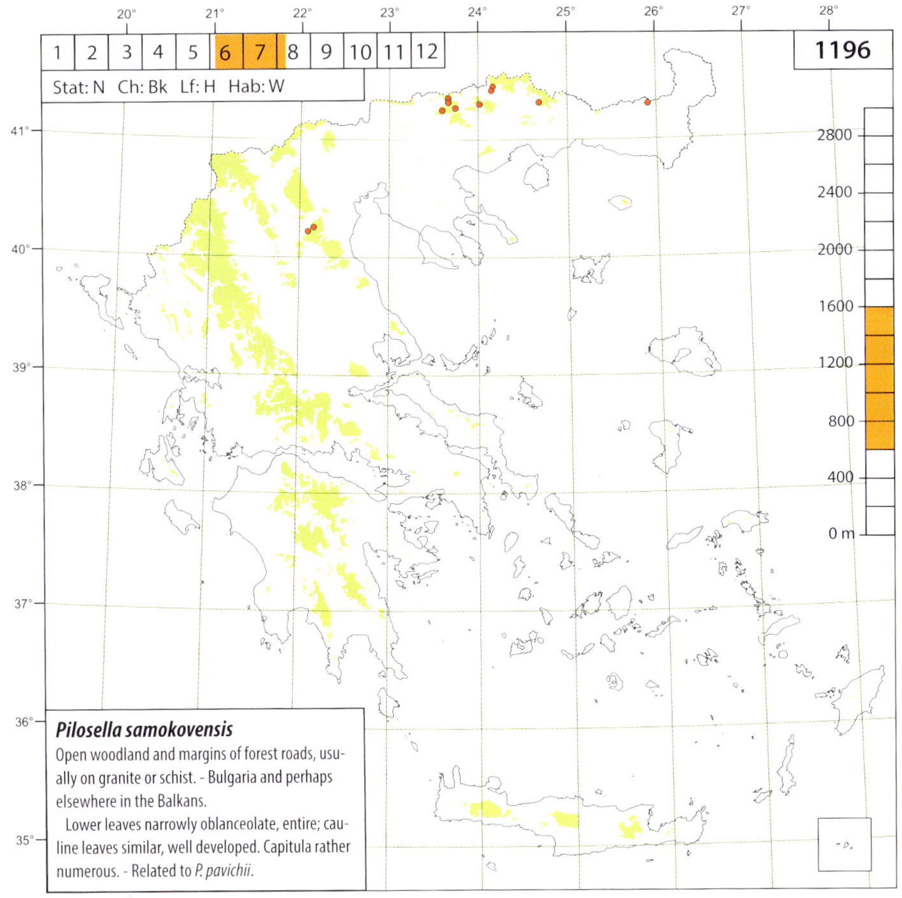

Pilosella samokovensis

Open woodland and margins of forest roads, usually on granite or schist. - Bulgaria and perhaps elsewhere in the Balkans.

Lower leaves narrowly oblanceolate, entire; cauline leaves similar, well developed. Capitula rather numerous. - Related to *P. pavichii*.

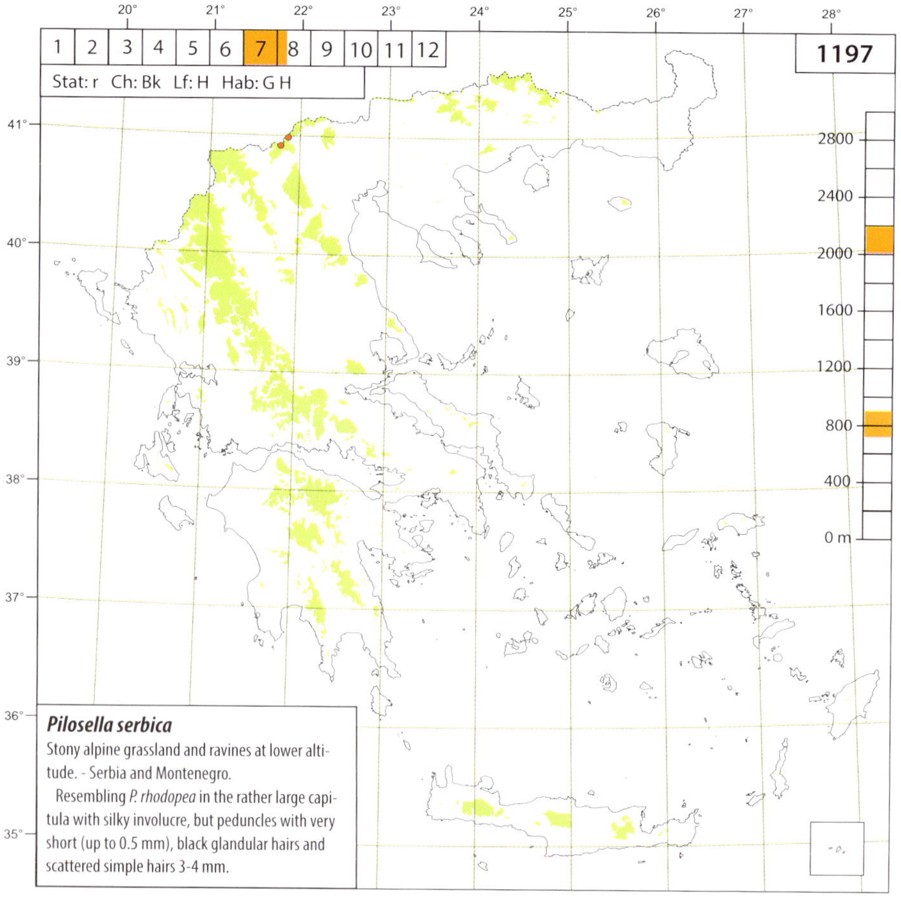

Pilosella serbica

Stony alpine grassland and ravines at lower altitude. - Serbia and Montenegro.

Resembling *P. rhodopea* in the rather large capitula with silky involucre, but peduncles with very short (up to 0.5 mm), black glandular hairs and scattered simple hairs 3-4 mm.

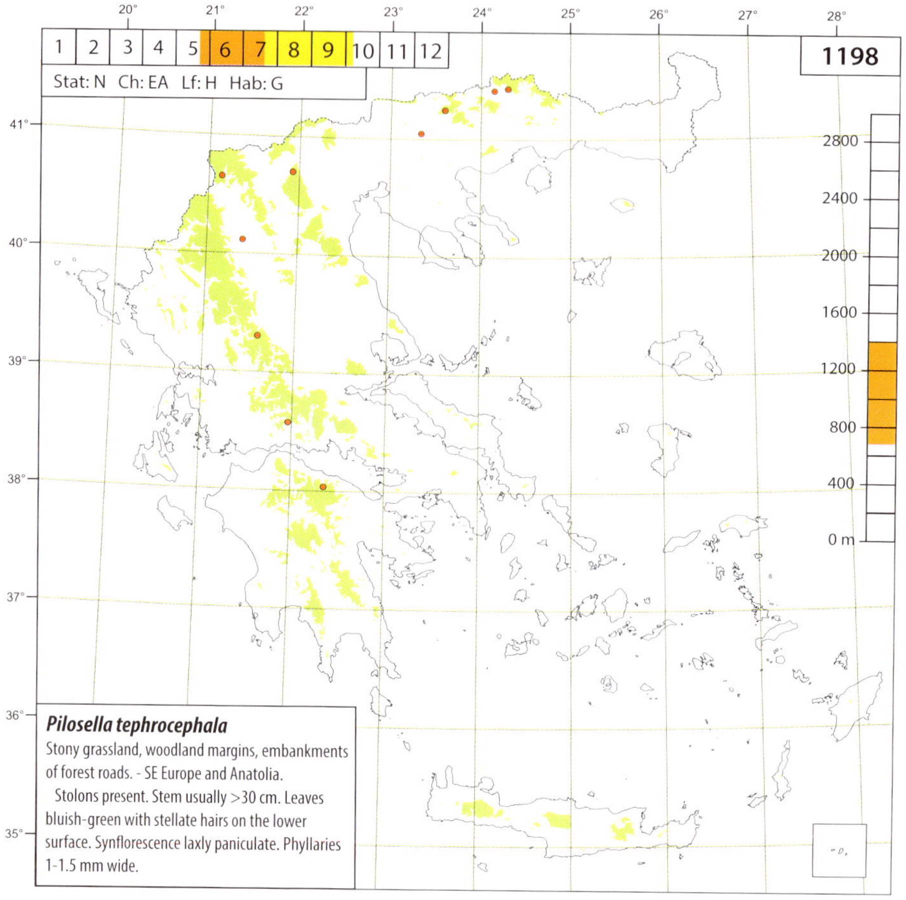

Pilosella tephrocephala

Stony grassland, woodland margins, embankments of forest roads. - SE Europe and Anatolia.

Stolons present. Stem usually >30 cm. Leaves bluish-green with stellate hairs on the lower surface. Synflorescence laxly paniculate. Phyllaries 1-1.5 mm wide.

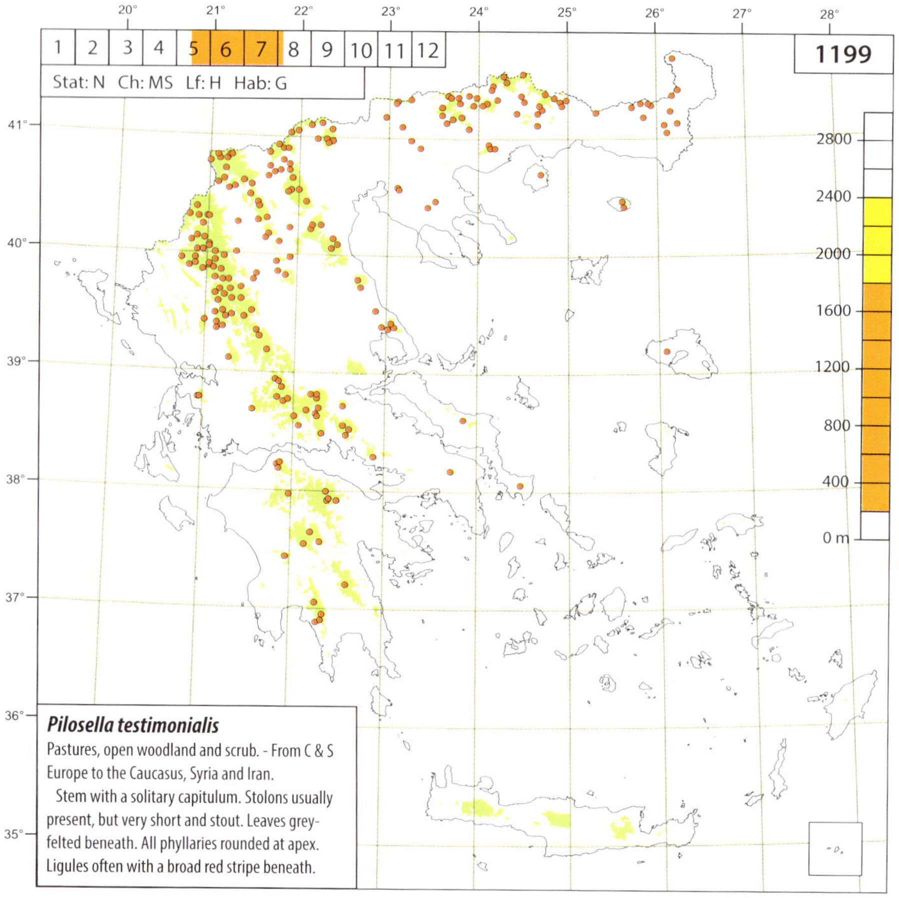

Pilosella testimonialis

Pastures, open woodland and scrub. - From C & S Europe to the Caucasus, Syria and Iran.

Stem with a solitary capitulum. Stolons usually present, but very short and stout. Leaves grey-felted beneath. All phyllaries rounded at apex. Ligules often with a broad red stripe beneath.

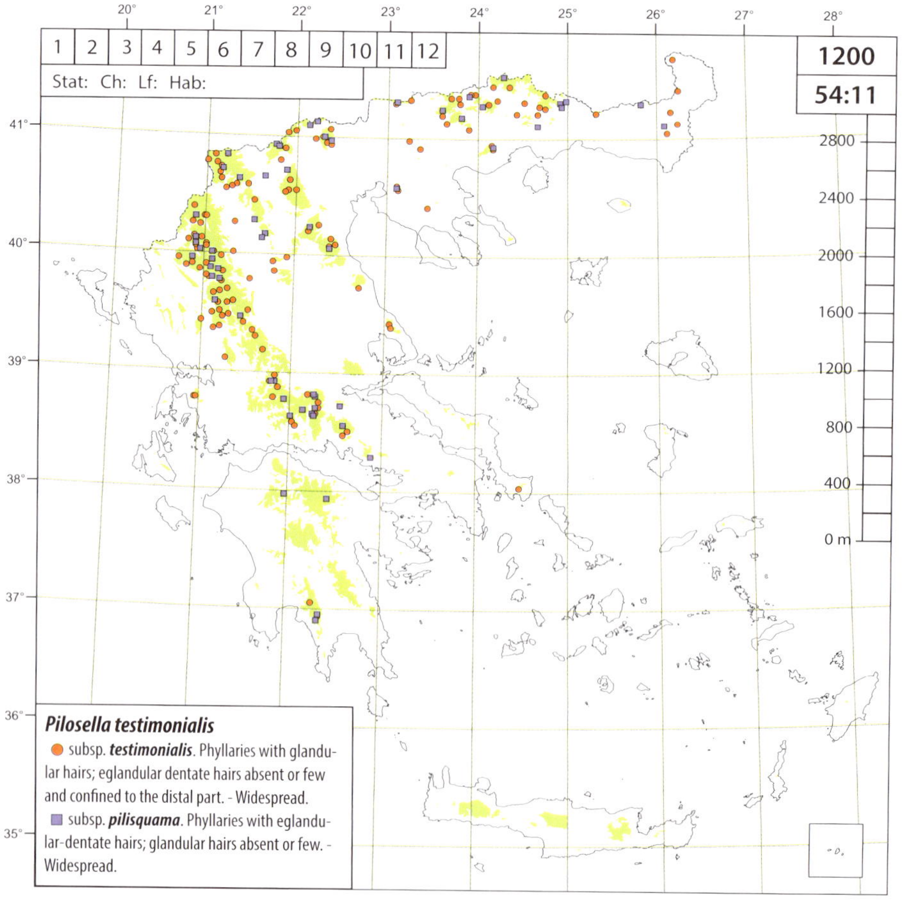

Pilosella testimonialis

● subsp. ***testimonialis***. Phyllaries with glandular hairs; eglandular dentate hairs absent or few and confined to the distal part. - Widespread.

■ subsp. ***pilisquama***. Phyllaries with eglandular-dentate hairs; glandular hairs absent or few. - Widespread.

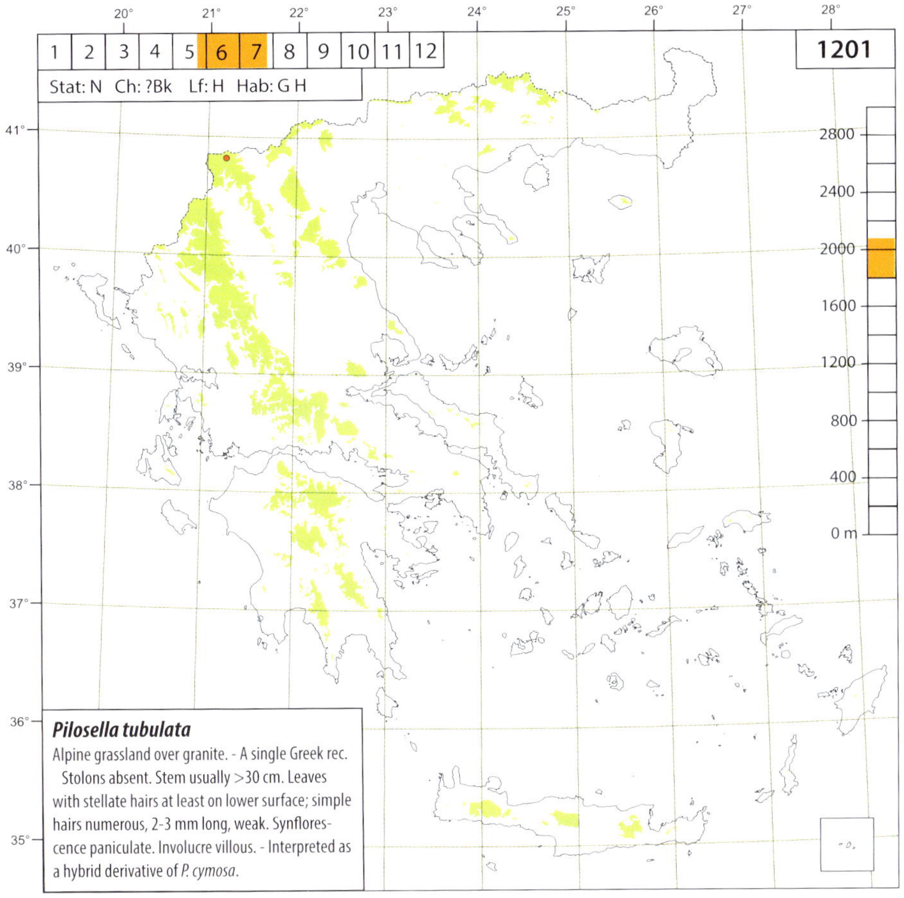

Pilosella tubulata

Alpine grassland over granite. - A single Greek rec.

Stolons absent. Stem usually >30 cm. Leaves with stellate hairs at least on lower surface; simple hairs numerous, 2-3 mm long, weak. Synflorescence paniculate. Involucre villous. - Interpreted as a hybrid derivative of *P. cymosa*.

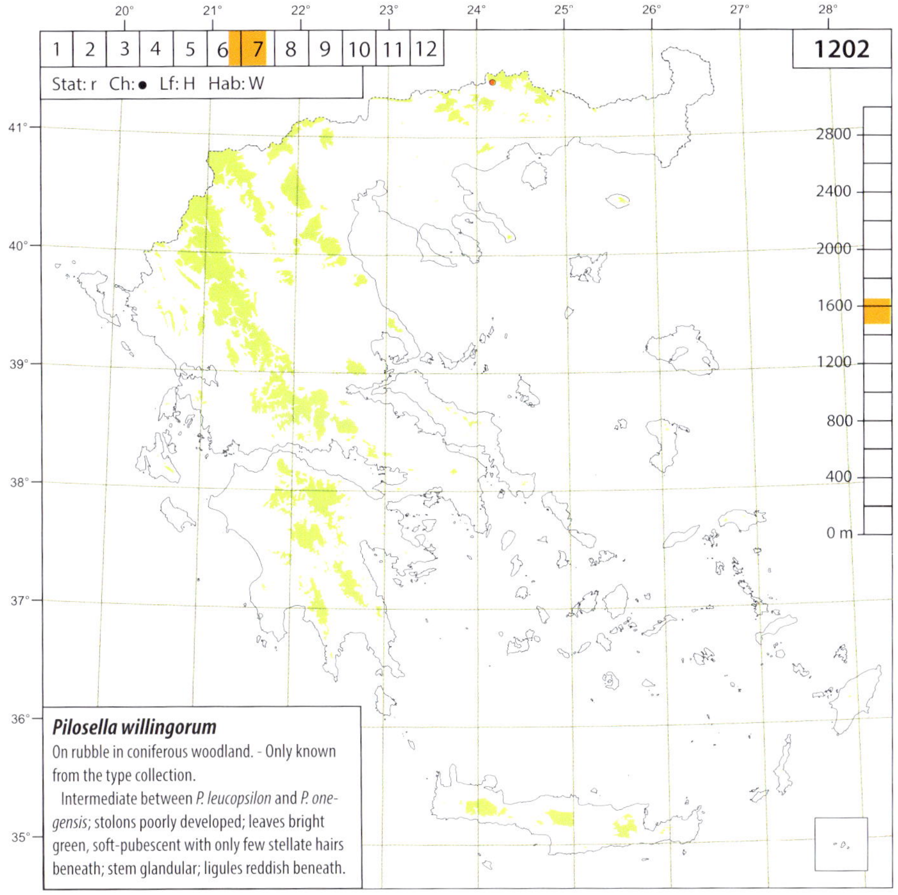

Pilosella willingorum

On rubble in coniferous woodland. - Only known from the type collection.

Intermediate between *P. leucopsilon* and *P. onegensis*; stolons poorly developed; leaves bright green, soft-pubescent with only few stellate hairs beneath; stem glandular; ligules reddish beneath.

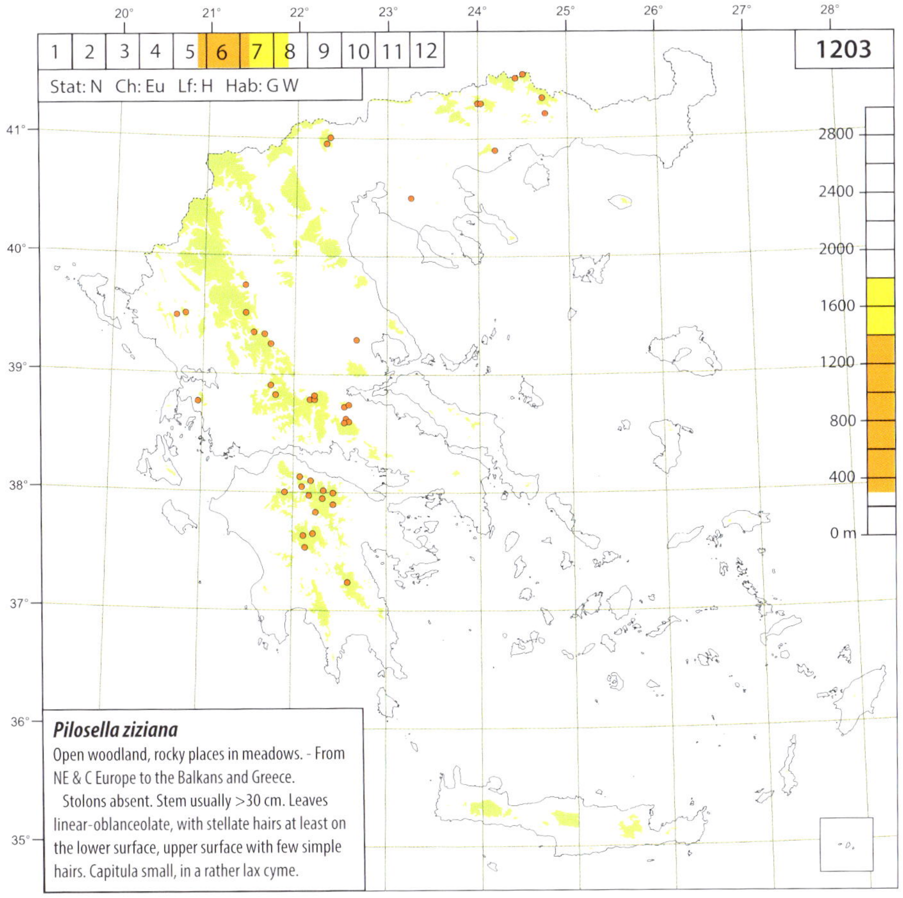

Pilosella ziziana

Open woodland, rocky places in meadows. - From NE & C Europe to the Balkans and Greece.

Stolons absent. Stem usually >30 cm. Leaves linear-oblanceolate, with stellate hairs at least on the lower surface, upper surface with few simple hairs. Capitula small, in a rather lax cyme.

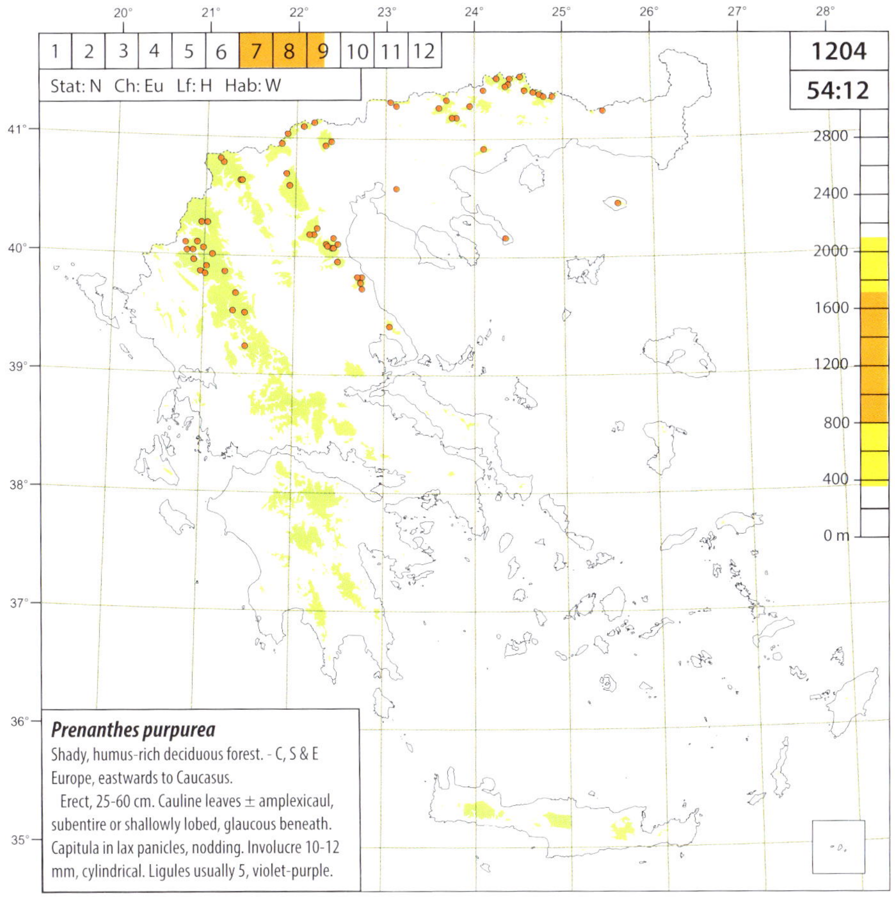

Prenanthes purpurea

Shady, humus-rich deciduous forest. - C, S & E Europe, eastwards to Caucasus.

Erect, 25-60 cm. Cauline leaves ± amplexicaul, subentire or shallowly lobed, glaucous beneath. Capitula in lax panicles, nodding. Involucre 10-12 mm, cylindrical. Ligules usually 5, violet-purple.

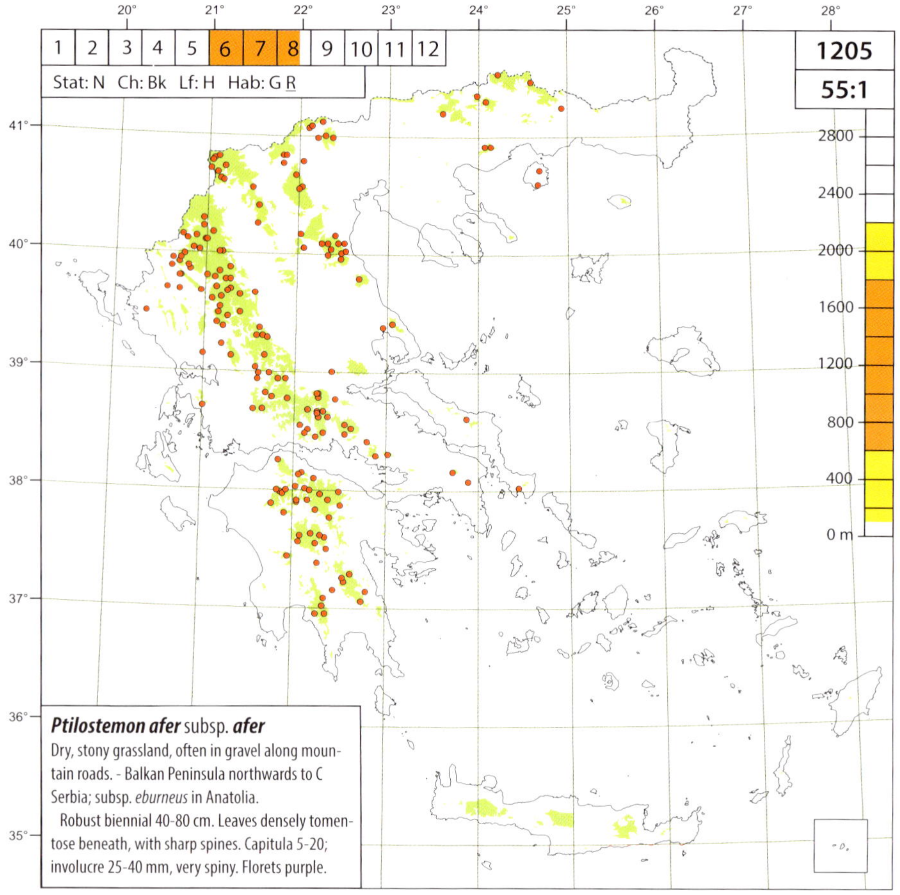

Ptilostemon afer subsp. _afer_
Dry, stony grassland, often in gravel along mountain roads. - Balkan Peninsula northwards to C Serbia; subsp. *eburneus* in Anatolia.
Robust biennial 40-80 cm. Leaves densely tomentose beneath, with sharp spines. Capitula 5-20; involucre 25-40 mm, very spiny. Florets purple.

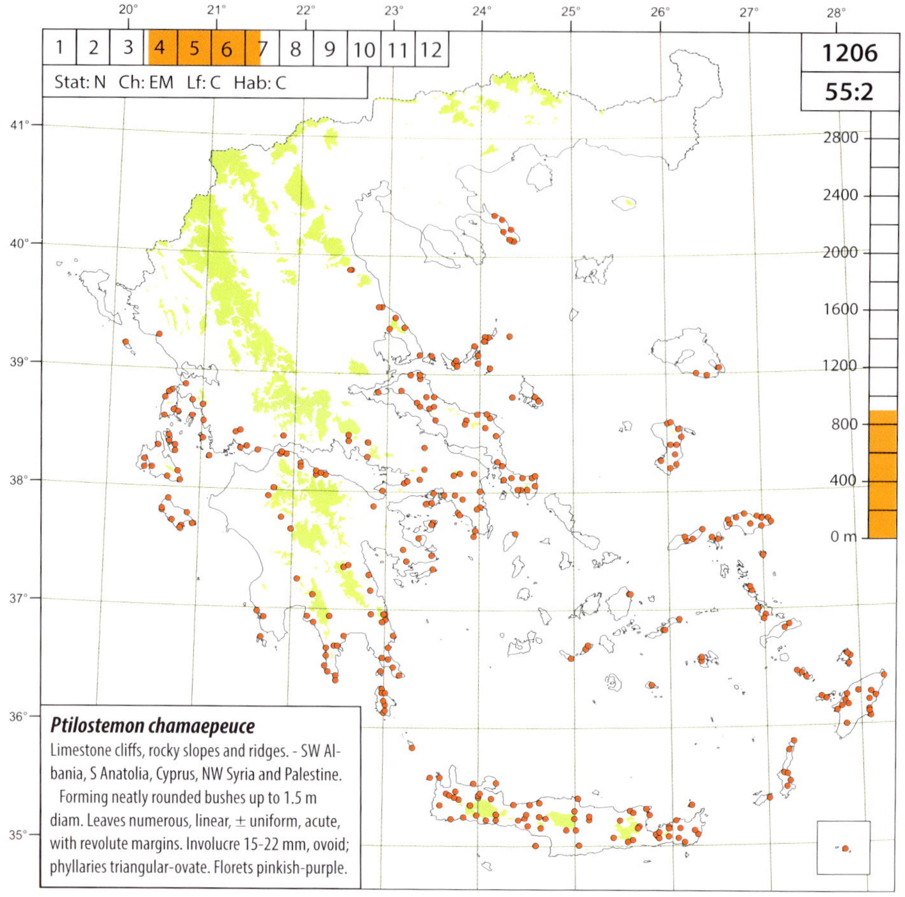

Ptilostemon chamaepeuce
Limestone cliffs, rocky slopes and ridges. - SW Albania, S Anatolia, Cyprus, NW Syria and Palestine.
Forming neatly rounded bushes up to 1.5 m diam. Leaves numerous, linear, ± uniform, acute, with revolute margins. Involucre 15-22 mm, ovoid; phyllaries triangular-ovate. Florets pinkish-purple.

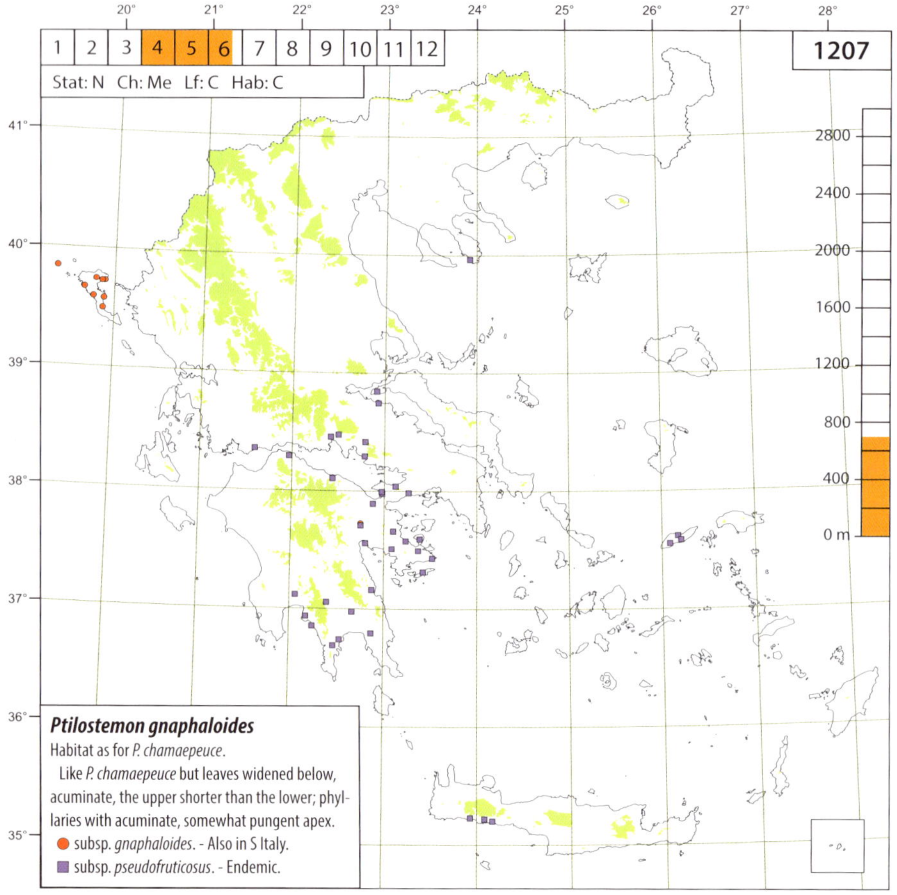

Ptilostemon gnaphaloides
Habitat as for *P. chamaepeuce*.
Like *P. chamaepeuce* but leaves widened below, acuminate, the upper shorter than the lower; phyllaries with acuminate, somewhat pungent apex.
● subsp. *gnaphaloides*. - Also in S Italy.
■ subsp. *pseudofruticosus*. - Endemic.

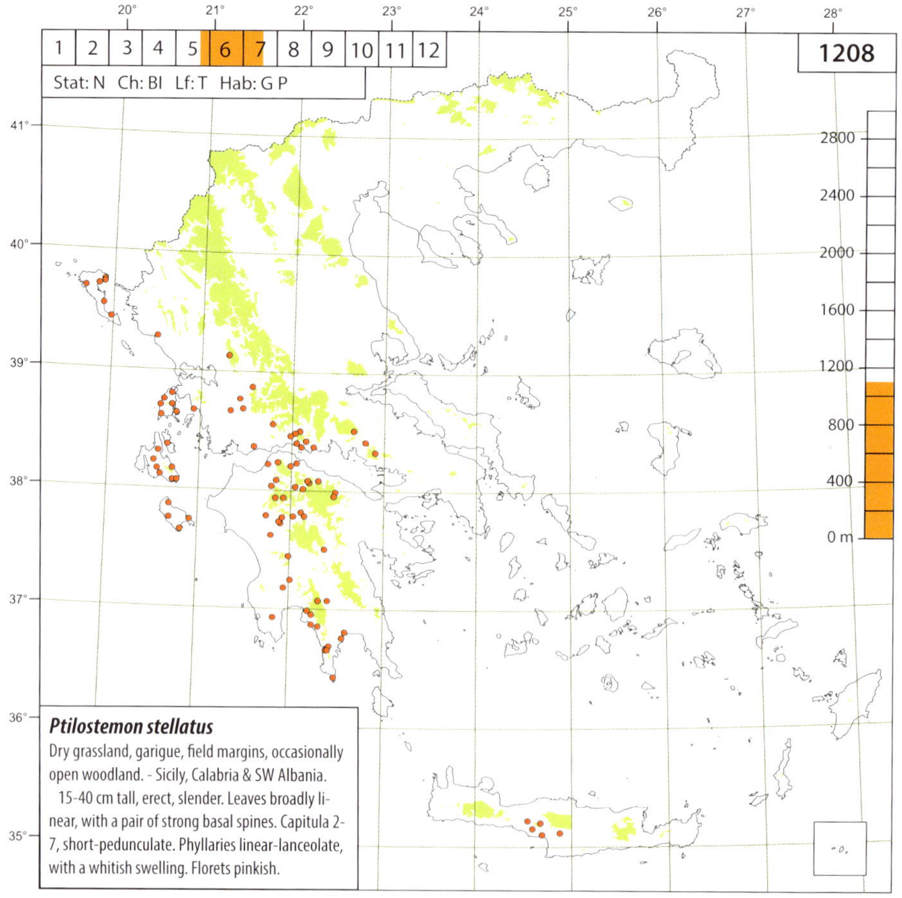

Ptilostemon stellatus
Dry grassland, garigue, field margins, occasionally open woodland. - Sicily, Calabria & SW Albania.
15-40 cm tall, erect, slender. Leaves broadly linear, with a pair of strong basal spines. Capitula 2-7, short-pedunculate. Phyllaries linear-lanceolate, with a whitish swelling. Florets pinkish.

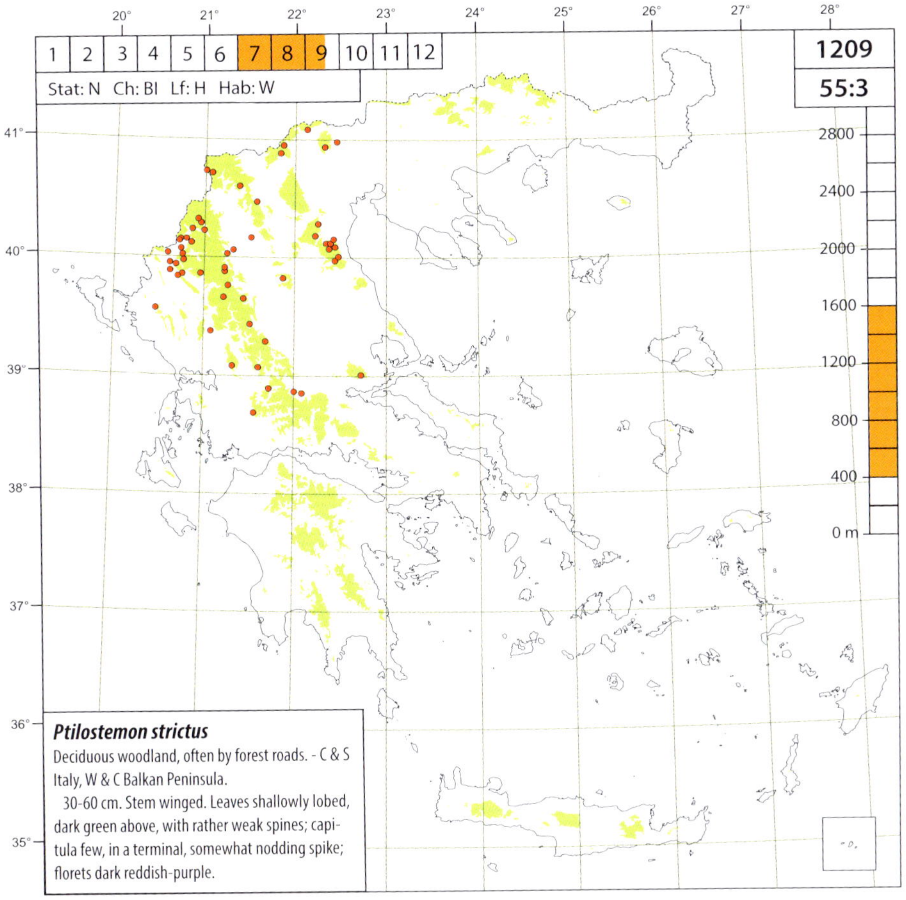

Ptilostemon strictus
Deciduous woodland, often by forest roads. - C & S Italy, W & C Balkan Peninsula.
30-60 cm. Stem winged. Leaves shallowly lobed, dark green above, with rather weak spines; capitula few, in a terminal, somewhat nodding spike; florets dark reddish-purple.

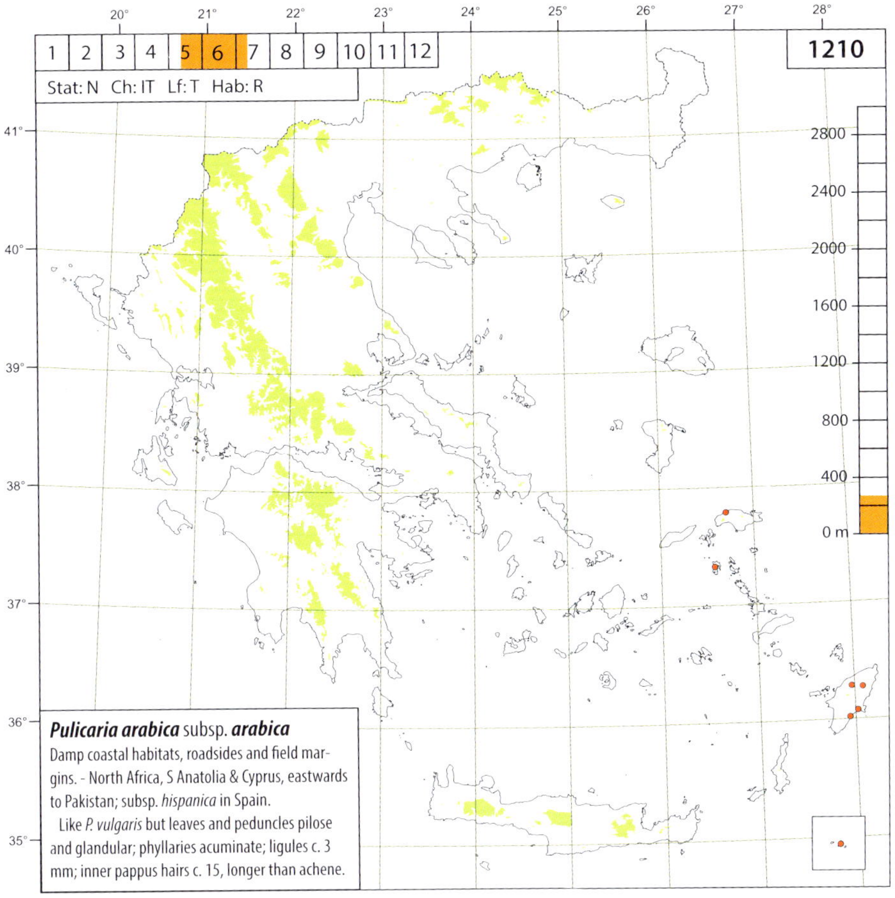

Pulicaria arabica subsp. ***arabica***
Damp coastal habitats, roadsides and field margins. - North Africa, S Anatolia & Cyprus, eastwards to Pakistan; subsp. *hispanica* in Spain.
Like *P. vulgaris* but leaves and peduncles pilose and glandular; phyllaries acuminate; ligules c. 3 mm; inner pappus hairs c. 15, longer than achene.

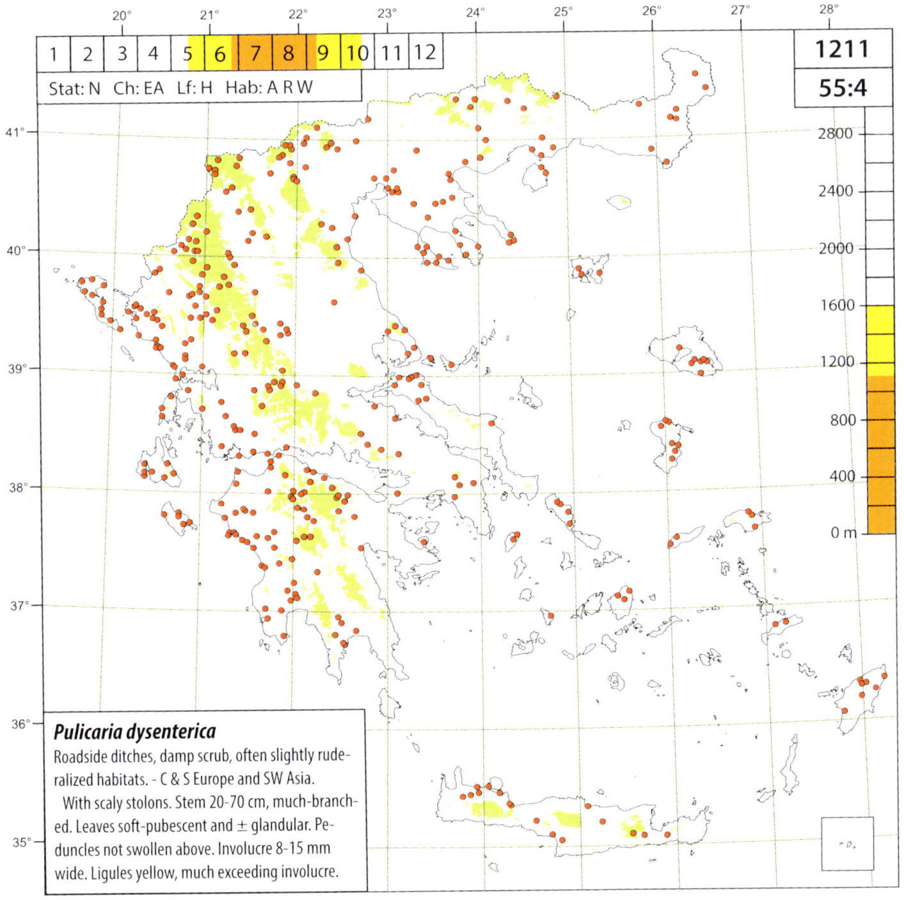

Pulicaria dysenterica
Roadside ditches, damp scrub, often slightly ruderalized habitats. - C & S Europe and SW Asia.
With scaly stolons. Stem 20-70 cm, much-branched. Leaves soft-pubescent and ± glandular. Peduncles not swollen above. Involucre 8-15 mm wide. Ligules yellow, much exceeding involucre.

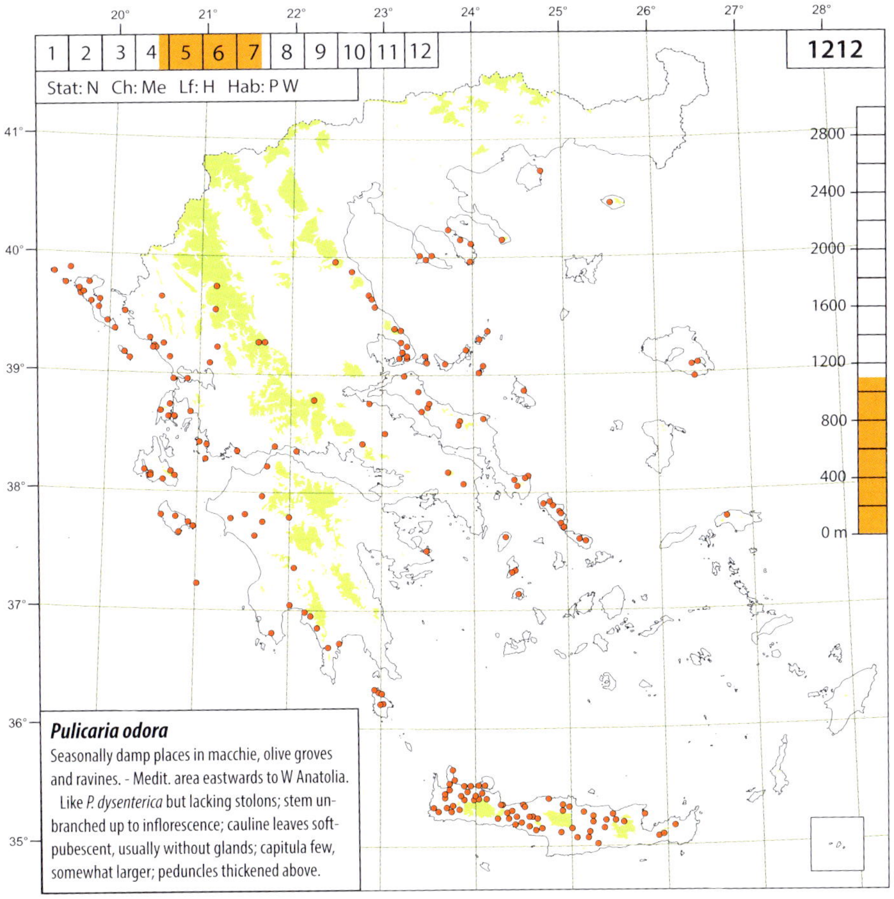

Pulicaria odora
Seasonally damp places in macchie, olive groves and ravines. - Medit. area eastwards to W Anatolia.
Like *P. dysenterica* but lacking stolons; stem unbranched up to inflorescence; cauline leaves soft-pubescent, usually without glands; capitula few, somewhat larger; peduncles thickened above.

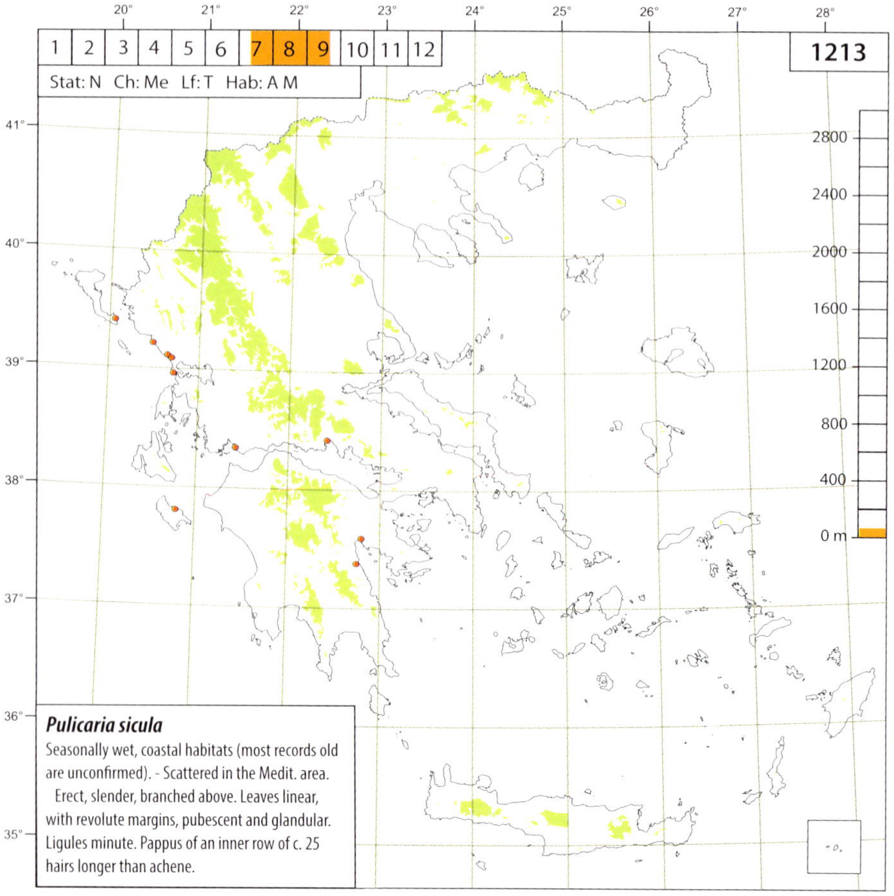

Pulicaria sicula

Seasonally wet, coastal habitats (most records old are unconfirmed). - Scattered in the Medit. area.

Erect, slender, branched above. Leaves linear, with revolute margins, pubescent and glandular. Ligules minute. Pappus of an inner row of c. 25 hairs longer than achene.

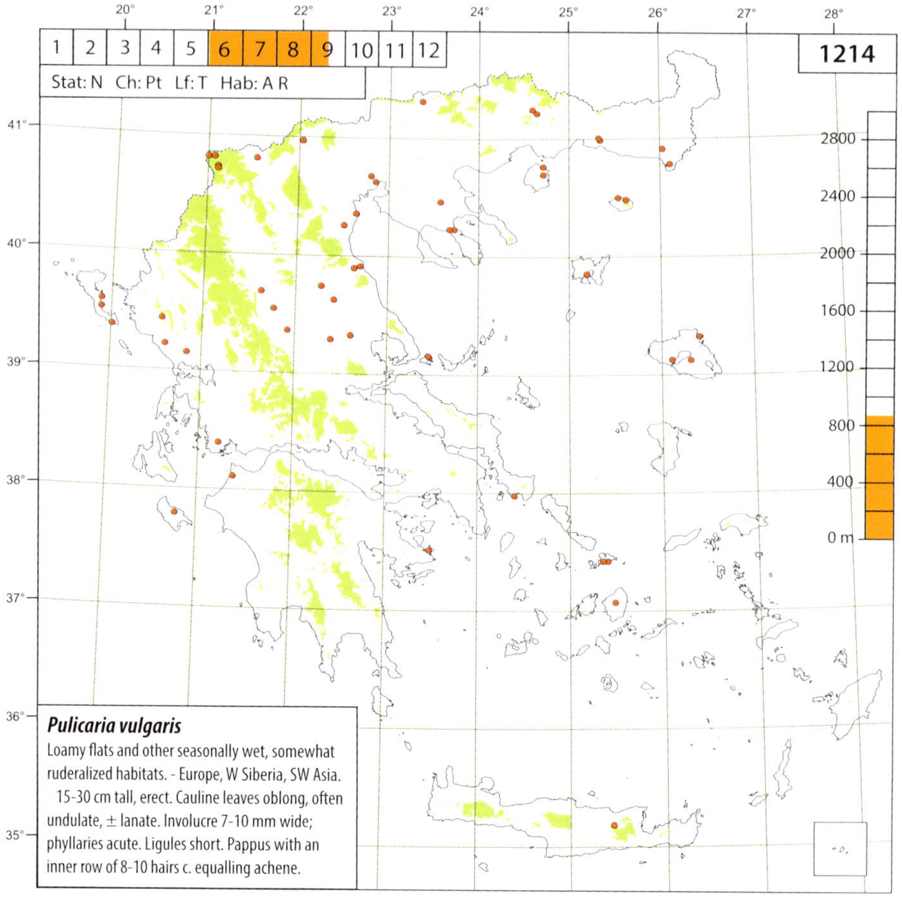

Pulicaria vulgaris

Loamy flats and other seasonally wet, somewhat ruderalized habitats. - Europe, W Siberia, SW Asia.

15-30 cm tall, erect. Cauline leaves oblong, often undulate, ± lanate. Involucre 7-10 mm wide; phyllaries acute. Ligules short. Pappus with an inner row of 8-10 hairs c. equalling achene.

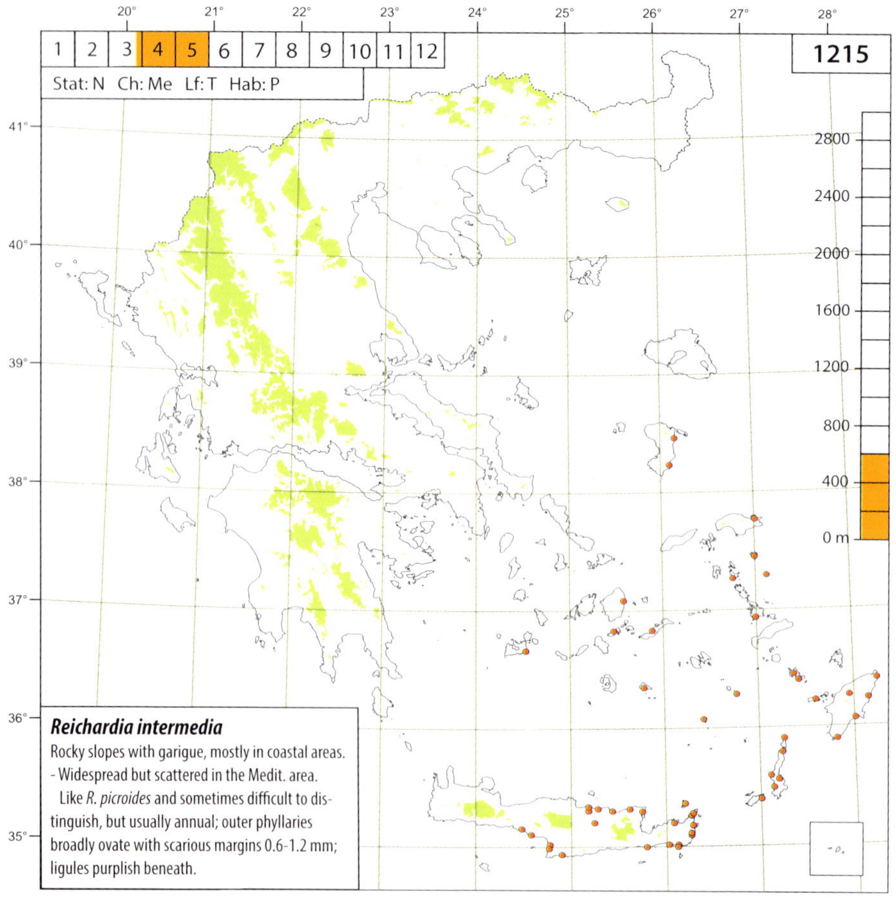

Reichardia intermedia

Rocky slopes with garigue, mostly in coastal areas. - Widespread but scattered in the Medit. area.

Like *R. picroides* and sometimes difficult to distinguish, but usually annual; outer phyllaries broadly ovate with scarious margins 0.6-1.2 mm; ligules purplish beneath.

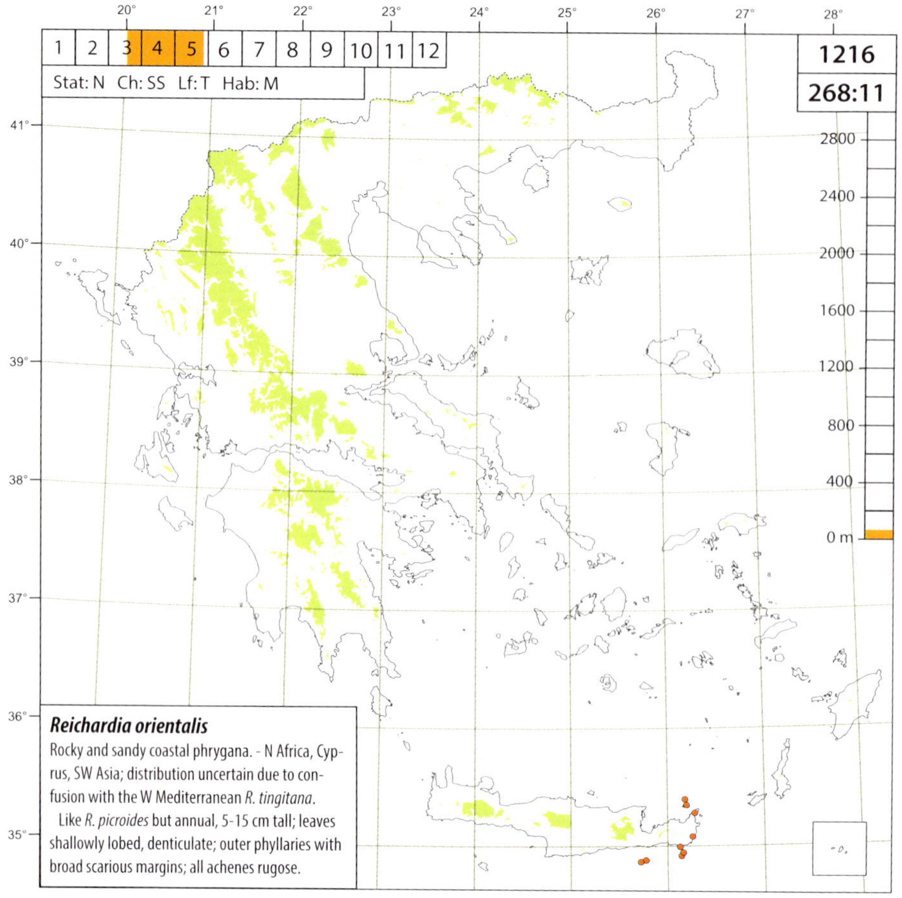

Reichardia orientalis

Rocky and sandy coastal phrygana. - N Africa, Cyprus, SW Asia; distribution uncertain due to confusion with the W Mediterranean *R. tingitana*.

Like *R. picroides* but annual, 5-15 cm tall; leaves shallowly lobed, denticulate; outer phyllaries with broad scarious margins; all achenes rugose.

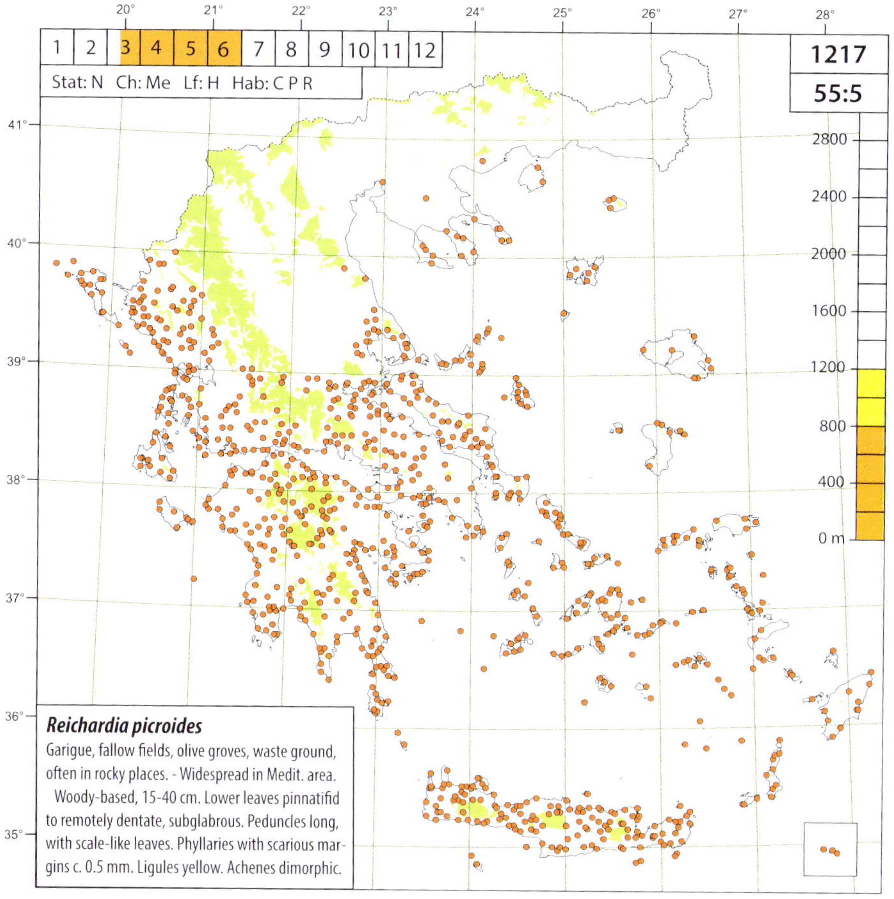

Reichardia picroides
Garigue, fallow fields, olive groves, waste ground, often in rocky places. - Widespread in Medit. area.
Woody-based, 15-40 cm. Lower leaves pinnatifid to remotely dentate, subglabrous. Peduncles long, with scale-like leaves. Phyllaries with scarious margins c. 0.5 mm. Ligules yellow. Achenes dimorphic.

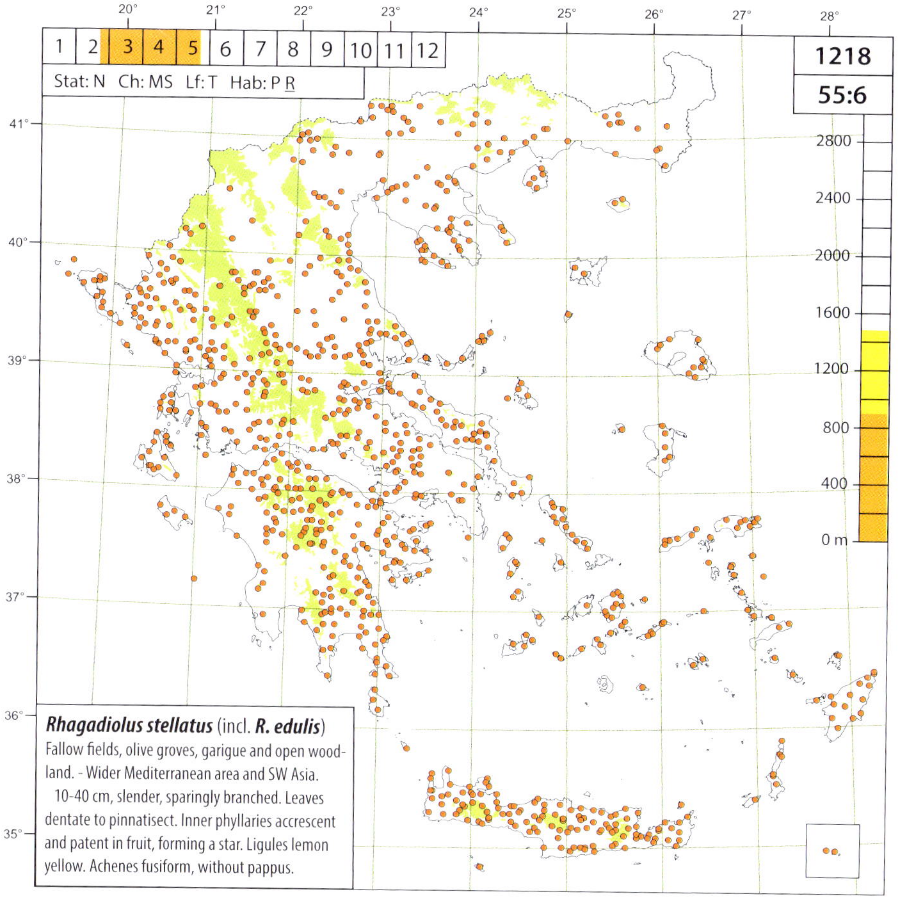

Rhagadiolus stellatus (incl. ***R. edulis***)
Fallow fields, olive groves, garigue and open woodland. - Wider Mediterranean area and SW Asia.
10-40 cm, slender, sparingly branched. Leaves dentate to pinnatisect. Inner phyllaries accrescent and patent in fruit, forming a star. Ligules lemon yellow. Achenes fusiform, without pappus.

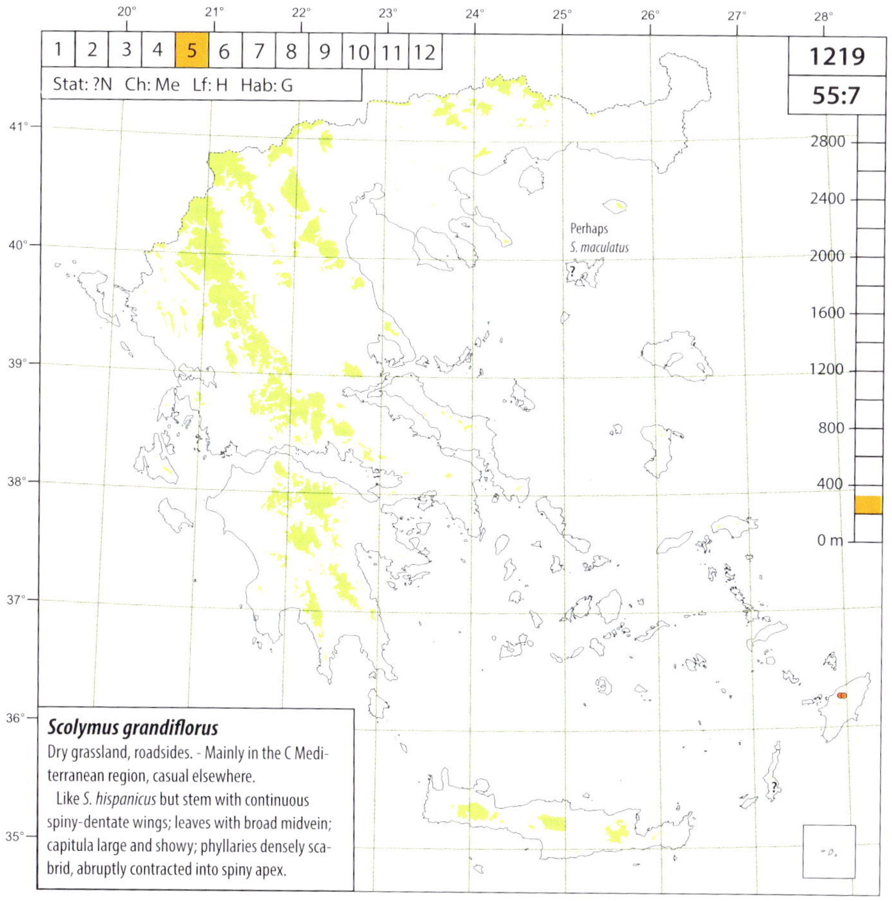

Scolymus grandiflorus
Dry grassland, roadsides. - Mainly in the C Mediterranean region, casual elsewhere.
Like *S. hispanicus* but stem with continuous spiny-dentate wings; leaves with broad midvein; capitula large and showy; phyllaries densely scabrid, abruptly contracted into spiny apex.

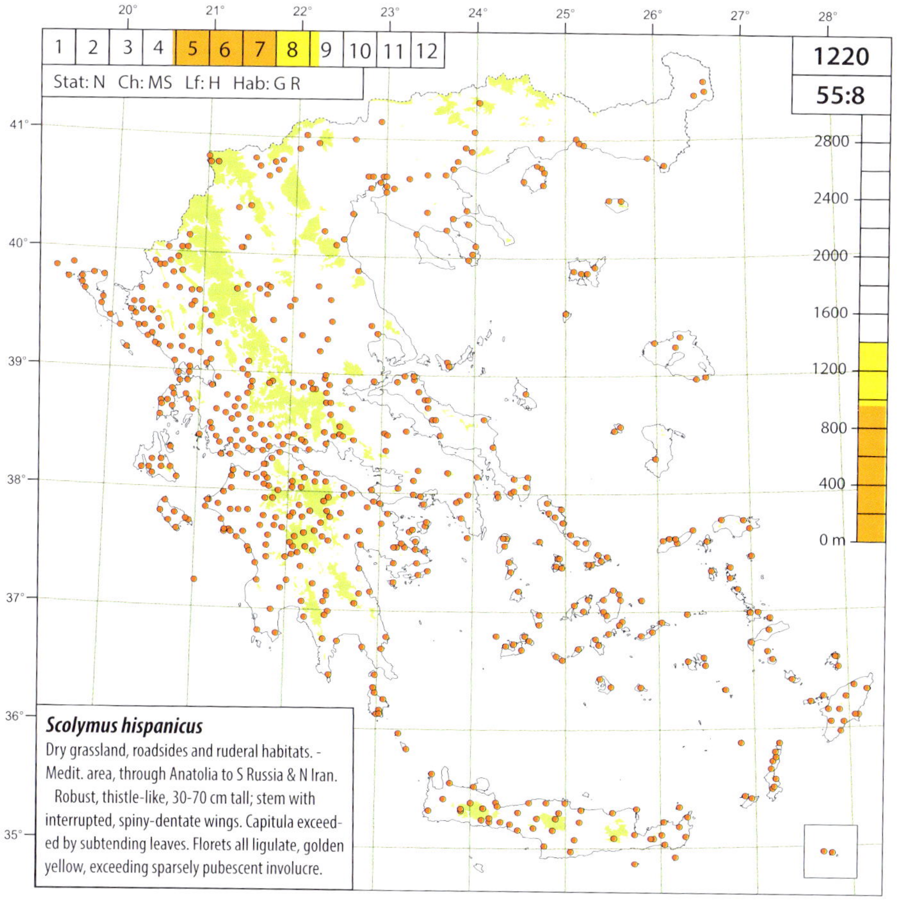

Scolymus hispanicus
Dry grassland, roadsides and ruderal habitats. - Medit. area, through Anatolia to S Russia & N Iran.
Robust, thistle-like, 30-70 cm tall; stem with interrupted, spiny-dentate wings. Capitula exceeded by subtending leaves. Florets all ligulate, golden yellow, exceeding sparsely pubescent involucre.

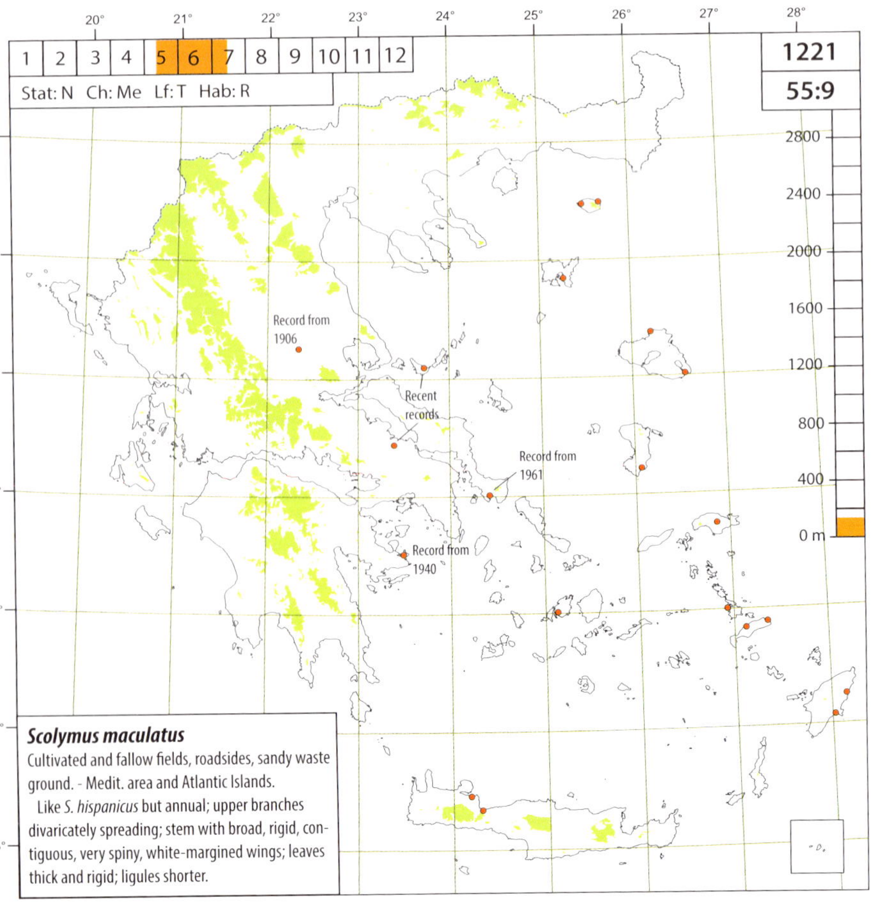

Scolymus maculatus

Cultivated and fallow fields, roadsides, sandy waste ground. - Medit. area and Atlantic Islands.

Like *S. hispanicus* but annual; upper branches divaricately spreading; stem with broad, rigid, contiguous, very spiny, white-margined wings; leaves thick and rigid; ligules shorter.

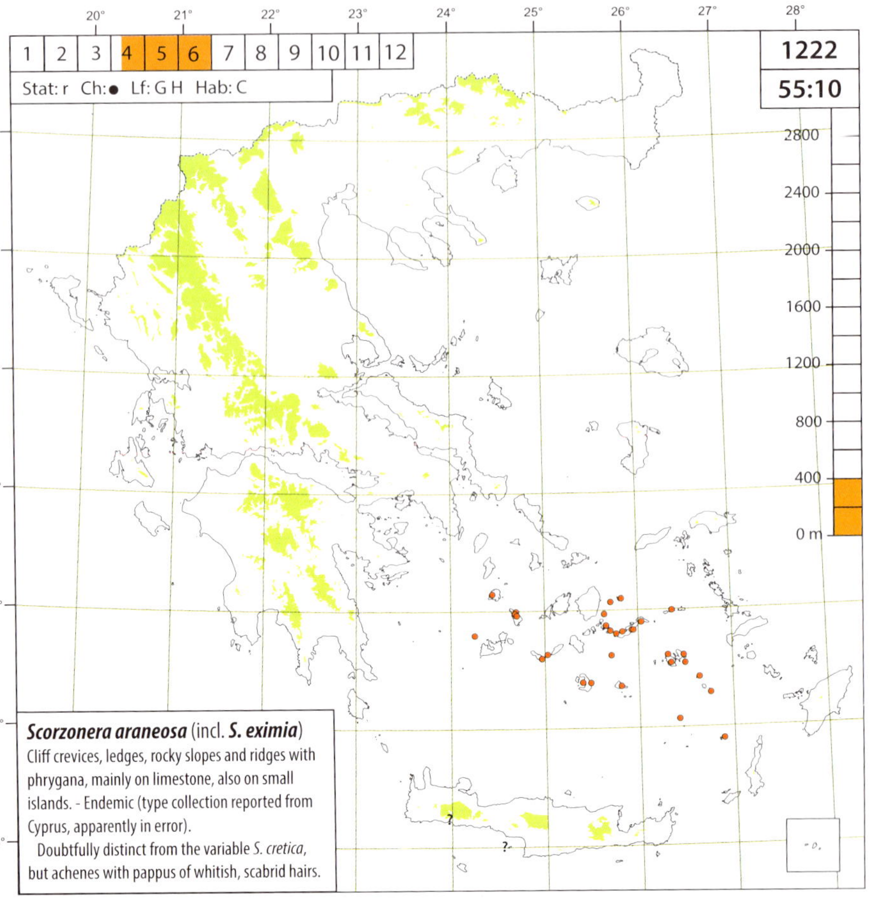

Scorzonera araneosa (incl. *S. eximia*)

Cliff crevices, ledges, rocky slopes and ridges with phrygana, mainly on limestone, also on small islands. - Endemic (type collection reported from Cyprus, apparently in error).

Doubtfully distinct from the variable *S. cretica*, but achenes with pappus of whitish, scabrid hairs.

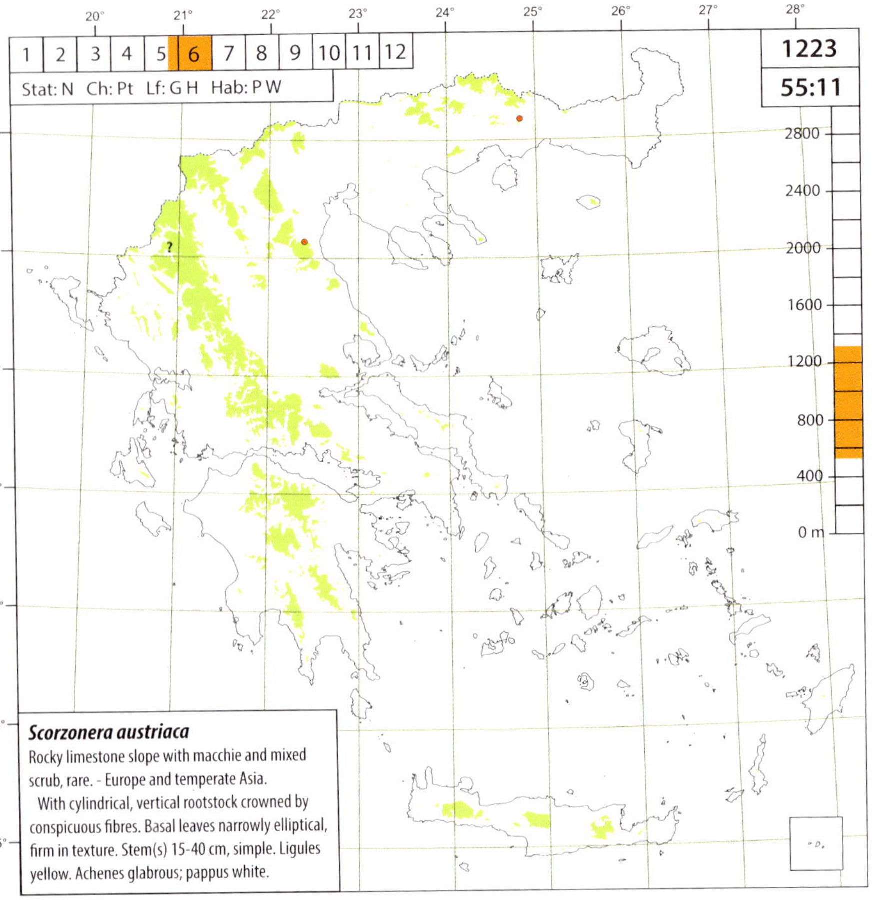

Scorzonera austriaca

Rocky limestone slope with macchie and mixed scrub, rare. - Europe and temperate Asia.

With cylindrical, vertical rootstock crowned by conspicuous fibres. Basal leaves narrowly elliptical, firm in texture. Stem(s) 15-40 cm, simple. Ligules yellow. Achenes glabrous; pappus white.

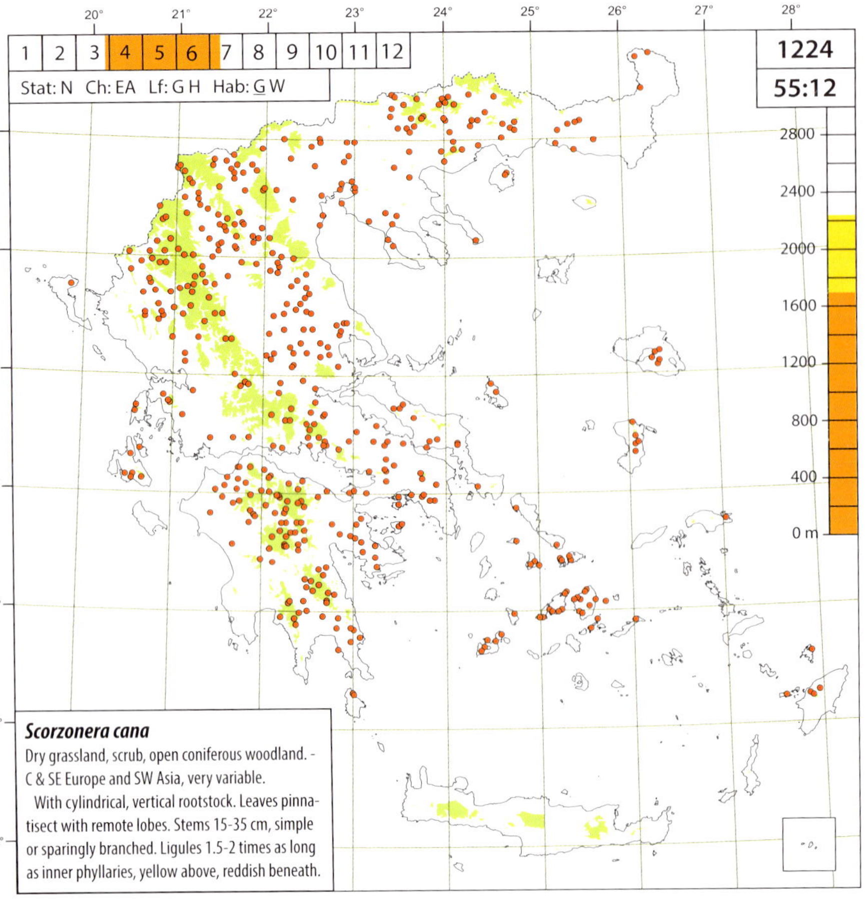

Scorzonera cana

Dry grassland, scrub, open coniferous woodland. - C & SE Europe and SW Asia, very variable.

With cylindrical, vertical rootstock. Leaves pinnatisect with remote lobes. Stems 15-35 cm, simple or sparingly branched. Ligules 1.5-2 times as long as inner phyllaries, yellow above, reddish beneath.

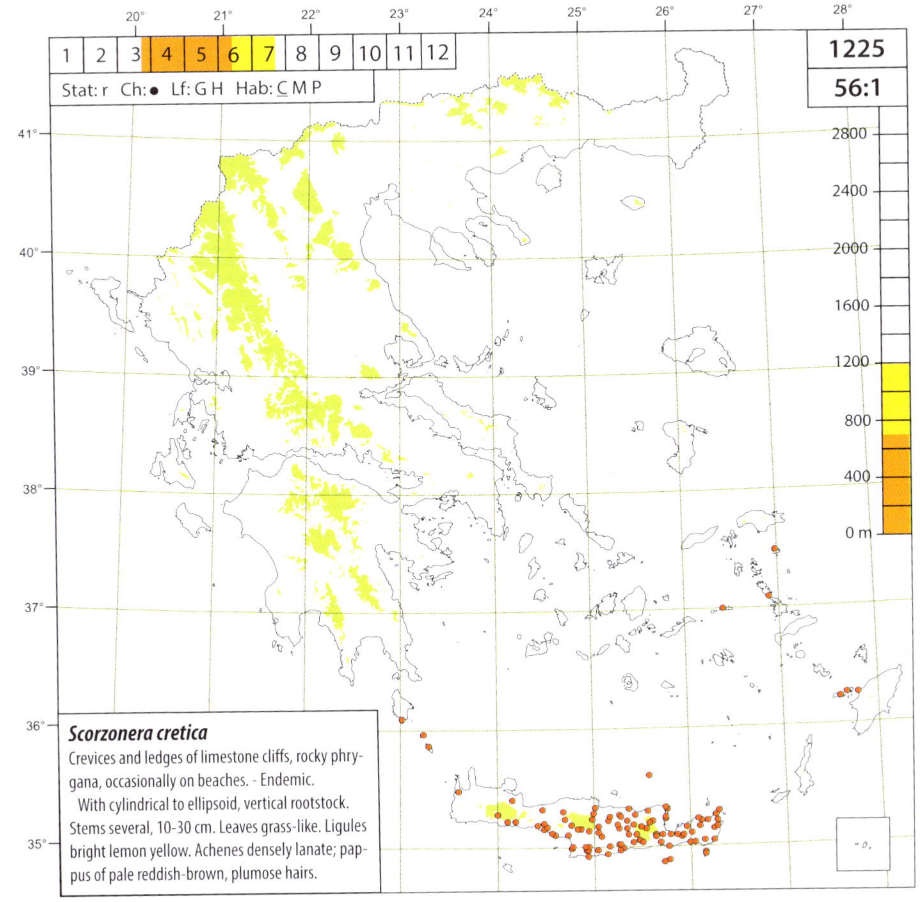

Scorzonera cretica
Crevices and ledges of limestone cliffs, rocky phrygana, occasionally on beaches. - Endemic.
With cylindrical to ellipsoid, vertical rootstock. Stems several, 10-30 cm. Leaves grass-like. Ligules bright lemon yellow. Achenes densely lanate; pappus of pale reddish-brown, plumose hairs.

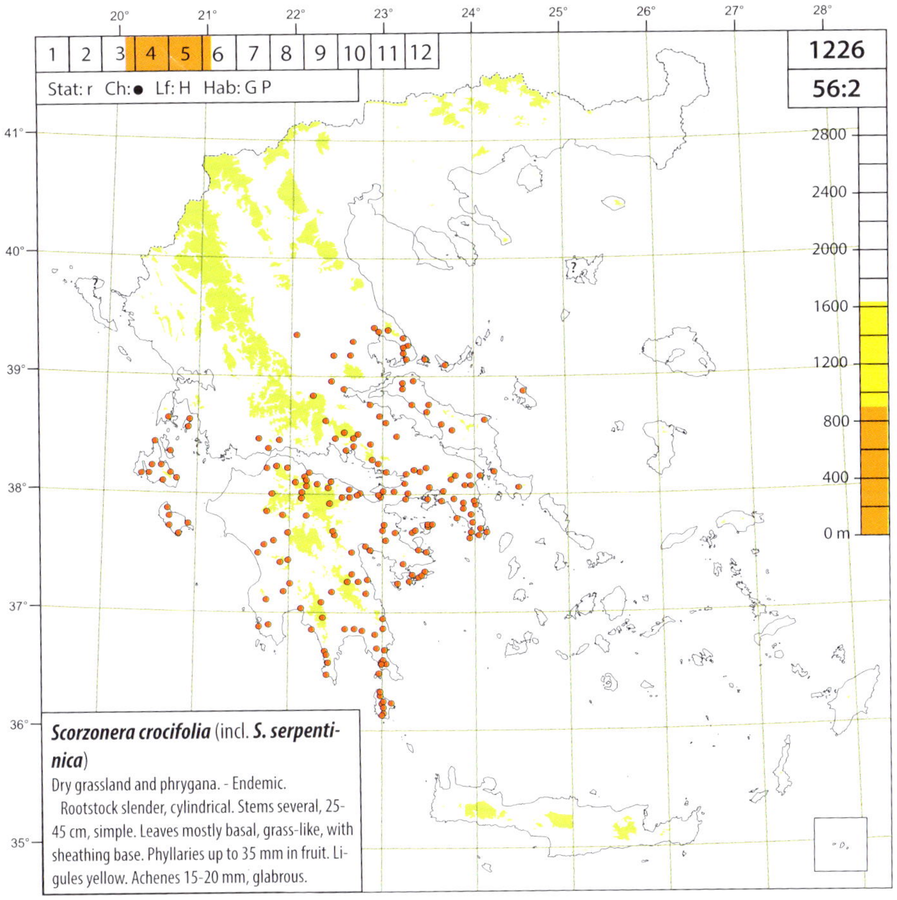

Scorzonera crocifolia (incl. ***S. serpentinica***)
Dry grassland and phrygana. - Endemic.
Rootstock slender, cylindrical. Stems several, 25-45 cm, simple. Leaves mostly basal, grass-like, with sheathing base. Phyllaries up to 35 mm in fruit. Ligules yellow. Achenes 15-20 mm, glabrous.

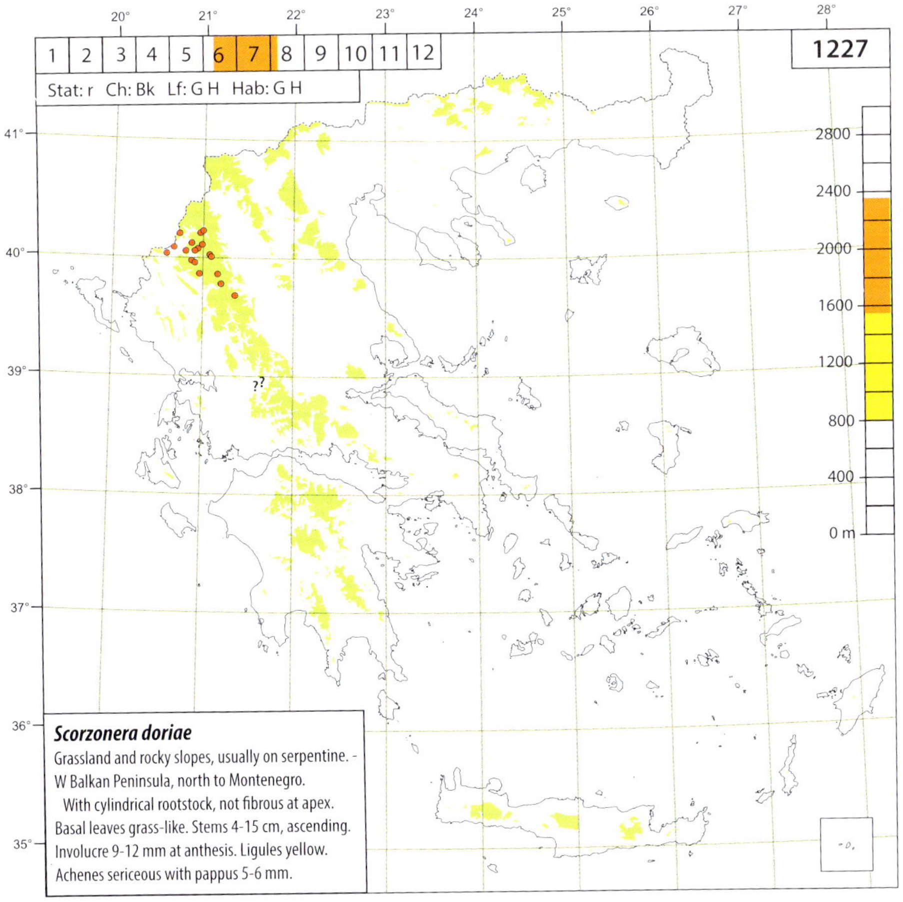

Scorzonera doriae
Grassland and rocky slopes, usually on serpentine. - W Balkan Peninsula, north to Montenegro.
With cylindrical rootstock, not fibrous at apex. Basal leaves grass-like. Stems 4-15 cm, ascending. Involucre 9-12 mm at anthesis. Ligules yellow. Achenes sericeous with pappus 5-6 mm.

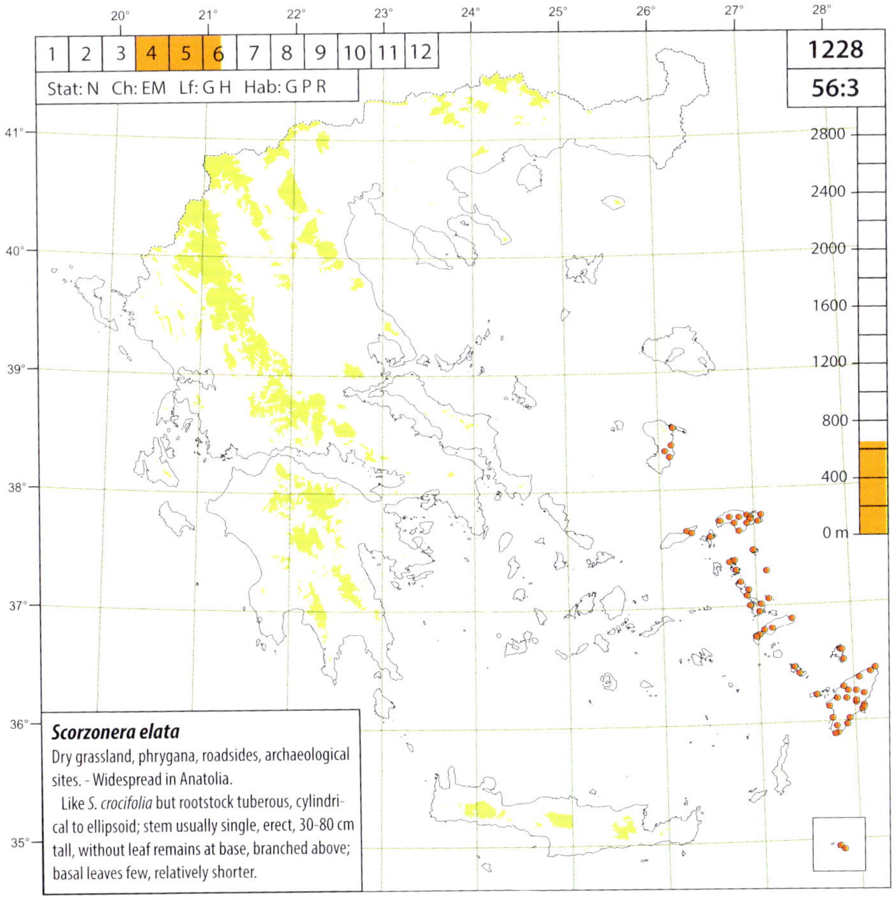

Scorzonera elata
Dry grassland, phrygana, roadsides, archaeological sites. - Widespread in Anatolia.
Like *S. crocifolia* but rootstock tuberous, cylindrical to ellipsoid; stem usually single, erect, 30-80 cm tall, without leaf remains at base, branched above; basal leaves few, relatively shorter.

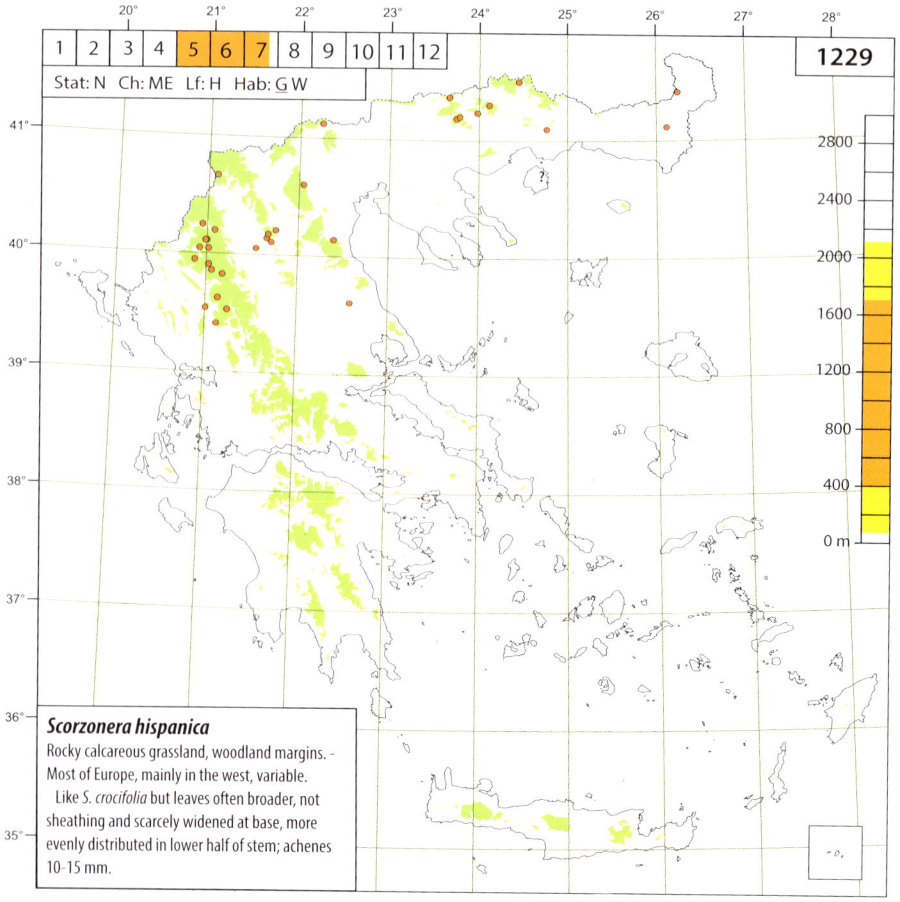

Scorzonera hispanica
Rocky calcareous grassland, woodland margins. - Most of Europe, mainly in the west, variable.
Like *S. crocifolia* but leaves often broader, not sheathing and scarcely widened at base, more evenly distributed in lower half of stem; achenes 10-15 mm.

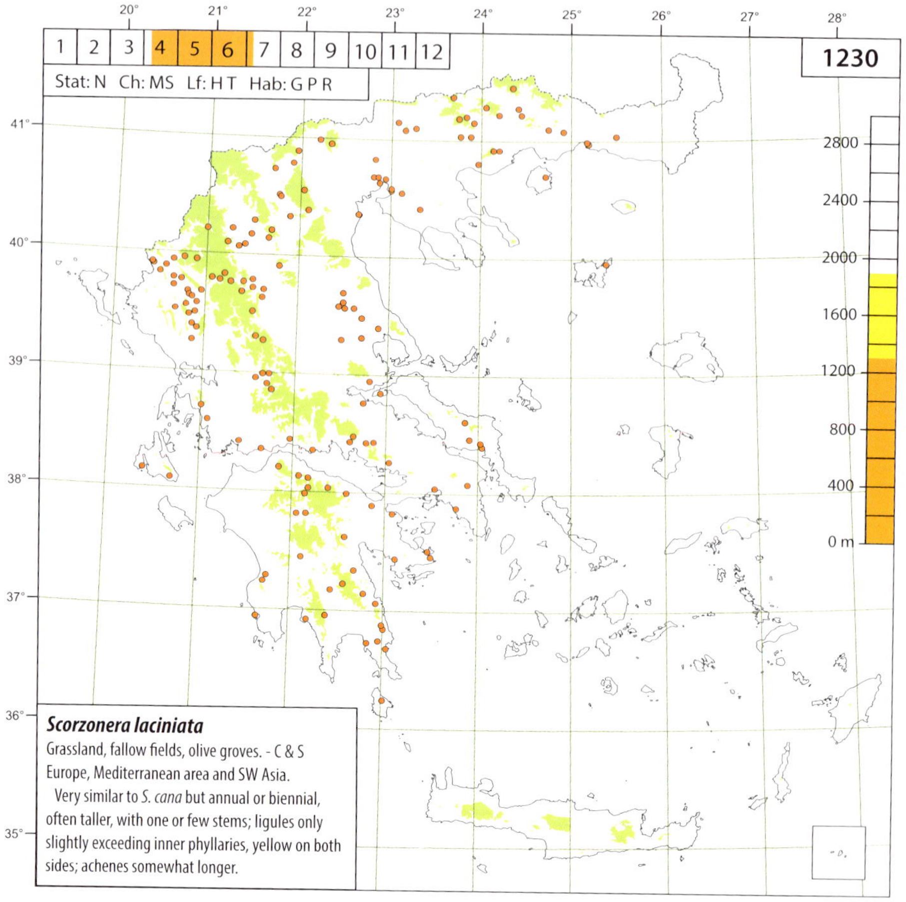

Scorzonera laciniata
Grassland, fallow fields, olive groves. - C & S Europe, Mediterranean area and SW Asia.
Very similar to *S. cana* but annual or biennial, often taller, with one or few stems; ligules only slightly exceeding inner phyllaries, yellow on both sides; achenes somewhat longer.

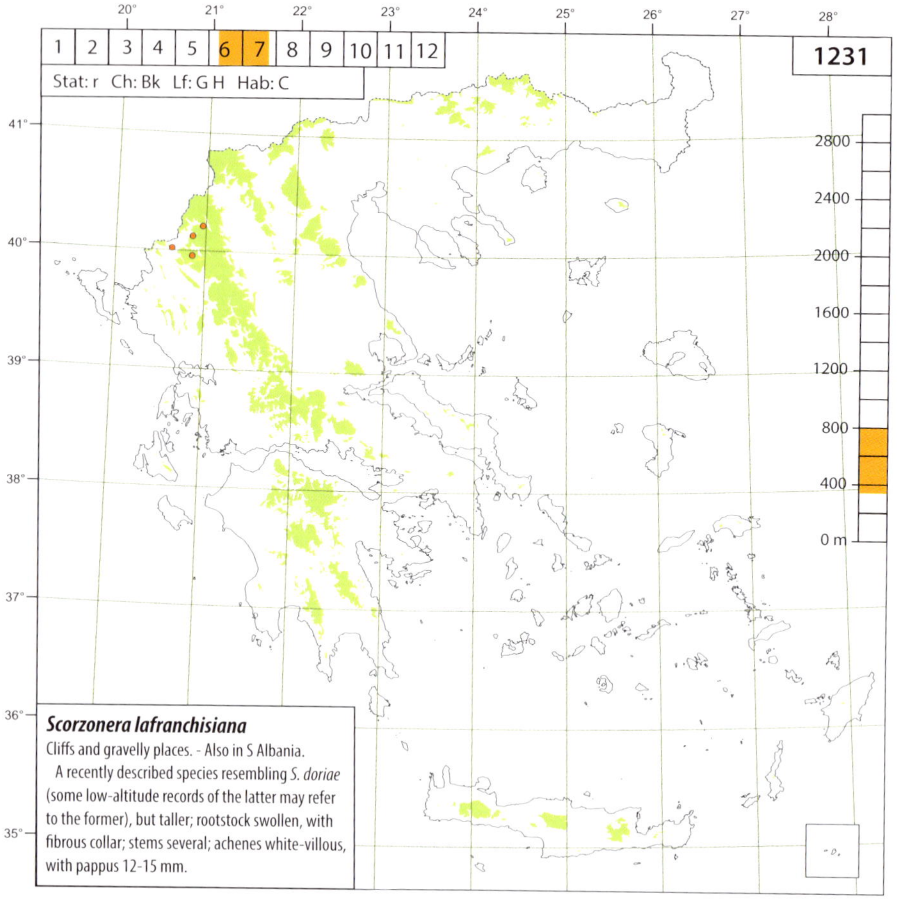

Scorzonera lafranchisiana
Cliffs and gravelly places. - Also in S Albania.
A recently described species resembling *S. doriae* (some low-altitude records of the latter may refer to the former), but taller; rootstock swollen, with fibrous collar; stems several; achenes white-villous, with pappus 12-15 mm.

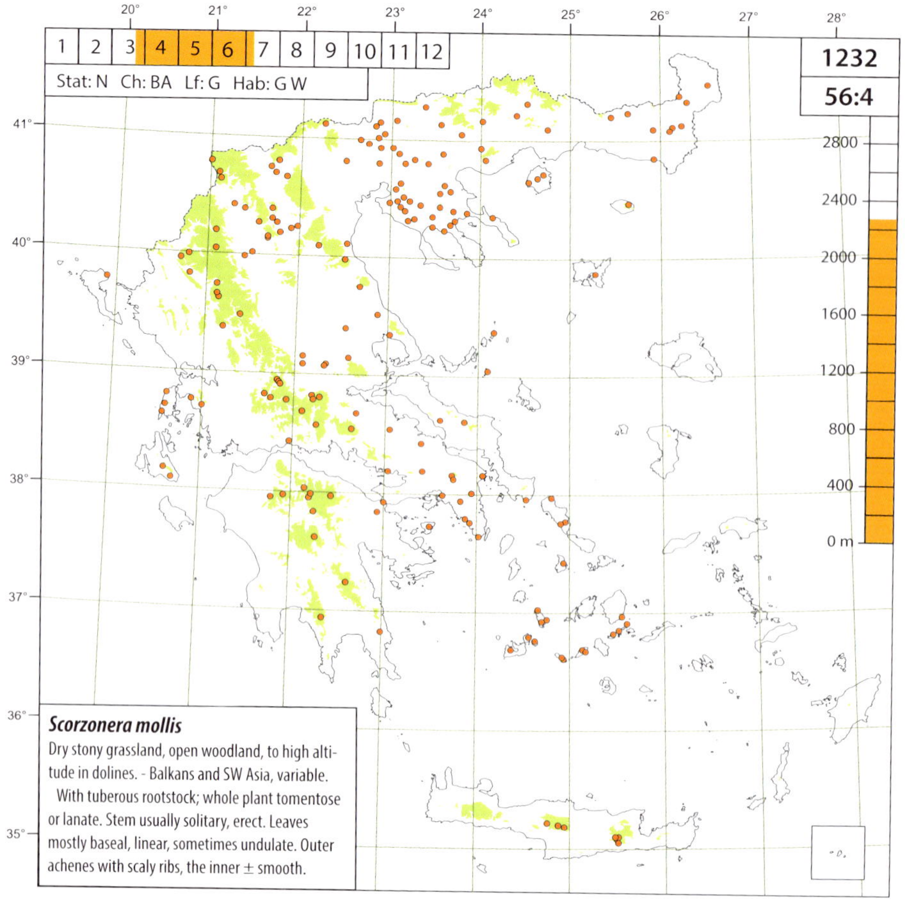

Scorzonera mollis
Dry stony grassland, open woodland, to high altitude in dolines. - Balkans and SW Asia, variable.
With tuberous rootstock; whole plant tomentose or lanate. Stem usually solitary, erect. Leaves mostly basal, linear, sometimes undulate. Outer achenes with scaly ribs, the inner ± smooth.

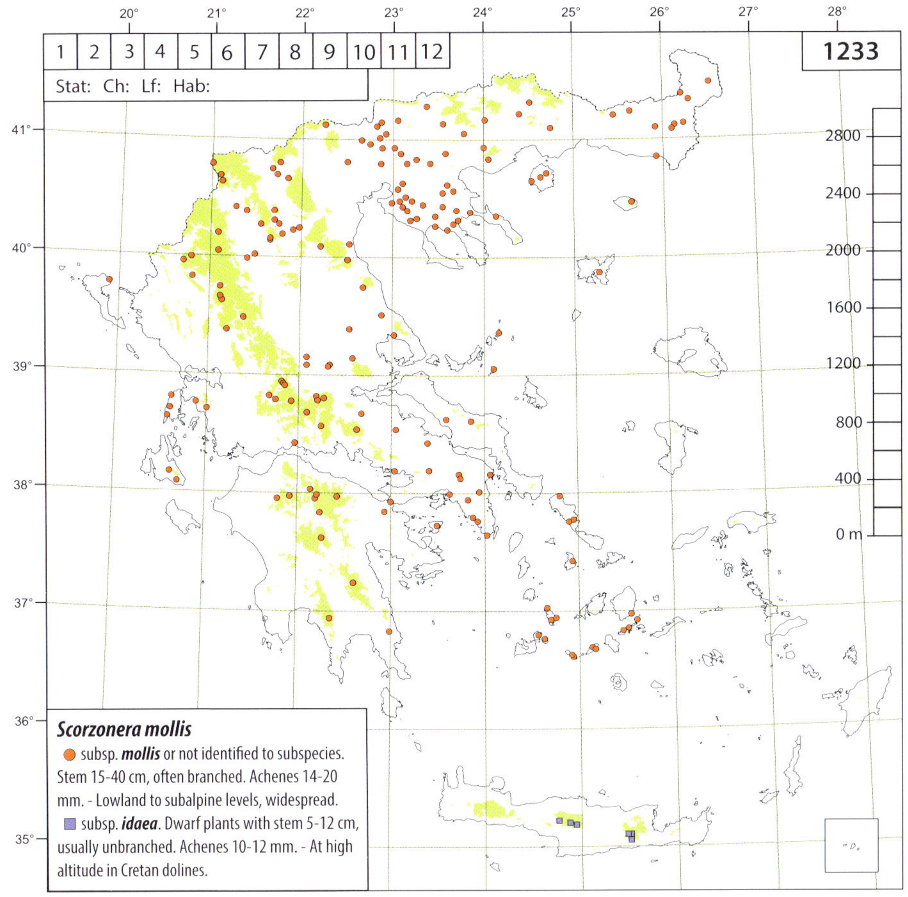

Scorzonera mollis
● subsp. ***mollis*** or not identified to subspecies. Stem 15-40 cm, often branched. Achenes 14-20 mm. - Lowland to subalpine levels, widespread.
■ subsp. ***idaea***. Dwarf plants with stem 5-12 cm, usually unbranched. Achenes 10-12 mm. - At high altitude in Cretan dolines.

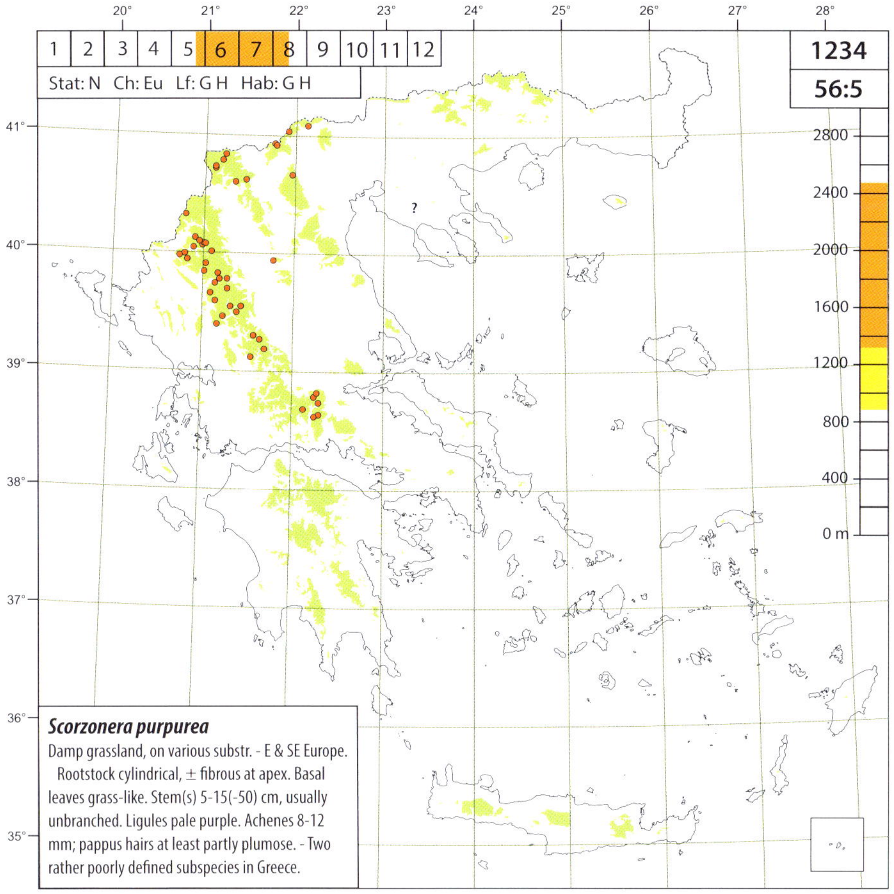

Scorzonera purpurea
Damp grassland, on various substr. - E & SE Europe.
Rootstock cylindrical, ± fibrous at apex. Basal leaves grass-like. Stem(s) 5-15(-50) cm, usually unbranched. Ligules pale purple. Achenes 8-12 mm; pappus hairs at least partly plumose. - Two rather poorly defined subspecies in Greece.

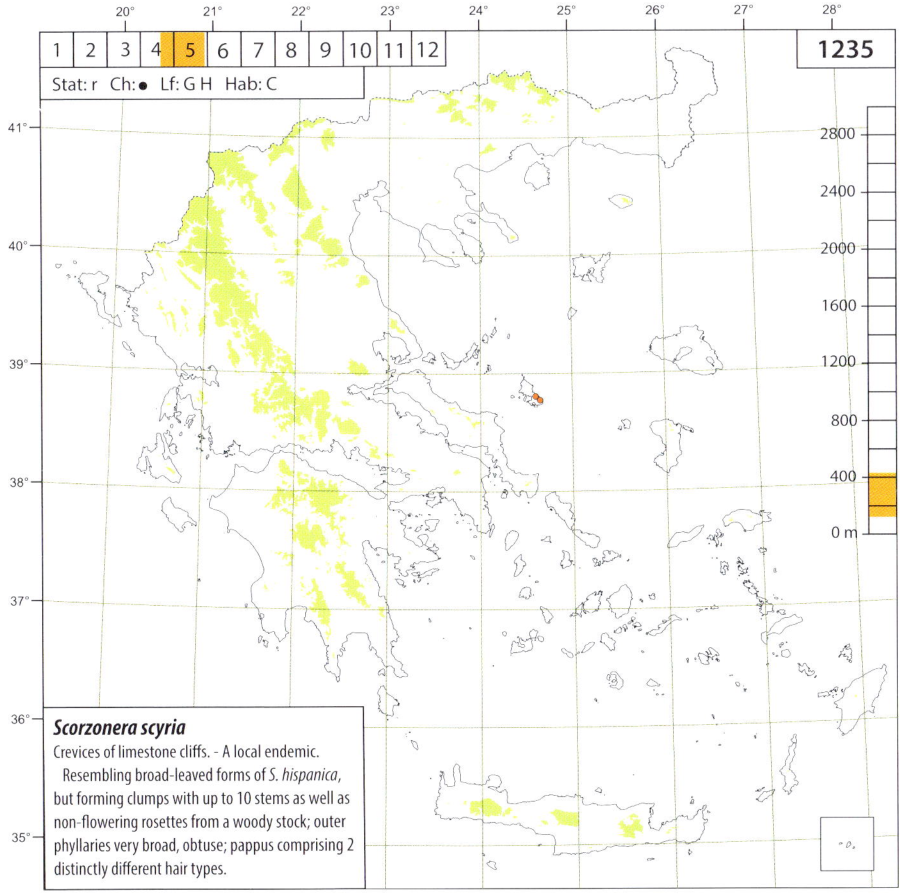

Scorzonera scyria
Crevices of limestone cliffs. - A local endemic.
Resembling broad-leaved forms of *S. hispanica*, but forming clumps with up to 10 stems as well as non-flowering rosettes from a woody stock; outer phyllaries very broad, obtuse; pappus comprising 2 distinctly different hair types.

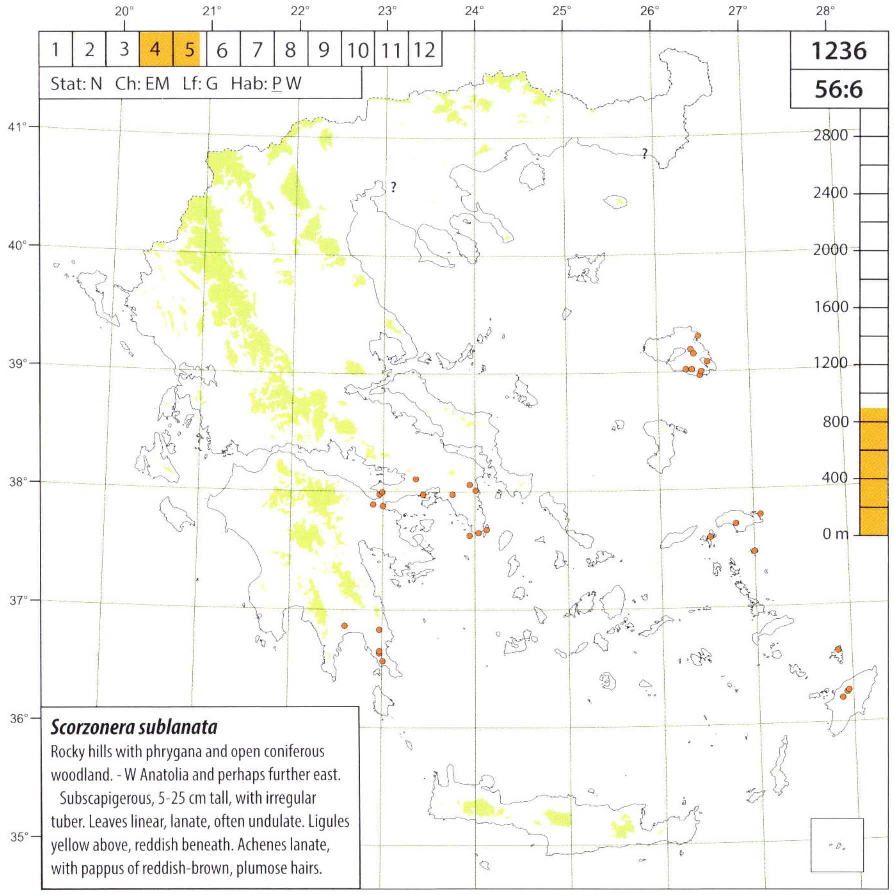

Scorzonera sublanata
Rocky hills with phrygana and open coniferous woodland. - W Anatolia and perhaps further east.
Subscapigerous, 5-25 cm tall, with irregular tuber. Leaves linear, lanate, often undulate. Ligules yellow above, reddish beneath. Achenes lanate, with pappus of reddish-brown, plumose hairs.

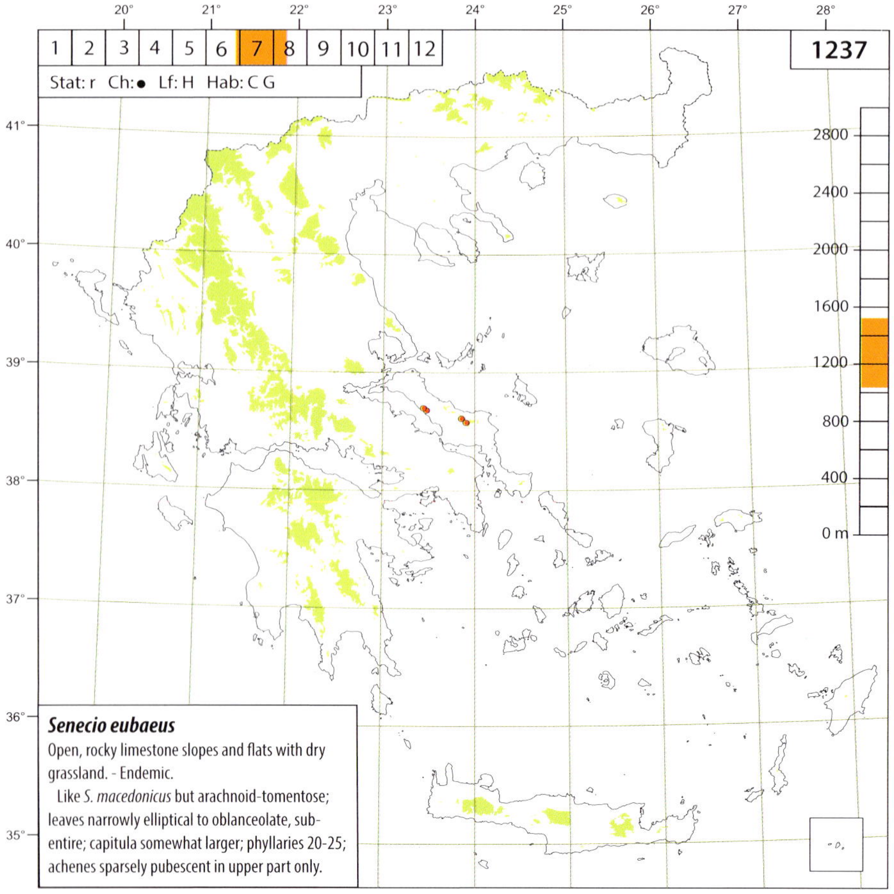

Senecio eubaeus
Open, rocky limestone slopes and flats with dry grassland. - Endemic.
Like *S. macedonicus* but arachnoid-tomentose; leaves narrowly elliptical to oblanceolate, subentire; capitula somewhat larger; phyllaries 20-25; achenes sparsely pubescent in upper part only.

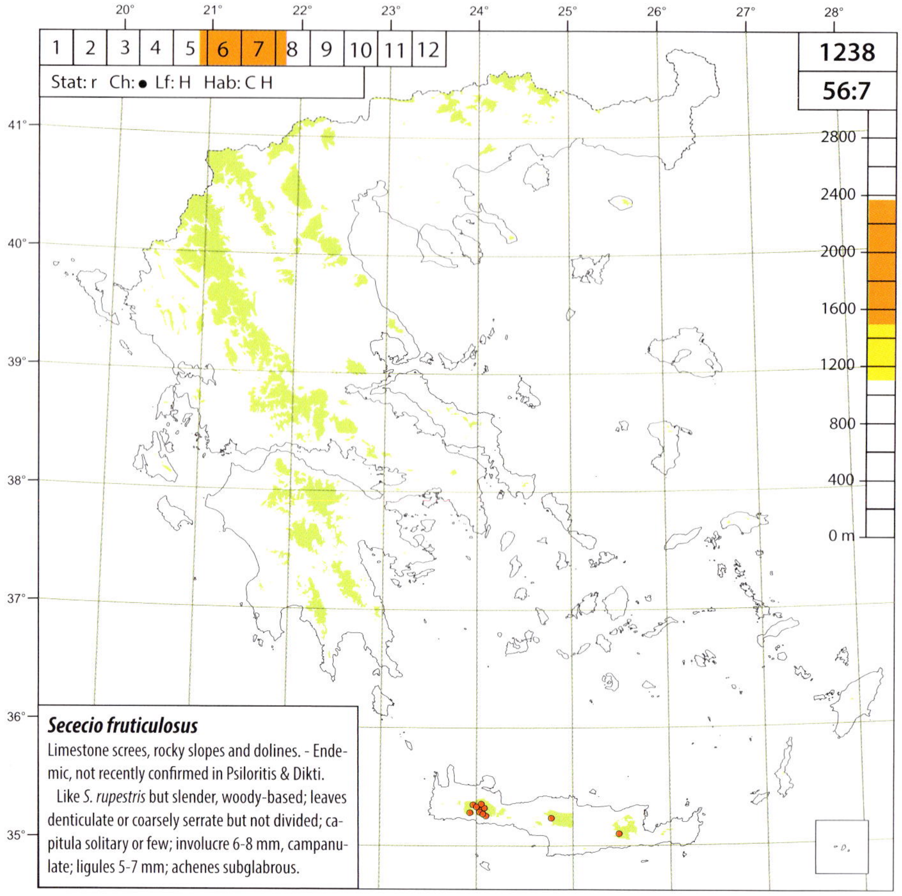

Sececio fruticulosus
Limestone screes, rocky slopes and dolines. - Endemic, not recently confirmed in Psiloritis & Dikti.
Like *S. rupestris* but slender, woody-based; leaves denticulate or coarsely serrate but not divided; capitula solitary or few; involucre 6-8 mm, campanulate; ligules 5-7 mm; achenes subglabrous.

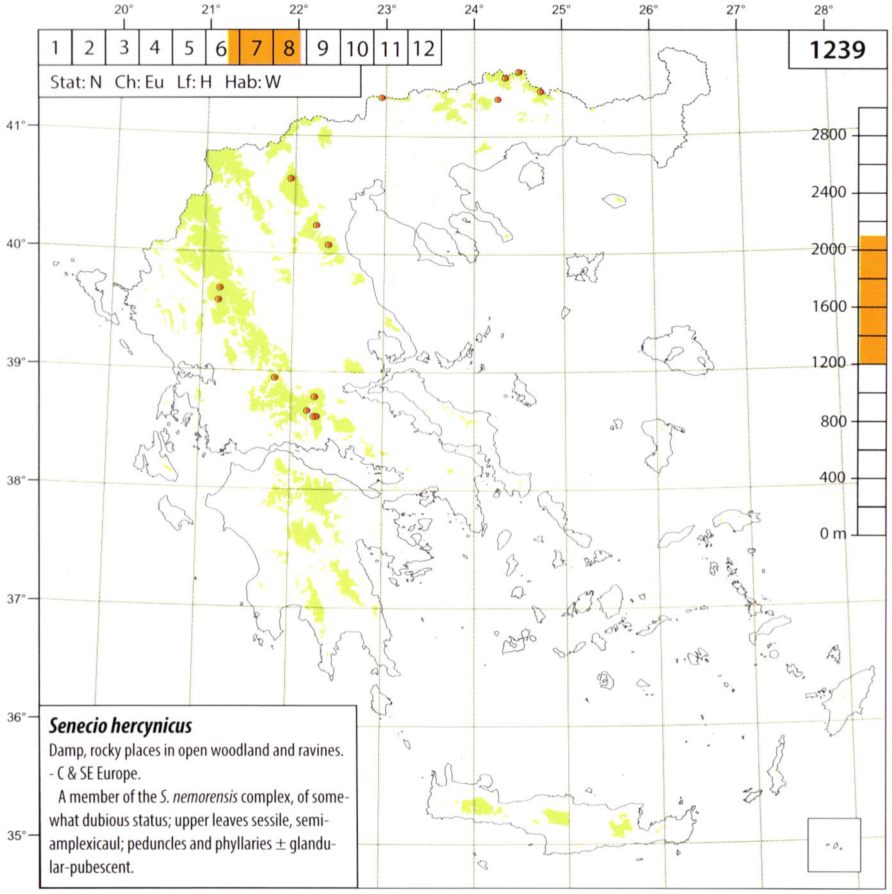

Senecio hercynicus
Damp, rocky places in open woodland and ravines. - C & SE Europe.
A member of the *S. nemorensis* complex, of somewhat dubious status; upper leaves sessile, semi-amplexicaul; peduncles and phyllaries ± glandular-pubescent.

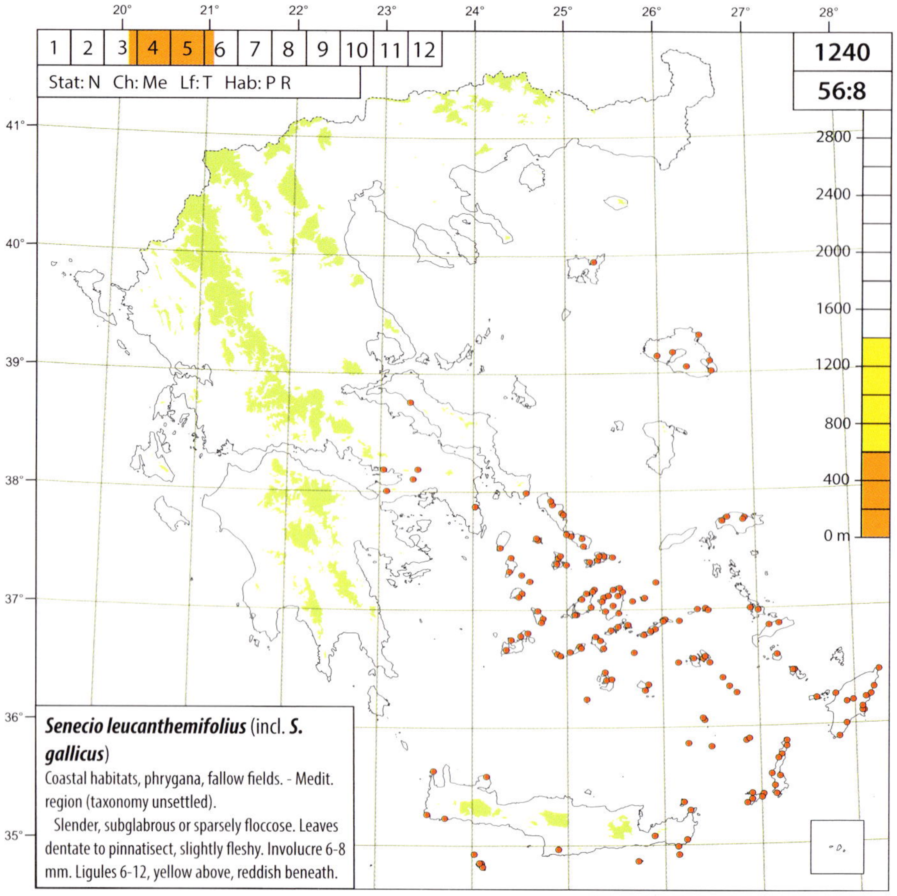

Senecio leucanthemifolius (incl. **S. gallicus**)
Coastal habitats, phrygana, fallow fields. - Medit. region (taxonomy unsettled).
Slender, subglabrous or sparsely floccose. Leaves dentate to pinnatisect, slightly fleshy. Involucre 6-8 mm. Ligules 6-12, yellow above, reddish beneath.

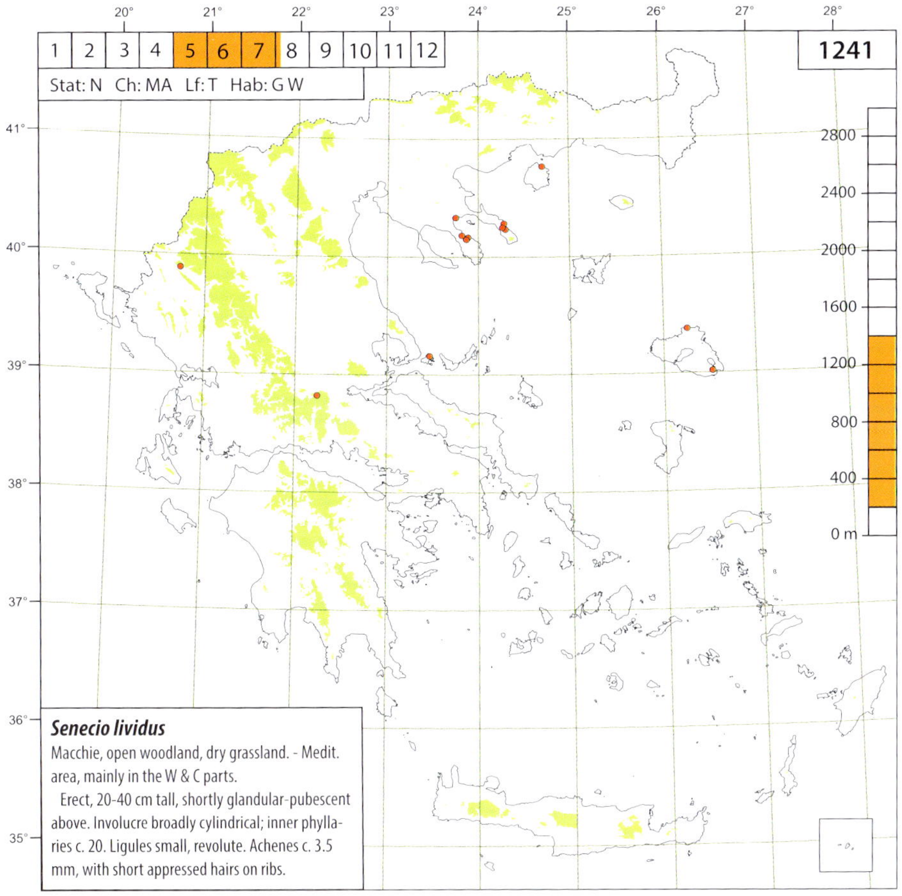

Senecio lividus
Macchie, open woodland, dry grassland. - Medit. area, mainly in the W & C parts.
Erect, 20-40 cm tall, shortly glandular-pubescent above. Involucre broadly cylindrical; inner phyllaries c. 20. Ligules small, revolute. Achenes c. 3.5 mm, with short appressed hairs on ribs.

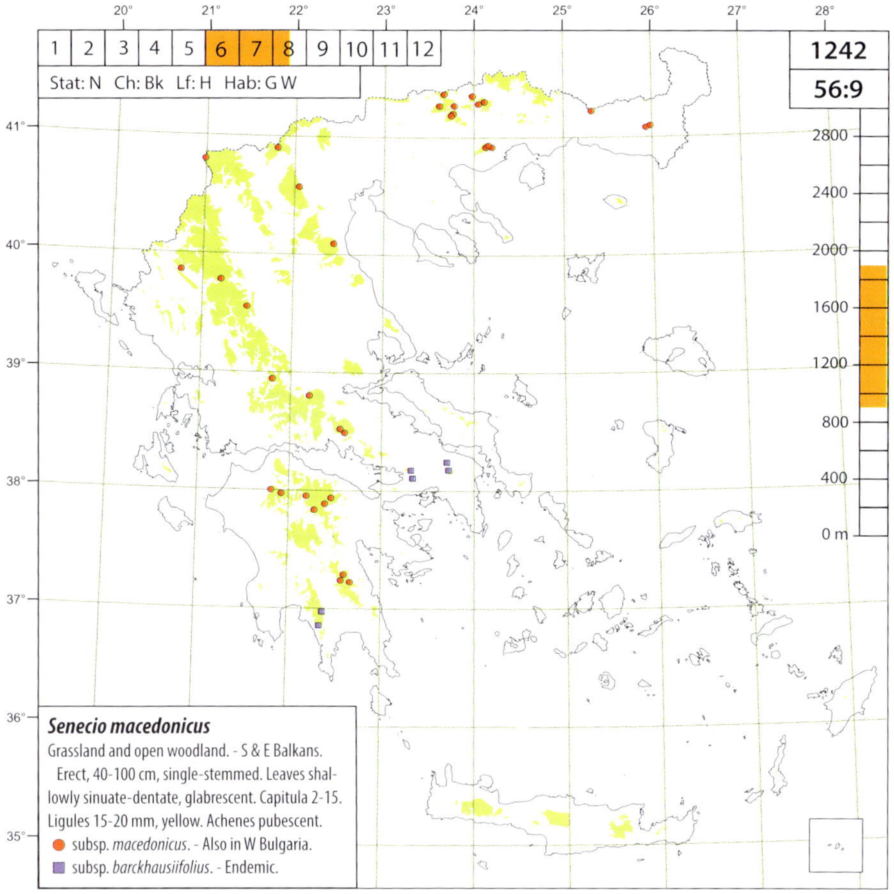

Senecio macedonicus
Grassland and open woodland. - S & E Balkans.
Erect, 40-100 cm, single-stemmed. Leaves shallowly sinuate-dentate, glabrescent. Capitula 2-15. Ligules 15-20 mm, yellow. Achenes pubescent.
● subsp. *macedonicus*. - Also in W Bulgaria.
■ subsp. *barckhausiifolius*. - Endemic.

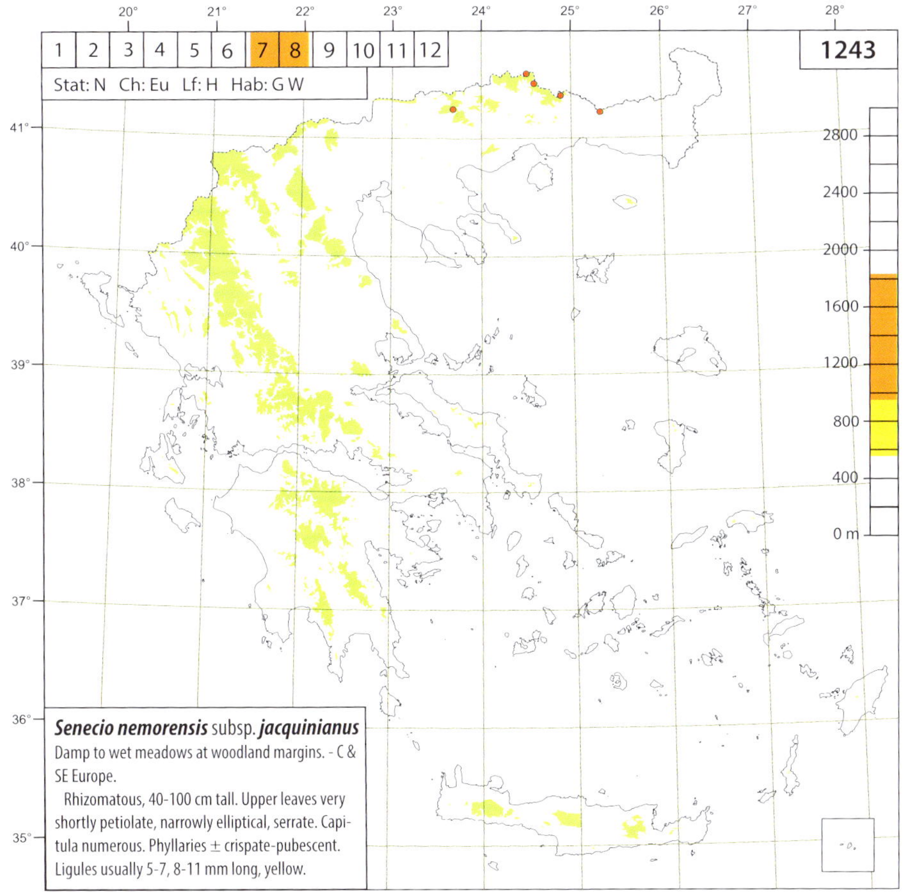

Senecio nemorensis subsp. ***jacquinianus***
Damp to wet meadows at woodland margins. - C & SE Europe.
Rhizomatous, 40-100 cm tall. Upper leaves very shortly petiolate, narrowly elliptical, serrate. Capitula numerous. Phyllaries ± crispate-pubescent. Ligules usually 5-7, 8-11 mm long, yellow.

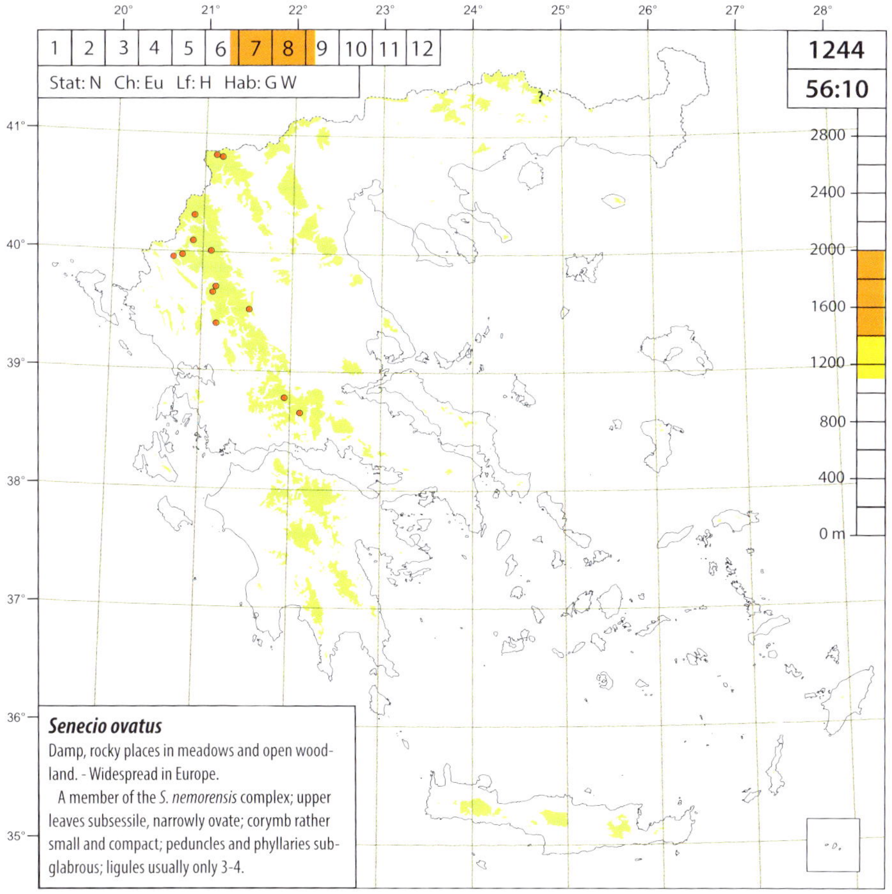

Senecio ovatus
Damp, rocky places in meadows and open woodland. - Widespread in Europe.
A member of the *S. nemorensis* complex; upper leaves subsessile, narrowly ovate; corymb rather small and compact; peduncles and phyllaries subglabrous; ligules usually only 3-4.

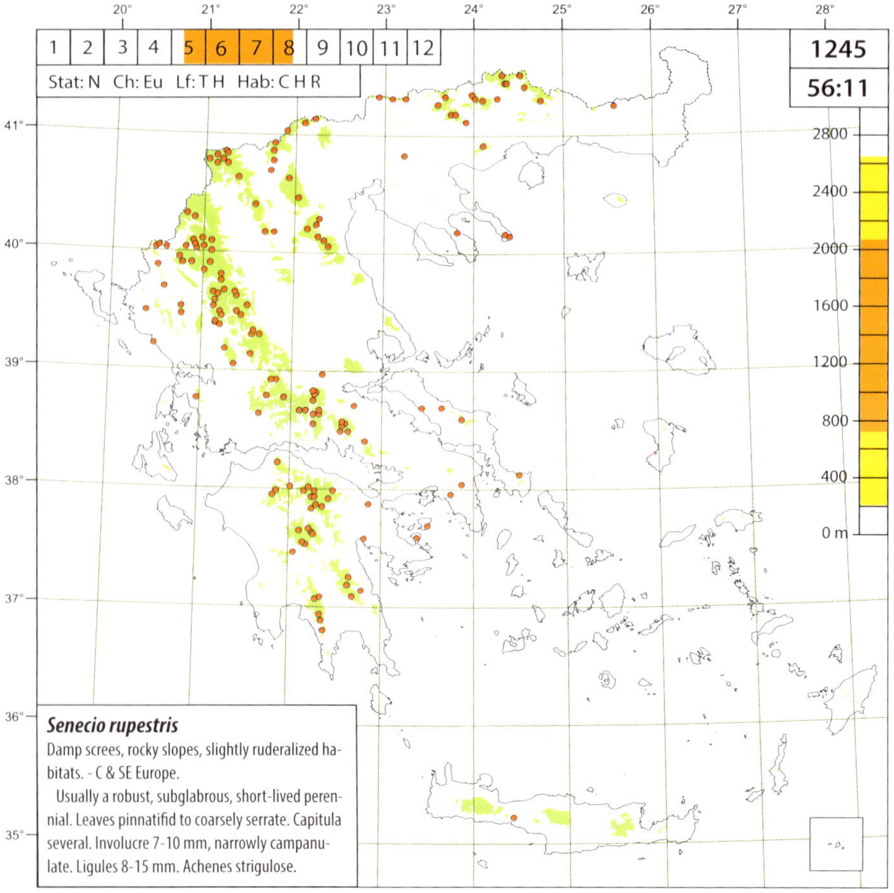

Senecio rupestris
Damp screes, rocky slopes, slightly ruderalized habitats. - C & SE Europe.
Usually a robust, subglabrous, short-lived perennial. Leaves pinnatifid to coarsely serrate. Capitula several. Involucre 7-10 mm, narrowly campanulate. Ligules 8-15 mm. Achenes strigulose.

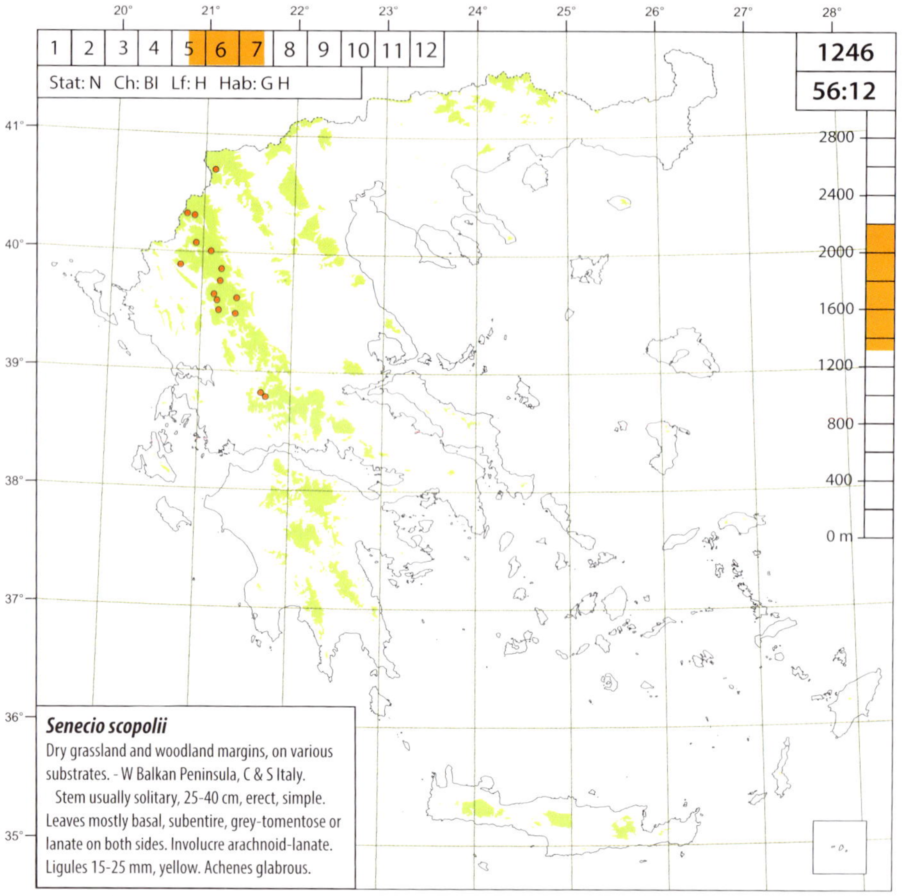

Senecio scopolii
Dry grassland and woodland margins, on various substrates. - W Balkan Peninsula, C & S Italy.
Stem usually solitary, 25-40 cm, erect, simple. Leaves mostly basal, subentire, grey-tomentose or lanate on both sides. Involucre arachnoid-lanate. Ligules 15-25 mm, yellow. Achenes glabrous.

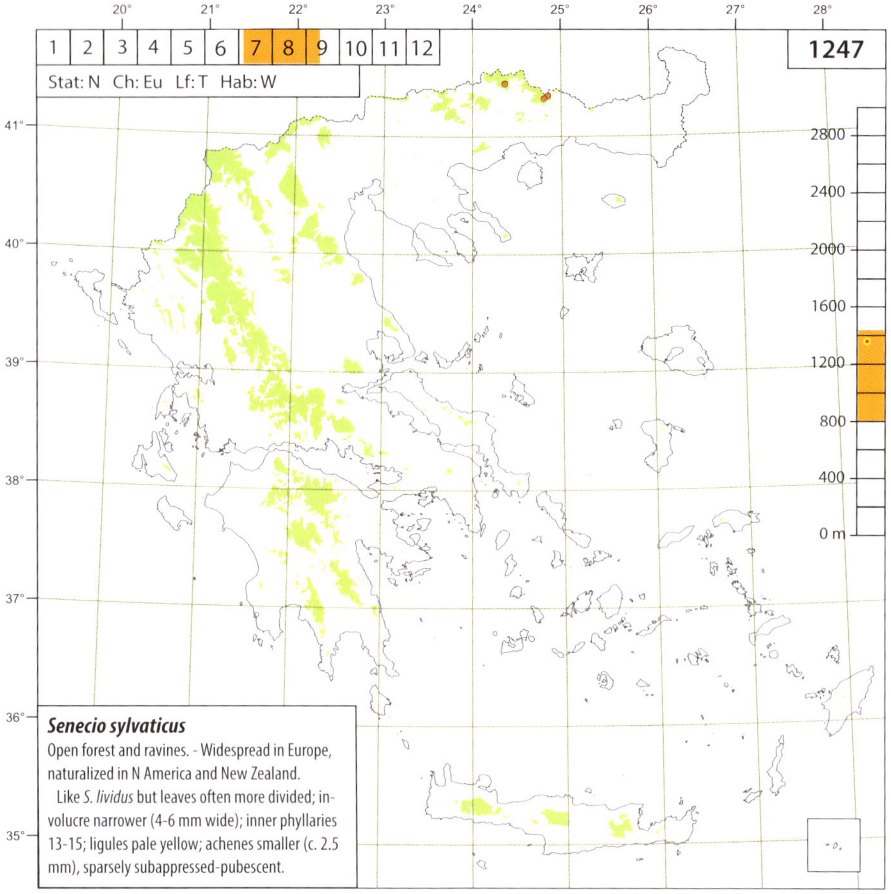

Senecio sylvaticus
Open forest and ravines. - Widespread in Europe, naturalized in N America and New Zealand.
Like *S. lividus* but leaves often more divided; involucre narrower (4-6 mm wide); inner phyllaries 13-15; ligules pale yellow; achenes smaller (c. 2.5 mm), sparsely subappressed-pubescent.

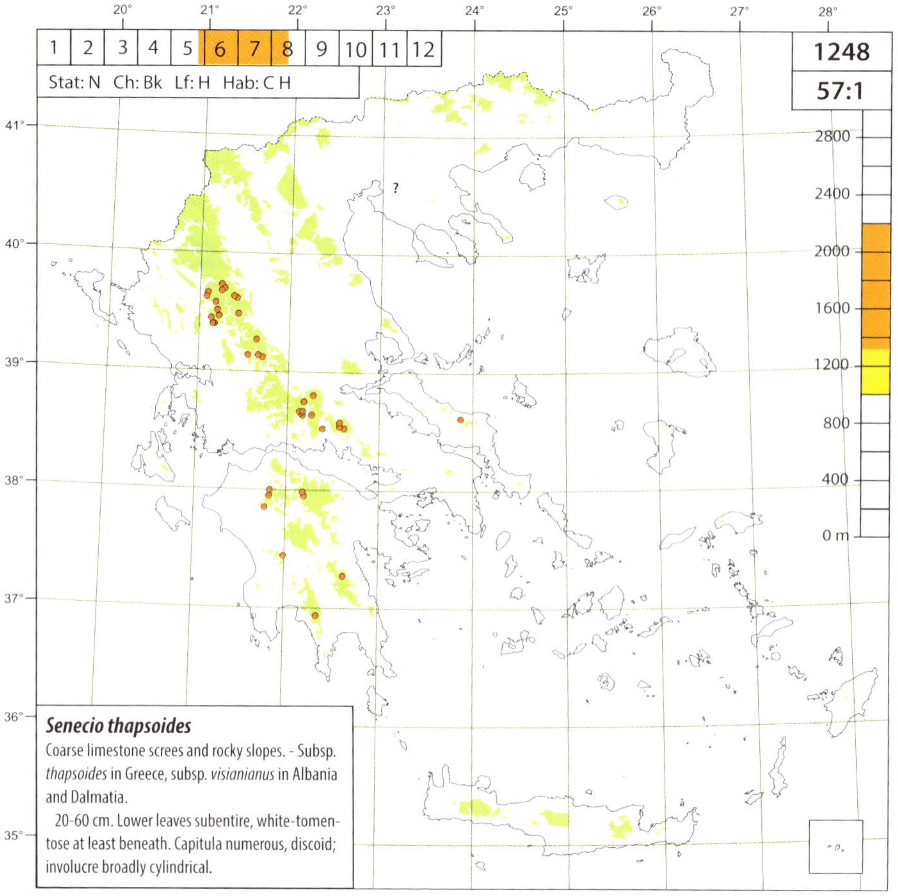

Senecio thapsoides
Coarse limestone screes and rocky slopes. - Subsp. *thapsoides* in Greece, subsp. *visianianus* in Albania and Dalmatia.
20-60 cm. Lower leaves subentire, white-tomentose at least beneath. Capitula numerous, discoid; involucre broadly cylindrical.

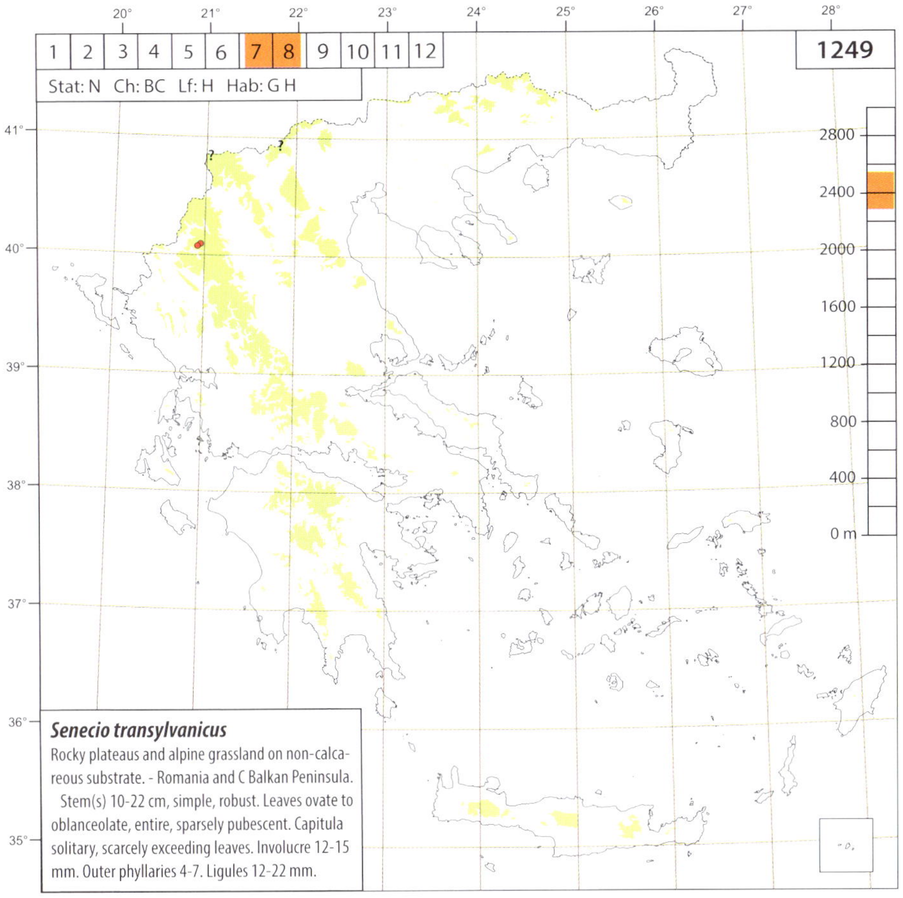

Senecio transylvanicus
Rocky plateaus and alpine grassland on non-calcareous substrate. - Romania and C Balkan Peninsula.
Stem(s) 10-22 cm, simple, robust. Leaves ovate to oblanceolate, entire, sparsely pubescent. Capitula solitary, scarcely exceeding leaves. Involucre 12-15 mm. Outer phyllaries 4-7. Ligules 12-22 mm.

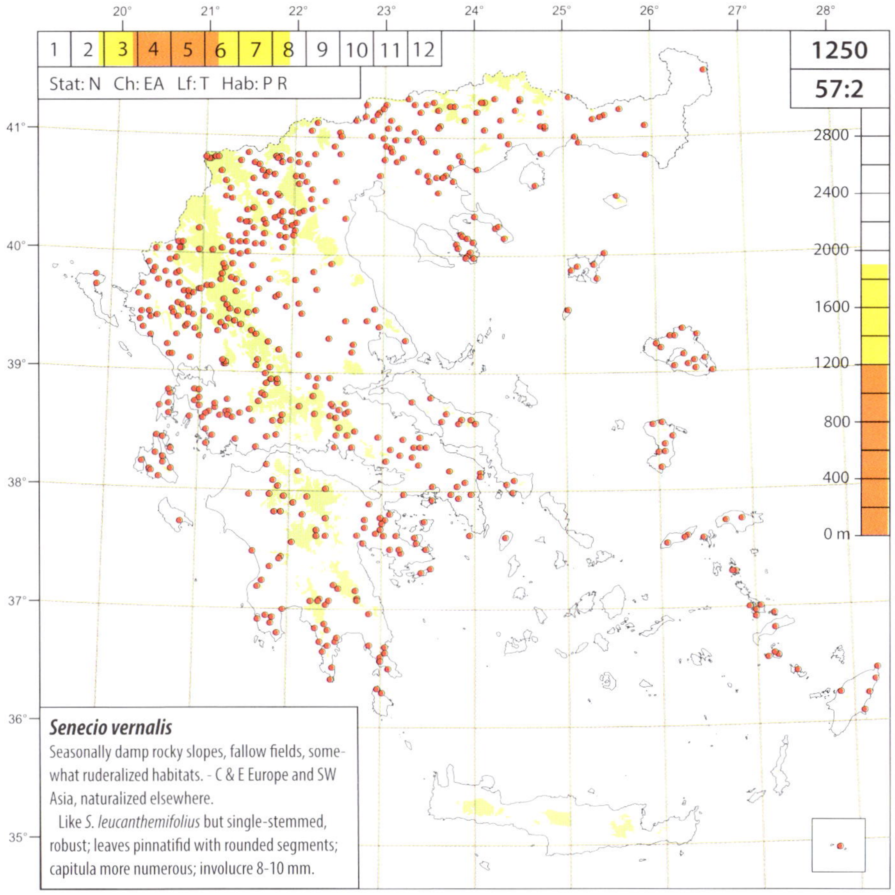

Senecio vernalis
Seasonally damp rocky slopes, fallow fields, somewhat ruderalized habitats. - C & E Europe and SW Asia, naturalized elsewhere.
Like *S. leucanthemifolius* but single-stemmed, robust; leaves pinnatifid with rounded segments; capitula more numerous; involucre 8-10 mm.

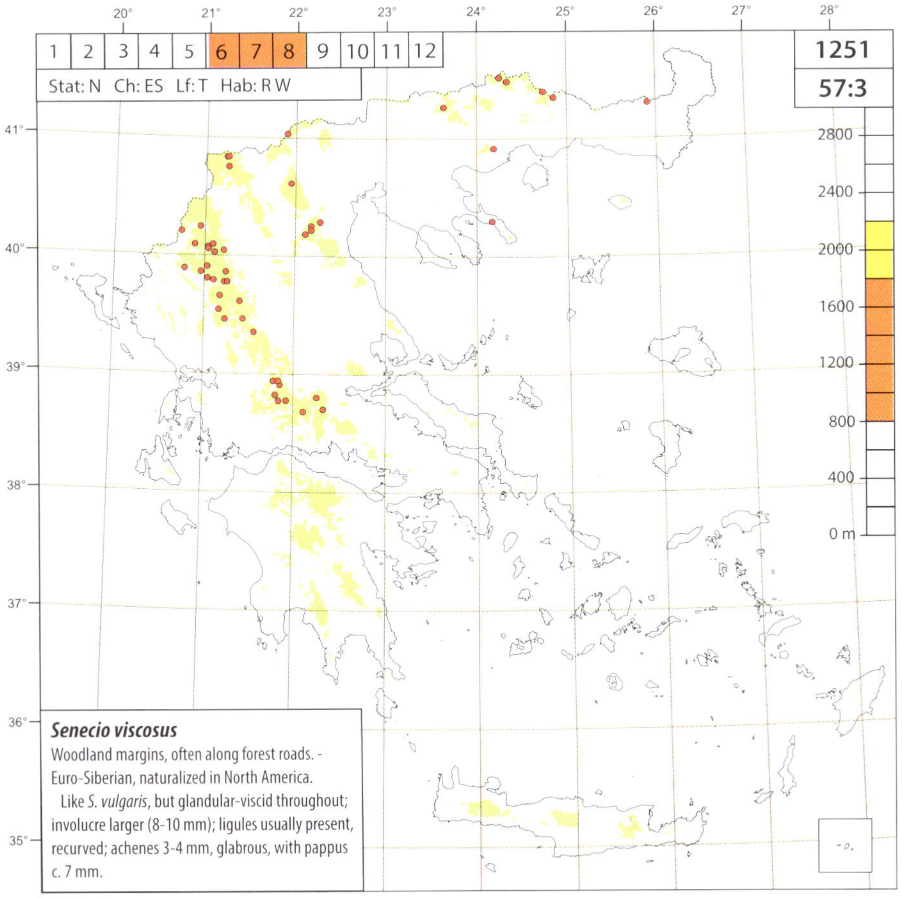

Senecio viscosus
Woodland margins, often along forest roads. - Euro-Siberian, naturalized in North America.
Like *S. vulgaris*, but glandular-viscid throughout; involucre larger (8-10 mm); ligules usually present, recurved; achenes 3-4 mm, glabrous, with pappus c. 7 mm.

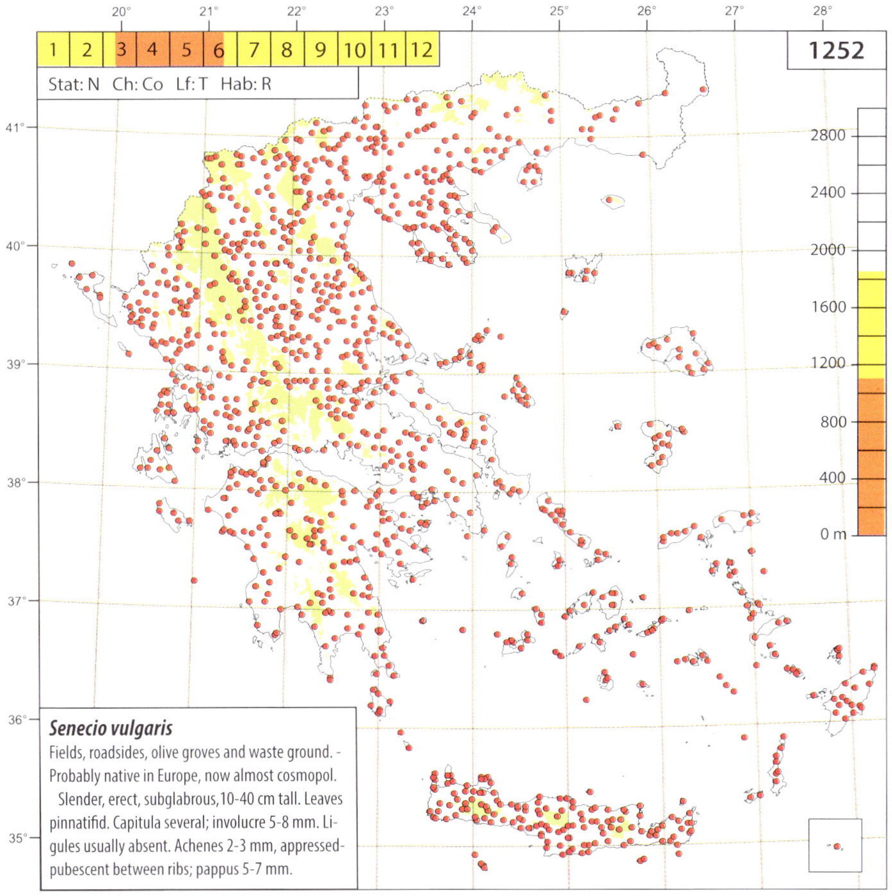

Senecio vulgaris
Fields, roadsides, olive groves and waste ground. - Probably native in Europe, now almost cosmopol.
Slender, erect, subglabrous, 10-40 cm tall. Leaves pinnatifid. Capitula several; involucre 5-8 mm. Ligules usually absent. Achenes 2-3 mm, appressed-pubescent between ribs; pappus 5-7 mm.

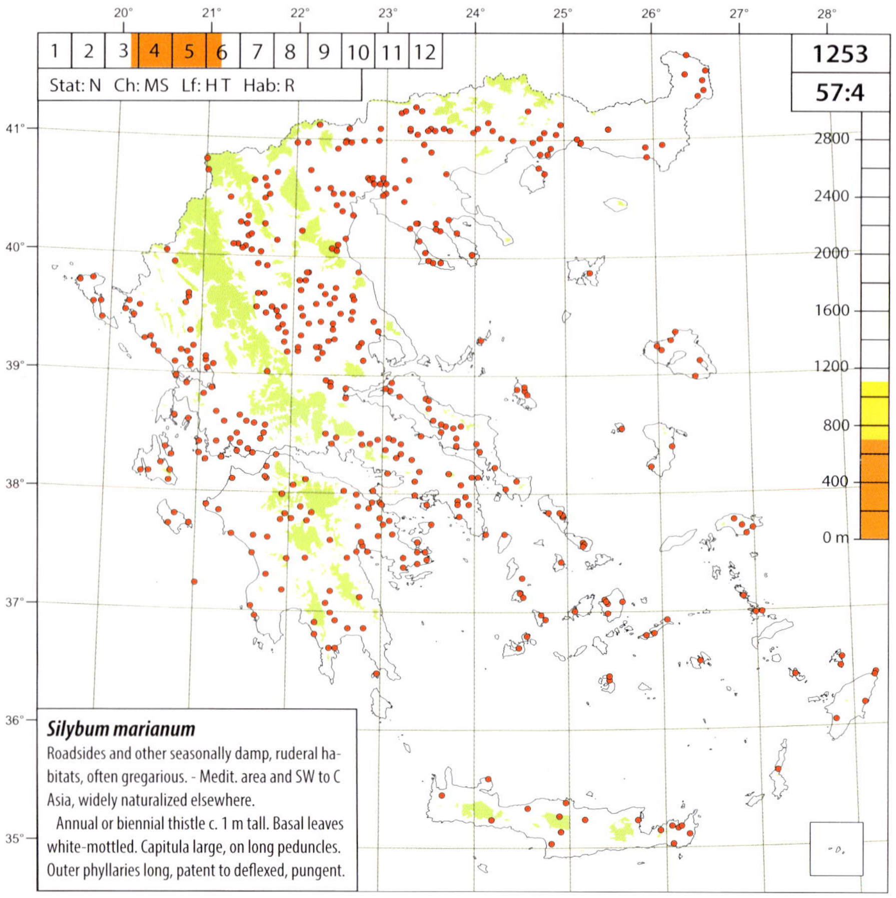

Silybum marianum
Roadsides and other seasonally damp, ruderal habitats, often gregarious. - Medit. area and SW to C Asia, widely naturalized elsewhere.
Annual or biennial thistle c. 1 m tall. Basal leaves white-mottled. Capitula large, on long peduncles. Outer phyllaries long, patent to deflexed, pungent.

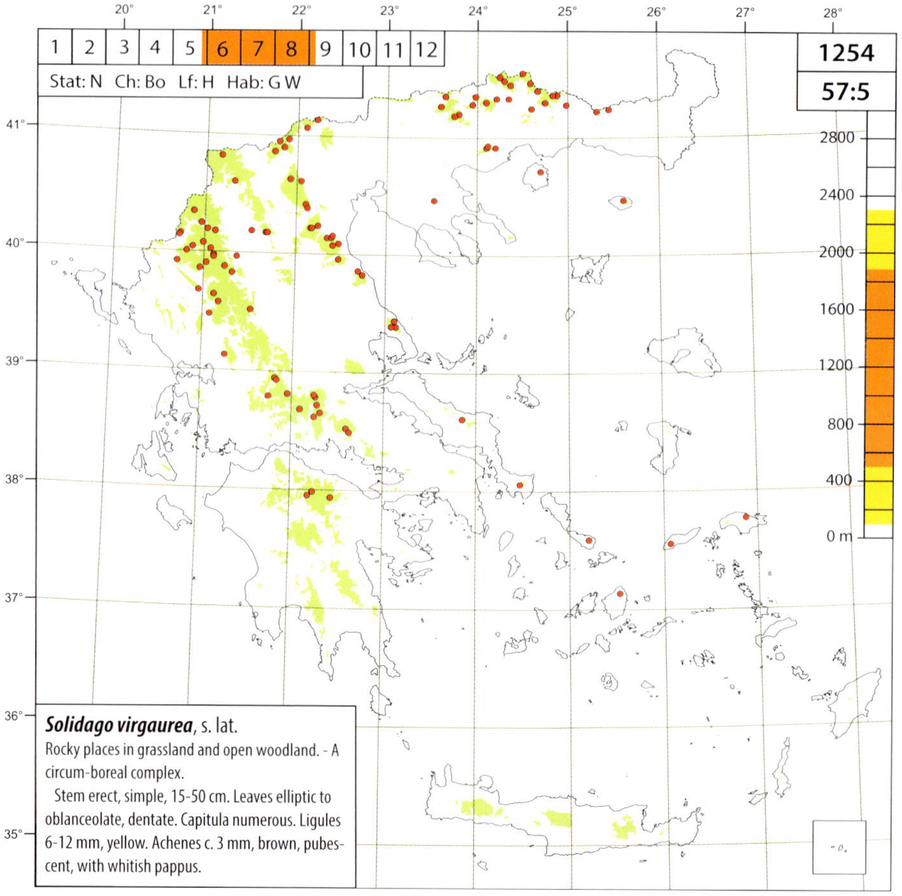

Solidago virgaurea, s. lat.
Rocky places in grassland and open woodland. - A circum-boreal complex.
Stem erect, simple, 15-50 cm. Leaves elliptic to oblanceolate, dentate. Capitula numerous. Ligules 6-12 mm, yellow. Achenes c. 3 mm, brown, pubescent, with whitish pappus.

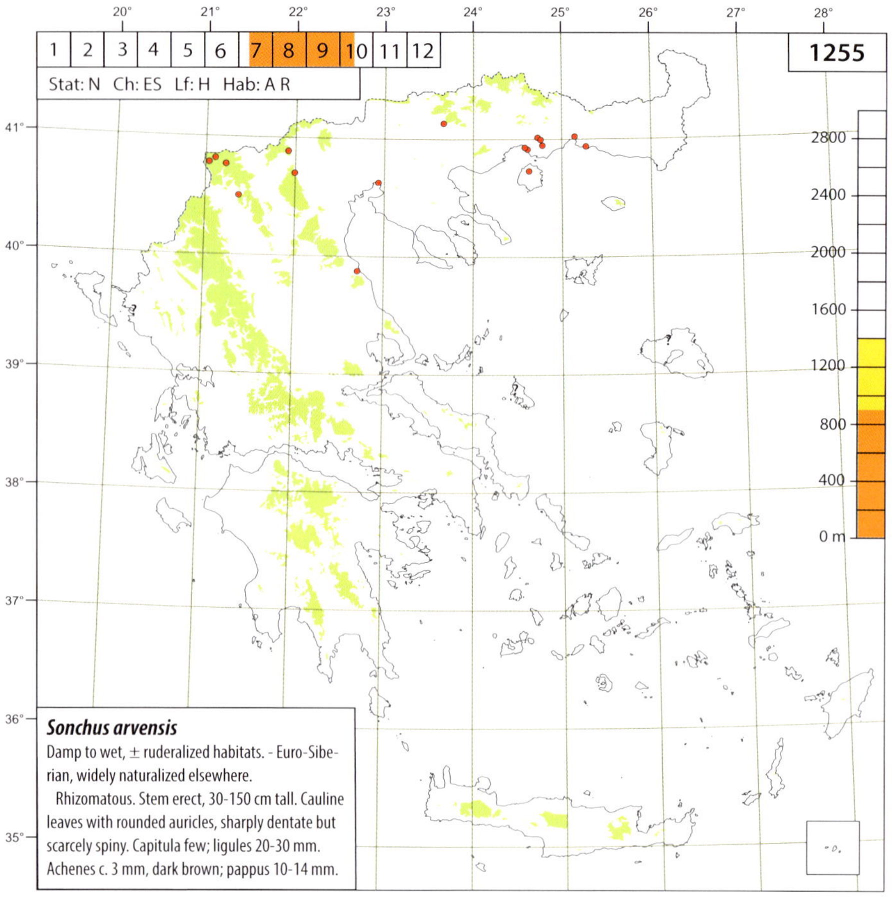

Sonchus arvensis
Damp to wet, ± ruderalized habitats. - Euro-Siberian, widely naturalized elsewhere.
Rhizomatous. Stem erect, 30-150 cm tall. Cauline leaves with rounded auricles, sharply dentate but scarcely spiny. Capitula few; ligules 20-30 mm. Achenes c. 3 mm, dark brown; pappus 10-14 mm.

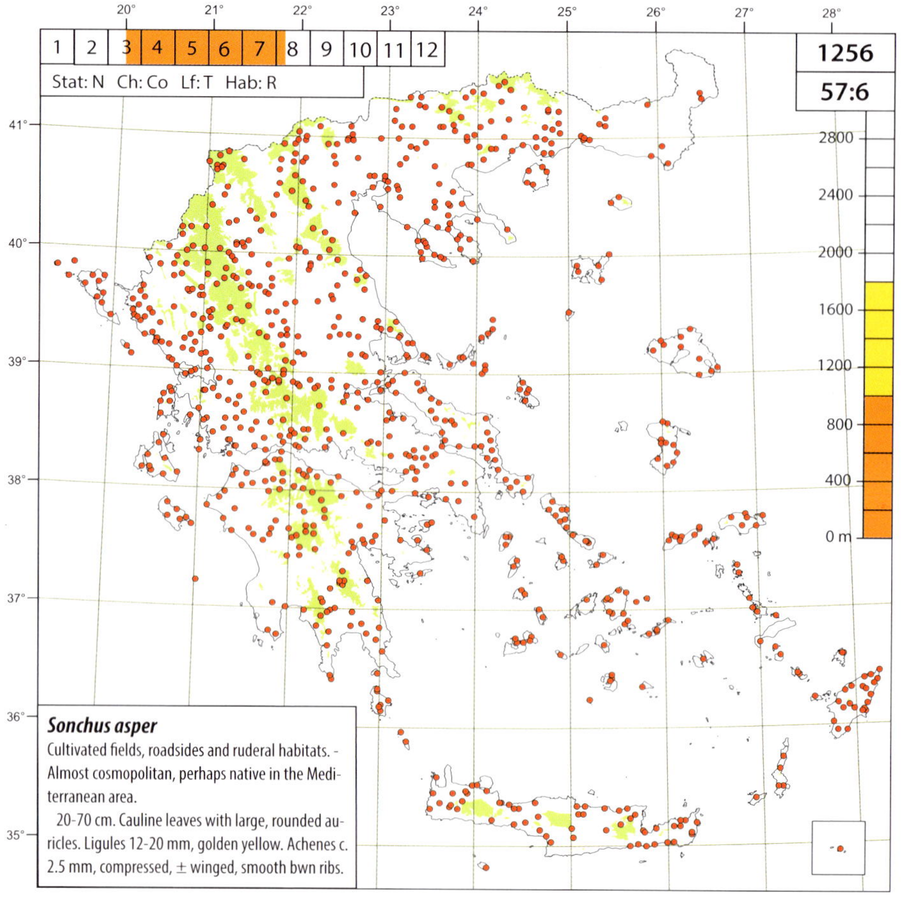

Sonchus asper
Cultivated fields, roadsides and ruderal habitats. - Almost cosmopolitan, perhaps native in the Mediterranean area.
20-70 cm. Cauline leaves with large, rounded auricles. Ligules 12-20 mm, golden yellow. Achenes c. 2.5 mm, compressed, ± winged, smooth bwn ribs.

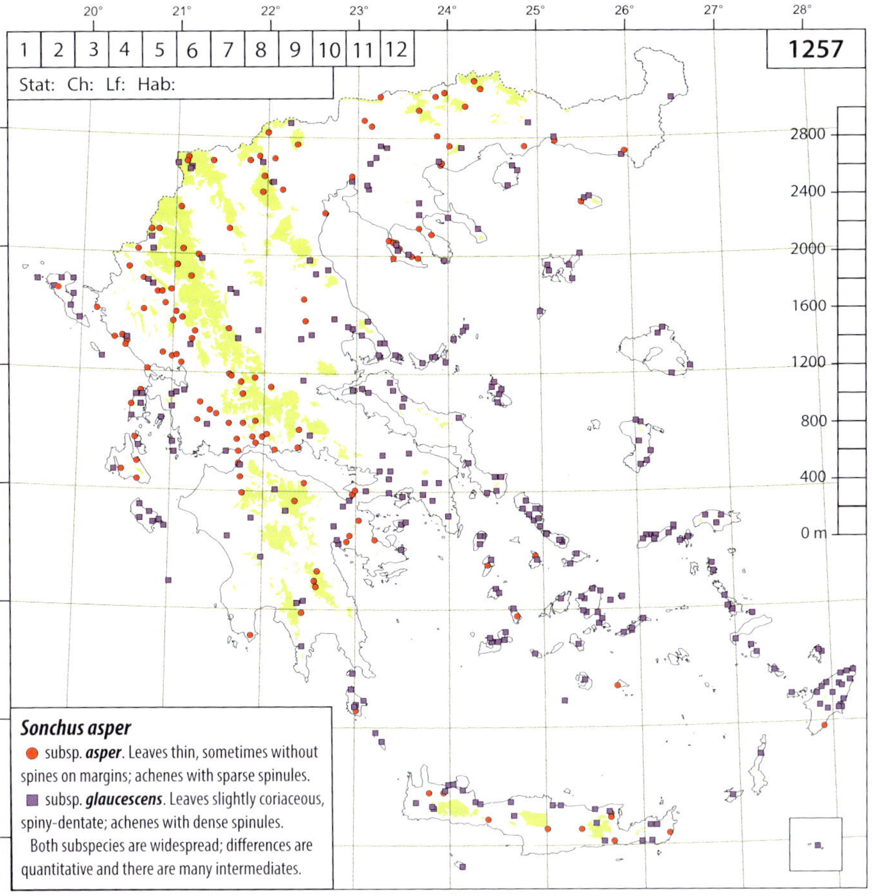

Sonchus asper
● subsp. ***asper***. Leaves thin, sometimes without spines on margins; achenes with sparse spinules.
■ subsp. ***glaucescens***. Leaves slightly coriaceous, spiny-dentate; achenes with dense spinules.
Both subspecies are widespread; differences are quantitative and there are many intermediates.

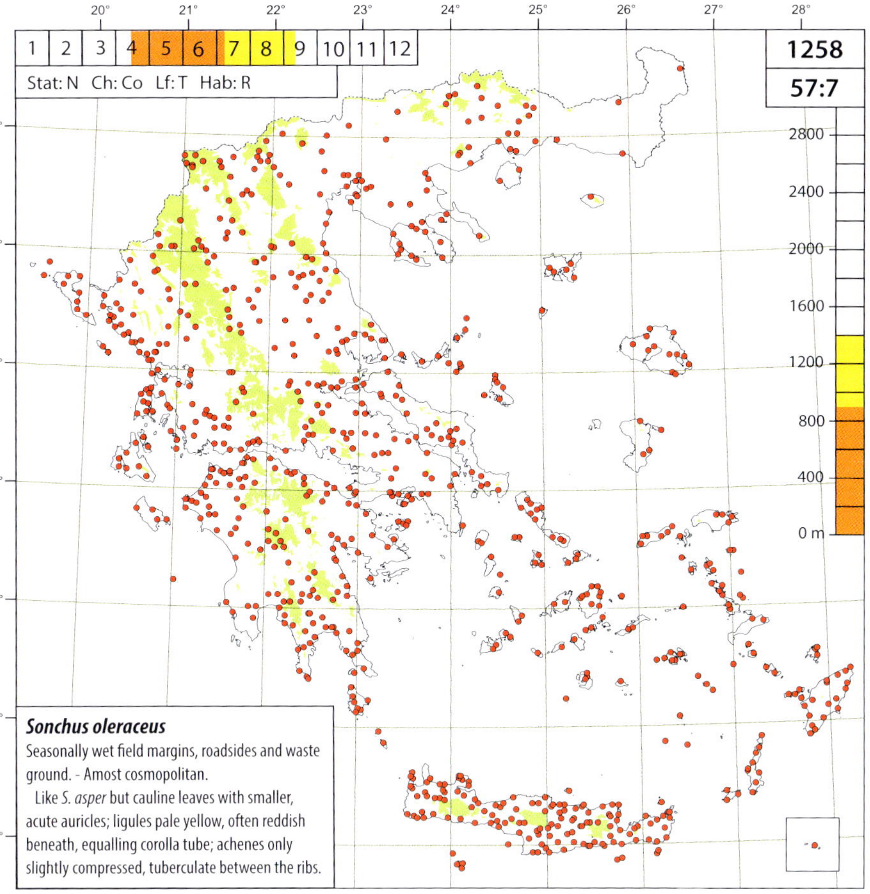

Sonchus oleraceus
Seasonally wet field margins, roadsides and waste ground. - Amost cosmopolitan.
Like *S. asper* but cauline leaves with smaller, acute auricles; ligules pale yellow, often reddish beneath, equalling corolla tube; achenes only slightly compressed, tuberculate between the ribs.

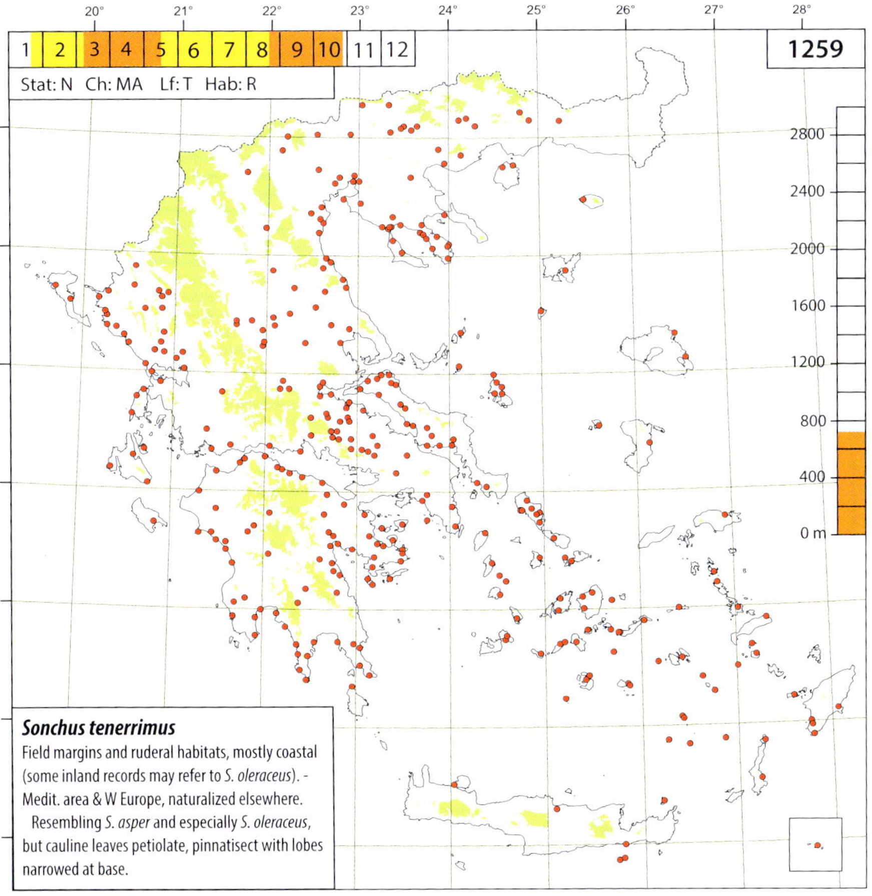

Sonchus tenerrimus
Field margins and ruderal habitats, mostly coastal (some inland records may refer to *S. oleraceus*). - Medit. area & W Europe, naturalized elsewhere.
Resembling *S. asper* and especially *S. oleraceus*, but cauline leaves petiolate, pinnatisect with lobes narrowed at base.

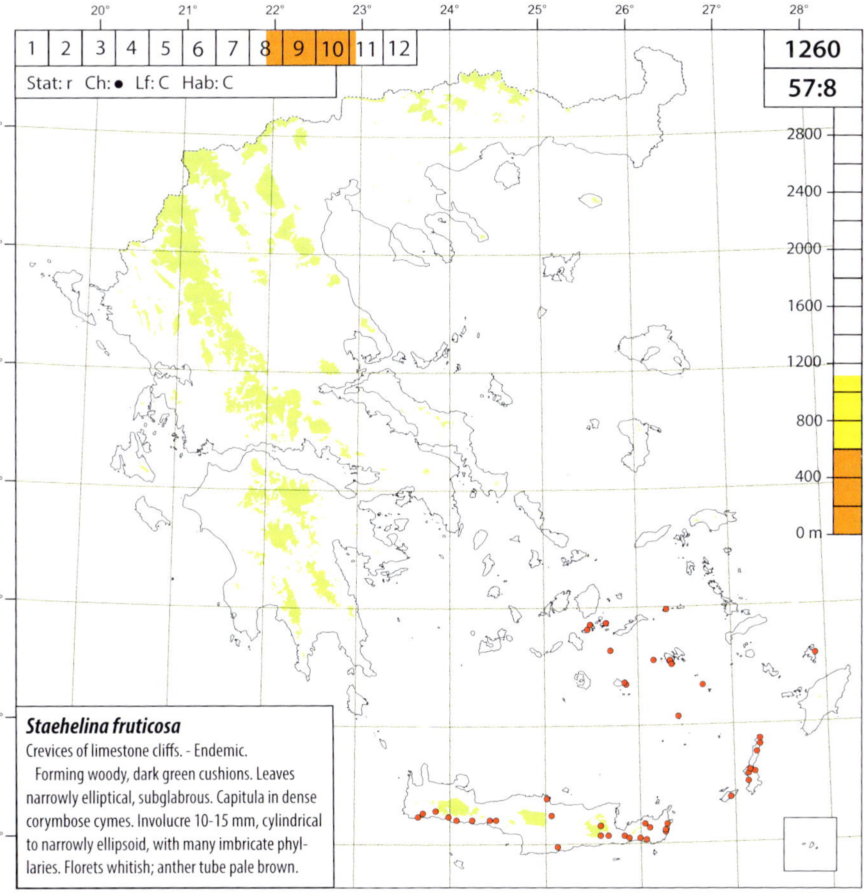

Staehelina fruticosa
Crevices of limestone cliffs. - Endemic.
Forming woody, dark green cushions. Leaves narrowly elliptical, subglabrous. Capitula in dense corymbose cymes. Involucre 10-15 mm, cylindrical to narrowly ellipsoid, with many imbricate phyllaries. Florets whitish; anther tube pale brown.

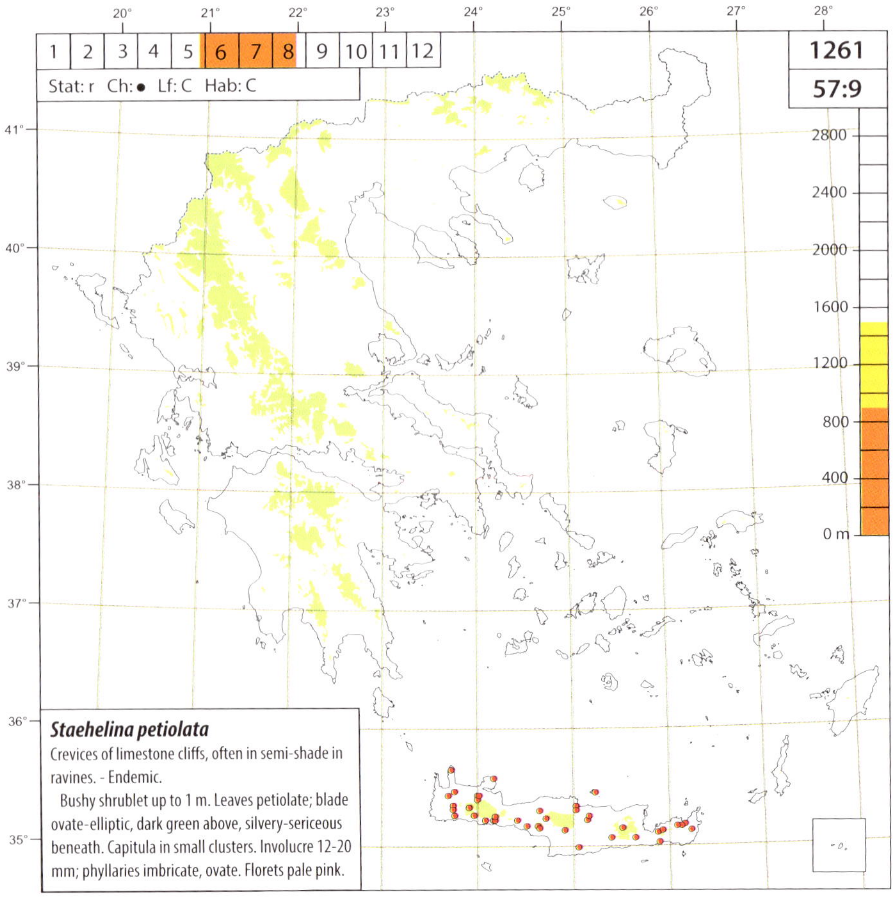

Staehelina petiolata

Crevices of limestone cliffs, often in semi-shade in ravines. - Endemic.

Bushy shrublet up to 1 m. Leaves petiolate; blade ovate-elliptic, dark green above, silvery-sericeous beneath. Capitula in small clusters. Involucre 12-20 mm; phyllaries imbricate, ovate. Florets pale pink.

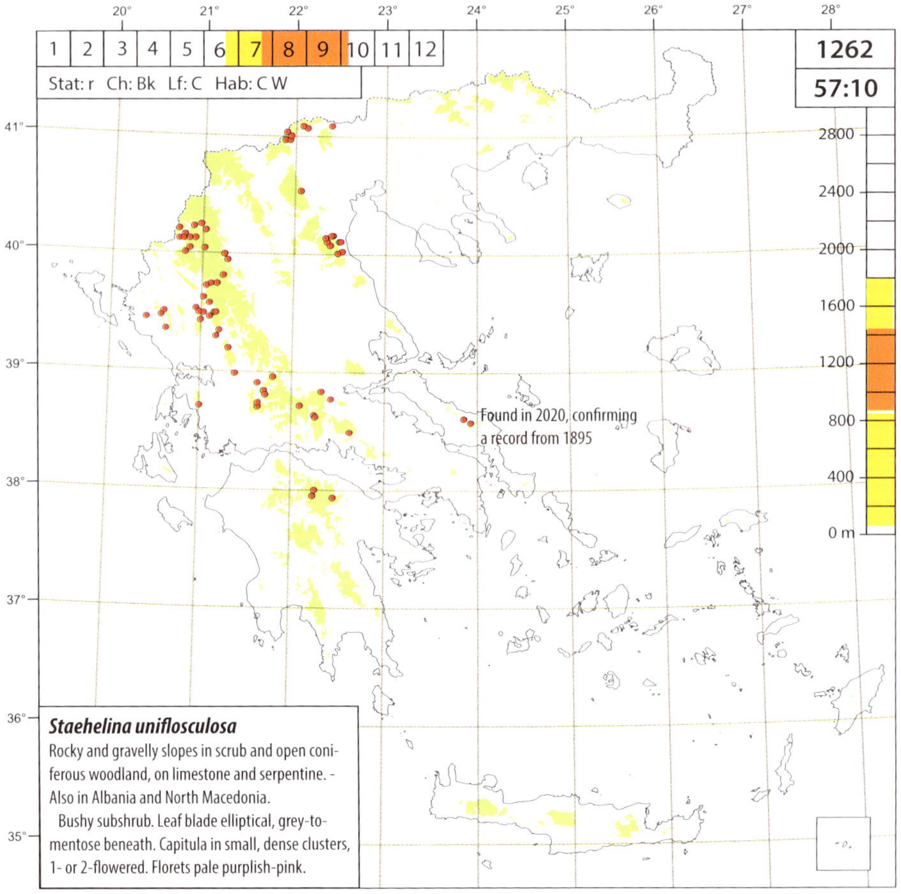

Staehelina uniflosculosa

Rocky and gravelly slopes in scrub and open coniferous woodland, on limestone and serpentine. - Also in Albania and North Macedonia.

Bushy subshrub. Leaf blade elliptical, grey-tomentose beneath. Capitula in small, dense clusters, 1- or 2-flowered. Florets pale purplish-pink.

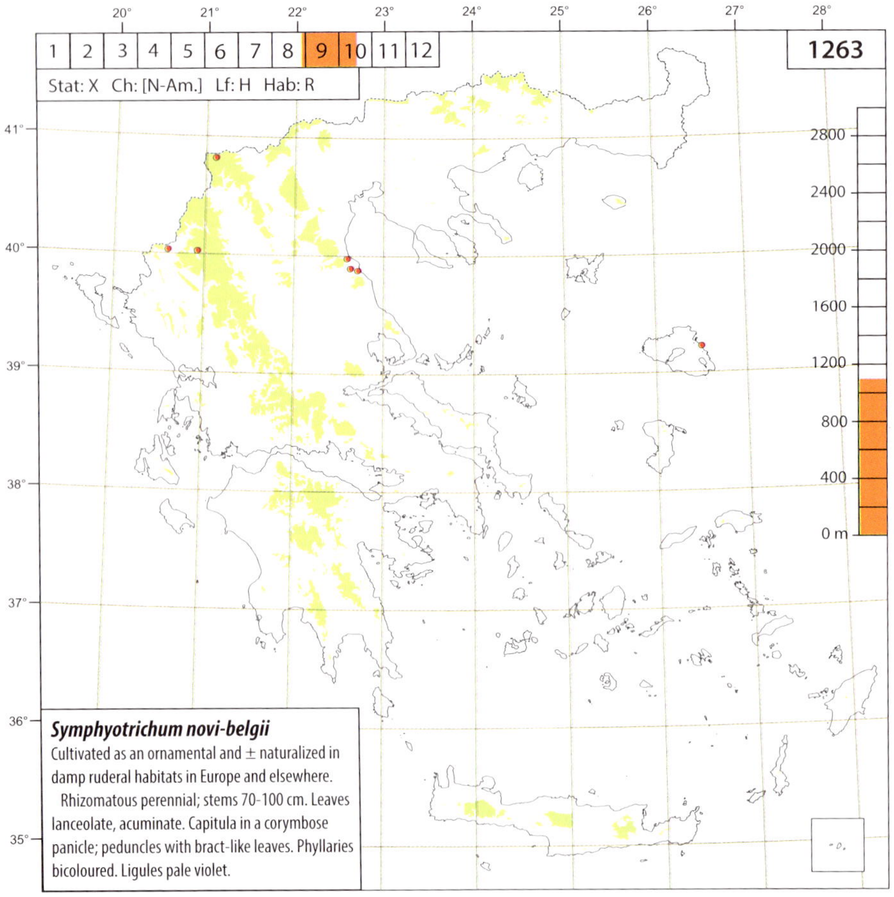

Symphyotrichum novi-belgii

Cultivated as an ornamental and ± naturalized in damp ruderal habitats in Europe and elsewhere.

Rhizomatous perennial; stems 70-100 cm. Leaves lanceolate, acuminate. Capitula in a corymbose panicle; peduncles with bract-like leaves. Phyllaries bicoloured. Ligules pale violet.

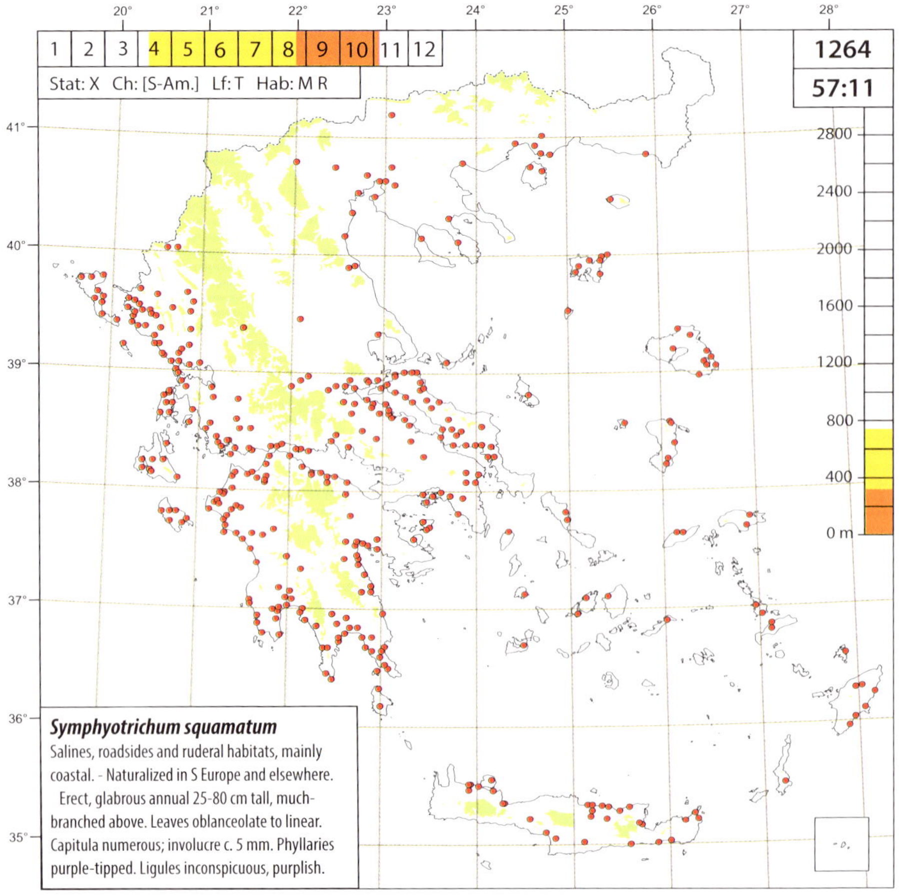

Symphyotrichum squamatum

Salines, roadsides and ruderal habitats, mainly coastal. - Naturalized in S Europe and elsewhere.

Erect, glabrous annual 25-80 cm tall, much-branched above. Leaves oblanceolate to linear. Capitula numerous; involucre c. 5 mm. Phyllaries purple-tipped. Ligules inconspicuous, purplish.

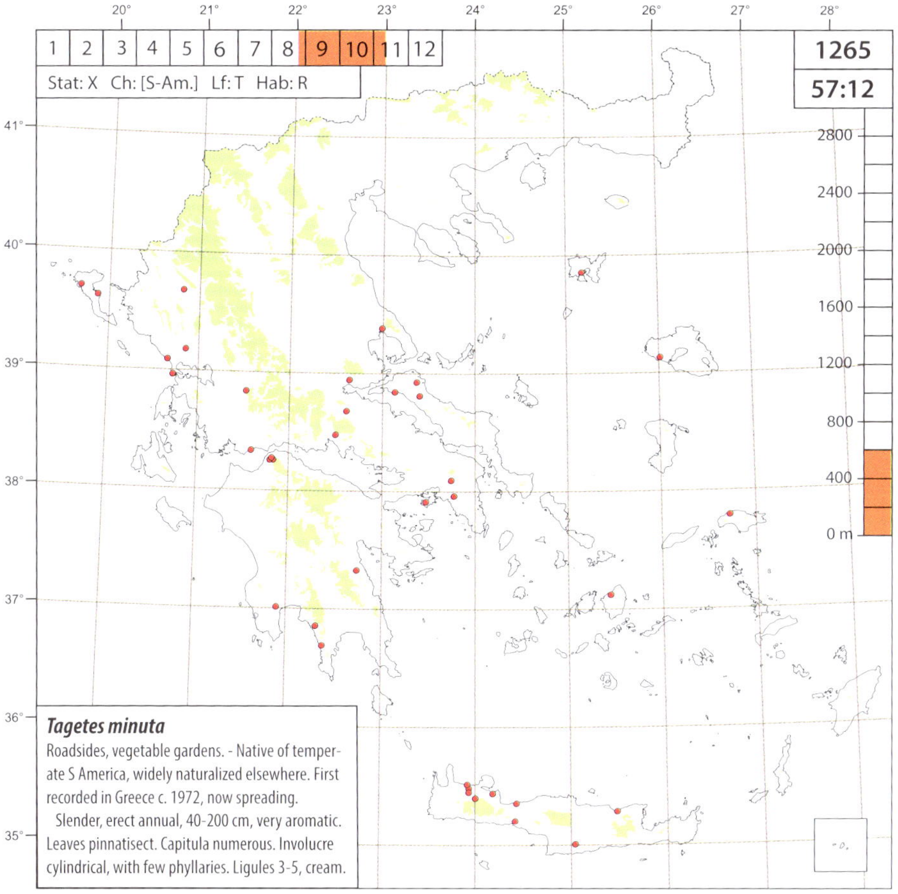

Tagetes minuta

Roadsides, vegetable gardens. - Native of temperate S America, widely naturalized elsewhere. First recorded in Greece c. 1972, now spreading.

Slender, erect annual, 40-200 cm, very aromatic. Leaves pinnatisect. Capitula numerous. Involucre cylindrical, with few phyllaries. Ligules 3-5, cream.

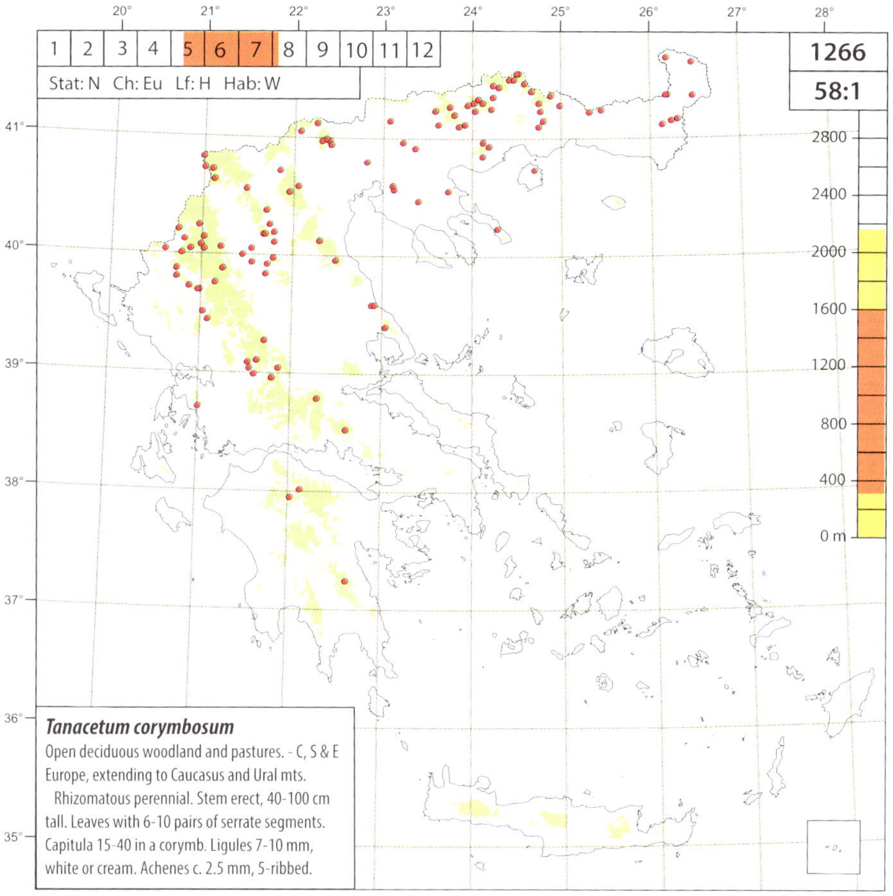

Tanacetum corymbosum

Open deciduous woodland and pastures. - C, S & E Europe, extending to Caucasus and Ural mts.

Rhizomatous perennial. Stem erect, 40-100 cm tall. Leaves with 6-10 pairs of serrate segments. Capitula 15-40 in a corymb. Ligules 7-10 mm, white or cream. Achenes c. 2.5 mm, 5-ribbed.

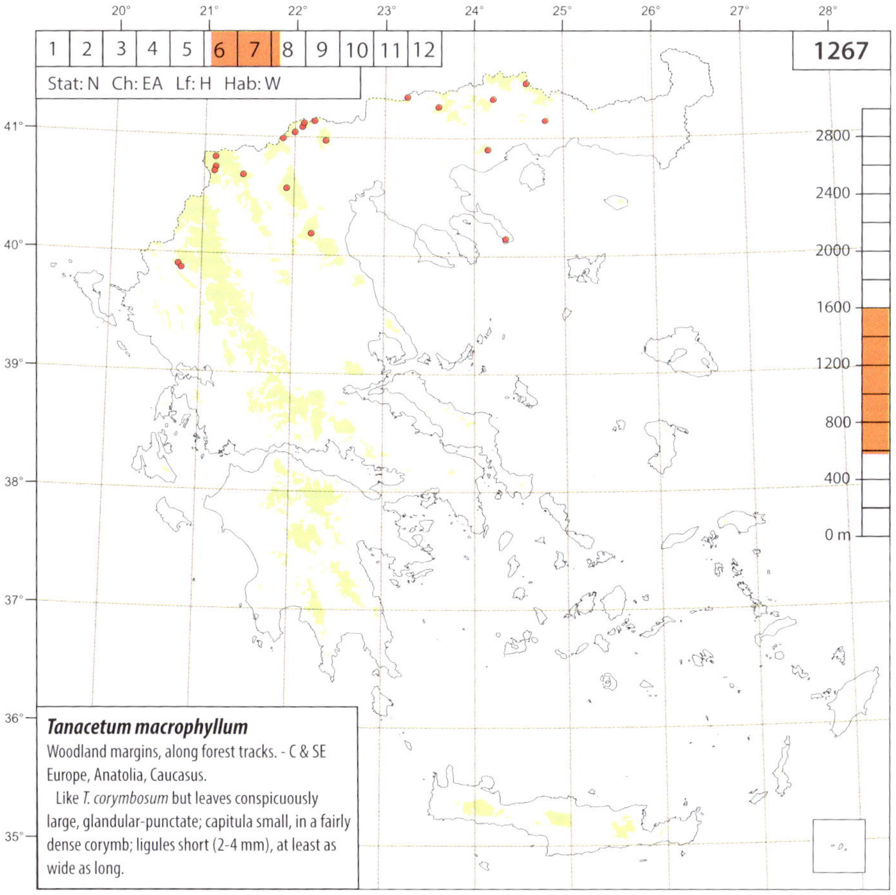

Tanacetum macrophyllum

Woodland margins, along forest tracks. - C & SE Europe, Anatolia, Caucasus.

Like *T. corymbosum* but leaves conspicuously large, glandular-punctate; capitula small, in a fairly dense corymb; ligules short (2-4 mm), at least as wide as long.

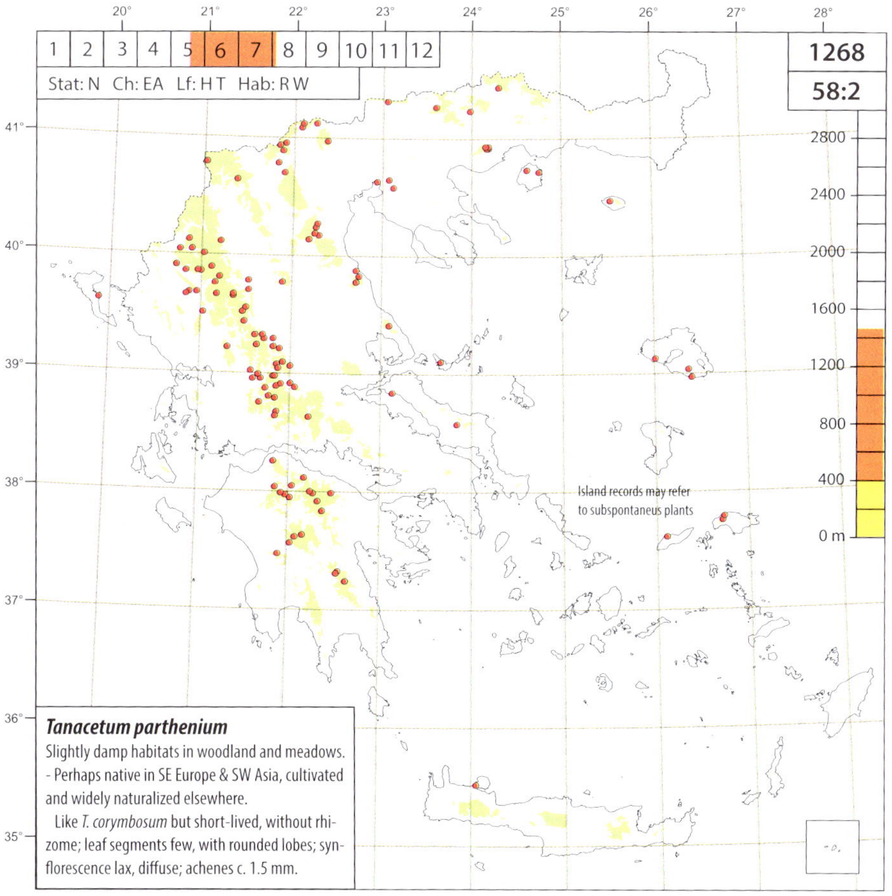

Tanacetum parthenium

Slightly damp habitats in woodland and meadows. - Perhaps native in SE Europe & SW Asia, cultivated and widely naturalized elsewhere.

Like *T. corymbosum* but short-lived, without rhizome; leaf segments few, with rounded lobes; synflorescence lax, diffuse; achenes c. 1.5 mm.

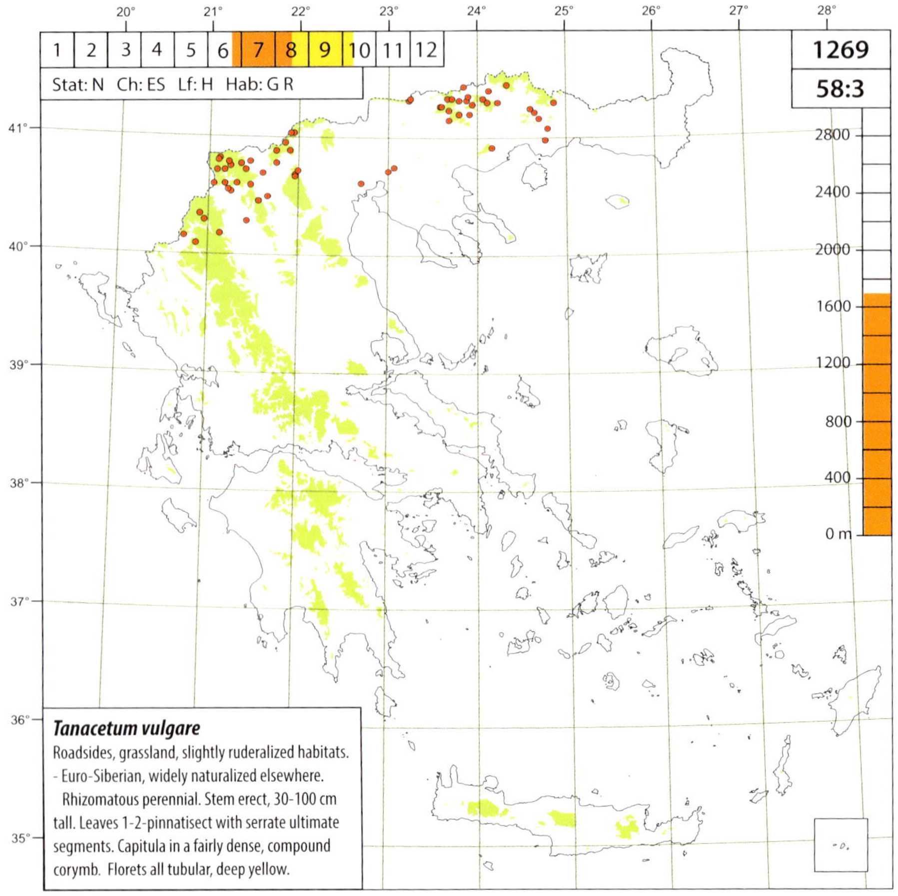

Tanacetum vulgare
Roadsides, grassland, slightly ruderalized habitats. - Euro-Siberian, widely naturalized elsewhere.
Rhizomatous perennial. Stem erect, 30-100 cm tall. Leaves 1-2-pinnatisect with serrate ultimate segments. Capitula in a fairly dense, compound corymb. Florets all tubular, deep yellow.

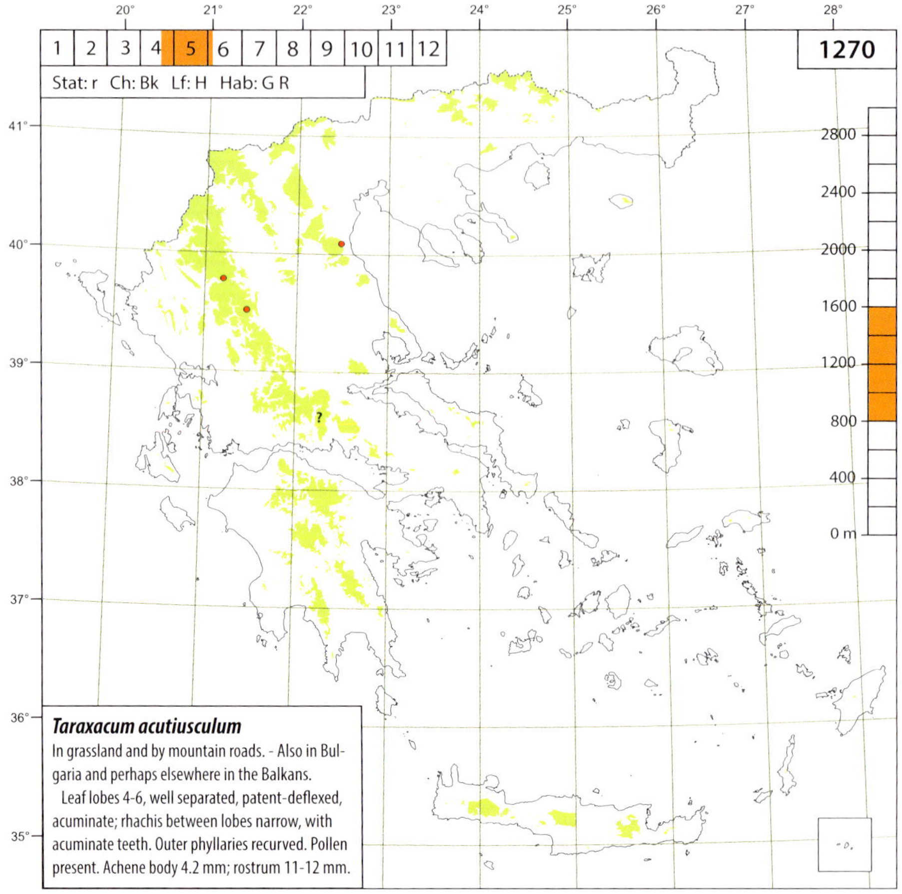

Taraxacum acutiusculum
In grassland and by mountain roads. - Also in Bulgaria and perhaps elsewhere in the Balkans.
Leaf lobes 4-6, well separated, patent-deflexed, acuminate; rhachis between lobes narrow, with acuminate teeth. Outer phyllaries recurved. Pollen present. Achene body 4.2 mm; rostrum 11-12 mm.

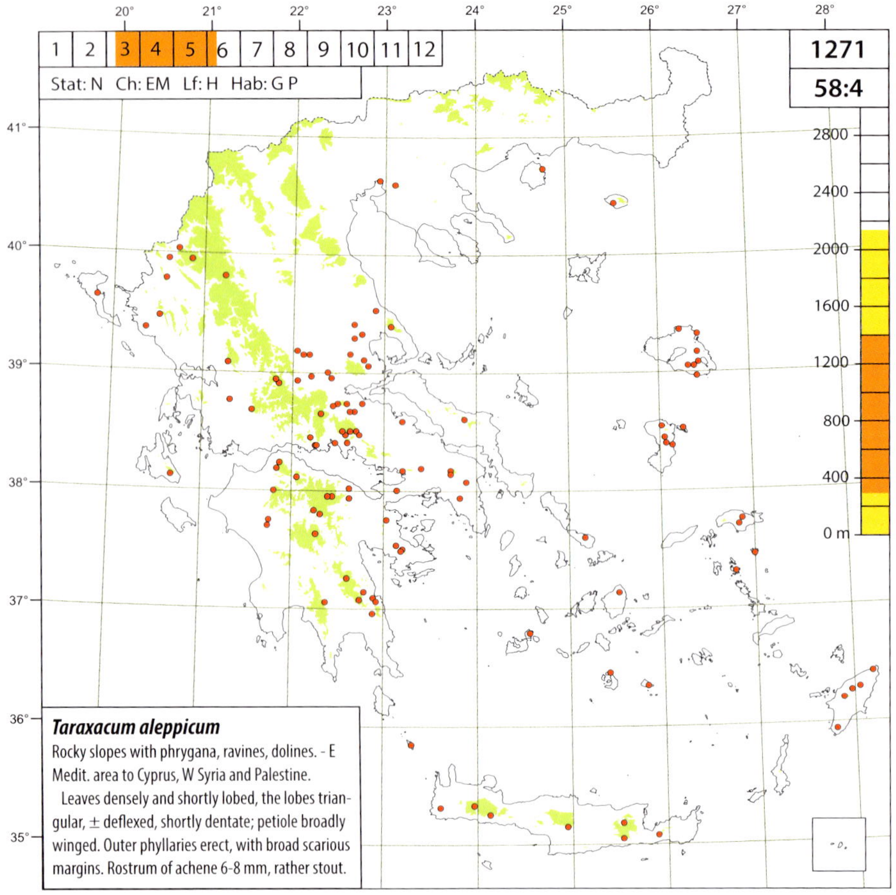

Taraxacum aleppicum
Rocky slopes with phrygana, ravines, dolines. - E Medit. area to Cyprus, W Syria and Palestine.
Leaves densely and shortly lobed, the lobes triangular, ± deflexed, shortly dentate; petiole broadly winged. Outer phyllaries erect, with broad scarious margins. Rostrum of achene 6-8 mm, rather stout.

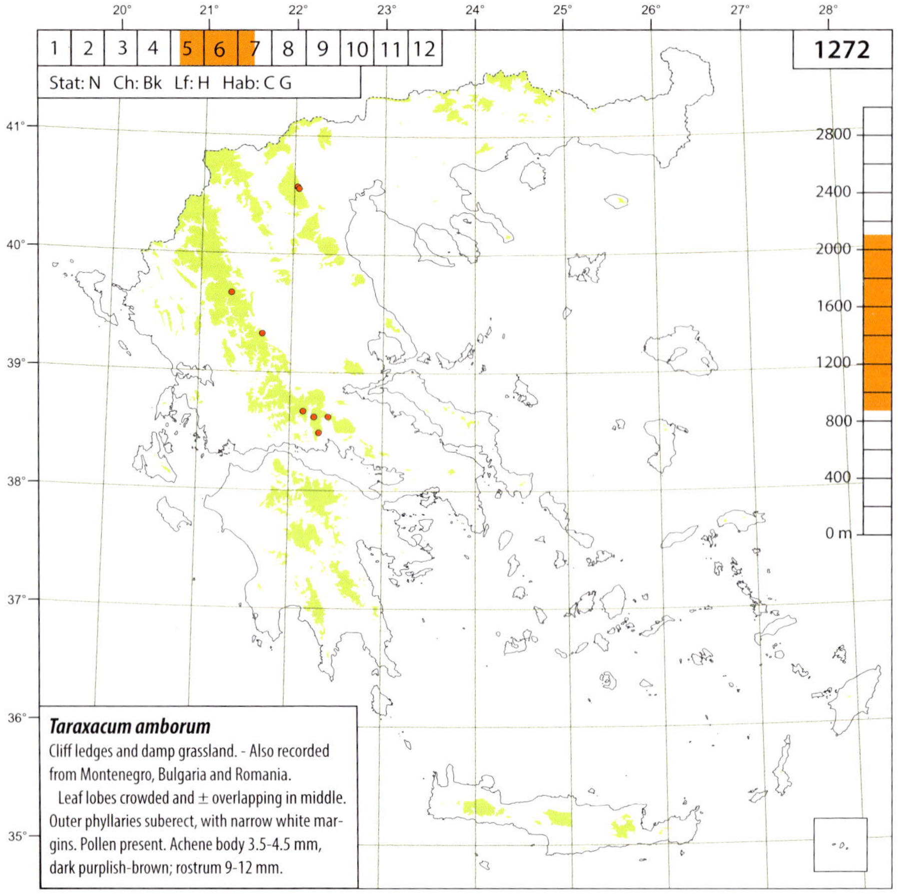

Taraxacum amborum
Cliff ledges and damp grassland. - Also recorded from Montenegro, Bulgaria and Romania.
Leaf lobes crowded and ± overlapping in middle. Outer phyllaries suberect, with narrow white margins. Pollen present. Achene body 3.5-4.5 mm, dark purplish-brown; rostrum 9-12 mm.

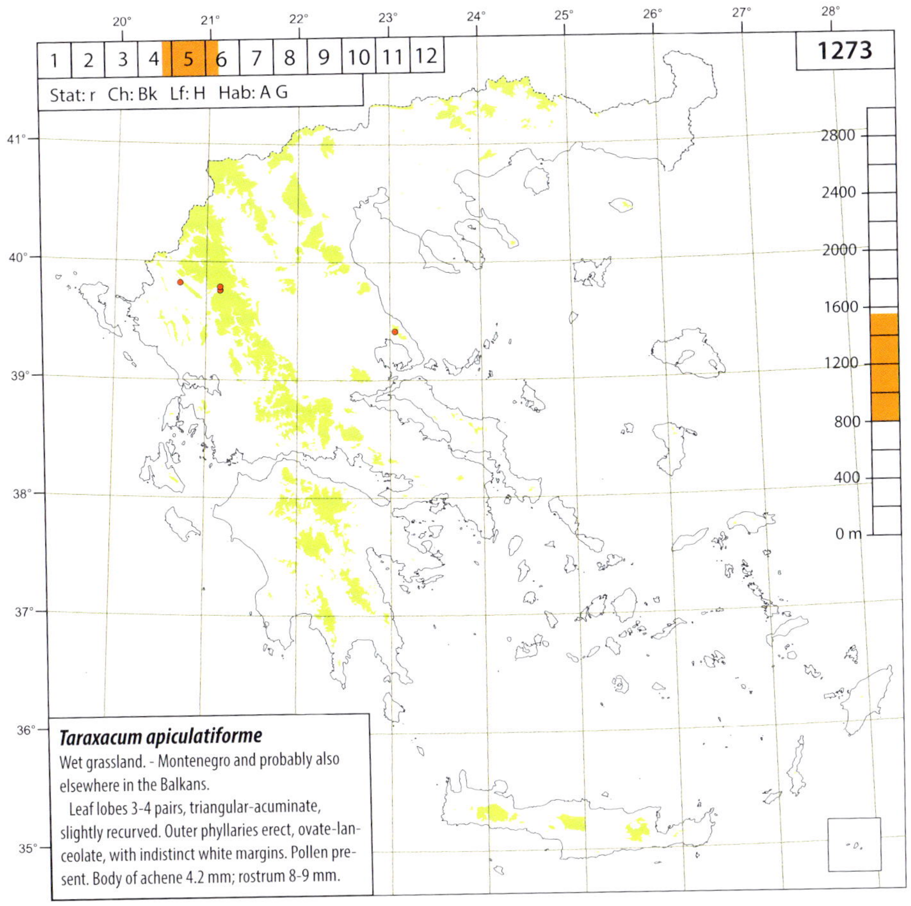

Taraxacum apiculatiforme

Wet grassland. - Montenegro and probably also elsewhere in the Balkans.

Leaf lobes 3-4 pairs, triangular-acuminate, slightly recurved. Outer phyllaries erect, ovate-lanceolate, with indistinct white margins. Pollen present. Body of achene 4.2 mm; rostrum 8-9 mm.

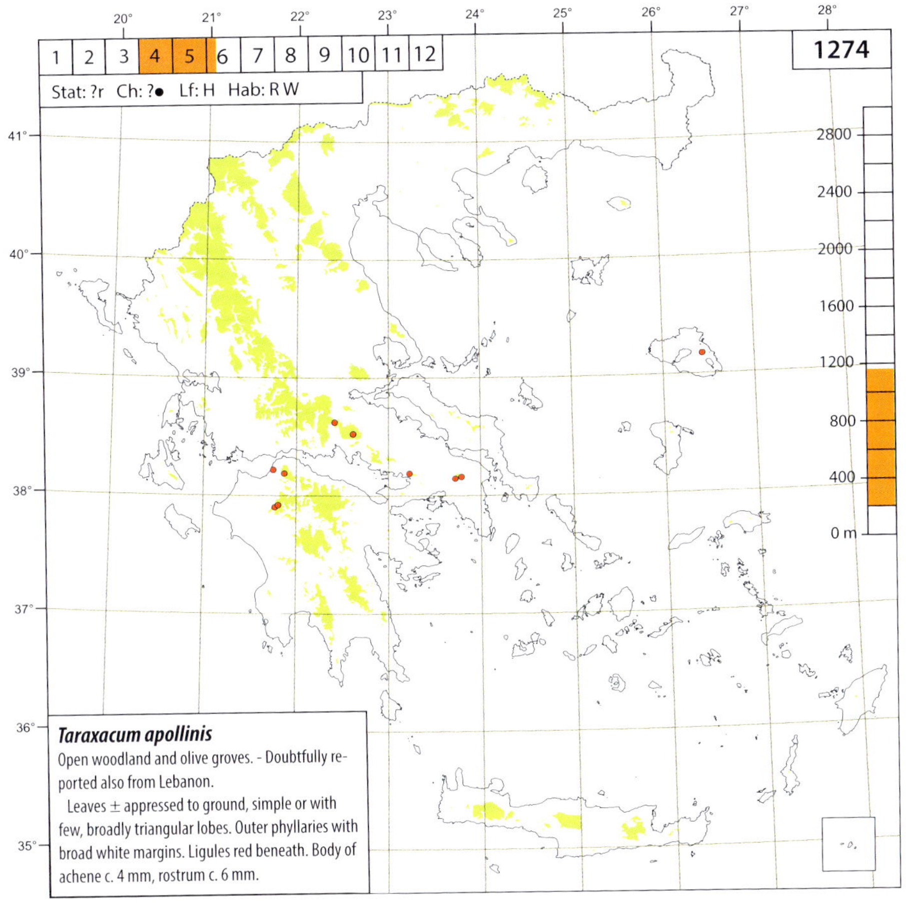

Taraxacum apollinis

Open woodland and olive groves. - Doubtfully reported also from Lebanon.

Leaves ± appressed to ground, simple or with few, broadly triangular lobes. Outer phyllaries with broad white margins. Ligules red beneath. Body of achene c. 4 mm, rostrum c. 6 mm.

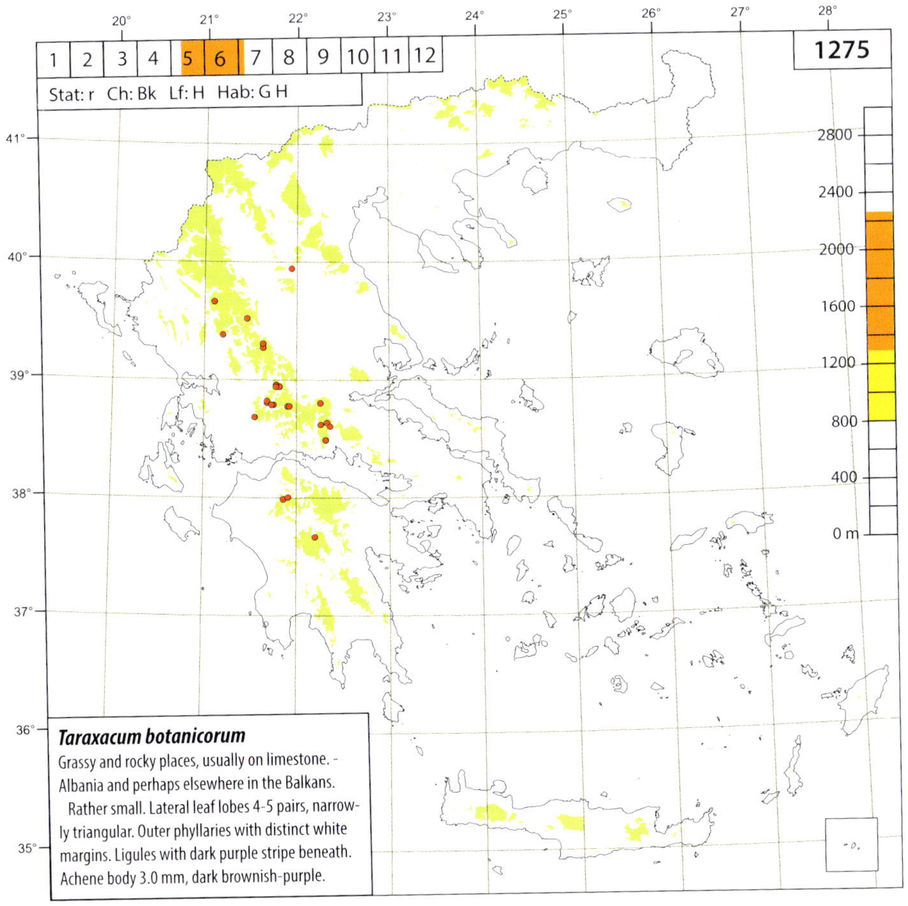

Taraxacum botanicorum

Grassy and rocky places, usually on limestone. - Albania and perhaps elsewhere in the Balkans.

Rather small. Lateral leaf lobes 4-5 pairs, narrowly triangular. Outer phyllaries with distinct white margins. Ligules with dark purple stripe beneath. Achene body 3.0 mm, dark brownish-purple.

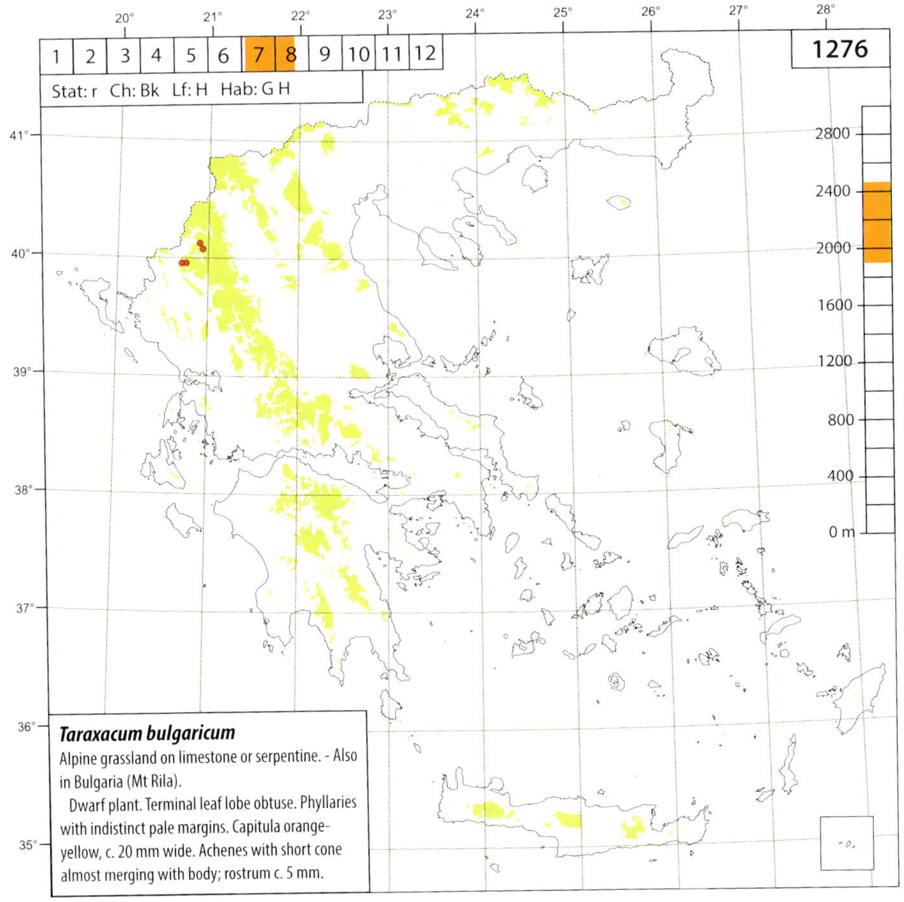

Taraxacum bulgaricum

Alpine grassland on limestone or serpentine. - Also in Bulgaria (Mt Rila).

Dwarf plant. Terminal leaf lobe obtuse. Phyllaries with indistinct pale margins. Capitula orange-yellow, c. 20 mm wide. Achenes with short cone almost merging with body; rostrum c. 5 mm.

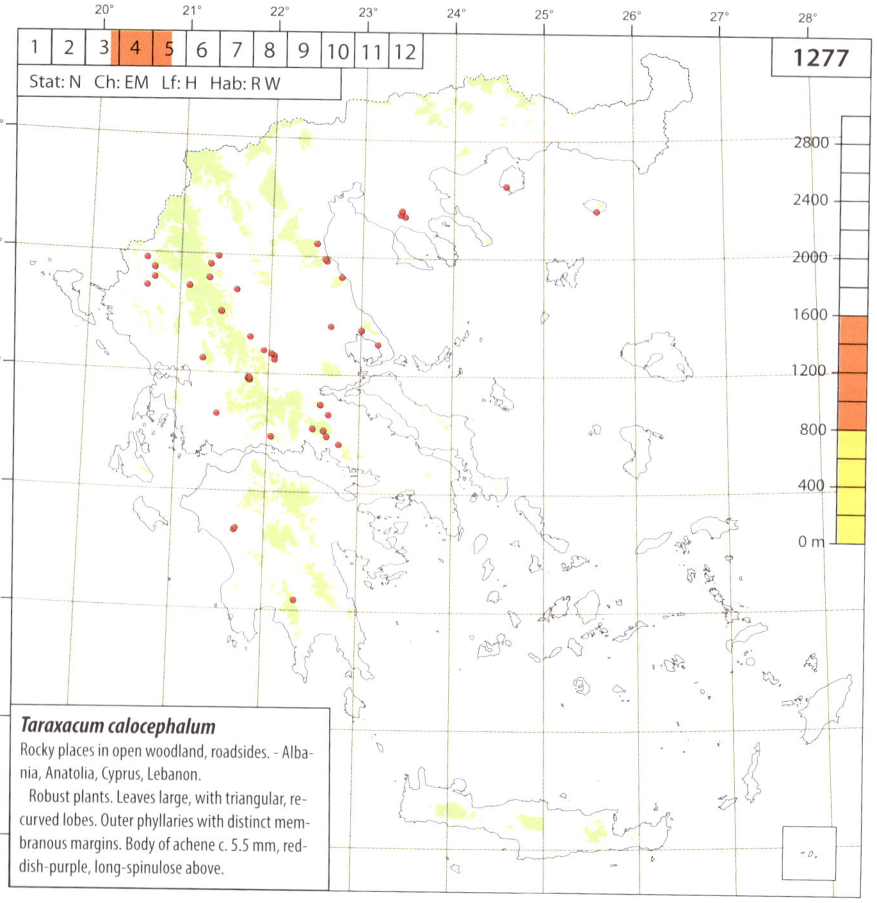

Taraxacum calocephalum

Rocky places in open woodland, roadsides. - Albania, Anatolia, Cyprus, Lebanon.

Robust plants. Leaves large, with triangular, recurved lobes. Outer phyllaries with distinct membranous margins. Body of achene c. 5.5 mm, red-dish-purple, long-spinulose above.

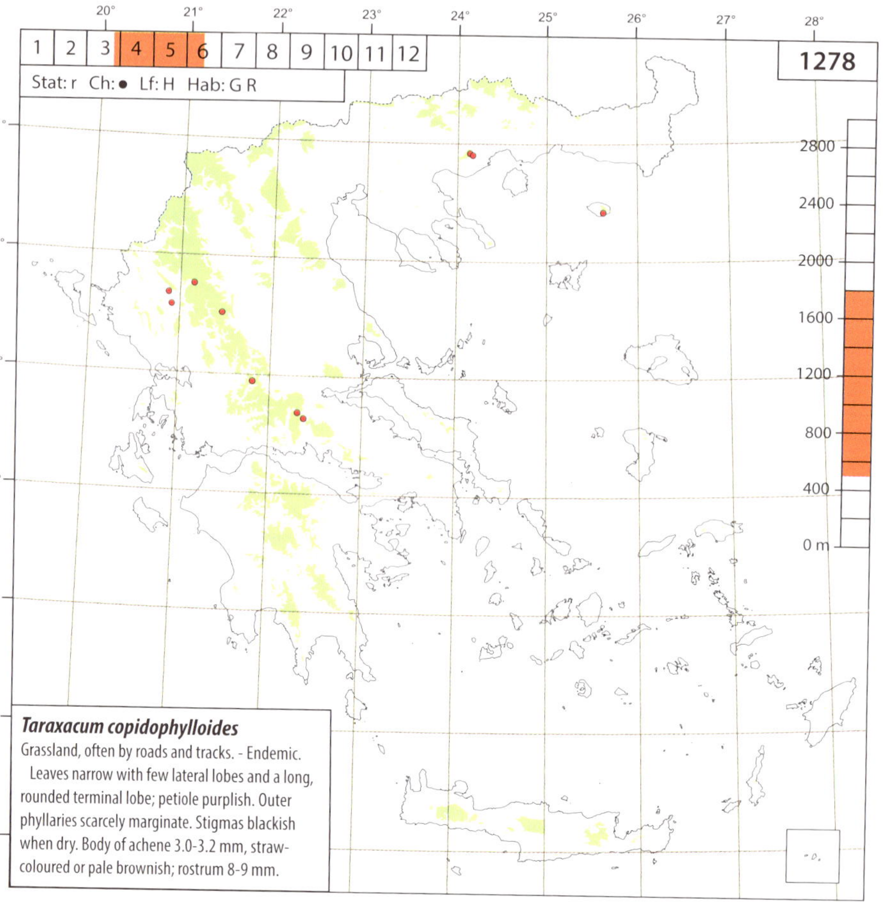

Taraxacum copidophylloides

Grassland, often by roads and tracks. - Endemic.

Leaves narrow with few lateral lobes and a long, rounded terminal lobe; petiole purplish. Outer phyllaries scarcely marginate. Stigmas blackish when dry. Body of achene 3.0-3.2 mm, straw-coloured or pale brownish; rostrum 8-9 mm.

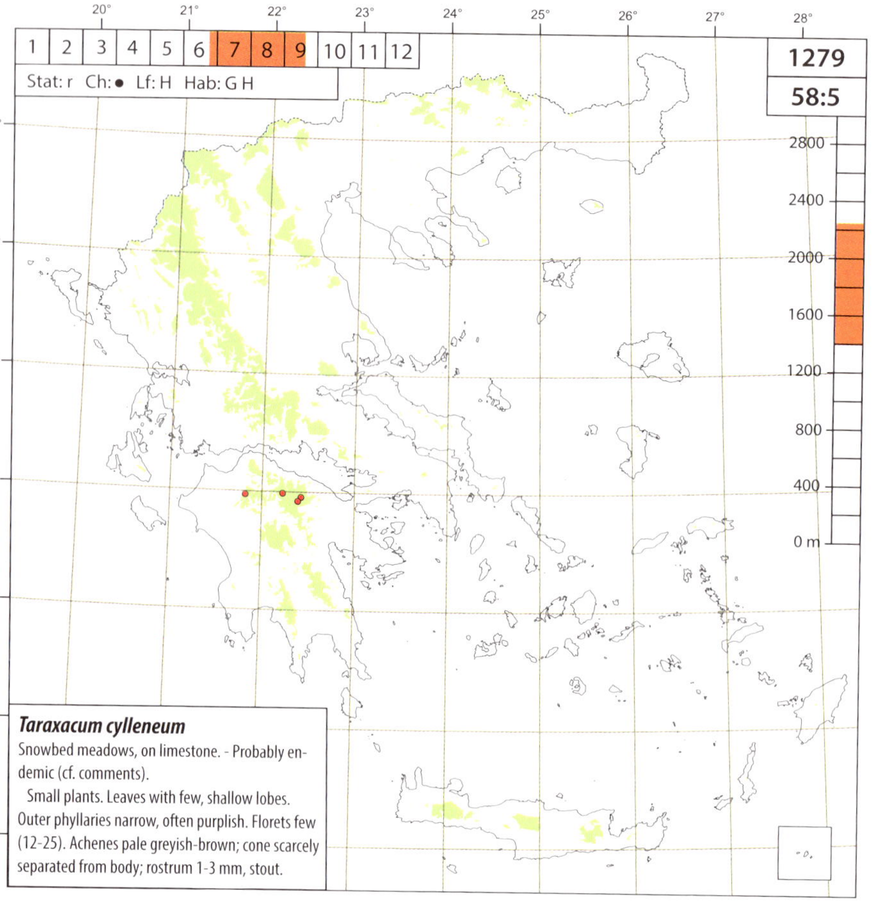

Taraxacum cylleneum

Snowbed meadows, on limestone. - Probably endemic (cf. comments).

Small plants. Leaves with few, shallow lobes. Outer phyllaries narrow, often purplish. Florets few (12-25). Achenes pale greyish-brown; cone scarcely separated from body; rostrum 1-3 mm, stout.

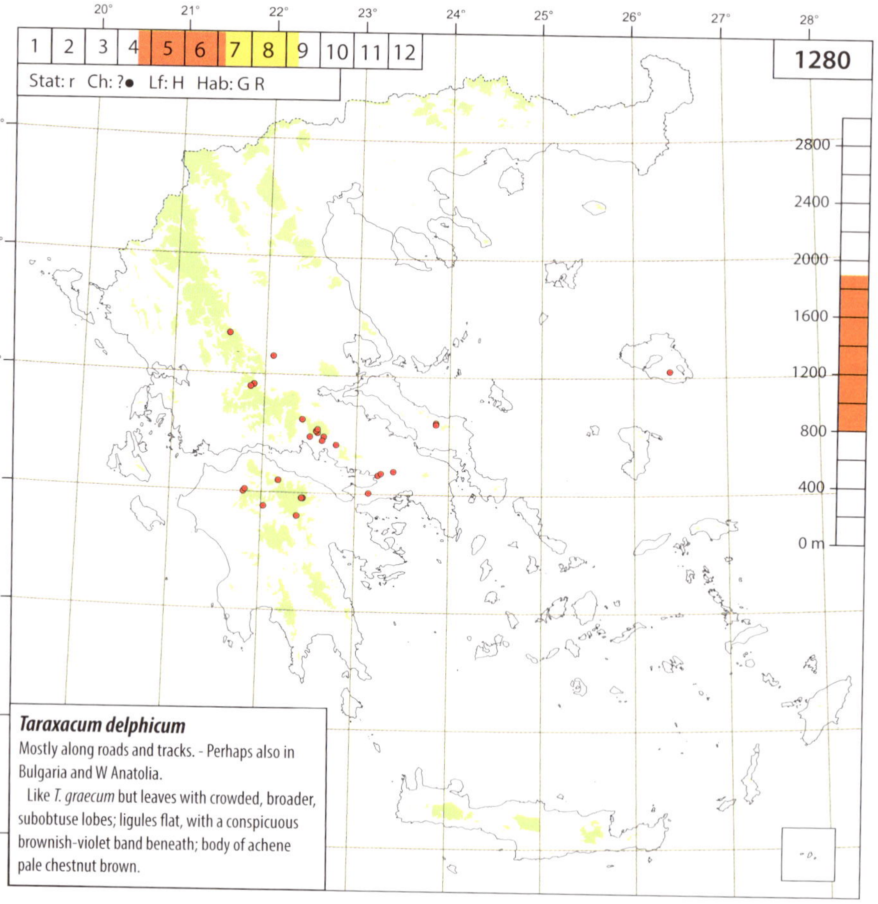

Taraxacum delphicum

Mostly along roads and tracks. - Perhaps also in Bulgaria and W Anatolia.

Like *T. graecum* but leaves with crowded, broader, subobtuse lobes; ligules flat, with a conspicuous brownish-violet band beneath; body of achene pale chestnut brown.

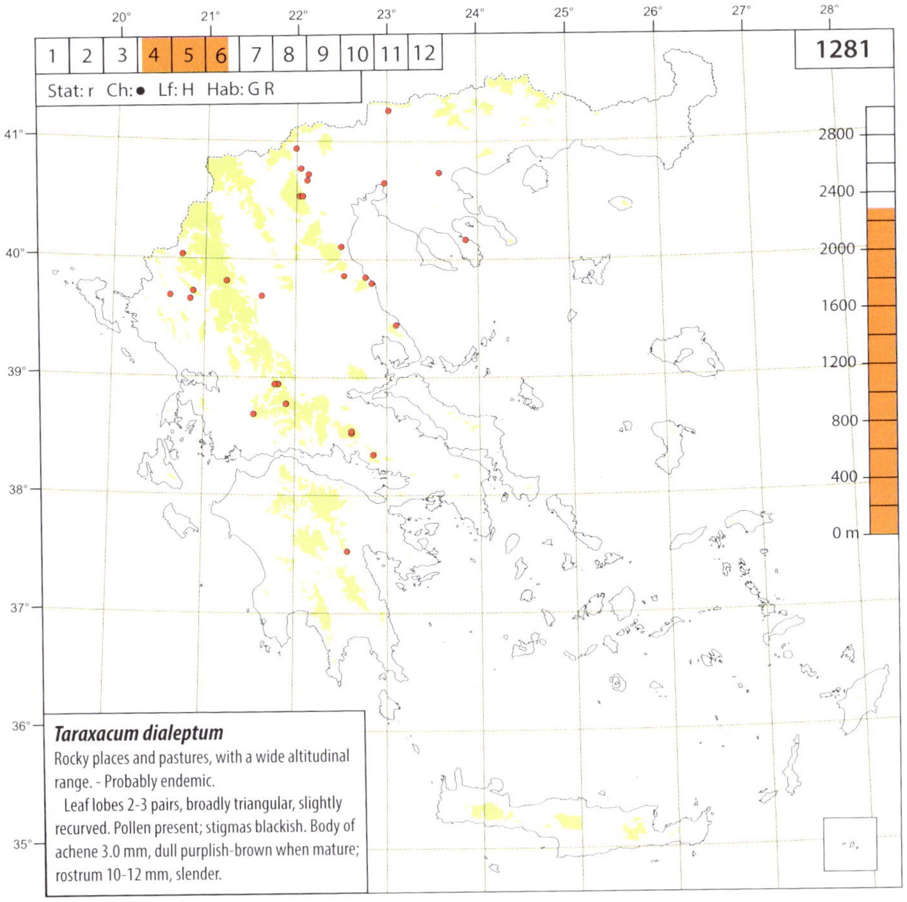

Taraxacum dialeptum
Rocky places and pastures, with a wide altitudinal range. - Probably endemic.
Leaf lobes 2-3 pairs, broadly triangular, slightly recurved. Pollen present; stigmas blackish. Body of achene 3.0 mm, dull purplish-brown when mature; rostrum 10-12 mm, slender.

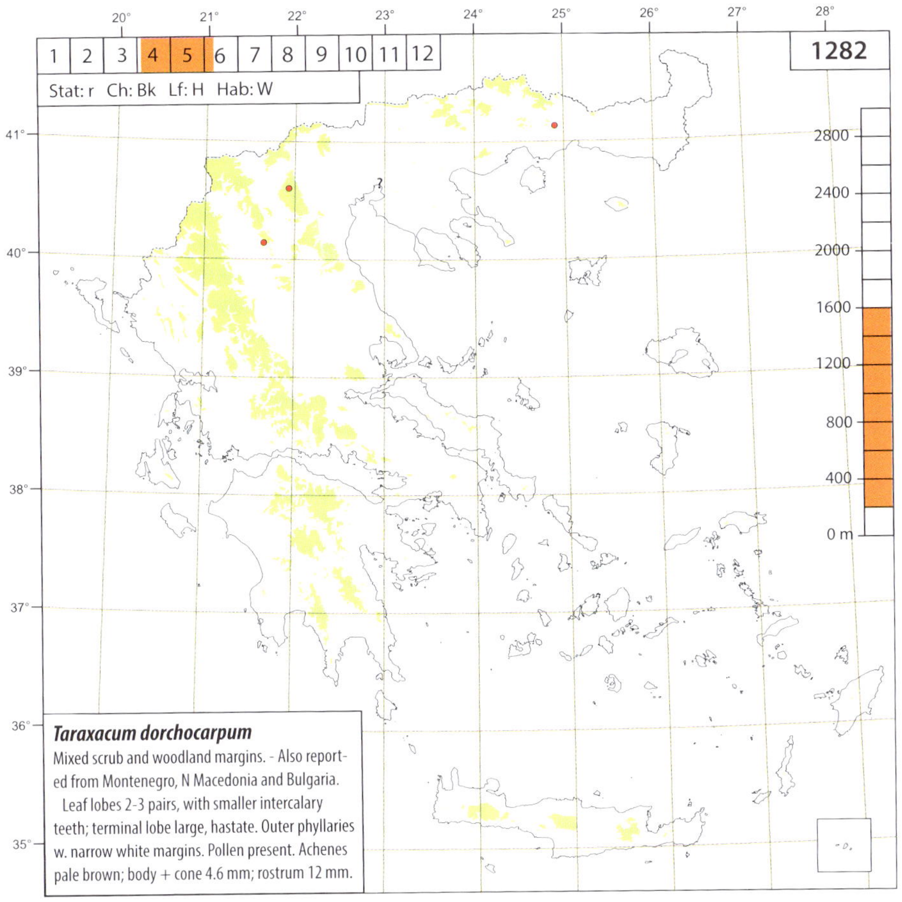

Taraxacum dorchocarpum
Mixed scrub and woodland margins. - Also reported from Montenegro, N Macedonia and Bulgaria.
Leaf lobes 2-3 pairs, with smaller intercalary teeth; terminal lobe large, hastate. Outer phyllaries w. narrow white margins. Pollen present. Achenes pale brown; body + cone 4.6 mm; rostrum 12 mm.

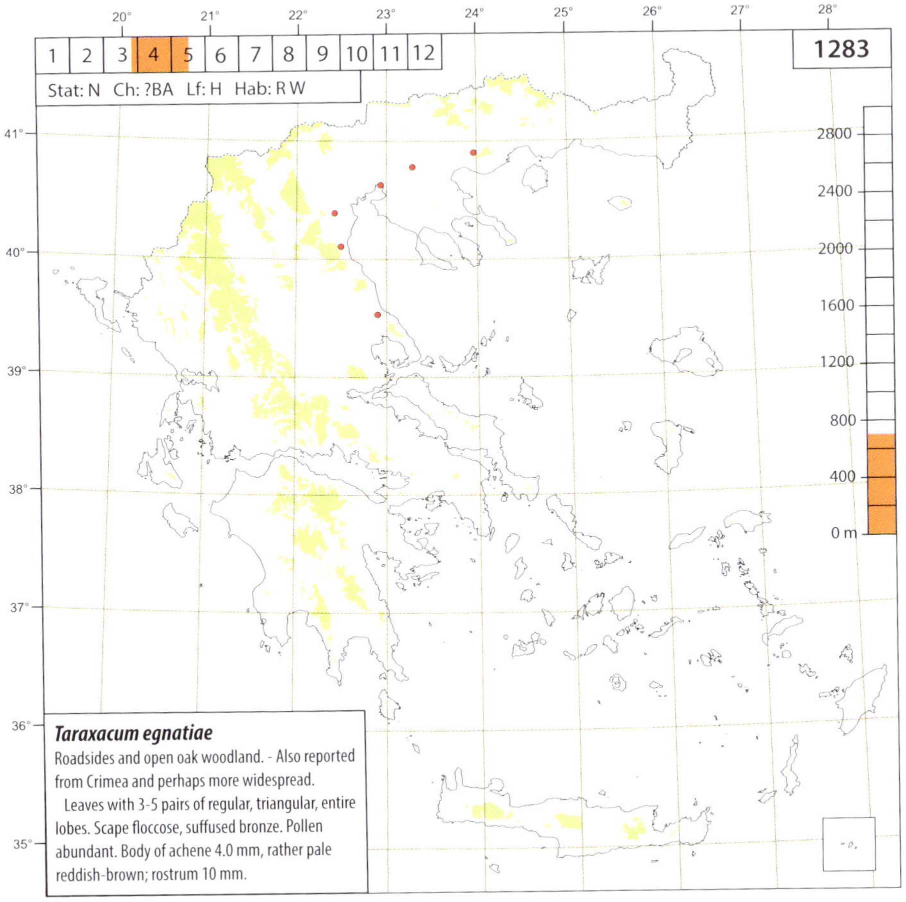

Taraxacum egnatiae
Roadsides and open oak woodland. - Also reported from Crimea and perhaps more widespread.
Leaves with 3-5 pairs of regular, triangular, entire lobes. Scape floccose, suffused bronze. Pollen abundant. Body of achene 4.0 mm, rather pale reddish-brown; rostrum 10 mm.

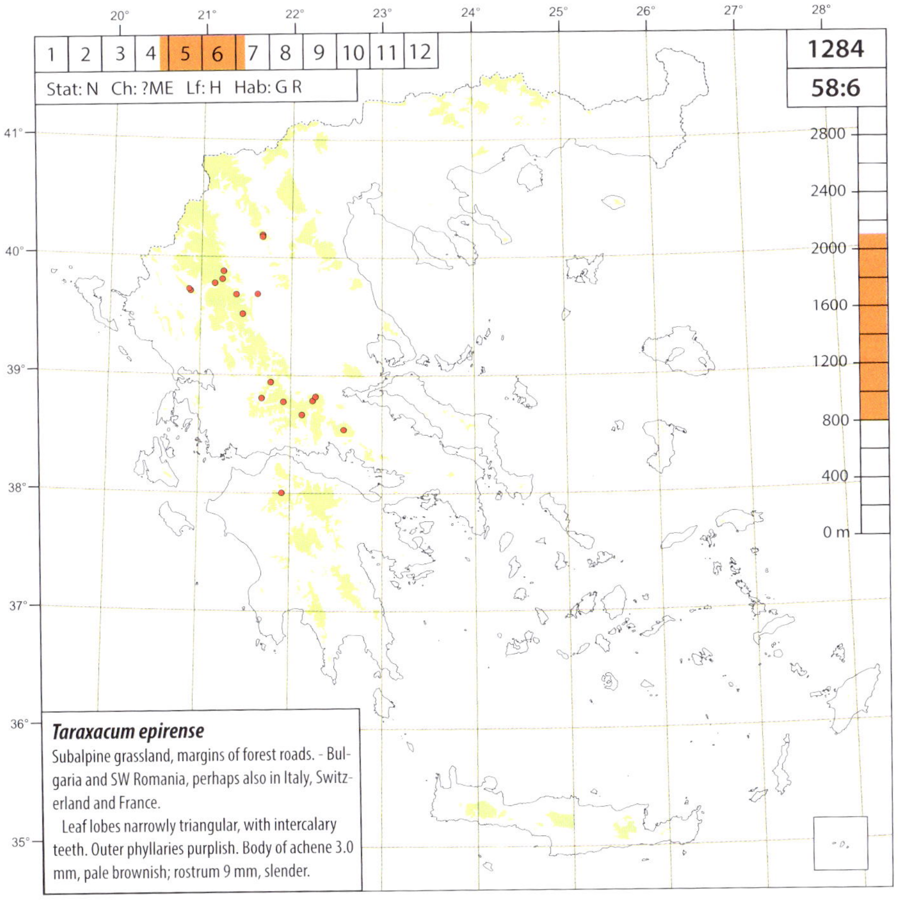

Taraxacum epirense
Subalpine grassland, margins of forest roads. - Bulgaria and SW Romania, perhaps also in Italy, Switzerland and France.
Leaf lobes narrowly triangular, with intercalary teeth. Outer phyllaries purplish. Body of achene 3.0 mm, pale brownish; rostrum 9 mm, slender.

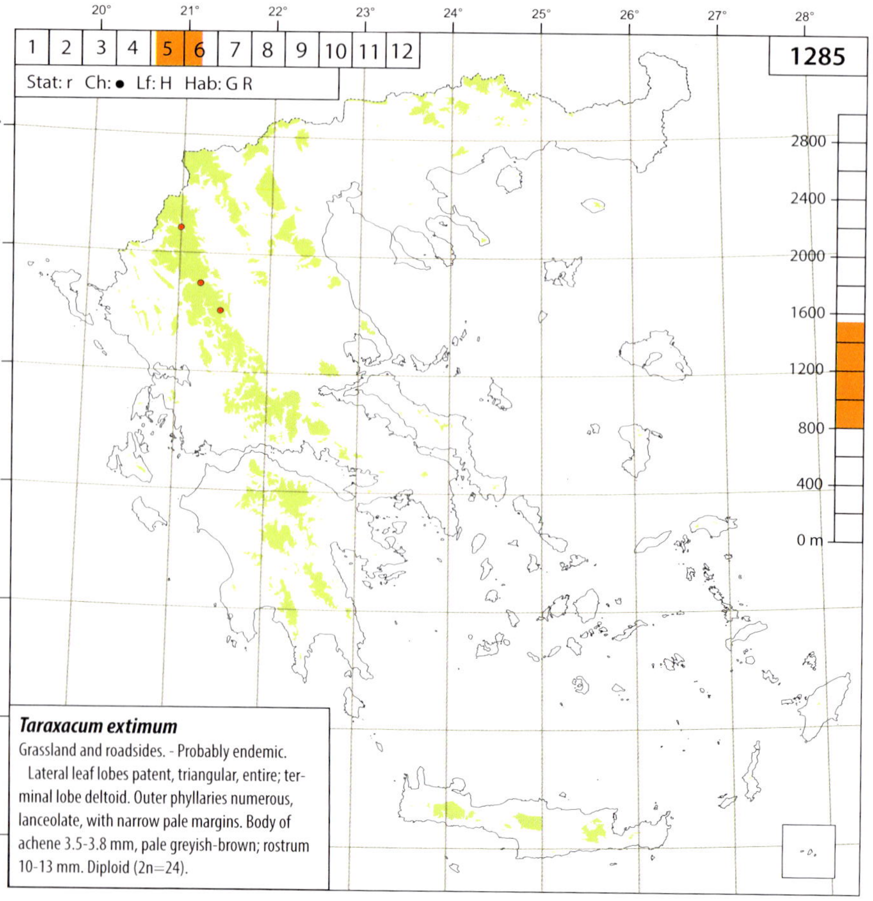

Taraxacum extimum

Grassland and roadsides. - Probably endemic.

Lateral leaf lobes patent, triangular, entire; terminal lobe deltoid. Outer phyllaries numerous, lanceolate, with narrow pale margins. Body of achene 3.5-3.8 mm, pale greyish-brown; rostrum 10-13 mm. Diploid (2n=24).

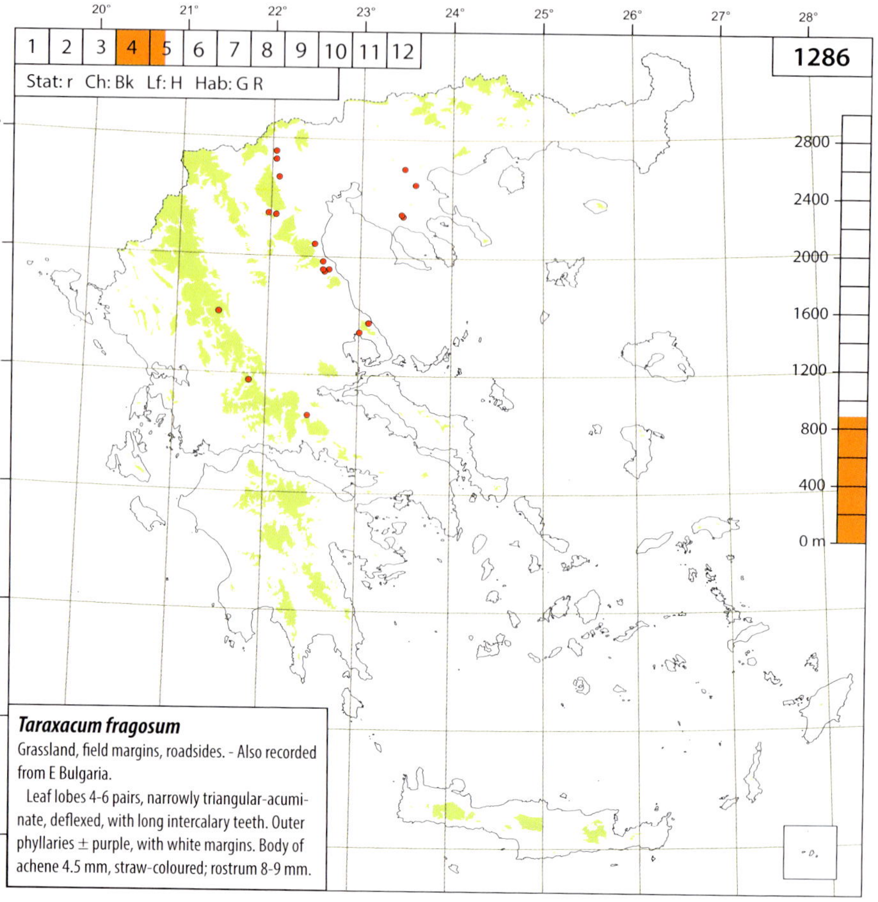

Taraxacum fragosum

Grassland, field margins, roadsides. - Also recorded from E Bulgaria.

Leaf lobes 4-6 pairs, narrowly triangular-acuminate, deflexed, with long intercalary teeth. Outer phyllaries ± purple, with white margins. Body of achene 4.5 mm, straw-coloured; rostrum 8-9 mm.

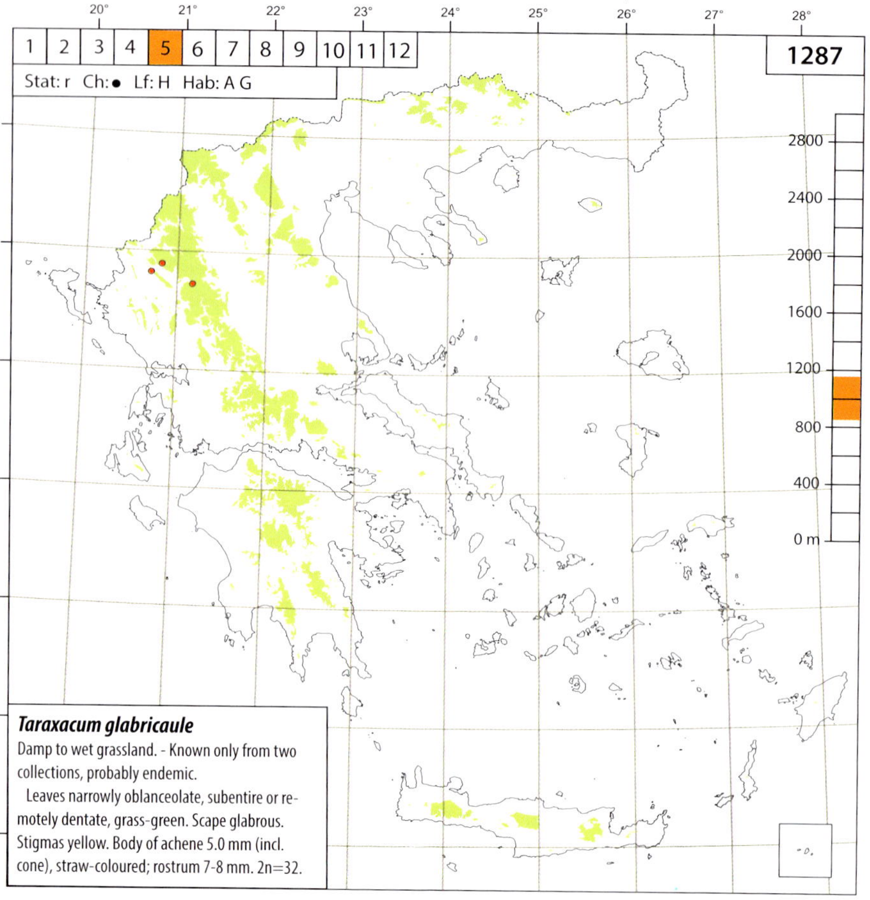

Taraxacum glabricaule

Damp to wet grassland. - Known only from two collections, probably endemic.

Leaves narrowly oblanceolate, subentire or remotely dentate, grass-green. Scape glabrous. Stigmas yellow. Body of achene 5.0 mm (incl. cone), straw-coloured; rostrum 7-8 mm. 2n=32.

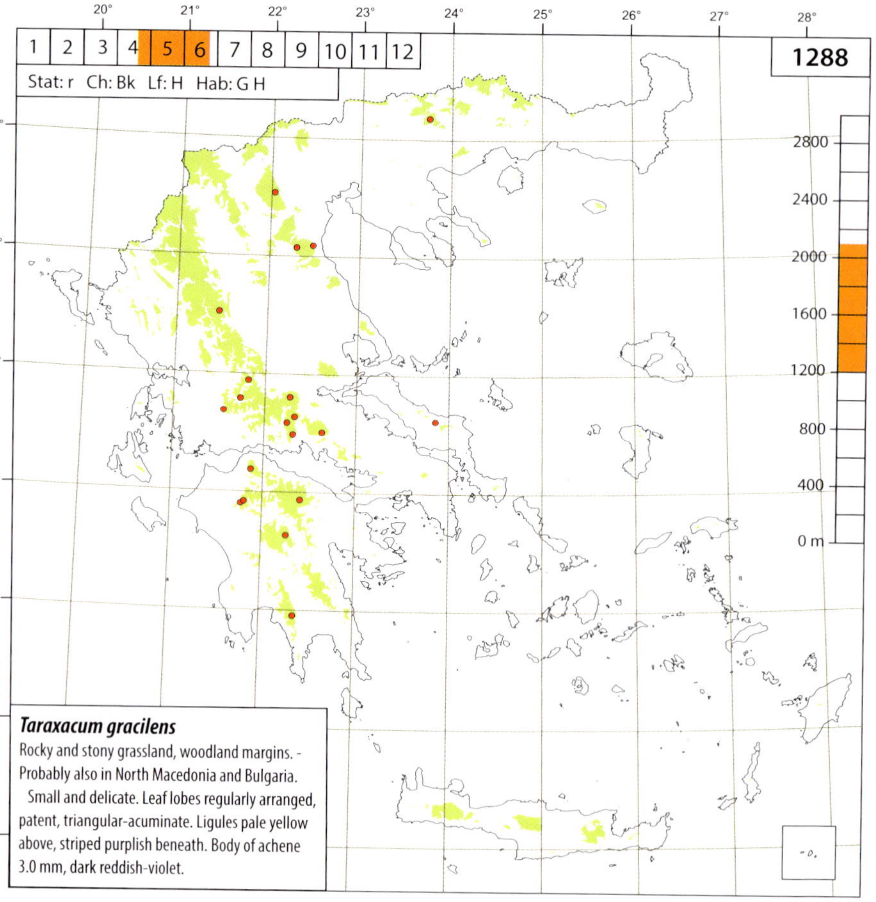

Taraxacum gracilens

Rocky and stony grassland, woodland margins. - Probably also in North Macedonia and Bulgaria.

Small and delicate. Leaf lobes regularly arranged, patent, triangular-acuminate. Ligules pale yellow above, striped purplish beneath. Body of achene 3.0 mm, dark reddish-violet.

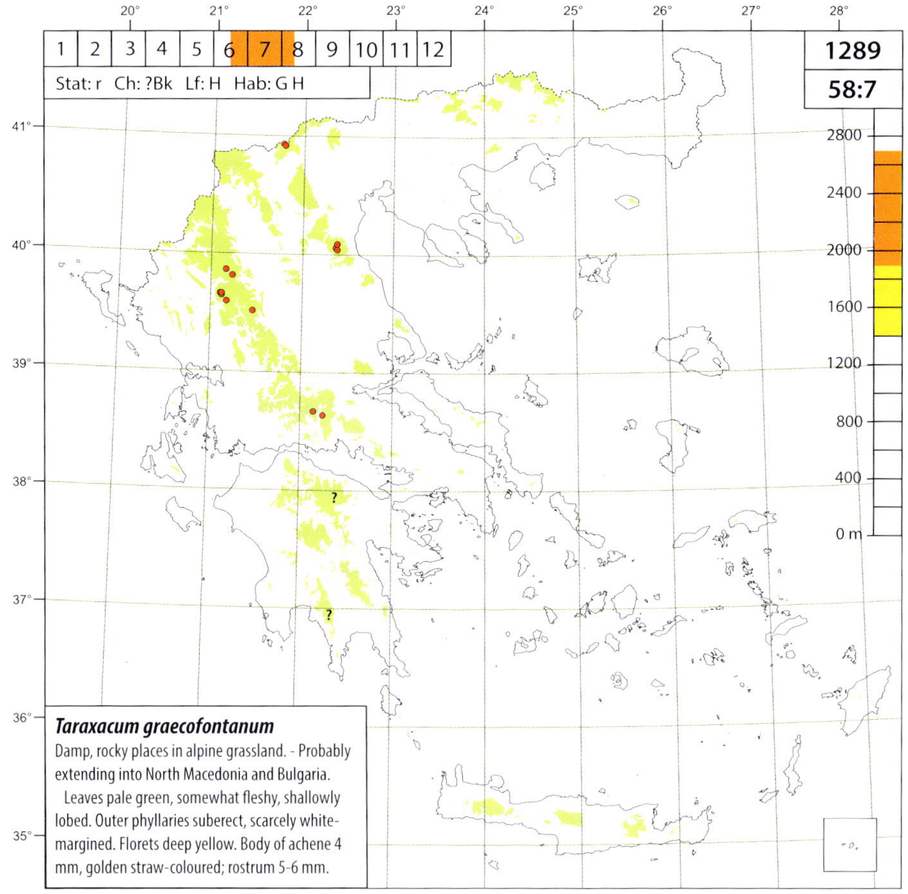

Taraxacum graecofontanum

Damp, rocky places in alpine grassland. - Probably extending into North Macedonia and Bulgaria.

Leaves pale green, somewhat fleshy, shallowly lobed. Outer phyllaries suberect, scarcely white-margined. Florets deep yellow. Body of achene 4 mm, golden straw-coloured; rostrum 5-6 mm.

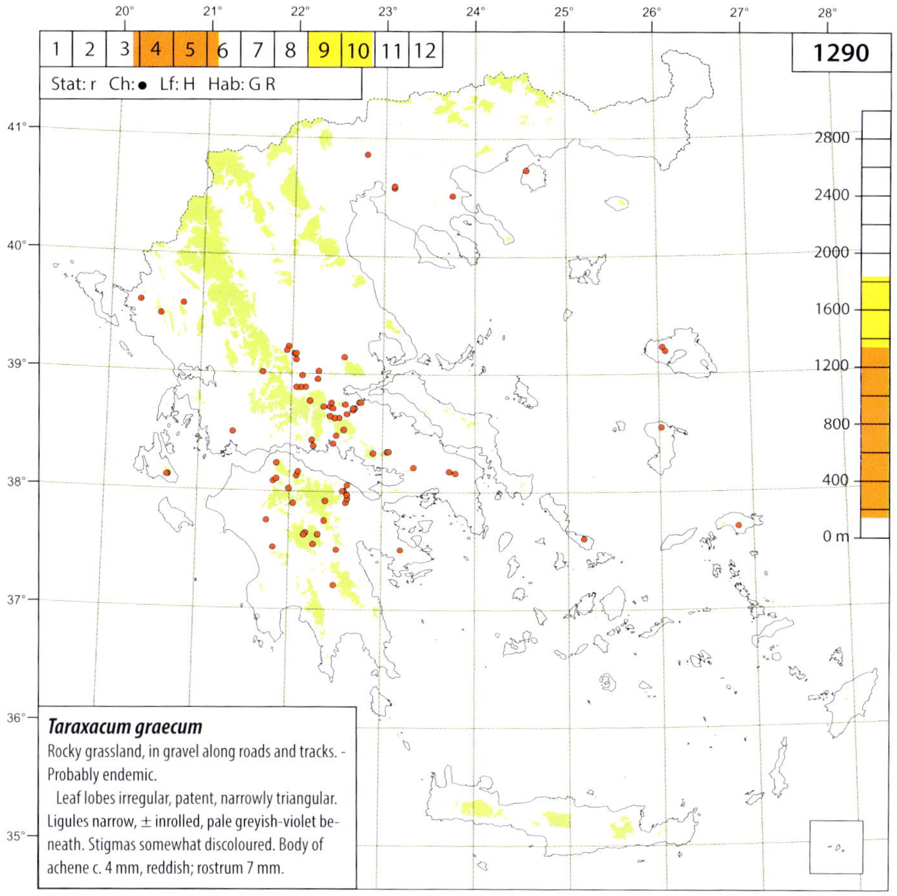

Taraxacum graecum

Rocky grassland, in gravel along roads and tracks. - Probably endemic.

Leaf lobes irregular, patent, narrowly triangular. Ligules narrow, ± inrolled, pale greyish-violet beneath. Stigmas somewhat discoloured. Body of achene c. 4 mm, reddish; rostrum 7 mm.

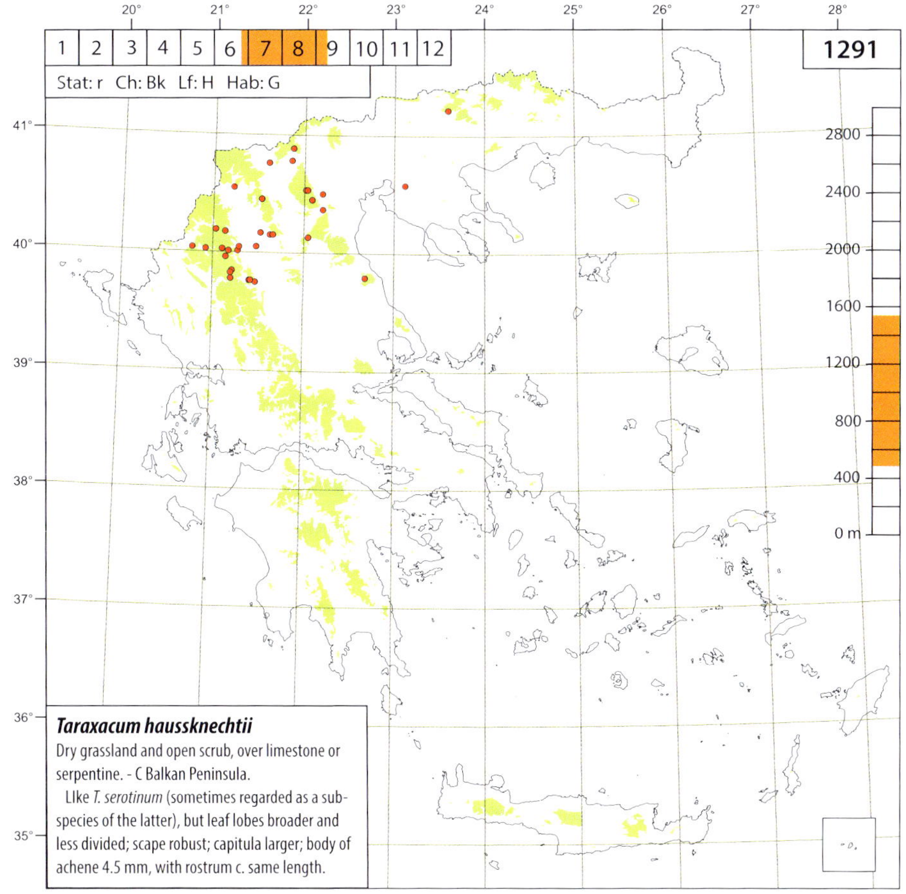

Taraxacum haussknechtii

Dry grassland and open scrub, over limestone or serpentine. - C Balkan Peninsula.

Like *T. serotinum* (sometimes regarded as a subspecies of the latter), but leaf lobes broader and less divided; scape robust; capitula larger; body of achene 4.5 mm, with rostrum c. same length.

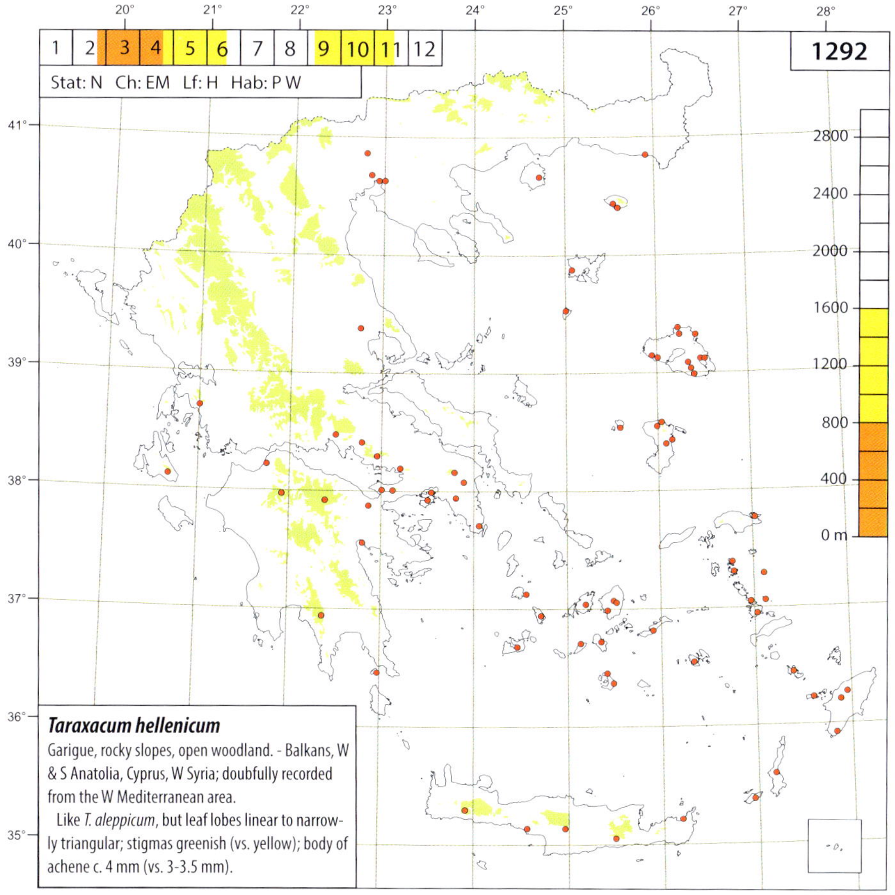

Taraxacum hellenicum

Garigue, rocky slopes, open woodland. - Balkans, W & S Anatolia, Cyprus, W Syria; doubtfully recorded from the W Mediterranean area.

Like *T. aleppicum*, but leaf lobes linear to narrowly triangular; stigmas greenish (vs. yellow); body of achene c. 4 mm (vs. 3-3.5 mm).

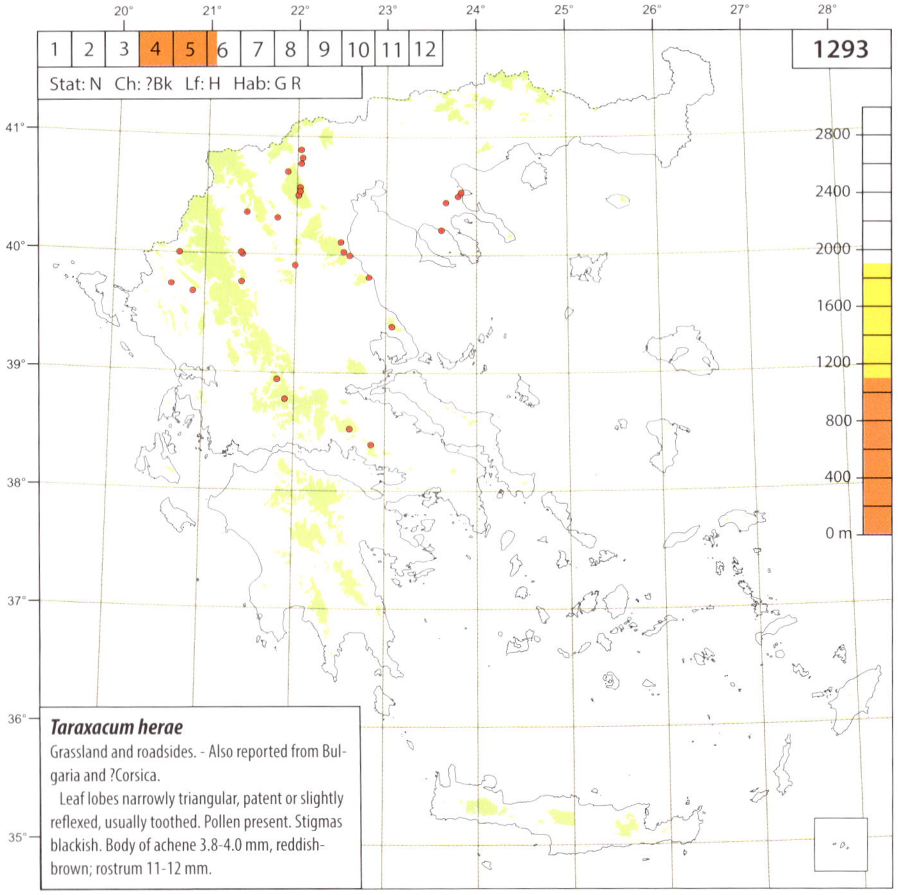

Taraxacum herae

Grassland and roadsides. - Also reported from Bulgaria and ?Corsica.

Leaf lobes narrowly triangular, patent or slightly reflexed, usually toothed. Pollen present. Stigmas blackish. Body of achene 3.8-4.0 mm, reddish-brown; rostrum 11-12 mm.

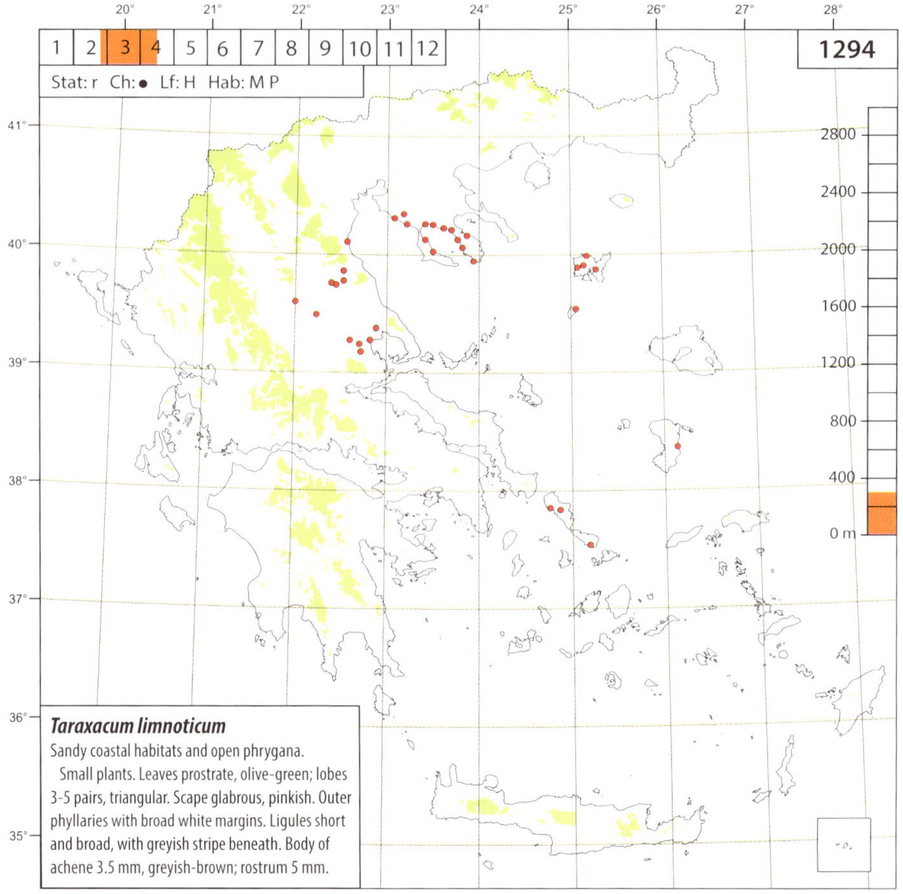

Taraxacum limnoticum

Sandy coastal habitats and open phrygana.

Small plants. Leaves prostrate, olive-green; lobes 3-5 pairs, triangular. Scape glabrous, pinkish. Outer phyllaries with broad white margins. Ligules short and broad, with greyish stripe beneath. Body of achene 3.5 mm, greyish-brown; rostrum 5 mm.

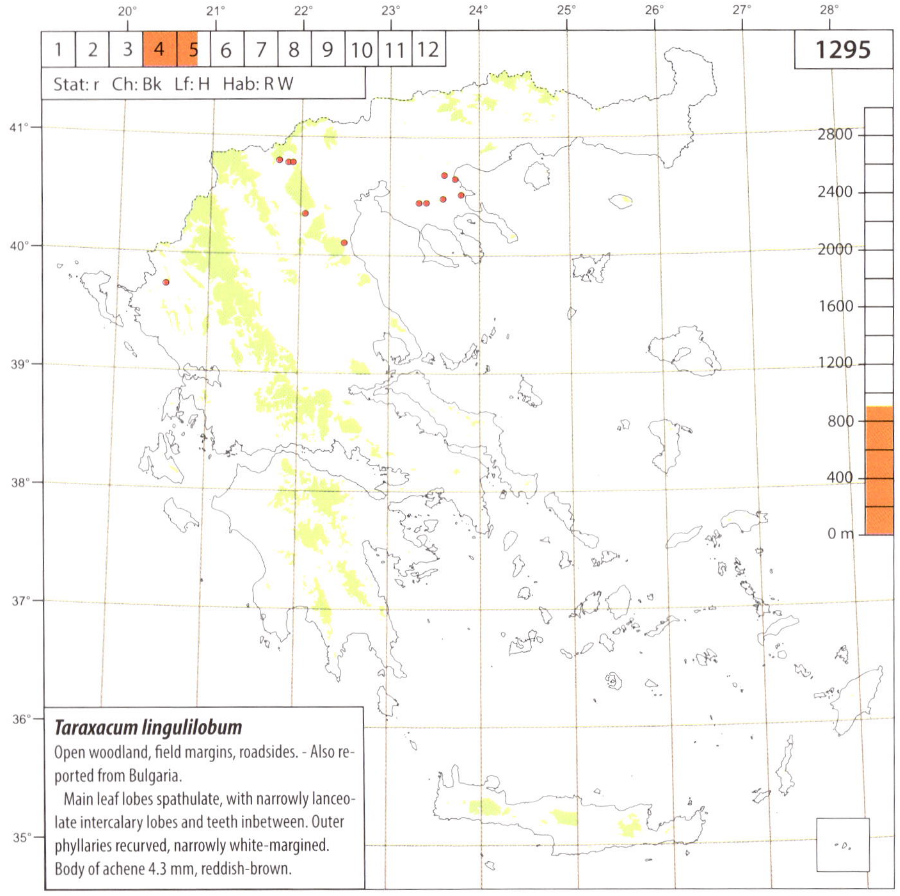

Taraxacum lingulilobum

Open woodland, field margins, roadsides. - Also reported from Bulgaria.

Main leaf lobes spathulate, with narrowly lanceolate intercalary lobes and teeth inbetween. Outer phyllaries recurved, narrowly white-margined. Body of achene 4.3 mm, reddish-brown.

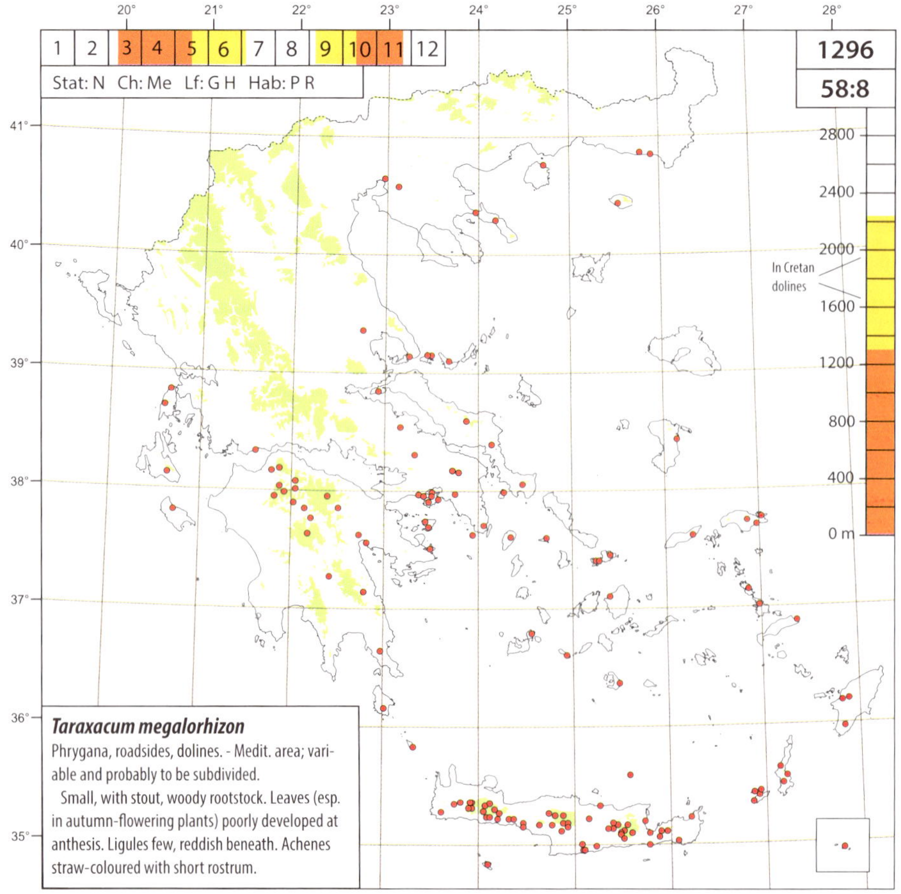

Taraxacum megalorhizon

Phrygana, roadsides, dolines. - Medit. area; variable and probably to be subdivided.

Small, with stout, woody rootstock. Leaves (esp. in autumn-flowering plants) poorly developed at anthesis. Ligules few, reddish beneath. Achenes straw-coloured with short rostrum.

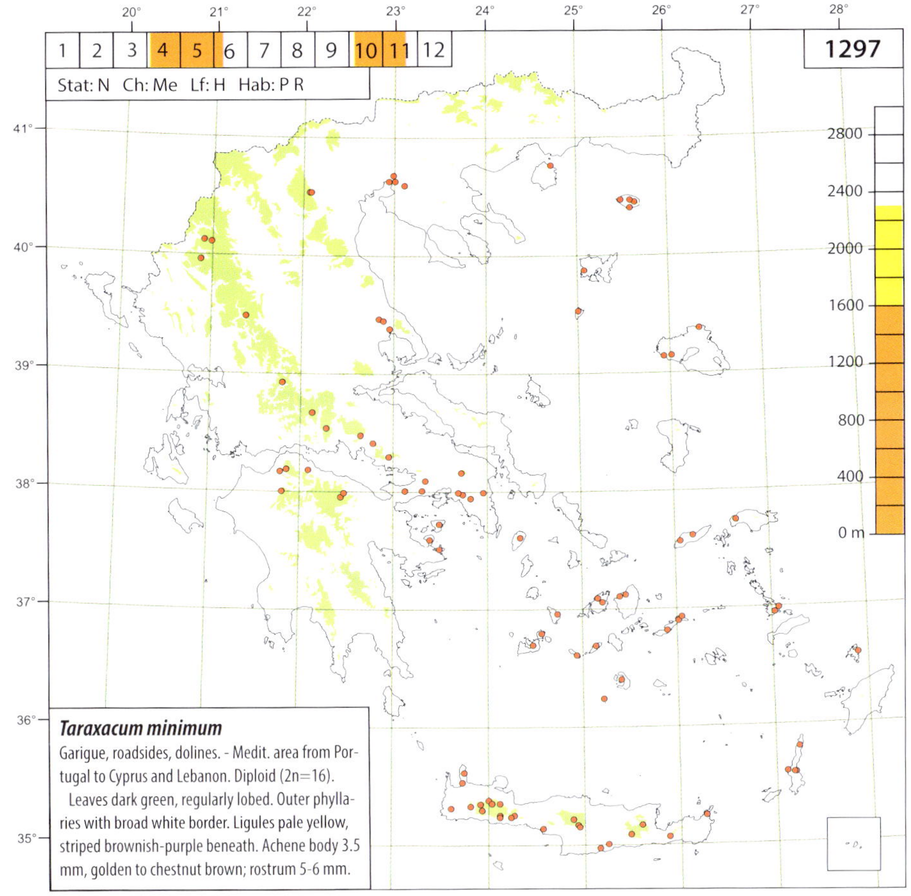

Taraxacum minimum

Garigue, roadsides, dolines. - Medit. area from Portugal to Cyprus and Lebanon. Diploid (2n=16).

Leaves dark green, regularly lobed. Outer phyllaries with broad white border. Ligules pale yellow, striped brownish-purple beneath. Achene body 3.5 mm, golden to chestnut brown; rostrum 5-6 mm.

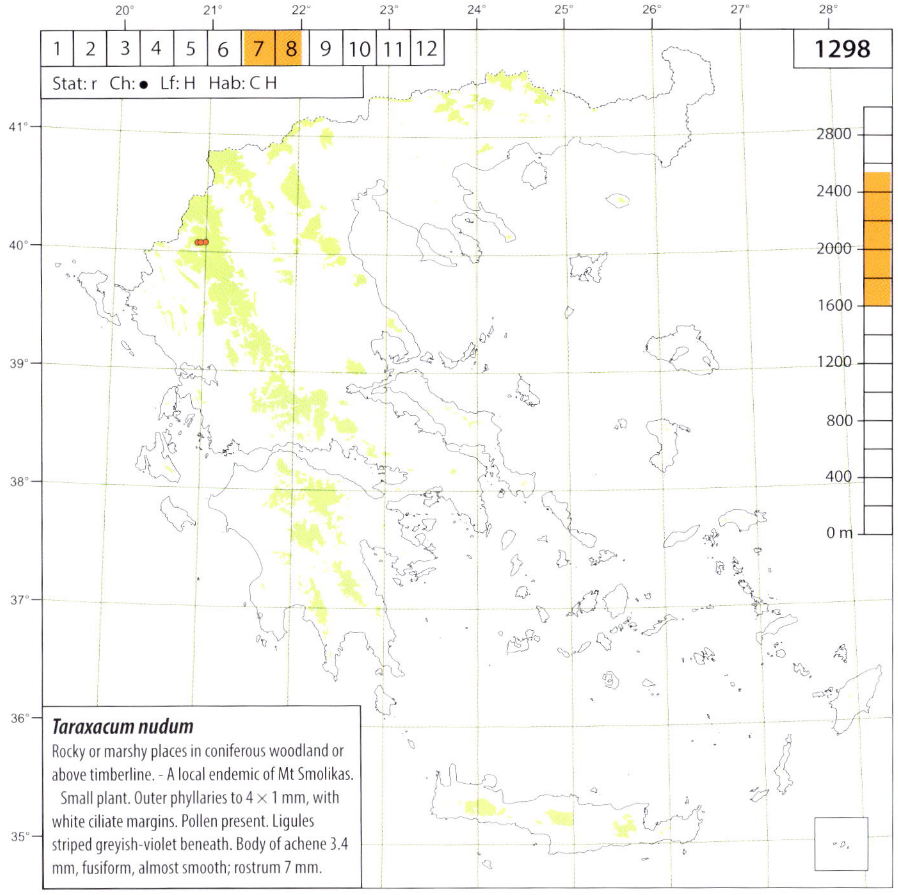

Taraxacum nudum

Rocky or marshy places in coniferous woodland or above timberline. - A local endemic of Mt Smolikas.

Small plant. Outer phyllaries to 4 × 1 mm, with white ciliate margins. Pollen present. Ligules striped greyish-violet beneath. Body of achene 3.4 mm, fusiform, almost smooth; rostrum 7 mm.

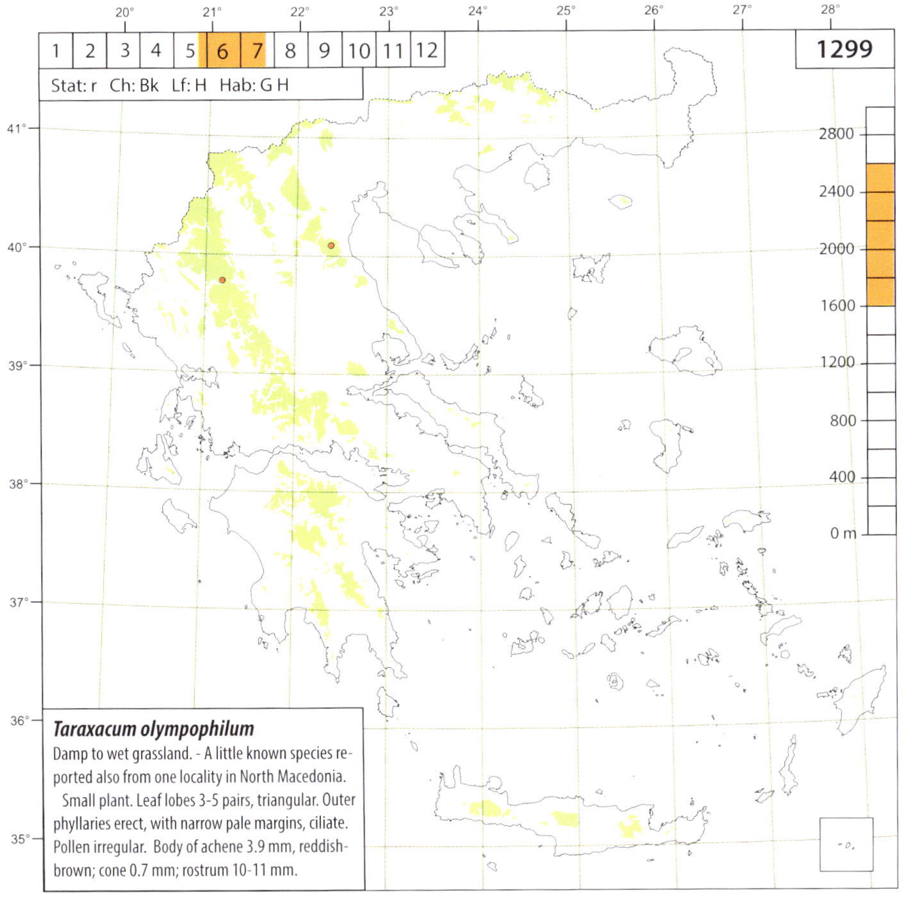

Taraxacum olympophilum

Damp to wet grassland. - A little known species reported also from one locality in North Macedonia.

Small plant. Leaf lobes 3-5 pairs, triangular. Outer phyllaries erect, with narrow pale margins, ciliate. Pollen irregular. Body of achene 3.9 mm, reddish-brown; cone 0.7 mm; rostrum 10-11 mm.

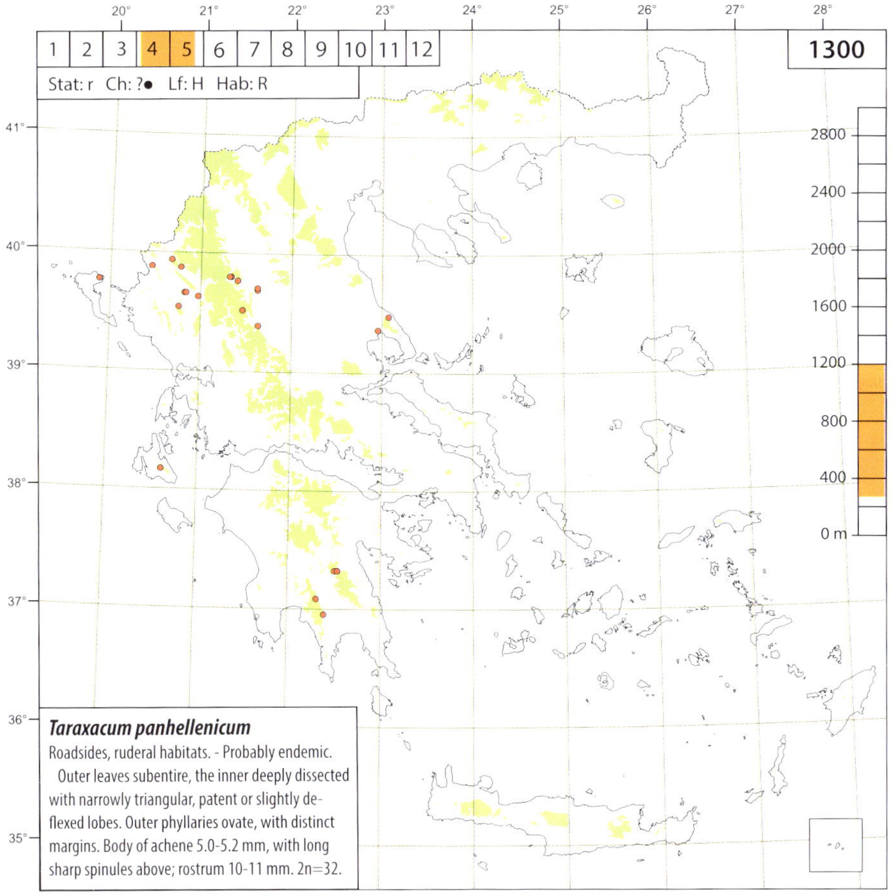

Taraxacum panhellenicum

Roadsides, ruderal habitats. - Probably endemic.

Outer leaves subentire, the inner deeply dissected with narrowly triangular, patent or slightly deflexed lobes. Outer phyllaries ovate, with distinct margins. Body of achene 5.0-5.2 mm, with long sharp spinules above; rostrum 10-11 mm. 2n=32.

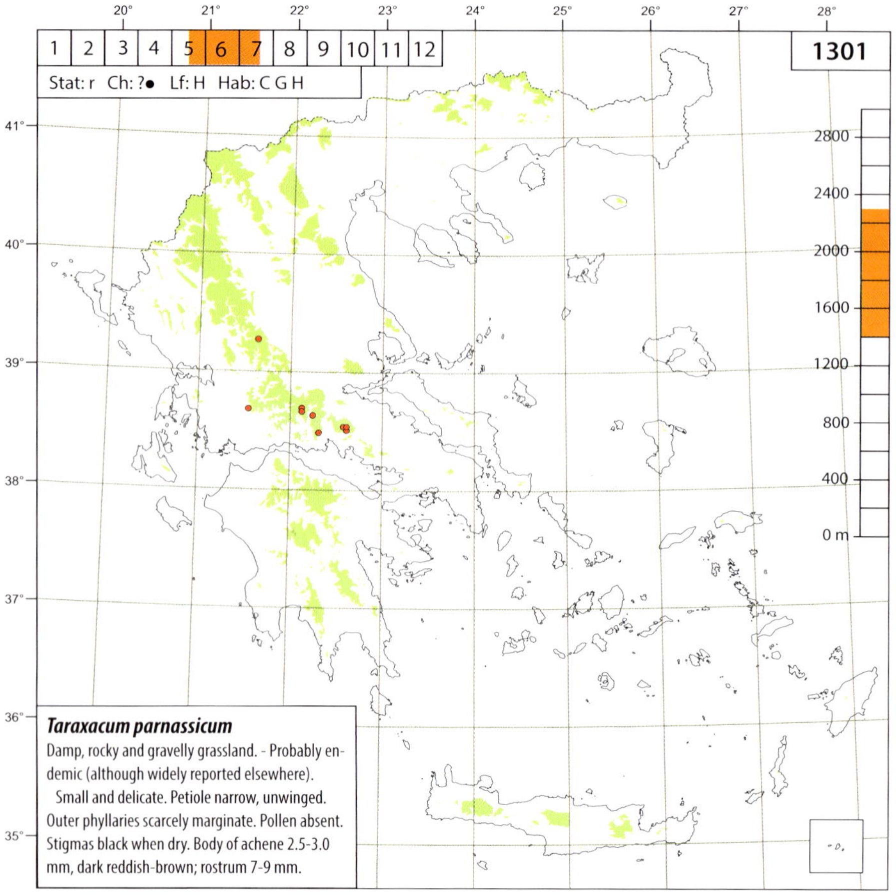

Taraxacum parnassicum

Damp, rocky and gravelly grassland. - Probably endemic (although widely reported elsewhere).

Small and delicate. Petiole narrow, unwinged. Outer phyllaries scarcely marginate. Pollen absent. Stigmas black when dry. Body of achene 2.5-3.0 mm, dark reddish-brown; rostrum 7-9 mm.

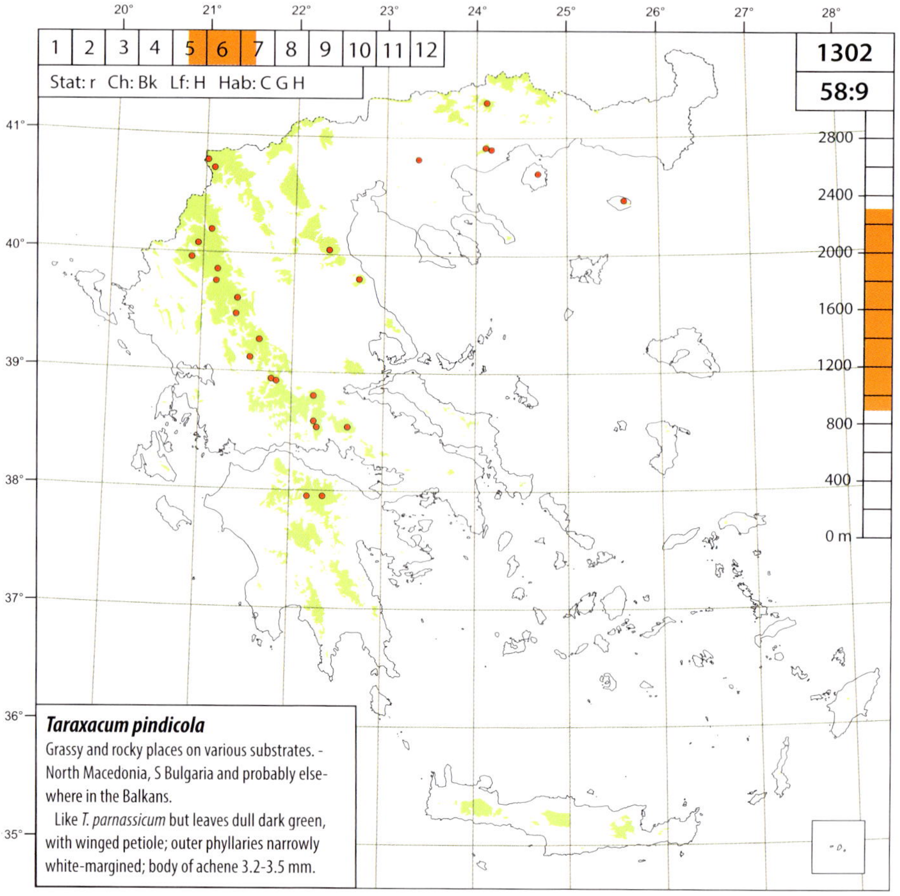

Taraxacum pindicola

Grassy and rocky places on various substrates. - North Macedonia, S Bulgaria and probably elsewhere in the Balkans.

Like *T. parnassicum* but leaves dull dark green, with winged petiole; outer phyllaries narrowly white-margined; body of achene 3.2-3.5 mm.

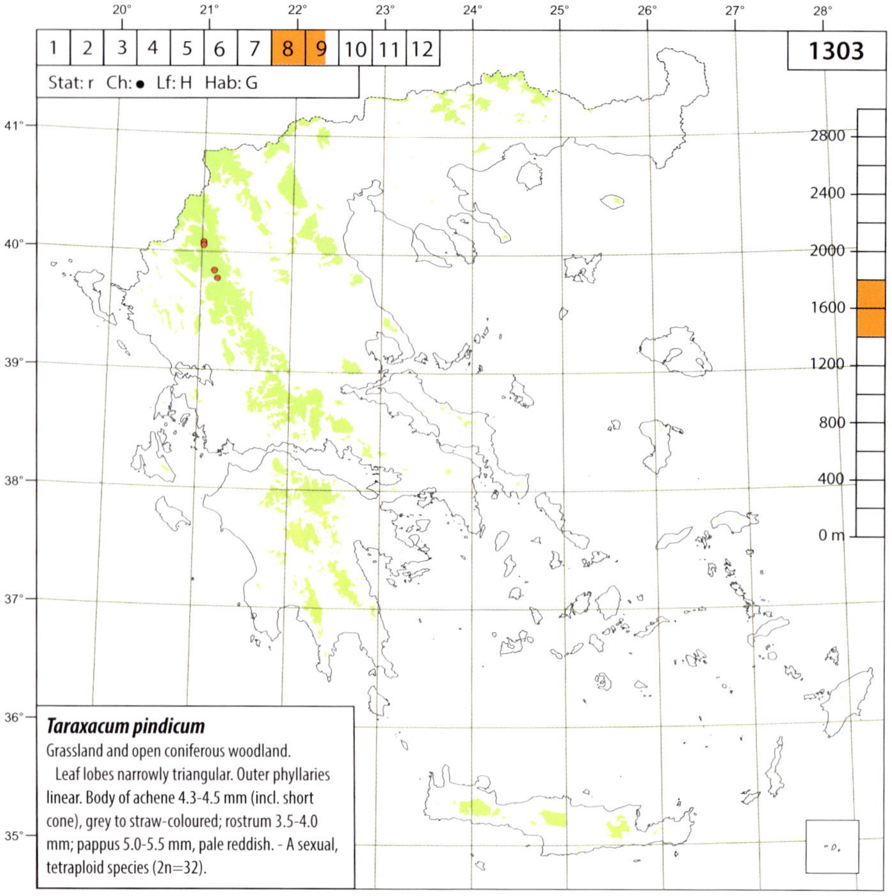

Taraxacum pindicum

Grassland and open coniferous woodland.

Leaf lobes narrowly triangular. Outer phyllaries linear. Body of achene 4.3-4.5 mm (incl. short cone), grey to straw-coloured; rostrum 3.5-4.0 mm; pappus 5.0-5.5 mm, pale reddish. - A sexual, tetraploid species (2n=32).

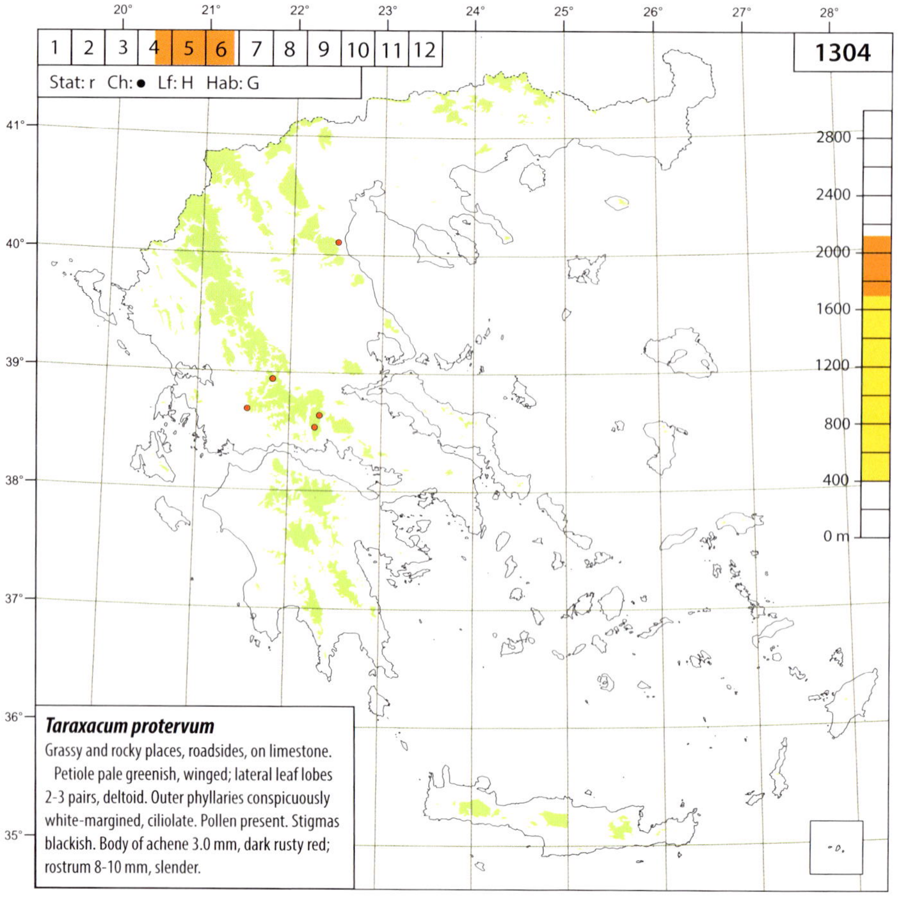

Taraxacum protervum

Grassy and rocky places, roadsides, on limestone.

Petiole pale greenish, winged; lateral leaf lobes 2-3 pairs, deltoid. Outer phyllaries conspicuously white-margined, ciliolate. Pollen present. Stigmas blackish. Body of achene 3.0 mm, dark rusty red; rostrum 8-10 mm, slender.

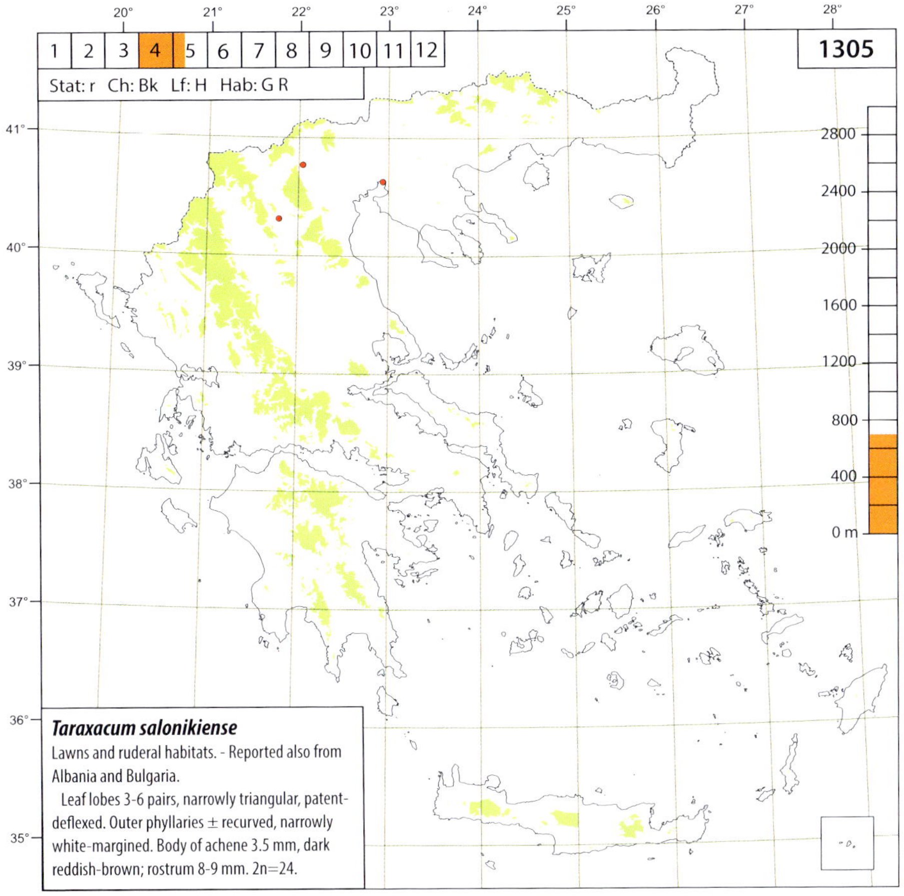

Taraxacum salonikiense
Lawns and ruderal habitats. - Reported also from Albania and Bulgaria.
Leaf lobes 3-6 pairs, narrowly triangular, patent-deflexed. Outer phyllaries ± recurved, narrowly white-margined. Body of achene 3.5 mm, dark reddish-brown; rostrum 8-9 mm. 2n=24.

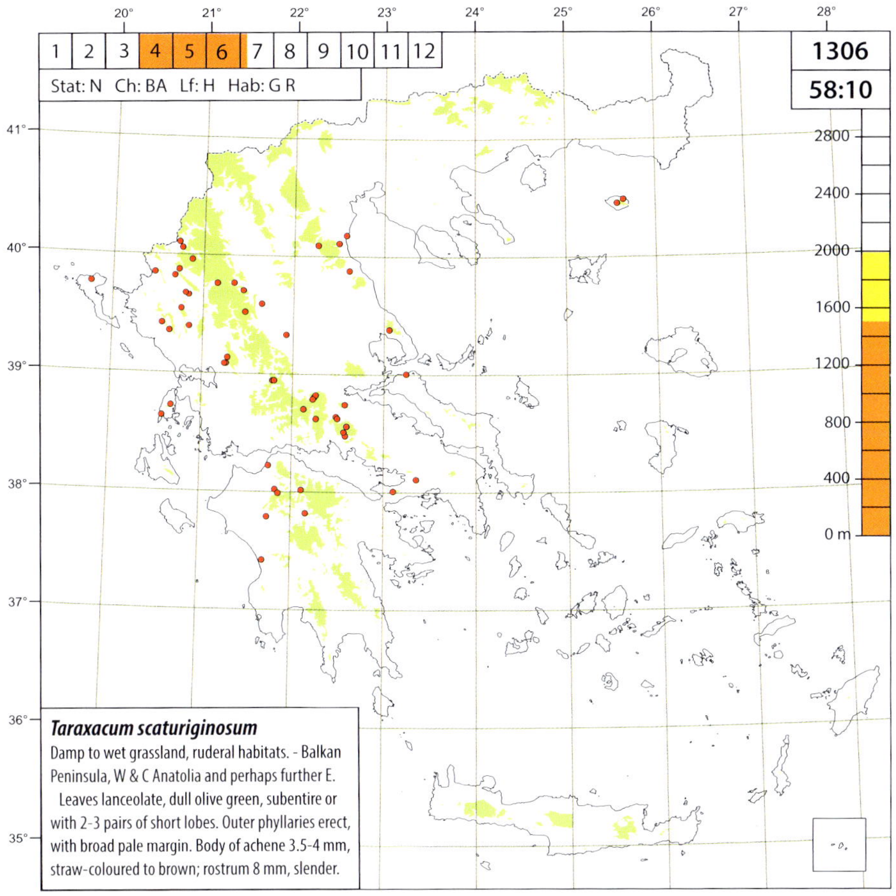

Taraxacum scaturiginosum
Damp to wet grassland, ruderal habitats. - Balkan Peninsula, W & C Anatolia and perhaps further E.
Leaves lanceolate, dull olive green, subentire or with 2-3 pairs of short lobes. Outer phyllaries erect, with broad pale margin. Body of achene 3.5-4 mm, straw-coloured to brown; rostrum 8 mm, slender.

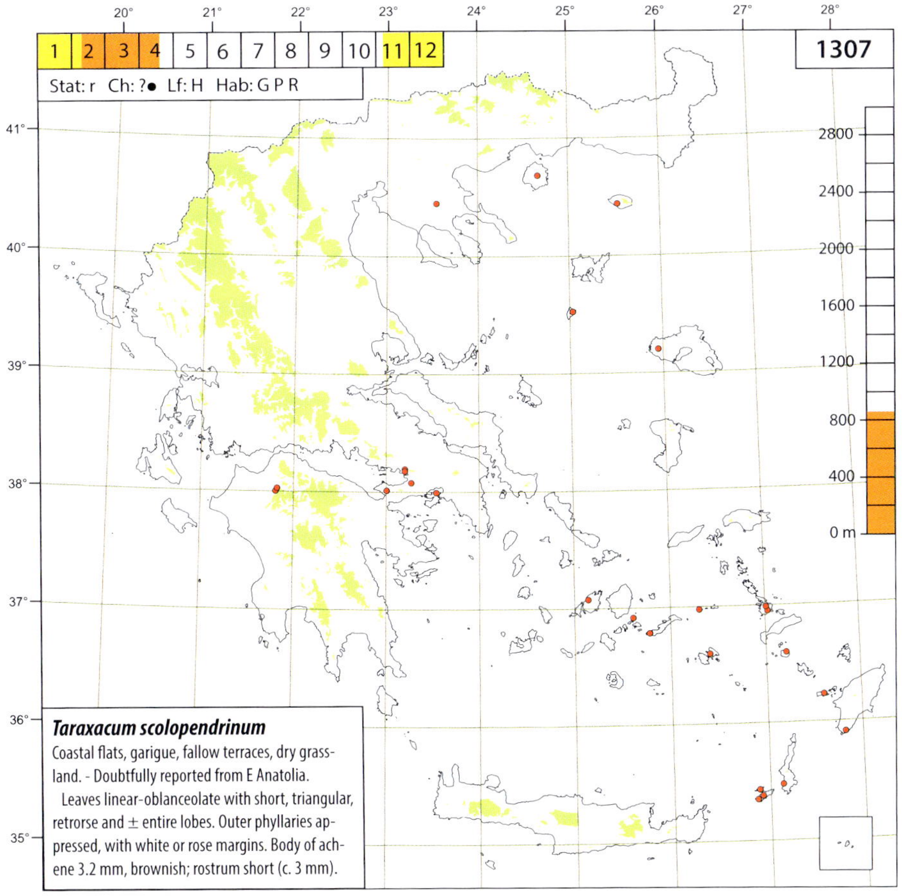

Taraxacum scolopendrinum
Coastal flats, garigue, fallow terraces, dry grassland. - Doubtfully reported from E Anatolia.
Leaves linear-oblanceolate with short, triangular, retrorse and ± entire lobes. Outer phyllaries appressed, with white or rose margins. Body of achene 3.2 mm, brownish; rostrum short (c. 3 mm).

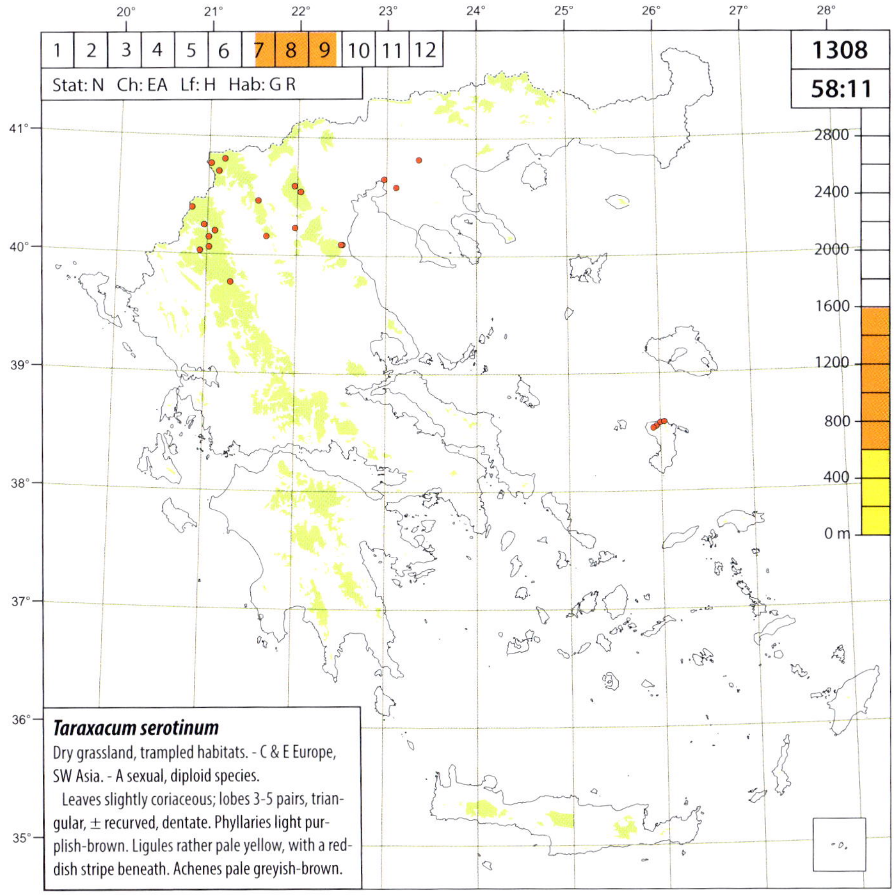

Taraxacum serotinum
Dry grassland, trampled habitats. - C & E Europe, SW Asia. - A sexual, diploid species.
Leaves slightly coriaceous; lobes 3-5 pairs, triangular, ± recurved, dentate. Phyllaries light purplish-brown. Ligules rather pale yellow, with a reddish stripe beneath. Achenes pale greyish-brown.

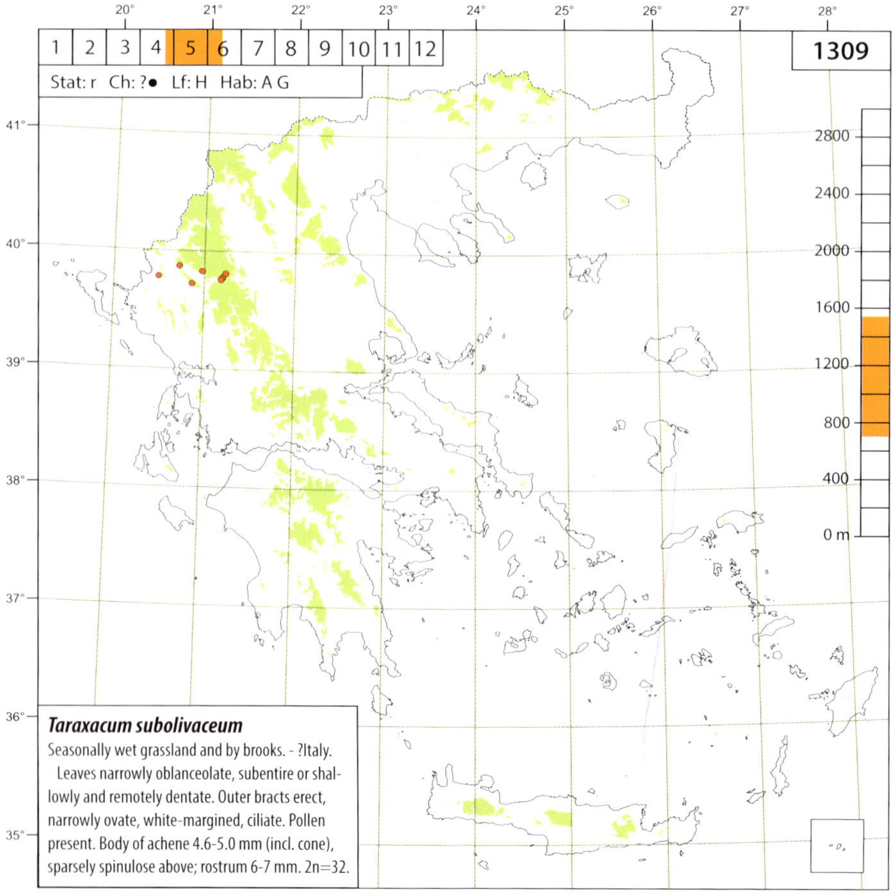

Taraxacum subolivaceum

Seasonally wet grassland and by brooks. - ?Italy.

Leaves narrowly oblanceolate, subentire or shallowly and remotely dentate. Outer bracts erect, narrowly ovate, white-margined, ciliate. Pollen present. Body of achene 4.6-5.0 mm (incl. cone), sparsely spinulose above; rostrum 6-7 mm. 2n=32.

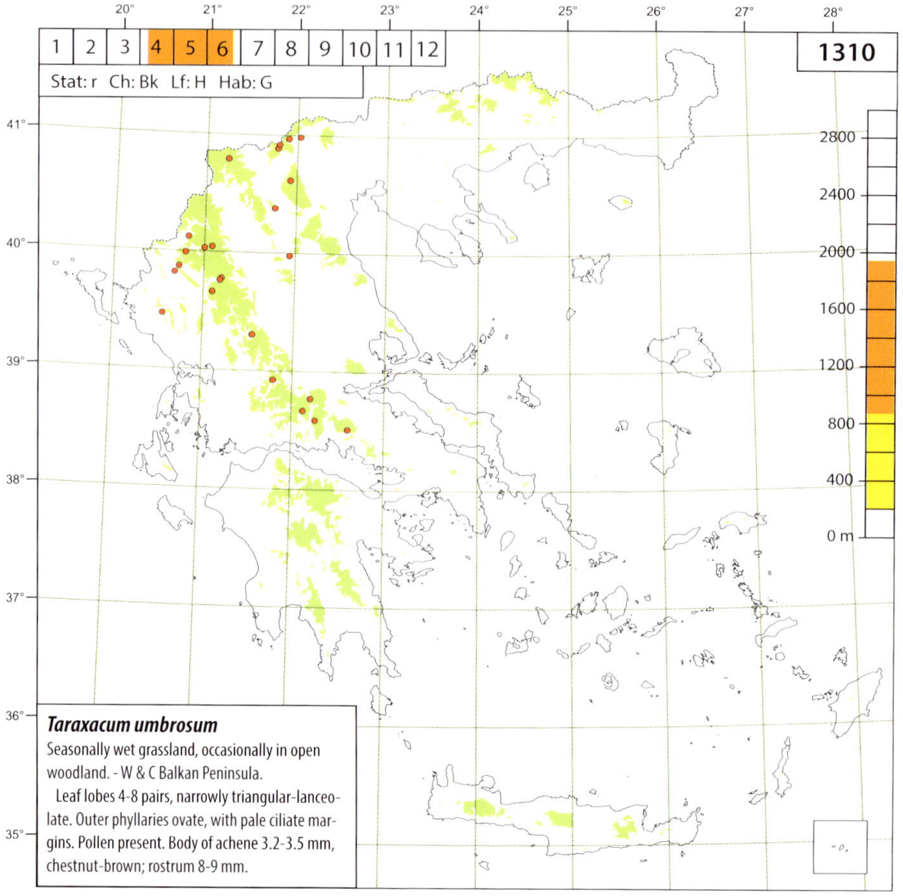

Taraxacum umbrosum

Seasonally wet grassland, occasionally in open woodland. - W & C Balkan Peninsula.

Leaf lobes 4-8 pairs, narrowly triangular-lanceolate. Outer phyllaries ovate, with pale ciliate margins. Pollen present. Body of achene 3.2-3.5 mm, chestnut-brown; rostrum 8-9 mm.

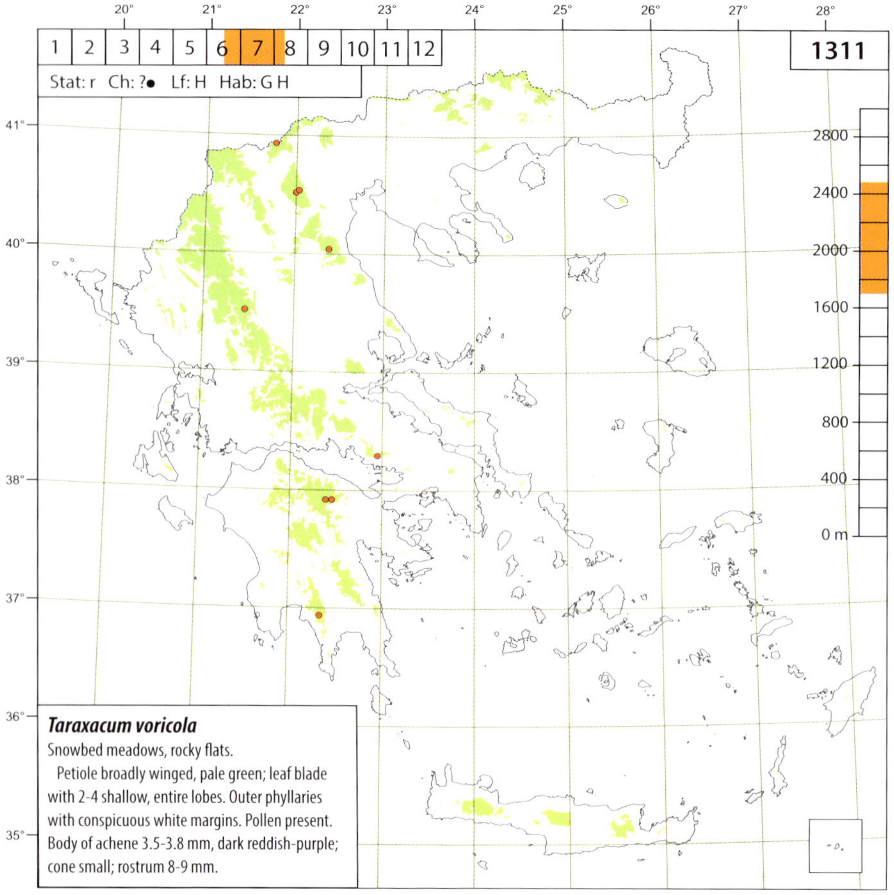

Taraxacum voricola

Snowbed meadows, rocky flats.

Petiole broadly winged, pale green; leaf blade with 2-4 shallow, entire lobes. Outer phyllaries with conspicuous white margins. Pollen present. Body of achene 3.5-3.8 mm, dark reddish-purple; cone small; rostrum 8-9 mm.

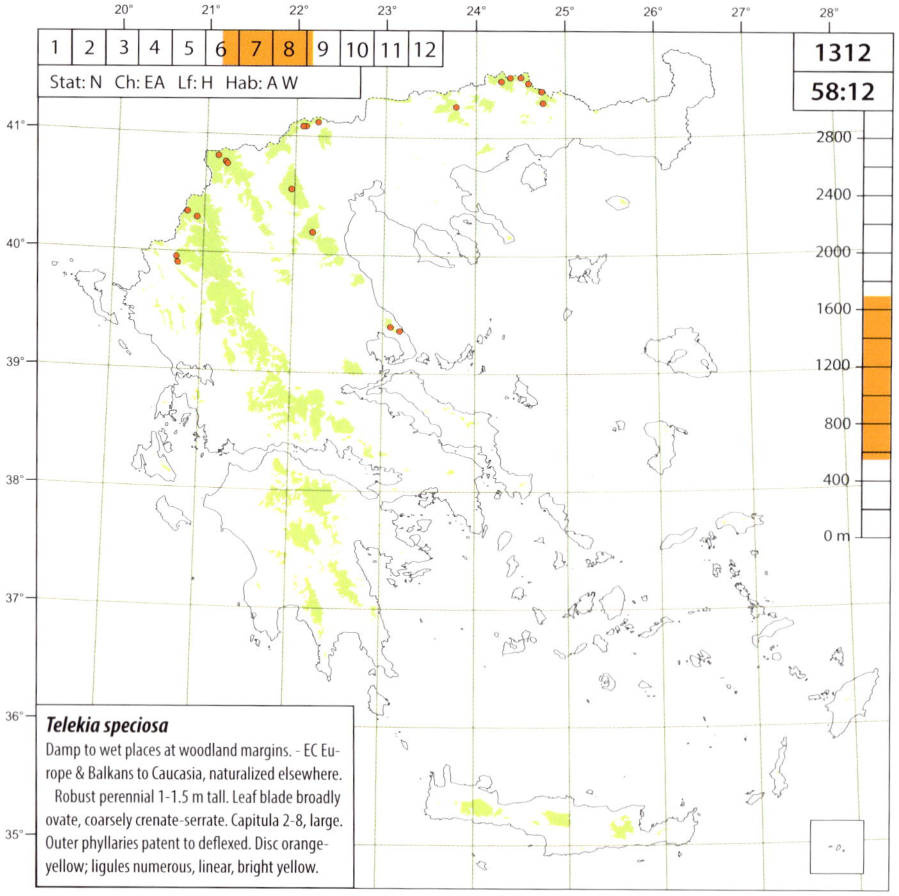

Telekia speciosa

Damp to wet places at woodland margins. - EC Europe & Balkans to Caucasia, naturalized elsewhere.

Robust perennial 1-1.5 m tall. Leaf blade broadly ovate, coarsely crenate-serrate. Capitula 2-8, large. Outer phyllaries patent to deflexed. Disc orange-yellow; ligules numerous, linear, bright yellow.

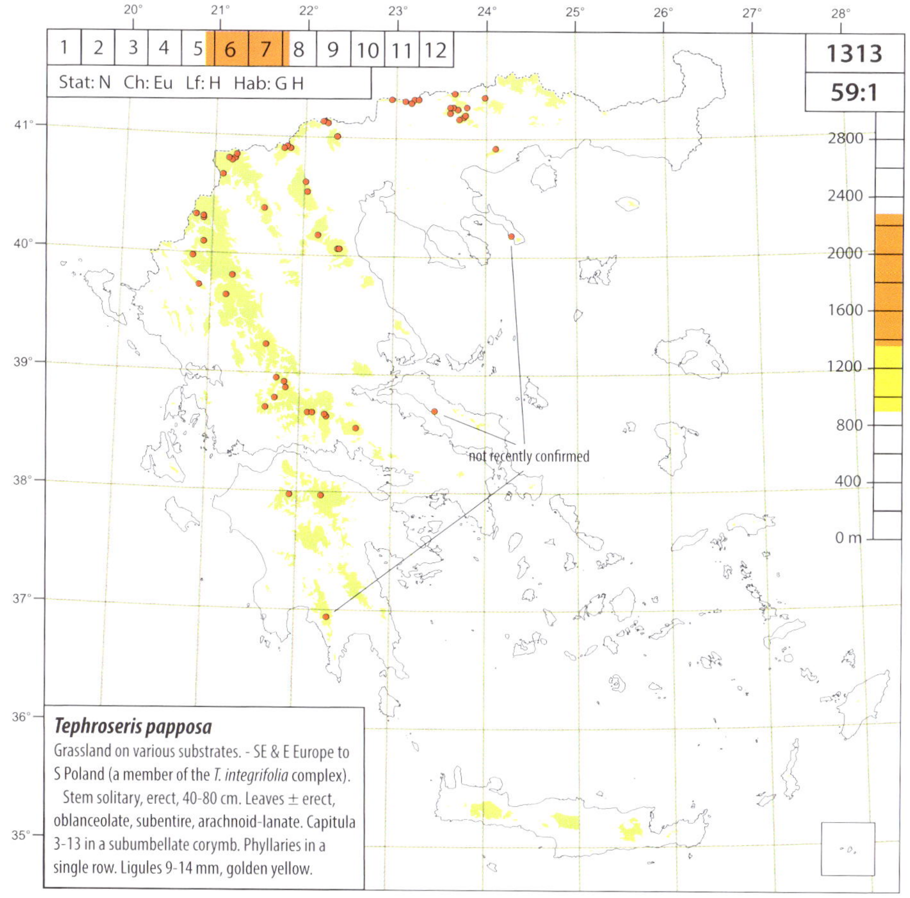

Tephroseris papposa

Grassland on various substrates. - SE & E Europe to S Poland (a member of the *T. integrifolia* complex).

Stem solitary, erect, 40-80 cm. Leaves ± erect, oblanceolate, subentire, arachnoid-lanate. Capitula 3-13 in a subumbellate corymb. Phyllaries in a single row. Ligules 9-14 mm, golden yellow.

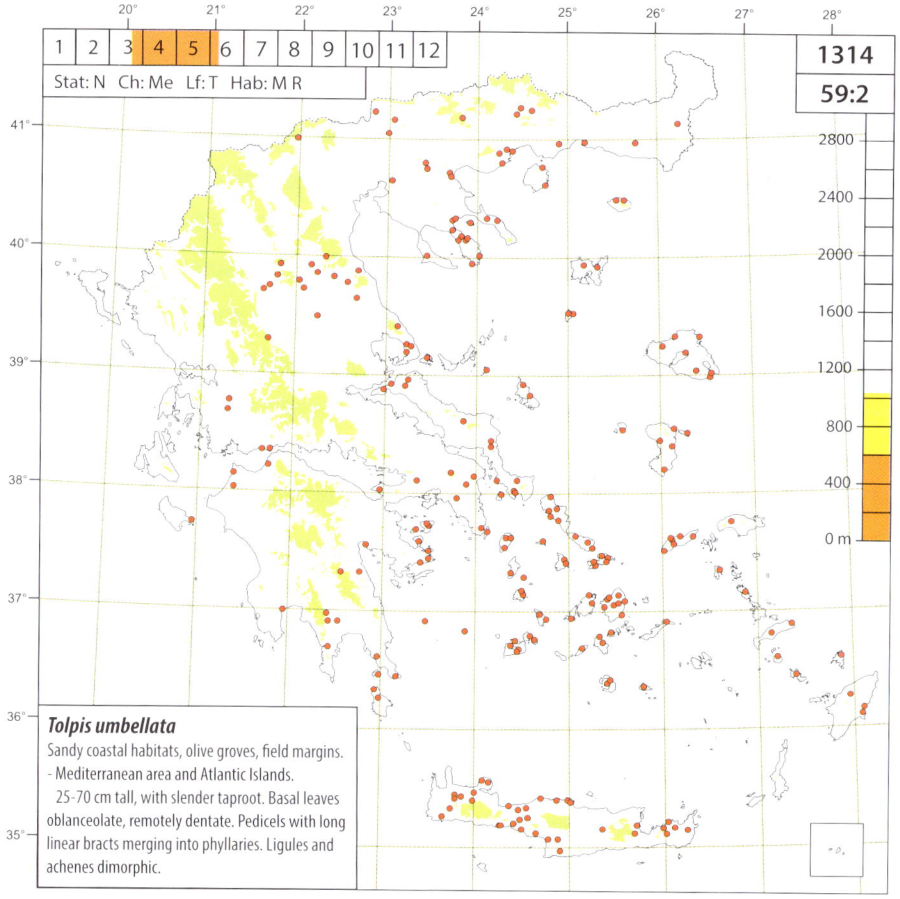

Tolpis umbellata

Sandy coastal habitats, olive groves, field margins. - Mediterranean area and Atlantic Islands.

25-70 cm tall, with slender taproot. Basal leaves oblanceolate, remotely dentate. Pedicels with long linear bracts merging into phyllaries. Ligules and achenes dimorphic.

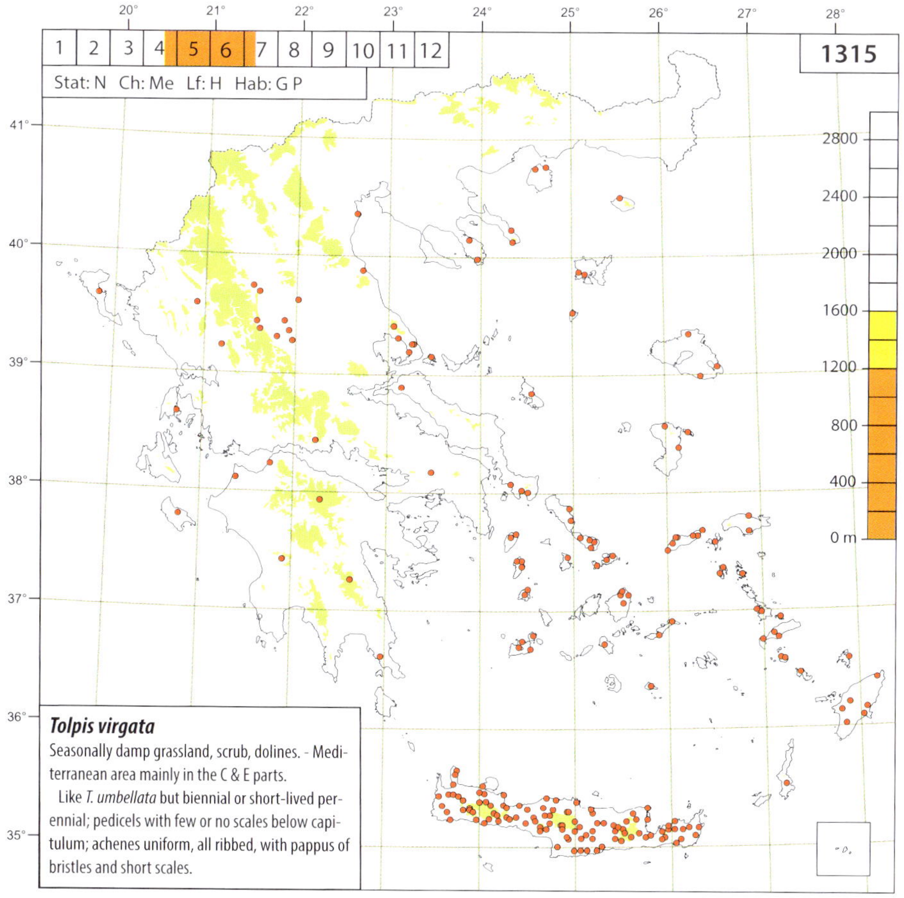

Tolpis virgata

Seasonally damp grassland, scrub, dolines. - Mediterranean area mainly in the C & E parts.

Like *T. umbellata* but biennial or short-lived perennial; pedicels with few or no scales below capitulum; achenes uniform, all ribbed, with pappus of bristles and short scales.

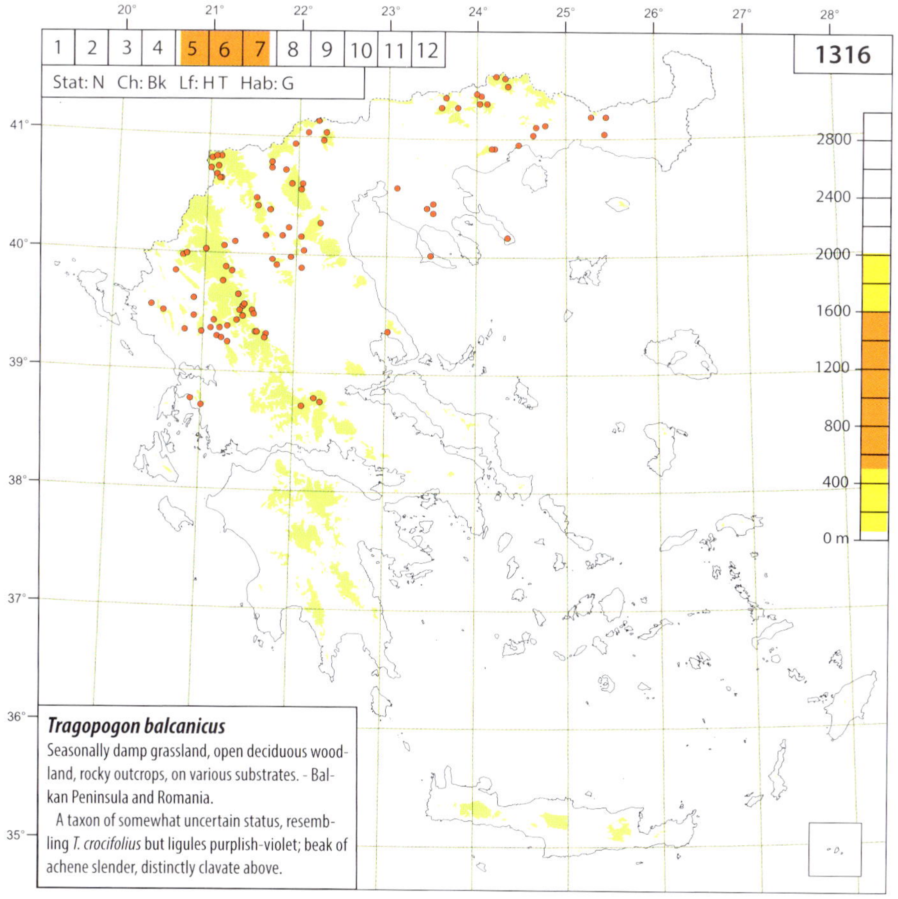

Tragopogon balcanicus

Seasonally damp grassland, open deciduous woodland, rocky outcrops, on various substrates. - Balkan Peninsula and Romania.

A taxon of somewhat uncertain status, resembling *T. crocifolius* but ligules purplish-violet; beak of achene slender, distinctly clavate above.

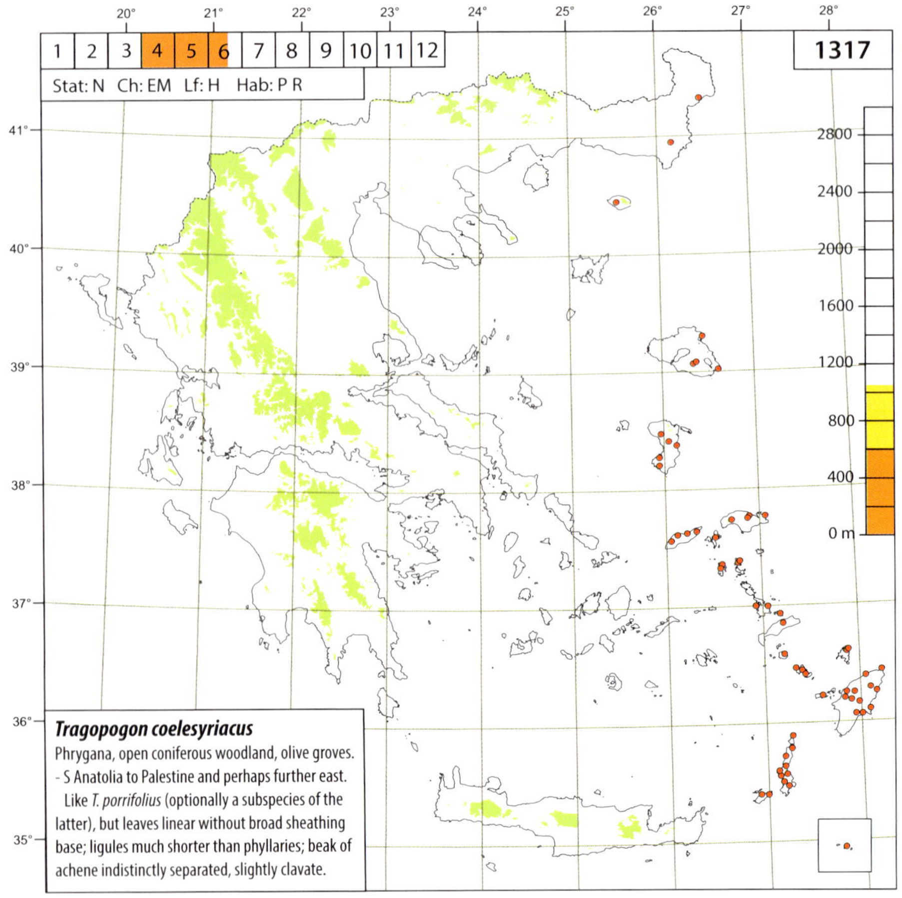

Tragopogon coelesyriacus
Phrygana, open coniferous woodland, olive groves. - S Anatolia to Palestine and perhaps further east.
Like *T. porrifolius* (optionally a subspecies of the latter), but leaves linear without broad sheathing base; ligules much shorter than phyllaries; beak of achene indistinctly separated, slightly clavate.

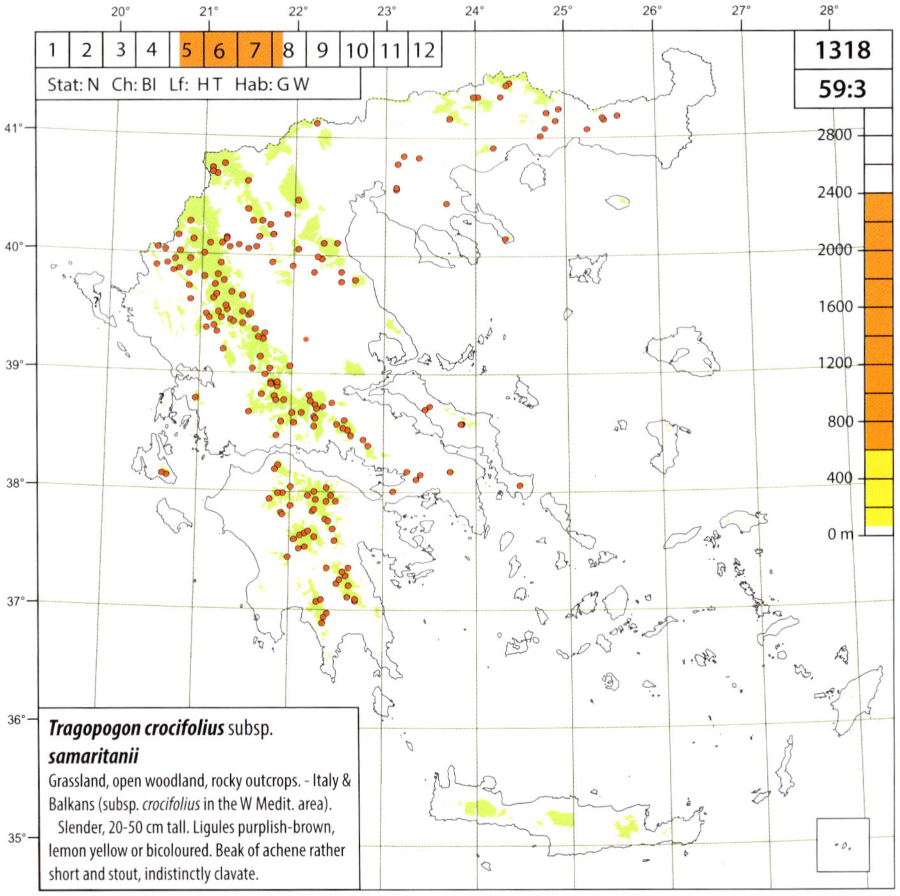

Tragopogon crocifolius subsp. ***samaritanii***
Grassland, open woodland, rocky outcrops. - Italy & Balkans (subsp. *crocifolius* in the W Medit. area).
Slender, 20-50 cm tall. Ligules purplish-brown, lemon yellow or bicoloured. Beak of achene rather short and stout, indistinctly clavate.

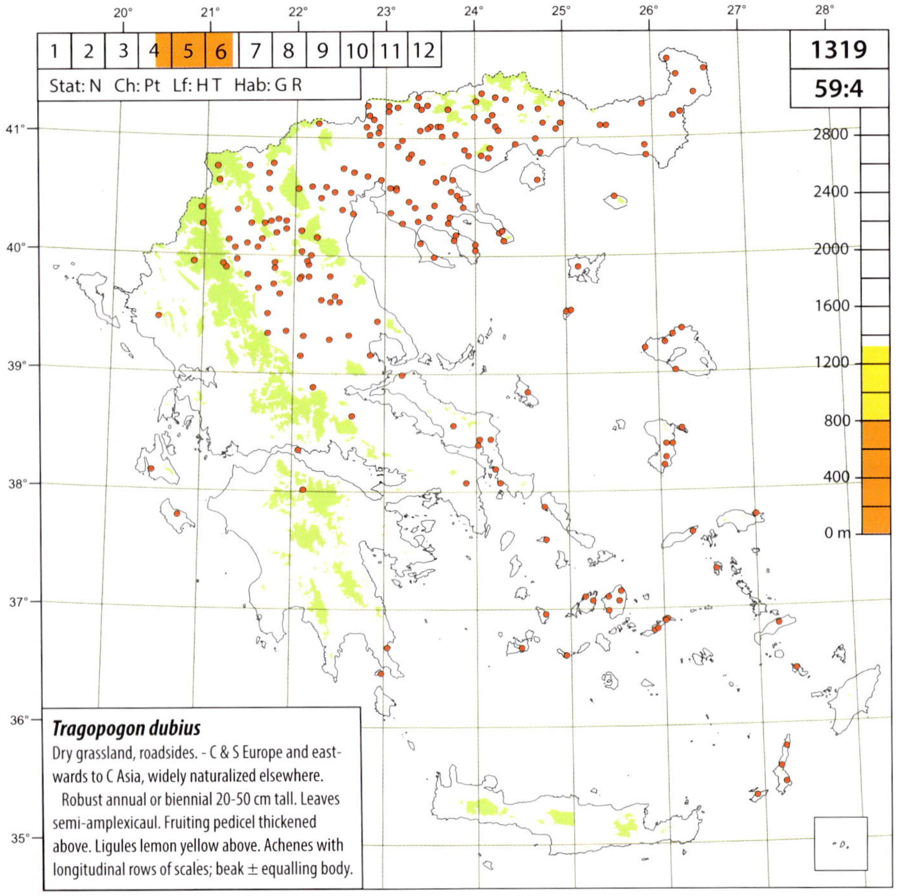

Tragopogon dubius
Dry grassland, roadsides. - C & S Europe and eastwards to C Asia, widely naturalized elsewhere.
Robust annual or biennial 20-50 cm tall. Leaves semi-amplexicaul. Fruiting pedicel thickened above. Ligules lemon yellow above. Achenes with longitudinal rows of scales; beak ± equalling body.

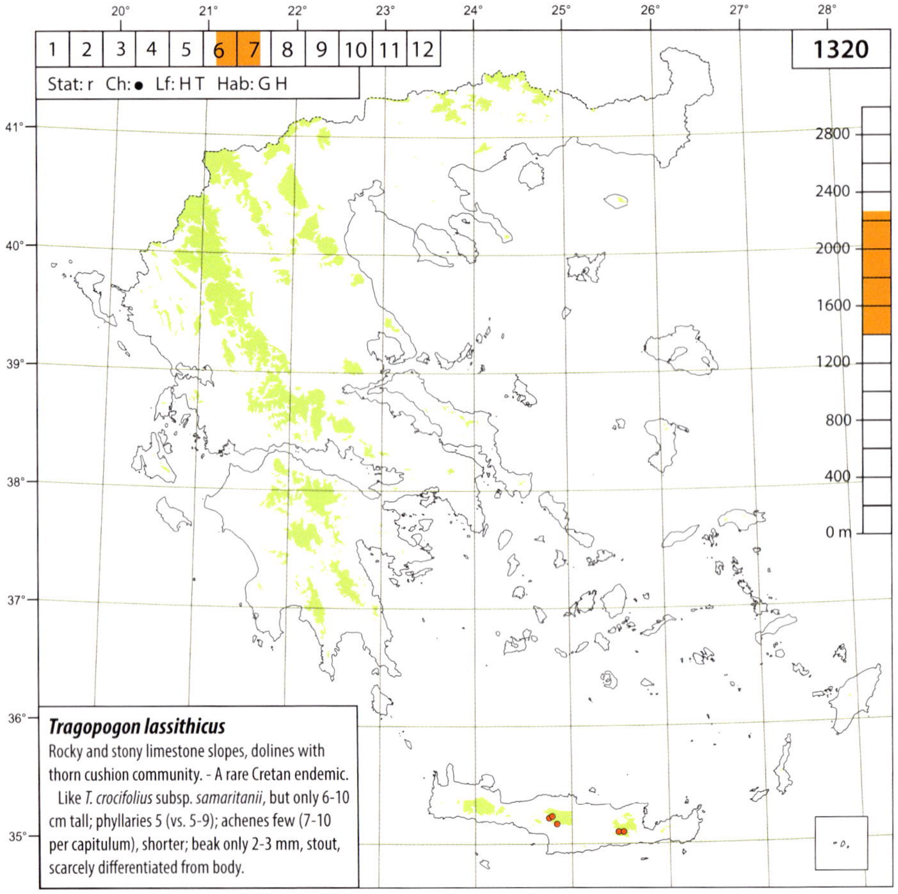

Tragopogon lassithicus
Rocky and stony limestone slopes, dolines with thorn cushion community. - A rare Cretan endemic.
Like *T. crocifolius* subsp. *samaritanii*, but only 6-10 cm tall; phyllaries 5 (vs. 5-9); achenes few (7-10 per capitulum), shorter; beak only 2-3 mm, stout, scarcely differentiated from body.

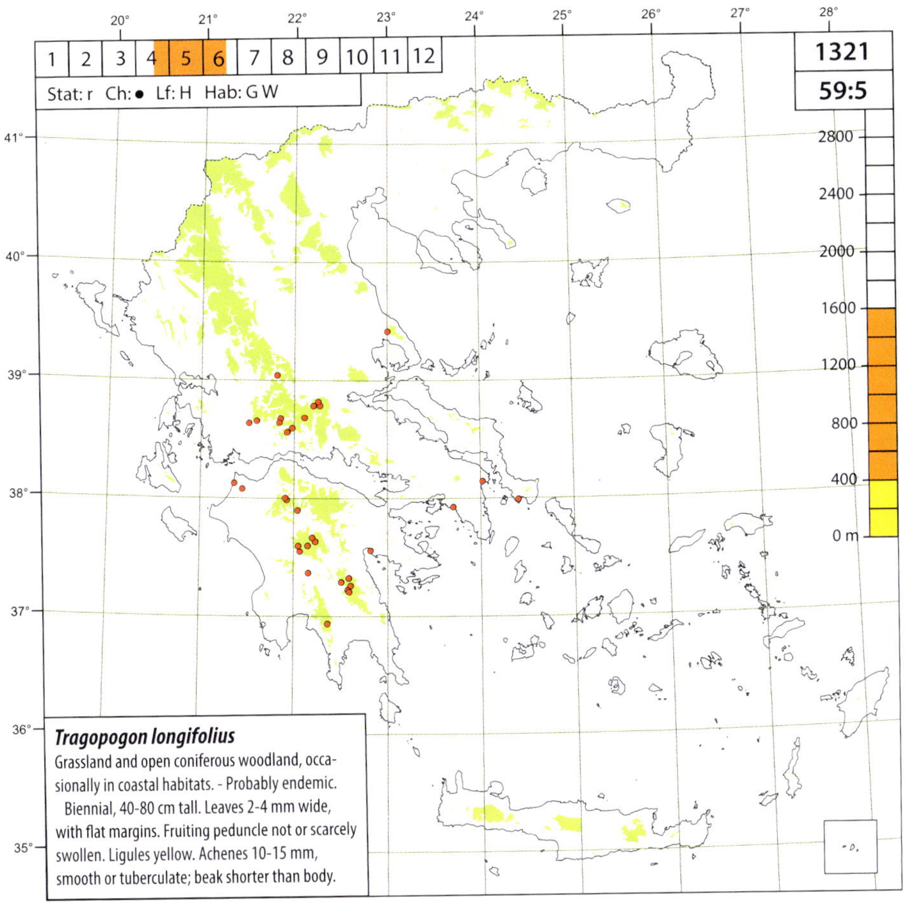

Tragopogon longifolius

Grassland and open coniferous woodland, occasionally in coastal habitats. - Probably endemic.
Biennial, 40-80 cm tall. Leaves 2-4 mm wide, with flat margins. Fruiting peduncle not or scarcely swollen. Ligules yellow. Achenes 10-15 mm, smooth or tuberculate; beak shorter than body.

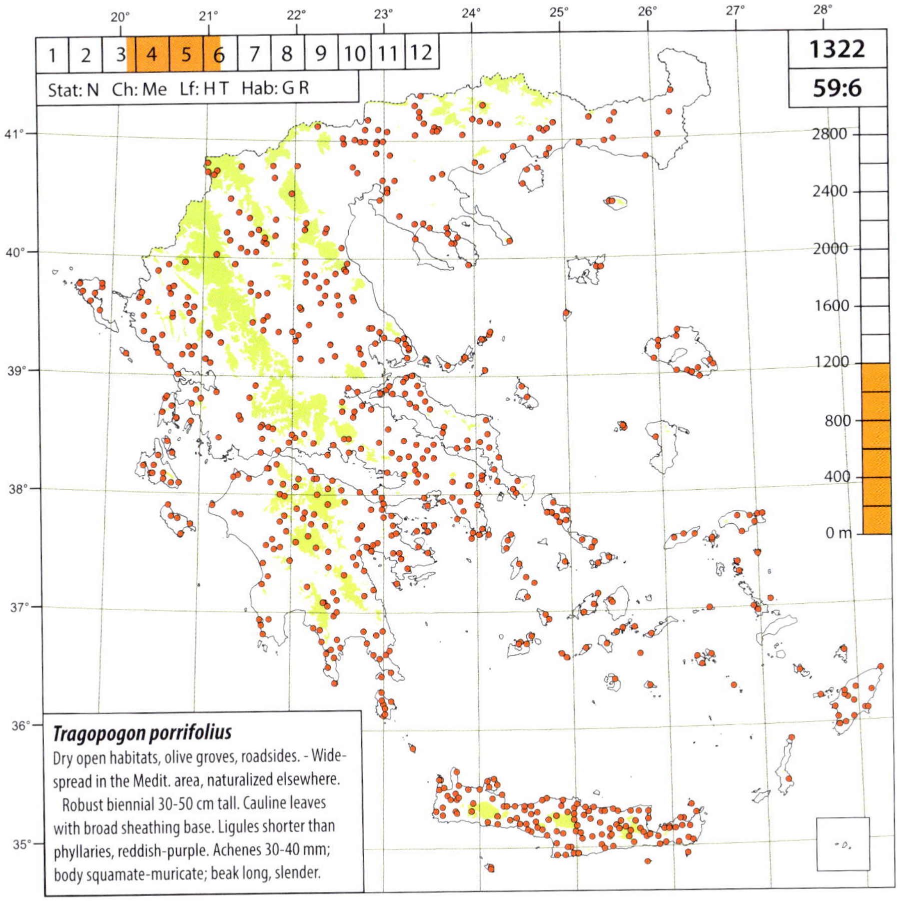

Tragopogon porrifolius

Dry open habitats, olive groves, roadsides. - Widespread in the Medit. area, naturalized elsewhere.
Robust biennial 30-50 cm tall. Cauline leaves with broad sheathing base. Ligules shorter than phyllaries, reddish-purple. Achenes 30-40 mm; body squamate-muricate; beak long, slender.

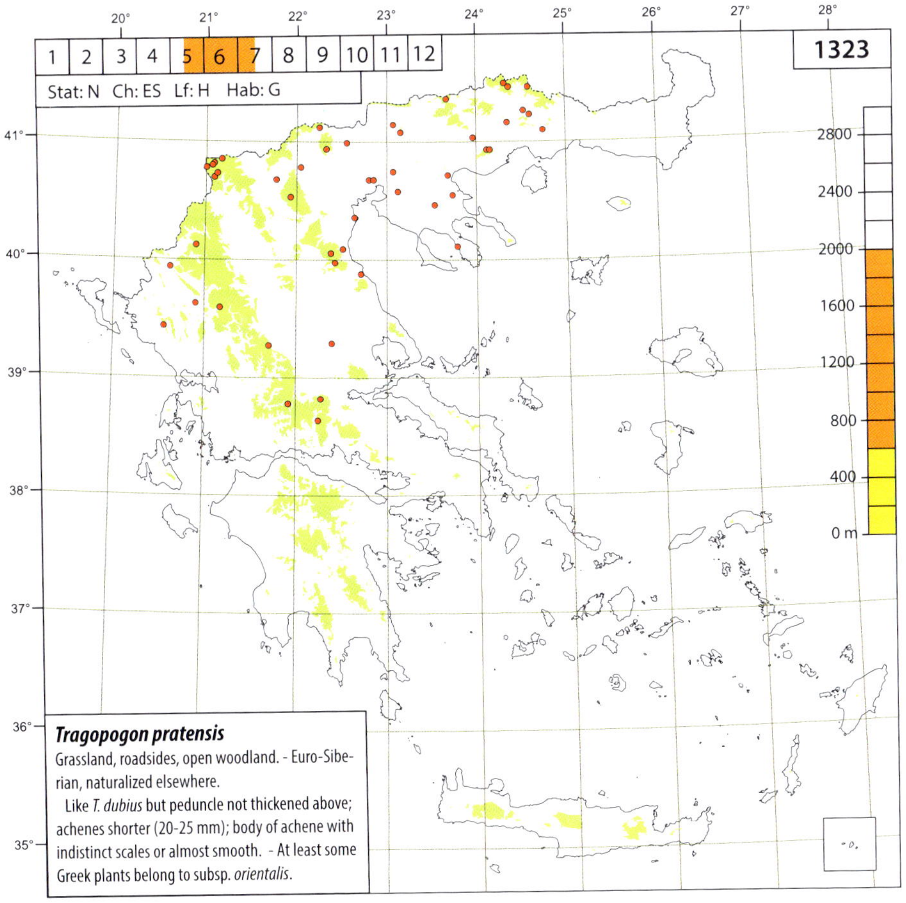

Tragopogon pratensis

Grassland, roadsides, open woodland. - Euro-Siberian, naturalized elsewhere.
Like *T. dubius* but peduncle not thickened above; achenes shorter (20-25 mm); body of achene with indistinct scales or almost smooth. - At least some Greek plants belong to subsp. *orientalis*.

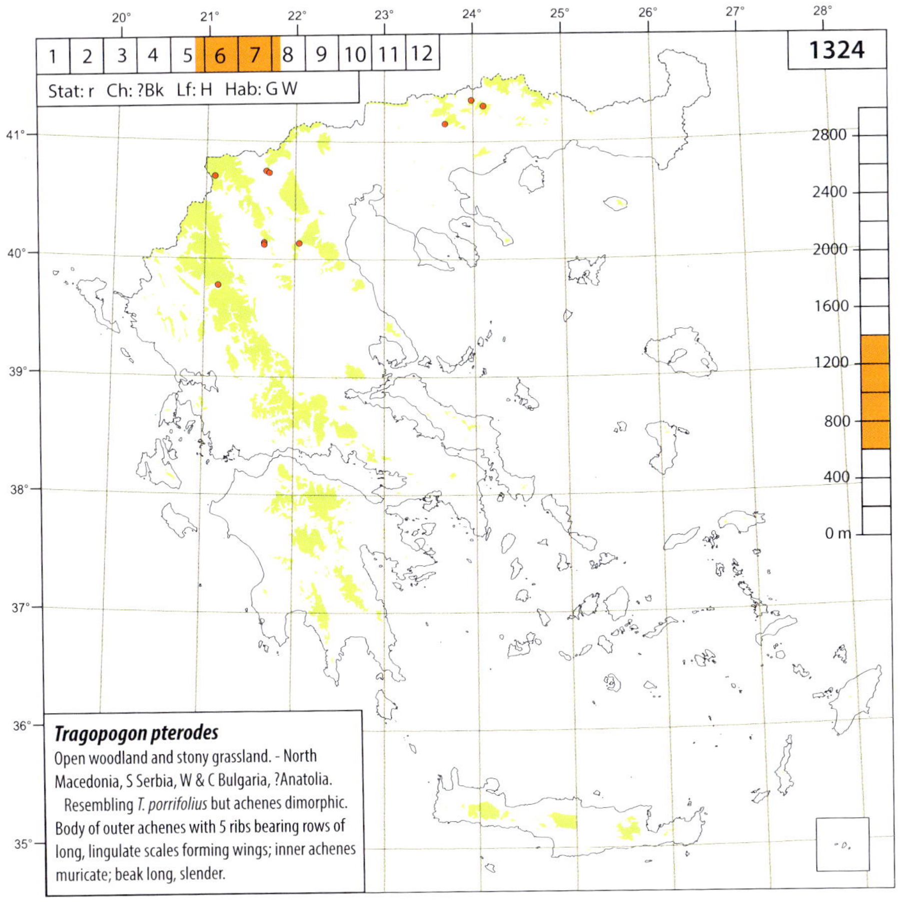

Tragopogon pterodes

Open woodland and stony grassland. - North Macedonia, S Serbia, W & C Bulgaria, ?Anatolia.
Resembling *T. porrifolius* but achenes dimorphic. Body of outer achenes with 5 ribs bearing rows of long, lingulate scales forming wings; inner achenes muricate; beak long, slender.

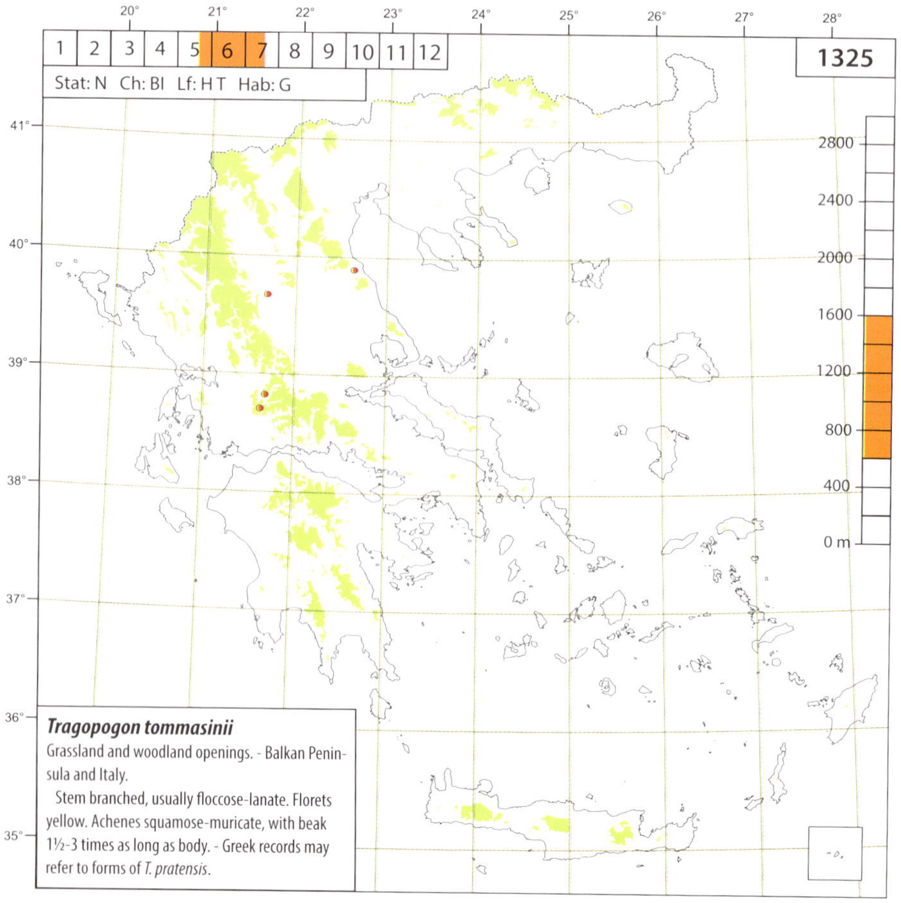

Tragopogon tommasinii
Grassland and woodland openings. - Balkan Peninsula and Italy.

Stem branched, usually floccose-lanate. Florets yellow. Achenes squamose-muricate, with beak 1½-3 times as long as body. - Greek records may refer to forms of *T. pratensis*.

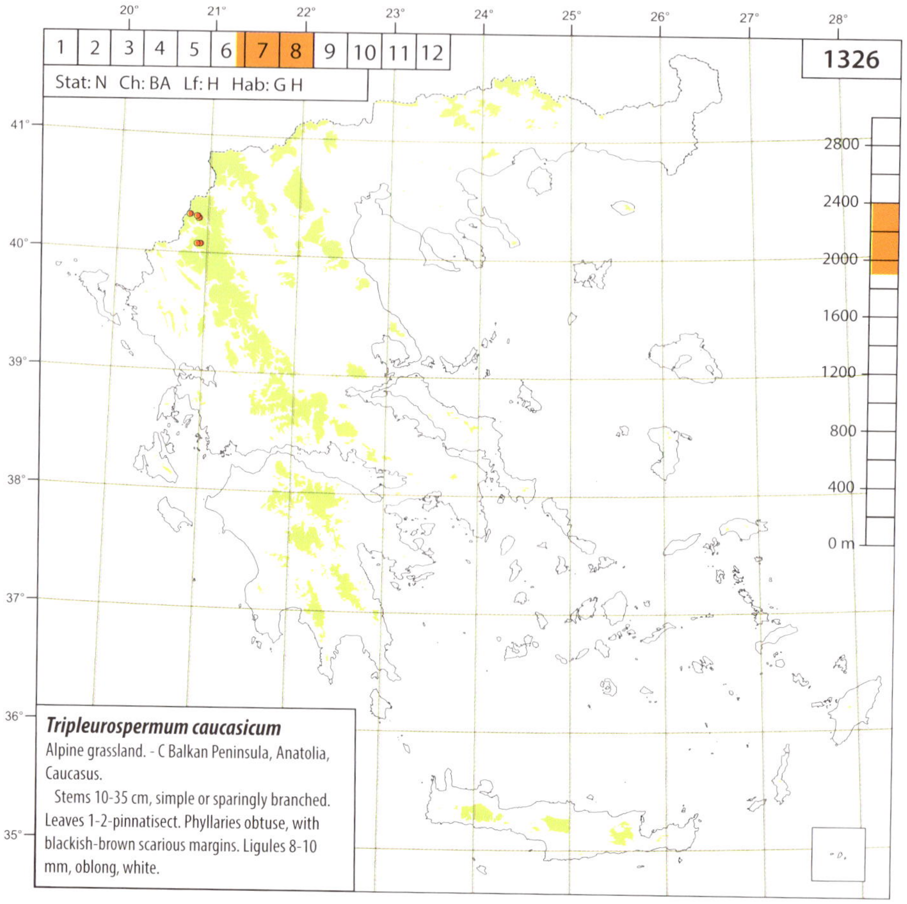

Tripleurospermum caucasicum
Alpine grassland. - C Balkan Peninsula, Anatolia, Caucasus.

Stems 10-35 cm, simple or sparingly branched. Leaves 1-2-pinnatisect. Phyllaries obtuse, with blackish-brown scarious margins. Ligules 8-10 mm, oblong, white.

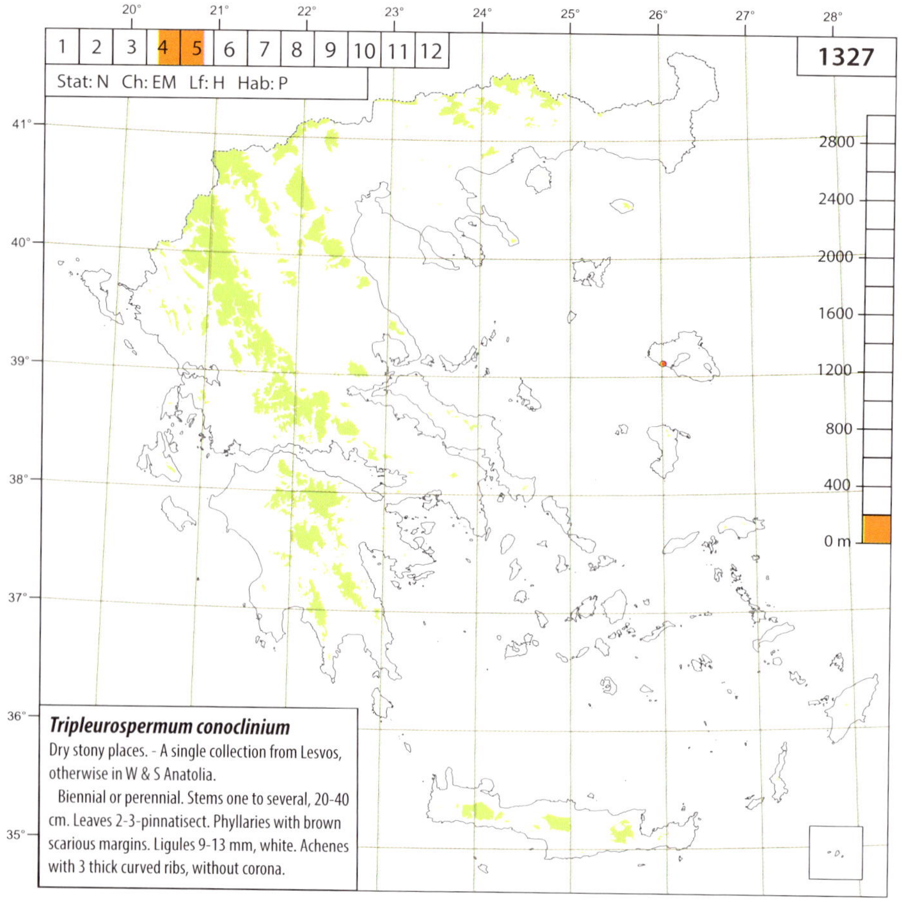

Tripleurospermum conoclinium
Dry stony places. - A single collection from Lesvos, otherwise in W & S Anatolia.

Biennial or perennial. Stems one to several, 20-40 cm. Leaves 2-3-pinnatisect. Phyllaries with brown scarious margins. Ligules 9-13 mm, white. Achenes with 3 thick curved ribs, without corona.

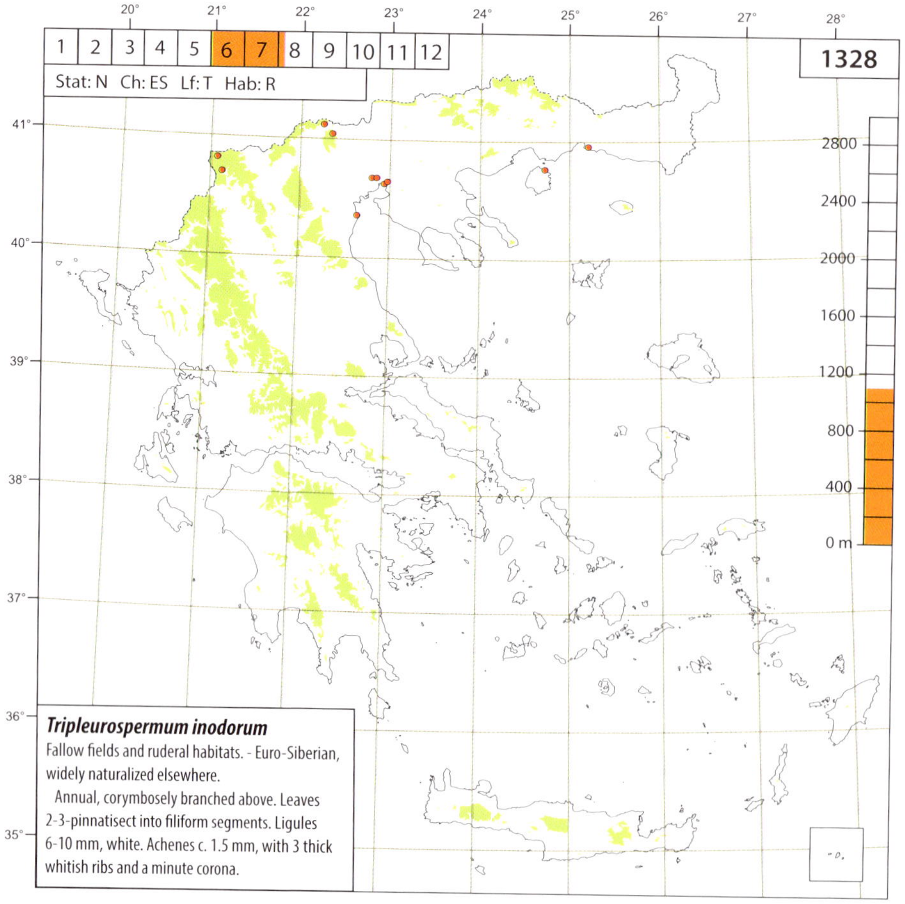

Tripleurospermum inodorum
Fallow fields and ruderal habitats. - Euro-Siberian, widely naturalized elsewhere.

Annual, corymbosely branched above. Leaves 2-3-pinnatisect into filiform segments. Ligules 6-10 mm, white. Achenes c. 1.5 mm, with 3 thick whitish ribs and a minute corona.

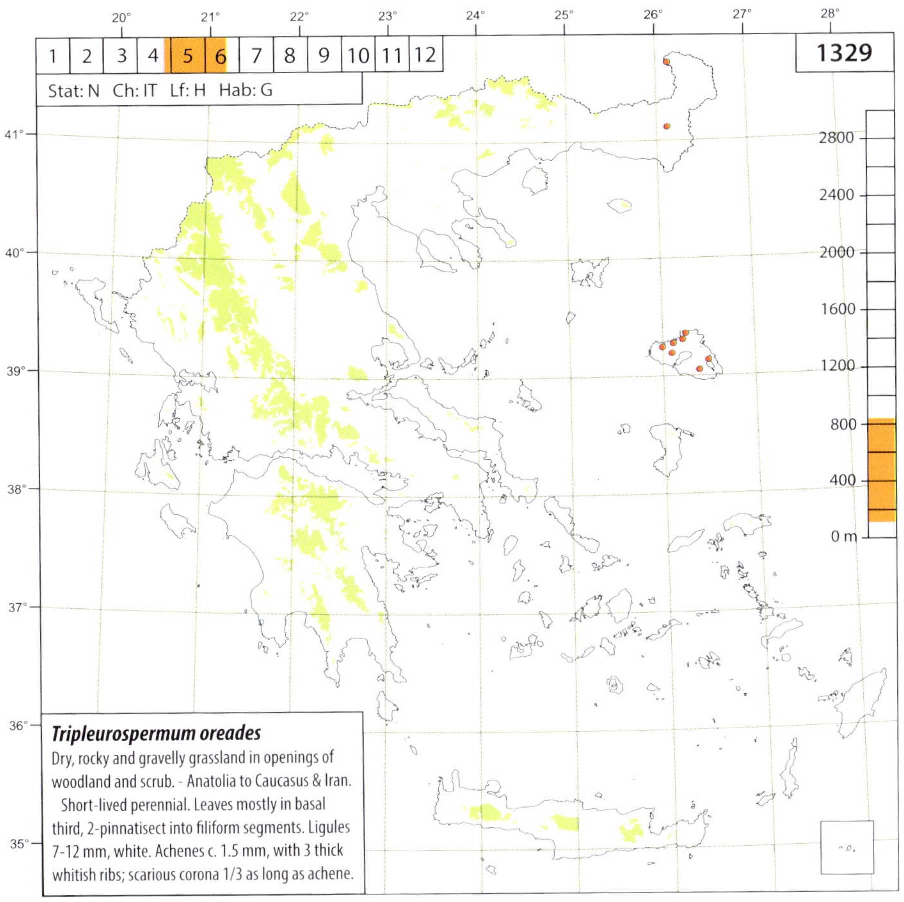

1329

Stat: N Ch: IT Lf: H Hab: G

Tripleurospermum oreades
Dry, rocky and gravelly grassland in openings of woodland and scrub. - Anatolia to Caucasus & Iran.
Short-lived perennial. Leaves mostly in basal third, 2-pinnatisect into filiform segments. Ligules 7-12 mm, white. Achenes c. 1.5 mm, with 3 thick whitish ribs; scarious corona 1/3 as long as achene.

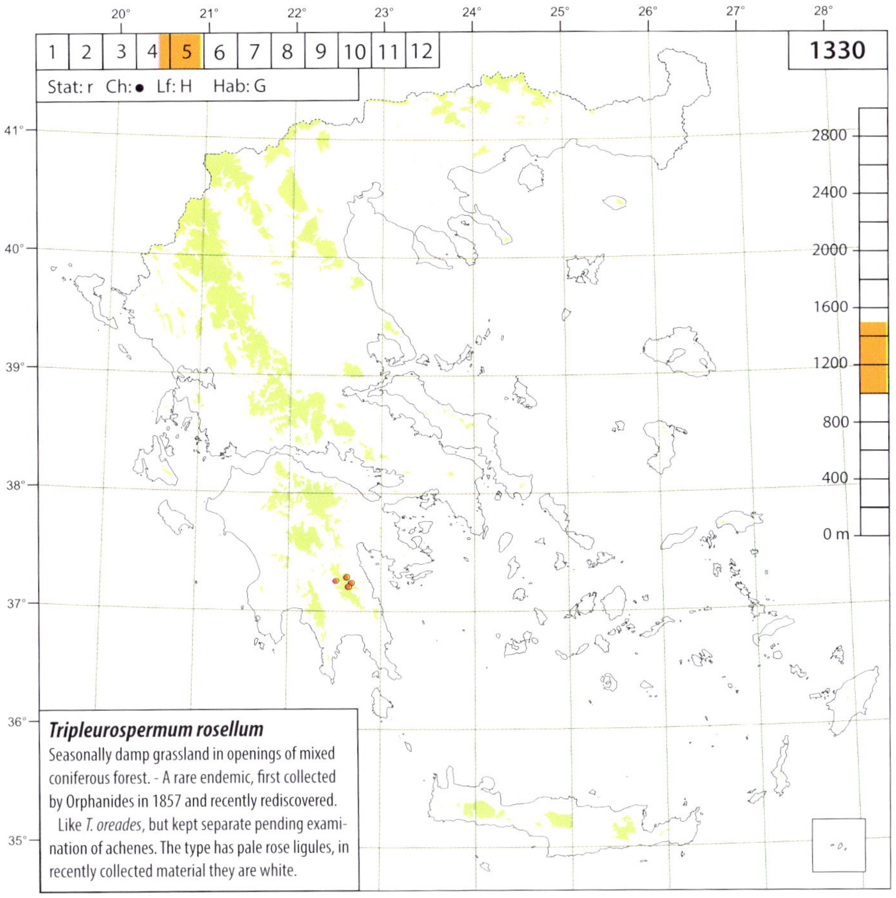

1330

Stat: r Ch: ● Lf: H Hab: G

Tripleurospermum rosellum
Seasonally damp grassland in openings of mixed coniferous forest. - A rare endemic, first collected by Orphanides in 1857 and recently rediscovered.
Like *T. oreades*, but kept separate pending examination of achenes. The type has pale rose ligules, in recently collected material they are white.

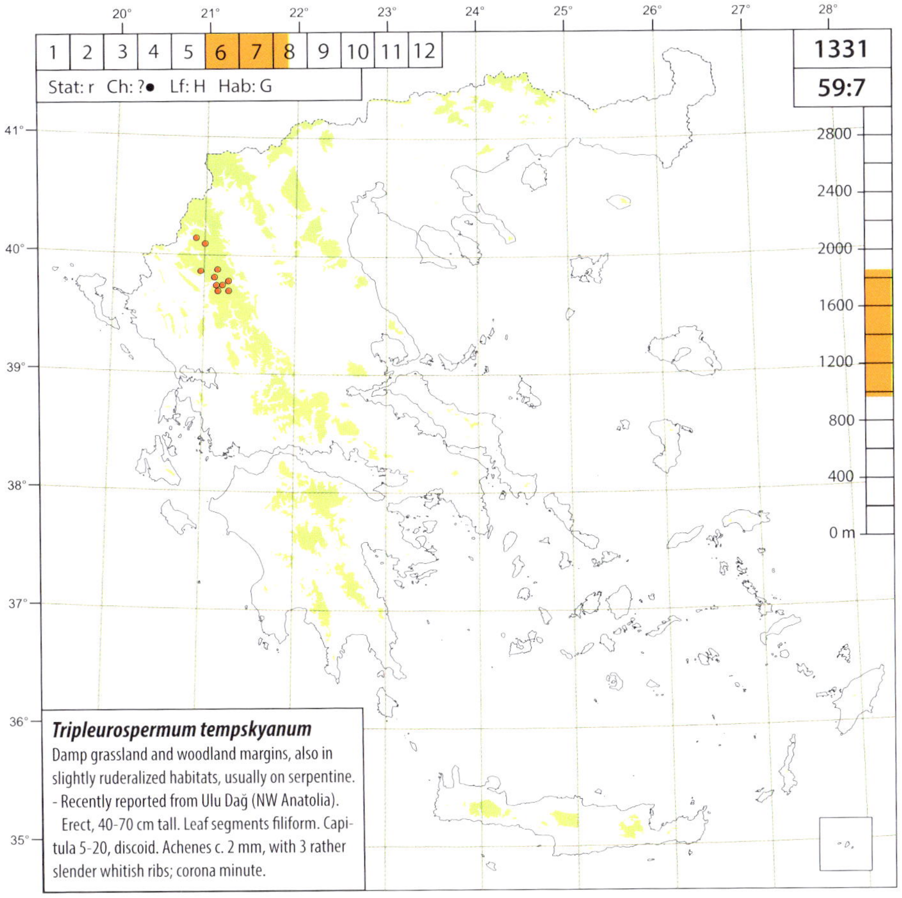

1331
59:7

Stat: r Ch: ?● Lf: H Hab: G

Tripleurospermum tempskyanum
Damp grassland and woodland margins, also in slightly ruderalized habitats, usually on serpentine. - Recently reported from Ulu Dağ (NW Anatolia).
Erect, 40-70 cm tall. Leaf segments filiform. Capitula 5-20, discoid. Achenes c. 2 mm, with 3 rather slender whitish ribs; corona minute.

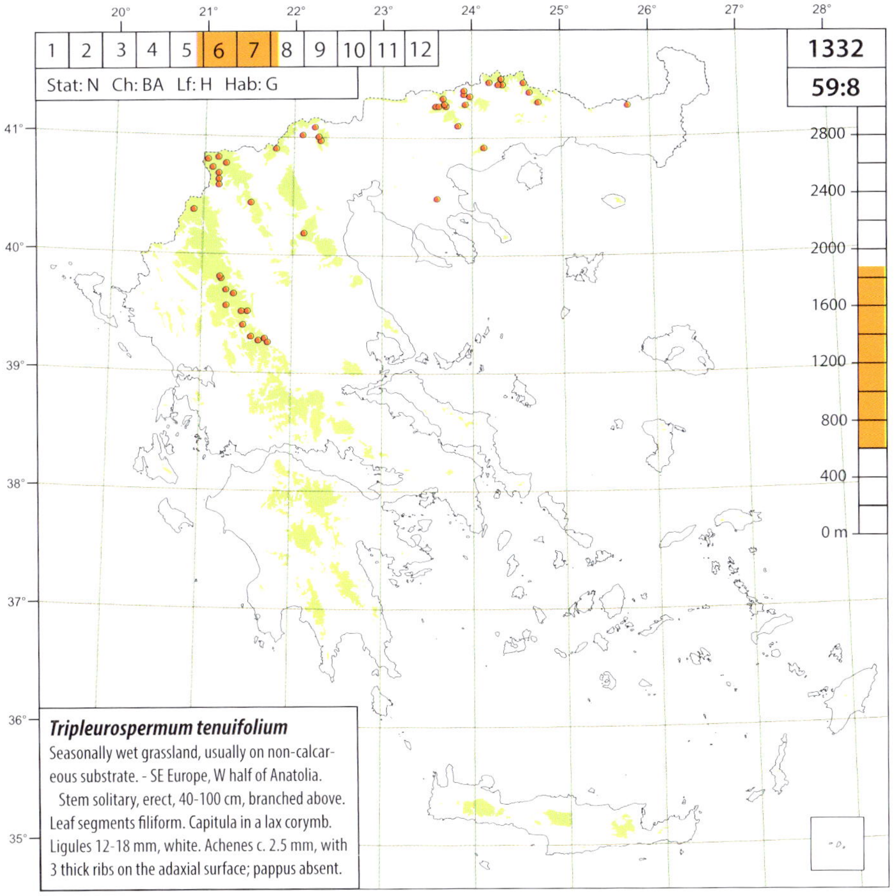

1332
59:8

Stat: N Ch: BA Lf: H Hab: G

Tripleurospermum tenuifolium
Seasonally wet grassland, usually on non-calcareous substrate. - SE Europe, W half of Anatolia.
Stem solitary, erect, 40-100 cm, branched above. Leaf segments filiform. Capitula in a lax corymb. Ligules 12-18 mm, white. Achenes c. 2.5 mm, with 3 thick ribs on the adaxial surface; pappus absent.

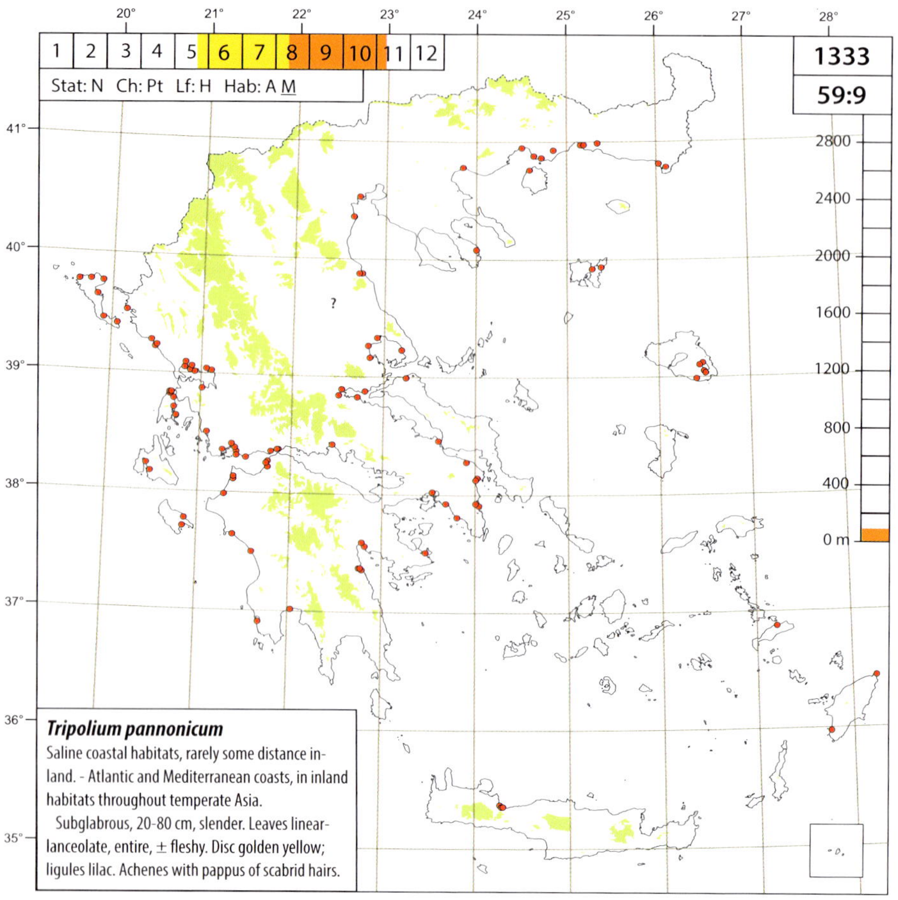

Tripolium pannonicum
Saline coastal habitats, rarely some distance inland. - Atlantic and Mediterranean coasts, in inland habitats throughout temperate Asia.
Subglabrous, 20-80 cm, slender. Leaves linear-lanceolate, entire, ± fleshy. Disc golden yellow; ligules lilac. Achenes with pappus of scabrid hairs.

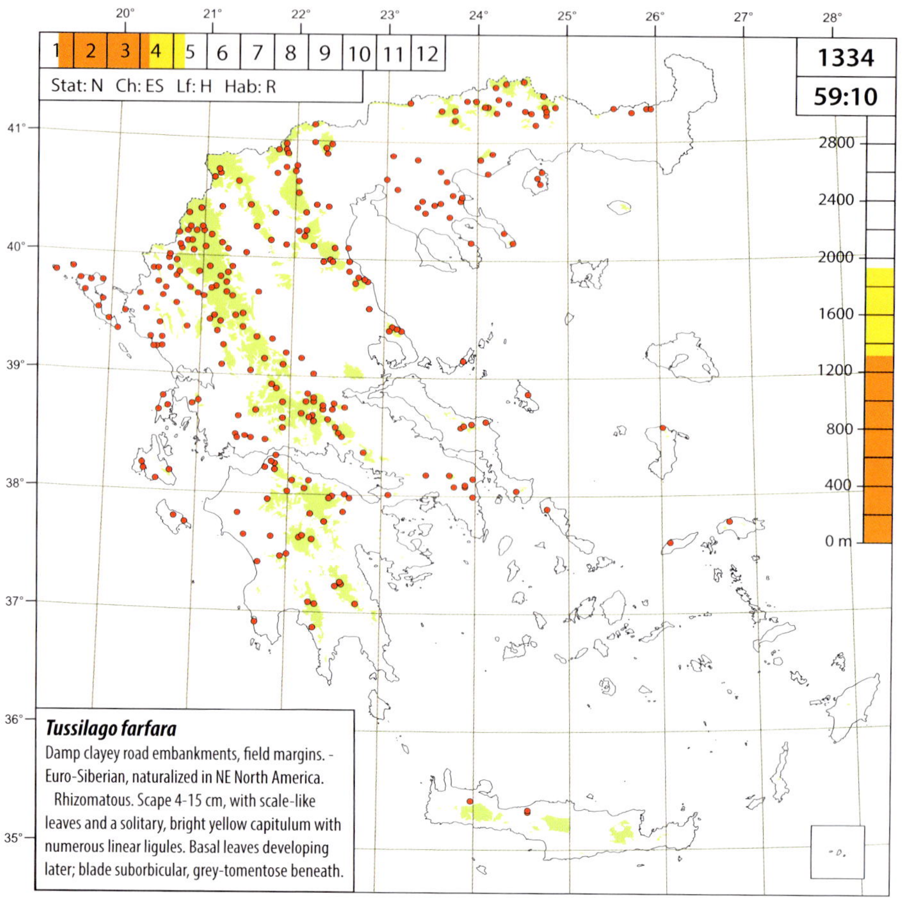

Tussilago farfara
Damp clayey road embankments, field margins. - Euro-Siberian, naturalized in NE North America.
Rhizomatous. Scape 4-15 cm, with scale-like leaves and a solitary, bright yellow capitulum with numerous linear ligules. Basal leaves developing later; blade suborbicular, grey-tomentose beneath.

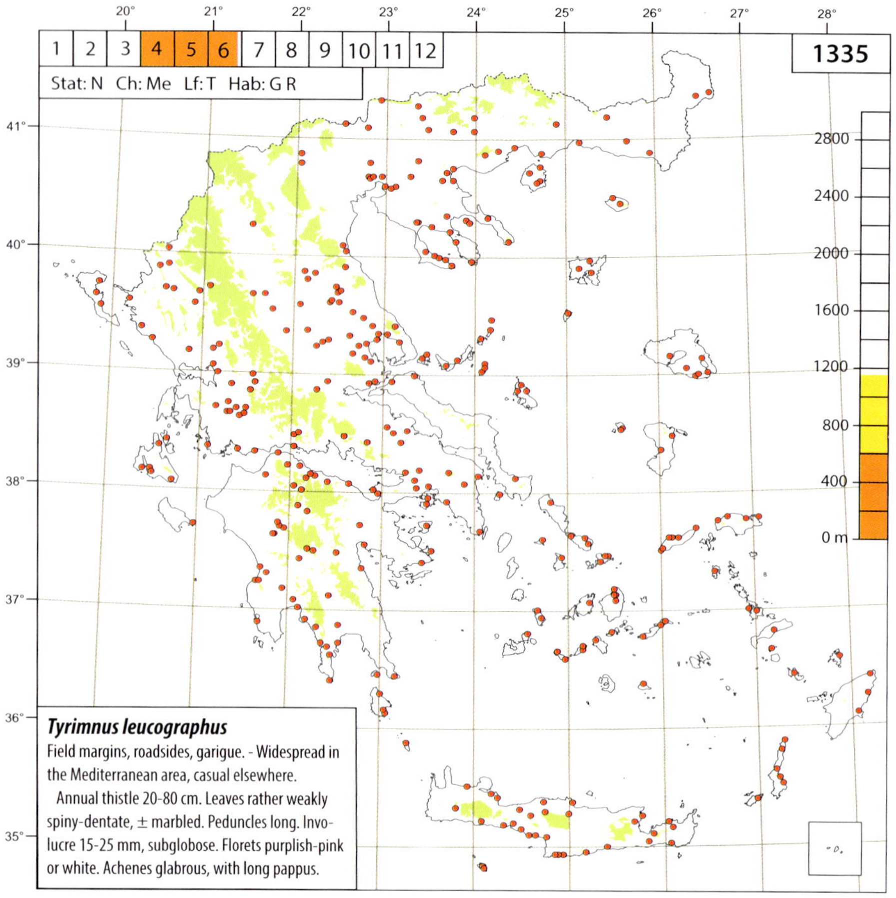

Tyrimnus leucographus
Field margins, roadsides, garigue. - Widespread in the Mediterranean area, casual elsewhere.
Annual thistle 20-80 cm. Leaves rather weakly spiny-dentate, ± marbled. Peduncles long. Involucre 15-25 mm, subglobose. Florets purplish-pink or white. Achenes glabrous, with long pappus.

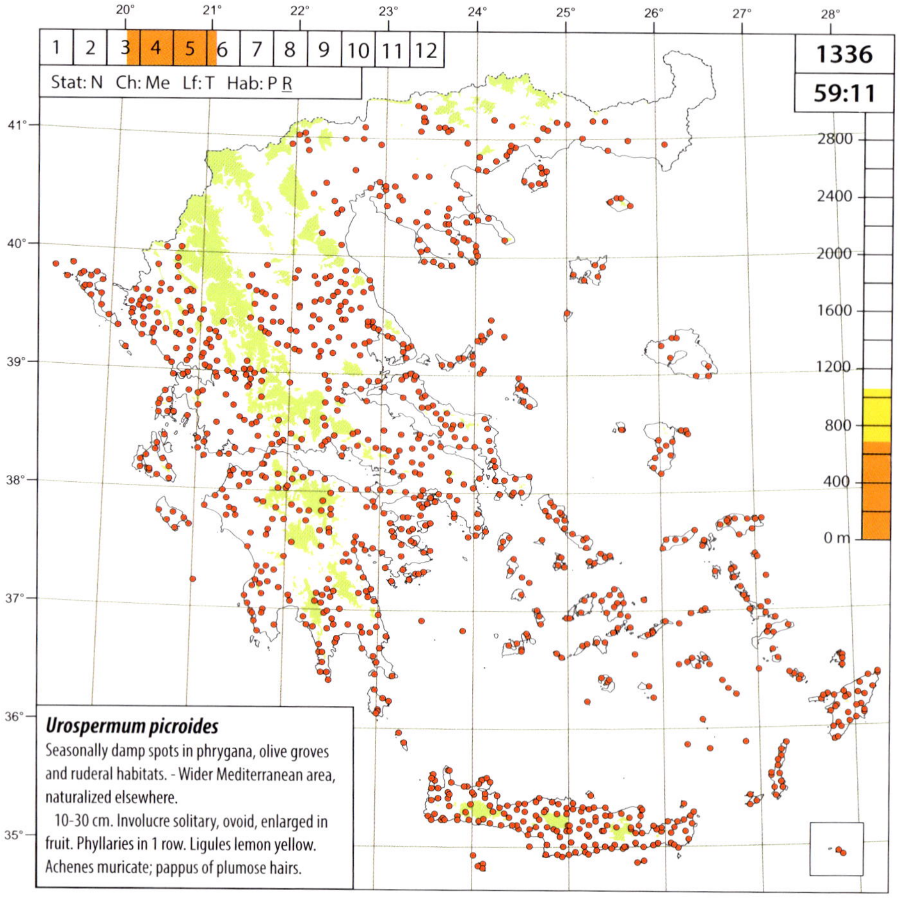

Urospermum picroides
Seasonally damp spots in phrygana, olive groves and ruderal habitats. - Wider Mediterranean area, naturalized elsewhere.
10-30 cm. Involucre solitary, ovoid, enlarged in fruit. Phyllaries in 1 row. Ligules lemon yellow. Achenes muricate; pappus of plumose hairs.

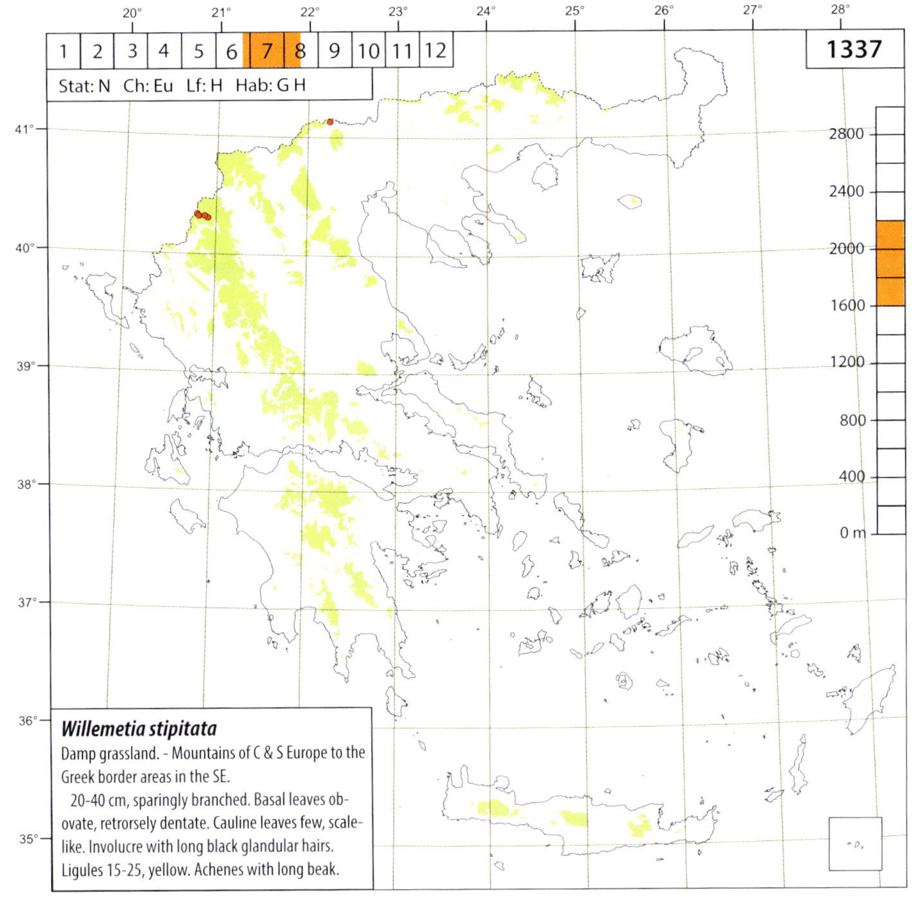

Willemetia stipitata

Damp grassland. - Mountains of C & S Europe to the Greek border areas in the SE.

20-40 cm, sparingly branched. Basal leaves obovate, retrorsely dentate. Cauline leaves few, scale-like. Involucre with long black glandular hairs. Ligules 15-25, yellow. Achenes with long beak.

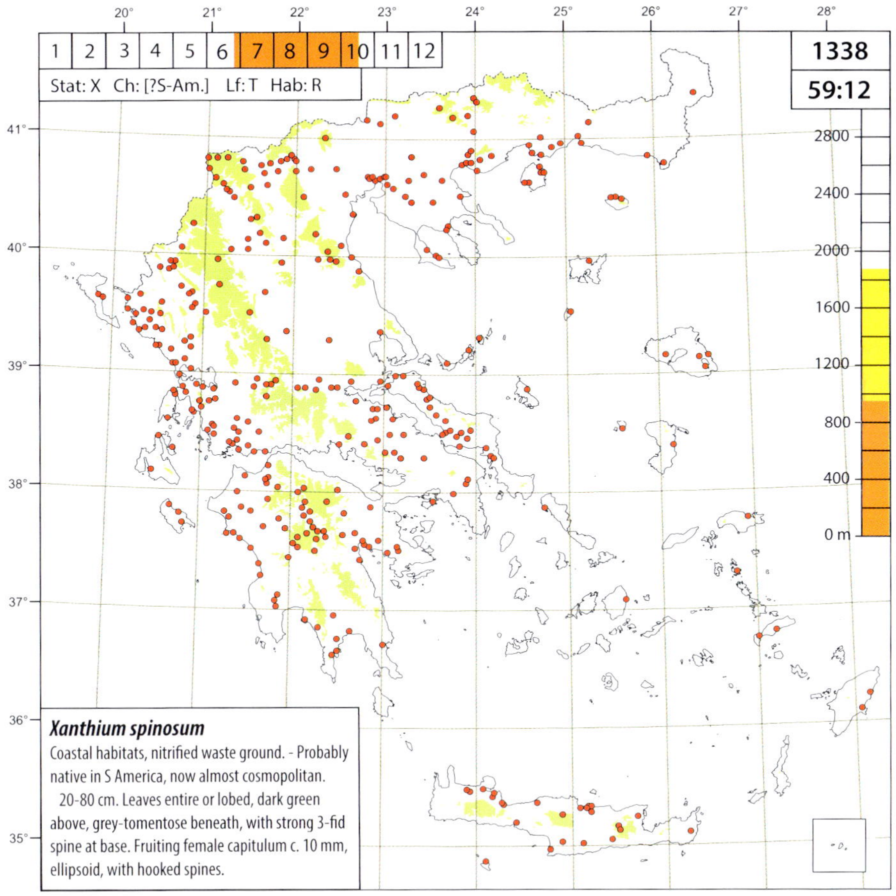

Xanthium spinosum

Coastal habitats, nitrified waste ground. - Probably native in S America, now almost cosmopolitan.

20-80 cm. Leaves entire or lobed, dark green above, grey-tomentose beneath, with strong 3-fid spine at base. Fruiting female capitulum c. 10 mm, ellipsoid, with hooked spines.

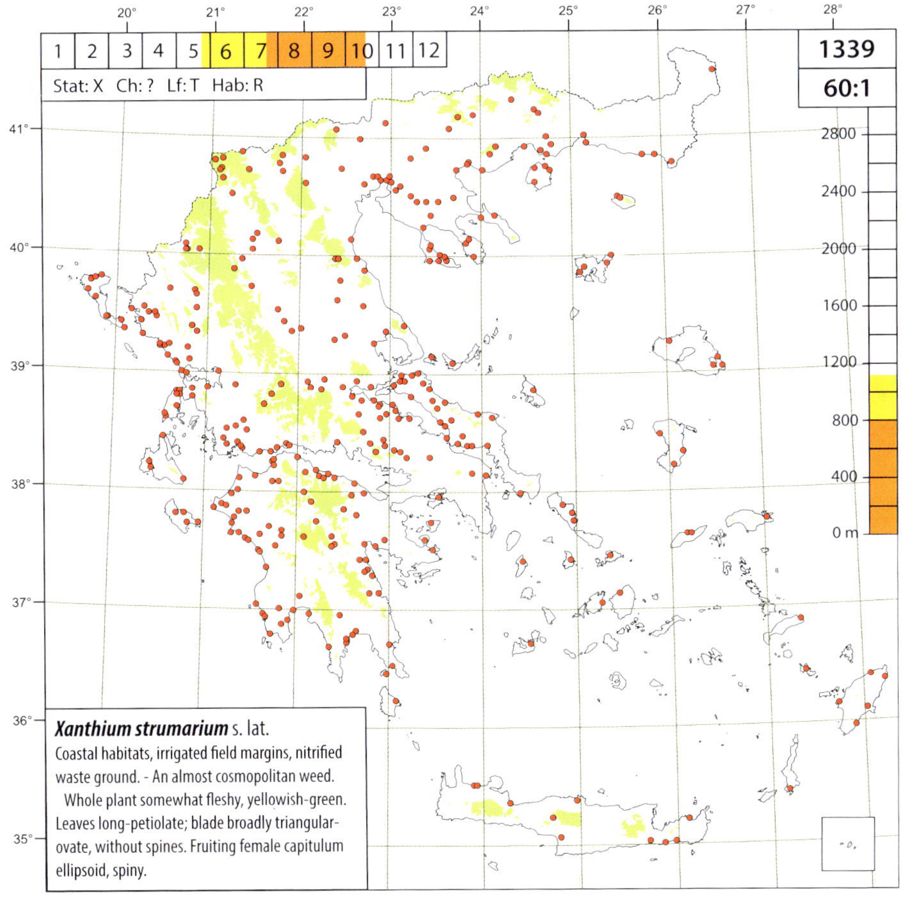

Xanthium strumarium s. lat.

Coastal habitats, irrigated field margins, nitrified waste ground. - An almost cosmopolitan weed.

Whole plant somewhat fleshy, yellowish-green. Leaves long-petiolate; blade broadly triangular-ovate, without spines. Fruiting female capitulum ellipsoid, spiny.

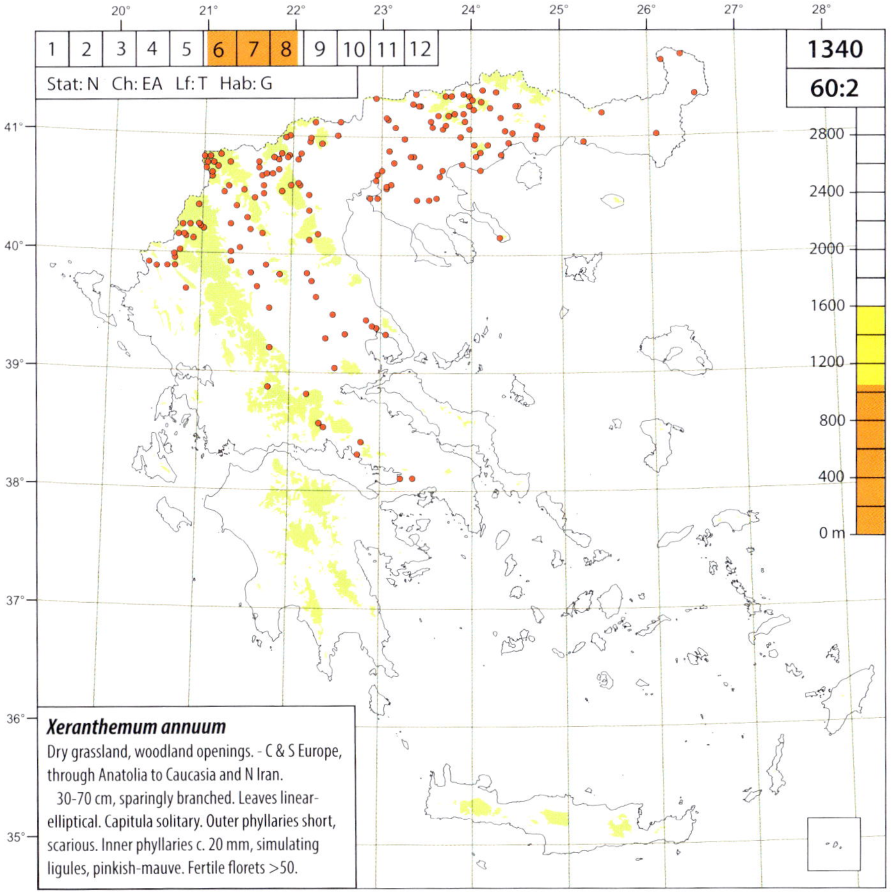

Xeranthemum annuum

Dry grassland, woodland openings. - C & S Europe, through Anatolia to Caucasia and N Iran.

30-70 cm, sparingly branched. Leaves linear-elliptical. Capitula solitary. Outer phyllaries short, scarious. Inner phyllaries c. 20 mm, simulating ligules, pinkish-mauve. Fertile florets >50.

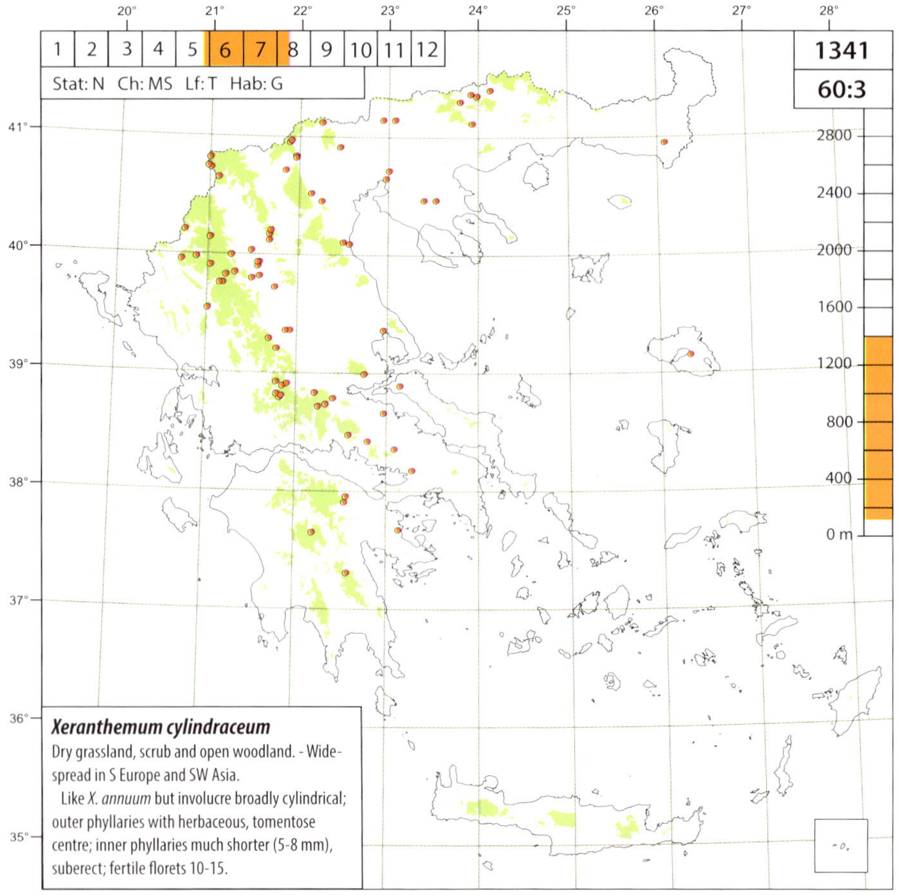

Xeranthemum cylindraceum
Dry grassland, scrub and open woodland. - Wide-spread in S Europe and SW Asia.
Like *X. annuum* but involucre broadly cylindrical; outer phyllaries with herbaceous, tomentose centre; inner phyllaries much shorter (5-8 mm), suberect; fertile florets 10-15.

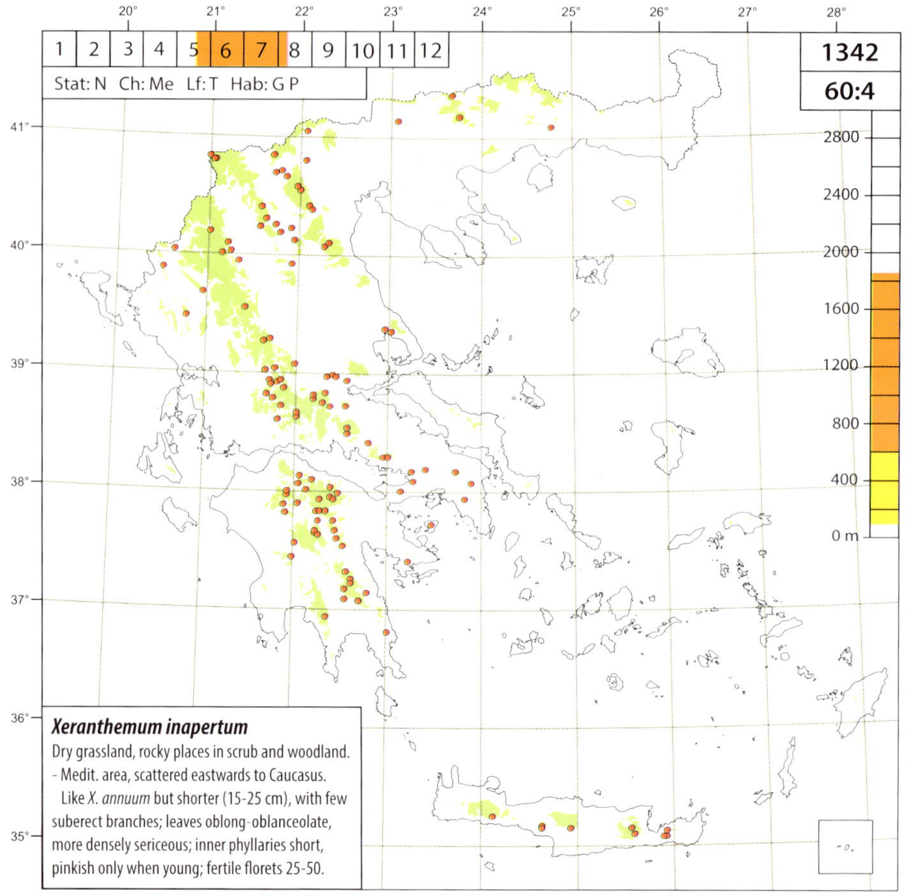

Xeranthemum inapertum
Dry grassland, rocky places in scrub and woodland. - Medit. area, scattered eastwards to Caucasus.
Like *X. annuum* but shorter (15-25 cm), with few suberect branches; leaves oblong-oblanceolate, more densely sericeous; inner phyllaries short, pinkish only when young; fertile florets 25-50.

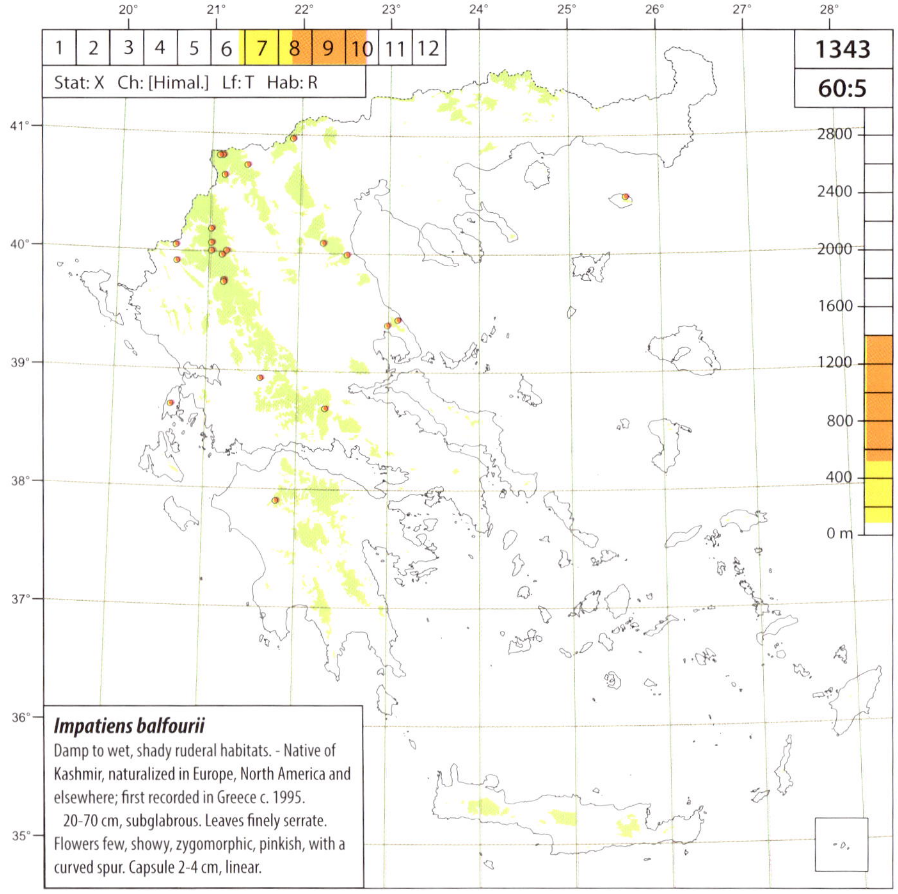

Impatiens balfourii
Damp to wet, shady ruderal habitats. - Native of Kashmir, naturalized in Europe, North America and elsewhere; first recorded in Greece c. 1995.
20-70 cm, subglabrous. Leaves finely serrate. Flowers few, showy, zygomorphic, pinkish, with a curved spur. Capsule 2-4 cm, linear.

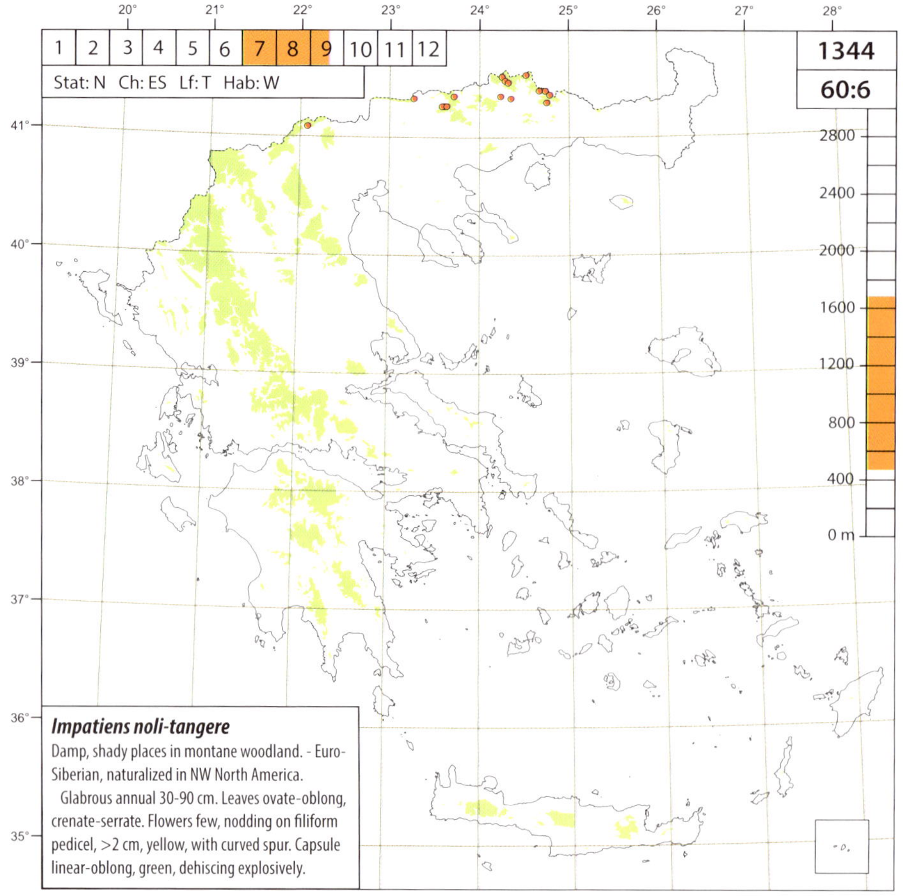

Impatiens noli-tangere
Damp, shady places in montane woodland. - Euro-Siberian, naturalized in NW North America.
Glabrous annual 30-90 cm. Leaves ovate-oblong, crenate-serrate. Flowers few, nodding on filiform pedicel, >2 cm, yellow, with curved spur. Capsule linear-oblong, green, dehiscing explosively.

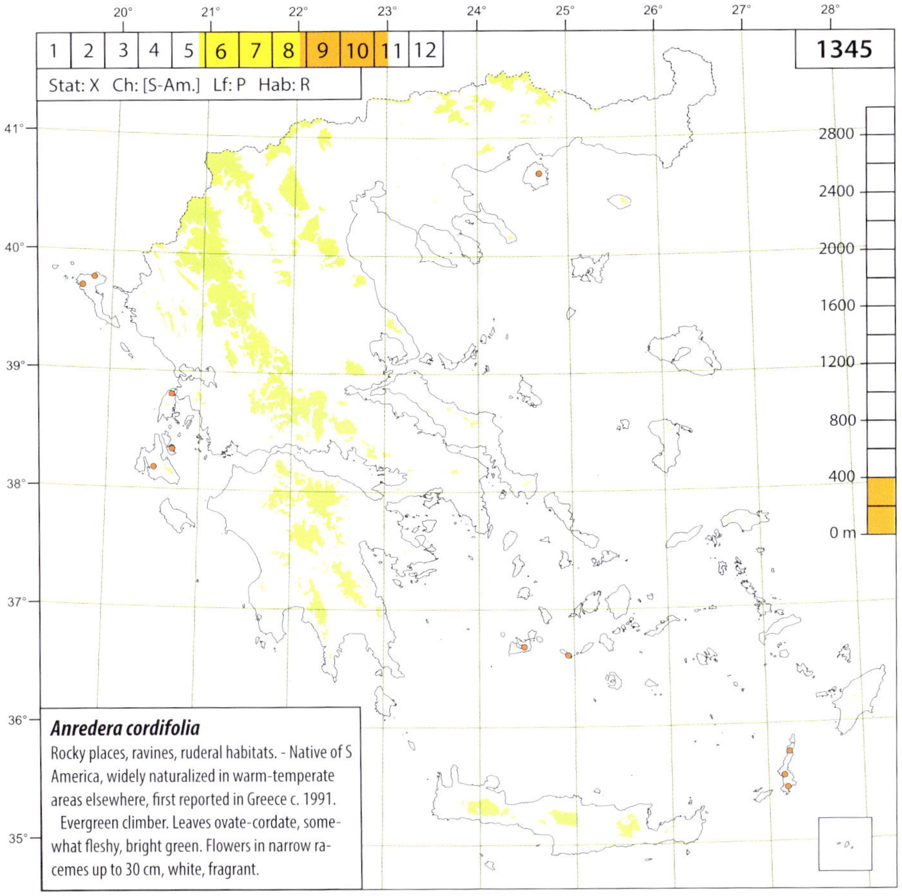

Anredera cordifolia

Rocky places, ravines, ruderal habitats. - Native of S America, widely naturalized in warm-temperate areas elsewhere, first reported in Greece c. 1991.

Evergreen climber. Leaves ovate-cordate, somewhat fleshy, bright green. Flowers in narrow racemes up to 30 cm, white, fragrant.

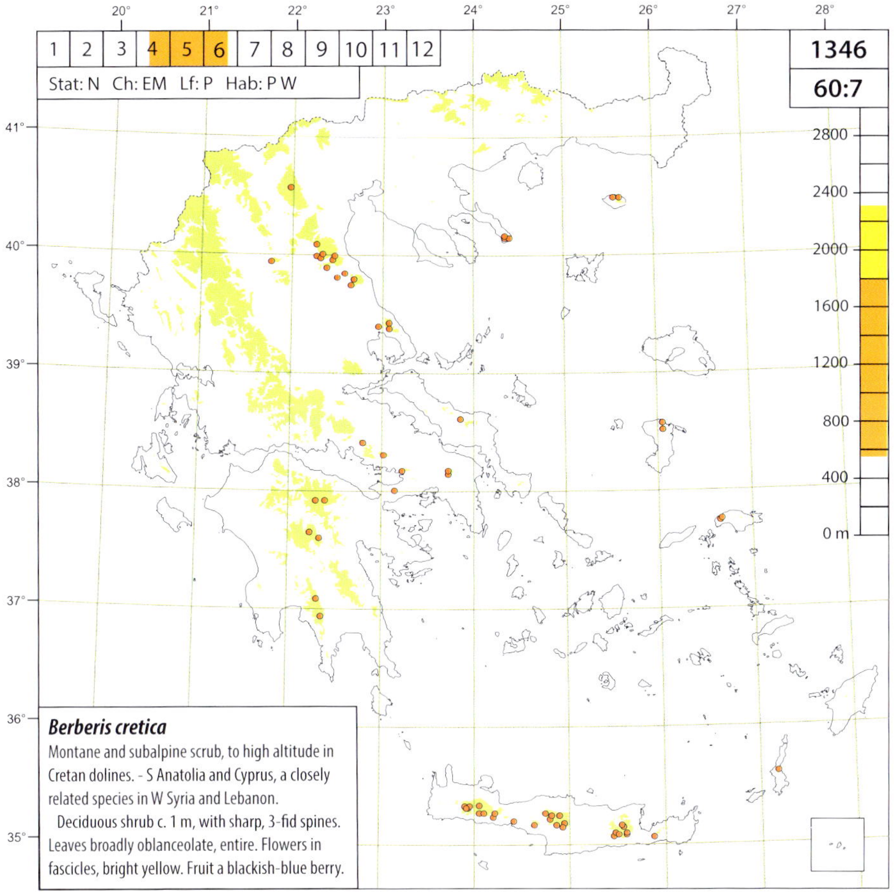

Berberis cretica

Montane and subalpine scrub, to high altitude in Cretan dolines. - S Anatolia and Cyprus, a closely related species in W Syria and Lebanon.

Deciduous shrub c. 1 m, with sharp, 3-fid spines. Leaves broadly oblanceolate, entire. Flowers in fascicles, bright yellow. Fruit a blackish-blue berry.

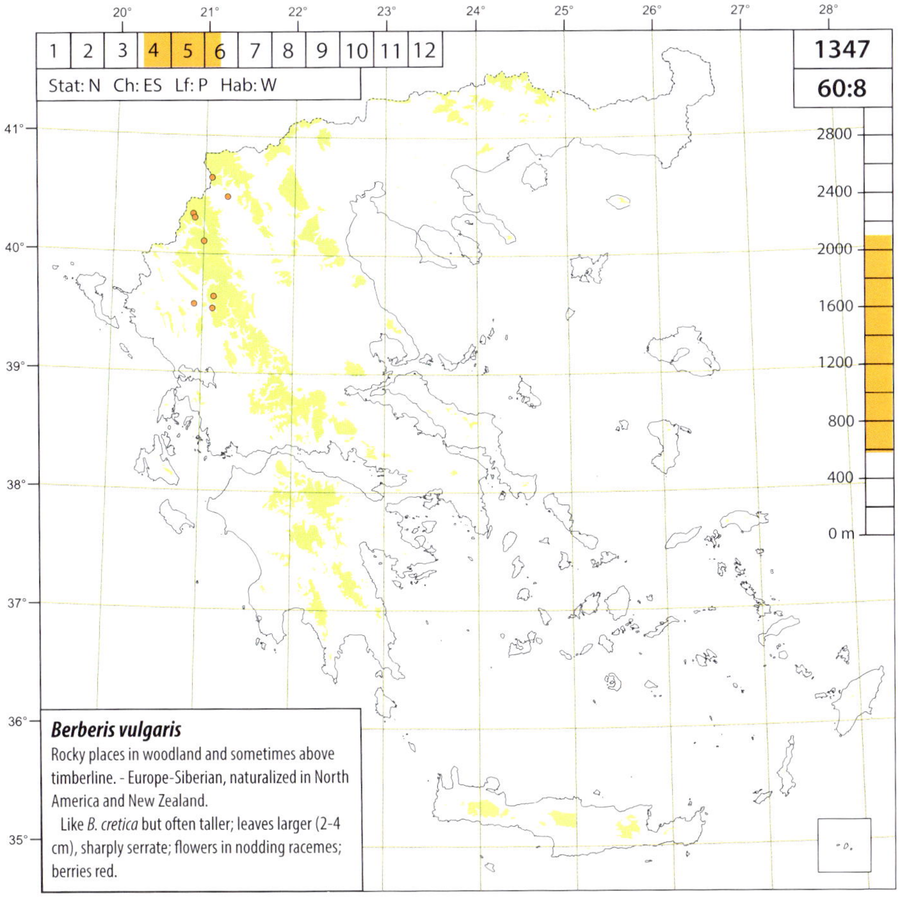

Berberis vulgaris

Rocky places in woodland and sometimes above timberline. - Europe-Siberian, naturalized in North America and New Zealand.

Like *B. cretica* but often taller; leaves larger (2-4 cm), sharply serrate; flowers in nodding racemes; berries red.

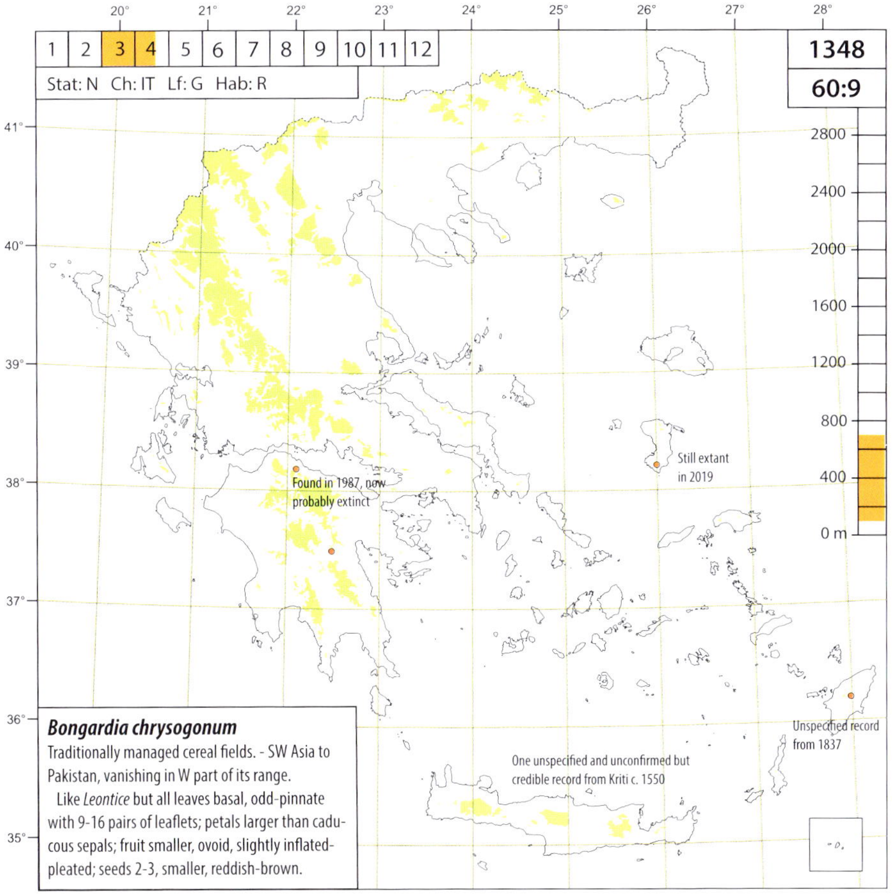

Bongardia chrysogonum

Traditionally managed cereal fields. - SW Asia to Pakistan, vanishing in W part of its range.

Like *Leontice* but all leaves basal, odd-pinnate with 9-16 pairs of leaflets; petals larger than caducous sepals; fruit smaller, ovoid, slightly inflated-pleated; seeds 2-3, smaller, reddish-brown.

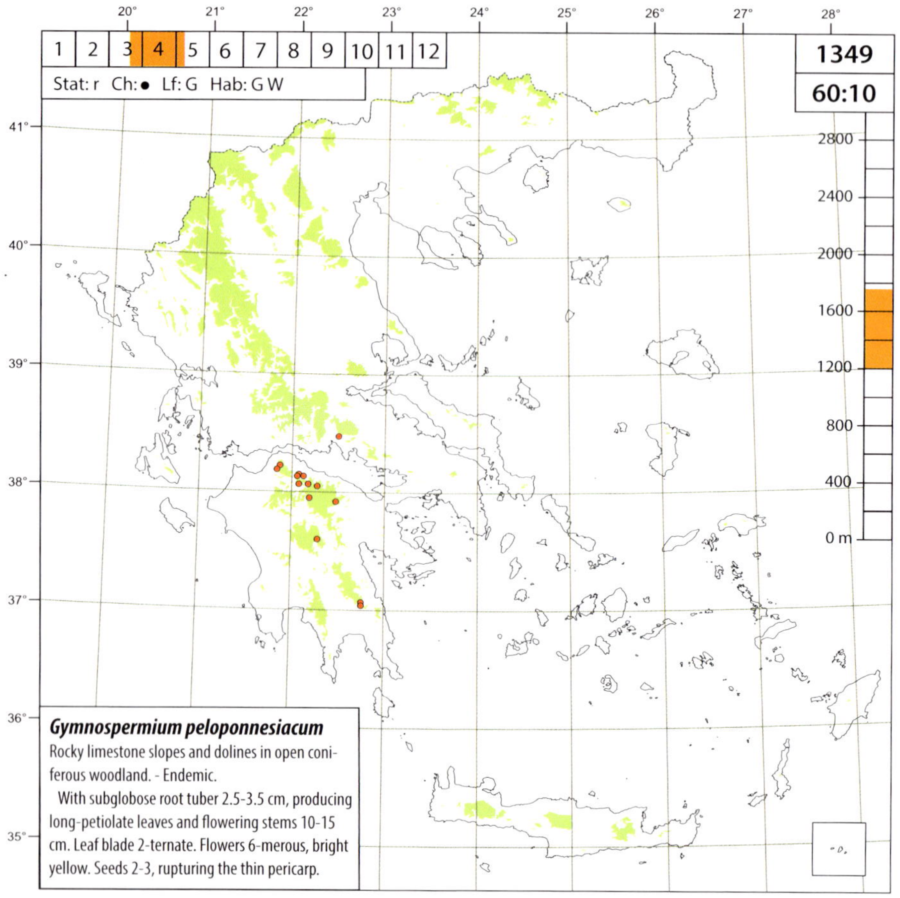

Gymnospermium peloponnesiacum
Rocky limestone slopes and dolines in open coniferous woodland. - Endemic.
With subglobose root tuber 2.5-3.5 cm, producing long-petiolate leaves and flowering stems 10-15 cm. Leaf blade 2-ternate. Flowers 6-merous, bright yellow. Seeds 2-3, rupturing the thin pericarp.

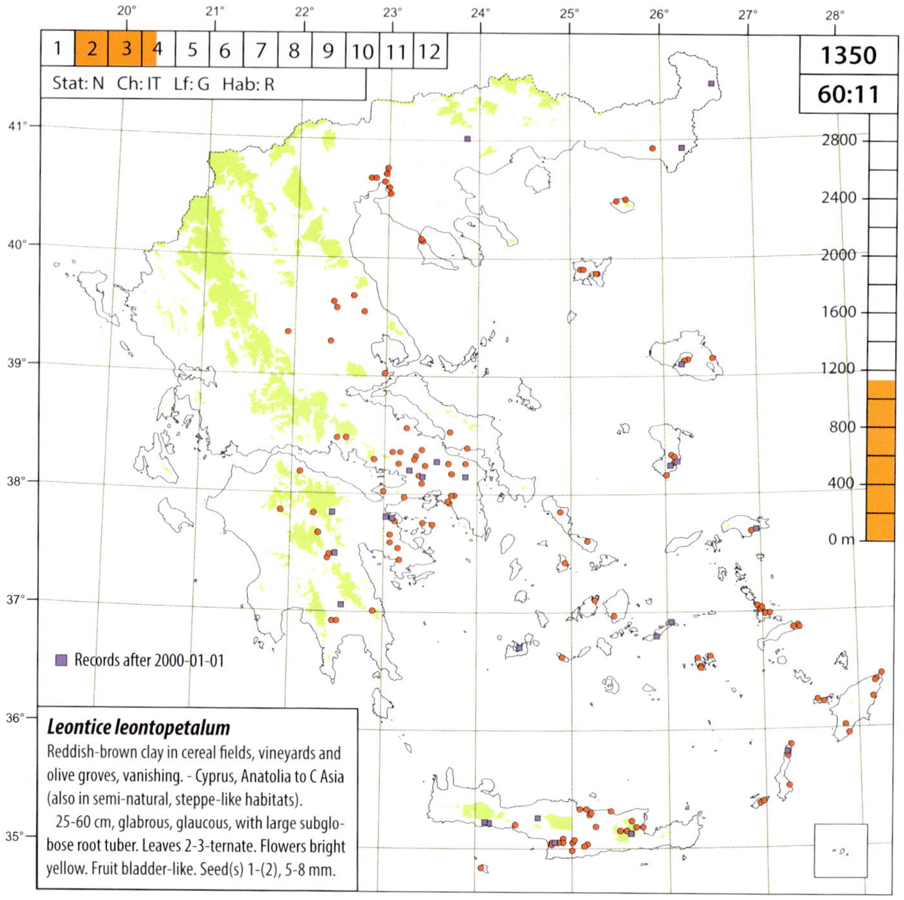

Leontice leontopetalum
Reddish-brown clay in cereal fields, vineyards and olive groves, vanishing. - Cyprus, Anatolia to C Asia (also in semi-natural, steppe-like habitats).
25-60 cm, glabrous, glaucous, with large subglobose root tuber. Leaves 2-3-ternate. Flowers bright yellow. Fruit bladder-like. Seed(s) 1-(2), 5-8 mm.

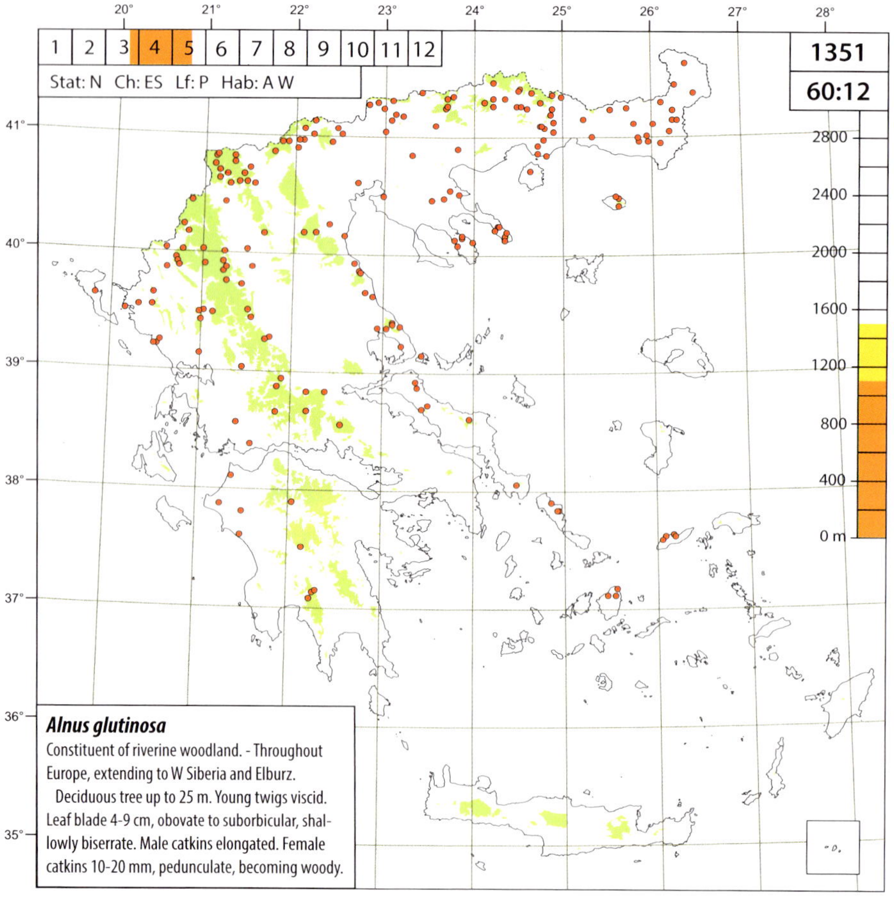

Alnus glutinosa
Constituent of riverine woodland. - Throughout Europe, extending to W Siberia and Elburz.
Deciduous tree up to 25 m. Young twigs viscid. Leaf blade 4-9 cm, obovate to suborbicular, shallowly biserrate. Male catkins elongated. Female catkins 10-20 mm, pedunculate, becoming woody.

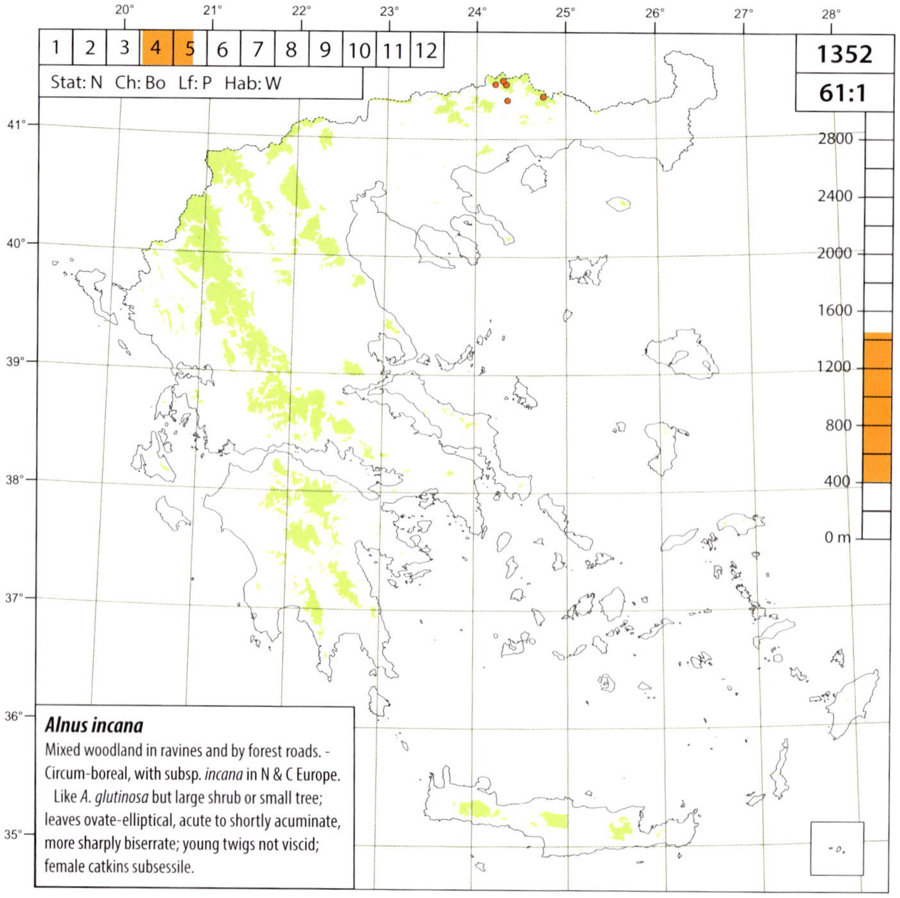

Alnus incana
Mixed woodland in ravines and by forest roads. - Circum-boreal, with subsp. *incana* in N & C Europe.
Like *A. glutinosa* but large shrub or small tree; leaves ovate-elliptical, acute to shortly acuminate, more sharply biserrate; young twigs not viscid; female catkins subsessile.

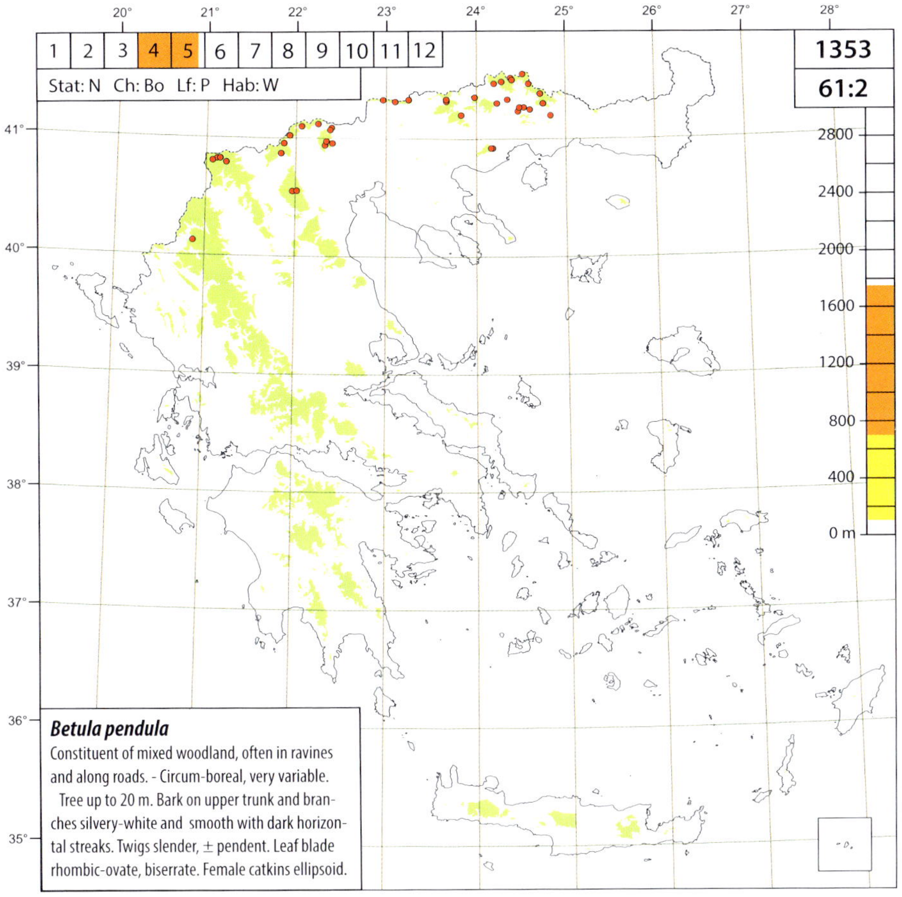

Betula pendula

Constituent of mixed woodland, often in ravines and along roads. - Circum-boreal, very variable.

Tree up to 20 m. Bark on upper trunk and branches silvery-white and smooth with dark horizontal streaks. Twigs slender, ± pendent. Leaf blade rhombic-ovate, biserrate. Female catkins ellipsoid.

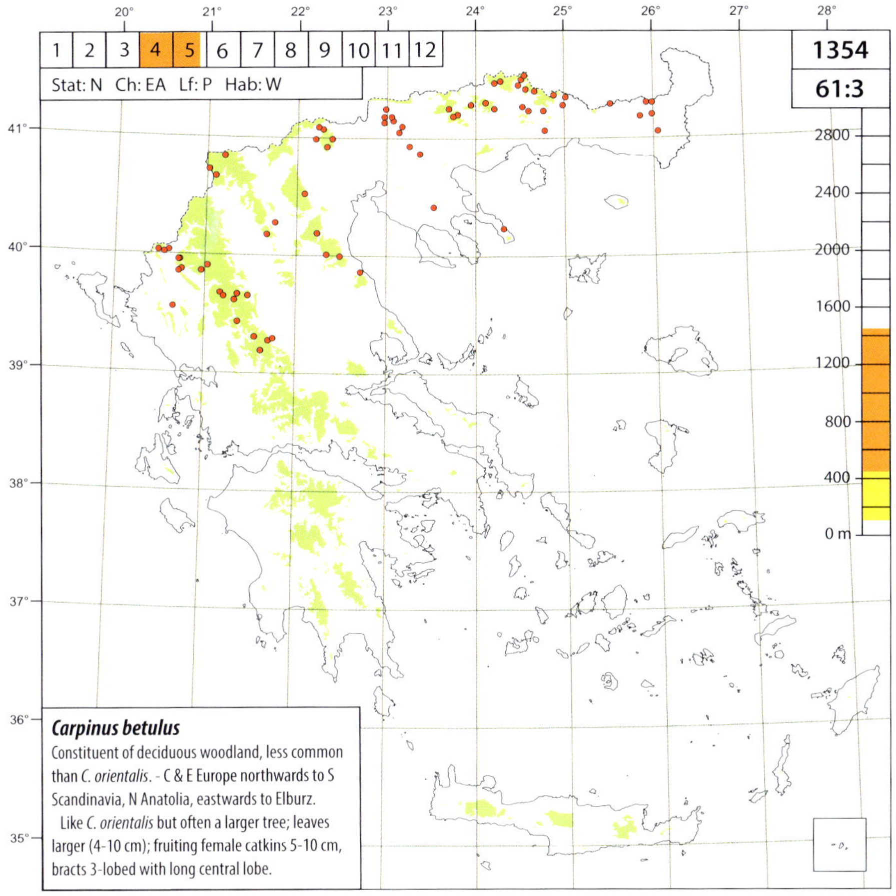

Carpinus betulus

Constituent of deciduous woodland, less common than *C. orientalis*. - C & E Europe northwards to S Scandinavia, N Anatolia, eastwards to Elburz.

Like *C. orientalis* but often a larger tree; leaves larger (4-10 cm); fruiting female catkins 5-10 cm, bracts 3-lobed with long central lobe.

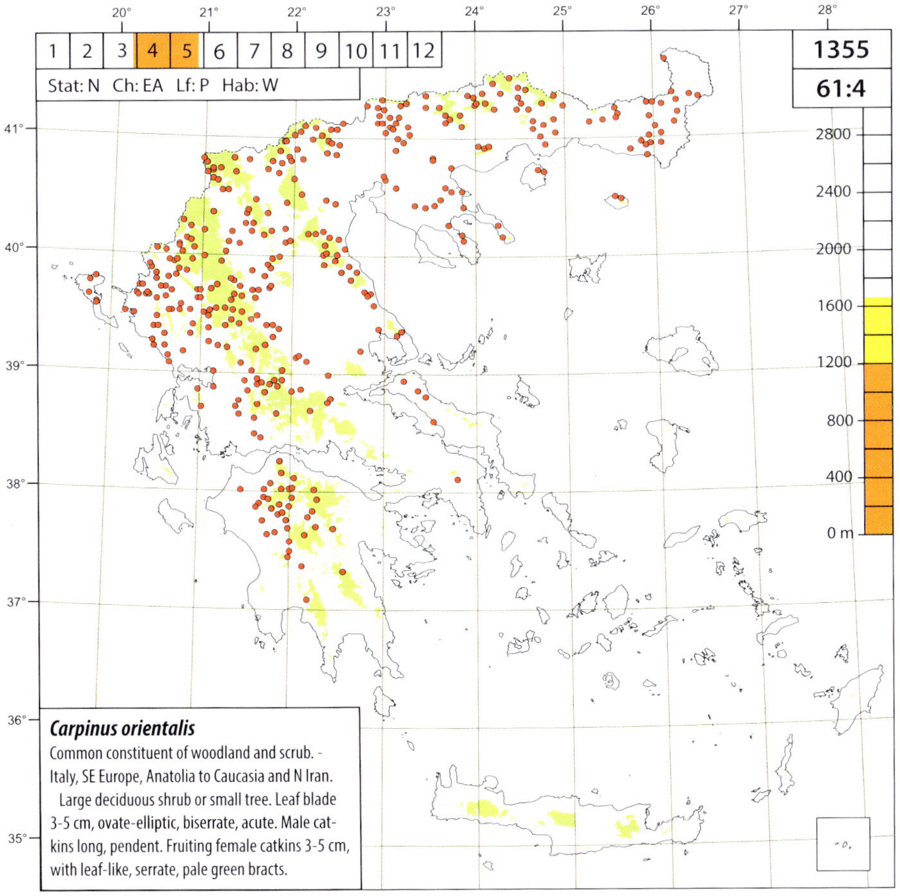

Carpinus orientalis

Common constituent of woodland and scrub. - Italy, SE Europe, Anatolia to Caucasia and N Iran.

Large deciduous shrub or small tree. Leaf blade 3-5 cm, ovate-elliptic, biserrate, acute. Male catkins long, pendent. Fruiting female catkins 3-5 cm, with leaf-like, serrate, pale green bracts.

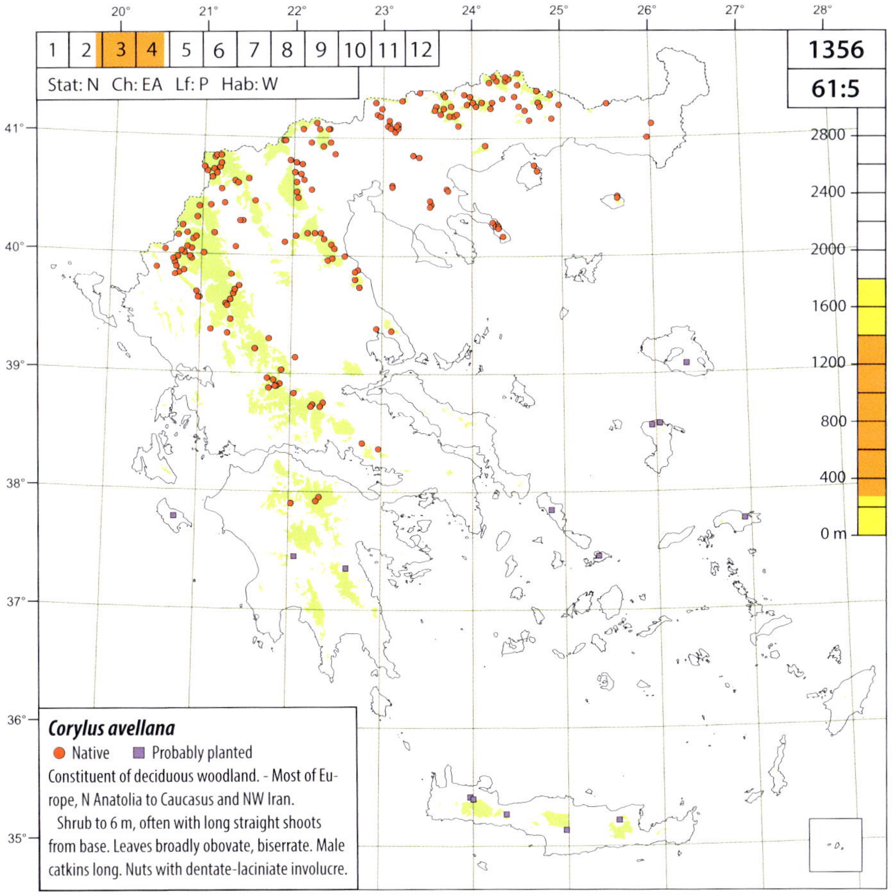

Corylus avellana

● Native ■ Probably planted

Constituent of deciduous woodland. - Most of Europe, N Anatolia to Caucasus and NW Iran.

Shrub to 6 m, often with long straight shoots from base. Leaves broadly obovate, biserrate. Male catkins long. Nuts with dentate-laciniate involucre.

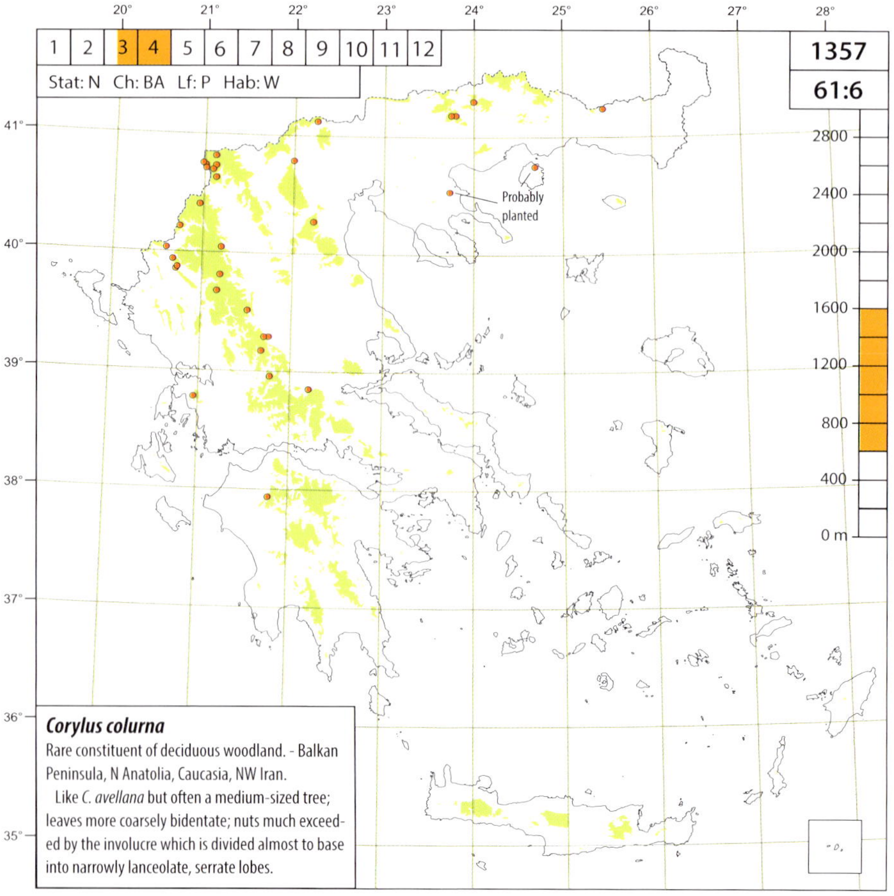

Corylus colurna
Rare constituent of deciduous woodland. - Balkan Peninsula, N Anatolia, Caucasia, NW Iran.
Like *C. avellana* but often a medium-sized tree; leaves more coarsely bidentate; nuts much exceeded by the involucre which is divided almost to base into narrowly lanceolate, serrate lobes.

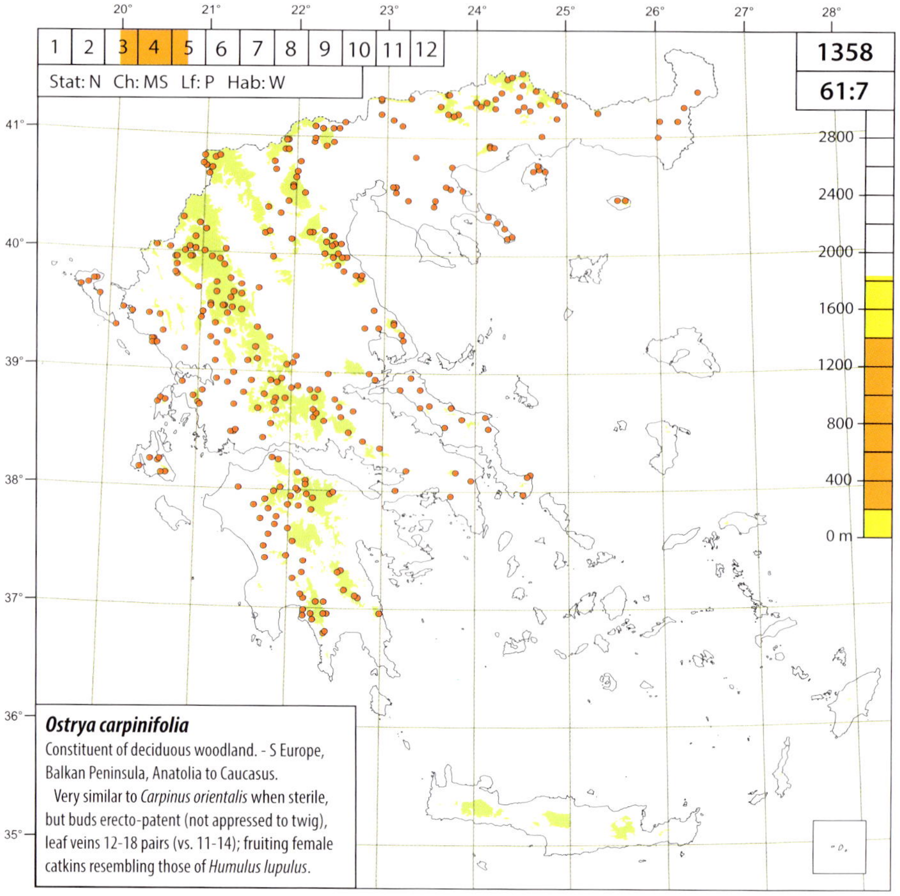

Ostrya carpinifolia
Constituent of deciduous woodland. - S Europe, Balkan Peninsula, Anatolia to Caucasus.
Very similar to *Carpinus orientalis* when sterile, but buds erecto-patent (not appressed to twig), leaf veins 12-18 pairs (vs. 11-14); fruiting female catkins resembling those of *Humulus lupulus*.

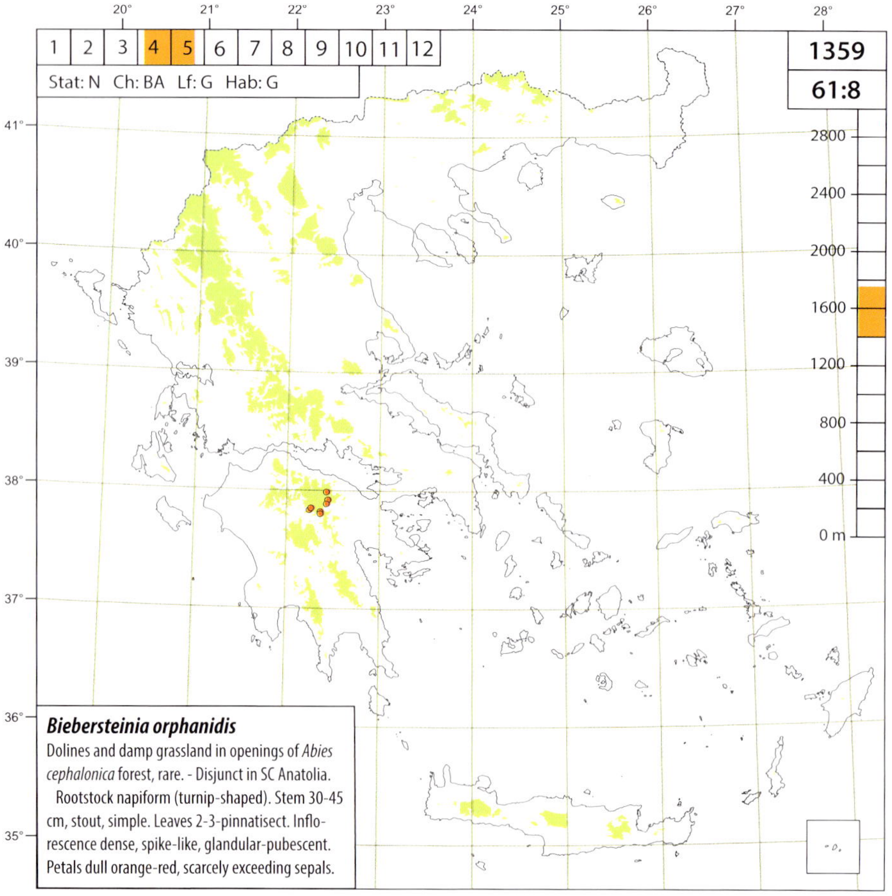

Biebersteinia orphanidis
Dolines and damp grassland in openings of *Abies cephalonica* forest, rare. - Disjunct in SC Anatolia.
Rootstock napiform (turnip-shaped). Stem 30-45 cm, stout, simple. Leaves 2-3-pinnatisect. Inflorescence dense, spike-like, glandular-pubescent. Petals dull orange-red, scarcely exceeding sepals.

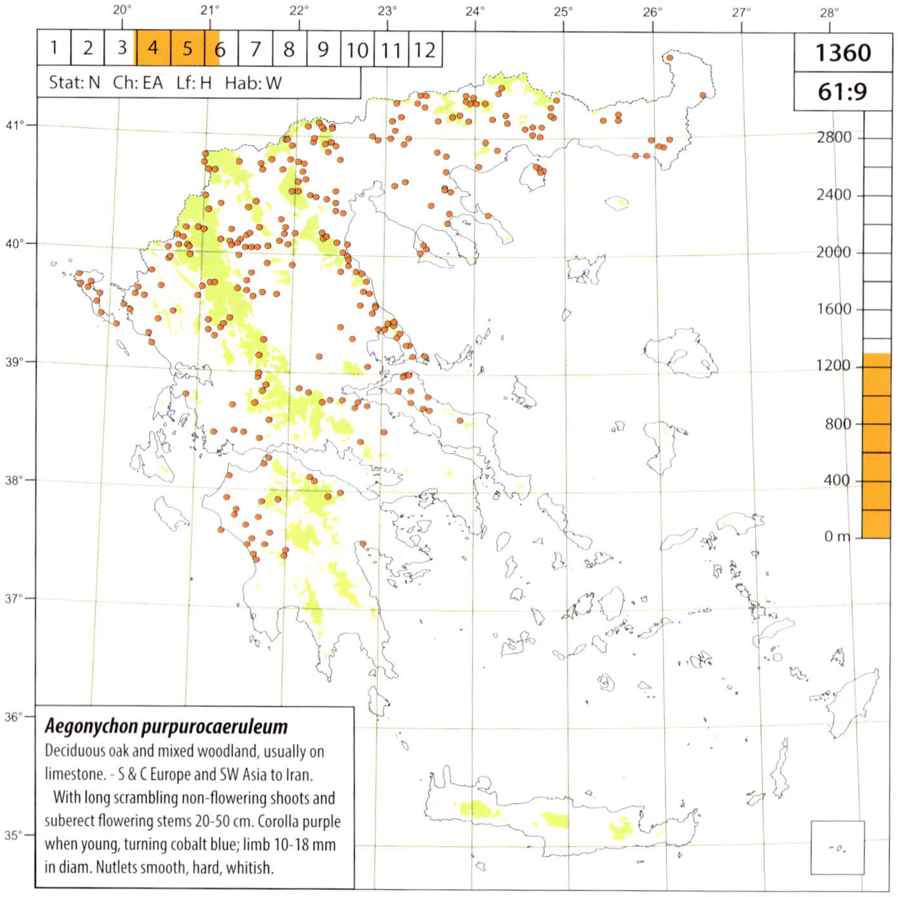

Aegonychon purpurocaeruleum
Deciduous oak and mixed woodland, usually on limestone. - S & C Europe and SW Asia to Iran.
With long scrambling non-flowering shoots and suberect flowering stems 20-50 cm. Corolla purple when young, turning cobalt blue; limb 10-18 mm in diam. Nutlets smooth, hard, whitish.

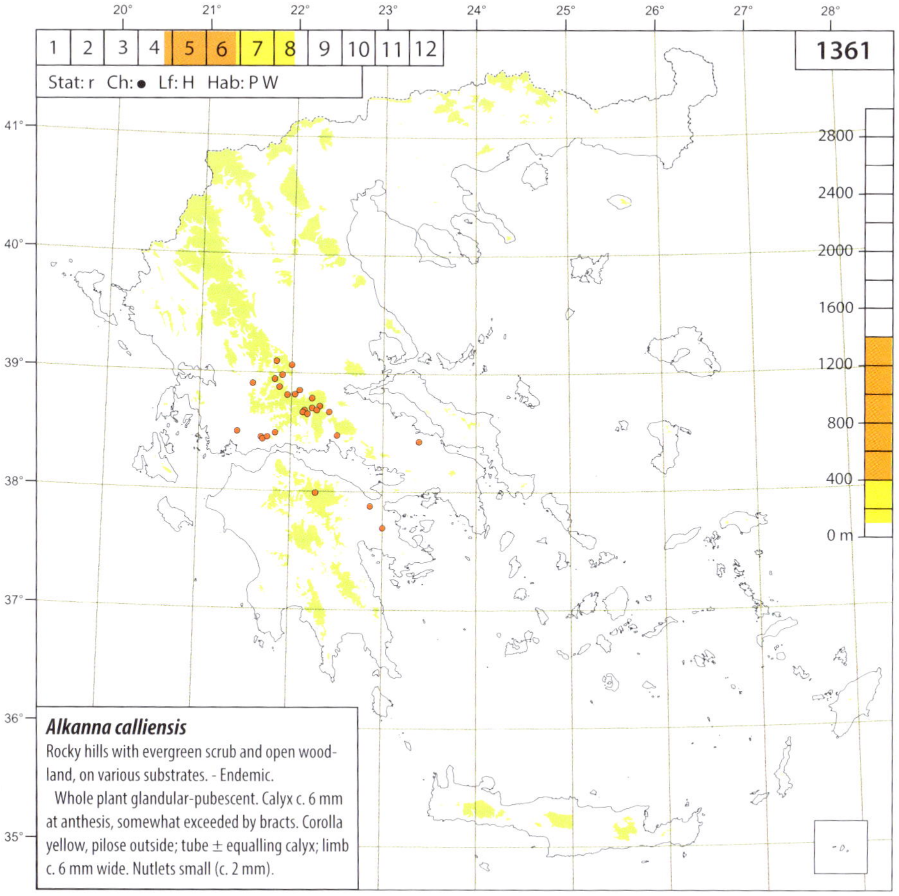

Alkanna calliensis

Rocky hills with evergreen scrub and open woodland, on various substrates. - Endemic.

Whole plant glandular-pubescent. Calyx c. 6 mm at anthesis, somewhat exceeded by bracts. Corolla yellow, pilose outside; tube ± equalling calyx; limb c. 6 mm wide. Nutlets small (c. 2 mm).

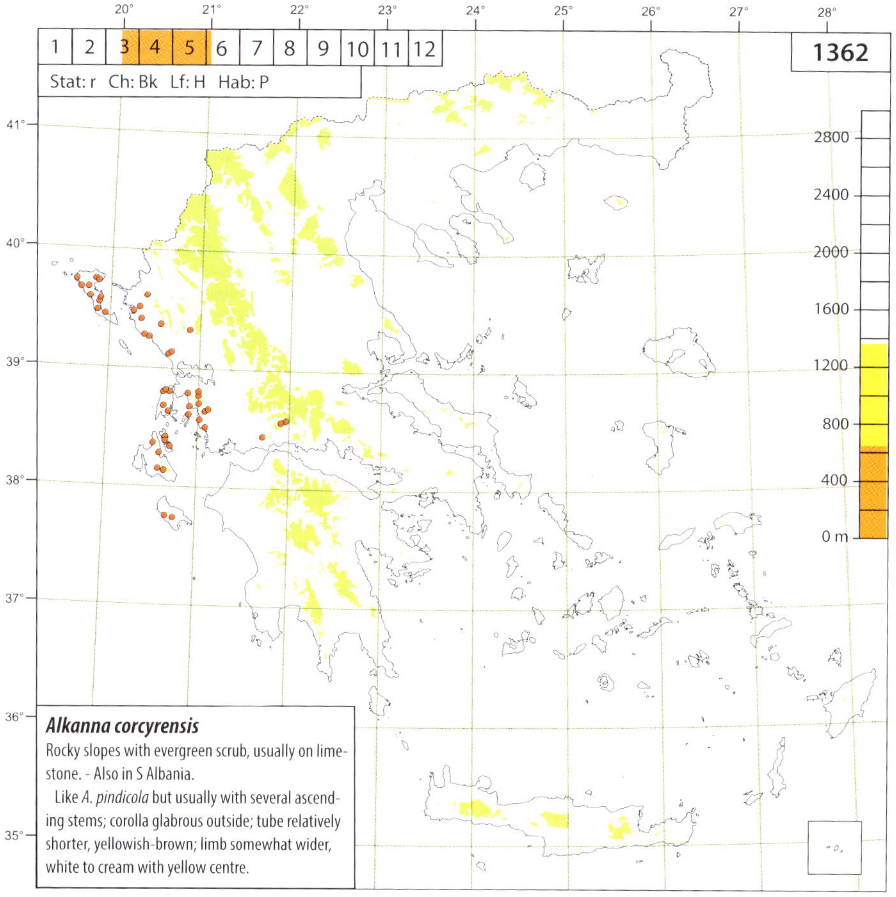

Alkanna corcyrensis

Rocky slopes with evergreen scrub, usually on limestone. - Also in S Albania.

Like *A. pindicola* but usually with several ascending stems; corolla glabrous outside; tube relatively shorter, yellowish-brown; limb somewhat wider, white to cream with yellow centre.

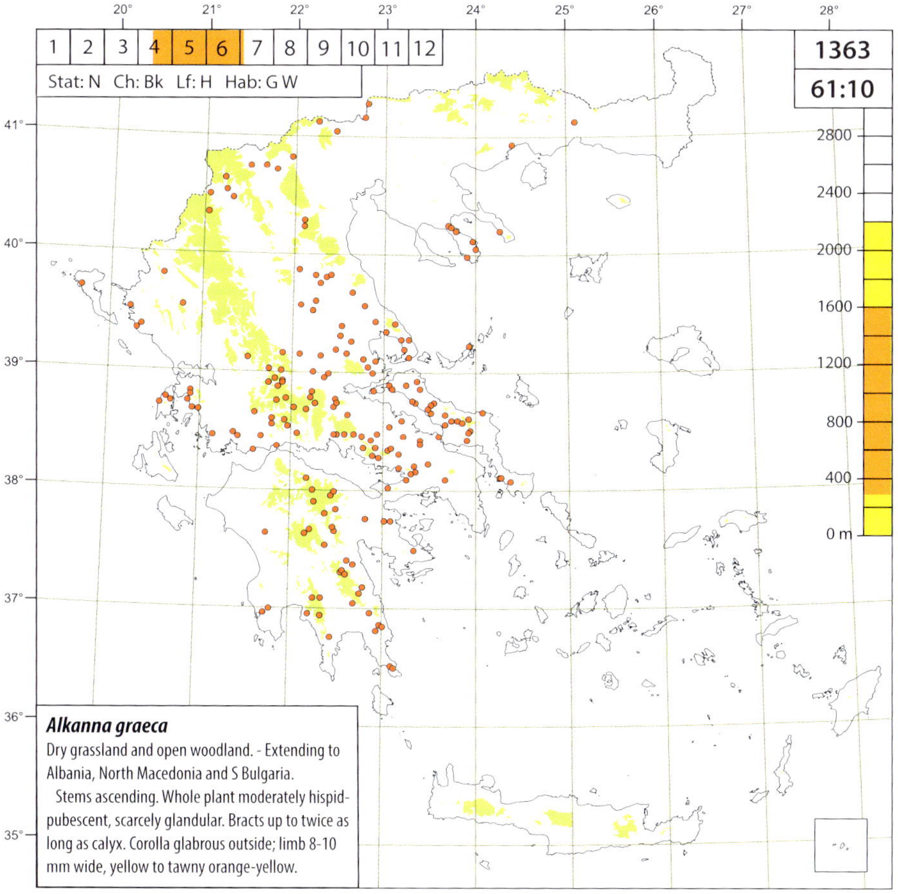

Alkanna graeca

Dry grassland and open woodland. - Extending to Albania, North Macedonia and S Bulgaria.

Stems ascending. Whole plant moderately hispid-pubescent, scarcely glandular. Bracts up to twice as long as calyx. Corolla glabrous outside; limb 8-10 mm wide, yellow to tawny orange-yellow.

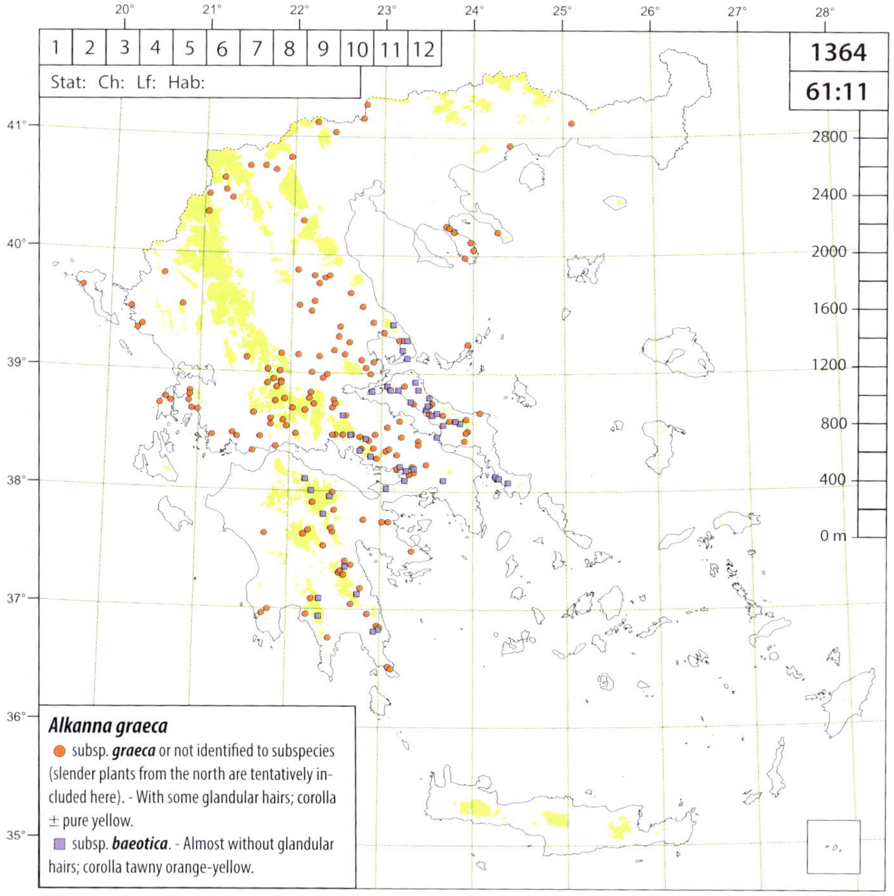

Alkanna graeca

● subsp. ***graeca*** or not identified to subspecies (slender plants from the north are tentatively included here). - With some glandular hairs; corolla ± pure yellow.

■ subsp. ***baeotica***. - Almost without glandular hairs; corolla tawny orange-yellow.

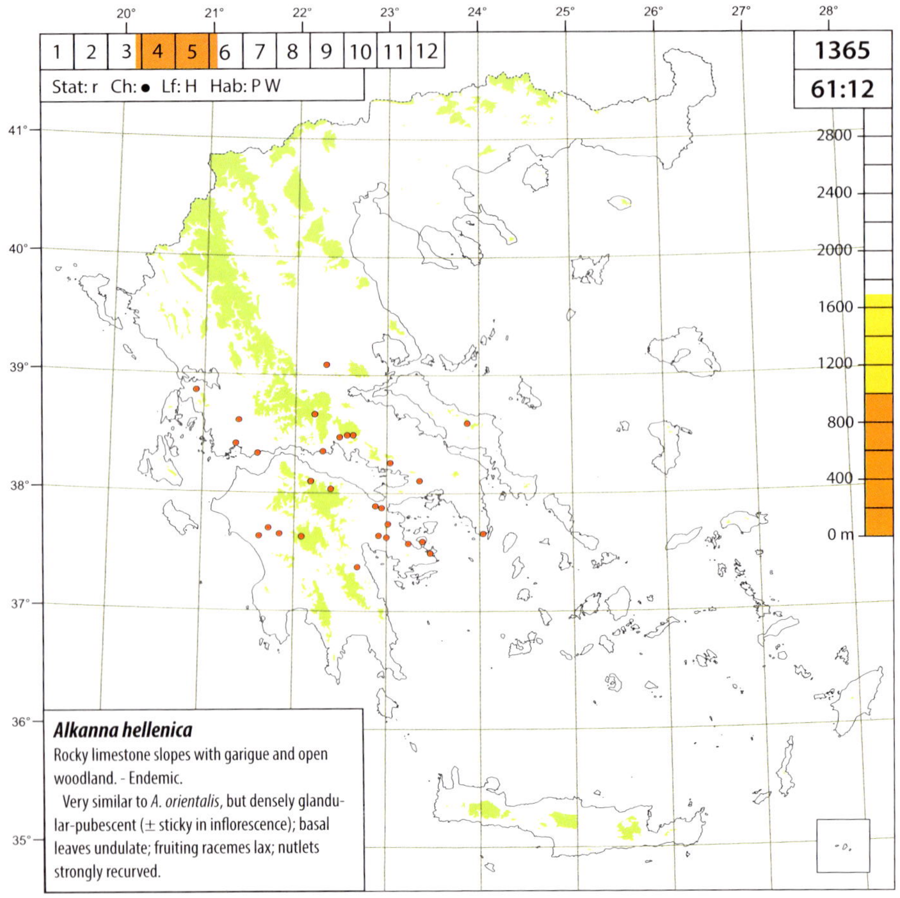

Alkanna hellenica
Rocky limestone slopes with garigue and open woodland. - Endemic.
Very similar to *A. orientalis*, but densely glandular-pubescent (± sticky in inflorescence); basal leaves undulate; fruiting racemes lax; nutlets strongly recurved.

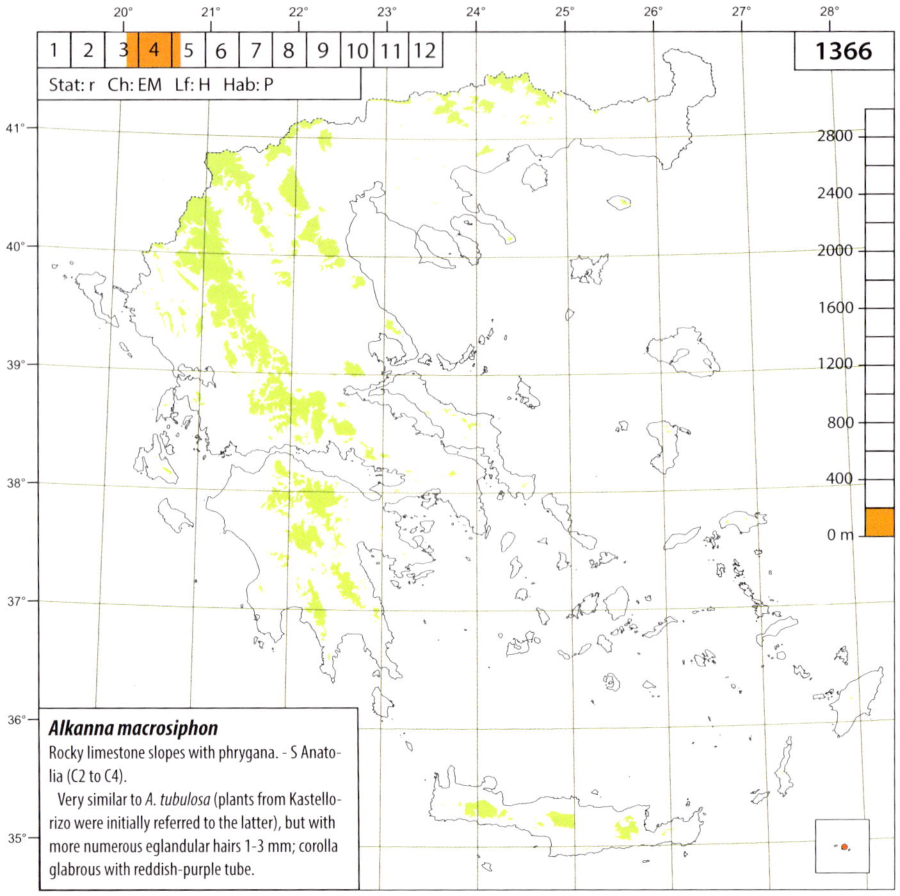

Alkanna macrosiphon
Rocky limestone slopes with phrygana. - S Anatolia (C2 to C4).
Very similar to *A. tubulosa* (plants from Kastellorizo were initially referred to the latter), but with more numerous eglandular hairs 1-3 mm; corolla glabrous with reddish-purple tube.

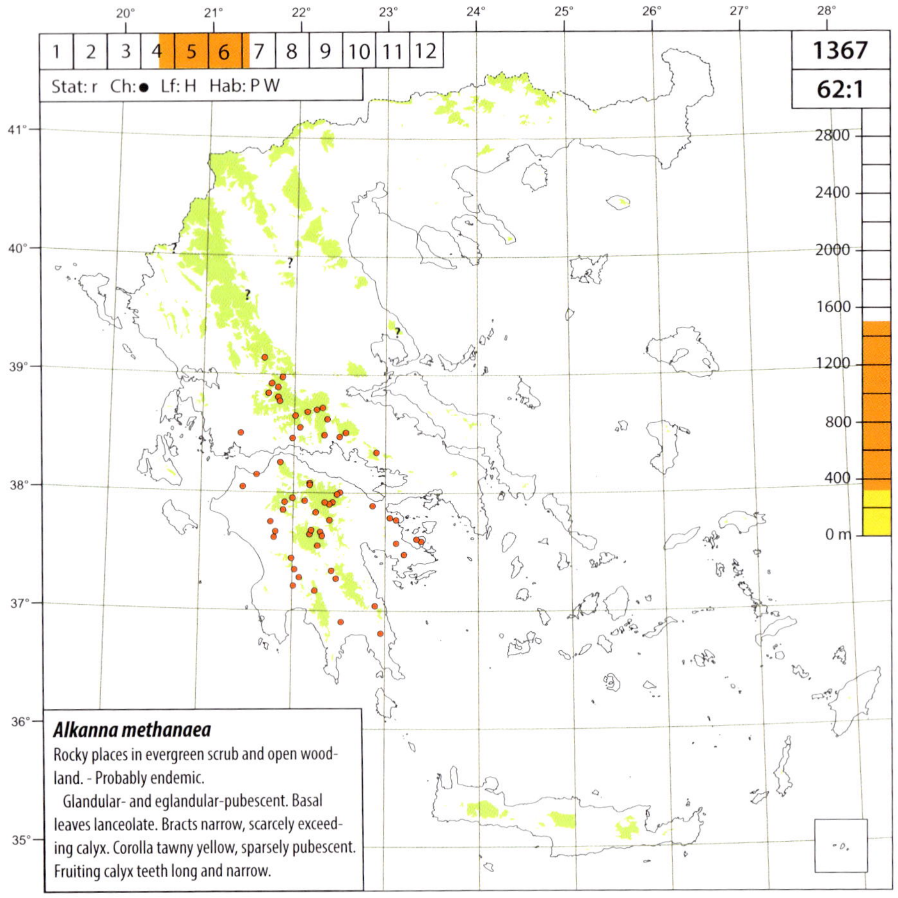

Alkanna methanaea
Rocky places in evergreen scrub and open woodland. - Probably endemic.
Glandular- and eglandular-pubescent. Basal leaves lanceolate. Bracts narrow, scarcely exceeding calyx. Corolla tawny yellow, sparsely pubescent. Fruiting calyx teeth long and narrow.

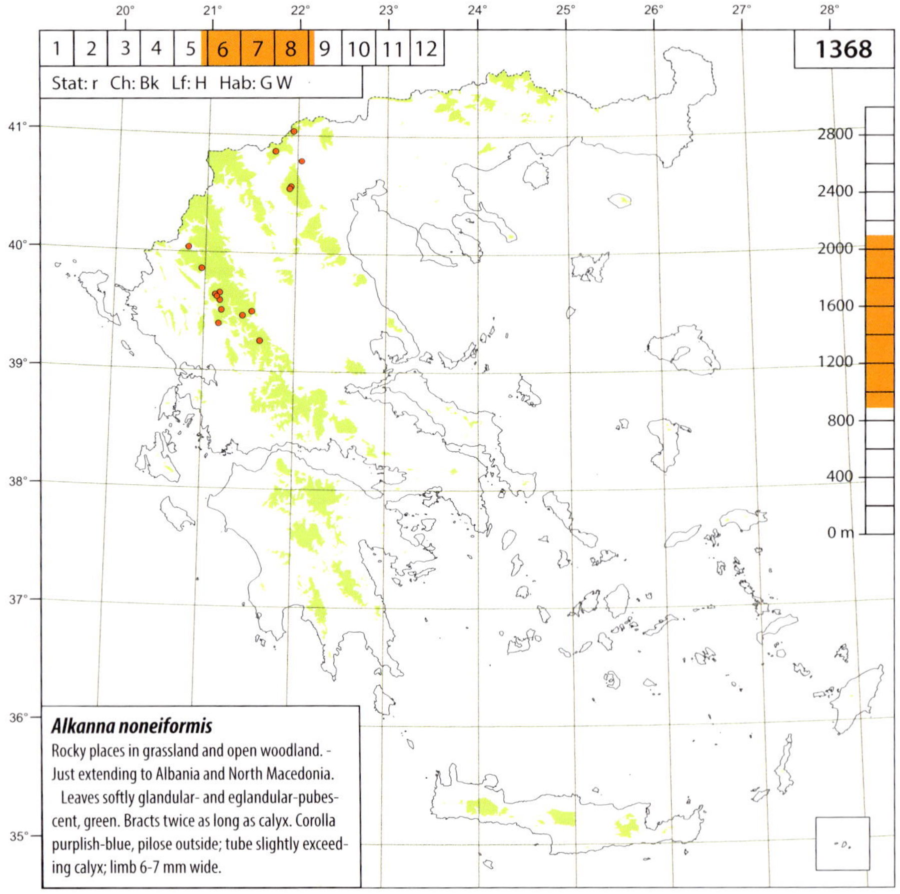

Alkanna noneiformis
Rocky places in grassland and open woodland. - Just extending to Albania and North Macedonia.
Leaves softly glandular- and eglandular-pubescent, green. Bracts twice as long as calyx. Corolla purplish-blue, pilose outside; tube slightly exceeding calyx; limb 6-7 mm wide.

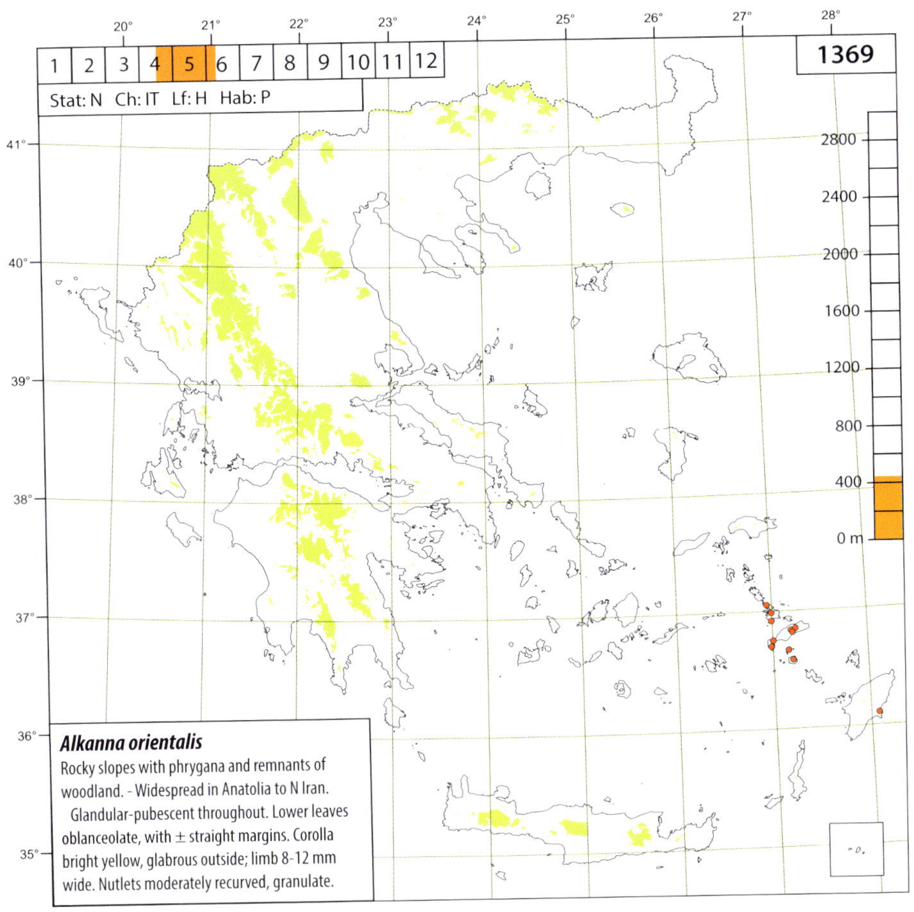

Alkanna orientalis

Rocky slopes with phrygana and remnants of woodland. - Widespread in Anatolia to N Iran.

Glandular-pubescent throughout. Lower leaves oblanceolate, with ± straight margins. Corolla bright yellow, glabrous outside; limb 8-12 mm wide. Nutlets moderately recurved, granulate.

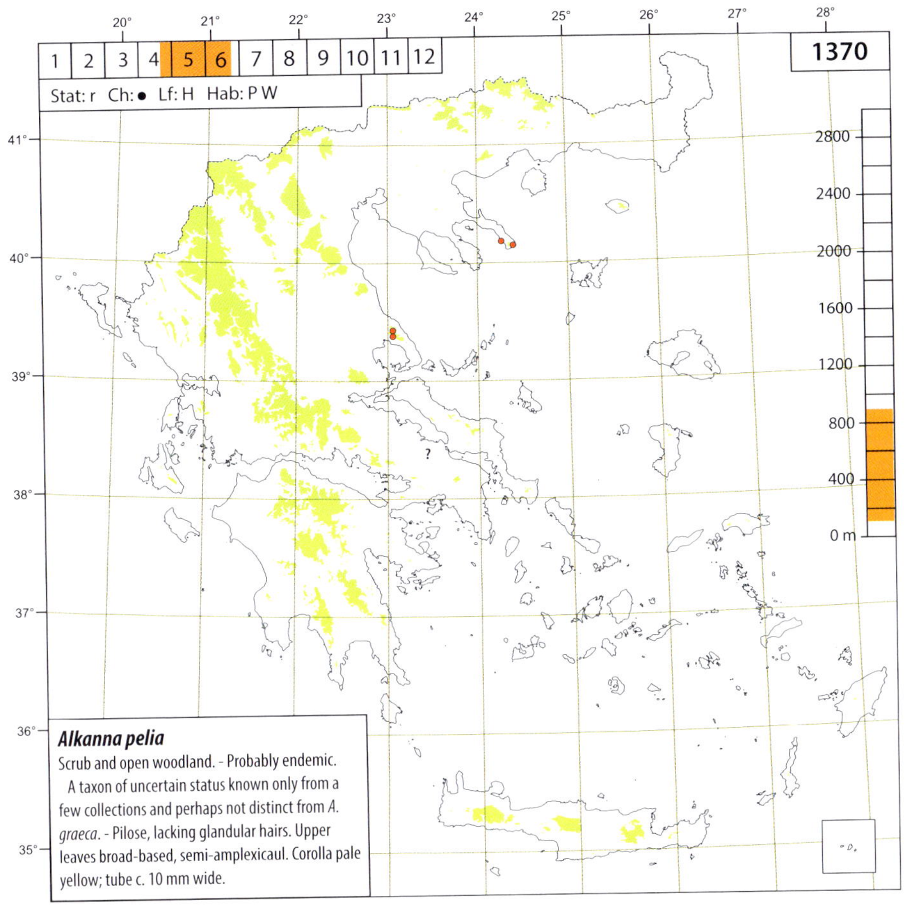

Alkanna pelia

Scrub and open woodland. - Probably endemic.

A taxon of uncertain status known only from a few collections and perhaps not distinct from *A. graeca*. - Pilose, lacking glandular hairs. Upper leaves broad-based, semi-amplexicaul. Corolla pale yellow; tube c. 10 mm wide.

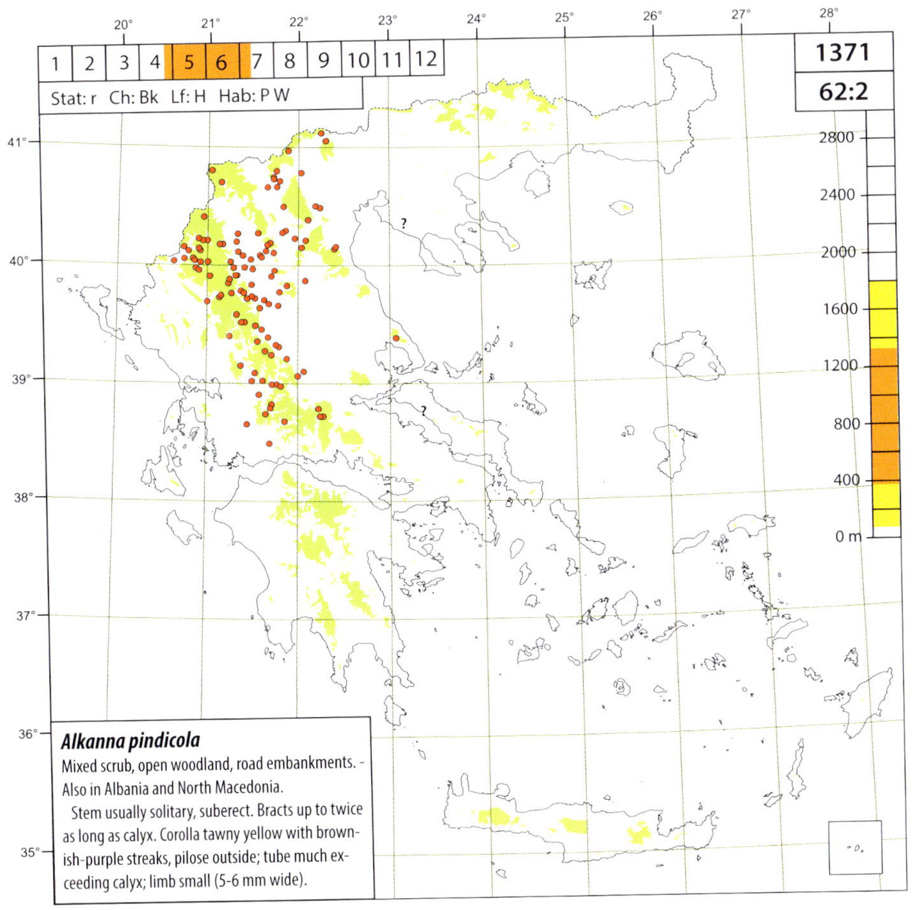

Alkanna pindicola

Mixed scrub, open woodland, road embankments. - Also in Albania and North Macedonia.

Stem usually solitary, suberect. Bracts up to twice as long as calyx. Corolla tawny yellow with brownish-purple streaks, pilose outside; tube much exceeding calyx; limb small (5-6 mm wide).

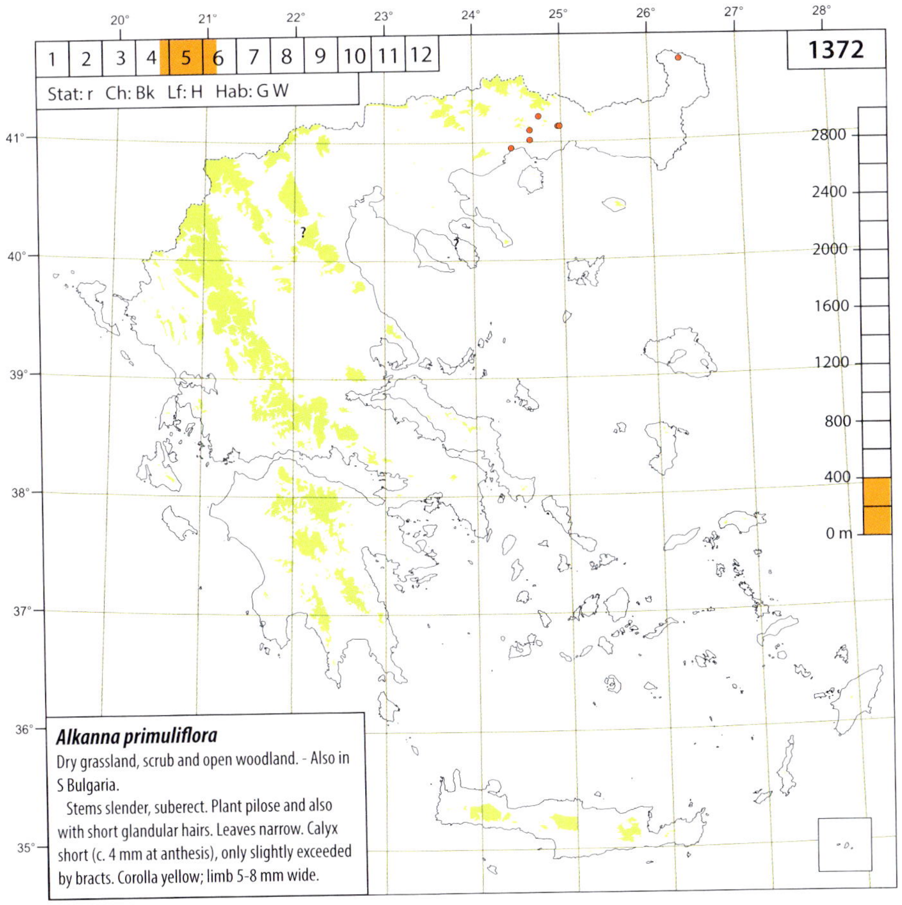

Alkanna primuliflora

Dry grassland, scrub and open woodland. - Also in S Bulgaria.

Stems slender, suberect. Plant pilose and also with short glandular hairs. Leaves narrow. Calyx short (c. 4 mm at anthesis), only slightly exceeded by bracts. Corolla yellow; limb 5-8 mm wide.

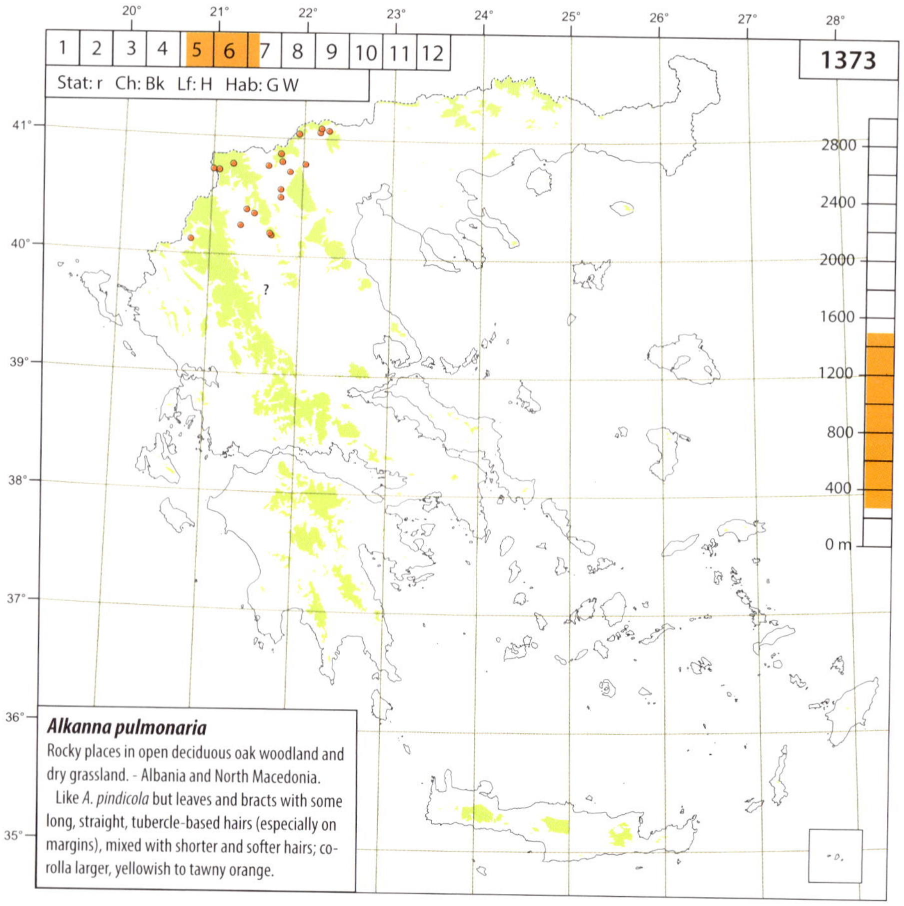

Alkanna pulmonaria

Rocky places in open deciduous oak woodland and dry grassland. - Albania and North Macedonia.

Like *A. pindicola* but leaves and bracts with some long, straight, tubercle-based hairs (especially on margins), mixed with shorter and softer hairs; corolla larger, yellowish to tawny orange.

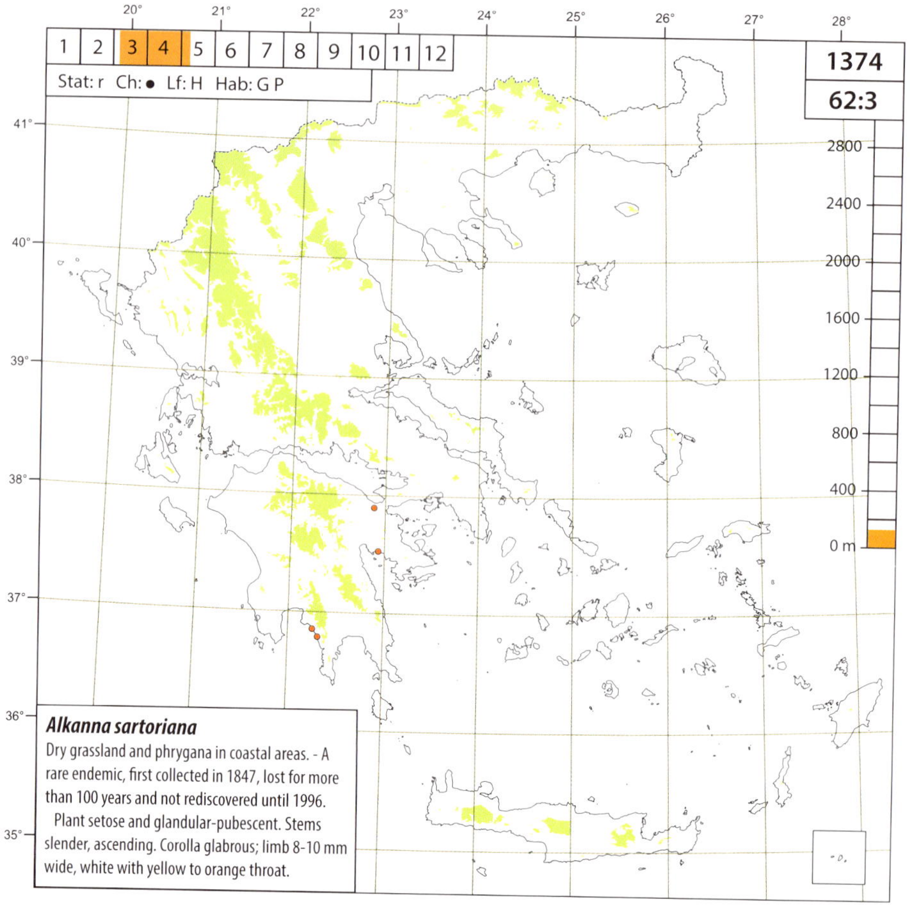

Alkanna sartoriana

Dry grassland and phrygana in coastal areas. - A rare endemic, first collected in 1847, lost for more than 100 years and not rediscovered until 1996.

Plant setose and glandular-pubescent. Stems slender, ascending. Corolla glabrous; limb 8-10 mm wide, white with yellow to orange throat.

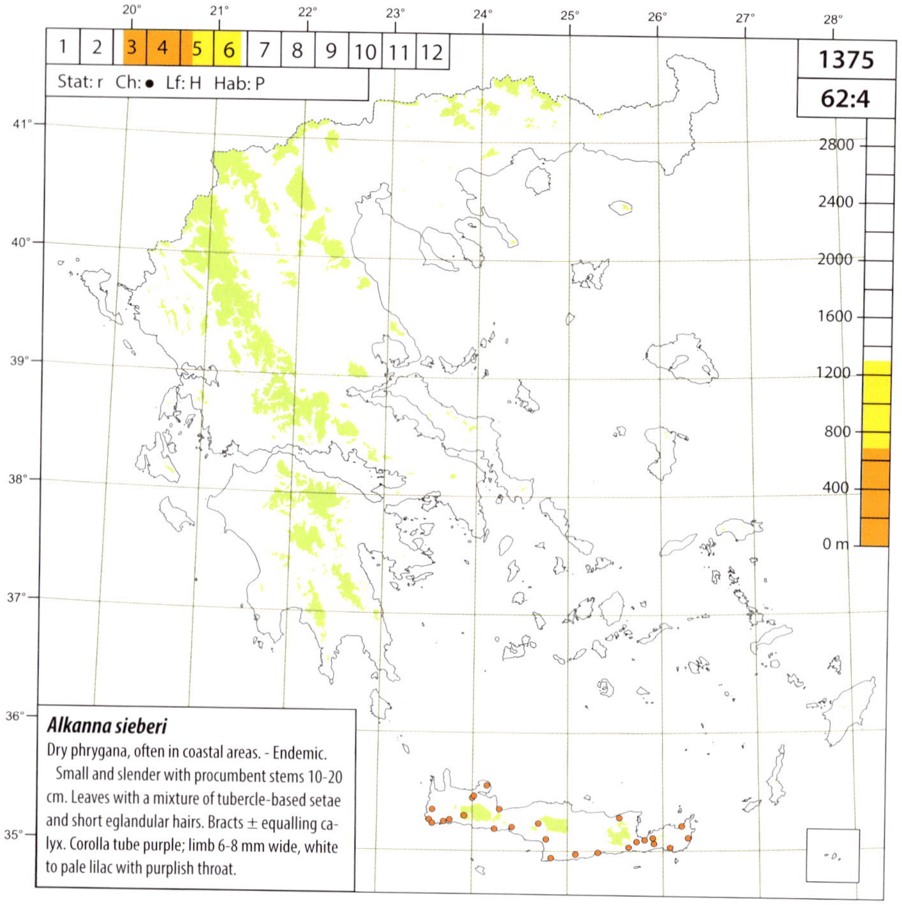

Alkanna sieberi

Dry phrygana, often in coastal areas. - Endemic.

Small and slender with procumbent stems 10-20 cm. Leaves with a mixture of tubercle-based setae and short eglandular hairs. Bracts ± equalling calyx. Corolla tube purple; limb 6-8 mm wide, white to pale lilac with purplish throat.

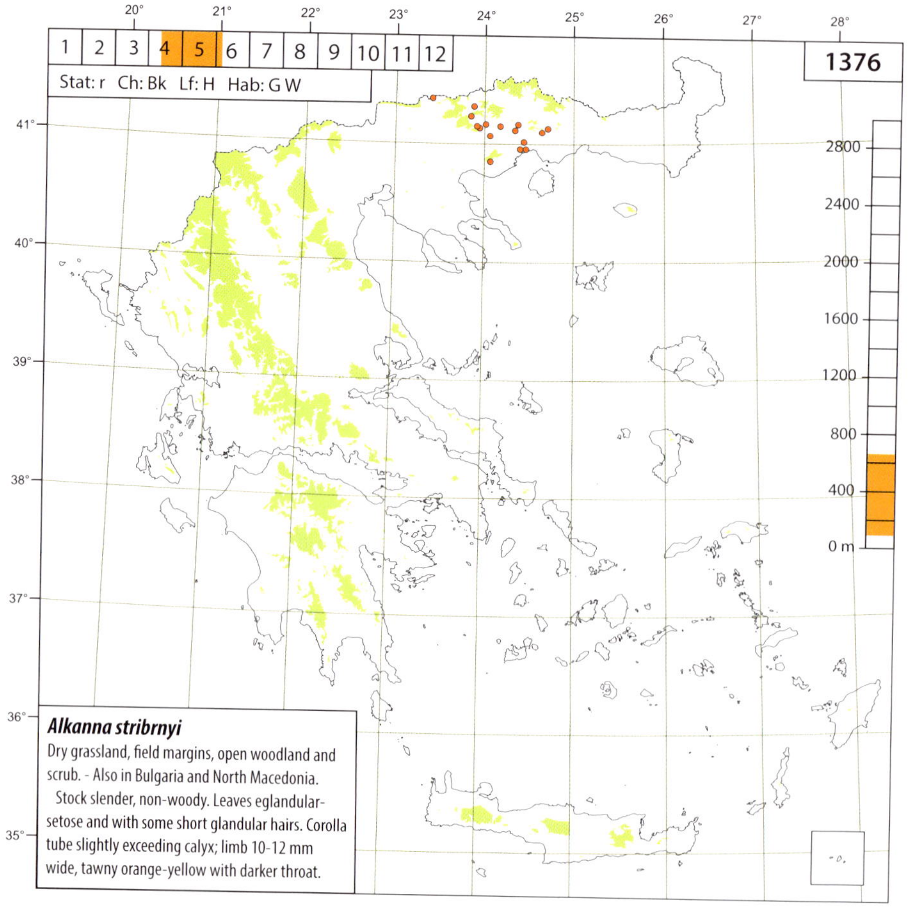

Alkanna stribrnyi

Dry grassland, field margins, open woodland and scrub. - Also in Bulgaria and North Macedonia.

Stock slender, non-woody. Leaves eglandular-setose and with some short glandular hairs. Corolla tube slightly exceeding calyx; limb 10-12 mm wide, tawny orange-yellow with darker throat.

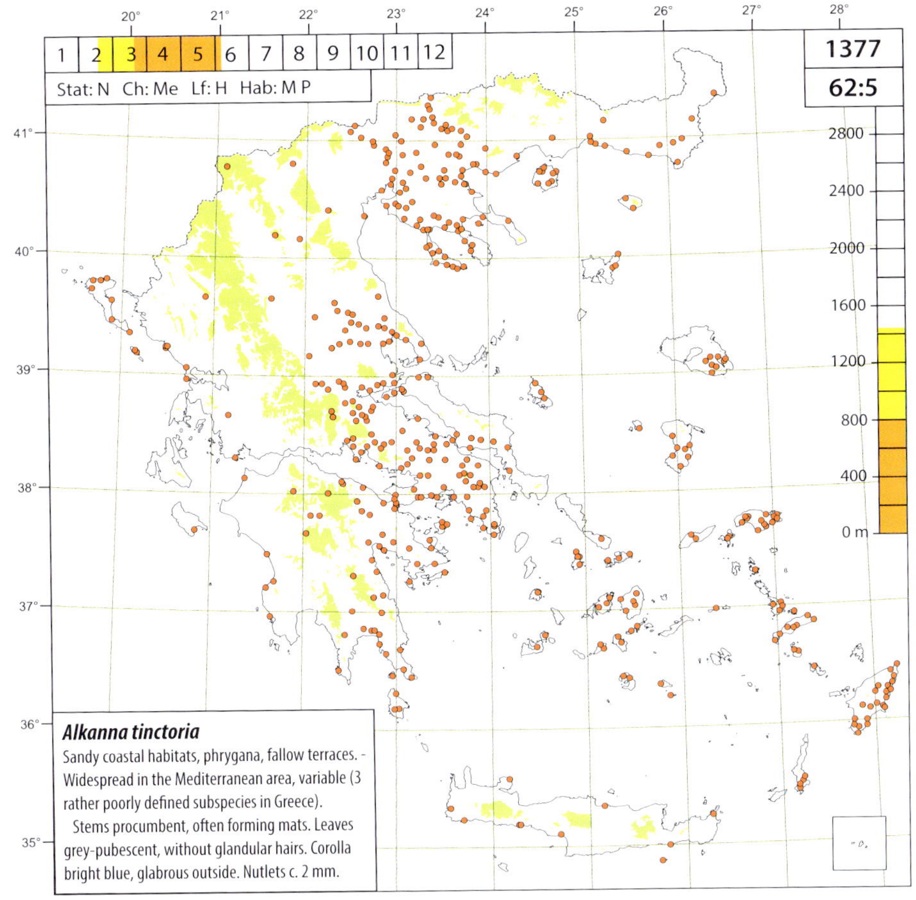

Alkanna tinctoria

Sandy coastal habitats, phrygana, fallow terraces. - Widespread in the Mediterranean area, variable (3 rather poorly defined subspecies in Greece).

Stems procumbent, often forming mats. Leaves grey-pubescent, without glandular hairs. Corolla bright blue, glabrous outside. Nutlets c. 2 mm.

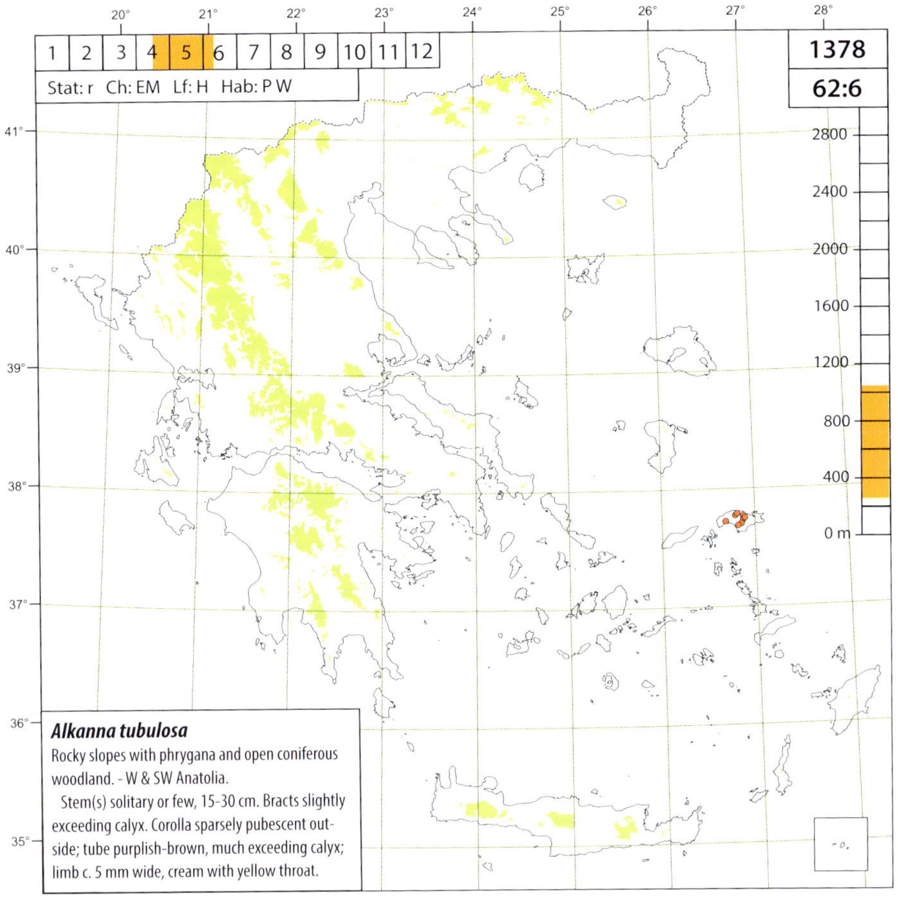

Alkanna tubulosa

Rocky slopes with phrygana and open coniferous woodland. - W & SW Anatolia.

Stem(s) solitary or few, 15-30 cm. Bracts slightly exceeding calyx. Corolla sparsely pubescent outside; tube purplish-brown, much exceeding calyx; limb c. 5 mm wide, cream with yellow throat.

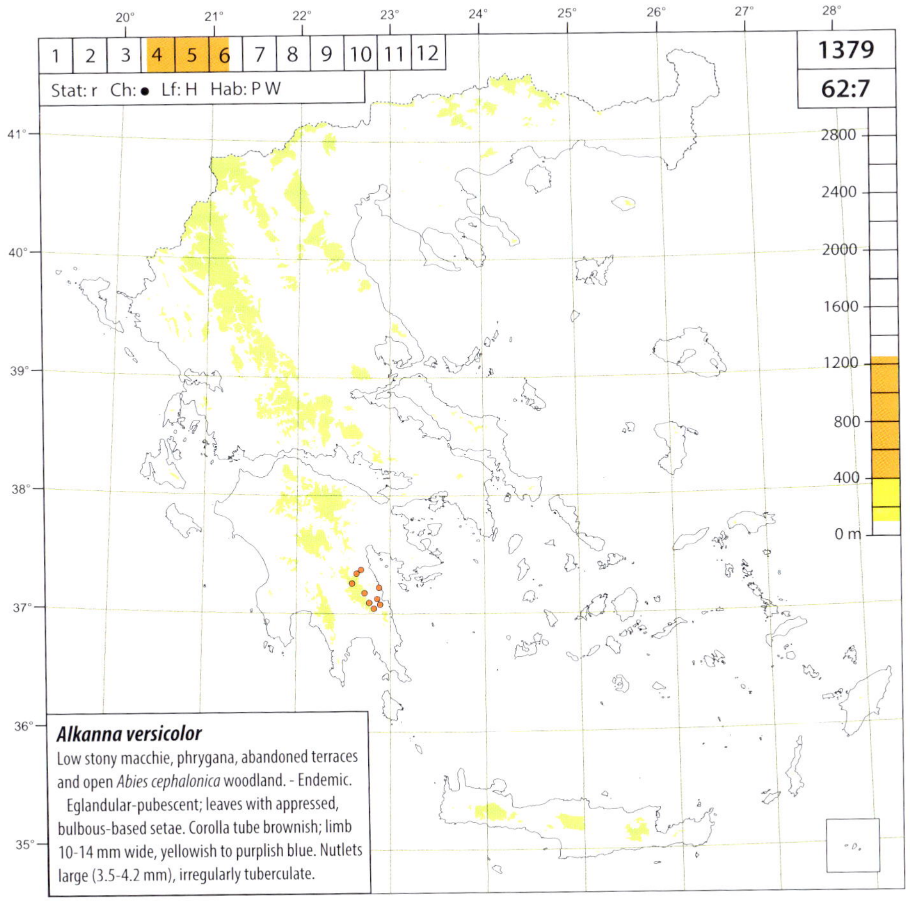

Alkanna versicolor

Low stony macchie, phrygana, abandoned terraces and open *Abies cephalonica* woodland. - Endemic.

Eglandular-pubescent; leaves with appressed, bulbous-based setae. Corolla tube brownish; limb 10-14 mm wide, yellowish to purplish blue. Nutlets large (3.5-4.2 mm), irregularly tuberculate.

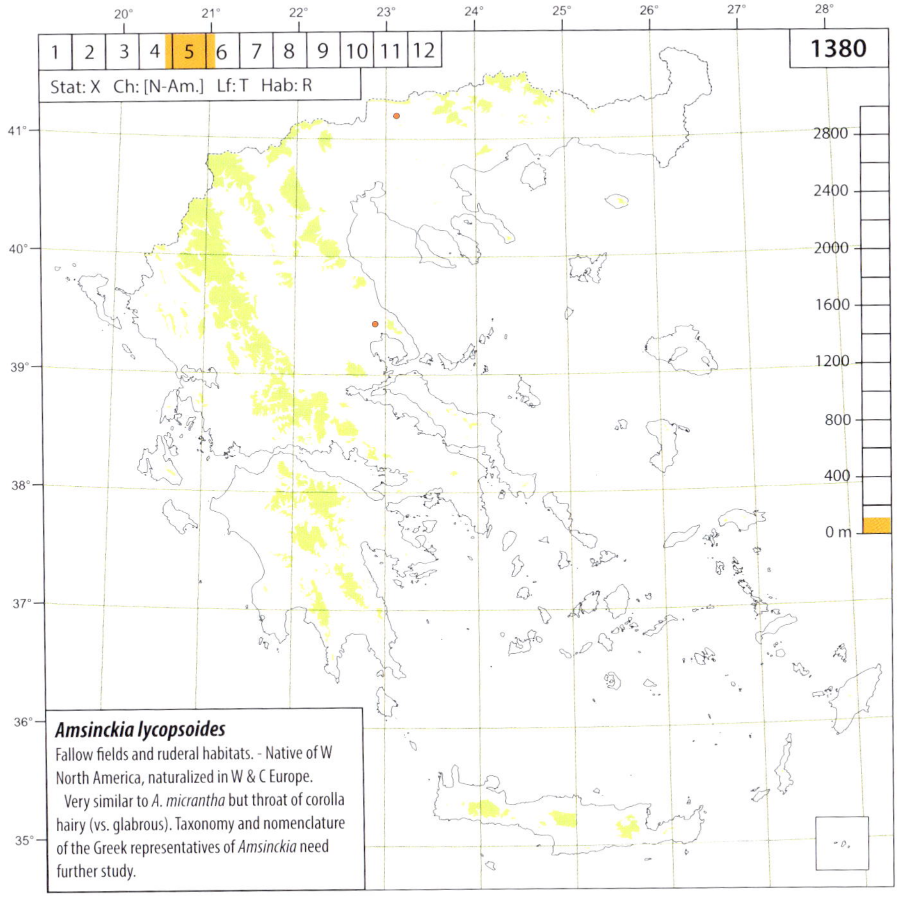

Amsinckia lycopsoides

Fallow fields and ruderal habitats. - Native of W North America, naturalized in W & C Europe.

Very similar to *A. micrantha* but throat of corolla hairy (vs. glabrous). Taxonomy and nomenclature of the Greek representatives of *Amsinckia* need further study.

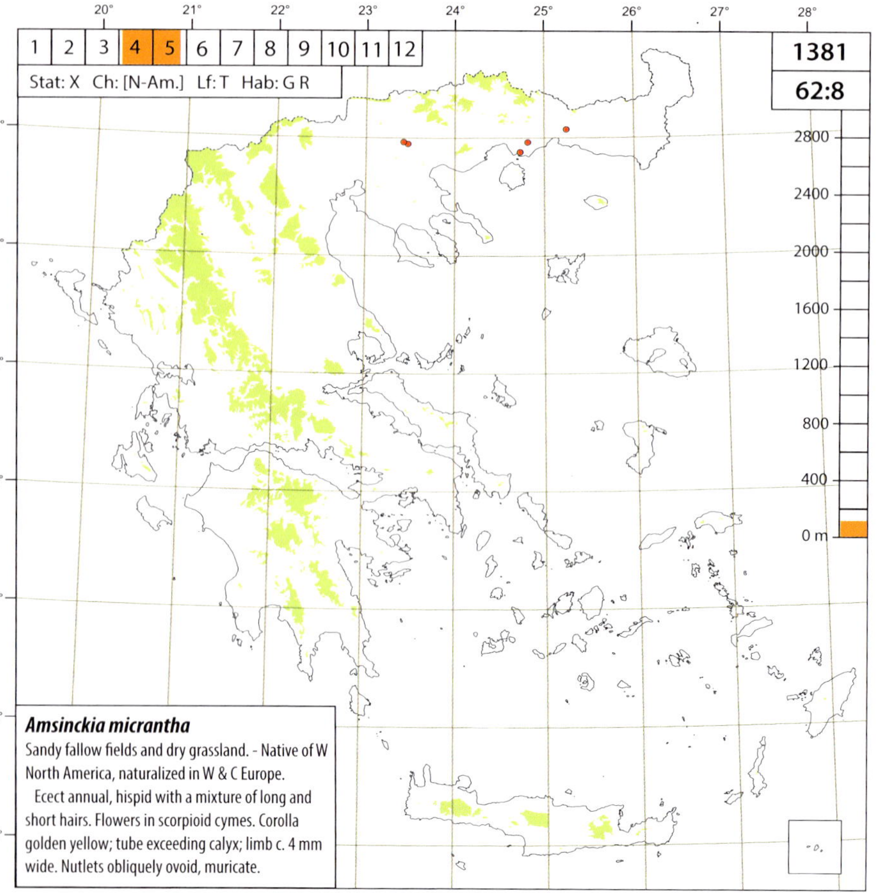

Amsinckia micrantha

Sandy fallow fields and dry grassland. - Native of W North America, naturalized in W & C Europe.

Ecect annual, hispid with a mixture of long and short hairs. Flowers in scorpioid cymes. Corolla golden yellow; tube exceeding calyx; limb c. 4 mm wide. Nutlets obliquely ovoid, muricate.

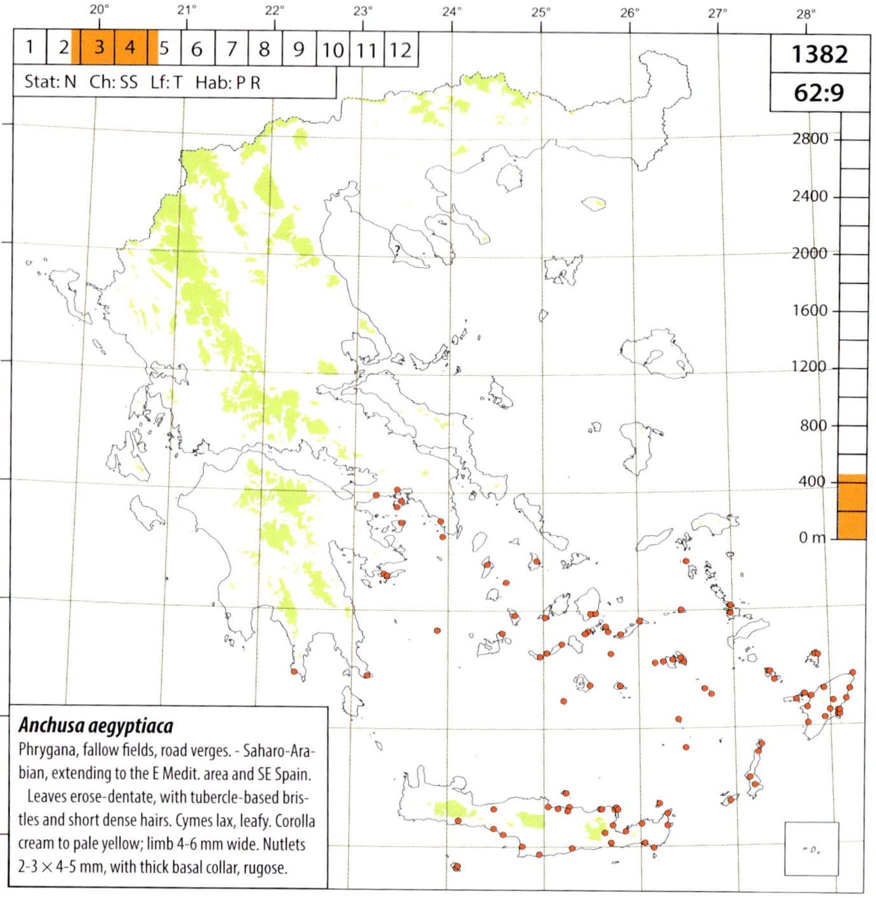

Anchusa aegyptiaca

Phrygana, fallow fields, road verges. - Saharo-Arabian, extending to the E Medit. area and SE Spain.

Leaves erose-dentate, with tubercle-based bristles and short dense hairs. Cymes lax, leafy. Corolla cream to pale yellow; limb 4-6 mm wide. Nutlets 2-3 × 4-5 mm, with thick basal collar, rugose.

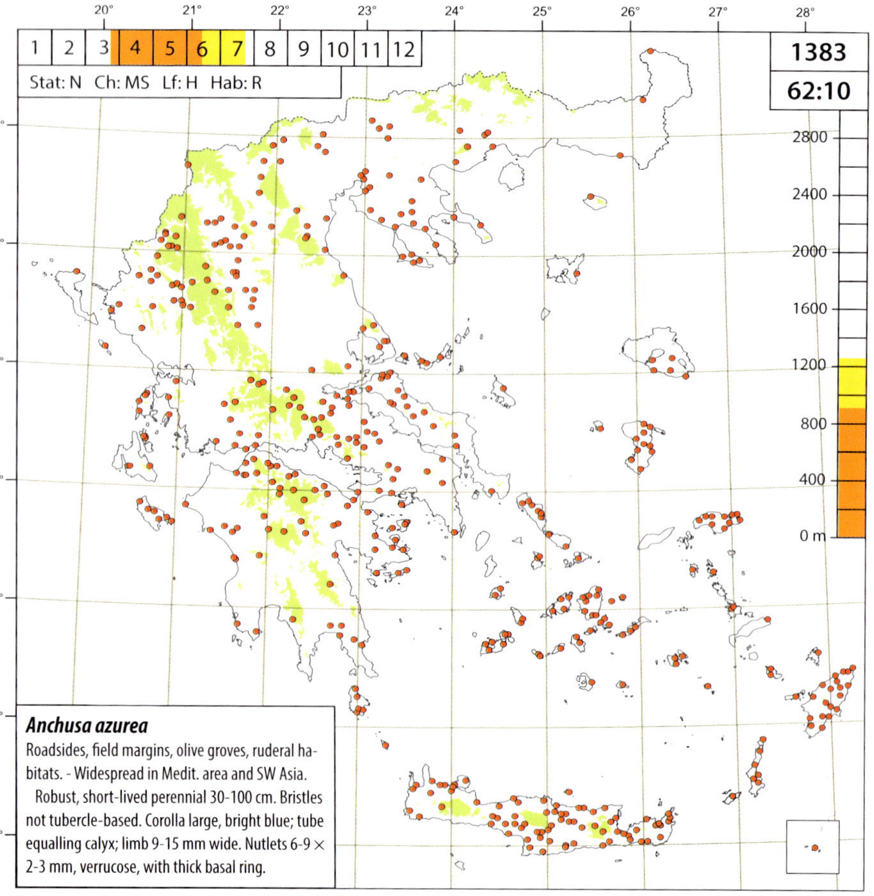

Anchusa azurea

Roadsides, field margins, olive groves, ruderal habitats. - Widespread in Medit. area and SW Asia.

Robust, short-lived perennial 30-100 cm. Bristles not tubercle-based. Corolla large, bright blue; tube equalling calyx; limb 9-15 mm wide. Nutlets 6-9 × 2-3 mm, verrucose, with thick basal ring.

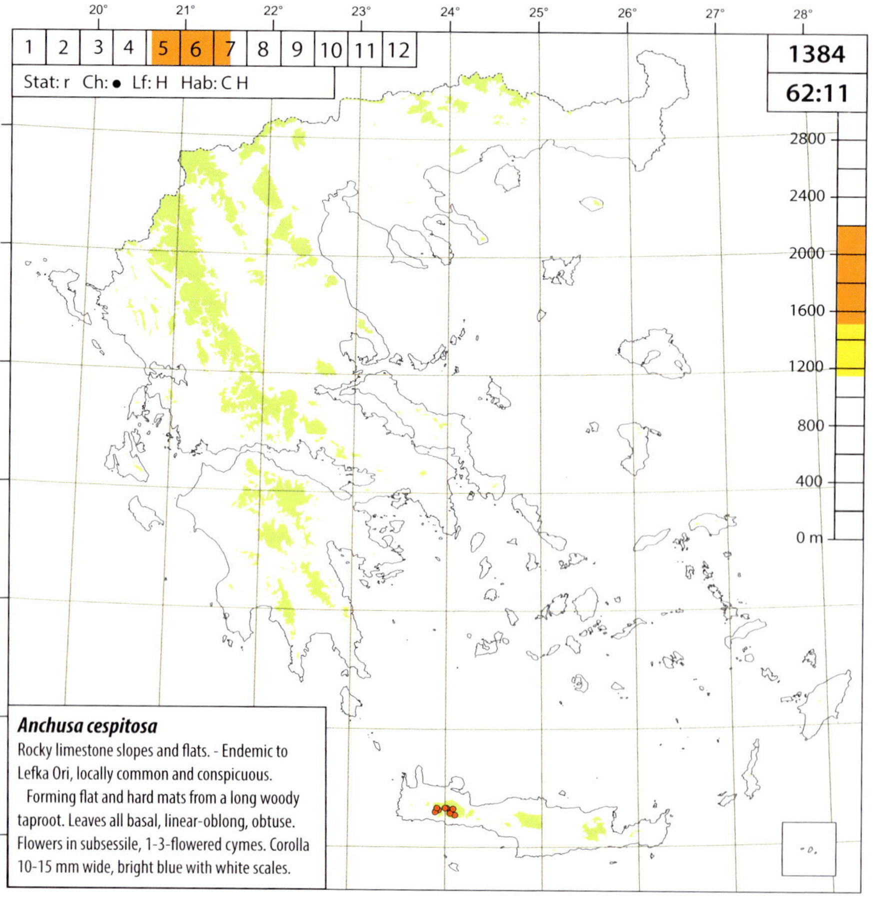

Anchusa cespitosa

Rocky limestone slopes and flats. - Endemic to Lefka Ori, locally common and conspicuous.

Forming flat and hard mats from a long woody taproot. Leaves all basal, linear-oblong, obtuse. Flowers in subsessile, 1-3-flowered cymes. Corolla 10-15 mm wide, bright blue with white scales.

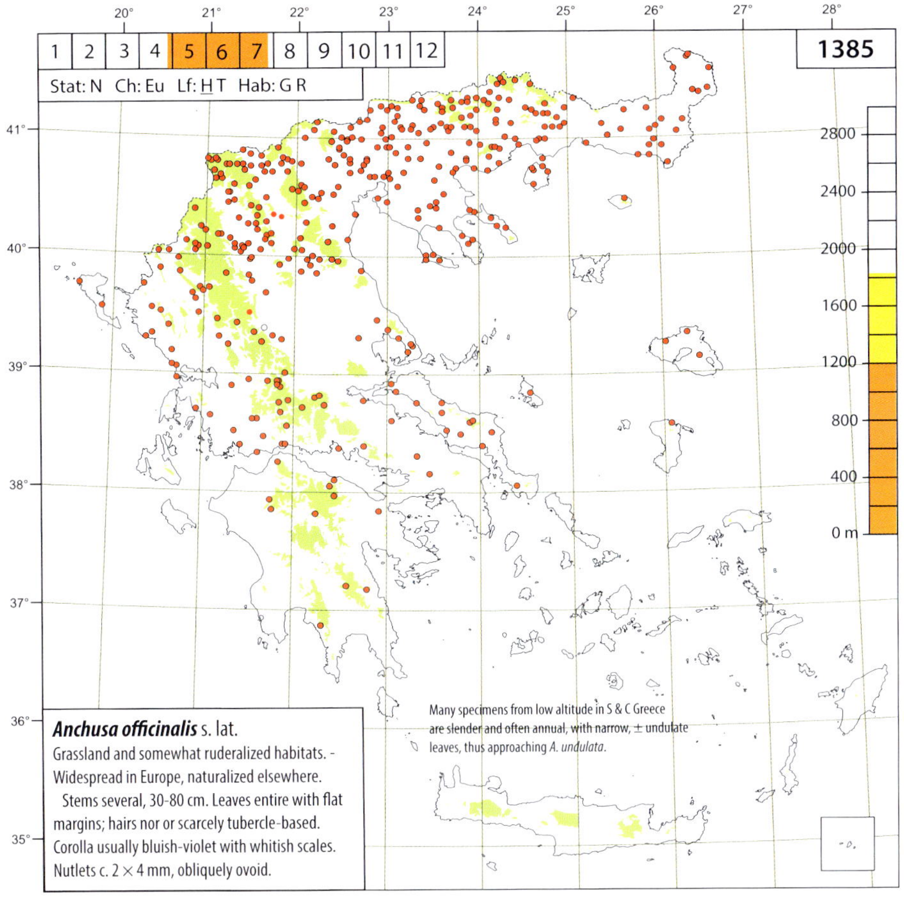

Anchusa officinalis s. lat.
Grassland and somewhat ruderalized habitats. - Widespread in Europe, naturalized elsewhere.
Stems several, 30-80 cm. Leaves entire with flat margins; hairs nor or scarcely tubercle-based. Corolla usually bluish-violet with whitish scales. Nutlets c. 2 × 4 mm, obliquely ovoid.

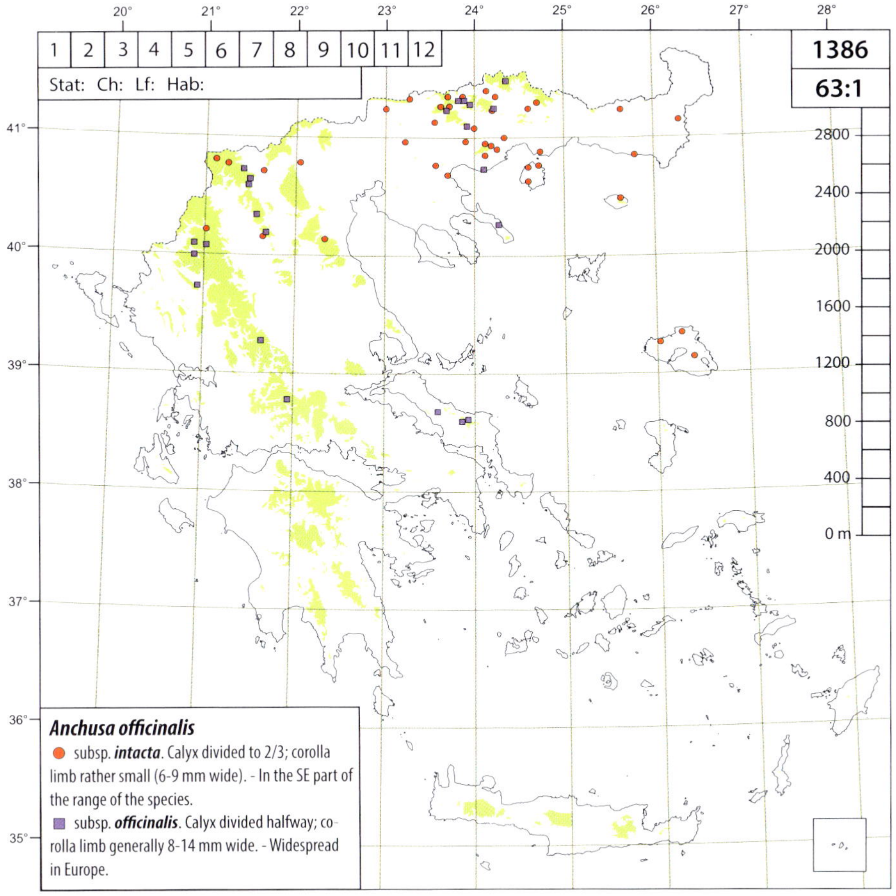

Anchusa officinalis
● subsp. ***intacta***. Calyx divided to 2/3; corolla limb rather small (6-9 mm wide). - In the SE part of the range of the species.
■ subsp. ***officinalis***. Calyx divided halfway; corolla limb generally 8-14 mm wide. - Widespread in Europe.

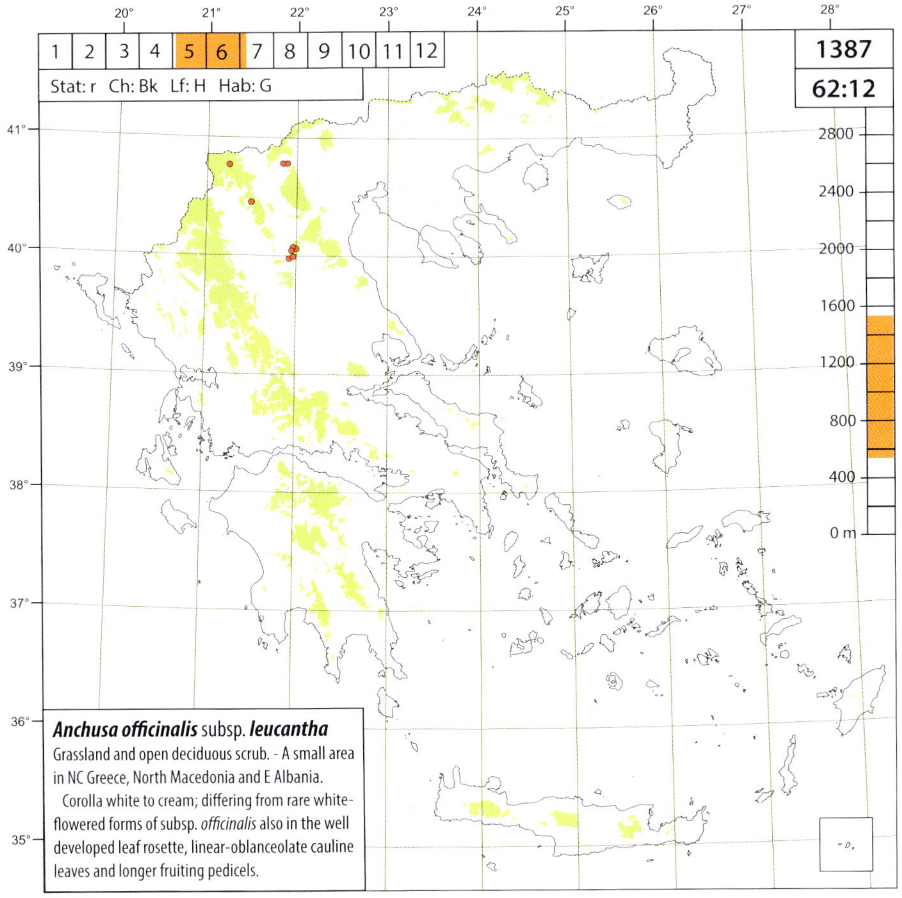

Anchusa officinalis subsp. ***leucantha***
Grassland and open deciduous scrub. - A small area in NC Greece, North Macedonia and E Albania.
Corolla white to cream; differing from rare white-flowered forms of subsp. *officinalis* also in the well developed leaf rosette, linear-oblanceolate cauline leaves and longer fruiting pedicels.

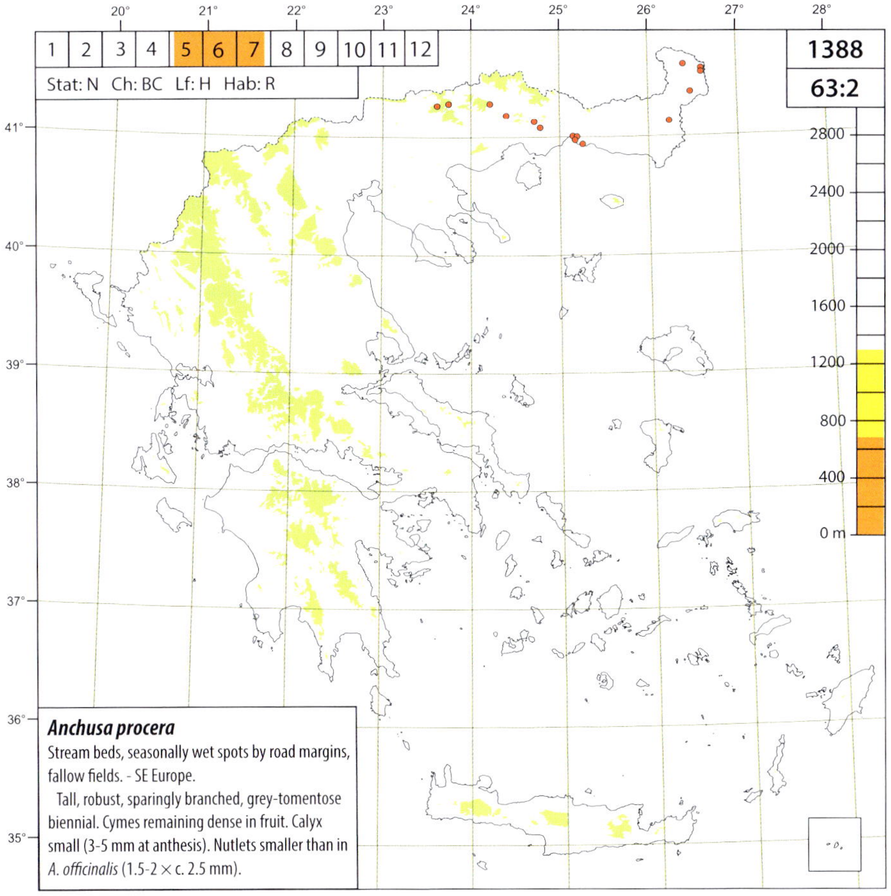

Anchusa procera
Stream beds, seasonally wet spots by road margins, fallow fields. - SE Europe.
Tall, robust, sparingly branched, grey-tomentose biennial. Cymes remaining dense in fruit. Calyx small (3-5 mm at anthesis). Nutlets smaller than in *A. officinalis* (1.5-2 × c. 2.5 mm).

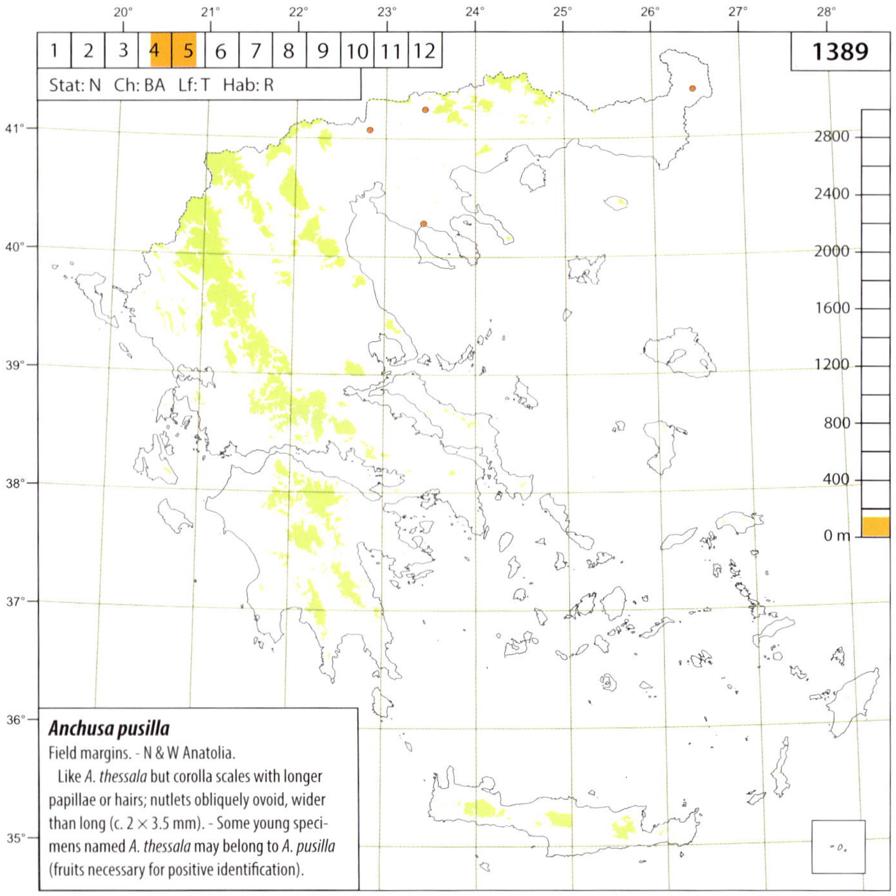

Anchusa pusilla

Field margins. - N & W Anatolia.

Like *A. thessala* but corolla scales with longer papillae or hairs; nutlets obliquely ovoid, wider than long (c. 2 × 3.5 mm). - Some young specimens named *A. thessala* may belong to *A. pusilla* (fruits necessary for positive identification).

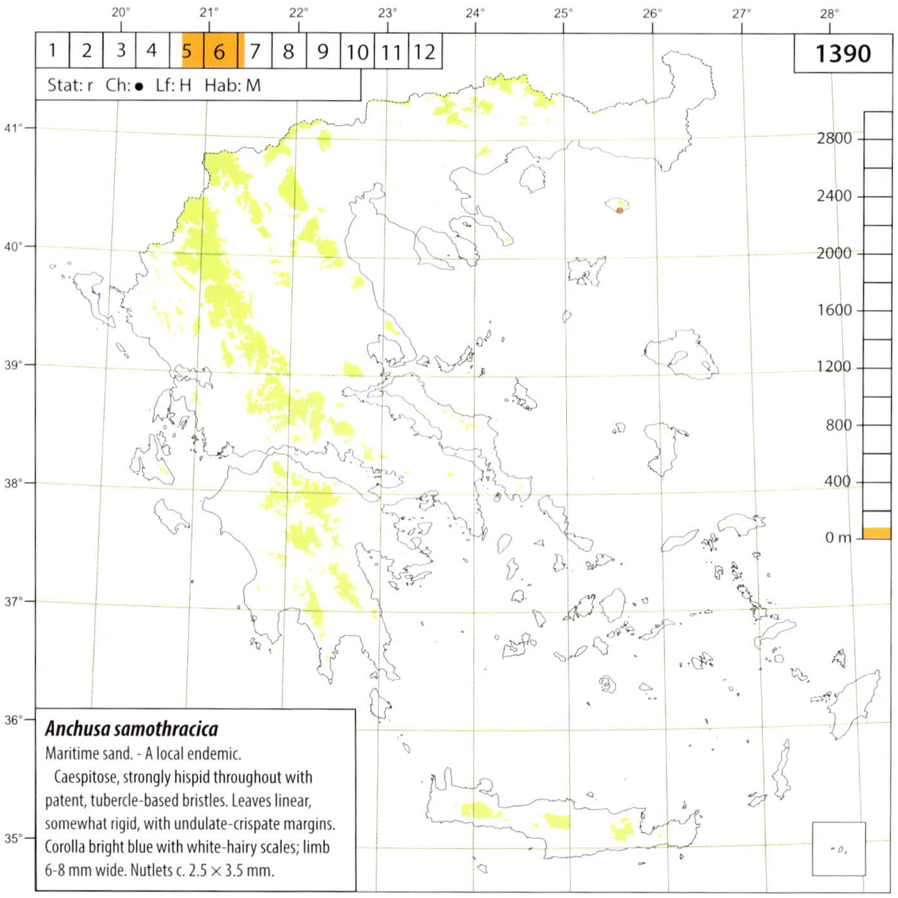

Anchusa samothracica

Maritime sand. - A local endemic.

Caespitose, strongly hispid throughout with patent, tubercle-based bristles. Leaves linear, somewhat rigid, with undulate-crispate margins. Corolla bright blue with white-hairy scales; limb 6-8 mm wide. Nutlets c. 2.5 × 3.5 mm.

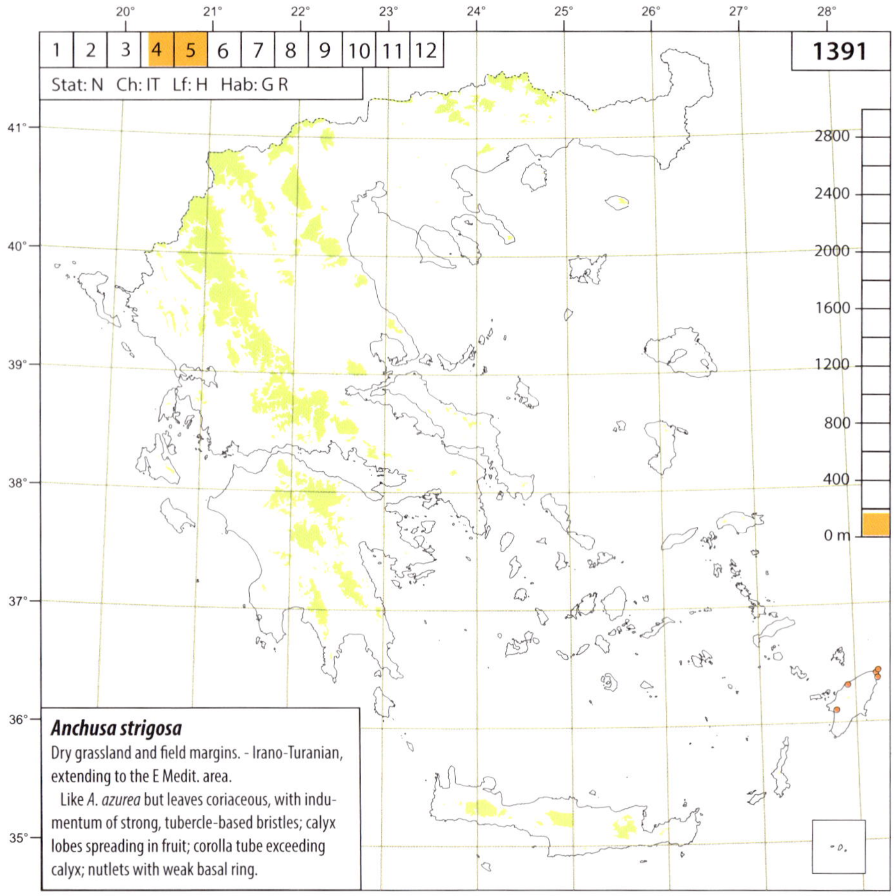

Anchusa strigosa

Dry grassland and field margins. - Irano-Turanian, extending to the E Medit. area.

Like *A. azurea* but leaves coriaceous, with indumentum of strong, tubercle-based bristles; calyx lobes spreading in fruit; corolla tube exceeding calyx; nutlets with weak basal ring.

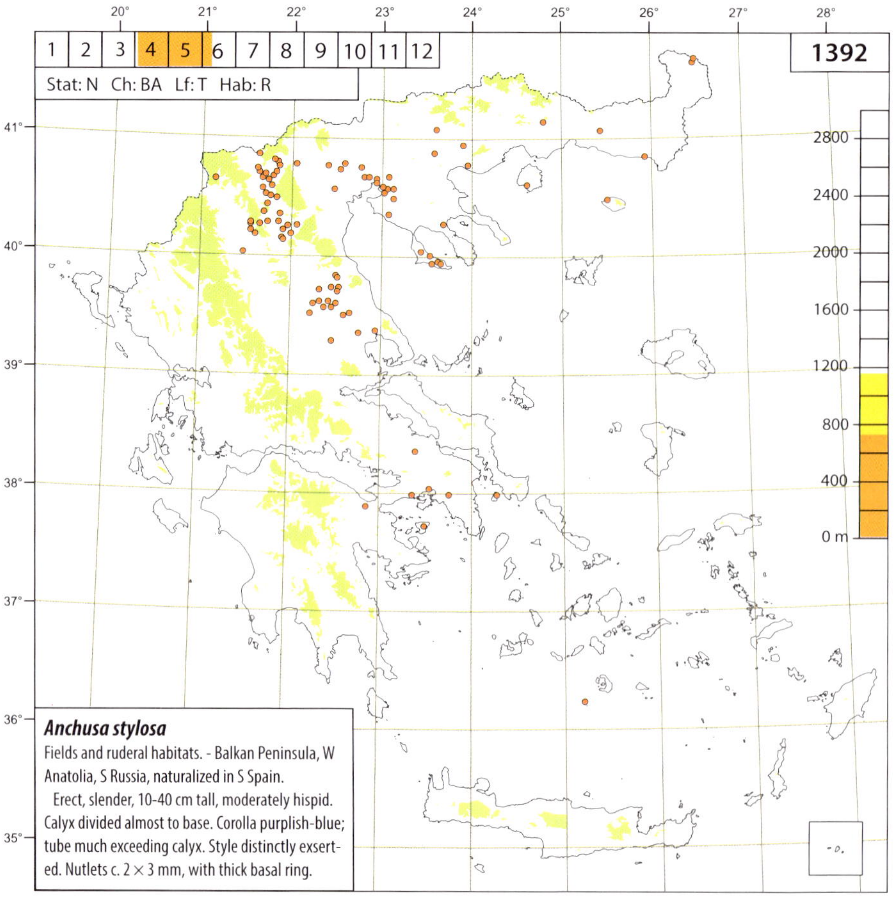

Anchusa stylosa

Fields and ruderal habitats. - Balkan Peninsula, W Anatolia, S Russia, naturalized in S Spain.

Erect, slender, 10-40 cm tall, moderately hispid. Calyx divided almost to base. Corolla purplish-blue; tube much exceeding calyx. Style distinctly exserted. Nutlets c. 2 × 3 mm, with thick basal ring.

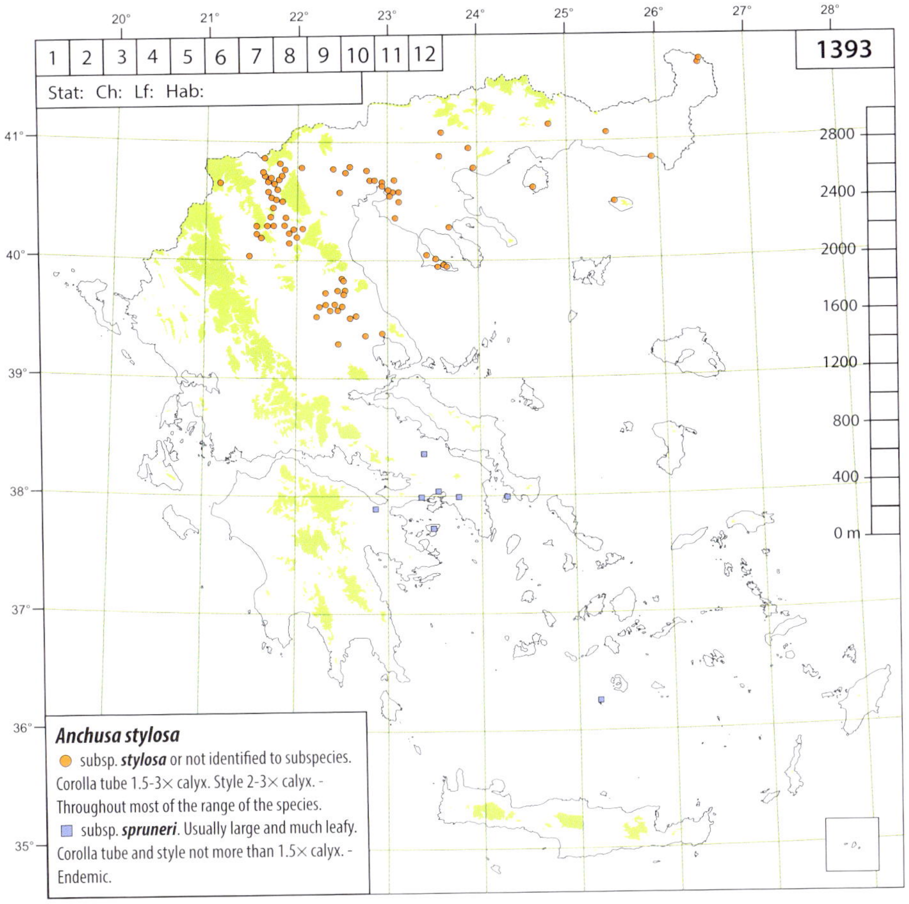

Anchusa stylosa
subsp. ***stylosa*** or not identified to subspecies. Corolla tube 1.5-3× calyx. Style 2-3× calyx. - Throughout most of the range of the species.
subsp. ***spruneri***. Usually large and much leafy. Corolla tube and style not more than 1.5× calyx. - Endemic.

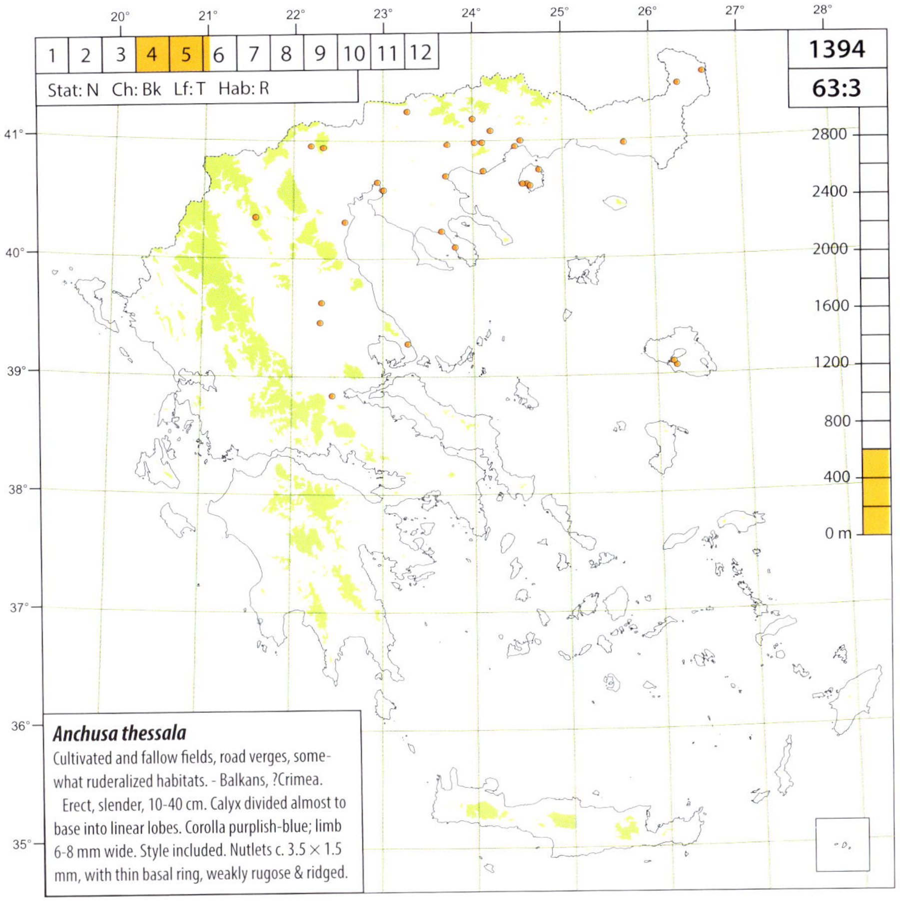

Anchusa thessala
Cultivated and fallow fields, road verges, somewhat ruderalized habitats. - Balkans, ?Crimea.
Erect, slender, 10-40 cm. Calyx divided almost to base into linear lobes. Corolla purplish-blue; limb 6-8 mm wide. Style included. Nutlets c. 3.5 × 1.5 mm, with thin basal ring, weakly rugose & ridged.

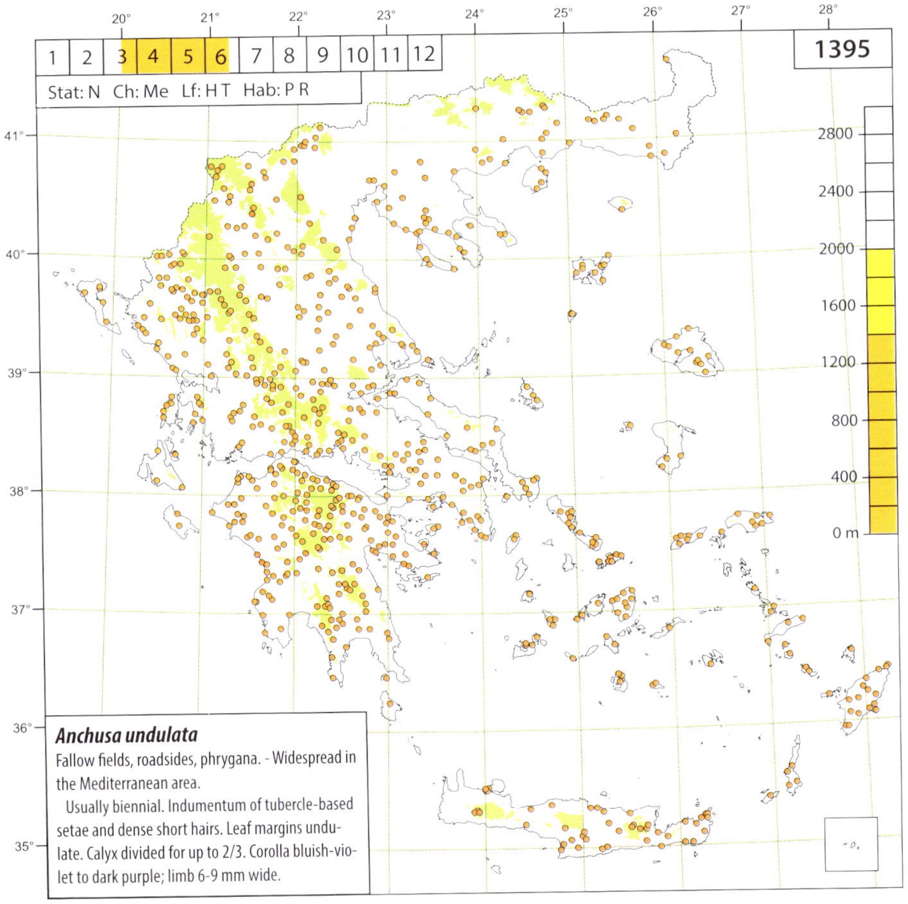

Anchusa undulata
Fallow fields, roadsides, phrygana. - Widespread in the Mediterranean area.
Usually biennial. Indumentum of tubercle-based setae and dense short hairs. Leaf margins undulate. Calyx divided for up to 2/3. Corolla bluish-violet to dark purple; limb 6-9 mm wide.

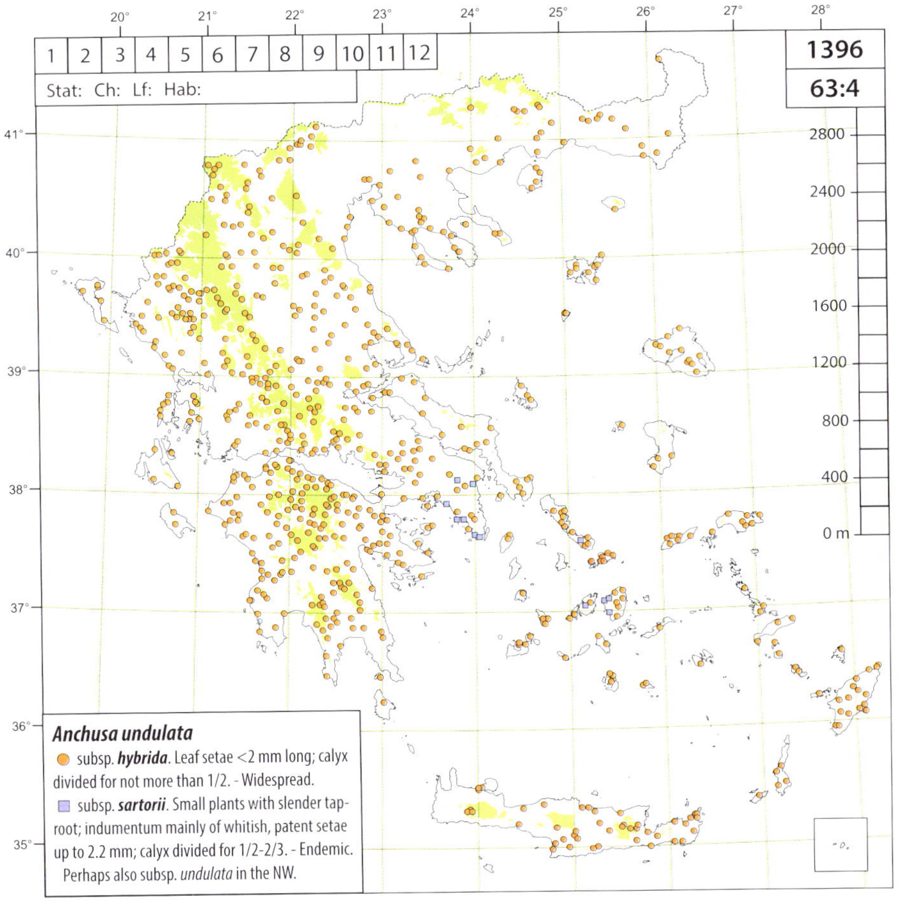

Anchusa undulata
subsp. ***hybrida***. Leaf setae <2 mm long; calyx divided for not more than 1/2. - Widespread.
subsp. ***sartorii***. Small plants with slender taproot; indumentum mainly of whitish, patent setae up to 2.2 mm; calyx divided for 1/2-2/3. - Endemic. Perhaps also subsp. *undulata* in the NW.

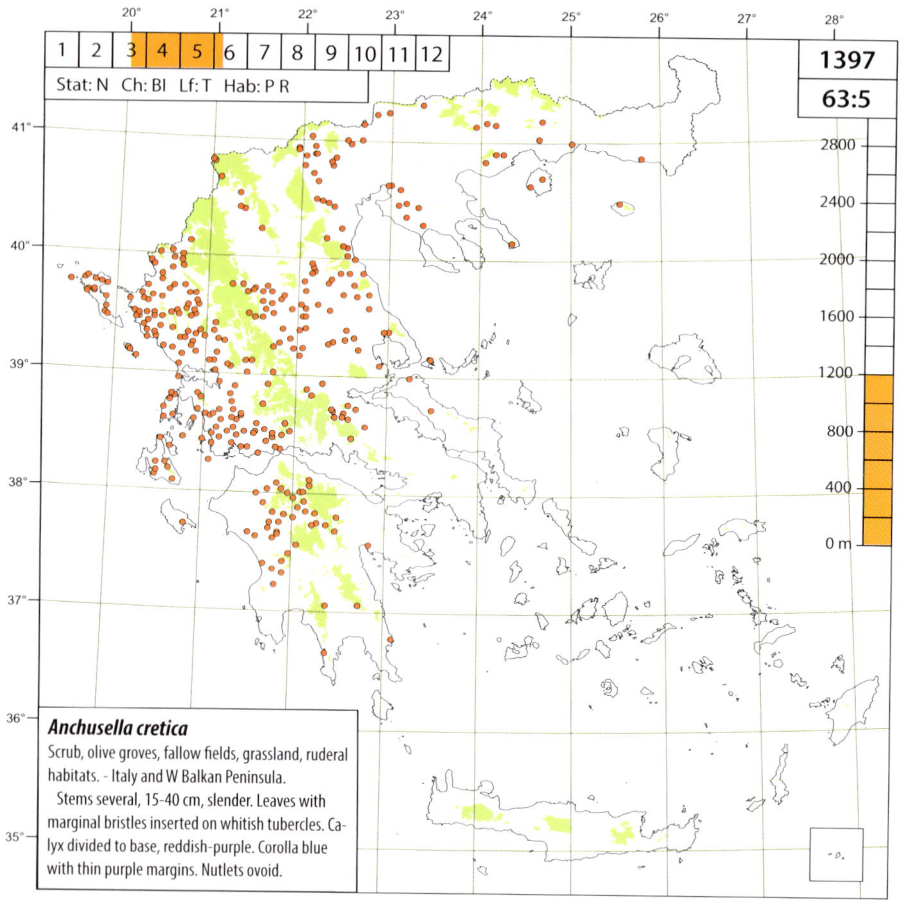

Anchusella cretica
Scrub, olive groves, fallow fields, grassland, ruderal habitats. - Italy and W Balkan Peninsula.
Stems several, 15-40 cm, slender. Leaves with marginal bristles inserted on whitish tubercles. Calyx divided to base, reddish-purple. Corolla blue with thin purple margins. Nutlets ovoid.

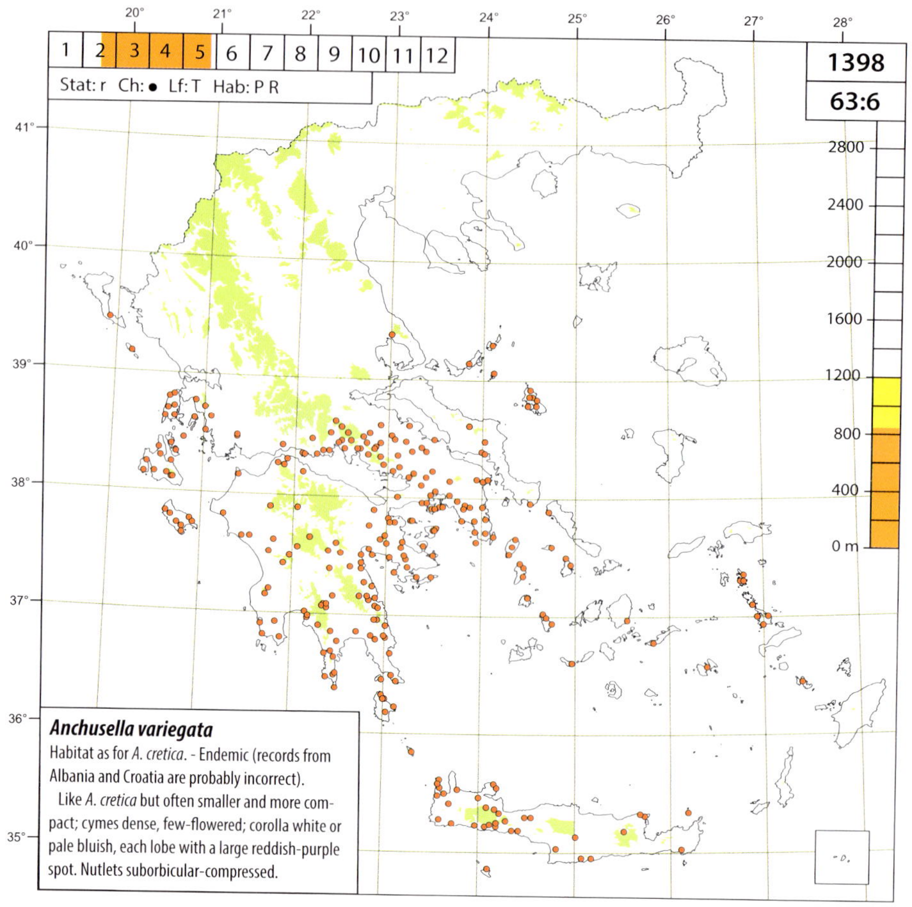

Anchusella variegata
Habitat as for *A. cretica*. - Endemic (records from Albania and Croatia are probably incorrect).
Like *A. cretica* but often smaller and more compact; cymes dense, few-flowered; corolla white or pale bluish, each lobe with a large reddish-purple spot. Nutlets suborbicular-compressed.

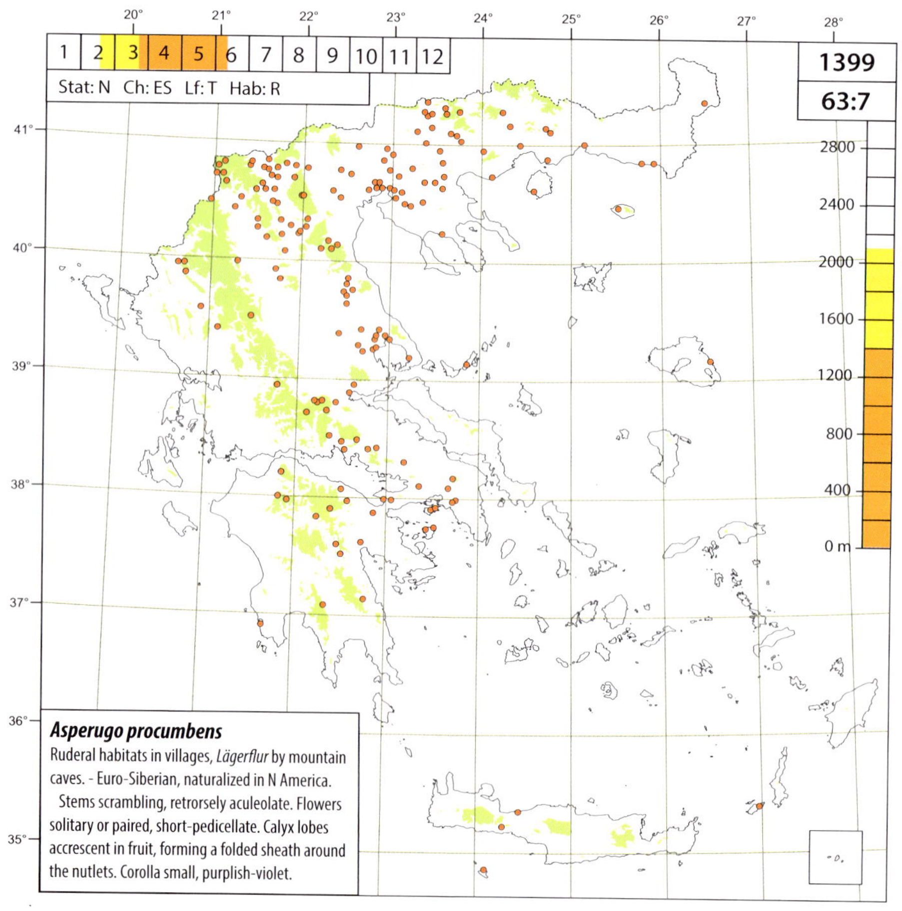

Asperugo procumbens
Ruderal habitats in villages, *Lägerflur* by mountain caves. - Euro-Siberian, naturalized in N America.
Stems scrambling, retrorsely aculeolate. Flowers solitary or paired, short-pedicellate. Calyx lobes accrescent in fruit, forming a folded sheath around the nutlets. Corolla small, purplish-violet.

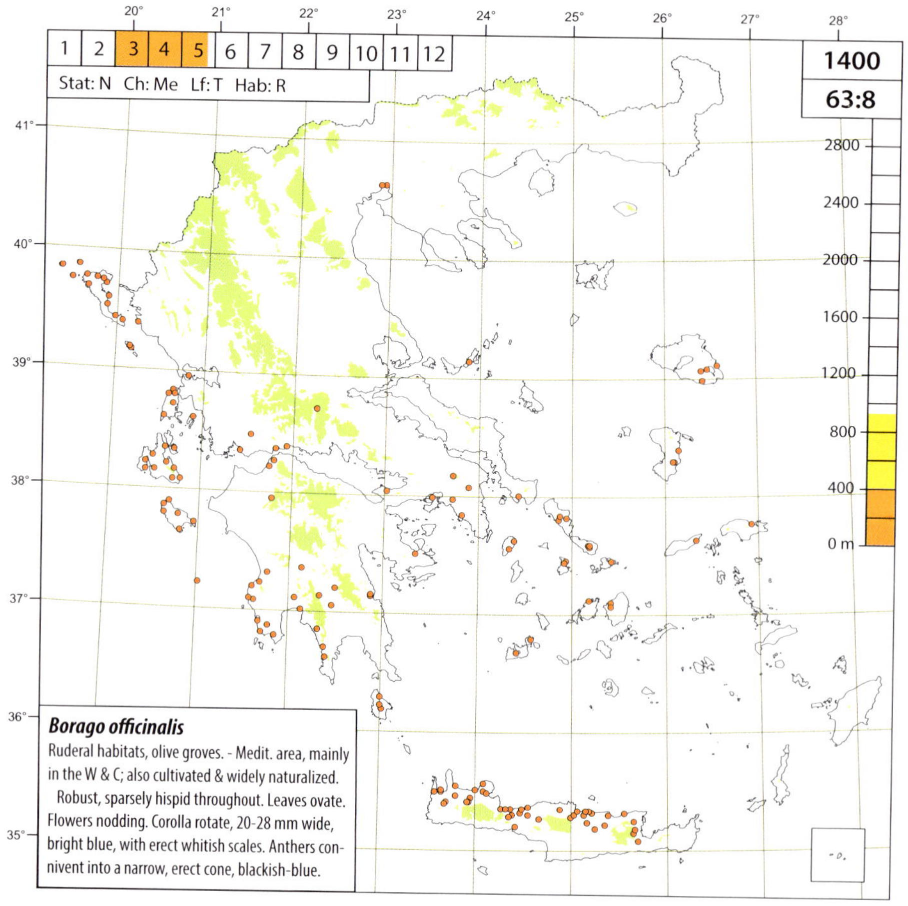

Borago officinalis
Ruderal habitats, olive groves. - Medit. area, mainly in the W & C; also cultivated & widely naturalized.
Robust, sparsely hispid throughout. Leaves ovate. Flowers nodding. Corolla rotate, 20-28 mm wide, bright blue, with erect whitish scales. Anthers connivent into a narrow, erect cone, blackish-blue.

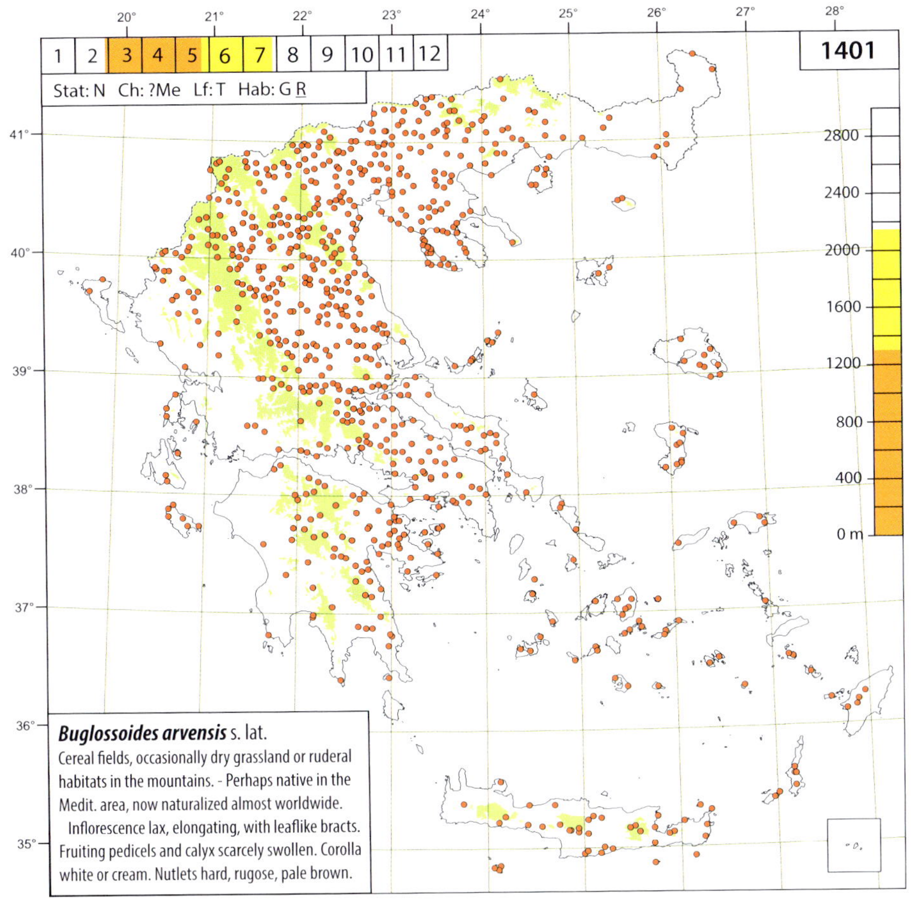

Buglossoides arvensis s. lat.
Cereal fields, occasionally dry grassland or ruderal habitats in the mountains. - Perhaps native in the Medit. area, now naturalized almost worldwide.
Inflorescence lax, elongating, with leaflike bracts. Fruiting pedicels and calyx scarcely swollen. Corolla white or cream. Nutlets hard, rugose, pale brown.

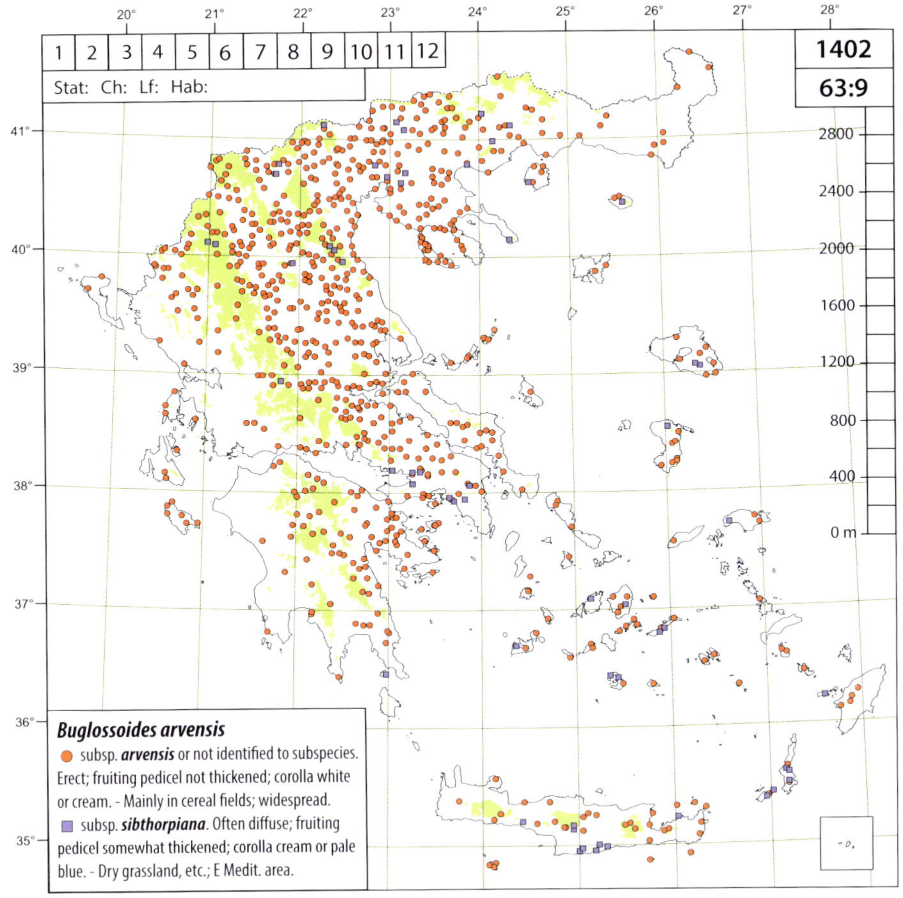

Buglossoides arvensis
● subsp. ***arvensis*** or not identified to subspecies. Erect; fruiting pedicel not thickened; corolla white or cream. - Mainly in cereal fields; widespread.
■ subsp. ***sibthorpiana***. Often diffuse; fruiting pedicel somewhat thickened; corolla cream or pale blue. - Dry grassland, etc.; E Medit. area.

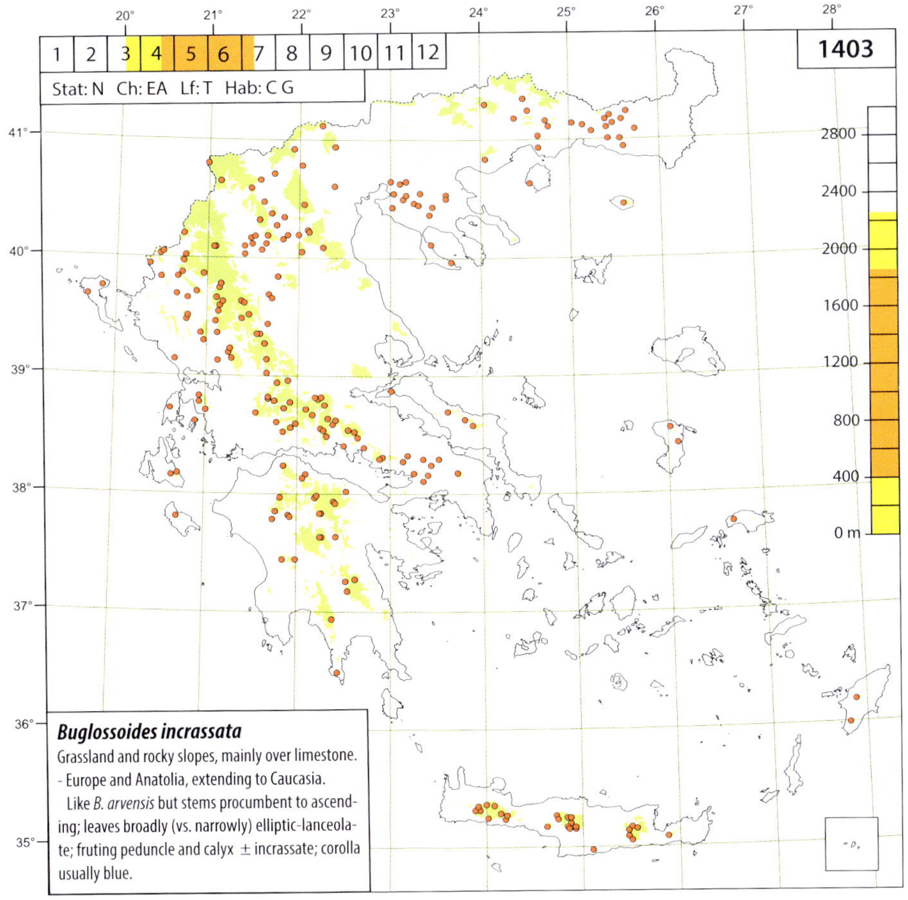

Buglossoides incrassata
Grassland and rocky slopes, mainly over limestone. - Europe and Anatolia, extending to Caucasia.
Like *B. arvensis* but stems procumbent to ascending; leaves broadly (vs. narrowly) elliptic-lanceolate; fruting peduncle and calyx ± incrassate; corolla usually blue.

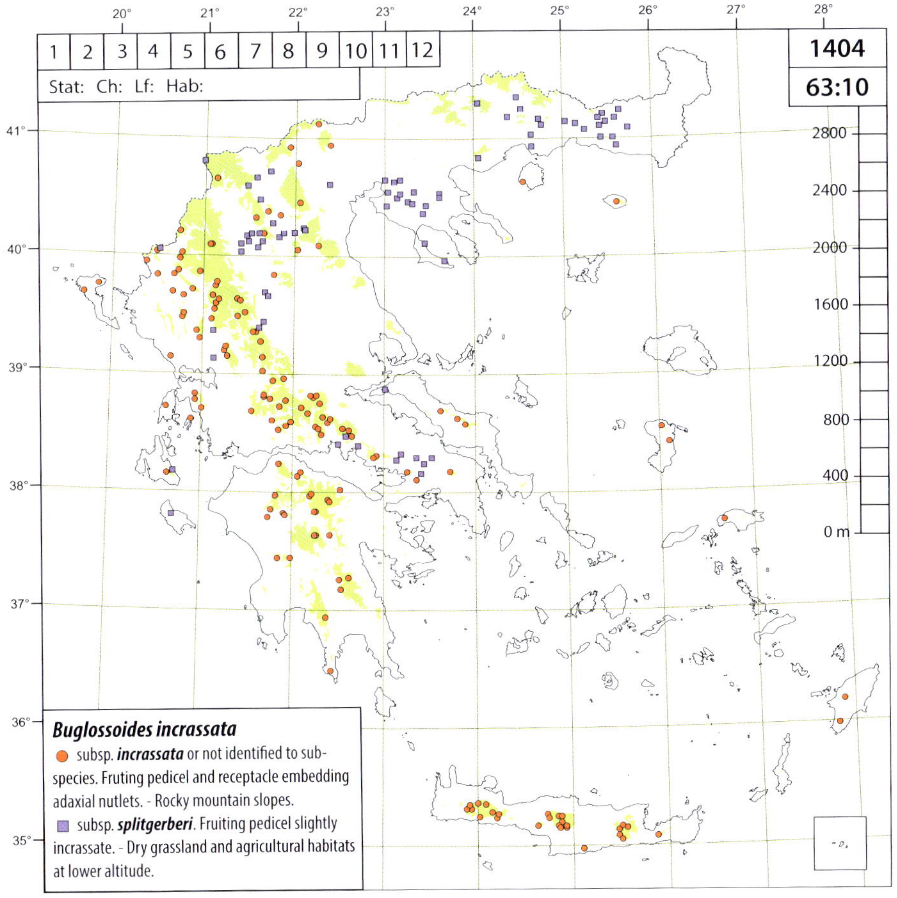

Buglossoides incrassata
● subsp. ***incrassata*** or not identified to subspecies. Fruting pedicel and receptacle embedding adaxial nutlets. - Rocky mountain slopes.
■ subsp. ***splitgerberi***. Fruiting pedicel slightly incrassate. - Dry grassland and agricultural habitats at lower altitude.

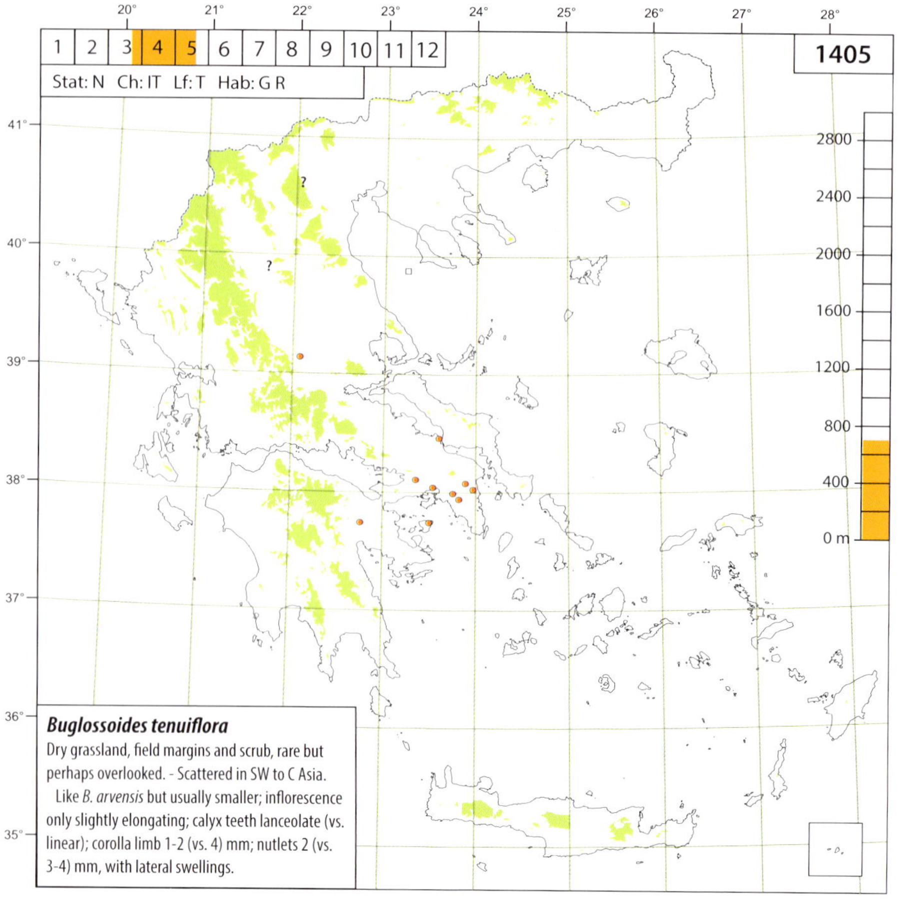

Buglossoides tenuiflora
Dry grassland, field margins and scrub, rare but perhaps overlooked. - Scattered in SW to C Asia.
Like *B. arvensis* but usually smaller; inflorescence only slightly elongating; calyx teeth lanceolate (vs. linear); corolla limb 1-2 (vs. 4) mm; nutlets 2 (vs. 3-4) mm, with lateral swellings.

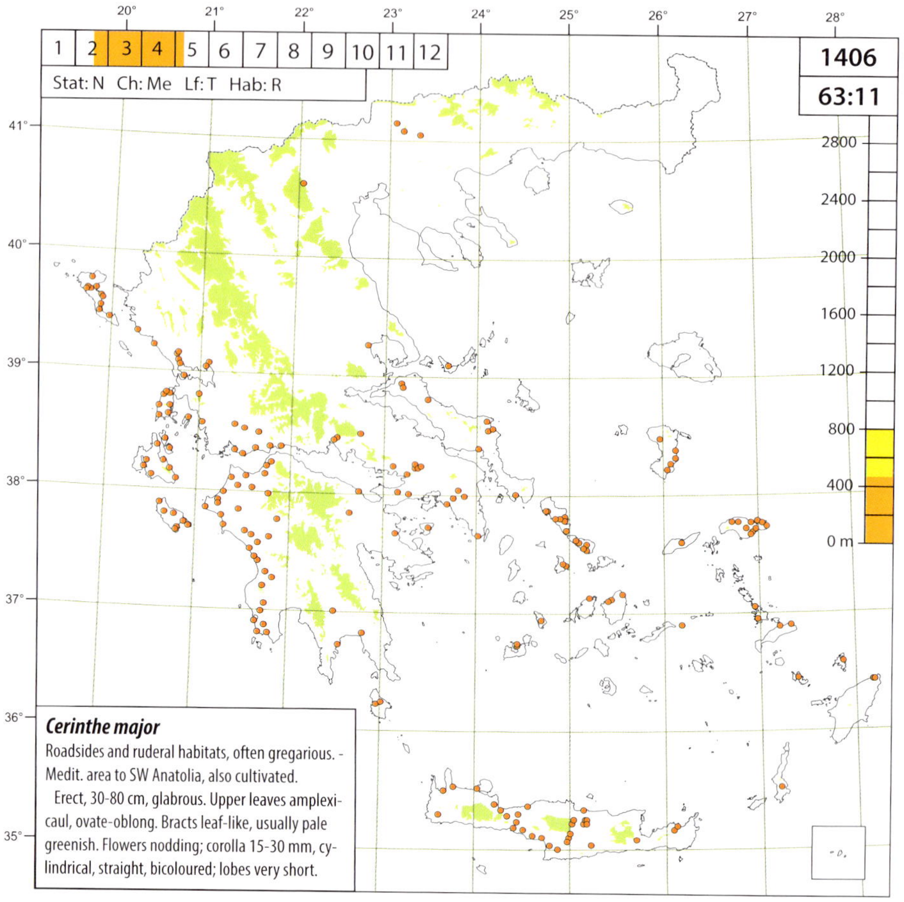

Cerinthe major
Roadsides and ruderal habitats, often gregarious. - Medit. area to SW Anatolia, also cultivated.
Erect, 30-80 cm, glabrous. Upper leaves amplexicaul, ovate-oblong. Bracts leaf-like, usually pale greenish. Flowers nodding; corolla 15-30 mm, cylindrical, straight, bicoloured; lobes very short.

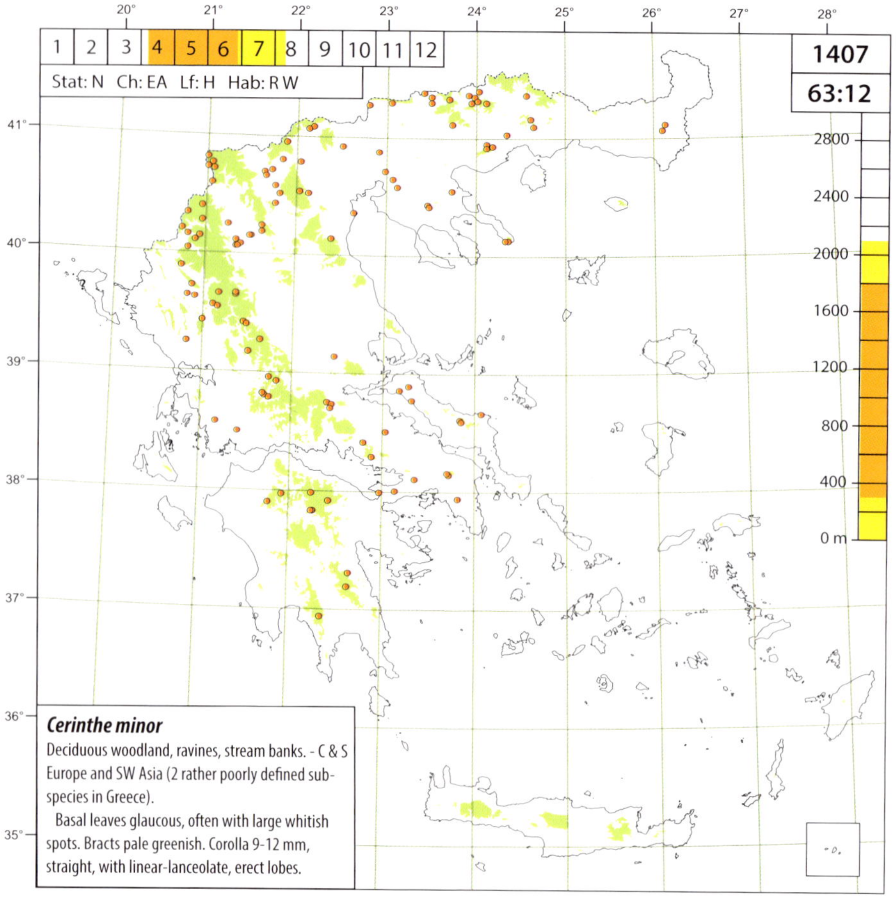

Cerinthe minor
Deciduous woodland, ravines, stream banks. - C & S Europe and SW Asia (2 rather poorly defined subspecies in Greece).
Basal leaves glaucous, often with large whitish spots. Bracts pale greenish. Corolla 9-12 mm, straight, with linear-lanceolate, erect lobes.

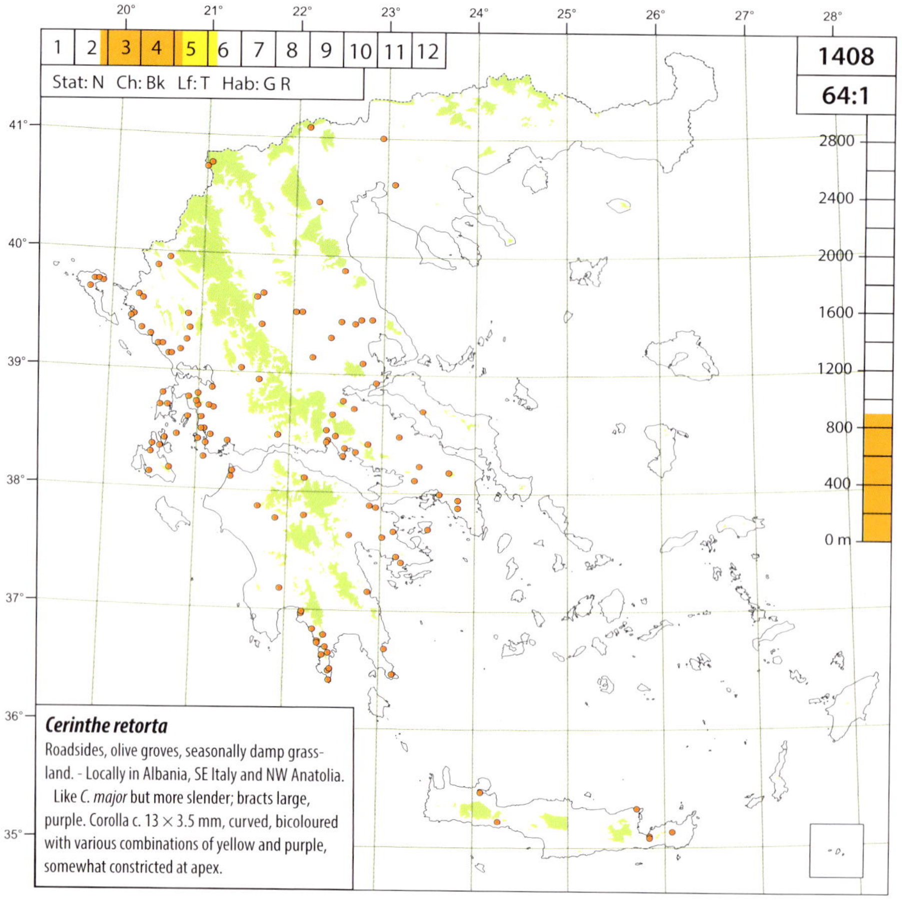

Cerinthe retorta
Roadsides, olive groves, seasonally damp grassland. - Locally in Albania, SE Italy and NW Anatolia.
Like *C. major* but more slender; bracts large, purple. Corolla c. 13 × 3.5 mm, curved, bicoloured with various combinations of yellow and purple, somewhat constricted at apex.

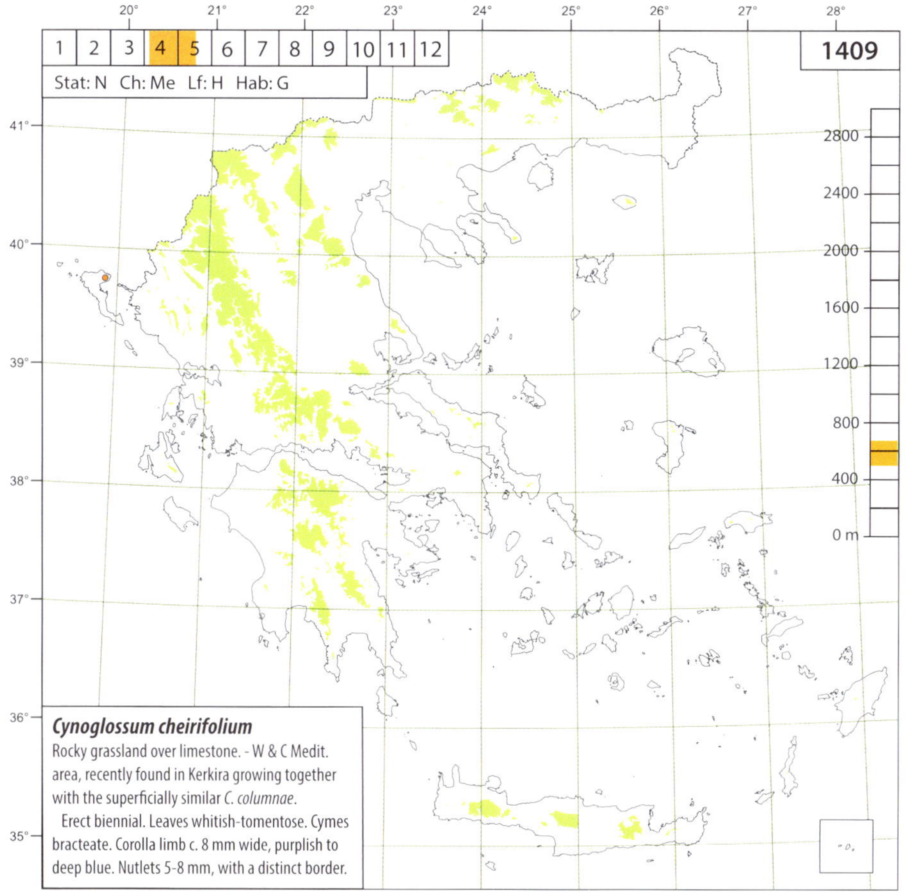

Cynoglossum cheirifolium
Rocky grassland over limestone. - W & C Medit. area, recently found in Kerkira growing together with the superficially similar *C. columnae*.
Erect biennial. Leaves whitish-tomentose. Cymes bracteate. Corolla limb c. 8 mm wide, purplish to deep blue. Nutlets 5-8 mm, with a distinct border.

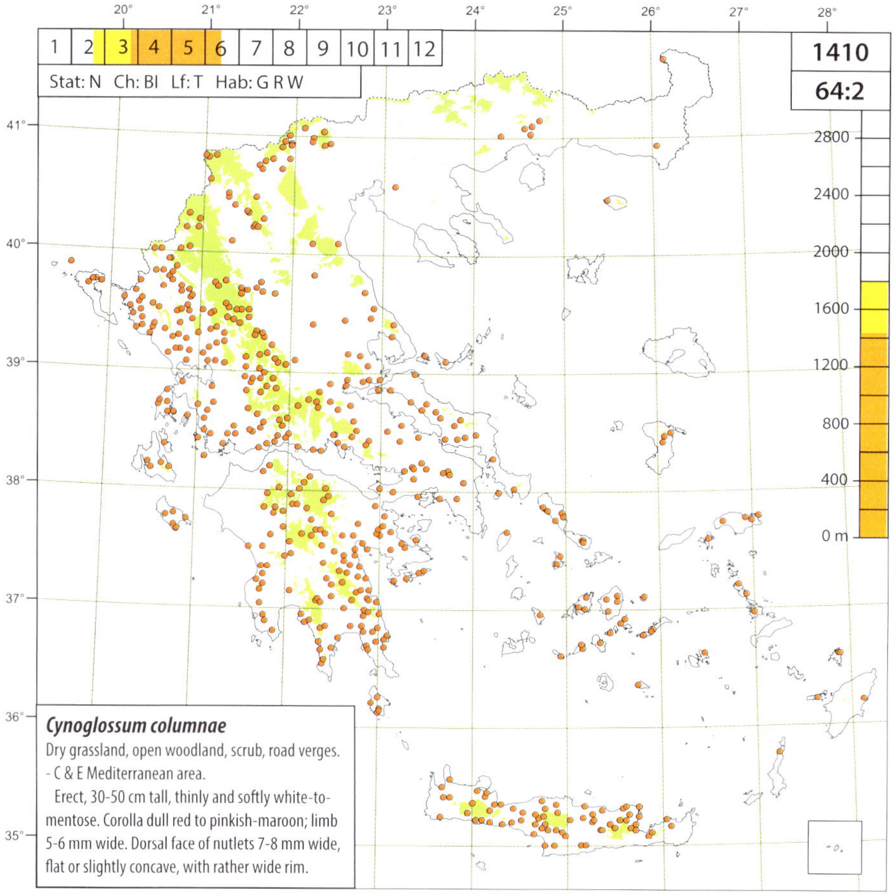

Cynoglossum columnae
Dry grassland, open woodland, scrub, road verges. - C & E Mediterranean area.
Erect, 30-50 cm tall, thinly and softly white-tomentose. Corolla dull red to pinkish-maroon; limb 5-6 mm wide. Dorsal face of nutlets 7-8 mm wide, flat or slightly concave, with rather wide rim.

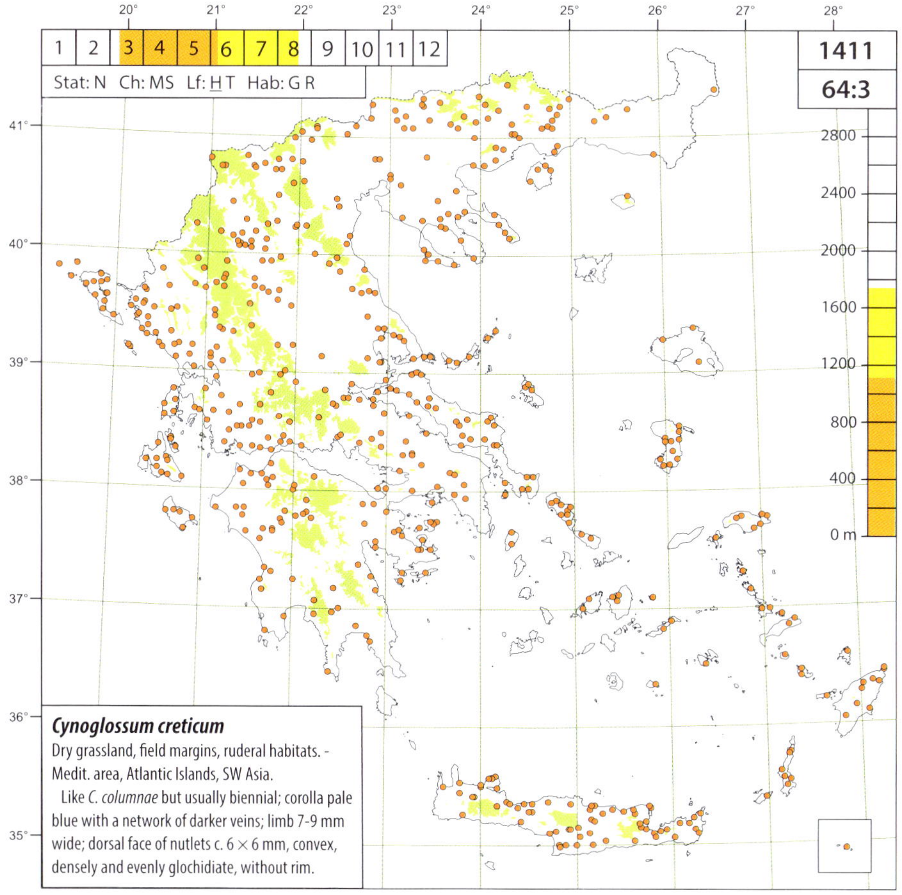

Cynoglossum creticum
Dry grassland, field margins, ruderal habitats. - Medit. area, Atlantic Islands, SW Asia.
Like *C. columnae* but usually biennial; corolla pale blue with a network of darker veins; limb 7-9 mm wide; dorsal face of nutlets c. 6 × 6 mm, convex, densely and evenly glochidiate, without rim.

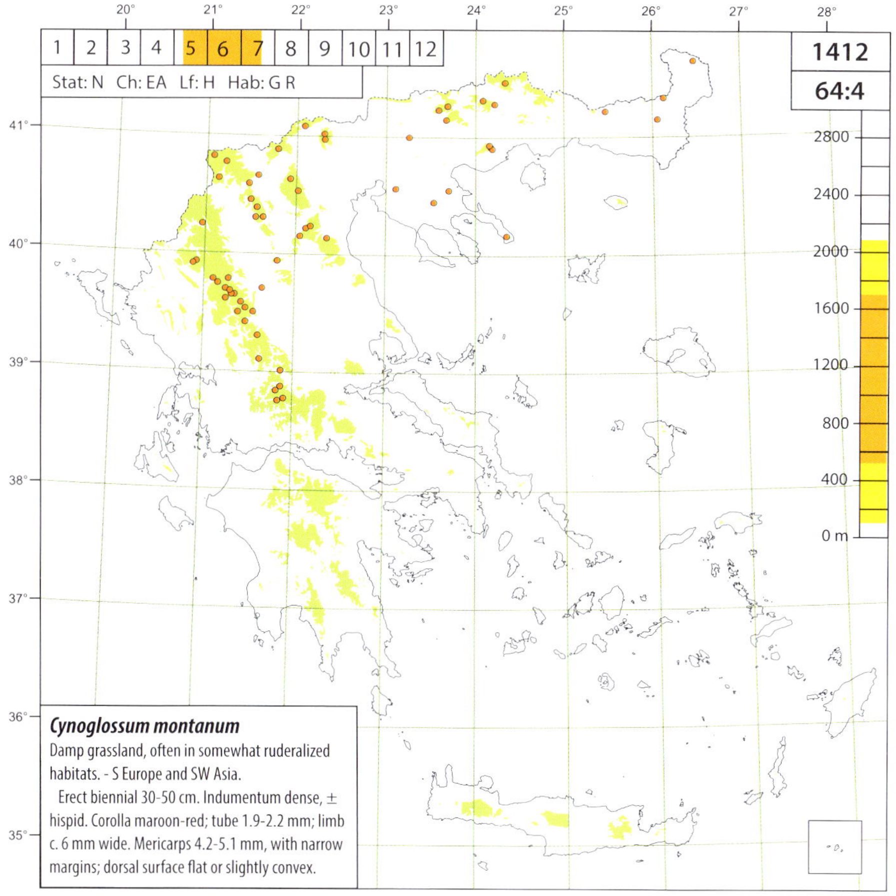

Cynoglossum montanum
Damp grassland, often in somewhat ruderalized habitats. - S Europe and SW Asia.
Erect biennial 30-50 cm. Indumentum dense, ± hispid. Corolla maroon-red; tube 1.9-2.2 mm; limb c. 6 mm wide. Mericarps 4.2-5.1 mm, with narrow margins; dorsal surface flat or slightly convex.

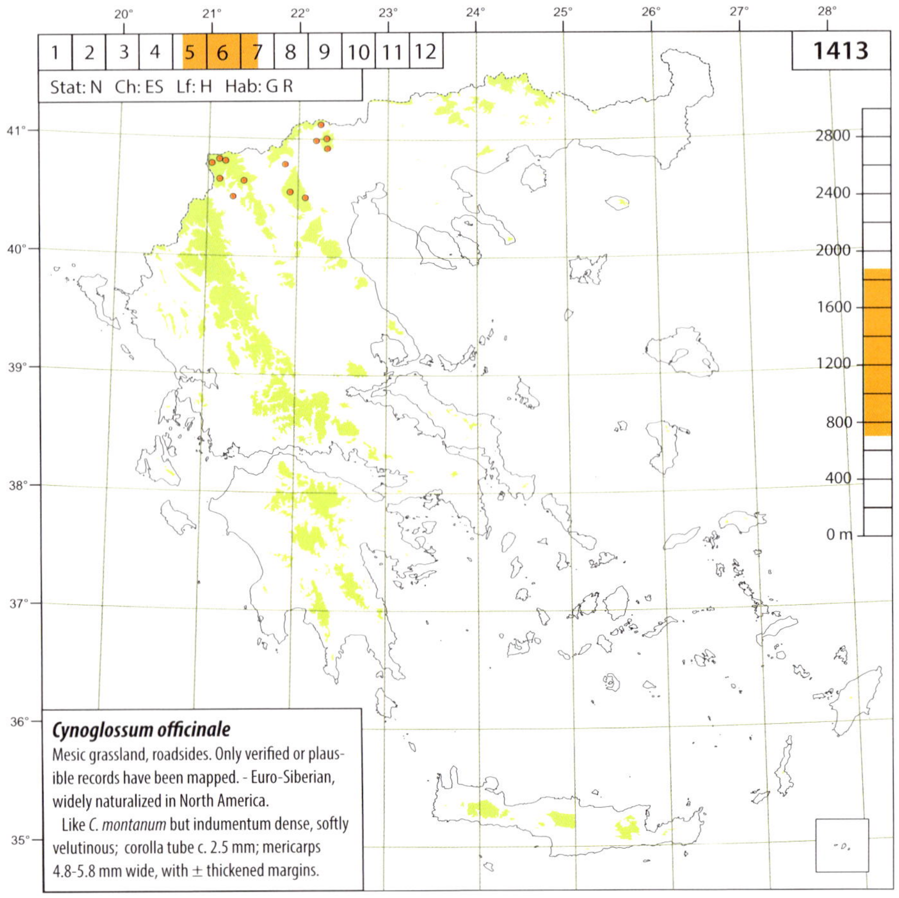

Cynoglossum officinale
Mesic grassland, roadsides. Only verified or plausible records have been mapped. - Euro-Siberian, widely naturalized in North America.
Like *C. montanum* but indumentum dense, softly velutinous; corolla tube c. 2.5 mm; mericarps 4.8-5.8 mm wide, with ± thickened margins.

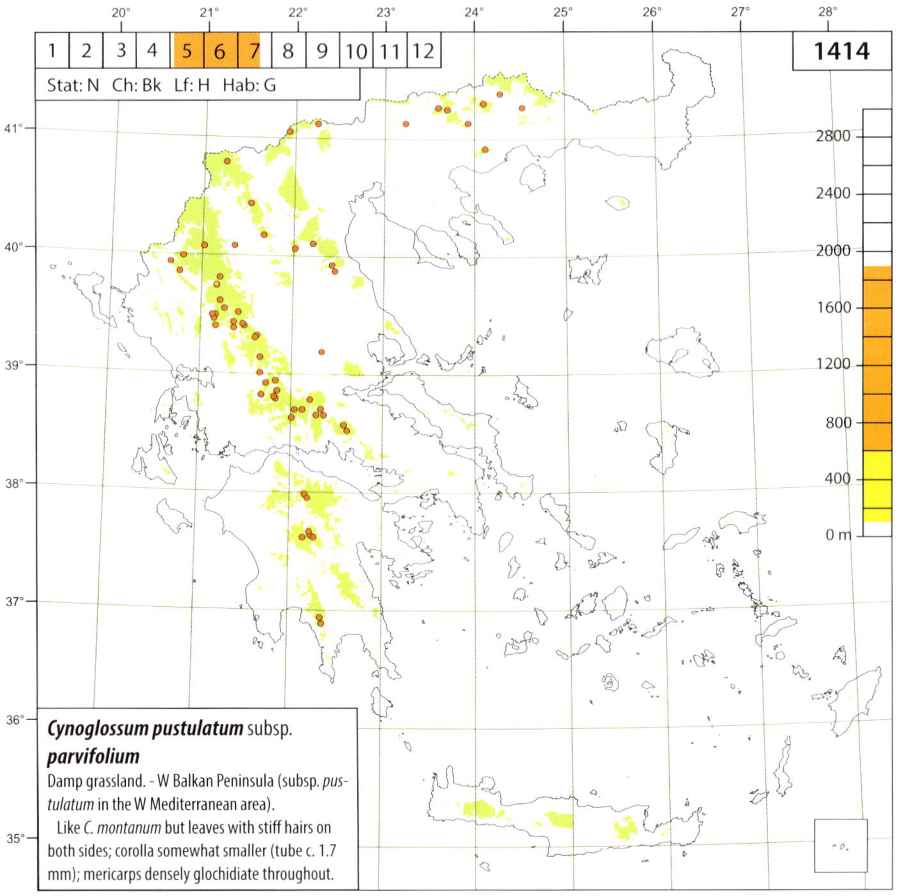

Cynoglossum pustulatum* subsp. *parvifolium
Damp grassland. - W Balkan Peninsula (subsp. *pustulatum* in the W Mediterranean area).
Like *C. montanum* but leaves with stiff hairs on both sides; corolla somewhat smaller (tube c. 1.7 mm); mericarps densely glochidiate throughout.

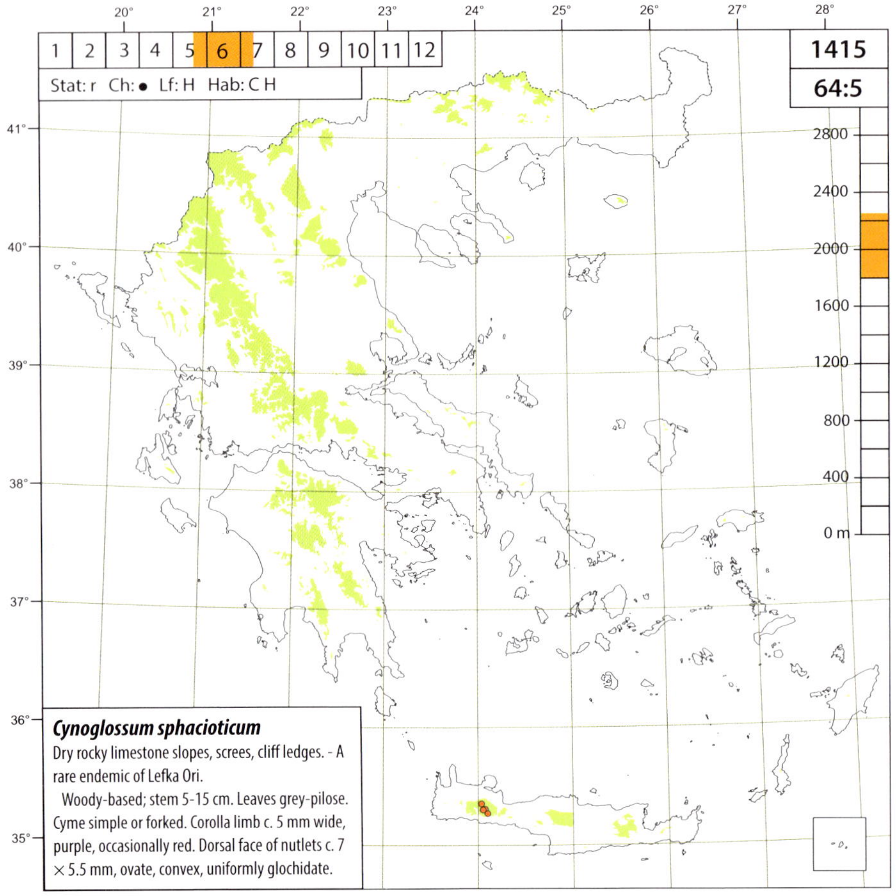

Cynoglossum sphacioticum
Dry rocky limestone slopes, screes, cliff ledges. - A rare endemic of Lefka Ori.
Woody-based; stem 5-15 cm. Leaves grey-pilose. Cyme simple or forked. Corolla limb c. 5 mm wide, purple, occasionally red. Dorsal face of nutlets c. 7 × 5.5 mm, ovate, convex, uniformly glochidate.

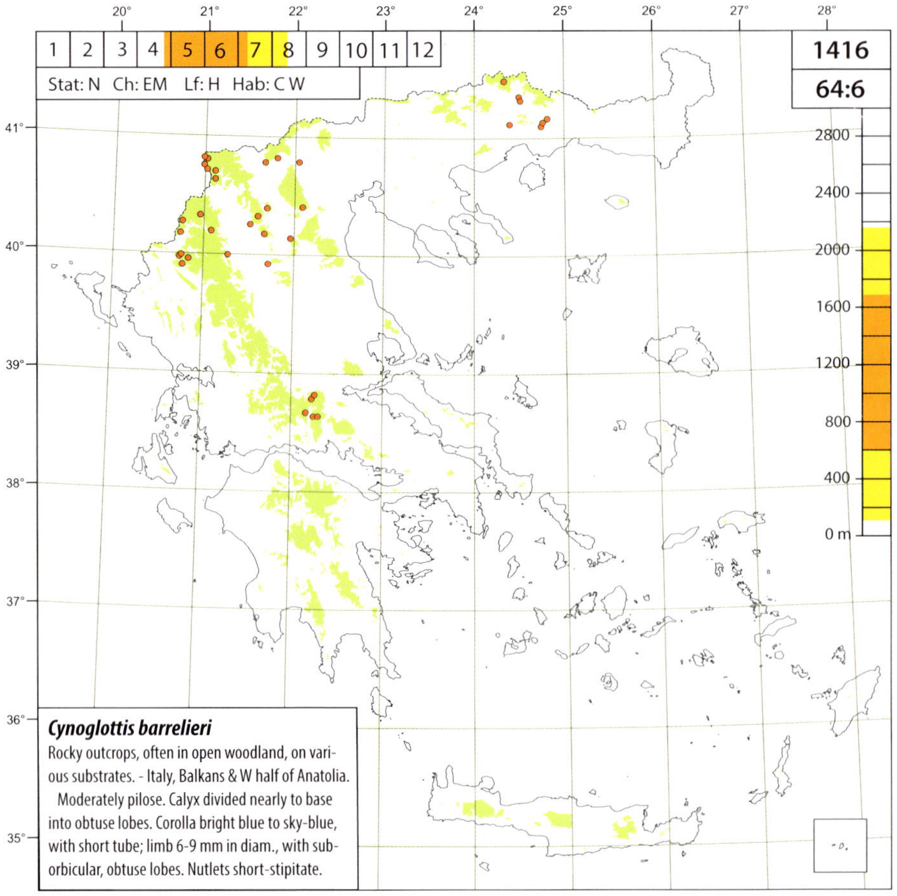

Cynoglottis barrelieri
Rocky outcrops, often in open woodland, on various substrates. - Italy, Balkans & W half of Anatolia.
Moderately pilose. Calyx divided nearly to base into obtuse lobes. Corolla bright blue to sky-blue, with short tube; limb 6-9 mm in diam., with suborbicular, obtuse lobes. Nutlets short-stipitate.

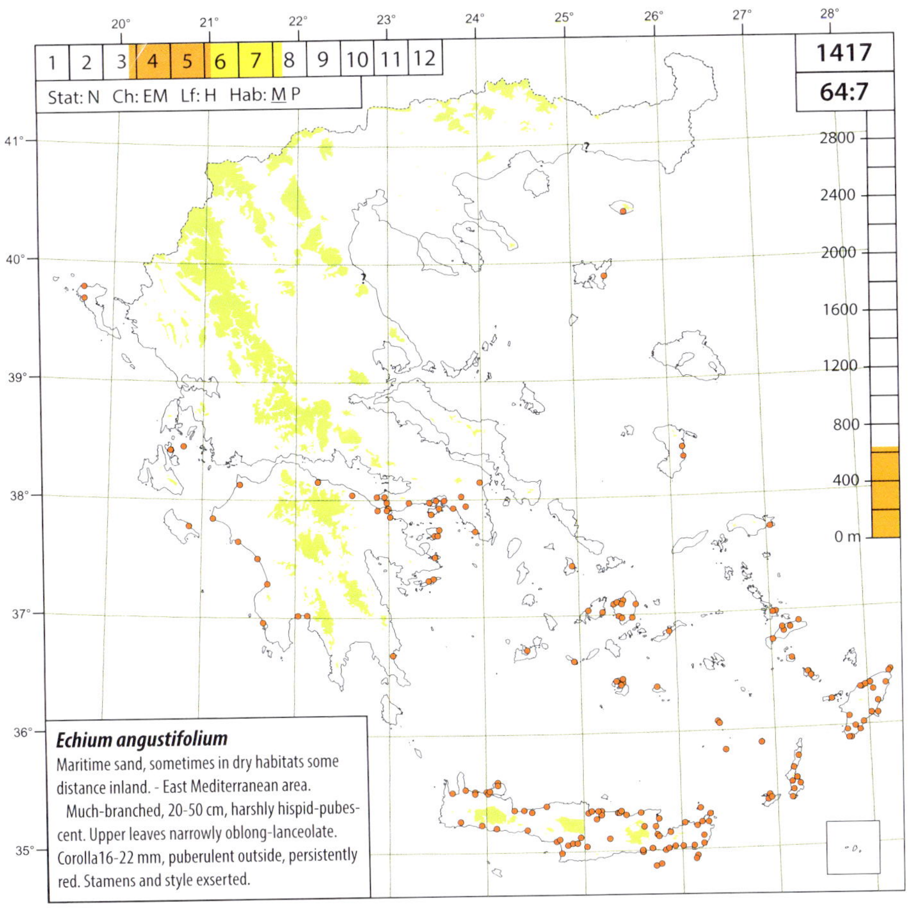

Echium angustifolium

Maritime sand, sometimes in dry habitats some distance inland. - East Mediterranean area.

Much-branched, 20-50 cm, harshly hispid-pubescent. Upper leaves narrowly oblong-lanceolate. Corolla16-22 mm, puberulent outside, persistently red. Stamens and style exserted.

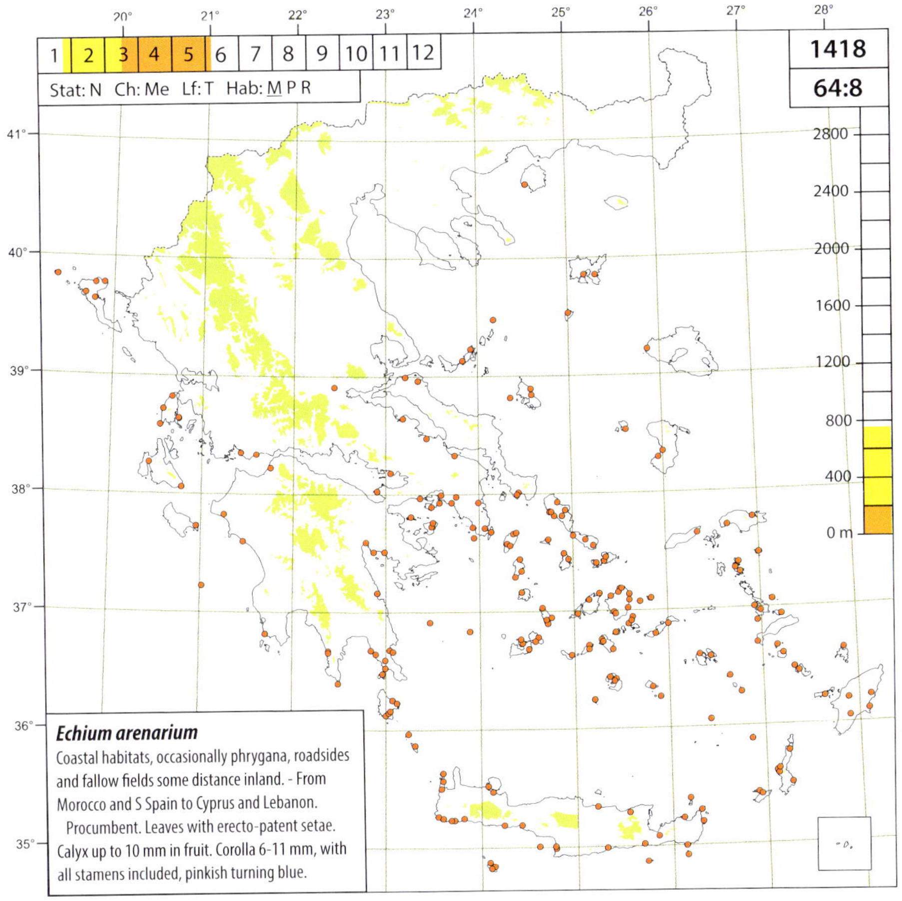

Echium arenarium

Coastal habitats, occasionally phrygana, roadsides and fallow fields some distance inland. - From Morocco and S Spain to Cyprus and Lebanon.

Procumbent. Leaves with erecto-patent setae. Calyx up to 10 mm in fruit. Corolla 6-11 mm, with all stamens included, pinkish turning blue.

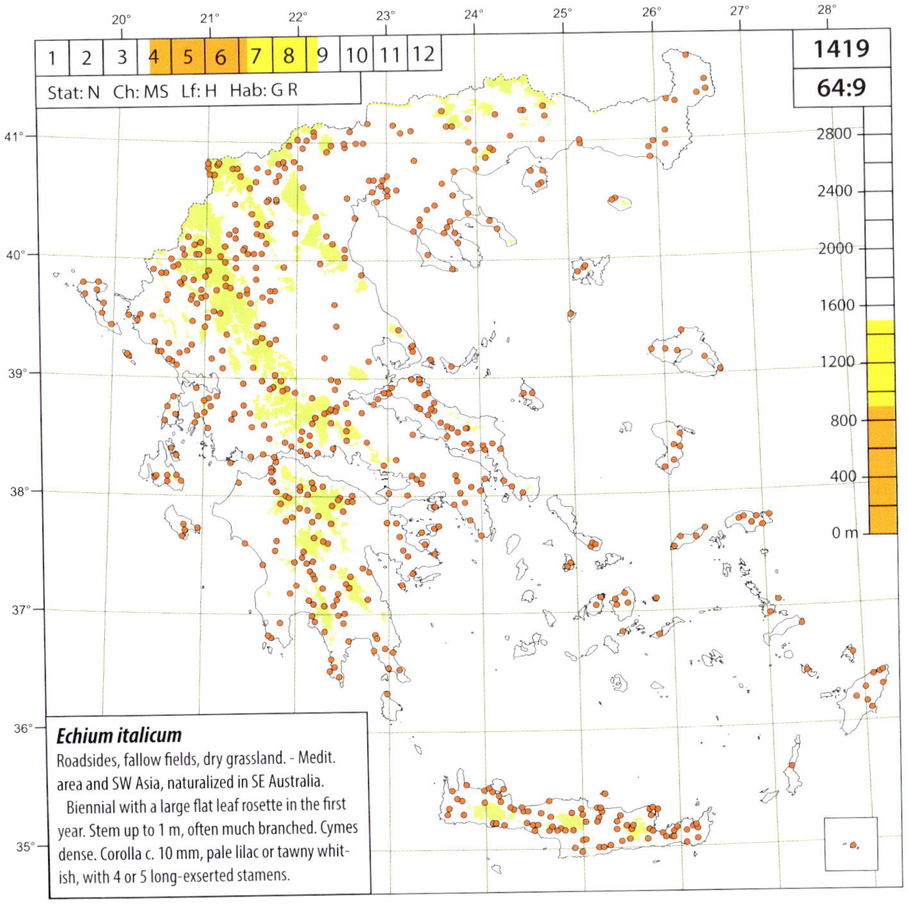

Echium italicum

Roadsides, fallow fields, dry grassland. - Medit. area and SW Asia, naturalized in SE Australia.

Biennial with a large flat leaf rosette in the first year. Stem up to 1 m, often much branched. Cymes dense. Corolla c. 10 mm, pale lilac or tawny whitish, with 4 or 5 long-exserted stamens.

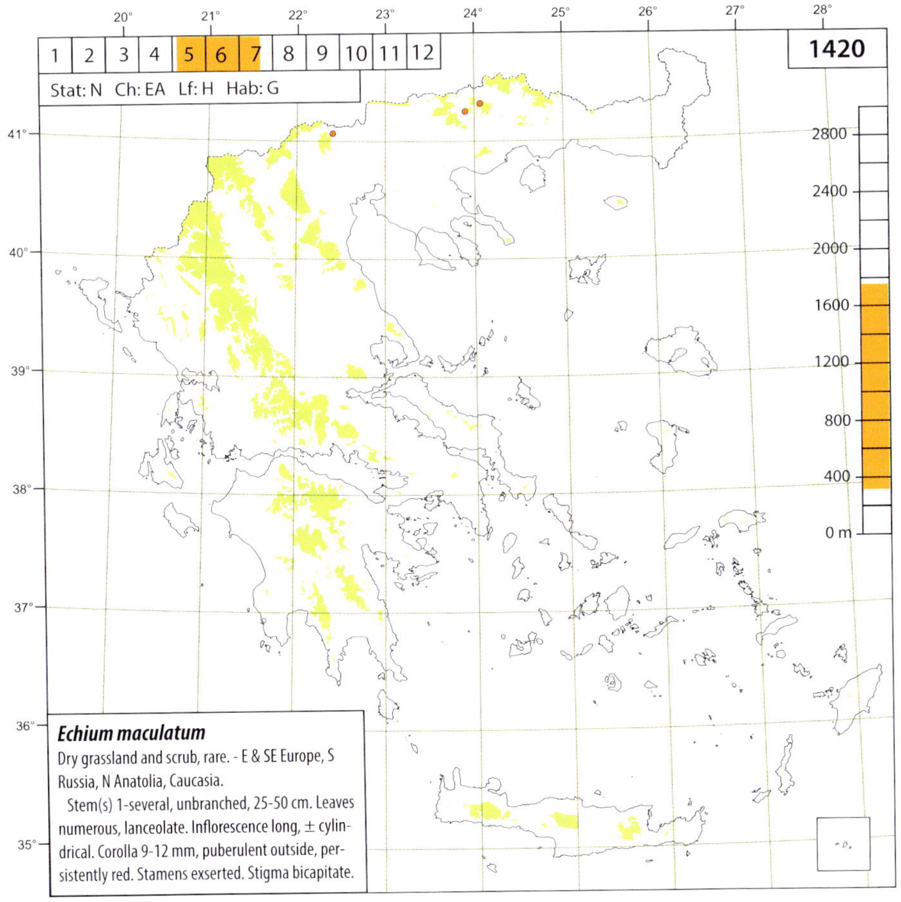

Echium maculatum

Dry grassland and scrub, rare. - E & SE Europe, S Russia, N Anatolia, Caucasia.

Stem(s) 1-several, unbranched, 25-50 cm. Leaves numerous, lanceolate. Inflorescence long, ± cylindrical. Corolla 9-12 mm, puberulent outside, persistently red. Stamens exserted. Stigma bicapitate.

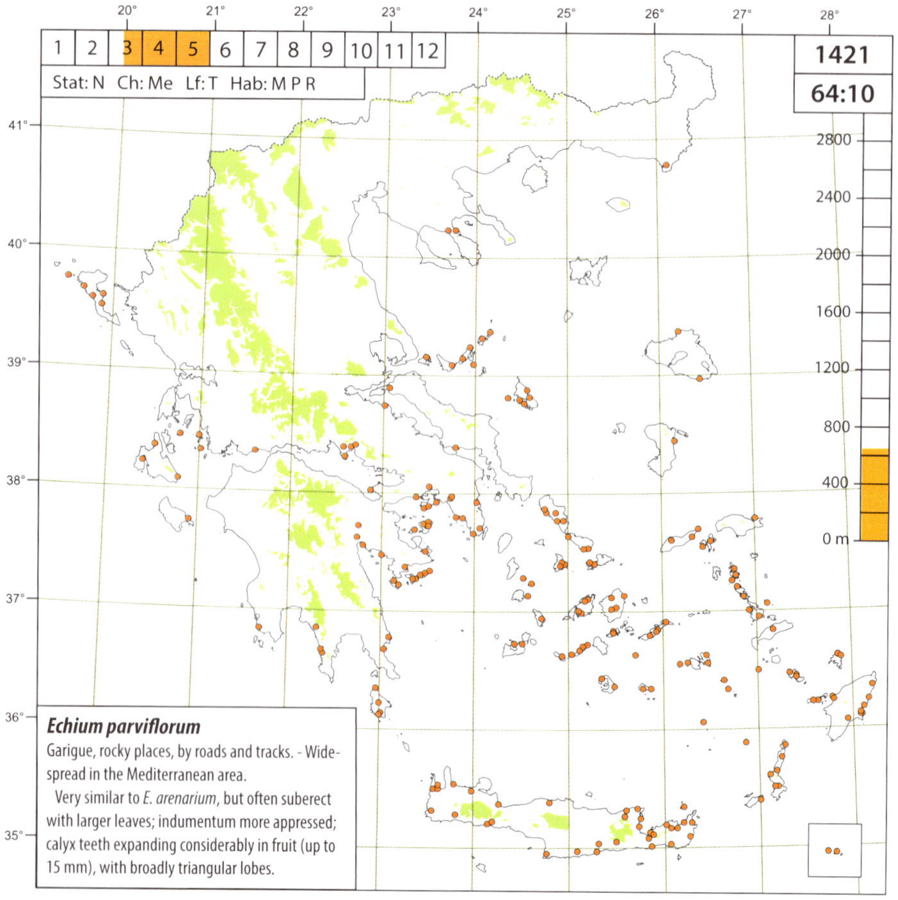

Echium parviflorum

Garigue, rocky places, by roads and tracks. - Widespread in the Mediterranean area.

Very similar to *E. arenarium*, but often suberect with larger leaves; indumentum more appressed; calyx teeth expanding considerably in fruit (up to 15 mm), with broadly triangular lobes.

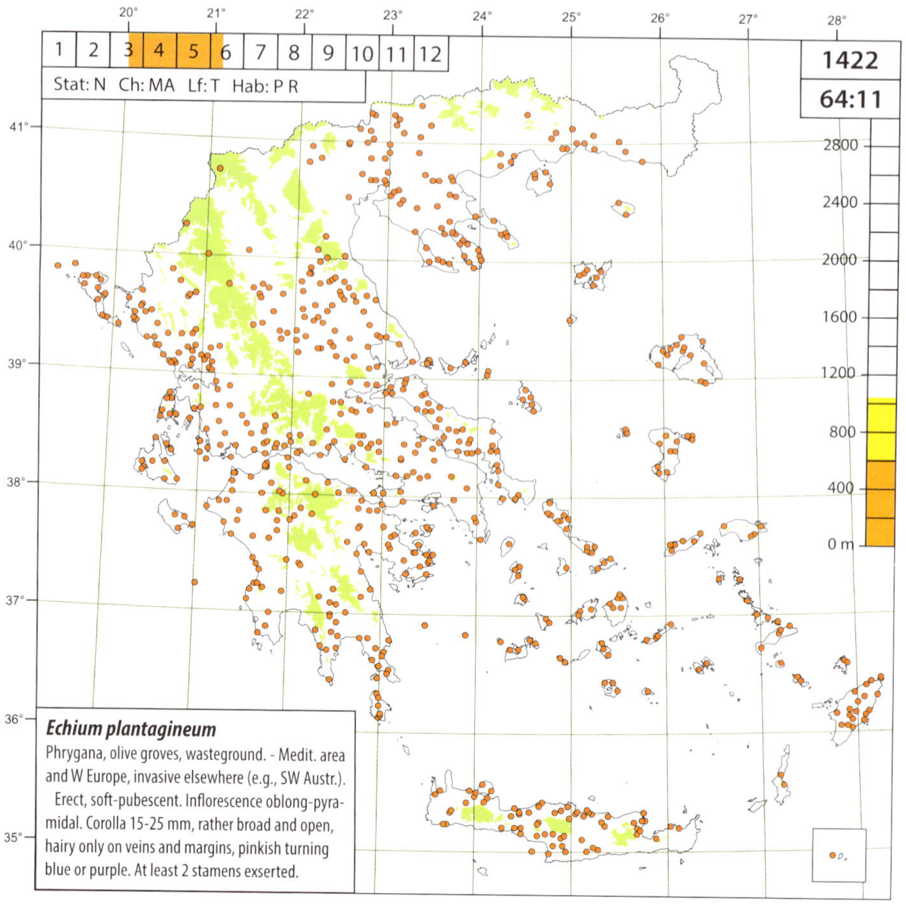

Echium plantagineum

Phrygana, olive groves, wasteground. - Medit. area and W Europe, invasive elsewhere (e.g., SW Austr.).

Erect, soft-pubescent. Inflorescence oblong-pyramidal. Corolla 15-25 mm, rather broad and open, hairy only on veins and margins, pinkish turning blue or purple. At least 2 stamens exserted.

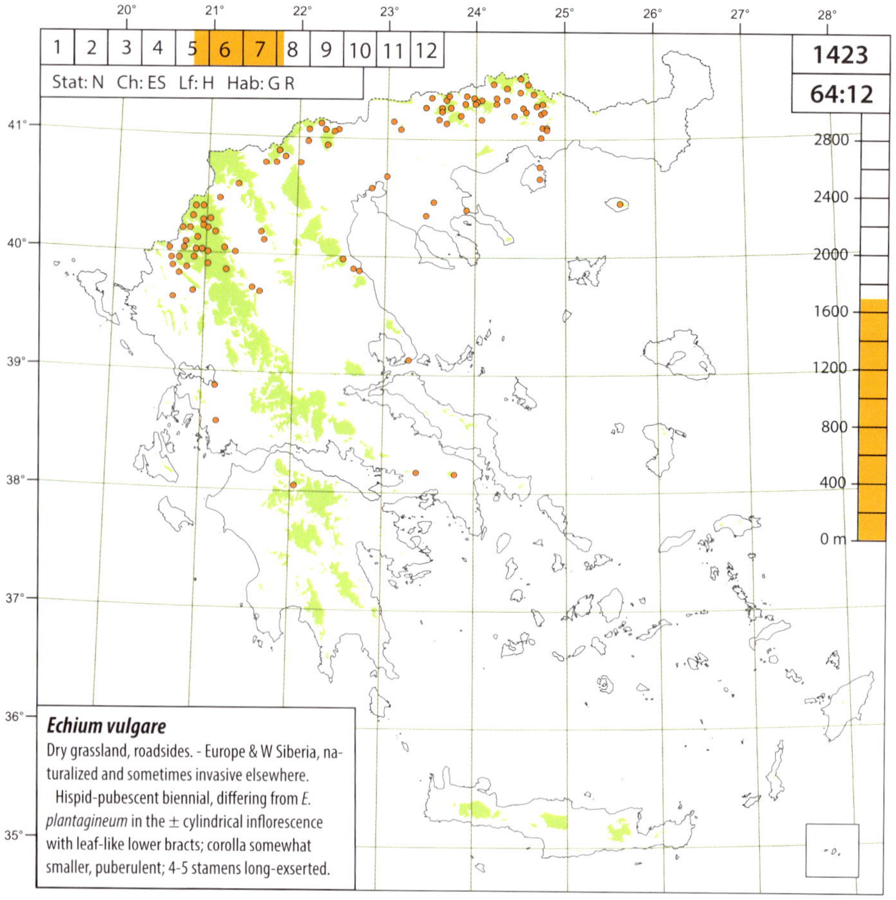

Echium vulgare

Dry grassland, roadsides. - Europe & W Siberia, naturalized and sometimes invasive elsewhere.

Hispid-pubescent biennial, differing from *E. plantagineum* in the ± cylindrical inflorescence with leaf-like lower bracts; corolla somewhat smaller, puberulent; 4-5 stamens long-exserted.

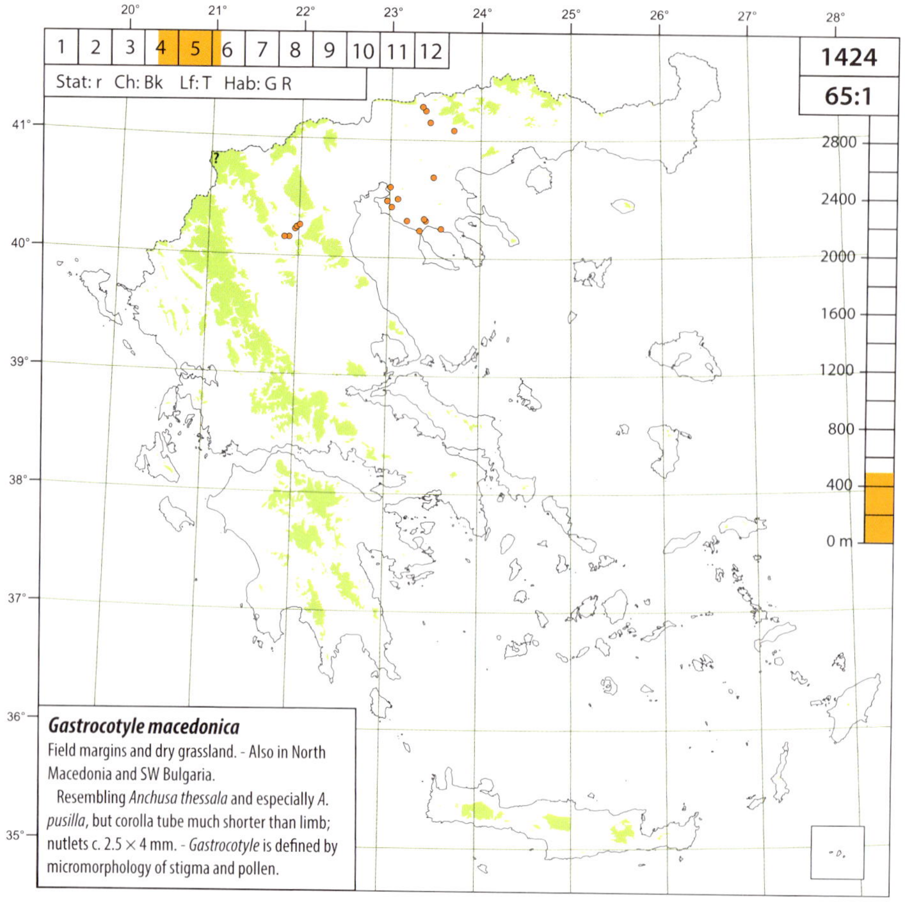

Gastrocotyle macedonica

Field margins and dry grassland. - Also in North Macedonia and SW Bulgaria.

Resembling *Anchusa thessala* and especially *A. pusilla*, but corolla tube much shorter than limb; nutlets c. 2.5 × 4 mm. - *Gastrocotyle* is defined by micromorphology of stigma and pollen.

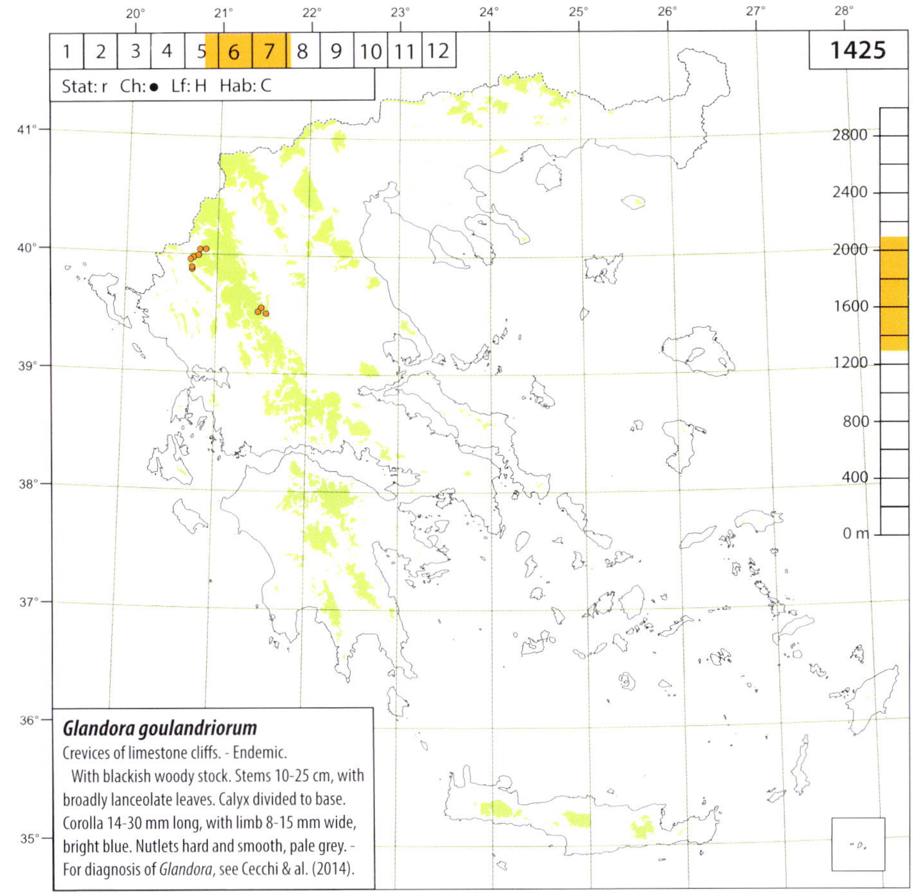

Glandora goulandriorum

Crevices of limestone cliffs. - Endemic.

With blackish woody stock. Stems 10-25 cm, with broadly lanceolate leaves. Calyx divided to base. Corolla 14-30 mm long, with limb 8-15 mm wide, bright blue. Nutlets hard and smooth, pale grey. - For diagnosis of *Glandora*, see Cecchi & al. (2014).

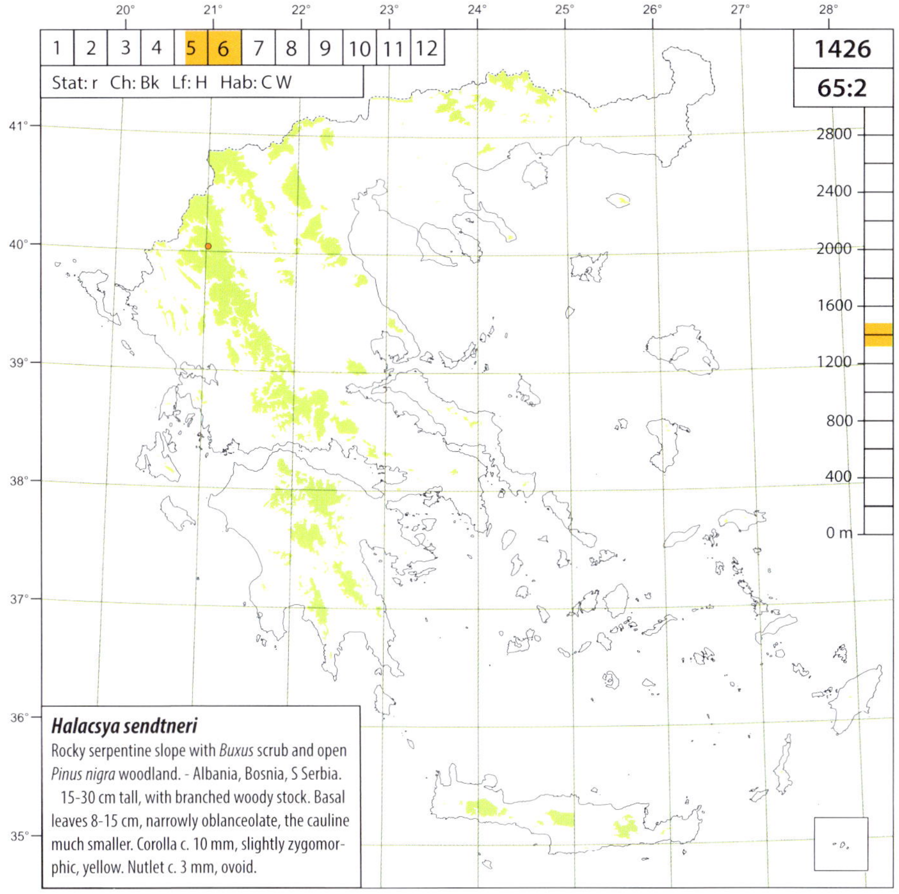

Halacsya sendtneri

Rocky serpentine slope with *Buxus* scrub and open *Pinus nigra* woodland. - Albania, Bosnia, S Serbia.

15-30 cm tall, with branched woody stock. Basal leaves 8-15 cm, narrowly oblanceolate, the cauline much smaller. Corolla c. 10 mm, slightly zygomorphic, yellow. Nutlet c. 3 mm, ovoid.

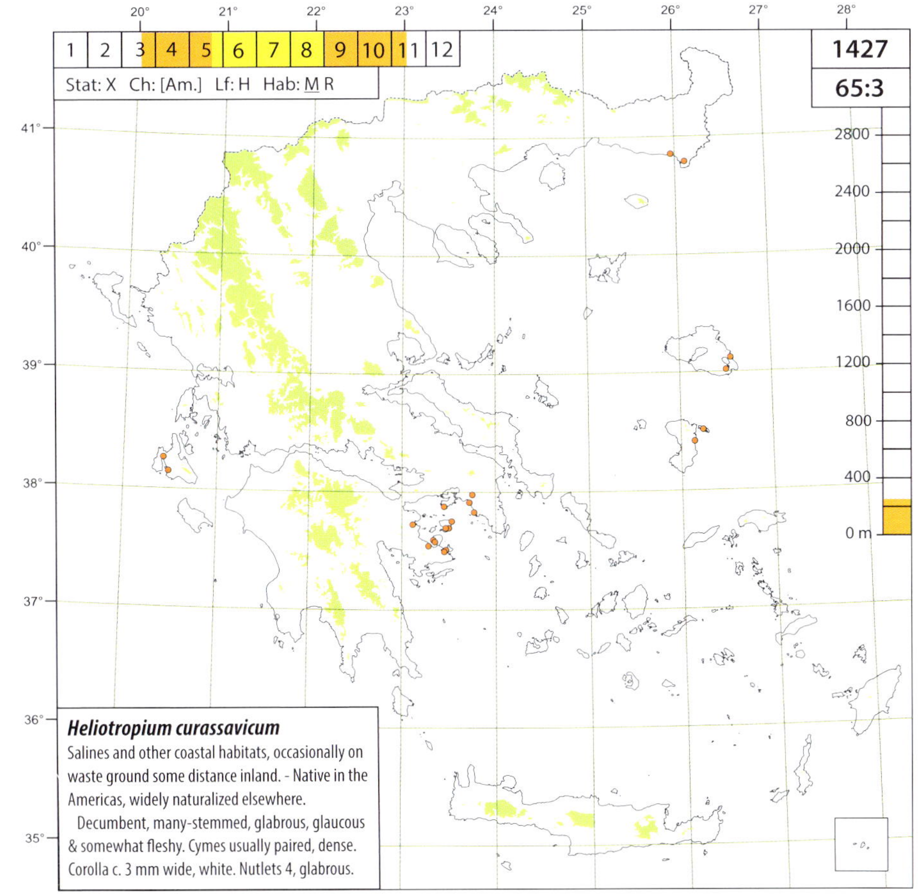

Heliotropium curassavicum

Salines and other coastal habitats, occasionally on waste ground some distance inland. - Native in the Americas, widely naturalized elsewhere.

Decumbent, many-stemmed, glabrous, glaucous & somewhat fleshy. Cymes usually paired, dense. Corolla c. 3 mm wide, white. Nutlets 4, glabrous.

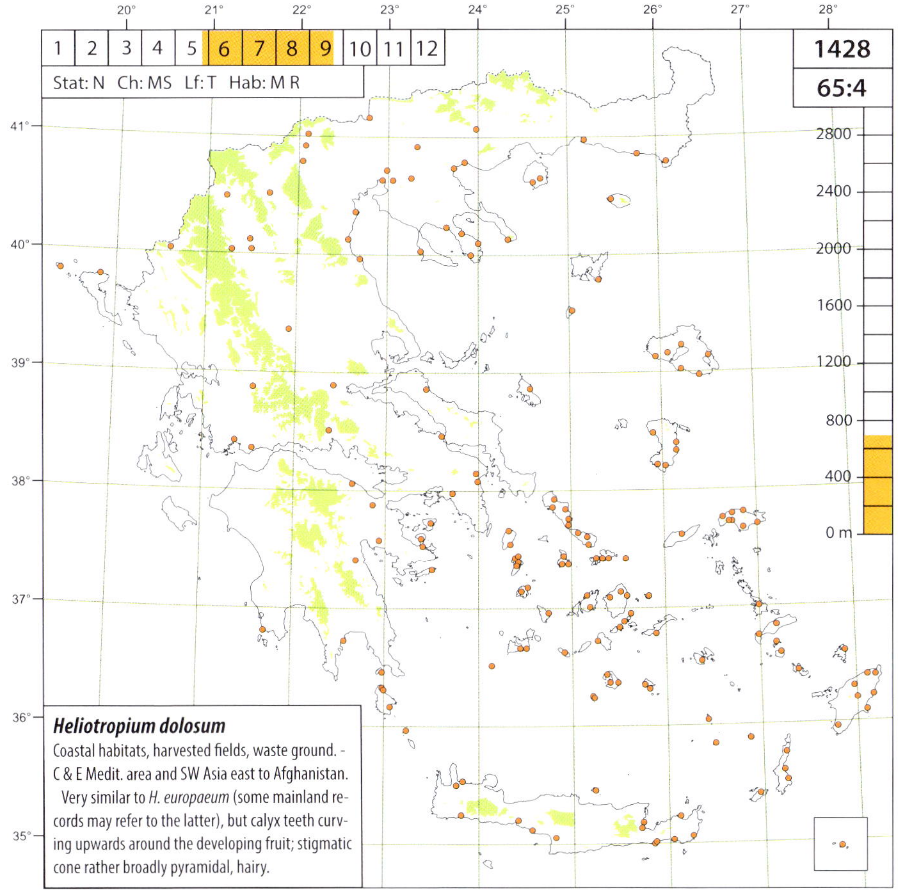

Heliotropium dolosum

Coastal habitats, harvested fields, waste ground. - C & E Medit. area and SW Asia east to Afghanistan.

Very similar to *H. europaeum* (some mainland records may refer to the latter), but calyx teeth curving upwards around the developing fruit; stigmatic cone rather broadly pyramidal, hairy.

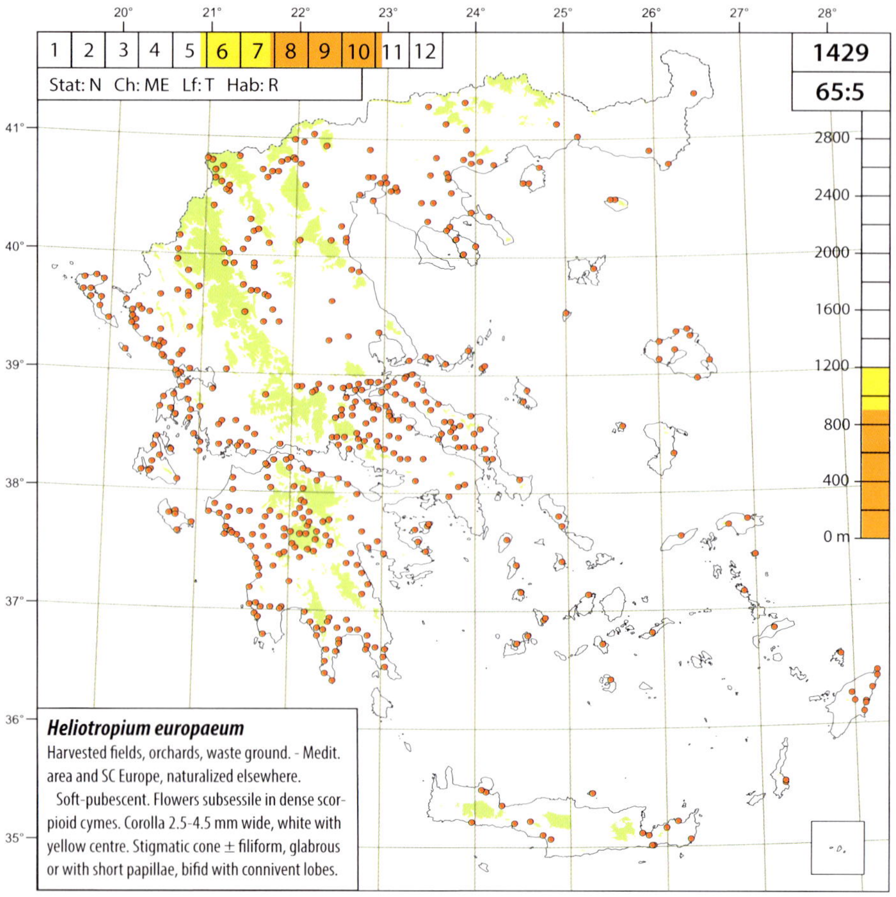

Heliotropium europaeum
Harvested fields, orchards, waste ground. - Medit. area and SC Europe, naturalized elsewhere.
Soft-pubescent. Flowers subsessile in dense scorpioid cymes. Corolla 2.5-4.5 mm wide, white with yellow centre. Stigmatic cone ± filiform, glabrous or with short papillae, bifid with connivent lobes.

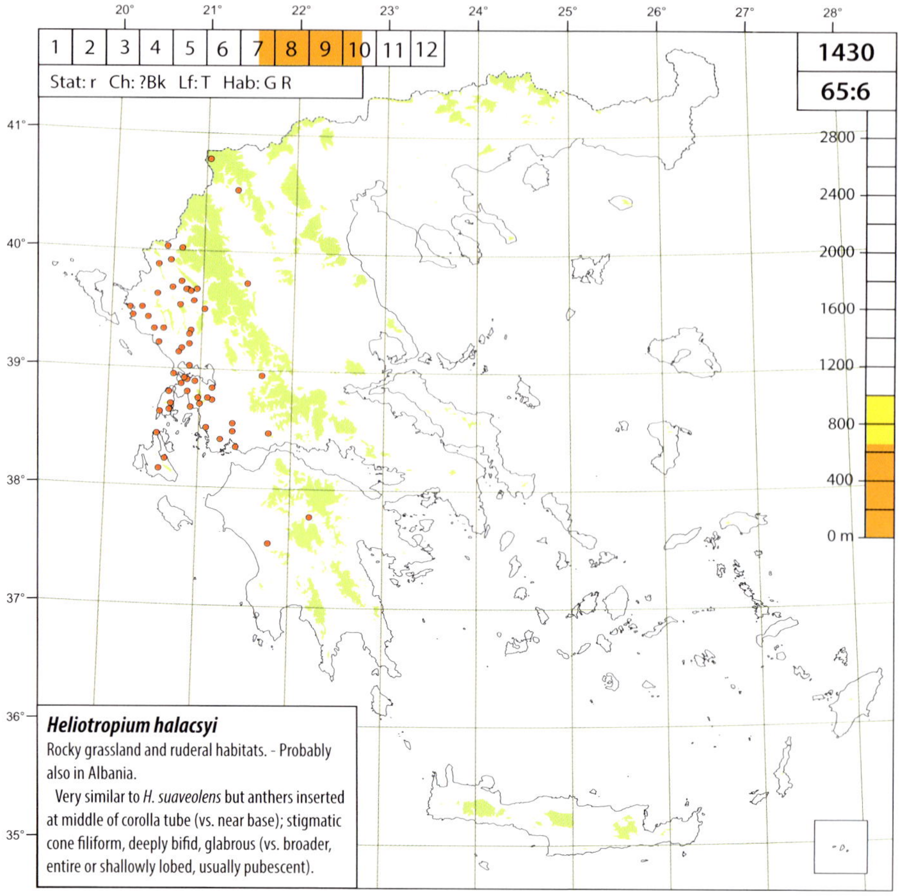

Heliotropium halacsyi
Rocky grassland and ruderal habitats. - Probably also in Albania.
Very similar to *H. suaveolens* but anthers inserted at middle of corolla tube (vs. near base); stigmatic cone filiform, deeply bifid, glabrous (vs. broader, entire or shallowly lobed, usually pubescent).

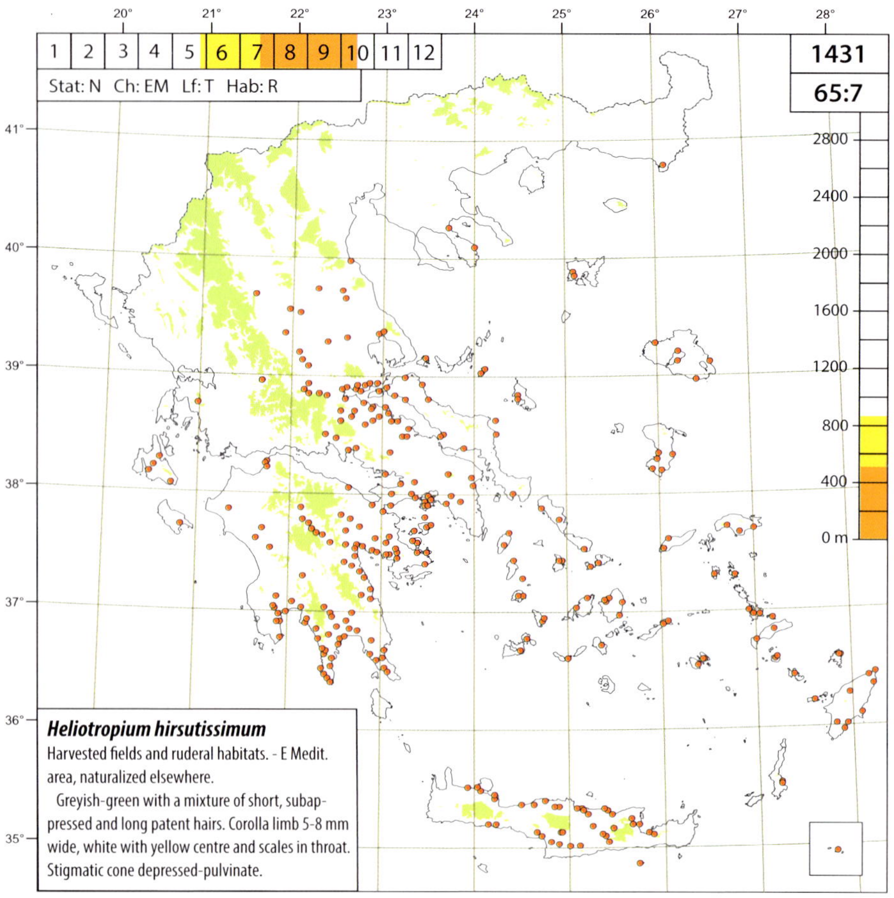

Heliotropium hirsutissimum
Harvested fields and ruderal habitats. - E Medit. area, naturalized elsewhere.
Greyish-green with a mixture of short, subappressed and long patent hairs. Corolla limb 5-8 mm wide, white with yellow centre and scales in throat. Stigmatic cone depressed-pulvinate.

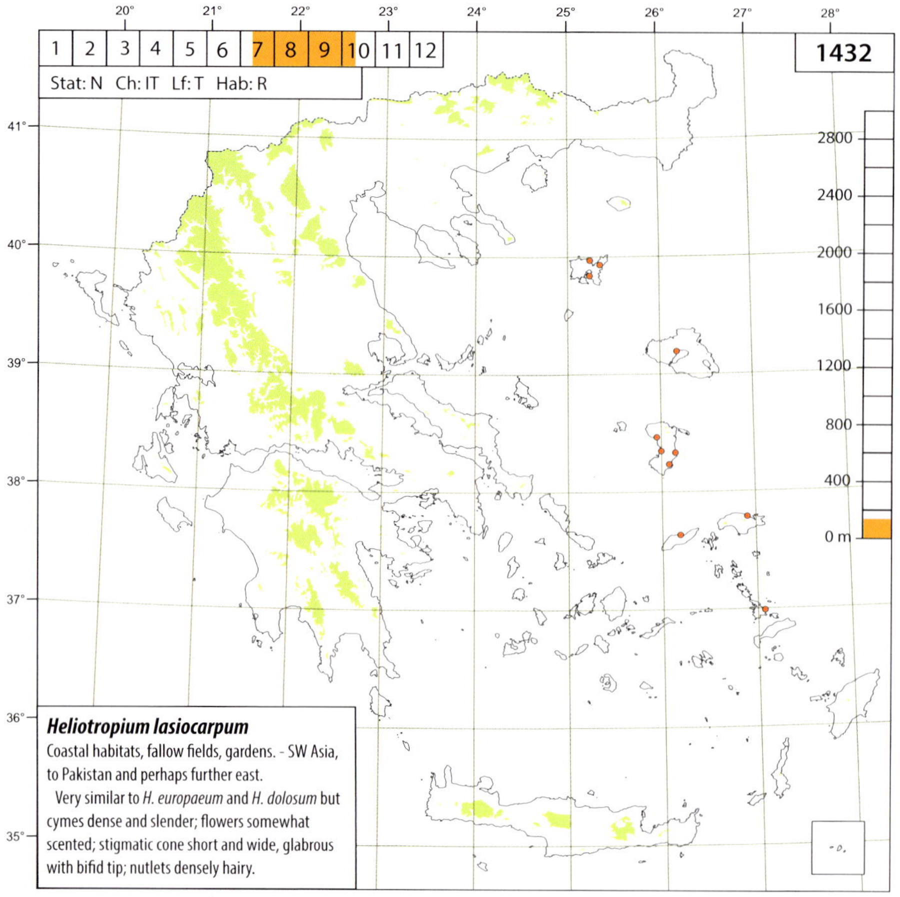

Heliotropium lasiocarpum
Coastal habitats, fallow fields, gardens. - SW Asia, to Pakistan and perhaps further east.
Very similar to *H. europaeum* and *H. dolosum* but cymes dense and slender; flowers somewhat scented; stigmatic cone short and wide, glabrous with bifid tip; nutlets densely hairy.

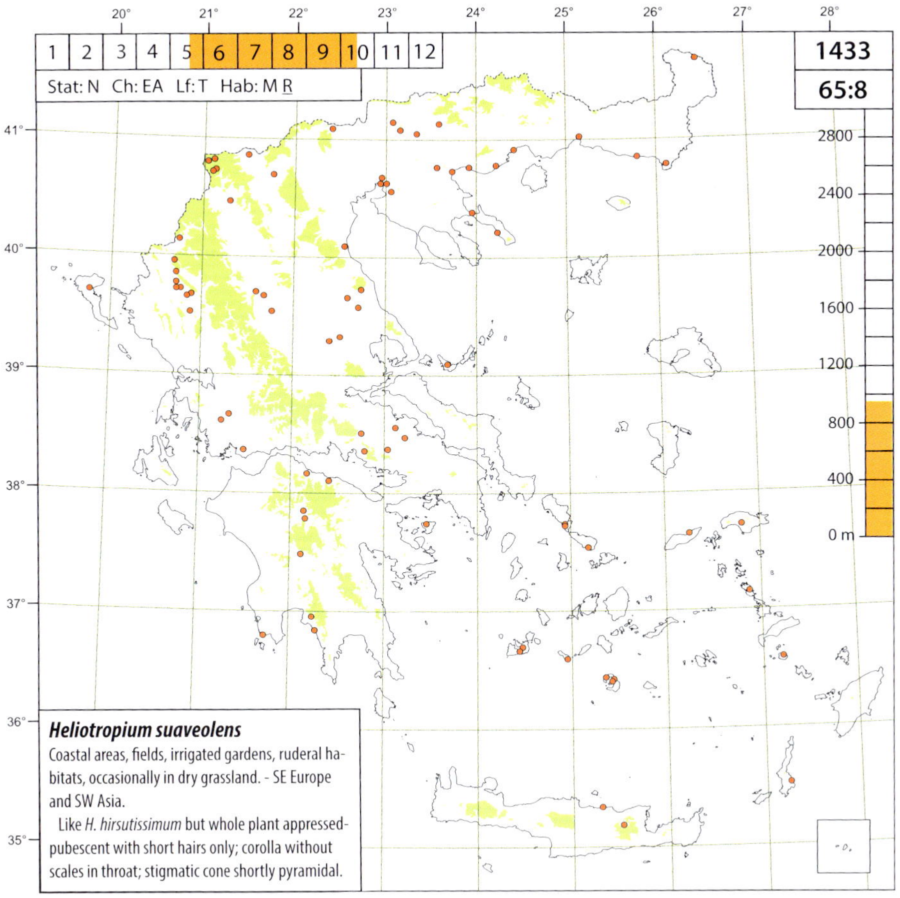

Heliotropium suaveolens

Coastal areas, fields, irrigated gardens, ruderal habitats, occasionally in dry grassland. - SE Europe and SW Asia.

Like *H. hirsutissimum* but whole plant appressed-pubescent with short hairs only; corolla without scales in throat; stigmatic cone shortly pyramidal.

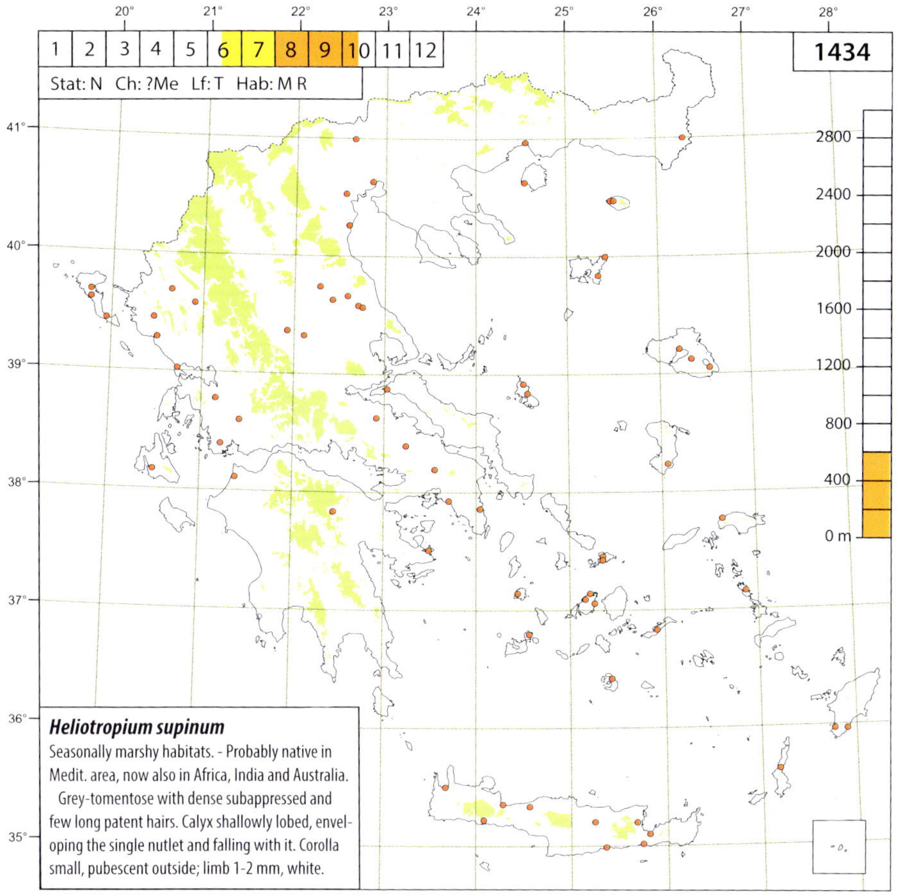

Heliotropium supinum

Seasonally marshy habitats. - Probably native in Medit. area, now also in Africa, India and Australia.

Grey-tomentose with dense subappressed and few long patent hairs. Calyx shallowly lobed, enveloping the single nutlet and falling with it. Corolla small, pubescent outside; limb 1-2 mm, white.

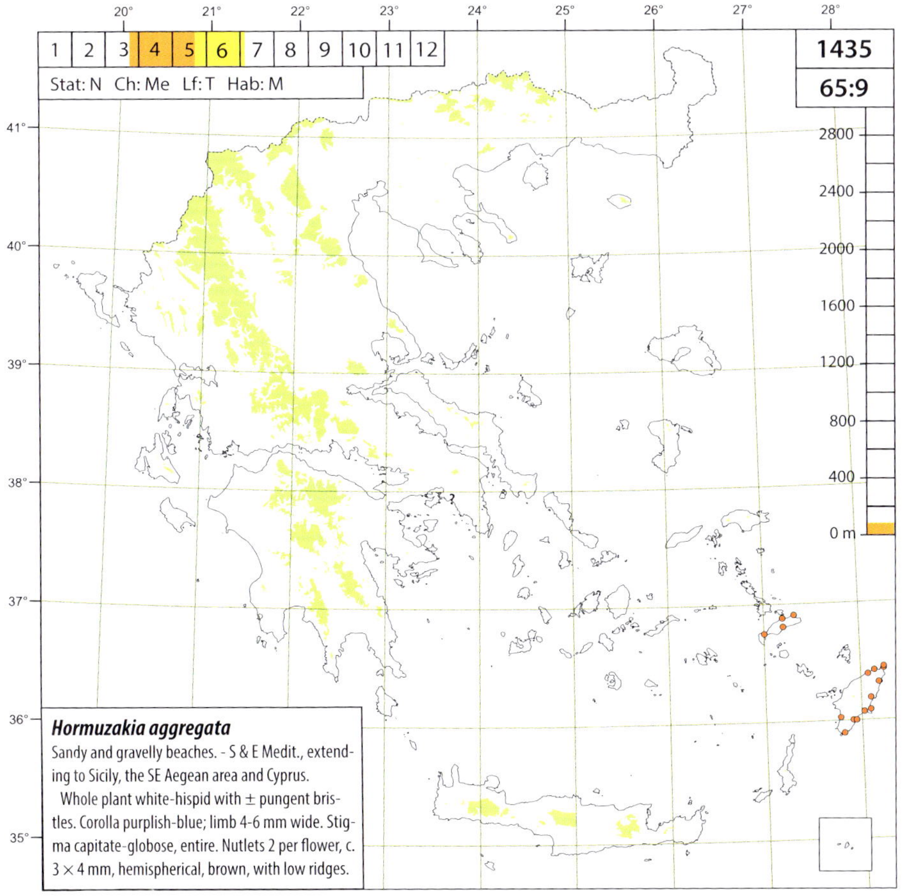

Hormuzakia aggregata

Sandy and gravelly beaches. - S & E Medit., extending to Sicily, the SE Aegean area and Cyprus.

Whole plant white-hispid with ± pungent bristles. Corolla purplish-blue; limb 4-6 mm wide. Stigma capitate-globose, entire. Nutlets 2 per flower, c. 3 × 4 mm, hemispherical, brown, with low ridges.

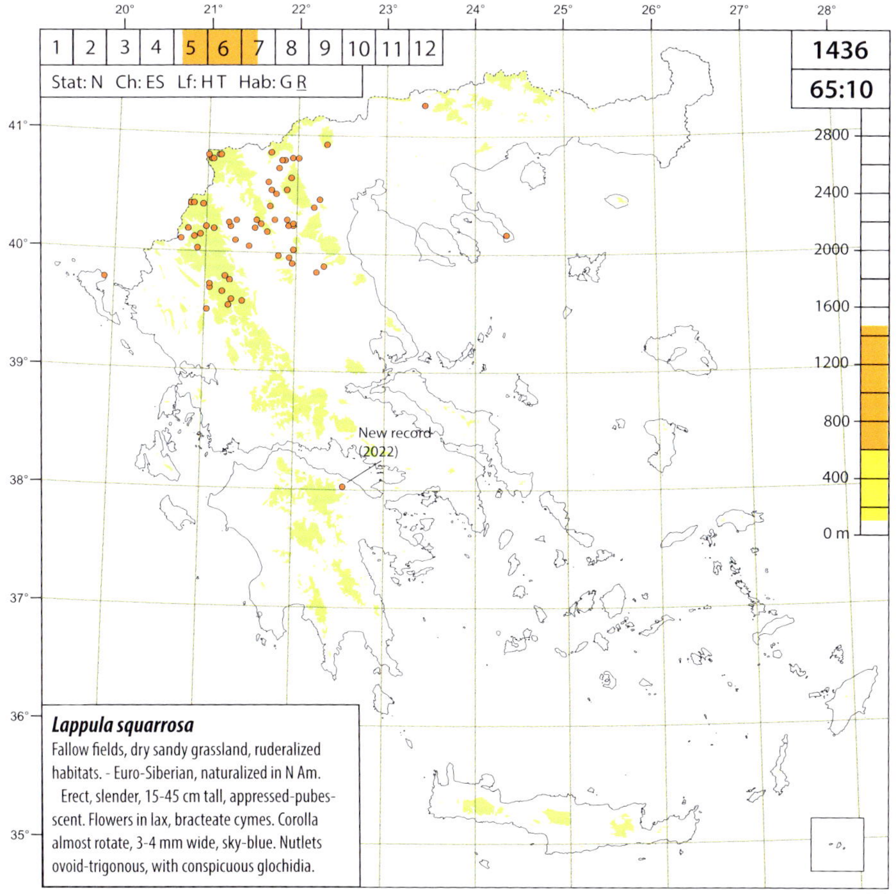

Lappula squarrosa

Fallow fields, dry sandy grassland, ruderalized habitats. - Euro-Siberian, naturalized in N Am.

Erect, slender, 15-45 cm tall, appressed-pubescent. Flowers in lax, bracteate cymes. Corolla almost rotate, 3-4 mm wide, sky-blue. Nutlets ovoid-trigonous, with conspicuous glochidia.

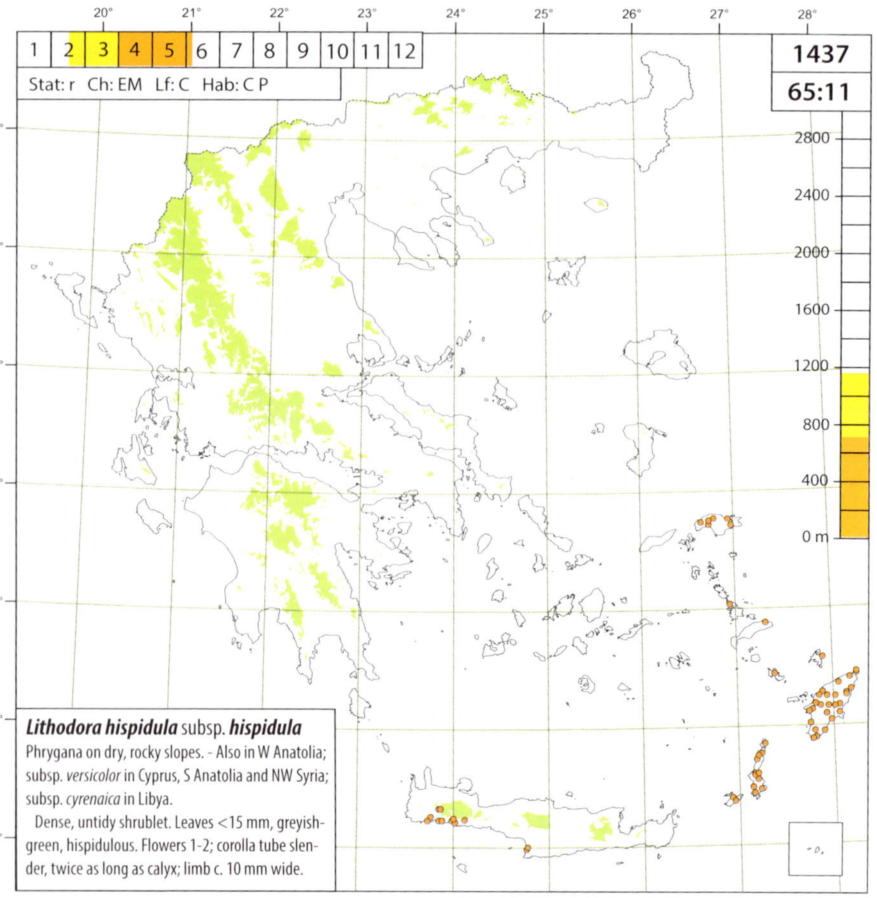

Lithodora hispidula* subsp. *hispidula

Phrygana on dry, rocky slopes. - Also in W Anatolia; subsp. *versicolor* in Cyprus, S Anatolia and NW Syria; subsp. *cyrenaica* in Libya.

Dense, untidy shrublet. Leaves <15 mm, greyish-green, hispidulous. Flowers 1-2; corolla tube slender, twice as long as calyx; limb c. 10 mm wide.

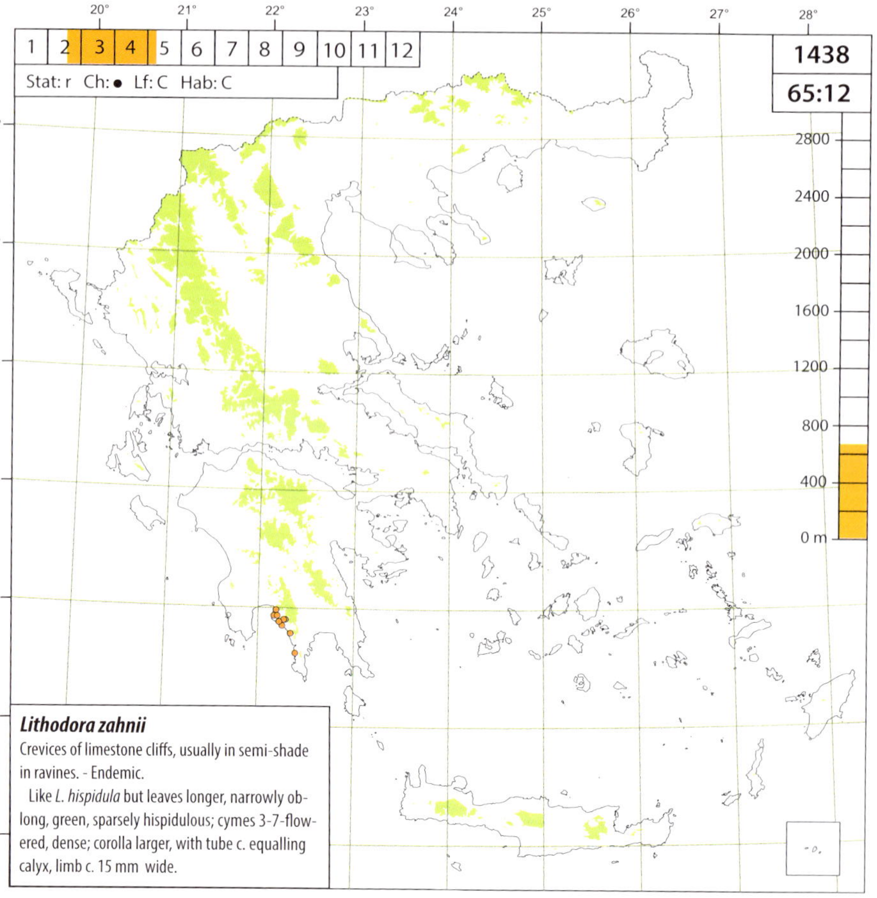

Lithodora zahnii

Crevices of limestone cliffs, usually in semi-shade in ravines. - Endemic.

Like *L. hispidula* but leaves longer, narrowly oblong, green, sparsely hispidulous; cymes 3-7-flowered, dense; corolla larger, with tube c. equalling calyx, limb c. 15 mm wide.

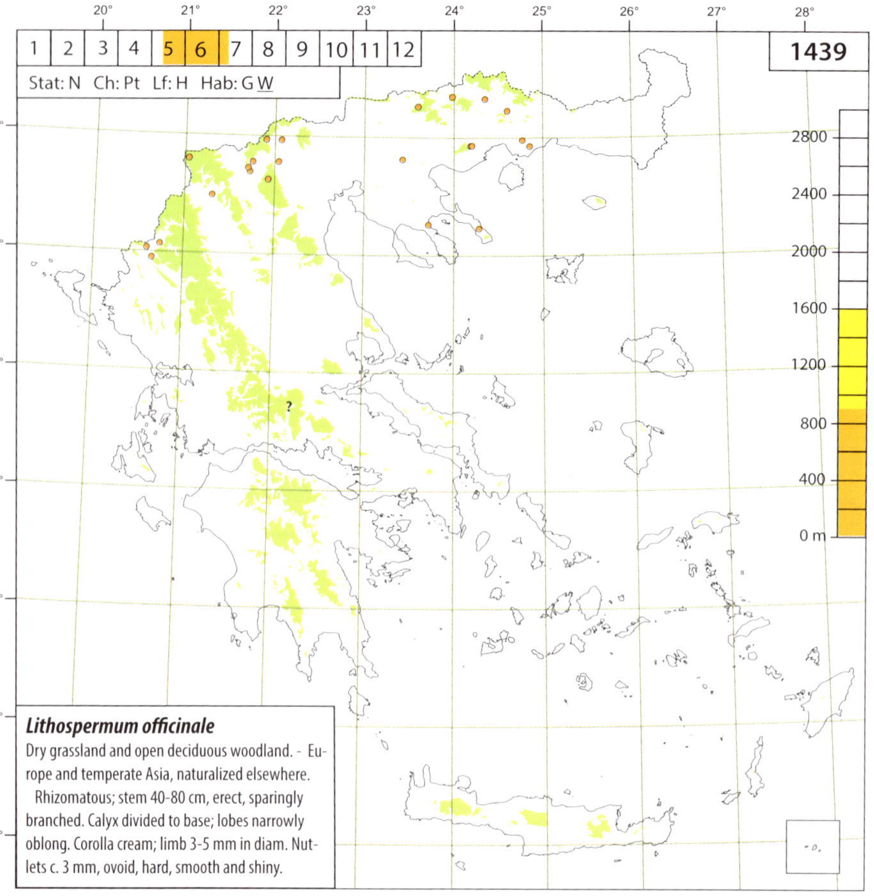

Lithospermum officinale

Dry grassland and open deciduous woodland. - Europe and temperate Asia, naturalized elsewhere.

Rhizomatous; stem 40-80 cm, erect, sparingly branched. Calyx divided to base; lobes narrowly oblong. Corolla cream; limb 3-5 mm in diam. Nutlets c. 3 mm, ovoid, hard, smooth and shiny.

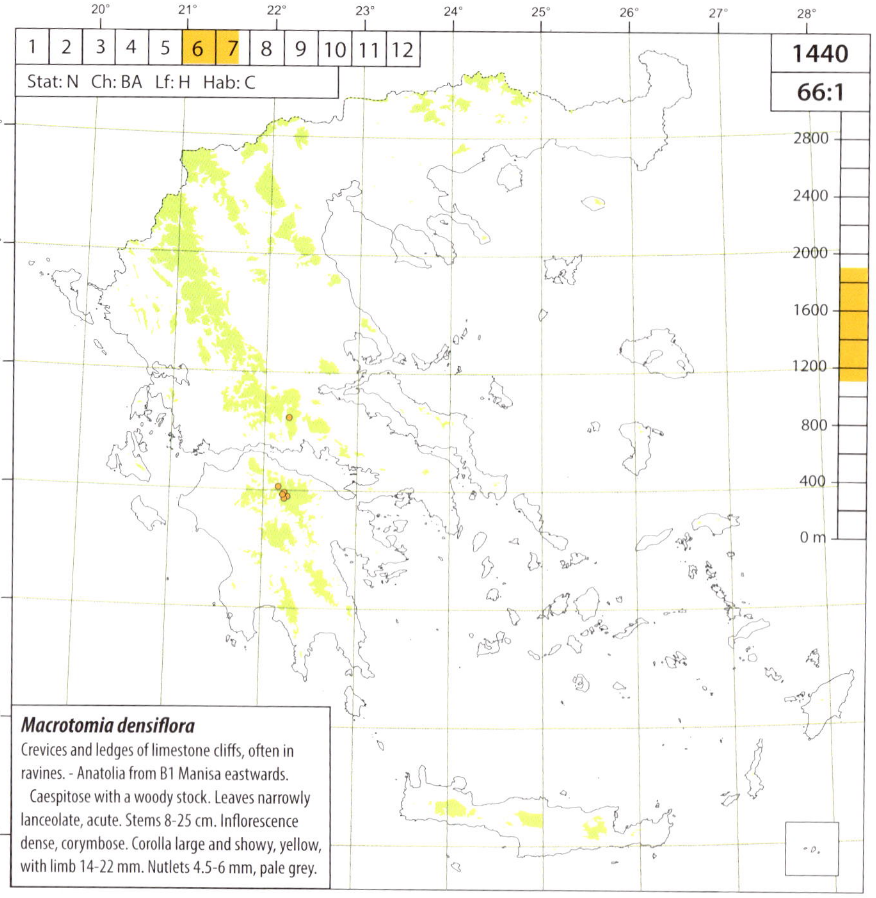

Macrotomia densiflora

Crevices and ledges of limestone cliffs, often in ravines. - Anatolia from B1 Manisa eastwards.

Caespitose with a woody stock. Leaves narrowly lanceolate, acute. Stems 8-25 cm. Inflorescence dense, corymbose. Corolla large and showy, yellow, with limb 14-22 mm. Nutlets 4.5-6 mm, pale grey.

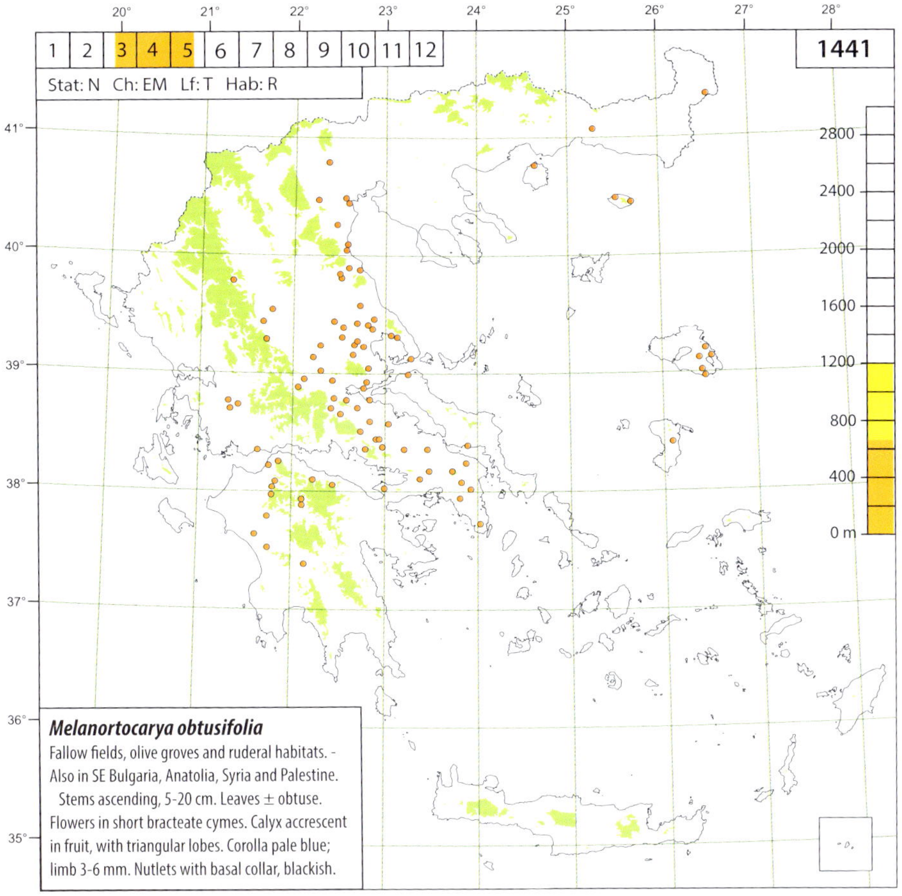

Melanortocarya obtusifolia
Fallow fields, olive groves and ruderal habitats. - Also in SE Bulgaria, Anatolia, Syria and Palestine.
Stems ascending, 5-20 cm. Leaves ± obtuse. Flowers in short bracteate cymes. Calyx accrescent in fruit, with triangular lobes. Corolla pale blue; limb 3-6 mm. Nutlets with basal collar, blackish.

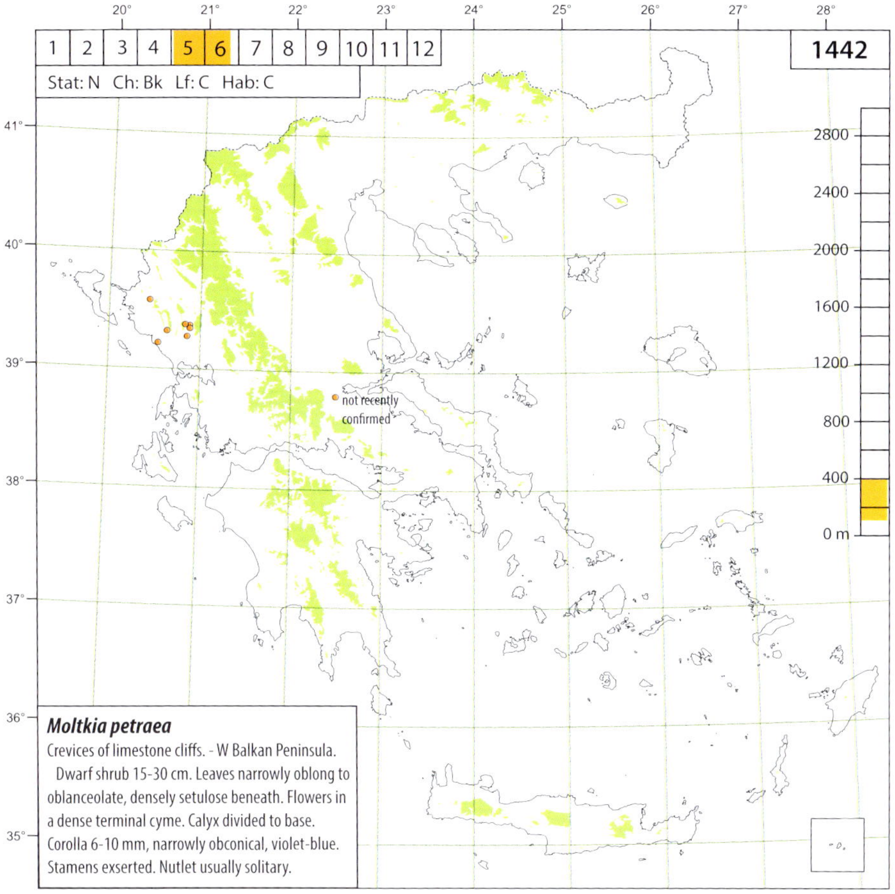

Moltkia petraea
Crevices of limestone cliffs. - W Balkan Peninsula.
Dwarf shrub 15-30 cm. Leaves narrowly oblong to oblanceolate, densely setulose beneath. Flowers in a dense terminal cyme. Calyx divided to base. Corolla 6-10 mm, narrowly obconical, violet-blue. Stamens exserted. Nutlet usually solitary.

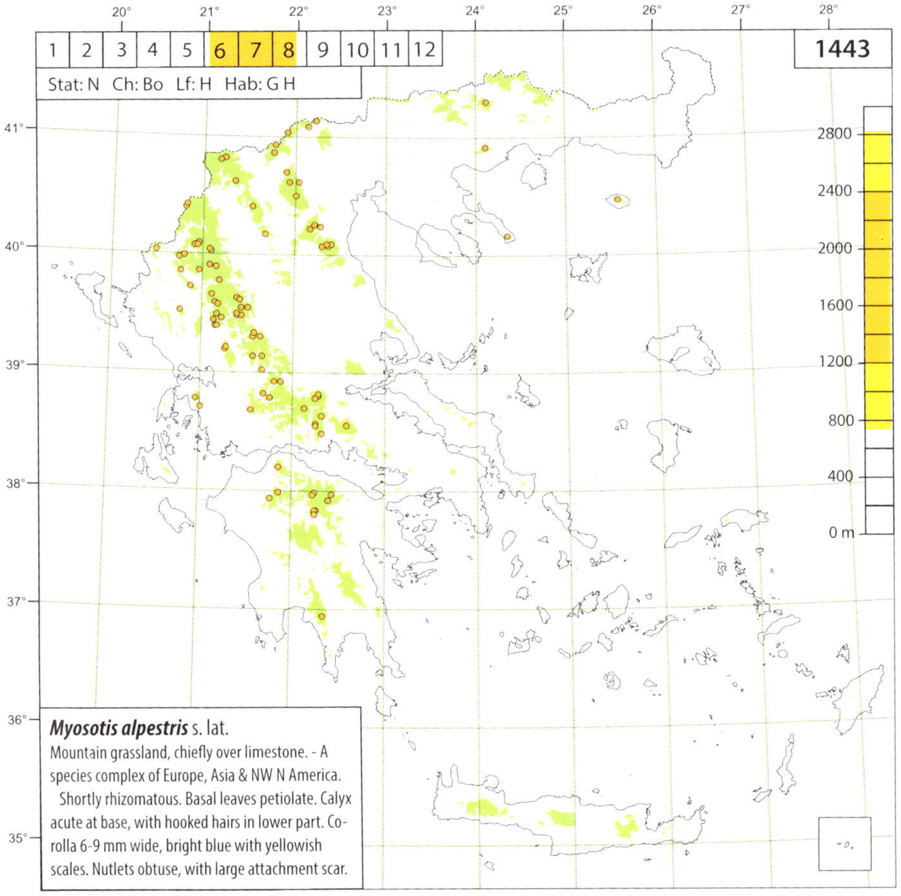

Myosotis alpestris s. lat.
Mountain grassland, chiefly over limestone. - A species complex of Europe, Asia & NW N America.
Shortly rhizomatous. Basal leaves petiolate. Calyx acute at base, with hooked hairs in lower part. Corolla 6-9 mm wide, bright blue with yellowish scales. Nutlets obtuse, with large attachment scar.

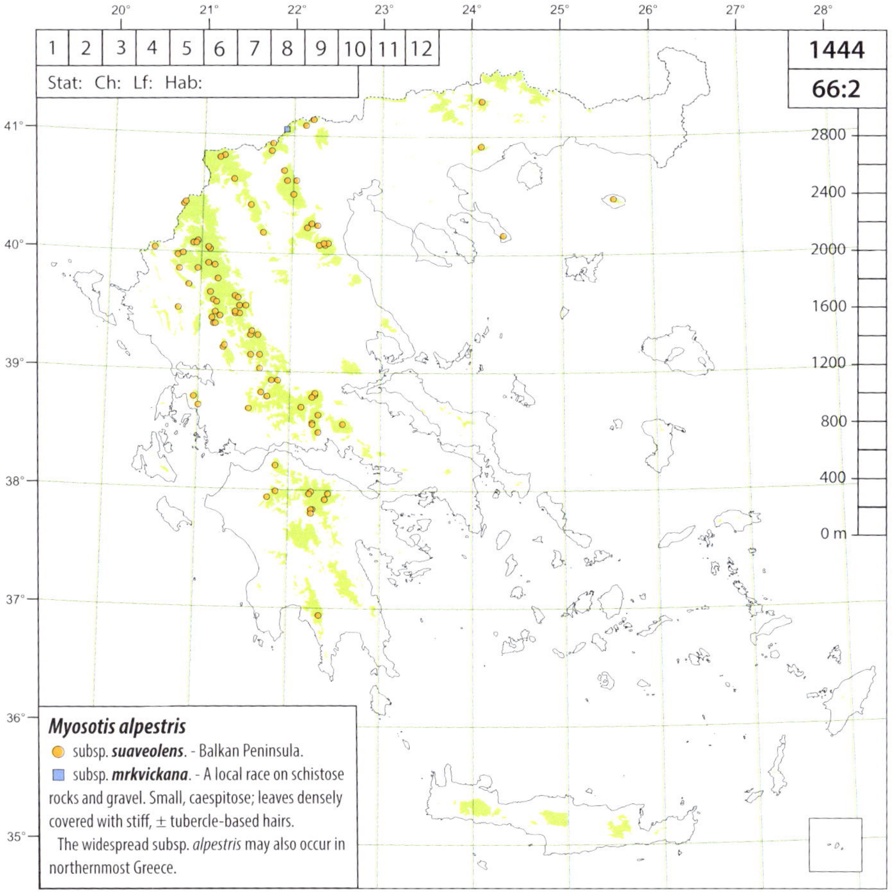

Myosotis alpestris
● subsp. ***suaveolens***. - Balkan Peninsula.
■ subsp. ***mrkvickana***. - A local race on schistose rocks and gravel. Small, caespitose; leaves densely covered with stiff, ± tubercle-based hairs.
The widespread subsp. *alpestris* may also occur in northernmost Greece.

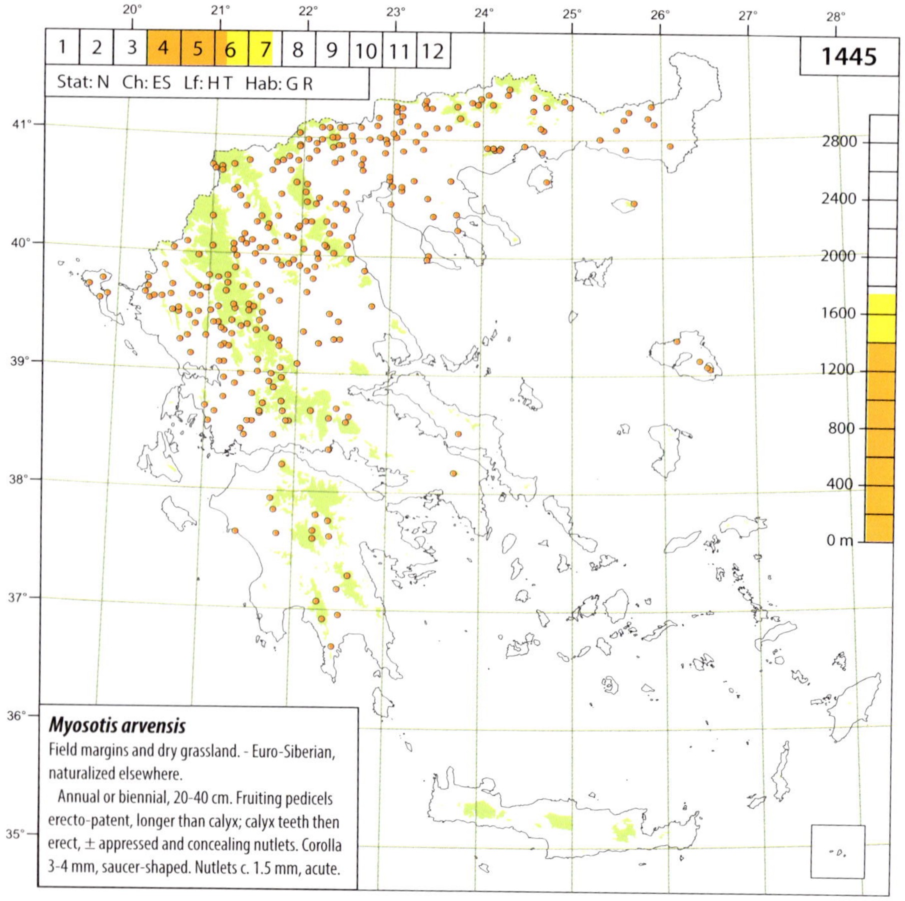

Myosotis arvensis

Field margins and dry grassland. - Euro-Siberian, naturalized elsewhere.

Annual or biennial, 20-40 cm. Fruiting pedicels erecto-patent, longer than calyx; calyx teeth then erect, ± appressed and concealing nutlets. Corolla 3-4 mm, saucer-shaped. Nutlets c. 1.5 mm, acute.

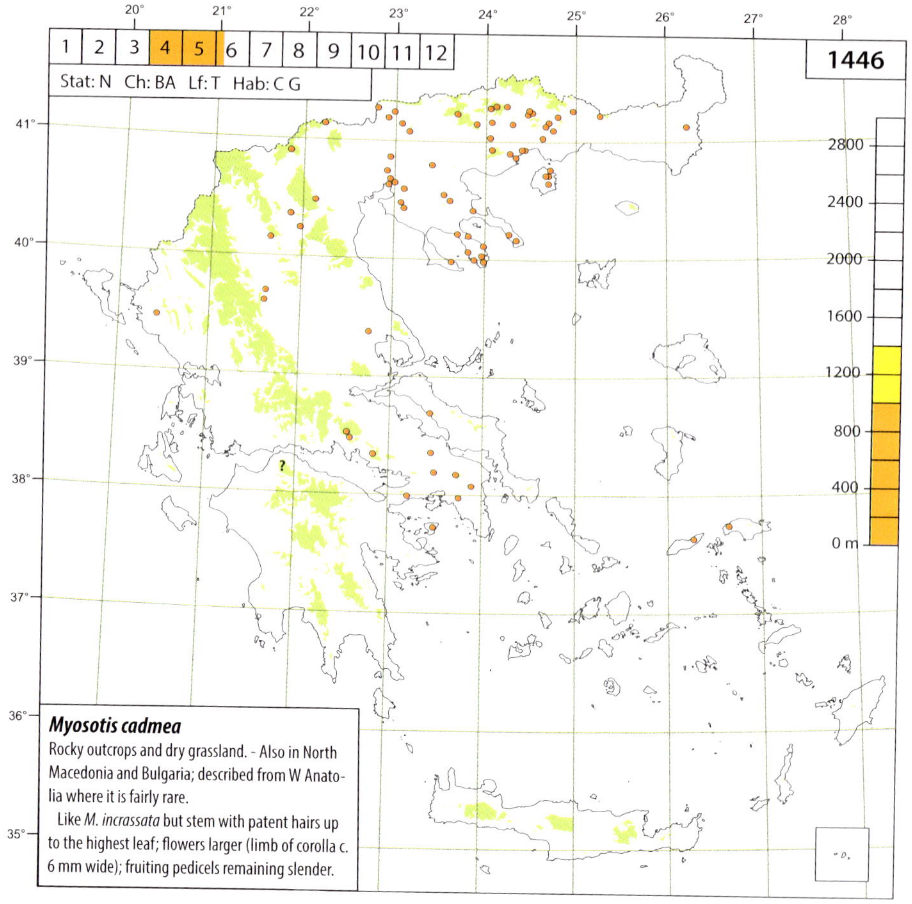

Myosotis cadmea

Rocky outcrops and dry grassland. - Also in North Macedonia and Bulgaria; described from W Anatolia where it is fairly rare.

Like *M. incrassata* but stem with patent hairs up to the highest leaf; flowers larger (limb of corolla c. 6 mm wide); fruiting pedicels remaining slender.

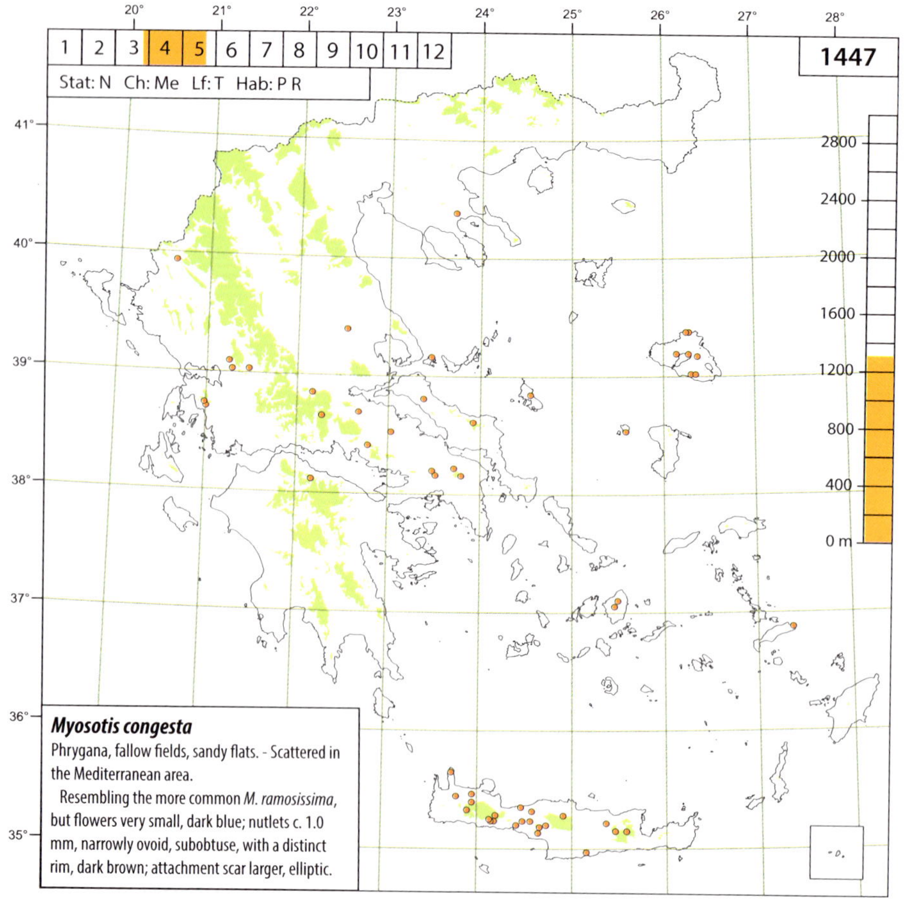

Myosotis congesta

Phrygana, fallow fields, sandy flats. - Scattered in the Mediterranean area.

Resembling the more common *M. ramosissima*, but flowers very small, dark blue; nutlets c. 1.0 mm, narrowly ovoid, subobtuse, with a distinct rim, dark brown; attachment scar larger, elliptic.

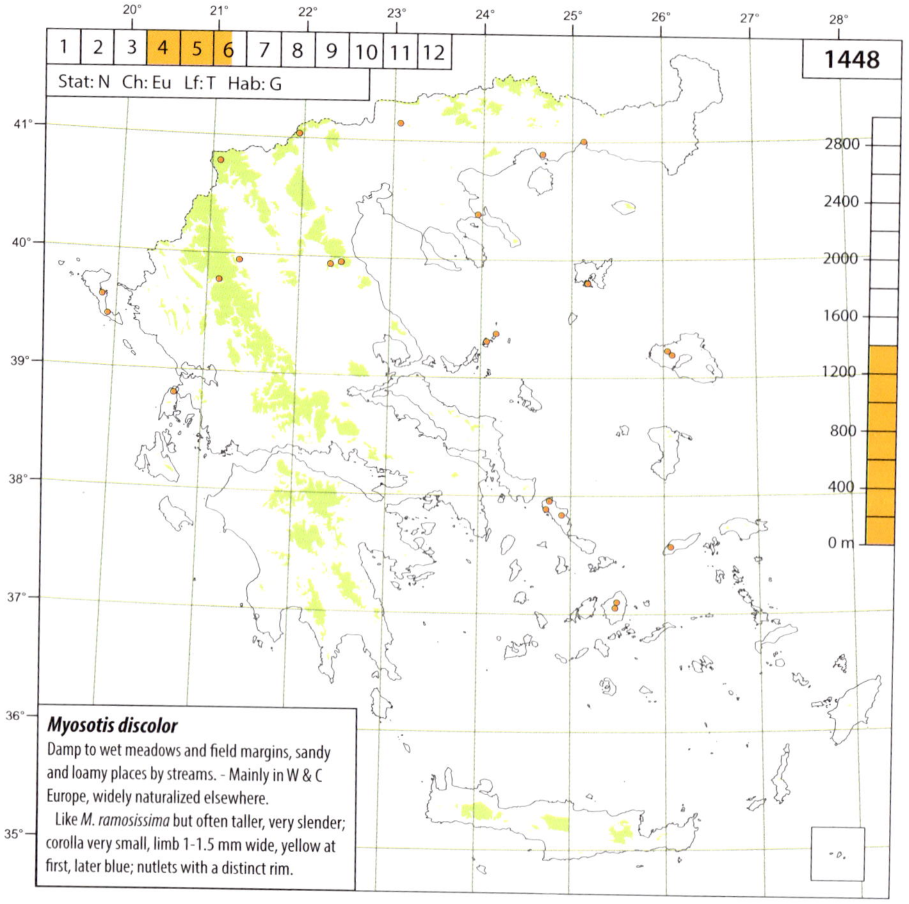

Myosotis discolor

Damp to wet meadows and field margins, sandy and loamy places by streams. - Mainly in W & C Europe, widely naturalized elsewhere.

Like *M. ramosissima* but often taller, very slender; corolla very small, limb 1-1.5 mm wide, yellow at first, later blue; nutlets with a distinct rim.

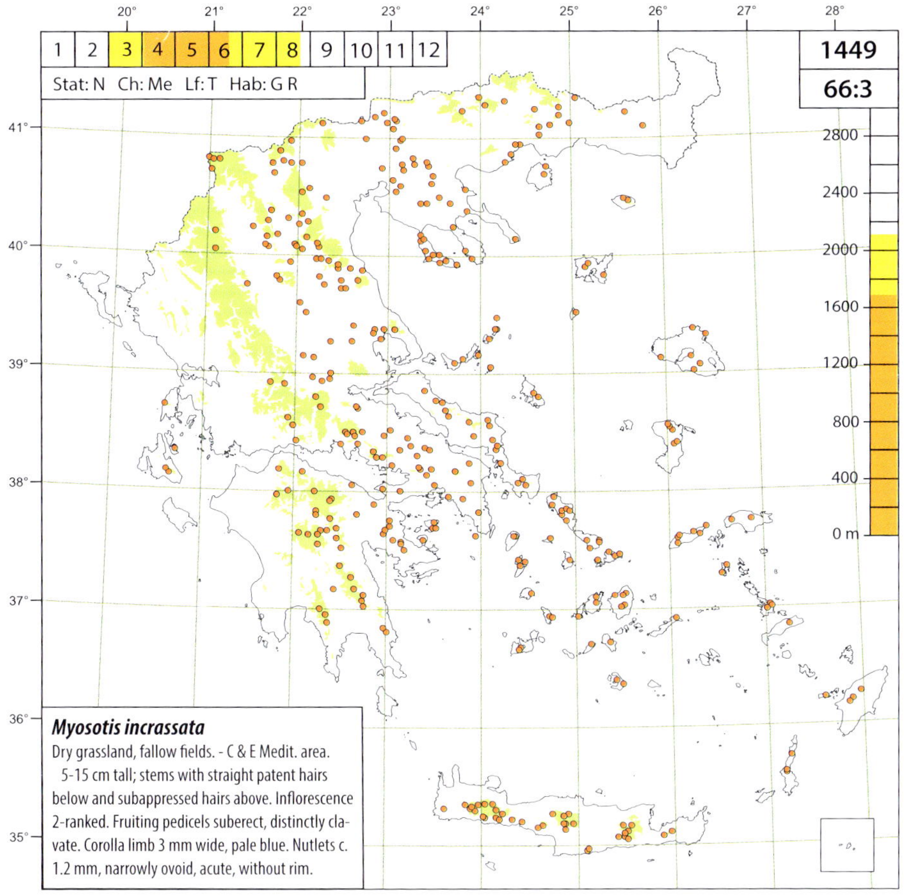

Myosotis incrassata
Dry grassland, fallow fields. - C & E Medit. area.
5-15 cm tall; stems with straight patent hairs below and subappressed hairs above. Inflorescence 2-ranked. Fruiting pedicels suberect, distinctly clavate. Corolla limb 3 mm wide, pale blue. Nutlets c. 1.2 mm, narrowly ovoid, acute, without rim.

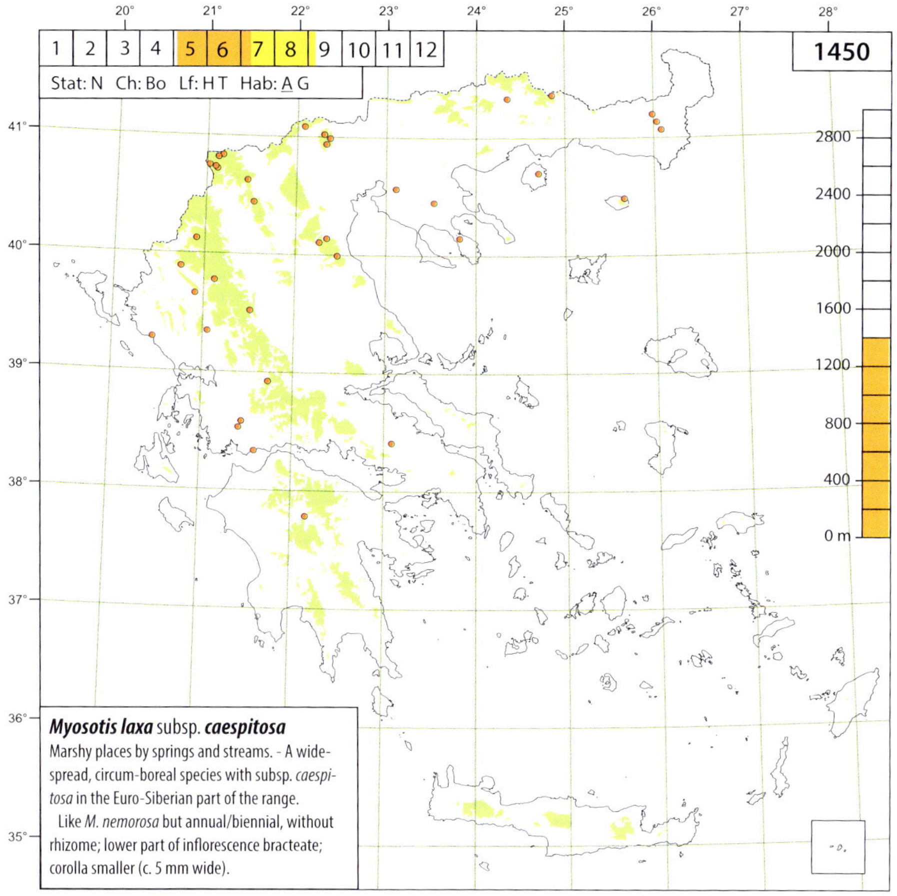

Myosotis laxa **subsp.** ***caespitosa***
Marshy places by springs and streams. - A widespread, circum-boreal species with subsp. *caespitosa* in the Euro-Siberian part of the range.
Like *M. nemorosa* but annual/biennial, without rhizome; lower part of inflorescence bracteate; corolla smaller (c. 5 mm wide).

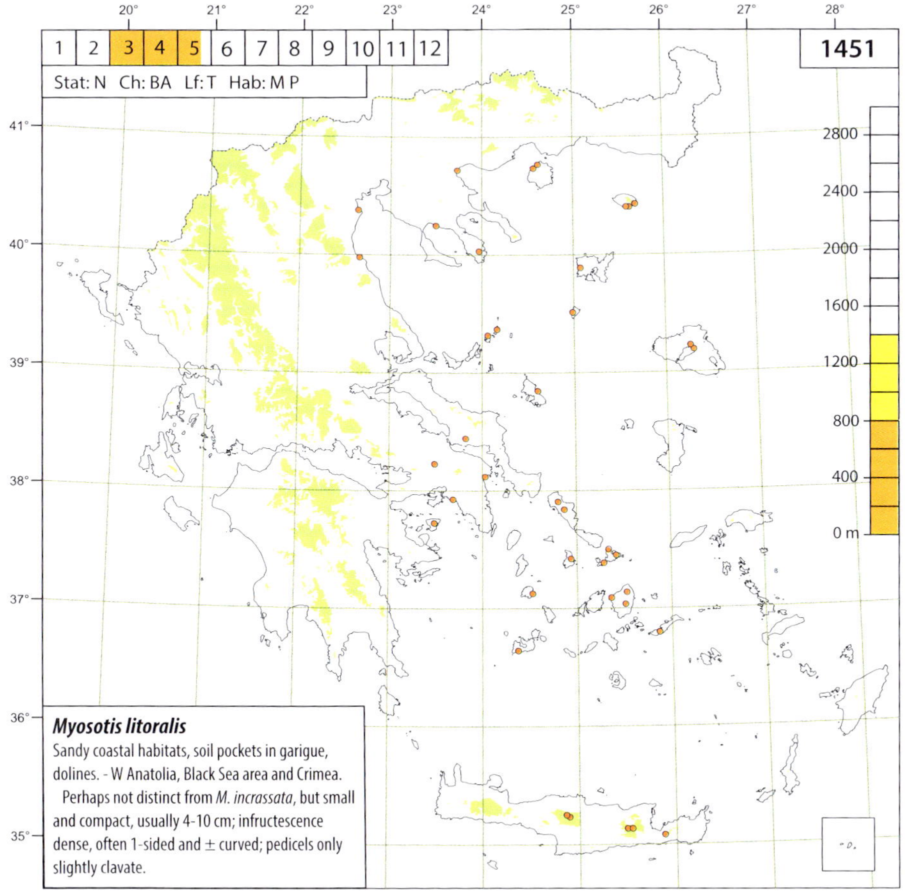

Myosotis litoralis
Sandy coastal habitats, soil pockets in garigue, dolines. - W Anatolia, Black Sea area and Crimea.
Perhaps not distinct from *M. incrassata*, but small and compact, usually 4-10 cm; infructescence dense, often 1-sided and ± curved; pedicels only slightly clavate.

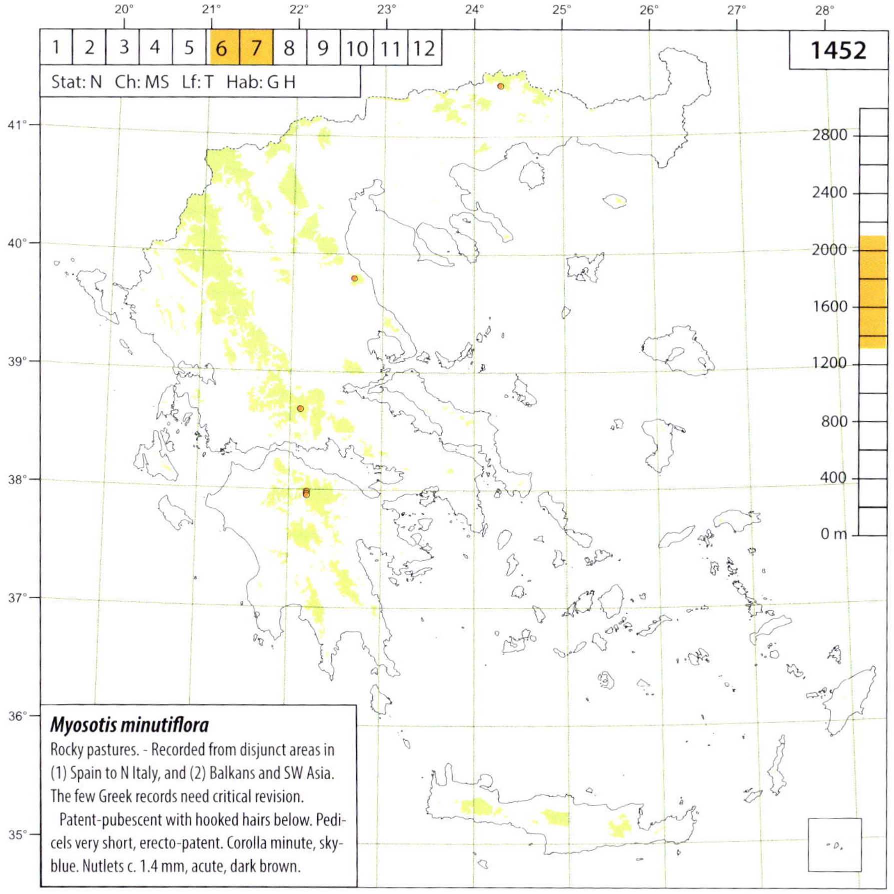

Myosotis minutiflora
Rocky pastures. - Recorded from disjunct areas in (1) Spain to N Italy, and (2) Balkans and SW Asia. The few Greek records need critical revision.
Patent-pubescent with hooked hairs below. Pedicels very short, erecto-patent. Corolla minute, sky-blue. Nutlets c. 1.4 mm, acute, dark brown.

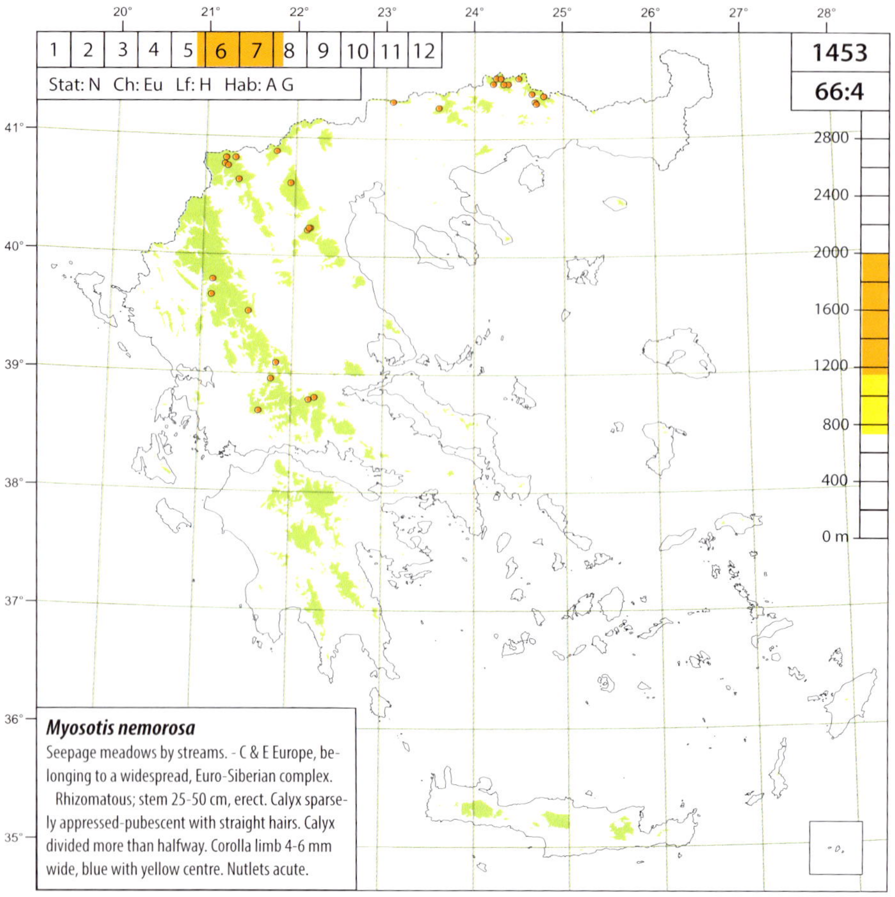

Myosotis nemorosa
Seepage meadows by streams. - C & E Europe, belonging to a widespread, Euro-Siberian complex.
Rhizomatous; stem 25-50 cm, erect. Calyx sparsely appressed-pubescent with straight hairs. Calyx divided more than halfway. Corolla limb 4-6 mm wide, blue with yellow centre. Nutlets acute.

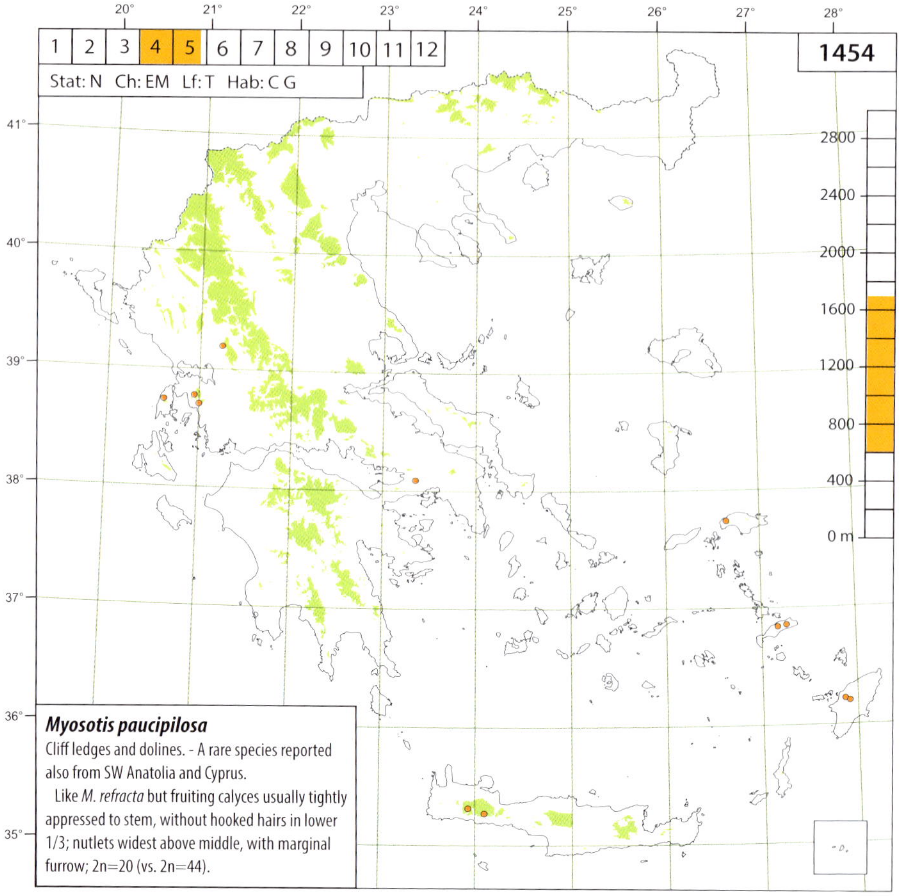

Myosotis paucipilosa
Cliff ledges and dolines. - A rare species reported also from SW Anatolia and Cyprus.
Like *M. refracta* but fruiting calyces usually tightly appressed to stem, without hooked hairs in lower 1/3; nutlets widest above middle, with marginal furrow; 2n=20 (vs. 2n=44).

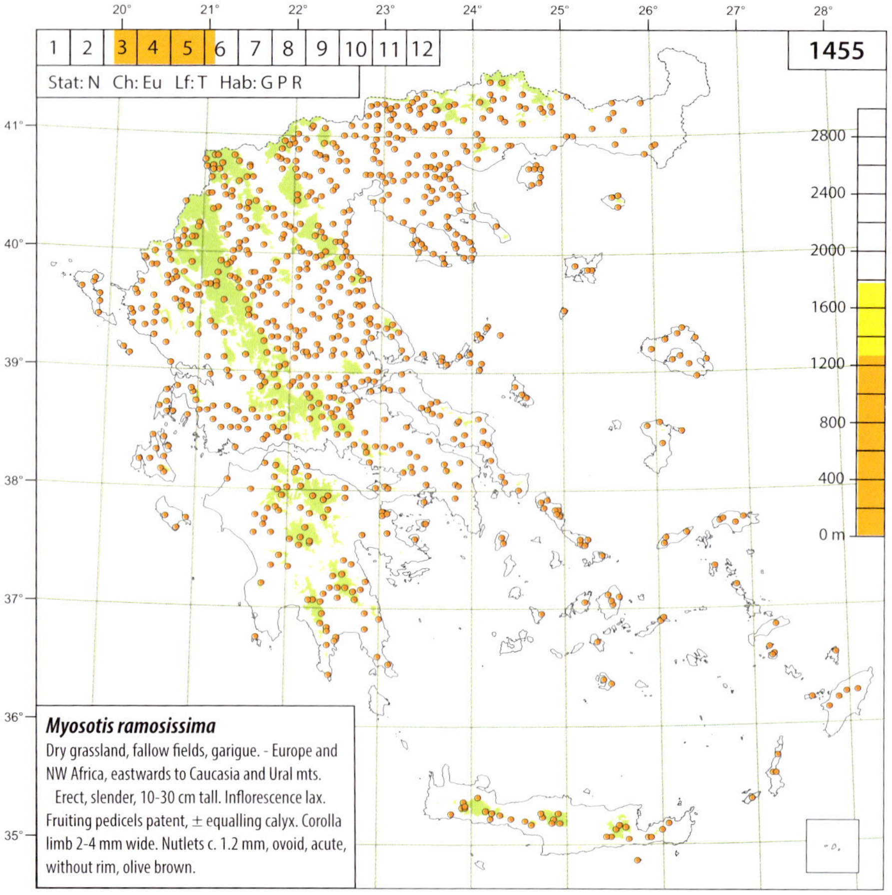

Myosotis ramosissima
Dry grassland, fallow fields, garigue. - Europe and NW Africa, eastwards to Caucasia and Ural mts.
Erect, slender, 10-30 cm tall. Inflorescence lax. Fruiting pedicels patent, ± equalling calyx. Corolla limb 2-4 mm wide. Nutlets c. 1.2 mm, ovoid, acute, without rim, olive brown.

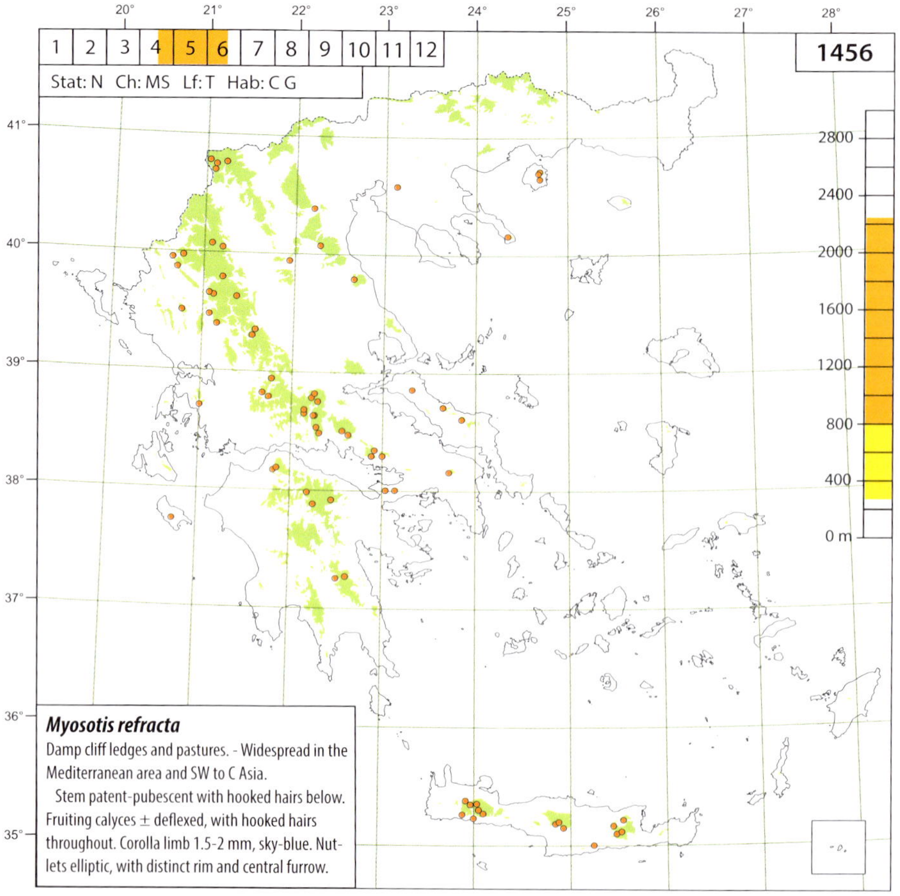

Myosotis refracta
Damp cliff ledges and pastures. - Widespread in the Mediterranean area and SW to C Asia.
Stem patent-pubescent with hooked hairs below. Fruiting calyces ± deflexed, with hooked hairs throughout. Corolla limb 1.5-2 mm, sky-blue. Nutlets elliptic, with distinct rim and central furrow.

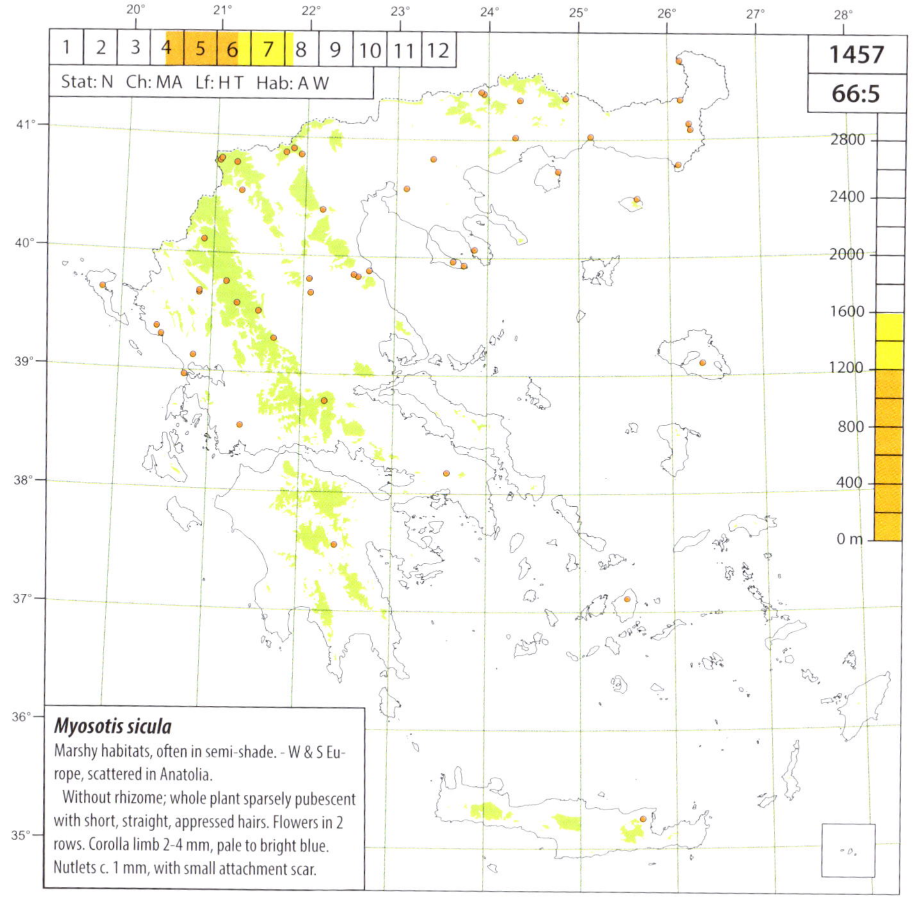

Myosotis sicula

Marshy habitats, often in semi-shade. - W & S Europe, scattered in Anatolia.

Without rhizome; whole plant sparsely pubescent with short, straight, appressed hairs. Flowers in 2 rows. Corolla limb 2-4 mm, pale to bright blue. Nutlets c. 1 mm, with small attachment scar.

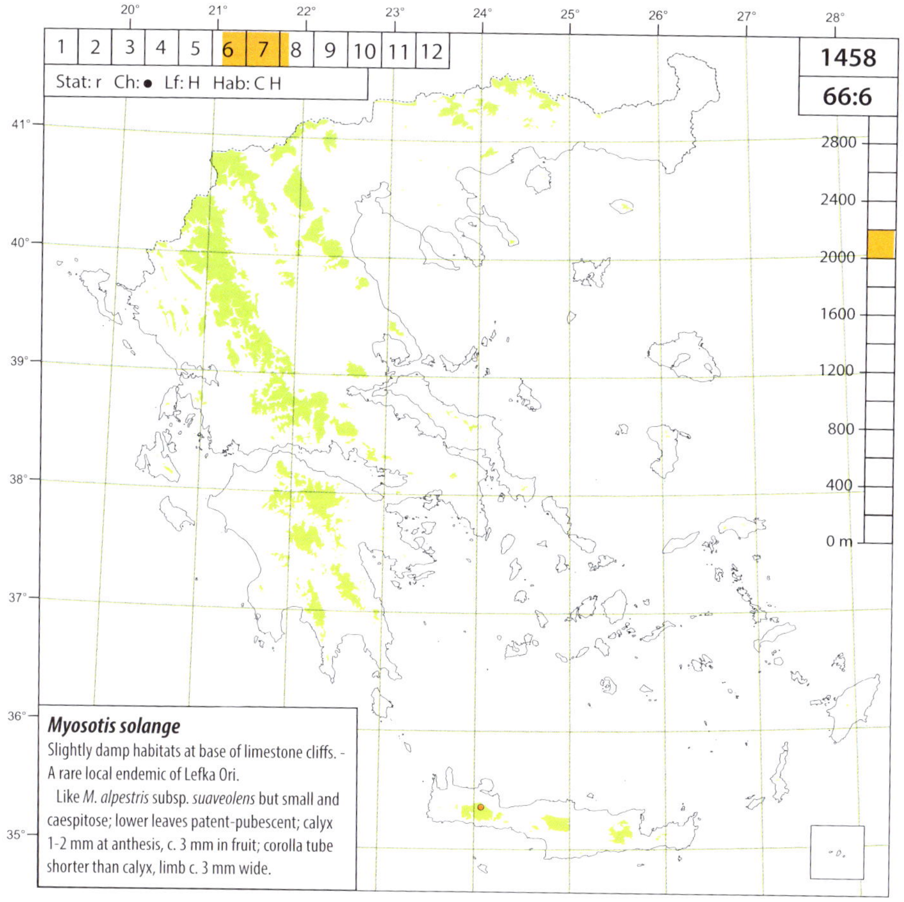

Myosotis solange

Slightly damp habitats at base of limestone cliffs. - A rare local endemic of Lefka Ori.

Like *M. alpestris* subsp. *suaveolens* but small and caespitose; lower leaves patent-pubescent; calyx 1-2 mm at anthesis, c. 3 mm in fruit; corolla tube shorter than calyx, limb c. 3 mm wide.

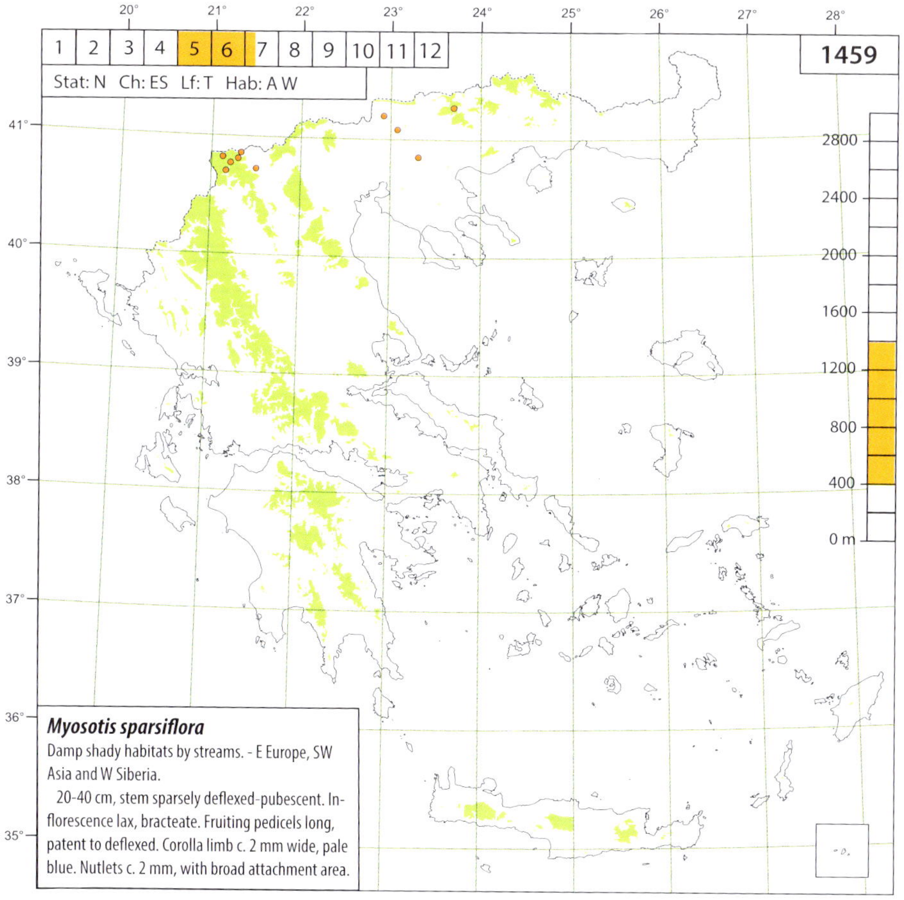

Myosotis sparsiflora

Damp shady habitats by streams. - E Europe, SW Asia and W Siberia.

20-40 cm, stem sparsely deflexed-pubescent. Inflorescence lax, bracteate. Fruiting pedicels long, patent to deflexed. Corolla limb c. 2 mm wide, pale blue. Nutlets c. 2 mm, with broad attachment area.

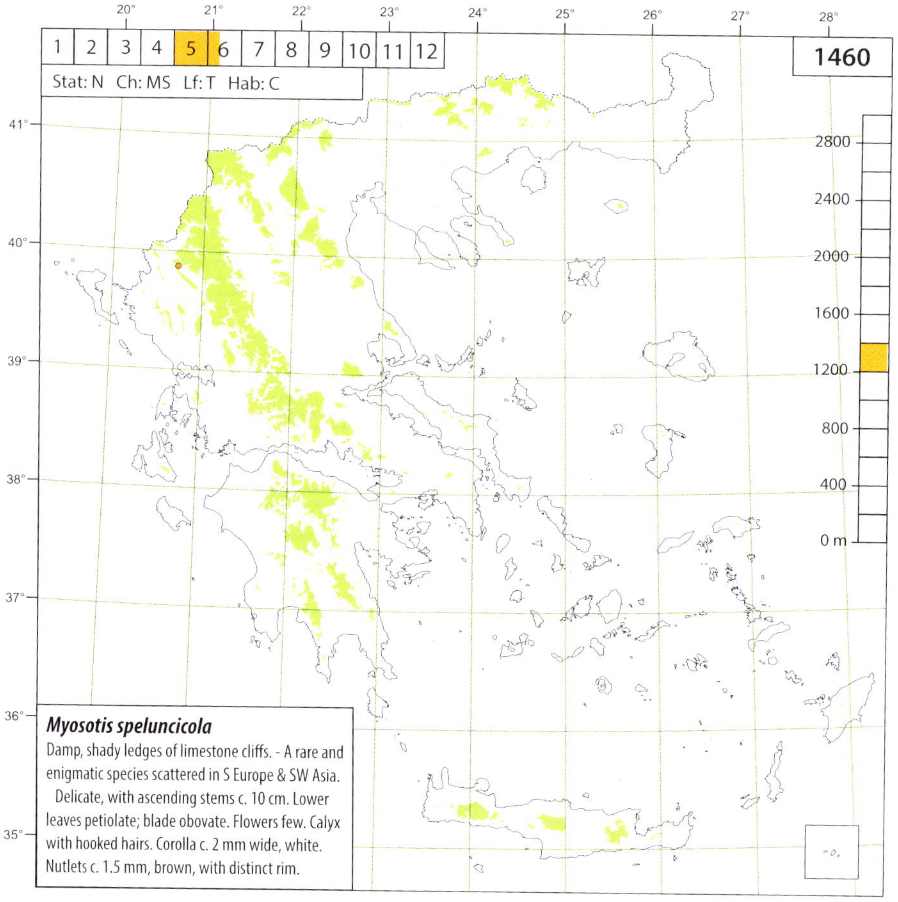

Myosotis speluncicola

Damp, shady ledges of limestone cliffs. - A rare and enigmatic species scattered in S Europe & SW Asia.

Delicate, with ascending stems c. 10 cm. Lower leaves petiolate; blade obovate. Flowers few. Calyx with hooked hairs. Corolla c. 2 mm wide, white. Nutlets c. 1.5 mm, brown, with distinct rim.

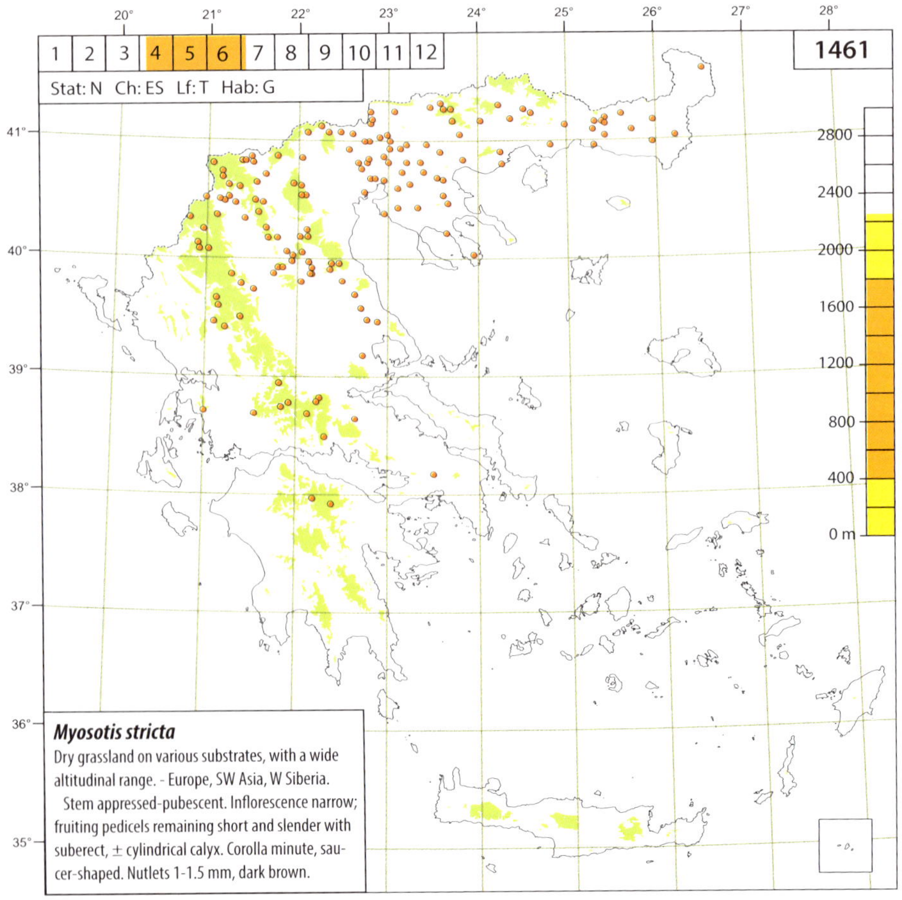

Myosotis stricta
Dry grassland on various substrates, with a wide altitudinal range. - Europe, SW Asia, W Siberia.
Stem appressed-pubescent. Inflorescence narrow; fruiting pedicels remaining short and slender with suberect, ± cylindrical calyx. Corolla minute, saucer-shaped. Nutlets 1-1.5 mm, dark brown.

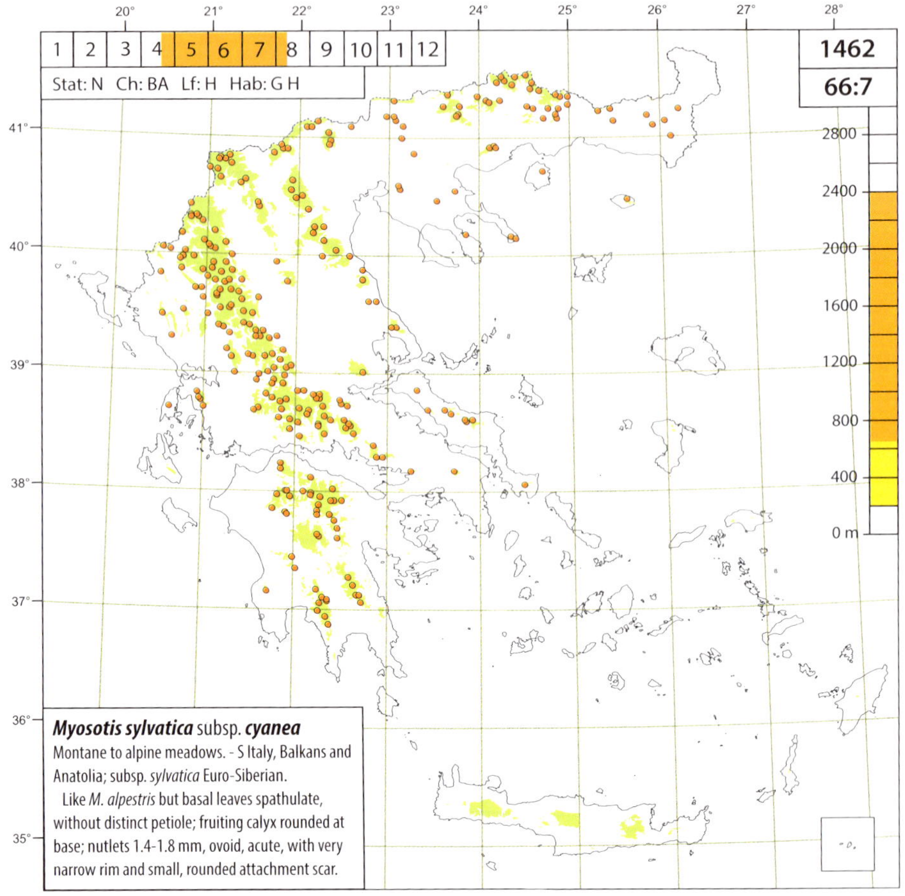

Myosotis sylvatica subsp. ***cyanea***
Montane to alpine meadows. - S Italy, Balkans and Anatolia; subsp. *sylvatica* Euro-Siberian.
Like *M. alpestris* but basal leaves spathulate, without distinct petiole; fruiting calyx rounded at base; nutlets 1.4-1.8 mm, ovoid, acute, with very narrow rim and small, rounded attachment scar.

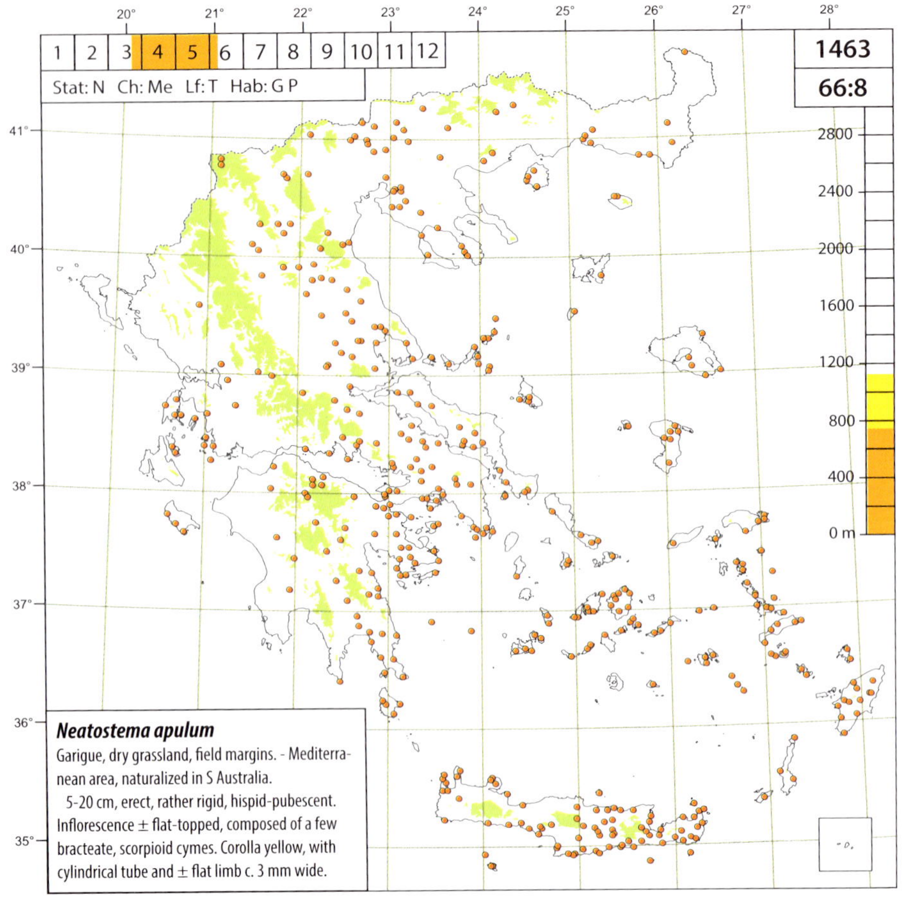

Neatostema apulum
Garigue, dry grassland, field margins. - Mediterranean area, naturalized in S Australia.
5-20 cm, erect, rather rigid, hispid-pubescent. Inflorescence ± flat-topped, composed of a few bracteate, scorpioid cymes. Corolla yellow, with cylindrical tube and ± flat limb c. 3 mm wide.

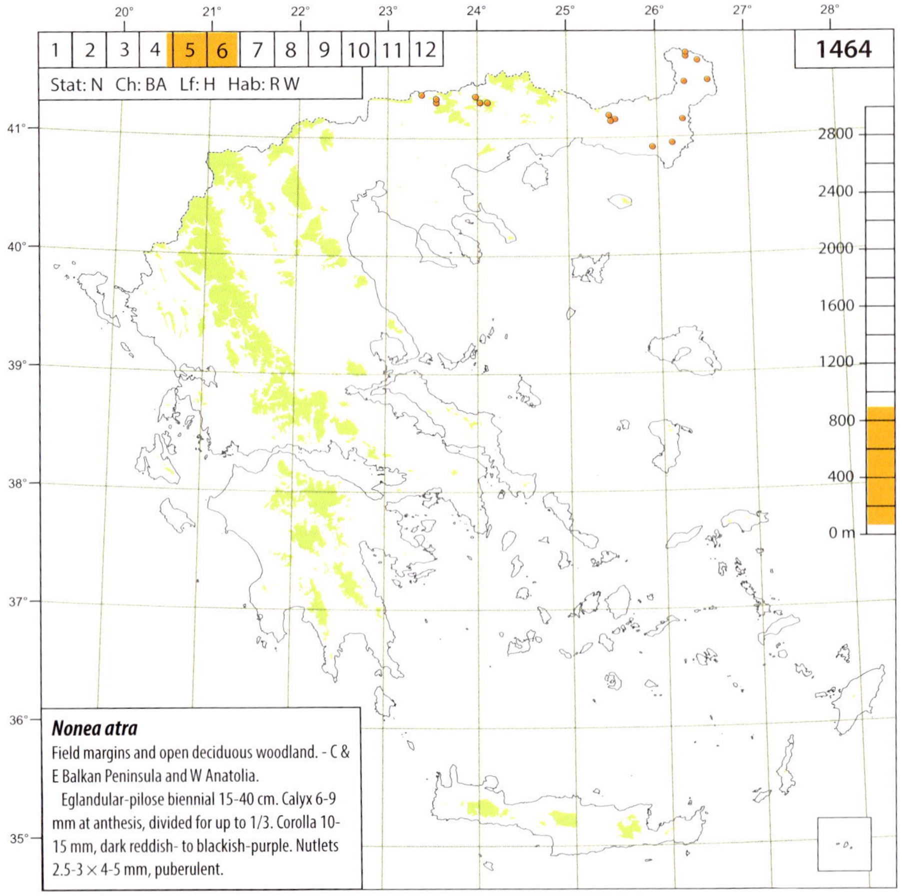

Nonea atra
Field margins and open deciduous woodland. - C & E Balkan Peninsula and W Anatolia.
Eglandular-pilose biennial 15-40 cm. Calyx 6-9 mm at anthesis, divided for up to 1/3. Corolla 10-15 mm, dark reddish- to blackish-purple. Nutlets 2.5-3 × 4-5 mm, puberulent.

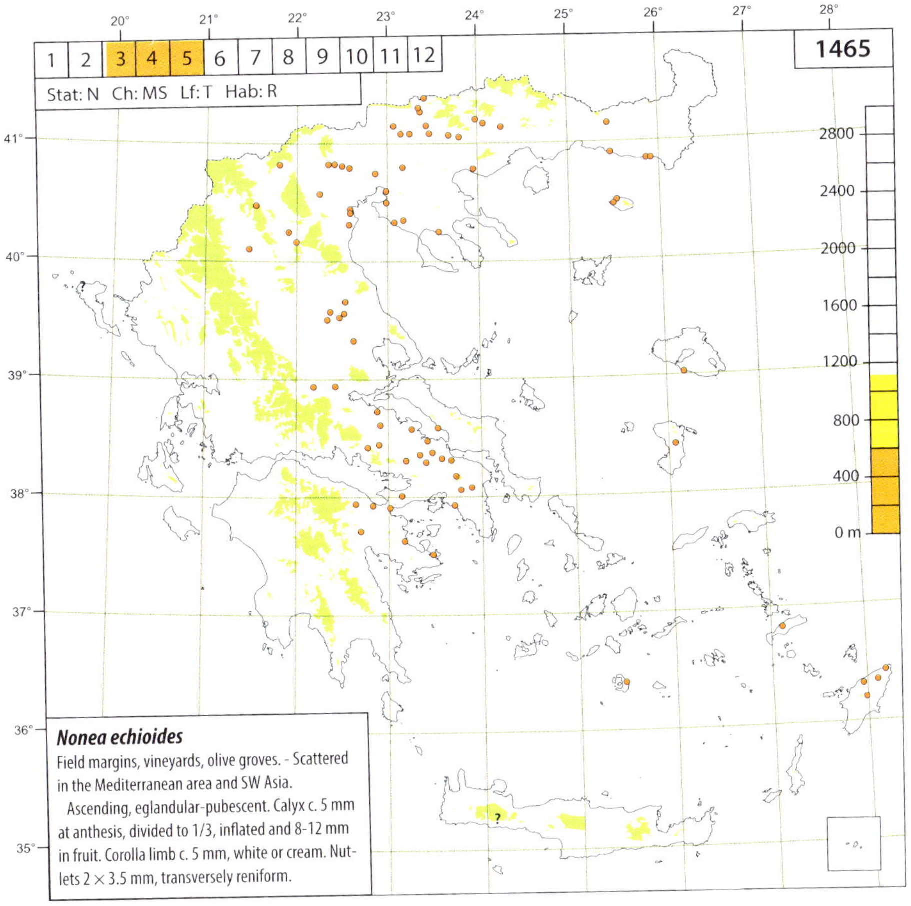

Nonea echioides
Field margins, vineyards, olive groves. - Scattered in the Mediterranean area and SW Asia.
Ascending, eglandular-pubescent. Calyx c. 5 mm at anthesis, divided to 1/3, inflated and 8-12 mm in fruit. Corolla limb c. 5 mm, white or cream. Nutlets 2 × 3.5 mm, transversely reniform.

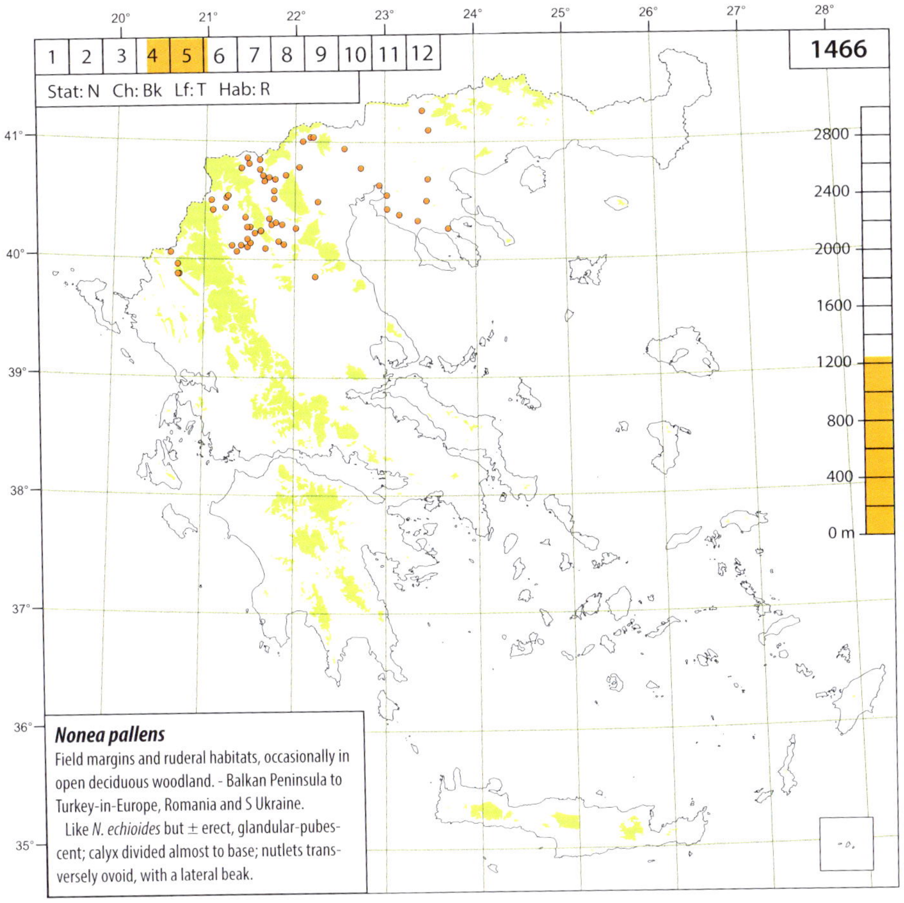

Nonea pallens
Field margins and ruderal habitats, occasionally in open deciduous woodland. - Balkan Peninsula to Turkey-in-Europe, Romania and S Ukraine.
Like *N. echioides* but ± erect, glandular-pubescent; calyx divided almost to base; nutlets transversely ovoid, with a lateral beak.

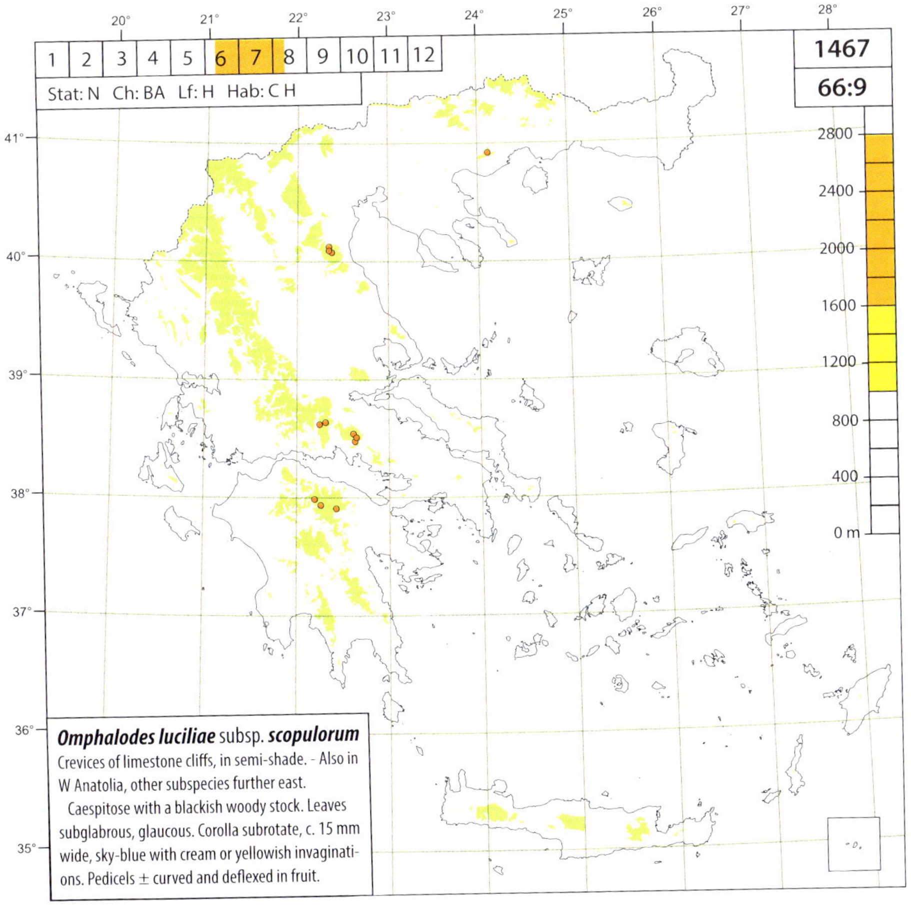

Omphalodes luciliae subsp. ***scopulorum***
Crevices of limestone cliffs, in semi-shade. - Also in W Anatolia, other subspecies further east.
Caespitose with a blackish woody stock. Leaves subglabrous, glaucous. Corolla subrotate, c. 15 mm wide, sky-blue with cream or yellowish invaginations. Pedicels ± curved and deflexed in fruit.

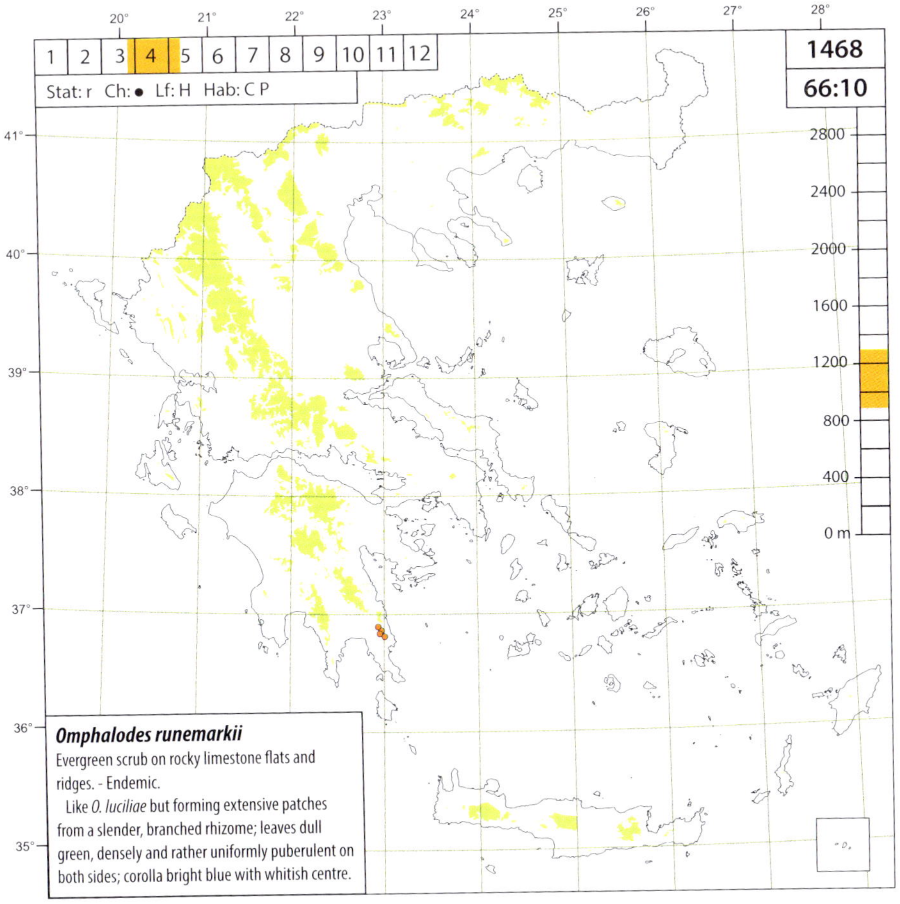

Omphalodes runemarkii
Evergreen scrub on rocky limestone flats and ridges. - Endemic.
Like *O. luciliae* but forming extensive patches from a slender, branched rhizome; leaves dull green, densely and rather uniformly puberulent on both sides; corolla bright blue with whitish centre.

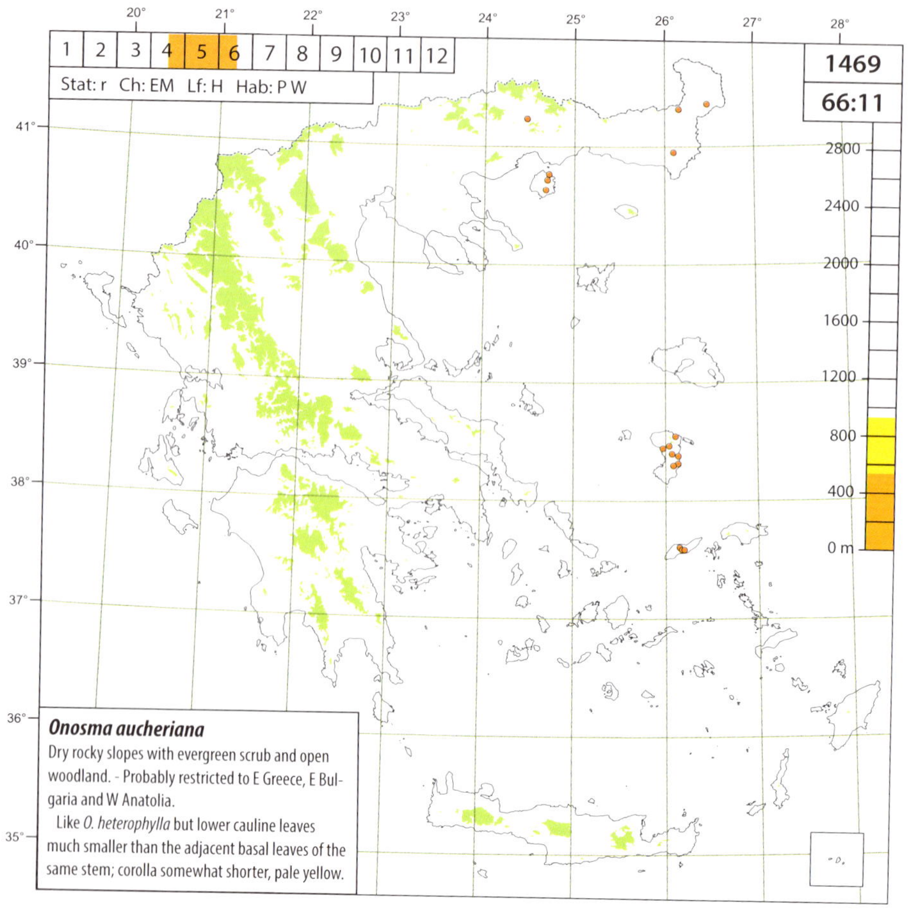

Onosma aucheriana

Dry rocky slopes with evergreen scrub and open woodland. - Probably restricted to E Greece, E Bulgaria and W Anatolia.

Like *O. heterophylla* but lower cauline leaves much smaller than the adjacent basal leaves of the same stem; corolla somewhat shorter, pale yellow.

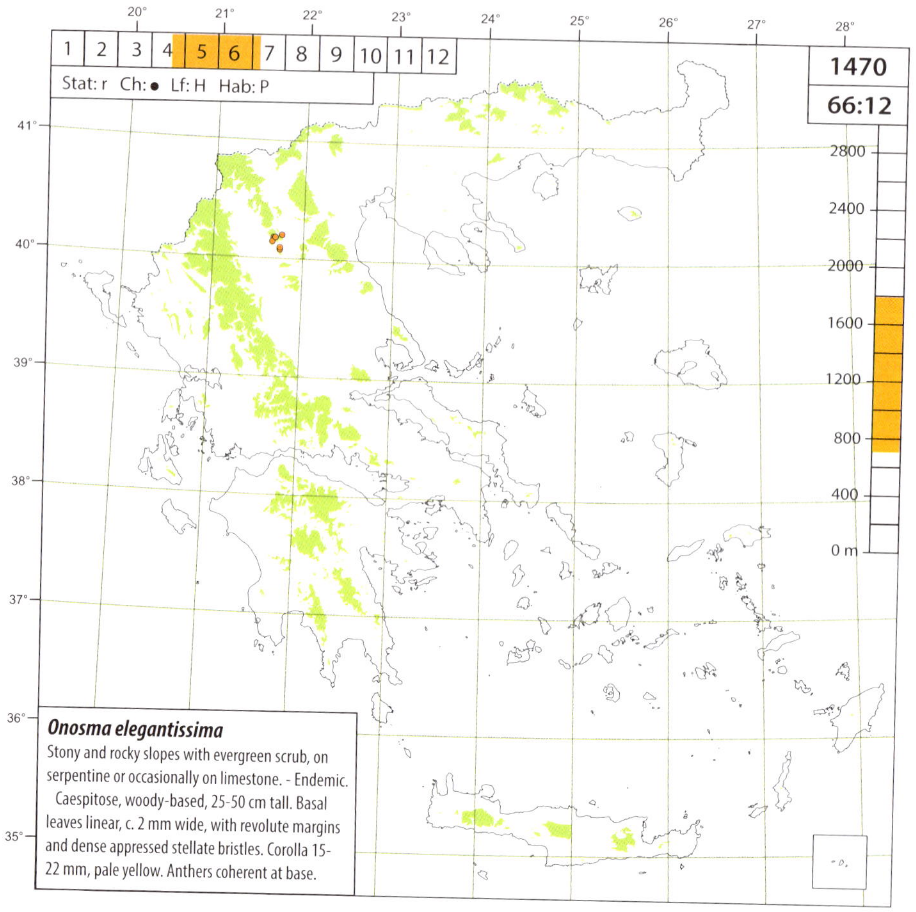

Onosma elegantissima

Stony and rocky slopes with evergreen scrub, on serpentine or occasionally on limestone. - Endemic.

Caespitose, woody-based, 25-50 cm tall. Basal leaves linear, c. 2 mm wide, with revolute margins and dense appressed stellate bristles. Corolla 15-22 mm, pale yellow. Anthers coherent at base.

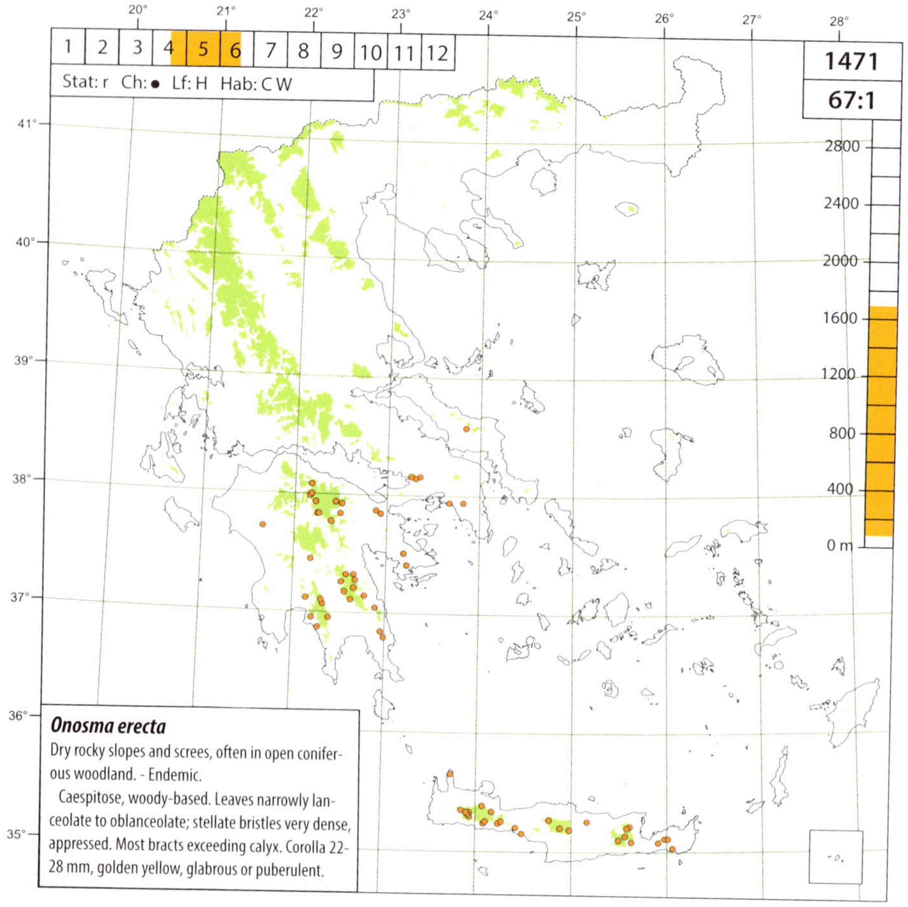

Onosma erecta

Dry rocky slopes and screes, often in open coniferous woodland. - Endemic.

Caespitose, woody-based. Leaves narrowly lanceolate to oblanceolate; stellate bristles very dense, appressed. Most bracts exceeding calyx. Corolla 22-28 mm, golden yellow, glabrous or puberulent.

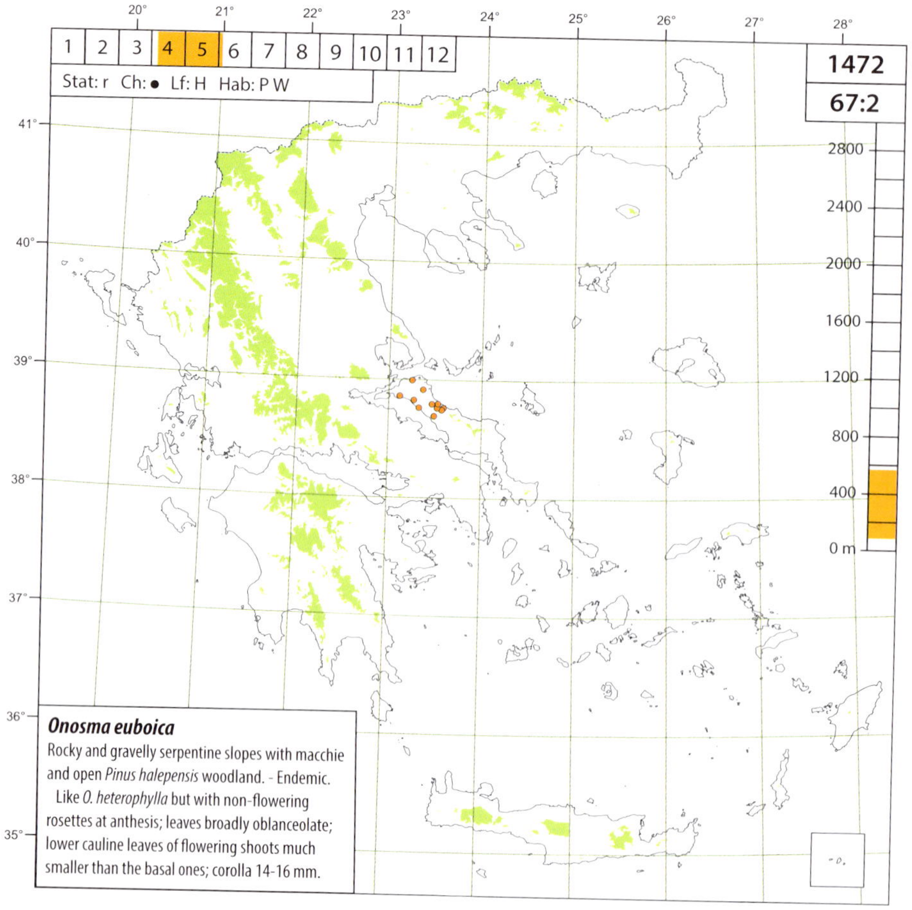

Onosma euboica

Rocky and gravelly serpentine slopes with macchie and open *Pinus halepensis* woodland. - Endemic.

Like *O. heterophylla* but with non-flowering rosettes at anthesis; leaves broadly oblanceolate; lower cauline leaves of flowering shoots much smaller than the basal ones; corolla 14-16 mm.

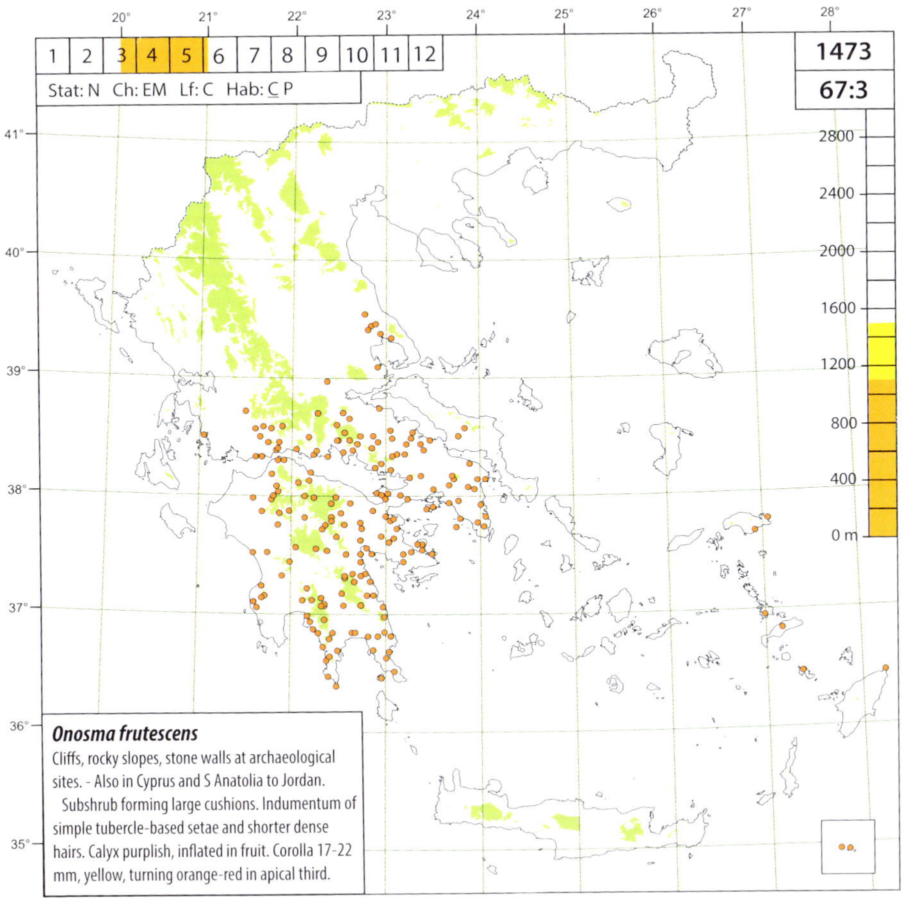

Onosma frutescens

Cliffs, rocky slopes, stone walls at archaeological sites. - Also in Cyprus and S Anatolia to Jordan.

Subshrub forming large cushions. Indumentum of simple tubercle-based setae and shorter dense hairs. Calyx purplish, inflated in fruit. Corolla 17-22 mm, yellow, turning orange-red in apical third.

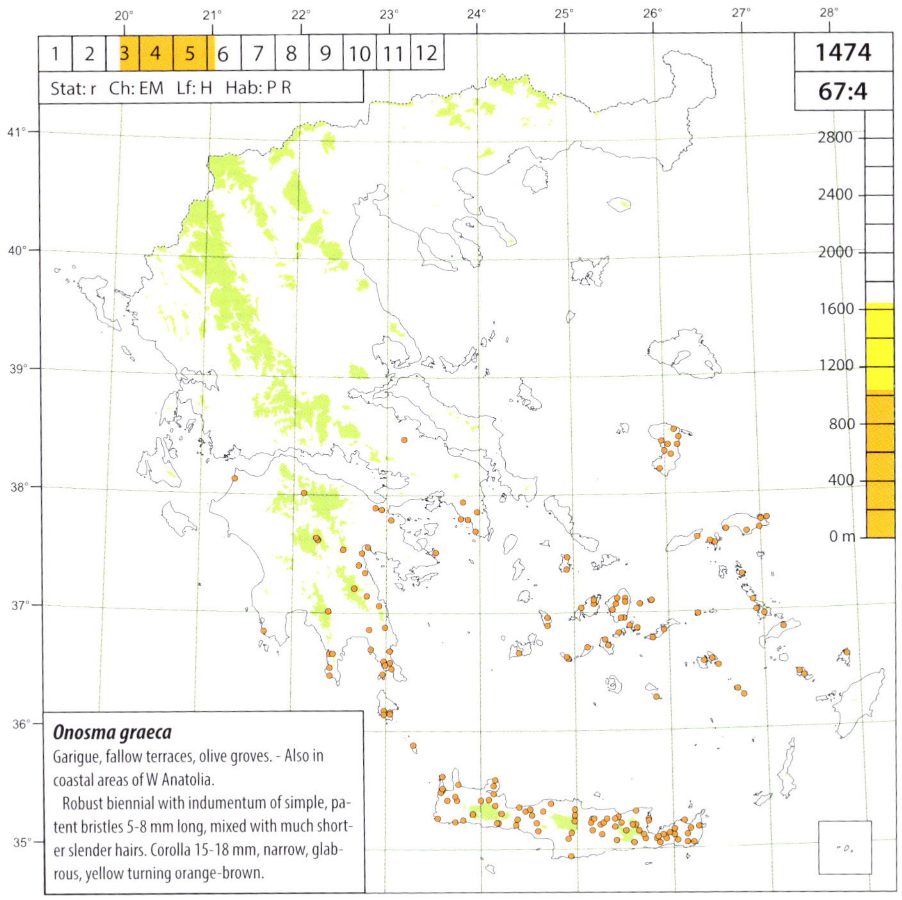

Onosma graeca

Garigue, fallow terraces, olive groves. - Also in coastal areas of W Anatolia.

Robust biennial with indumentum of simple, patent bristles 5-8 mm long, mixed with much shorter slender hairs. Corolla 15-18 mm, narrow, glabrous, yellow turning orange-brown.

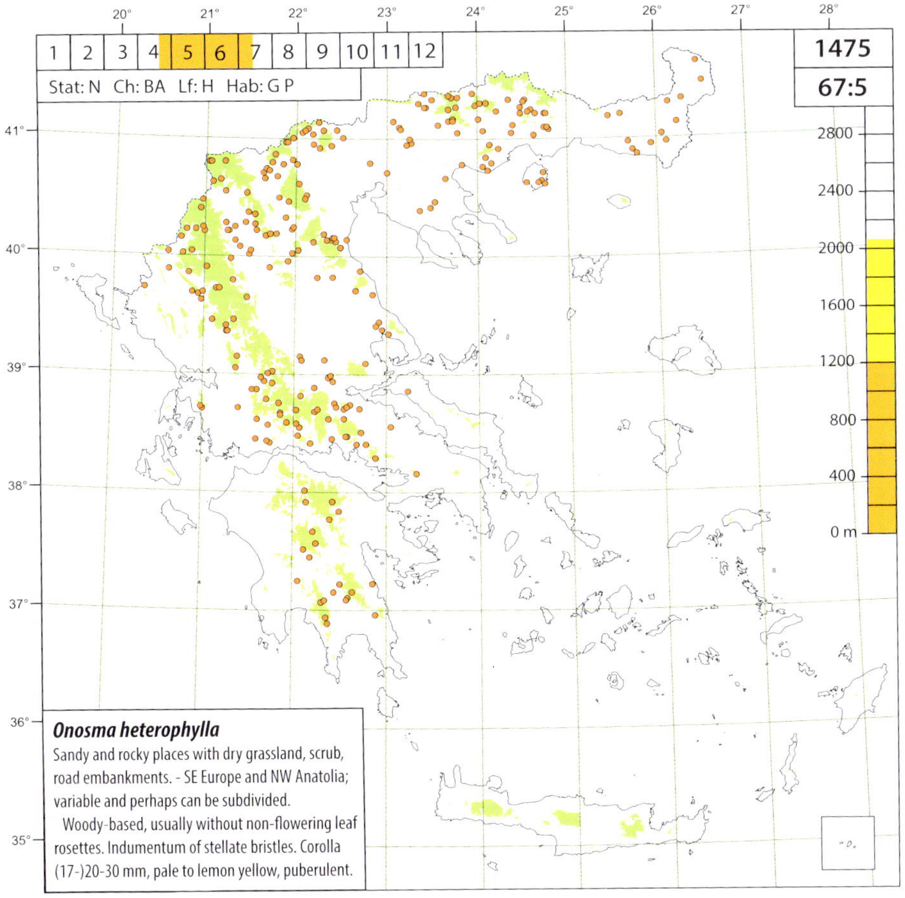

Onosma heterophylla

Sandy and rocky places with dry grassland, scrub, road embankments. - SE Europe and NW Anatolia; variable and perhaps can be subdivided.

Woody-based, usually without non-flowering leaf rosettes. Indumentum of stellate bristles. Corolla (17-)20-30 mm, pale to lemon yellow, puberulent.

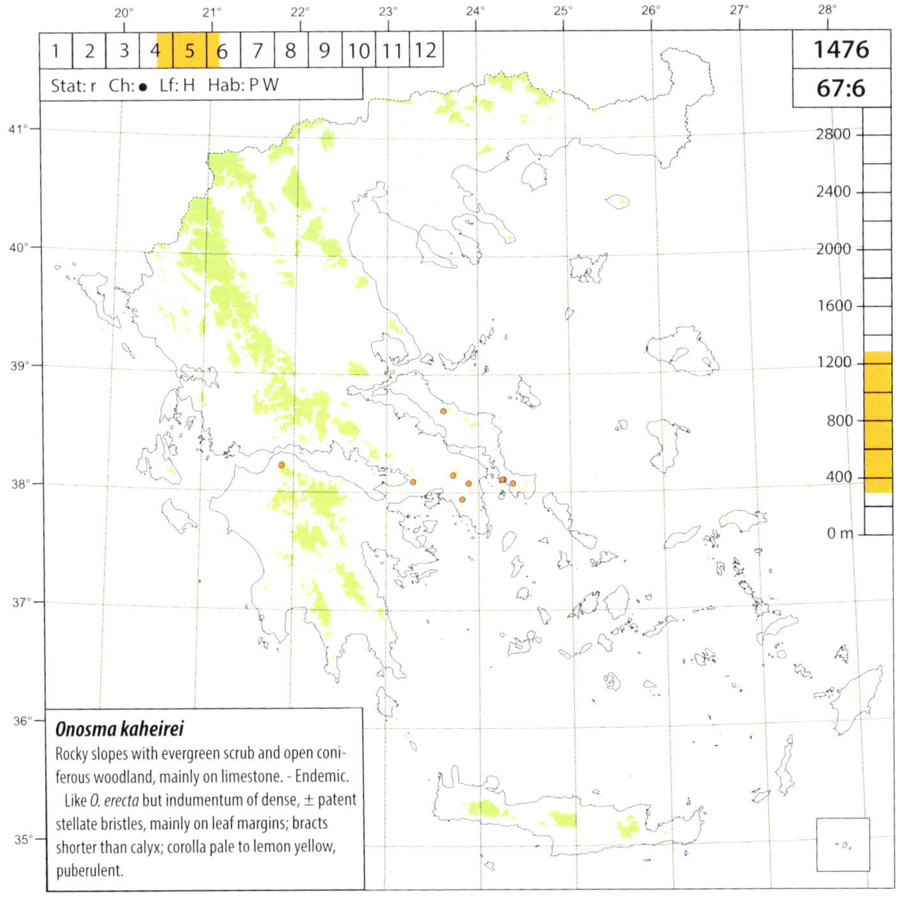

Onosma kaheirei

Rocky slopes with evergreen scrub and open coniferous woodland, mainly on limestone. - Endemic.

Like *O. erecta* but indumentum of dense, ± patent stellate bristles, mainly on leaf margins; bracts shorter than calyx; corolla pale to lemon yellow, puberulent.

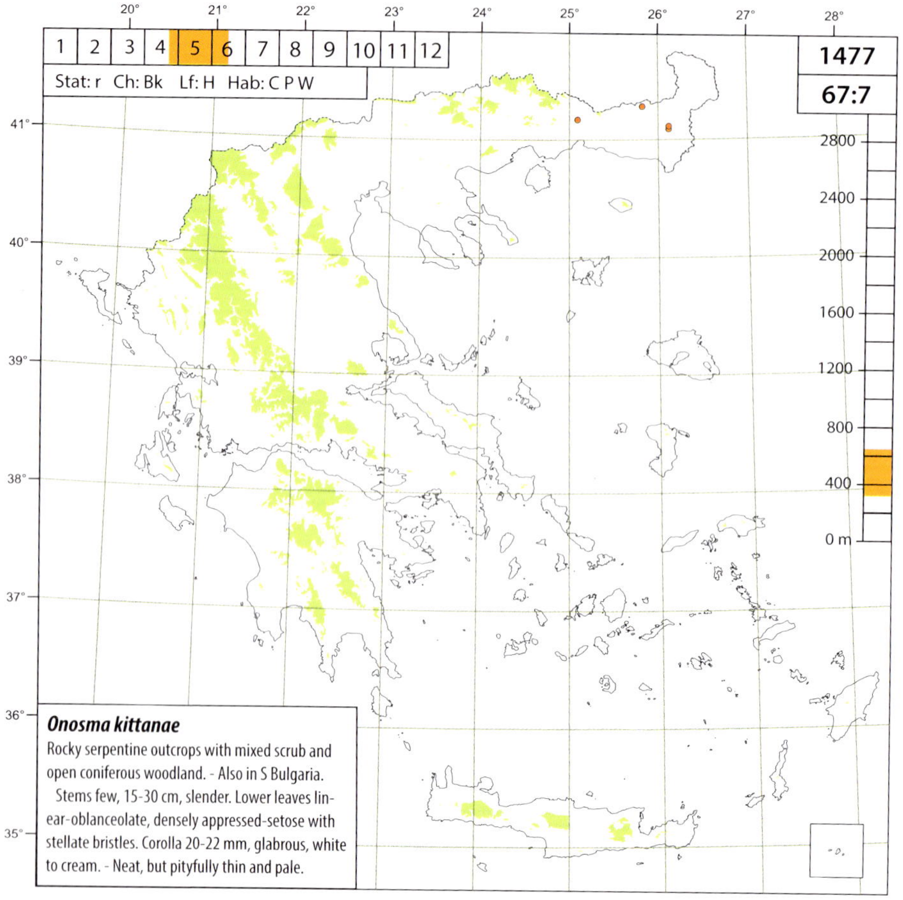

Onosma kittanae
Rocky serpentine outcrops with mixed scrub and open coniferous woodland. - Also in S Bulgaria.
Stems few, 15-30 cm, slender. Lower leaves linear-oblanceolate, densely appressed-setose with stellate bristles. Corolla 20-22 mm, glabrous, white to cream. - Neat, but pityfully thin and pale.

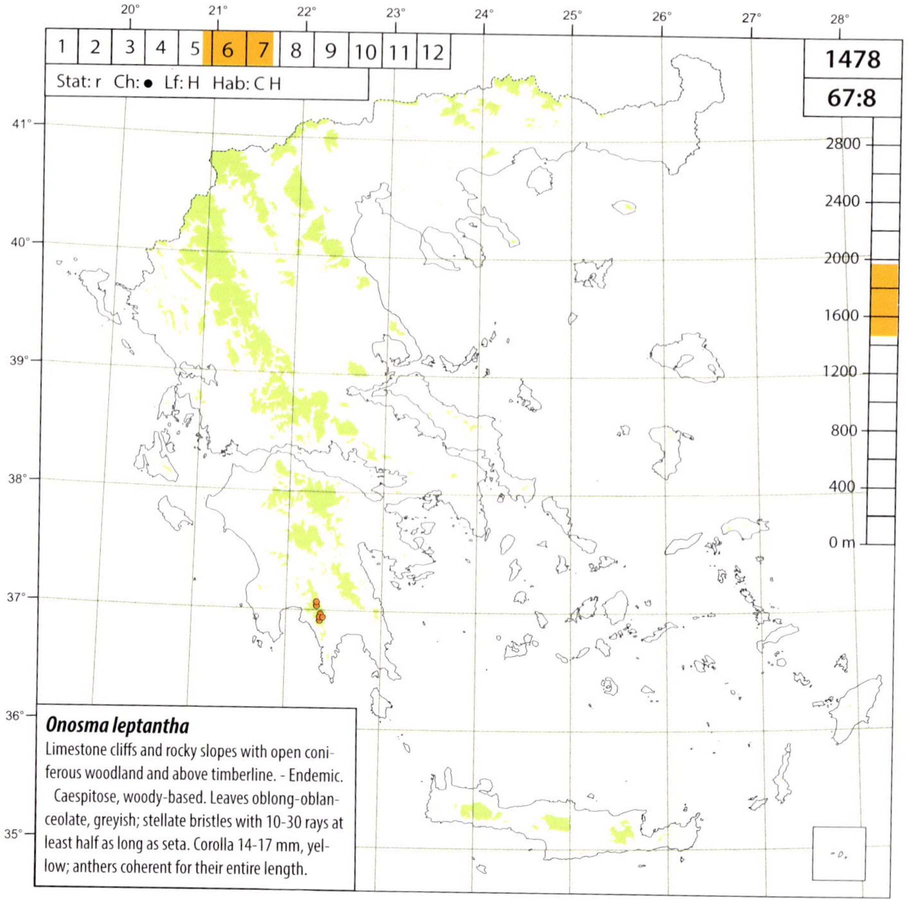

Onosma leptantha
Limestone cliffs and rocky slopes with open coniferous woodland and above timberline. - Endemic.
Caespitose, woody-based. Leaves oblong-oblanceolate, greyish; stellate bristles with 10-30 rays at least half as long as seta. Corolla 14-17 mm, yellow; anthers coherent for their entire length.

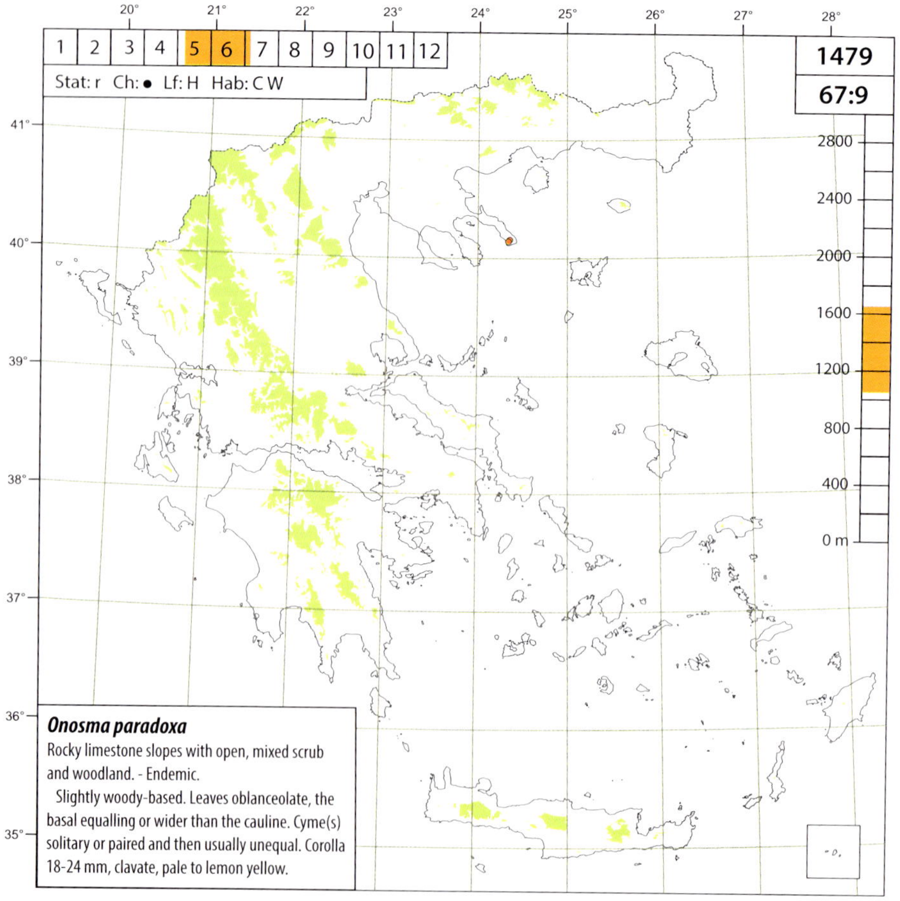

Onosma paradoxa
Rocky limestone slopes with open, mixed scrub and woodland. - Endemic.
Slightly woody-based. Leaves oblanceolate, the basal equalling or wider than the cauline. Cyme(s) solitary or paired and then usually unequal. Corolla 18-24 mm, clavate, pale to lemon yellow.

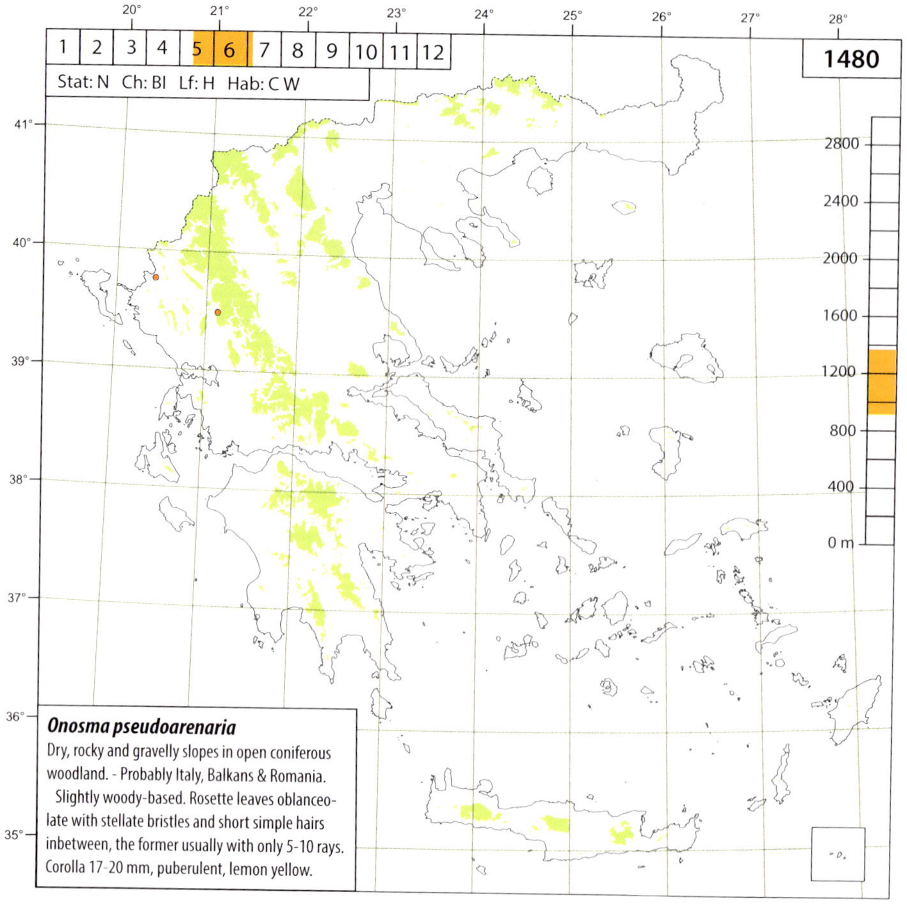

Onosma pseudoarenaria
Dry, rocky and gravelly slopes in open coniferous woodland. - Probably Italy, Balkans & Romania.
Slightly woody-based. Rosette leaves oblanceolate with stellate bristles and short simple hairs inbetween, the former usually with only 5-10 rays. Corolla 17-20 mm, puberulent, lemon yellow.

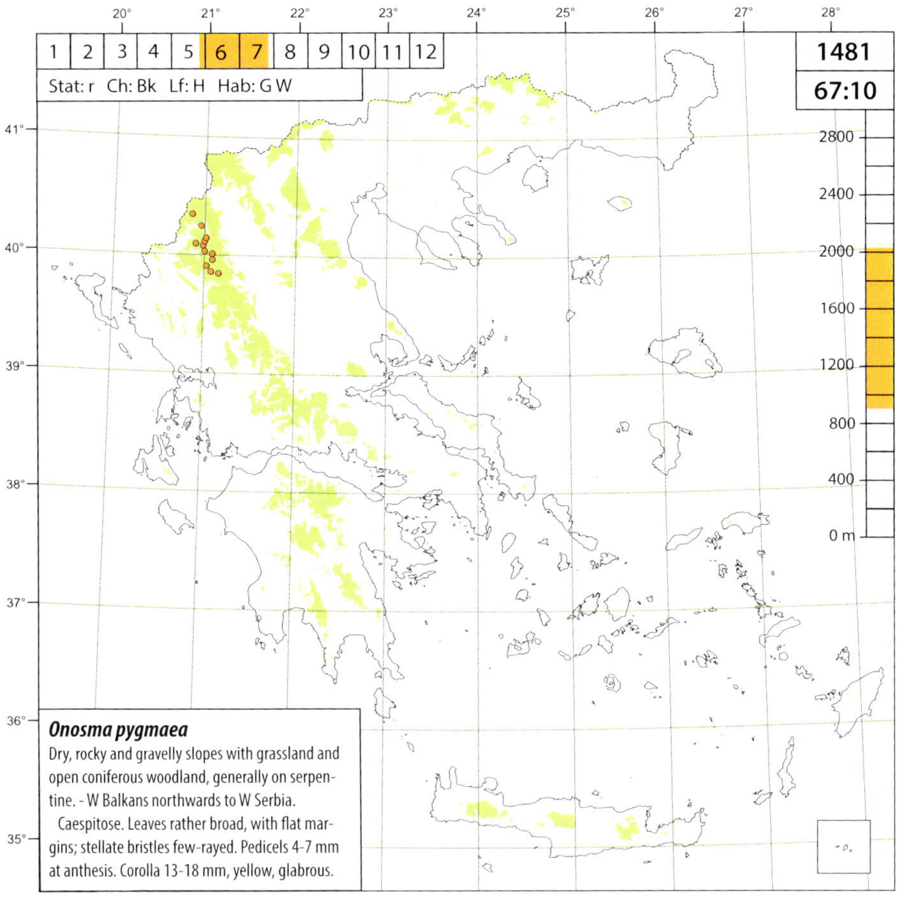

Onosma pygmaea
Dry, rocky and gravelly slopes with grassland and open coniferous woodland, generally on serpentine. - W Balkans northwards to W Serbia.
Caespitose. Leaves rather broad, with flat margins; stellate bristles few-rayed. Pedicels 4-7 mm at anthesis. Corolla 13-18 mm, yellow, glabrous.

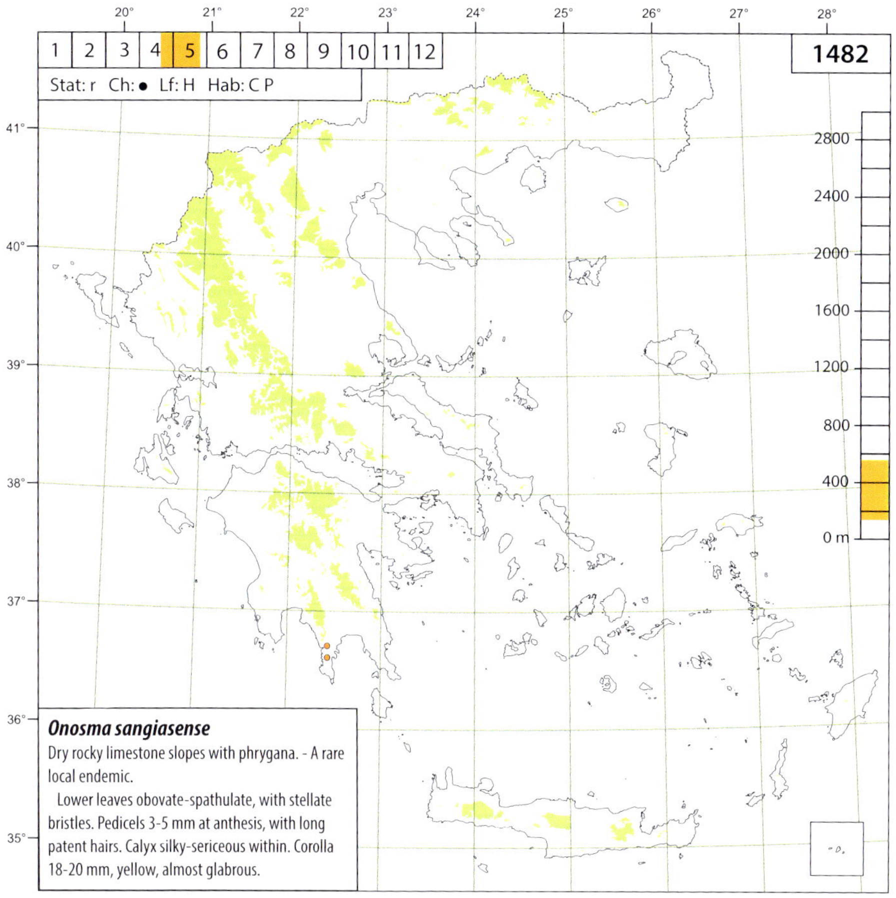

Onosma sangiasense
Dry rocky limestone slopes with phrygana. - A rare local endemic.
Lower leaves obovate-spathulate, with stellate bristles. Pedicels 3-5 mm at anthesis, with long patent hairs. Calyx silky-sericeous within. Corolla 18-20 mm, yellow, almost glabrous.

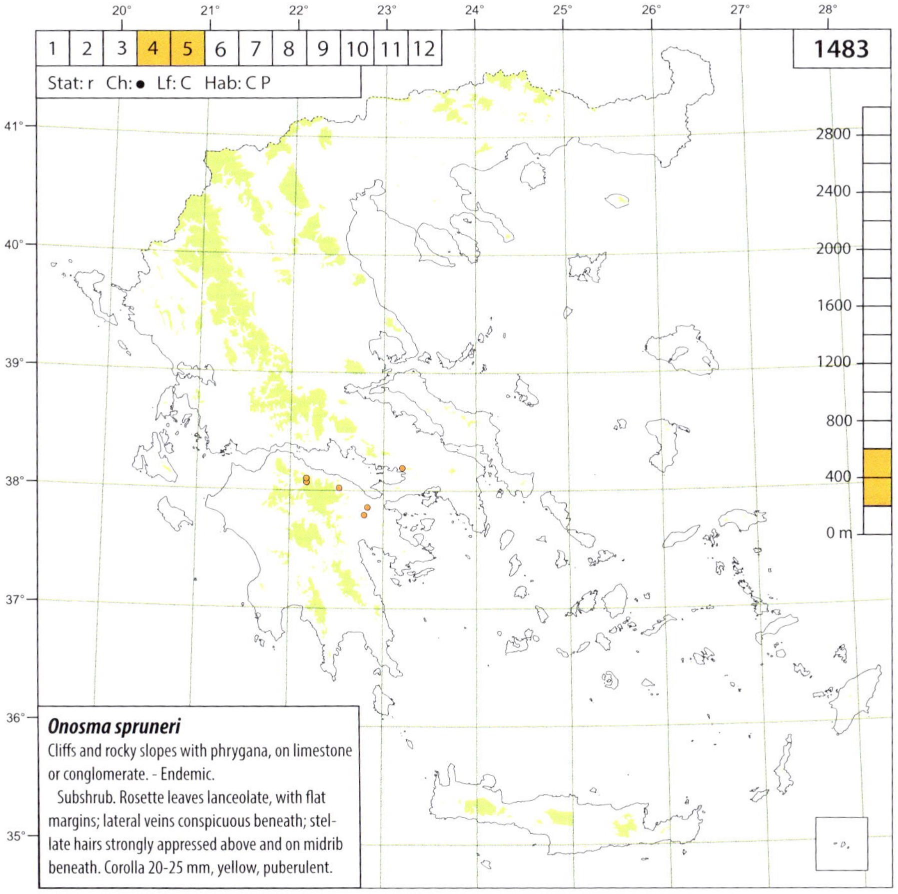

Onosma spruneri
Cliffs and rocky slopes with phrygana, on limestone or conglomerate. - Endemic.
Subshrub. Rosette leaves lanceolate, with flat margins; lateral veins conspicuous beneath; stellate hairs strongly appressed above and on midrib beneath. Corolla 20-25 mm, yellow, puberulent.

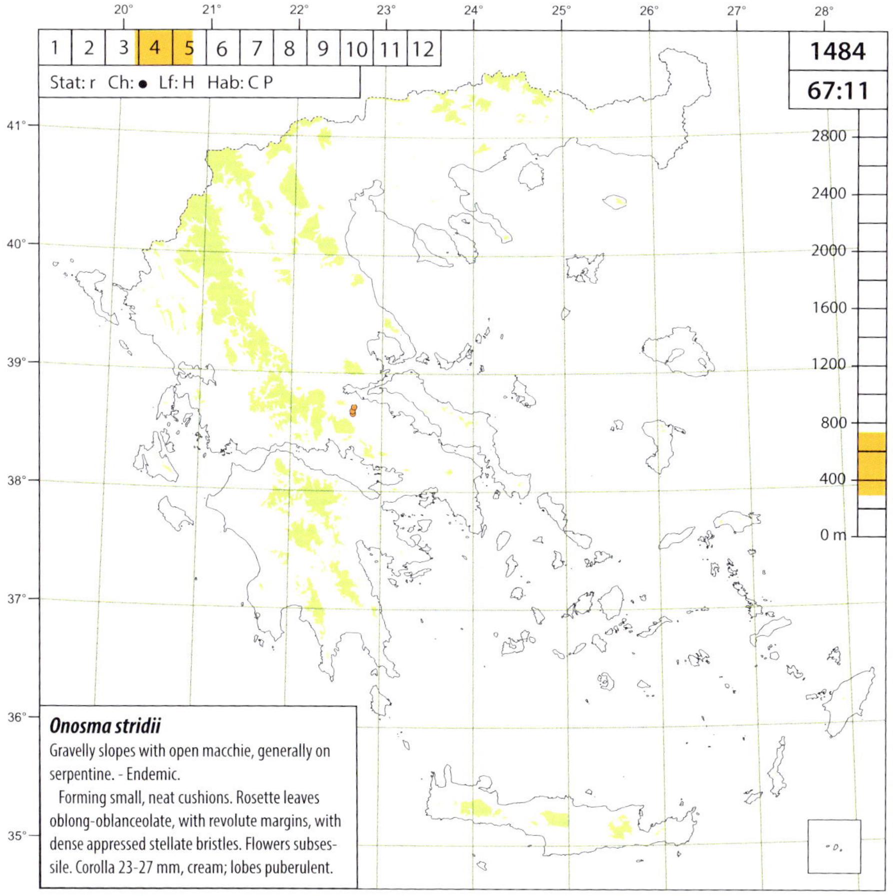

Onosma stridii
Gravelly slopes with open macchie, generally on serpentine. - Endemic.
Forming small, neat cushions. Rosette leaves oblong-oblanceolate, with revolute margins, with dense appressed stellate bristles. Flowers subsessile. Corolla 23-27 mm, cream; lobes puberulent.

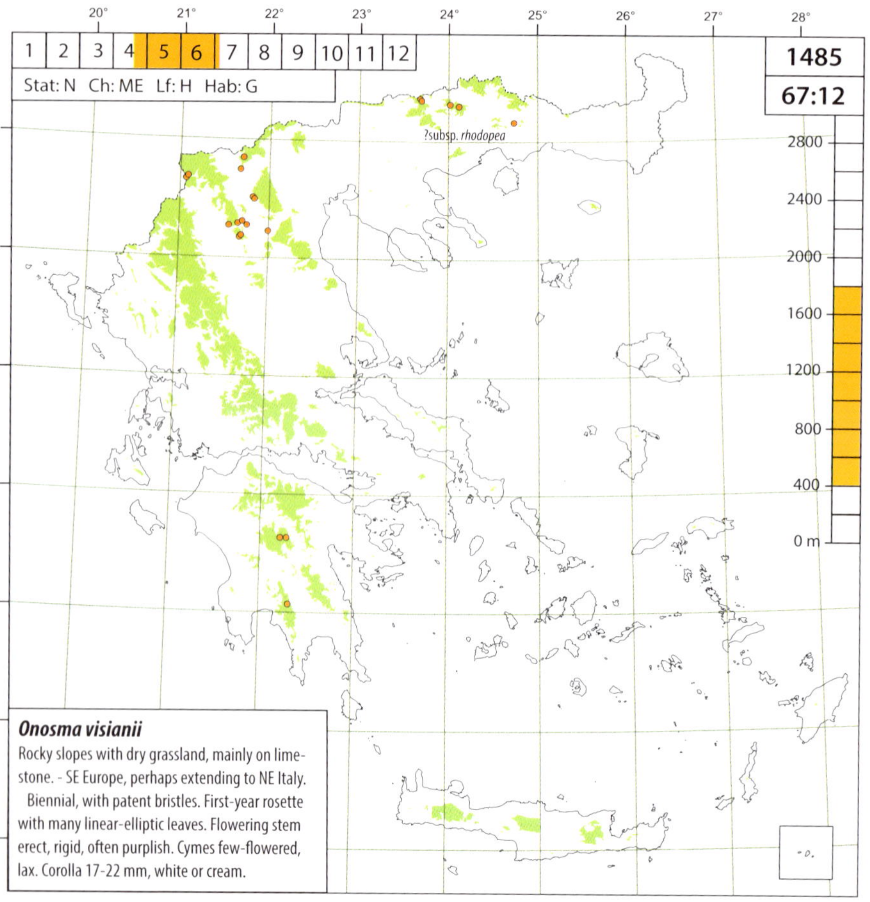

Onosma visianii
Rocky slopes with dry grassland, mainly on limestone. - SE Europe, perhaps extending to NE Italy.
Biennial, with patent bristles. First-year rosette with many linear-elliptic leaves. Flowering stem erect, rigid, often purplish. Cymes few-flowered, lax. Corolla 17-22 mm, white or cream.

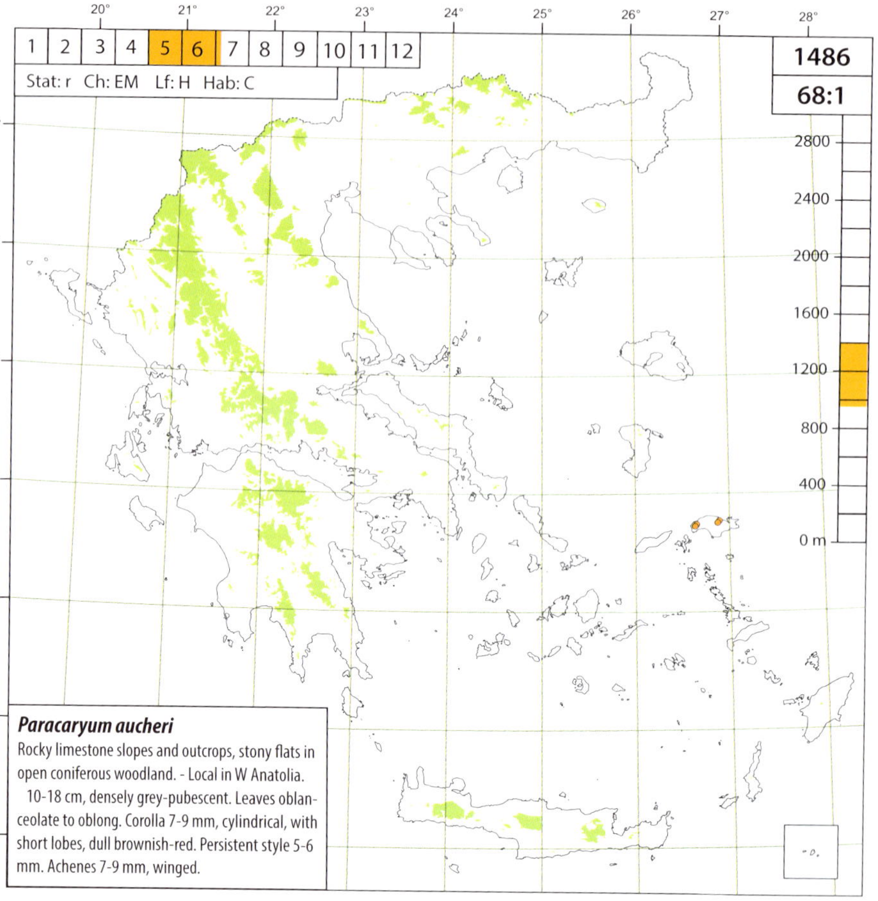

Paracaryum aucheri
Rocky limestone slopes and outcrops, stony flats in open coniferous woodland. - Local in W Anatolia.
10-18 cm, densely grey-pubescent. Leaves oblanceolate to oblong. Corolla 7-9 mm, cylindrical, with short lobes, dull brownish-red. Persistent style 5-6 mm. Achenes 7-9 mm, winged.

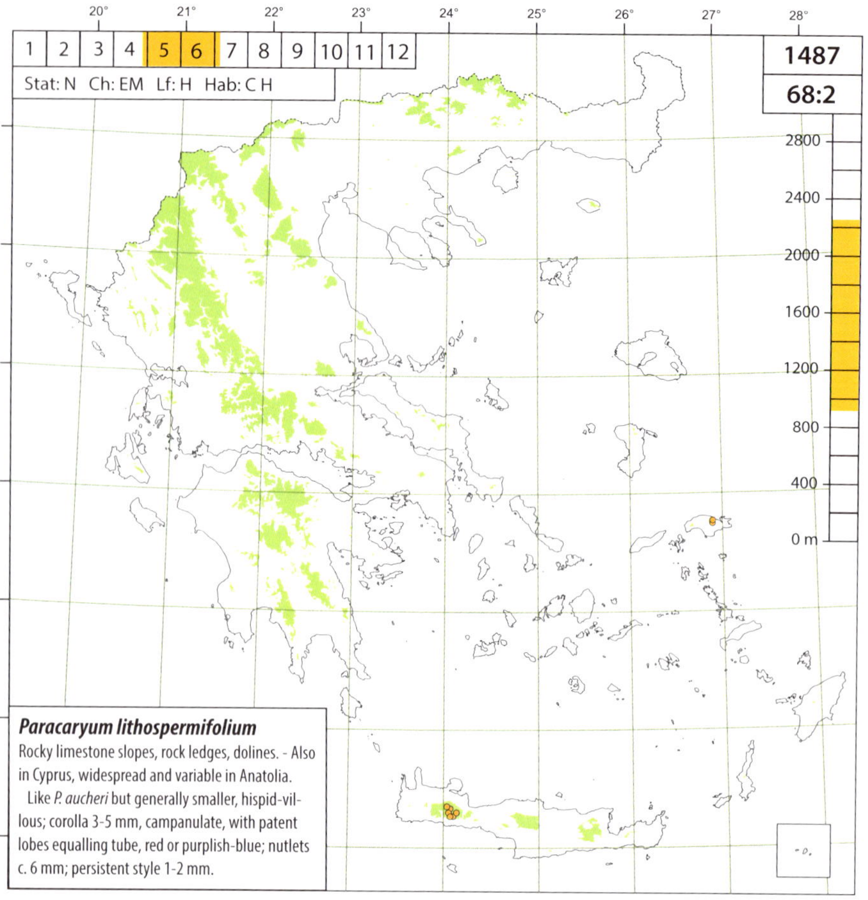

Paracaryum lithospermifolium
Rocky limestone slopes, rock ledges, dolines. - Also in Cyprus, widespread and variable in Anatolia.
Like *P. aucheri* but generally smaller, hispid-villous; corolla 3-5 mm, campanulate, with patent lobes equalling tube, red or purplish-blue; nutlets c. 6 mm; persistent style 1-2 mm.

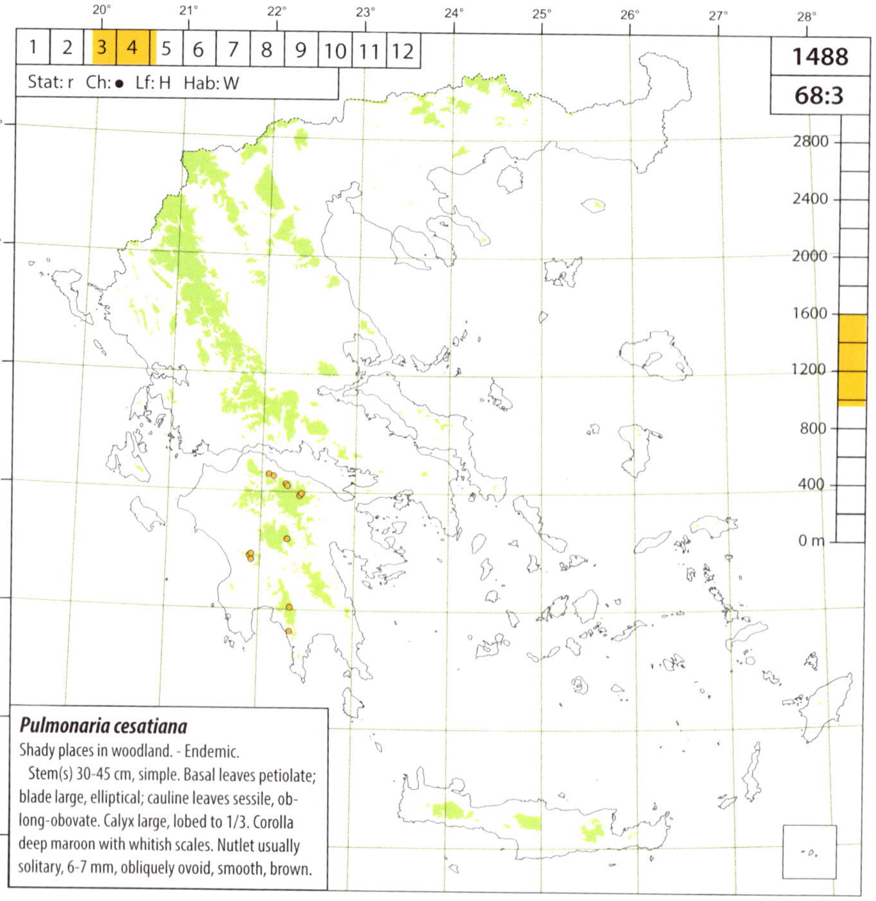

Pulmonaria cesatiana
Shady places in woodland. - Endemic.
Stem(s) 30-45 cm, simple. Basal leaves petiolate; blade large, elliptical; cauline leaves sessile, oblong-obovate. Calyx large, lobed to 1/3. Corolla deep maroon with whitish scales. Nutlet usually solitary, 6-7 mm, obliquely ovoid, smooth, brown.

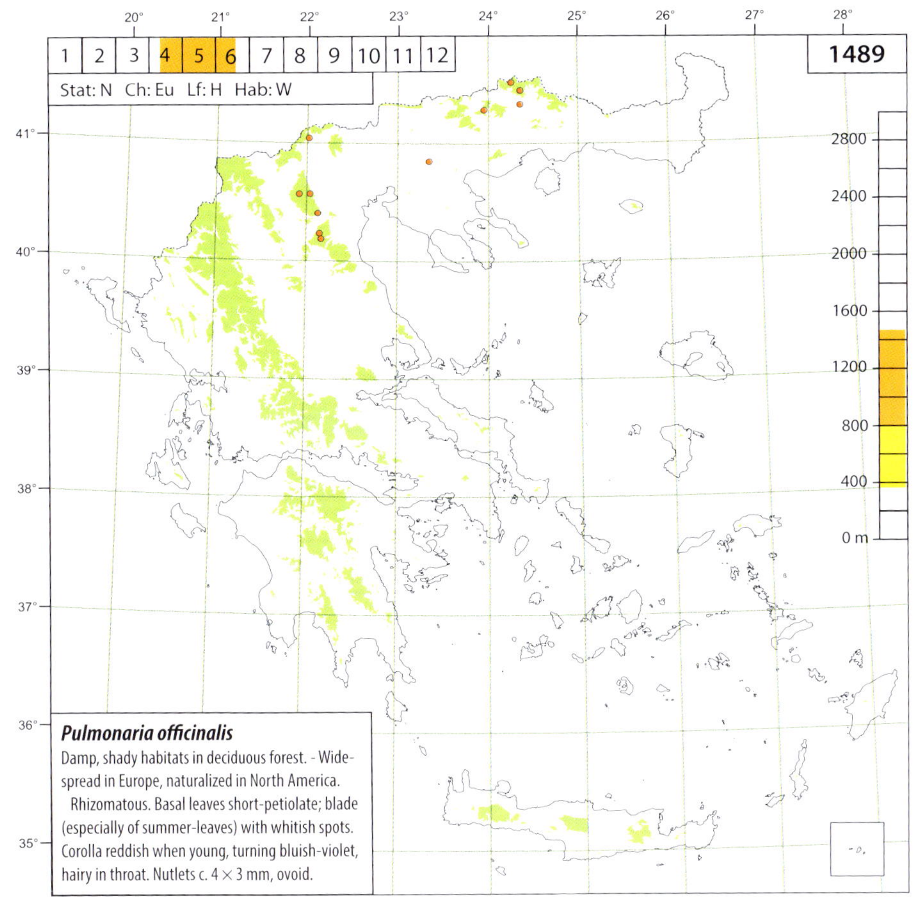

Pulmonaria officinalis

Damp, shady habitats in deciduous forest. - Widespread in Europe, naturalized in North America.

Rhizomatous. Basal leaves short-petiolate; blade (especially of summer-leaves) with whitish spots. Corolla reddish when young, turning bluish-violet, hairy in throat. Nutlets c. 4 × 3 mm, ovoid.

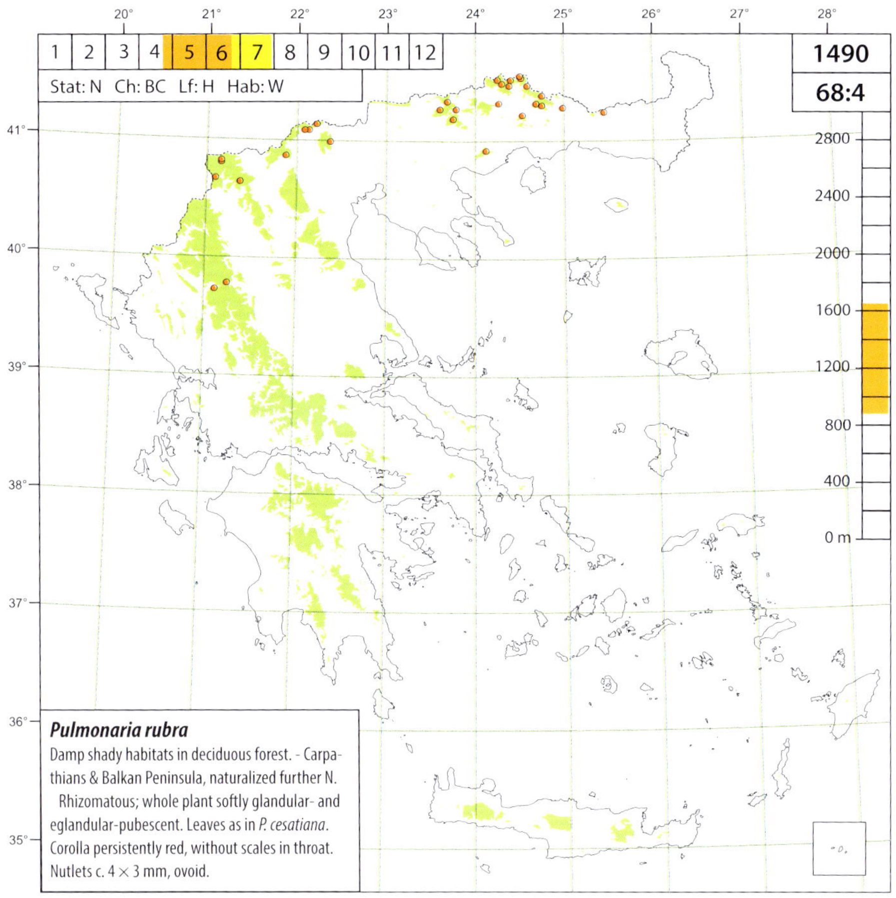

Pulmonaria rubra

Damp shady habitats in deciduous forest. - Carpathians & Balkan Peninsula, naturalized further N.

Rhizomatous; whole plant softly glandular- and eglandular-pubescent. Leaves as in *P. cesatiana*. Corolla persistently red, without scales in throat. Nutlets c. 4 × 3 mm, ovoid.

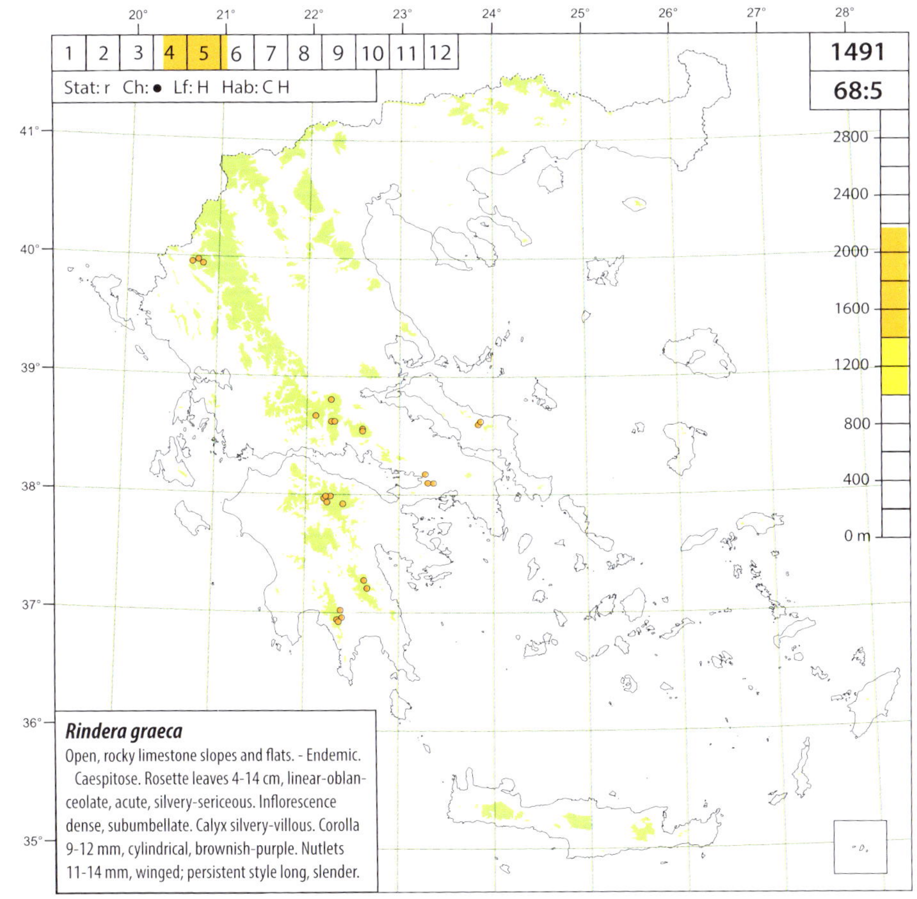

Rindera graeca

Open, rocky limestone slopes and flats. - Endemic.

Caespitose. Rosette leaves 4-14 cm, linear-oblanceolate, acute, silvery-sericeous. Inflorescence dense, subumbellate. Calyx silvery-villous. Corolla 9-12 mm, cylindrical, brownish-purple. Nutlets 11-14 mm, winged; persistent style long, slender.

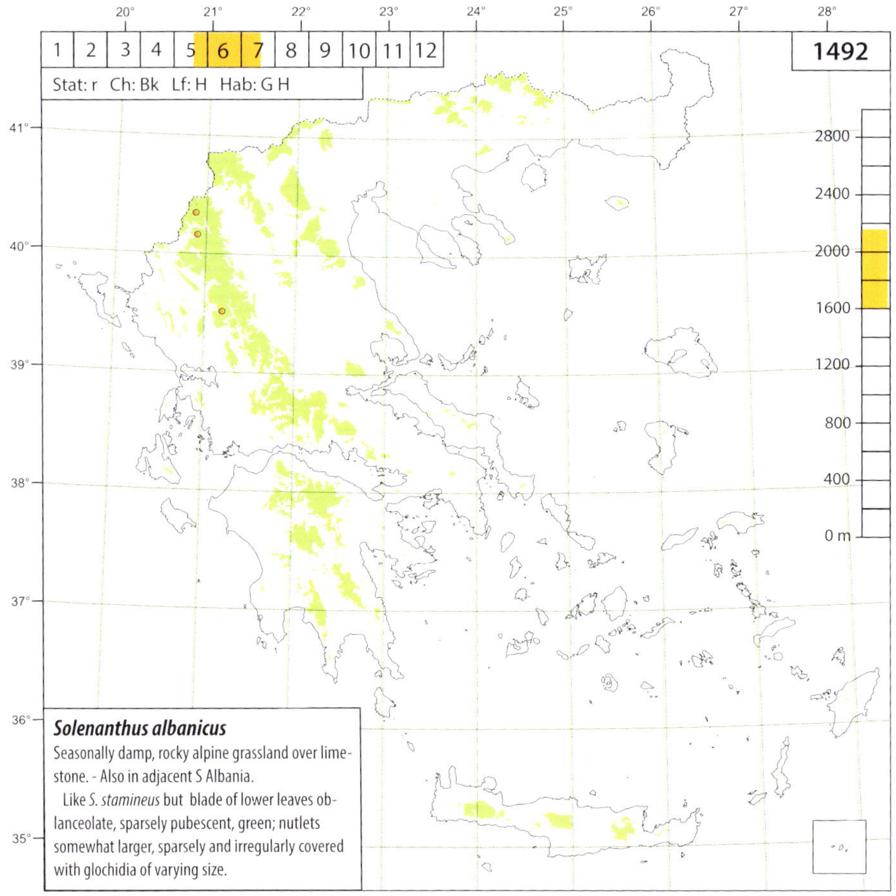

Solenanthus albanicus

Seasonally damp, rocky alpine grassland over limestone. - Also in adjacent S Albania.

Like *S. stamineus* but blade of lower leaves oblanceolate, sparsely pubescent, green; nutlets somewhat larger, sparsely and irregularly covered with glochidia of varying size.

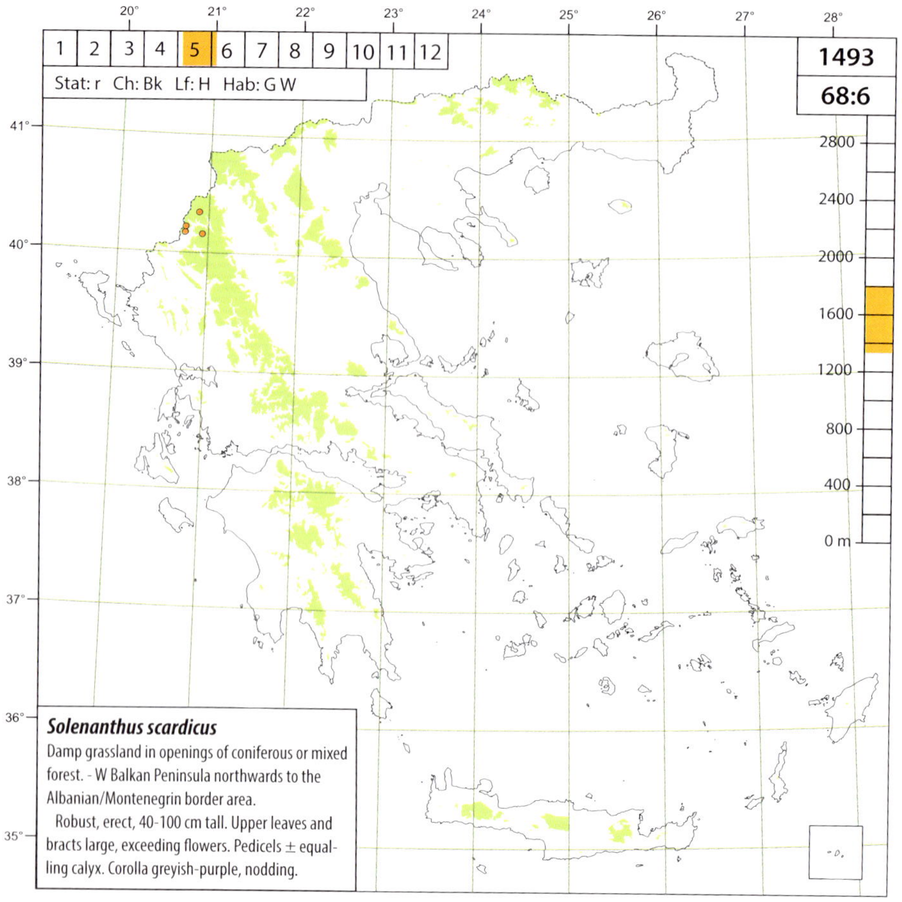

Solenanthus scardicus

Damp grassland in openings of coniferous or mixed forest. - W Balkan Peninsula northwards to the Albanian/Montenegrin border area.

Robust, erect, 40-100 cm tall. Upper leaves and bracts large, exceeding flowers. Pedicels ± equalling calyx. Corolla greyish-purple, nodding.

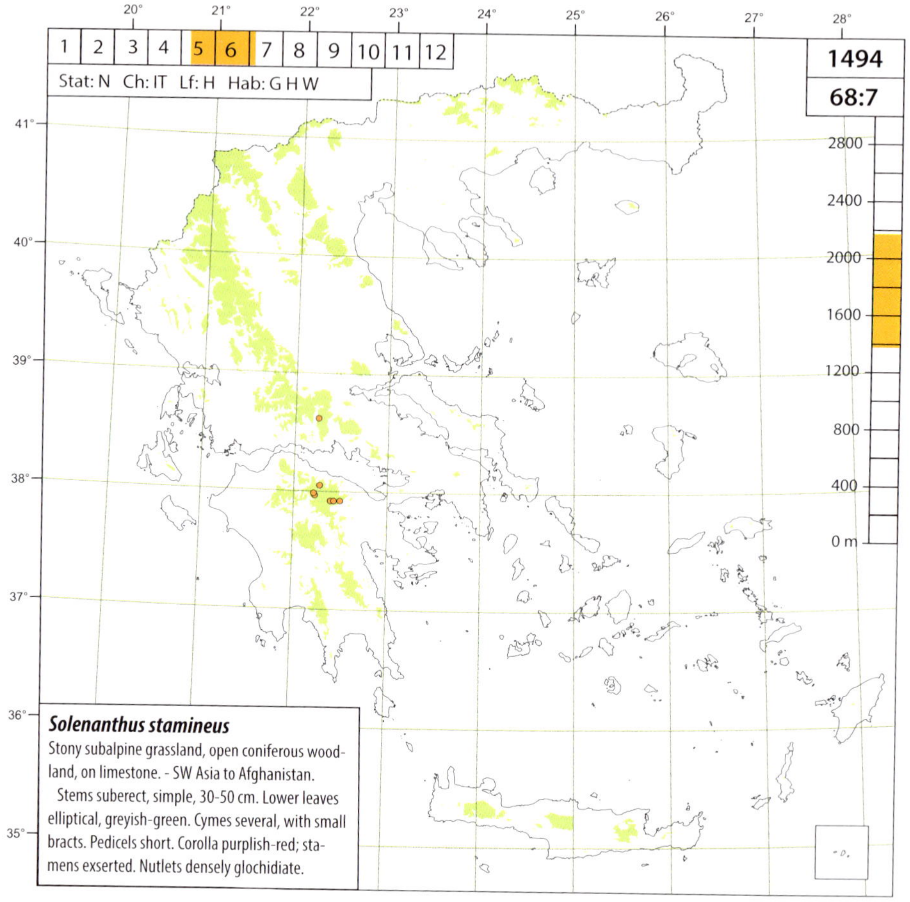

Solenanthus stamineus

Stony subalpine grassland, open coniferous woodland, on limestone. - SW Asia to Afghanistan.

Stems suberect, simple, 30-50 cm. Lower leaves elliptical, greyish-green. Cymes several, with small bracts. Pedicels short. Corolla purplish-red; stamens exserted. Nutlets densely glochidiate.

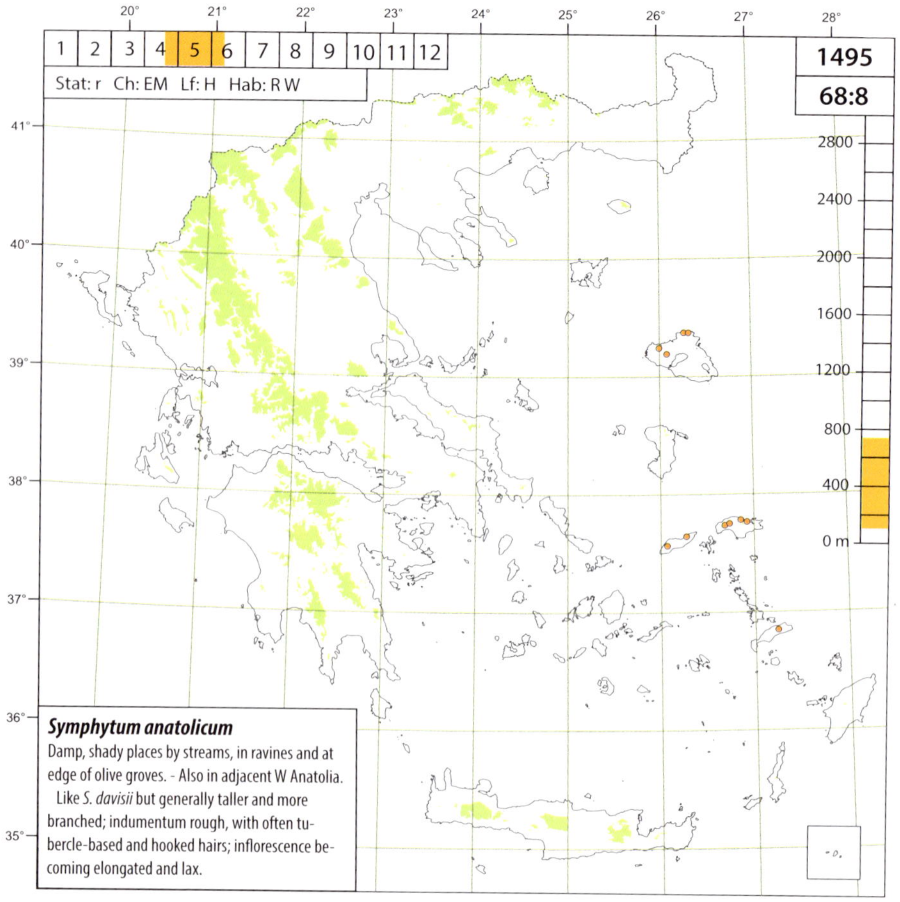

Symphytum anatolicum

Damp, shady places by streams, in ravines and at edge of olive groves. - Also in adjacent W Anatolia.

Like *S. davisii* but generally taller and more branched; indumentum rough, with often tubercle-based and hooked hairs; inflorescence becoming elongated and lax.

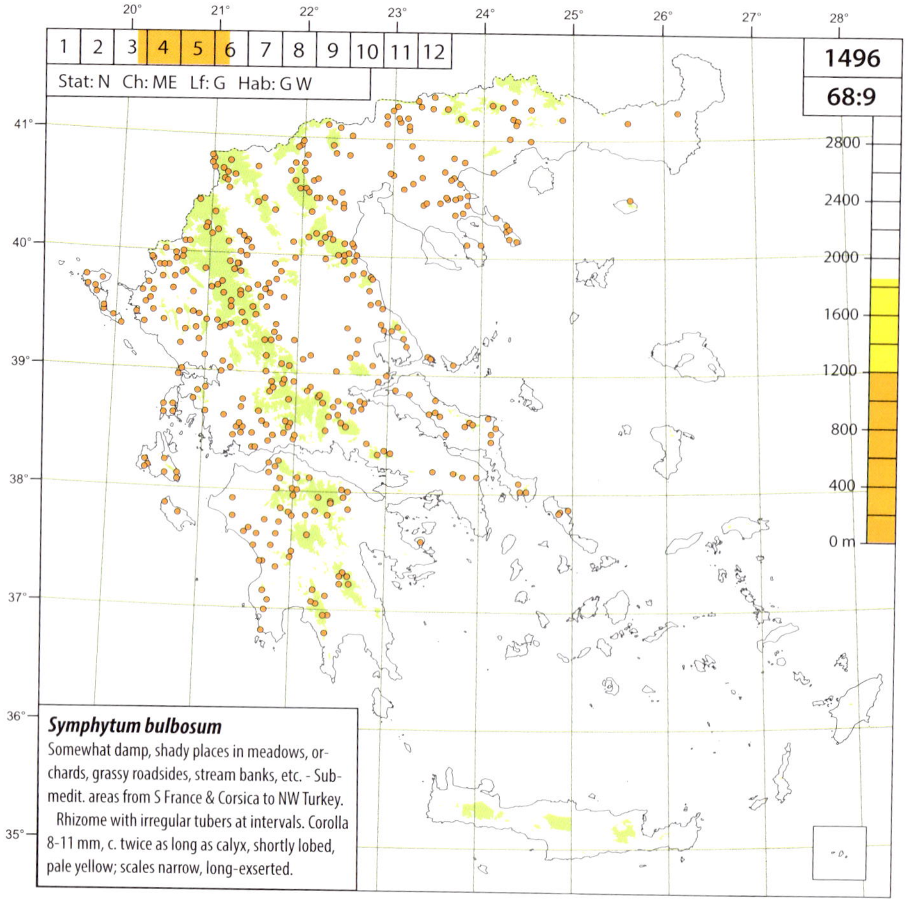

Symphytum bulbosum

Somewhat damp, shady places in meadows, orchards, grassy roadsides, stream banks, etc. - Submedit. areas from S France & Corsica to NW Turkey.

Rhizome with irregular tubers at intervals. Corolla 8-11 mm, c. twice as long as calyx, shortly lobed, pale yellow; scales narrow, long-exserted.

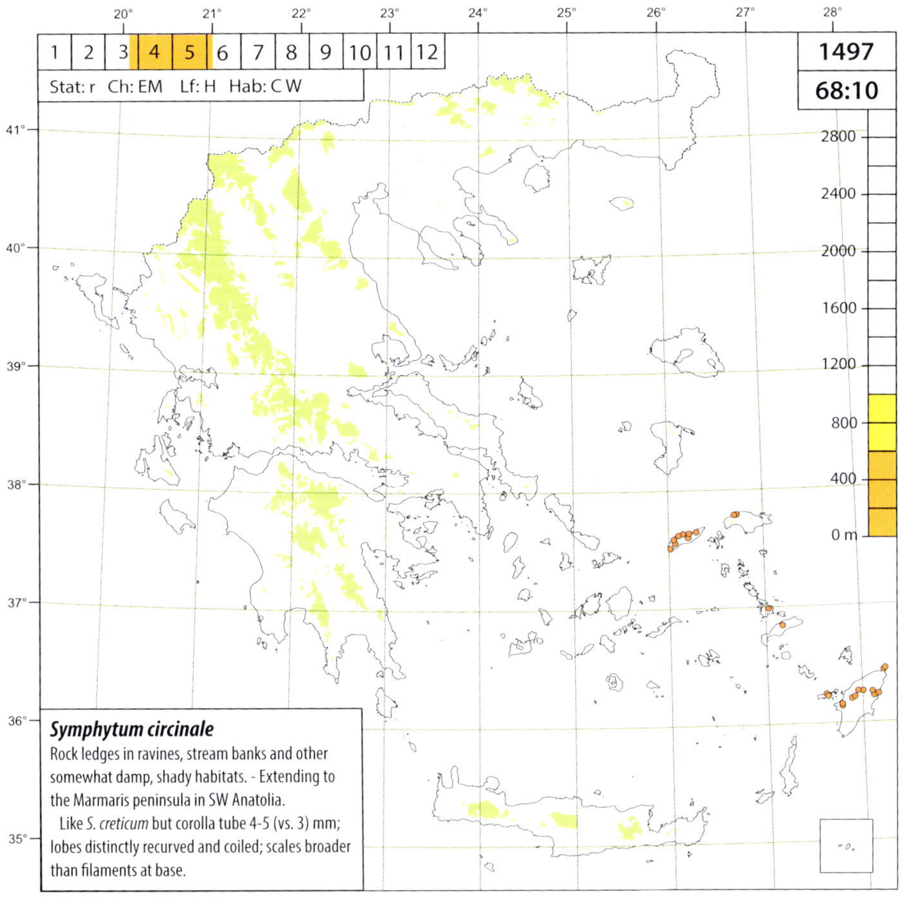

Symphytum circinale

Rock ledges in ravines, stream banks and other somewhat damp, shady habitats. - Extending to the Marmaris peninsula in SW Anatolia.

Like *S. creticum* but corolla tube 4-5 (vs. 3) mm; lobes distinctly recurved and coiled; scales broader than filaments at base.

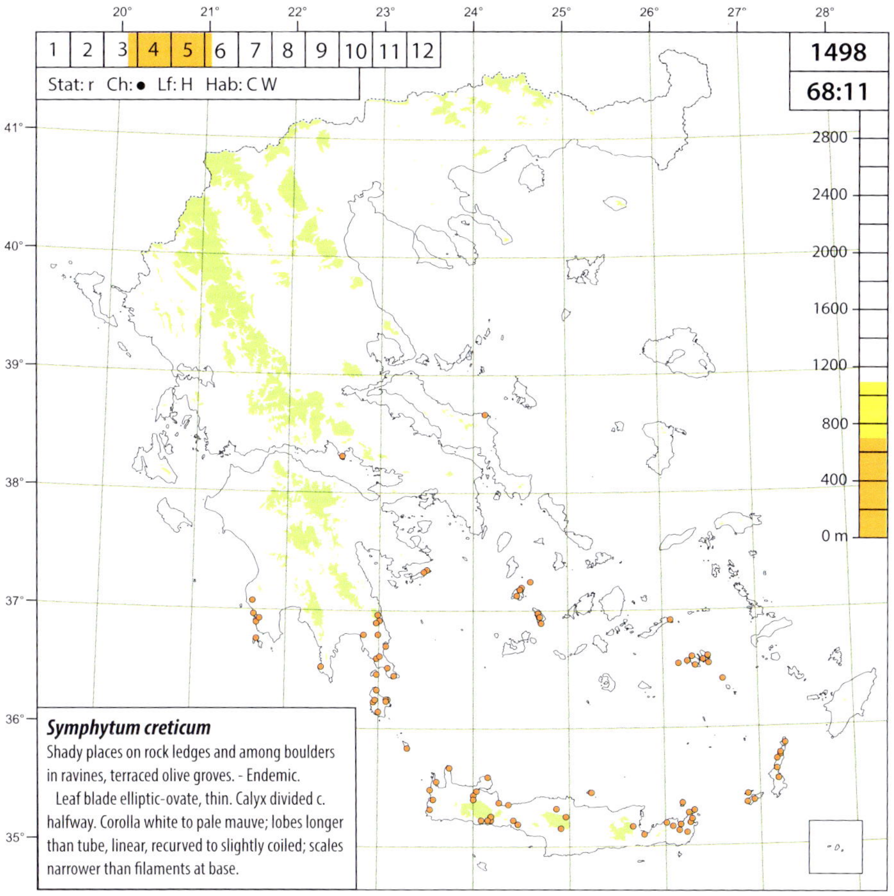

Symphytum creticum

Shady places on rock ledges and among boulders in ravines, terraced olive groves. - Endemic.

Leaf blade elliptic-ovate, thin. Calyx divided c. halfway. Corolla white to pale mauve; lobes longer than tube, linear, recurved to slightly coiled; scales narrower than filaments at base.

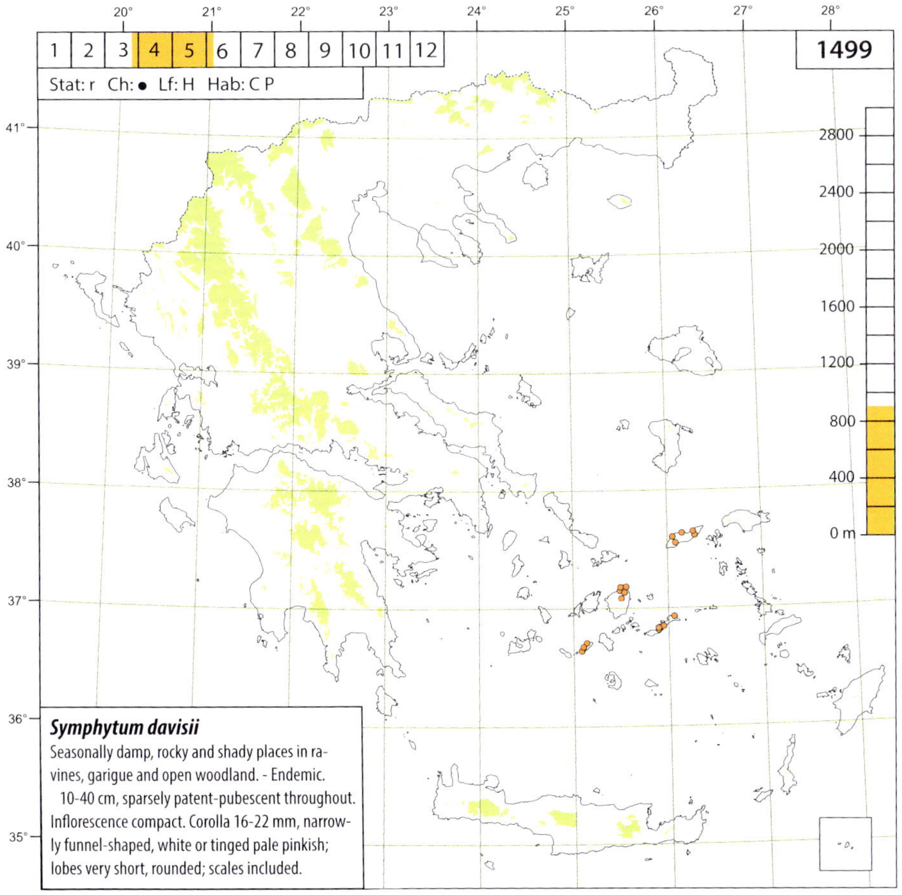

Symphytum davisii

Seasonally damp, rocky and shady places in ravines, garigue and open woodland. - Endemic.

10-40 cm, sparsely patent-pubescent throughout. Inflorescence compact. Corolla 16-22 mm, narrowly funnel-shaped, white or tinged pale pinkish; lobes very short, rounded; scales included.

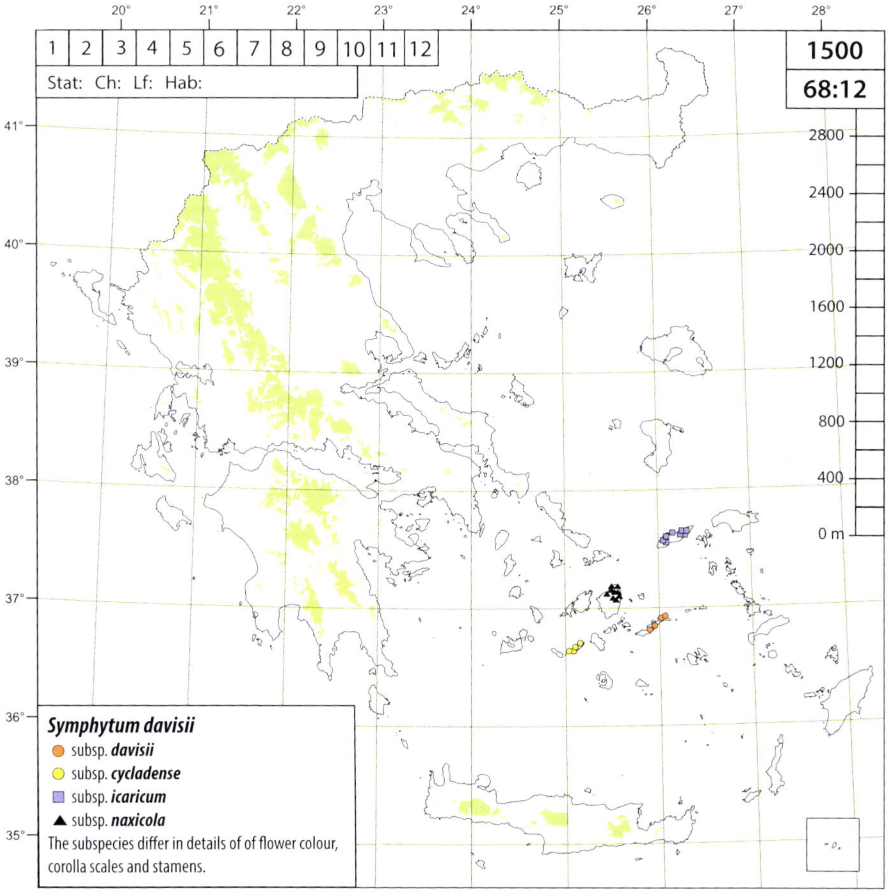

Symphytum davisii

- ● subsp. ***davisii***
- ● subsp. ***cycladense***
- ■ subsp. ***icaricum***
- ▲ subsp. ***naxicola***

The subspecies differ in details of of flower colour, corolla scales and stamens.

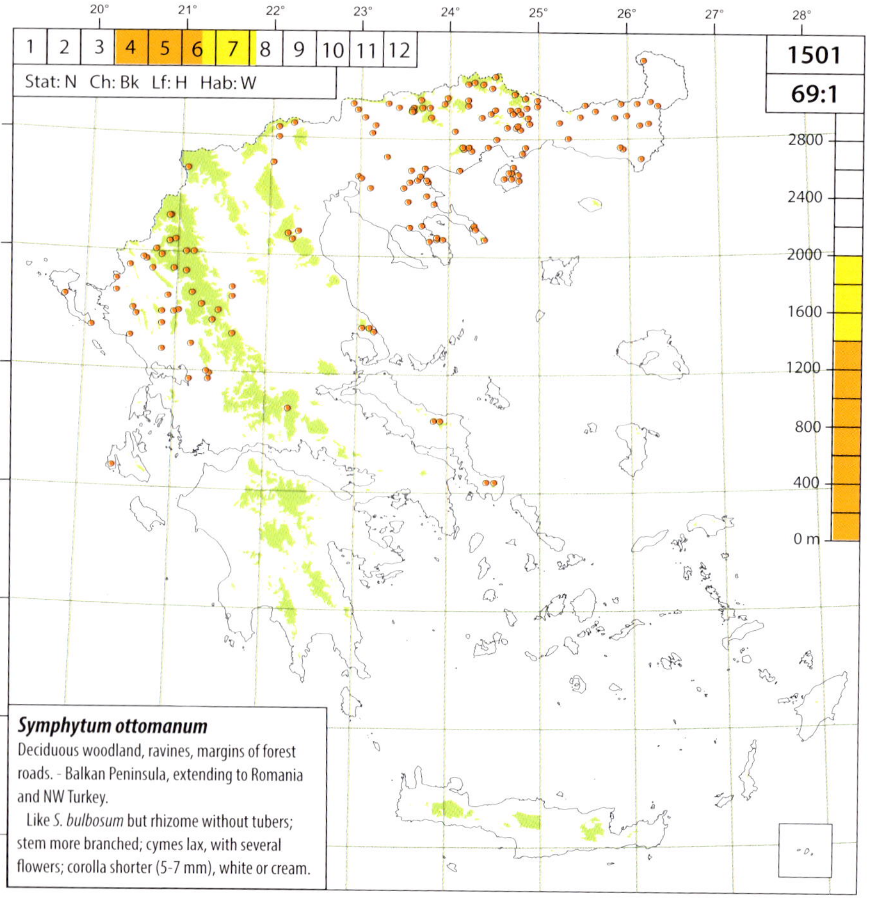

Symphytum ottomanum
Deciduous woodland, ravines, margins of forest roads. - Balkan Peninsula, extending to Romania and NW Turkey.
Like *S. bulbosum* but rhizome without tubers; stem more branched; cymes lax, with several flowers; corolla shorter (5-7 mm), white or cream.

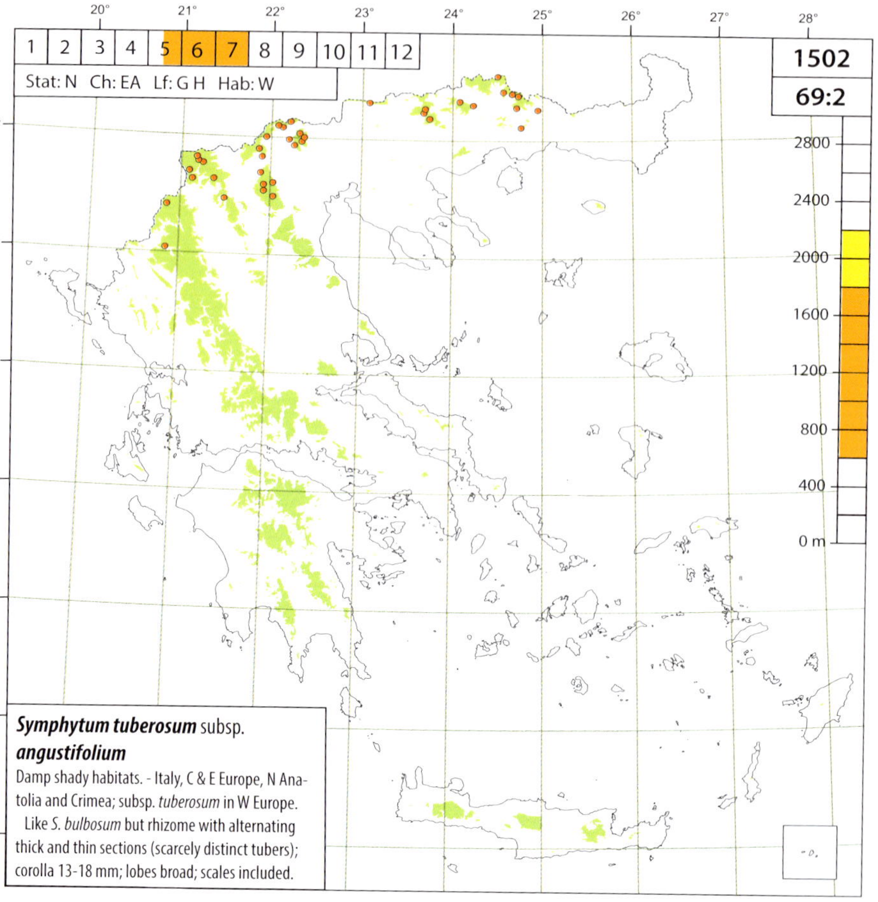

Symphytum tuberosum subsp. ***angustifolium***
Damp shady habitats. - Italy, C & E Europe, N Anatolia and Crimea; subsp. *tuberosum* in W Europe.
Like *S. bulbosum* but rhizome with alternating thick and thin sections (scarcely distinct tubers); corolla 13-18 mm; lobes broad; scales included.

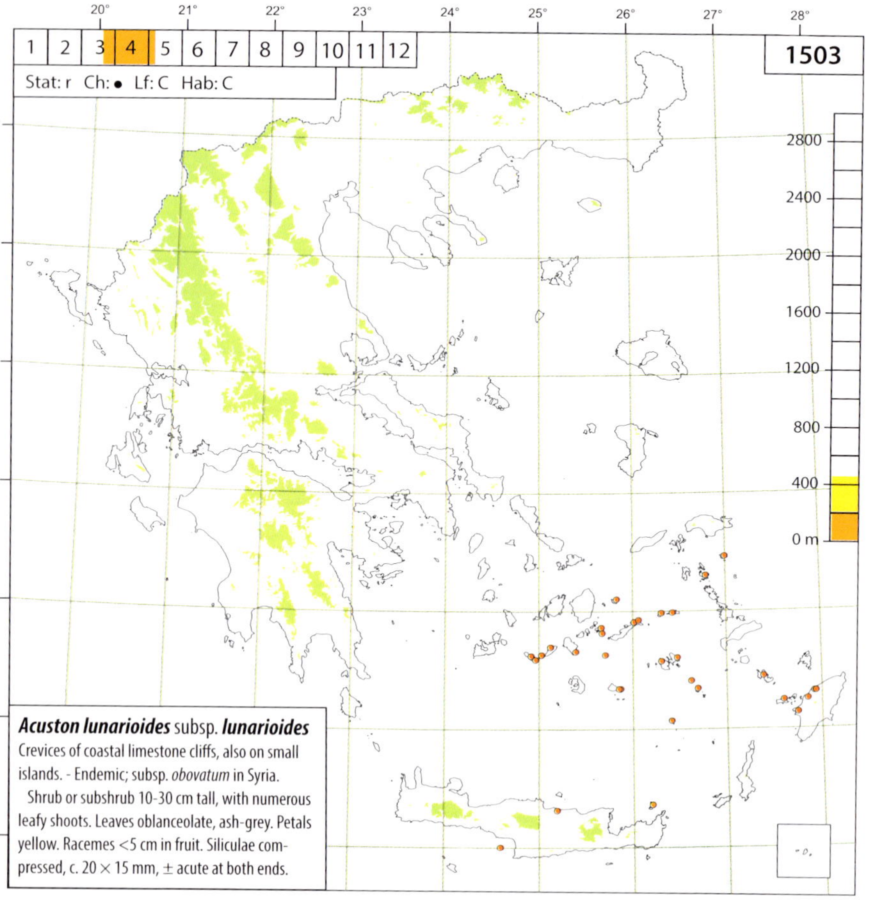

Acuston lunarioides subsp. ***lunarioides***
Crevices of coastal limestone cliffs, also on small islands. - Endemic; subsp. *obovatum* in Syria.
Shrub or subshrub 10-30 cm tall, with numerous leafy shoots. Leaves oblanceolate, ash-grey. Petals yellow. Racemes <5 cm in fruit. Siliculae compressed, c. 20 × 15 mm, ± acute at both ends.

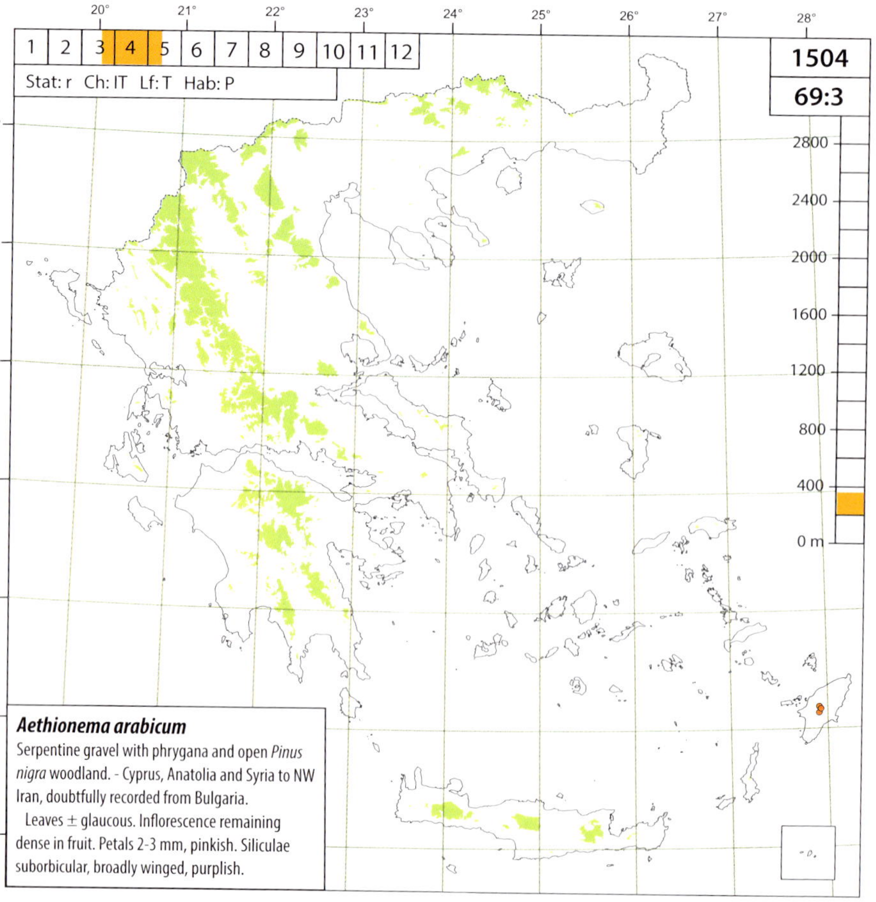

Aethionema arabicum
Serpentine gravel with phrygana and open *Pinus nigra* woodland. - Cyprus, Anatolia and Syria to NW Iran, doubtfully recorded from Bulgaria.
Leaves ± glaucous. Inflorescence remaining dense in fruit. Petals 2-3 mm, pinkish. Silicule suborbicular, broadly winged, purplish.

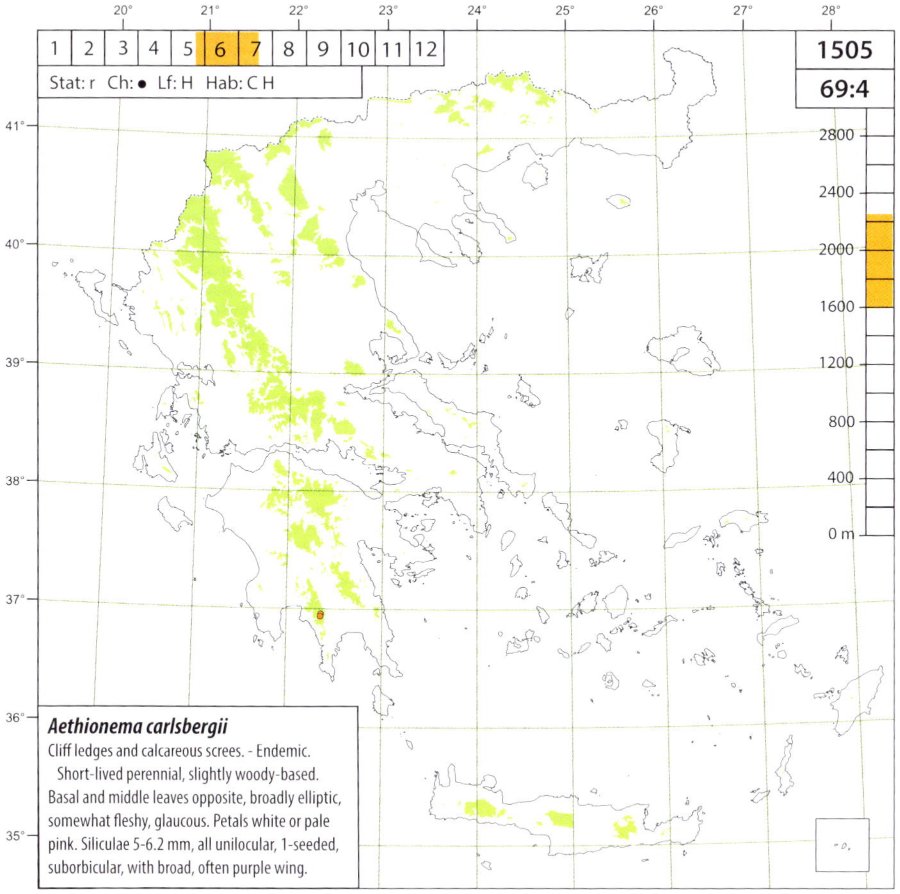

Aethionema carlsbergii

Cliff ledges and calcareous screes. - Endemic.

Short-lived perennial, slightly woody-based. Basal and middle leaves opposite, broadly elliptic, somewhat fleshy, glaucous. Petals white or pale pink. Siliculae 5-6.2 mm, all unilocular, 1-seeded, suborbicular, with broad, often purple wing.

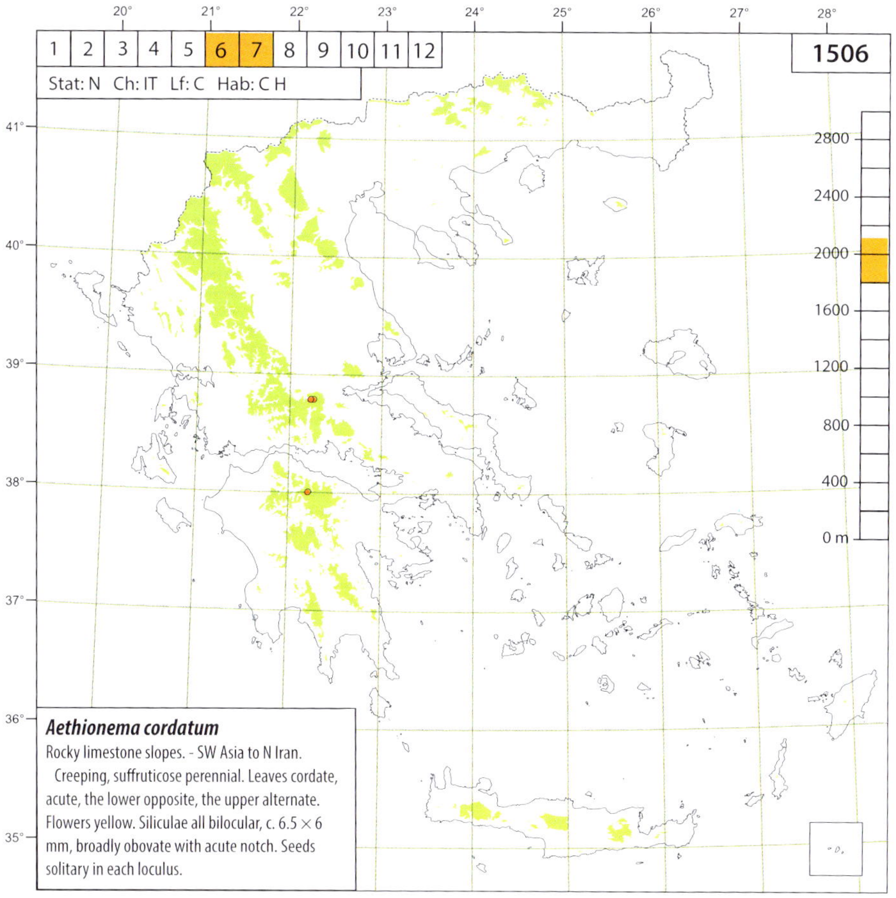

Aethionema cordatum

Rocky limestone slopes. - SW Asia to N Iran.

Creeping, suffruticose perennial. Leaves cordate, acute, the lower opposite, the upper alternate. Flowers yellow. Siliculae all bilocular, c. 6.5 × 6 mm, broadly obovate with acute notch. Seeds solitary in each loculus.

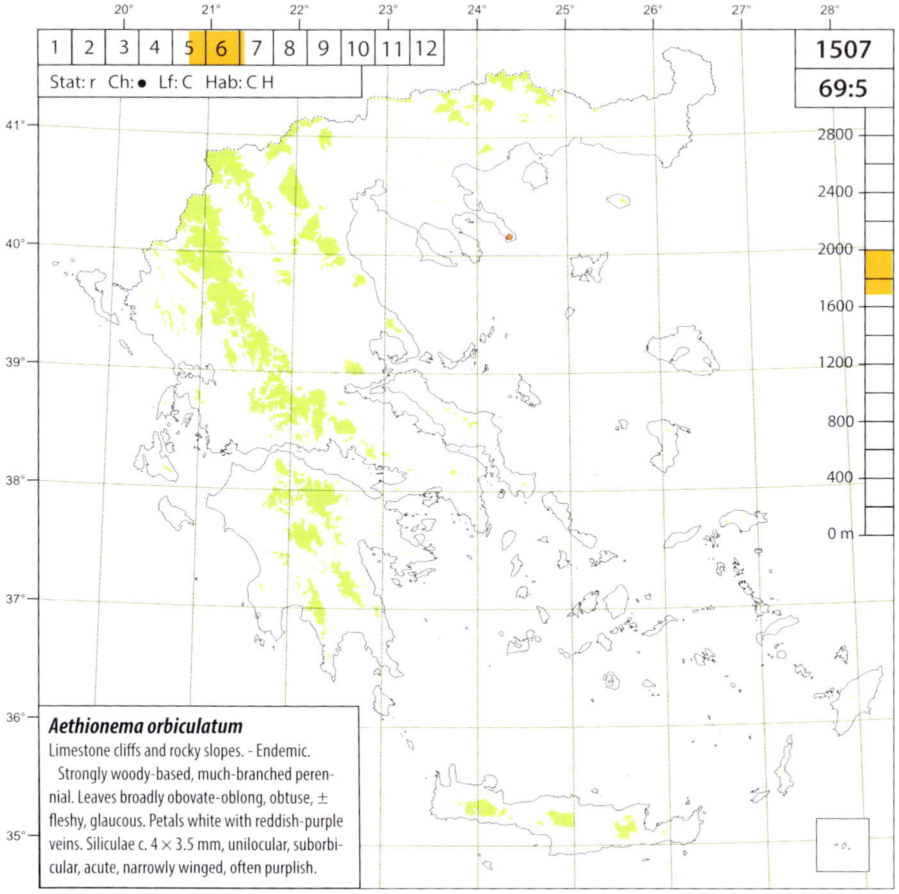

Aethionema orbiculatum

Limestone cliffs and rocky slopes. - Endemic.

Strongly woody-based, much-branched perennial. Leaves broadly obovate-oblong, obtuse, ± fleshy, glaucous. Petals white with reddish-purple veins. Siliculae c. 4 × 3.5 mm, unilocular, suborbicular, acute, narrowly winged, often purplish.

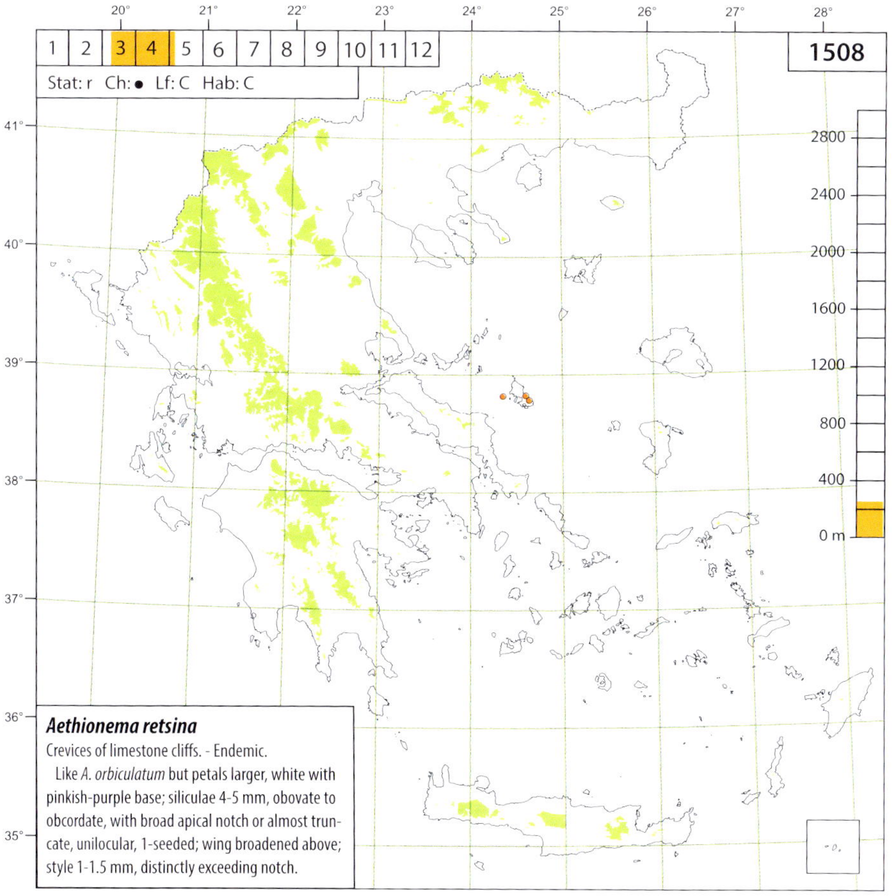

Aethionema retsina

Crevices of limestone cliffs. - Endemic.

Like *A. orbiculatum* but petals larger, white with pinkish-purple base; siliculae 4-5 mm, obovate to obcordate, with broad apical notch or almost truncate, unilocular, 1-seeded; wing broadened above; style 1-1.5 mm, distinctly exceeding notch.

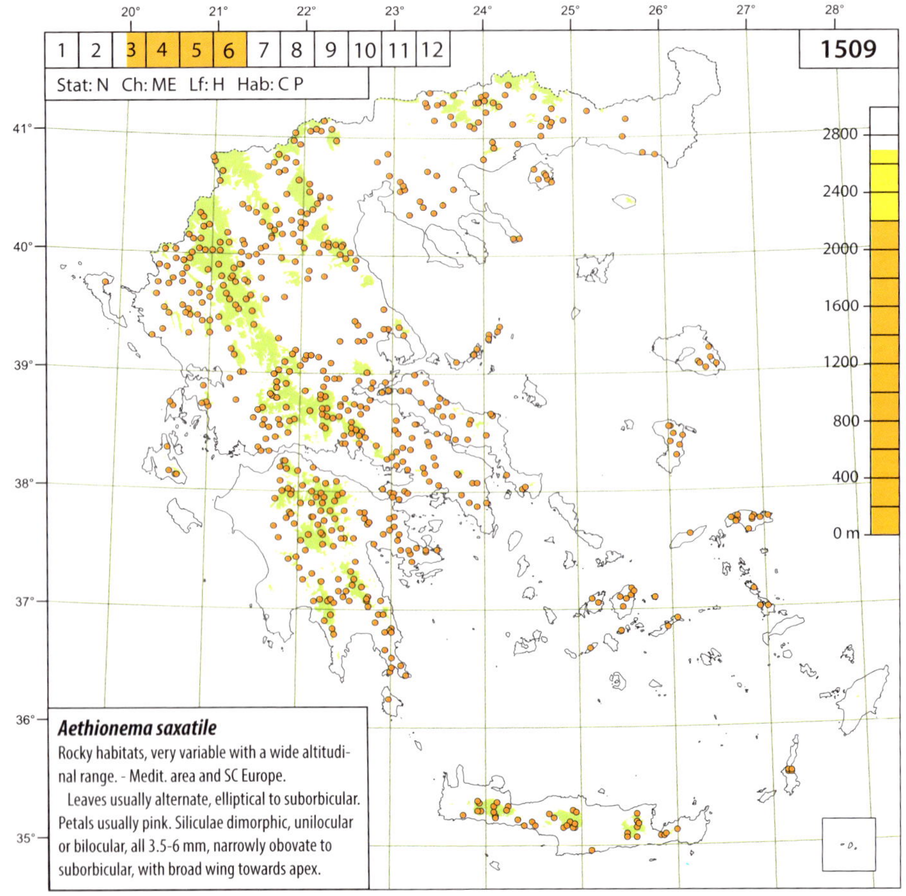

Aethionema saxatile
Rocky habitats, very variable with a wide altitudinal range. - Medit. area and SC Europe.
Leaves usually alternate, elliptical to suborbicular. Petals usually pink. Siliculae dimorphic, unilocular or bilocular, all 3.5-6 mm, narrowly obovate to suborbicular, with broad wing towards apex.

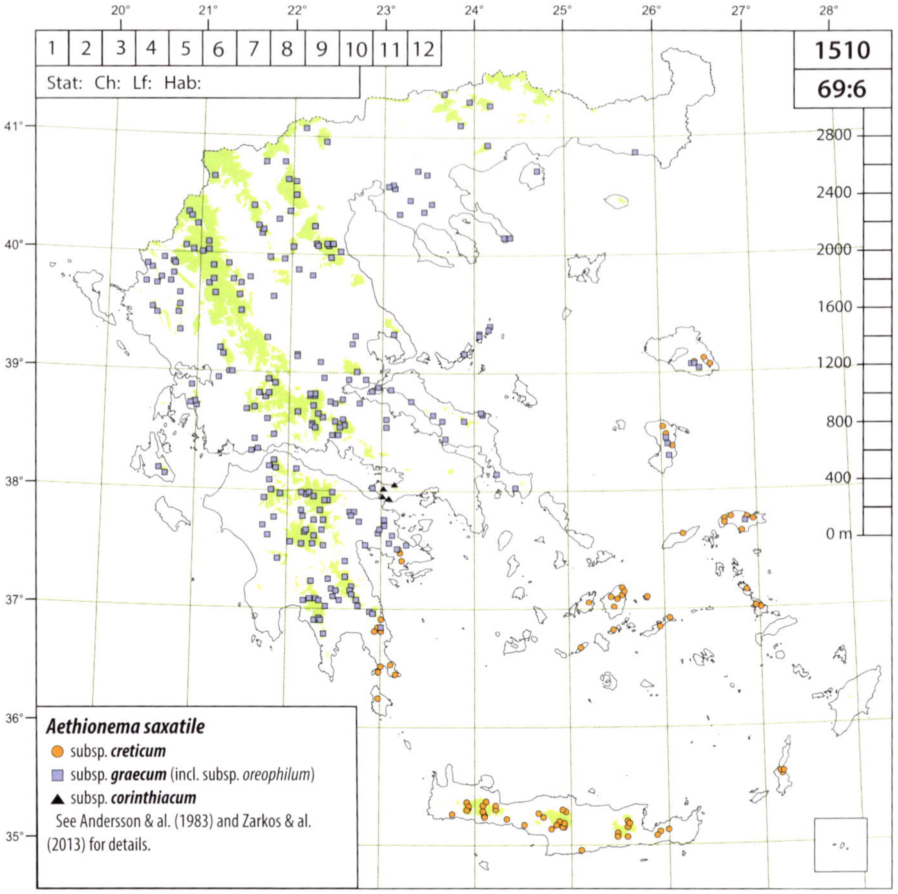

Aethionema saxatile
subsp. ***creticum***
subsp. ***graecum*** (incl. subsp. *oreophilum*)
subsp. ***corinthiacum***
See Andersson & al. (1983) and Zarkos & al. (2013) for details.

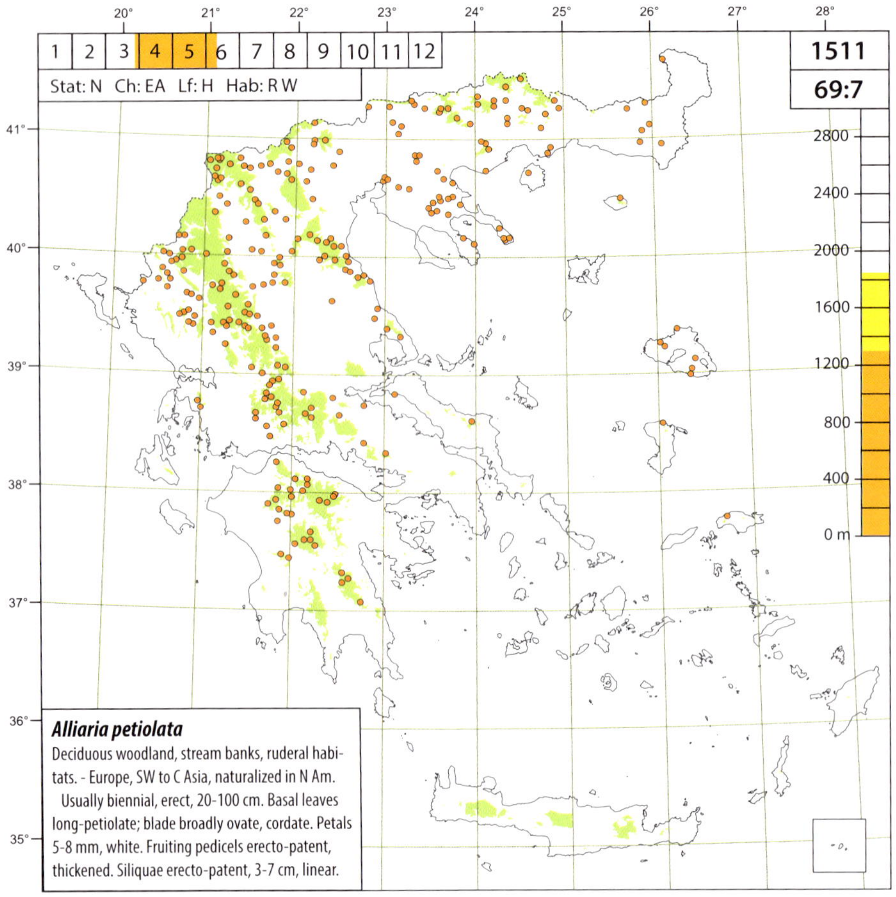

Alliaria petiolata
Deciduous woodland, stream banks, ruderal habitats. - Europe, SW to C Asia, naturalized in N Am.
Usually biennial, erect, 20-100 cm. Basal leaves long-petiolate; blade broadly ovate, cordate. Petals 5-8 mm, white. Fruiting pedicels erecto-patent, thickened. Siliquae erecto-patent, 3-7 cm, linear.

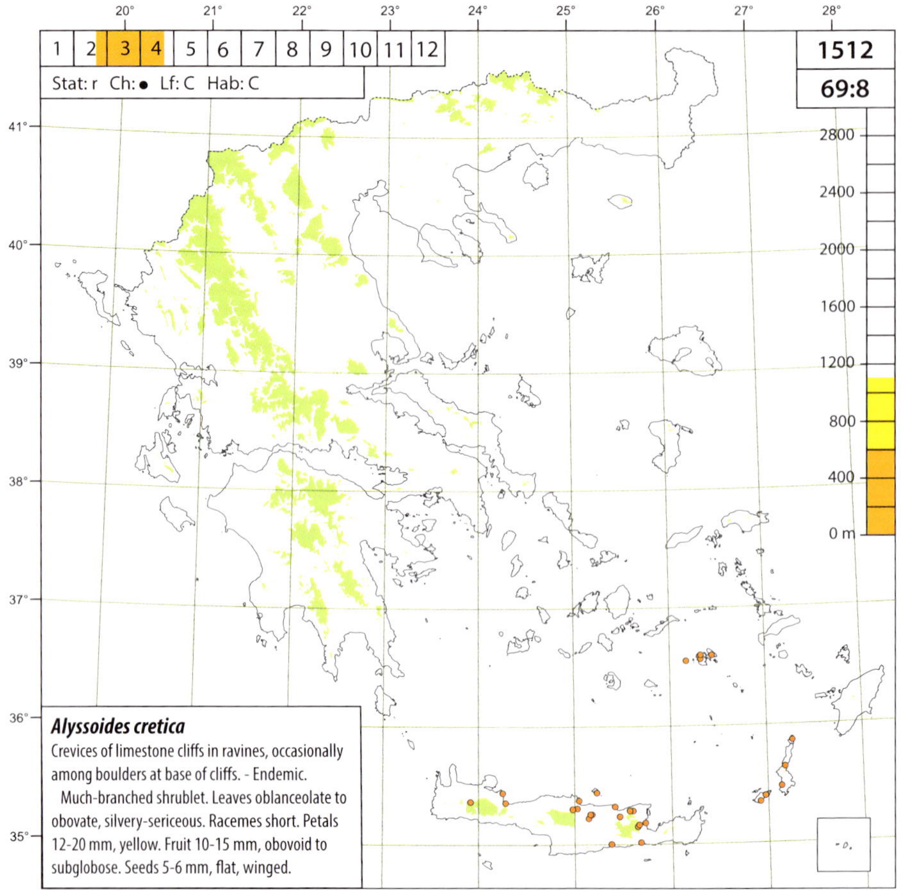

Alyssoides cretica
Crevices of limestone cliffs in ravines, occasionally among boulders at base of cliffs. - Endemic.
Much-branched shrublet. Leaves oblanceolate to obovate, silvery-sericeous. Racemes short. Petals 12-20 mm, yellow. Fruit 10-15 mm, obovoid to subglobose. Seeds 5-6 mm, flat, winged.

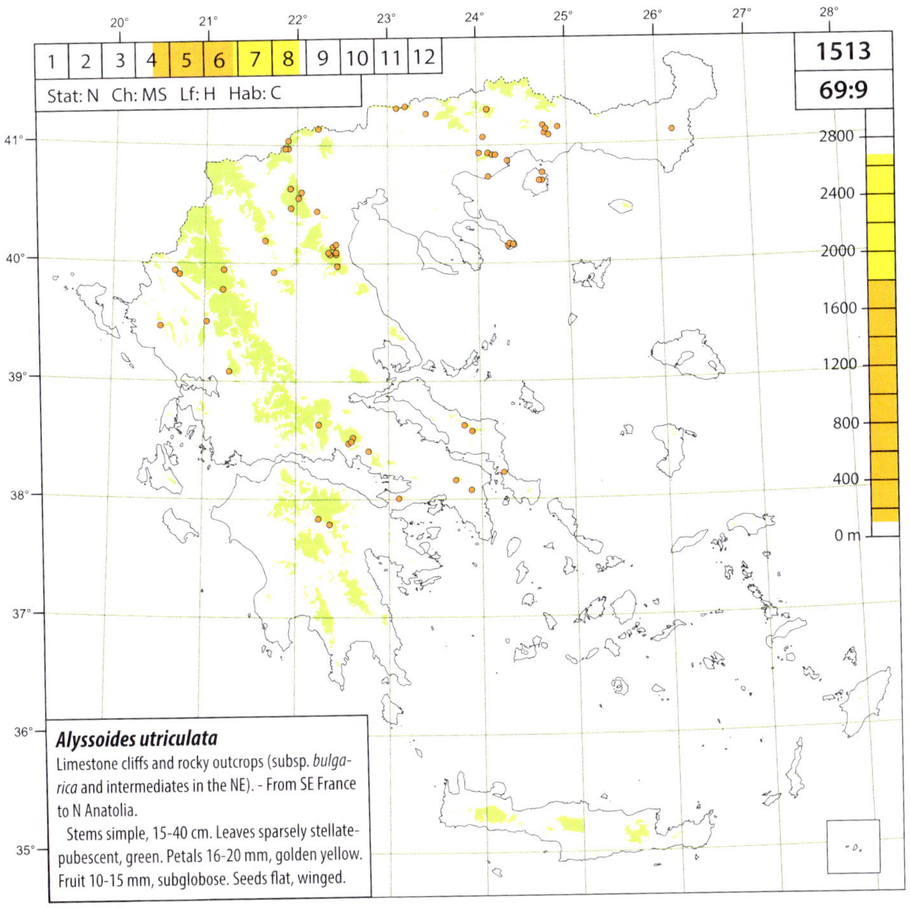

Alyssoides utriculata

Limestone cliffs and rocky outcrops (subsp. *bulgarica* and intermediates in the NE). - From SE France to N Anatolia.

Stems simple, 15-40 cm. Leaves sparsely stellate-pubescent, green. Petals 16-20 mm, golden yellow. Fruit 10-15 mm, subglobose. Seeds flat, winged.

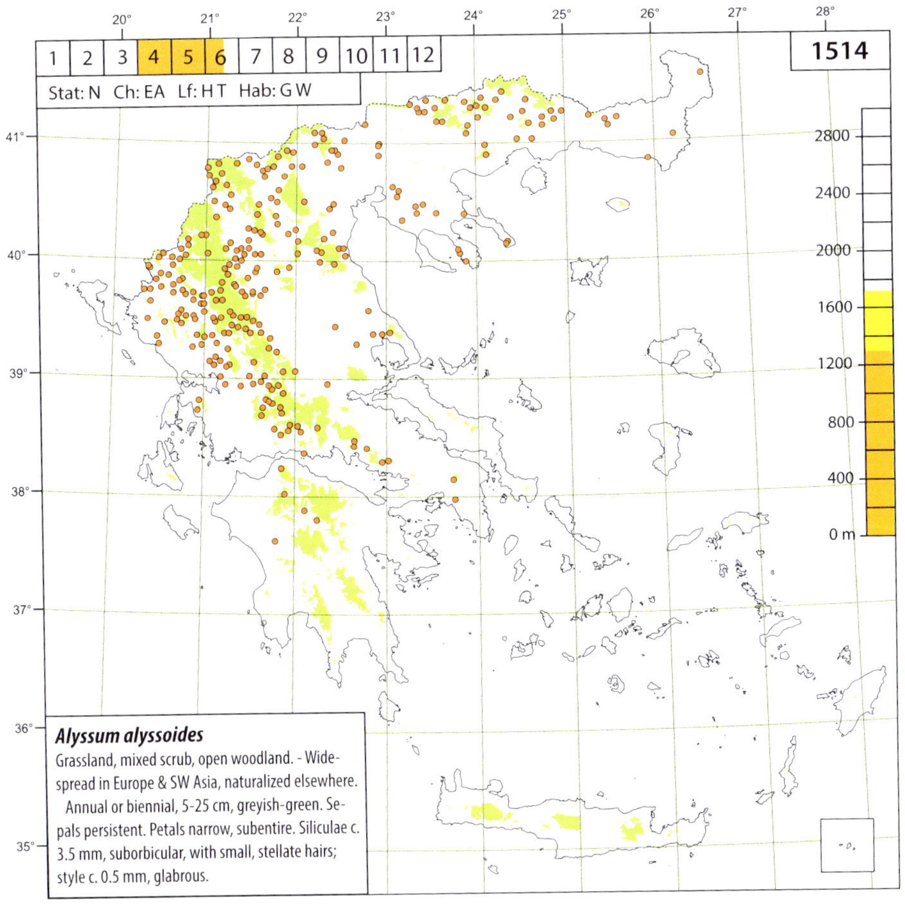

Alyssum alyssoides

Grassland, mixed scrub, open woodland. - Widespread in Europe & SW Asia, naturalized elsewhere.

Annual or biennial, 5-25 cm, greyish-green. Sepals persistent. Petals narrow, subentire. Siliculae c. 3.5 mm, suborbicular, with small, stellate hairs; style c. 0.5 mm, glabrous.

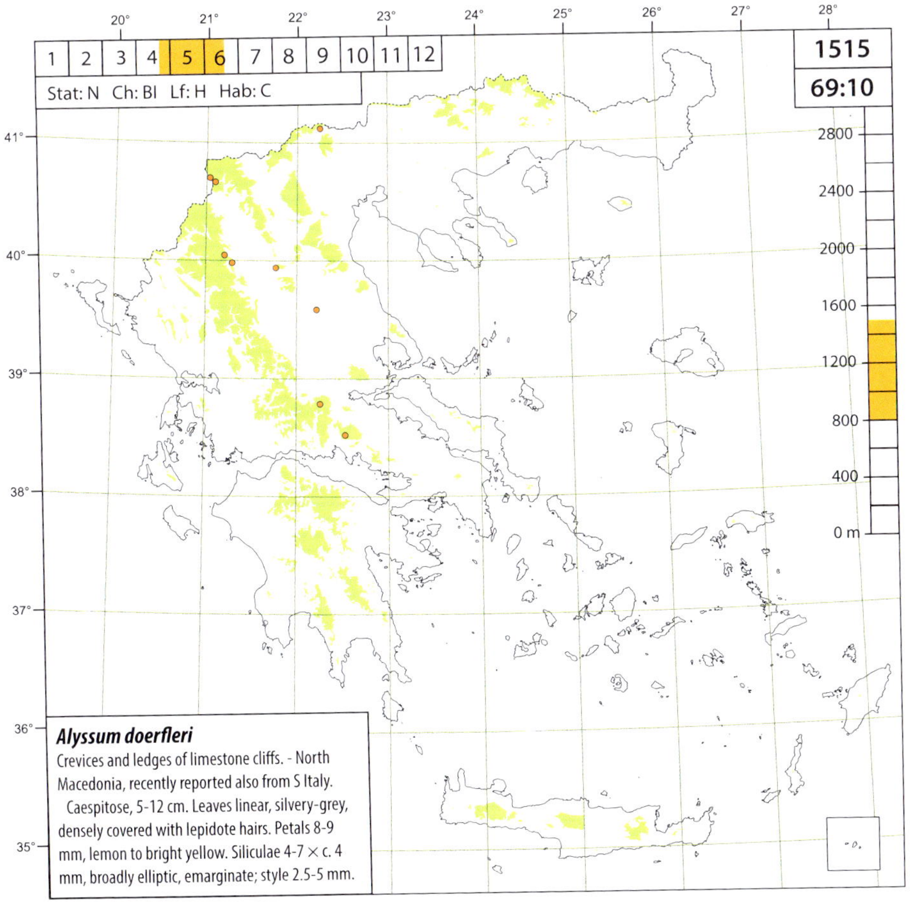

Alyssum doerfleri

Crevices and ledges of limestone cliffs. - North Macedonia, recently reported also from S Italy.

Caespitose, 5-12 cm. Leaves linear, silvery-grey, densely covered with lepidote hairs. Petals 8-9 mm, lemon to bright yellow. Siliculae 4-7 × c. 4 mm, broadly elliptic, emarginate; style 2.5-5 mm.

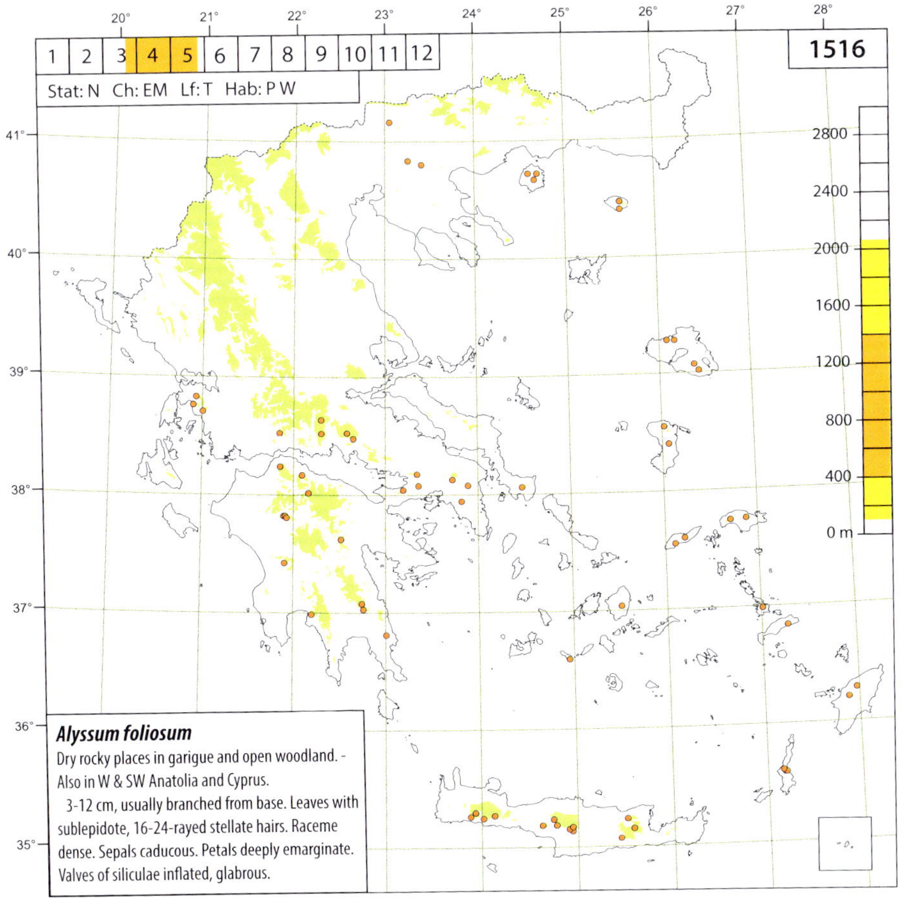

Alyssum foliosum

Dry rocky places in garigue and open woodland. - Also in W & SW Anatolia and Cyprus.

3-12 cm, usually branched from base. Leaves with sublepidote, 16-24-rayed stellate hairs. Raceme dense. Sepals caducous. Petals deeply emarginate. Valves of siliculae inflated, glabrous.

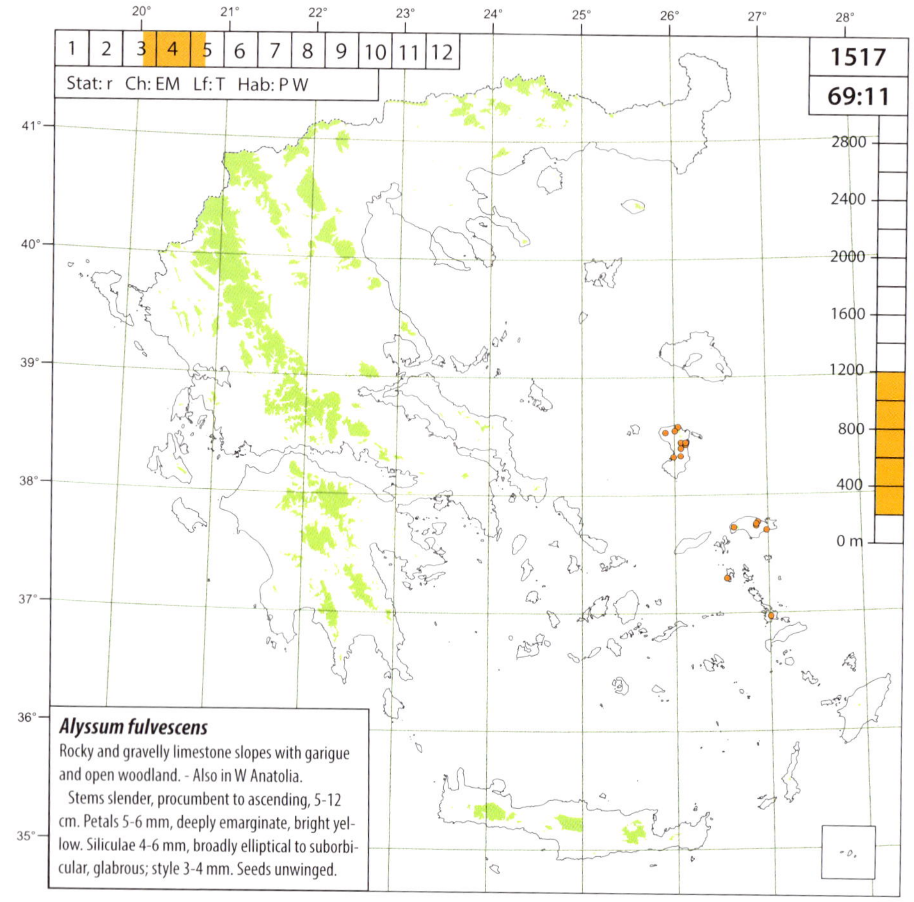

Alyssum fulvescens

Rocky and gravelly limestone slopes with garigue and open woodland. - Also in W Anatolia.

Stems slender, procumbent to ascending, 5-12 cm. Petals 5-6 mm, deeply emarginate, bright yellow. Siliculae 4-6 mm, broadly elliptical to suborbicular, glabrous; style 3-4 mm. Seeds unwinged.

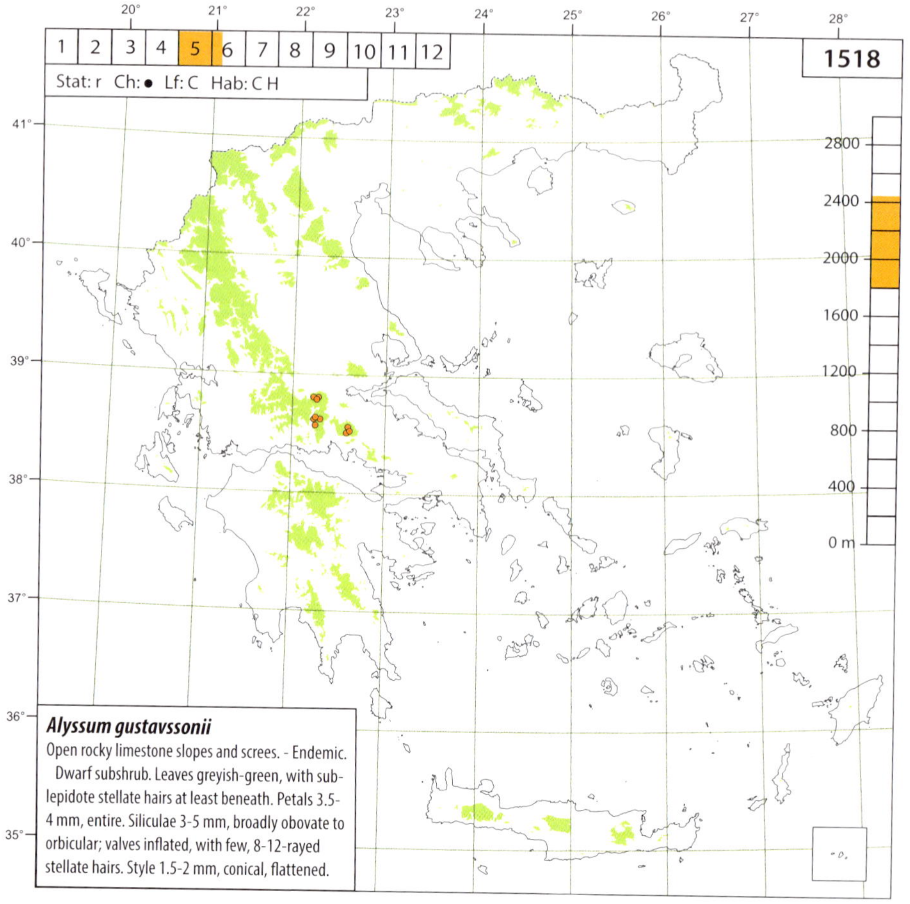

Alyssum gustavssonii

Open rocky limestone slopes and screes. - Endemic.

Dwarf subshrub. Leaves greyish-green, with sub-lepidote stellate hairs at least beneath. Petals 3.5-4 mm, entire. Siliculae 3-5 mm, broadly obovate to orbicular; valves inflated, with few, 8-12-rayed stellate hairs. Style 1.5-2 mm, conical, flattened.

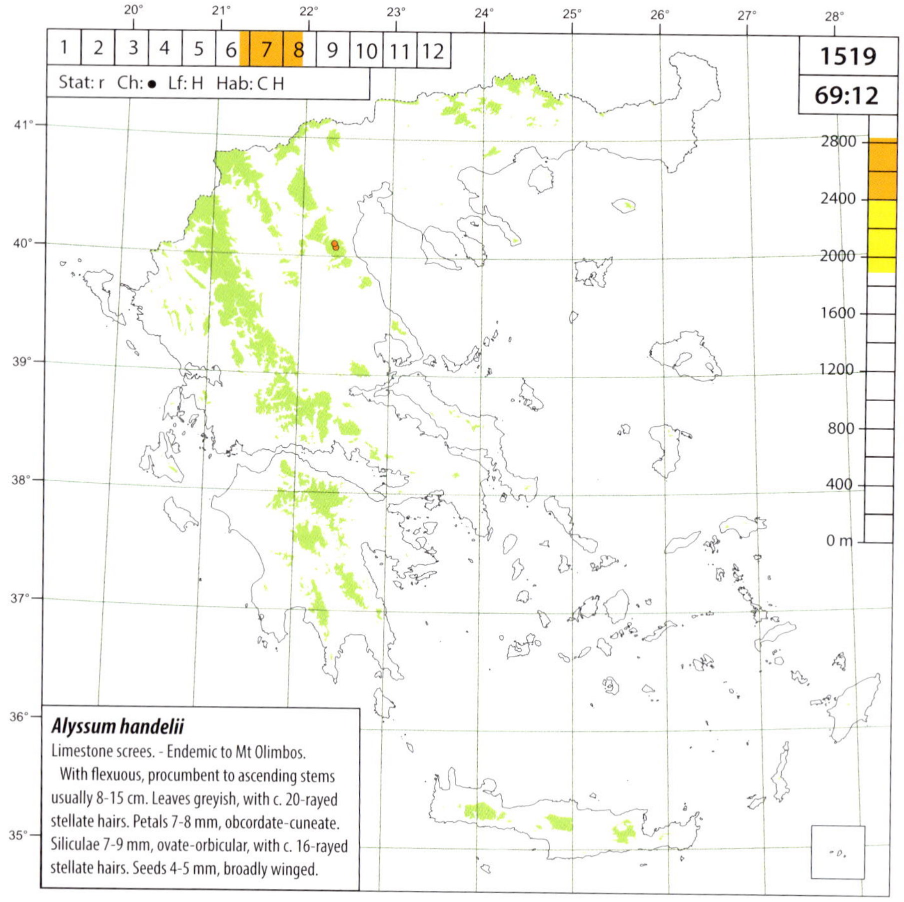

Alyssum handelii

Limestone screes. - Endemic to Mt Olimbos.

With flexuous, procumbent to ascending stems usually 8-15 cm. Leaves greyish, with c. 20-rayed stellate hairs. Petals 7-8 mm, obcordate-cuneate. Siliculae 7-9 mm, ovate-orbicular, with c. 16-rayed stellate hairs. Seeds 4-5 mm, broadly winged.

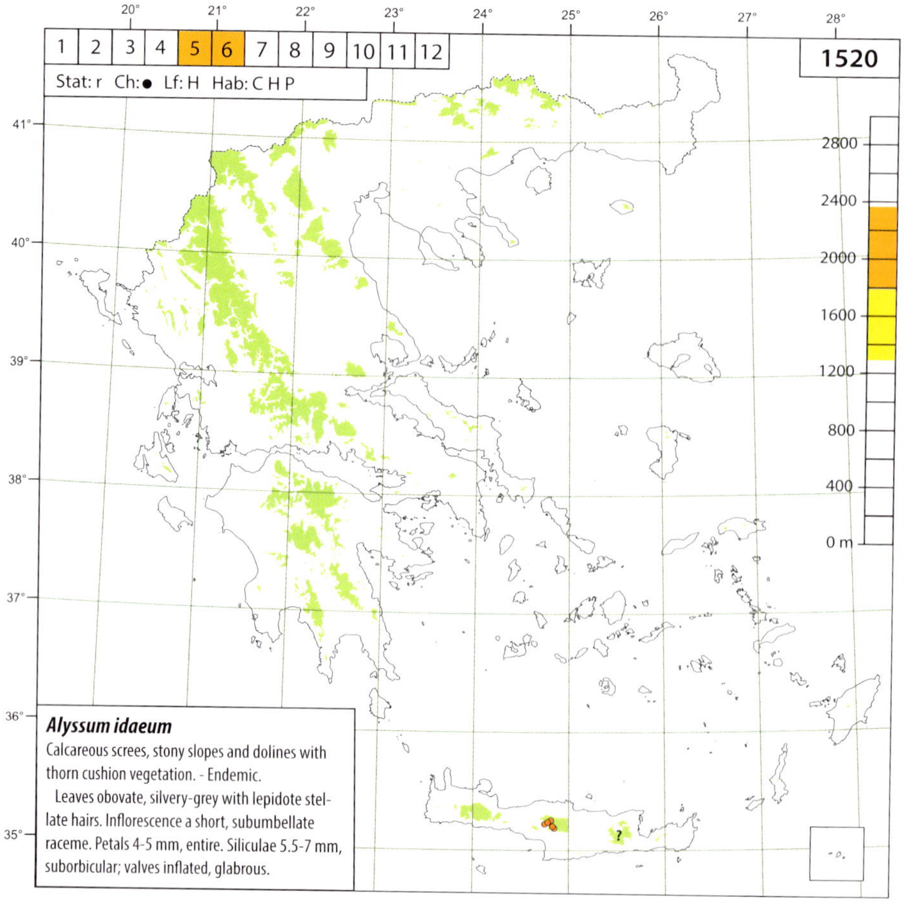

Alyssum idaeum

Calcareous screes, stony slopes and dolines with thorn cushion vegetation. - Endemic.

Leaves obovate, silvery-grey with lepidote stellate hairs. Inflorescence a short, subumbellate raceme. Petals 4-5 mm, entire. Siliculae 5.5-7 mm, suborbicular; valves inflated, glabrous.

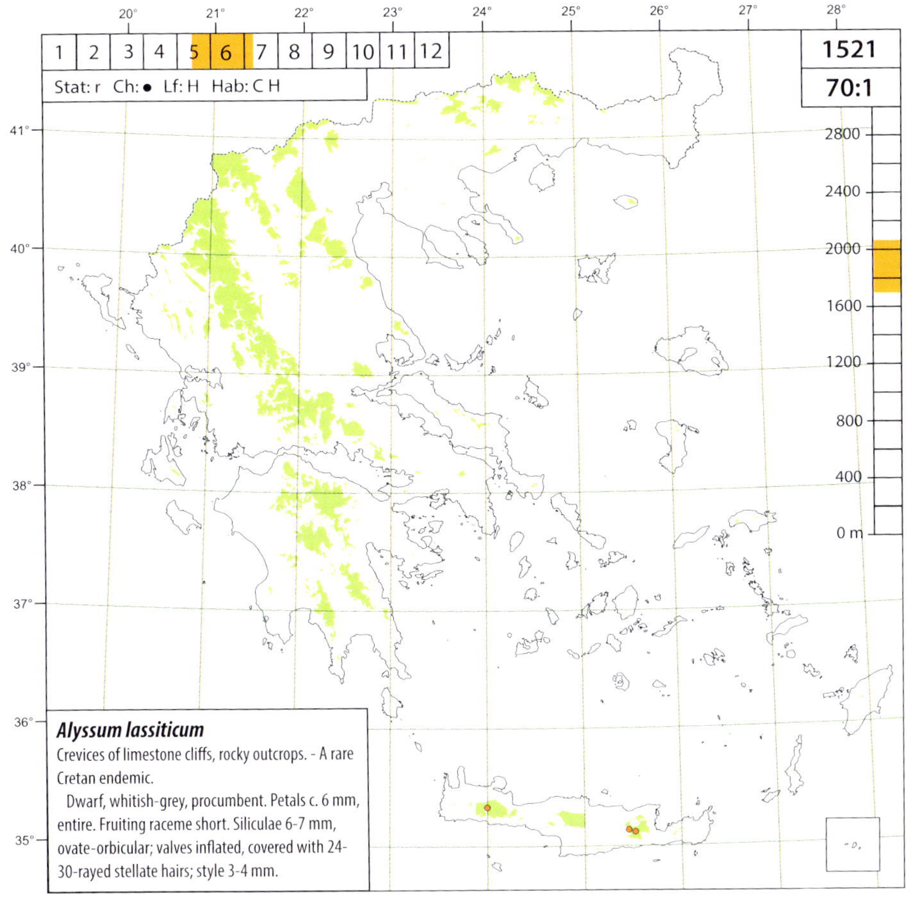

Alyssum lassiticum

Crevices of limestone cliffs, rocky outcrops. - A rare Cretan endemic.

Dwarf, whitish-grey, procumbent. Petals c. 6 mm, entire. Fruiting raceme short. Siliculae 6-7 mm, ovate-orbicular; valves inflated, covered with 24-30-rayed stellate hairs; style 3-4 mm.

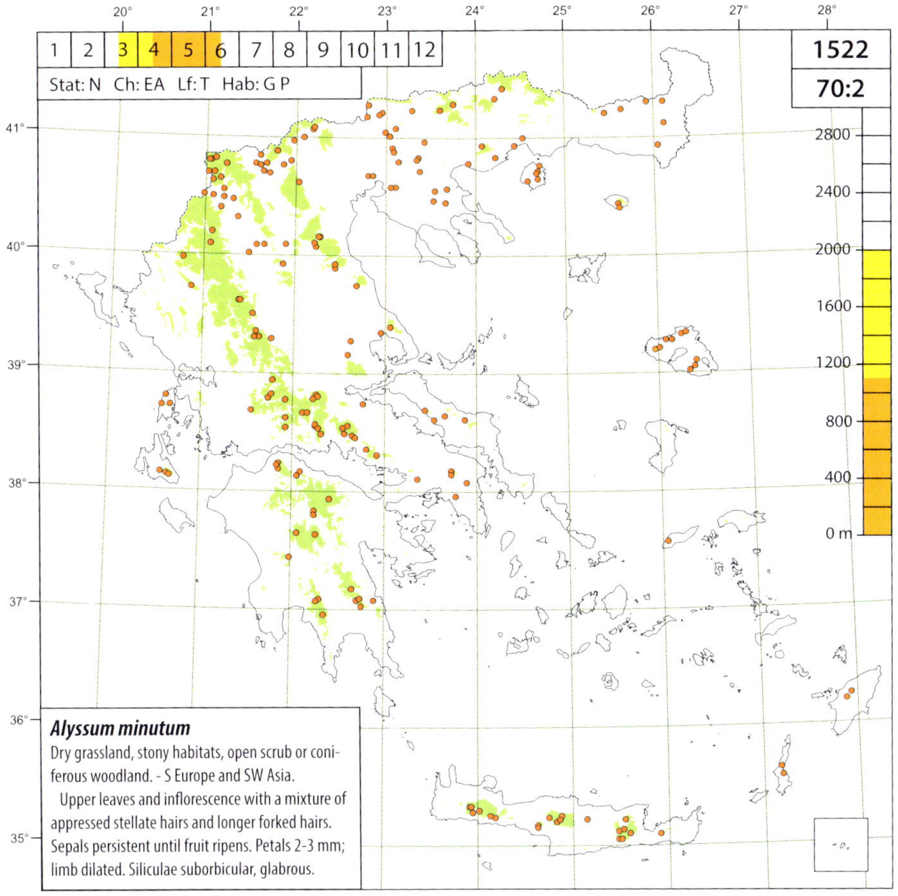

Alyssum minutum

Dry grassland, stony habitats, open scrub or coniferous woodland. - S Europe and SW Asia.

Upper leaves and inflorescence with a mixture of appressed stellate hairs and longer forked hairs. Sepals persistent until fruit ripens. Petals 2-3 mm; limb dilated. Siliculae suborbicular, glabrous.

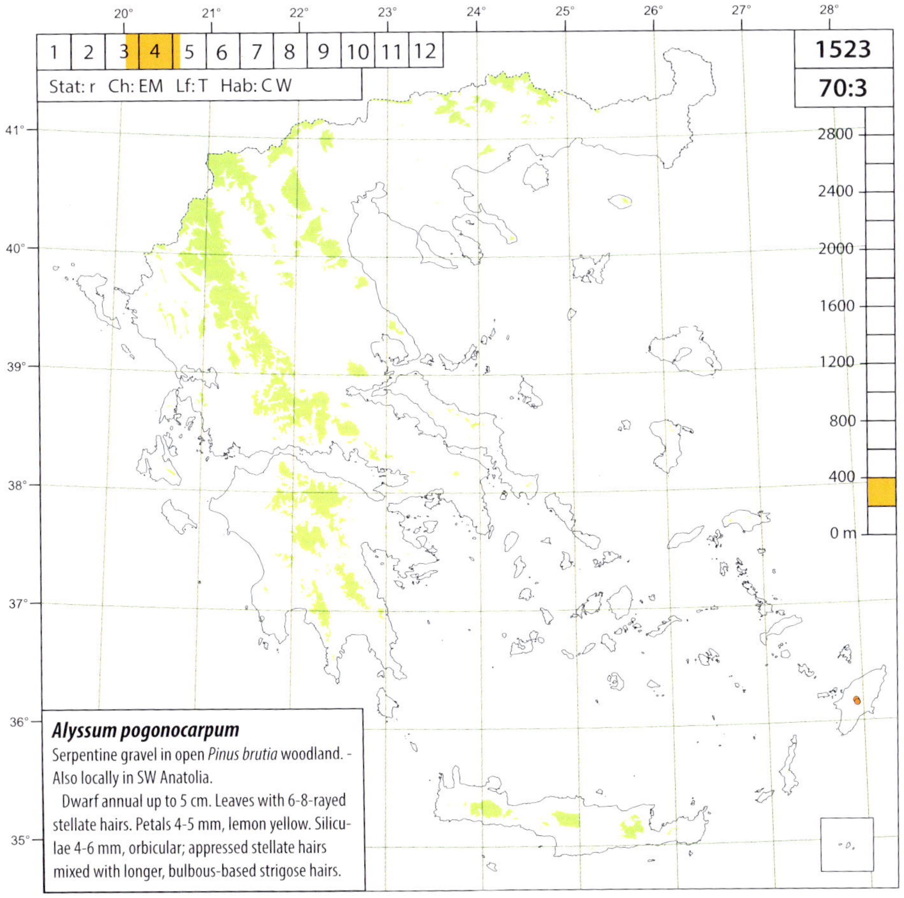

Alyssum pogonocarpum

Serpentine gravel in open *Pinus brutia* woodland. - Also locally in SW Anatolia.

Dwarf annual up to 5 cm. Leaves with 6-8-rayed stellate hairs. Petals 4-5 mm, lemon yellow. Siliculae 4-6 mm, orbicular; appressed stellate hairs mixed with longer, bulbous-based strigose hairs.

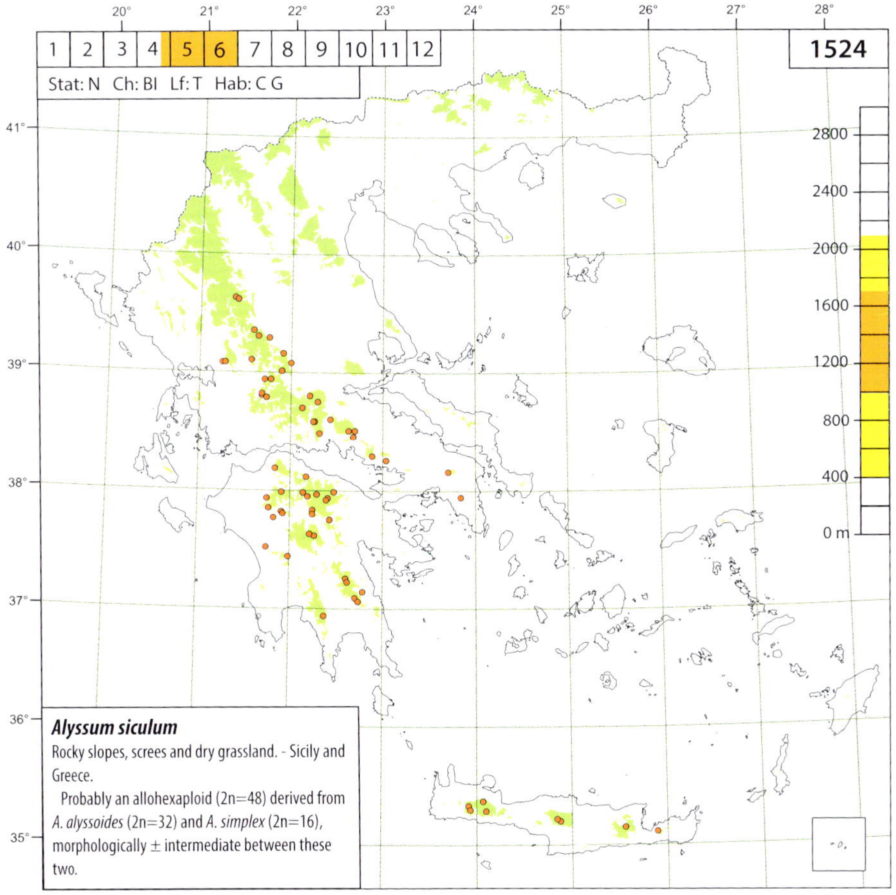

Alyssum siculum

Rocky slopes, screes and dry grassland. - Sicily and Greece.

Probably an allohexaploid (2n=48) derived from *A. alyssoides* (2n=32) and *A. simplex* (2n=16), morphologically ± intermediate between these two.

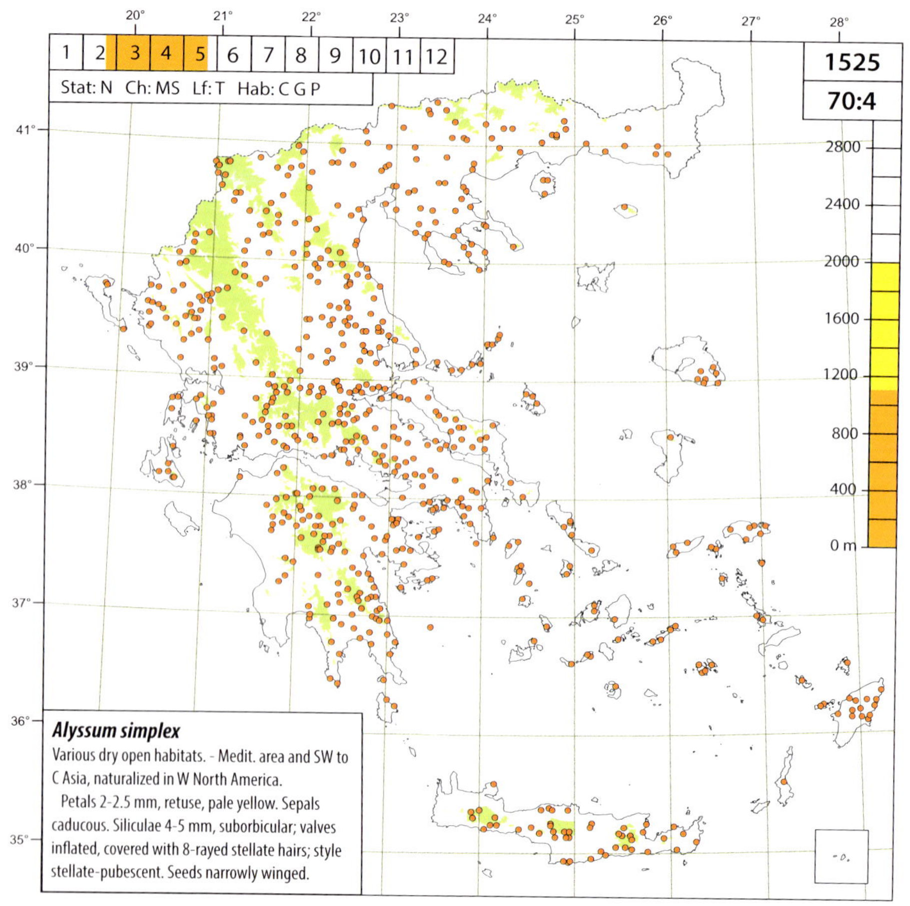

Alyssum simplex
Various dry open habitats. - Medit. area and SW to C Asia, naturalized in W North America.
Petals 2-2.5 mm, retuse, pale yellow. Sepals caducous. Siliculae 4-5 mm, suborbicular; valves inflated, covered with 8-rayed stellate hairs; style stellate-pubescent. Seeds narrowly winged.

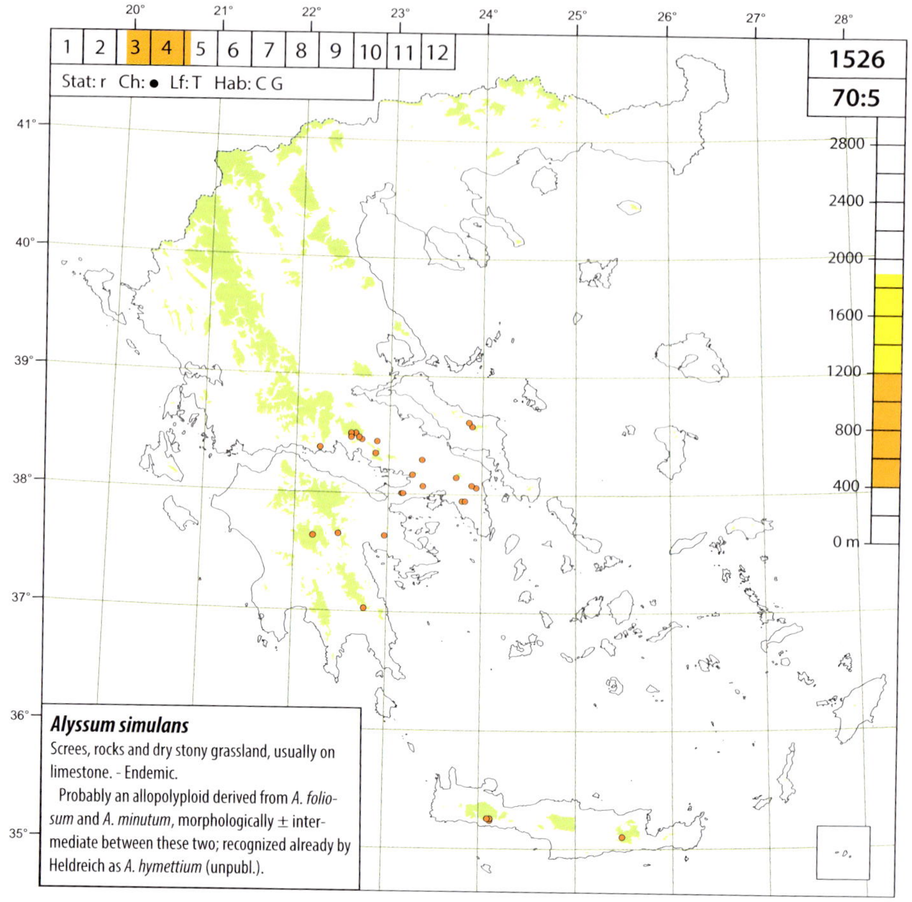

Alyssum simulans
Screes, rocks and dry stony grassland, usually on limestone. - Endemic.
Probably an allopolyploid derived from *A. foliosum* and *A. minutum*, morphologically ± intermediate between these two; recognized already by Heldreich as *A. hymettium* (unpubl.).

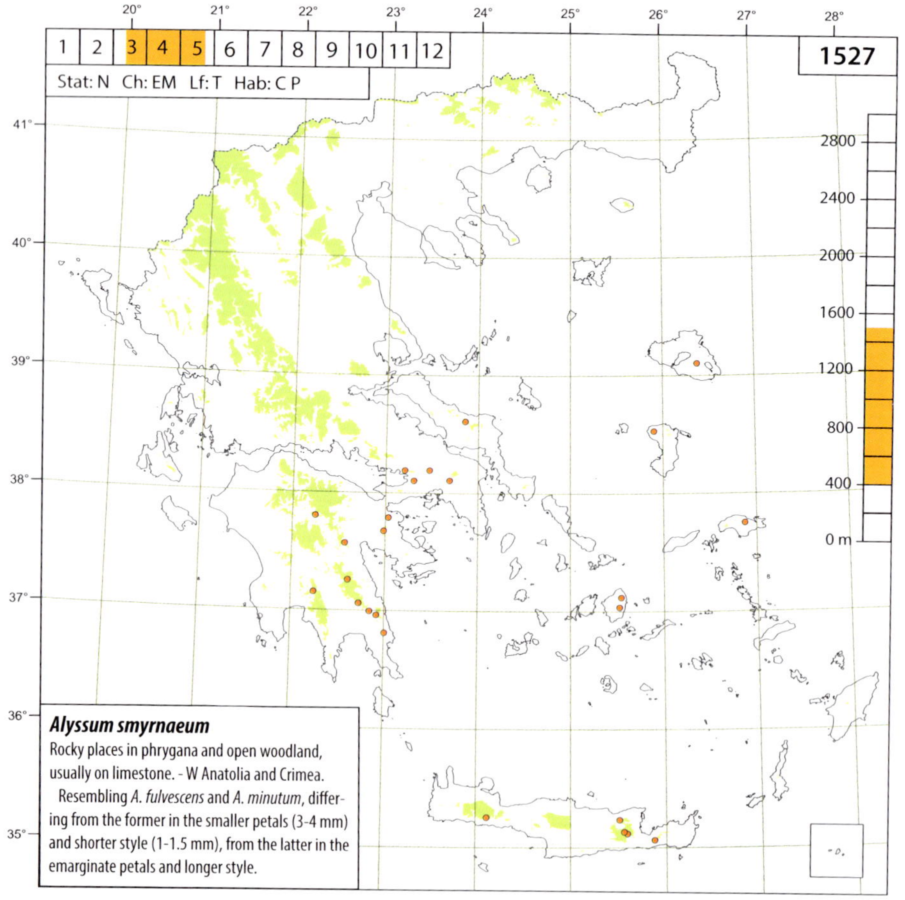

Alyssum smyrnaeum
Rocky places in phrygana and open woodland, usually on limestone. - W Anatolia and Crimea.
Resembling *A. fulvescens* and *A. minutum*, differing from the former in the smaller petals (3-4 mm) and shorter style (1-1.5 mm), from the latter in the emarginate petals and longer style.

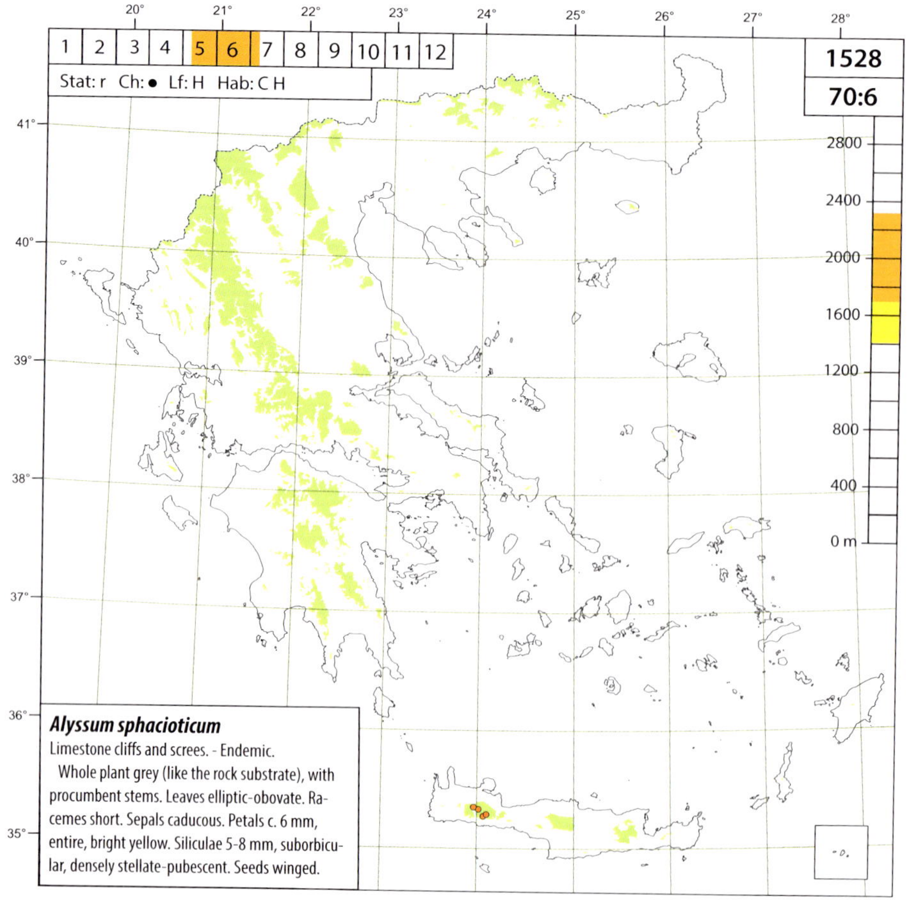

Alyssum sphacioticum
Limestone cliffs and screes. - Endemic.
Whole plant grey (like the rock substrate), with procumbent stems. Leaves elliptic-obovate. Racemes short. Sepals caducous. Petals c. 6 mm, entire, bright yellow. Siliculae 5-8 mm, suborbicular, densely stellate-pubescent. Seeds winged.

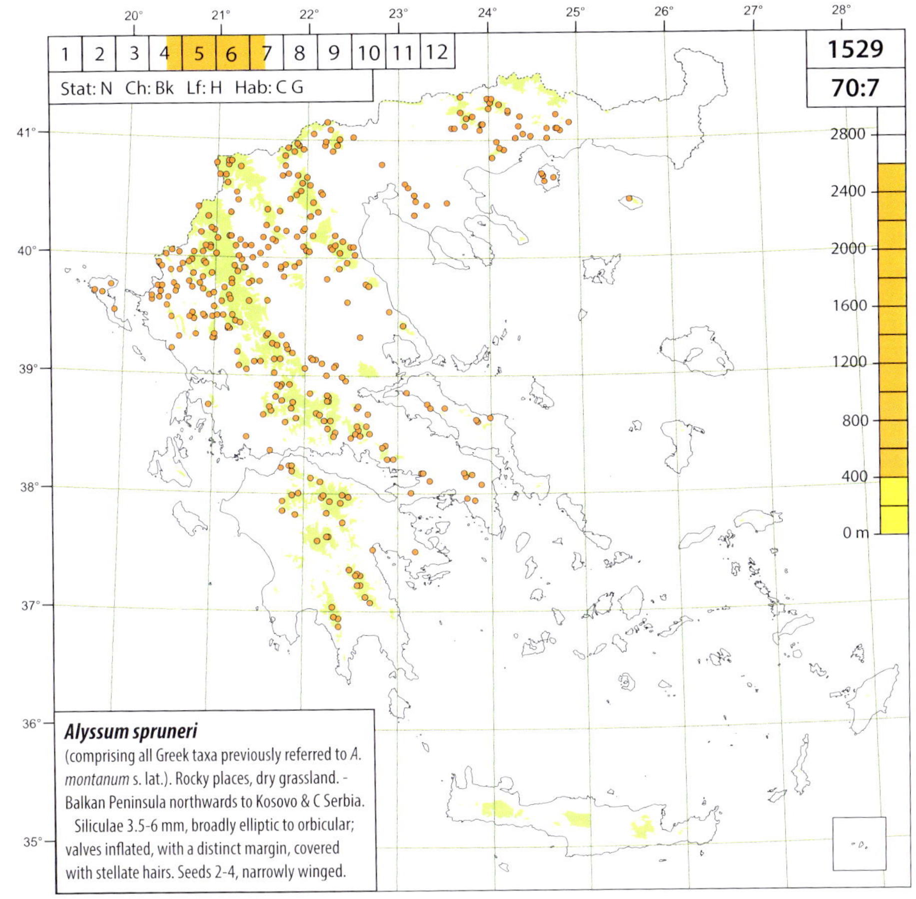

Alyssum spruneri

(comprising all Greek taxa previously referred to *A. montanum* s. lat.). Rocky places, dry grassland. - Balkan Peninsula northwards to Kosovo & C Serbia.

Siliculae 3.5-6 mm, broadly elliptic to orbicular; valves inflated, with a distinct margin, covered with stellate hairs. Seeds 2-4, narrowly winged.

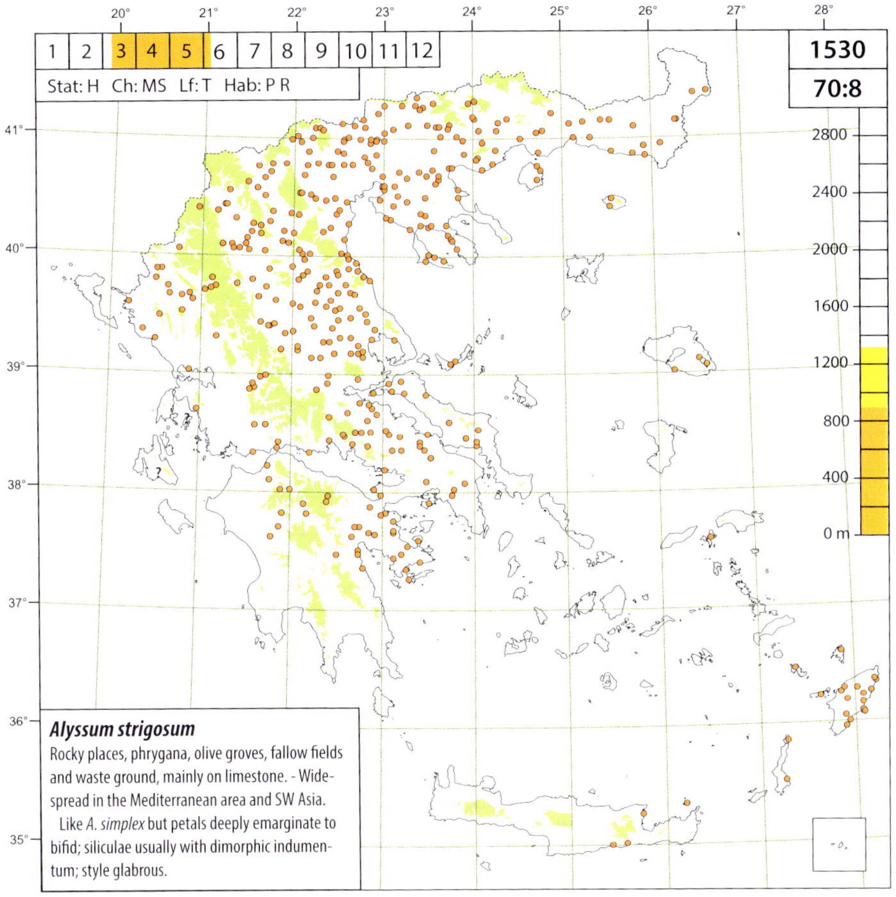

Alyssum strigosum

Rocky places, phrygana, olive groves, fallow fields and waste ground, mainly on limestone. - Widespread in the Mediterranean area and SW Asia.

Like *A. simplex* but petals deeply emarginate to bifid; siliculae usually with dimorphic indumentum; style glabrous.

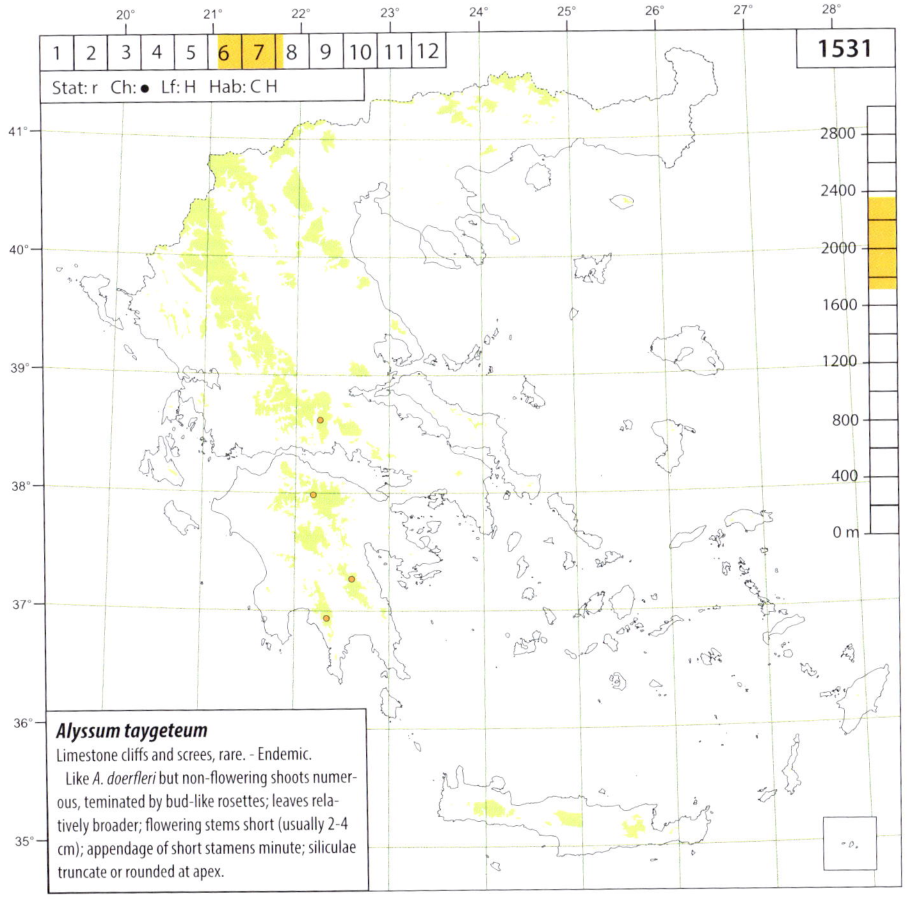

Alyssum taygeteum

Limestone cliffs and screes, rare. - Endemic.

Like *A. doerfleri* but non-flowering shoots numerous, teminated by bud-like rosettes; leaves relatively broader; flowering stems short (usually 2-4 cm); appendage of short stamens minute; siliculae truncate or rounded at apex.

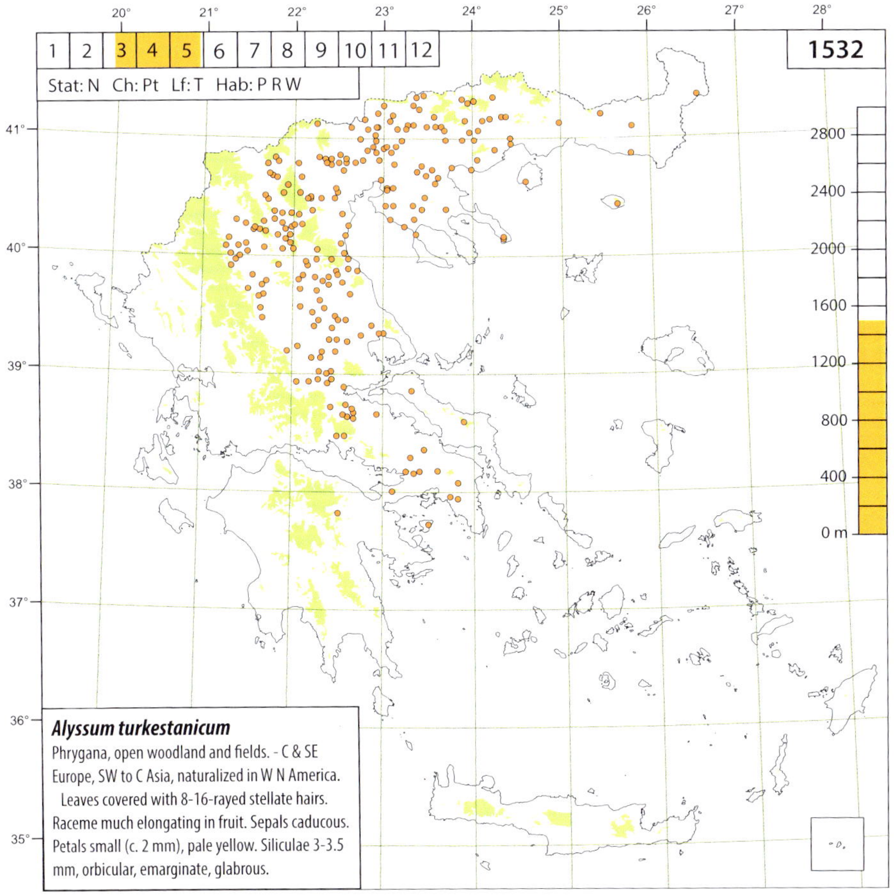

Alyssum turkestanicum

Phrygana, open woodland and fields. - C & SE Europe, SW to C Asia, naturalized in W N America.

Leaves covered with 8-16-rayed stellate hairs. Raceme much elongating in fruit. Sepals caducous. Petals small (c. 2 mm), pale yellow. Siliculae 3-3.5 mm, orbicular, emarginate, glabrous.

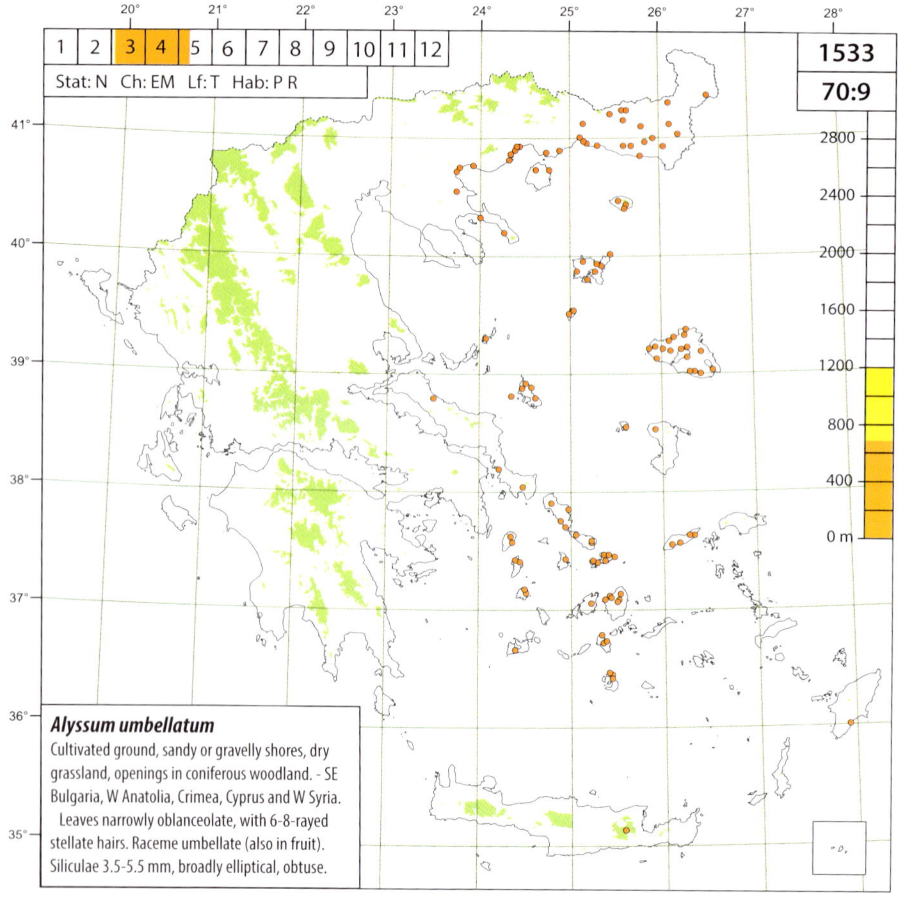

Alyssum umbellatum

Cultivated ground, sandy or gravelly shores, dry grassland, openings in coniferous woodland. - SE Bulgaria, W Anatolia, Crimea, Cyprus and W Syria.

Leaves narrowly oblanceolate, with 6-8-rayed stellate hairs. Raceme umbellate (also in fruit). Siliculae 3.5-5.5 mm, broadly elliptical, obtuse.

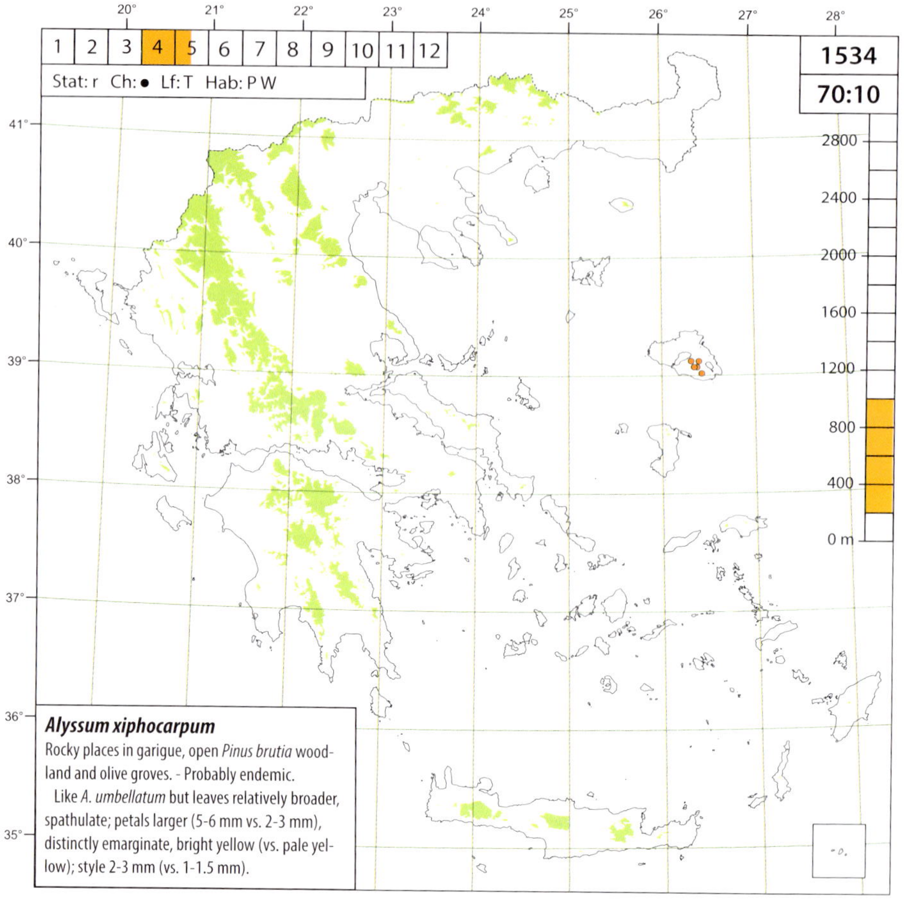

Alyssum xiphocarpum

Rocky places in garigue, open *Pinus brutia* woodland and olive groves. - Probably endemic.

Like *A. umbellatum* but leaves relatively broader, spathulate; petals larger (5-6 mm vs. 2-3 mm), distinctly emarginate, bright yellow (vs. pale yellow); style 2-3 mm (vs. 1-1.5 mm).

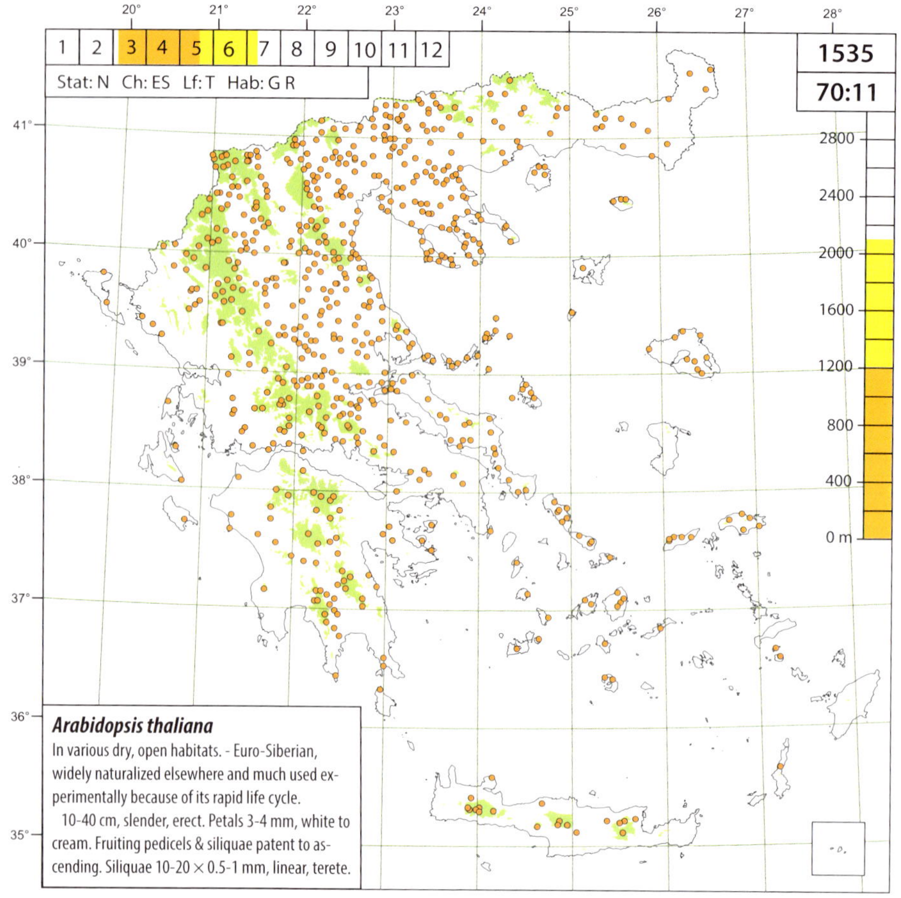

Arabidopsis thaliana

In various dry, open habitats. - Euro-Siberian, widely naturalized elsewhere and much used experimentally because of its rapid life cycle.

10-40 cm, slender, erect. Petals 3-4 mm, white to cream. Fruiting pedicels & siliquae patent to ascending. Siliquae 10-20 × 0.5-1 mm, linear, terete.

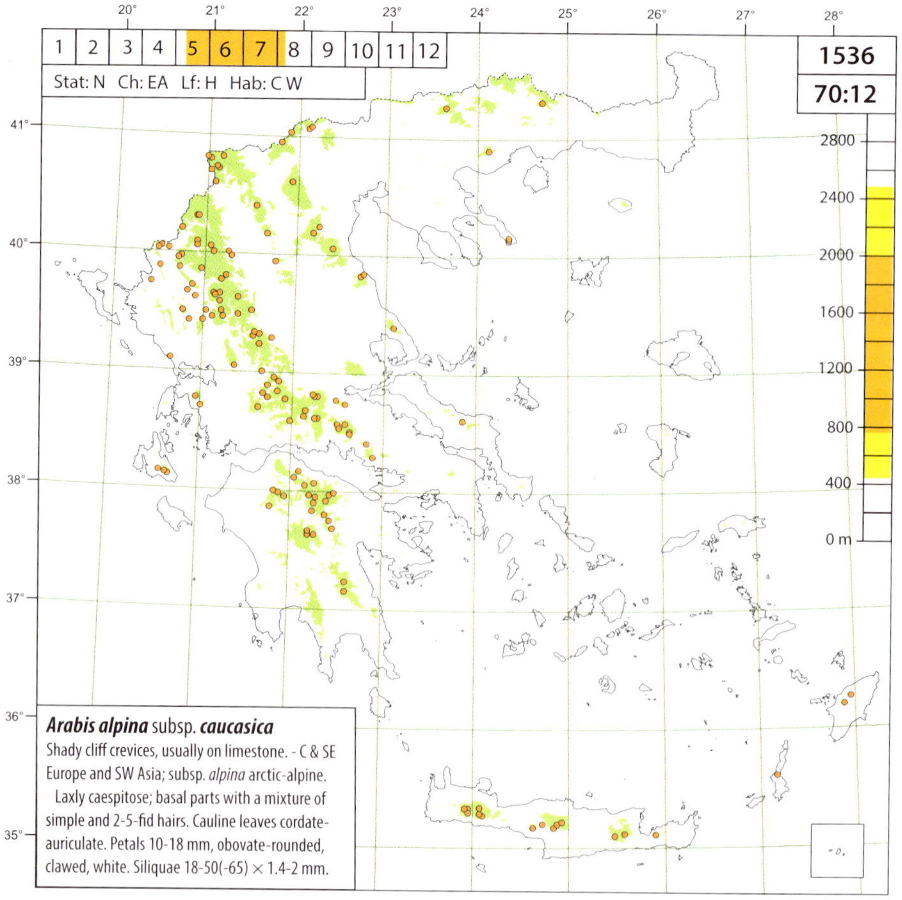

Arabis alpina subsp. ***caucasica***

Shady cliff crevices, usually on limestone. - C & SE Europe and SW Asia; subsp. *alpina* arctic-alpine.

Laxly caespitose; basal parts with a mixture of simple and 2-5-fid hairs. Cauline leaves cordate-auriculate. Petals 10-18 mm, obovate-rounded, clawed, white. Siliquae 18-50(-65) × 1.4-2 mm.

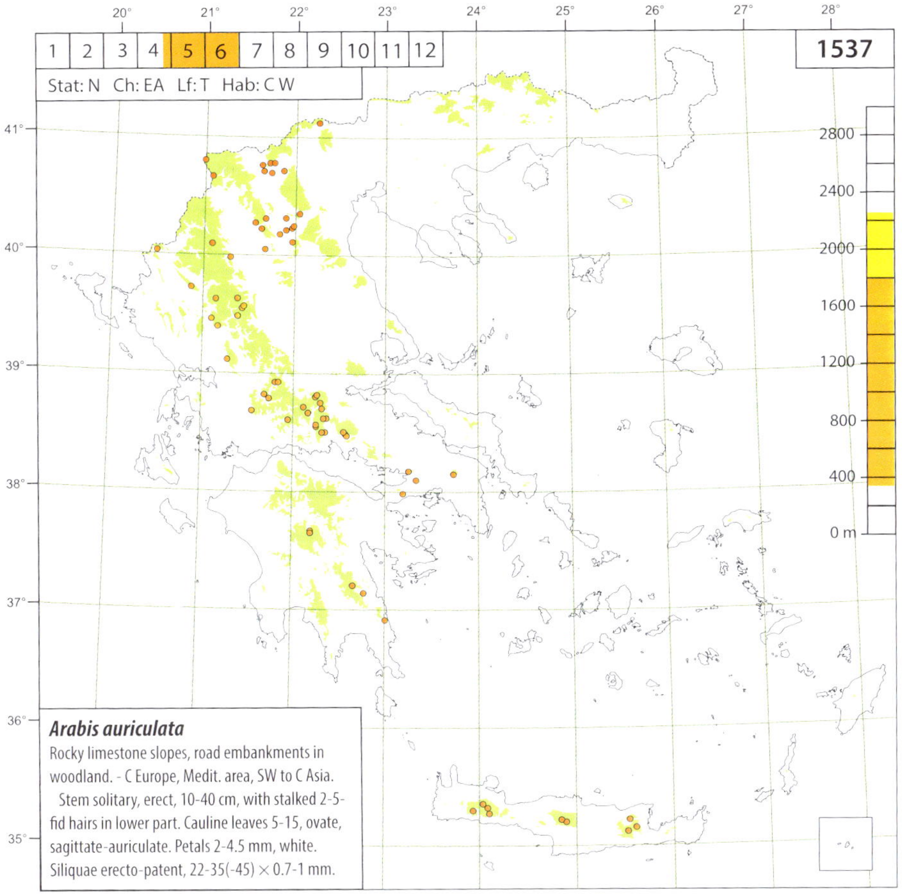

Arabis auriculata

Rocky limestone slopes, road embankments in woodland. - C Europe, Medit. area, SW to C Asia.

Stem solitary, erect, 10-40 cm, with stalked 2-5-fid hairs in lower part. Cauline leaves 5-15, ovate, sagittate-auriculate. Petals 2-4.5 mm, white. Siliquae erecto-patent, 22-35(-45) × 0.7-1 mm.

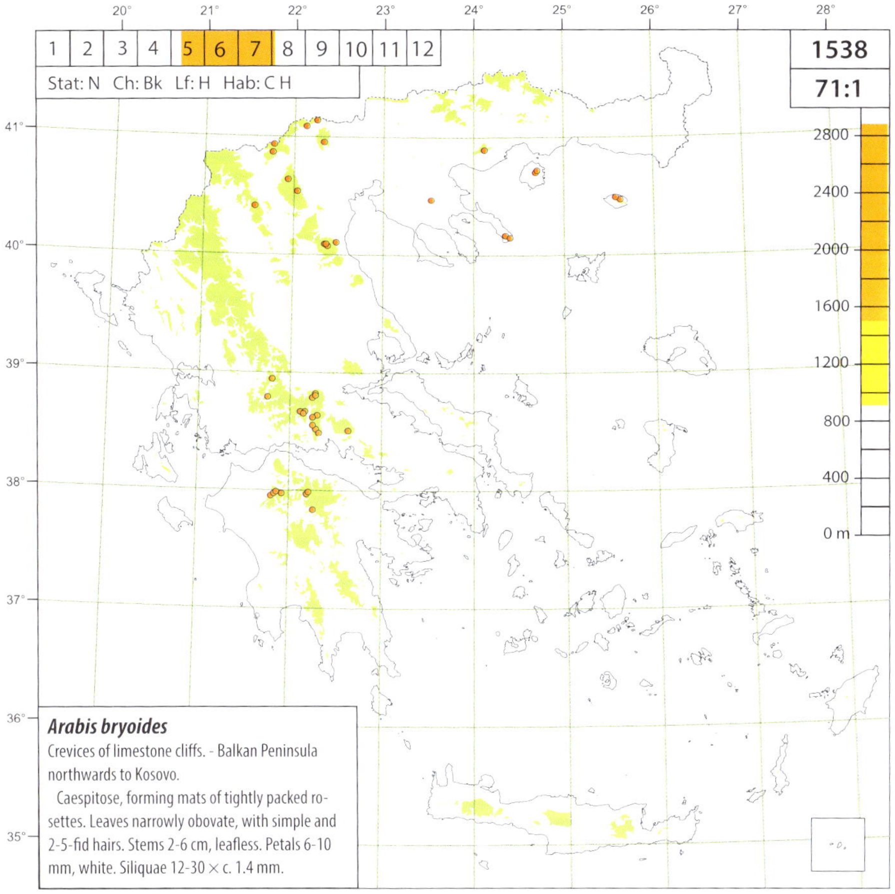

Arabis bryoides

Crevices of limestone cliffs. - Balkan Peninsula northwards to Kosovo.

Caespitose, forming mats of tightly packed rosettes. Leaves narrowly obovate, with simple and 2-5-fid hairs. Stems 2-6 cm, leafless. Petals 6-10 mm, white. Siliquae 12-30 × c. 1.4 mm.

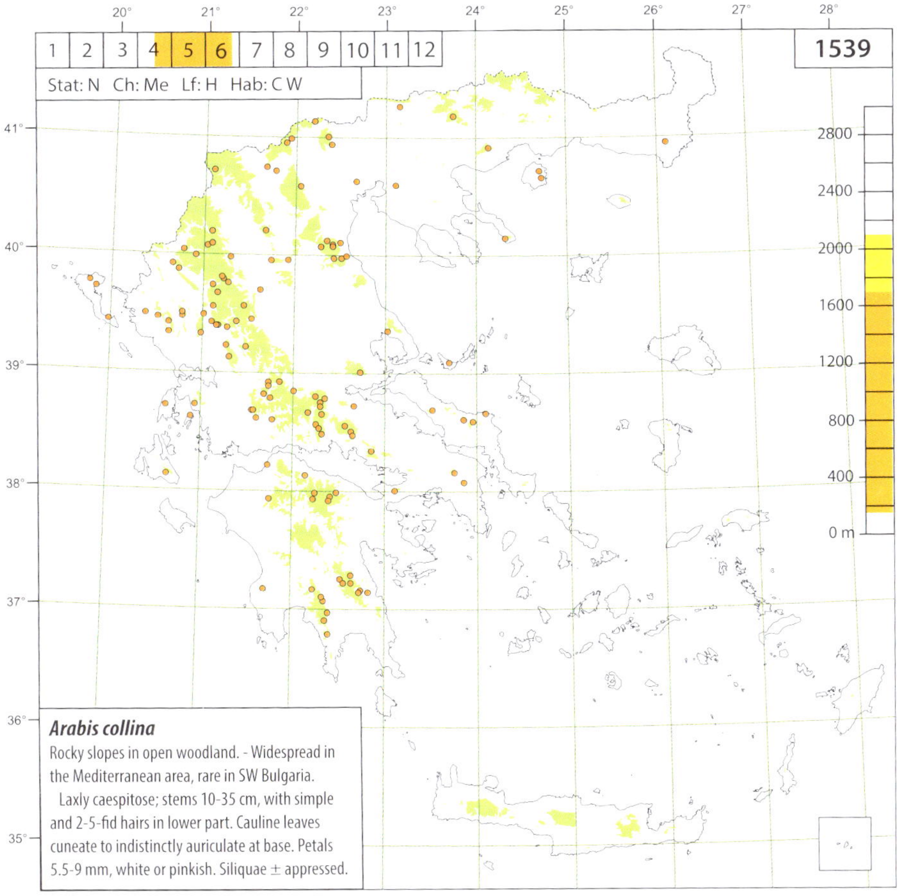

Arabis collina

Rocky slopes in open woodland. - Widespread in the Mediterranean area, rare in SW Bulgaria.

Laxly caespitose; stems 10-35 cm, with simple and 2-5-fid hairs in lower part. Cauline leaves cuneate to indistinctly auriculate at base. Petals 5.5-9 mm, white or pinkish. Siliquae ± appressed.

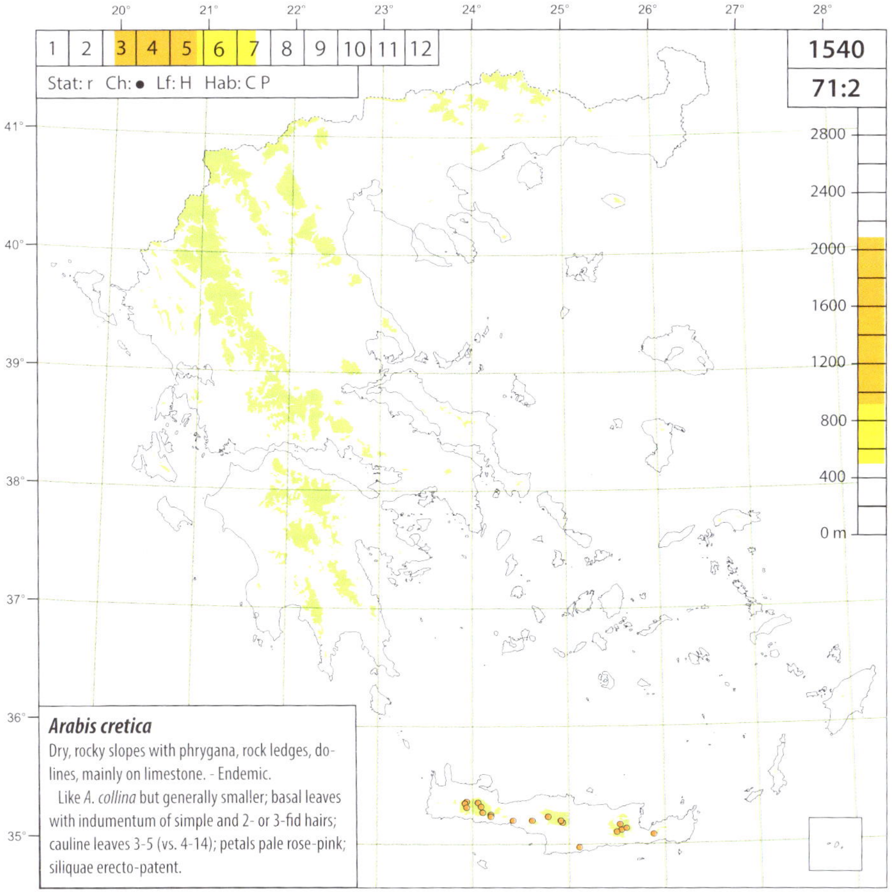

Arabis cretica

Dry, rocky slopes with phrygana, rock ledges, dolines, mainly on limestone. - Endemic.

Like *A. collina* but generally smaller; basal leaves with indumentum of simple and 2- or 3-fid hairs; cauline leaves 3-5 (vs. 4-14); petals pale rose-pink; siliquae erecto-patent.

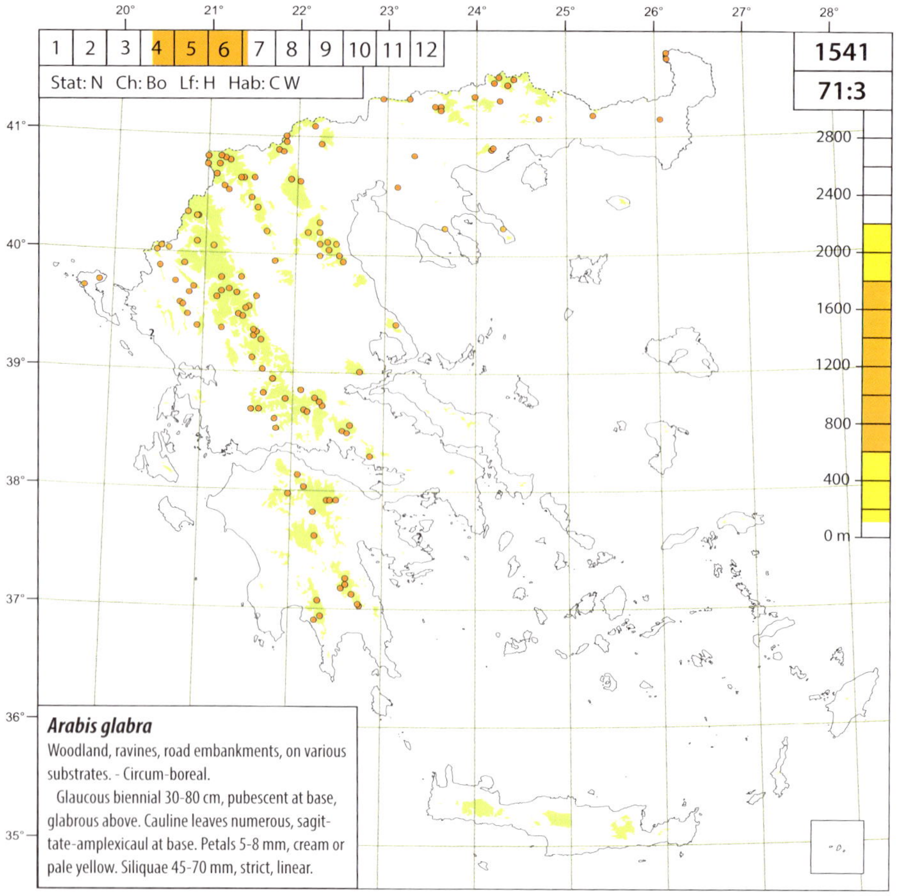

Arabis glabra

Woodland, ravines, road embankments, on various substrates. - Circum-boreal.

Glaucous biennial 30-80 cm, pubescent at base, glabrous above. Cauline leaves numerous, sagittate-amplexicaul at base. Petals 5-8 mm, cream or pale yellow. Siliquae 45-70 mm, strict, linear.

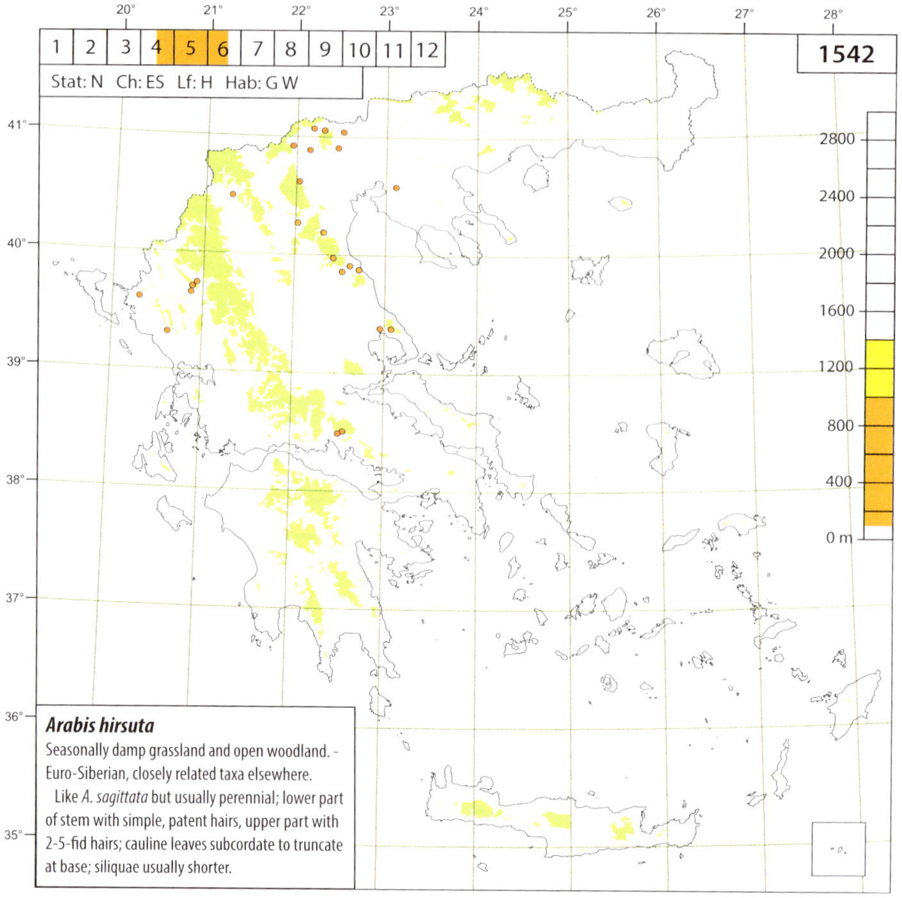

Arabis hirsuta

Seasonally damp grassland and open woodland. - Euro-Siberian, closely related taxa elsewhere.

Like *A. sagittata* but usually perennial; lower part of stem with simple, patent hairs, upper part with 2-5-fid hairs; cauline leaves subcordate to truncate at base; siliquae usually shorter.

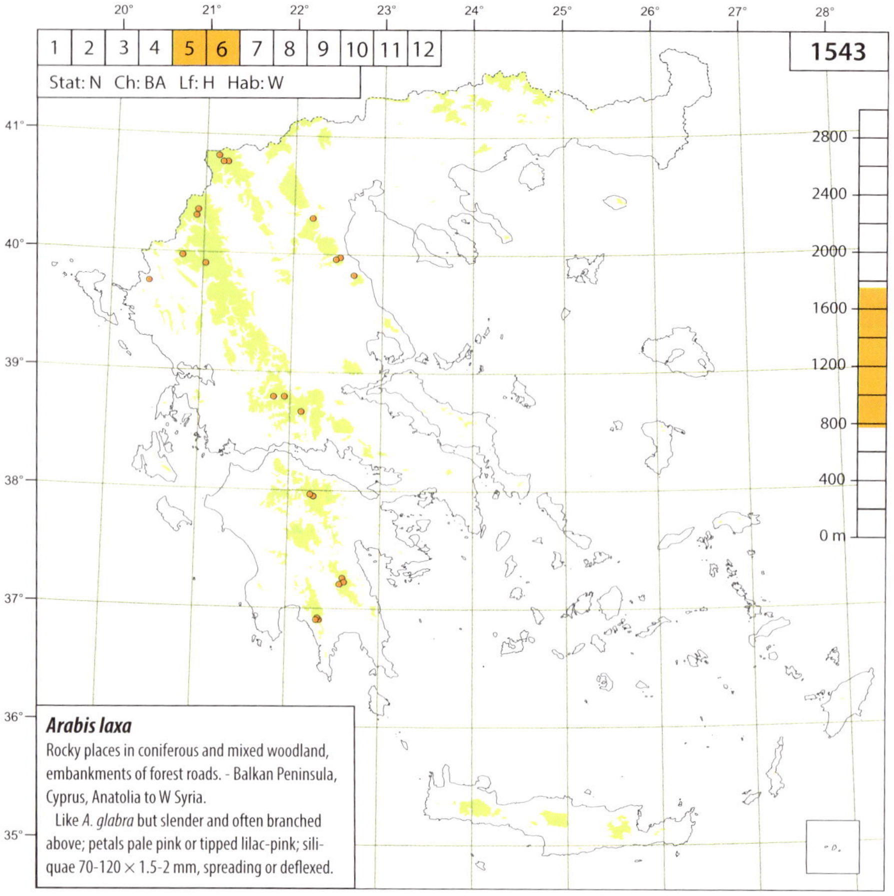

Arabis laxa

Rocky places in coniferous and mixed woodland, embankments of forest roads. - Balkan Peninsula, Cyprus, Anatolia to W Syria.

Like *A. glabra* but slender and often branched above; petals pale pink or tipped lilac-pink; siliquae 70-120 × 1.5-2 mm, spreading or deflexed.

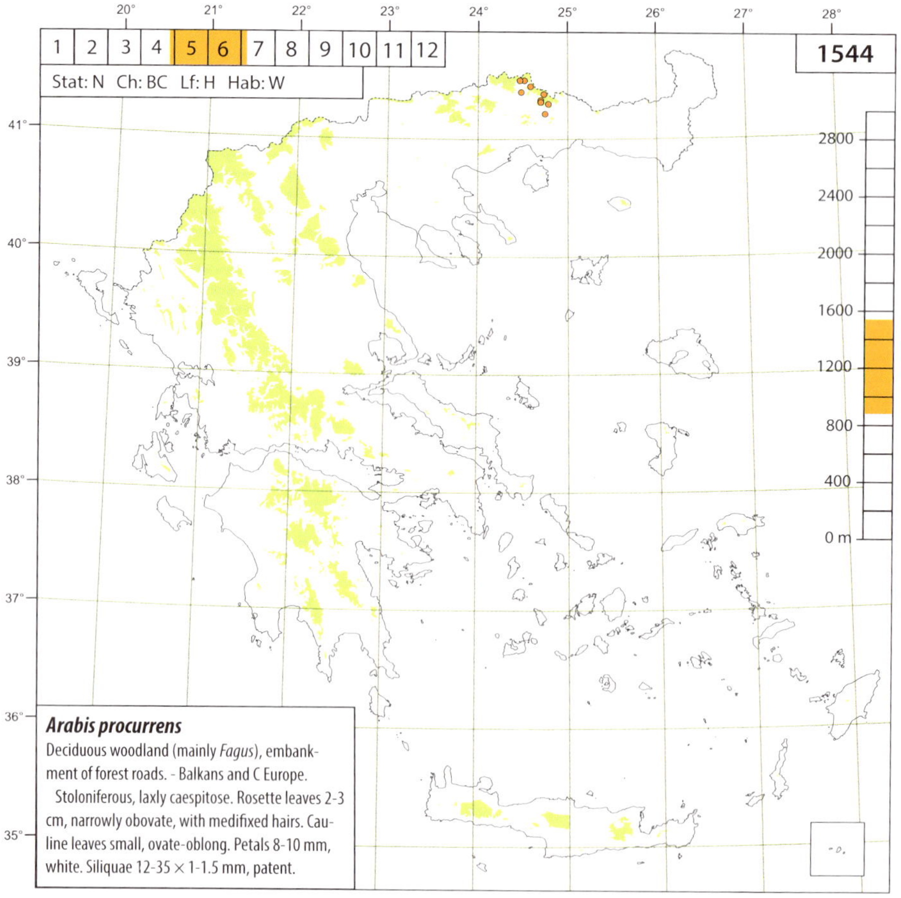

Arabis procurrens

Deciduous woodland (mainly *Fagus*), embankment of forest roads. - Balkans and C Europe.

Stoloniferous, laxly caespitose. Rosette leaves 2-3 cm, narrowly obovate, with medifixed hairs. Cauline leaves small, ovate-oblong. Petals 8-10 mm, white. Siliquae 12-35 × 1-1.5 mm, patent.

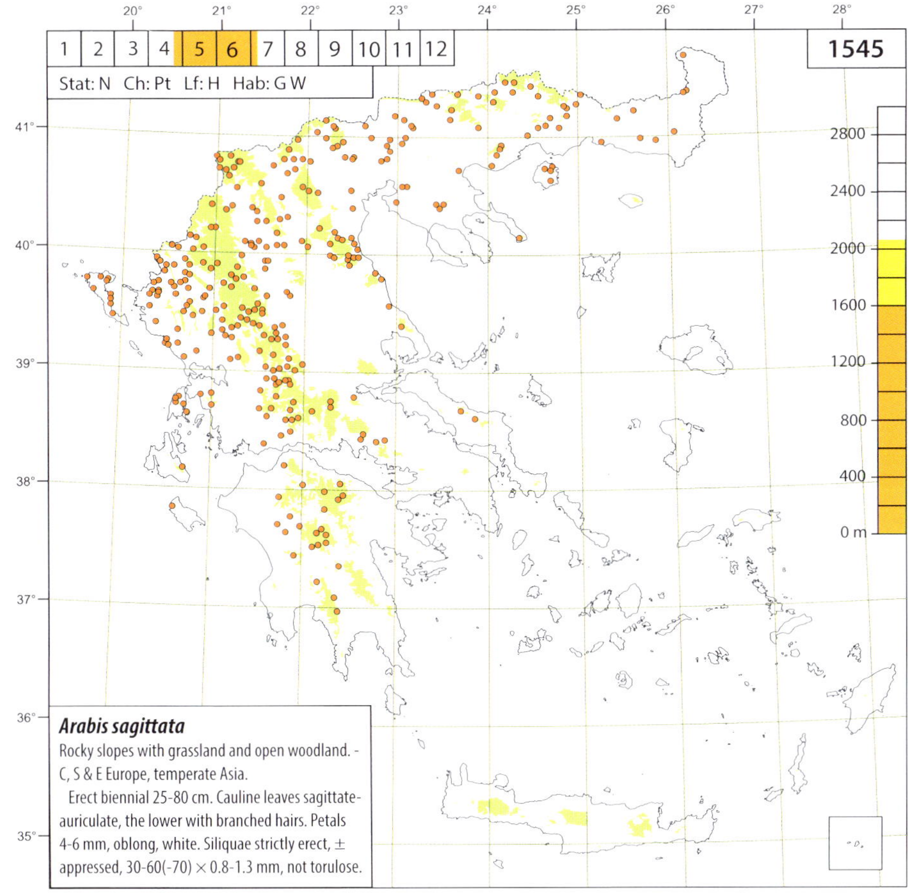

Arabis sagittata

Rocky slopes with grassland and open woodland. - C, S & E Europe, temperate Asia.

Erect biennial 25-80 cm. Cauline leaves sagittate-auriculate, the lower with branched hairs. Petals 4-6 mm, oblong, white. Siliquae strictly erect, ± appressed, 30-60(-70) × 0.8-1.3 mm, not torulose.

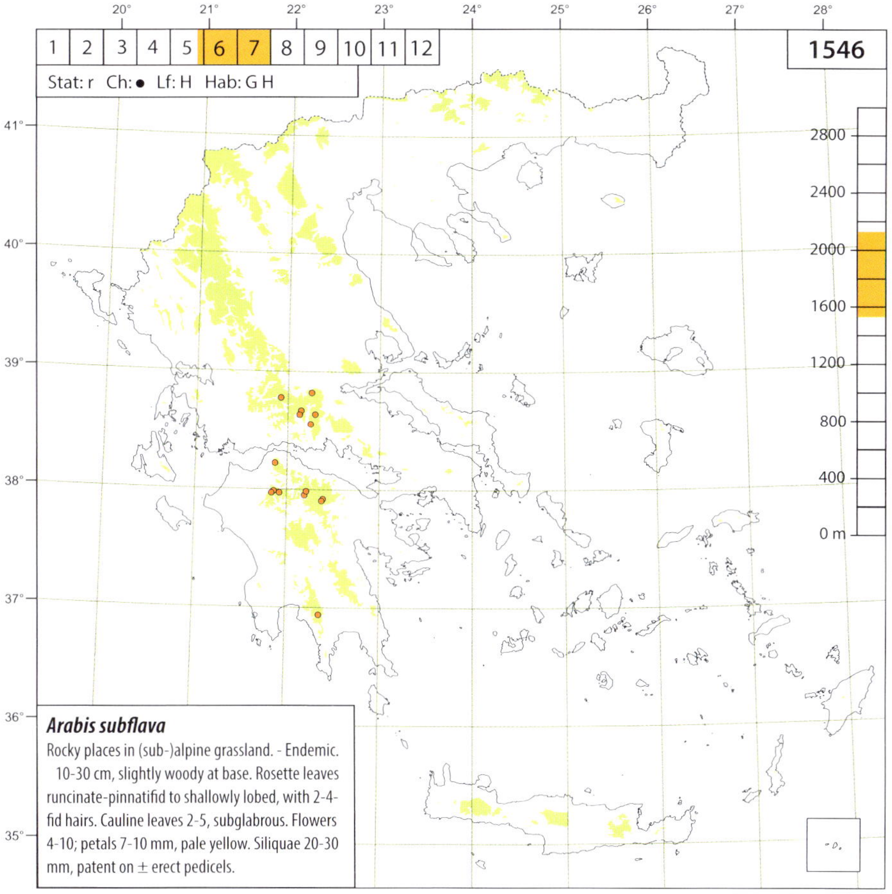

Arabis subflava

Rocky places in (sub-)alpine grassland. - Endemic.

10-30 cm, slightly woody at base. Rosette leaves runcinate-pinnatifid to shallowly lobed, with 2-4-fid hairs. Cauline leaves 2-5, subglabrous. Flowers 4-10; petals 7-10 mm, pale yellow. Siliquae 20-30 mm, patent on ± erect pedicels.

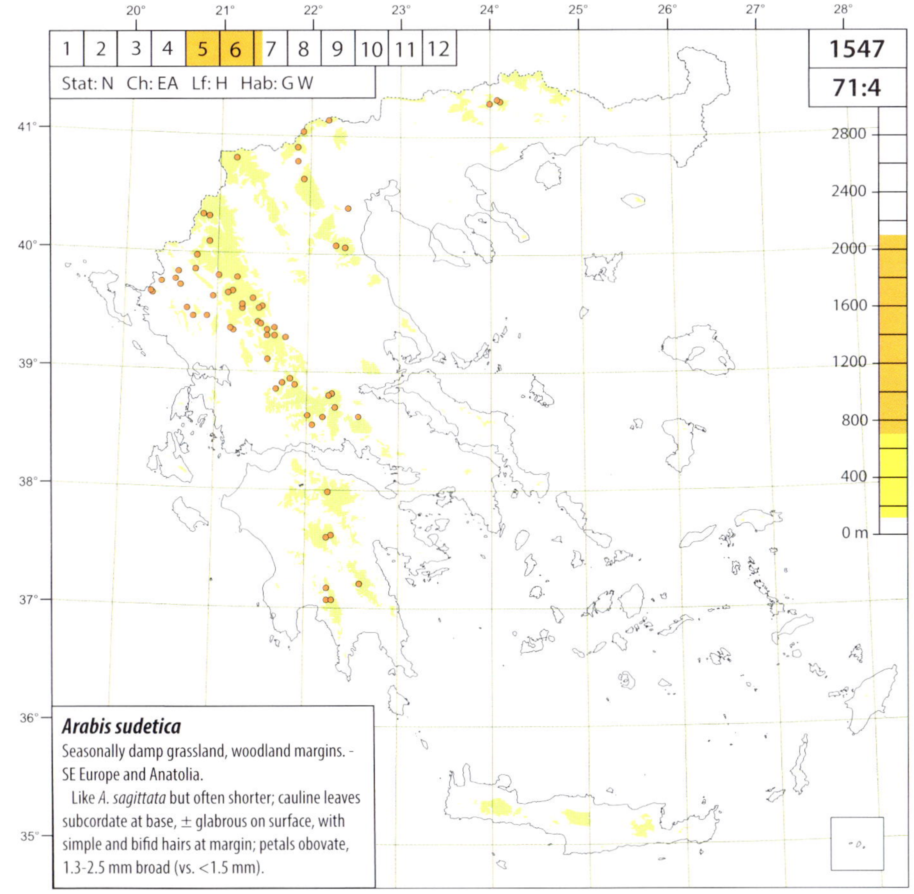

Arabis sudetica

Seasonally damp grassland, woodland margins. - SE Europe and Anatolia.

Like *A. sagittata* but often shorter; cauline leaves subcordate at base, ± glabrous on surface, with simple and bifid hairs at margin; petals obovate, 1.3-2.5 mm broad (vs. <1.5 mm).

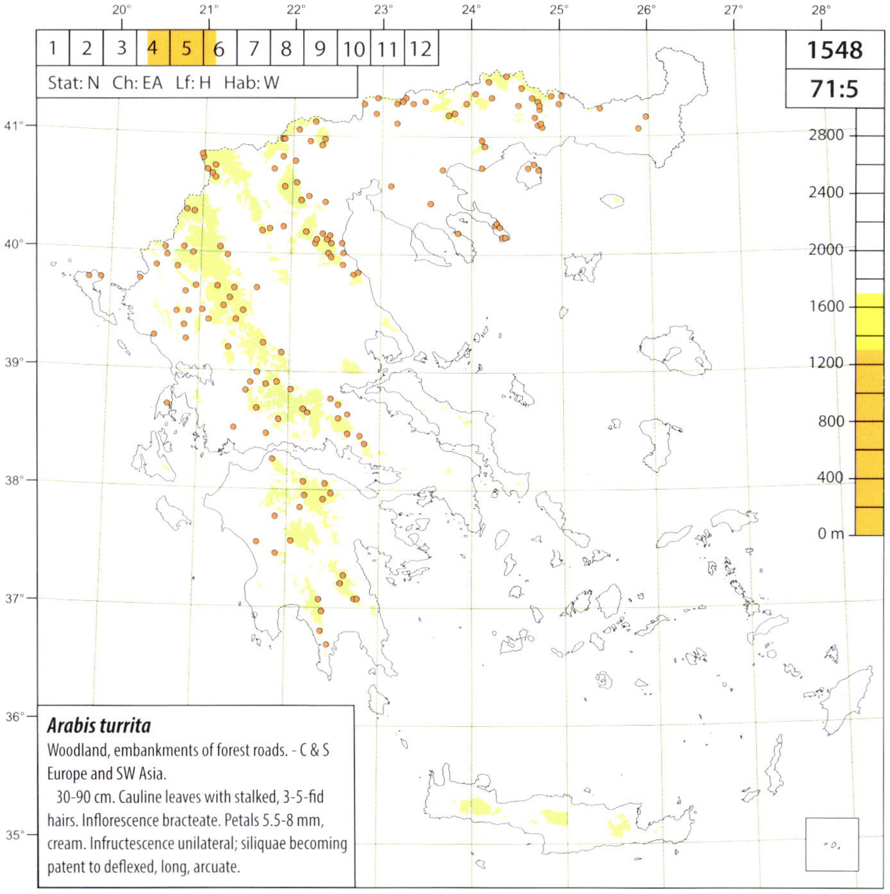

Arabis turrita

Woodland, embankments of forest roads. - C & S Europe and SW Asia.

30-90 cm. Cauline leaves with stalked, 3-5-fid hairs. Inflorescence bracteate. Petals 5.5-8 mm, cream. Infructescence unilateral; siliquae becoming patent to deflexed, long, arcuate.

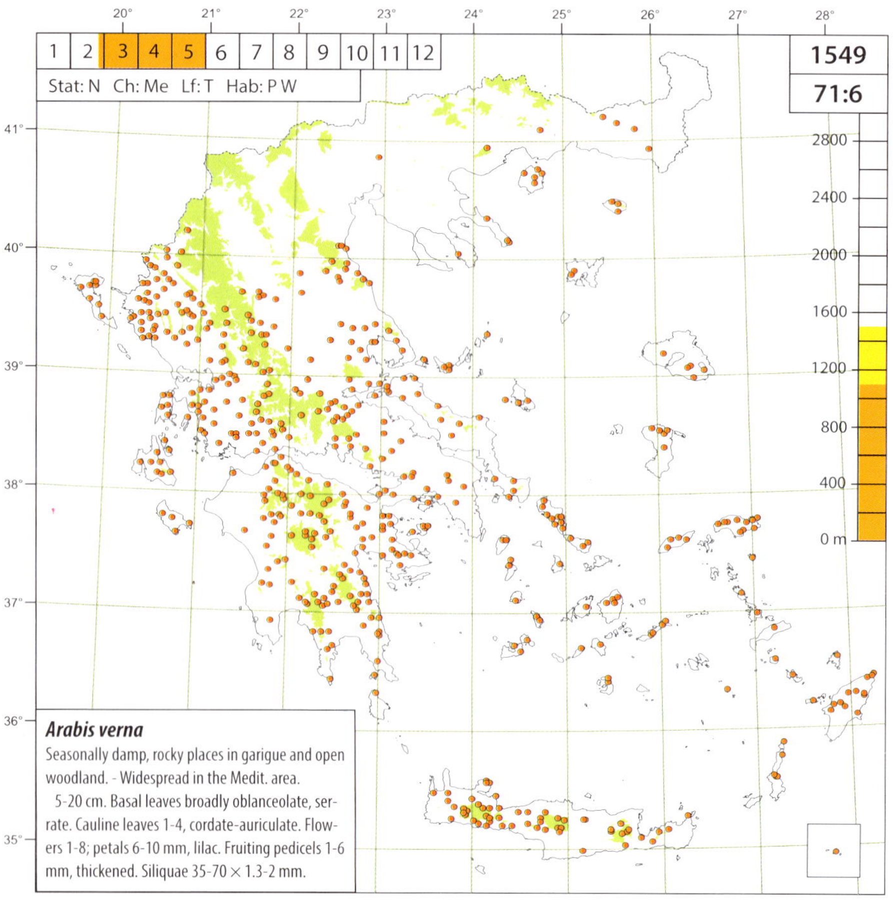

Arabis verna

Seasonally damp, rocky places in garigue and open woodland. - Widespread in the Medit. area.

5-20 cm. Basal leaves broadly oblanceolate, serrate. Cauline leaves 1-4, cordate-auriculate. Flowers 1-8; petals 6-10 mm, lilac. Fruiting pedicels 1-6 mm, thickened. Siliquae 35-70 × 1.3-2 mm.

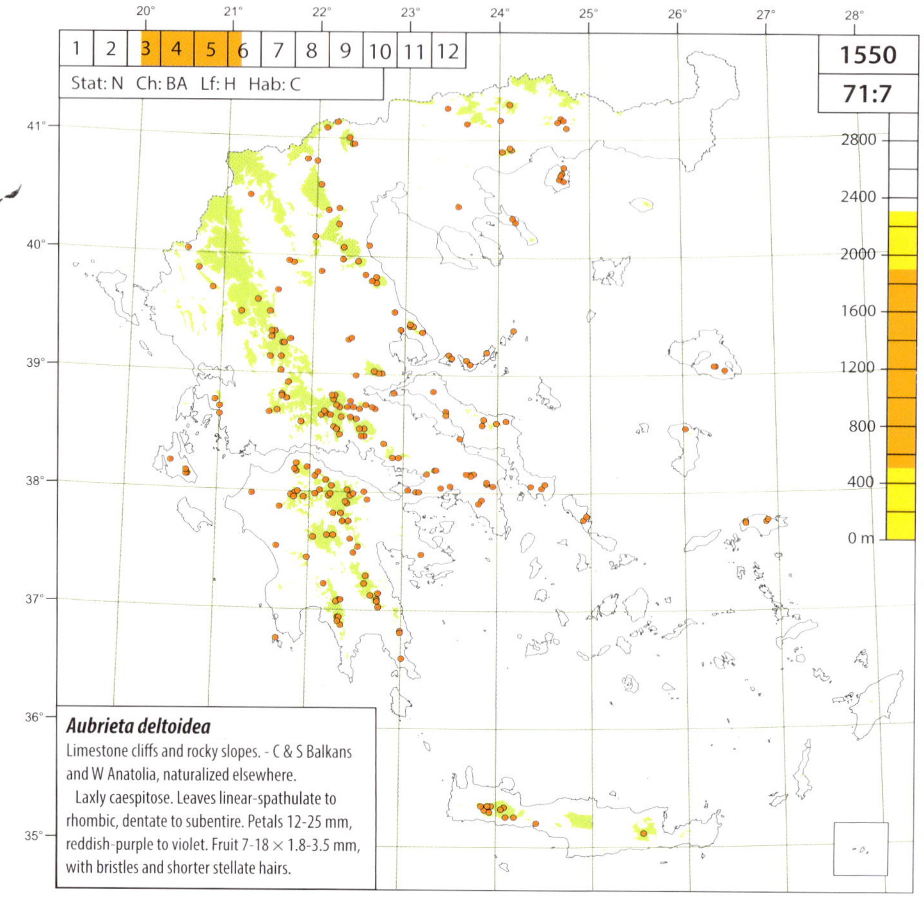

Aubrieta deltoidea

Limestone cliffs and rocky slopes. - C & S Balkans and W Anatolia, naturalized elsewhere.

Laxly caespitose. Leaves linear-spathulate to rhombic, dentate to subentire. Petals 12-25 mm, reddish-purple to violet. Fruit 7-18 × 1.8-3.5 mm, with bristles and shorter stellate hairs.

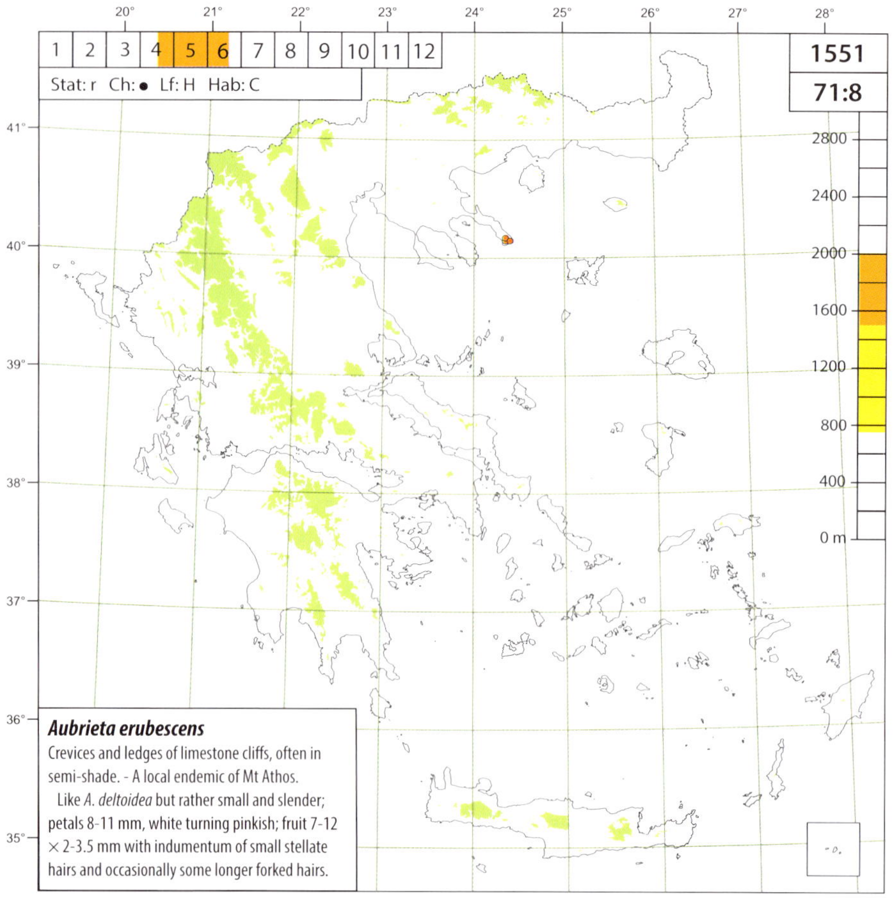

Aubrieta erubescens

Crevices and ledges of limestone cliffs, often in semi-shade. - A local endemic of Mt Athos.

Like *A. deltoidea* but rather small and slender; petals 8-11 mm, white turning pinkish; fruit 7-12 × 2-3.5 mm with indumentum of small stellate hairs and occasionally some longer forked hairs.

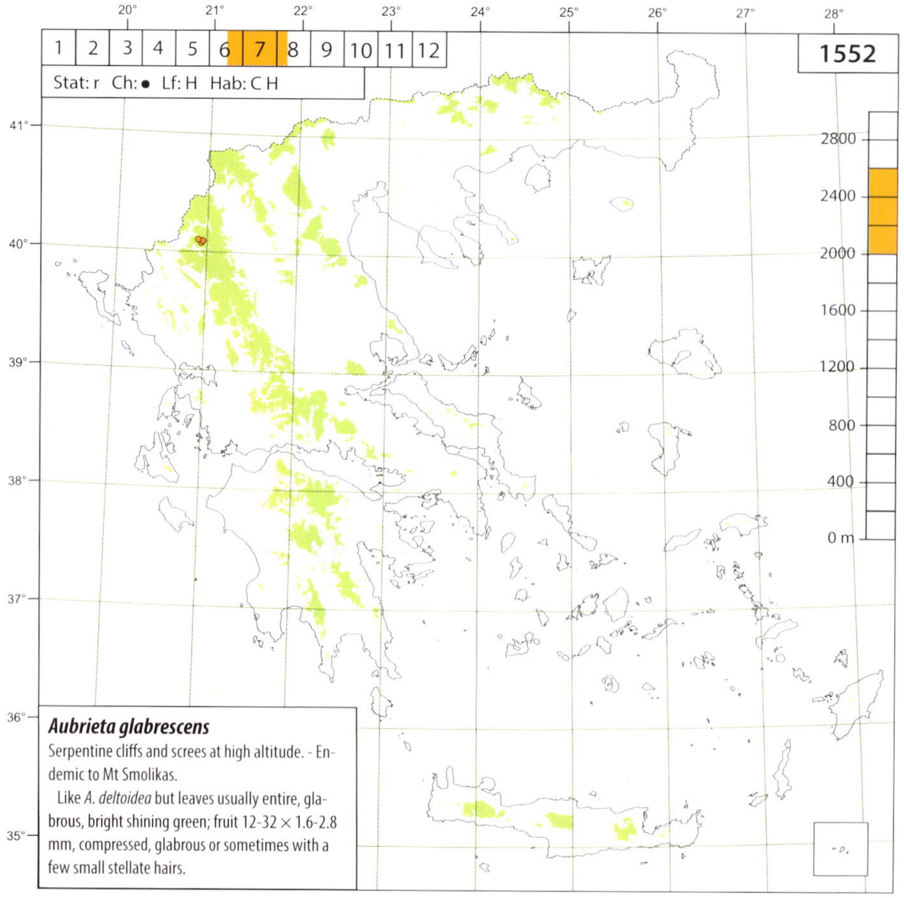

Aubrieta glabrescens

Serpentine cliffs and screes at high altitude. - Endemic to Mt Smolikas.

Like *A. deltoidea* but leaves usually entire, glabrous, bright shining green; fruit 12-32 × 1.6-2.8 mm, compressed, glabrous or sometimes with a few small stellate hairs.

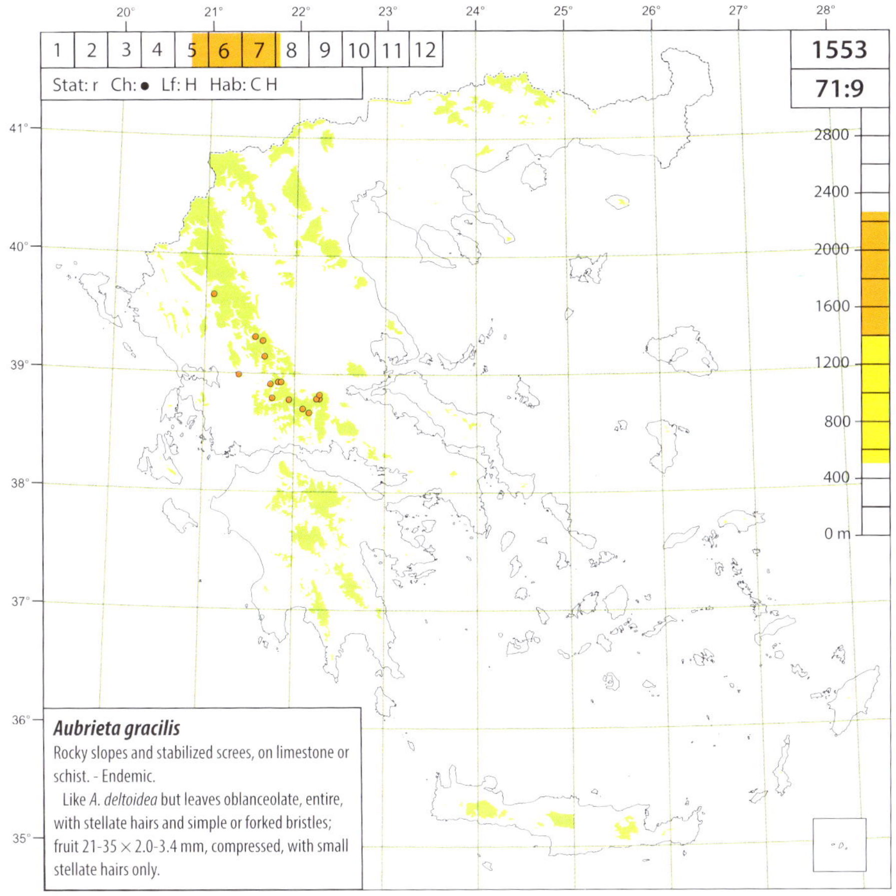

Aubrieta gracilis

Rocky slopes and stabilized screes, on limestone or schist. - Endemic.

Like *A. deltoidea* but leaves oblanceolate, entire, with stellate hairs and simple or forked bristles; fruit 21-35 × 2.0-3.4 mm, compressed, with small stellate hairs only.

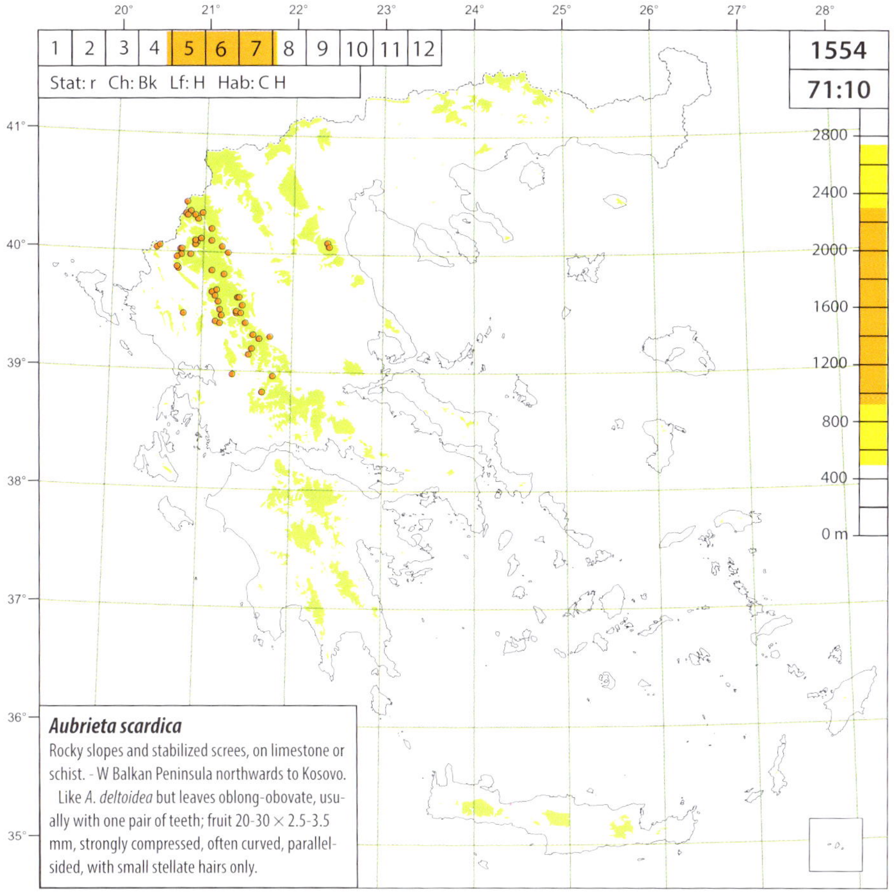

Aubrieta scardica

Rocky slopes and stabilized screes, on limestone or schist. - W Balkan Peninsula northwards to Kosovo.

Like *A. deltoidea* but leaves oblong-obovate, usually with one pair of teeth; fruit 20-30 × 2.5-3.5 mm, strongly compressed, often curved, parallel-sided, with small stellate hairs only.

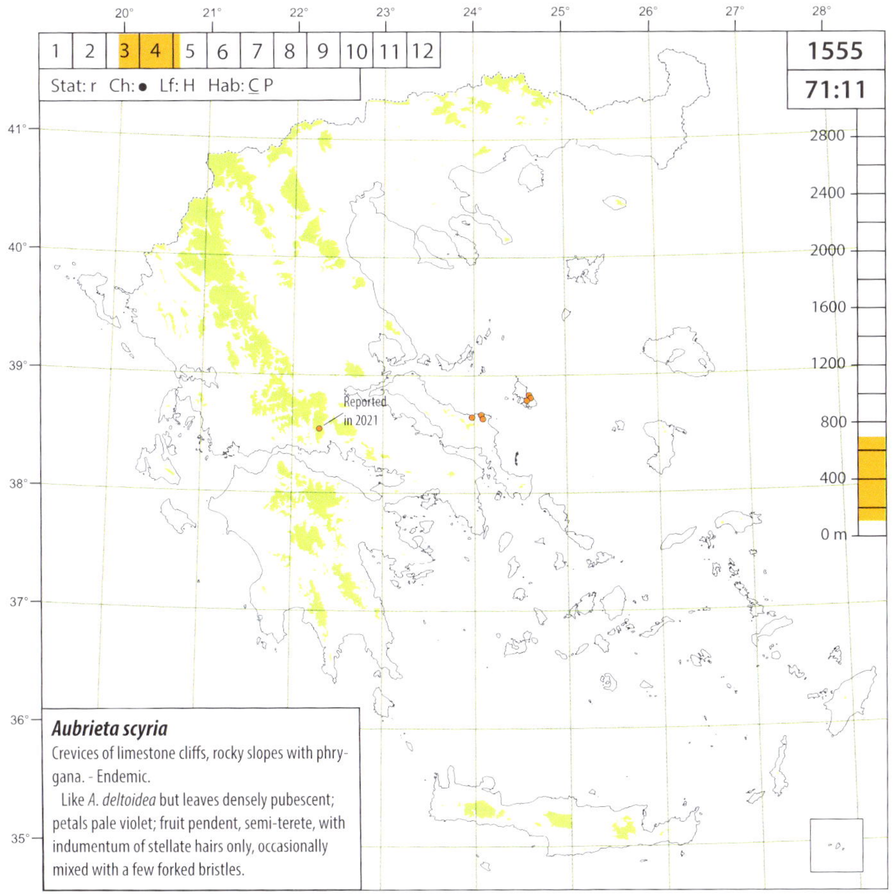

Aubrieta scyria

Crevices of limestone cliffs, rocky slopes with phrygana. - Endemic.

Like *A. deltoidea* but leaves densely pubescent; petals pale violet; fruit pendent, semi-terete, with indumentum of stellate hairs only, occasionally mixed with a few forked bristles.

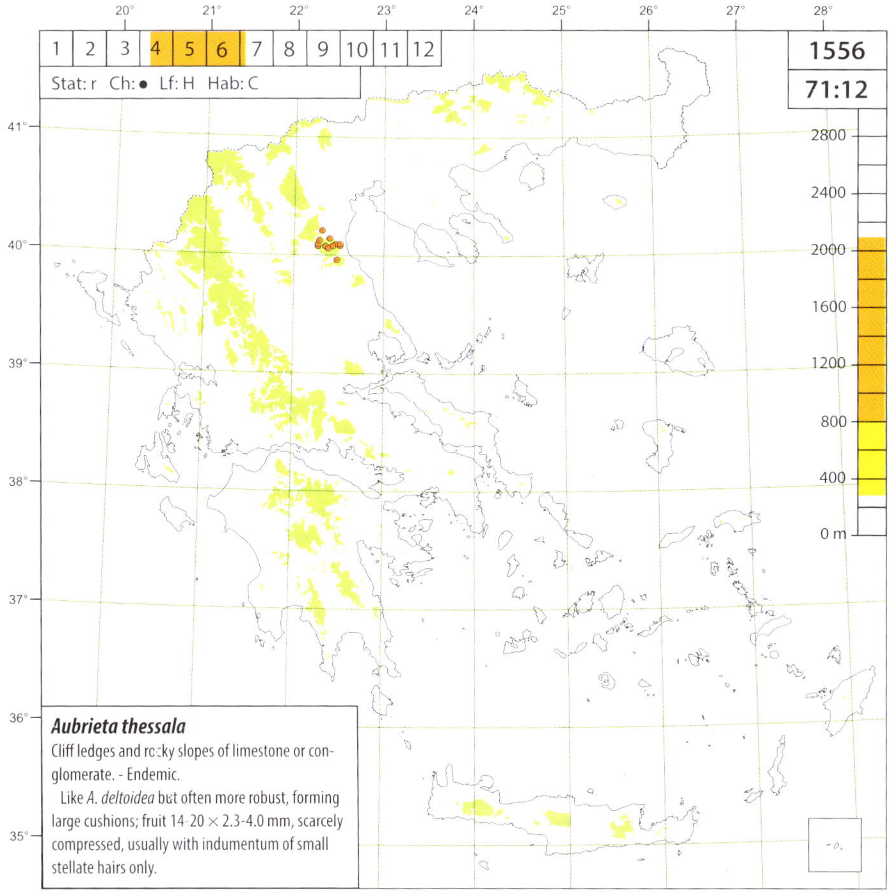

Aubrieta thessala

Cliff ledges and rocky slopes of limestone or conglomerate. - Endemic.

Like *A. deltoidea* but often more robust, forming large cushions; fruit 14-20 × 2.3-4.0 mm, scarcely compressed, usually with indumentum of small stellate hairs only.

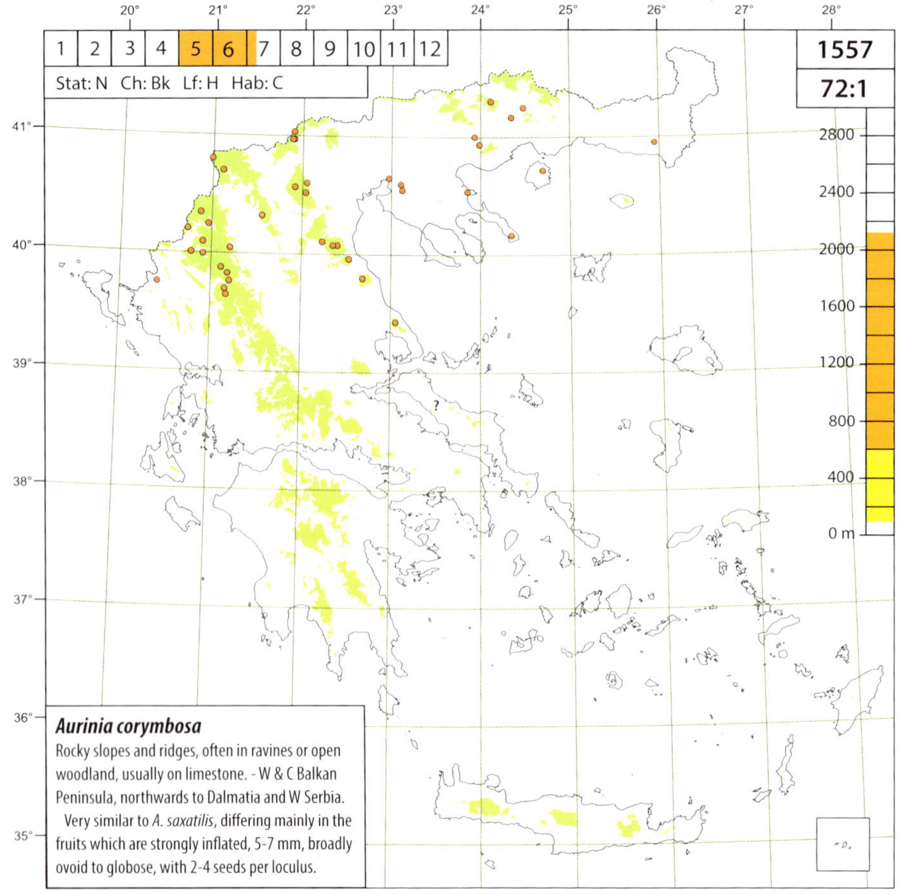

Aurinia corymbosa

Rocky slopes and ridges, often in ravines or open woodland, usually on limestone. - W & C Balkan Peninsula, northwards to Dalmatia and W Serbia.

Very similar to *A. saxatilis*, differing mainly in the fruits which are strongly inflated, 5-7 mm, broadly ovoid to globose, with 2-4 seeds per loculus.

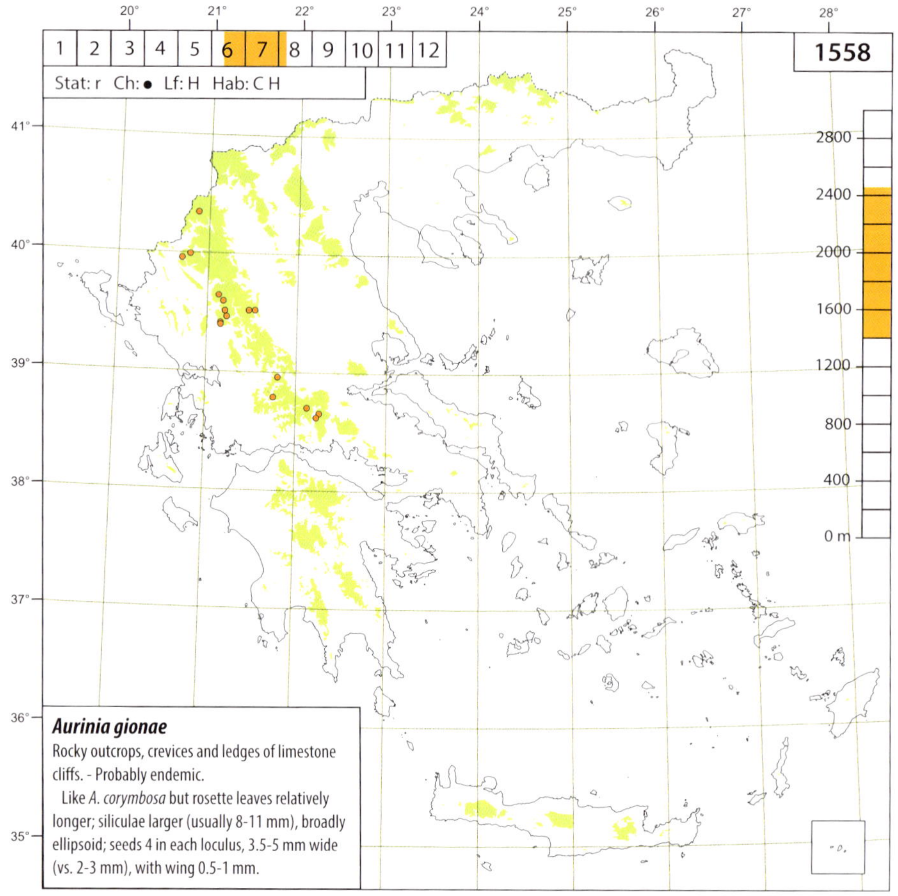

Aurinia gionae

Rocky outcrops, crevices and ledges of limestone cliffs. - Probably endemic.

Like *A. corymbosa* but rosette leaves relatively longer; siliculae larger (usually 8-11 mm), broadly ellipsoid; seeds 4 in each loculus, 3.5-5 mm wide (vs. 2-3 mm), with wing 0.5-1 mm.

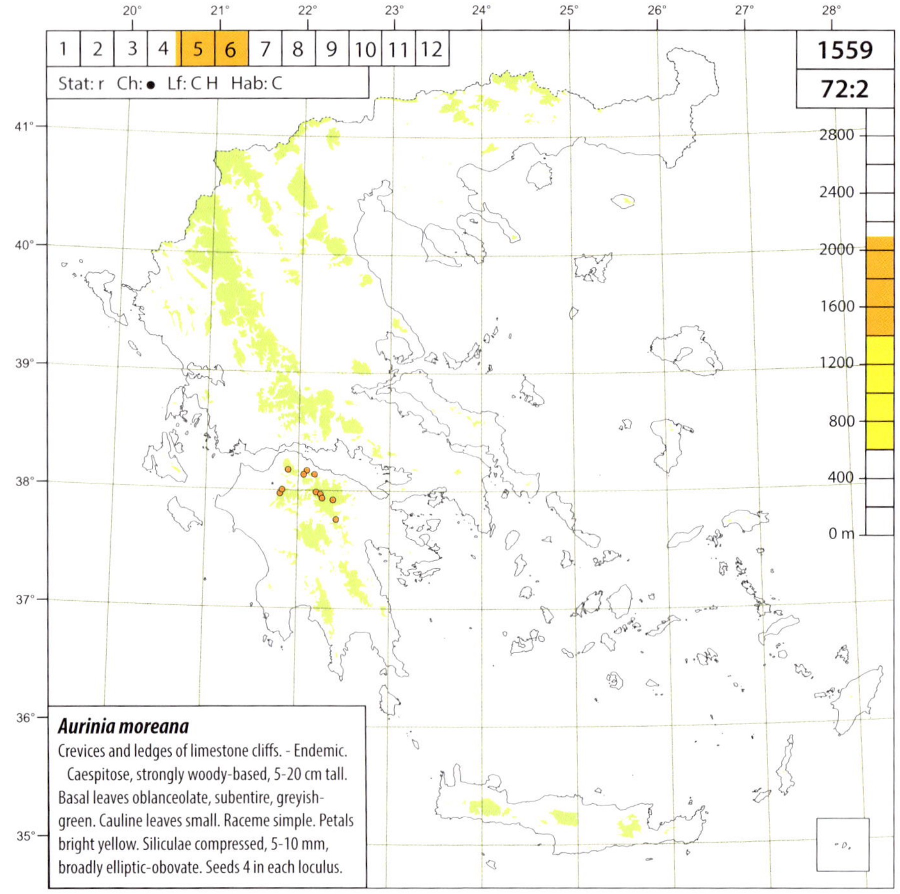

Aurinia moreana

Crevices and ledges of limestone cliffs. - Endemic.

Caespitose, strongly woody-based, 5-20 cm tall. Basal leaves oblanceolate, subentire, greyish-green. Cauline leaves small. Raceme simple. Petals bright yellow. Siliculae compressed, 5-10 mm, broadly elliptic-obovate. Seeds 4 in each loculus.

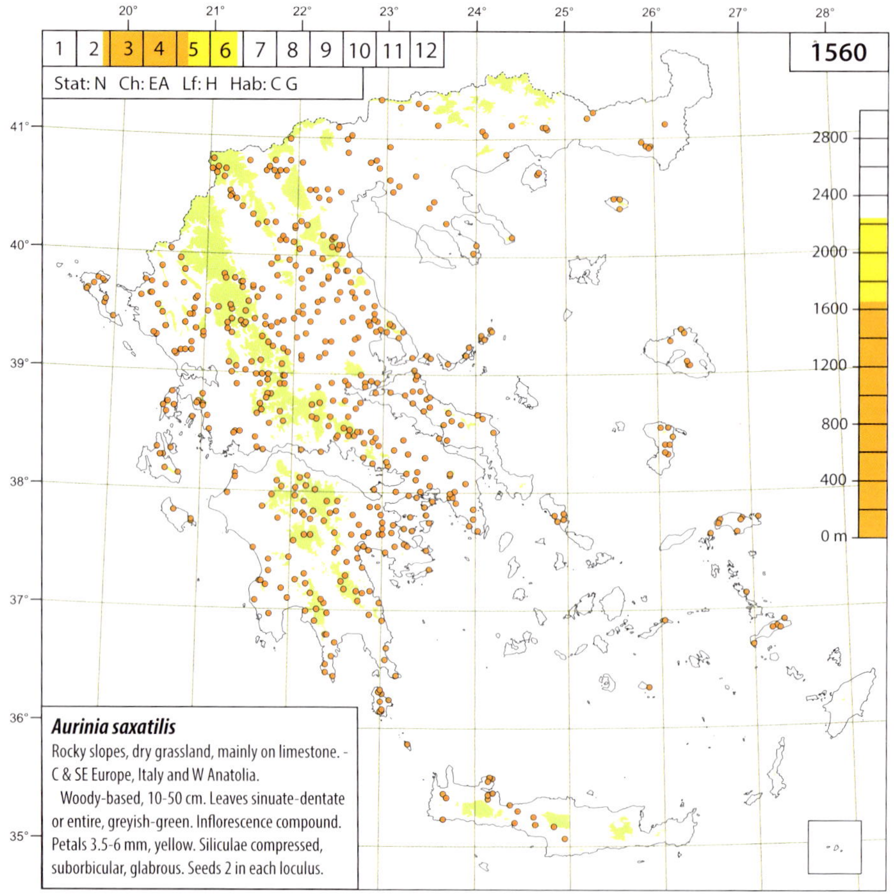

Aurinia saxatilis

Rocky slopes, dry grassland, mainly on limestone. - C & SE Europe, Italy and W Anatolia.

Woody-based, 10-50 cm. Leaves sinuate-dentate or entire, greyish-green. Inflorescence compound. Petals 3.5-6 mm, yellow. Siliculae compressed, suborbicular, glabrous. Seeds 2 in each loculus.

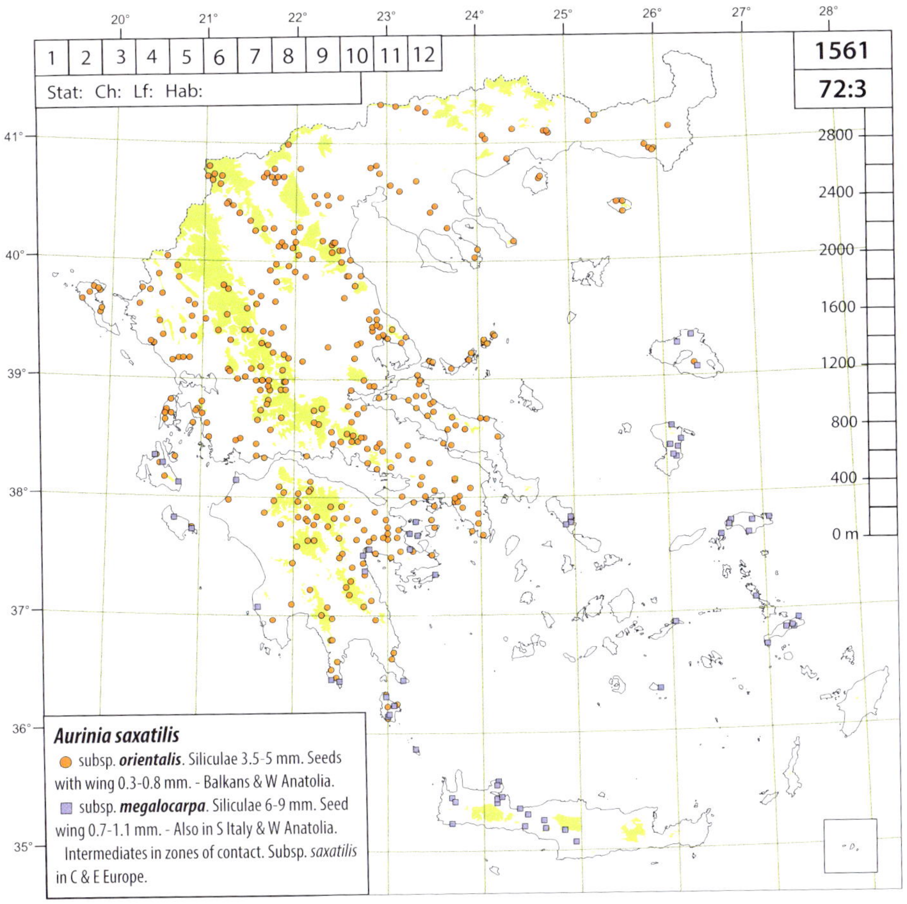

Aurinia saxatilis
subsp. ***orientalis***. Siliculae 3.5-5 mm. Seeds with wing 0.3-0.8 mm. - Balkans & W Anatolia.
subsp. ***megalocarpa***. Siliculae 6-9 mm. Seed wing 0.7-1.1 mm. - Also in S Italy & W Anatolia.
Intermediates in zones of contact. Subsp. *saxatilis* in C & E Europe.

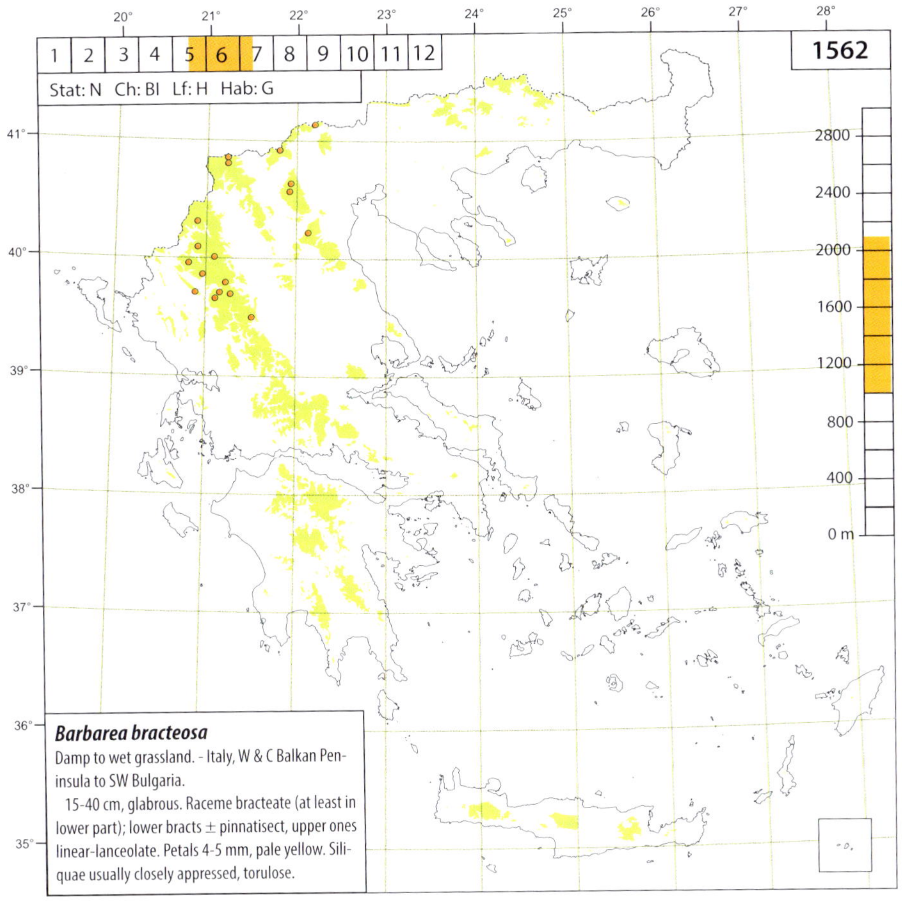

Barbarea bracteosa
Damp to wet grassland. - Italy, W & C Balkan Peninsula to SW Bulgaria.
15-40 cm, glabrous. Raceme bracteate (at least in lower part); lower bracts ± pinnatisect, upper ones linear-lanceolate. Petals 4-5 mm, pale yellow. Siliquae usually closely appressed, torulose.

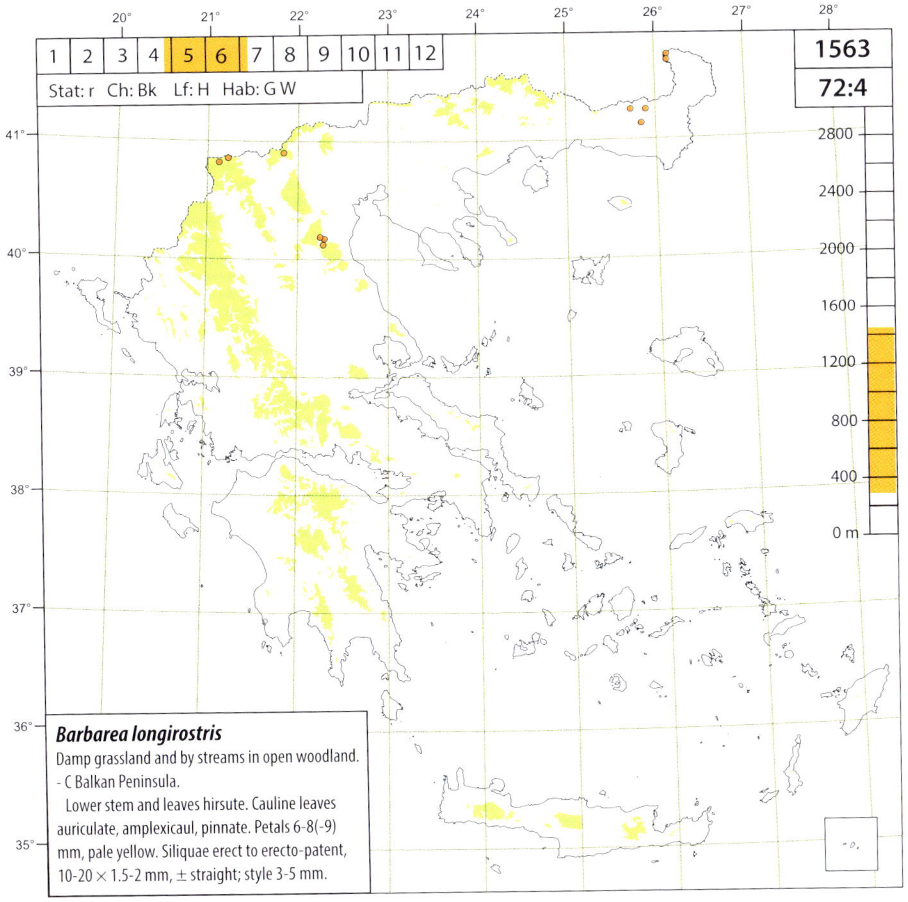

Barbarea longirostris
Damp grassland and by streams in open woodland. - C Balkan Peninsula.
Lower stem and leaves hirsute. Cauline leaves auriculate, amplexicaul, pinnate. Petals 6-8(-9) mm, pale yellow. Siliquae erect to erecto-patent, 10-20 × 1.5-2 mm, ± straight; style 3-5 mm.

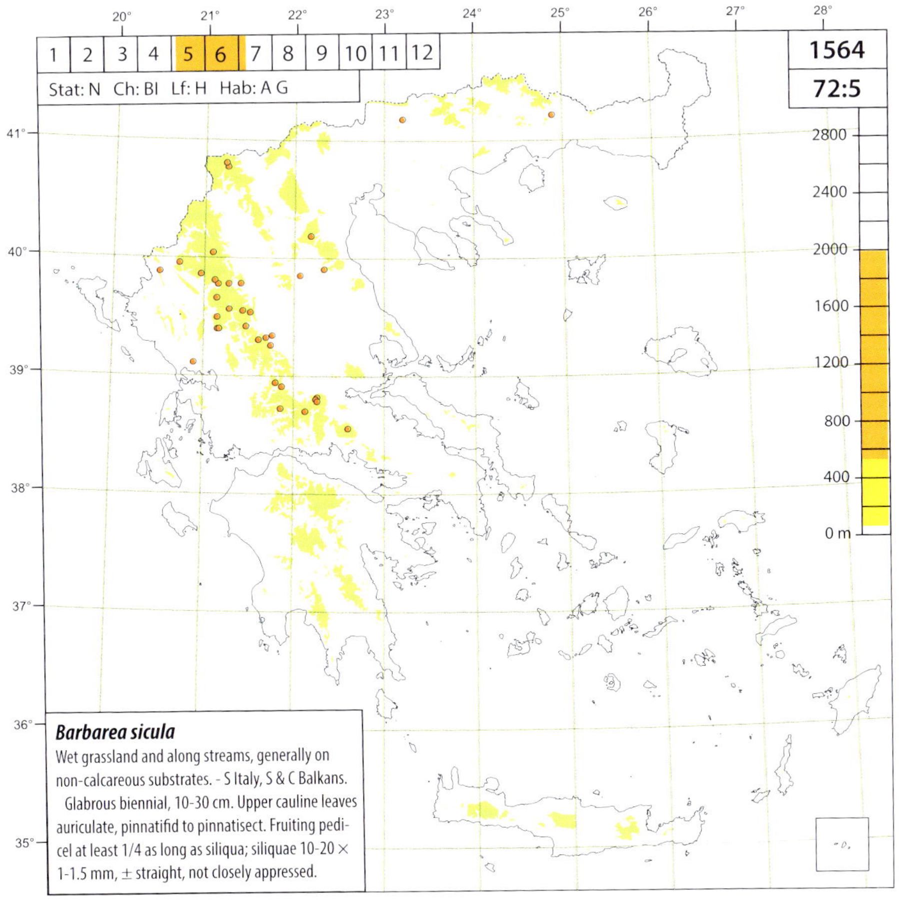

Barbarea sicula
Wet grassland and along streams, generally on non-calcareous substrates. - S Italy, S & C Balkans.
Glabrous biennial, 10-30 cm. Upper cauline leaves auriculate, pinnatifid to pinnatisect. Fruiting pedicel at least 1/4 as long as siliqua; siliquae 10-20 × 1-1.5 mm, ± straight, not closely appressed.

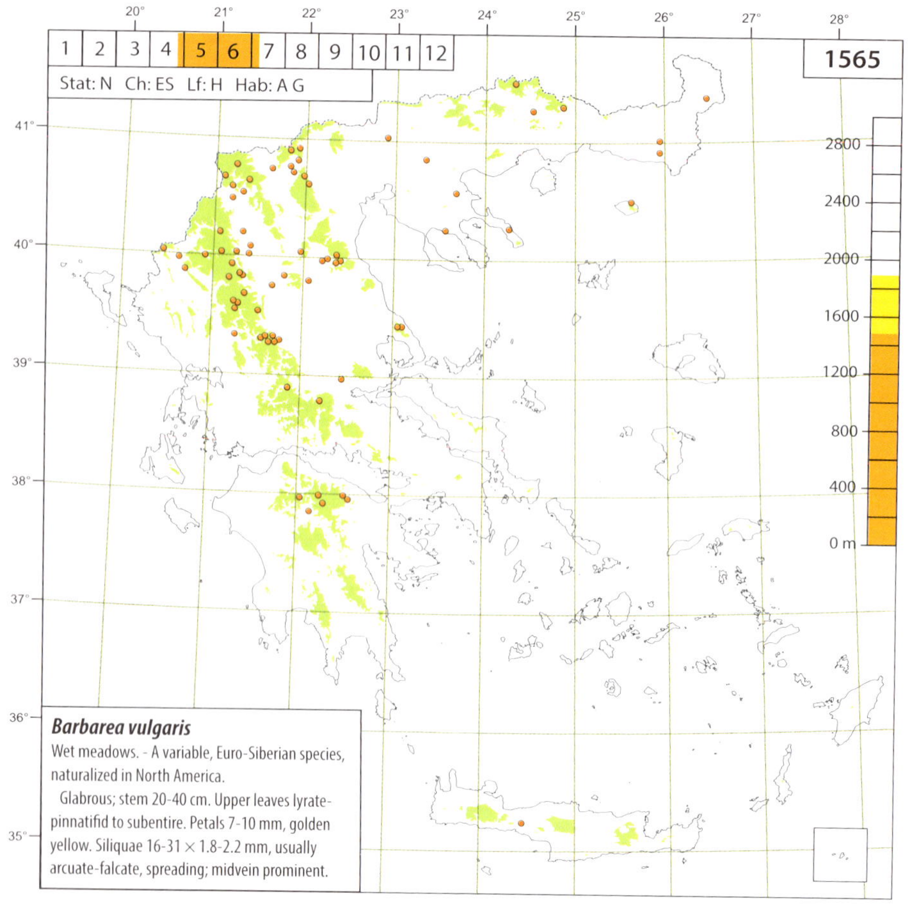

Barbarea vulgaris

Wet meadows. - A variable, Euro-Siberian species, naturalized in North America.

Glabrous; stem 20-40 cm. Upper leaves lyrate-pinnatifid to subentire. Petals 7-10 mm, golden yellow. Siliquae 16-31 × 1.8-2.2 mm, usually arcuate-falcate, spreading; midvein prominent.

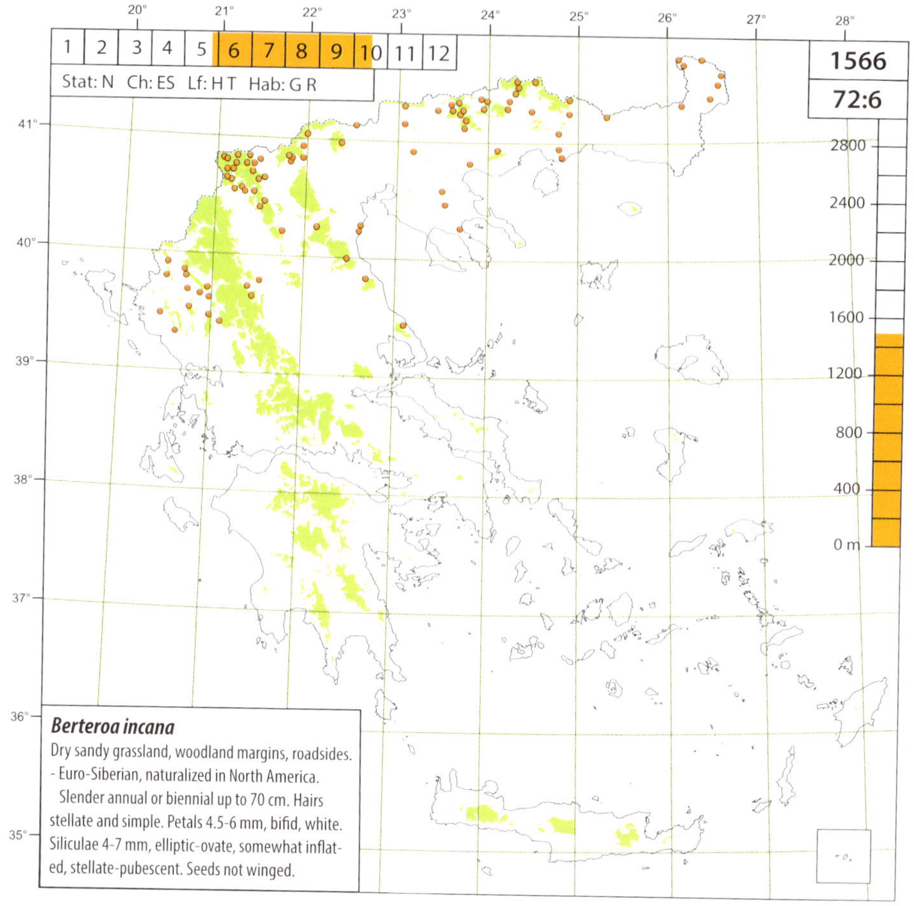

Berteroa incana

Dry sandy grassland, woodland margins, roadsides. - Euro-Siberian, naturalized in North America.

Slender annual or biennial up to 70 cm. Hairs stellate and simple. Petals 4.5-6 mm, bifid, white. Siliculae 4-7 mm, elliptic-ovate, somewhat inflated, stellate-pubescent. Seeds not winged.

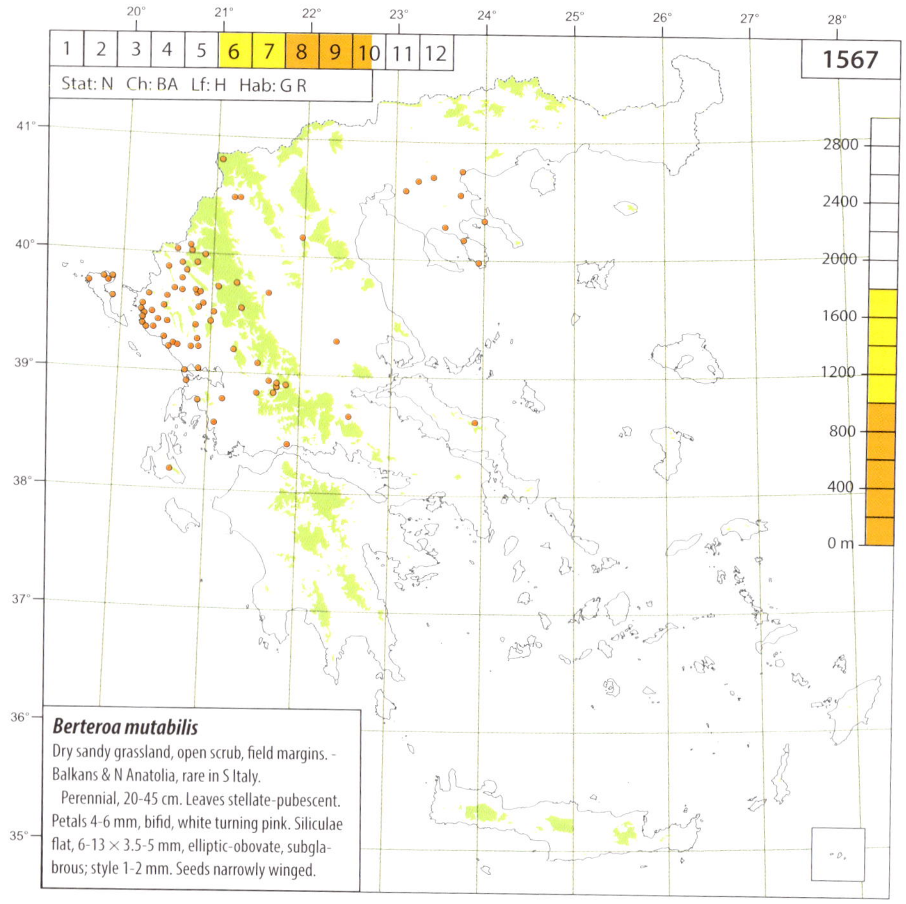

Berteroa mutabilis

Dry sandy grassland, open scrub, field margins. - Balkans & N Anatolia, rare in S Italy.

Perennial, 20-45 cm. Leaves stellate-pubescent. Petals 4-6 mm, bifid, white turning pink. Siliculae flat, 6-13 × 3.5-5 mm, elliptic-obovate, subglabrous; style 1-2 mm. Seeds narrowly winged.

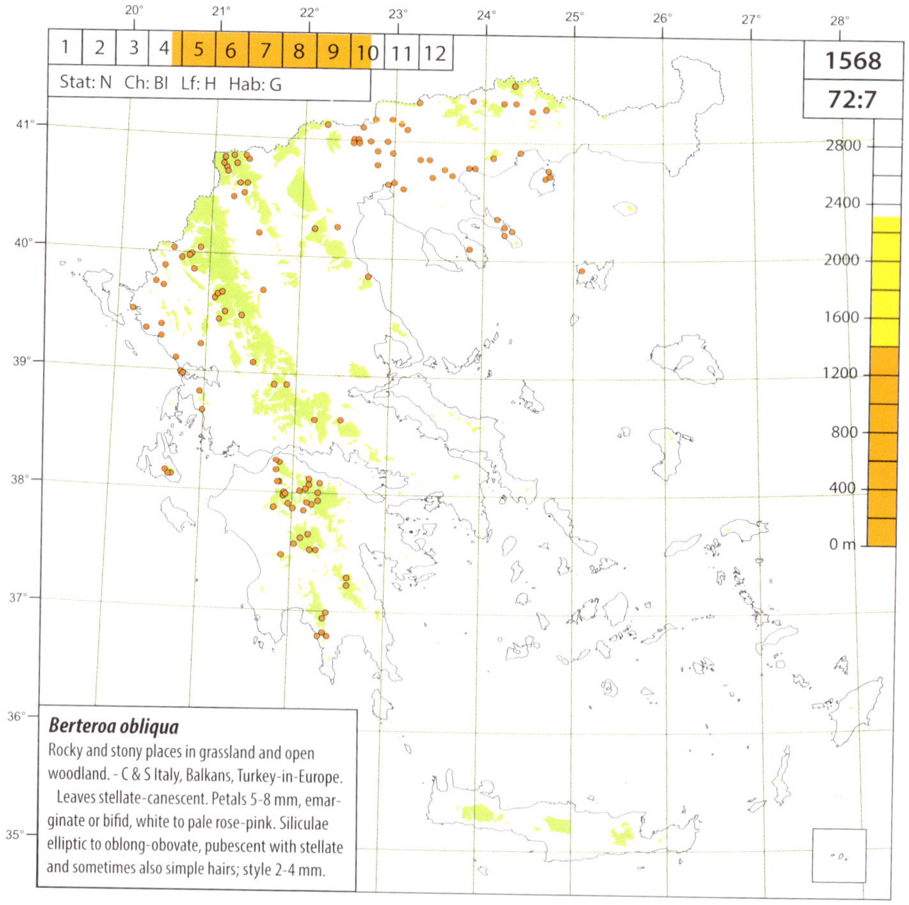

Berteroa obliqua

Rocky and stony places in grassland and open woodland. - C & S Italy, Balkans, Turkey-in-Europe.

Leaves stellate-canescent. Petals 5-8 mm, emarginate or bifid, white to pale rose-pink. Siliculae elliptic to oblong-obovate, pubescent with stellate and sometimes also simple hairs; style 2-4 mm.

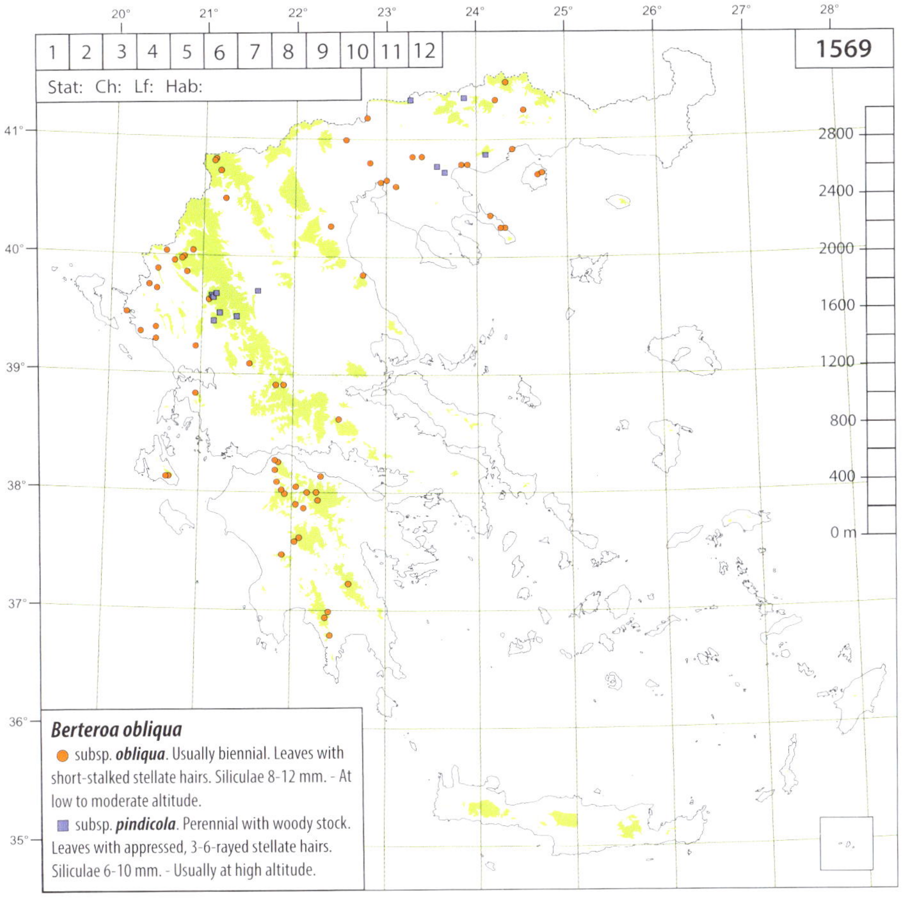

Berteroa obliqua
● subsp. ***obliqua***. Usually biennial. Leaves with short-stalked stellate hairs. Siliculae 8-12 mm. - At low to moderate altitude.
■ subsp. ***pindicola***. Perennial with woody stock. Leaves with appressed, 3-6-rayed stellate hairs. Siliculae 6-10 mm. - Usually at high altitude.

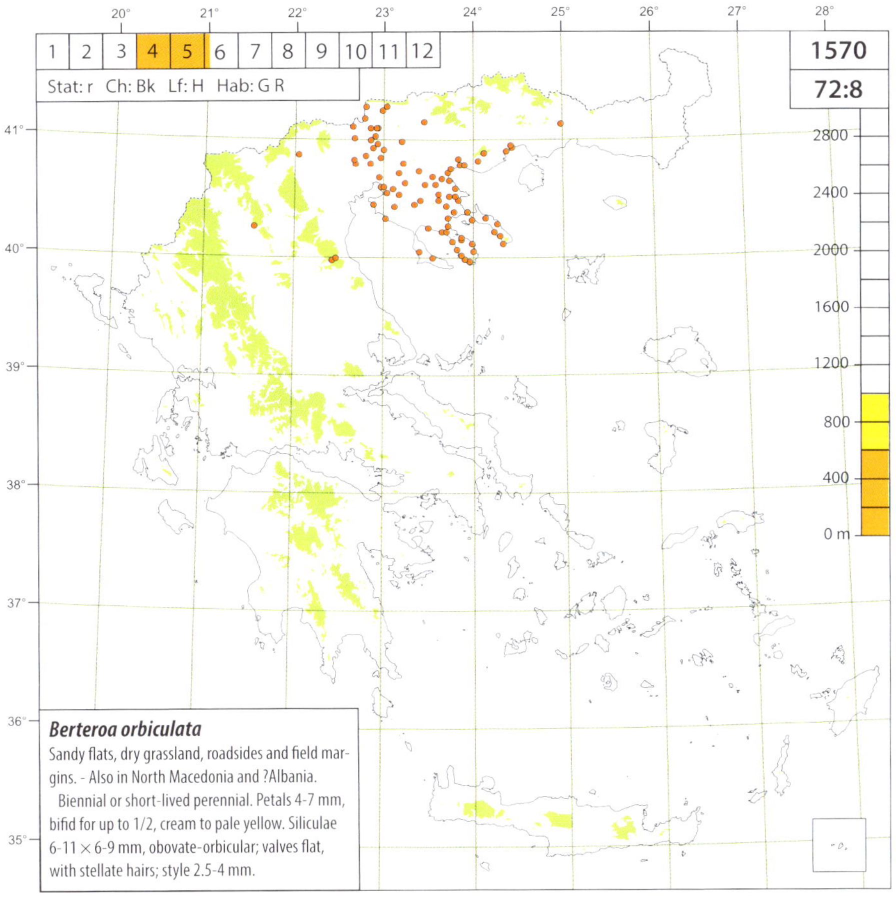

Berteroa orbiculata
Sandy flats, dry grassland, roadsides and field margins. - Also in North Macedonia and ?Albania.
Biennial or short-lived perennial. Petals 4-7 mm, bifid for up to 1/2, cream to pale yellow. Siliculae 6-11 × 6-9 mm, obovate-orbicular; valves flat, with stellate hairs; style 2.5-4 mm.

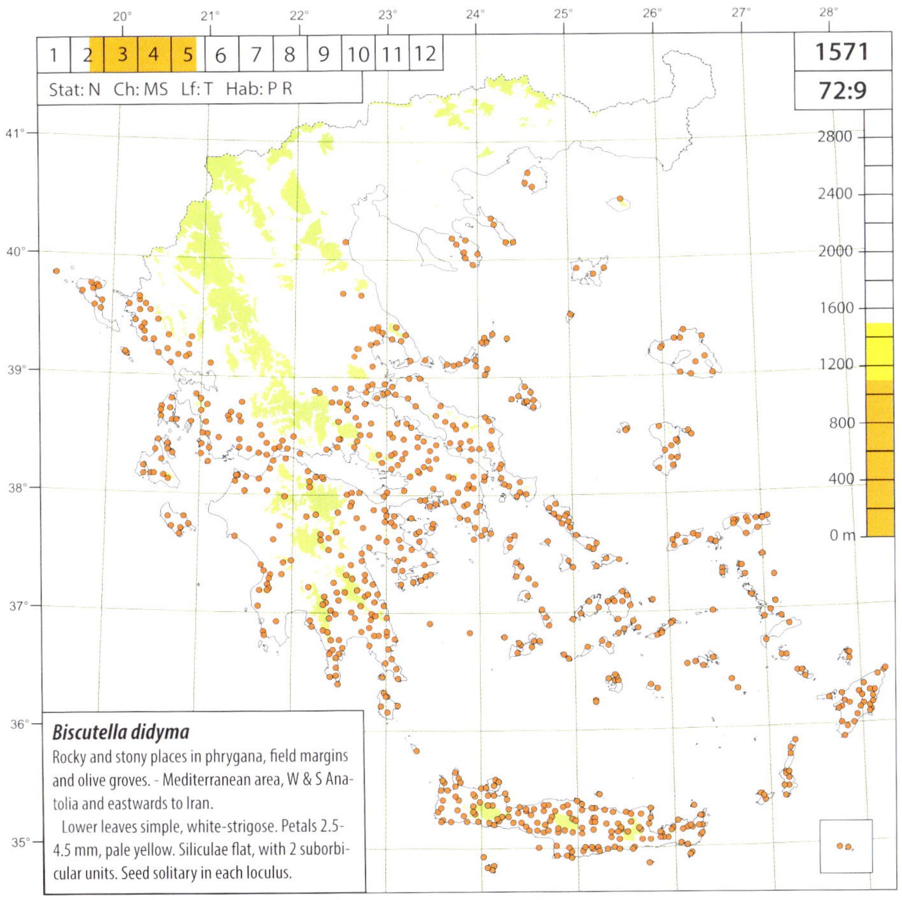

Biscutella didyma
Rocky and stony places in phrygana, field margins and olive groves. - Mediterranean area, W & S Anatolia and eastwards to Iran.
Lower leaves simple, white-strigose. Petals 2.5-4.5 mm, pale yellow. Siliculae flat, with 2 suborbicular units. Seed solitary in each loculus.

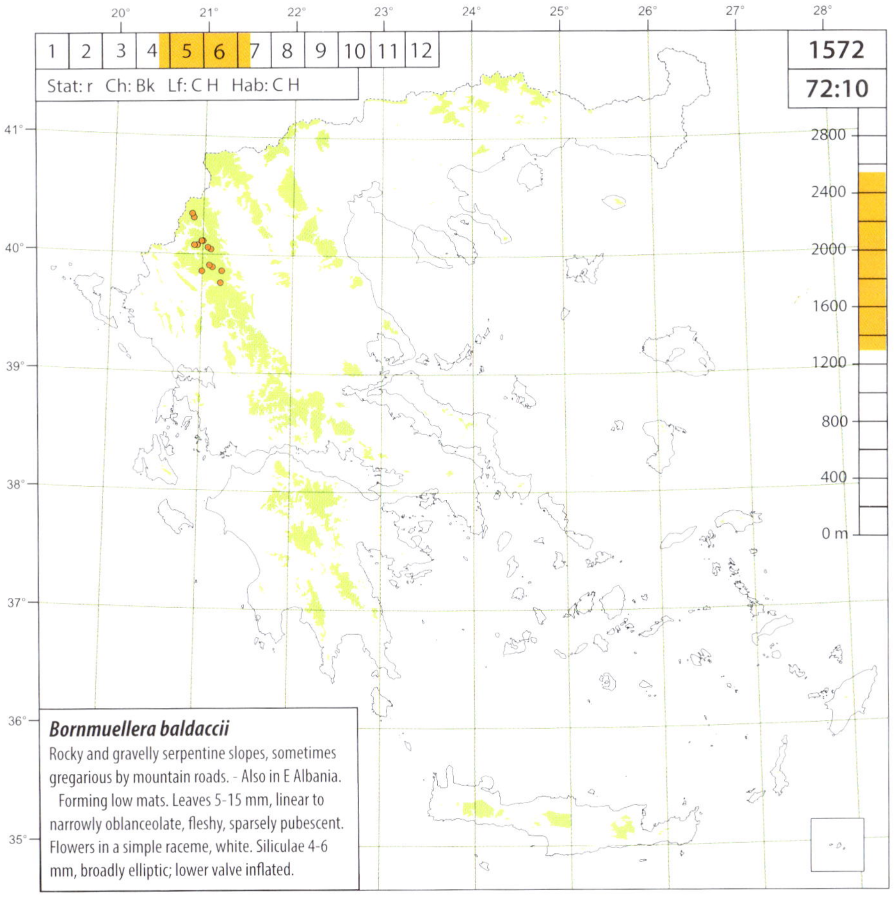

Bornmuellera baldaccii
Rocky and gravelly serpentine slopes, sometimes gregarious by mountain roads. - Also in E Albania.
Forming low mats. Leaves 5-15 mm, linear to narrowly oblanceolate, fleshy, sparsely pubescent. Flowers in a simple raceme, white. Siliculae 4-6 mm, broadly elliptic; lower valve inflated.

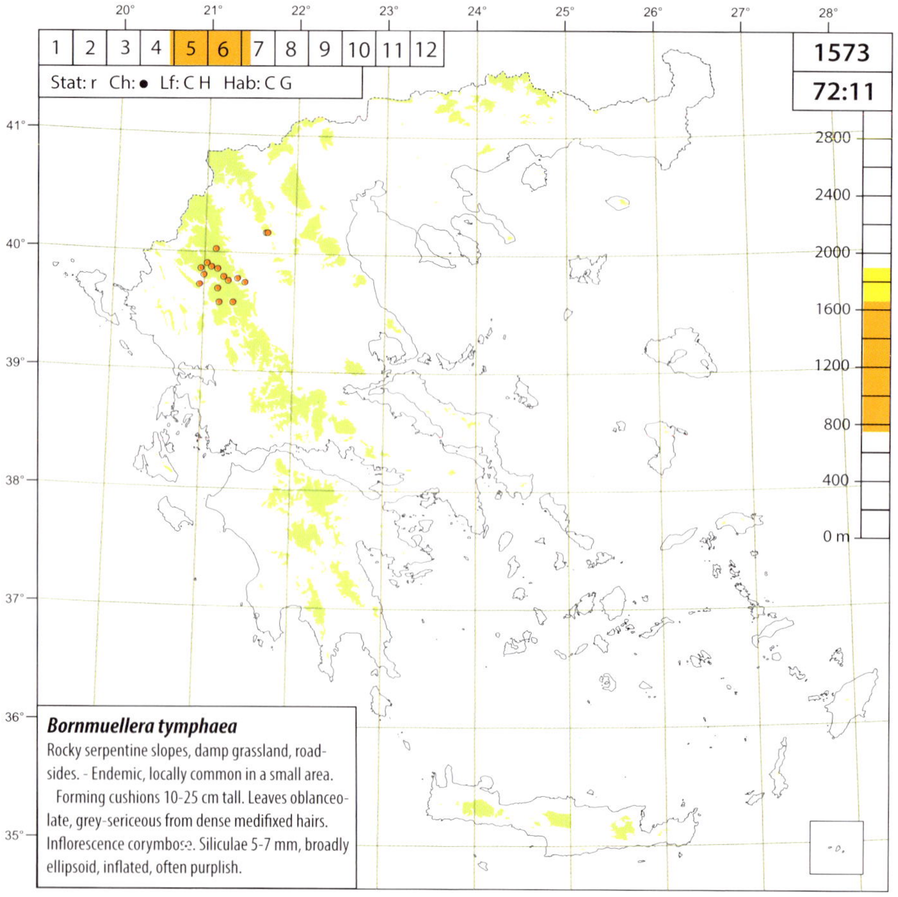

Bornmuellera tymphaea

Rocky serpentine slopes, damp grassland, roadsides. - Endemic, locally common in a small area.

Forming cushions 10-25 cm tall. Leaves oblanceolate, grey-sericeous from dense medifixed hairs. Inflorescence corymbose. Siliculae 5-7 mm, broadly ellipsoid, inflated, often purplish.

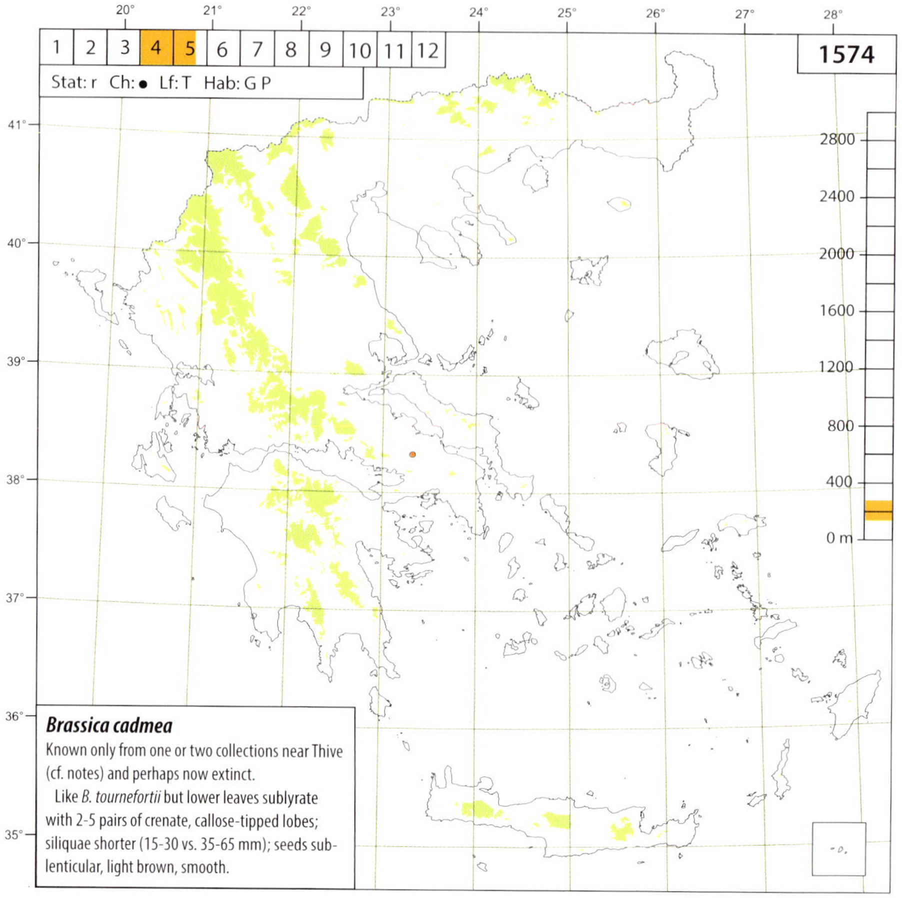

Brassica cadmea

Known only from one or two collections near Thive (cf. notes) and perhaps now extinct.

Like *B. tournefortii* but lower leaves sublyrate with 2-5 pairs of crenate, callose-tipped lobes; siliquae shorter (15-30 vs. 35-65 mm); seeds sublenticular, light brown, smooth.

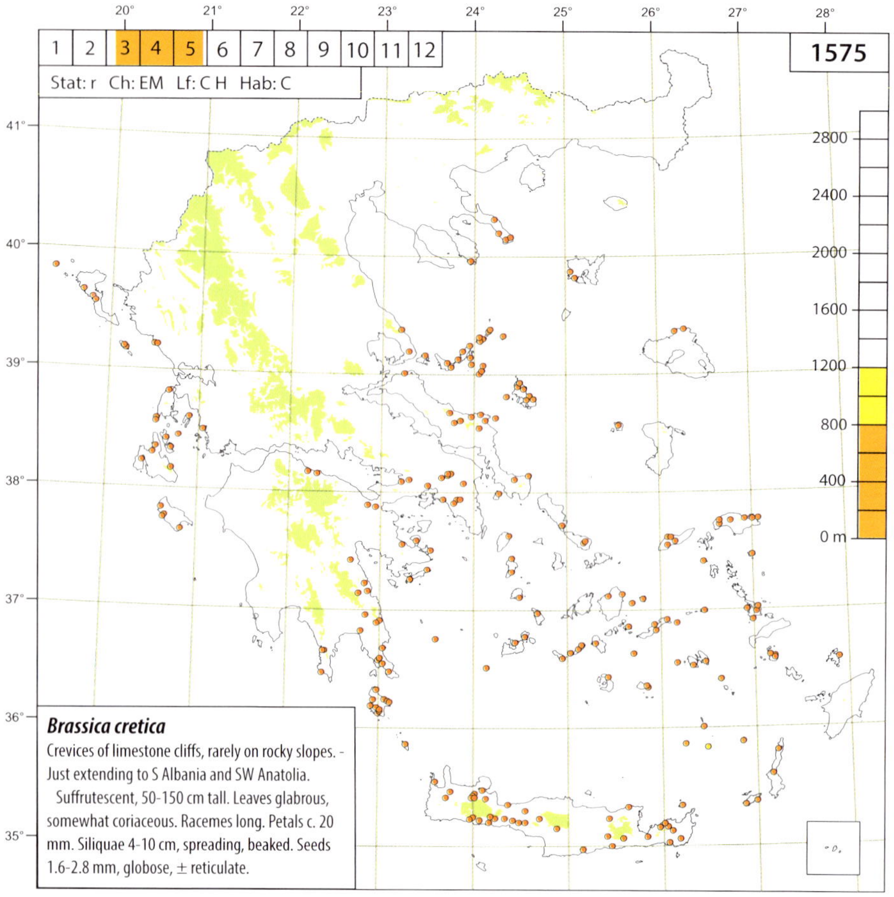

Brassica cretica

Crevices of limestone cliffs, rarely on rocky slopes. - Just extending to S Albania and SW Anatolia.

Suffrutescent, 50-150 cm tall. Leaves glabrous, somewhat coriaceous. Racemes long. Petals c. 20 mm. Siliquae 4-10 cm, spreading, beaked. Seeds 1.6-2.8 mm, globose, ± reticulate.

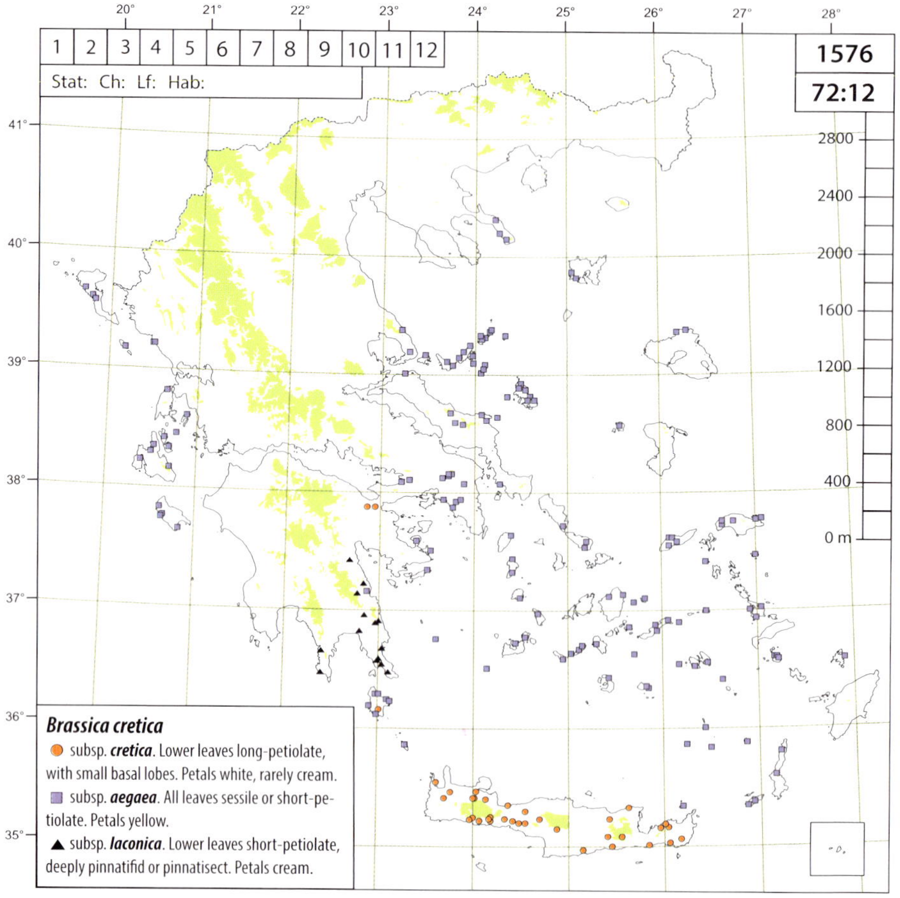

Brassica cretica

● subsp. ***cretica***. Lower leaves long-petiolate, with small basal lobes. Petals white, rarely cream.

■ subsp. ***aegaea***. All leaves sessile or short-petiolate. Petals yellow.

▲ subsp. ***laconica***. Lower leaves short-petiolate, deeply pinnatifid or pinnatisect. Petals cream.

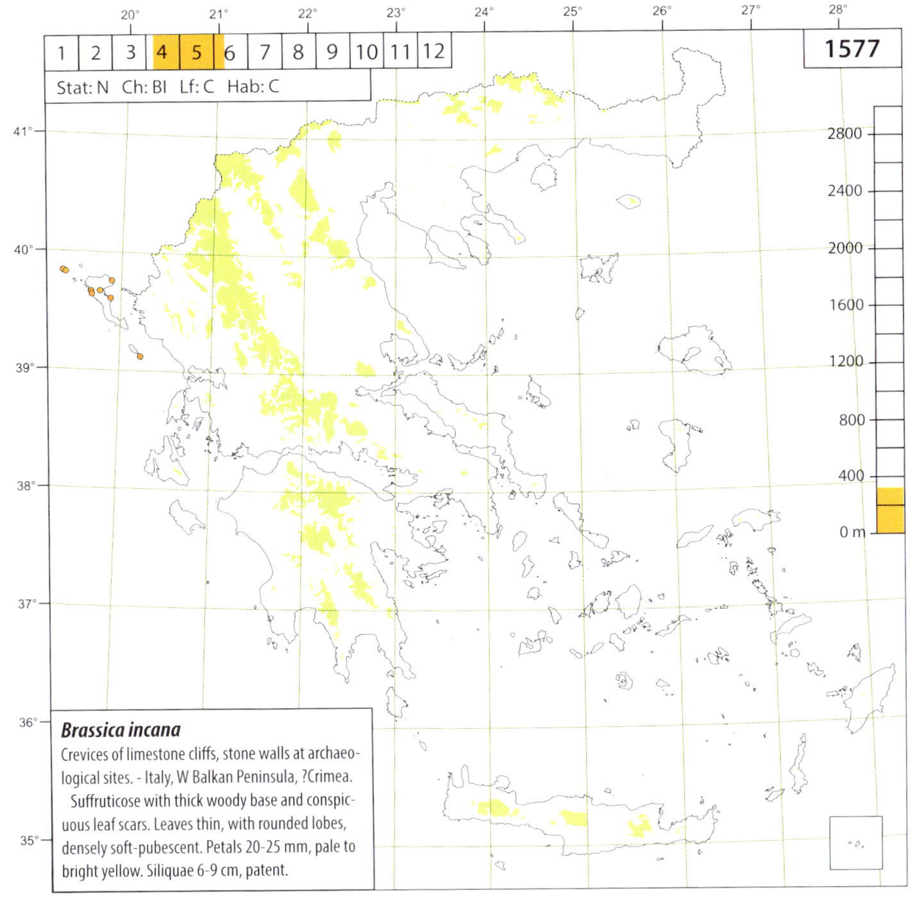

Brassica incana
Crevices of limestone cliffs, stone walls at archaeological sites. - Italy, W Balkan Peninsula, ?Crimea.
Suffruticose with thick woody base and conspicuous leaf scars. Leaves thin, with rounded lobes, densely soft-pubescent. Petals 20-25 mm, pale to bright yellow. Siliquae 6-9 cm, patent.

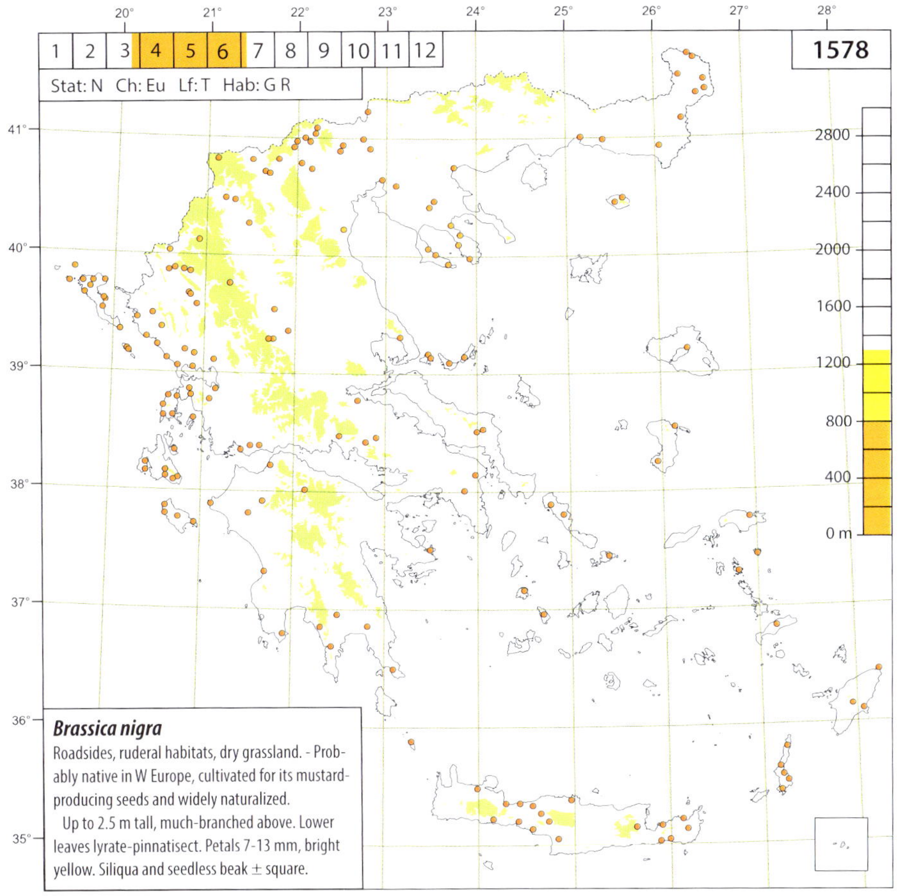

Brassica nigra
Roadsides, ruderal habitats, dry grassland. - Probably native in W Europe, cultivated for its mustard-producing seeds and widely naturalized.
Up to 2.5 m tall, much-branched above. Lower leaves lyrate-pinnatisect. Petals 7-13 mm, bright yellow. Siliqua and seedless beak ± square.

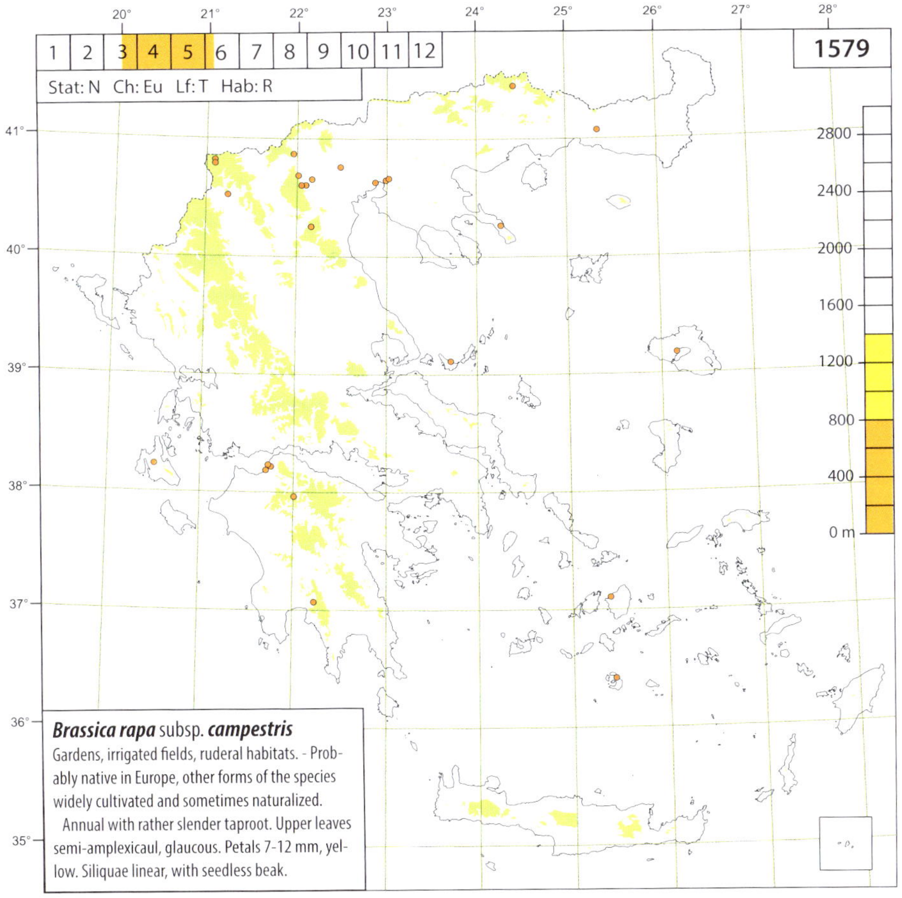

Brassica rapa* subsp. *campestris
Gardens, irrigated fields, ruderal habitats. - Probably native in Europe, other forms of the species widely cultivated and sometimes naturalized.
Annual with rather slender taproot. Upper leaves semi-amplexicaul, glaucous. Petals 7-12 mm, yellow. Siliquae linear, with seedless beak.

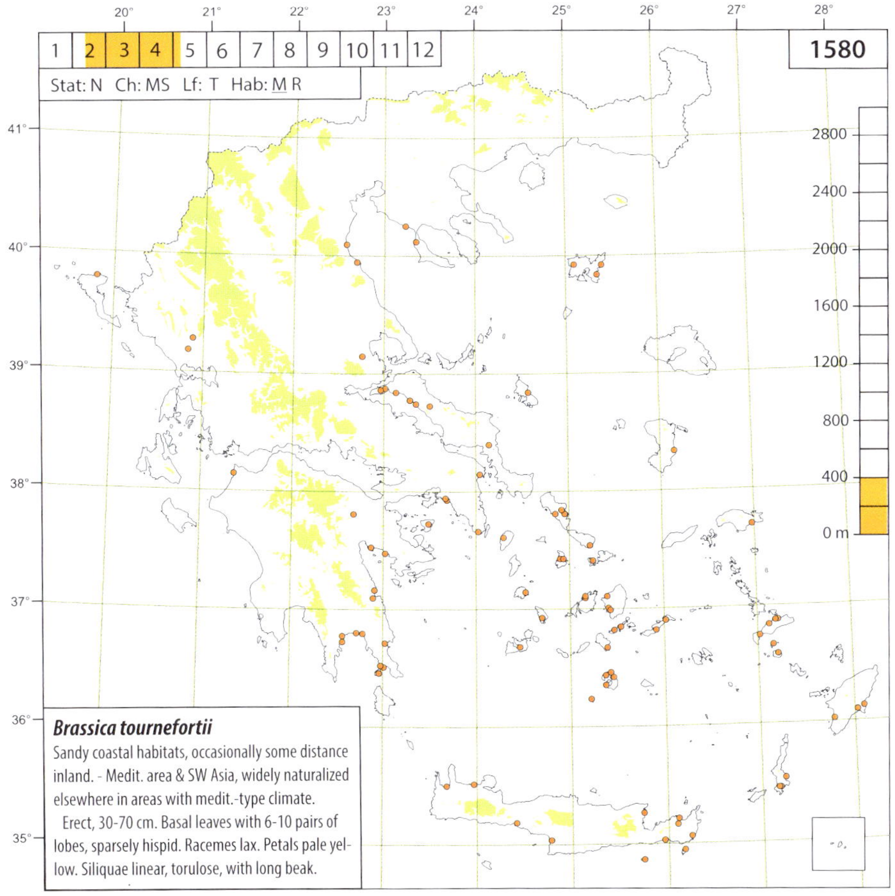

Brassica tournefortii
Sandy coastal habitats, occasionally some distance inland. - Medit. area & SW Asia, widely naturalized elsewhere in areas with medit.-type climate.
Erect, 30-70 cm. Basal leaves with 6-10 pairs of lobes, sparsely hispid. Racemes lax. Petals pale yellow. Siliquae linear, torulose, with long beak.

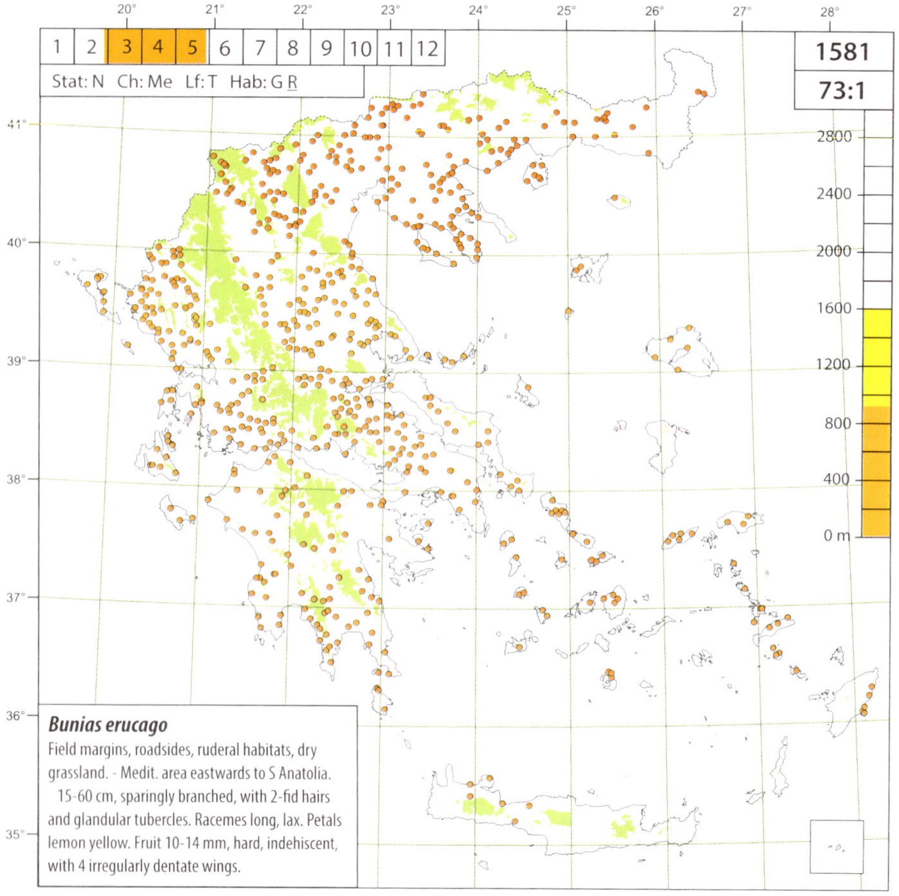

Bunias erucago
Field margins, roadsides, ruderal habitats, dry grassland. - Medit. area eastwards to S Anatolia.
15-60 cm, sparingly branched, with 2-fid hairs and glandular tubercles. Racemes long, lax. Petals lemon yellow. Fruit 10-14 mm, hard, indehiscent, with 4 irregularly dentate wings.

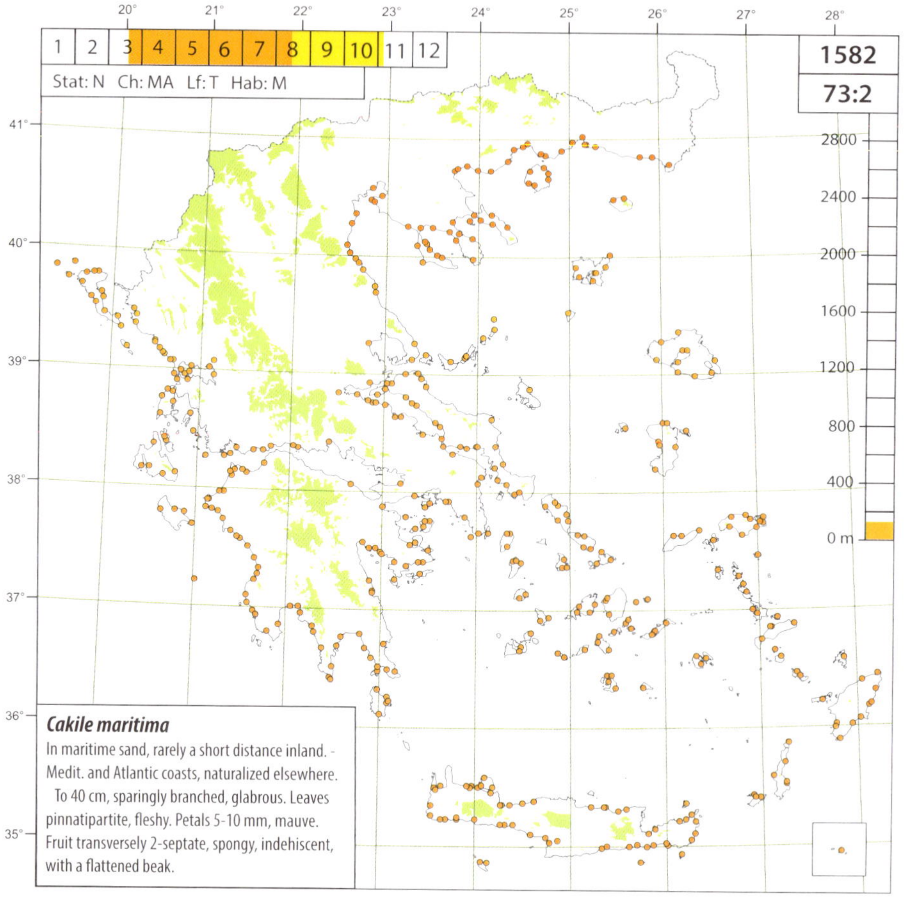

Cakile maritima
In maritime sand, rarely a short distance inland. - Medit. and Atlantic coasts, naturalized elsewhere.
To 40 cm, sparingly branched, glabrous. Leaves pinnatipartite, fleshy. Petals 5-10 mm, mauve. Fruit transversely 2-septate, spongy, indehiscent, with a flattened beak.

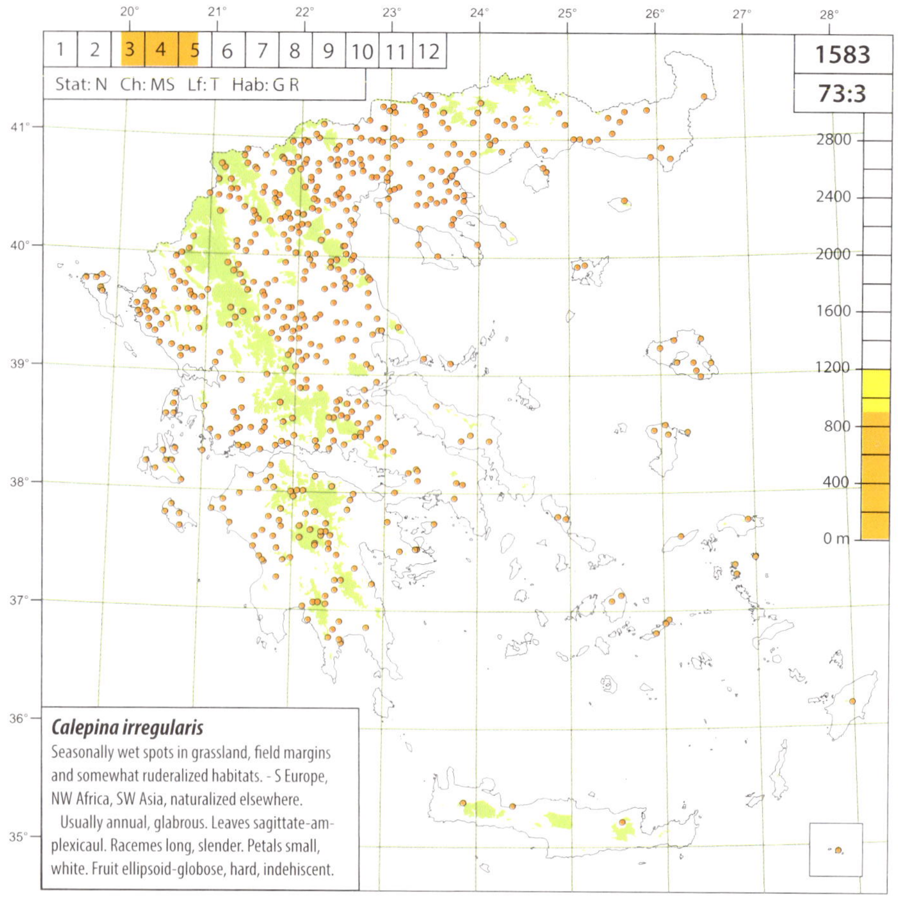

Calepina irregularis
Seasonally wet spots in grassland, field margins and somewhat ruderalized habitats. - S Europe, NW Africa, SW Asia, naturalized elsewhere.
Usually annual, glabrous. Leaves sagittate-amplexicaul. Racemes long, slender. Petals small, white. Fruit ellipsoid-globose, hard, indehiscent.

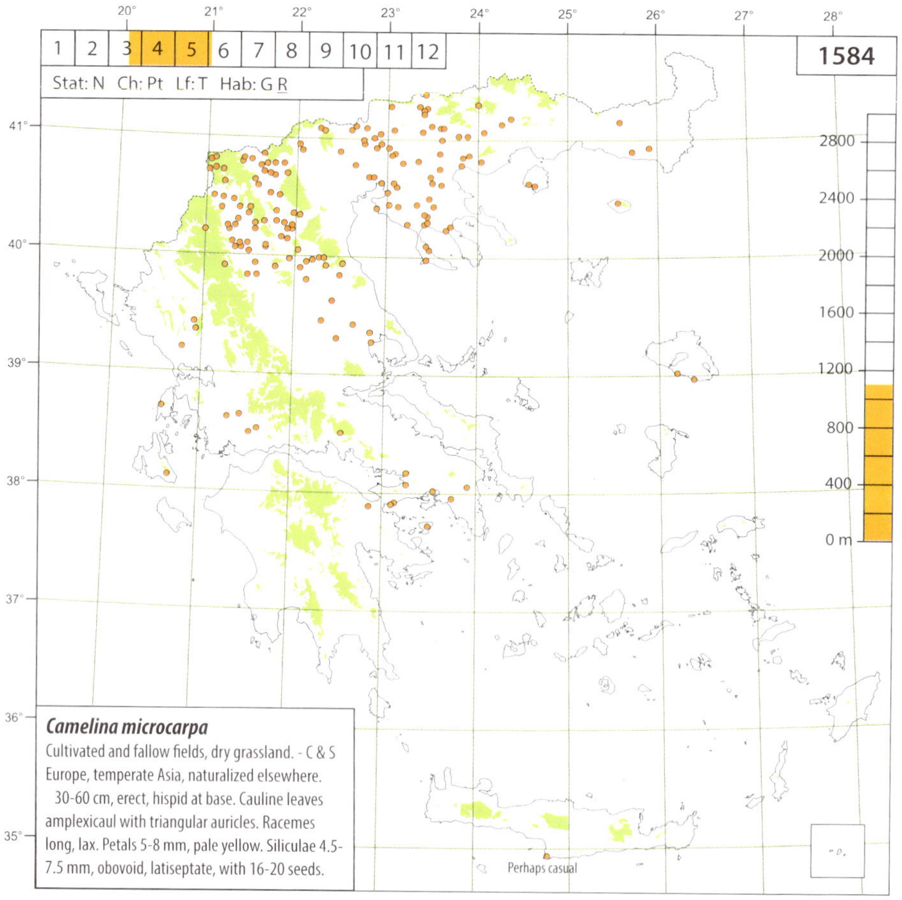

Camelina microcarpa
Cultivated and fallow fields, dry grassland. - C & S Europe, temperate Asia, naturalized elsewhere.
30-60 cm, erect, hispid at base. Cauline leaves amplexicaul with triangular auricles. Racemes long, lax. Petals 5-8 mm, pale yellow. Siliculae 4.5-7.5 mm, obovoid, latiseptate, with 16-20 seeds.

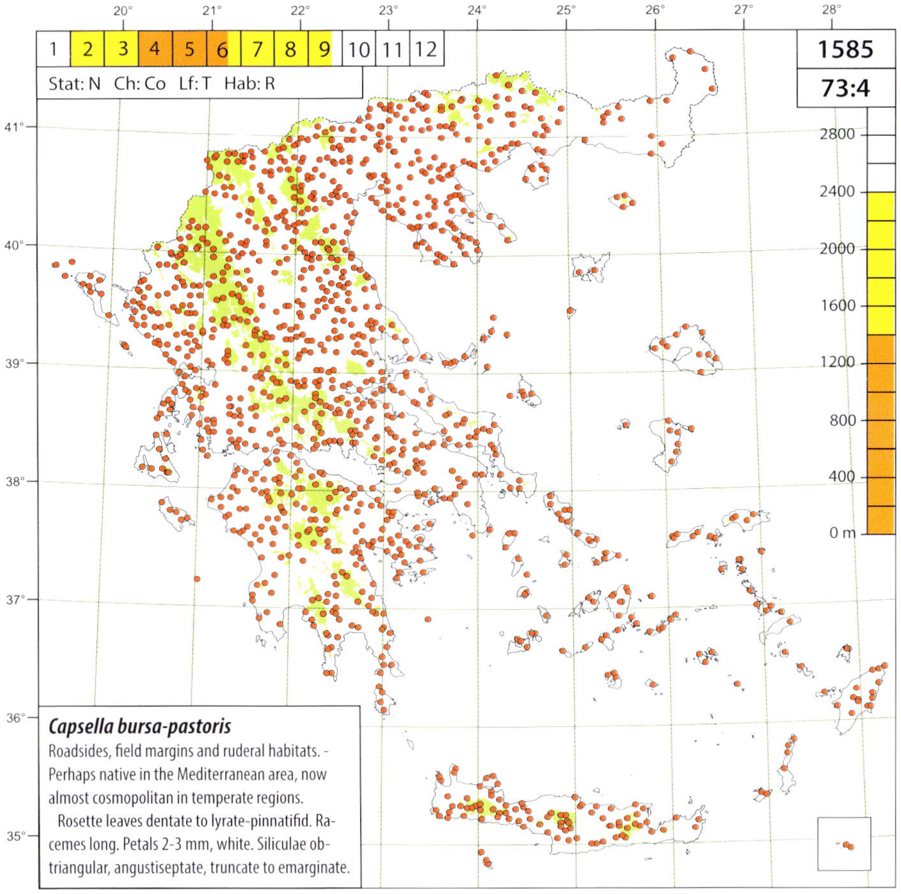

Capsella bursa-pastoris
Roadsides, field margins and ruderal habitats. - Perhaps native in the Mediterranean area, now almost cosmopolitan in temperate regions.
Rosette leaves dentate to lyrate-pinnatifid. Racemes long. Petals 2-3 mm, white. Siliculae obtriangular, angustiseptate, truncate to emarginate.

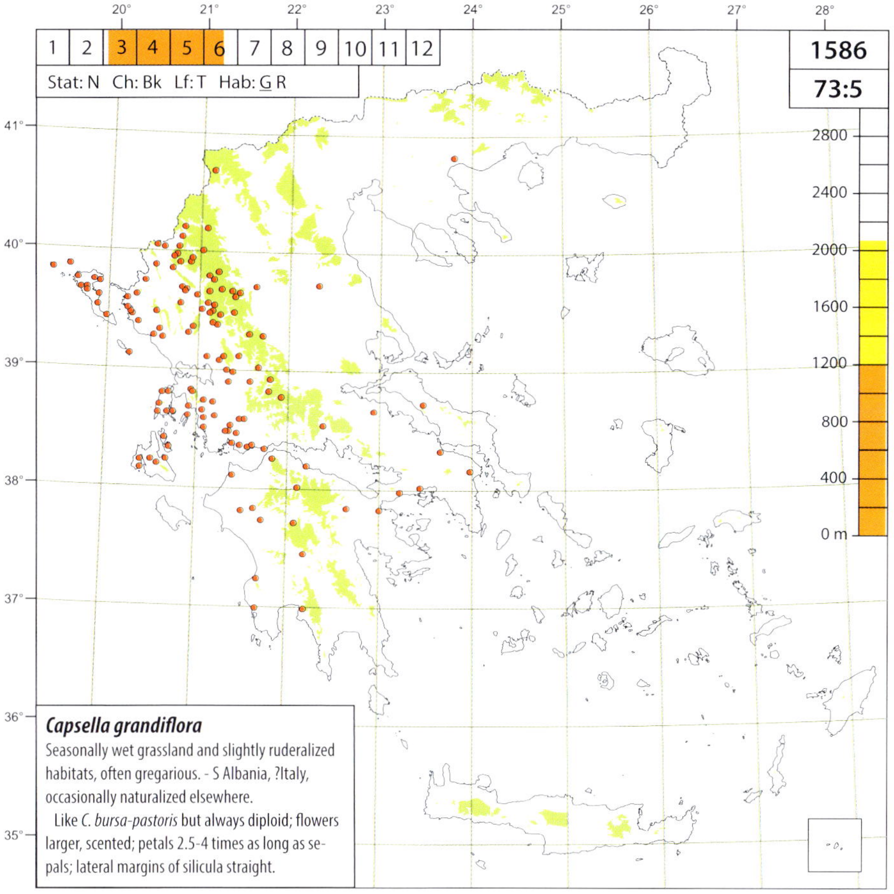

Capsella grandiflora
Seasonally wet grassland and slightly ruderalized habitats, often gregarious. - S Albania, ?Italy, occasionally naturalized elsewhere.
Like *C. bursa-pastoris* but always diploid; flowers larger, scented; petals 2.5-4 times as long as sepals; lateral margins of silicula straight.

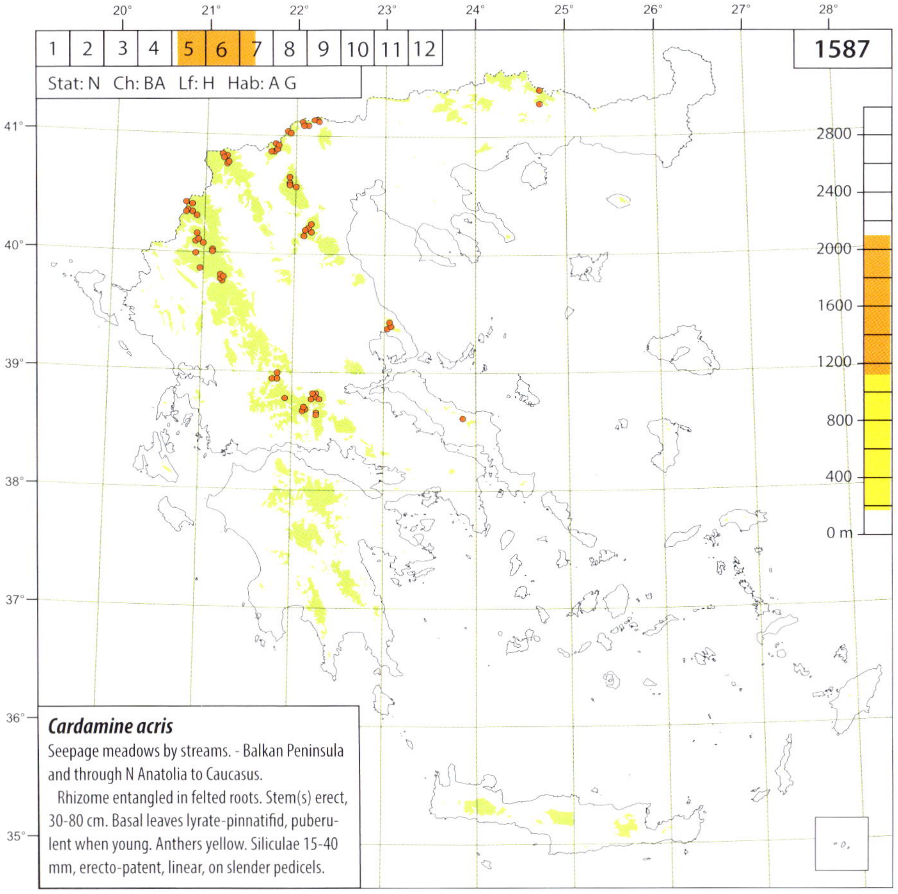

Cardamine acris
Seepage meadows by streams. - Balkan Peninsula and through N Anatolia to Caucasus.
Rhizome entangled in felted roots. Stem(s) erect, 30-80 cm. Basal leaves lyrate-pinnatifid, puberulent when young. Anthers yellow. Siliculae 15-40 mm, erecto-patent, linear, on slender pedicels.

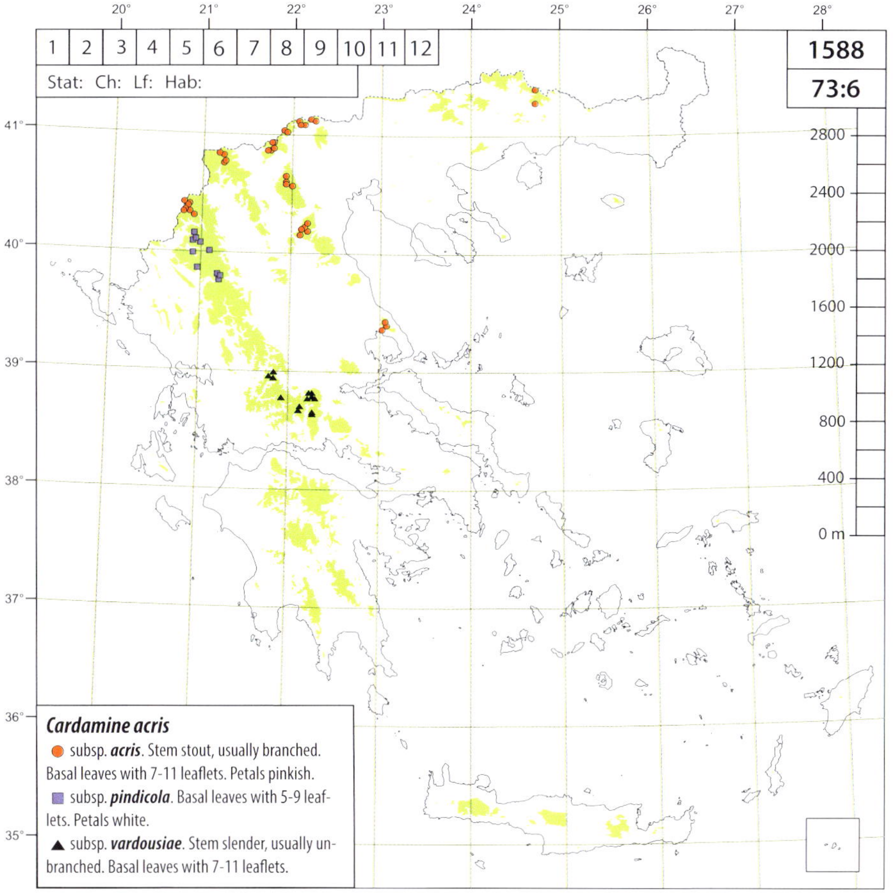

Cardamine acris
● subsp. ***acris***. Stem stout, usually branched. Basal leaves with 7-11 leaflets. Petals pinkish.
■ subsp. ***pindicola***. Basal leaves with 5-9 leaflets. Petals white.
▲ subsp. ***vardousiae***. Stem slender, usually unbranched. Basal leaves with 7-11 leaflets.

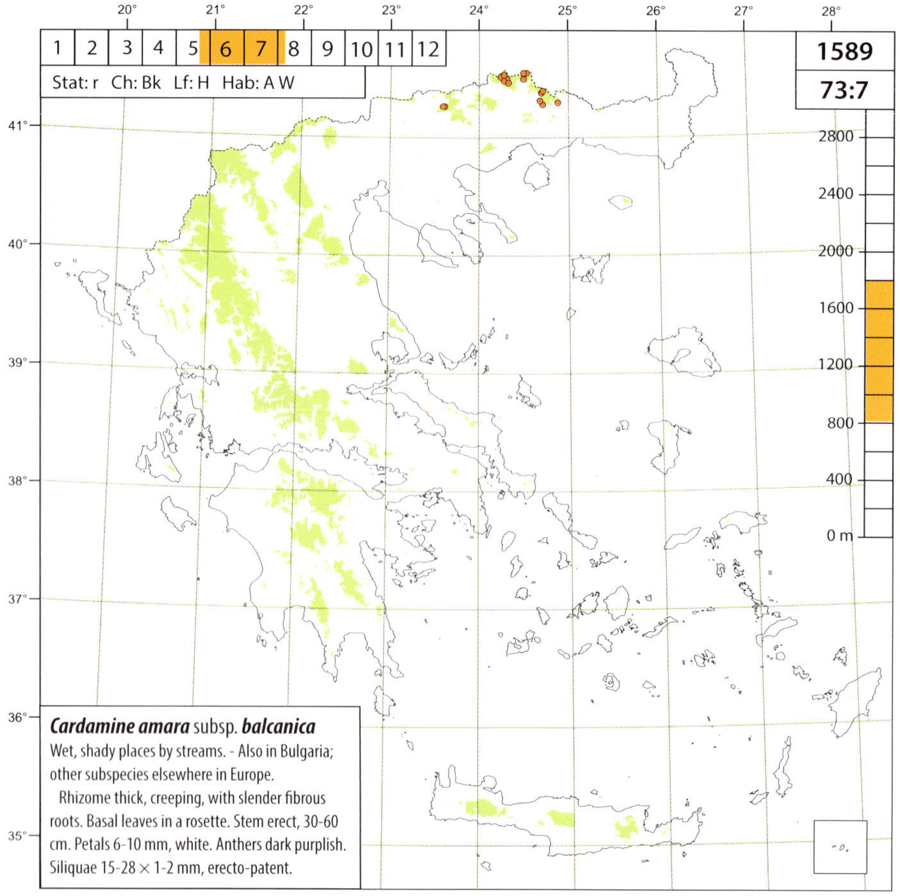

Cardamine amara subsp. **balcanica**
Wet, shady places by streams. - Also in Bulgaria; other subspecies elsewhere in Europe.
Rhizome thick, creeping, with slender fibrous roots. Basal leaves in a rosette. Stem erect, 30-60 cm. Petals 6-10 mm, white. Anthers dark purplish. Siliquae 15-28 × 1-2 mm, erecto-patent.

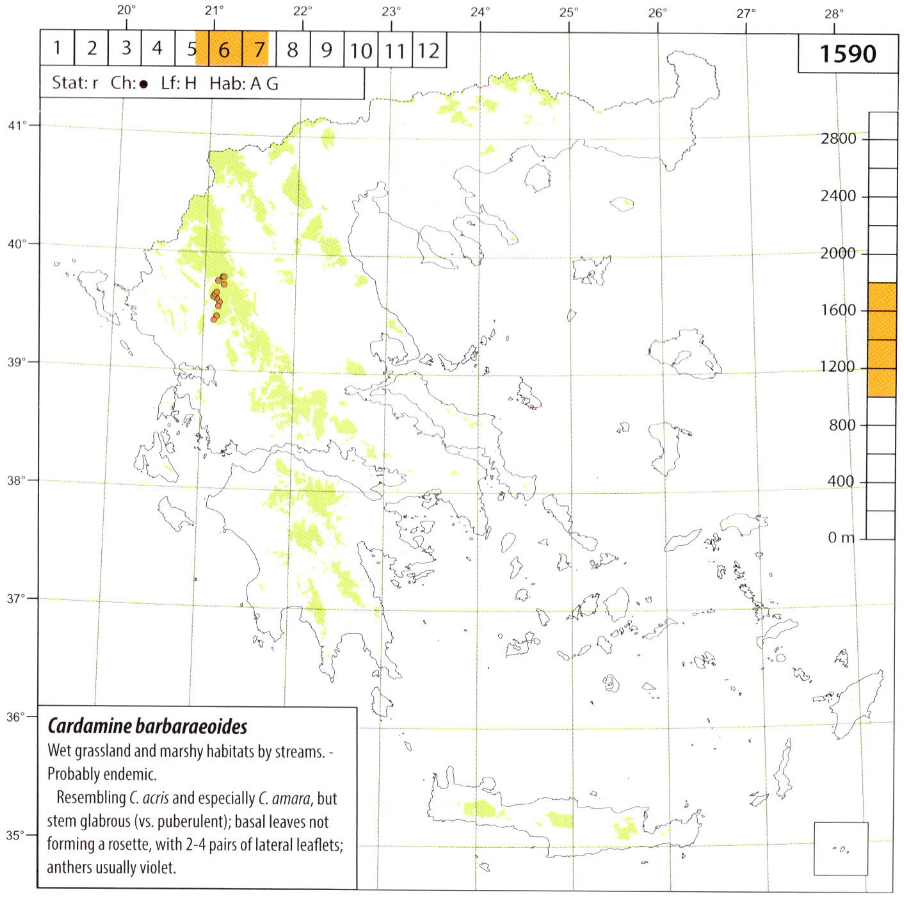

Cardamine barbaraeoides
Wet grassland and marshy habitats by streams. - Probably endemic.
Resembling *C. acris* and especially *C. amara*, but stem glabrous (vs. puberulent); basal leaves not forming a rosette, with 2-4 pairs of lateral leaflets; anthers usually violet.

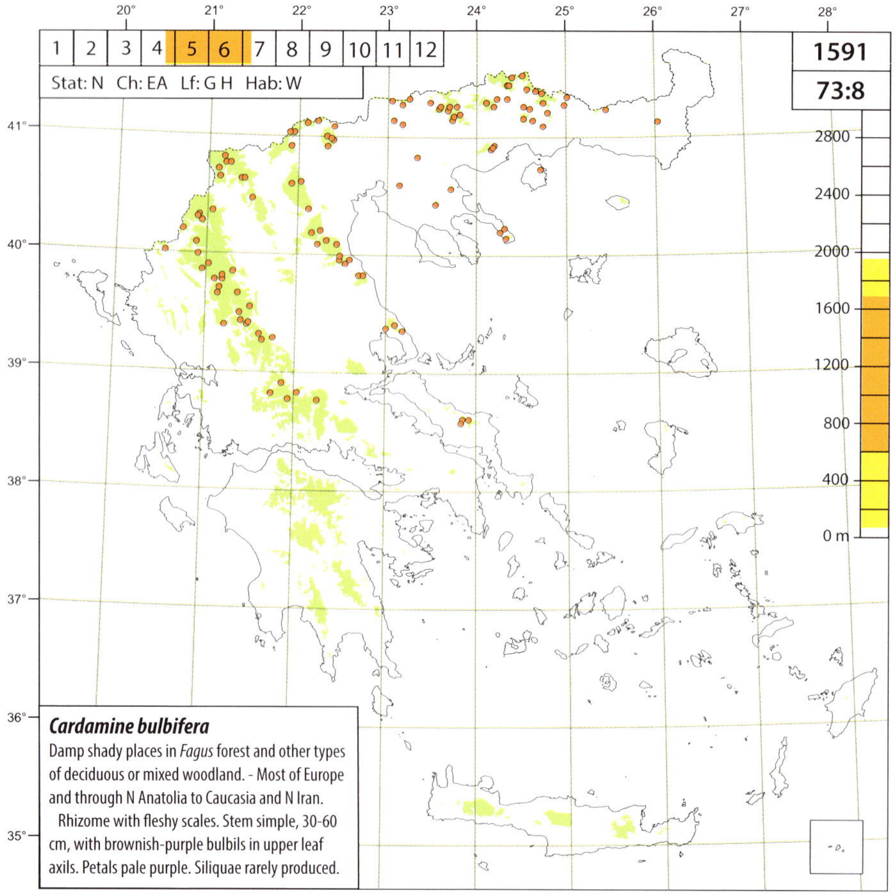

Cardamine bulbifera
Damp shady places in *Fagus* forest and other types of deciduous or mixed woodland. - Most of Europe and through N Anatolia to Caucasia and N Iran.
Rhizome with fleshy scales. Stem simple, 30-60 cm, with brownish-purple bulbils in upper leaf axils. Petals pale purple. Siliquae rarely produced.

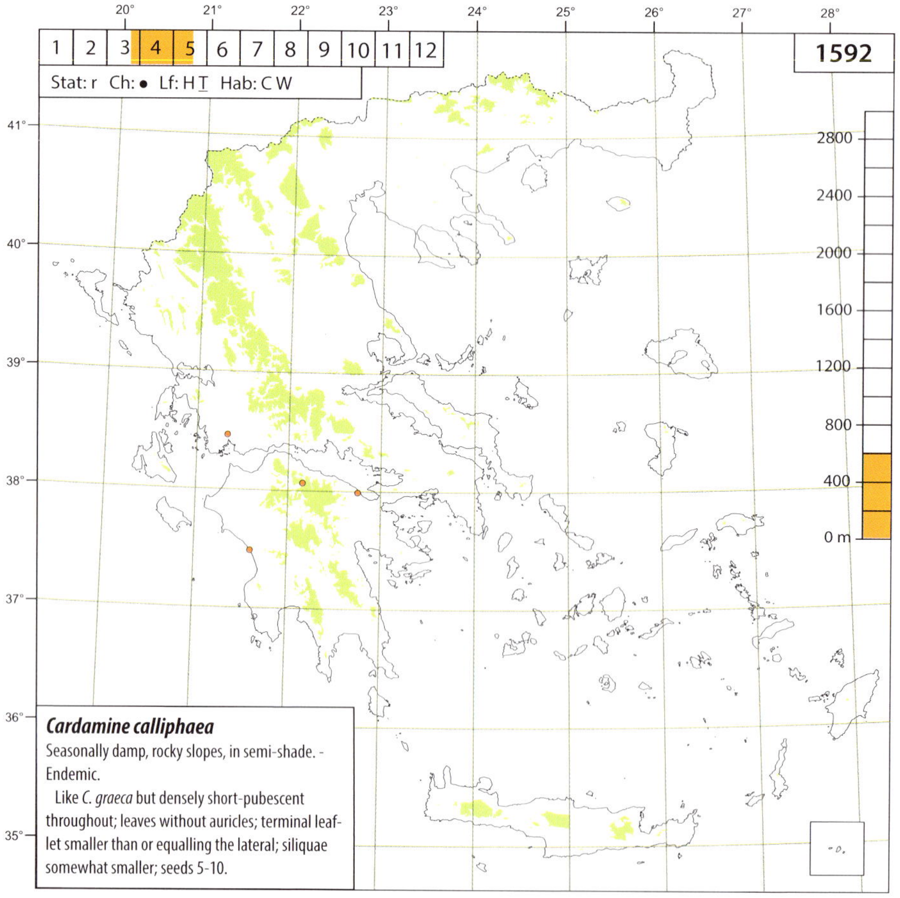

Cardamine calliphaea
Seasonally damp, rocky slopes, in semi-shade. - Endemic.
Like *C. graeca* but densely short-pubescent throughout; leaves without auricles; terminal leaflet smaller than or equalling the lateral; siliquae somewhat smaller; seeds 5-10.

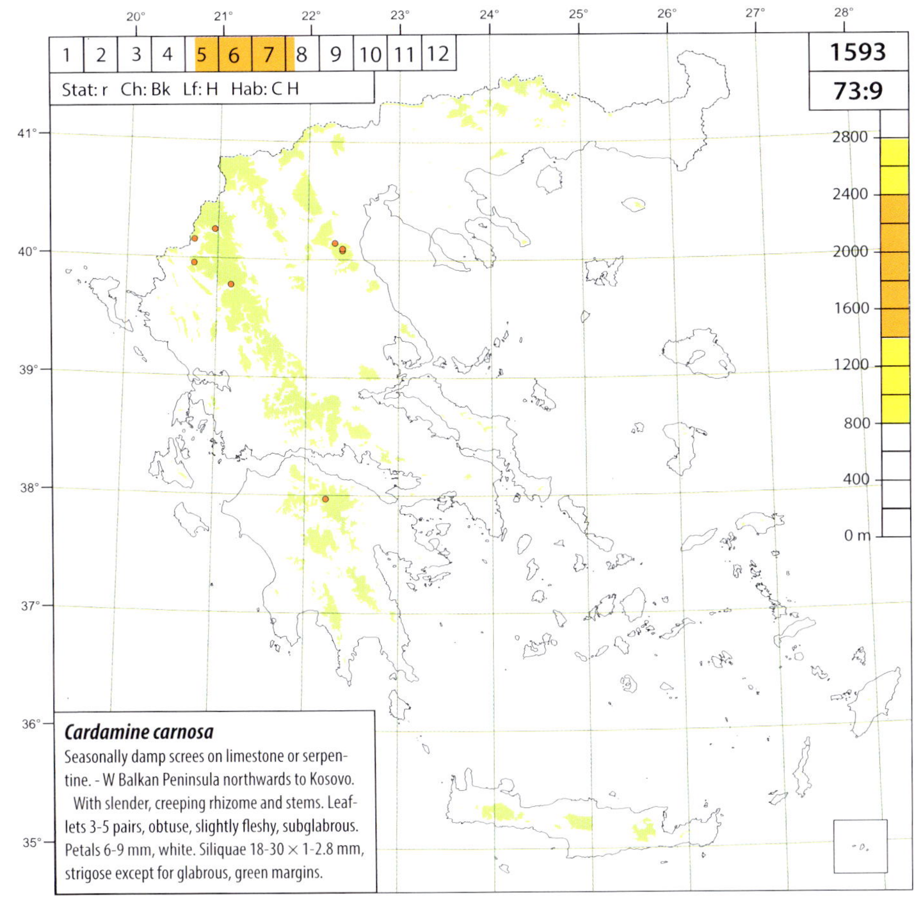

Cardamine carnosa

Seasonally damp screes on limestone or serpentine. - W Balkan Peninsula northwards to Kosovo.

With slender, creeping rhizome and stems. Leaflets 3-5 pairs, obtuse, slightly fleshy, subglabrous. Petals 6-9 mm, white. Siliquae 18-30 × 1-2.8 mm, strigose except for glabrous, green margins.

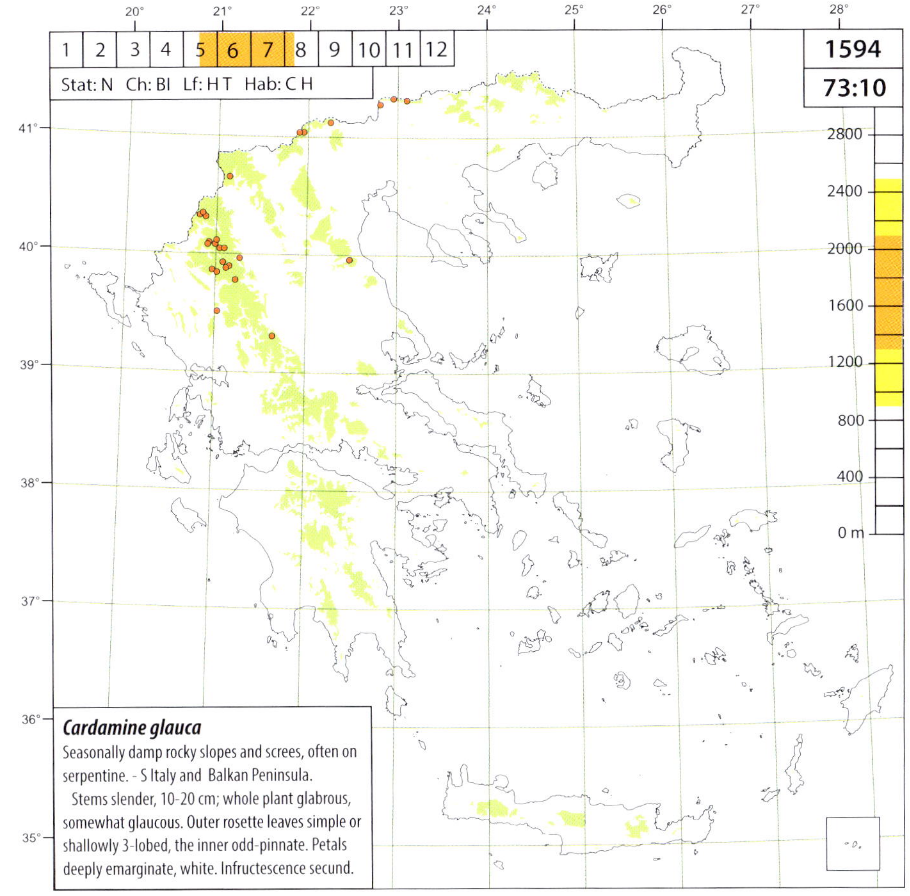

Cardamine glauca

Seasonally damp rocky slopes and screes, often on serpentine. - S Italy and Balkan Peninsula.

Stems slender, 10-20 cm; whole plant glabrous, somewhat glaucous. Outer rosette leaves simple or shallowly 3-lobed, the inner odd-pinnate. Petals deeply emarginate, white. Infructescence secund.

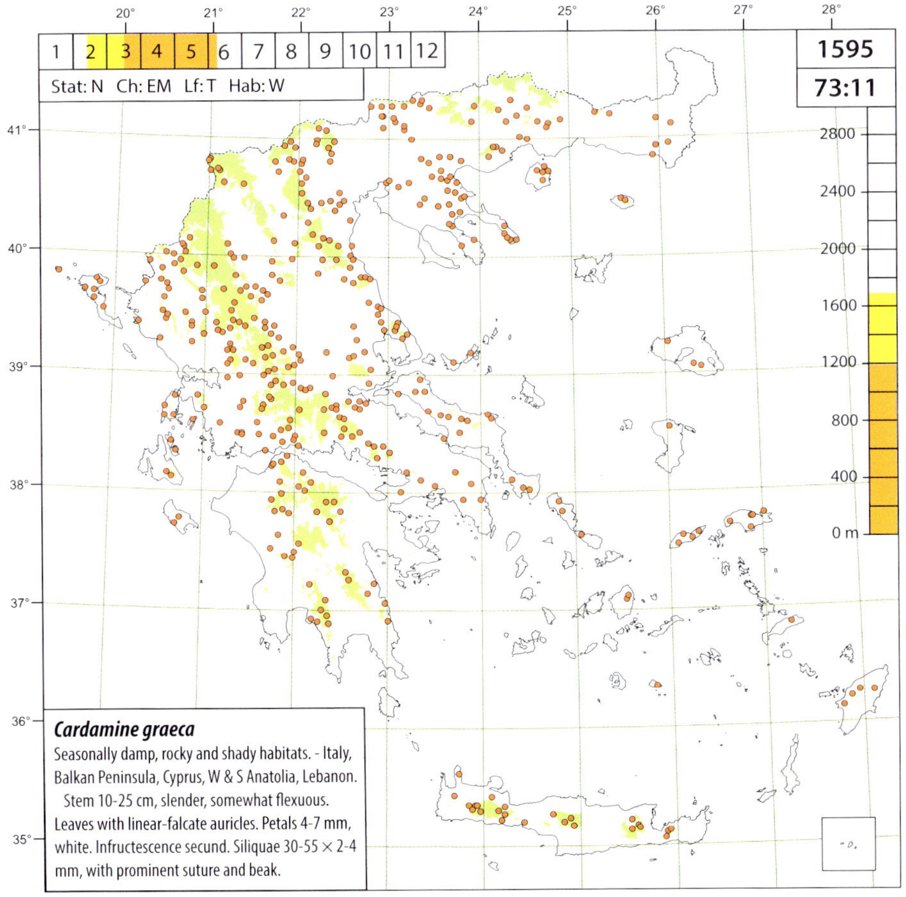

Cardamine graeca

Seasonally damp, rocky and shady habitats. - Italy, Balkan Peninsula, Cyprus, W & S Anatolia, Lebanon.

Stem 10-25 cm, slender, somewhat flexuous. Leaves with linear-falcate auricles. Petals 4-7 mm, white. Infructescence secund. Siliquae 30-55 × 2-4 mm, with prominent suture and beak.

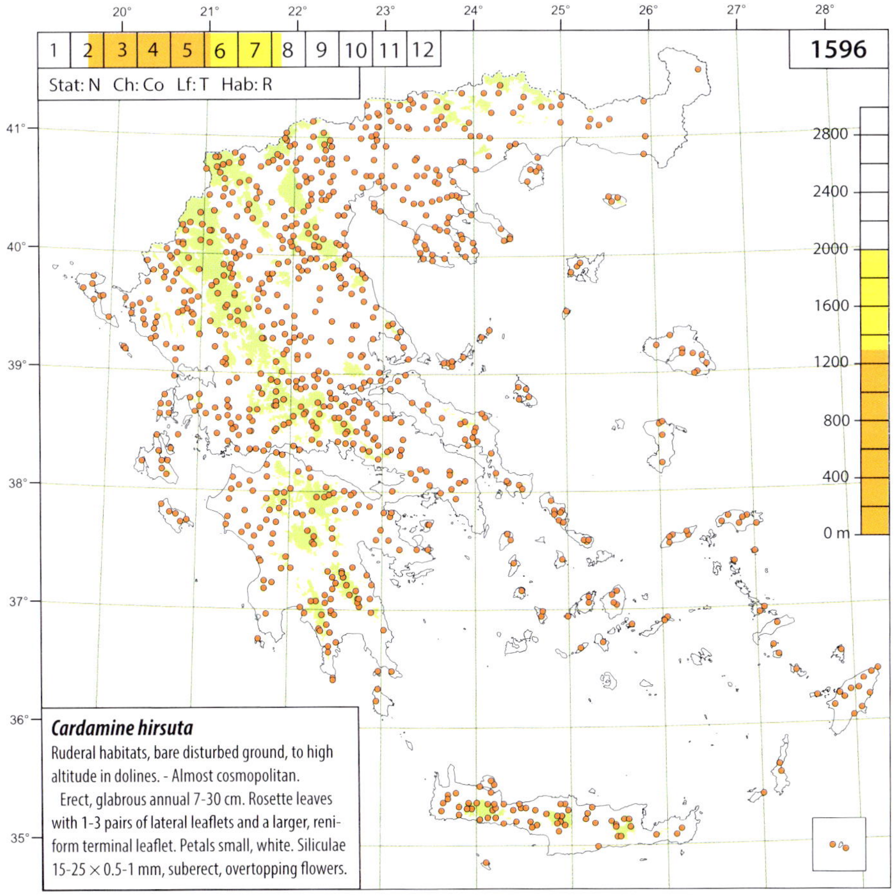

Cardamine hirsuta

Ruderal habitats, bare disturbed ground, to high altitude in dolines. - Almost cosmopolitan.

Erect, glabrous annual 7-30 cm. Rosette leaves with 1-3 pairs of lateral leaflets and a larger, reniform terminal leaflet. Petals small, white. Siliculae 15-25 × 0.5-1 mm, suberect, overtopping flowers.

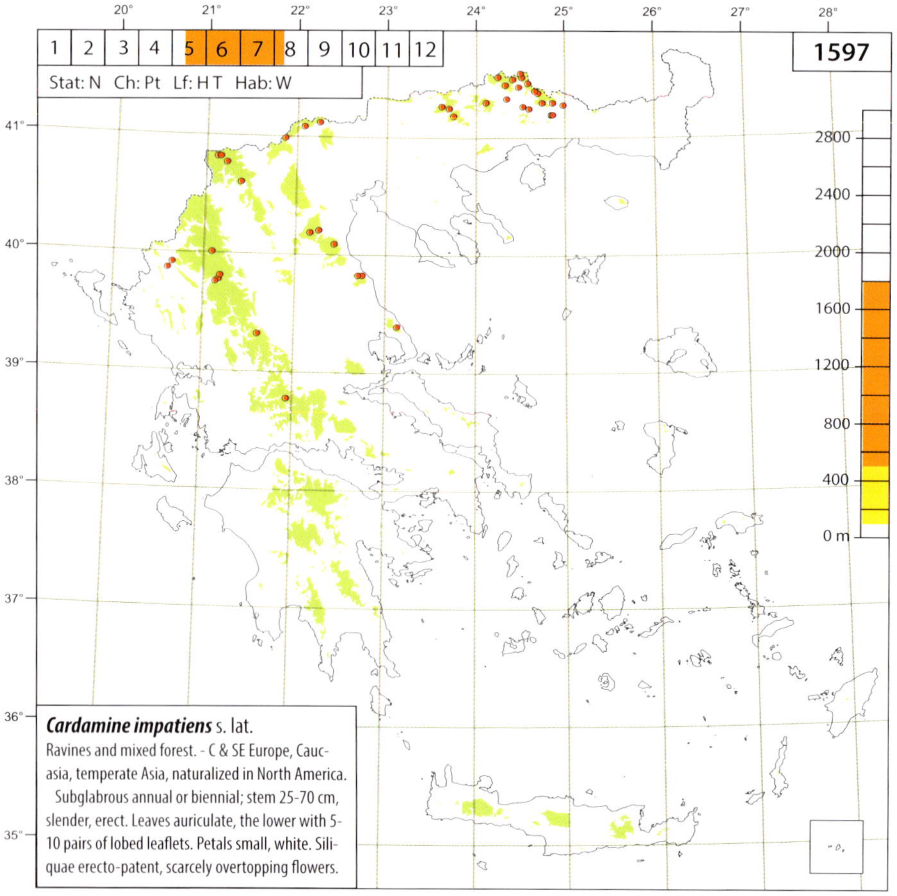

Cardamine impatiens s. lat.
Ravines and mixed forest. - C & SE Europe, Caucasia, temperate Asia, naturalized in North America.
Subglabrous annual or biennial; stem 25-70 cm, slender, erect. Leaves auriculate, the lower with 5-10 pairs of lobed leaflets. Petals small, white. Siliquae erecto-patent, scarcely overtopping flowers.

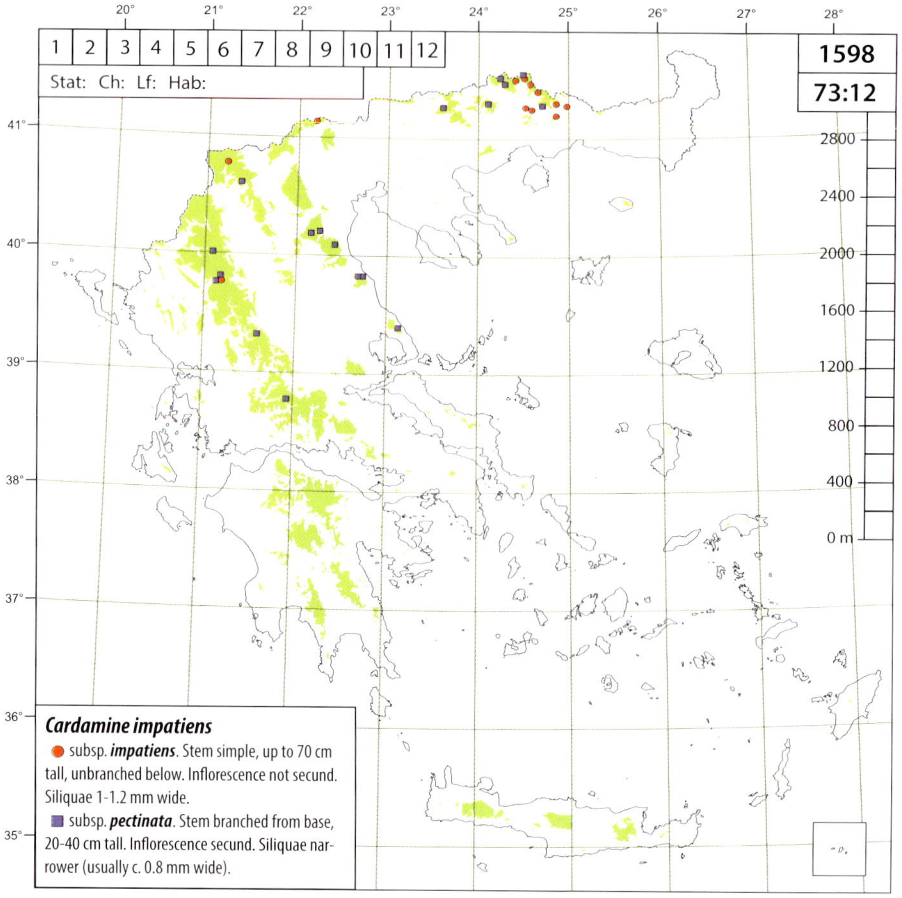

Cardamine impatiens
● subsp. ***impatiens***. Stem simple, up to 70 cm tall, unbranched below. Inflorescence not secund. Siliquae 1-1.2 mm wide.
■ subsp. ***pectinata***. Stem branched from base, 20-40 cm tall. Inflorescence secund. Siliquae narrower (usually c. 0.8 mm wide).

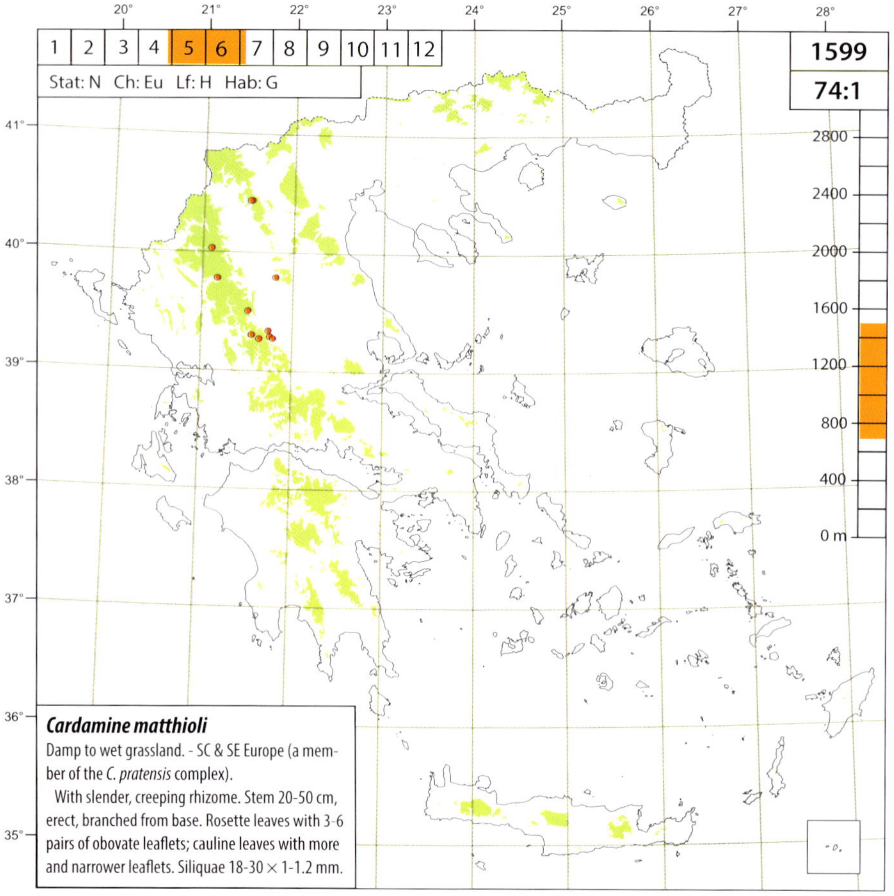

Cardamine matthioli
Damp to wet grassland. - SC & SE Europe (a member of the *C. pratensis* complex).
With slender, creeping rhizome. Stem 20-50 cm, erect, branched from base. Rosette leaves with 3-6 pairs of obovate leaflets; cauline leaves with more and narrower leaflets. Siliquae 18-30 × 1-1.2 mm.

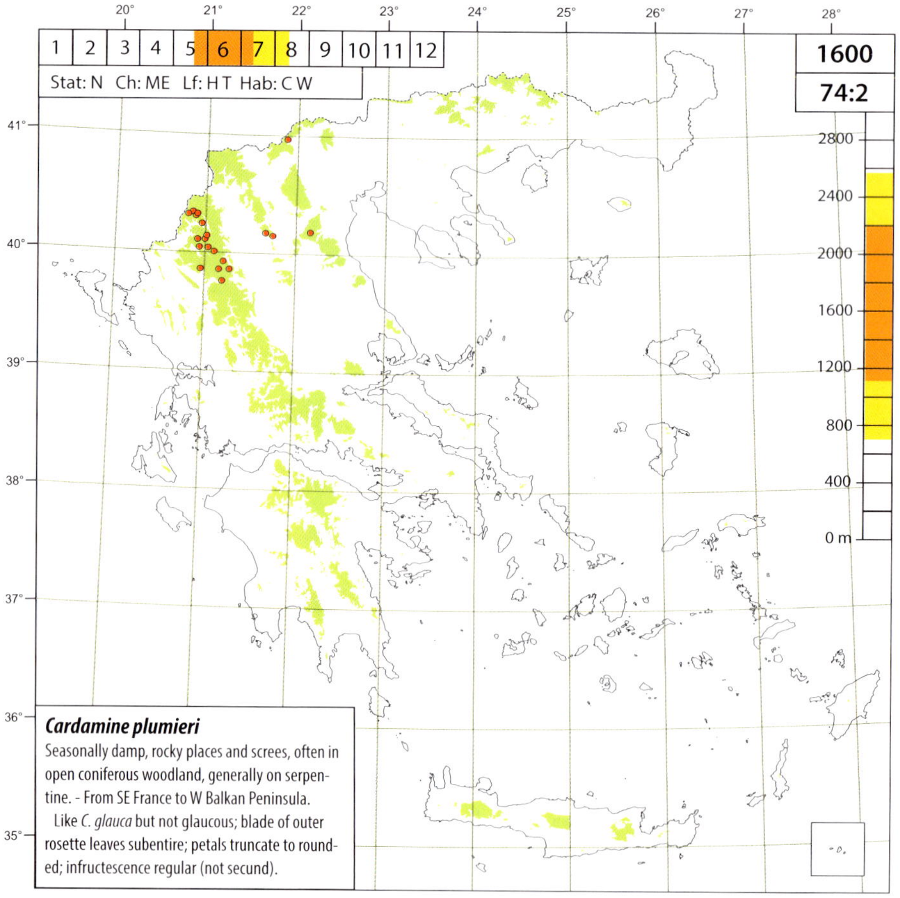

Cardamine plumieri
Seasonally damp, rocky places and screes, often in open coniferous woodland, generally on serpentine. - From SE France to W Balkan Peninsula.
Like *C. glauca* but not glaucous; blade of outer rosette leaves subentire; petals truncate to rounded; infructescence regular (not secund).

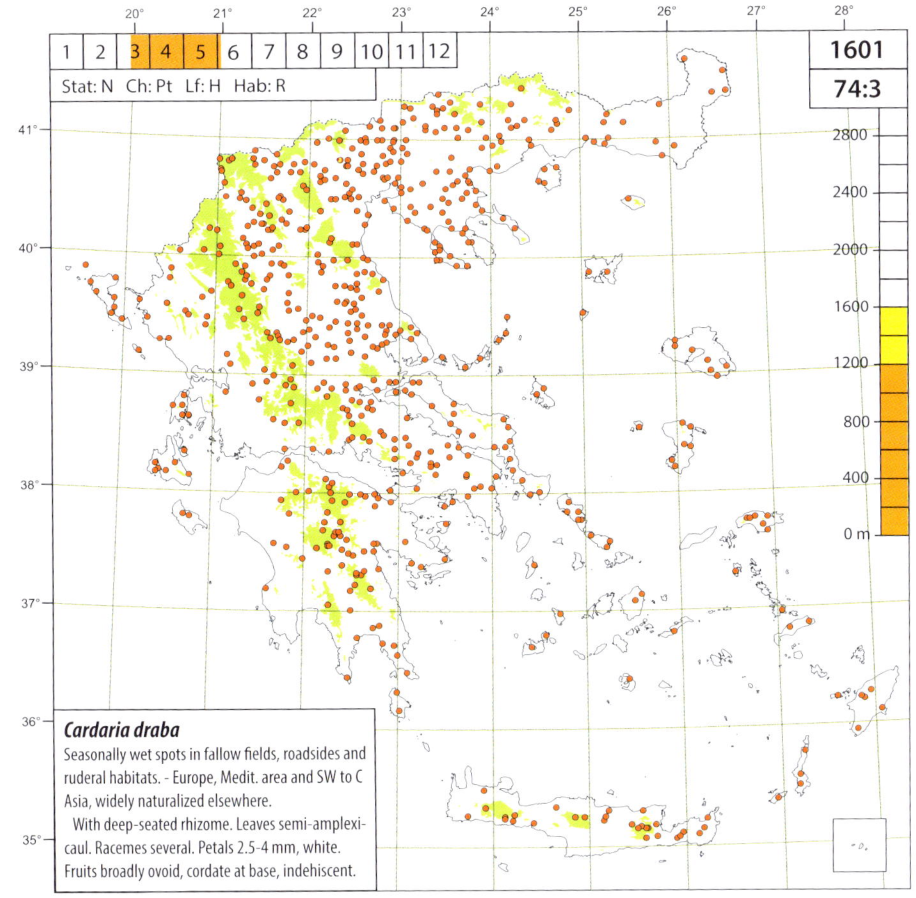

Cardaria draba

Seasonally wet spots in fallow fields, roadsides and ruderal habitats. - Europe, Medit. area and SW to C Asia, widely naturalized elsewhere.

With deep-seated rhizome. Leaves semi-amplexicaul. Racemes several. Petals 2.5-4 mm, white. Fruits broadly ovoid, cordate at base, indehiscent.

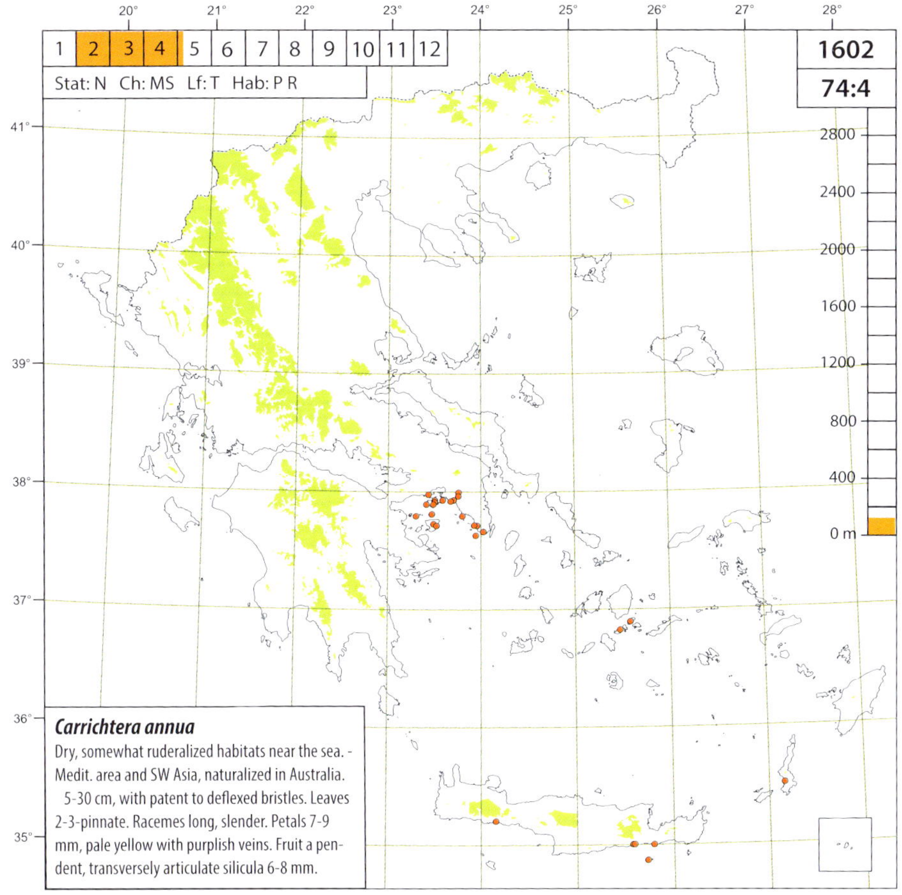

Carrichtera annua

Dry, somewhat ruderalized habitats near the sea. - Medit. area and SW Asia, naturalized in Australia.

5-30 cm, with patent to deflexed bristles. Leaves 2-3-pinnate. Racemes long, slender. Petals 7-9 mm, pale yellow with purplish veins. Fruit a pendent, transversely articulate silicula 6-8 mm.

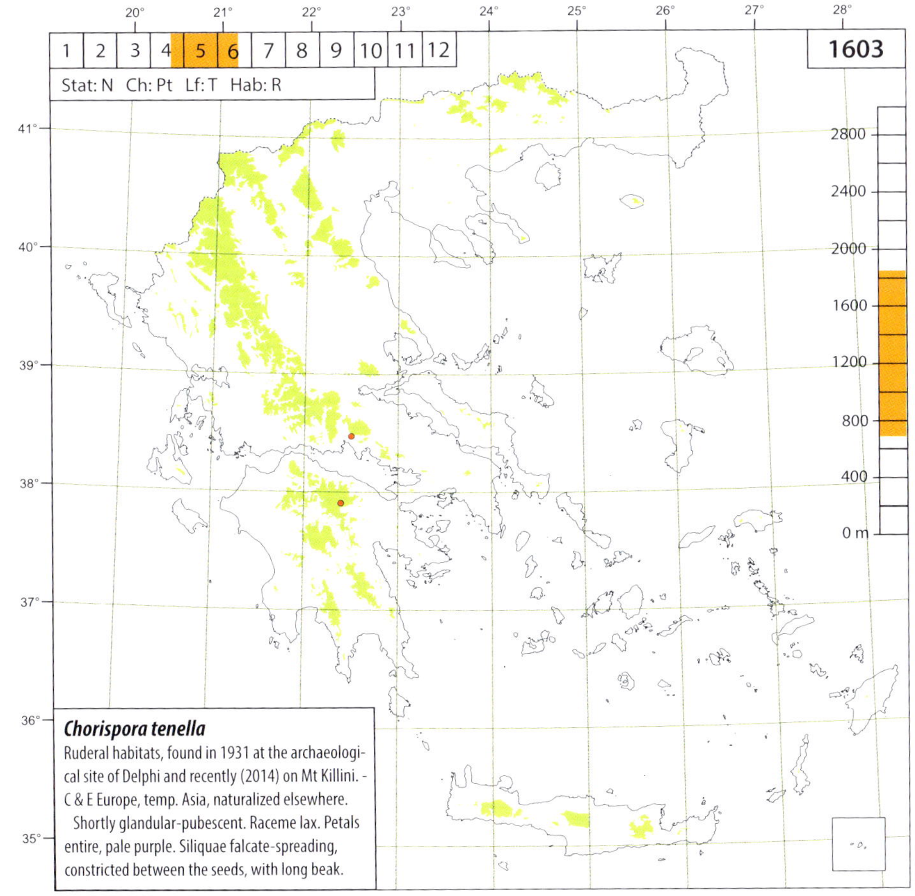

Chorispora tenella

Ruderal habitats, found in 1931 at the archaeological site of Delphi and recently (2014) on Mt Killini. - C & E Europe, temp. Asia, naturalized elsewhere.

Shortly glandular-pubescent. Raceme lax. Petals pale purple. Siliquae falcate-spreading, constricted between the seeds, with long beak.

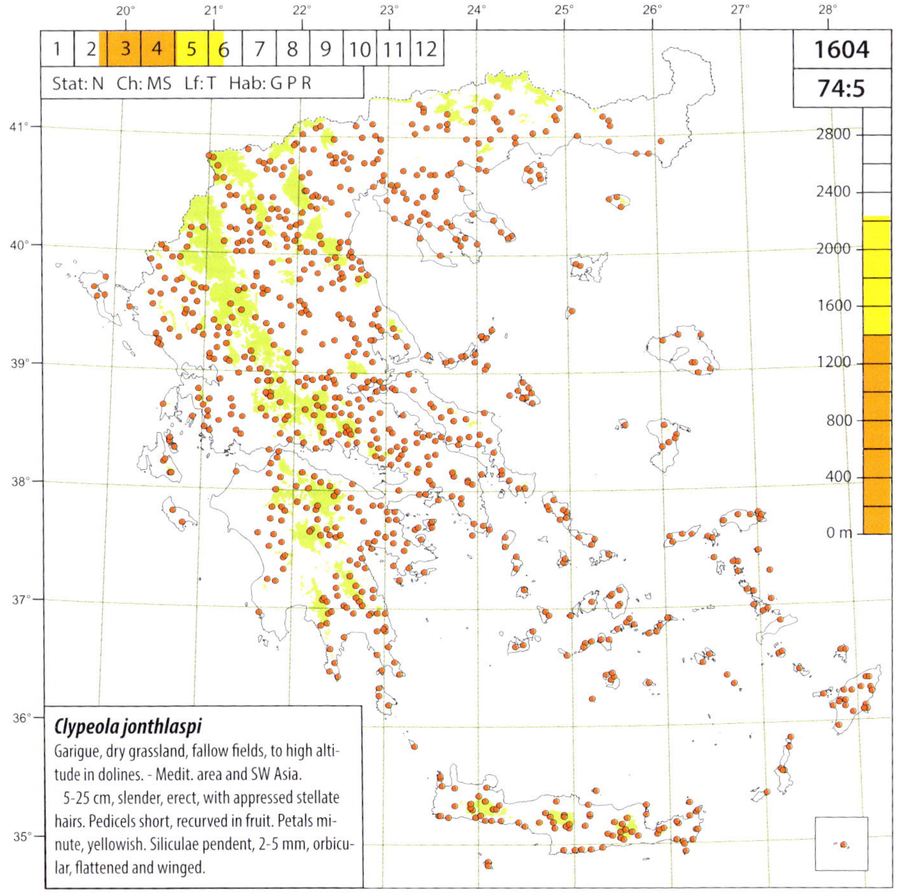

Clypeola jonthlaspi

Garigue, dry grassland, fallow fields, to high altitude in dolines. - Medit. area and SW Asia.

5-25 cm, slender, erect, with appressed stellate hairs. Pedicels short, recurved in fruit. Petals minute, yellowish. Siliculae pendent, 2-5 mm, orbicular, flattened and winged.

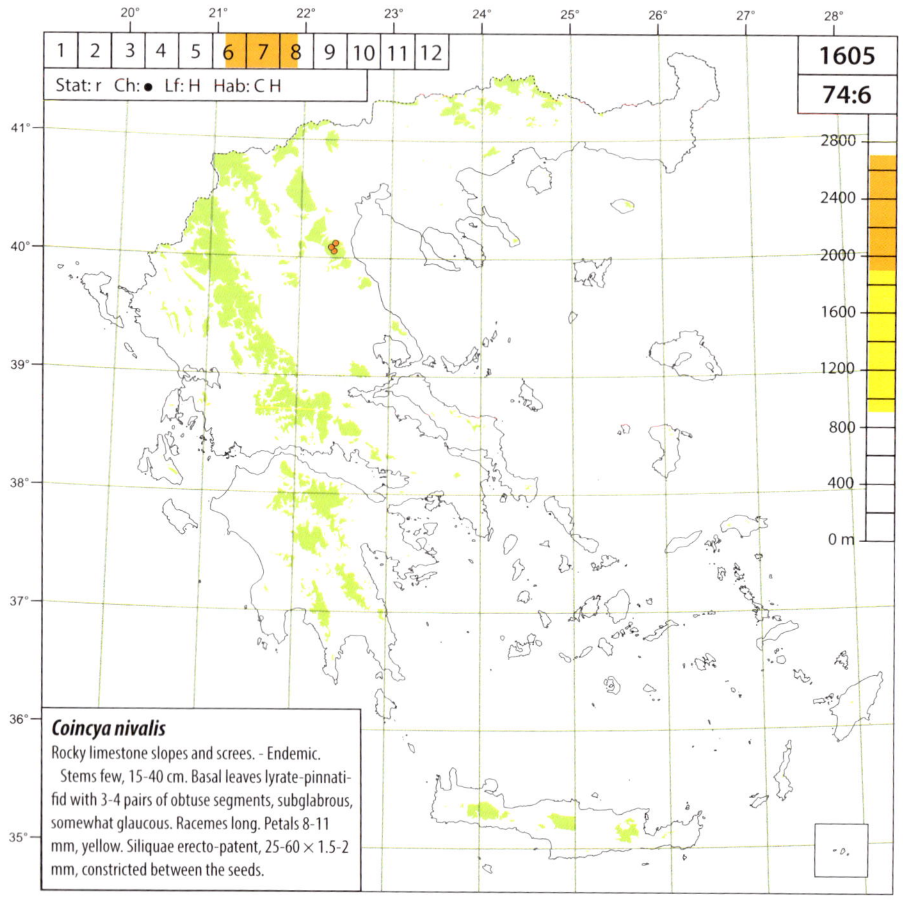

Stat: r Ch: ● Lf: H Hab: C H

Coincya nivalis

Rocky limestone slopes and screes. - Endemic.

Stems few, 15-40 cm. Basal leaves lyrate-pinnatifid with 3-4 pairs of obtuse segments, subglabrous, somewhat glaucous. Racemes long. Petals 8-11 mm, yellow. Siliquae erecto-patent, 25-60 × 1.5-2 mm, constricted between the seeds.

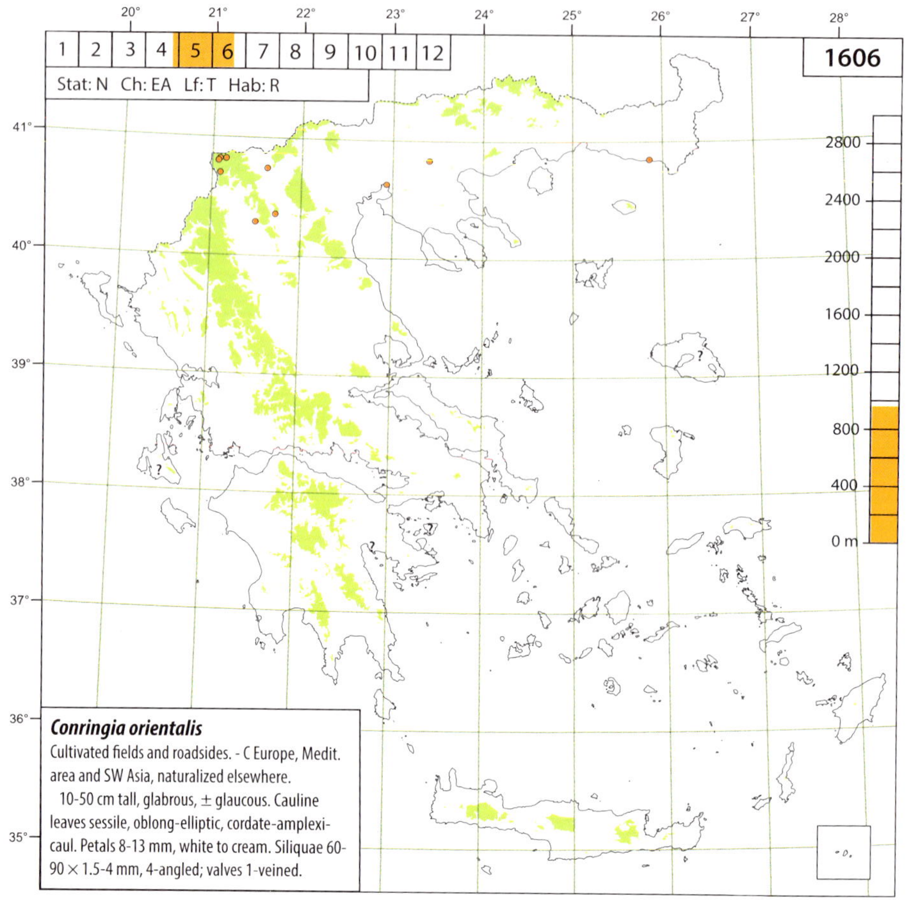

Stat: N Ch: EA Lf: T Hab: R

Conringia orientalis

Cultivated fields and roadsides. - C Europe, Medit. area and SW Asia, naturalized elsewhere.

10-50 cm tall, glabrous, ± glaucous. Cauline leaves sessile, oblong-elliptic, cordate-amplexicaul. Petals 8-13 mm, white to cream. Siliquae 60-90 × 1.5-4 mm, 4-angled; valves 1-veined.

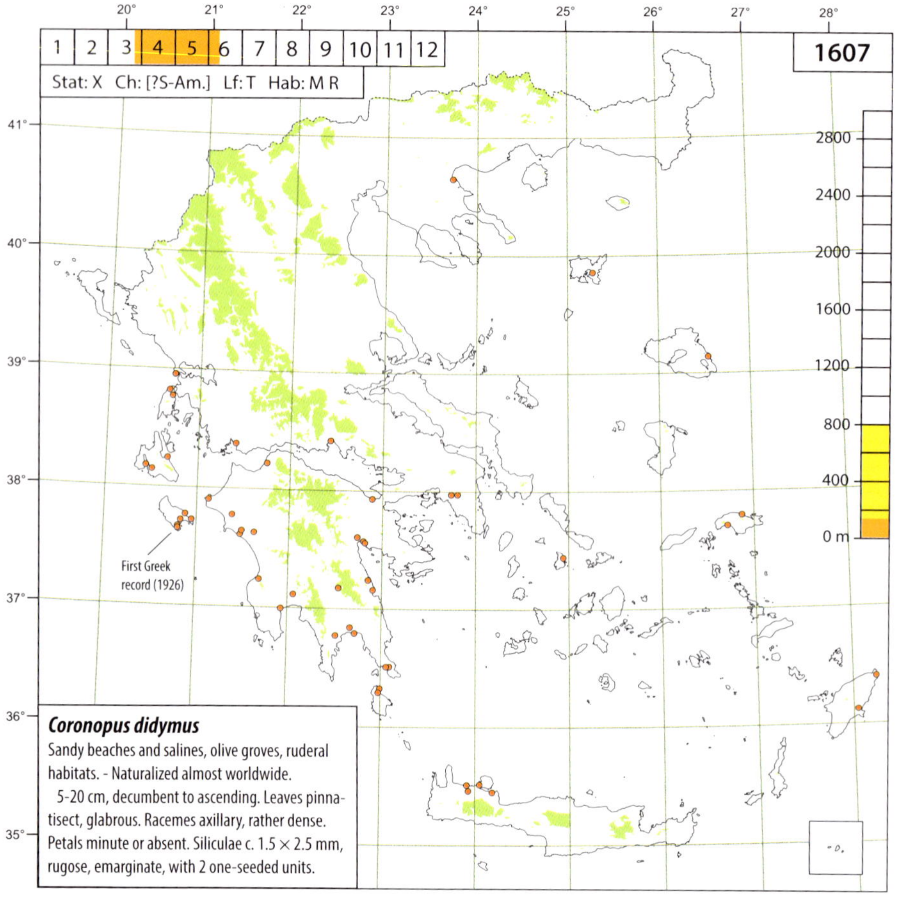

Stat: X Ch: [?S-Am.] Lf: T Hab: M R

Coronopus didymus

Sandy beaches and salines, olive groves, ruderal habitats. - Naturalized almost worldwide.

5-20 cm, decumbent to ascending. Leaves pinnatisect, glabrous. Racemes axillary, rather dense. Petals minute or absent. Siliculae c. 1.5 × 2.5 mm, rugose, emarginate, with 2 one-seeded units.

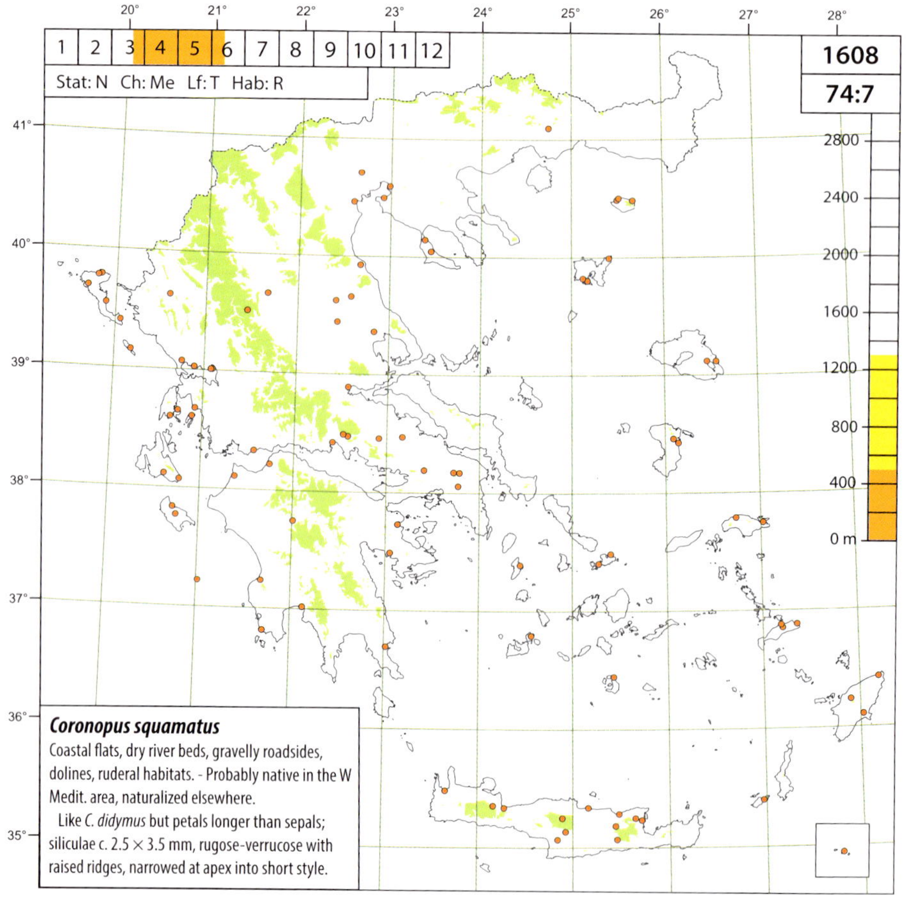

Stat: N Ch: Me Lf: T Hab: R

Coronopus squamatus

Coastal flats, dry river beds, gravelly roadsides, dolines, ruderal habitats. - Probably native in the W Medit. area, naturalized elsewhere.

Like *C. didymus* but petals longer than sepals; siliculae c. 2.5 × 3.5 mm, rugose-verrucose with raised ridges, narrowed at apex into short style.

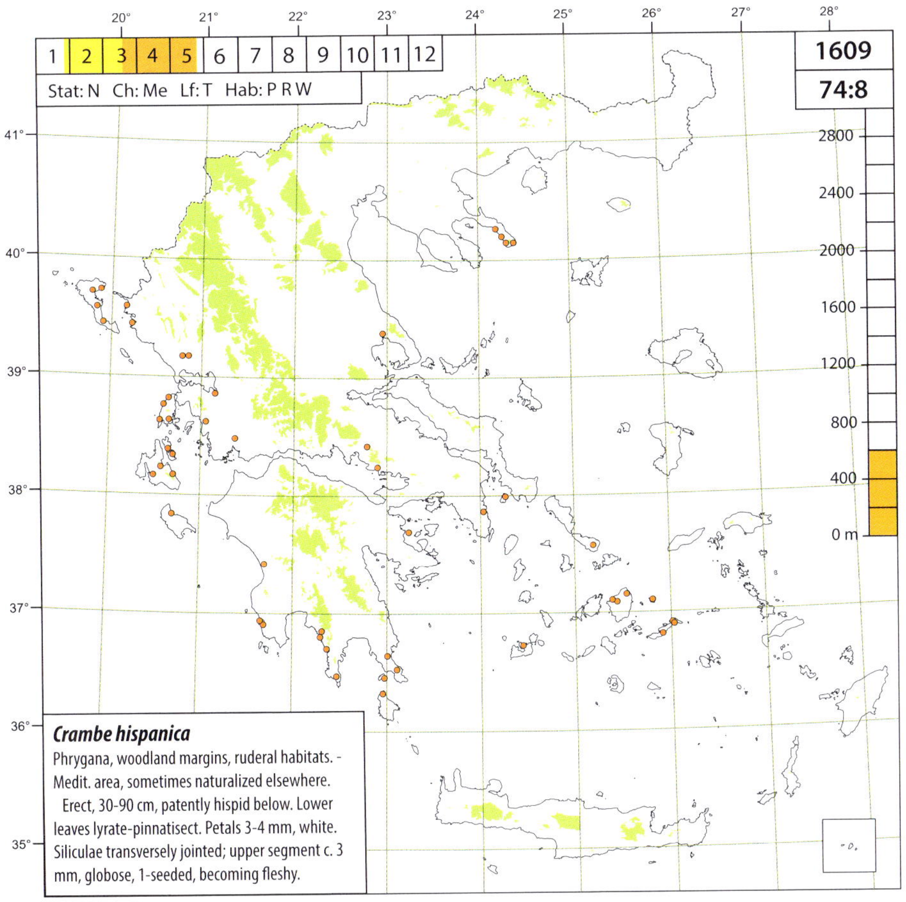

Crambe hispanica
Phrygana, woodland margins, ruderal habitats. - Medit. area, sometimes naturalized elsewhere.
Erect, 30-90 cm, patently hispid below. Lower leaves lyrate-pinnatisect. Petals 3-4 mm, white. Siliculae transversely jointed; upper segment c. 3 mm, globose, 1-seeded, becoming fleshy.

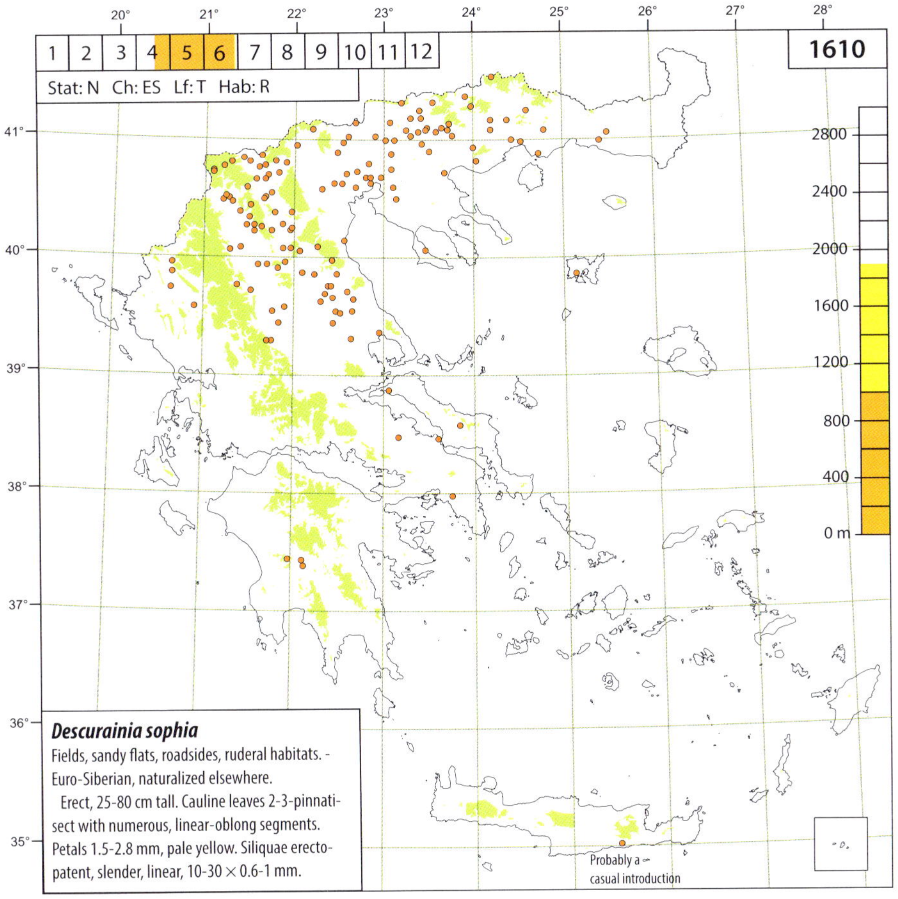

Descurainia sophia
Fields, sandy flats, roadsides, ruderal habitats. - Euro-Siberian, naturalized elsewhere.
Erect, 25-80 cm tall. Cauline leaves 2-3-pinnatisect with numerous, linear-oblong segments. Petals 1.5-2.8 mm, pale yellow. Siliquae erecto-patent, slender, linear, 10-30 × 0.6-1 mm.

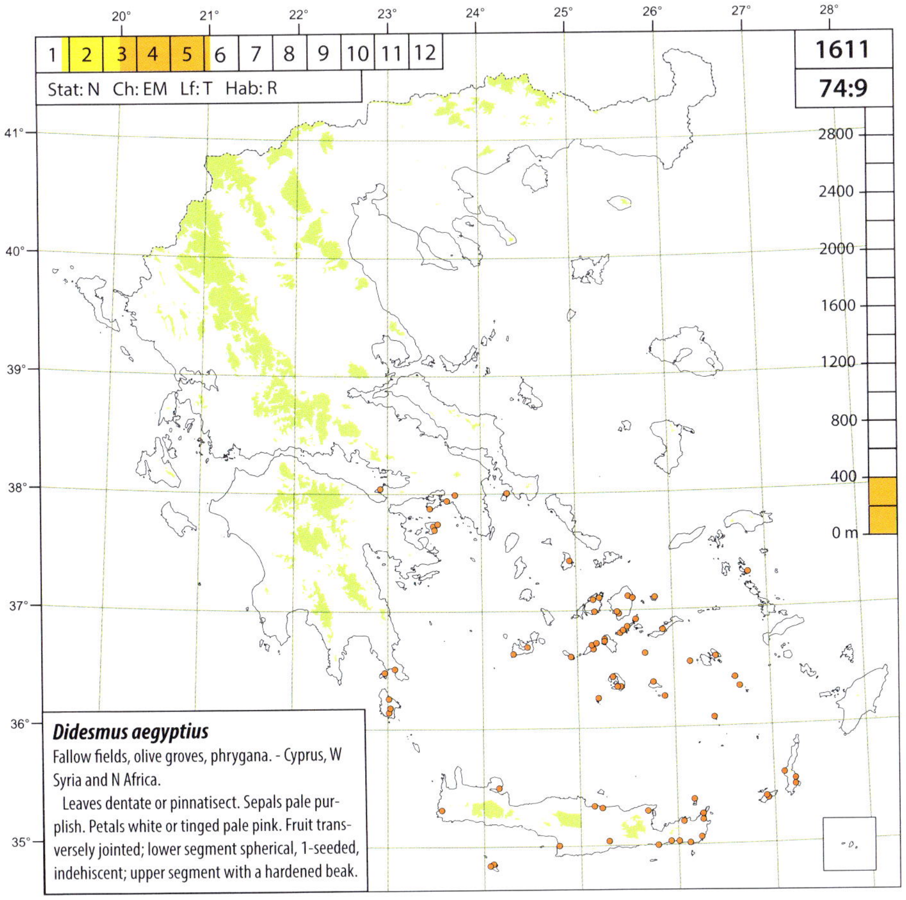

Didesmus aegyptius
Fallow fields, olive groves, phrygana. - Cyprus, W Syria and N Africa.
Leaves dentate or pinnatisect. Sepals pale purplish. Petals white or tinged pale pink. Fruit transversely jointed; lower segment spherical, 1-seeded, indehiscent; upper segment with a hardened beak.

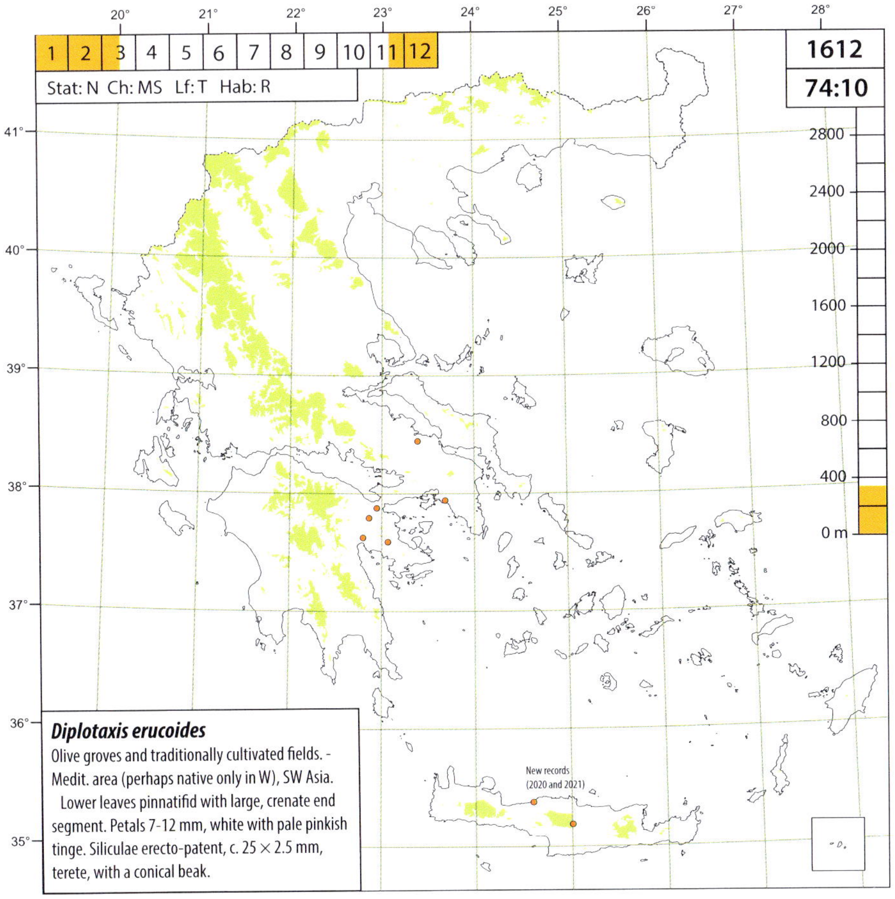

Diplotaxis erucoides
Olive groves and traditionally cultivated fields. - Medit. area (perhaps native only in W), SW Asia.
Lower leaves pinnatifid with large, crenate end segment. Petals 7-12 mm, white with pale pinkish tinge. Siliculae erecto-patent, c. 25 × 2.5 mm, terete, with a conical beak.

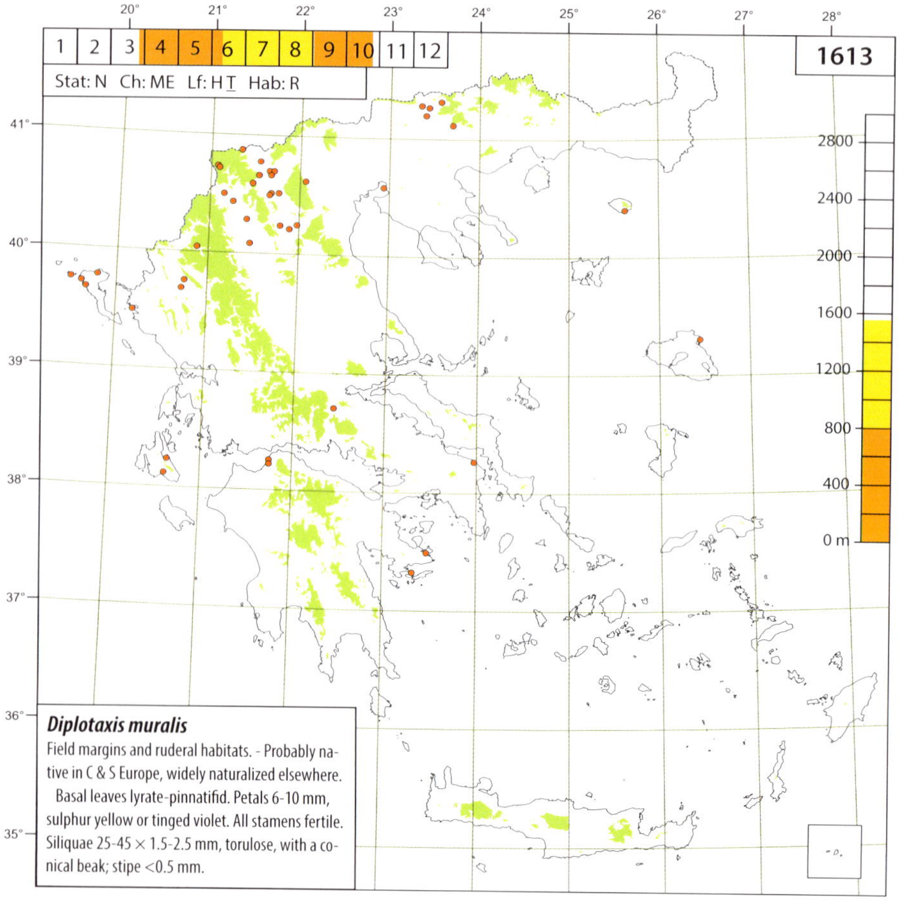

Diplotaxis muralis
Field margins and ruderal habitats. - Probably native in C & S Europe, widely naturalized elsewhere.
Basal leaves lyrate-pinnatifid. Petals 6-10 mm, sulphur yellow or tinged violet. All stamens fertile. Siliquae 25-45 × 1.5-2.5 mm, torulose, with a conical beak; stipe <0.5 mm.

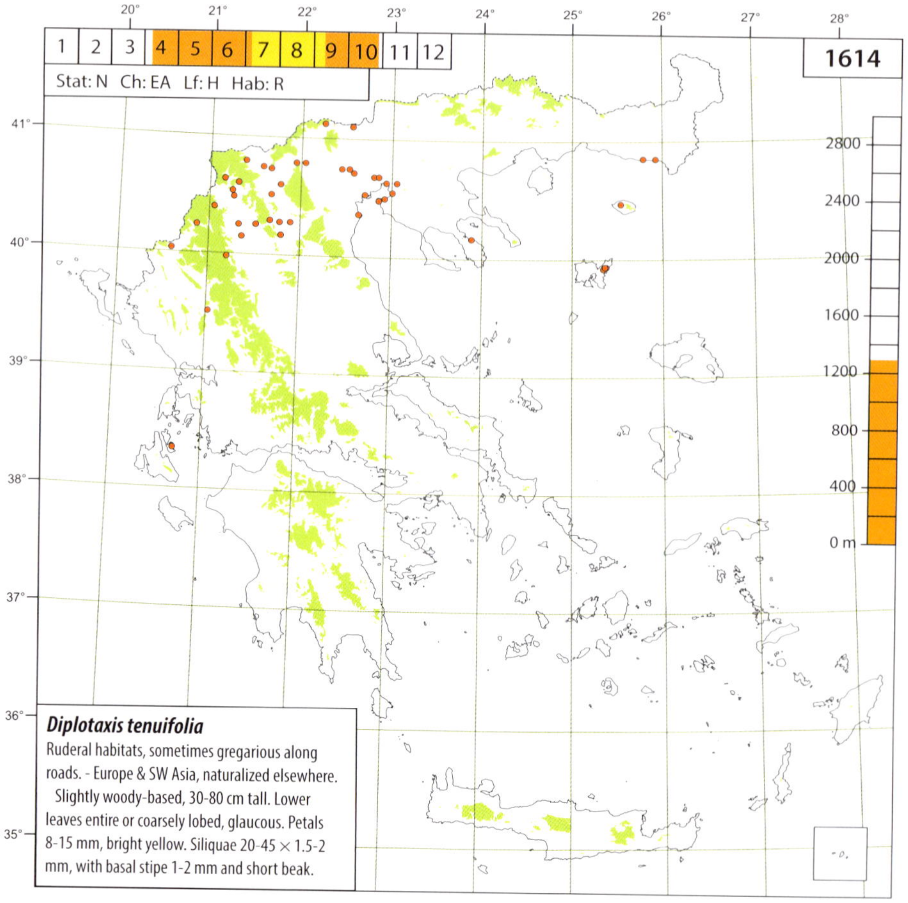

Diplotaxis tenuifolia
Ruderal habitats, sometimes gregarious along roads. - Europe & SW Asia, naturalized elsewhere.
Slightly woody-based, 30-80 cm tall. Lower leaves entire or coarsely lobed, glaucous. Petals 8-15 mm, bright yellow. Siliquae 20-45 × 1.5-2 mm, with basal stipe 1-2 mm and short beak.

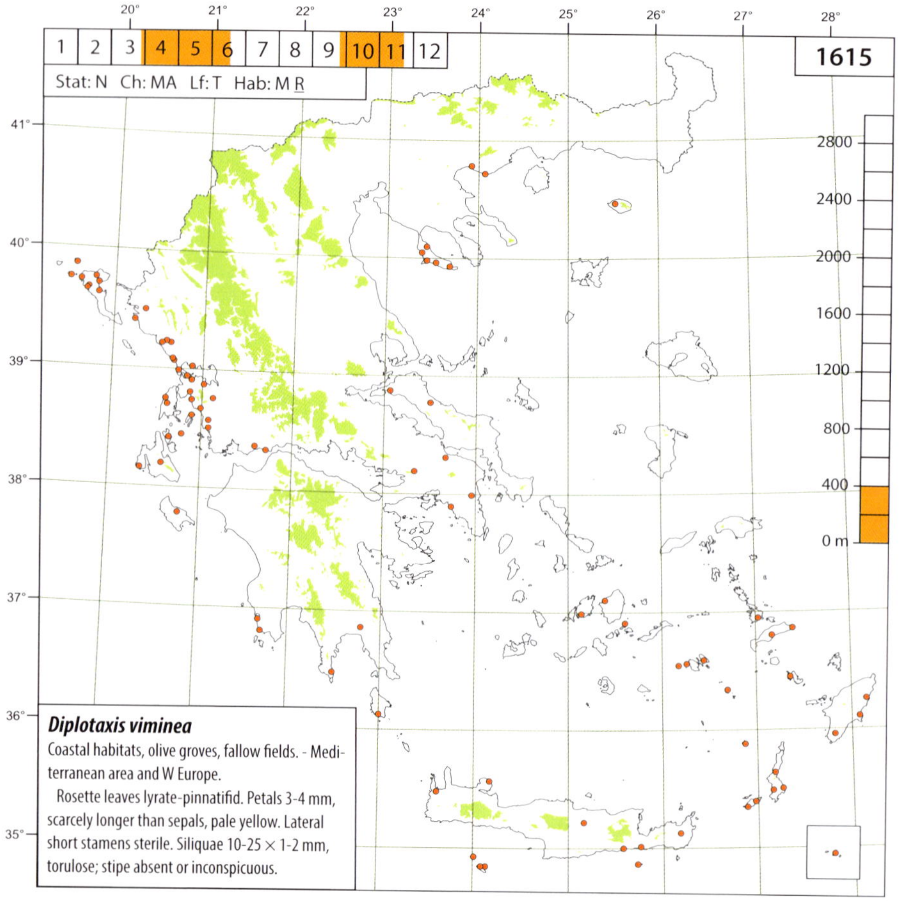

Diplotaxis viminea
Coastal habitats, olive groves, fallow fields. - Mediterranean area and W Europe.
Rosette leaves lyrate-pinnatifid. Petals 3-4 mm, scarcely longer than sepals, pale yellow. Lateral short stamens sterile. Siliquae 10-25 × 1-2 mm, torulose; stipe absent or inconspicuous.

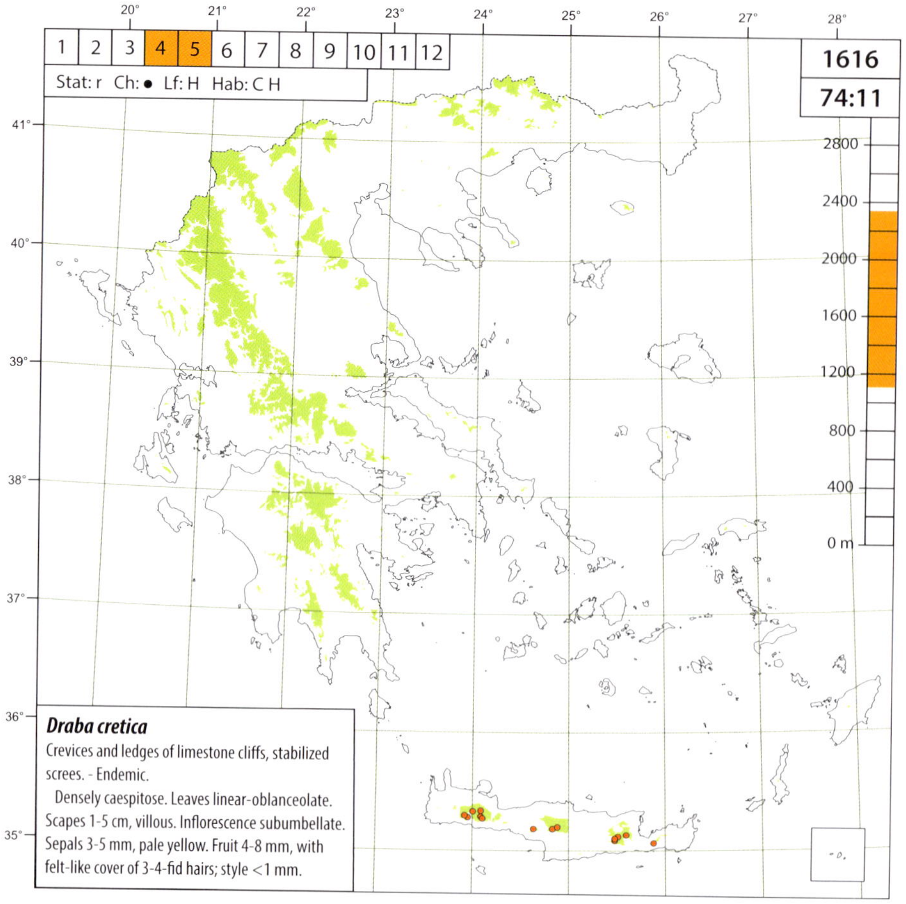

Draba cretica
Crevices and ledges of limestone cliffs, stabilized screes. - Endemic.
Densely caespitose. Leaves linear-oblanceolate. Scapes 1-5 cm, villous. Inflorescence subumbellate. Sepals 3-5 mm, pale yellow. Fruit 4-8 mm, with felt-like cover of 3-4-fid hairs; style <1 mm.

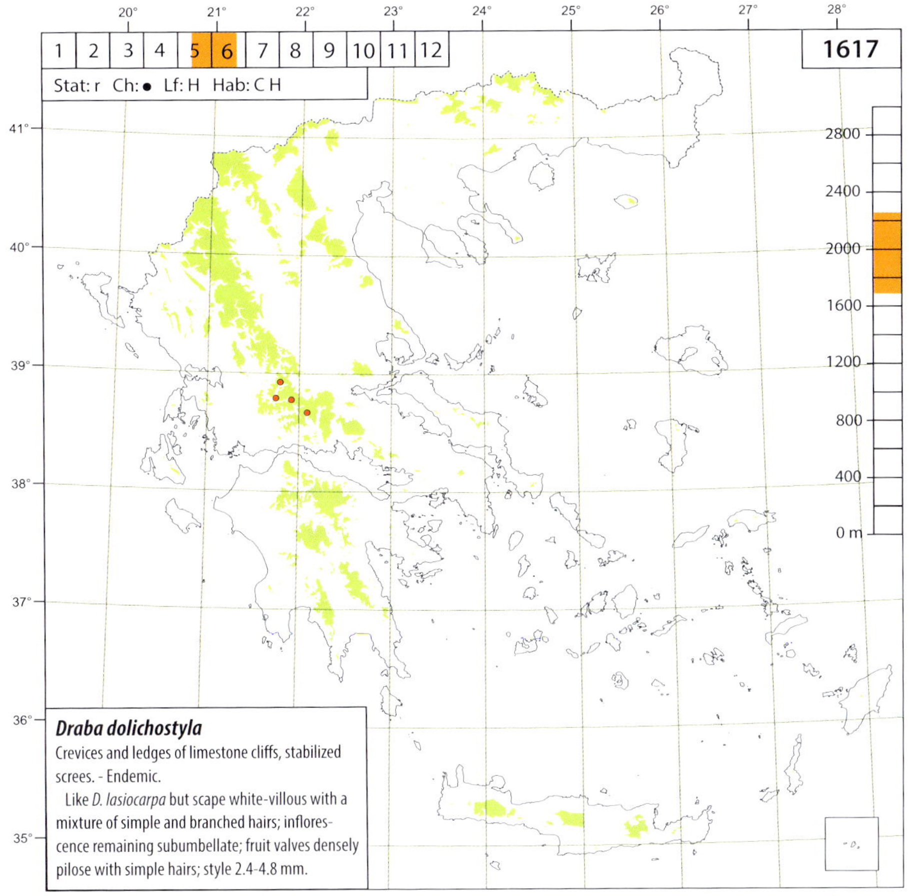

1617

Stat: r Ch: ● Lf: H Hab: C H

Draba dolichostyla
Crevices and ledges of limestone cliffs, stabilized screes. - Endemic.
Like *D. lasiocarpa* but scape white-villous with a mixture of simple and branched hairs; inflorescence remaining subumbellate; fruit valves densely pilose with simple hairs; style 2.4-4.8 mm.

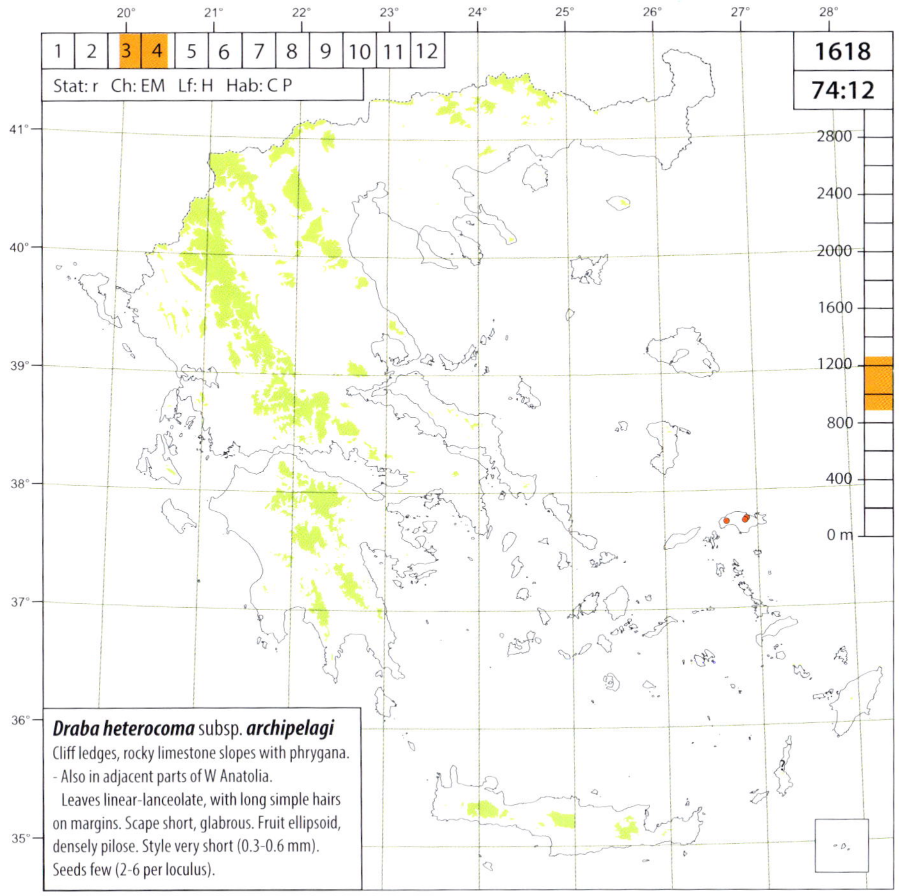

1618

74:12

Stat: r Ch: EM Lf: H Hab: C P

Draba heterocoma subsp. ***archipelagi***
Cliff ledges, rocky limestone slopes with phrygana. - Also in adjacent parts of W Anatolia.
Leaves linear-lanceolate, with long simple hairs on margins. Scape short, glabrous. Fruit ellipsoid, densely pilose. Style very short (0.3-0.6 mm). Seeds few (2-6 per loculus).

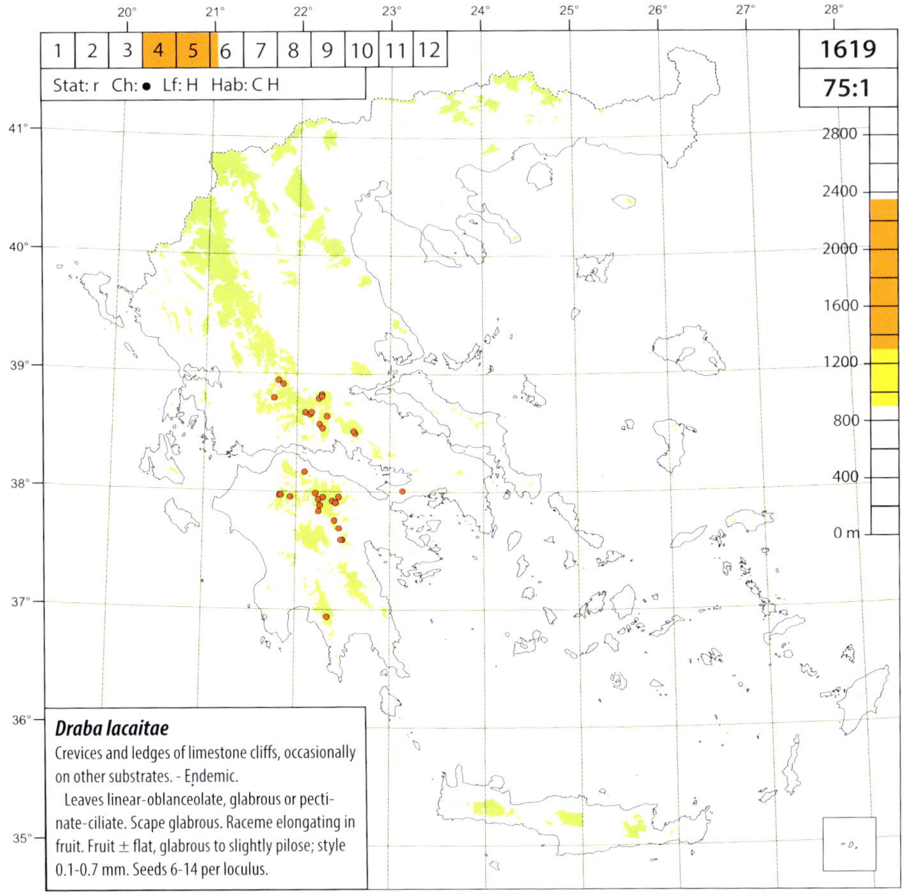

1619

75:1

Stat: r Ch: ● Lf: H Hab: C H

Draba lacaitae
Crevices and ledges of limestone cliffs, occasionally on other substrates. - Endemic.
Leaves linear-oblanceolate, glabrous or pectinate-ciliate. Scape glabrous. Raceme elongating in fruit. Fruit ± flat, glabrous to slightly pilose; style 0.1-0.7 mm. Seeds 6-14 per loculus.

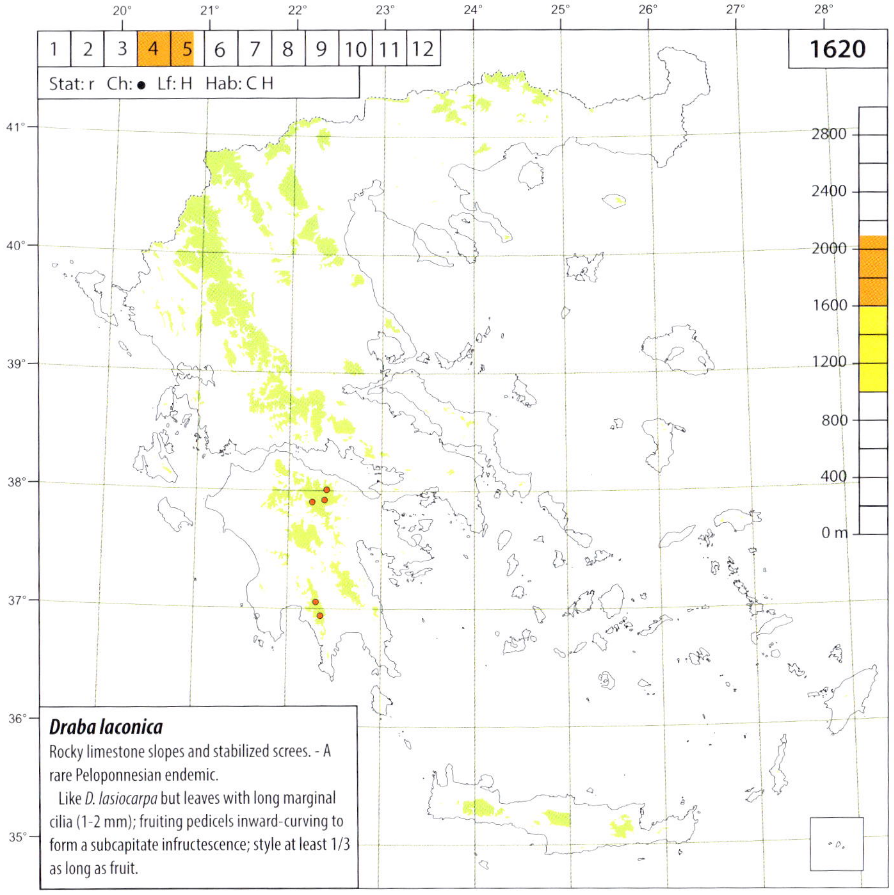

1620

Stat: r Ch: ● Lf: H Hab: C H

Draba laconica
Rocky limestone slopes and stabilized screes. - A rare Peloponnesian endemic.
Like *D. lasiocarpa* but leaves with long marginal cilia (1-2 mm); fruiting pedicels inward-curving to form a subcapitate infructescence; style at least 1/3 as long as fruit.

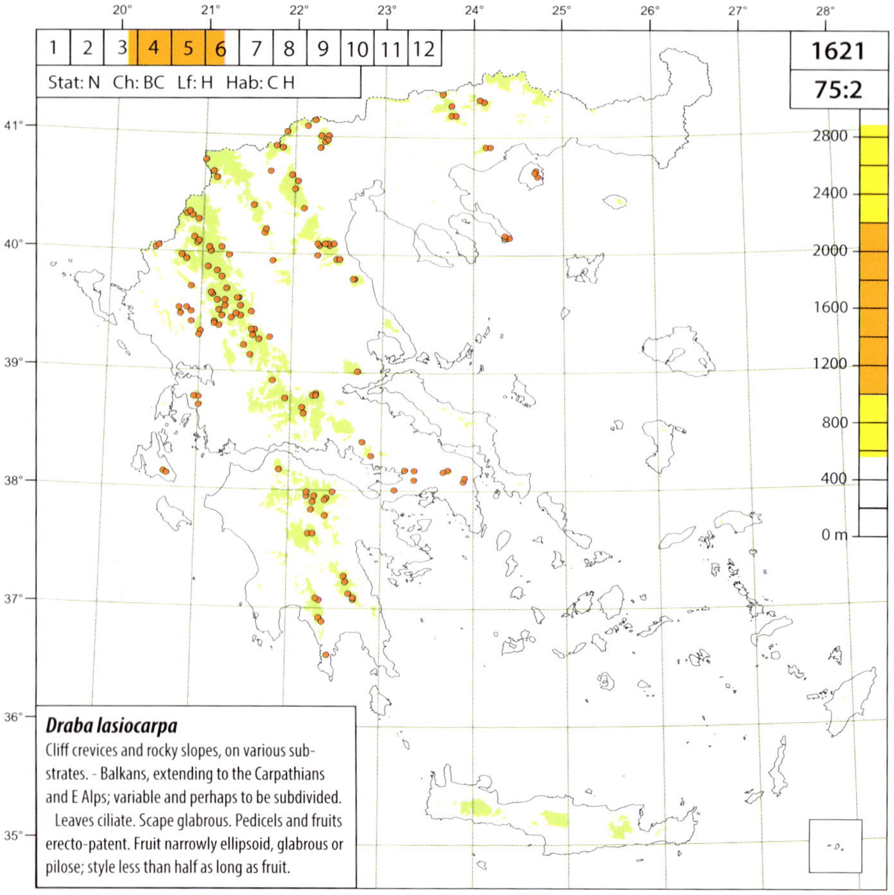

Draba lasiocarpa
Cliff crevices and rocky slopes, on various substrates. - Balkans, extending to the Carpathians and E Alps; variable and perhaps to be subdivided.
Leaves ciliate. Scape glabrous. Pedicels and fruits erecto-patent. Fruit narrowly ellipsoid, glabrous or pilose; style less than half as long as fruit.

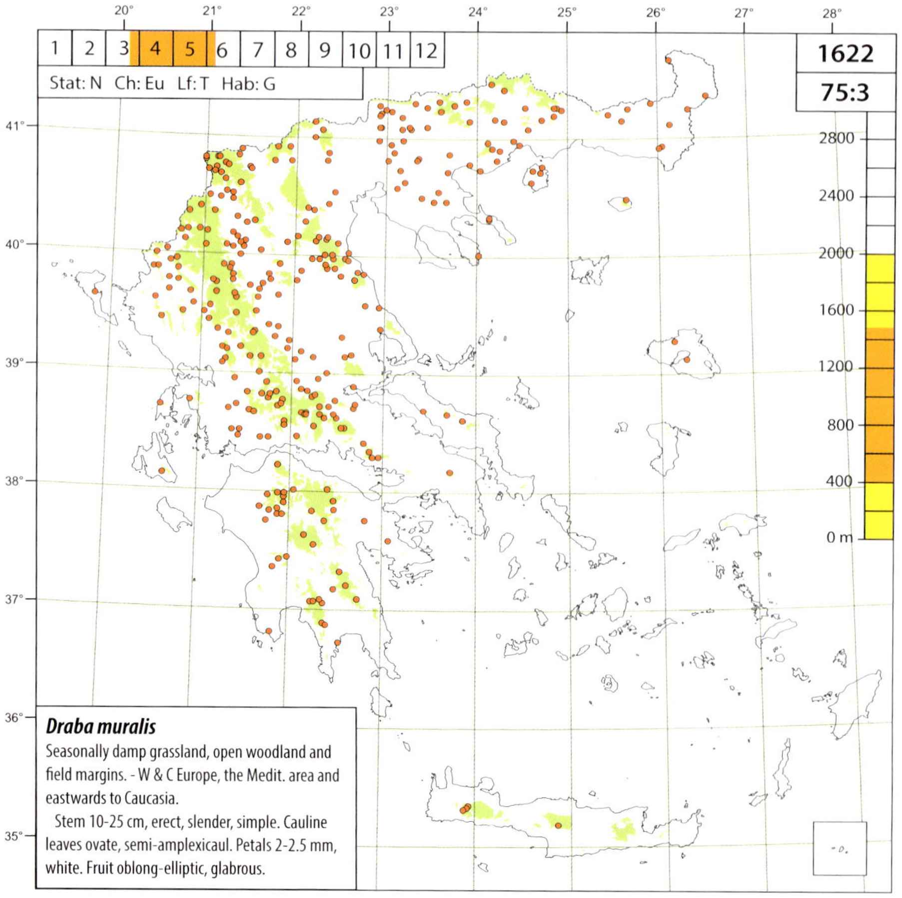

Draba muralis
Seasonally damp grassland, open woodland and field margins. - W & C Europe, the Medit. area and eastwards to Caucasia.
Stem 10-25 cm, erect, slender, simple. Cauline leaves ovate, semi-amplexicaul. Petals 2-2.5 mm, white. Fruit oblong-elliptic, glabrous.

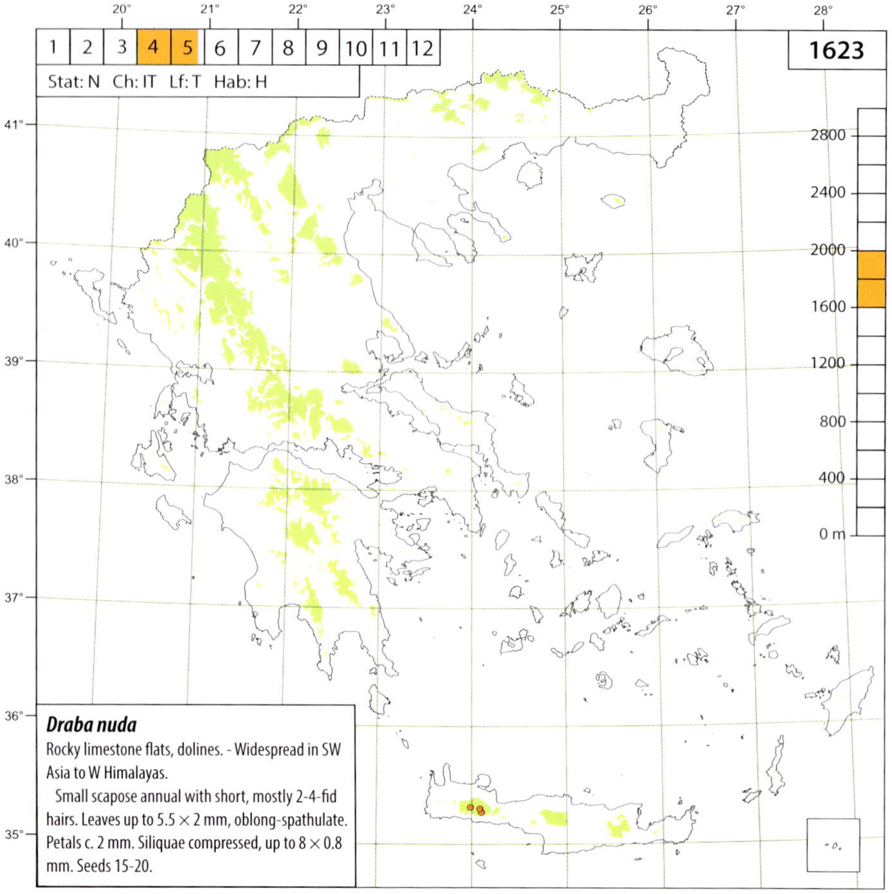

Draba nuda
Rocky limestone flats, dolines. - Widespread in SW Asia to W Himalayas.
Small scapose annual with short, mostly 2-4-fid hairs. Leaves up to 5.5 × 2 mm, oblong-spathulate. Petals c. 2 mm. Siliquae compressed, up to 8 × 0.8 mm. Seeds 15-20.

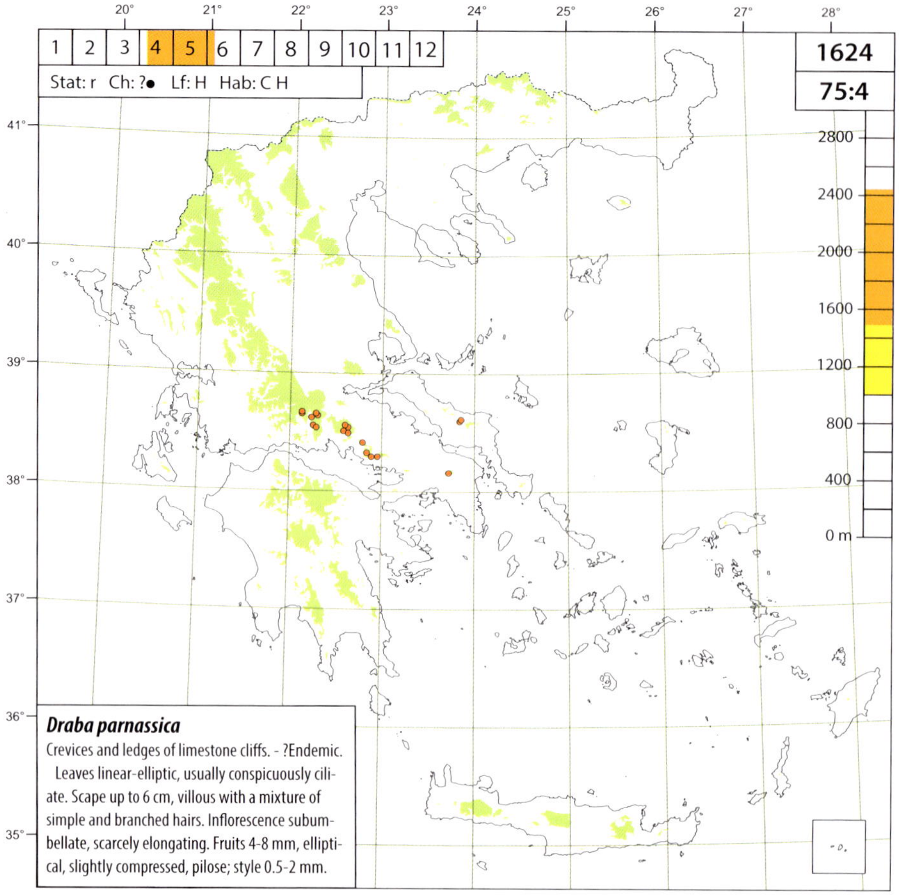

Draba parnassica
Crevices and ledges of limestone cliffs. - ?Endemic.
Leaves linear-elliptic, usually conspicuously ciliate. Scape up to 6 cm, villous with a mixture of simple and branched hairs. Inflorescence subumbellate, scarcely elongating. Fruits 4-8 mm, elliptical, slightly compressed, pilose; style 0.5-2 mm.

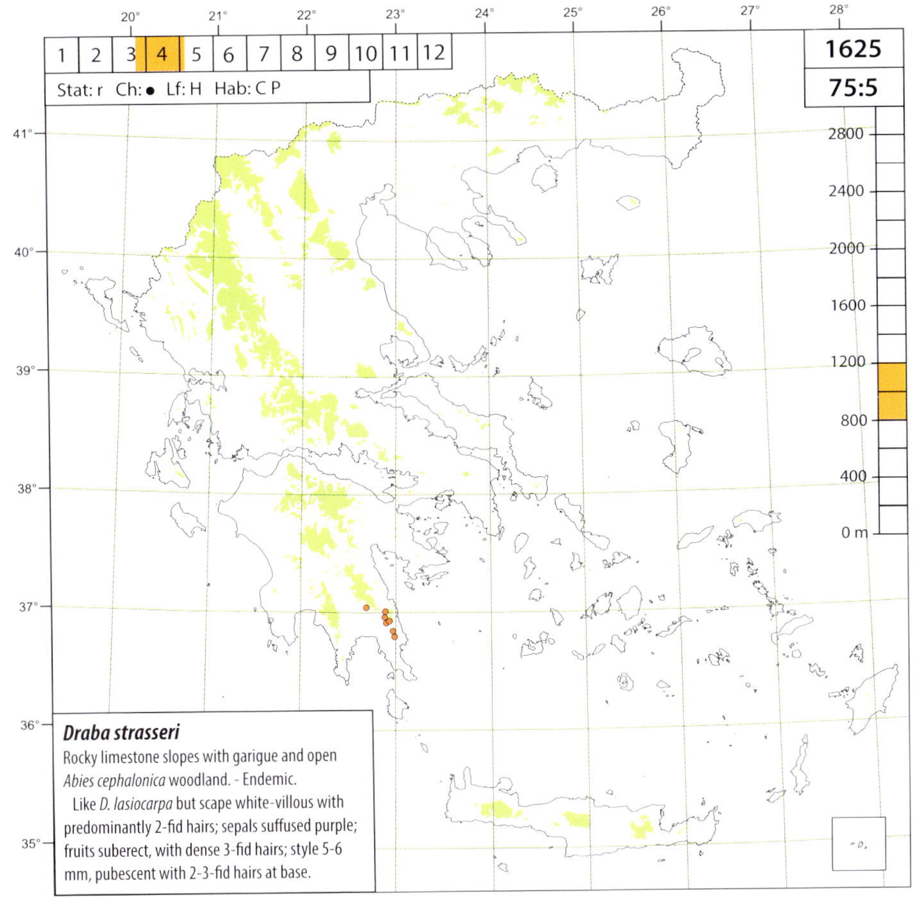

Draba strasseri
Rocky limestone slopes with garigue and open *Abies cephalonica* woodland. - Endemic.
Like *D. lasiocarpa* but scape white-villous with predominantly 2-fid hairs; sepals suffused purple; fruits suberect, with dense 3-fid hairs; style 5-6 mm, pubescent with 2-3-fid hairs at base.

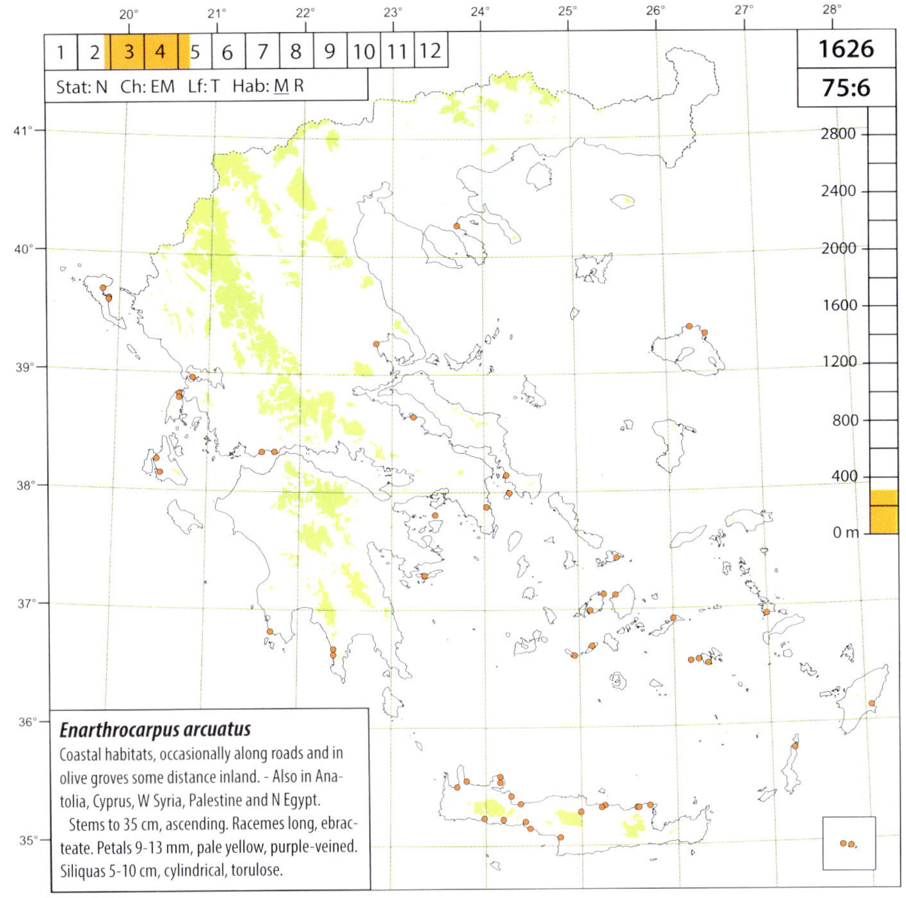

Enarthrocarpus arcuatus
Coastal habitats, occasionally along roads and in olive groves some distance inland. - Also in Anatolia, Cyprus, W Syria, Palestine and N Egypt.
Stems to 35 cm, ascending. Racemes long, ebracteate. Petals 9-13 mm, pale yellow, purple-veined. Siliquas 5-10 cm, cylindrical, torulose.

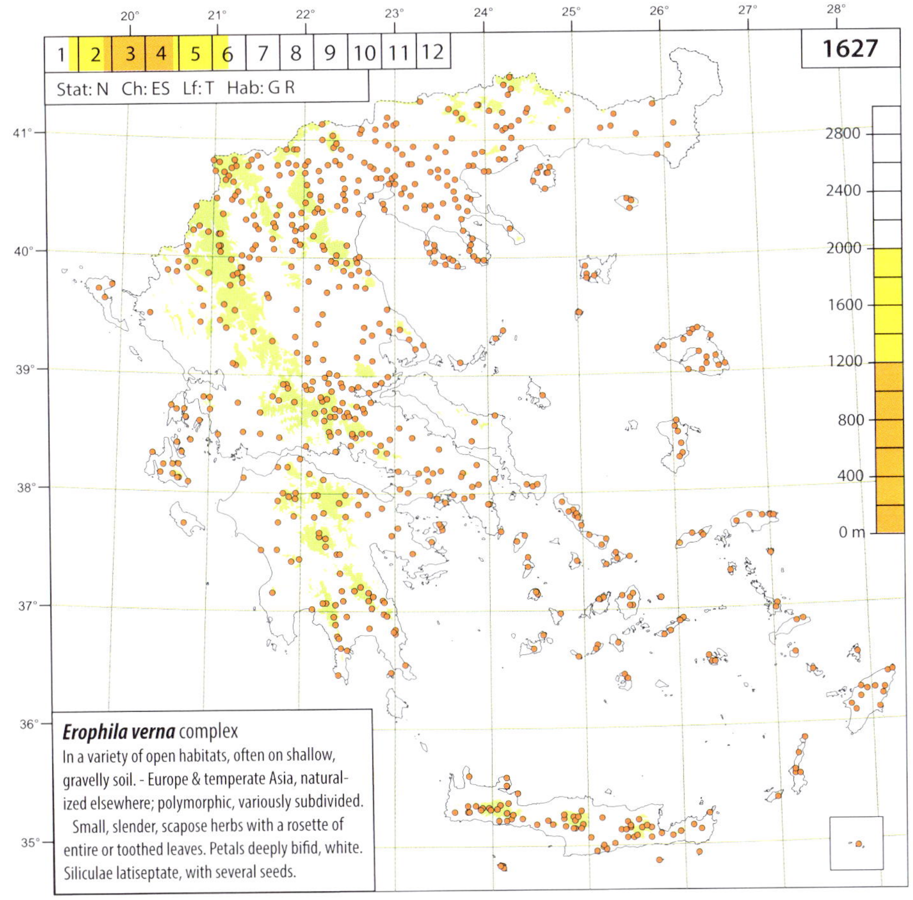

***Erophila verna* complex**
In a variety of open habitats, often on shallow, gravelly soil. - Europe & temperate Asia, naturalized elsewhere; polymorphic, variously subdivided.
Small, slender, scapose herbs with a rosette of entire or toothed leaves. Petals deeply bifid, white. Siliculae latiseptate, with several seeds.

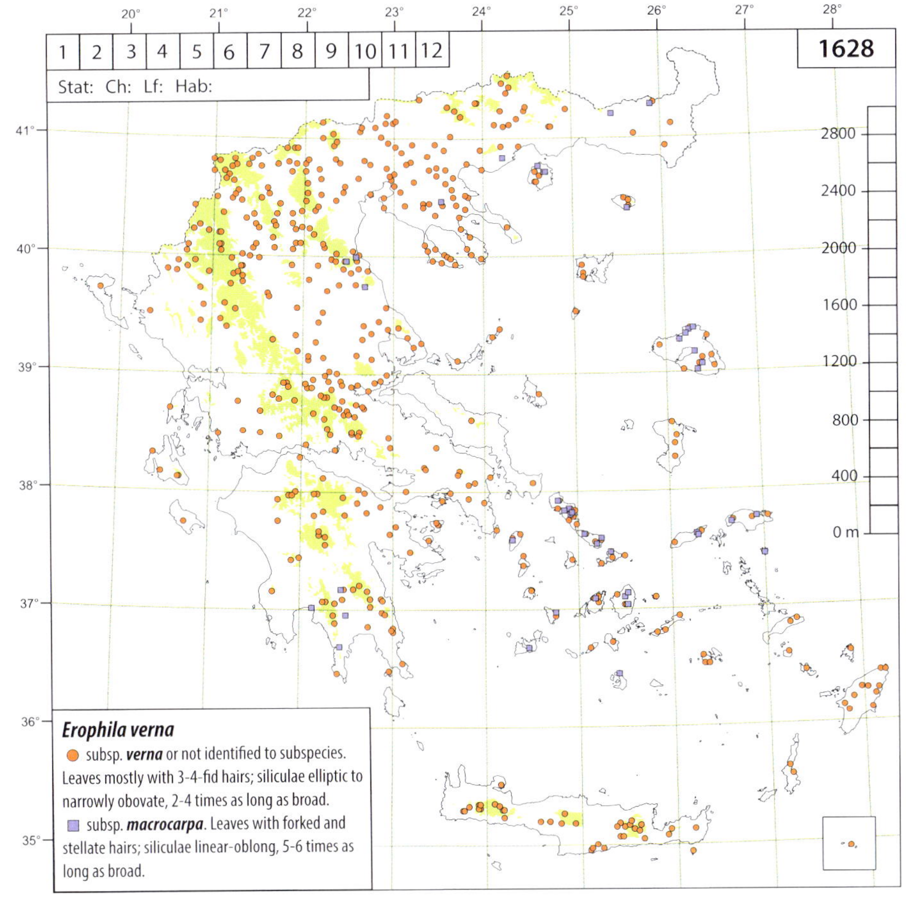

Erophila verna
● subsp. ***verna*** or not identified to subspecies. Leaves mostly with 3-4-fid hairs; siliculae elliptic to narrowly obovate, 2-4 times as long as broad.
■ subsp. ***macrocarpa***. Leaves with forked and stellate hairs; siliculae linear-oblong, 5-6 times as long as broad.

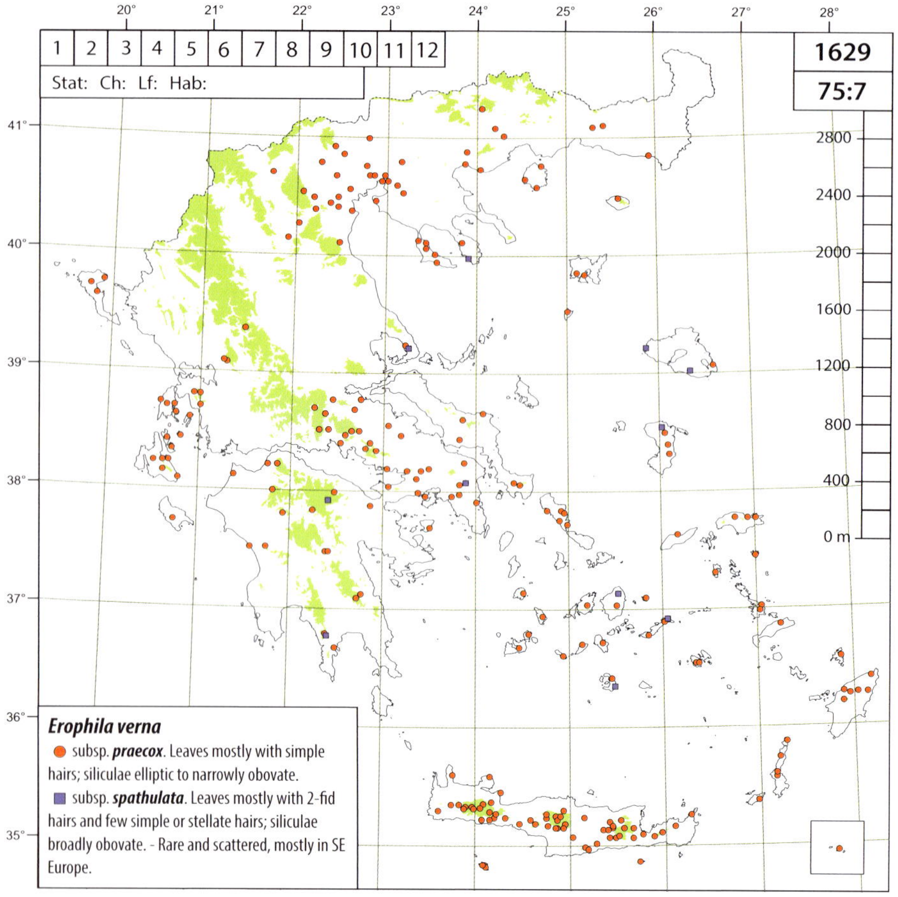

Erophila verna
● subsp. ***praecox***. Leaves mostly with simple hairs; siliculae elliptic to narrowly obovate.
■ subsp. ***spathulata***. Leaves mostly with 2-fid hairs and few simple or stellate hairs; siliculae broadly obovate. - Rare and scattered, mostly in SE Europe.

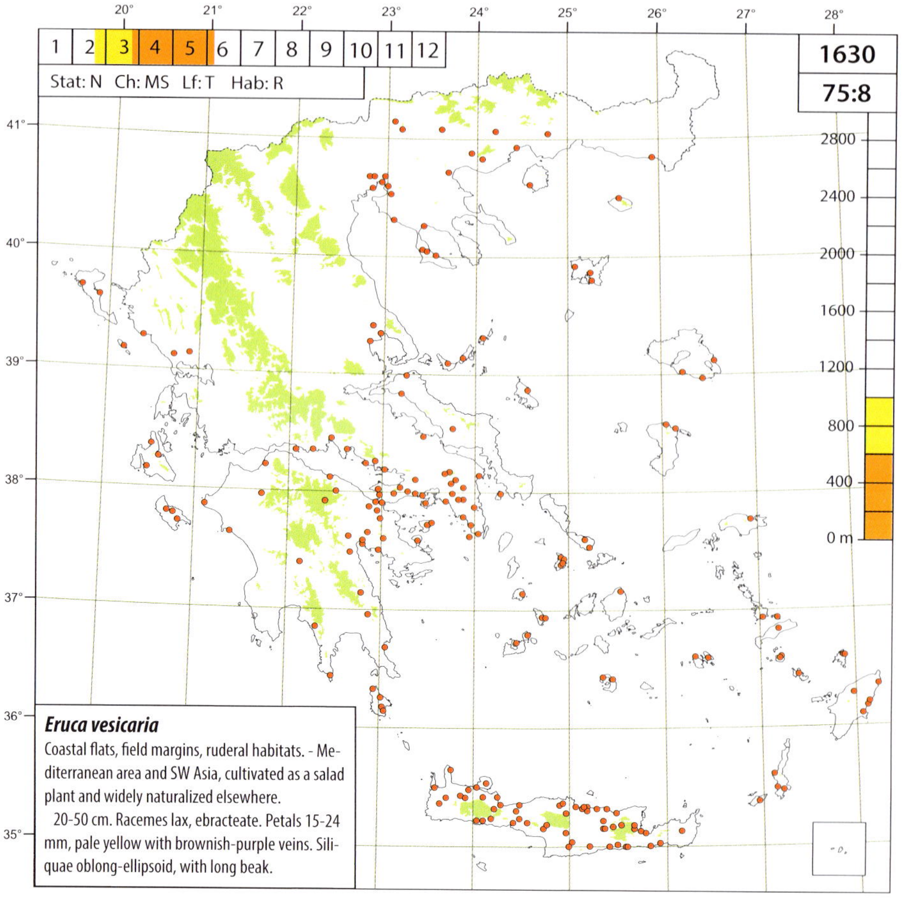

Eruca vesicaria
Coastal flats, field margins, ruderal habitats. - Mediterranean area and SW Asia, cultivated as a salad plant and widely naturalized elsewhere.
20-50 cm. Racemes lax, ebracteate. Petals 15-24 mm, pale yellow with brownish-purple veins. Siliquae oblong-ellipsoid, with long beak.

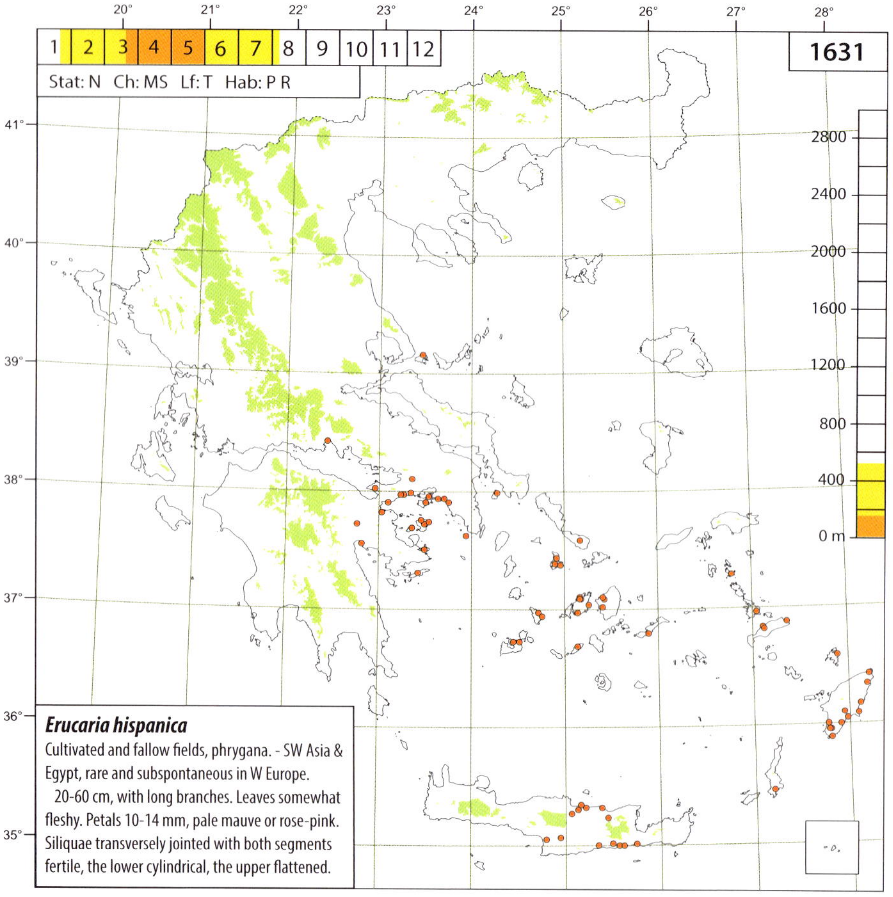

Erucaria hispanica
Cultivated and fallow fields, phrygana. - SW Asia & Egypt, rare and subspontaneous in W Europe.
20-60 cm, with long branches. Leaves somewhat fleshy. Petals 10-14 mm, pale mauve or rose-pink. Siliquae transversely jointed with both segments fertile, the lower cylindrical, the upper flattened.

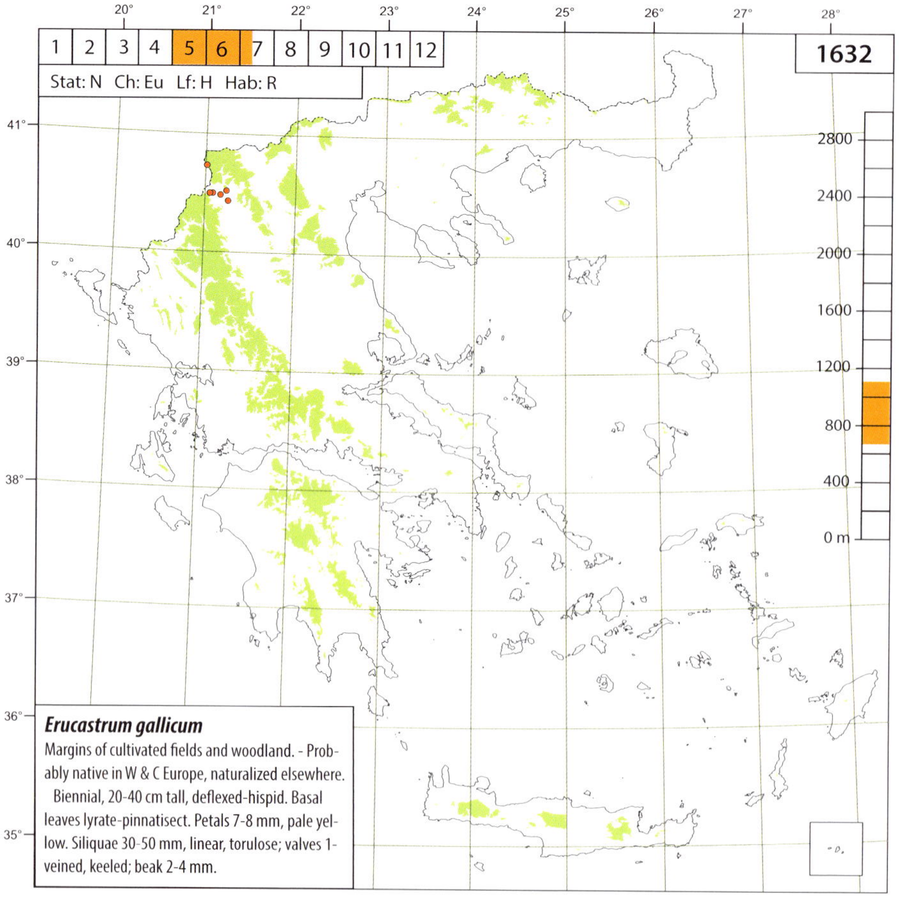

Erucastrum gallicum
Margins of cultivated fields and woodland. - Probably native in W & C Europe, naturalized elsewhere.
Biennial, 20-40 cm tall, deflexed-hispid. Basal leaves lyrate-pinnatisect. Petals 7-8 mm, pale yellow. Siliquae 30-50 mm, linear, torulose; valves 1-veined, keeled; beak 2-4 mm.

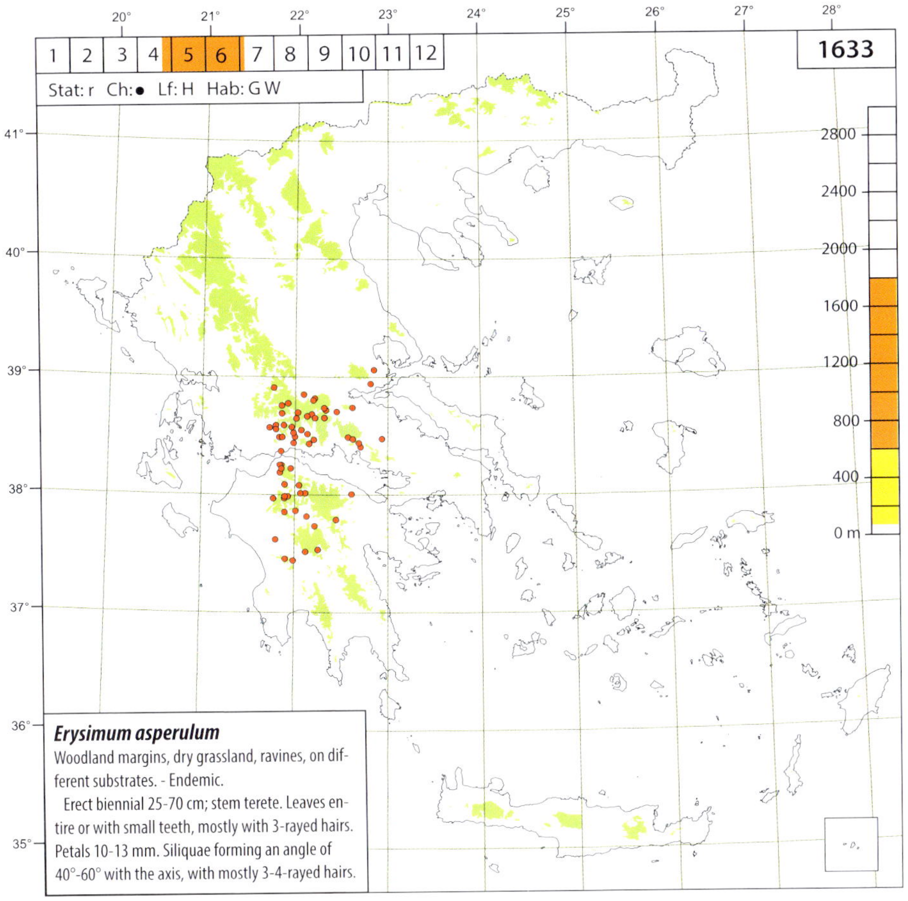

Erysimum asperulum
Woodland margins, dry grassland, ravines, on different substrates. - Endemic.
Erect biennial 25-70 cm; stem terete. Leaves entire or with small teeth, mostly with 3-rayed hairs. Petals 10-13 mm. Siliquae forming an angle of 40°-60° with the axis, with mostly 3-4-rayed hairs.

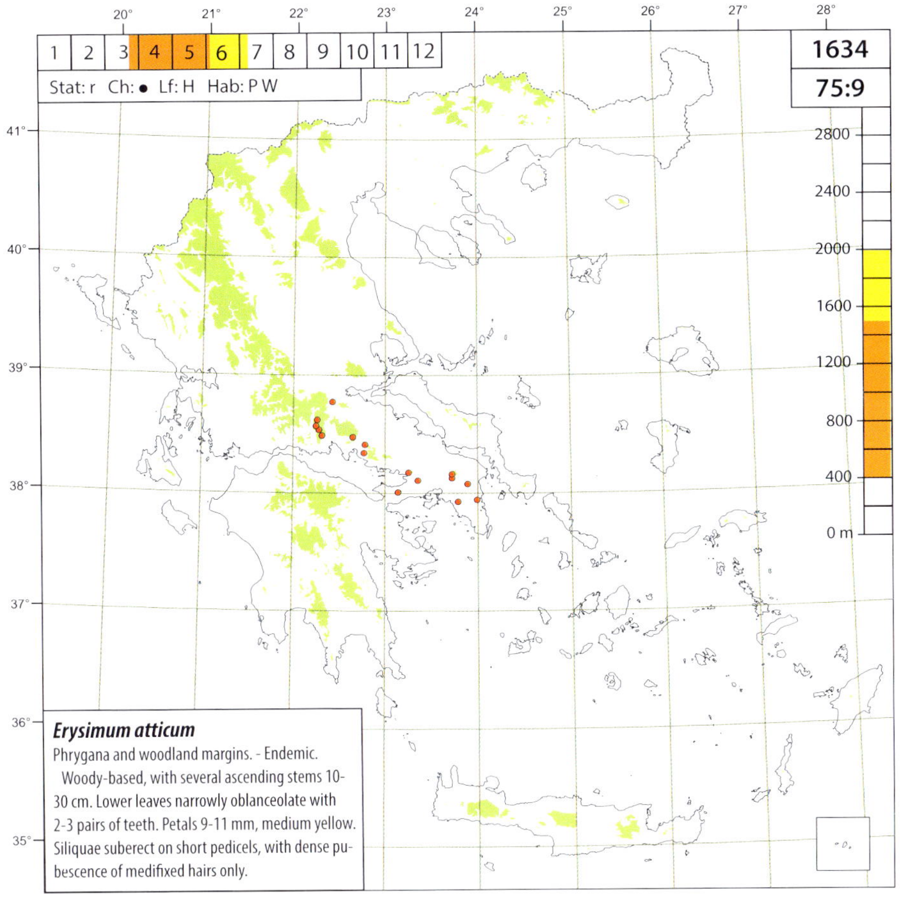

Erysimum atticum
Phrygana and woodland margins. - Endemic.
Woody-based, with several ascending stems 10-30 cm. Lower leaves narrowly oblanceolate with 2-3 pairs of teeth. Petals 9-11 mm, medium yellow. Siliquae suberect on short pedicels, with dense pubescence of medifixed hairs only.

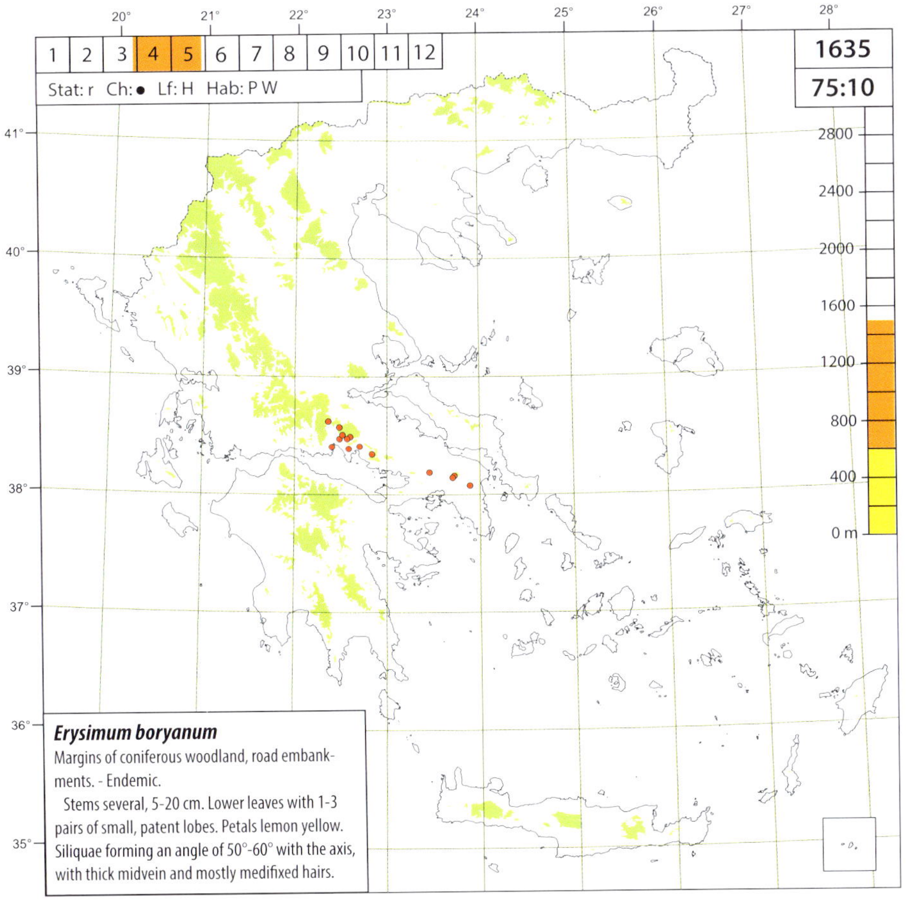

Erysimum boryanum
Margins of coniferous woodland, road embankments. - Endemic.
Stems several, 5-20 cm. Lower leaves with 1-3 pairs of small, patent lobes. Petals lemon yellow. Siliquae forming an angle of 50°-60° with the axis, with thick midvein and mostly medifixed hairs.

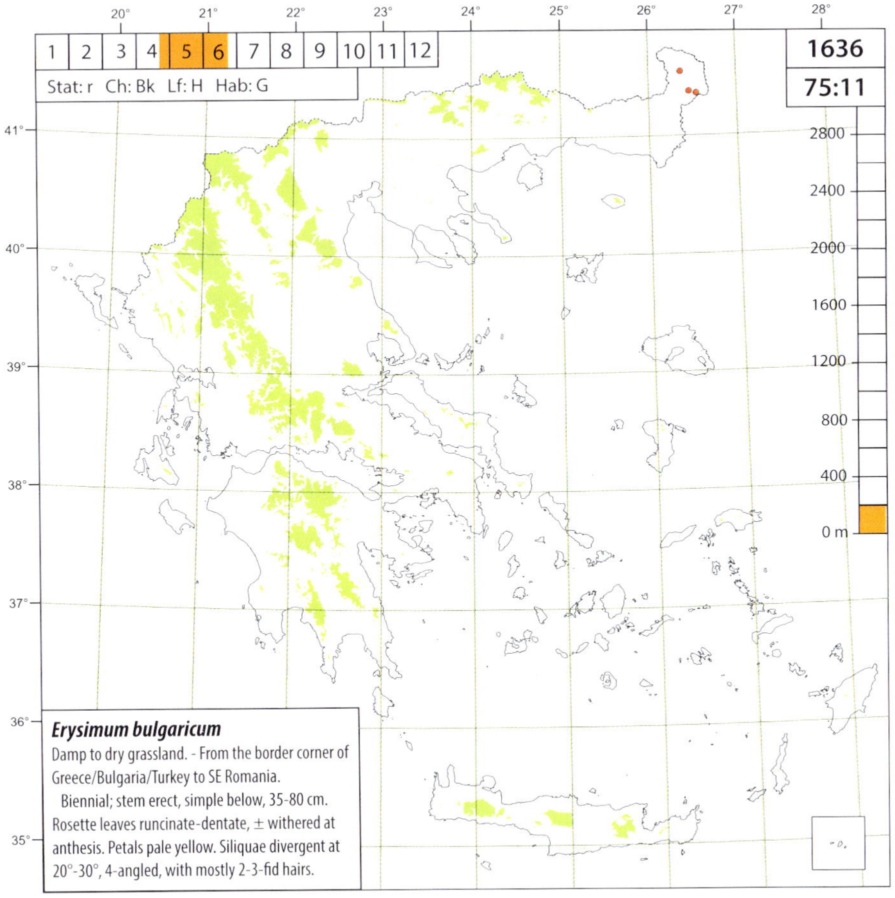

Erysimum bulgaricum
Damp to dry grassland. - From the border corner of Greece/Bulgaria/Turkey to SE Romania.
Biennial; stem erect, simple below, 35-80 cm. Rosette leaves runcinate-dentate, ± withered at anthesis. Petals pale yellow. Siliquae divergent at 20°-30°, 4-angled, with mostly 2-3-fid hairs.

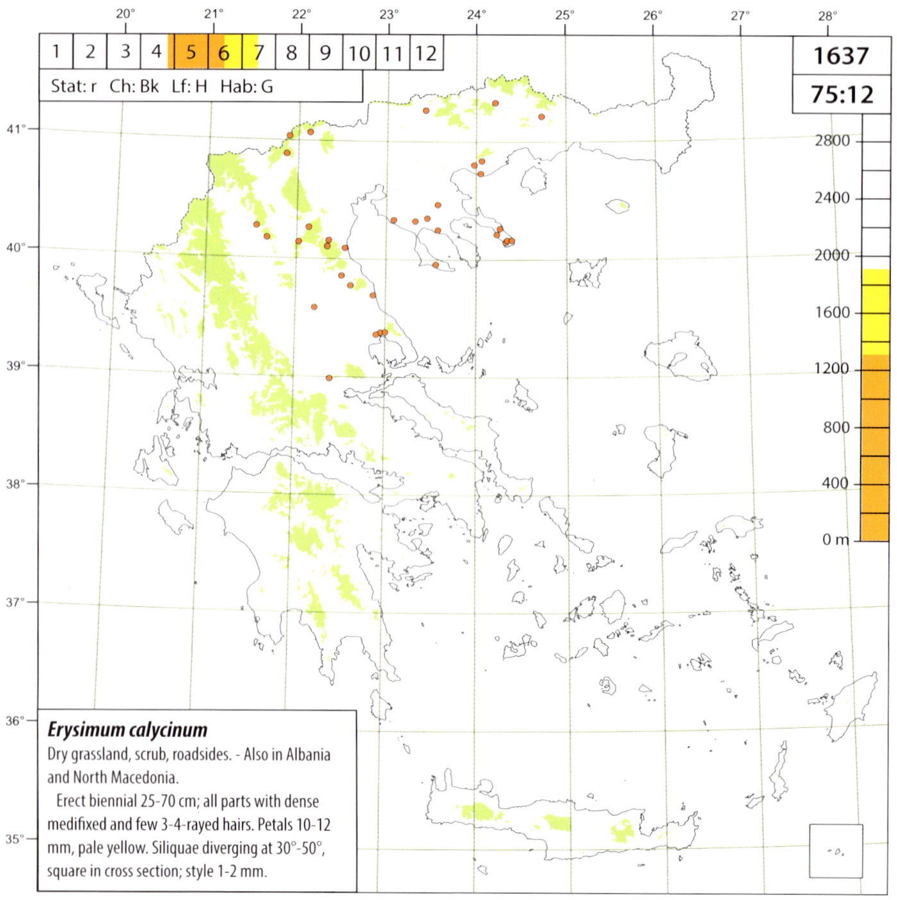

Erysimum calycinum
Dry grassland, scrub, roadsides. - Also in Albania and North Macedonia.
Erect biennial 25-70 cm; all parts with dense medifixed and few 3-4-rayed hairs. Petals 10-12 mm, pale yellow. Siliquae diverging at 30°-50°, square in cross section; style 1-2 mm.

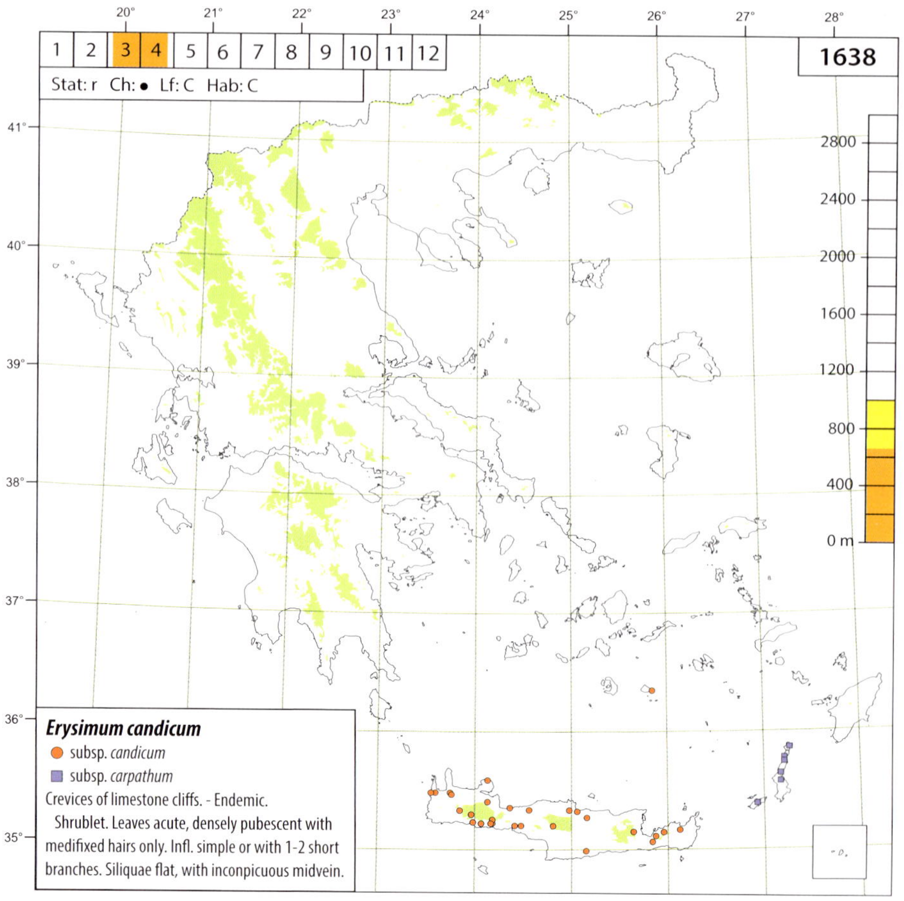

Erysimum candicum
● subsp. *candicum*
■ subsp. *carpathum*
Crevices of limestone cliffs. - Endemic.
Shrublet. Leaves acute, densely pubescent with medifixed hairs only. Infl. simple or with 1-2 short branches. Siliquae flat, with inconpicuous midvein.

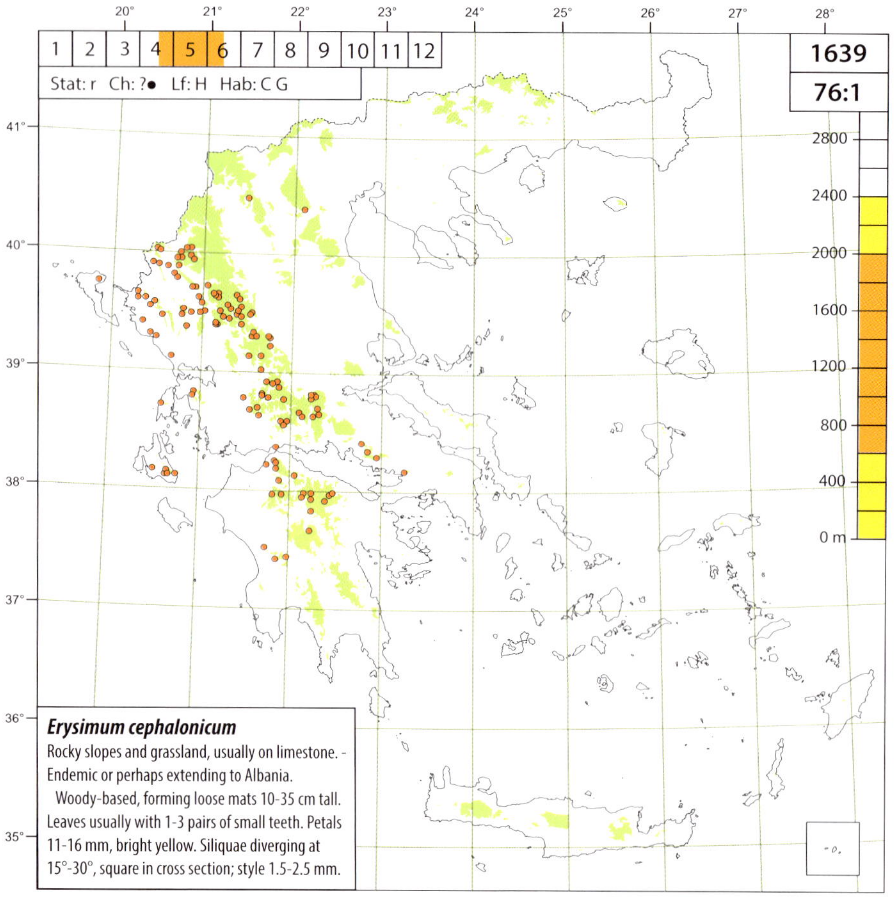

Erysimum cephalonicum
Rocky slopes and grassland, usually on limestone. - Endemic or perhaps extending to Albania.
Woody-based, forming loose mats 10-35 cm tall. Leaves usually with 1-3 pairs of small teeth. Petals 11-16 mm, bright yellow. Siliquae diverging at 15°-30°, square in cross section; style 1.5-2.5 mm.

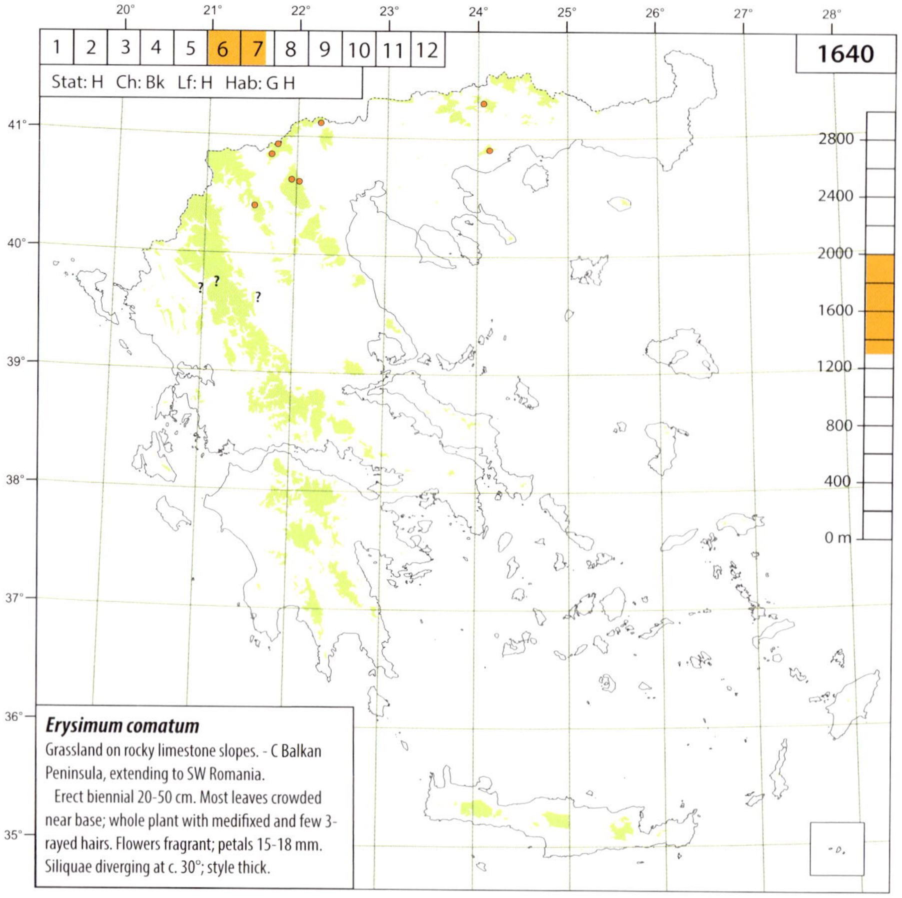

Erysimum comatum
Grassland on rocky limestone slopes. - C Balkan Peninsula, extending to SW Romania.
Erect biennial 20-50 cm. Most leaves crowded near base; whole plant with medifixed and few 3-rayed hairs. Flowers fragrant; petals 15-18 mm. Siliquae diverging at c. 30°; style thick.

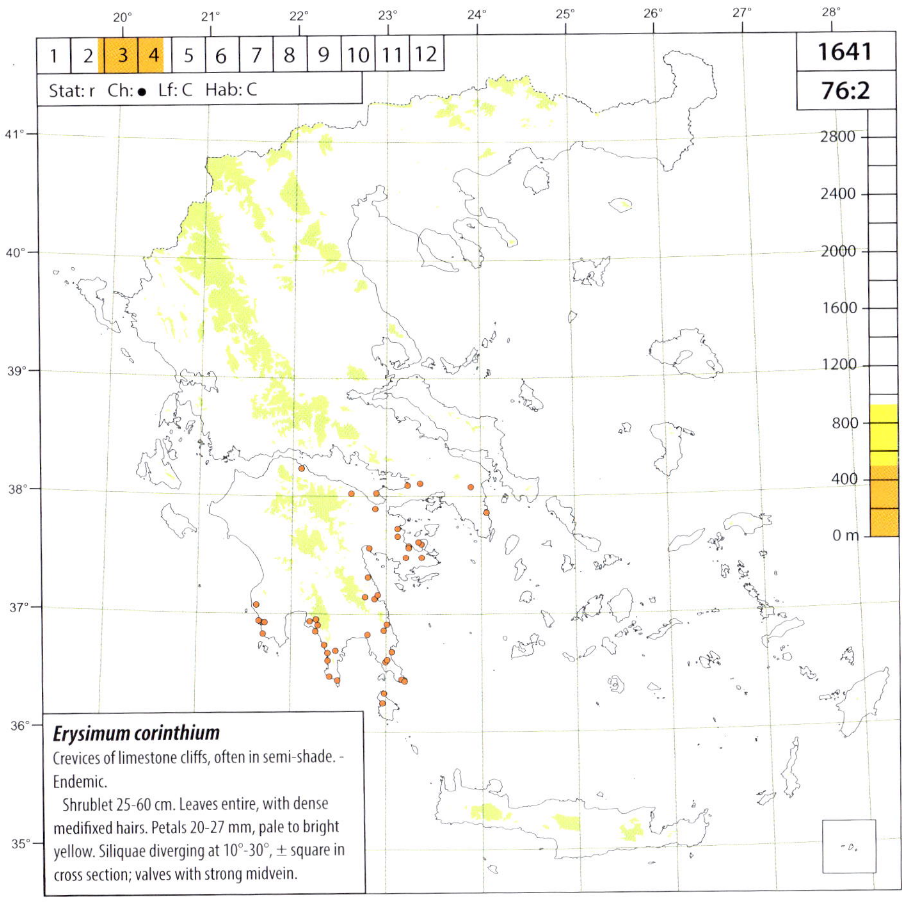

Erysimum corinthium
Crevices of limestone cliffs, often in semi-shade. - Endemic.
Shrublet 25-60 cm. Leaves entire, with dense medifixed hairs. Petals 20-27 mm, pale to bright yellow. Siliquae diverging at 10°-30°, ± square in cross section; valves with strong midvein.

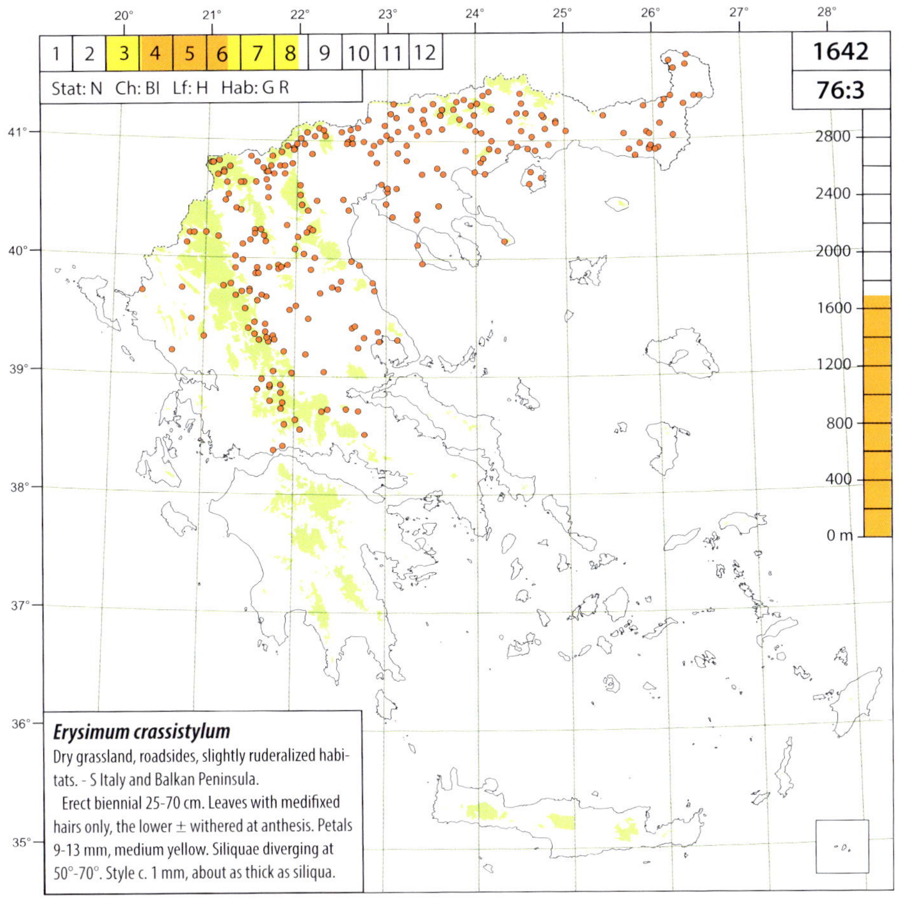

Erysimum crassistylum
Dry grassland, roadsides, slightly ruderalized habitats. - S Italy and Balkan Peninsula.
Erect biennial 25-70 cm. Leaves with medifixed hairs only, the lower ± withered at anthesis. Petals 9-13 mm, medium yellow. Siliquae diverging at 50°-70°. Style c. 1 mm, about as thick as siliqua.

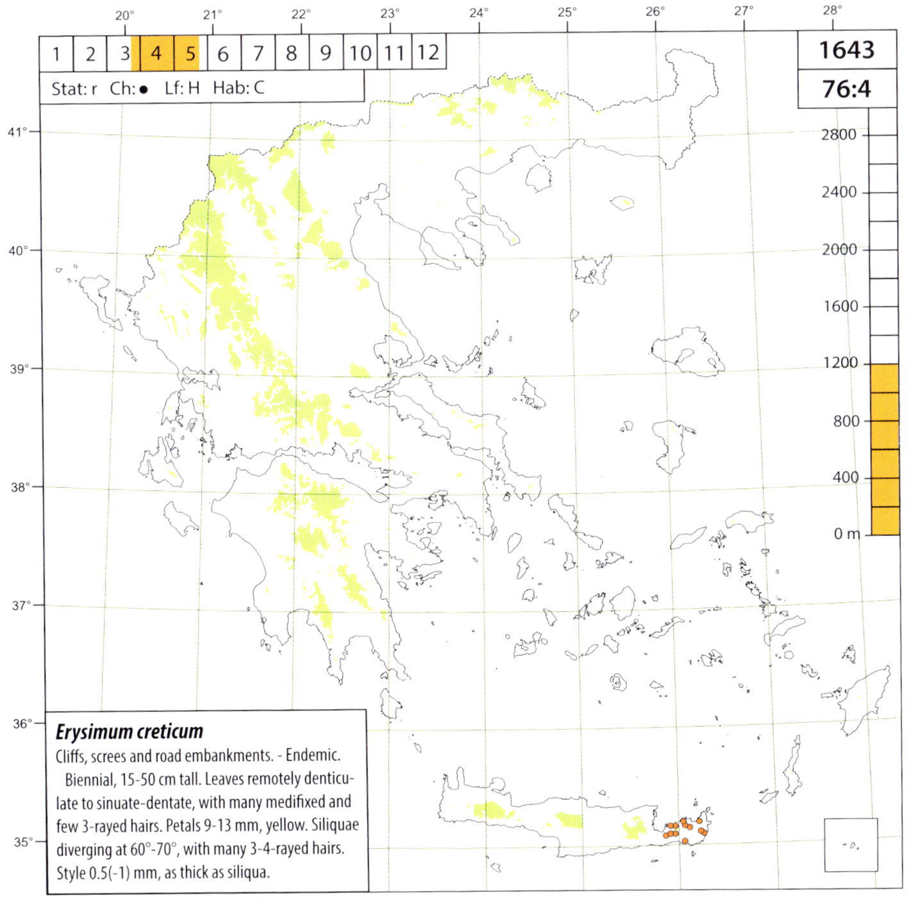

Erysimum creticum
Cliffs, screes and road embankments. - Endemic.
Biennial, 15-50 cm tall. Leaves remotely denticulate to sinuate-dentate, with many medifixed and few 3-rayed hairs. Petals 9-13 mm, yellow. Siliquae diverging at 60°-70°, with many 3-4-rayed hairs. Style 0.5(-1) mm, as thick as siliqua.

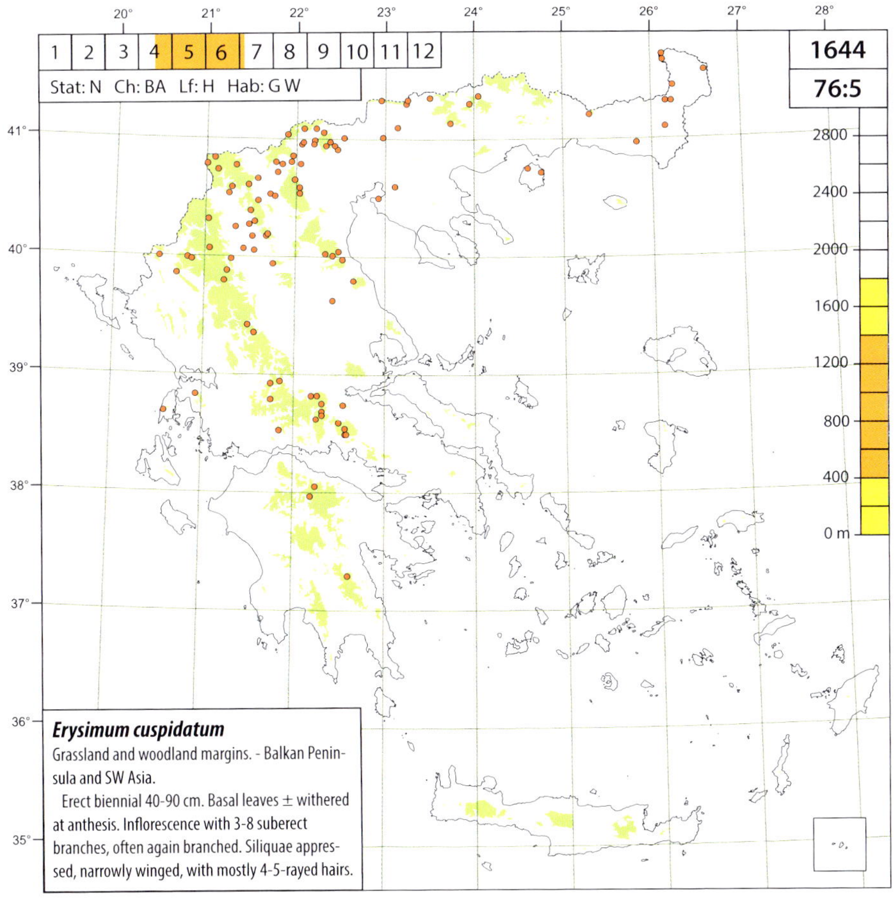

Erysimum cuspidatum
Grassland and woodland margins. - Balkan Peninsula and SW Asia.
Erect biennial 40-90 cm. Basal leaves ± withered at anthesis. Inflorescence with 3-8 suberect branches, often again branched. Siliquae appressed, narrowly winged, with mostly 4-5-rayed hairs.

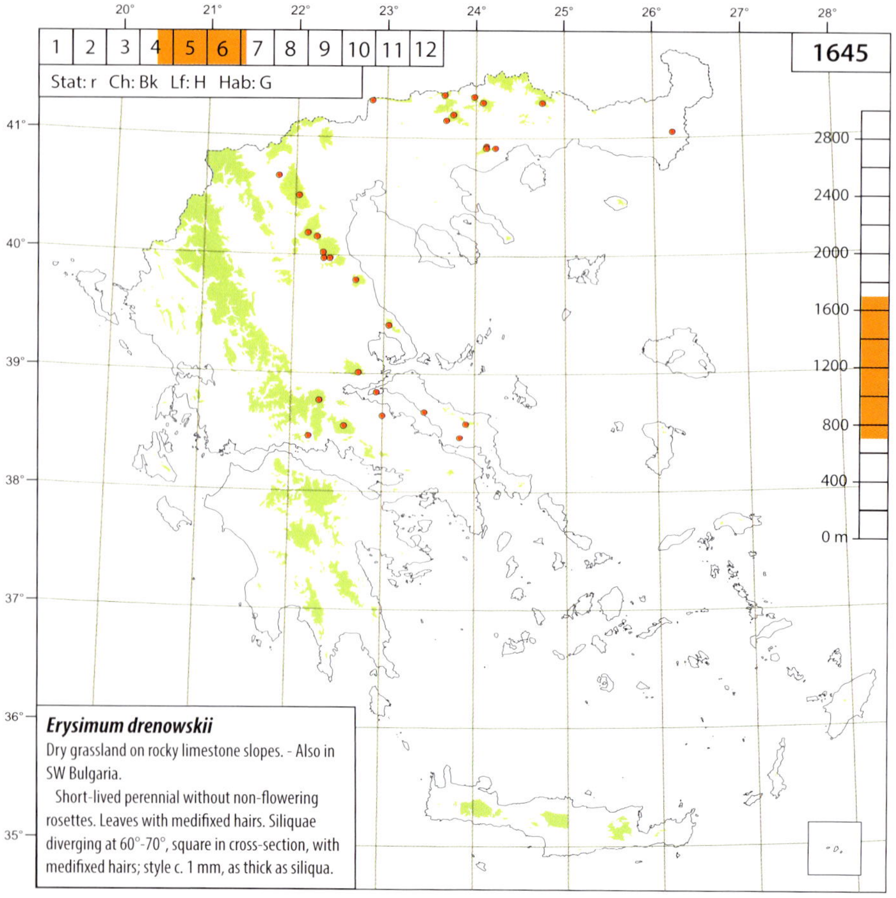

Erysimum drenowskii

Dry grassland on rocky limestone slopes. - Also in SW Bulgaria.

Short-lived perennial without non-flowering rosettes. Leaves with medifixed hairs. Siliquae diverging at 60°-70°, square in cross-section, with medifixed hairs; style c. 1 mm, as thick as siliqua.

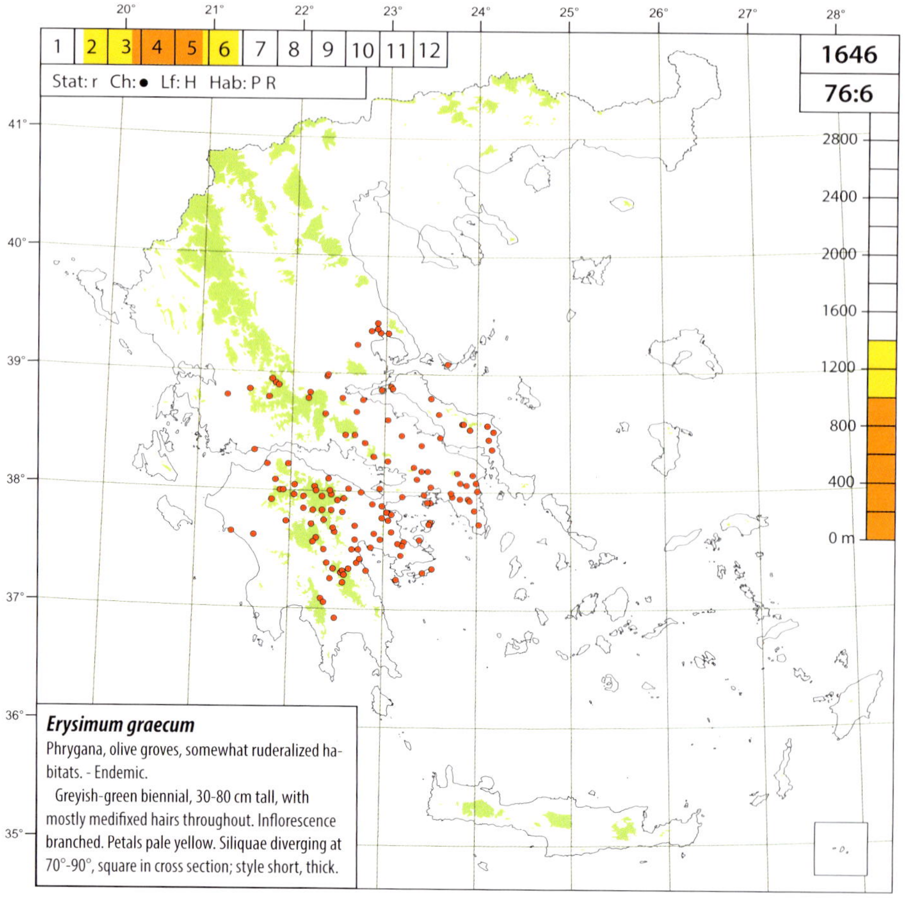

Erysimum graecum

Phrygana, olive groves, somewhat ruderalized habitats. - Endemic.

Greyish-green biennial, 30-80 cm tall, with mostly medifixed hairs throughout. Inflorescence branched. Petals pale yellow. Siliquae diverging at 70°-90°, square in cross section; style short, thick.

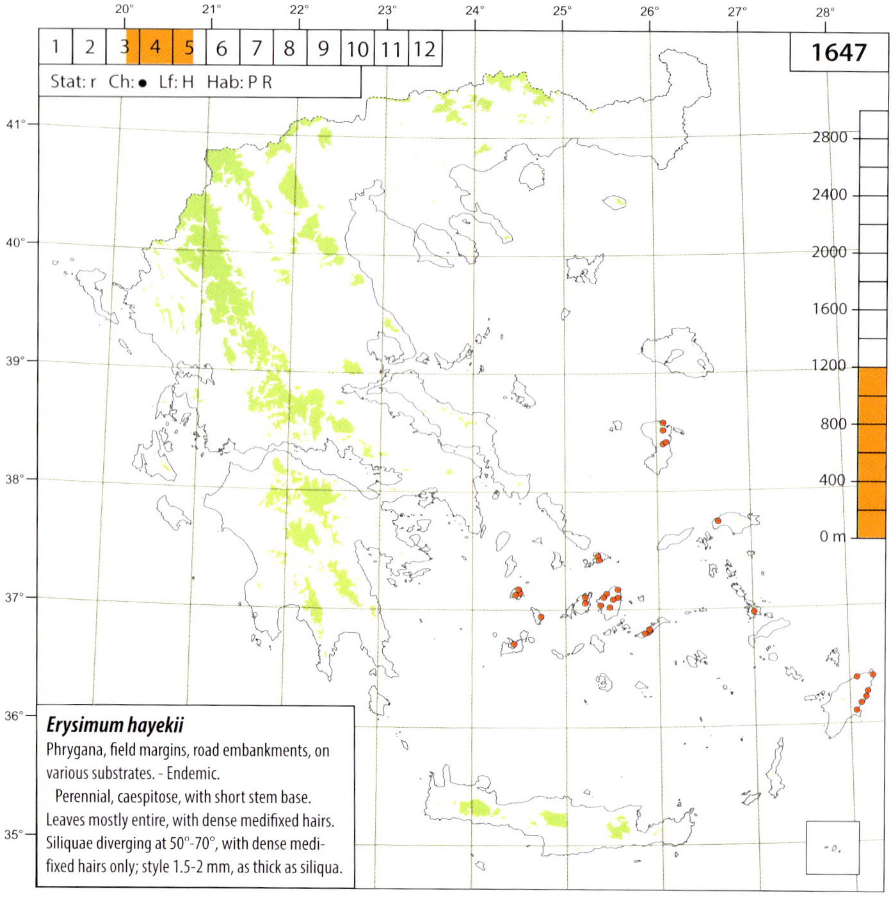

Erysimum hayekii

Phrygana, field margins, road embankments, on various substrates. - Endemic.

Perennial, caespitose, with short stem base. Leaves mostly entire, with dense medifixed hairs. Siliquae diverging at 50°-70°, with dense medifixed hairs only; style 1.5-2 mm, as thick as siliqua.

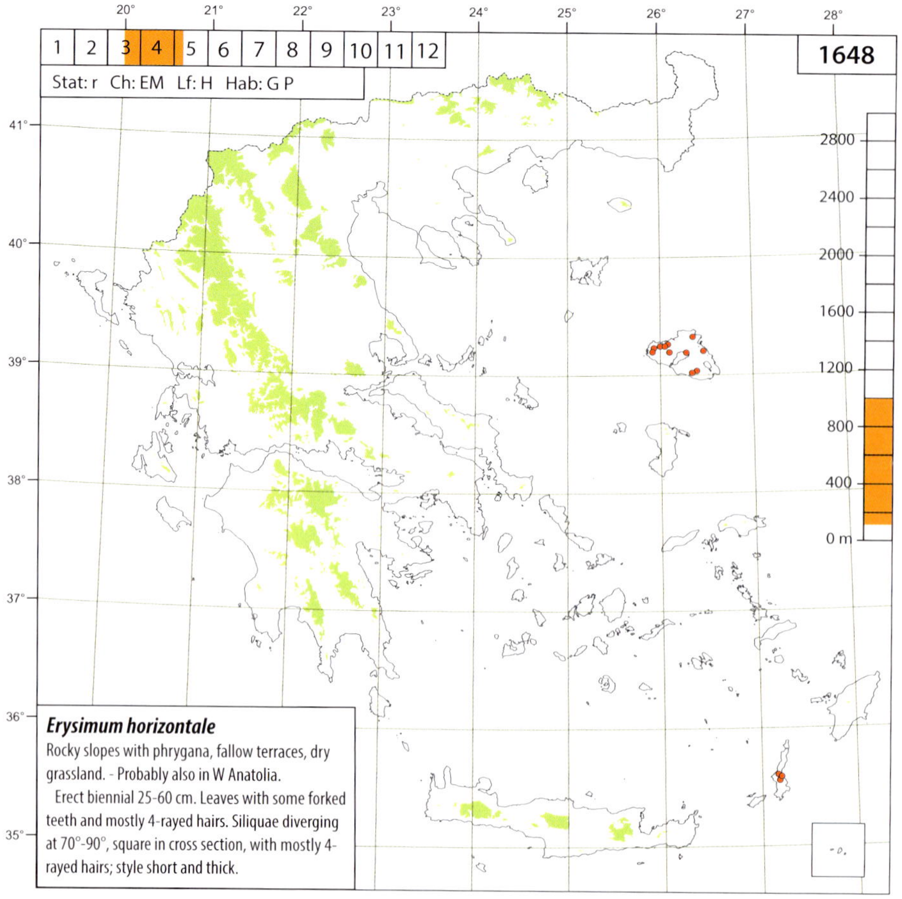

Erysimum horizontale

Rocky slopes with phrygana, fallow terraces, dry grassland. - Probably also in W Anatolia.

Erect biennial 25-60 cm. Leaves with some forked teeth and mostly 4-rayed hairs. Siliquae diverging at 70°-90°, square in cross section, with mostly 4-rayed hairs; style short and thick.

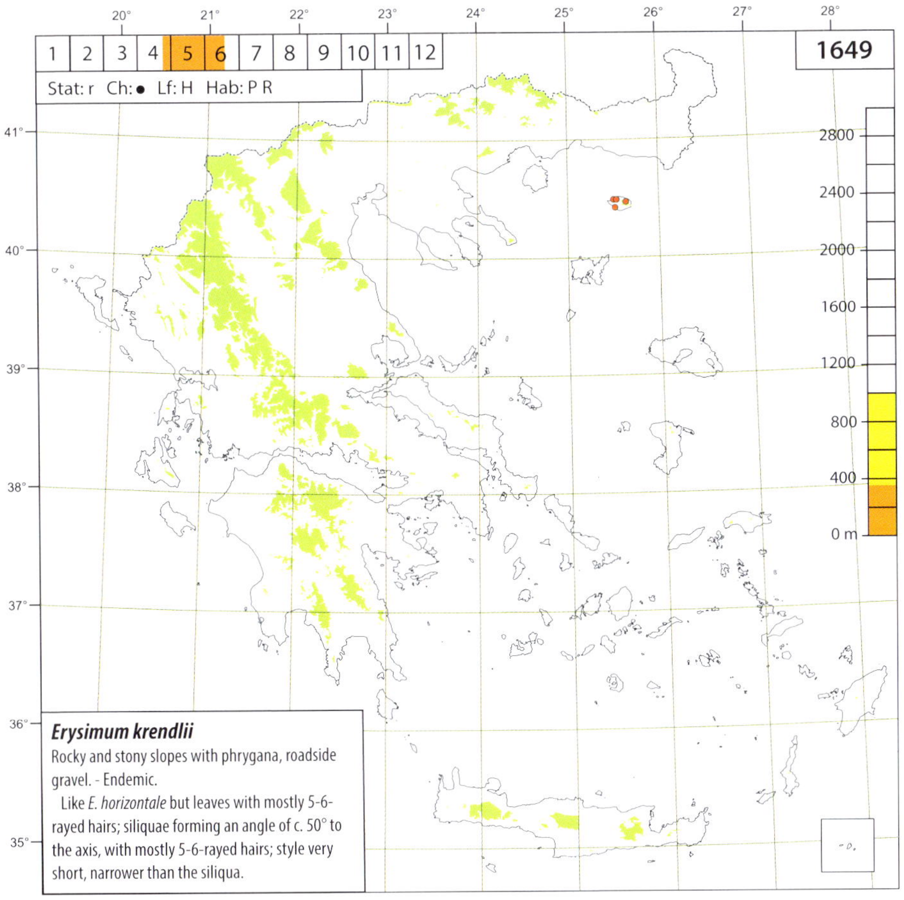

Erysimum krendlii
Rocky and stony slopes with phrygana, roadside gravel. - Endemic.
Like *E. horizontale* but leaves with mostly 5-6-rayed hairs; siliquae forming an angle of c. 50° to the axis, with mostly 5-6-rayed hairs; style very short, narrower than the siliqua.

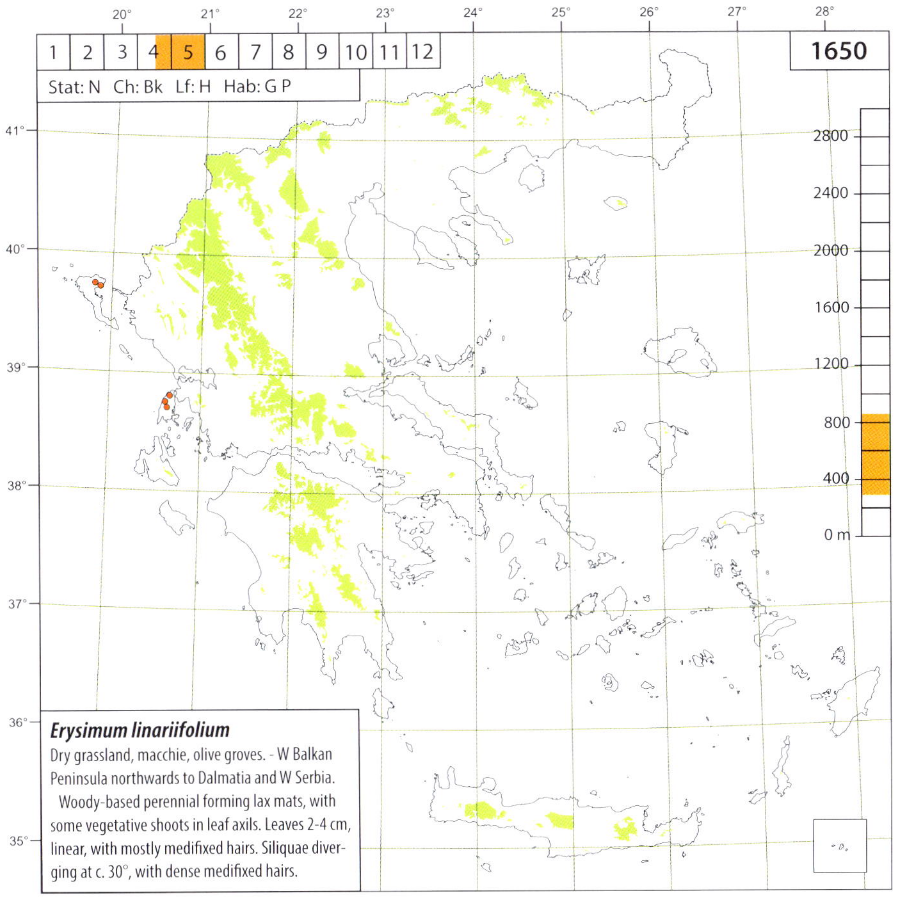

Erysimum linariifolium
Dry grassland, macchie, olive groves. - W Balkan Peninsula northwards to Dalmatia and W Serbia.
Woody-based perennial forming lax mats, with some vegetative shoots in leaf axils. Leaves 2-4 cm, linear, with mostly medifixed hairs. Siliquae diverging at c. 30°, with dense medifixed hairs.

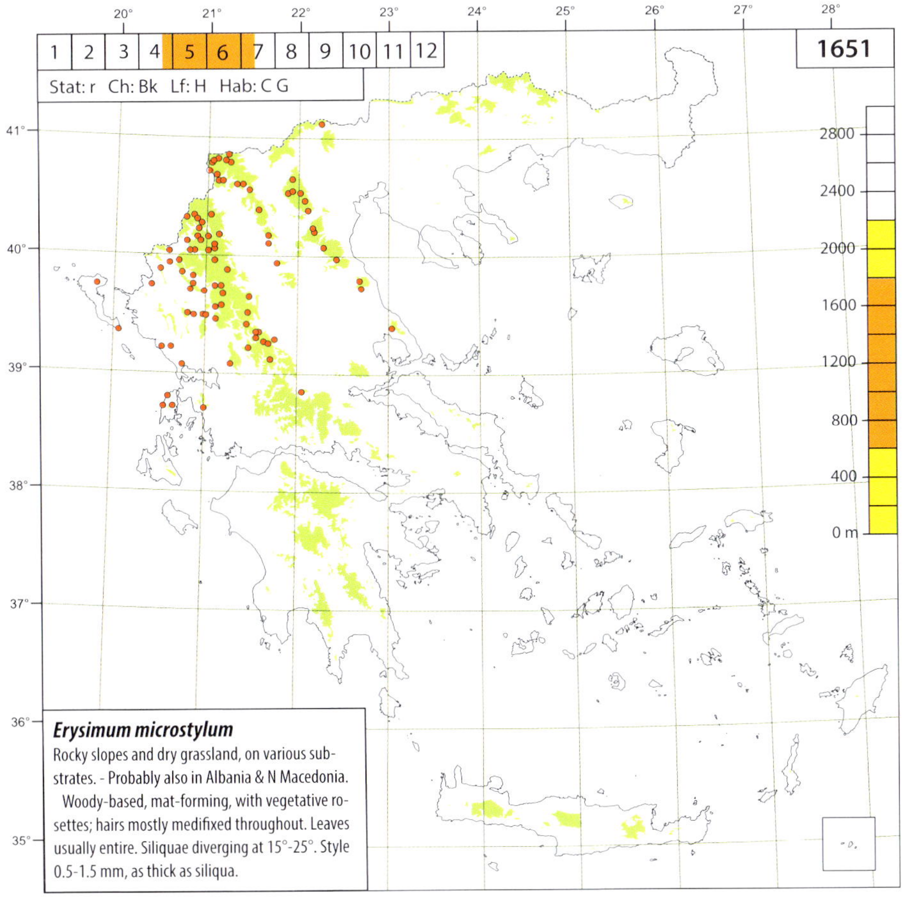

Erysimum microstylum
Rocky slopes and dry grassland, on various substrates. - Probably also in Albania & N Macedonia.
Woody-based, mat-forming, with vegetative rosettes; hairs mostly medifixed throughout. Leaves usually entire. Siliquae diverging at 15°-25°. Style 0.5-1.5 mm, as thick as siliqua.

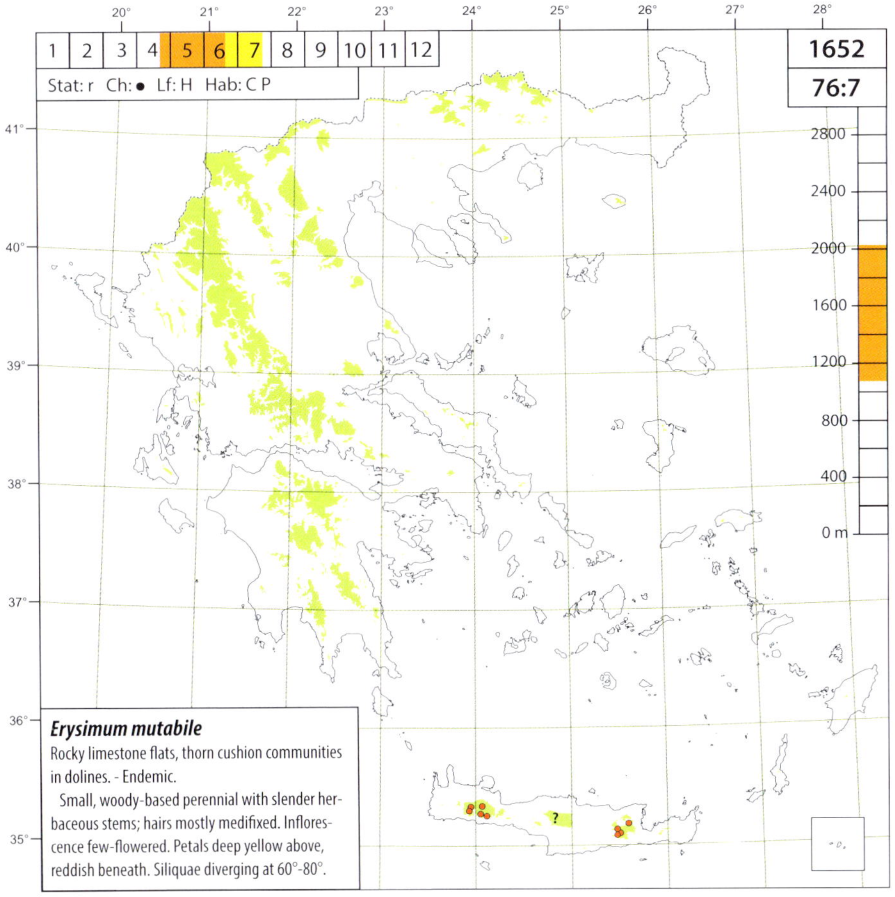

Erysimum mutabile
Rocky limestone flats, thorn cushion communities in dolines. - Endemic.
Small, woody-based perennial with slender herbaceous stems; hairs mostly medifixed. Inflorescence few-flowered. Petals deep yellow above, reddish beneath. Siliquae diverging at 60°-80°.

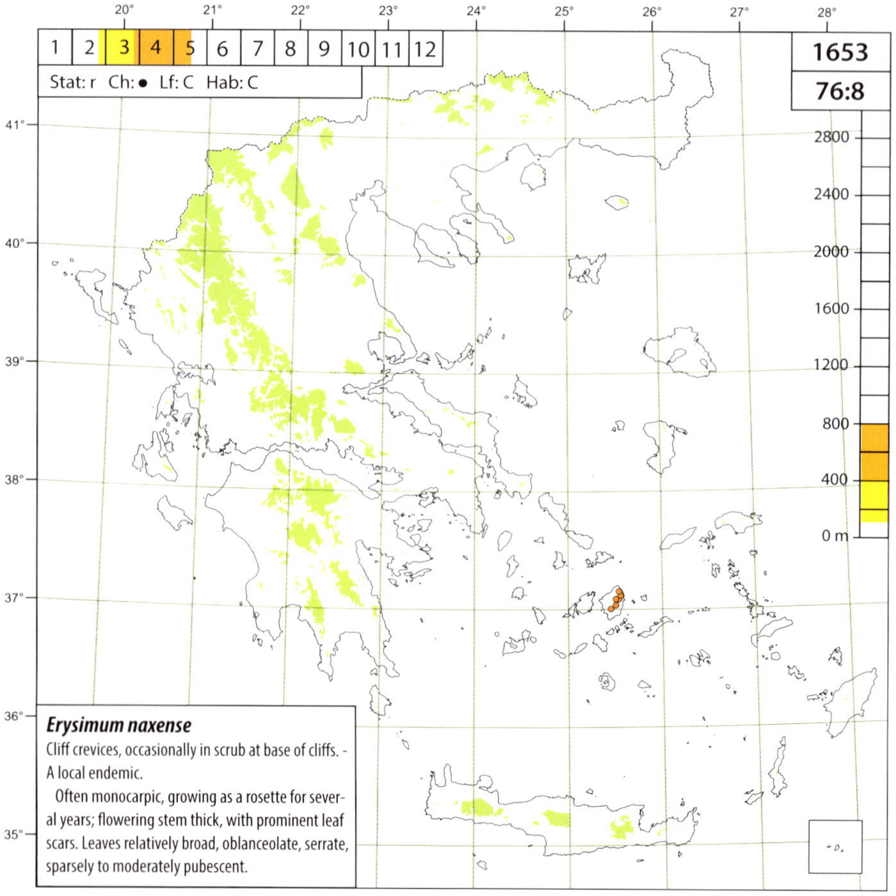

Erysimum naxense

Cliff crevices, occasionally in scrub at base of cliffs. - A local endemic.

Often monocarpic, growing as a rosette for several years; flowering stem thick, with prominent leaf scars. Leaves relatively broad, oblanceolate, serrate, sparsely to moderately pubescent.

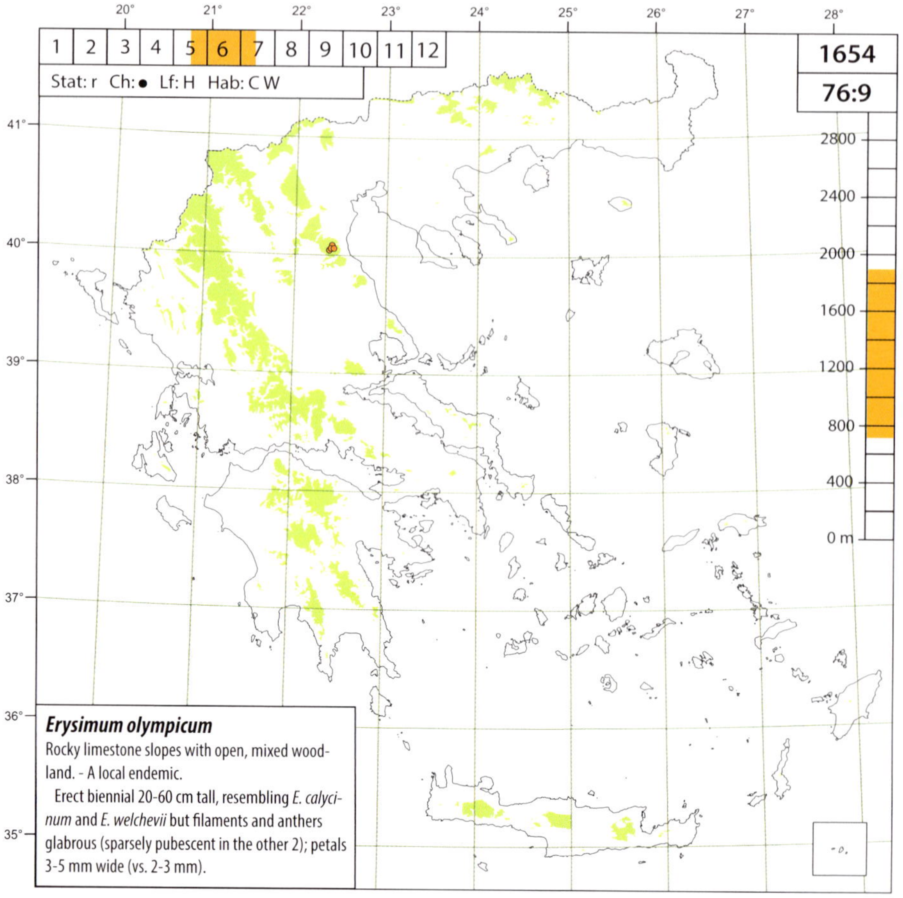

Erysimum olympicum

Rocky limestone slopes with open, mixed woodland. - A local endemic.

Erect biennial 20-60 cm tall, resembling *E. calycinum* and *E. welchevii* but filaments and anthers glabrous (sparsely pubescent in the other 2); petals 3-5 mm wide (vs. 2-3 mm).

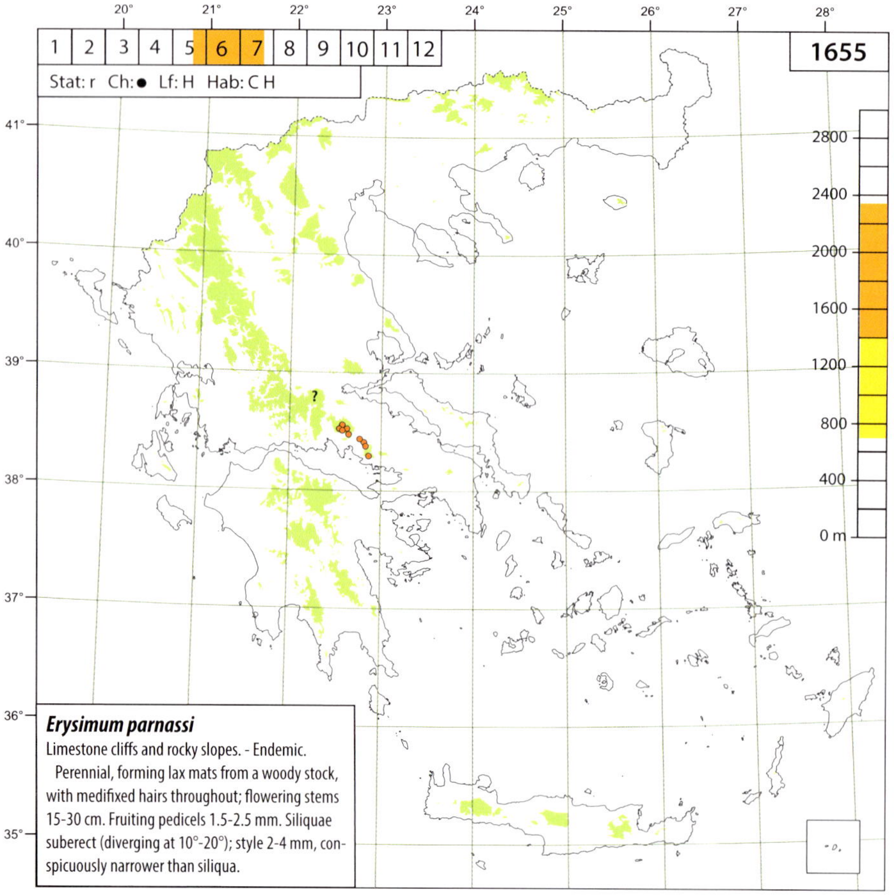

Erysimum parnassi

Limestone cliffs and rocky slopes. - Endemic.

Perennial, forming lax mats from a woody stock, with medifixed hairs throughout; flowering stems 15-30 cm. Fruiting pedicels 1.5-2.5 mm. Siliquae suberect (diverging at 10°-20°); style 2-4 mm, conspicuously narrower than siliqua.

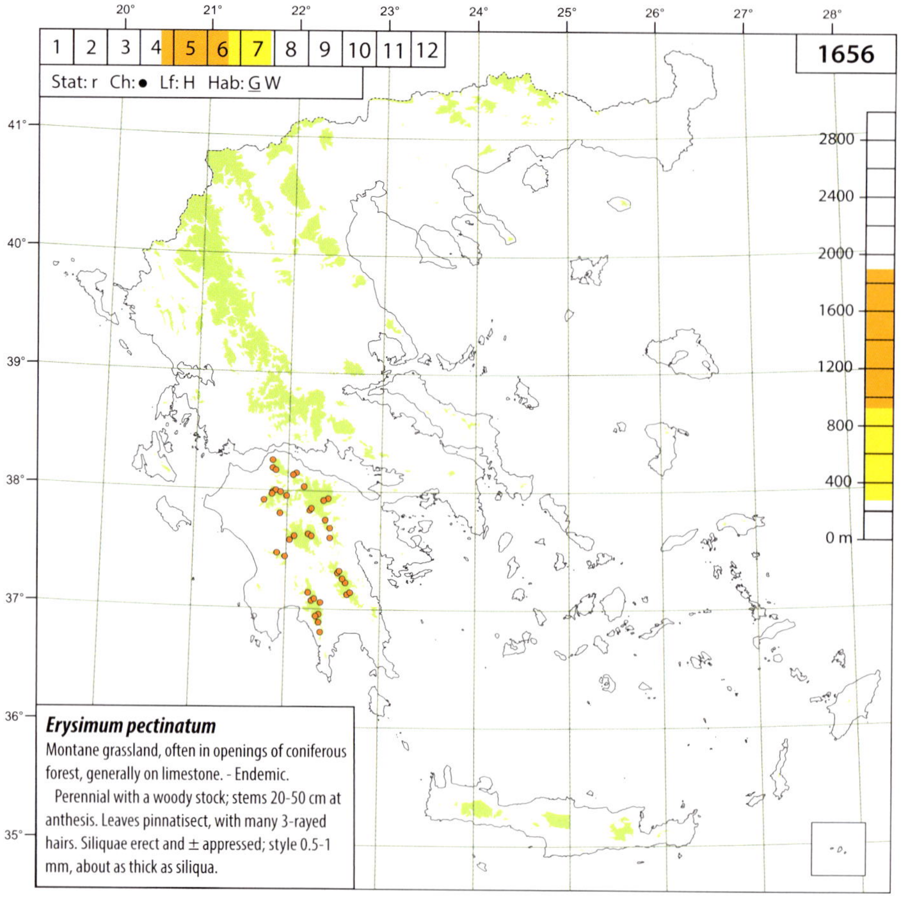

Erysimum pectinatum

Montane grassland, often in openings of coniferous forest, generally on limestone. - Endemic.

Perennial with a woody stock; stems 20-50 cm at anthesis. Leaves pinnatisect, with many 3-rayed hairs. Siliquae erect and ± appressed; style 0.5-1 mm, about as thick as siliqua.

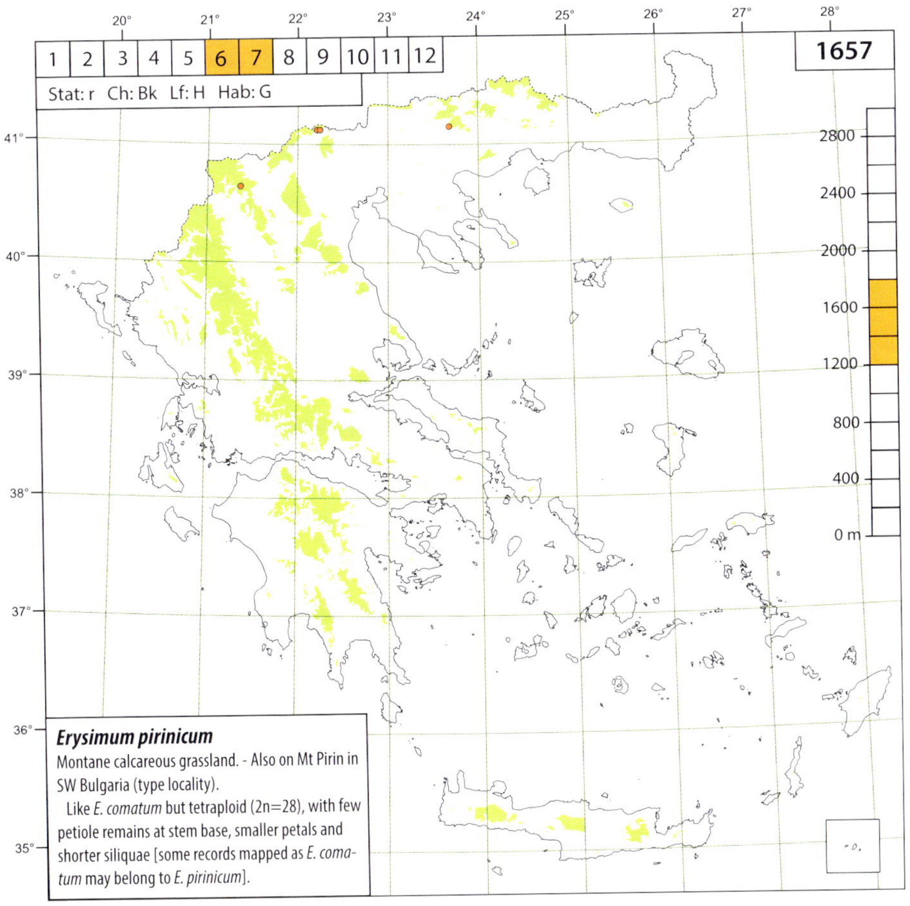

Erysimum pirinicum
Montane calcareous grassland. - Also on Mt Pirin in SW Bulgaria (type locality).
Like *E. comatum* but tetraploid (2n=28), with few petiole remains at stem base, smaller petals and shorter siliquae [some records mapped as *E. comatum* may belong to *E. pirinicum*].

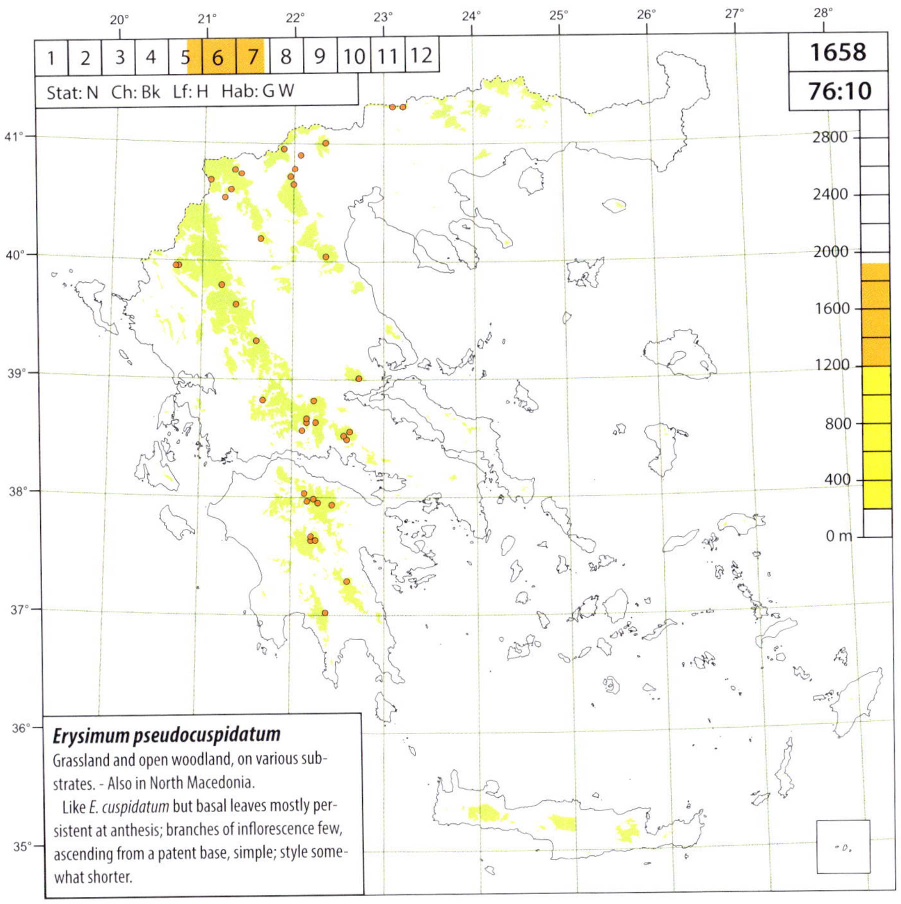

Erysimum pseudocuspidatum
Grassland and open woodland, on various substrates. - Also in North Macedonia.
Like *E. cuspidatum* but basal leaves mostly persistent at anthesis; branches of inflorescence few, ascending from a patent base, simple; style somewhat shorter.

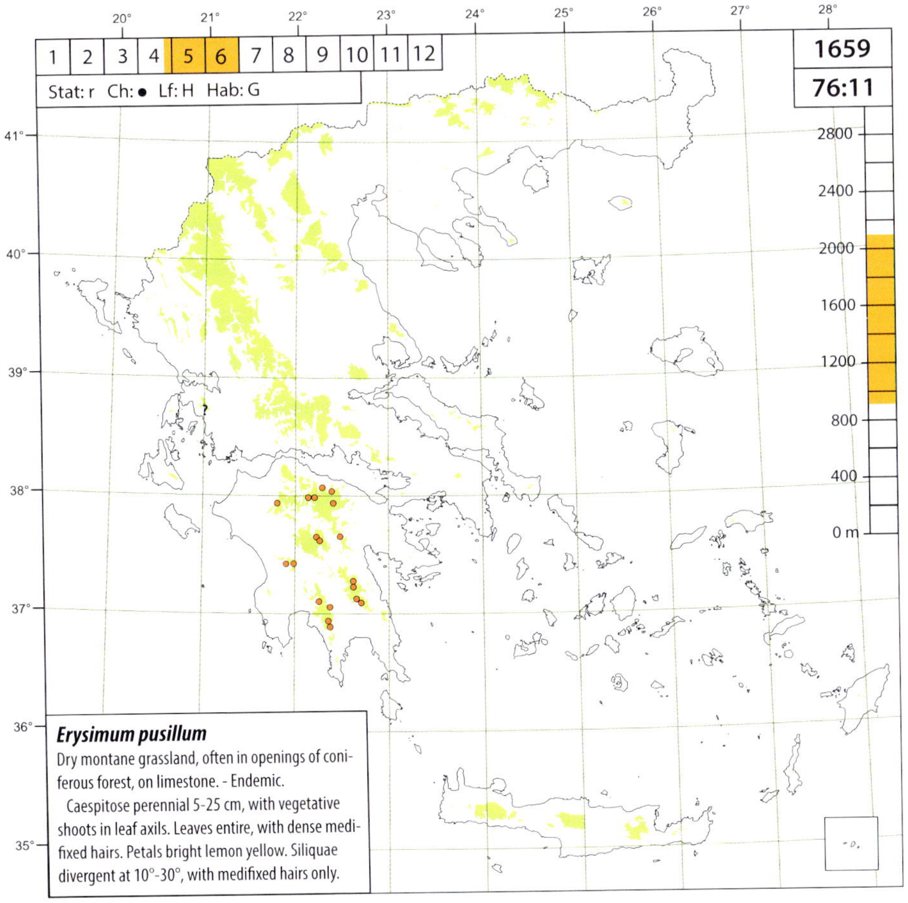

Erysimum pusillum
Dry montane grassland, often in openings of coniferous forest, on limestone. - Endemic.
Caespitose perennial 5-25 cm, with vegetative shoots in leaf axils. Leaves entire, with dense medifixed hairs. Petals bright lemon yellow. Siliquae divergent at 10°-30°, with medifixed hairs only.

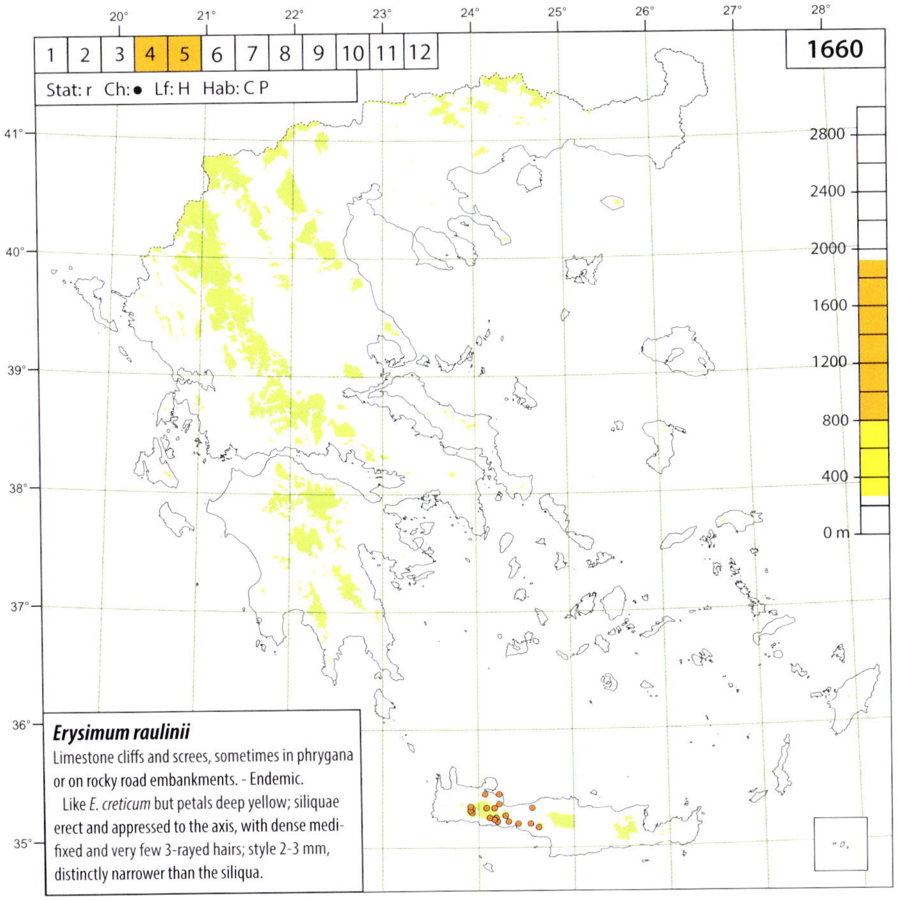

Erysimum raulinii
Limestone cliffs and screes, sometimes in phrygana or on rocky road embankments. - Endemic.
Like *E. creticum* but petals deep yellow; siliquae erect and appressed to the axis, with dense medifixed and very few 3-rayed hairs; style 2-3 mm, distinctly narrower than the siliqua.

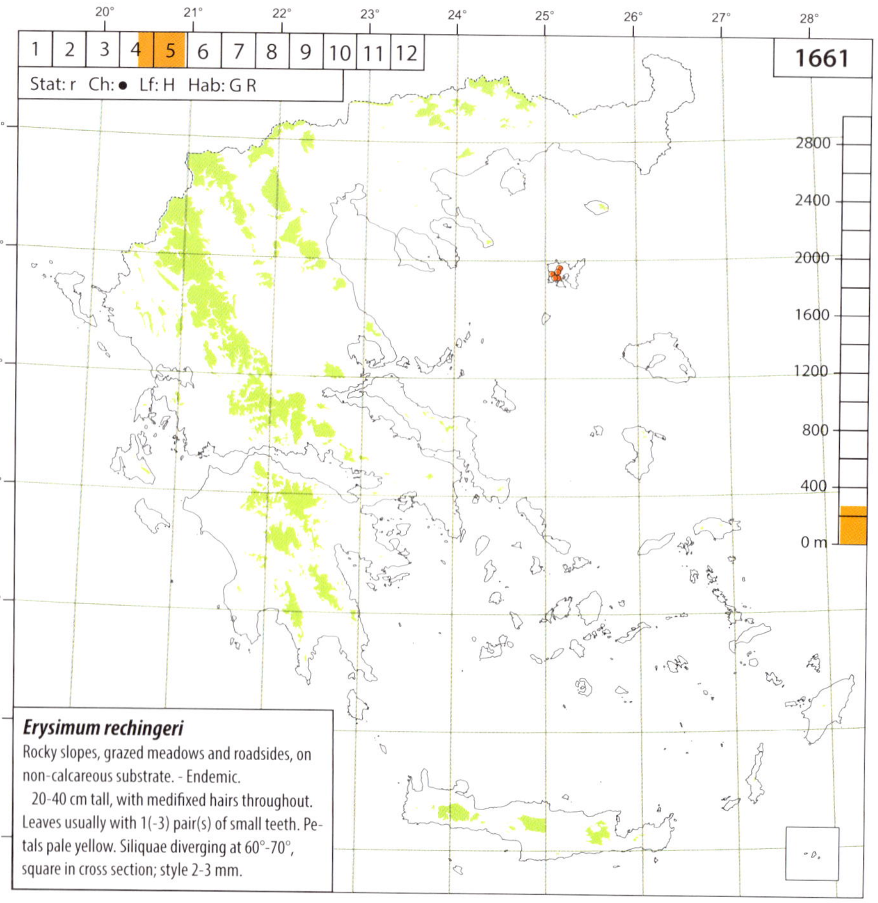

Stat: r Ch: ● Lf: H Hab: G R

Erysimum rechingeri

Rocky slopes, grazed meadows and roadsides, on non-calcareous substrate. - Endemic.

20-40 cm tall, with medifixed hairs throughout. Leaves usually with 1(-3) pair(s) of small teeth. Petals pale yellow. Siliquae diverging at 60°-70°, square in cross section; style 2-3 mm.

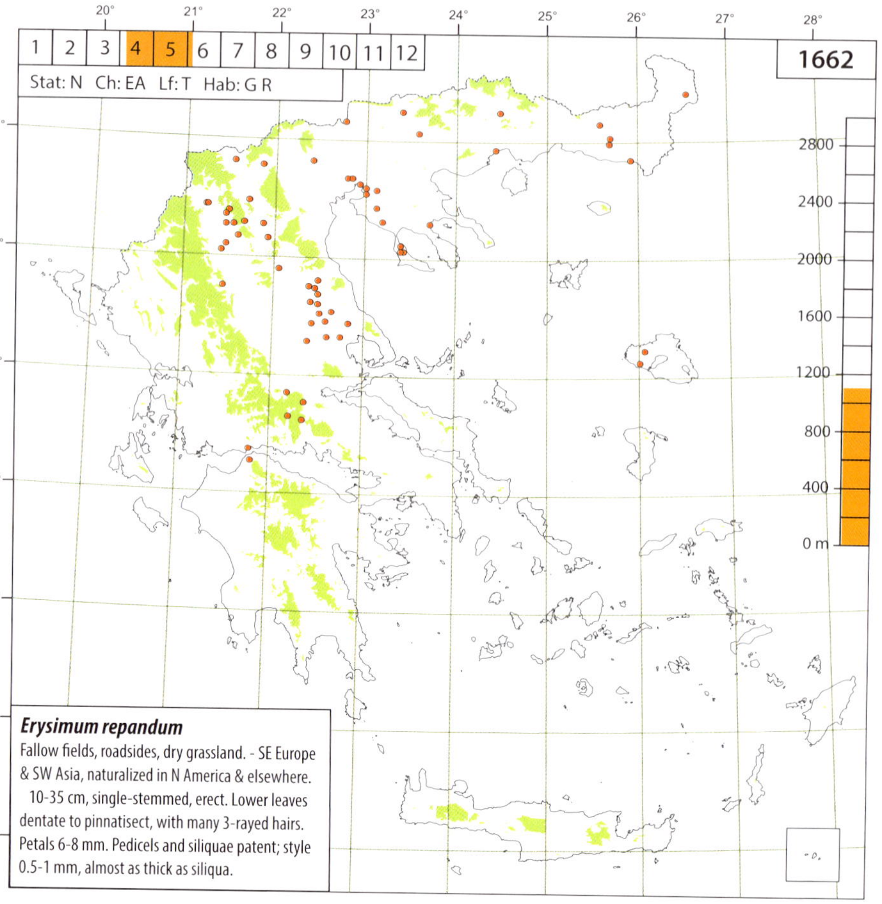

Stat: N Ch: EA Lf: T Hab: G R

Erysimum repandum

Fallow fields, roadsides, dry grassland. - SE Europe & SW Asia, naturalized in N America & elsewhere.

10-35 cm, single-stemmed, erect. Lower leaves dentate to pinnatisect, with many 3-rayed hairs. Petals 6-8 mm. Pedicels and siliquae patent; style 0.5-1 mm, almost as thick as siliqua.

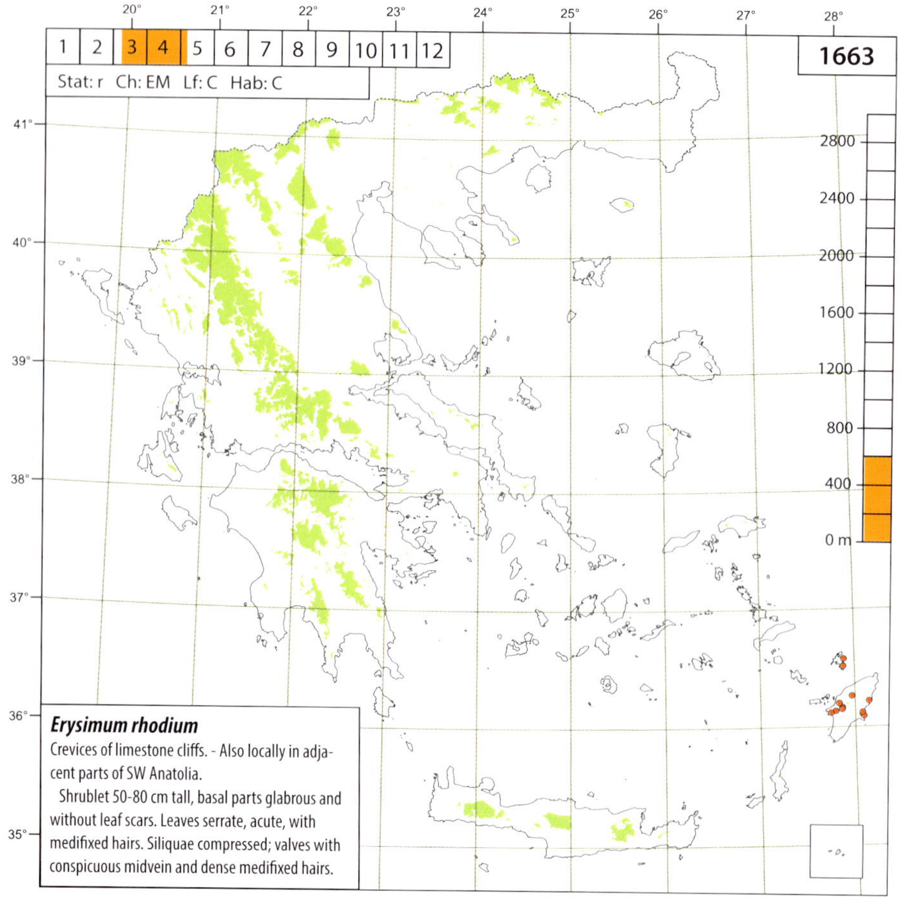

Stat: r Ch: EM Lf: C Hab: C

Erysimum rhodium

Crevices of limestone cliffs. - Also locally in adjacent parts of SW Anatolia.

Shrublet 50-80 cm tall, basal parts glabrous and without leaf scars. Leaves serrate, acute, with medifixed hairs. Siliquae compressed; valves with conspicuous midvein and dense medifixed hairs.

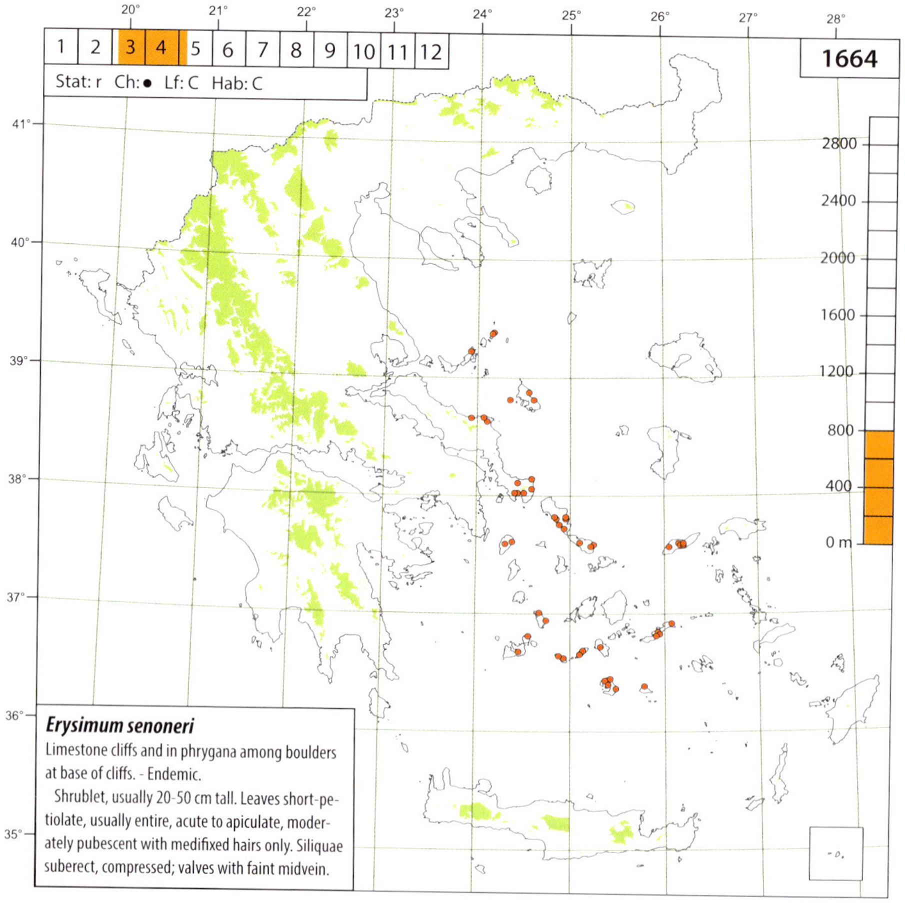

Stat: r Ch: ● Lf: C Hab: C

Erysimum senoneri

Limestone cliffs and in phrygana among boulders at base of cliffs. - Endemic.

Shrublet, usually 20-50 cm tall. Leaves short-petiolate, usually entire, acute to apiculate, moderately pubescent with medifixed hairs only. Siliquae suberect, compressed; valves with faint midvein.

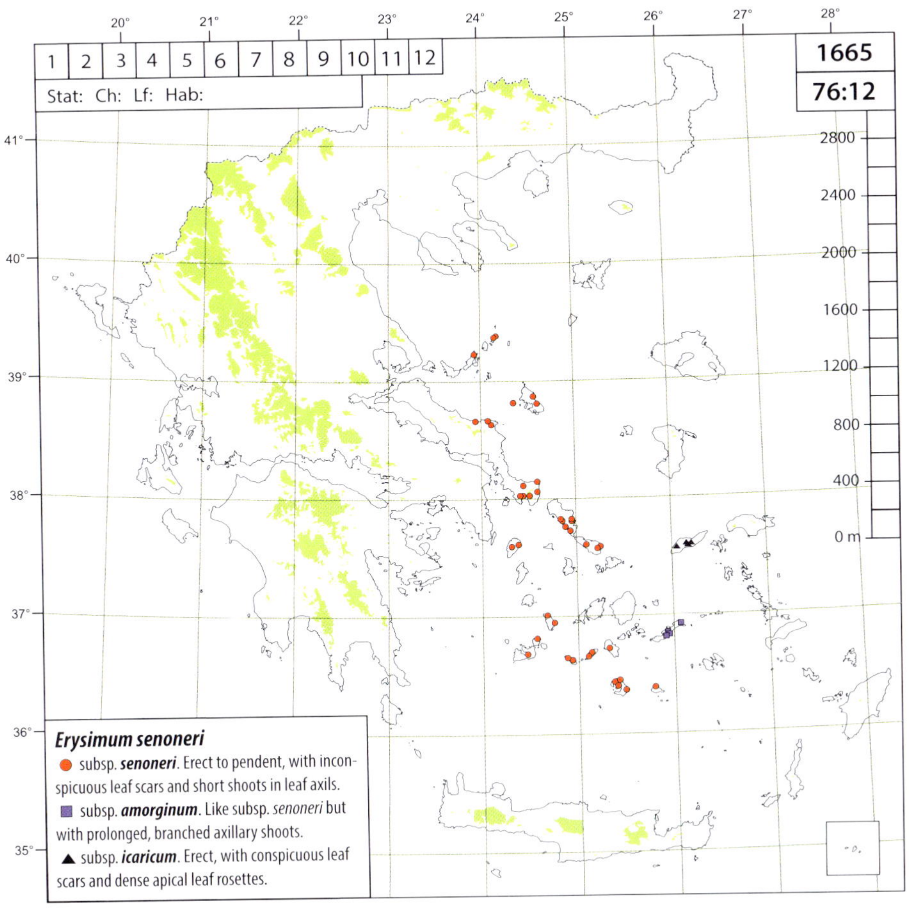

Erysimum senoneri

● subsp. ***senoneri***. Erect to pendent, with inconspicuous leaf scars and short shoots in leaf axils.

■ subsp. ***amorginum***. Like subsp. *senoneri* but with prolonged, branched axillary shoots.

▲ subsp. ***icaricum***. Erect, with conspicuous leaf scars and dense apical leaf rosettes.

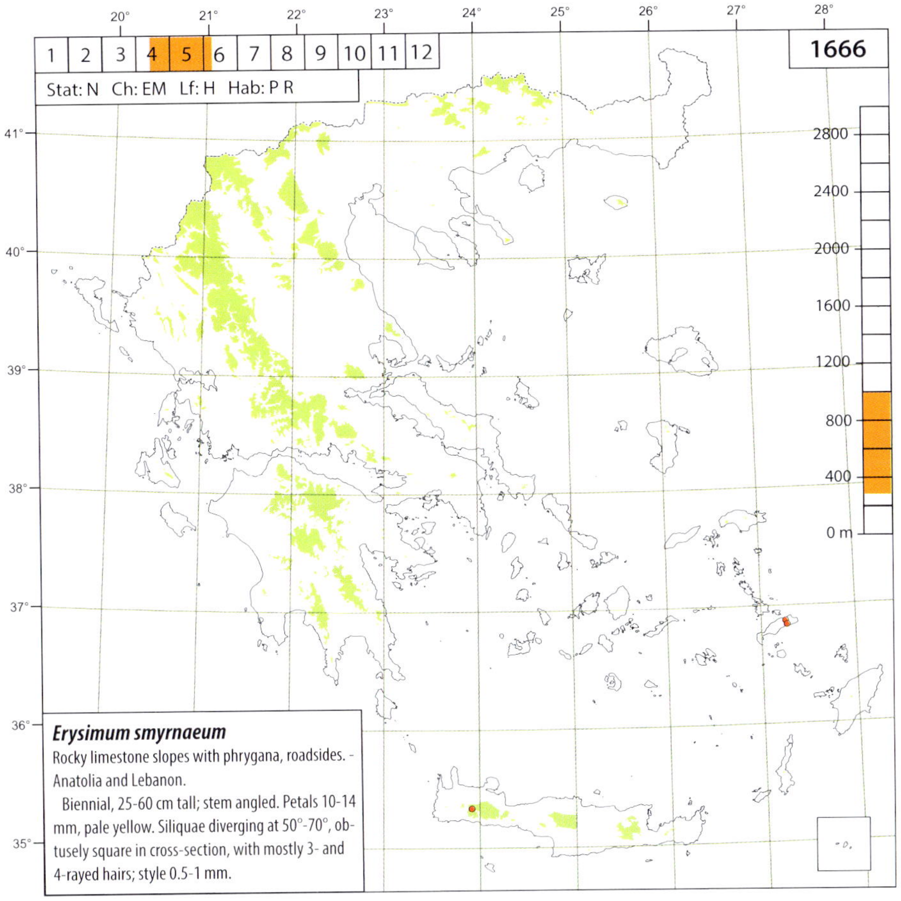

Erysimum smyrnaeum

Rocky limestone slopes with phrygana, roadsides. - Anatolia and Lebanon.

Biennial, 25-60 cm tall; stem angled. Petals 10-14 mm, pale yellow. Siliquae diverging at 50°-70°, obtusely square in cross-section, with mostly 3- and 4-rayed hairs; style 0.5-1 mm.

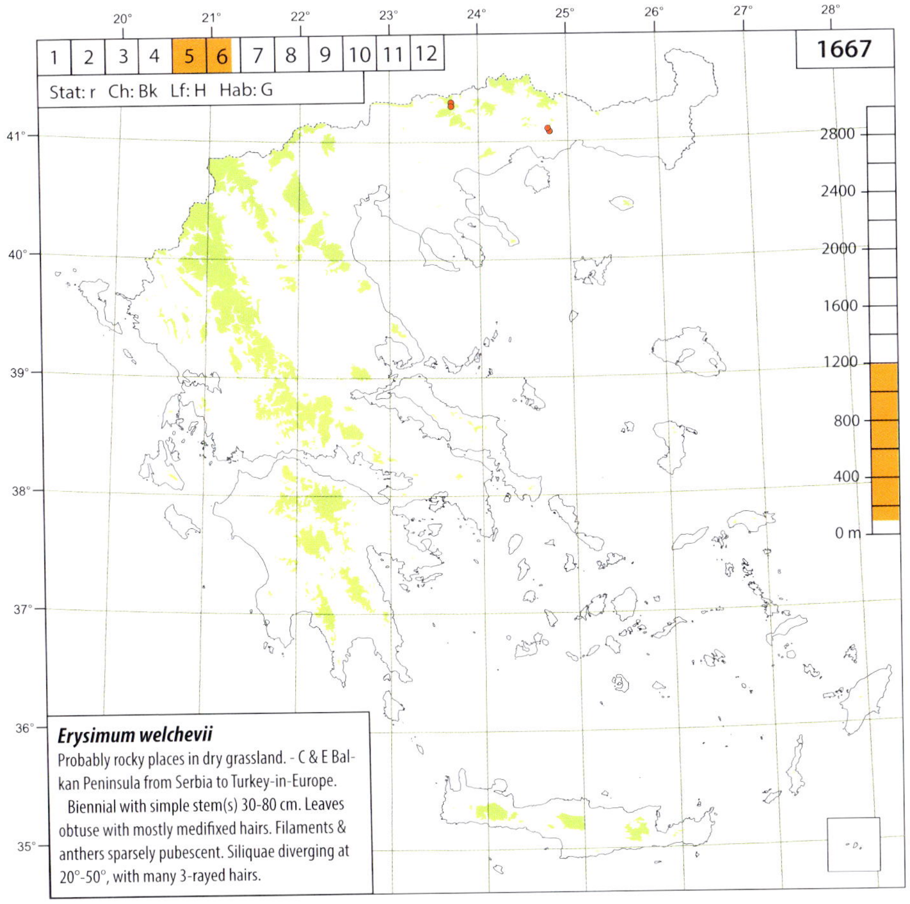

Erysimum welchevii

Probably rocky places in dry grassland. - C & E Balkan Peninsula from Serbia to Turkey-in-Europe.

Biennial with simple stem(s) 30-80 cm. Leaves obtuse with mostly medifixed hairs. Filaments & anthers sparsely pubescent. Siliquae diverging at 20°-50°, with many 3-rayed hairs.

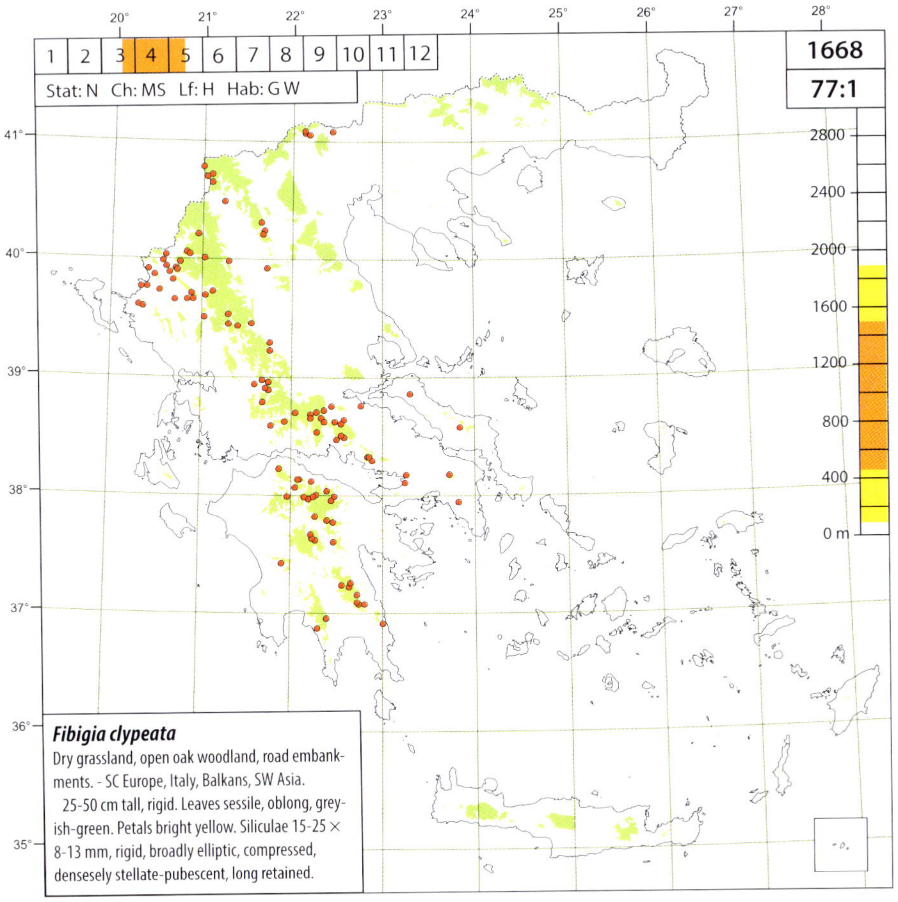

Fibigia clypeata

Dry grassland, open oak woodland, road embankments. - SC Europe, Italy, Balkans, SW Asia.

25-50 cm tall, rigid. Leaves sessile, oblong, greyish-green. Petals bright yellow. Siliculae 15-25 × 8-13 mm, rigid, broadly elliptic, compressed, densesely stellate-pubescent, long retained.

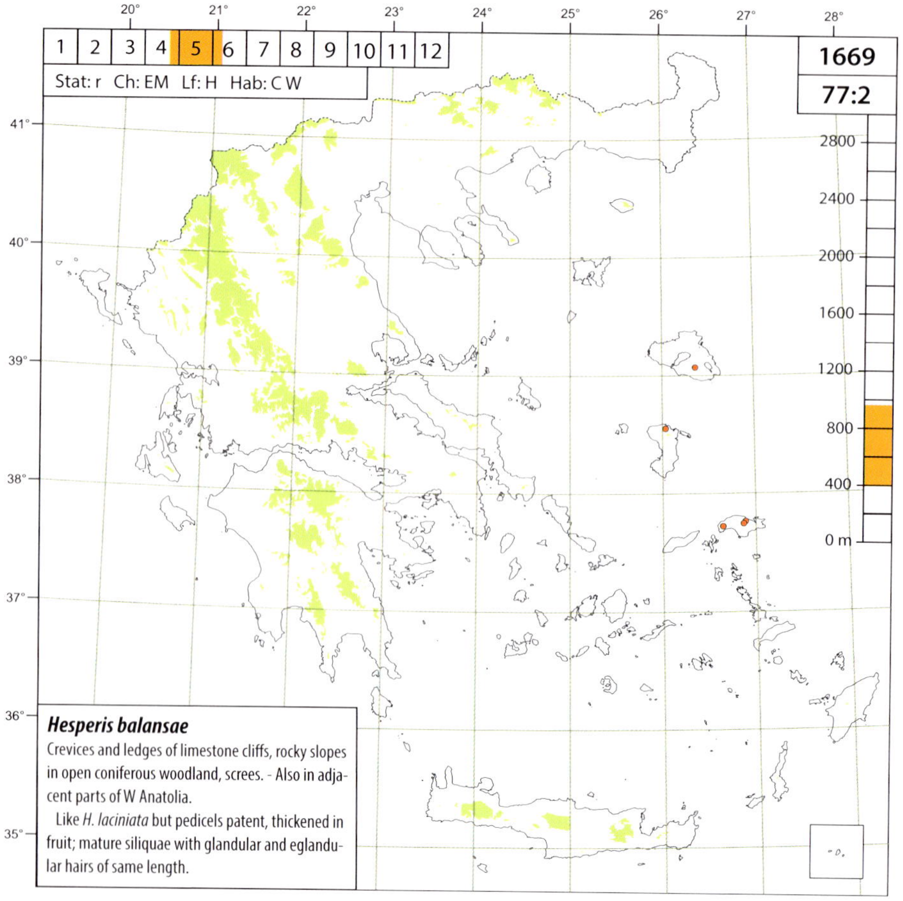

Hesperis balansae
Crevices and ledges of limestone cliffs, rocky slopes in open coniferous woodland, screes. - Also in adjacent parts of W Anatolia.
Like *H. laciniata* but pedicels patent, thickened in fruit; mature siliquae with glandular and eglandular hairs of same length.

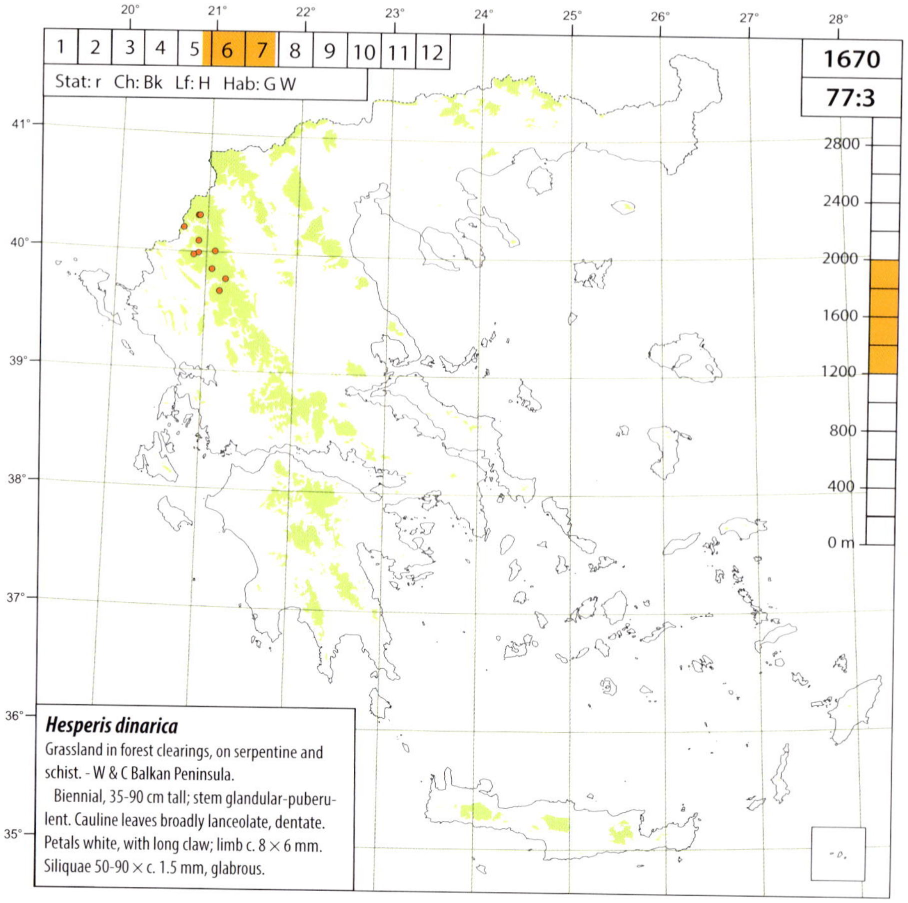

Hesperis dinarica
Grassland in forest clearings, on serpentine and schist. - W & C Balkan Peninsula.
Biennial, 35-90 cm tall; stem glandular-puberulent. Cauline leaves broadly lanceolate, dentate. Petals white, with long claw; limb c. 8 × 6 mm. Siliquae 50-90 × c. 1.5 mm, glabrous.

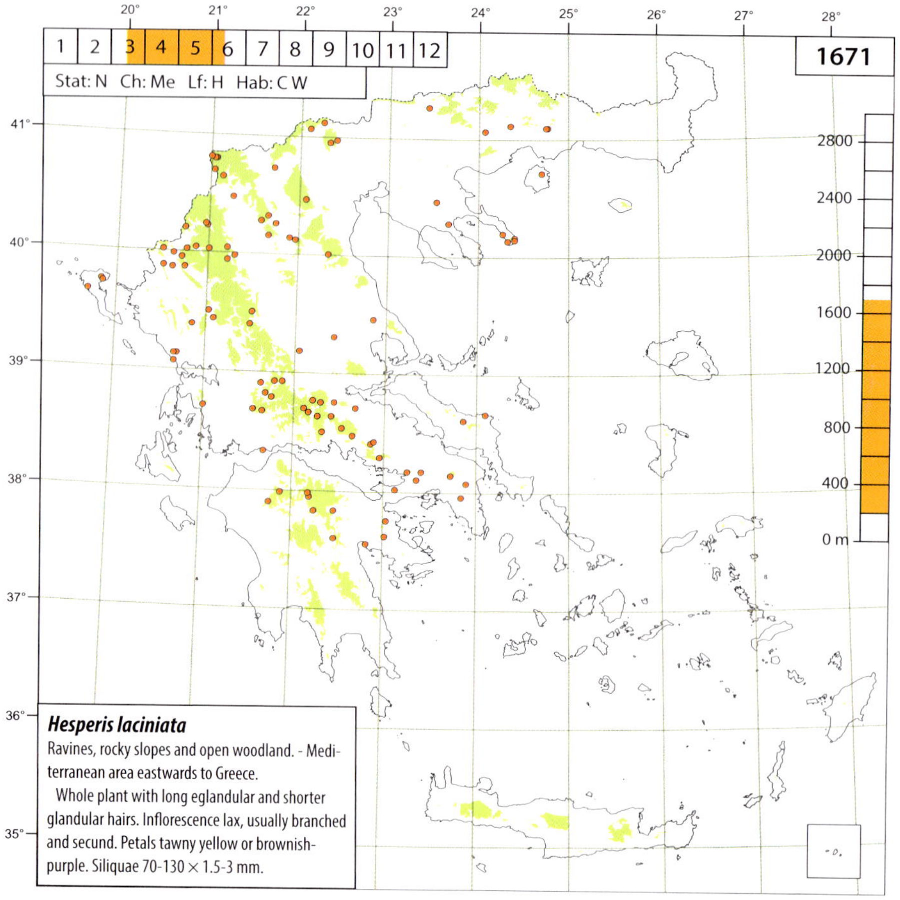

Hesperis laciniata
Ravines, rocky slopes and open woodland. - Mediterranean area eastwards to Greece.
Whole plant with long glandular and shorter glandular hairs. Inflorescence lax, usually branched and secund. Petals tawny yellow or brownish-purple. Siliqua 70-130 × 1.5-3 mm.

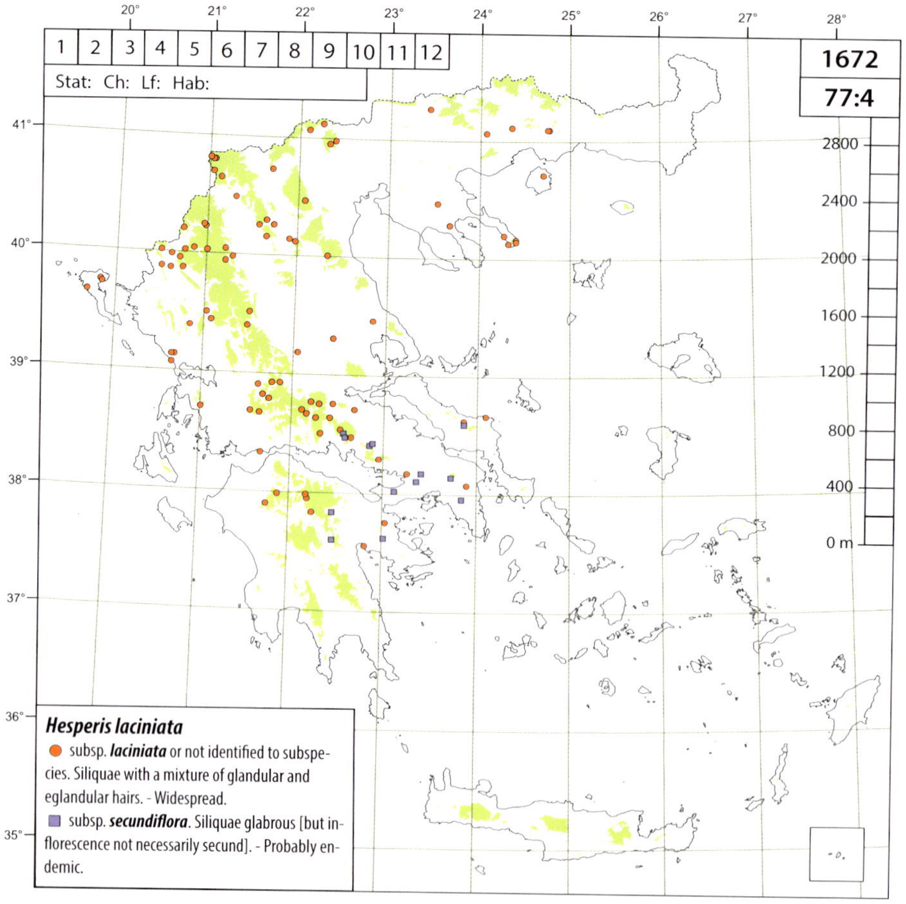

Hesperis laciniata
● subsp. ***laciniata*** or not identified to subspecies. Siliquae with a mixture of glandular and eglandular hairs. - Widespread.
■ subsp. ***secundiflora***. Siliquae glabrous [but inflorescence not necessarily secund]. - Probably endemic.

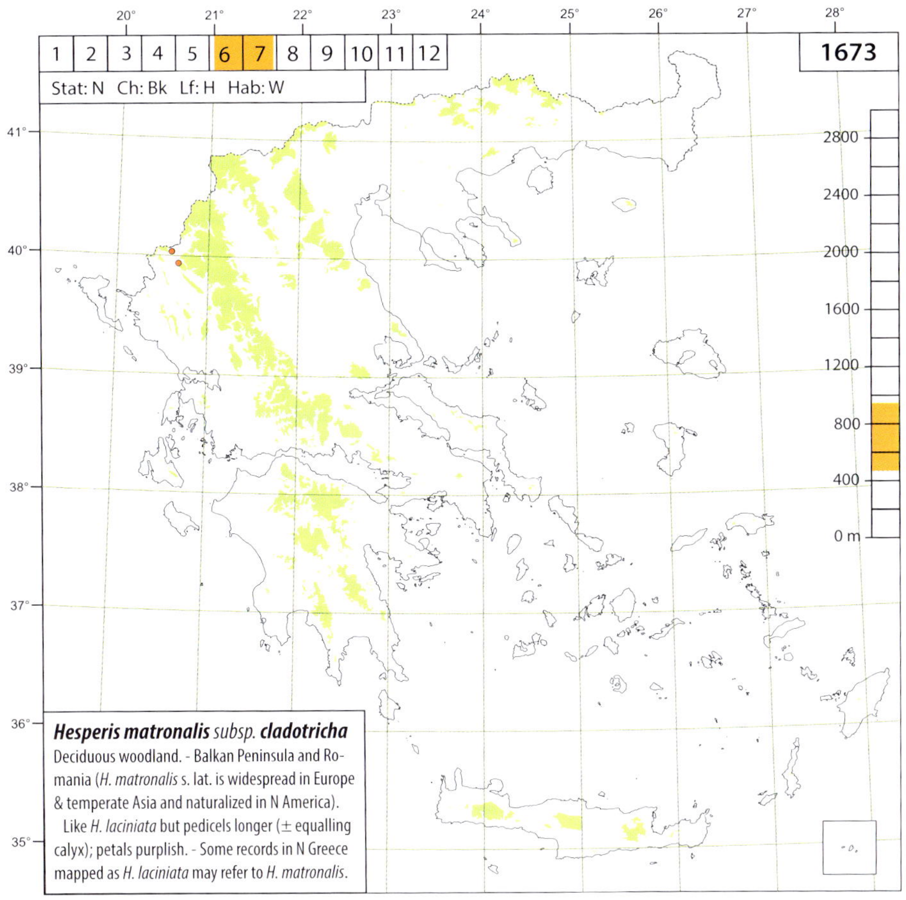

Hesperis matronalis *subsp.* ***cladotricha***
Deciduous woodland. - Balkan Peninsula and Romania (*H. matronalis* s. lat. is widespread in Europe & temperate Asia and naturalized in N America).
Like *H. laciniata* but pedicels longer (± equalling calyx); petals purplish. - Some records in N Greece mapped as *H. laciniata* may refer to *H. matronalis*.

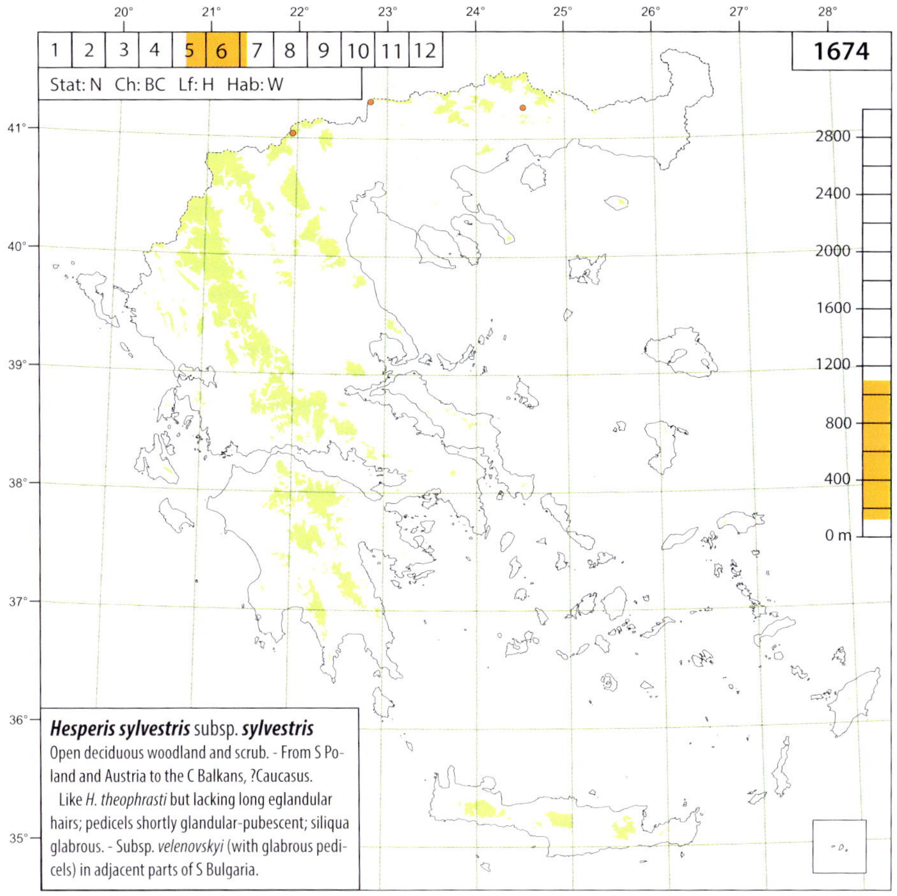

Hesperis sylvestris subsp. ***sylvestris***
Open deciduous woodland and scrub. - From S Poland and Austria to the C Balkans, ?Caucasus.
Like *H. theophrasti* but lacking long eglandular hairs; pedicels shortly glandular-pubescent; siliqua glabrous. - Subsp. *velenovskyi* (with glabrous pedicels) in adjacent parts of S Bulgaria.

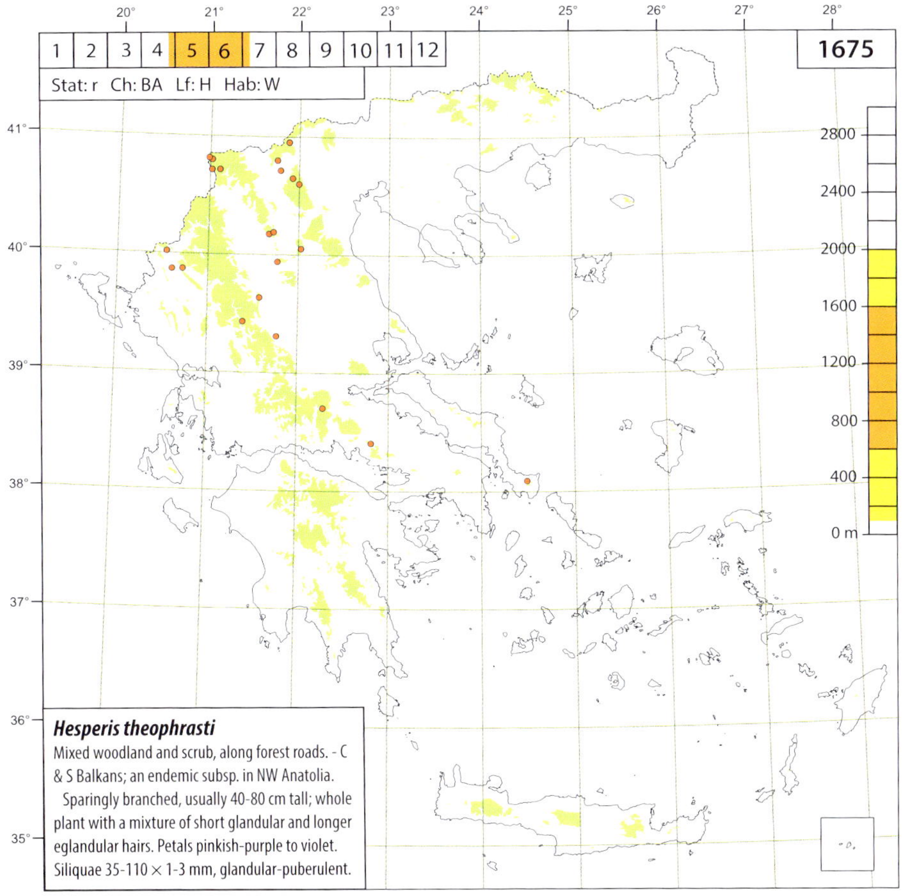

Hesperis theophrasti
Mixed woodland and scrub, along forest roads. - C & S Balkans; an endemic subsp. in NW Anatolia.
Sparingly branched, usually 40-80 cm tall; whole plant with a mixture of short glandular and longer eglandular hairs. Petals pinkish-purple to violet. Siliquae 35-110 × 1-3 mm, glandular-puberulent.

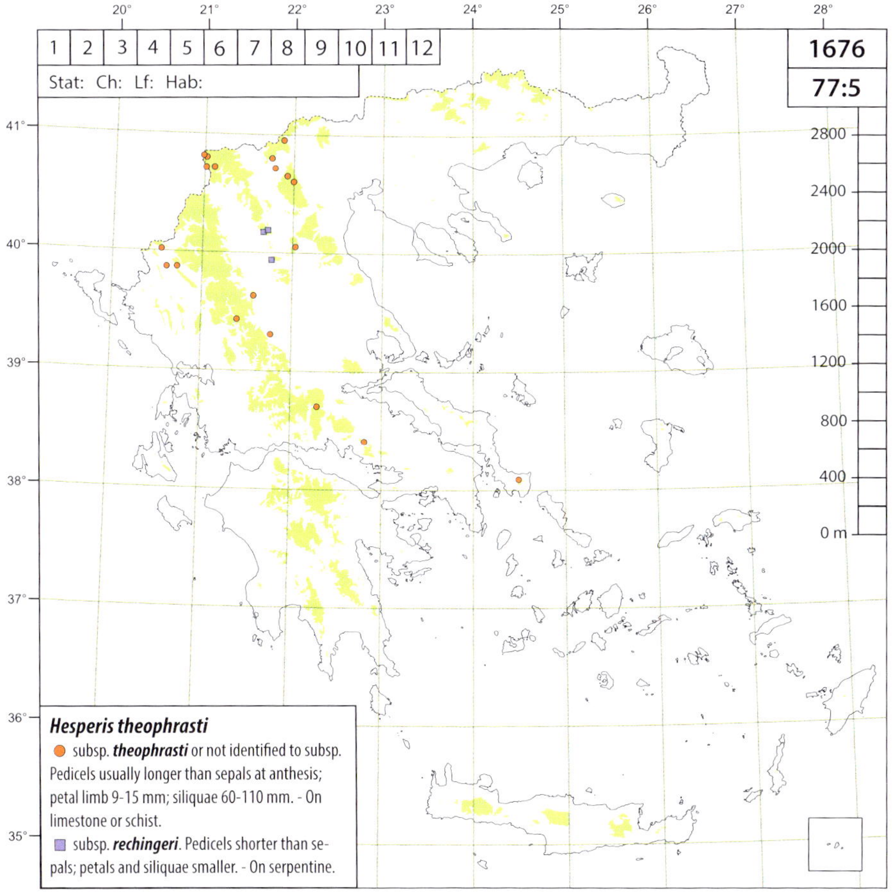

Hesperis theophrasti
● subsp. ***theophrasti*** or not identified to subsp. Pedicels usually longer than sepals at anthesis; petal limb 9-15 mm; siliquae 60-110 mm. - On limestone or schist.
■ subsp. ***rechingeri***. Pedicels shorter than sepals; petals and siliquae smaller. - On serpentine.

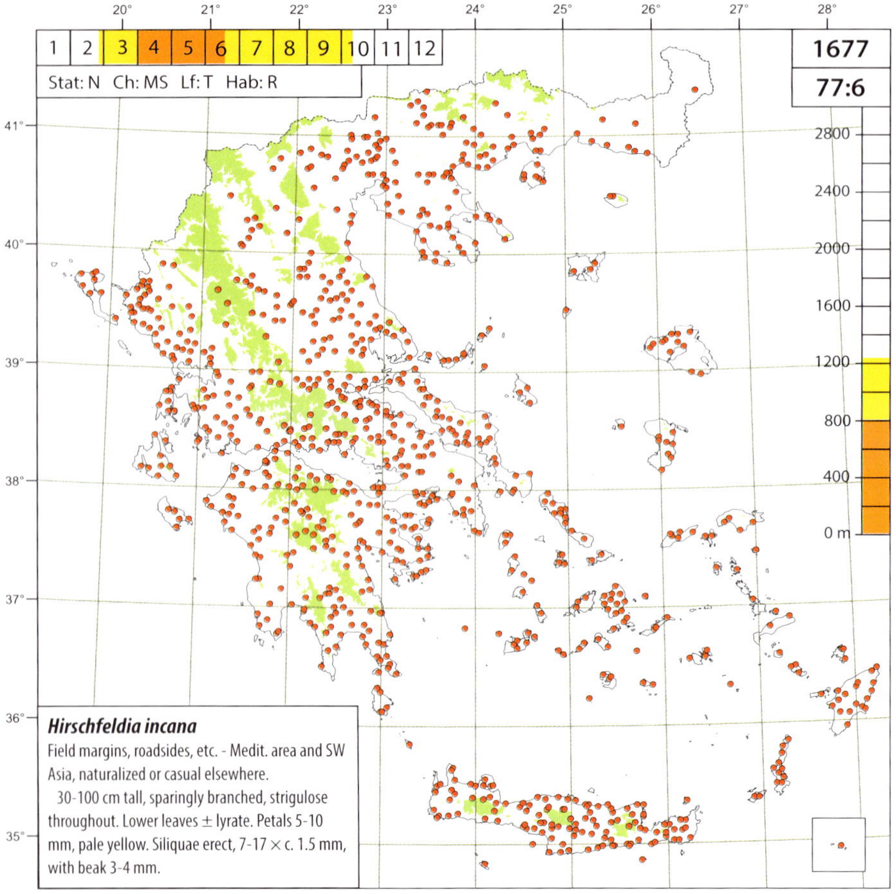

Hirschfeldia incana
Field margins, roadsides, etc. - Medit. area and SW Asia, naturalized or casual elsewhere.
30-100 cm tall, sparingly branched, strigulose throughout. Lower leaves ± lyrate. Petals 5-10 mm, pale yellow. Siliquae erect, 7-17 × c. 1.5 mm, with beak 3-4 mm.

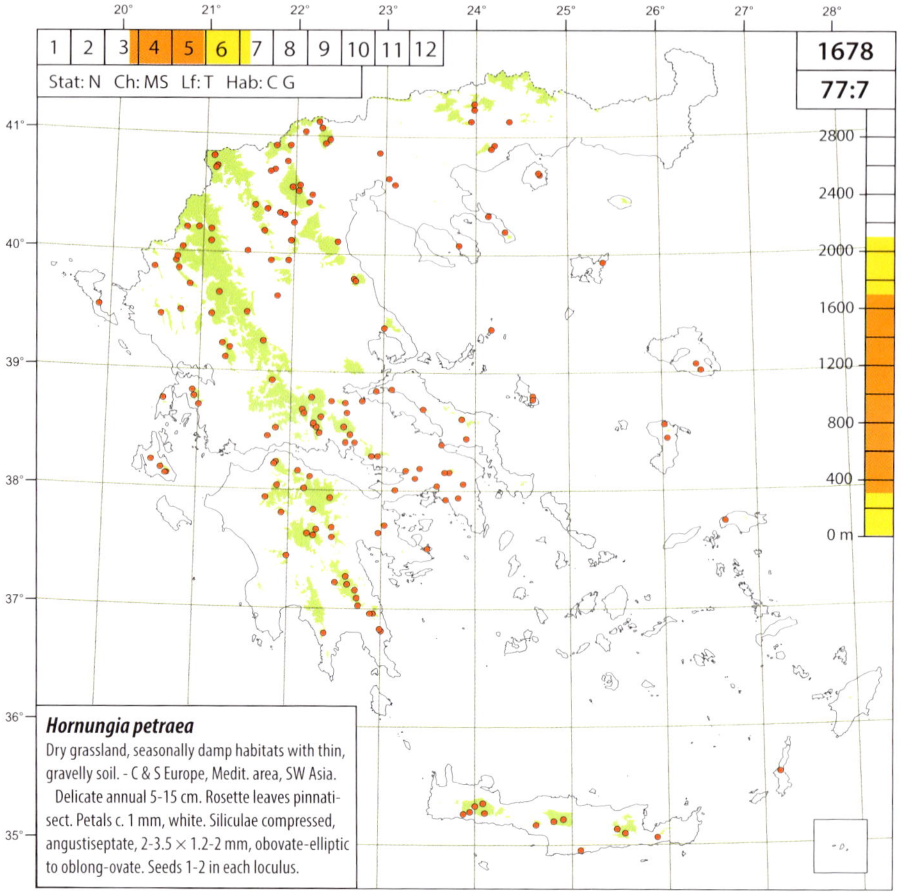

Hornungia petraea
Dry grassland, seasonally damp habitats with thin, gravelly soil. - C & S Europe, Medit. area, SW Asia.
Delicate annual 5-15 cm. Rosette leaves pinnatisect. Petals c. 1 mm, white. Siliculae compressed, angustiseptate, 2-3.5 × 1.2-2 mm, obovate-elliptic to oblong-ovate. Seeds 1-2 in each loculus.

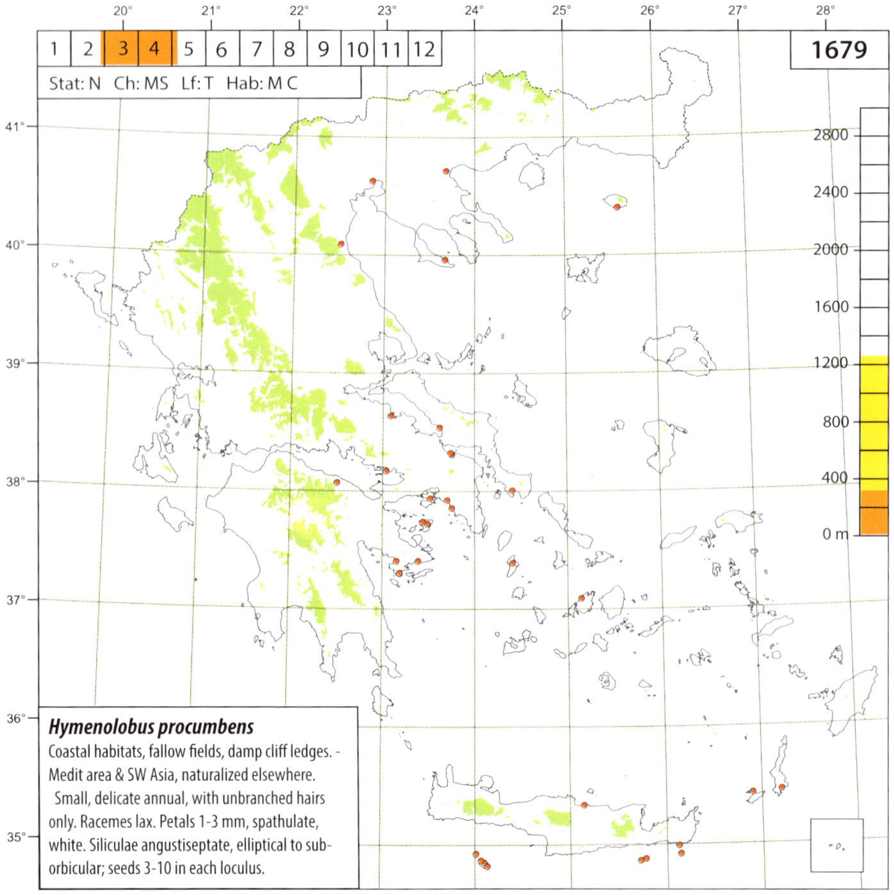

Hymenolobus procumbens
Coastal habitats, fallow fields, damp cliff ledges. - Medit area & SW Asia, naturalized elsewhere.
Small, delicate annual, with unbranched hairs only. Racemes lax. Petals 1-3 mm, spathulate, white. Siliculae angustiseptate, elliptical to suborbicular; seeds 3-10 in each loculus.

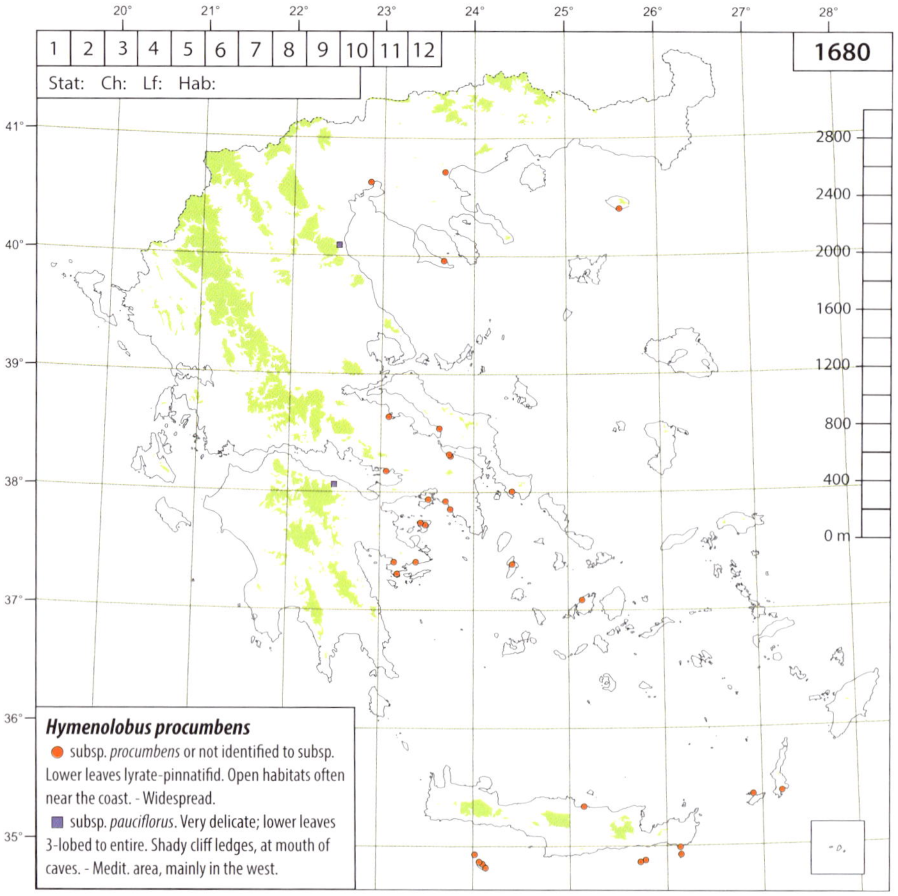

Hymenolobus procumbens
● subsp. *procumbens* or not identified to subsp. Lower leaves lyrate-pinnatifid. Open habitats often near the coast. - Widespread.
■ subsp. *pauciflorus*. Very delicate; lower leaves 3-lobed to entire. Shady cliff ledges, at mouth of caves. - Medit. area, mainly in the west.

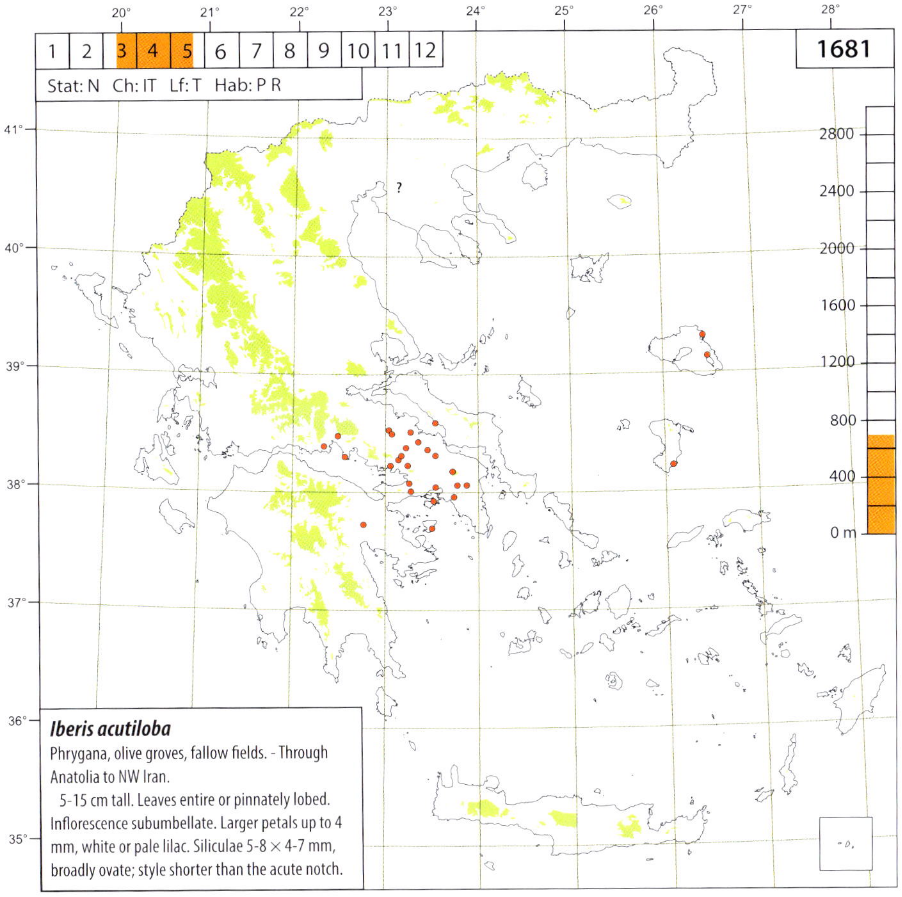

Iberis acutiloba
Phrygana, olive groves, fallow fields. - Through Anatolia to NW Iran.
5-15 cm tall. Leaves entire or pinnately lobed. Inflorescence subumbellate. Larger petals up to 4 mm, white or pale lilac. Siliculae 5-8 × 4-7 mm, broadly ovate; style shorter than the acute notch.

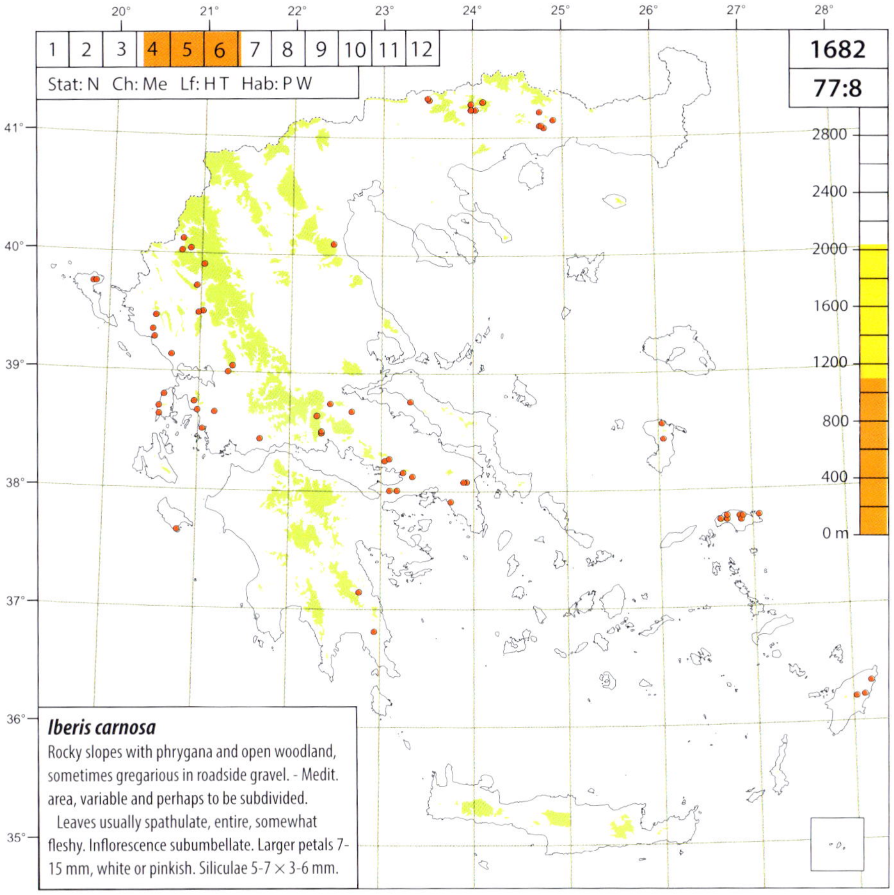

Iberis carnosa
Rocky slopes with phrygana and open woodland, sometimes gregarious in roadside gravel. - Medit. area, variable and perhaps to be subdivided.
Leaves usually spathulate, entire, somewhat fleshy. Inflorescence subumbellate. Larger petals 7-15 mm, white or pinkish. Siliculae 5-7 × 3-6 mm.

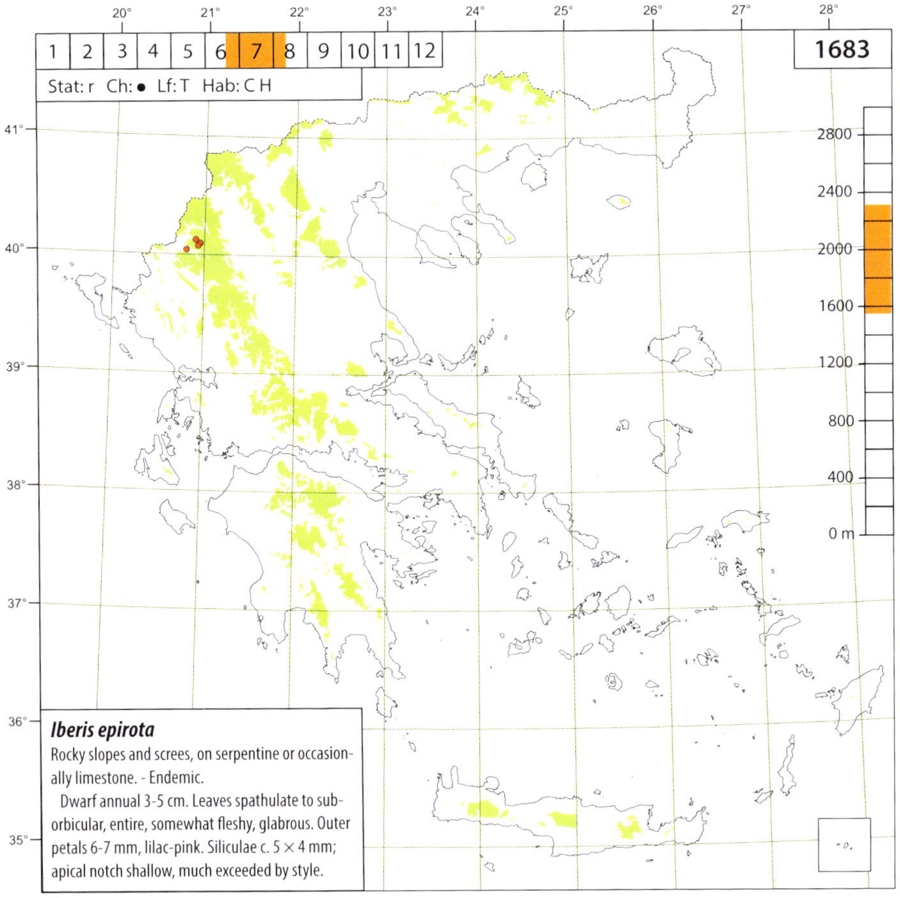

Iberis epirota
Rocky slopes and screes, on serpentine or occasionally limestone. - Endemic.
Dwarf annual 3-5 cm. Leaves spathulate to suborbicular, entire, somewhat fleshy, glabrous. Outer petals 6-7 mm, lilac-pink. Siliculae c. 5 × 4 mm; apical notch shallow, much exceeded by style.

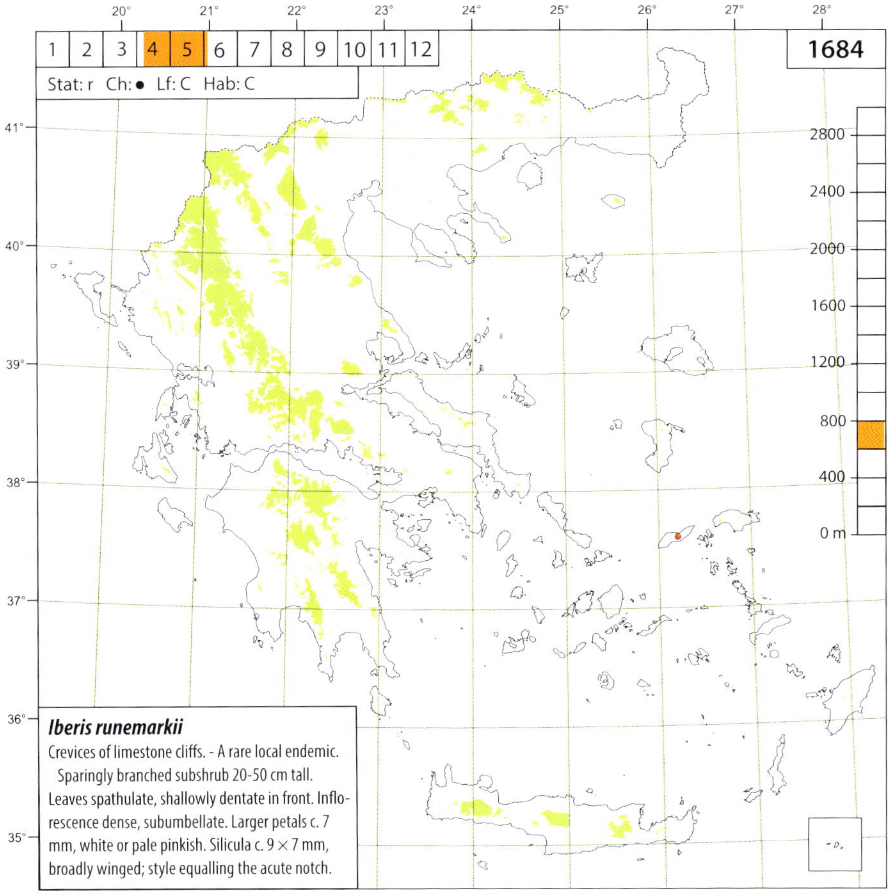

Iberis runemarkii
Crevices of limestone cliffs. - A rare local endemic.
Sparingly branched subshrub 20-50 cm tall. Leaves spathulate, shallowly dentate in front. Inflorescence dense, subumbellate. Larger petals c. 7 mm, white or pale pinkish. Silicula c. 9 × 7 mm, broadly winged; style equalling the acute notch.

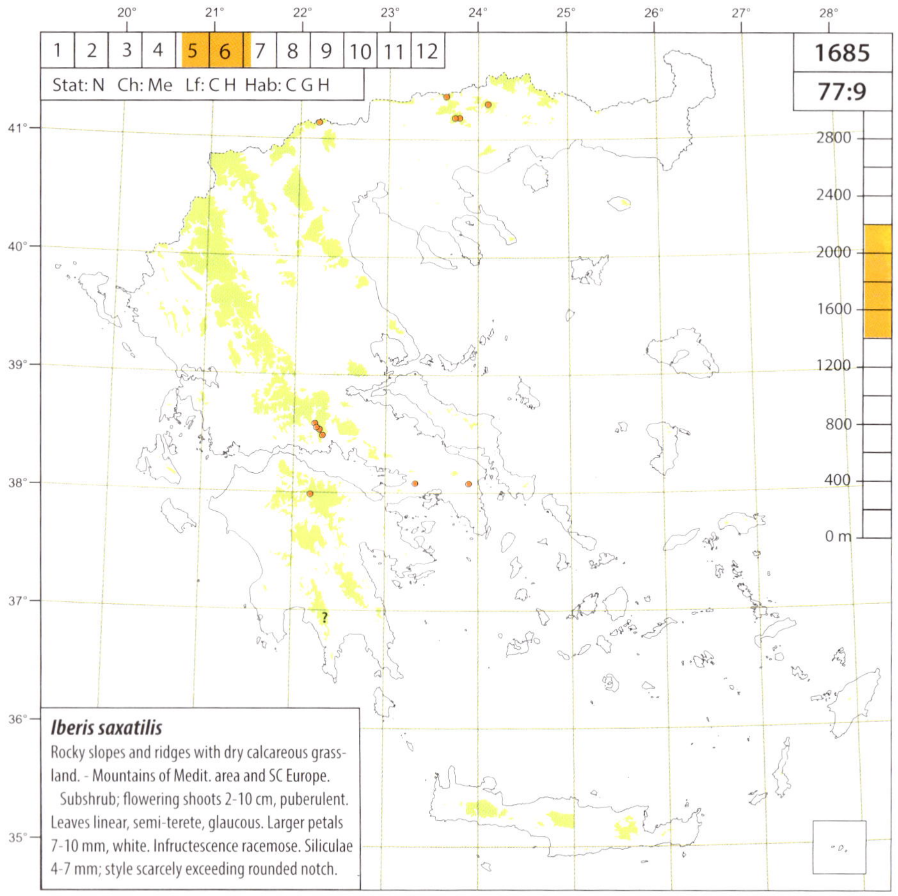

Iberis saxatilis

Rocky slopes and ridges with dry calcareous grassland. - Mountains of Medit. area and SC Europe.

Subshrub; flowering shoots 2-10 cm, puberulent. Leaves linear, semi-terete, glaucous. Larger petals 7-10 mm, white. Infructescence racemose. Siliculae 4-7 mm; style scarcely exceeding rounded notch.

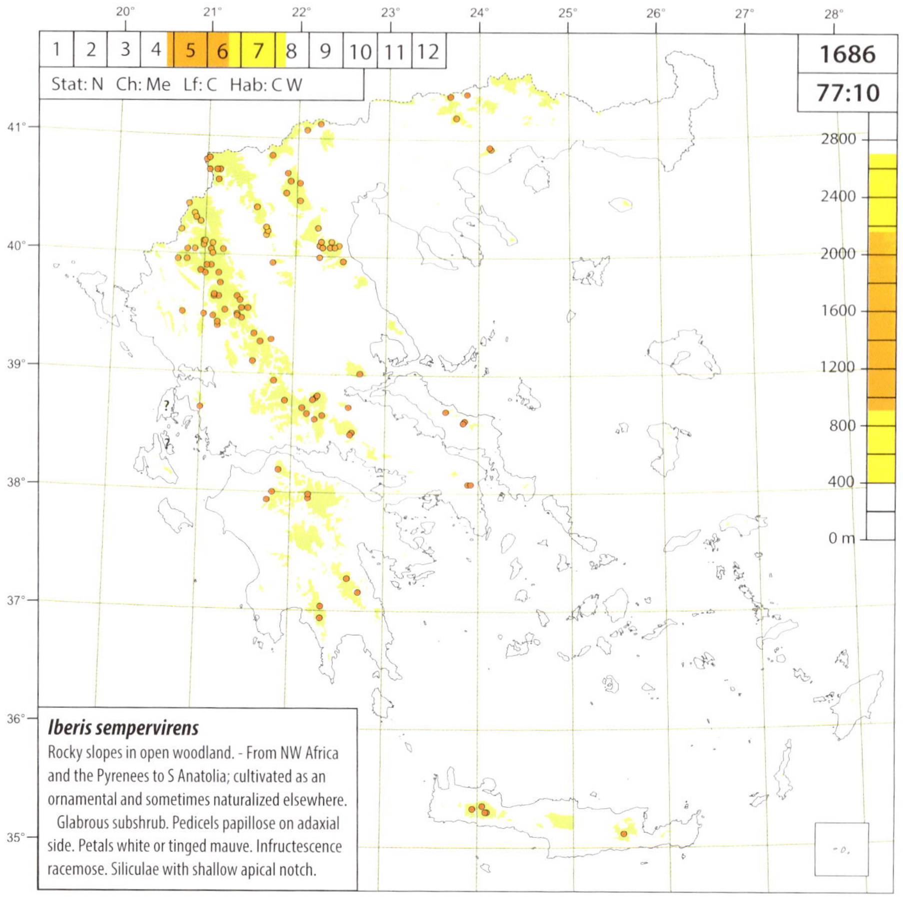

Iberis sempervirens

Rocky slopes in open woodland. - From NW Africa and the Pyrenees to S Anatolia; cultivated as an ornamental and sometimes naturalized elsewhere.

Glabrous subshrub. Pedicels papillose on adaxial side. Petals white or tinged mauve. Infructescence racemose. Siliculae with shallow apical notch.

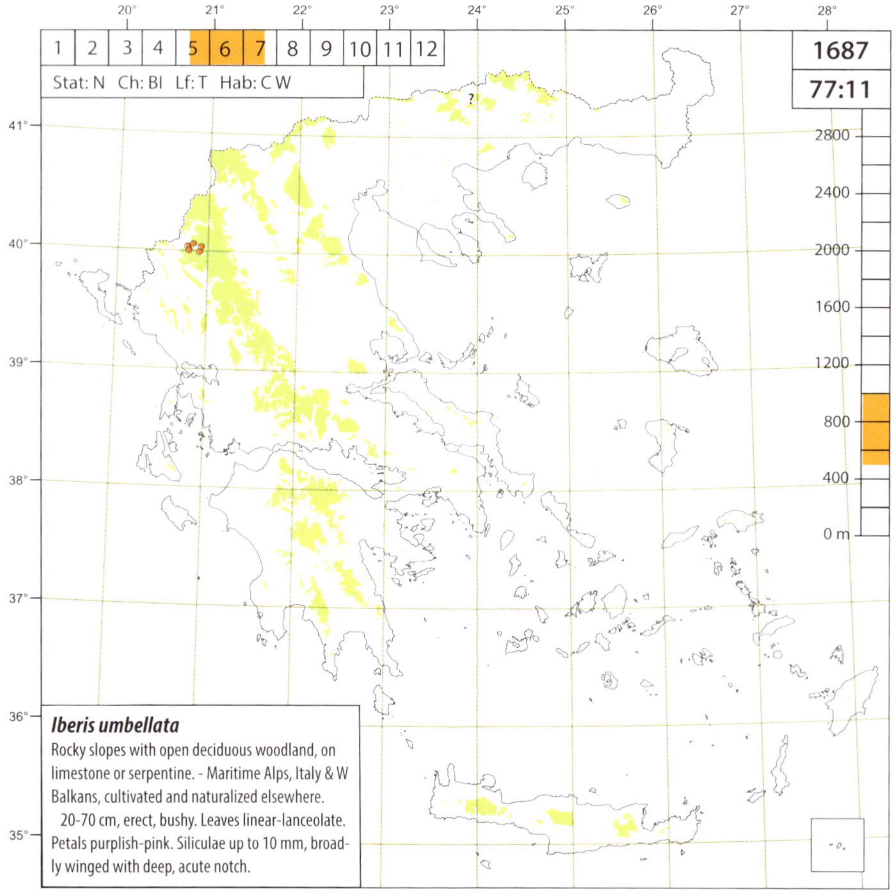

Iberis umbellata

Rocky slopes with open deciduous woodland, on limestone or serpentine. - Maritime Alps, Italy & W Balkans, cultivated and naturalized elsewhere.

20-70 cm, erect, bushy. Leaves linear-lanceolate. Petals purplish-pink. Siliculae up to 10 mm, broadly winged with deep, acute notch.

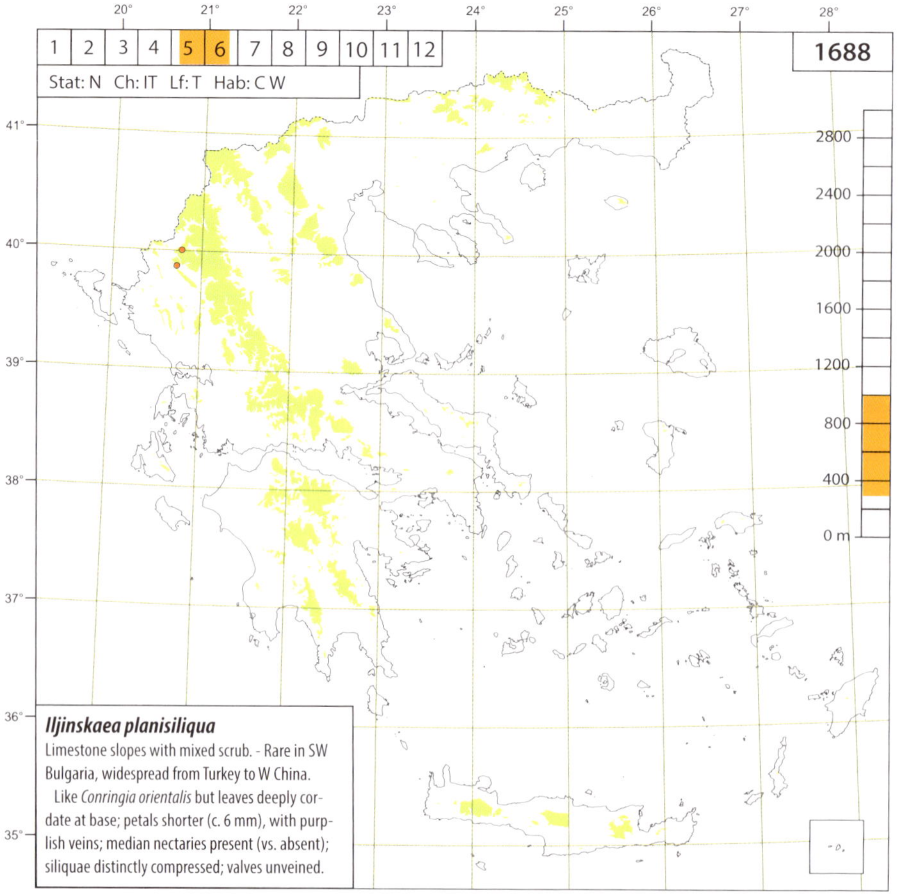

Iljinskaea planisiliqua

Limestone slopes with mixed scrub. - Rare in SW Bulgaria, widespread from Turkey to W China.

Like *Conringia orientalis* but leaves deeply cordate at base; petals shorter (c. 6 mm), with purplish veins; median nectaries present (vs. absent); siliquae distinctly compressed; valves unveined.

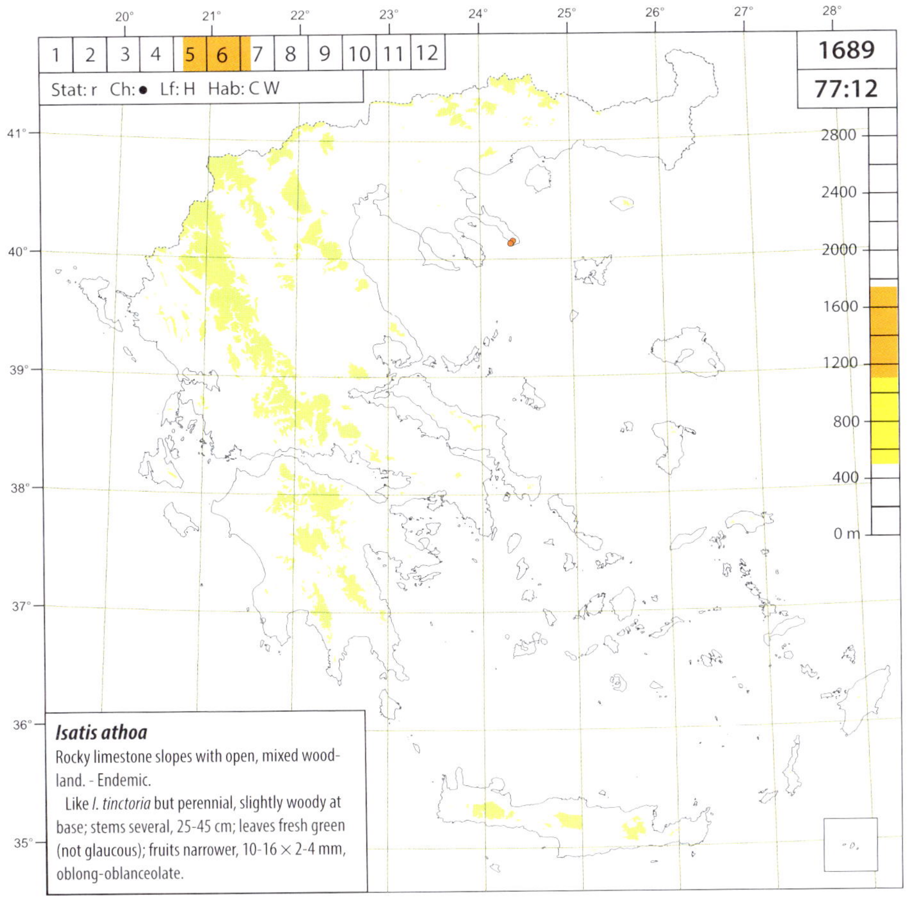

Isatis athoa

Rocky limestone slopes with open, mixed woodland. - Endemic.

Like *I. tinctoria* but perennial, slightly woody at base; stems several, 25-45 cm; leaves fresh green (not glaucous); fruits narrower, 10-16 × 2-4 mm, oblong-oblanceolate.

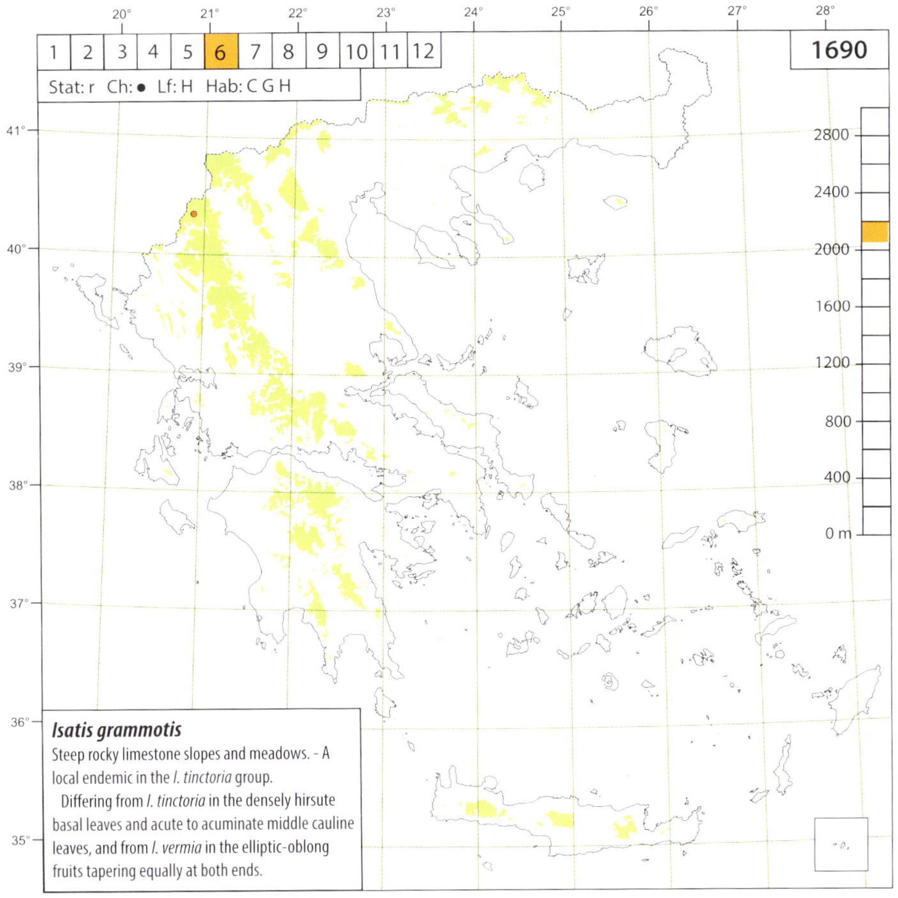

Isatis grammotis

Steep rocky limestone slopes and meadows. - A local endemic in the *I. tinctoria* group.

Differing from *I. tinctoria* in the densely hirsute basal leaves and acute to acuminate middle cauline leaves, and from *I. vermia* in the elliptic-oblong fruits tapering equally at both ends.

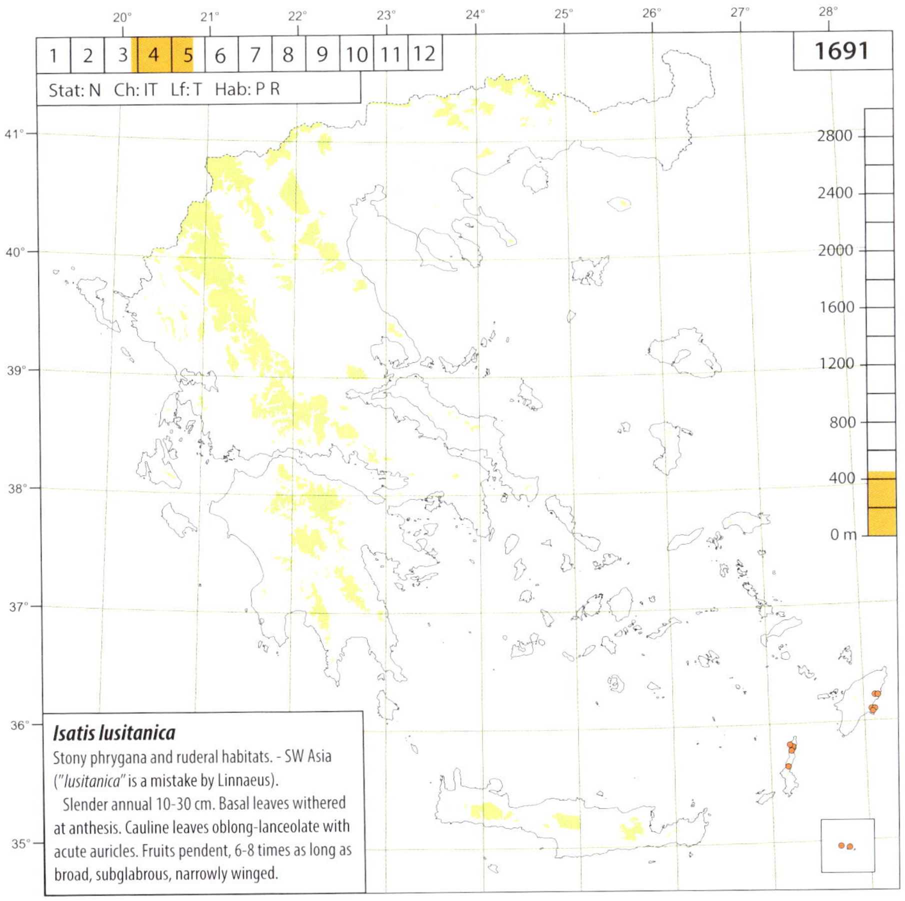

Isatis lusitanica

Stony phrygana and ruderal habitats. - SW Asia ("*lusitanica*" is a mistake by Linnaeus).

Slender annual 10-30 cm. Basal leaves withered at anthesis. Cauline leaves oblong-lanceolate with acute auricles. Fruits pendent, 6-8 times as long as broad, subglabrous, narrowly winged.

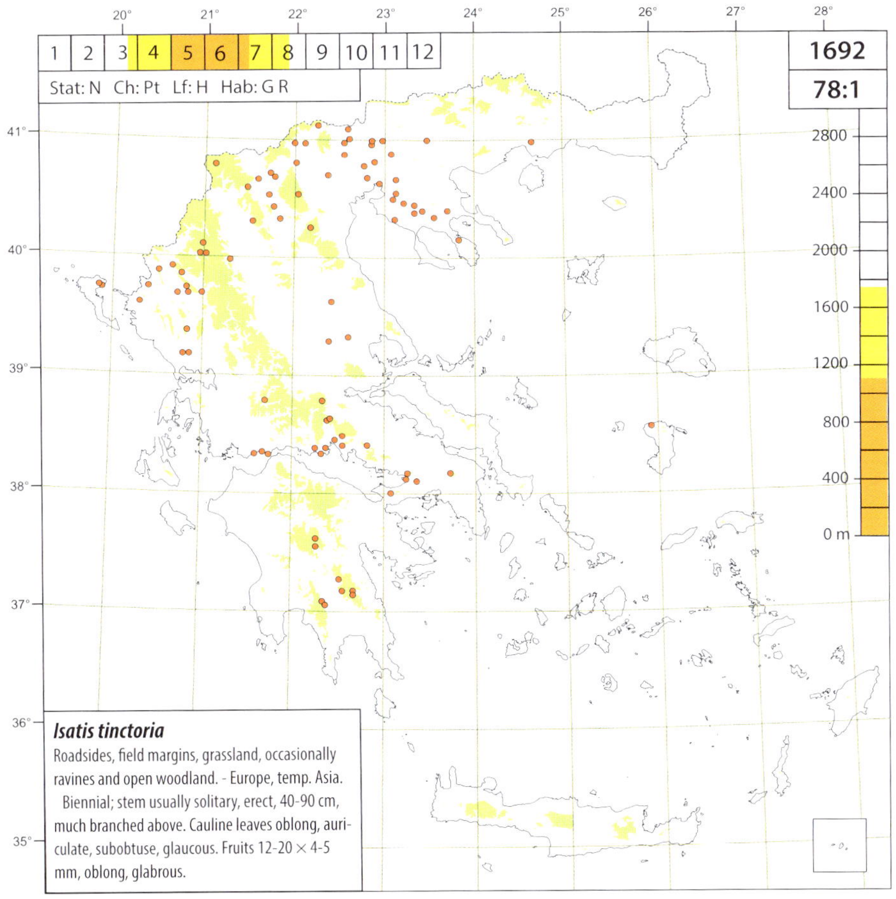

Isatis tinctoria

Roadsides, field margins, grassland, occasionally ravines and open woodland. - Europe, temp. Asia.

Biennial; stem usually solitary, erect, 40-90 cm, much branched above. Cauline leaves oblong, auriculate, subobtuse, glaucous. Fruits 12-20 × 4-5 mm, oblong, glabrous.

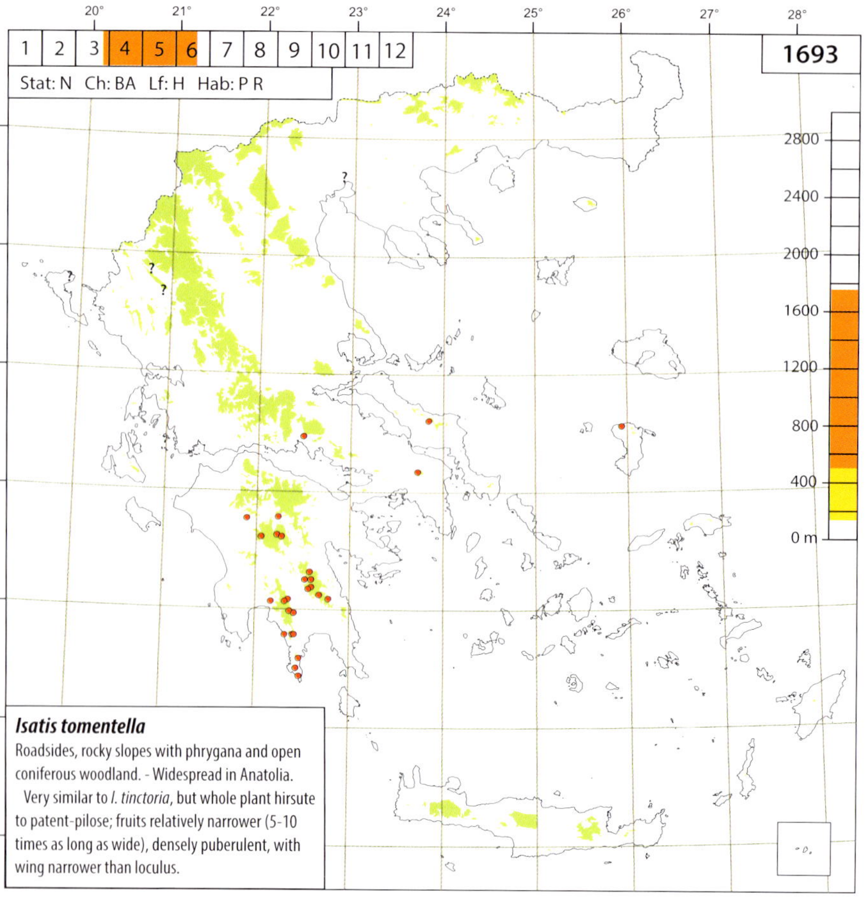

Isatis tomentella
Roadsides, rocky slopes with phrygana and open coniferous woodland. - Widespread in Anatolia.
Very similar to *I. tinctoria*, but whole plant hirsute to patent-pilose; fruits relatively narrower (5-10 times as long as wide), densely puberulent, with wing narrower than loculus.

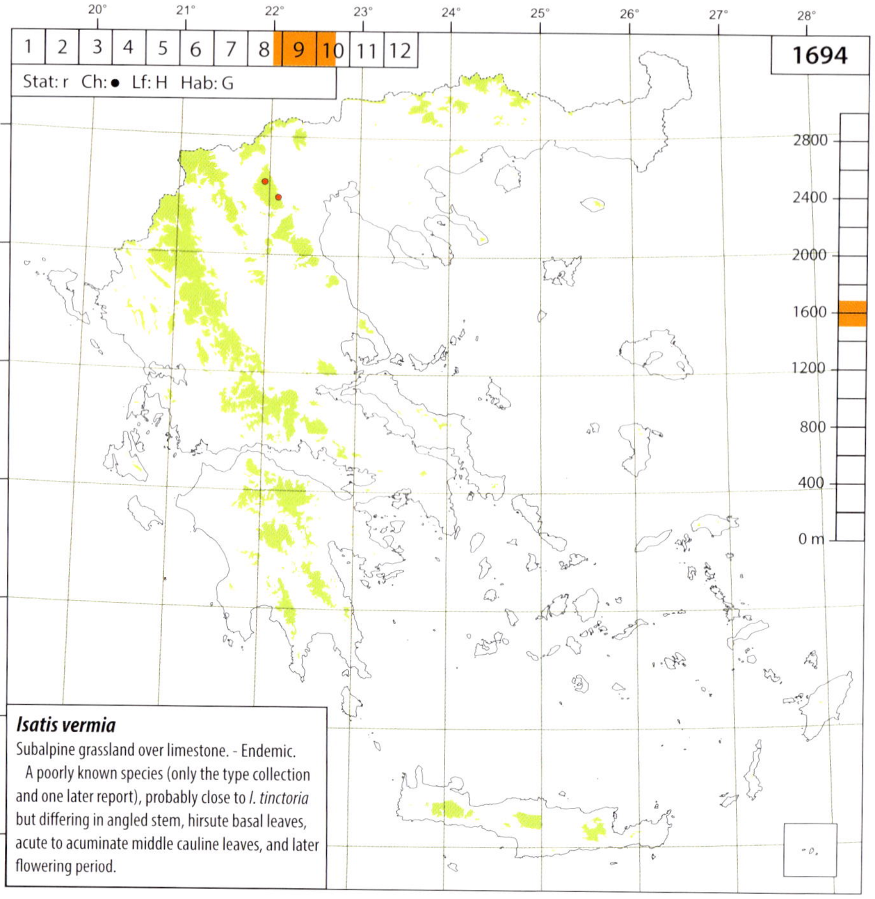

Isatis vermia
Subalpine grassland over limestone. - Endemic.
A poorly known species (only the type collection and one later report), probably close to *I. tinctoria* but differing in angled stem, hirsute basal leaves, acute to acuminate middle cauline leaves, and later flowering period.

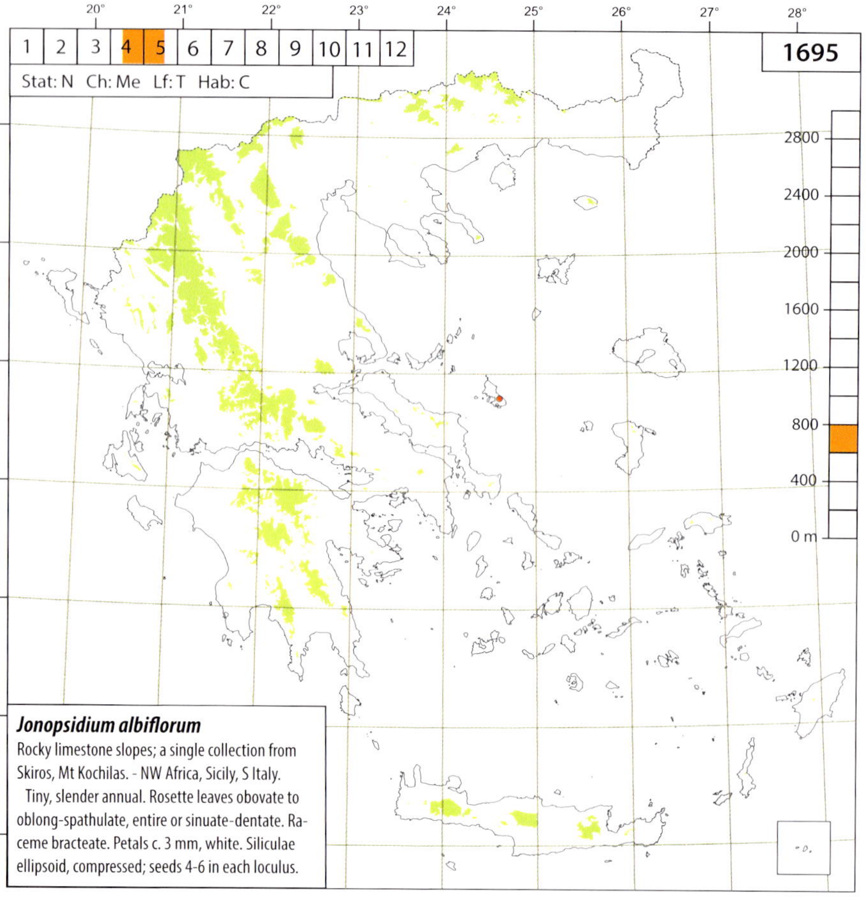

Jonopsidium albiflorum
Rocky limestone slopes; a single collection from Skiros, Mt Kochilas. - NW Africa, Sicily, S Italy.
Tiny, slender annual. Rosette leaves obovate to oblong-spathulate, entire or sinuate-dentate. Raceme bracteate. Petals c. 3 mm, white. Siliculae ellipsoid, compressed; seeds 4-6 in each loculus.

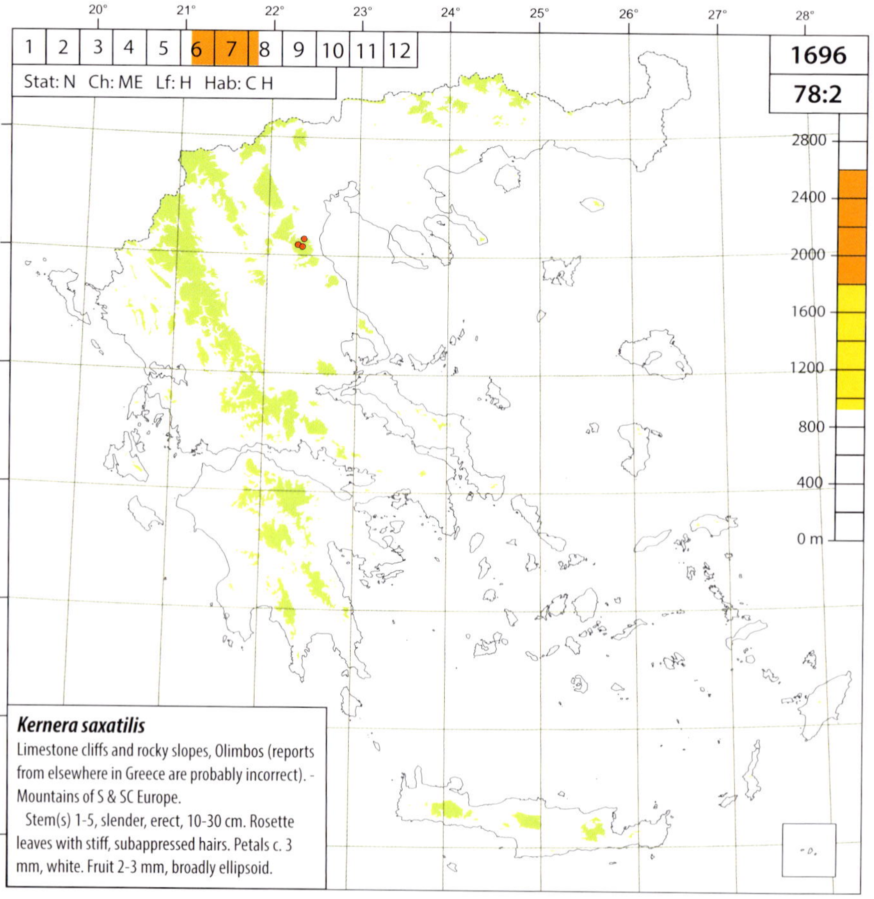

Kernera saxatilis
Limestone cliffs and rocky slopes, Olimbos (reports from elsewhere in Greece are probably incorrect). - Mountains of S & SC Europe.
Stem(s) 1-5, slender, erect, 10-30 cm. Rosette leaves with stiff, subappressed hairs. Petals c. 3 mm, white. Fruit 2-3 mm, broadly ellipsoid.

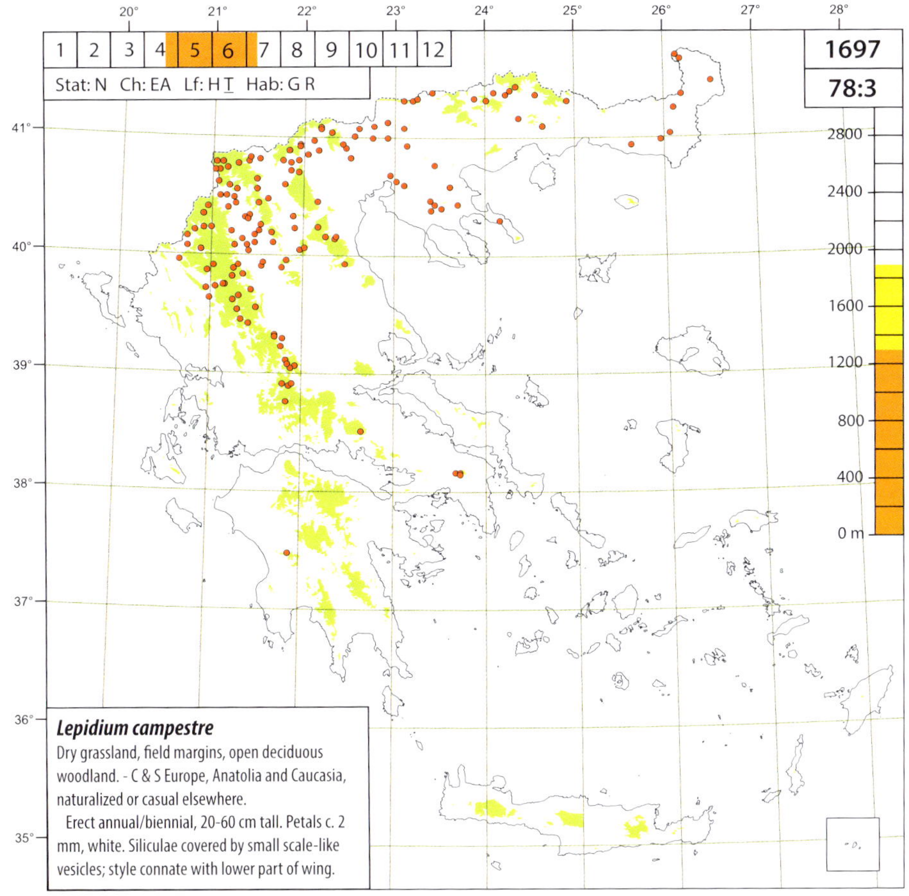

Lepidium campestre

Dry grassland, field margins, open deciduous woodland. - C & S Europe, Anatolia and Caucasia, naturalized or casual elsewhere.

Erect annual/biennial, 20-60 cm tall. Petals c. 2 mm, white. Siliculae covered by small scale-like vesicles; style connate with lower part of wing.

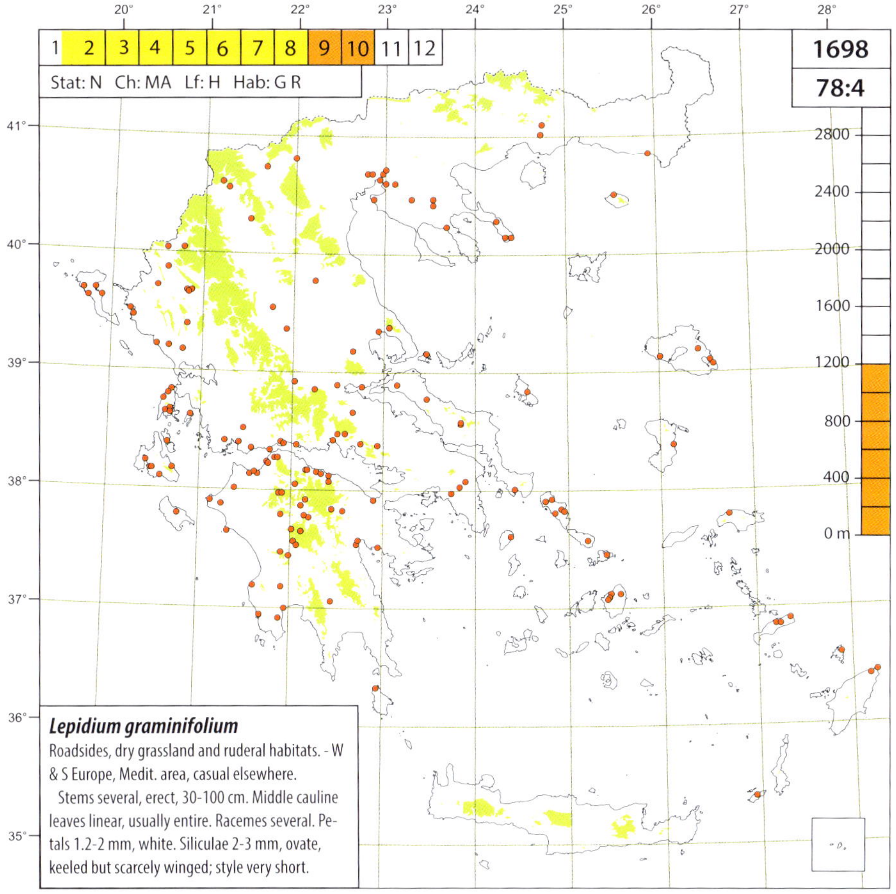

Lepidium graminifolium

Roadsides, dry grassland and ruderal habitats. - W & S Europe, Medit. area, casual elsewhere.

Stems several, erect, 30-100 cm. Middle cauline leaves linear, usually entire. Racemes several. Petals 1.2-2 mm, white. Siliculae 2-3 mm, ovate, keeled but scarcely winged; style very short.

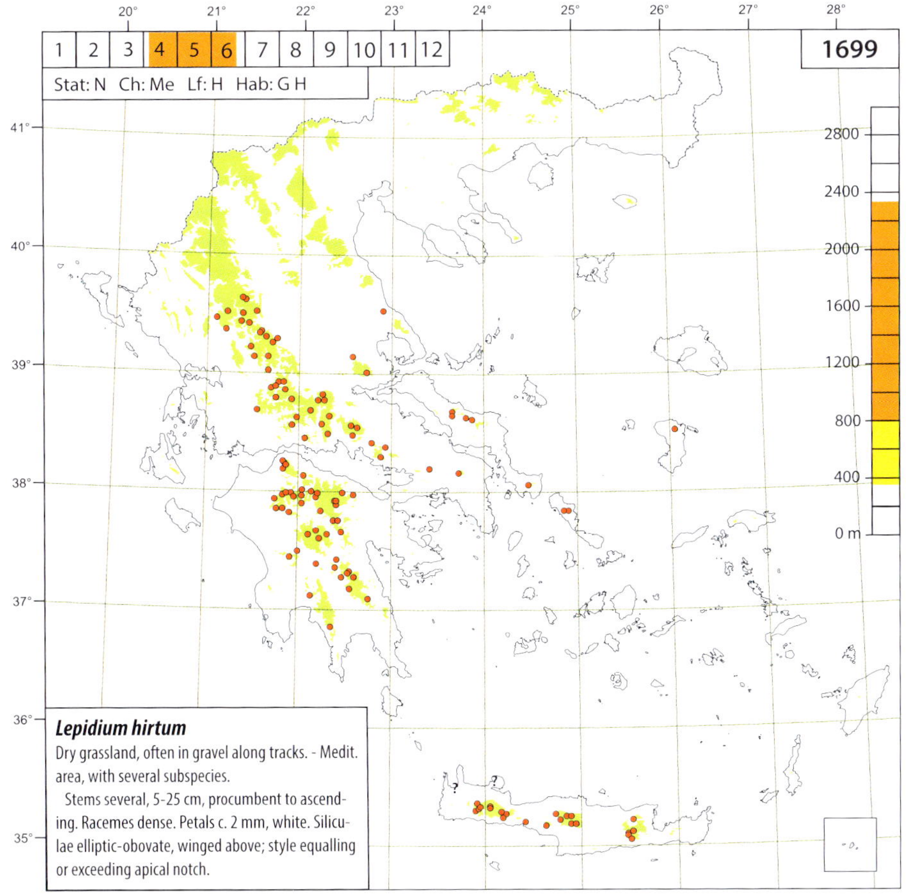

Lepidium hirtum

Dry grassland, often in gravel along tracks. - Medit. area, with several subspecies.

Stems several, 5-25 cm, procumbent to ascending. Racemes dense. Petals c. 2 mm, white. Siliculae elliptic-obovate, winged above; style equalling or exceeding apical notch.

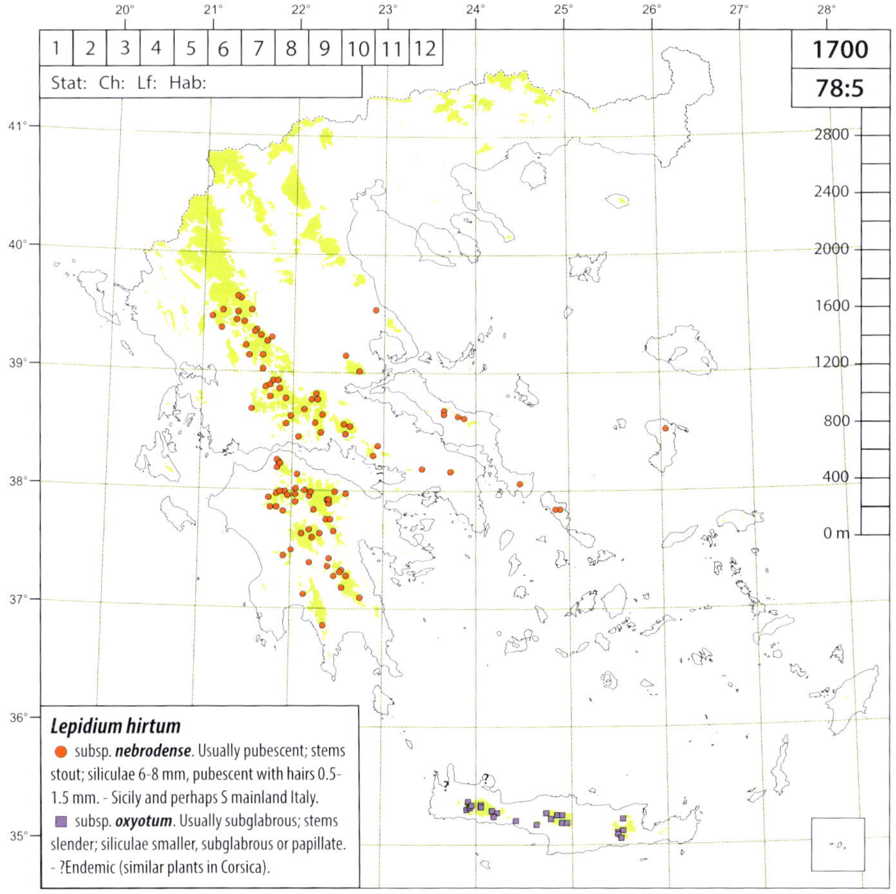

Lepidium hirtum

● subsp. ***nebrodense***. Usually pubescent; stems stout; siliculae 6-8 mm, pubescent with hairs 0.5-1.5 mm. - Sicily and perhaps S mainland Italy.

■ subsp. ***oxyotum***. Usually subglabrous; stems slender; siliculae smaller, subglabrous or papillate. - ?Endemic (similar plants in Corsica).

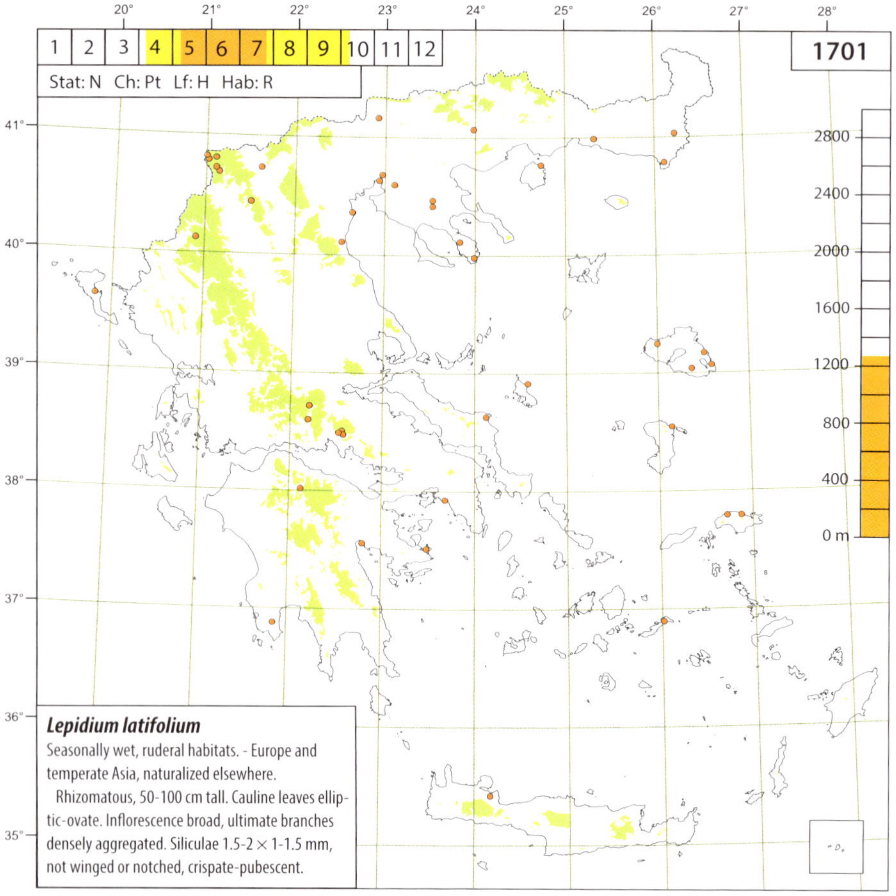

Lepidium latifolium

Seasonally wet, ruderal habitats. - Europe and temperate Asia, naturalized elsewhere.

Rhizomatous, 50-100 cm tall. Cauline leaves elliptic-ovate. Inflorescence broad, ultimate branches densely aggregated. Siliculae 1.5-2 × 1-1.5 mm, not winged or notched, crispate-pubescent.

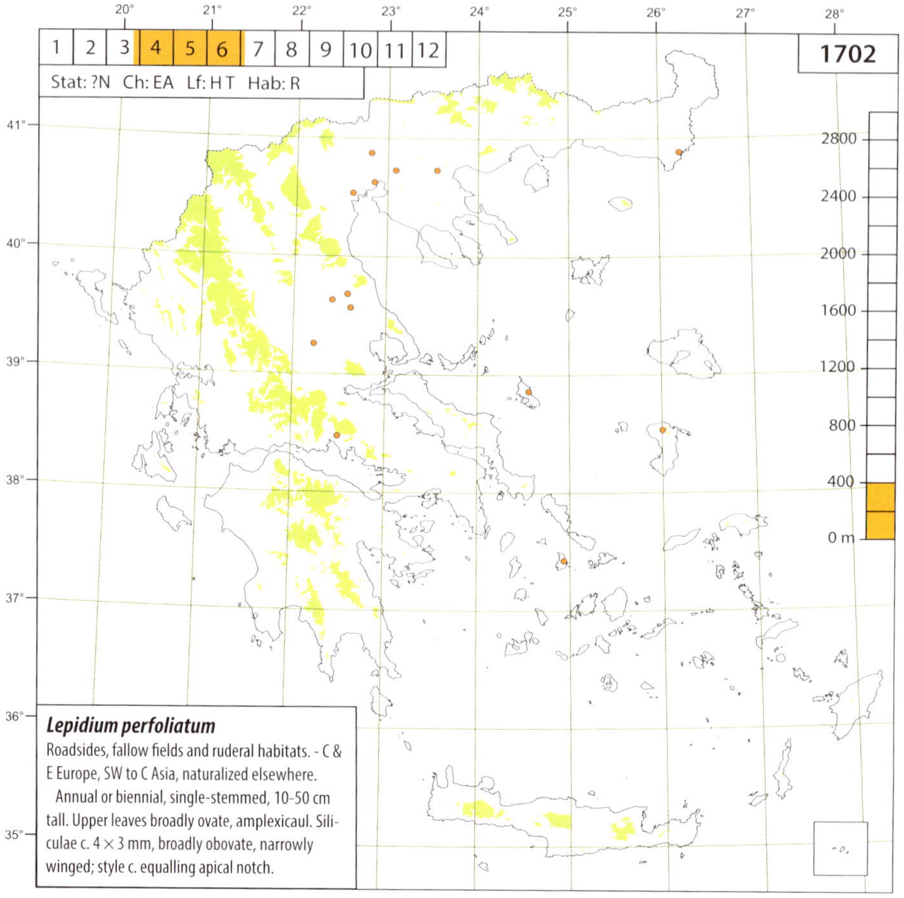

Lepidium perfoliatum

Roadsides, fallow fields and ruderal habitats. - C & E Europe, SW to C Asia, naturalized elsewhere.

Annual or biennial, single-stemmed, 10-50 cm tall. Upper leaves broadly ovate, amplexicaul. Siliculae c. 4 × 3 mm, broadly obovate, narrowly winged; style c. equalling apical notch.

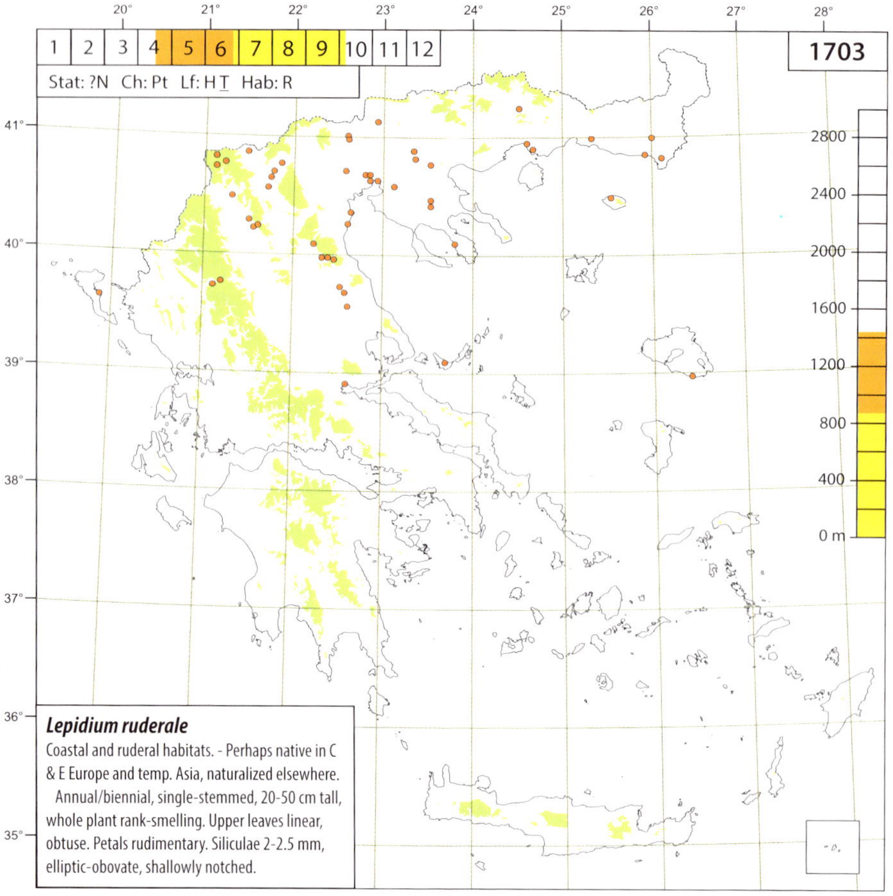

Lepidium ruderale

Coastal and ruderal habitats. - Perhaps native in C & E Europe and temp. Asia, naturalized elsewhere.

Annual/biennial, single-stemmed, 20-50 cm tall, whole plant rank-smelling. Upper leaves linear, obtuse. Petals rudimentary. Siliculae 2-2.5 mm, elliptic-obovate, shallowly notched.

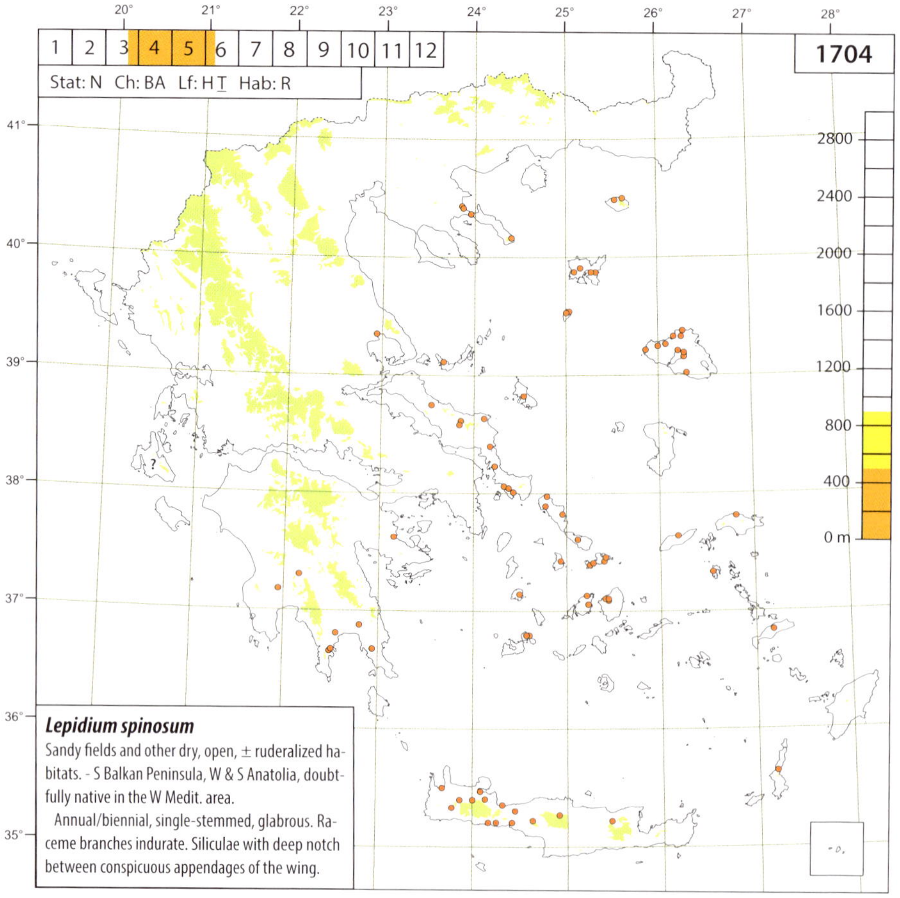

Lepidium spinosum

Sandy fields and other dry, open, ± ruderalized habitats. - S Balkan Peninsula, W & S Anatolia, doubtfully native in the W Medit. area.

Annual/biennial, single-stemmed, glabrous. Raceme branches indurate. Siliculae with deep notch between conspicuous appendages of the wing.

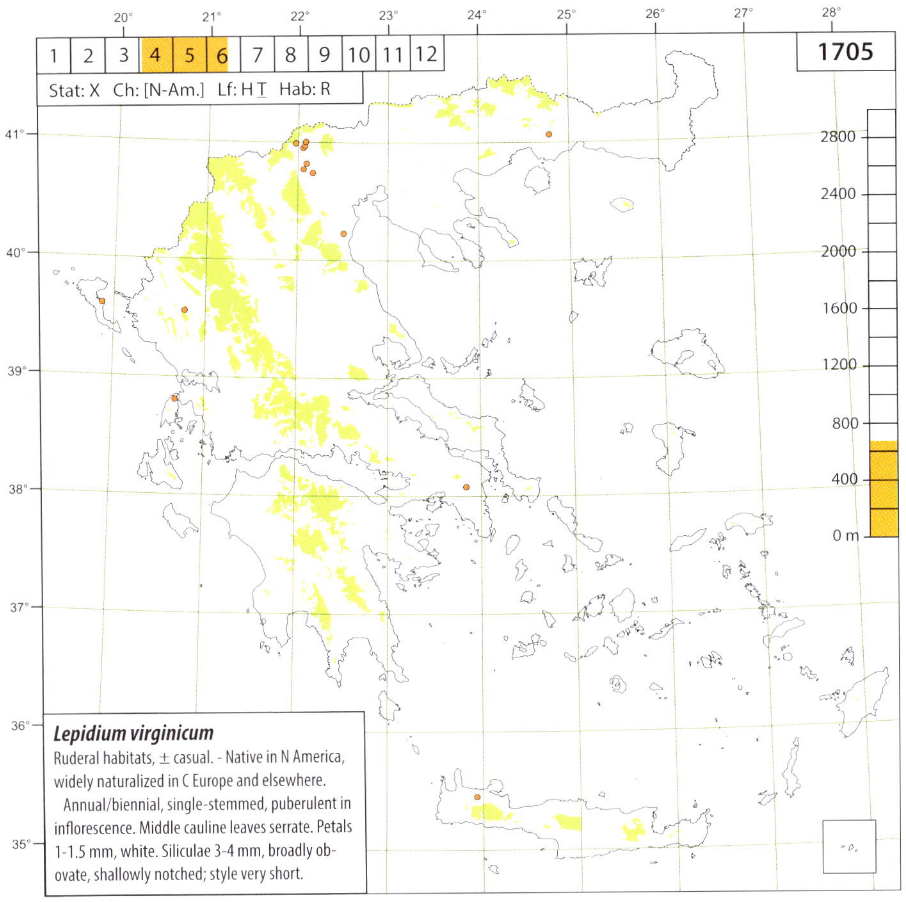

Lepidium virginicum

Ruderal habitats, ± casual. - Native in N America, widely naturalized in C Europe and elsewhere.

Annual/biennial, single-stemmed, puberulent in inflorescence. Middle cauline leaves serrate. Petals 1-1.5 mm, white. Siliculae 3-4 mm, broadly obovate, shallowly notched; style very short.

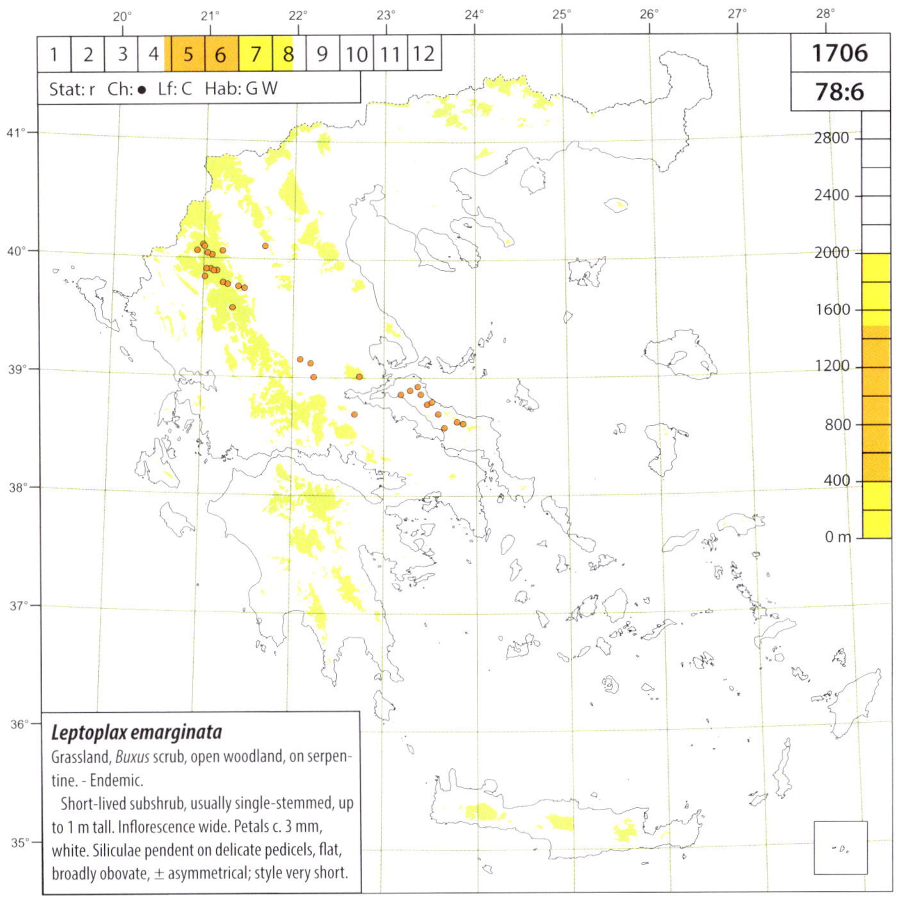

Leptoplax emarginata

Grassland, *Buxus* scrub, open woodland, on serpentine. - Endemic.

Short-lived subshrub, usually single-stemmed, up to 1 m tall. Inflorescence wide. Petals c. 3 mm, white. Siliculae pendent on delicate pedicels, flat, broadly obovate, ± asymmetrical; style very short.

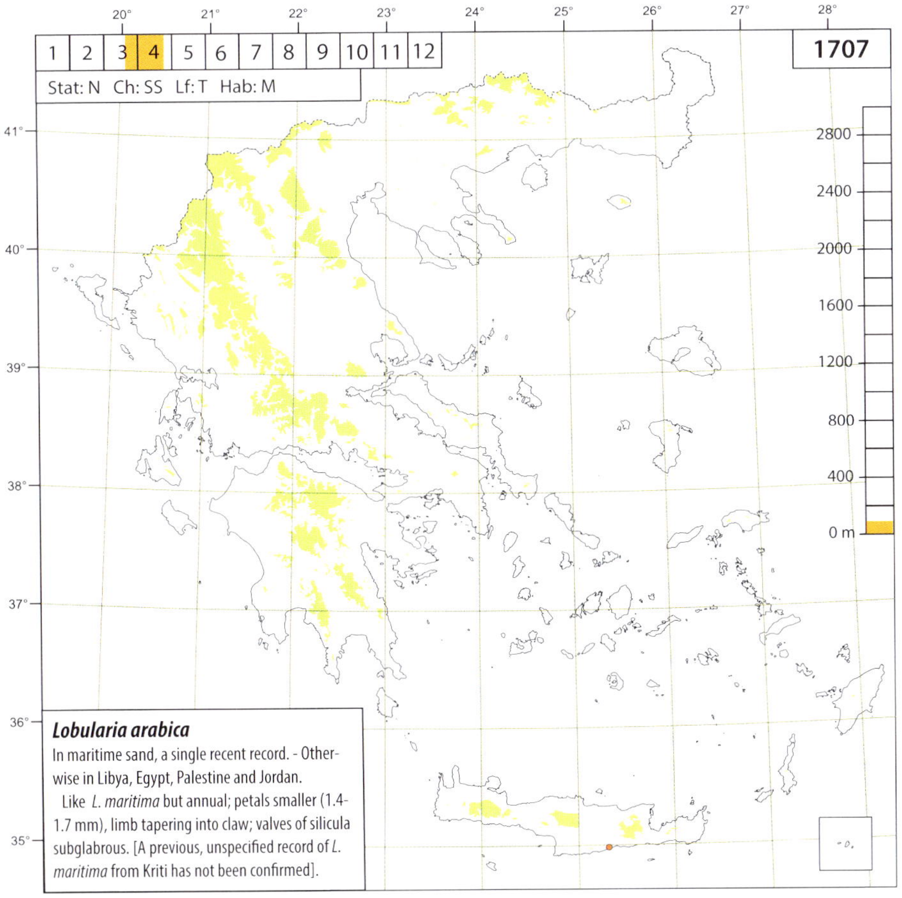

Lobularia arabica

In maritime sand, a single recent record. - Otherwise in Libya, Egypt, Palestine and Jordan.

Like *L. maritima* but annual; petals smaller (1.4-1.7 mm), limb tapering into claw; valves of silicula subglabrous. [A previous, unspecified record of *L. maritima* from Kriti has not been confirmed].

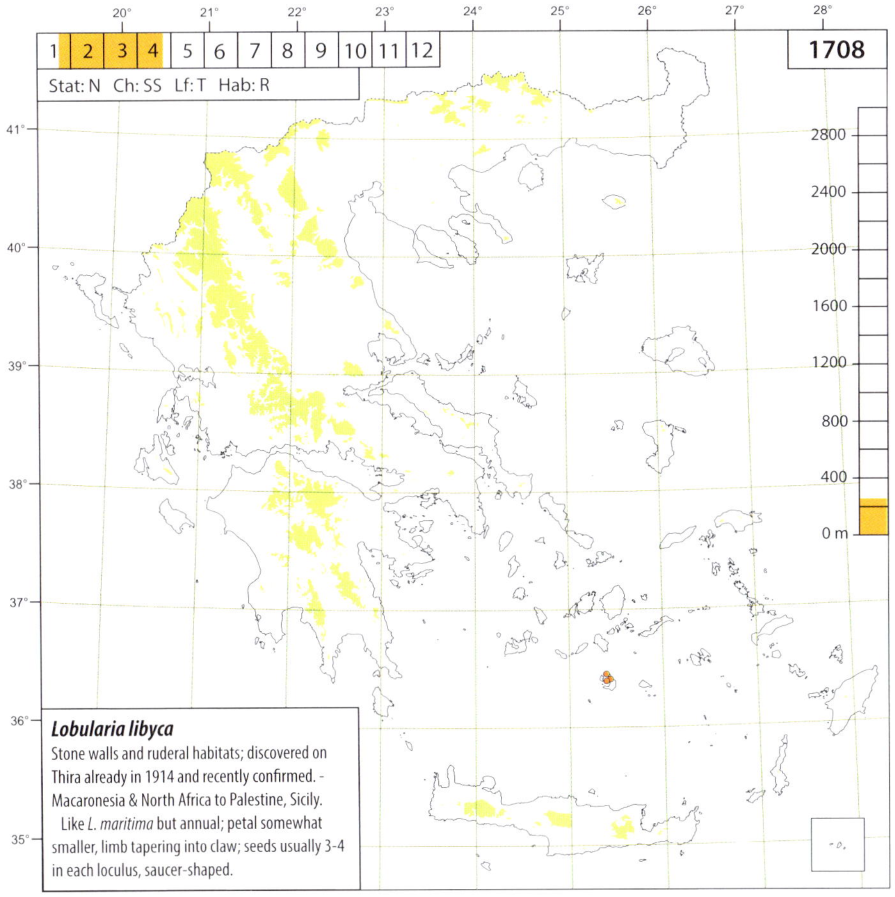

Lobularia libyca

Stone walls and ruderal habitats; discovered on Thira already in 1914 and recently confirmed. - Macaronesia & North Africa to Palestine, Sicily.

Like *L. maritima* but annual; petal somewhat smaller, limb tapering into claw; seeds usually 3-4 in each loculus, saucer-shaped.

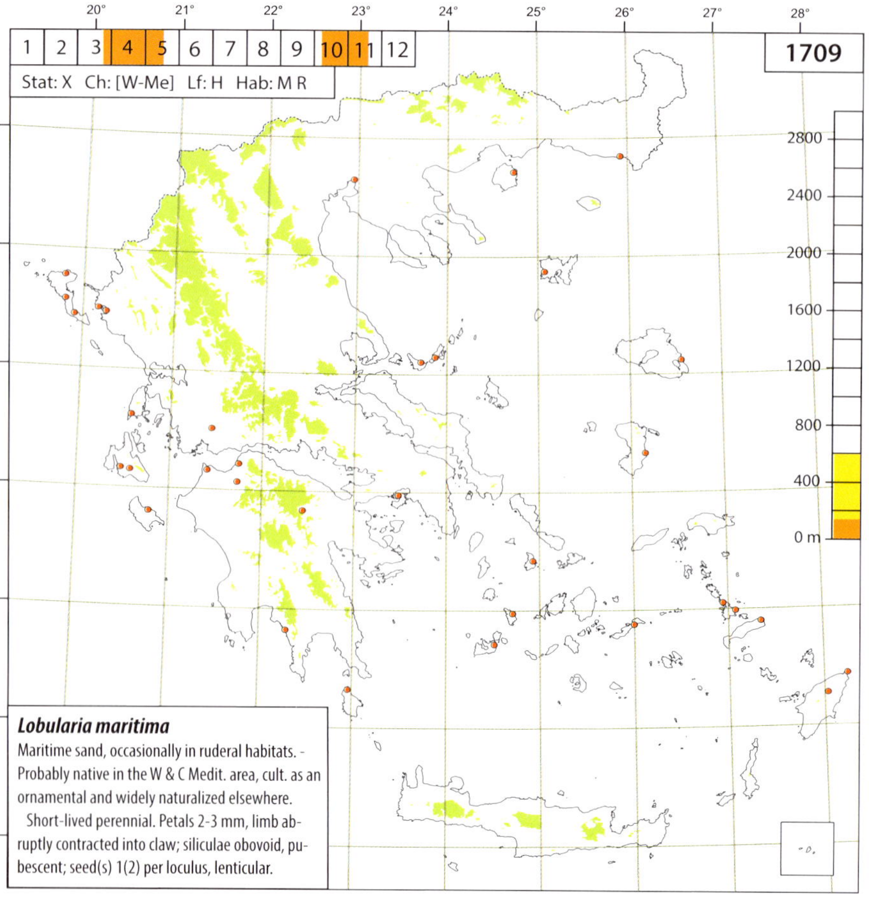

Lobularia maritima

Maritime sand, occasionally in ruderal habitats. - Probably native in the W & C Medit. area, cult. as an ornamental and widely naturalized elsewhere.

Short-lived perennial. Petals 2-3 mm, limb abruptly contracted into claw; siliculae obovoid, pubescent; seed(s) 1(2) per loculus, lenticular.

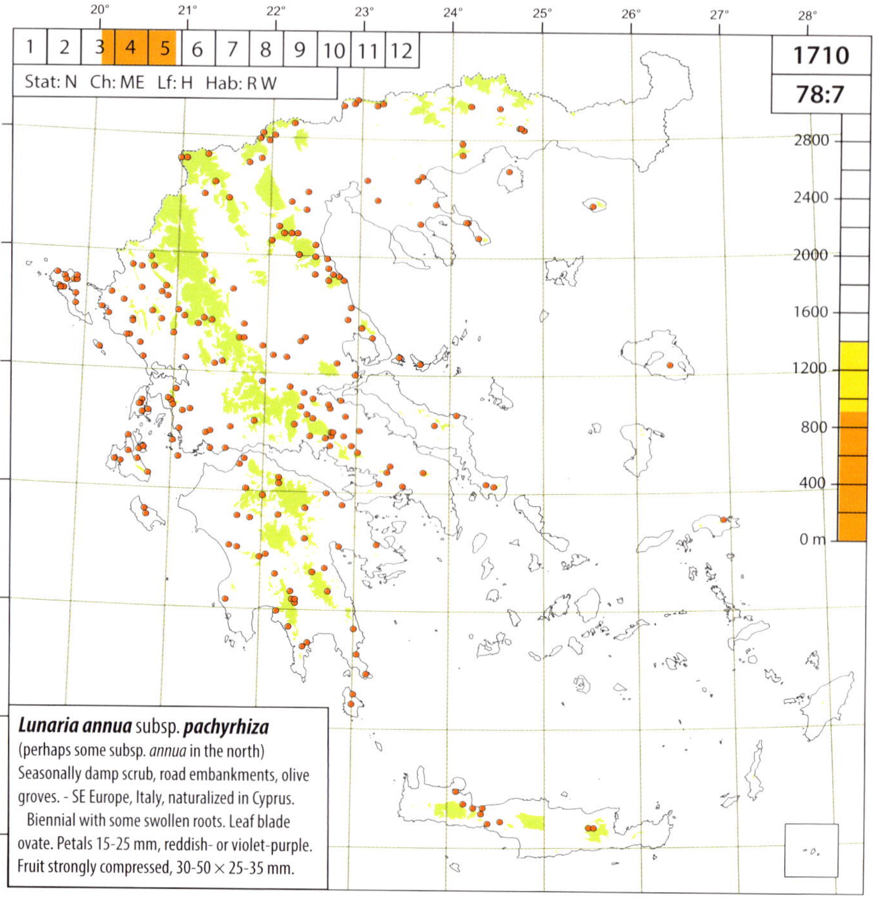

Lunaria annua subsp. **pachyrhiza**

(perhaps some subsp. *annua* in the north)

Seasonally damp scrub, road embankments, olive groves. - SE Europe, Italy, naturalized in Cyprus.

Biennial with some swollen roots. Leaf blade ovate. Petals 15-25 mm, reddish- or violet-purple. Fruit strongly compressed, 30-50 × 25-35 mm.

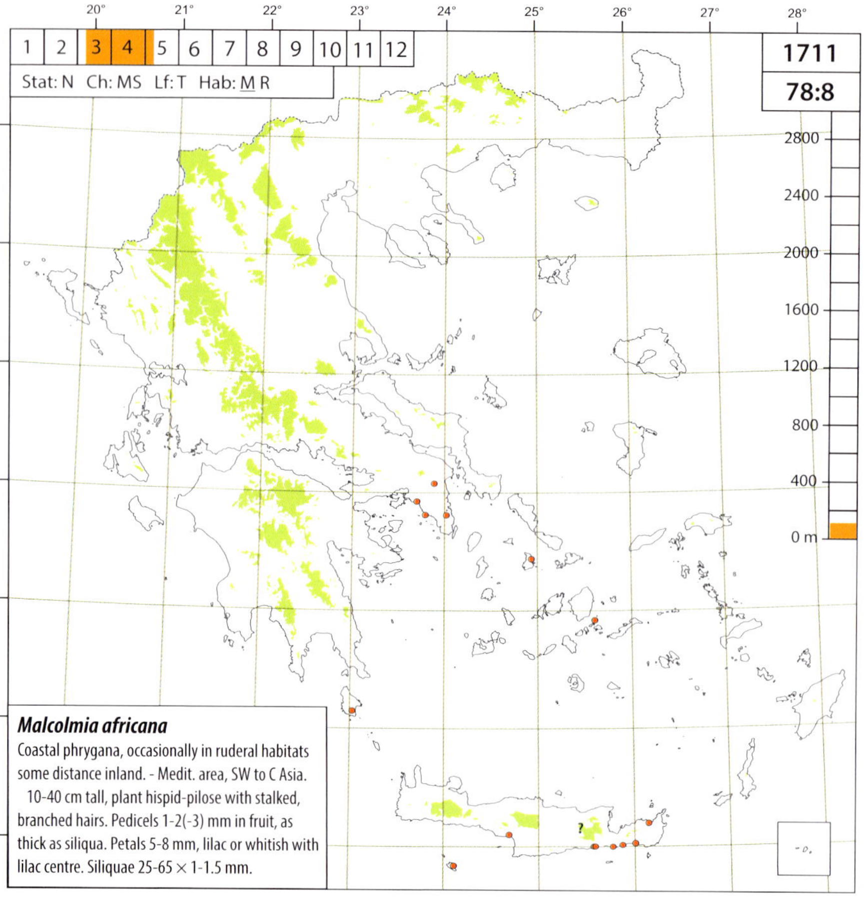

Malcolmia africana

Coastal phrygana, occasionally in ruderal habitats some distance inland. - Medit. area, SW to C Asia.

10-40 cm tall, plant hispid-pilose with stalked, branched hairs. Pedicels 1-2(-3) mm in fruit, as thick as siliqua. Petals 5-8 mm, lilac or whitish with lilac centre. Siliquae 25-65 × 1-1.5 mm.

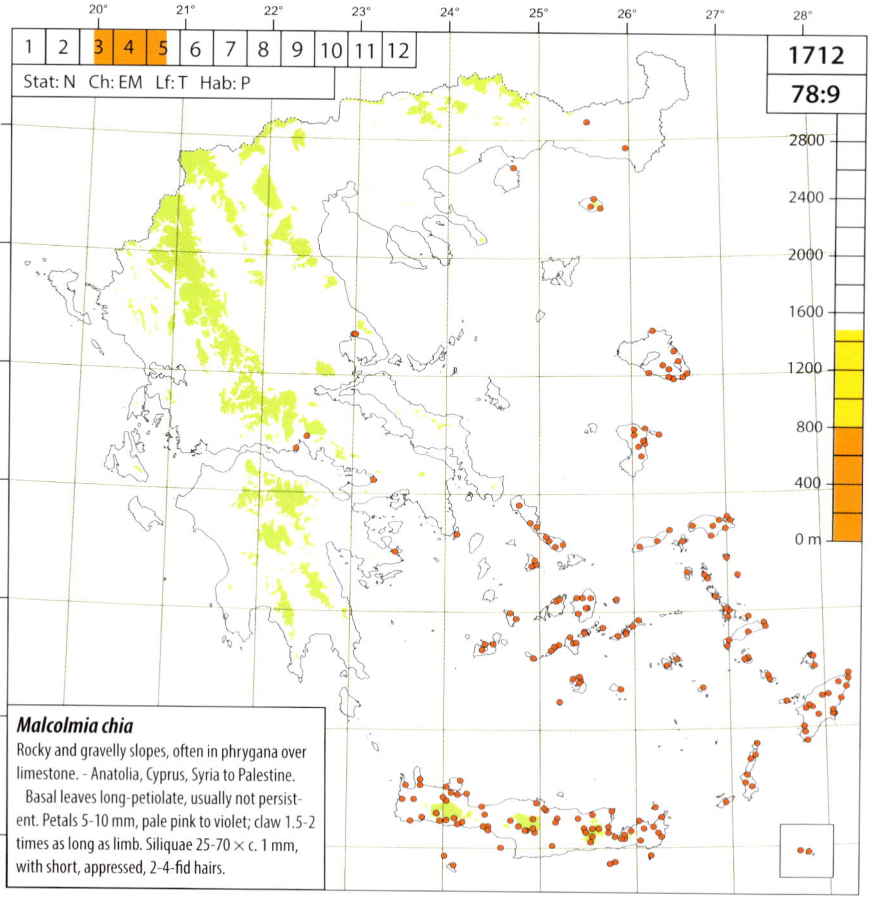

Malcolmia chia

Rocky and gravelly slopes, often in phrygana over limestone. - Anatolia, Cyprus, Syria to Palestine.

Basal leaves long-petiolate, usually not persistent. Petals 5-10 mm, pale pink to violet; claw 1.5-2 times as long as limb. Siliquae 25-70 × c. 1 mm, with short, appressed, 2-4-fid hairs.

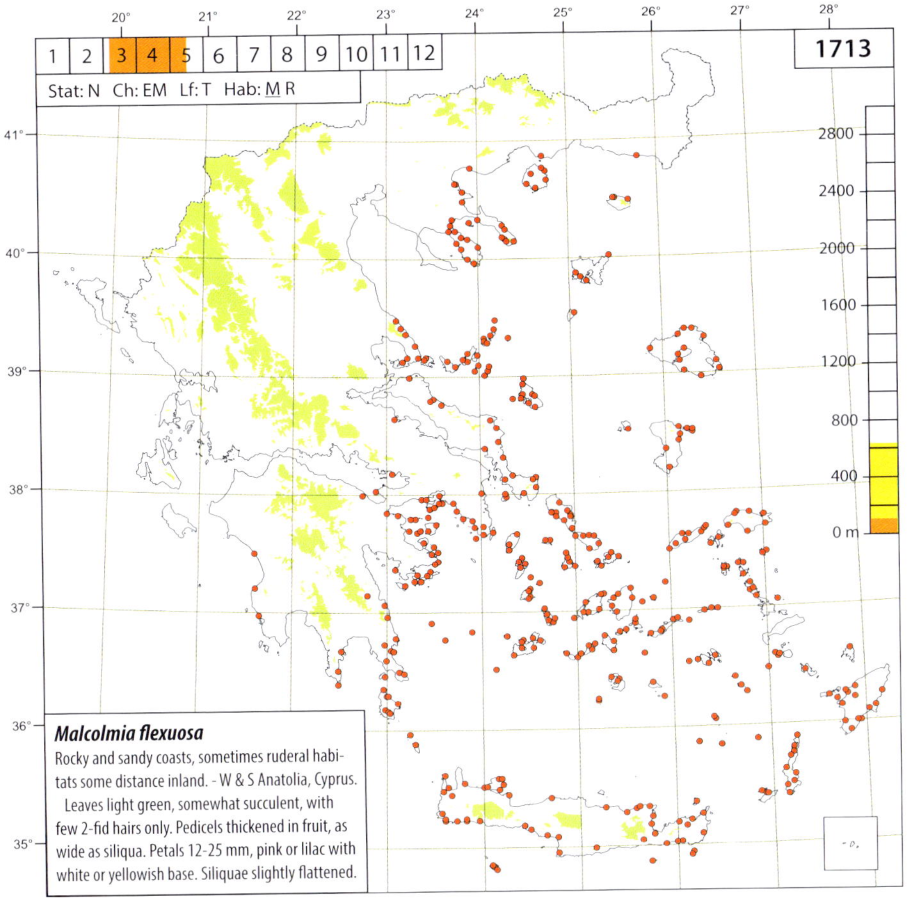

Malcolmia flexuosa
Rocky and sandy coasts, sometimes ruderal habitats some distance inland. - W & S Anatolia, Cyprus.
Leaves light green, somewhat succulent, with few 2-fid hairs only. Pedicels thickened in fruit, as wide as siliqua. Petals 12-25 mm, pink or lilac with white or yellowish base. Siliquae slightly flattened.

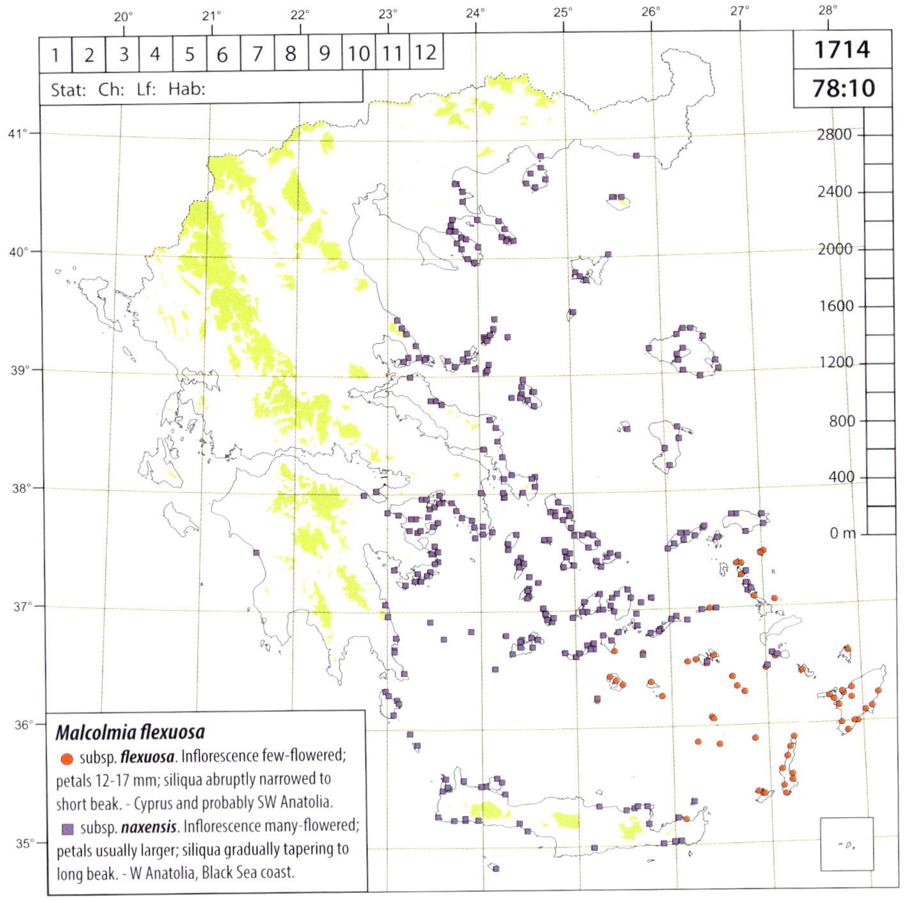

Malcolmia flexuosa
● subsp. ***flexuosa***. Inflorescence few-flowered; petals 12-17 mm; siliqua abruptly narrowed to short beak. - Cyprus and probably SW Anatolia.
■ subsp. ***naxensis***. Inflorescence many-flowered; petals usually larger; siliqua gradually tapering to long beak. - W Anatolia, Black Sea coast.

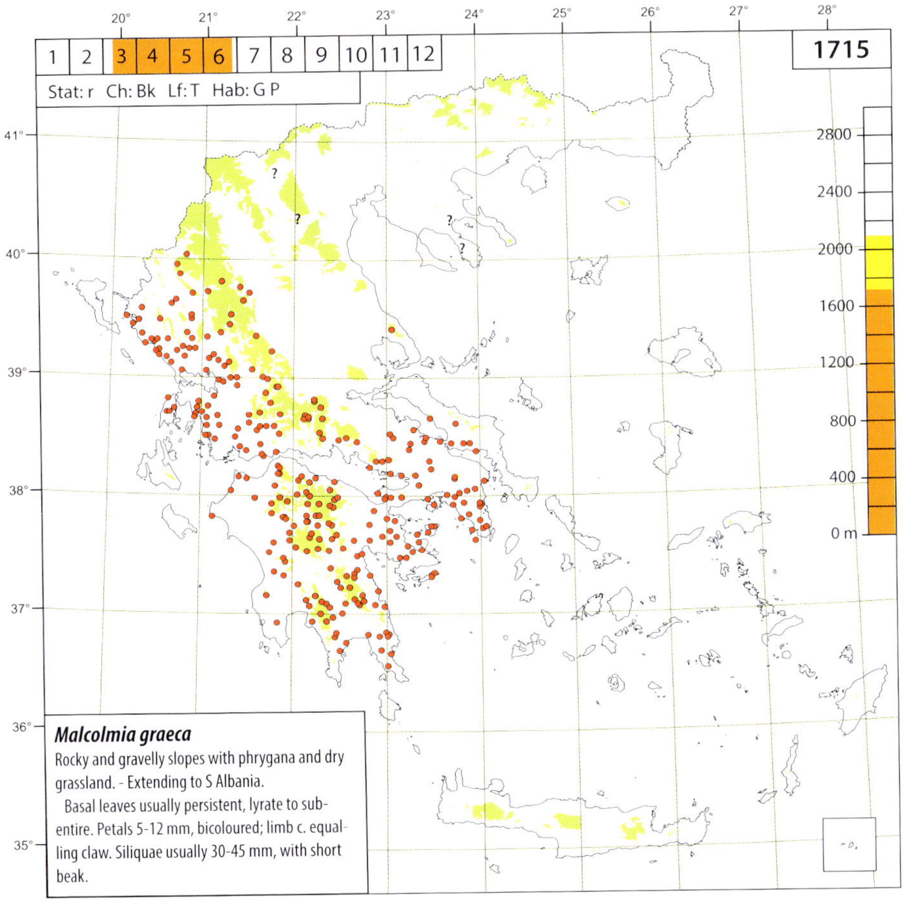

Malcolmia graeca
Rocky and gravelly slopes with phrygana and dry grassland. - Extending to S Albania.
Basal leaves usually persistent, lyrate to subentire. Petals 5-12 mm, bicoloured; limb c. equalling claw. Siliquae usually 30-45 mm, with short beak.

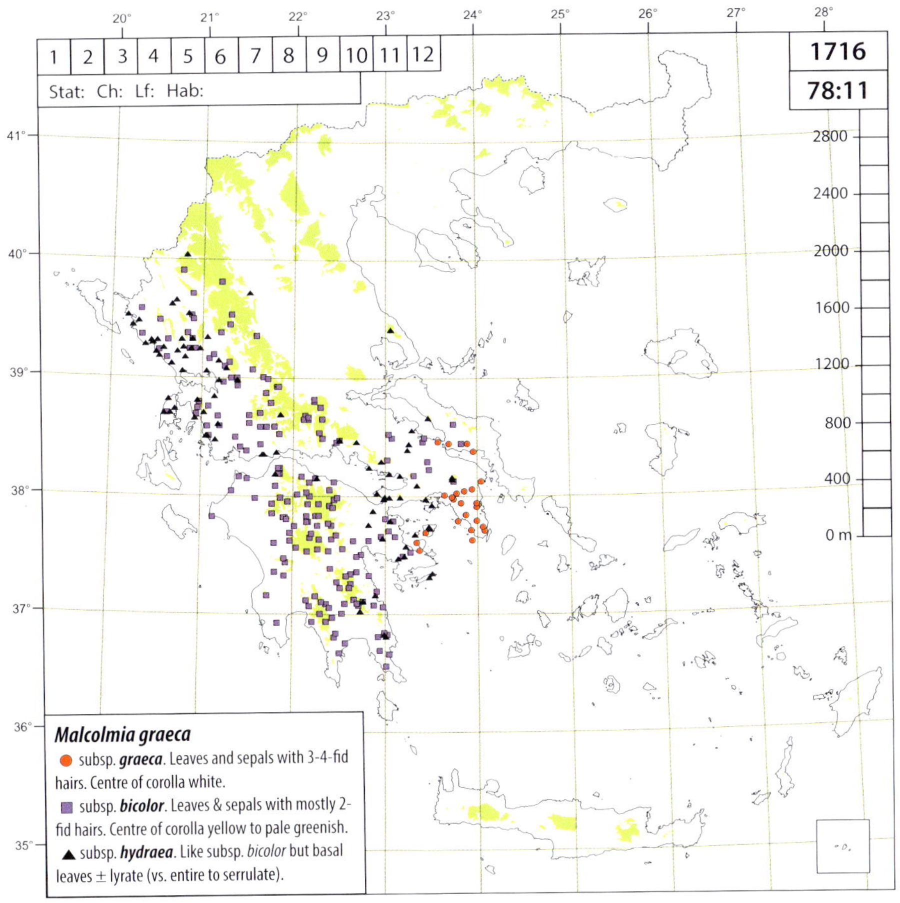

Malcolmia graeca
● subsp. ***graeca***. Leaves and sepals with 3-4-fid hairs. Centre of corolla white.
■ subsp. ***bicolor***. Leaves & sepals with mostly 2-fid hairs. Centre of corolla yellow to pale greenish.
▲ subsp. ***hydraea***. Like subsp. *bicolor* but basal leaves ± lyrate (vs. entire to serrulate).

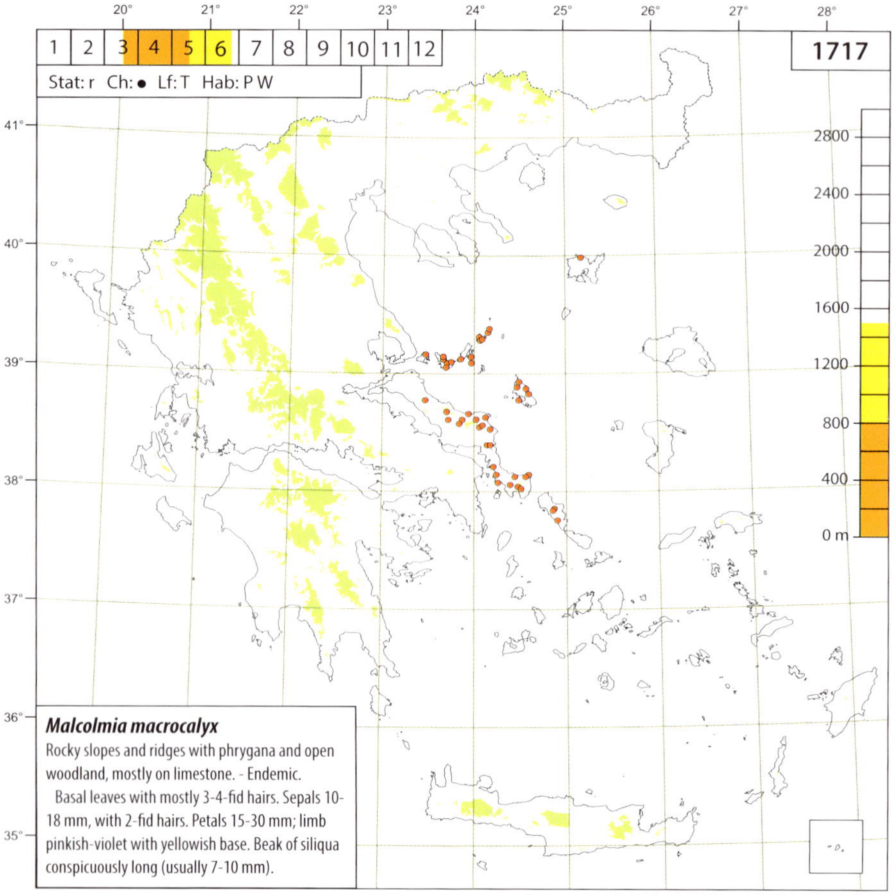

Malcolmia macrocalyx
Rocky slopes and ridges with phrygana and open woodland, mostly on limestone. - Endemic.
Basal leaves with mostly 3-4-fid hairs. Sepals 10-18 mm, with 2-fid hairs. Petals 15-30 mm; limb pinkish-violet with yellowish base. Beak of siliqua conspicuously long (usually 7-10 mm).

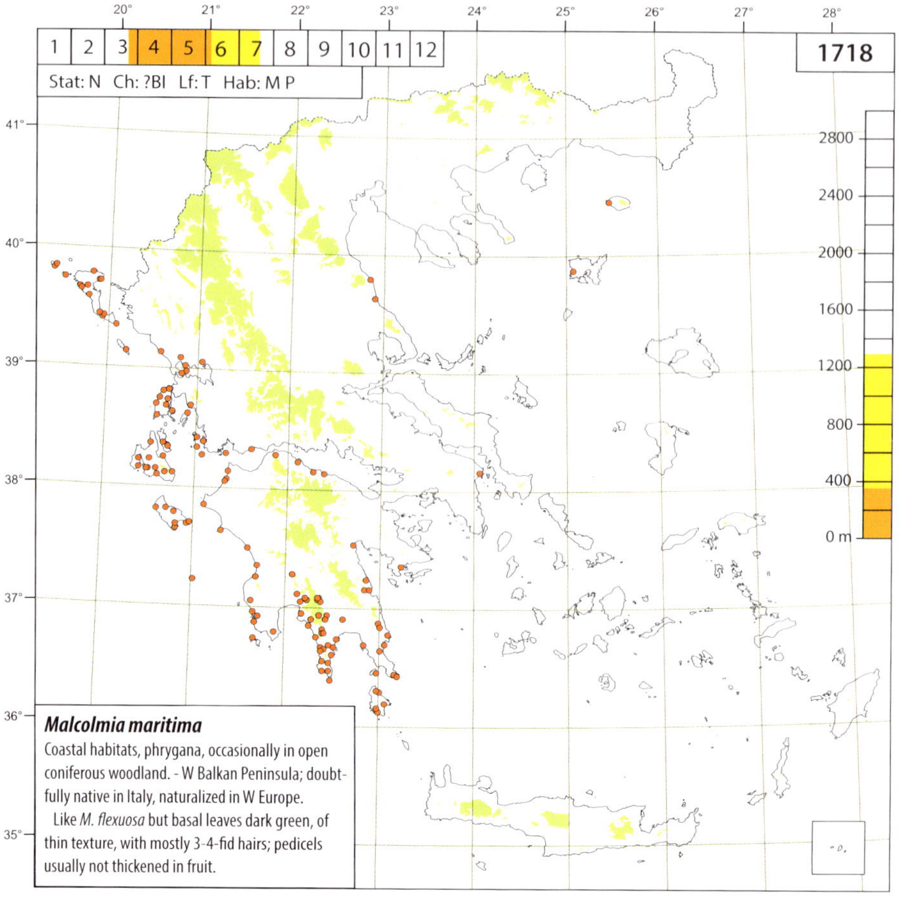

Malcolmia maritima
Coastal habitats, phrygana, occasionally in open coniferous woodland. - W Balkan Peninsula; doubtfully native in Italy, naturalized in W Europe.
Like *M. flexuosa* but basal leaves dark green, of thin texture, with mostly 3-4-fid hairs; pedicels usually not thickened in fruit.

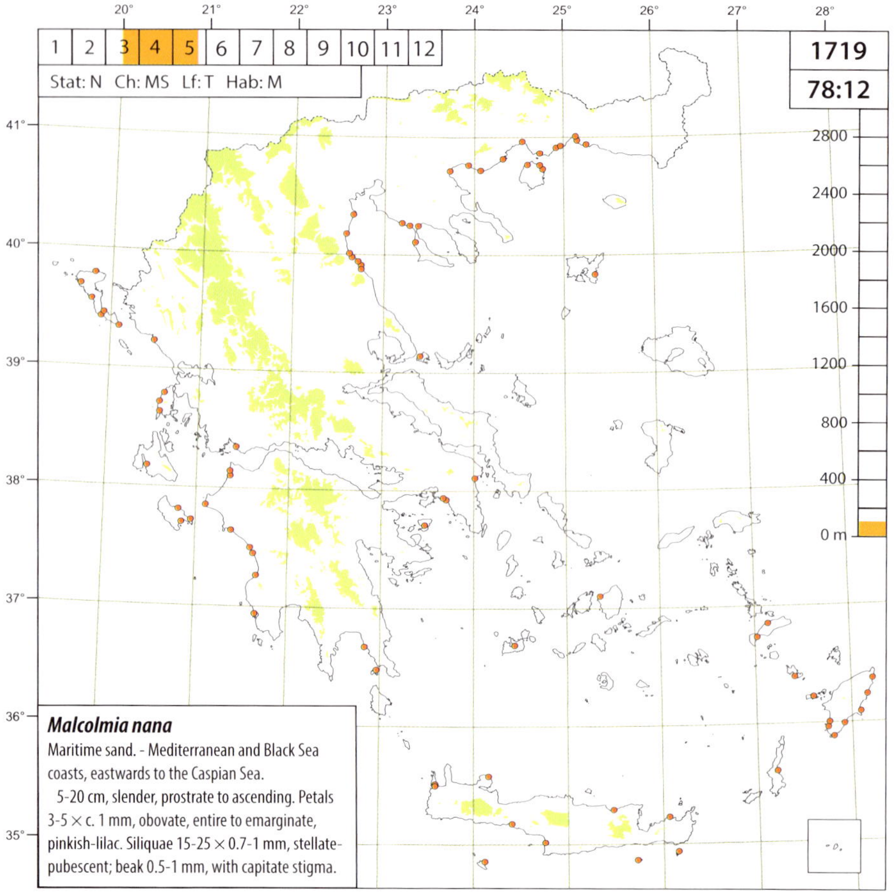

Malcolmia nana
Maritime sand. - Mediterranean and Black Sea coasts, eastwards to the Caspian Sea.
5-20 cm, slender, prostrate to ascending. Petals 3-5 × c. 1 mm, obovate, entire to emarginate, pinkish-lilac. Siliquae 15-25 × 0.7-1 mm, stellate-pubescent; beak 0.5-1 mm, with capitate stigma.

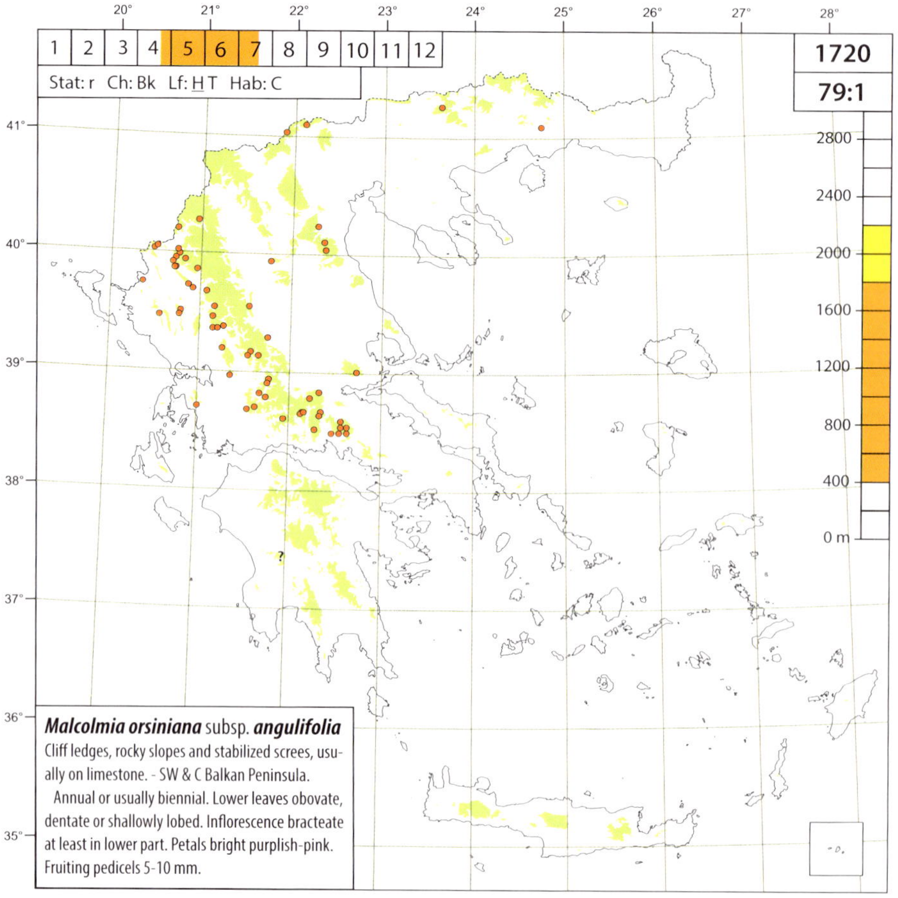

Malcolmia orsiniana subsp. _angulifolia_
Cliff ledges, rocky slopes and stabilized screes, usually on limestone. - SW & C Balkan Peninsula.
Annual or usually biennial. Lower leaves obovate, dentate or shallowly lobed. Inflorescence bracteate at least in lower part. Petals bright purplish-pink. Fruiting pedicels 5-10 mm.

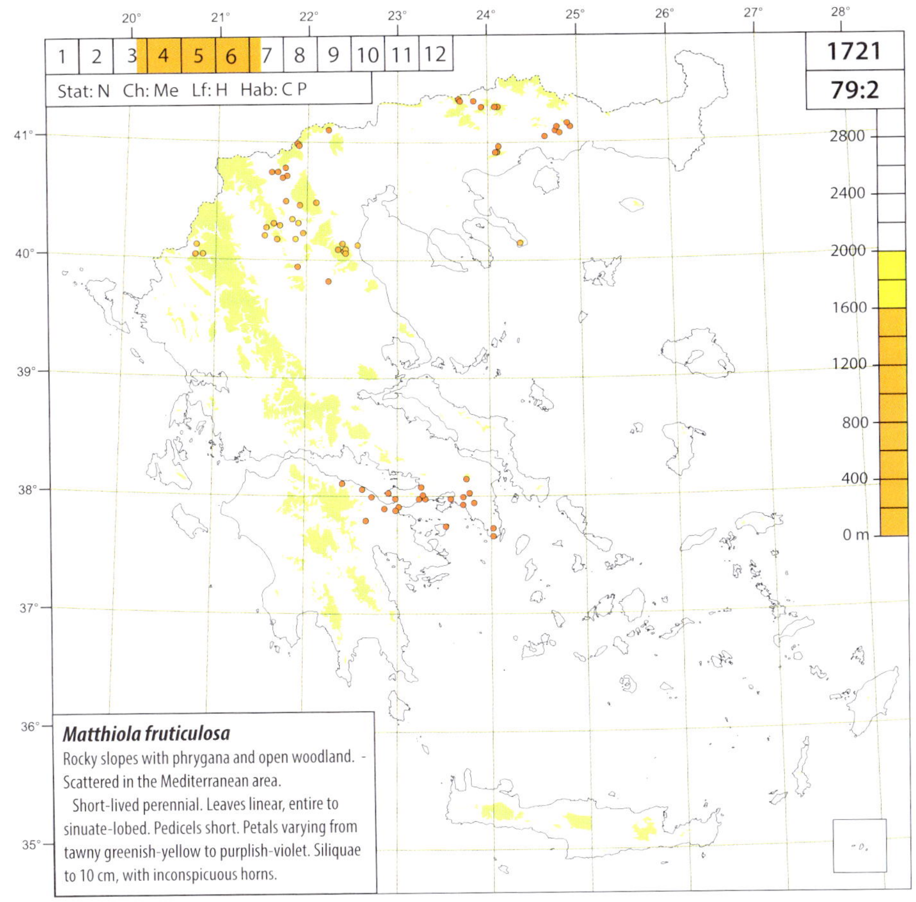

Matthiola fruticulosa

Rocky slopes with phrygana and open woodland. - Scattered in the Mediterranean area.

Short-lived perennial. Leaves linear, entire to sinuate-lobed. Pedicels short. Petals varying from tawny greenish-yellow to purplish-violet. Siliquae to 10 cm, with inconspicuous horns.

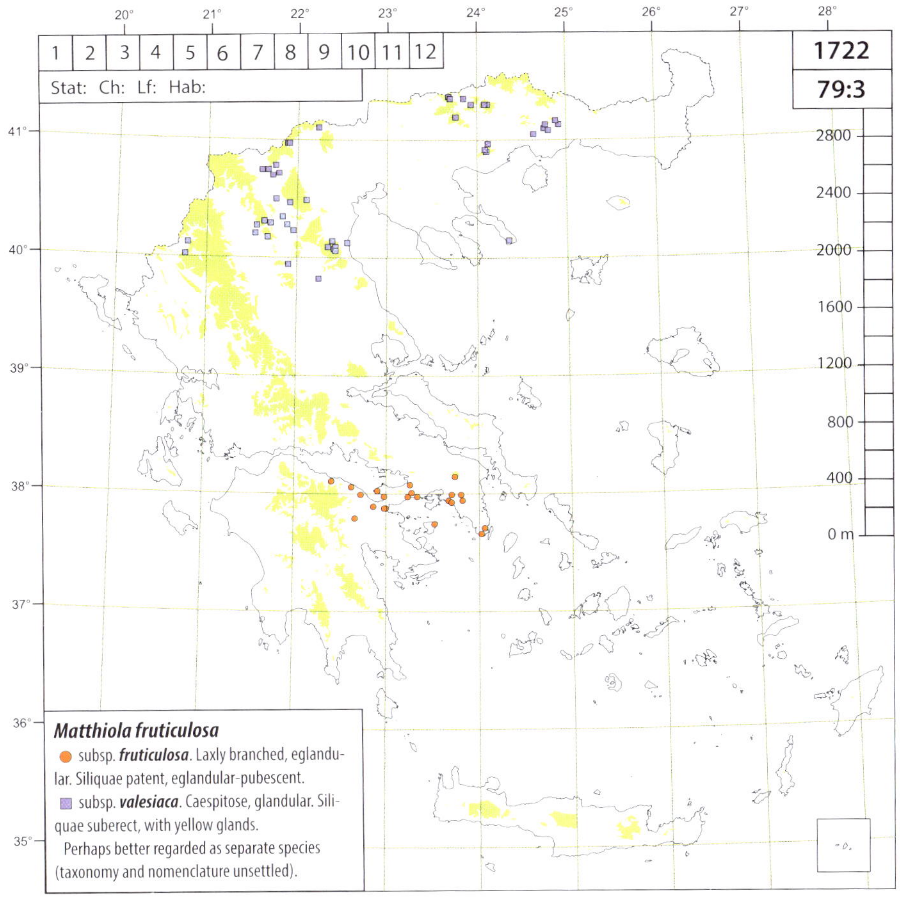

Matthiola fruticulosa

● subsp. ***fruticulosa***. Laxly branched, eglandular. Siliquae patent, eglandular-pubescent.

■ subsp. ***valesiaca***. Caespitose, glandular. Siliquae suberect, with yellow glands.

Perhaps better regarded as separate species (taxonomy and nomenclature unsettled).

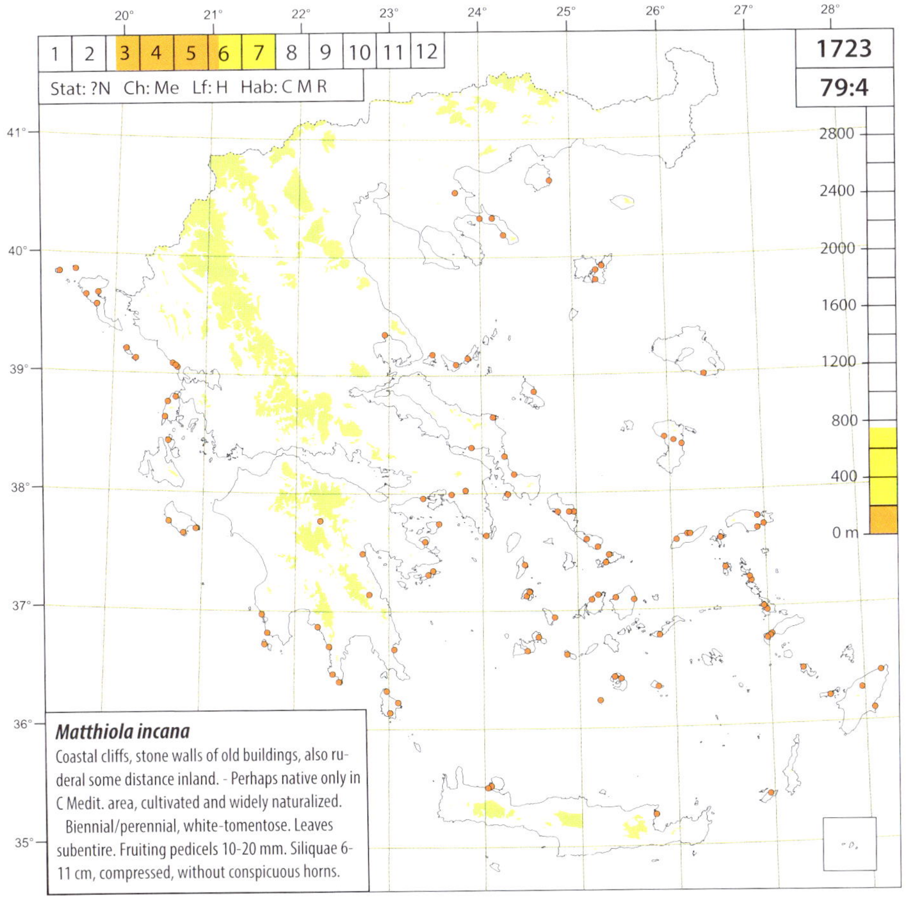

Matthiola incana

Coastal cliffs, stone walls of old buildings, also ruderal some distance inland. - Perhaps native only in C Medit. area, cultivated and widely naturalized.

Biennial/perennial, white-tomentose. Leaves subentire. Fruiting pedicels 10-20 mm. Siliquae 6-11 cm, compressed, without conspicuous horns.

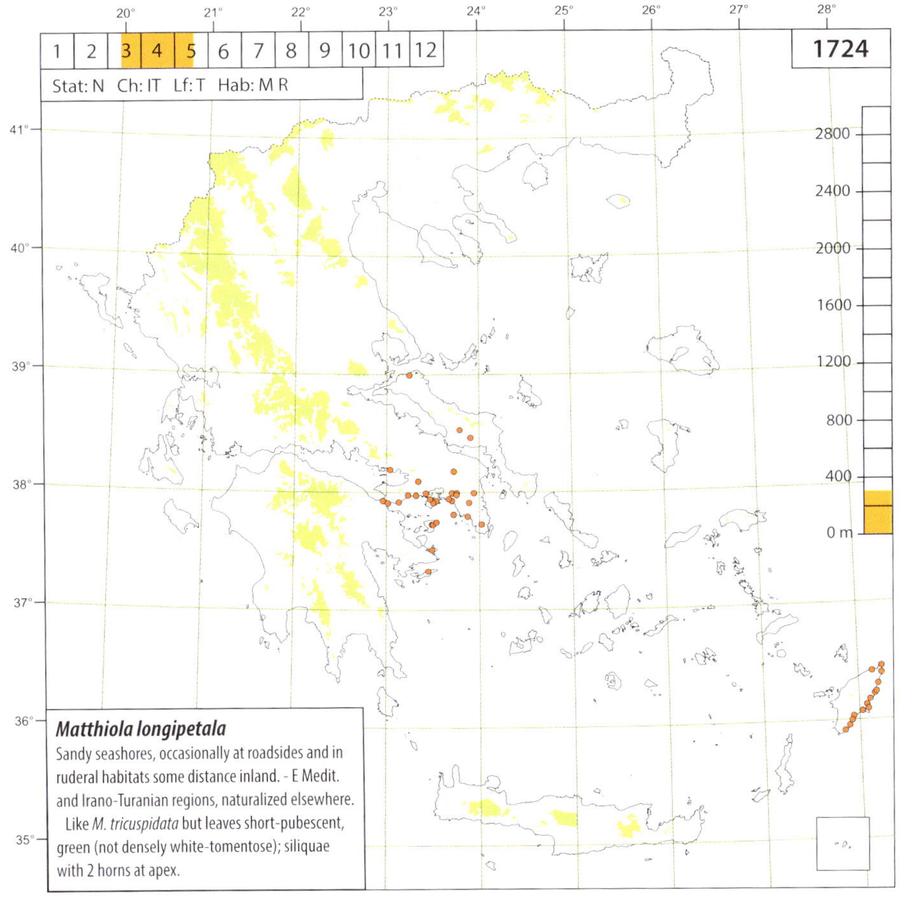

Matthiola longipetala

Sandy seashores, occasionally at roadsides and in ruderal habitats some distance inland. - E Medit. and Irano-Turanian regions, naturalized elsewhere.

Like *M. tricuspidata* but leaves short-pubescent, green (not densely white-tomentose); siliquae with 2 horns at apex.

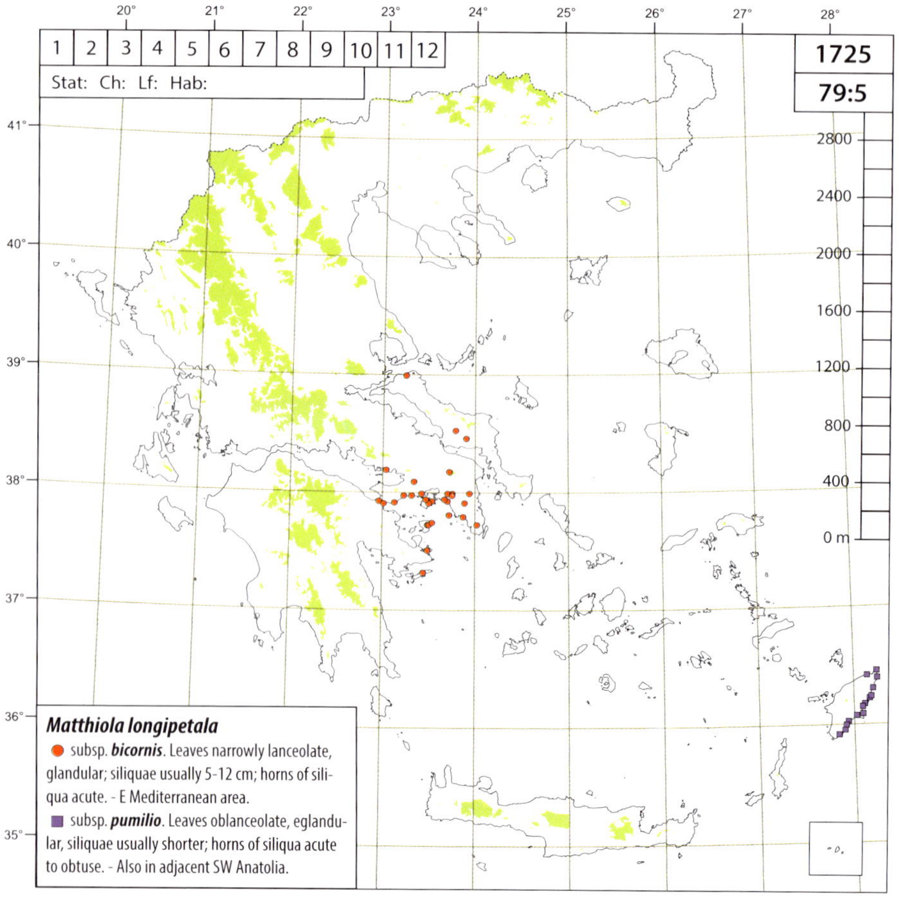

Matthiola longipetala

● subsp. ***bicornis***. Leaves narrowly lanceolate, glandular; siliquae usually 5-12 cm; horns of siliqua acute. - E Mediterranean area.

■ subsp. ***pumilio***. Leaves oblanceolate, eglandular, siliquae usually shorter; horns of siliqua acute to obtuse. - Also in adjacent SW Anatolia.

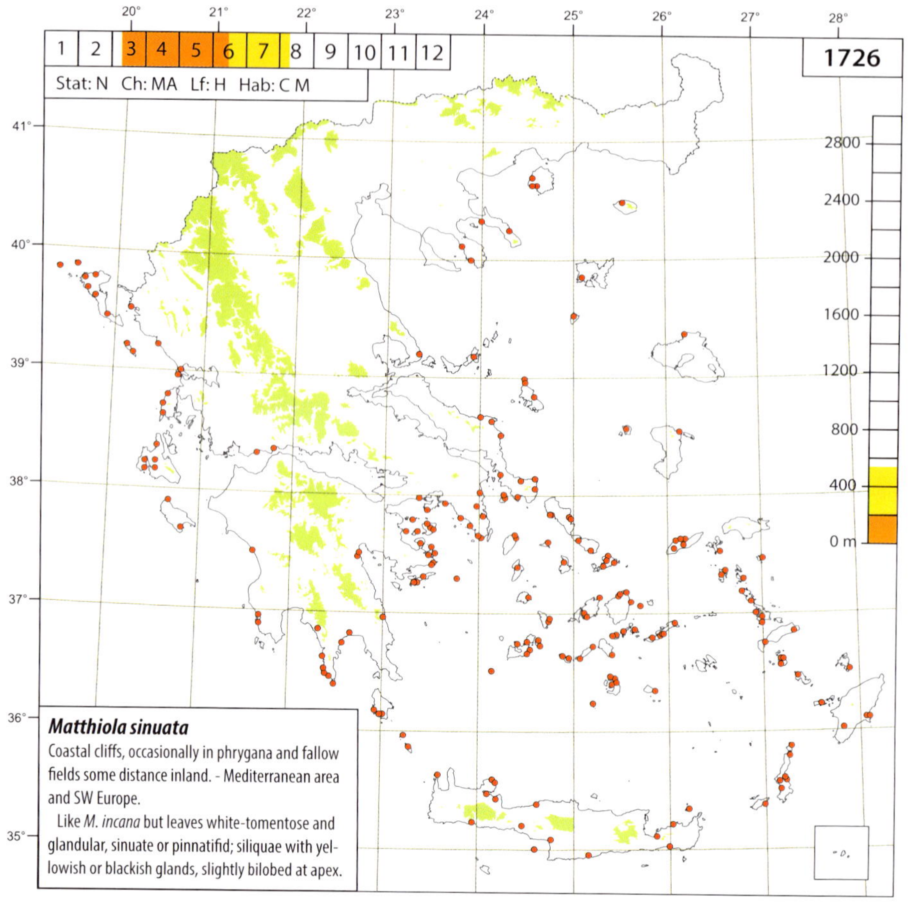

Matthiola sinuata

Coastal cliffs, occasionally in phrygana and fallow fields some distance inland. - Mediterranean area and SW Europe.

Like *M. incana* but leaves white-tomentose and glandular, sinuate or pinnatifid; siliquae with yellowish or blackish glands, slightly bilobed at apex.

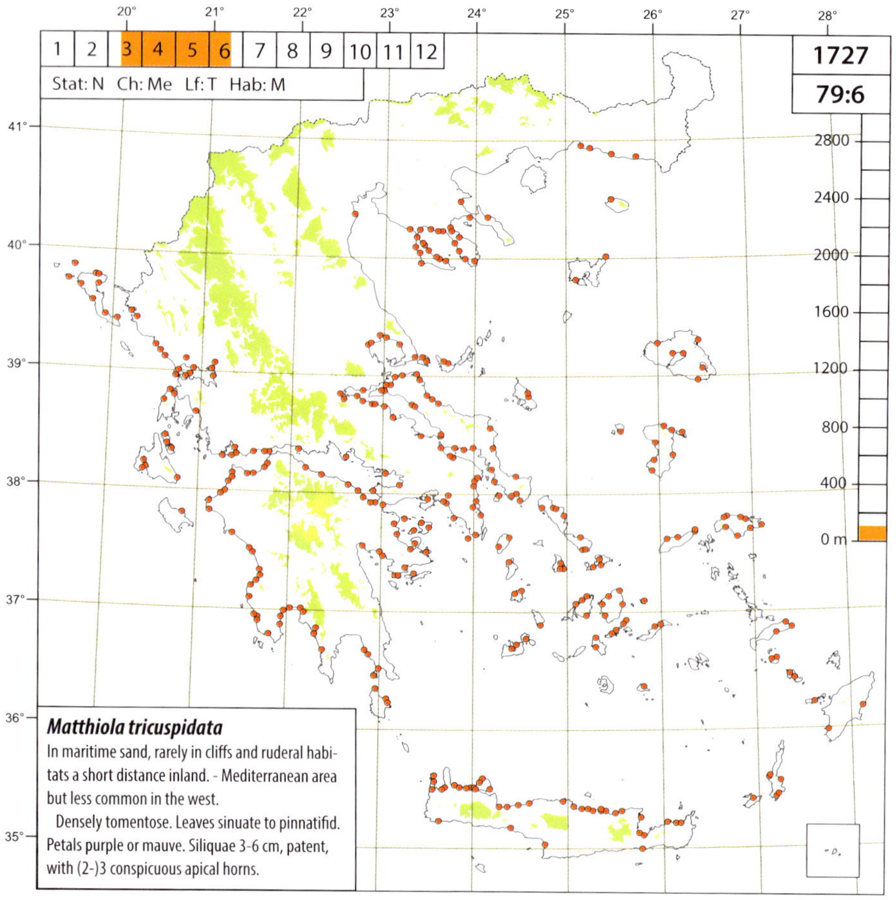

Matthiola tricuspidata

In maritime sand, rarely in cliffs and ruderal habitats a short distance inland. - Mediterranean area but less common in the west.

Densely tomentose. Leaves sinuate to pinnatifid. Petals purple or mauve. Siliquae 3-6 cm, patent, with (2-)3 conspicuous apical horns.

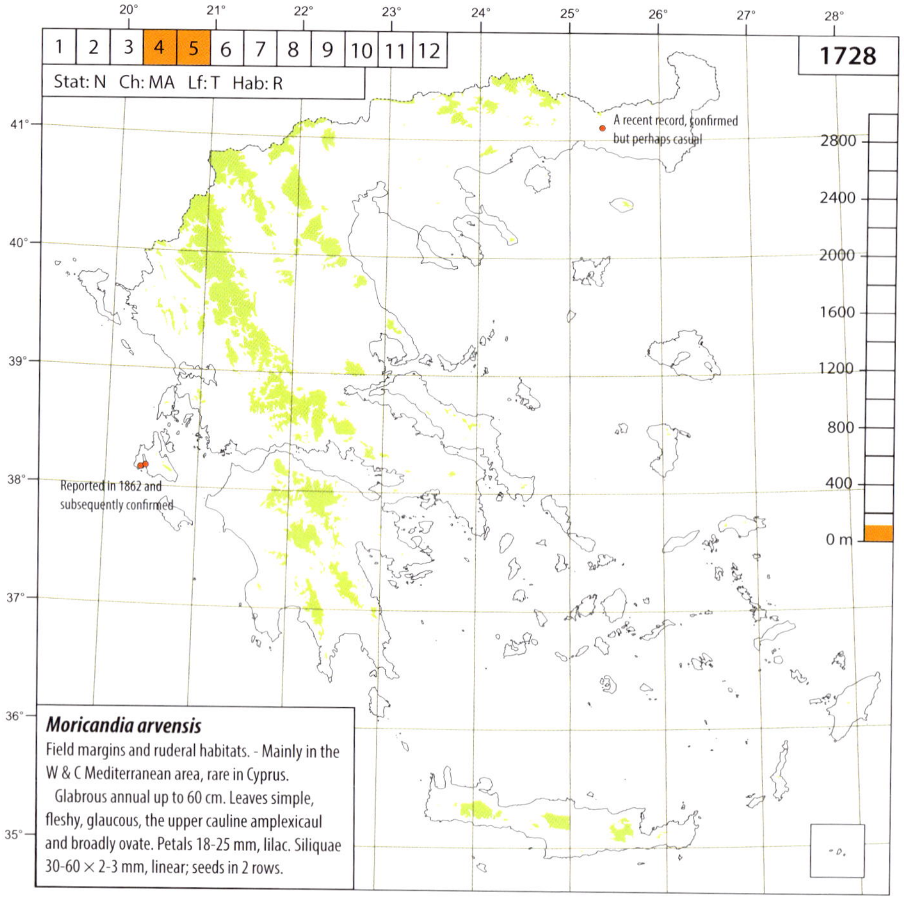

Moricandia arvensis

Field margins and ruderal habitats. - Mainly in the W & C Mediterranean area, rare in Cyprus.

Glabrous annual up to 60 cm. Leaves simple, fleshy, glaucous, the upper cauline amplexicaul and broadly ovate. Petals 18-25 mm, lilac. Siliquae 30-60 × 2-3 mm, linear; seeds in 2 rows.

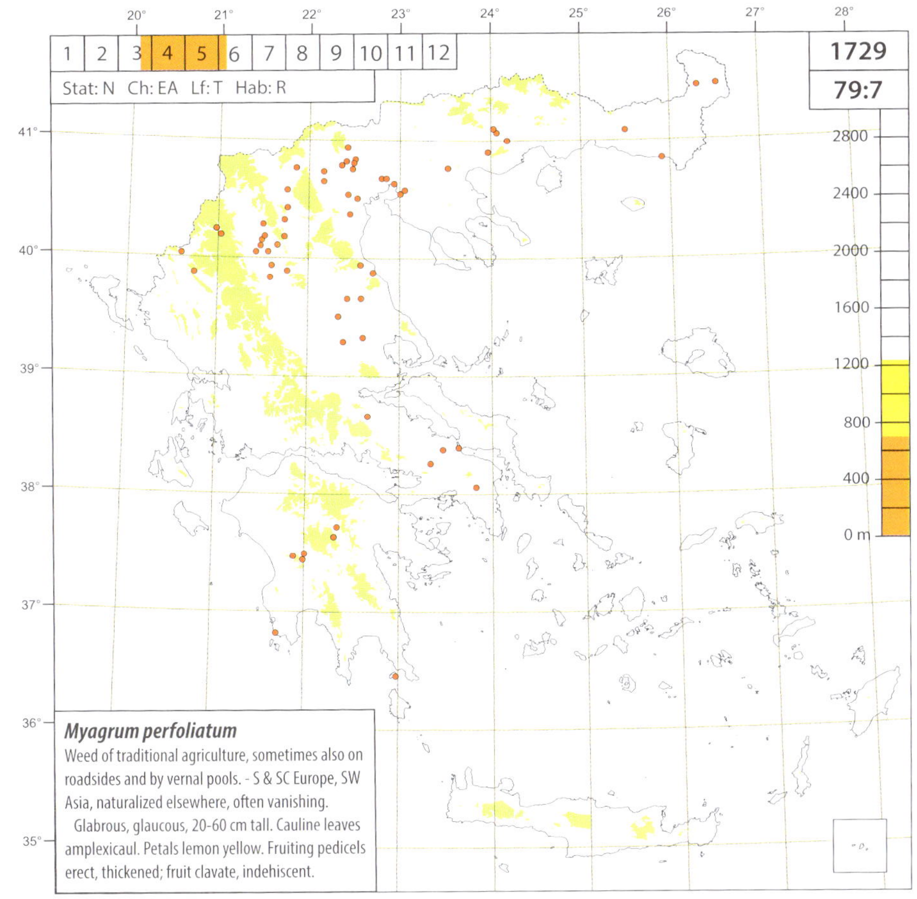

Myagrum perfoliatum
Weed of traditional agriculture, sometimes also on roadsides and by vernal pools. - S & SC Europe, SW Asia, naturalized elsewhere, often vanishing.
Glabrous, glaucous, 20-60 cm tall. Cauline leaves amplexicaul. Petals lemon yellow. Fruiting pedicels erect, thickened; fruit clavate, indehiscent.

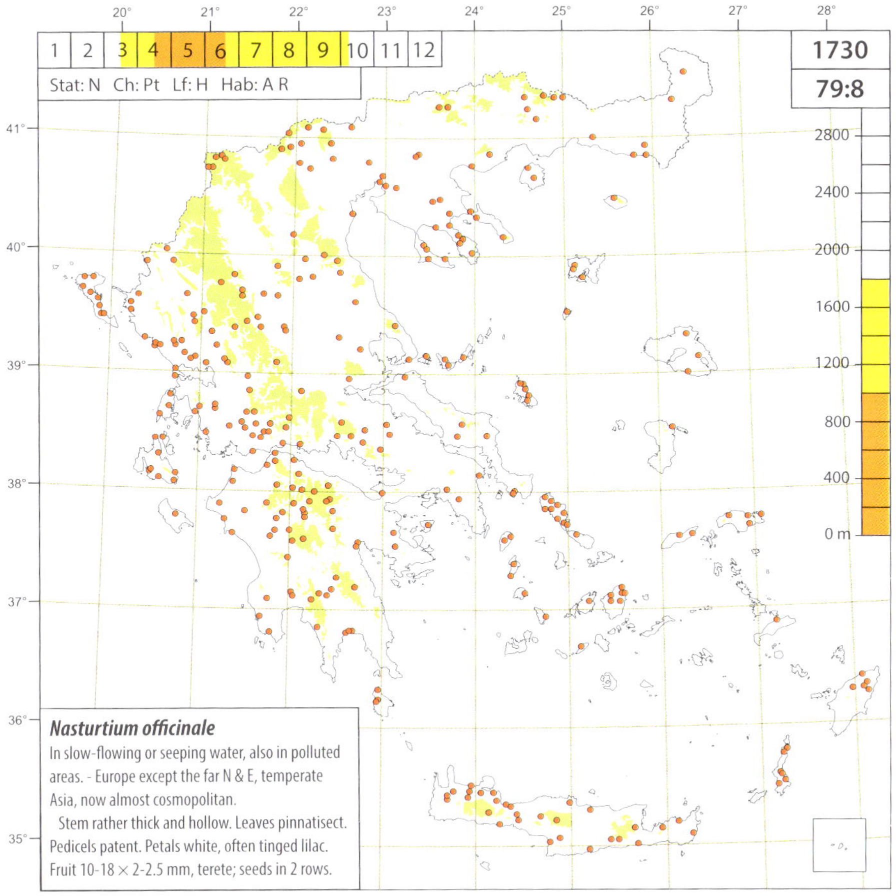

Nasturtium officinale
In slow-flowing or seeping water, also in polluted areas. - Europe except the far N & E, temperate Asia, now almost cosmopolitan.
Stem rather thick and hollow. Leaves pinnatisect. Pedicels patent. Petals white, often tinged lilac. Fruit 10-18 × 2-2.5 mm, terete; seeds in 2 rows.

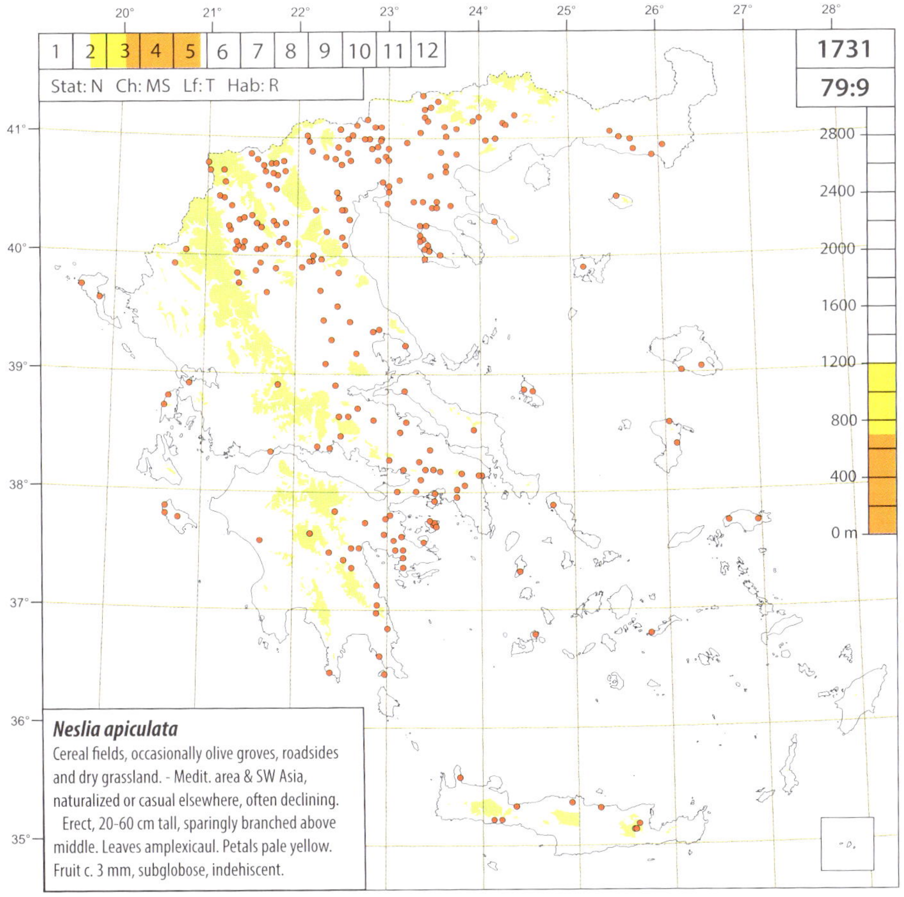

Neslia apiculata
Cereal fields, occasionally olive groves, roadsides and dry grassland. - Medit. area & SW Asia, naturalized or casual elsewhere, often declining.
Erect, 20-60 cm tall, sparingly branched above middle. Leaves amplexicaul. Petals pale yellow. Fruit c. 3 mm, subglobose, indehiscent.

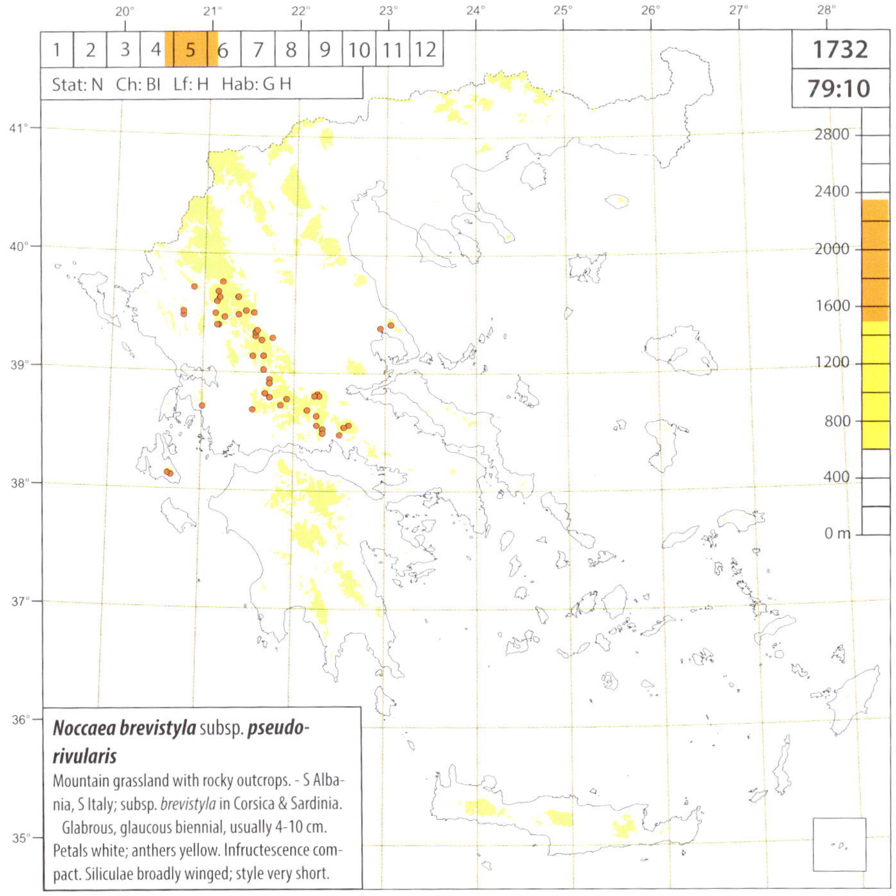

Noccaea brevistyla subsp. ***pseudo-rivularis***
Mountain grassland with rocky outcrops. - S Albania, S Italy; subsp. *brevistyla* in Corsica & Sardinia.
Glabrous, glaucous biennial, usually 4-10 cm. Petals white; anthers yellow. Infructescence compact. Siliculae broadly winged; style very short.

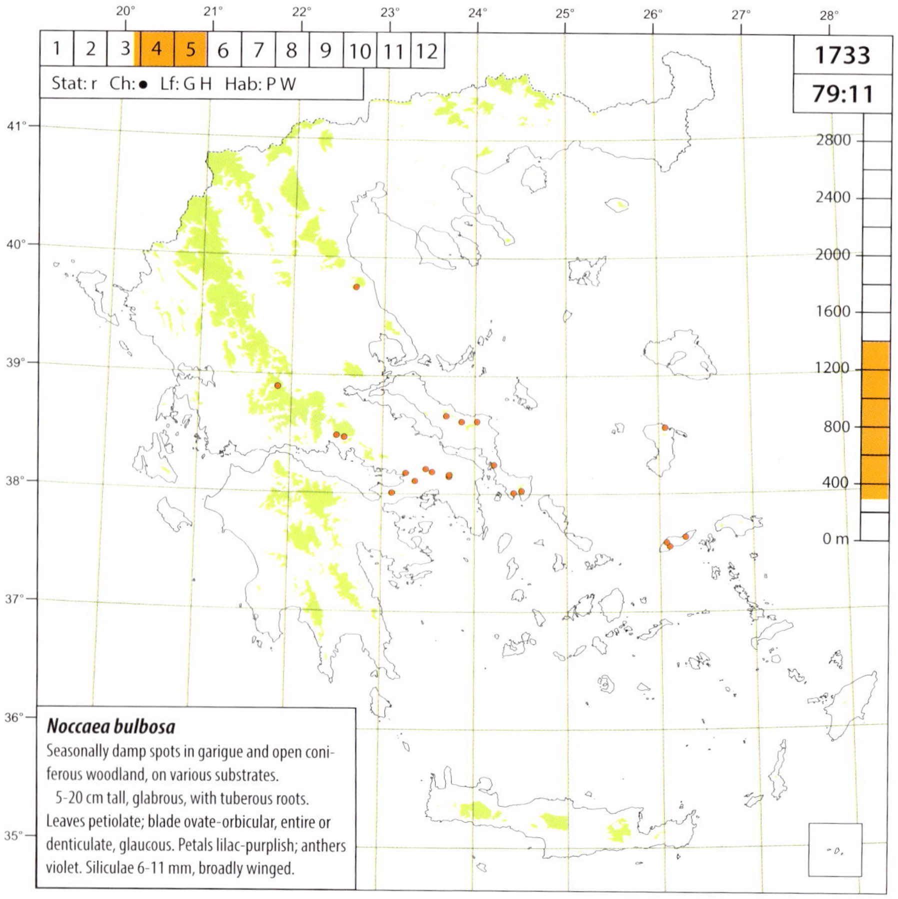

Noccaea bulbosa
Seasonally damp spots in garigue and open coniferous woodland, on various substrates.
5-20 cm tall, glabrous, with tuberous roots. Leaves petiolate; blade ovate-orbicular, entire or denticulate, glaucous. Petals lilac-purplish; anthers violet. Siliculae 6-11 mm, broadly winged.

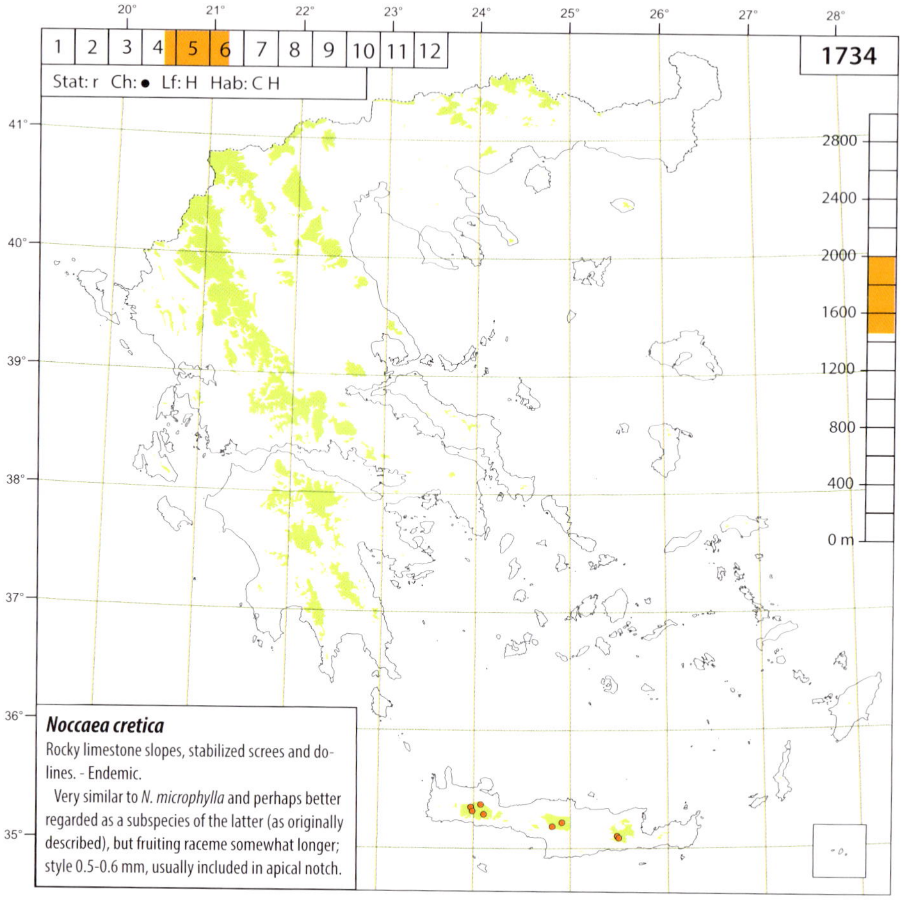

Noccaea cretica
Rocky limestone slopes, stabilized screes and dolines. - Endemic.
Very similar to *N. microphylla* and perhaps better regarded as a subspecies of the latter (as originally described), but fruiting raceme somewhat longer; style 0.5-0.6 mm, usually included in apical notch.

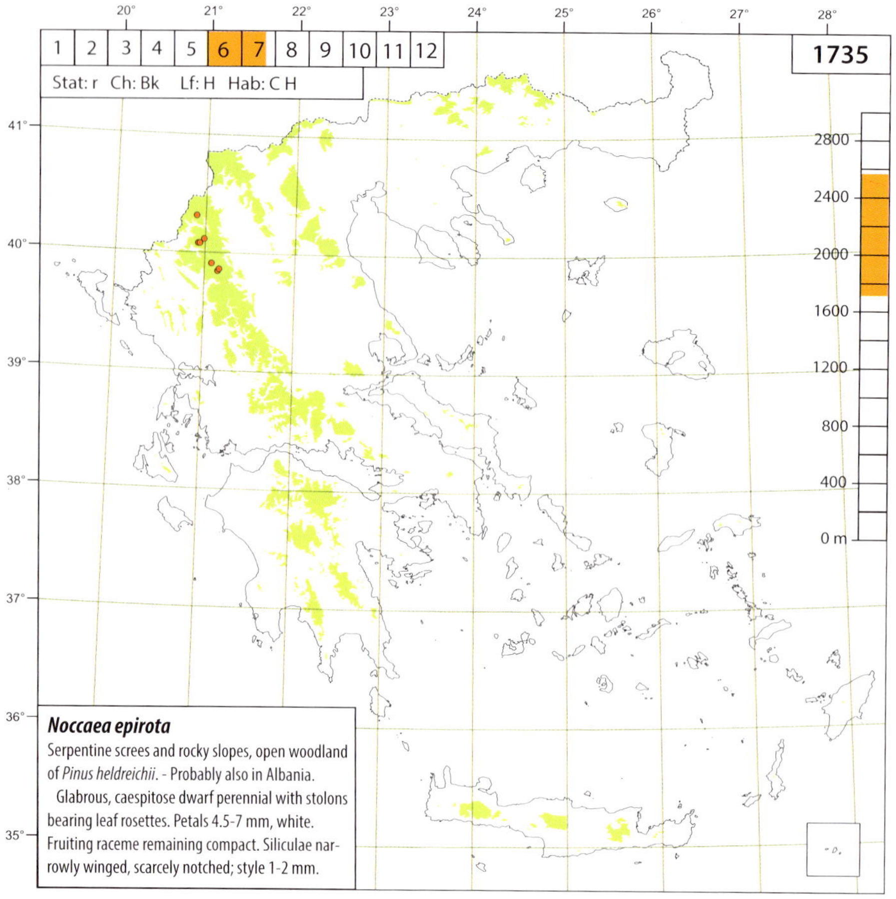

Noccaea epirota
Serpentine screes and rocky slopes, open woodland of *Pinus heldreichii*. - Probably also in Albania.
Glabrous, caespitose dwarf perennial with stolons bearing leaf rosettes. Petals 4.5-7 mm, white. Fruiting raceme remaining compact. Siliculae narrowly winged, scarcely notched; style 1-2 mm.

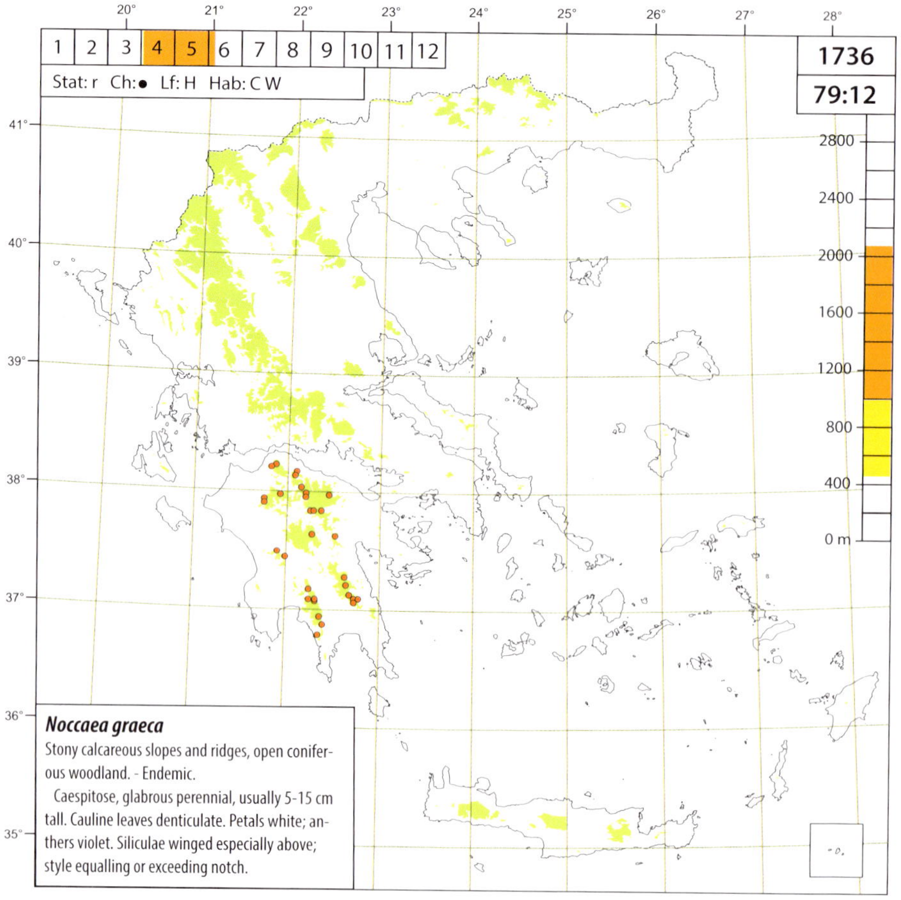

Noccaea graeca
Stony calcareous slopes and ridges, open coniferous woodland. - Endemic.
Caespitose, glabrous perennial, usually 5-15 cm tall. Cauline leaves denticulate. Petals white; anthers violet. Siliculae winged especially above; style equalling or exceeding notch.

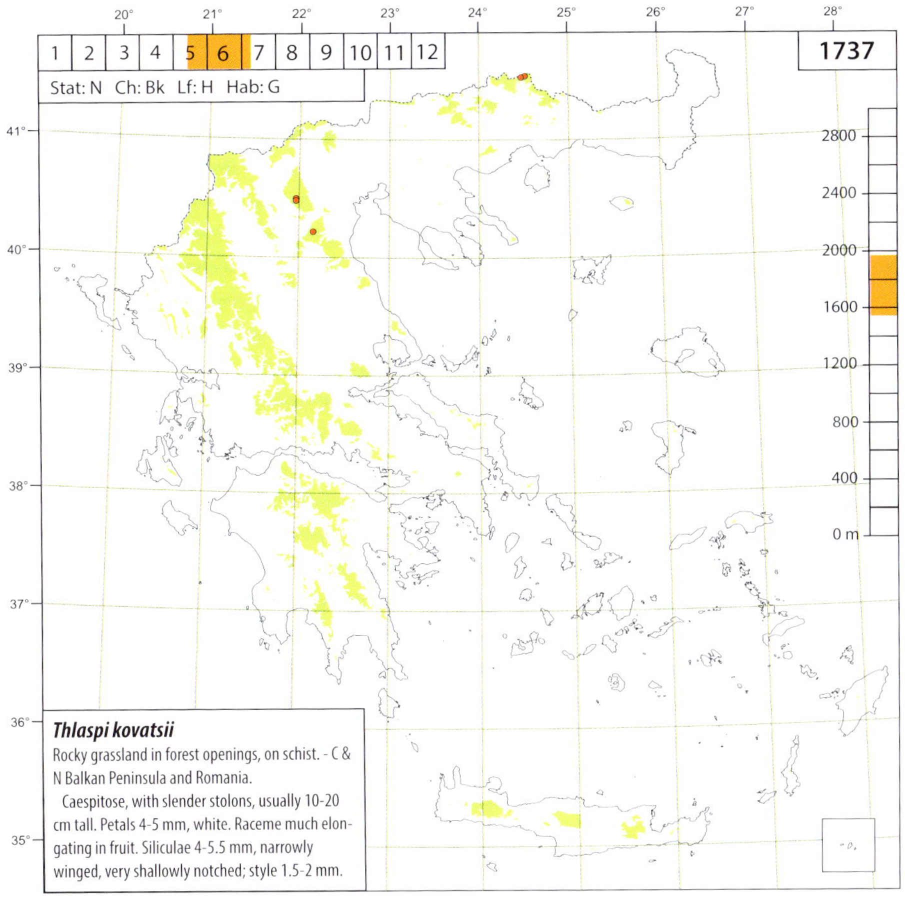

Thlaspi kovatsii
Rocky grassland in forest openings, on schist. - C & N Balkan Peninsula and Romania.
Caespitose, with slender stolons, usually 10-20 cm tall. Petals 4-5 mm, white. Raceme much elongating in fruit. Siliculae 4-5.5 mm, narrowly winged, very shallowly notched; style 1.5-2 mm.

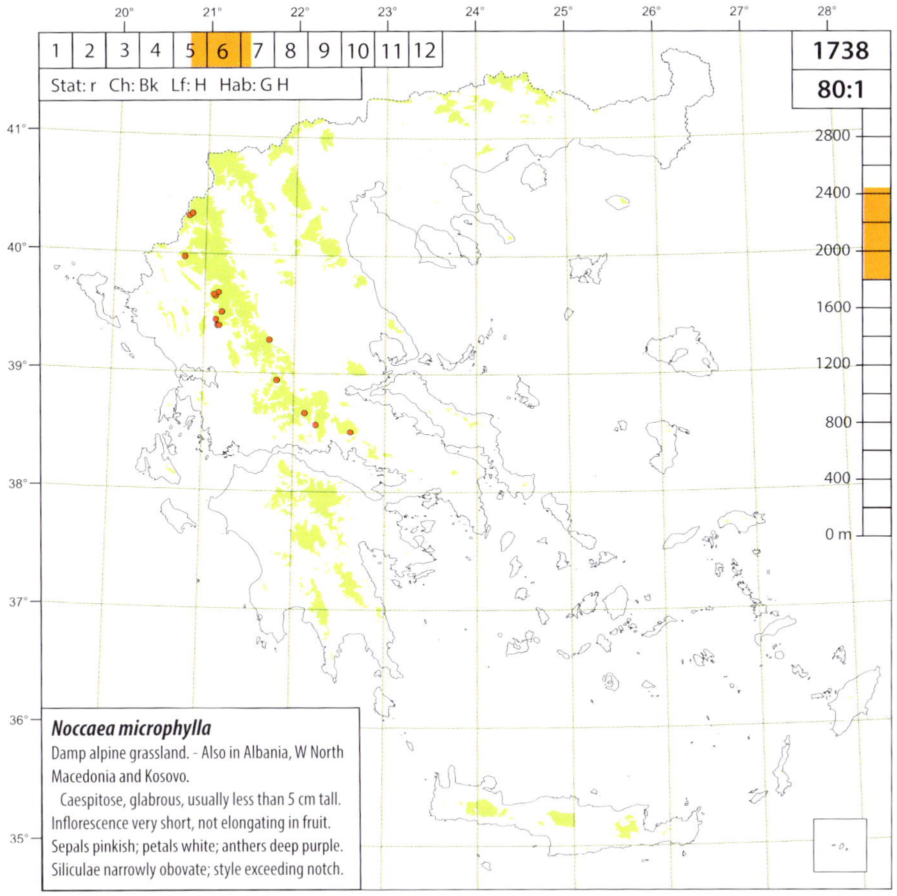

Noccaea microphylla
Damp alpine grassland. - Also in Albania, W North Macedonia and Kosovo.
Caespitose, glabrous, usually less than 5 cm tall. Inflorescence very short, not elongating in fruit. Sepals pinkish; petals white; anthers deep purple. Siliculae narrowly obovate; style exceeding notch.

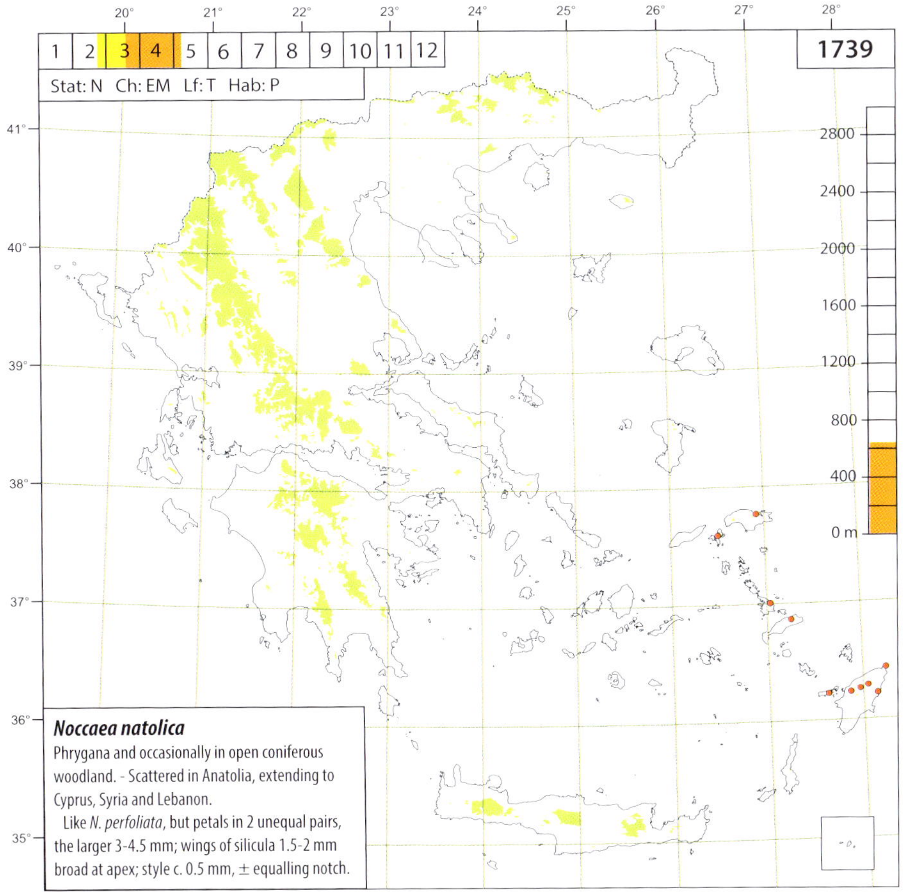

Noccaea natolica
Phrygana and occasionally in open coniferous woodland. - Scattered in Anatolia, extending to Cyprus, Syria and Lebanon.
Like *N. perfoliata*, but petals in 2 unequal pairs, the larger 3-4.5 mm; wings of silicula 1.5-2 mm broad at apex; style c. 0.5 mm, ± equalling notch.

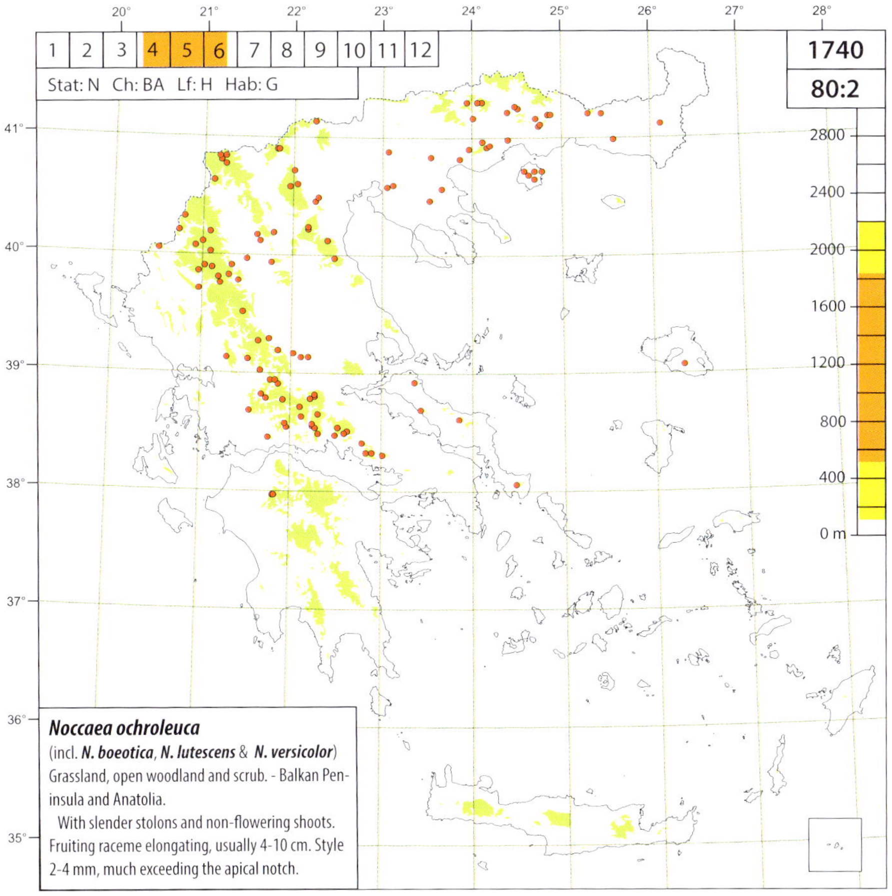

Noccaea ochroleuca
(incl. ***N. boeotica***, ***N. lutescens*** & ***N. versicolor***)
Grassland, open woodland and scrub. - Balkan Peninsula and Anatolia.
With slender stolons and non-flowering shoots. Fruiting raceme elongating, usually 4-10 cm. Style 2-4 mm, much exceeding the apical notch.

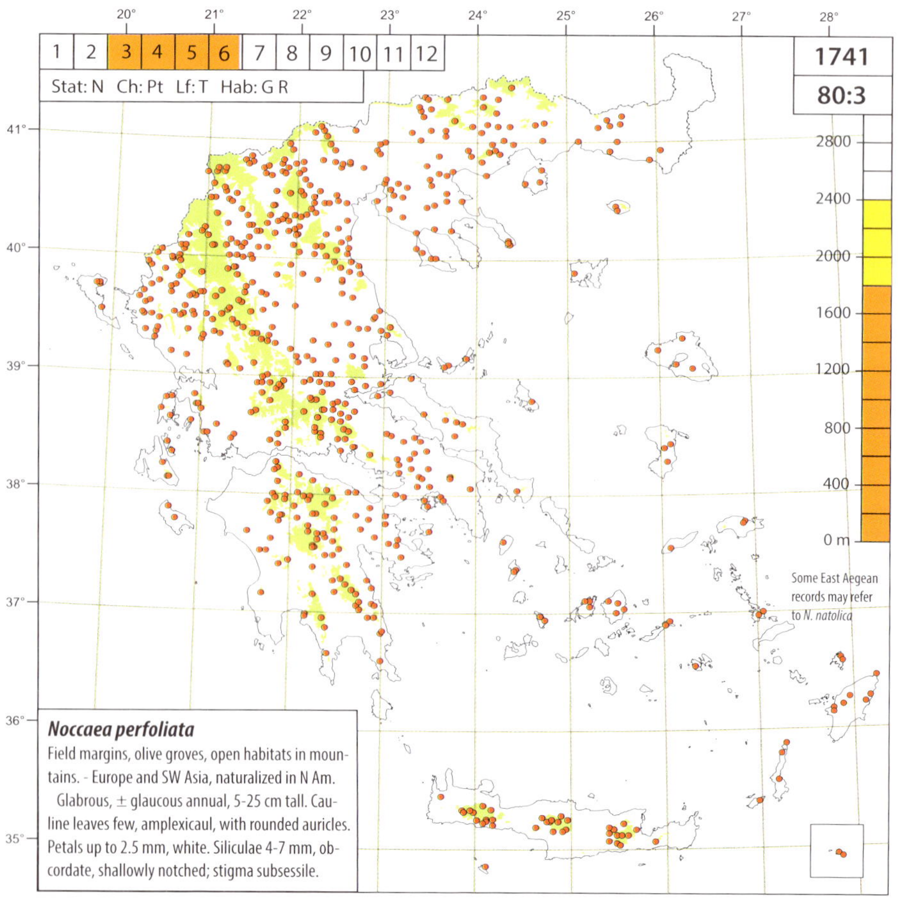

Noccaea perfoliata
Field margins, olive groves, open habitats in mountains. - Europe and SW Asia, naturalized in N Am.
Glabrous, ± glaucous annual, 5-25 cm tall. Cauline leaves few, amplexicaul, with rounded auricles. Petals up to 2.5 mm, white. Siliculae 4-7 mm, obcordate, shallowly notched; stigma subsessile.

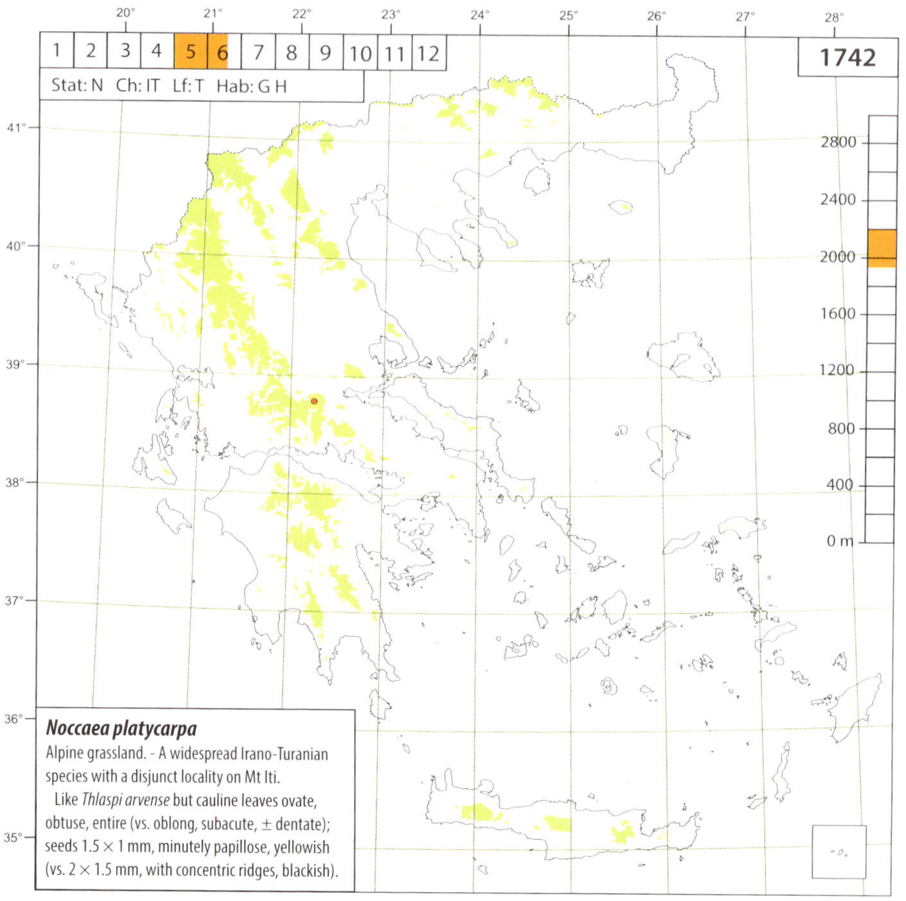

Noccaea platycarpa
Alpine grassland. - A widespread Irano-Turanian species with a disjunct locality on Mt Iti.
Like *Thlaspi arvense* but cauline leaves ovate, obtuse, entire (vs. oblong, subacute, ± dentate); seeds 1.5 × 1 mm, minutely papillose, yellowish (vs. 2 × 1.5 mm, with concentric ridges, blackish).

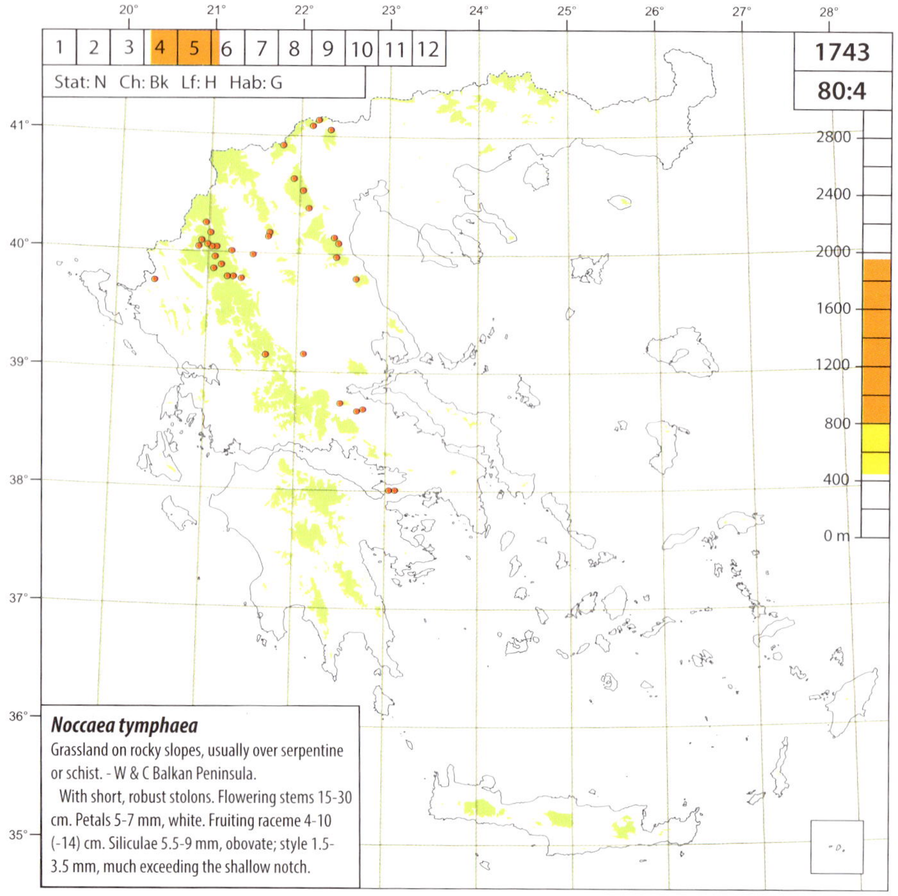

Noccaea tymphaea
Grassland on rocky slopes, usually over serpentine or schist. - W & C Balkan Peninsula.
With short, robust stolons. Flowering stems 15-30 cm. Petals 5-7 mm, white. Fruiting raceme 4-10 (-14) cm. Siliculae 5.5-9 mm, obovate; style 1.5-3.5 mm, much exceeding the shallow notch.

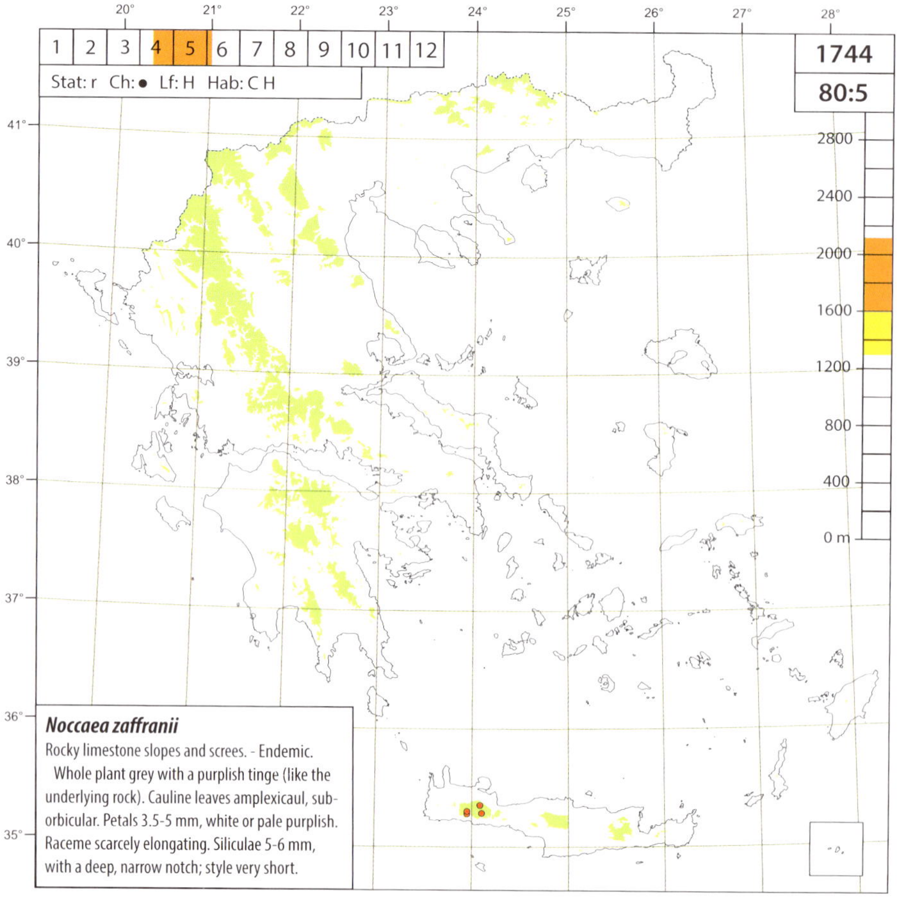

Noccaea zaffranii
Rocky limestone slopes and screes. - Endemic.
Whole plant grey with a purplish tinge (like the underlying rock). Cauline leaves amplexicaul, suborbicular. Petals 3.5-5 mm, white or pale purplish. Raceme scarcely elongating. Siliculae 5-6 mm, with a deep, narrow notch; style very short.

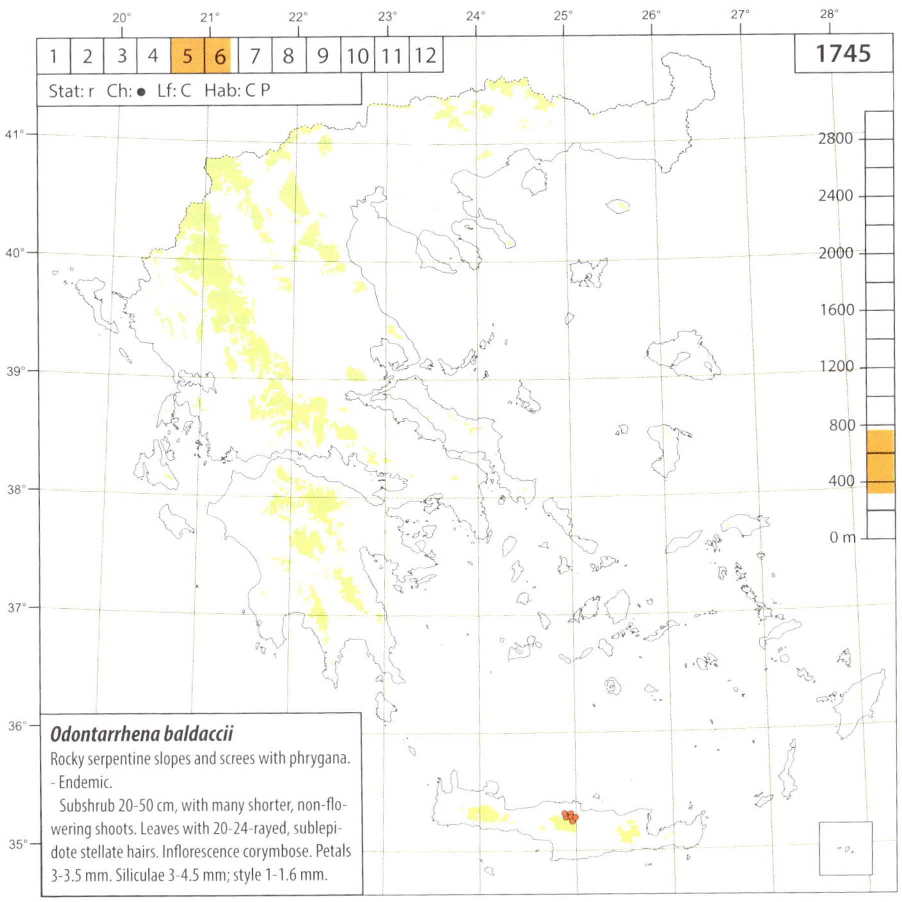

Odontarrhena baldaccii

Rocky serpentine slopes and screes with phrygana. - Endemic.

Subshrub 20-50 cm, with many shorter, non-flowering shoots. Leaves with 20-24-rayed, sublepidote stellate hairs. Inflorescence corymbose. Petals 3-3.5 mm. Siliculae 3-4.5 mm; style 1-1.6 mm.

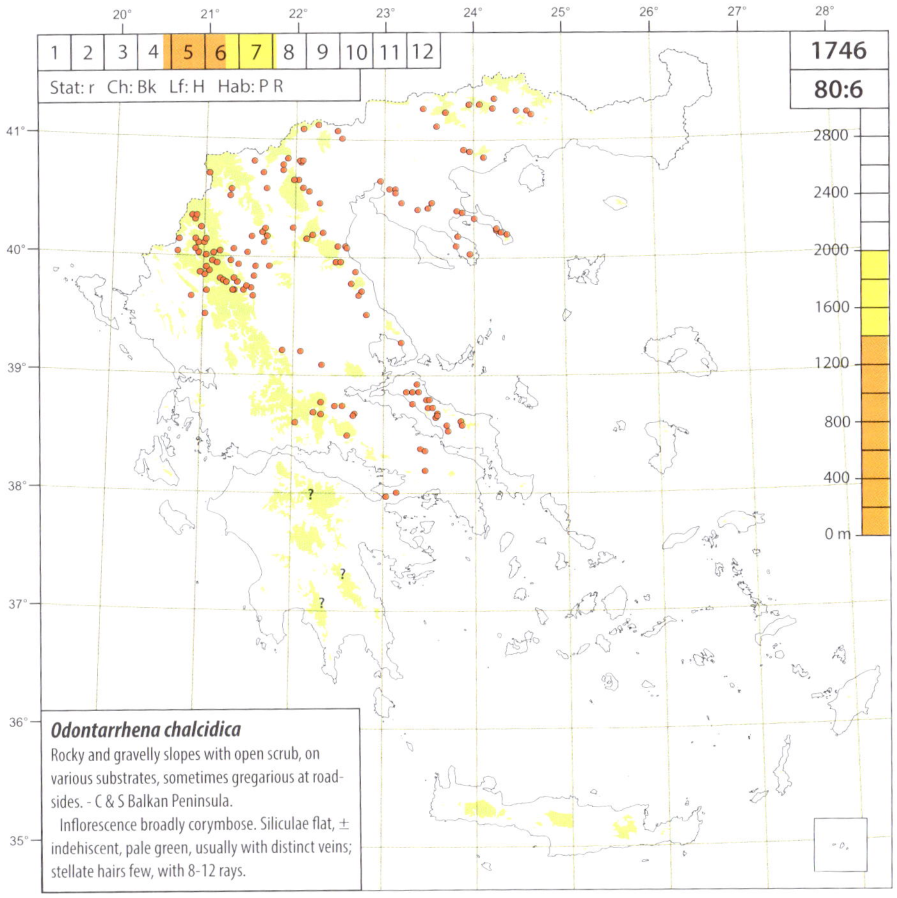

Odontarrhena chalcidica

Rocky and gravelly slopes with open scrub, on various substrates, sometimes gregarious at roadsides. - C & S Balkan Peninsula.

Inflorescence broadly corymbose. Siliculae flat, ± indehiscent, pale green, usually with distinct veins; stellate hairs few, with 8-12 rays.

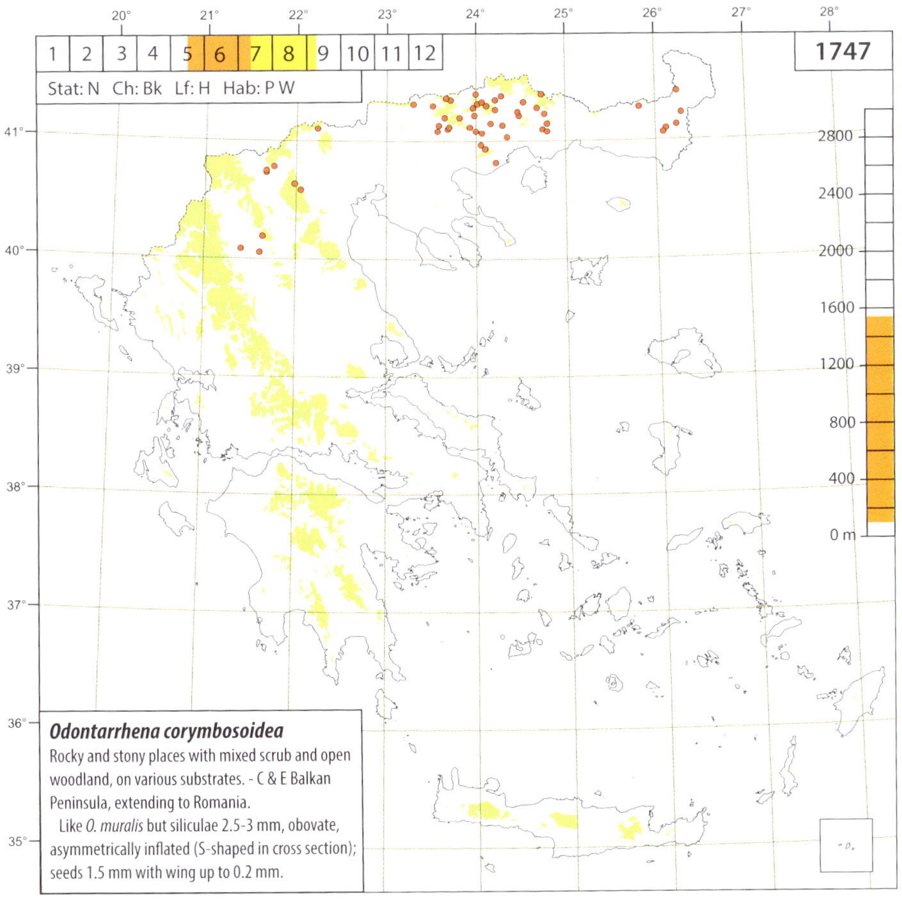

Odontarrhena corymbosoidea

Rocky and stony places with mixed scrub and open woodland, on various substrates. - C & E Balkan Peninsula, extending to Romania.

Like *O. muralis* but siliculae 2.5-3 mm, obovate, asymmetrically inflated (S-shaped in cross section); seeds 1.5 mm with wing up to 0.2 mm.

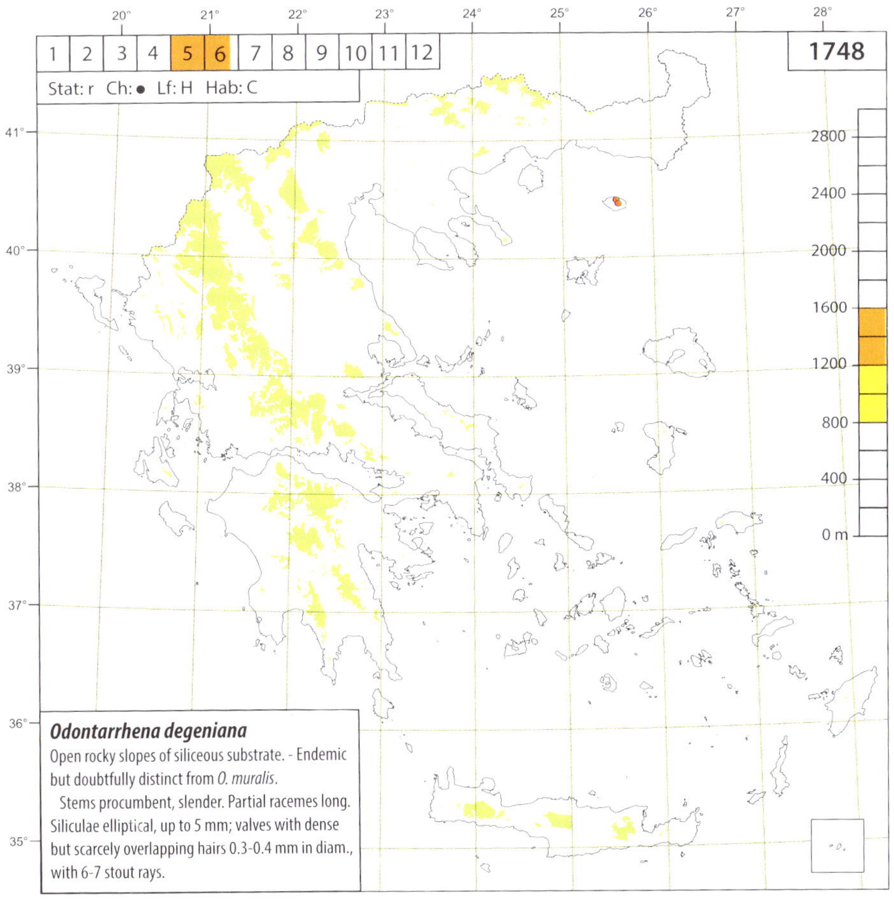

Odontarrhena degeniana

Open rocky slopes of siliceous substrate. - Endemic but doubtfully distinct from *O. muralis*.

Stems procumbent, slender. Partial racemes long. Siliculae elliptical, up to 5 mm; valves with dense but scarcely overlapping hairs 0.3-0.4 mm in diam., with 6-7 stout rays.

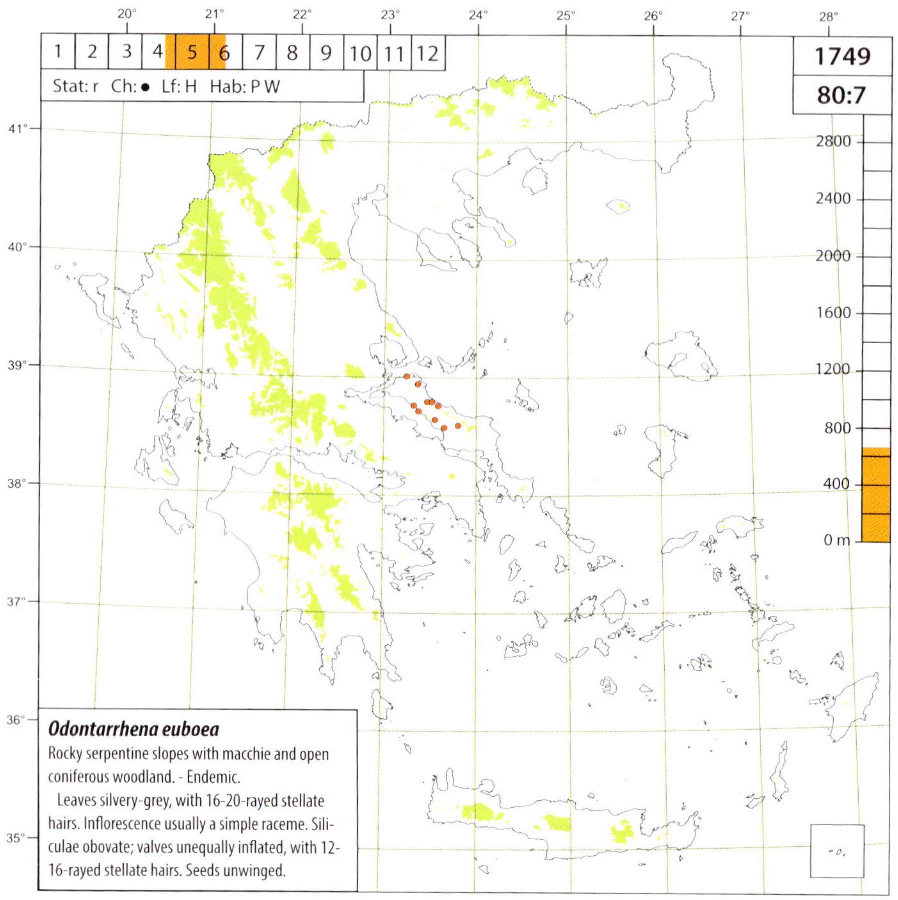

Odontarrhena euboea
Rocky serpentine slopes with macchie and open coniferous woodland. - Endemic.
Leaves silvery-grey, with 16-20-rayed stellate hairs. Inflorescence usually a simple raceme. Siliculae obovate; valves unequally inflated, with 12-16-rayed stellate hairs. Seeds unwinged.

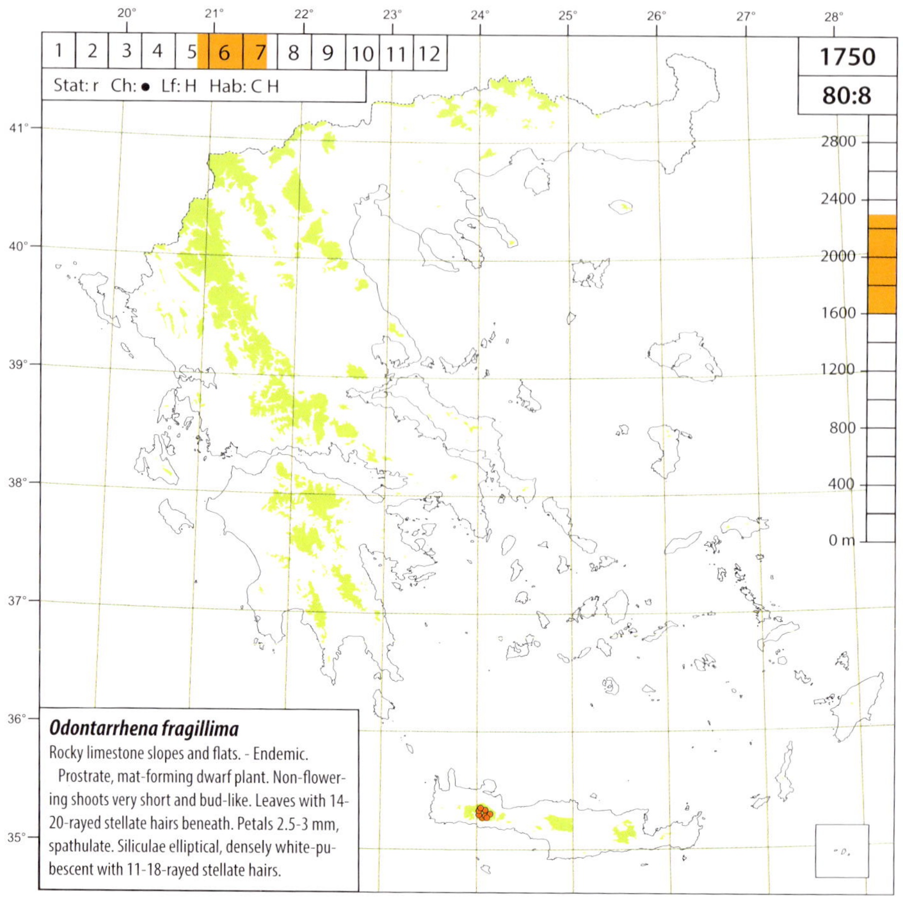

Odontarrhena fragillima
Rocky limestone slopes and flats. - Endemic.
Prostrate, mat-forming dwarf plant. Non-flowering shoots very short and bud-like. Leaves with 14-20-rayed stellate hairs beneath. Petals 2.5-3 mm, spathulate. Siliculae elliptical, densely white-pubescent with 11-18-rayed stellate hairs.

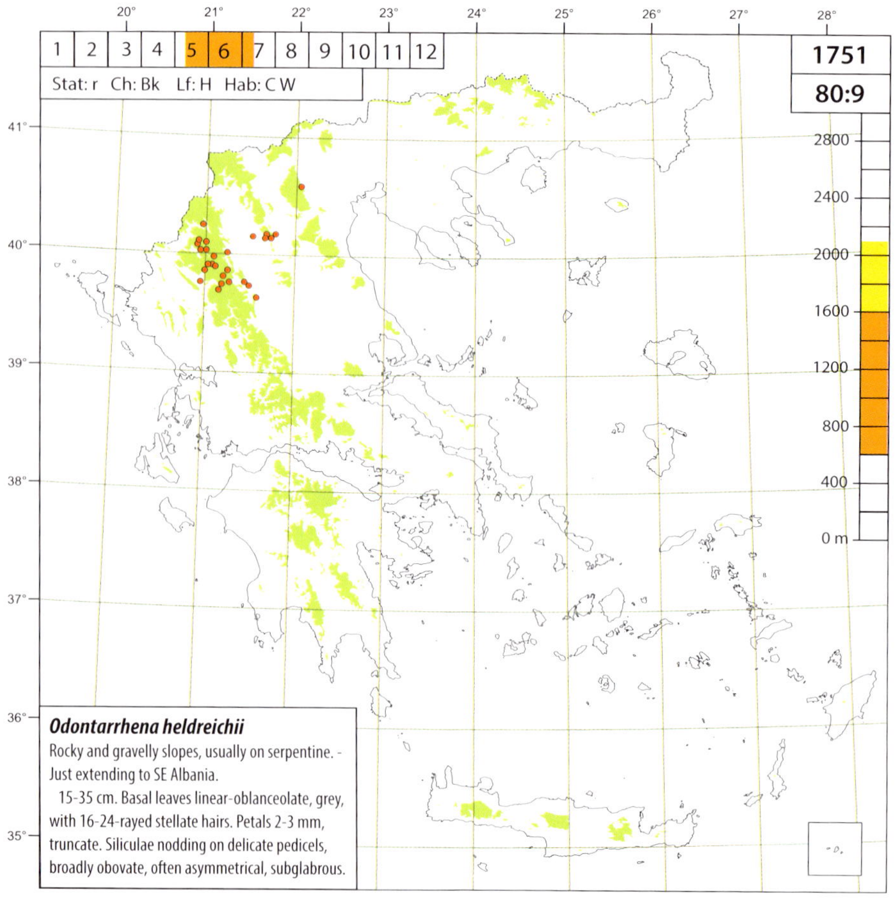

Odontarrhena heldreichii
Rocky and gravelly slopes, usually on serpentine. - Just extending to SE Albania.
15-35 cm. Basal leaves linear-oblanceolate, grey, with 16-24-rayed stellate hairs. Petals 2-3 mm, truncate. Siliculae nodding on delicate pedicels, broadly obovate, often asymmetrical, subglabrous.

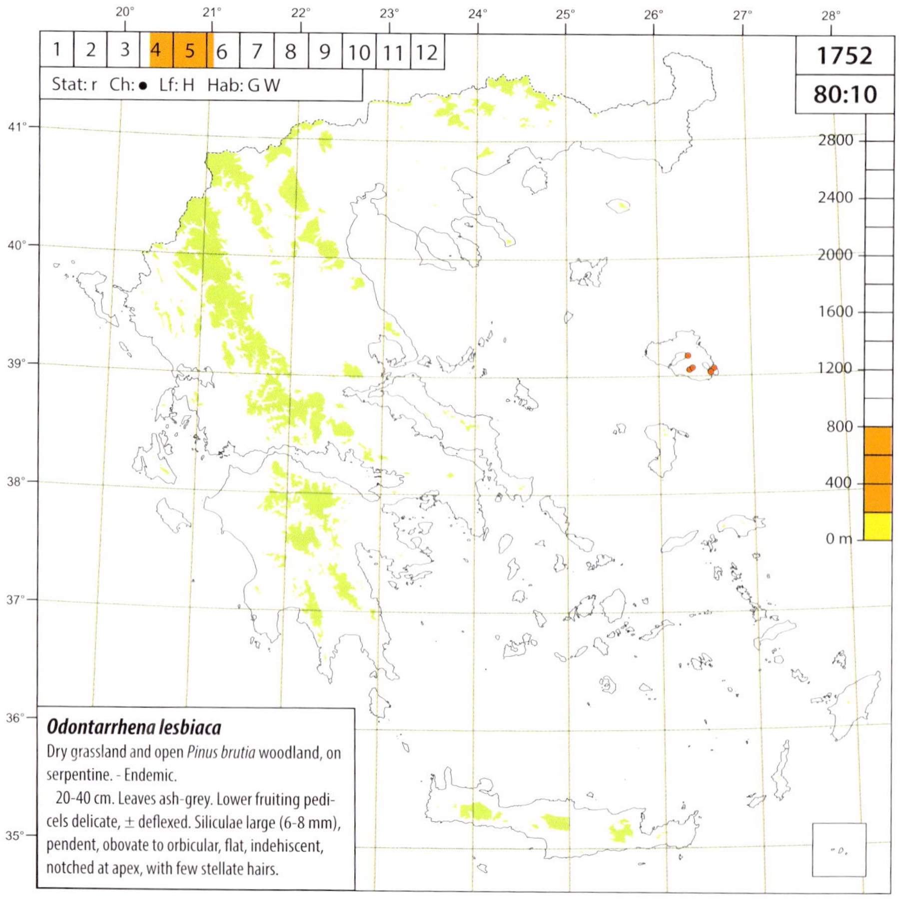

Odontarrhena lesbiaca
Dry grassland and open *Pinus brutia* woodland, on serpentine. - Endemic.
20-40 cm. Leaves ash-grey. Lower fruiting pedicels delicate, ± deflexed. Siliculae large (6-8 mm), pendent, obovate to orbicular, flat, indehiscent, notched at apex, with few stellate hairs.

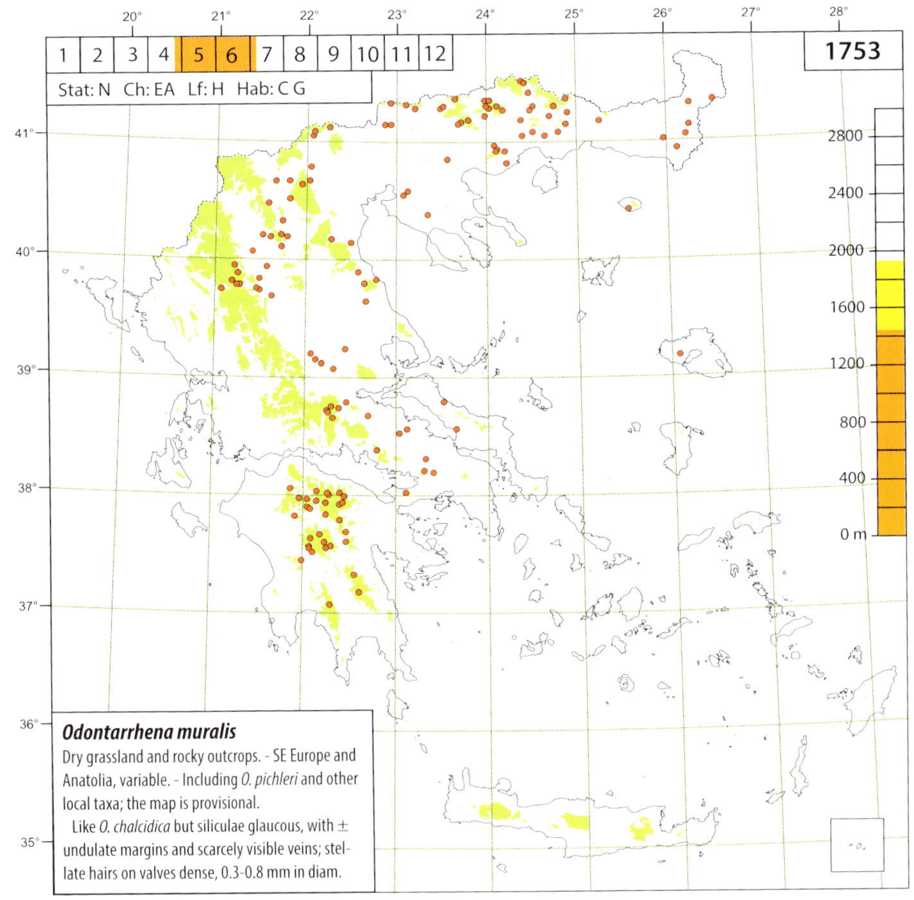

Odontarrhena muralis
Dry grassland and rocky outcrops. - SE Europe and Anatolia, variable. - Including *O. pichleri* and other local taxa; the map is provisional.
Like *O. chalcidica* but siliculae glaucous, with ± undulate margins and scarcely visible veins; stellate hairs on valves dense, 0.3-0.8 mm in diam.

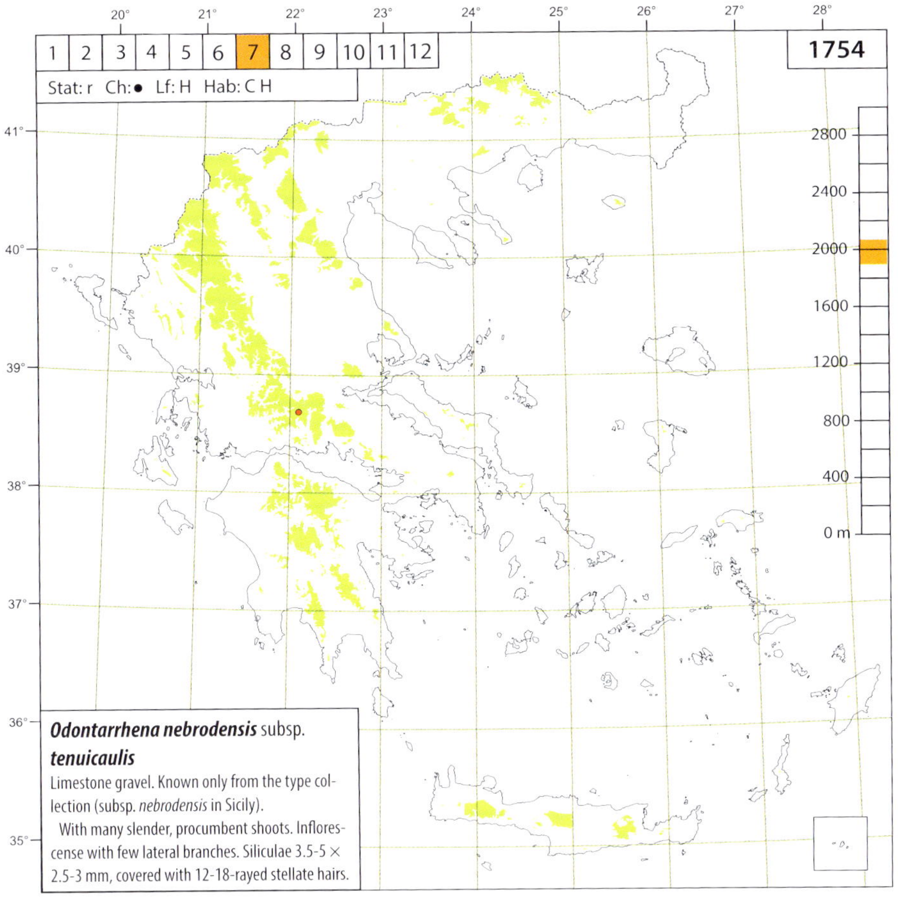

Odontarrhena nebrodensis **subsp. *tenuicaulis***
Limestone gravel. Known only from the type collection (subsp. *nebrodensis* in Sicily).
With many slender, procumbent shoots. Inflorescence with few lateral branches. Siliculae 3.5-5 × 2.5-3 mm, covered with 12-18-rayed stellate hairs.

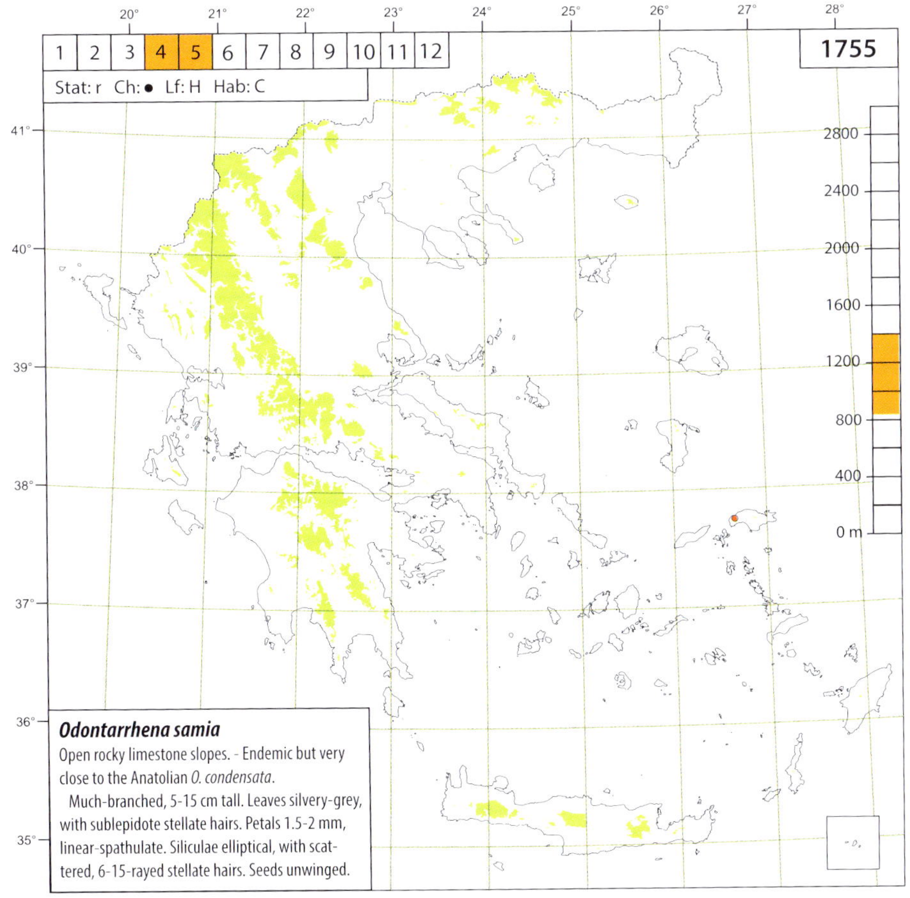

Odontarrhena samia
Open rocky limestone slopes. - Endemic but very close to the Anatolian *O. condensata*.
Much-branched, 5-15 cm tall. Leaves silvery-grey, with sublepidote stellate hairs. Petals 1.5-2 mm, linear-spathulate. Siliculae elliptical, with scattered, 6-15-rayed stellate hairs. Seeds unwinged.

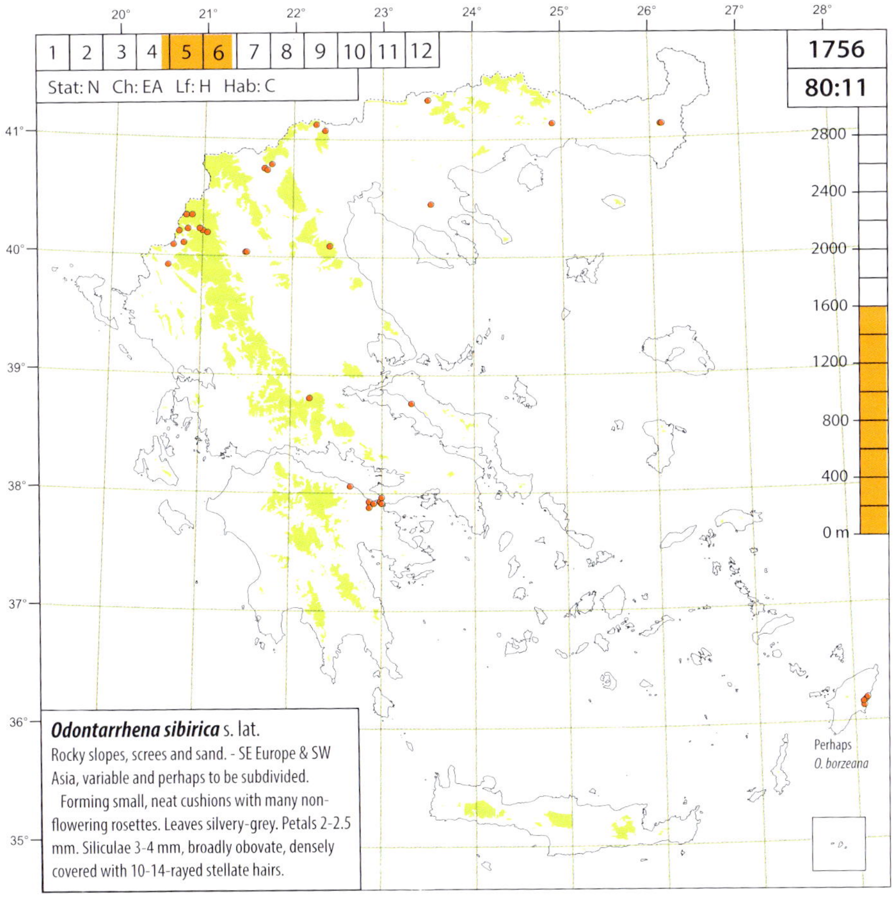

Odontarrhena sibirica s. lat.
Rocky slopes, screes and sand. - SE Europe & SW Asia, variable and perhaps to be subdivided.
Forming small, neat cushions with many non-flowering rosettes. Leaves silvery-grey. Petals 2-2.5 mm. Siliculae 3-4 mm, broadly obovate, densely covered with 10-14-rayed stellate hairs.

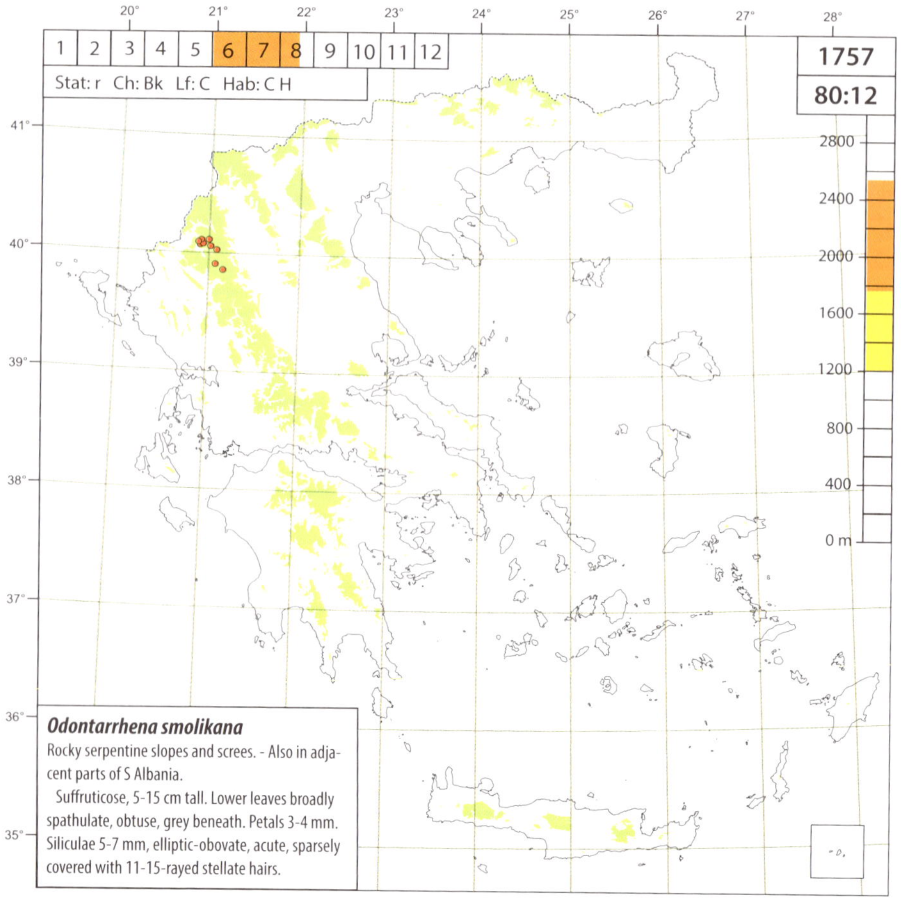

Odontarrhena smolikana

Rocky serpentine slopes and screes. - Also in adjacent parts of S Albania.

Suffruticose, 5-15 cm tall. Lower leaves broadly spathulate, obtuse, grey beneath. Petals 3-4 mm. Siliculae 5-7 mm, elliptic-obovate, acute, sparsely covered with 11-15-rayed stellate hairs.

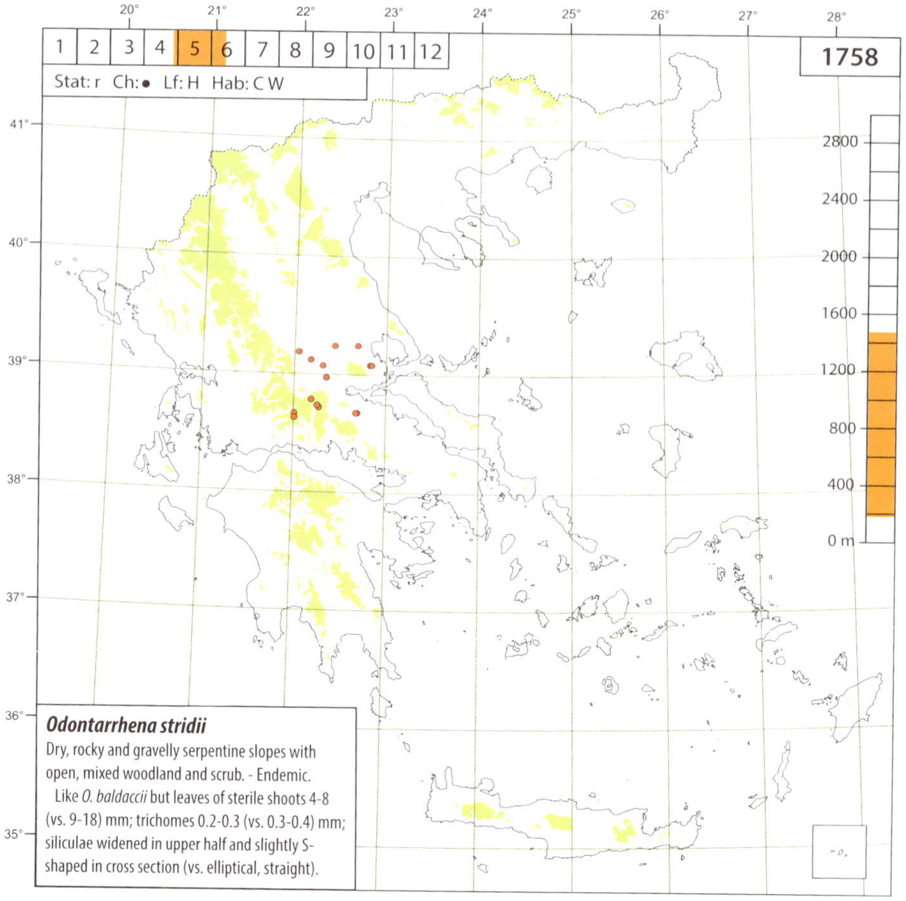

Odontarrhena stridii

Dry, rocky and gravelly serpentine slopes with open, mixed woodland and scrub. - Endemic.

Like *O. baldaccii* but leaves of sterile shoots 4-8 (vs. 9-18) mm; trichomes 0.2-0.3 (vs. 0.3-0.4) mm; siliculae widened in upper half and slightly S-shaped in cross section (vs. elliptical, straight).

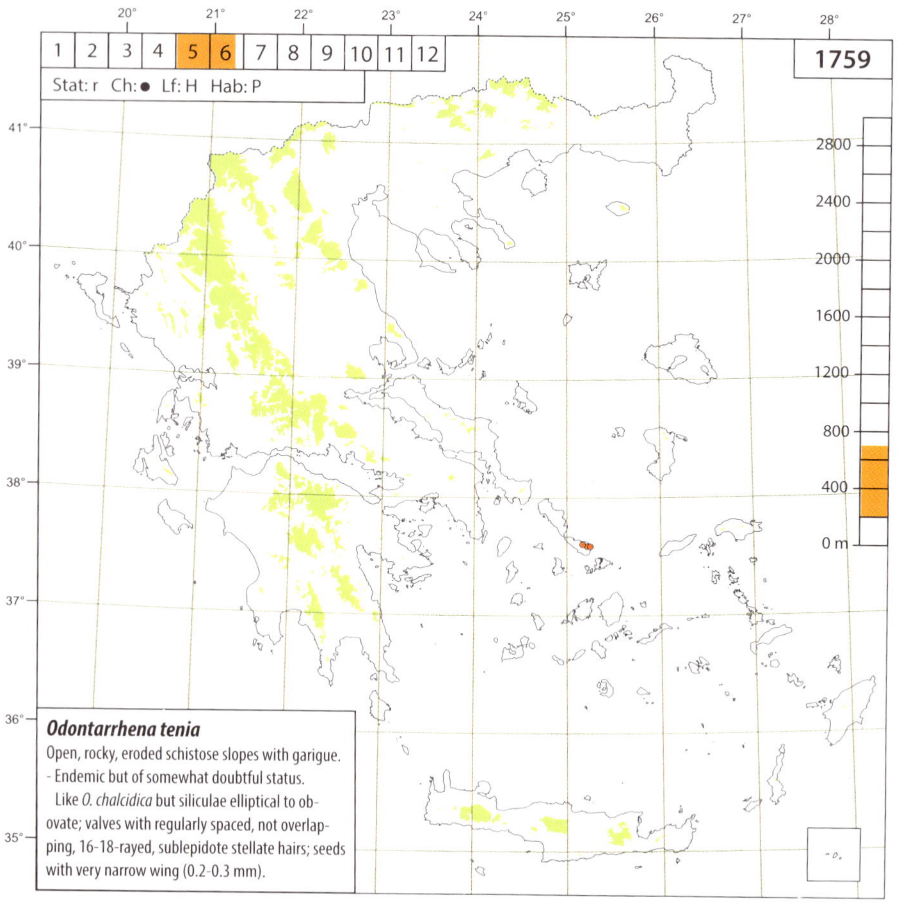

Odontarrhena tenia

Open, rocky, eroded schistose slopes with garigue. - Endemic but of somewhat doubtful status.

Like *O. chalcidica* but siliculae elliptical to obovate; valves with regularly spaced, not overlapping, 16-18-rayed, sublepidote stellate hairs; seeds with very narrow wing (0.2-0.3 mm).

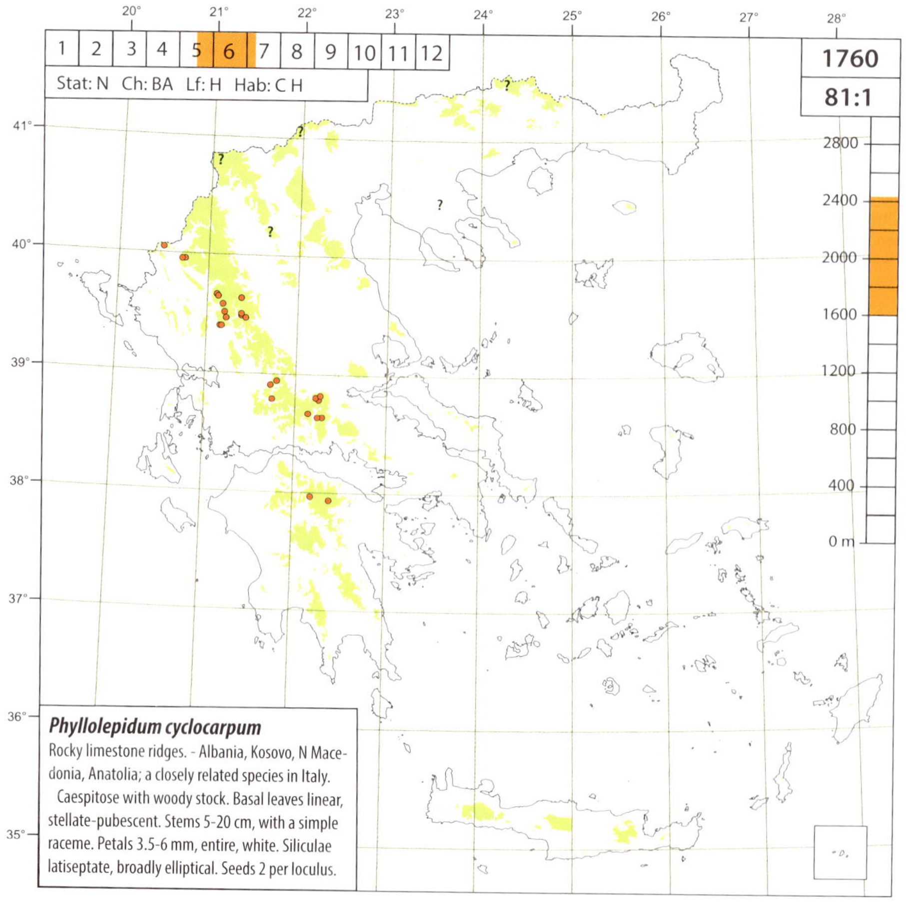

Phyllolepidum cyclocarpum

Rocky limestone ridges. - Albania, Kosovo, N Macedonia, Anatolia; a closely related species in Italy.

Caespitose with woody stock. Basal leaves linear, stellate-pubescent. Stems 5-20 cm, with a simple raceme. Petals 3.5-6 mm, entire, white. Siliculae latiseptate, broadly elliptical. Seeds 2 per loculus.

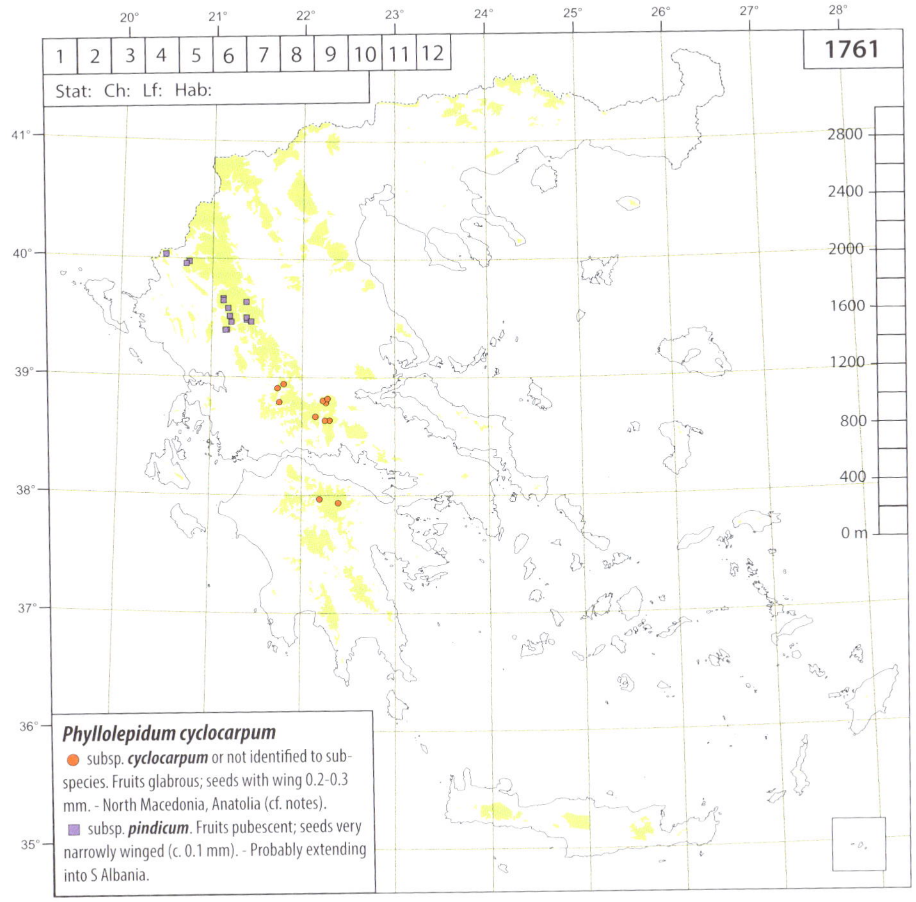

Phyllolepidum cyclocarpum
● subsp. ***cyclocarpum*** or not identified to subspecies. Fruits glabrous; seeds with wing 0.2-0.3 mm. - North Macedonia, Anatolia (cf. notes).
■ subsp. ***pindicum***. Fruits pubescent; seeds very narrowly winged (c. 0.1 mm). - Probably extending into S Albania.

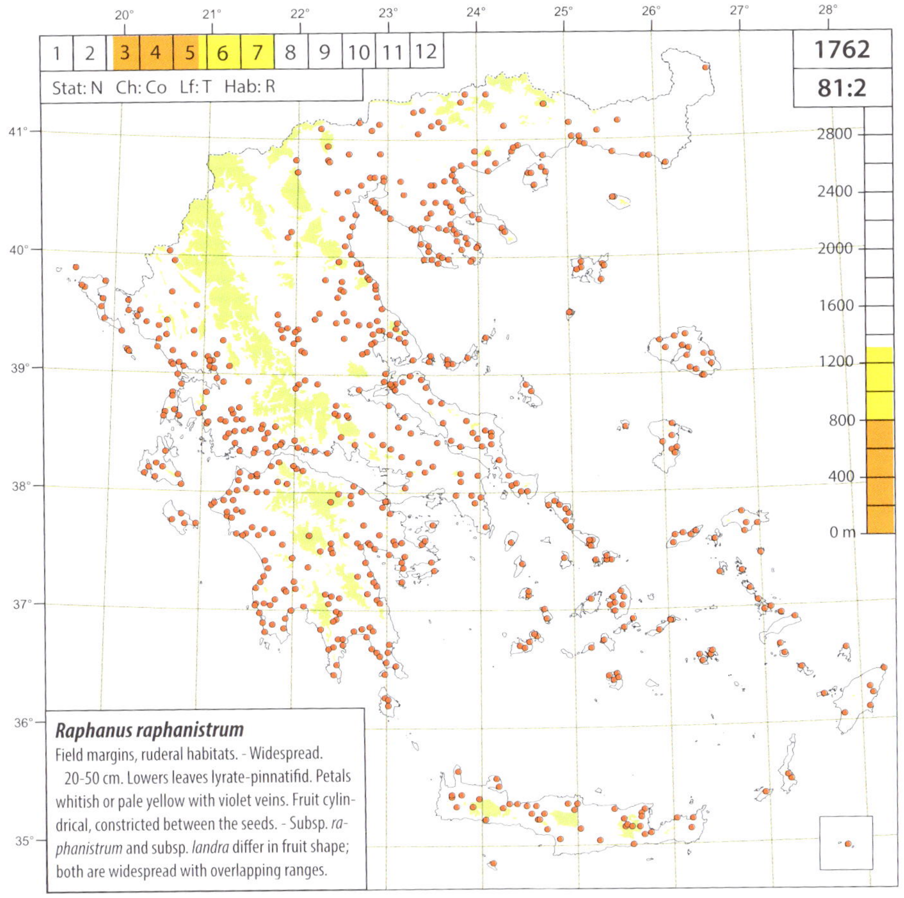

Raphanus raphanistrum
Field margins, ruderal habitats. - Widespread.
20-50 cm. Lowers leaves lyrate-pinnatifid. Petals whitish or pale yellow with violet veins. Fruit cylindrical, constricted between the seeds. - Subsp. *raphanistrum* and subsp. *landra* differ in fruit shape; both are widespread with overlapping ranges.

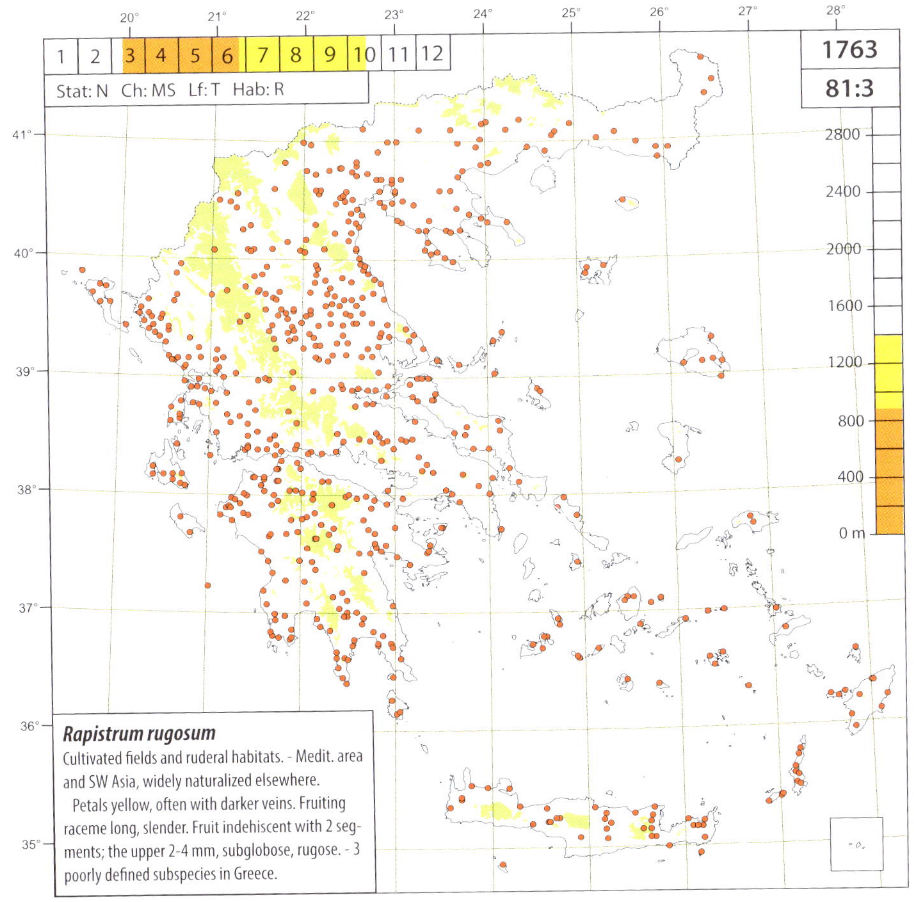

Rapistrum rugosum
Cultivated fields and ruderal habitats. - Medit. area and SW Asia, widely naturalized elsewhere.
Petals yellow, often with darker veins. Fruiting raceme long, slender. Fruit indehiscent with 2 segments; the upper 2-4 mm, subglobose, rugose. - 3 poorly defined subspecies in Greece.

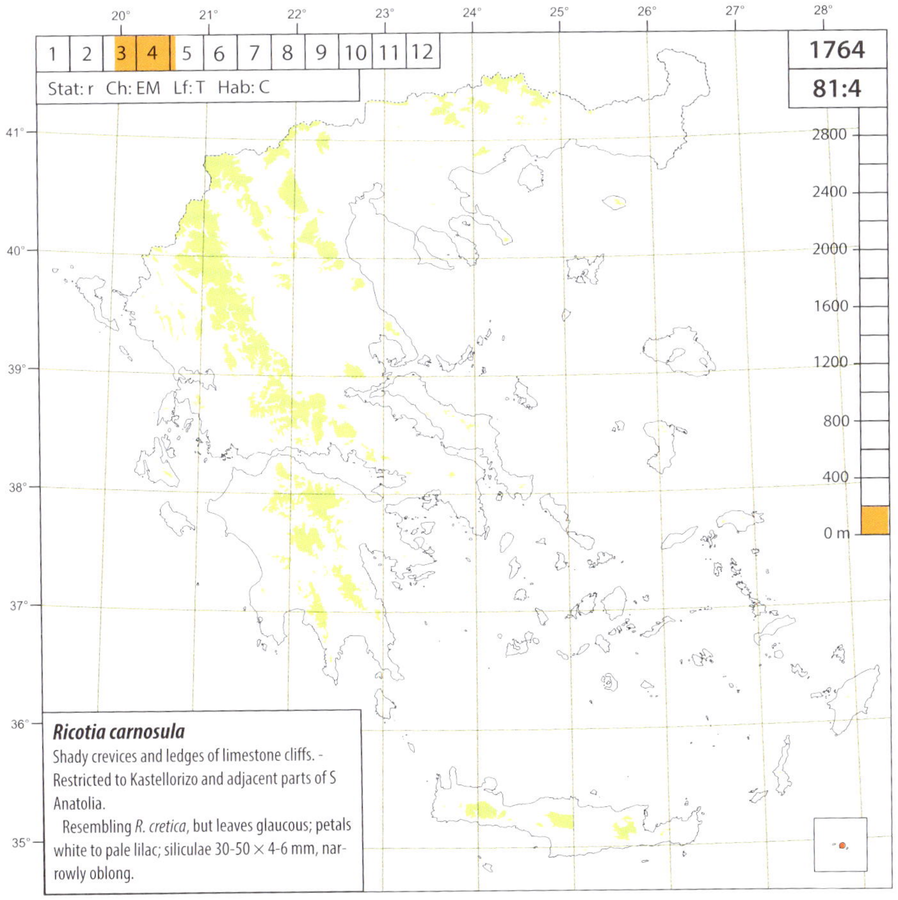

Ricotia carnosula
Shady crevices and ledges of limestone cliffs. - Restricted to Kastellorizo and adjacent parts of S Anatolia.
Resembling *R. cretica*, but leaves glaucous; petals white to pale lilac; siliculae 30-50 × 4-6 mm, narrowly oblong.

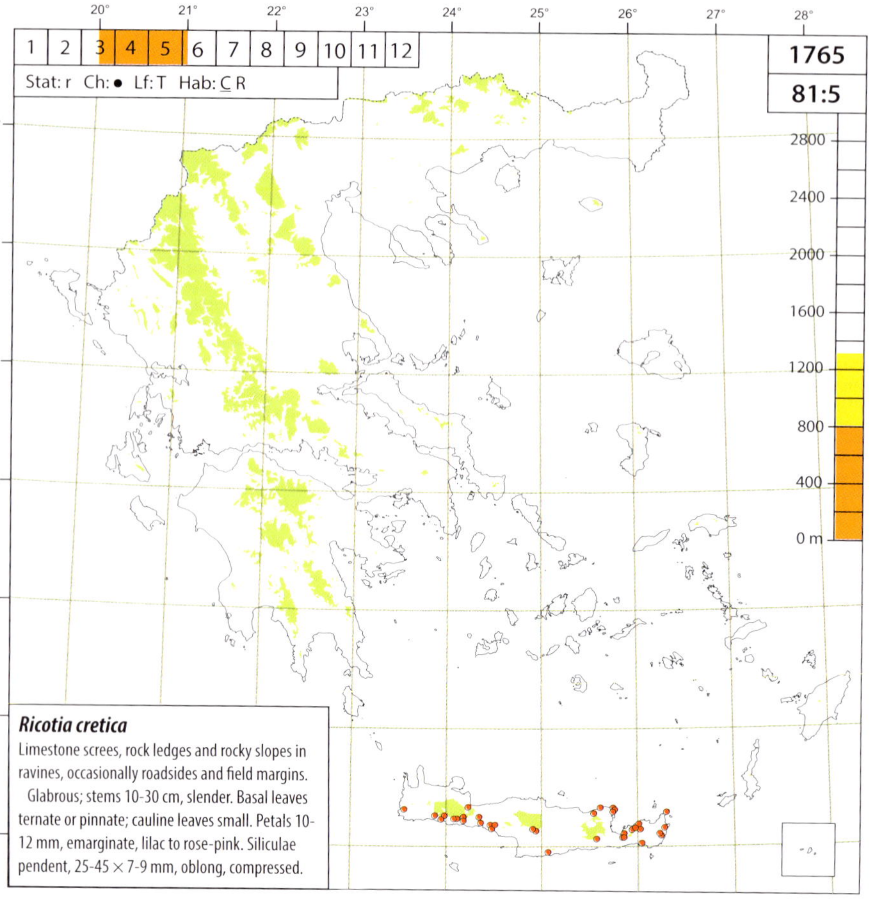

Ricotia cretica
Limestone screes, rock ledges and rocky slopes in ravines, occasionally roadsides and field margins.
Glabrous; stems 10-30 cm, slender. Basal leaves ternate or pinnate; cauline leaves small. Petals 10-12 mm, emarginate, lilac to rose-pink. Siliculae pendent, 25-45 × 7-9 mm, oblong, compressed.

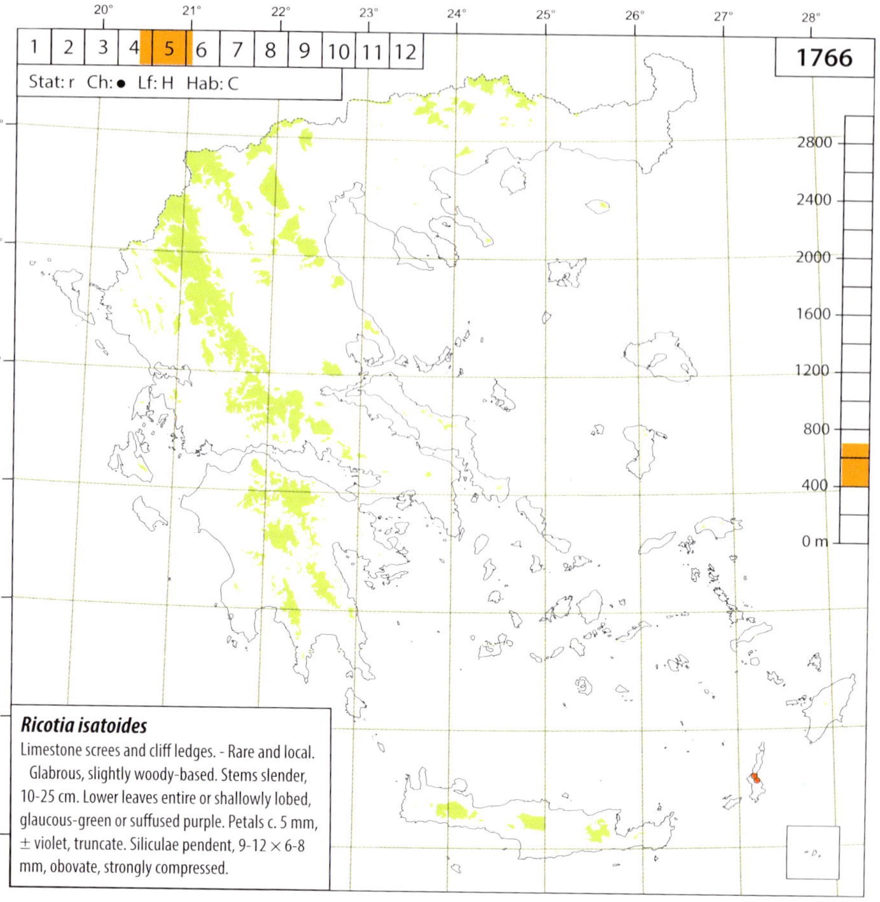

Ricotia isatoides
Limestone screes and cliff ledges. - Rare and local.
Glabrous, slightly woody-based. Stems slender, 10-25 cm. Lower leaves entire or shallowly lobed, glaucous-green or suffused purple. Petals c. 5 mm, ± violet, truncate. Siliculae pendent, 9-12 × 6-8 mm, obovate, strongly compressed.

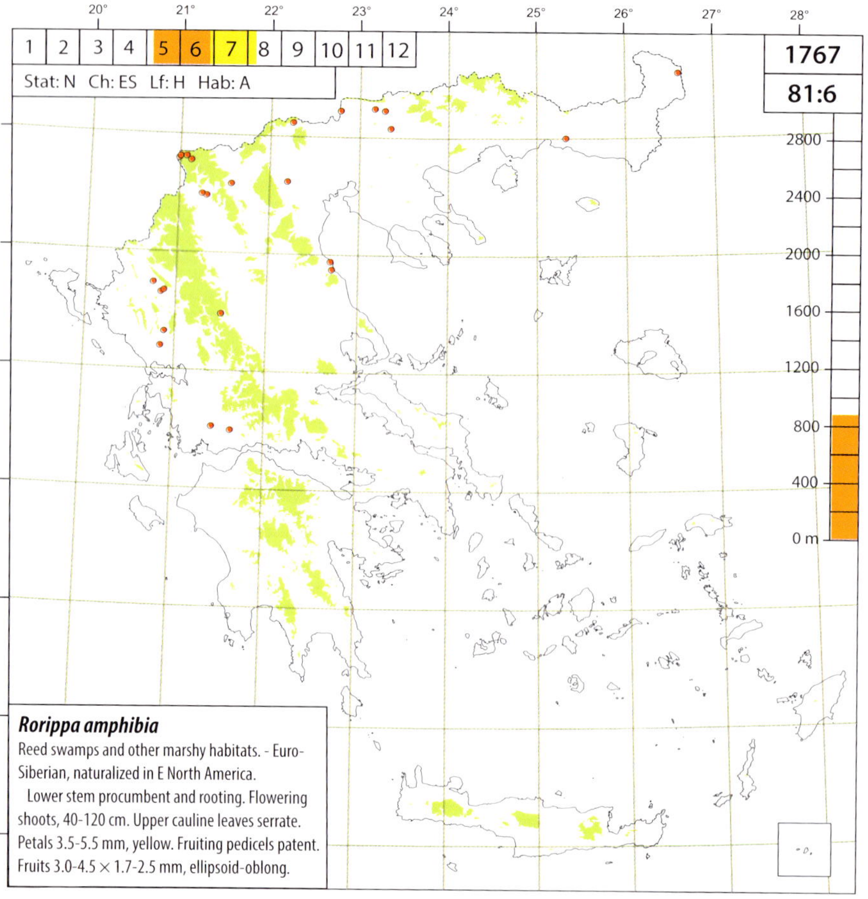

Rorippa amphibia
Reed swamps and other marshy habitats. - Euro-Siberian, naturalized in E North America.
Lower stem procumbent and rooting. Flowering shoots, 40-120 cm. Upper cauline leaves serrate. Petals 3.5-5.5 mm, yellow. Fruiting pedicels patent. Fruits 3.0-4.5 × 1.7-2.5 mm, ellipsoid-oblong.

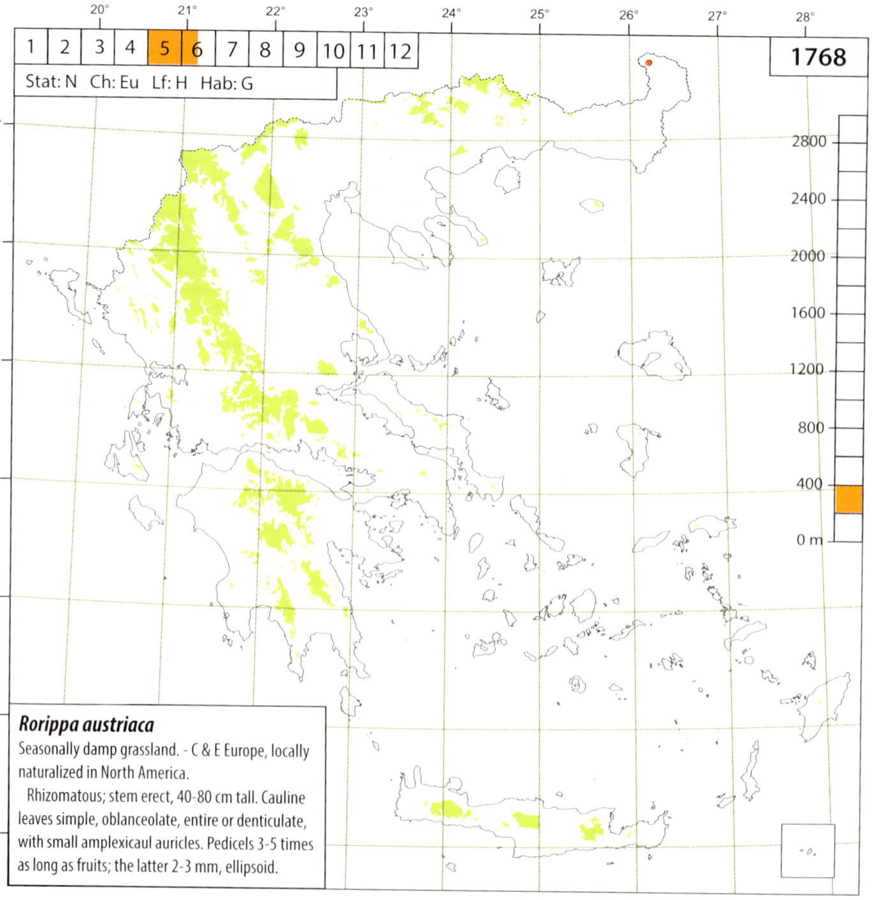

Rorippa austriaca
Seasonally damp grassland. - C & E Europe, locally naturalized in North America.
Rhizomatous; stem erect, 40-80 cm tall. Cauline leaves simple, oblanceolate, entire or denticulate, with small amplexicaul auricles. Pedicels 3-5 times as long as fruits; the latter 2-3 mm, ellipsoid.

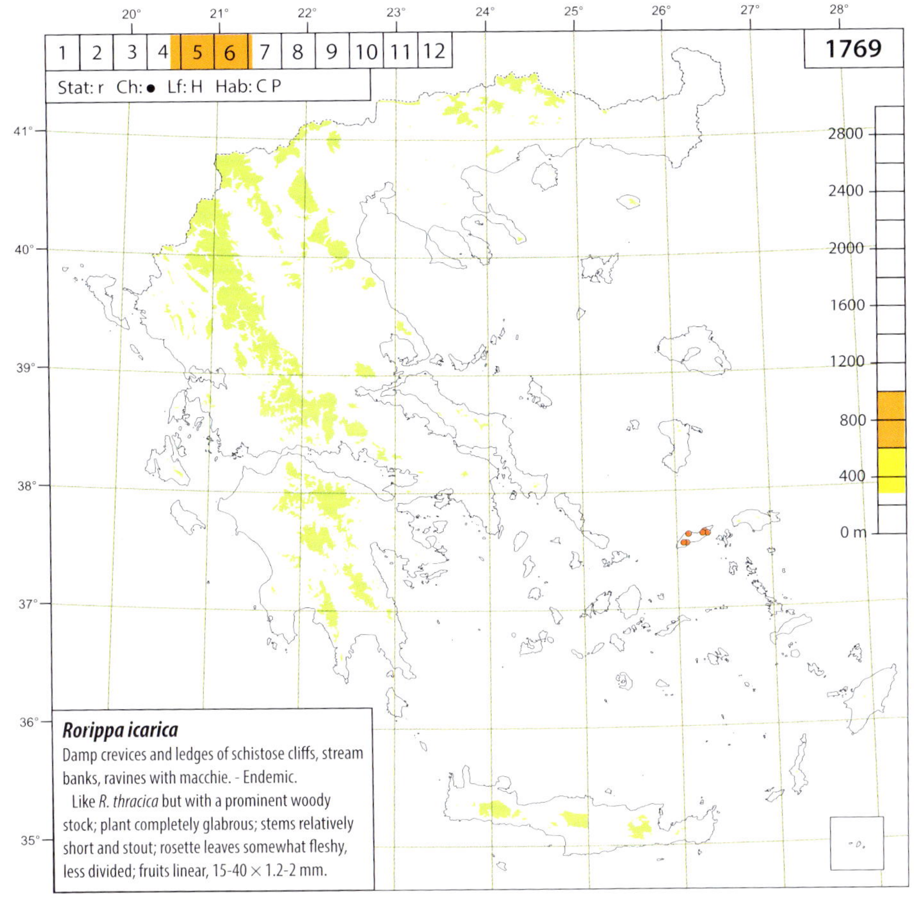

Stat: r Ch: ● Lf: H Hab: C P

Rorippa icarica

Damp crevices and ledges of schistose cliffs, stream banks, ravines with macchie. - Endemic.

Like *R. thracica* but with a prominent woody stock; plant completely glabrous; stems relatively short and stout; rosette leaves somewhat fleshy, less divided; fruits linear, 15-40 × 1.2-2 mm.

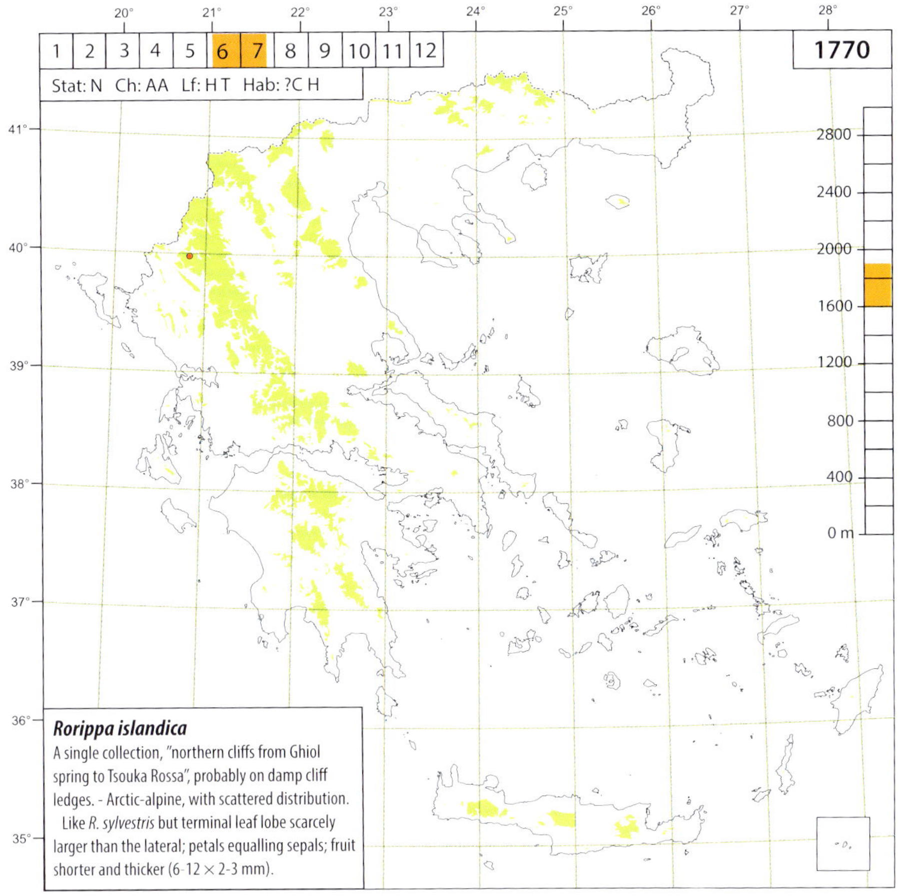

Stat: N Ch: AA Lf: H T Hab: ?C H

Rorippa islandica

A single collection, "northern cliffs from Ghiol spring to Tsouka Rossa", probably on damp cliff ledges. - Arctic-alpine, with scattered distribution.

Like *R. sylvestris* but terminal leaf lobe scarcely larger than the lateral; petals equalling sepals; fruit shorter and thicker (6-12 × 2-3 mm).

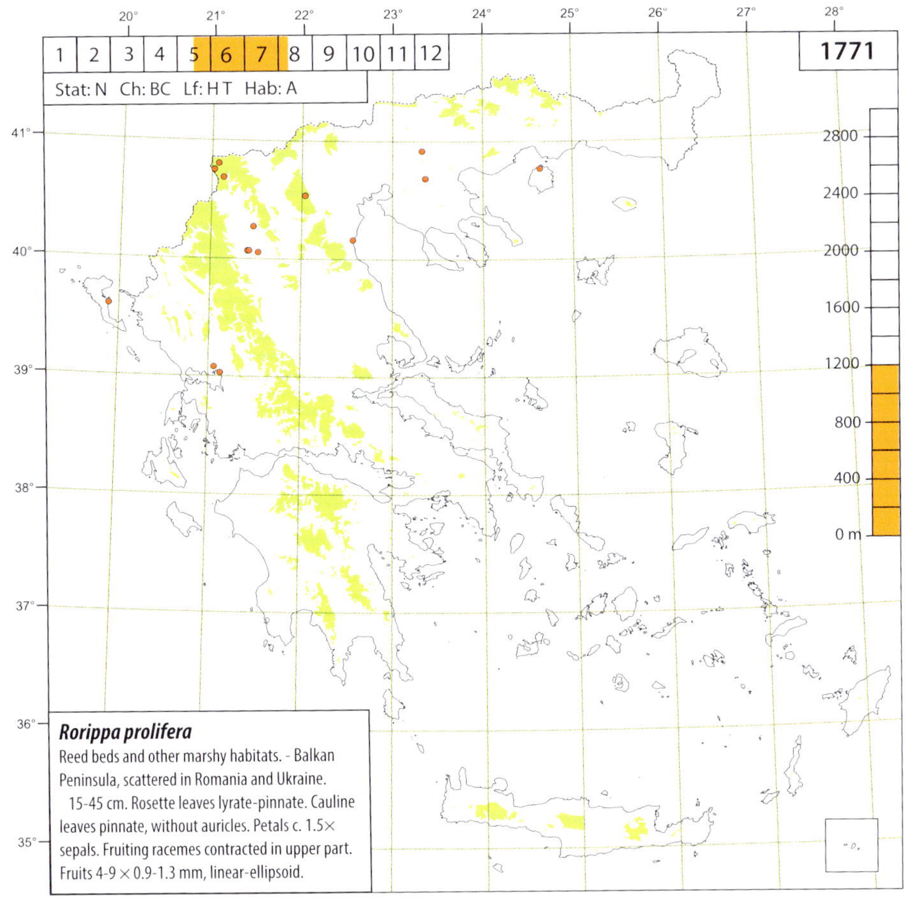

Stat: N Ch: BC Lf: H T Hab: A

Rorippa prolifera

Reed beds and other marshy habitats. - Balkan Peninsula, scattered in Romania and Ukraine.

15-45 cm. Rosette leaves lyrate-pinnate. Cauline leaves pinnate, without auricles. Petals c. 1.5× sepals. Fruiting racemes contracted in upper part. Fruits 4-9 × 0.9-1.3 mm, linear-ellipsoid.

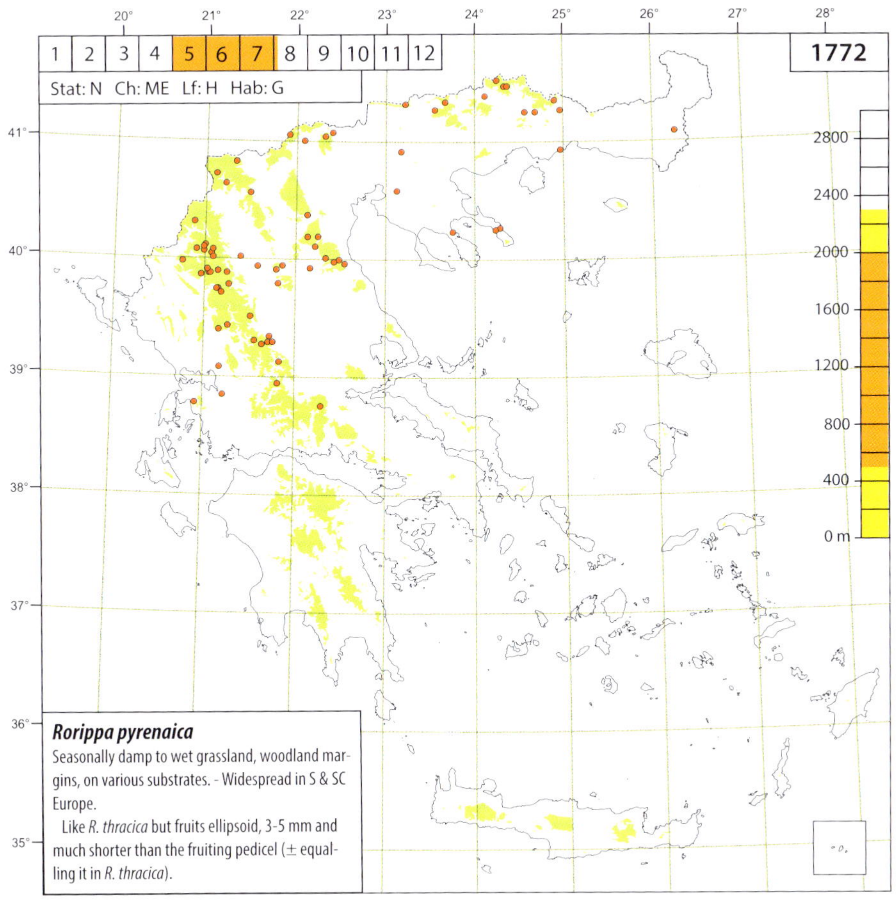

Stat: N Ch: ME Lf: H Hab: G

Rorippa pyrenaica

Seasonally damp to wet grassland, woodland margins, on various substrates. - Widespread in S & SC Europe.

Like *R. thracica* but fruits ellipsoid, 3-5 mm and much shorter than the fruiting pedicel (± equalling it in *R. thracica*).

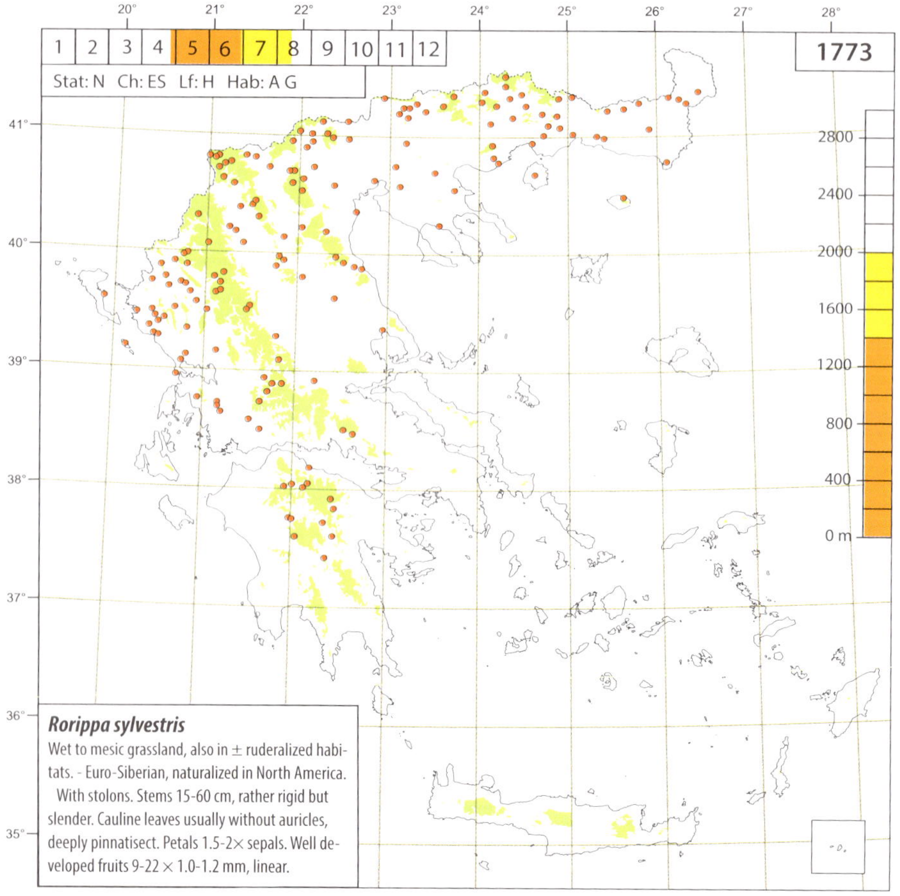

Rorippa sylvestris
Wet to mesic grassland, also in ± ruderalized habitats. - Euro-Siberian, naturalized in North America.
With stolons. Stems 15-60 cm, rather rigid but slender. Cauline leaves usually without auricles, deeply pinnatisect. Petals 1.5-2× sepals. Well developed fruits 9-22 × 1.0-1.2 mm, linear.

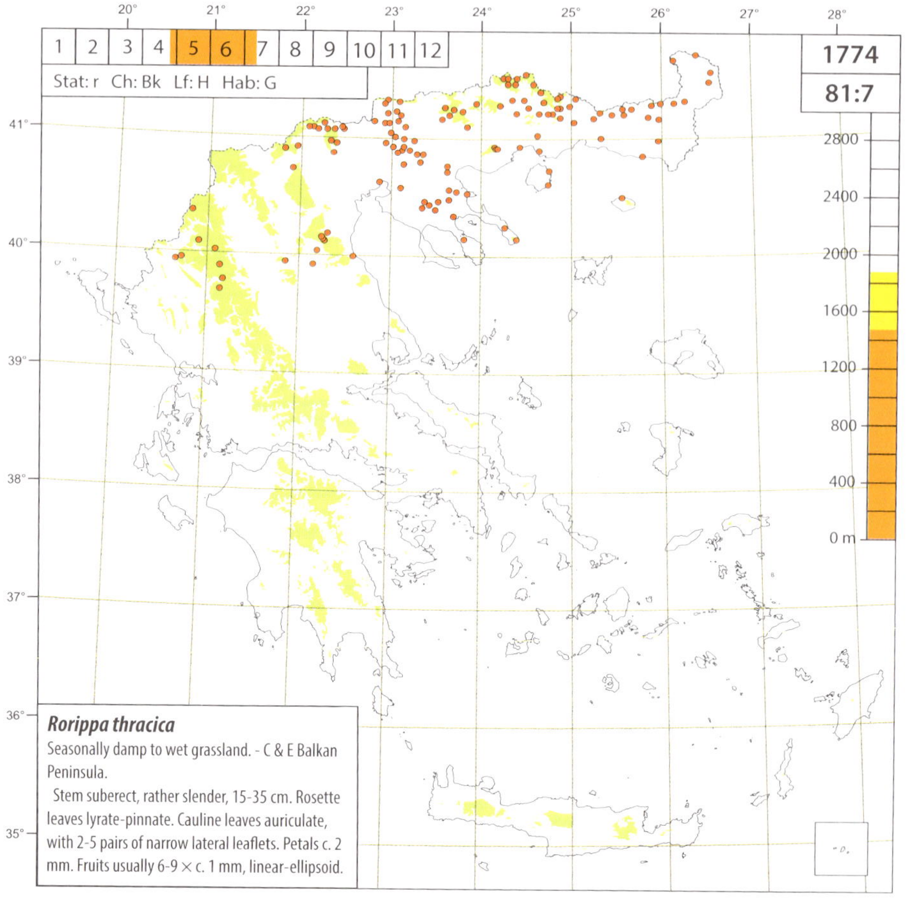

Rorippa thracica
Seasonally damp to wet grassland. - C & E Balkan Peninsula.
Stem suberect, rather slender, 15-35 cm. Rosette leaves lyrate-pinnate. Cauline leaves auriculate, with 2-5 pairs of narrow lateral leaflets. Petals c. 2 mm. Fruits usually 6-9 × c. 1 mm, linear-ellipsoid.

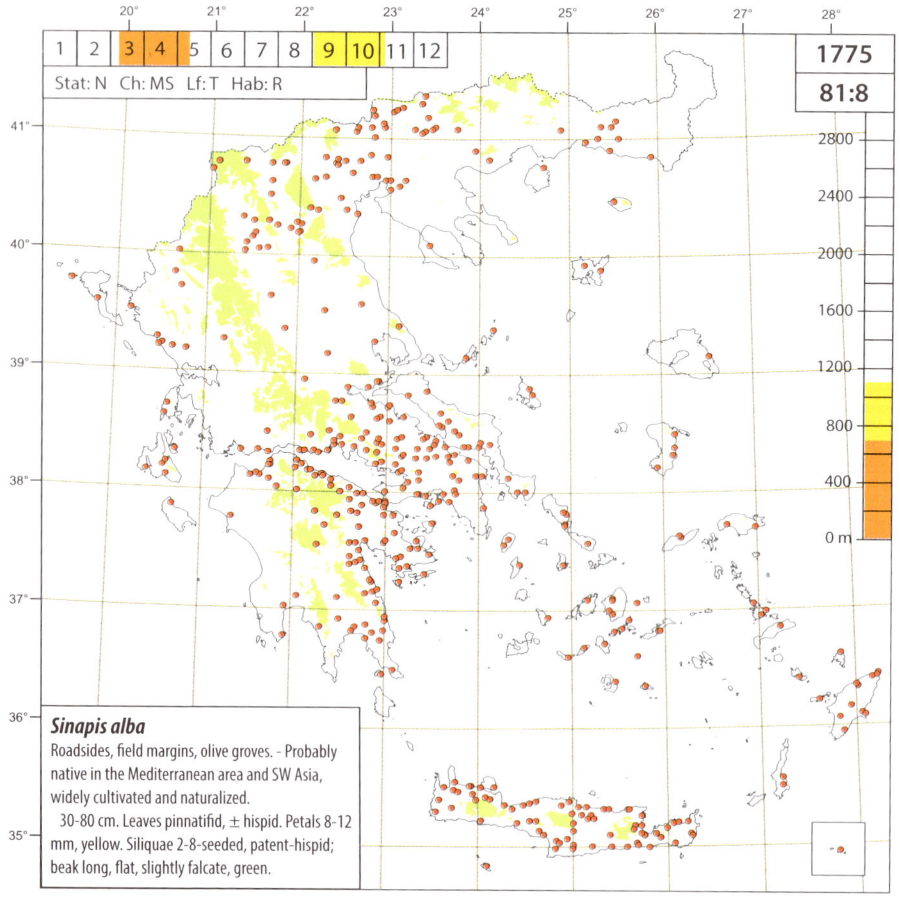

Sinapis alba
Roadsides, field margins, olive groves. - Probably native in the Mediterranean area and SW Asia, widely cultivated and naturalized.
30-80 cm. Leaves pinnatifid, ± hispid. Petals 8-12 mm, yellow. Siliquae 2-8-seeded, patent-hispid; beak long, flat, slightly falcate, green.

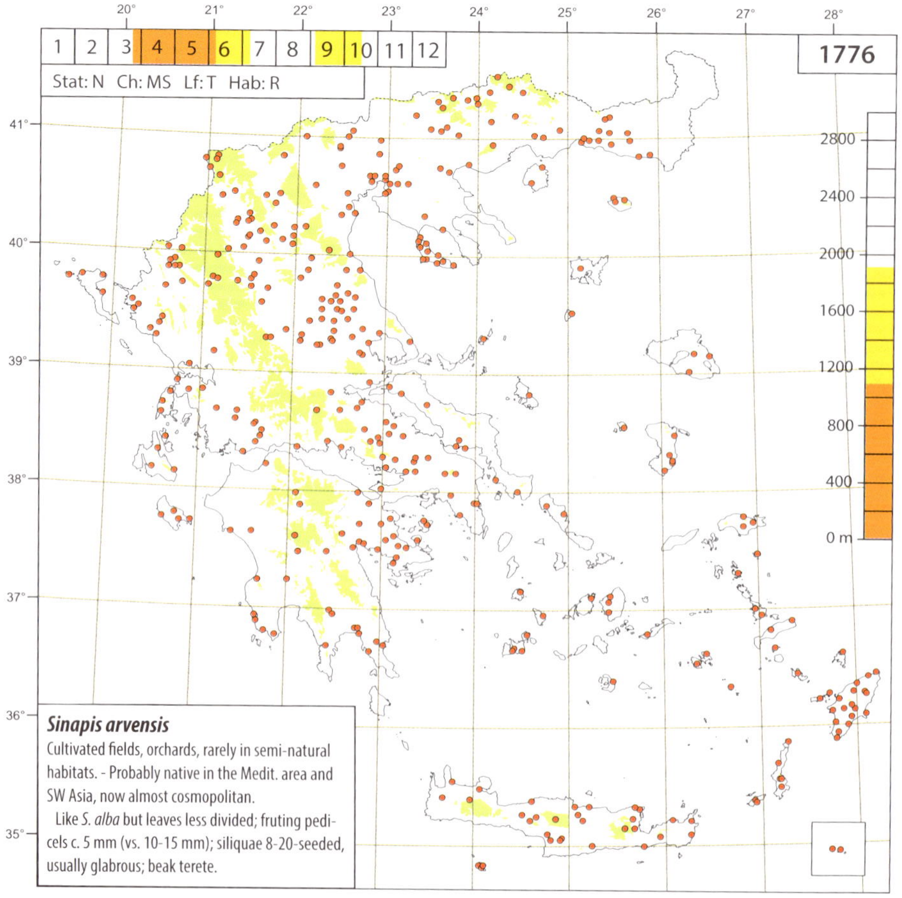

Sinapis arvensis
Cultivated fields, orchards, rarely in semi-natural habitats. - Probably native in the Medit. area and SW Asia, now almost cosmopolitan.
Like *S. alba* but leaves less divided; fruting pedicels c. 5 mm (vs. 10-15 mm); siliquae 8-20-seeded, usually glabrous; beak terete.

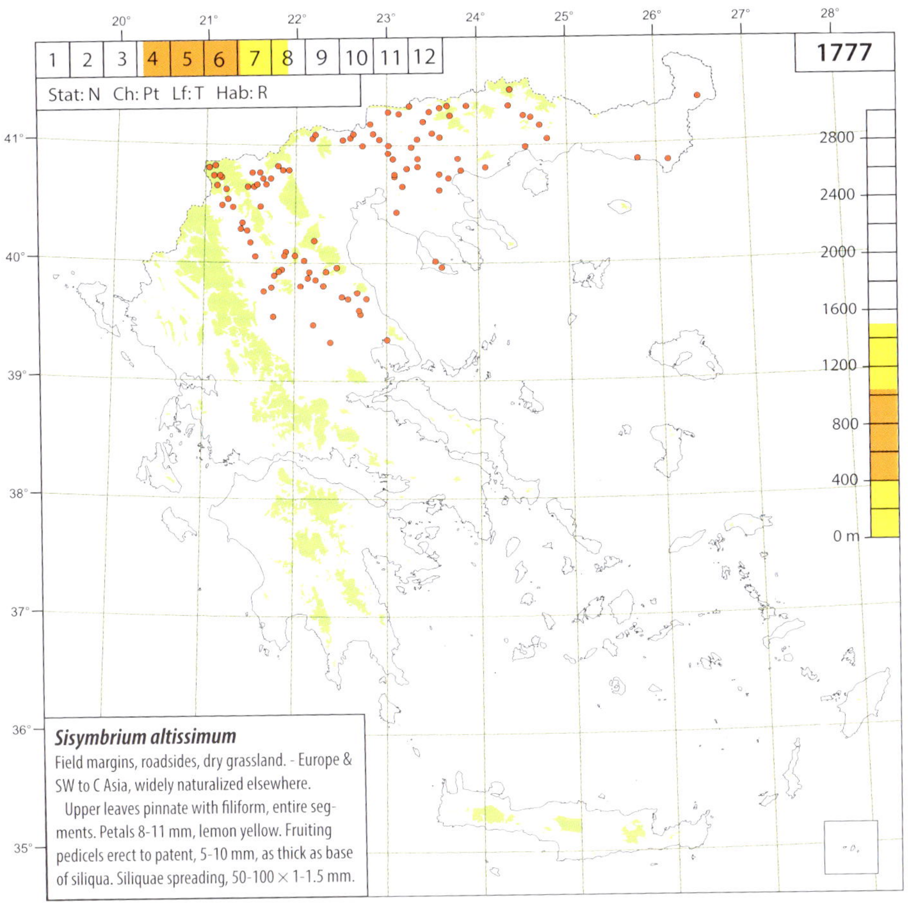

Sisymbrium altissimum
Field margins, roadsides, dry grassland. - Europe & SW to C Asia, widely naturalized elsewhere.
Upper leaves pinnate with filiform, entire segments. Petals 8-11 mm, lemon yellow. Fruiting pedicels erect to patent, 5-10 mm, as thick as base of siliqua. Siliquae spreading, 50-100 × 1-1.5 mm.

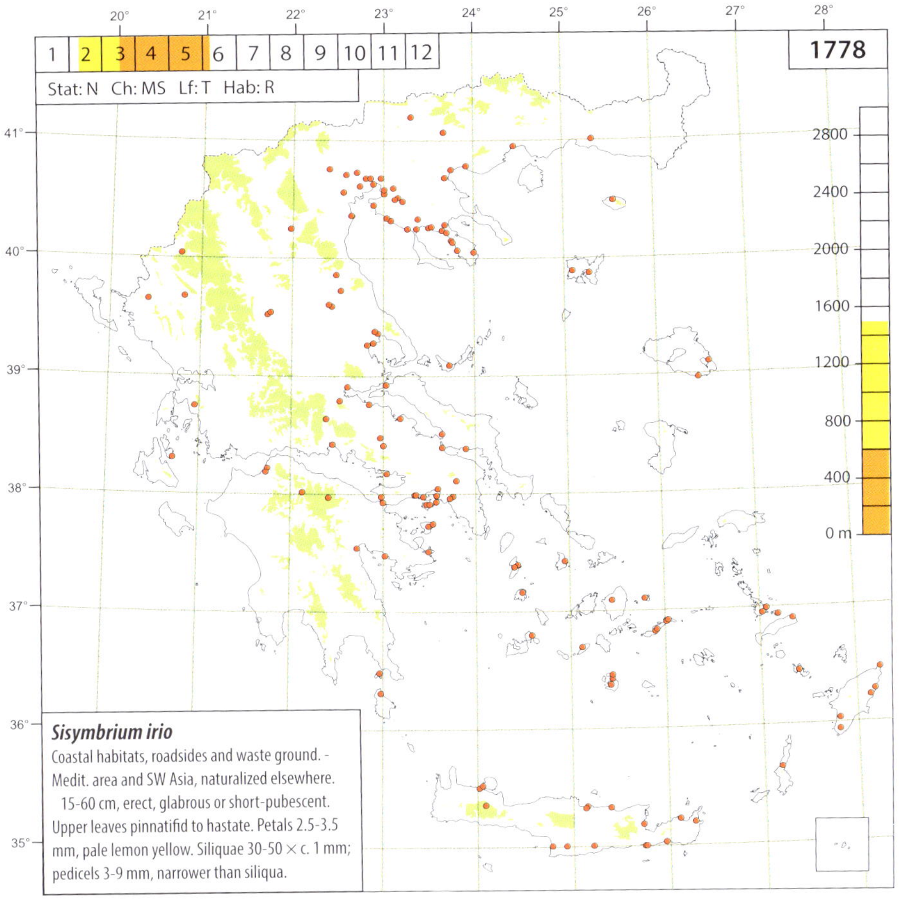

Sisymbrium irio
Coastal habitats, roadsides and waste ground. - Medit. area and SW Asia, naturalized elsewhere.
15-60 cm, erect, glabrous or short-pubescent. Upper leaves pinnatifid to hastate. Petals 2.5-3.5 mm, pale lemon yellow. Siliquae 30-50 × c. 1 mm; pedicels 3-9 mm, narrower than siliqua.

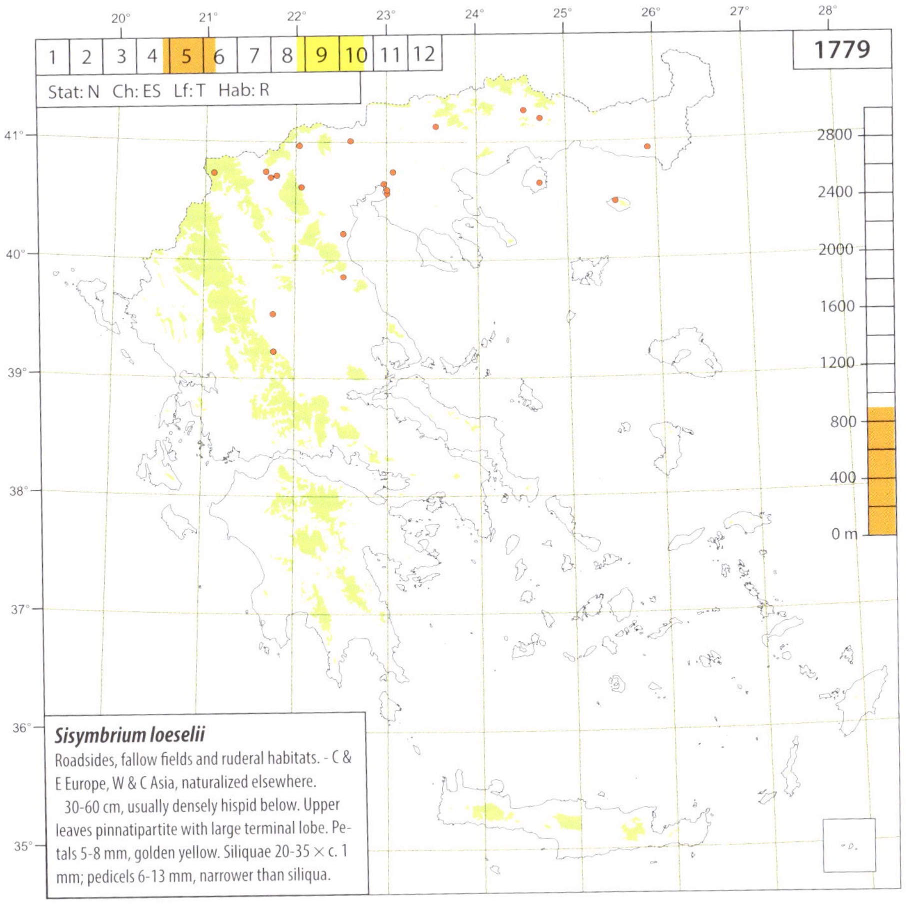

Sisymbrium loeselii
Roadsides, fallow fields and ruderal habitats. - C & E Europe, W & C Asia, naturalized elsewhere.
30-60 cm, usually densely hispid below. Upper leaves pinnatipartite with large terminal lobe. Petals 5-8 mm, golden yellow. Siliquae 20-35 × c. 1 mm; pedicels 6-13 mm, narrower than siliqua.

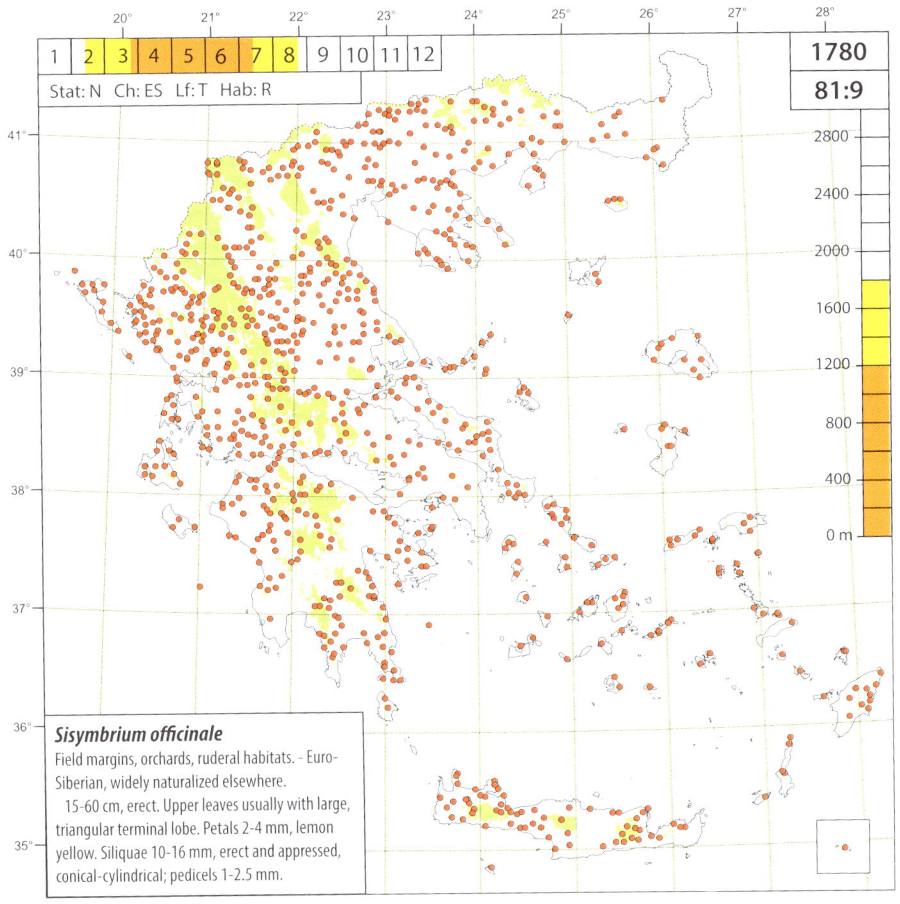

Sisymbrium officinale
Field margins, orchards, ruderal habitats. - Euro-Siberian, widely naturalized elsewhere.
15-60 cm, erect. Upper leaves usually with large, triangular terminal lobe. Petals 2-4 mm, lemon yellow. Siliquae 10-16 mm, erect and appressed, conical-cylindrical; pedicels 1-2.5 mm.

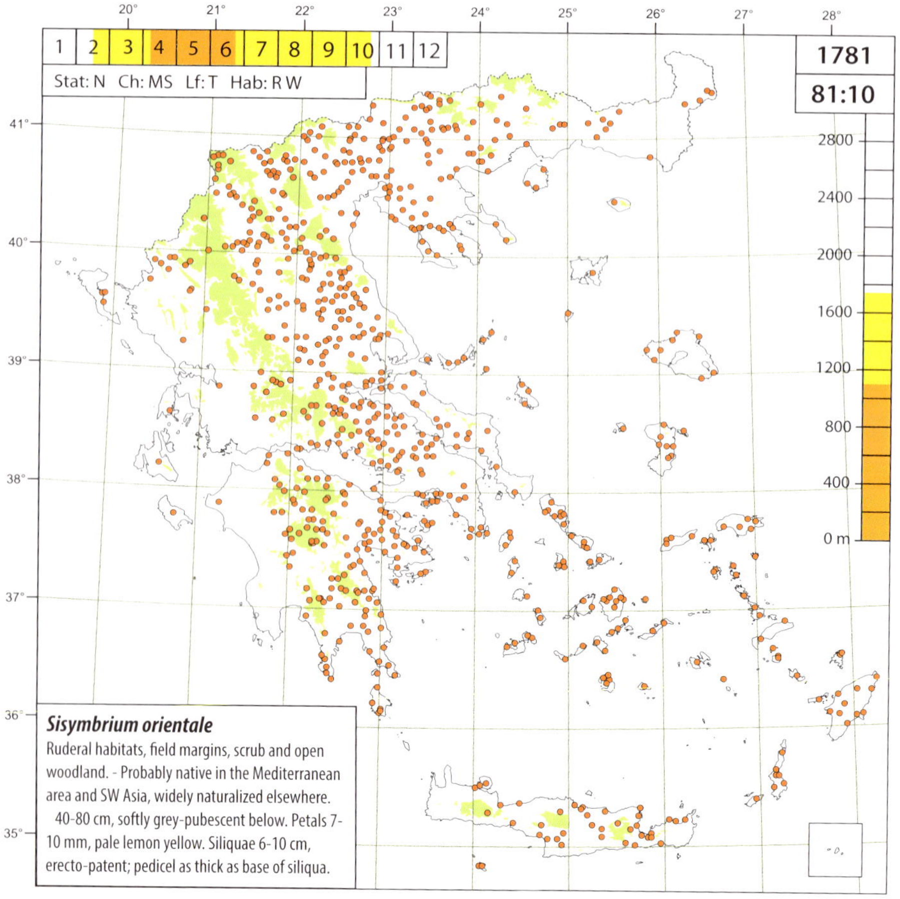

Sisymbrium orientale
Ruderal habitats, field margins, scrub and open woodland. - Probably native in the Mediterranean area and SW Asia, widely naturalized elsewhere.
40-80 cm, softly grey-pubescent below. Petals 7-10 mm, pale lemon yellow. Siliquae 6-10 cm, erecto-patent; pedicel as thick as base of siliqua.

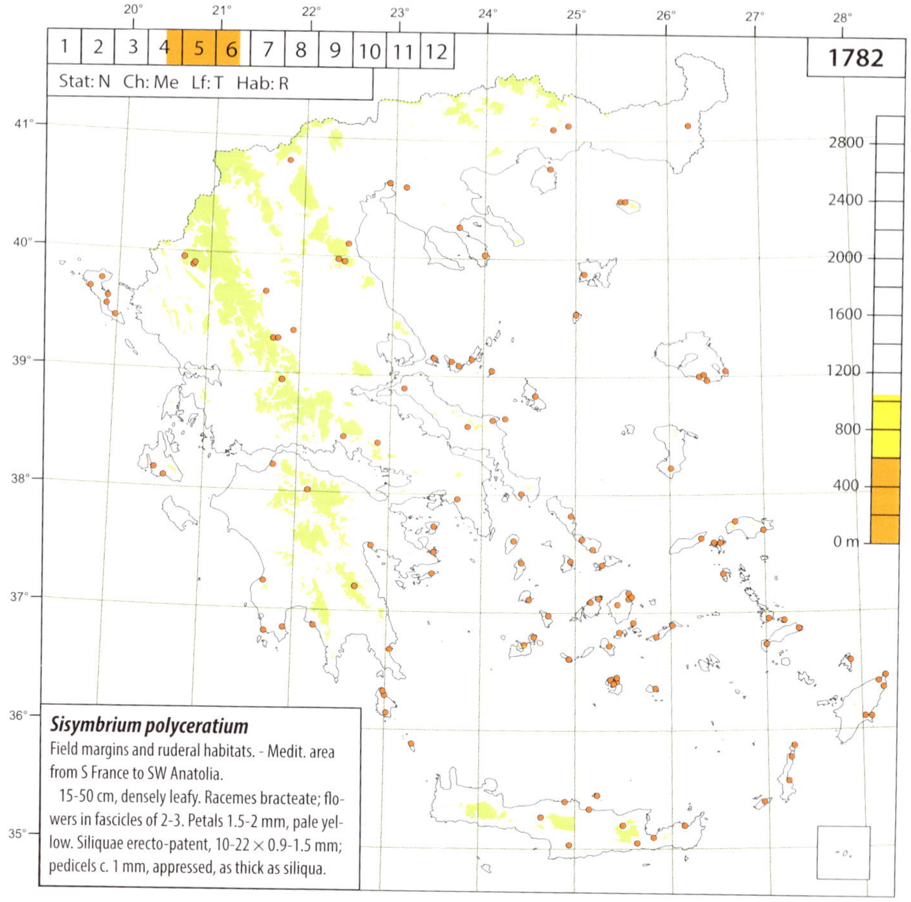

Sisymbrium polyceratium
Field margins and ruderal habitats. - Medit. area from S France to SW Anatolia.
15-50 cm, densely leafy. Racemes bracteate; flowers in fascicles of 2-3. Petals 1.5-2 mm, pale yellow. Siliquae erecto-patent, 10-22 × 0.9-1.5 mm; pedicels c. 1 mm, appressed, as thick as siliqua.

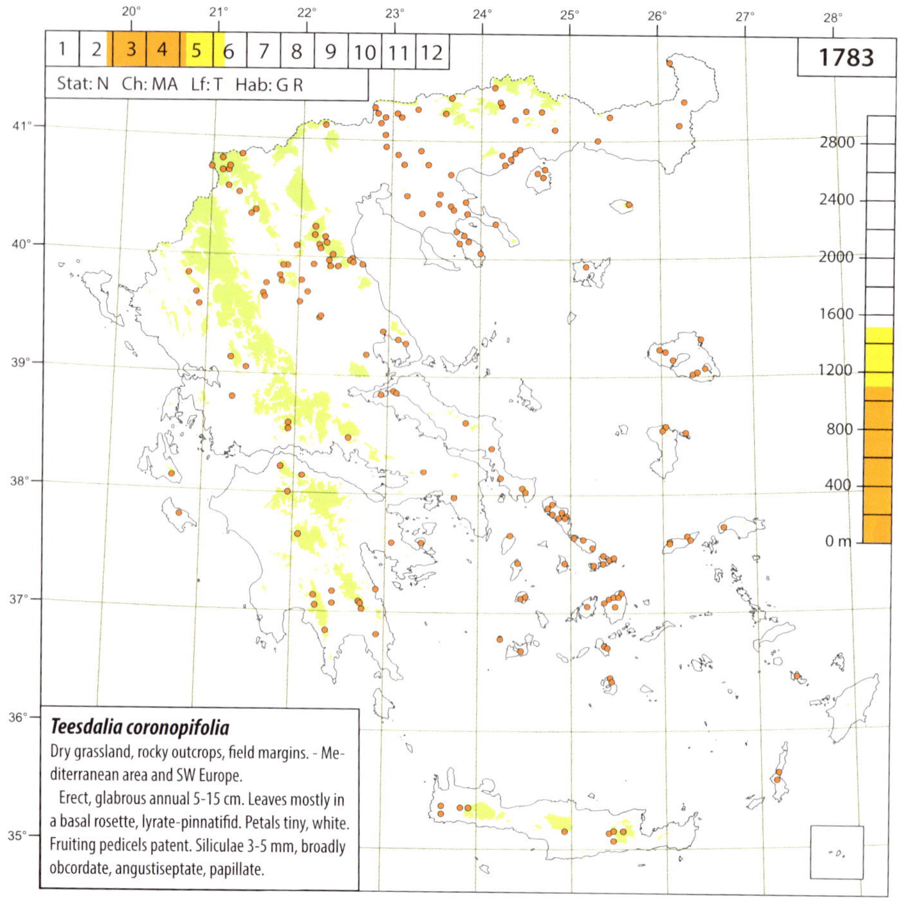

Teesdalia coronopifolia
Dry grassland, rocky outcrops, field margins. - Mediterranean area and SW Europe.
Erect, glabrous annual 5-15 cm. Leaves mostly in a basal rosette, lyrate-pinnatifid. Petals tiny, white. Fruiting pedicels patent. Siliculae 3-5 mm, broadly obcordate, angustiseptate, papillate.

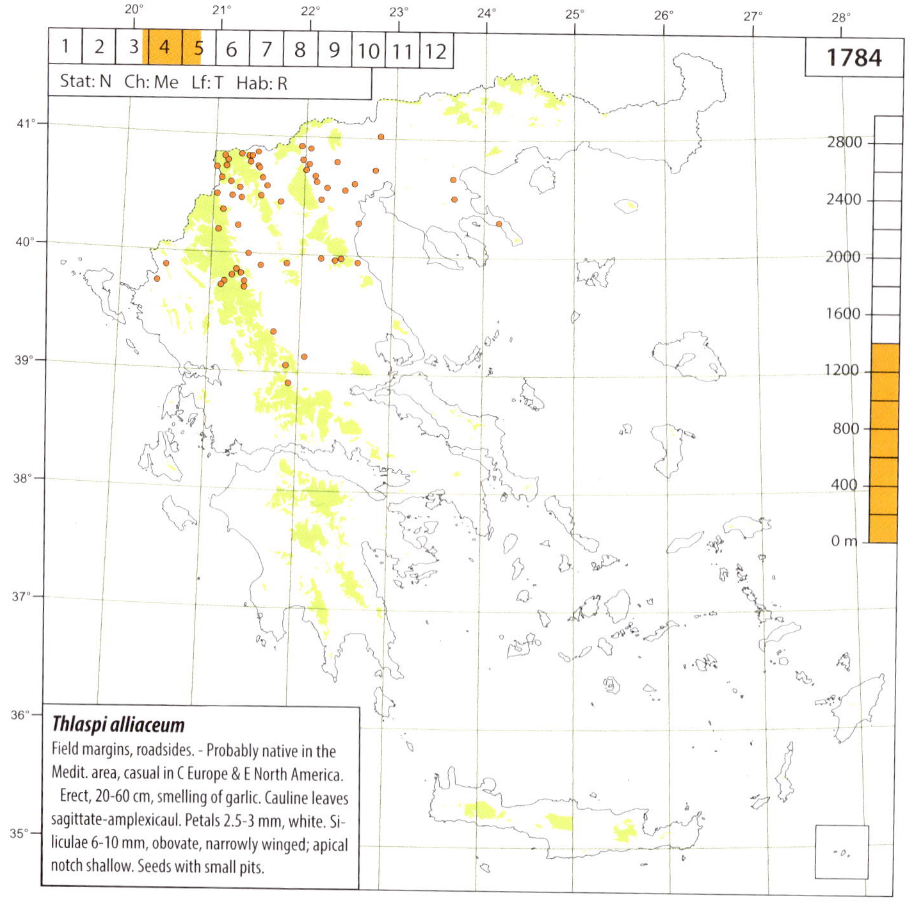

Thlaspi alliaceum
Field margins, roadsides. - Probably native in the Medit. area, casual in C Europe & E North America.
Erect, 20-60 cm, smelling of garlic. Cauline leaves sagittate-amplexicaul. Petals 2.5-3 mm, white. Siliculae 6-10 mm, obovate, narrowly winged; apical notch shallow. Seeds with small pits.

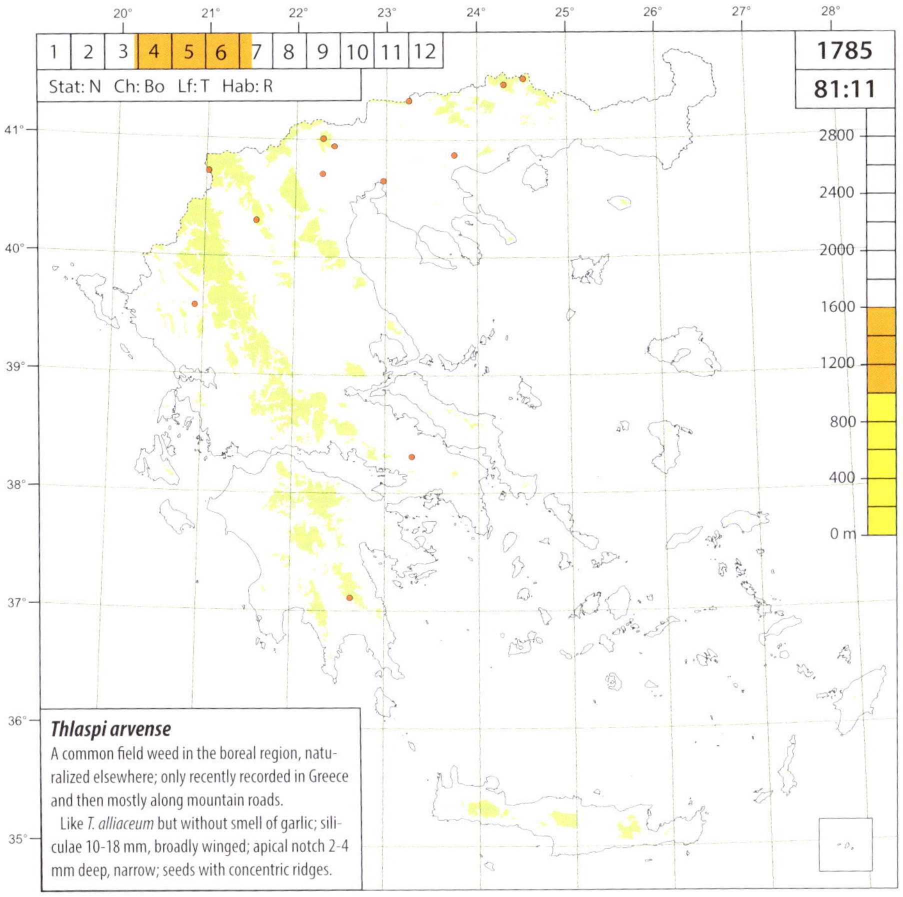

Thlaspi arvense

A common field weed in the boreal region, naturalized elsewhere; only recently recorded in Greece and then mostly along mountain roads.

Like *T. alliaceum* but without smell of garlic; siliculae 10-18 mm, broadly winged; apical notch 2-4 mm deep, narrow; seeds with concentric ridges.

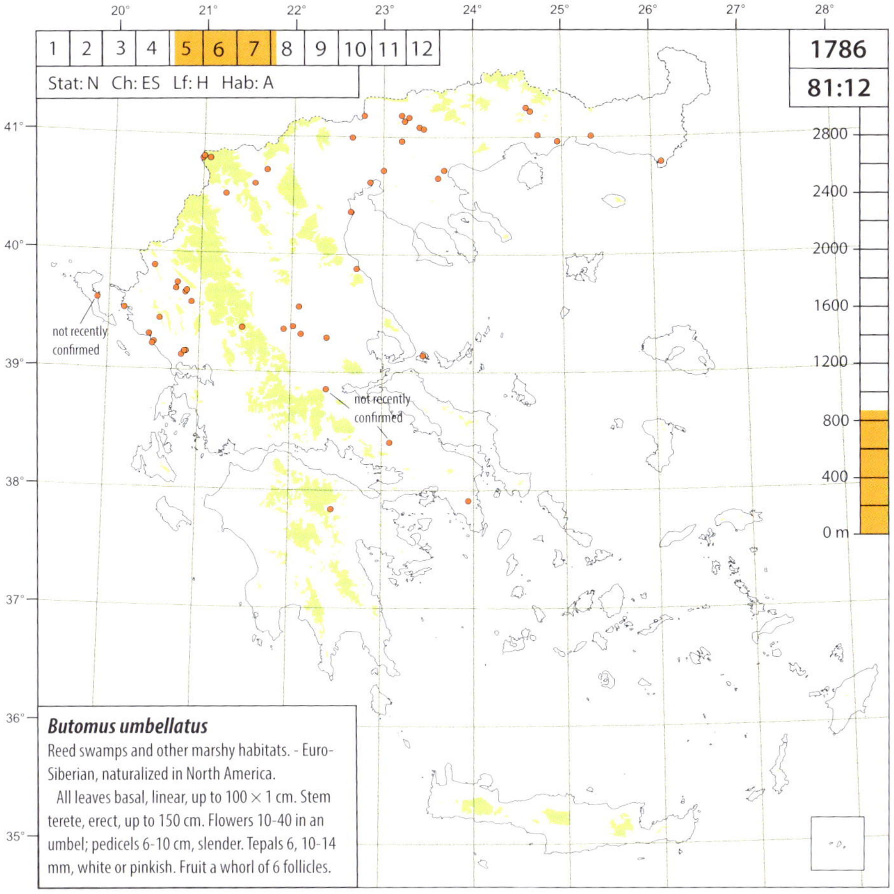

Butomus umbellatus

Reed swamps and other marshy habitats. - Euro-Siberian, naturalized in North America.

All leaves basal, linear, up to 100 × 1 cm. Stem terete, erect, up to 150 cm. Flowers 10-40 in an umbel; pedicels 6-10 cm, slender. Tepals 6, 10-14 mm, white or pinkish. Fruit a whorl of 6 follicles.

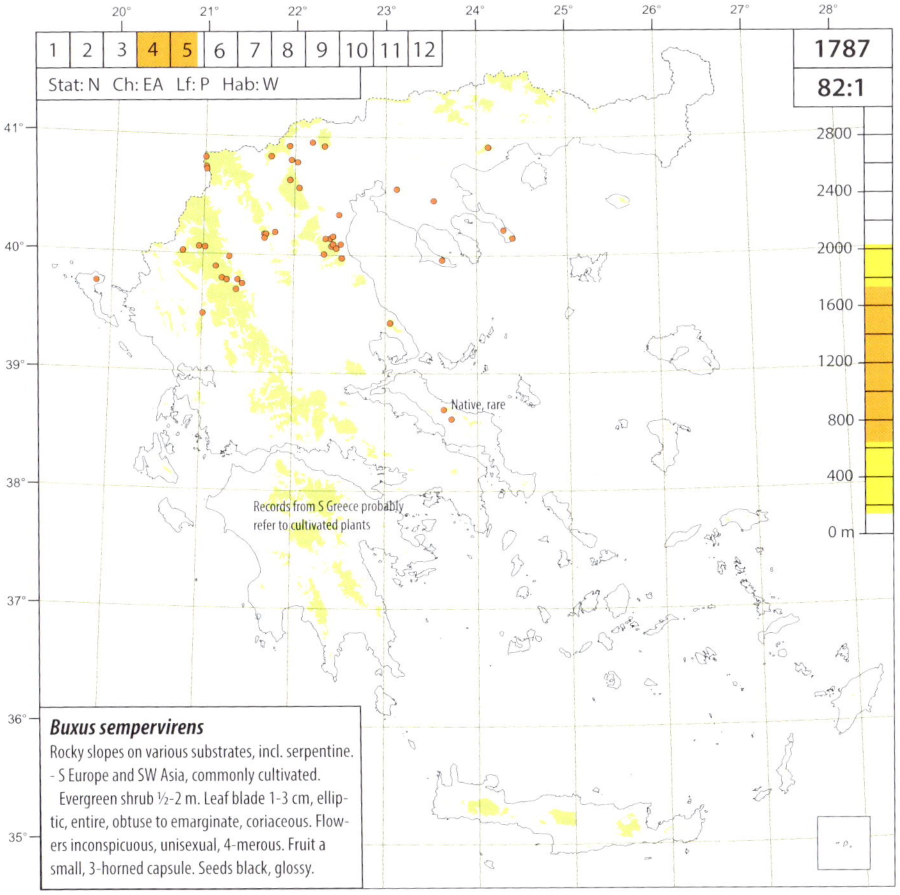

Buxus sempervirens

Rocky slopes on various substrates, incl. serpentine. - S Europe and SW Asia, commonly cultivated.

Evergreen shrub ½-2 m. Leaf blade 1-3 cm, elliptic, entire, obtuse to emarginate, coriaceous. Flowers inconspicuous, unisexual, 4-merous. Fruit a small, 3-horned capsule. Seeds black, glossy.

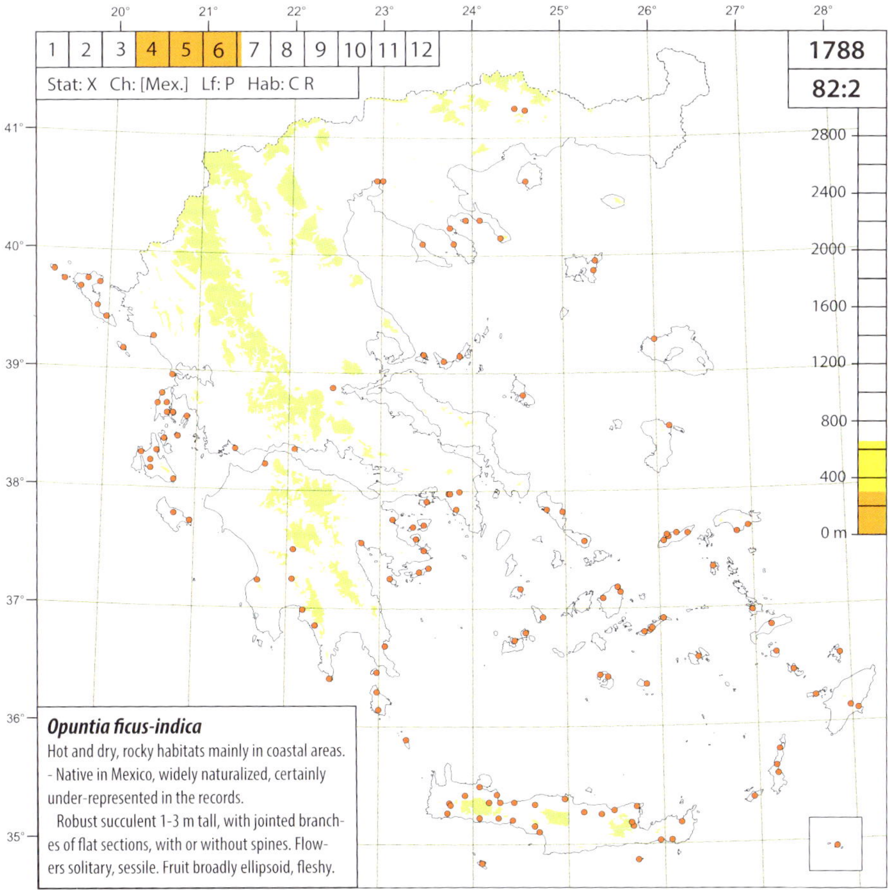

Opuntia ficus-indica

Hot and dry, rocky habitats mainly in coastal areas. - Native in Mexico, widely naturalized, certainly under-represented in the records.

Robust succulent 1-3 m tall, with jointed branches of flat sections, with or without spines. Flowers solitary, sessile. Fruit broadly ellipsoid, fleshy.

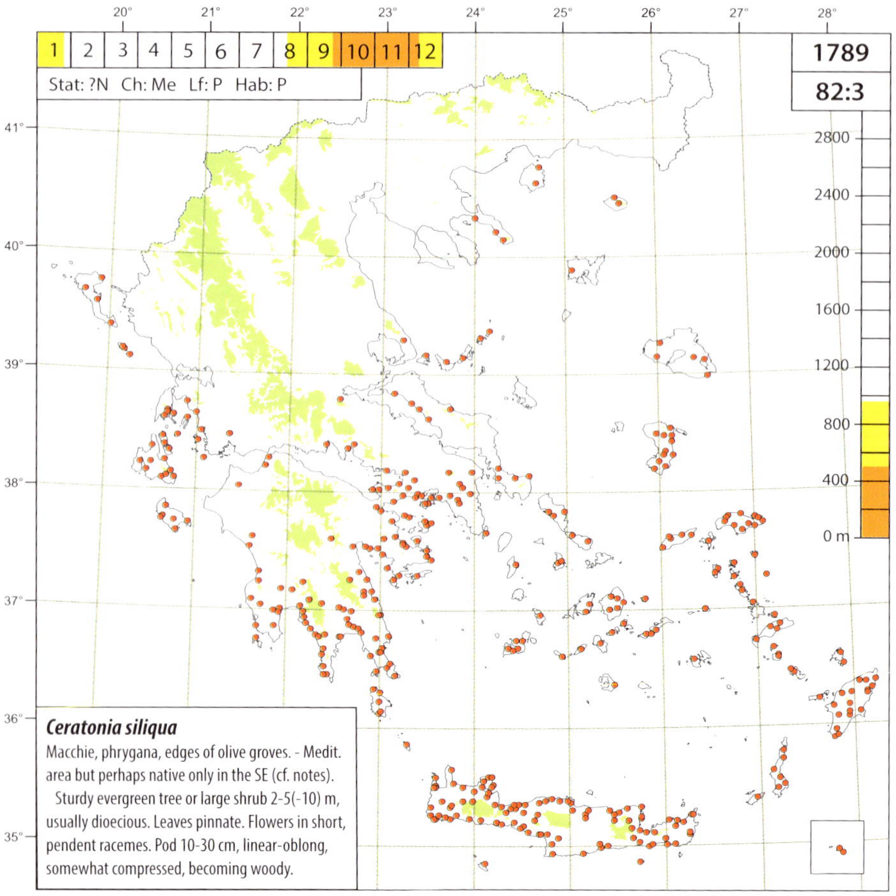

Ceratonia siliqua

Macchie, phrygana, edges of olive groves. - Medit. area but perhaps native only in the SE (cf. notes).

Sturdy evergreen tree or large shrub 2-5(-10) m, usually dioecious. Leaves pinnate. Flowers in short, pendent racemes. Pod 10-30 cm, linear-oblong, somewhat compressed, becoming woody.

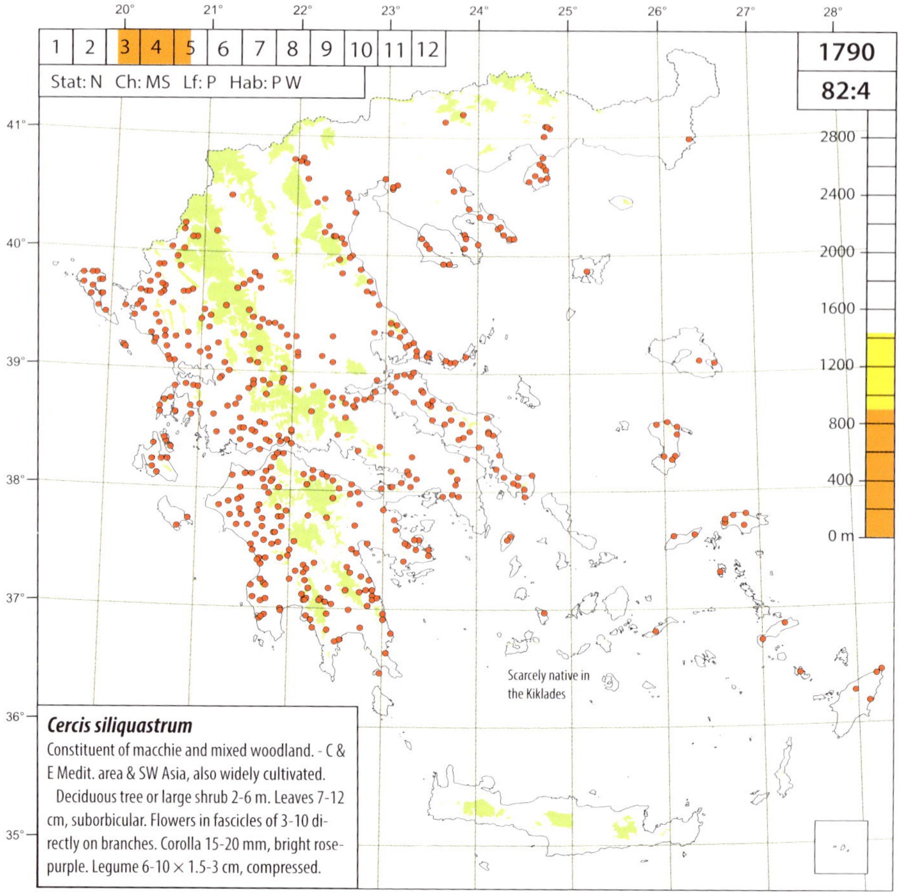

Cercis siliquastrum

Constituent of macchie and mixed woodland. - C & E Medit. area & SW Asia, also widely cultivated.

Deciduous tree or large shrub 2-6 m. Leaves 7-12 cm, suborbicular. Flowers in fascicles of 3-10 directly on branches. Corolla 15-20 mm, bright rose-purple. Legume 6-10 × 1.5-3 cm, compressed.

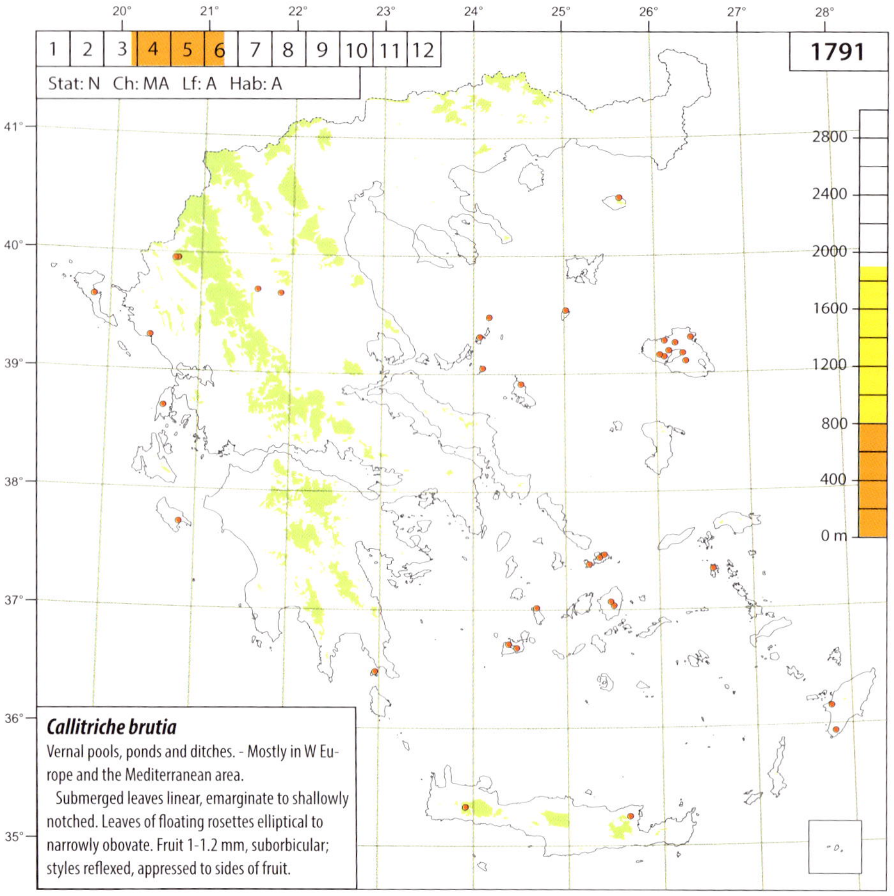

Callitriche brutia

Vernal pools, ponds and ditches. - Mostly in W Europe and the Mediterranean area.

Submerged leaves linear, emarginate to shallowly notched. Leaves of floating rosettes elliptical to narrowly obovate. Fruit 1-1.2 mm, suborbicular; styles reflexed, appressed to sides of fruit.

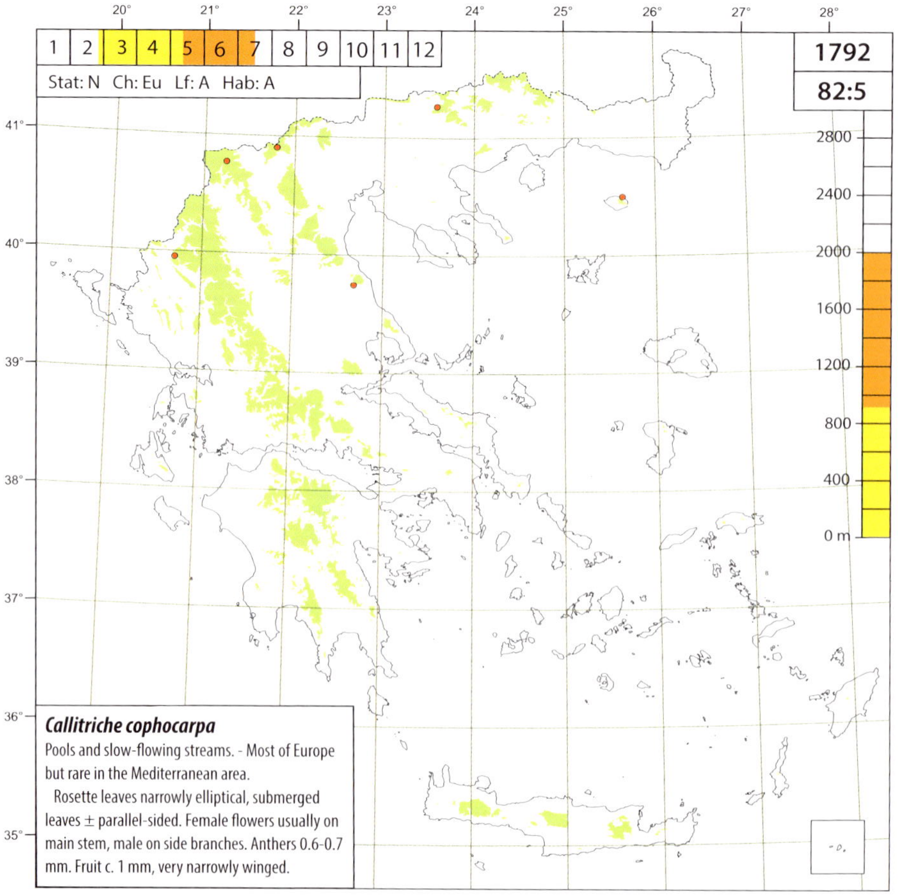

Callitriche cophocarpa

Pools and slow-flowing streams. - Most of Europe but rare in the Mediterranean area.

Rosette leaves narrowly elliptical, submerged leaves ± parallel-sided. Female flowers usually on main stem, male on side branches. Anthers 0.6-0.7 mm. Fruit c. 1 mm, very narrowly winged.

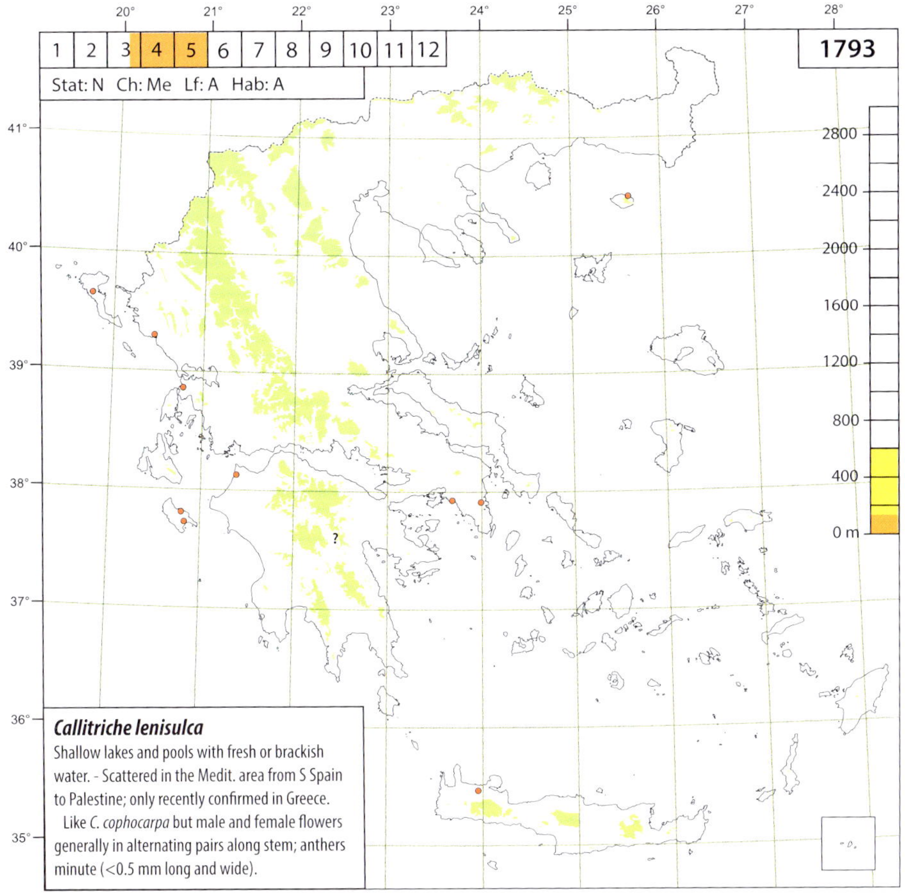

Callitriche lenisulca
Shallow lakes and pools with fresh or brackish water. - Scattered in the Medit. area from S Spain to Palestine; only recently confirmed in Greece.
Like *C. cophocarpa* but male and female flowers generally in alternating pairs along stem; anthers minute (<0.5 mm long and wide).

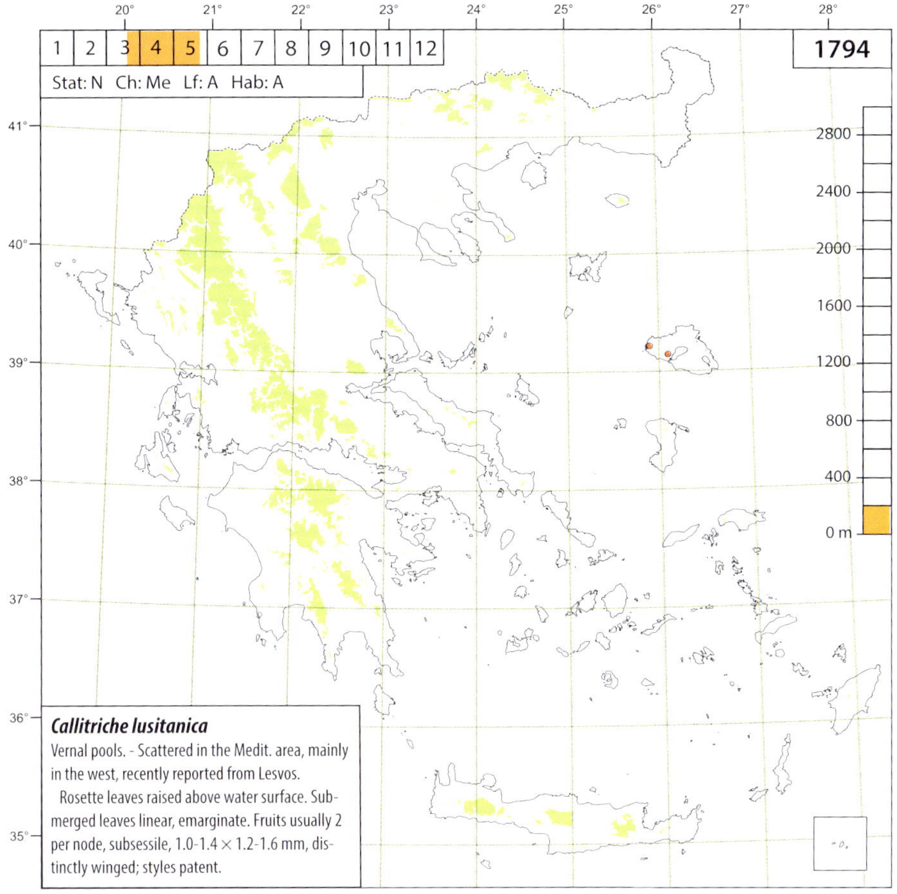

Callitriche lusitanica
Vernal pools. - Scattered in the Medit. area, mainly in the west, recently reported from Lesvos.
Rosette leaves raised above water surface. Submerged leaves linear, emarginate. Fruits usually 2 per node, subsessile, 1.0-1.4 × 1.2-1.6 mm, distinctly winged; styles patent.

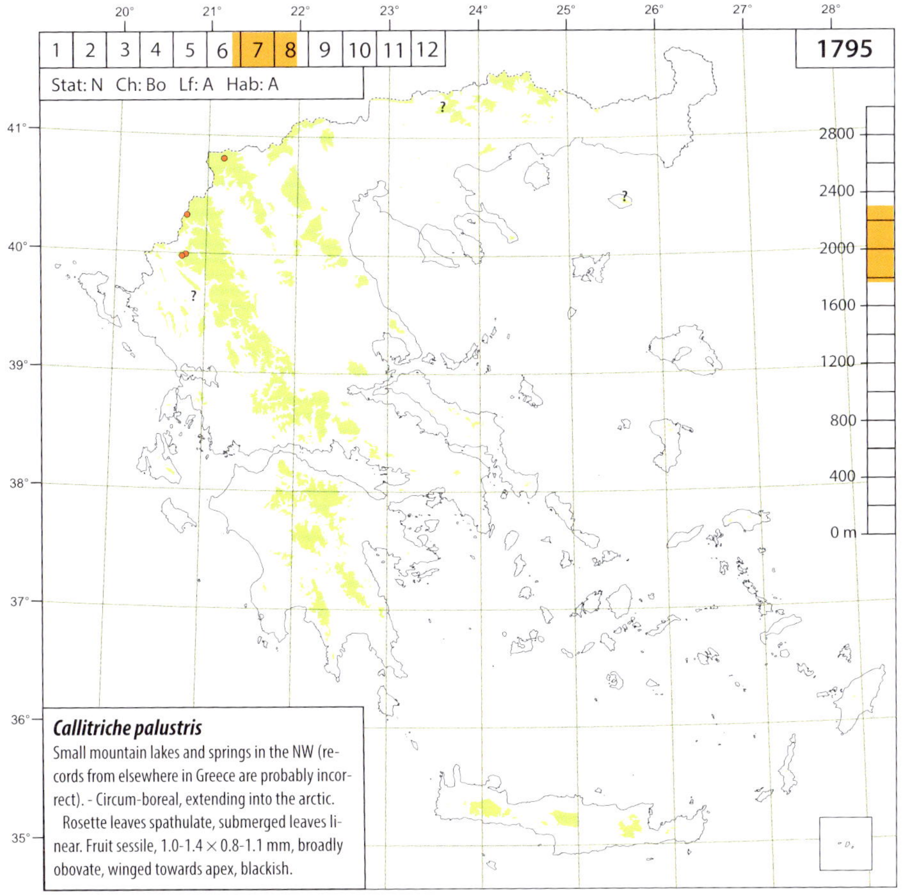

Callitriche palustris
Small mountain lakes and springs in the NW (records from elsewhere in Greece are probably incorrect). - Circum-boreal, extending into the arctic.
Rosette leaves spathulate, submerged leaves linear. Fruit sessile, 1.0-1.4 × 0.8-1.1 mm, broadly obovate, winged towards apex, blackish.

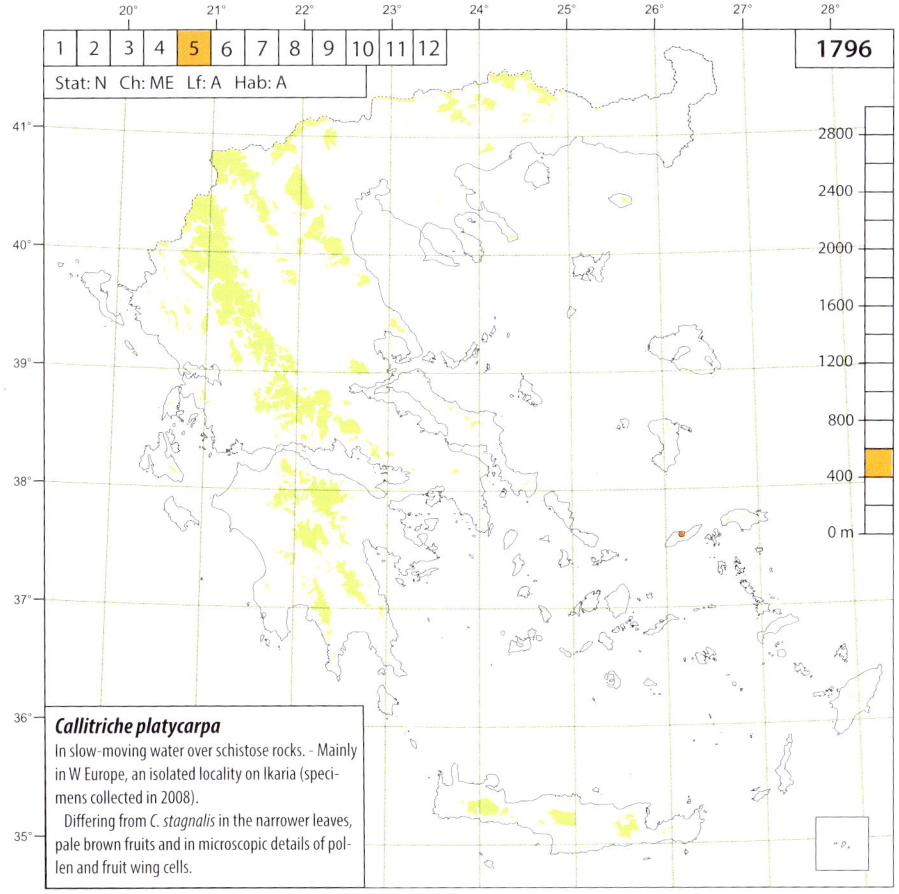

Callitriche platycarpa
In slow-moving water over schistose rocks. - Mainly in W Europe, an isolated locality on Ikaria (specimens collected in 2008).
Differing from *C. stagnalis* in the narrower leaves, pale brown fruits and in microscopic details of pollen and fruit wing cells.

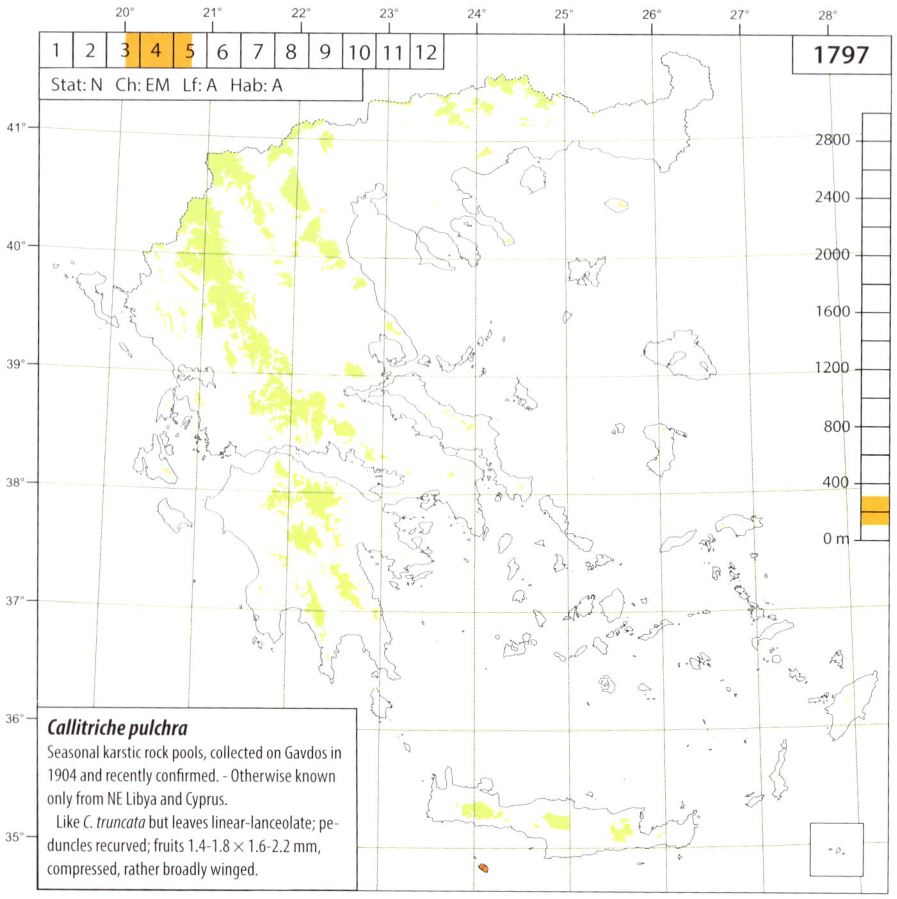

Callitriche pulchra

Seasonal karstic rock pools, collected on Gavdos in 1904 and recently confirmed. - Otherwise known only from NE Libya and Cyprus.

Like *C. truncata* but leaves linear-lanceolate; peduncles recurved; fruits 1.4-1.8 × 1.6-2.2 mm, compressed, rather broadly winged.

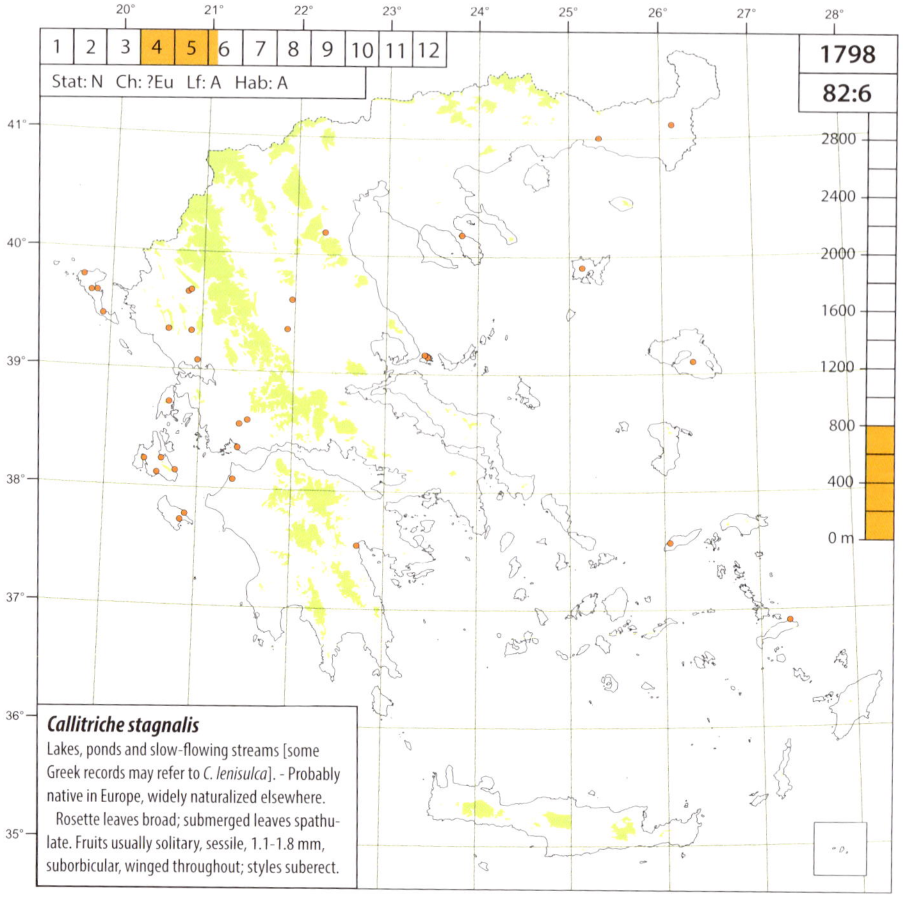

Callitriche stagnalis

Lakes, ponds and slow-flowing streams [some Greek records may refer to *C. lenisulca*]. - Probably native in Europe, widely naturalized elsewhere.

Rosette leaves broad; submerged leaves spathulate. Fruits usually solitary, sessile, 1.1-1.8 mm, suborbicular, winged throughout; styles suberect.

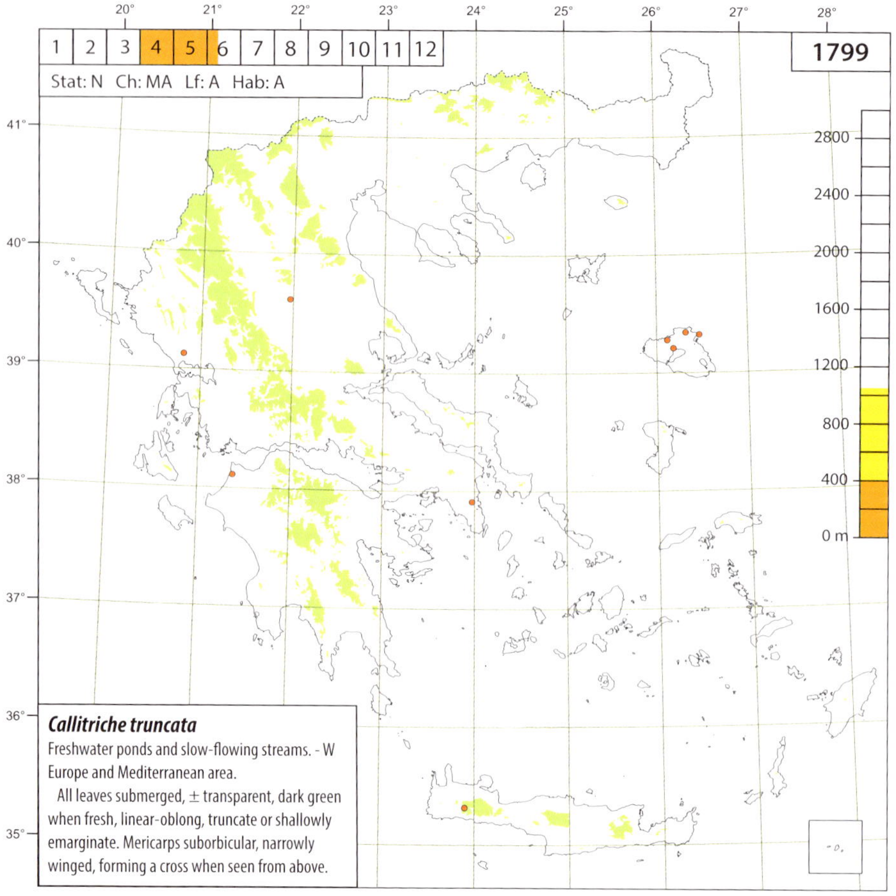

Callitriche truncata

Freshwater ponds and slow-flowing streams. - W Europe and Mediterranean area.

All leaves submerged, ± transparent, dark green when fresh, linear-oblong, truncate or shallowly emarginate. Mericarps suborbicular, narrowly winged, forming a cross when seen from above.

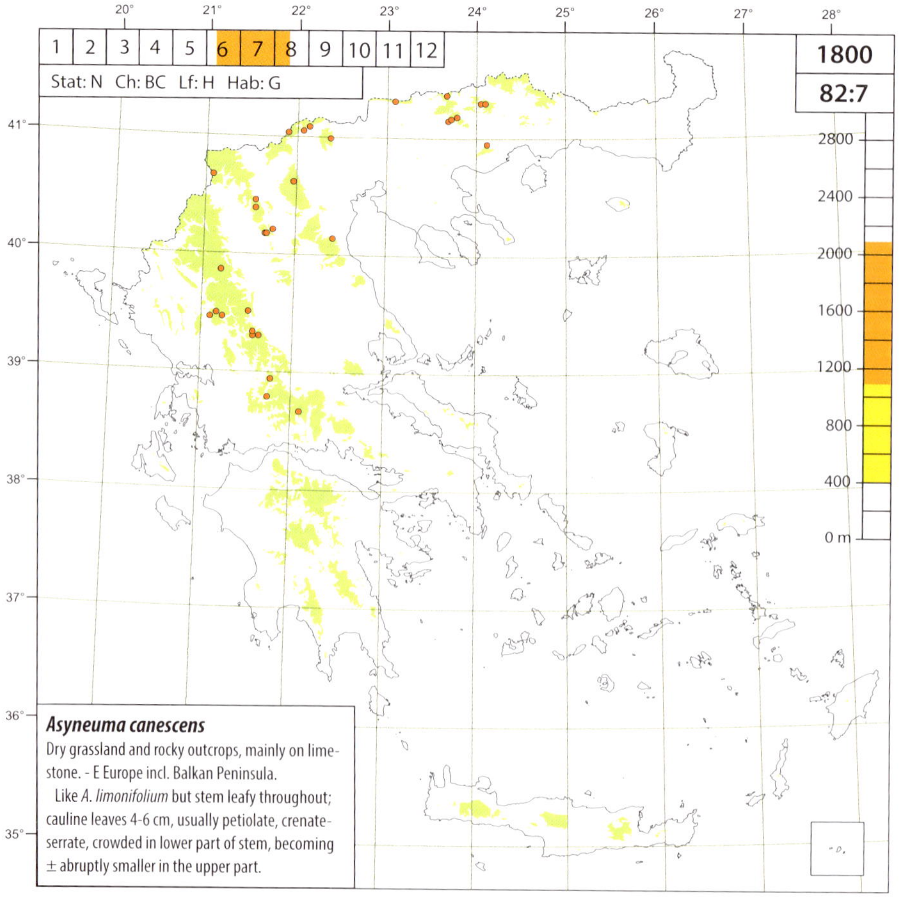

Asyneuma canescens

Dry grassland and rocky outcrops, mainly on limestone. - E Europe incl. Balkan Peninsula.

Like *A. limonifolium* but stem leafy throughout; cauline leaves 4-6 cm, usually petiolate, crenate-serrate, crowded in lower part of stem, becoming ± abruptly smaller in the upper part.

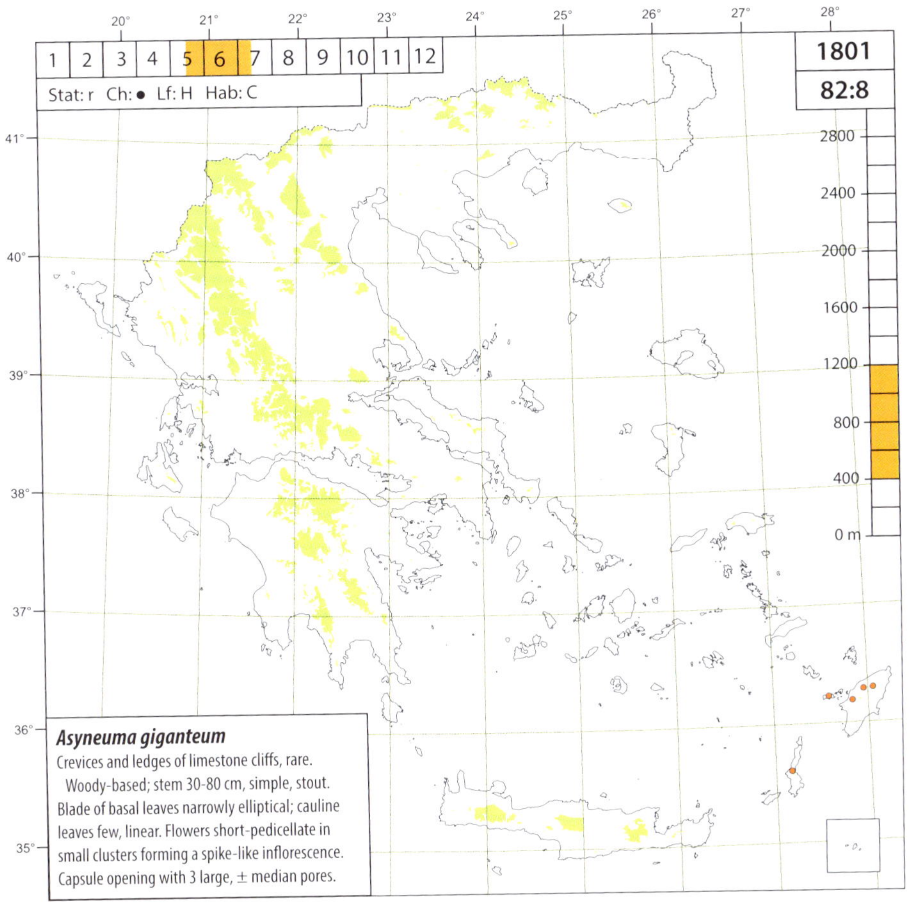

Asyneuma giganteum

Crevices and ledges of limestone cliffs, rare.

Woody-based; stem 30-80 cm, simple, stout. Blade of basal leaves narrowly elliptical; cauline leaves few, linear. Flowers short-pedicellate in small clusters forming a spike-like inflorescence. Capsule opening with 3 large, ± median pores.

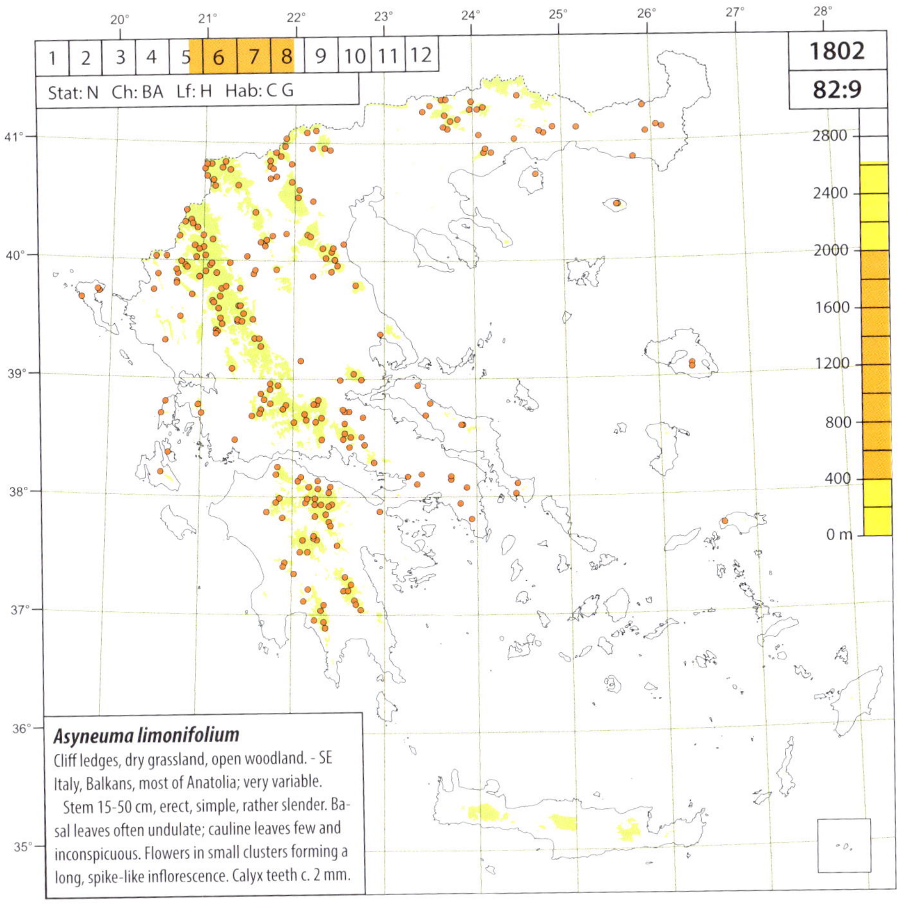

Asyneuma limonifolium

Cliff ledges, dry grassland, open woodland. - SE Italy, Balkans, most of Anatolia; very variable.

Stem 15-50 cm, erect, simple, rather slender. Basal leaves often undulate; cauline leaves few and inconspicuous. Flowers in small clusters forming a long, spike-like inflorescence. Calyx teeth c. 2 mm.

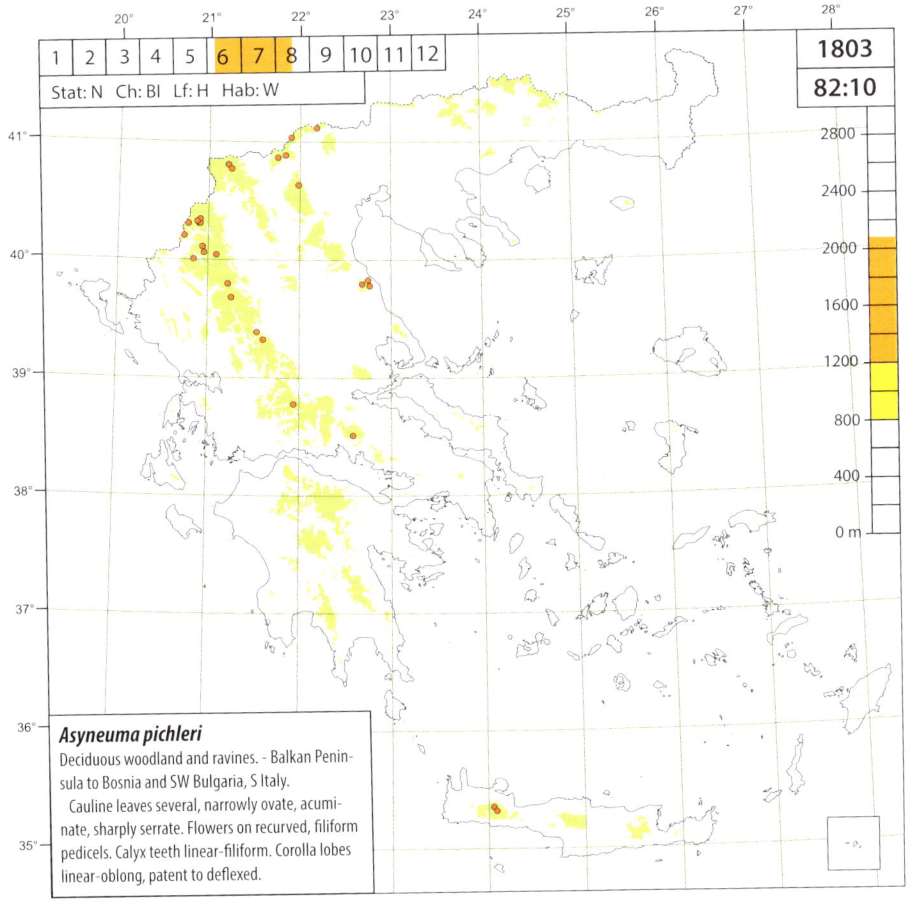

Asyneuma pichleri

Deciduous woodland and ravines. - Balkan Peninsula to Bosnia and SW Bulgaria, S Italy.

Cauline leaves several, narrowly ovate, acuminate, sharply serrate. Flowers on recurved, filiform pedicels. Calyx teeth linear-filiform. Corolla lobes linear-oblong, patent to deflexed.

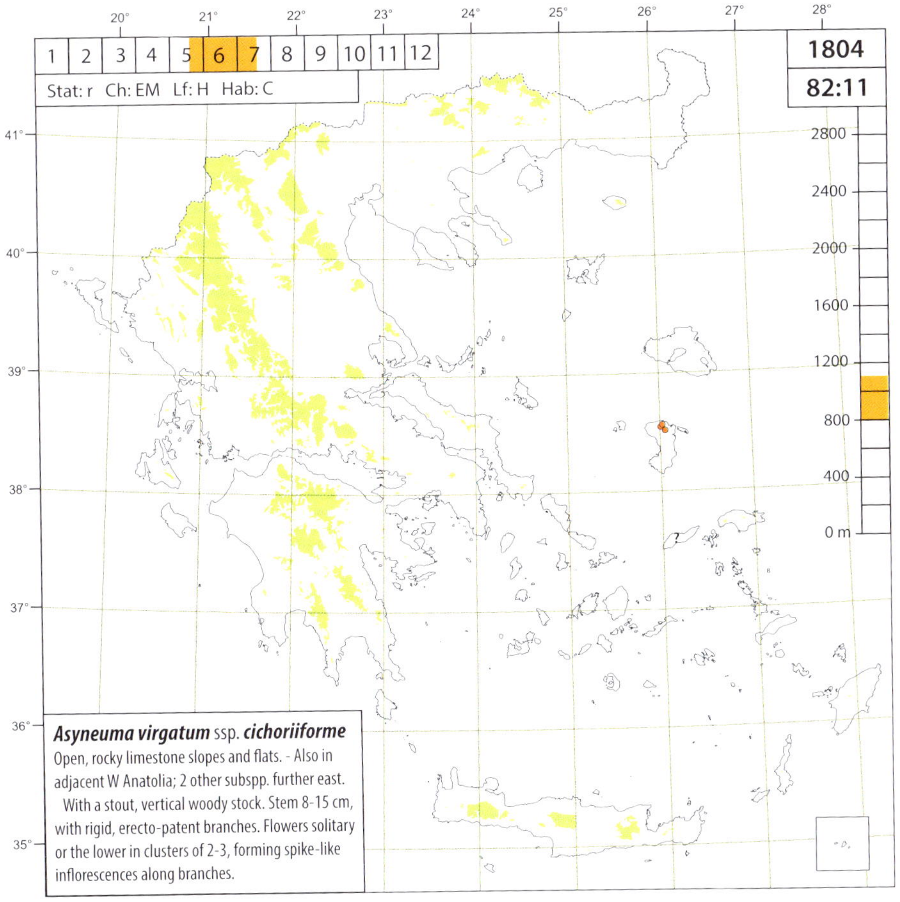

Asyneuma virgatum ssp. _cichoriiforme_

Open, rocky limestone slopes and flats. - Also in adjacent W Anatolia; 2 other subspp. further east.

With a stout, vertical woody stock. Stem 8-15 cm, with rigid, erecto-patent branches. Flowers solitary or the lower in clusters of 2-3, forming spike-like inflorescences along branches.

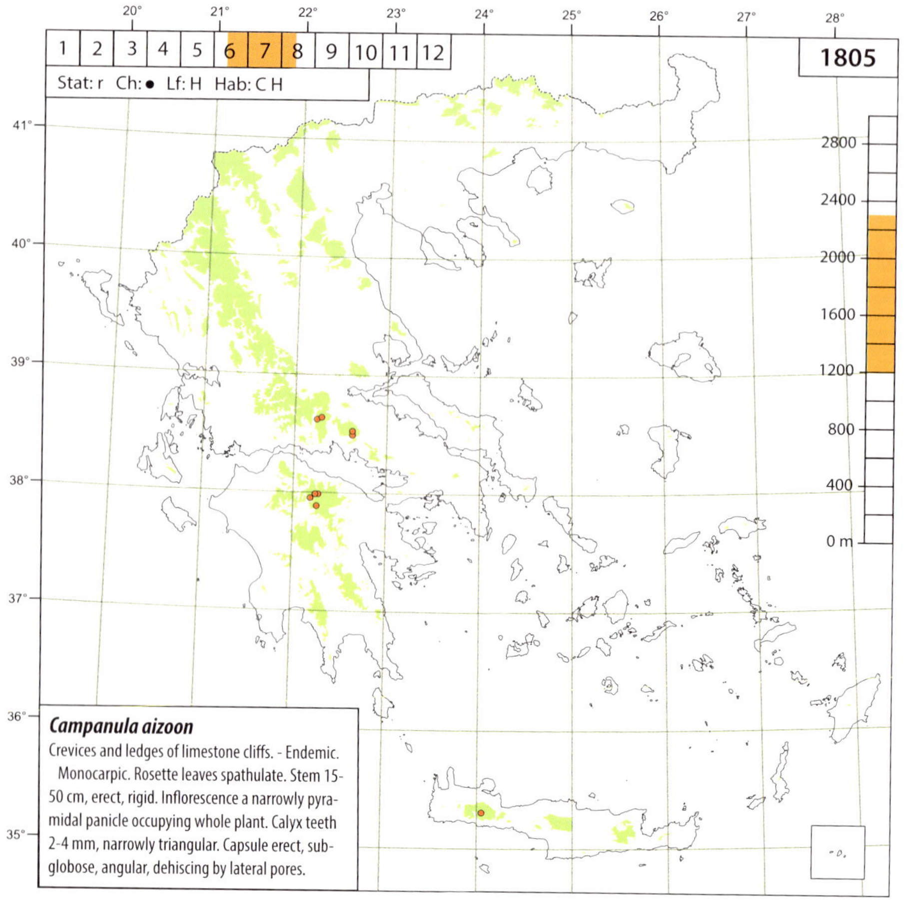

Campanula aizoon
Crevices and ledges of limestone cliffs. - Endemic.
Monocarpic. Rosette leaves spathulate. Stem 15-50 cm, erect, rigid. Inflorescence a narrowly pyramidal panicle occupying whole plant. Calyx teeth 2-4 mm, narrowly triangular. Capsule erect, subglobose, angular, dehiscing by lateral pores.

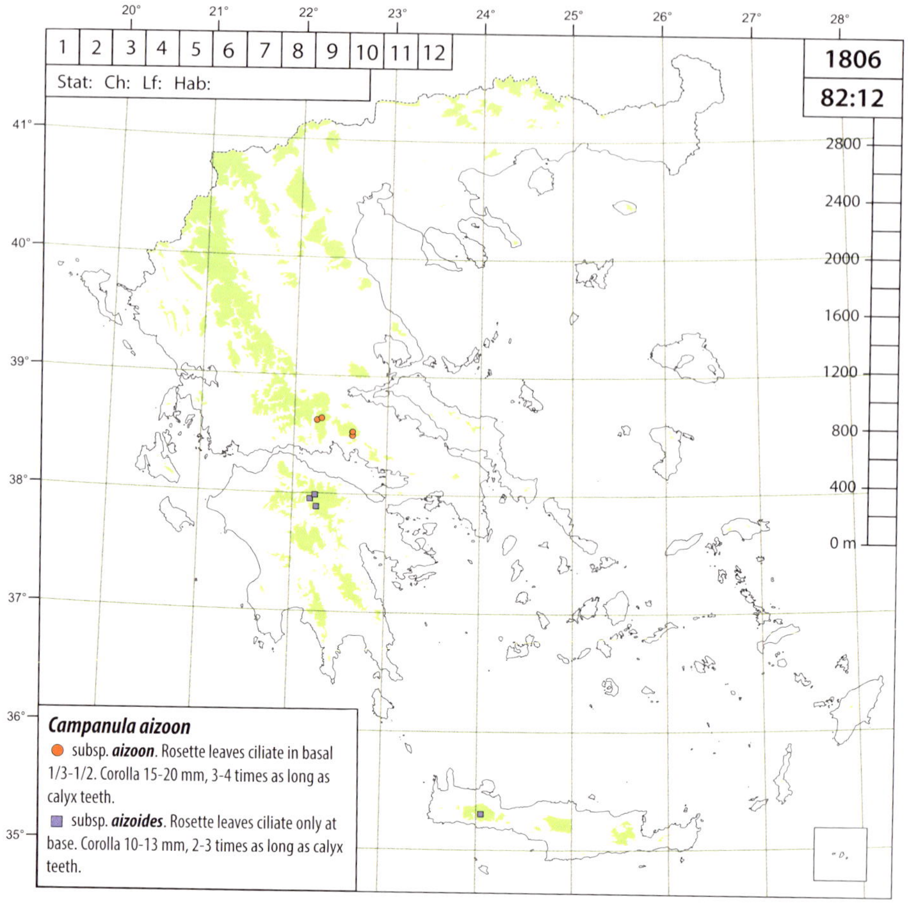

Campanula aizoon
● subsp. ***aizoon***. Rosette leaves ciliate in basal 1/3-1/2. Corolla 15-20 mm, 3-4 times as long as calyx teeth.
■ subsp. ***aizoides***. Rosette leaves ciliate only at base. Corolla 10-13 mm, 2-3 times as long as calyx teeth.

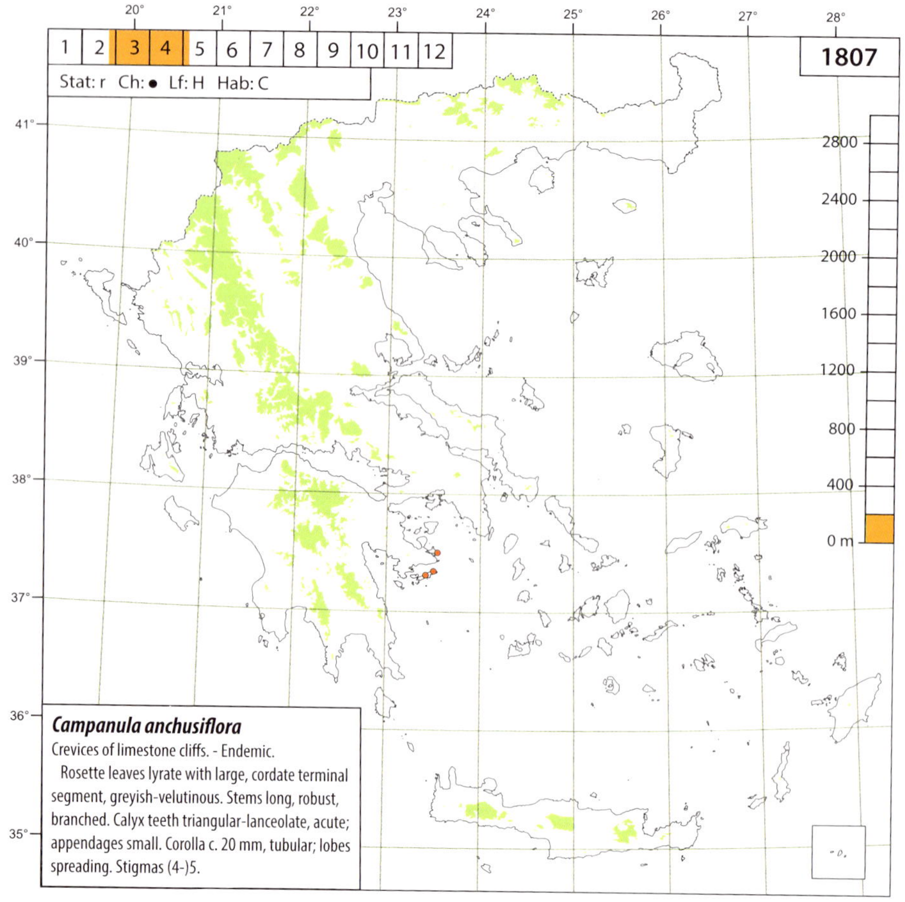

Campanula anchusiflora
Crevices of limestone cliffs. - Endemic.
Rosette leaves lyrate with large, cordate terminal segment, greyish-velutinous. Stems long, robust, branched. Calyx teeth triangular-lanceolate, acute; appendages small. Corolla c. 20 mm, tubular; lobes spreading. Stigmas (4-)5.

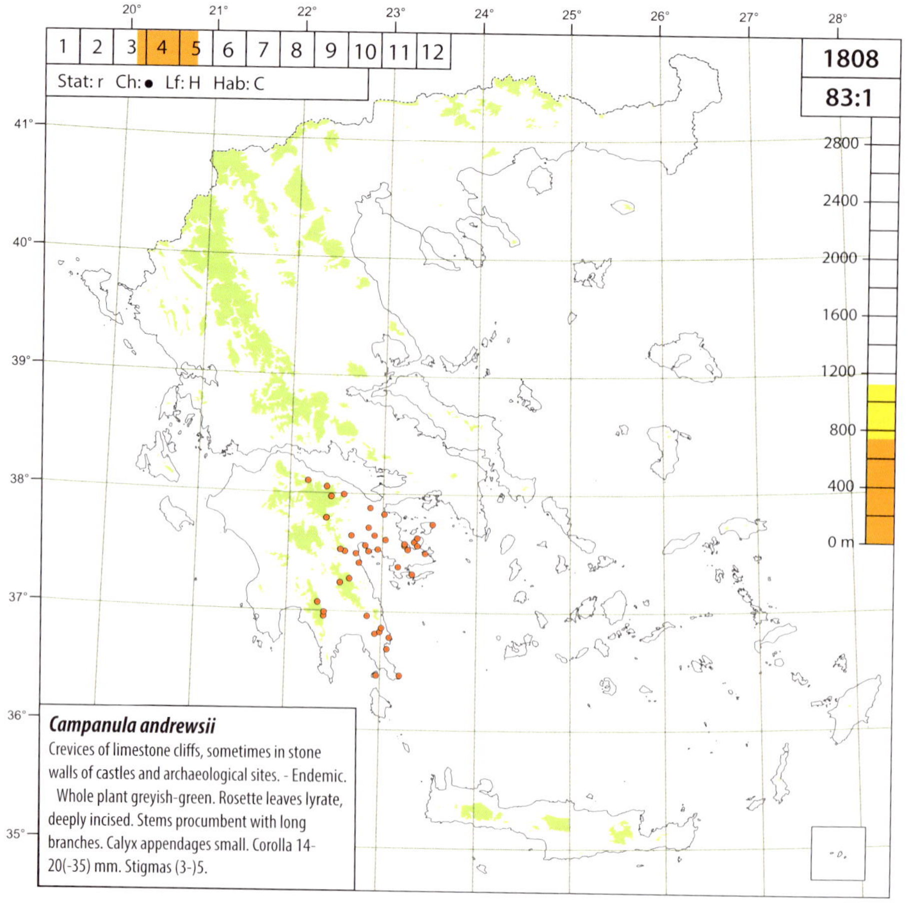

Campanula andrewsii
Crevices of limestone cliffs, sometimes in stone walls of castles and archaeological sites. - Endemic.
Whole plant greyish-green. Rosette leaves lyrate, deeply incised. Stems procumbent with long branches. Calyx appendages small. Corolla 14-20(-35) mm. Stigmas (3-)5.

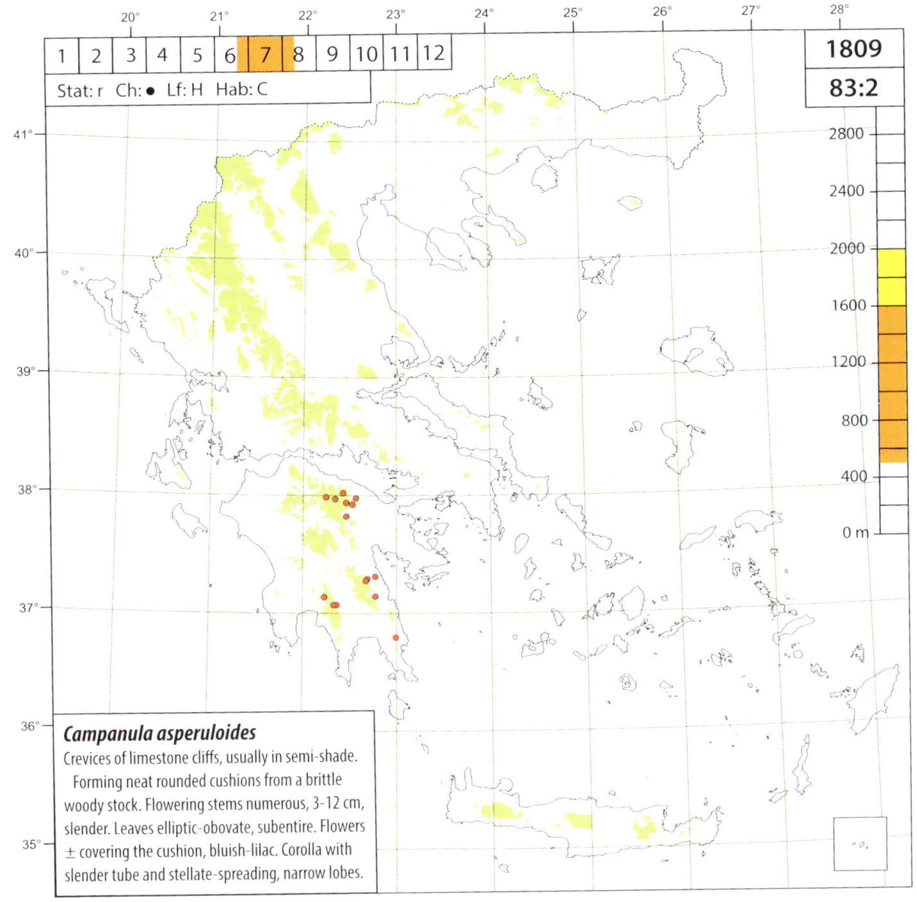

Campanula asperuloides

Crevices of limestone cliffs, usually in semi-shade.

Forming neat rounded cushions from a brittle woody stock. Flowering stems numerous, 3-12 cm, slender. Leaves elliptic-obovate, subentire. Flowers ± covering the cushion, bluish-lilac. Corolla with slender tube and stellate-spreading, narrow lobes.

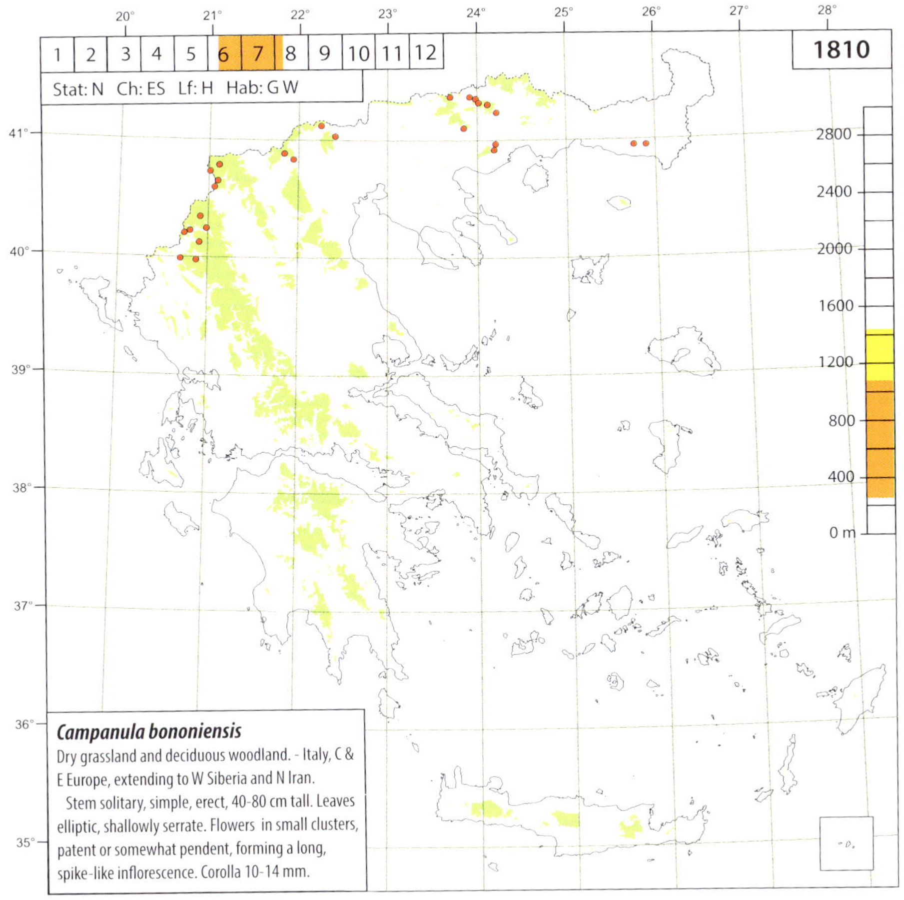

Campanula bononiensis

Dry grassland and deciduous woodland. - Italy, C & E Europe, extending to W Siberia and N Iran.

Stem solitary, simple, erect, 40-80 cm tall. Leaves elliptic, shallowly serrate. Flowers in small clusters, patent or somewhat pendent, forming a long, spike-like inflorescence. Corolla 10-14 mm.

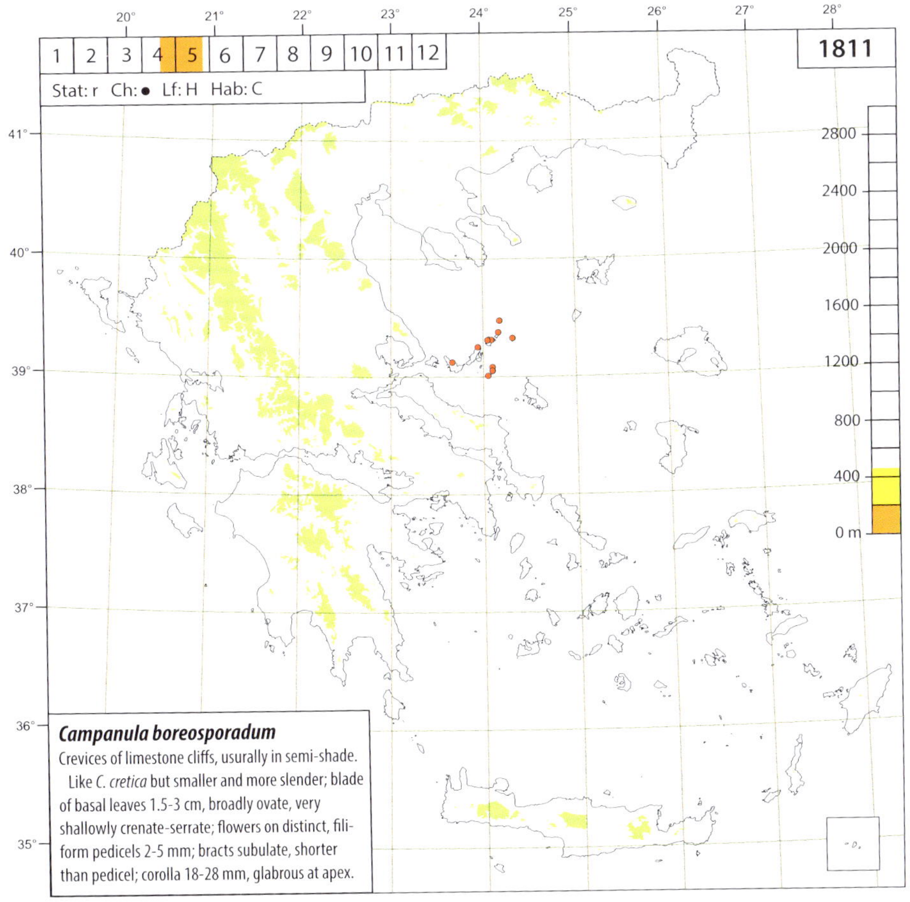

Campanula boreosporadum

Crevices of limestone cliffs, usurally in semi-shade.

Like *C. cretica* but smaller and more slender; blade of basal leaves 1.5-3 cm, broadly ovate, very shallowly crenate-serrate; flowers on distinct, filiform pedicels 2-5 mm; bracts subulate, shorter than pedicel; corolla 18-28 mm, glabrous at apex.

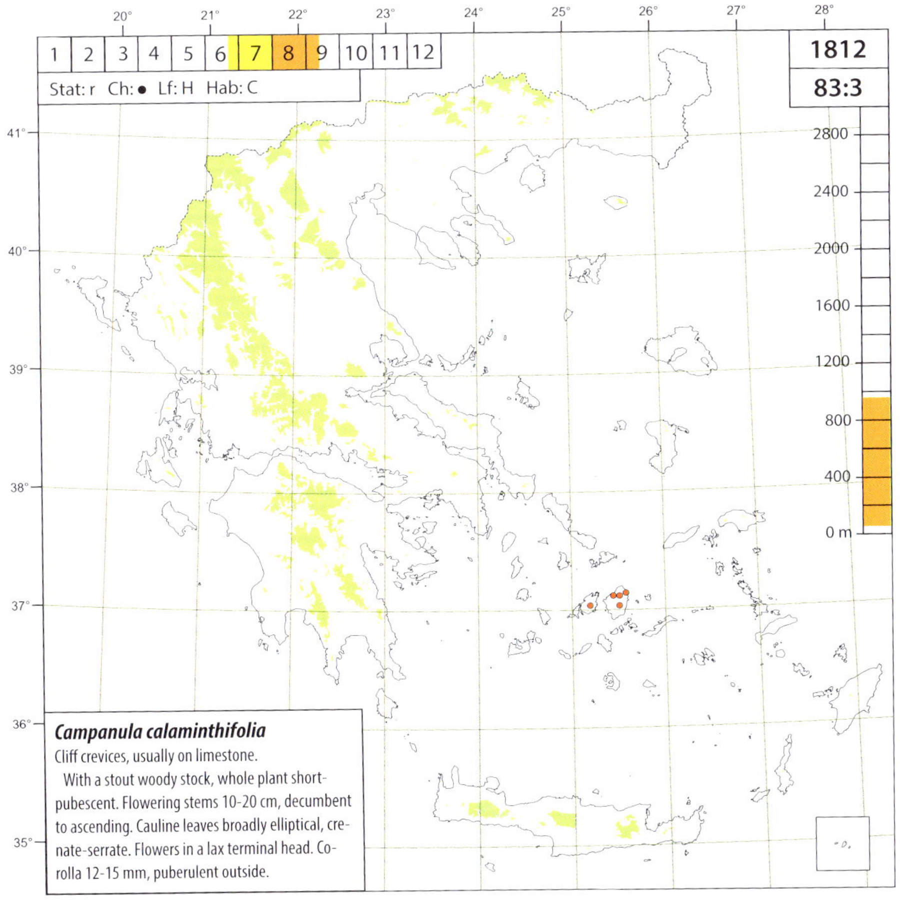

Campanula calaminthifolia

Cliff crevices, usually on limestone.

With a stout woody stock, whole plant short-pubescent. Flowering stems 10-20 cm, decumbent to ascending. Cauline leaves broadly elliptical, crenate-serrate. Flowers in a lax terminal head. Corolla 12-15 mm, puberulent outside.

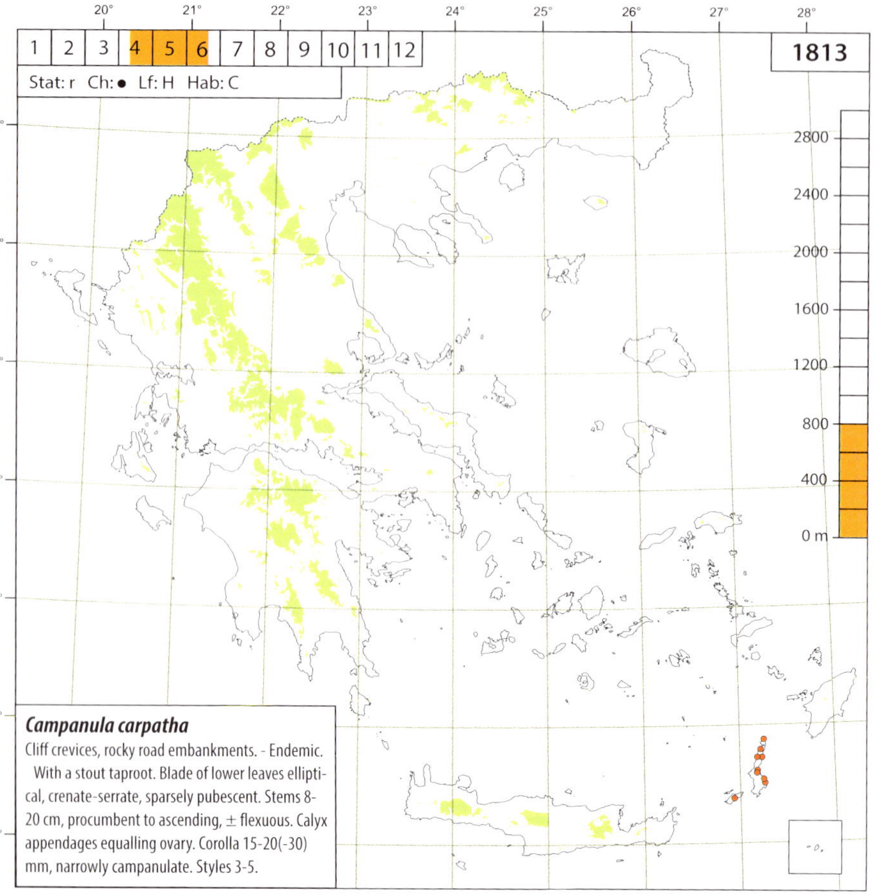

Campanula carpatha
Cliff crevices, rocky road embankments. - Endemic.
With a stout taproot. Blade of lower leaves elliptical, crenate-serrate, sparsely pubescent. Stems 8-20 cm, procumbent to ascending, ± flexuous. Calyx appendages equalling ovary. Corolla 15-20(-30) mm, narrowly campanulate. Styles 3-5.

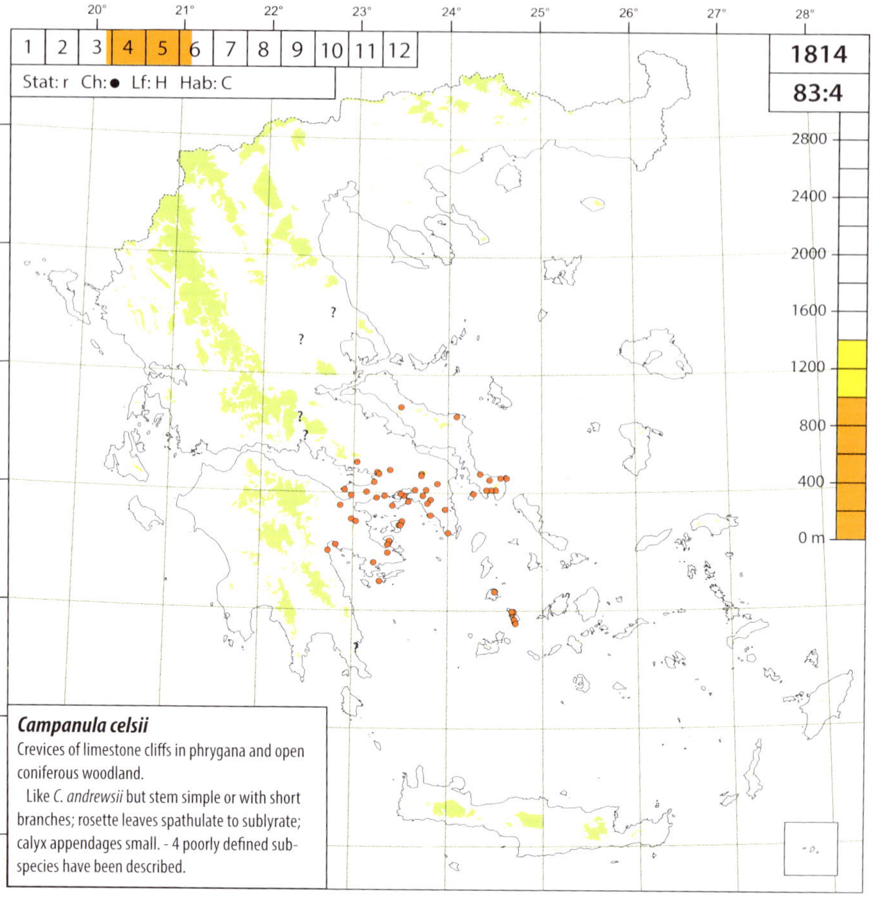

Campanula celsii
Crevices of limestone cliffs in phrygana and open coniferous woodland.
Like *C. andrewsii* but stem simple or with short branches; rosette leaves spathulate to sublyrate; calyx appendages small. - 4 poorly defined subspecies have been described.

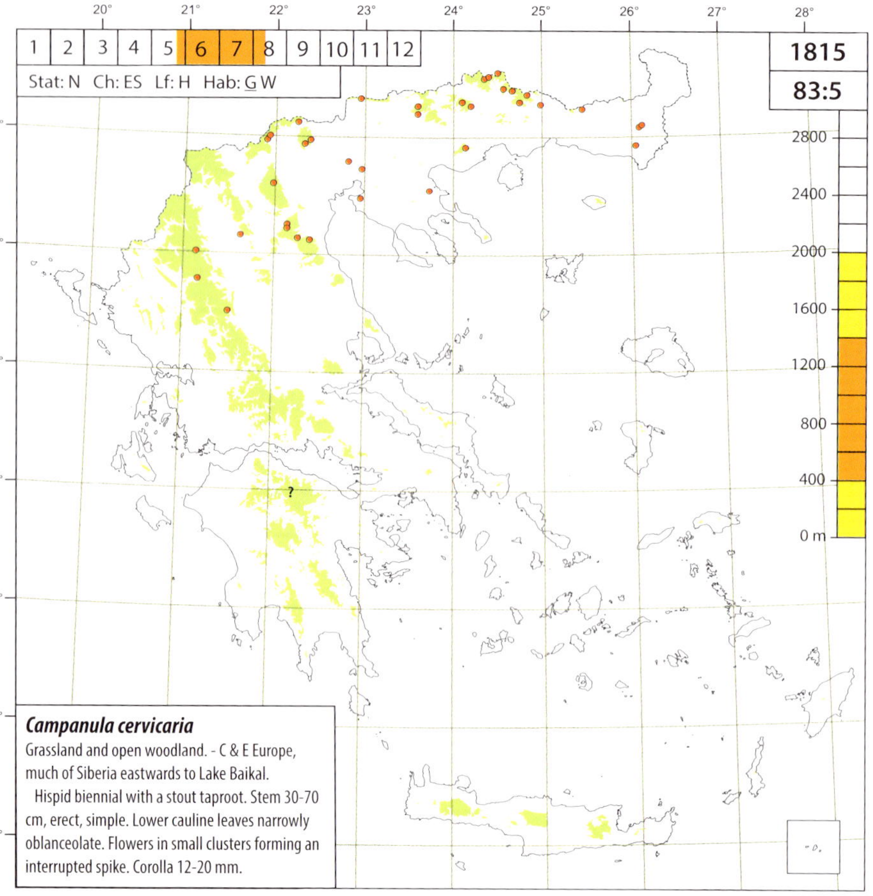

Campanula cervicaria
Grassland and open woodland. - C & E Europe, much of Siberia eastwards to Lake Baikal.
Hispid biennial with a stout taproot. Stem 30-70 cm, erect, simple. Lower cauline leaves narrowly oblanceolate. Flowers in small clusters forming an interrupted spike. Corolla 12-20 mm.

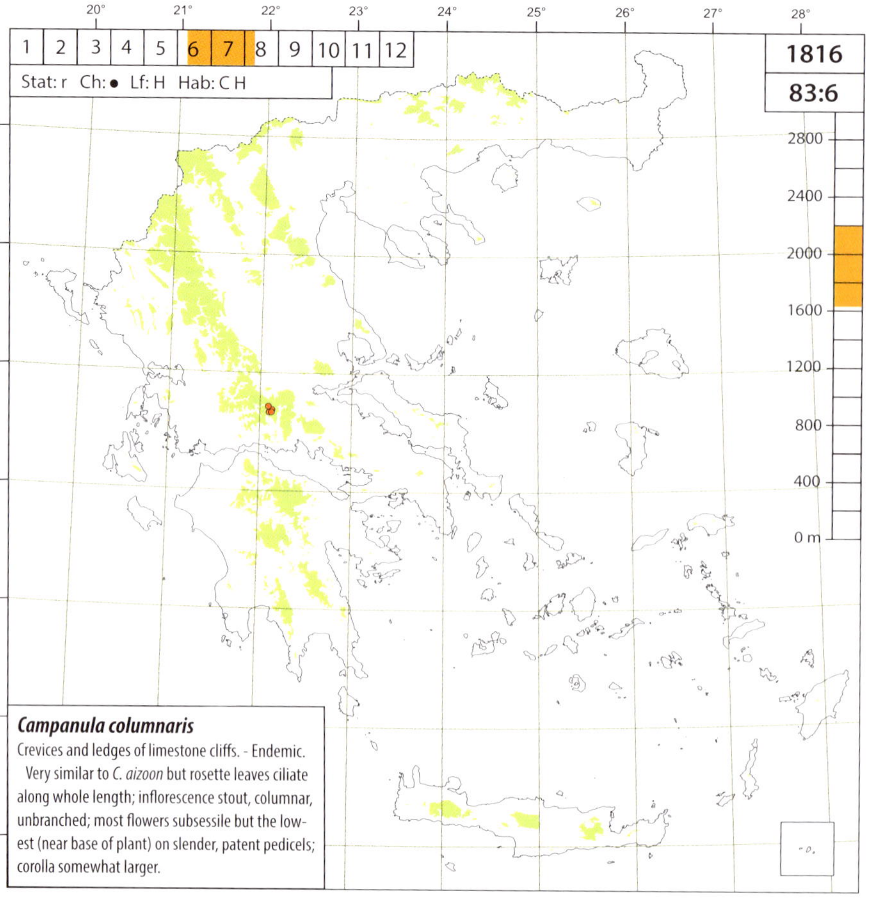

Campanula columnaris
Crevices and ledges of limestone cliffs. - Endemic.
Very similar to *C. aizoon* but rosette leaves ciliate along whole length; inflorescence stout, columnar, unbranched; most flowers subsessile but the lowest (near base of plant) on slender, patent pedicels; corolla somewhat larger.

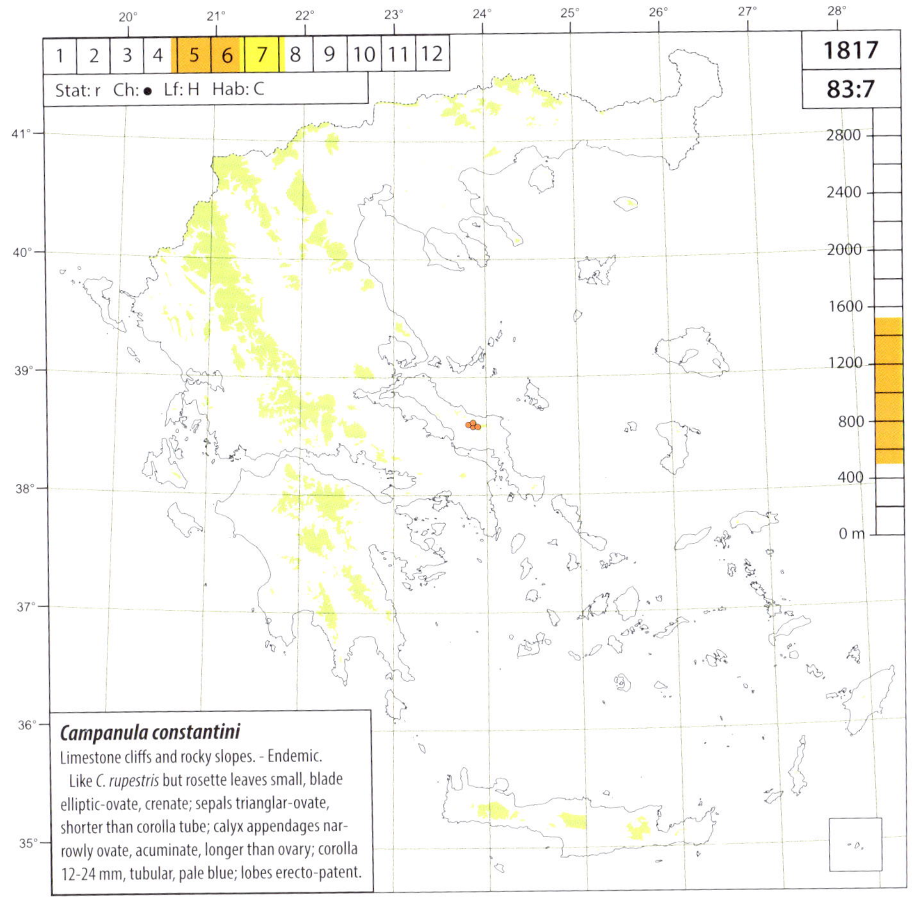

Campanula constantini

Limestone cliffs and rocky slopes. - Endemic.

Like *C. rupestris* but rosette leaves small, blade elliptic-ovate, crenate; sepals trianglar-ovate, shorter than corolla tube; calyx appendages narrowly ovate, acuminate, longer than ovary; corolla 12-24 mm, tubular, pale blue; lobes erecto-patent.

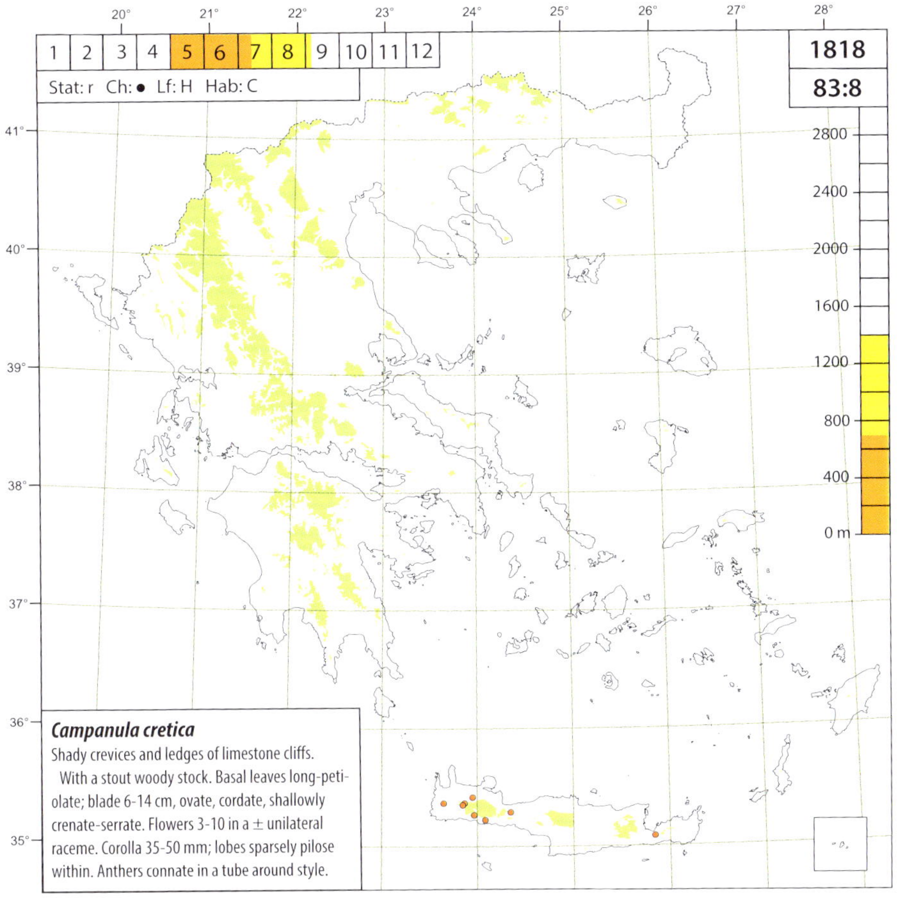

Campanula cretica

Shady crevices and ledges of limestone cliffs.

With a stout woody stock. Basal leaves long-petiolate; blade 6-14 cm, ovate, cordate, shallowly crenate-serrate. Flowers 3-10 in a ± unilateral raceme. Corolla 35-50 mm; lobes sparsely pilose within. Anthers connate in a tube around style.

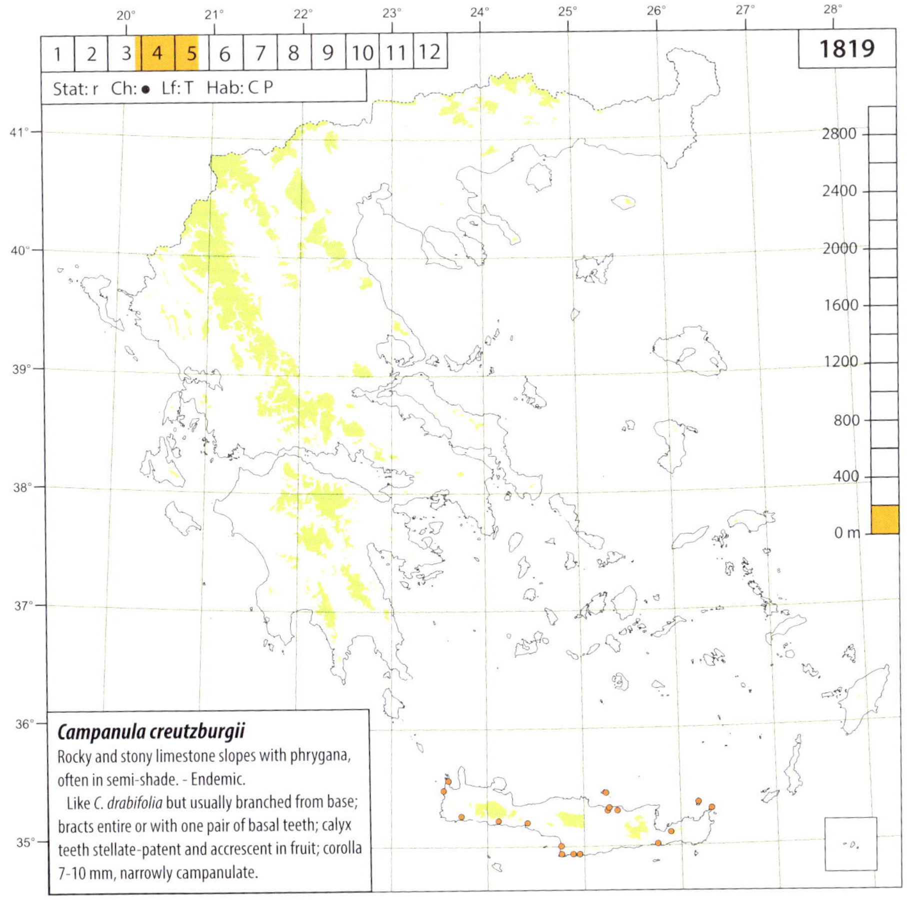

Campanula creutzburgii

Rocky and stony limestone slopes with phrygana, often in semi-shade. - Endemic.

Like *C. drabifolia* but usually branched from base; bracts entire or with one pair of basal teeth; calyx teeth stellate-patent and accrescent in fruit; corolla 7-10 mm, narrowly campanulate.

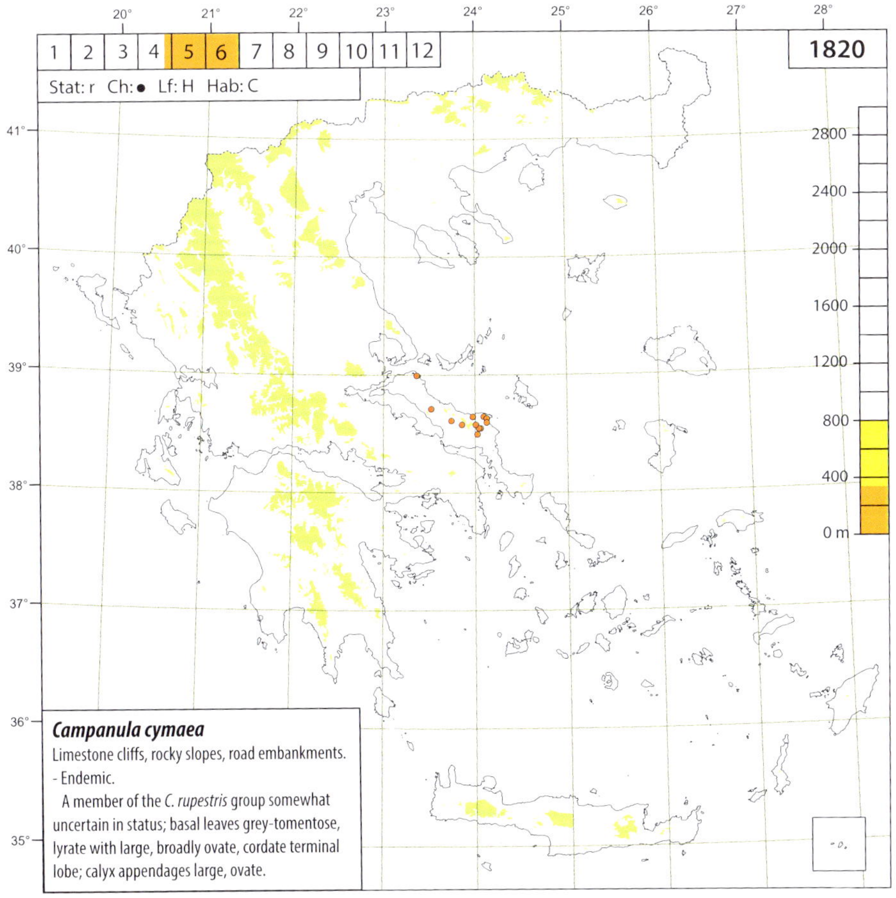

Campanula cymaea

Limestone cliffs, rocky slopes, road embankments. - Endemic.

A member of the *C. rupestris* group somewhat uncertain in status; basal leaves grey-tomentose, lyrate with large, broadly ovate, cordate terminal lobe; calyx appendages large, ovate.

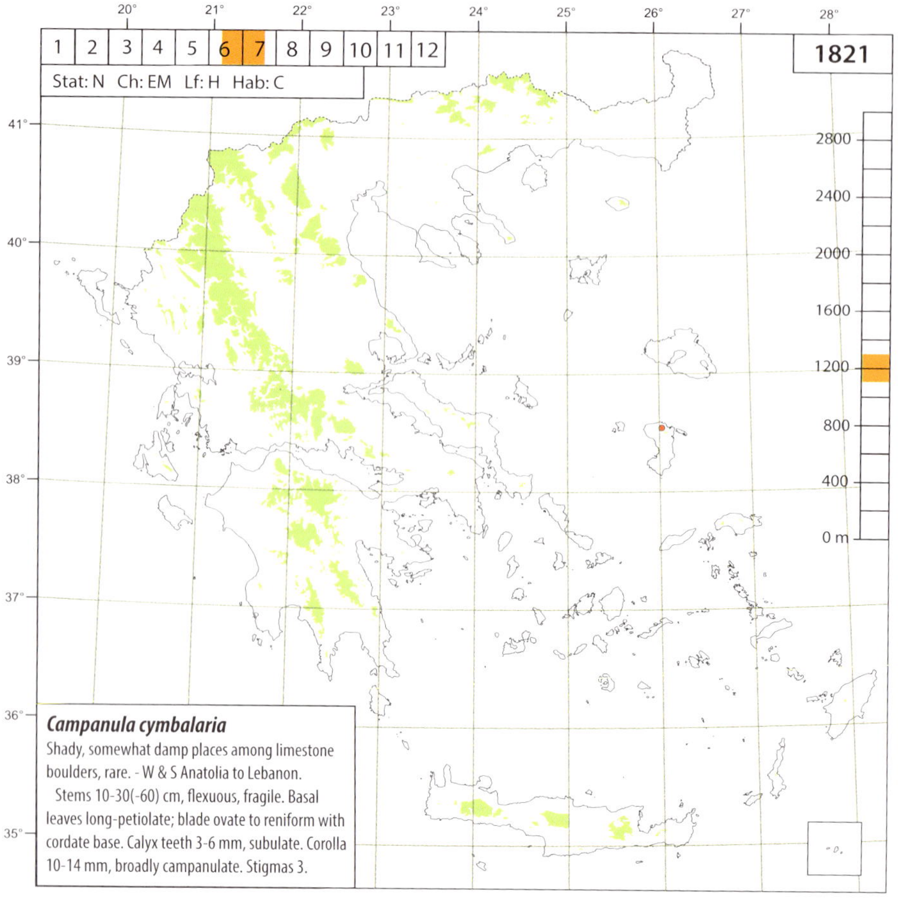

Campanula cymbalaria

Shady, somewhat damp places among limestone boulders, rare. - W & S Anatolia to Lebanon.

Stems 10-30(-60) cm, flexuous, fragile. Basal leaves long-petiolate; blade ovate to reniform with cordate base. Calyx teeth 3-6 mm, subulate. Corolla 10-14 mm, broadly campanulate. Stigmas 3.

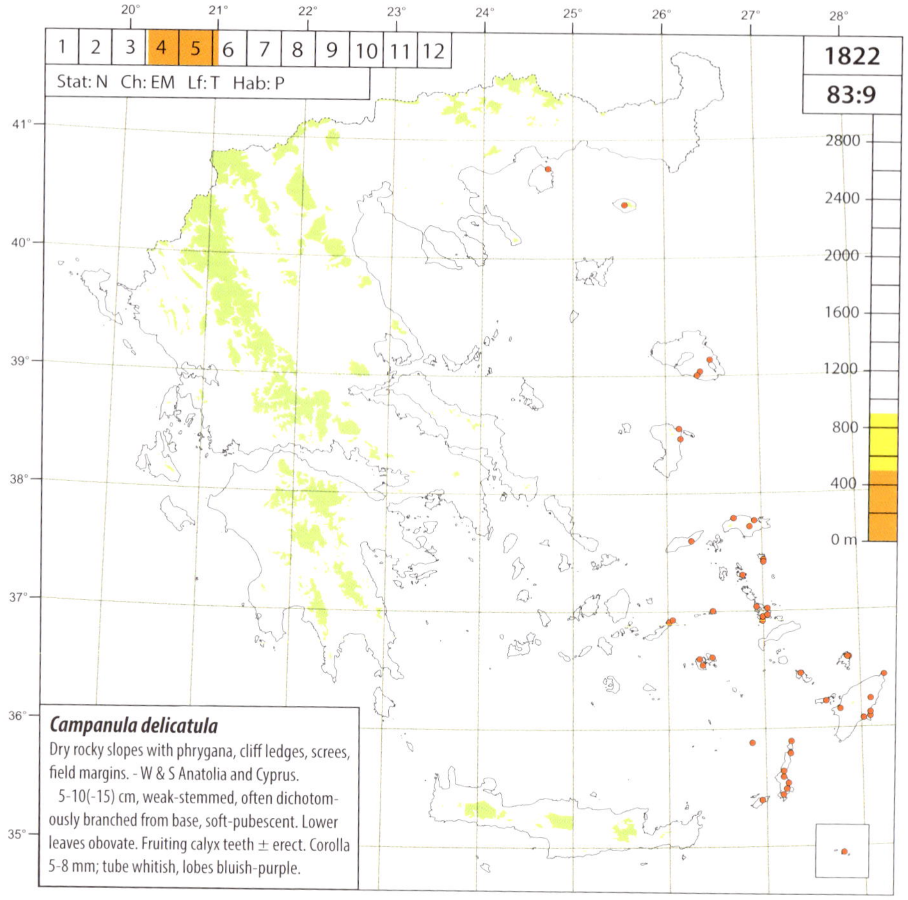

Campanula delicatula

Dry rocky slopes with phrygana, cliff ledges, screes, field margins. - W & S Anatolia and Cyprus.

5-10(-15) cm, weak-stemmed, often dichotomously branched from base, soft-pubescent. Lower leaves obovate. Fruiting calyx teeth ± erect. Corolla 5-8 mm; tube whitish, lobes bluish-purple.

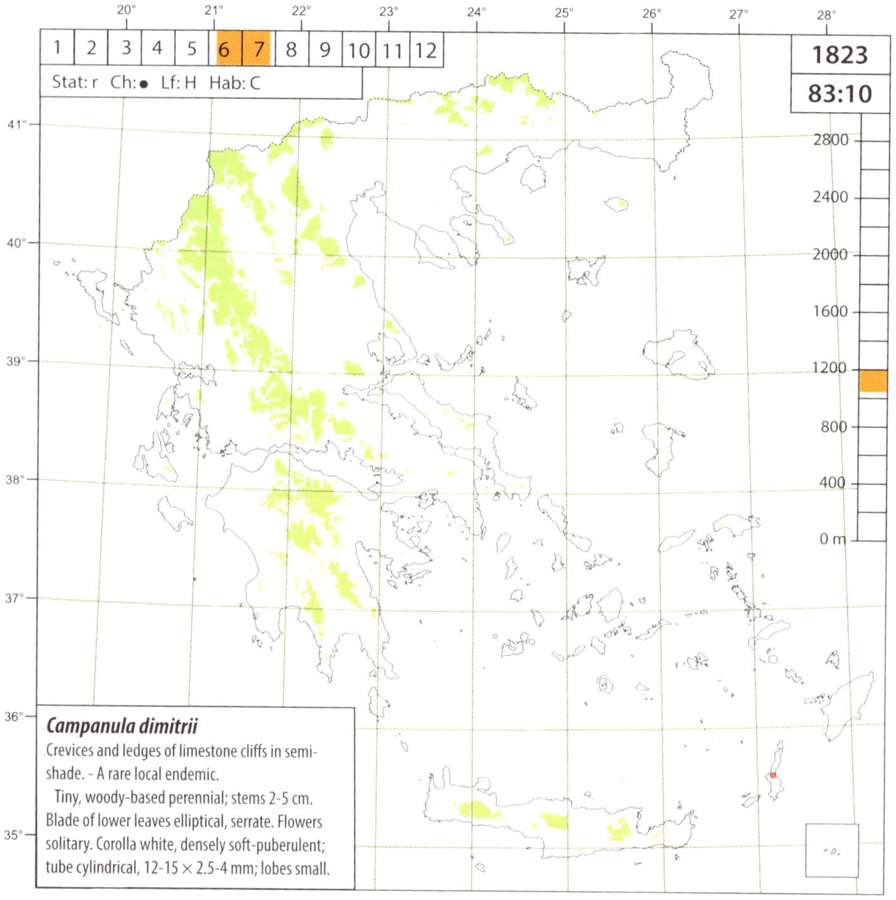

Campanula dimitrii

Crevices and ledges of limestone cliffs in semi-shade. - A rare local endemic.

Tiny, woody-based perennial; stems 2-5 cm. Blade of lower leaves elliptical, serrate. Flowers solitary. Corolla white, densely soft-puberulent; tube cylindrical, 12-15 × 2.5-4 mm; lobes small.

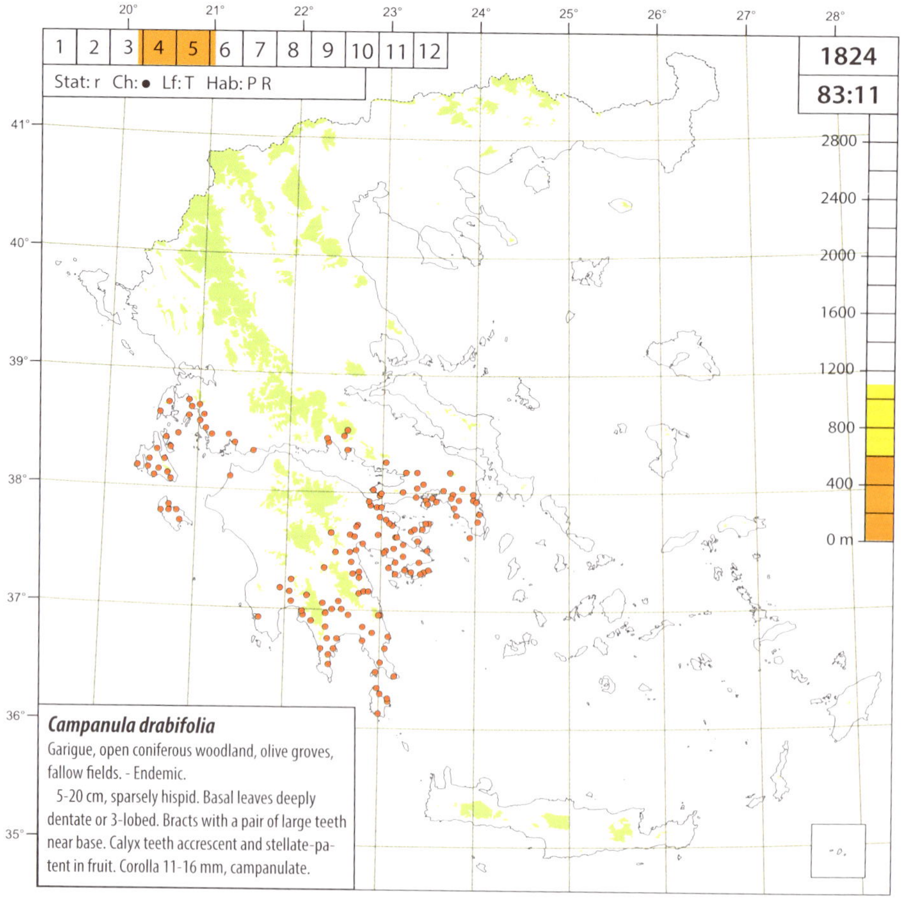

Campanula drabifolia

Garigue, open coniferous woodland, olive groves, fallow fields. - Endemic.

5-20 cm, sparsely hispid. Basal leaves deeply dentate or 3-lobed. Bracts with a pair of large teeth near base. Calyx teeth accrescent and stellate-patent in fruit. Corolla 11-16 mm, campanulate.

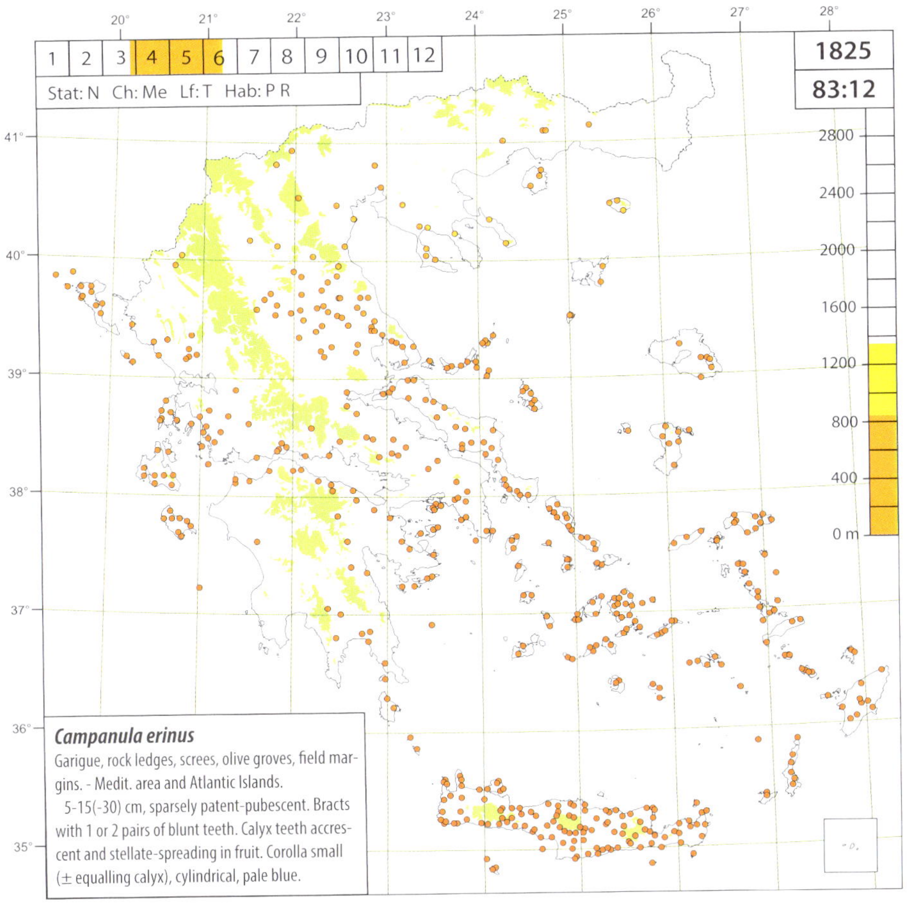

Campanula erinus

Garigue, rock ledges, screes, olive groves, field margins. - Medit. area and Atlantic Islands.

5-15(-30) cm, sparsely patent-pubescent. Bracts with 1 or 2 pairs of blunt teeth. Calyx teeth accrescent and stellate-spreading in fruit. Corolla small (± equalling calyx), cylindrical, pale blue.

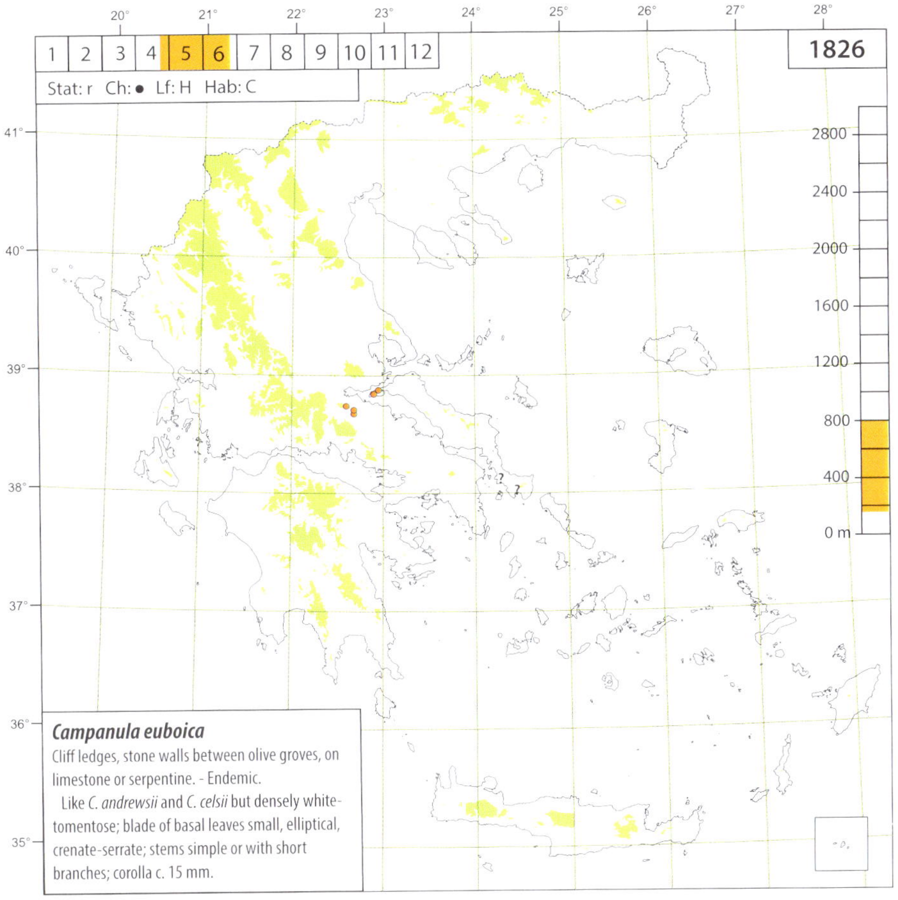

Campanula euboica

Cliff ledges, stone walls between olive groves, on limestone or serpentine. - Endemic.

Like *C. andrewsii* and *C. celsii* but densely white-tomentose; blade of basal leaves small, elliptical, crenate-serrate; stems simple or with short branches; corolla c. 15 mm.

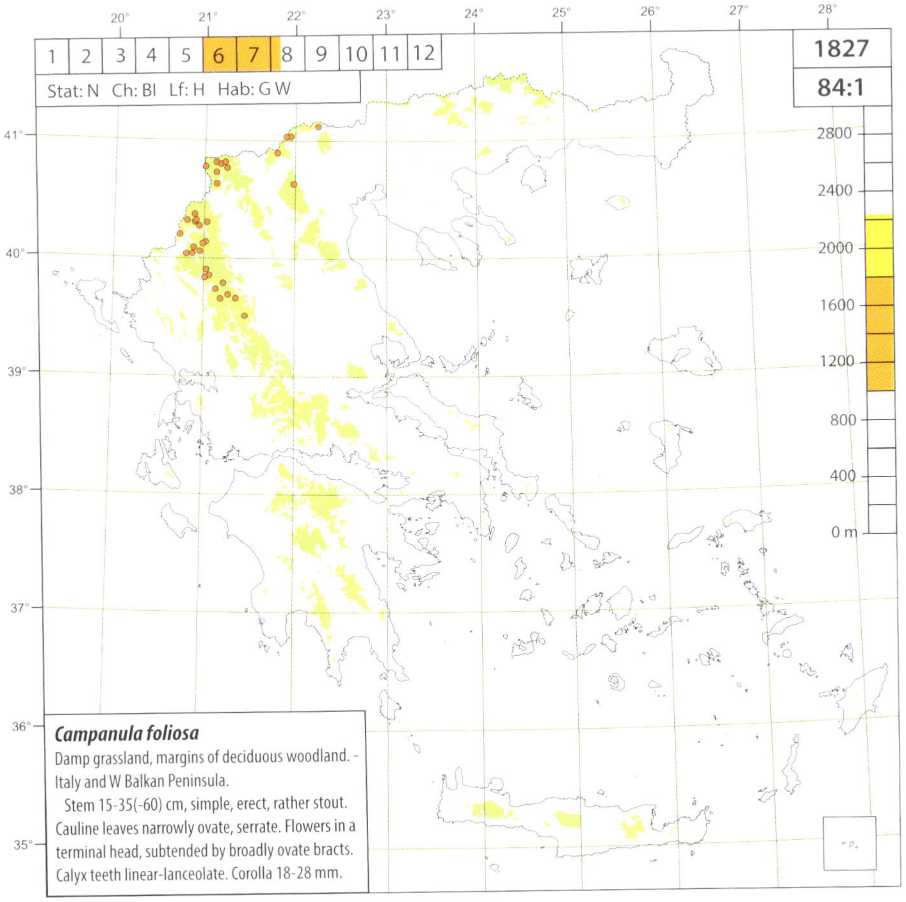

Campanula foliosa

Damp grassland, margins of deciduous woodland. - Italy and W Balkan Peninsula.

Stem 15-35(-60) cm, simple, erect, rather stout. Cauline leaves narrowly ovate, serrate. Flowers in a terminal head, subtended by broadly ovate bracts. Calyx teeth linear-lanceolate. Corolla 18-28 mm.

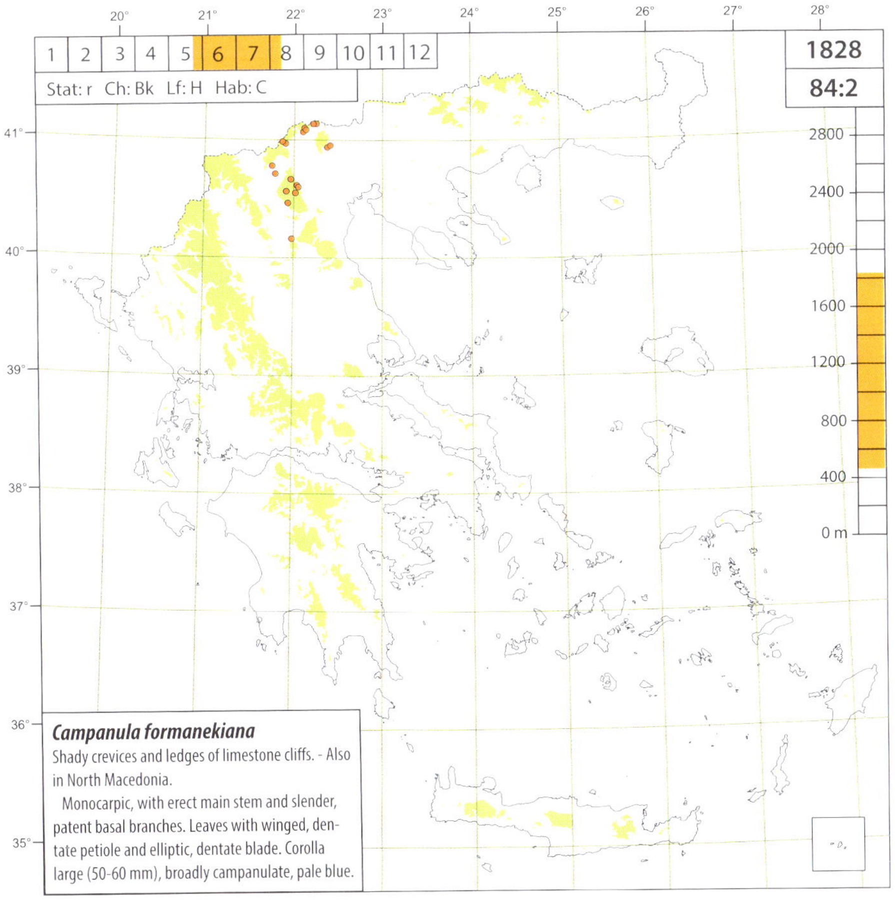

Campanula formanekiana

Shady crevices and ledges of limestone cliffs. - Also in North Macedonia.

Monocarpic, with erect main stem and slender, patent basal branches. Leaves with winged, dentate petiole and elliptic, dentate blade. Corolla large (50-60 mm), broadly campanulate, pale blue.

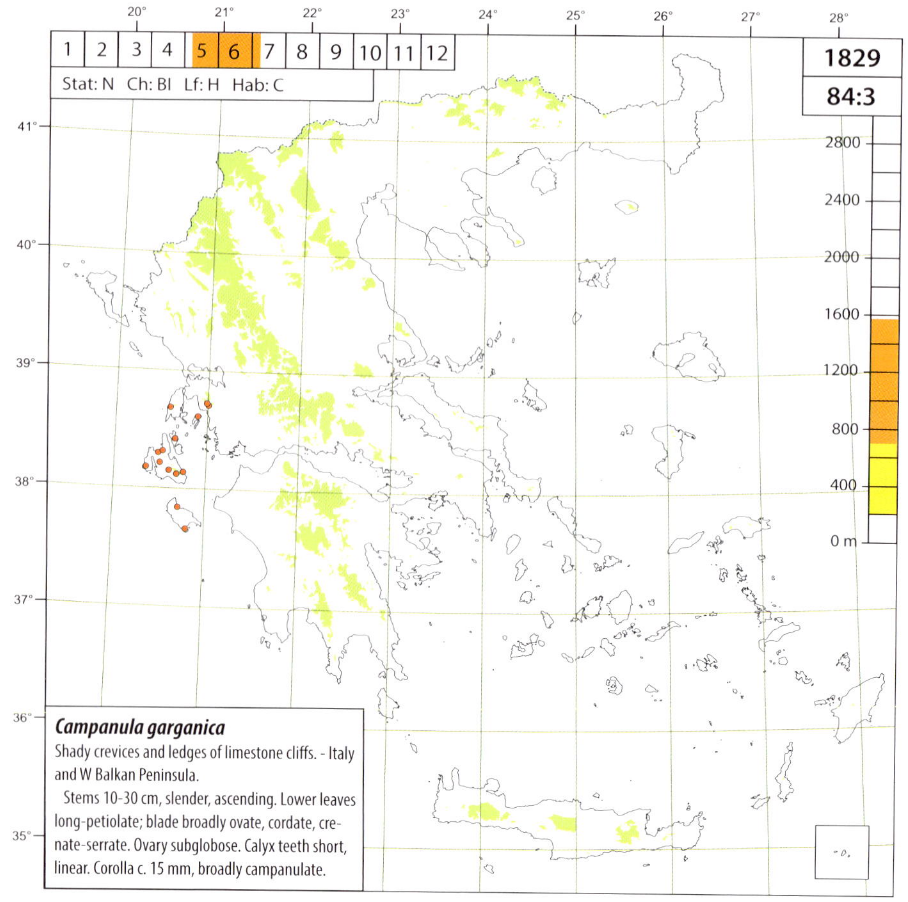

Campanula garganica

Shady crevices and ledges of limestone cliffs. - Italy and W Balkan Peninsula.

Stems 10-30 cm, slender, ascending. Lower leaves long-petiolate; blade broadly ovate, cordate, crenate-serrate. Ovary subglobose. Calyx teeth short, linear. Corolla c. 15 mm, broadly campanulate.

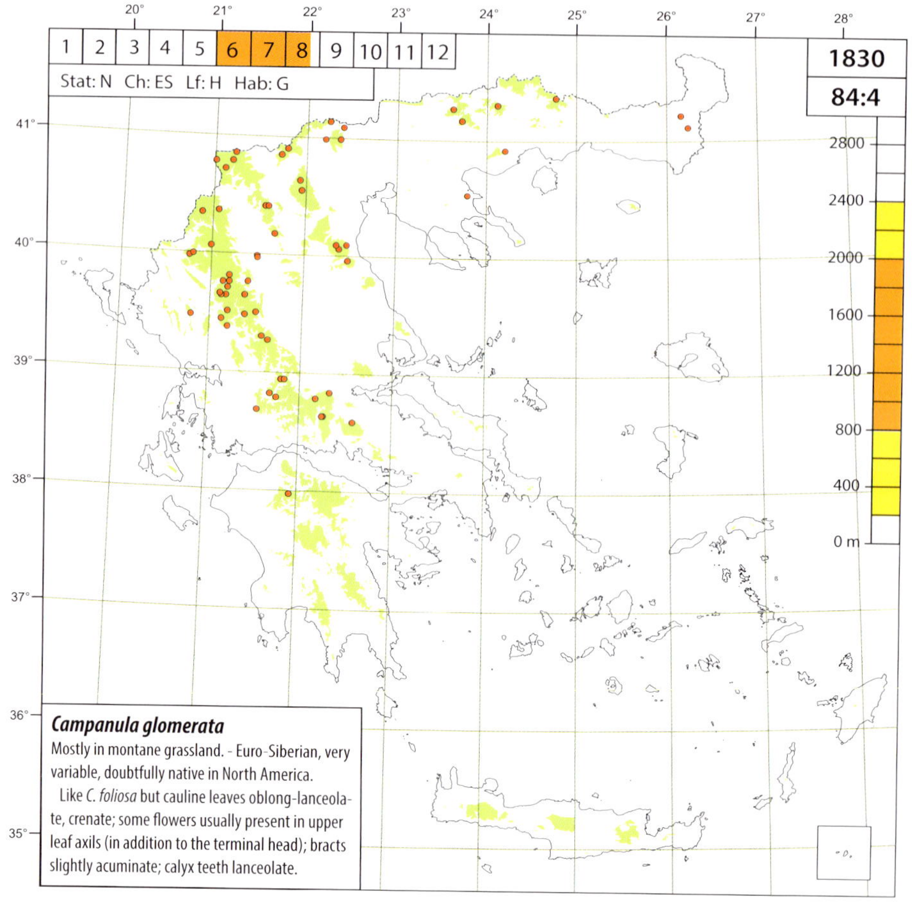

Campanula glomerata

Mostly in montane grassland. - Euro-Siberian, very variable, doubtfully native in North America.

Like *C. foliosa* but cauline leaves oblong-lanceolate, crenate; some flowers usually present in upper leaf axils (in addition to the terminal head); bracts slightly acuminate; calyx teeth lanceolate.

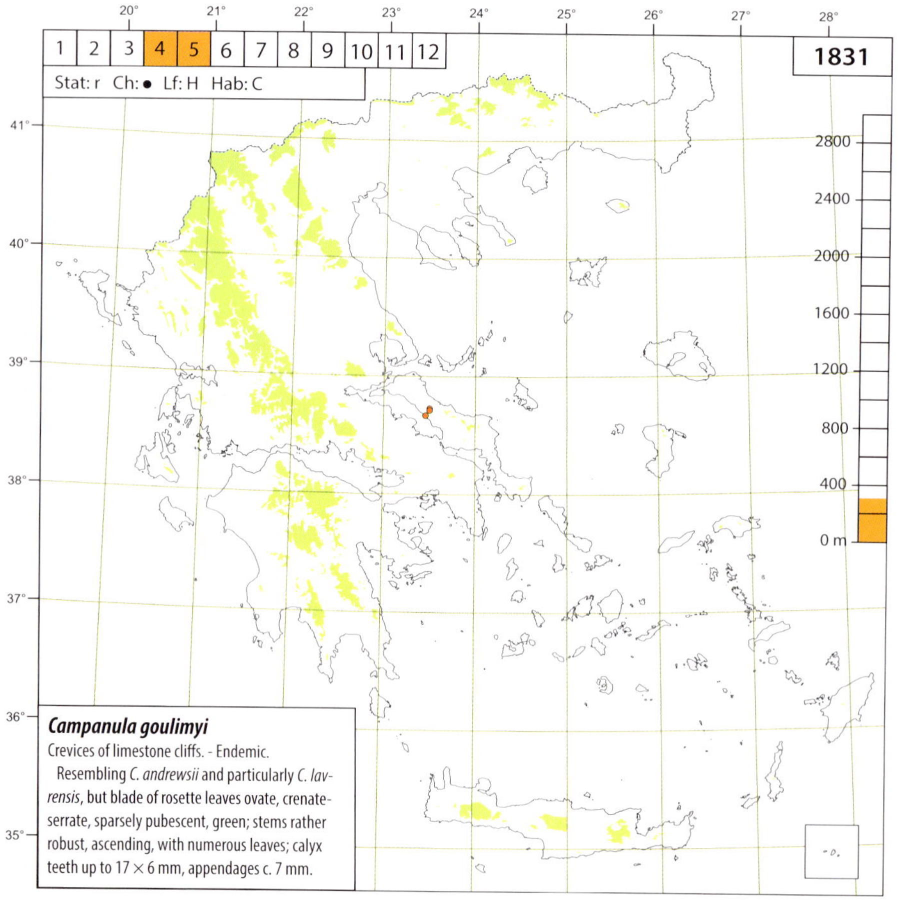

Campanula goulimyi

Crevices of limestone cliffs. - Endemic.

Resembling *C. andrewsii* and particularly *C. lavrensis*, but blade of rosette leaves ovate, crenate-serrate, sparsely pubescent, green; stems rather robust, ascending, with numerous leaves; calyx teeth up to 17 × 6 mm, appendages c. 7 mm.

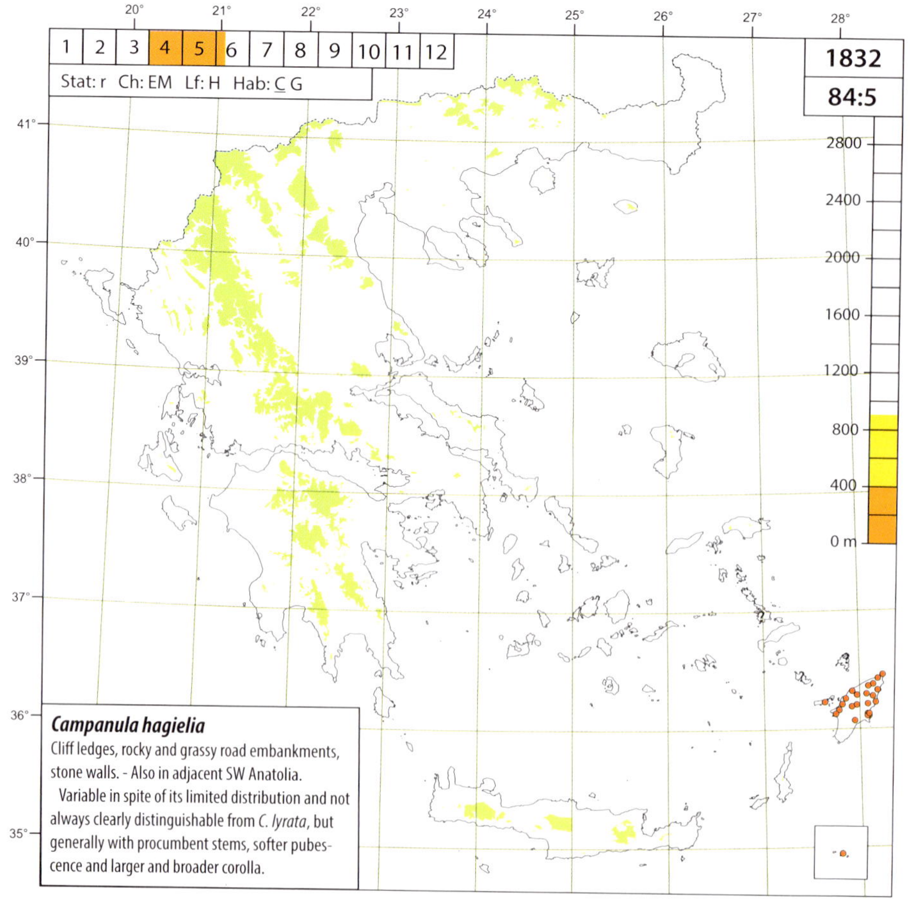

Campanula hagielia

Cliff ledges, rocky and grassy road embankments, stone walls. - Also in adjacent SW Anatolia.

Variable in spite of its limited distribution and not always clearly distinguishable from *C. lyrata*, but generally with procumbent stems, softer pubescence and larger and broader corolla.

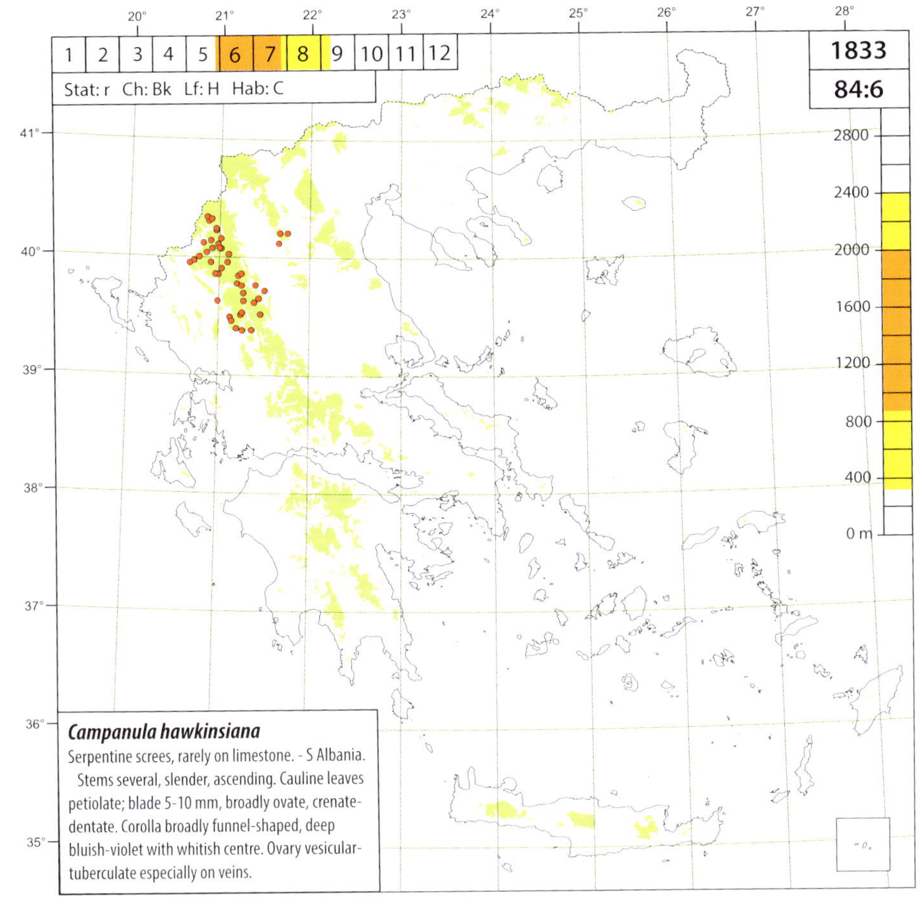

Campanula hawkinsiana

Serpentine screes, rarely on limestone. - S Albania.

Stems several, slender, ascending. Cauline leaves petiolate; blade 5-10 mm, broadly ovate, crenate-dentate. Corolla broadly funnel-shaped, deep bluish-violet with whitish centre. Ovary vesicular-tuberculate especially on veins.

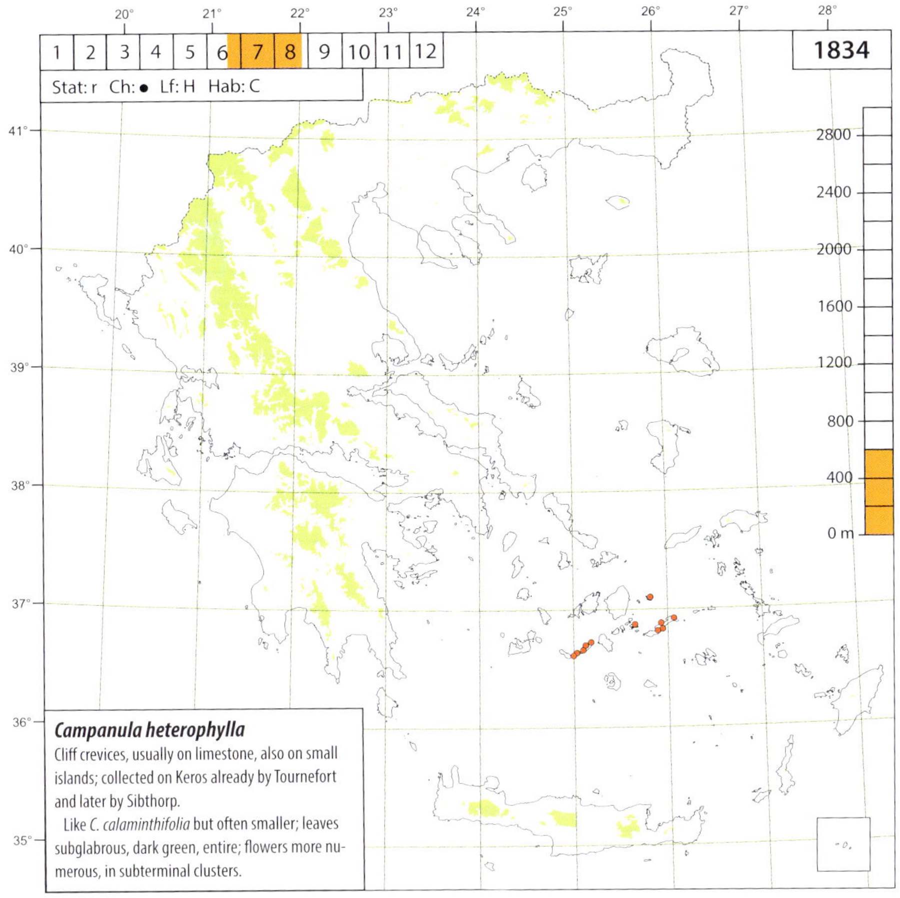

Campanula heterophylla

Cliff crevices, usually on limestone, also on small islands; collected on Keros already by Tournefort and later by Sibthorp.

Like *C. calaminthifolia* but often smaller; leaves subglabrous, dark green, entire; flowers more numerous, in subterminal clusters.

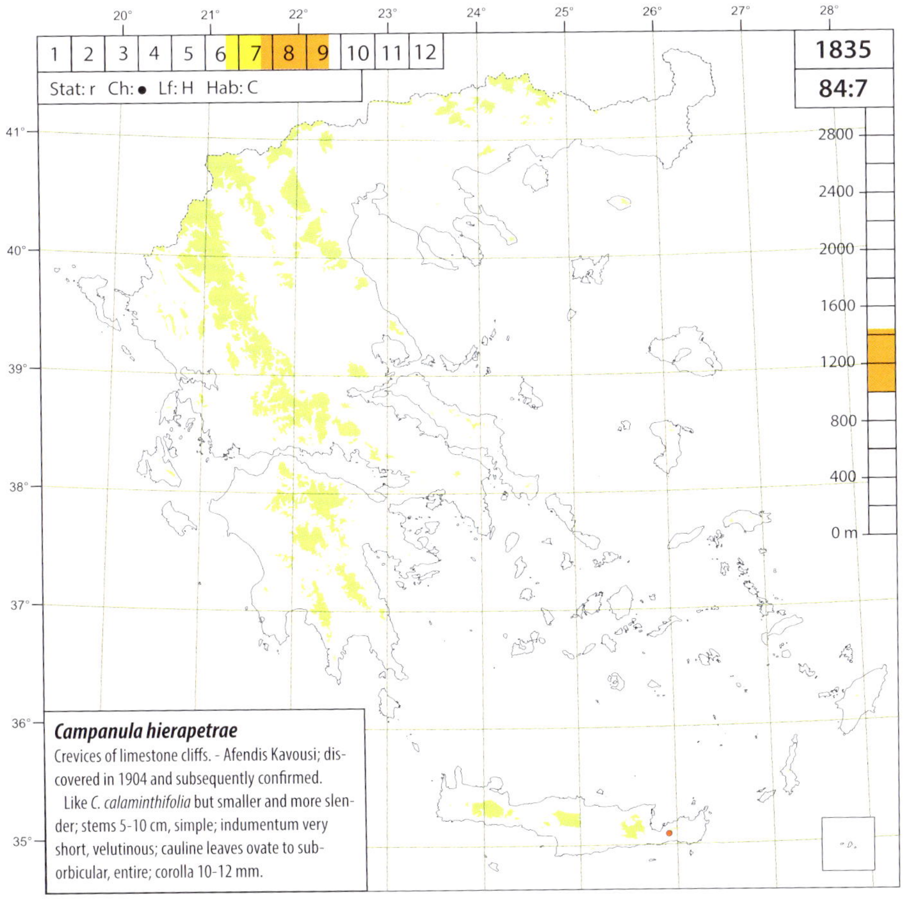

Campanula hierapetrae

Crevices of limestone cliffs. - Afendis Kavousi; discovered in 1904 and subsequently confirmed.

Like *C. calaminthifolia* but smaller and more slender; stems 5-10 cm, simple; indumentum very short, velutinous; cauline leaves ovate to sub-orbicular, entire; corolla 10-12 mm.

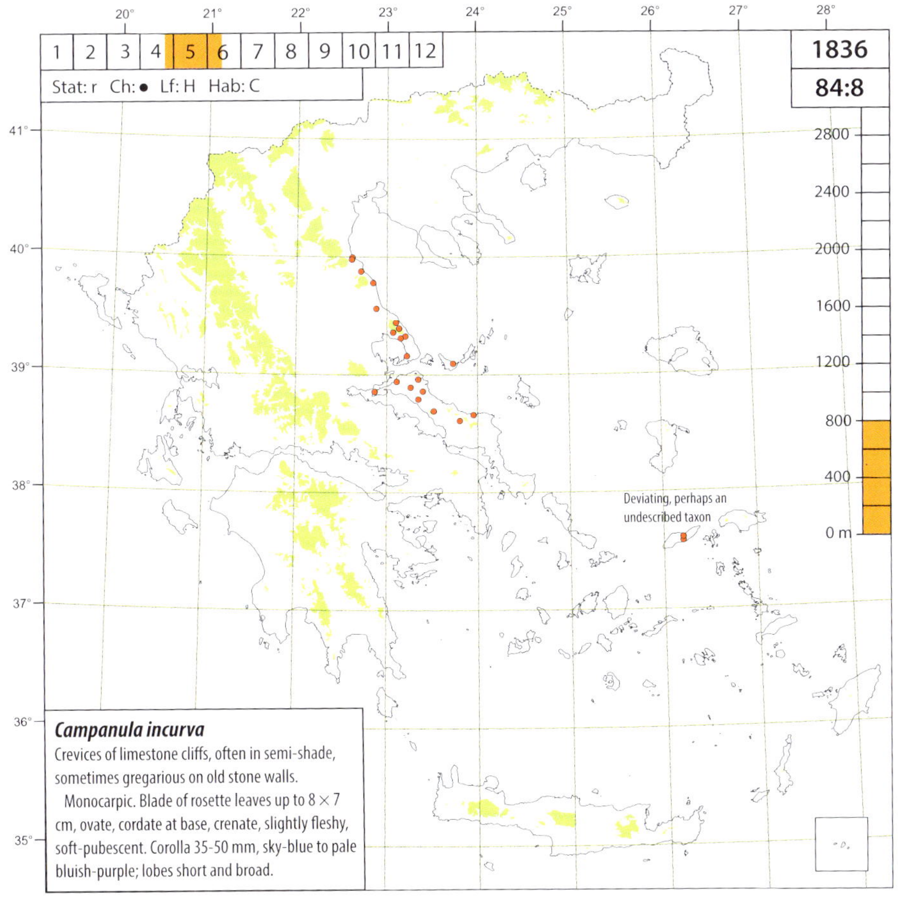

Campanula incurva

Crevices of limestone cliffs, often in semi-shade, sometimes gregarious on old stone walls.

Monocarpic. Blade of rosette leaves up to 8 × 7 cm, ovate, cordate at base, crenate, slightly fleshy, soft-pubescent. Corolla 35-50 mm, sky-blue to pale bluish-purple; lobes short and broad.

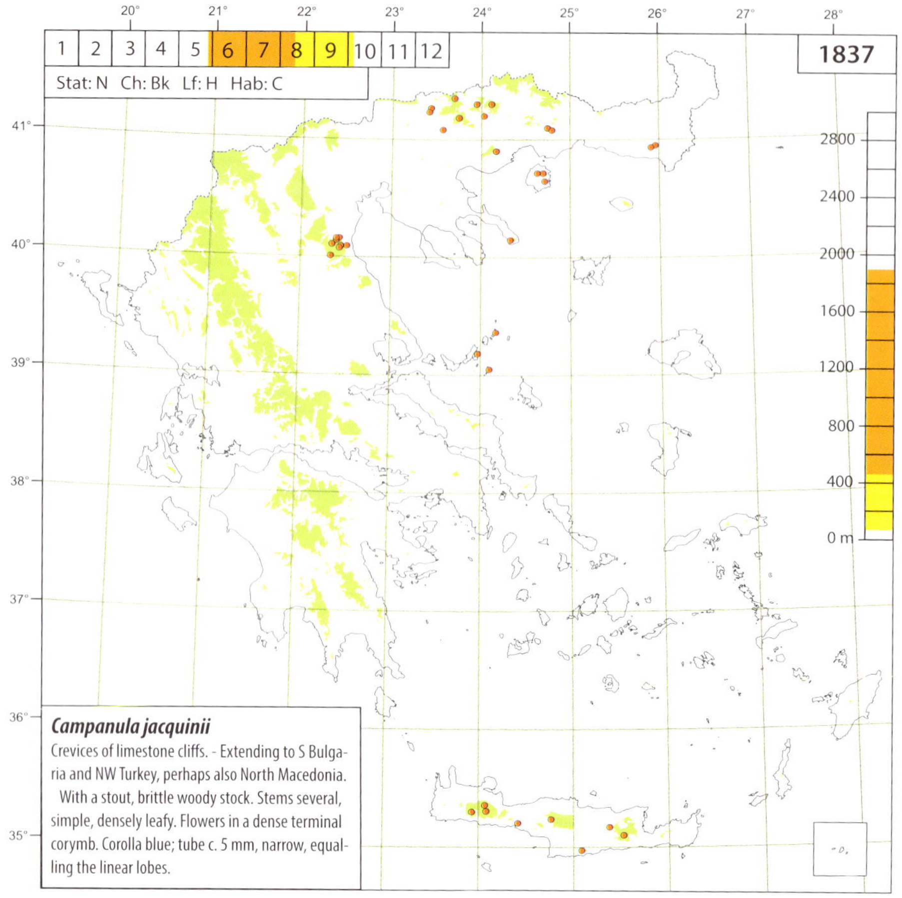

Campanula jacquinii
Crevices of limestone cliffs. - Extending to S Bulgaria and NW Turkey, perhaps also North Macedonia.
With a stout, brittle woody stock. Stems several, simple, densely leafy. Flowers in a dense terminal corymb. Corolla blue; tube c. 5 mm, narrow, equalling the linear lobes.

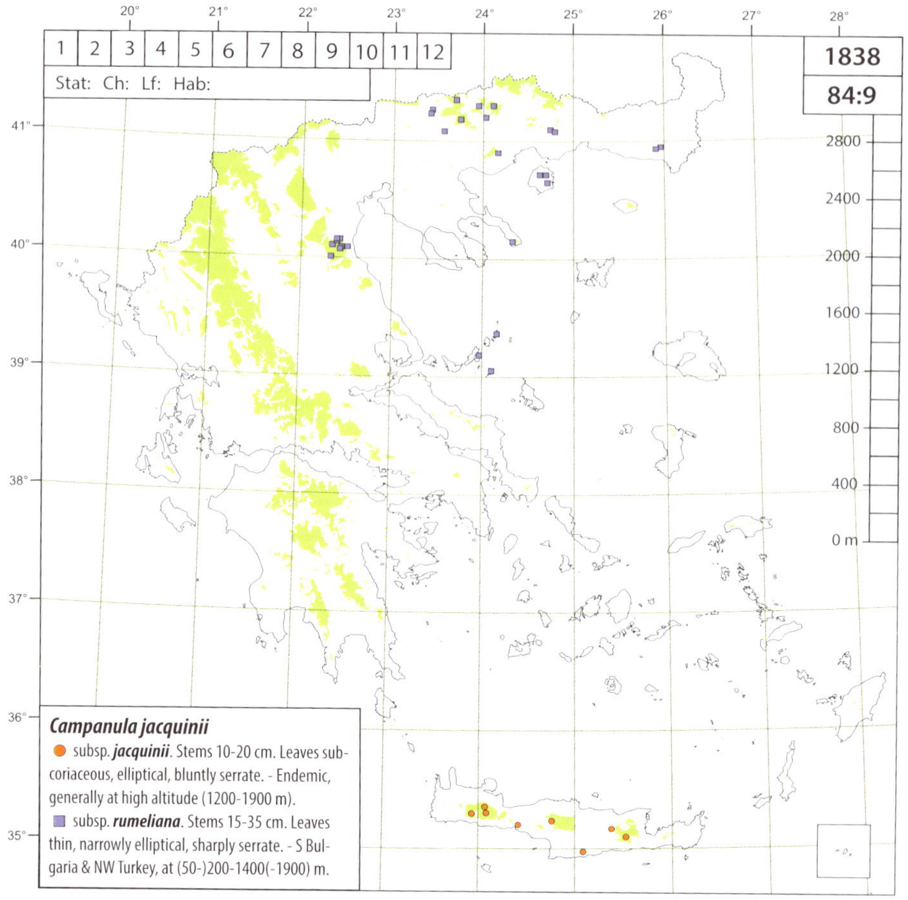

Campanula jacquinii
● subsp. ***jacquinii***. Stems 10-20 cm. Leaves subcoriaceous, elliptical, bluntly serrate. - Endemic, generally at high altitude (1200-1900 m).
■ subsp. ***rumeliana***. Stems 15-35 cm. Leaves thin, narrowly elliptical, sharply serrate. - S Bulgaria & NW Turkey, at (50-)200-1400(-1900) m.

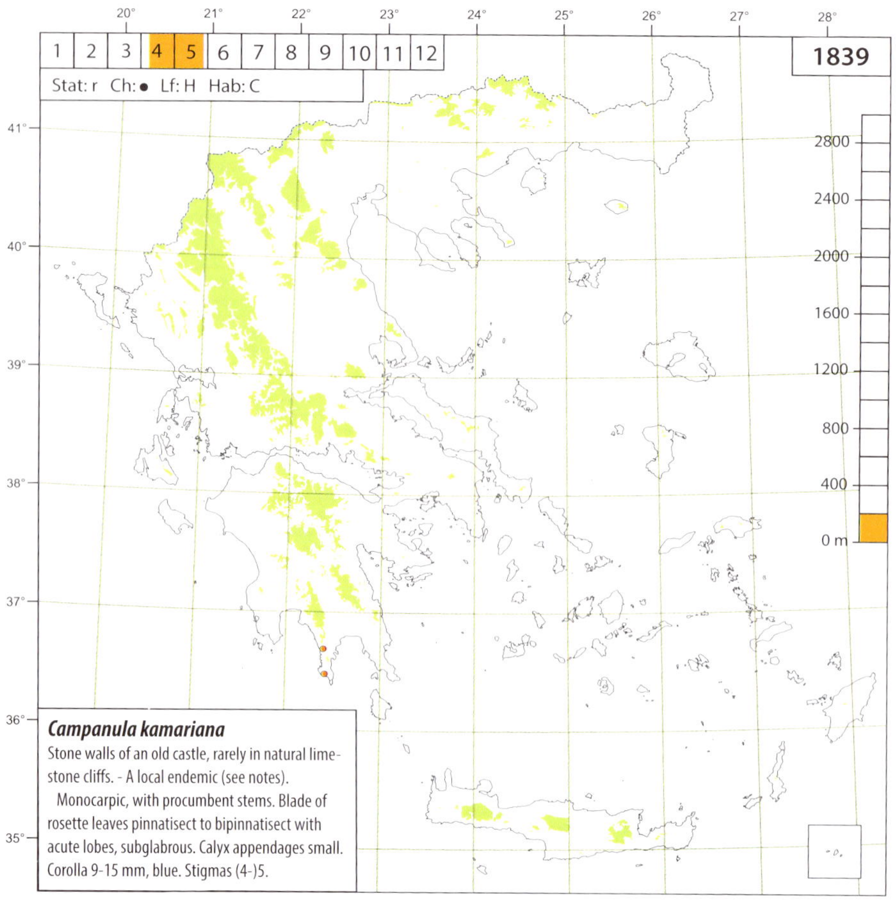

Campanula kamariana
Stone walls of an old castle, rarely in natural limestone cliffs. - A local endemic (see notes).
Monocarpic, with procumbent stems. Blade of rosette leaves pinnatisect to bipinnatisect with acute lobes, subglabrous. Calyx appendages small. Corolla 9-15 mm, blue. Stigmas (4-)5.

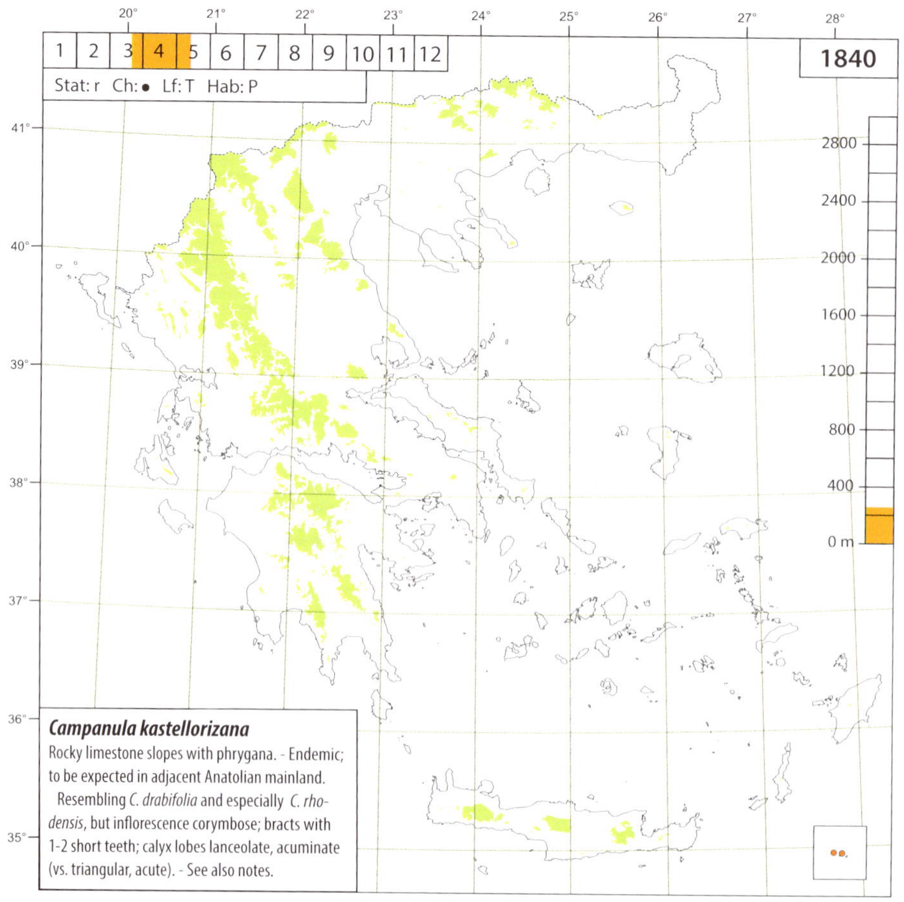

Campanula kastellorizana
Rocky limestone slopes with phrygana. - Endemic; to be expected in adjacent Anatolian mainland.
Resembling *C. drabifolia* and especially *C. rhodensis*, but inflorescence corymbose; bracts with 1-2 short teeth; calyx lobes lanceolate, acuminate (vs. triangular, acute). - See also notes.

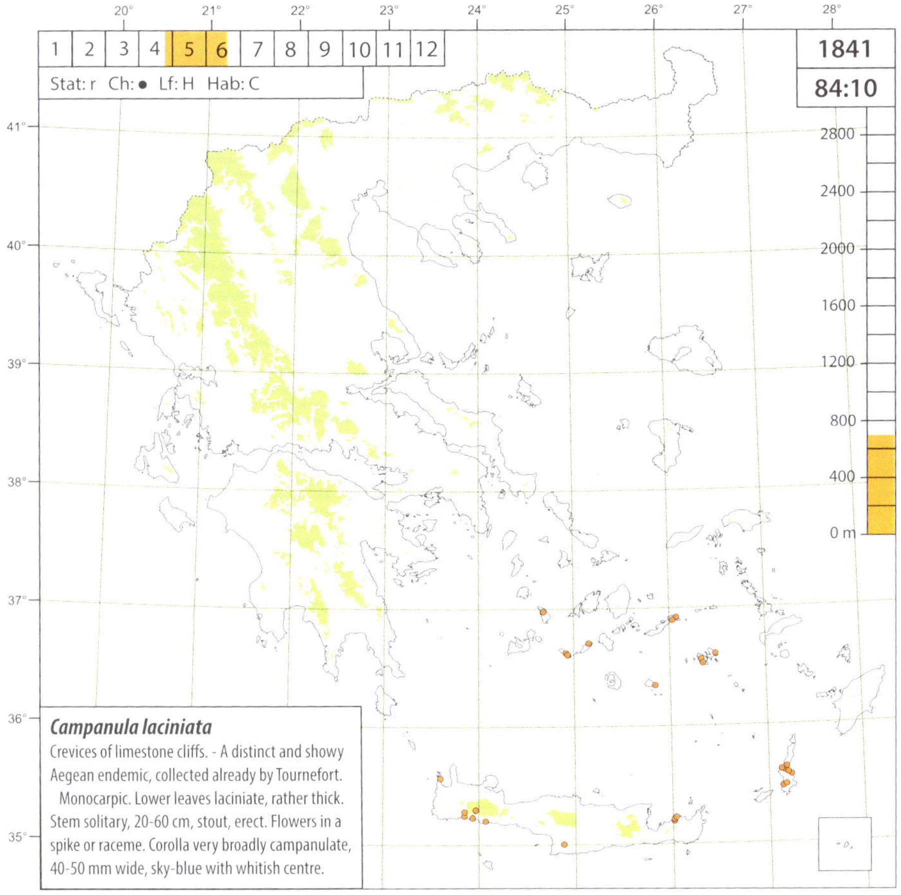

Campanula laciniata

Crevices of limestone cliffs. - A distinct and showy Aegean endemic, collected already by Tournefort.

Monocarpic. Lower leaves laciniate, rather thick. Stem solitary, 20-60 cm, stout, erect. Flowers in a spike or raceme. Corolla very broadly campanulate, 40-50 mm wide, sky-blue with whitish centre.

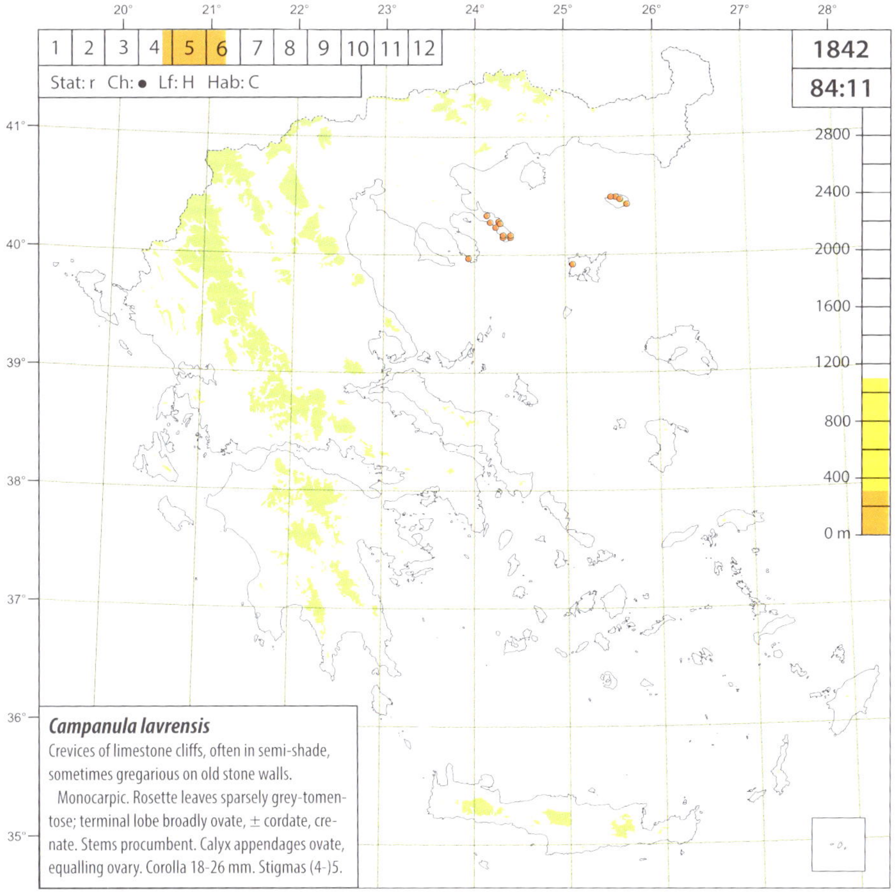

Campanula lavrensis

Crevices of limestone cliffs, often in semi-shade, sometimes gregarious on old stone walls.

Monocarpic. Rosette leaves sparsely grey-tomentose; terminal lobe broadly ovate, ± cordate, crenate. Stems procumbent. Calyx appendages ovate, equalling ovary. Corolla 18-26 mm. Stigmas (4-)5.

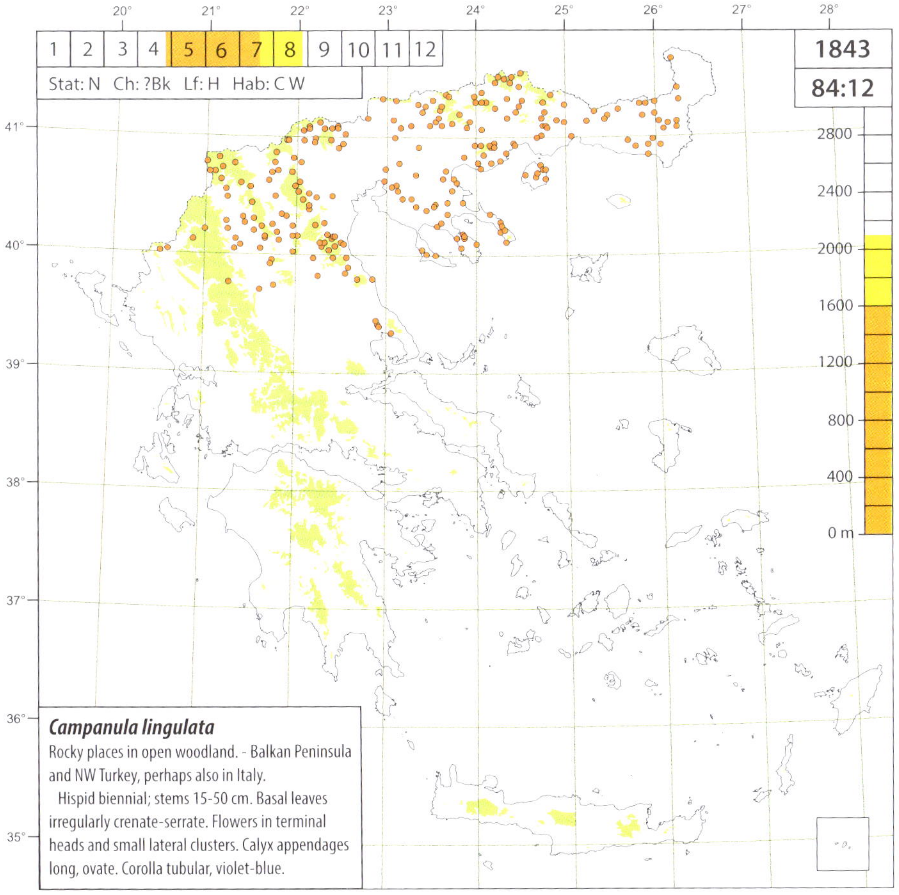

Campanula lingulata

Rocky places in open woodland. - Balkan Peninsula and NW Turkey, perhaps also in Italy.

Hispid biennial; stems 15-50 cm. Basal leaves irregularly crenate-serrate. Flowers in terminal heads and small lateral clusters. Calyx appendages long, ovate. Corolla tubular, violet-blue.

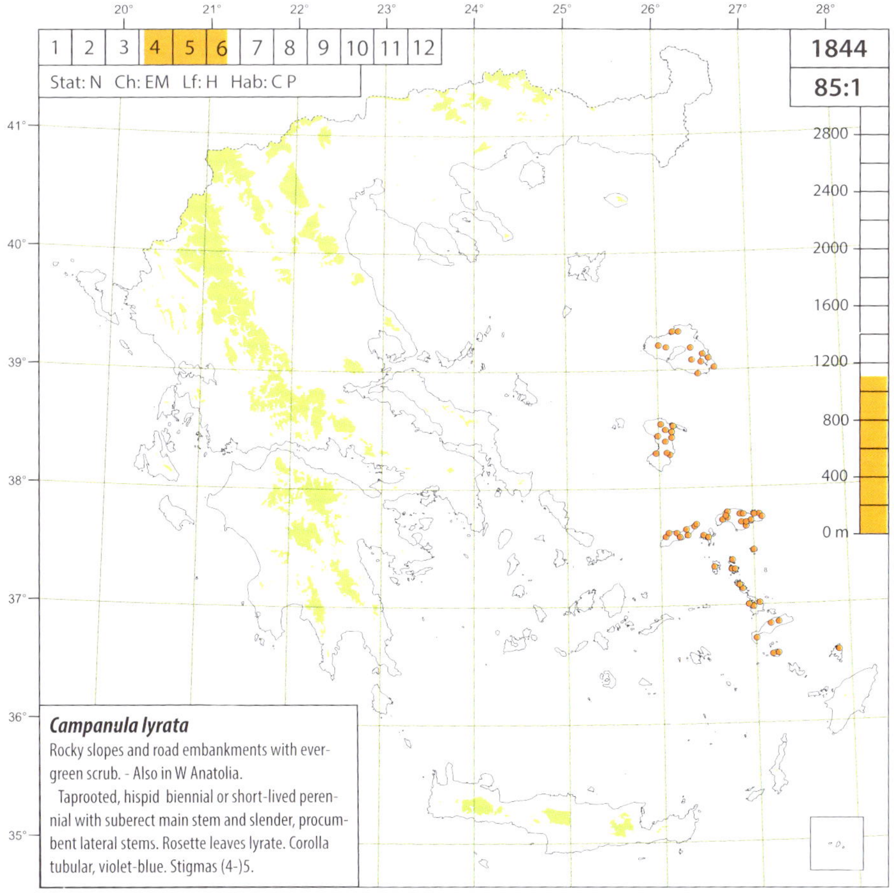

Campanula lyrata

Rocky slopes and road embankments with evergreen scrub. - Also in W Anatolia.

Taprooted, hispid biennial or short-lived perennial with suberect main stem and slender, procumbent lateral stems. Rosette leaves lyrate. Corolla tubular, violet-blue. Stigmas (4-)5.

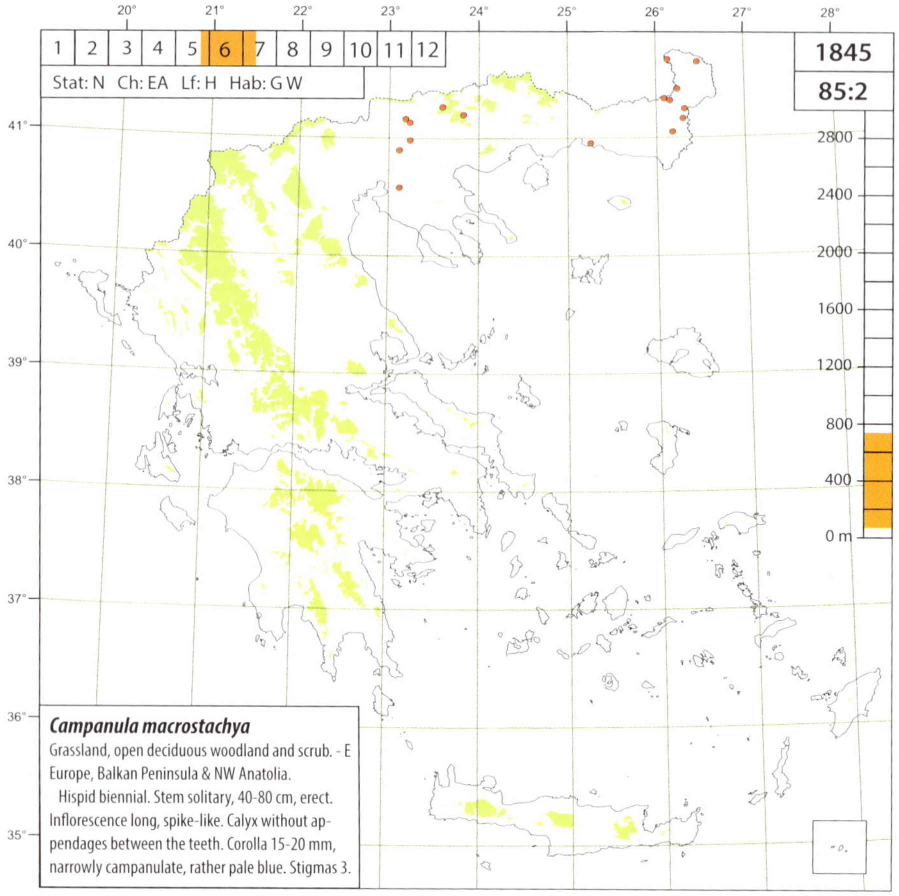

Campanula macrostachya

Grassland, open deciduous woodland and scrub. - E Europe, Balkan Peninsula & NW Anatolia.

Hispid biennial. Stem solitary, 40-80 cm, erect. Inflorescence long, spike-like. Calyx without appendages between the teeth. Corolla 15-20 mm, narrowly campanulate, rather pale blue. Stigmas 3.

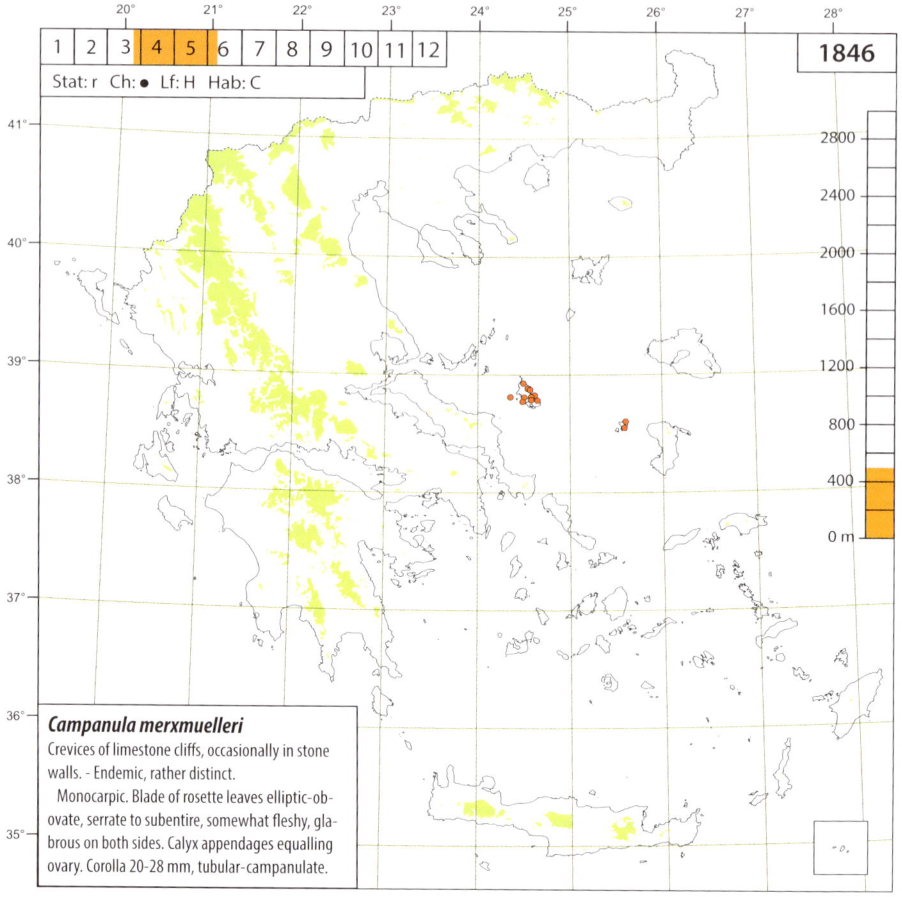

Campanula merxmuelleri

Crevices of limestone cliffs, occasionally in stone walls. - Endemic, rather distinct.

Monocarpic. Blade of rosette leaves elliptic-obovate, serrate to subentire, somewhat fleshy, glabrous on both sides. Calyx appendages equalling ovary. Corolla 20-28 mm, tubular-campanulate.

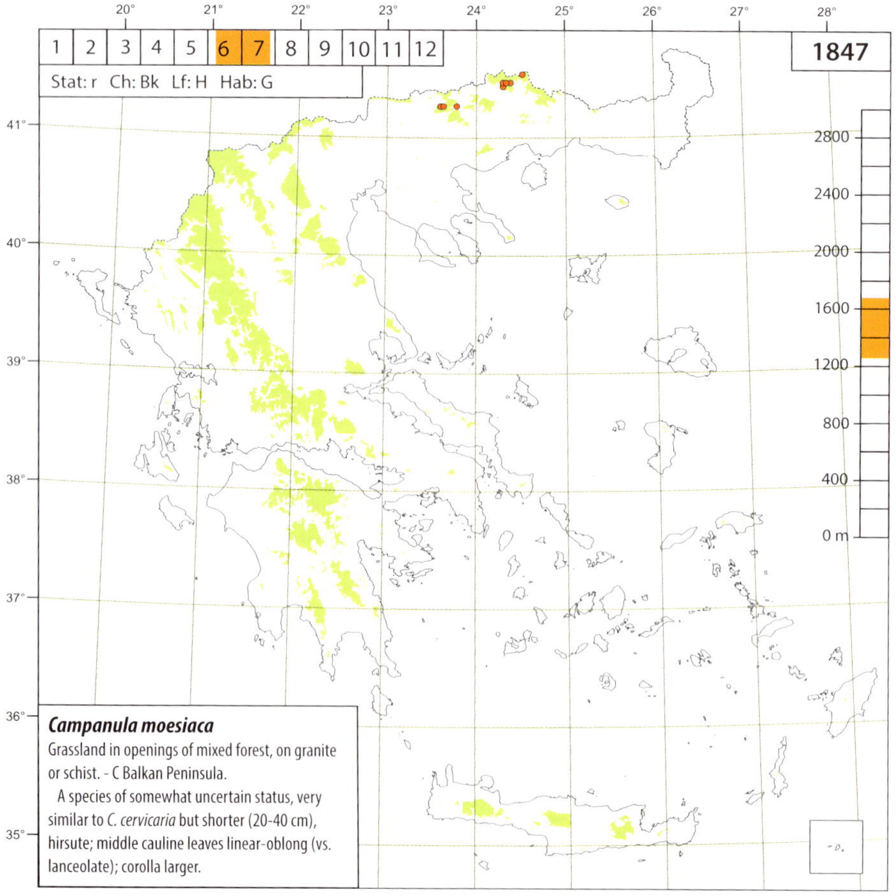

Campanula moesiaca

Grassland in openings of mixed forest, on granite or schist. - C Balkan Peninsula.

A species of somewhat uncertain status, very similar to *C. cervicaria* but shorter (20-40 cm), hirsute; middle cauline leaves linear-oblong (vs. lanceolate); corolla larger.

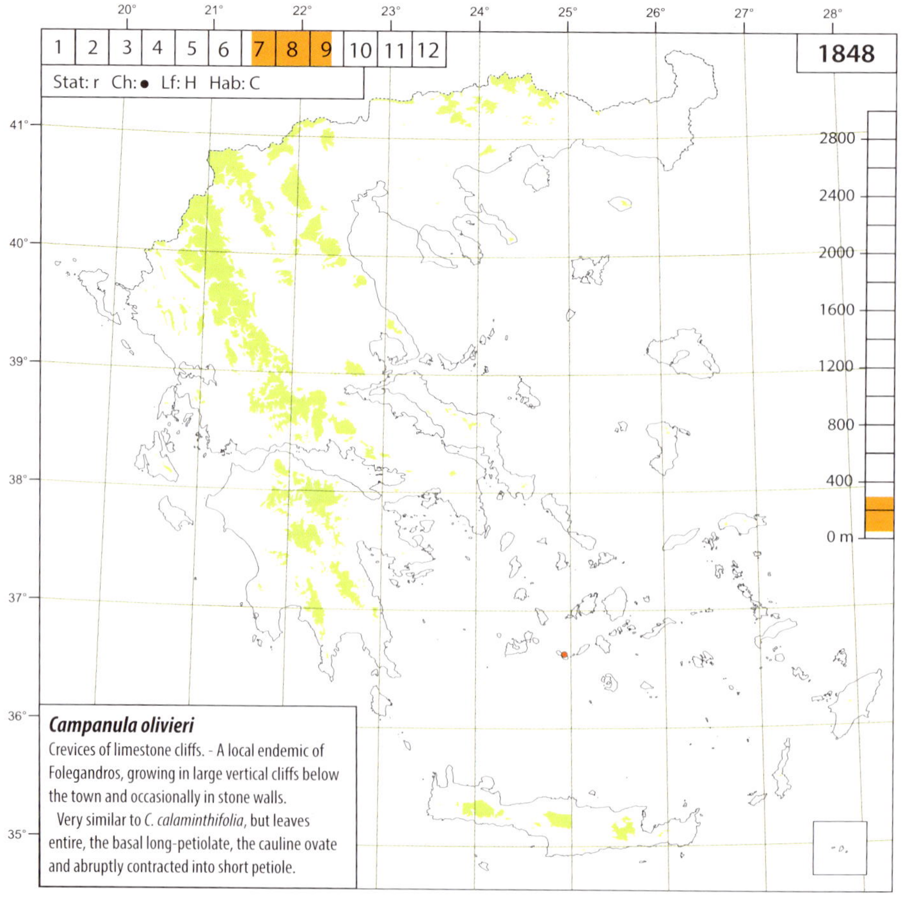

Campanula olivieri

Crevices of limestone cliffs. - A local endemic of Folegandros, growing in large vertical cliffs below the town and occasionally in stone walls.

Very similar to *C. calaminthifolia*, but leaves entire, the basal long-petiolate, the cauline ovate and abruptly contracted into short petiole.

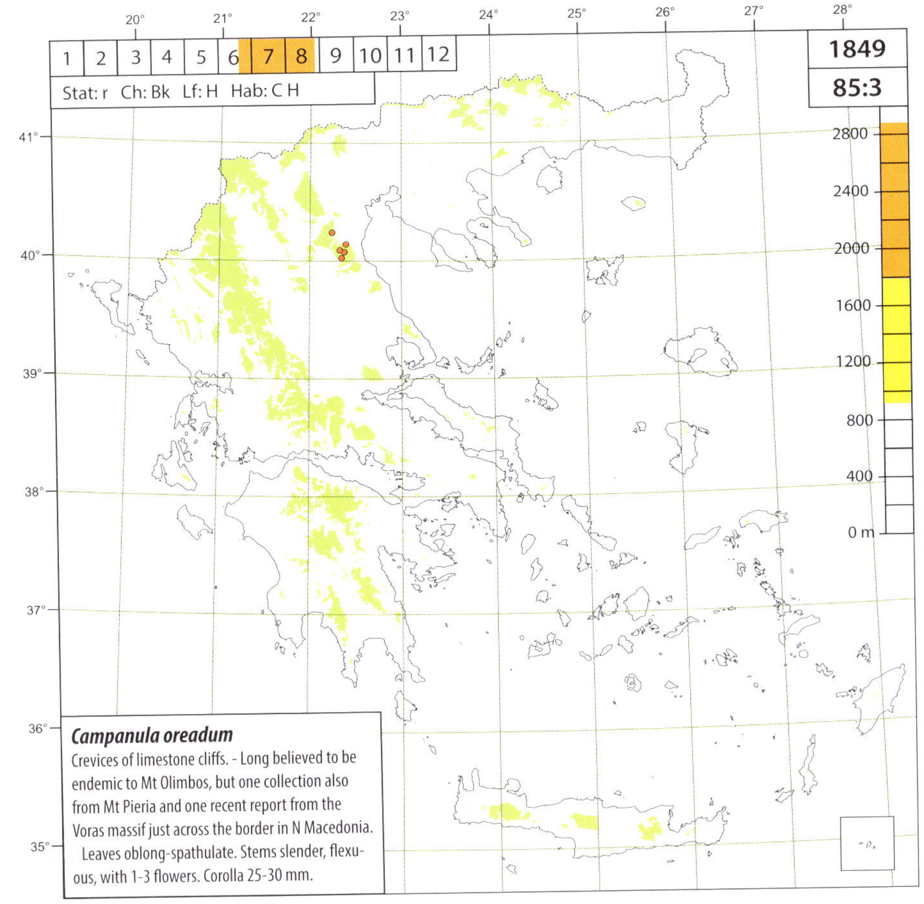

Campanula oreadum

Crevices of limestone cliffs. - Long believed to be endemic to Mt Olimbos, but one collection also from Mt Pieria and one recent report from the Voras massif just across the border in N Macedonia.

Leaves oblong-spathulate. Stems slender, flexuous, with 1-3 flowers. Corolla 25-30 mm.

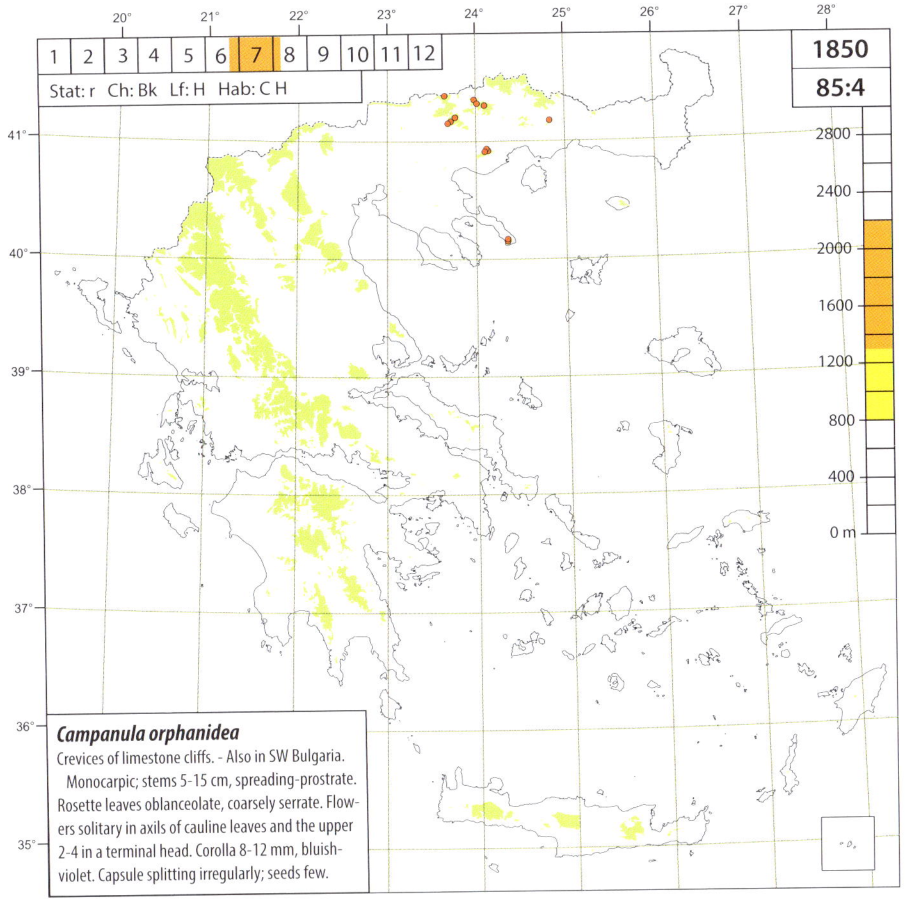

Campanula orphanidea

Crevices of limestone cliffs. - Also in SW Bulgaria.

Monocarpic; stems 5-15 cm, spreading-prostrate. Rosette leaves oblanceolate, coarsely serrate. Flowers solitary in axils of cauline leaves and the upper 2-4 in a terminal head. Corolla 8-12 mm, bluish-violet. Capsule splitting irregularly; seeds few.

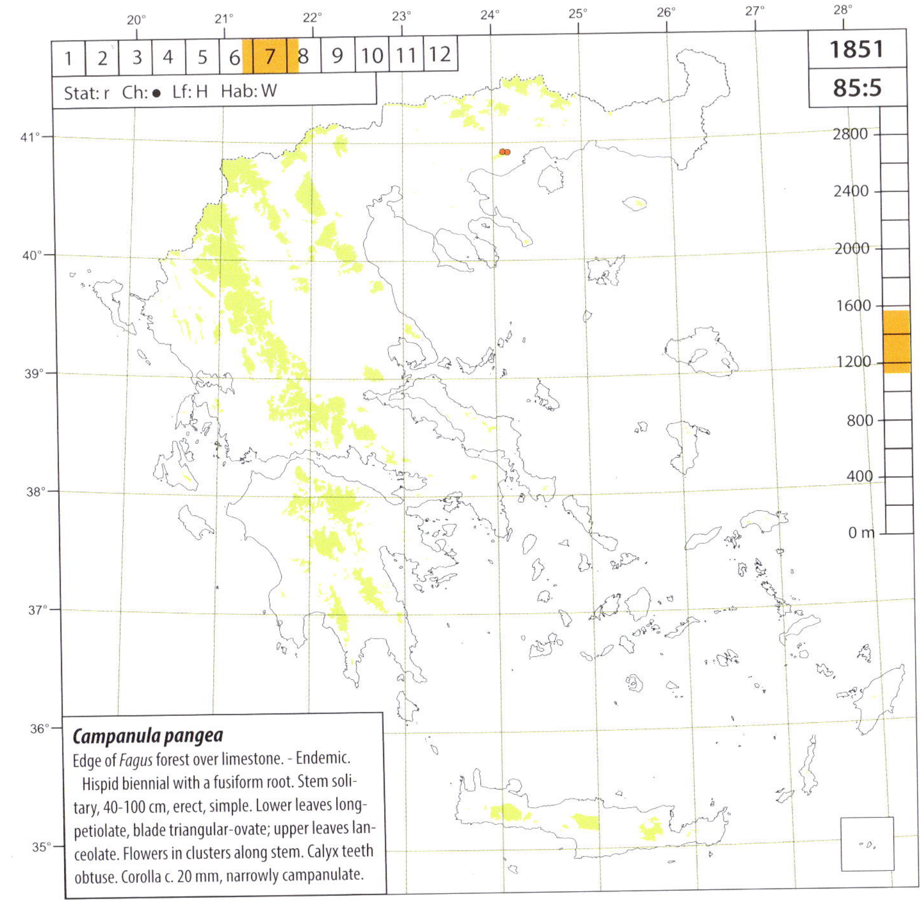

Campanula pangea

Edge of *Fagus* forest over limestone. - Endemic.

Hispid biennial with a fusiform root. Stem solitary, 40-100 cm, erect, simple. Lower leaves long-petiolate, blade triangular-ovate; upper leaves lanceolate. Flowers in clusters along stem. Calyx teeth obtuse. Corolla c. 20 mm, narrowly campanulate.

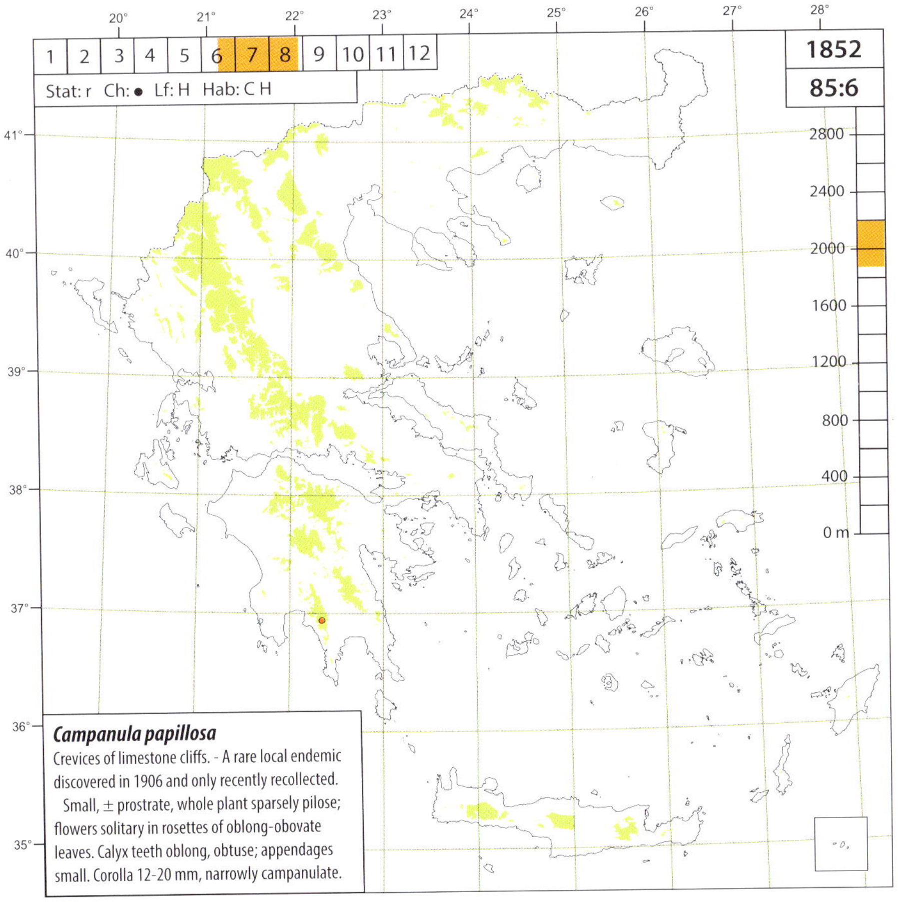

Campanula papillosa

Crevices of limestone cliffs. - A rare local endemic discovered in 1906 and only recently recollected.

Small, ± prostrate, whole plant sparsely pilose; flowers solitary in rosettes of oblong-obovate leaves. Calyx teeth oblong, obtuse; appendages small. Corolla 12-20 mm, narrowly campanulate.

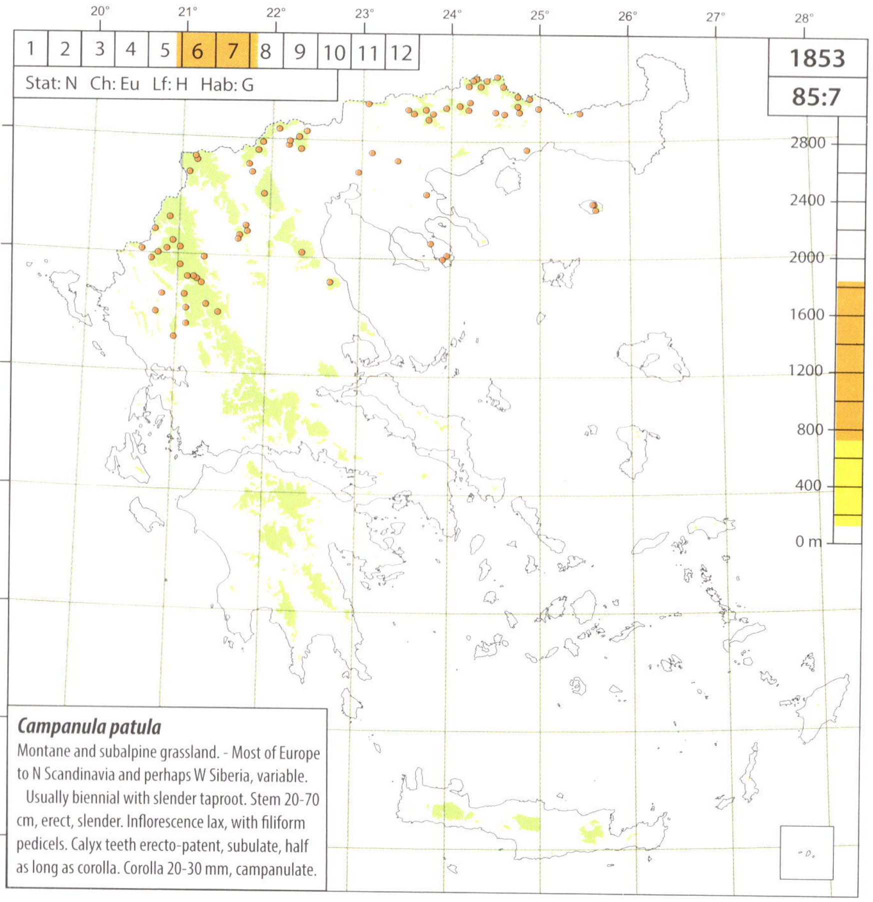

Campanula patula

Montane and subalpine grassland. - Most of Europe to N Scandinavia and perhaps W Siberia, variable.

Usually biennial with slender taproot. Stem 20-70 cm, erect, slender. Inflorescence lax, with filiform pedicels. Calyx teeth erecto-patent, subulate, half as long as corolla. Corolla 20-30 mm, campanulate.

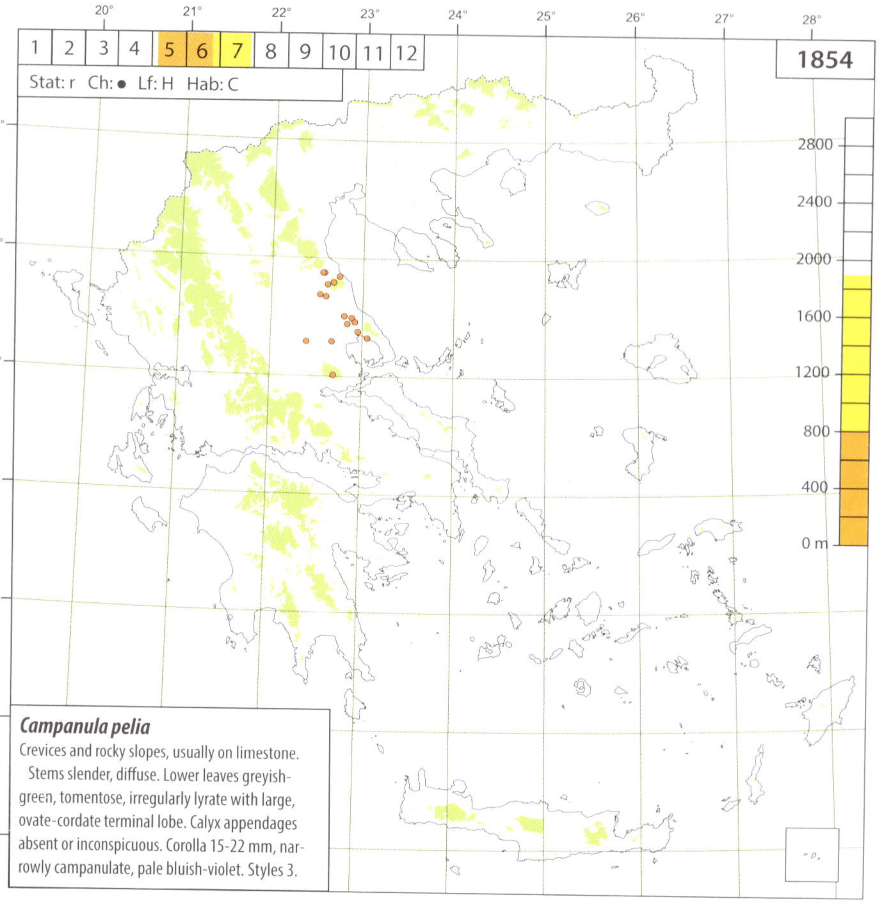

Campanula pelia

Crevices and rocky slopes, usually on limestone.

Stems slender, diffuse. Lower leaves greyish-green, tomentose, irregularly lyrate with large, ovate-cordate terminal lobe. Calyx appendages absent or inconspicuous. Corolla 15-22 mm, narrowly campanulate, pale bluish-violet. Styles 3.

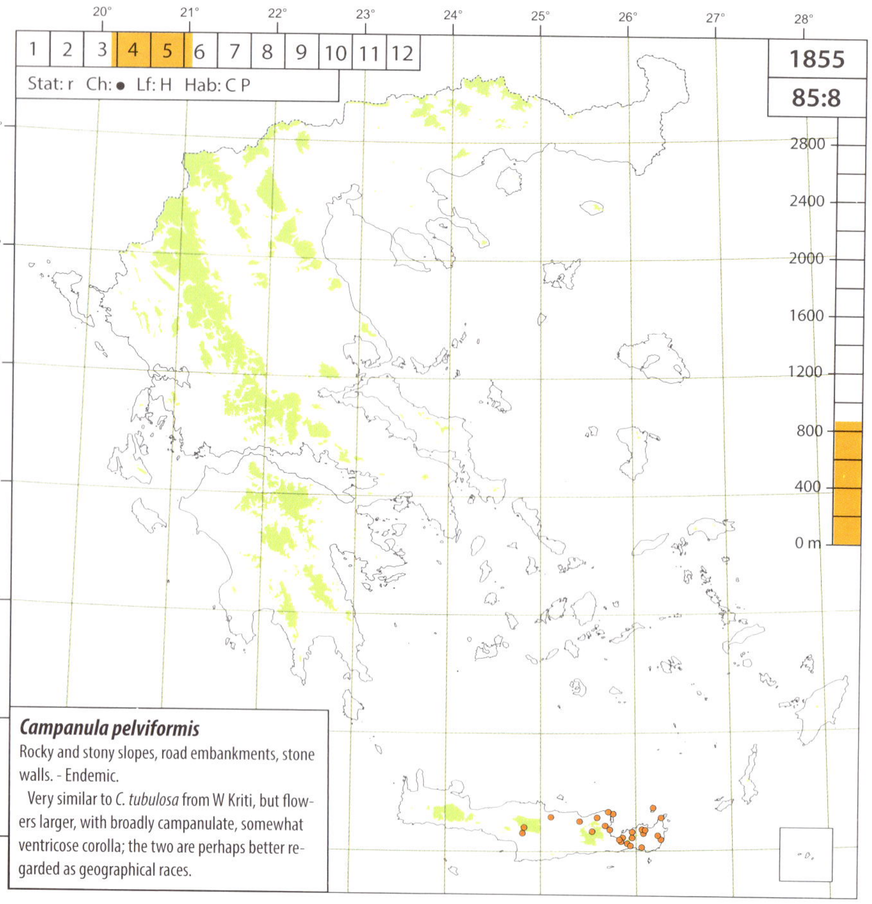

Campanula pelviformis

Rocky and stony slopes, road embankments, stone walls. - Endemic.

Very similar to *C. tubulosa* from W Kriti, but flowers larger, with broadly campanulate, somewhat ventricose corolla; the two are perhaps better regarded as geographical races.

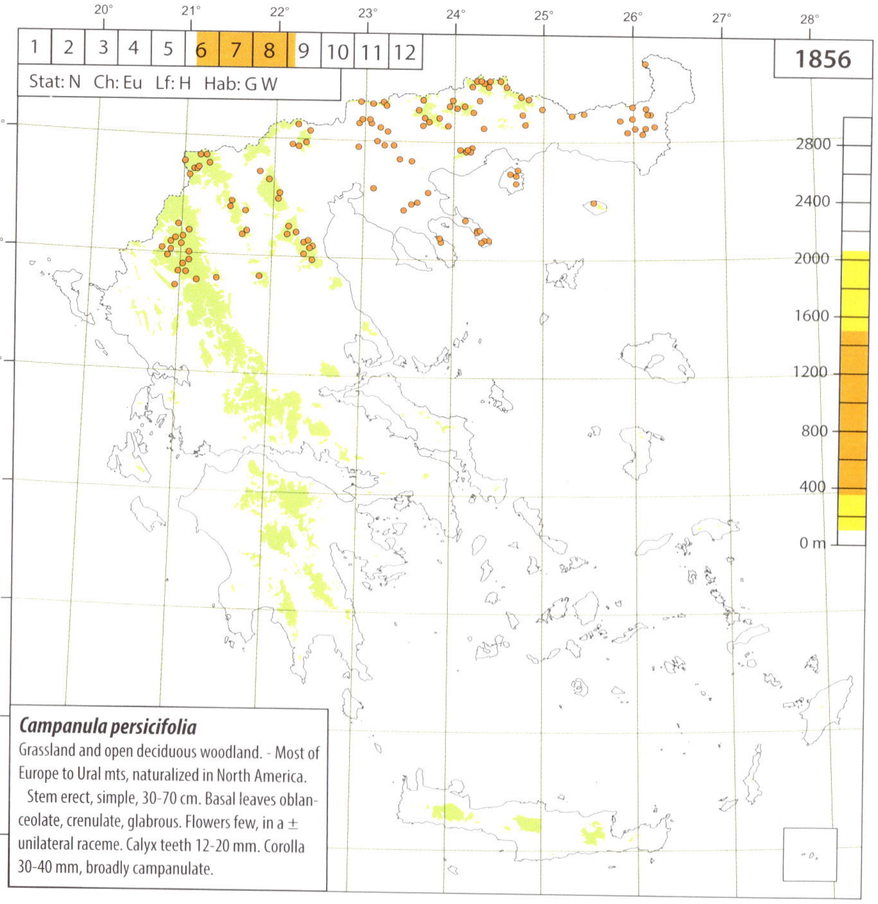

Campanula persicifolia

Grassland and open deciduous woodland. - Most of Europe to Ural mts, naturalized in North America.

Stem erect, simple, 30-70 cm. Basal leaves oblanceolate, crenulate, glabrous. Flowers few, in a ± unilateral raceme. Calyx teeth 12-20 mm. Corolla 30-40 mm, broadly campanulate.

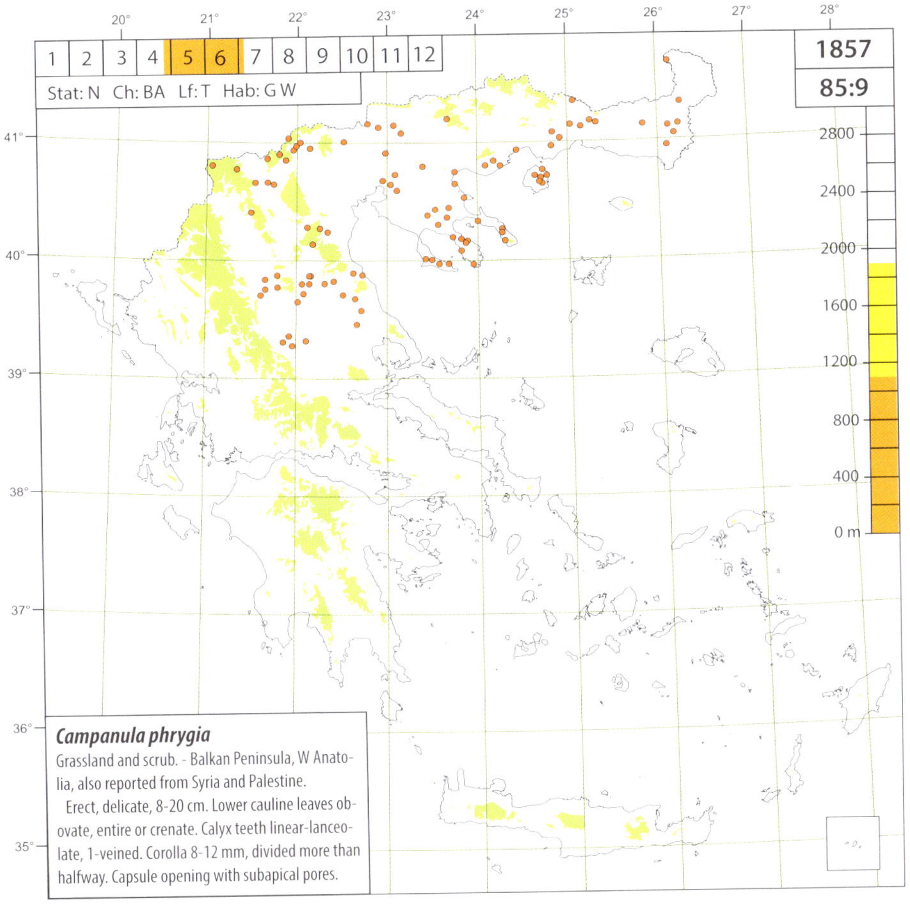

Campanula phrygia
Grassland and scrub. - Balkan Peninsula, W Anatolia, also reported from Syria and Palestine.
Erect, delicate, 8-20 cm. Lower cauline leaves obovate, entire or crenate. Calyx teeth linear-lanceolate, 1-veined. Corolla 8-12 mm, divided more than halfway. Capsule opening with subapical pores.

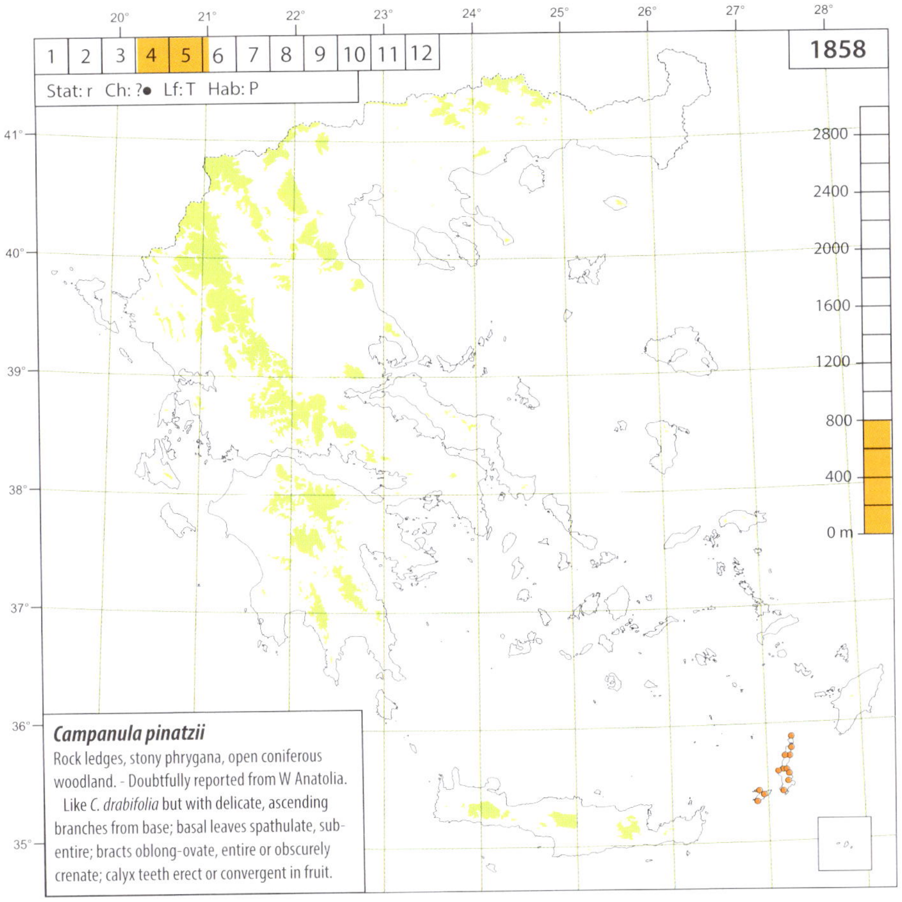

Campanula pinatzii
Rock ledges, stony phrygana, open coniferous woodland. - Doubtfully reported from W Anatolia.
Like *C. drabifolia* but with delicate, ascending branches from base; basal leaves spathulate, subentire; bracts oblong-ovate, entire or obscurely crenate; calyx teeth erect or convergent in fruit.

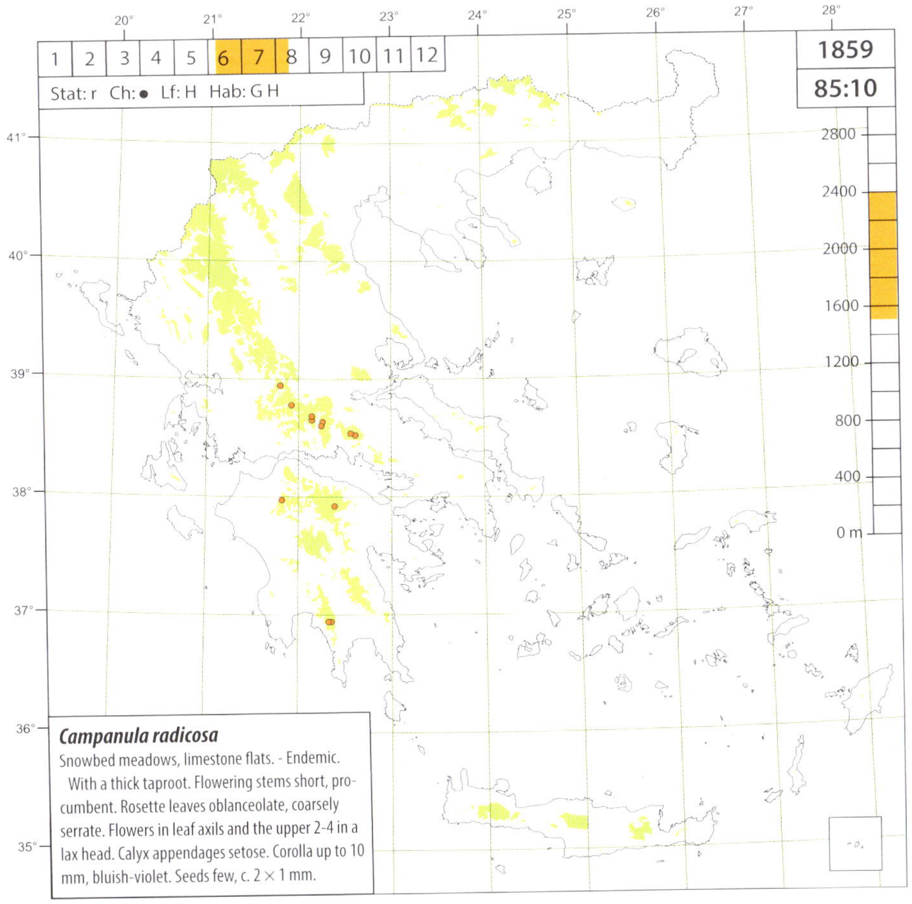

Campanula radicosa
Snowbed meadows, limestone flats. - Endemic.
With a thick taproot. Flowering stems short, procumbent. Rosette leaves oblanceolate, coarsely serrate. Flowers in leaf axils and the upper 2-4 in a lax head. Calyx appendages setose. Corolla up to 10 mm, bluish-violet. Seeds few, c. 2 × 1 mm.

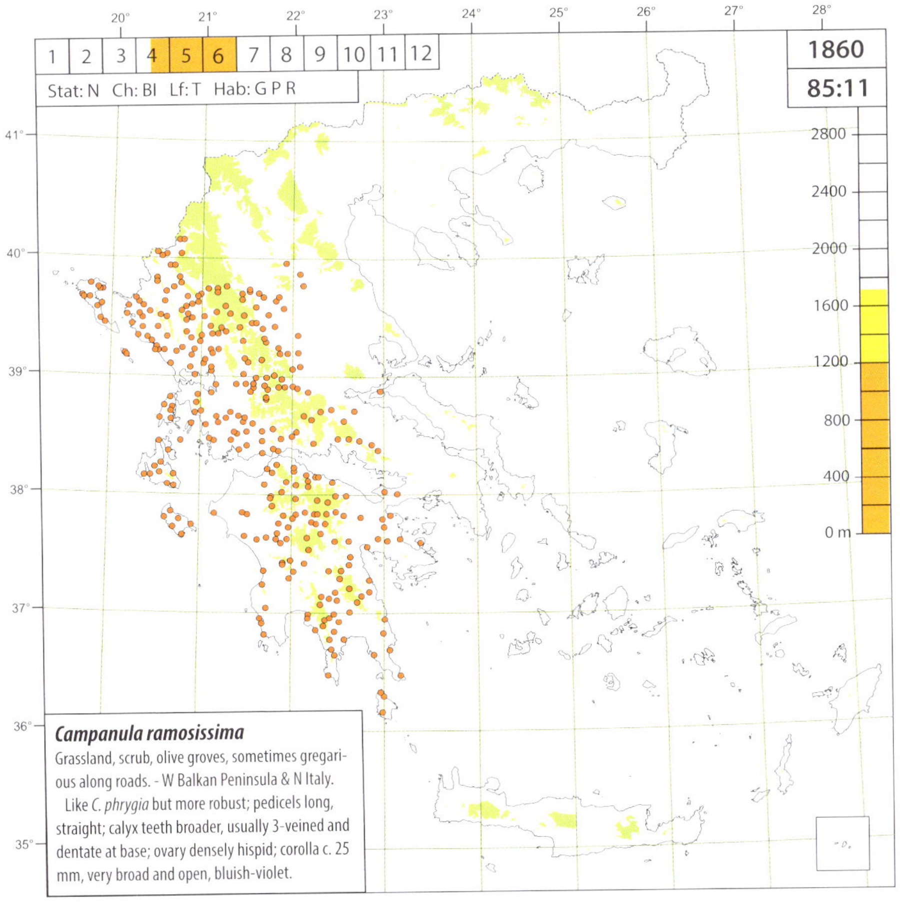

Campanula ramosissima
Grassland, scrub, olive groves, sometimes gregarious along roads. - W Balkan Peninsula & N Italy.
Like *C. phrygia* but more robust; pedicels long, straight; calyx teeth broader, usually 3-veined and dentate at base; ovary densely hispid; corolla c. 25 mm, very broad and open, bluish-violet.

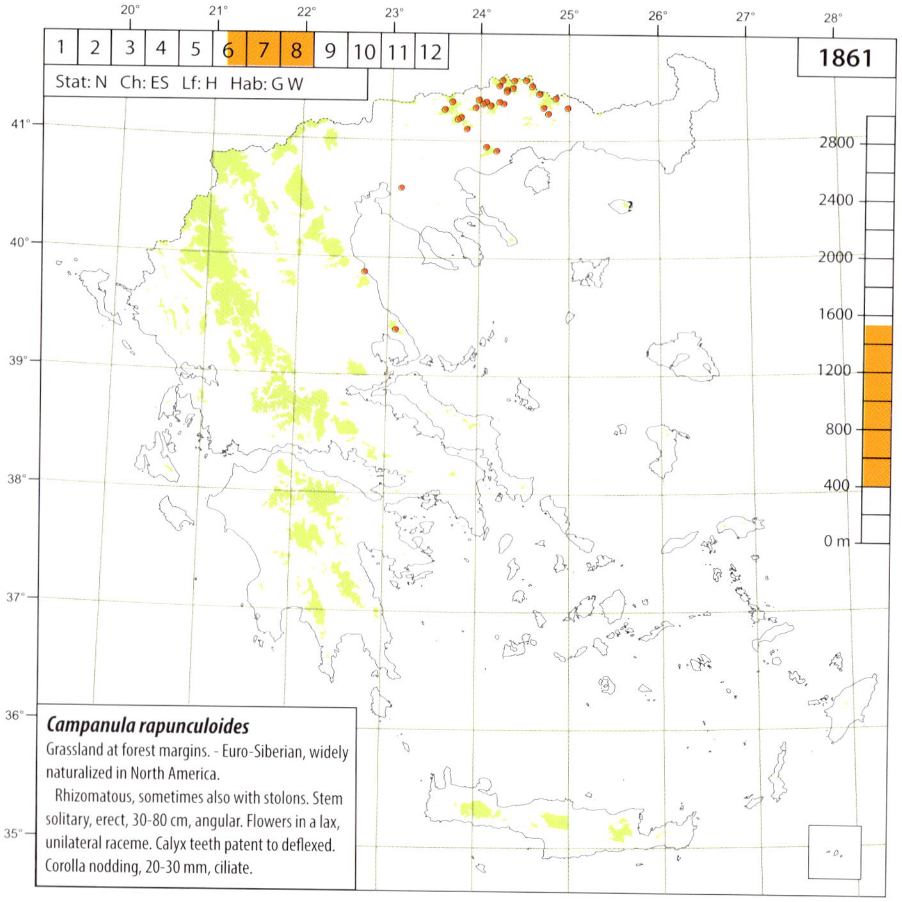

Campanula rapunculoides

Grassland at forest margins. - Euro-Siberian, widely naturalized in North America.

Rhizomatous, sometimes also with stolons. Stem solitary, erect, 30-80 cm, angular. Flowers in a lax, unilateral raceme. Calyx teeth patent to deflexed. Corolla nodding, 20-30 mm, ciliate.

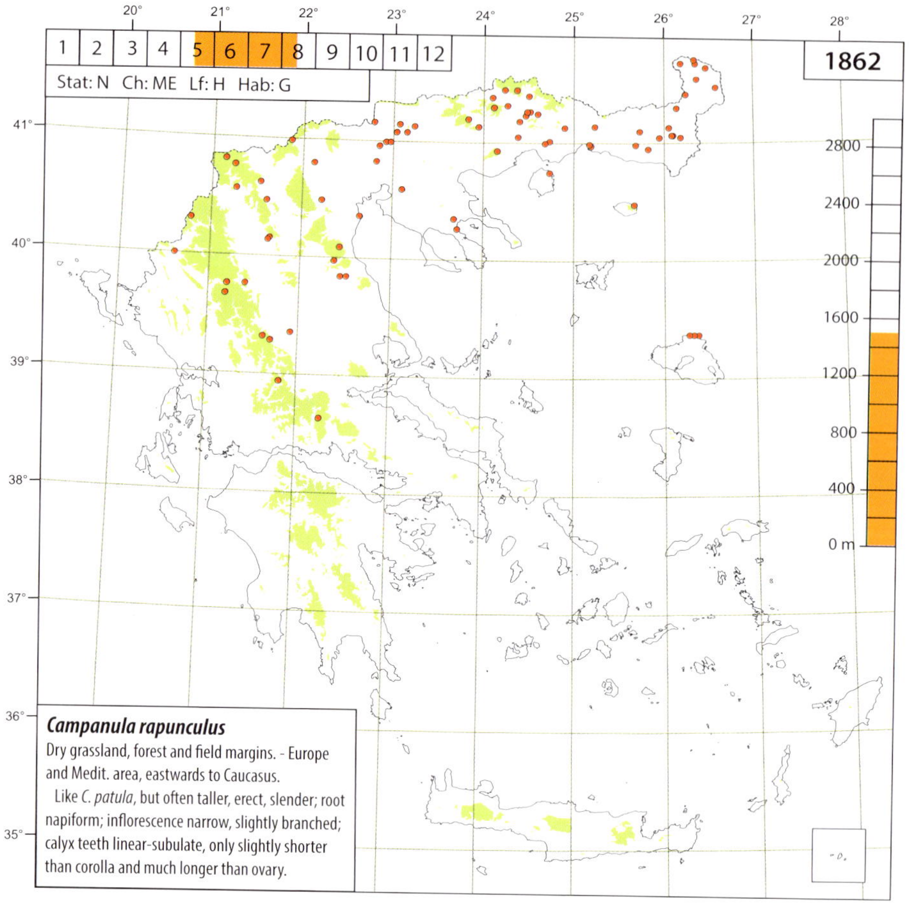

Campanula rapunculus

Dry grassland, forest and field margins. - Europe and Medit. area, eastwards to Caucasus.

Like *C. patula*, but often taller, erect, slender; root napiform; inflorescence narrow, slightly branched; calyx teeth linear-subulate, only slightly shorter than corolla and much longer than ovary.

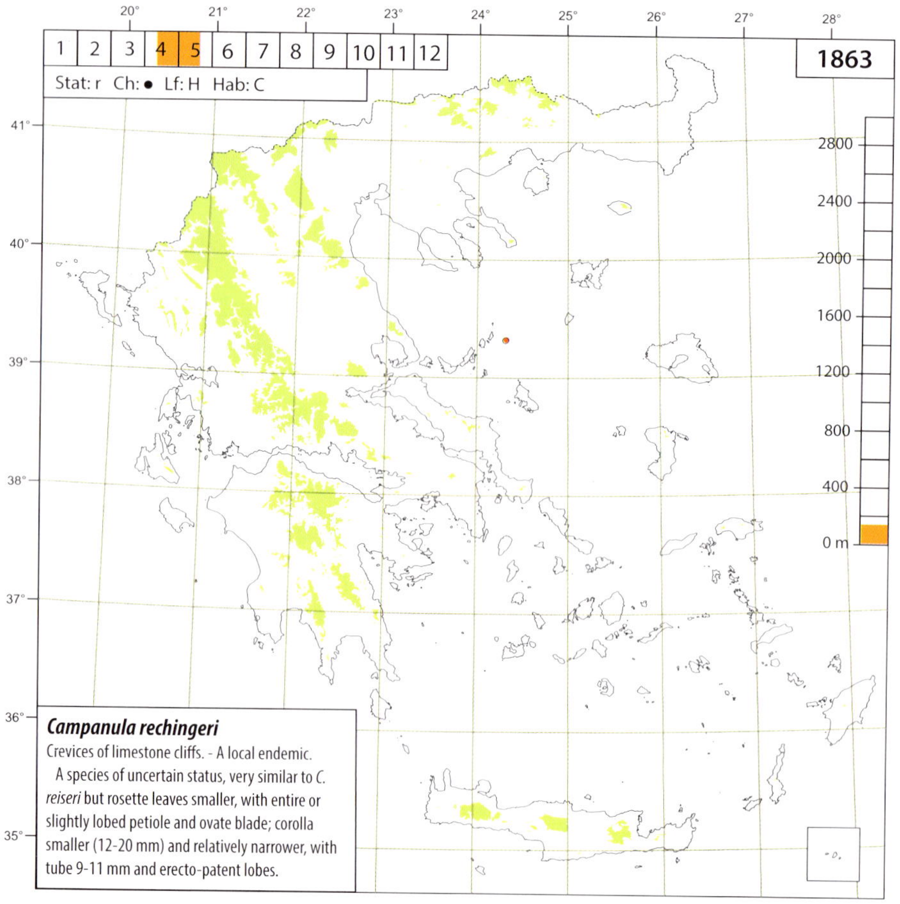

Campanula rechingeri

Crevices of limestone cliffs. - A local endemic.

A species of uncertain status, very similar to *C. reiseri* but rosette leaves smaller, with entire or slightly lobed petiole and ovate blade; corolla smaller (12-20 mm) and relatively narrower, with tube 9-11 mm and erecto-patent lobes.

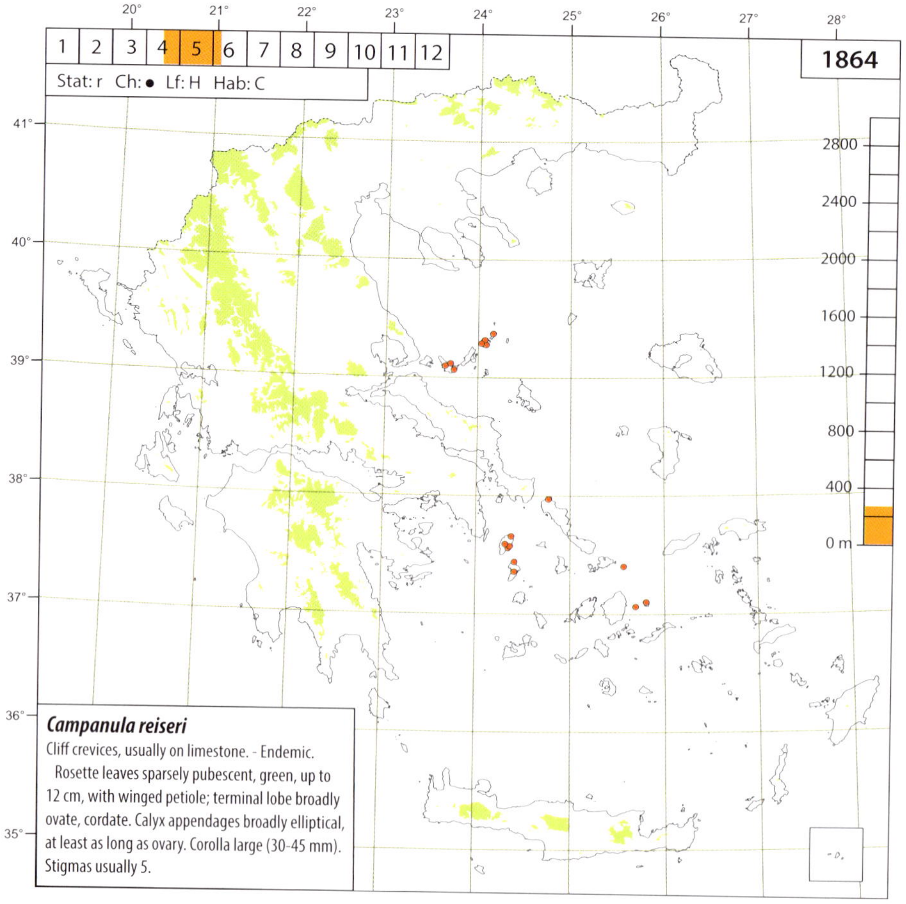

Campanula reiseri

Cliff crevices, usually on limestone. - Endemic.

Rosette leaves sparsely pubescent, green, up to 12 cm, with winged petiole; terminal lobe broadly ovate, cordate. Calyx appendages broadly elliptical, at least as long as ovary. Corolla large (30-45 mm). Stigmas usually 5.

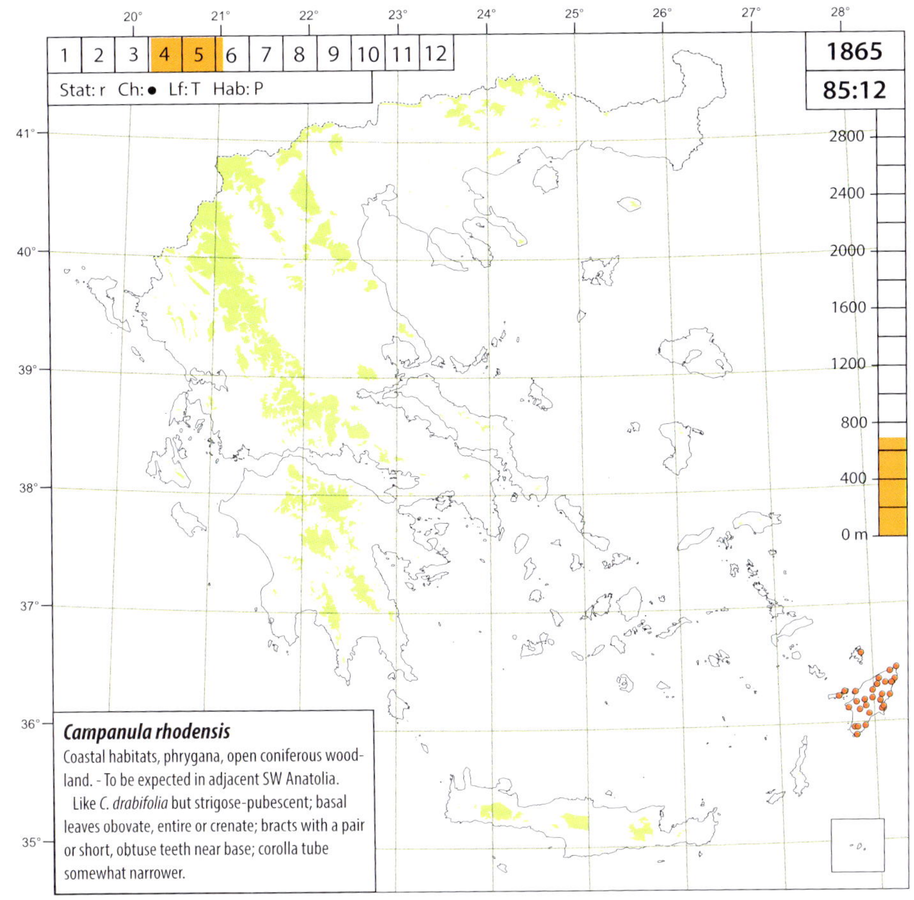

Campanula rhodensis
Coastal habitats, phrygana, open coniferous woodland. - To be expected in adjacent SW Anatolia.
Like *C. drabifolia* but strigose-pubescent; basal leaves obovate, entire or crenate; bracts with a pair or short, obtuse teeth near base; corolla tube somewhat narrower.

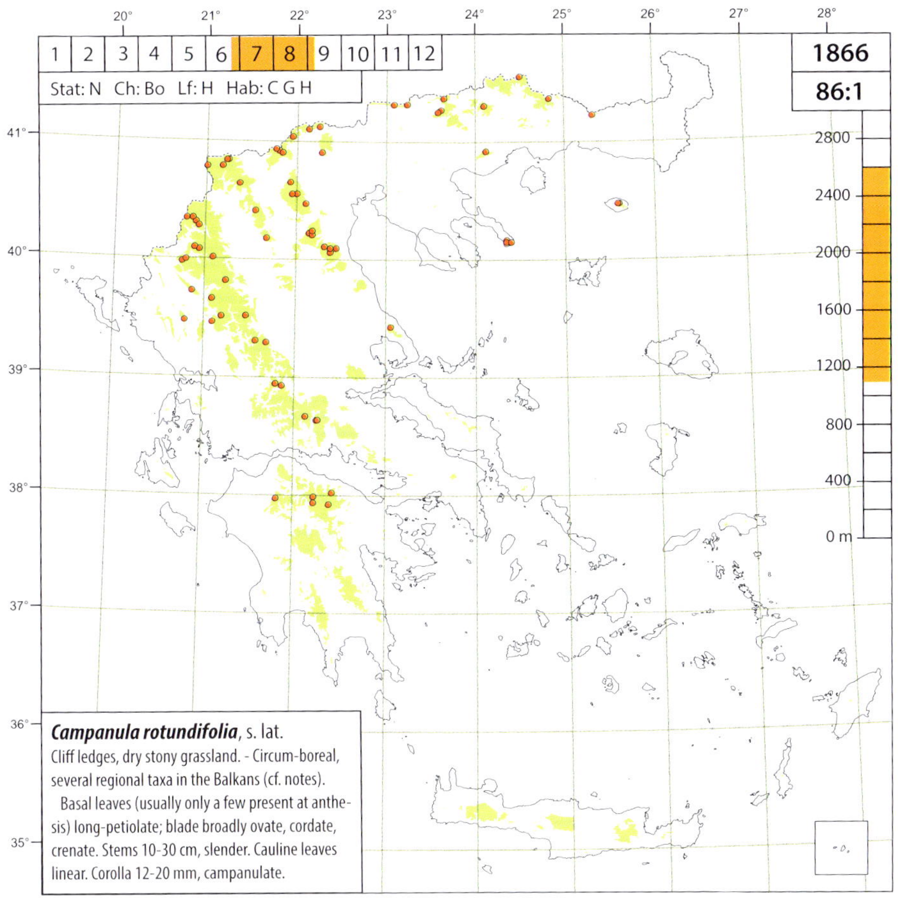

Campanula rotundifolia, s. lat.
Cliff ledges, dry stony grassland. - Circum-boreal, several regional taxa in the Balkans (cf. notes).
Basal leaves (usually only a few present at anthesis) long-petiolate; blade broadly ovate, cordate, crenate. Stems 10-30 cm, slender. Cauline leaves linear. Corolla 12-20 mm, campanulate.

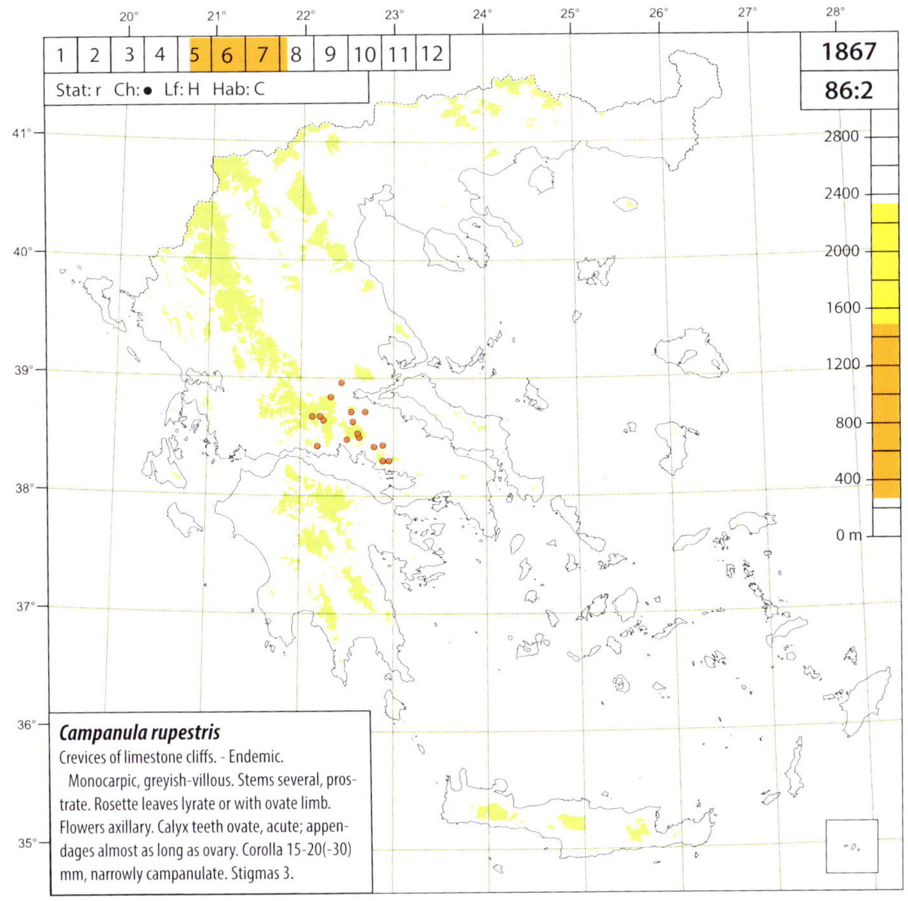

Campanula rupestris
Crevices of limestone cliffs. - Endemic.
Monocarpic, greyish-villous. Stems several, prostrate. Rosette leaves lyrate or with ovate limb. Flowers axillary. Calyx teeth ovate, acute; appendages almost as long as ovary. Corolla 15-20(-30) mm, narrowly campanulate. Stigmas 3.

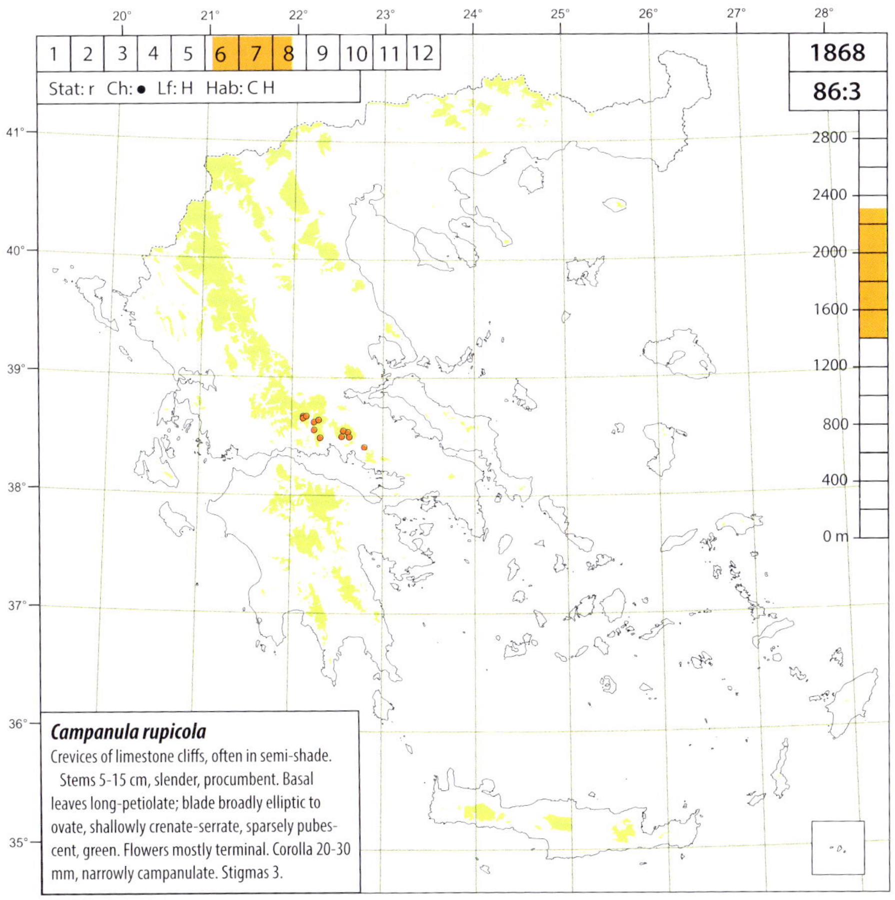

Campanula rupicola
Crevices of limestone cliffs, often in semi-shade.
Stems 5-15 cm, slender, procumbent. Basal leaves long-petiolate; blade broadly elliptic to ovate, shallowly crenate-serrate, sparsely pubescent, green. Flowers mostly terminal. Corolla 20-30 mm, narrowly campanulate. Stigmas 3.

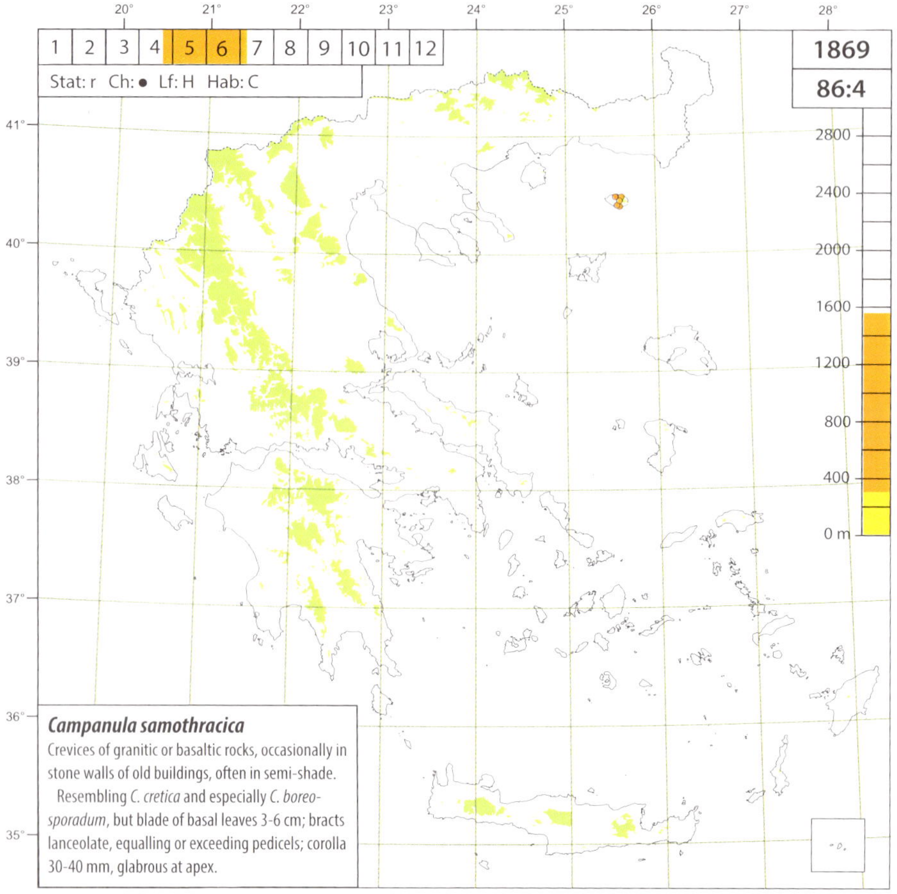

Campanula samothracica

Crevices of granitic or basaltic rocks, occasionally in stone walls of old buildings, often in semi-shade.

Resembling *C. cretica* and especially *C. boreo-sporadum*, but blade of basal leaves 3-6 cm; bracts lanceolate, equalling or exceeding pedicels; corolla 30-40 mm, glabrous at apex.

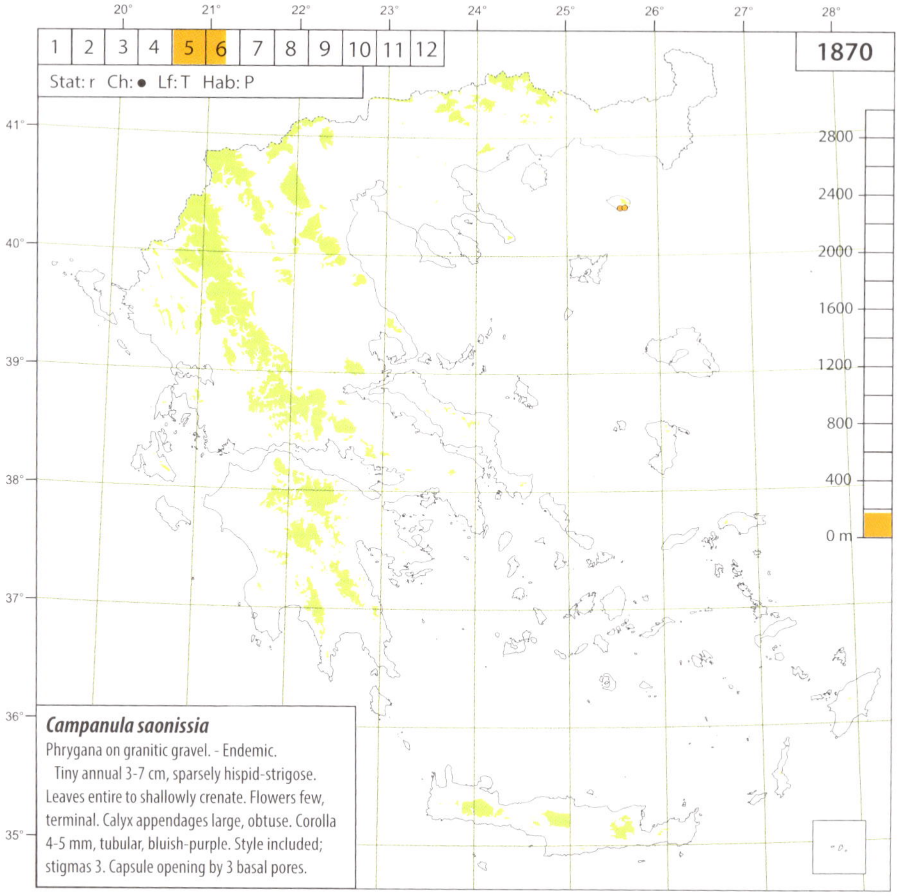

Campanula saonissia

Phrygana on granitic gravel. - Endemic.

Tiny annual 3-7 cm, sparsely hispid-strigose. Leaves entire to shallowly crenate. Flowers few, terminal. Calyx appendages large, obtuse. Corolla 4-5 mm, tubular, bluish-purple. Style included; stigmas 3. Capsule opening by 3 basal pores.

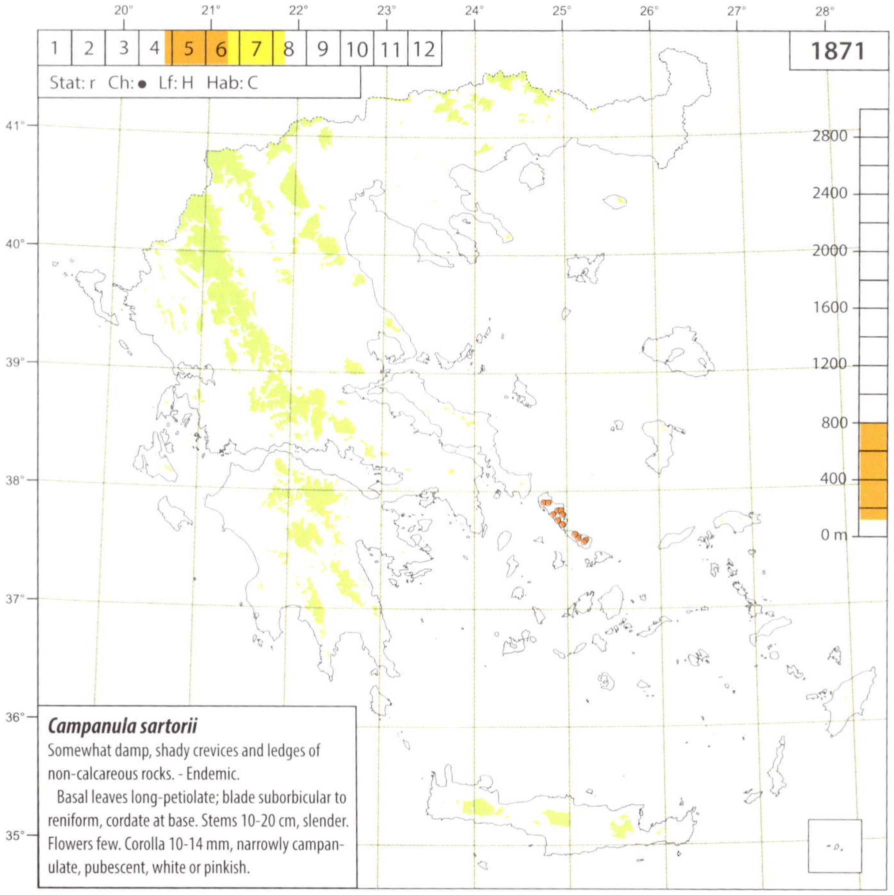

Campanula sartorii

Somewhat damp, shady crevices and ledges of non-calcareous rocks. - Endemic.

Basal leaves long-petiolate; blade suborbicular to reniform, cordate at base. Stems 10-20 cm, slender. Flowers few. Corolla 10-14 mm, narrowly campanulate, pubescent, white or pinkish.

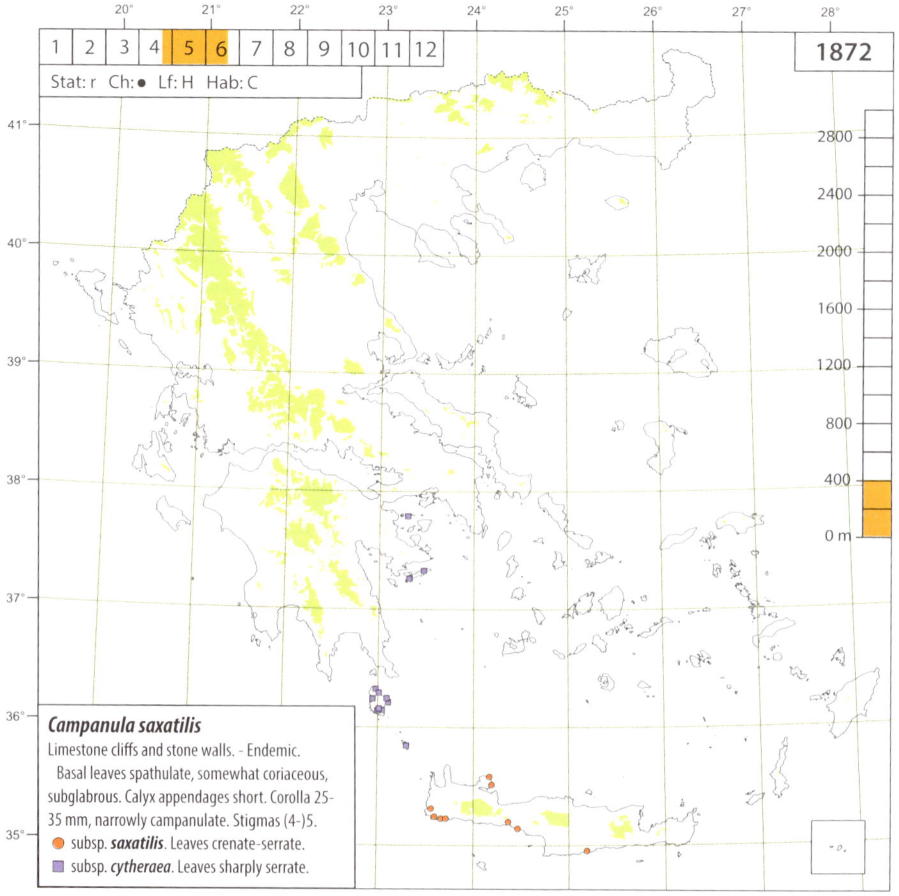

Campanula saxatilis

Limestone cliffs and stone walls. - Endemic.

Basal leaves spathulate, somewhat coriaceous, subglabrous. Calyx appendages short. Corolla 25-35 mm, narrowly campanulate. Stigmas (4-)5.

● subsp. ***saxatilis***. Leaves crenate-serrate.

■ subsp. ***cytheraea***. Leaves sharply serrate.

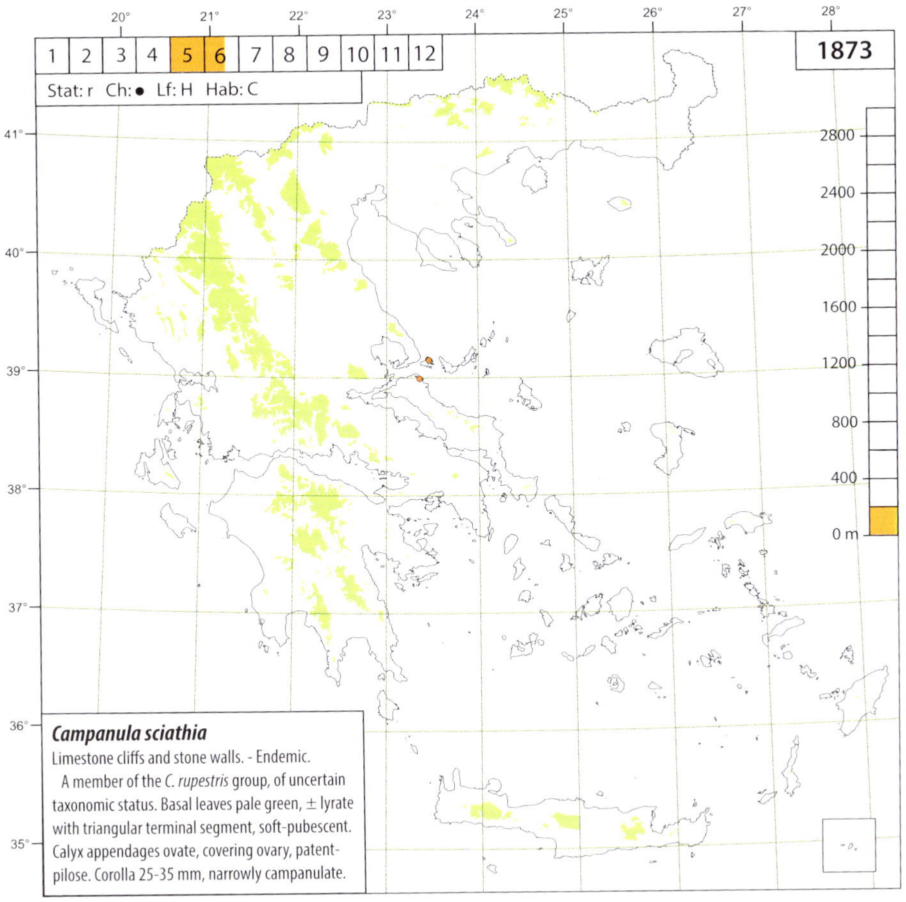

Campanula sciathia
Limestone cliffs and stone walls. - Endemic.
A member of the *C. rupestris* group, of uncertain taxonomic status. Basal leaves pale green, ± lyrate with triangular terminal segment, soft-pubescent. Calyx appendages ovate, covering ovary, patent-pilose. Corolla 25-35 mm, narrowly campanulate.

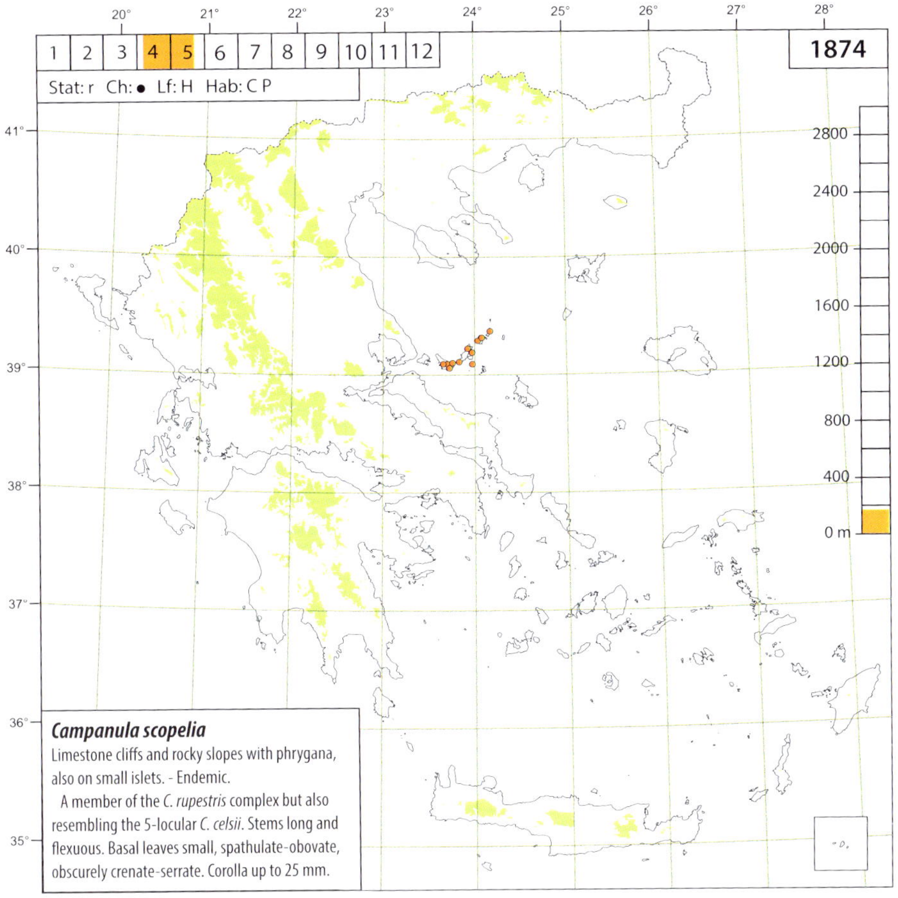

Campanula scopelia
Limestone cliffs and rocky slopes with phrygana, also on small islets. - Endemic.
A member of the *C. rupestris* complex but also resembling the 5-locular *C. celsii*. Stems long and flexuous. Basal leaves small, spathulate-obovate, obscurely crenate-serrate. Corolla up to 25 mm.

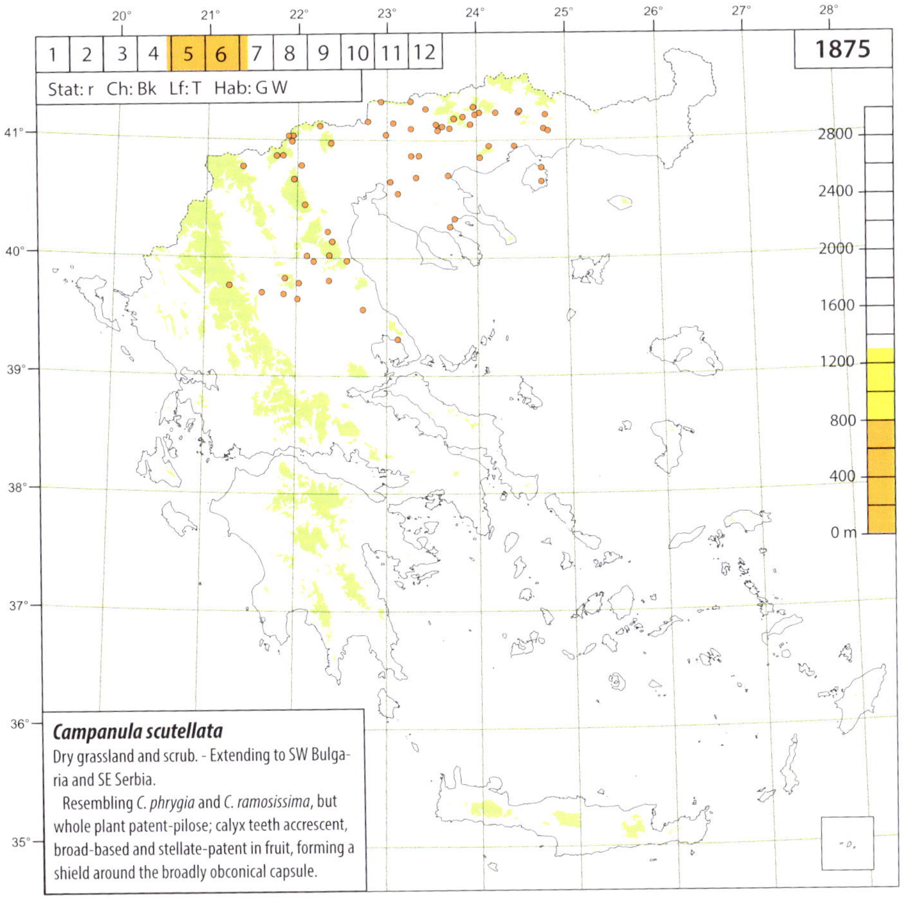

Campanula scutellata
Dry grassland and scrub. - Extending to SW Bulgaria and SE Serbia.
Resembling *C. phrygia* and *C. ramosissima*, but whole plant patent-pilose; calyx teeth accrescent, broad-based and stellate-patent in fruit, forming a shield around the broadly obconical capsule.

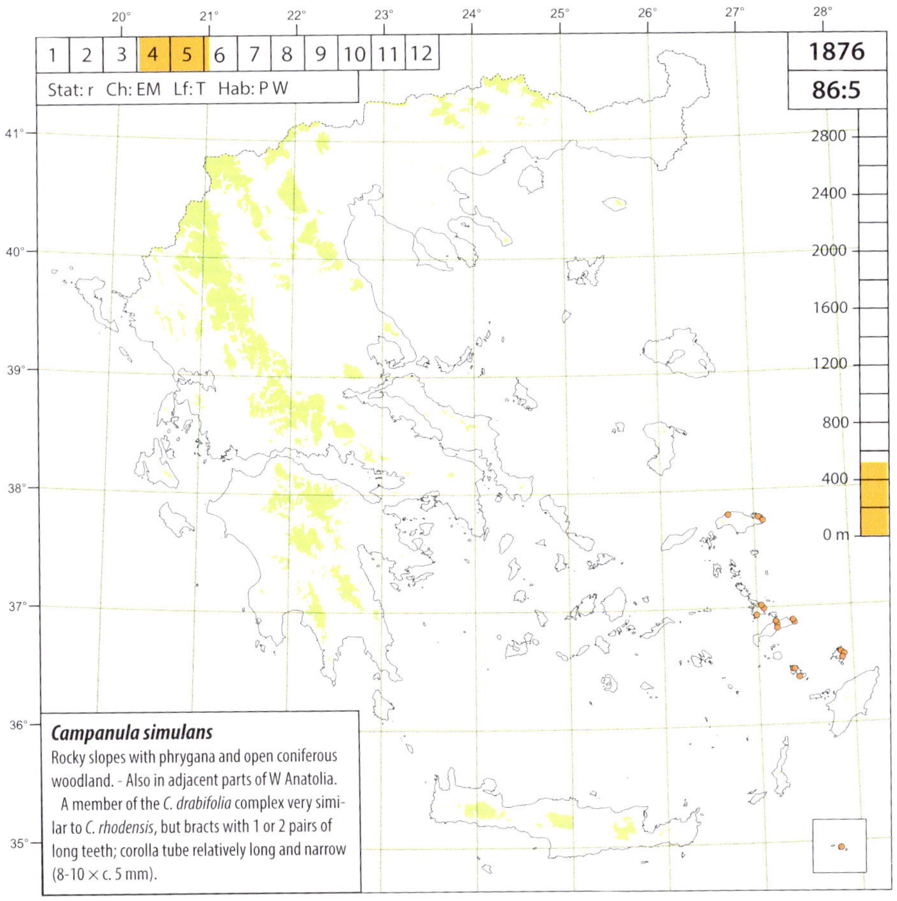

Campanula simulans
Rocky slopes with phrygana and open coniferous woodland. - Also in adjacent parts of W Anatolia.
A member of the *C. drabifolia* complex very similar to *C. rhodensis*, but bracts with 1 or 2 pairs of long teeth; corolla tube relatively long and narrow (8-10 × c. 5 mm).

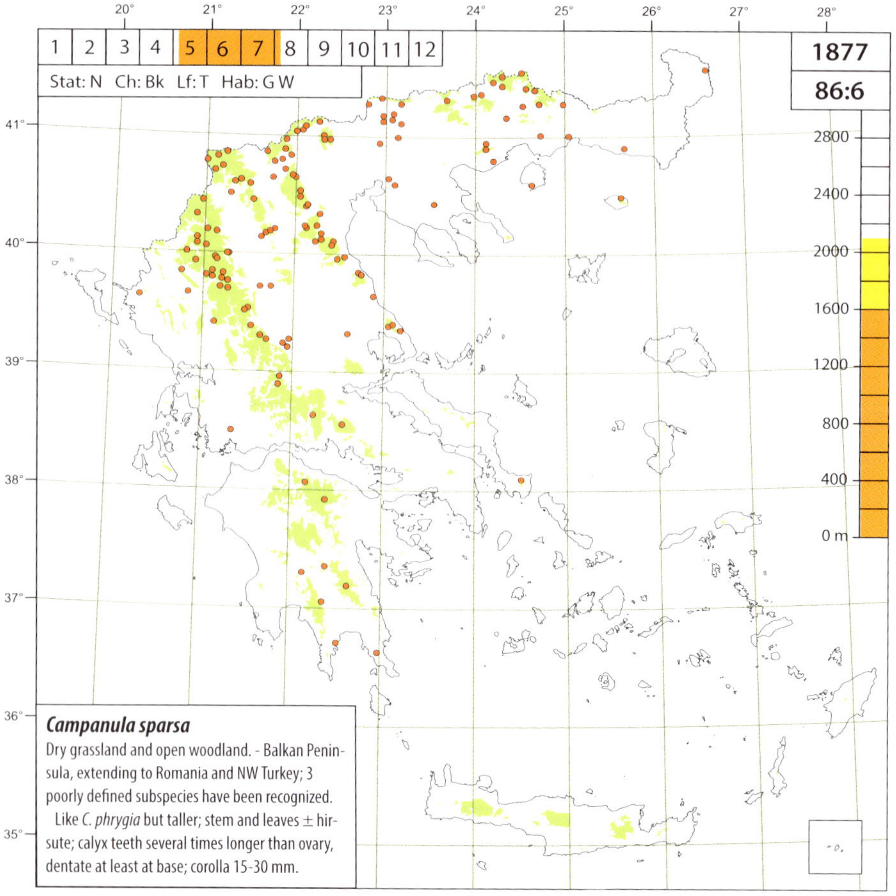

Campanula sparsa
Dry grassland and open woodland. - Balkan Peninsula, extending to Romania and NW Turkey; 3 poorly defined subspecies have been recognized.
Like *C. phrygia* but taller; stem and leaves ± hirsute; calyx teeth several times longer than ovary, dentate at least at base; corolla 15-30 mm.

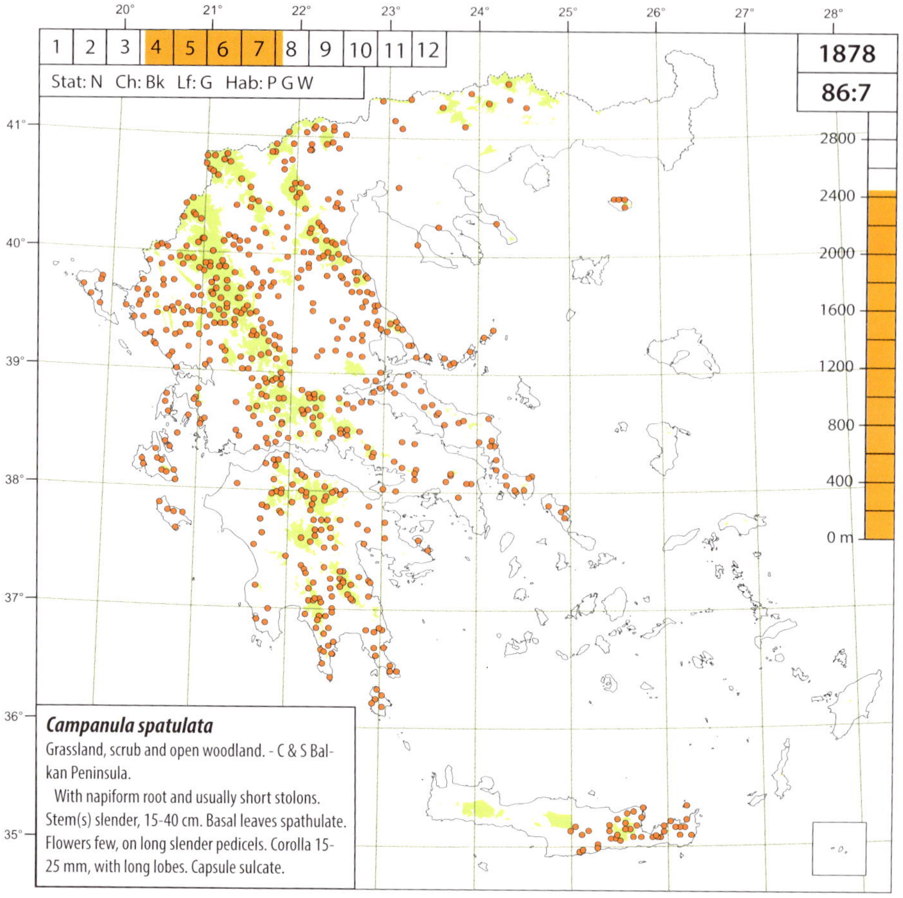

Campanula spatulata
Grassland, scrub and open woodland. - C & S Balkan Peninsula.
With napiform root and usually short stolons. Stem(s) slender, 15-40 cm. Basal leaves spathulate. Flowers few, on long slender pedicels. Corolla 15-25 mm, with long lobes. Capsule sulcate.

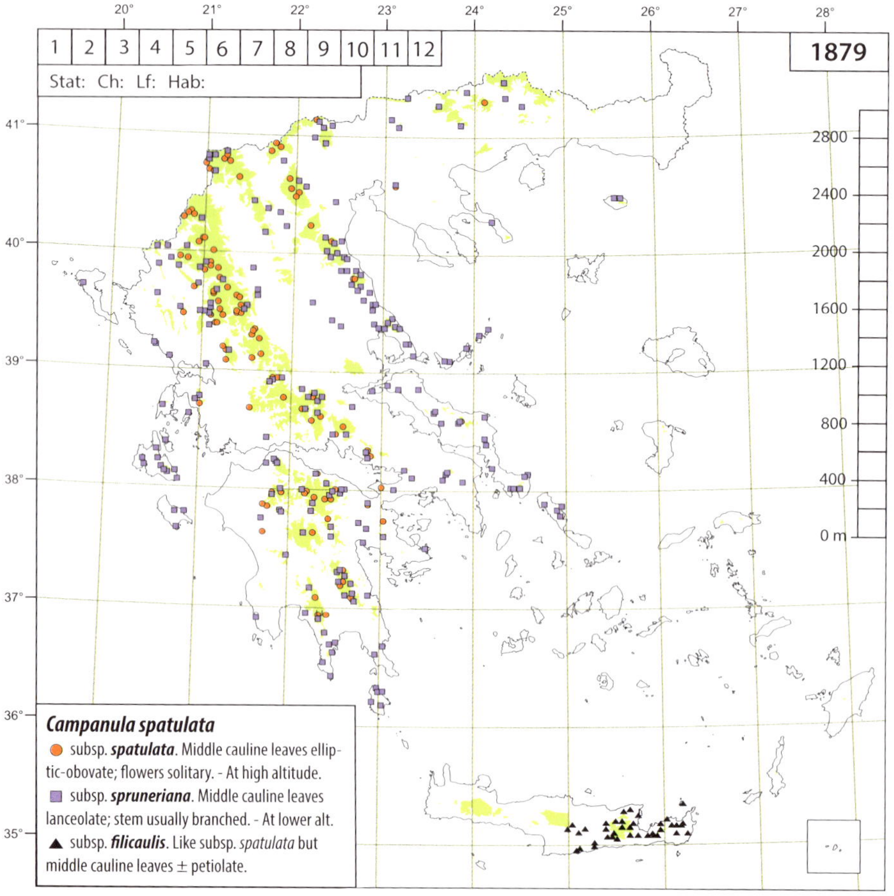

Campanula spatulata
● subsp. ***spatulata***. Middle cauline leaves elliptic-obovate; flowers solitary. - At high altitude.
■ subsp. ***spruneriana***. Middle cauline leaves lanceolate; stem usually branched. - At lower alt.
▲ subsp. ***filicaulis***. Like subsp. *spatulata* but middle cauline leaves ± petiolate.

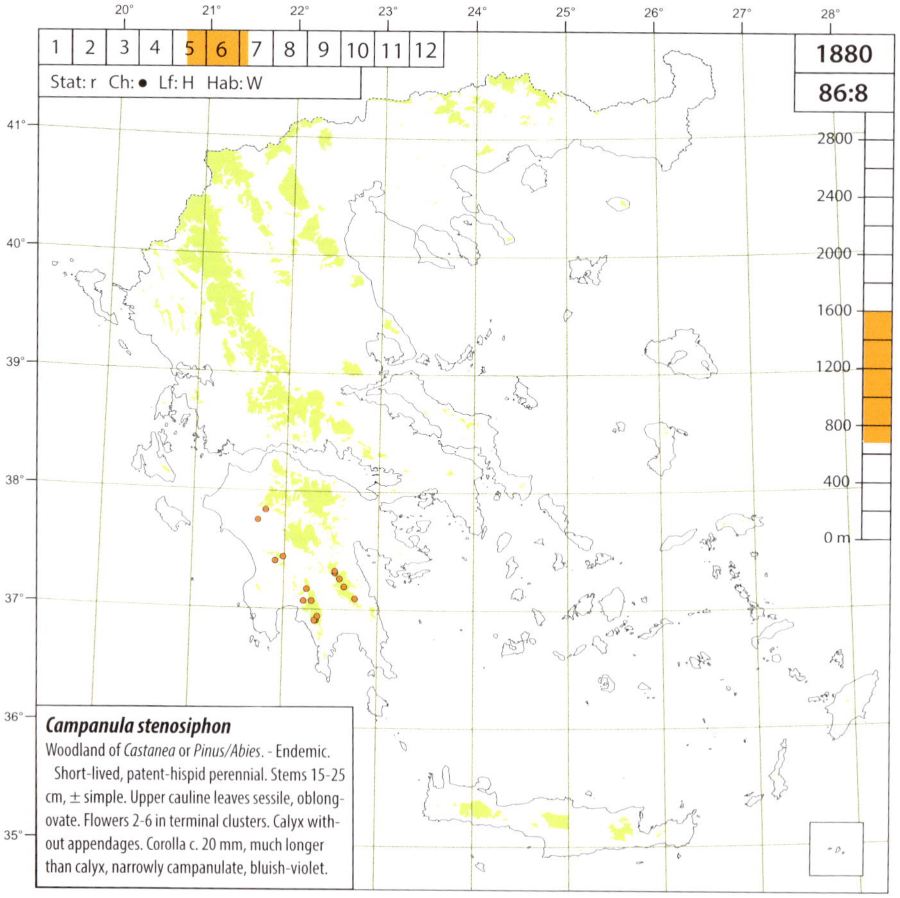

Campanula stenosiphon
Woodland of *Castanea* or *Pinus/Abies*. - Endemic.
Short-lived, patent-hispid perennial. Stems 15-25 cm, ± simple. Upper cauline leaves sessile, oblong-ovate. Flowers 2-6 in terminal clusters. Calyx without appendages. Corolla c. 20 mm, much longer than calyx, narrowly campanulate, bluish-violet.

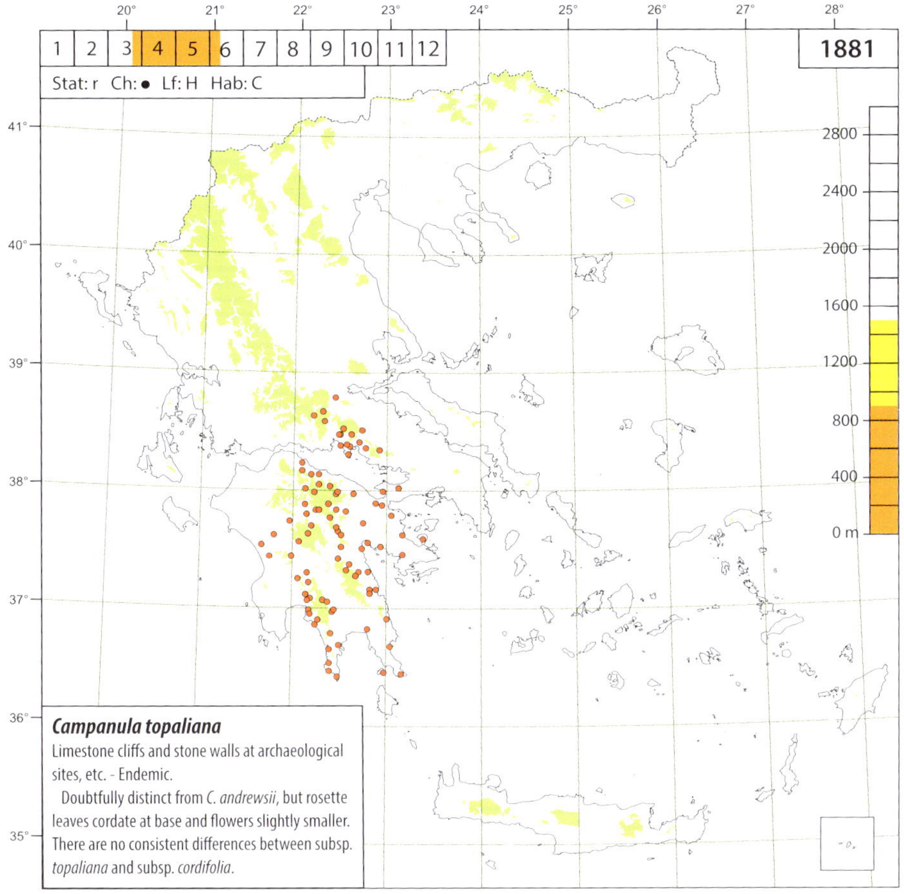

Campanula topaliana
Limestone cliffs and stone walls at archaeological sites, etc. - Endemic.
Doubtfully distinct from *C. andrewsii*, but rosette leaves cordate at base and flowers slightly smaller. There are no consistent differences between subsp. *topaliana* and subsp. *cordifolia*.

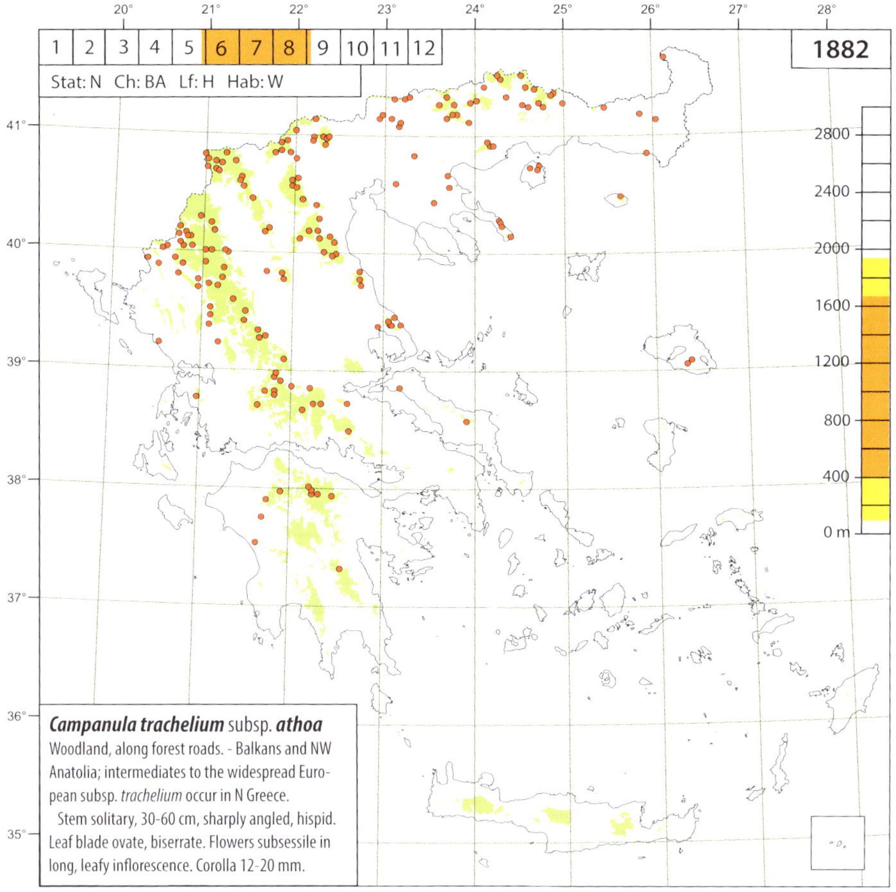

Campanula trachelium* subsp. *athoa
Woodland, along forest roads. - Balkans and NW Anatolia; intermediates to the widespread European subsp. *trachelium* occur in N Greece.
Stem solitary, 30-60 cm, sharply angled, hispid. Leaf blade ovate, biserrate. Flowers subsessile in long, leafy inflorescence. Corolla 12-20 mm.

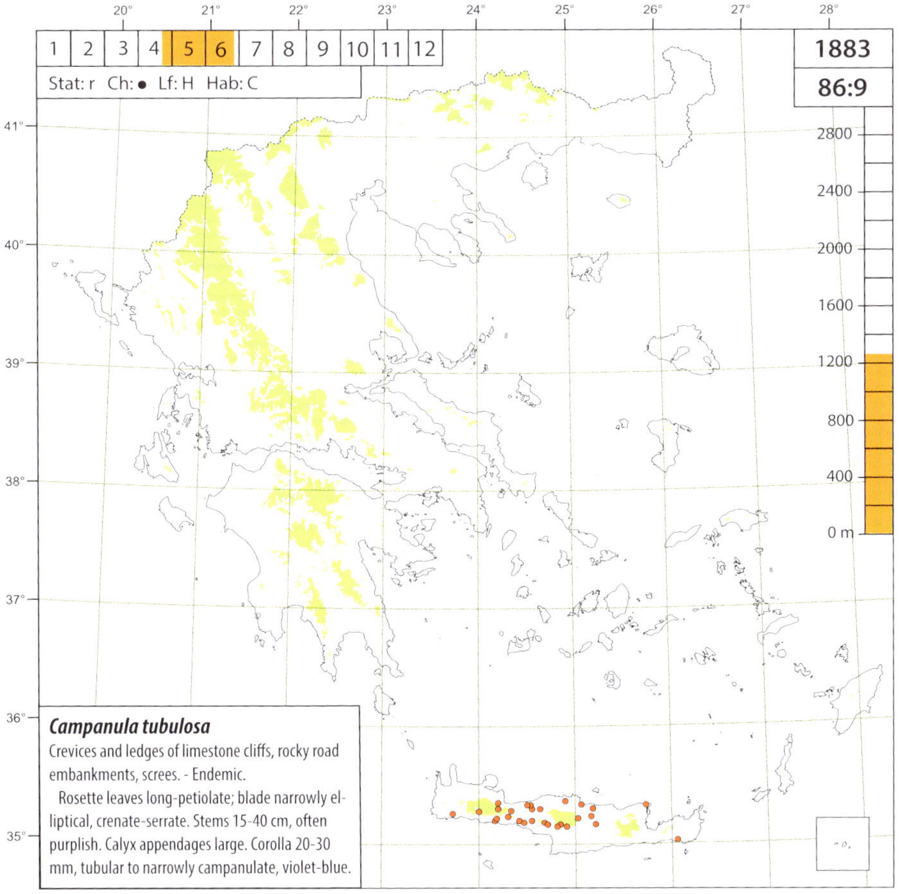

Campanula tubulosa
Crevices and ledges of limestone cliffs, rocky road embankments, screes. - Endemic.
Rosette leaves long-petiolate; blade narrowly elliptical, crenate-serrate. Stems 15-40 cm, often purplish. Calyx appendages large. Corolla 20-30 mm, tubular to narrowly campanulate, violet-blue.

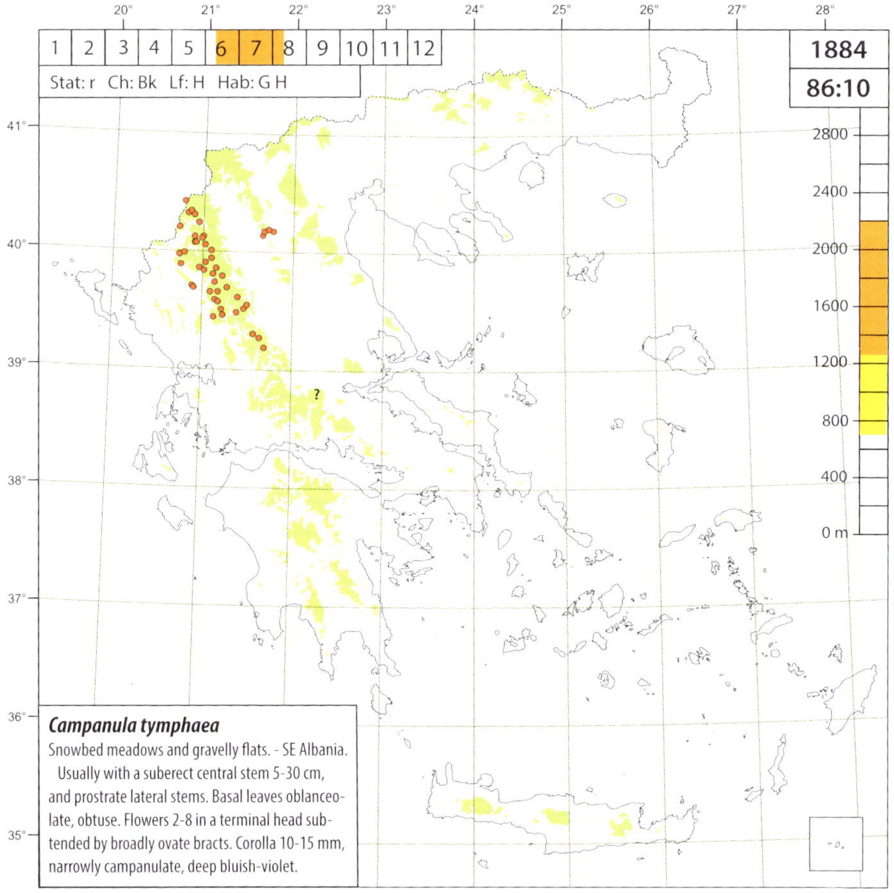

Campanula tymphaea
Snowbed meadows and gravelly flats. - SE Albania.
Usually with a suberect central stem 5-30 cm, and prostrate lateral stems. Basal leaves oblanceolate, obtuse. Flowers 2-8 in a terminal head subtended by broadly ovate bracts. Corolla 10-15 mm, narrowly campanulate, deep bluish-violet.

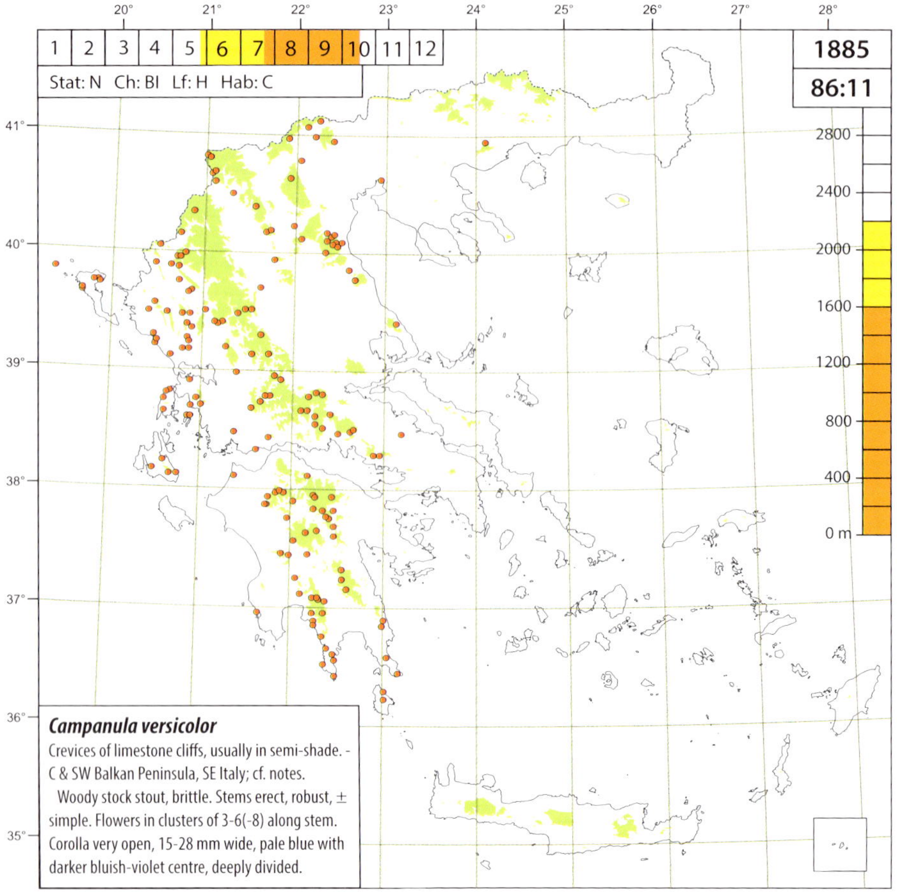

Campanula versicolor
Crevices of limestone cliffs, usually in semi-shade. - C & SW Balkan Peninsula, SE Italy; cf. notes.
Woody stock stout, brittle. Stems erect, robust, ± simple. Flowers in clusters of 3-6(-8) along stem. Corolla very open, 15-28 mm wide, pale blue with darker bluish-violet centre, deeply divided.

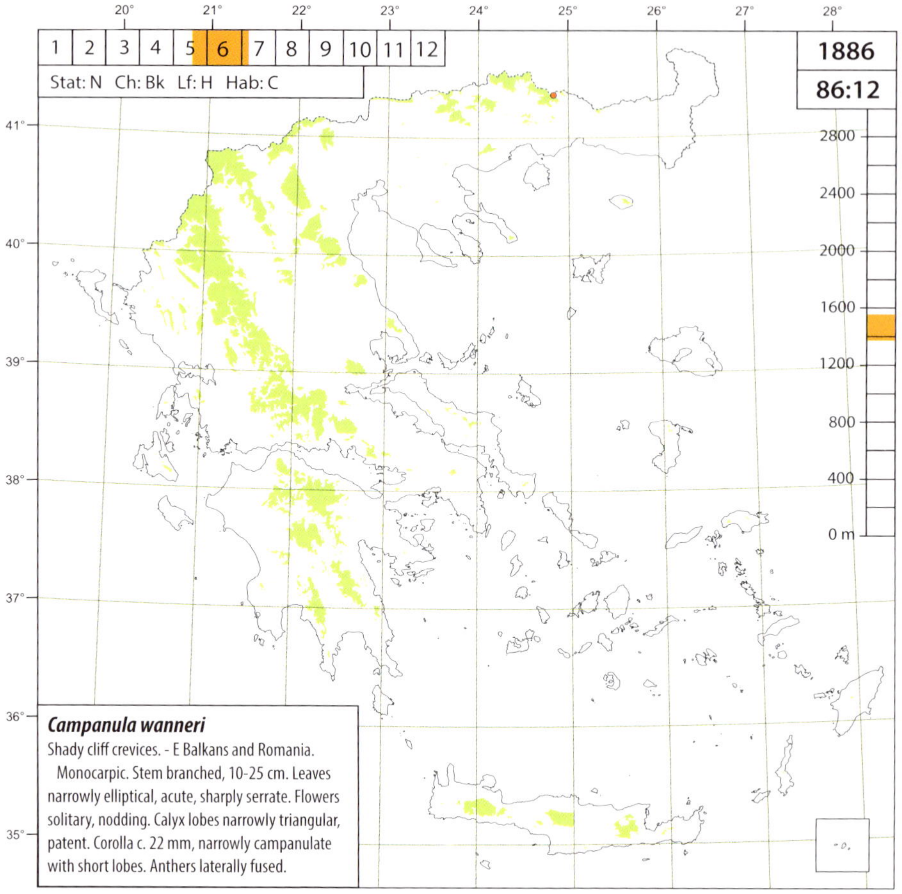

Campanula wanneri
Shady cliff crevices. - E Balkans and Romania.
Monocarpic. Stem branched, 10-25 cm. Leaves narrowly elliptical, acute, sharply serrate. Flowers solitary, nodding. Calyx lobes narrowly triangular, patent. Corolla c. 22 mm, narrowly campanulate with short lobes. Anthers laterally fused.

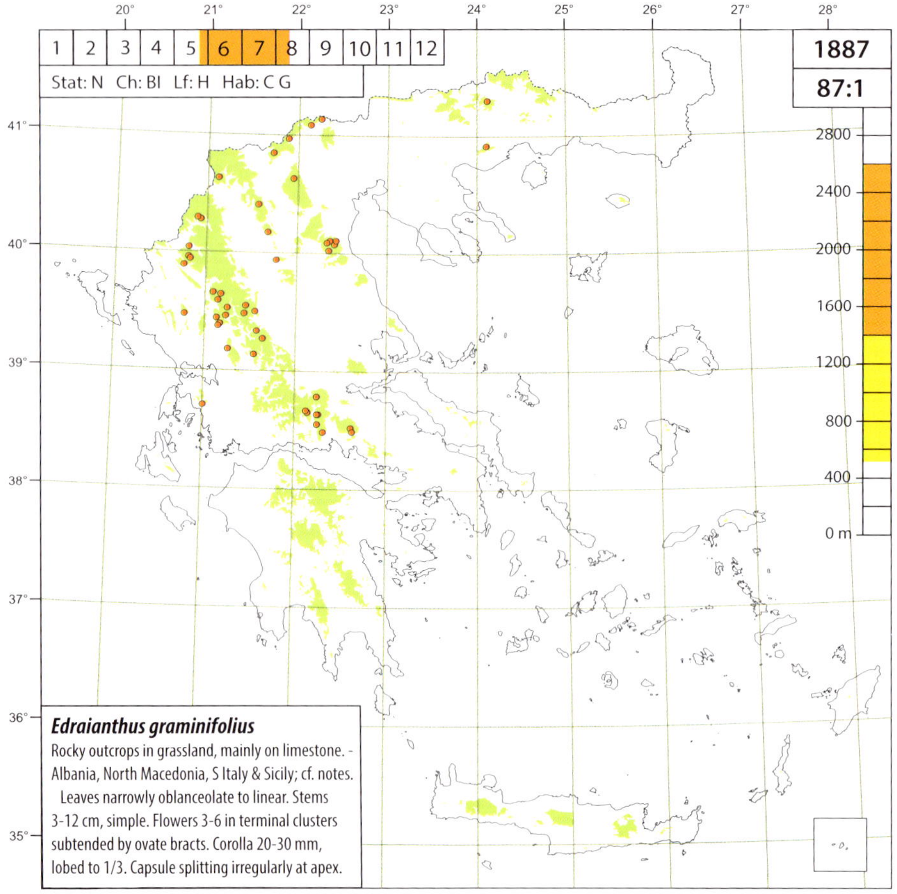

Edraianthus graminifolius
Rocky outcrops in grassland, mainly on limestone. - Albania, North Macedonia, S Italy & Sicily; cf. notes.
Leaves narrowly oblanceolate to linear. Stems 3-12 cm, simple. Flowers 3-6 in terminal clusters subtended by ovate bracts. Corolla 20-30 mm, lobed to 1/3. Capsule splitting irregularly at apex.

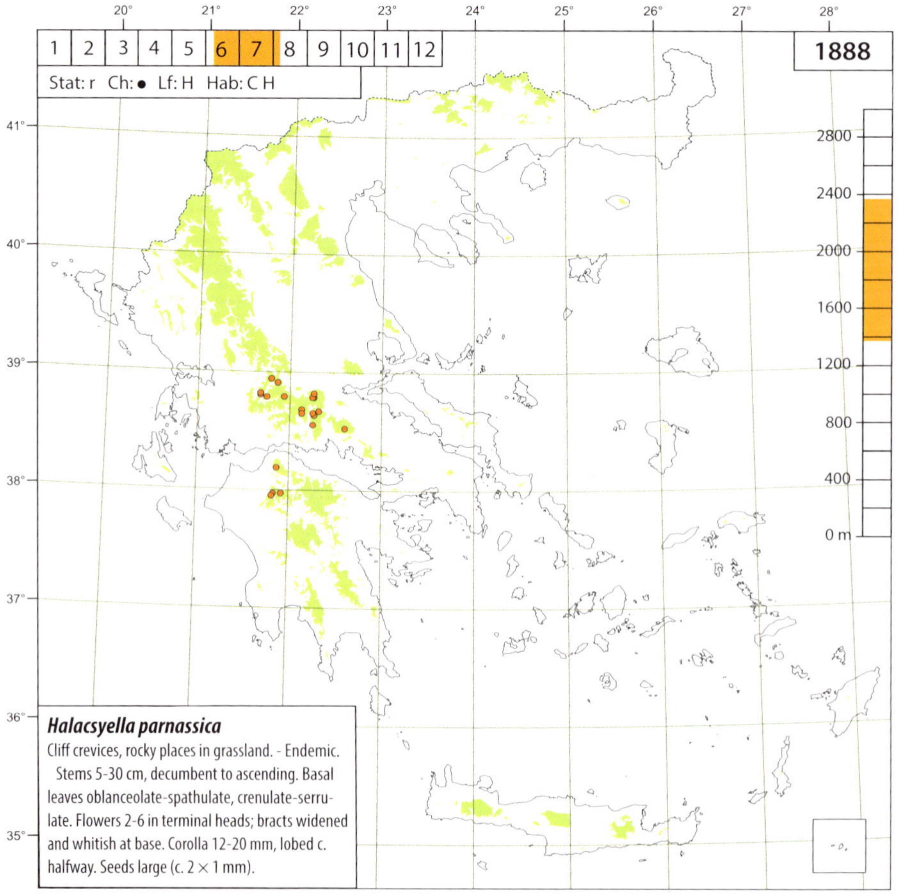

Halacsyella parnassica
Cliff crevices, rocky places in grassland. - Endemic.
Stems 5-30 cm, decumbent to ascending. Basal leaves oblanceolate-spathulate, crenulate-serrulate. Flowers 2-6 in terminal heads; bracts widened and whitish at base. Corolla 12-20 mm, lobed c. halfway. Seeds large (c. 2 × 1 mm).

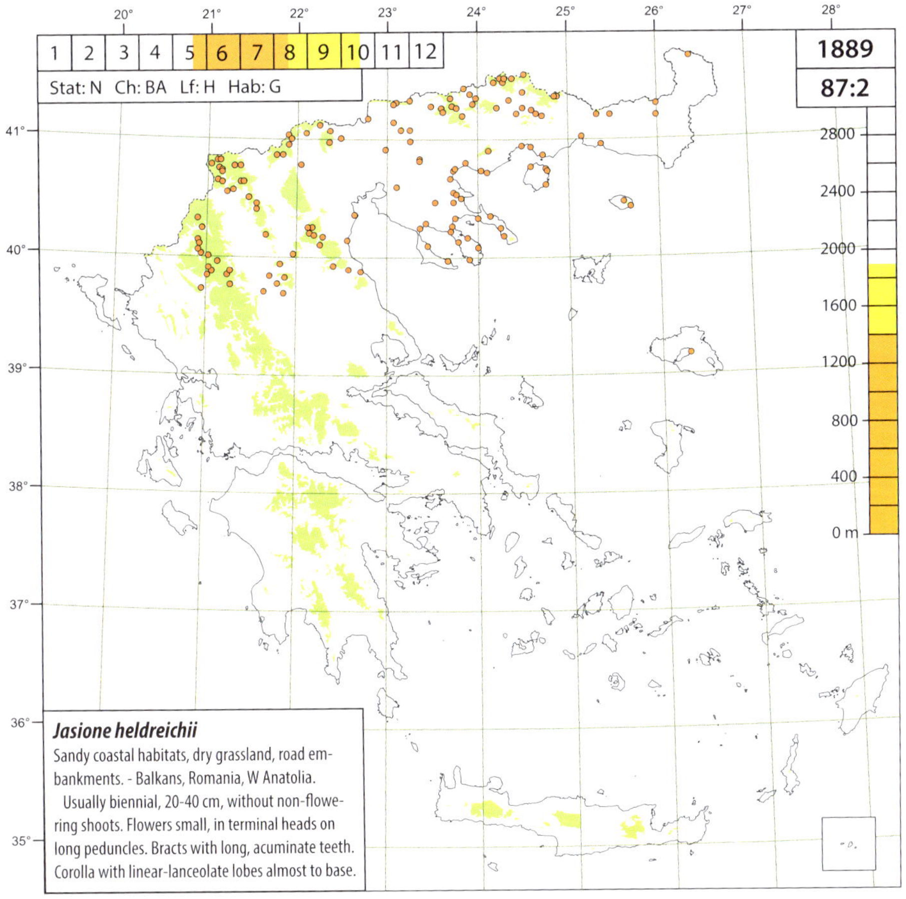

Jasione heldreichii

Sandy coastal habitats, dry grassland, road embankments. - Balkans, Romania, W Anatolia.

Usually biennial, 20-40 cm, without non-flowering shoots. Flowers small, in terminal heads on long peduncles. Bracts with long, acuminate teeth. Corolla with linear-lanceolate lobes almost to base.

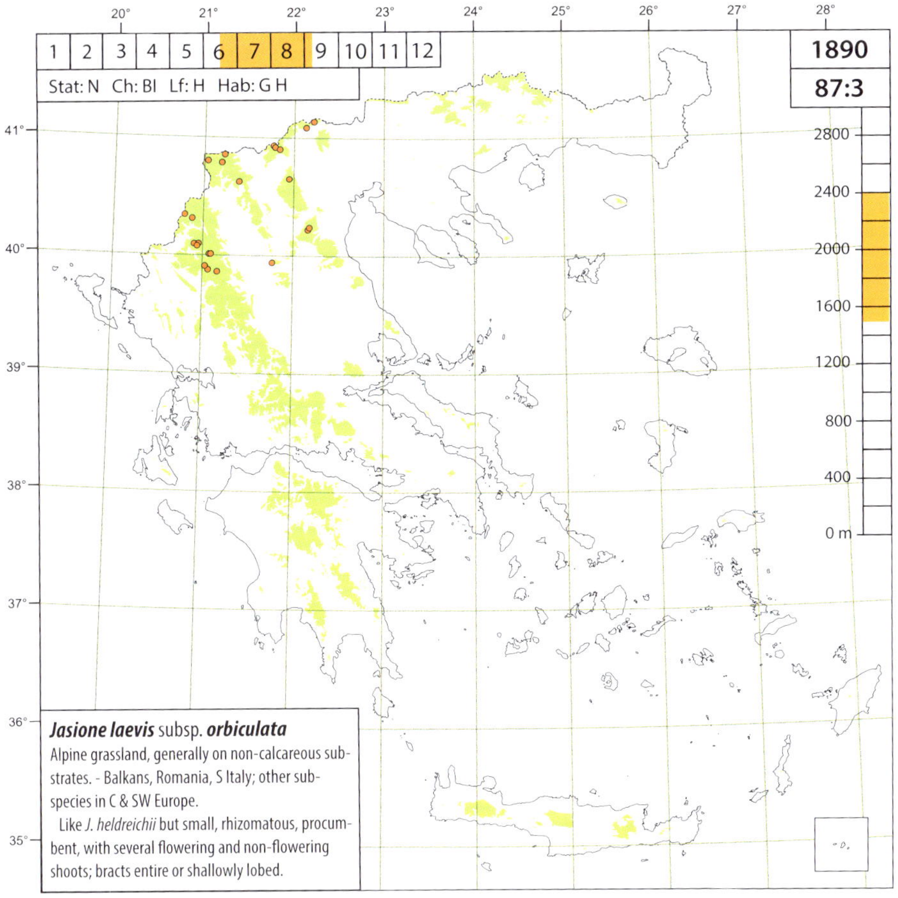

Jasione laevis* subsp. *orbiculata

Alpine grassland, generally on non-calcareous substrates. - Balkans, Romania, S Italy; other subspecies in C & SW Europe.

Like *J. heldreichii* but small, rhizomatous, procumbent, with several flowering and non-flowering shoots; bracts entire or shallowly lobed.

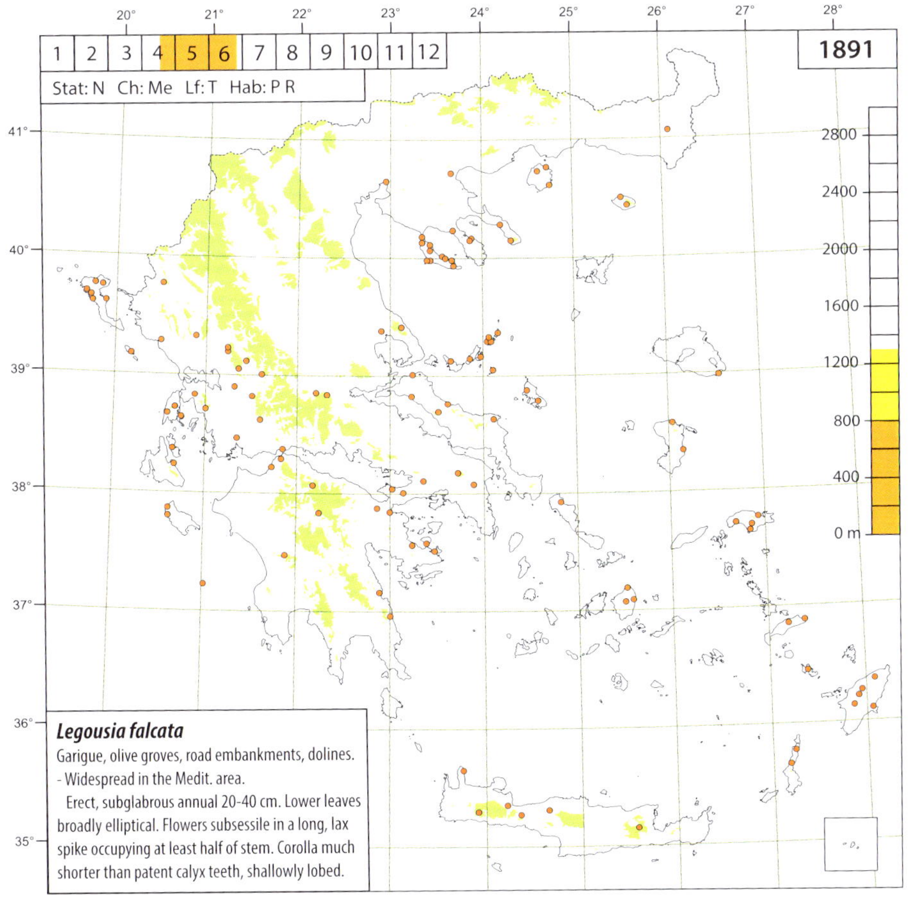

Legousia falcata

Garigue, olive groves, road embankments, dolines. - Widespread in the Medit. area.

Erect, subglabrous annual 20-40 cm. Lower leaves broadly elliptical. Flowers subsessile in a long, lax spike occupying at least half of stem. Corolla much shorter than patent calyx teeth, shallowly lobed.

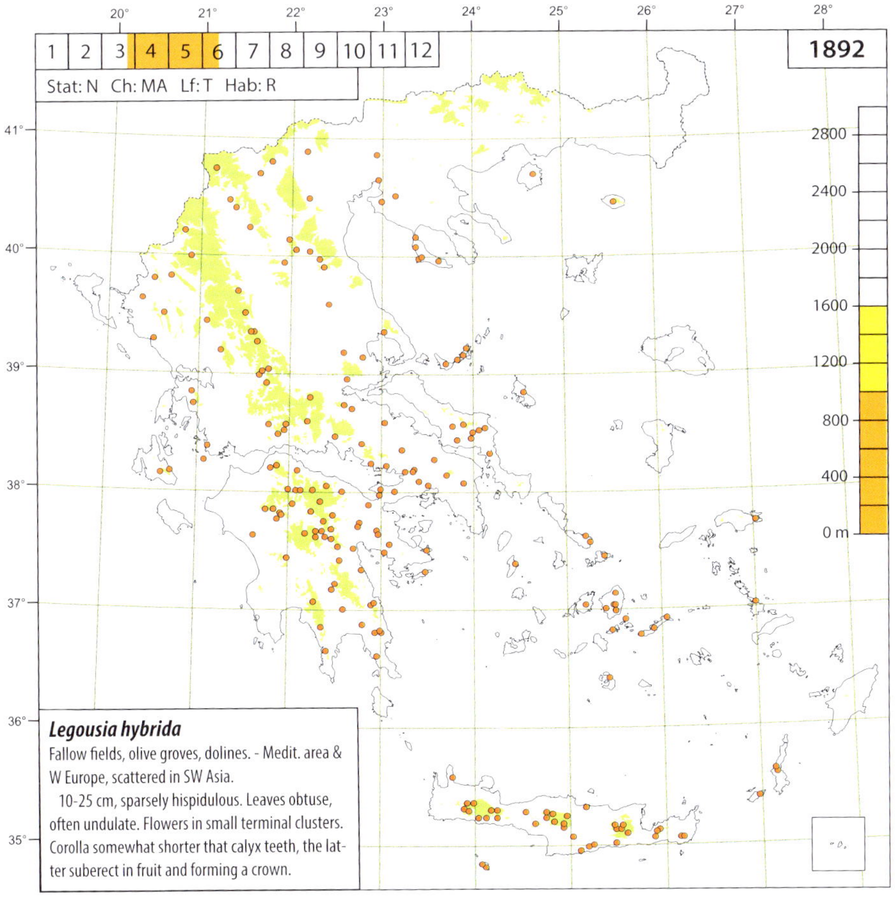

Legousia hybrida

Fallow fields, olive groves, dolines. - Medit. area & W Europe, scattered in SW Asia.

10-25 cm, sparsely hispidulous. Leaves obtuse, often undulate. Flowers in small terminal clusters. Corolla somewhat shorter that calyx teeth, the latter suberect in fruit and forming a crown.

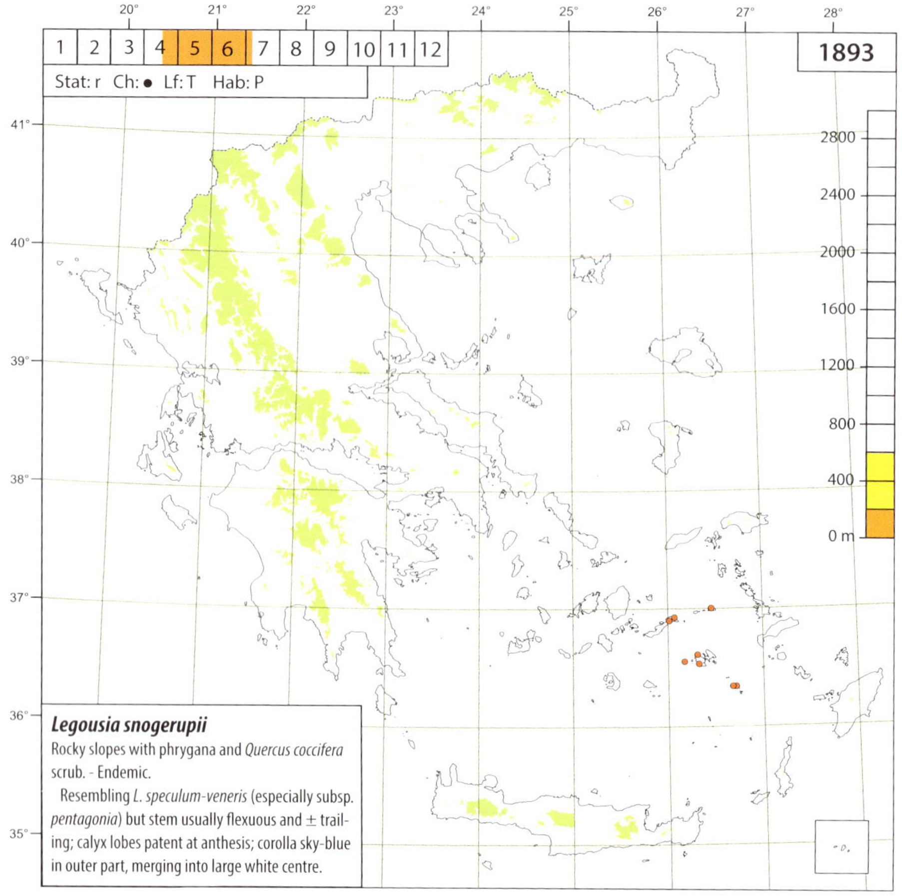

Legousia snogerupii
Rocky slopes with phrygana and *Quercus coccifera* scrub. - Endemic.
Resembling *L. speculum-veneris* (especially subsp. *pentagonia*) but stem usually flexuous and ± trailing; calyx lobes patent at anthesis; corolla sky-blue in outer part, merging into large white centre.

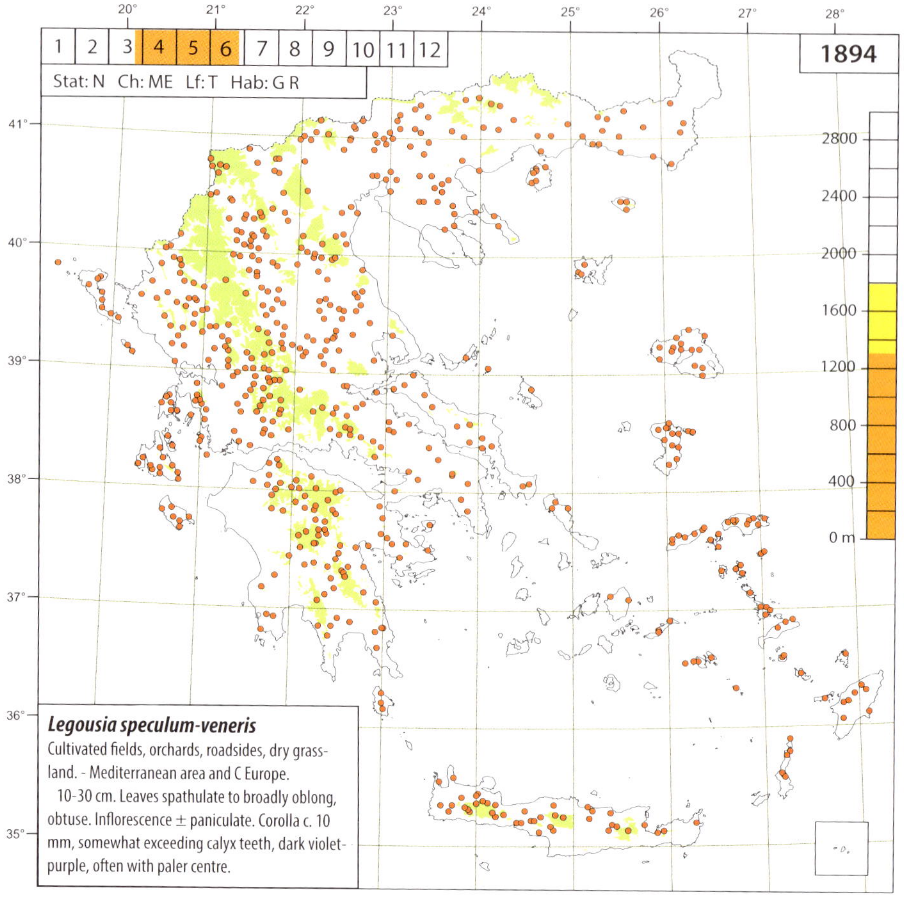

Legousia speculum-veneris
Cultivated fields, orchards, roadsides, dry grassland. - Mediterranean area and C Europe.
10-30 cm. Leaves spathulate to broadly oblong, obtuse. Inflorescence ± paniculate. Corolla c. 10 mm, somewhat exceeding calyx teeth, dark violet-purple, often with paler centre.

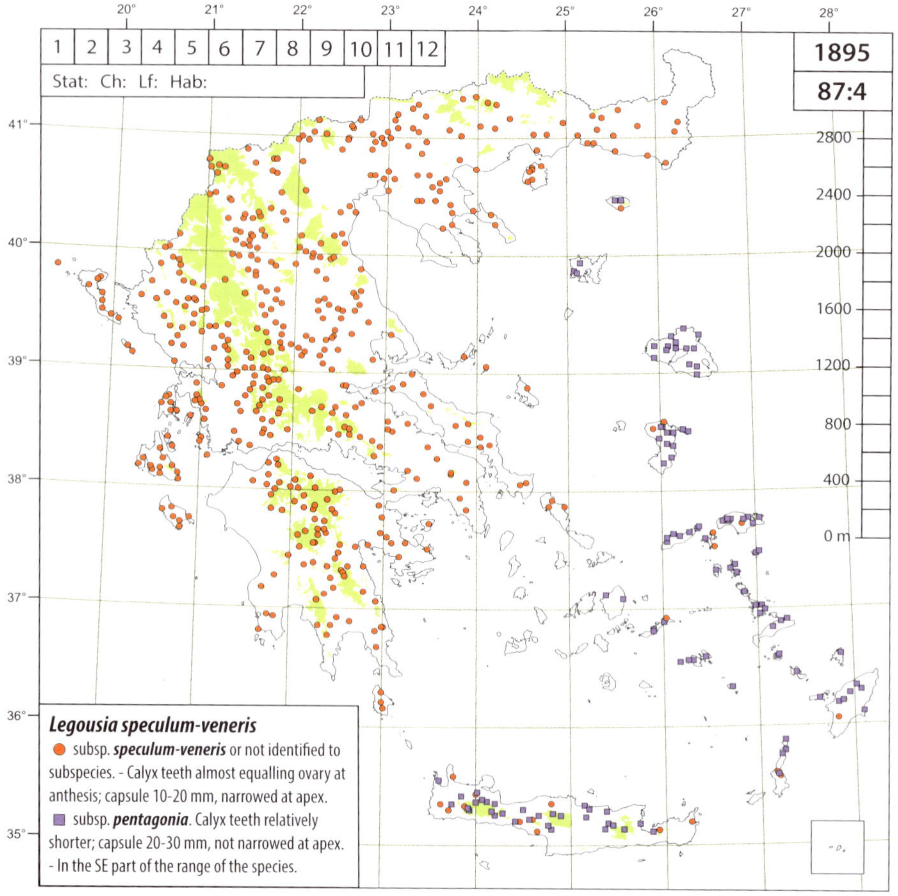

Legousia speculum-veneris
● subsp. ***speculum-veneris*** or not identified to subspecies. - Calyx teeth almost equalling ovary at anthesis; capsule 10-20 mm, narrowed at apex.
■ subsp. ***pentagonia***. Calyx teeth relatively shorter; capsule 20-30 mm, not narrowed at apex. - In the SE part of the range of the species.

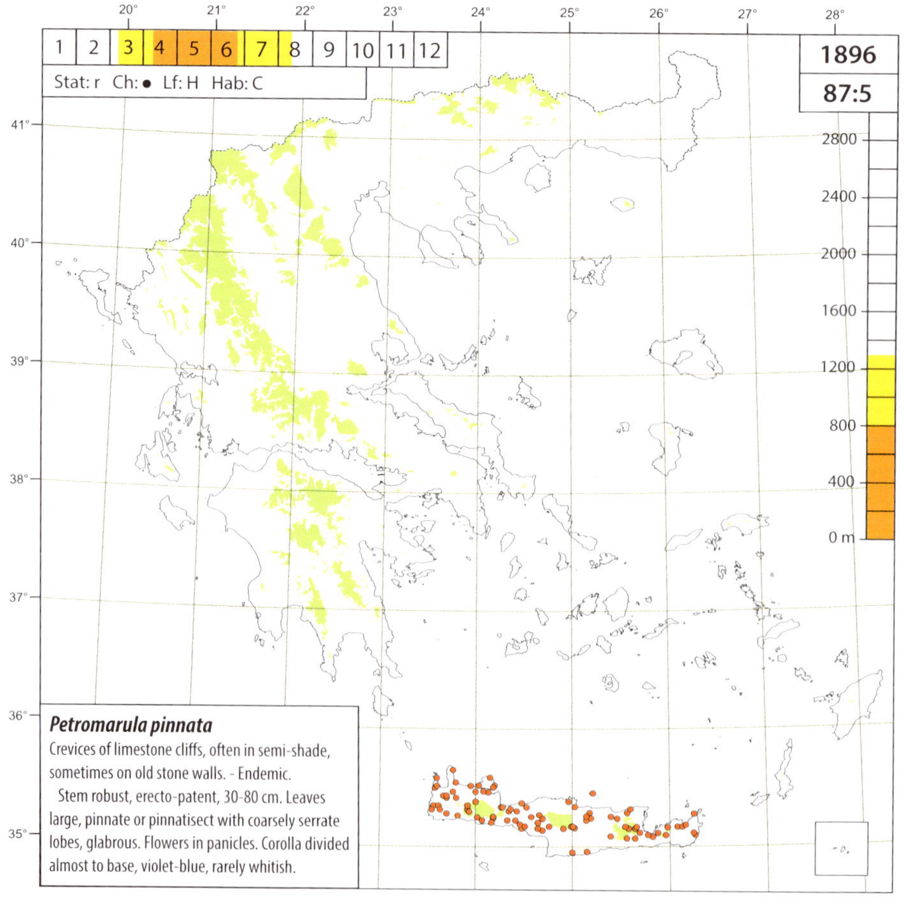

Petromarula pinnata
Crevices of limestone cliffs, often in semi-shade, sometimes on old stone walls. - Endemic.
Stem robust, erecto-patent, 30-80 cm. Leaves large, pinnate or pinnatisect with coarsely serrate lobes, glabrous. Flowers in panicles. Corolla divided almost to base, violet-blue, rarely whitish.

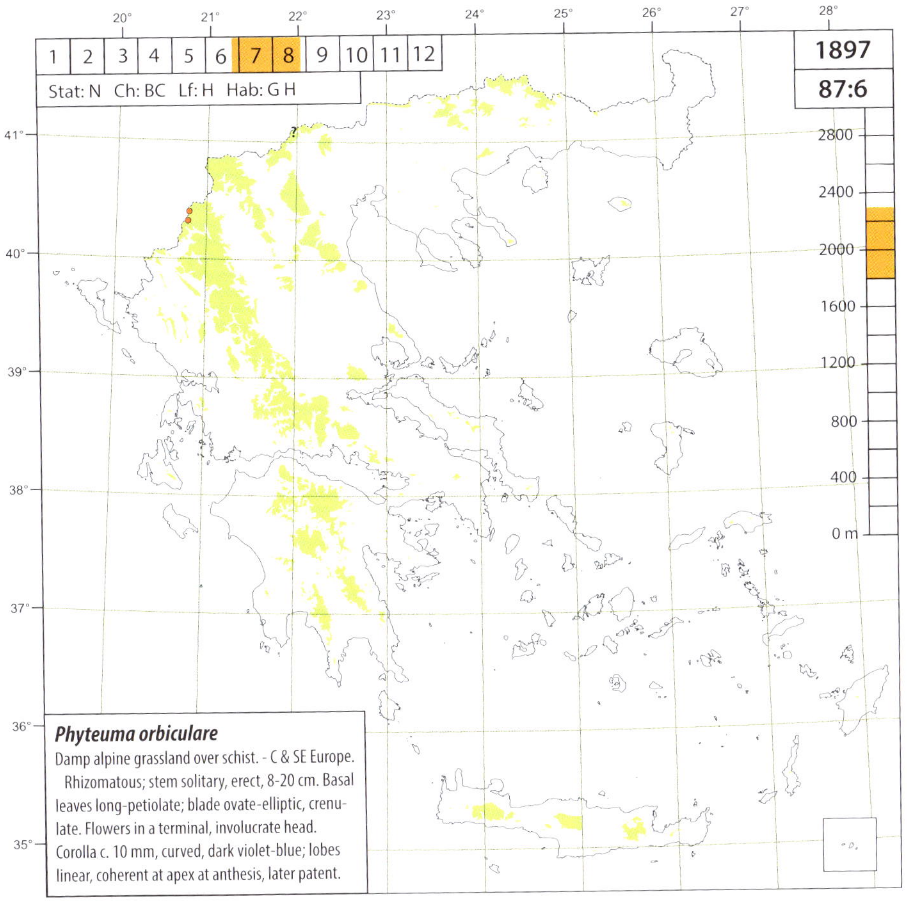

Phyteuma orbiculare
Damp alpine grassland over schist. - C & SE Europe.
Rhizomatous; stem solitary, erect, 8-20 cm. Basal leaves long-petiolate; blade ovate-elliptic, crenulate. Flowers in a terminal, involucrate head. Corolla c. 10 mm, curved, dark violet-blue; lobes linear, coherent at apex at anthesis, later patent.

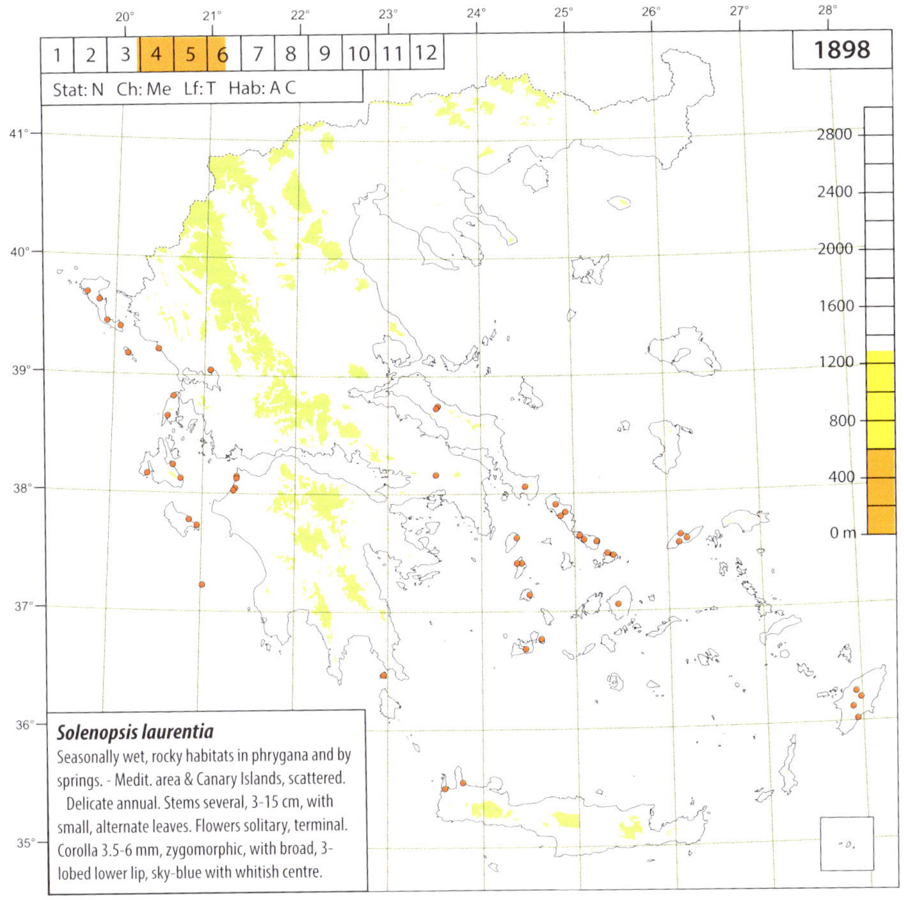

Solenopsis laurentia
Seasonally wet, rocky habitats in phrygana and by springs. - Medit. area & Canary Islands, scattered.
Delicate annual. Stems several, 3-15 cm, with small, alternate leaves. Flowers solitary, terminal. Corolla 3.5-6 mm, zygomorphic, with broad, 3-lobed lower lip, sky-blue with whitish centre.

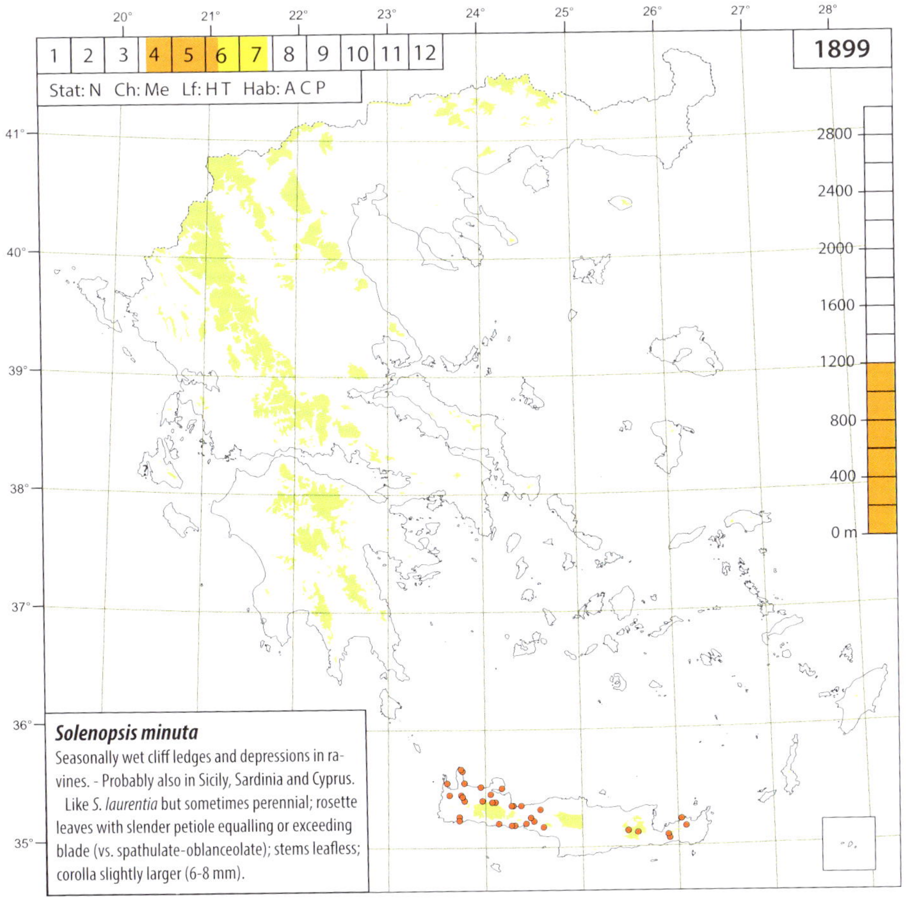

Solenopsis minuta
Seasonally wet cliff ledges and depressions in ravines. - Probably also in Sicily, Sardinia and Cyprus.
Like *S. laurentia* but sometimes perennial; rosette leaves with slender petiole equalling or exceeding blade (vs. spathulate-oblanceolate); stems leafless; corolla slightly larger (6-8 mm).

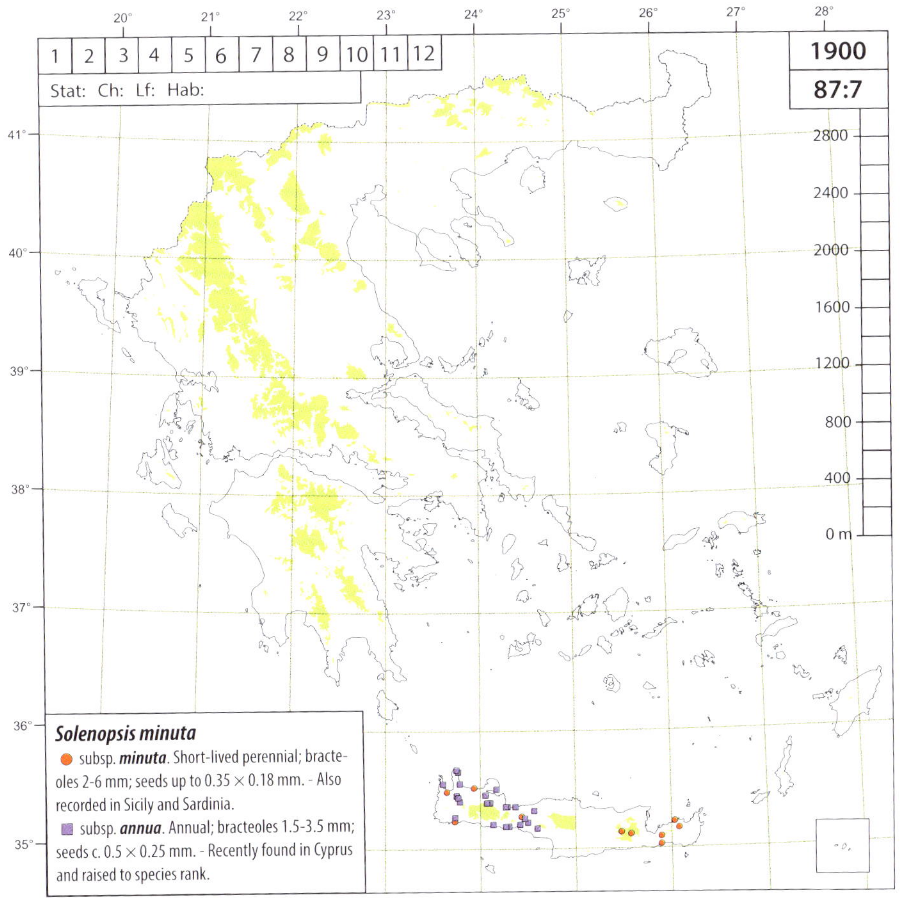

Solenopsis minuta
● subsp. ***minuta***. Short-lived perennial; bracteoles 2-6 mm; seeds up to 0.35 × 0.18 mm. - Also recorded in Sicily and Sardinia.
■ subsp. ***annua***. Annual; bracteoles 1.5-3.5 mm; seeds c. 0.5 × 0.25 mm. - Recently found in Cyprus and raised to species rank.

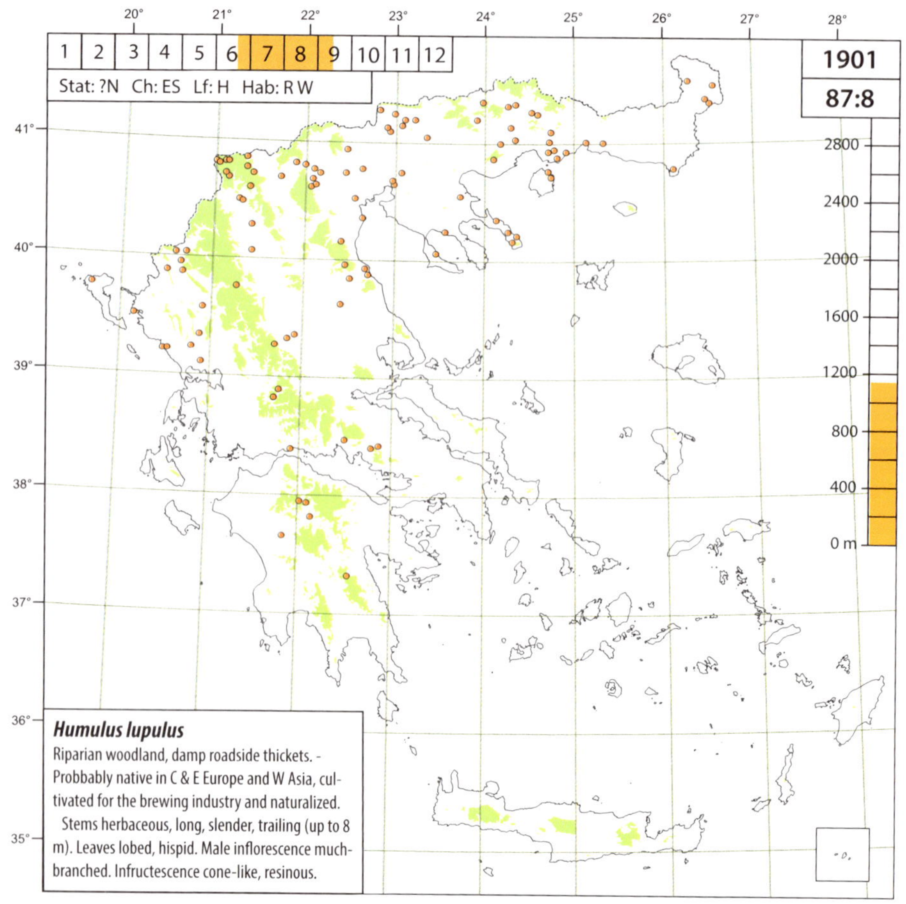

Humulus lupulus
Riparian woodland, damp roadside thickets. - Probbably native in C & E Europe and W Asia, cultivated for the brewing industry and naturalized.
Stems herbaceous, long, slender, trailing (up to 8 m). Leaves lobed, hispid. Male inflorescence much-branched. Infructescence cone-like, resinous.

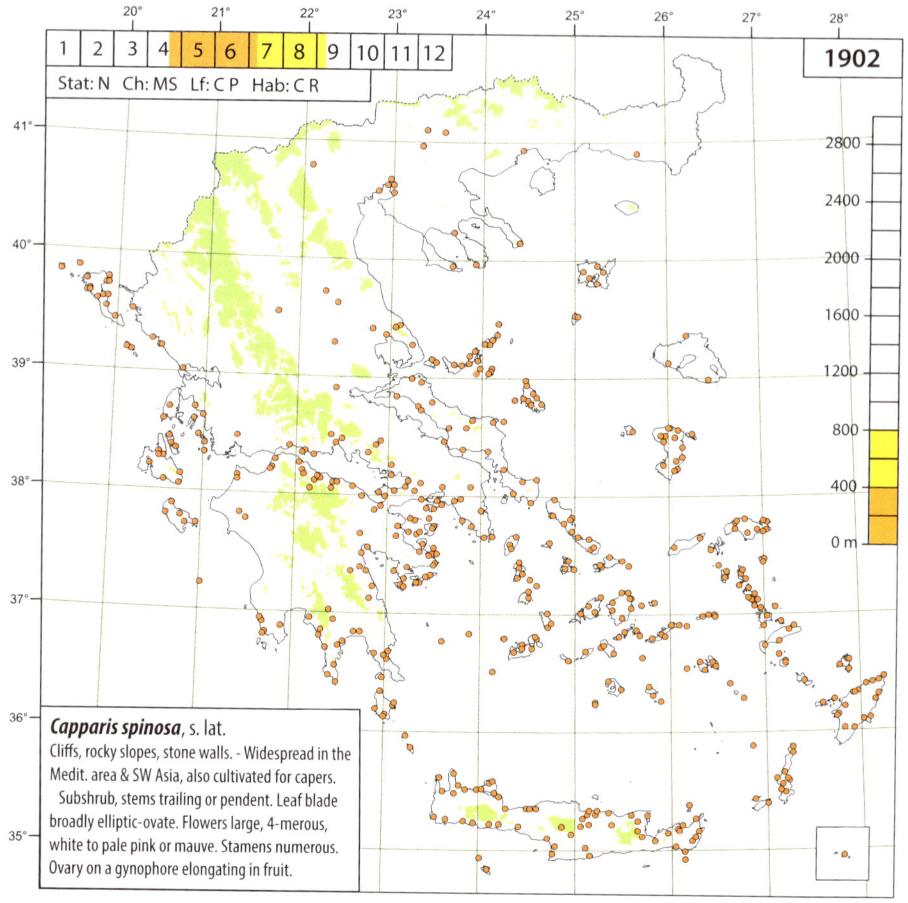

Capparis spinosa, s. lat.
Cliffs, rocky slopes, stone walls. - Widespread in the Medit. area & SW Asia, also cultivated for capers.
Subshrub, stems trailing or pendent. Leaf blade broadly elliptic-ovate. Flowers large, 4-merous, white to pale pink or mauve. Stamens numerous. Ovary on a gynophore elongating in fruit.

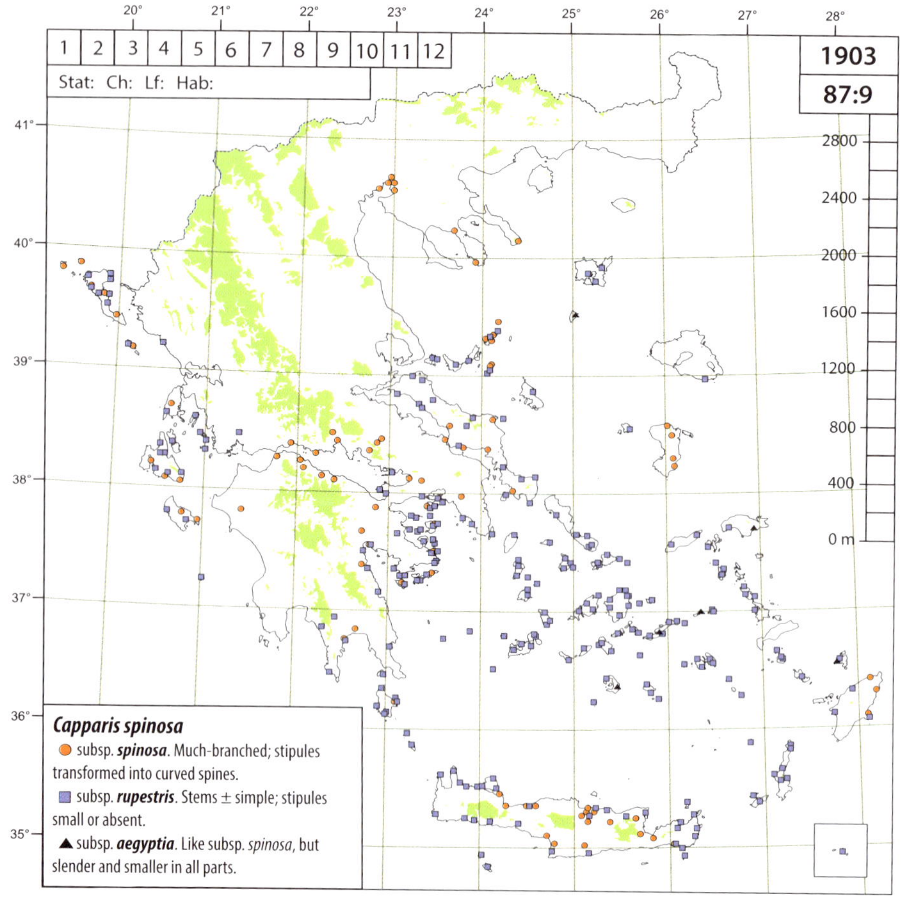

Capparis spinosa
● subsp. ***spinosa***. Much-branched; stipules transformed into curved spines.
■ subsp. ***rupestris***. Stems ± simple; stipules small or absent.
▲ subsp. ***aegyptia***. Like subsp. *spinosa*, but slender and smaller in all parts.

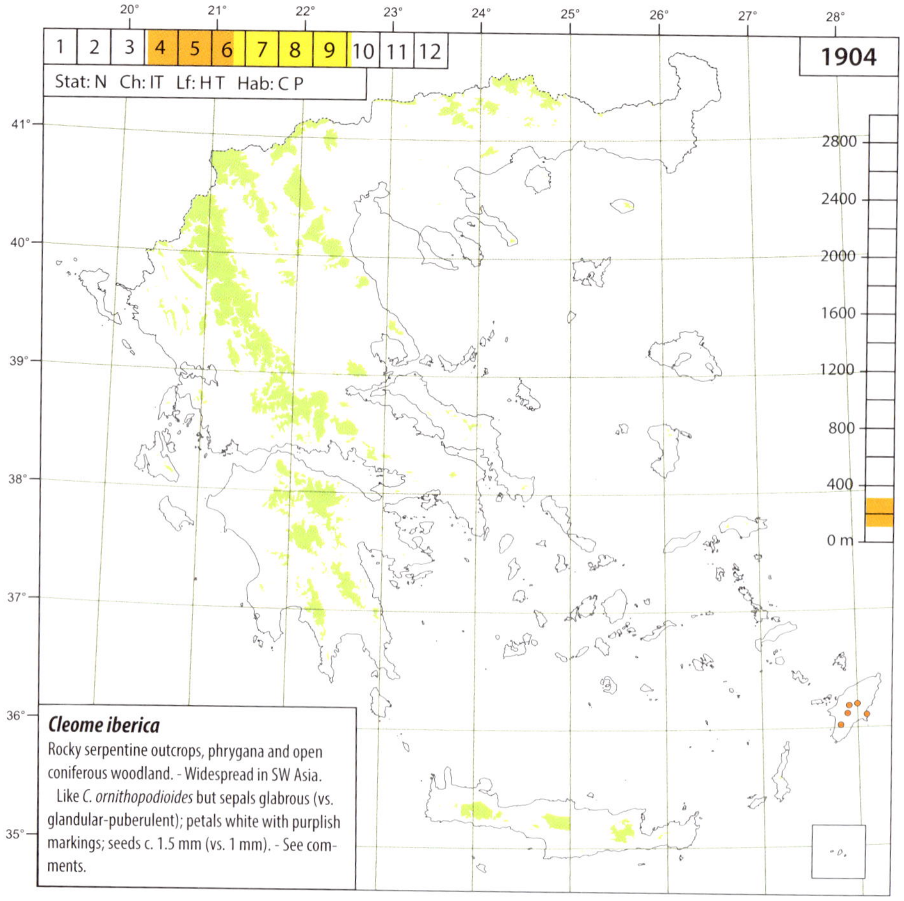

Cleome iberica
Rocky serpentine outcrops, phrygana and open coniferous woodland. - Widespread in SW Asia.
Like *C. ornithopodioides* but sepals glabrous (vs. glandular-puberulent); petals white with purplish markings; seeds c. 1.5 mm (vs. 1 mm). - See comments.

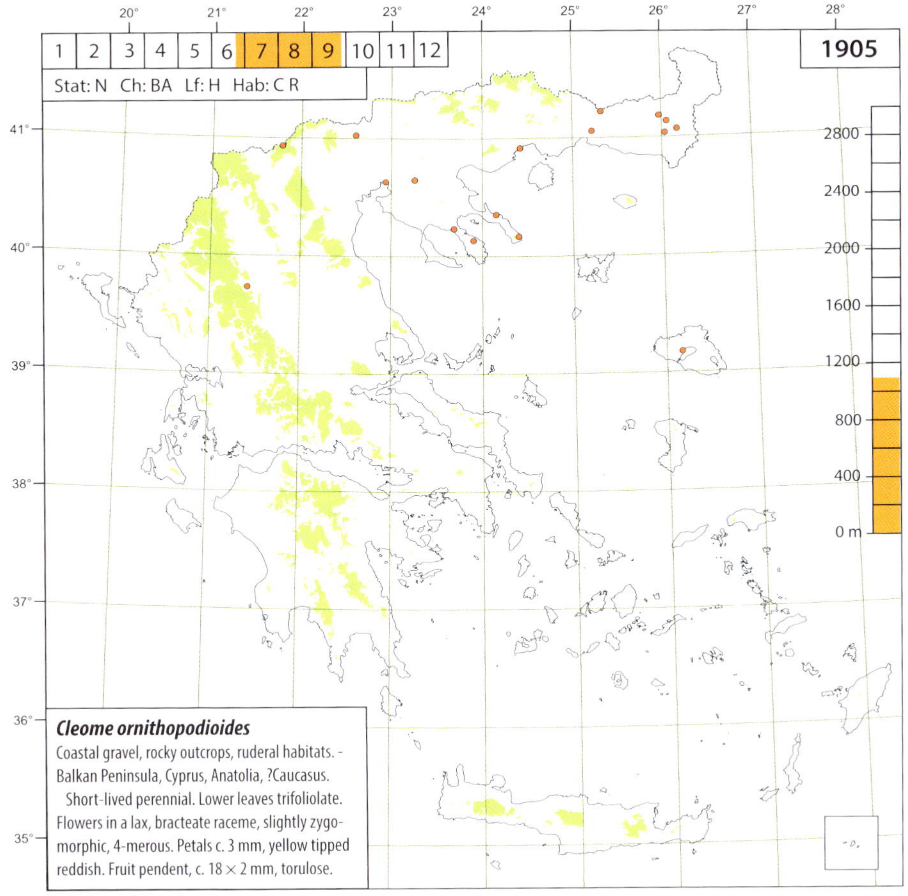

Cleome ornithopodioides
Coastal gravel, rocky outcrops, ruderal habitats. - Balkan Peninsula, Cyprus, Anatolia, ?Caucasus.
Short-lived perennial. Lower leaves trifoliolate. Flowers in a lax, bracteate raceme, slightly zygomorphic, 4-merous. Petals c. 3 mm, yellow tipped reddish. Fruit pendent, c. 18 × 2 mm, torulose.

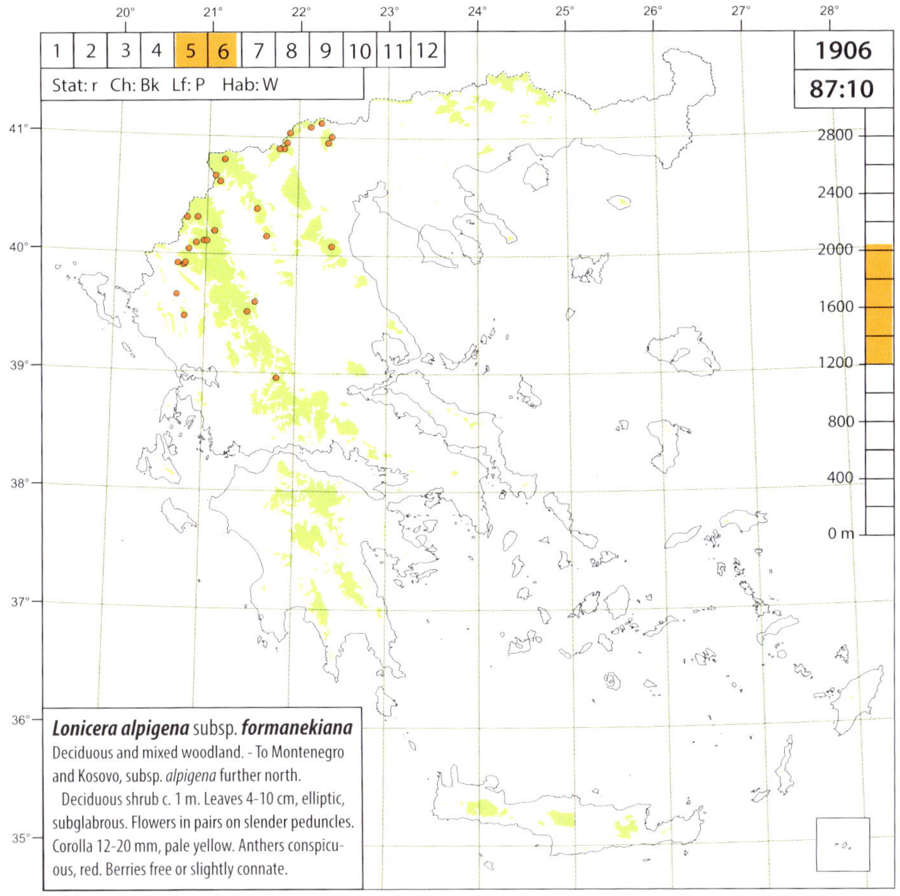

Lonicera alpigena **subsp. *formanekiana***
Deciduous and mixed woodland. - To Montenegro and Kosovo, subsp. *alpigena* further north.
Deciduous shrub c. 1 m. Leaves 4-10 cm, elliptic, subglabrous. Flowers in pairs on slender peduncles. Corolla 12-20 mm, pale yellow. Anthers conspicuous, red. Berries free or slightly connate.

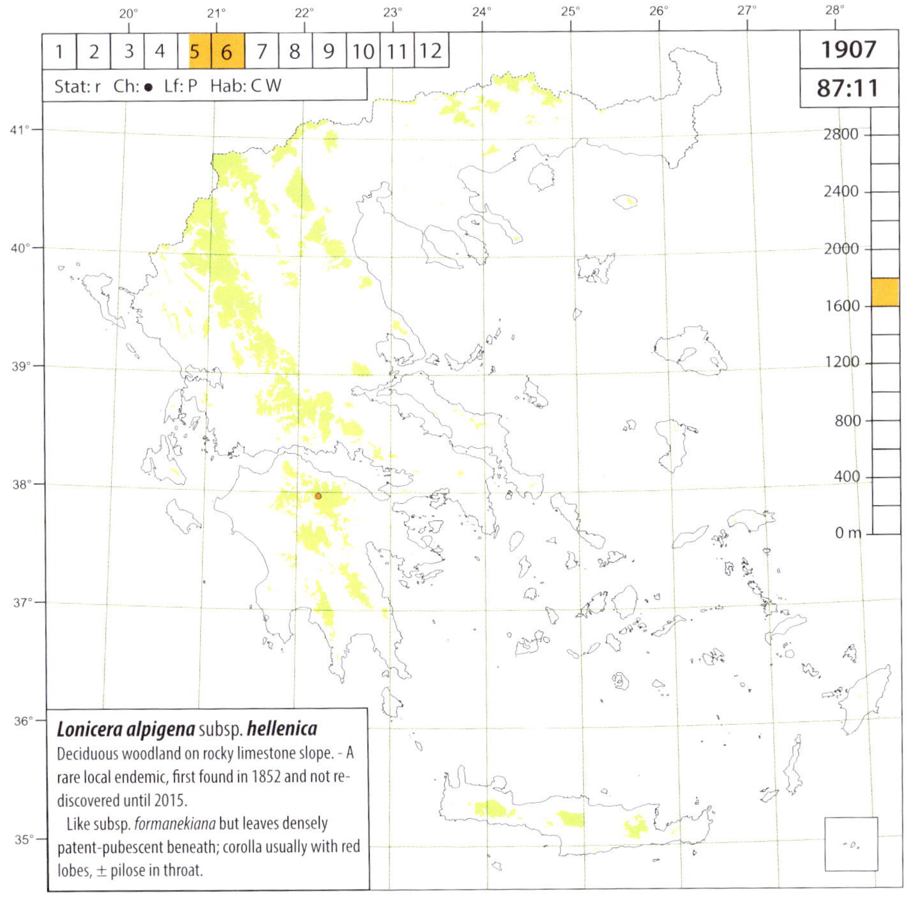

Lonicera alpigena **subsp. *hellenica***
Deciduous woodland on rocky limestone slope. - A rare local endemic, first found in 1852 and not rediscovered until 2015.
Like subsp. *formanekiana* but leaves densely patent-pubescent beneath; corolla usually with red lobes, ± pilose in throat.

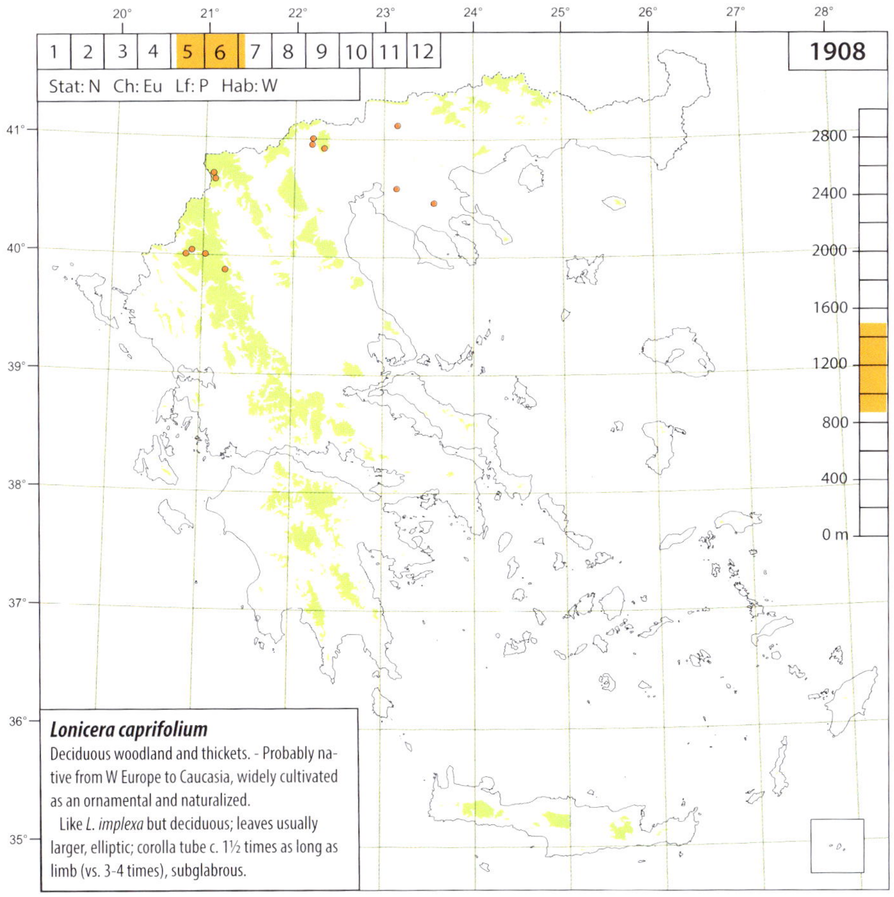

Lonicera caprifolium
Deciduous woodland and thickets. - Probably native from W Europe to Caucasia, widely cultivated as an ornamental and naturalized.
Like *L. implexa* but deciduous; leaves usually larger, elliptic; corolla tube c. 1½ times as long as limb (vs. 3-4 times), subglabrous.

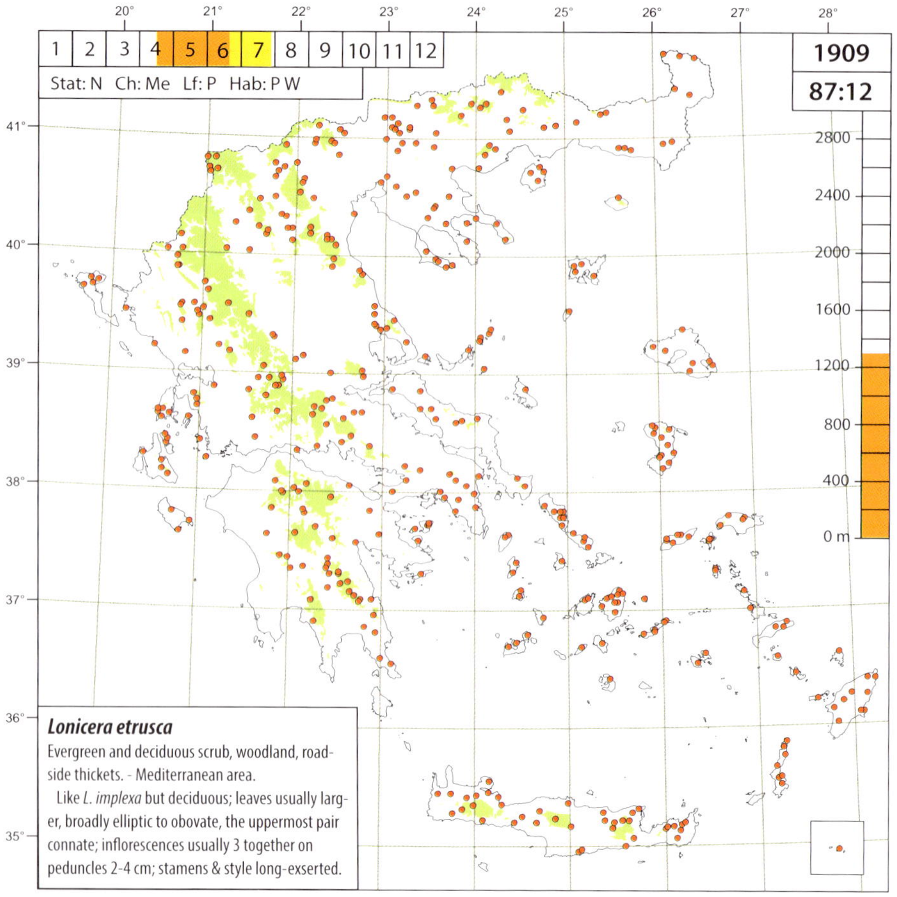

Lonicera etrusca

Evergreen and deciduous scrub, woodland, roadside thickets. - Mediterranean area.

Like *L. implexa* but deciduous; leaves usually larger, broadly elliptic to obovate, the uppermost pair connate; inflorescences usually 3 together on peduncles 2-4 cm; stamens & style long-exserted.

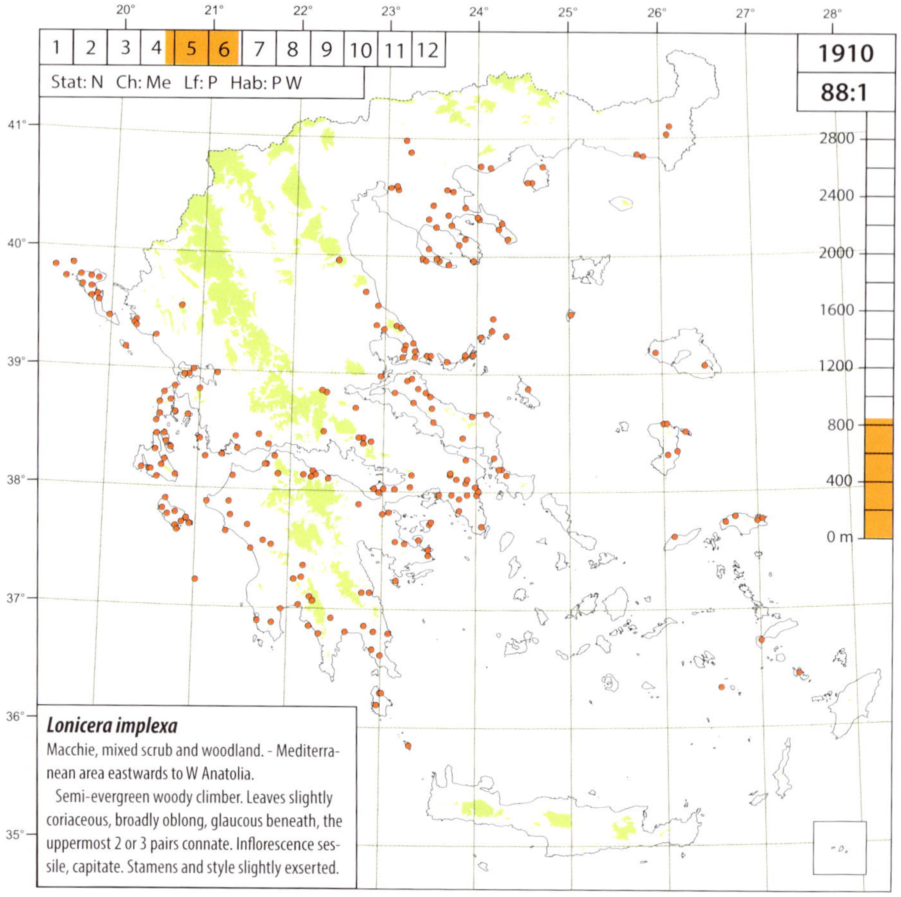

Lonicera implexa

Macchie, mixed scrub and woodland. - Mediterranean area eastwards to W Anatolia.

Semi-evergreen woody climber. Leaves slightly coriaceous, broadly oblong, glaucous beneath, the uppermost 2 or 3 pairs connate. Inflorescence sessile, capitate. Stamens and style slightly exserted.

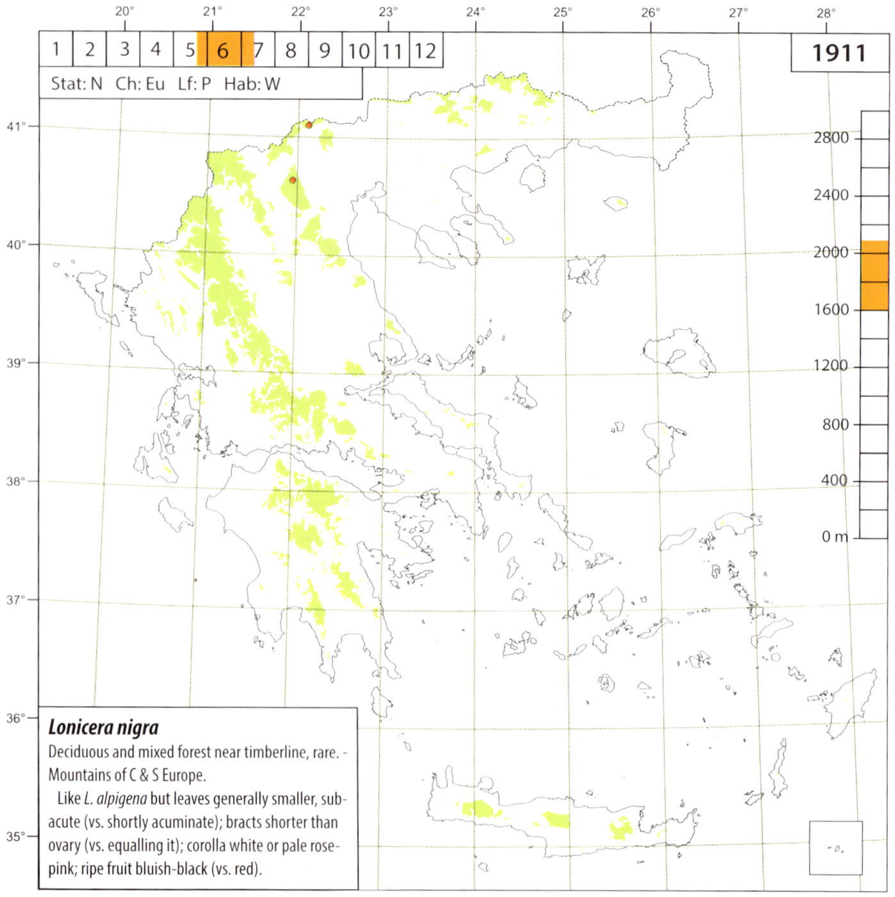

Lonicera nigra

Deciduous and mixed forest near timberline, rare. - Mountains of C & S Europe.

Like *L. alpigena* but leaves generally smaller, subacute (vs. shortly acuminate); bracts shorter than ovary (vs. equalling it); corolla white or pale rose-pink; ripe fruit bluish-black (vs. red).

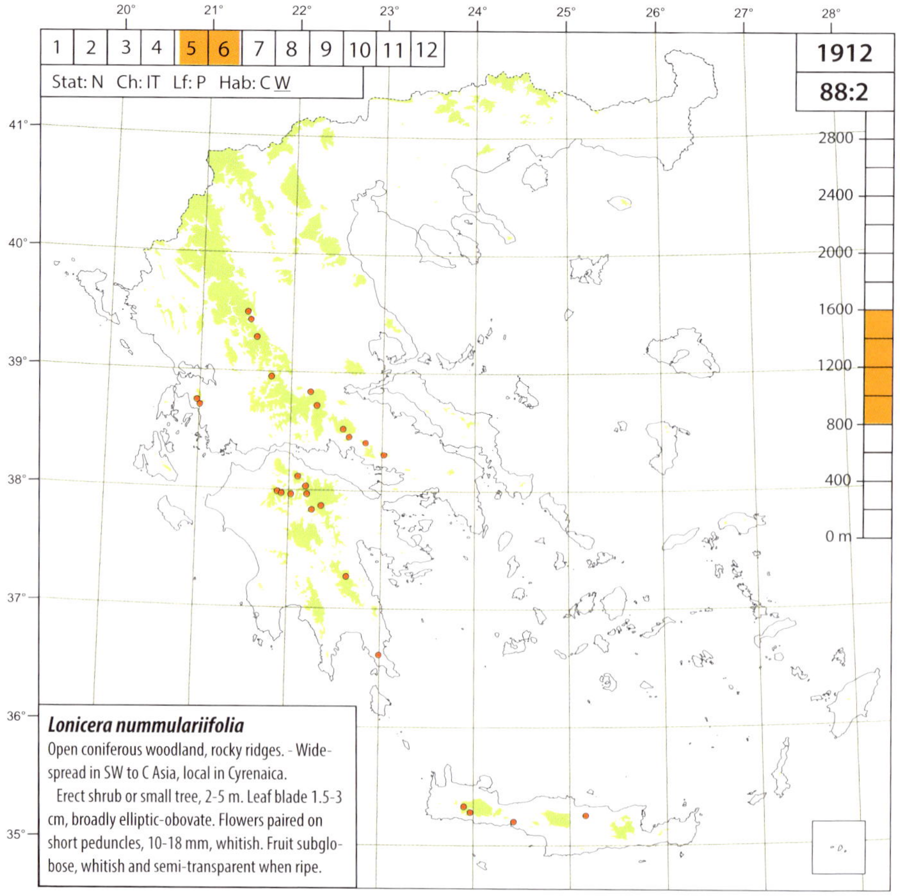

Lonicera nummulariifolia

Open coniferous woodland, rocky ridges. - Widespread in SW to C Asia, local in Cyrenaica.

Erect shrub or small tree, 2-5 m. Leaf blade 1.5-3 cm, broadly elliptic-obovate. Flowers paired on short peduncles, 10-18 mm, whitish. Fruit subglobose, whitish and semi-transparent when ripe.

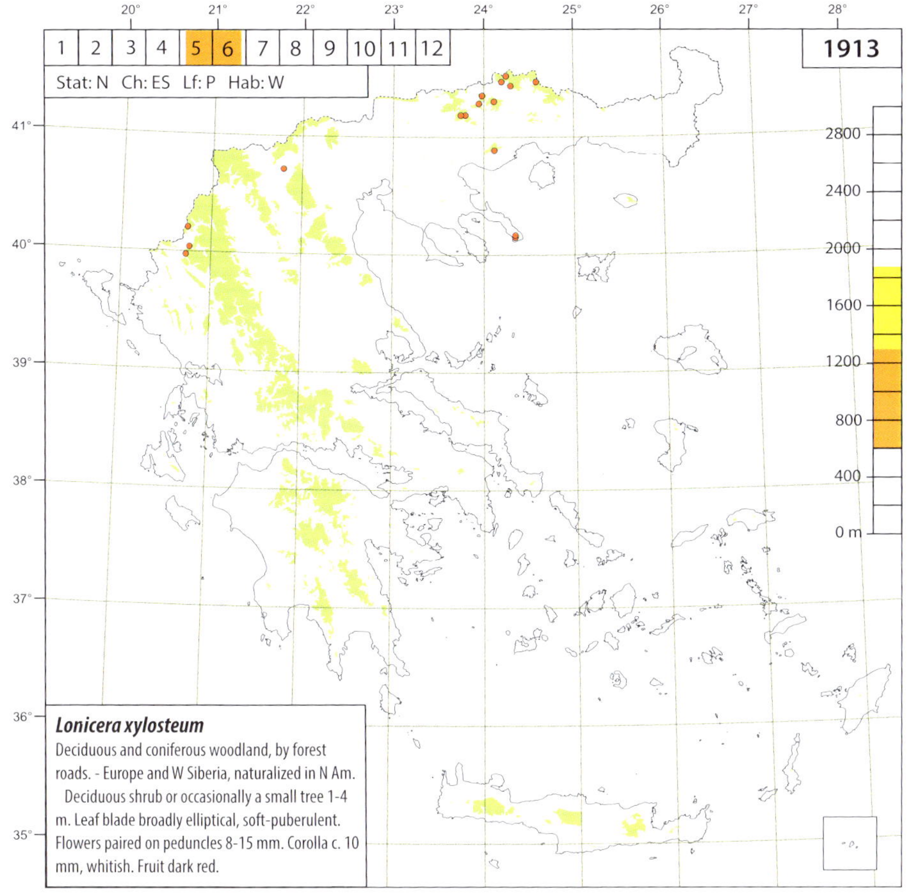

Lonicera xylosteum
Deciduous and coniferous woodland, by forest roads. - Europe and W Siberia, naturalized in N Am.
Deciduous shrub or occasionally a small tree 1-4 m. Leaf blade broadly elliptical, soft-puberulent. Flowers paired on peduncles 8-15 mm. Corolla c. 10 mm, whitish. Fruit dark red.

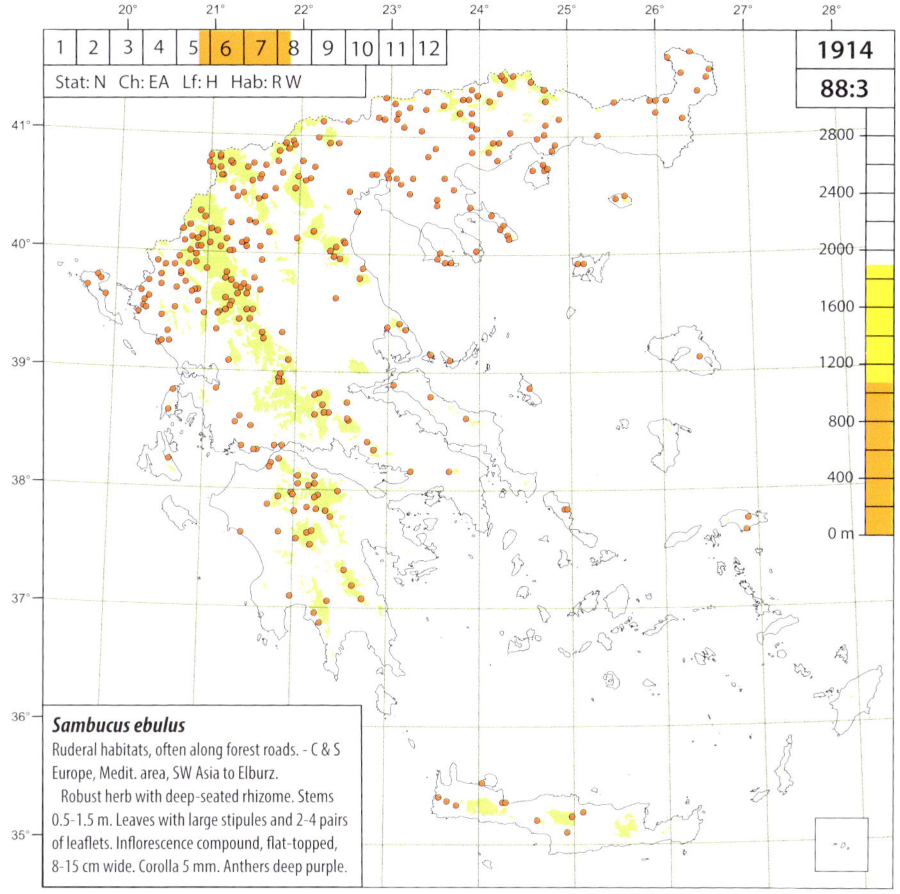

Sambucus ebulus
Ruderal habitats, often along forest roads. - C & S Europe, Medit. area, SW Asia to Elburz.
Robust herb with deep-seated rhizome. Stems 0.5-1.5 m. Leaves with large stipules and 2-4 pairs of leaflets. Inflorescence compound, flat-topped, 8-15 cm wide. Corolla 5 mm. Anthers deep purple.

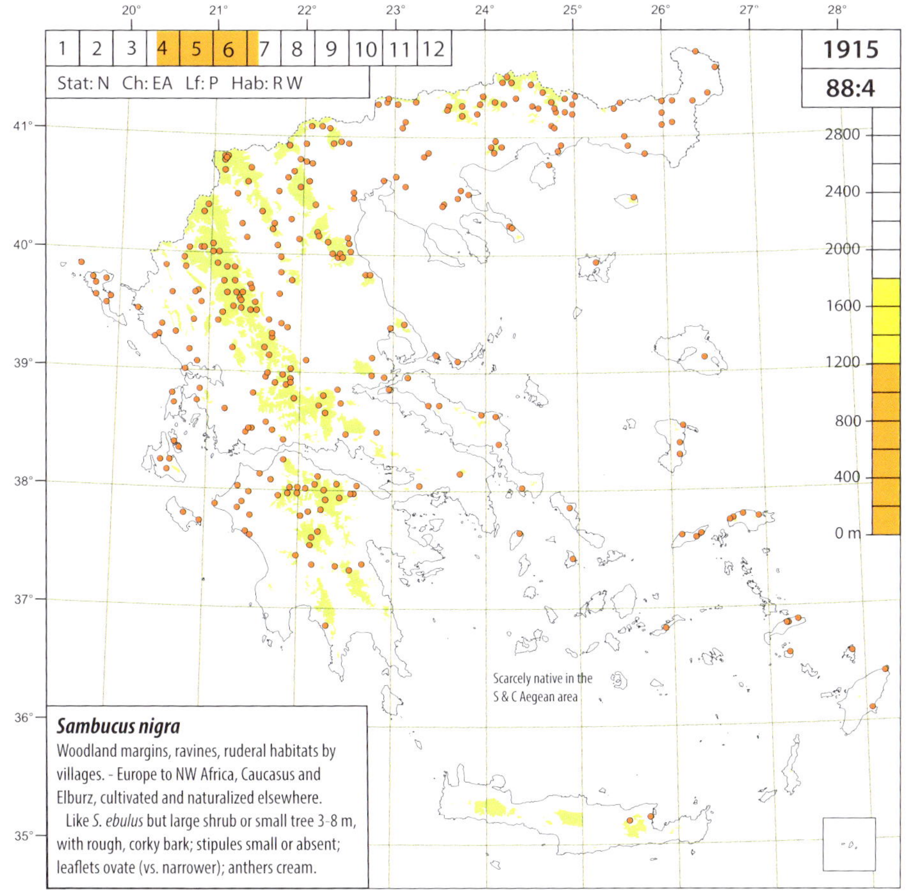

Sambucus nigra
Woodland margins, ravines, ruderal habitats by villages. - Europe to NW Africa, Caucasus and Elburz, cultivated and naturalized elsewhere.
Like *S. ebulus* but large shrub or small tree 3-8 m, with rough, corky bark; stipules small or absent; leaflets ovate (vs. narrower); anthers cream.

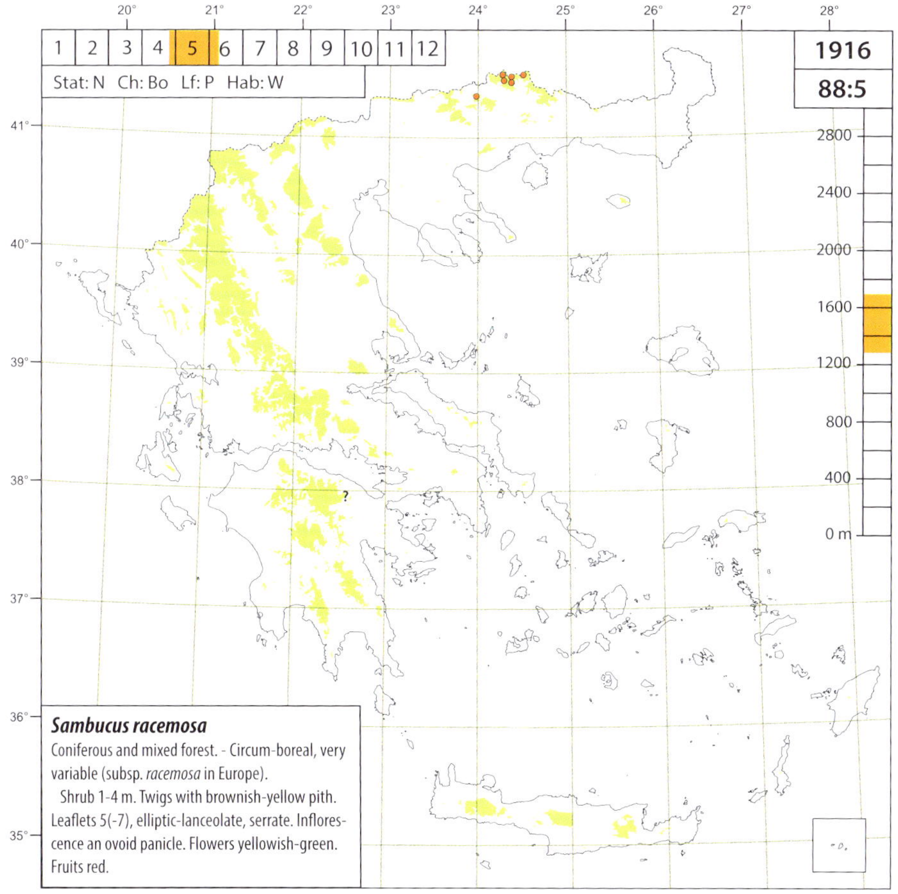

Sambucus racemosa
Coniferous and mixed forest. - Circum-boreal, very variable (subsp. *racemosa* in Europe).
Shrub 1-4 m. Twigs with brownish-yellow pith. Leaflets 5(-7), elliptic-lanceolate, serrate. Inflorescence an ovoid panicle. Flowers yellowish-green. Fruits red.

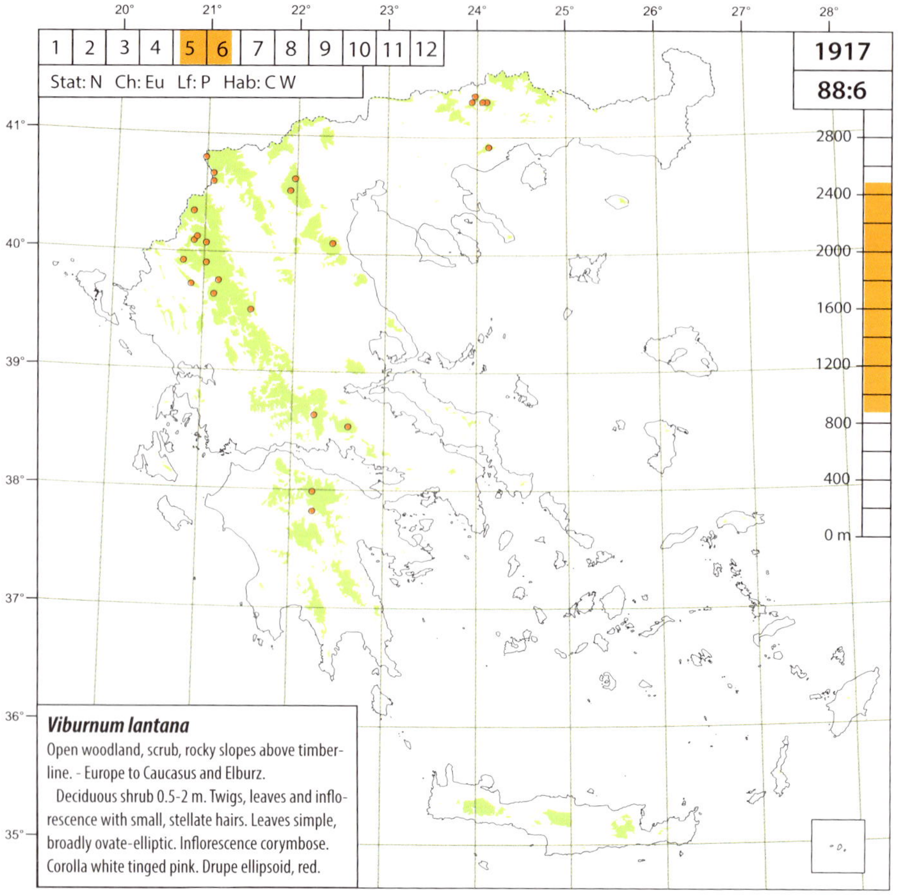

Viburnum lantana
Open woodland, scrub, rocky slopes above timberline. - Europe to Caucasus and Elburz.
Deciduous shrub 0.5-2 m. Twigs, leaves and inflorescence with small, stellate hairs. Leaves simple, broadly ovate-elliptic. Inflorescence corymbose. Corolla white tinged pink. Drupe ellipsoid, red.

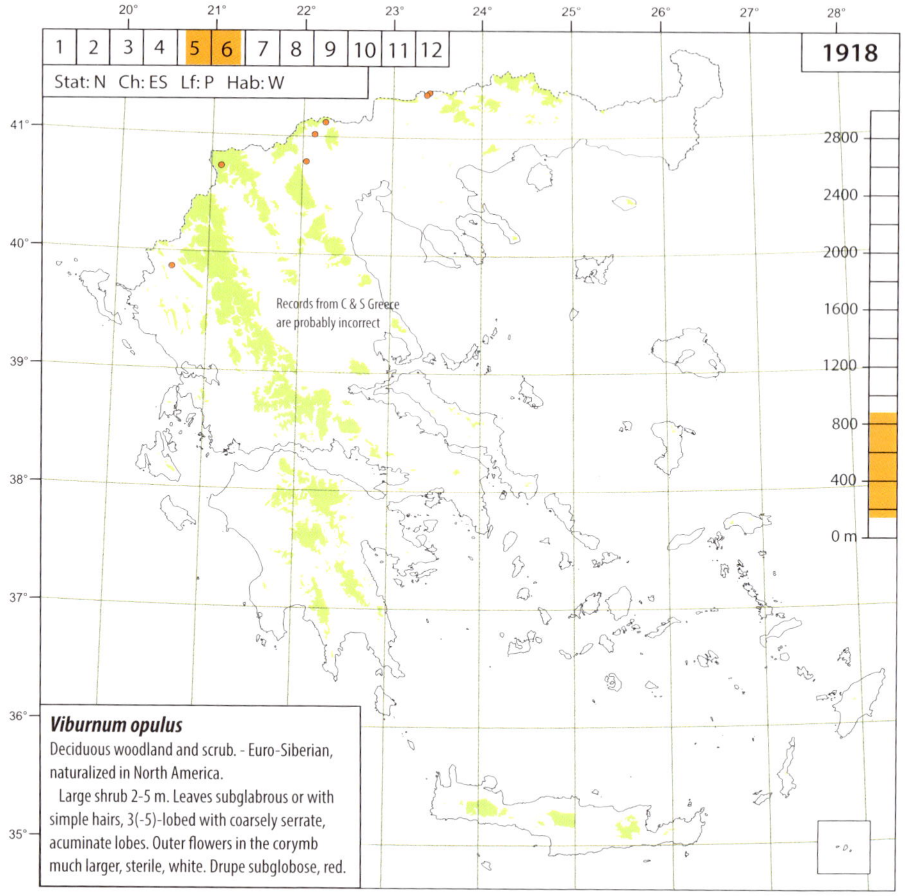

Viburnum opulus
Deciduous woodland and scrub. - Euro-Siberian, naturalized in North America.
Large shrub 2-5 m. Leaves subglabrous or with simple hairs, 3(-5)-lobed with coarsely serrate, acuminate lobes. Outer flowers in the corymb much larger, sterile, white. Drupe subglobose, red.

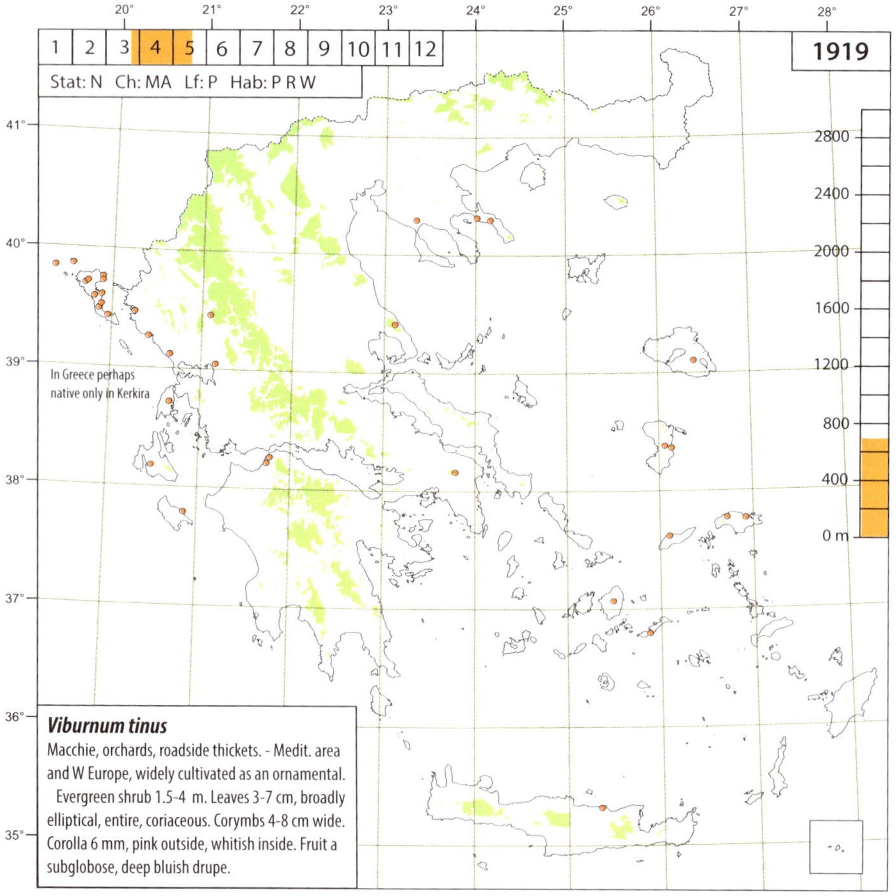

Viburnum tinus
Macchie, orchards, roadside thickets. - Medit. area and W Europe, widely cultivated as an ornamental.
Evergreen shrub 1.5-4 m. Leaves 3-7 cm, broadly elliptical, entire, coriaceous. Corymbs 4-8 cm wide. Corolla 6 mm, pink outside, whitish inside. Fruit a subglobose, deep bluish drupe.

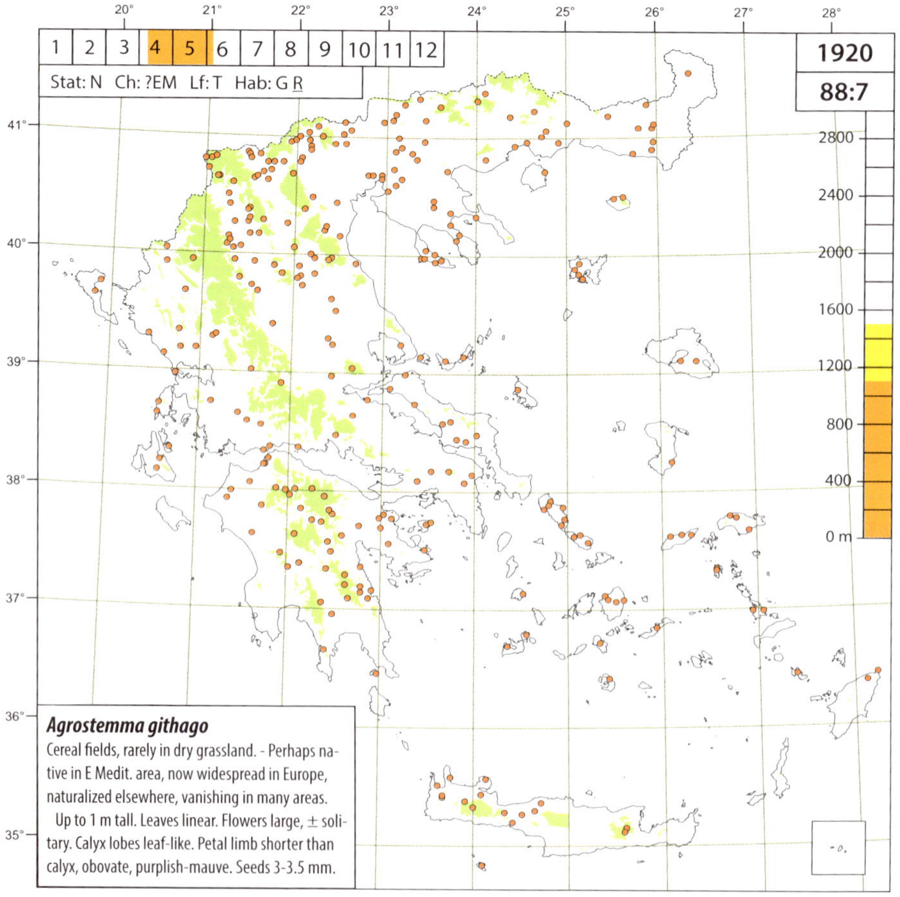

Agrostemma githago
Cereal fields, rarely in dry grassland. - Perhaps native in E Medit. area, now widespread in Europe, naturalized elsewhere, vanishing in many areas.
Up to 1 m tall. Leaves linear. Flowers large, ± solitary. Calyx lobes leaf-like. Petal limb shorter than calyx, obovate, purplish-mauve. Seeds 3-3.5 mm.

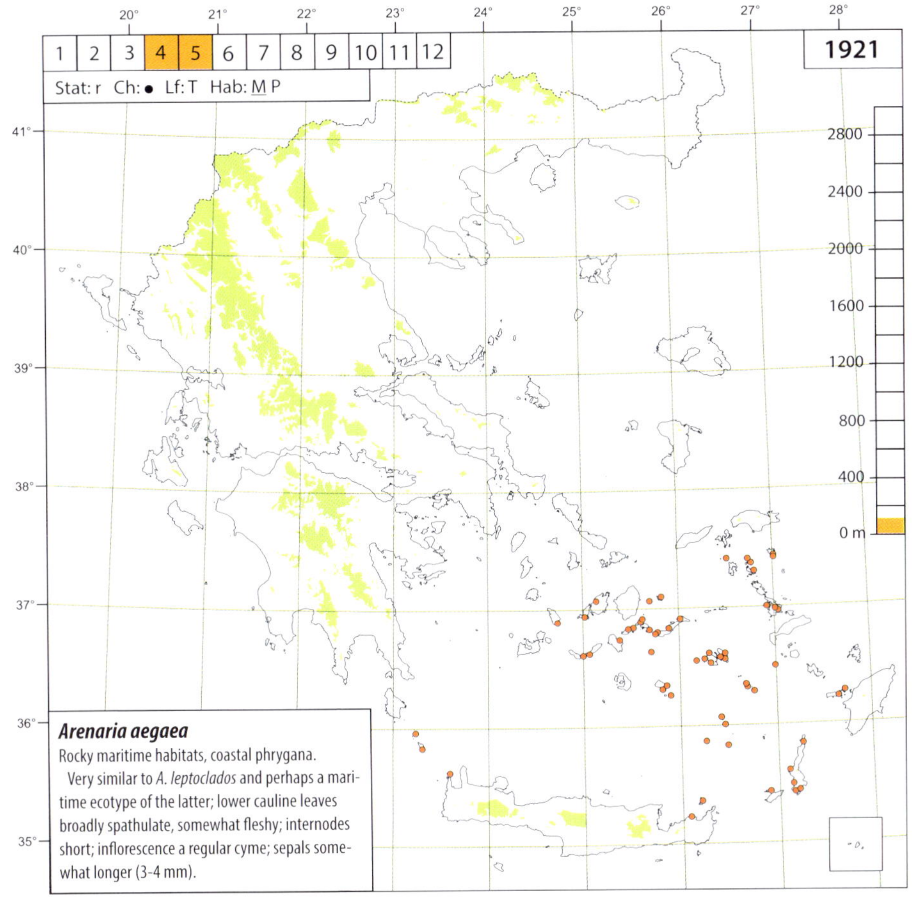

Arenaria aegaea

Rocky maritime habitats, coastal phrygana.

Very similar to *A. leptoclados* and perhaps a maritime ecotype of the latter; lower cauline leaves broadly spathulate, somewhat fleshy; internodes short; inflorescence a regular cyme; sepals somewhat longer (3-4 mm).

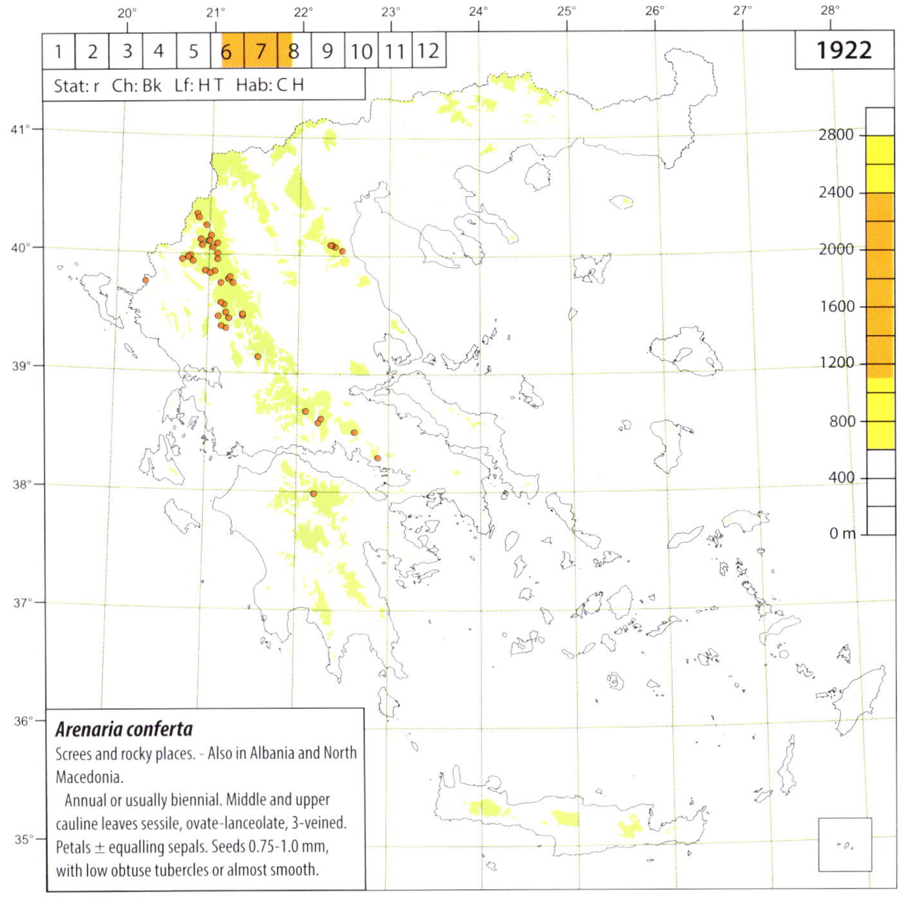

Arenaria conferta

Screes and rocky places. - Also in Albania and North Macedonia.

Annual or usually biennial. Middle and upper cauline leaves sessile, ovate-lanceolate, 3-veined. Petals ± equalling sepals. Seeds 0.75-1.0 mm, with low obtuse tubercles or almost smooth.

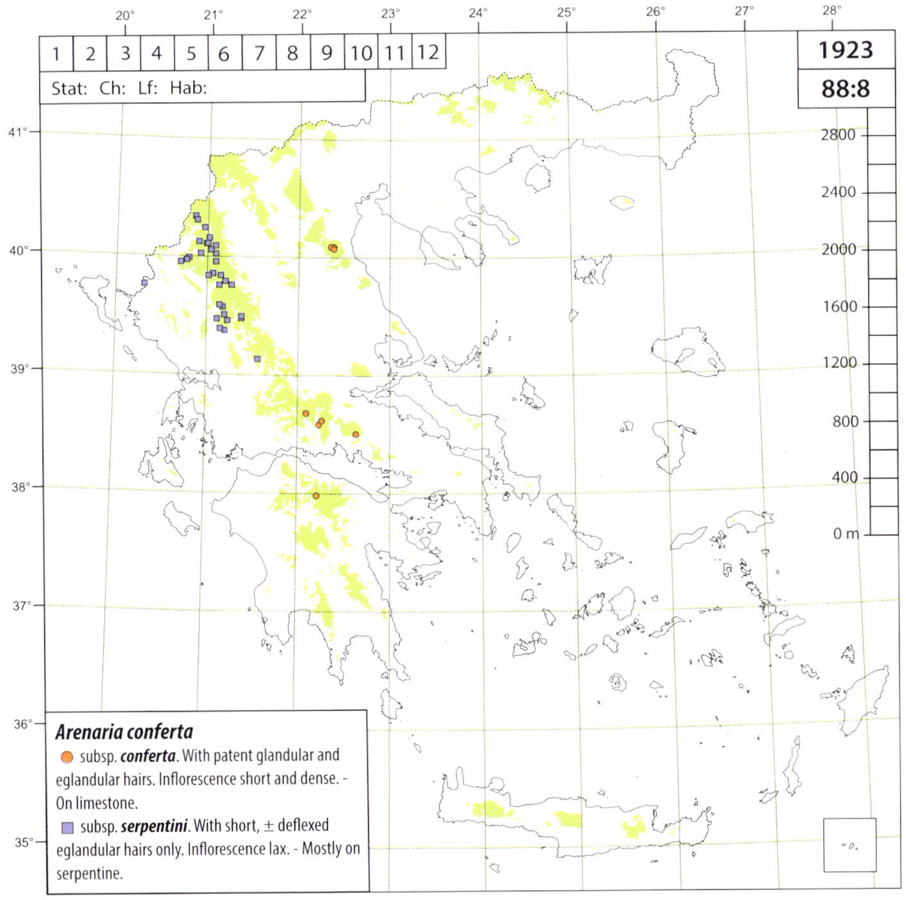

Arenaria conferta

● subsp. ***conferta***. With patent glandular and eglandular hairs. Inflorescence short and dense. - On limestone.

■ subsp. ***serpentini***. With short, ± deflexed eglandular hairs only. Inflorescence lax. - Mostly on serpentine.

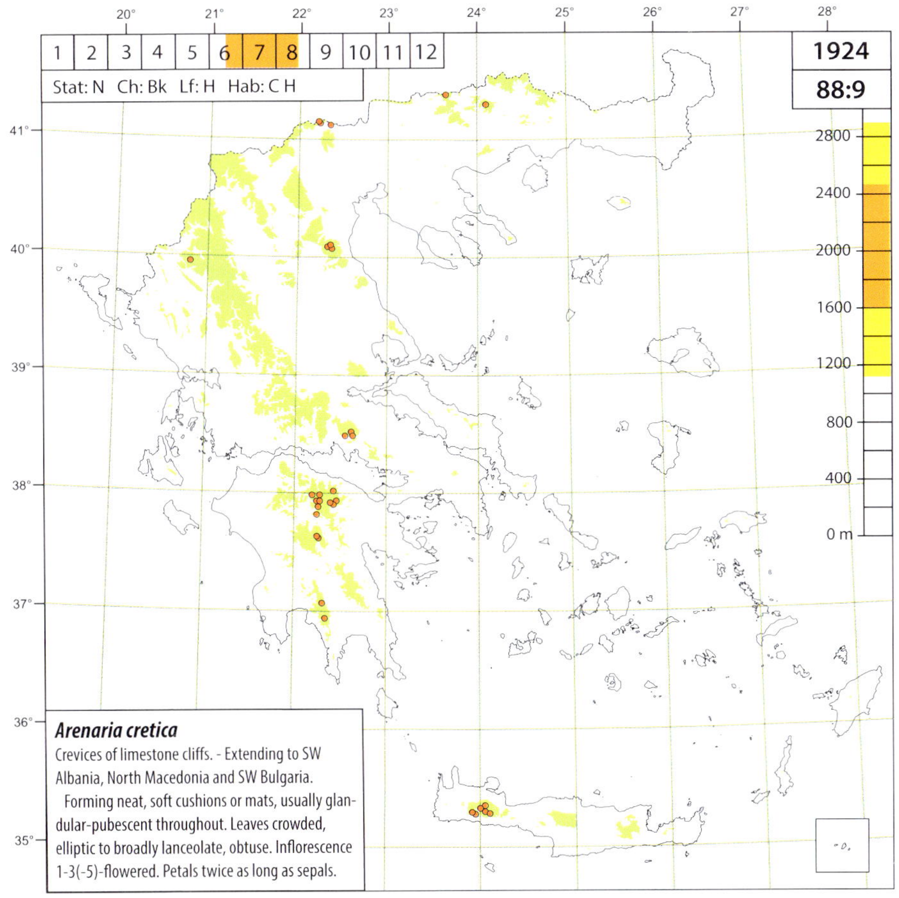

Arenaria cretica

Crevices of limestone cliffs. - Extending to SW Albania, North Macedonia and SW Bulgaria.

Forming neat, soft cushions or mats, usually glandular-pubescent throughout. Leaves crowded, elliptic to broadly lanceolate, obtuse. Inflorescence 1-3(-5)-flowered. Petals twice as long as sepals.

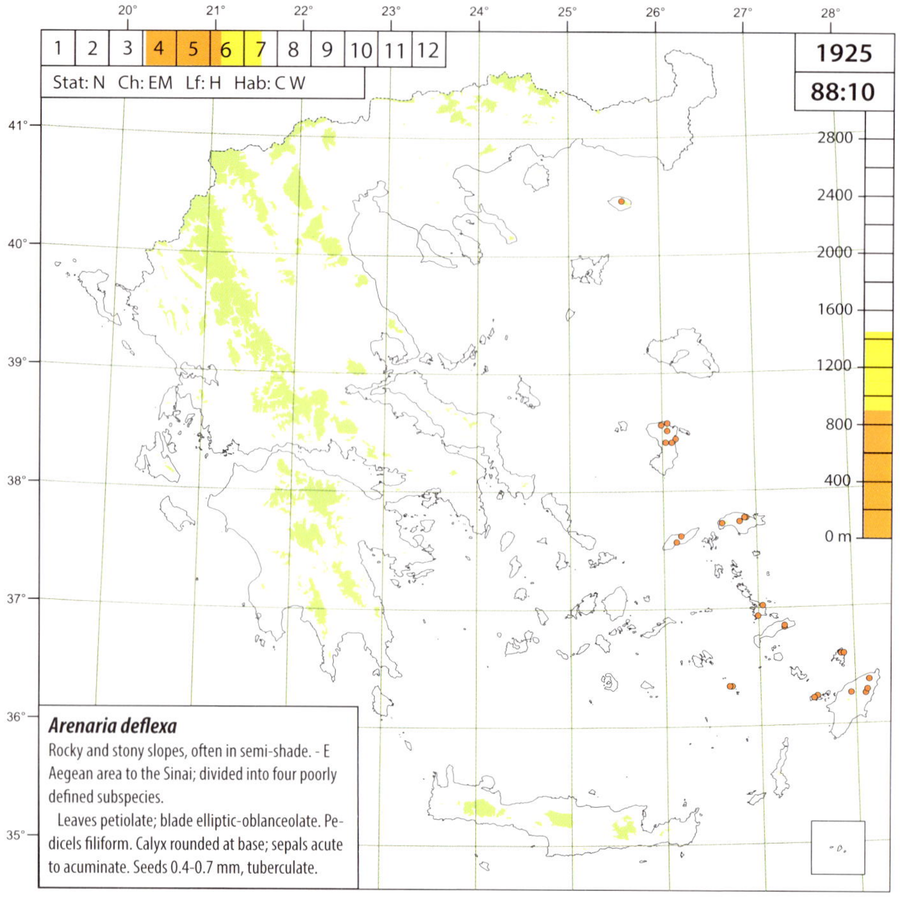

Arenaria deflexa

Rocky and stony slopes, often in semi-shade. - E Aegean area to the Sinai; divided into four poorly defined subspecies.

Leaves petiolate; blade elliptic-oblanceolate. Pedicels filiform. Calyx rounded at base; sepals acute to acuminate. Seeds 0.4-0.7 mm, tuberculate.

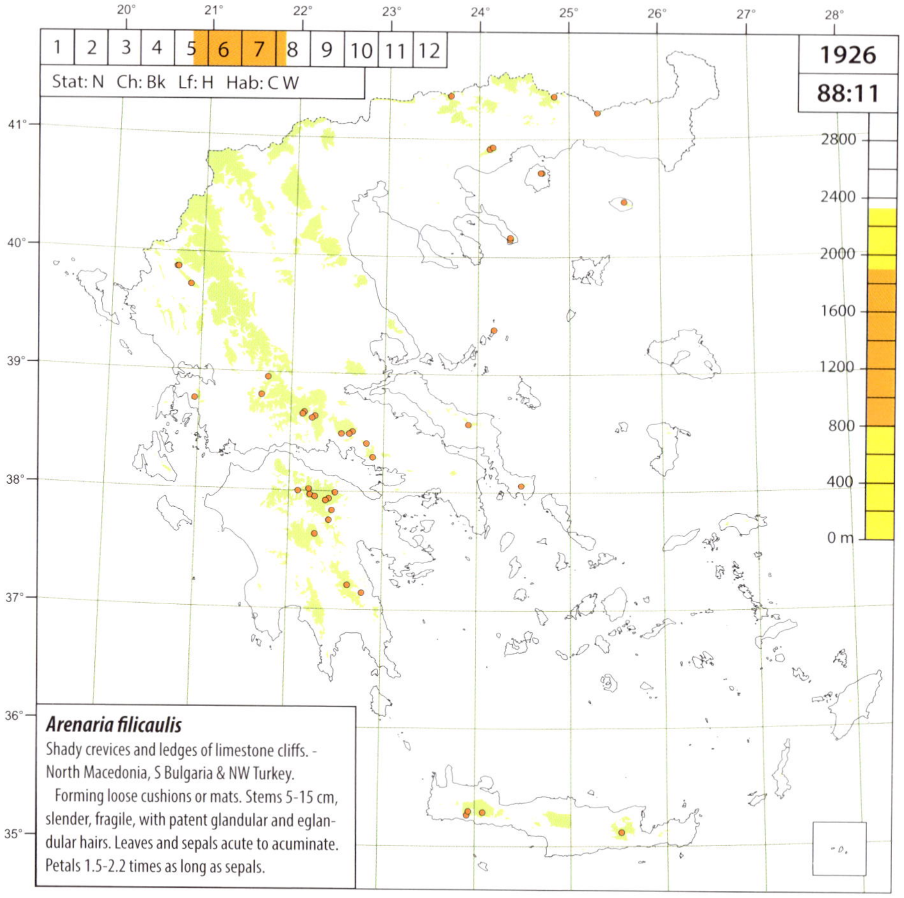

Arenaria filicaulis

Shady crevices and ledges of limestone cliffs. - North Macedonia, S Bulgaria & NW Turkey.

Forming loose cushions or mats. Stems 5-15 cm, slender, fragile, with patent glandular and eglandular hairs. Leaves and sepals acute to acuminate. Petals 1.5-2.2 times as long as sepals.

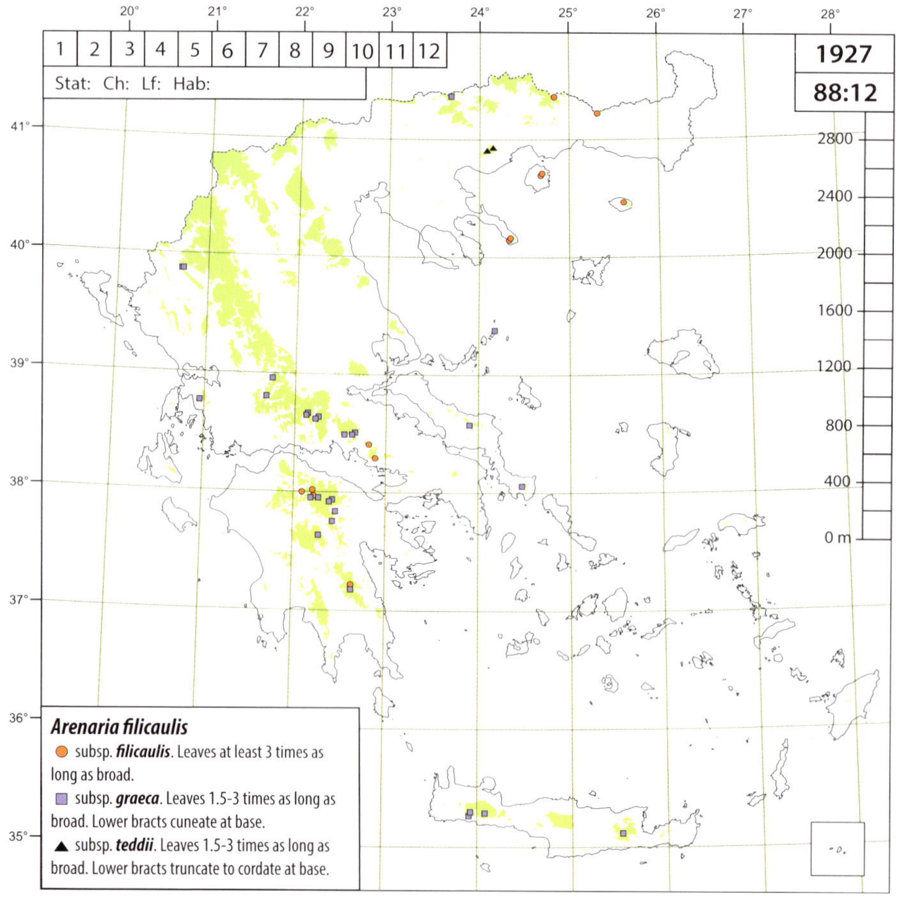

Arenaria filicaulis

● subsp. ***filicaulis***. Leaves at least 3 times as long as broad.

■ subsp. ***graeca***. Leaves 1.5-3 times as long as broad. Lower bracts cuneate at base.

▲ subsp. ***teddii***. Leaves 1.5-3 times as long as broad. Lower bracts truncate to cordate at base.

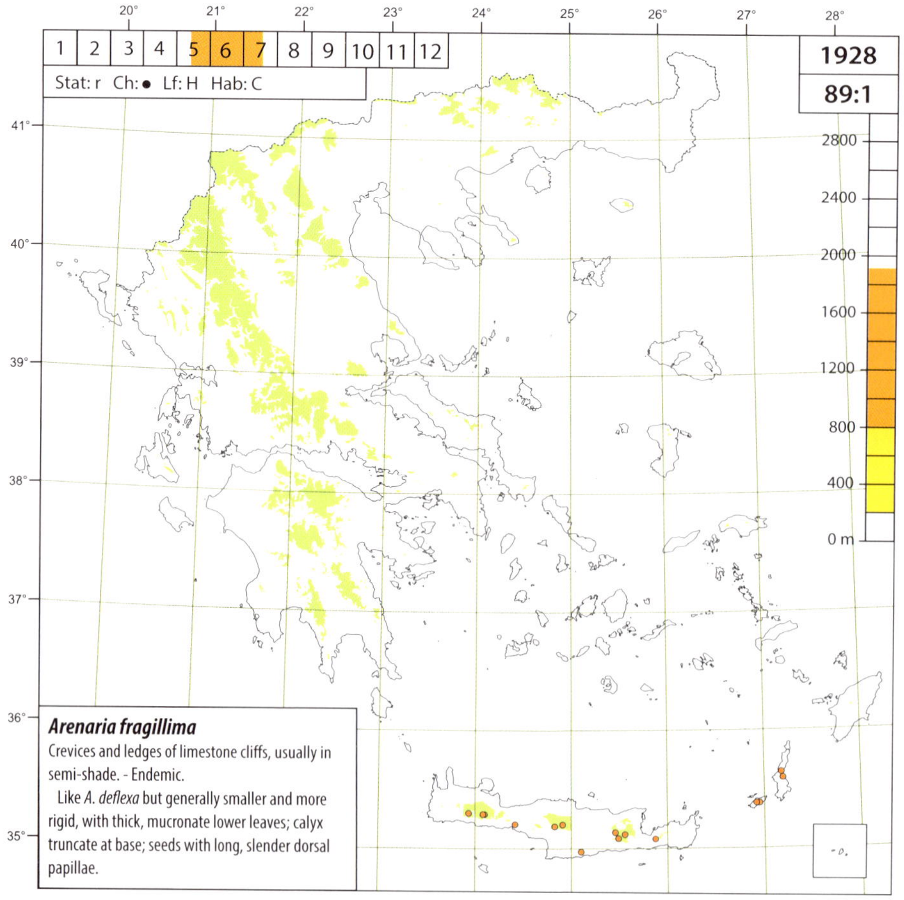

Arenaria fragillima

Crevices and ledges of limestone cliffs, usually in semi-shade. - Endemic.

Like *A. deflexa* but generally smaller and more rigid, with thick, mucronate lower leaves; calyx truncate at base; seeds with long, slender dorsal papillae.

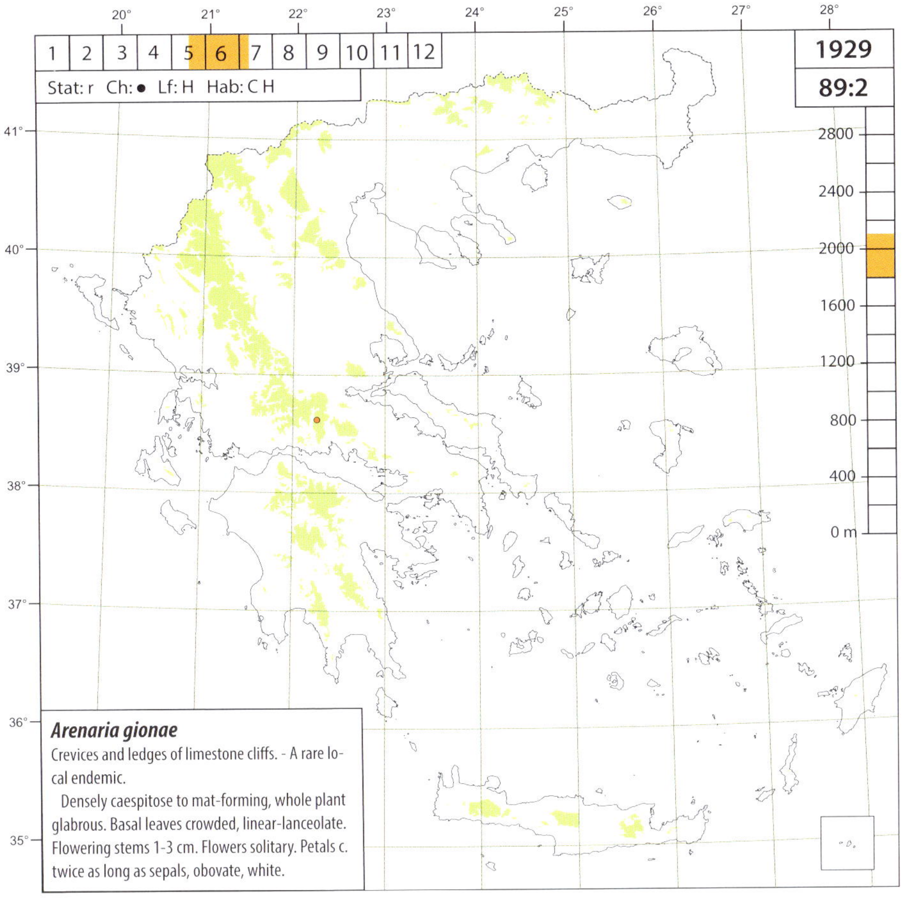

Arenaria gionae

Crevices and ledges of limestone cliffs. - A rare local endemic.

Densely caespitose to mat-forming, whole plant glabrous. Basal leaves crowded, linear-lanceolate. Flowering stems 1-3 cm. Flowers solitary. Petals c. twice as long as sepals, obovate, white.

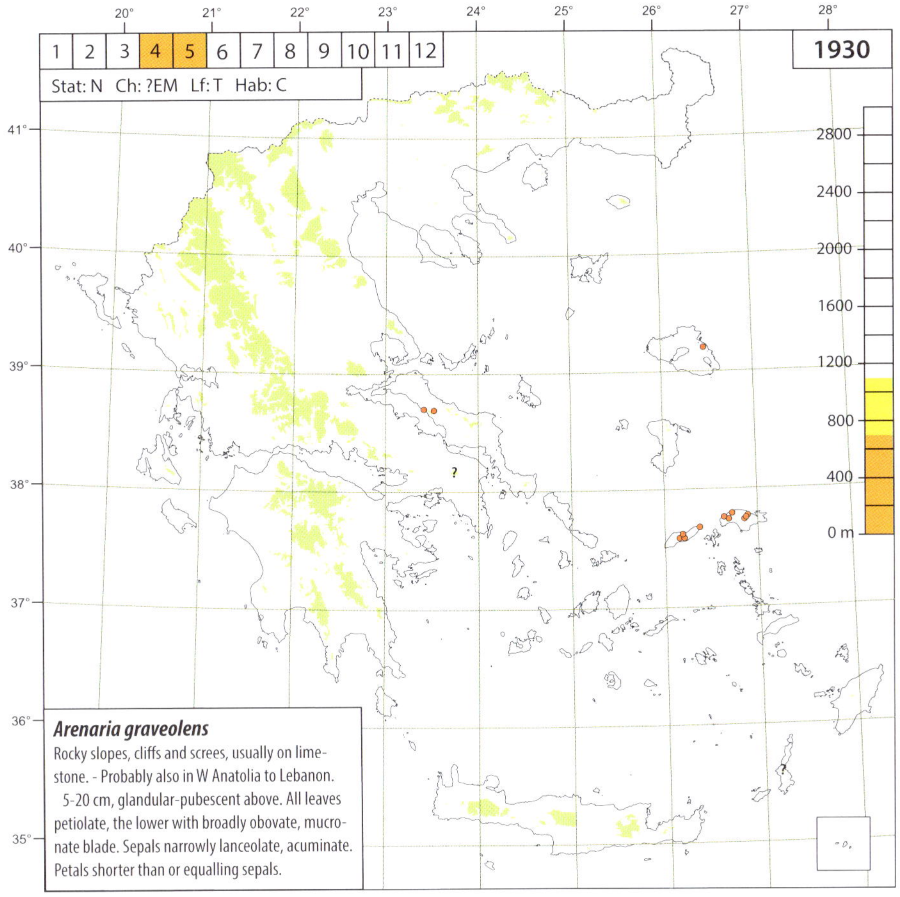

Arenaria graveolens

Rocky slopes, cliffs and screes, usually on limestone. - Probably also in W Anatolia to Lebanon.

5-20 cm, glandular-pubescent above. All leaves petiolate, the lower with broadly obovate, mucronate blade. Sepals narrowly lanceolate, acuminate. Petals shorter than or equalling sepals.

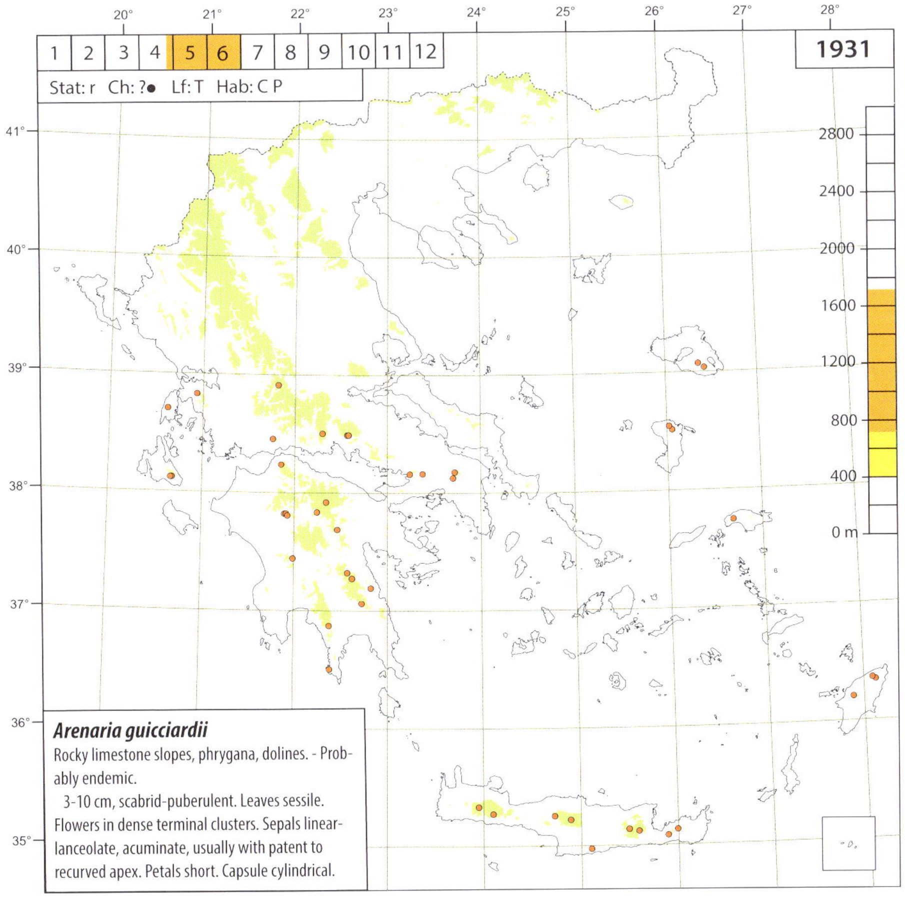

Arenaria guicciardii

Rocky limestone slopes, phrygana, dolines. - Probably endemic.

3-10 cm, scabrid-puberulent. Leaves sessile. Flowers in dense terminal clusters. Sepals linear-lanceolate, acuminate, usually with patent to recurved apex. Petals short. Capsule cylindrical.

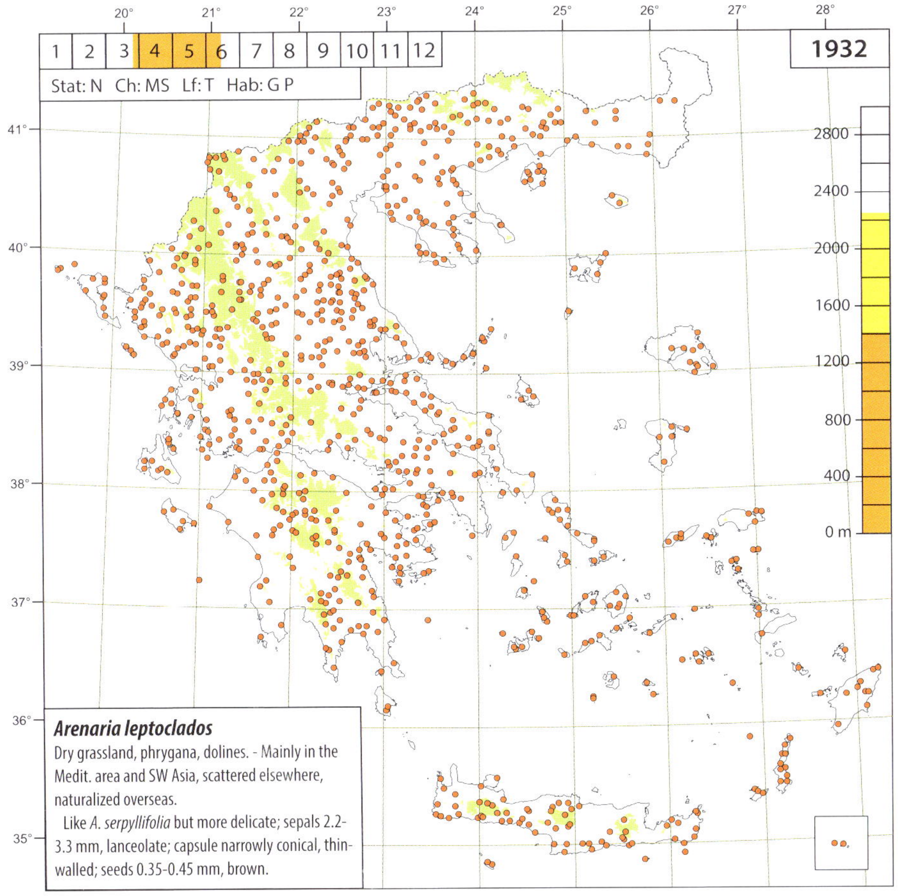

Arenaria leptoclados

Dry grassland, phrygana, dolines. - Mainly in the Medit. area and SW Asia, scattered elsewhere, naturalized overseas.

Like *A. serpyllifolia* but more delicate; sepals 2.2-3.3 mm, lanceolate; capsule narrowly conical, thin-walled; seeds 0.35-0.45 mm, brown.

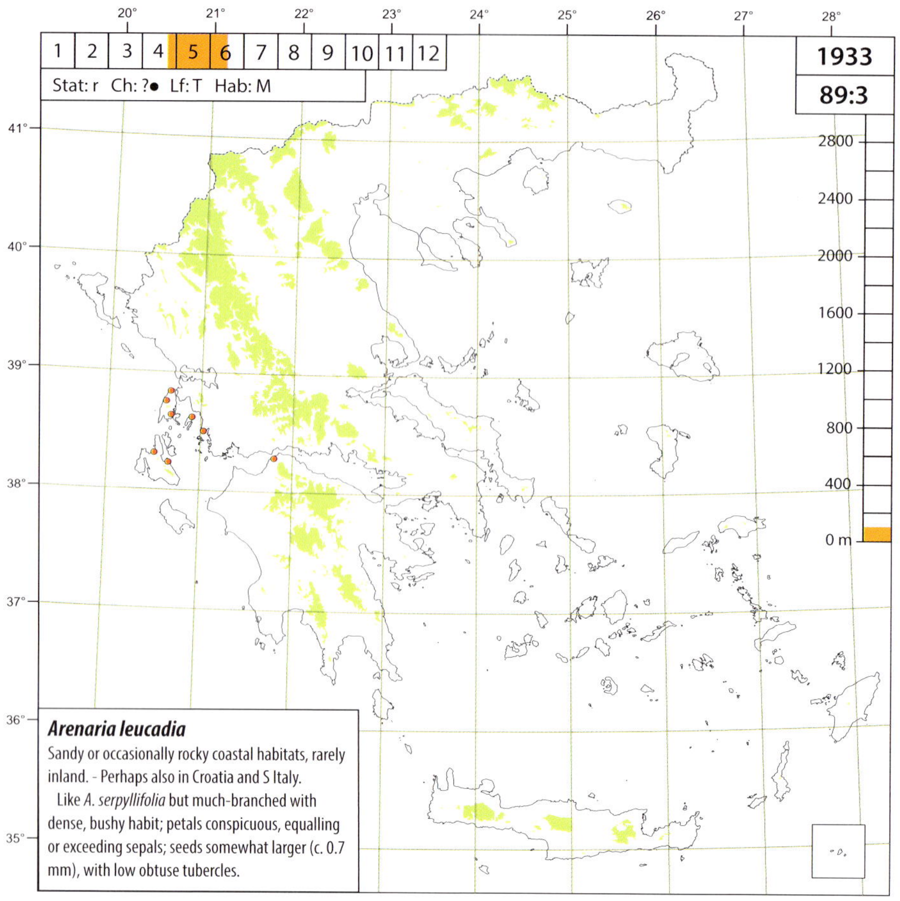

Arenaria leucadia
Sandy or occasionally rocky coastal habitats, rarely inland. - Perhaps also in Croatia and S Italy.
Like *A. serpyllifolia* but much-branched with dense, bushy habit; petals conspicuous, equalling or exceeding sepals; seeds somewhat larger (c. 0.7 mm), with low obtuse tubercles.

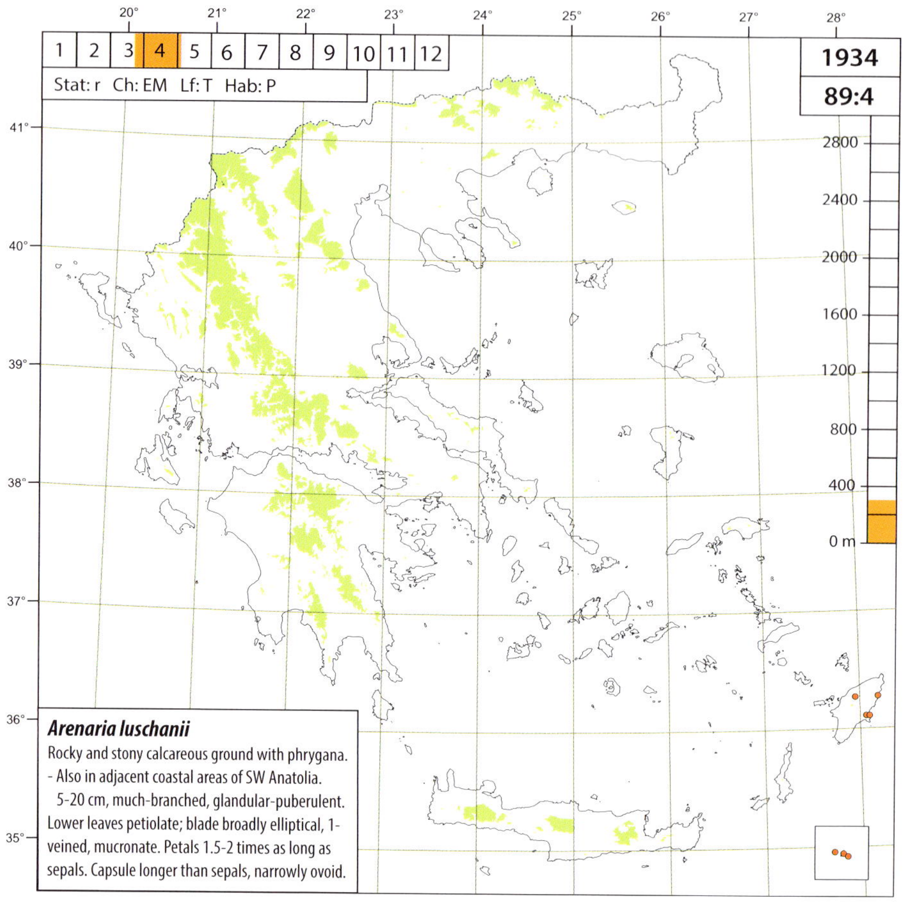

Arenaria luschanii
Rocky and stony calcareous ground with phrygana. - Also in adjacent coastal areas of SW Anatolia.
5-20 cm, much-branched, glandular-puberulent. Lower leaves petiolate; blade broadly elliptical, 1-veined, mucronate. Petals 1.5-2 times as long as sepals. Capsule longer than sepals, narrowly ovoid.

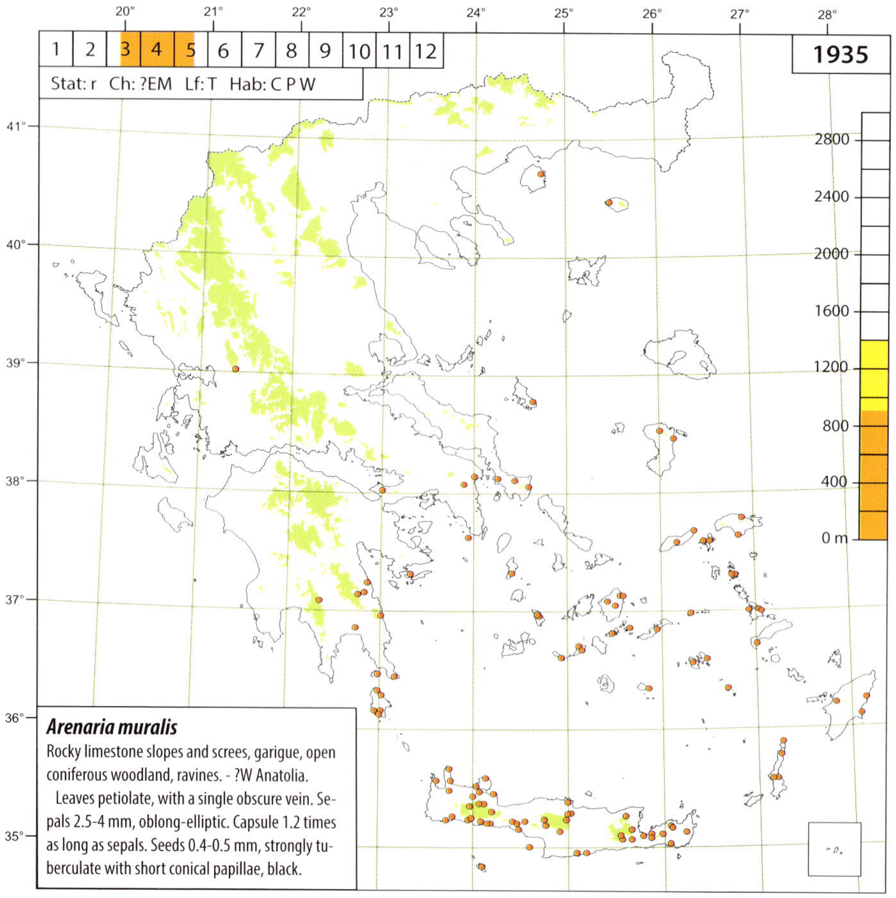

Arenaria muralis
Rocky limestone slopes and screes, garigue, open coniferous woodland, ravines. - ?W Anatolia.
Leaves petiolate, with a single obscure vein. Sepals 2.5-4 mm, oblong-elliptic. Capsule 1.2 times as long as sepals. Seeds 0.4-0.5 mm, strongly tuberculate with short conical papillae, black.

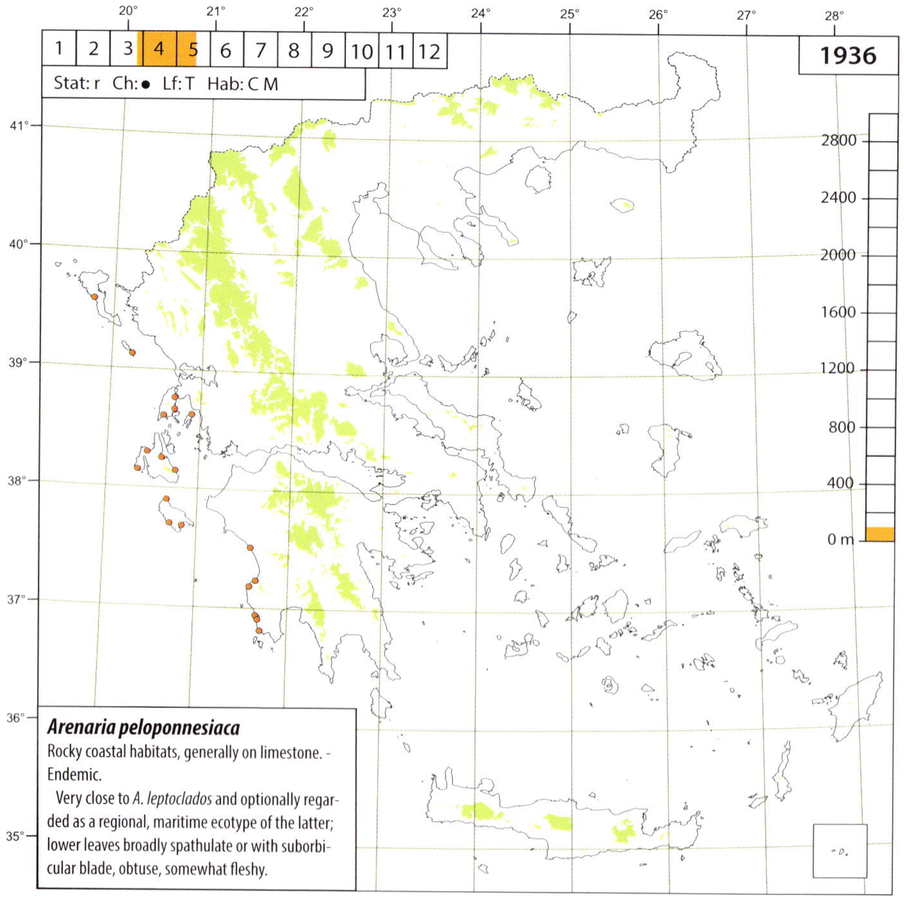

Arenaria peloponnesiaca
Rocky coastal habitats, generally on limestone. - Endemic.
Very close to *A. leptoclados* and optionally regarded as a regional, maritime ecotype of the latter; lower leaves broadly spathulate or with suborbicular blade, obtuse, somewhat fleshy.

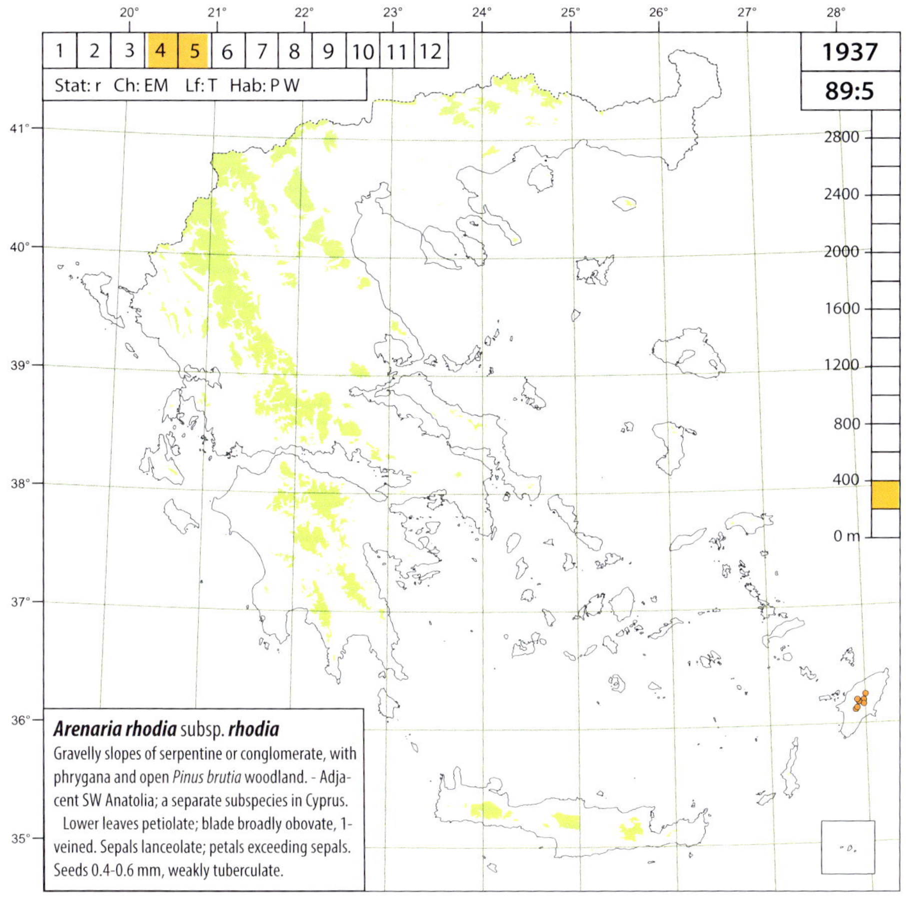

Arenaria rhodia subsp. _rhodia_
Gravelly slopes of serpentine or conglomerate, with phrygana and open *Pinus brutia* woodland. - Adjacent SW Anatolia; a separate subspecies in Cyprus.
Lower leaves petiolate; blade broadly obovate, 1-veined. Sepals lanceolate; petals exceeding sepals. Seeds 0.4-0.6 mm, weakly tuberculate.

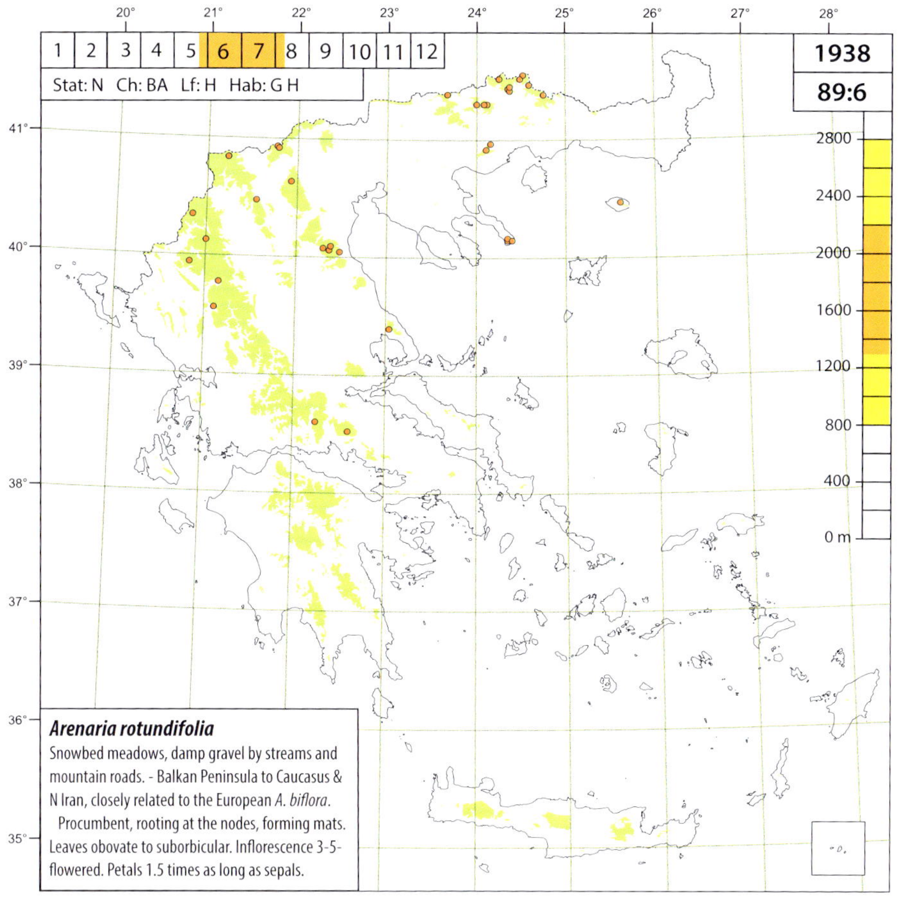

Arenaria rotundifolia
Snowbed meadows, damp gravel by streams and mountain roads. - Balkan Peninsula to Caucasus & N Iran, closely related to the European *A. biflora*.
Procumbent, rooting at the nodes, forming mats. Leaves obovate to suborbicular. Inflorescence 3-5-flowered. Petals 1.5 times as long as sepals.

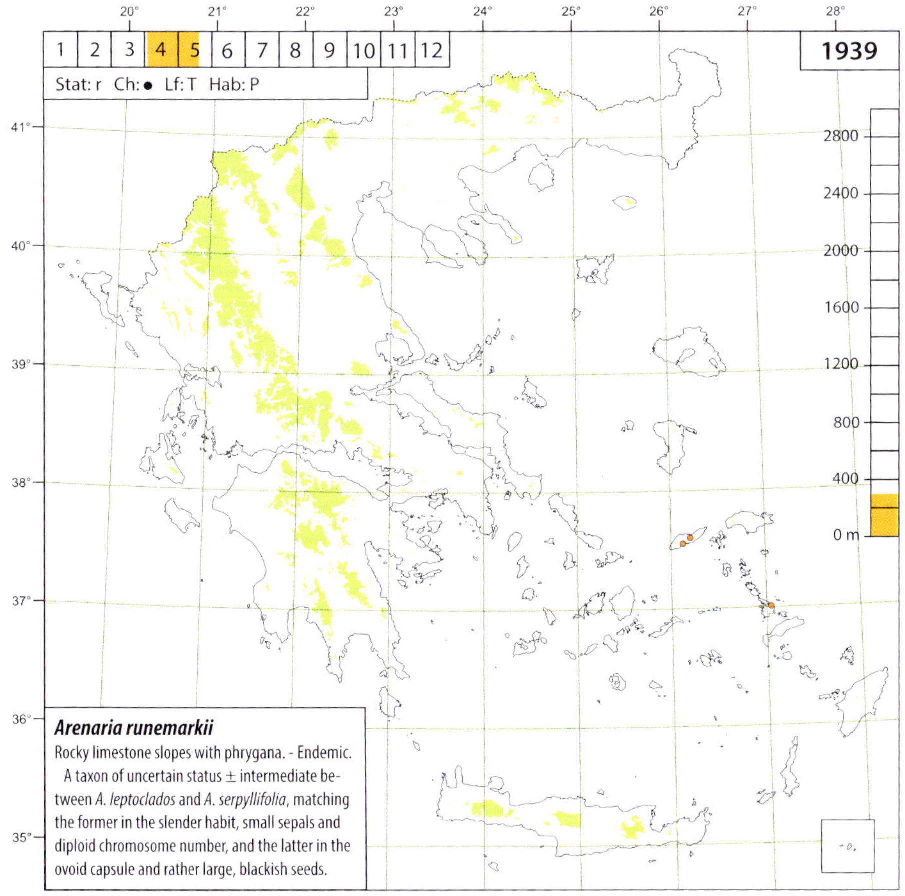

Arenaria runemarkii
Rocky limestone slopes with phrygana. - Endemic.
A taxon of uncertain status ± intermediate between *A. leptoclados* and *A. serpyllifolia*, matching the former in the slender habit, small sepals and diploid chromosome number, and the latter in the ovoid capsule and rather large, blackish seeds.

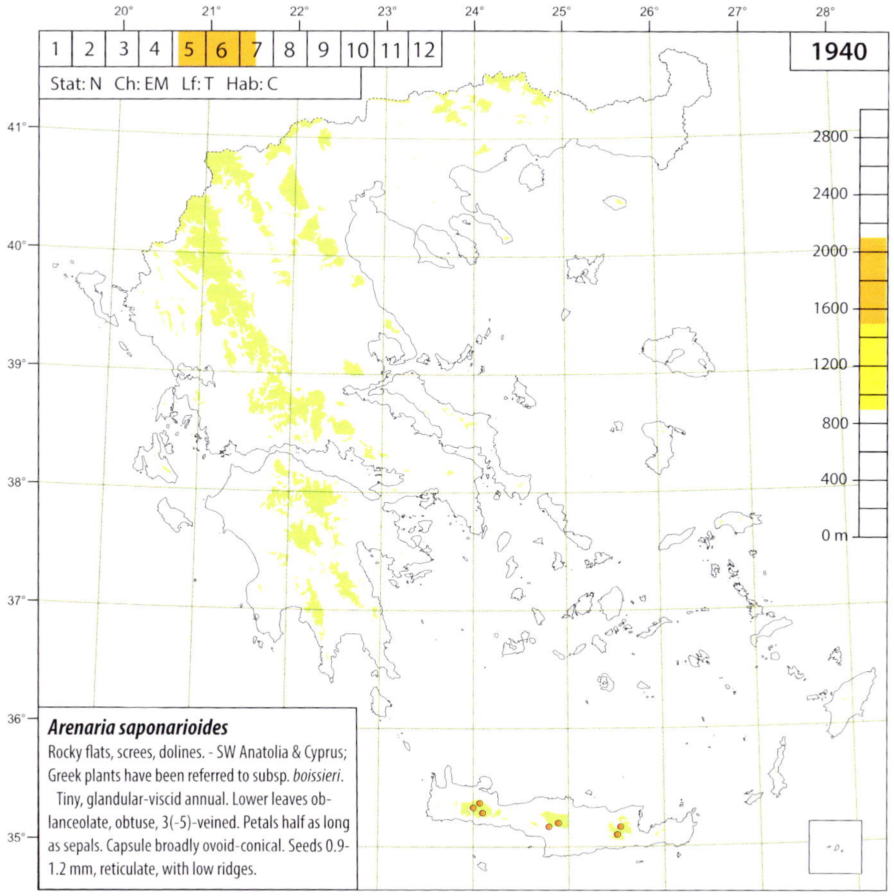

Arenaria saponarioides
Rocky flats, screes, dolines. - SW Anatolia & Cyprus; Greek plants have been referred to subsp. *boissieri*.
Tiny, glandular-viscid annual. Lower leaves oblanceolate, obtuse, 3(-5)-veined. Petals half as long as sepals. Capsule broadly ovoid-conical. Seeds 0.9-1.2 mm, reticulate, with low ridges.

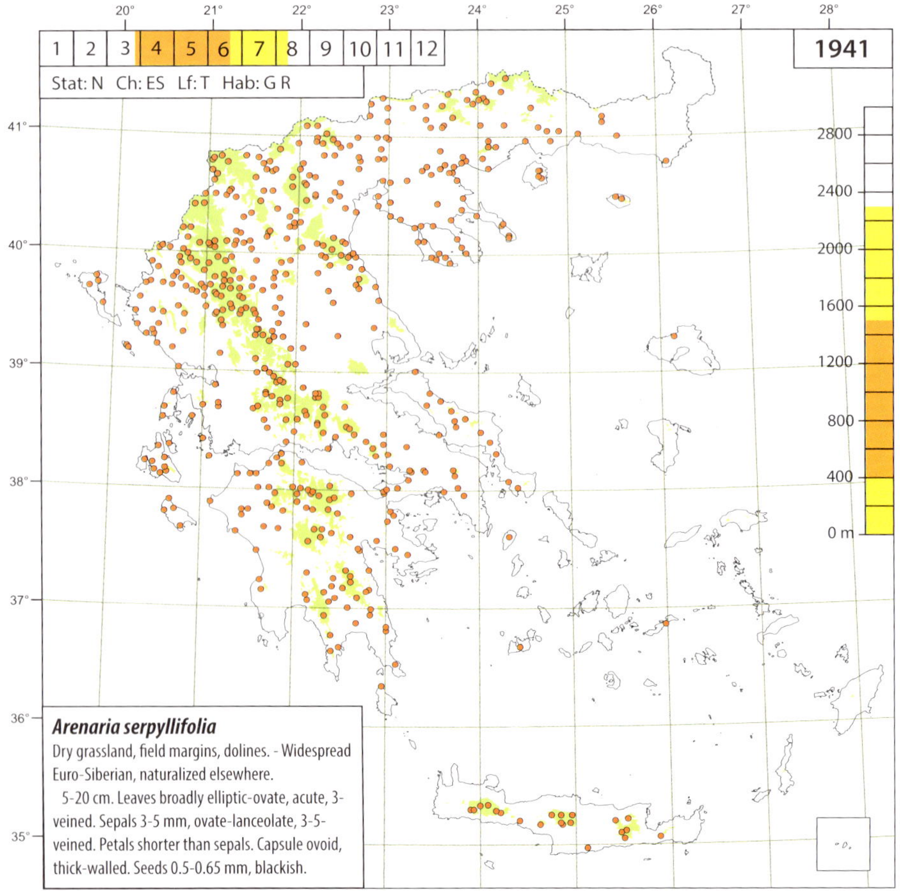

Arenaria serpyllifolia
Dry grassland, field margins, dolines. - Widespread Euro-Siberian, naturalized elsewhere.
5-20 cm. Leaves broadly elliptic-ovate, acute, 3-veined. Sepals 3-5 mm, ovate-lanceolate, 3-5-veined. Petals shorter than sepals. Capsule ovoid, thick-walled. Seeds 0.5-0.65 mm, blackish.

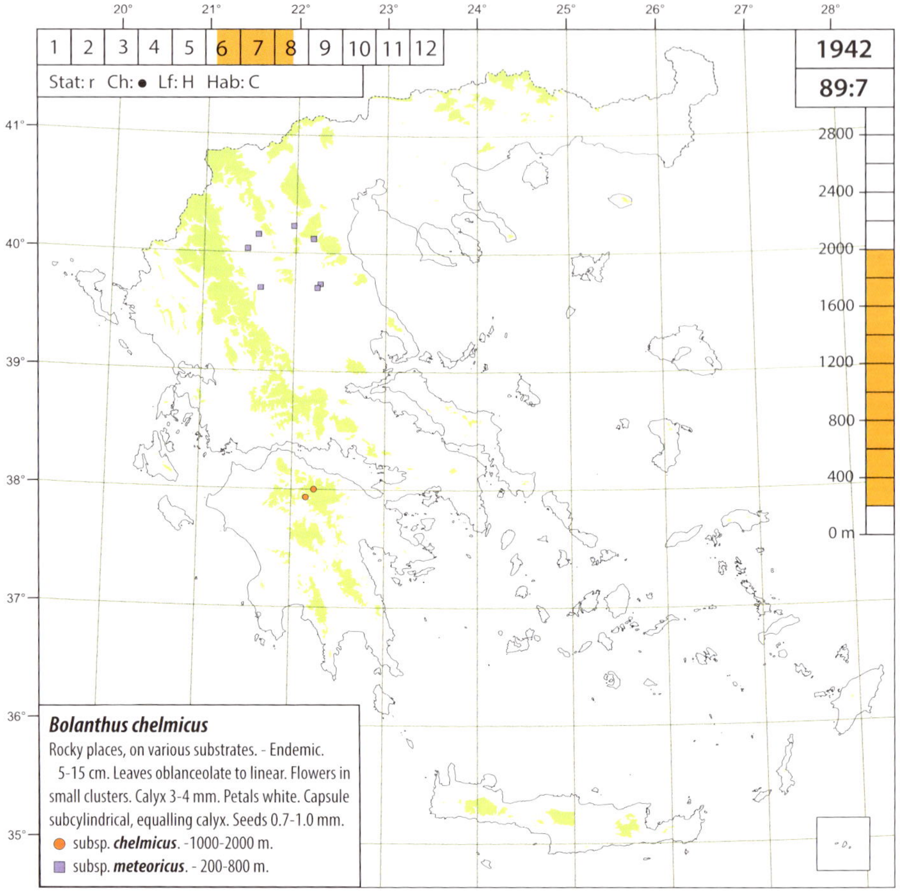

Bolanthus chelmicus
Rocky places, on various substrates. - Endemic.
5-15 cm. Leaves oblanceolate to linear. Flowers in small clusters. Calyx 3-4 mm. Petals white. Capsule subcylindrical, equalling calyx. Seeds 0.7-1.0 mm.
● subsp. ***chelmicus***. - 1000-2000 m.
■ subsp. ***meteoricus***. - 200-800 m.

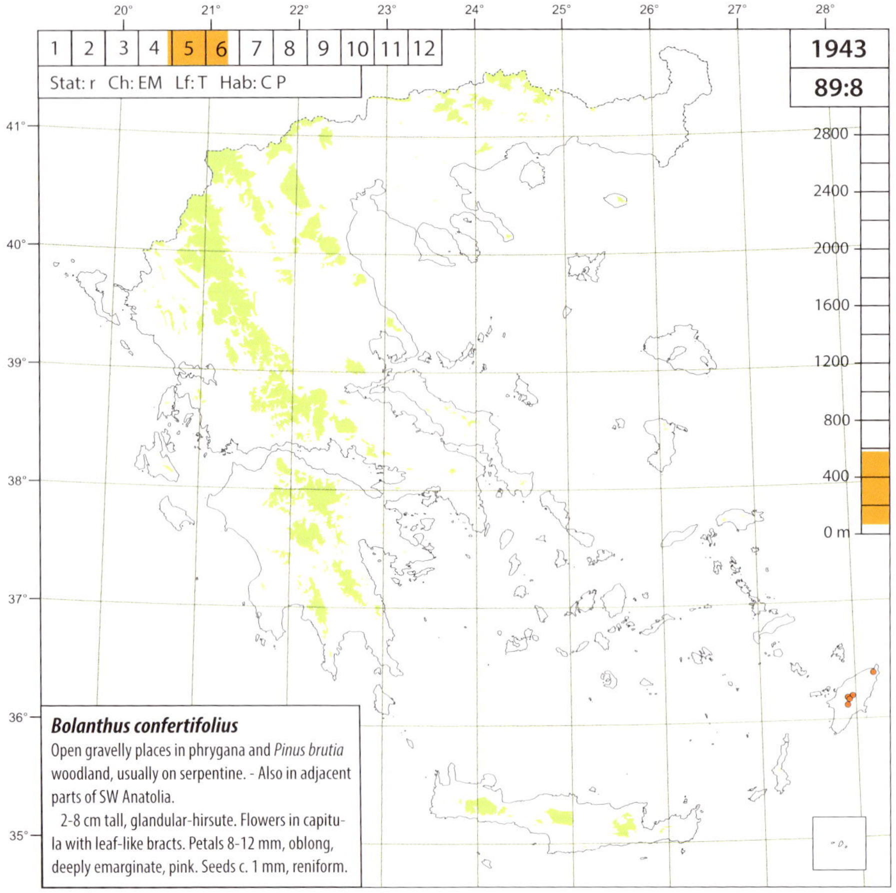

Bolanthus confertifolius
Open gravelly places in phrygana and *Pinus brutia* woodland, usually on serpentine. - Also in adjacent parts of SW Anatolia.
2-8 cm tall, glandular-hirsute. Flowers in capitula with leaf-like bracts. Petals 8-12 mm, oblong, deeply emarginate, pink. Seeds c. 1 mm, reniform.

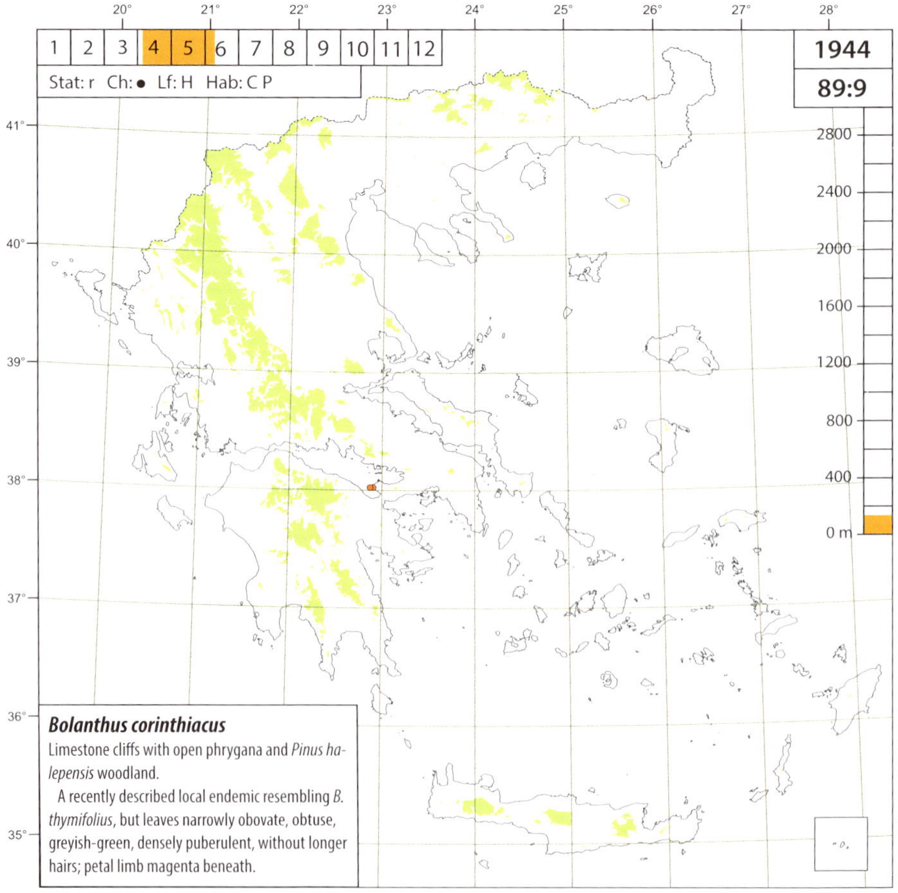

Bolanthus corinthiacus
Limestone cliffs with open phrygana and *Pinus halepensis* woodland.
A recently described local endemic resembling *B. thymifolius*, but leaves narrowly obovate, obtuse, greyish-green, densely puberulent, without longer hairs; petal limb magenta beneath.

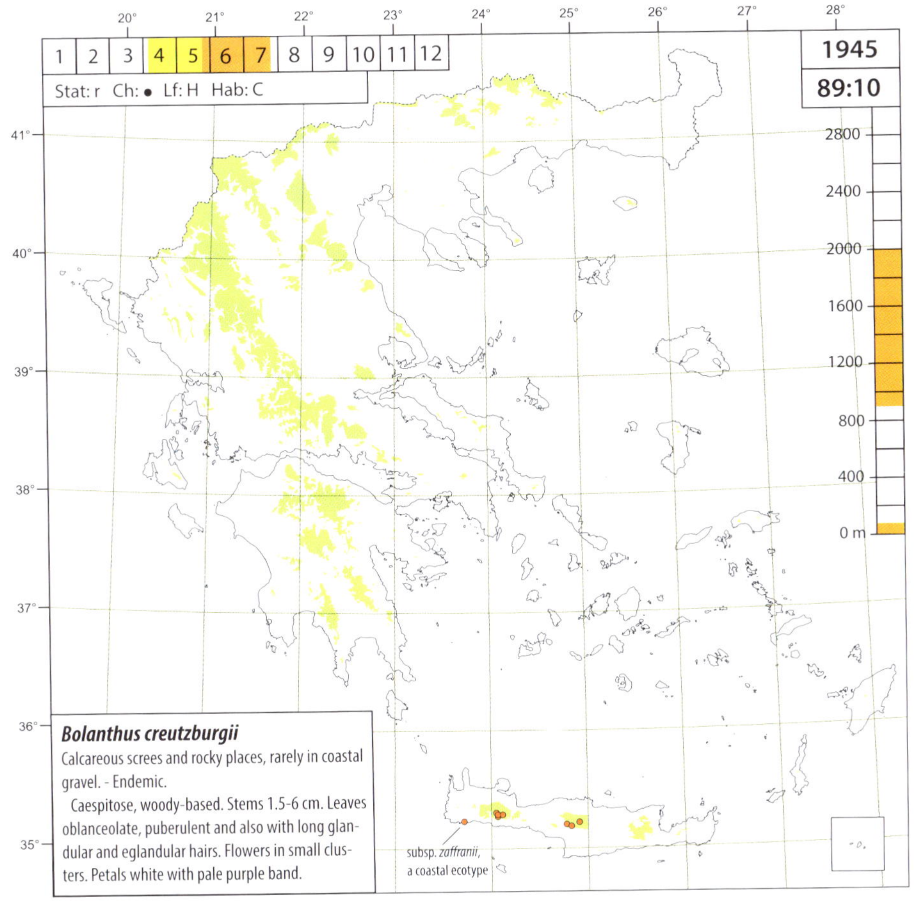

Bolanthus creutzburgii

Calcareous screes and rocky places, rarely in coastal gravel. - Endemic.

Caespitose, woody-based. Stems 1.5-6 cm. Leaves oblanceolate, puberulent and also with long glandular and eglandular hairs. Flowers in small clusters. Petals white with pale purple band.

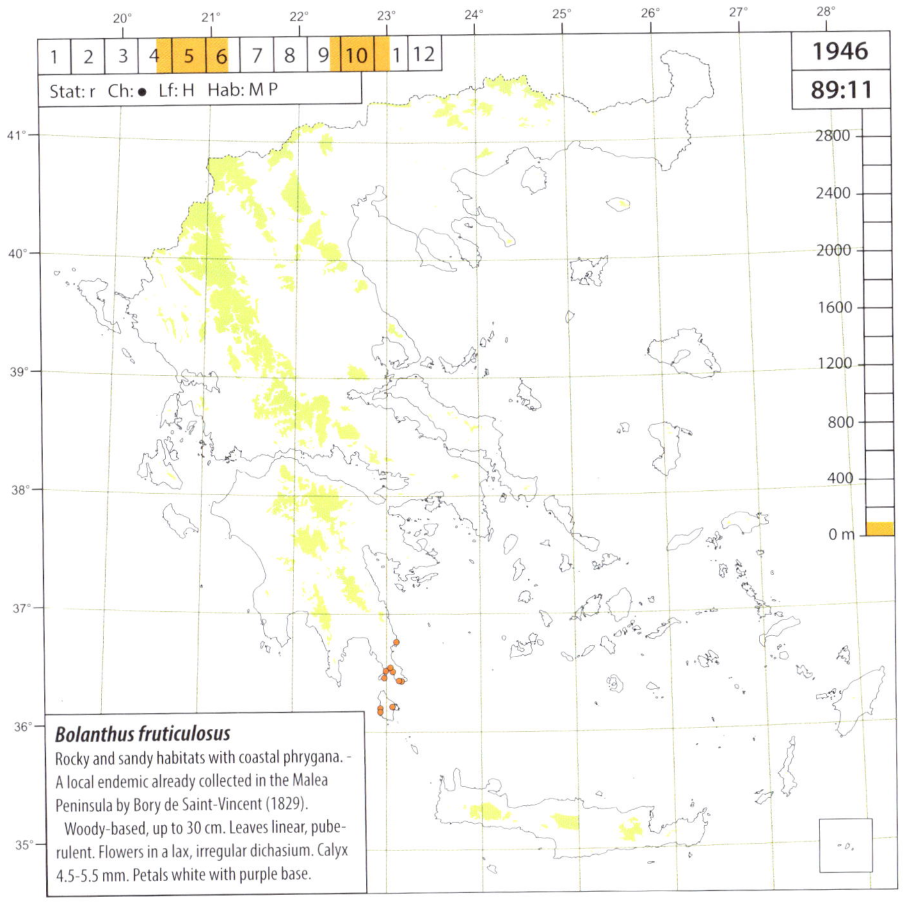

Bolanthus fruticulosus

Rocky and sandy habitats with coastal phrygana. - A local endemic already collected in the Malea Peninsula by Bory de Saint-Vincent (1829).

Woody-based, up to 30 cm. Leaves linear, puberulent. Flowers in a lax, irregular dichasium. Calyx 4.5-5.5 mm. Petals white with purple base.

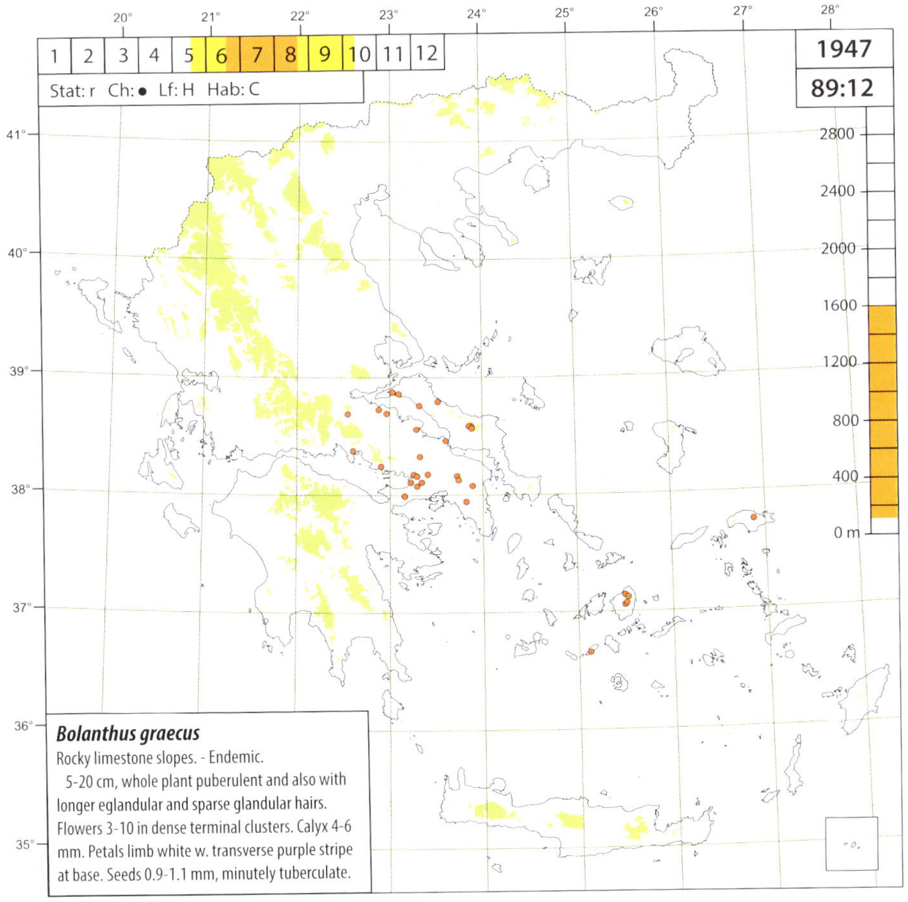

Bolanthus graecus

Rocky limestone slopes. - Endemic.

5-20 cm, whole plant puberulent and also with longer eglandular and sparse glandular hairs. Flowers 3-10 in dense terminal clusters. Calyx 4-6 mm. Petals limb white w. transverse purple stripe at base. Seeds 0.9-1.1 mm, minutely tuberculate.

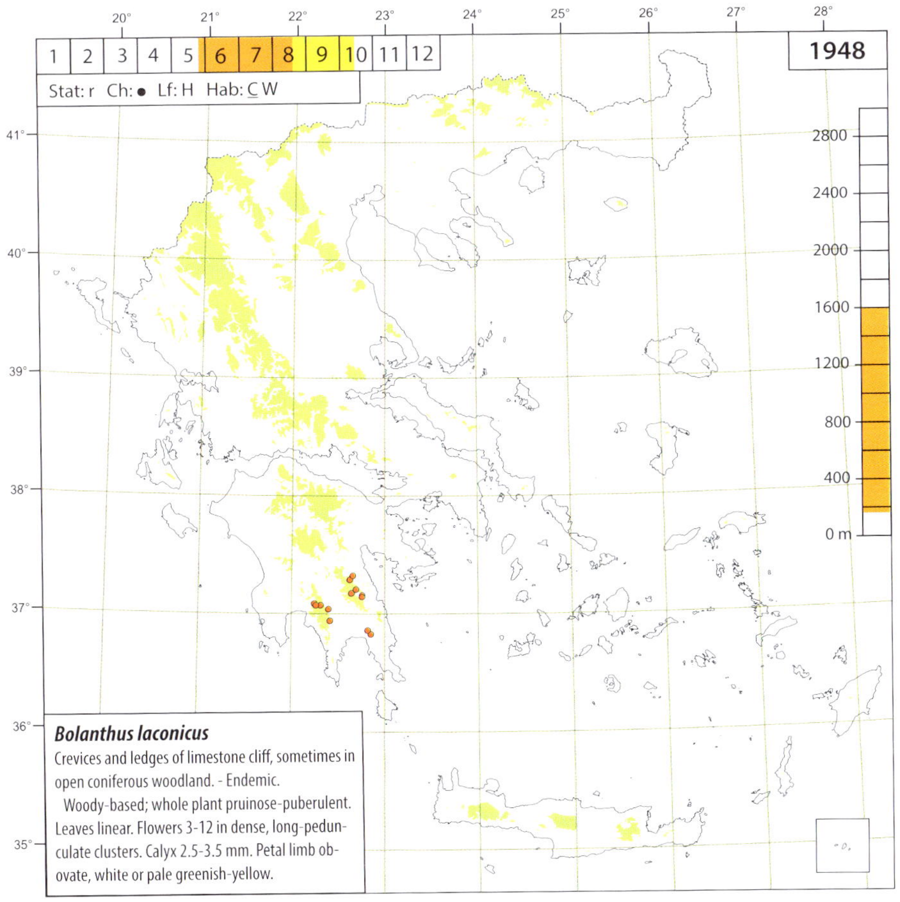

Bolanthus laconicus

Crevices and ledges of limestone cliff, sometimes in open coniferous woodland. - Endemic.

Woody-based; whole plant pruinose-puberulent. Leaves linear. Flowers 3-12 in dense, long-pedunculate clusters. Calyx 2.5-3.5 mm. Petal limb obovate, white or pale greenish-yellow.

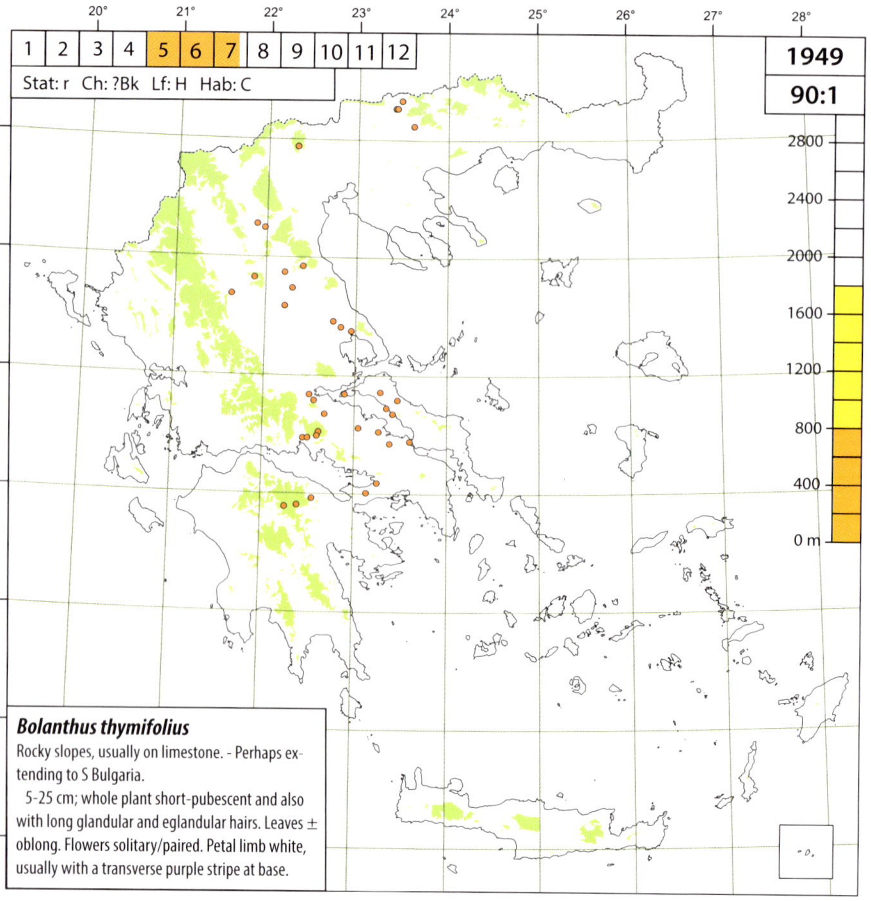

Bolanthus thymifolius
Rocky slopes, usually on limestone. - Perhaps extending to S Bulgaria.
5-25 cm; whole plant short-pubescent and also with long glandular and eglandular hairs. Leaves ± oblong. Flowers solitary/paired. Petal limb white, usually with a transverse purple stripe at base.

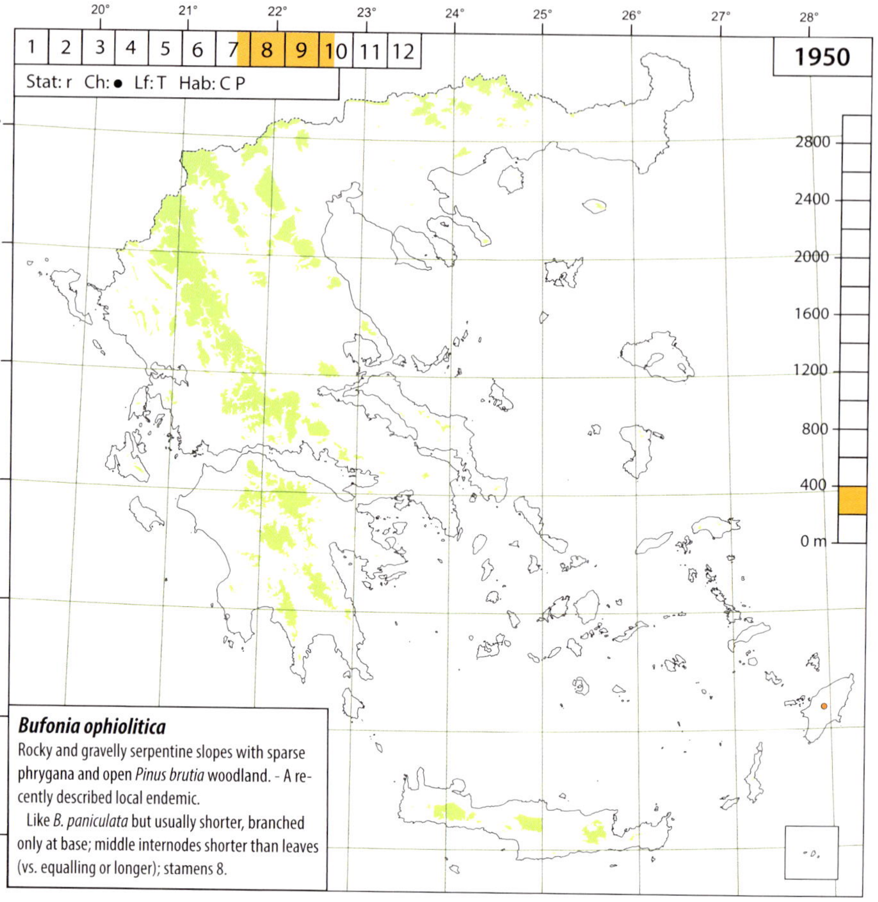

Bufonia ophiolitica
Rocky and gravelly serpentine slopes with sparse phrygana and open *Pinus brutia* woodland. - A recently described local endemic.
Like *B. paniculata* but usually shorter, branched only at base; middle internodes shorter than leaves (vs. equalling or longer); stamens 8.

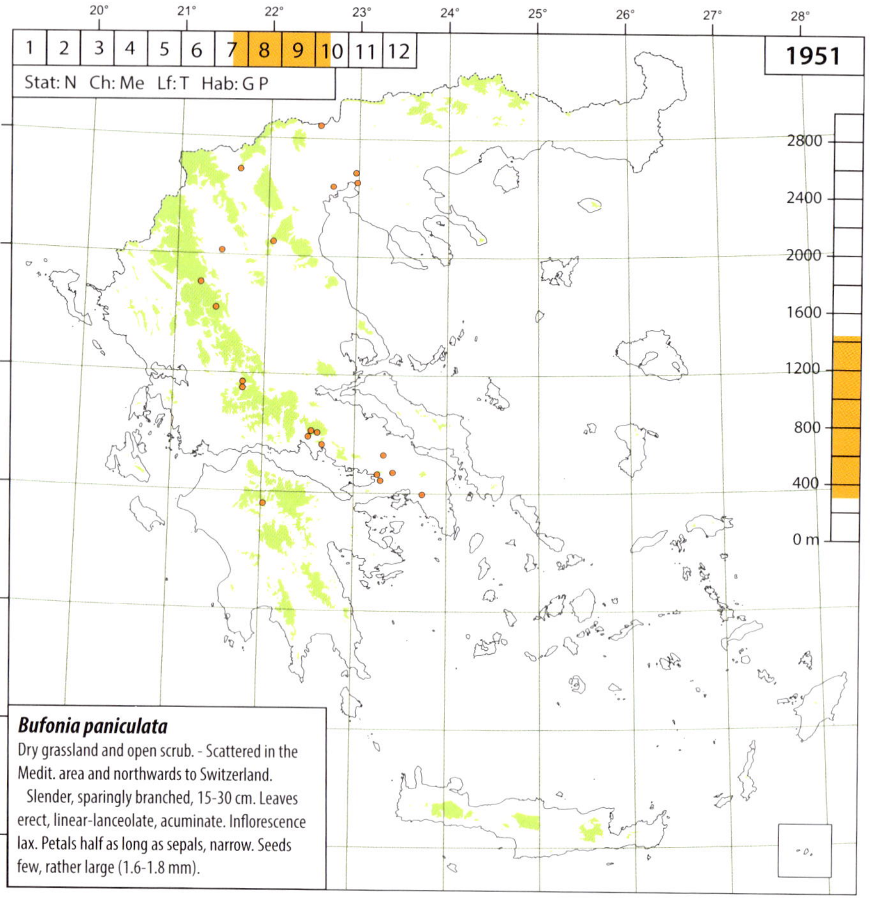

Bufonia paniculata
Dry grassland and open scrub. - Scattered in the Medit. area and northwards to Switzerland.
Slender, sparingly branched, 15-30 cm. Leaves erect, linear-lanceolate, acuminate. Inflorescence lax. Petals half as long as sepals, narrow. Seeds few, rather large (1.6-1.8 mm).

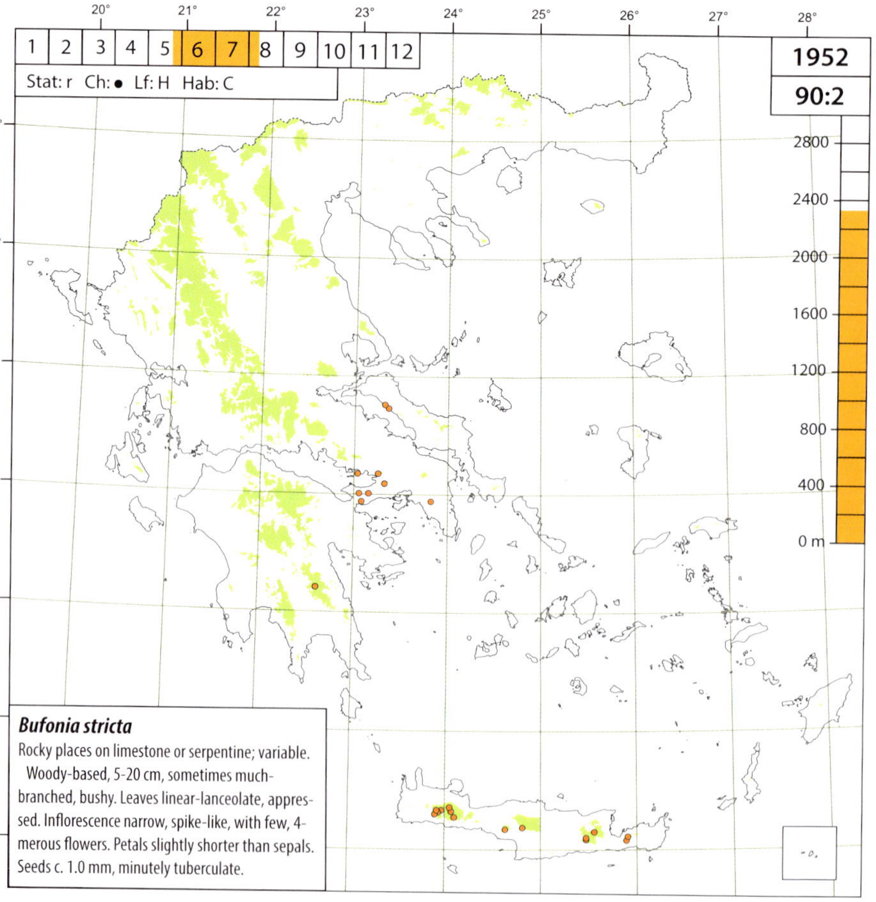

Bufonia stricta
Rocky places on limestone or serpentine; variable.
Woody-based, 5-20 cm, sometimes much-branched, bushy. Leaves linear-lanceolate, appressed. Inflorescence narrow, spike-like, with few, 4-merous flowers. Petals slightly shorter than sepals. Seeds c. 1.0 mm, minutely tuberculate.

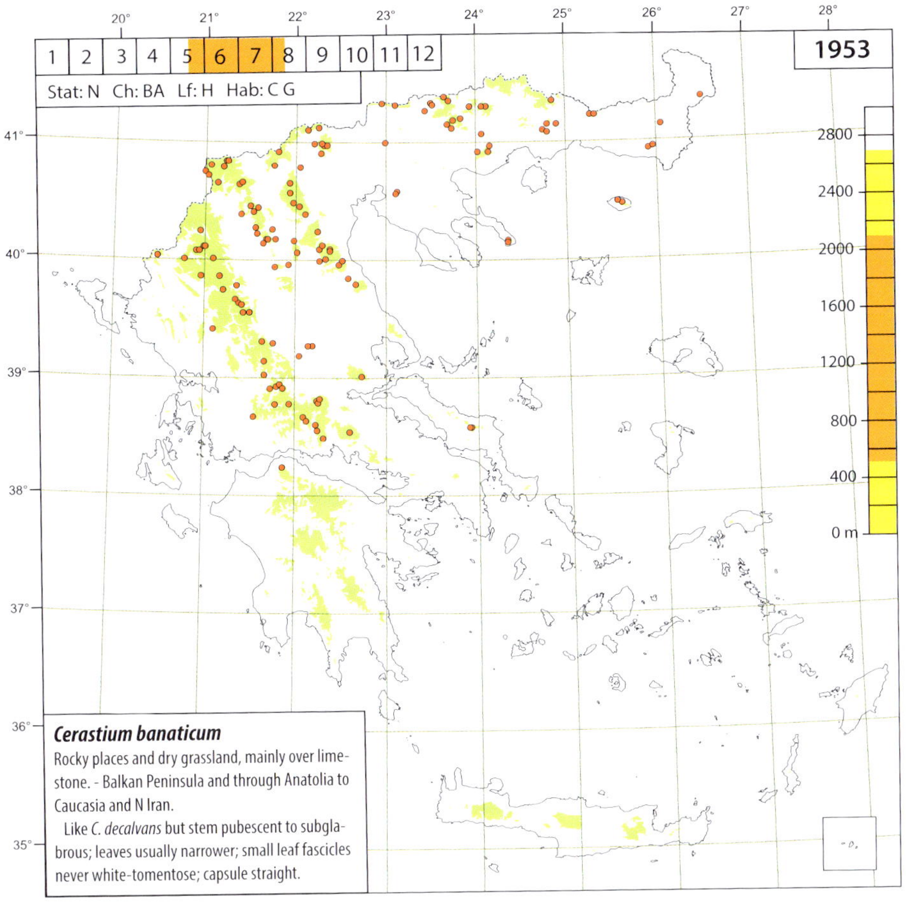

Cerastium banaticum

Rocky places and dry grassland, mainly over limestone. - Balkan Peninsula and through Anatolia to Caucasia and N Iran.

Like *C. decalvans* but stem pubescent to subglabrous; leaves usually narrower; small leaf fascicles never white-tomentose; capsule straight.

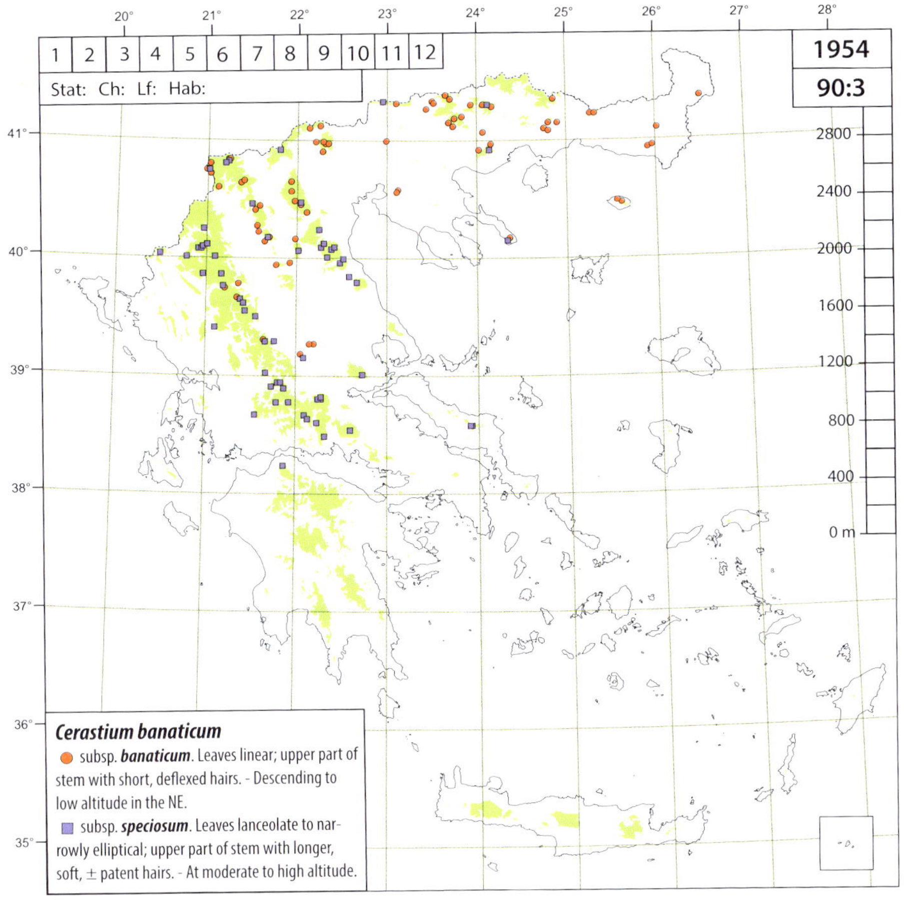

Cerastium banaticum

● subsp. ***banaticum***. Leaves linear; upper part of stem with short, deflexed hairs. - Descending to low altitude in the NE.

■ subsp. ***speciosum***. Leaves lanceolate to narrowly elliptical; upper part of stem with longer, soft, ± patent hairs. - At moderate to high altitude.

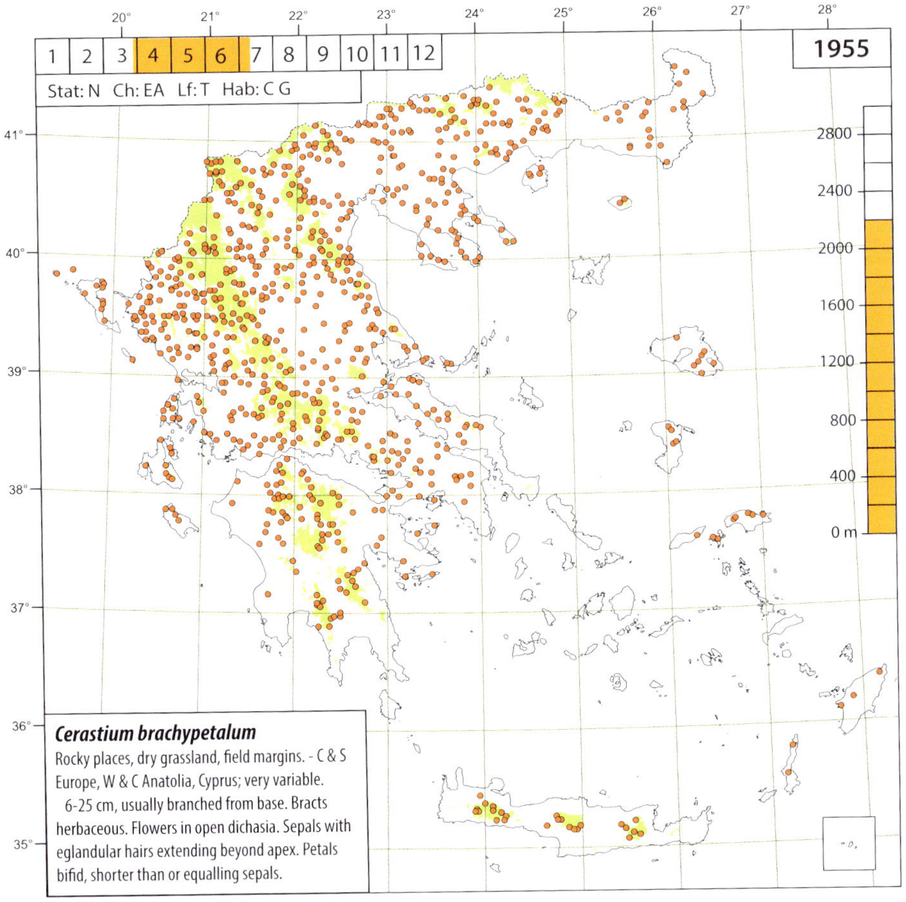

Cerastium brachypetalum

Rocky places, dry grassland, field margins. - C & S Europe, W & C Anatolia, Cyprus; very variable.

6-25 cm, usually branched from base. Bracts herbaceous. Flowers in open dichasia. Sepals with eglandular hairs extending beyond apex. Petals bifid, shorter than or equalling sepals.

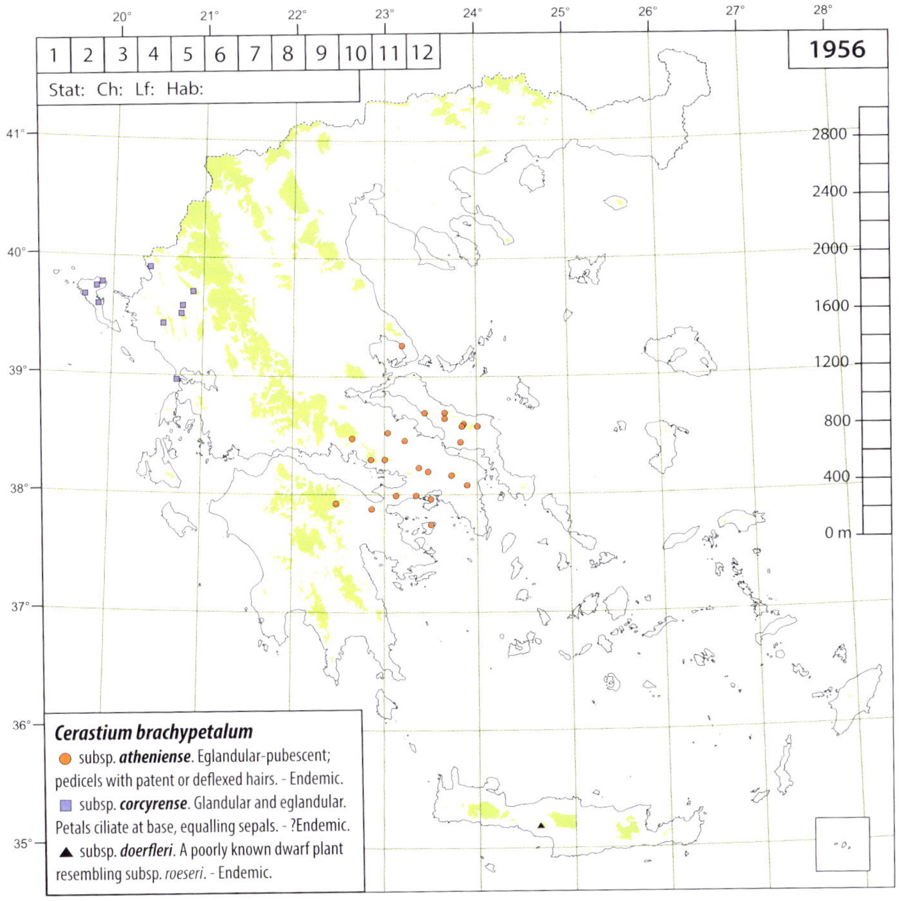

Cerastium brachypetalum

● subsp. ***atheniense***. Eglandular-pubescent; pedicels with patent or deflexed hairs. - Endemic.

■ subsp. ***corcyrense***. Glandular and eglandular. Petals ciliate at base, equalling sepals. - ?Endemic.

▲ subsp. ***doerfleri***. A poorly known dwarf plant resembling subsp. *roeseri*. - Endemic.

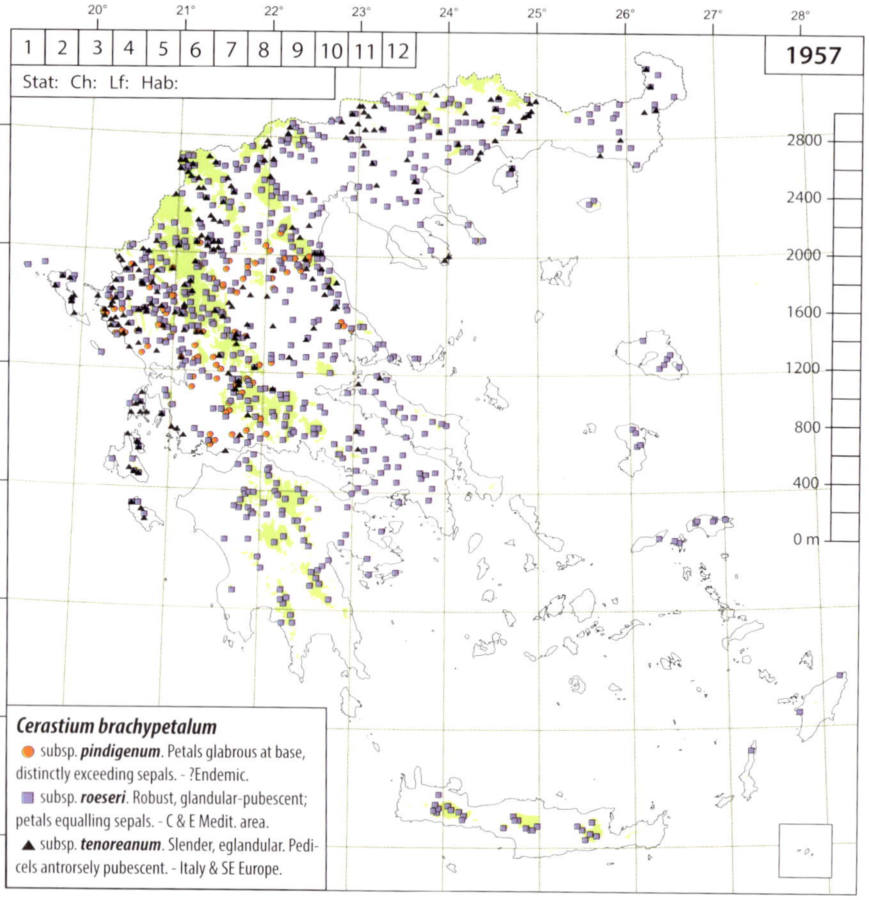

Cerastium brachypetalum
● subsp. ***pindigenum***. Petals glabrous at base, distinctly exceeding sepals. - ?Endemic.
■ subsp. ***roeseri***. Robust, glandular-pubescent; petals equalling sepals. - C & E Medit. area.
▲ subsp. ***tenoreanum***. Slender, eglandular. Pedicels antrorsely pubescent. - Italy & SE Europe.

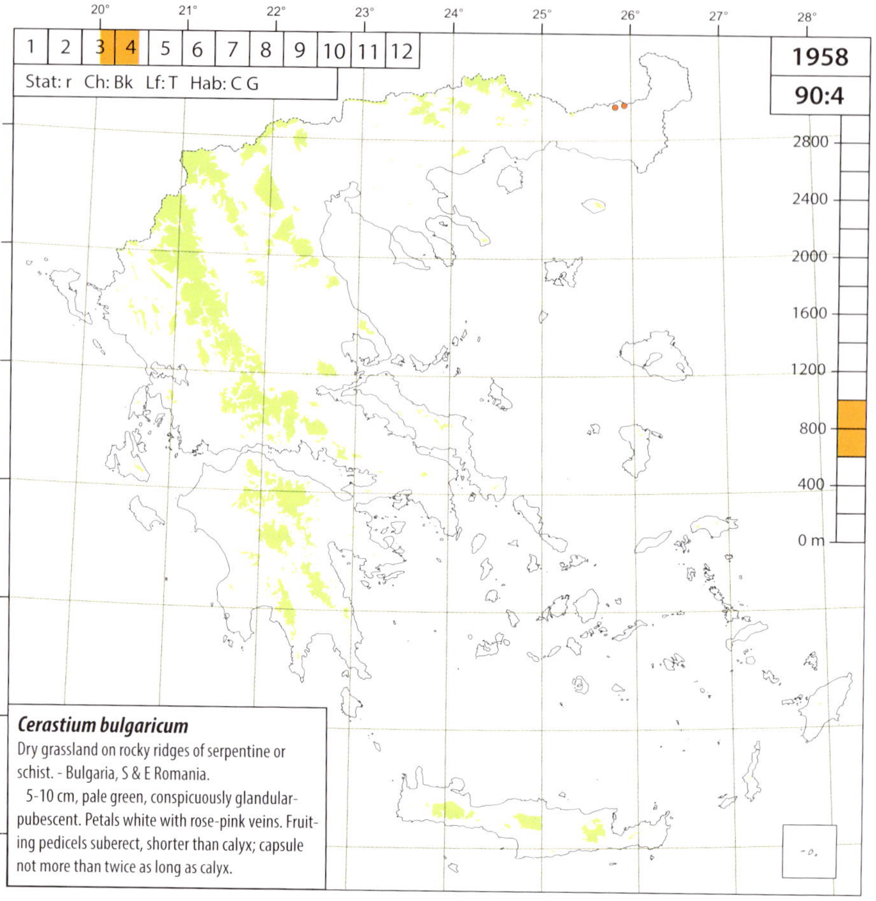

Cerastium bulgaricum
Dry grassland on rocky ridges of serpentine or schist. - Bulgaria, S & E Romania.
5-10 cm, pale green, conspicuously glandular-pubescent. Petals white with rose-pink veins. Fruiting pedicels suberect, shorter than calyx; capsule not more than twice as long as calyx.

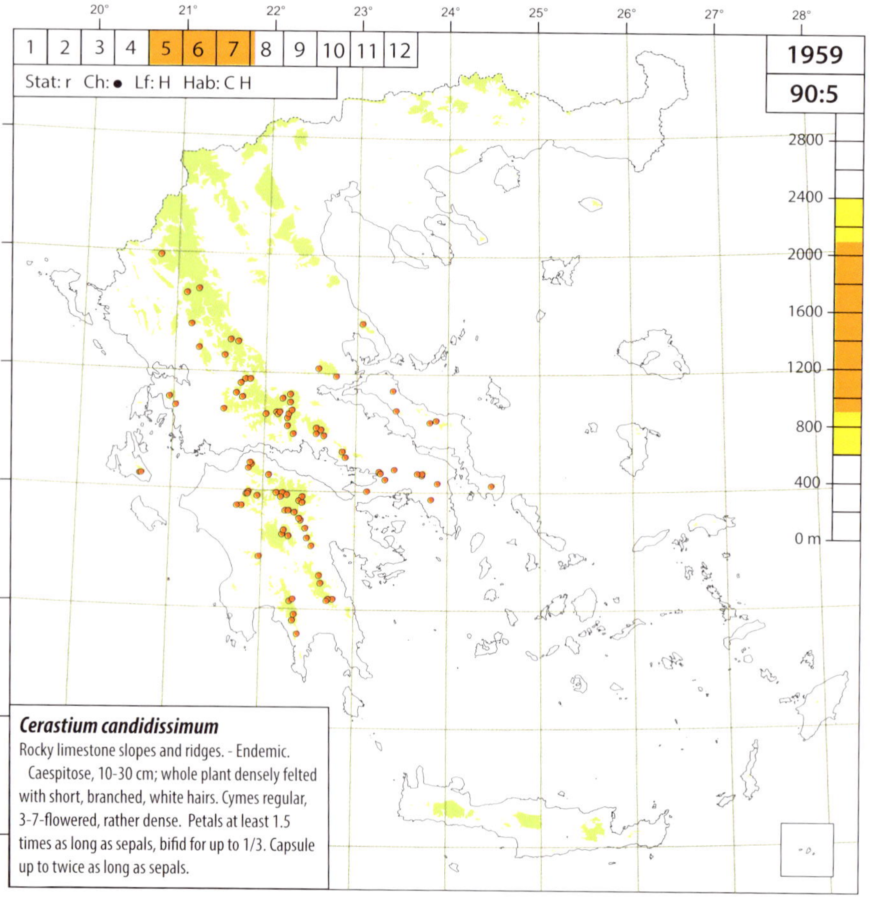

Cerastium candidissimum
Rocky limestone slopes and ridges. - Endemic.
Caespitose, 10-30 cm; whole plant densely felted with short, branched, white hairs. Cymes regular, 3-7-flowered, rather dense. Petals at least 1.5 times as long as sepals, bifid for up to 1/3. Capsule up to twice as long as sepals.

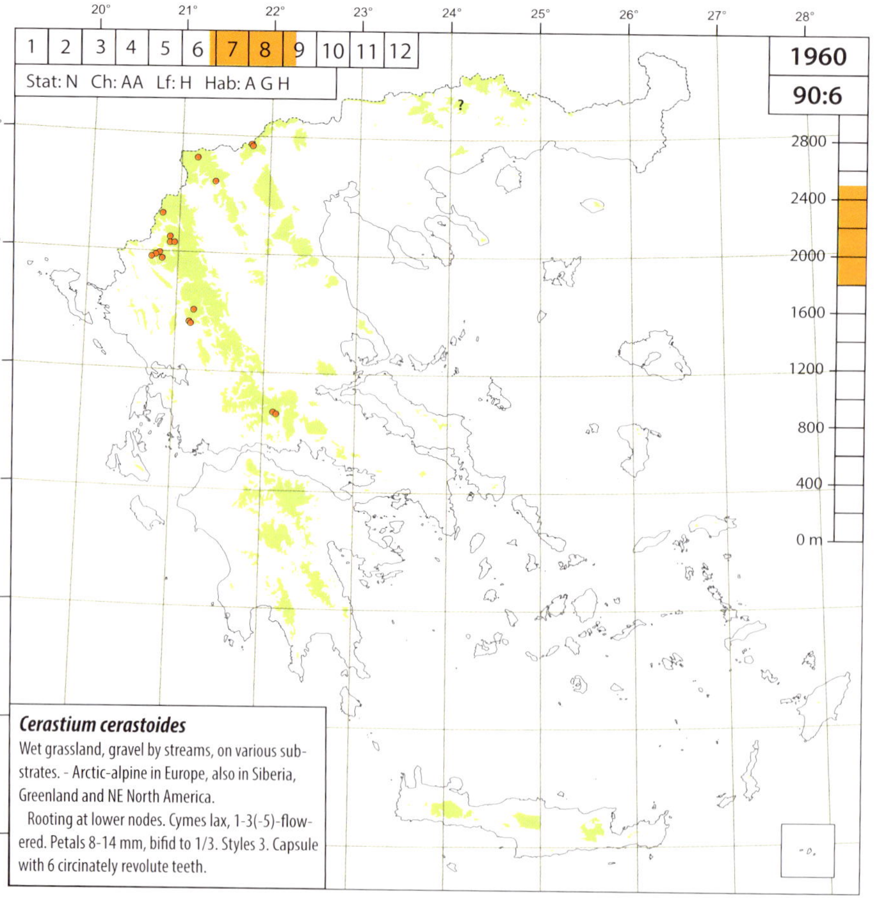

Cerastium cerastoides
Wet grassland, gravel by streams, on various substrates. - Arctic-alpine in Europe, also in Siberia, Greenland and NE North America.
Rooting at lower nodes. Cymes lax, 1-3(-5)-flowered. Petals 8-14 mm, bifid to 1/3. Styles 3. Capsule with 6 circinately revolute teeth.

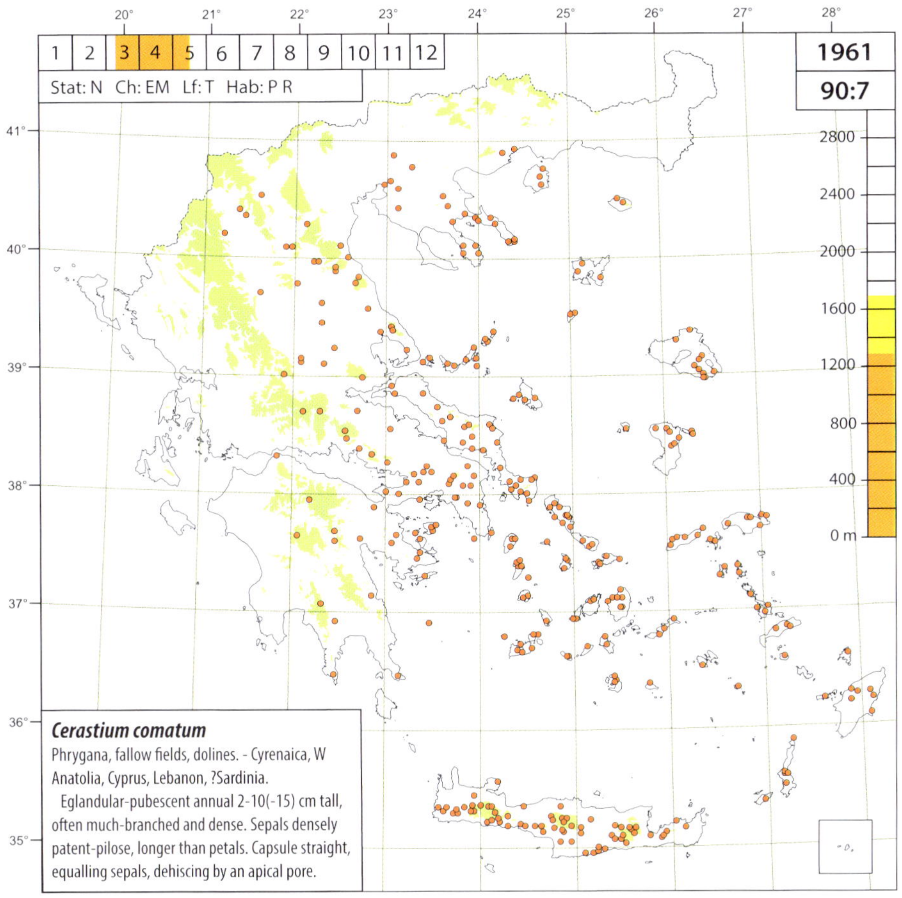

Cerastium comatum

Phrygana, fallow fields, dolines. - Cyrenaica, W Anatolia, Cyprus, Lebanon, ?Sardinia.

Eglandular-pubescent annual 2-10(-15) cm tall, often much-branched and dense. Sepals densely patent-pilose, longer than petals. Capsule straight, equalling sepals, dehiscing by an apical pore.

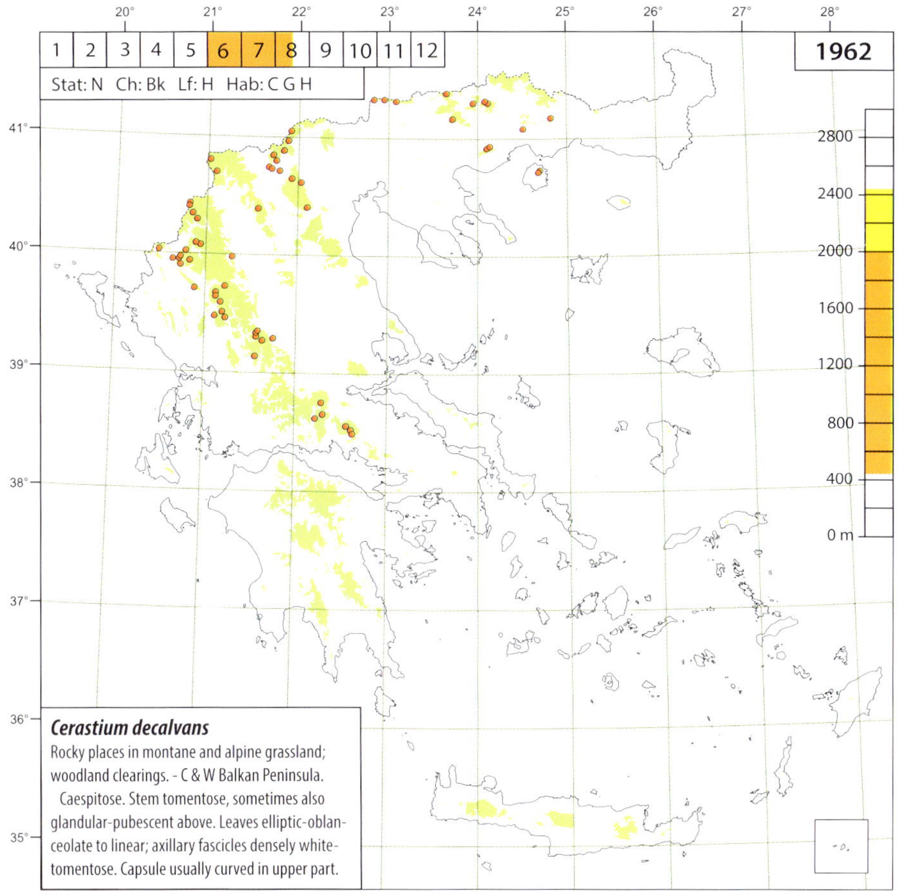

Cerastium decalvans

Rocky places in montane and alpine grassland; woodland clearings. - C & W Balkan Peninsula.

Caespitose. Stem tomentose, sometimes also glandular-pubescent above. Leaves elliptic-oblanceolate to linear; axillary fascicles densely white-tomentose. Capsule usually curved in upper part.

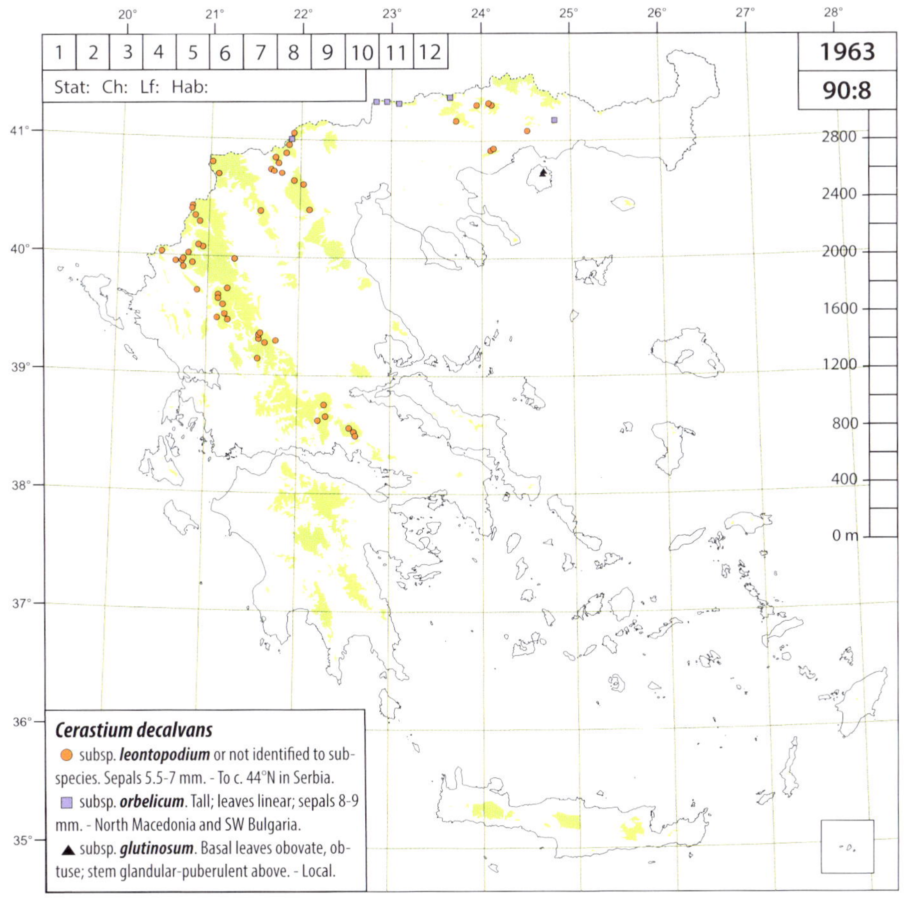

Cerastium decalvans

● subsp. ***leontopodium*** or not identified to subspecies. Sepals 5.5-7 mm. - To c. 44°N in Serbia.

■ subsp. ***orbelicum***. Tall; leaves linear; sepals 8-9 mm. - North Macedonia and SW Bulgaria.

▲ subsp. ***glutinosum***. Basal leaves obovate, obtuse; stem glandular-puberulent above. - Local.

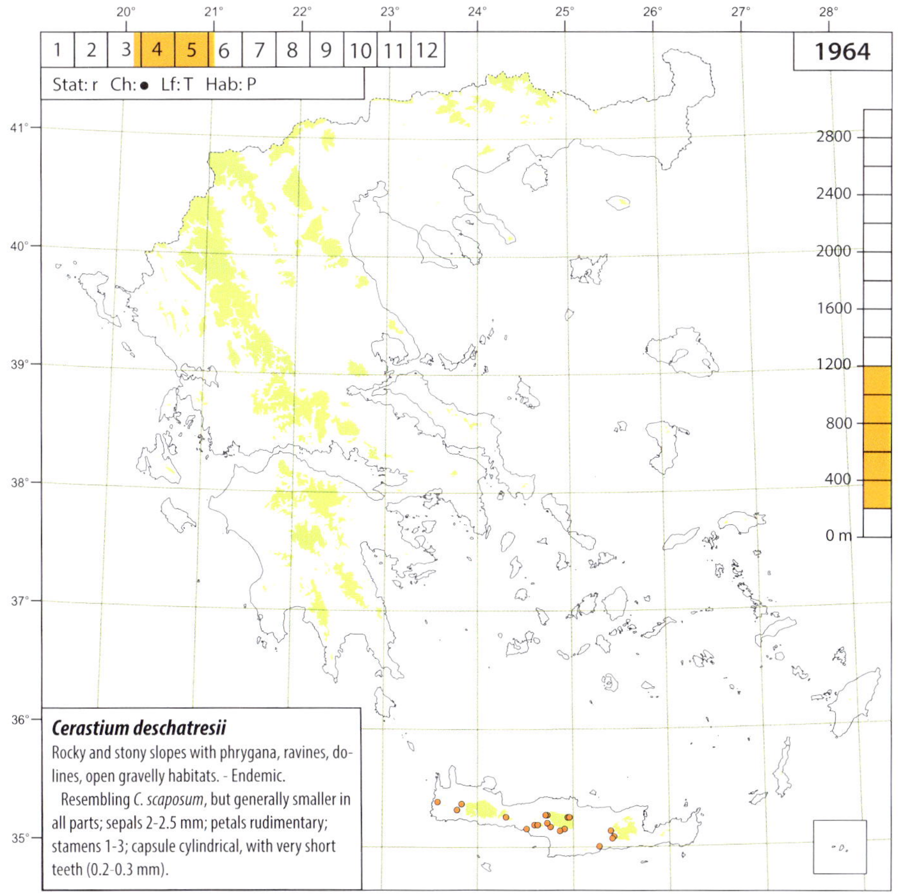

Cerastium deschatresii

Rocky and stony slopes with phrygana, ravines, dolines, open gravelly habitats. - Endemic.

Resembling *C. scaposum*, but generally smaller in all parts; sepals 2-2.5 mm; petals rudimentary; stamens 1-3; capsule cylindrical, with very short teeth (0.2-0.3 mm).

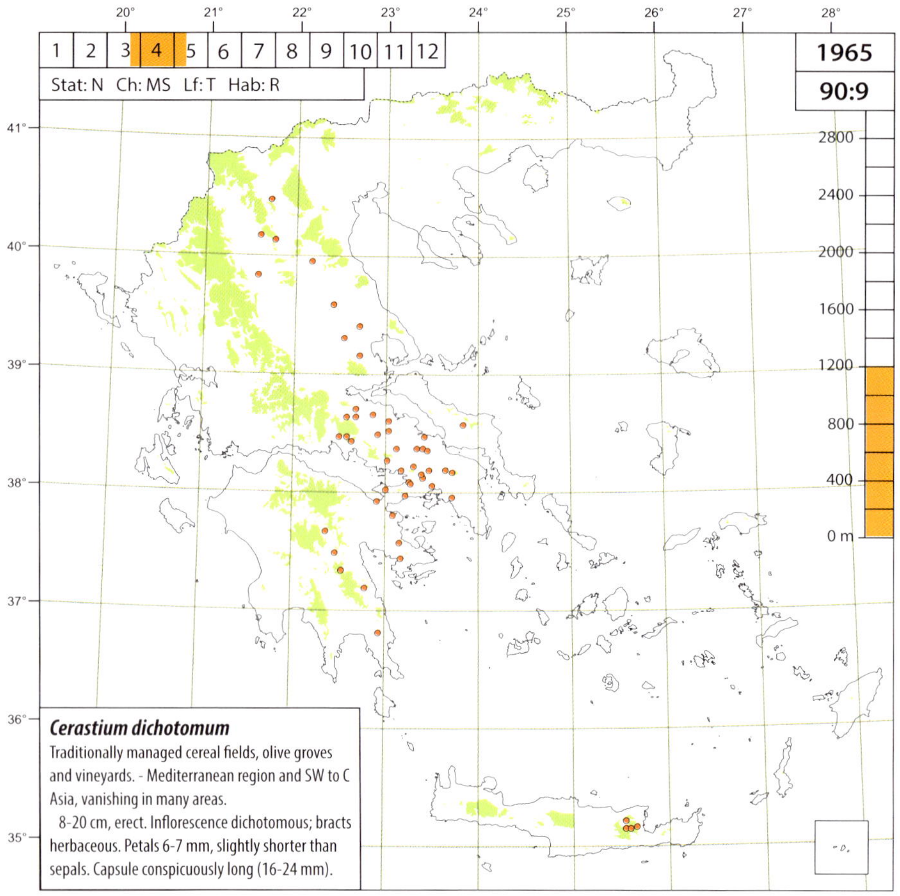

Cerastium dichotomum

Traditionally managed cereal fields, olive groves and vineyards. - Mediterranean region and SW to C Asia, vanishing in many areas.

8-20 cm, erect. Inflorescence dichotomous; bracts herbaceous. Petals 6-7 mm, slightly shorter than sepals. Capsule conspicuously long (16-24 mm).

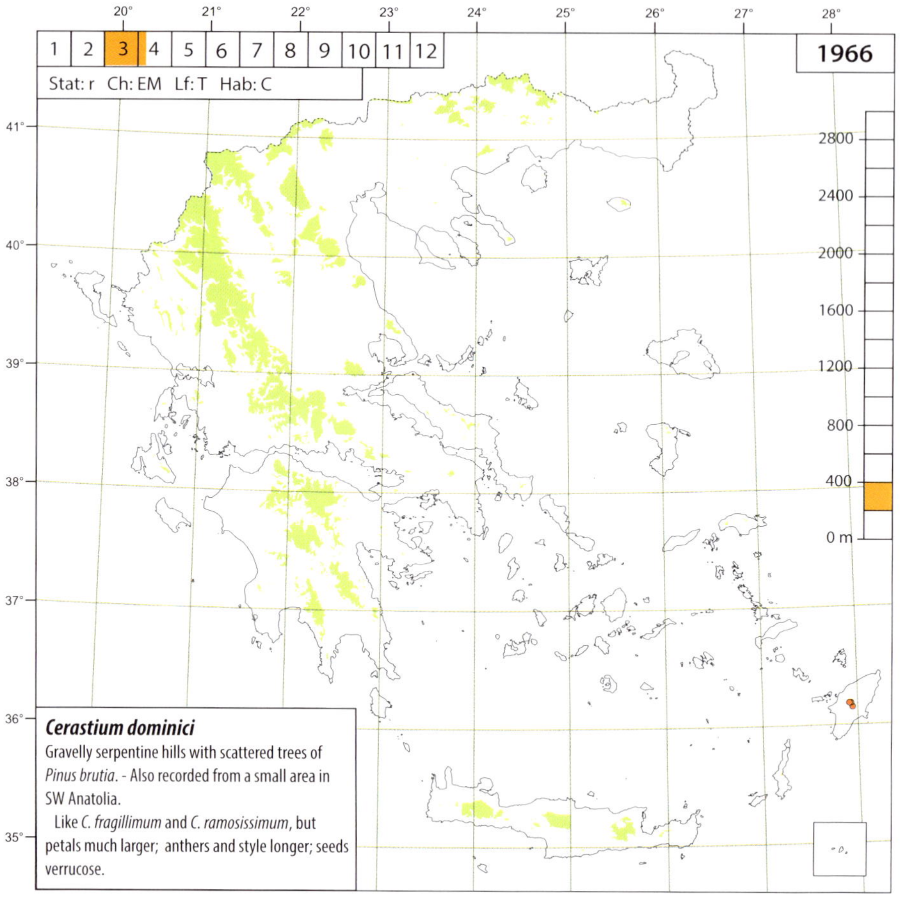

Cerastium dominici

Gravelly serpentine hills with scattered trees of *Pinus brutia*. - Also recorded from a small area in SW Anatolia.

Like *C. fragillimum* and *C. ramosissimum*, but petals much larger; anthers and style longer; seeds verrucose.

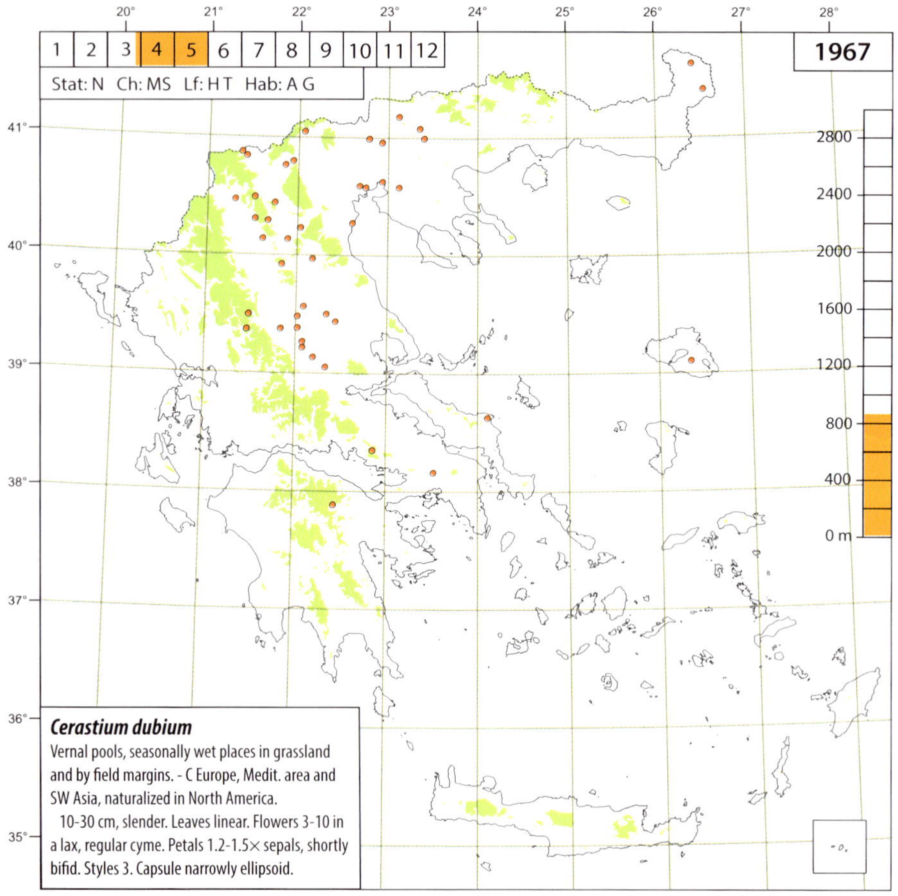

Cerastium dubium

Vernal pools, seasonally wet places in grassland and by field margins. - C Europe, Medit. area and SW Asia, naturalized in North America.

10-30 cm, slender. Leaves linear. Flowers 3-10 in a lax, regular cyme. Petals 1.2-1.5× sepals, shortly bifid. Styles 3. Capsule narrowly ellipsoid.

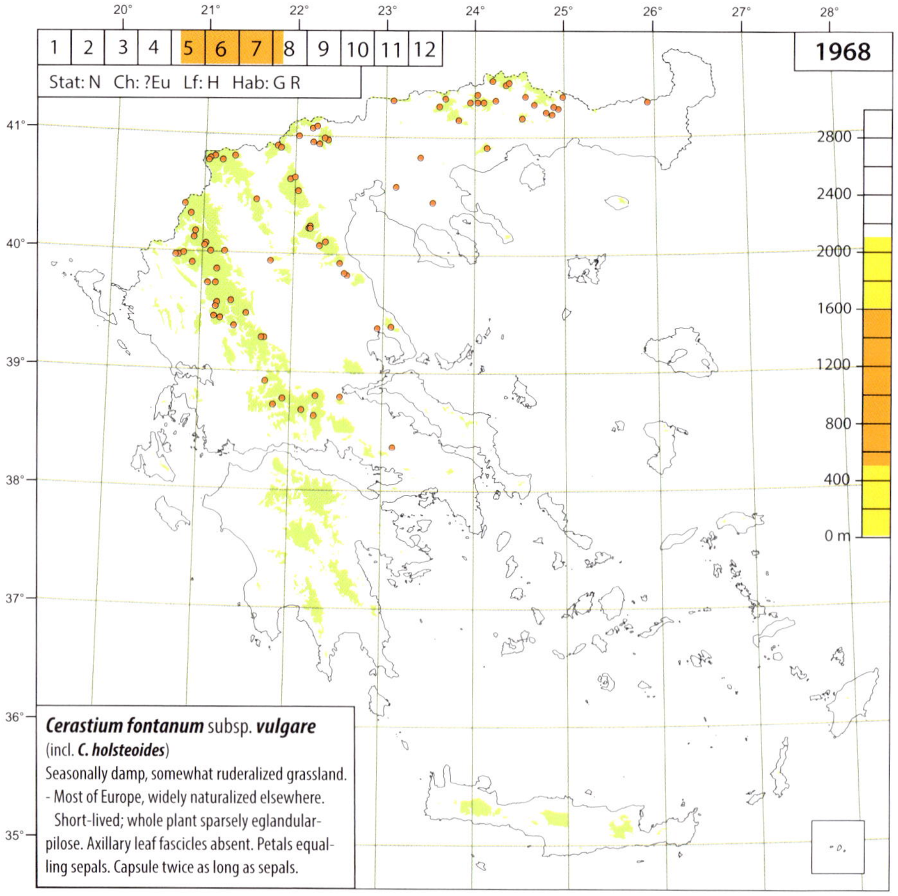

Cerastium fontanum subsp. ***vulgare***

(incl. ***C. holsteoides***)

Seasonally damp, somewhat ruderalized grassland. - Most of Europe, widely naturalized elsewhere.

Short-lived; whole plant sparsely eglandular-pilose. Axillary leaf fascicles absent. Petals equalling sepals. Capsule twice as long as sepals.

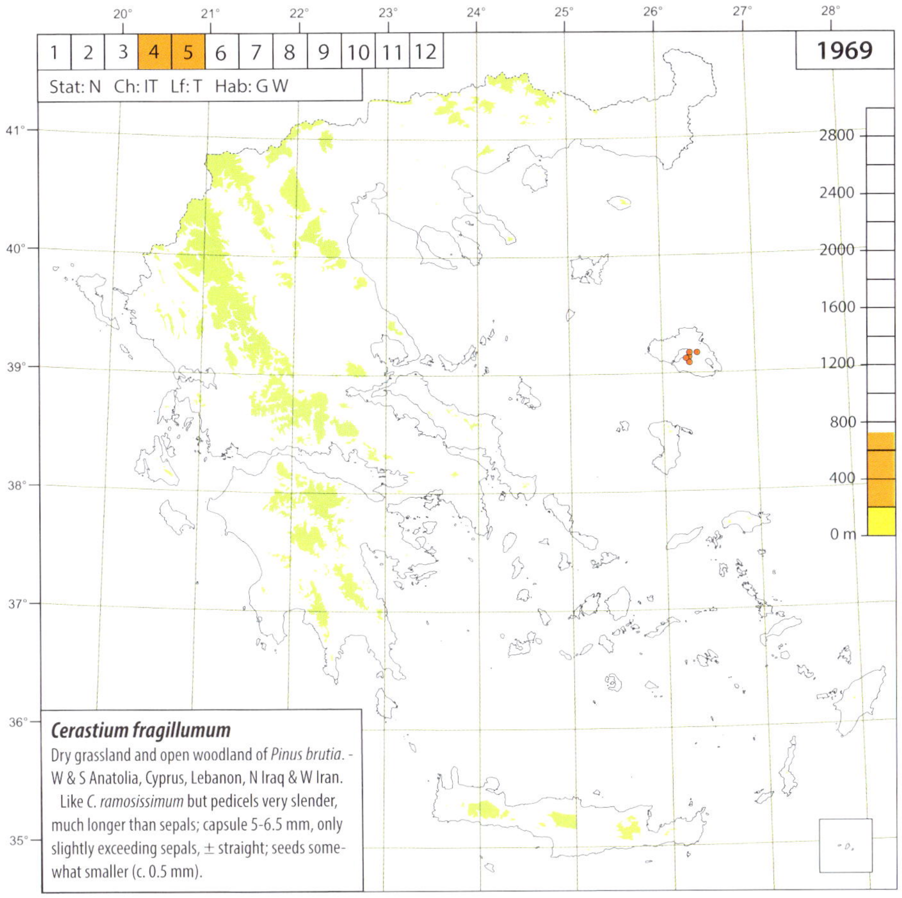

Cerastium fragillumum

Dry grassland and open woodland of *Pinus brutia*. - W & S Anatolia, Cyprus, Lebanon, N Iraq & W Iran.

Like *C. ramosissimum* but pedicels very slender, much longer than sepals; capsule 5-6.5 mm, only slightly exceeding sepals, ± straight; seeds somewhat smaller (c. 0.5 mm).

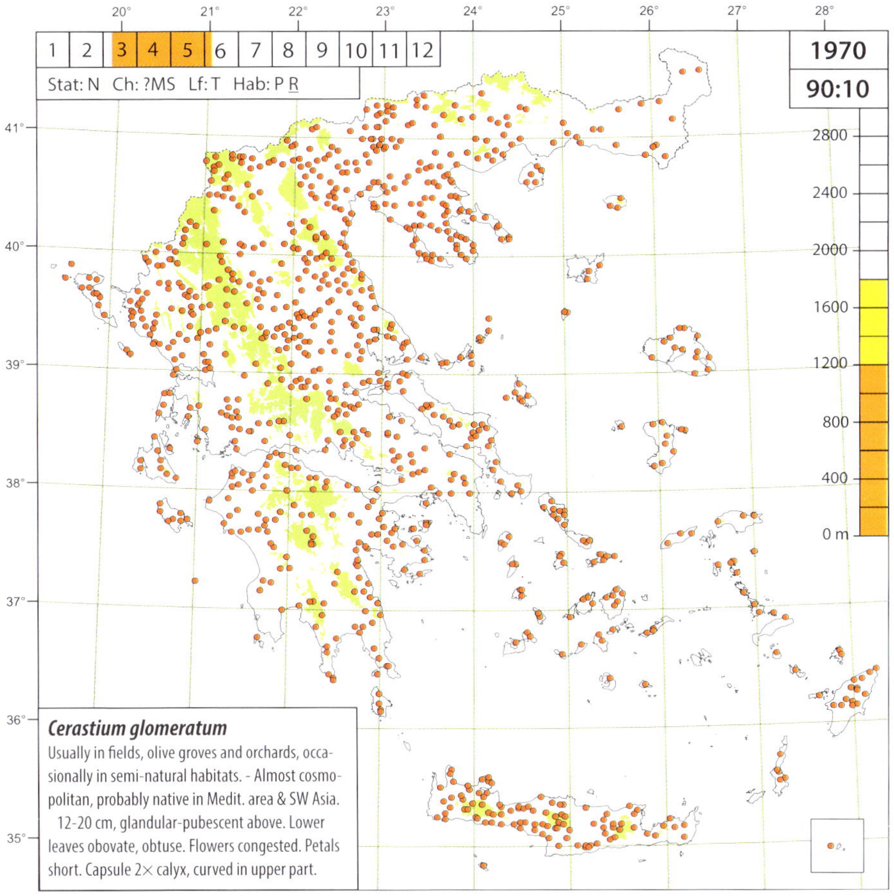

Cerastium glomeratum

Usually in fields, olive groves and orchards, occasionally in semi-natural habitats. - Almost cosmopolitan, probably native in Medit. area & SW Asia.

12-20 cm, glandular-pubescent above. Lower leaves obovate, obtuse. Flowers congested. Petals short. Capsule 2× calyx, curved in upper part.

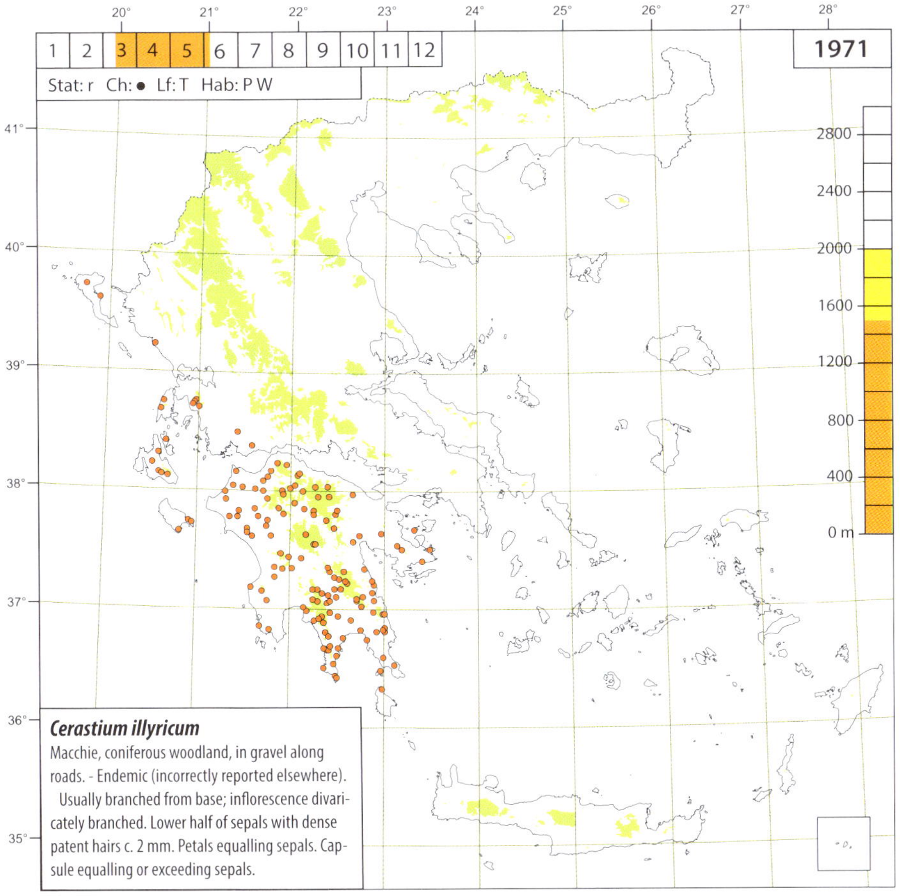

Cerastium illyricum

Macchie, coniferous woodland, in gravel along roads. - Endemic (incorrectly reported elsewhere).

Usually branched from base; inflorescence divaricately branched. Lower half of sepals with dense patent hairs c. 2 mm. Petals equalling sepals. Capsule equalling or exceeding sepals.

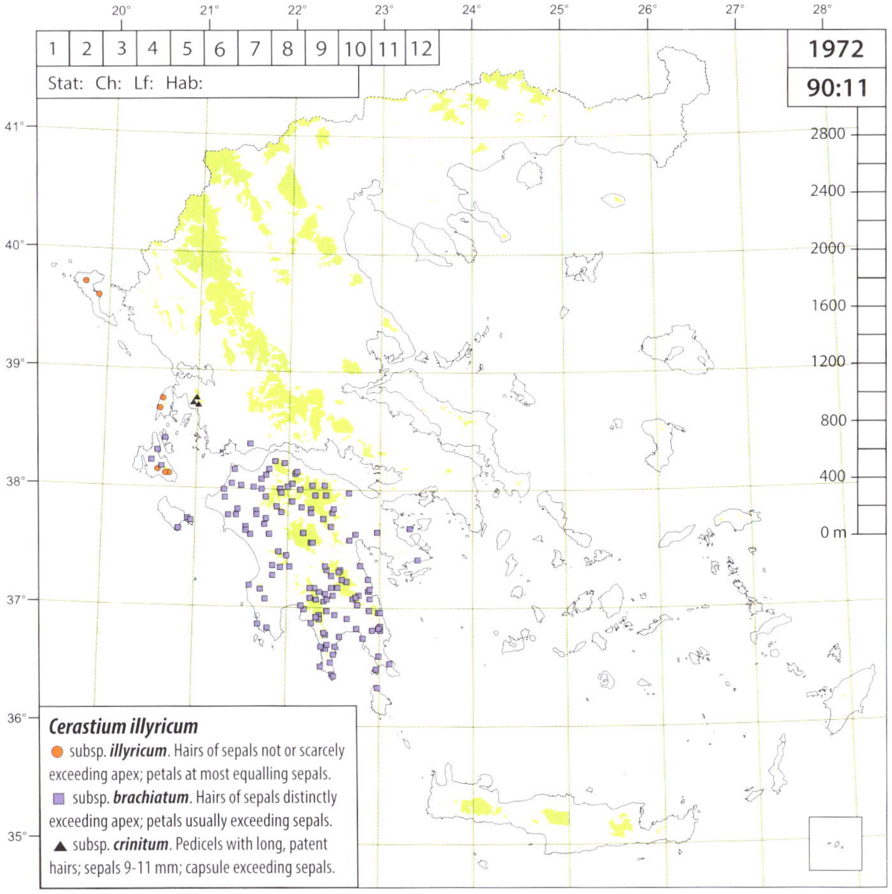

Cerastium illyricum

● subsp. ***illyricum***. Hairs of sepals not or scarcely exceeding apex; petals at most equalling sepals.

■ subsp. ***brachiatum***. Hairs of sepals distinctly exceeding apex; petals usually exceeding sepals.

▲ subsp. ***crinitum***. Pedicels with long, patent hairs; sepals 9-11 mm; capsule exceeding sepals.

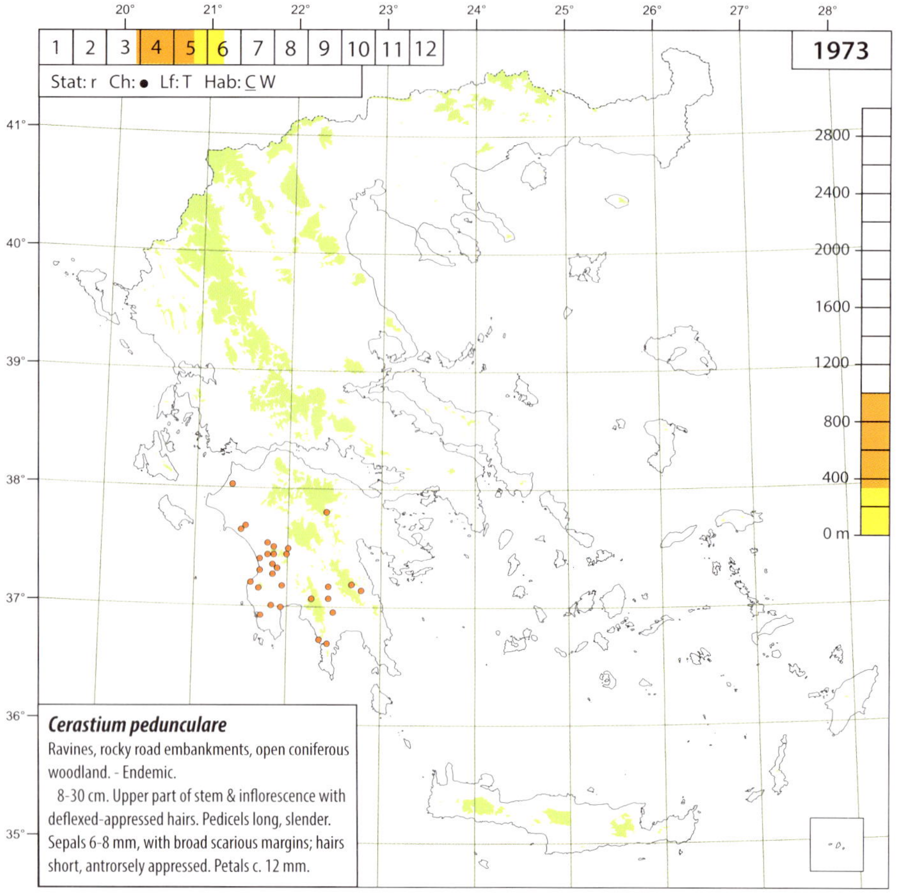

Cerastium pedunculare
Ravines, rocky road embankments, open coniferous woodland. - Endemic.
8-30 cm. Upper part of stem & inflorescence with deflexed-appressed hairs. Pedicels long, slender. Sepals 6-8 mm, with broad scarious margins; hairs short, antrorsely appressed. Petals c. 12 mm.

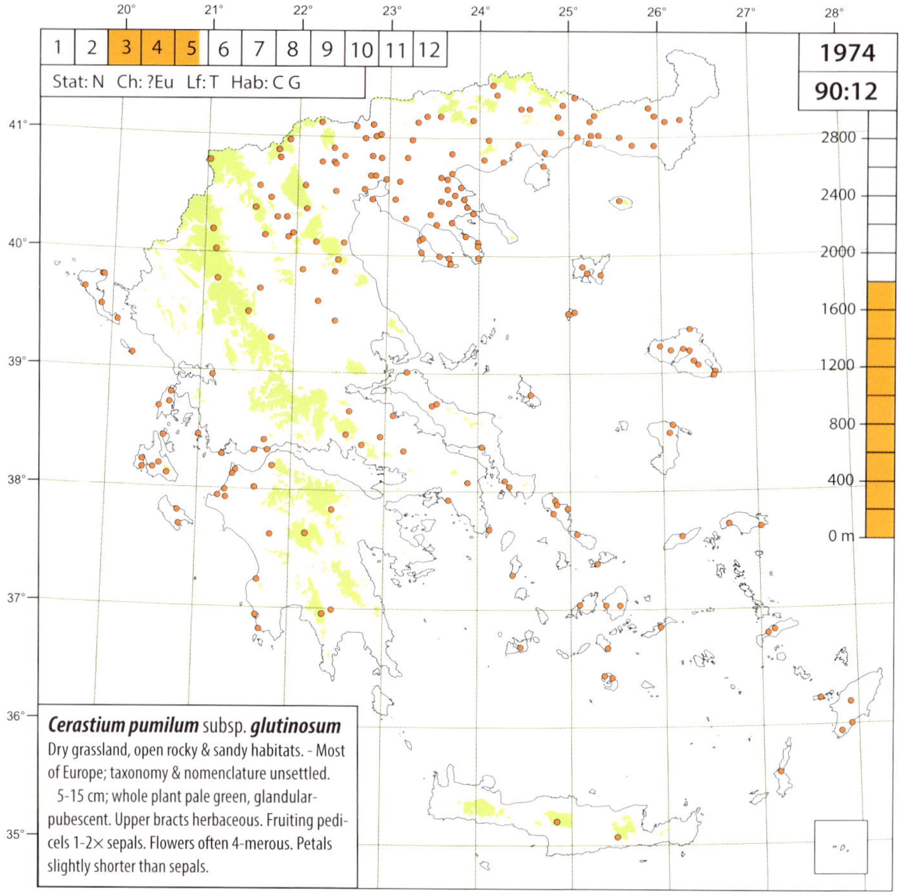

Cerastium pumilum subsp. ***glutinosum***
Dry grassland, open rocky & sandy habitats. - Most of Europe; taxonomy & nomenclature unsettled.
5-15 cm; whole plant pale green, glandular-pubescent. Upper bracts herbaceous. Fruiting pedicels 1-2× sepals. Flowers often 4-merous. Petals slightly shorter than sepals.

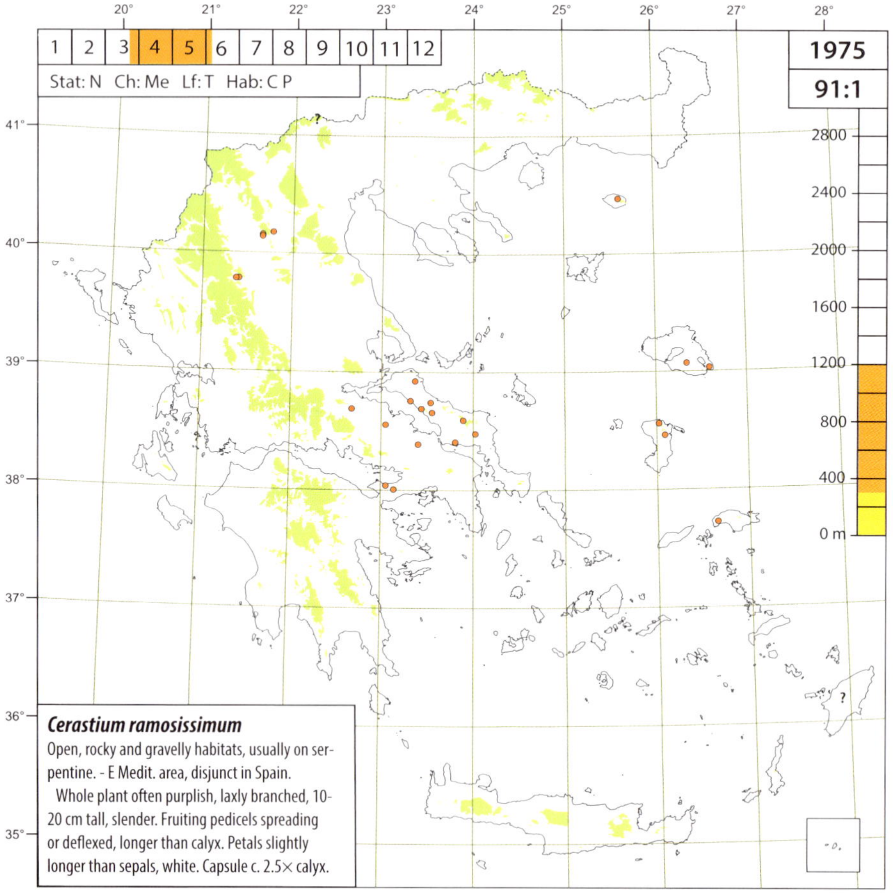

Cerastium ramosissimum
Open, rocky and gravelly habitats, usually on serpentine. - E Medit. area, disjunct in Spain.
Whole plant often purplish, laxly branched, 10-20 cm tall, slender. Fruiting pedicels spreading or deflexed, longer than calyx. Petals slightly longer than sepals, white. Capsule c. 2.5× calyx.

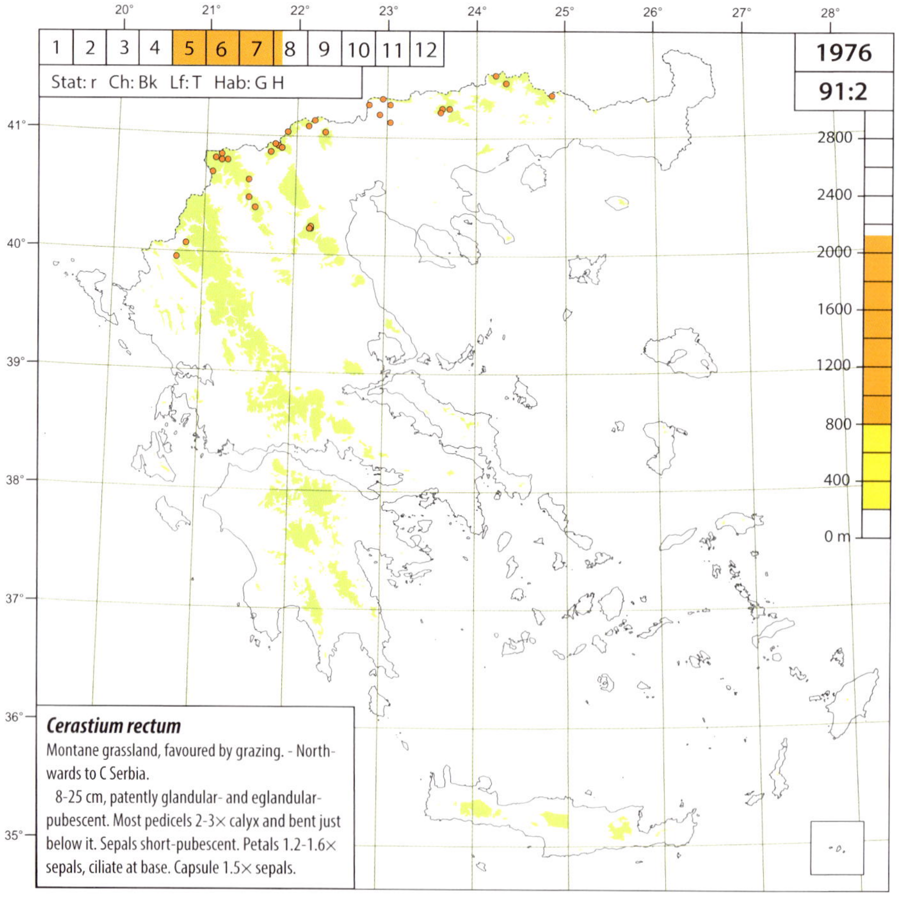

Cerastium rectum
Montane grassland, favoured by grazing. - Northwards to C Serbia.
8-25 cm, patently glandular- and eglandular-pubescent. Most pedicels 2-3× calyx and bent just below it. Sepals short-pubescent. Petals 1.2-1.6× sepals, ciliate at base. Capsule 1.5× sepals.

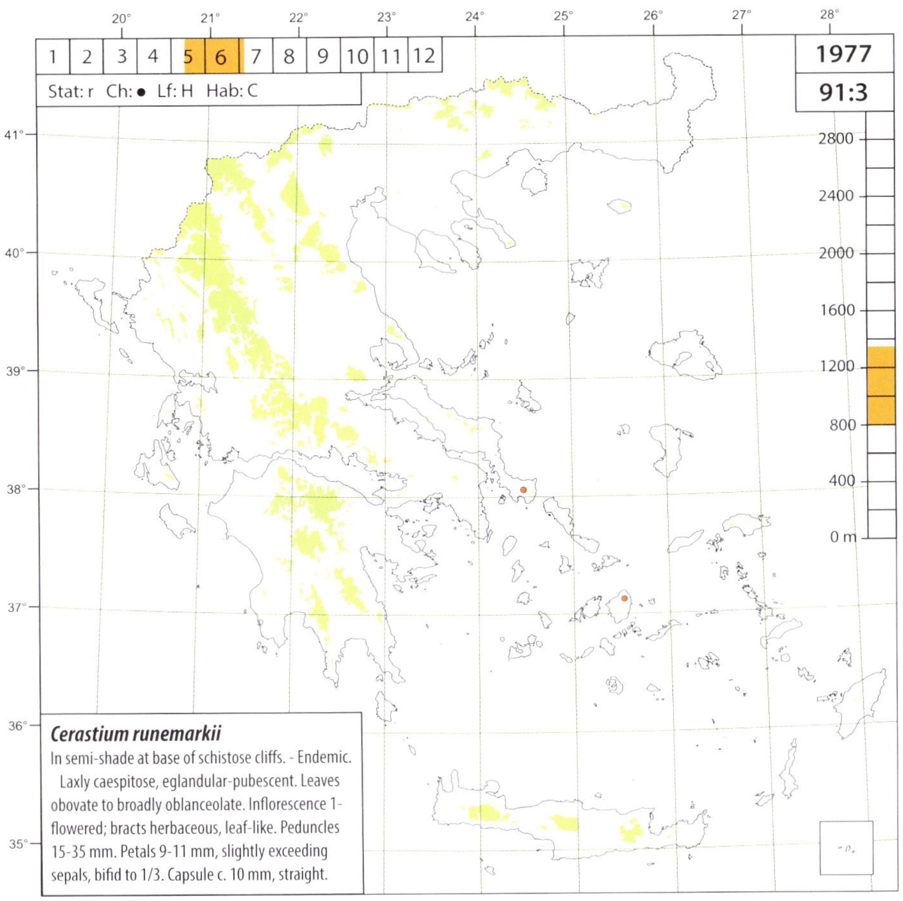

Cerastium runemarkii
In semi-shade at base of schistose cliffs. - Endemic.
Laxly caespitose, eglandular-pubescent. Leaves obovate to broadly oblanceolate. Inflorescence 1-flowered; bracts herbaceous, leaf-like. Peduncles 15-35 mm. Petals 9-11 mm, slightly exceeding sepals, bifid to 1/3. Capsule c. 10 mm, straight.

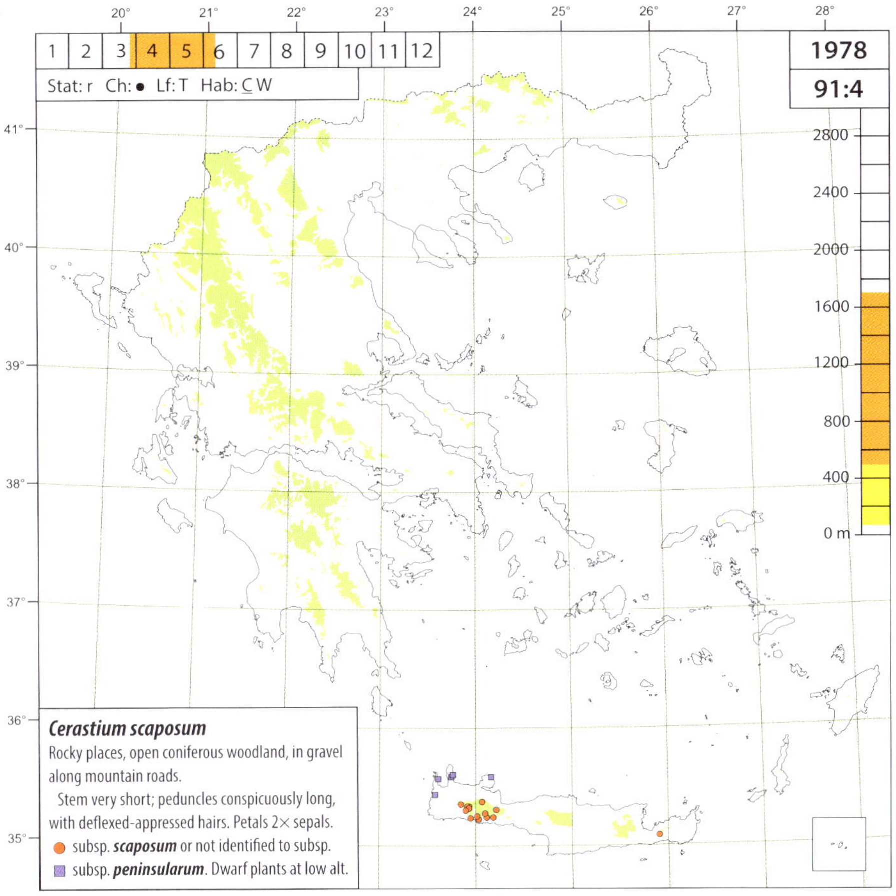

Cerastium scaposum
Rocky places, open coniferous woodland, in gravel along mountain roads.
Stem very short; peduncles conspicuously long, with deflexed-appressed hairs. Petals 2× sepals.
● subsp. ***scaposum*** or not identified to subsp.
■ subsp. ***peninsularum***. Dwarf plants at low alt.

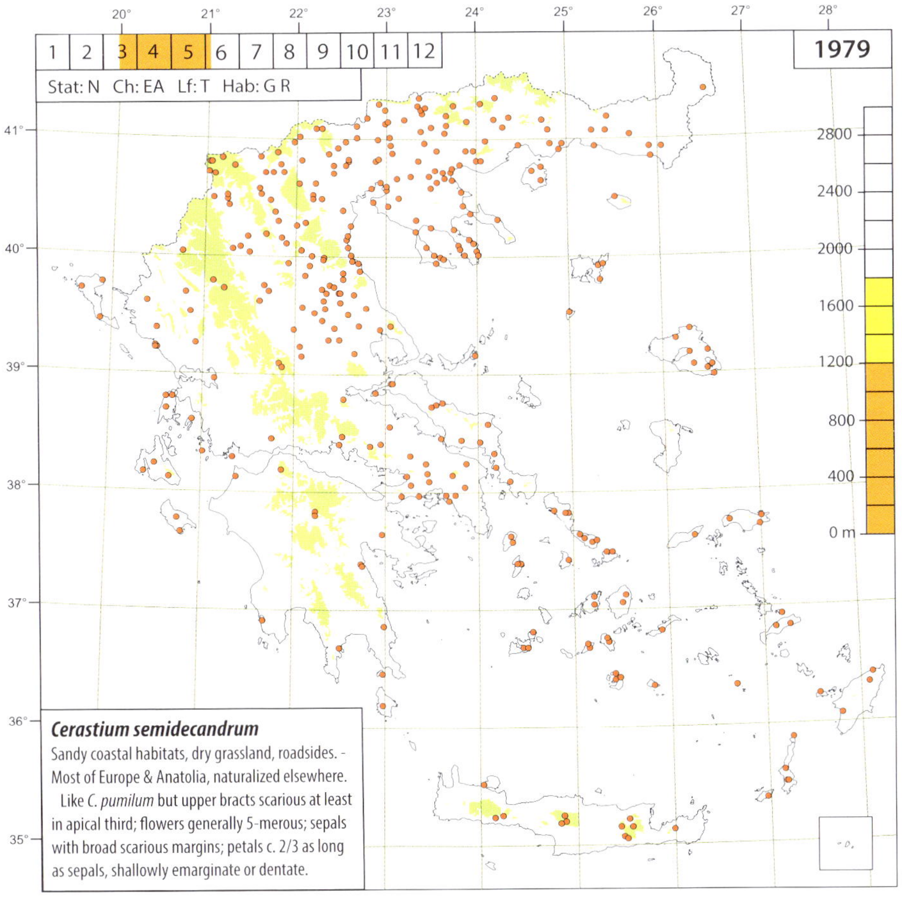

Cerastium semidecandrum
Sandy coastal habitats, dry grassland, roadsides. - Most of Europe & Anatolia, naturalized elsewhere.
Like *C. pumilum* but upper bracts scarious at least in apical third; flowers generally 5-merous; sepals with broad scarious margins; petals c. 2/3 as long as sepals, shallowly emarginate or dentate.

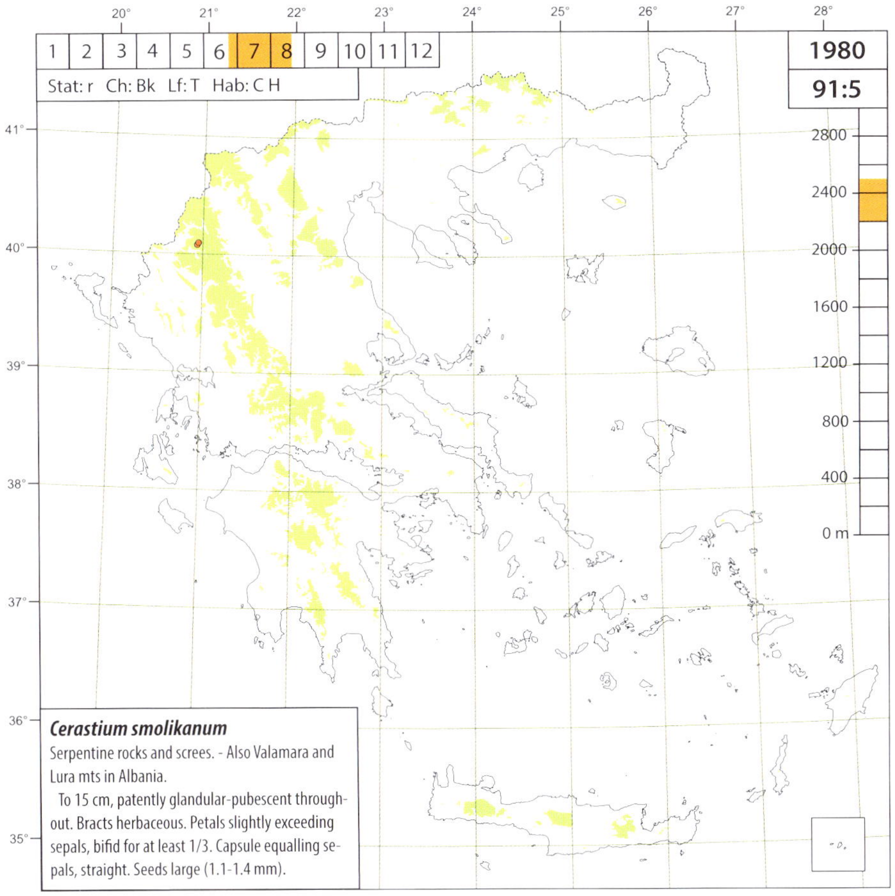

Cerastium smolikanum
Serpentine rocks and screes. - Also Valamara and Lura mts in Albania.
To 15 cm, patently glandular-pubescent throughout. Bracts herbaceous. Petals slightly exceeding sepals, bifid for at least 1/3. Capsule equalling sepals, straight. Seeds large (1.1-1.4 mm).

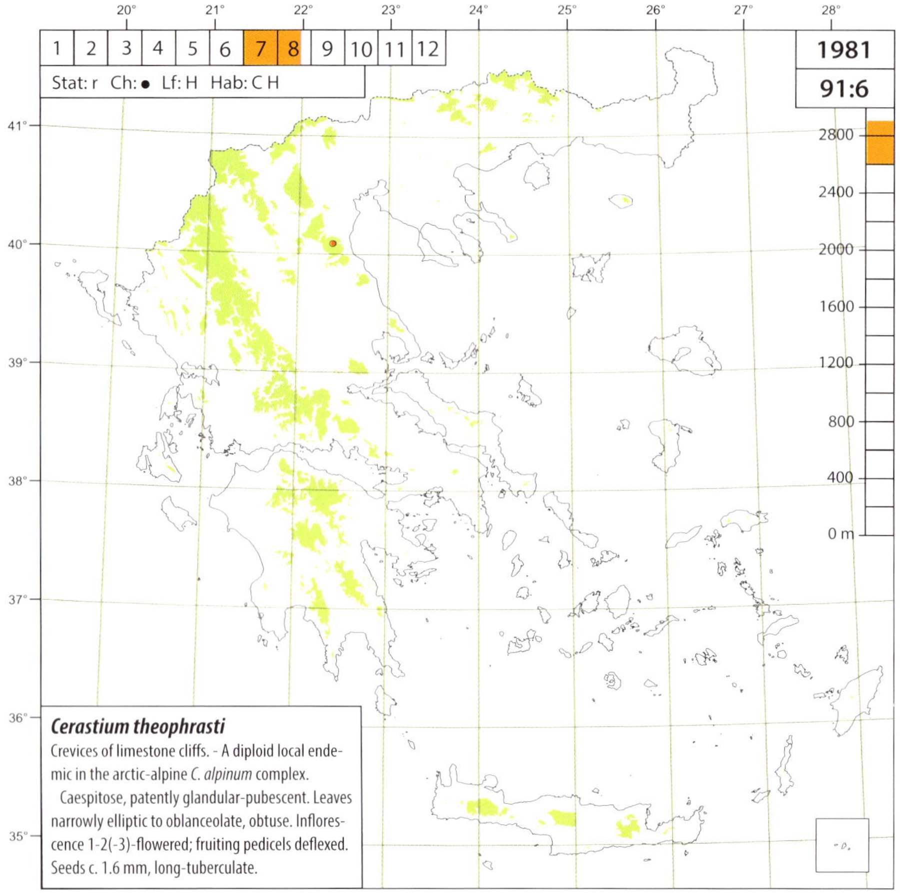

Cerastium theophrasti
Crevices of limestone cliffs. - A diploid local endemic in the arctic-alpine *C. alpinum* complex.
Caespitose, patently glandular-pubescent. Leaves narrowly elliptic to oblanceolate, obtuse. Inflorescence 1-2(-3)-flowered; fruiting pedicels deflexed. Seeds c. 1.6 mm, long-tuberculate.

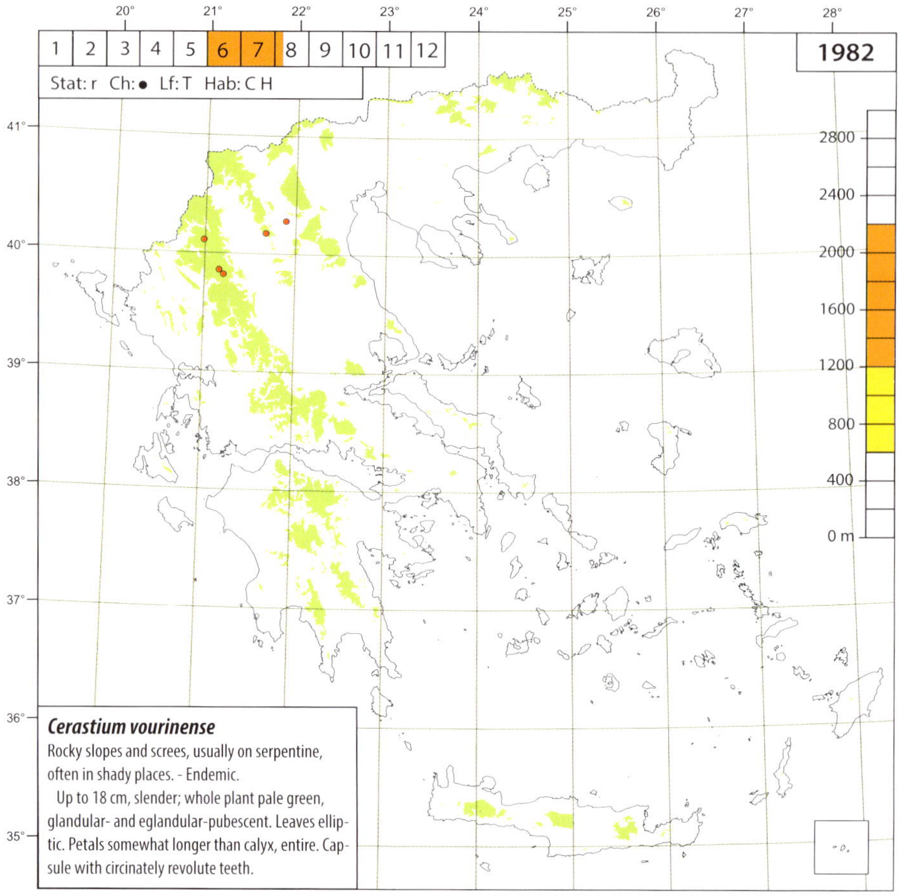

Cerastium vourinense
Rocky slopes and screes, usually on serpentine, often in shady places. - Endemic.
Up to 18 cm, slender; whole plant pale green, glandular- and eglandular-pubescent. Leaves elliptic. Petals somewhat longer than calyx, entire. Capsule with circinately revolute teeth.

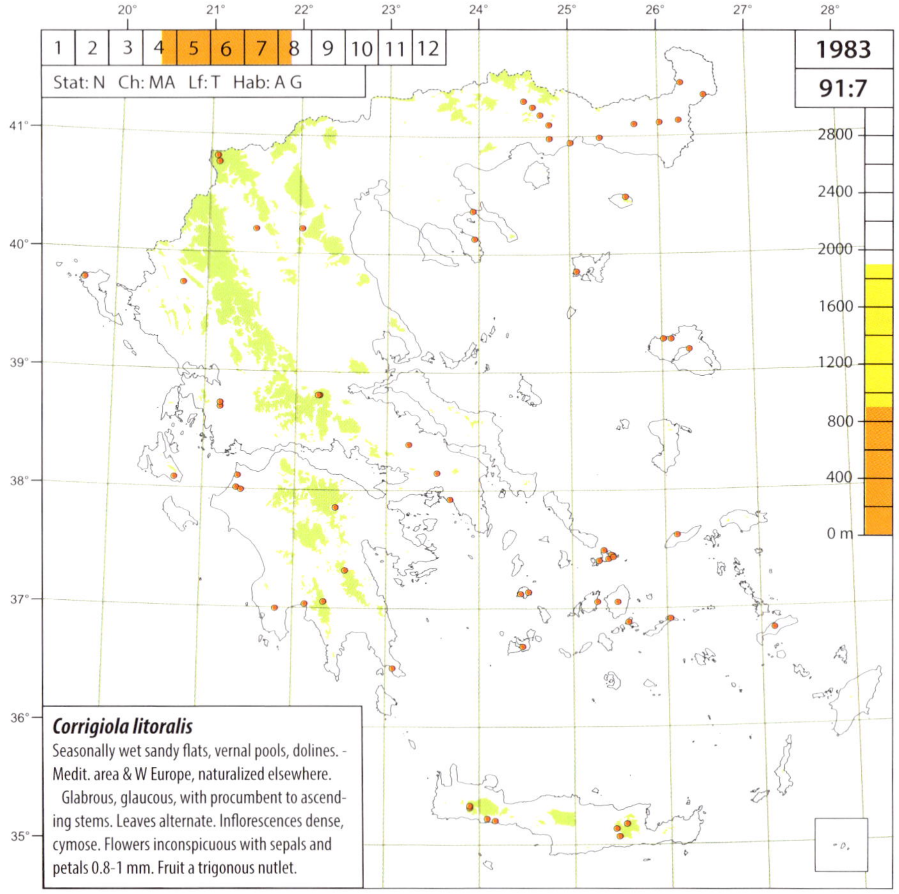

Corrigiola litoralis
Seasonally wet sandy flats, vernal pools, dolines. - Medit. area & W Europe, naturalized elsewhere.
Glabrous, glaucous, with procumbent to ascending stems. Leaves alternate. Inflorescences dense, cymose. Flowers inconspicuous with sepals and petals 0.8-1 mm. Fruit a trigonous nutlet.

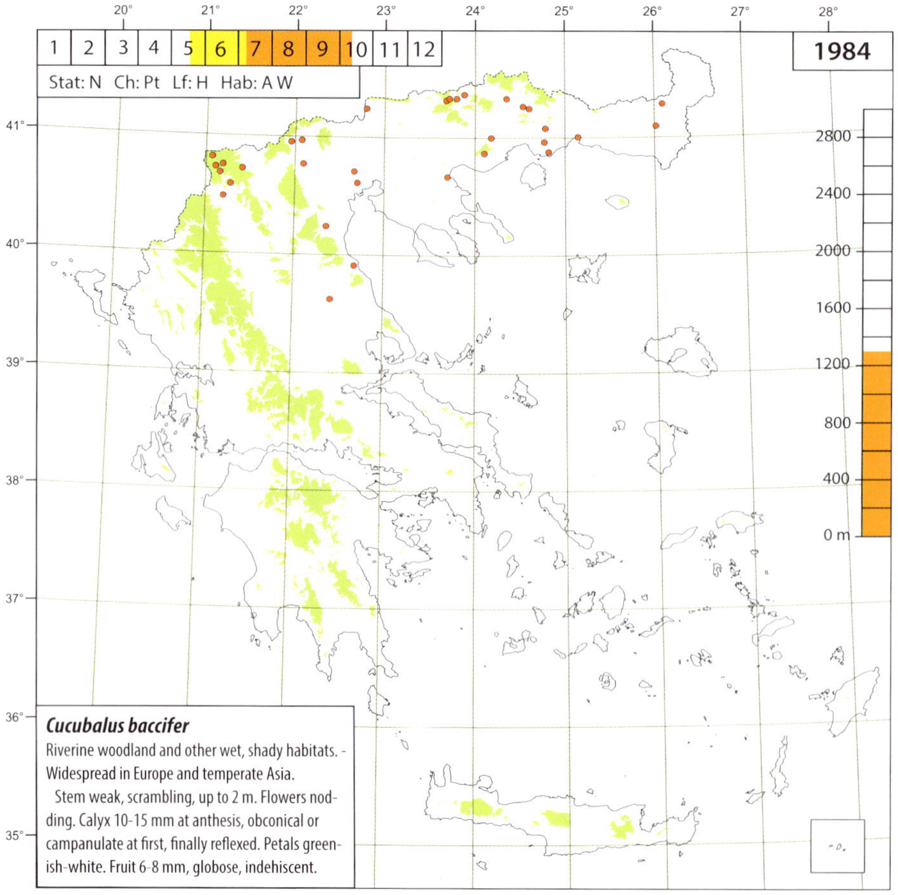

Cucubalus baccifer
Riverine woodland and other wet, shady habitats. - Widespread in Europe and temperate Asia.
Stem weak, scrambling, up to 2 m. Flowers nodding. Calyx 10-15 mm at anthesis, obconical or campanulate at first, finally reflexed. Petals greenish-white. Fruit 6-8 mm, globose, indehiscent.

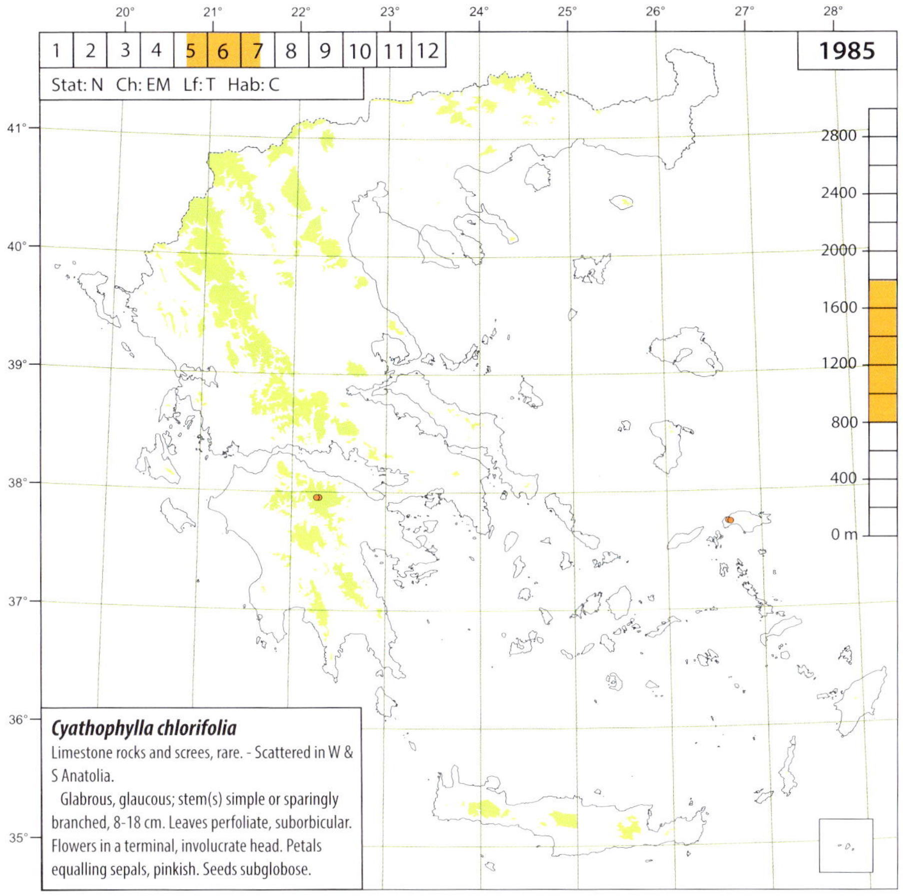

Cyathophylla chlorifolia
Limestone rocks and screes, rare. - Scattered in W & S Anatolia.

Glabrous, glaucous; stem(s) simple or sparingly branched, 8-18 cm. Leaves perfoliate, suborbicular. Flowers in a terminal, involucrate head. Petals equalling sepals, pinkish. Seeds subglobose.

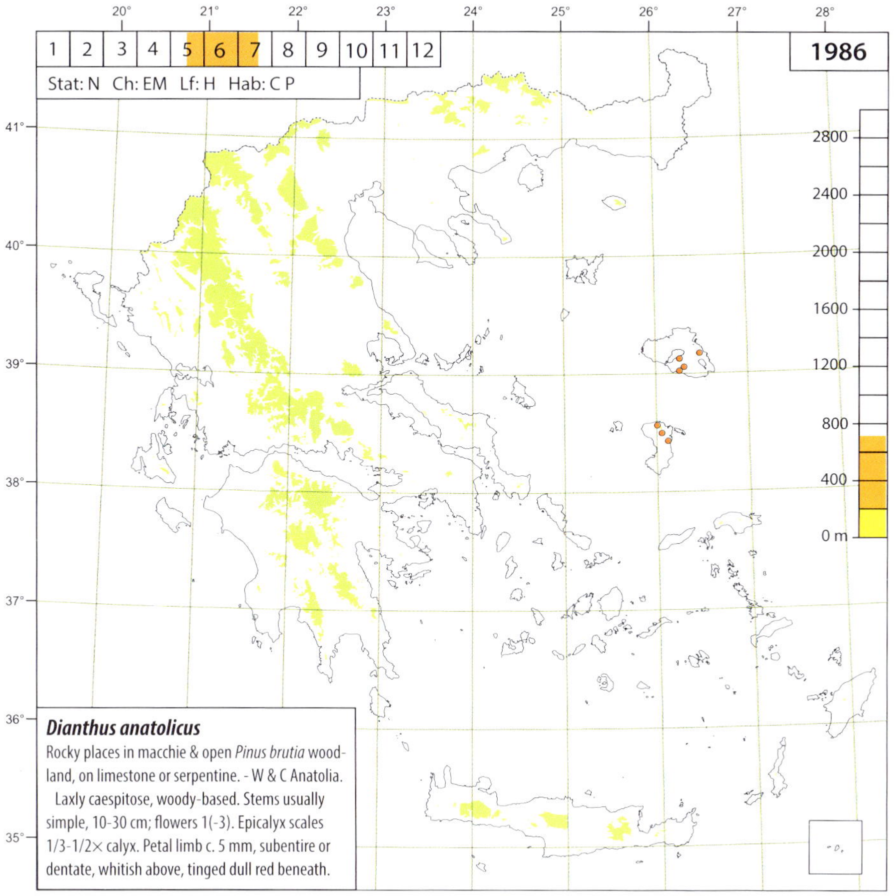

Dianthus anatolicus
Rocky places in macchie & open *Pinus brutia* woodland, on limestone or serpentine. - W & C Anatolia.

Laxly caespitose, woody-based. Stems usually simple, 10-30 cm; flowers 1(-3). Epicalyx scales 1/3-1/2× calyx. Petal limb c. 5 mm, subentire or dentate, whitish above, tinged dull red beneath.

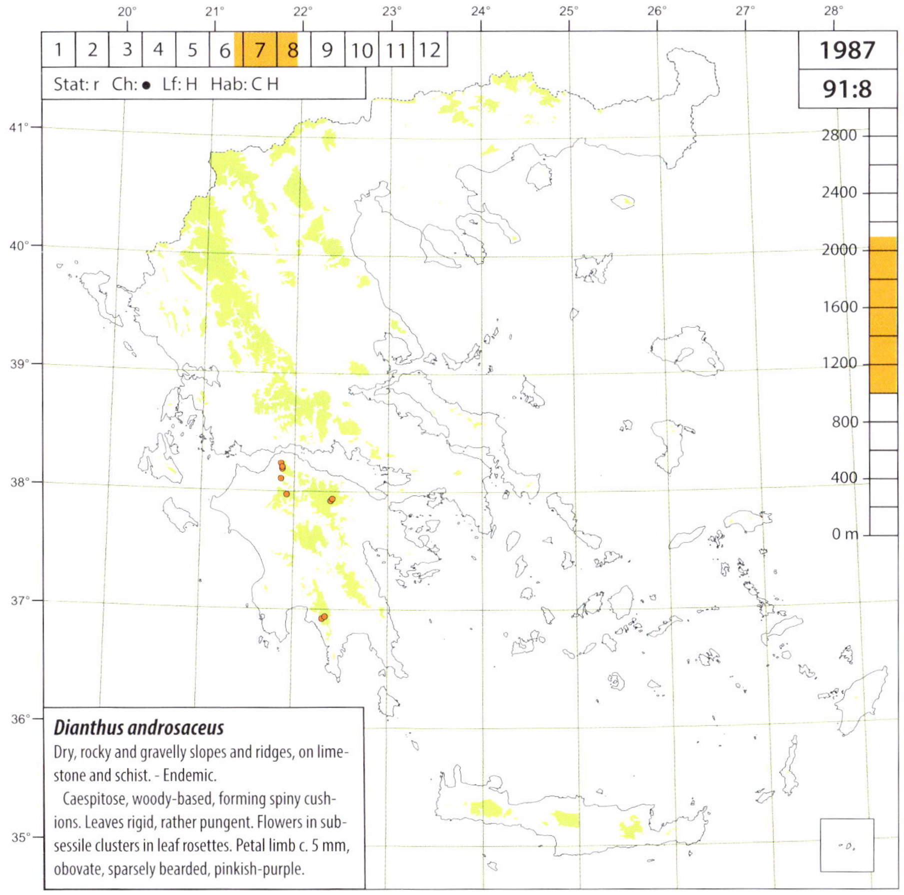

Dianthus androsaceus
Dry, rocky and gravelly slopes and ridges, on limestone and schist. - Endemic.

Caespitose, woody-based, forming spiny cushions. Leaves rigid, rather pungent. Flowers in subsessile clusters in leaf rosettes. Petal limb c. 5 mm, obovate, sparsely bearded, pinkish-purple.

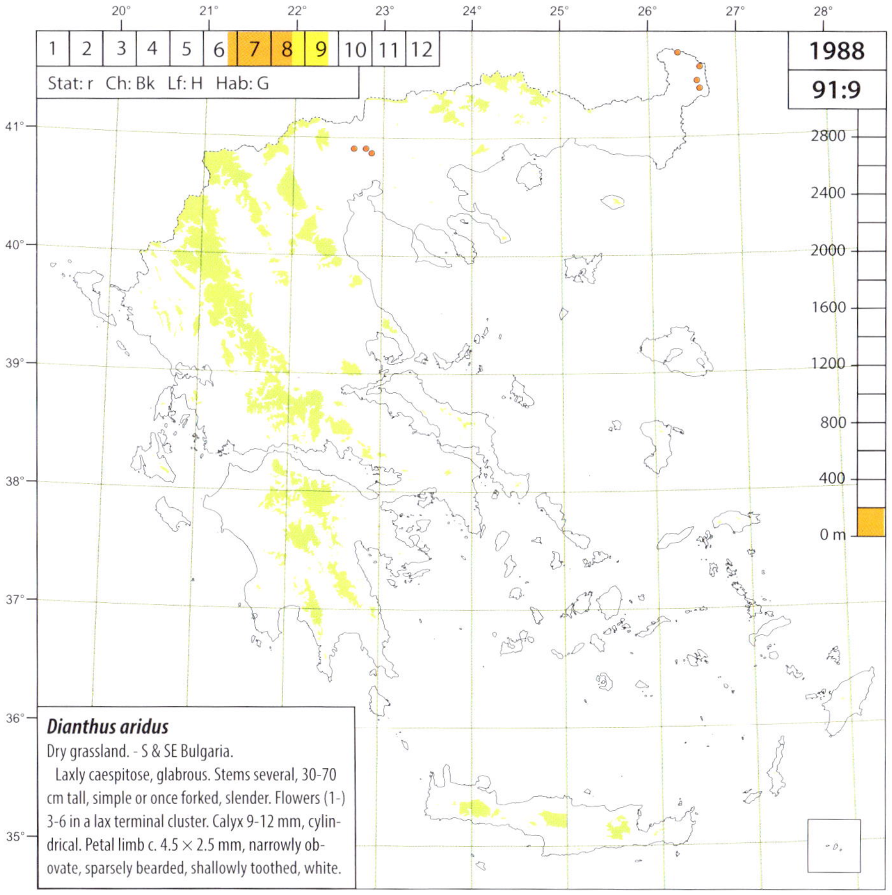

Dianthus aridus
Dry grassland. - S & SE Bulgaria.

Laxly caespitose, glabrous. Stems several, 30-70 cm tall, simple or once forked, slender. Flowers (1-) 3-6 in a lax terminal cluster. Calyx 9-12 mm, cylindrical. Petal limb c. 4.5 × 2.5 mm, narrowly obovate, sparsely bearded, shallowly toothed, white.

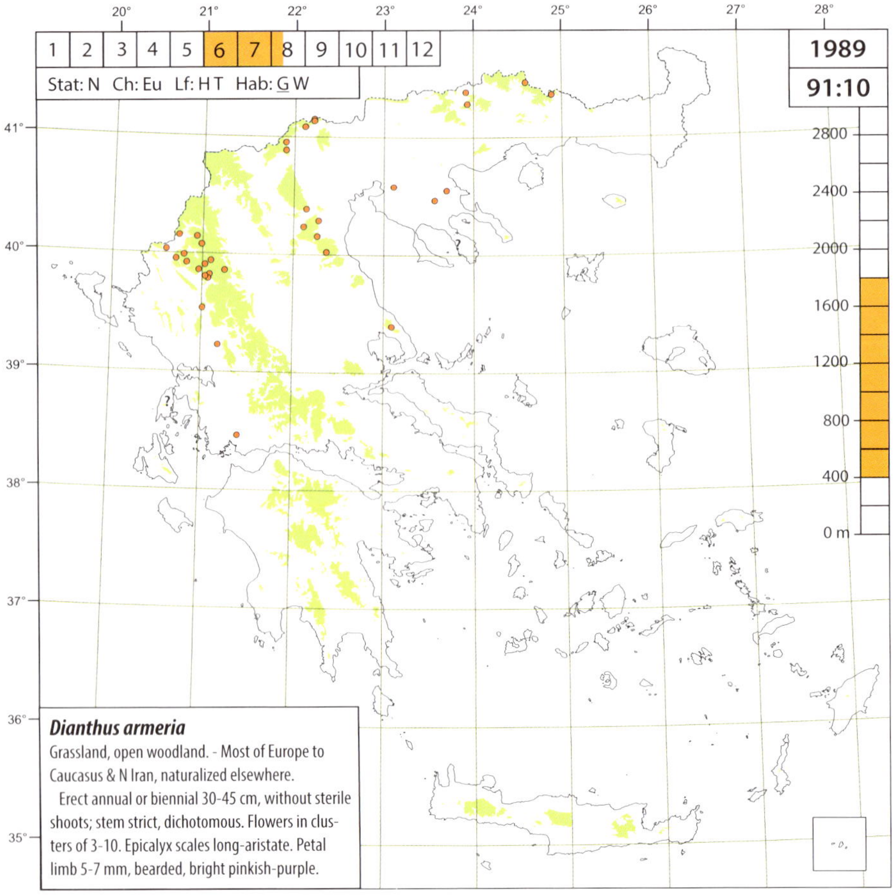

Dianthus armeria
Grassland, open woodland. - Most of Europe to Caucasus & N Iran, naturalized elsewhere.
Erect annual or biennial 30-45 cm, without sterile shoots; stem strict, dichotomous. Flowers in clusters of 3-10. Epicalyx scales long-aristate. Petal limb 5-7 mm, bearded, bright pinkish-purple.

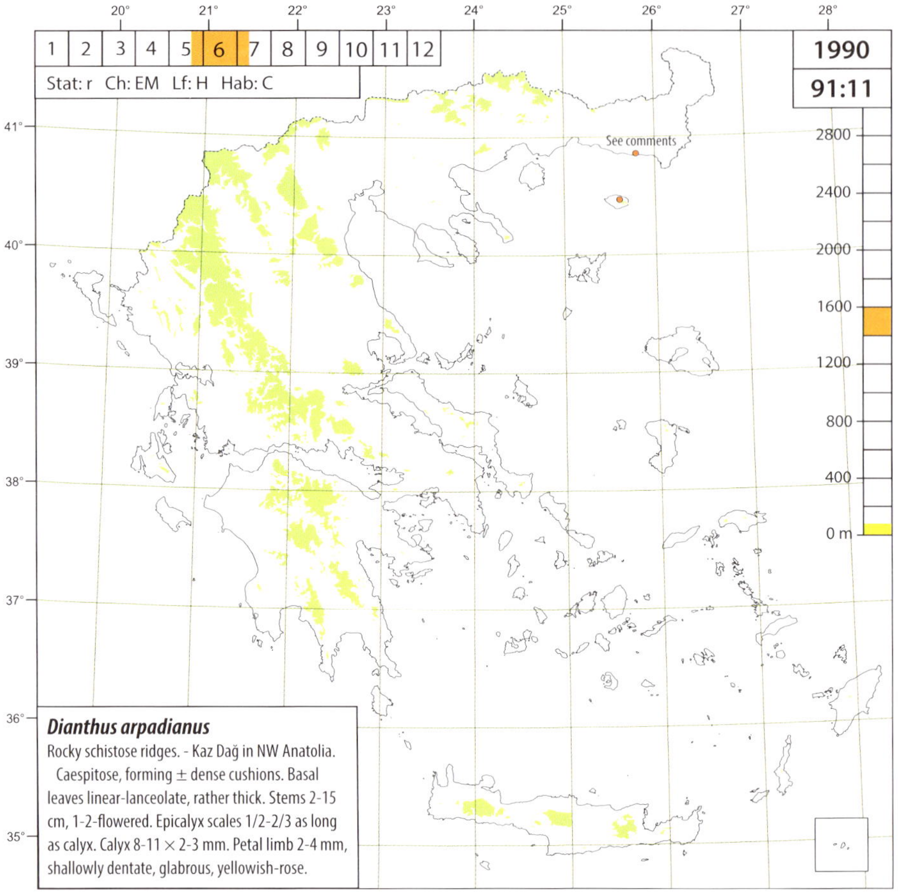

Dianthus arpadianus
Rocky schistose ridges. - Kaz Dağ in NW Anatolia.
Caespitose, forming ± dense cushions. Basal leaves linear-lanceolate, rather thick. Stems 2-15 cm, 1-2-flowered. Epicalyx scales 1/2-2/3 as long as calyx. Calyx 8-11 × 2-3 mm. Petal limb 2-4 mm, shallowly dentate, glabrous, yellowish-rose.

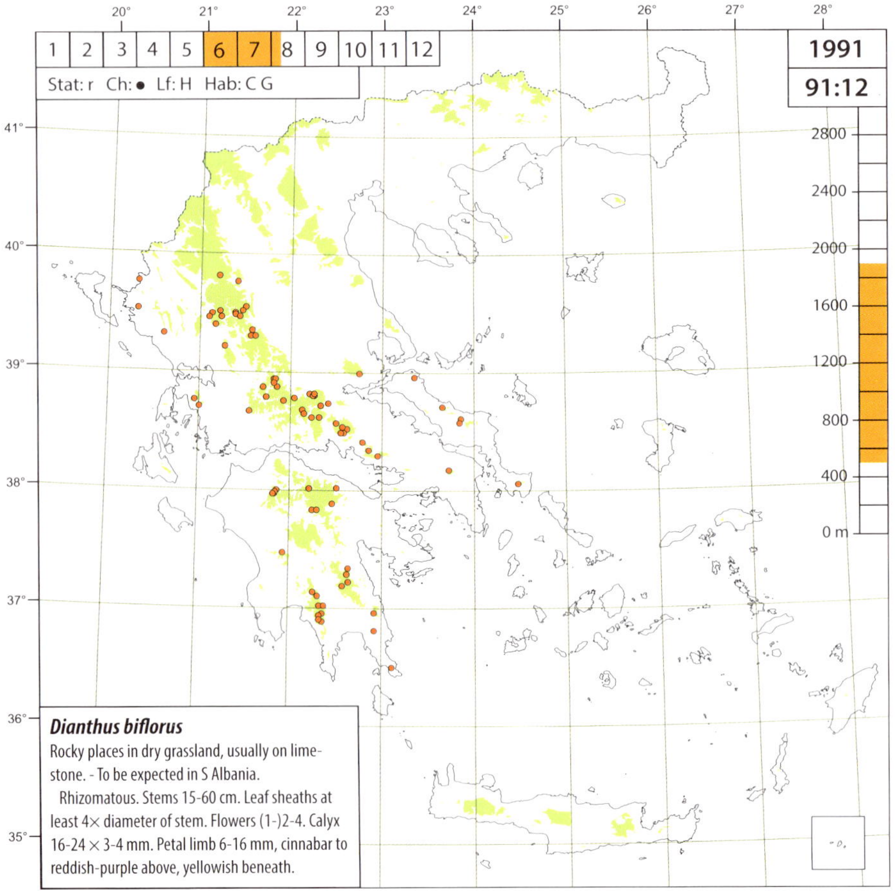

Dianthus biflorus
Rocky places in dry grassland, usually on limestone. - To be expected in S Albania.
Rhizomatous. Stems 15-60 cm. Leaf sheaths at least 4× diameter of stem. Flowers (1-)2-4. Calyx 16-24 × 3-4 mm. Petal limb 6-16 mm, cinnabar to reddish-purple above, yellowish beneath.

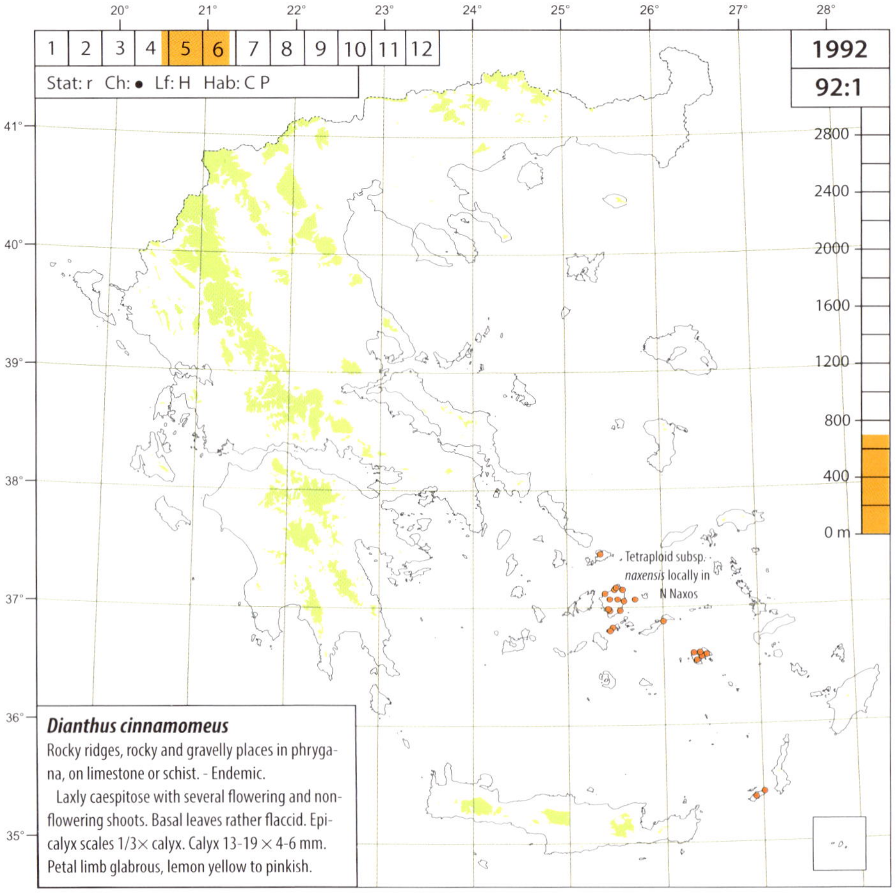

Dianthus cinnamomeus
Rocky ridges, rocky and gravelly places in phrygana, on limestone or schist. - Endemic.
Laxly caespitose with several flowering and non-flowering shoots. Basal leaves rather flaccid. Epicalyx scales 1/3× calyx. Calyx 13-19 × 4-6 mm. Petal limb glabrous, lemon yellow to pinkish.

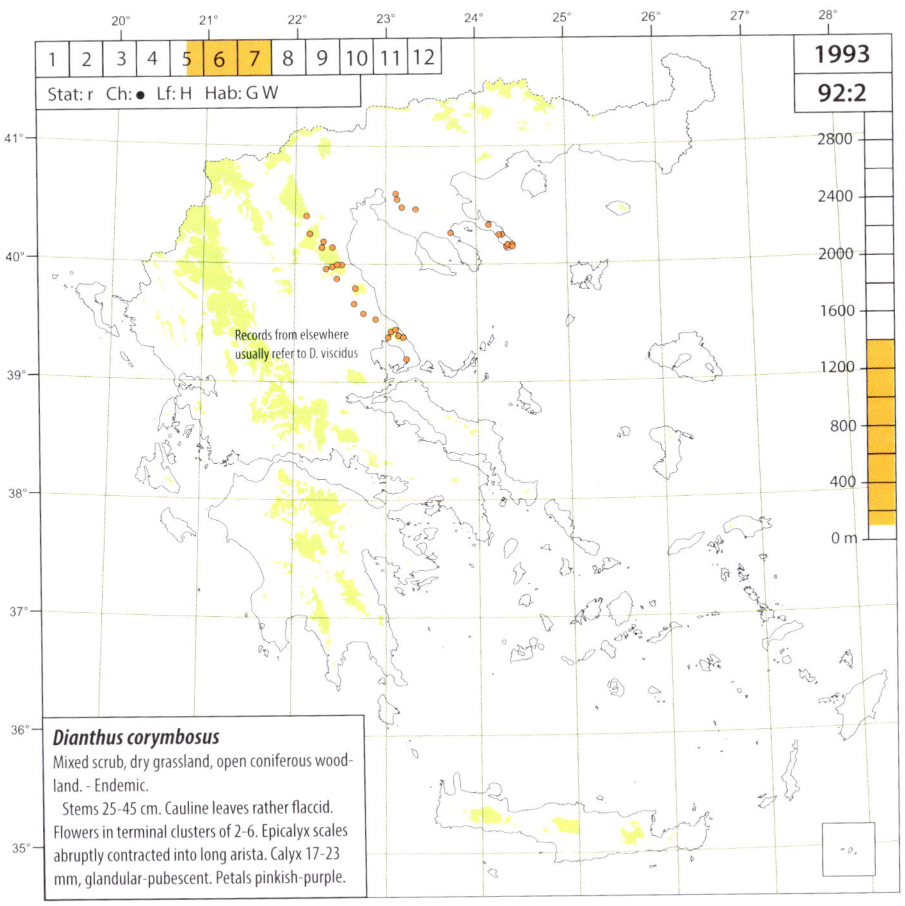

Dianthus corymbosus

Mixed scrub, dry grassland, open coniferous woodland. - Endemic.

Stems 25-45 cm. Cauline leaves rather flaccid. Flowers in terminal clusters of 2-6. Epicalyx scales abruptly contracted into long arista. Calyx 17-23 mm, glandular-pubescent. Petals pinkish-purple.

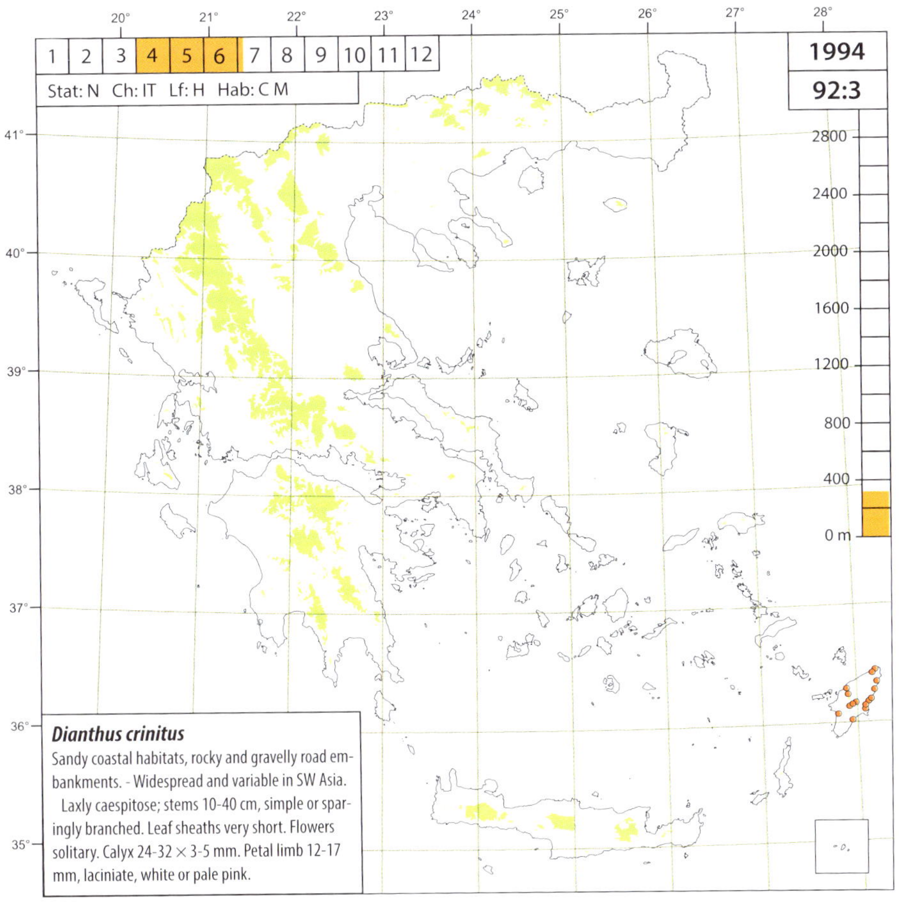

Dianthus crinitus

Sandy coastal habitats, rocky and gravelly road embankments. - Widespread and variable in SW Asia.

Laxly caespitose; stems 10-40 cm, simple or sparingly branched. Leaf sheaths very short. Flowers solitary. Calyx 24-32 × 3-5 mm. Petal limb 12-17 mm, laciniate, white or pale pink.

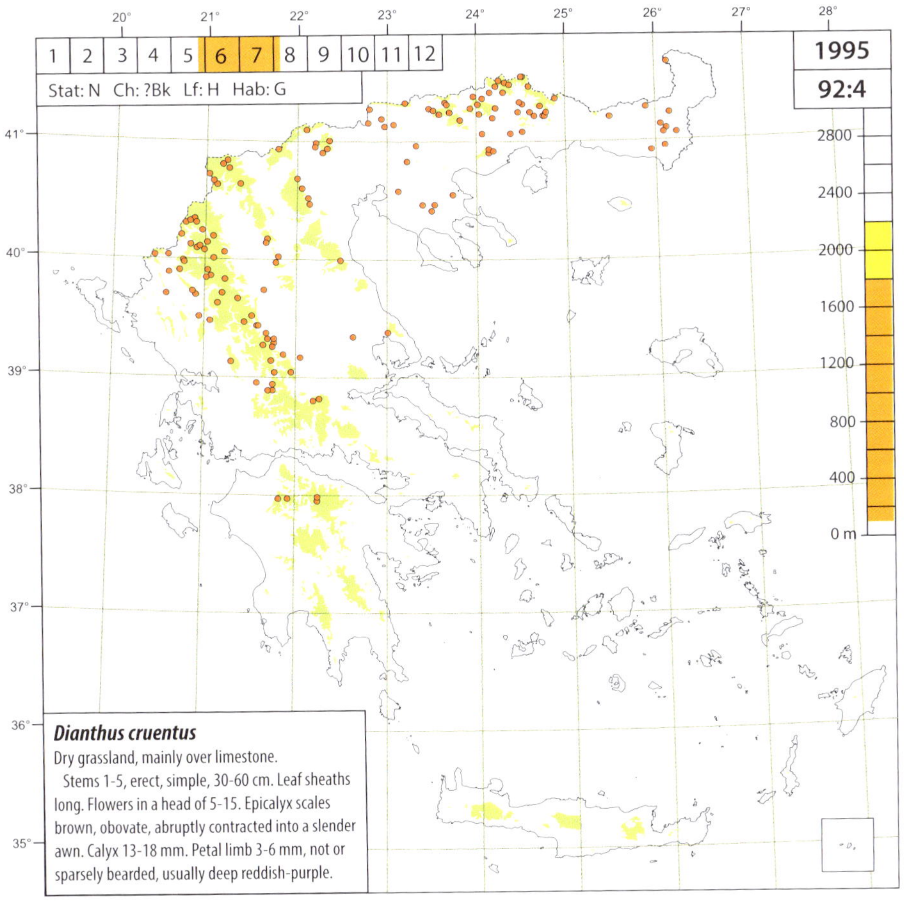

Dianthus cruentus

Dry grassland, mainly over limestone.

Stems 1-5, erect, simple, 30-60 cm. Leaf sheaths long. Flowers in a head of 5-15. Epicalyx scales brown, obovate, abruptly contracted into a slender awn. Calyx 13-18 mm. Petal limb 3-6 mm, not or sparsely bearded, usually deep reddish-purple.

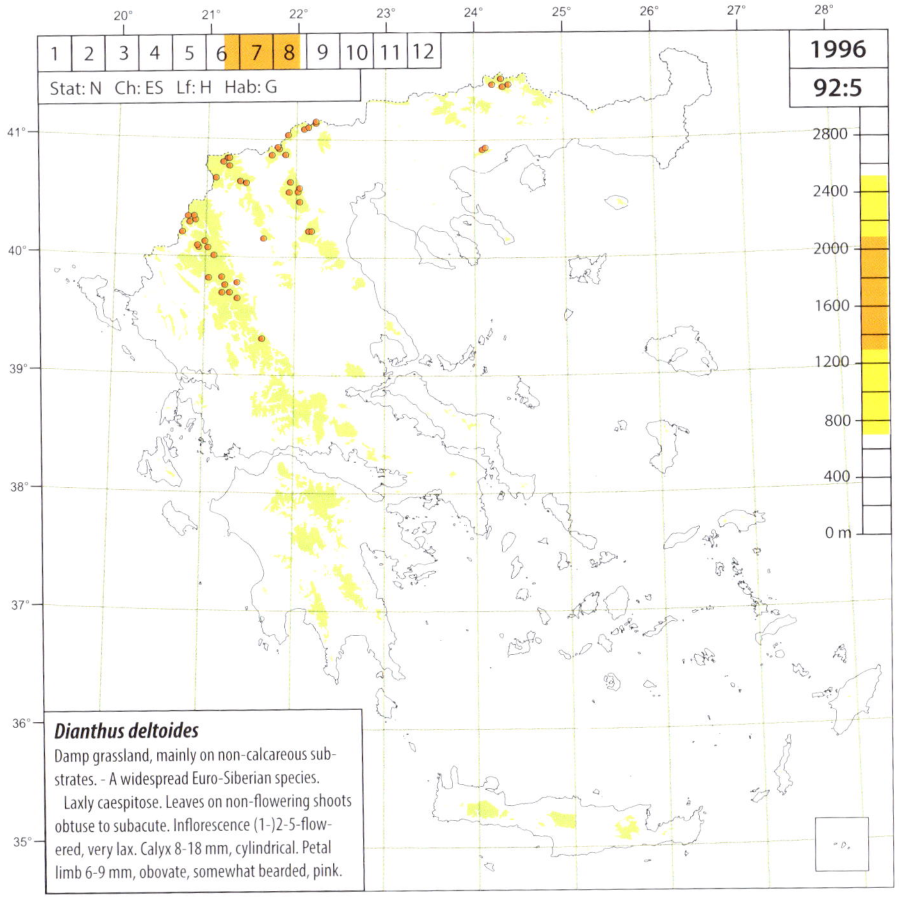

Dianthus deltoides

Damp grassland, mainly on non-calcareous substrates. - A widespread Euro-Siberian species.

Laxly caespitose. Leaves on non-flowering shoots obtuse to subacute. Inflorescence (1-)2-5-flowered, very lax. Calyx 8-18 mm, cylindrical. Petal limb 6-9 mm, obovate, somewhat bearded, pink.

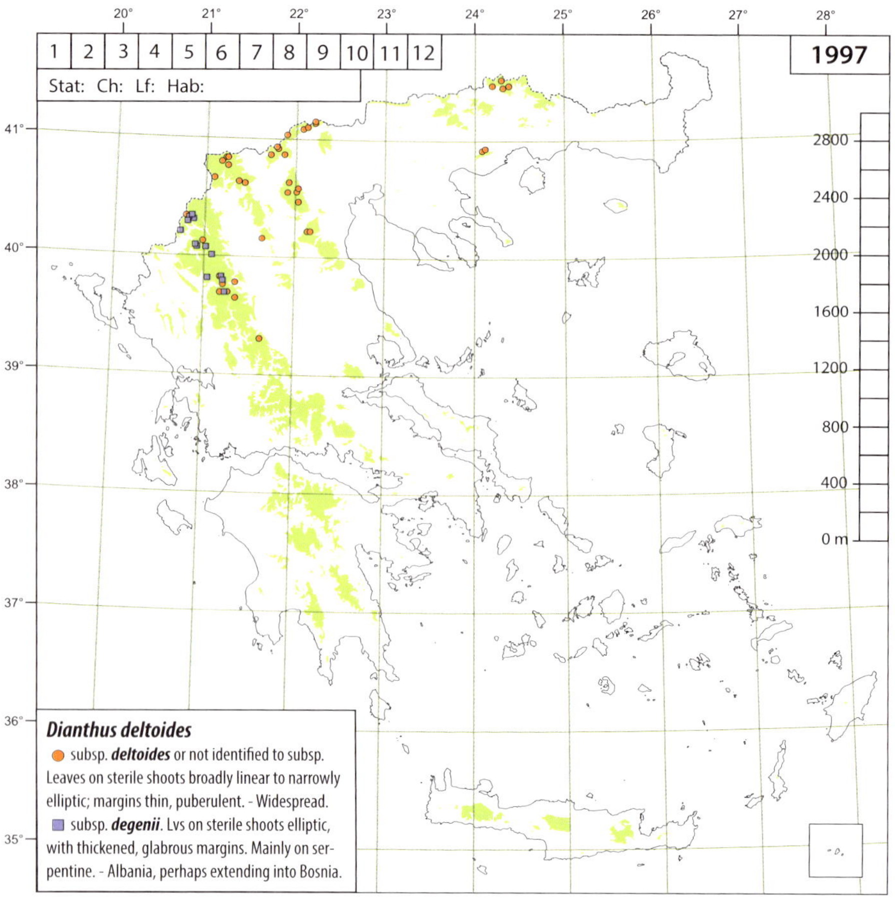

Dianthus deltoides
subsp. ***deltoides*** or not identified to subsp. Leaves on sterile shoots broadly linear to narrowly elliptic; margins thin, puberulent. - Widespread.
subsp. ***degenii***. Lvs on sterile shoots elliptic, with thickened, glabrous margins. Mainly on serpentine. - Albania, perhaps extending into Bosnia.

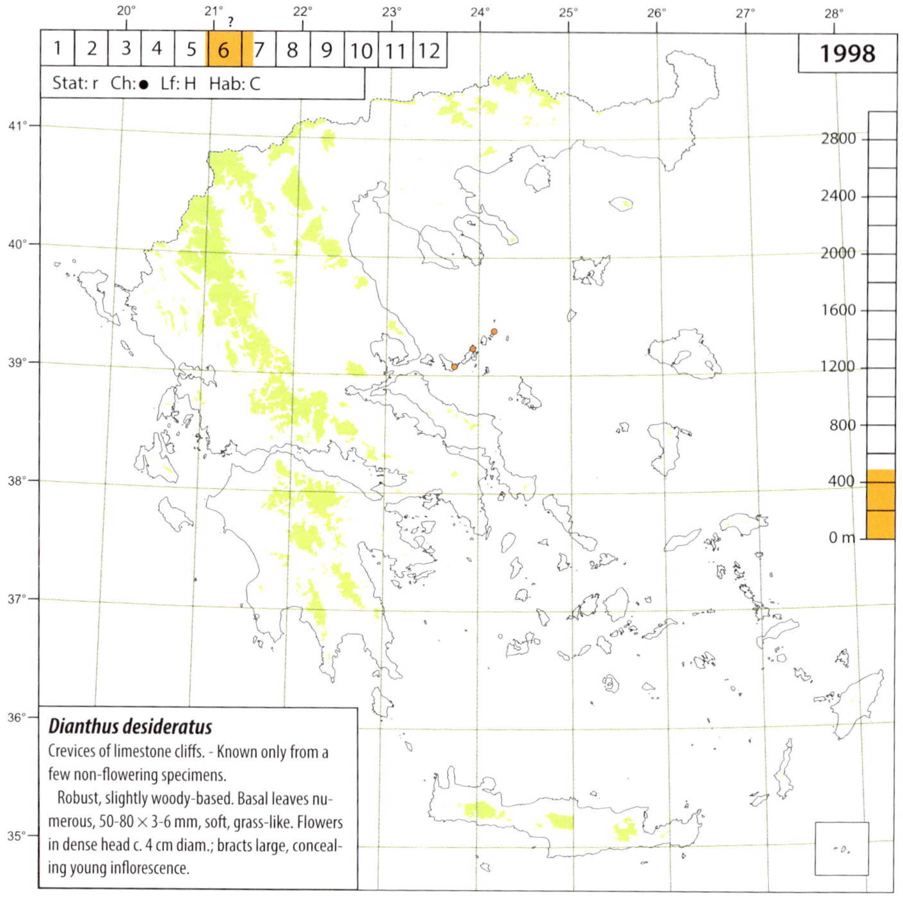

Dianthus desideratus
Crevices of limestone cliffs. - Known only from a few non-flowering specimens.
Robust, slightly woody-based. Basal leaves numerous, 50-80 × 3-6 mm, soft, grass-like. Flowers in dense head c. 4 cm diam.; bracts large, concealing young inflorescence.

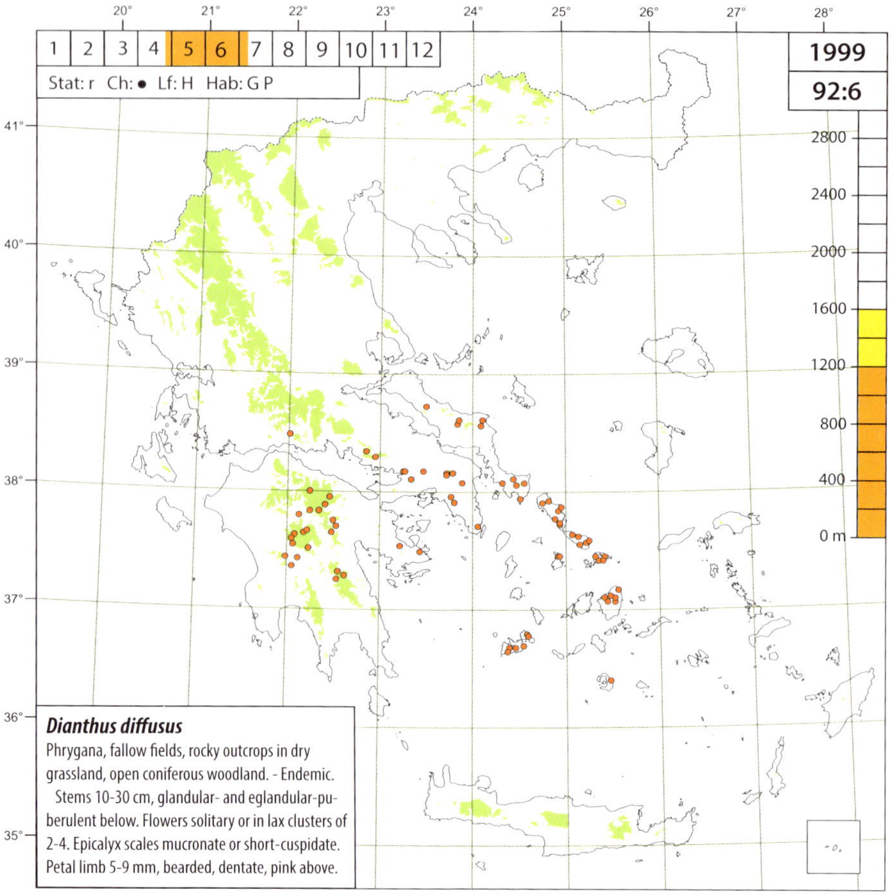

Dianthus diffusus
Phrygana, fallow fields, rocky outcrops in dry grassland, open coniferous woodland. - Endemic.
Stems 10-30 cm, glandular- and eglandular-puberulent below. Flowers solitary or in lax clusters of 2-4. Epicalyx scales mucronate or short-cuspidate. Petal limb 5-9 mm, bearded, dentate, pink above.

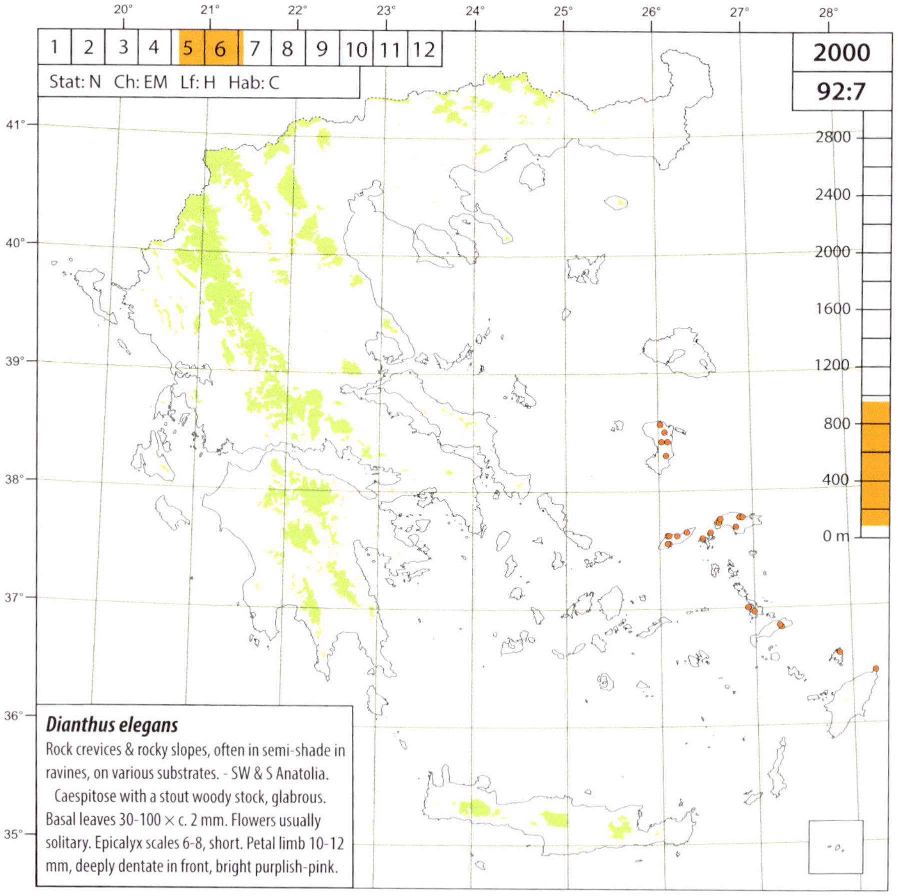

Dianthus elegans
Rock crevices & rocky slopes, often in semi-shade in ravines, on various substrates. - SW & S Anatolia.
Caespitose with a stout woody stock, glabrous. Basal leaves 30-100 × c. 2 mm. Flowers usually solitary. Epicalyx scales 6-8, short. Petal limb 10-12 mm, deeply dentate in front, bright purplish-pink.

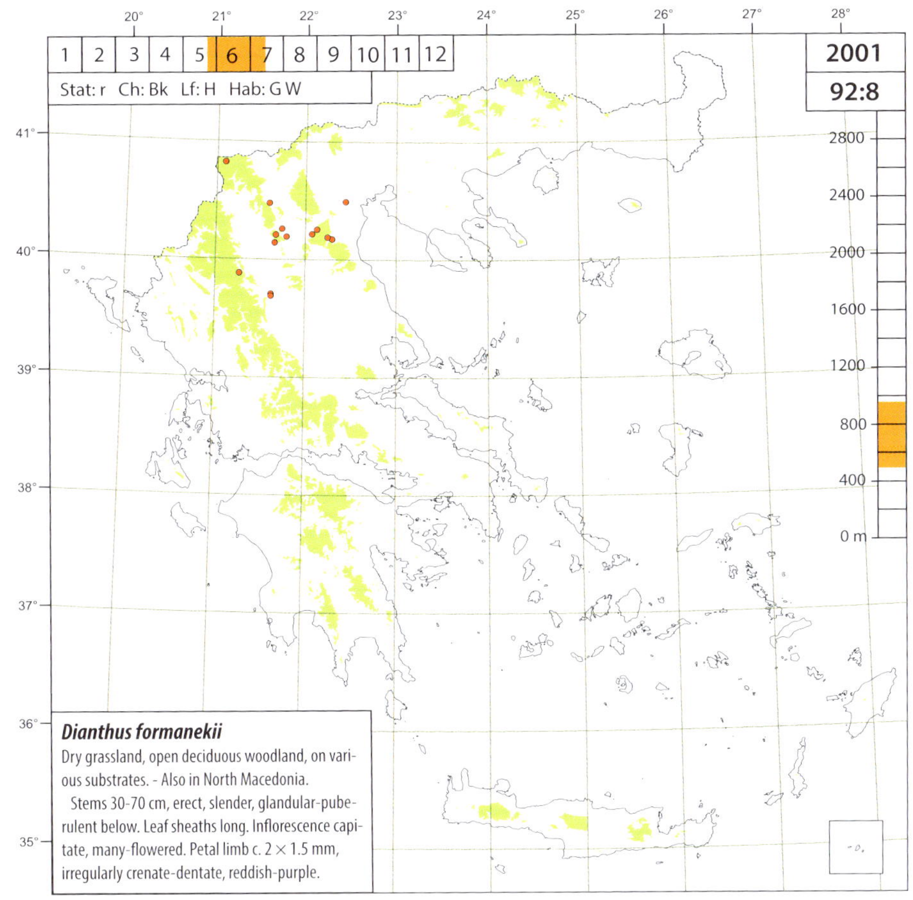

Dianthus formanekii

Dry grassland, open deciduous woodland, on various substrates. - Also in North Macedonia.

Stems 30-70 cm, erect, slender, glandular-puberulent below. Leaf sheaths long. Inflorescence capitate, many-flowered. Petal limb c. 2 × 1.5 mm, irregularly crenate-dentate, reddish-purple.

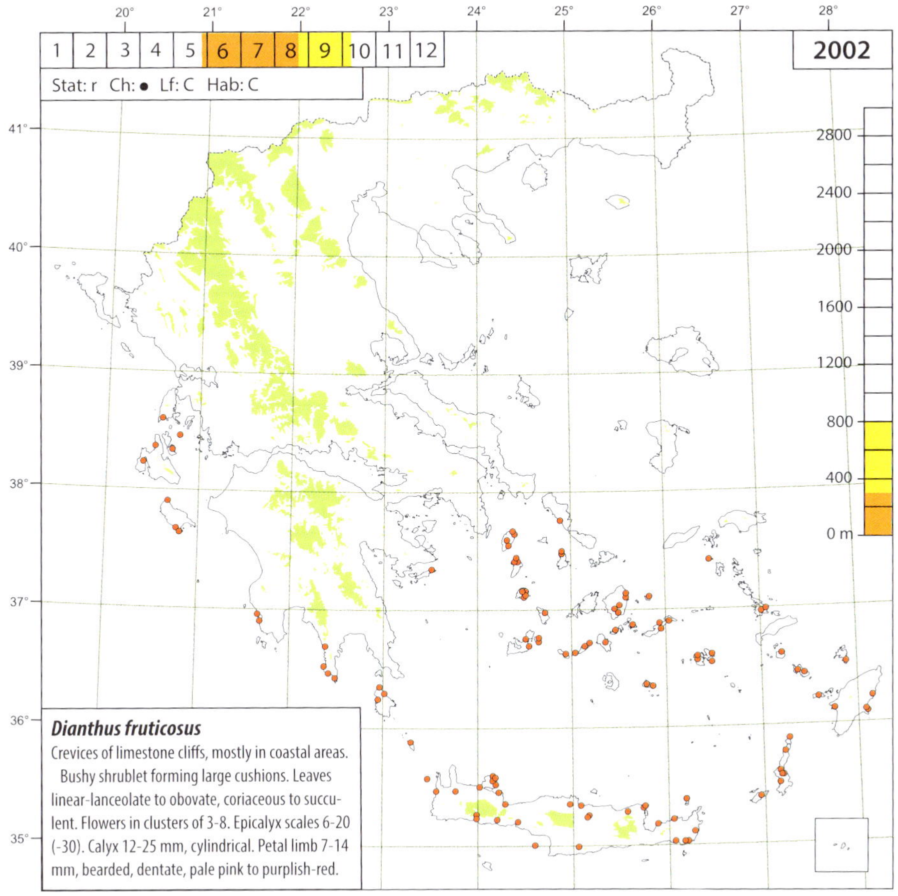

Dianthus fruticosus

Crevices of limestone cliffs, mostly in coastal areas.

Bushy shrublet forming large cushions. Leaves linear-lanceolate to obovate, coriaceous to succulent. Flowers in clusters of 3-8. Epicalyx scales 6-20 (-30). Calyx 12-25 mm, cylindrical. Petal limb 7-14 mm, bearded, dentate, pale pink to purplish-red.

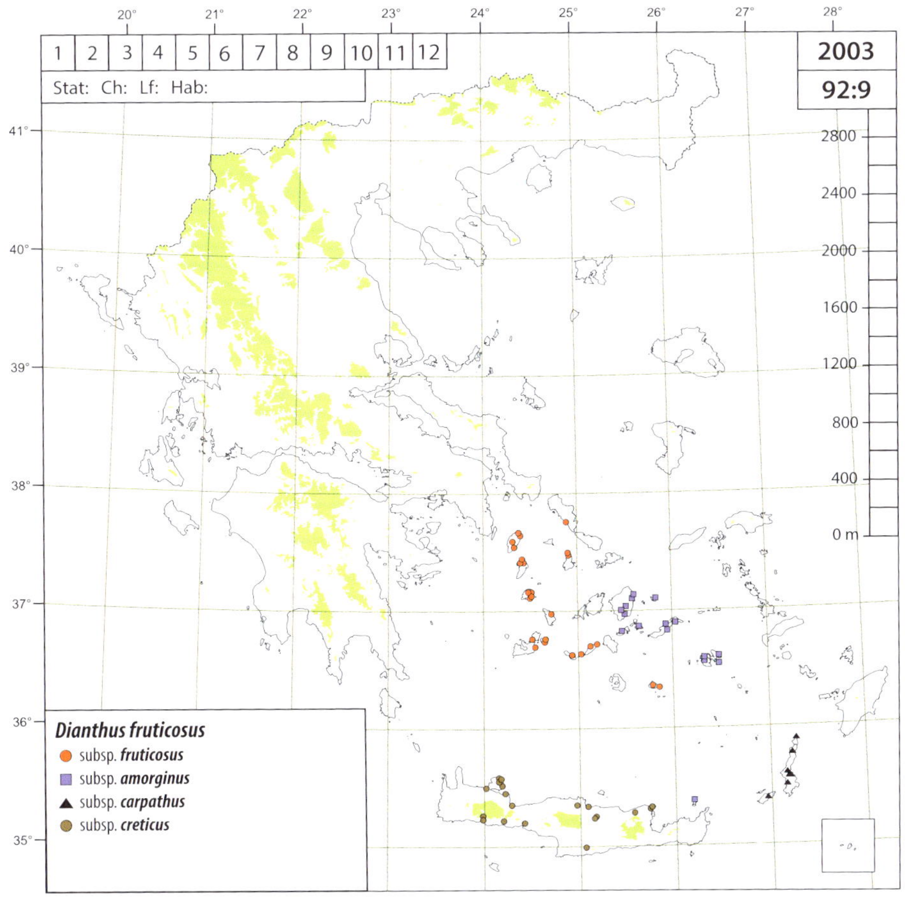

Dianthus fruticosus

- subsp. ***fruticosus***
- subsp. ***amorginus***
- subsp. ***carpathus***
- subsp. ***creticus***

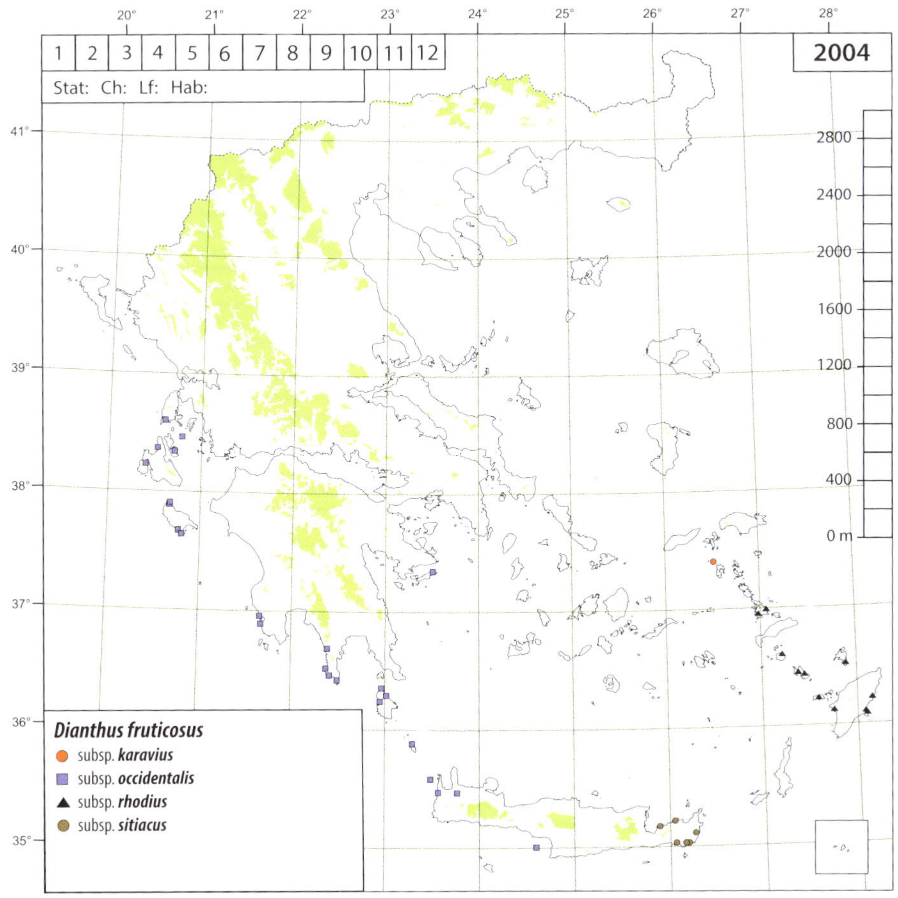

Dianthus fruticosus

- subsp. ***karavius***
- subsp. ***occidentalis***
- subsp. ***rhodius***
- subsp. ***sitiacus***

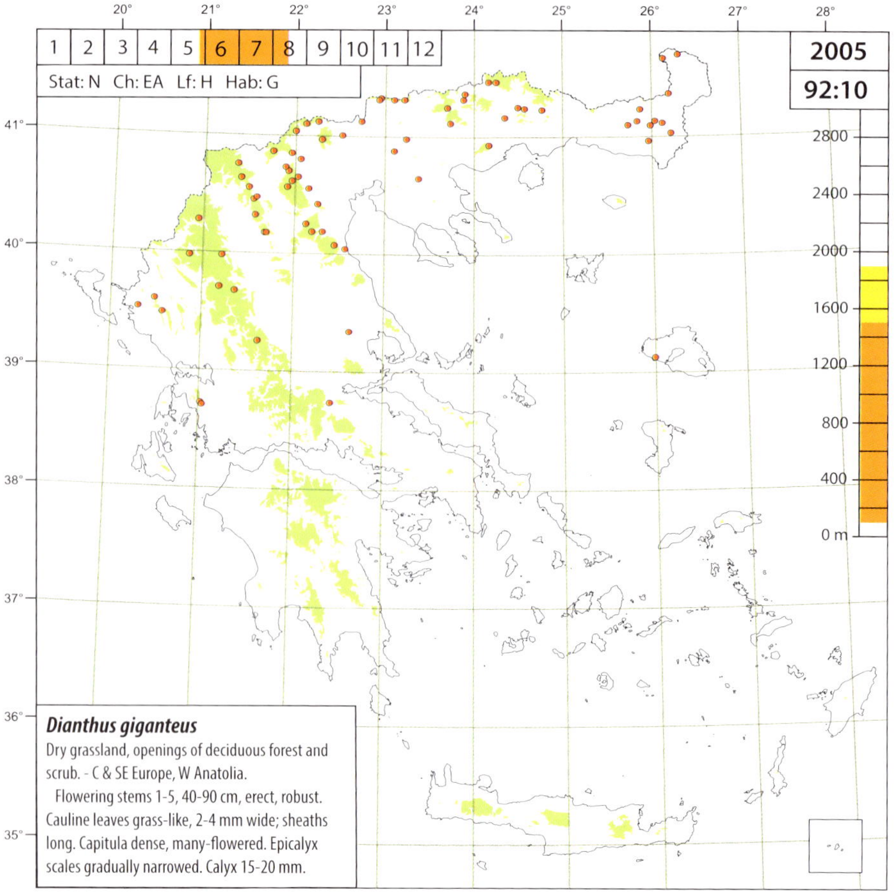

Dianthus giganteus

Dry grassland, openings of deciduous forest and scrub. - C & SE Europe, W Anatolia.

Flowering stems 1-5, 40-90 cm, erect, robust. Cauline leaves grass-like, 2-4 mm wide; sheaths long. Capitula dense, many-flowered. Epicalyx scales gradually narrowed. Calyx 15-20 mm.

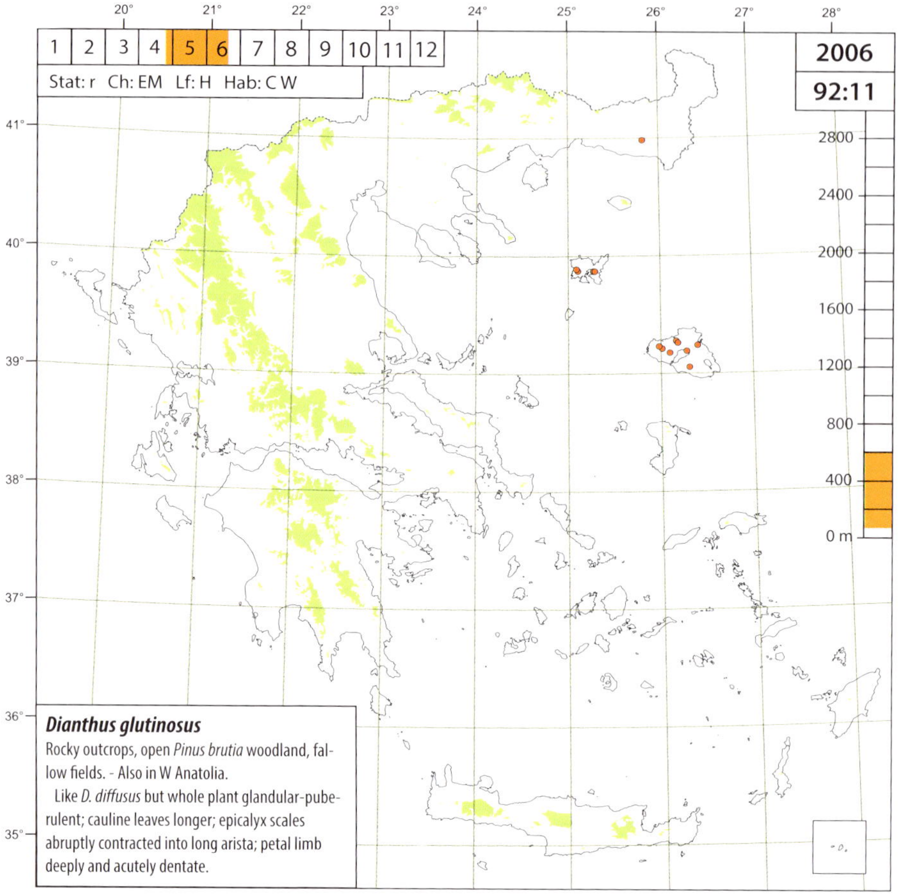

Dianthus glutinosus

Rocky outcrops, open *Pinus brutia* woodland, fallow fields. - Also in W Anatolia.

Like *D. diffusus* but whole plant glandular-pubescent; cauline leaves longer; epicalyx scales abruptly contracted into long arista; petal limb deeply and acutely dentate.

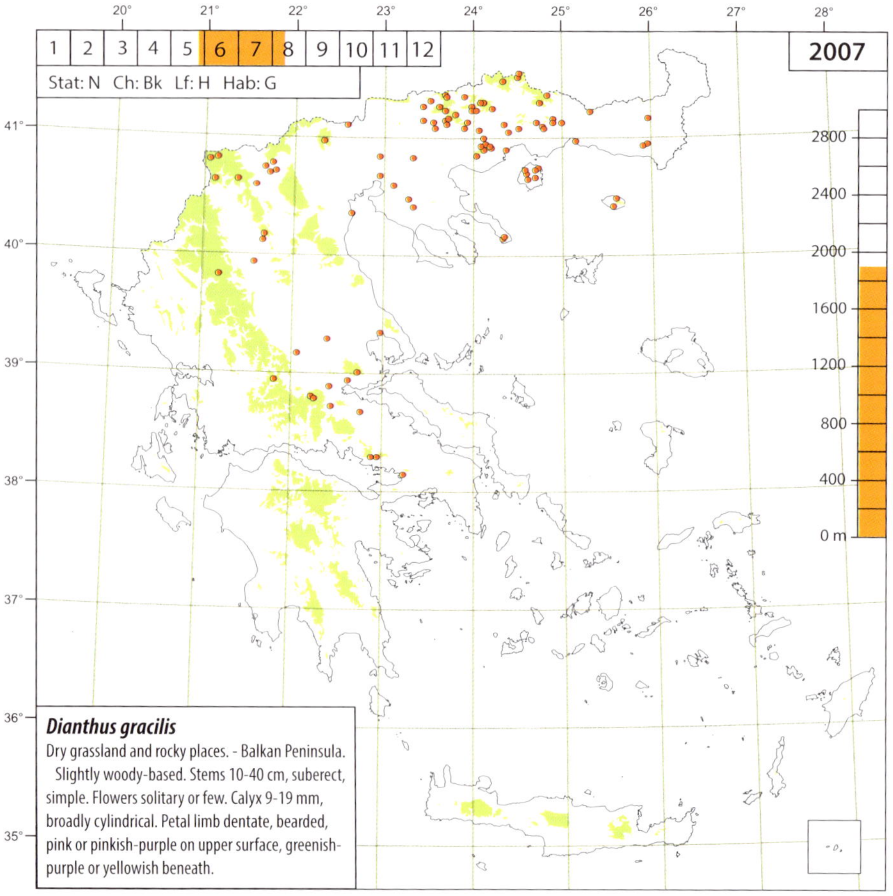

Dianthus gracilis

Dry grassland and rocky places. - Balkan Peninsula.

Slightly woody-based. Stems 10-40 cm, suberect, simple. Flowers solitary or few. Calyx 9-19 mm, broadly cylindrical. Petal limb dentate, bearded, pink or pinkish-purple on upper surface, greenish-purple or yellowish beneath.

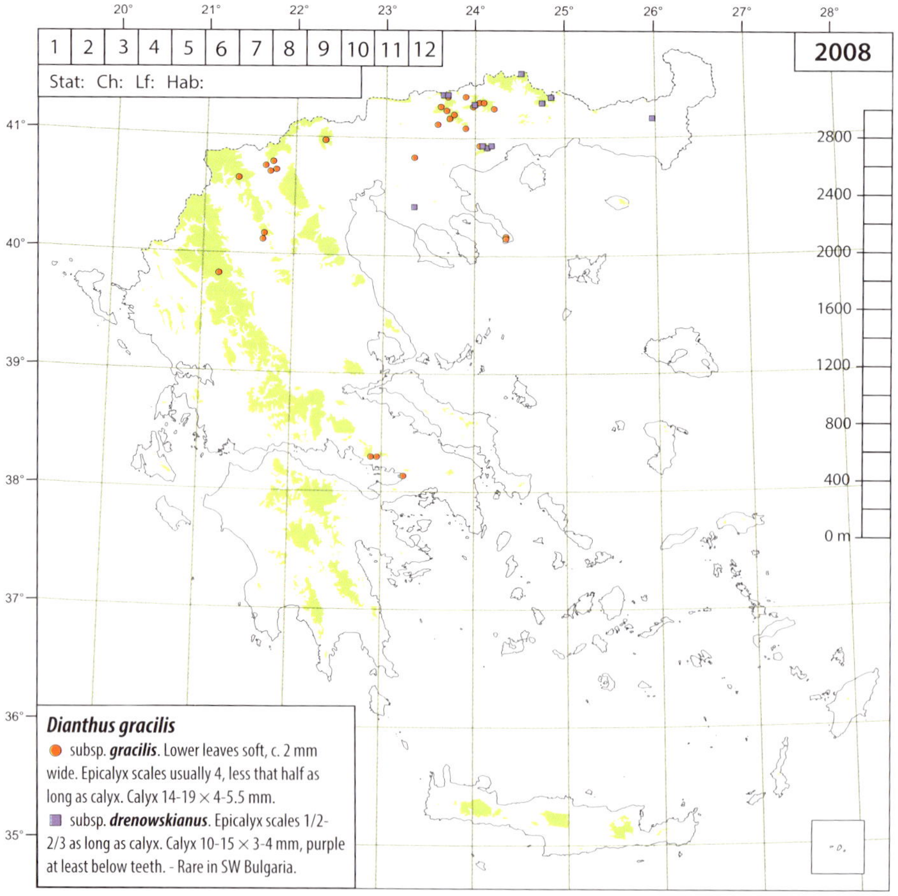

Dianthus gracilis

● subsp. ***gracilis***. Lower leaves soft, c. 2 mm wide. Epicalyx scales usually 4, less that half as long as calyx. Calyx 14-19 × 4-5.5 mm.

■ subsp. ***drenowskianus***. Epicalyx scales 1/2-2/3 as long as calyx. Calyx 10-15 × 3-4 mm, purple at least below teeth. - Rare in SW Bulgaria.

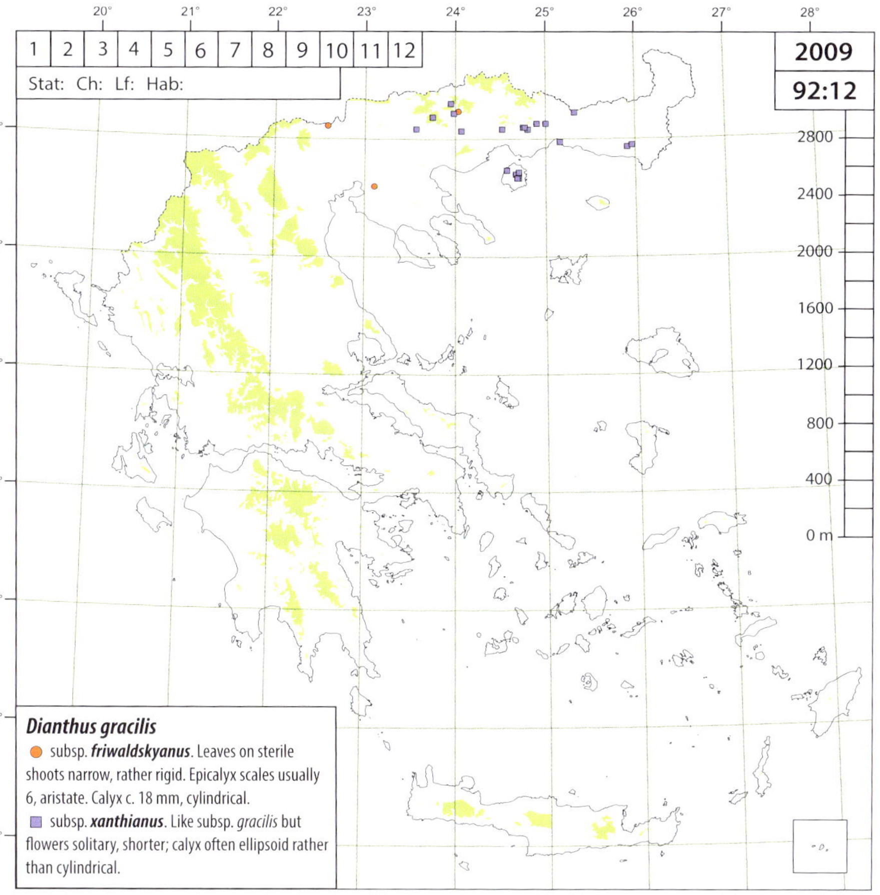

Dianthus gracilis

● subsp. ***friwaldskyanus***. Leaves on sterile shoots narrow, rather rigid. Epicalyx scales usually 6, aristate. Calyx c. 18 mm, cylindrical.

■ subsp. ***xanthianus***. Like subsp. *gracilis* but flowers solitary, shorter; calyx often ellipsoid rather than cylindrical.

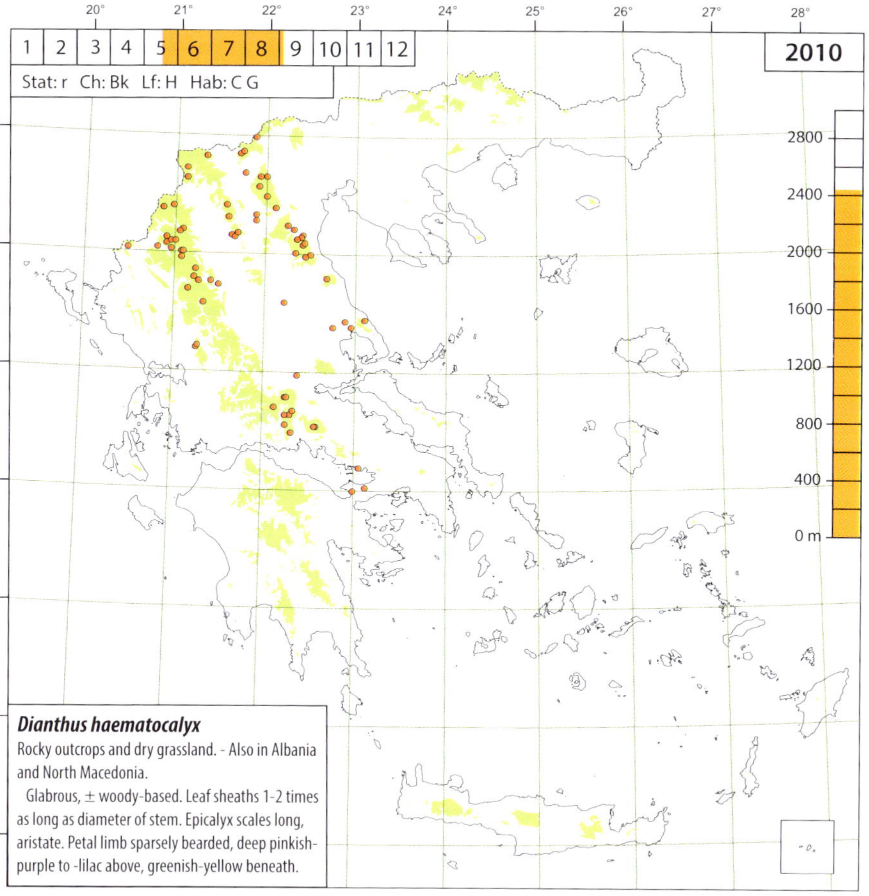

Dianthus haematocalyx

Rocky outcrops and dry grassland. - Also in Albania and North Macedonia.

Glabrous, ± woody-based. Leaf sheaths 1-2 times as long as diameter of stem. Epicalyx scales long, aristate. Petal limb sparsely bearded, deep pinkish-purple to -lilac above, greenish-yellow beneath.

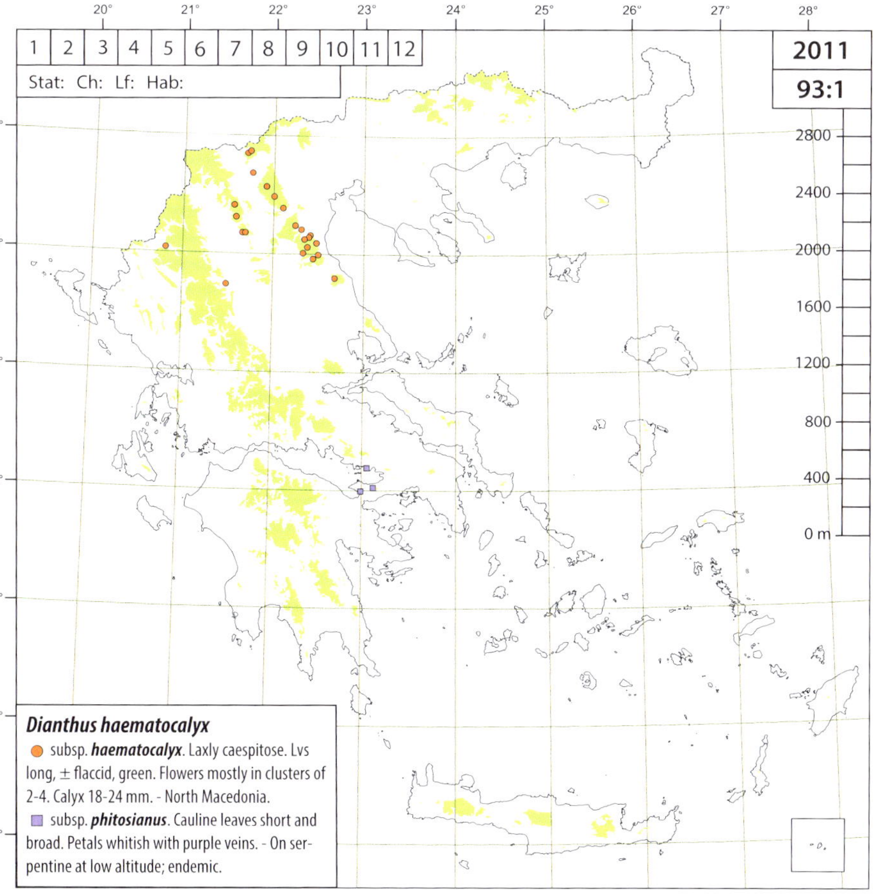

Dianthus haematocalyx

● subsp. ***haematocalyx***. Laxly caespitose. Lvs long, ± flaccid, green. Flowers mostly in clusters of 2-4. Calyx 18-24 mm. - North Macedonia.

■ subsp. ***phitosianus***. Cauline leaves short and broad. Petals whitish with purple veins. - On serpentine at low altitude; endemic.

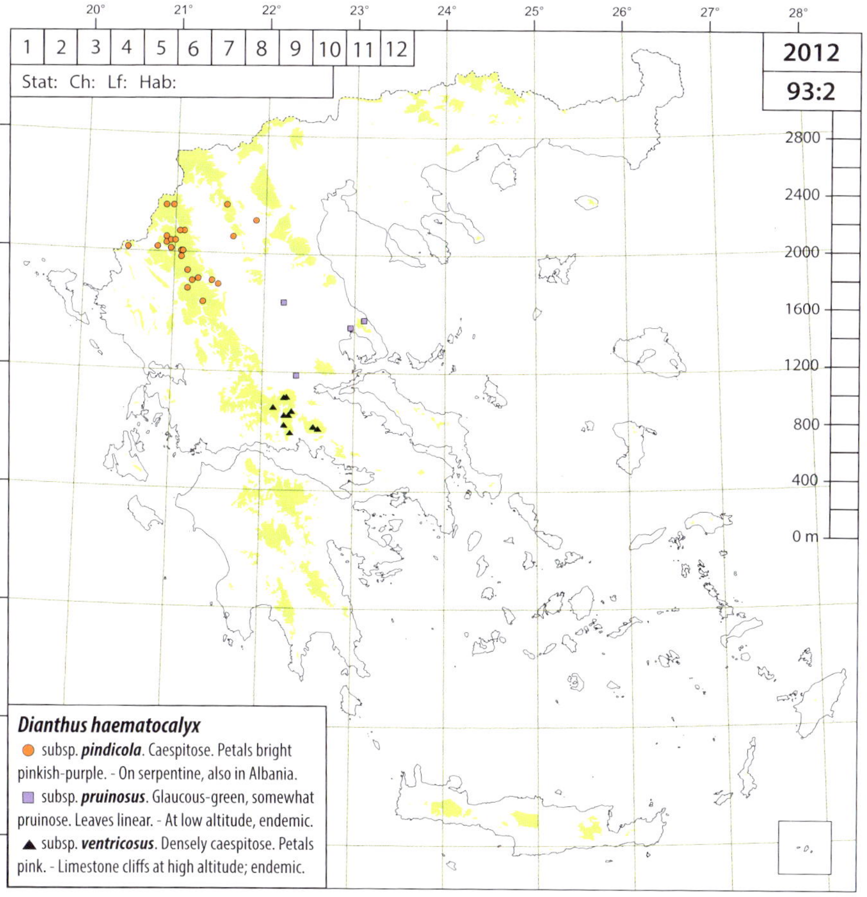

Dianthus haematocalyx

● subsp. ***pindicola***. Caespitose. Petals bright pinkish-purple. - On serpentine, also in Albania.

■ subsp. ***pruinosus***. Glaucous-green, somewhat pruinose. Leaves linear. - At low altitude, endemic.

▲ subsp. ***ventricosus***. Densely caespitose. Petals pink. - Limestone cliffs at high altitude; endemic.

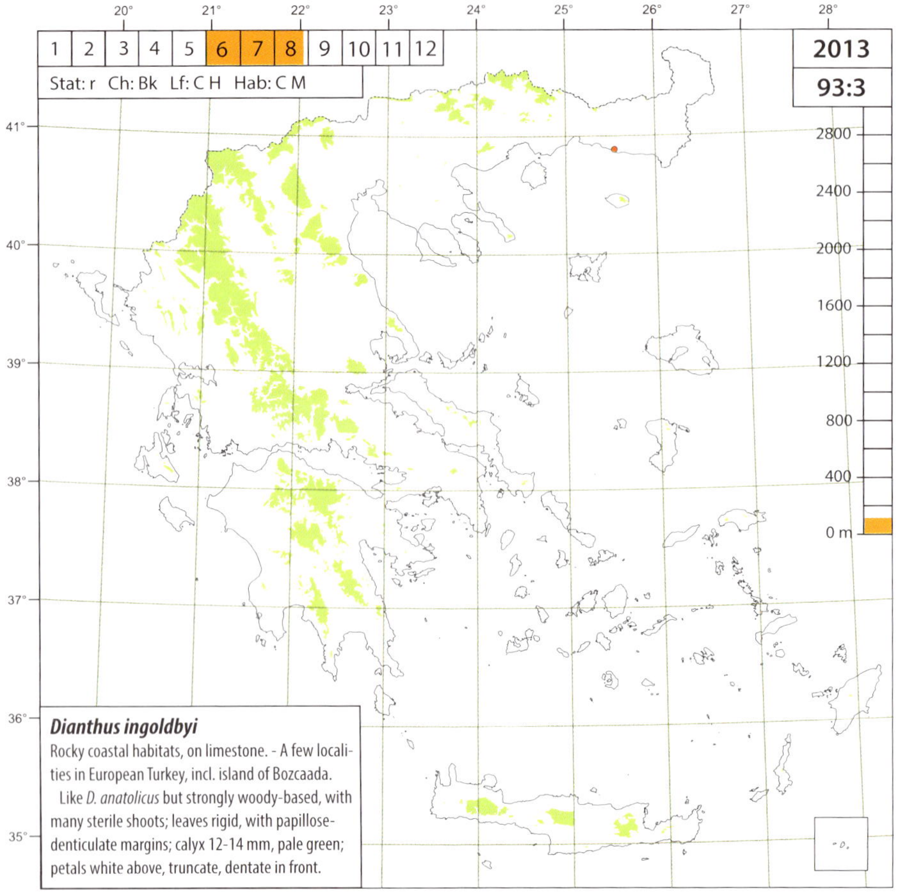

Dianthus ingoldbyi

Rocky coastal habitats, on limestone. - A few localities in European Turkey, incl. island of Bozcaada.

Like *D. anatolicus* but strongly woody-based, with many sterile shoots; leaves rigid, with papillose-denticulate margins; calyx 12-14 mm, pale green; petals white above, truncate, dentate in front.

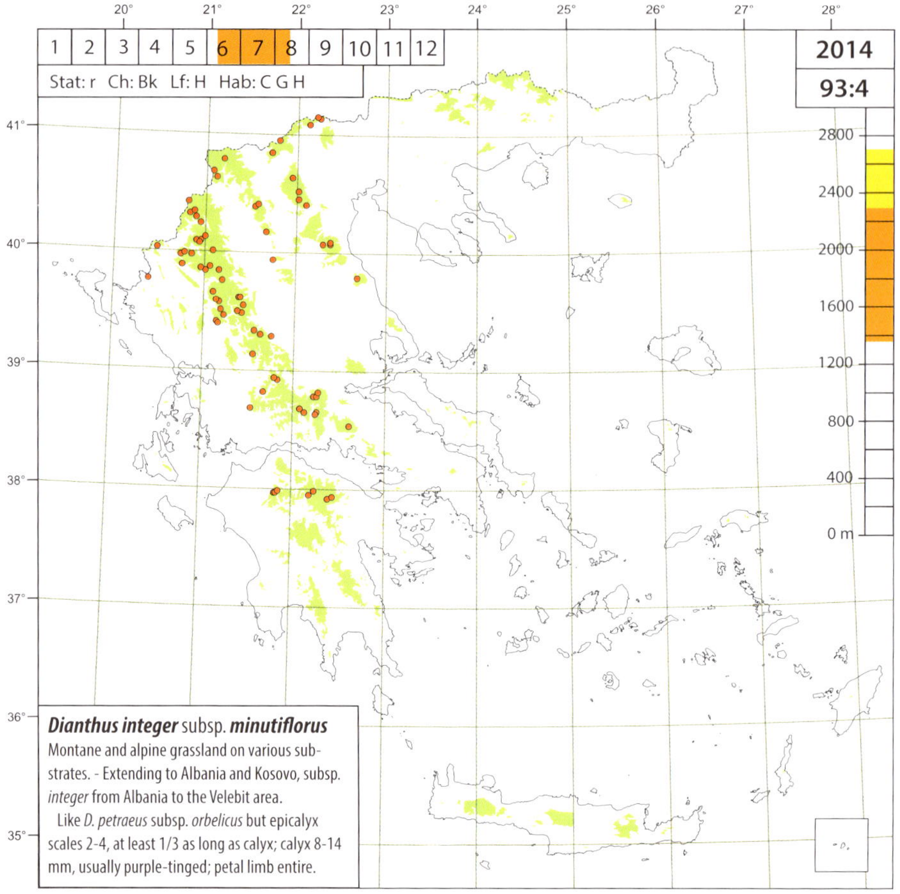

Dianthus integer subsp. *minutiflorus*

Montane and alpine grassland on various substrates. - Extending to Albania and Kosovo, subsp. *integer* from Albania to the Velebit area.

Like *D. petraeus* subsp. *orbelicus* but epicalyx scales 2-4, at least 1/3 as long as calyx; calyx 8-14 mm, usually purple-tinged; petal limb entire.

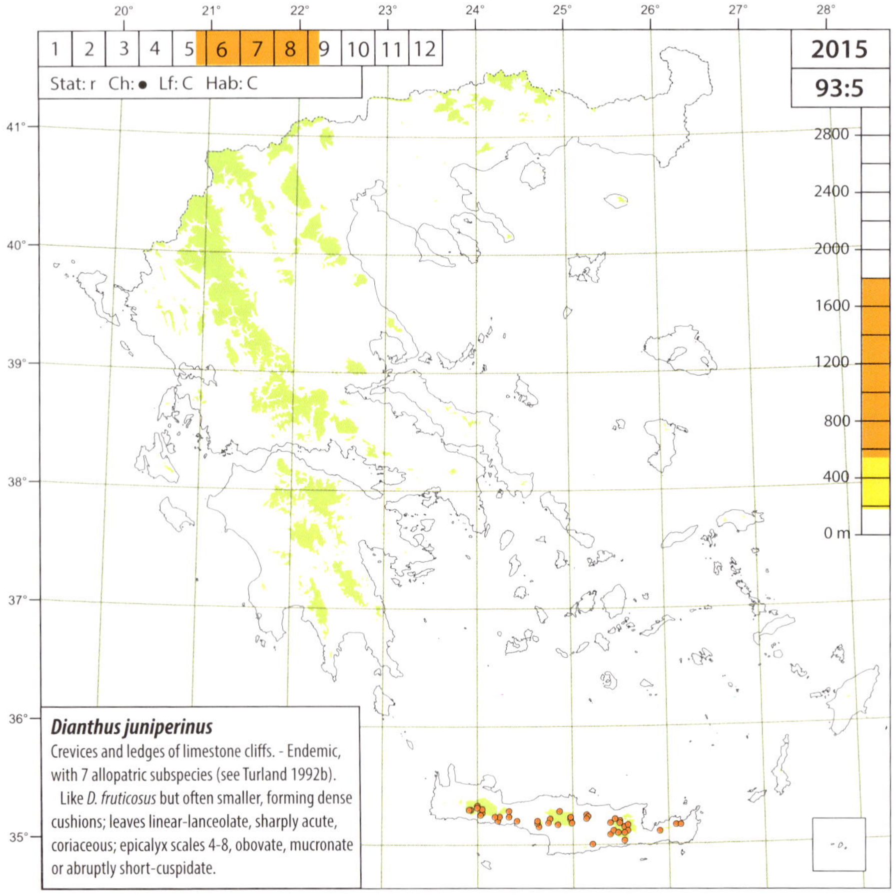

Dianthus juniperinus

Crevices and ledges of limestone cliffs. - Endemic, with 7 allopatric subspecies (see Turland 1992b).

Like *D. fruticosus* but often smaller, forming dense cushions; leaves linear-lanceolate, sharply acute, coriaceous; epicalyx scales 4-8, obovate, mucronate or abruptly short-cuspidate.

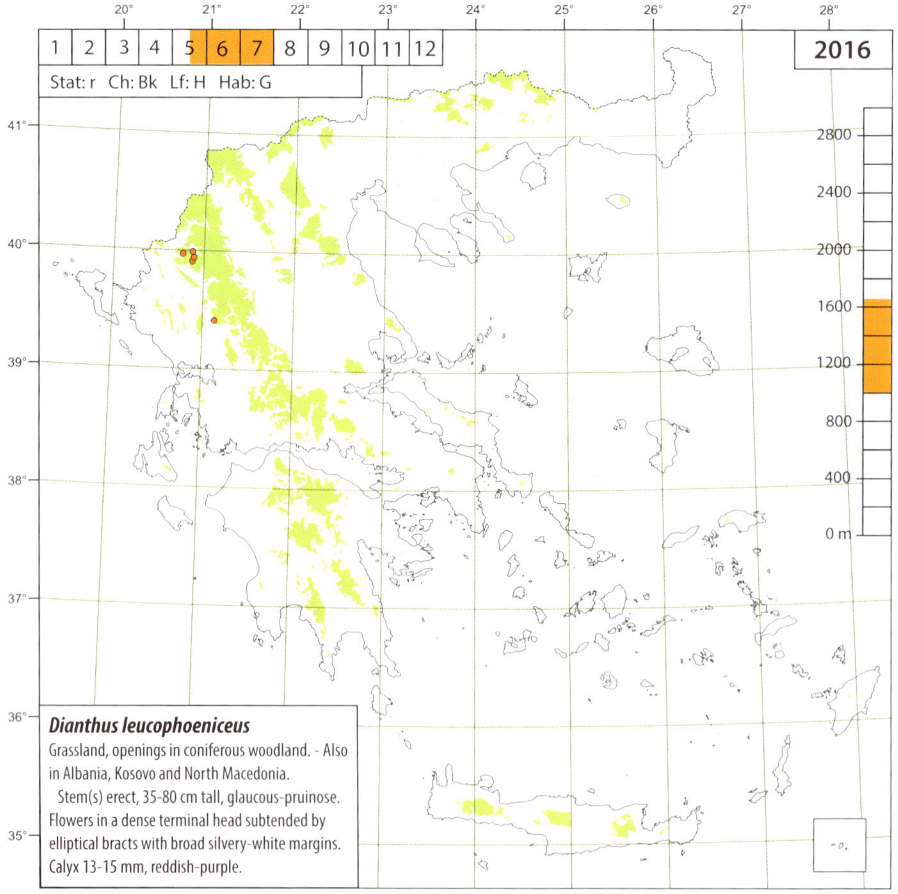

Dianthus leucophoeniceus

Grassland, openings in coniferous woodland. - Also in Albania, Kosovo and North Macedonia.

Stem(s) erect, 35-80 cm tall, glaucous-pruinose. Flowers in a dense terminal head subtended by elliptical bracts with broad silvery-white margins. Calyx 13-15 mm, reddish-purple.

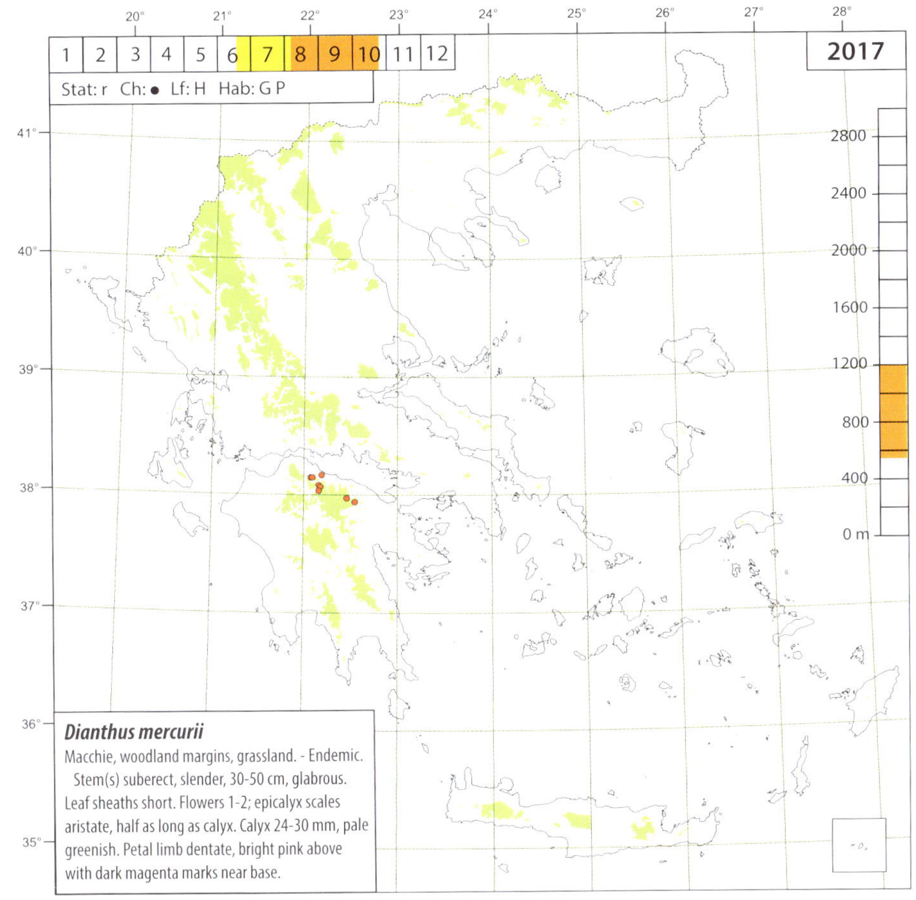

Dianthus mercurii
Macchie, woodland margins, grassland. - Endemic.
Stem(s) suberect, slender, 30-50 cm, glabrous. Leaf sheaths short. Flowers 1-2; epicalyx scales aristate, half as long as calyx. Calyx 24-30 mm, pale greenish. Petal limb dentate, bright pink above with dark magenta marks near base.

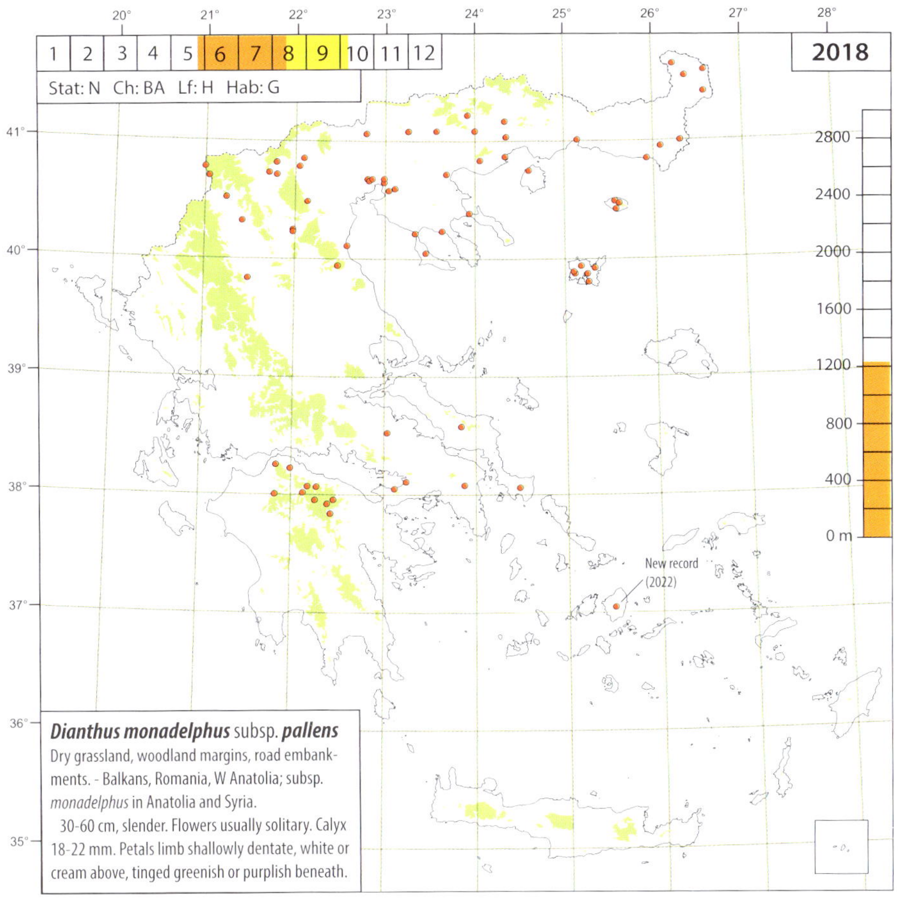

Dianthus monadelphus subsp. ***pallens***
Dry grassland, woodland margins, road embankments. - Balkans, Romania, W Anatolia; subsp. *monadelphus* in Anatolia and Syria.
30-60 cm, slender. Flowers usually solitary. Calyx 18-22 mm. Petals limb shallowly dentate, white or cream above, tinged greenish or purplish beneath.

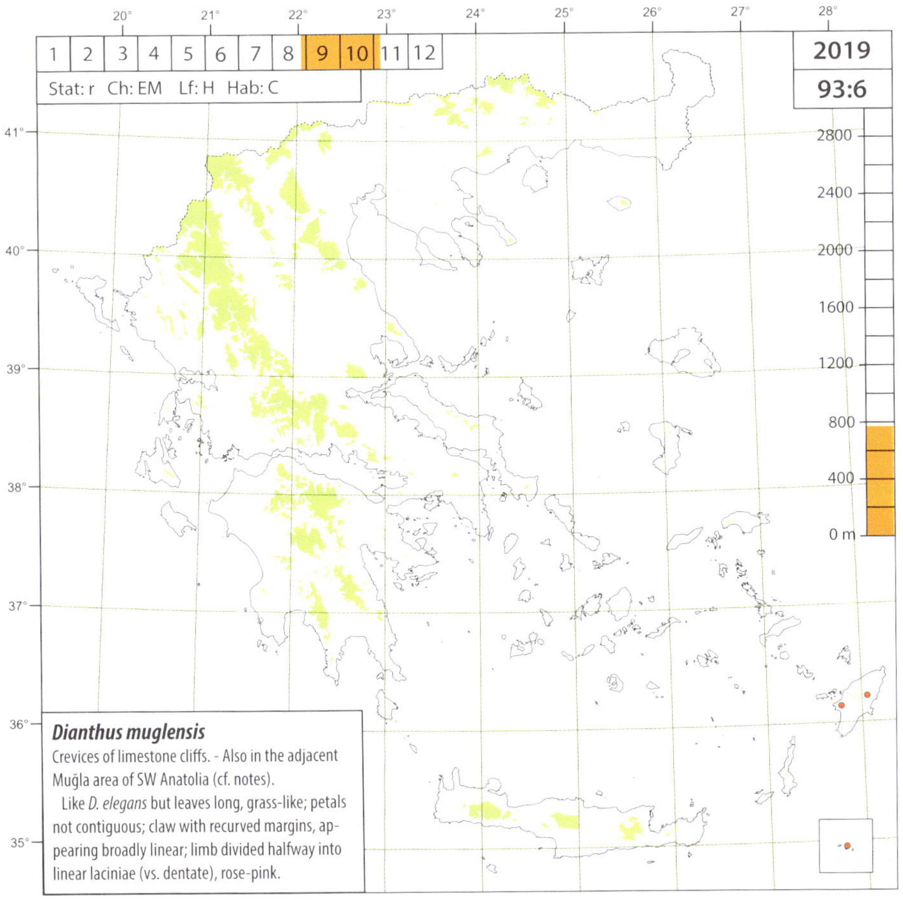

Dianthus muglensis
Crevices of limestone cliffs. - Also in the adjacent Muğla area of SW Anatolia (cf. notes).
Like *D. elegans* but leaves long, grass-like; petals not contiguous; claw with recurved margins, appearing broadly linear; limb divided halfway into linear laciniae (vs. dentate), rose-pink.

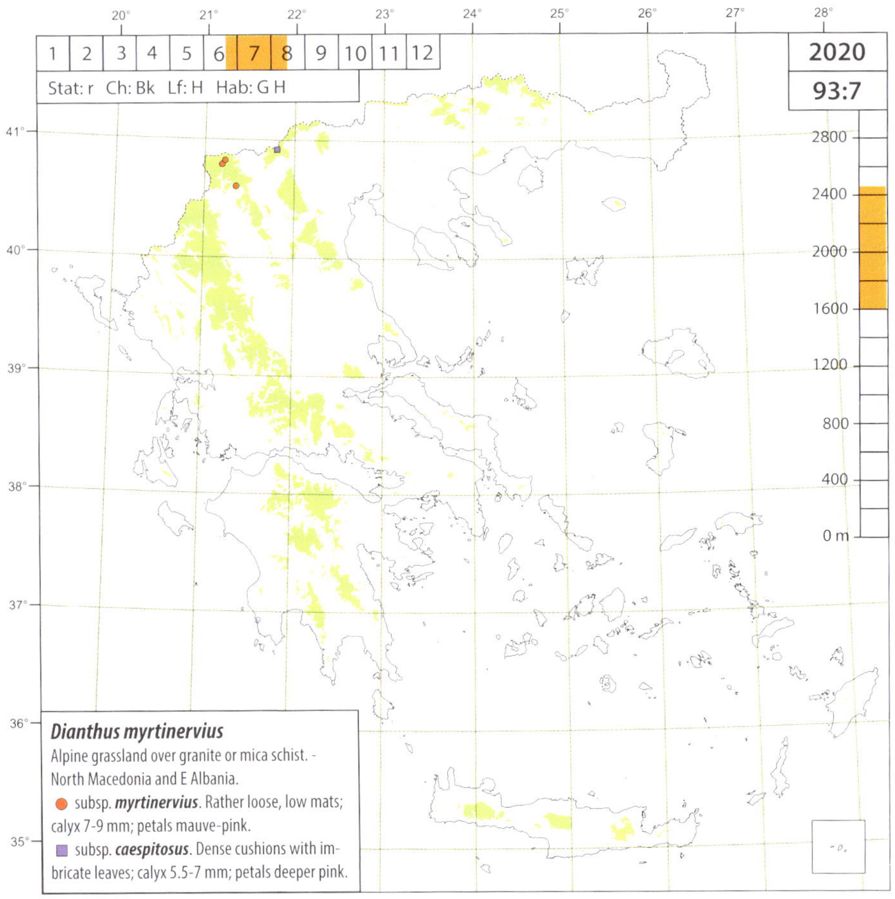

Dianthus myrtinervius
Alpine grassland over granite or mica schist. - North Macedonia and E Albania.
● subsp. ***myrtinervius***. Rather loose, low mats; calyx 7-9 mm; petals mauve-pink.
■ subsp. ***caespitosus***. Dense cushions with imbricate leaves; calyx 5.5-7 mm; petals deeper pink.

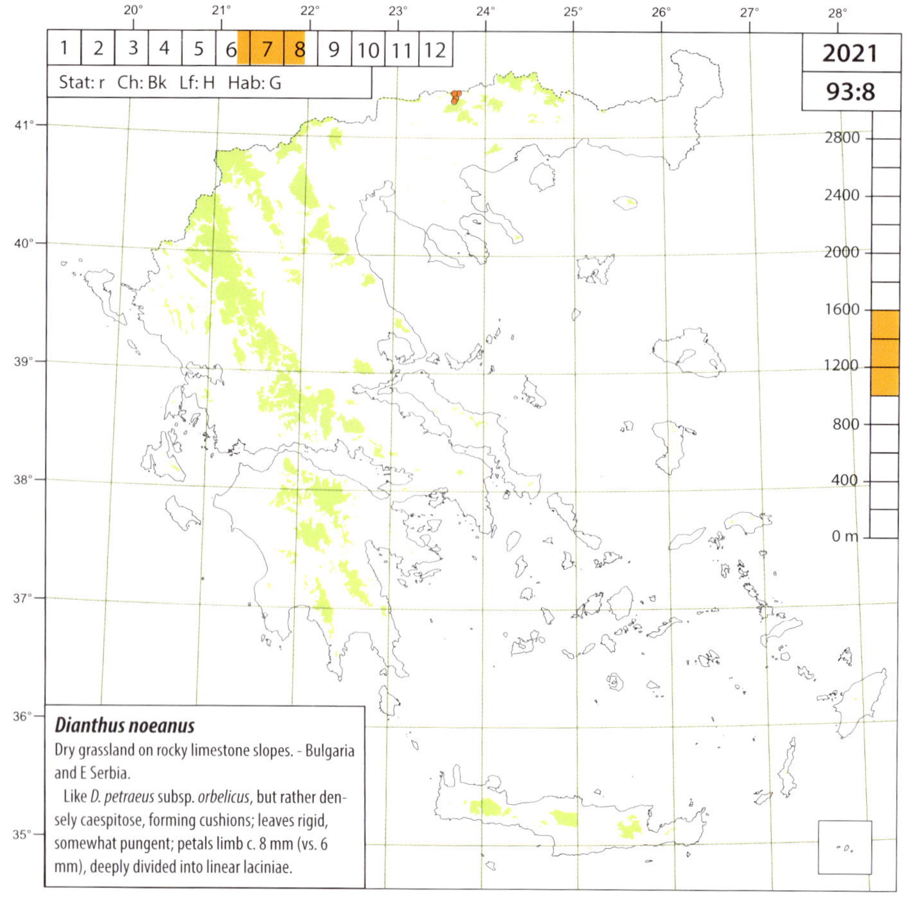

Dianthus noeanus
Dry grassland on rocky limestone slopes. - Bulgaria and E Serbia.
Like *D. petraeus* subsp. *orbelicus*, but rather densely caespitose, forming cushions; leaves rigid, somewhat pungent; petals limb c. 8 mm (vs. 6 mm), deeply divided into linear laciniae.

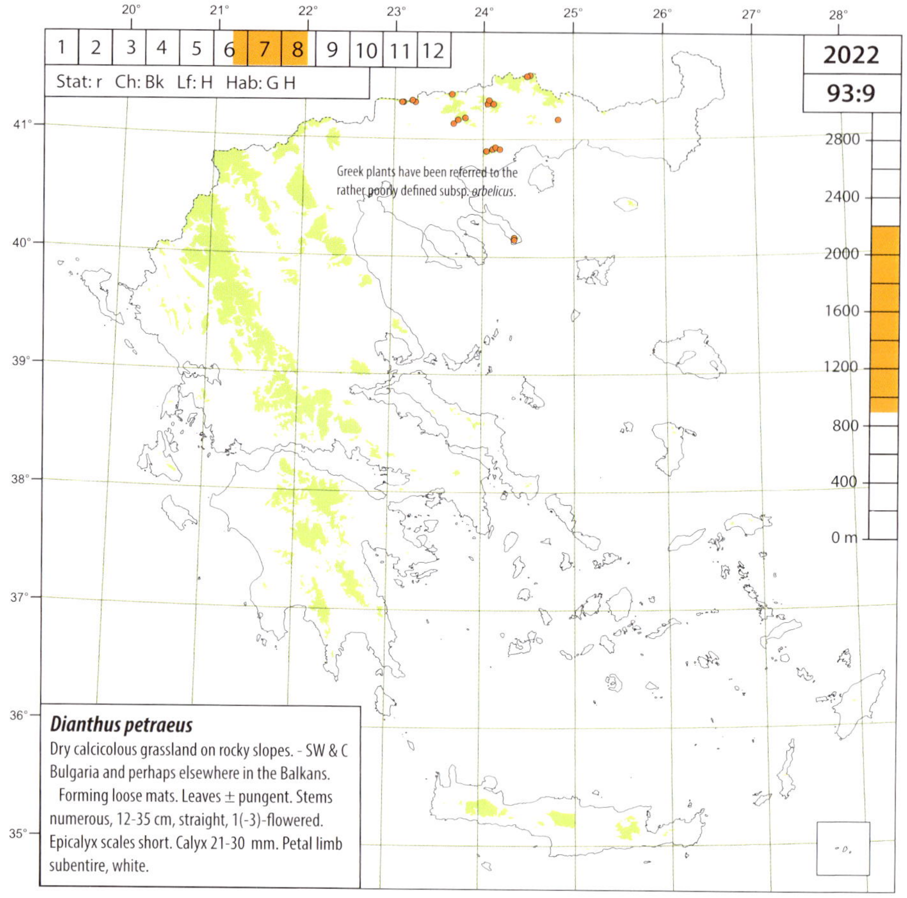

Dianthus petraeus
Dry calcicolous grassland on rocky slopes. - SW & C Bulgaria and perhaps elsewhere in the Balkans.
Forming loose mats. Leaves ± pungent. Stems numerous, 12-35 cm, straight, 1(-3)-flowered. Epicalyx scales short. Calyx 21-30 mm. Petal limb subentire, white.

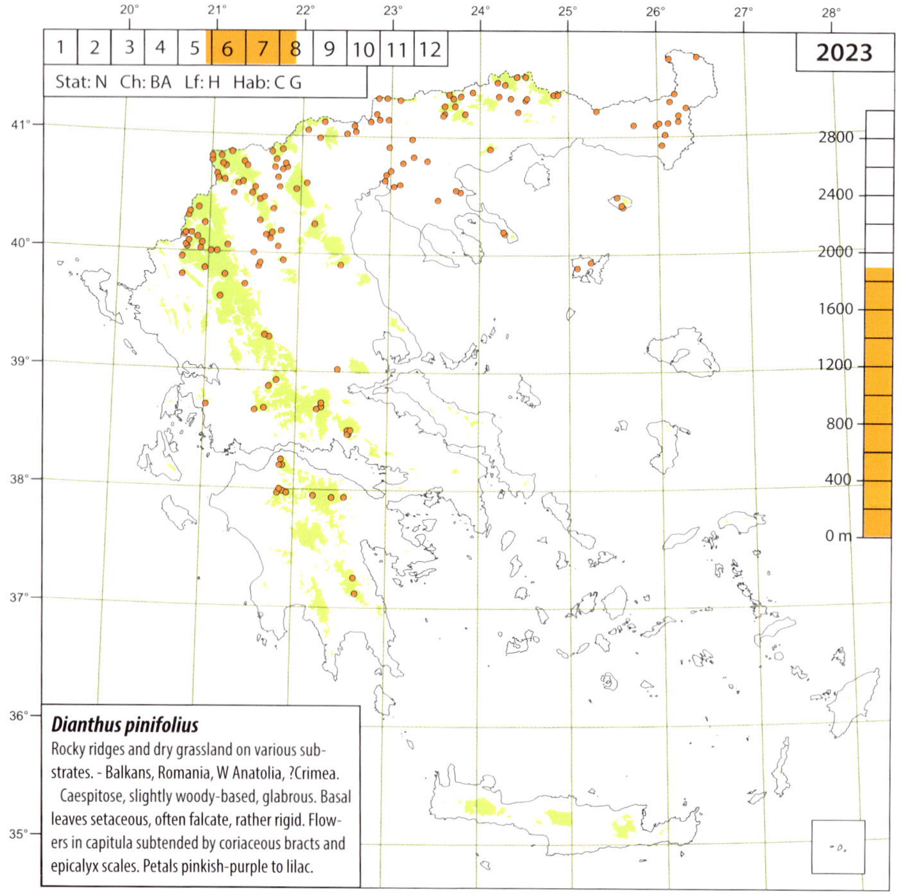

Dianthus pinifolius
Rocky ridges and dry grassland on various substrates. - Balkans, Romania, W Anatolia, ?Crimea.
Caespitose, slightly woody-based, glabrous. Basal leaves setaceous, often falcate, rather rigid. Flowers in capitula subtended by coriaceous bracts and epicalyx scales. Petals pinkish-purple to lilac.

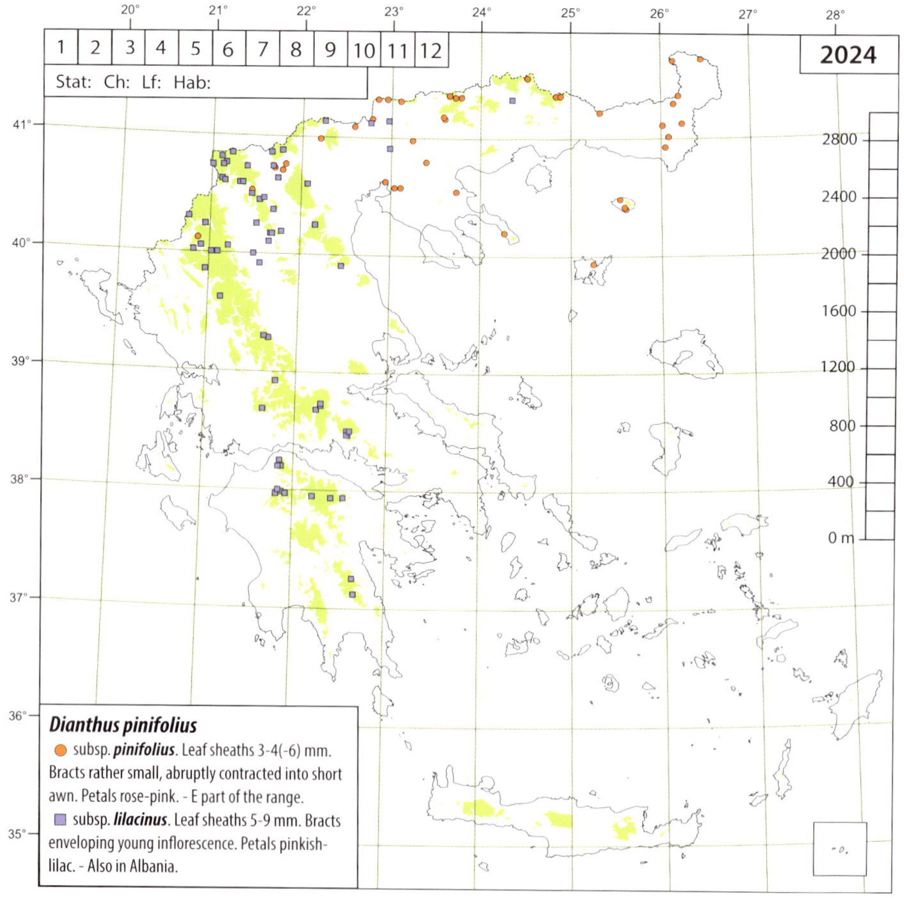

Dianthus pinifolius
● subsp. ***pinifolius***. Leaf sheaths 3-4(-6) mm. Bracts rather small, abruptly contracted into short awn. Petals rose-pink. - E part of the range.
■ subsp. ***lilacinus***. Leaf sheaths 5-9 mm. Bracts enveloping young inflorescence. Petals pinkish-lilac. - Also in Albania.

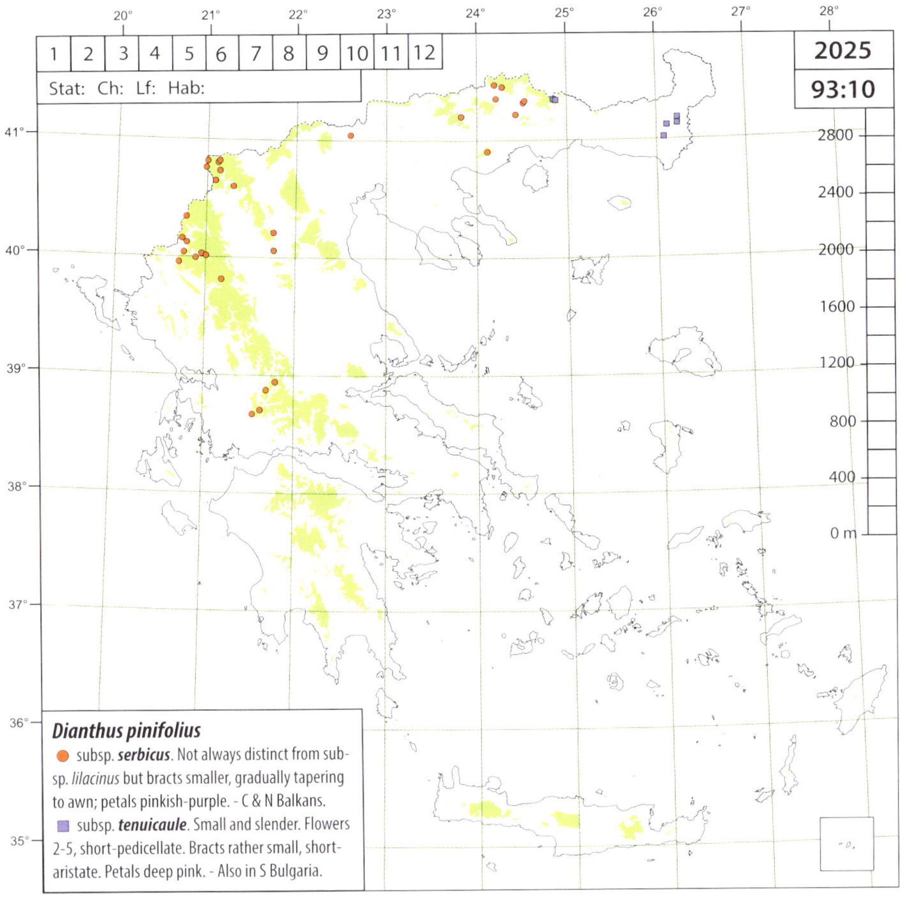

Dianthus pinifolius

● subsp. ***serbicus***. Not always distinct from subsp. *lilacinus* but bracts smaller, gradually tapering to awn; petals pinkish-purple. - C & N Balkans.

■ subsp. ***tenuicaule***. Small and slender. Flowers 2-5, short-pedicellate. Bracts rather small, short-aristate. Petals deep pink. - Also in S Bulgaria.

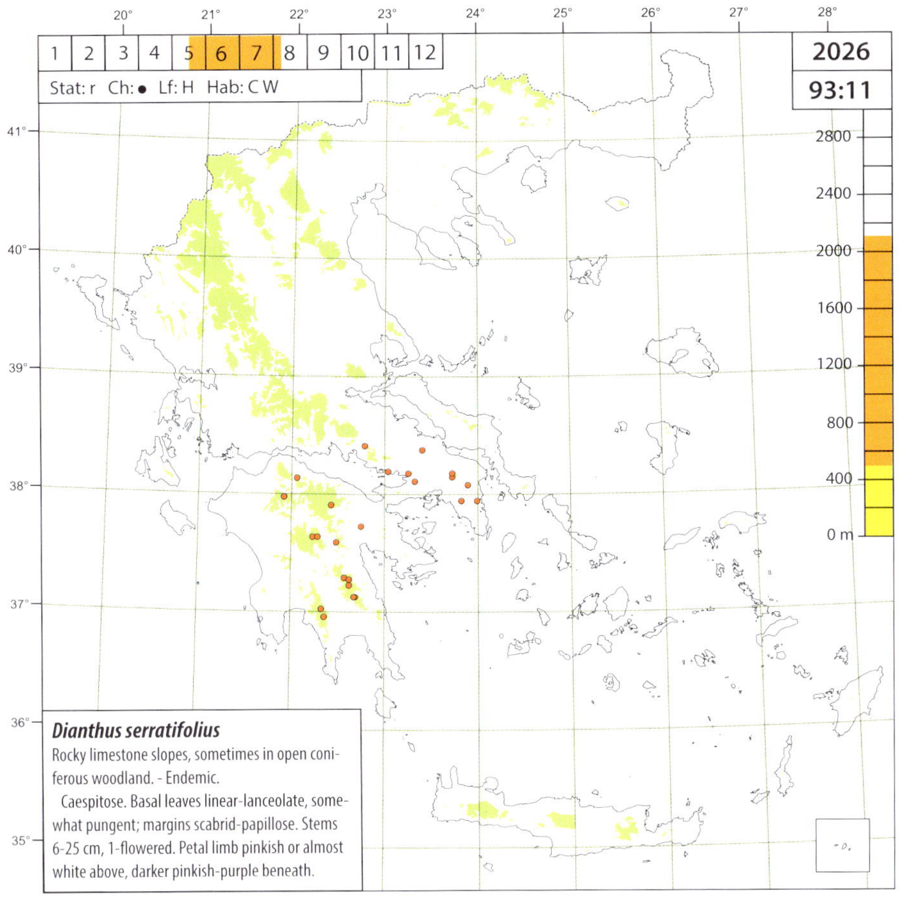

Dianthus serratifolius

Rocky limestone slopes, sometimes in open coniferous woodland. - Endemic.

Caespitose. Basal leaves linear-lanceolate, somewhat pungent; margins scabrid-papillose. Stems 6-25 cm, 1-flowered. Petal limb pinkish or almost white above, darker pinkish-purple beneath.

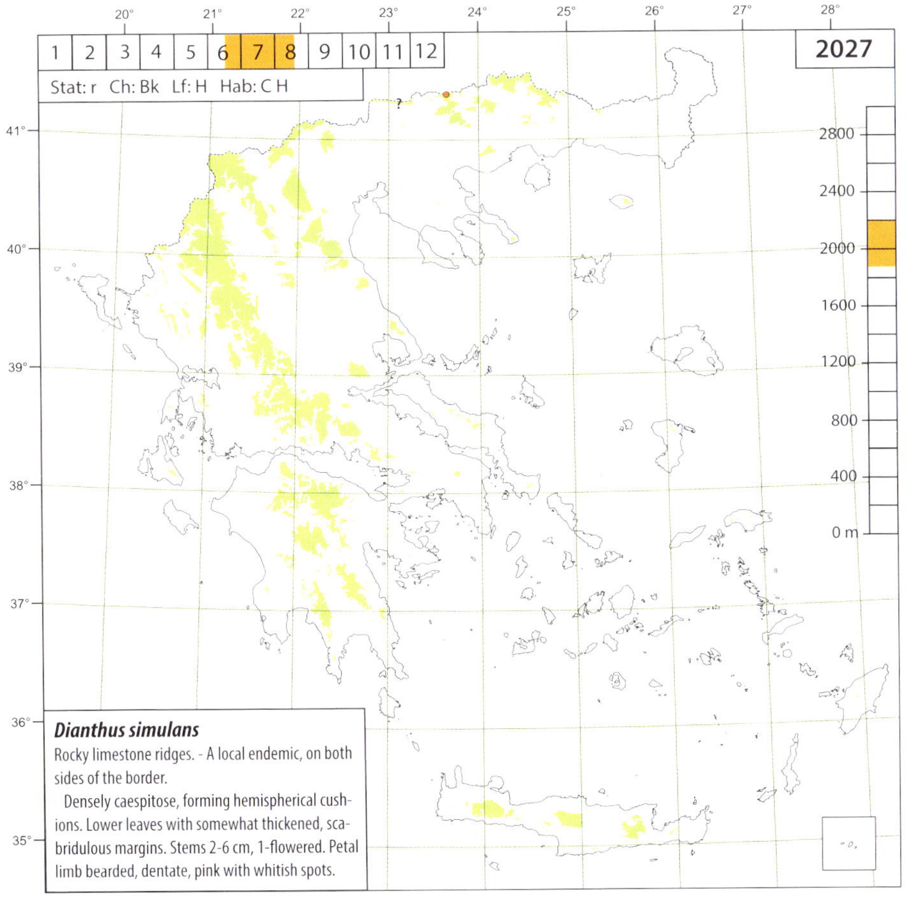

Dianthus simulans

Rocky limestone ridges. - A local endemic, on both sides of the border.

Densely caespitose, forming hemispherical cushions. Lower leaves with somewhat thickened, scabridulous margins. Stems 2-6 cm, 1-flowered. Petal limb bearded, dentate, pink with whitish spots.

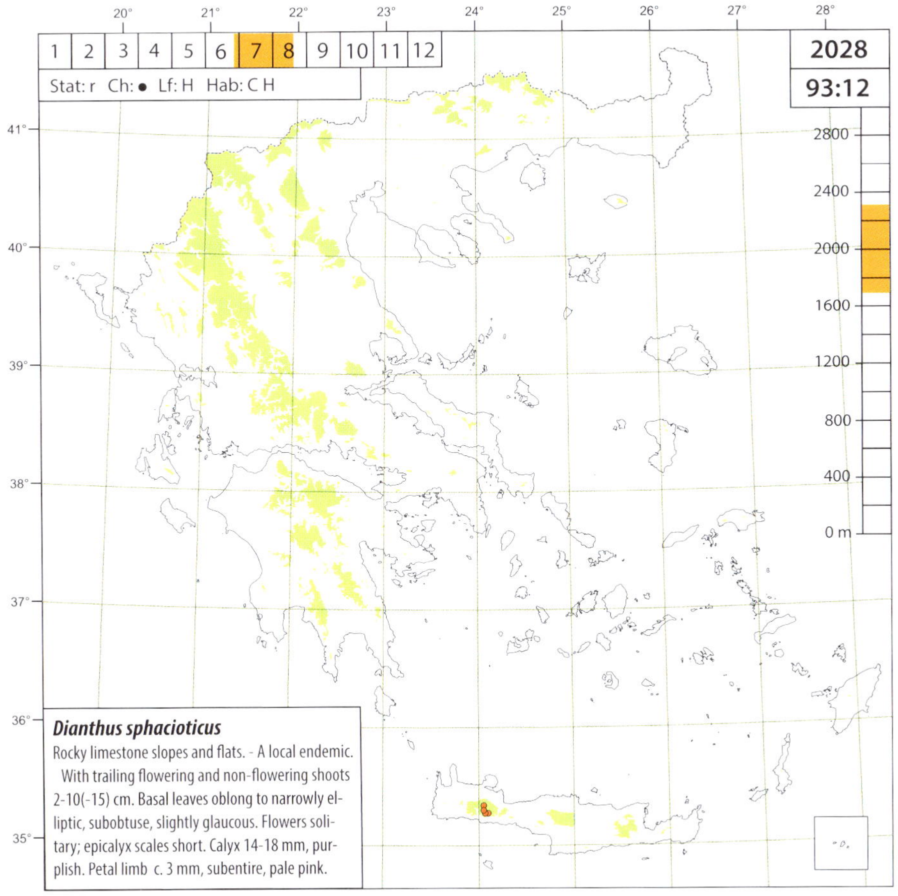

Dianthus sphacioticus

Rocky limestone slopes and flats. - A local endemic.

With trailing flowering and non-flowering shoots 2-10(-15) cm. Basal leaves oblong to narrowly elliptic, subobtuse, slightly glaucous. Flowers solitary; epicalyx scales short. Calyx 14-18 mm, purplish. Petal limb c. 3 mm, subentire, pale pink.

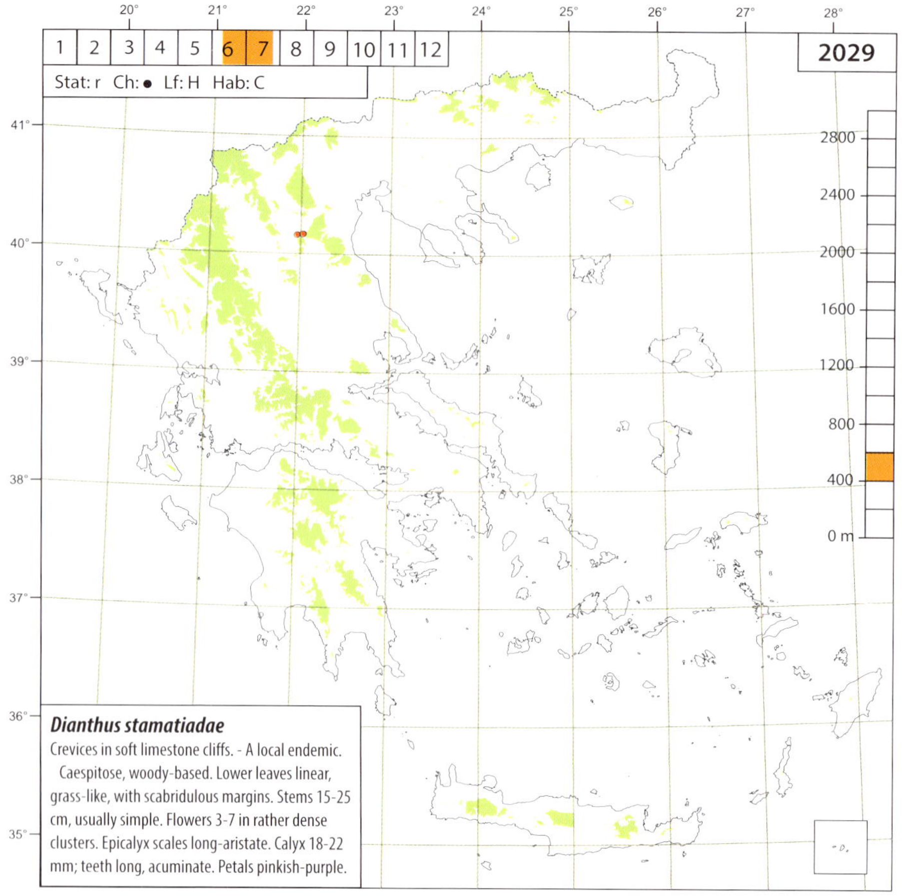

Dianthus stamatiadae

Crevices in soft limestone cliffs. - A local endemic.

Caespitose, woody-based. Lower leaves linear, grass-like, with scabridulous margins. Stems 15-25 cm, usually simple. Flowers 3-7 in rather dense clusters. Epicalyx scales long-aristate. Calyx 18-22 mm; teeth long, acuminate. Petals pinkish-purple.

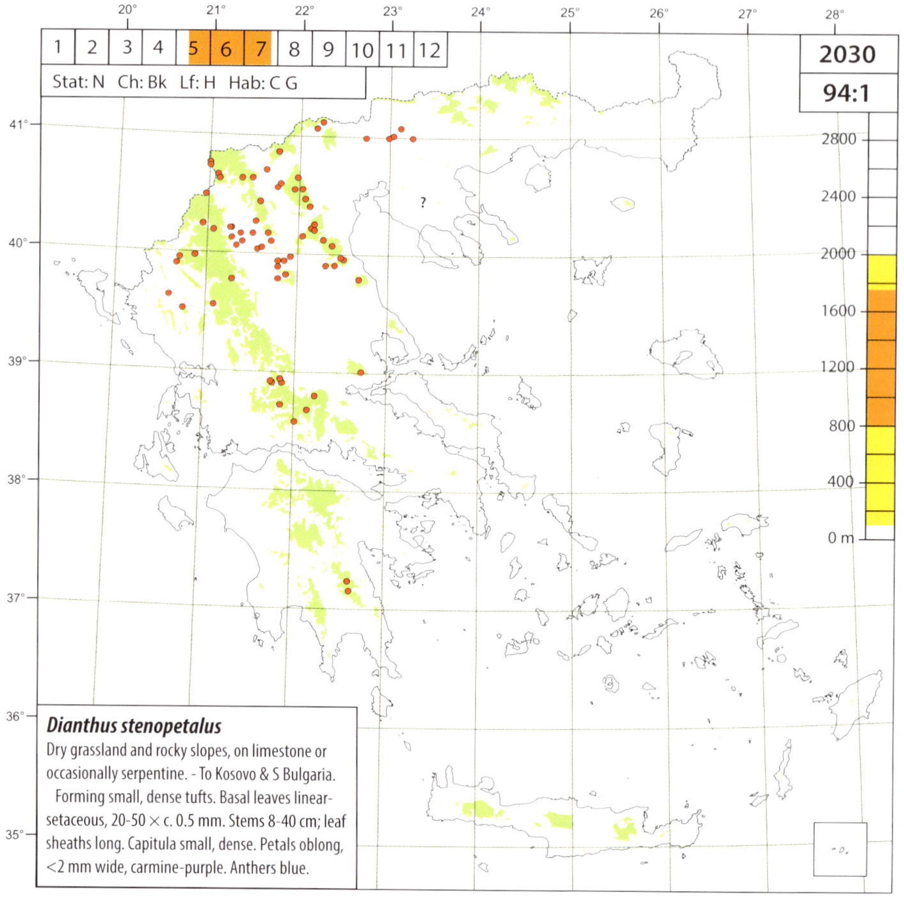

Dianthus stenopetalus

Dry grassland and rocky slopes, on limestone or occasionally serpentine. - To Kosovo & S Bulgaria.

Forming small, dense tufts. Basal leaves linear-setaceous, 20-50 × c. 0.5 mm. Stems 8-40 cm; leaf sheaths long. Capitula small, dense. Petals oblong, <2 mm wide, carmine-purple. Anthers blue.

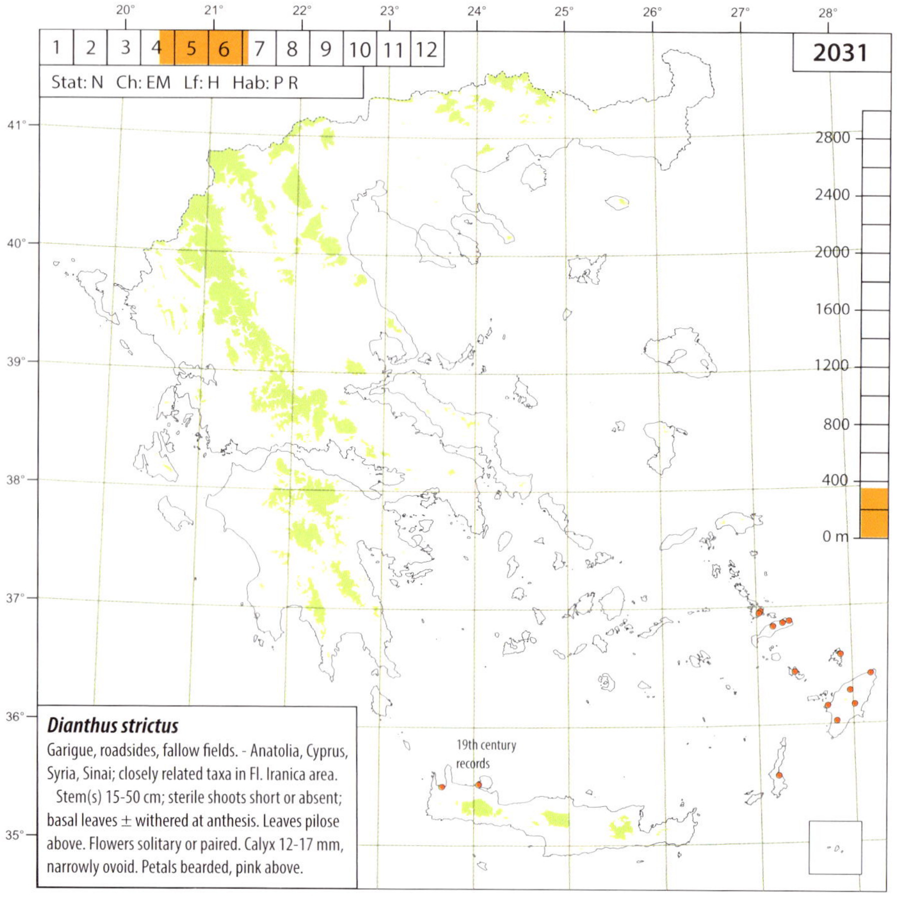

Dianthus strictus

Garigue, roadsides, fallow fields. - Anatolia, Cyprus, Syria, Sinai; closely related taxa in Fl. Iranica area.

Stem(s) 15-50 cm; sterile shoots short or absent; basal leaves ± withered at anthesis. Leaves pilose above. Flowers solitary or paired. Calyx 12-17 mm, narrowly ovoid. Petals bearded, pink above.

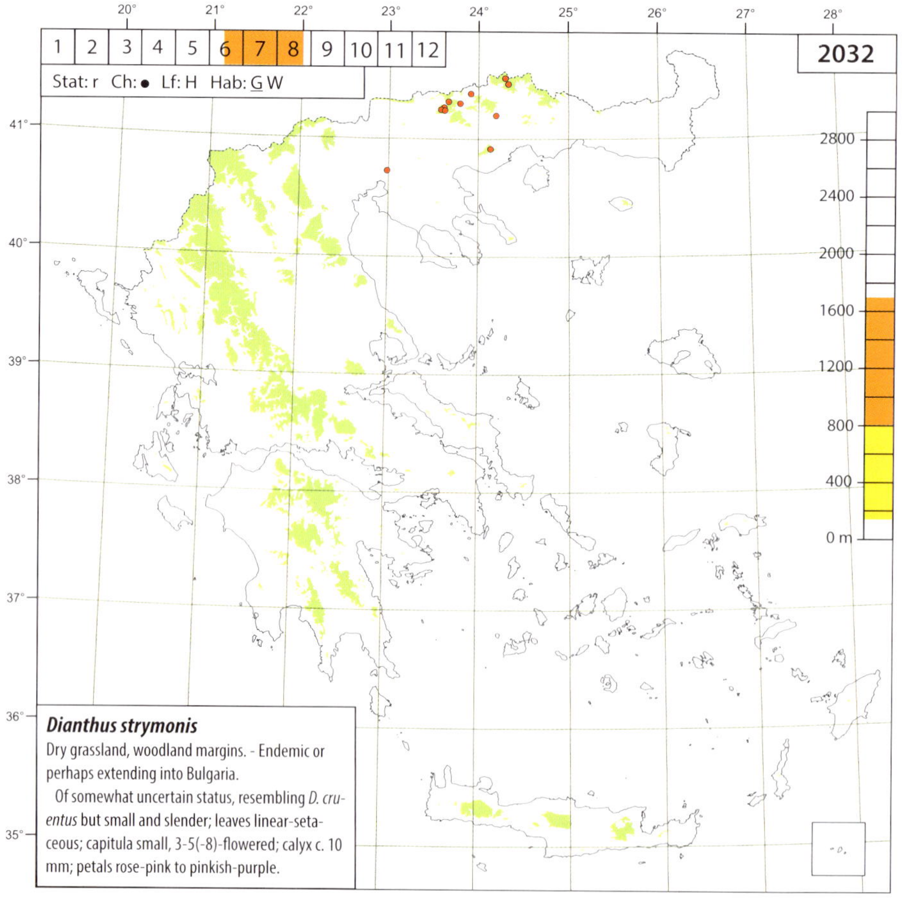

Dianthus strymonis

Dry grassland, woodland margins. - Endemic or perhaps extending into Bulgaria.

Of somewhat uncertain status, resembling *D. cruentus* but small and slender; leaves linear-setaceous; capitula small, 3-5(-8)-flowered; calyx c. 10 mm; petals rose-pink to pinkish-purple.

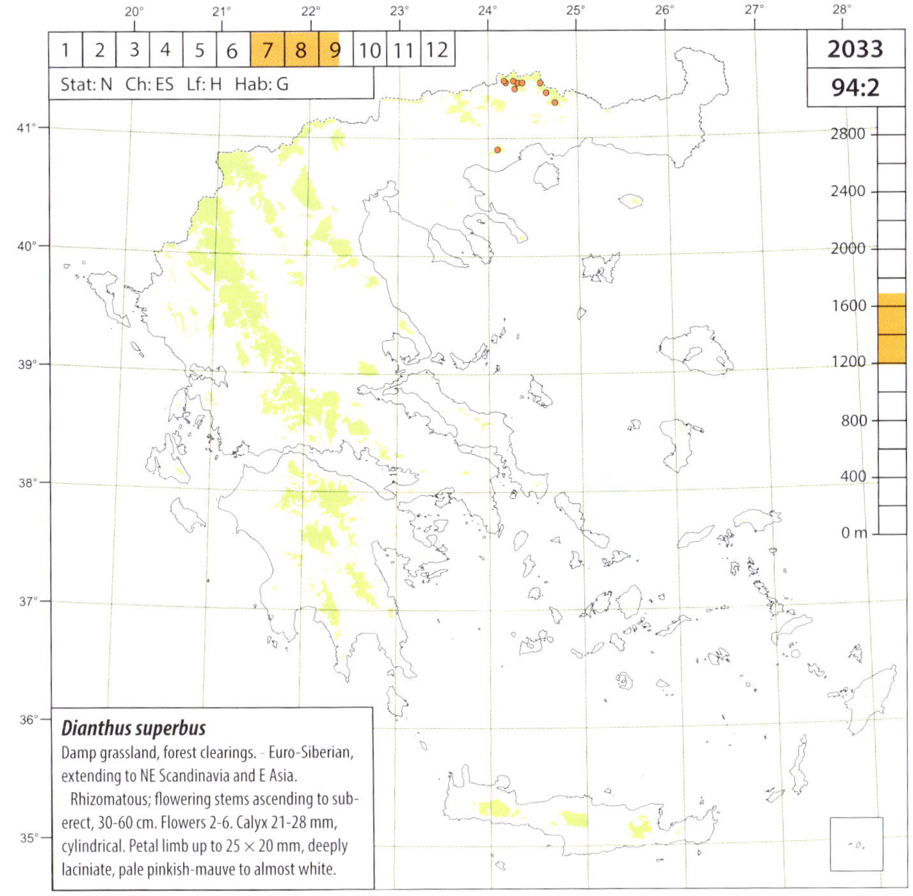

Stat: N Ch: ES Lf: H Hab: G

Dianthus superbus

Damp grassland, forest clearings. - Euro-Siberian, extending to NE Scandinavia and E Asia.

Rhizomatous; flowering stems ascending to sub-erect, 30-60 cm. Flowers 2-6. Calyx 21-28 mm, cylindrical. Petal limb up to 25 × 20 mm, deeply laciniate, pale pinkish-mauve to almost white.

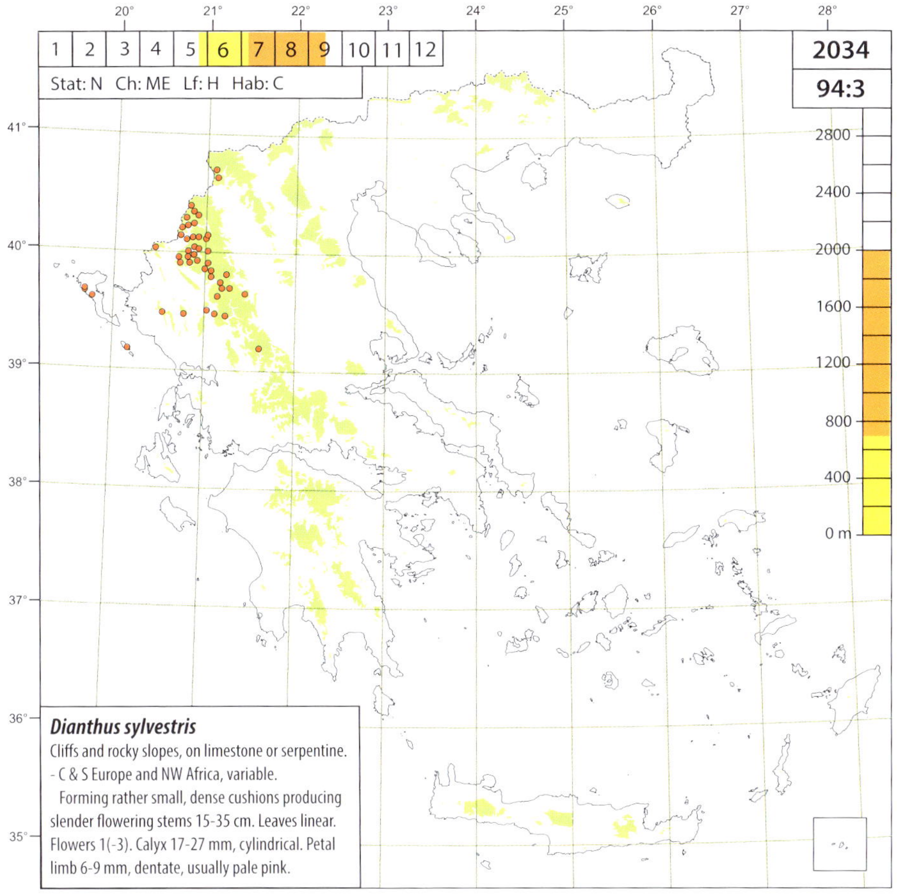

Stat: N Ch: ME Lf: H Hab: C

Dianthus sylvestris

Cliffs and rocky slopes, on limestone or serpentine. - C & S Europe and NW Africa, variable.

Forming rather small, dense cushions producing slender flowering stems 15-35 cm. Leaves linear. Flowers 1(-3). Calyx 17-27 mm, cylindrical. Petal limb 6-9 mm, dentate, usually pale pink.

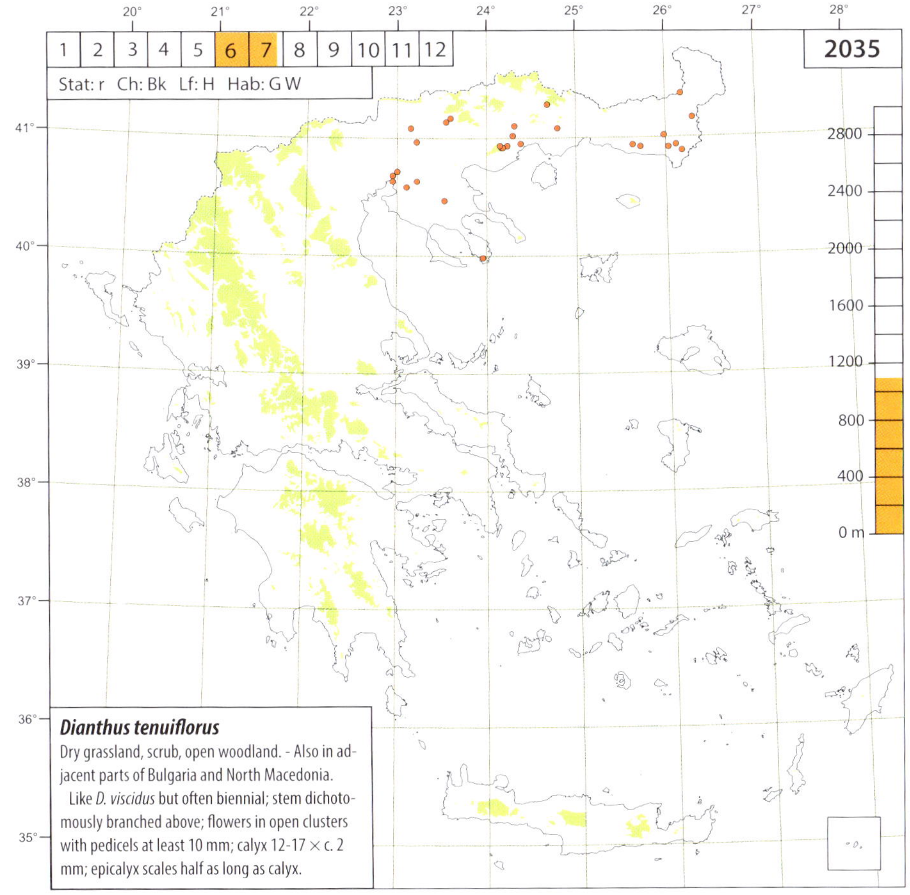

Stat: r Ch: Bk Lf: H Hab: G W

Dianthus tenuiflorus

Dry grassland, scrub, open woodland. - Also in adjacent parts of Bulgaria and North Macedonia.

Like *D. viscidus* but often biennial; stem dichotomously branched above; flowers in open clusters with pedicels at least 10 mm; calyx 12-17 × c. 2 mm; epicalyx scales half as long as calyx.

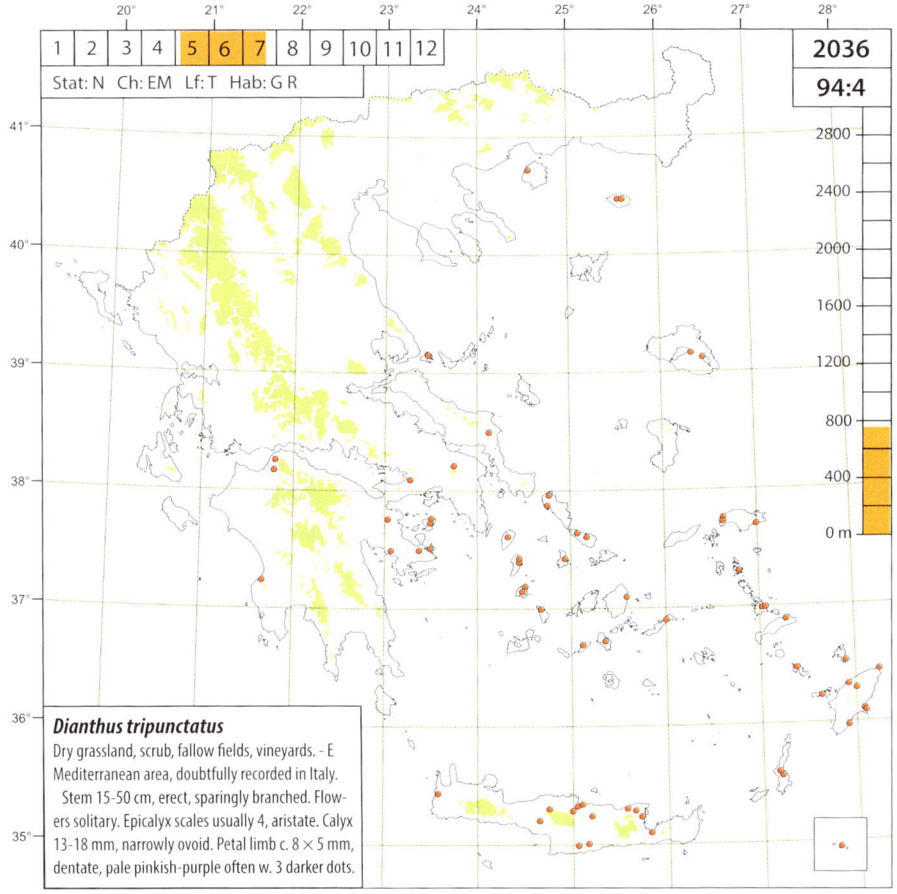

Stat: N Ch: EM Lf: T Hab: G R

Dianthus tripunctatus

Dry grassland, scrub, fallow fields, vineyards. - E Mediterranean area, doubtfully recorded in Italy.

Stem 15-50 cm, erect, sparingly branched. Flowers solitary. Epicalyx scales usually 4, aristate. Calyx 13-18 mm, narrowly ovoid. Petal limb c. 8 × 5 mm, dentate, pale pinkish-purple often w. 3 darker dots.

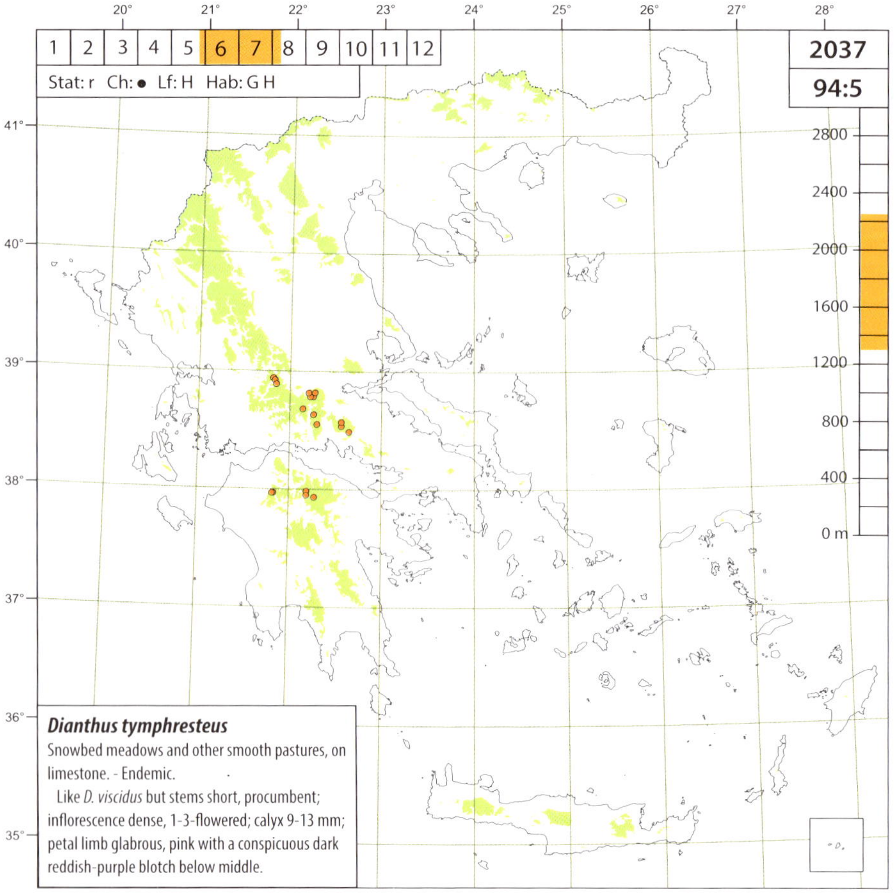

Dianthus tymphresteus

Snowbed meadows and other smooth pastures, on limestone. - Endemic.

Like *D. viscidus* but stems short, procumbent; inflorescence dense, 1-3-flowered; calyx 9-13 mm; petal limb glabrous, pink with a conspicuous dark reddish-purple blotch below middle.

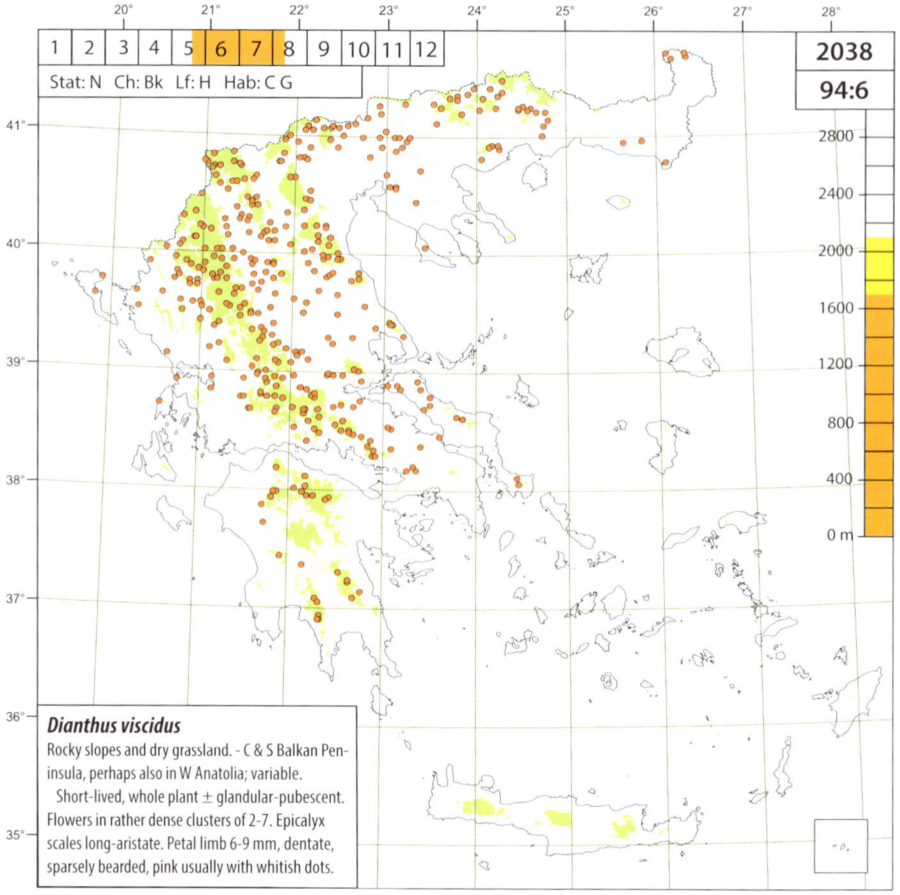

Dianthus viscidus

Rocky slopes and dry grassland. - C & S Balkan Peninsula, perhaps also in W Anatolia; variable.

Short-lived, whole plant ± glandular-pubescent. Flowers in rather dense clusters of 2-7. Epicalyx scales long-aristate. Petal limb 6-9 mm, dentate, sparsely bearded, pink usually with whitish dots.

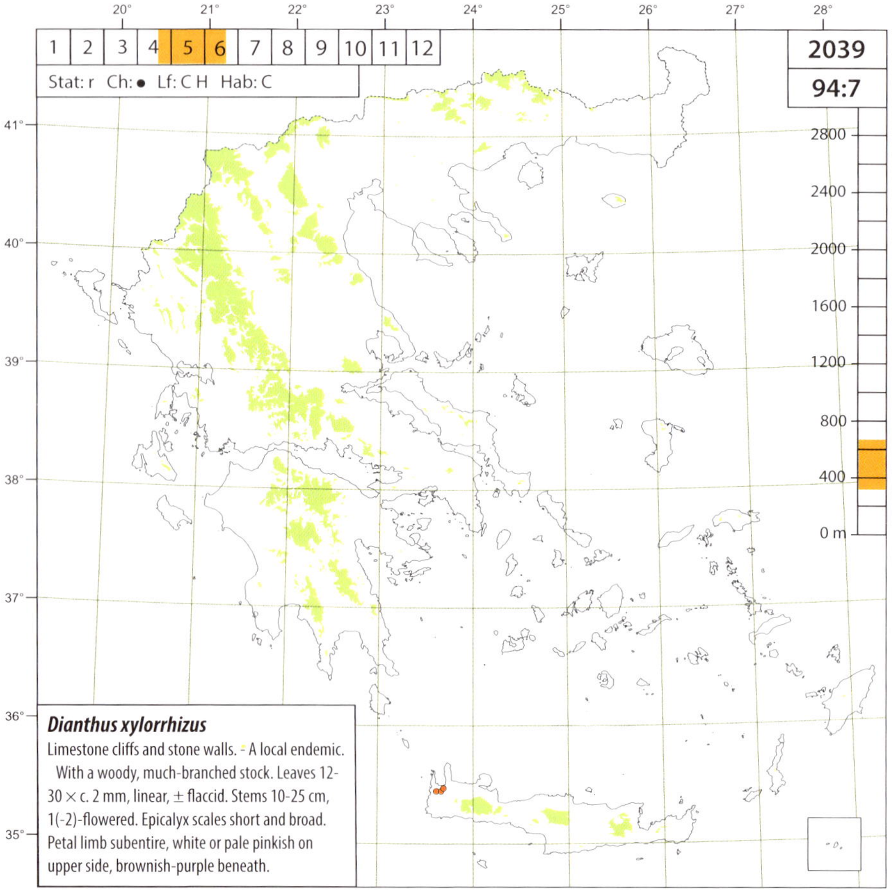

Dianthus xylorrhizus

Limestone cliffs and stone walls. - A local endemic.

With a woody, much-branched stock. Leaves 12-30 × c. 2 mm, linear, ± flaccid. Stems 10-25 cm, 1(-2)-flowered. Epicalyx scales short and broad. Petal limb subentire, white or pale pinkish on upper side, brownish-purple beneath.

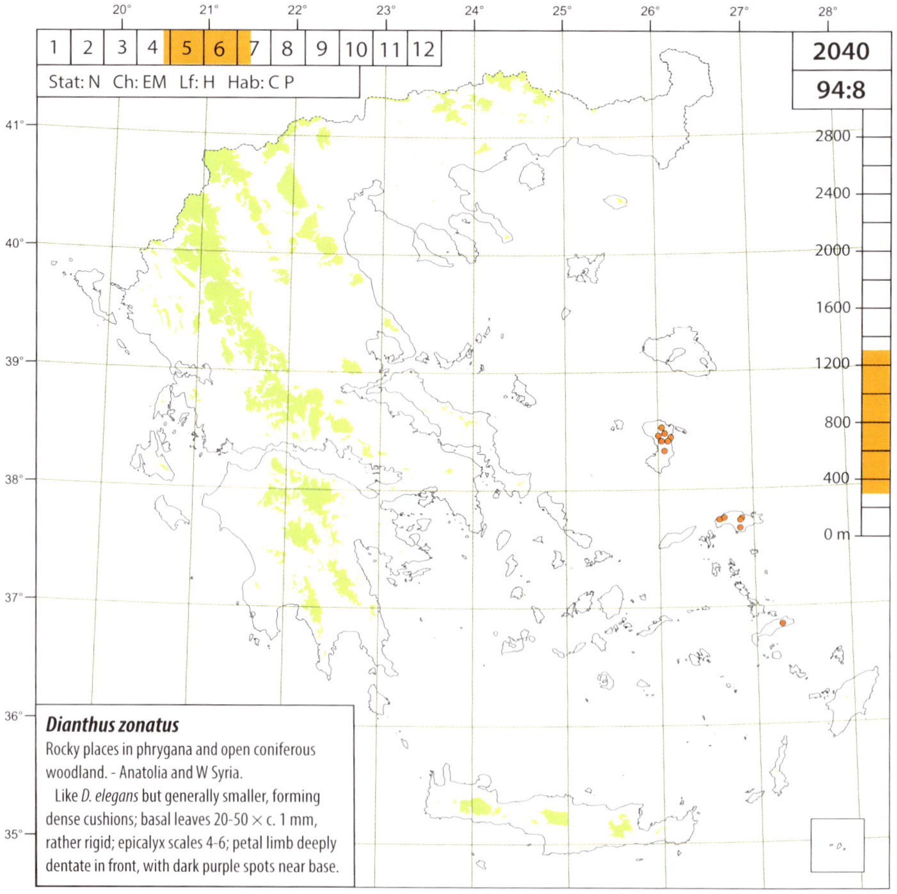

Dianthus zonatus

Rocky places in phrygana and open coniferous woodland. - Anatolia and W Syria.

Like *D. elegans* but generally smaller, forming dense cushions; basal leaves 20-50 × c. 1 mm, rather rigid; epicalyx scales 4-6; petal limb deeply dentate in front, with dark purple spots near base.

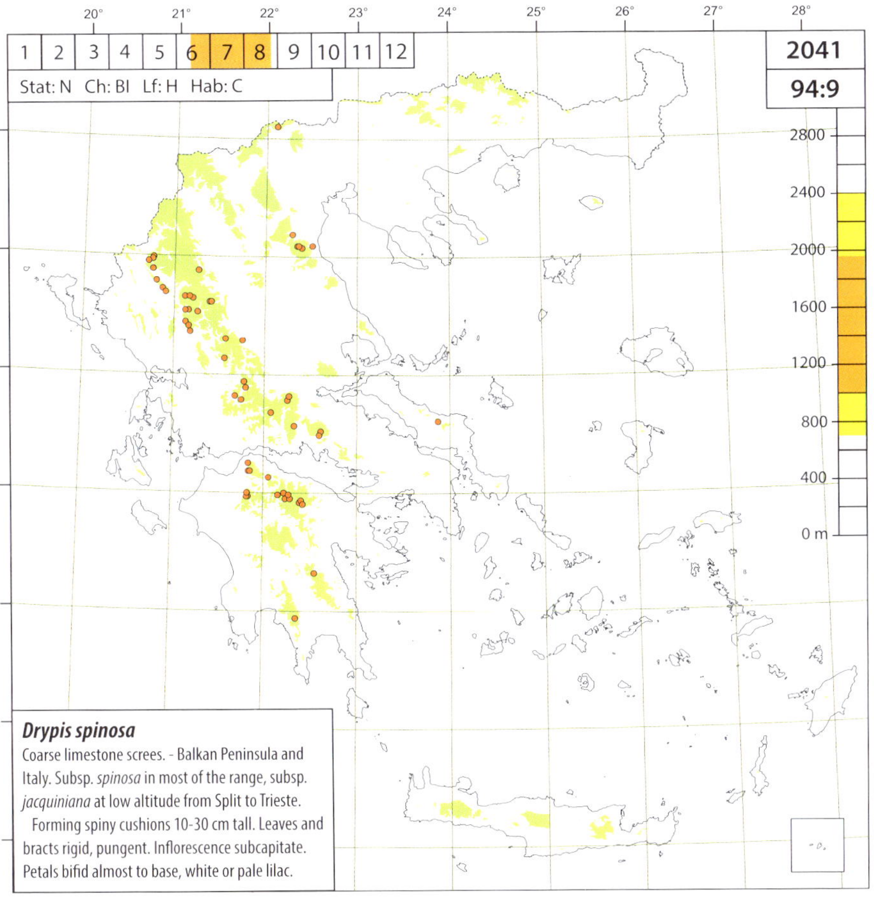

Drypis spinosa

Coarse limestone screes. - Balkan Peninsula and Italy. Subsp. *spinosa* in most of the range, subsp. *jacquiniana* at low altitude from Split to Trieste.

Forming spiny cushions 10-30 cm tall. Leaves and bracts rigid, pungent. Inflorescence subcapitate. Petals bifid almost to base, white or pale lilac.

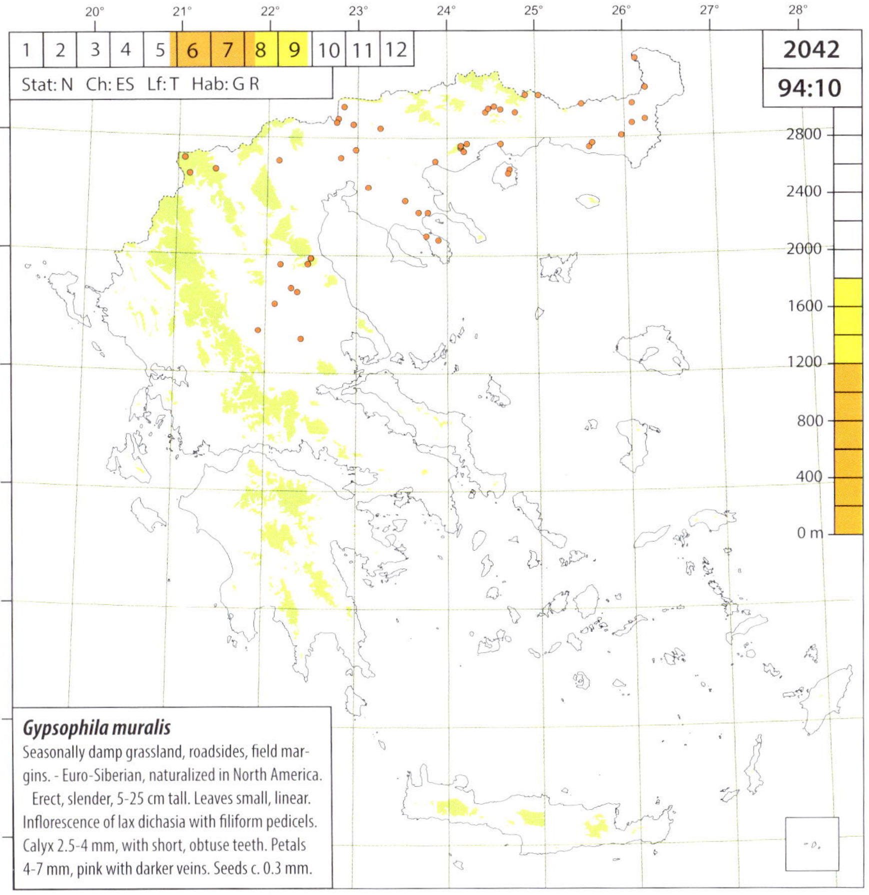

Gypsophila muralis

Seasonally damp grassland, roadsides, field margins. - Euro-Siberian, naturalized in North America.

Erect, slender, 5-25 cm tall. Leaves small, linear. Inflorescence of lax dichasia with filiform pedicels. Calyx 2.5-4 mm, with short, obtuse teeth. Petals 4-7 mm, pink with darker veins. Seeds c. 0.3 mm.

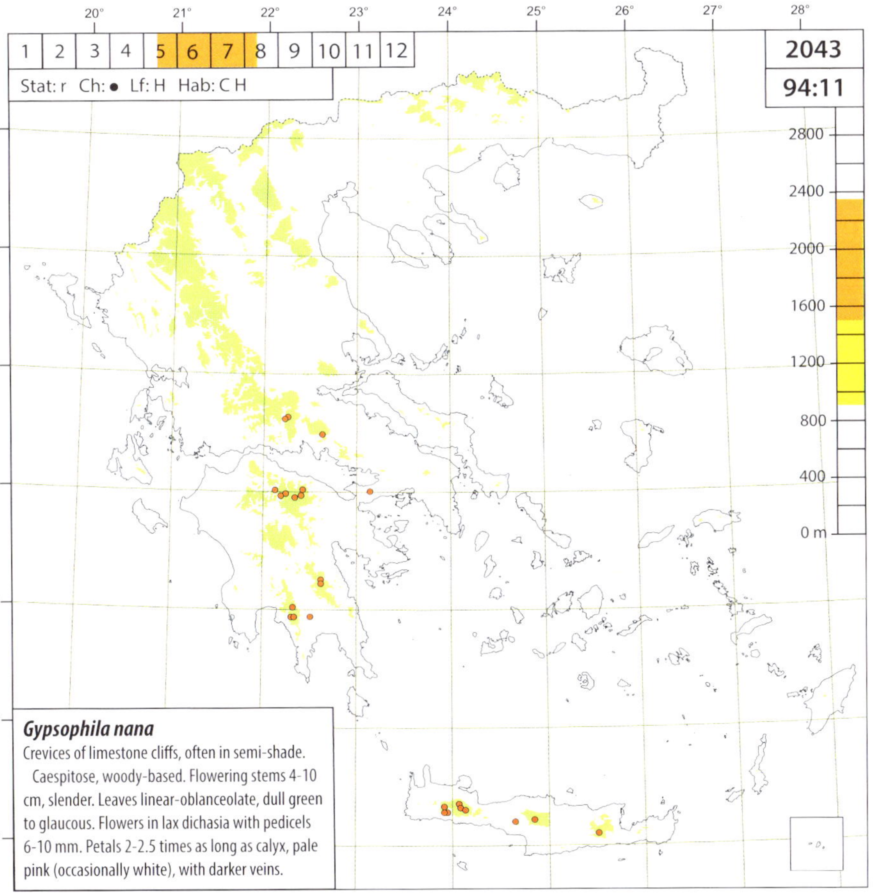

Gypsophila nana

Crevices of limestone cliffs, often in semi-shade.

Caespitose, woody-based. Flowering stems 4-10 cm, slender. Leaves linear-oblanceolate, dull green to glaucous. Flowers in lax dichasia with pedicels 6-10 mm. Petals 2-2.5 times as long as calyx, pale pink (occasionally white), with darker veins.

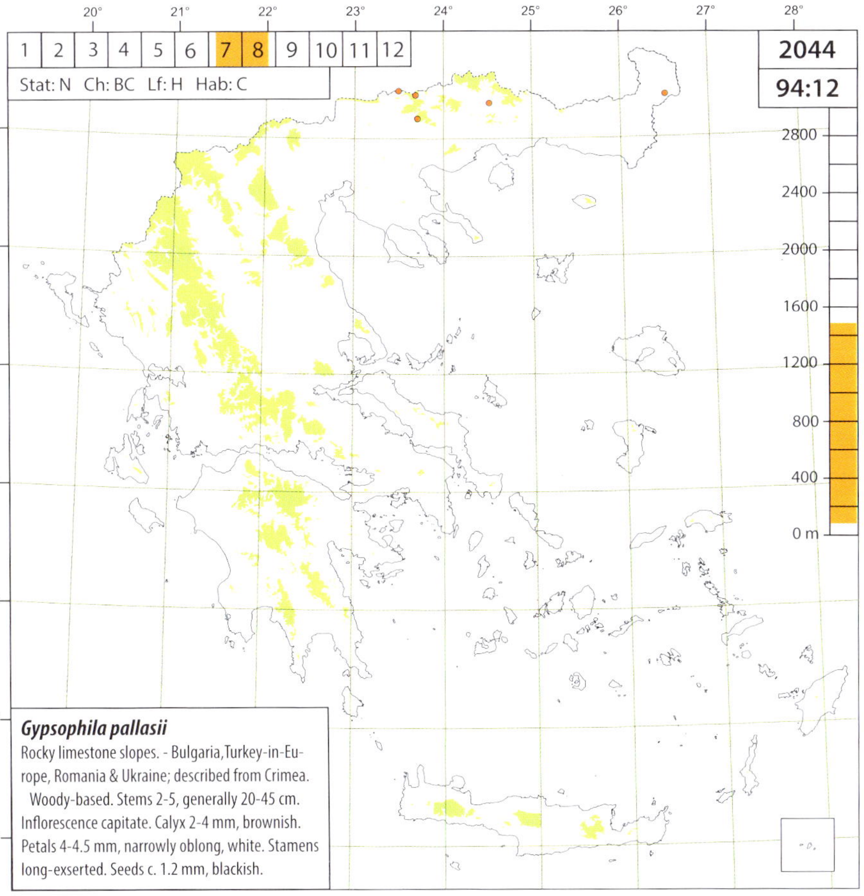

Gypsophila pallasii

Rocky limestone slopes. - Bulgaria, Turkey-in-Europe, Romania & Ukraine; described from Crimea.

Woody-based. Stems 2-5, generally 20-45 cm. Inflorescence capitate. Calyx 2-4 mm, brownish. Petals 4-4.5 mm, narrowly oblong, white. Stamens long-exserted. Seeds c. 1.2 mm, blackish.

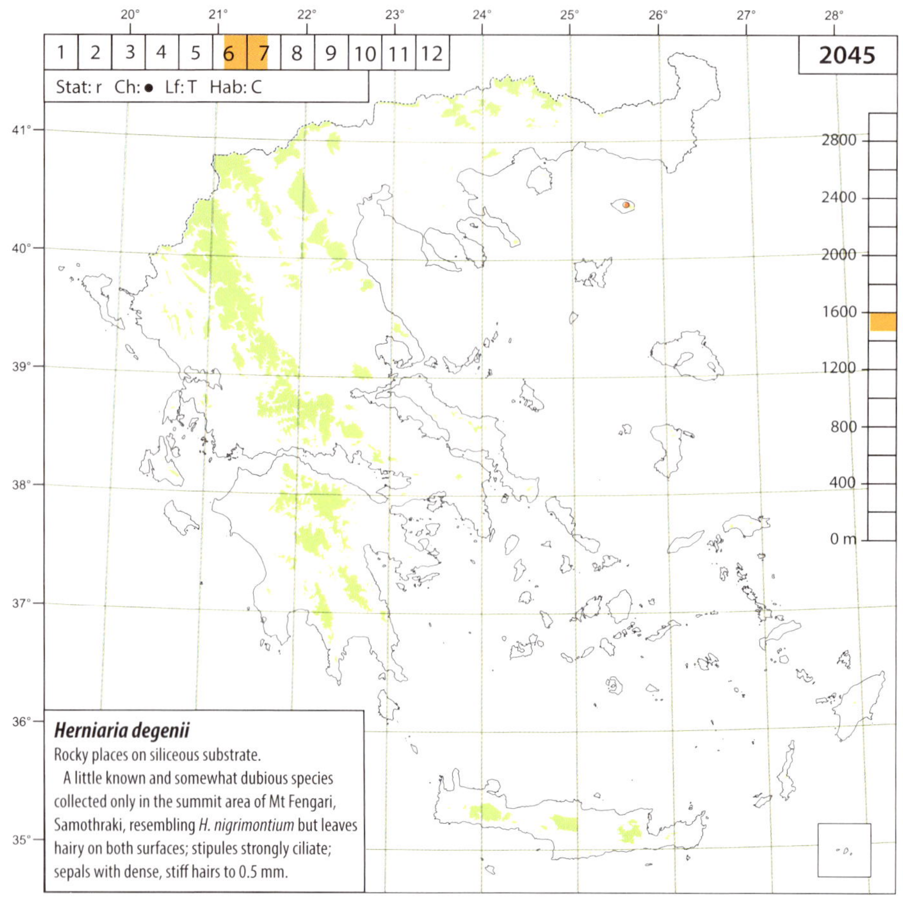

Herniaria degenii
Rocky places on siliceous substrate.
A little known and somewhat dubious species collected only in the summit area of Mt Fengari, Samothraki, resembling *H. nigrimontium* but leaves hairy on both surfaces; stipules strongly ciliate; sepals with dense, stiff hairs to 0.5 mm.

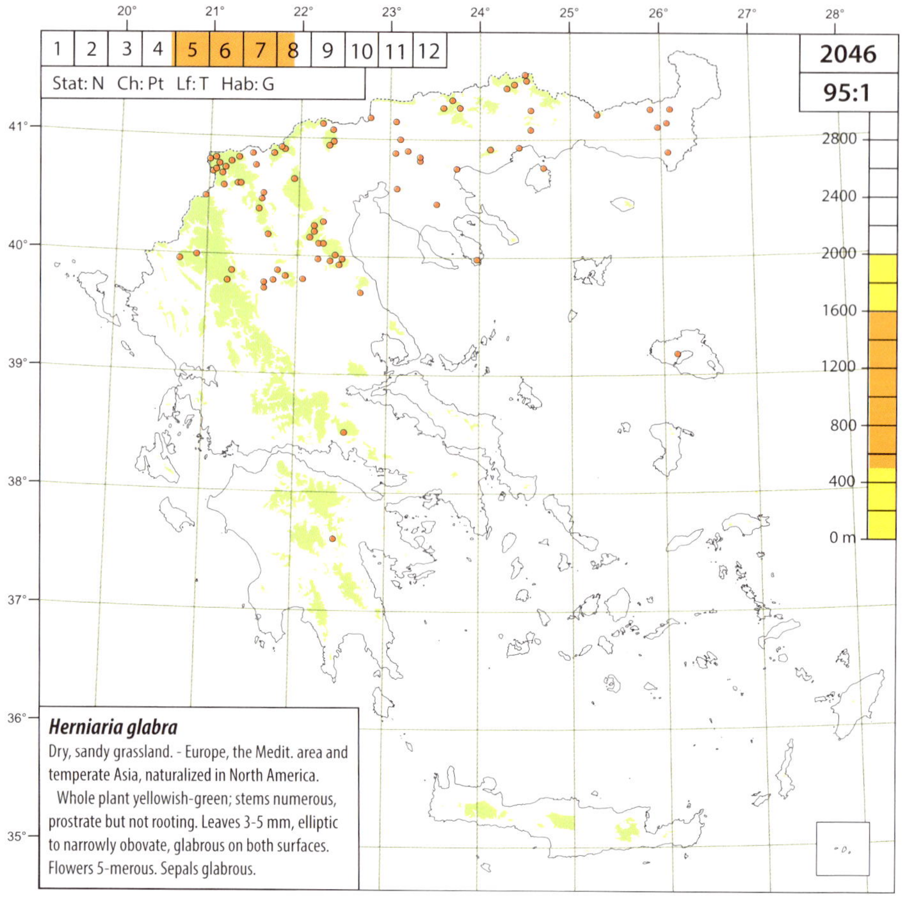

Herniaria glabra
Dry, sandy grassland. - Europe, the Medit. area and temperate Asia, naturalized in North America.
Whole plant yellowish-green; stems numerous, prostrate but not rooting. Leaves 3-5 mm, elliptic to narrowly obovate, glabrous on both surfaces. Flowers 5-merous. Sepals glabrous.

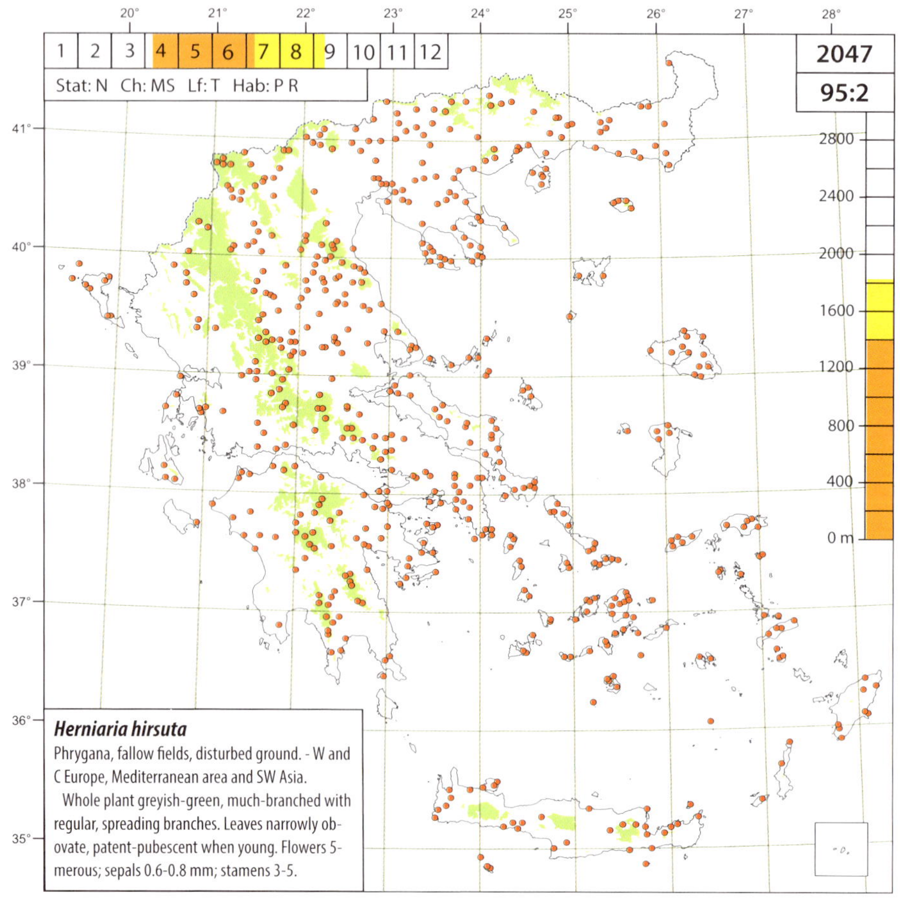

Herniaria hirsuta
Phrygana, fallow fields, disturbed ground. - W and C Europe, Mediterranean area and SW Asia.
Whole plant greyish-green, much-branched with regular, spreading branches. Leaves narrowly ob-ovate, patent-pubescent when young. Flowers 5-merous; sepals 0.6-0.8 mm; stamens 3-5.

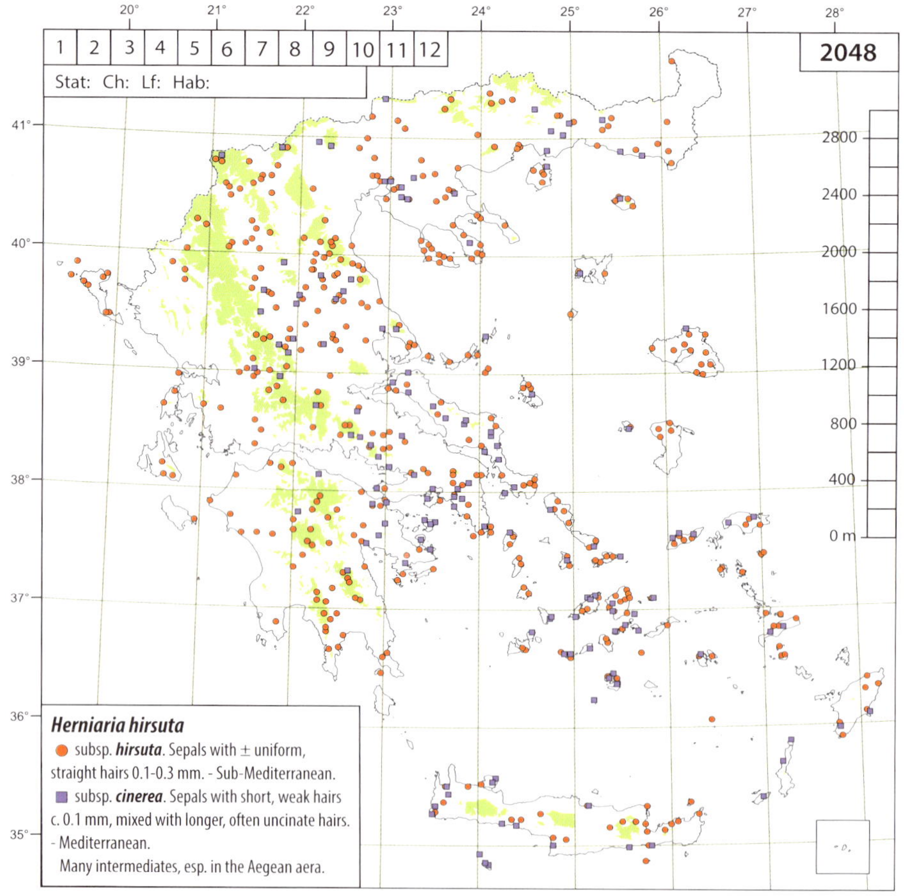

Herniaria hirsuta
● subsp. ***hirsuta***. Sepals with ± uniform, straight hairs 0.1-0.3 mm. - Sub-Mediterranean.
■ subsp. ***cinerea***. Sepals with short, weak hairs c. 0.1 mm, mixed with longer, often uncinate hairs. - Mediterranean.
Many intermediates, esp. in the Aegean aera.

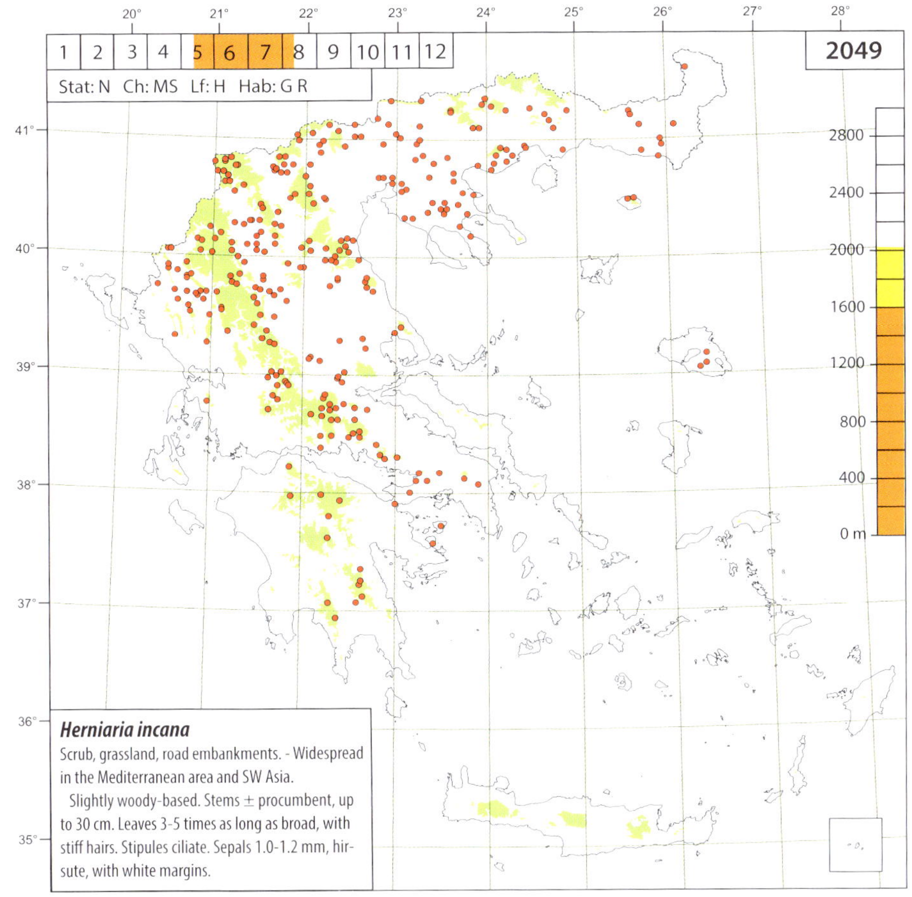

Herniaria incana

Scrub, grassland, road embankments. - Widespread in the Mediterranean area and SW Asia.

Slightly woody-based. Stems ± procumbent, up to 30 cm. Leaves 3-5 times as long as broad, with stiff hairs. Stipules ciliate. Sepals 1.0-1.2 mm, hirsute, with white margins.

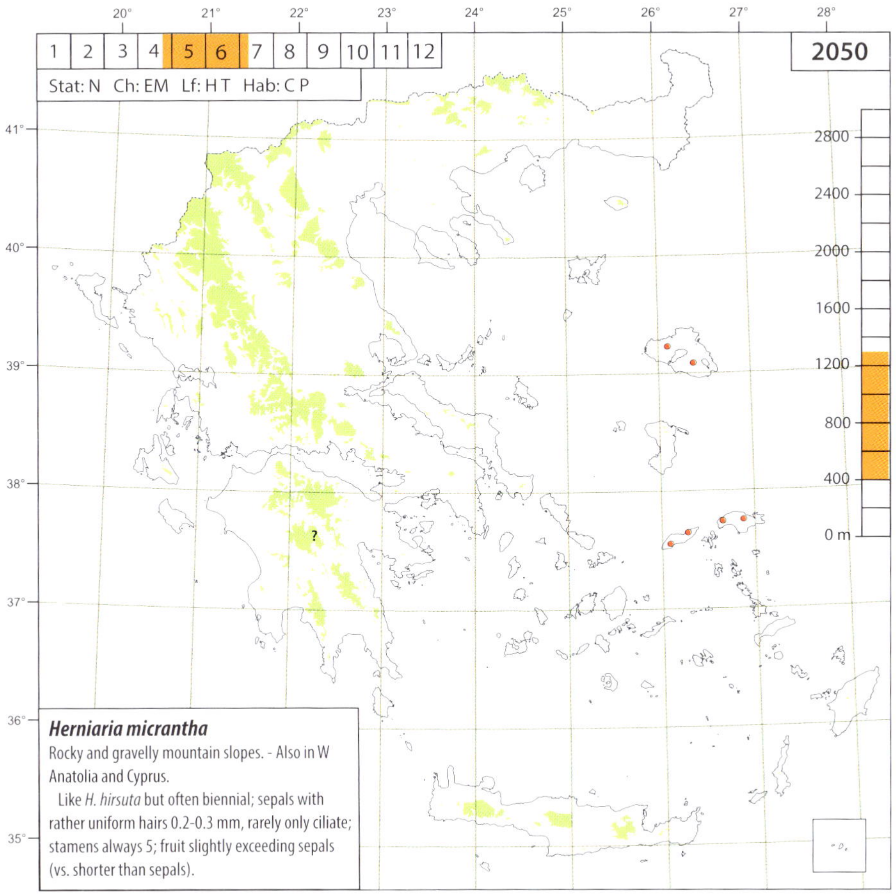

Herniaria micrantha

Rocky and gravelly mountain slopes. - Also in W Anatolia and Cyprus.

Like *H. hirsuta* but often biennial; sepals with rather uniform hairs 0.2-0.3 mm, rarely only ciliate; stamens always 5; fruit slightly exceeding sepals (vs. shorter than sepals).

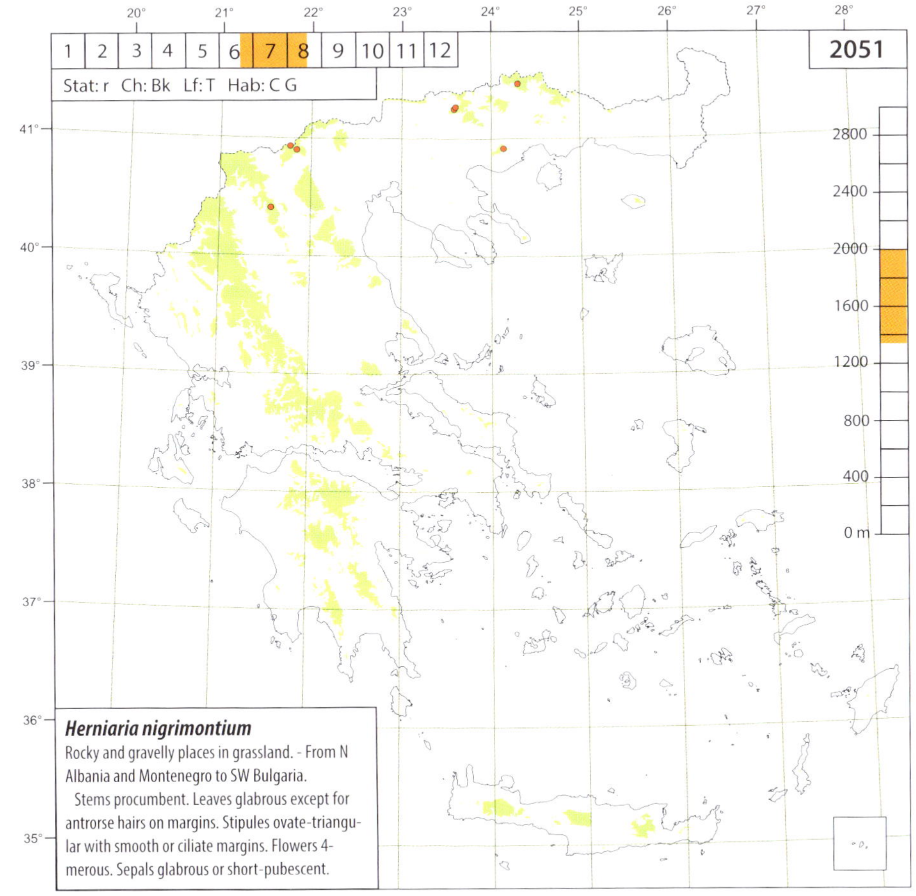

Herniaria nigrimontium

Rocky and gravelly places in grassland. - From N Albania and Montenegro to SW Bulgaria.

Stems procumbent. Leaves glabrous except for antrorse hairs on margins. Stipules ovate-triangular with smooth or ciliate margins. Flowers 4-merous. Sepals glabrous or short-pubescent.

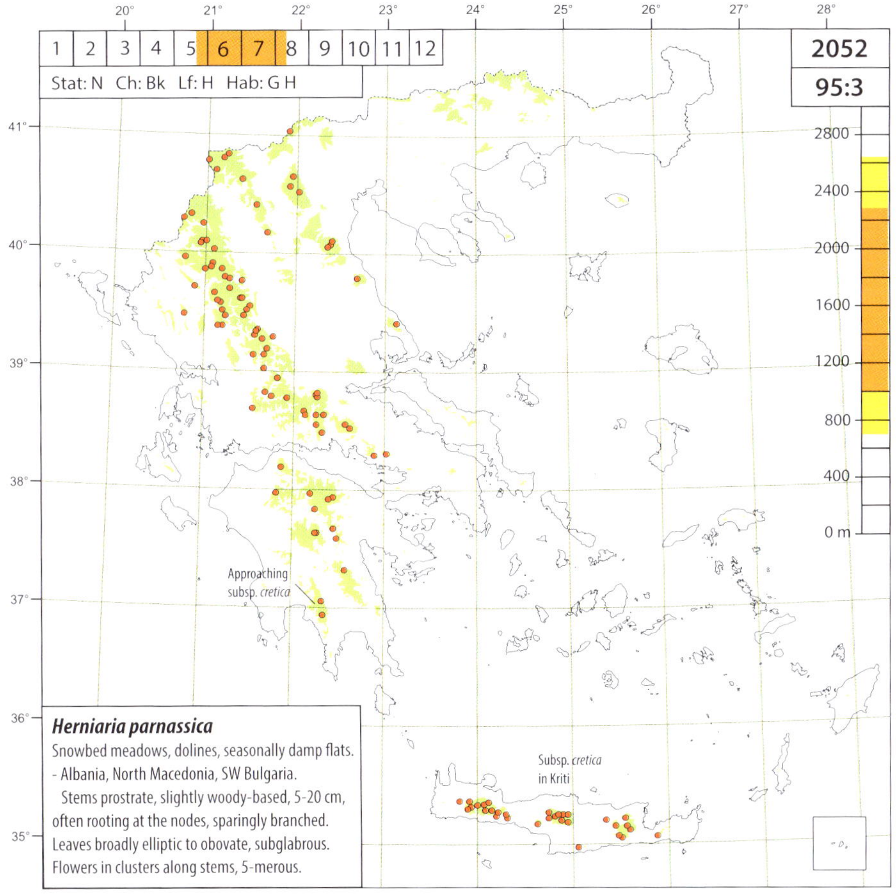

Herniaria parnassica

Snowbed meadows, dolines, seasonally damp flats. - Albania, North Macedonia, SW Bulgaria.

Stems prostrate, slightly woody-based, 5-20 cm, often rooting at the nodes, sparingly branched. Leaves broadly elliptic to obovate, subglabrous. Flowers in clusters along stems, 5-merous.

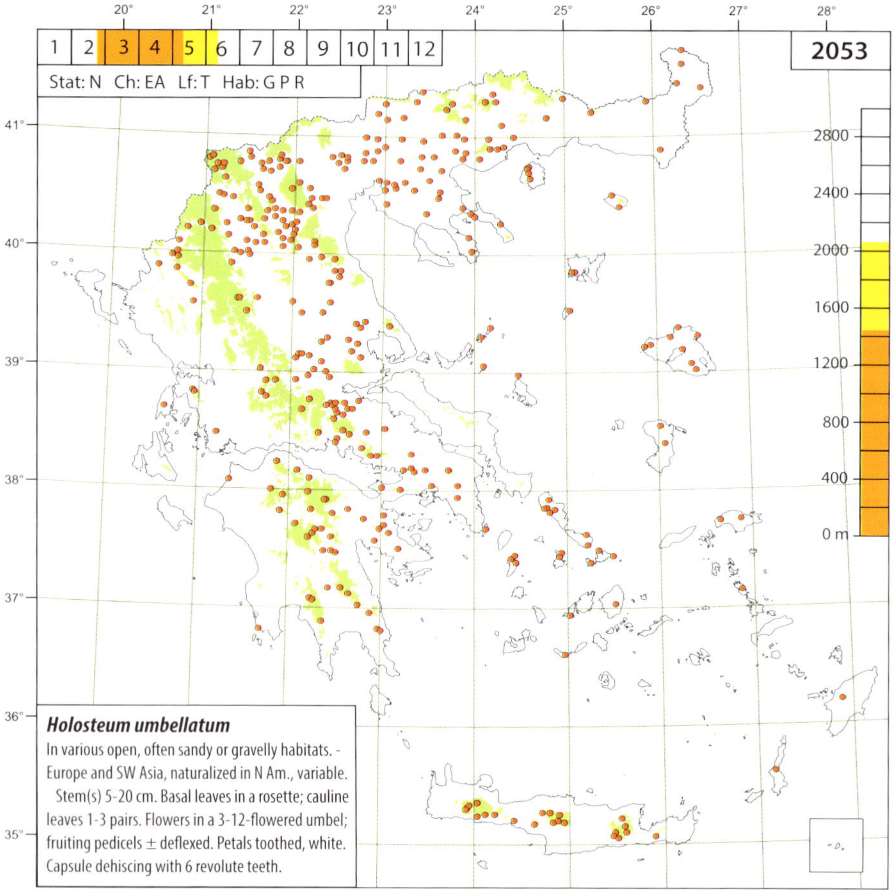

Holosteum umbellatum

In various open, often sandy or gravelly habitats. - Europe and SW Asia, naturalized in N Am., variable.

Stem(s) 5-20 cm. Basal leaves in a rosette; cauline leaves 1-3 pairs. Flowers in a 3-12-flowered umbel; fruiting pedicels ± deflexed. Petals toothed, white. Capsule dehiscing with 6 revolute teeth.

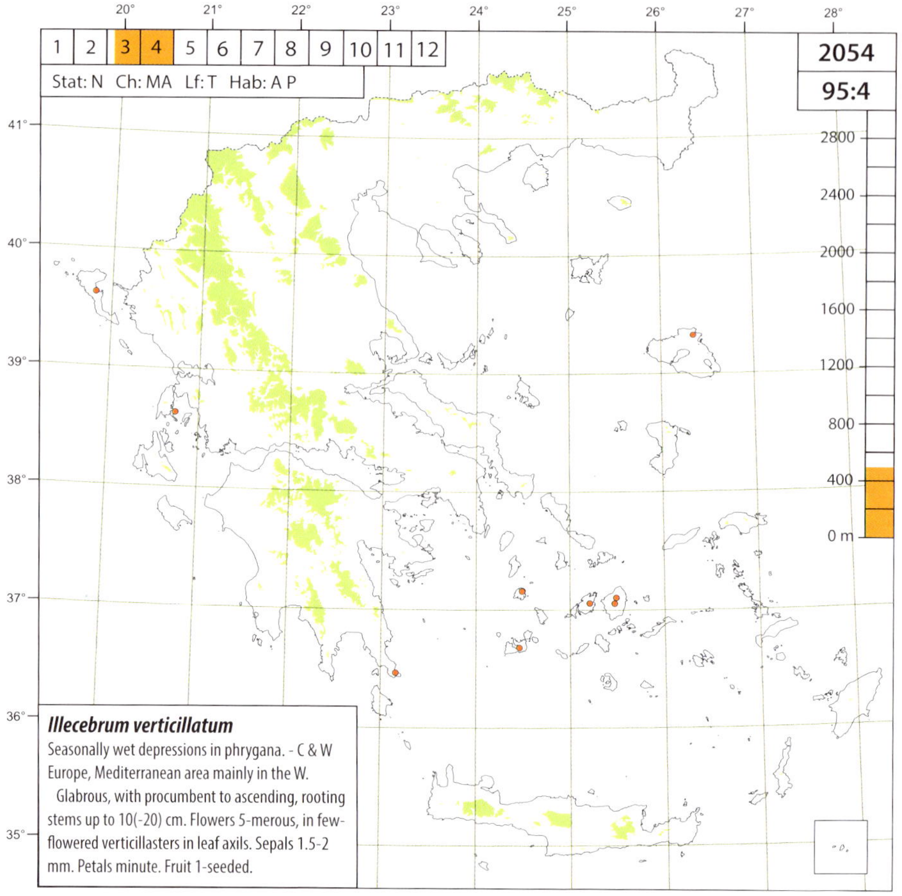

Illecebrum verticillatum

Seasonally wet depressions in phrygana. - C & W Europe, Mediterranean area mainly in the W.

Glabrous, with procumbent to ascending, rooting stems up to 10(-20) cm. Flowers 5-merous, in few-flowered verticillasters in leaf axils. Sepals 1.5-2 mm. Petals minute. Fruit 1-seeded.

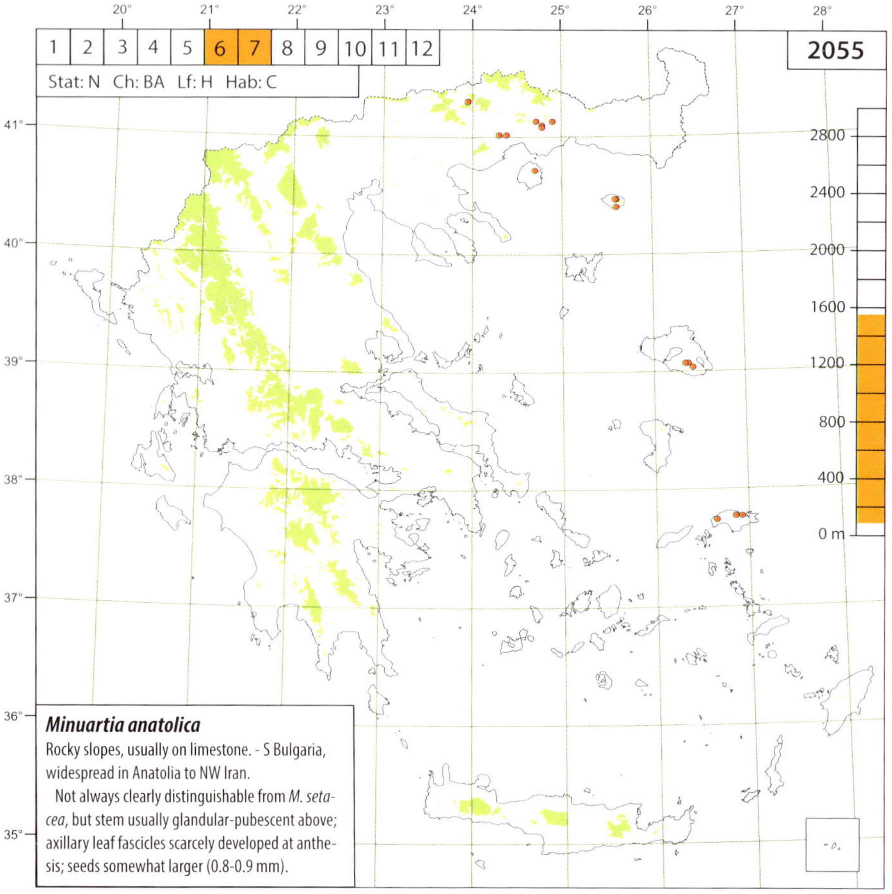

Minuartia anatolica

Rocky slopes, usually on limestone. - S Bulgaria, widespread in Anatolia to NW Iran.

Not always clearly distinguishable from *M. setacea*, but stem usually glandular-pubescent above; axillary leaf fascicles scarcely developed at anthesis; seeds somewhat larger (0.8-0.9 mm).

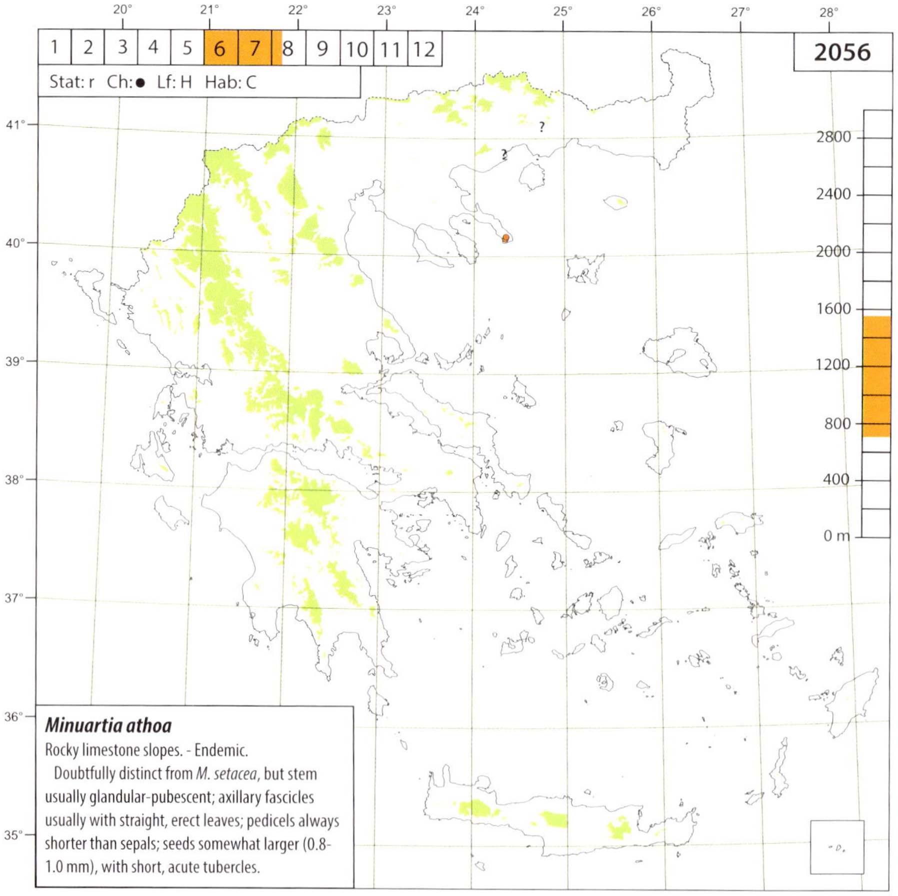

Minuartia athoa

Rocky limestone slopes. - Endemic.

Doubtfully distinct from *M. setacea*, but stem usually glandular-pubescent; axillary fascicles usually with straight, erect leaves; pedicels always shorter than sepals; seeds somewhat larger (0.8-1.0 mm), with short, acute tubercles.

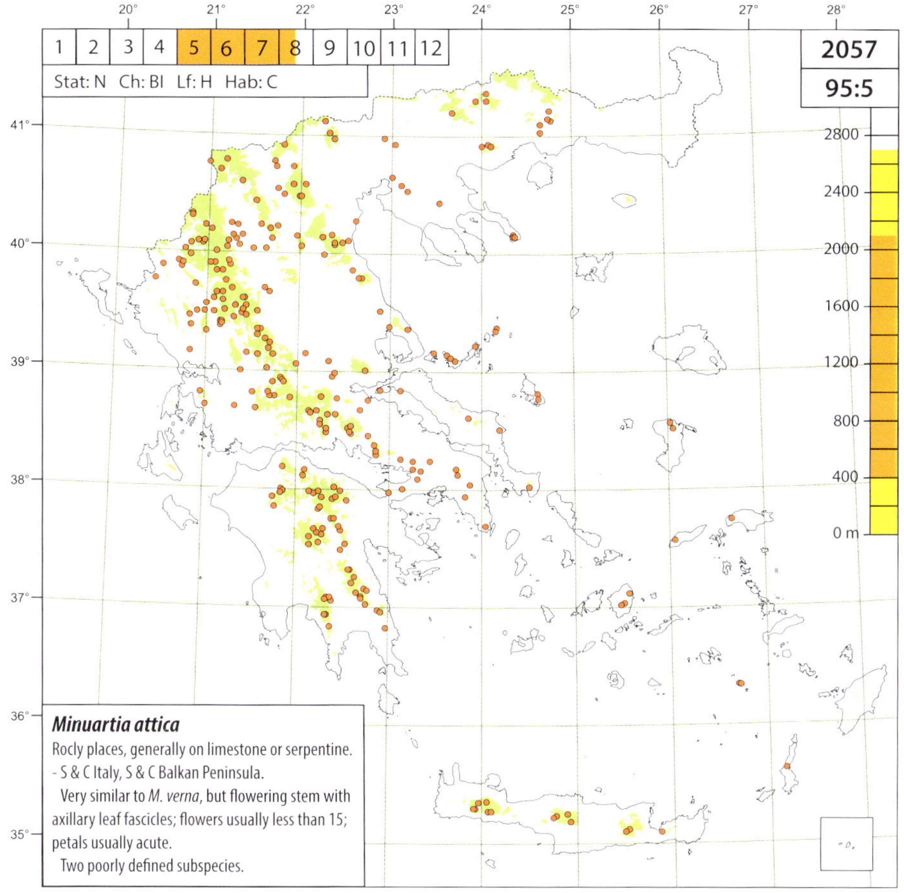

Minuartia attica

Rocly places, generally on limestone or serpentine. - S & C Italy, S & C Balkan Peninsula.

Very similar to *M. verna*, but flowering stem with axillary leaf fascicles; flowers usually less than 15; petals usually acute.

Two poorly defined subspecies.

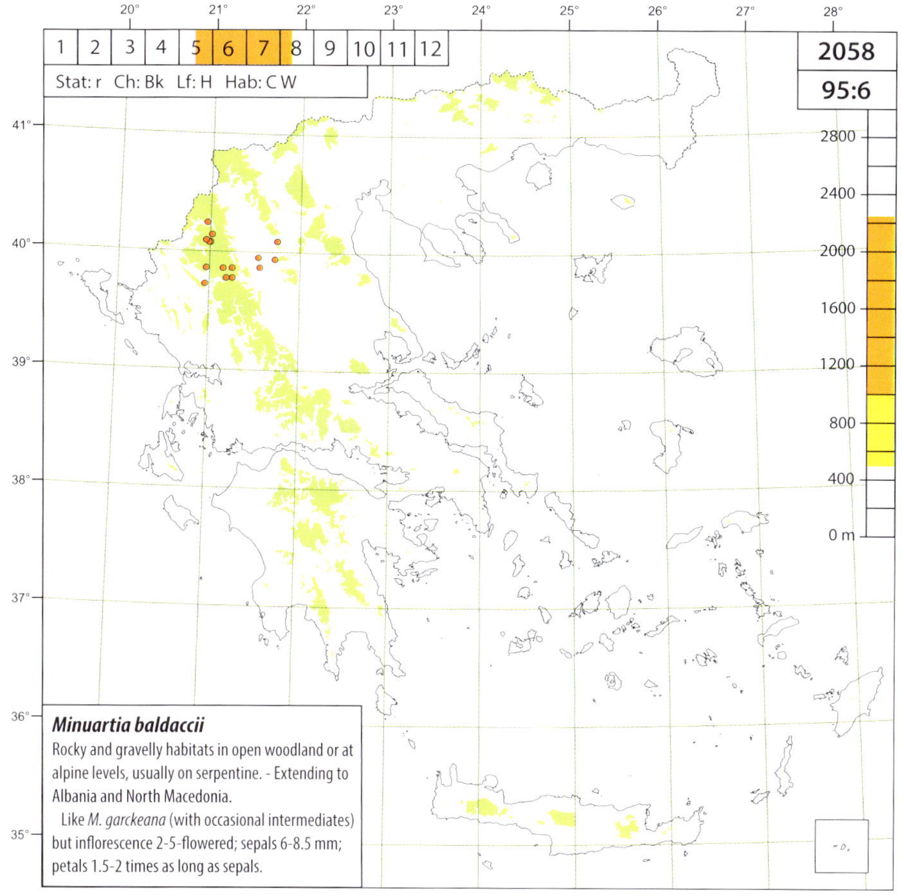

Minuartia baldaccii

Rocky and gravelly habitats in open woodland or at alpine levels, usually on serpentine. - Extending to Albania and North Macedonia.

Like *M. garckeana* (with occasional intermediates) but inflorescence 2-5-flowered; sepals 6-8.5 mm; petals 1.5-2 times as long as sepals.

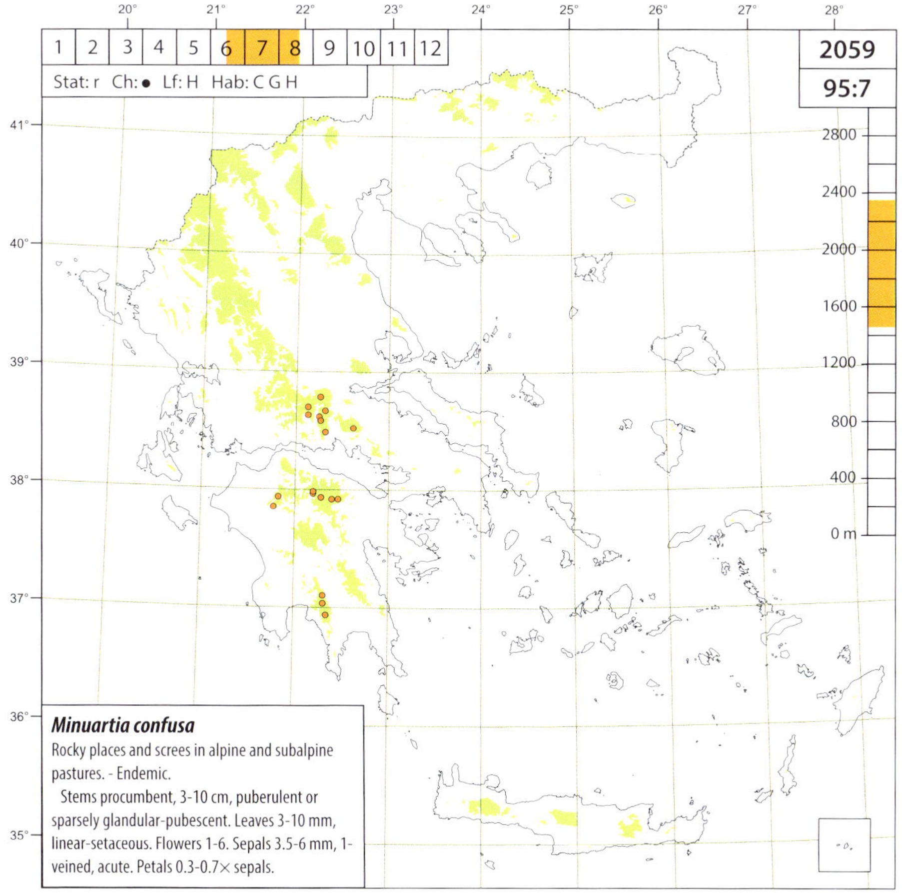

Minuartia confusa

Rocky places and screes in alpine and subalpine pastures. - Endemic.

Stems procumbent, 3-10 cm, puberulent or sparsely glandular-pubescent. Leaves 3-10 mm, linear-setaceous. Flowers 1-6. Sepals 3.5-6 mm, 1-veined, acute. Petals 0.3-0.7× sepals.

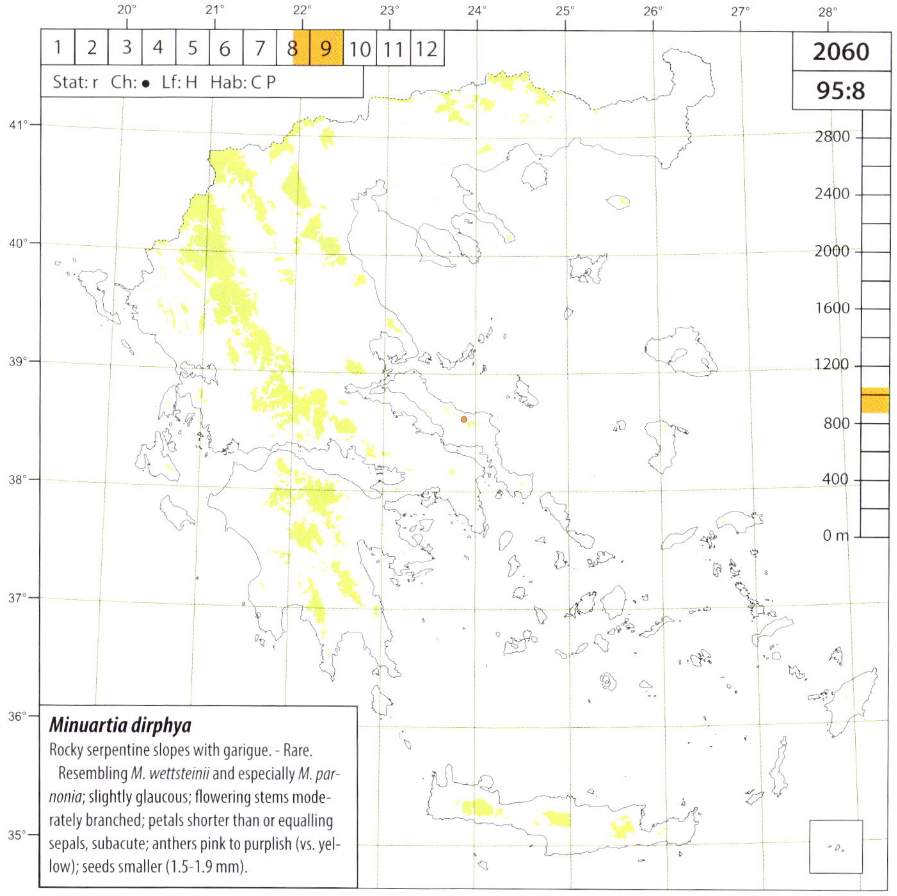

Minuartia dirphya

Rocky serpentine slopes with garigue. - Rare.

Resembling *M. wettsteinii* and especially *M. parnonia*; slightly glaucous; flowering stems moderately branched; petals shorter than or equalling sepals, subacute; anthers pink to purplish (vs. yellow); seeds smaller (1.5-1.9 mm).

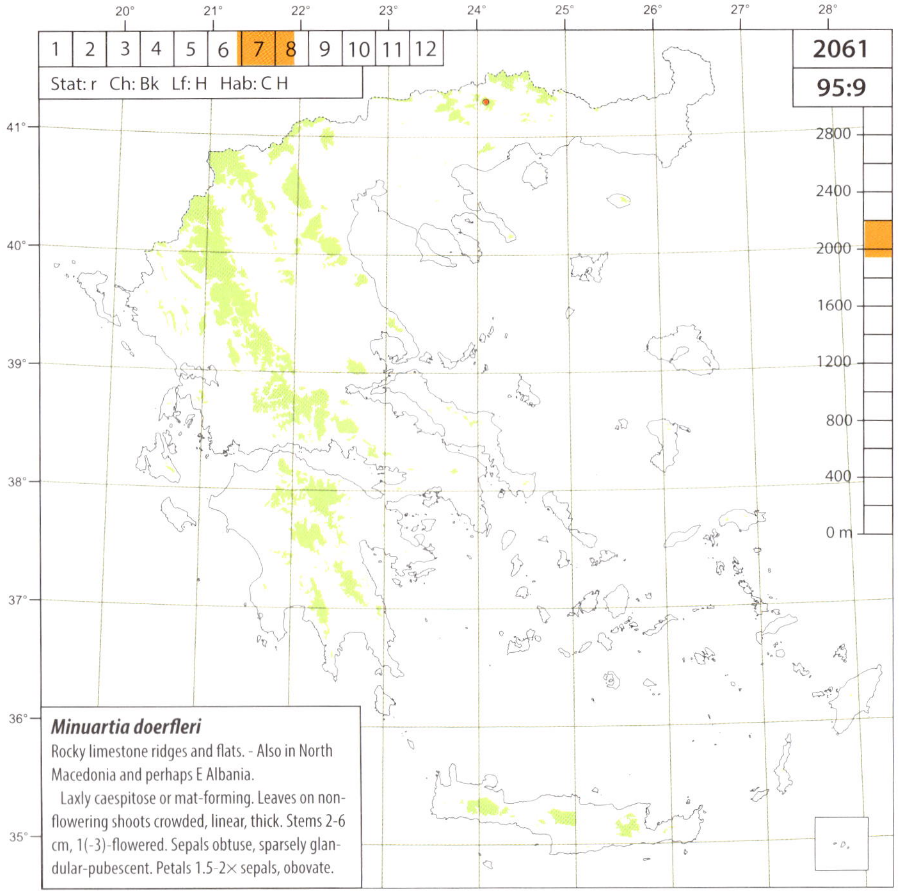

Minuartia doerfleri
Rocky limestone ridges and flats. - Also in North Macedonia and perhaps E Albania.
Laxly caespitose or mat-forming. Leaves on non-flowering shoots crowded, linear, thick. Stems 2-6 cm, 1(-3)-flowered. Sepals obtuse, sparsely glandular-pubescent. Petals 1.5-2× sepals, obovate.

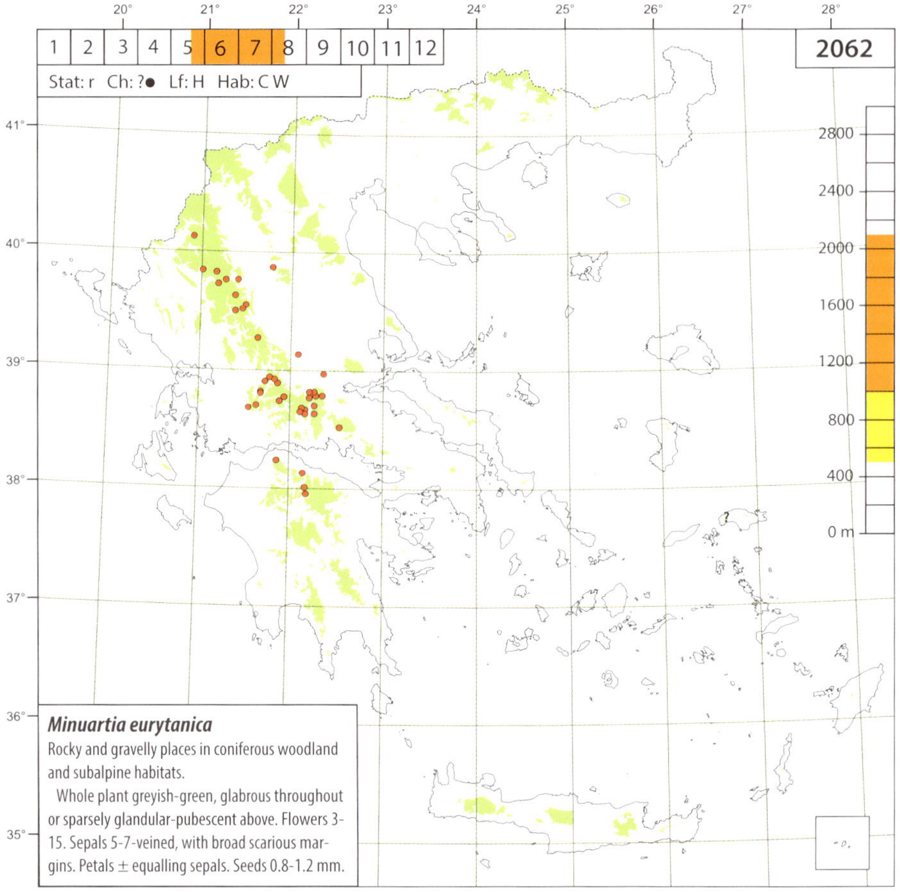

Minuartia eurytanica
Rocky and gravelly places in coniferous woodland and subalpine habitats.
Whole plant greyish-green, glabrous throughout or sparsely glandular-pubescent above. Flowers 3-15. Sepals 5-7-veined, with broad scarious margins. Petals ± equalling sepals. Seeds 0.8-1.2 mm.

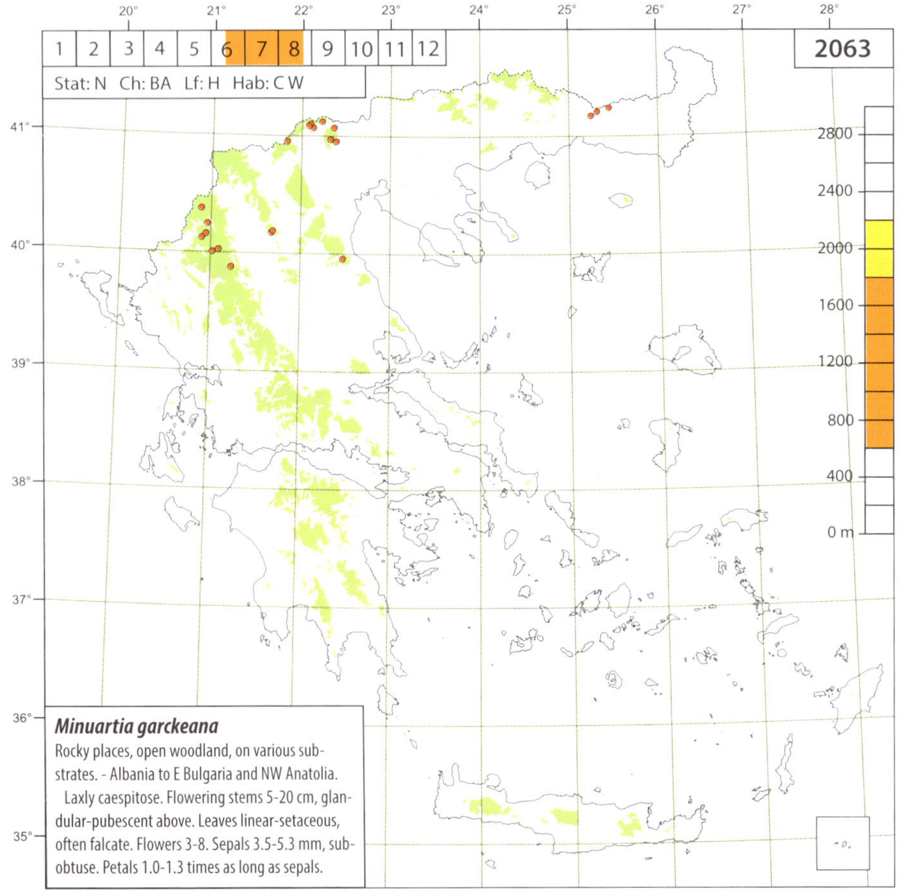

Minuartia garckeana
Rocky places, open woodland, on various substrates. - Albania to E Bulgaria and NW Anatolia.
Laxly caespitose. Flowering stems 5-20 cm, glandular-pubescent above. Leaves linear-setaceous, often falcate. Flowers 3-8. Sepals 3.5-5.3 mm, subobtuse. Petals 1.0-1.3 times as long as sepals.

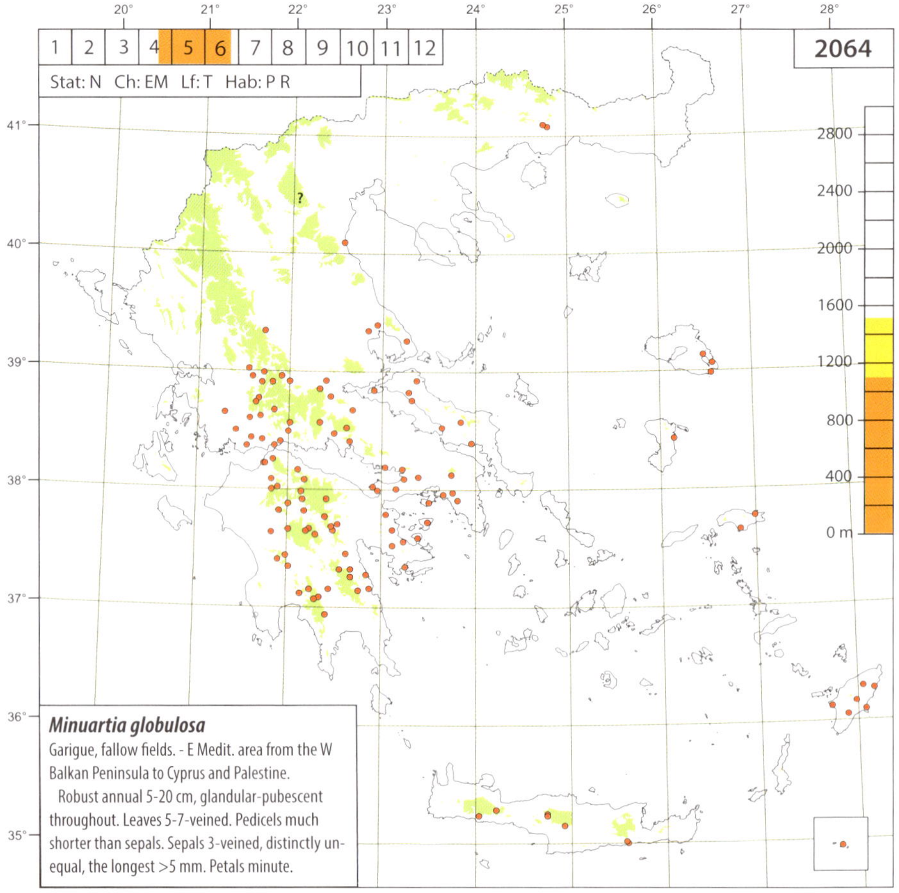

Minuartia globulosa
Garigue, fallow fields. - E Medit. area from the W Balkan Peninsula to Cyprus and Palestine.
Robust annual 5-20 cm, glandular-pubescent throughout. Leaves 5-7-veined. Pedicels much shorter than sepals. Sepals 3-veined, distinctly unequal, the longest >5 mm. Petals minute.

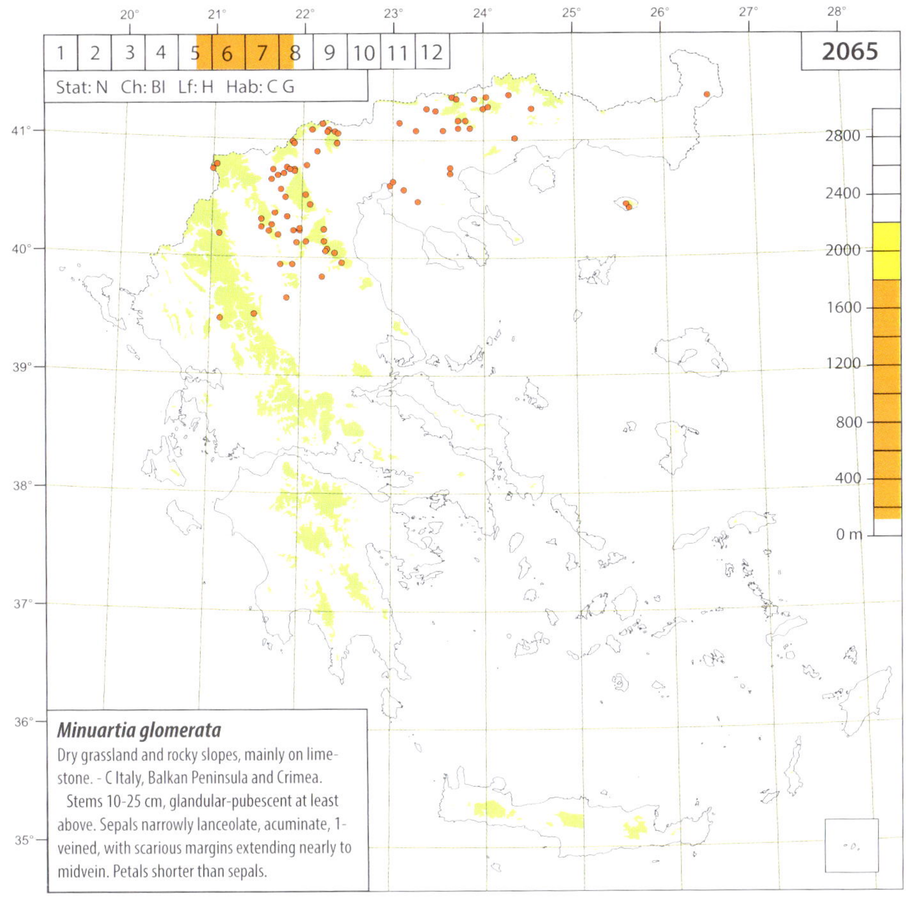

Minuartia glomerata
Dry grassland and rocky slopes, mainly on limestone. - C Italy, Balkan Peninsula and Crimea.
Stems 10-25 cm, glandular-pubescent at least above. Sepals narrowly lanceolate, acuminate, 1-veined, with scarious margins extending nearly to midvein. Petals shorter than sepals.

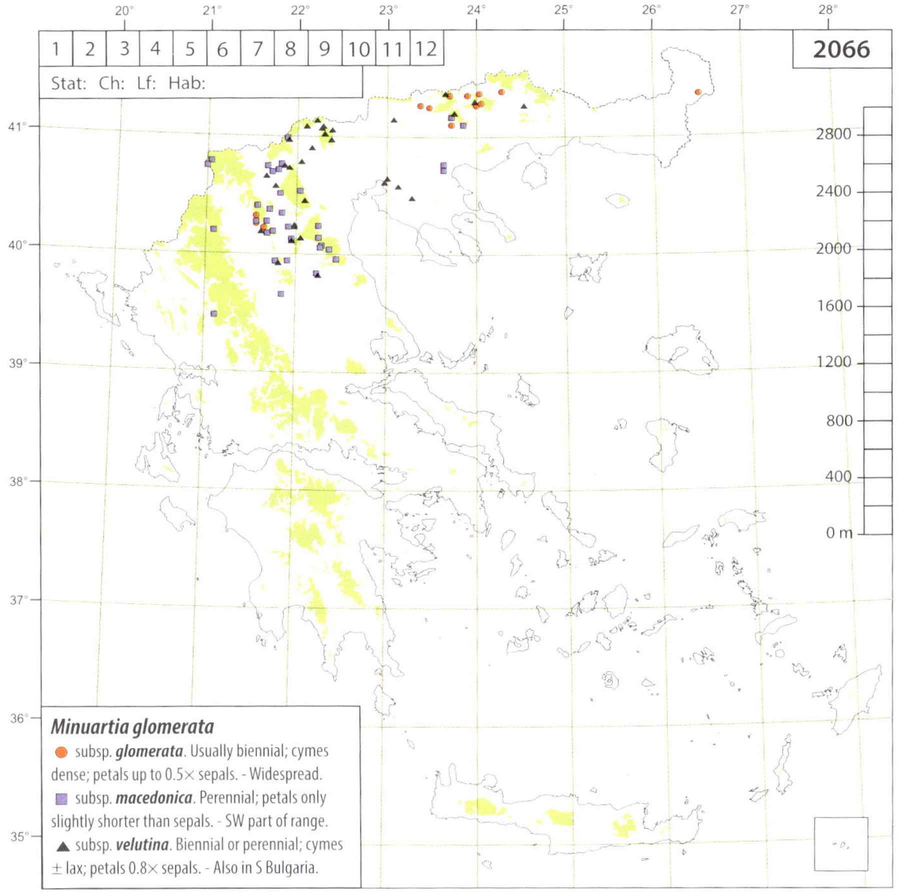

Minuartia glomerata
● subsp. ***glomerata***. Usually biennial; cymes dense; petals up to 0.5× sepals. - Widespread.
■ subsp. ***macedonica***. Perennial; petals only slightly shorter than sepals. - SW part of range.
▲ subsp. ***velutina***. Biennial or perennial; cymes ± lax; petals 0.8× sepals. - Also in S Bulgaria.

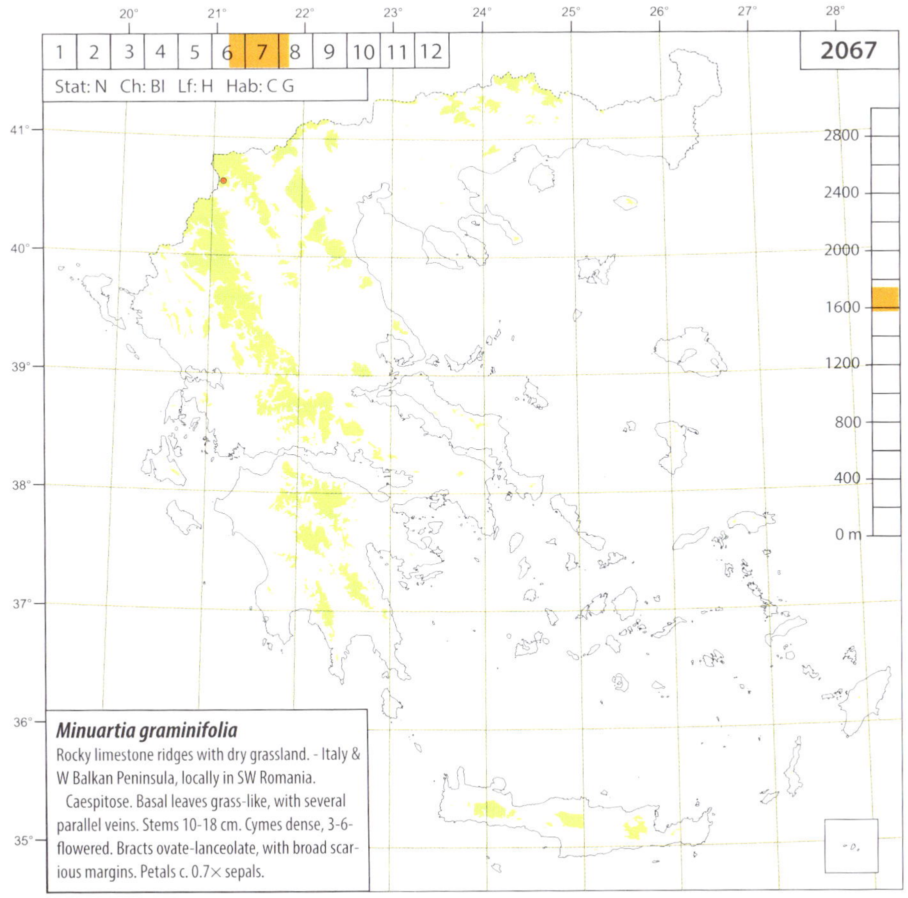

Minuartia graminifolia
Rocky limestone ridges with dry grassland. - Italy & W Balkan Peninsula, locally in SW Romania.
Caespitose. Basal leaves grass-like, with several parallel veins. Stems 10-18 cm. Cymes dense, 3-6-flowered. Bracts ovate-lanceolate, with broad scarious margins. Petals c. 0.7× sepals.

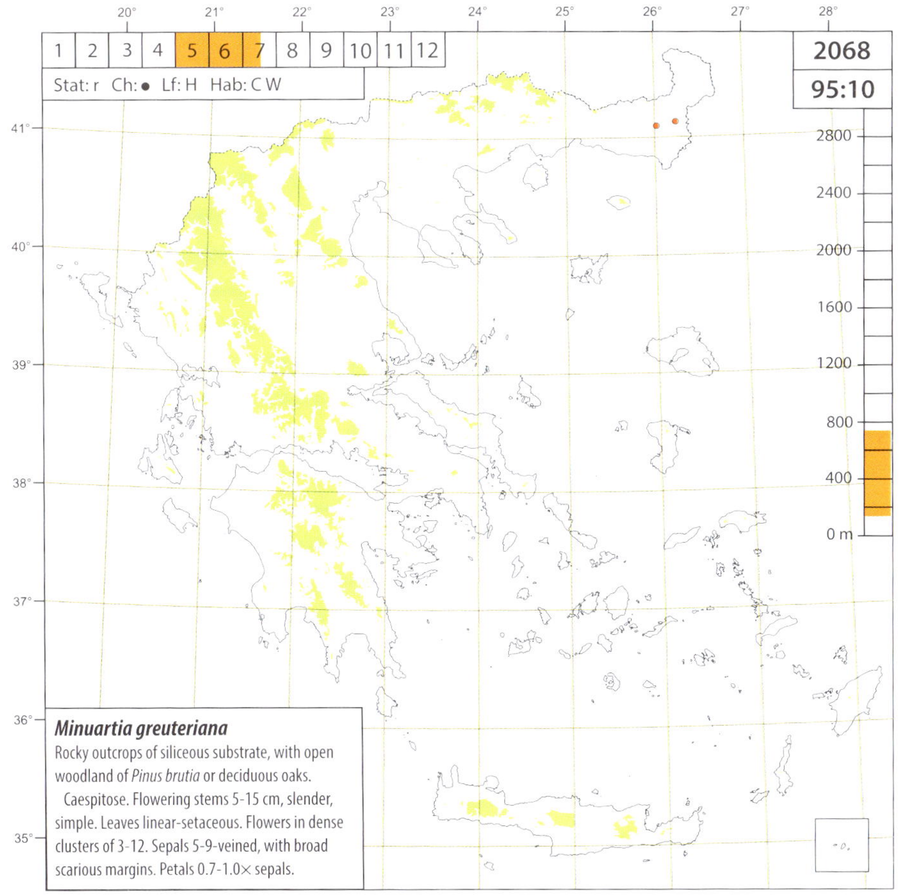

Minuartia greuteriana
Rocky outcrops of siliceous substrate, with open woodland of *Pinus brutia* or deciduous oaks.
Caespitose. Flowering stems 5-15 cm, slender, simple. Leaves linear-setaceous. Flowers in dense clusters of 3-12. Sepals 5-9-veined, with broad scarious margins. Petals 0.7-1.0× sepals.

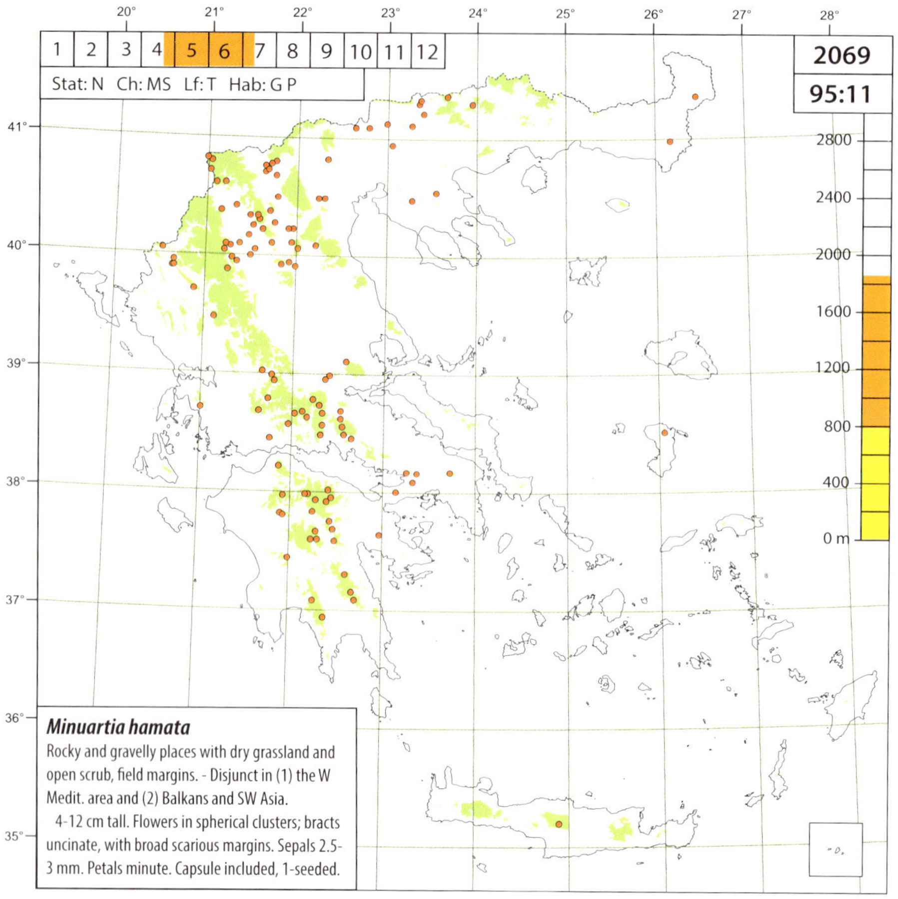

Minuartia hamata

Rocky and gravelly places with dry grassland and open scrub, field margins. - Disjunct in (1) the W Medit. area and (2) Balkans and SW Asia.

4-12 cm tall. Flowers in spherical clusters; bracts uncinate, with broad scarious margins. Sepals 2.5-3 mm. Petals minute. Capsule included, 1-seeded.

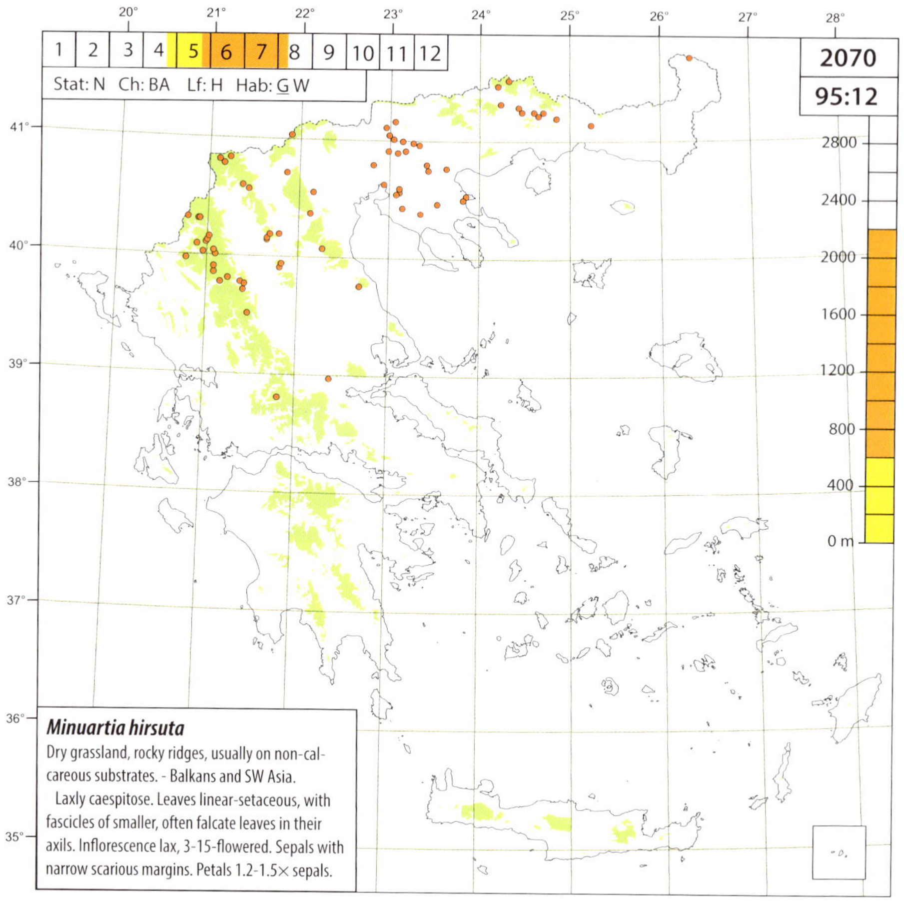

Minuartia hirsuta

Dry grassland, rocky ridges, usually on non-calcareous substrates. - Balkans and SW Asia.

Laxly caespitose. Leaves linear-setaceous, with fascicles of smaller, often falcate leaves in their axils. Inflorescence lax, 3-15-flowered. Sepals with narrow scarious margins. Petals 1.2-1.5× sepals.

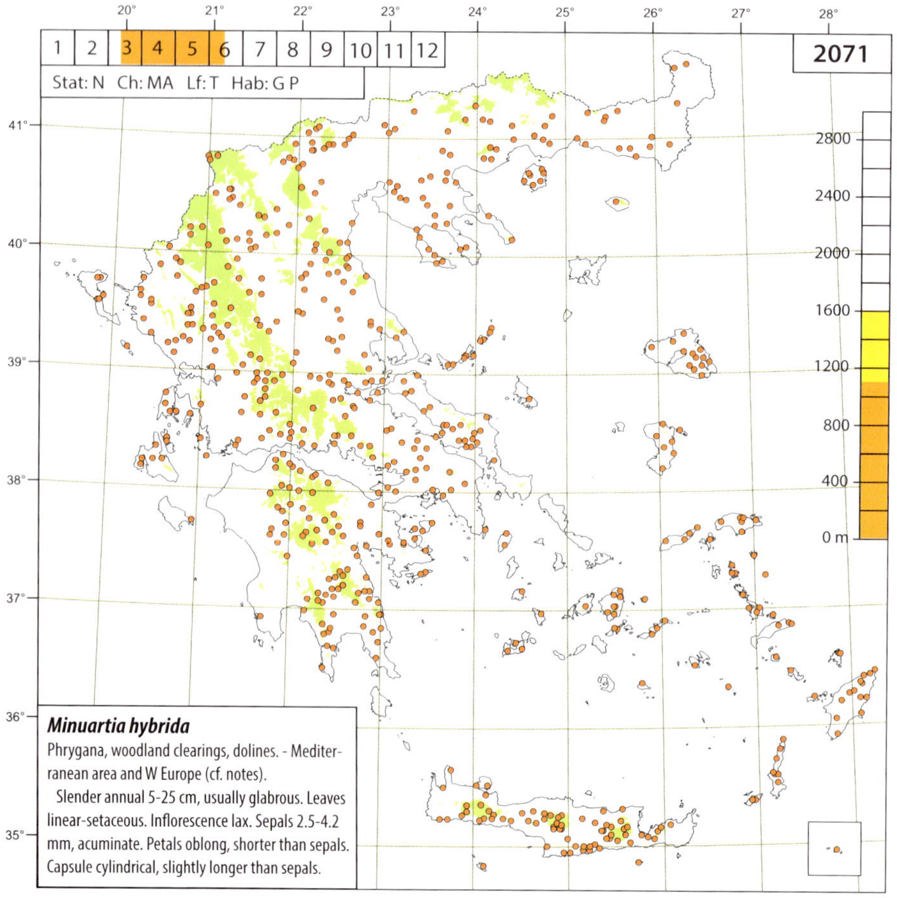

Minuartia hybrida

Phrygana, woodland clearings, dolines. - Mediterranean area and W Europe (cf. notes).

Slender annual 5-25 cm, usually glabrous. Leaves linear-setaceous. Inflorescence lax. Sepals 2.5-4.2 mm, acuminate. Petals oblong, shorter than sepals. Capsule cylindrical, slightly longer than sepals.

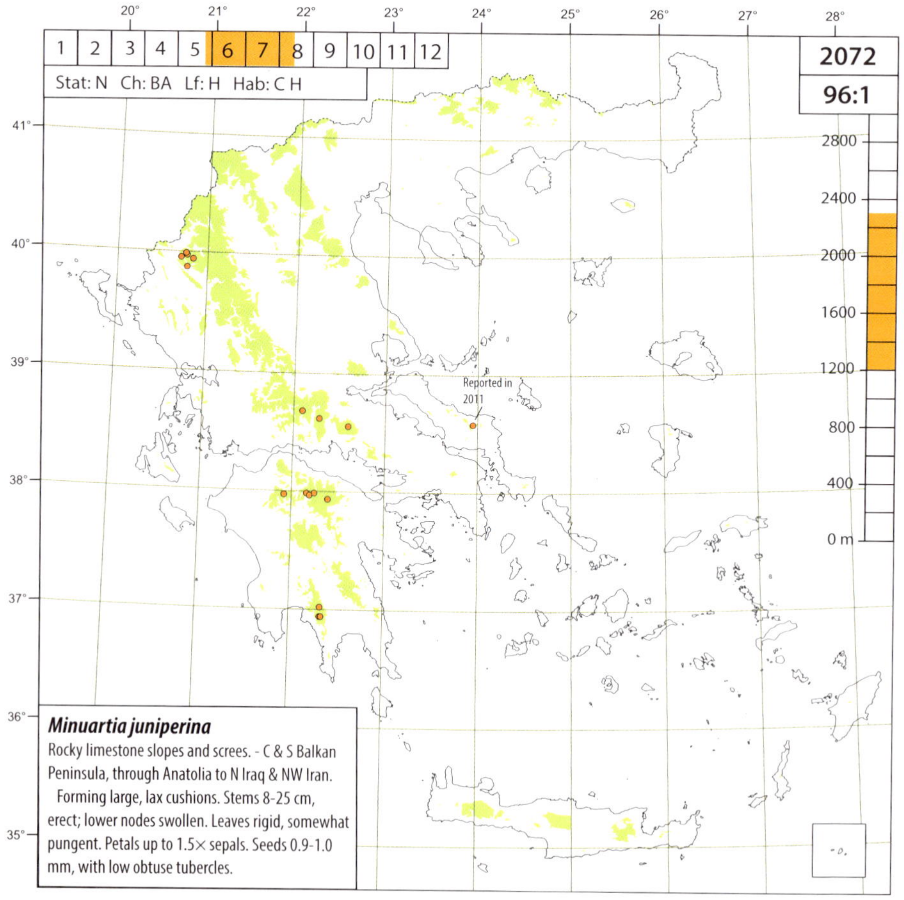

Minuartia juniperina

Rocky limestone slopes and screes. - C & S Balkan Peninsula, through Anatolia to N Iraq & NW Iran.

Forming large, lax cushions. Stems 8-25 cm, erect; lower nodes swollen. Leaves rigid, somewhat pungent. Petals up to 1.5× sepals. Seeds 0.9-1.0 mm, with low obtuse tubercles.

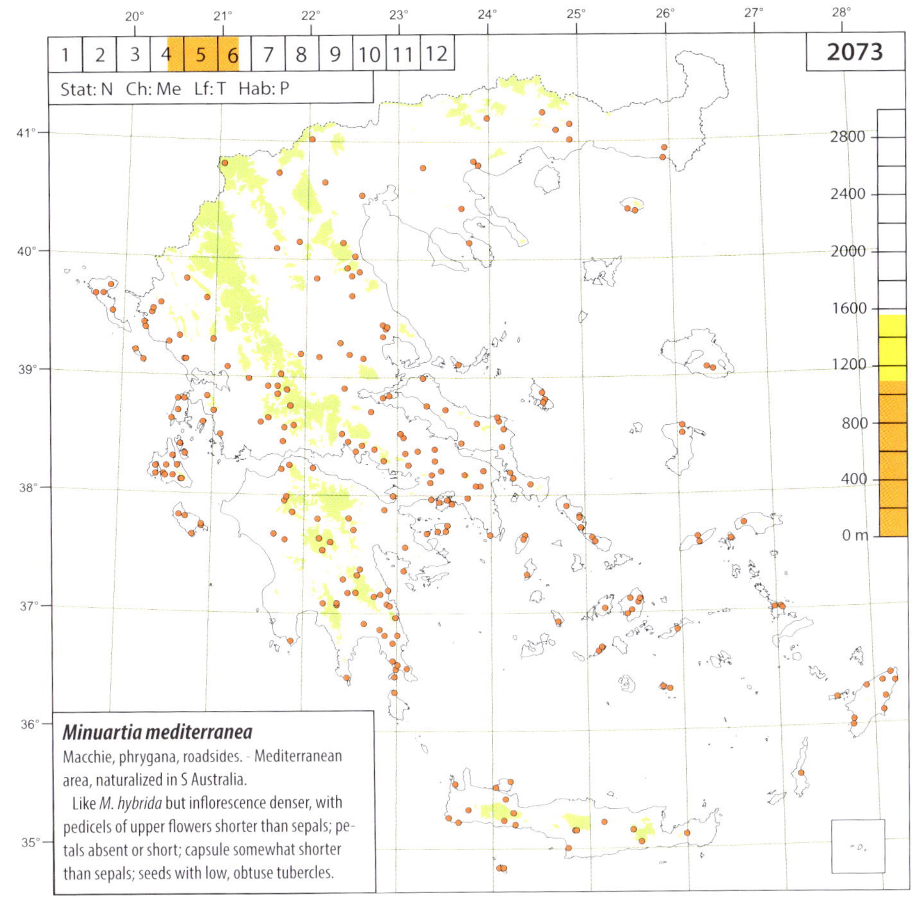

Minuartia mediterranea

Macchie, phrygana, roadsides. - Mediterranean area, naturalized in S Australia.

Like *M. hybrida* but inflorescence denser, with pedicels of upper flowers shorter than sepals; petals absent or short; capsule somewhat shorter than sepals; seeds with low, obtuse tubercles.

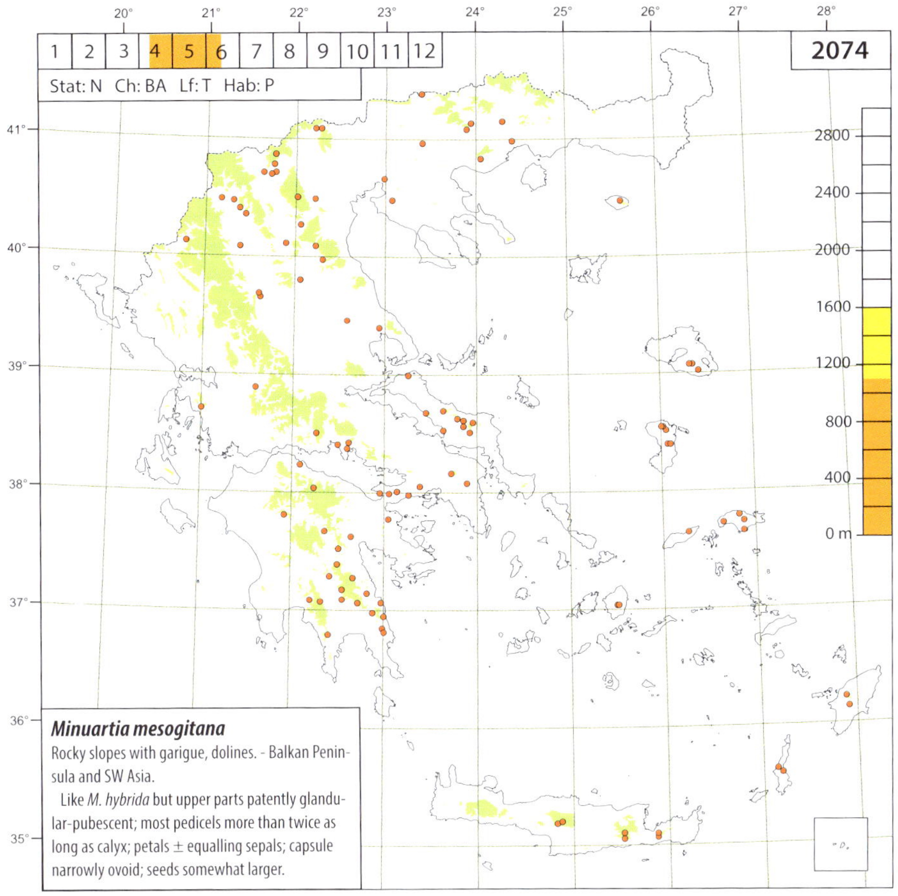

Minuartia mesogitana

Rocky slopes with garigue, dolines. - Balkan Peninsula and SW Asia.

Like *M. hybrida* but upper parts patently glandular-pubescent; most pedicels more than twice as long as calyx; petals ± equalling sepals; capsule narrowly ovoid; seeds somewhat larger.

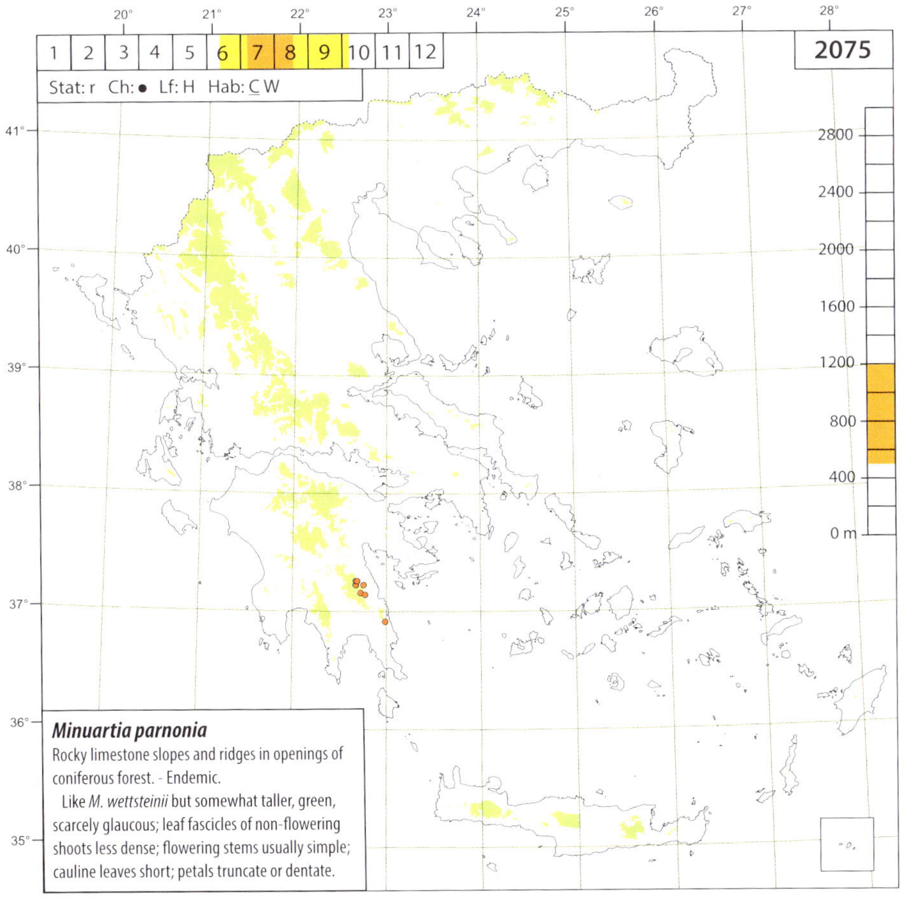

Minuartia parnonia

Rocky limestone slopes and ridges in openings of coniferous forest. - Endemic.

Like *M. wettsteinii* but somewhat taller, green, scarcely glaucous; leaf fascicles of non-flowering shoots less dense; flowering stems usually simple; cauline leaves short; petals truncate or dentate.

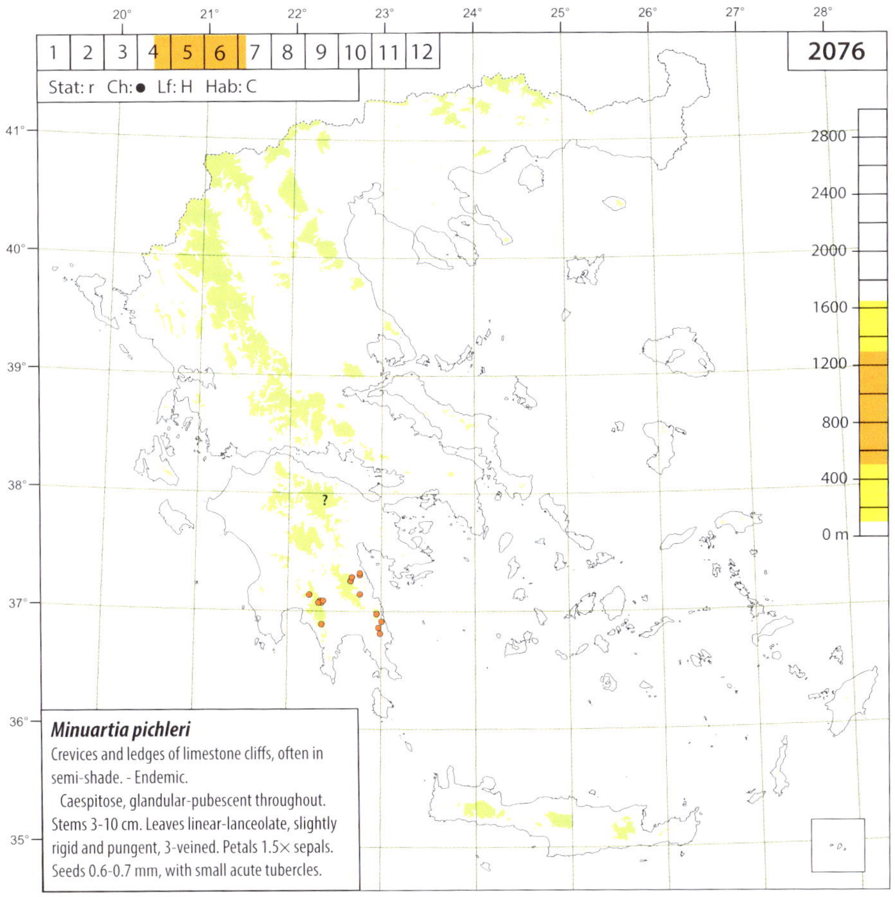

Minuartia pichleri

Crevices and ledges of limestone cliffs, often in semi-shade. - Endemic.

Caespitose, glandular-pubescent throughout. Stems 3-10 cm. Leaves linear-lanceolate, slightly rigid and pungent, 3-veined. Petals 1.5× sepals. Seeds 0.6-0.7 mm, with small acute tubercles.

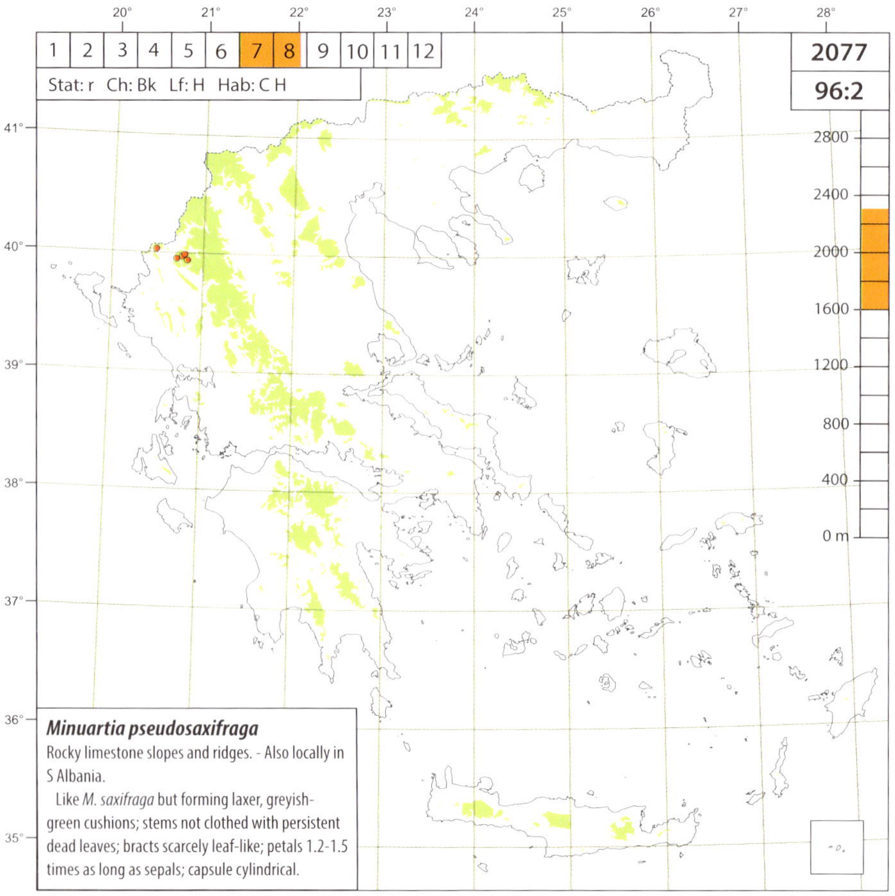

Minuartia pseudosaxifraga

Rocky limestone slopes and ridges. - Also locally in S Albania.

Like *M. saxifraga* but forming laxer, greyish-green cushions; stems not clothed with persistent dead leaves; bracts scarcely leaf-like; petals 1.2-1.5 times as long as sepals; capsule cylindrical.

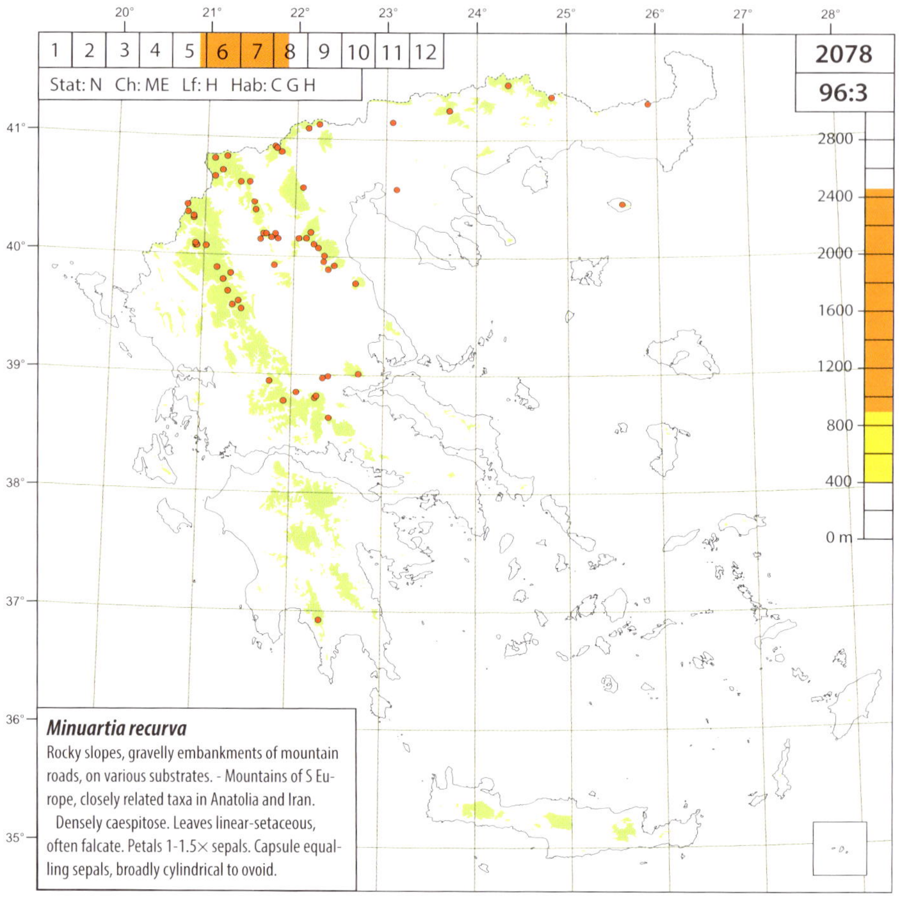

Minuartia recurva

Rocky slopes, gravelly embankments of mountain roads, on various substrates. - Mountains of S Europe, closely related taxa in Anatolia and Iran.

Densely caespitose. Leaves linear-setaceous, often falcate. Petals 1-1.5× sepals. Capsule equalling sepals, broadly cylindrical to ovoid.

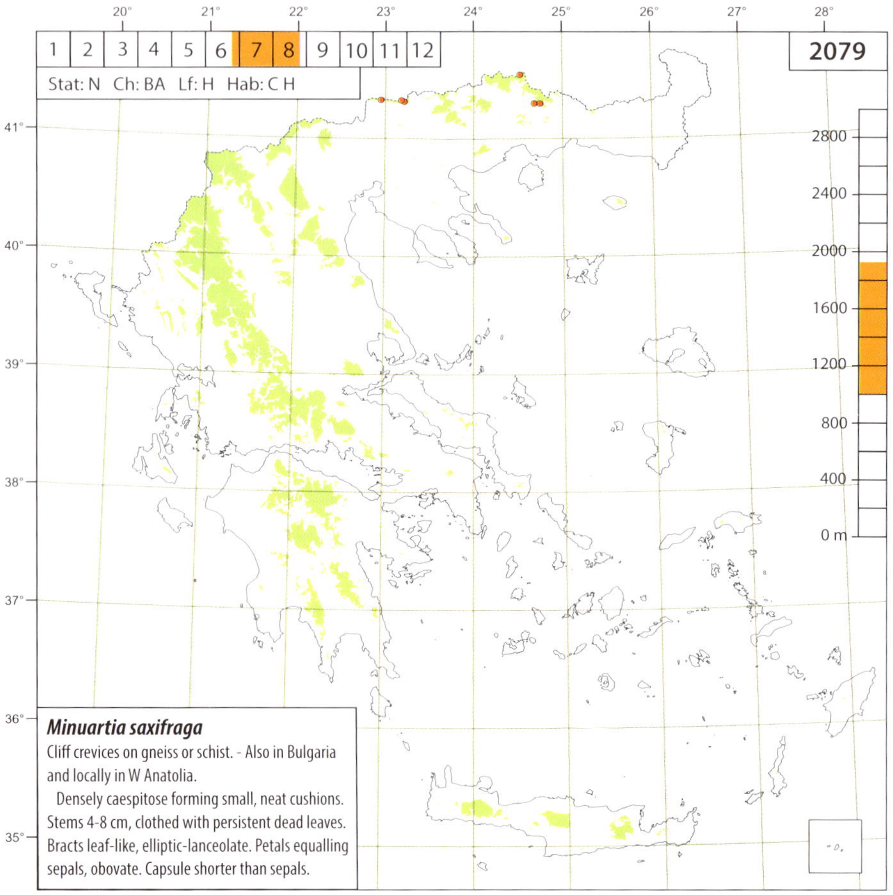

Minuartia saxifraga

Cliff crevices on gneiss or schist. - Also in Bulgaria and locally in W Anatolia.

Densely caespitose forming small, neat cushions. Stems 4-8 cm, clothed with persistent dead leaves. Bracts leaf-like, elliptic-lanceolate. Petals equalling sepals, obovate. Capsule shorter than sepals.

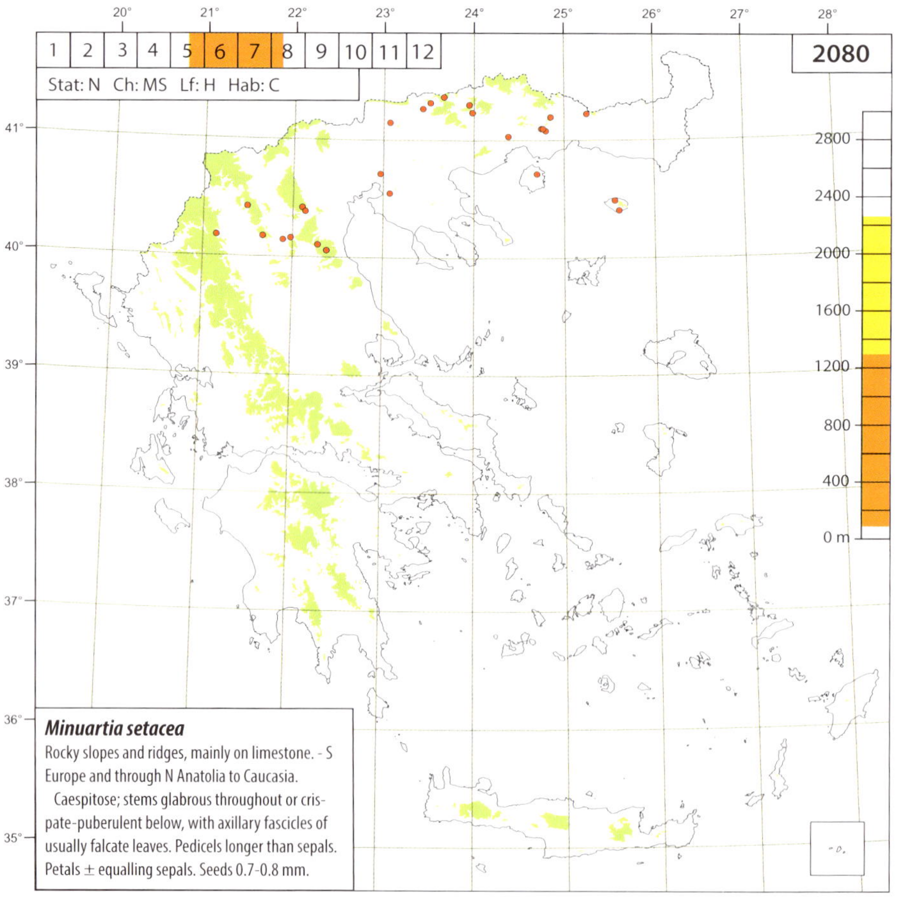

Minuartia setacea

Rocky slopes and ridges, mainly on limestone. - S Europe and through N Anatolia to Caucasia.

Caespitose; stems glabrous throughout or crispate-puberulent below, with axillary fascicles of usually falcate leaves. Pedicels longer than sepals. Petals ± equalling sepals. Seeds 0.7-0.8 mm.

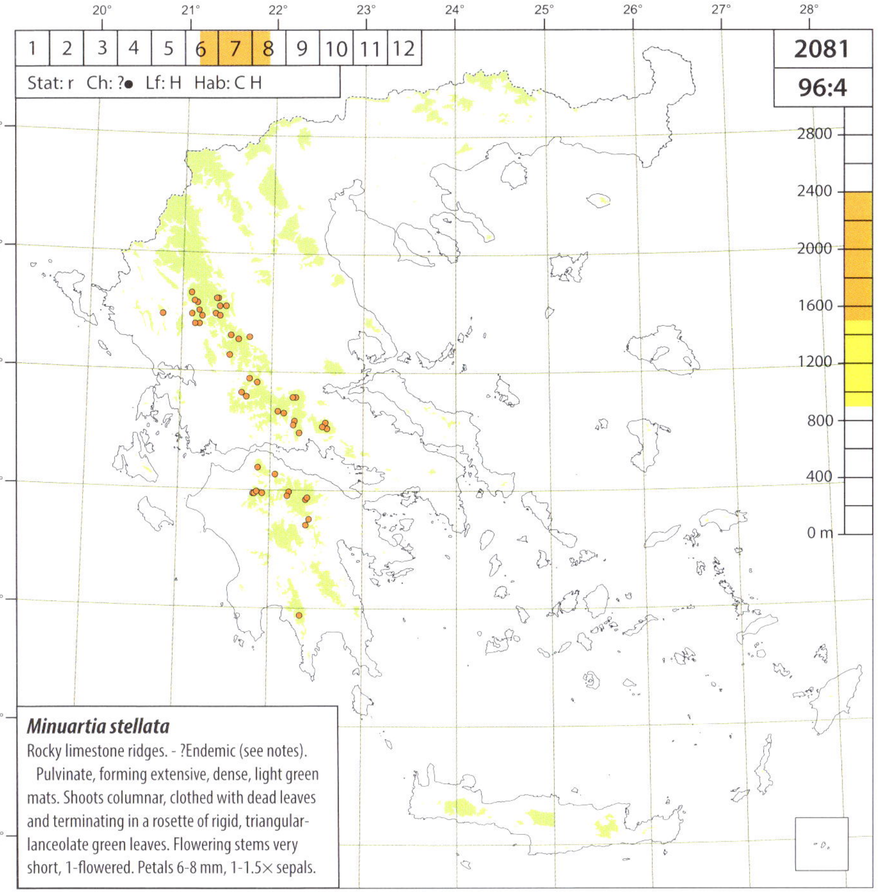

Minuartia stellata

Rocky limestone ridges. - ?Endemic (see notes).

Pulvinate, forming extensive, dense, light green mats. Shoots columnar, clothed with dead leaves and terminating in a rosette of rigid, triangular-lanceolate green leaves. Flowering stems very short, 1-flowered. Petals 6-8 mm, 1-1.5× sepals.

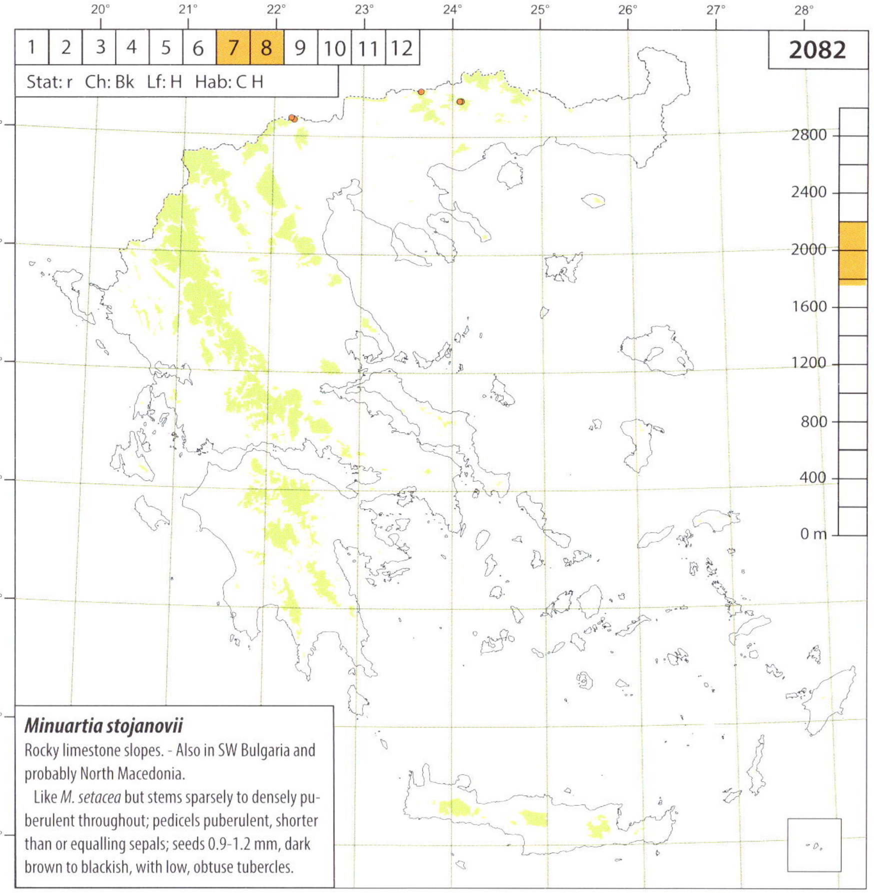

Minuartia stojanovii

Rocky limestone slopes. - Also in SW Bulgaria and probably North Macedonia.

Like *M. setacea* but stems sparsely to densely puberulent throughout; pedicels puberulent, shorter than or equalling sepals; seeds 0.9-1.2 mm, dark brown to blackish, with low, obtuse tubercles.

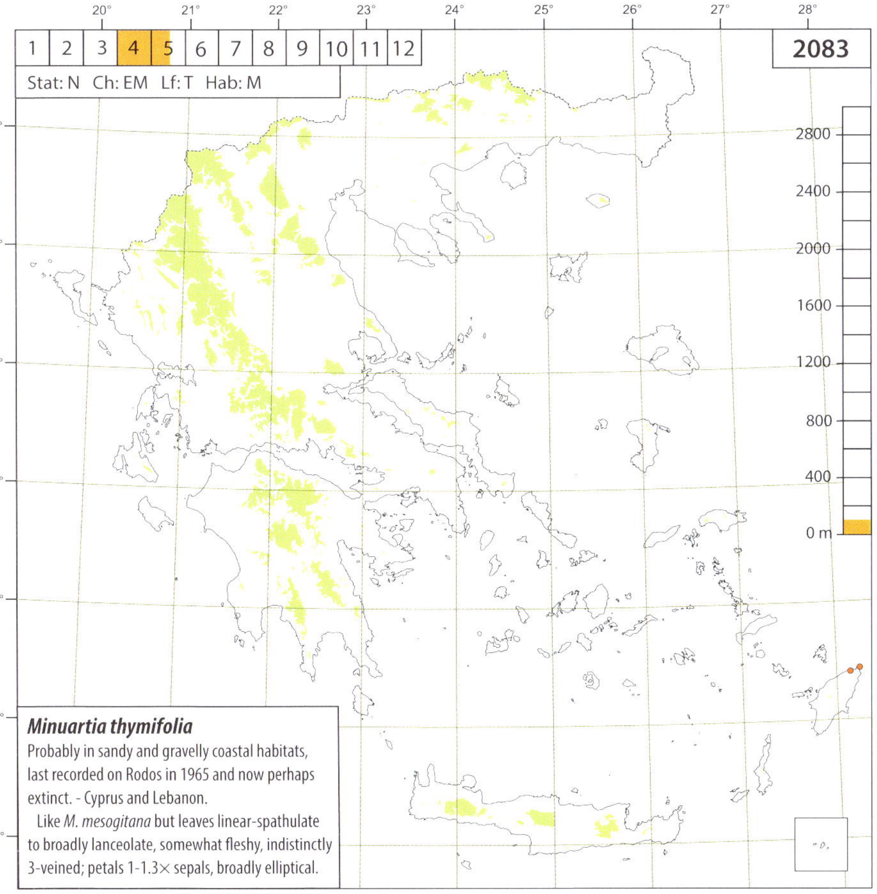

Minuartia thymifolia

Probably in sandy and gravelly coastal habitats, last recorded on Rodos in 1965 and now perhaps extinct. - Cyprus and Lebanon.

Like *M. mesogitana* but leaves linear-spathulate to broadly lanceolate, somewhat fleshy, indistinctly 3-veined; petals 1-1.3× sepals, broadly elliptical.

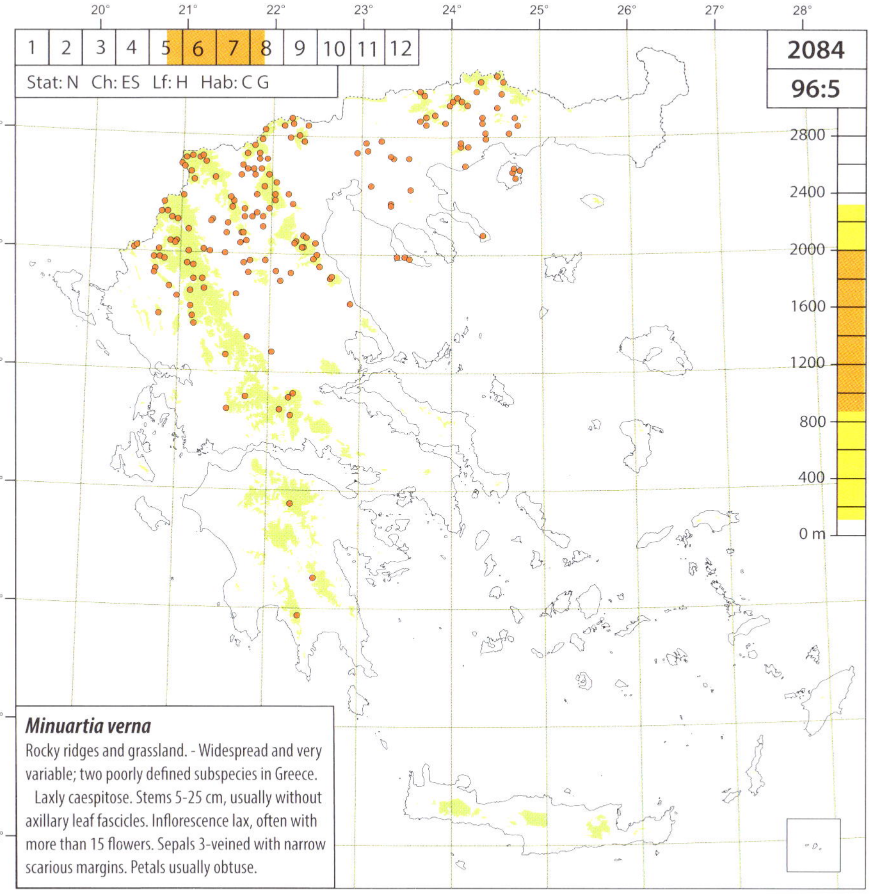

Minuartia verna

Rocky ridges and grassland. - Widespread and very variable; two poorly defined subspecies in Greece.

Laxly caespitose. Stems 5-25 cm, usually without axillary leaf fascicles. Inflorescence lax, often with more than 15 flowers. Sepals 3-veined with narrow scarious margins. Petals usually obtuse.

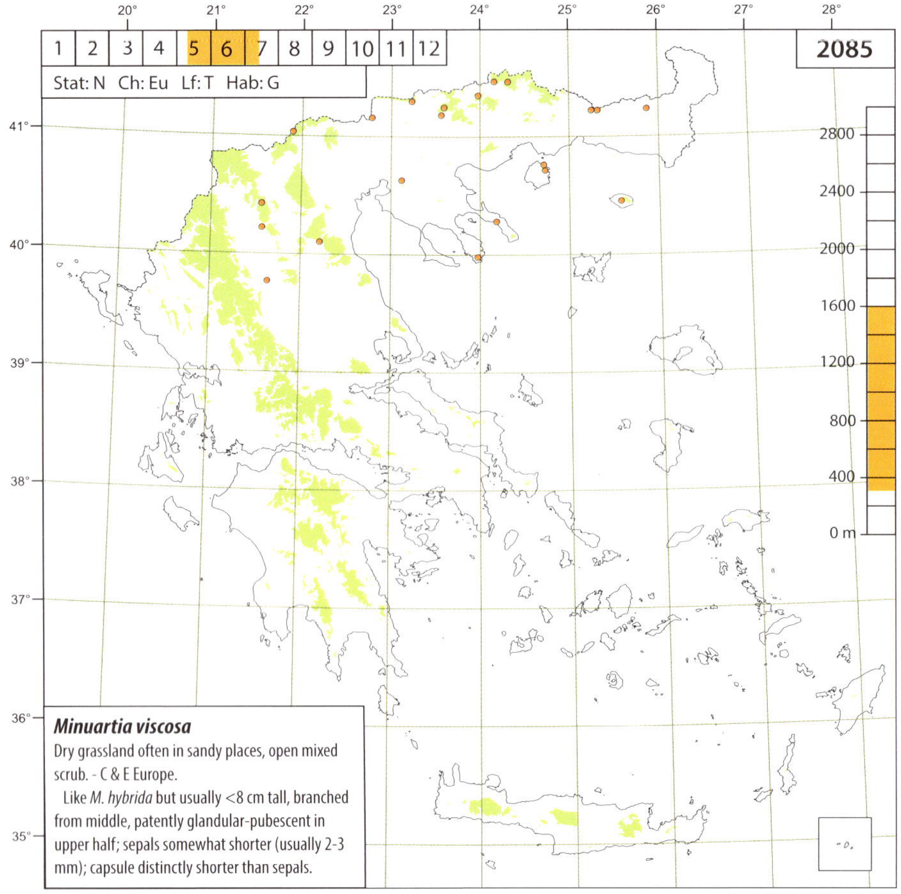

Minuartia viscosa

Dry grassland often in sandy places, open mixed scrub. - C & E Europe.

Like *M. hybrida* but usually <8 cm tall, branched from middle, patently glandular-pubescent in upper half; sepals somewhat shorter (usually 2-3 mm); capsule distinctly shorter than sepals.

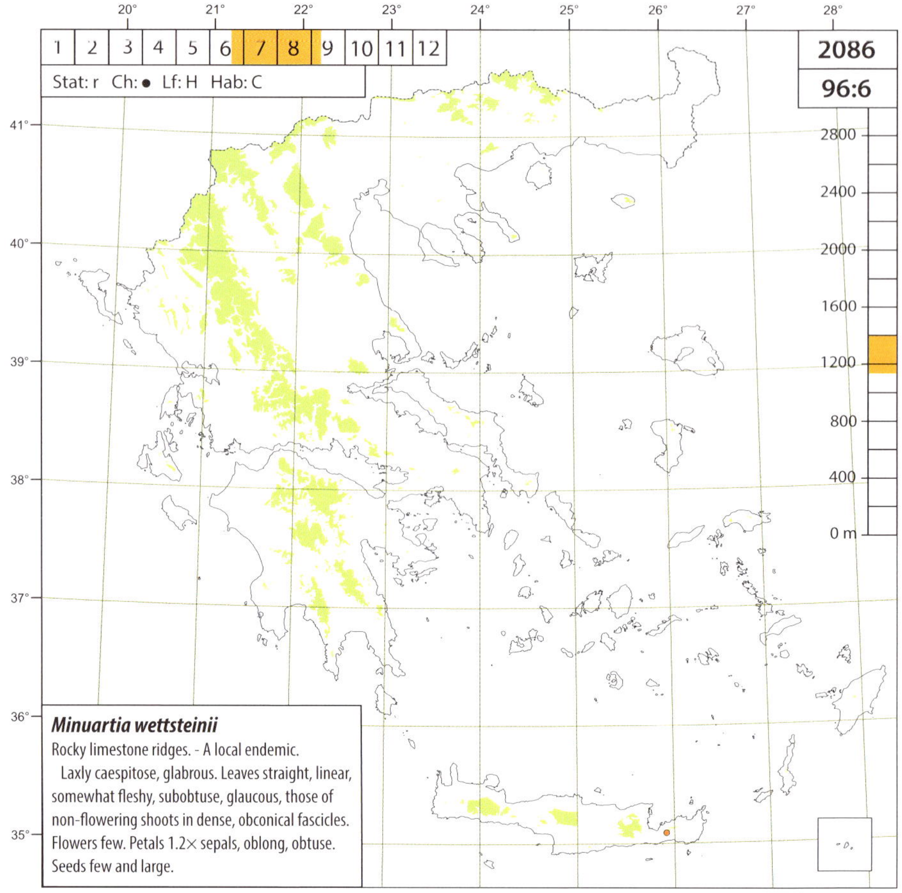

Minuartia wettsteinii

Rocky limestone ridges. - A local endemic.

Laxly caespitose, glabrous. Leaves straight, linear, somewhat fleshy, subobtuse, glaucous, those of non-flowering shoots in dense, obconical fascicles. Flowers few. Petals 1.2× sepals, oblong, obtuse. Seeds few and large.

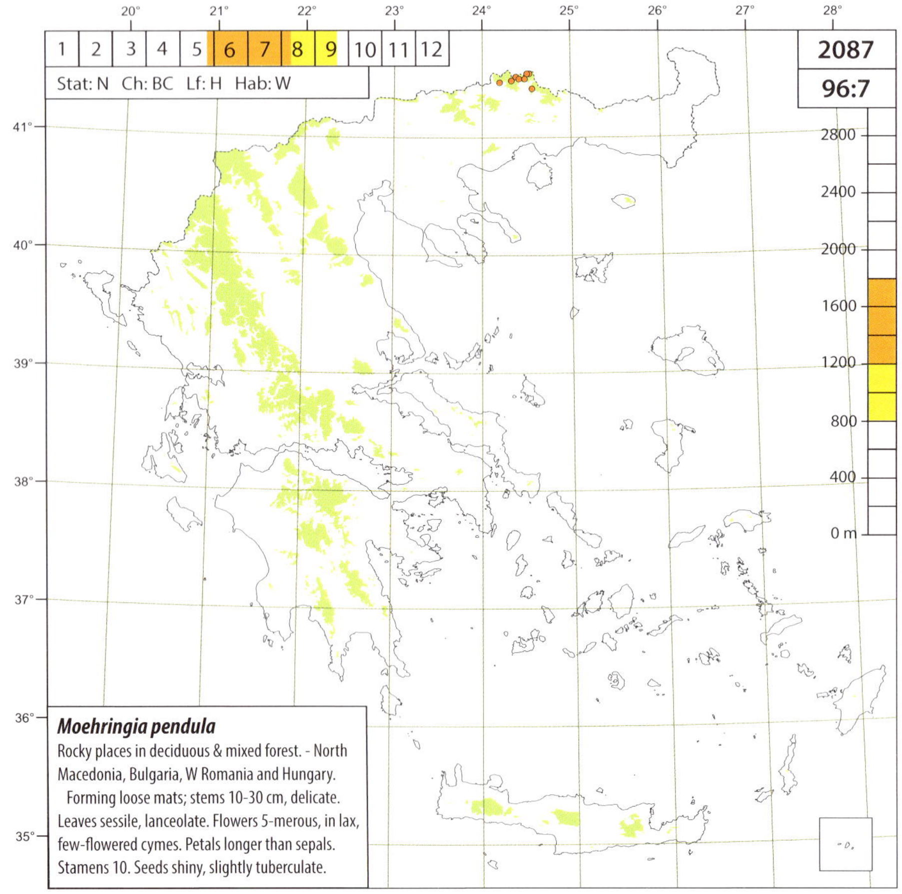

Moehringia pendula

Rocky places in deciduous & mixed forest. - North Macedonia, Bulgaria, W Romania and Hungary.

Forming loose mats; stems 10-30 cm, delicate. Leaves sessile, lanceolate. Flowers 5-merous, in lax, few-flowered cymes. Petals longer than sepals. Stamens 10. Seeds shiny, slightly tuberculate.

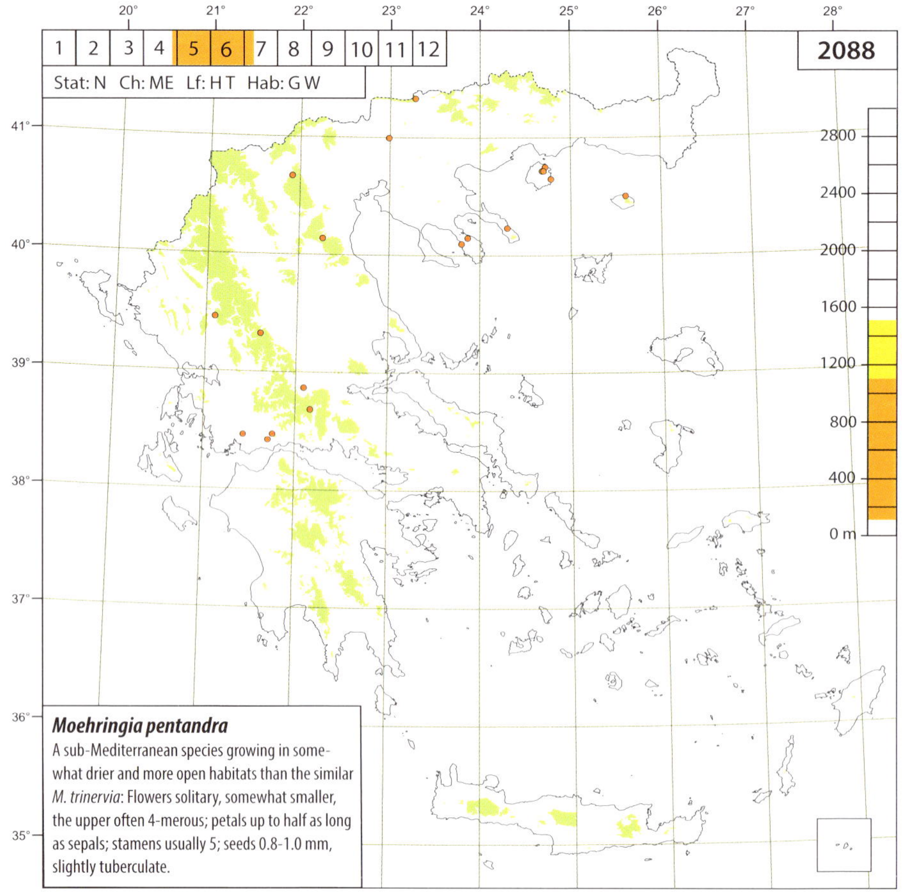

Moehringia pentandra

A sub-Mediterranean species growing in somewhat drier and more open habitats than the similar *M. trinervia*: Flowers solitary, somewhat smaller, the upper often 4-merous; petals up to half as long as sepals; stamens usually 5; seeds 0.8-1.0 mm, slightly tuberculate.

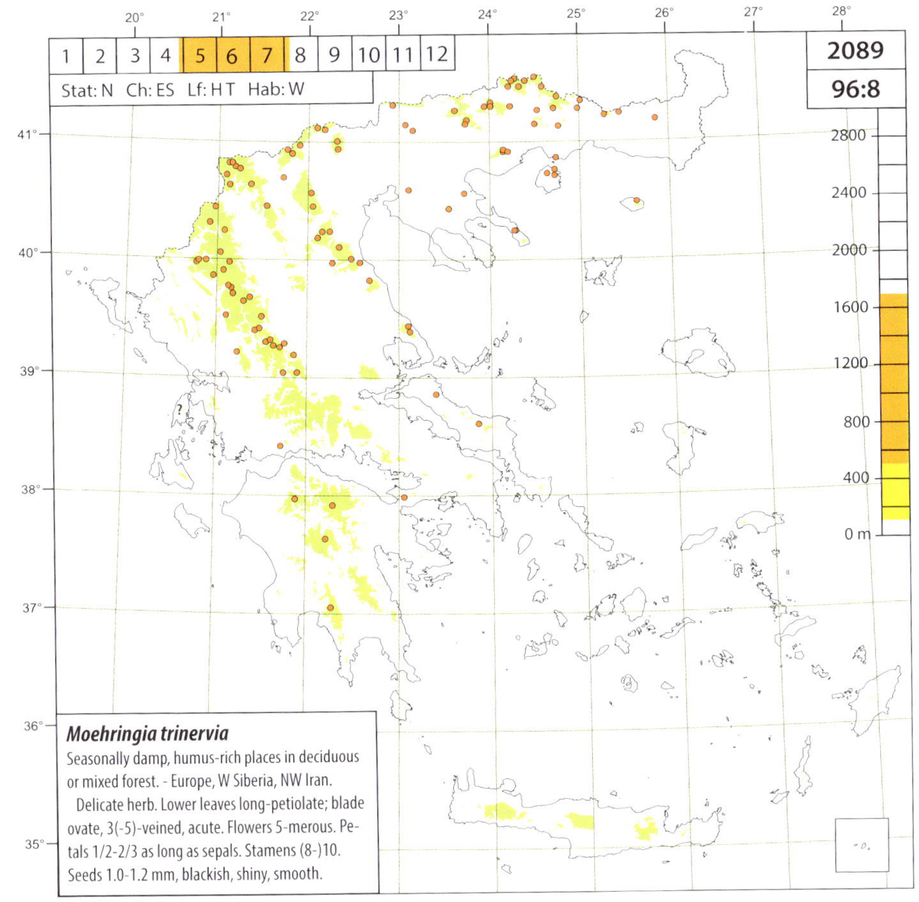

Moehringia trinervia

Seasonally damp, humus-rich places in deciduous or mixed forest. - Europe, W Siberia, NW Iran.

Delicate herb. Lower leaves long-petiolate; blade ovate, 3(-5)-veined, acute. Flowers 5-merous. Petals 1/2-2/3 as long as sepals. Stamens (8-)10. Seeds 1.0-1.2 mm, blackish, shiny, smooth.

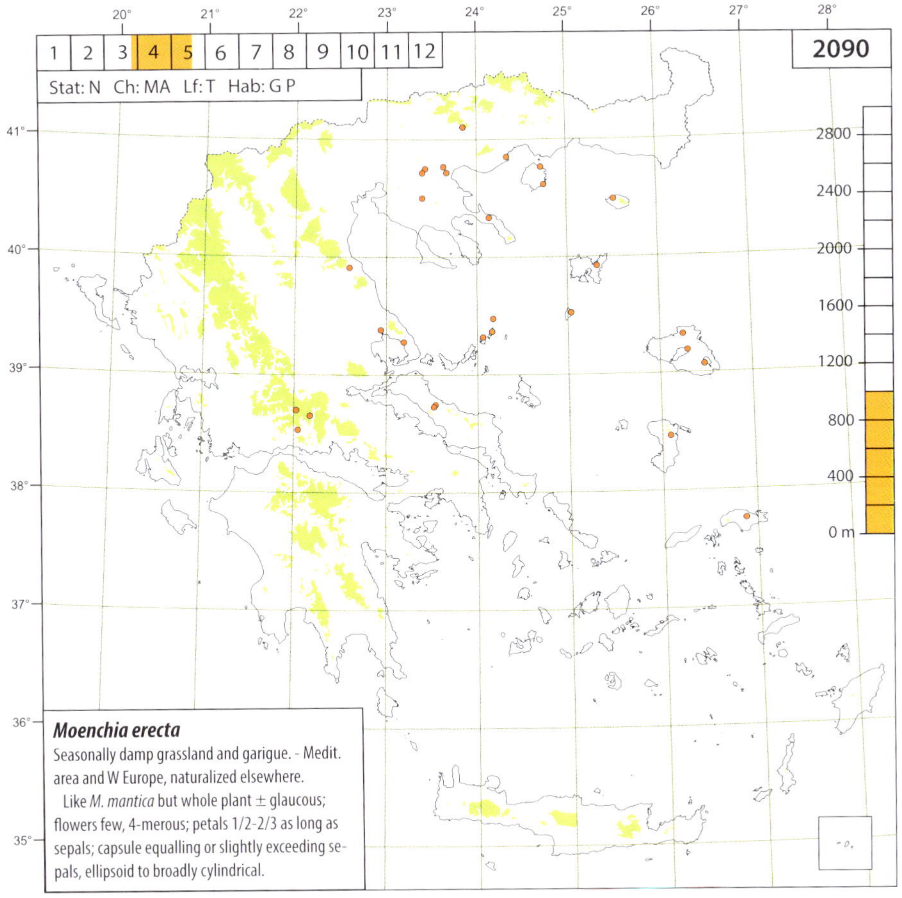

Moenchia erecta

Seasonally damp grassland and garigue. - Medit. area and W Europe, naturalized elsewhere.

Like *M. mantica* but whole plant ± glaucous; flowers few, 4-merous; petals 1/2-2/3 as long as sepals; capsule equalling or slightly exceeding sepals, ellipsoid to broadly cylindrical.

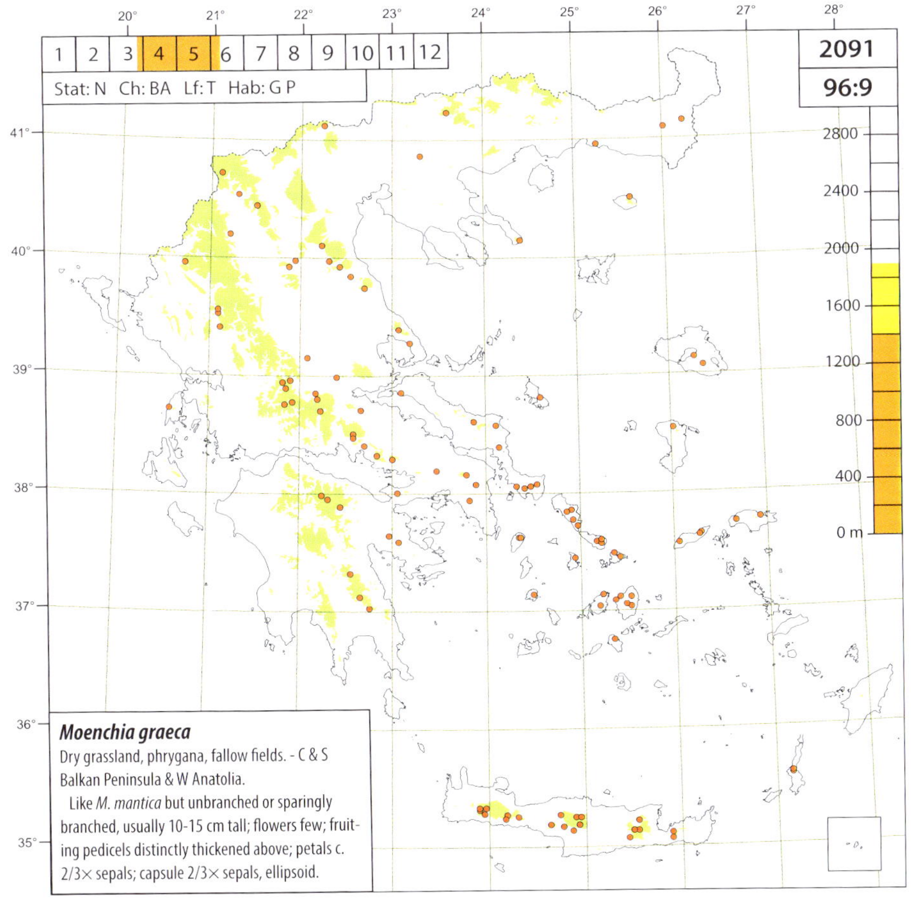

Moenchia graeca

Dry grassland, phrygana, fallow fields. - C & S Balkan Peninsula & W Anatolia.

Like *M. mantica* but unbranched or sparingly branched, usually 10-15 cm tall; flowers few; fruiting pedicels distinctly thickened above; petals c. 2/3× sepals; capsule 2/3× sepals, ellipsoid.

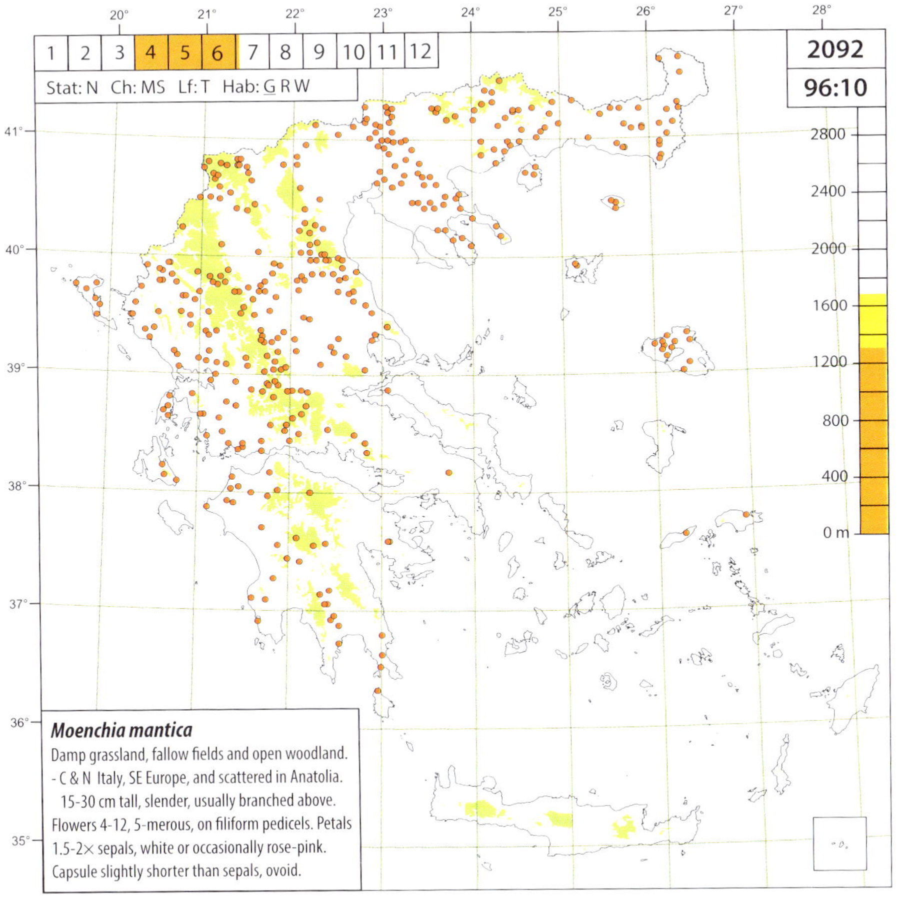

Moenchia mantica

Damp grassland, fallow fields and open woodland. - C & N Italy, SE Europe, and scattered in Anatolia.

15-30 cm tall, slender, usually branched above. Flowers 4-12, 5-merous, on filiform pedicels. Petals 1.5-2× sepals, white or occasionally rose-pink. Capsule slightly shorter than sepals, ovoid.

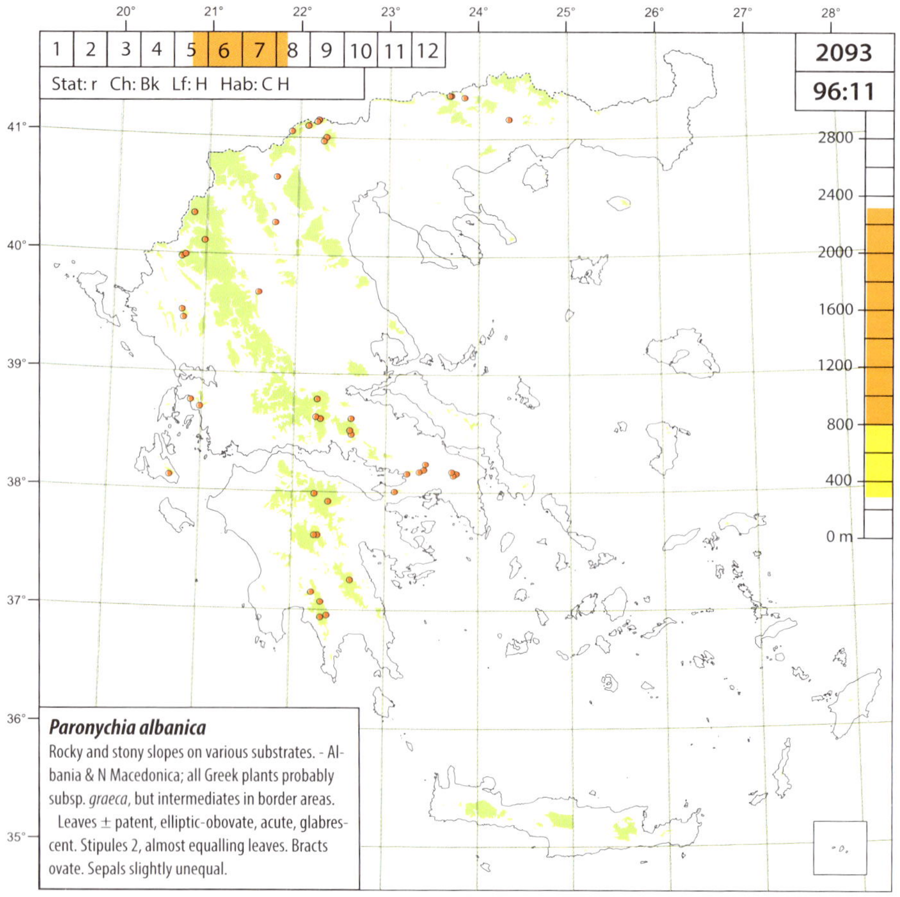

Paronychia albanica
Rocky and stony slopes on various substrates. - Albania & N Macedonica; all Greek plants probably subsp. *graeca*, but intermediates in border areas.
Leaves ± patent, elliptic-obovate, acute, glabrescent. Stipules 2, almost equalling leaves. Bracts ovate. Sepals slightly unequal.

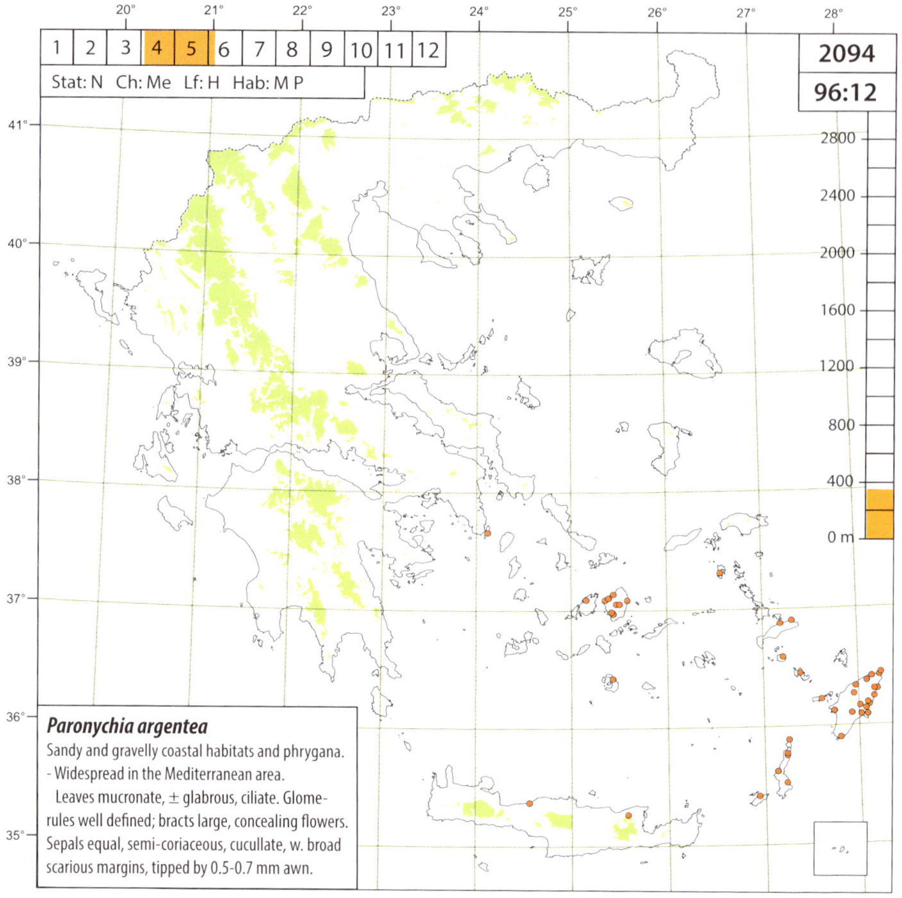

Paronychia argentea
Sandy and gravelly coastal habitats and phrygana. - Widespread in the Mediterranean area.
Leaves mucronate, ± glabrous, ciliate. Glomerules well defined; bracts large, concealing flowers. Sepals equal, semi-coriaceous, cucullate, w. broad scarious margins, tipped by 0.5-0.7 mm awn.

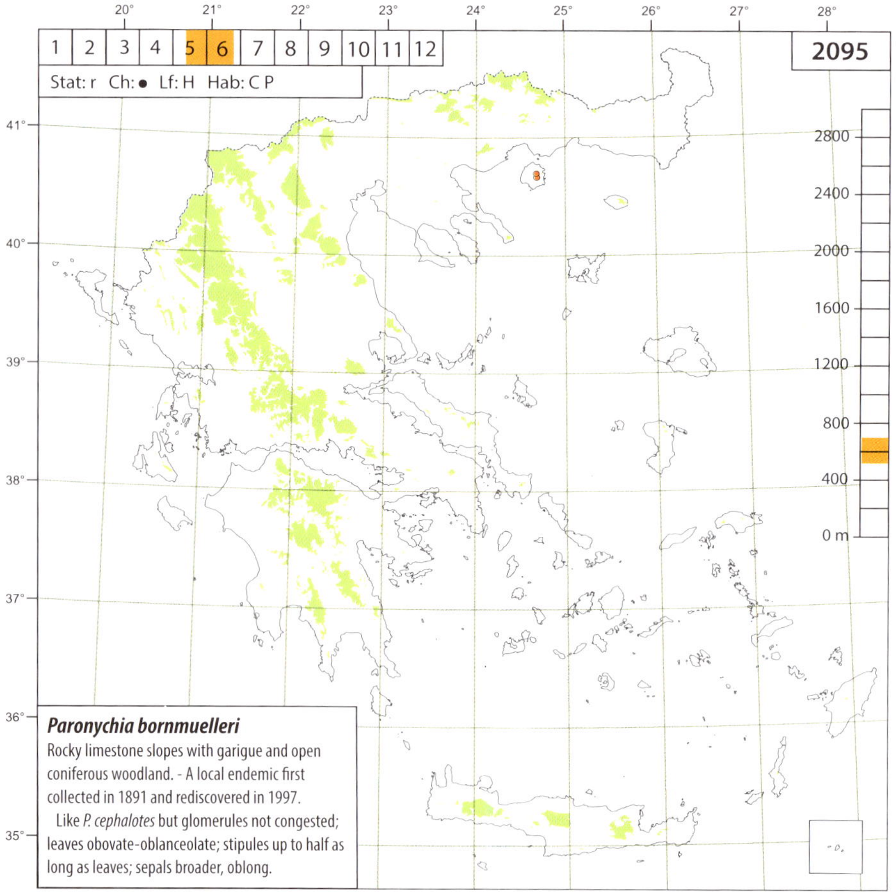

Paronychia bornmuelleri
Rocky limestone slopes with garigue and open coniferous woodland. - A local endemic first collected in 1891 and rediscovered in 1997.
Like *P. cephalotes* but glomerules not congested; leaves obovate-oblanceolate; stipules up to half as long as leaves; sepals broader, oblong.

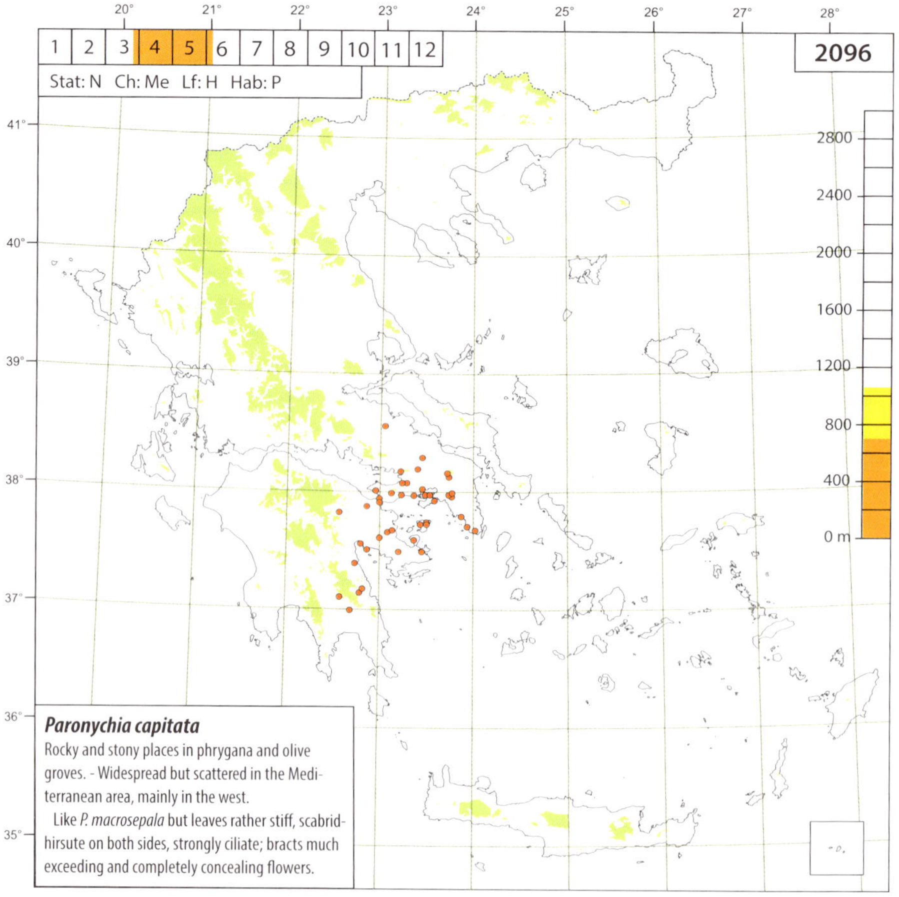

Paronychia capitata
Rocky and stony places in phrygana and olive groves. - Widespread but scattered in the Mediterranean area, mainly in the west.
Like *P. macrosepala* but leaves rather stiff, scabrid-hirsute on both sides, strongly ciliate; bracts much exceeding and completely concealing flowers.

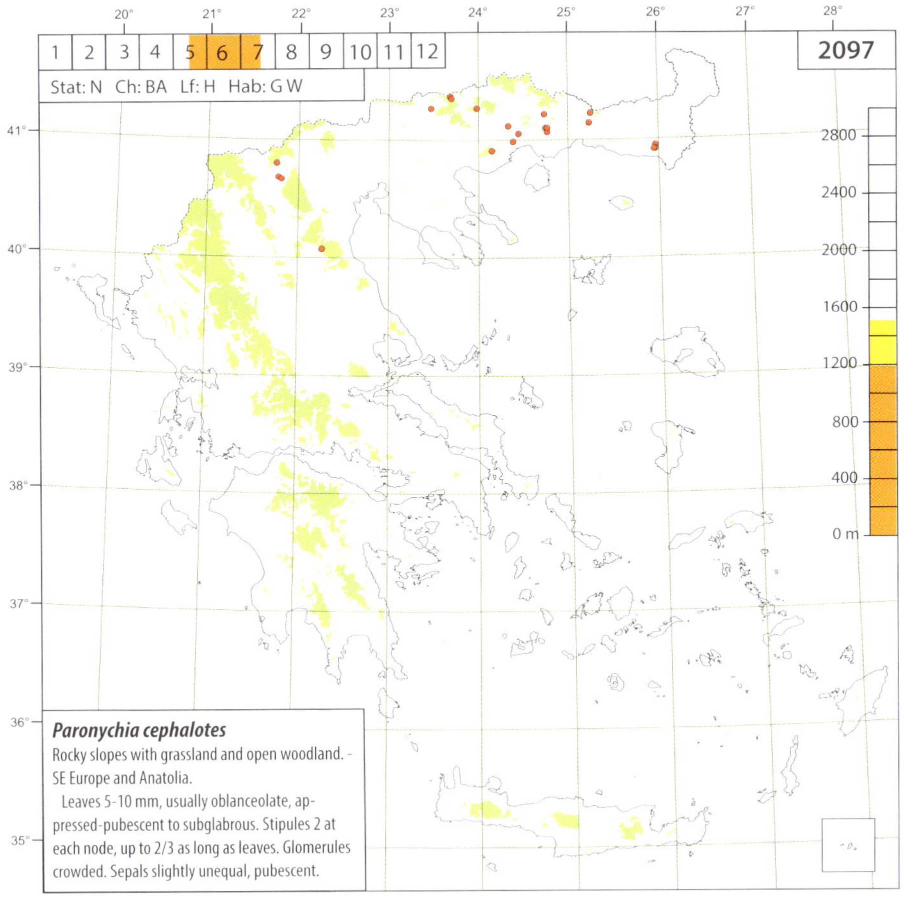

Paronychia cephalotes
Rocky slopes with grassland and open woodland. - SE Europe and Anatolia.
Leaves 5-10 mm, usually oblanceolate, appressed-pubescent to subglabrous. Stipules 2 at each node, up to 2/3 as long as leaves. Glomerules crowded. Sepals slightly unequal, pubescent.

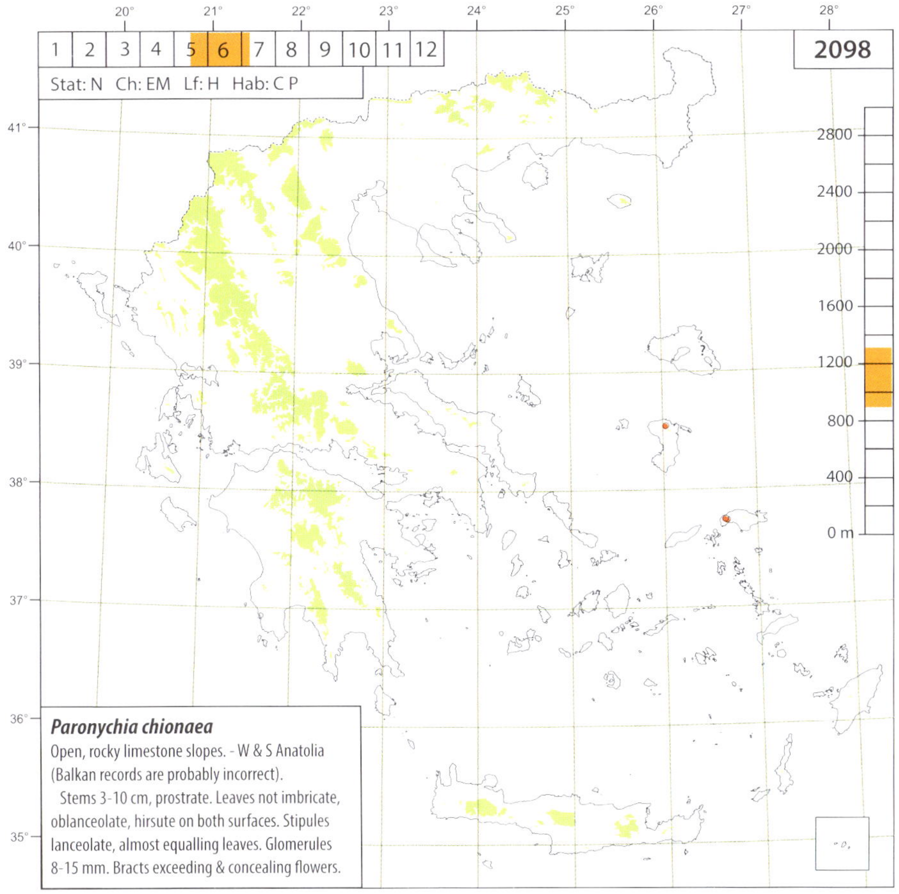

Paronychia chionaea
Open, rocky limestone slopes. - W & S Anatolia (Balkan records are probably incorrect).
Stems 3-10 cm, prostrate. Leaves not imbricate, oblanceolate, hirsute on both surfaces. Stipules lanceolate, almost equalling leaves. Glomerules 8-15 mm. Bracts exceeding & concealing flowers.

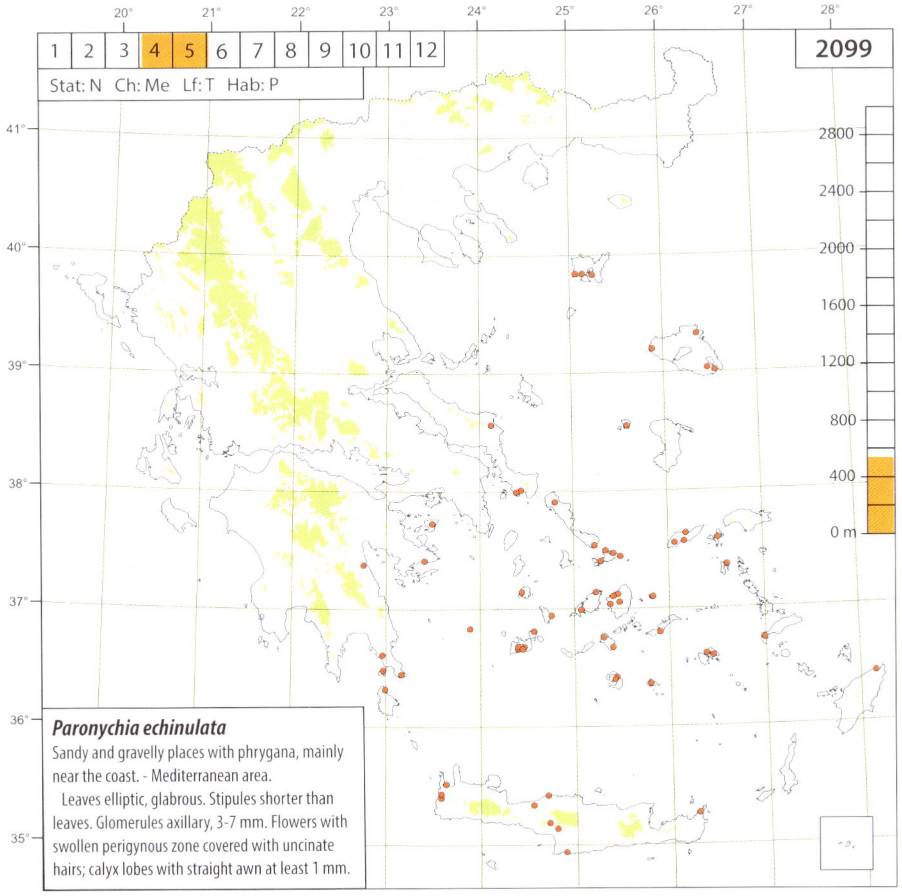

Paronychia echinulata
Sandy and gravelly places with phrygana, mainly near the coast. - Mediterranean area.
Leaves elliptic, glabrous. Stipules shorter than leaves. Glomerules axillary, 3-7 mm. Flowers with swollen perigynous zone covered with uncinate hairs; calyx lobes with straight awn at least 1 mm.

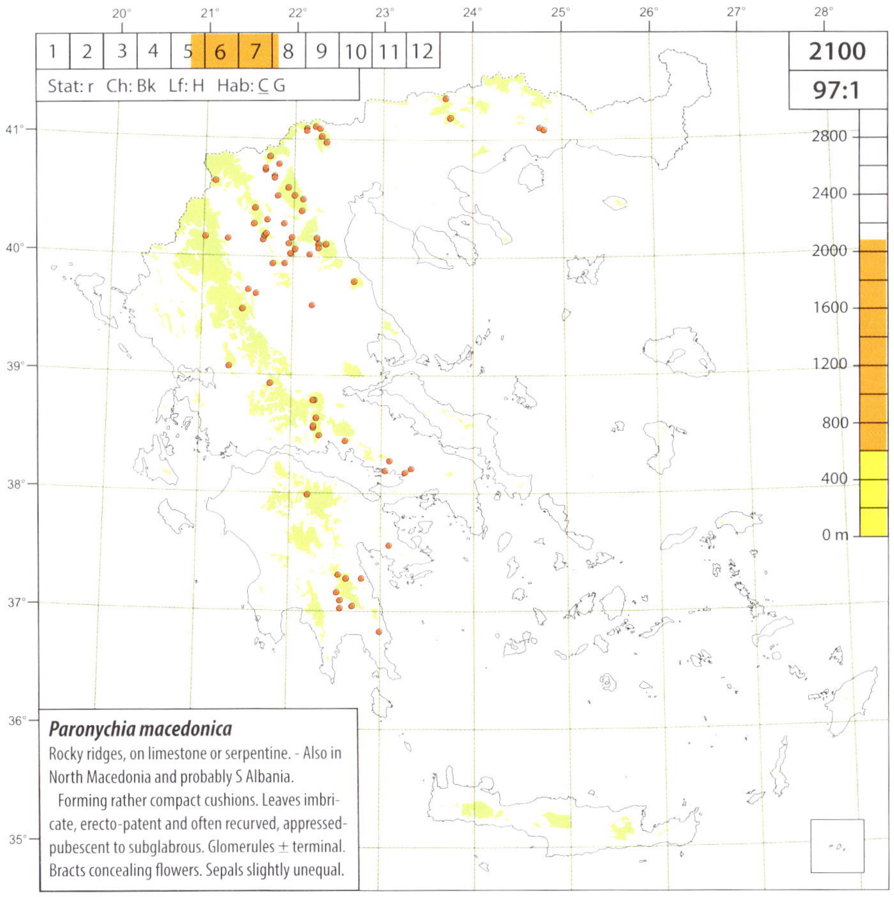

Paronychia macedonica
Rocky ridges, on limestone or serpentine. - Also in North Macedonia and probably S Albania.
Forming rather compact cushions. Leaves imbricate, erecto-patent and often recurved, appressed-pubescent to subglabrous. Glomerules ± terminal. Bracts concealing flowers. Sepals slightly unequal.

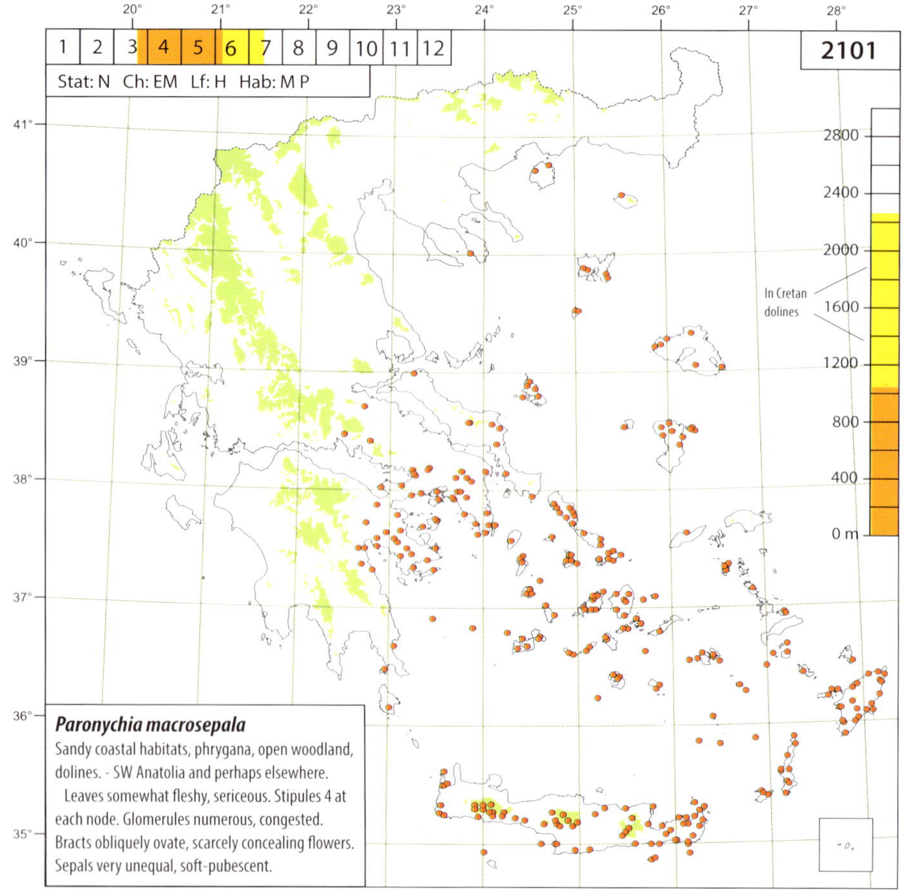

Paronychia macrosepala

Sandy coastal habitats, phrygana, open woodland, dolines. - SW Anatolia and perhaps elsewhere.

Leaves somewhat fleshy, sericeous. Stipules 4 at each node. Glomerules numerous, congested. Bracts obliquely ovate, scarcely concealing flowers. Sepals very unequal, soft-pubescent.

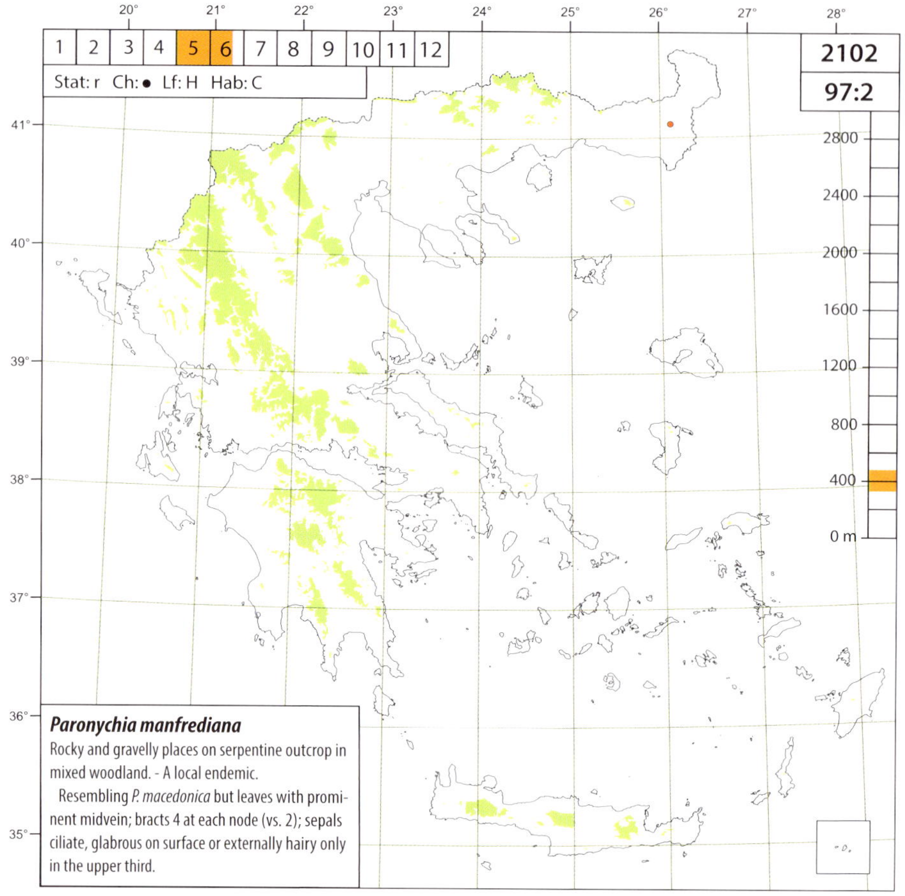

Paronychia manfrediana

Rocky and gravelly places on serpentine outcrop in mixed woodland. - A local endemic.

Resembling *P. macedonica* but leaves with prominent midvein; bracts 4 at each node (vs. 2); sepals ciliate, glabrous on surface or externally hairy only in the upper third.

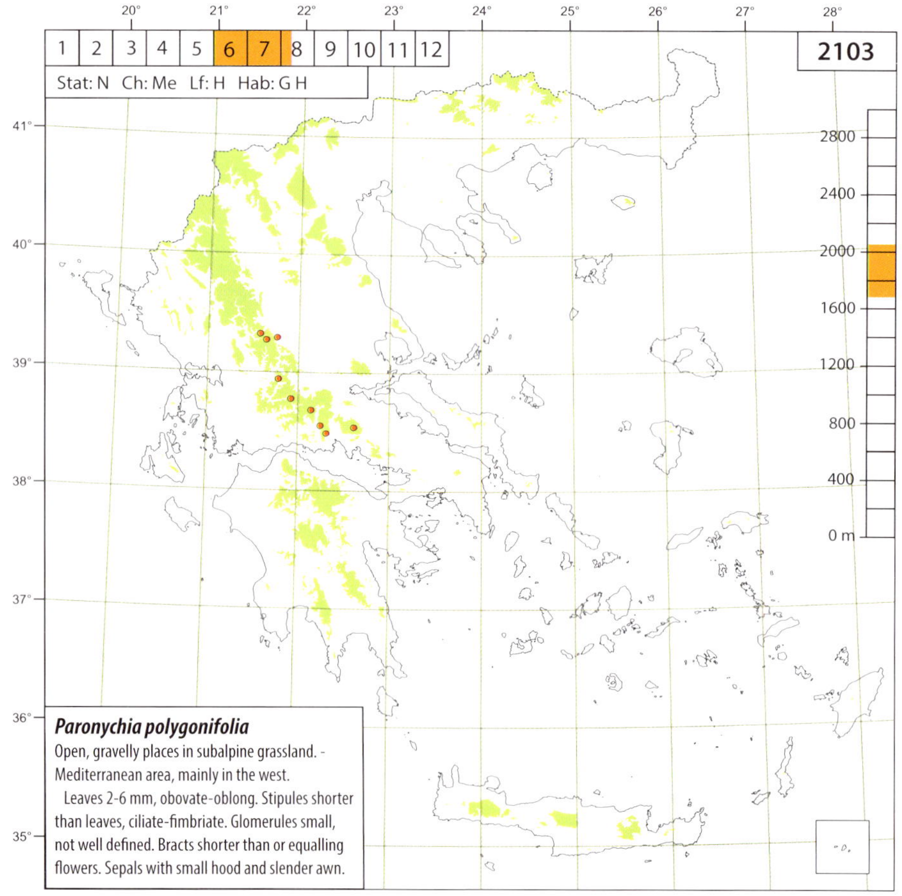

Paronychia polygonifolia

Open, gravelly places in subalpine grassland. - Mediterranean area, mainly in the west.

Leaves 2-6 mm, obovate-oblong. Stipules shorter than leaves, ciliate-fimbriate. Glomerules small, not well defined. Bracts shorter than or equalling flowers. Sepals with small hood and slender awn.

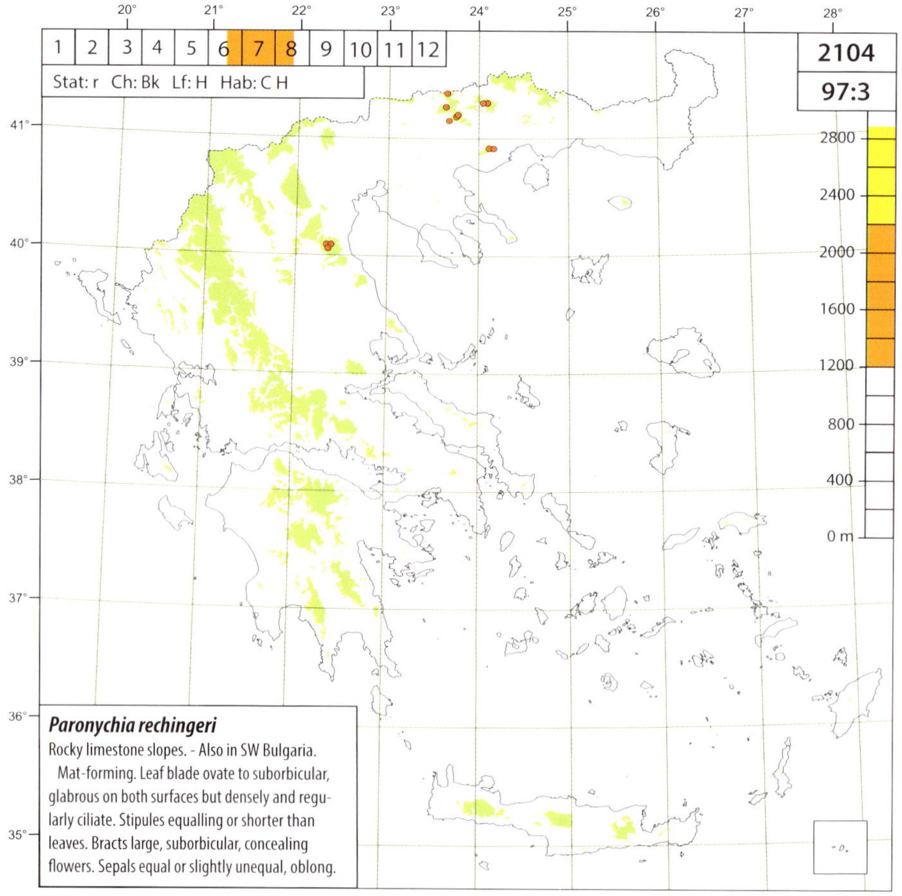

Paronychia rechingeri

Rocky limestone slopes. - Also in SW Bulgaria.

Mat-forming. Leaf blade ovate to suborbicular, glabrous on both surfaces but densely and regularly ciliate. Stipules equalling or shorter than leaves. Bracts large, suborbicular, concealing flowers. Sepals equal or slightly unequal, oblong.

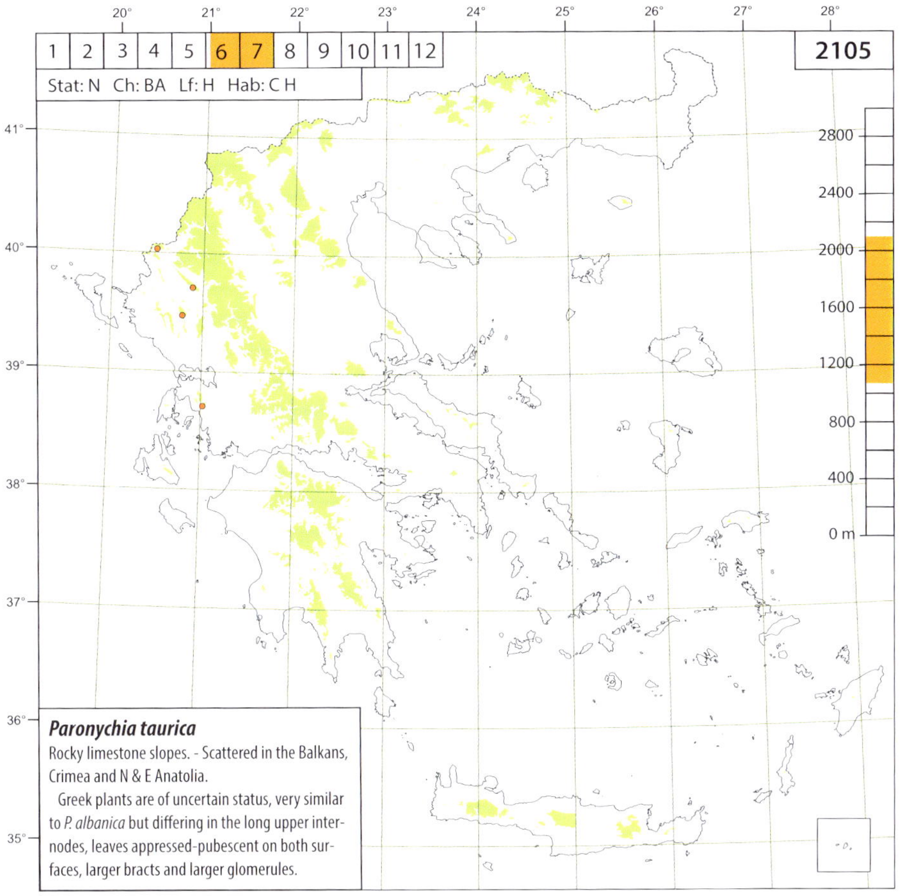

Paronychia taurica
Rocky limestone slopes. - Scattered in the Balkans, Crimea and N & E Anatolia.
Greek plants are of uncertain status, very similar to *P. albanica* but differing in the long upper internodes, leaves appressed-pubescent on both surfaces, larger bracts and larger glomerules.

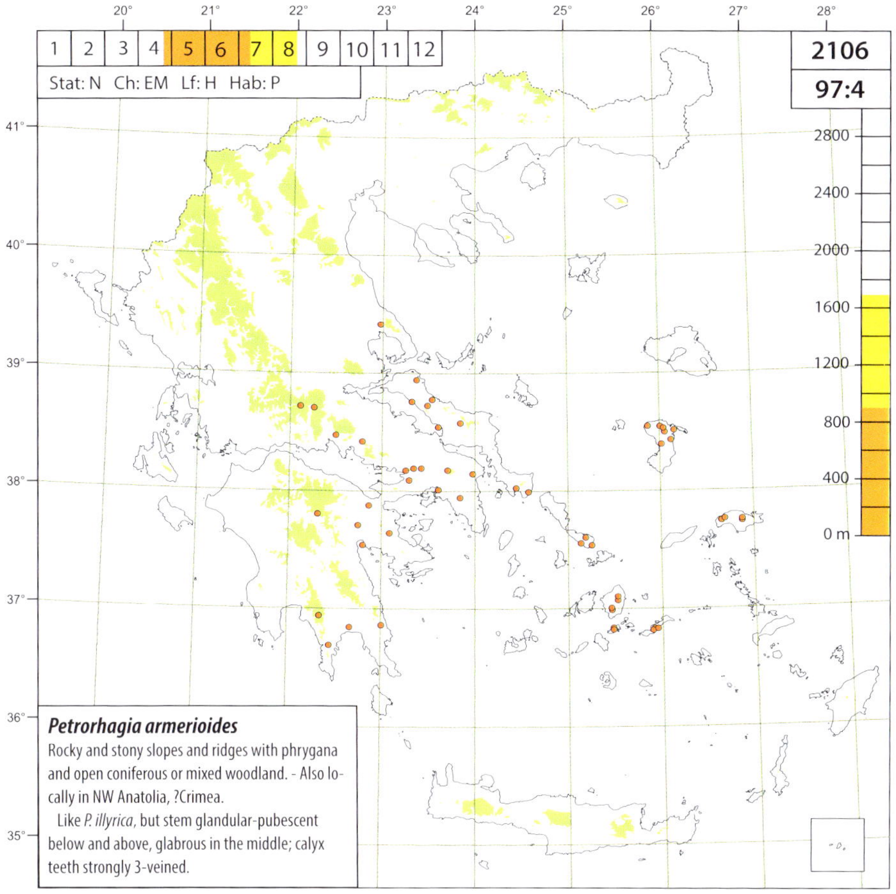

Petrorhagia armerioides
Rocky and stony slopes and ridges with phrygana and open coniferous or mixed woodland. - Also locally in NW Anatolia, ?Crimea.
Like *P. illyrica*, but stem glandular-pubescent below and above, glabrous in the middle; calyx teeth strongly 3-veined.

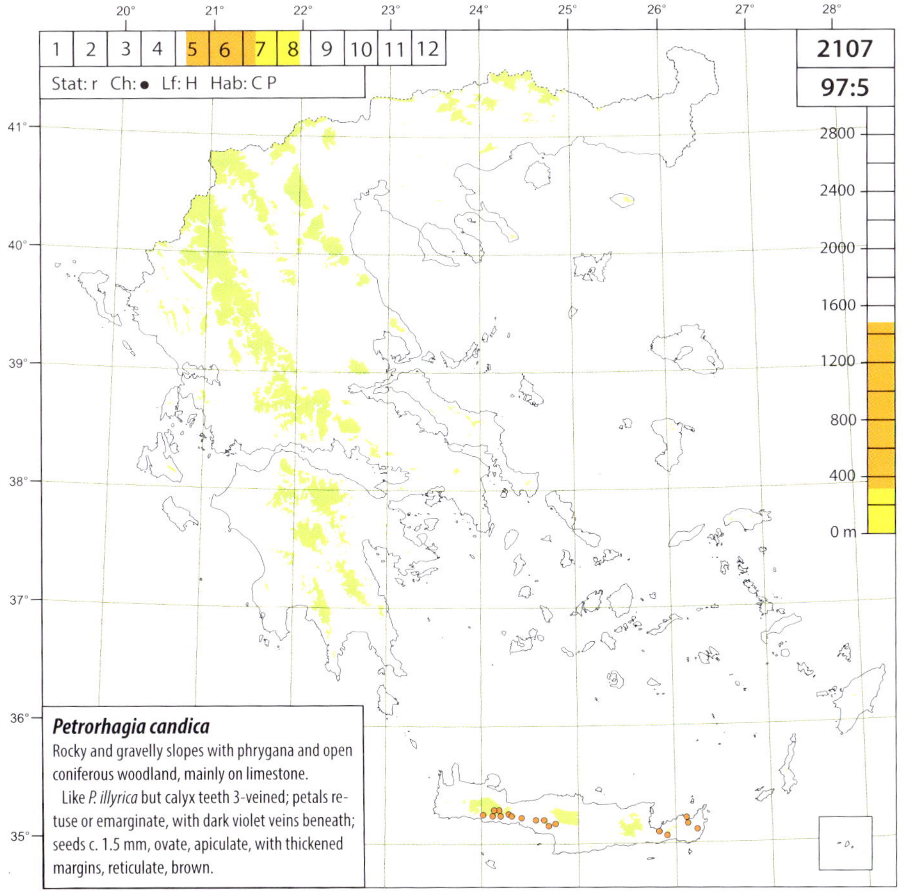

Petrorhagia candica
Rocky and gravelly slopes with phrygana and open coniferous woodland, mainly on limestone.
Like *P. illyrica* but calyx teeth 3-veined; petals retuse or emarginate, with dark violet veins beneath; seeds c. 1.5 mm, ovate, apiculate, with thickened margins, reticulate, brown.

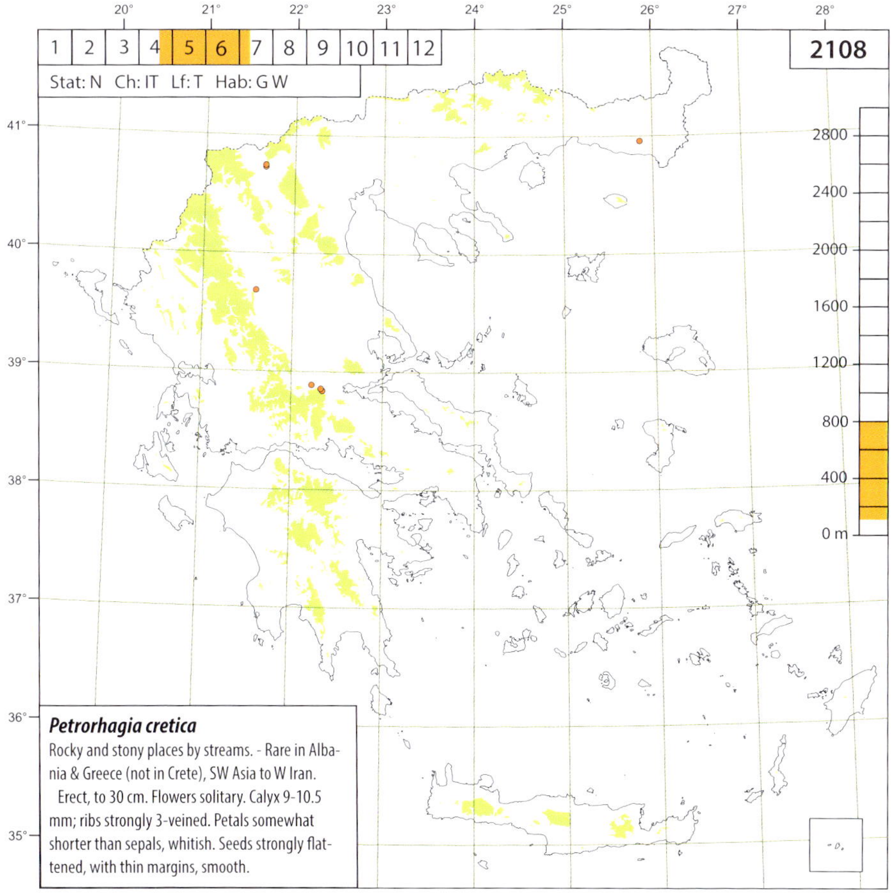

Petrorhagia cretica
Rocky and stony places by streams. - Rare in Albania & Greece (not in Crete), SW Asia to W Iran.
Erect, to 30 cm. Flowers solitary. Calyx 9-10.5 mm; ribs strongly 3-veined. Petals somewhat shorter than sepals, whitish. Seeds strongly flattened, with thin margins, smooth.

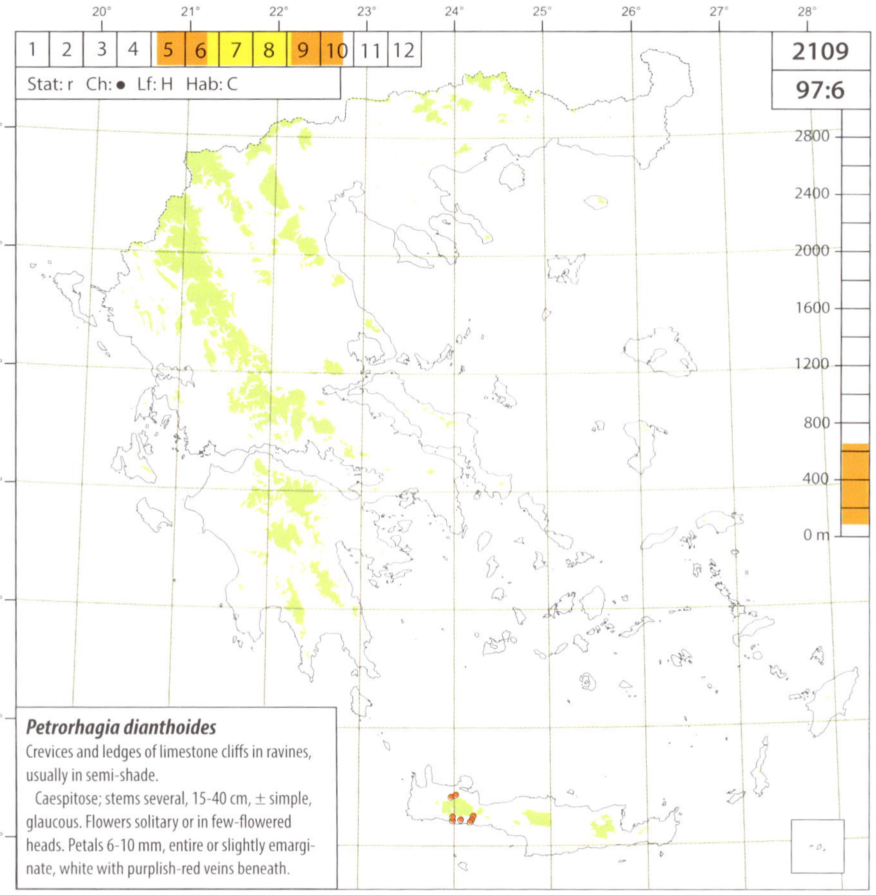

Petrorhagia dianthoides

Crevices and ledges of limestone cliffs in ravines, usually in semi-shade.

Caespitose; stems several, 15-40 cm, ± simple, glaucous. Flowers solitary or in few-flowered heads. Petals 6-10 mm, entire or slightly emarginate, white with purplish-red veins beneath.

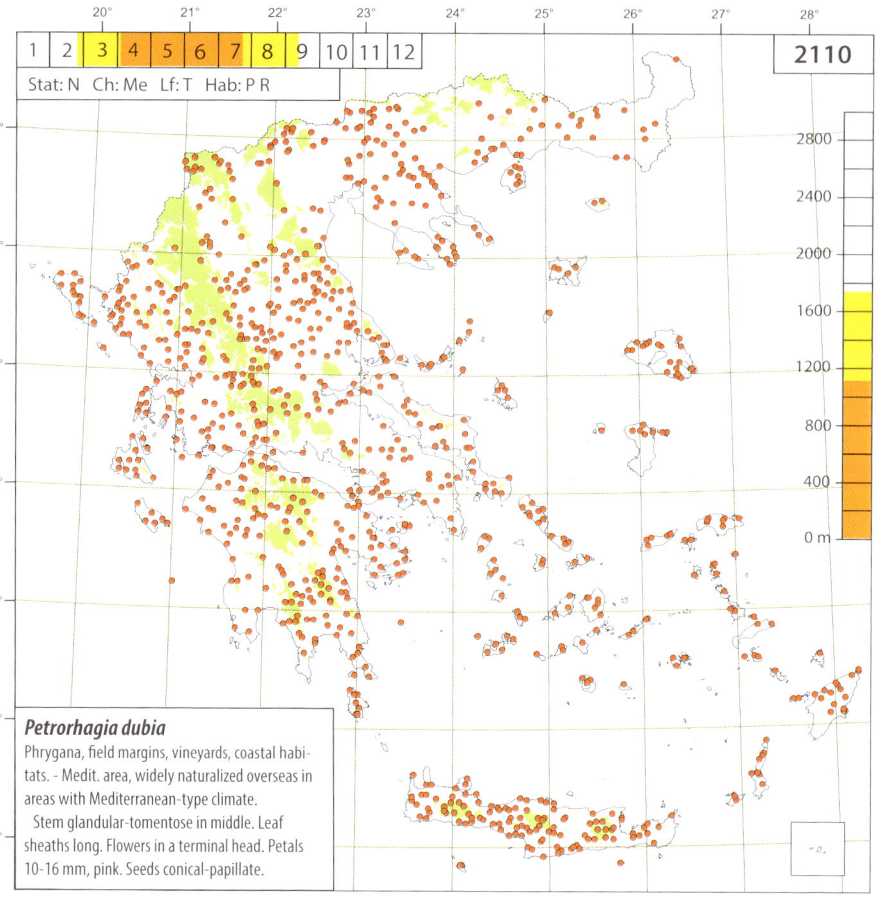

Petrorhagia dubia

Phrygana, field margins, vineyards, coastal habitats. - Medit. area, widely naturalized overseas in areas with Mediterranean-type climate.

Stem glandular-tomentose in middle. Leaf sheaths long. Flowers in a terminal head. Petals 10-16 mm, pink. Seeds conical-papillate.

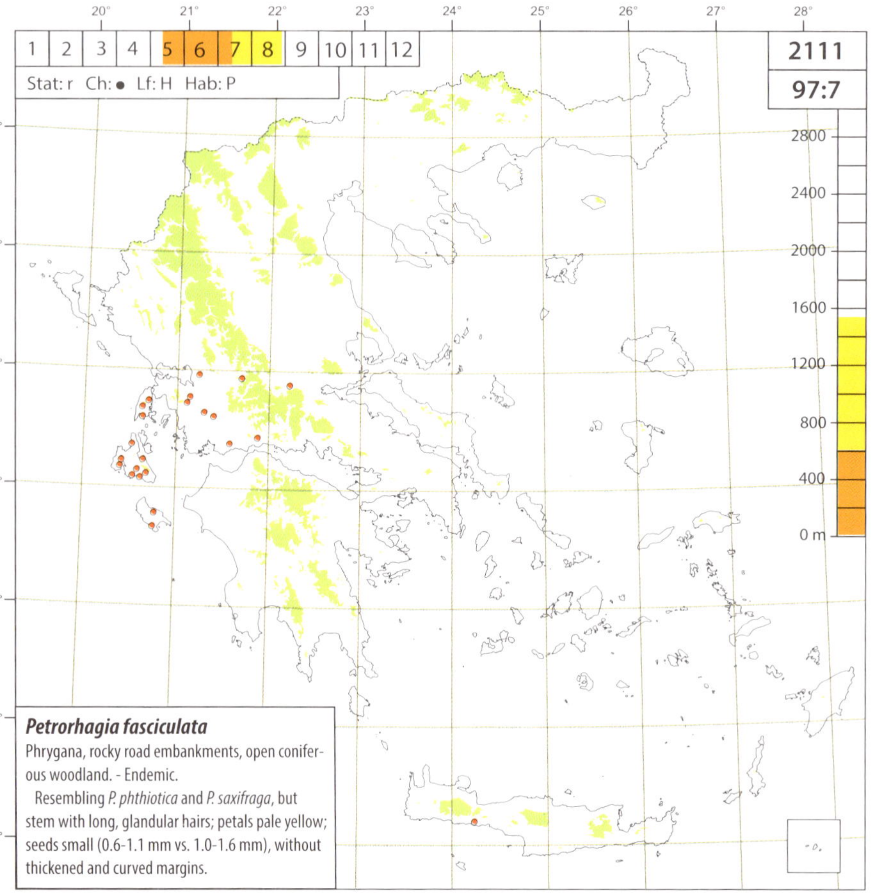

Petrorhagia fasciculata

Phrygana, rocky road embankments, open coniferous woodland. - Endemic.

Resembling *P. phthiotica* and *P. saxifraga*, but stem with long, glandular hairs; petals pale yellow; seeds small (0.6-1.1 mm vs. 1.0-1.6 mm), without thickened and curved margins.

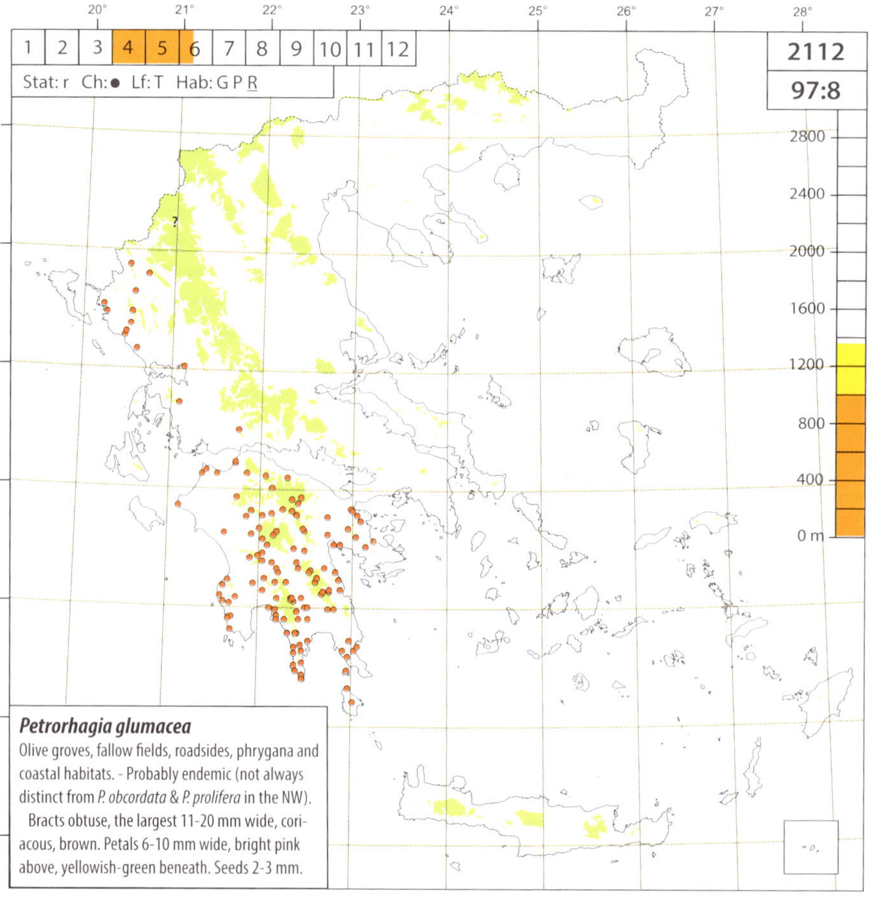

Petrorhagia glumacea

Olive groves, fallow fields, roadsides, phrygana and coastal habitats. - Probably endemic (not always distinct from *P. obcordata* & *P. prolifera* in the NW).

Bracts obtuse, the largest 11-20 mm wide, coriaceous, brown. Petals 6-10 mm wide, bright pink above, yellowish-green beneath. Seeds 2-3 mm.

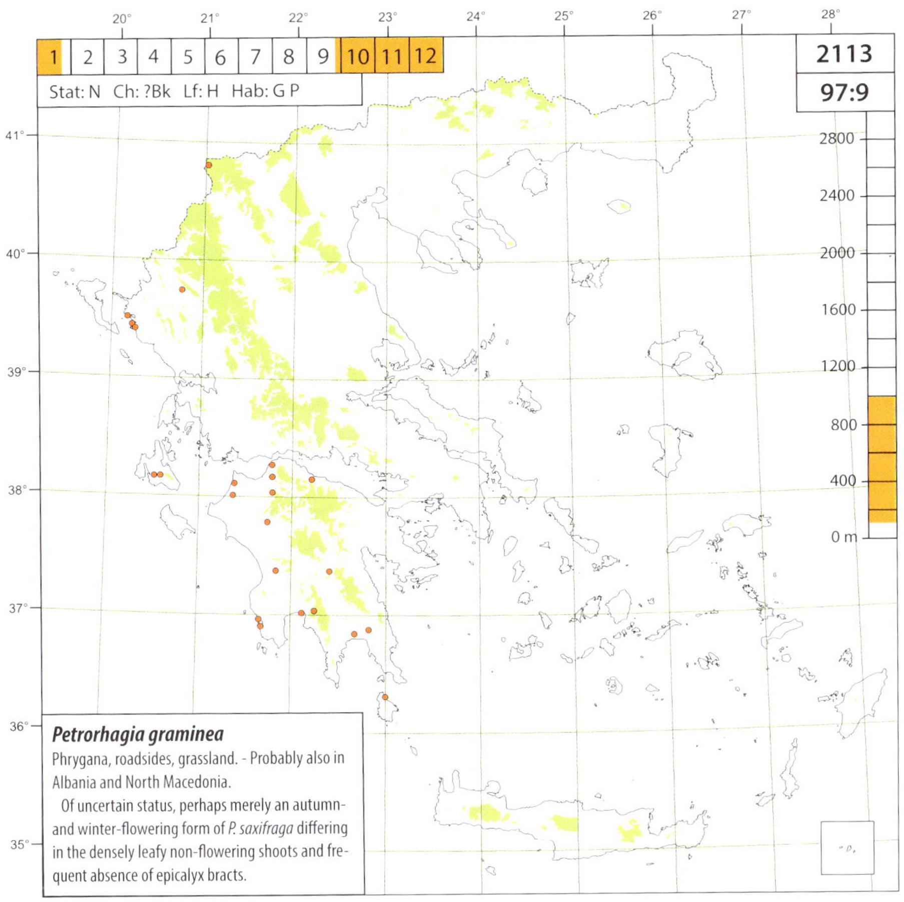

Petrorhagia graminea

Phrygana, roadsides, grassland. - Probably also in Albania and North Macedonia.

Of uncertain status, perhaps merely an autumn- and winter-flowering form of *P. saxifraga* differing in the densely leafy non-flowering shoots and frequent absence of epicalyx bracts.

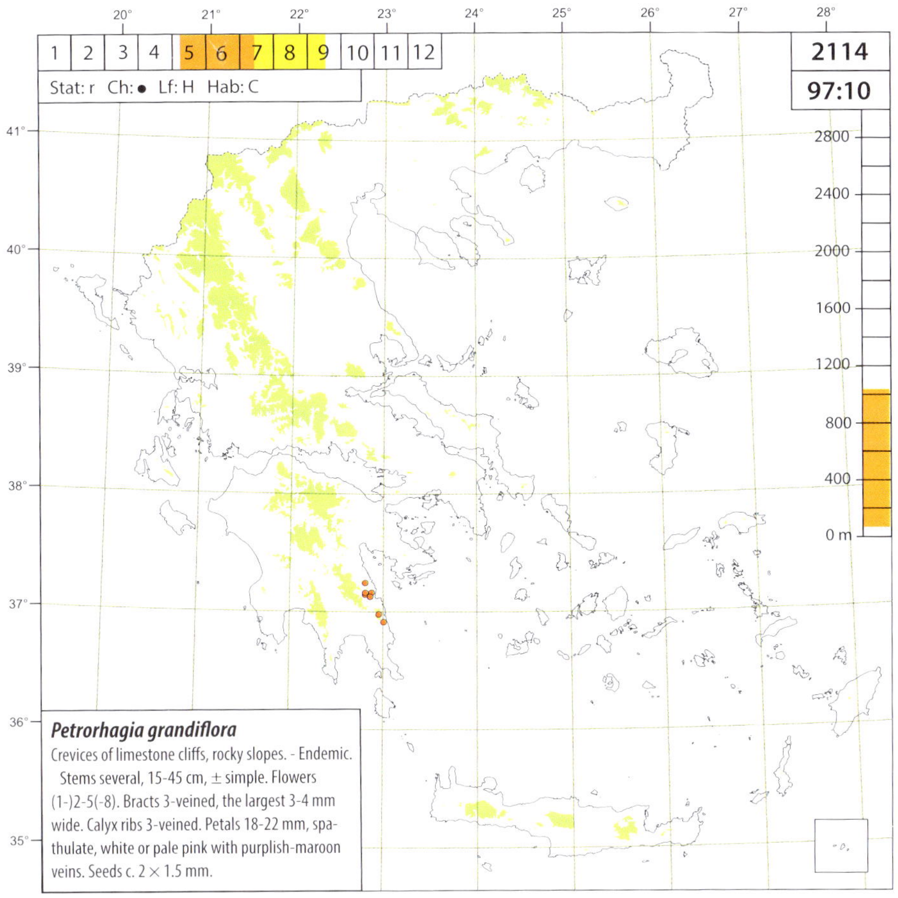

Petrorhagia grandiflora

Crevices of limestone cliffs, rocky slopes. - Endemic.

Stems several, 15-45 cm, ± simple. Flowers (1-)2-5(-8). Bracts 3-veined, the largest 3-4 mm wide. Calyx ribs 3-veined. Petals 18-22 mm, spathulate, white or pale pink with purplish-maroon veins. Seeds c. 2 × 1.5 mm.

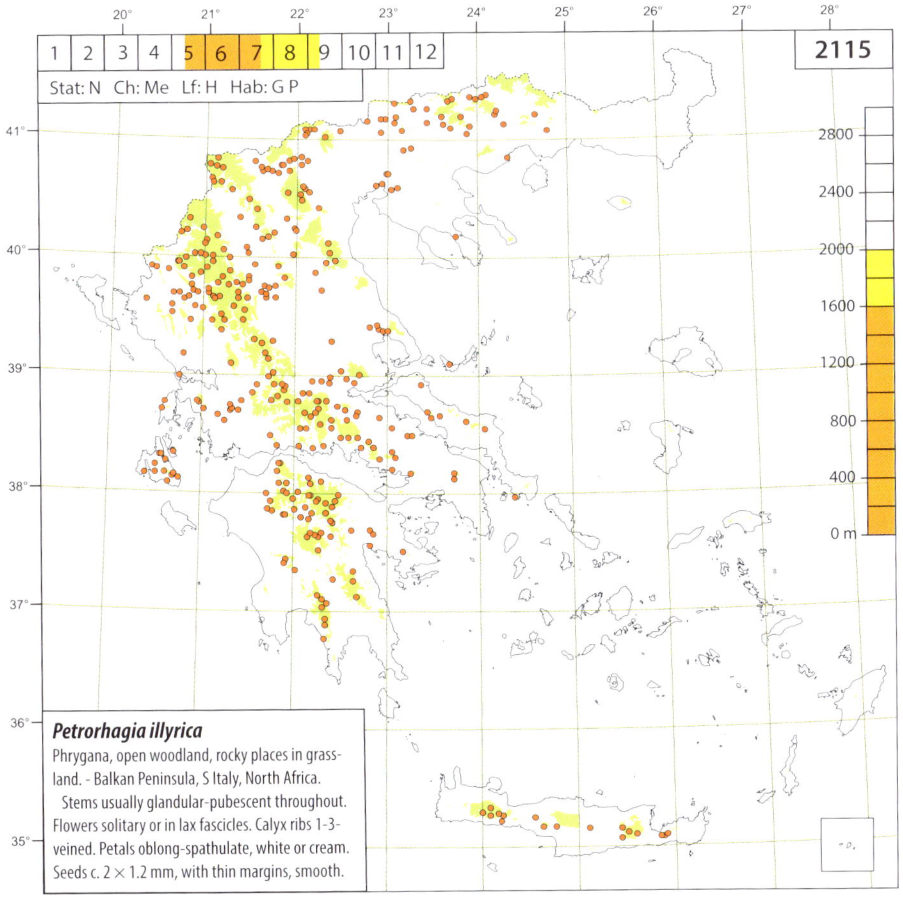

Petrorhagia illyrica

Phrygana, open woodland, rocky places in grassland. - Balkan Peninsula, S Italy, North Africa.

Stems usually glandular-pubescent throughout. Flowers solitary or in lax fascicles. Calyx ribs 1-3-veined. Petals oblong-spathulate, white or cream. Seeds c. 2 × 1.2 mm, with thin margins, smooth.

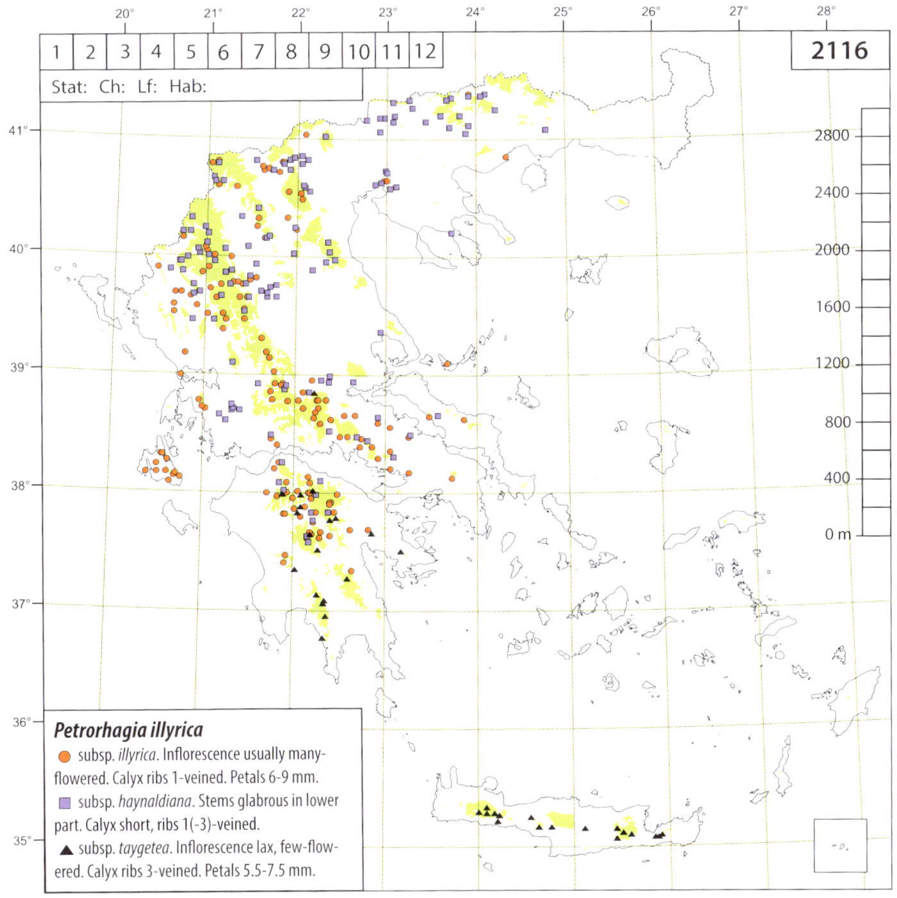

Petrorhagia illyrica

● subsp. *illyrica*. Inflorescence usually many-flowered. Calyx ribs 1-veined. Petals 6-9 mm.

■ subsp. *haynaldiana*. Stems glabrous in lower part. Calyx short, ribs 1(-3)-veined.

▲ subsp. *taygetea*. Inflorescence lax, few-flowered. Calyx ribs 3-veined. Petals 5.5-7.5 mm.

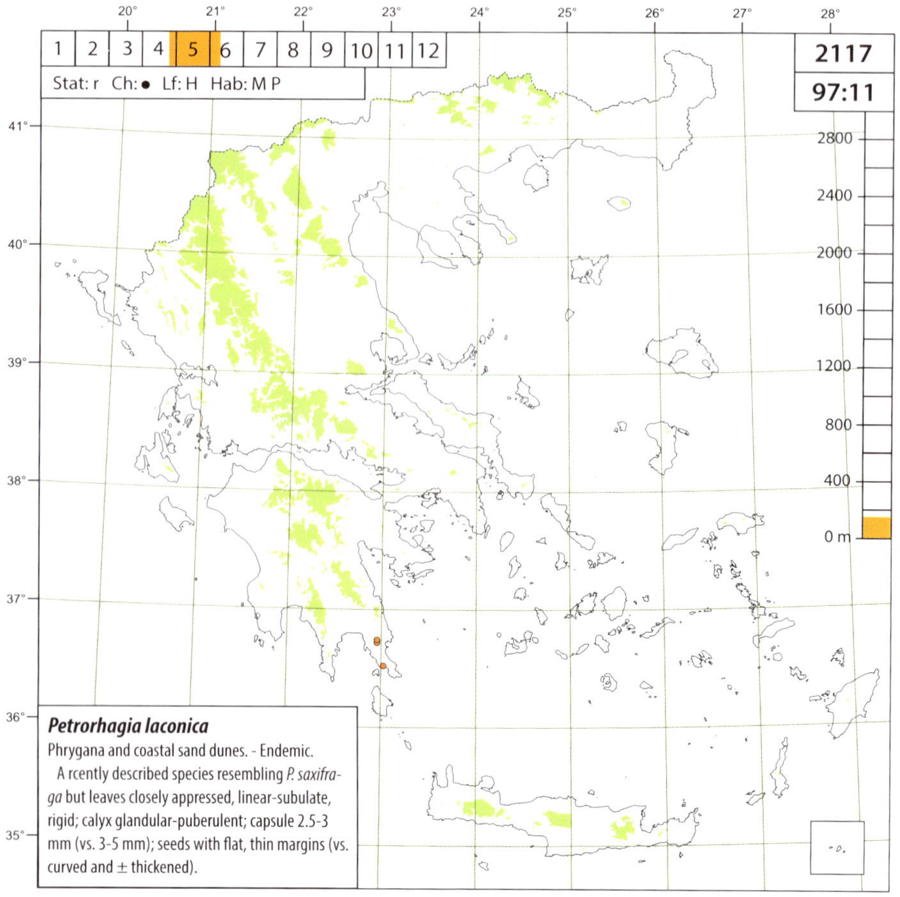

Petrorhagia laconica
Phrygana and coastal sand dunes. - Endemic.
A rcently described species resembling *P. saxifraga* but leaves closely appressed, linear-subulate, rigid; calyx glandular-puberulent; capsule 2.5-3 mm (vs. 3-5 mm); seeds with flat, thin margins (vs. curved and ± thickened).

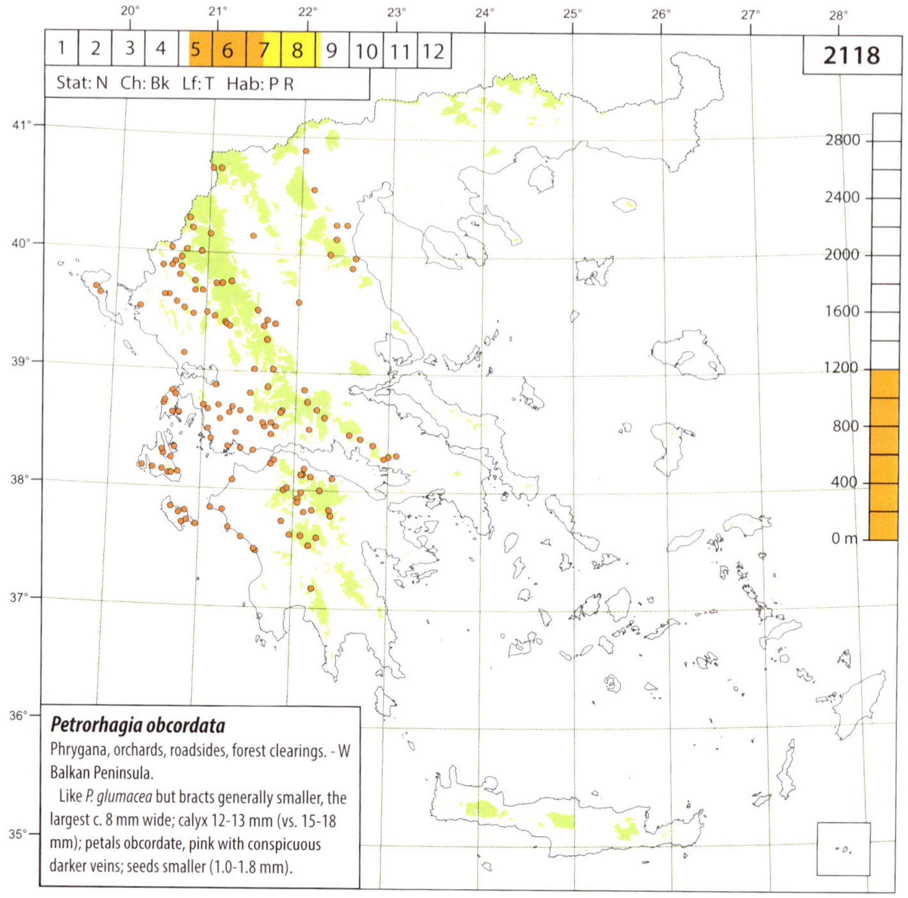

Petrorhagia obcordata
Phrygana, orchards, roadsides, forest clearings. - W Balkan Peninsula.
Like *P. glumacea* but bracts generally smaller, the largest c. 8 mm wide; calyx 12-13 mm (vs. 15-18 mm); petals obcordate, pink with conspicuous darker veins; seeds smaller (1.0-1.8 mm).

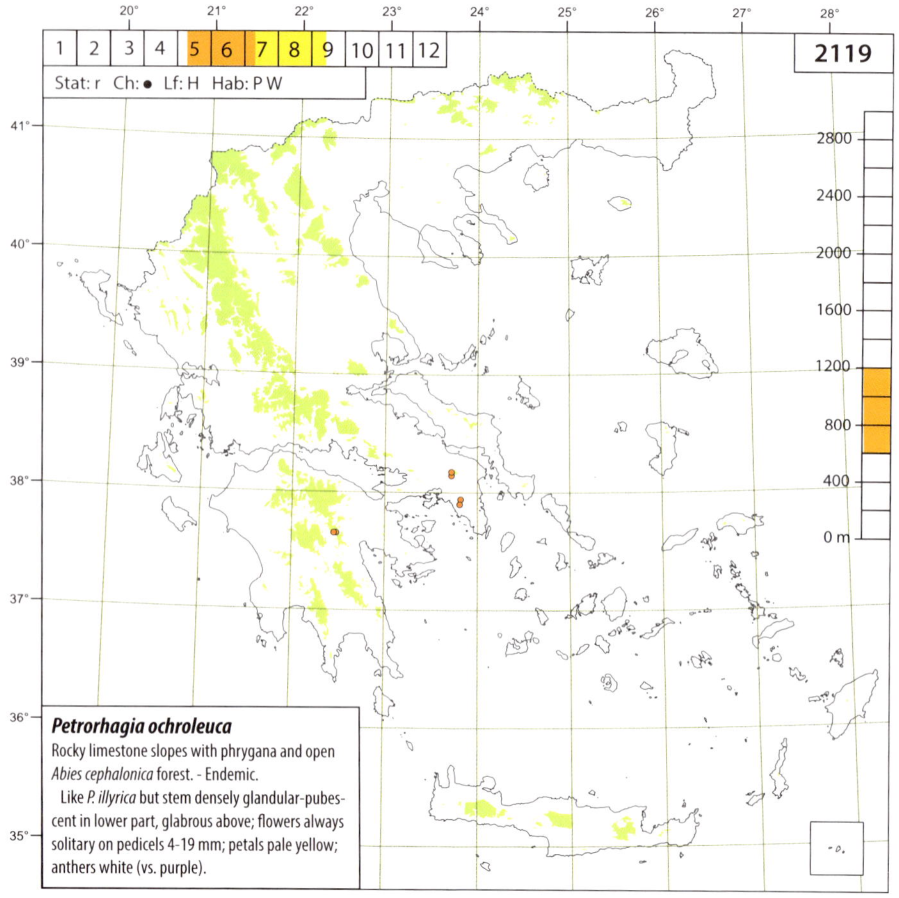

Petrorhagia ochroleuca
Rocky limestone slopes with phrygana and open *Abies cephalonica* forest. - Endemic.
Like *P. illyrica* but stem densely glandular-pubescent in lower part, glabrous above; flowers always solitary on pedicels 4-19 mm; petals pale yellow; anthers white (vs. purple).

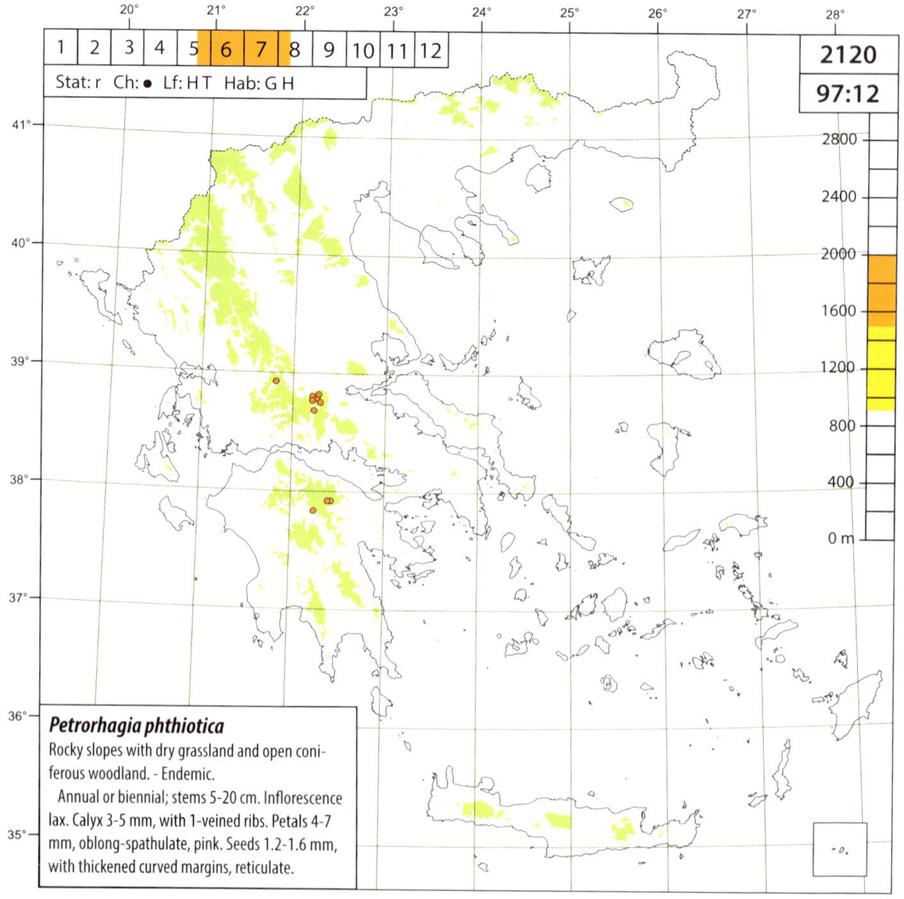

Petrorhagia phthiotica
Rocky slopes with dry grassland and open coniferous woodland. - Endemic.
Annual or biennial; stems 5-20 cm. Inflorescence lax. Calyx 3-5 mm, with 1-veined ribs. Petals 4-7 mm, oblong-spathulate, pink. Seeds 1.2-1.6 mm, with thickened curved margins, reticulate.

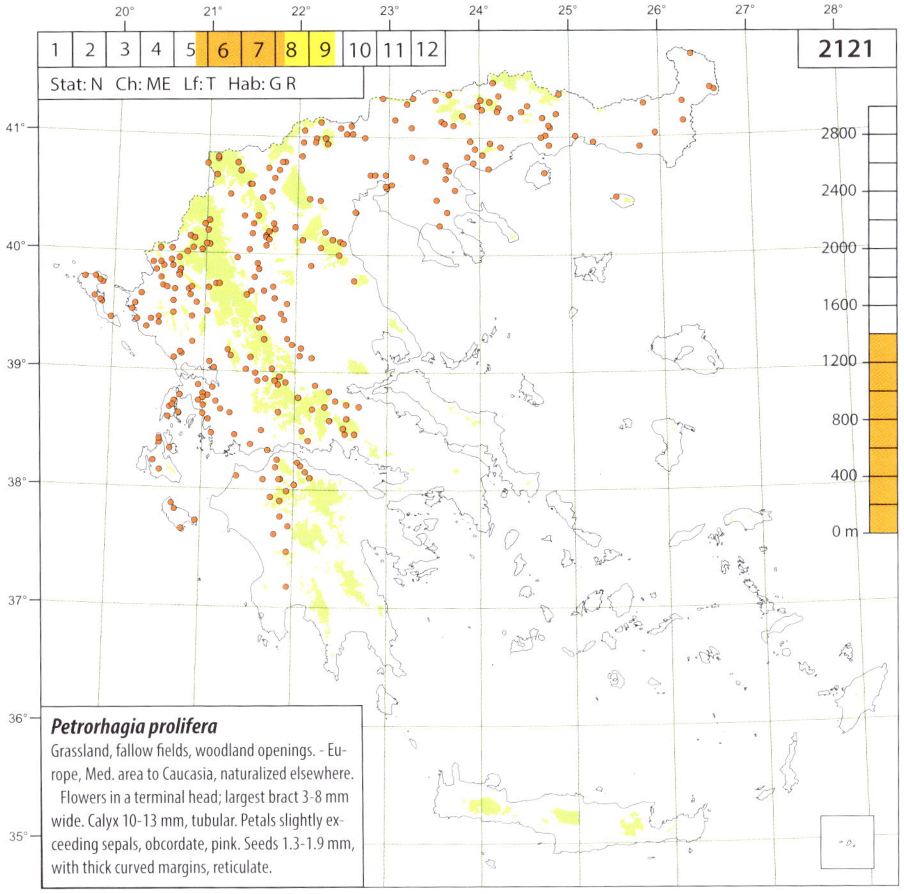

Petrorhagia prolifera

Grassland, fallow fields, woodland openings. - Europe, Med. area to Caucasia, naturalized elsewhere. Flowers in a terminal head; largest bract 3-8 mm wide. Calyx 10-13 mm, tubular. Petals slightly exceeding sepals, obcordate, pink. Seeds 1.3-1.9 mm, with thick curved margins, reticulate.

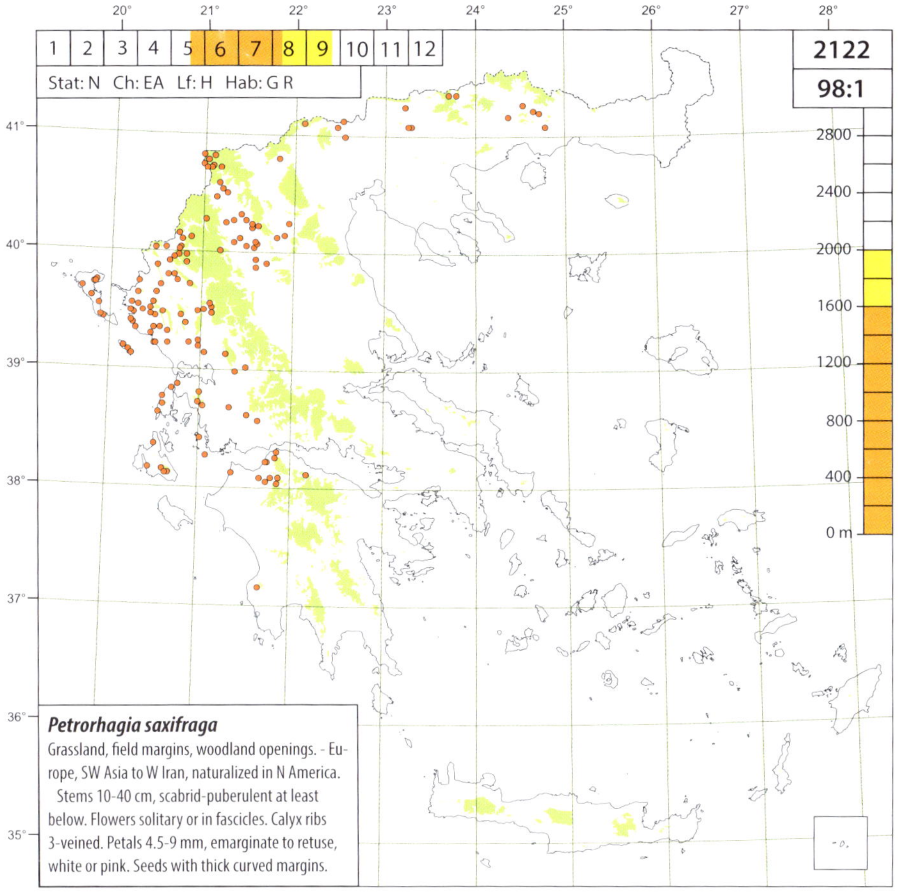

Petrorhagia saxifraga

Grassland, field margins, woodland openings. - Europe, SW Asia to W Iran, naturalized in N America. Stems 10-40 cm, scabrid-puberulent at least below. Flowers solitary or in fascicles. Calyx ribs 3-veined. Petals 4.5-9 mm, emarginate to retuse, white or pink. Seeds with thick curved margins.

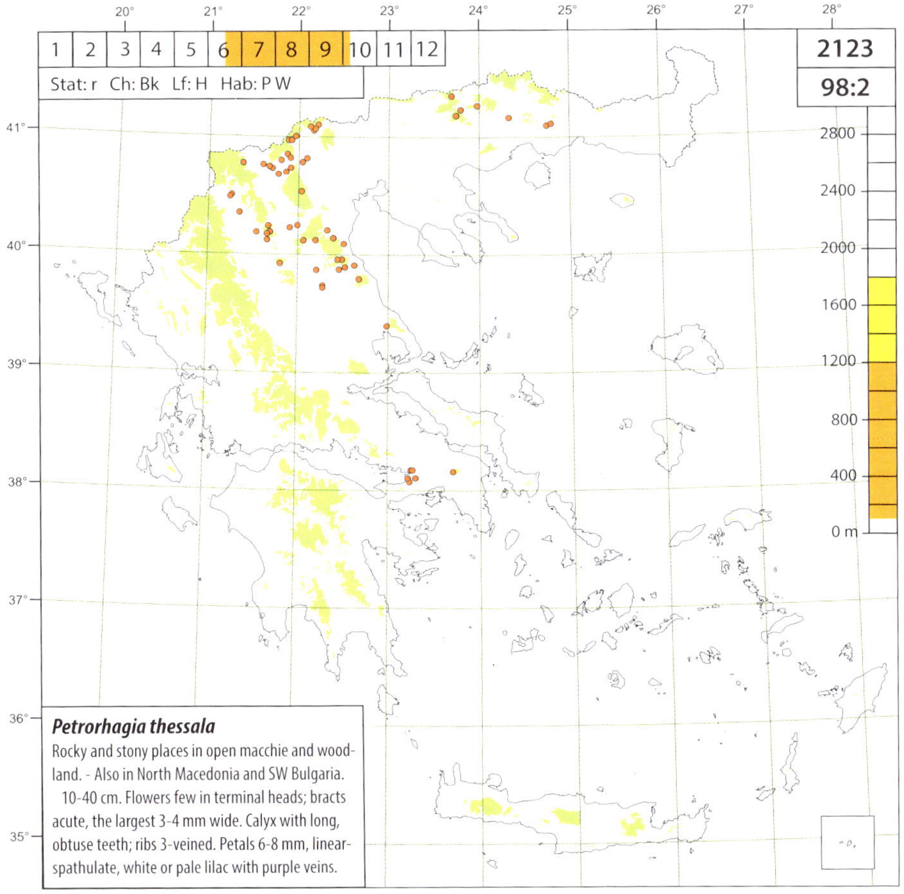

Petrorhagia thessala

Rocky and stony places in open macchie and woodland. - Also in North Macedonia and SW Bulgaria. 10-40 cm. Flowers few in terminal heads; bracts acute, the largest 3-4 mm wide. Calyx with long, obtuse teeth; ribs 3-veined. Petals 6-8 mm, linear-spathulate, white or pale lilac with purple veins.

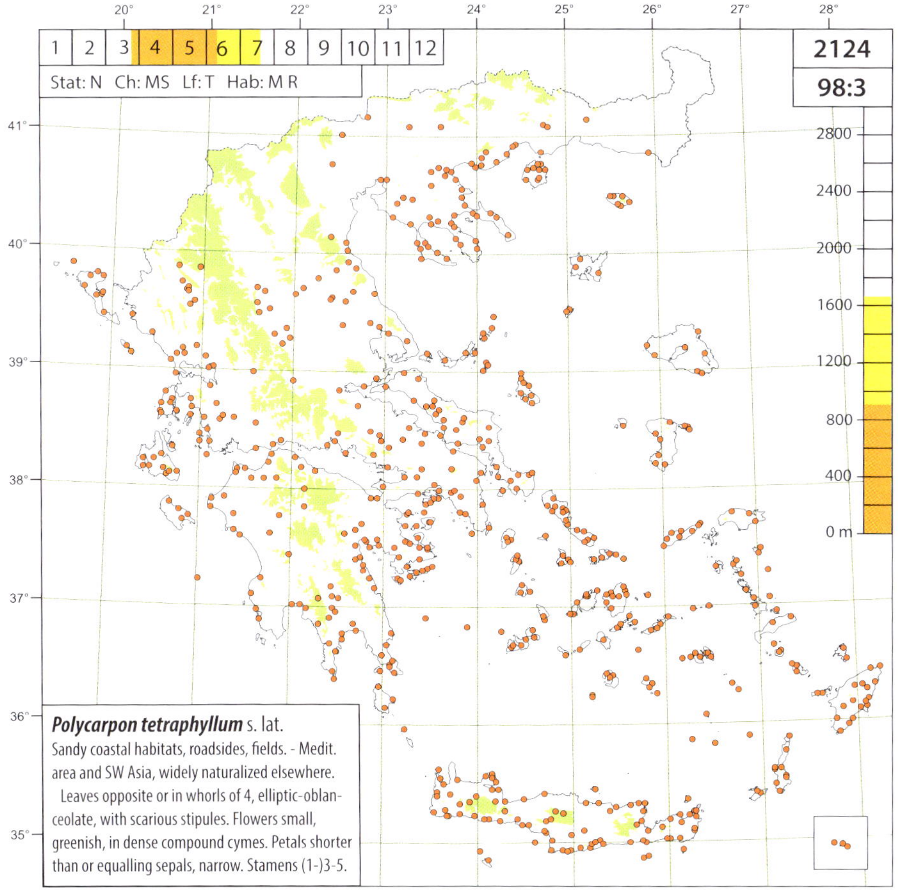

Polycarpon tetraphyllum s. lat.

Sandy coastal habitats, roadsides, fields. - Medit. area and SW Asia, widely naturalized elsewhere. Leaves opposite or in whorls of 4, elliptic-oblanceolate, with scarious stipules. Flowers small, greenish, in dense compound cymes. Petals shorter than or equalling sepals, narrow. Stamens (1-)3-5.

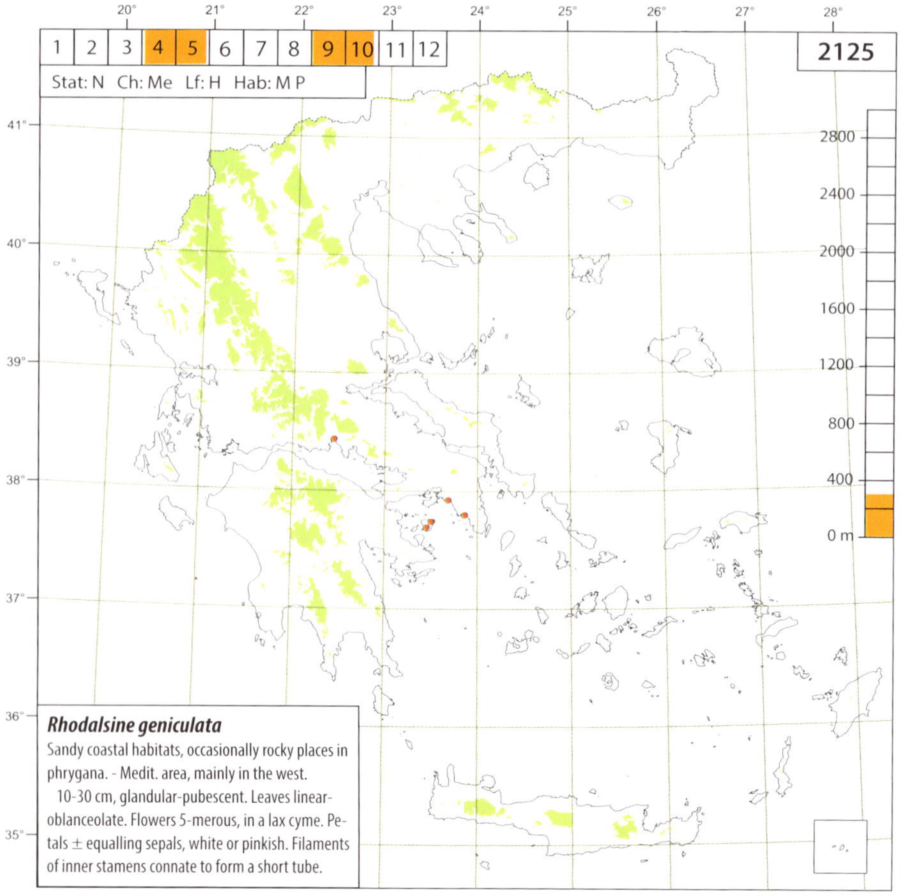

Rhodalsine geniculata
Sandy coastal habitats, occasionally rocky places in phrygana. - Medit. area, mainly in the west.
10-30 cm, glandular-pubescent. Leaves linear-oblanceolate. Flowers 5-merous, in a lax cyme. Petals ± equalling sepals, white or pinkish. Filaments of inner stamens connate to form a short tube.

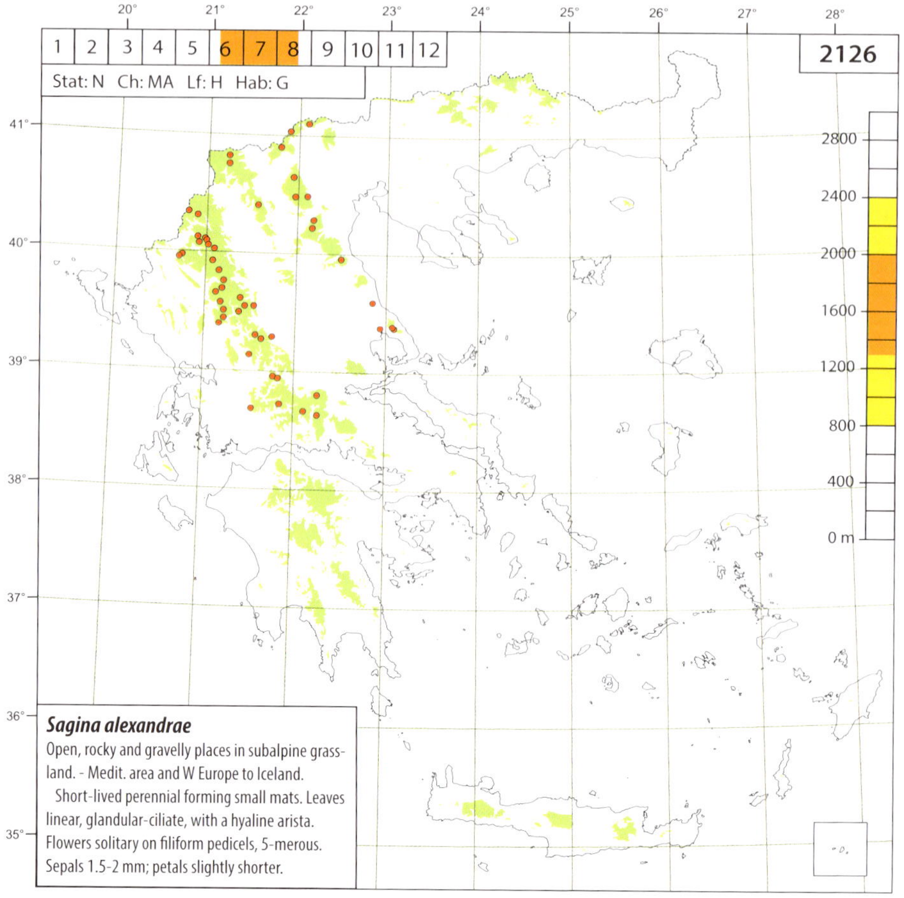

Sagina alexandrae
Open, rocky and gravelly places in subalpine grassland. - Medit. area and W Europe to Iceland.
Short-lived perennial forming small mats. Leaves linear, glandular-ciliate, with a hyaline arista. Flowers solitary on filiform pedicels, 5-merous. Sepals 1.5-2 mm; petals slightly shorter.

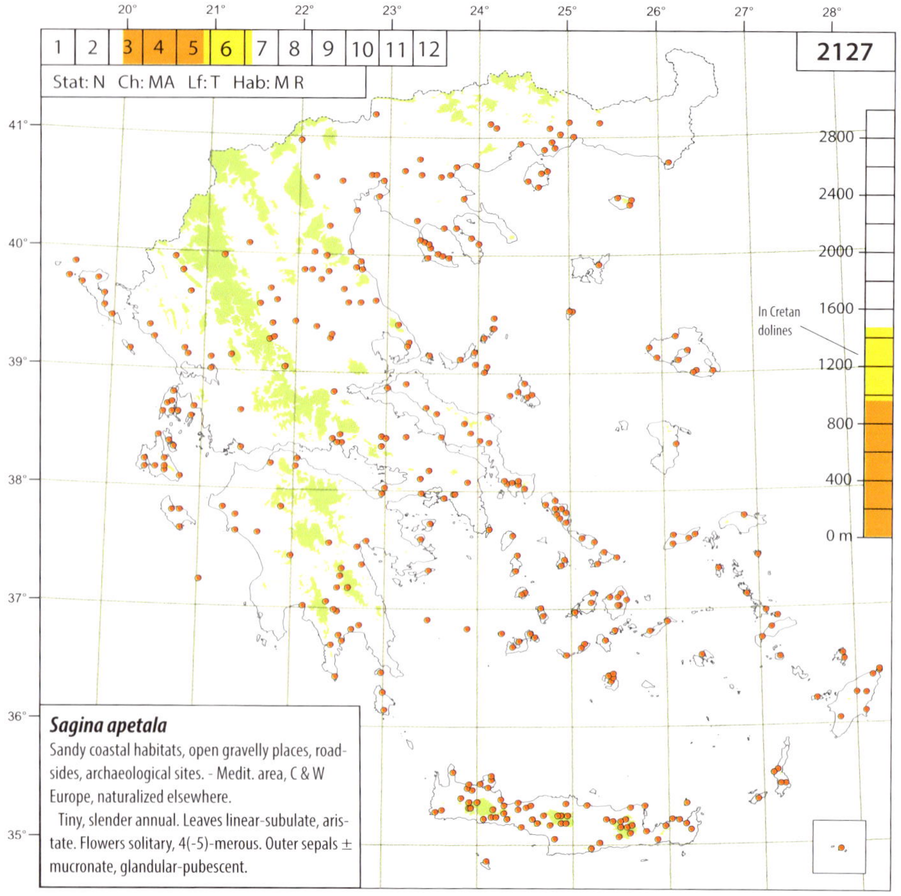

Sagina apetala
Sandy coastal habitats, open gravelly places, roadsides, archaeological sites. - Medit. area, C & W Europe, naturalized elsewhere.
Tiny, slender annual. Leaves linear-subulate, aristate. Flowers solitary, 4(-5)-merous. Outer sepals ± mucronate, glandular-pubescent.

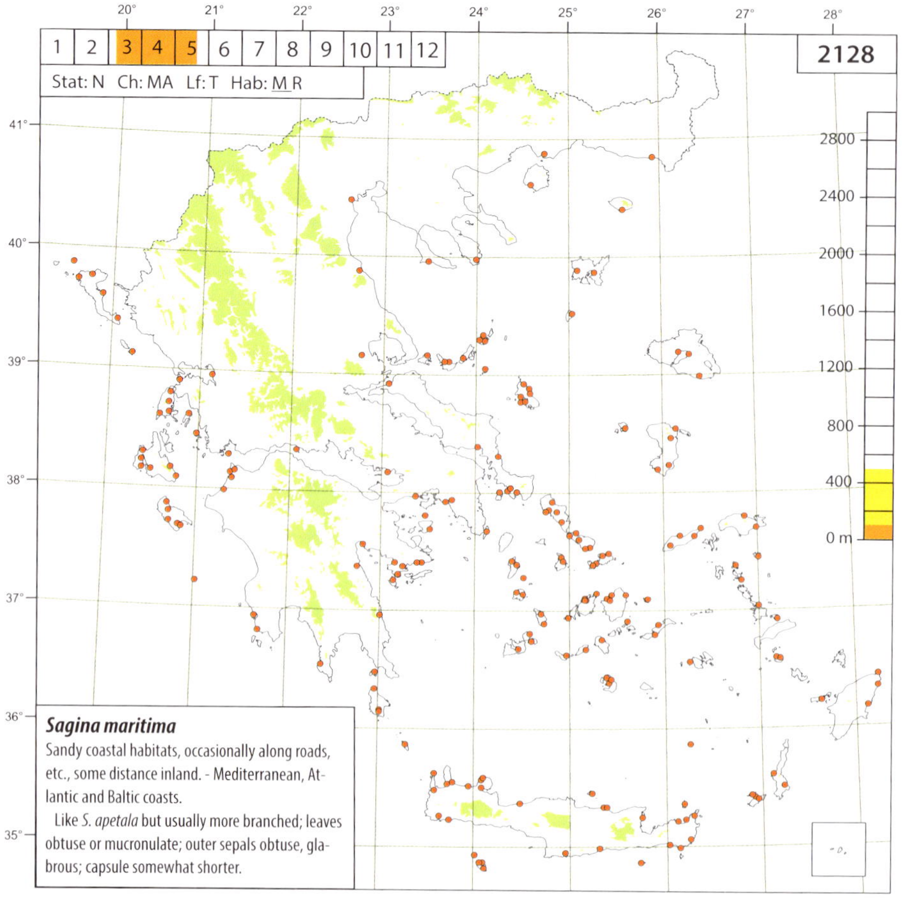

Sagina maritima
Sandy coastal habitats, occasionally along roads, etc., some distance inland. - Mediterranean, Atlantic and Baltic coasts.
Like *S. apetala* but usually more branched; leaves obtuse or mucronulate; outer sepals obtuse, glabrous; capsule somewhat shorter.

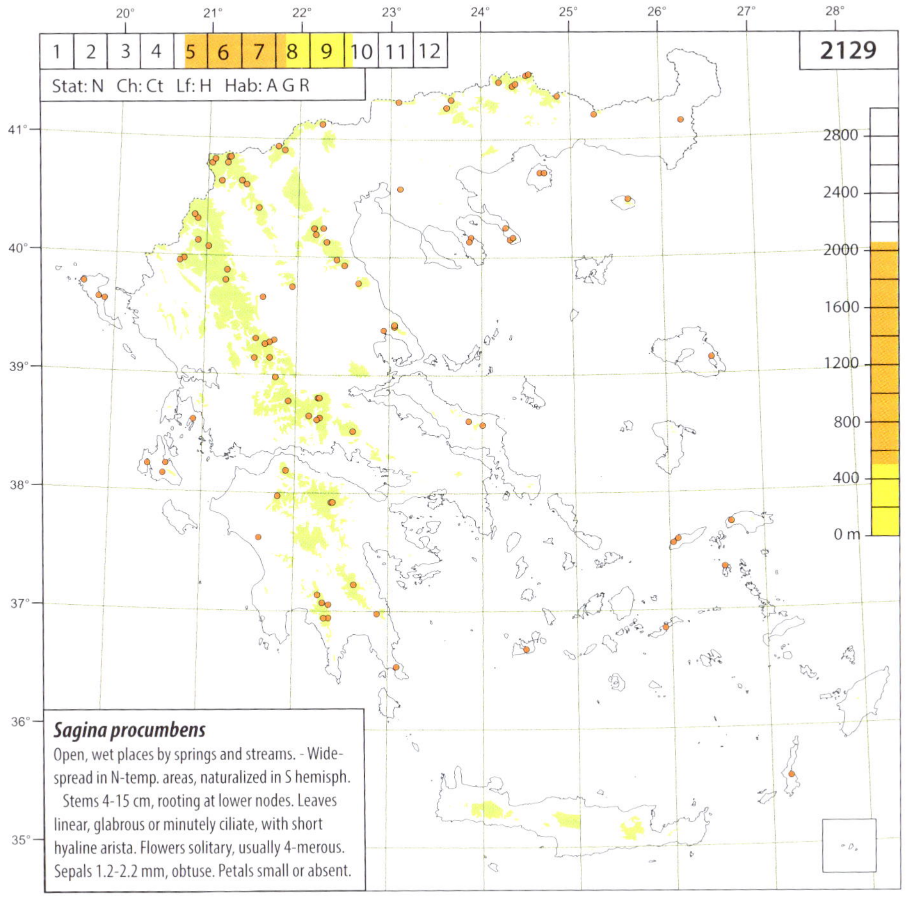

Sagina procumbens

Open, wet places by springs and streams. - Widespread in N-temp. areas, naturalized in S hemisph.

Stems 4-15 cm, rooting at lower nodes. Leaves linear, glabrous or minutely ciliate, with short hyaline arista. Flowers solitary, usually 4-merous. Sepals 1.2-2.2 mm, obtuse. Petals small or absent.

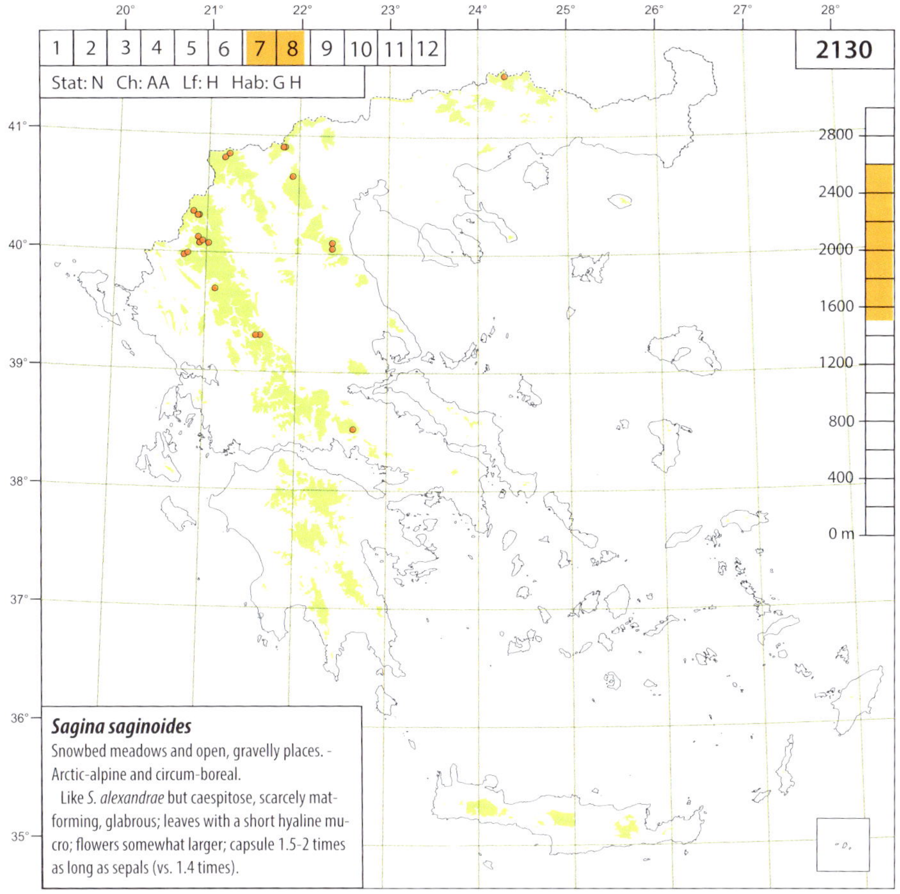

Sagina saginoides

Snowbed meadows and open, gravelly places. - Arctic-alpine and circum-boreal.

Like *S. alexandrae* but caespitose, scarcely mat-forming, glabrous; leaves with a short hyaline mucro; flowers somewhat larger; capsule 1.5-2 times as long as sepals (vs. 1.4 times).

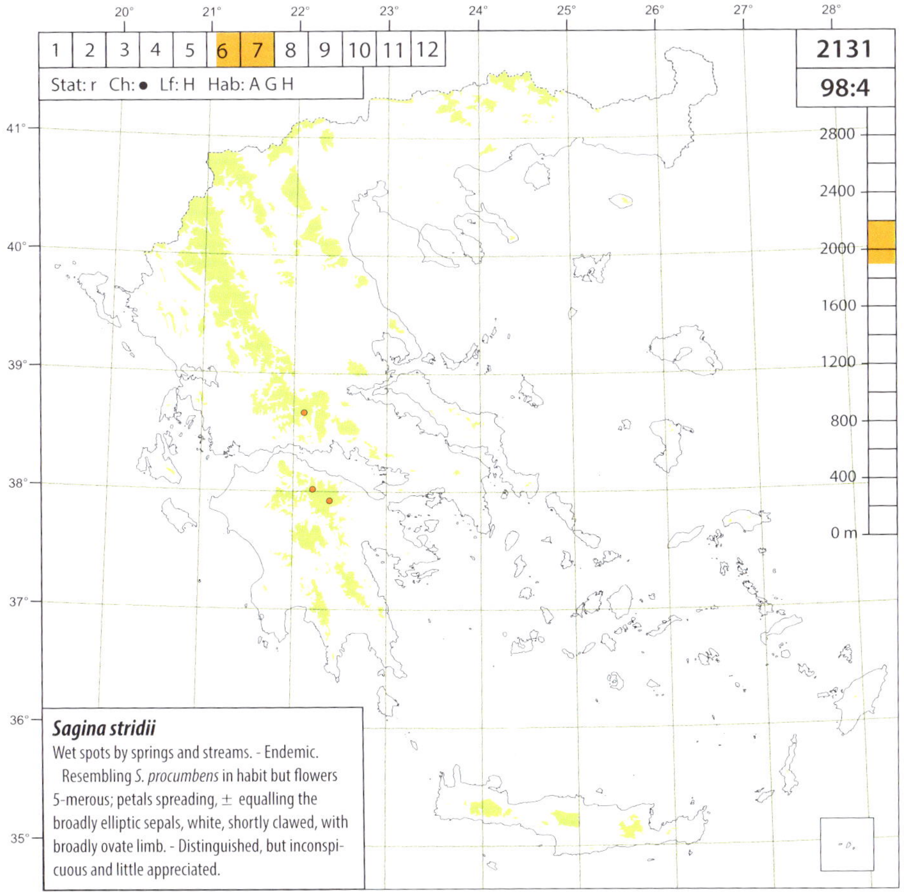

Sagina stridii

Wet spots by springs and streams. - Endemic.

Resembling *S. procumbens* in habit but flowers 5-merous; petals spreading, ± equalling the broadly elliptic sepals, white, shortly clawed, with broadly ovate limb. - Distinguished, but inconspicuous and little appreciated.

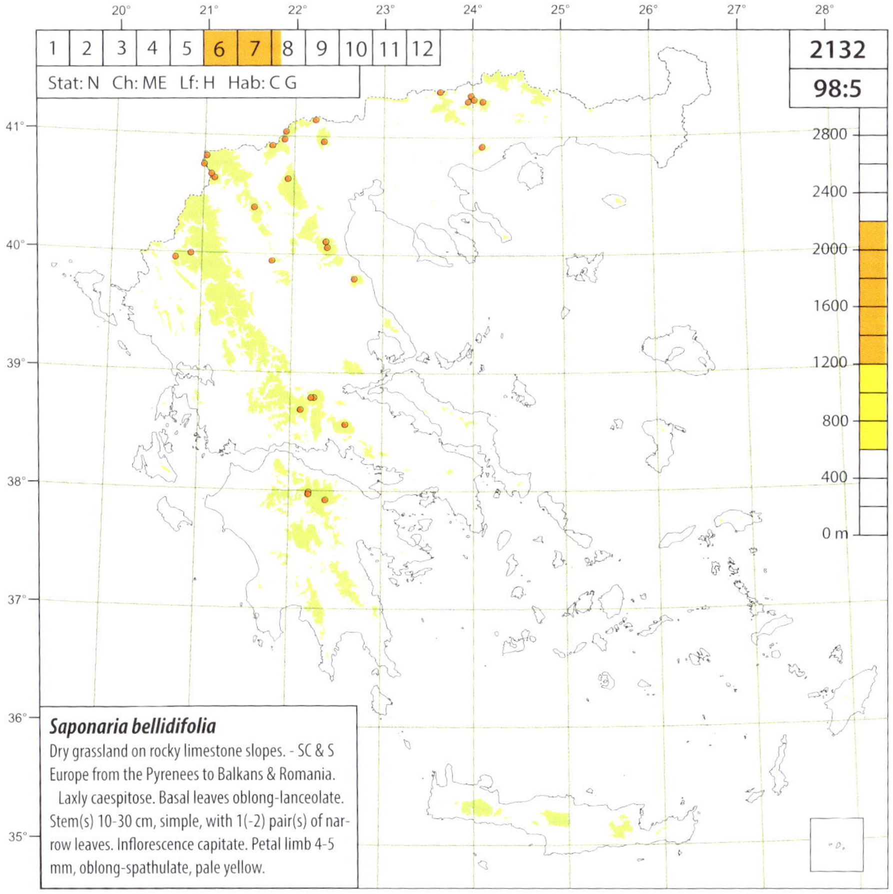

Saponaria bellidifolia

Dry grassland on rocky limestone slopes. - SC & S Europe from the Pyrenees to Balkans & Romania.

Laxly caespitose. Basal leaves oblong-lanceolate. Stem(s) 10-30 cm, simple, with 1(-2) pair(s) of narrow leaves. Inflorescence capitate. Petal limb 4-5 mm, oblong-spathulate, pale yellow.

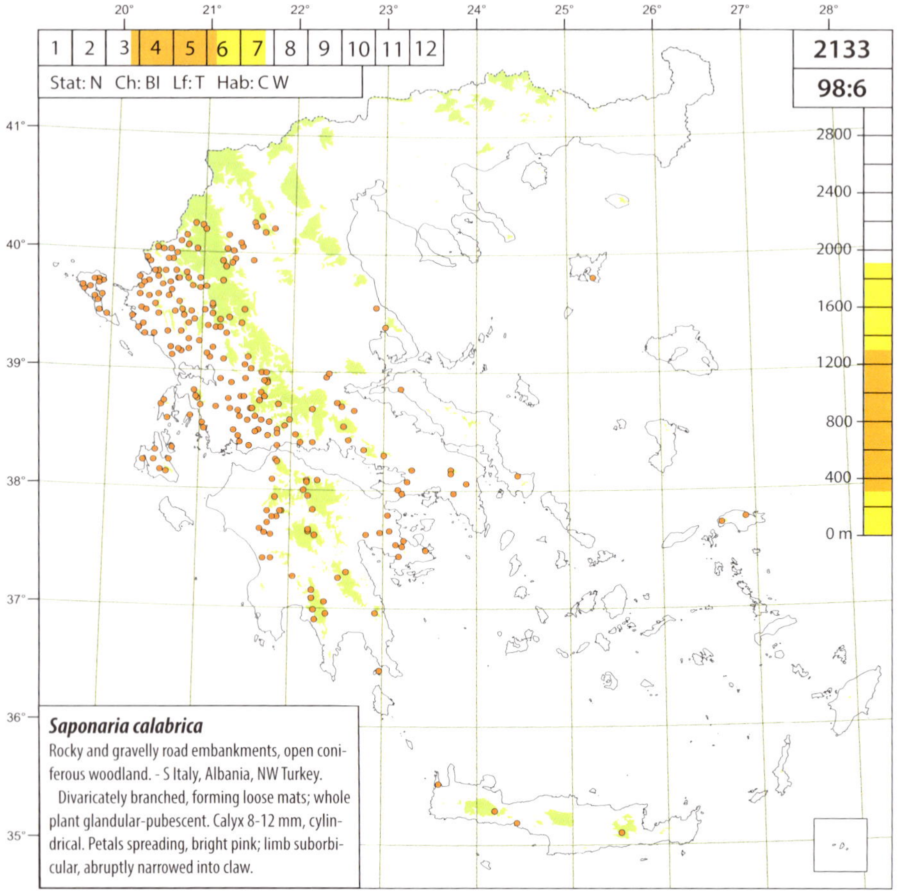

Saponaria calabrica

Rocky and gravelly road embankments, open coniferous woodland. - S Italy, Albania, NW Turkey.

Divaricately branched, forming loose mats; whole plant glandular-pubescent. Calyx 8-12 mm, cylindrical. Petals spreading, bright pink; limb suborbicular, abruptly narrowed into claw.

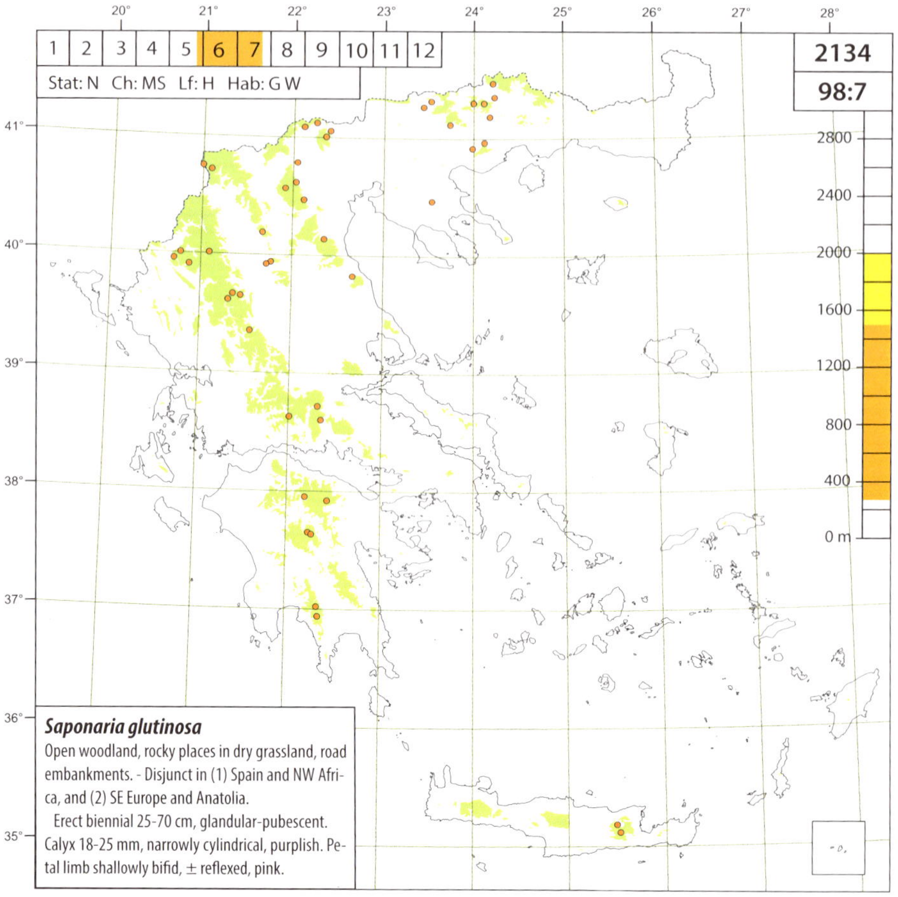

Saponaria glutinosa

Open woodland, rocky places in dry grassland, road embankments. - Disjunct in (1) Spain and NW Africa, and (2) SE Europe and Anatolia.

Erect biennial 25-70 cm, glandular-pubescent. Calyx 18-25 mm, narrowly cylindrical, purplish. Petal limb shallowly bifid, ± reflexed, pink.

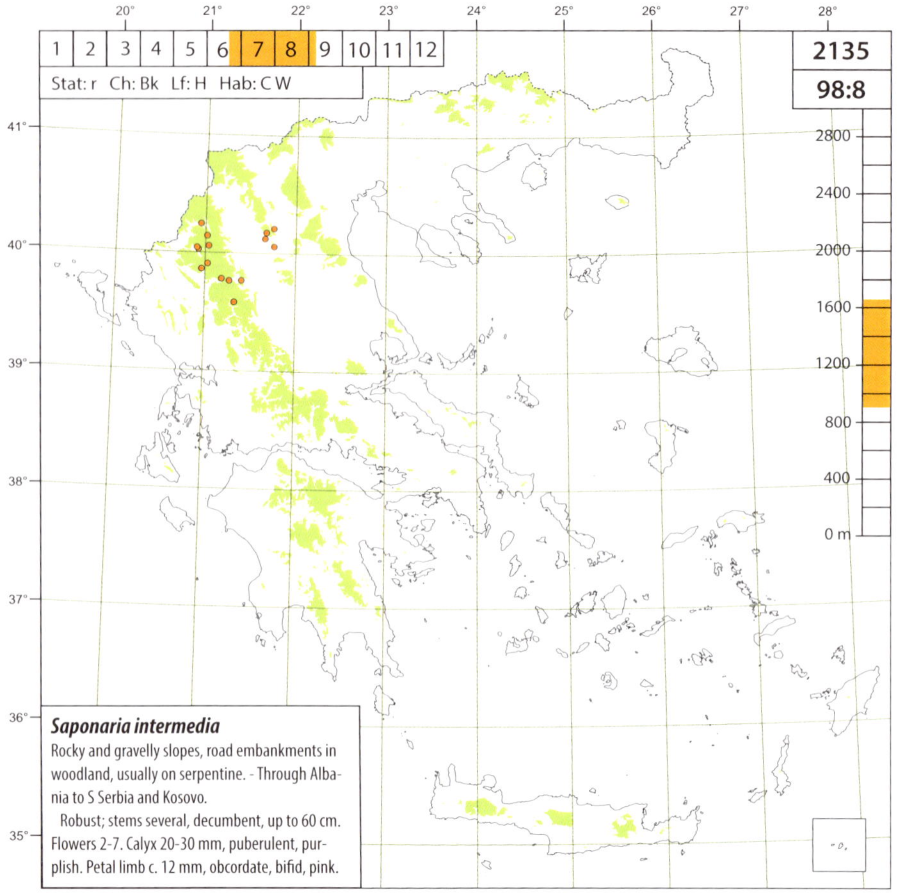

Saponaria intermedia

Rocky and gravelly slopes, road embankments in woodland, usually on serpentine. - Through Albania to S Serbia and Kosovo.

Robust; stems several, decumbent, up to 60 cm. Flowers 2-7. Calyx 20-30 mm, puberulent, purplish. Petal limb c. 12 mm, obcordate, bifid, pink.

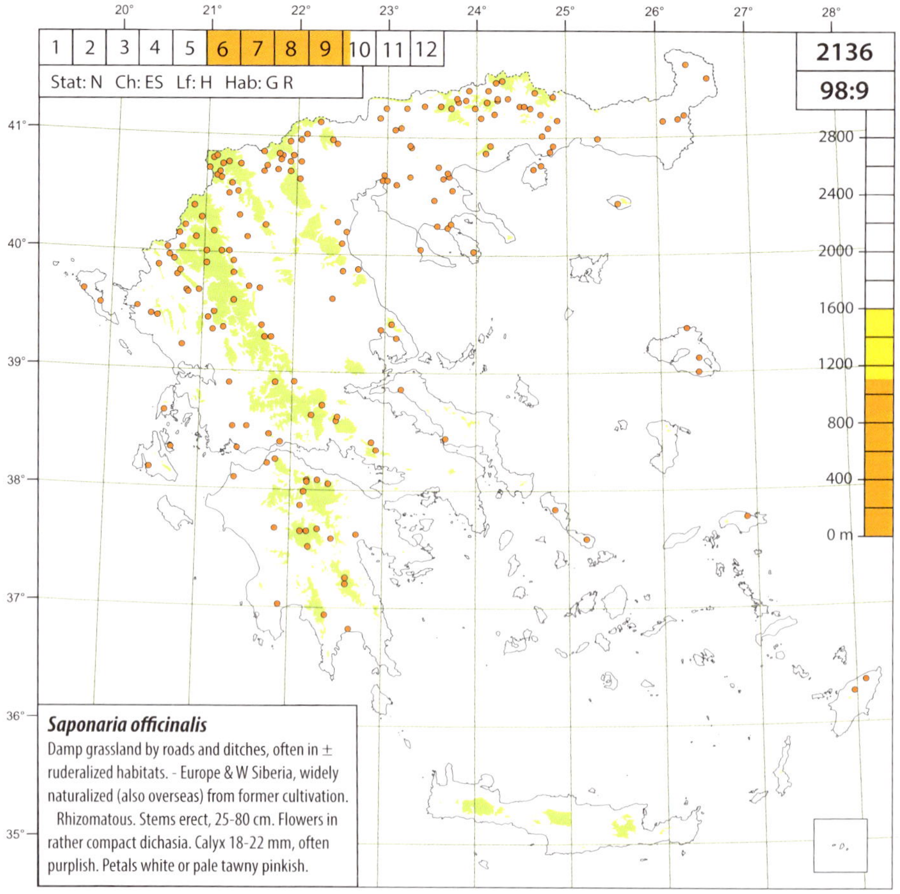

Saponaria officinalis

Damp grassland by roads and ditches, often in ± ruderalized habitats. - Europe & W Siberia, widely naturalized (also overseas) from former cultivation.

Rhizomatous. Stems erect, 25-80 cm. Flowers in rather compact dichasia. Calyx 18-22 mm, often purplish. Petals white or pale tawny pinkish.

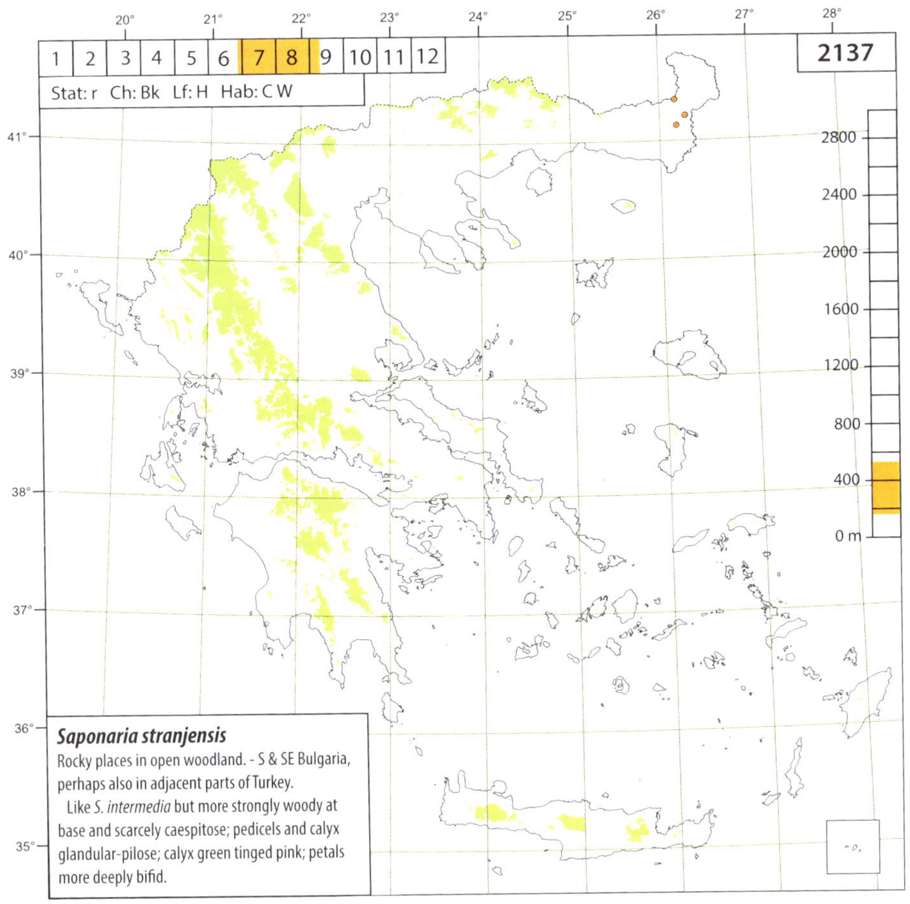

Saponaria stranjensis

Rocky places in open woodland. - S & SE Bulgaria, perhaps also in adjacent parts of Turkey.

Like *S. intermedia* but more strongly woody at base and scarcely caespitose; pedicels and calyx glandular-pilose; calyx green tinged pink; petals more deeply bifid.

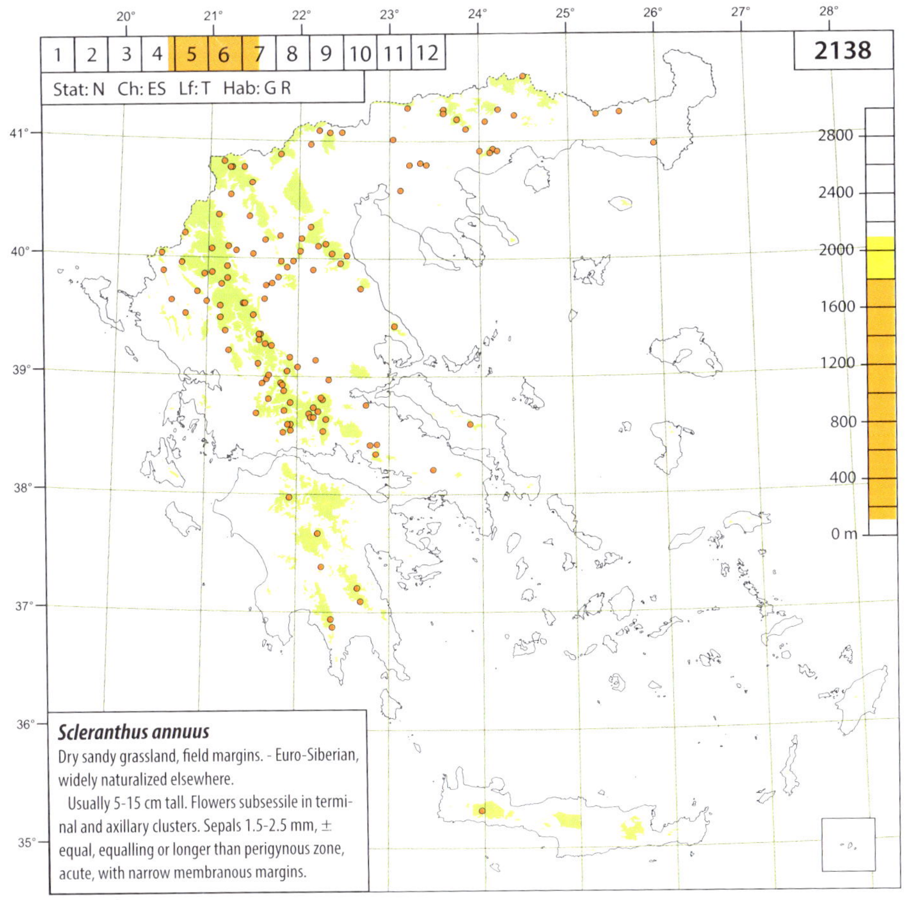

Scleranthus annuus

Dry sandy grassland, field margins. - Euro-Siberian, widely naturalized elsewhere.

Usually 5-15 cm tall. Flowers subsessile in terminal and axillary clusters. Sepals 1.5-2.5 mm, ± equal, equalling or longer than perigynous zone, acute, with narrow membranous margins.

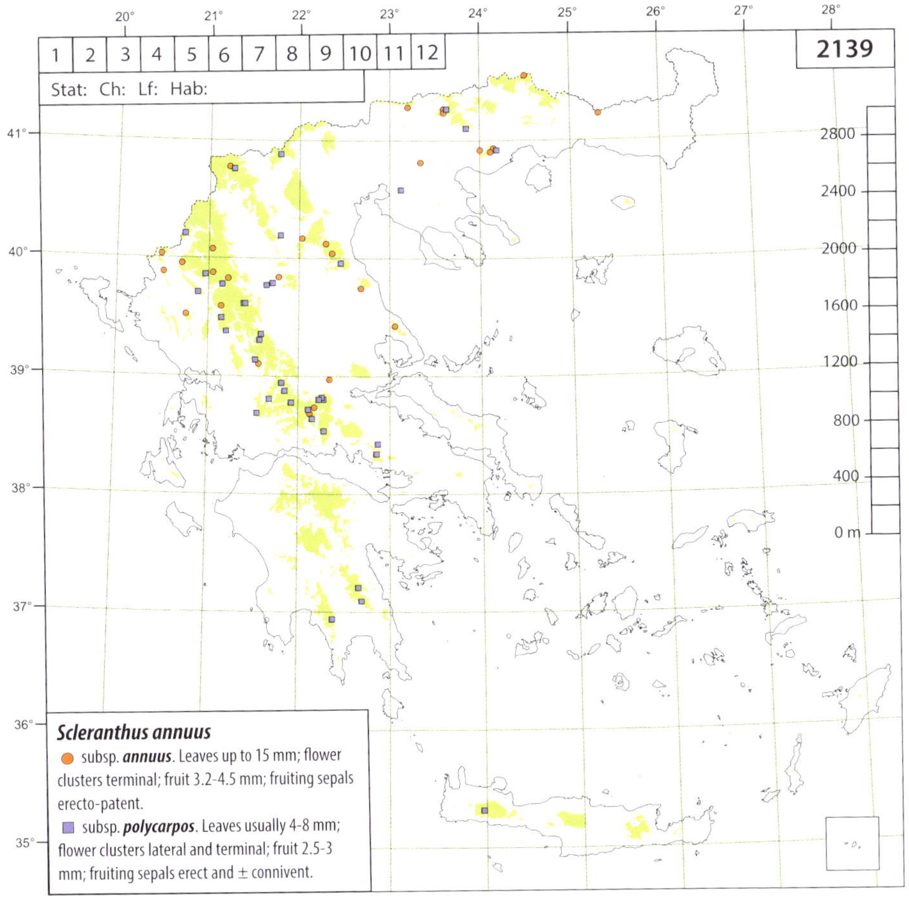

Scleranthus annuus

● subsp. ***annuus***. Leaves up to 15 mm; flower clusters terminal; fruit 3.2-4.5 mm; fruiting sepals erecto-patent.

■ subsp. ***polycarpos***. Leaves usually 4-8 mm; flower clusters lateral and terminal; fruit 2.5-3 mm; fruiting sepals erect and ± connivent.

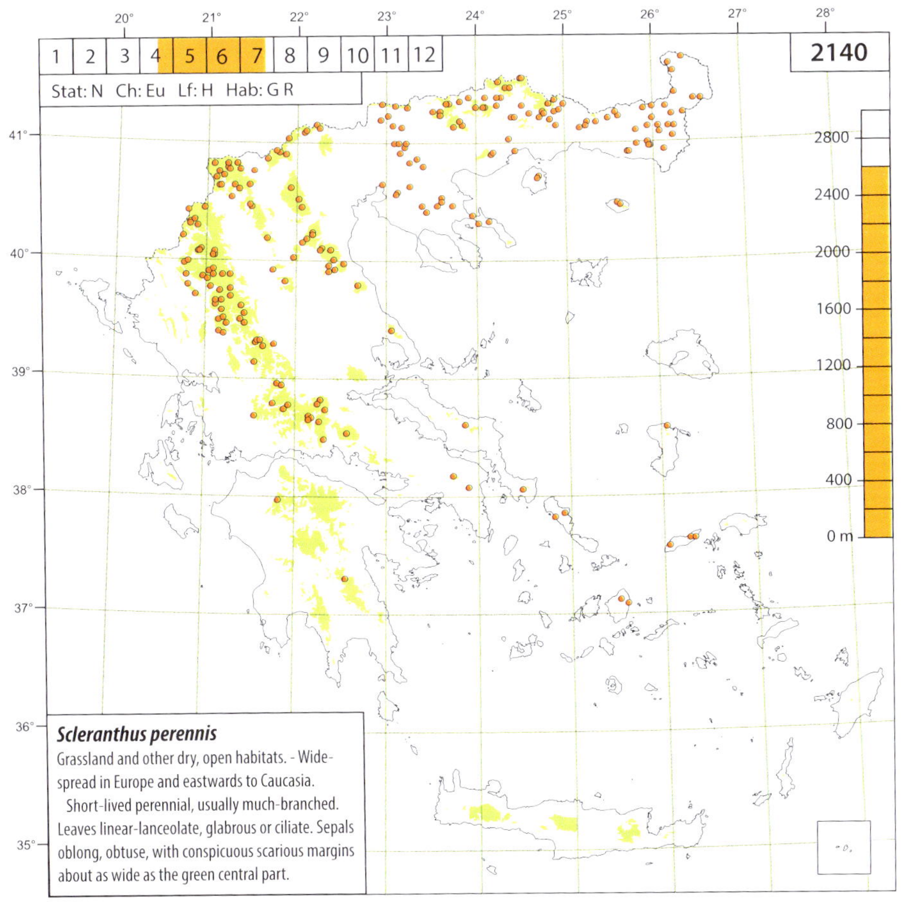

Scleranthus perennis

Grassland and other dry, open habitats. - Widespread in Europe and eastwards to Caucasia.

Short-lived perennial, usually much-branched. Leaves linear-lanceolate, glabrous or ciliate. Sepals oblong, obtuse, with conspicuous scarious margins about as wide as the green central part.

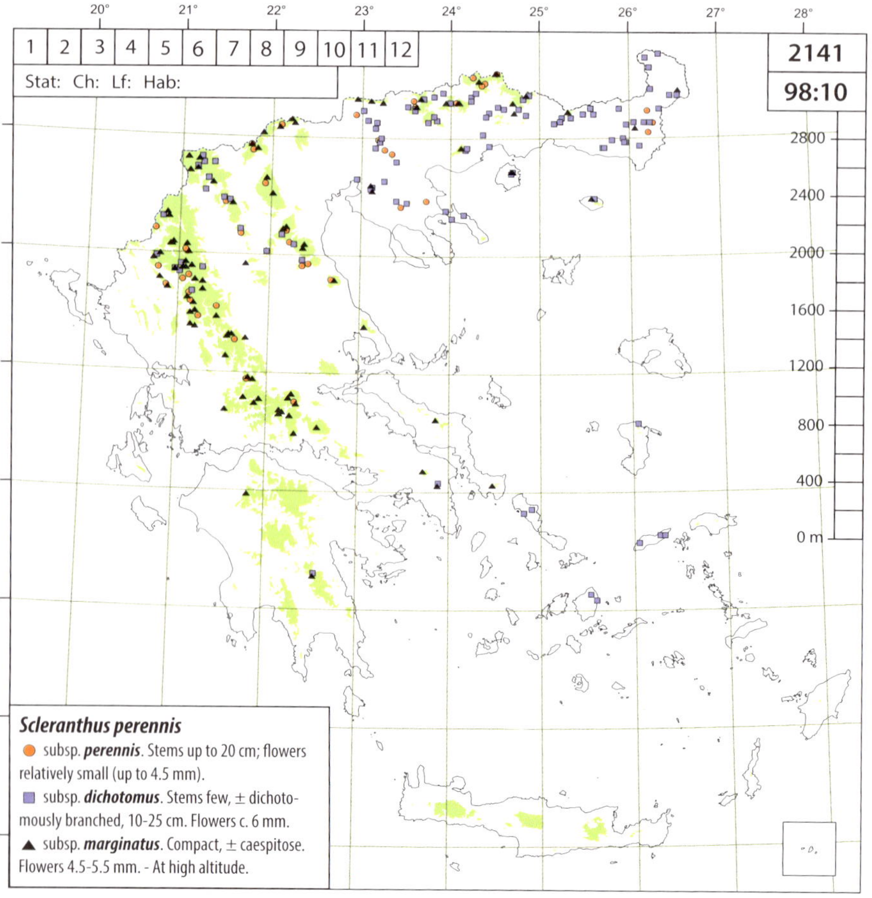

Scleranthus perennis

● subsp. ***perennis***. Stems up to 20 cm; flowers relatively small (up to 4.5 mm).

■ subsp. ***dichotomus***. Stems few, ± dichotomously branched, 10-25 cm. Flowers c. 6 mm.

▲ subsp. ***marginatus***. Compact, ± caespitose. Flowers 4.5-5.5 mm. - At high altitude.

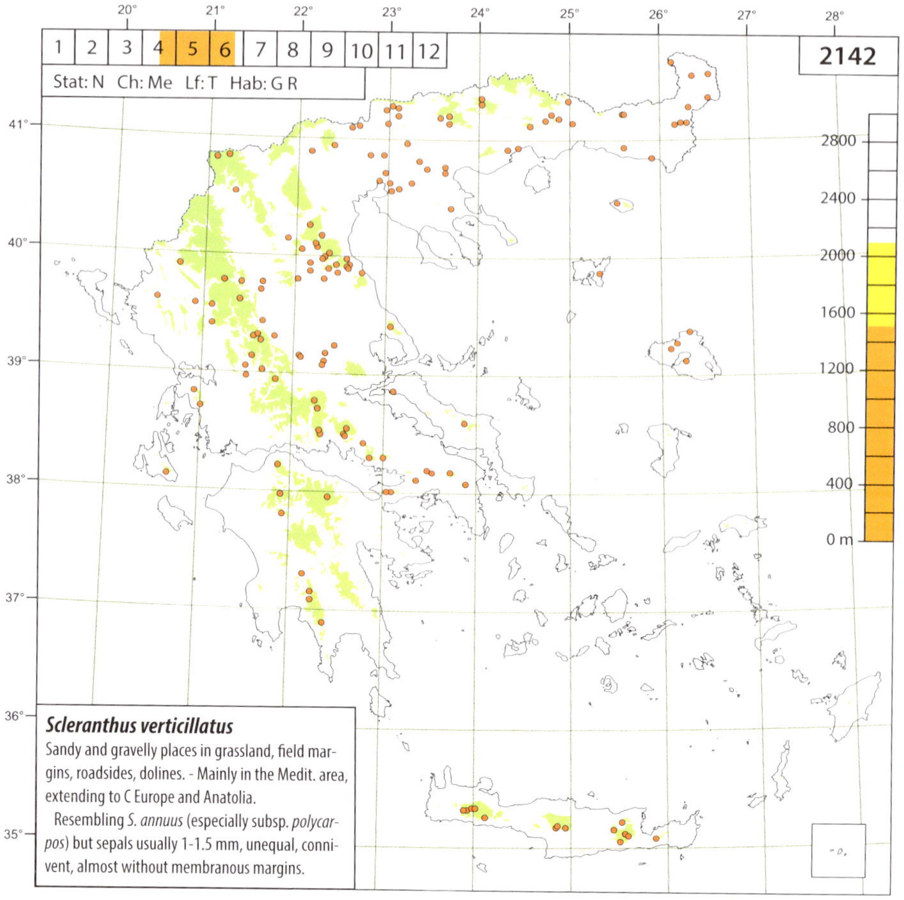

Scleranthus verticillatus

Sandy and gravelly places in grassland, field margins, roadsides, dolines. - Mainly in the Medit. area, extending to C Europe and Anatolia.

Resembling *S. annuus* (especially subsp. *polycarpos*) but sepals usually 1-1.5 mm, unequal, connivent, almost without membranous margins.

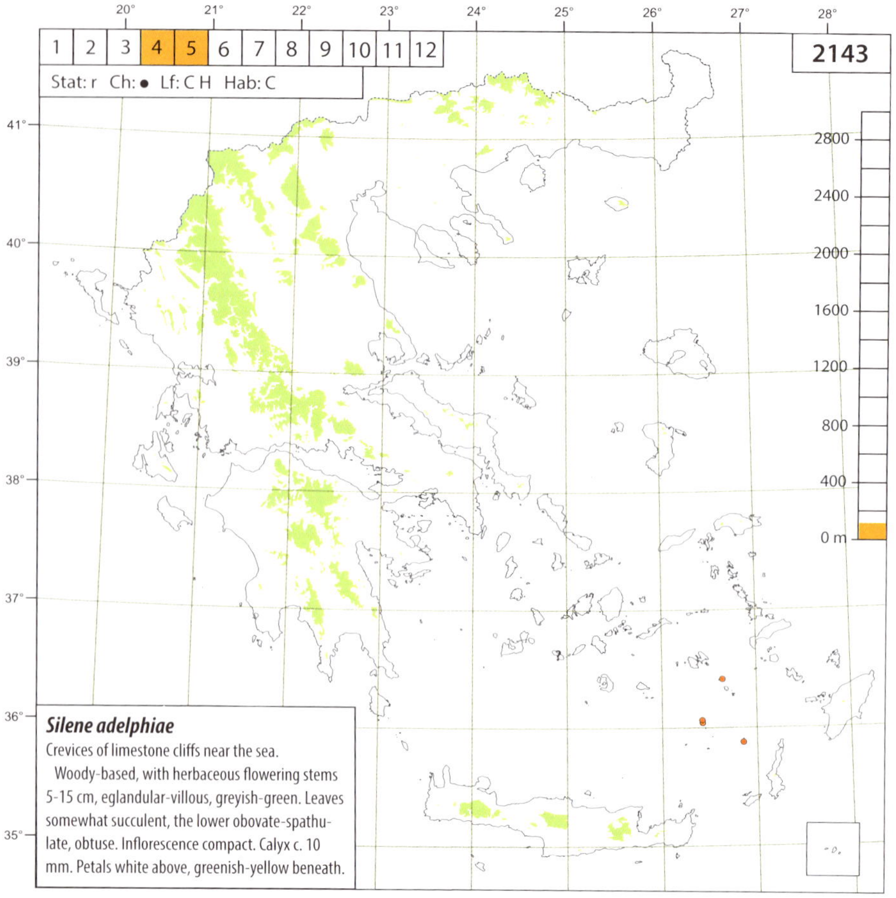

Silene adelphiae

Crevices of limestone cliffs near the sea.

Woody-based, with herbaceous flowering stems 5-15 cm, eglandular-villous, greyish-green. Leaves somewhat succulent, the lower obovate-spathulate, obtuse. Inflorescence compact. Calyx c. 10 mm. Petals white above, greenish-yellow beneath.

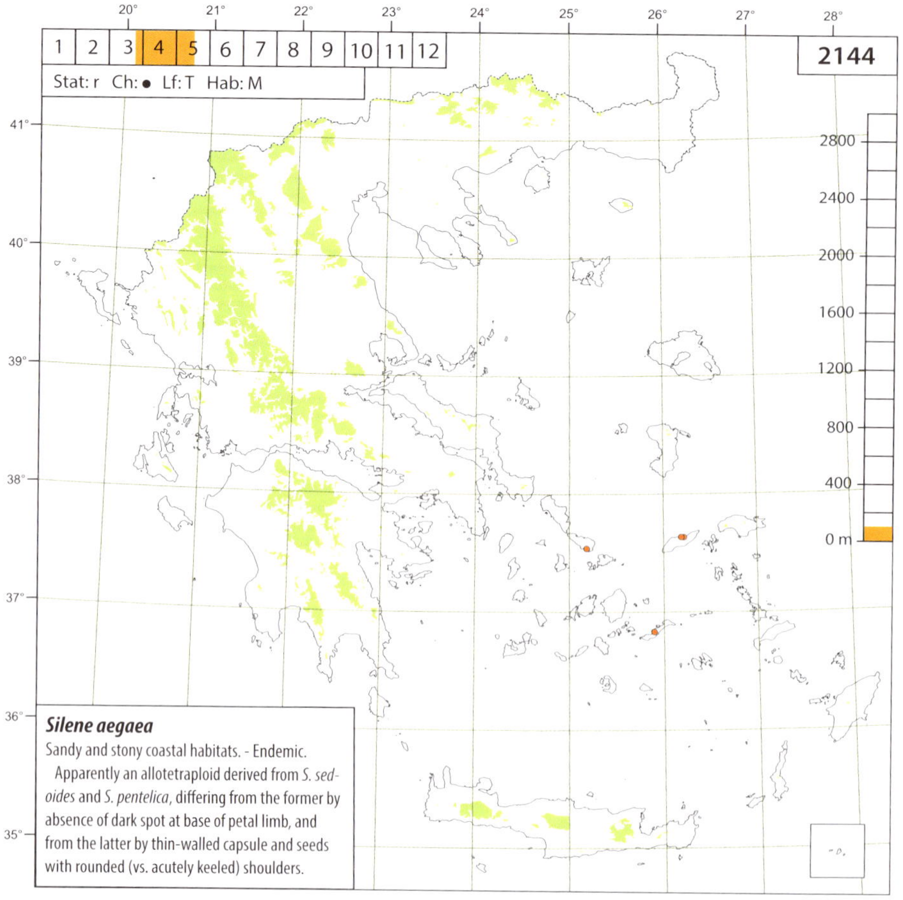

Silene aegaea

Sandy and stony coastal habitats. - Endemic.

Apparently an allotetraploid derived from *S. sedoides* and *S. pentelica*, differing from the former by absence of dark spot at base of petal limb, and from the latter by thin-walled capsule and seeds with rounded (vs. acutely keeled) shoulders.

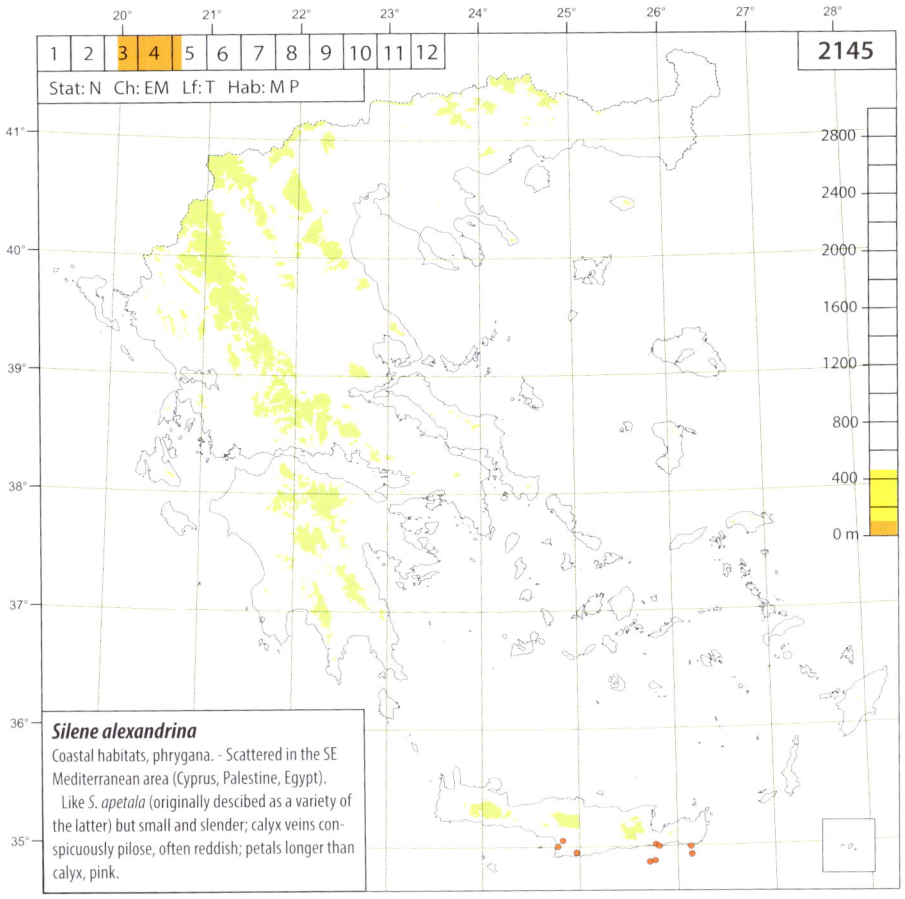

Silene alexandrina

Coastal habitats, phrygana. - Scattered in the SE Mediterranean area (Cyprus, Palestine, Egypt).

Like *S. apetala* (originally descibed as a variety of the latter) but small and slender; calyx veins conspicuously pilose, often reddish; petals longer than calyx, pink.

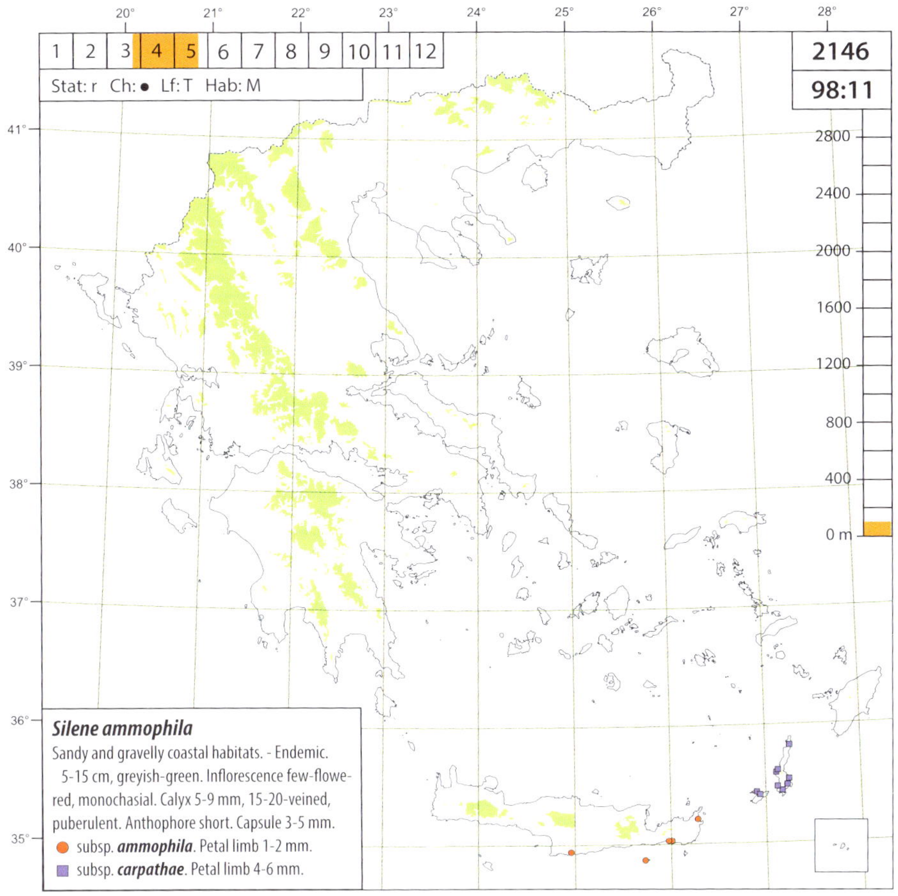

Silene ammophila

Sandy and gravelly coastal habitats. - Endemic.

5-15 cm, greyish-green. Inflorescence few-flowered, monochasial. Calyx 5-9 mm, 15-20-veined, puberulent. Anthophore short. Capsule 3-5 mm.

● subsp. ***ammophila***. Petal limb 1-2 mm.

■ subsp. ***carpathae***. Petal limb 4-6 mm.

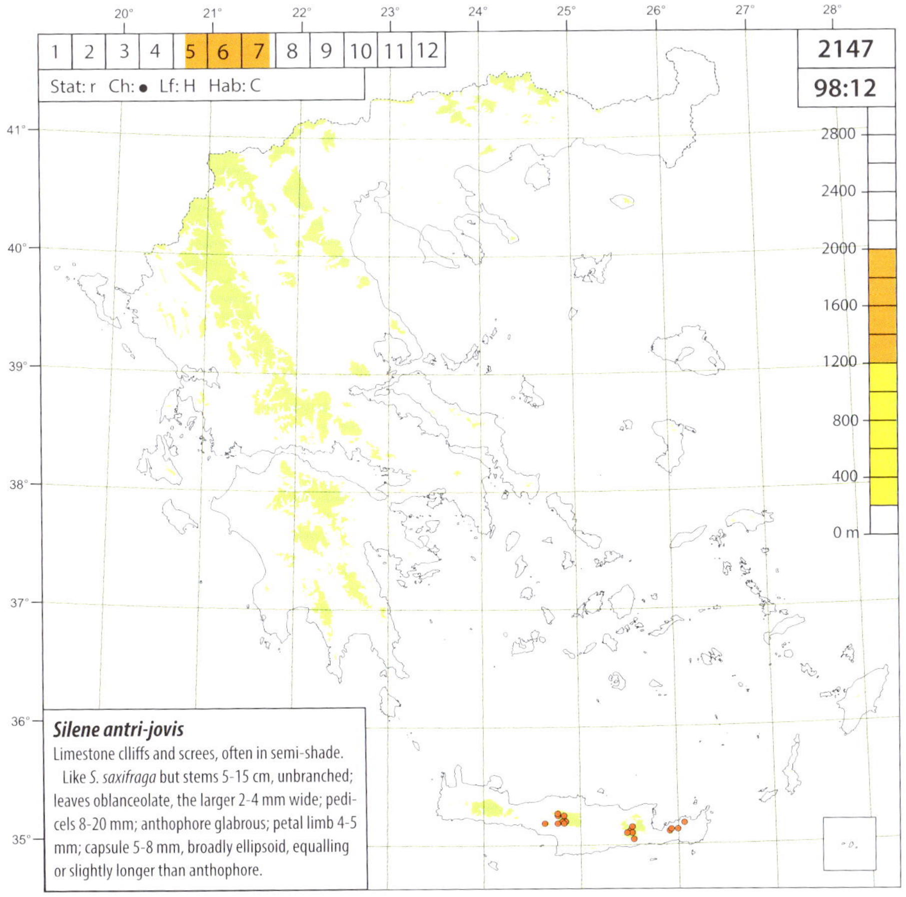

Silene antri-jovis

Limestone clliffs and screes, often in semi-shade.

Like *S. saxifraga* but stems 5-15 cm, unbranched; leaves oblanceolate, the larger 2-4 mm wide; pedicels 8-20 mm; anthophore glabrous; petal limb 4-5 mm; capsule 5-8 mm, broadly ellipsoid, equalling or slightly longer than anthophore.

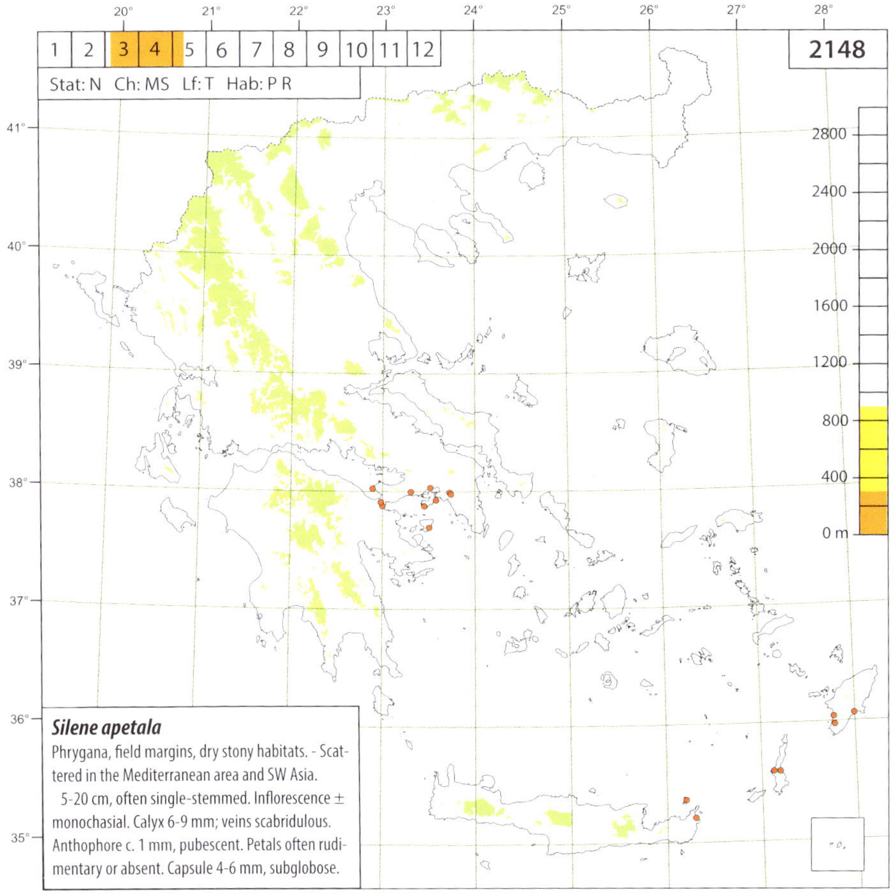

Silene apetala

Phrygana, field margins, dry stony habitats. - Scattered in the Mediterranean area and SW Asia.

5-20 cm, often single-stemmed. Inflorescence ± monochasial. Calyx 6-9 mm; veins scabridulous. Anthophore c. 1 mm, pubescent. Petals often rudimentary or absent. Capsule 4-6 mm, subglobose.

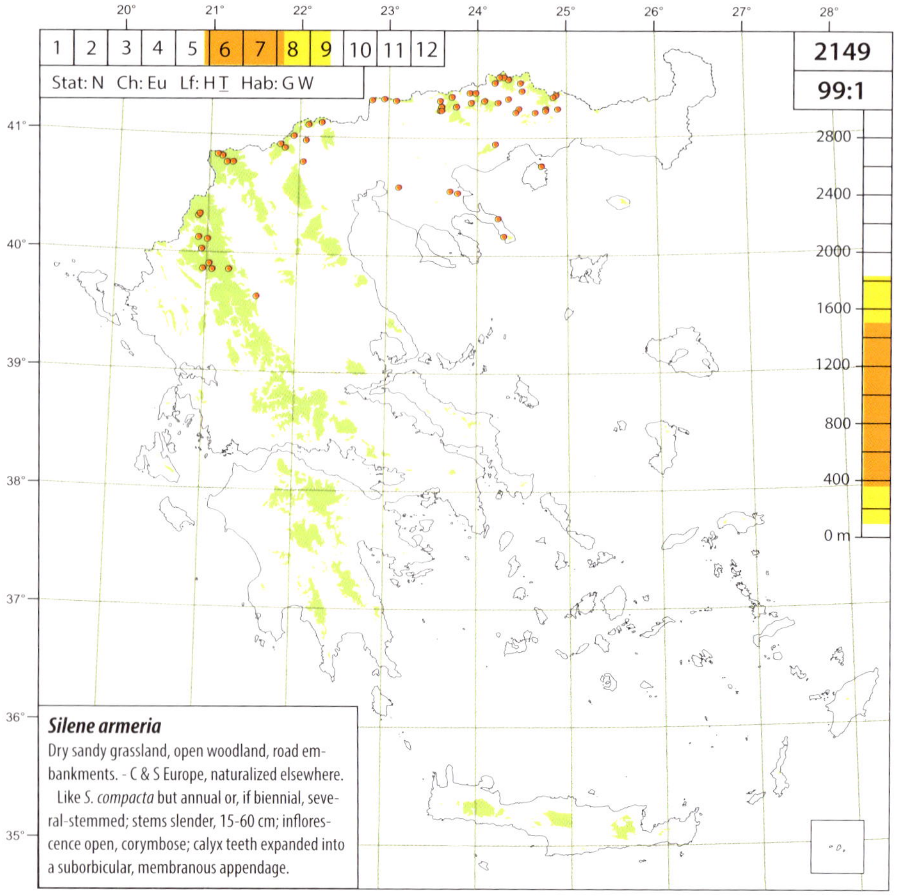

Silene armeria
Dry sandy grassland, open woodland, road embankments. - C & S Europe, naturalized elsewhere.
Like *S. compacta* but annual or, if biennial, several-stemmed; stems slender, 15-60 cm; inflorescence open, corymbose; calyx teeth expanded into a suborbicular, membranous appendage.

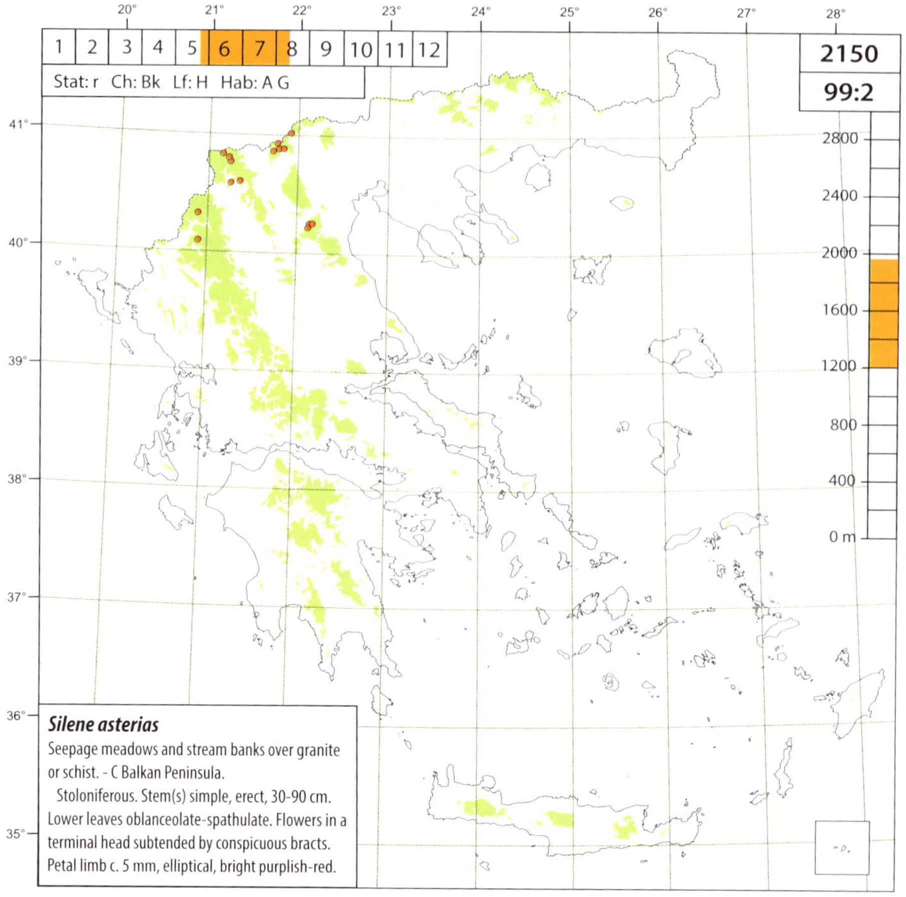

Silene asterias
Seepage meadows and stream banks over granite or schist. - C Balkan Peninsula.
Stoloniferous. Stem(s) simple, erect, 30-90 cm. Lower leaves oblanceolate-spathulate. Flowers in a terminal head subtended by conspicuous bracts. Petal limb c. 5 mm, elliptical, bright purplish-red.

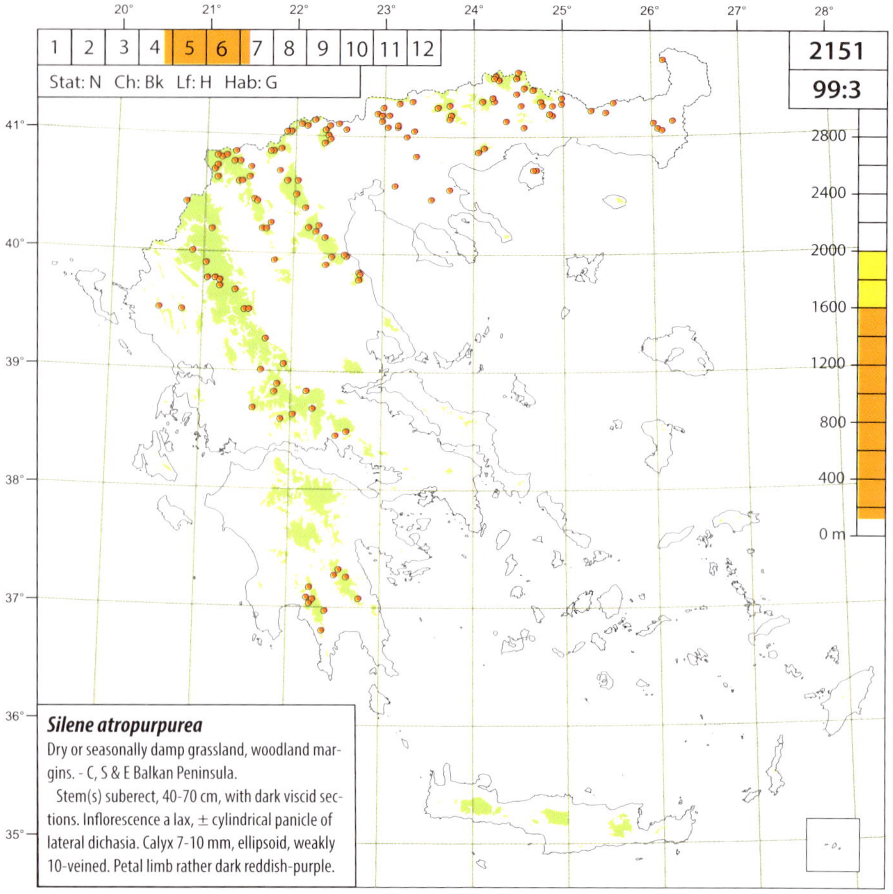

Silene atropurpurea
Dry or seasonally damp grassland, woodland margins. - C, S & E Balkan Peninsula.
Stem(s) suberect, 40-70 cm, with dark viscid sections. Inflorescence a lax, ± cylindrical panicle of lateral dichasia. Calyx 7-10 mm, ellipsoid, weakly 10-veined. Petal limb rather dark reddish-purple.

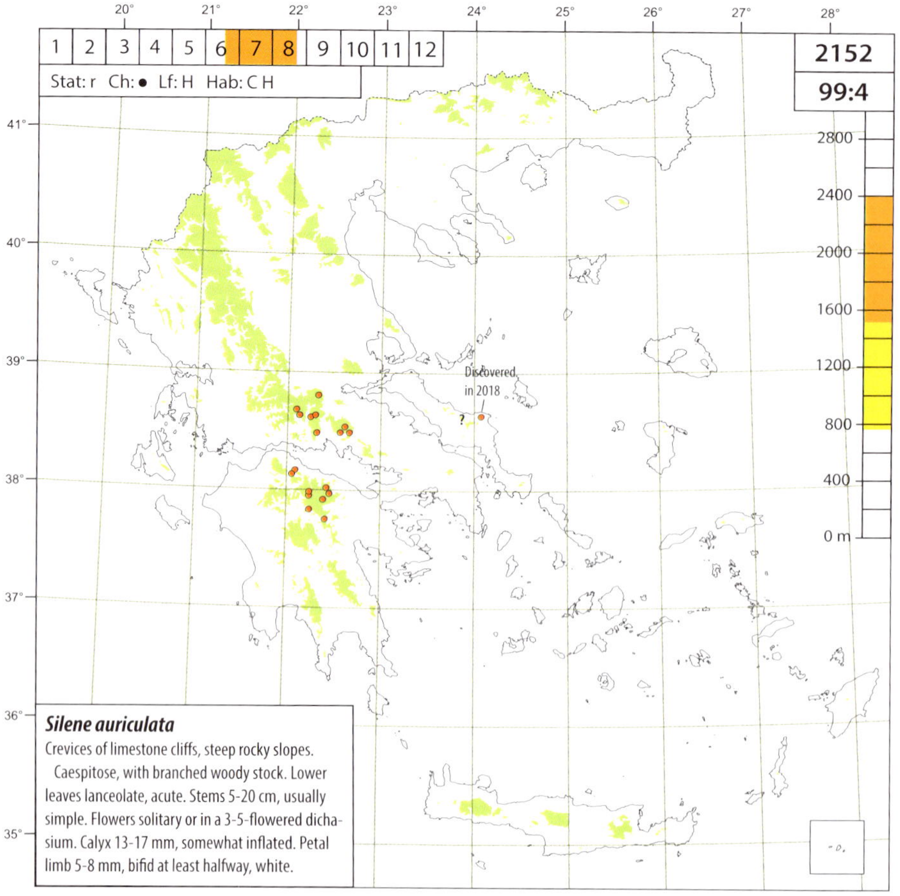

Silene auriculata
Crevices of limestone cliffs, steep rocky slopes.
Caespitose, with branched woody stock. Lower leaves lanceolate, acute. Stems 5-20 cm, usually simple. Flowers solitary or in a 3-5-flowered dichasium. Calyx 13-17 mm, somewhat inflated. Petal limb 5-8 mm, bifid at least halfway, white.

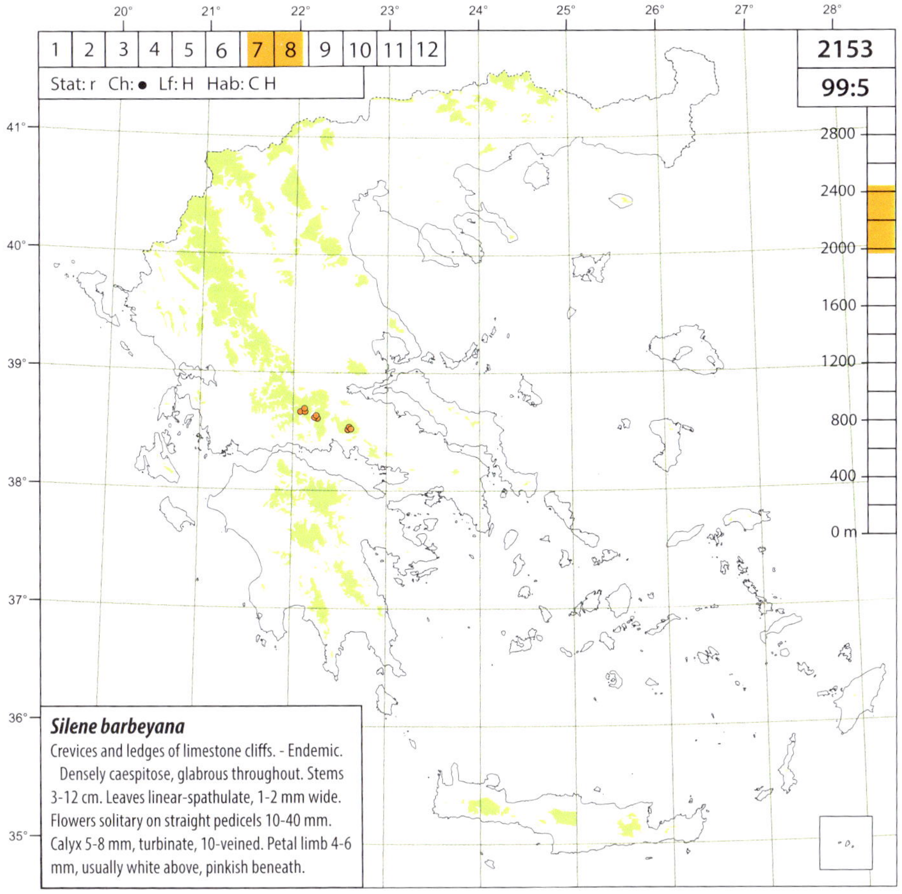

Silene barbeyana
Crevices and ledges of limestone cliffs. - Endemic.
Densely caespitose, glabrous throughout. Stems 3-12 cm. Leaves linear-spathulate, 1-2 mm wide. Flowers solitary on straight pedicels 10-40 mm. Calyx 5-8 mm, turbinate, 10-veined. Petal limb 4-6 mm, usually white above, pinkish beneath.

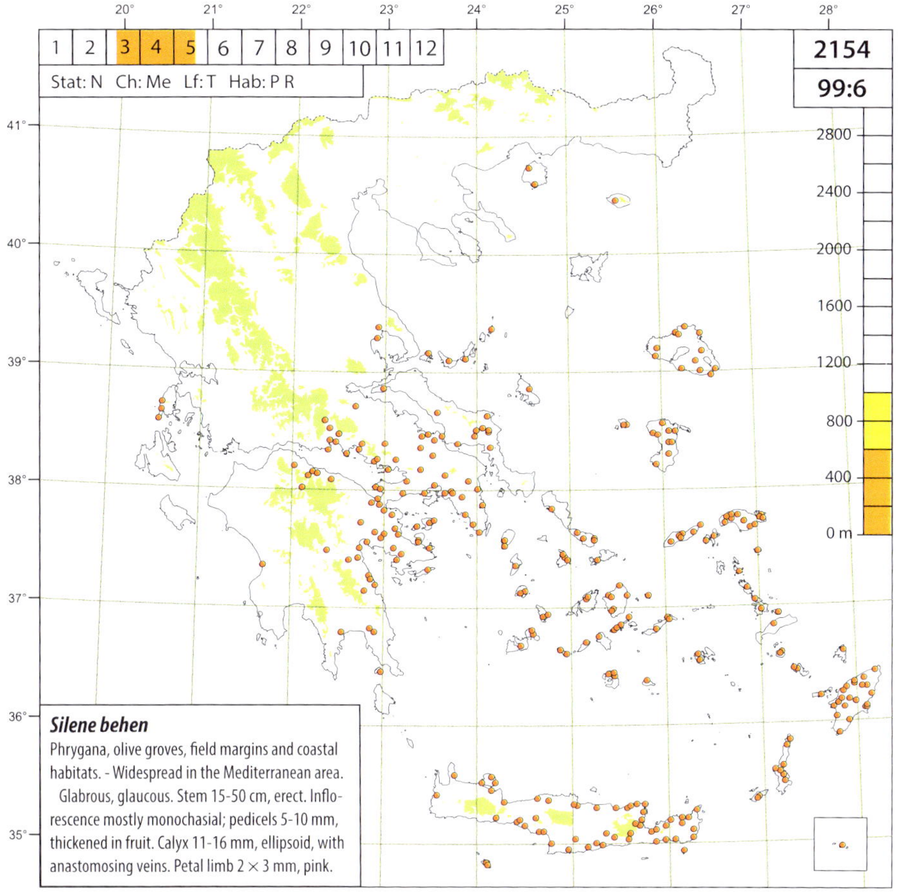

Silene behen
Phrygana, olive groves, field margins and coastal habitats. - Widespread in the Mediterranean area.
Glabrous, glaucous. Stem 15-50 cm, erect. Inflorescence mostly monochasial; pedicels 5-10 mm, thickened in fruit. Calyx 11-16 mm, ellipsoid, with anastomosing veins. Petal limb 2 × 3 mm, pink.

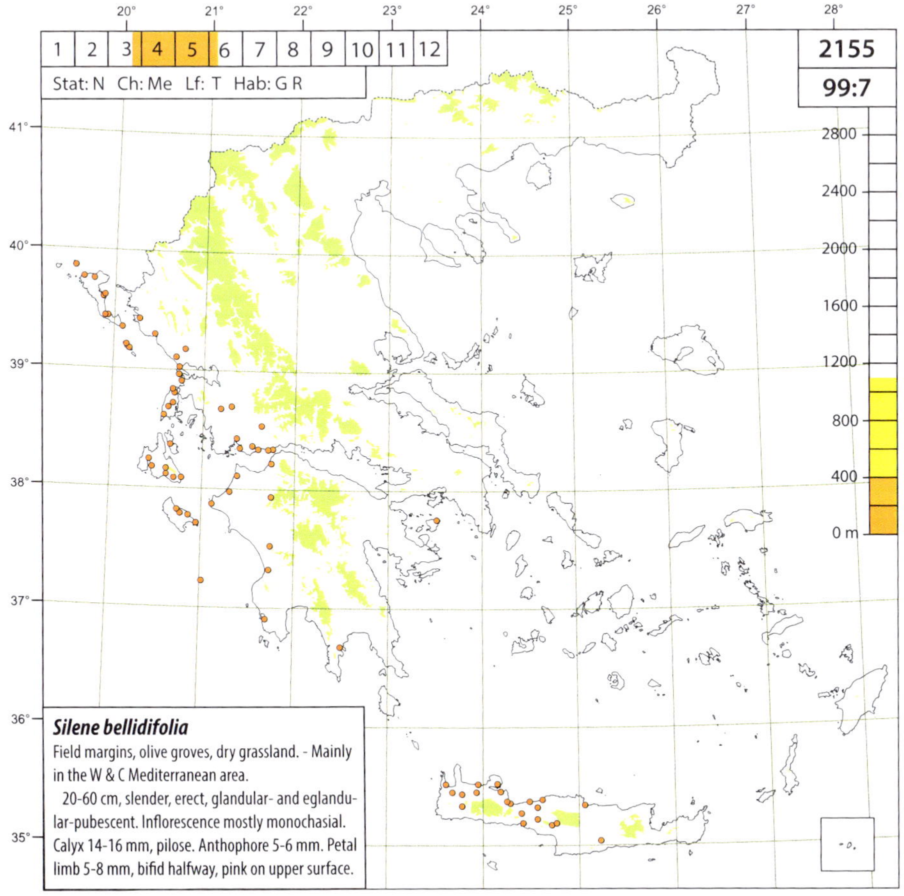

Silene bellidifolia
Field margins, olive groves, dry grassland. - Mainly in the W & C Mediterranean area.
20-60 cm, slender, erect, glandular- and eglandular-pubescent. Inflorescence mostly monochasial. Calyx 14-16 mm, pilose. Anthophore 5-6 mm. Petal limb 5-8 mm, bifid halfway, pink on upper surface.

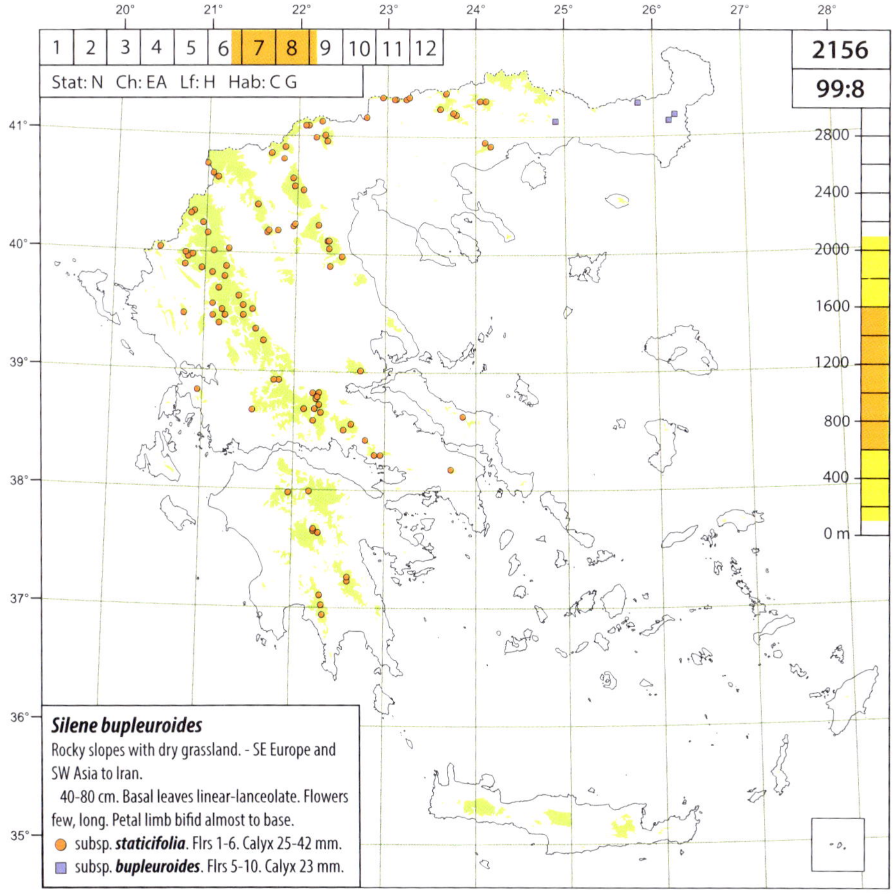

Silene bupleuroides
Rocky slopes with dry grassland. - SE Europe and SW Asia to Iran.
40-80 cm. Basal leaves linear-lanceolate. Flowers few, long. Petal limb bifid almost to base.
● subsp. ***staticifolia***. Flrs 1-6. Calyx 25-42 mm.
■ subsp. ***bupleuroides***. Flrs 5-10. Calyx 23 mm.

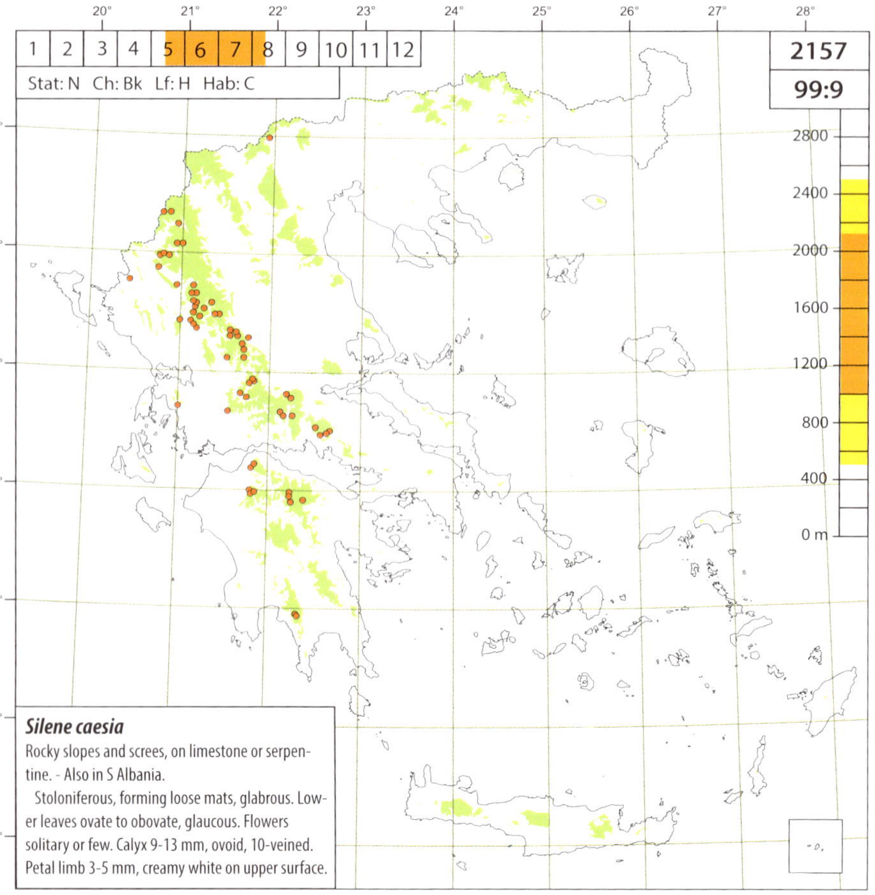

Silene caesia

Rocky slopes and screes, on limestone or serpentine. - Also in S Albania.

Stoloniferous, forming loose mats, glabrous. Lower leaves ovate to obovate, glaucous. Flowers solitary or few. Calyx 9-13 mm, ovoid, 10-veined. Petal limb 3-5 mm, creamy white on upper surface.

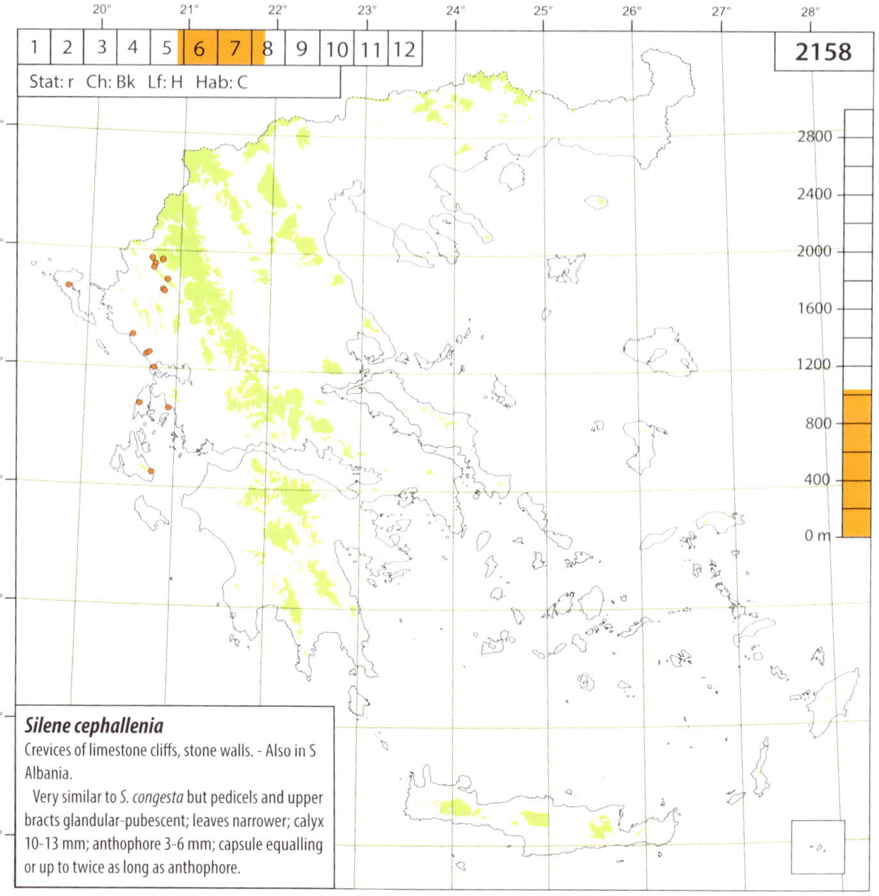

Silene cephallenia

Crevices of limestone cliffs, stone walls. - Also in S Albania.

Very similar to *S. congesta* but pedicels and upper bracts glandular-pubescent; leaves narrower; calyx 10-13 mm; anthophore 3-6 mm; capsule equalling or up to twice as long as anthophore.

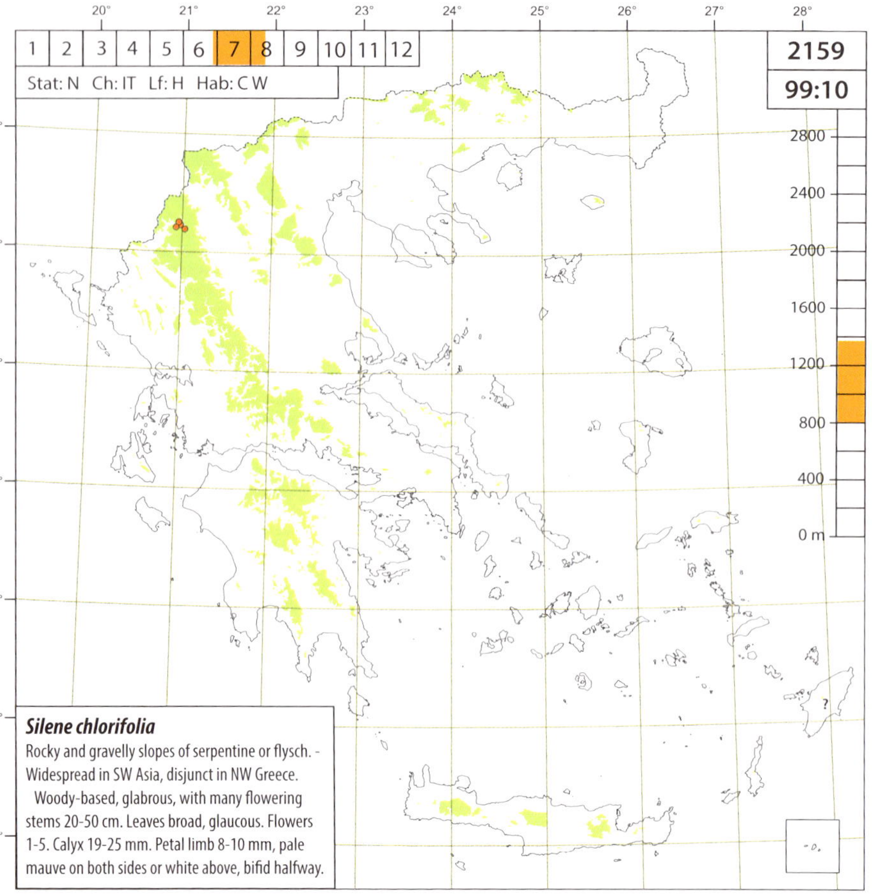

Silene chlorifolia

Rocky and gravelly slopes of serpentine or flysch. - Widespread in SW Asia, disjunct in NW Greece.

Woody-based, glabrous, with many flowering stems 20-50 cm. Leaves broad, glaucous. Flowers 1-5. Calyx 19-25 mm. Petal limb 8-10 mm, pale mauve on both sides or white above, bifid halfway.

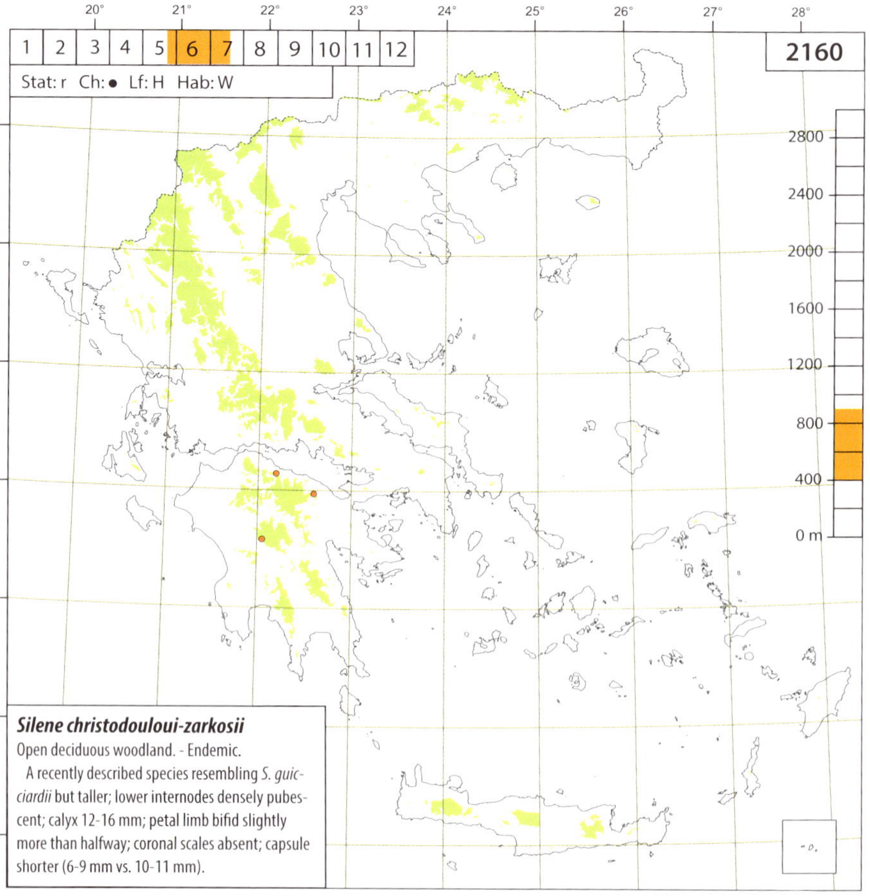

Silene christodouloui-zarkosii

Open deciduous woodland. - Endemic.

A recently described species resembling *S. guicciardii* but taller; lower internodes densely pubescent; calyx 12-16 mm; petal limb bifid slightly more than halfway; coronal scales absent; capsule shorter (6-9 mm vs. 10-11 mm).

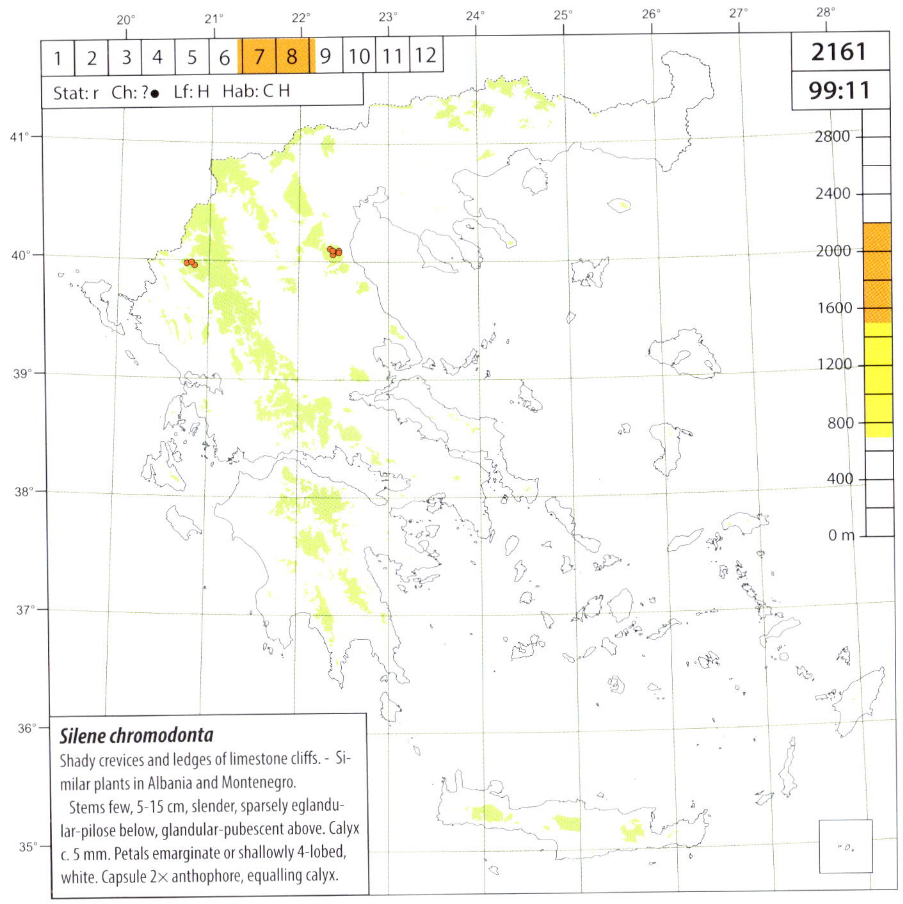

Silene chromodonta
Shady crevices and ledges of limestone cliffs. - Similar plants in Albania and Montenegro.
Stems few, 5-15 cm, slender, sparsely eglandular-pilose below, glandular-pubescent above. Calyx c. 5 mm. Petals emarginate or shallowly 4-lobed, white. Capsule 2× anthophore, equalling calyx.

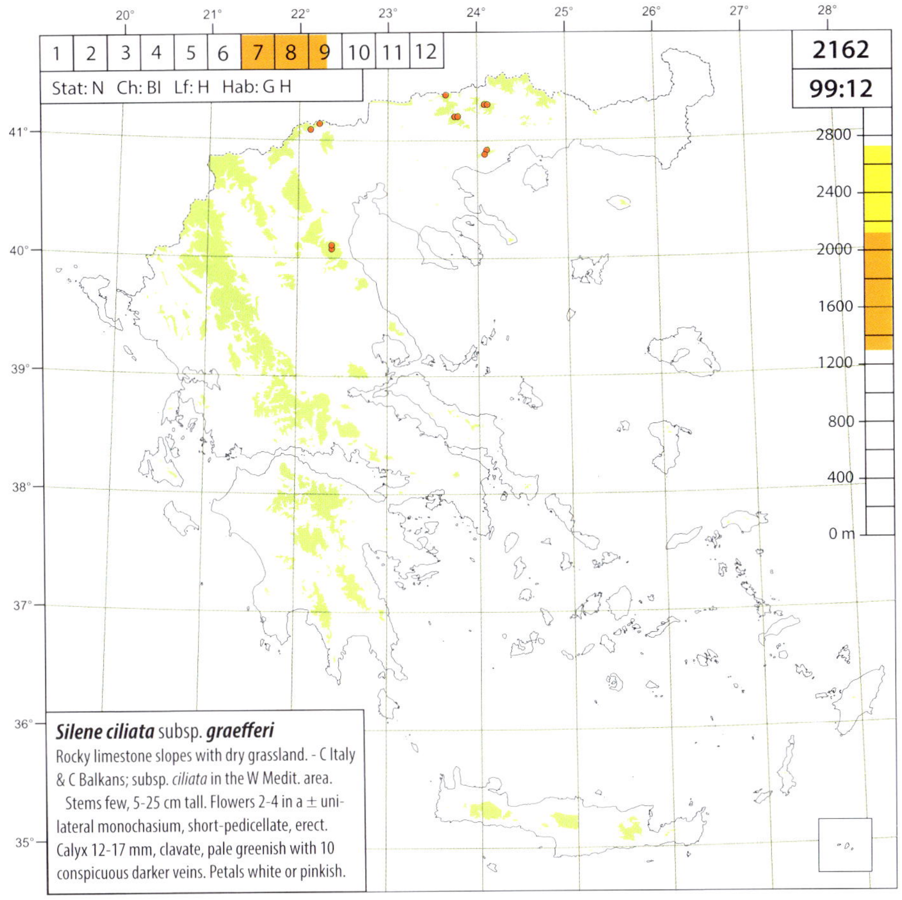

Silene ciliata* subsp. *graefferi
Rocky limestone slopes with dry grassland. - C Italy & C Balkans; subsp. *ciliata* in the W Medit. area.
Stems few, 5-25 cm tall. Flowers 2-4 in a ± unilateral monochasium, short-pedicellate, erect. Calyx 12-17 mm, clavate, pale greenish with 10 conspicuous darker veins. Petals white or pinkish.

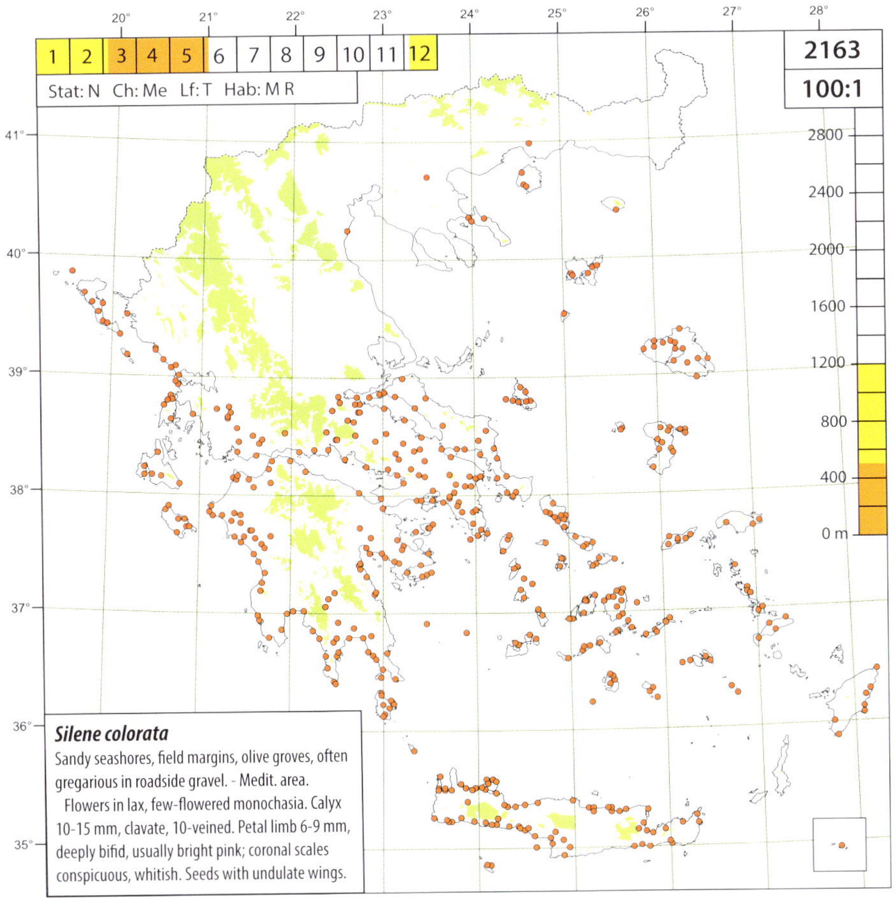

Silene colorata
Sandy seashores, field margins, olive groves, often gregarious in roadside gravel. - Medit. area.
Flowers in lax, few-flowered monochasia. Calyx 10-15 mm, clavate, 10-veined. Petal limb 6-9 mm, deeply bifid, usually bright pink; coronal scales conspicuous, whitish. Seeds with undulate wings.

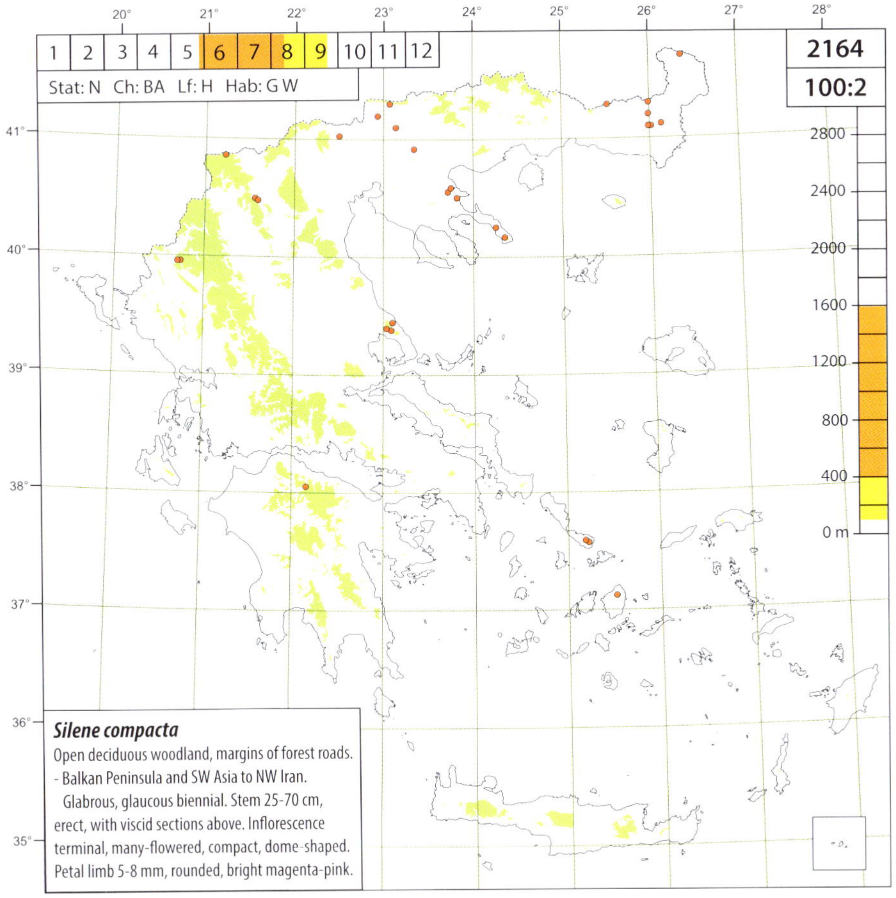

Silene compacta
Open deciduous woodland, margins of forest roads. - Balkan Peninsula and SW Asia to NW Iran.
Glabrous, glaucous biennial. Stem 25-70 cm, erect, with viscid sections above. Inflorescence terminal, many-flowered, compact, dome-shaped. Petal limb 5-8 mm, rounded, bright magenta-pink.

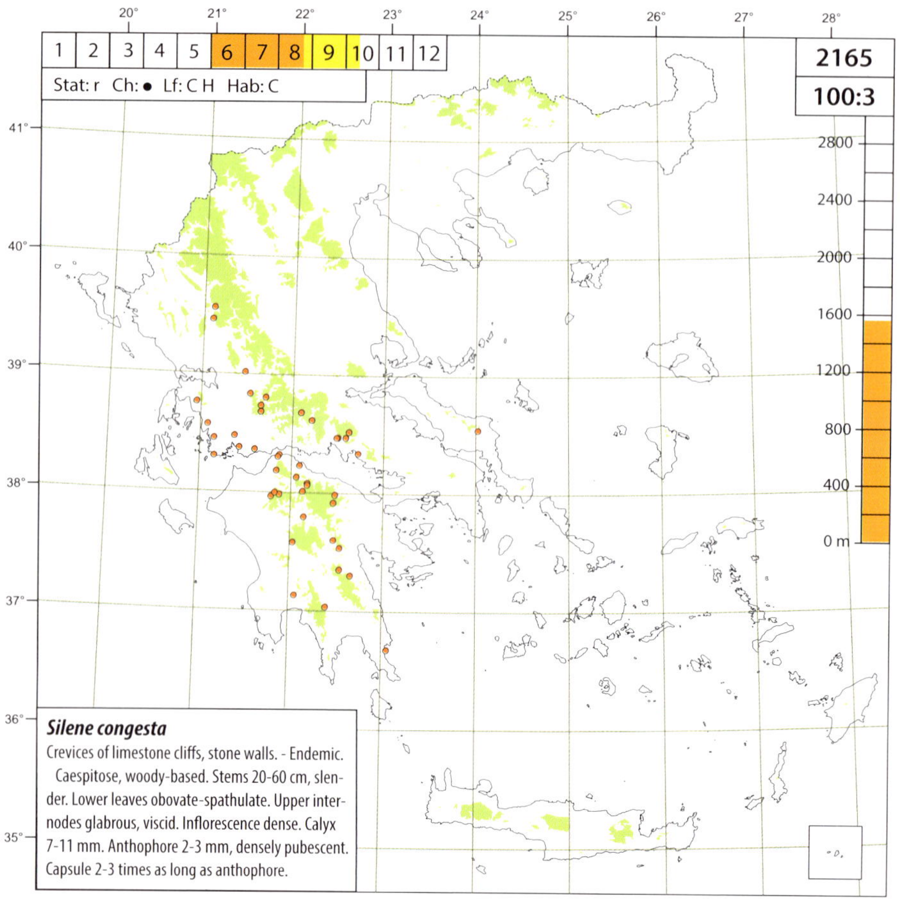

Silene congesta
Crevices of limestone cliffs, stone walls. - Endemic.
Caespitose, woody-based. Stems 20-60 cm, slender. Lower leaves obovate-spathulate. Upper internodes glabrous, viscid. Inflorescence dense. Calyx 7-11 mm. Anthophore 2-3 mm, densely pubescent. Capsule 2-3 times as long as anthophore.

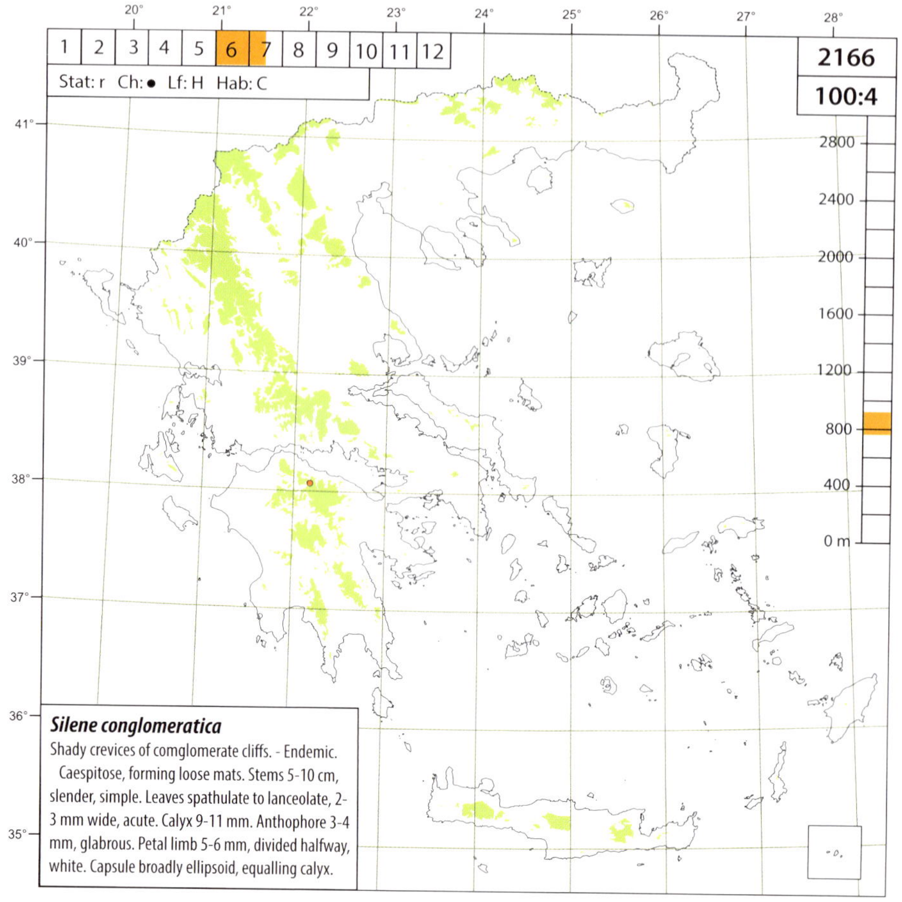

Silene conglomeratica
Shady crevices of comglomerate cliffs. - Endemic.
Caespitose, forming loose mats. Stems 5-10 cm, slender, simple. Leaves spathulate to lanceolate, 2-3 mm wide, acute. Calyx 9-11 mm. Anthophore 3-4 mm, glabrous. Petal limb 5-6 mm, divided halfway, white. Capsule broadly ellipsoid, equalling calyx.

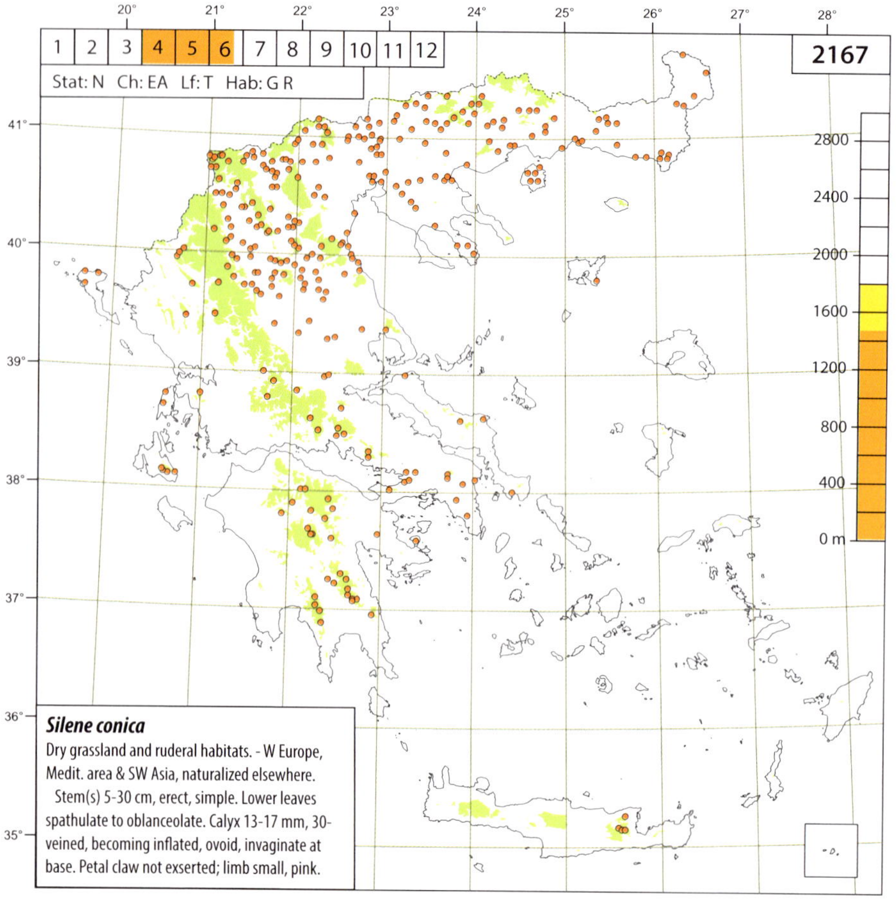

Silene conica
Dry grassland and ruderal habitats. - W Europe, Medit. area & SW Asia, naturalized elsewhere.
Stem(s) 5-30 cm, erect, simple. Lower leaves spathulate to oblanceolate. Calyx 13-17 mm, 30-veined, becoming inflated, ovoid, invaginate at base. Petal claw not exserted; limb small, pink.

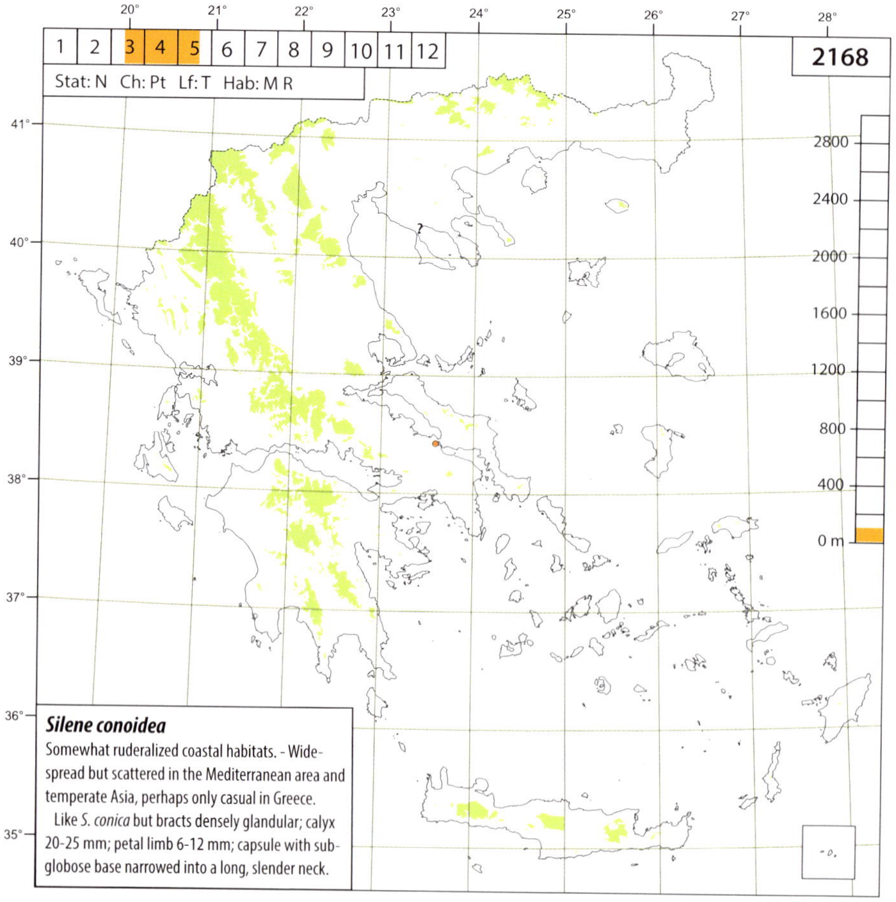

Silene conoidea
Somewhat ruderalized coastal habitats. - Widespread but scattered in the Mediterranean area and temperate Asia, perhaps only casual in Greece.
Like *S. conica* but bracts densely glandular; calyx 20-25 mm; petal limb 6-12 mm; capsule with subglobose base narrowed into a long, slender neck.

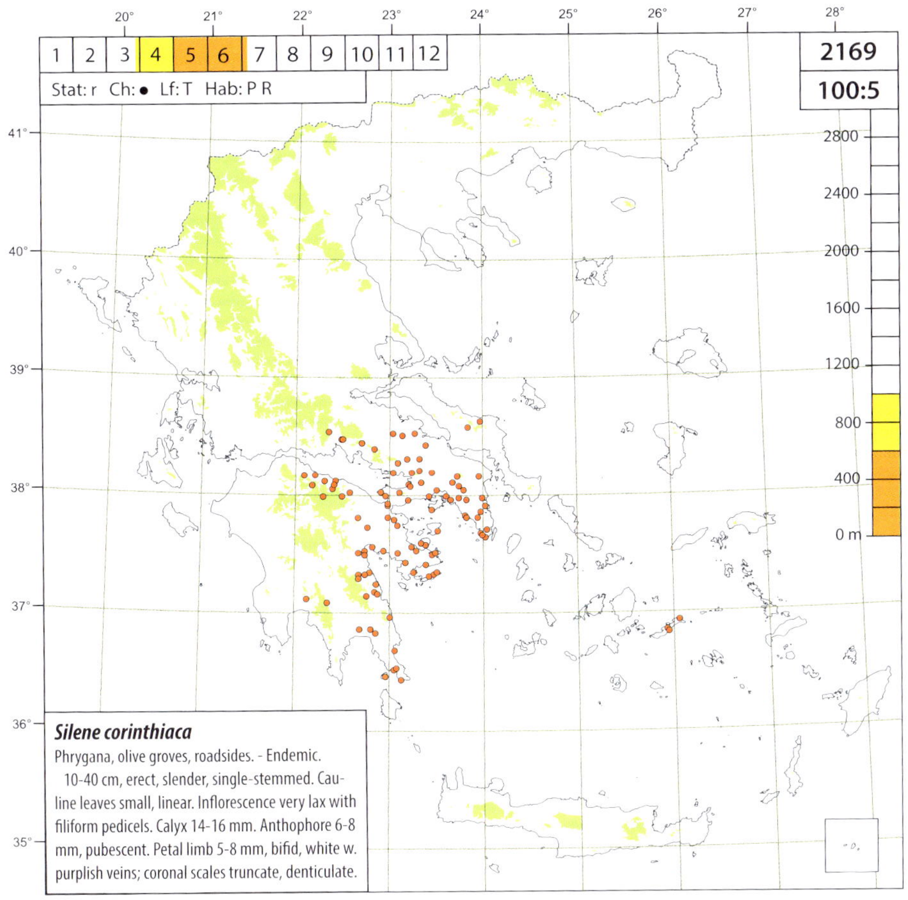

Silene corinthiaca

Phrygana, olive groves, roadsides. - Endemic.

10-40 cm, erect, slender, single-stemmed. Cauline leaves small, linear. Inflorescence very lax with filiform pedicels. Calyx 14-16 mm. Anthophore 6-8 mm, pubescent. Petal limb 5-8 mm, bifid, white w. purplish veins; coronal scales truncate, denticulate.

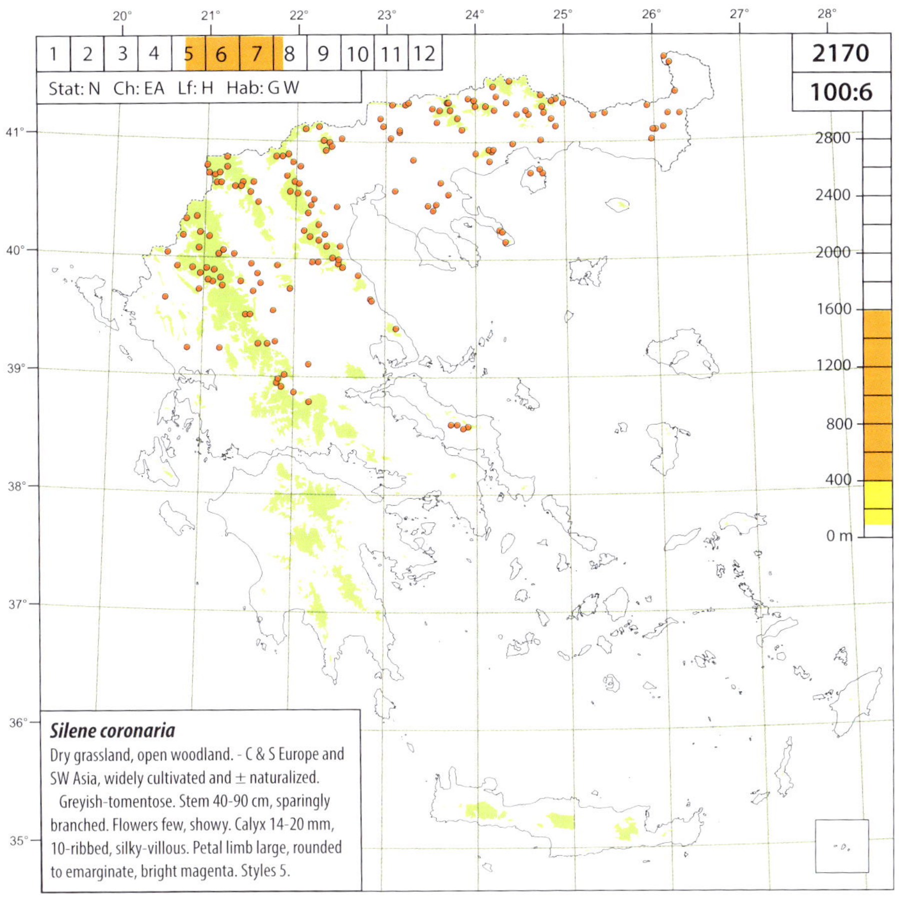

Silene coronaria

Dry grassland, open woodland. - C & S Europe and SW Asia, widely cultivated and ± naturalized.

Greyish-tomentose. Stem 40-90 cm, sparingly branched. Flowers few, showy. Calyx 14-20 mm, 10-ribbed, silky-villous. Petal limb large, rounded to emarginate, bright magenta. Styles 5.

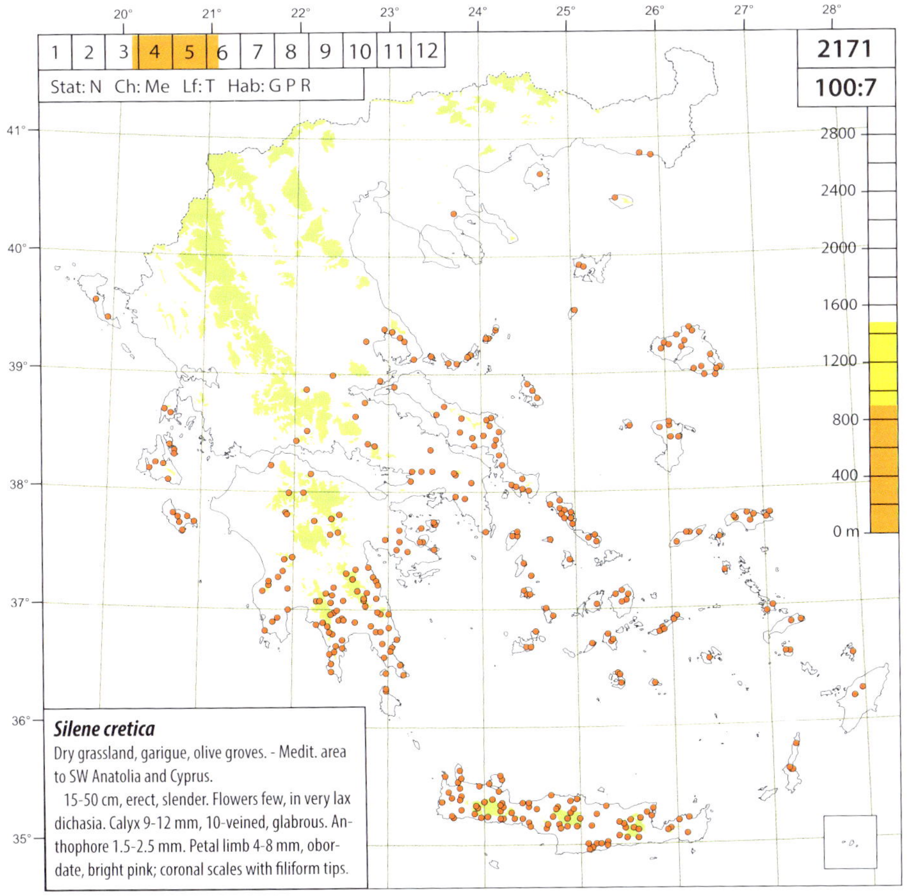

Silene cretica

Dry grassland, garigue, olive groves. - Medit. area to SW Anatolia and Cyprus.

15-50 cm, erect, slender. Flowers few, in very lax dichasia. Calyx 9-12 mm, 10-veined, glabrous. Anthophore 1.5-2.5 mm. Petal limb 4-8 mm, obordate, bright pink; coronal scales with filiform tips.

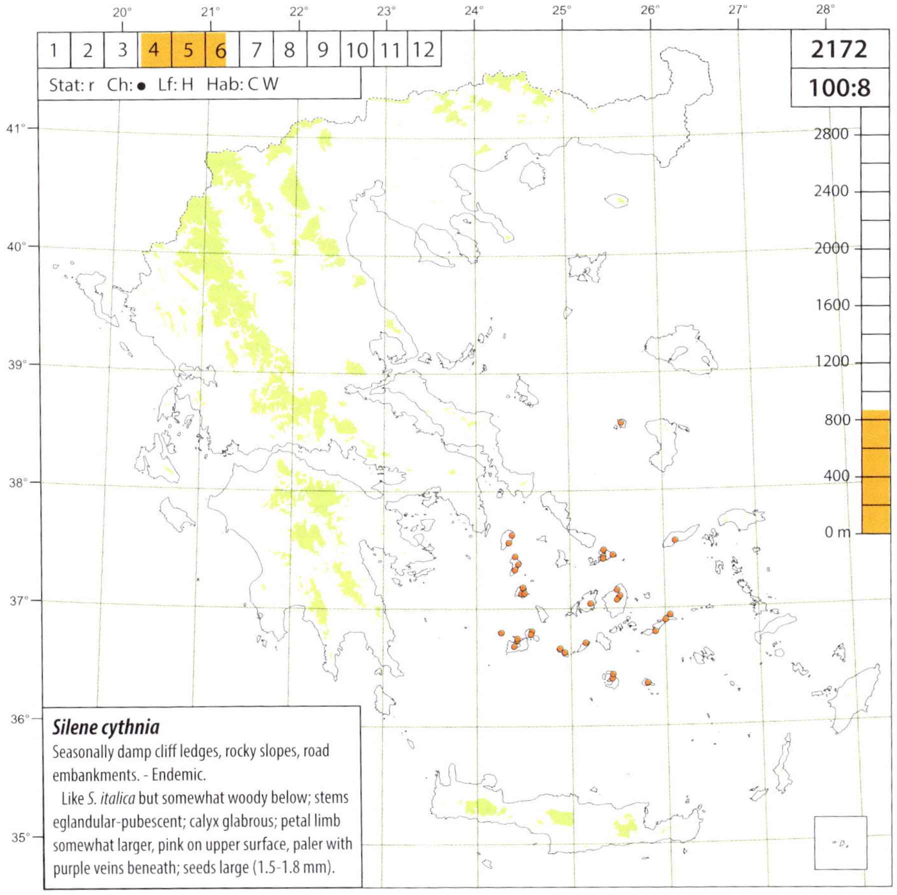

Silene cythnia

Seasonally damp cliff ledges, rocky slopes, road embankments. - Endemic.

Like *S. italica* but somewhat woody below; stems eglandular-pubescent; calyx glabrous; petal limb somewhat larger, pink on upper surface, paler with purple veins beneath; seeds large (1.5-1.8 mm).

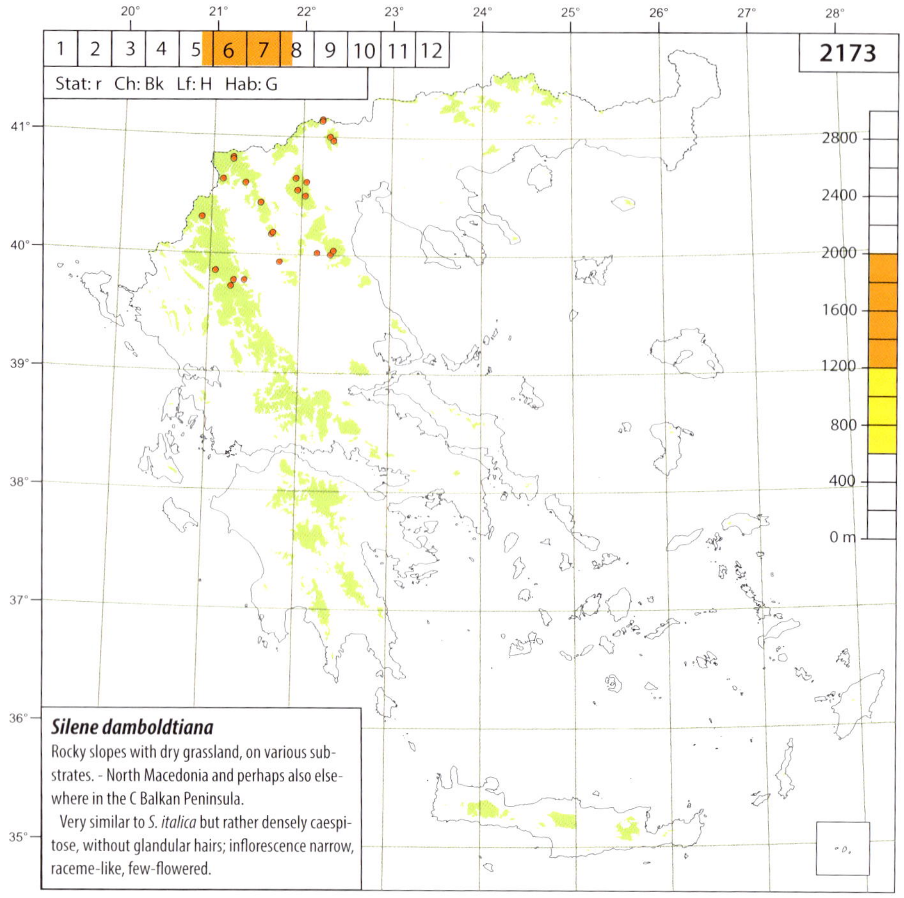

Silene damboldtiana

Rocky slopes with dry grassland, on various substrates. - North Macedonia and perhaps also elsewhere in the C Balkan Peninsula.

Very similar to *S. italica* but rather densely caespitose, without glandular hairs; inflorescence narrow, raceme-like, few-flowered.

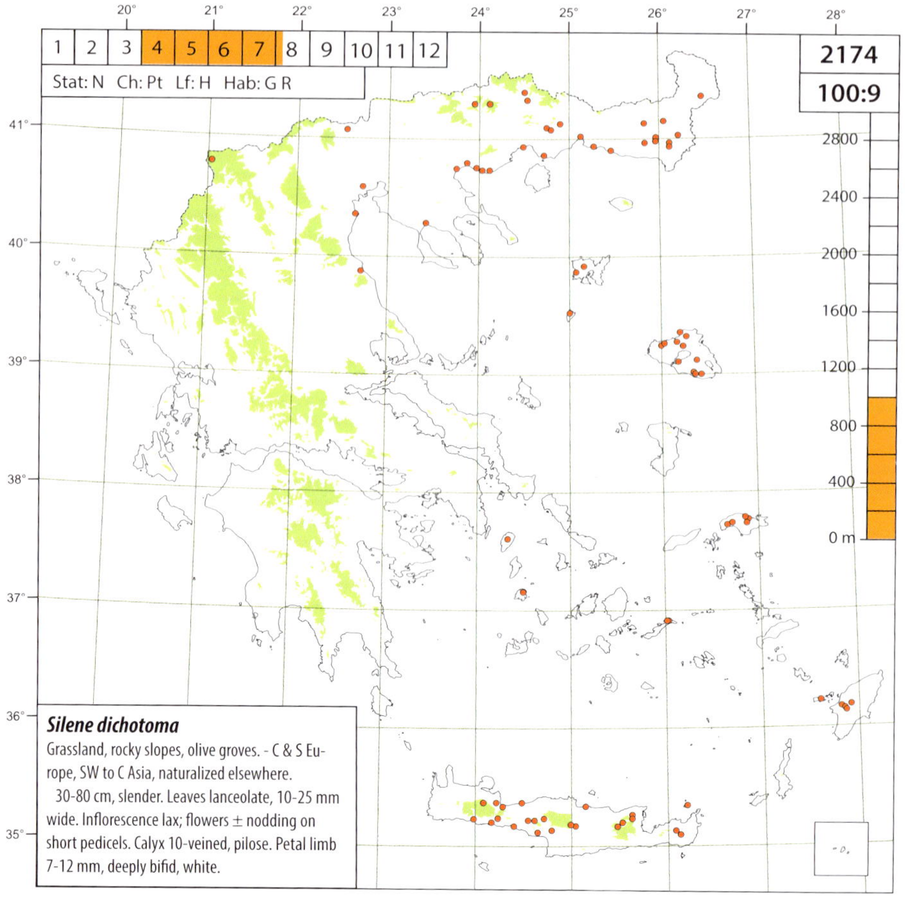

Silene dichotoma

Grassland, rocky slopes, olive groves. - C & S Europe, SW to C Asia, naturalized elsewhere.

30-80 cm, slender. Leaves lanceolate, 10-25 mm wide. Inflorescence lax; flowers ± nodding on short pedicels. Calyx 10-veined, pilose. Petal limb 7-12 mm, deeply bifid, white.

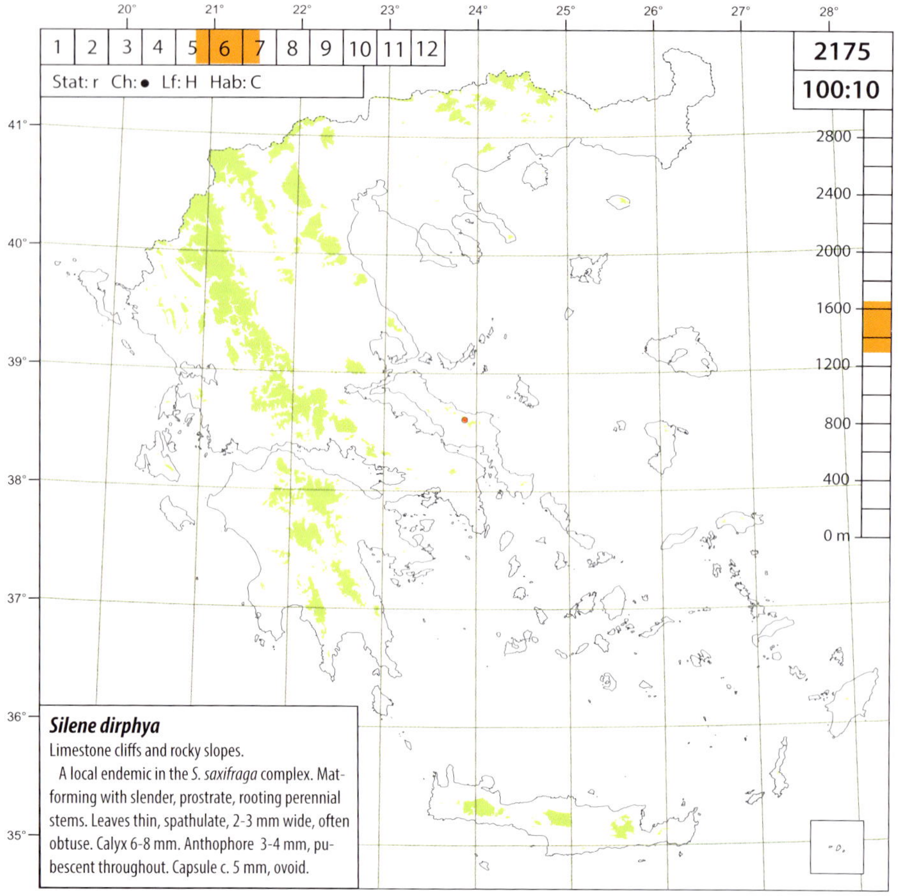

Silene dirphya

Limestone cliffs and rocky slopes.

A local endemic in the *S. saxifraga* complex. Mat-forming with slender, prostrate, rooting perennial stems. Leaves thin, spathulate, 2-3 mm wide, often obtuse. Calyx 6-8 mm. Anthophore 3-4 mm, pubescent throughout. Capsule c. 5 mm, ovoid.

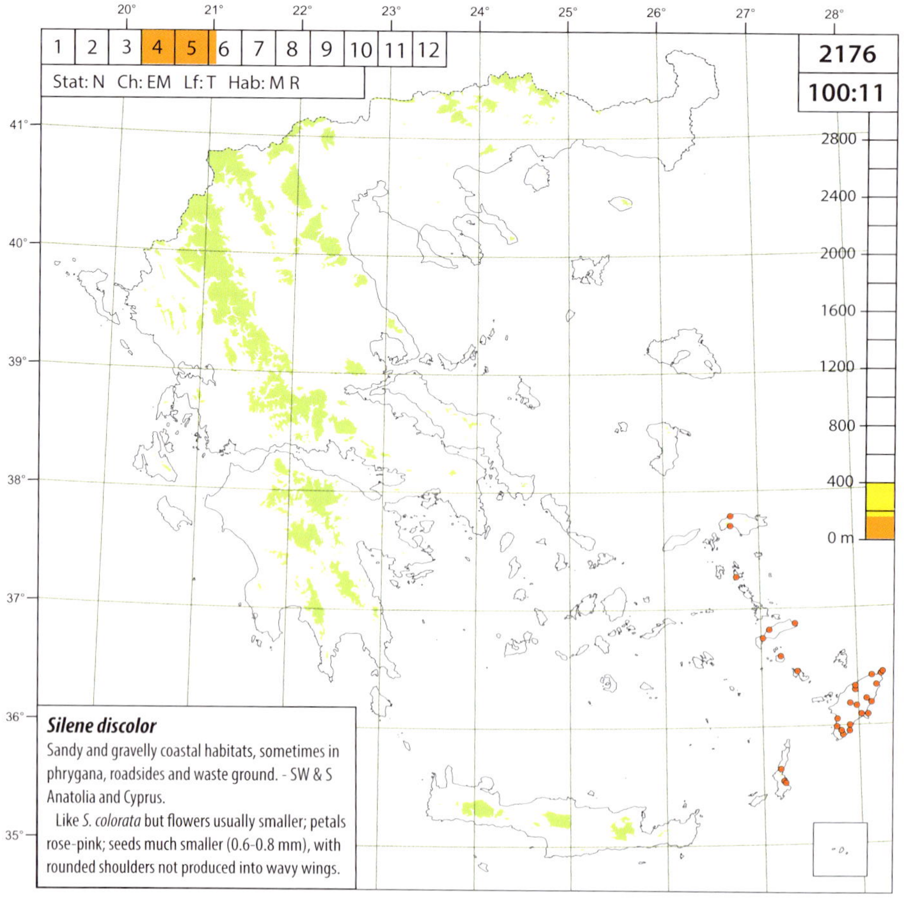

Silene discolor

Sandy and gravelly coastal habitats, sometimes in phrygana, roadsides and waste ground. - SW & S Anatolia and Cyprus.

Like *S. colorata* but flowers usually smaller; petals rose-pink; seeds much smaller (0.6-0.8 mm), with rounded shoulders not produced into wavy wings.

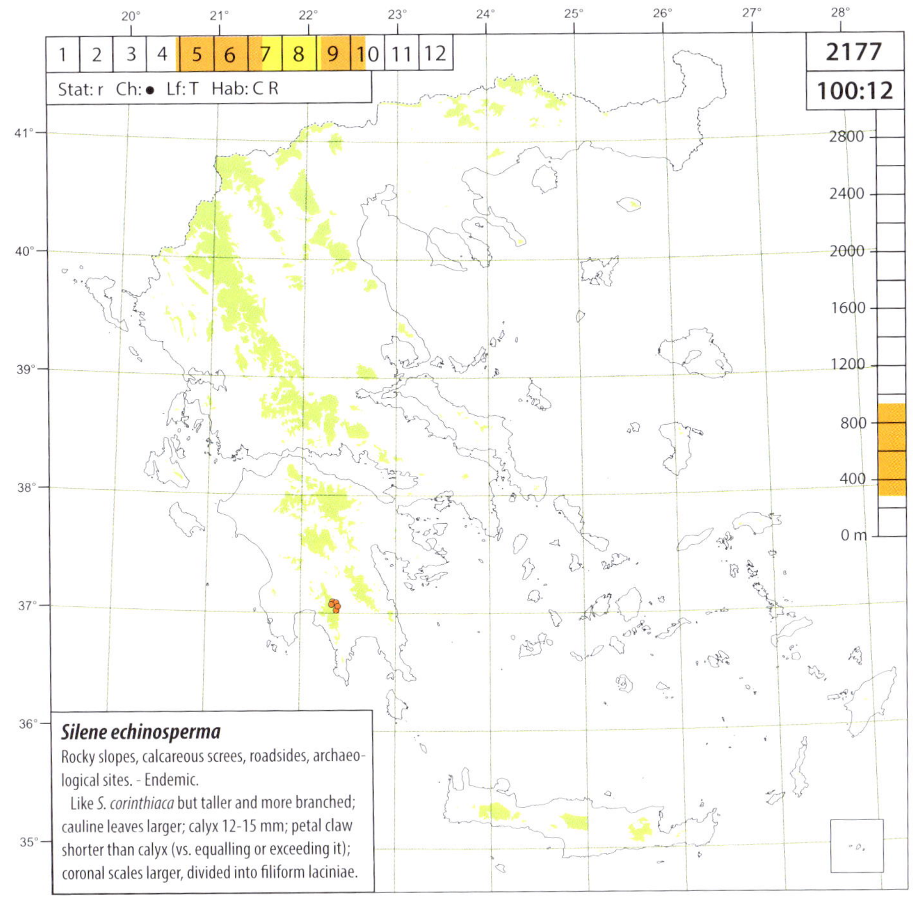

Silene echinosperma

Rocky slopes, calcareous screes, roadsides, archaeological sites. - Endemic.

Like *S. corinthiaca* but taller and more branched; cauline leaves larger; calyx 12-15 mm; petal claw shorter than calyx (vs. equalling or exceeding it); coronal scales larger, divided into filiform laciniae.

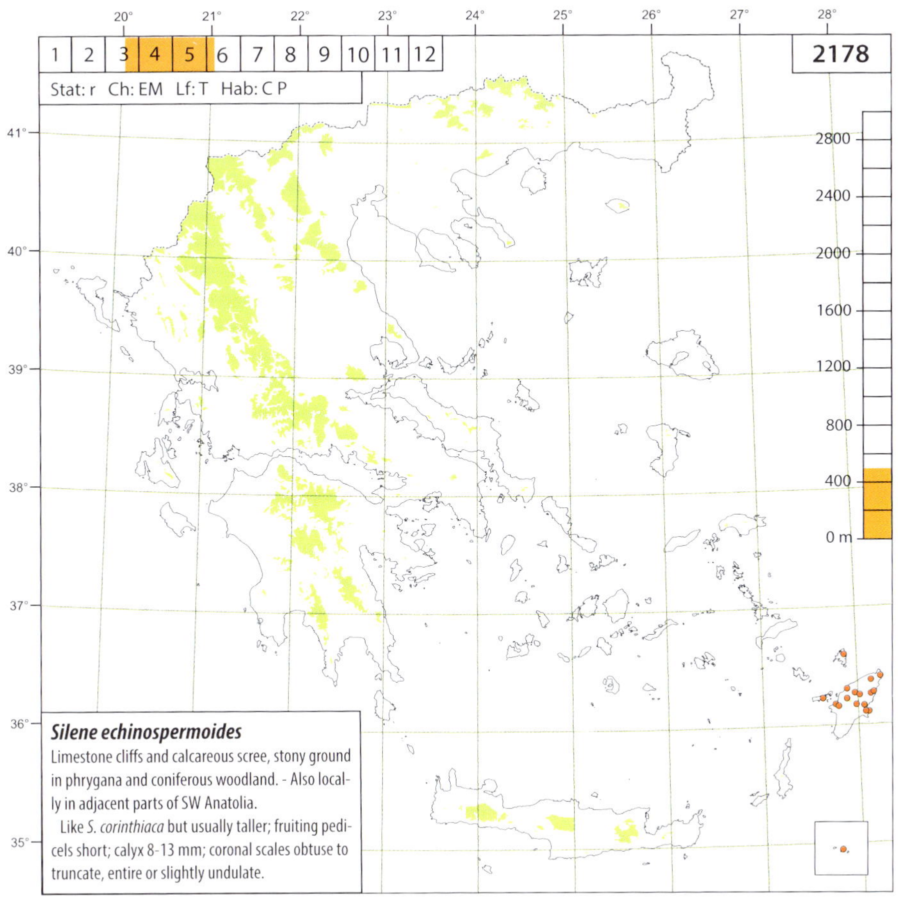

Silene echinospermoides

Limestone cliffs and calcareous scree, stony ground in phrygana and coniferous woodland. - Also locally in adjacent parts of SW Anatolia.

Like *S. corinthiaca* but usually taller; fruiting pedicels short; calyx 8-13 mm; coronal scales obtuse to truncate, entire or slightly undulate.

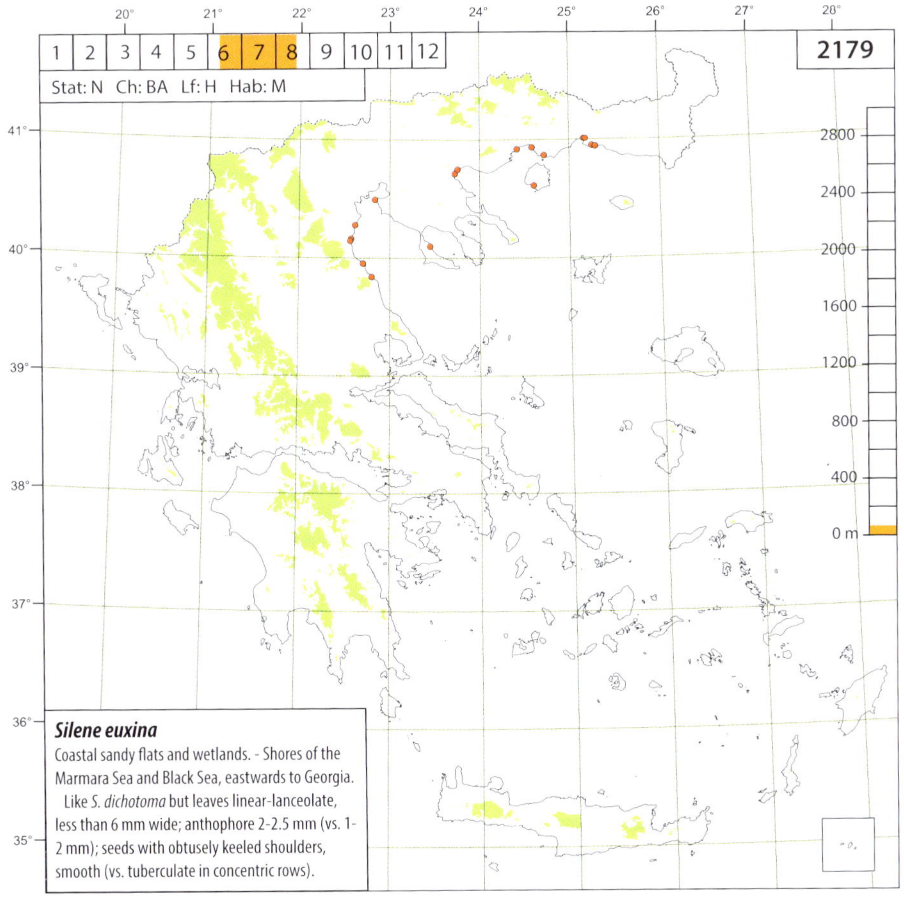

Silene euxina

Coastal sandy flats and wetlands. - Shores of the Marmara Sea and Black Sea, eastwards to Georgia.

Like *S. dichotoma* but leaves linear-lanceolate, less than 6 mm wide; anthophore 2-2.5 mm (vs. 1-2 mm); seeds with obtusely keeled shoulders, smooth (vs. tuberculate in concentric rows).

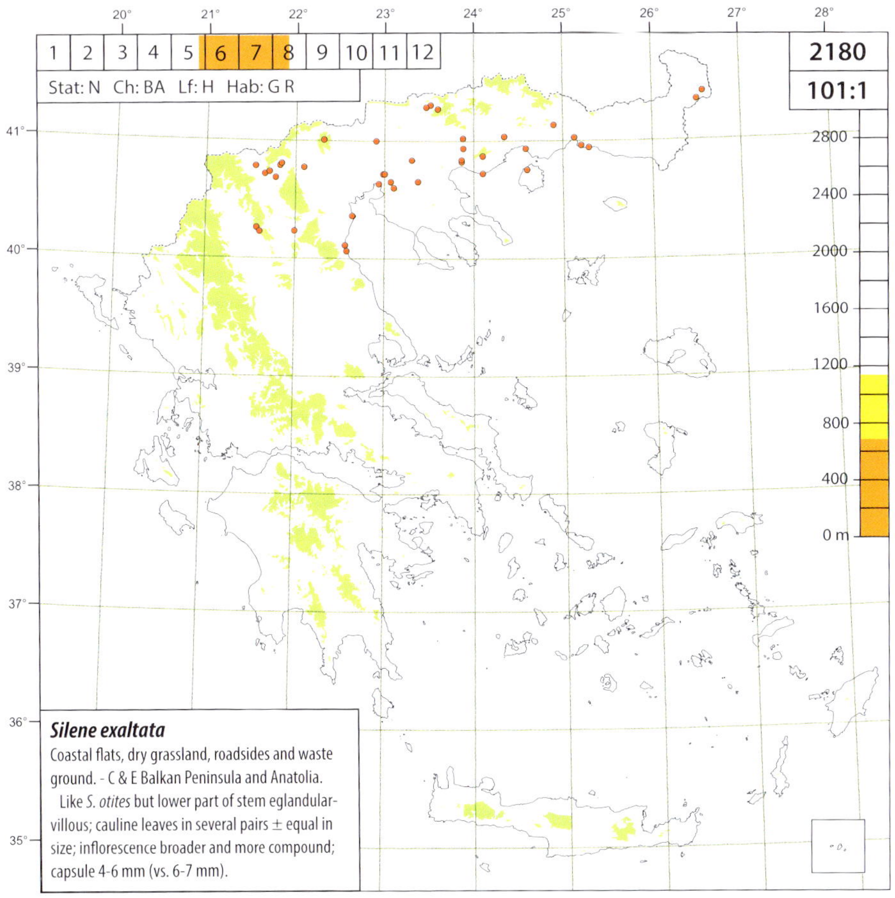

Silene exaltata

Coastal flats, dry grassland, roadsides and waste ground. - C & E Balkan Peninsula and Anatolia.

Like *S. otites* but lower part of stem eglandular-villous; cauline leaves in several pairs ± equal in size; inflorescence broader and more compound; capsule 4-6 mm (vs. 6-7 mm).

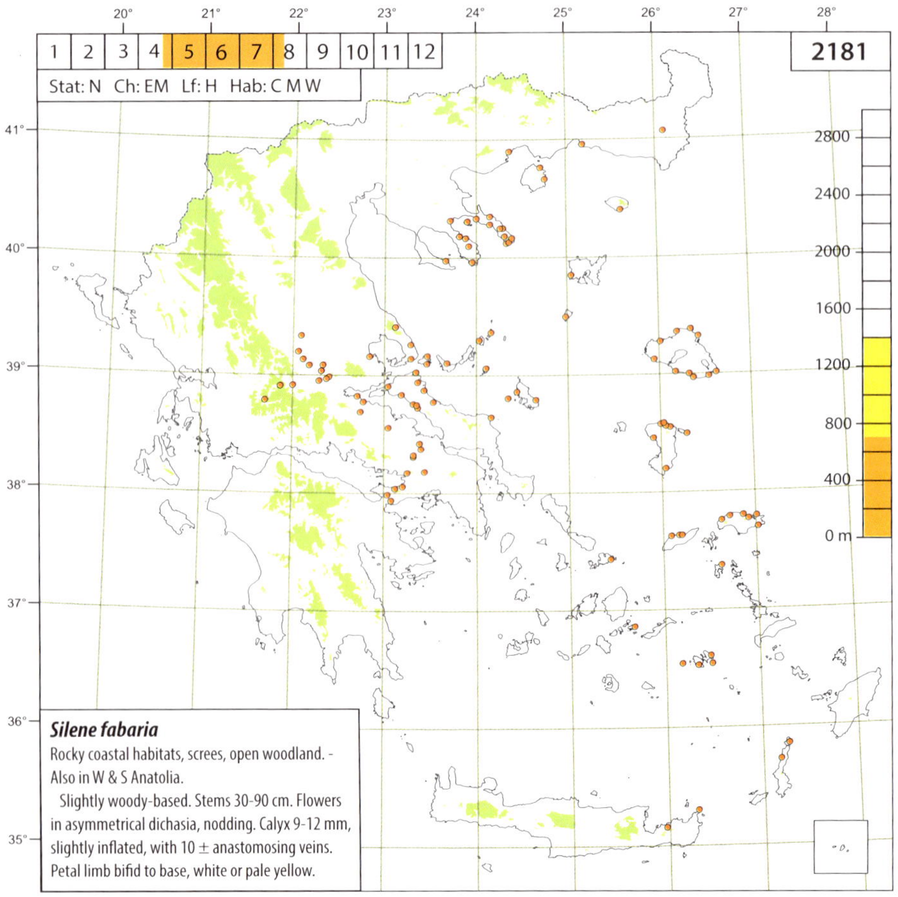

Silene fabaria

Rocky coastal habitats, screes, open woodland. - Also in W & S Anatolia.

Slightly woody-based. Stems 30-90 cm. Flowers in asymmetrical dichasia, nodding. Calyx 9-12 mm, slightly inflated, with 10 ± anastomosing veins. Petal limb bifid to base, white or pale yellow.

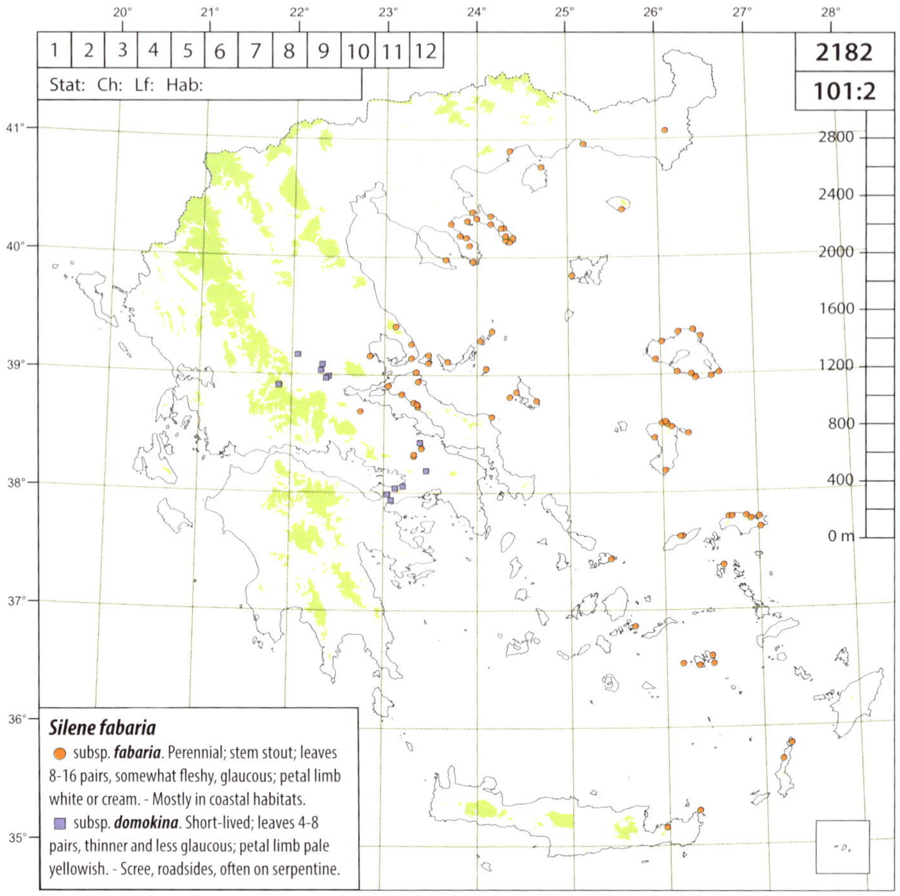

Silene fabaria

● subsp. ***fabaria***. Perennial; stem stout; leaves 8-16 pairs, somewhat fleshy, glaucous; petal limb white or cream. - Mostly in coastal habitats.

■ subsp. ***domokina***. Short-lived; leaves 4-8 pairs, thinner and less glaucous; petal limb pale yellowish. - Scree, roadsides, often on serpentine.

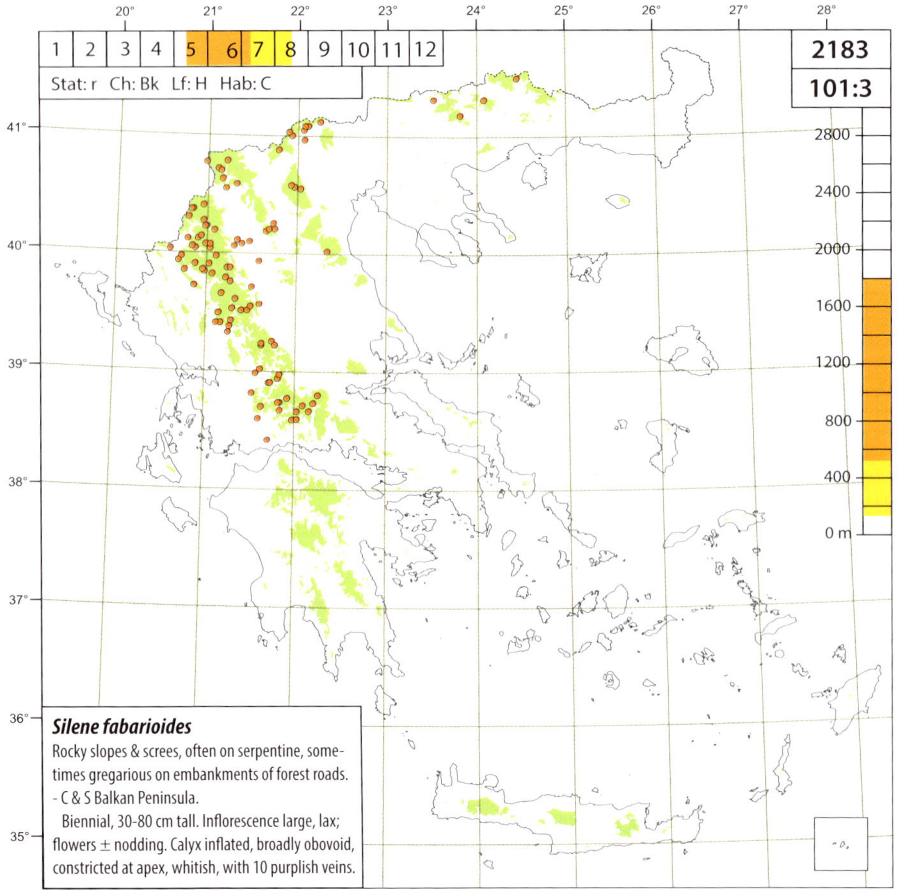

Silene fabarioides

Rocky slopes & screes, often on serpentine, sometimes gregarious on embankments of forest roads. - C & S Balkan Peninsula.

Biennial, 30-80 cm tall. Inflorescence large, lax; flowers ± nodding. Calyx inflated, broadly obovoid, constricted at apex, whitish, with 10 purplish veins.

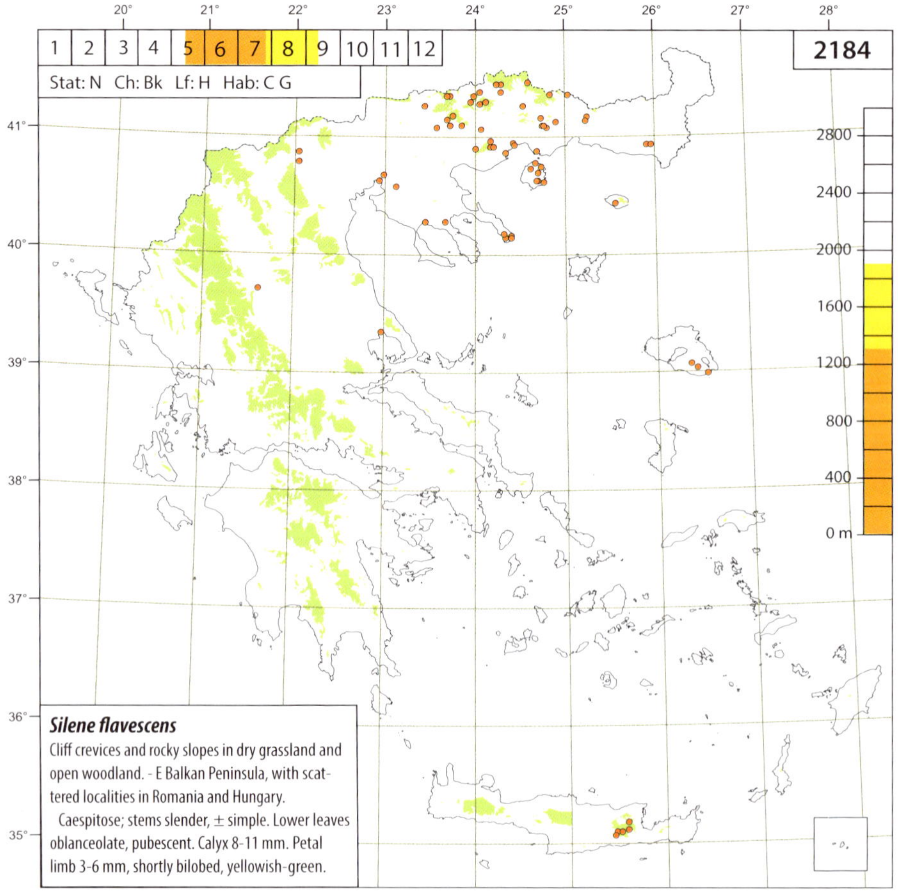

Silene flavescens

Cliff crevices and rocky slopes in dry grassland and open woodland. - E Balkan Peninsula, with scattered localities in Romania and Hungary.

Caespitose; stems slender, ± simple. Lower leaves oblanceolate, pubescent. Calyx 8-11 mm. Petal limb 3-6 mm, shortly bilobed, yellowish-green.

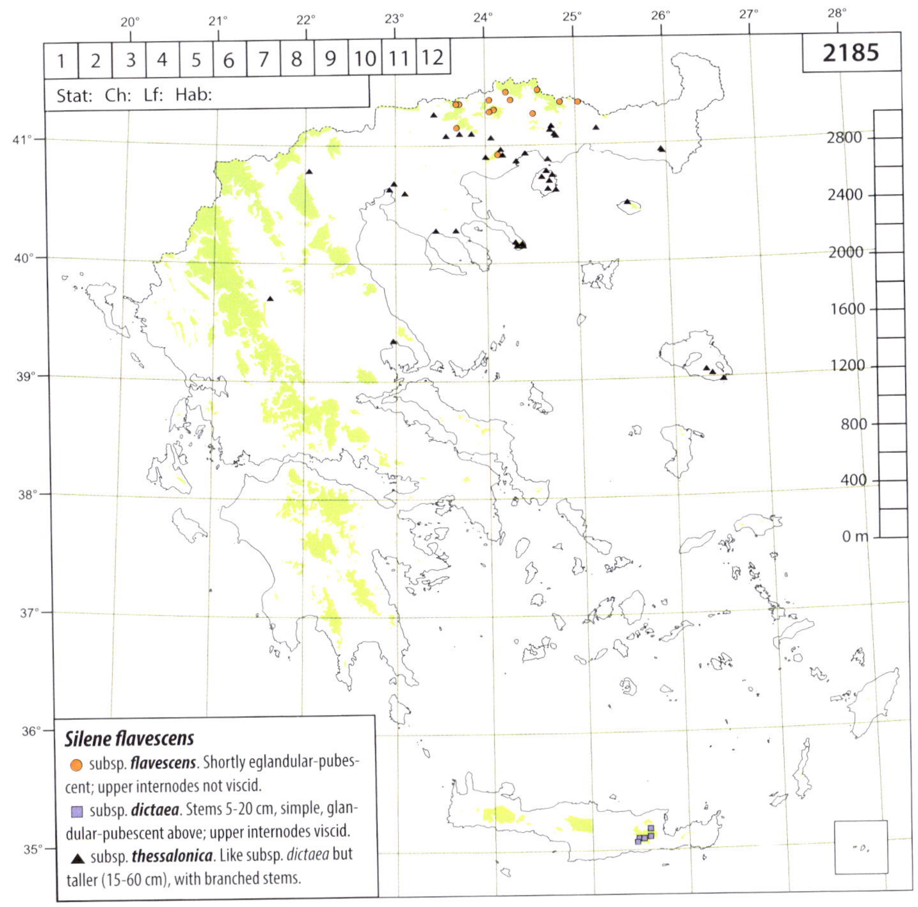

Silene flavescens

● subsp. ***flavescens***. Shortly eglandular-pubescent; upper internodes not viscid.

■ subsp. ***dictaea***. Stems 5-20 cm, simple, glandular-pubescent above; upper internodes viscid.

▲ subsp. ***thessalonica***. Like subsp. *dictaea* but taller (15-60 cm), with branched stems.

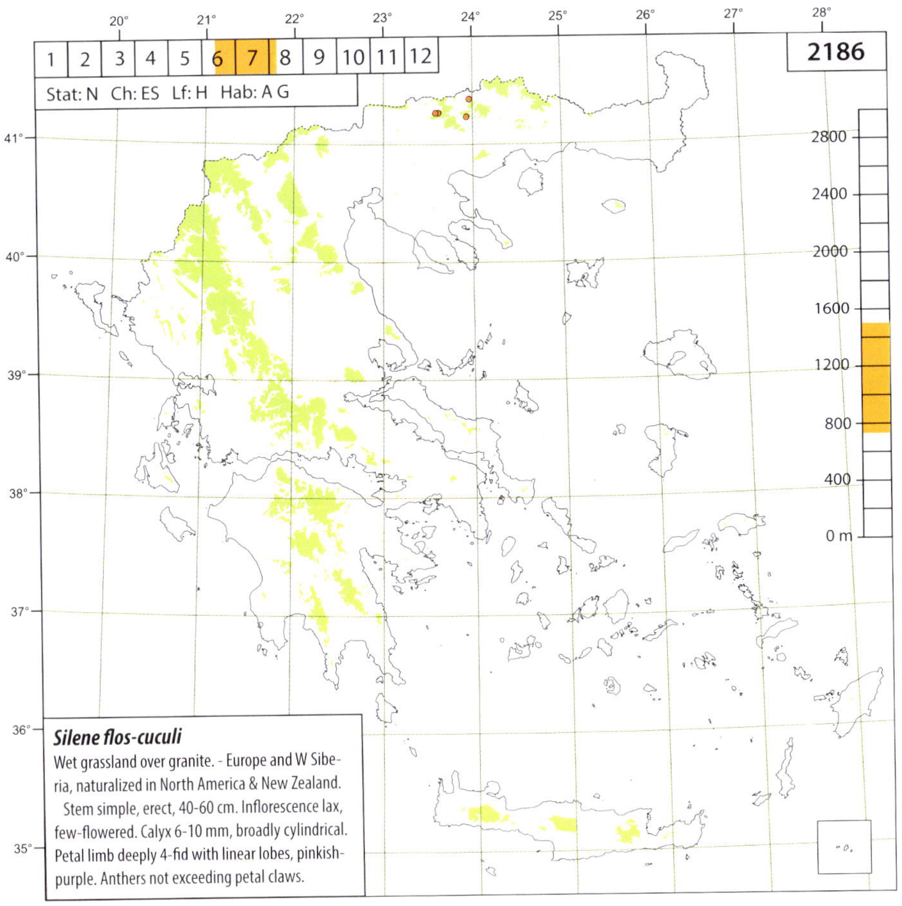

Silene flos-cuculi

Wet grassland over granite. - Europe and W Siberia, naturalized in North America & New Zealand.

Stem simple, erect, 40-60 cm. Inflorescence lax, few-flowered. Calyx 6-10 mm, broadly cylindrical. Petal limb deeply 4-fid with linear lobes, pinkish-purple. Anthers not exceeding petal claws.

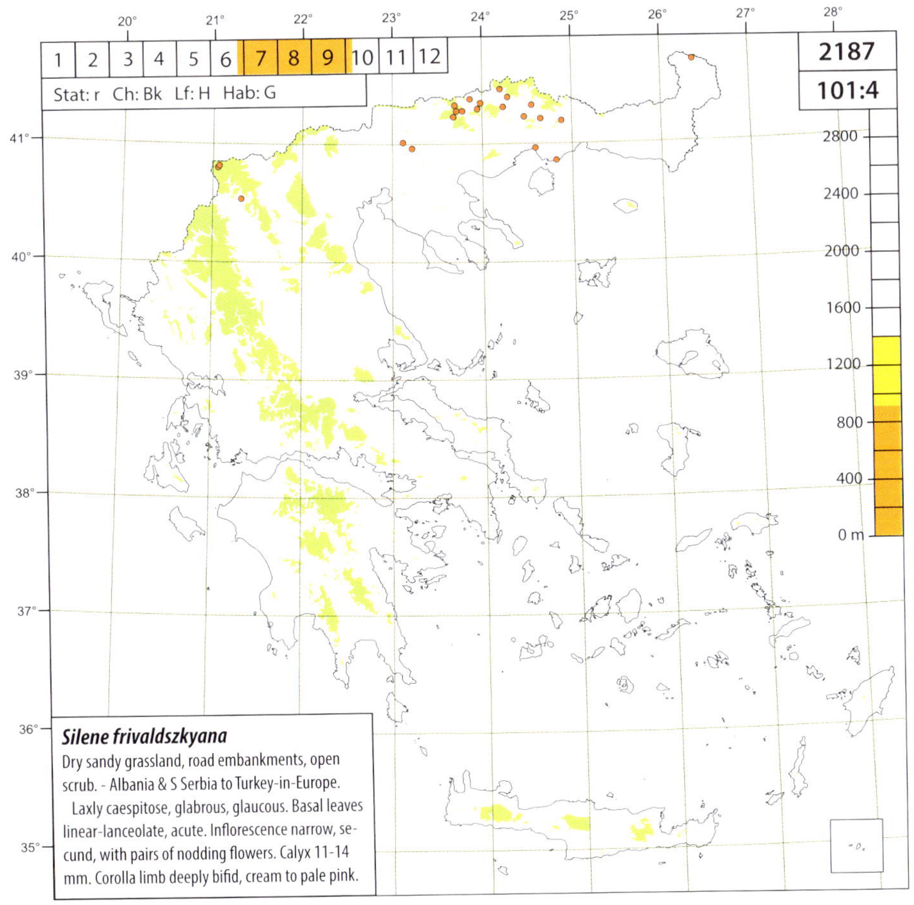

Silene frivaldszkyana

Dry sandy grassland, road embankments, open scrub. - Albania & S Serbia to Turkey-in-Europe.

Laxly caespitose, glabrous, glaucous. Basal leaves linear-lanceolate, acute. Inflorescence narrow, secund, with pairs of nodding flowers. Calyx 11-14 mm. Corolla limb deeply bifid, cream to pale pink.

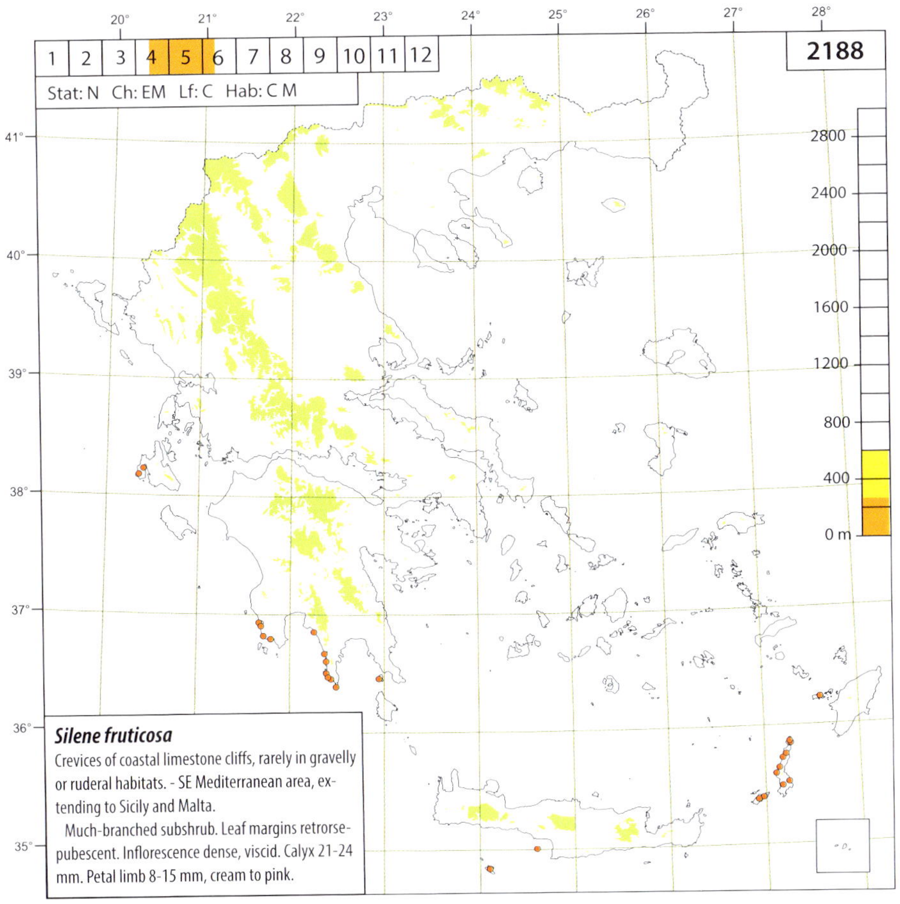

Silene fruticosa

Crevices of coastal limestone cliffs, rarely in gravelly or ruderal habitats. - SE Mediterranean area, extending to Sicily and Malta.

Much-branched subshrub. Leaf margins retrorse-pubescent. Inflorescence dense, viscid. Calyx 21-24 mm. Petal limb 8-15 mm, cream to pink.

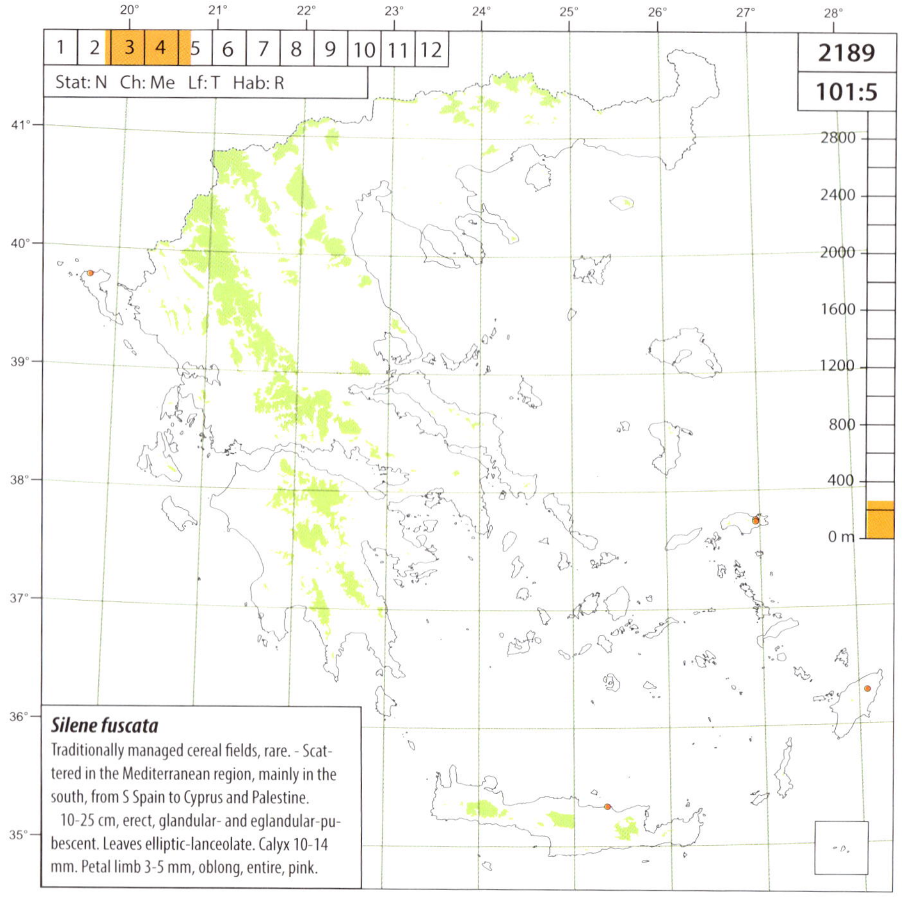

Silene fuscata
Traditionally managed cereal fields, rare. - Scattered in the Mediterranean region, mainly in the south, from S Spain to Cyprus and Palestine.
10-25 cm, erect, glandular- and eglandular-pubescent. Leaves elliptic-lanceolate. Calyx 10-14 mm. Petal limb 3-5 mm, oblong, entire, pink.

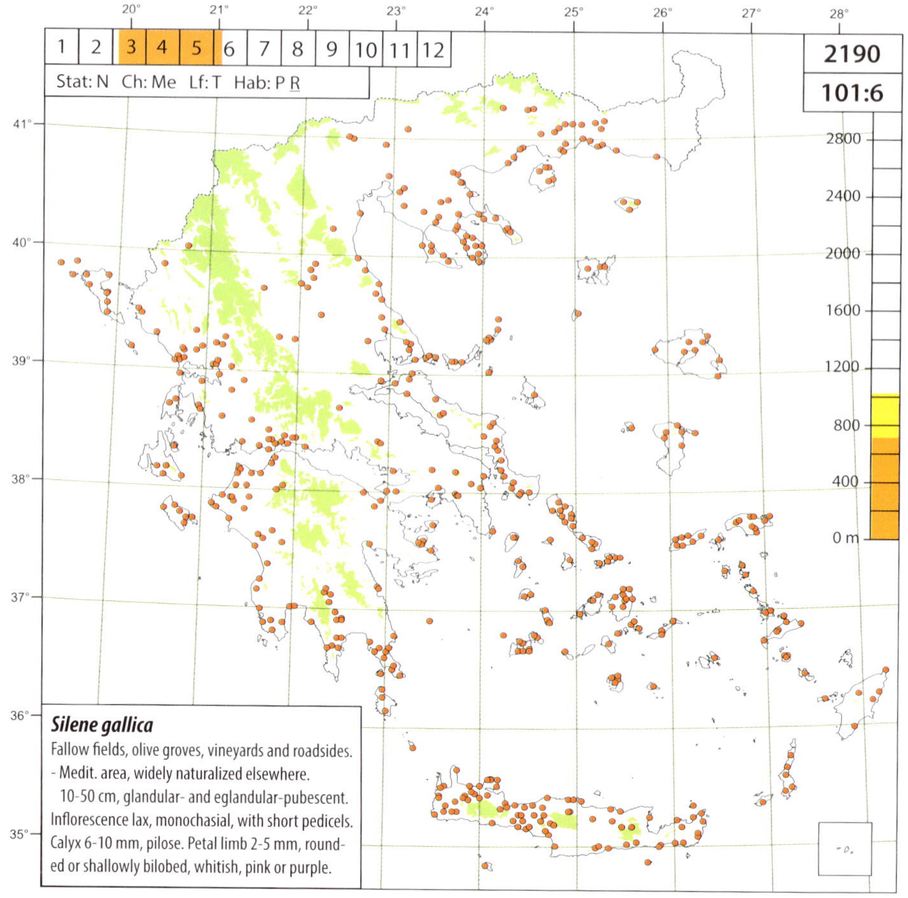

Silene gallica
Fallow fields, olive groves, vineyards and roadsides. - Medit. area, widely naturalized elsewhere.
10-50 cm, glandular- and eglandular-pubescent. Inflorescence lax, monochasial, with short pedicels. Calyx 6-10 mm, pilose. Petal limb 2-5 mm, rounded or shallowly bilobed, whitish, pink or purple.

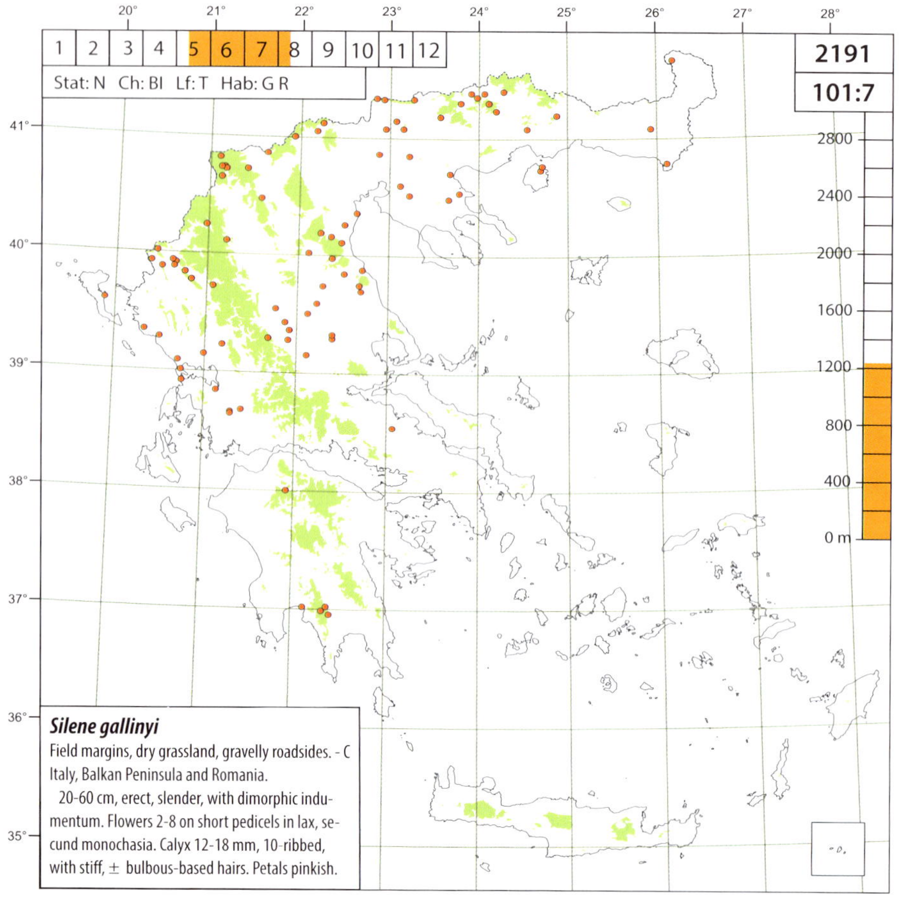

Silene gallinyi
Field margins, dry grassland, gravelly roadsides. - C Italy, Balkan Peninsula and Romania.
20-60 cm, erect, slender, with dimorphic indumentum. Flowers 2-8 on short pedicels in lax, secund monochasia. Calyx 12-18 mm, 10-ribbed, with stiff, ± bulbous-based hairs. Petals pinkish.

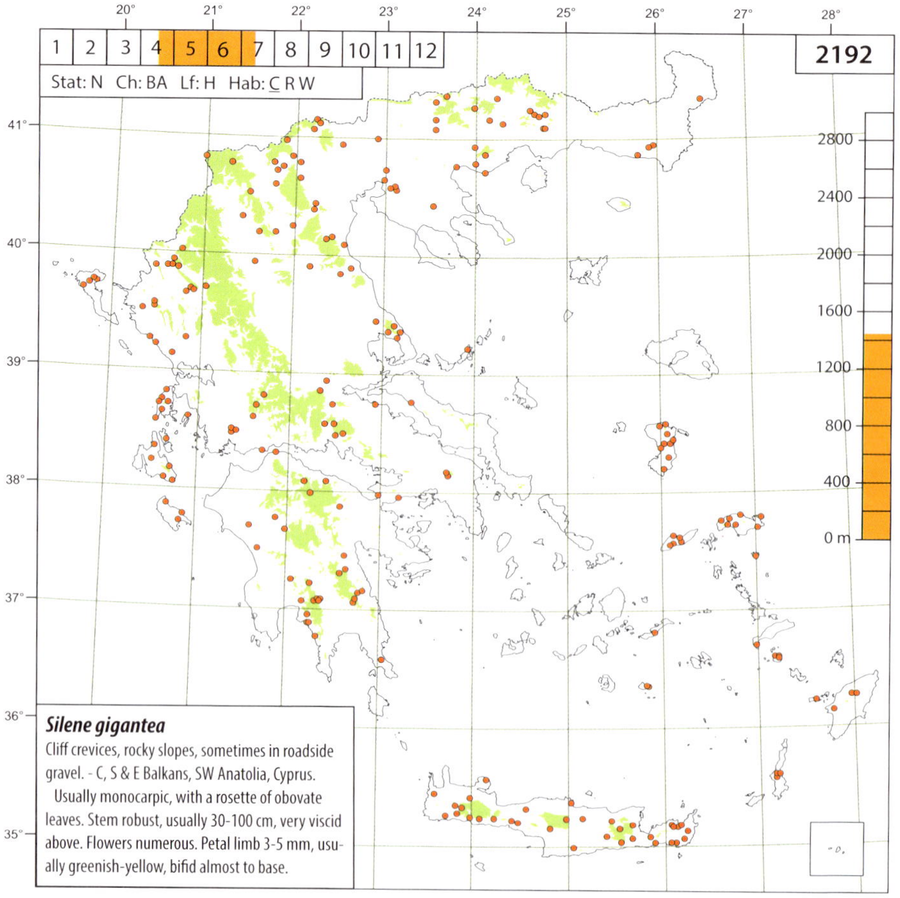

Silene gigantea
Cliff crevices, rocky slopes, sometimes in roadside gravel. - C, S & E Balkans, SW Anatolia, Cyprus.
Usually monocarpic, with a rosette of obovate leaves. Stem robust, usually 30-100 cm, very viscid above. Flowers numerous. Petal limb 3-5 mm, usually greenish-yellow, bifid almost to base.

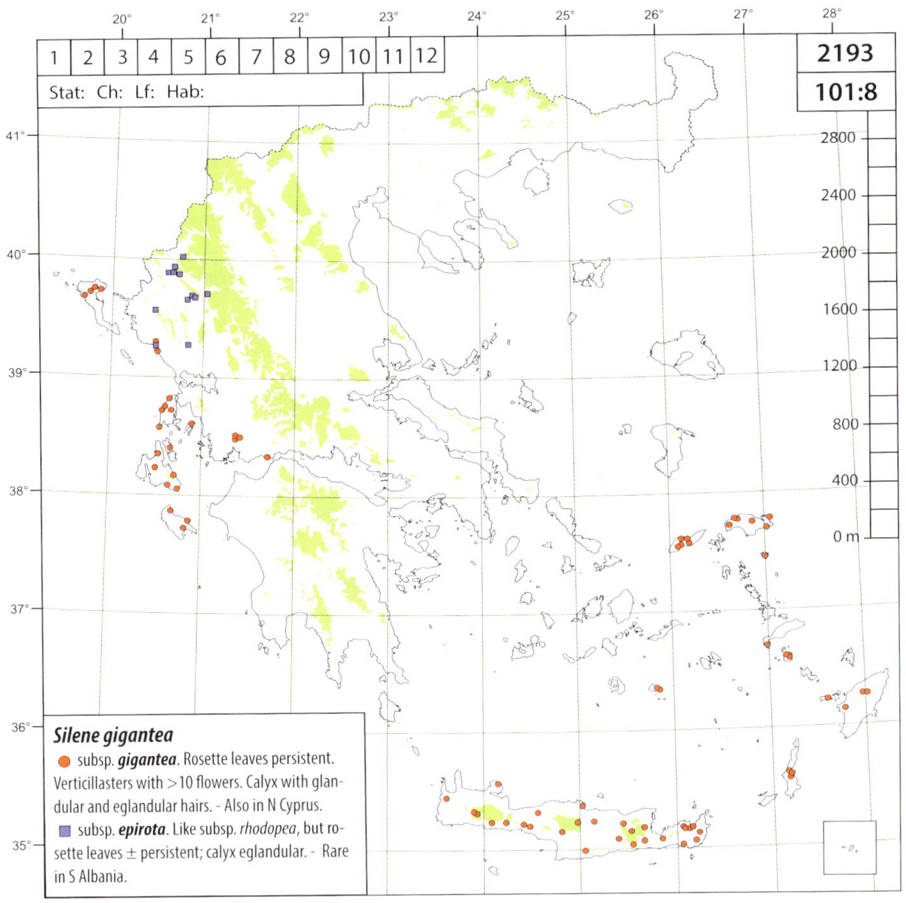

Silene gigantea
● subsp. ***gigantea***. Rosette leaves persistent. Verticillasters with >10 flowers. Calyx with glandular and eglandular hairs. - Also in N Cyprus.
■ subsp. ***epirota***. Like subsp. *rhodopea*, but rosette leaves ± persistent; calyx eglandular. - Rare in S Albania.

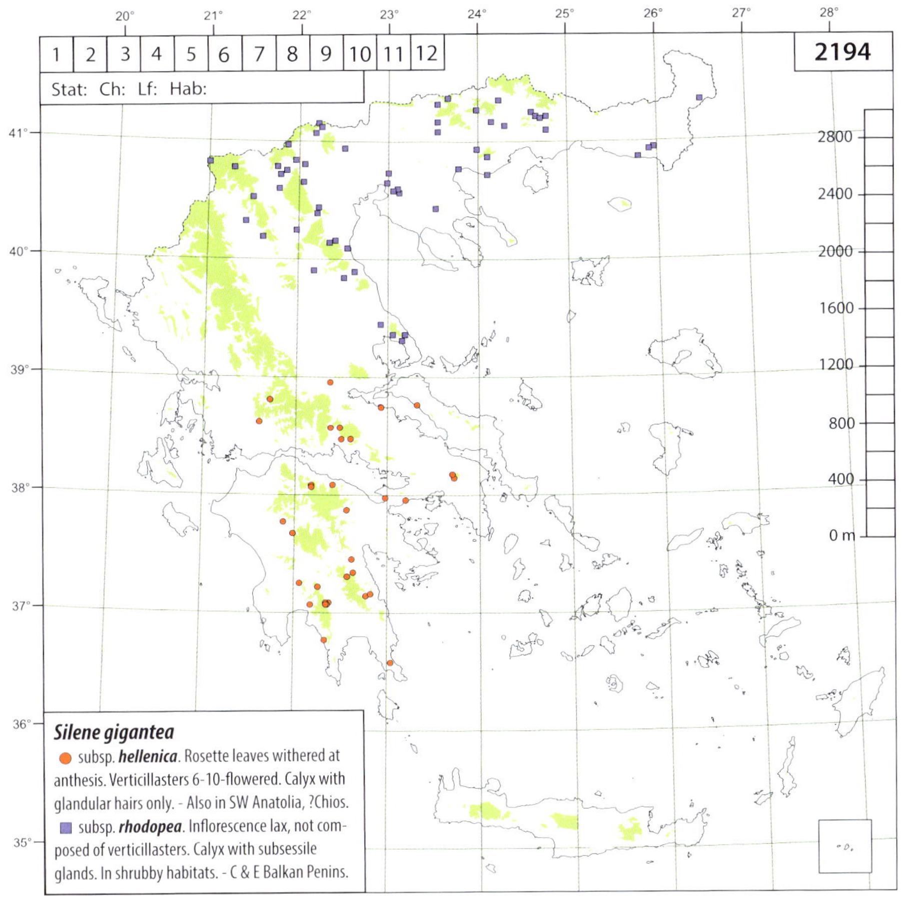

Silene gigantea
● subsp. ***hellenica***. Rosette leaves withered at anthesis. Verticillasters 6-10-flowered. Calyx with glandular hairs only. - Also in SW Anatolia, ?Chios.
■ subsp. ***rhodopea***. Inflorescence lax, not composed of verticillasters. Calyx with subsessile glands. In shrubby habitats. - C & E Balkan Penins.

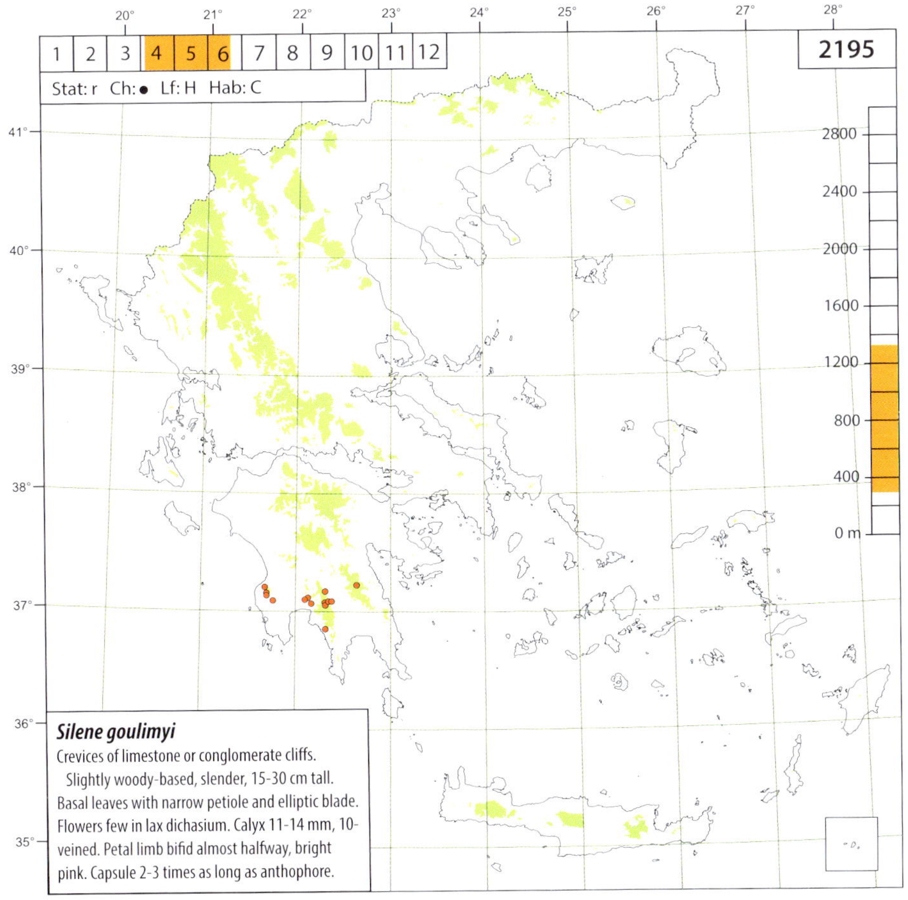

Silene goulimyi
Crevices of limestone or conglomerate cliffs.
Slightly woody-based, slender, 15-30 cm tall. Basal leaves with narrow petiole and elliptic blade. Flowers few in lax dichasium. Calyx 11-14 mm, 10-veined. Petal limb bifid almost halfway, bright pink. Capsule 2-3 times as long as anthophore.

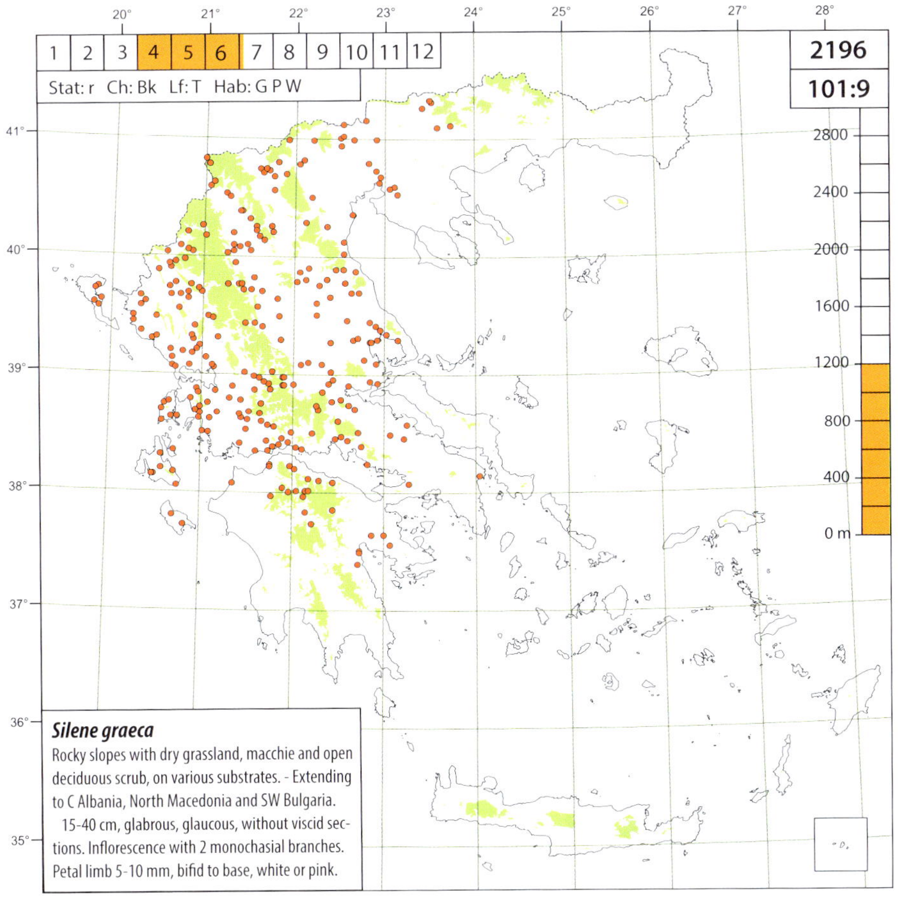

Silene graeca
Rocky slopes with dry grassland, macchie and open deciduous scrub, on various substrates. - Extending to C Albania, North Macedonia and SW Bulgaria.
15-40 cm, glabrous, glaucous, without viscid sections. Inflorescence with 2 monochasial branches. Petal limb 5-10 mm, bifid to base, white or pink.

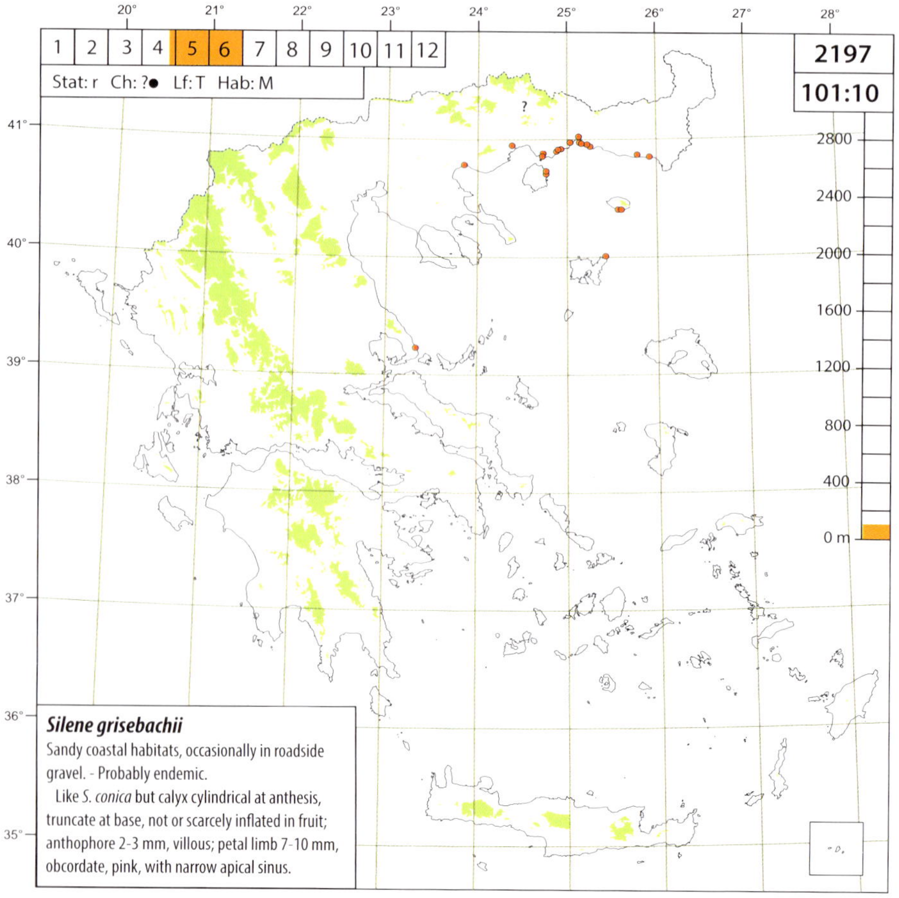

Silene grisebachii
Sandy coastal habitats, occasionally in roadside gravel. - Probably endemic.
Like *S. conica* but calyx cylindrical at anthesis, truncate at base, not or scarcely inflated in fruit; anthophore 2-3 mm, villous; petal limb 7-10 mm, obcordate, pink, with narrow apical sinus.

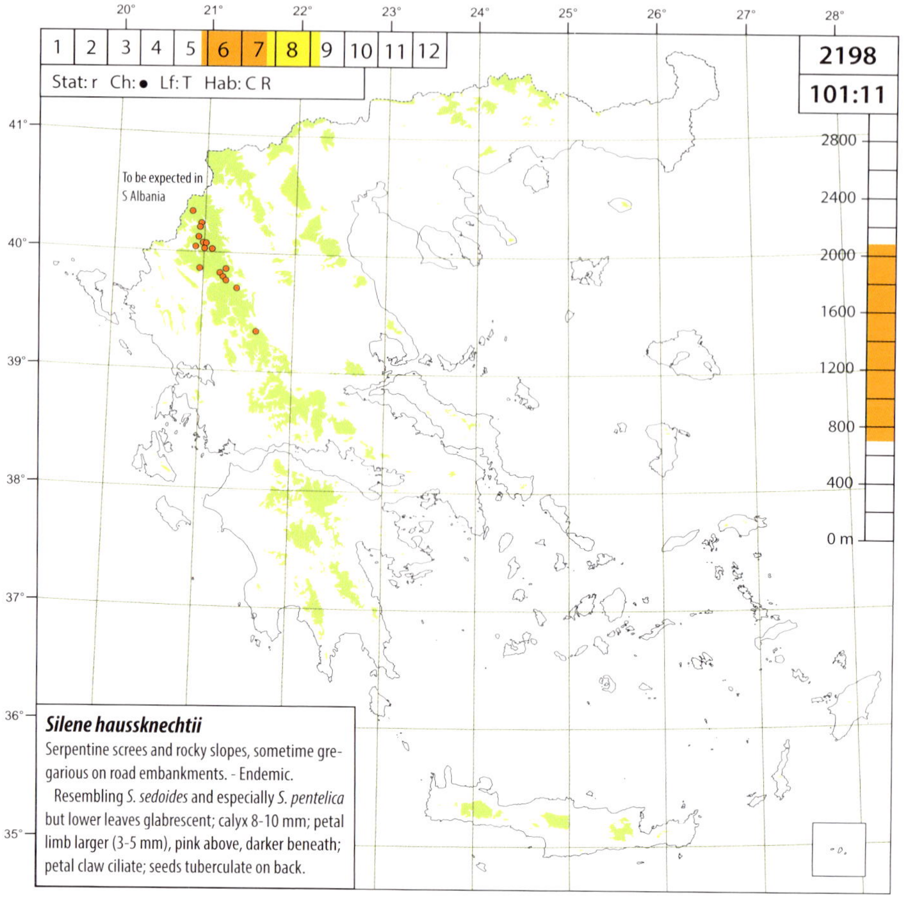

Silene haussknechtii
Serpentine screes and rocky slopes, sometime gregarious on road embankments. - Endemic.
Resembling *S. sedoides* and especially *S. pentelica* but lower leaves glabrescent; calyx 8-10 mm; petal limb larger (3-5 mm), pink above, darker beneath; petal claw ciliate; seeds tuberculate on back.

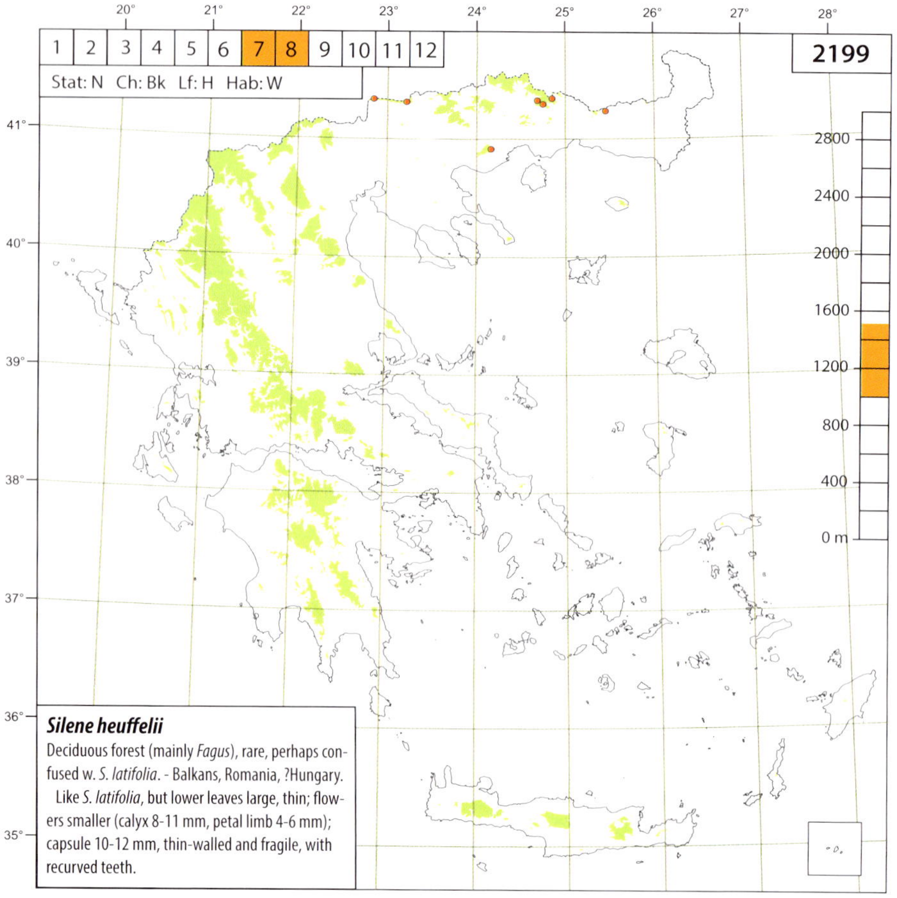

Silene heuffelii
Deciduous forest (mainly *Fagus*), rare, perhaps confused w. *S. latifolia*. - Balkans, Romania, ?Hungary.
Like *S. latifolia*, but lower leaves large, thin; flowers smaller (calyx 8-11 mm, petal limb 4-6 mm); capsule 10-12 mm, thin-walled and fragile, with recurved teeth.

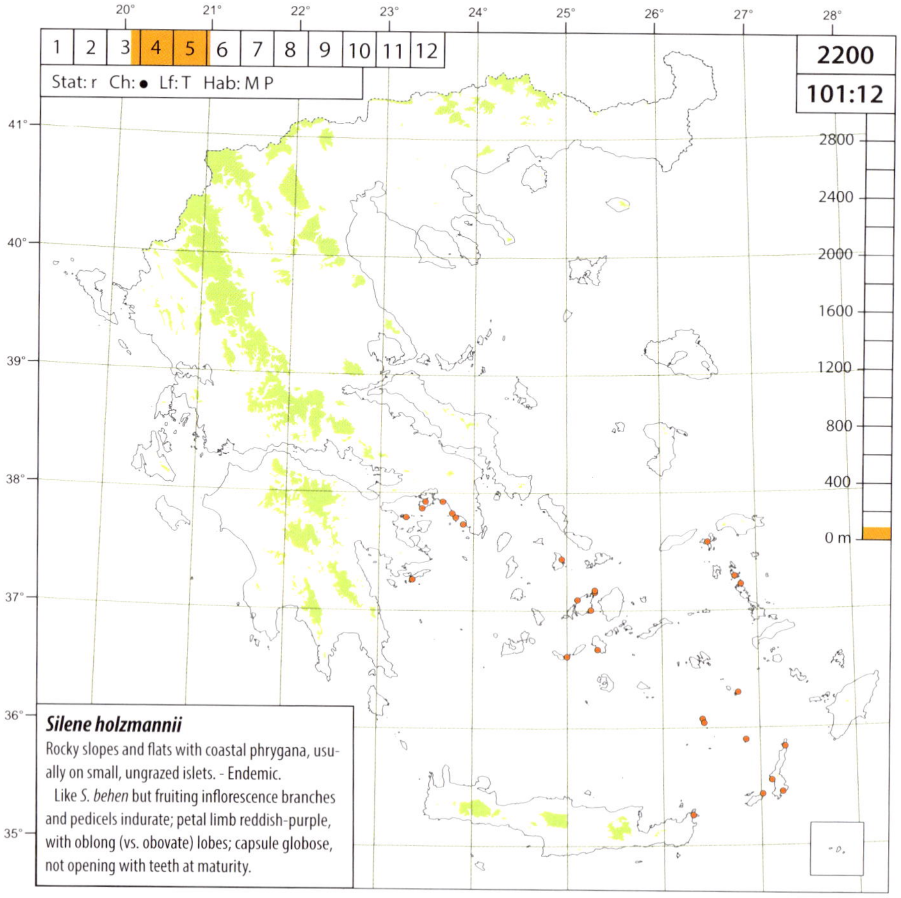

Silene holzmannii
Rocky slopes and flats with coastal phrygana, usually on small, ungrazed islets. - Endemic.
Like *S. behen* but fruiting inflorescence branches and pedicels indurate; petal limb reddish-purple, with oblong (vs. obovate) lobes; capsule globose, not opening with teeth at maturity.

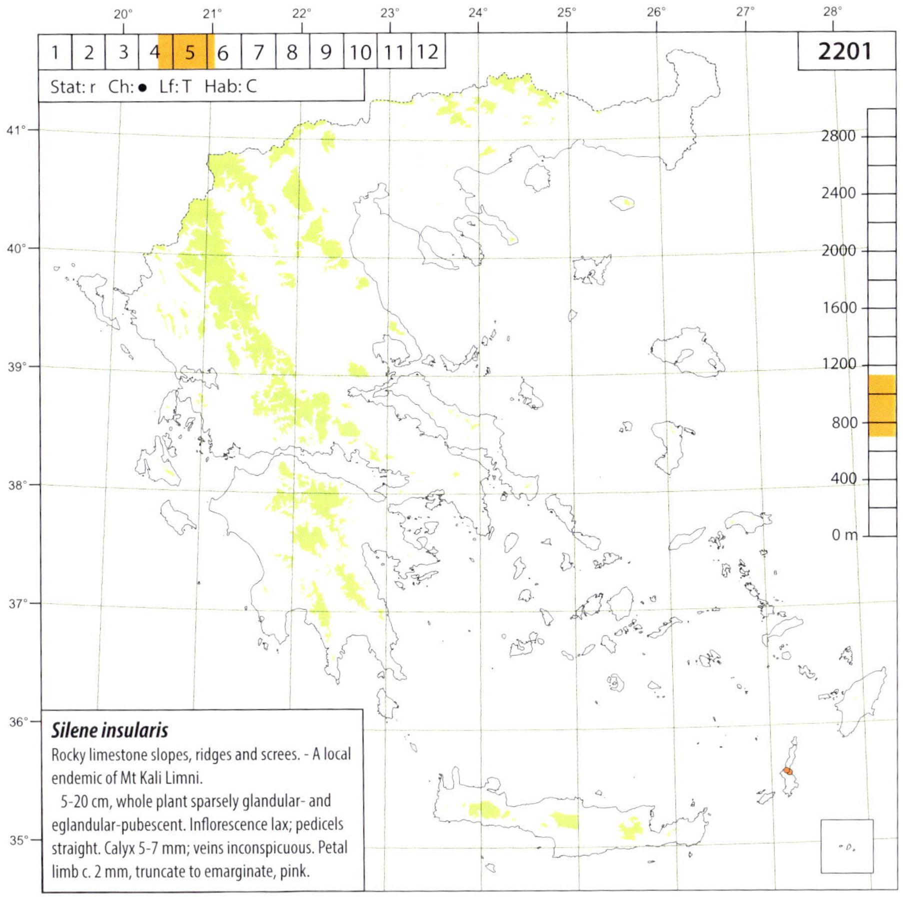

Silene insularis

Rocky limestone slopes, ridges and screes. - A local endemic of Mt Kali Limni.

5-20 cm, whole plant sparsely glandular- and eglandular-pubescent. Inflorescence lax; pedicels straight. Calyx 5-7 mm; veins inconspicuous. Petal limb c. 2 mm, truncate to emarginate, pink.

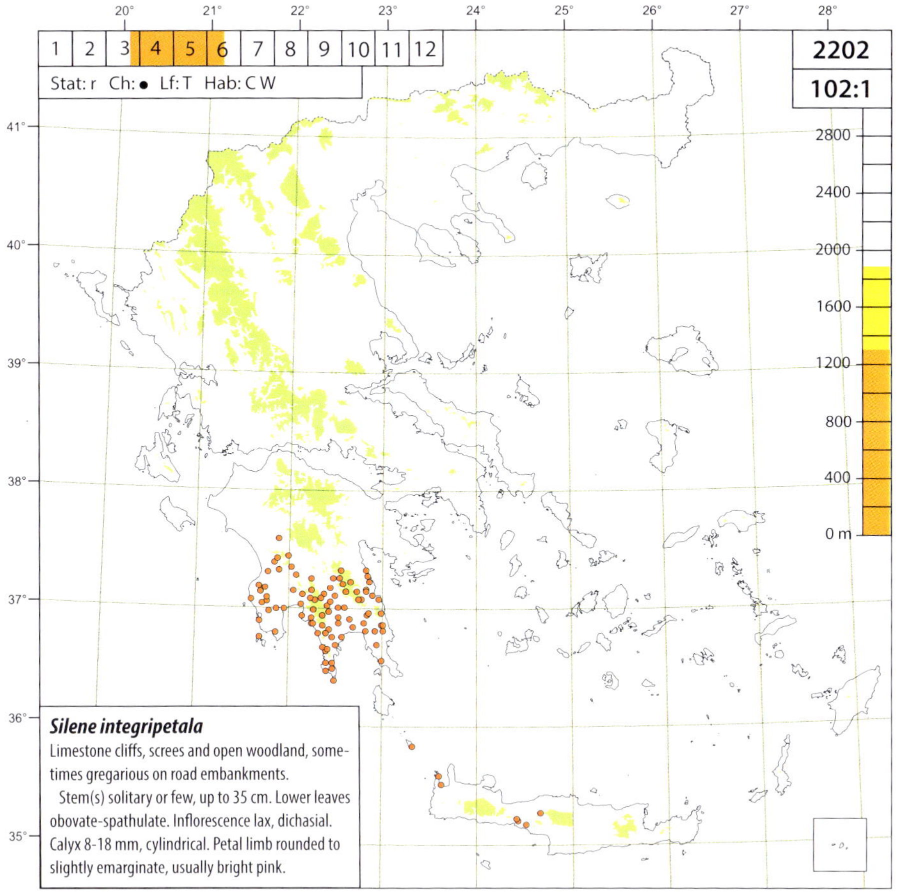

Silene integripetala

Limestone cliffs, screes and open woodland, sometimes gregarious on road embankments.

Stem(s) solitary or few, up to 35 cm. Lower leaves obovate-spathulate. Inflorescence lax, dichasial. Calyx 8-18 mm, cylindrical. Petal limb rounded to slightly emarginate, usually bright pink.

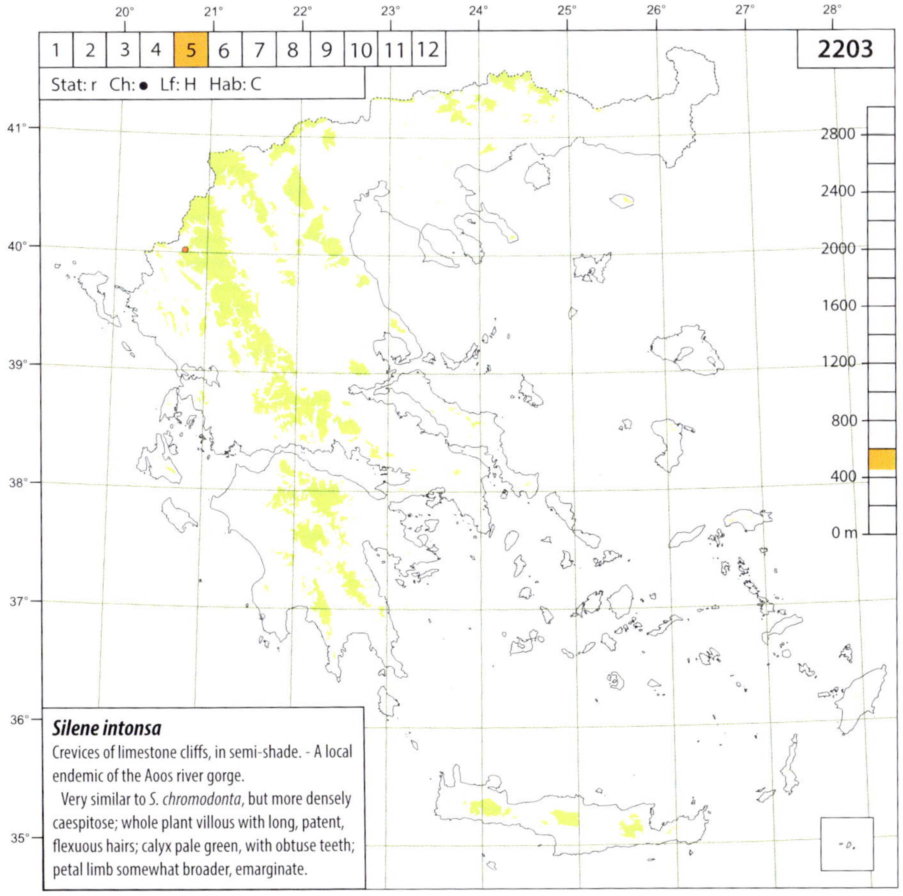

Silene intonsa

Crevices of limestone cliffs, in semi-shade. - A local endemic of the Aoos river gorge.

Very similar to *S. chromodonta*, but more densely caespitose; whole plant villous with long, patent, flexuous hairs; calyx pale green, with obtuse teeth; petal limb somewhat broader, emarginate.

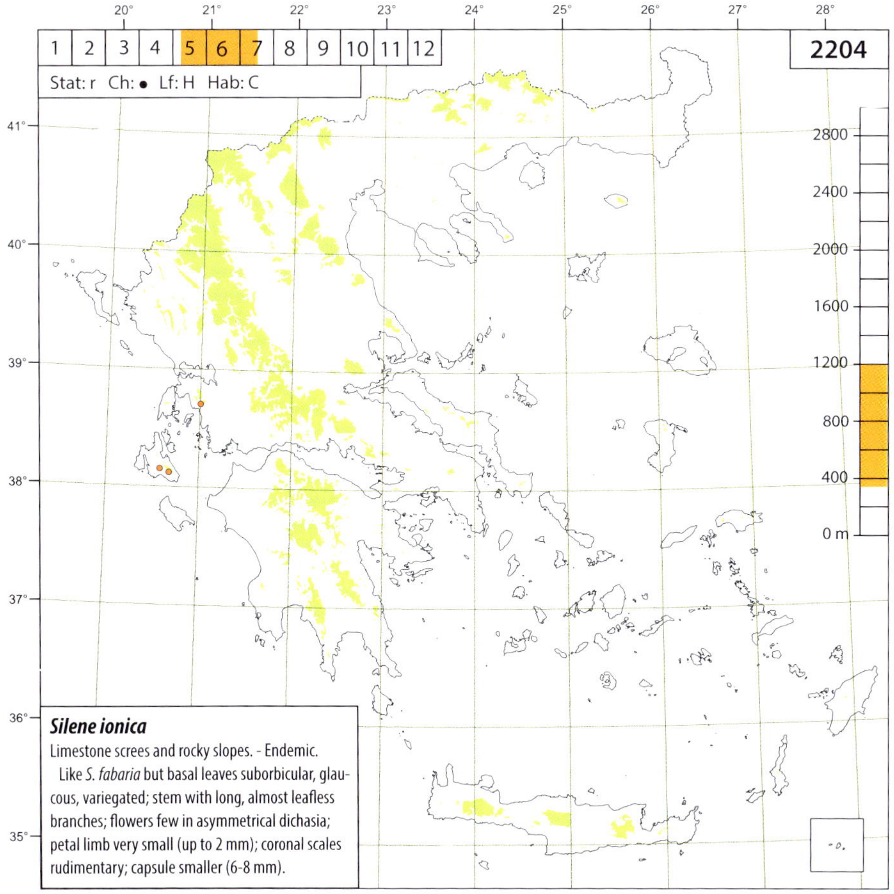

Silene ionica

Limestone screes and rocky slopes. - Endemic.

Like *S. fabaria* but basal leaves suborbicular, glaucous, variegated; stem with long, almost leafless branches; flowers few in asymmetrical dichasia; petal limb very small (up to 2 mm); coronal scales rudimentary; capsule smaller (6-8 mm).

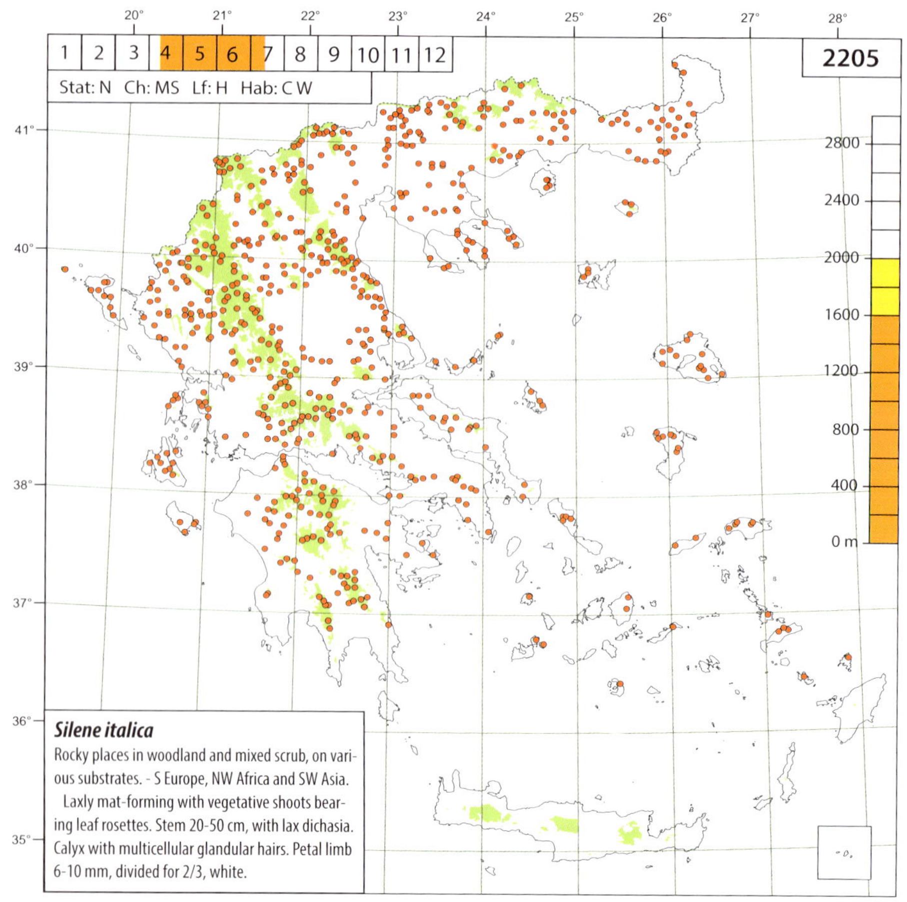

Silene italica

Rocky places in woodland and mixed scrub, on various substrates. - S Europe, NW Africa and SW Asia.

Laxly mat-forming with vegetative shoots bearing leaf rosettes. Stem 20-50 cm, with lax dichasia. Calyx with multicellular glandular hairs. Petal limb 6-10 mm, divided for 2/3, white.

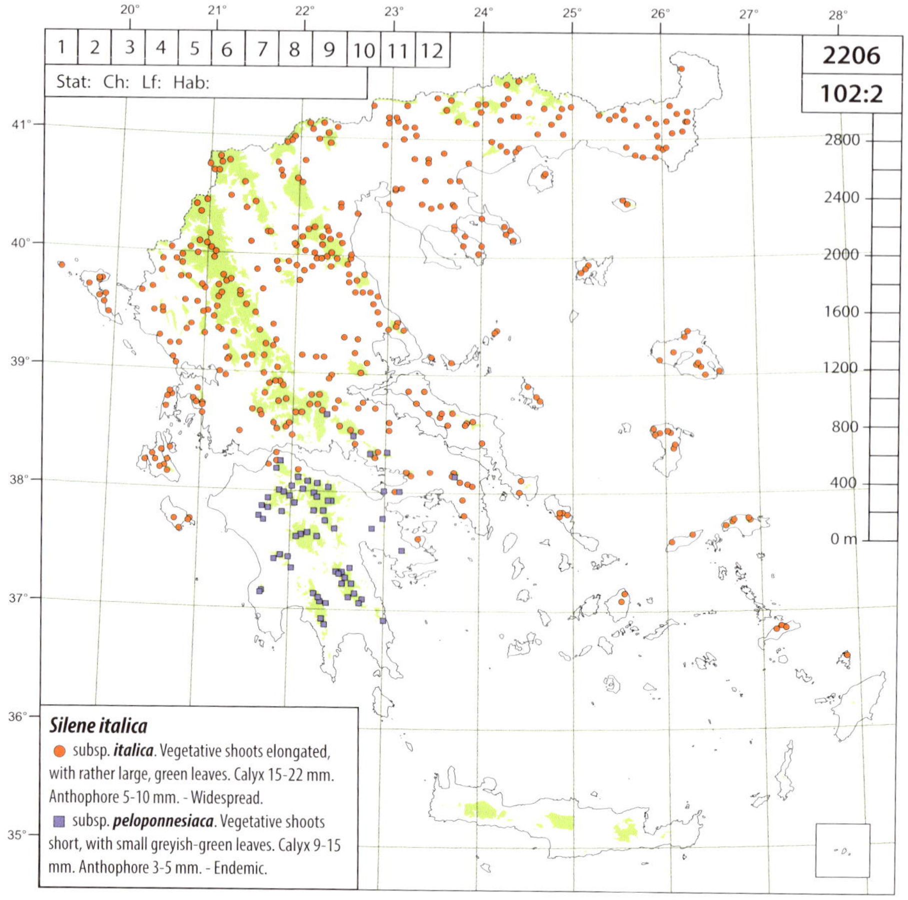

Silene italica

● subsp. ***italica***. Vegetative shoots elongated, with rather large, green leaves. Calyx 15-22 mm. Anthophore 5-10 mm. - Widespread.

■ subsp. ***peloponnesiaca***. Vegetative shoots short, with small greyish-green leaves. Calyx 9-15 mm. Anthophore 3-5 mm. - Endemic.

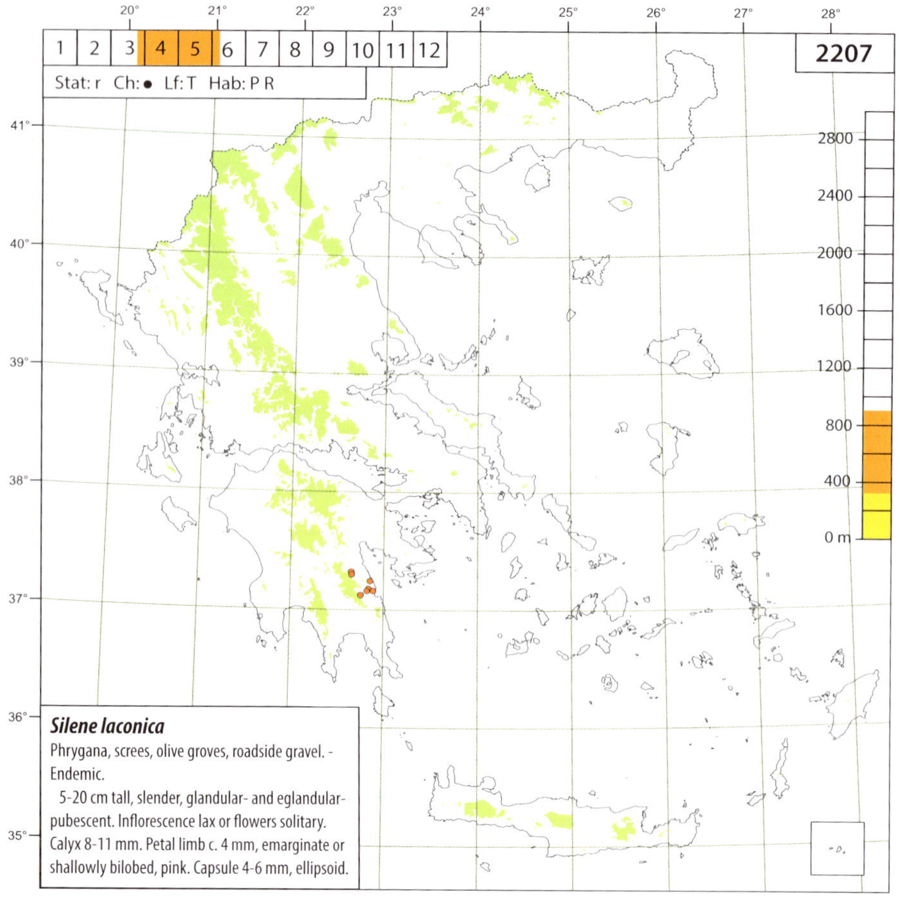

Silene laconica

Phrygana, screes, olive groves, roadside gravel. - Endemic.

5-20 cm tall, slender, glandular- and eglandular-pubescent. Inflorescence lax or flowers solitary. Calyx 8-11 mm. Petal limb c. 4 mm, emarginate or shallowly bilobed, pink. Capsule 4-6 mm, ellipsoid.

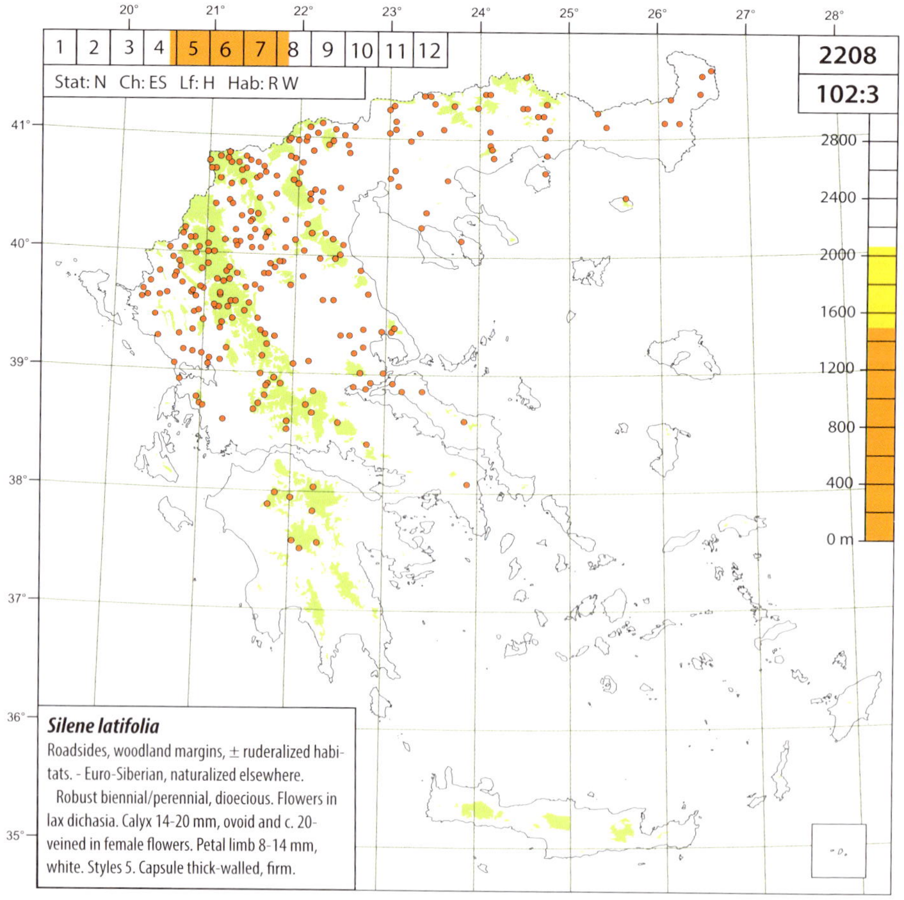

Silene latifolia

Roadsides, woodland margins, ± ruderalized habitats. - Euro-Siberian, naturalized elsewhere.

Robust biennial/perennial, dioecious. Flowers in lax dichasia. Calyx 14-20 mm, ovoid and c. 20-veined in female flowers. Petal limb 8-14 mm, white. Styles 5. Capsule thick-walled, firm.

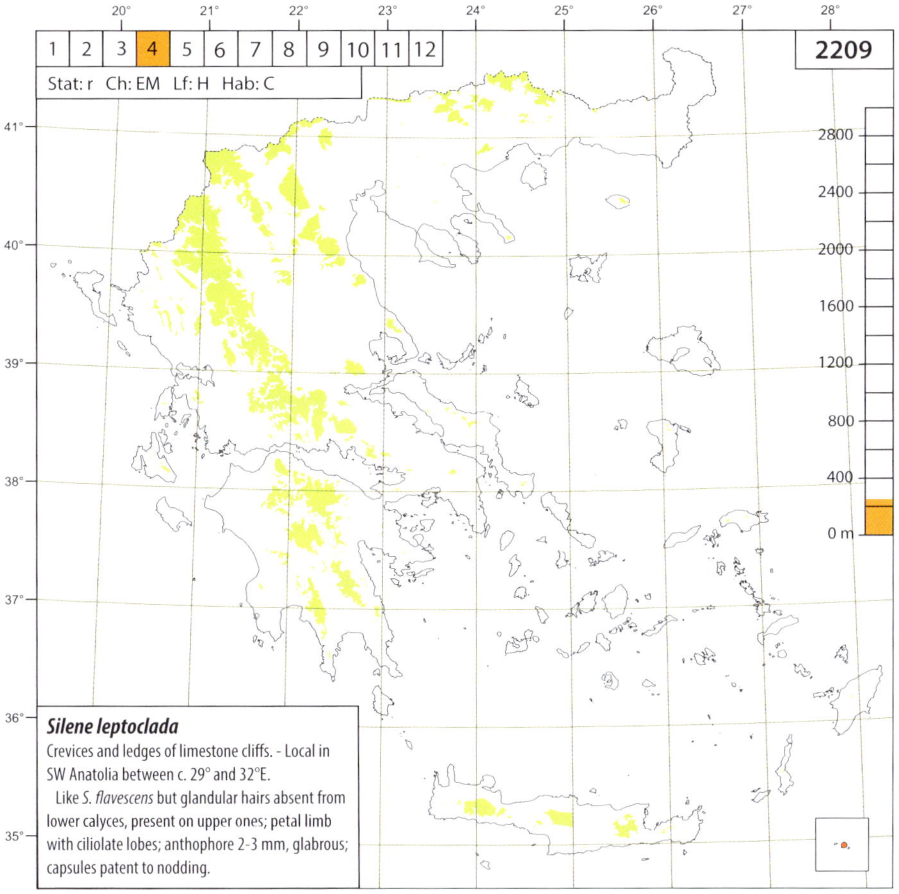

Silene leptoclada

Crevices and ledges of limestone cliffs. - Local in SW Anatolia between c. 29° and 32°E.

Like *S. flavescens* but glandular hairs absent from lower calyces, present on upper ones; petal limb with ciliolate lobes; anthophore 2-3 mm, glabrous; capsules patent to nodding.

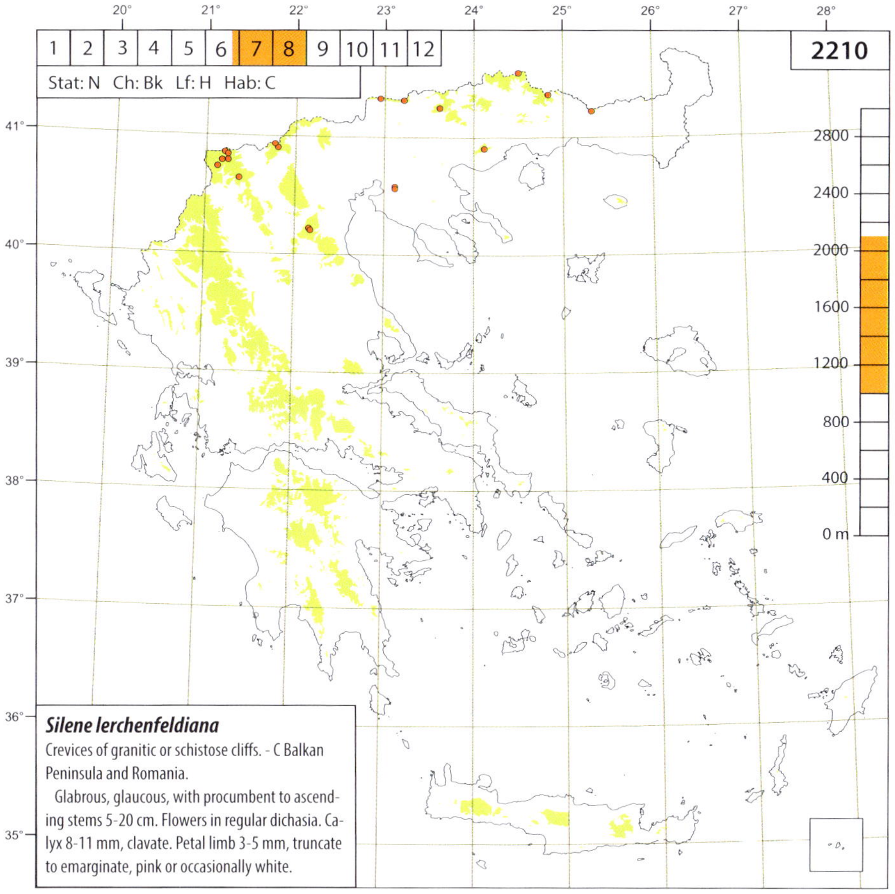

Silene lerchenfeldiana

Crevices of granitic or schistose cliffs. - C Balkan Peninsula and Romania.

Glabrous, glaucous, with procumbent to ascending stems 5-20 cm. Flowers in regular dichasia. Calyx 8-11 mm, clavate. Petal limb 3-5 mm, truncate to emarginate, pink or occasionally white.

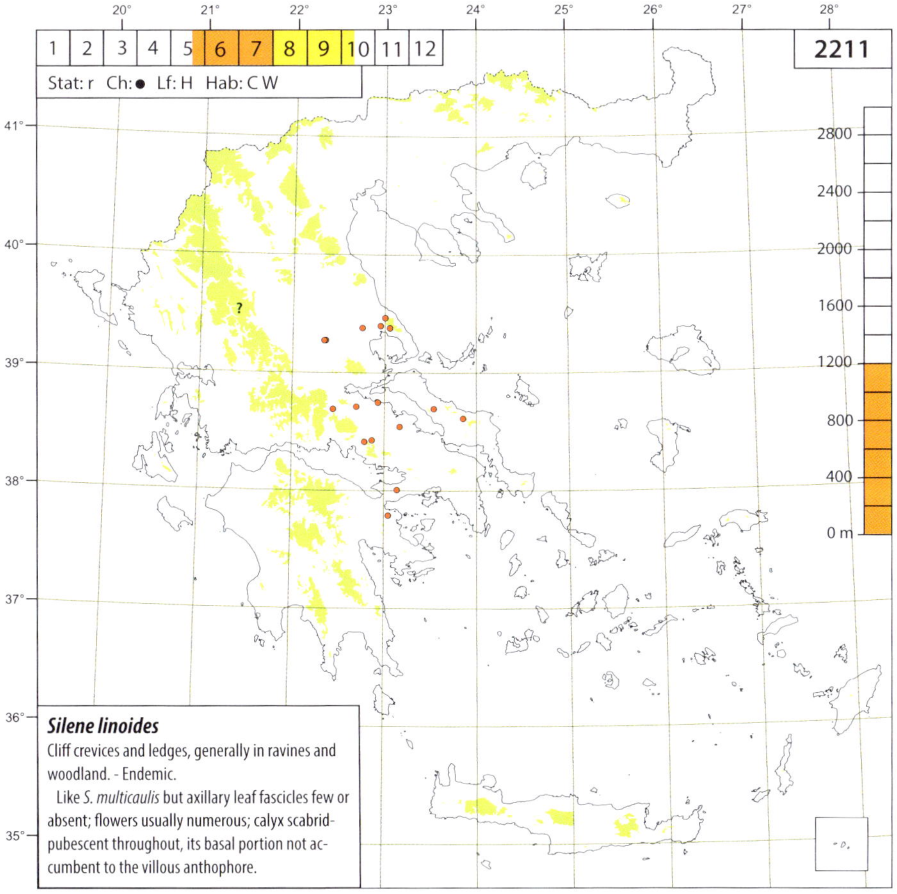

Silene linoides

Cliff crevices and ledges, generally in ravines and woodland. - Endemic.

Like *S. multicaulis* but axillary leaf fascicles few or absent; flowers usually numerous; calyx scabrid-pubescent throughout, its basal portion not accumbent to the villous anthophore.

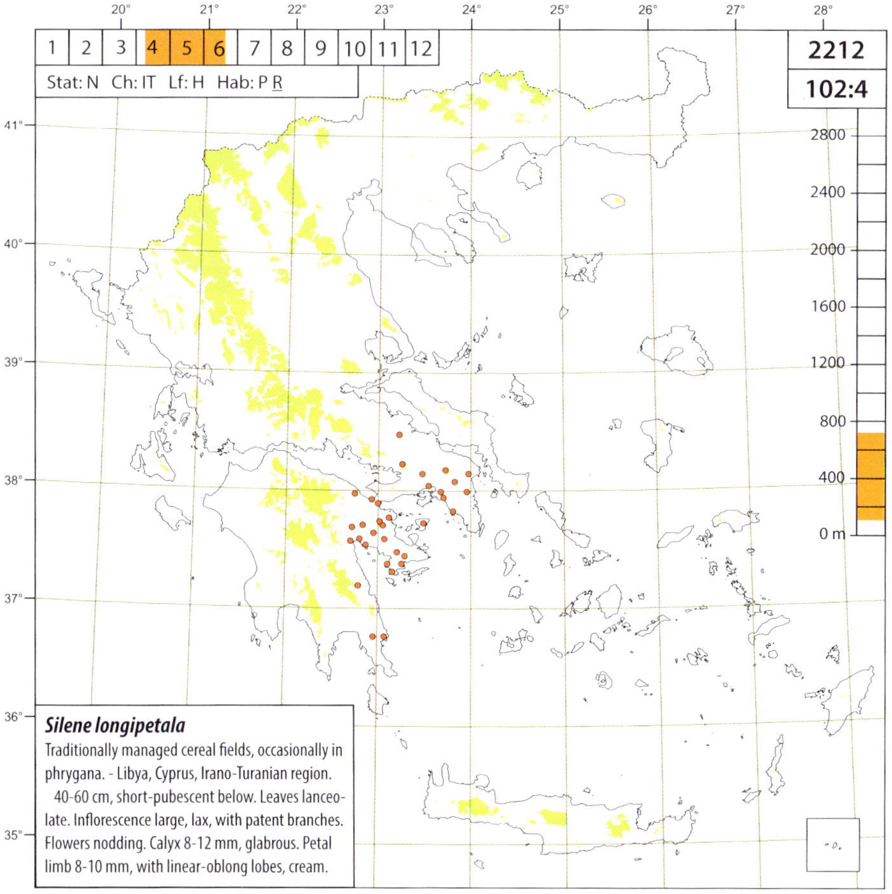

Silene longipetala

Traditionally managed cereal fields, occasionally in phrygana. - Libya, Cyprus, Irano-Turanian region.

40-60 cm, short-pubescent below. Leaves lanceolate. Inflorescence large, lax, with patent branches. Flowers nodding. Calyx 8-12 mm, glabrous. Petal limb 8-10 mm, with linear-oblong lobes, cream.

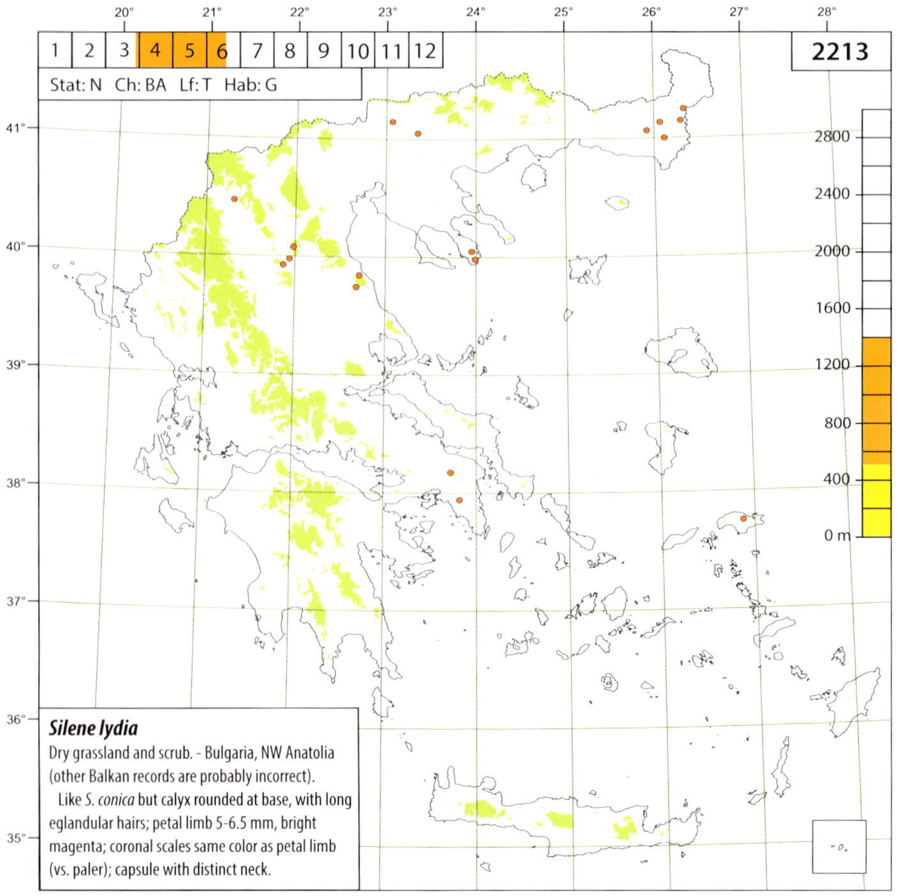

Silene lydia
Dry grassland and scrub. - Bulgaria, NW Anatolia (other Balkan records are probably incorrect).
Like *S. conica* but calyx rounded at base, with long eglandular hairs; petal limb 5-6.5 mm, bright magenta; coronal scales same color as petal limb (vs. paler); capsule with distinct neck.

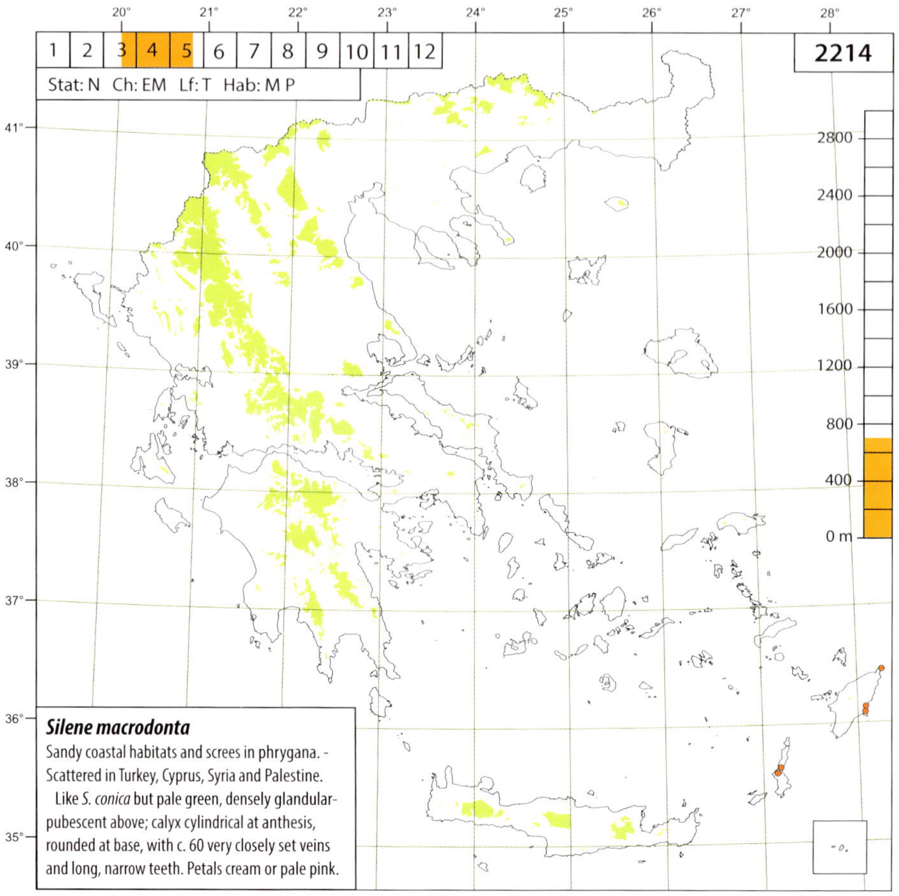

Silene macrodonta
Sandy coastal habitats and screes in phrygana. - Scattered in Turkey, Cyprus, Syria and Palestine.
Like *S. conica* but pale green, densely glandular-pubescent above; calyx cylindrical at anthesis, rounded at base, with c. 60 very closely set veins and long, narrow teeth. Petals cream or pale pink.

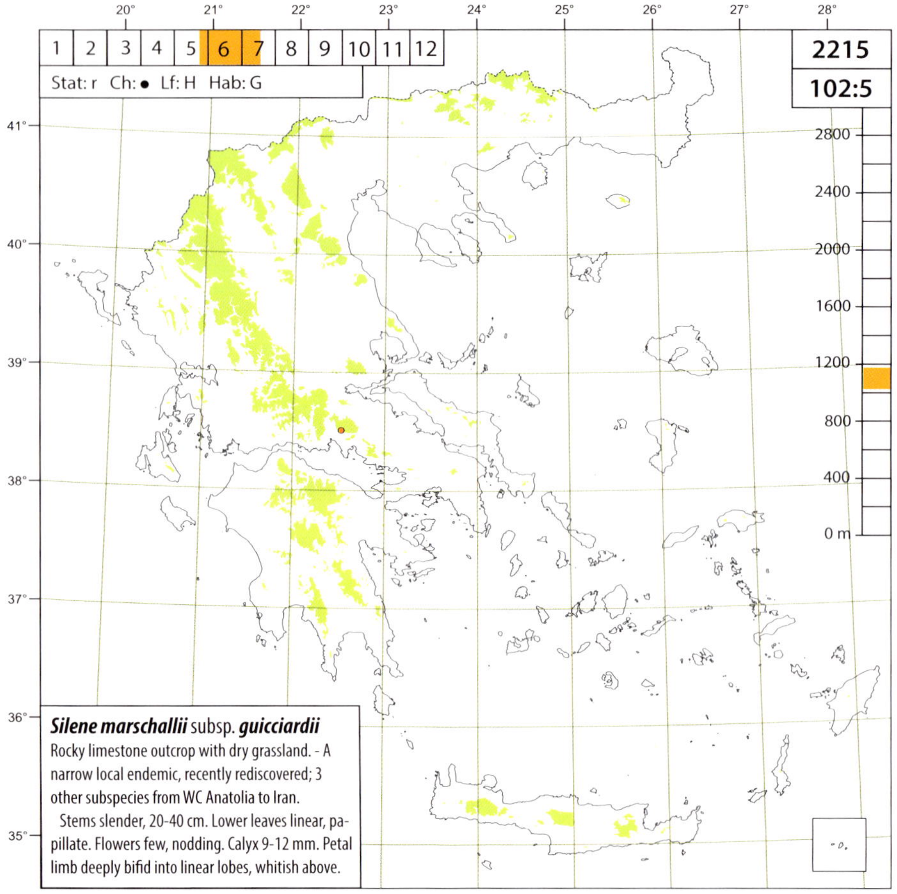

Silene marschallii* subsp. *guicciardii
Rocky limestone outcrop with dry grassland. - A narrow local endemic, recently rediscovered; 3 other subspecies from WC Anatolia to Iran.
Stems slender, 20-40 cm. Lower leaves linear, papillate. Flowers few, nodding. Calyx 9-12 mm. Petal limb deeply bifid into linear lobes, whitish above.

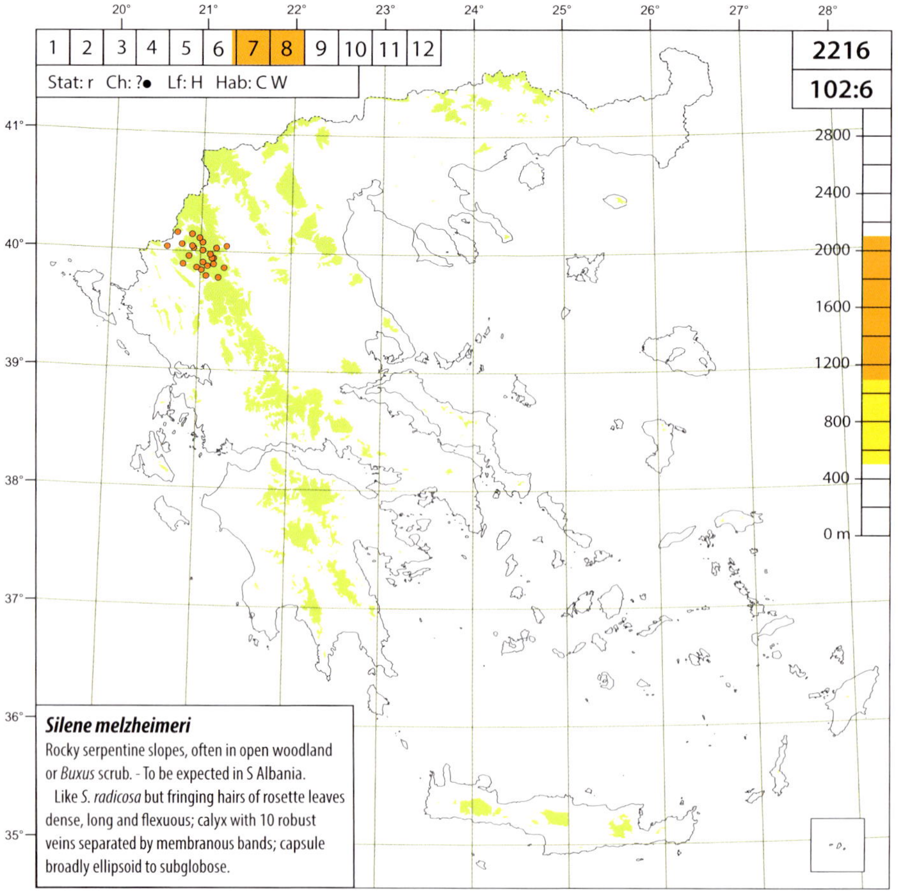

Silene melzheimeri
Rocky serpentine slopes, often in open woodland or *Buxus* scrub. - To be expected in S Albania.
Like *S. radicosa* but fringing hairs of rosette leaves dense, long and flexuous; calyx with 10 robust veins separated by membranous bands; capsule broadly ellipsoid to subglobose.

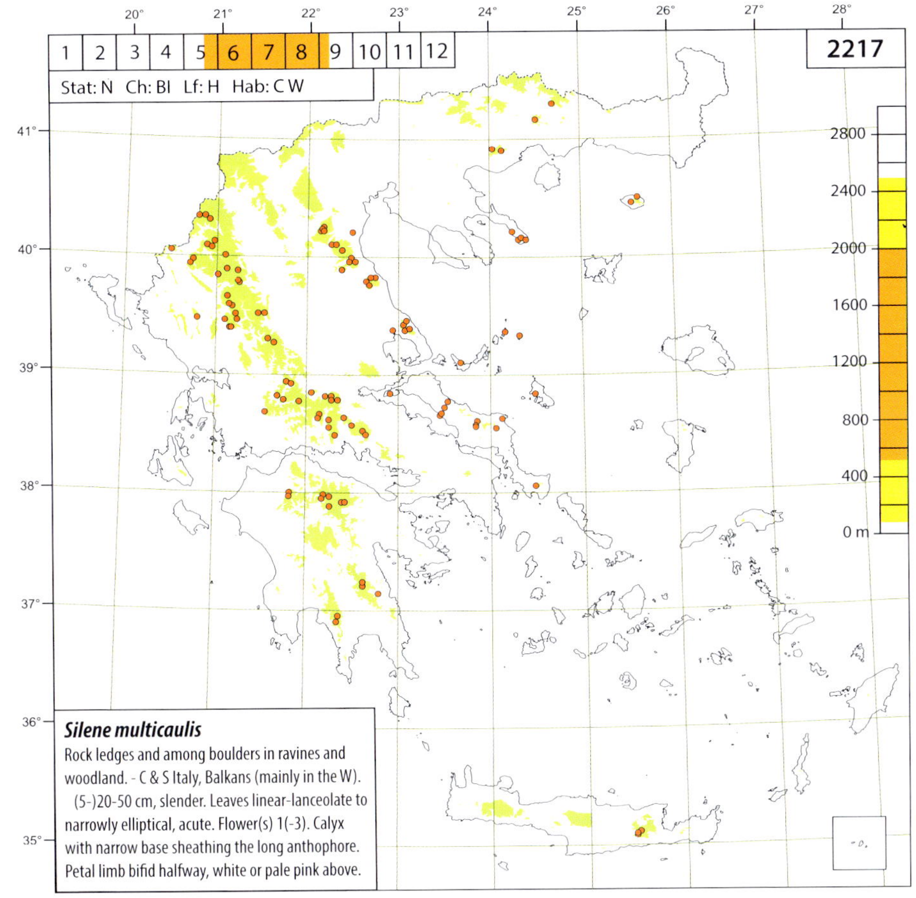

Silene multicaulis
Rock ledges and among boulders in ravines and woodland. - C & S Italy, Balkans (mainly in the W).
(5-)20-50 cm, slender. Leaves linear-lanceolate to narrowly elliptical, acute. Flower(s) 1(-3). Calyx with narrow base sheathing the long anthophore. Petal limb bifid halfway, white or pale pink above.

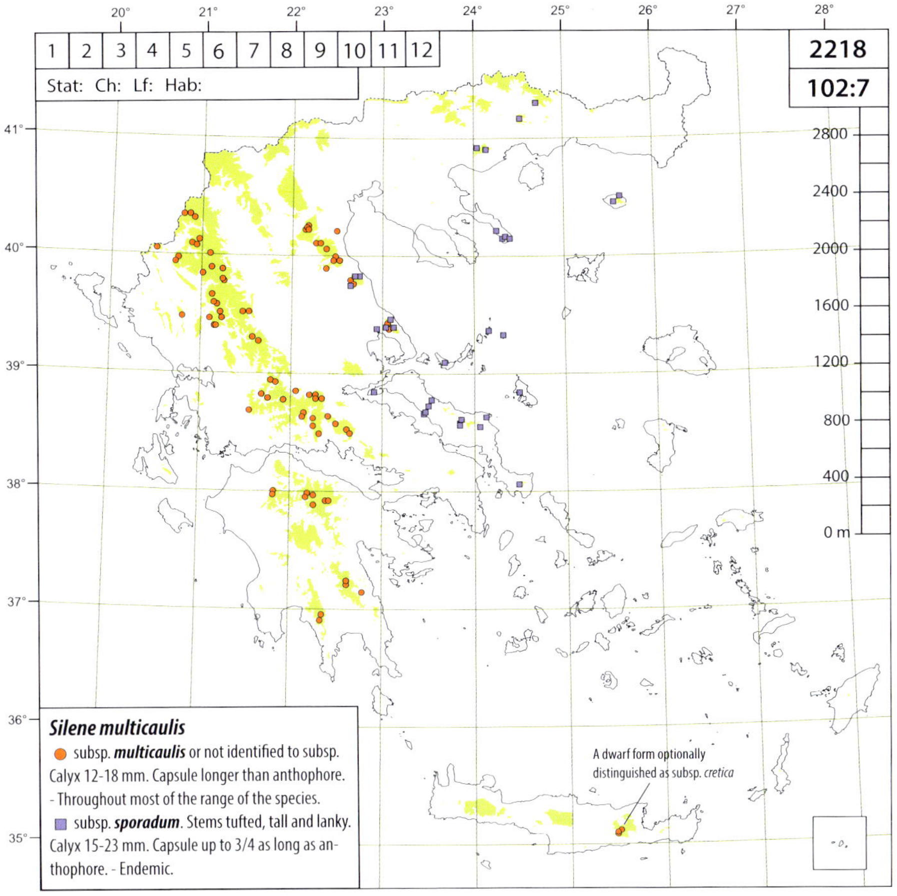

Silene multicaulis
● subsp. ***multicaulis*** or not identified to subsp. Calyx 12-18 mm. Capsule longer than anthophore. - Throughout most of the range of the species.
■ subsp. ***sporadum***. Stems tufted, tall and lanky. Calyx 15-23 mm. Capsule up to 3/4 as long as anthophore. - Endemic.

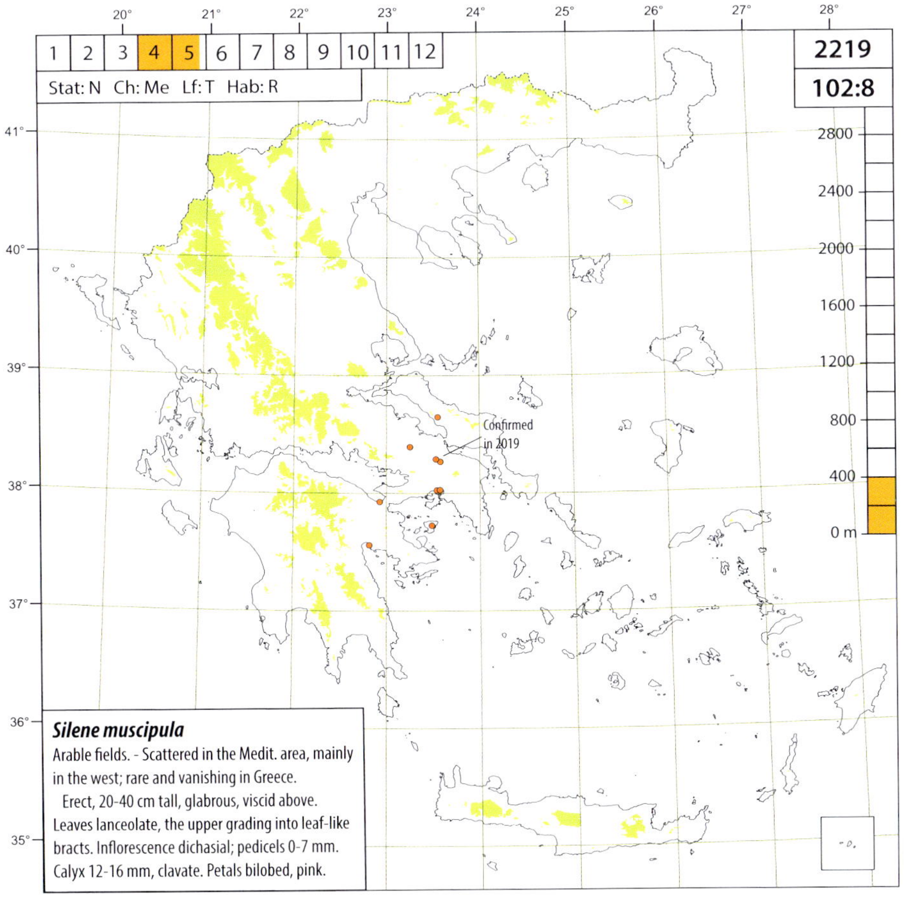

Silene muscipula
Arable fields. - Scattered in the Medit. area, mainly in the west; rare and vanishing in Greece.
Erect, 20-40 cm tall, glabrous, viscid above. Leaves lanceolate, the upper grading into leaf-like bracts. Inflorescence dichasial; pedicels 0-7 mm. Calyx 12-16 mm, clavate. Petals bilobed, pink.

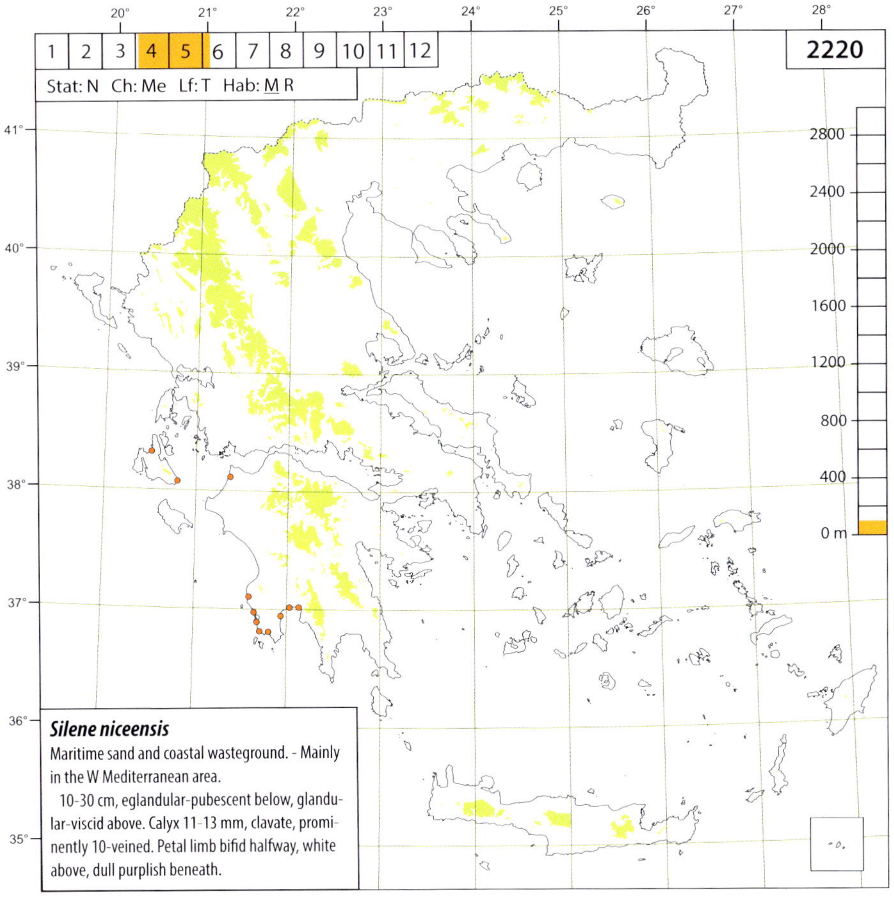

Silene niceensis
Maritime sand and coastal wasteground. - Mainly in the W Mediterranean area.
10-30 cm, eglandular-pubescent below, glandular-viscid above. Calyx 11-13 mm, clavate, prominently 10-veined. Petal limb bifid halfway, white above, dull purplish beneath.

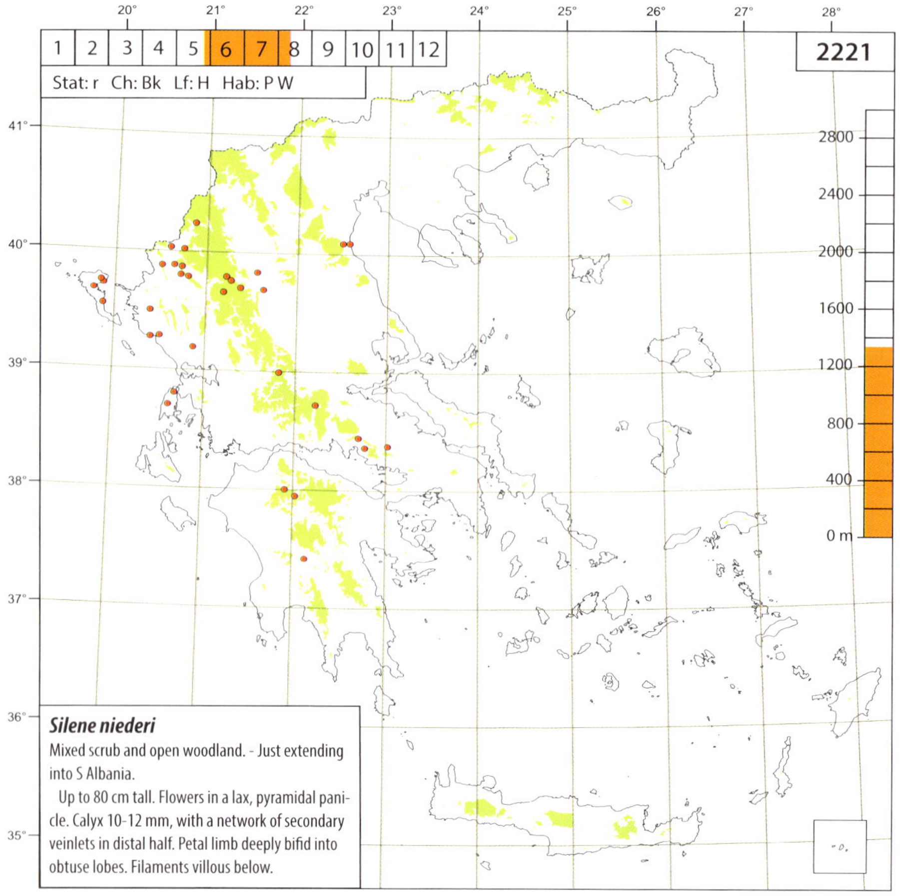

Silene niederi

Mixed scrub and open woodland. - Just extending into S Albania.

Up to 80 cm tall. Flowers in a lax, pyramidal panicle. Calyx 10-12 mm, with a network of secondary veinlets in distal half. Petal limb deeply bifid into obtuse lobes. Filaments villous below.

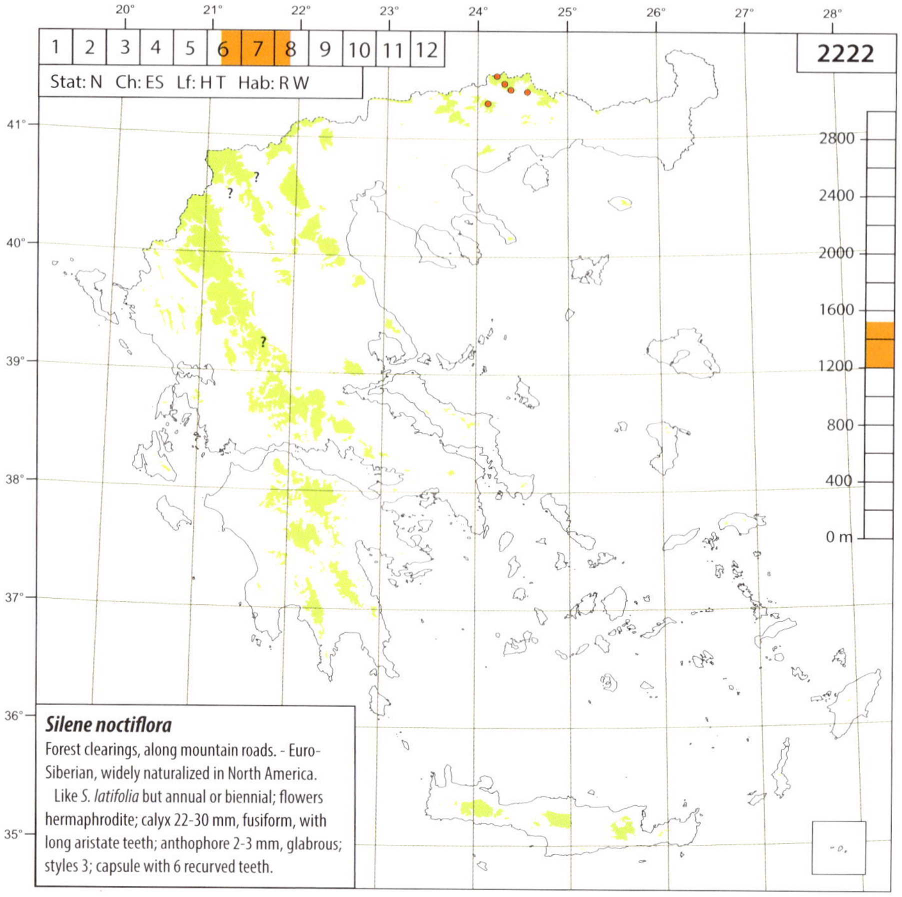

Silene noctiflora

Forest clearings, along mountain roads. - Euro-Siberian, widely naturalized in North America.

Like *S. latifolia* but annual or biennial; flowers hermaphrodite; calyx 22-30 mm, fusiform, with long aristate teeth; anthophore 2-3 mm, glabrous; styles 3; capsule with 6 recurved teeth.

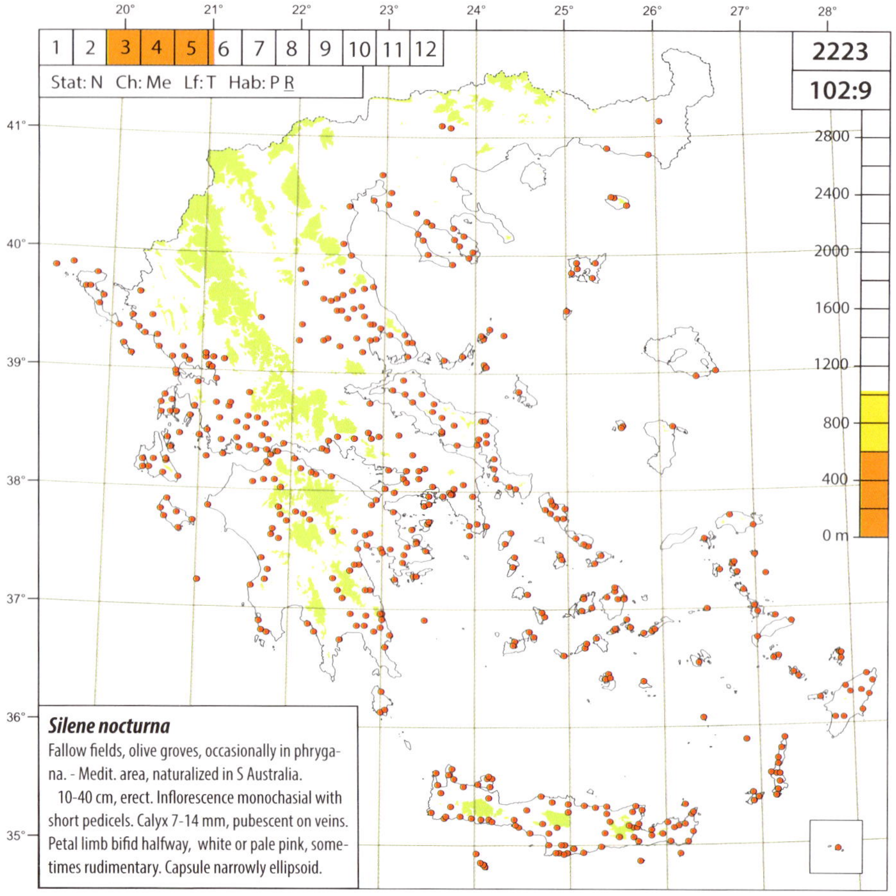

Silene nocturna

Fallow fields, olive groves, occasionally in phrygana. - Medit. area, naturalized in S Australia.

10-40 cm, erect. Inflorescence monochasial with short pedicels. Calyx 7-14 mm, pubescent on veins. Petal limb bifid halfway, white or pale pink, sometimes rudimentary. Capsule narrowly ellipsoid.

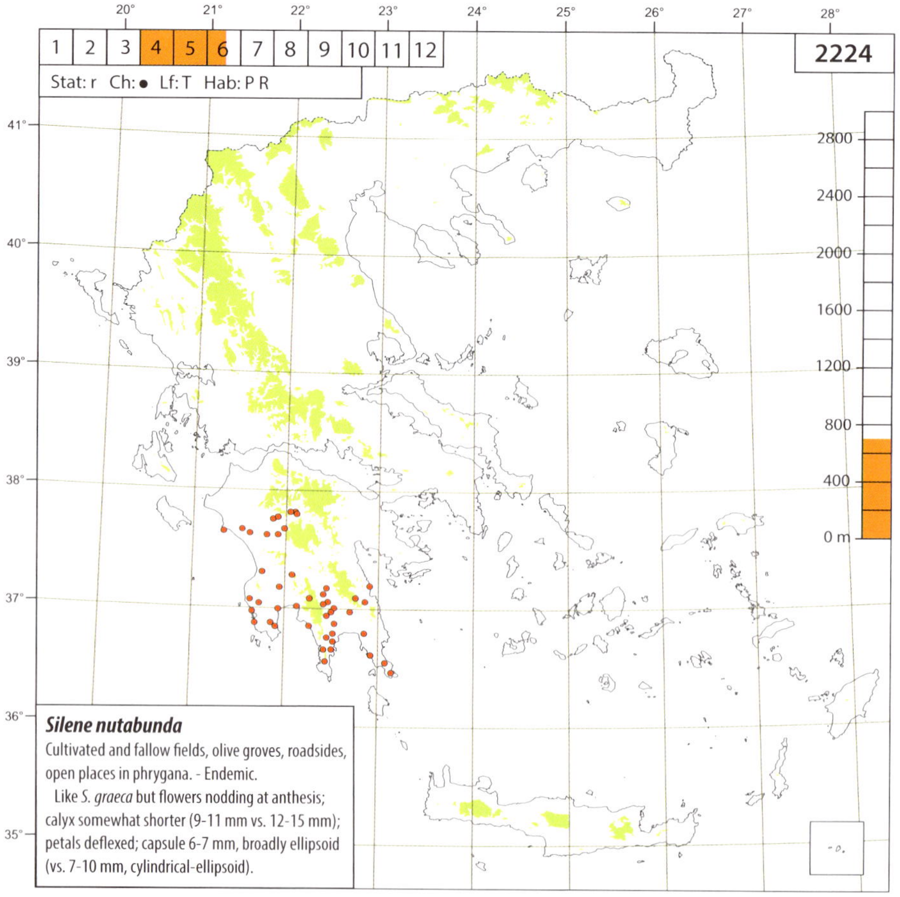

Silene nutabunda

Cultivated and fallow fields, olive groves, roadsides, open places in phrygana. - Endemic.

Like *S. graeca* but flowers nodding at anthesis; calyx somewhat shorter (9-11 mm vs. 12-15 mm); petals deflexed; capsule 6-7 mm, broadly ellipsoid (vs. 7-10 mm, cylindrical-ellipsoid).

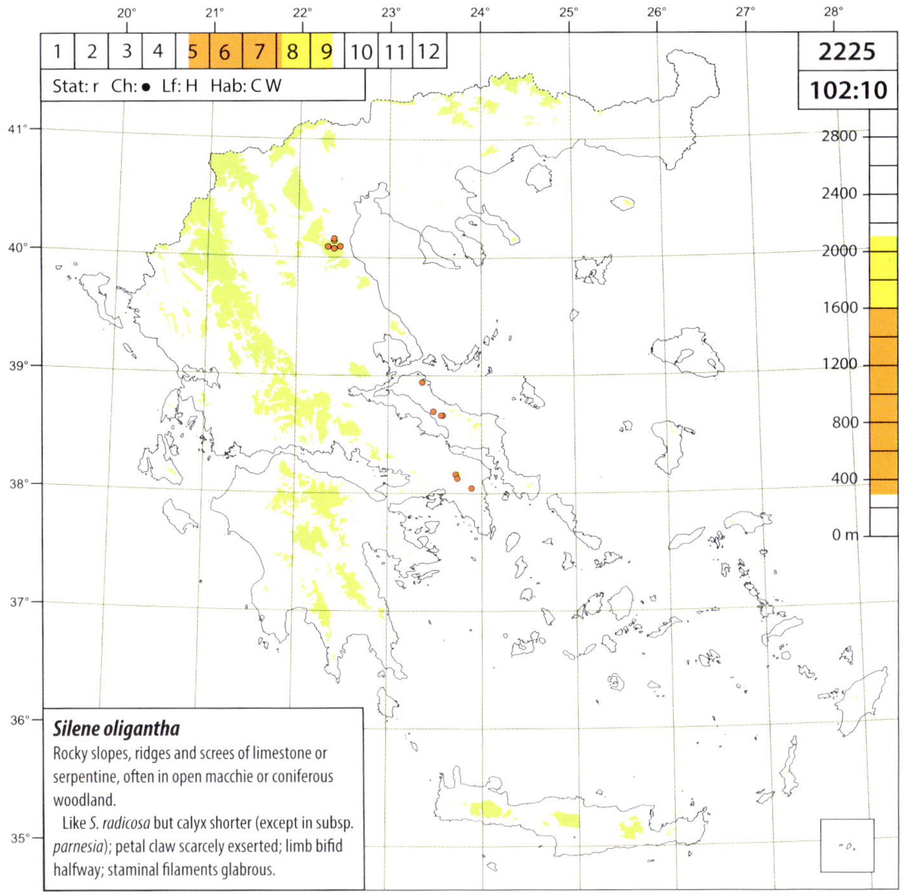

Silene oligantha

Rocky slopes, ridges and screes of limestone or serpentine, often in open macchie or coniferous woodland.

Like *S. radicosa* but calyx shorter (except in subsp. *parnesia*); petal claw scarcely exserted; limb bifid halfway; staminal filaments glabrous.

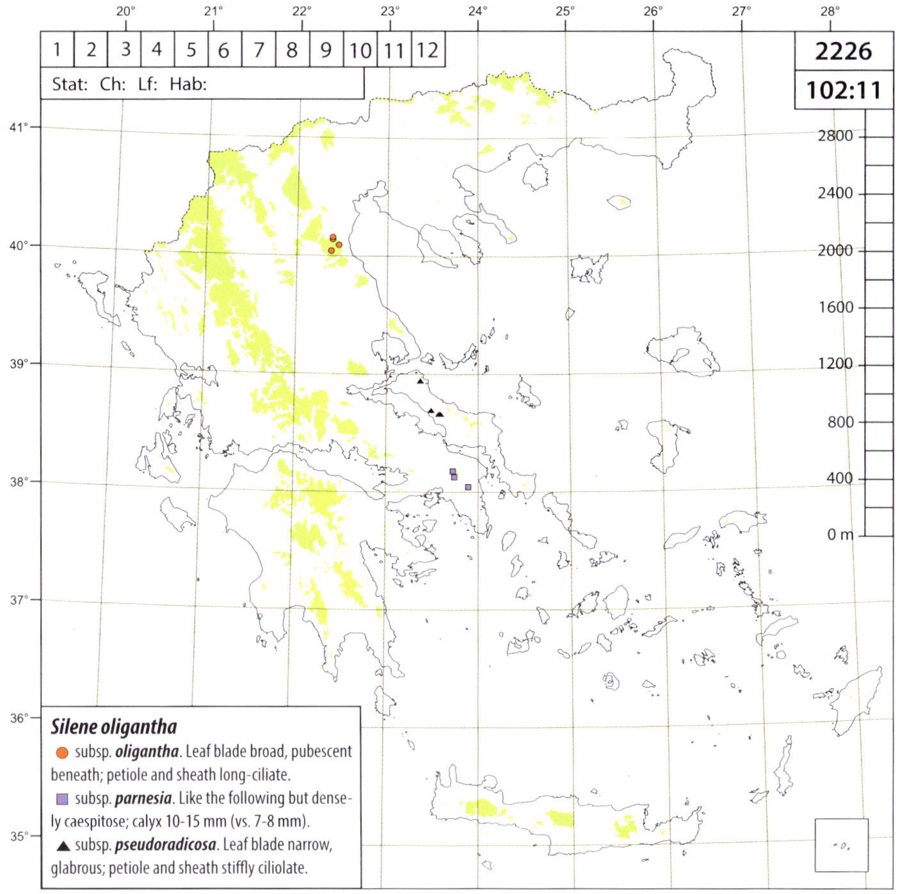

Silene oligantha

● subsp. ***oligantha***. Leaf blade broad, pubescent beneath; petiole and sheath long-ciliate.

■ subsp. ***parnesia***. Like the following but densely caespitose; calyx 10-15 mm (vs. 7-8 mm).

▲ subsp. ***pseudoradicosa***. Leaf blade narrow, glabrous; petiole and sheath stiffly ciliolate.

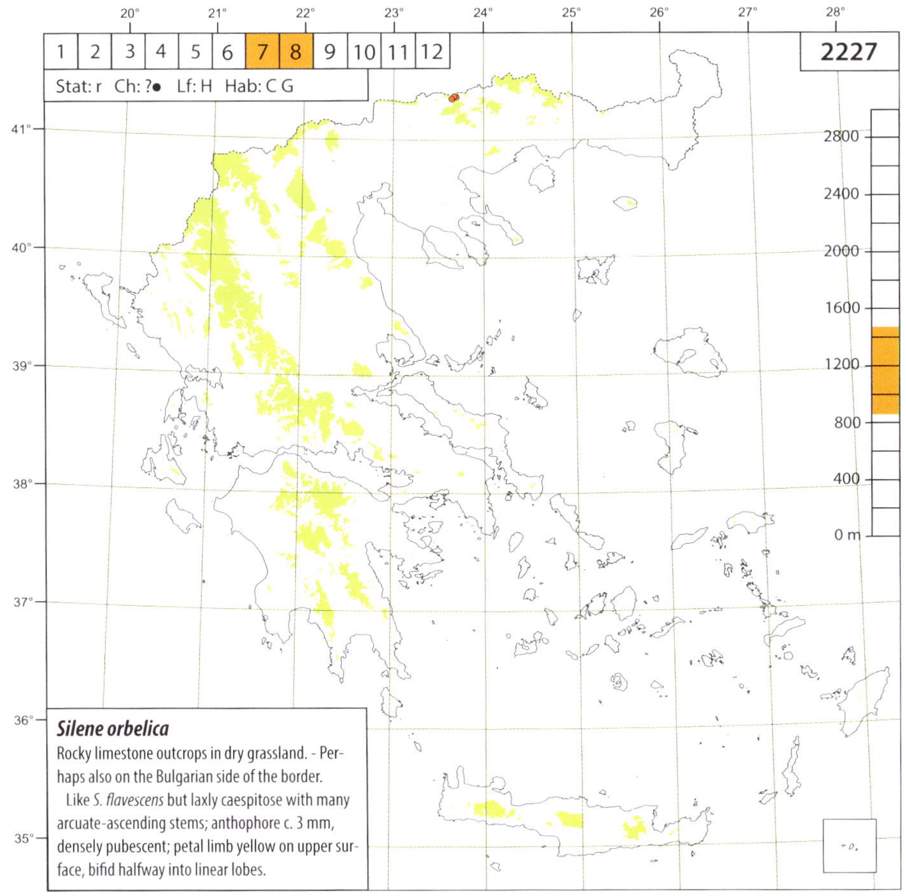

Silene orbelica

Rocky limestone outcrops in dry grassland. - Perhaps also on the Bulgarian side of the border.

Like *S. flavescens* but laxly caespitose with many arcuate-ascending stems; anthophore c. 3 mm, densely pubescent; petal limb yellow on upper surface, bifid halfway into linear lobes.

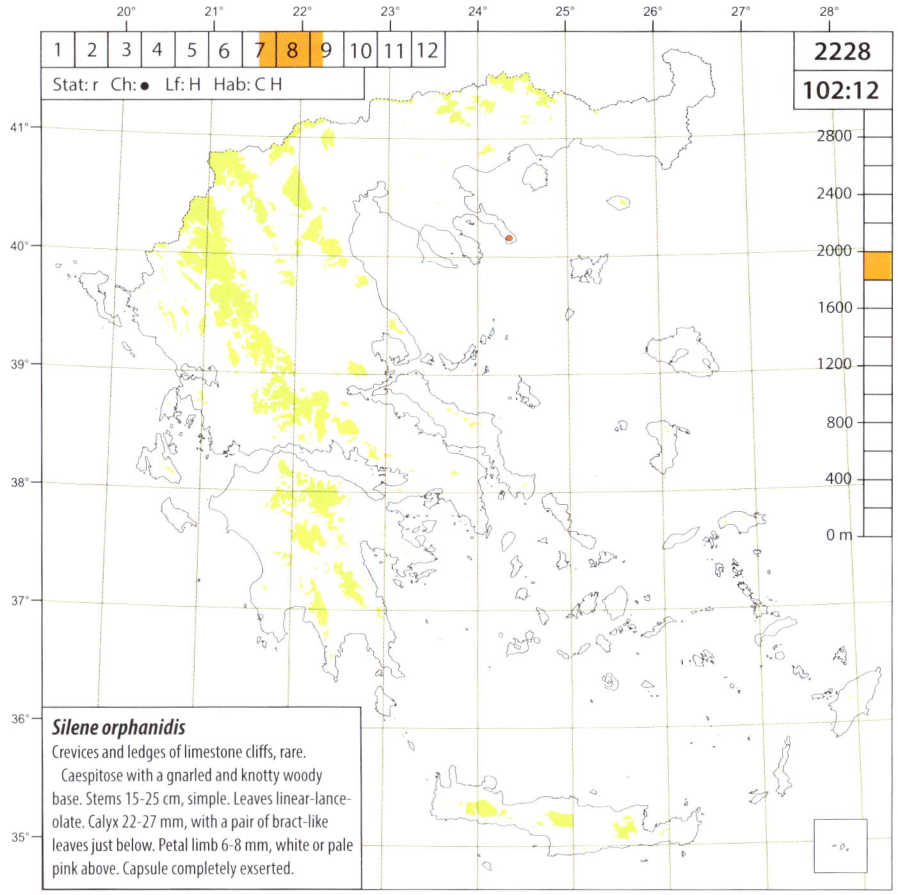

Silene orphanidis

Crevices and ledges of limestone cliffs, rare.

Caespitose with a gnarled and knotty woody base. Stems 15-25 cm, simple. Leaves linear-lanceolate. Calyx 22-27 mm, with a pair of bract-like leaves just below. Petal limb 6-8 mm, white or pale pink above. Capsule completely exserted.

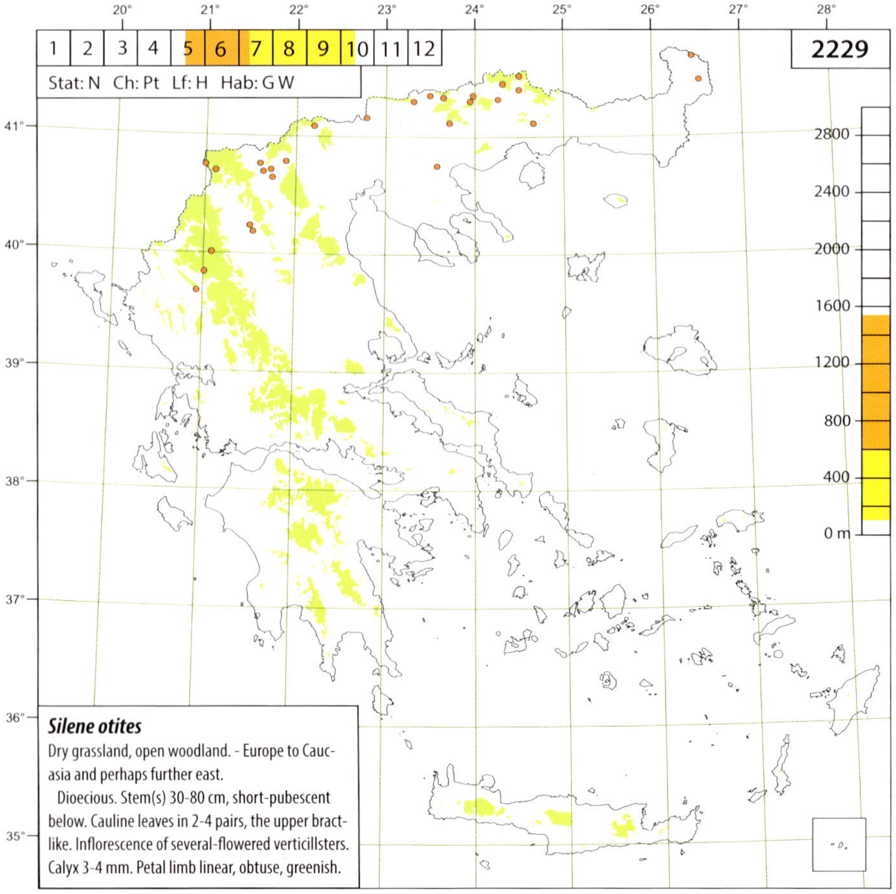

Silene otites

Dry grassland, open woodland. - Europe to Caucasia and perhaps further east.

Dioecious. Stem(s) 30-80 cm, short-pubescent below. Cauline leaves in 2-4 pairs, the upper bract-like. Inflorescence of several-flowered verticillsters. Calyx 3-4 mm. Petal limb linear, obtuse, greenish.

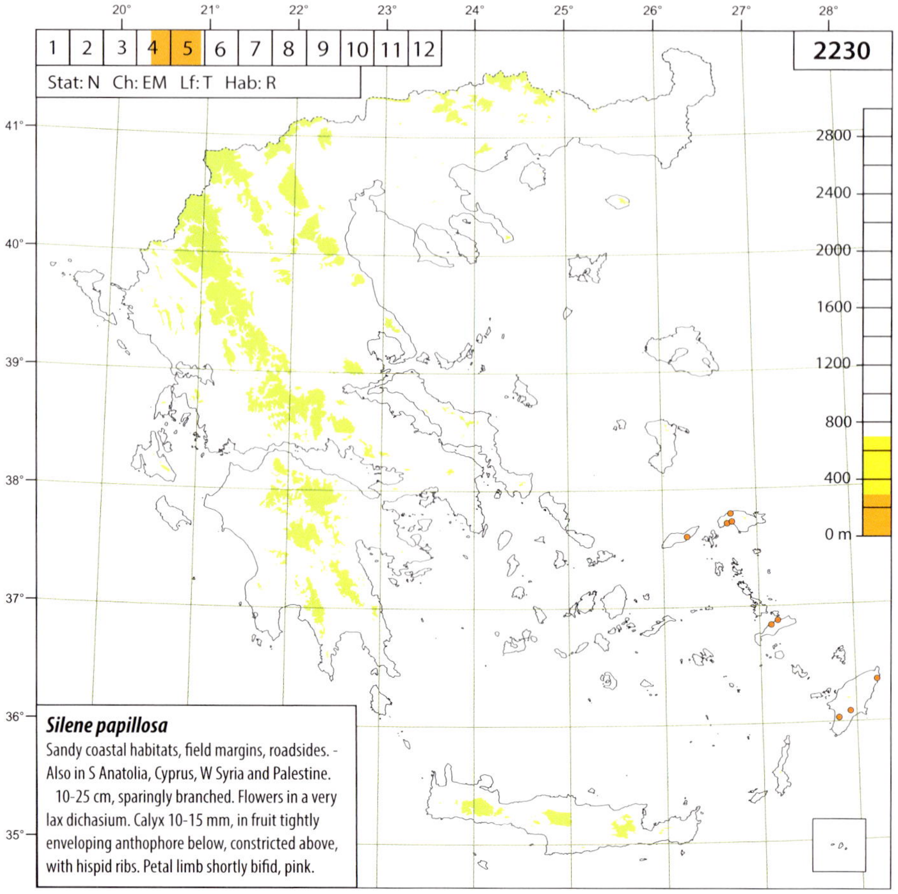

Silene papillosa

Sandy coastal habitats, field margins, roadsides. - Also in S Anatolia, Cyprus, W Syria and Palestine.

10-25 cm, sparingly branched. Flowers in a very lax dichasium. Calyx 10-15 mm, in fruit tightly enveloping anthophore below, constricted above, with hispid ribs. Petal limb shortly bifid, pink.

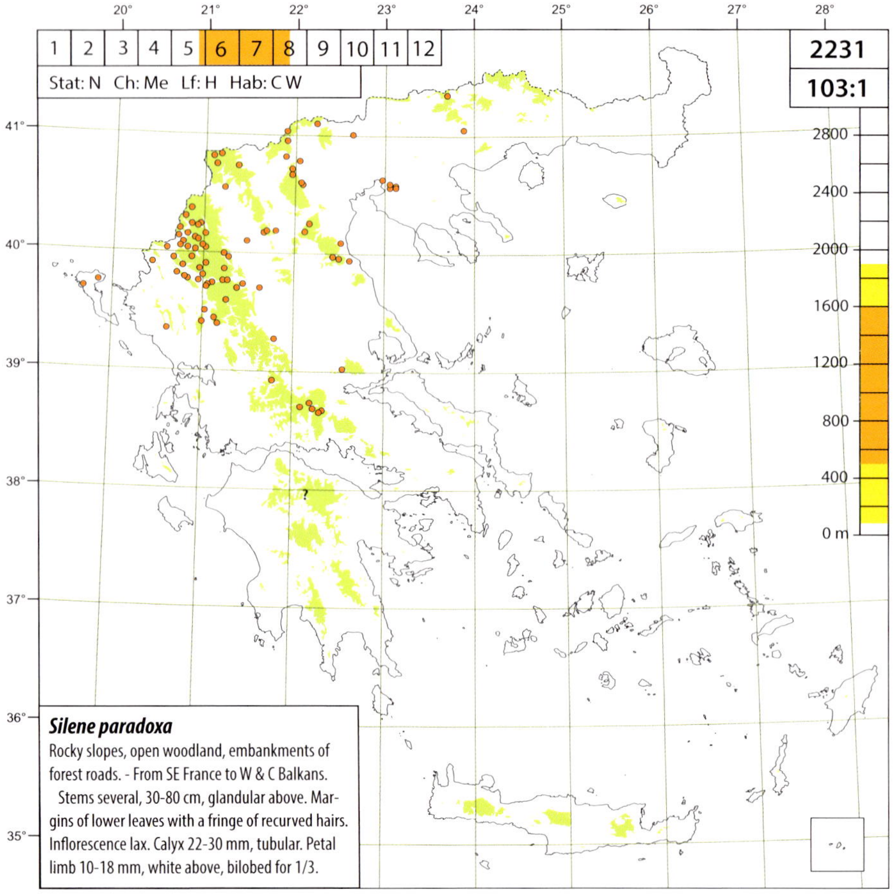

Silene paradoxa

Rocky slopes, open woodland, embankments of forest roads. - From SE France to W & C Balkans.

Stems several, 30-80 cm, glandular above. Margins of lower leaves with a fringe of recurved hairs. Inflorescence lax. Calyx 22-30 mm, tubular. Petal limb 10-18 mm, white above, bilobed for 1/3.

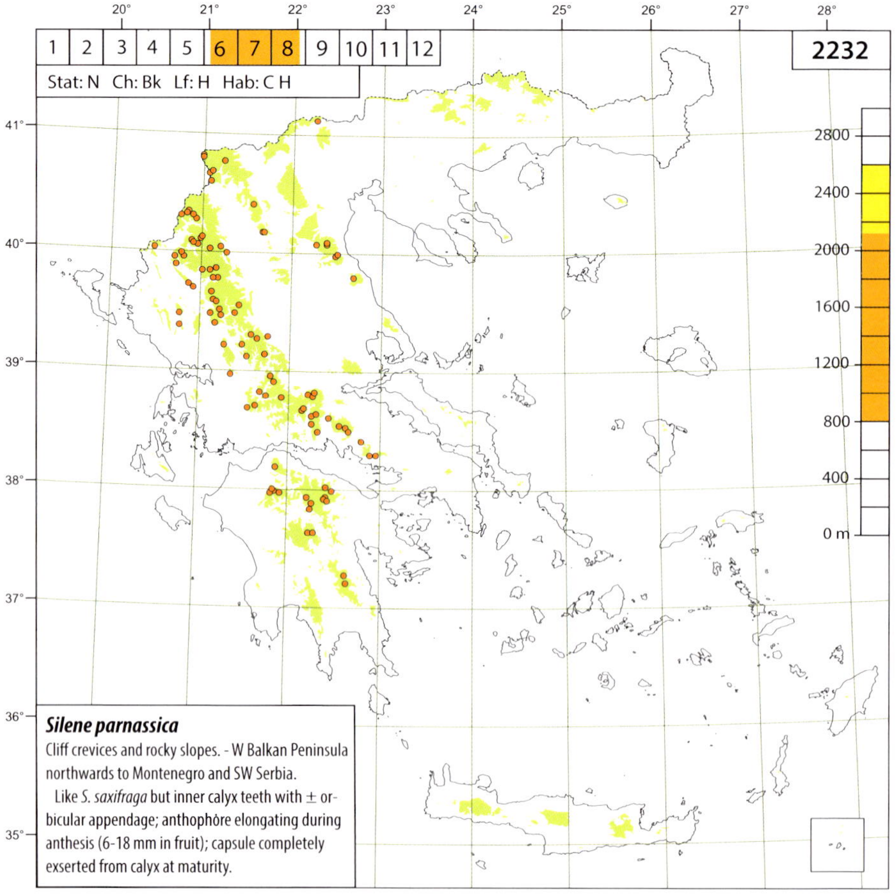

Silene parnassica

Cliff crevices and rocky slopes. - W Balkan Peninsula northwards to Montenegro and SW Serbia.

Like *S. saxifraga* but inner calyx teeth with ± orbicular appendage; anthophore elongating during anthesis (6-18 mm in fruit); capsule completely exserted from calyx at maturity.

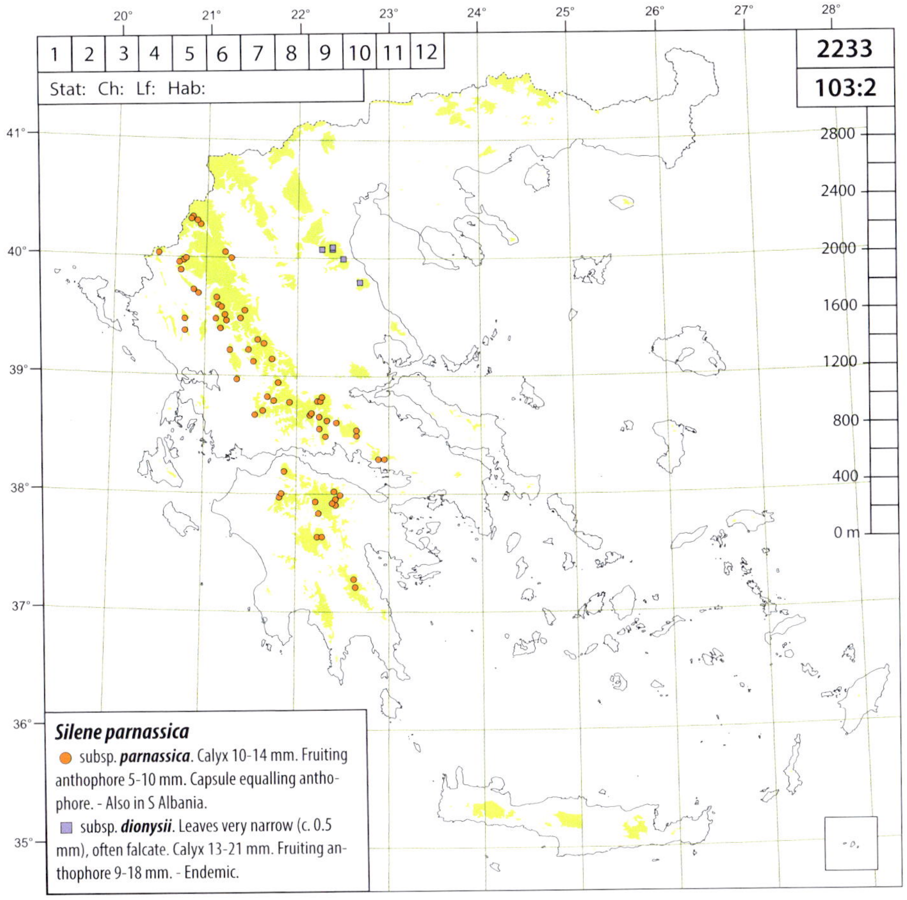

Silene parnassica

● subsp. ***parnassica***. Calyx 10-14 mm. Fruiting anthophore 5-10 mm. Capsule equalling anthophore. - Also in S Albania.

■ subsp. ***dionysii***. Leaves very narrow (c. 0.5 mm), often falcate. Calyx 13-21 mm. Fruiting anthophore 9-18 mm. - Endemic.

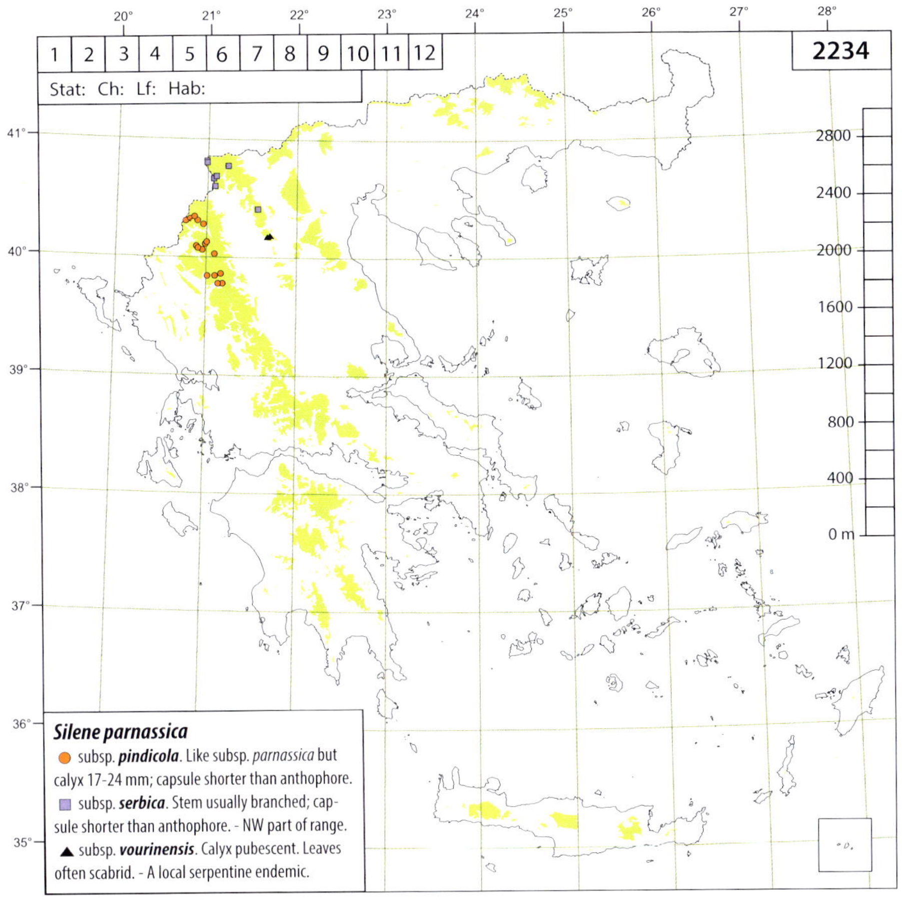

Silene parnassica

● subsp. ***pindicola***. Like subsp. *parnassica* but calyx 17-24 mm; capsule shorter than anthophore.

■ subsp. ***serbica***. Stem usually branched; capsule shorter than anthophore. - NW part of range.

▲ subsp. ***vourinensis***. Calyx pubescent. Leaves often scabrid. - A local serpentine endemic.

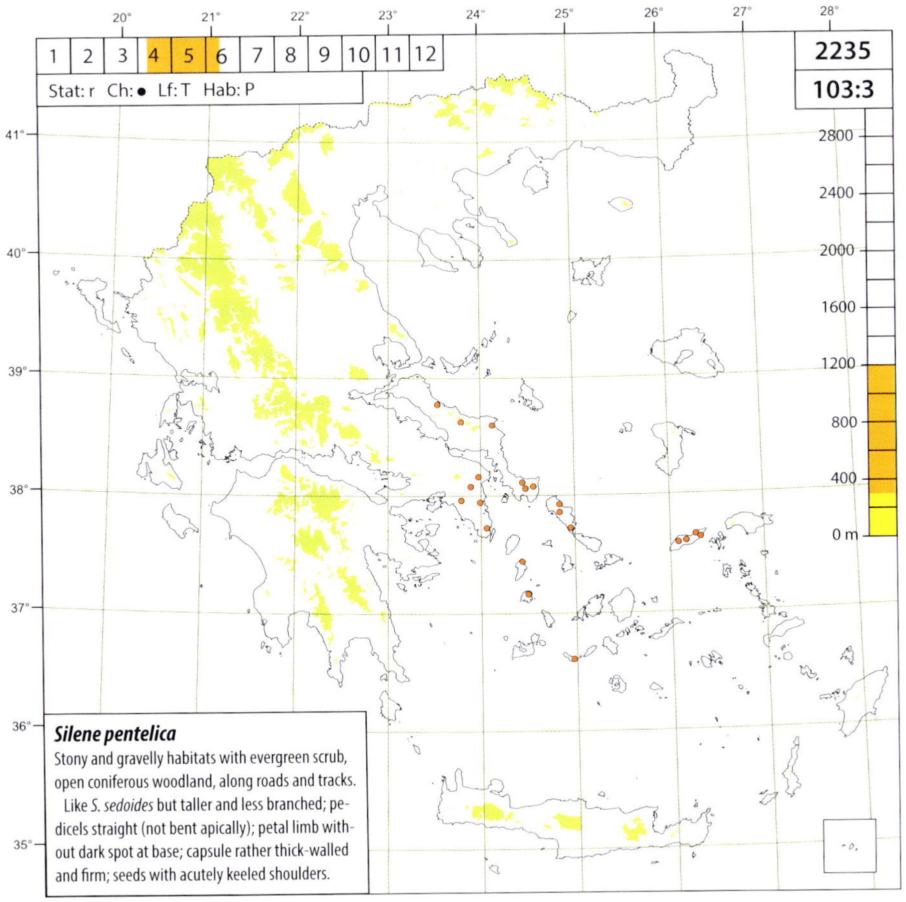

Silene pentelica

Stony and gravelly habitats with evergreen scrub, open coniferous woodland, along roads and tracks.

Like *S. sedoides* but taller and less branched; pedicels straight (not bent apically); petal limb without dark spot at base; capsule rather thick-walled and firm; seeds with acutely keeled shoulders.

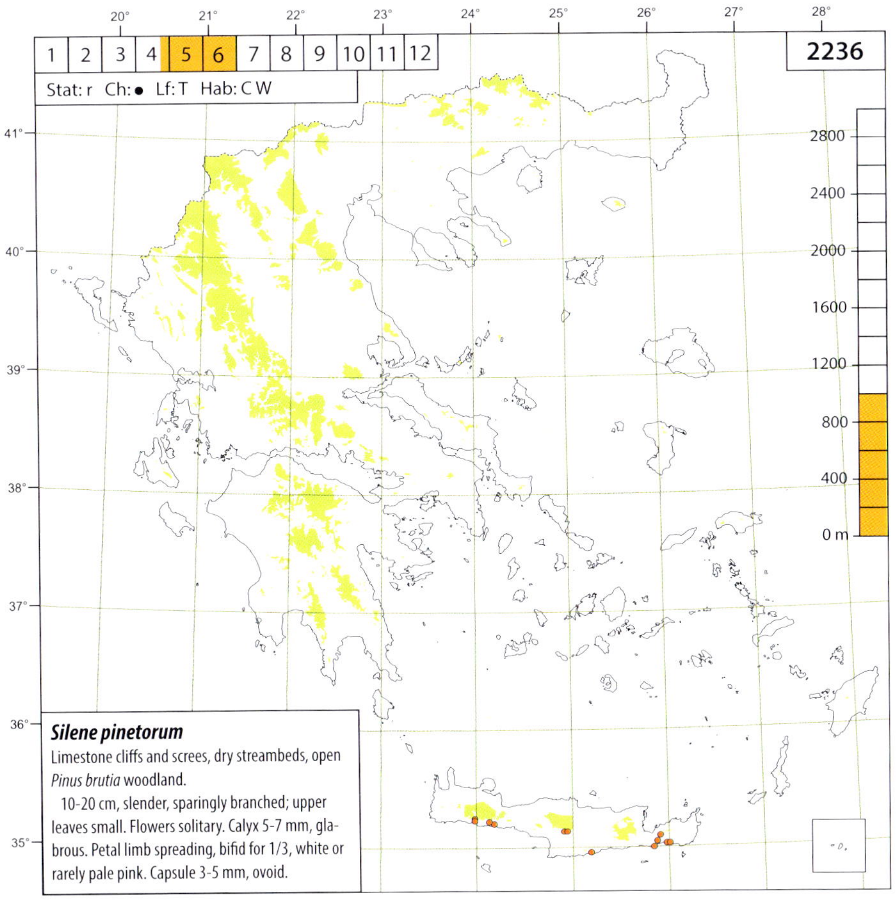

Silene pinetorum

Limestone cliffs and screes, dry streambeds, open *Pinus brutia* woodland.

10-20 cm, slender, sparingly branched; upper leaves small. Flowers solitary. Calyx 5-7 mm, glabrous. Petal limb spreading, bifid for 1/3, white or rarely pale pink. Capsule 3-5 mm, ovoid.

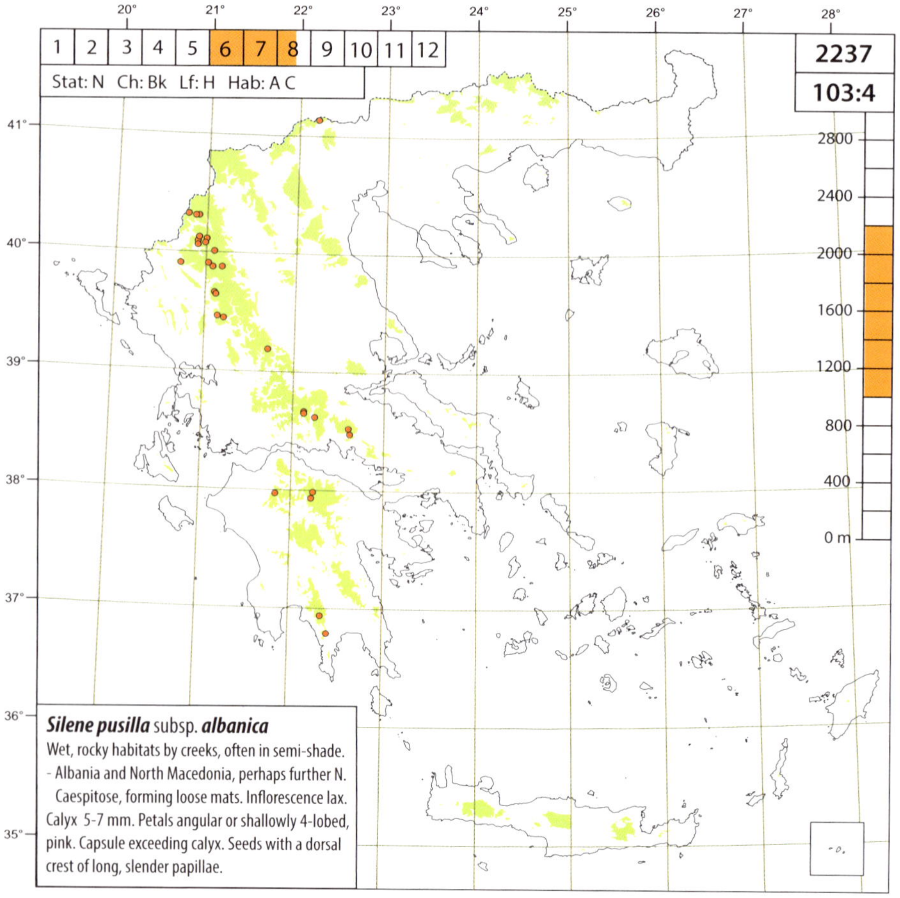

Silene pusilla subsp. ***albanica***
Wet, rocky habitats by creeks, often in semi-shade. - Albania and North Macedonia, perhaps further N.
Caespitose, forming loose mats. Inflorescence lax. Calyx 5-7 mm. Petals angular or shallowly 4-lobed, pink. Capsule exceeding calyx. Seeds with a dorsal crest of long, slender papillae.

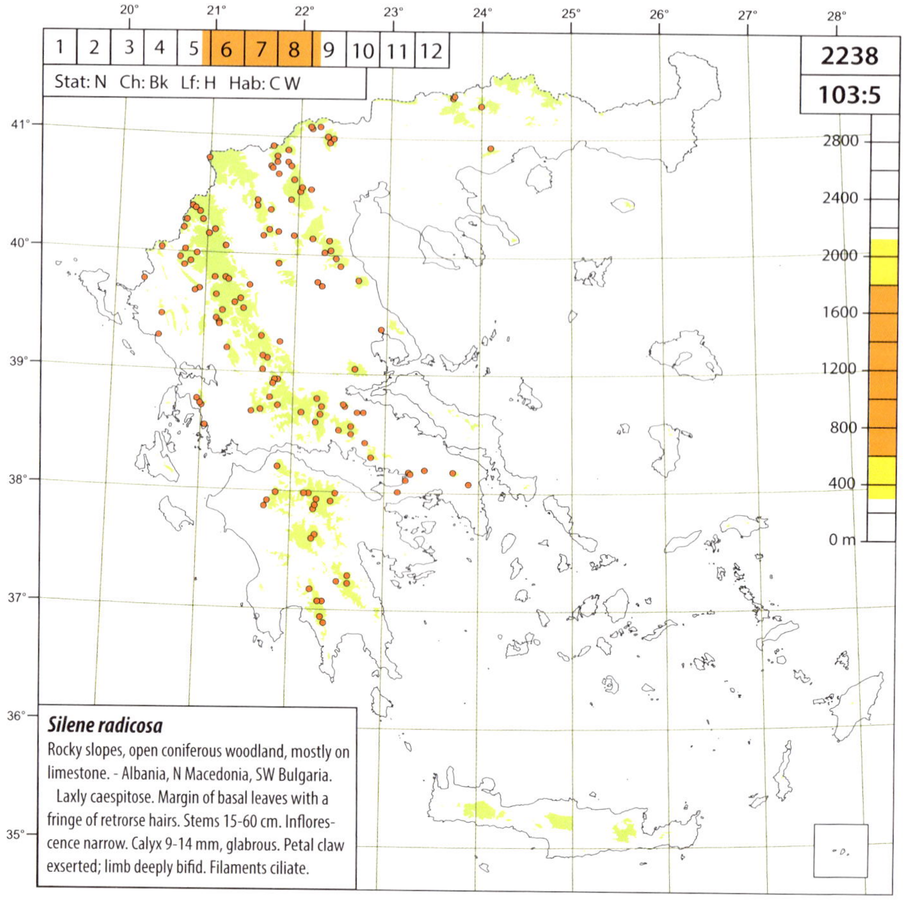

Silene radicosa
Rocky slopes, open coniferous woodland, mostly on limestone. - Albania, N Macedonia, SW Bulgaria.
Laxly caespitose. Margin of basal leaves with a fringe of retrorse hairs. Stems 15-60 cm. Inflorescence narrow. Calyx 9-14 mm, glabrous. Petal claw exserted; limb deeply bifid. Filaments ciliate.

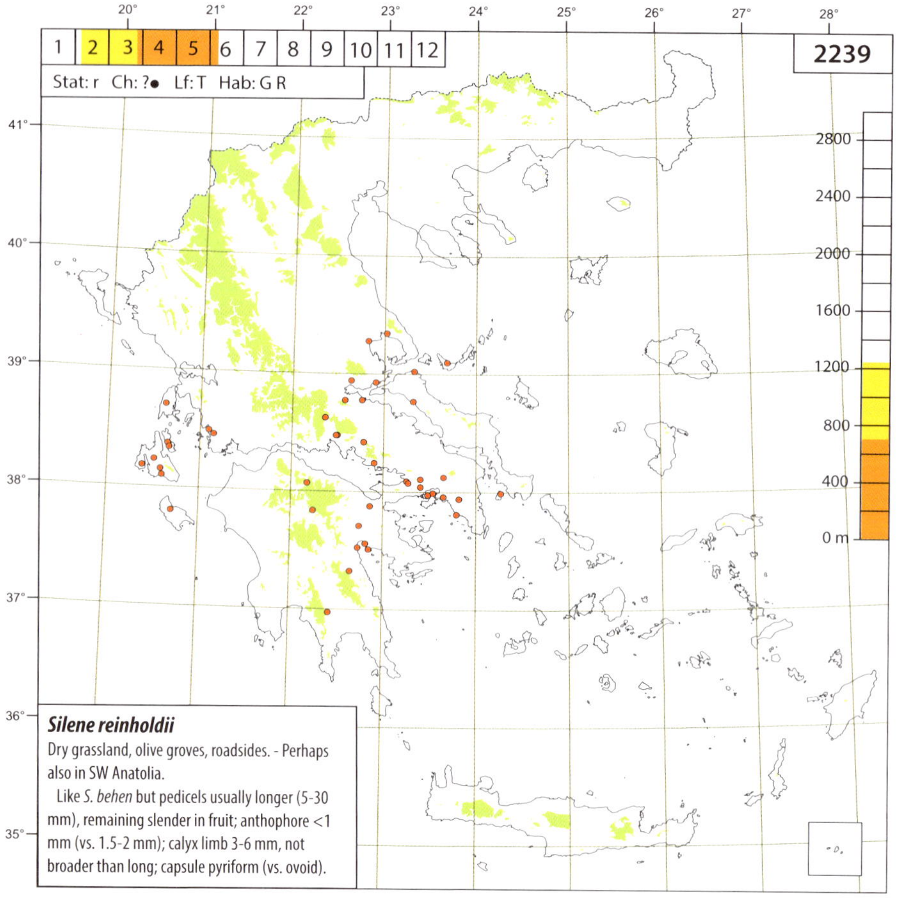

Silene reinholdii
Dry grassland, olive groves, roadsides. - Perhaps also in SW Anatolia.
Like *S. behen* but pedicels usually longer (5-30 mm), remaining slender in fruit; anthophore <1 mm (vs. 1.5-2 mm); calyx limb 3-6 mm, not broader than long; capsule pyriform (vs. ovoid).

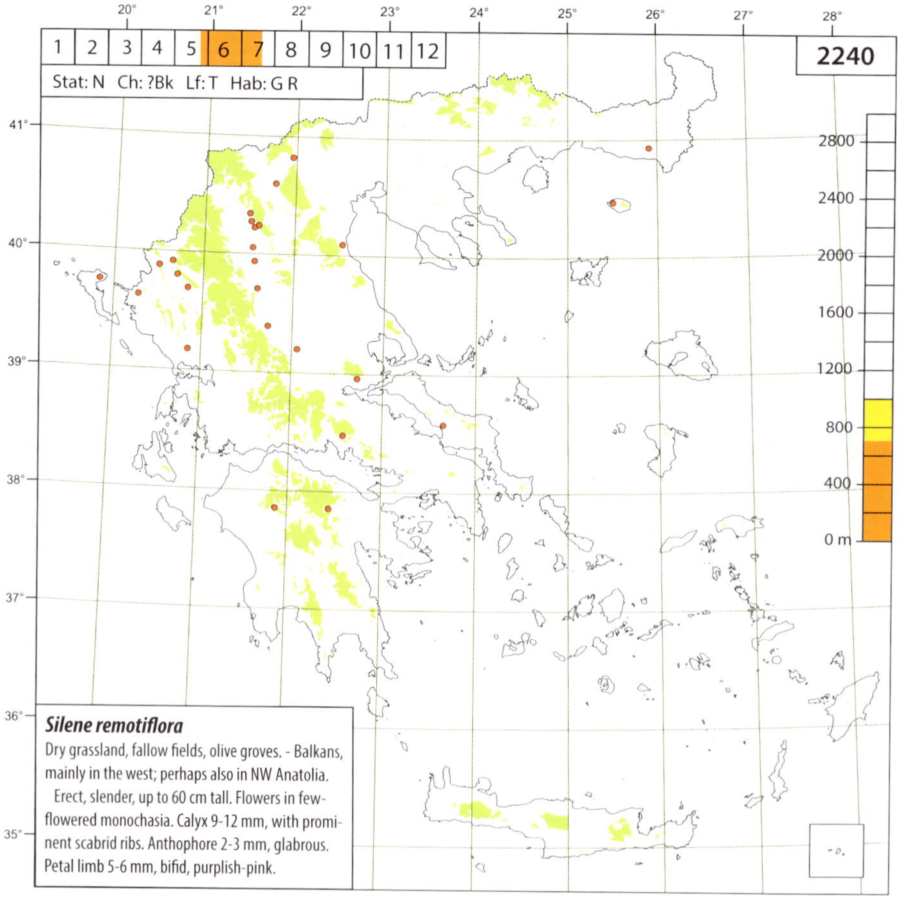

Silene remotiflora
Dry grassland, fallow fields, olive groves. - Balkans, mainly in the west; perhaps also in NW Anatolia.
Erect, slender, up to 60 cm tall. Flowers in few-flowered monochasia. Calyx 9-12 mm, with prominent scabrid ribs. Anthophore 2-3 mm, glabrous. Petal limb 5-6 mm, bifid, purplish-pink.

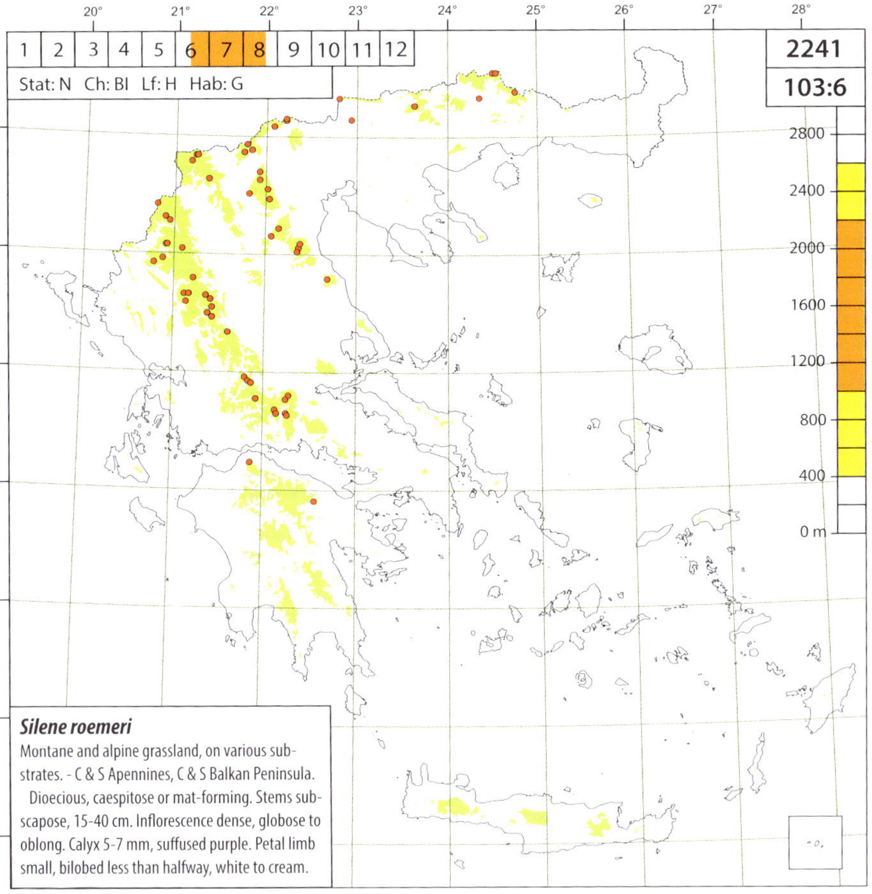

Silene roemeri

Montane and alpine grassland, on various substrates. - C & S Apennines, C & S Balkan Peninsula.

Dioecious, caespitose or mat-forming. Stems subscapose, 15-40 cm. Inflorescence dense, globose to oblong. Calyx 5-7 mm, suffused purple. Petal limb small, bilobed less than halfway, white to cream.

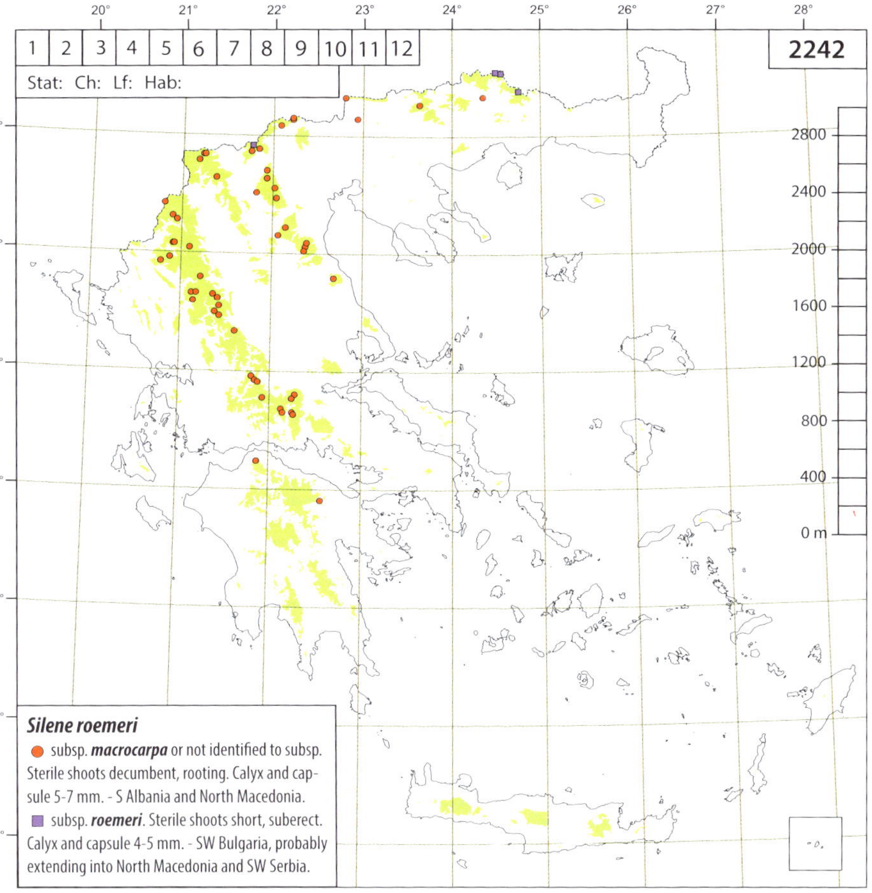

Silene roemeri

● subsp. ***macrocarpa*** or not identified to subsp. Sterile shoots decumbent, rooting. Calyx and capsule 5-7 mm. - S Albania and North Macedonia.

■ subsp. ***roemeri***. Sterile shoots short, suberect. Calyx and capsule 4-5 mm. - SW Bulgaria, probably extending into North Macedonia and SW Serbia.

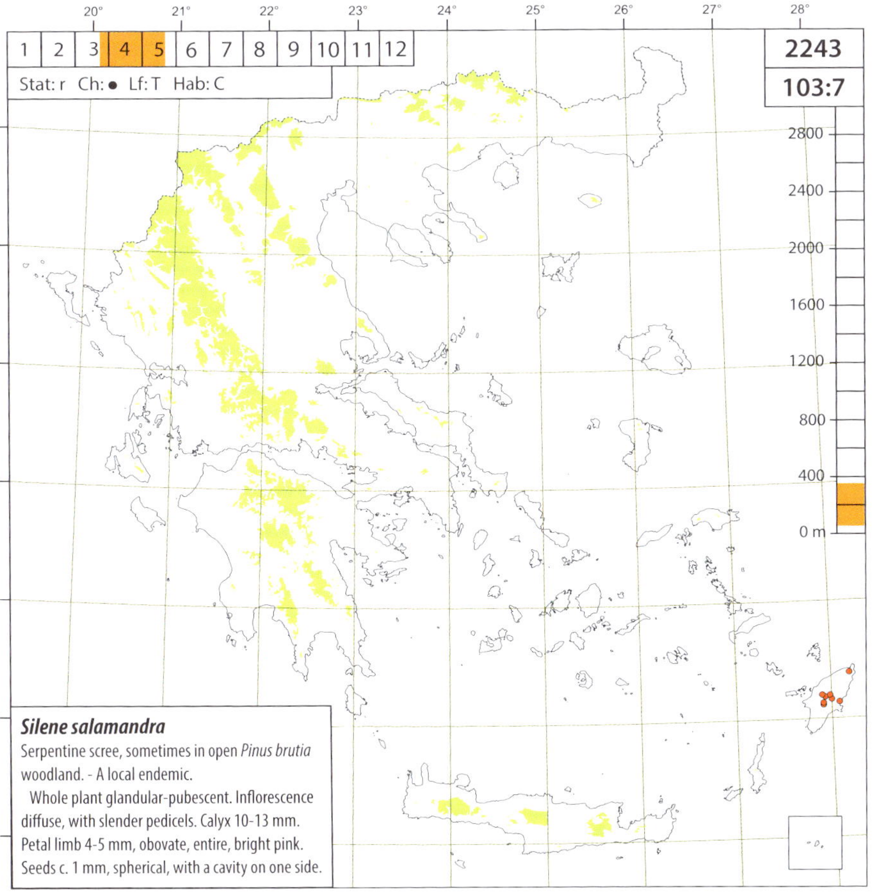

Silene salamandra

Serpentine scree, sometimes in open *Pinus brutia* woodland. - A local endemic.

Whole plant glandular-pubescent. Inflorescence diffuse, with slender pedicels. Calyx 10-13 mm. Petal limb 4-5 mm, obovate, entire, bright pink. Seeds c. 1 mm, spherical, with a cavity on one side.

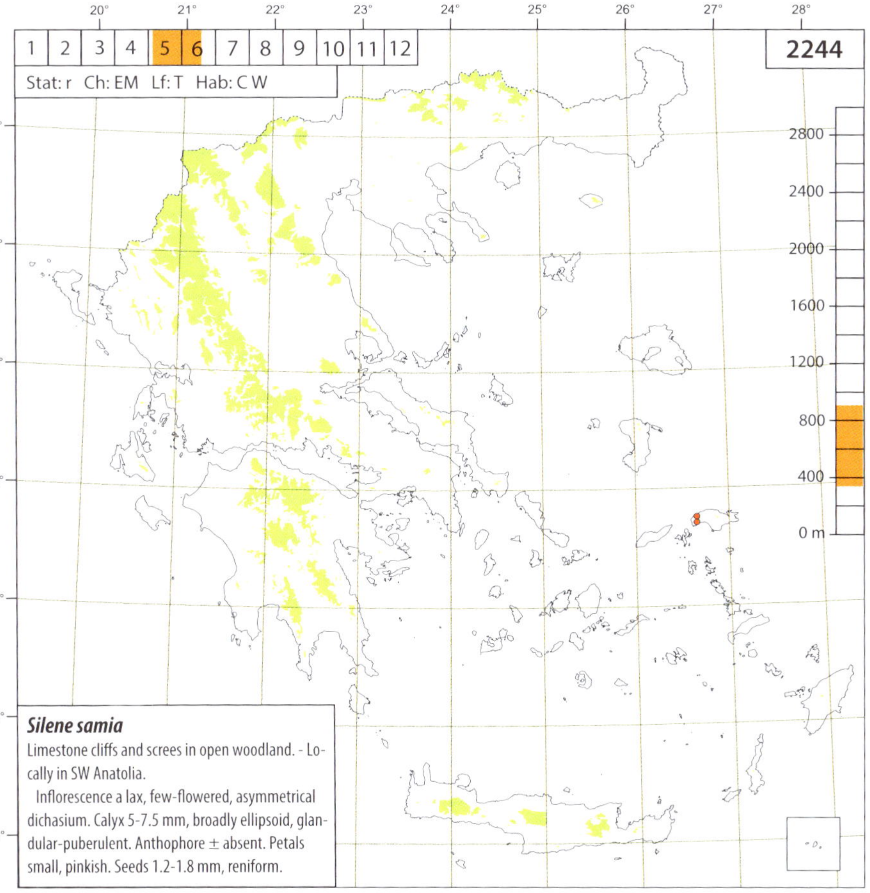

Silene samia

Limestone cliffs and screes in open woodland. - Locally in SW Anatolia.

Inflorescence a lax, few-flowered, asymmetrical dichasium. Calyx 5-7.5 mm, broadly ellipsoid, glandular-puberulent. Anthophore ± absent. Petals small, pinkish. Seeds 1.2-1.8 mm, reniform.

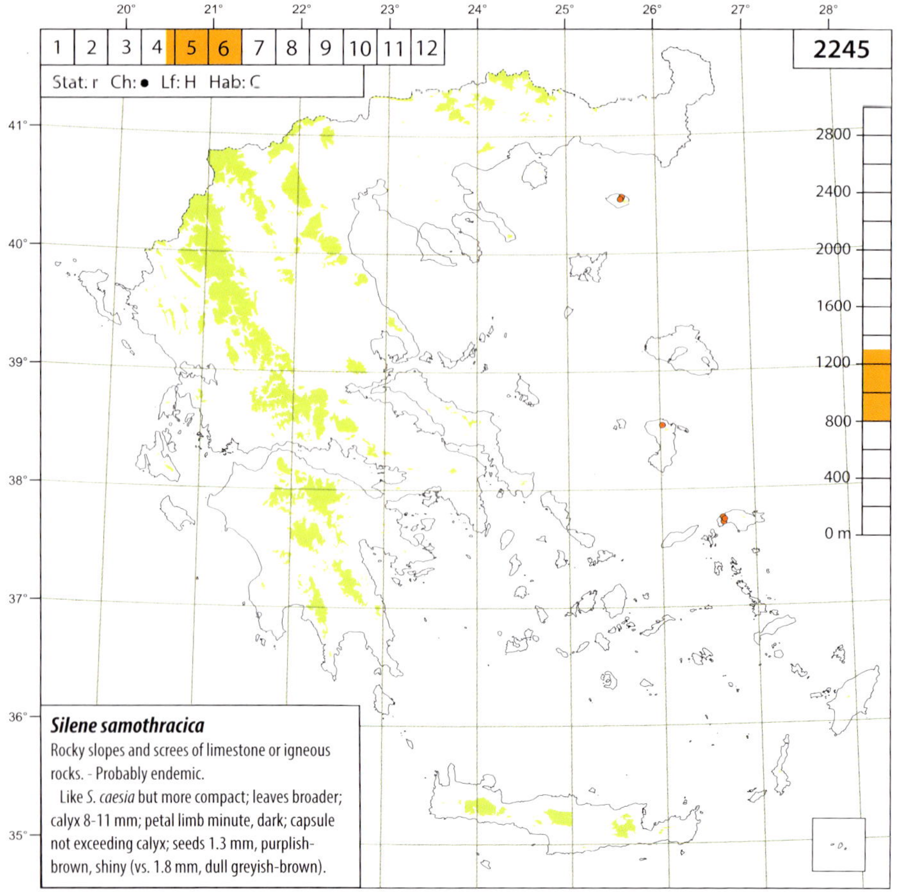

Silene samothracica
Rocky slopes and screes of limestone or igneous rocks. - Probably endemic.
Like *S. caesia* but more compact; leaves broader; calyx 8-11 mm; petal limb minute, dark; capsule not exceeding calyx; seeds 1.3 mm, purplish-brown, shiny (vs. 1.8 mm, dull greyish-brown).

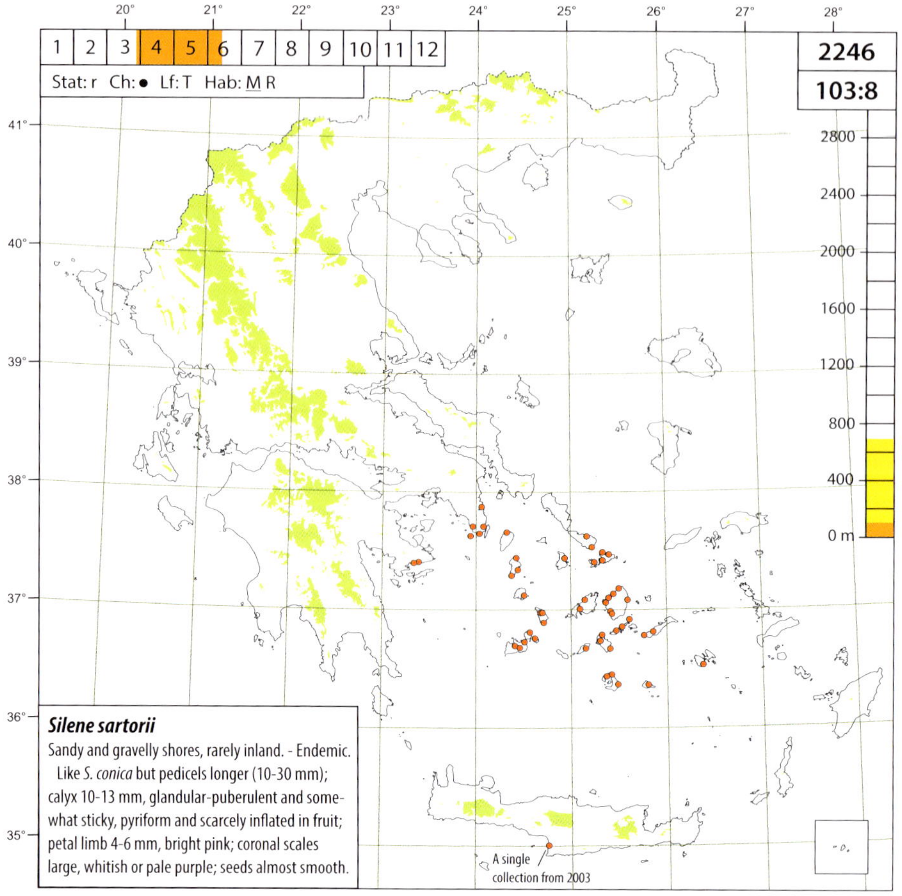

Silene sartorii
Sandy and gravelly shores, rarely inland. - Endemic.
Like *S. conica* but pedicels longer (10-30 mm); calyx 10-13 mm, glandular-puberulent and somewhat sticky, pyriform and scarcely inflated in fruit; petal limb 4-6 mm, bright pink; coronal scales large, whitish or pale purple; seeds almost smooth.

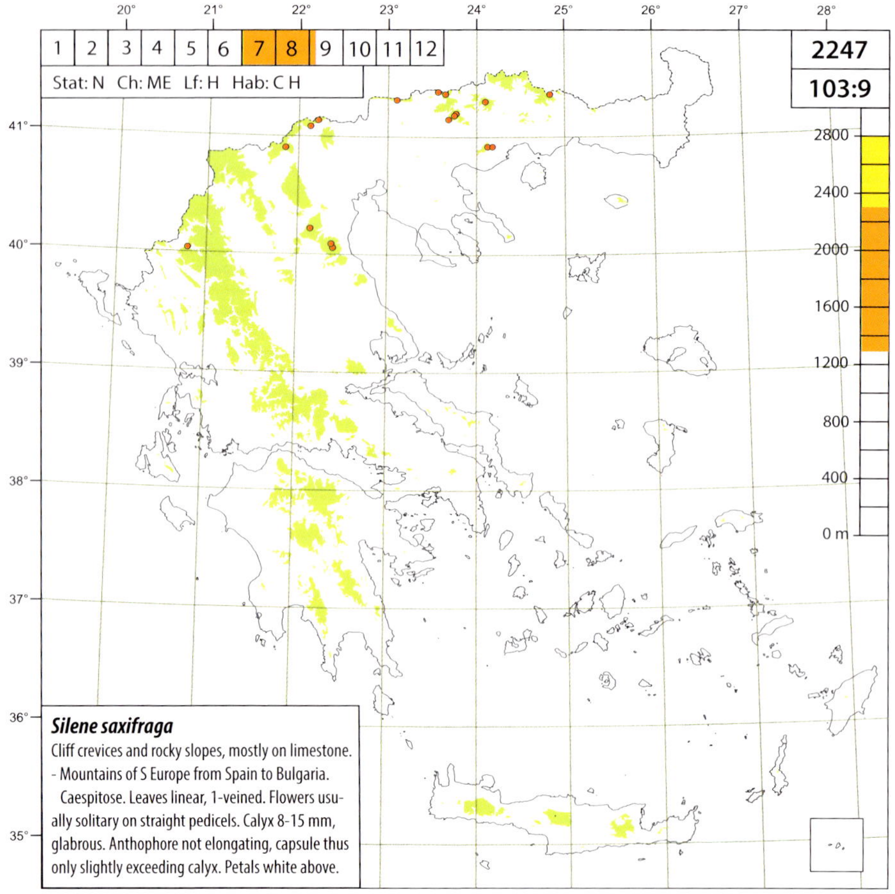

Silene saxifraga
Cliff crevices and rocky slopes, mostly on limestone. - Mountains of S Europe from Spain to Bulgaria.
Caespitose. Leaves linear, 1-veined. Flowers usually solitary on straight pedicels. Calyx 8-15 mm, glabrous. Anthophore not elongating, capsule thus only slightly exceeding calyx. Petals white above.

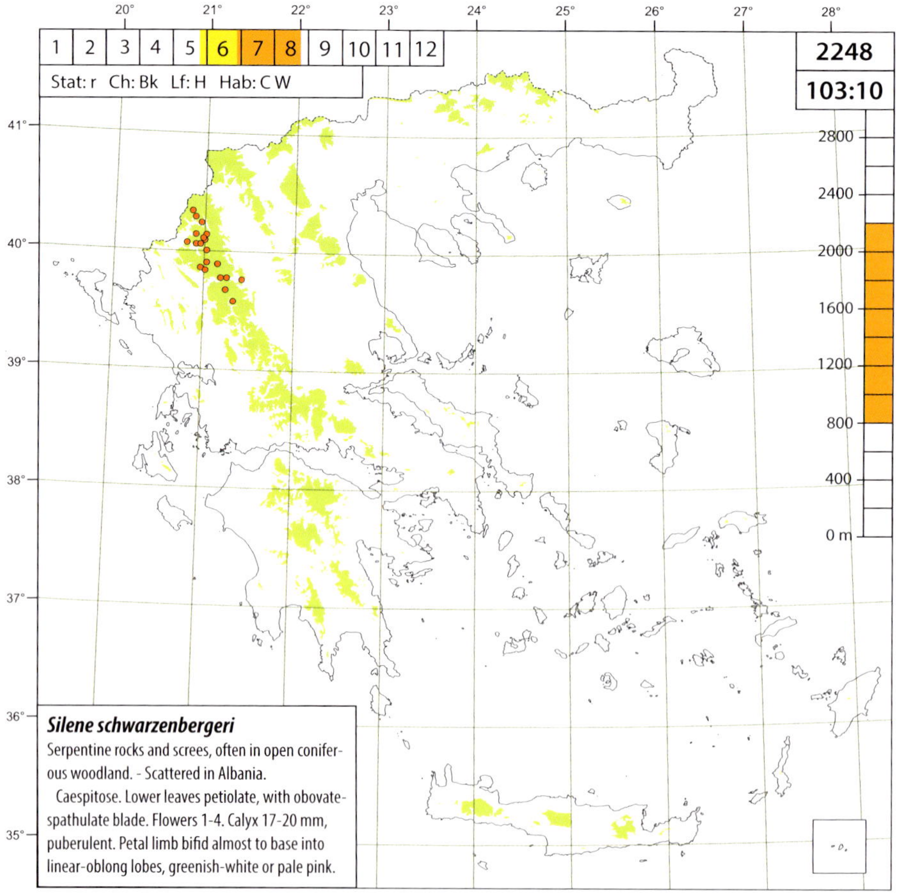

Silene schwarzenbergeri
Serpentine rocks and screes, often in open coniferous woodland. - Scattered in Albania.
Caespitose. Lower leaves petiolate, with obovate-spathulate blade. Flowers 1-4. Calyx 17-20 mm, puberulent. Petal limb bifid almost to base into linear-oblong lobes, greenish-white or pale pink.

2249 103:11

Stat: N Ch: SS Lf: T Hab: M P

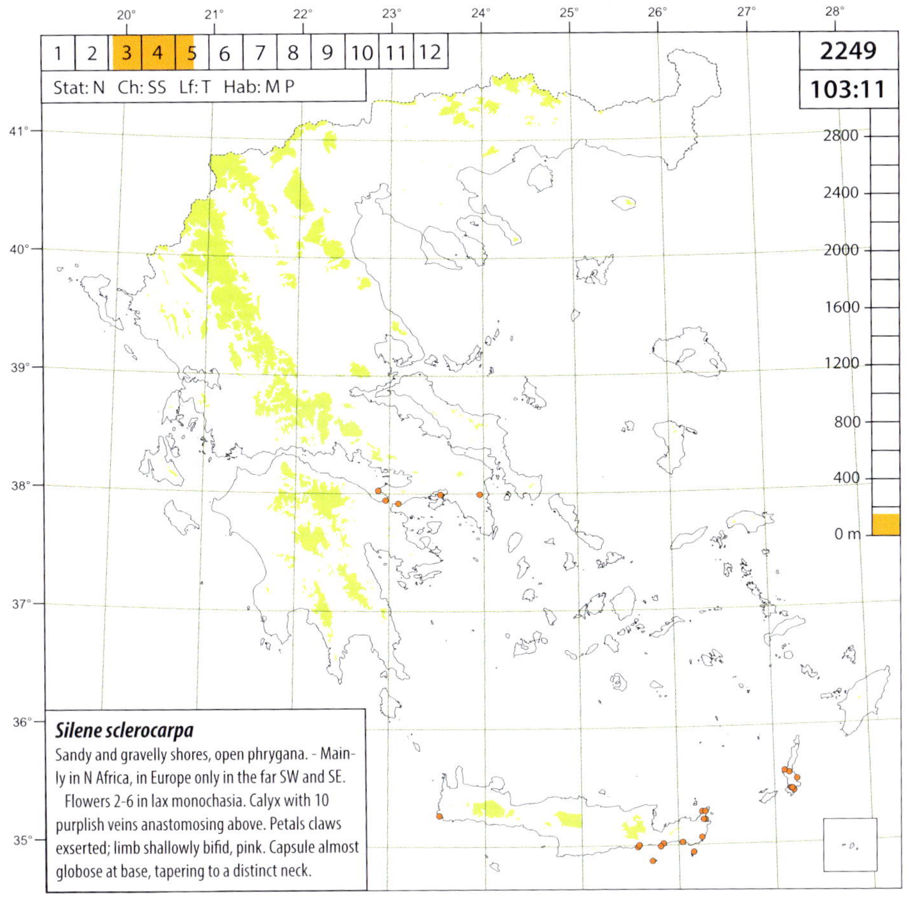

Silene sclerocarpa

Sandy and gravelly shores, open phrygana. - Mainly in N Africa, in Europe only in the far SW and SE.

Flowers 2-6 in lax monochasia. Calyx with 10 purplish veins anastomosing above. Petals claws exserted; limb shallowly bifid, pink. Capsule almost globose at base, tapering to a distinct neck.

2250 103:12

Stat: N Ch: Me Lf: T Hab: M P

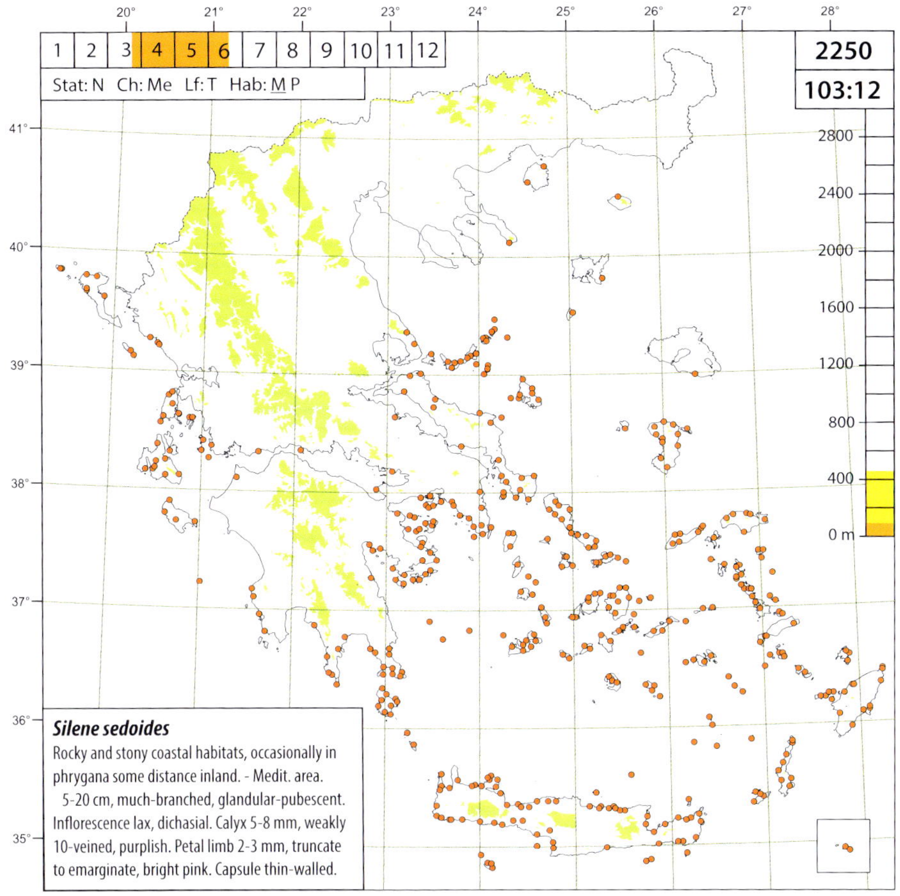

Silene sedoides

Rocky and stony coastal habitats, occasionally in phrygana some distance inland. - Medit. area.

5-20 cm, much-branched, glandular-pubescent. Inflorescence lax, dichasial. Calyx 5-8 mm, weakly 10-veined, purplish. Petal limb 2-3 mm, truncate to emarginate, bright pink. Capsule thin-walled.

2251

Stat: r Ch: Bk Lf: H Hab: G W

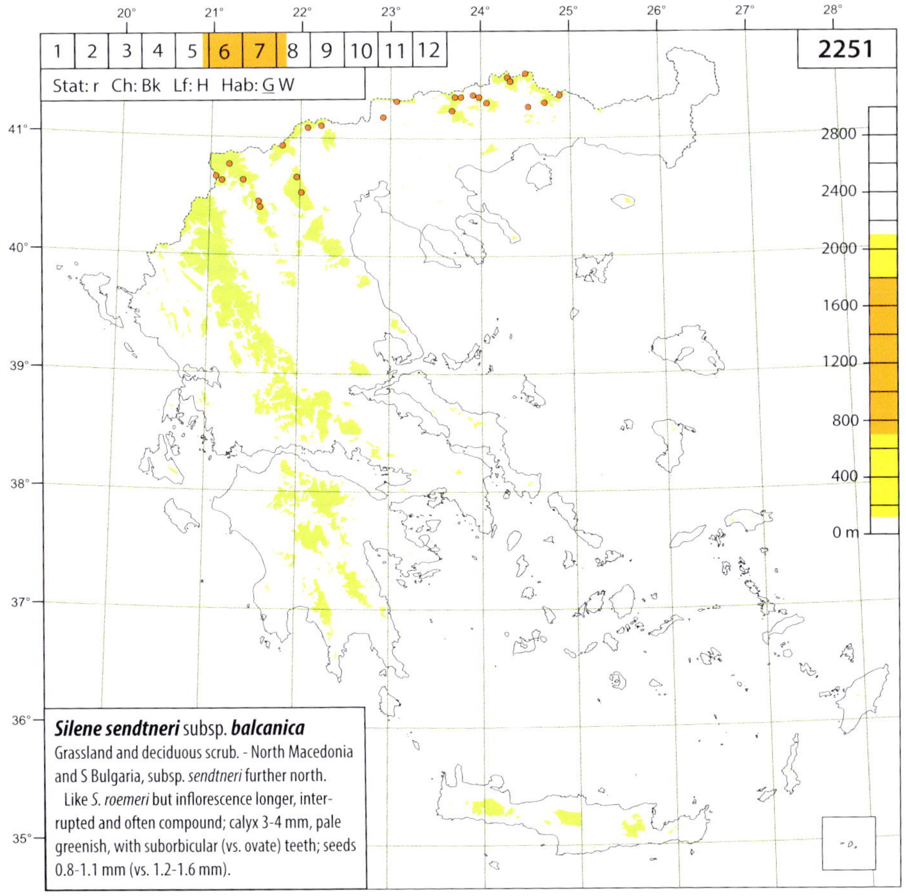

Silene sendtneri* subsp. *balcanica

Grassland and deciduous scrub. - North Macedonia and S Bulgaria, subsp. *sendtneri* further north.

Like *S. roemeri* but inflorescence longer, interrupted and often compound; calyx 3-4 mm, pale greenish, with suborbicular (vs. ovate) teeth; seeds 0.8-1.1 mm (vs. 1.2-1.6 mm).

2252

Stat: r Ch: ● Lf: H Hab: C P W

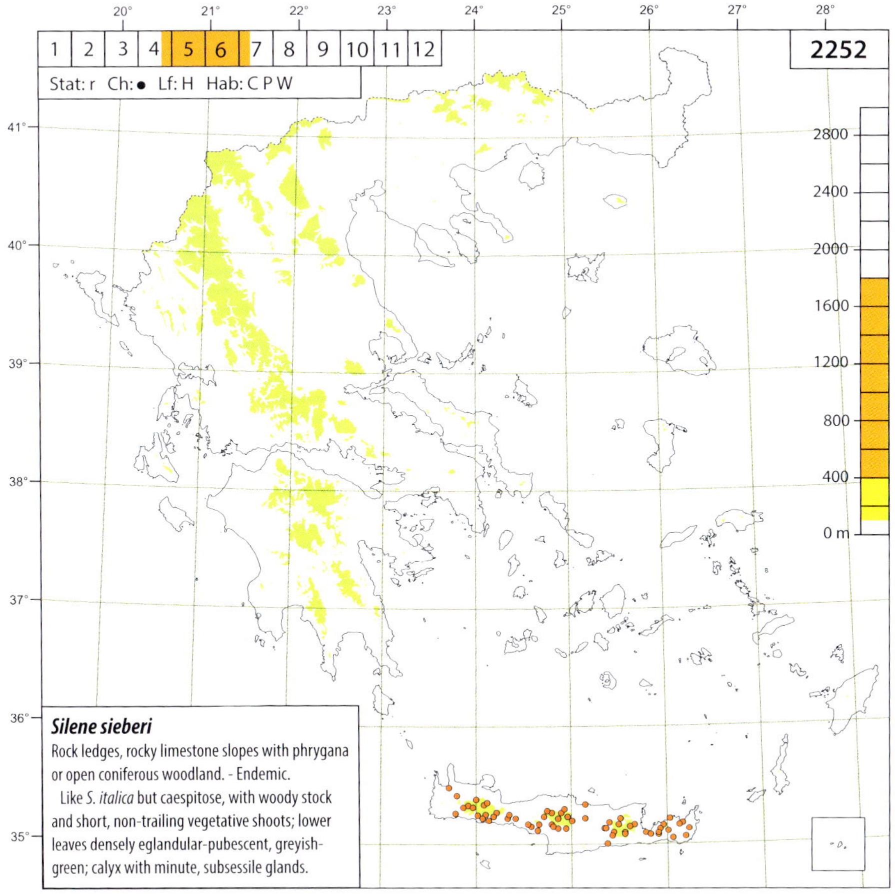

Silene sieberi

Rock ledges, rocky limestone slopes with phrygana or open coniferous woodland. - Endemic.

Like *S. italica* but caespitose, with woody stock and short, non-trailing vegetative shoots; lower leaves densely eglandular-pubescent, greyish-green; calyx with minute, subsessile glands.

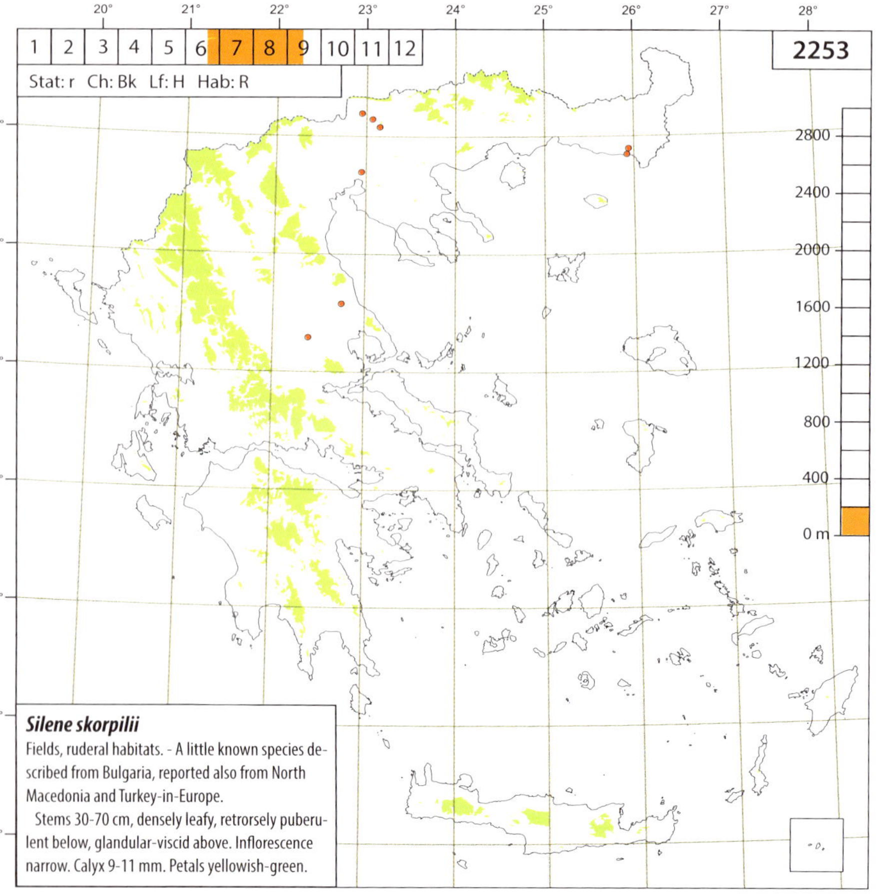

Silene skorpilii
Fields, ruderal habitats. - A little known species described from Bulgaria, reported also from North Macedonia and Turkey-in-Europe.
Stems 30-70 cm, densely leafy, retrorsely puberulent below, glandular-viscid above. Inflorescence narrow. Calyx 9-11 mm. Petals yellowish-green.

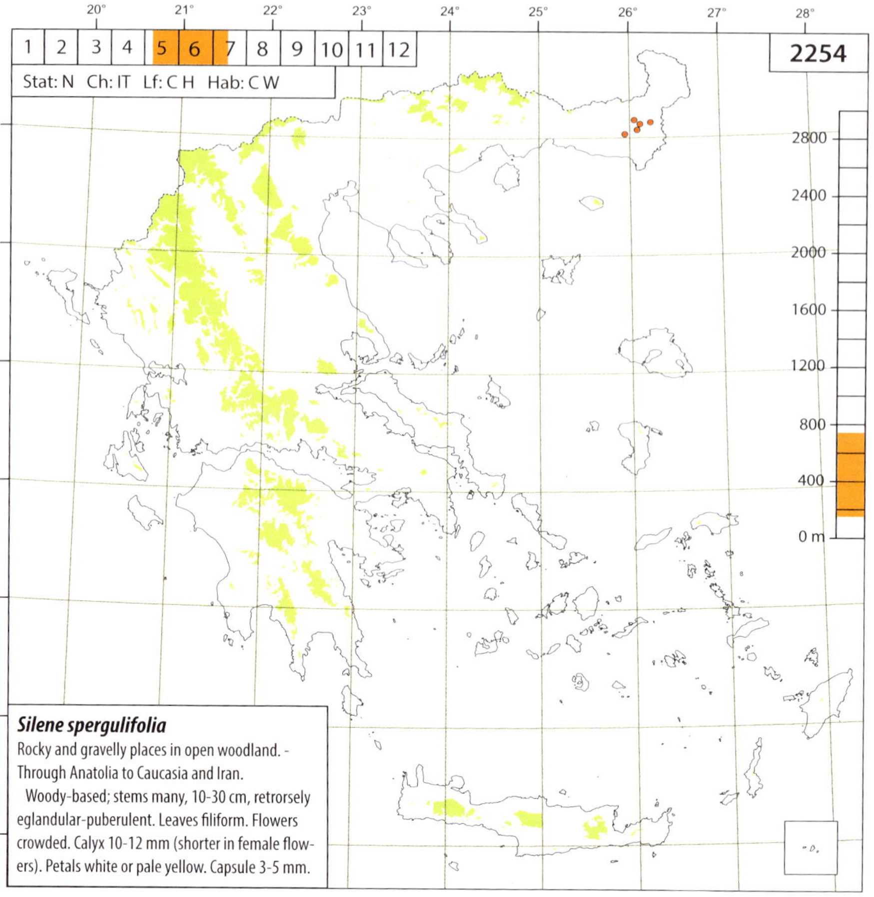

Silene spergulifolia
Rocky and gravelly places in open woodland. - Through Anatolia to Caucasia and Iran.
Woody-based; stems many, 10-30 cm, retrorsely eglandular-puberulent. Leaves filiform. Flowers crowded. Calyx 10-12 mm (shorter in female flowers). Petals white or pale yellow. Capsule 3-5 mm.

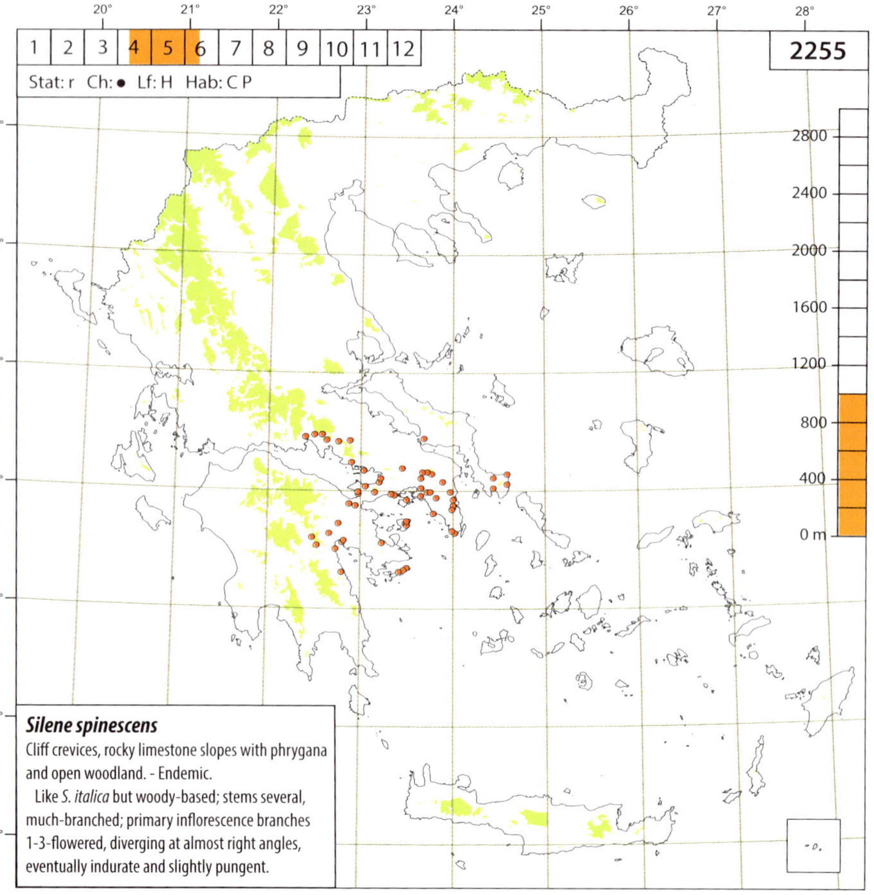

Silene spinescens
Cliff crevices, rocky limestone slopes with phrygana and open woodland. - Endemic.
Like *S. italica* but woody-based; stems several, much-branched; primary inflorescence branches 1-3-flowered, diverging at almost right angles, eventually indurate and slightly pungent.

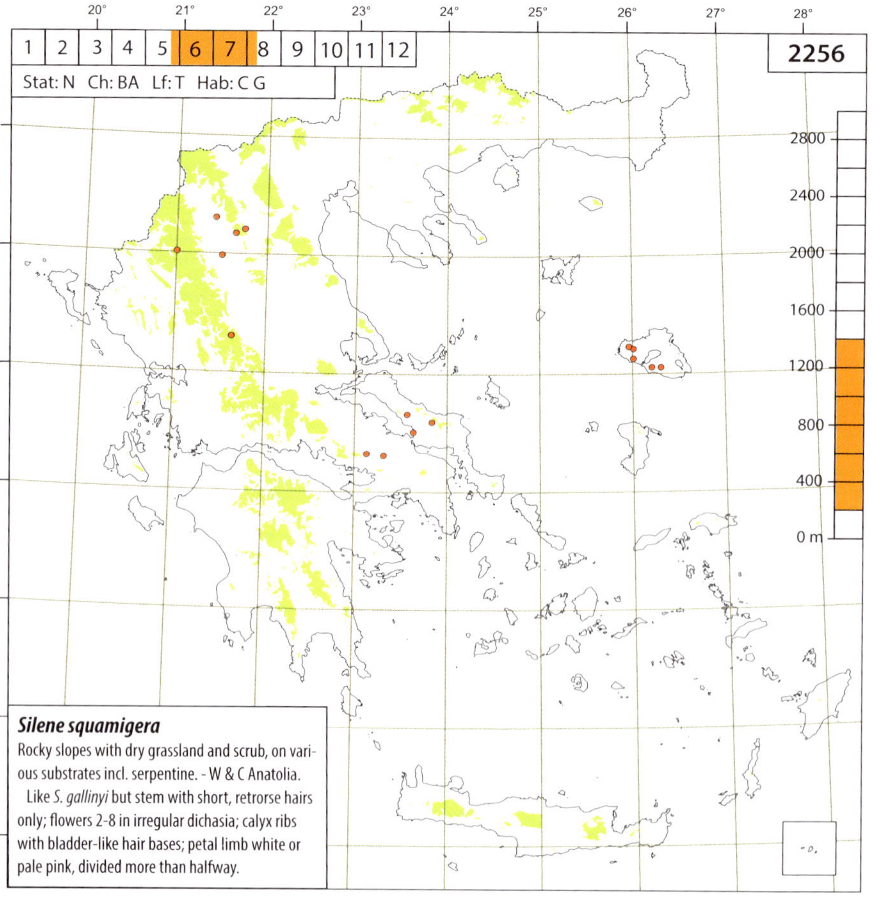

Silene squamigera
Rocky slopes with dry grassland and scrub, on various substrates incl. serpentine. - W & C Anatolia.
Like *S. gallinyi* but stem with short, retrorse hairs only; flowers 2-8 in irregular dichasia; calyx ribs with bladder-like hair bases; petal limb white or pale pink, divided more than halfway.

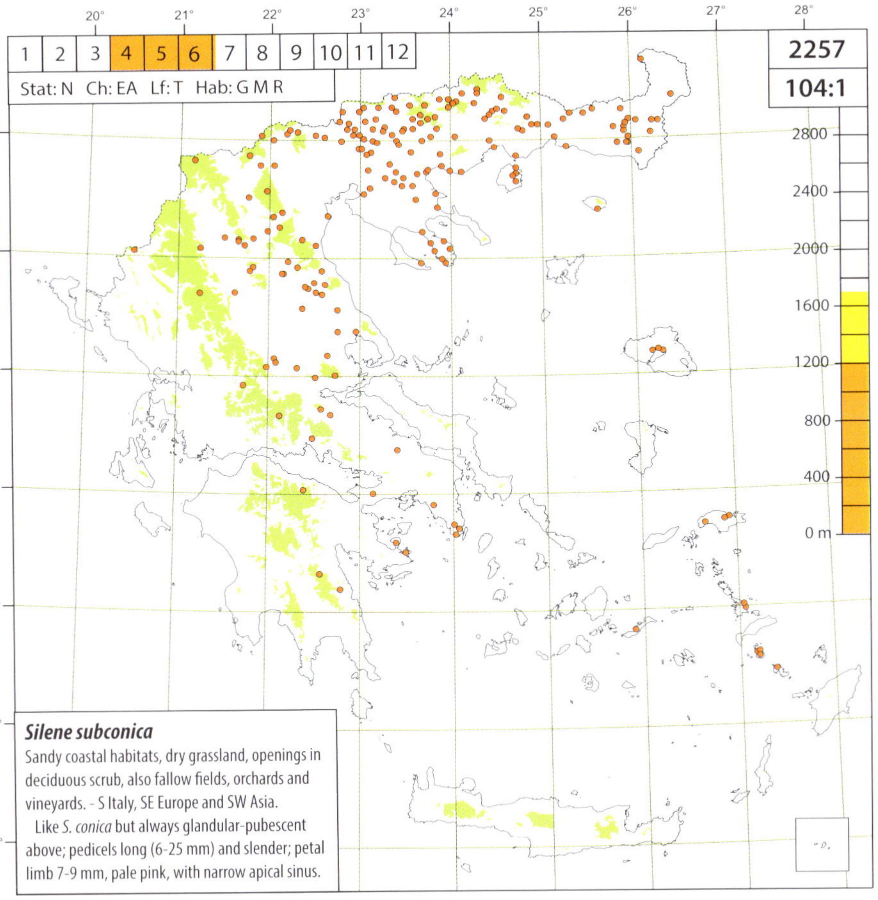

Silene subconica

Sandy coastal habitats, dry grassland, openings in deciduous scrub, also fallow fields, orchards and vineyards. - S Italy, SE Europe and SW Asia.

Like *S. conica* but always glandular-pubescent above; pedicels long (6-25 mm) and slender; petal limb 7-9 mm, pale pink, with narrow apical sinus.

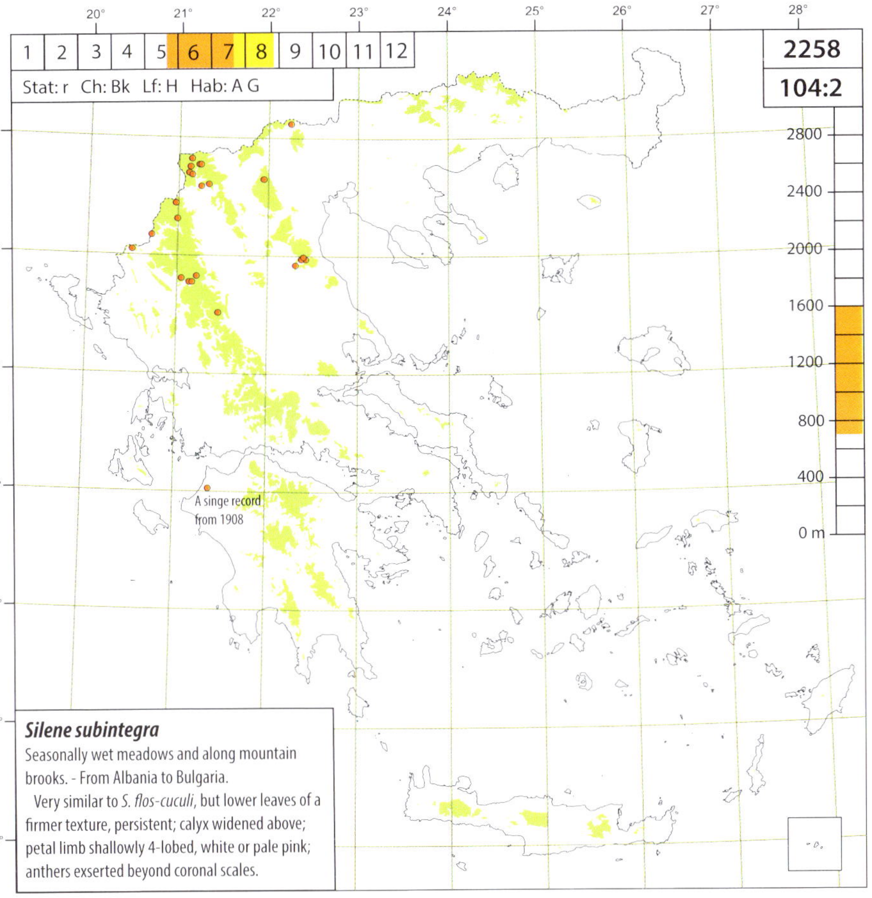

Silene subintegra

Seasonally wet meadows and along mountain brooks. - From Albania to Bulgaria.

Very similar to *S. flos-cuculi*, but lower leaves of a firmer texture, persistent; calyx widened above; petal limb shallowly 4-lobed, white or pale pink; anthers exserted beyond coronal scales.

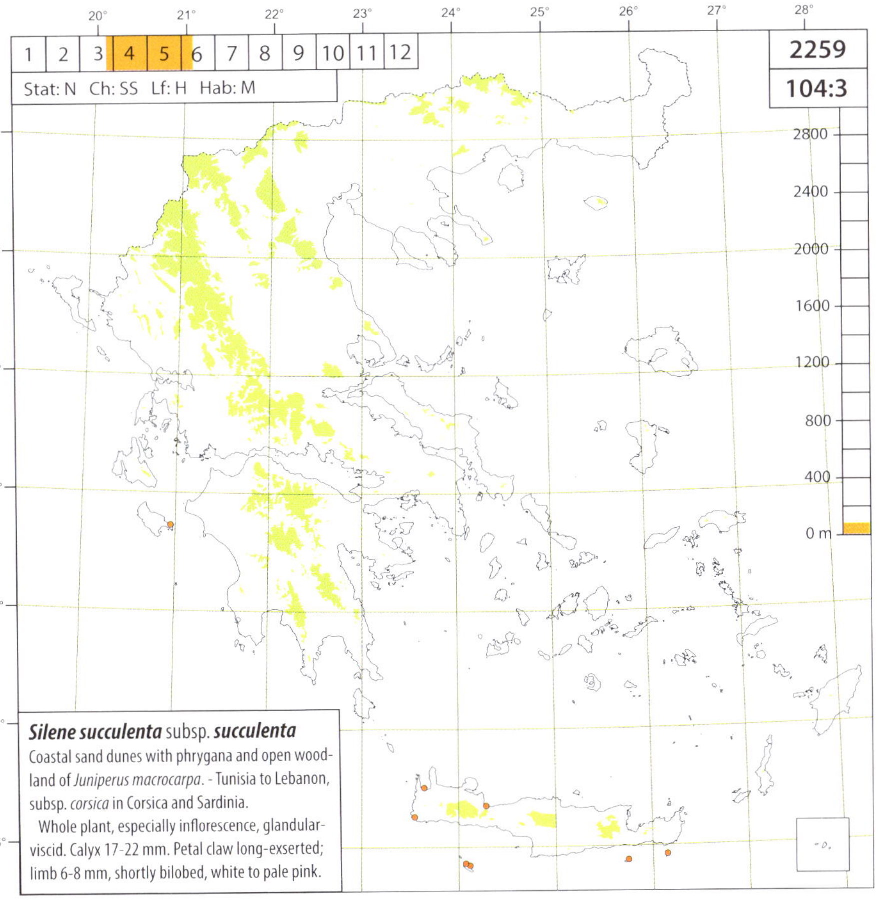

Silene succulenta subsp. *succulenta*

Coastal sand dunes with phrygana and open woodland of *Juniperus macrocarpa*. - Tunisia to Lebanon, subsp. *corsica* in Corsica and Sardinia.

Whole plant, especially inflorescence, glandular-viscid. Calyx 17-22 mm. Petal claw long-exserted; limb 6-8 mm, shortly bilobed, white to pale pink.

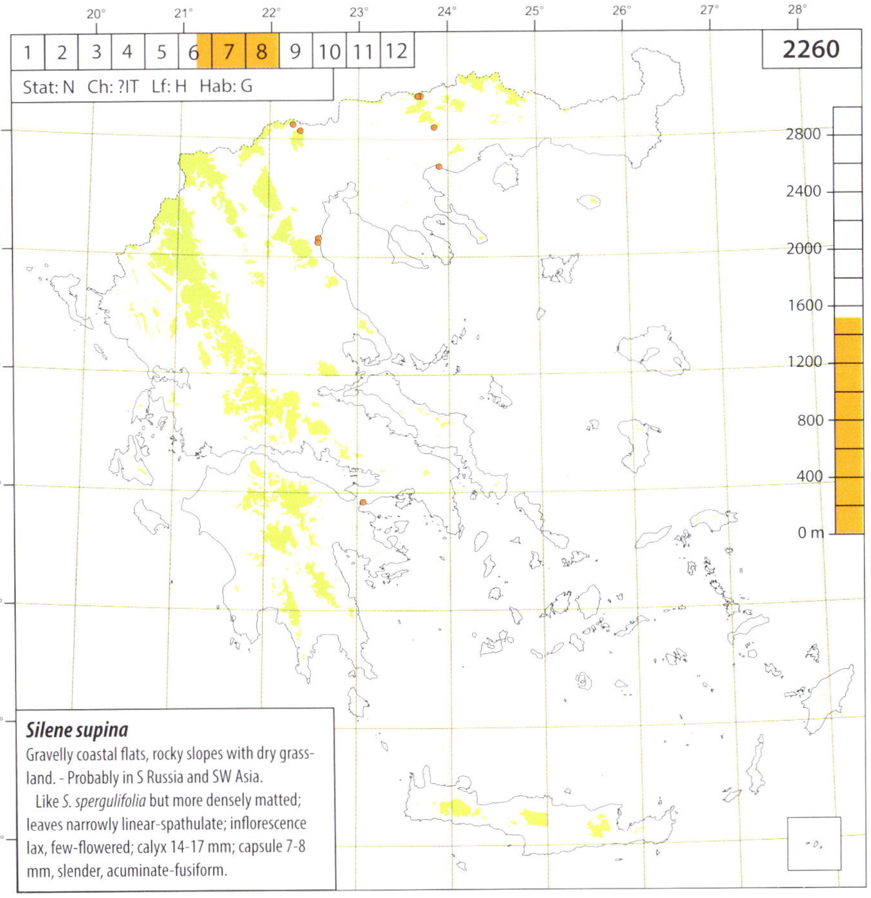

Silene supina

Gravelly coastal flats, rocky slopes with dry grassland. - Probably in S Russia and SW Asia.

Like *S. spergulifolia* but more densely matted; leaves narrowly linear-spathulate; inflorescence lax, few-flowered; calyx 14-17 mm; capsule 7-8 mm, slender, acuminate-fusiform.

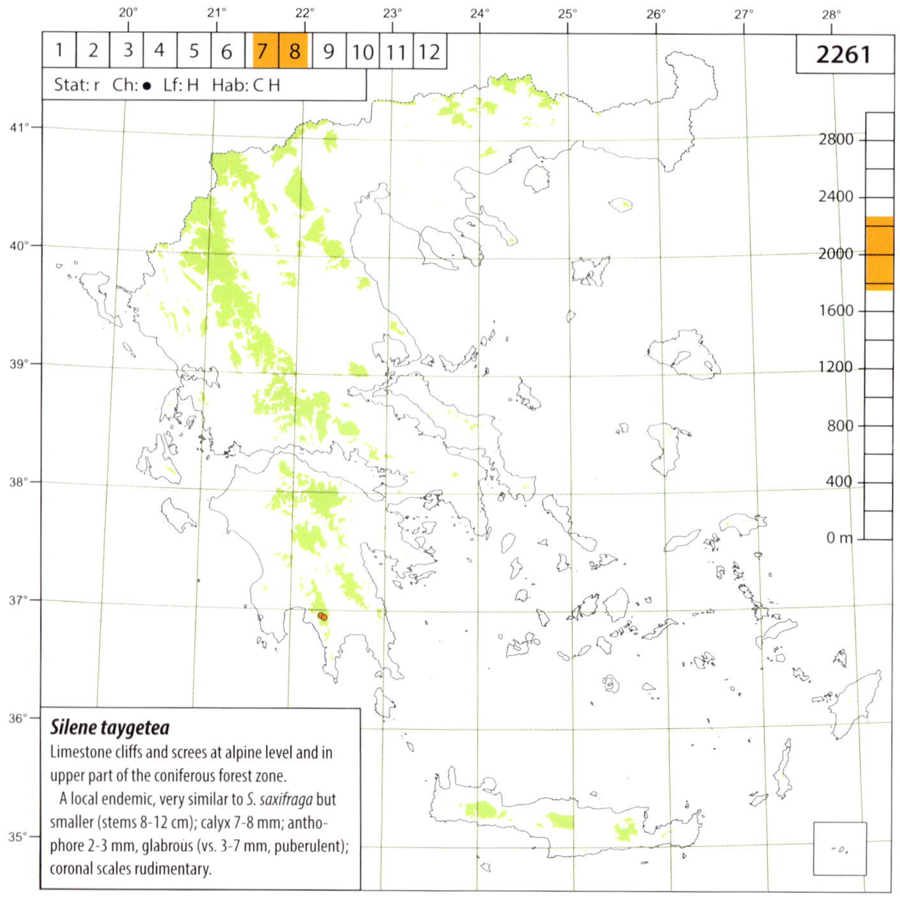

Silene taygetea

Limestone cliffs and screes at alpine level and in upper part of the coniferous forest zone.

A local endemic, very similar to *S. saxifraga* but smaller (stems 8-12 cm); calyx 7-8 mm; anthophore 2-3 mm, glabrous (vs. 3-7 mm, puberulent); coronal scales rudimentary.

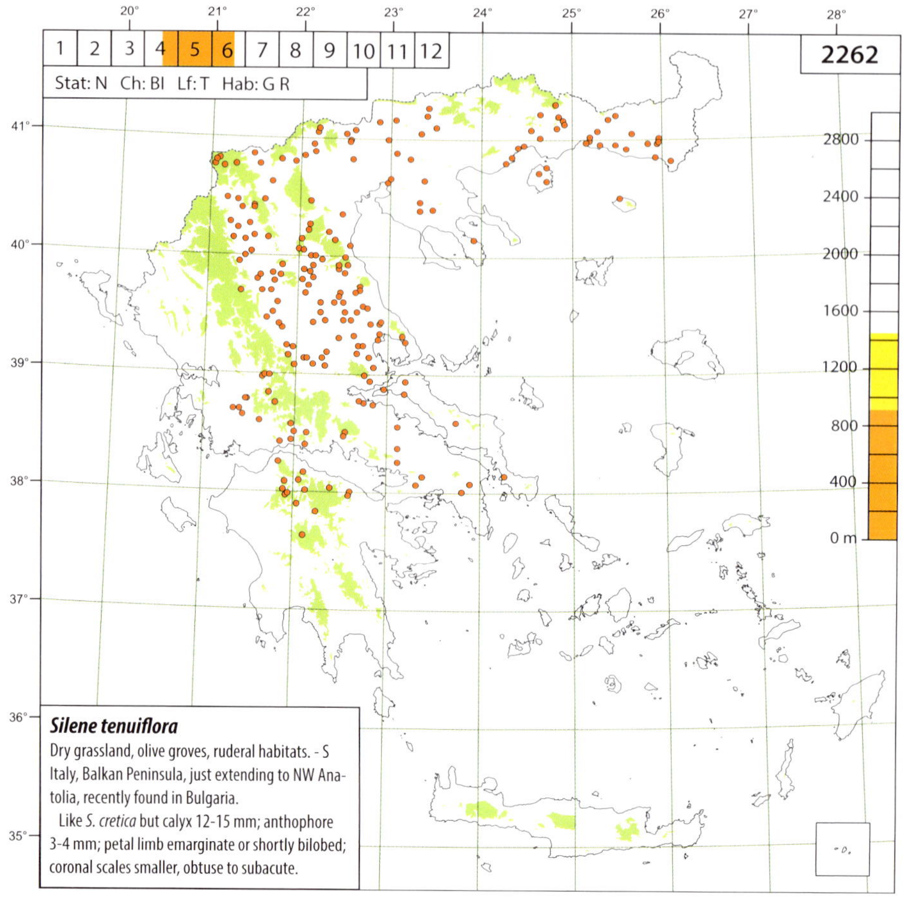

Silene tenuiflora

Dry grassland, olive groves, ruderal habitats. - S Italy, Balkan Peninsula, just extending to NW Anatolia, recently found in Bulgaria.

Like *S. cretica* but calyx 12-15 mm; anthophore 3-4 mm; petal limb emarginate or shortly bilobed; coronal scales smaller, obtuse to subacute.

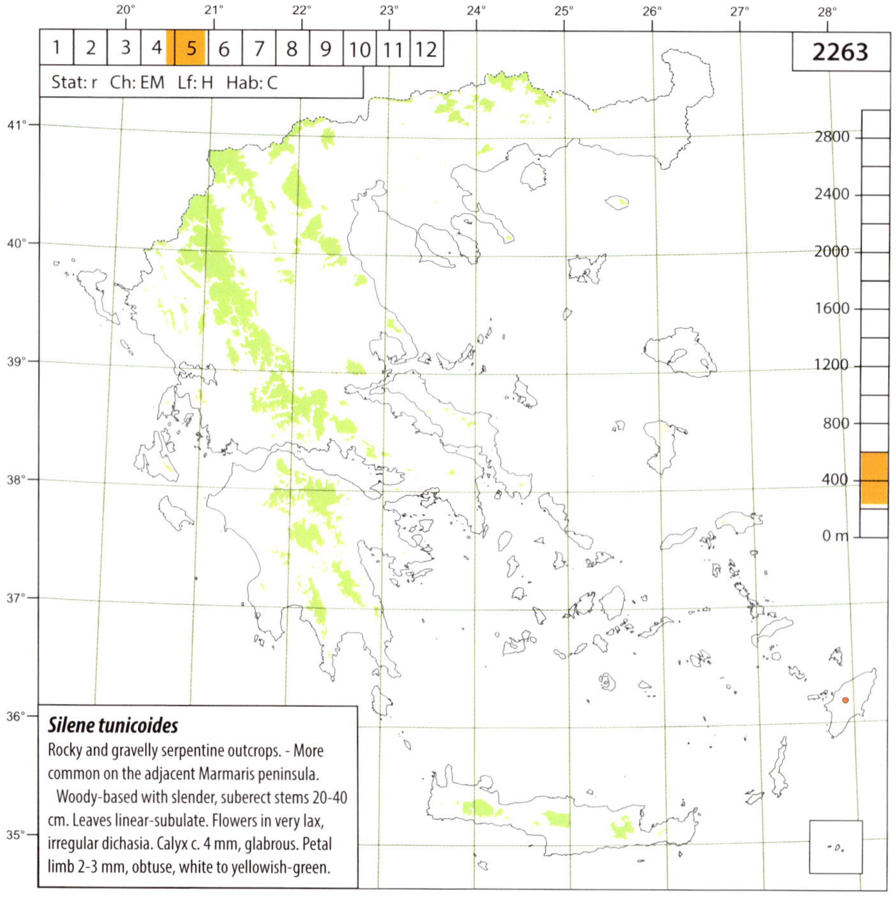

Silene tunicoides

Rocky and gravelly serpentine outcrops. - More common on the adjacent Marmaris peninsula.

Woody-based with slender, suberect stems 20-40 cm. Leaves linear-subulate. Flowers in very lax, irregular dichasia. Calyx c. 4 mm, glabrous. Petal limb 2-3 mm, obtuse, white to yellowish-green.

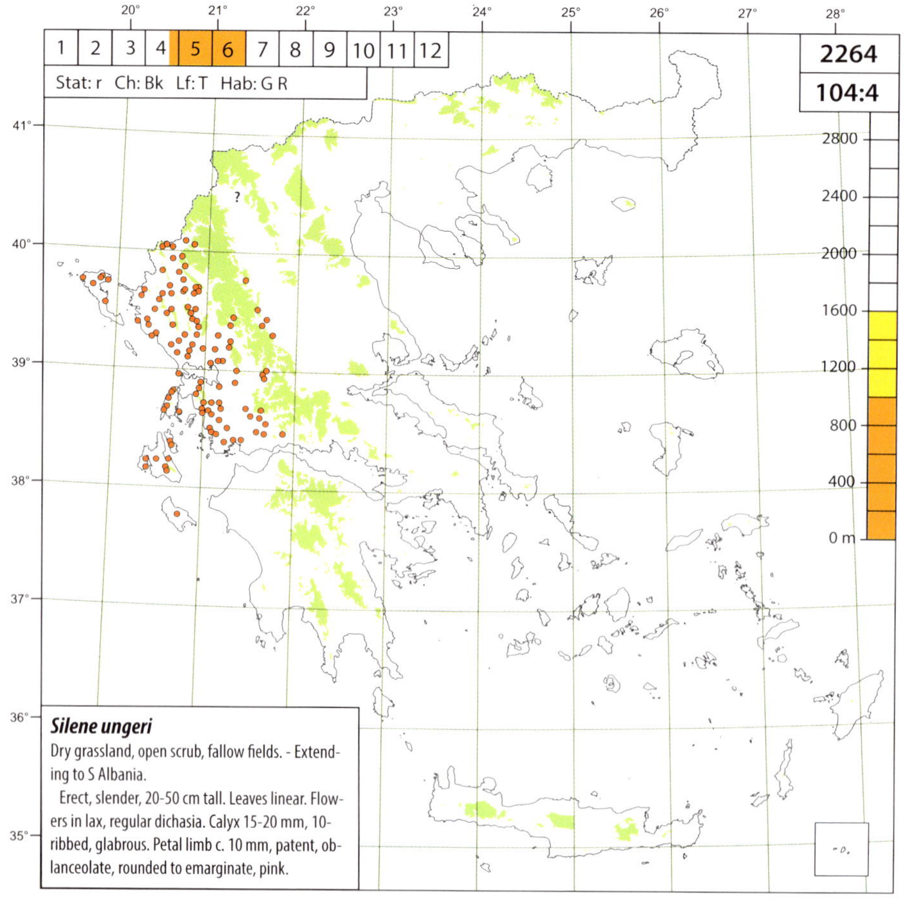

Silene ungeri

Dry grassland, open scrub, fallow fields. - Extending to S Albania.

Erect, slender, 20-50 cm tall. Leaves linear. Flowers in lax, regular dichasia. Calyx 15-20 mm, 10-ribbed, glabrous. Petal limb c. 10 mm, patent, oblanceolate, rounded to emarginate, pink.

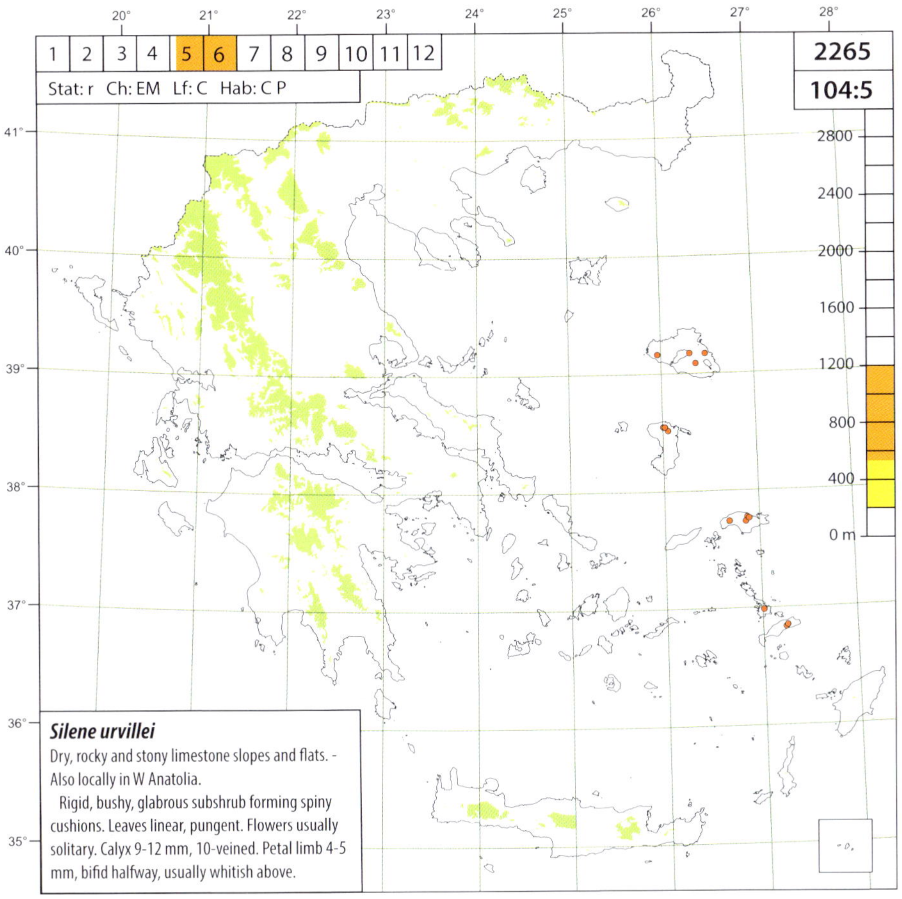

Silene urvillei

Dry, rocky and stony limestone slopes and flats. - Also locally in W Anatolia.

Rigid, bushy, glabrous subshrub forming spiny cushions. Leaves linear, pungent. Flowers usually solitary. Calyx 9-12 mm, 10-veined. Petal limb 4-5 mm, bifid halfway, usually whitish above.

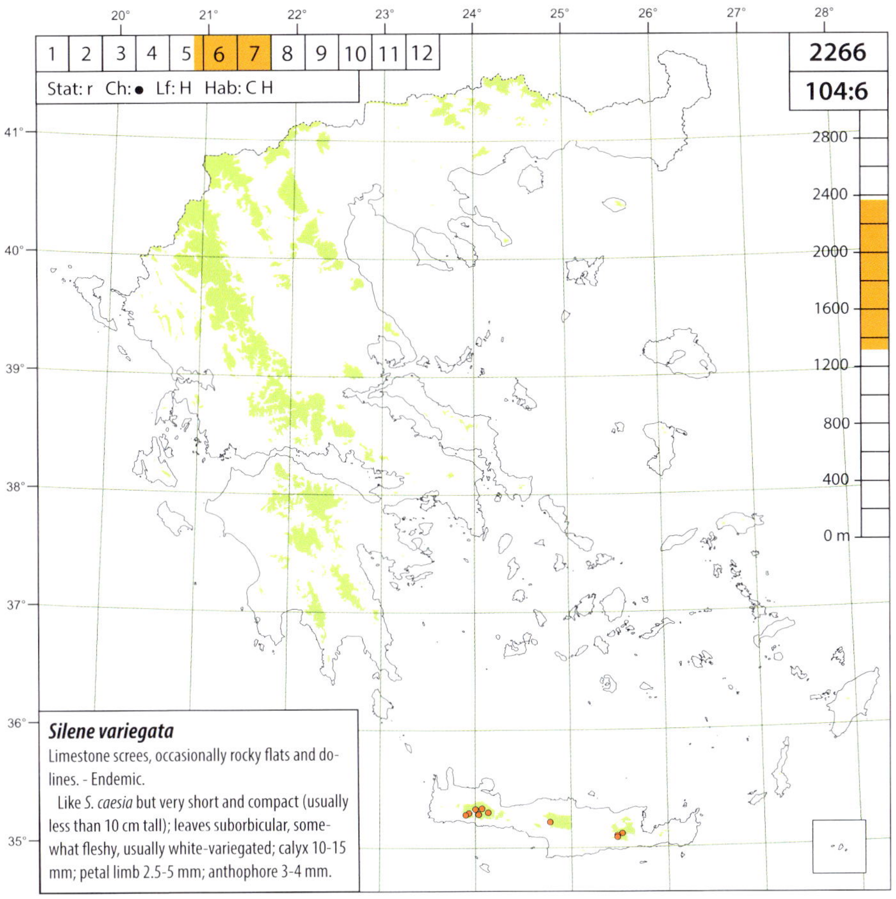

Silene variegata

Limestone screes, occasionally rocky flats and dolines. - Endemic.

Like *S. caesia* but very short and compact (usually less than 10 cm tall); leaves suborbicular, somewhat fleshy, usually white-variegated; calyx 10-15 mm; petal limb 2.5-5 mm; anthophore 3-4 mm.

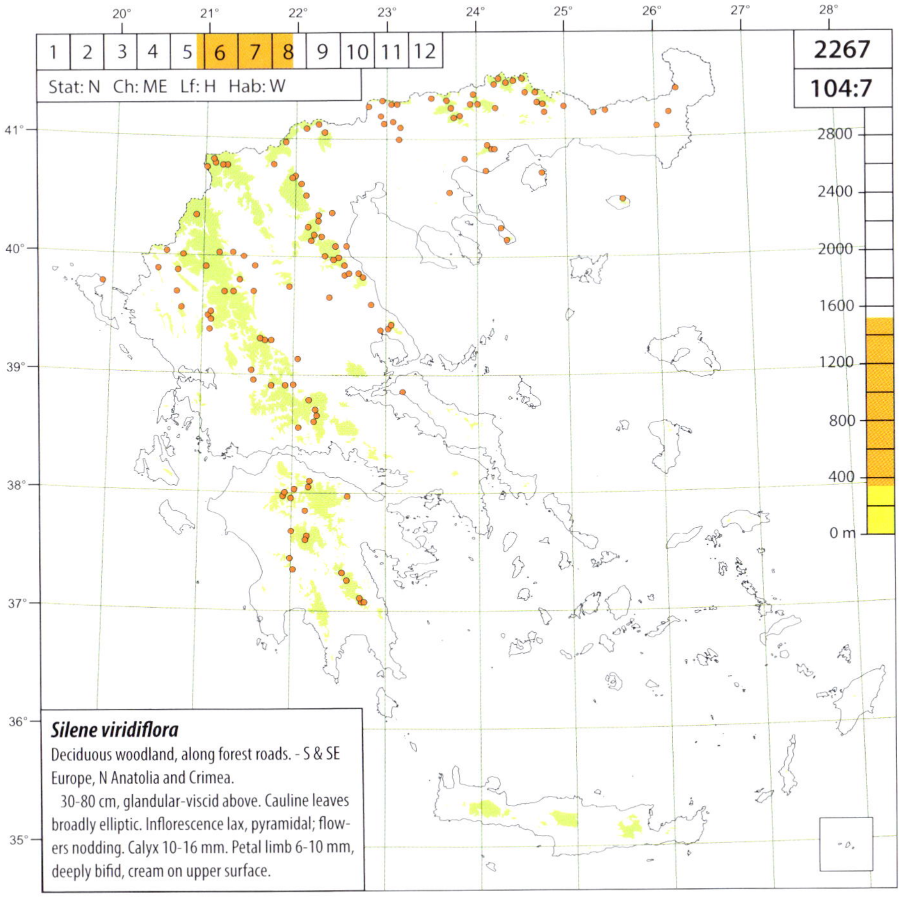

Silene viridiflora

Deciduous woodland, along forest roads. - S & SE Europe, N Anatolia and Crimea.

30-80 cm, glandular-viscid above. Cauline leaves broadly elliptic. Inflorescence lax, pyramidal; flowers nodding. Calyx 10-16 mm. Petal limb 6-10 mm, deeply bifid, cream on upper surface.

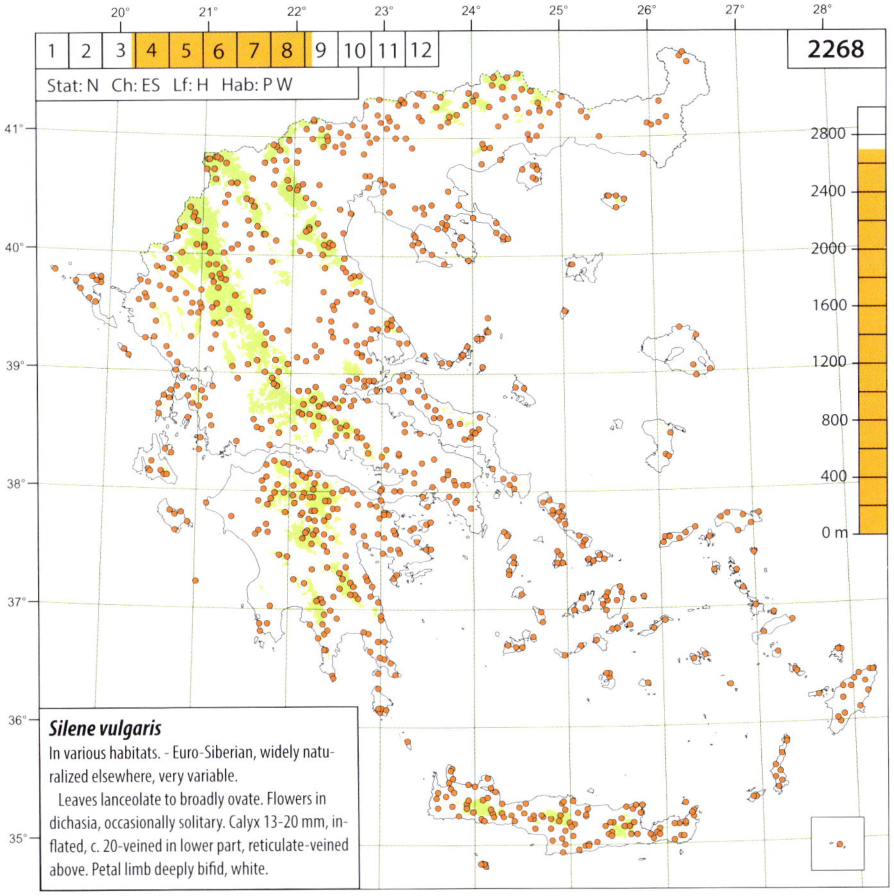

Silene vulgaris

In various habitats. - Euro-Siberian, widely naturalized elsewhere, very variable.

Leaves lanceolate to broadly ovate. Flowers in dichasia, occasionally solitary. Calyx 13-20 mm, inflated, c. 20-veined in lower part, reticulate-veined above. Petal limb deeply bifid, white.

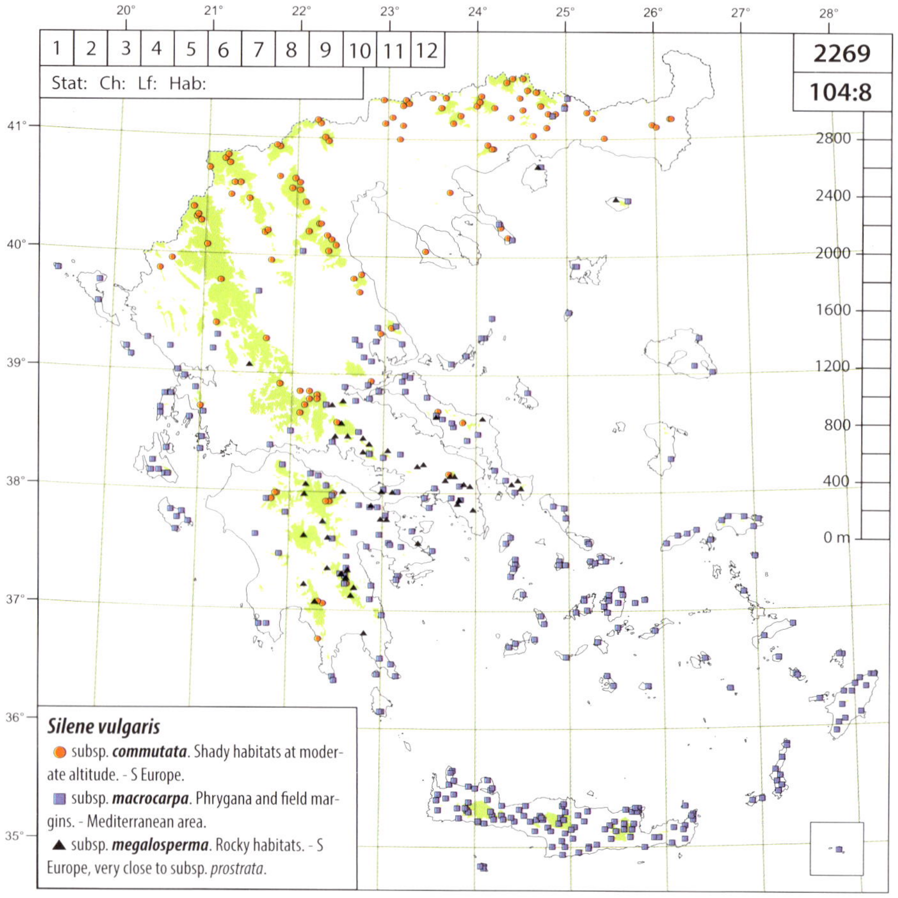

1 2 3 4 5 6 7 8 9 10 11 12
Stat: Ch: Lf: Hab:
2269
104:8
20° 21° 22° 23° 24° 25° 26° 27° 28°
41° 40° 39° 38° 37° 36° 35°
2800 2400 2000 1600 1200 800 400 0 m
Silene vulgaris
subsp. commutata. Shady habitats at moderate altitude. - S Europe.
subsp. macrocarpa. Phrygana and field margins. - Mediterranean area.
subsp. megalosperma. Rocky habitats. - S Europe, very close to subsp. prostrata.

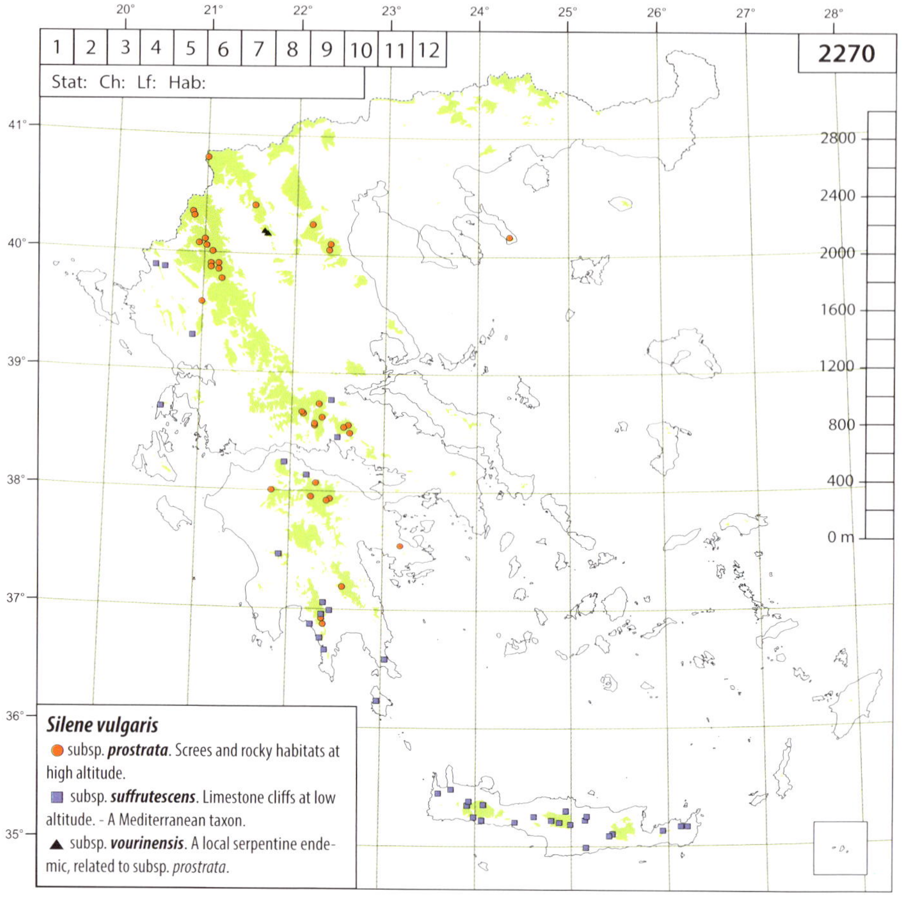

1 2 3 4 5 6 7 8 9 10 11 12
Stat: Ch: Lf: Hab:
2270
20° 21° 22° 23° 24° 25° 26° 27° 28°
41° 40° 39° 38° 37° 36° 35°
2800 2400 2000 1600 1200 800 400 0 m
Silene vulgaris
subsp. prostrata. Screes and rocky habitats at high altitude.
subsp. suffrutescens. Limestone cliffs at low altitude. - A Mediterranean taxon.
subsp. vourinensis. A local serpentine endemic, related to subsp. prostrata.

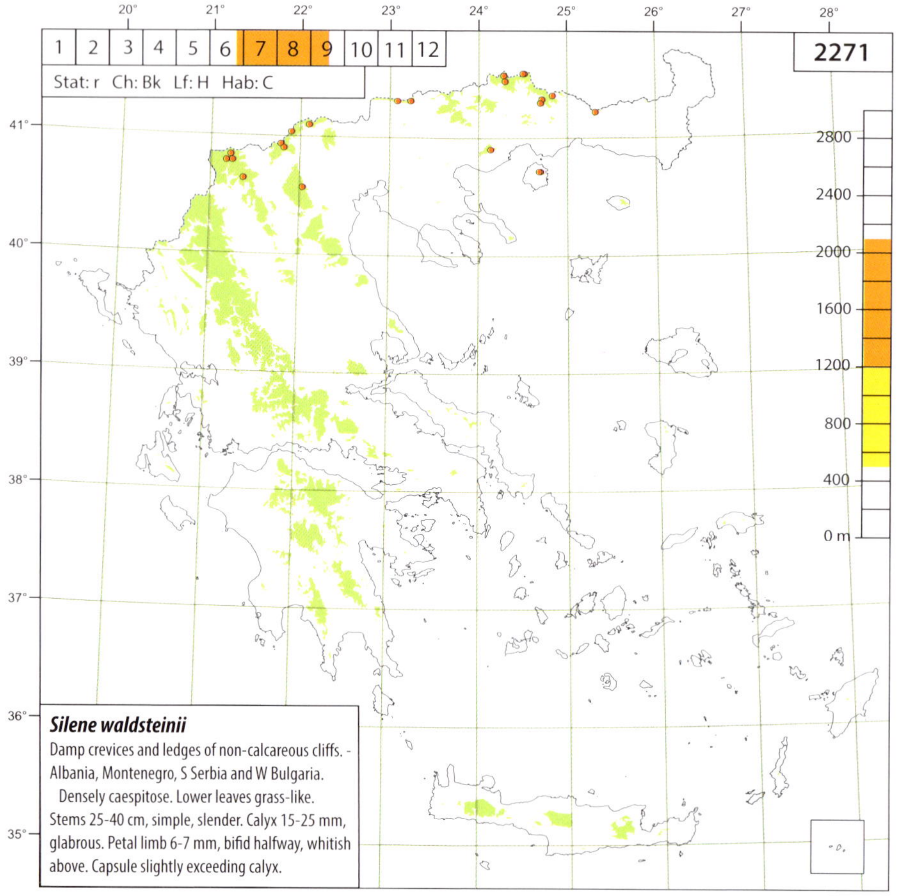

1 2 3 4 5 6 7 8 9 10 11 12
Stat: r Ch: Bk Lf: H Hab: C
2271
20° 21° 22° 23° 24° 25° 26° 27° 28°
41° 40° 39° 38° 37° 36° 35°
2800 2400 2000 1600 1200 800 400 0 m
Silene waldsteinii
Damp crevices and ledges of non-calcareous cliffs. - Albania, Montenegro, S Serbia and W Bulgaria.
Densely caespitose. Lower leaves grass-like. Stems 25-40 cm, simple, slender. Calyx 15-25 mm, glabrous. Petal limb 6-7 mm, bifid halfway, whitish above. Capsule slightly exceeding calyx.

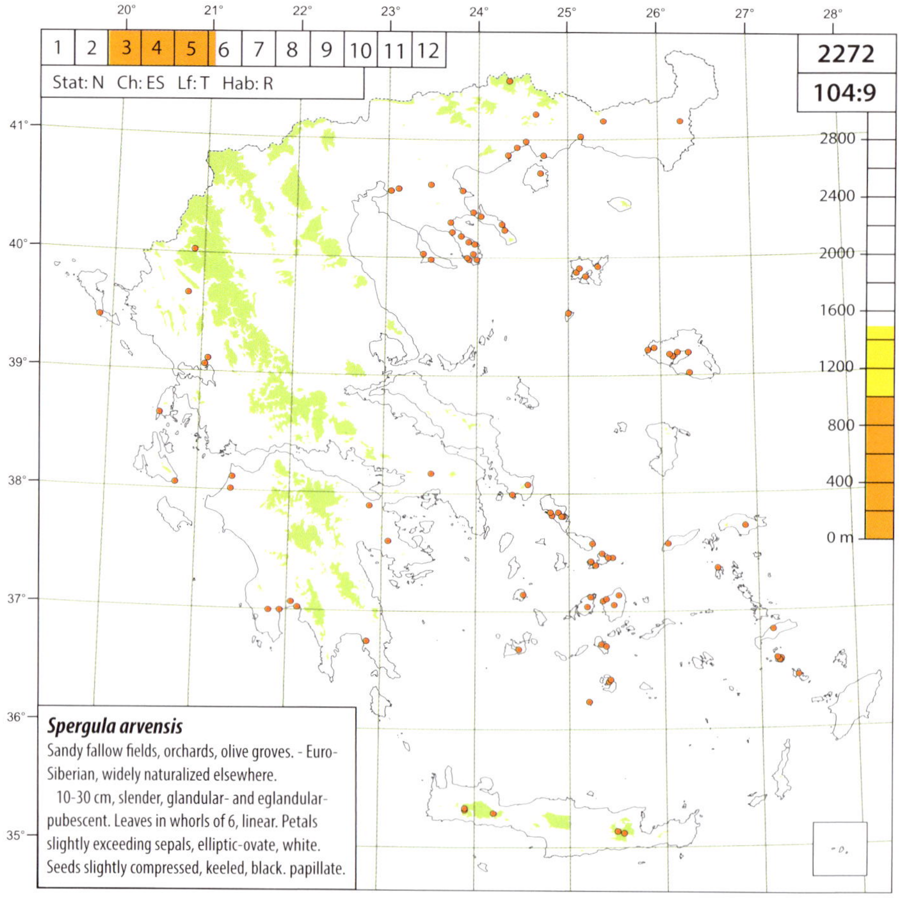

1 2 3 4 5 6 7 8 9 10 11 12
Stat: N Ch: ES Lf: T Hab: R
2272
104:9
20° 21° 22° 23° 24° 25° 26° 27° 28°
41° 40° 39° 38° 37° 36° 35°
2800 2400 2000 1600 1200 800 400 0 m
Spergula arvensis
Sandy fallow fields, orchards, olive groves. - Euro-Siberian, widely naturalized elsewhere.
10-30 cm, slender, glandular- and eglandular-pubescent. Leaves in whorls of 6, linear. Petals slightly exceeding sepals, elliptic-ovate, white. Seeds slightly compressed, keeled, black, papillate.

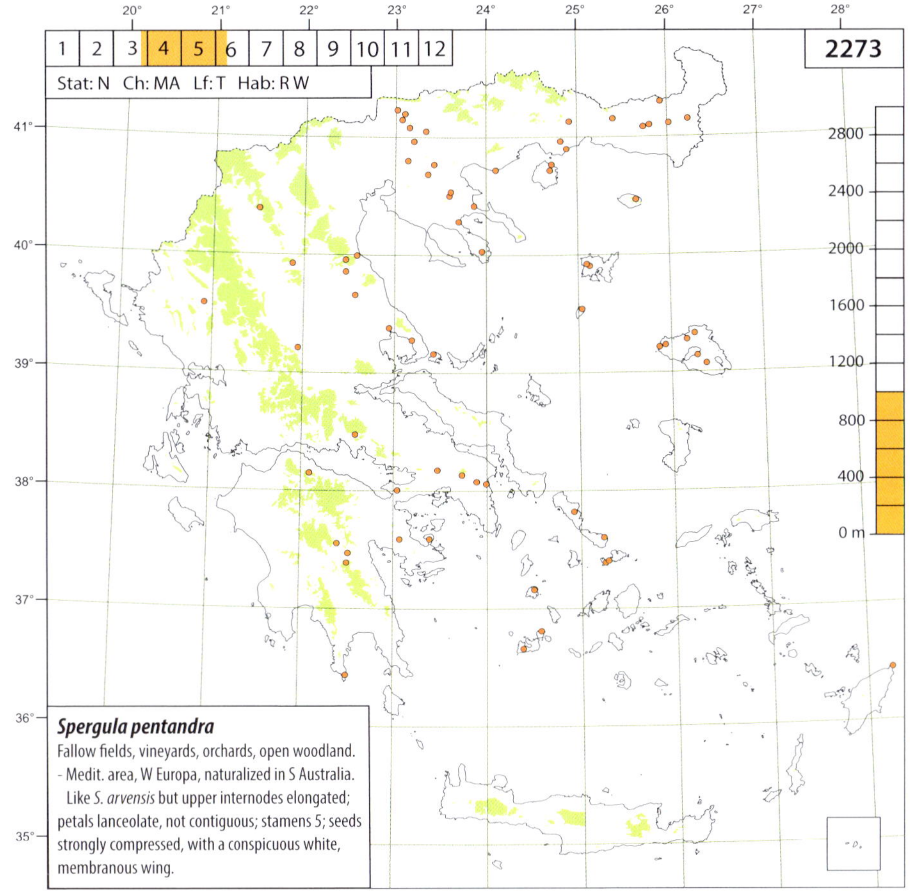

Spergula pentandra

Fallow fields, vineyards, orchards, open woodland. - Medit. area, W Europa, naturalized in S Australia.

Like *S. arvensis* but upper internodes elongated; petals lanceolate, not contiguous; stamens 5; seeds strongly compressed, with a conspicuous white, membranous wing.

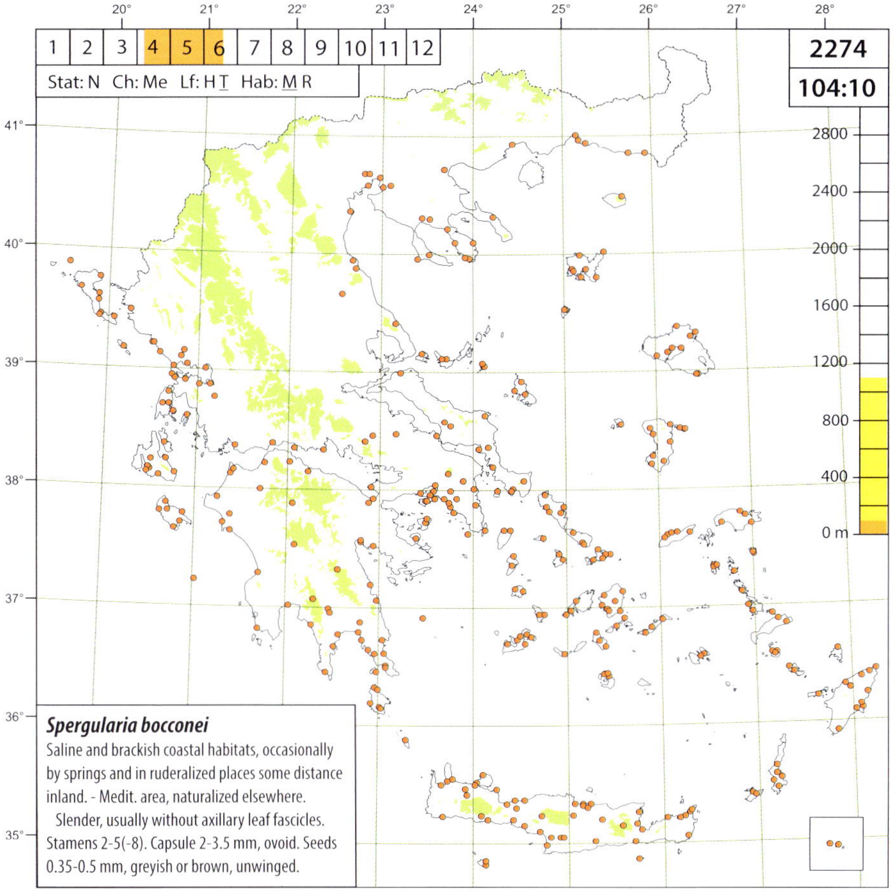

Spergularia bocconei

Saline and brackish coastal habitats, occasionally by springs and in ruderalized places some distance inland. - Medit. area, naturalized elsewhere.

Slender, usually without axillary leaf fascicles. Stamens 2-5(-8). Capsule 2-3.5 mm, ovoid. Seeds 0.35-0.5 mm, greyish or brown, unwinged.

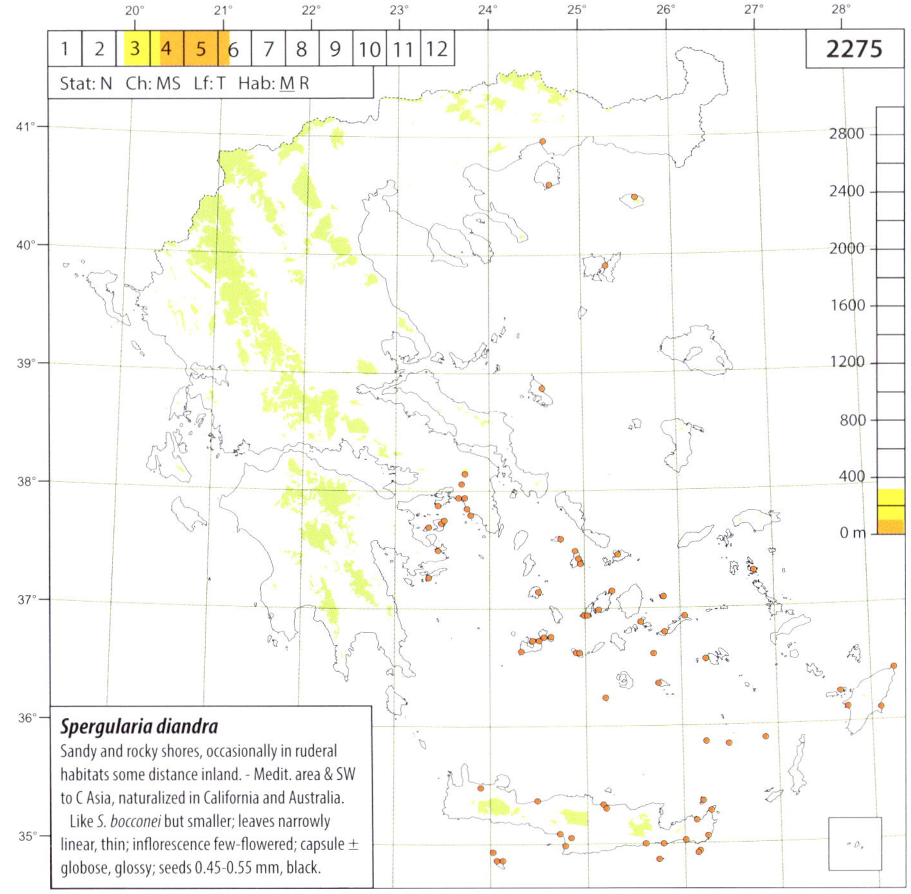

Spergularia diandra

Sandy and rocky shores, occasionally in ruderal habitats some distance inland. - Medit. area & SW to C Asia, naturalized in California and Australia.

Like *S. bocconei* but smaller; leaves narrowly linear, thin; inflorescence few-flowered; capsule ± globose, glossy; seeds 0.45-0.55 mm, black.

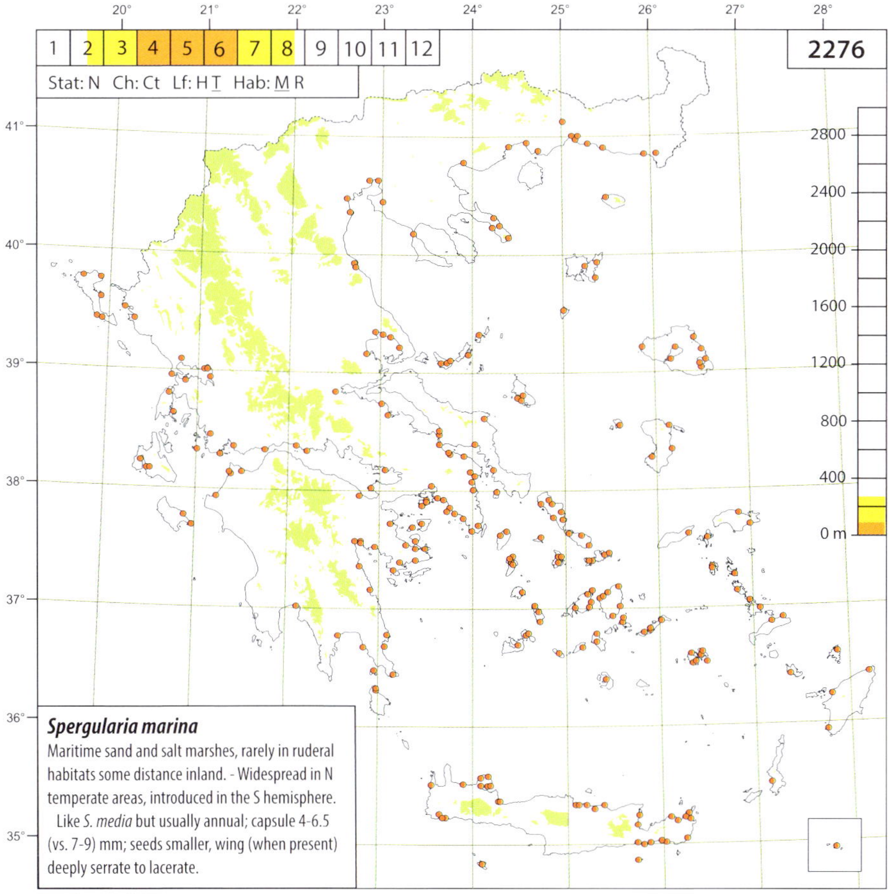

Spergularia marina

Maritime sand and salt marshes, rarely in ruderal habitats some distance inland. - Widespread in N temperate areas, introduced in the S hemisphere.

Like *S. media* but usually annual; capsule 4-6.5 (vs. 7-9) mm; seeds smaller, wing (when present) deeply serrate to lacerate.

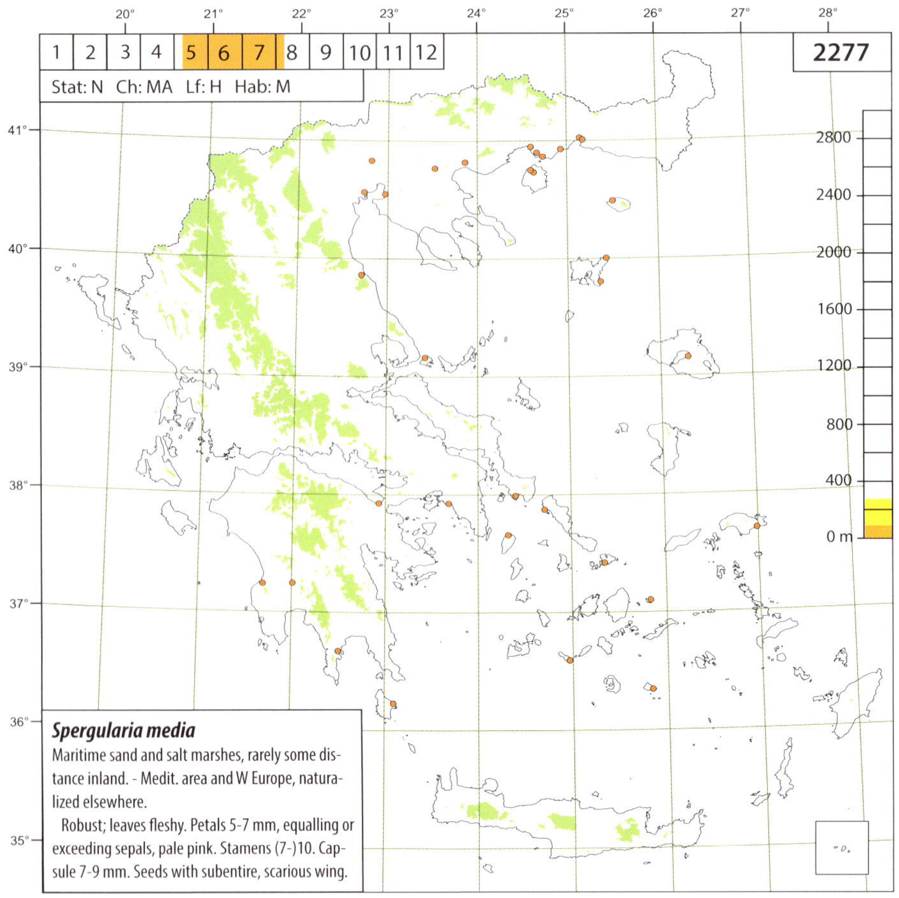

Spergularia media
Maritime sand and salt marshes, rarely some distance inland. - Medit. area and W Europe, naturalized elsewhere.
Robust; leaves fleshy. Petals 5-7 mm, equalling or exceeding sepals, pale pink. Stamens (7-)10. Capsule 7-9 mm. Seeds with subentire, scarious wing.

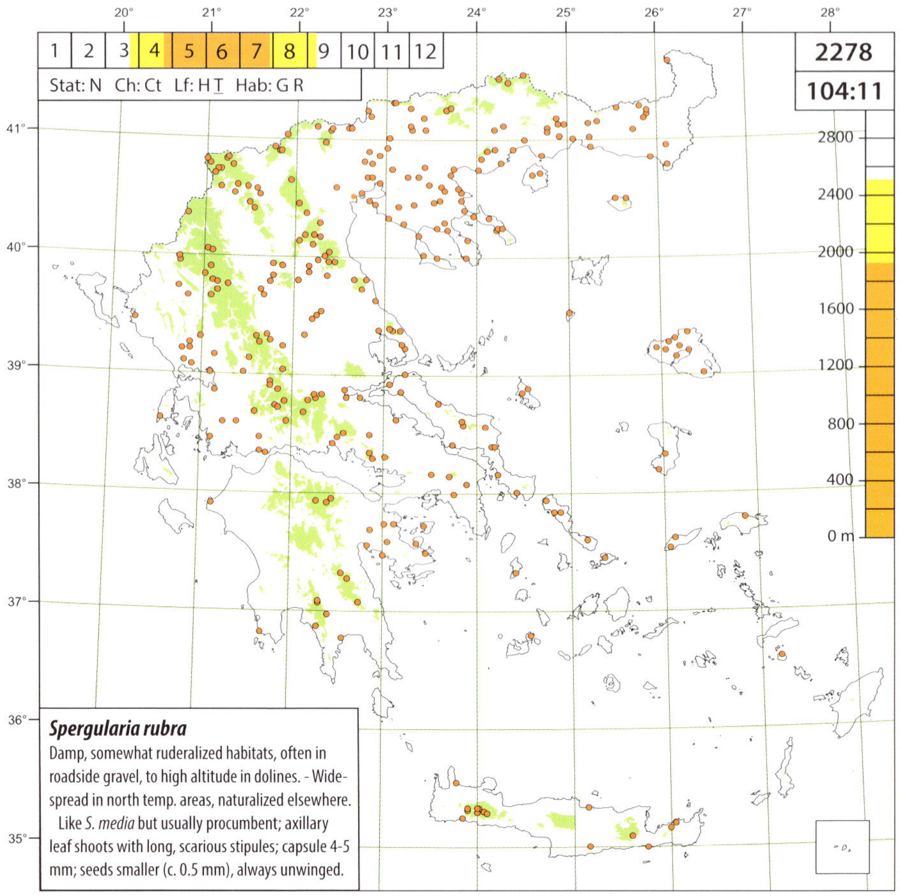

Spergularia rubra
Damp, somewhat ruderalized habitats, often in roadside gravel, to high altitude in dolines. - Widespread in north temp. areas, naturalized elsewhere.
Like *S. media* but usually procumbent; axillary leaf shoots with long, scarious stipules; capsule 4-5 mm; seeds smaller (c. 0.5 mm), always unwinged.

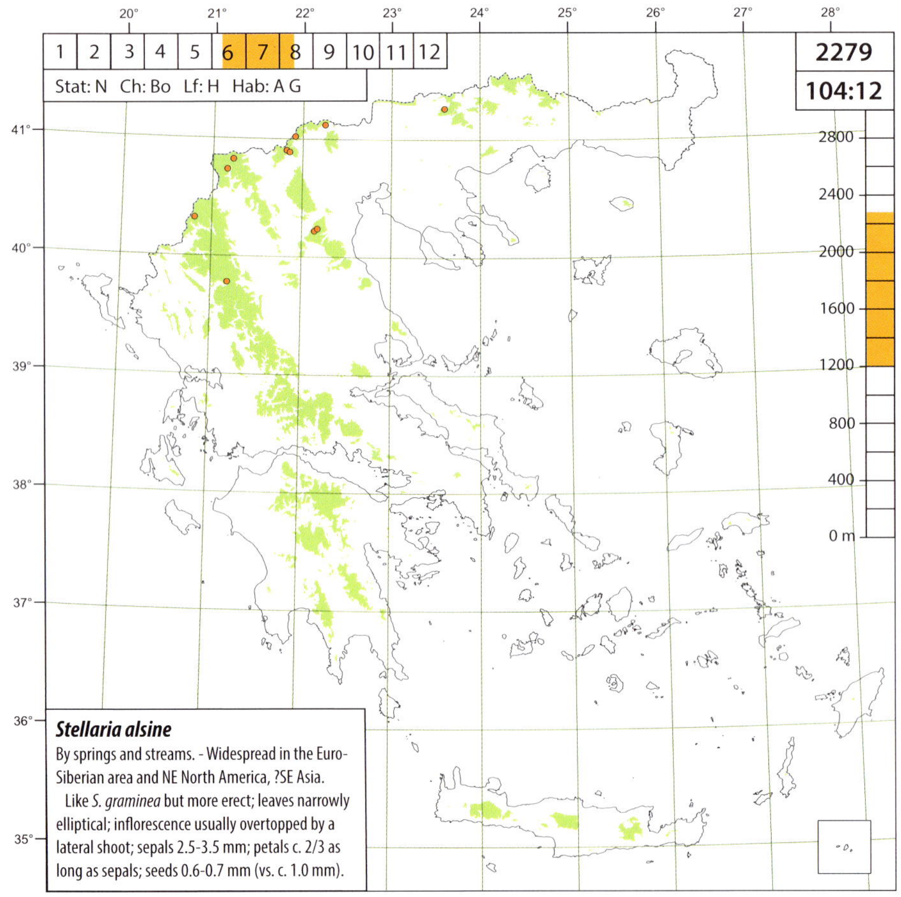

Stellaria alsine
By springs and streams. - Widespread in the Euro-Siberian area and NE North America, ?SE Asia.
Like *S. graminea* but more erect; leaves narrowly elliptical; inflorescence usually overtopped by a lateral shoot; sepals 2.5-3.5 mm; petals c. 2/3 as long as sepals; seeds 0.6-0.7 mm (vs. c. 1.0 mm).

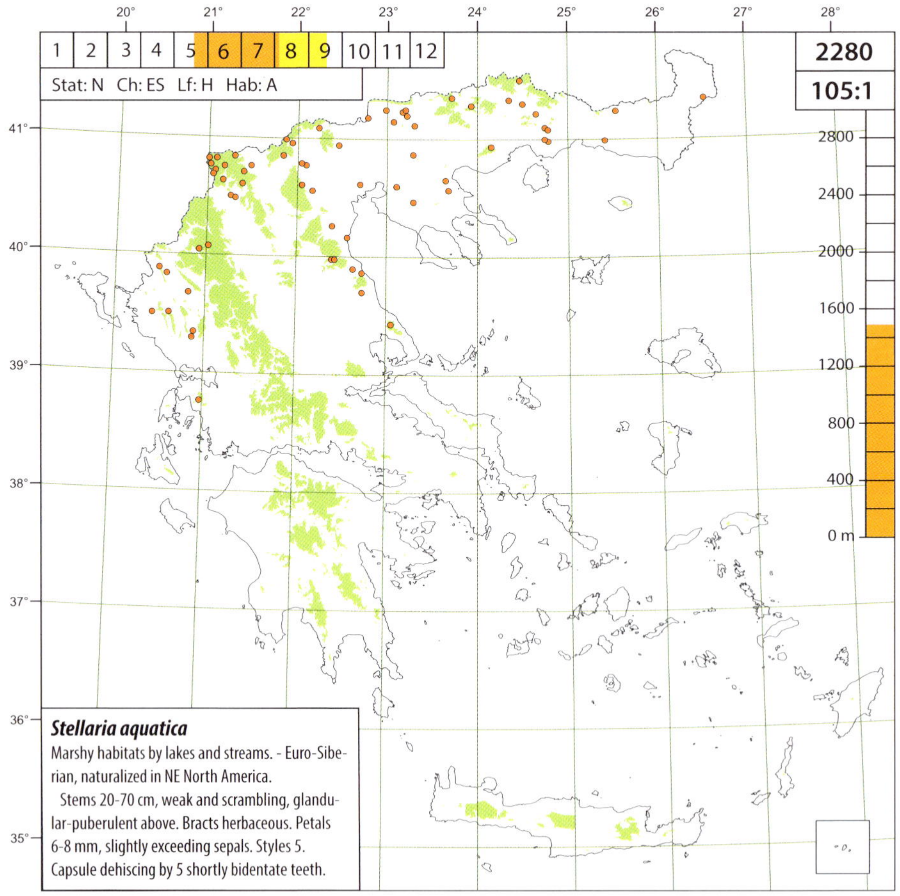

Stellaria aquatica
Marshy habitats by lakes and streams. - Euro-Siberian, naturalized in NE North America.
Stems 20-70 cm, weak and scrambling, glandular-puberulent above. Bracts herbaceous. Petals 6-8 mm, slightly exceeding sepals. Styles 5. Capsule dehiscing by 5 shortly bidentate teeth.

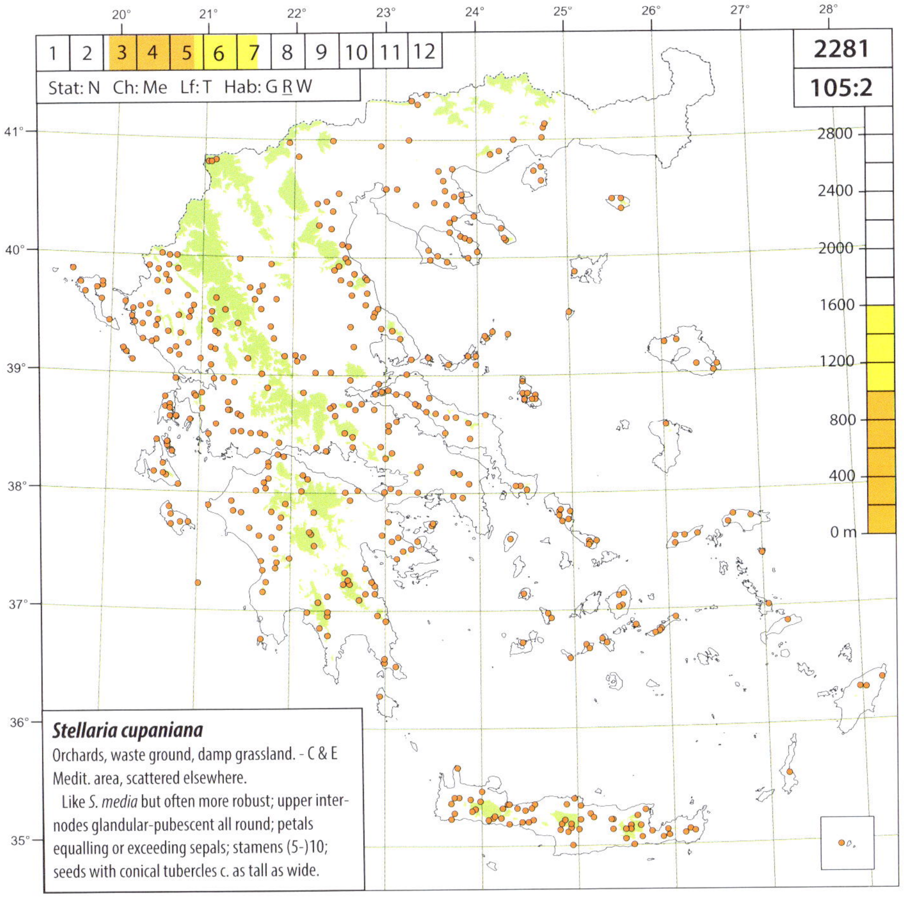

Stellaria cupaniana

Orchards, waste ground, damp grassland. - C & E Medit. area, scattered elsewhere.

Like *S. media* but often more robust; upper internodes glandular-pubescent all round; petals equalling or exceeding sepals; stamens (5-)10; seeds with conical tubercles c. as tall as wide.

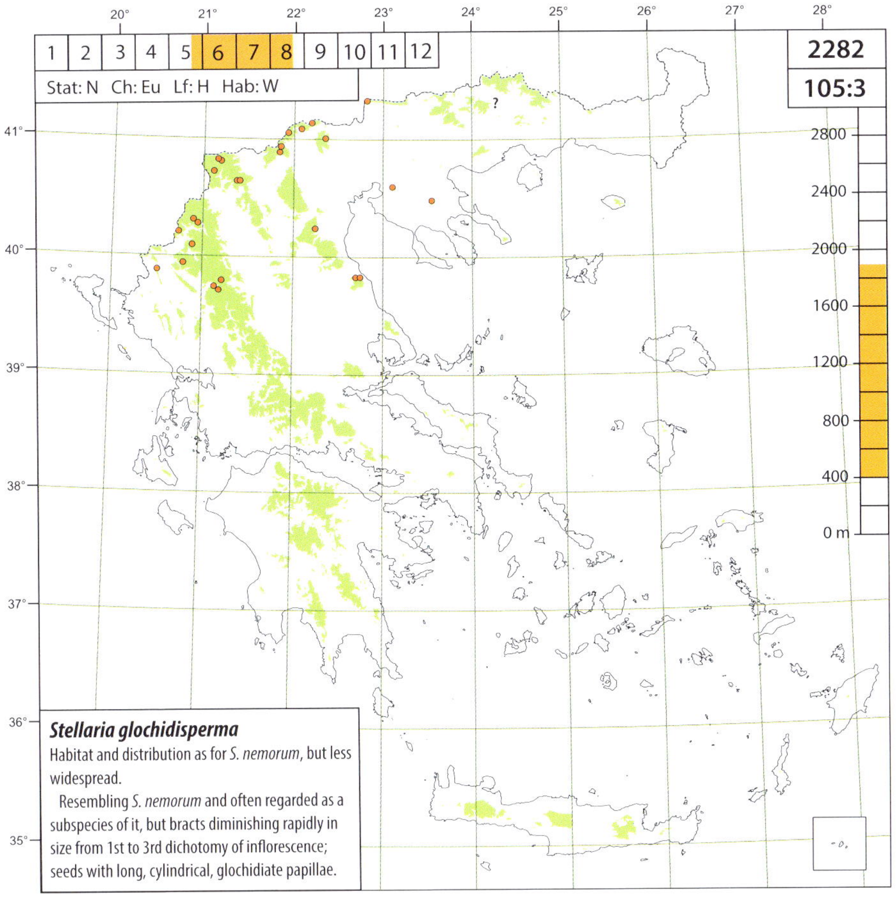

Stellaria glochidisperma

Habitat and distribution as for *S. nemorum*, but less widespread.

Resembling *S. nemorum* and often regarded as a subspecies of it, but bracts diminishing rapidly in size from 1st to 3rd dichotomy of inflorescence; seeds with long, cylindrical, glochidiate papillae.

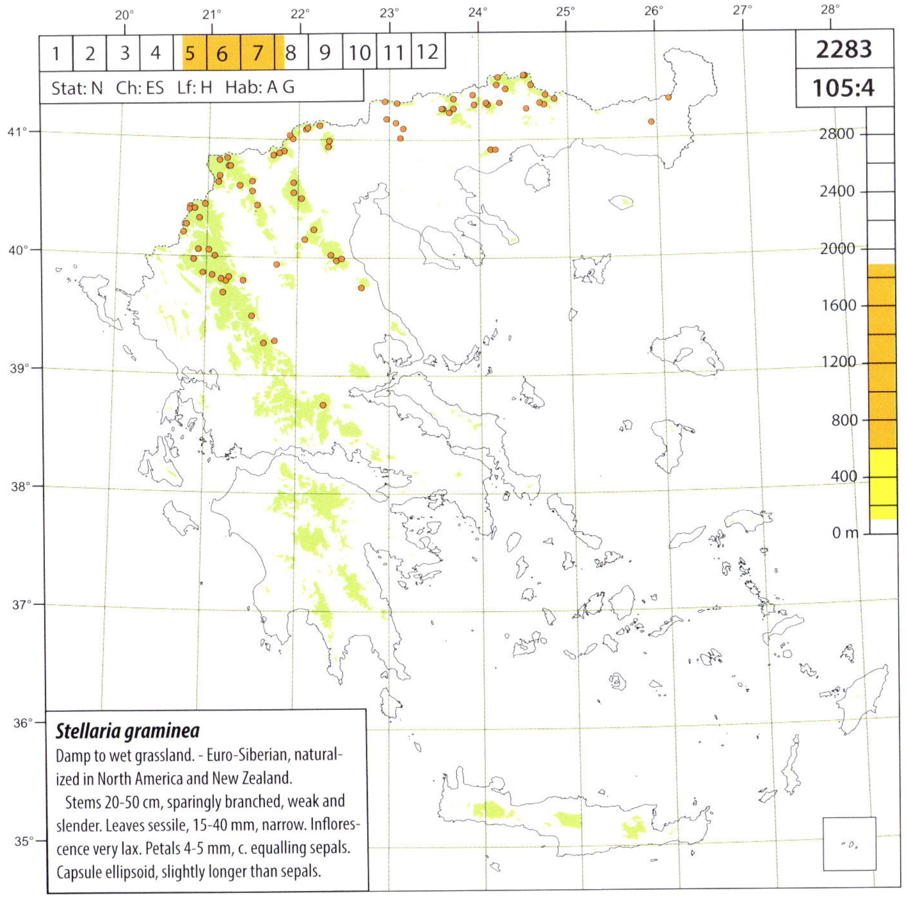

Stellaria graminea

Damp to wet grassland. - Euro-Siberian, naturalized in North America and New Zealand.

Stems 20-50 cm, sparingly branched, weak and slender. Leaves sessile, 15-40 mm, narrow. Inflorescence very lax. Petals 4-5 mm, c. equalling sepals. Capsule ellipsoid, slightly longer than sepals.

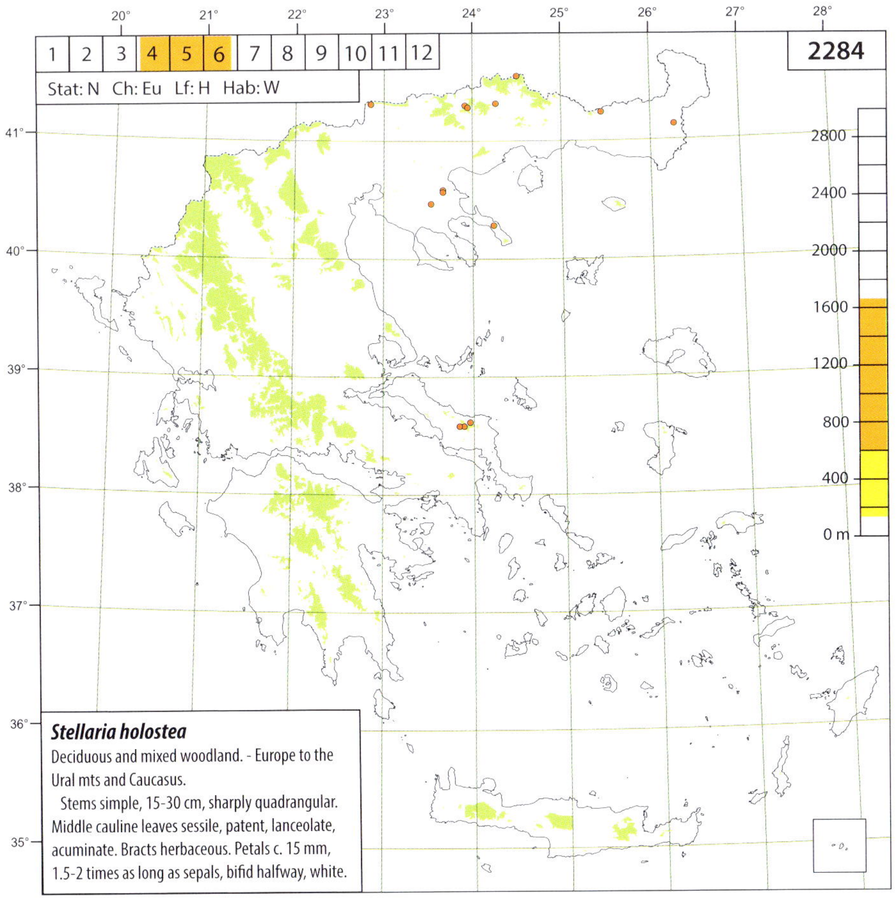

Stellaria holostea

Deciduous and mixed woodland. - Europe to the Ural mts and Caucasus.

Stems simple, 15-30 cm, sharply quadrangular. Middle cauline leaves sessile, patent, lanceolate, acuminate. Bracts herbaceous. Petals c. 15 mm, 1.5-2 times as long as sepals, bifid halfway, white.

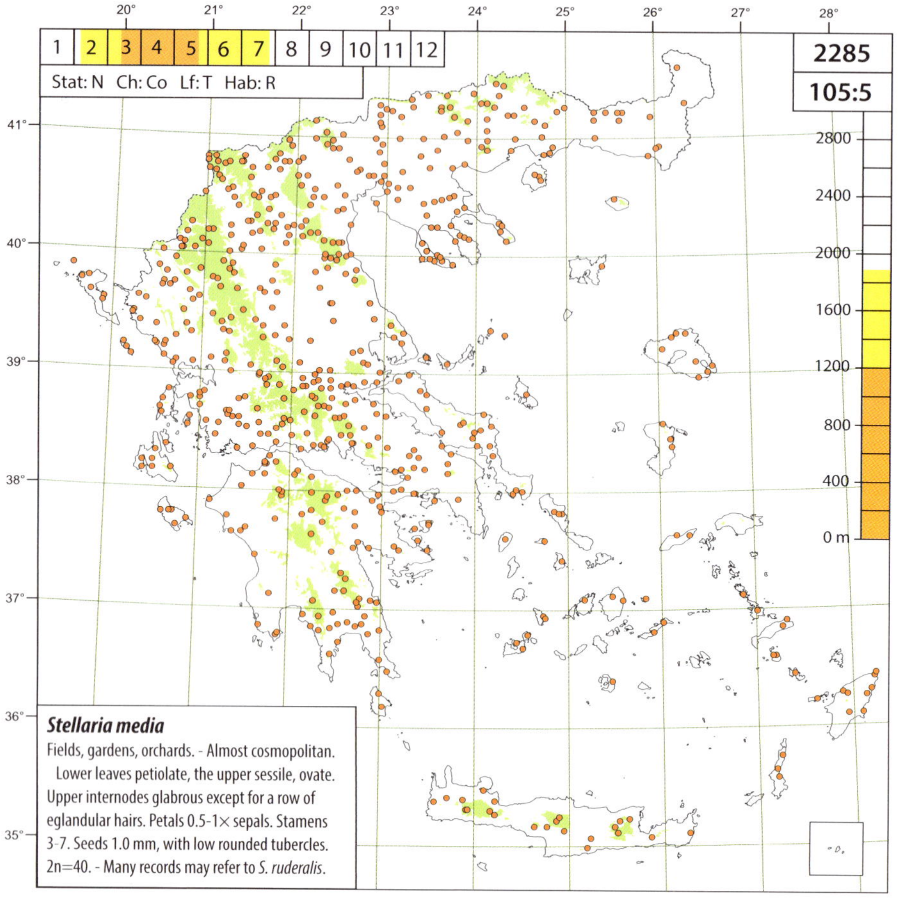

Stellaria media

Fields, gardens, orchards. - Almost cosmopolitan.

Lower leaves petiolate, the upper sessile, ovate. Upper internodes glabrous except for a row of eglandular hairs. Petals 0.5-1× sepals. Stamens 3-7. Seeds 1.0 mm, with low rounded tubercles. 2n=40. - Many records may refer to *S. ruderalis*.

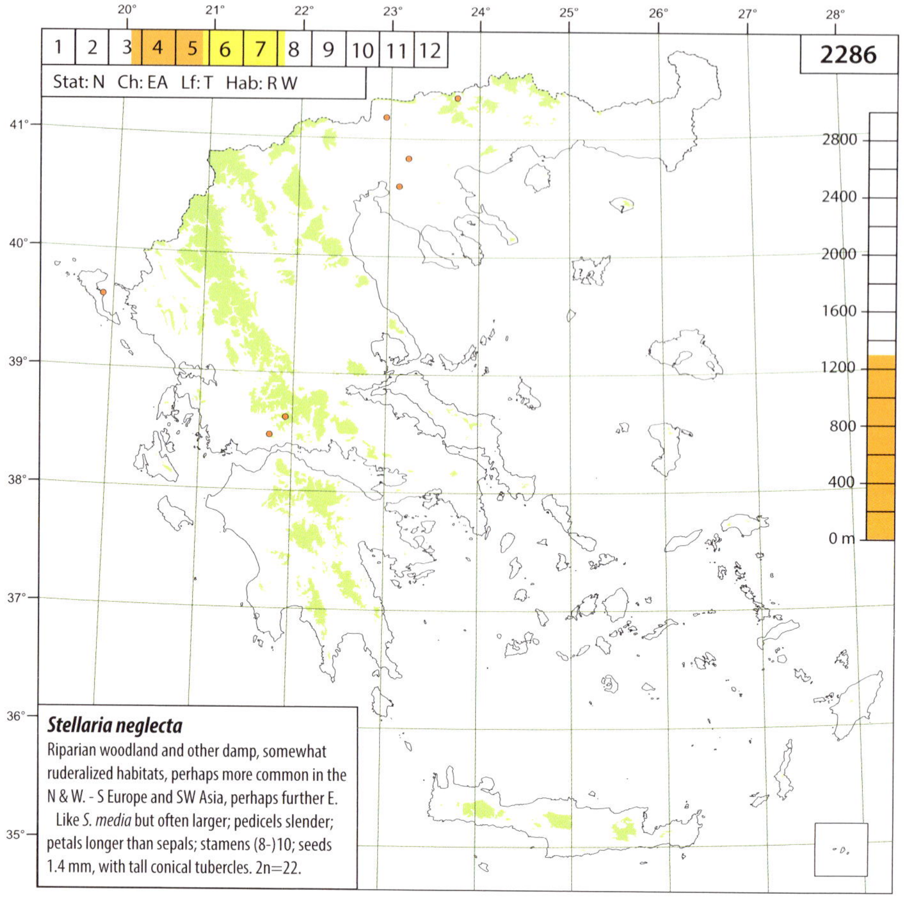

Stellaria neglecta

Riparian woodland and other damp, somewhat ruderalized habitats, perhaps more common in the N & W. - S Europe and SW Asia, perhaps further E.

Like *S. media* but often larger; pedicels slender; petals longer than sepals; stamens (8-)10; seeds 1.4 mm, with tall conical tubercles. 2n=22.

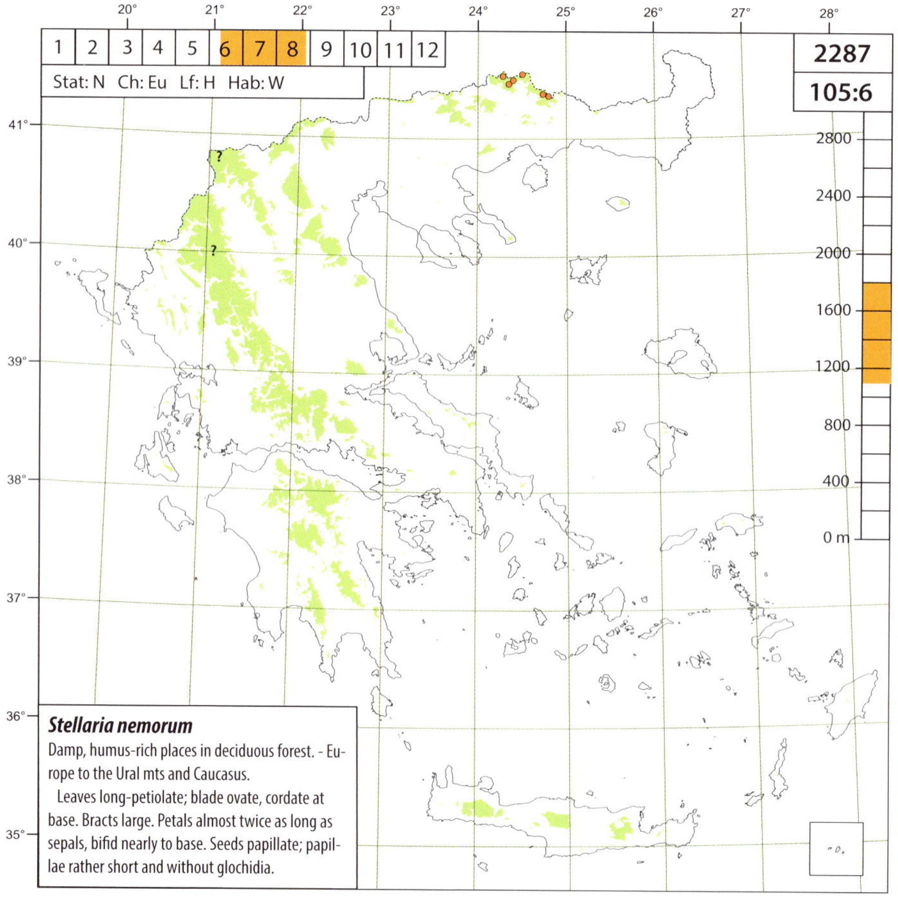

Stellaria nemorum

Damp, humus-rich places in deciduous forest. - Europe to the Ural mts and Caucasus.

Leaves long-petiolate; blade ovate, cordate at base. Bracts large. Petals almost twice as long as sepals, bifid nearly to base. Seeds papillate; papillae rather short and without glochidia.

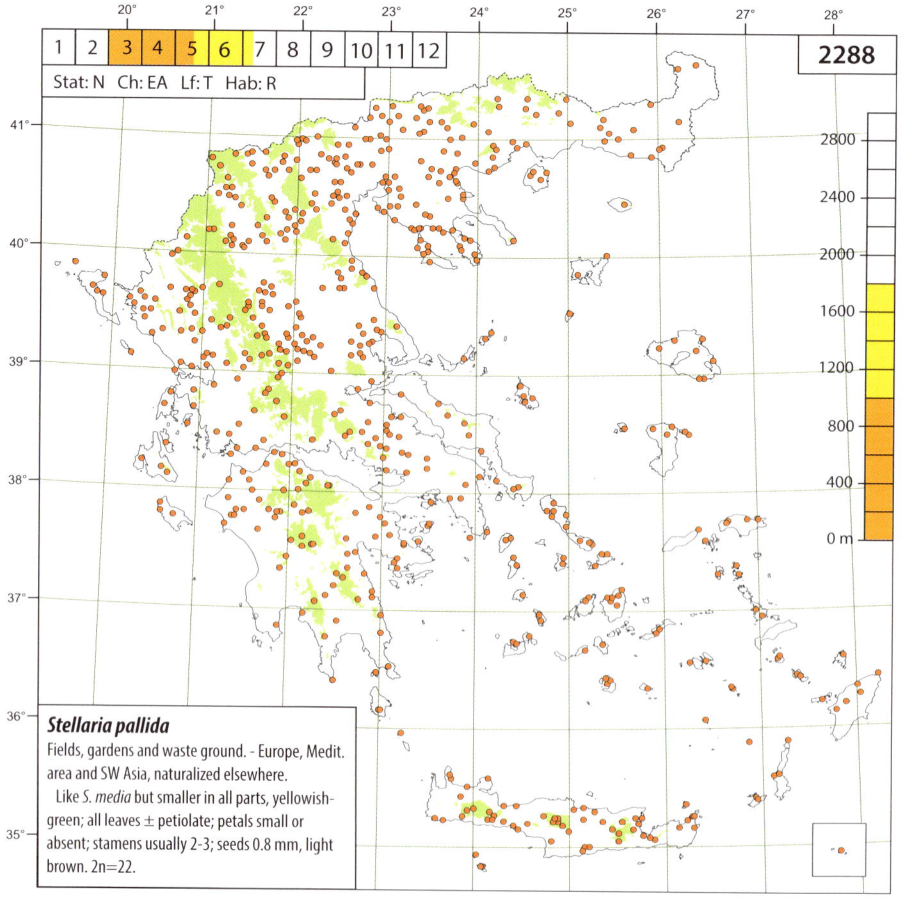

Stellaria pallida

Fields, gardens and waste ground. - Europe, Medit. area and SW Asia, naturalized elsewhere.

Like *S. media* but smaller in all parts, yellowish-green; all leaves ± petiolate; petals small or absent; stamens usually 2-3; seeds 0.8 mm, light brown. 2n=22.

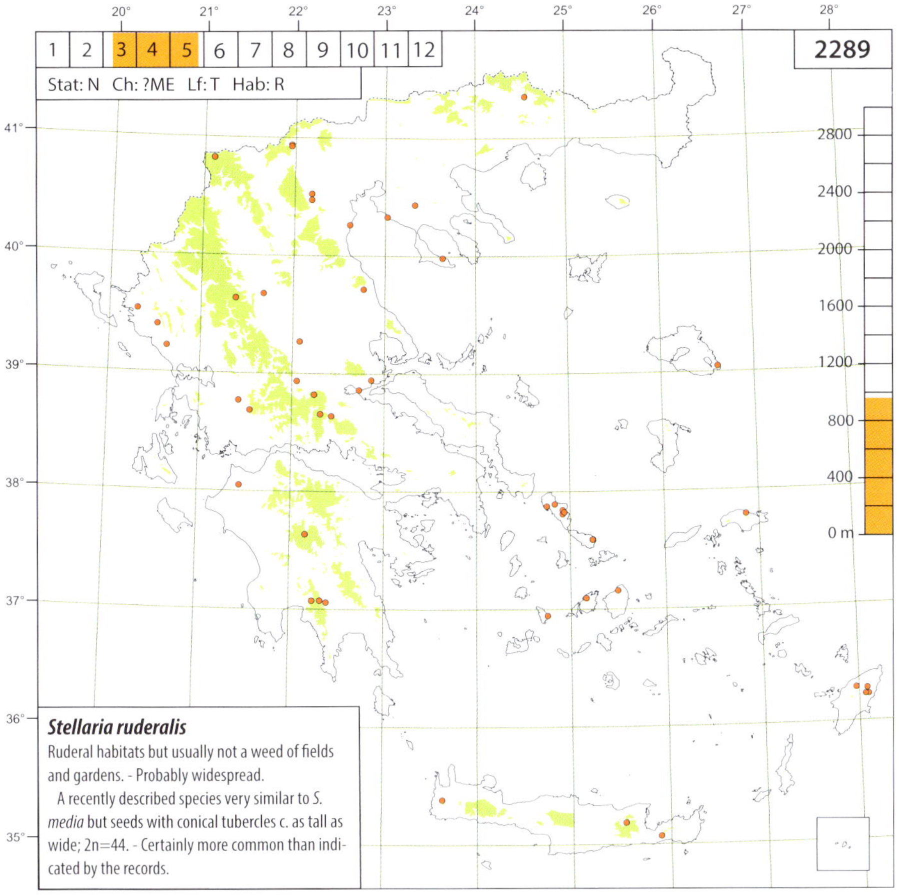

Stellaria ruderalis
Ruderal habitats but usually not a weed of fields and gardens. - Probably widespread.
A recently described species very similar to *S. media* but seeds with conical tubercles c. as tall as wide; 2n=44. - Certainly more common than indicated by the records.

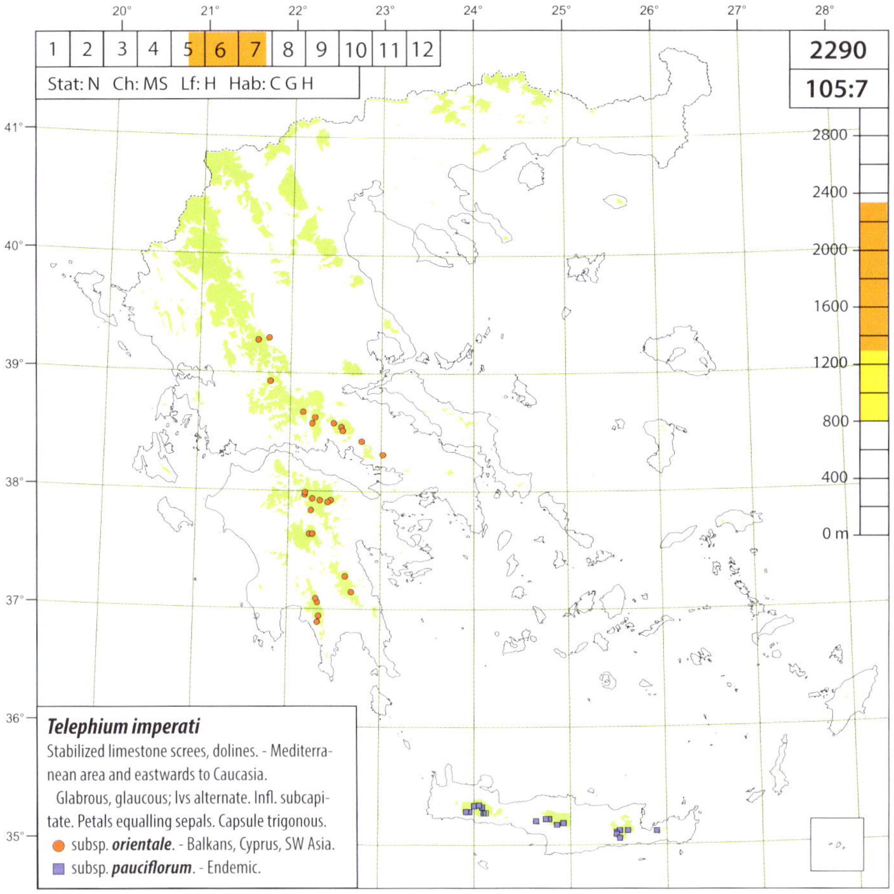

Telephium imperati
Stabilized limestone screes, dolines. - Mediterranean area and eastwards to Caucasia.
Glabrous, glaucous; lvs alternate. Infl. subcapitate. Petals equalling sepals. Capsule trigonous.
● subsp. ***orientale***. - Balkans, Cyprus, SW Asia.
■ subsp. ***pauciflorum***. - Endemic.

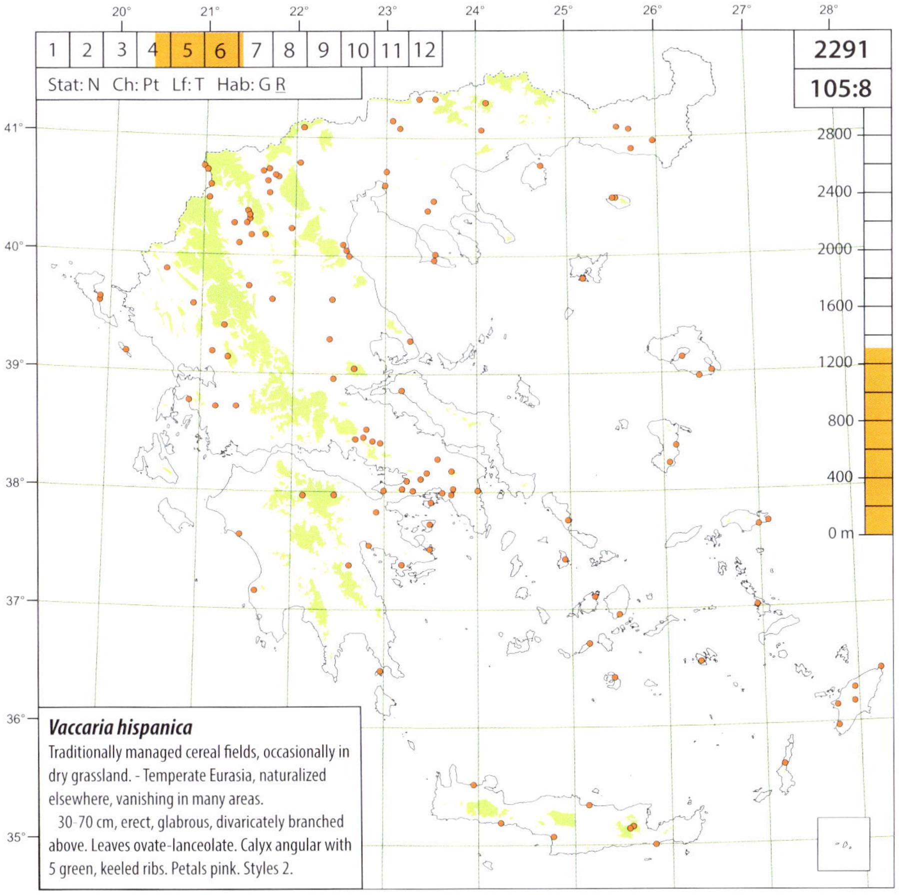

Vaccaria hispanica
Traditionally managed cereal fields, occasionally in dry grassland. - Temperate Eurasia, naturalized elsewhere, vanishing in many areas.
30-70 cm, erect, glabrous, divaricately branched above. Leaves ovate-lanceolate. Calyx angular with 5 green, keeled ribs. Petals pink. Styles 2.

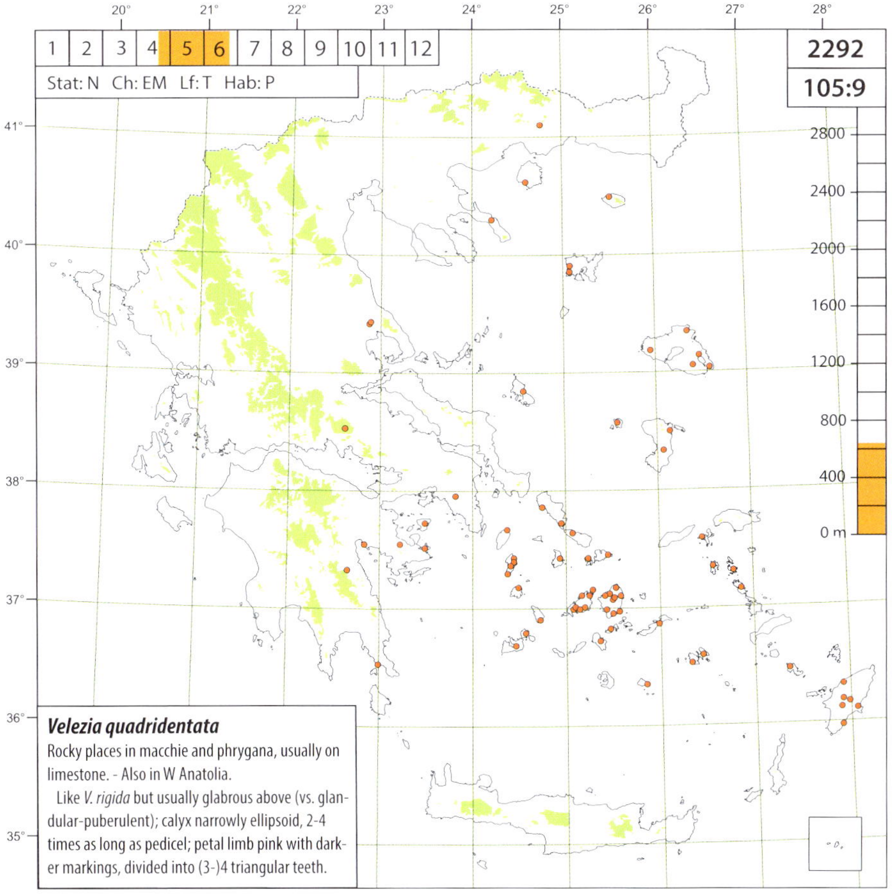

Velezia quadridentata
Rocky places in macchie and phrygana, usually on limestone. - Also in W Anatolia.
Like *V. rigida* but usually glabrous above (vs. glandular-puberulent); calyx narrowly ellipsoid, 2-4 times as long as pedicel; petal limb pink with darker markings, divided into (3-)4 triangular teeth.

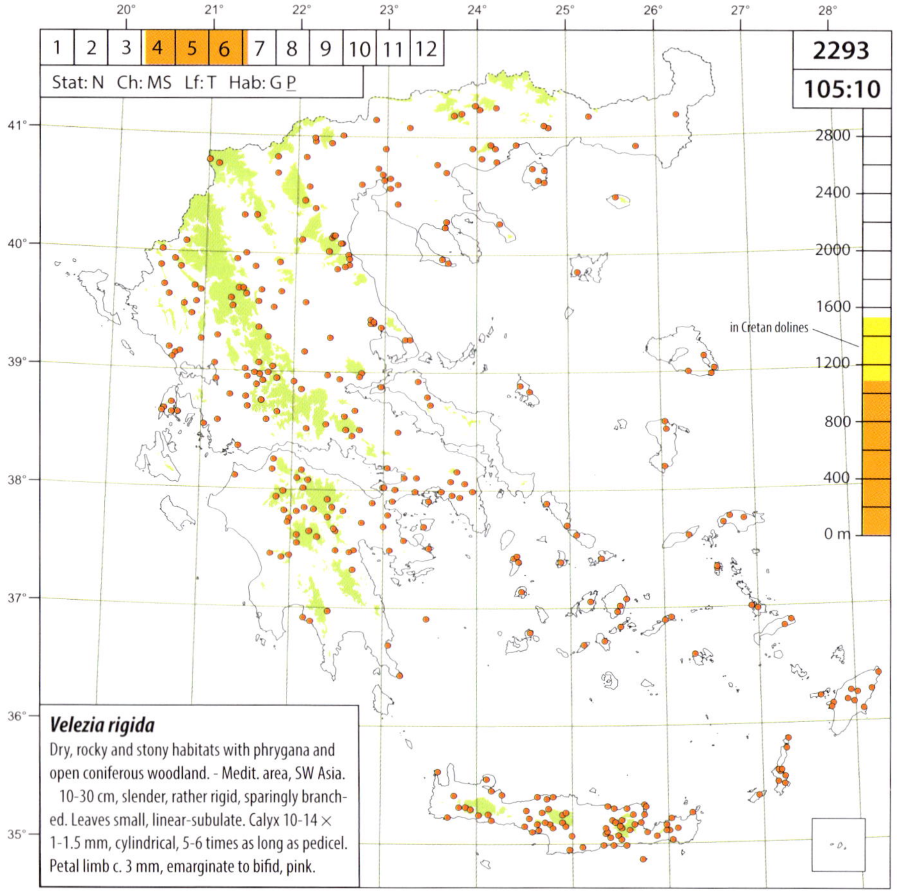

Velezia rigida
Dry, rocky and stony habitats with phrygana and open coniferous woodland. - Medit. area, SW Asia.
10-30 cm, slender, rather rigid, sparingly branched. Leaves small, linear-subulate. Calyx 10-14 × 1-1.5 mm, cylindrical, 5-6 times as long as pedicel. Petal limb c. 3 mm, emarginate to bifid, pink.

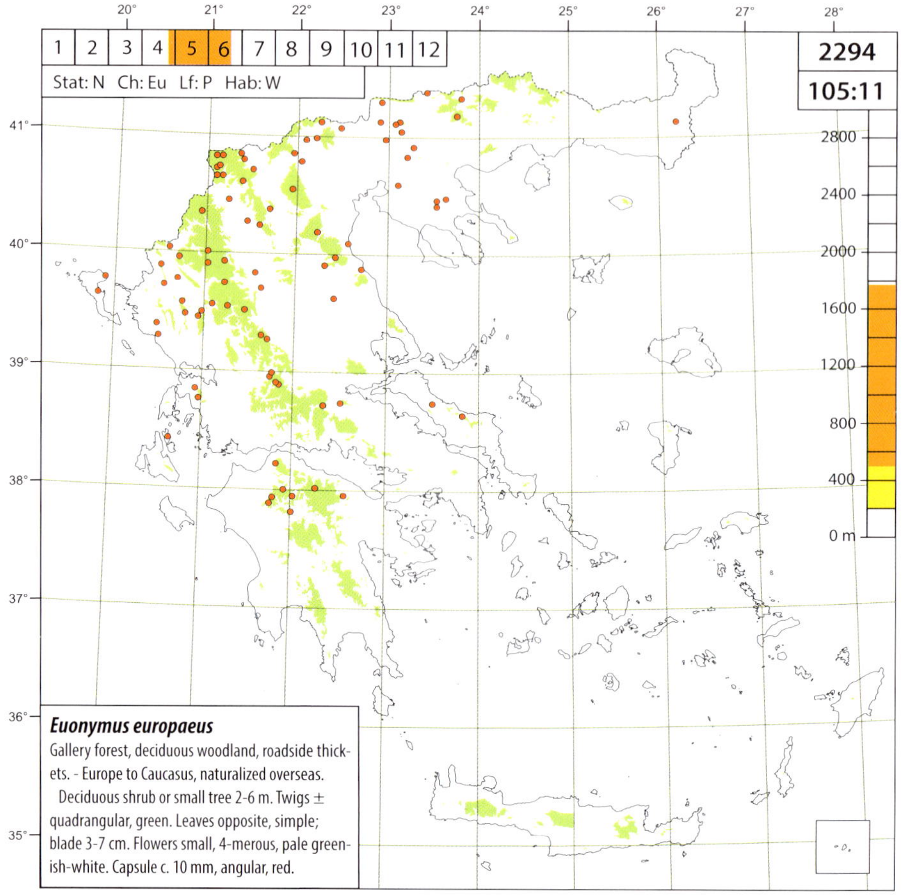

Euonymus europaeus
Gallery forest, deciduous woodland, roadside thickets. - Europe to Caucasus, naturalized overseas.
Deciduous shrub or small tree 2-6 m. Twigs ± quadrangular, green. Leaves opposite, simple; blade 3-7 cm. Flowers small, 4-merous, pale greenish-white. Capsule c. 10 mm, angular, red.

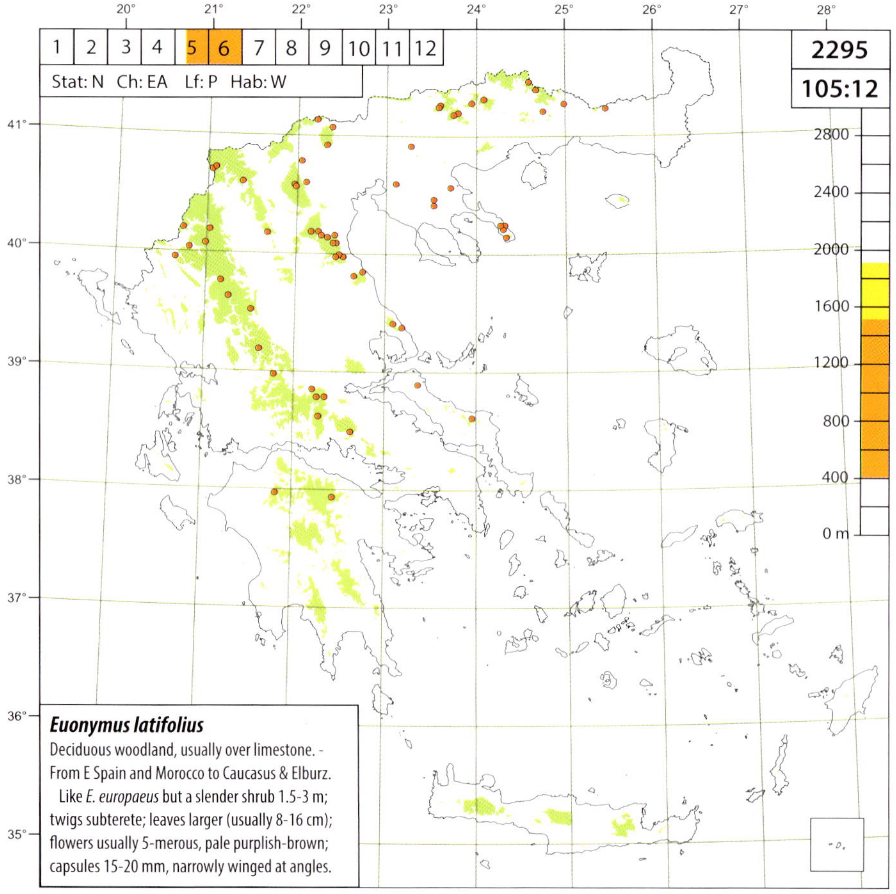

Euonymus latifolius
Deciduous woodland, usually over limestone. - From E Spain and Morocco to Caucasus & Elburz.
Like *E. europaeus* but a slender shrub 1.5-3 m; twigs subterete; leaves larger (usually 8-16 cm); flowers usually 5-merous, pale purplish-brown; capsules 15-20 mm, narrowly winged at angles.

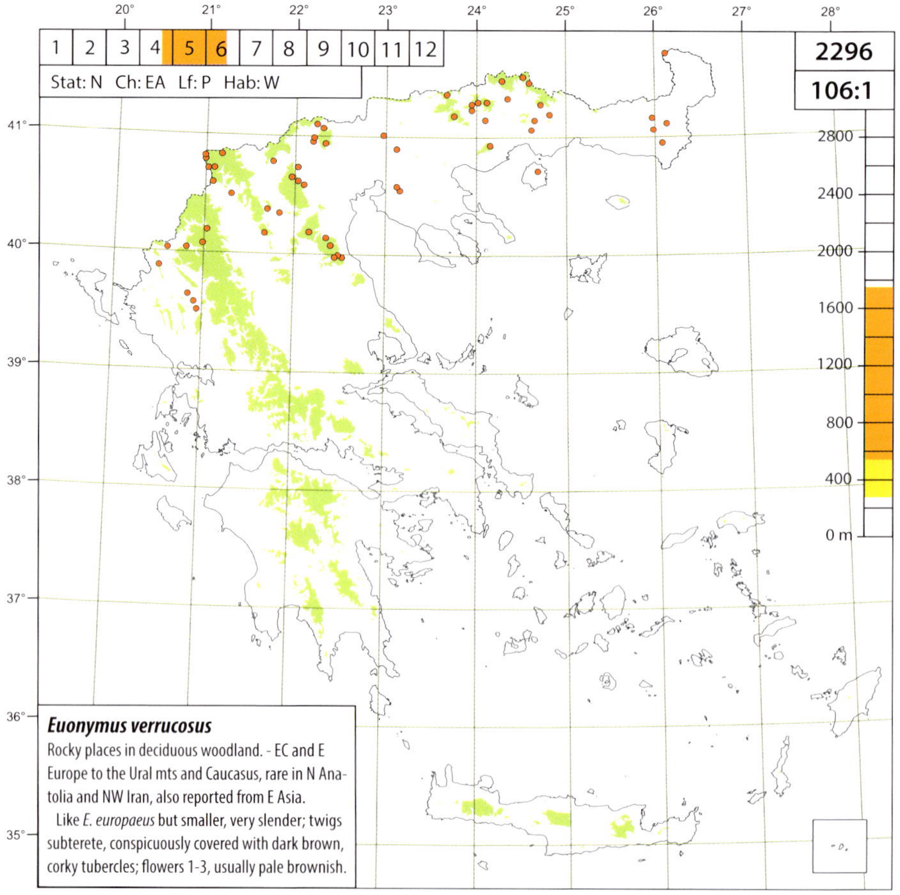

Euonymus verrucosus
Rocky places in deciduous woodland. - EC and E Europe to the Ural mts and Caucasus, rare in N Anatolia and NW Iran, also reported from E Asia.
Like *E. europaeus* but smaller, very slender; twigs subterete, conspicuously covered with dark brown, corky tubercles; flowers 1-3, usually pale brownish.

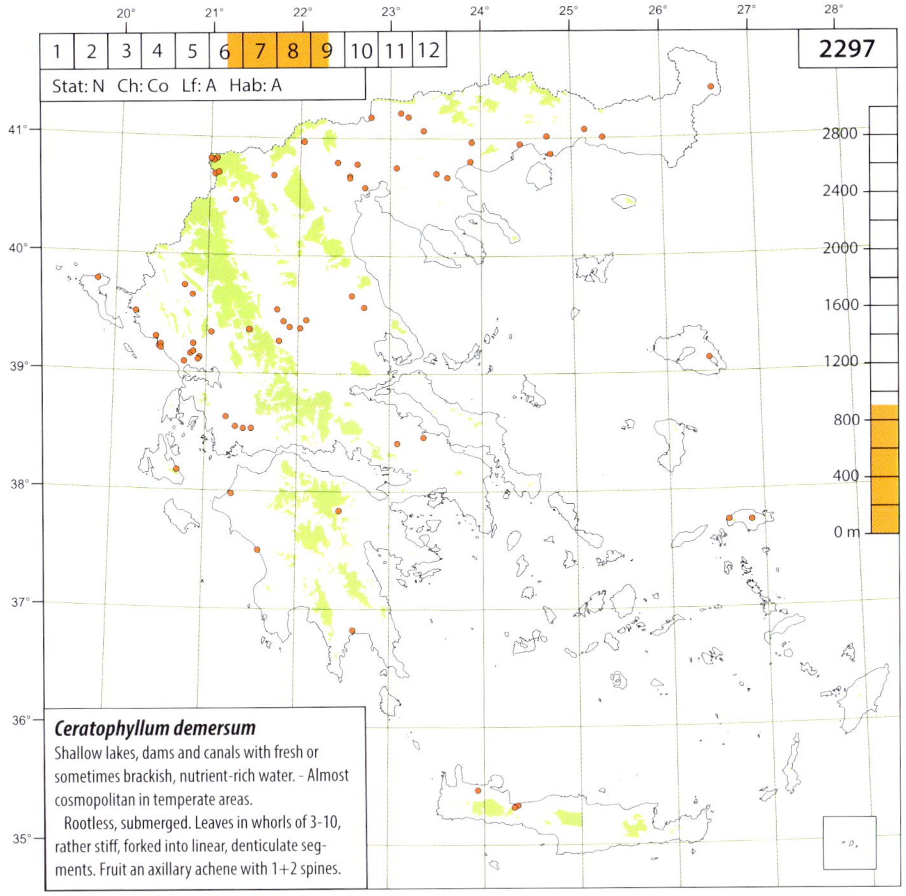

Ceratophyllum demersum

Shallow lakes, dams and canals with fresh or sometimes brackish, nutrient-rich water. - Almost cosmopolitan in temperate areas.

Rootless, submerged. Leaves in whorls of 3-10, rather stiff, forked into linear, denticulate segments. Fruit an axillary achene with 1+2 spines.

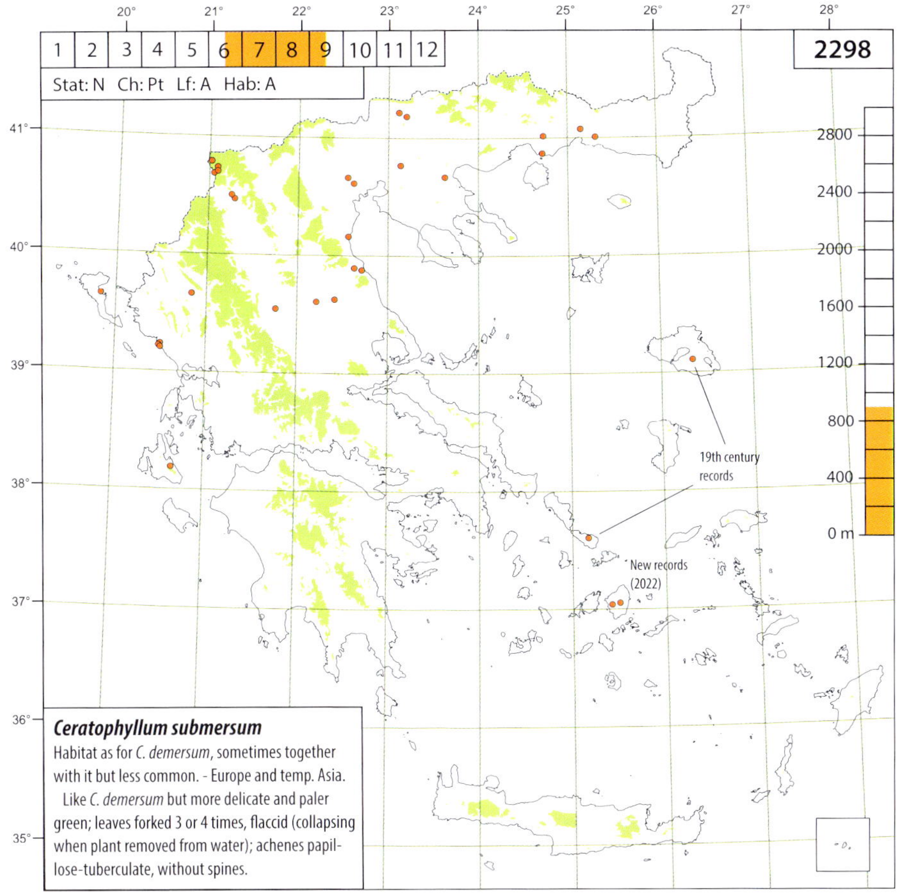

Ceratophyllum submersum

Habitat as for *C. demersum*, sometimes together with it but less common. - Europe and temp. Asia.

Like *C. demersum* but more delicate and paler green; leaves forked 3 or 4 times, flaccid (collapsing when plant removed from water); achenes papillose-tuberculate, without spines.

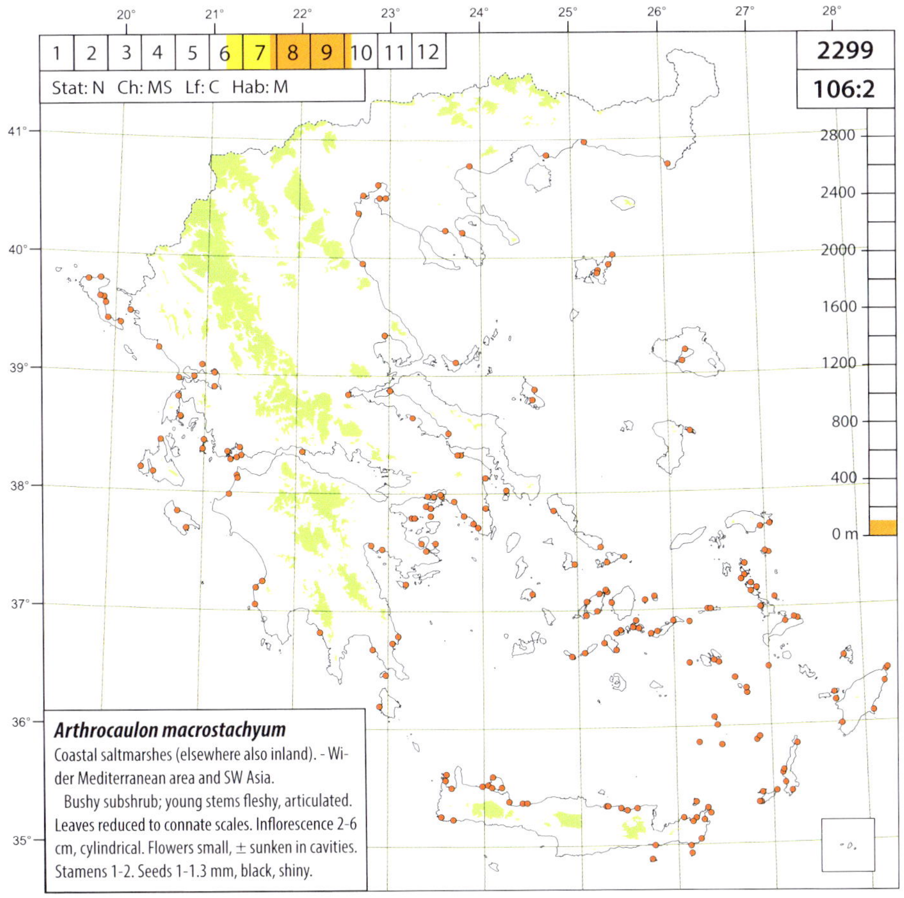

Arthrocaulon macrostachyum

Coastal saltmarshes (elsewhere also inland). - Wider Mediterranean area and SW Asia.

Bushy subshrub; young stems fleshy, articulated. Leaves reduced to connate scales. Inflorescence 2-6 cm, cylindrical. Flowers small, ± sunken in cavities. Stamens 1-2. Seeds 1-1.3 mm, black, shiny.

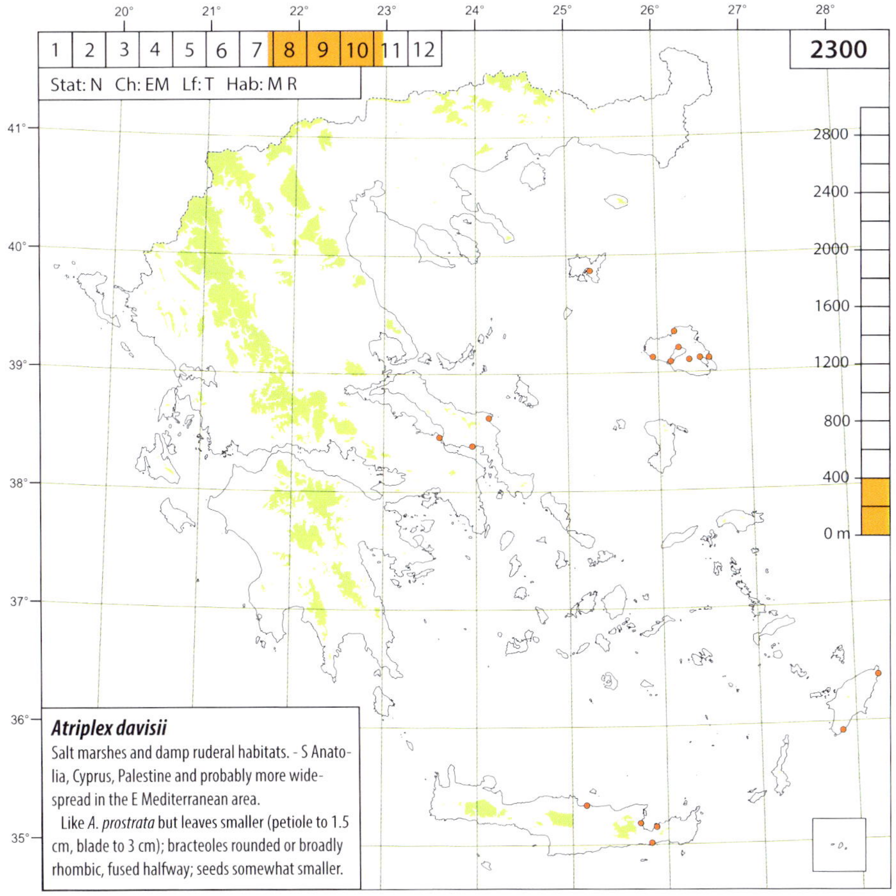

Atriplex davisii

Salt marshes and damp ruderal habitats. - S Anatolia, Cyprus, Palestine and probably more widespread in the E Mediterranean area.

Like *A. prostrata* but leaves smaller (petiole to 1.5 cm, blade to 3 cm); bracteoles rounded or broadly rhombic, fused halfway; seeds somewhat smaller.

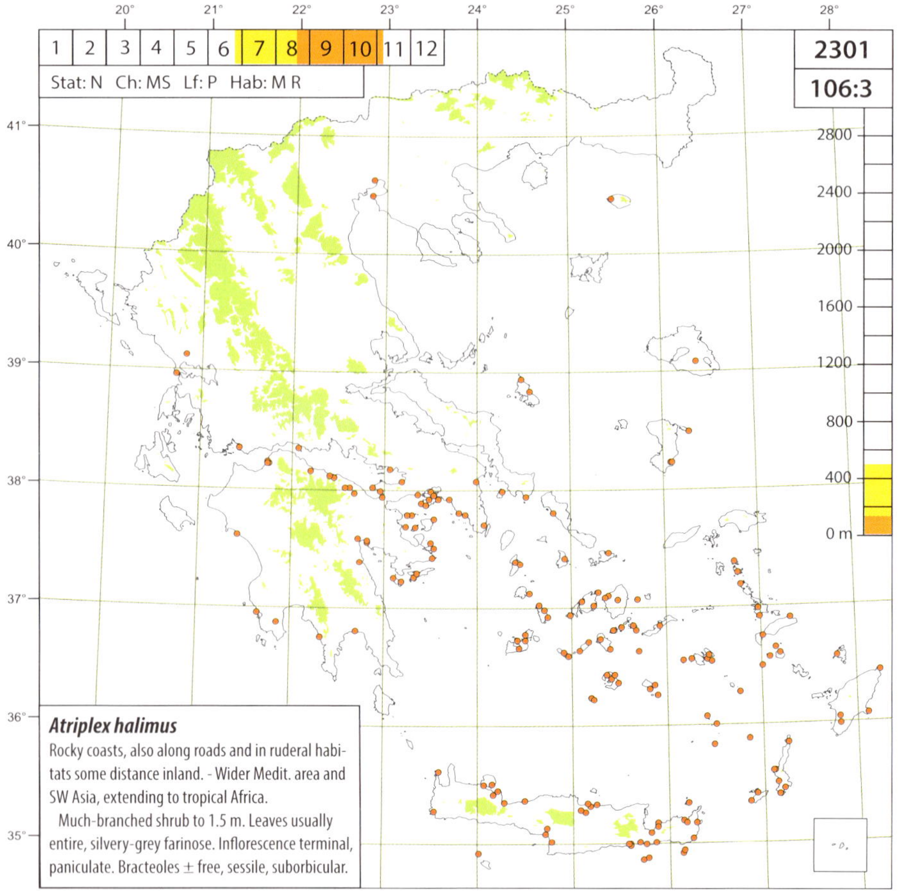

Atriplex halimus

Rocky coasts, also along roads and in ruderal habitats some distance inland. - Wider Medit. area and SW Asia, extending to tropical Africa.

Much-branched shrub to 1.5 m. Leaves usually entire, silvery-grey farinose. Inflorescence terminal, paniculate. Bracteoles ± free, sessile, suborbicular.

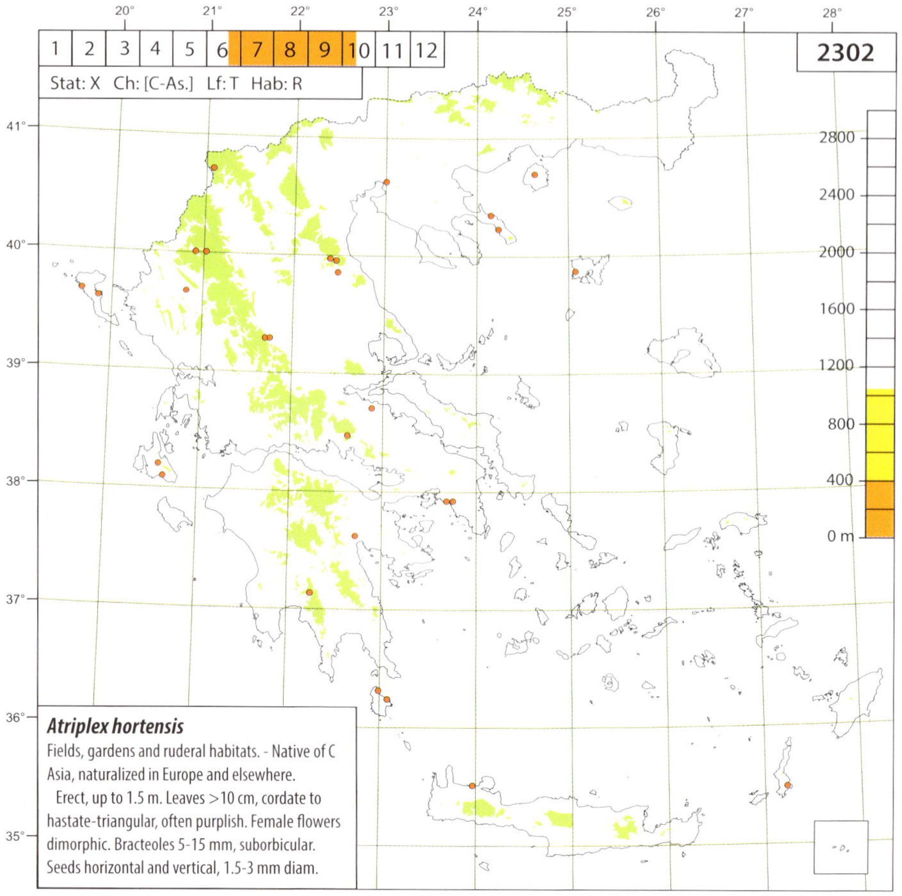

Atriplex hortensis

Fields, gardens and ruderal habitats. - Native of C Asia, naturalized in Europe and elsewhere.

Erect, up to 1.5 m. Leaves >10 cm, cordate to hastate-triangular, often purplish. Female flowers dimorphic. Bracteoles 5-15 mm, suborbicular. Seeds horizontal and vertical, 1.5-3 mm diam.

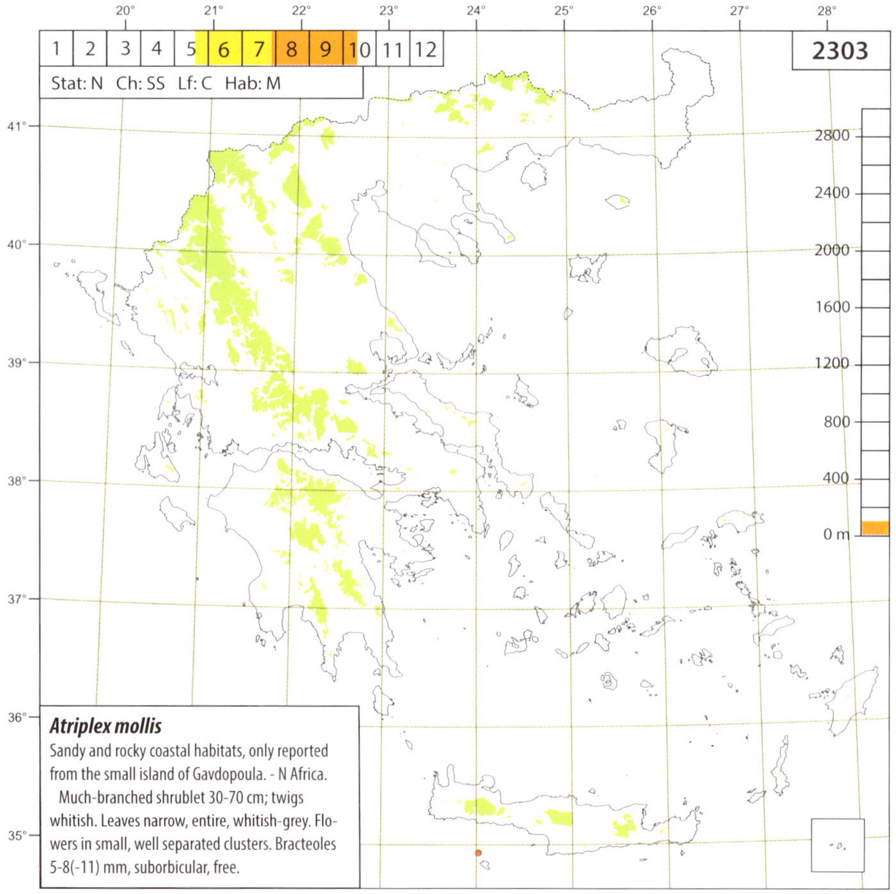

Atriplex mollis

Sandy and rocky coastal habitats, only reported from the small island of Gavdopoula. - N Africa.

Much-branched shrublet 30-70 cm; twigs whitish. Leaves narrow, entire, whitish-grey. Flowers in small, well separated clusters. Bracteoles 5-8(-11) mm, suborbicular, free.

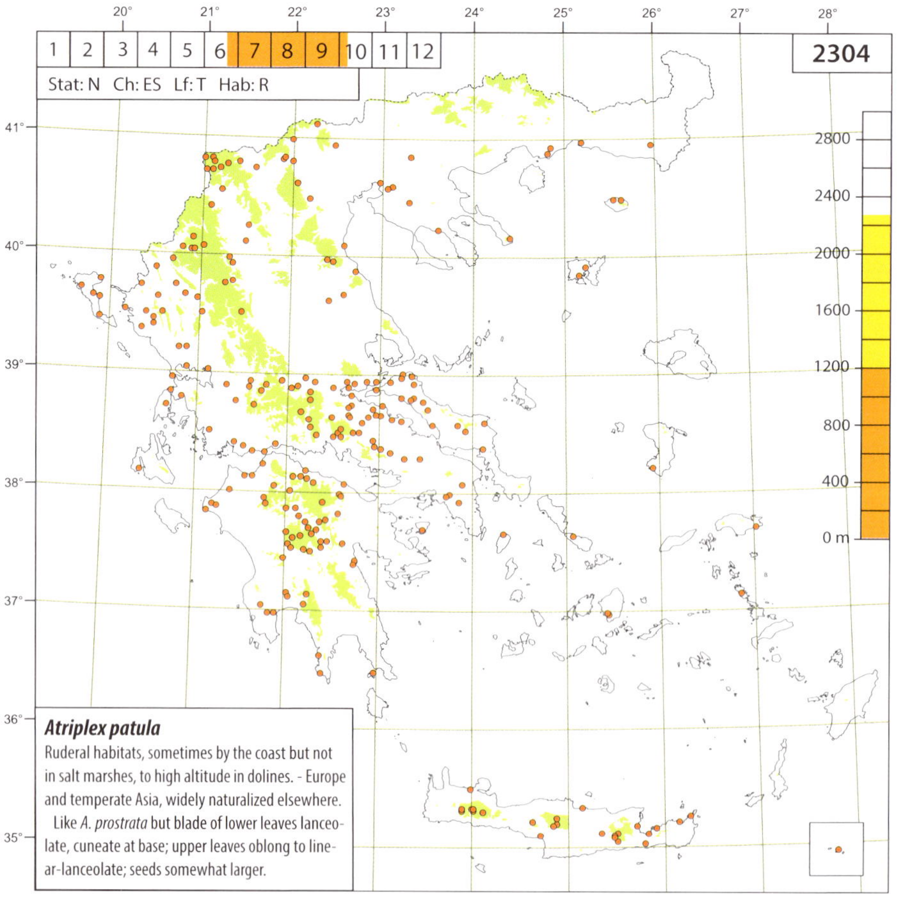

Atriplex patula

Ruderal habitats, sometimes by the coast but not in salt marshes, to high altitude in dolines. - Europe and temperate Asia, widely naturalized elsewhere.

Like *A. prostrata* but blade of lower leaves lanceolate, cuneate at base; upper leaves oblong to linear-lanceolate; seeds somewhat larger.

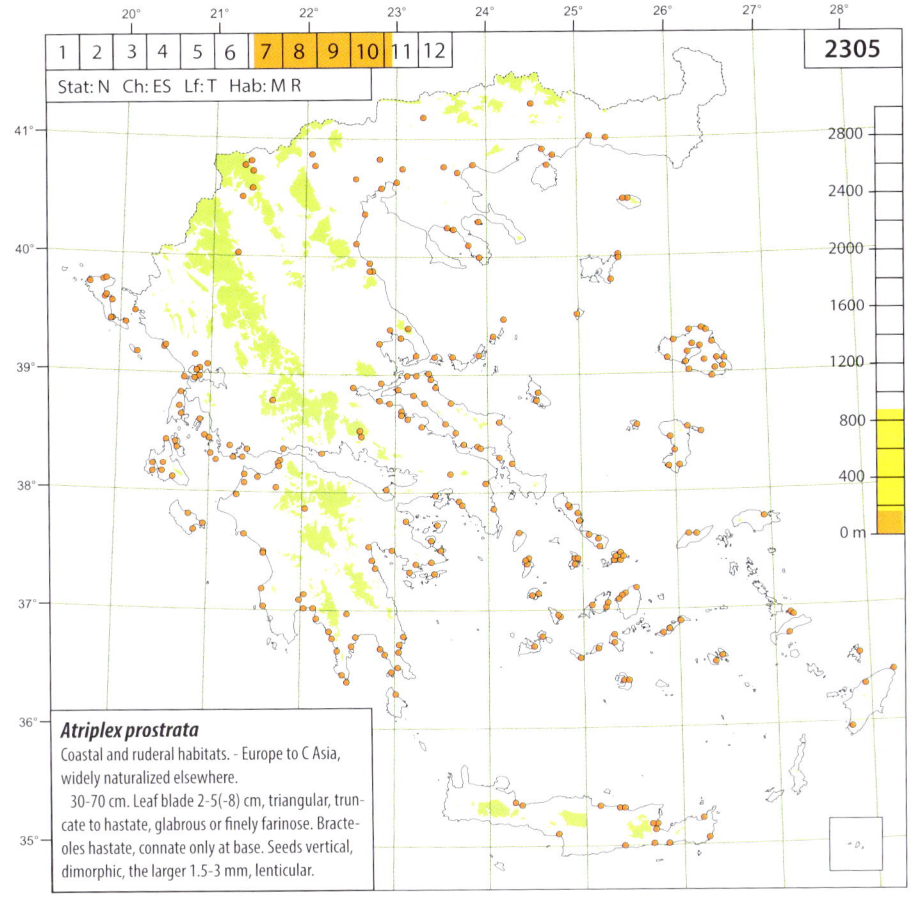

Atriplex prostrata
Coastal and ruderal habitats. - Europe to C Asia, widely naturalized elsewhere.
30-70 cm. Leaf blade 2-5(-8) cm, triangular, truncate to hastate, glabrous or finely farinose. Bracteoles hastate, connate only at base. Seeds vertical, dimorphic, the larger 1.5-3 mm, lenticular.

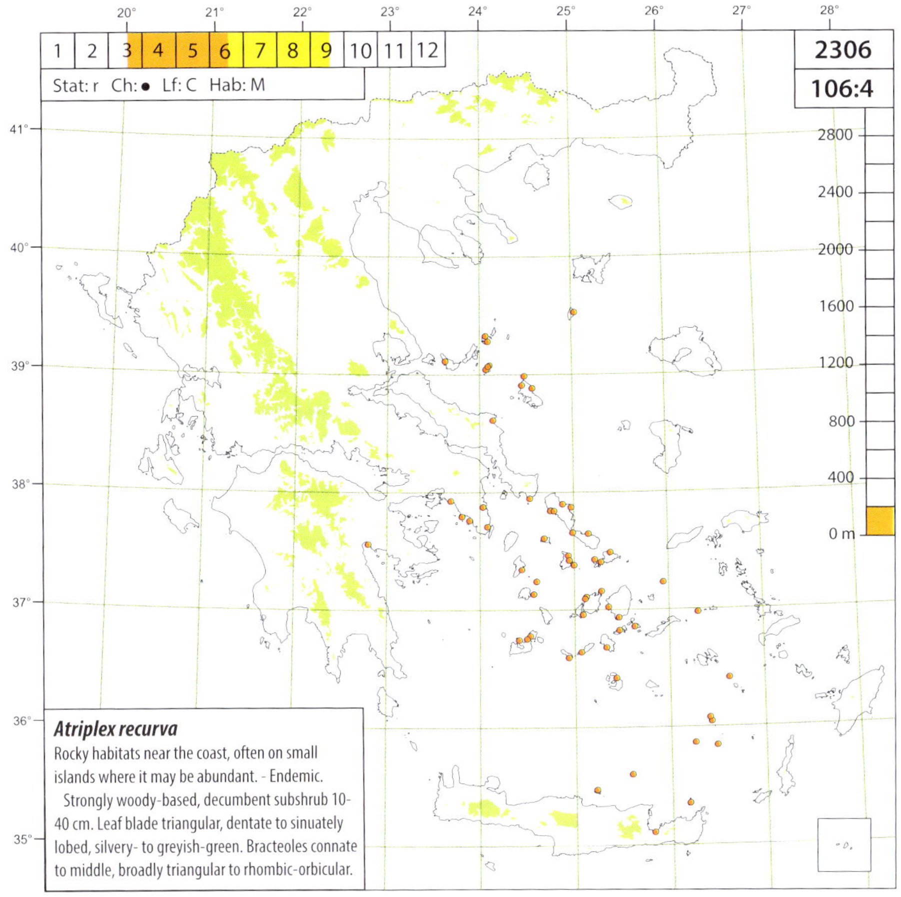

Atriplex recurva
Rocky habitats near the coast, often on small islands where it may be abundant. - Endemic.
Strongly woody-based, decumbent subshrub 10-40 cm. Leaf blade triangular, dentate to sinuately lobed, silvery- to greyish-green. Bracteoles connate to middle, broadly triangular to rhombic-orbicular.

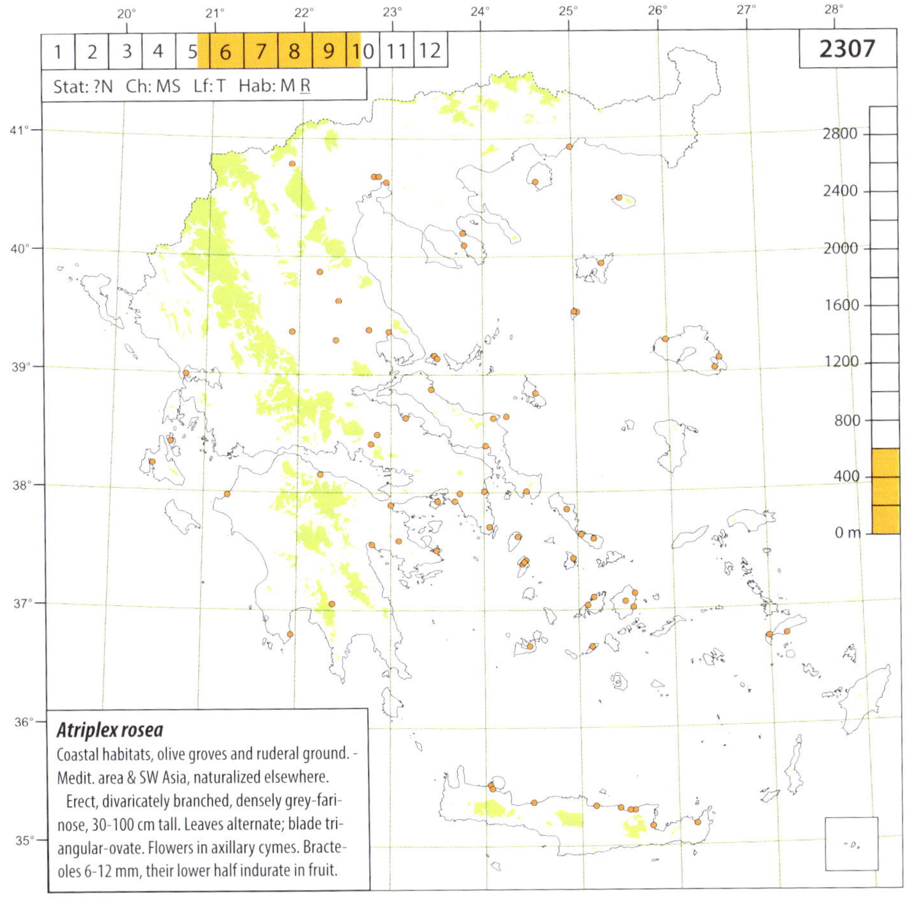

Atriplex rosea
Coastal habitats, olive groves and ruderal ground. - Medit. area & SW Asia, naturalized elsewhere.
Erect, divaricately branched, densely grey-farinose, 30-100 cm tall. Leaves alternate; blade triangular-ovate. Flowers in axillary cymes. Bracteoles 6-12 mm, their lower half indurate in fruit.

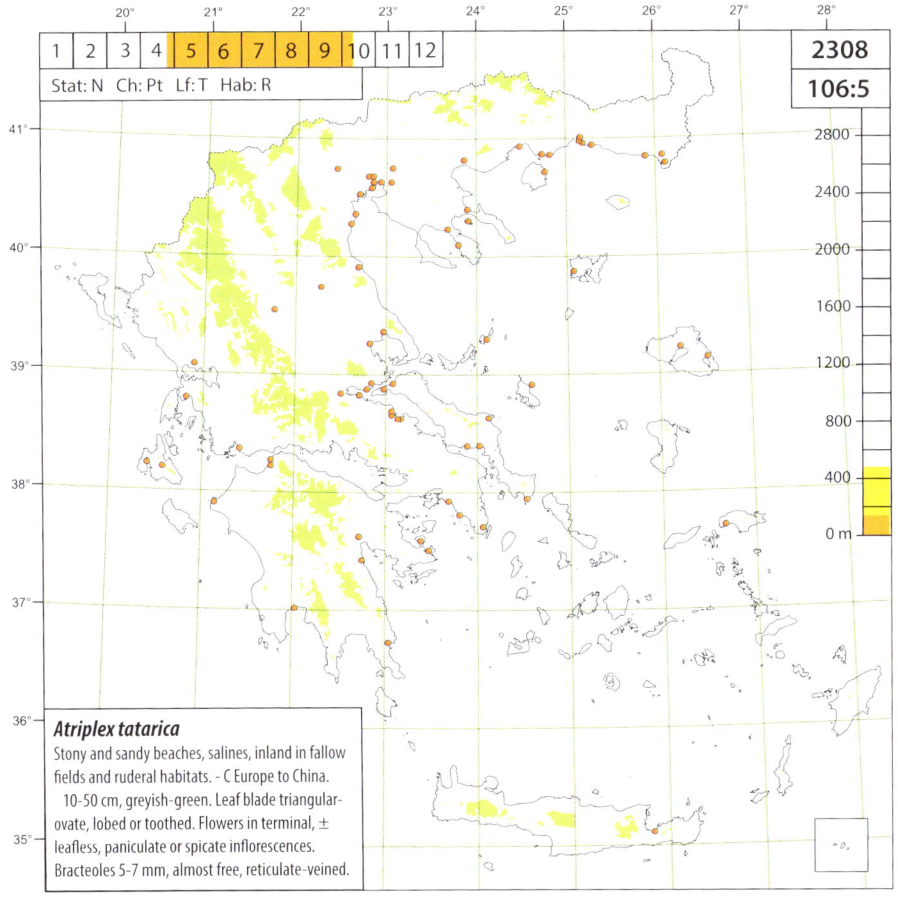

Atriplex tatarica
Stony and sandy beaches, salines, inland in fallow fields and ruderal habitats. - C Europe to China.
10-50 cm, greyish-green. Leaf blade triangular-ovate, lobed or toothed. Flowers in terminal, ± leafless, paniculate or spicate inflorescences. Bracteoles 5-7 mm, almost free, reticulate-veined.

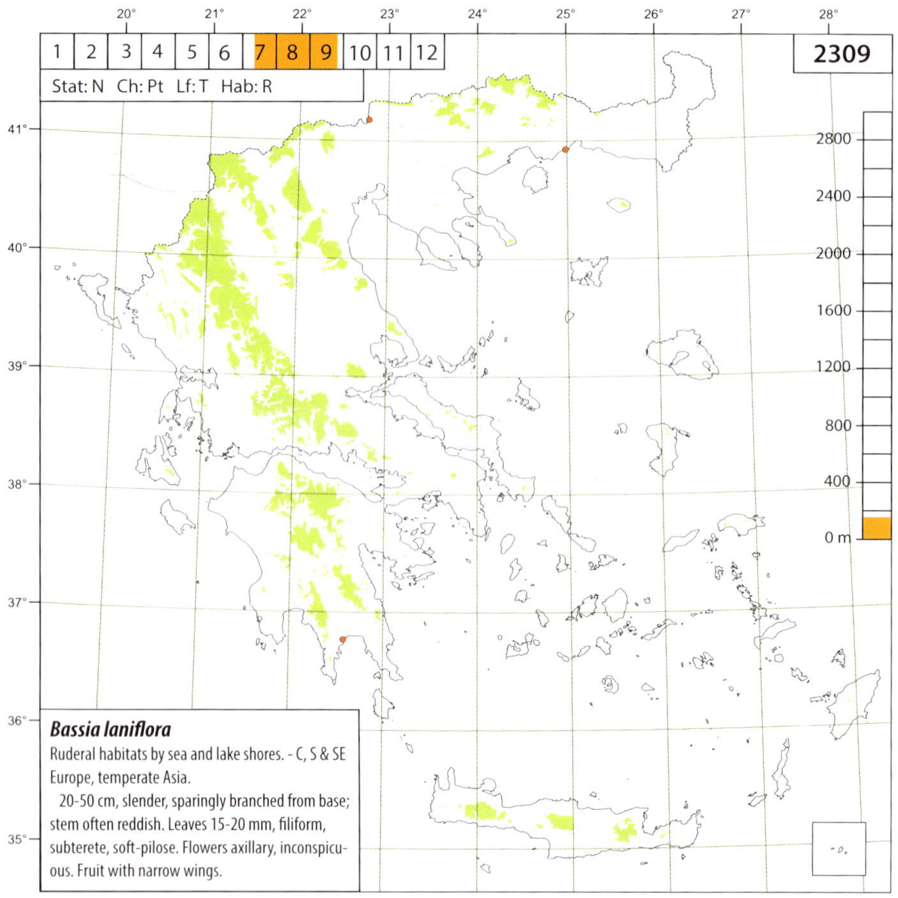

Bassia laniflora
Ruderal habitats by sea and lake shores. - C, S & SE Europe, temperate Asia.
20-50 cm, slender, sparingly branched from base; stem often reddish. Leaves 15-20 mm, filiform, subterete, soft-pilose. Flowers axillary, inconspicuous. Fruit with narrow wings.

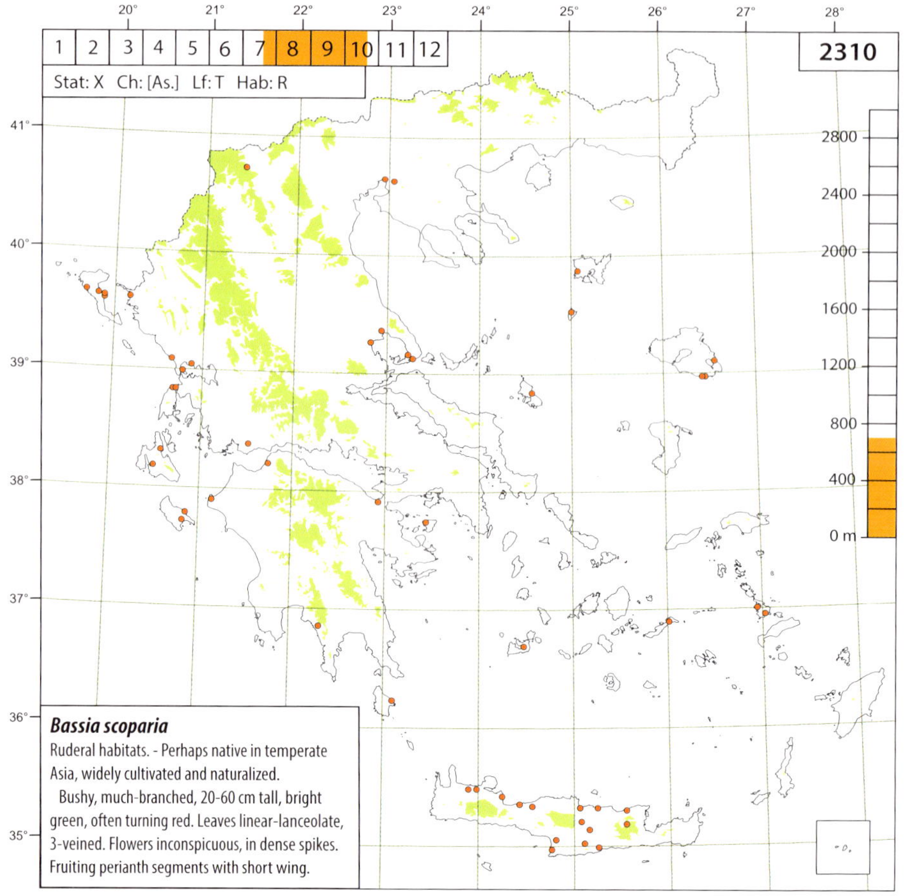

Bassia scoparia
Ruderal habitats. - Perhaps native in temperate Asia, widely cultivated and naturalized.
Bushy, much-branched, 20-60 cm tall, bright green, often turning red. Leaves linear-lanceolate, 3-veined. Flowers inconspicuous, in dense spikes. Fruiting perianth segments with short wing.

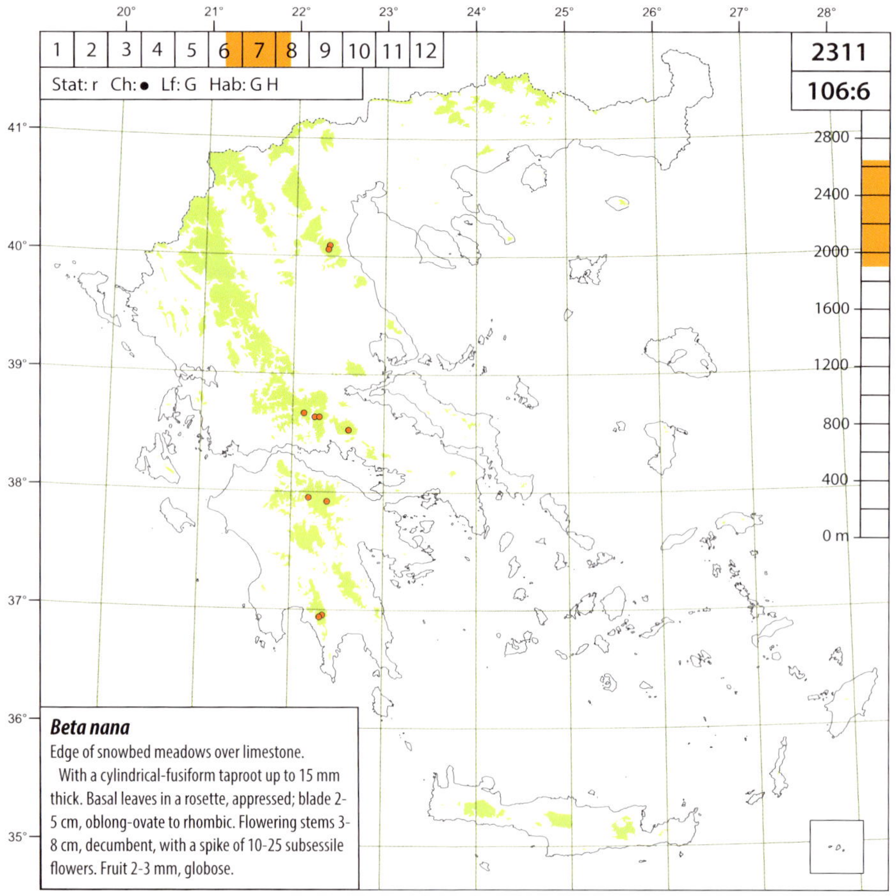

Beta nana
Edge of snowbed meadows over limestone.
With a cylindrical-fusiform taproot up to 15 mm thick. Basal leaves in a rosette, appressed; blade 2-5 cm, oblong-ovate to rhombic. Flowering stems 3-8 cm, decumbent, with a spike of 10-25 subsessile flowers. Fruit 2-3 mm, globose.

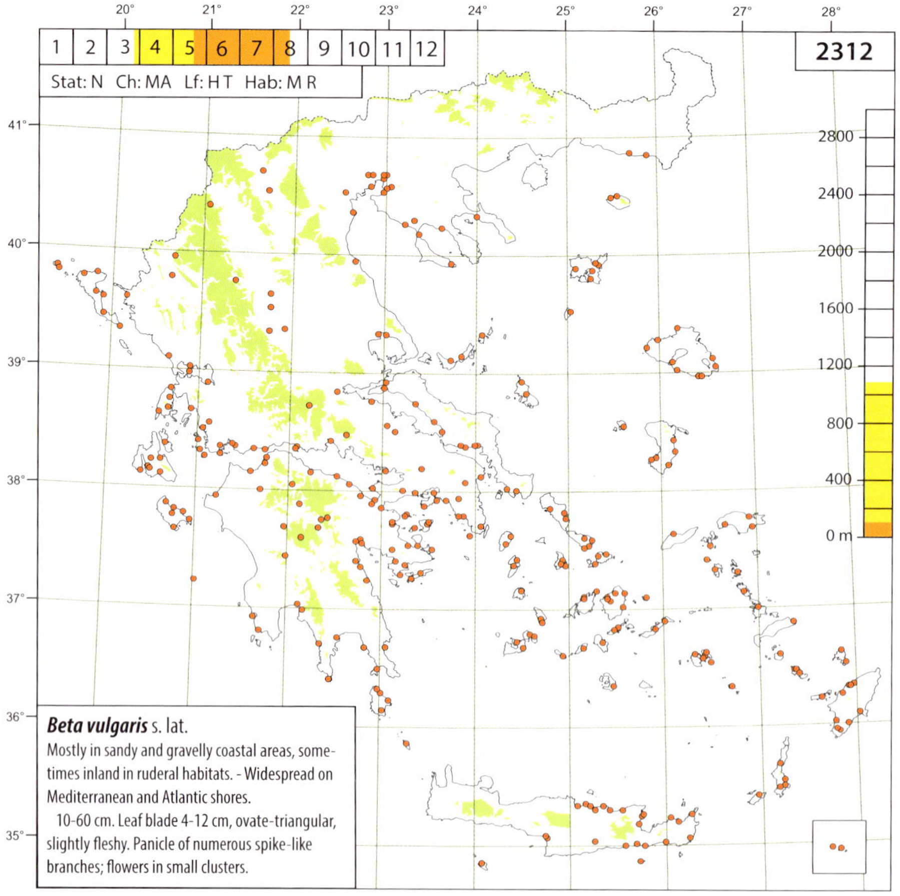

Beta vulgaris s. lat.
Mostly in sandy and gravelly coastal areas, sometimes inland in ruderal habitats. - Widespread on Mediterranean and Atlantic shores.
10-60 cm. Leaf blade 4-12 cm, ovate-triangular, slightly fleshy. Panicle of numerous spike-like branches; flowers in small clusters.

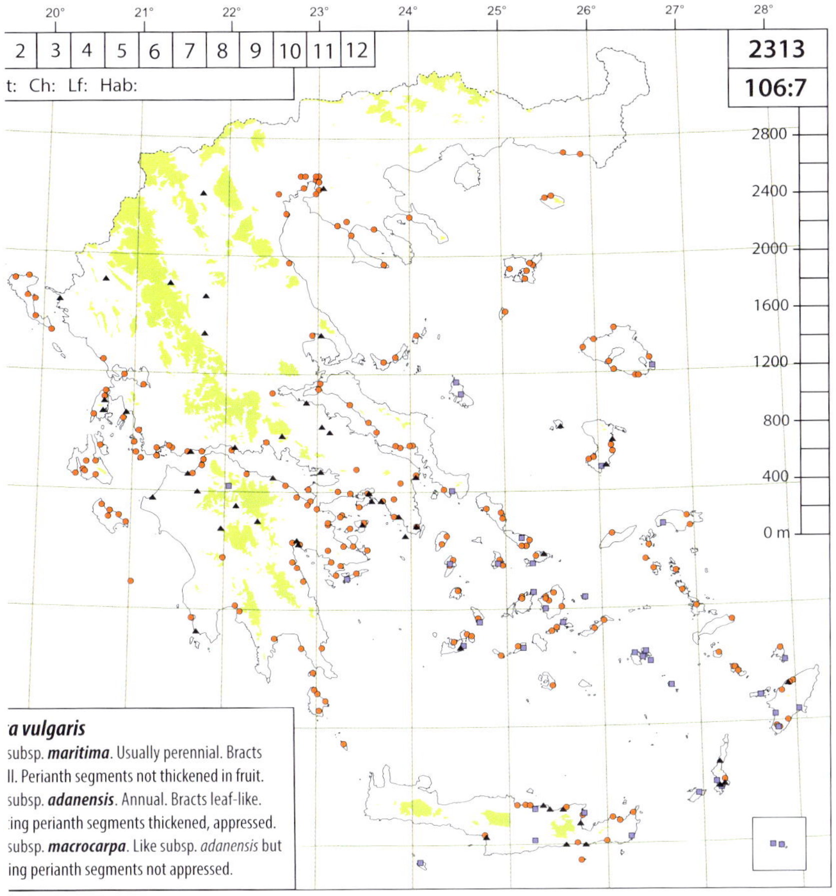

a vulgaris
subsp. ***maritima***. Usually perennial. Bracts ll. Perianth segments not thickened in fruit.
subsp. ***adanensis***. Annual. Bracts leaf-like. ing perianth segments thickened, appressed.
subsp. ***macrocarpa***. Like subsp. *adanensis* but ing perianth segments not appressed.

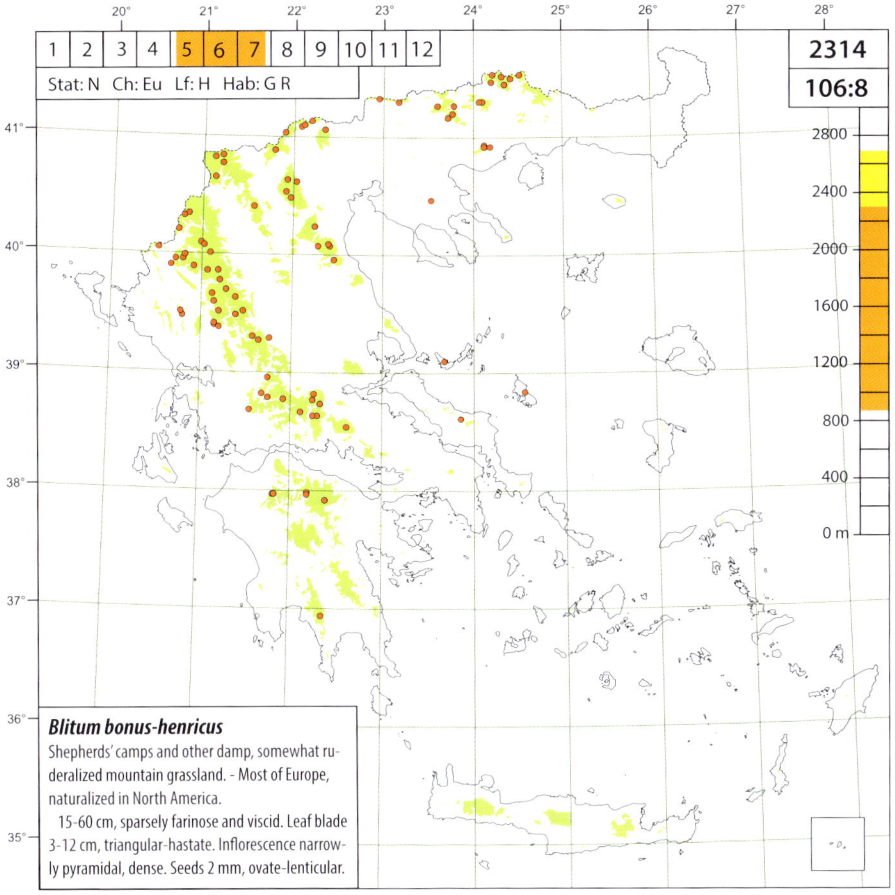

Blitum bonus-henricus
Shepherds' camps and other damp, somewhat ruderalized mountain grassland. - Most of Europe, naturalized in North America.
15-60 cm, sparsely farinose and viscid. Leaf blade 3-12 cm, triangular-hastate. Inflorescence narrowly pyramidal, dense. Seeds 2 mm, ovate-lenticular.

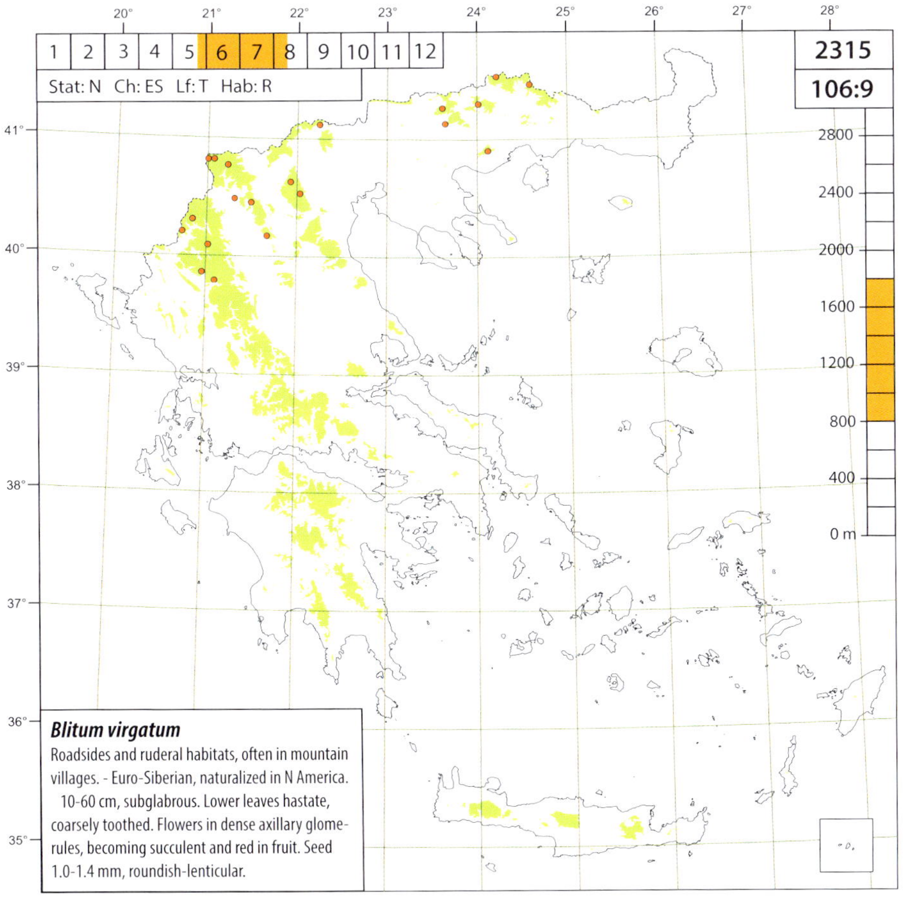

Blitum virgatum
Roadsides and ruderal habitats, often in mountain villages. - Euro-Siberian, naturalized in N America.
10-60 cm, subglabrous. Lower leaves hastate, coarsely toothed. Flowers in dense axillary glomerules, becoming succulent and red in fruit. Seed 1.0-1.4 mm, roundish-lenticular.

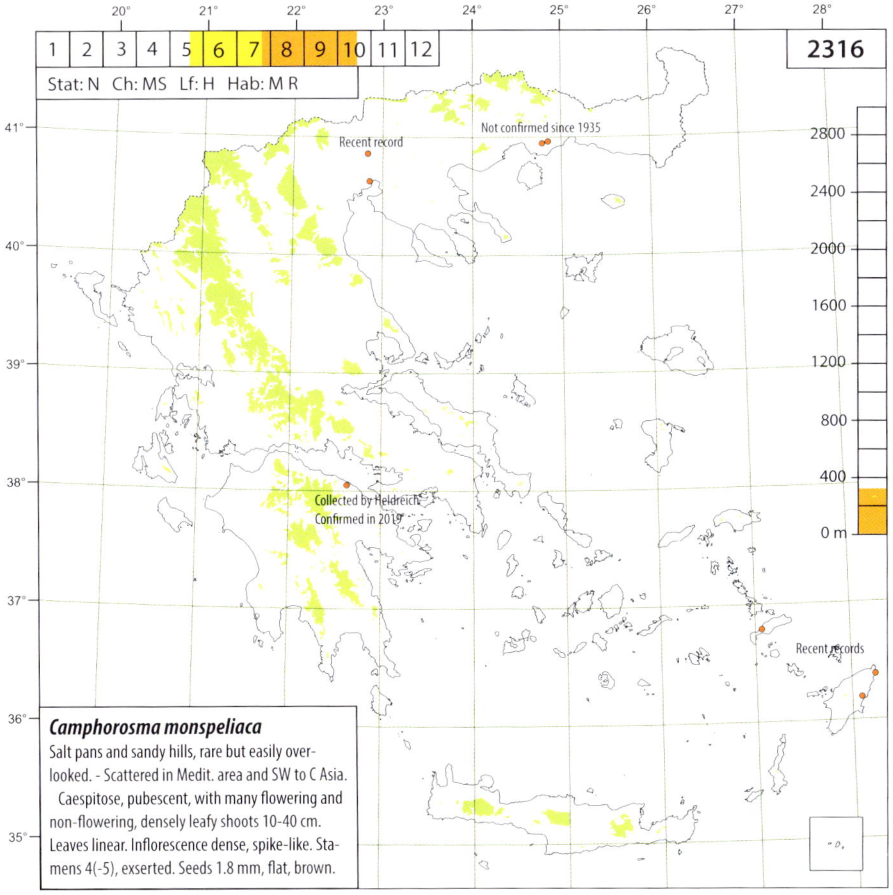

Camphorosma monspeliaca
Salt pans and sandy hills, rare but easily overlooked. - Scattered in Medit. area and SW to C Asia.
Caespitose, pubescent, with many flowering and non-flowering, densely leafy shoots 10-40 cm. Leaves linear. Inflorescence dense, spike-like. Stamens 4(-5), exserted. Seeds 1.8 mm, flat, brown.

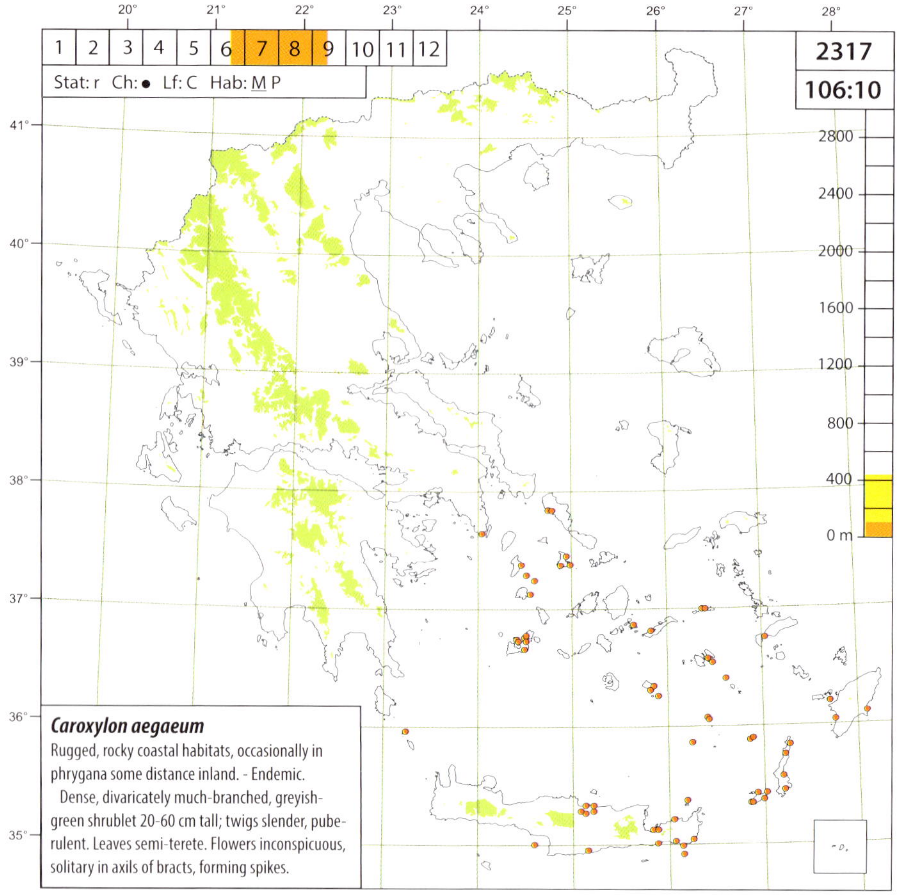

Caroxylon aegaeum

Rugged, rocky coastal habitats, occasionally in phrygana some distance inland. - Endemic.

Dense, divaricately much-branched, greyish-green shrublet 20-60 cm tall; twigs slender, puberulent. Leaves semi-terete. Flowers inconspicuous, solitary in axils of bracts, forming spikes.

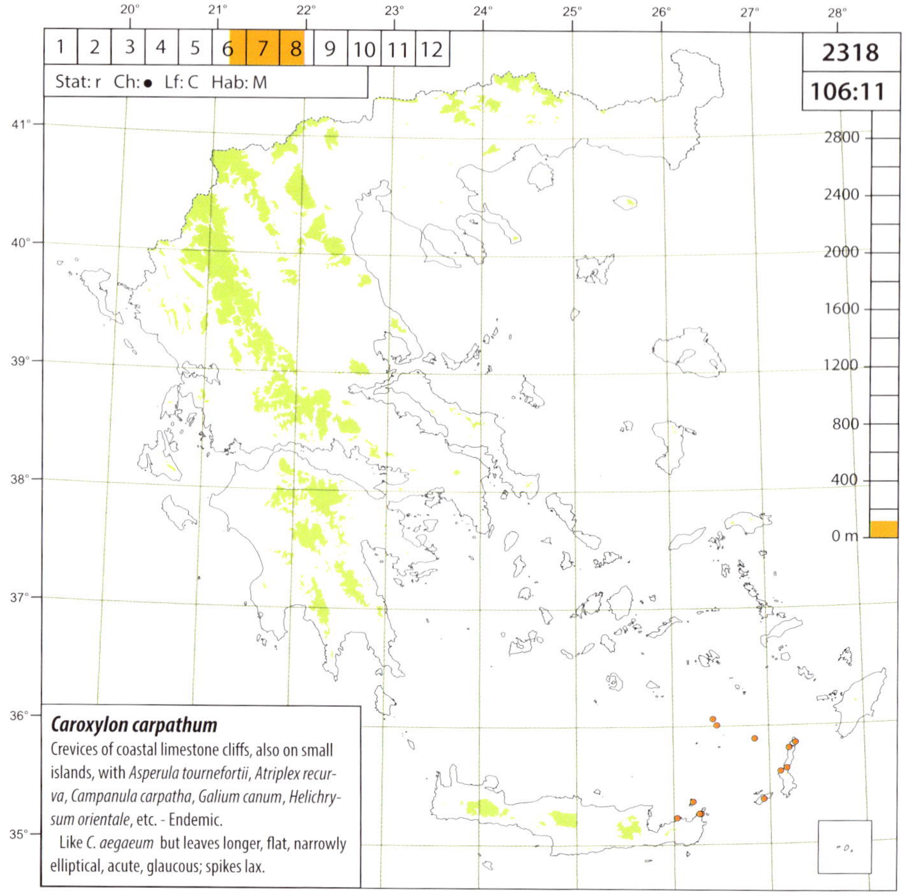

Caroxylon carpathum

Crevices of coastal limestone cliffs, also on small islands, with *Asperula tournefortii*, *Atriplex recurva*, *Campanula carpatha*, *Galium canum*, *Helichrysum orientale*, etc. - Endemic.

Like *C. aegaeum* but leaves longer, flat, narrowly elliptical, acute, glaucous; spikes lax.

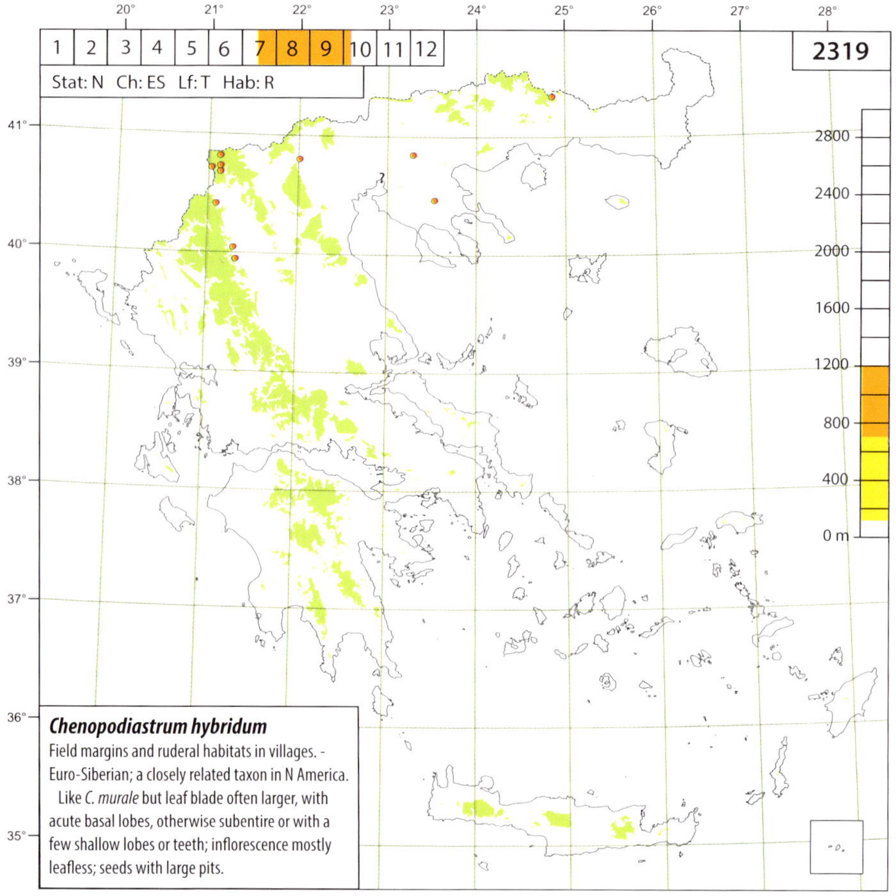

Chenopodiastrum hybridum

Field margins and ruderal habitats in villages. - Euro-Siberian; a closely related taxon in N America.

Like *C. murale* but leaf blade often larger, with acute basal lobes, otherwise subentire or with a few shallow lobes or teeth; inflorescence mostly leafless; seeds with large pits.

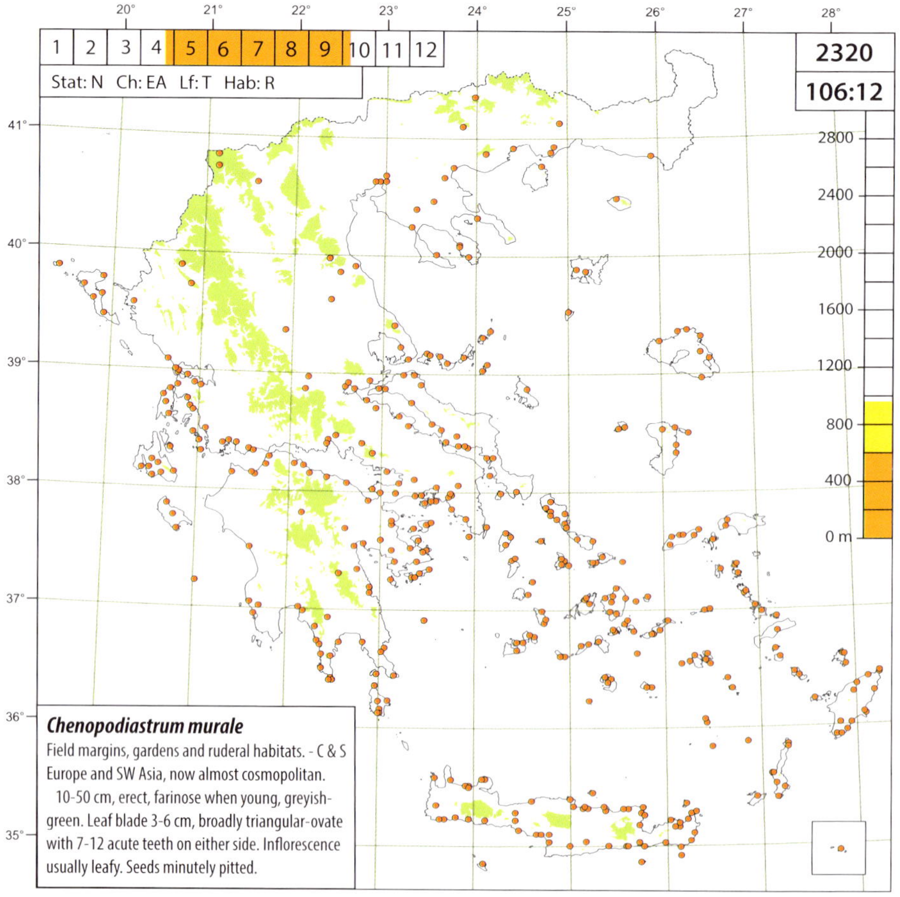

Chenopodiastrum murale

Field margins, gardens and ruderal habitats. - C & S Europe and SW Asia, now almost cosmopolitan.

10-50 cm, erect, farinose when young, greyish-green. Leaf blade 3-6 cm, broadly triangular-ovate with 7-12 acute teeth on either side. Inflorescence usually leafy. Seeds minutely pitted.

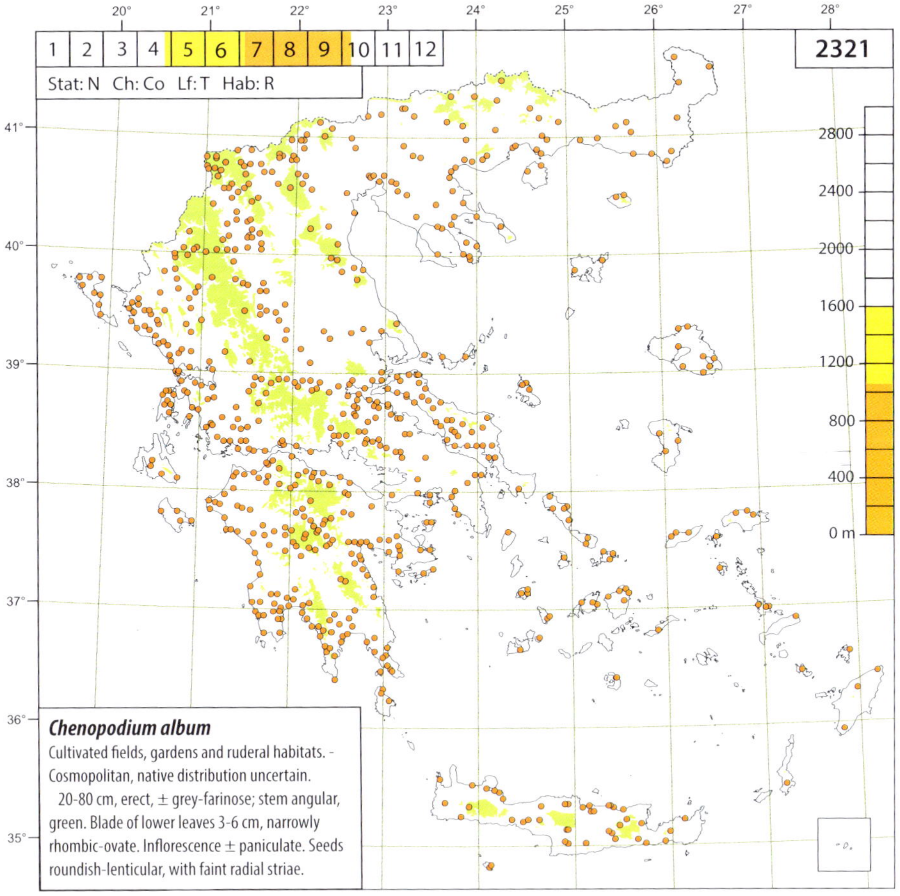

Chenopodium album
Cultivated fields, gardens and ruderal habitats. - Cosmopolitan, native distribution uncertain.
20-80 cm, erect, ± grey-farinose; stem angular, green. Blade of lower leaves 3-6 cm, narrowly rhombic-ovate. Inflorescence ± paniculate. Seeds roundish-lenticular, with faint radial striae.

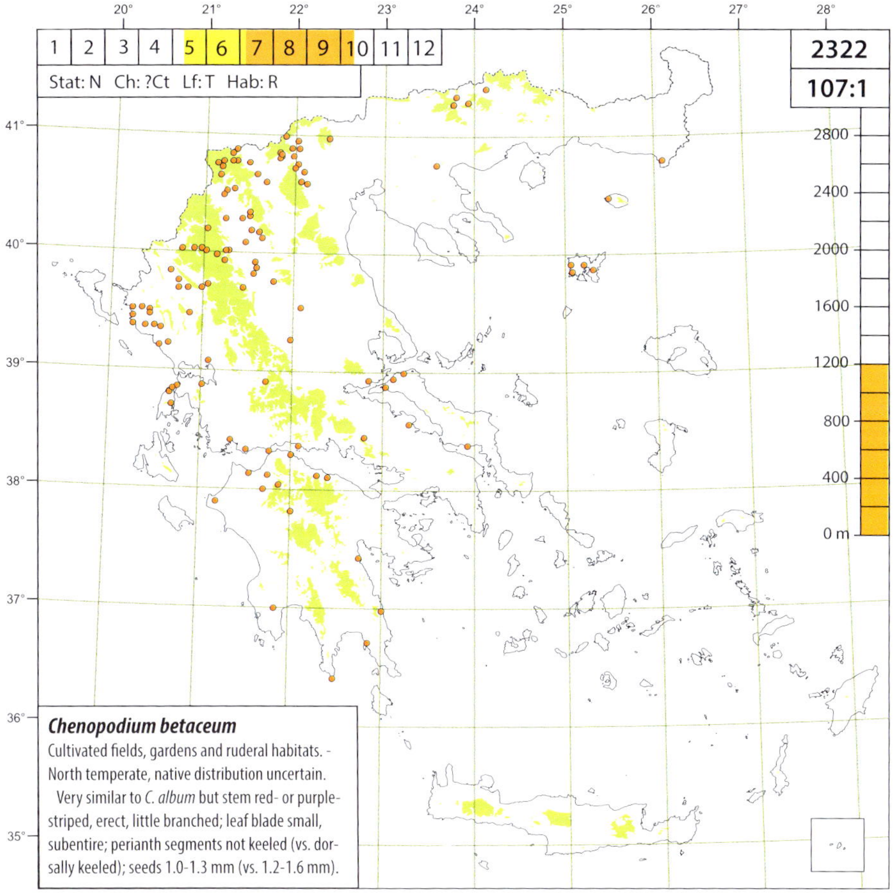

Chenopodium betaceum
Cultivated fields, gardens and ruderal habitats. - North temperate, native distribution uncertain.
Very similar to *C. album* but stem red- or purple-striped, erect, little branched; leaf blade small, subentire; perianth segments not keeled (vs. dorsally keeled); seeds 1.0-1.3 mm (vs. 1.2-1.6 mm).

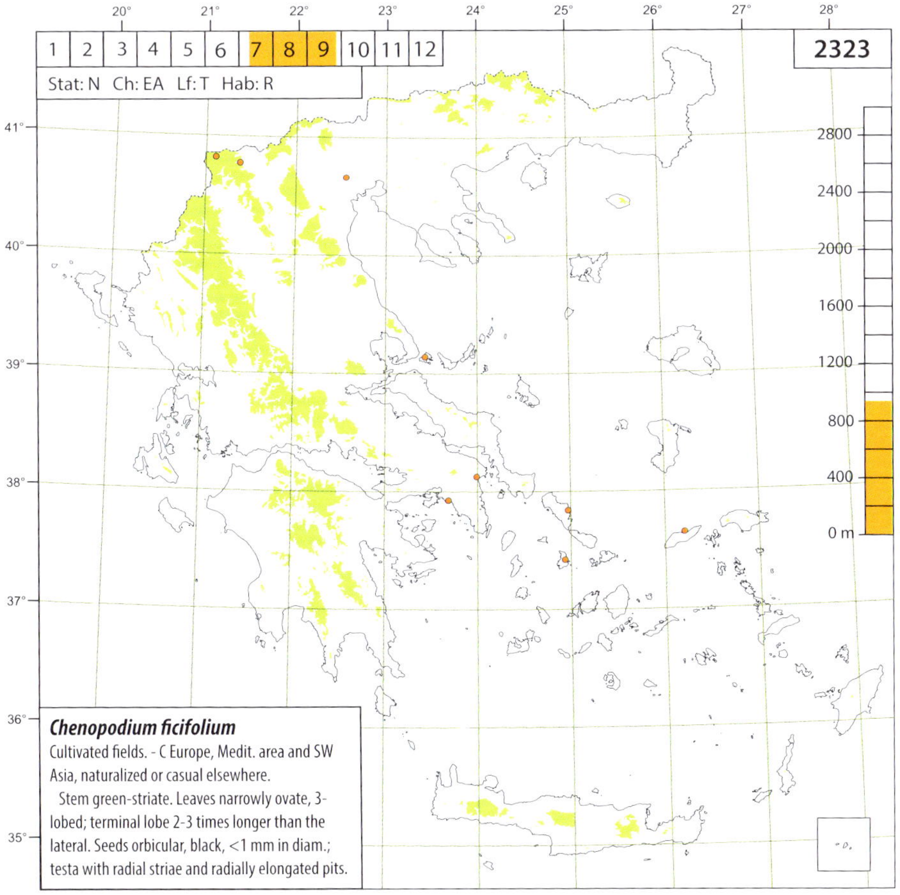

Chenopodium ficifolium
Cultivated fields. - C Europe, Medit. area and SW Asia, naturalized or casual elsewhere.
Stem green-striate. Leaves narrowly ovate, 3-lobed; terminal lobe 2-3 times longer than the lateral. Seeds orbicular, black, <1 mm in diam.; testa with radial striae and radially elongated pits.

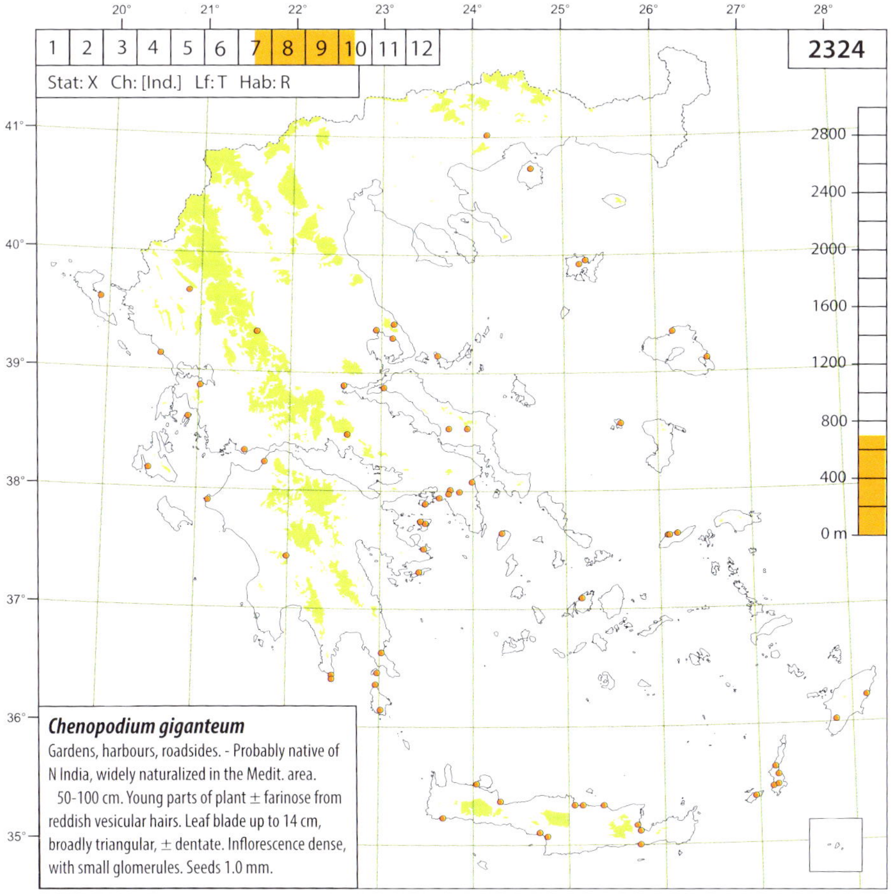

Chenopodium giganteum
Gardens, harbours, roadsides. - Probably native of N India, widely naturalized in the Medit. area.
50-100 cm. Young parts of plant ± farinose from reddish vesicular hairs. Leaf blade up to 14 cm, broadly triangular, ± dentate. Inflorescence dense, with small glomerules. Seeds 1.0 mm.

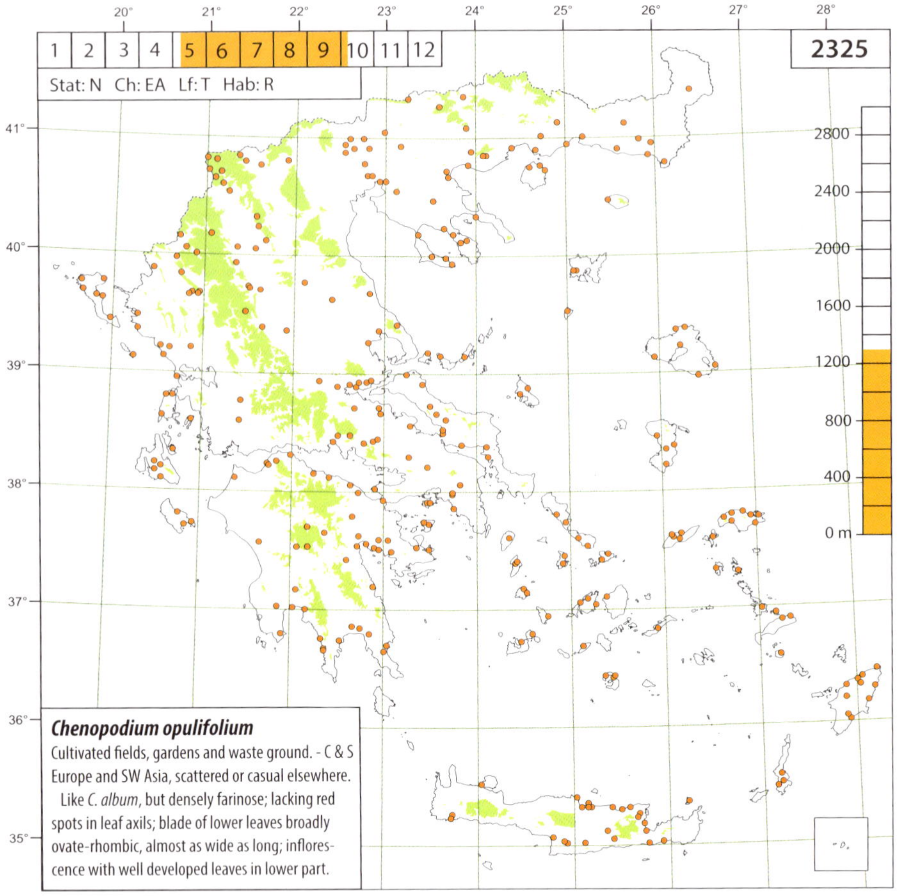

Chenopodium opulifolium

Cultivated fields, gardens and waste ground. - C & S Europe and SW Asia, scattered or casual elsewhere.

Like *C. album*, but densely farinose; lacking red spots in leaf axils; blade of lower leaves broadly ovate-rhombic, almost as wide as long; inflorescence with well developed leaves in lower part.

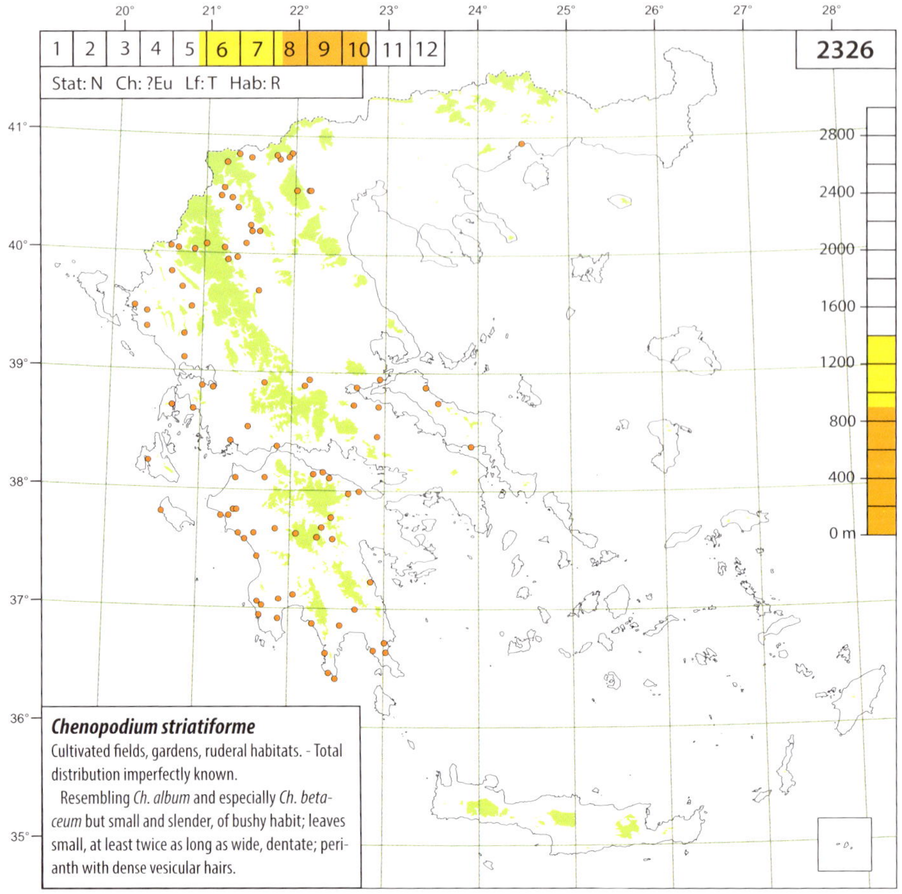

Chenopodium striatiforme

Cultivated fields, gardens, ruderal habitats. - Total distribution imperfectly known.

Resembling *Ch. album* and especially *Ch. betaceum* but small and slender, of bushy habit; leaves small, at least twice as long as wide, dentate; perianth with dense vesicular hairs.

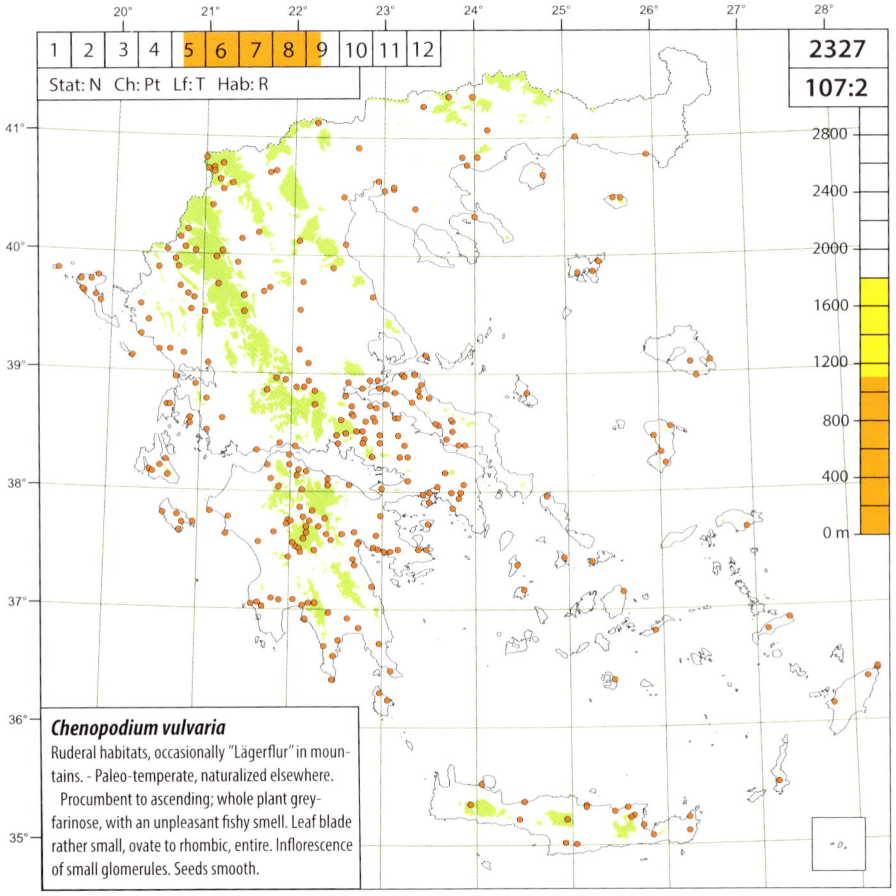

Chenopodium vulvaria

Ruderal habitats, occasionally "Lägerflur" in mountains. - Paleo-temperate, naturalized elsewhere.

Procumbent to ascending; whole plant grey-farinose, with an unpleasant fishy smell. Leaf blade rather small, ovate to rhombic, entire. Inflorescence of small glomerules. Seeds smooth.

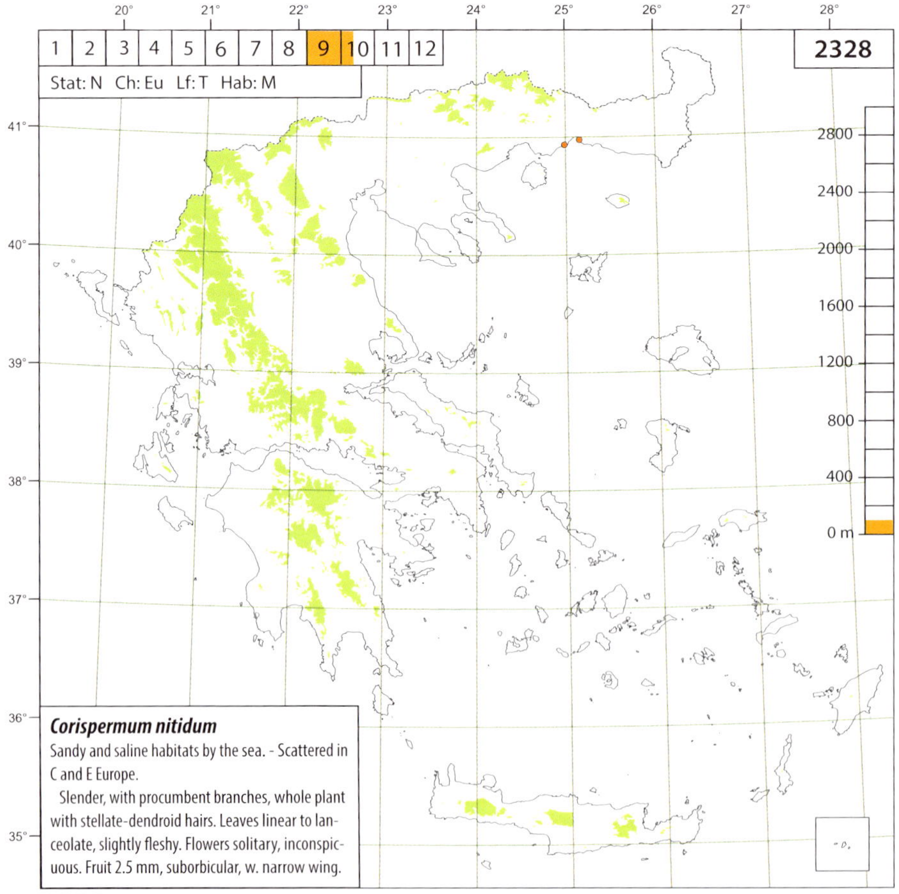

Corispermum nitidum

Sandy and saline habitats by the sea. - Scattered in C and E Europe.

Slender, with procumbent branches, whole plant with stellate-dendroid hairs. Leaves linear to lanceolate, slightly fleshy. Flowers solitary, inconspicuous. Fruit 2.5 mm, suborbicular, w. narrow wing.

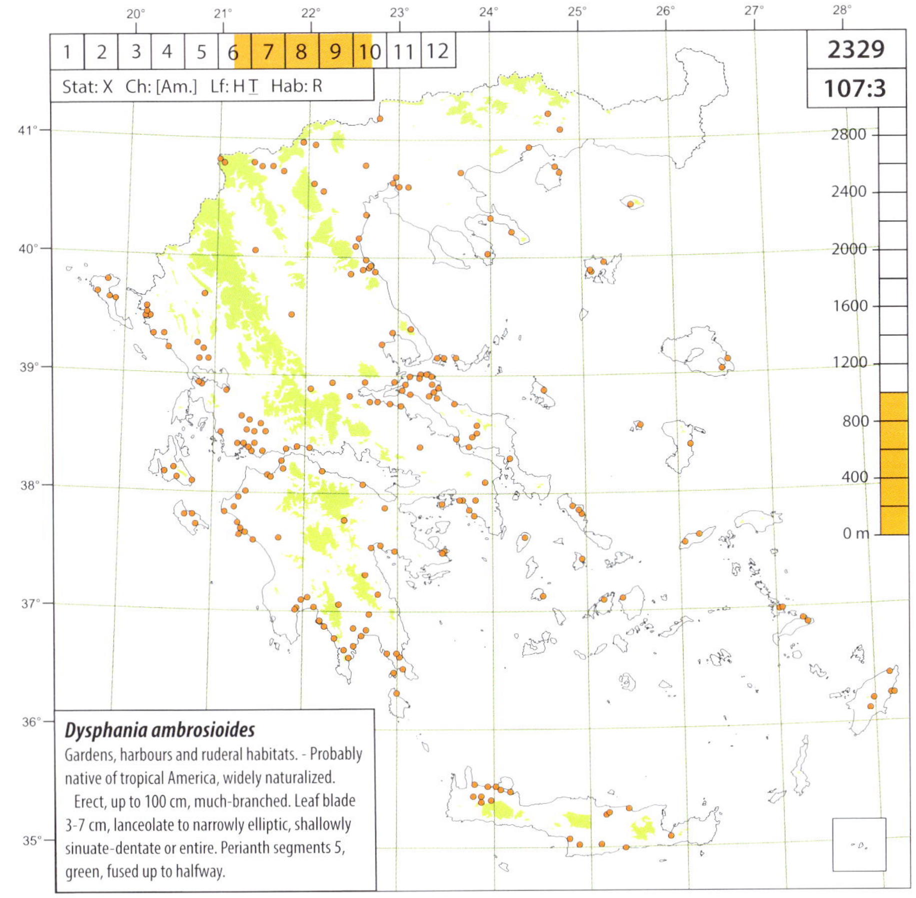

Dysphania ambrosioides

Gardens, harbours and ruderal habitats. - Probably native of tropical America, widely naturalized.

Erect, up to 100 cm, much-branched. Leaf blade 3-7 cm, lanceolate to narrowly elliptic, shallowly sinuate-dentate or entire. Perianth segments 5, green, fused up to halfway.

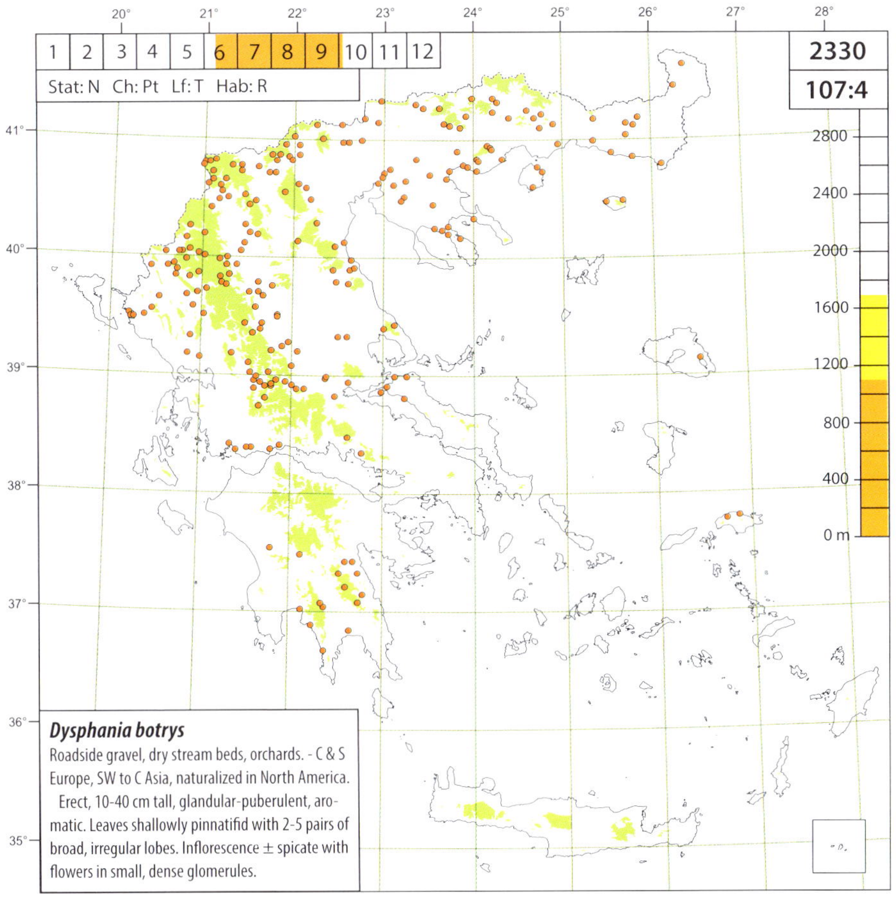

Dysphania botrys

Roadside gravel, dry stream beds, orchards. - C & S Europe, SW to C Asia, naturalized in North America.

Erect, 10-40 cm tall, glandular-puberulent, aromatic. Leaves shallowly pinnatifid with 2-5 pairs of broad, irregular lobes. Inflorescence ± spicate with flowers in small, dense glomerules.

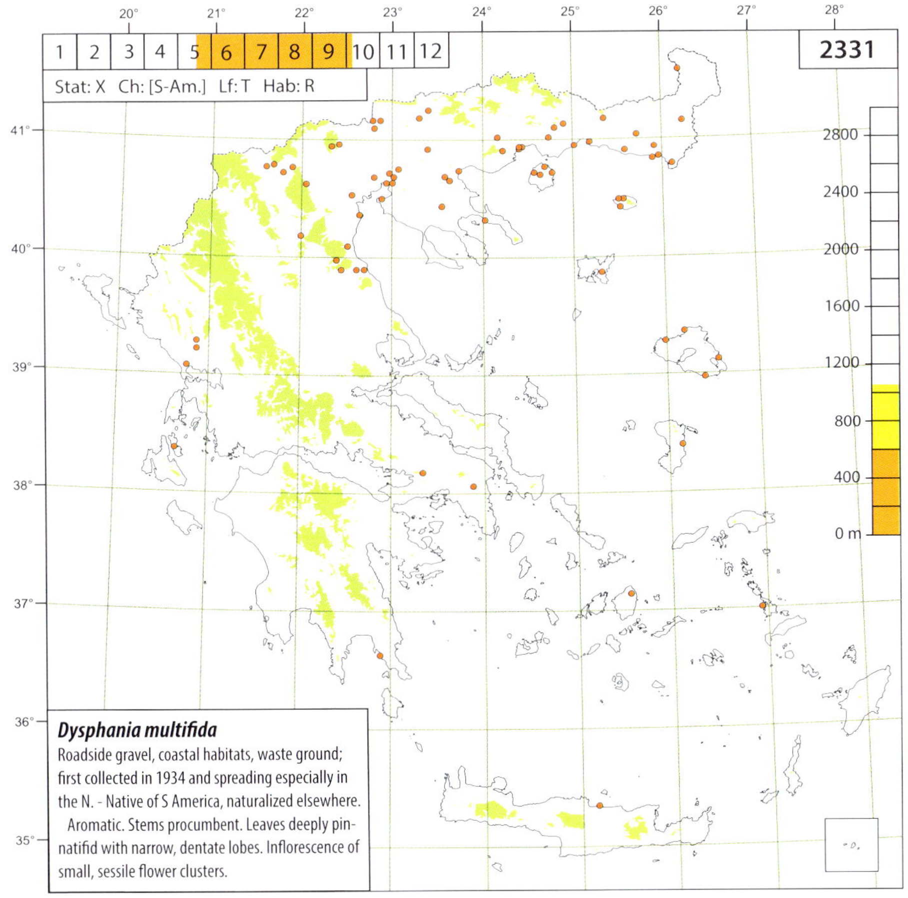

Dysphania multifida

Roadside gravel, coastal habitats, waste ground; first collected in 1934 and spreading especially in the N. - Native of S America, naturalized elsewhere.

Aromatic. Stems procumbent. Leaves deeply pinnatifid with narrow, dentate lobes. Inflorescence of small, sessile flower clusters.

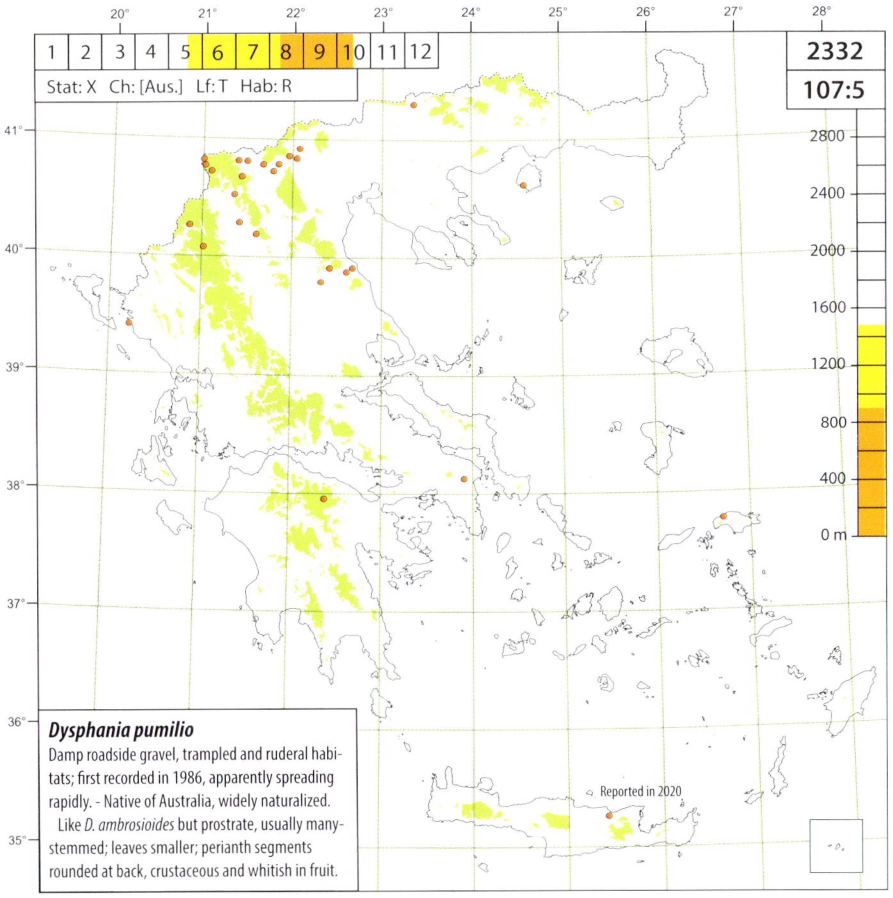

Dysphania pumilio

Damp roadside gravel, trampled and ruderal habitats; first recorded in 1986, apparently spreading rapidly. - Native of Australia, widely naturalized.

Like *D. ambrosioides* but prostrate, usually many-stemmed; leaves smaller; perianth segments rounded at back, crustaceous and whitish in fruit.

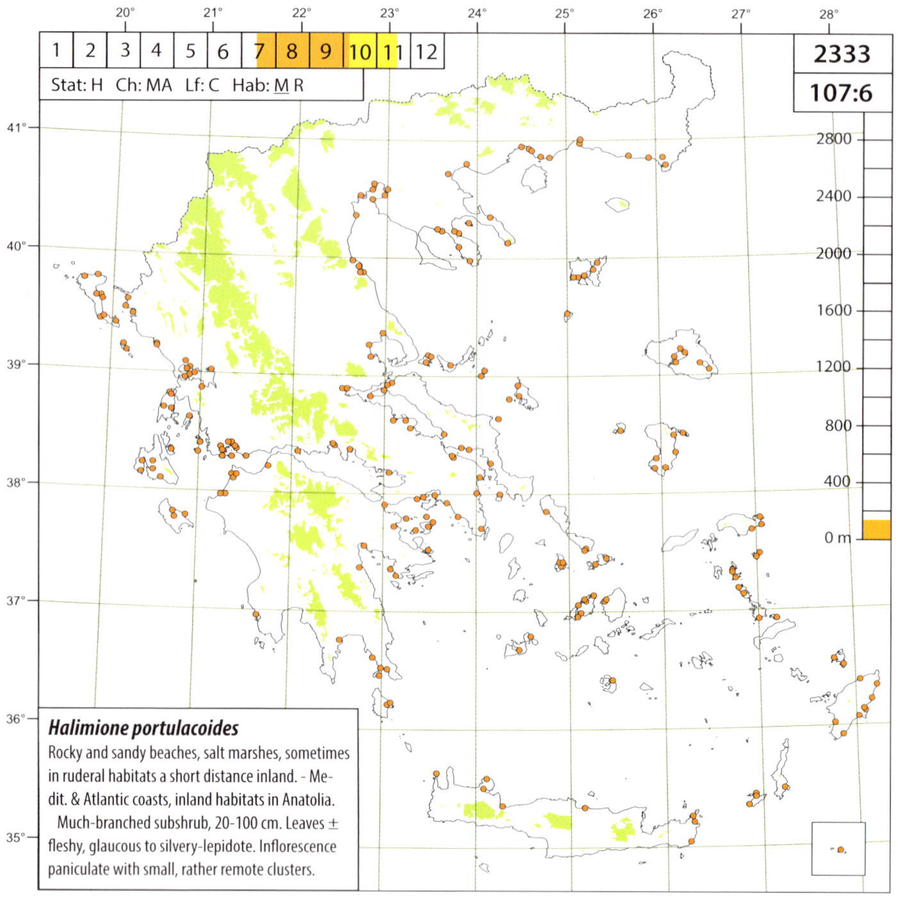

Halimione portulacoides

Rocky and sandy beaches, salt marshes, sometimes in ruderal habitats a short distance inland. - Medit. & Atlantic coasts, inland habitats in Anatolia.

Much-branched subshrub, 20-100 cm. Leaves ± fleshy, glaucous to silvery-lepidote. Inflorescence paniculate with small, rather remote clusters.

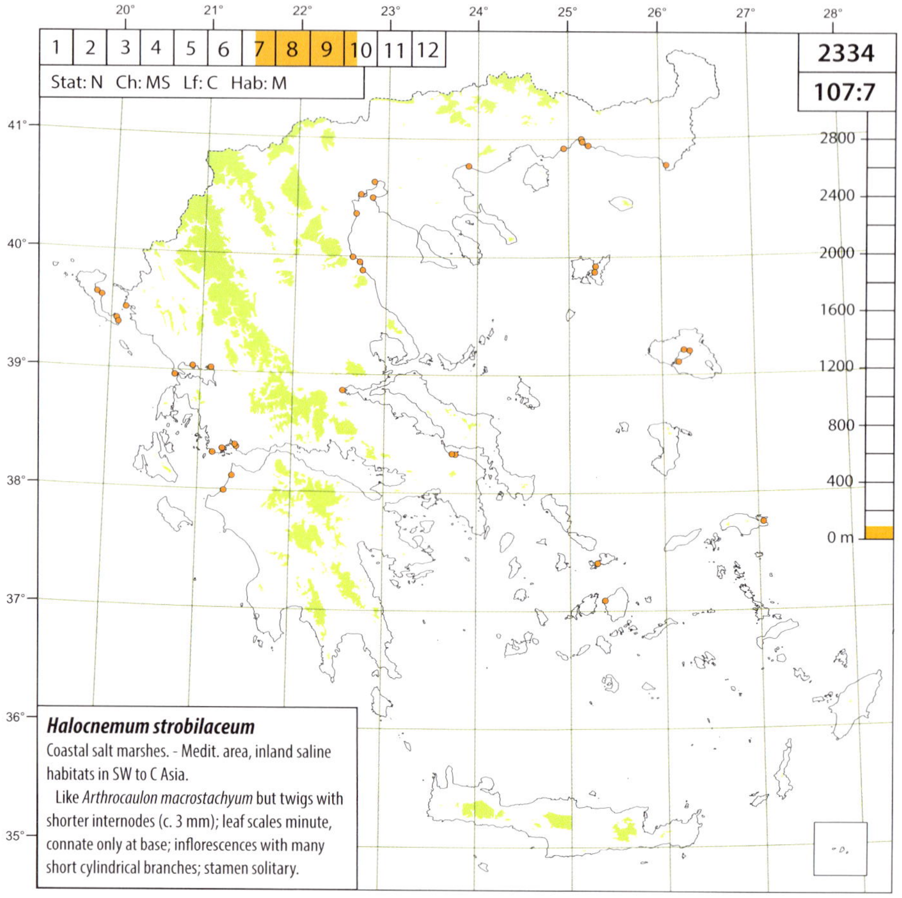

Halocnemum strobilaceum

Coastal salt marshes. - Medit. area, inland saline habitats in SW to C Asia.

Like *Arthrocaulon macrostachyum* but twigs with shorter internodes (c. 3 mm); leaf scales minute, connate only at base; inflorescences with many short cylindrical branches; stamen solitary.

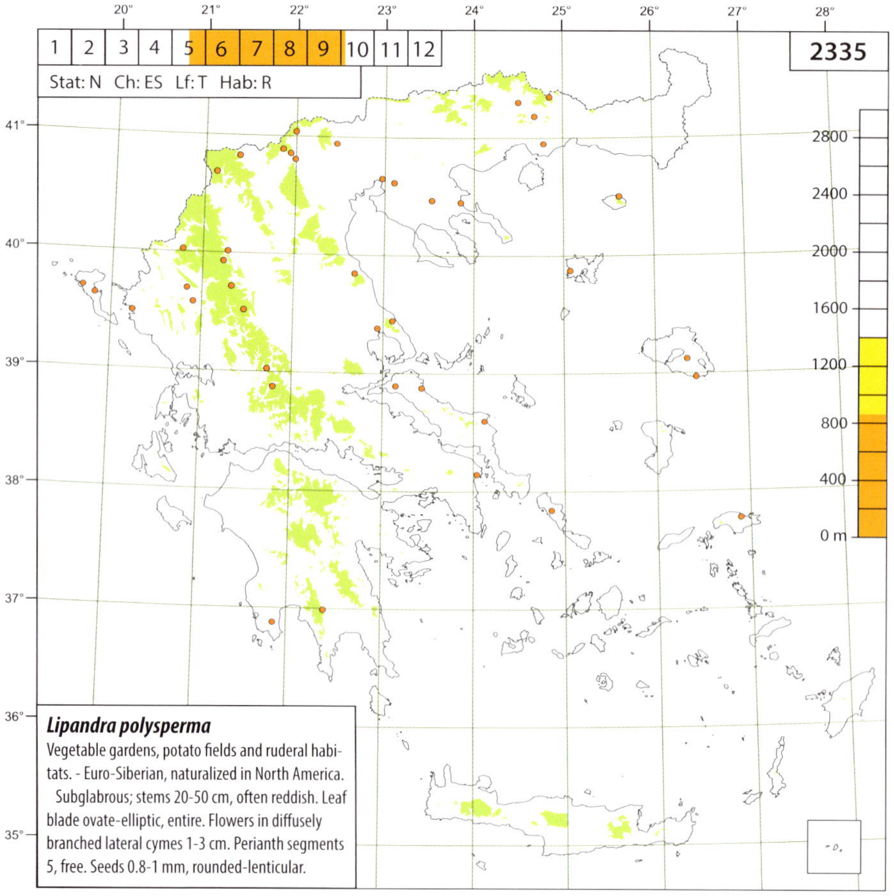

Lipandra polysperma

Vegetable gardens, potato fields and ruderal habitats. - Euro-Siberian, naturalized in North America.

Subglabrous; stems 20-50 cm, often reddish. Leaf blade ovate-elliptic, entire. Flowers in diffusely branched lateral cymes 1-3 cm. Perianth segments 5, free. Seeds 0.8-1 mm, rounded-lenticular.

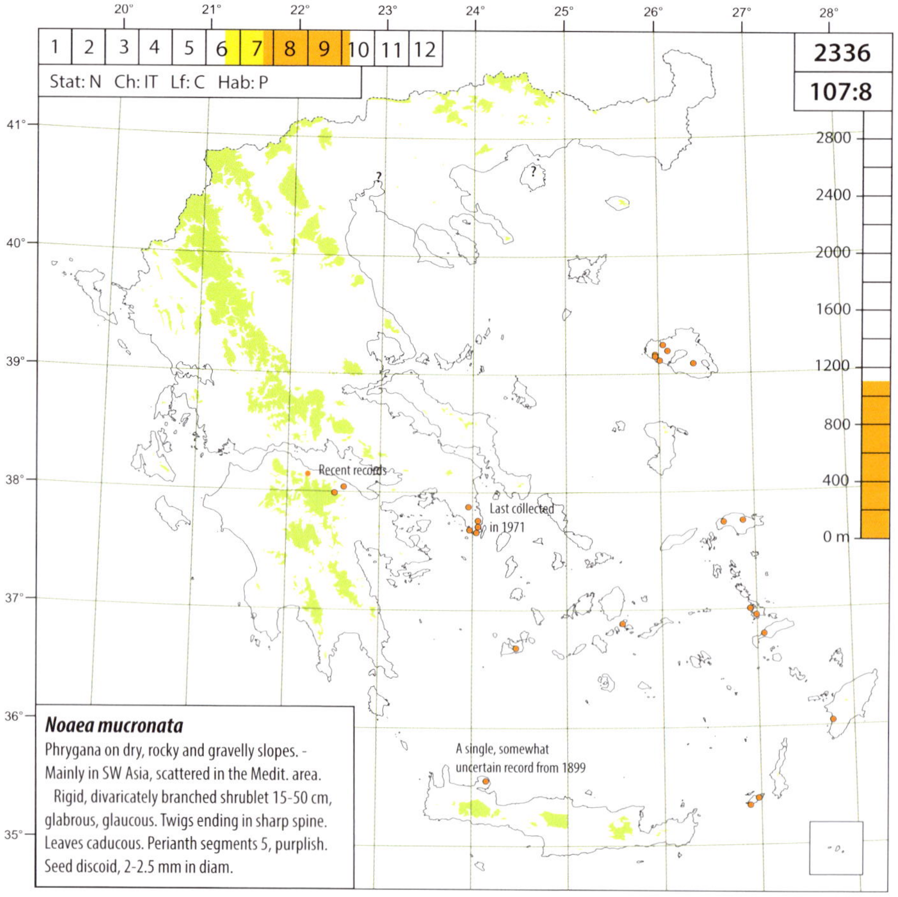

Noaea mucronata

Phrygana on dry, rocky and gravelly slopes. - Mainly in SW Asia, scattered in the Medit. area.

Rigid, divaricately branched shrublet 15-50 cm, glabrous, glaucous. Twigs ending in sharp spine. Leaves caducous. Perianth segments 5, purplish. Seed discoid, 2-2.5 mm in diam.

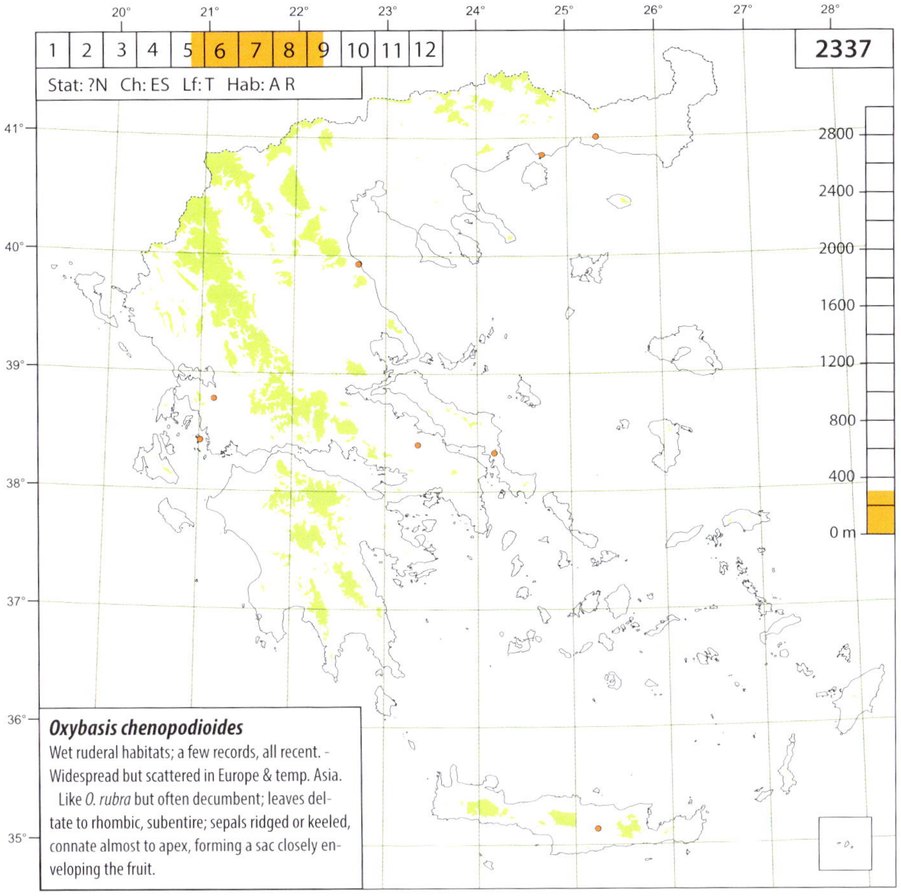

Oxybasis chenopodioides
Wet ruderal habitats; a few records, all recent. - Widespread but scattered in Europe & temp. Asia.
Like *O. rubra* but often decumbent; leaves deltate to rhombic, subentire; sepals ridged or keeled, connate almost to apex, forming a sac closely enveloping the fruit.

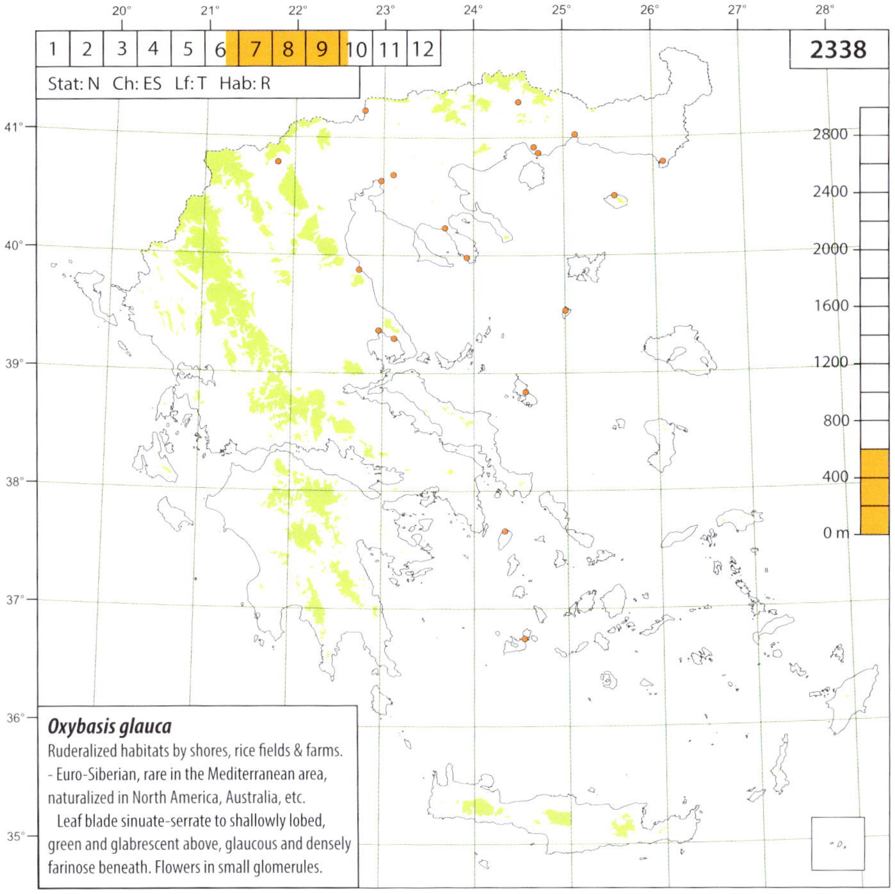

Oxybasis glauca
Ruderalized habitats by shores, rice fields & farms. - Euro-Siberian, rare in the Mediterranean area, naturalized in North America, Australia, etc.
Leaf blade sinuate-serrate to shallowly lobed, green and glabrescent above, glaucous and densely farinose beneath. Flowers in small glomerules.

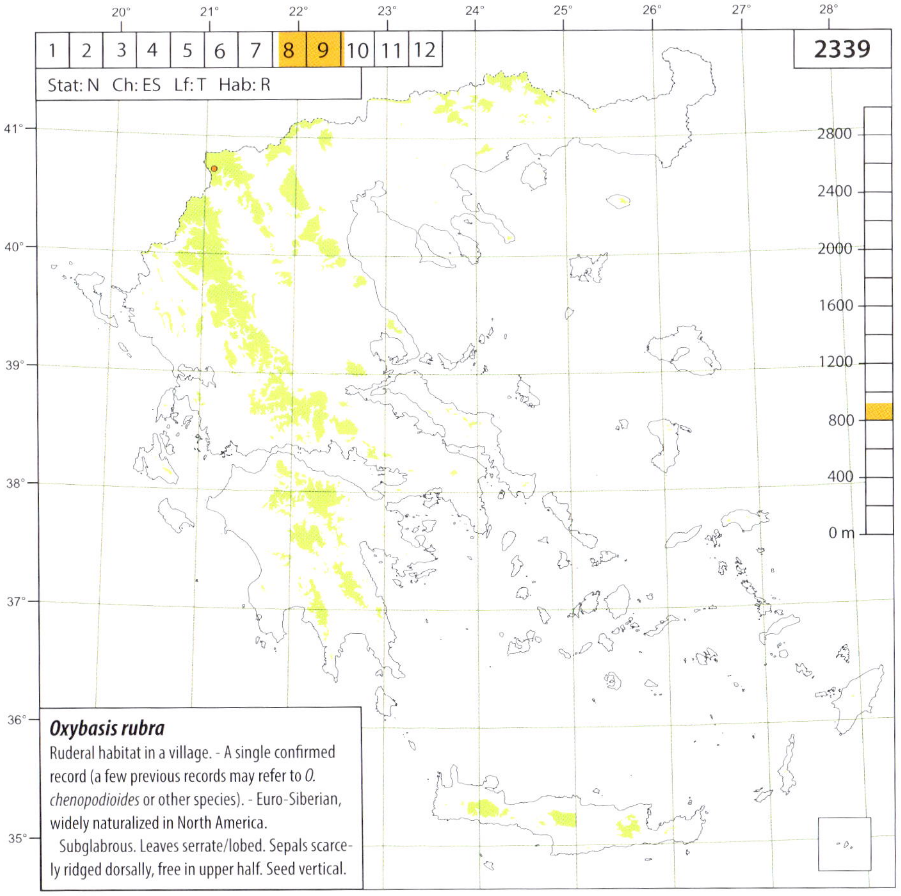

Oxybasis rubra
Ruderal habitat in a village. - A single confirmed record (a few previous records may refer to *O. chenopodioides* or other species). - Euro-Siberian, widely naturalized in North America.
Subglabrous. Leaves serrate/lobed. Sepals scarcely ridged dorsally, free in upper half. Seed vertical.

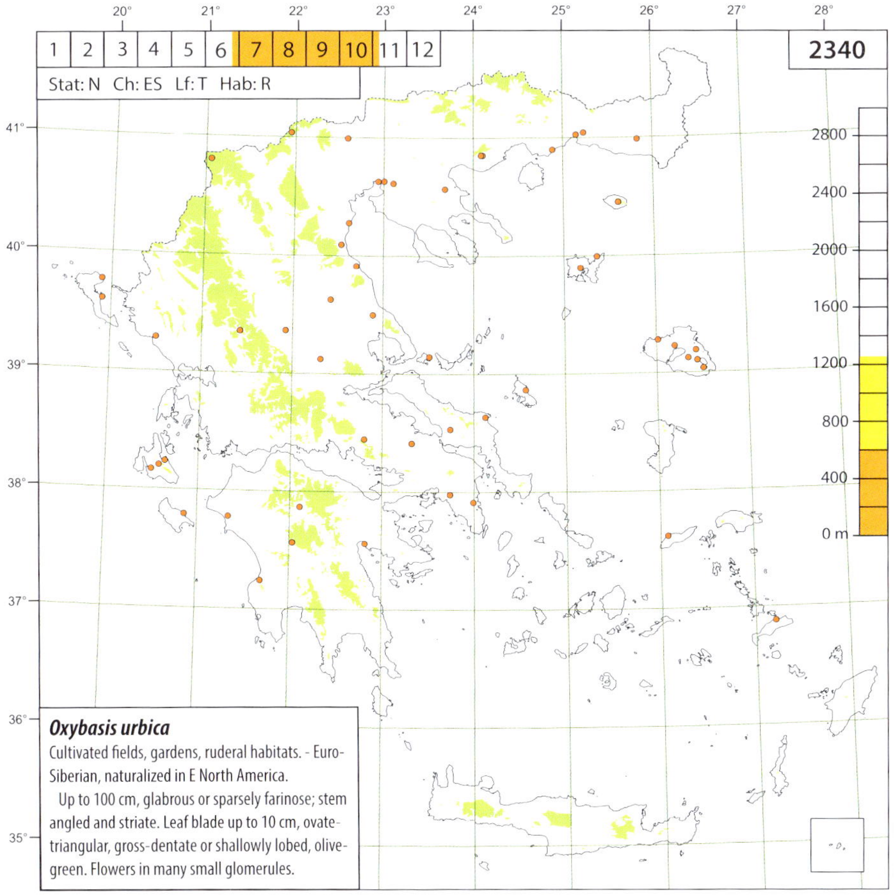

Oxybasis urbica
Cultivated fields, gardens, ruderal habitats. - Euro-Siberian, naturalized in E North America.
Up to 100 cm, glabrous or sparsely farinose; stem angled and striate. Leaf blade up to 10 cm, ovate-triangular, gross-dentate or shallowly lobed, olive-green. Flowers in many small glomerules.

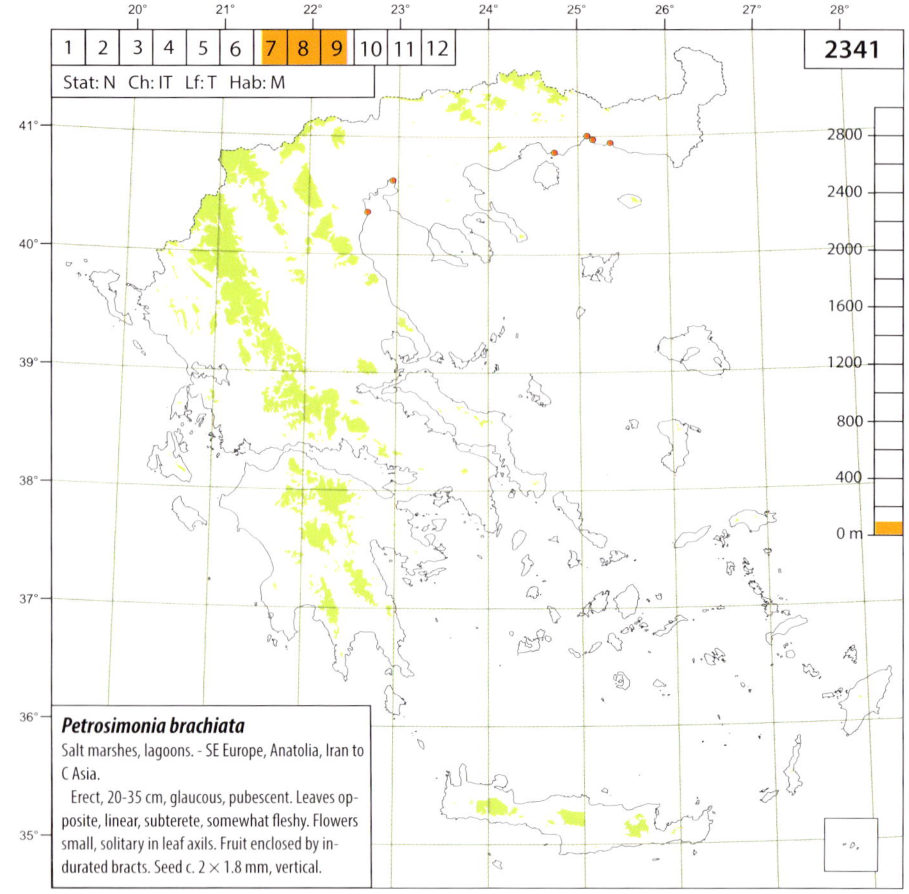

Petrosimonia brachiata
Salt marshes, lagoons. - SE Europe, Anatolia, Iran to C Asia.
Erect, 20-35 cm, glaucous, pubescent. Leaves opposite, linear, subterete, somewhat fleshy. Flowers small, solitary in leaf axils. Fruit enclosed by indurated bracts. Seed c. 2 × 1.8 mm, vertical.

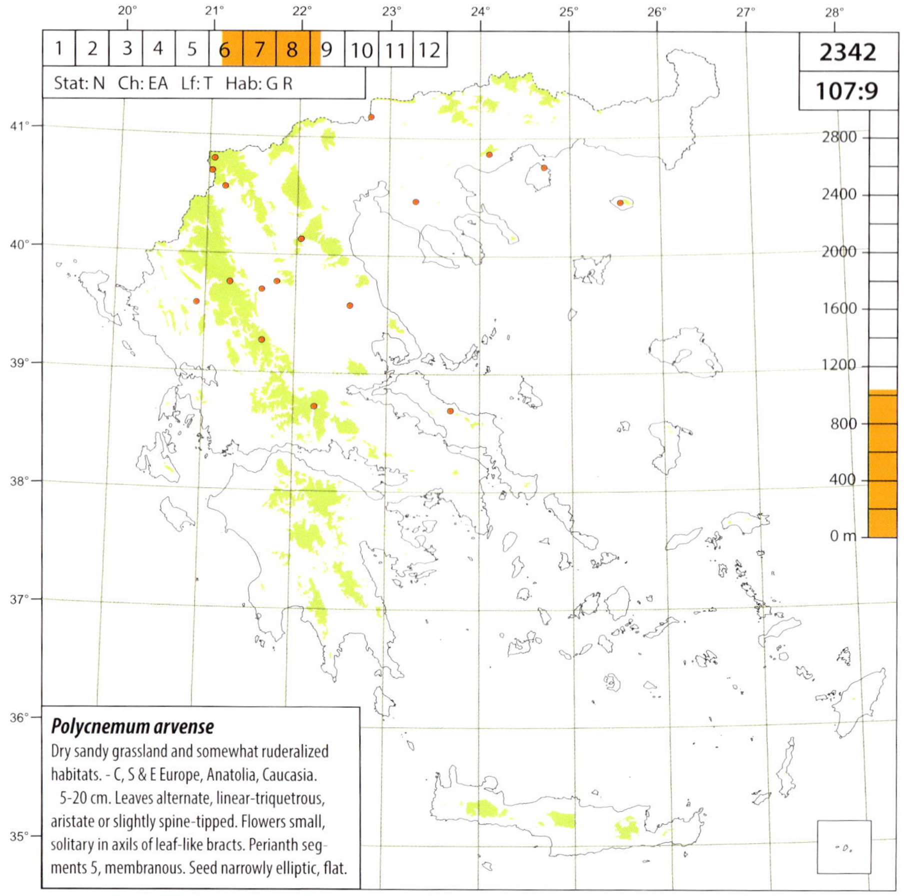

Polycnemum arvense
Dry sandy grassland and somewhat ruderalized habitats. - C, S & E Europe, Anatolia, Caucasia.
5-20 cm. Leaves alternate, linear-triquetrous, aristate or slightly spine-tipped. Flowers small, solitary in axils of leaf-like bracts. Perianth segments 5, membranous. Seed narrowly elliptic, flat.

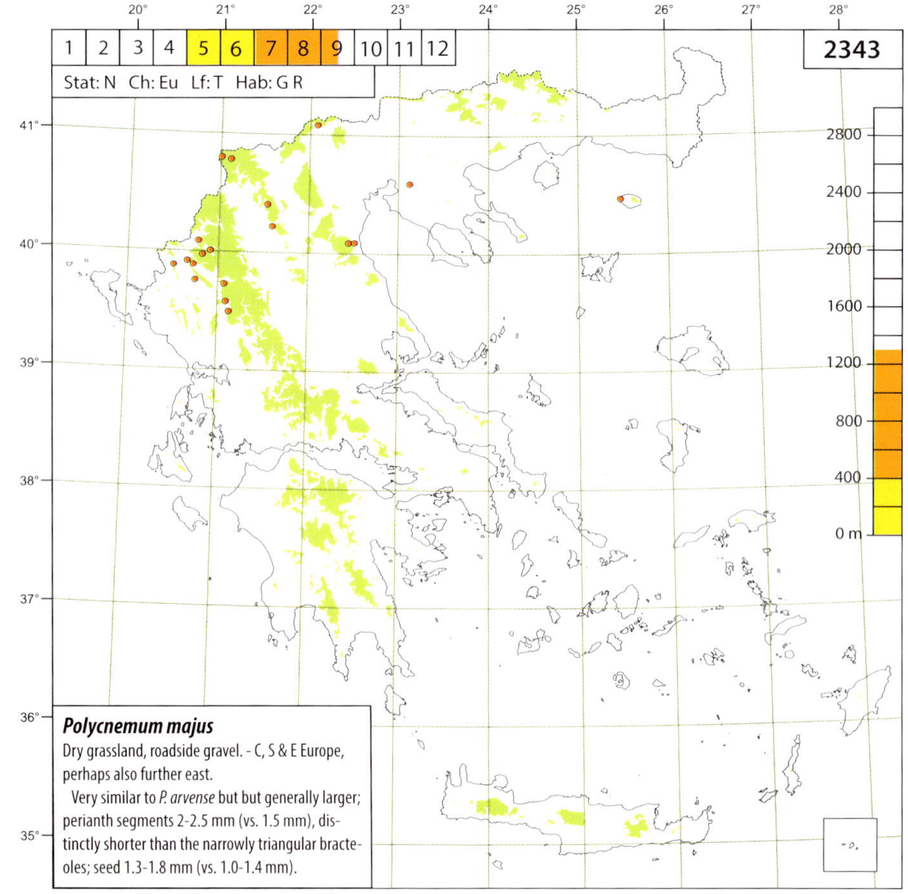

Polycnemum majus
Dry grassland, roadside gravel. - C, S & E Europe, perhaps also further east.
Very similar to *P. arvense* but but generally larger; perianth segments 2-2.5 mm (vs. 1.5 mm), distinctly shorter than the narrowly triangular bracteoles; seed 1.3-1.8 mm (vs. 1.0-1.4 mm).

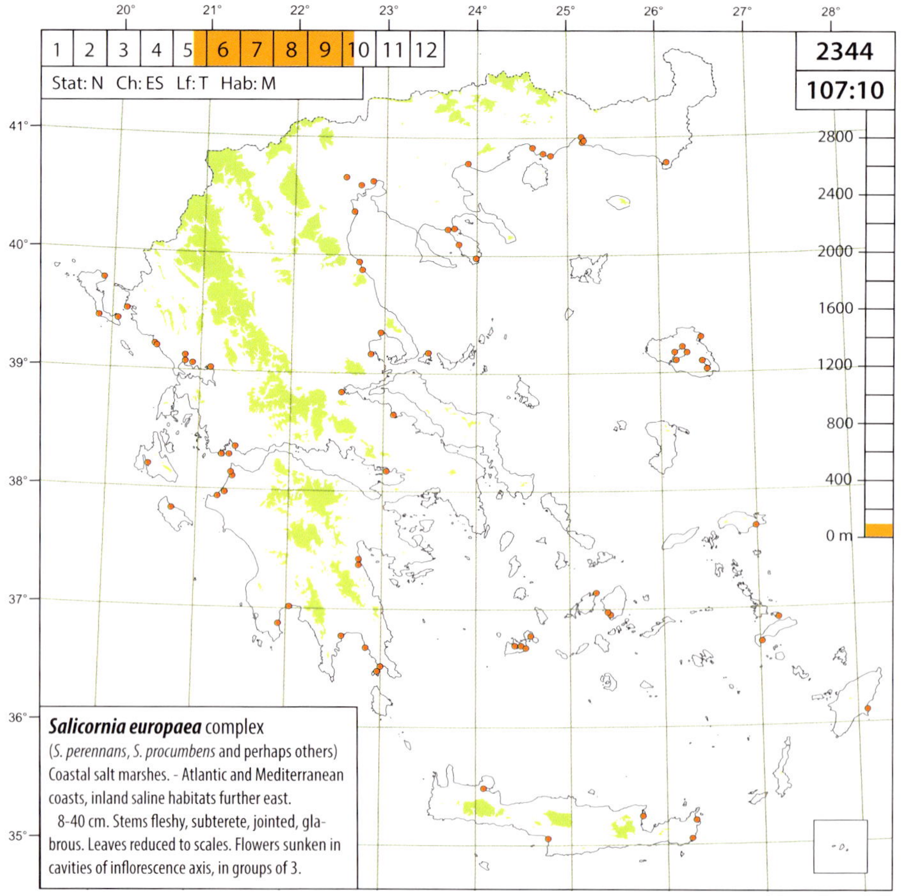

Salicornia europaea complex
(*S. perennans*, *S. procumbens* and perhaps others)
Coastal salt marshes. - Atlantic and Mediterranean coasts, inland saline habitats further east.
8-40 cm. Stems fleshy, subterete, jointed, glabrous. Leaves reduced to scales. Flowers sunken in cavities of inflorescence axis, in groups of 3.

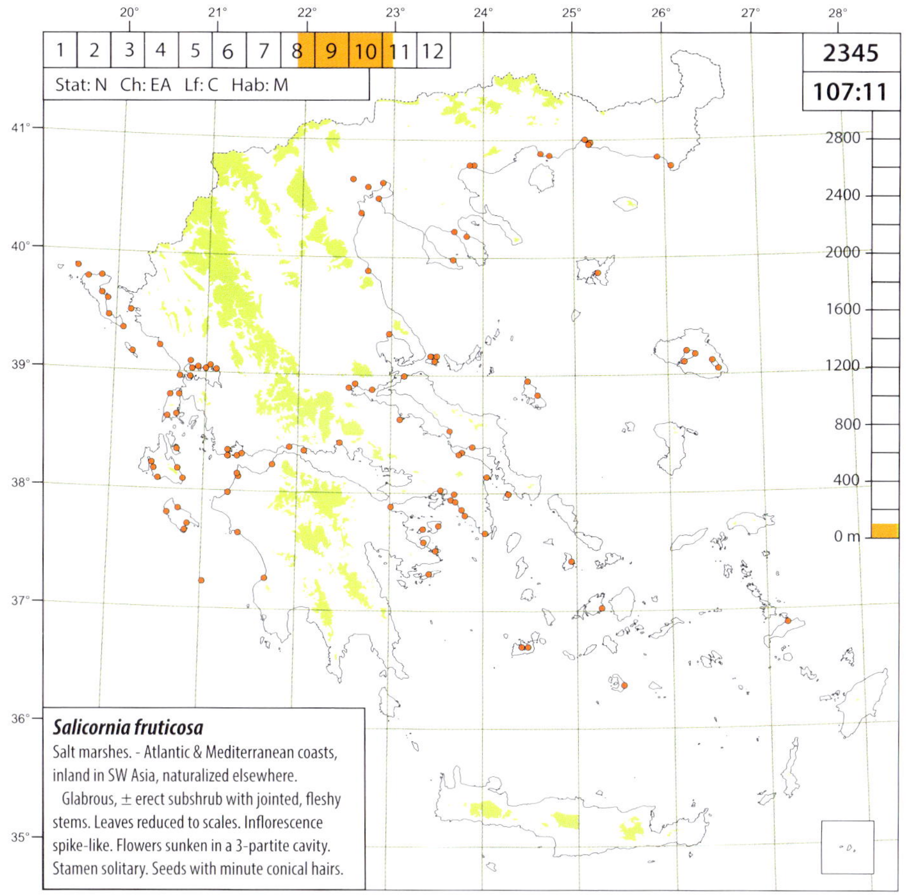

Salicornia fruticosa
Salt marshes. - Atlantic & Mediterranean coasts, inland in SW Asia, naturalized elsewhere.
Glabrous, ± erect subshrub with jointed, fleshy stems. Leaves reduced to scales. Inflorescence spike-like. Flowers sunken in a 3-partite cavity. Stamen solitary. Seeds with minute conical hairs.

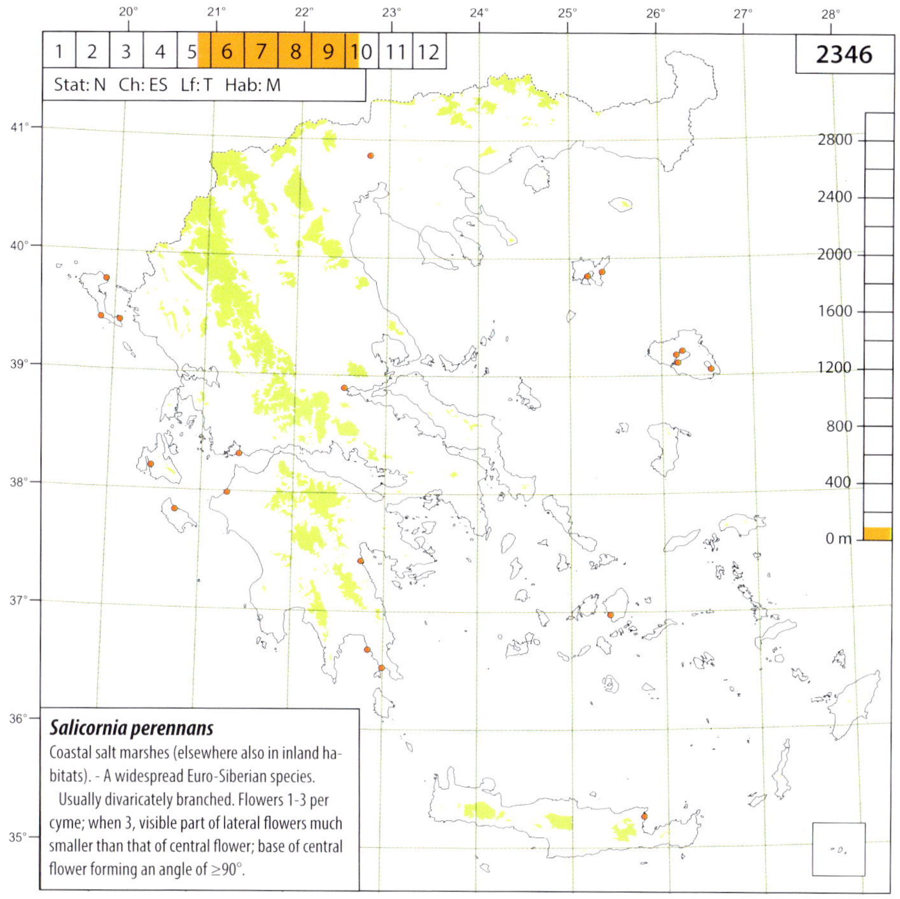

Salicornia perennans
Coastal salt marshes (elsewhere also in inland habitats). - A widespread Euro-Siberian species.
Usually divaricately branched. Flowers 1-3 per cyme; when 3, visible part of lateral flowers much smaller than that of central flower; base of central flower forming an angle of ≥90°.

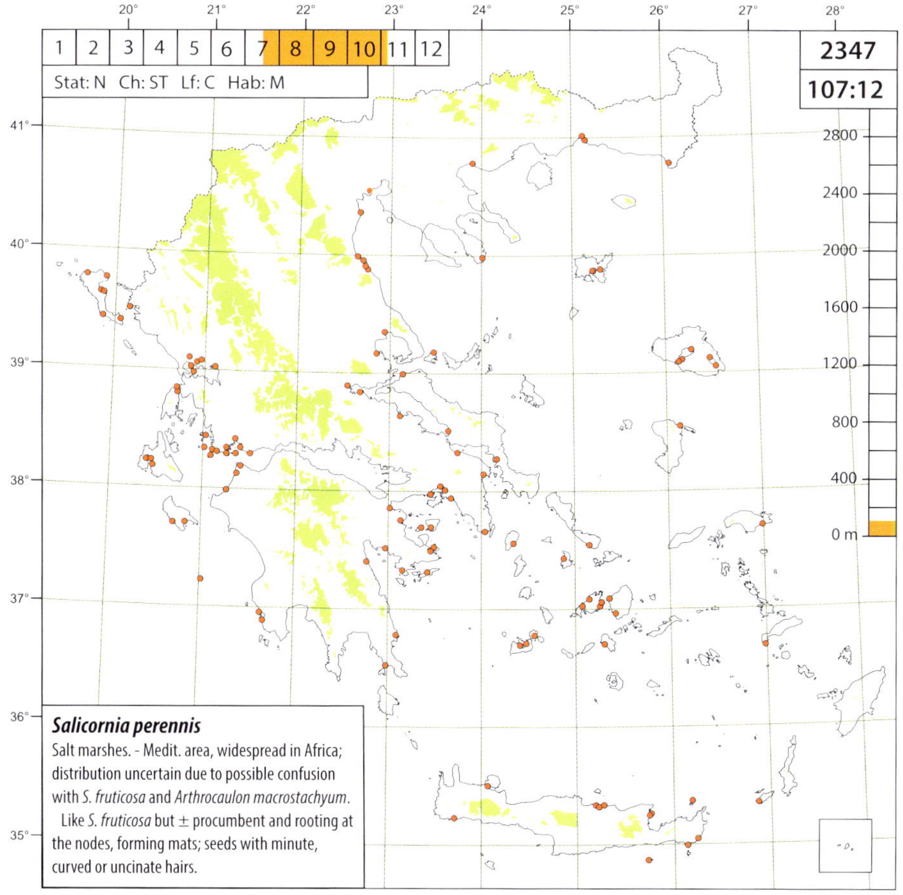

Salicornia perennis
Salt marshes. - Medit. area, widespread in Africa; distribution uncertain due to possible confusion with *S. fruticosa* and *Arthrocaulon macrostachyum*.
Like *S. fruticosa* but ± procumbent and rooting at the nodes, forming mats; seeds with minute, curved or uncinate hairs.

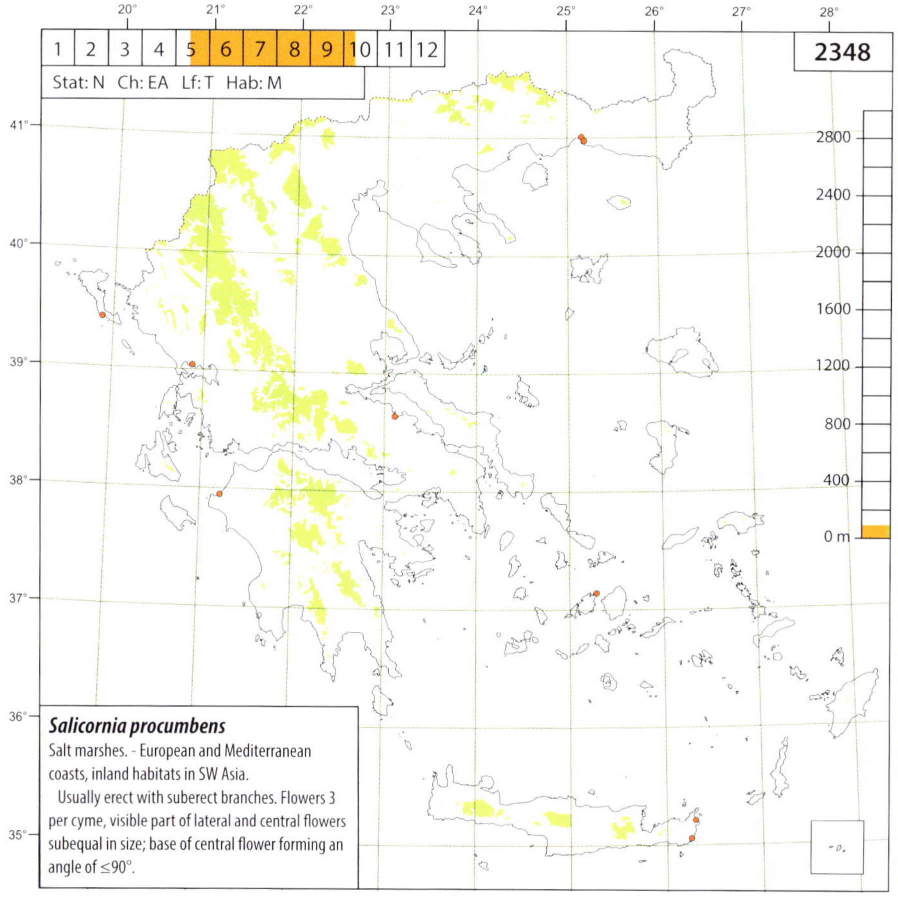

Salicornia procumbens
Salt marshes. - European and Mediterranean coasts, inland habitats in SW Asia.
Usually erect with suberect branches. Flowers 3 per cyme, visible part of lateral and central flowers subequal in size; base of central flower forming an angle of ≤90°.

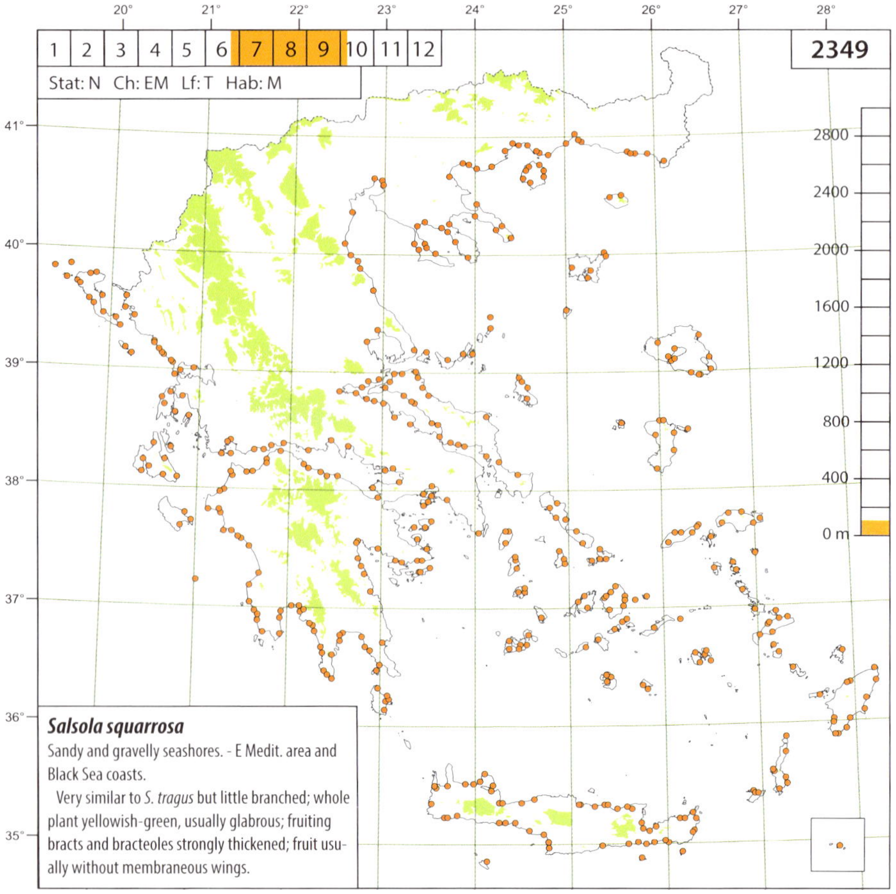

Salsola squarrosa

Sandy and gravelly seashores. - E Medit. area and Black Sea coasts.

Very similar to *S. tragus* but little branched; whole plant yellowish-green, usually glabrous; fruiting bracts and bracteoles strongly thickened; fruit usually without membraneous wings.

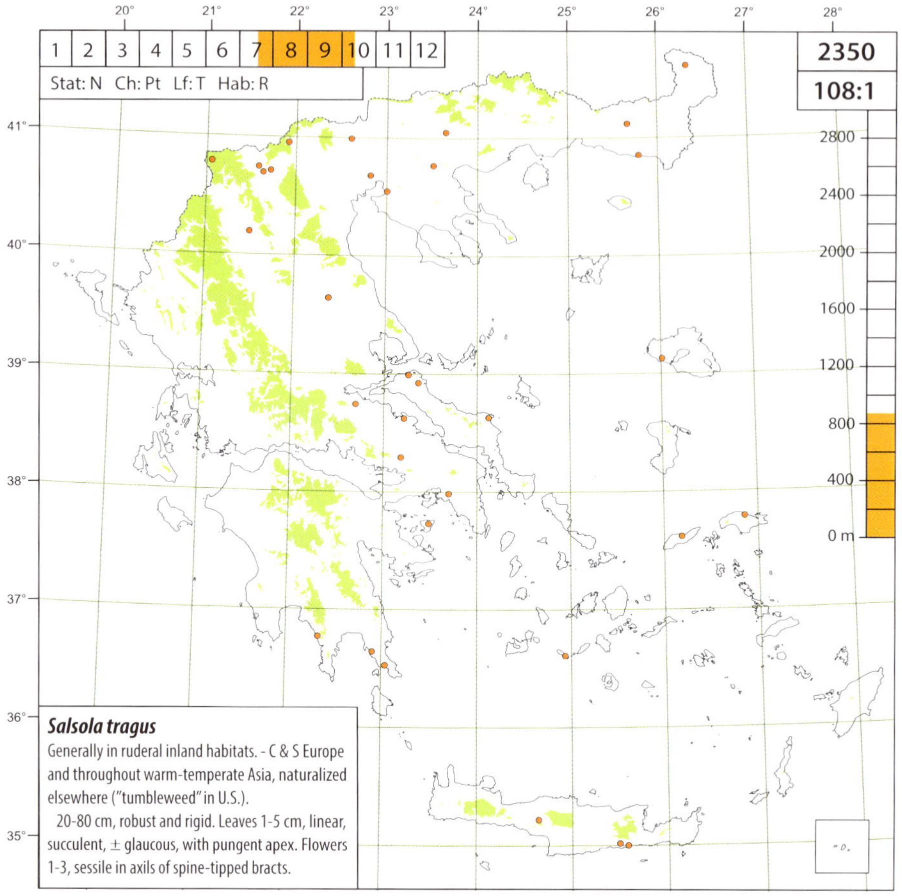

Salsola tragus

Generally in ruderal inland habitats. - C & S Europe and throughout warm-temperate Asia, naturalized elsewhere ("tumbleweed" in U.S.).

20-80 cm, robust and rigid. Leaves 1-5 cm, linear, succulent, ± glaucous, with pungent apex. Flowers 1-3, sessile in axils of spine-tipped bracts.

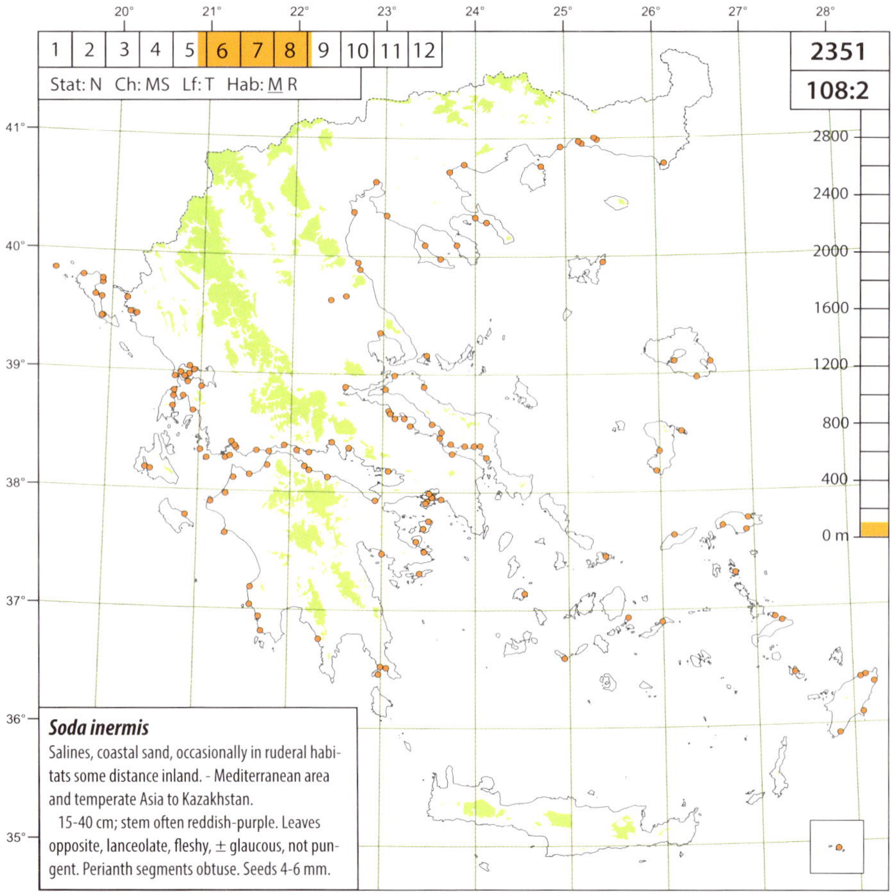

Soda inermis

Salines, coastal sand, occasionally in ruderal habitats some distance inland. - Mediterranean area and temperate Asia to Kazakhstan.

15-40 cm; stem often reddish-purple. Leaves opposite, lanceolate, fleshy, ± glaucous, not pungent. Perianth segments obtuse. Seeds 4-6 mm.

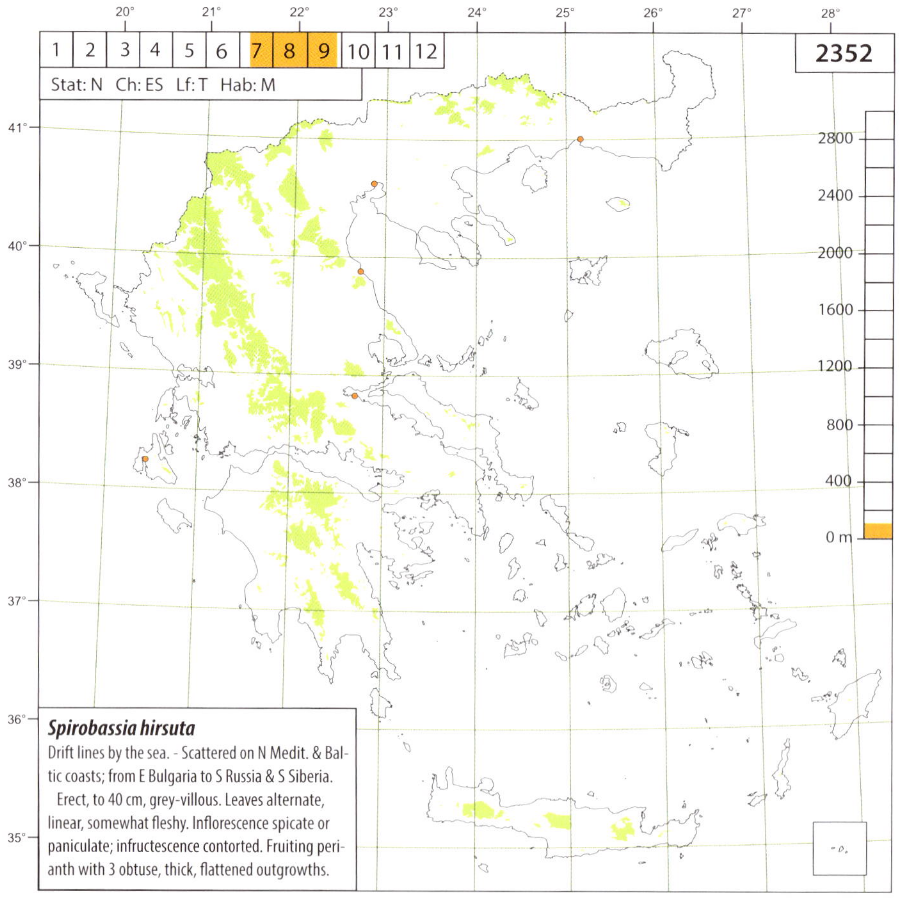

Spirobassia hirsuta

Drift lines by the sea. - Scattered on N Medit. & Baltic coasts; from E Bulgaria to S Russia & S Siberia.

Erect, to 40 cm, grey-villous. Leaves alternate, linear, somewhat fleshy. Inflorescence spicate or paniculate; infructescence contorted. Fruiting perianth with 3 obtuse, thick, flattened outgrowths.

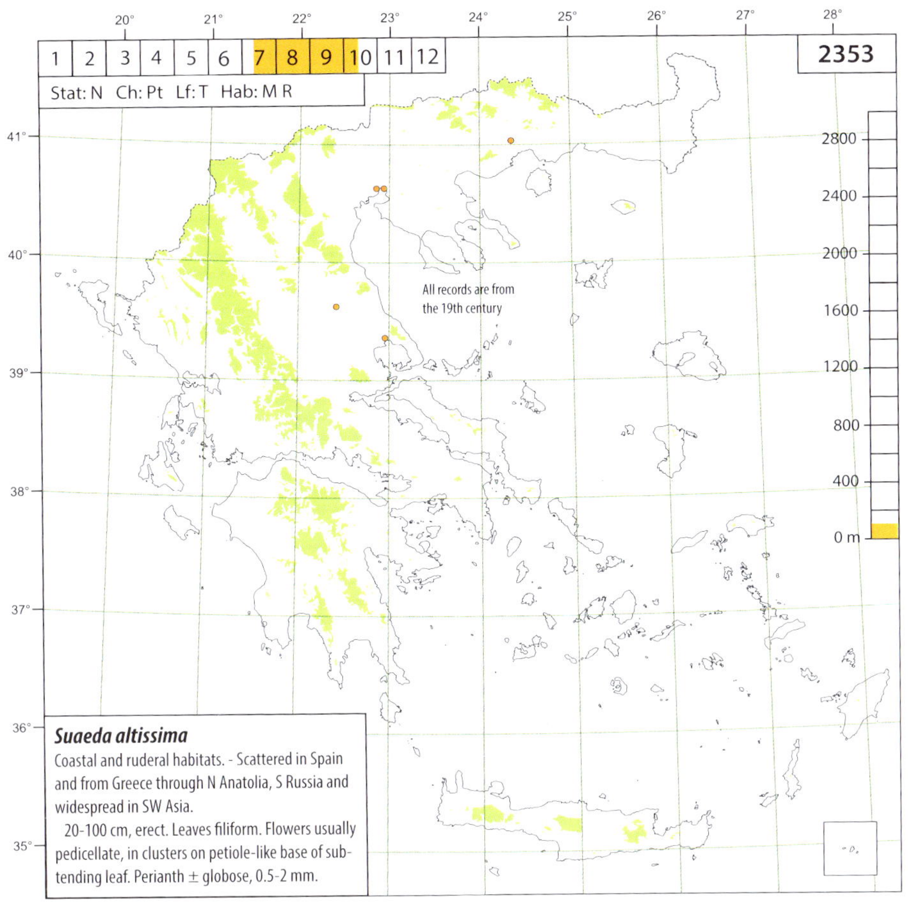

Suaeda altissima

Coastal and ruderal habitats. - Scattered in Spain and from Greece through N Anatolia, S Russia and widespread in SW Asia.

20-100 cm, erect. Leaves filiform. Flowers usually pedicellate, in clusters on petiole-like base of subtending leaf. Perianth ± globose, 0.5-2 mm.

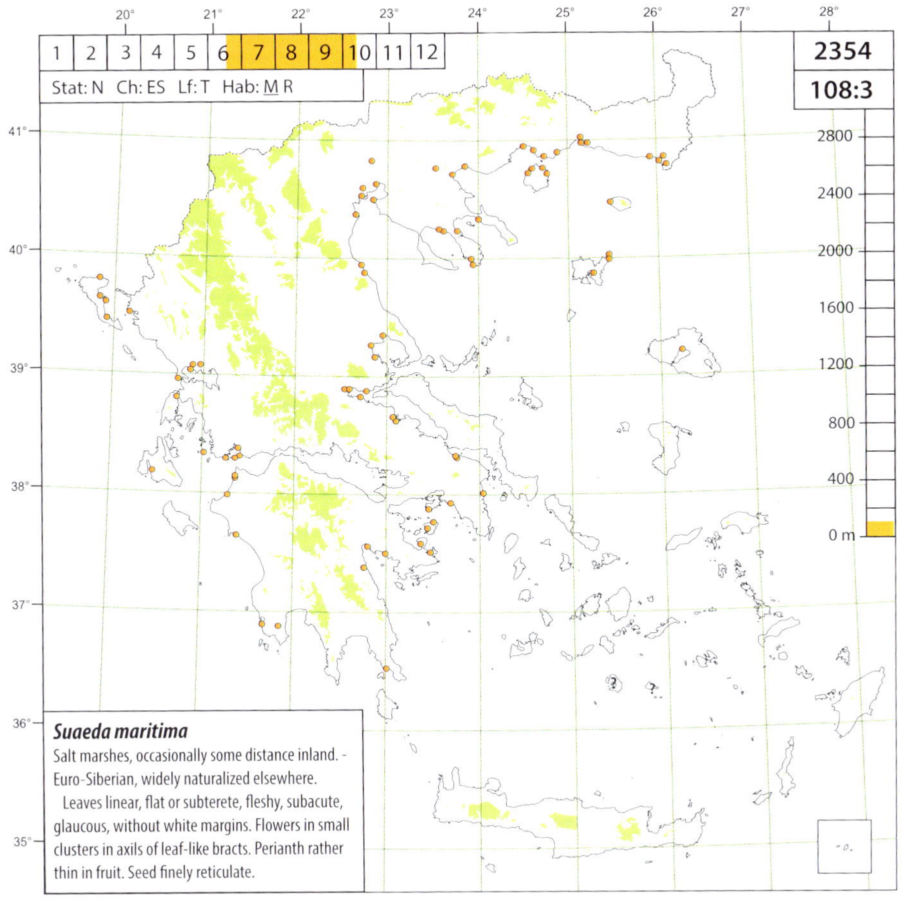

Suaeda maritima

Salt marshes, occasionally some distance inland. - Euro-Siberian, widely naturalized elsewhere.

Leaves linear, flat or subterete, fleshy, subacute, glaucous, without white margins. Flowers in small clusters in axils of leaf-like bracts. Perianth rather thin in fruit. Seed finely reticulate.

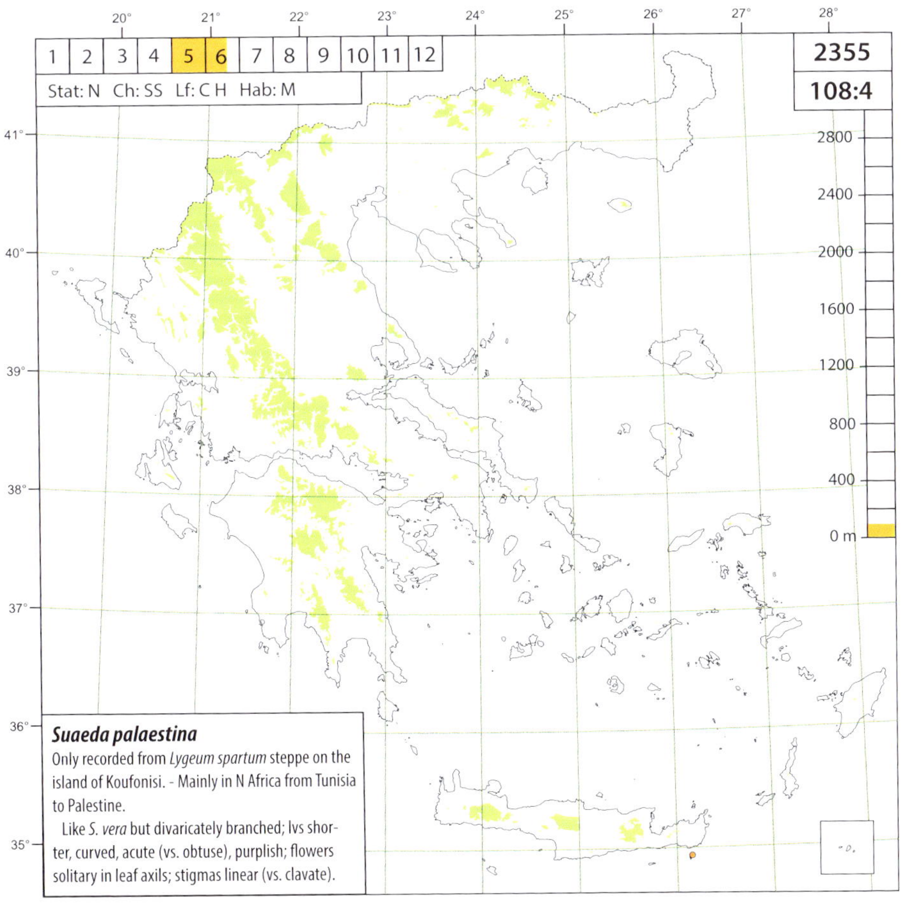

Suaeda palaestina

Only recorded from *Lygeum spartum* steppe on the island of Koufonisi. - Mainly in N Africa from Tunisia to Palestine.

Like *S. vera* but divaricately branched; lvs shorter, curved, acute (vs. obtuse), purplish; flowers solitary in leaf axils; stigmas linear (vs. clavate).

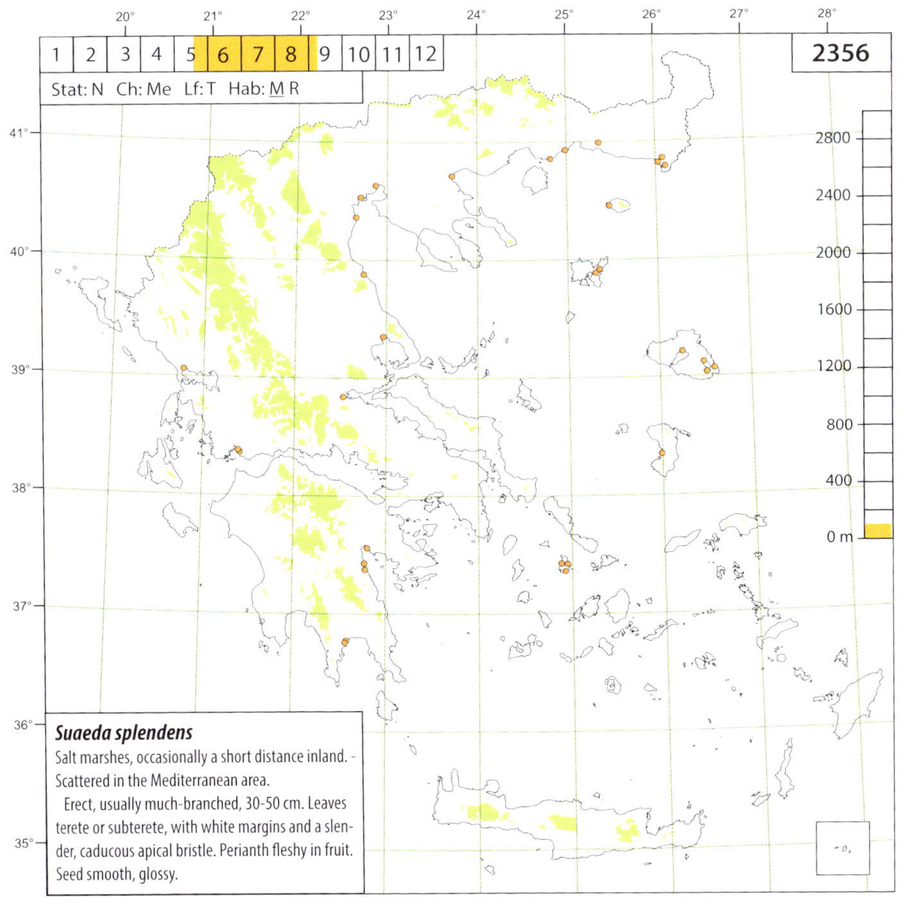

Suaeda splendens

Salt marshes, occasionally a short distance inland. - Scattered in the Mediterranean area.

Erect, usually much-branched, 30-50 cm. Leaves terete or subterete, with white margins and a slender, caducous apical bristle. Perianth fleshy in fruit. Seed smooth, glossy.

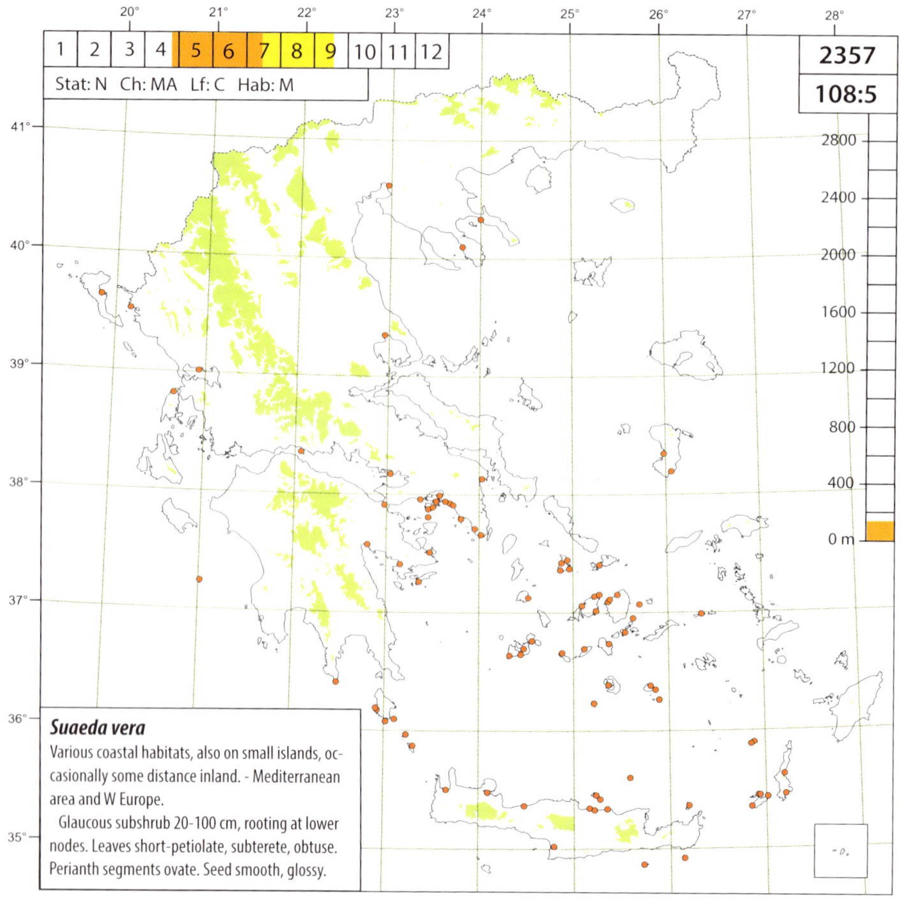

Suaeda vera
Various coastal habitats, also on small islands, occasionally some distance inland. - Mediterranean area and W Europe.
Glaucous subshrub 20-100 cm, rooting at lower nodes. Leaves short-petiolate, subterete, obtuse. Perianth segments ovate. Seed smooth, glossy.

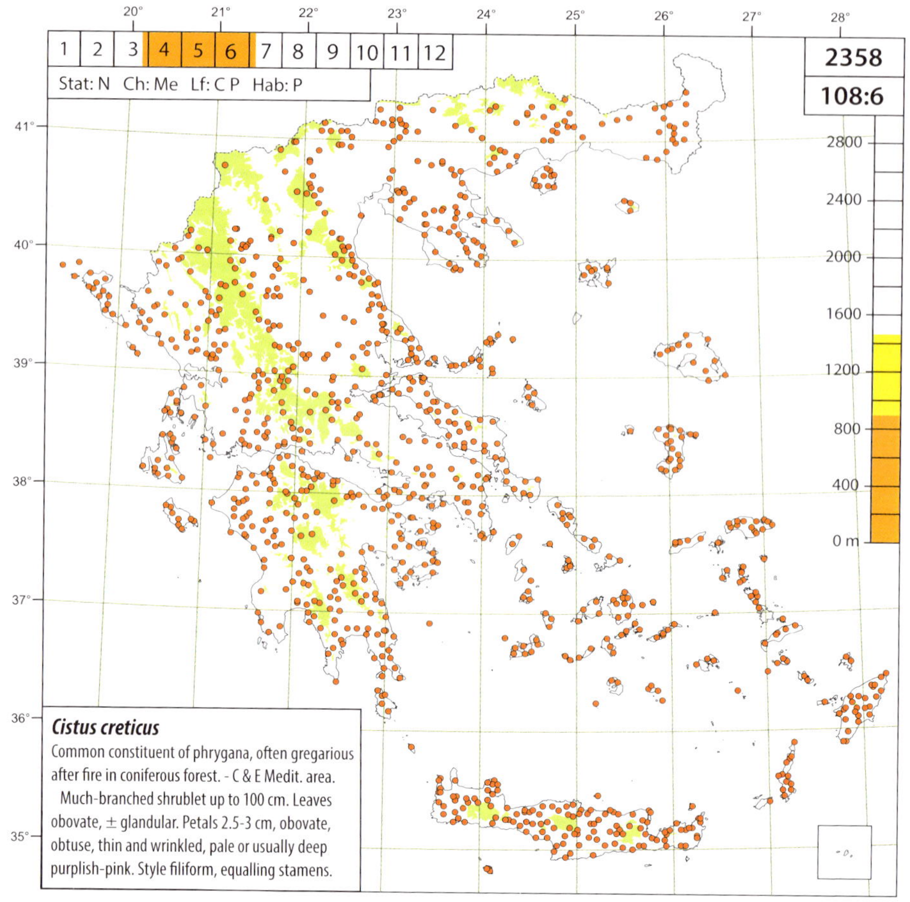

Cistus creticus
Common constituent of phrygana, often gregarious after fire in coniferous forest. - C & E Medit. area.
Much-branched shrublet up to 100 cm. Leaves obovate, ± glandular. Petals 2.5-3 cm, obovate, obtuse, thin and wrinkled, pale or usually deep purplish-pink. Style filiform, equalling stamens.

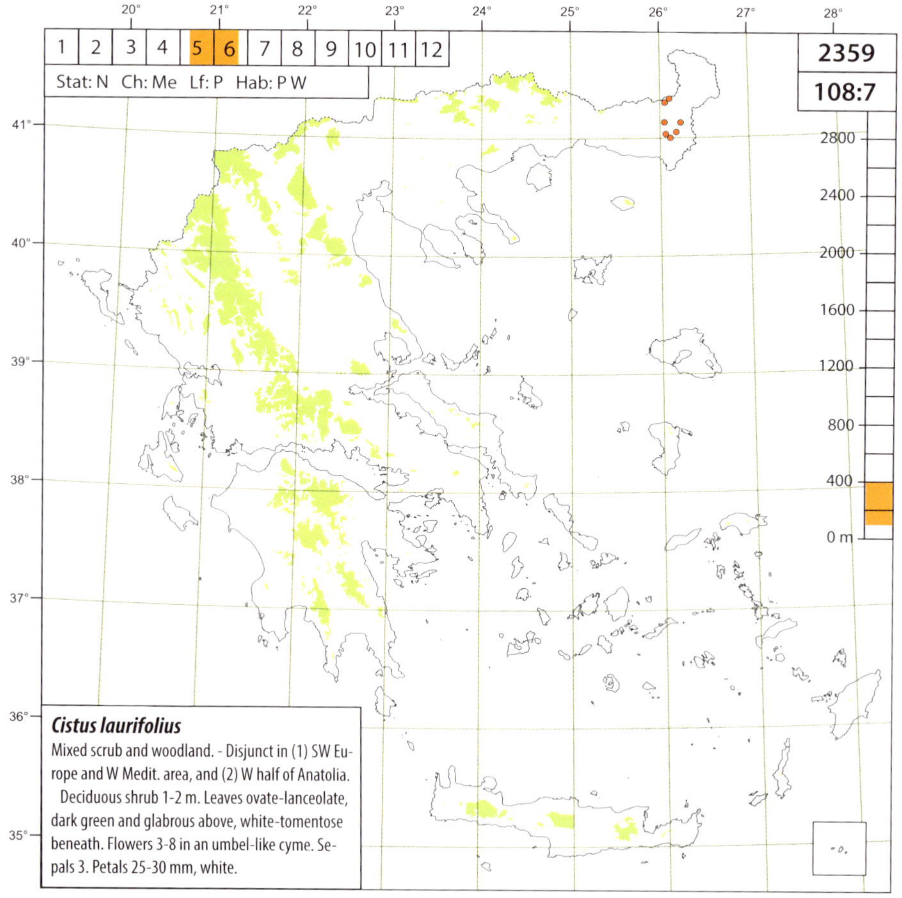

Cistus laurifolius
Mixed scrub and woodland. - Disjunct in (1) SW Europe and W Medit. area, and (2) W half of Anatolia.
Deciduous shrub 1-2 m. Leaves ovate-lanceolate, dark green and glabrous above, white-tomentose beneath. Flowers 3-8 in an umbel-like cyme. Sepals 3. Petals 25-30 mm, white.

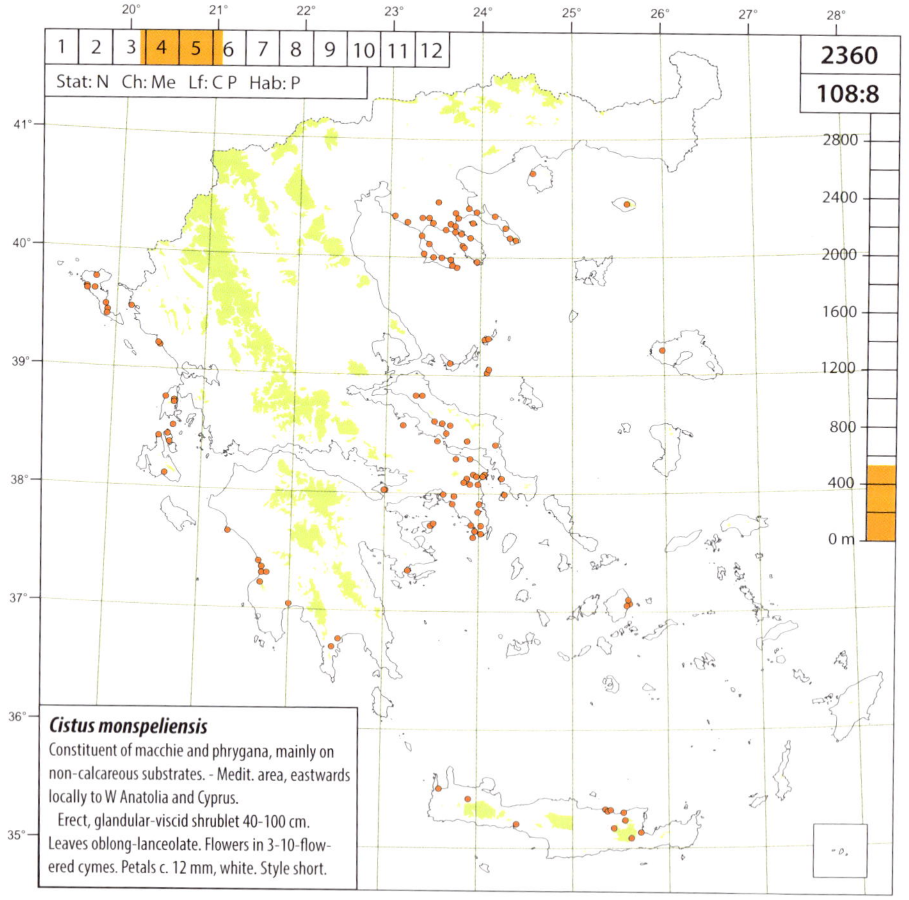

Cistus monspeliensis
Constituent of macchie and phrygana, mainly on non-calcareous substrates. - Medit. area, eastwards locally to W Anatolia and Cyprus.
Erect, glandular-viscid shrublet 40-100 cm. Leaves oblong-lanceolate. Flowers in 3-10-flowered cymes. Petals c. 12 mm, white. Style short.

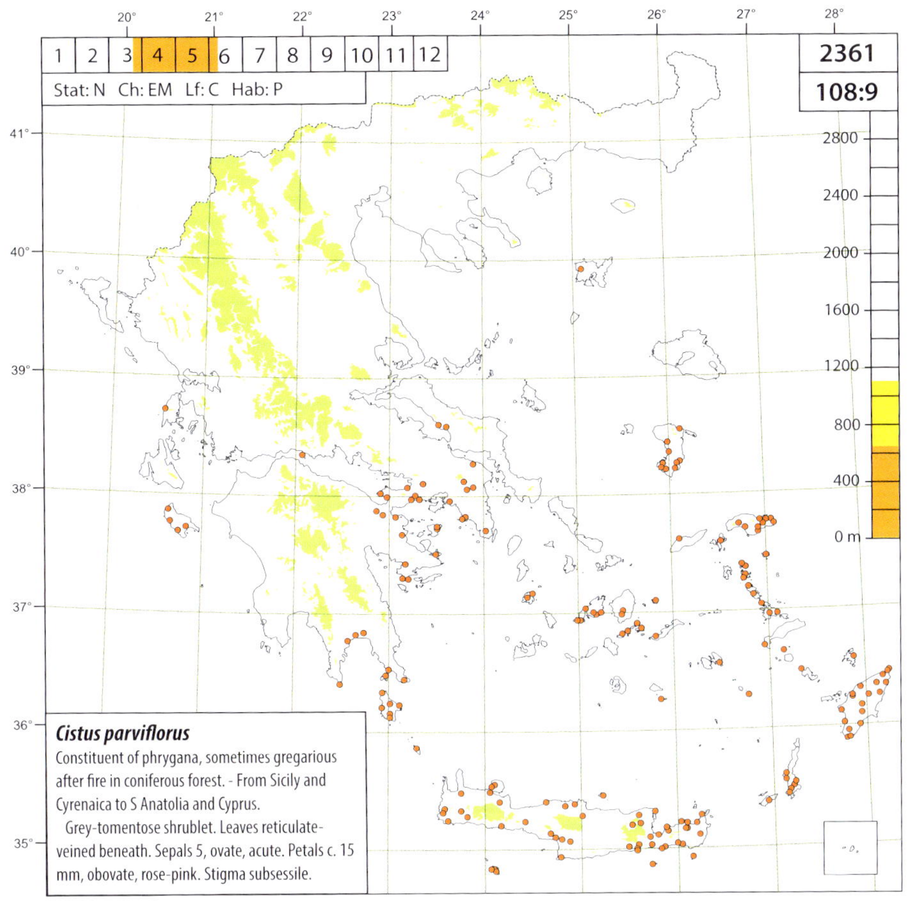

Cistus parviflorus

Constituent of phrygana, sometimes gregarious after fire in coniferous forest. - From Sicily and Cyrenaica to S Anatolia and Cyprus.

Grey-tomentose shrublet. Leaves reticulate-veined beneath. Sepals 5, ovate, acute. Petals c. 15 mm, obovate, rose-pink. Stigma subsessile.

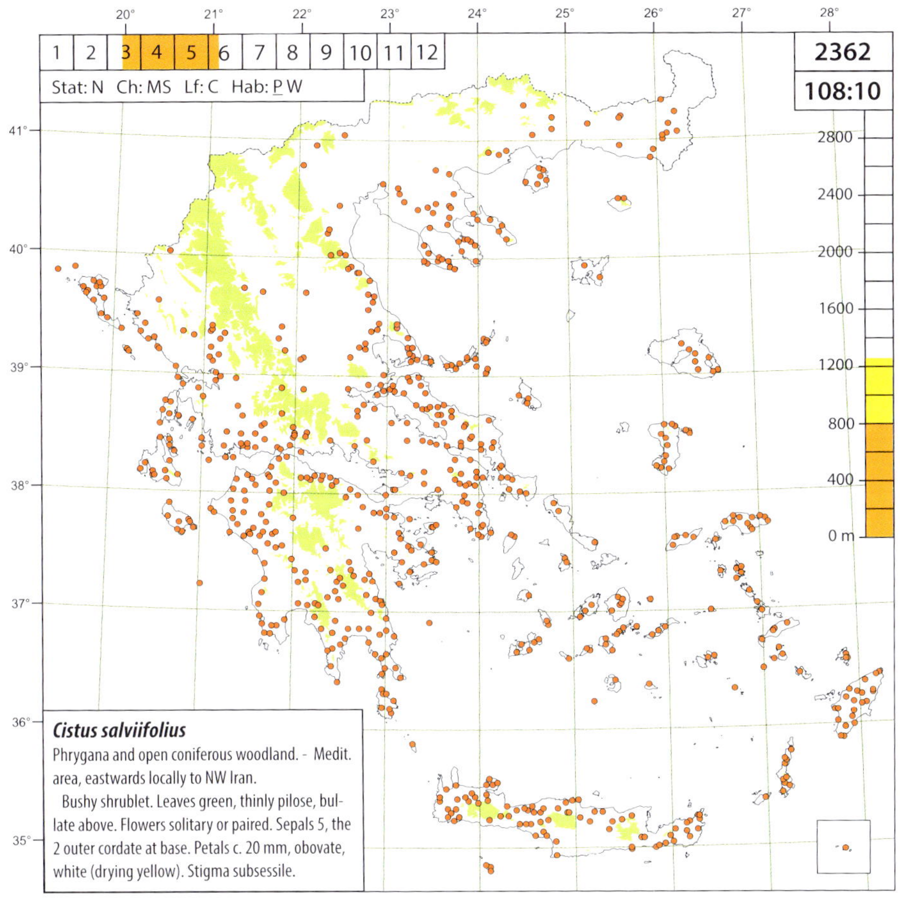

Cistus salviifolius

Phrygana and open coniferous woodland. - Medit. area, eastwards locally to NW Iran.

Bushy shrublet. Leaves green, thinly pilose, bullate above. Flowers solitary or paired. Sepals 5, the 2 outer cordate at base. Petals c. 20 mm, obovate, white (drying yellow). Stigma subsessile.

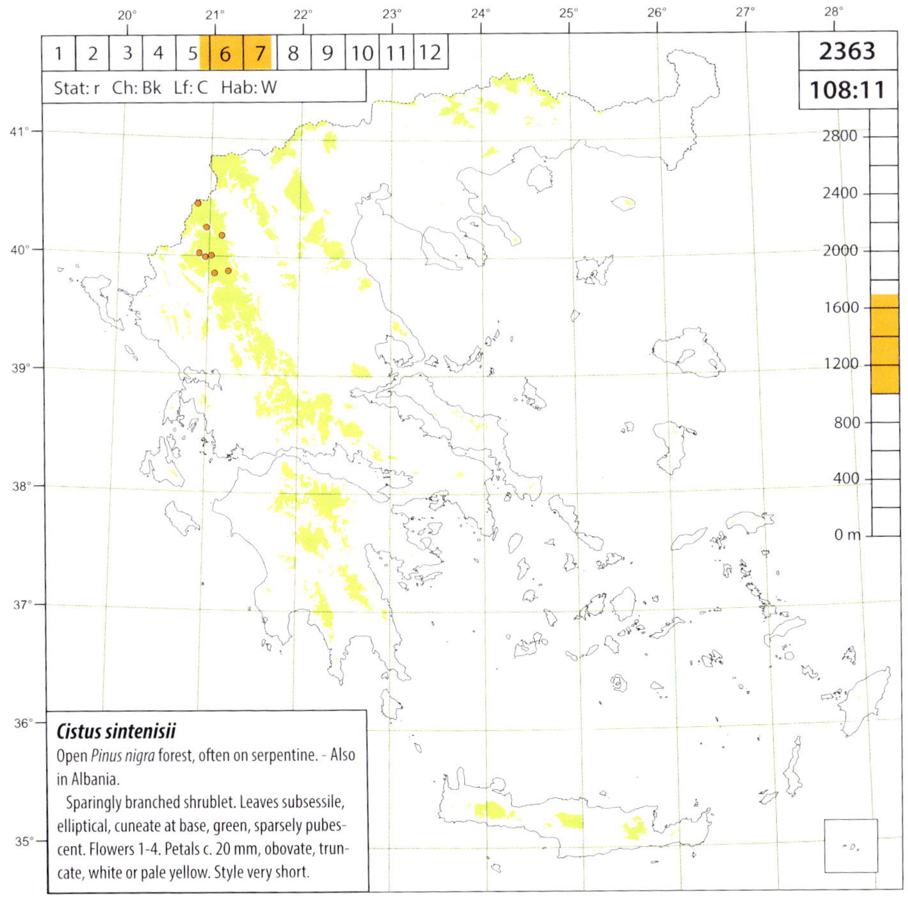

Cistus sintenisii

Open *Pinus nigra* forest, often on serpentine. - Also in Albania.

Sparingly branched shrublet. Leaves subsessile, elliptical, cuneate at base, green, sparsely pubescent. Flowers 1-4. Petals c. 20 mm, obovate, truncate, white or pale yellow. Style very short.

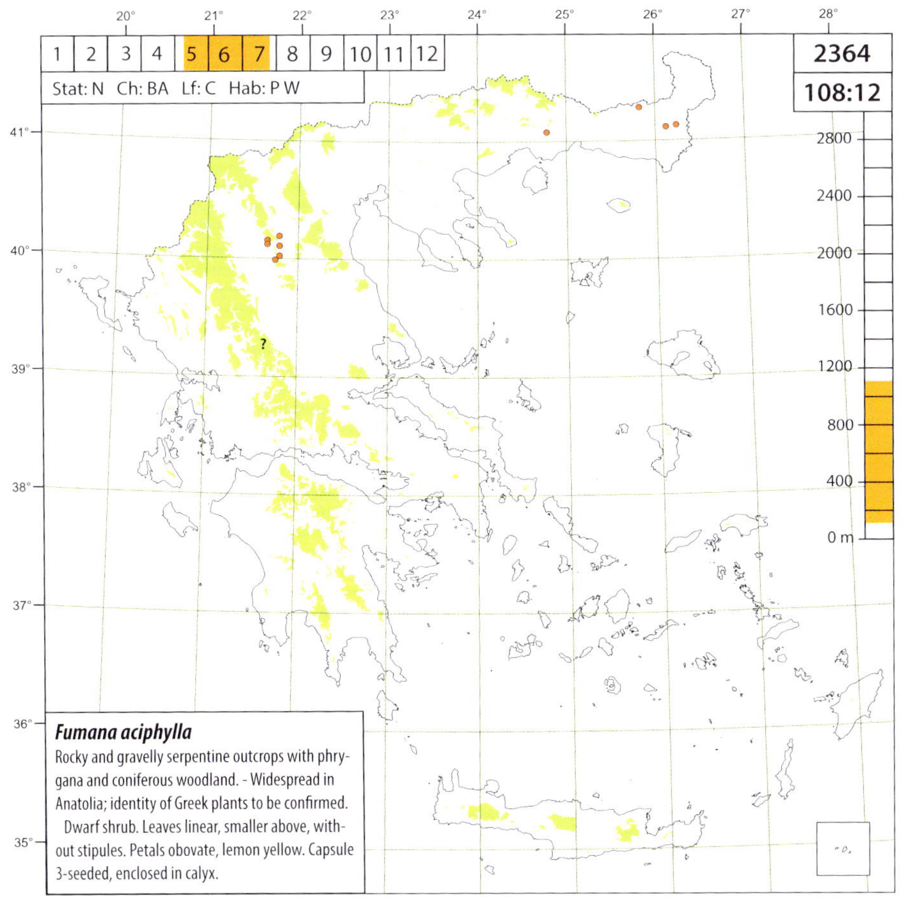

Fumana aciphylla

Rocky and gravelly serpentine outcrops with phrygana and coniferous woodland. - Widespread in Anatolia; identity of Greek plants to be confirmed.

Dwarf shrub. Leaves linear, smaller above, without stipules. Petals obovate, lemon yellow. Capsule 3-seeded, enclosed in calyx.

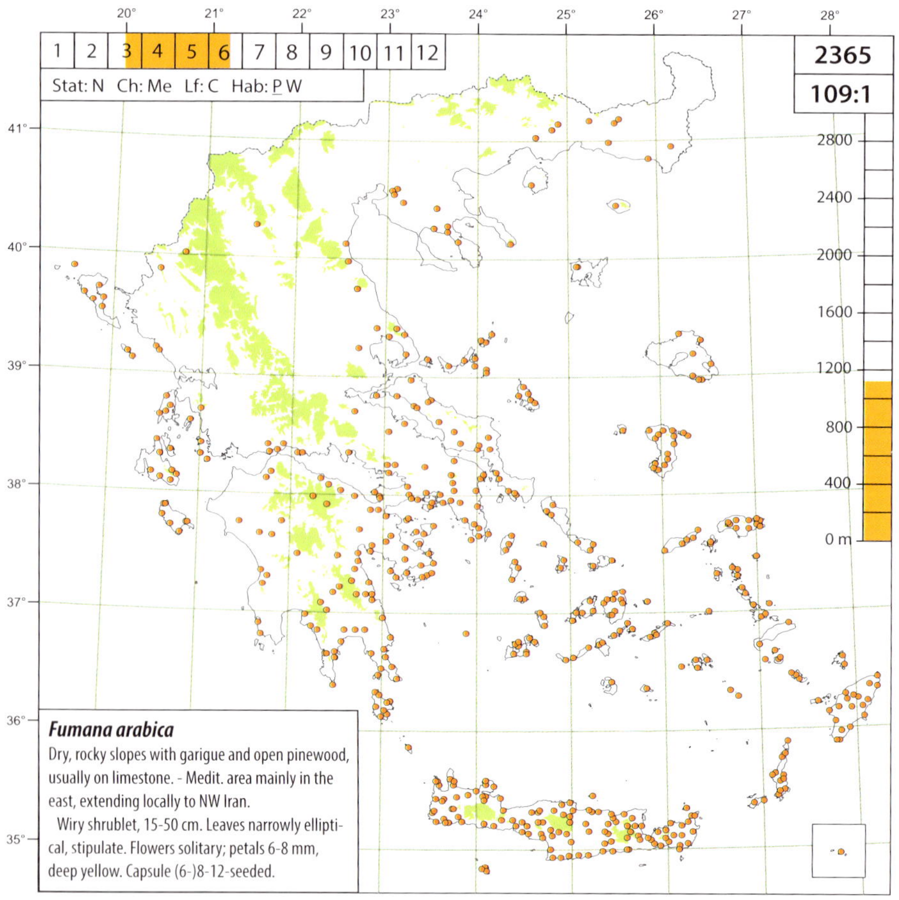

Fumana arabica
Dry, rocky slopes with garigue and open pinewood, usually on limestone. - Medit. area mainly in the east, extending locally to NW Iran.
Wiry shrublet, 15-50 cm. Leaves narrowly elliptical, stipulate. Flowers solitary; petals 6-8 mm, deep yellow. Capsule (6-)8-12-seeded.

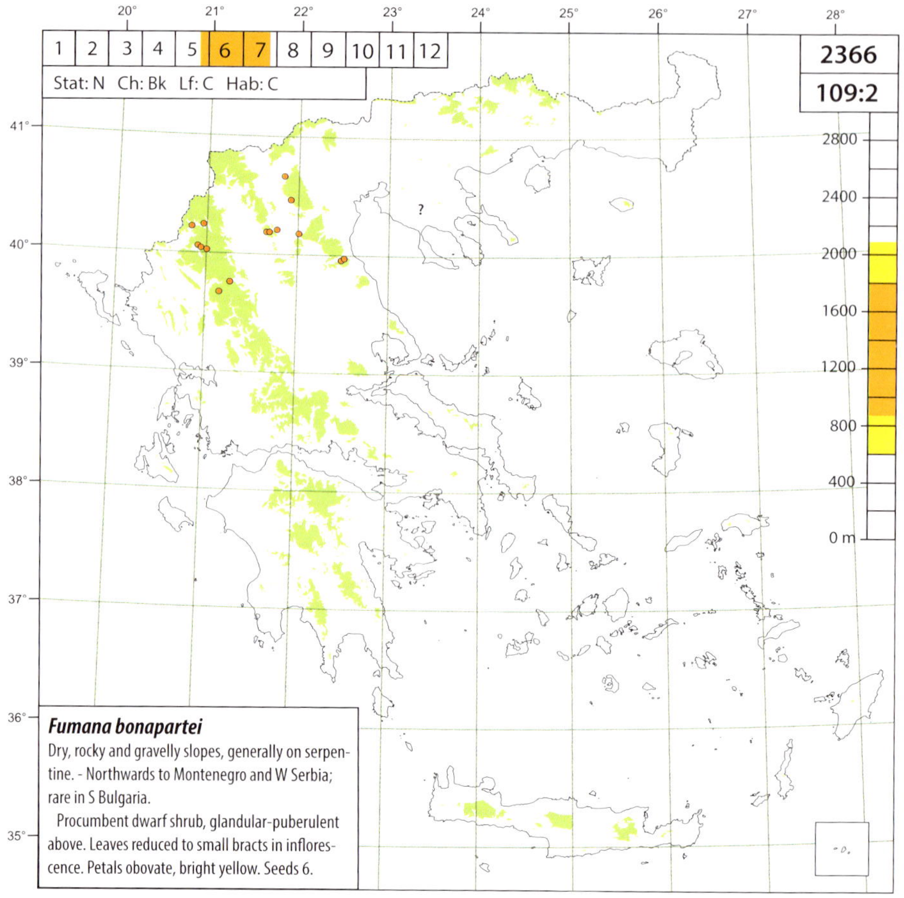

Fumana bonapartei
Dry, rocky and gravelly slopes, generally on serpentine. - Northwards to Montenegro and W Serbia; rare in S Bulgaria.
Procumbent dwarf shrub, glandular-puberulent above. Leaves reduced to small bracts in inflorescence. Petals obovate, bright yellow. Seeds 6.

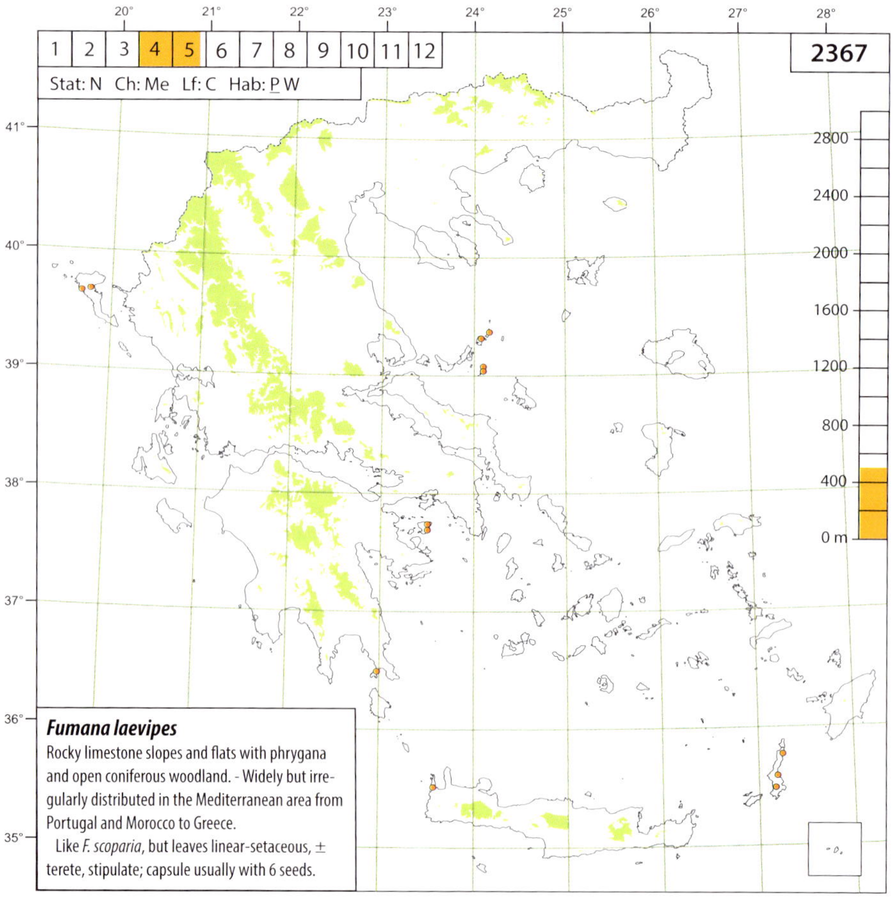

Fumana laevipes
Rocky limestone slopes and flats with phrygana and open coniferous woodland. - Widely but irregularly distributed in the Mediterranean area from Portugal and Morocco to Greece.
Like *F. scoparia*, but leaves linear-setaceous, ± terete, stipulate; capsule usually with 6 seeds.

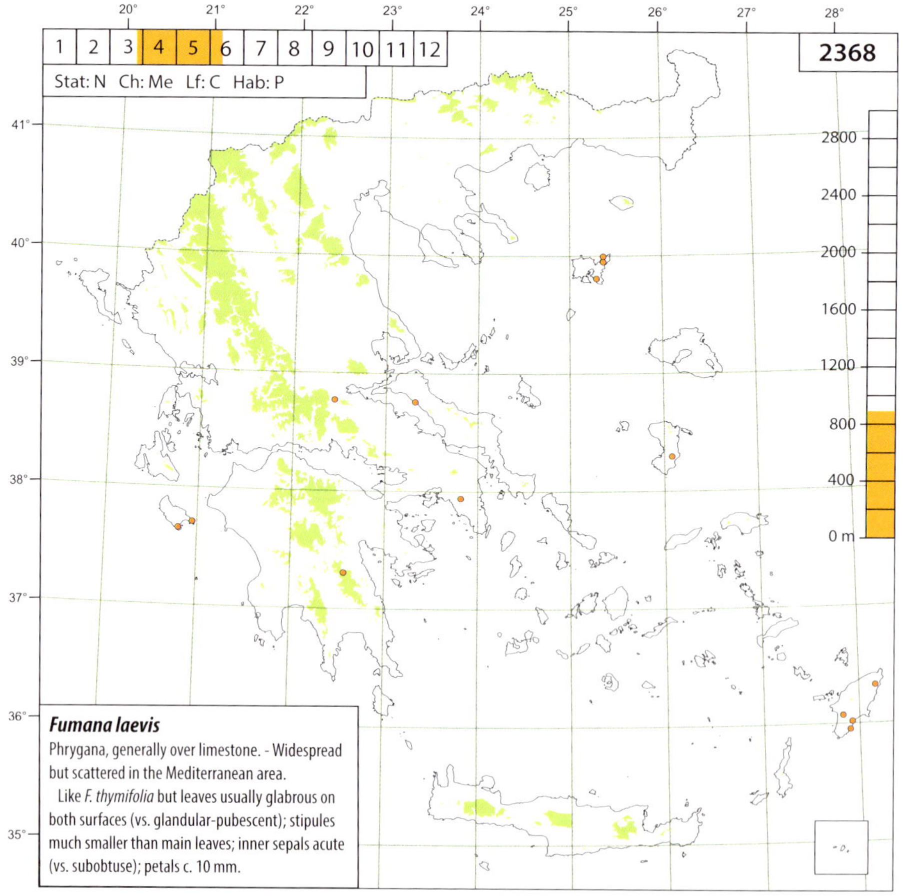

Fumana laevis
Phrygana, generally over limestone. - Widespread but scattered in the Mediterranean area.
Like *F. thymifolia* but leaves usually glabrous on both surfaces (vs. glandular-pubescent); stipules much smaller than main leaves; inner sepals acute (vs. subobtuse); petals c. 10 mm.

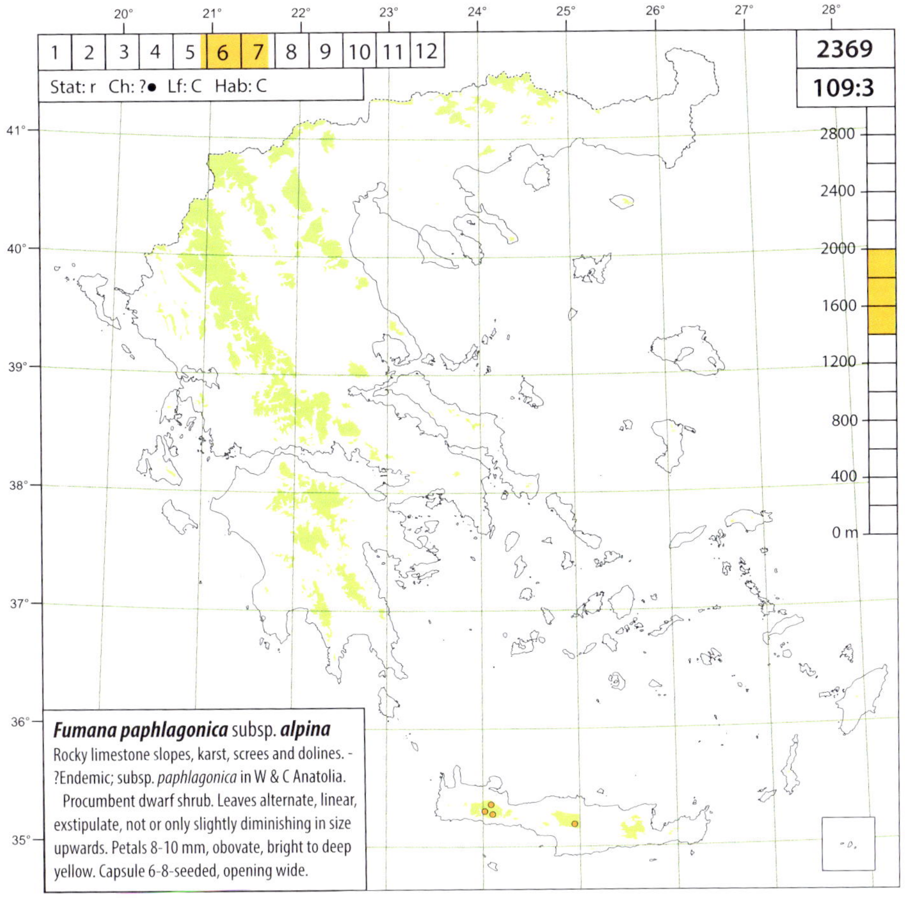

Fumana paphlagonica subsp. ***alpina***
Rocky limestone slopes, karst, screes and dolines. - ?Endemic; subsp. *paphlagonica* in W & C Anatolia.
Procumbent dwarf shrub. Leaves alternate, linear, exstipulate, not or only slightly diminishing in size upwards. Petals 8-10 mm, obovate, bright to deep yellow. Capsule 6-8-seeded, opening wide.

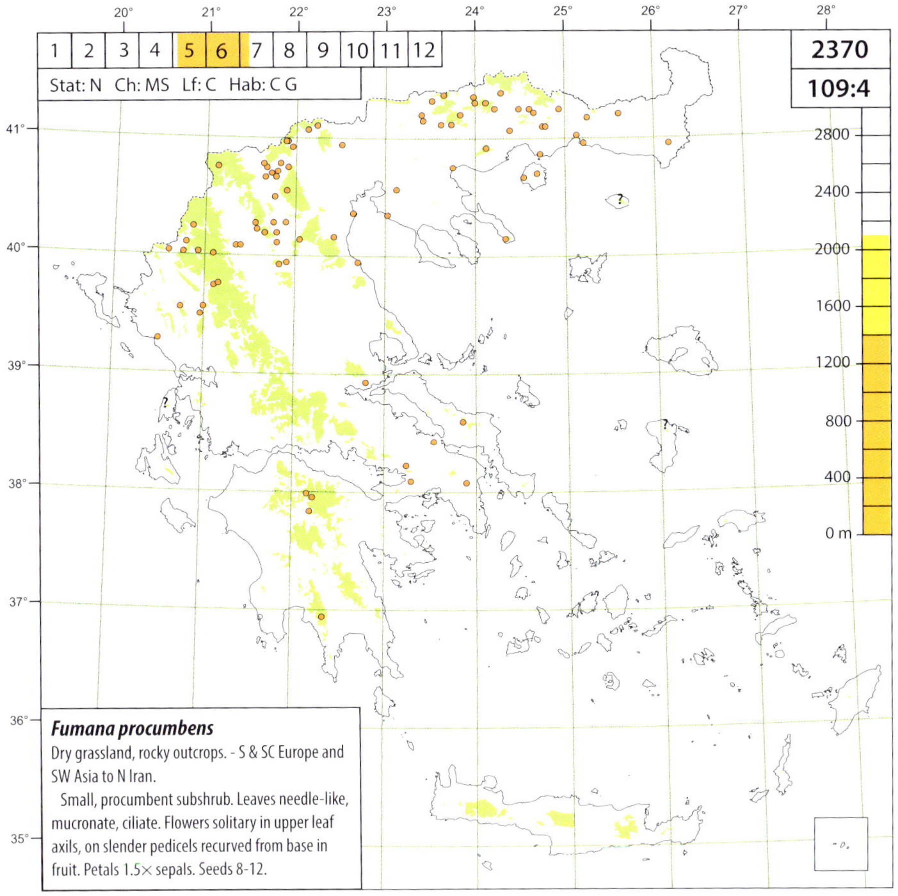

Fumana procumbens
Dry grassland, rocky outcrops. - S & SC Europe and SW Asia to N Iran.
Small, procumbent subshrub. Leaves needle-like, mucronate, ciliate. Flowers solitary in upper leaf axils, on slender pedicels recurved from base in fruit. Petals 1.5× sepals. Seeds 8-12.

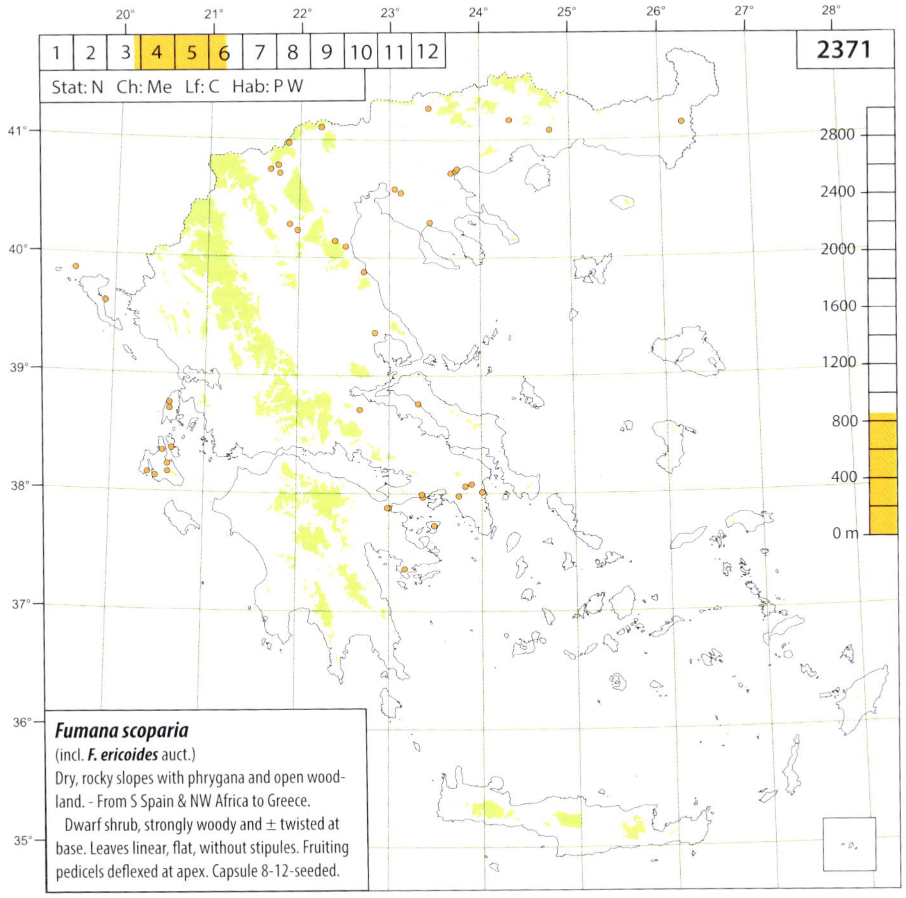

Fumana scoparia
(incl. ***F. ericoides*** auct.)
Dry, rocky slopes with phrygana and open woodland. - From S Spain & NW Africa to Greece.
Dwarf shrub, strongly woody and ± twisted at base. Leaves linear, flat, without stipules. Fruiting pedicels deflexed at apex. Capsule 8-12-seeded.

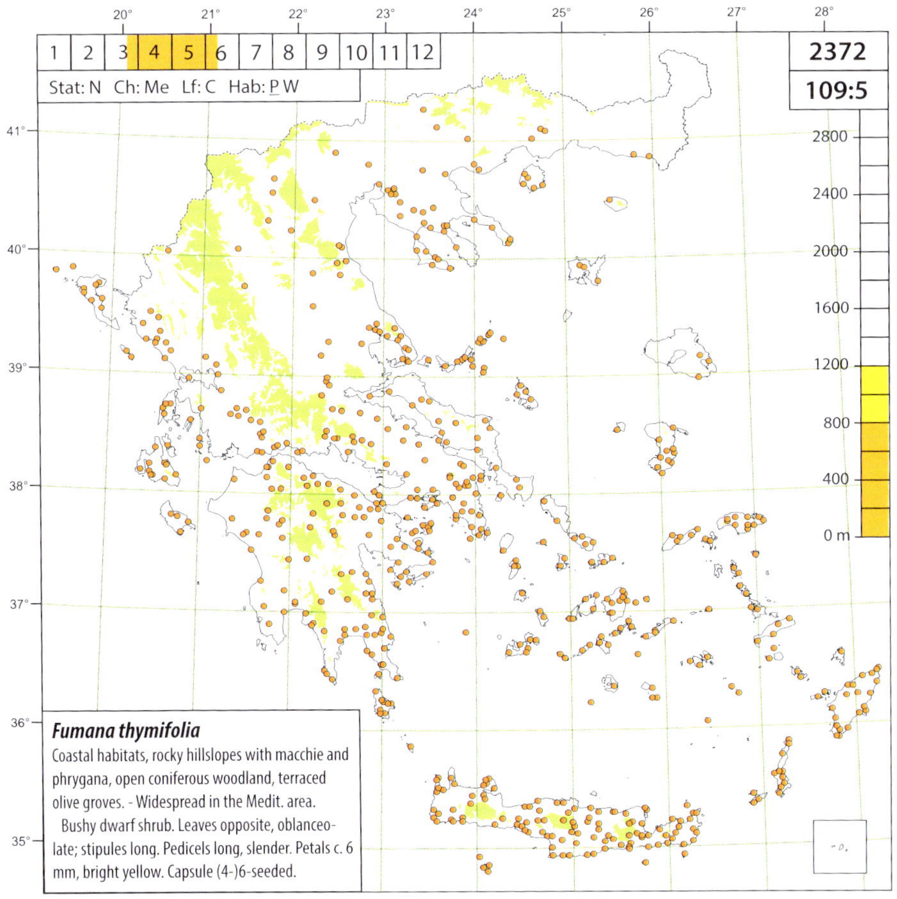

Fumana thymifolia
Coastal habitats, rocky hillslopes with macchie and phrygana, open coniferous woodland, terraced olive groves. - Widespread in the Medit. area.
Bushy dwarf shrub. Leaves opposite, oblanceolate; stipules long. Pedicels long, slender. Petals c. 6 mm, bright yellow. Capsule (4-)6-seeded.

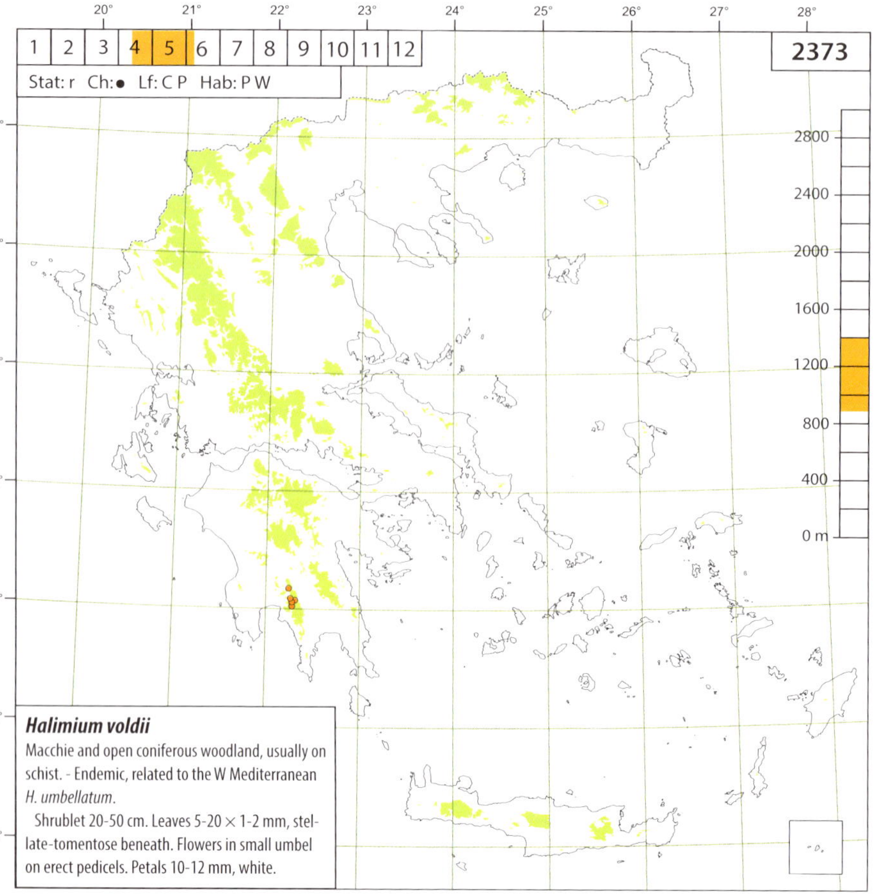

Halimium voldii

Macchie and open coniferous woodland, usually on schist. - Endemic, related to the W Mediterranean *H. umbellatum*.

Shrublet 20-50 cm. Leaves 5-20 × 1-2 mm, stellate-tomentose beneath. Flowers in small umbel on erect pedicels. Petals 10-12 mm, white.

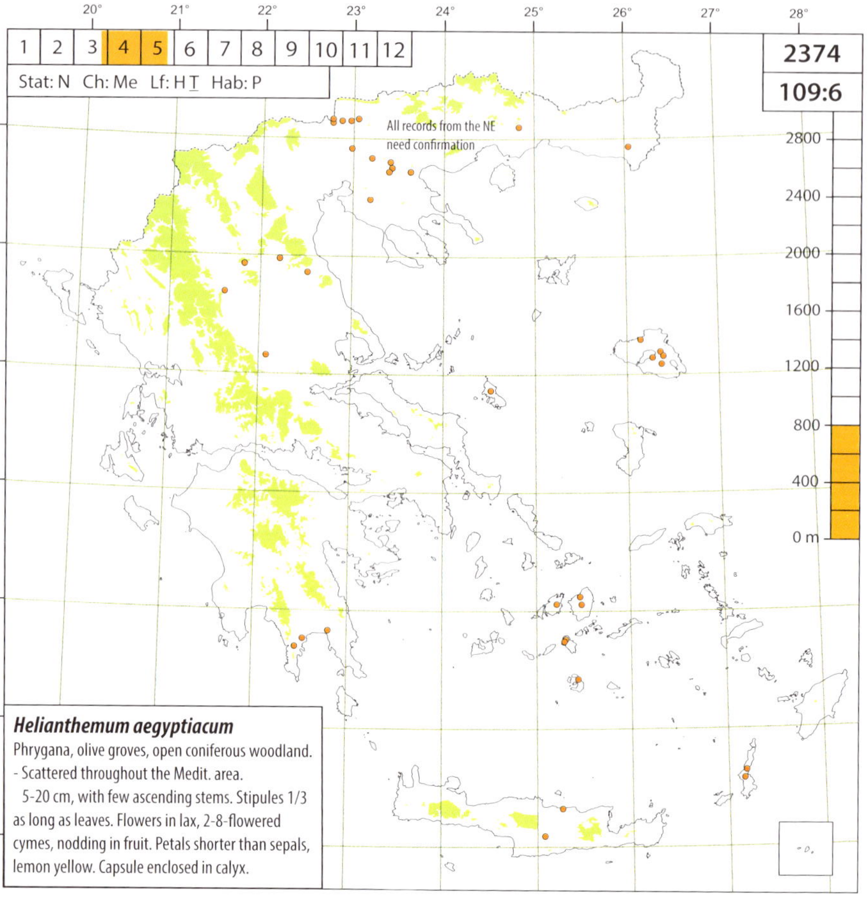

Helianthemum aegyptiacum

Phrygana, olive groves, open coniferous woodland. - Scattered throughout the Medit. area.

5-20 cm, with few ascending stems. Stipules 1/3 as long as leaves. Flowers in lax, 2-8-flowered cymes, nodding in fruit. Petals shorter than sepals, lemon yellow. Capsule enclosed in calyx.

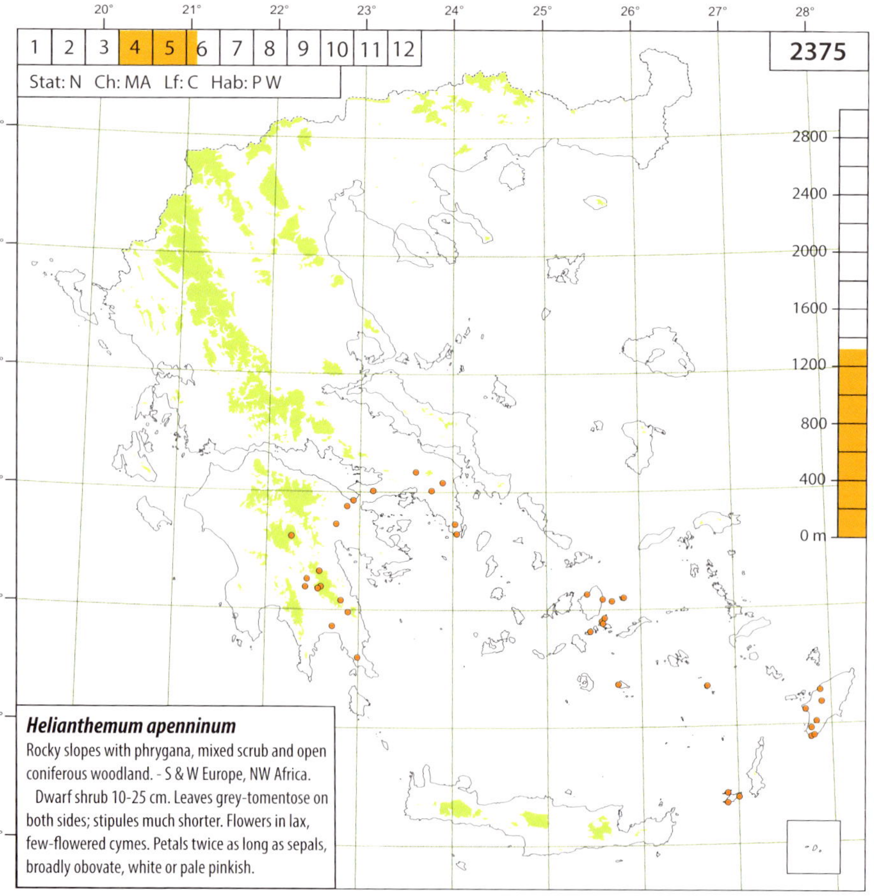

Helianthemum apenninum

Rocky slopes with phrygana, mixed scrub and open coniferous woodland. - S & W Europe, NW Africa.

Dwarf shrub 10-25 cm. Leaves grey-tomentose on both sides; stipules much shorter. Flowers in lax, few-flowered cymes. Petals twice as long as sepals, broadly obovate, white or pale pinkish.

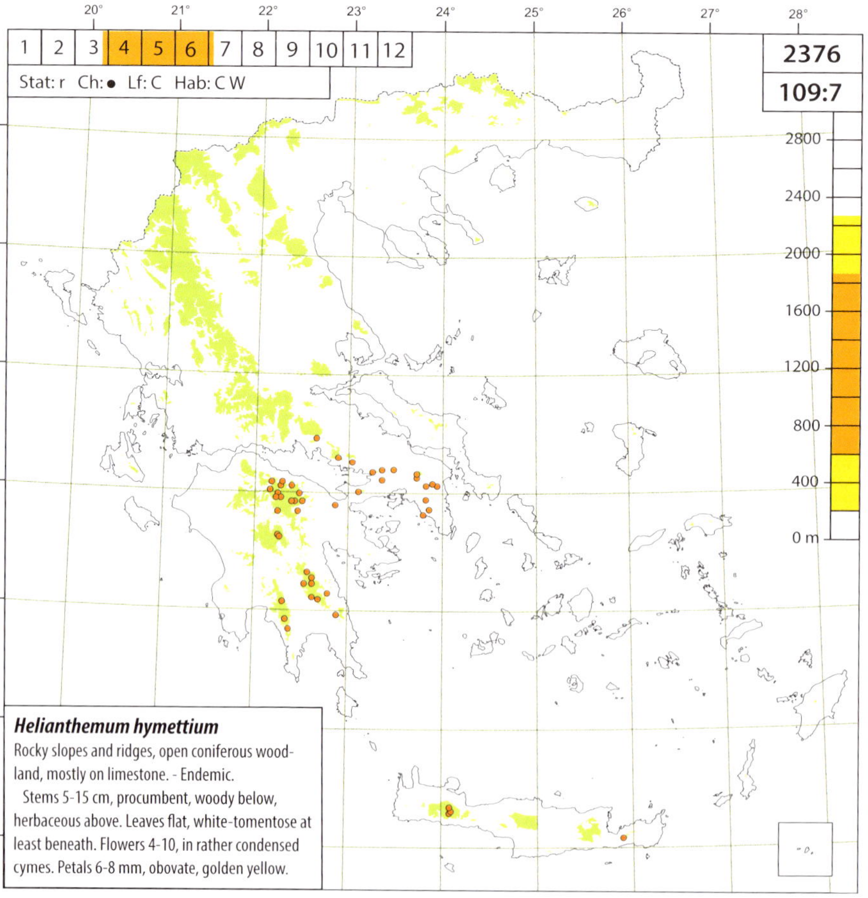

Helianthemum hymettium

Rocky slopes and ridges, open coniferous woodland, mostly on limestone. - Endemic.

Stems 5-15 cm, procumbent, woody below, herbaceous above. Leaves flat, white-tomentose at least beneath. Flowers 4-10, in rather condensed cymes. Petals 6-8 mm, obovate, golden yellow.

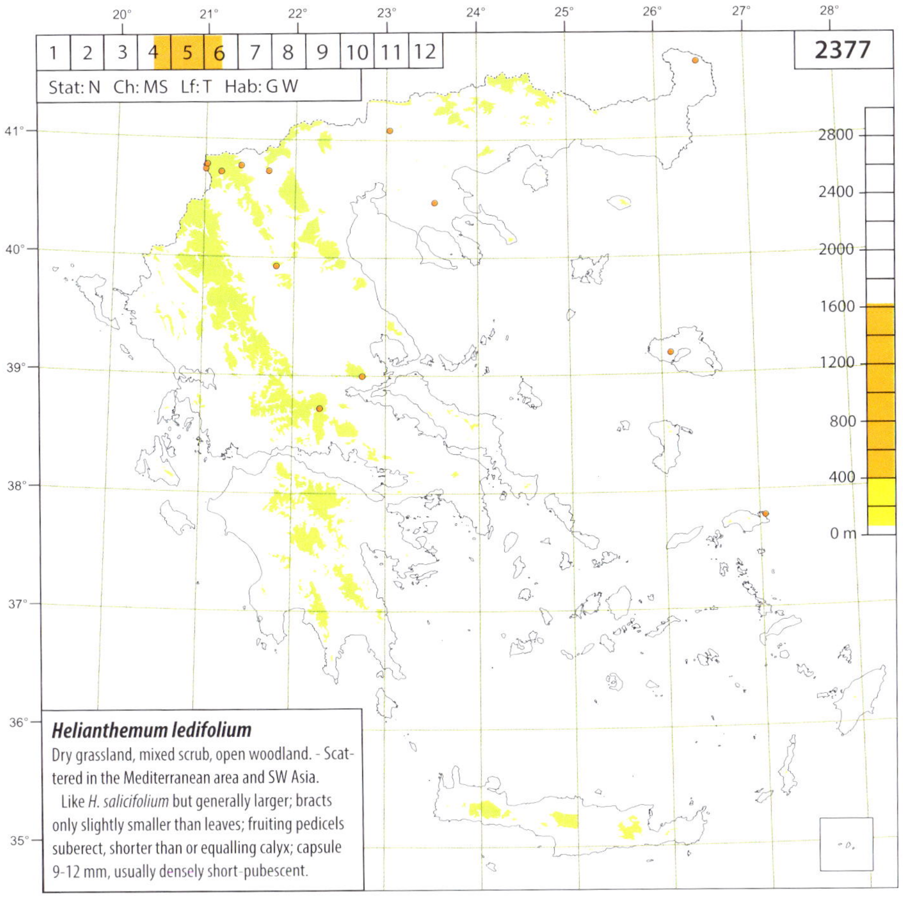

Helianthemum ledifolium
Dry grassland, mixed scrub, open woodland. - Scattered in the Mediterranean area and SW Asia.
Like *H. salicifolium* but generally larger; bracts only slightly smaller than leaves; fruiting pedicels suberect, shorter than or equalling calyx; capsule 9-12 mm, usually densely short-pubescent.

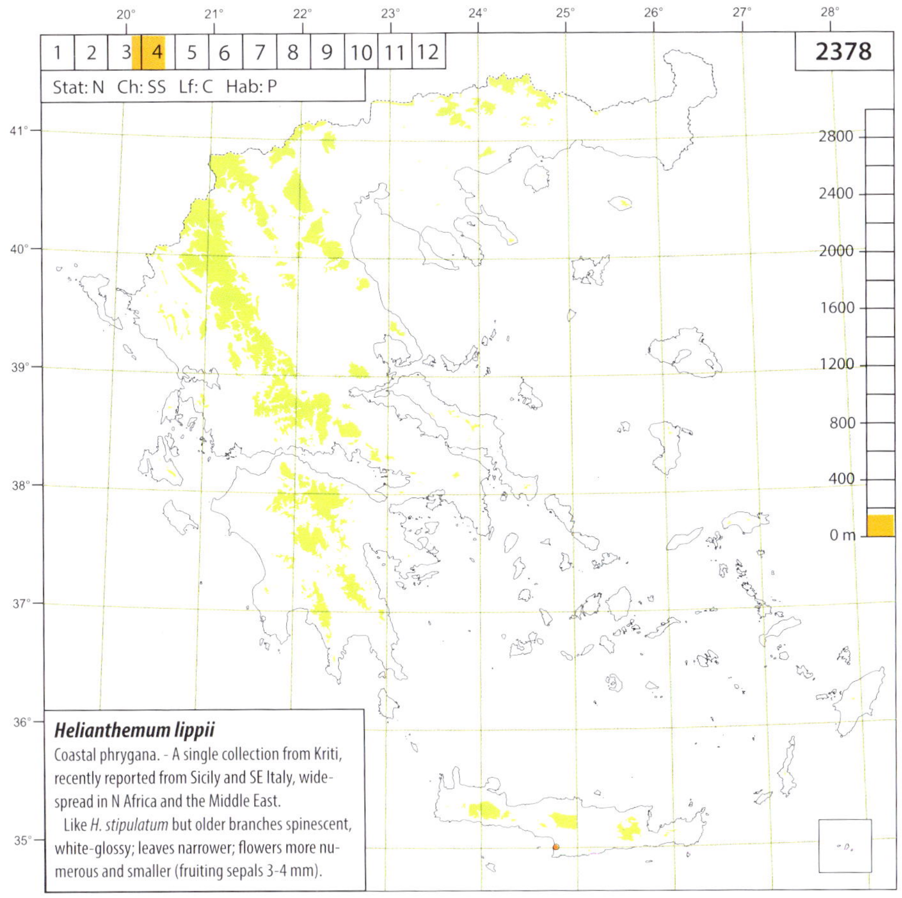

Helianthemum lippii
Coastal phrygana. - A single collection from Kriti, recently reported from Sicily and SE Italy, widespread in N Africa and the Middle East.
Like *H. stipulatum* but older branches spinescent, white-glossy; leaves narrower; flowers more numerous and smaller (fruiting sepals 3-4 mm).

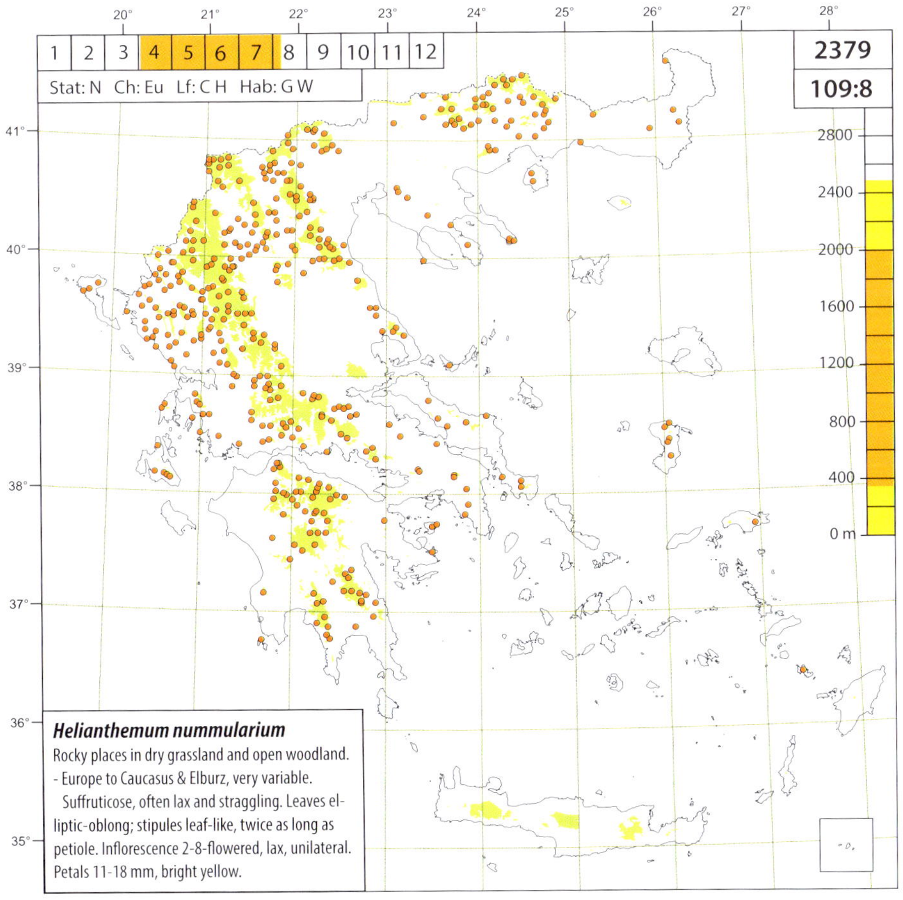

Helianthemum nummularium
Rocky places in dry grassland and open woodland. - Europe to Caucasus & Elburz, very variable.
Suffruticose, often lax and straggling. Leaves elliptic-oblong; stipules leaf-like, twice as long as petiole. Inflorescence 2-8-flowered, lax, unilateral. Petals 11-18 mm, bright yellow.

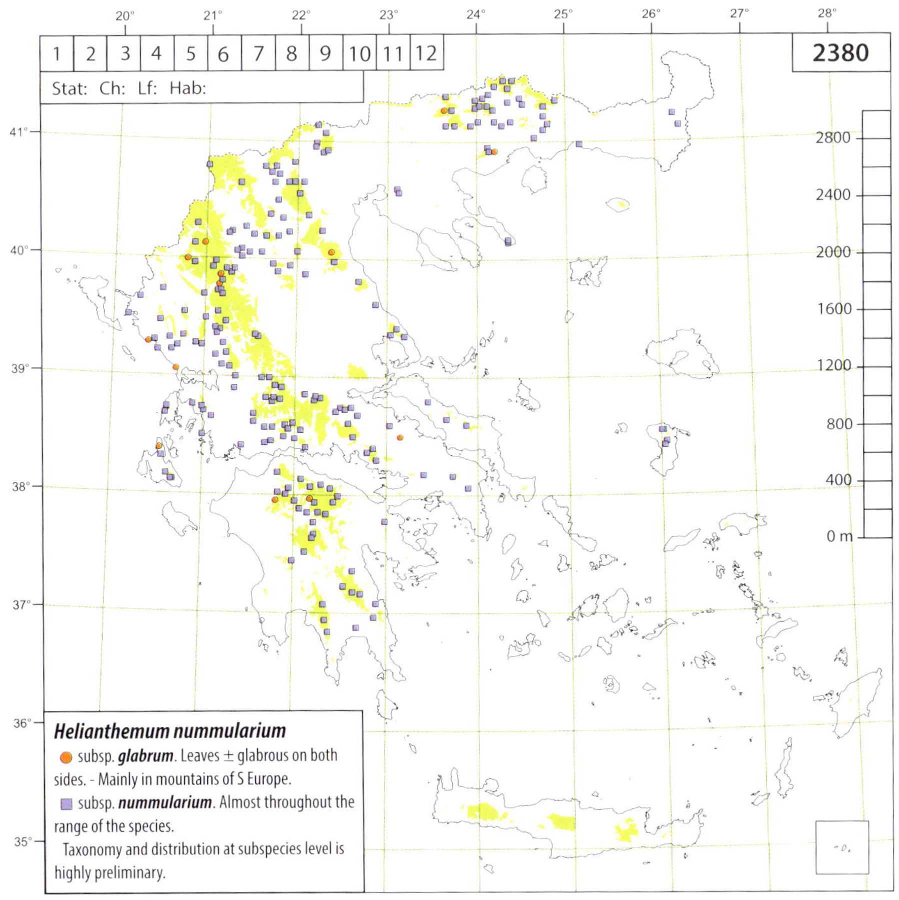

Helianthemum nummularium
● subsp. ***glabrum***. Leaves ± glabrous on both sides. - Mainly in mountains of S Europe.
■ subsp. ***nummularium***. Almost throughout the range of the species.
Taxonomy and distribution at subspecies level is highly preliminary.

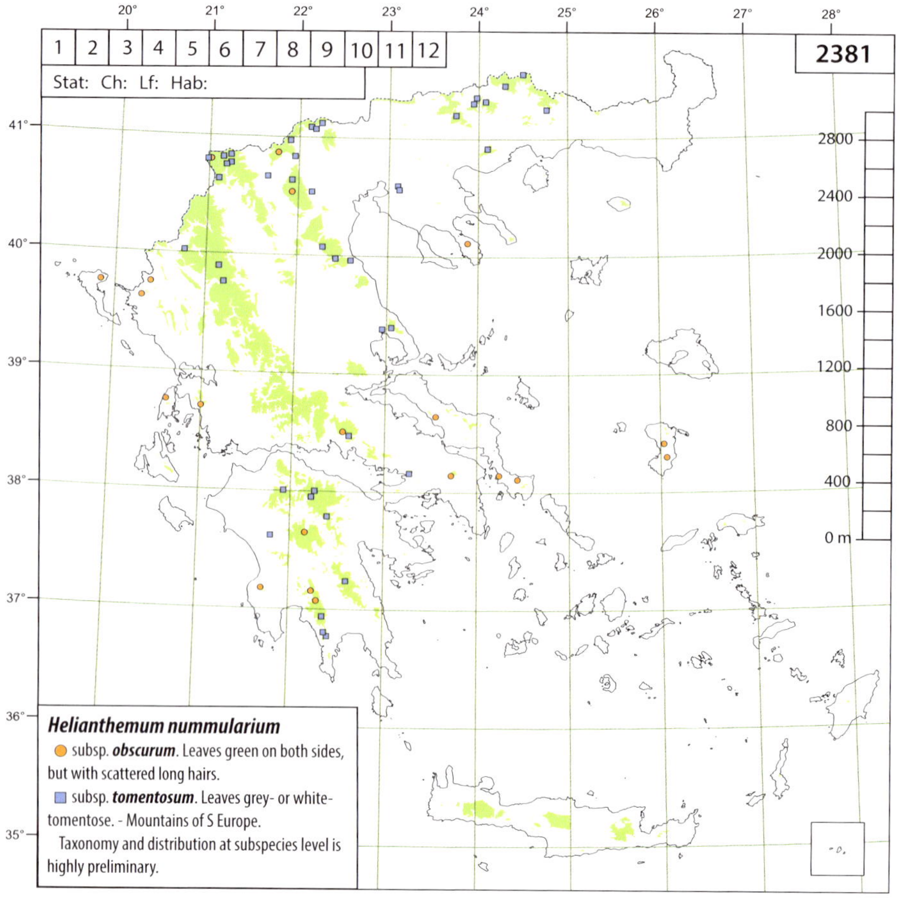

Helianthemum nummularium
● subsp. ***obscurum***. Leaves green on both sides, but with scattered long hairs.
■ subsp. ***tomentosum***. Leaves grey- or white-tomentose. - Mountains of S Europe.
Taxonomy and distribution at subspecies level is highly preliminary.

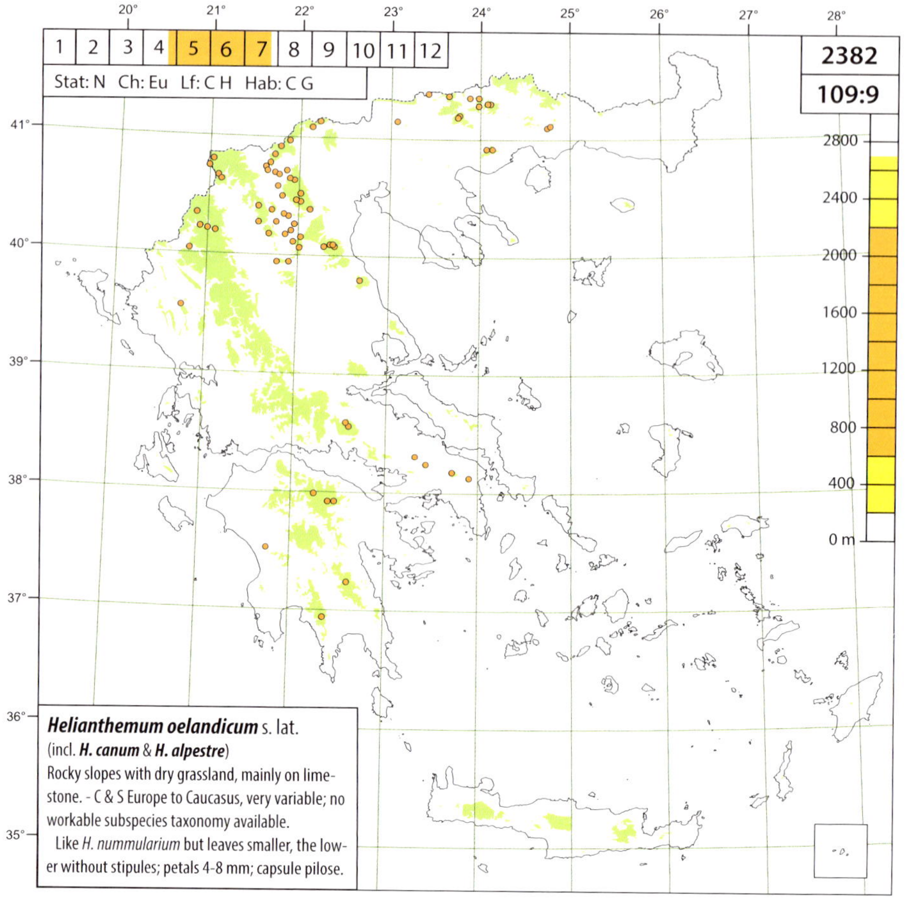

Helianthemum oelandicum s. lat.
(incl. ***H. canum*** & ***H. alpestre***)
Rocky slopes with dry grassland, mainly on limestone. - C & S Europe to Caucasus, very variable; no workable subspecies taxonomy available.
Like *H. nummularium* but leaves smaller, the lower without stipules; petals 4-8 mm; capsule pilose.

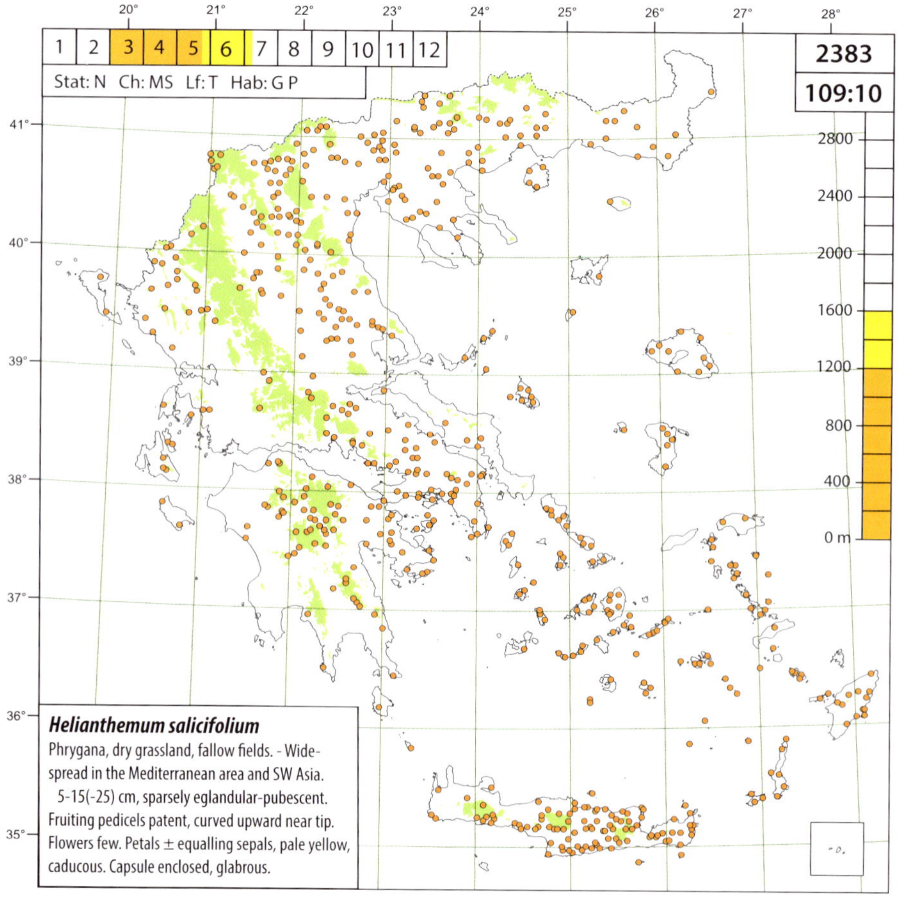

Helianthemum salicifolium
Phrygana, dry grassland, fallow fields. - Widespread in the Mediterranean area and SW Asia.
5-15(-25) cm, sparsely eglandular-pubescent. Fruiting pedicels patent, curved upward near tip. Flowers few. Petals ± equalling sepals, pale yellow, caducous. Capsule enclosed, glabrous.

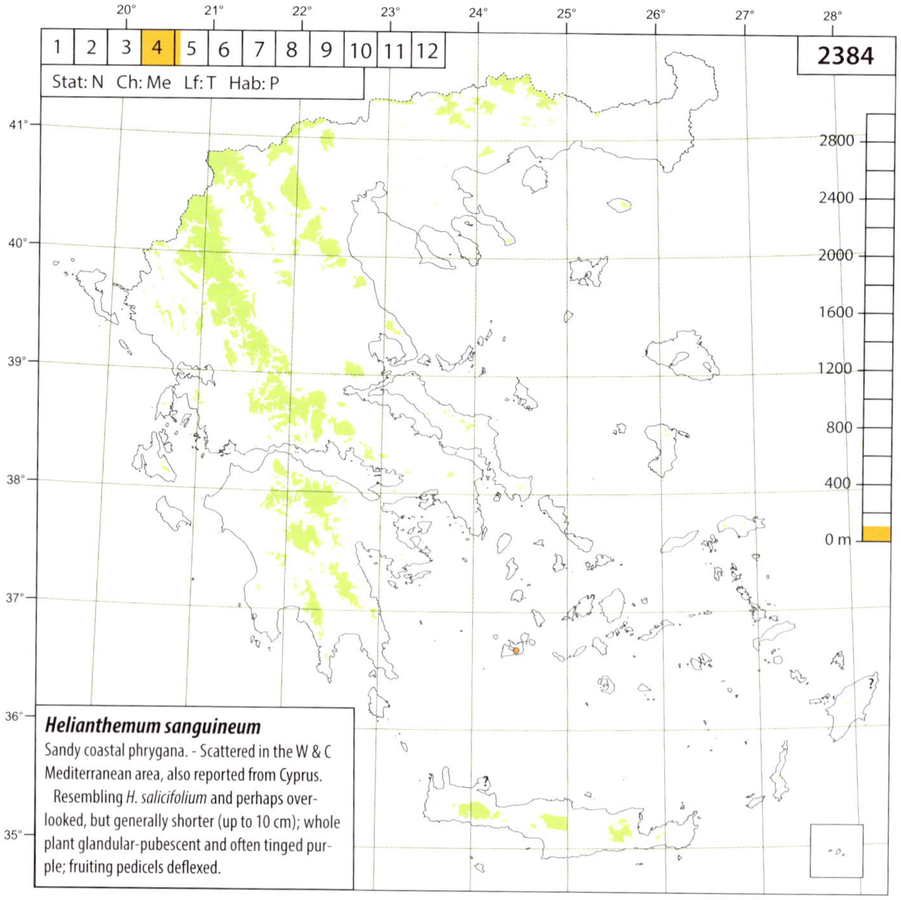

Helianthemum sanguineum
Sandy coastal phrygana. - Scattered in the W & C Mediterranean area, also reported from Cyprus.
Resembling *H. salicifolium* and perhaps overlooked, but generally shorter (up to 10 cm); whole plant glandular-pubescent and often tinged purple; fruiting pedicels deflexed.

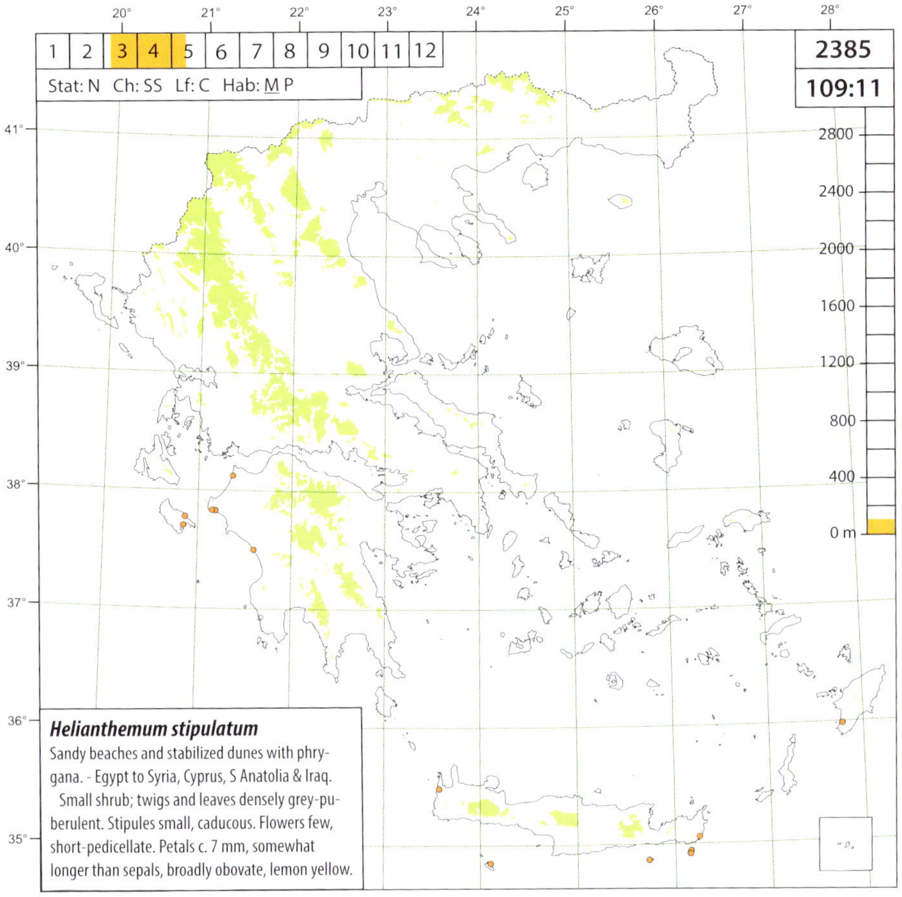

Helianthemum stipulatum
Sandy beaches and stabilized dunes with phrygana. - Egypt to Syria, Cyprus, S Anatolia & Iraq.
Small shrub; twigs and leaves densely grey-puberulent. Stipules small, caducous. Flowers few, short-pedicellate. Petals c. 7 mm, somewhat longer than sepals, broadly obovate, lemon yellow.

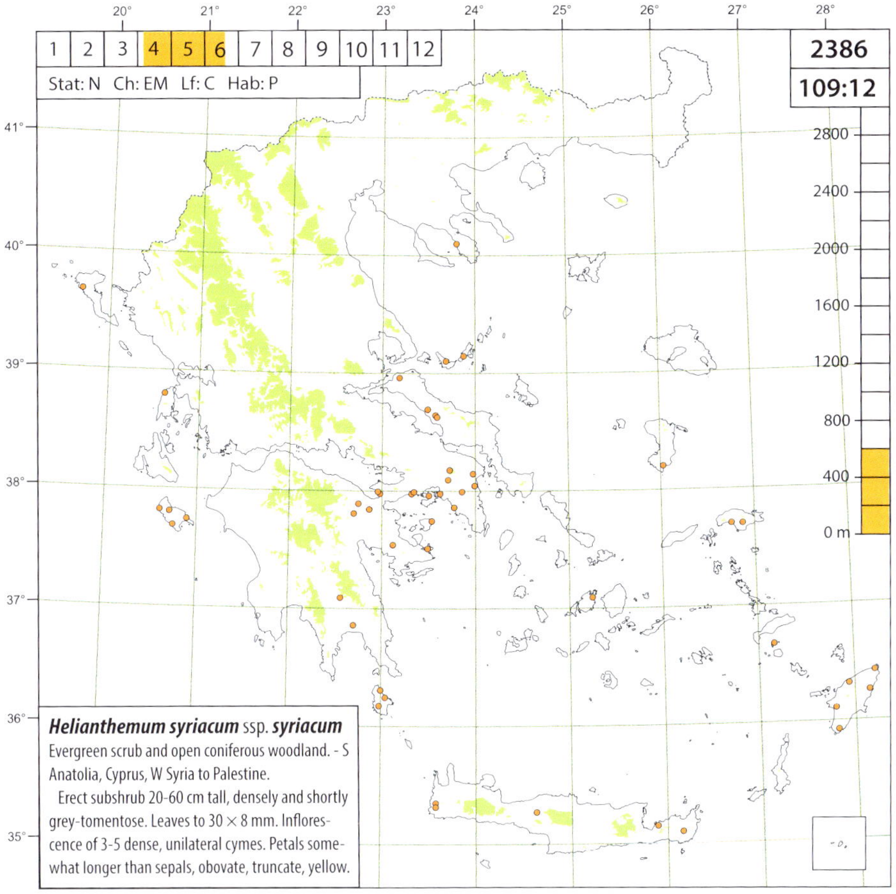

Helianthemum syriacum ssp. ***syriacum***
Evergreen scrub and open coniferous woodland. - S Anatolia, Cyprus, W Syria to Palestine.
Erect subshrub 20-60 cm tall, densely and shortly grey-tomentose. Leaves to 30 × 8 mm. Inflorescence of 3-5 dense, unilateral cymes. Petals somewhat longer than sepals, obovate, truncate, yellow.

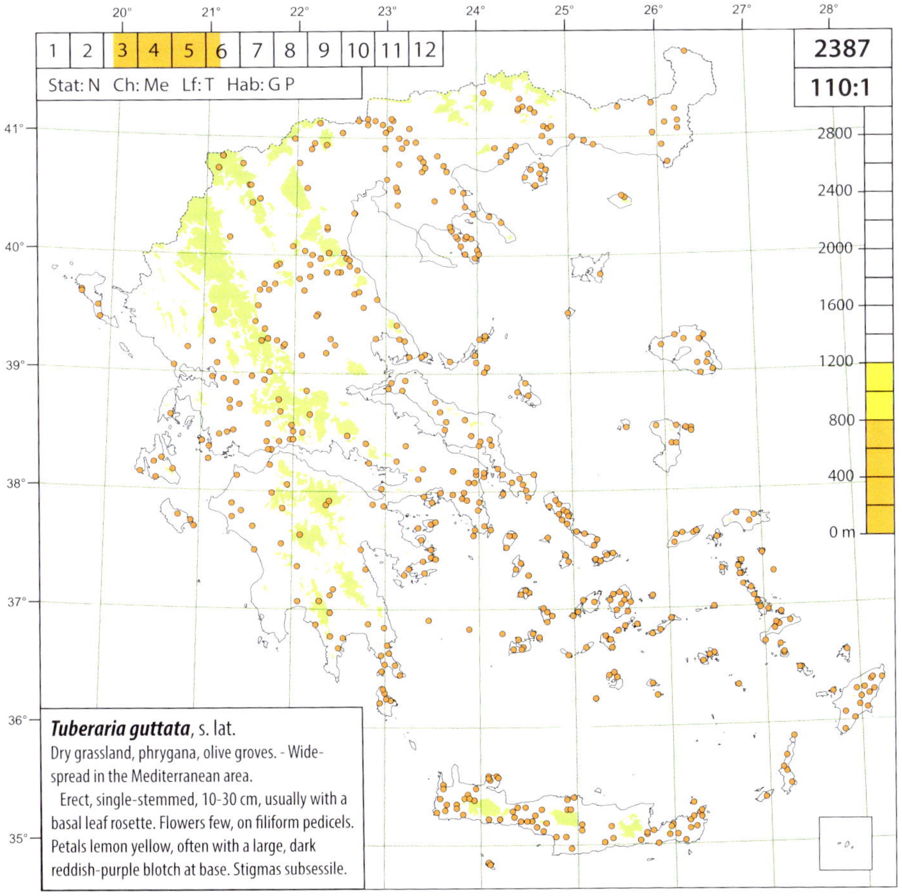

Tuberaria guttata, s. lat.
Dry grassland, phrygana, olive groves. - Widespread in the Mediterranean area.
Erect, single-stemmed, 10-30 cm, usually with a basal leaf rosette. Flowers few, on filiform pedicels. Petals lemon yellow, often with a large, dark reddish-purple blotch at base. Stigmas subsessile.

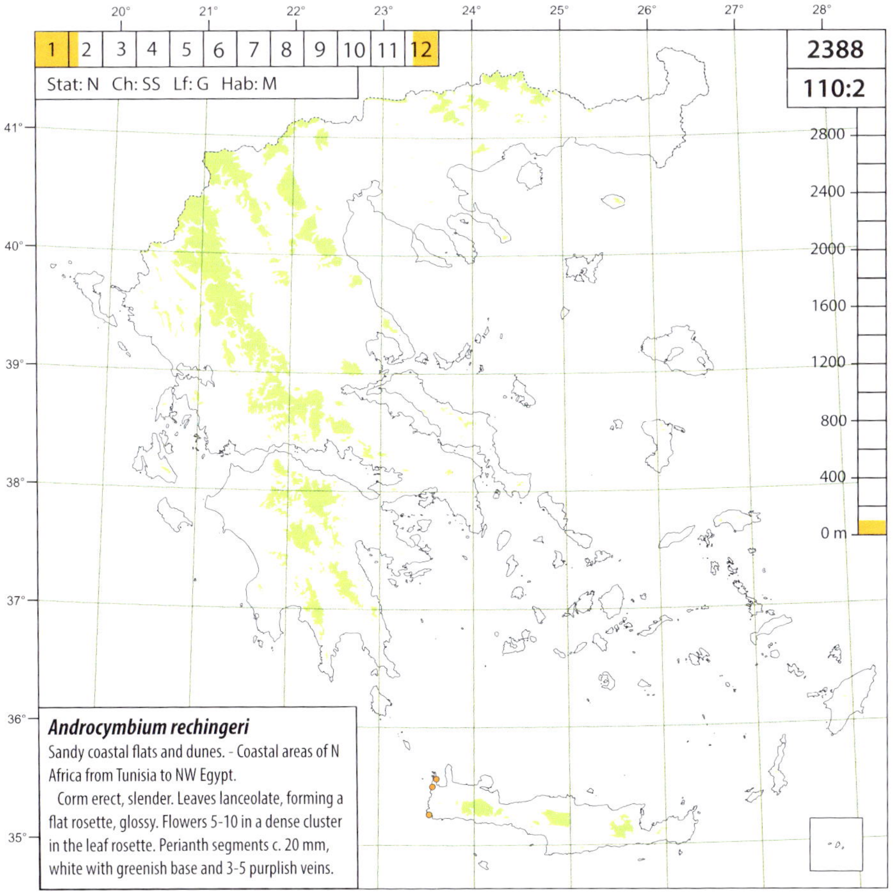

Androcymbium rechingeri
Sandy coastal flats and dunes. - Coastal areas of N Africa from Tunisia to NW Egypt.
Corm erect, slender. Leaves lanceolate, forming a flat rosette, glossy. Flowers 5-10 in a dense cluster in the leaf rosette. Perianth segments c. 20 mm, white with greenish base and 3-5 purplish veins.

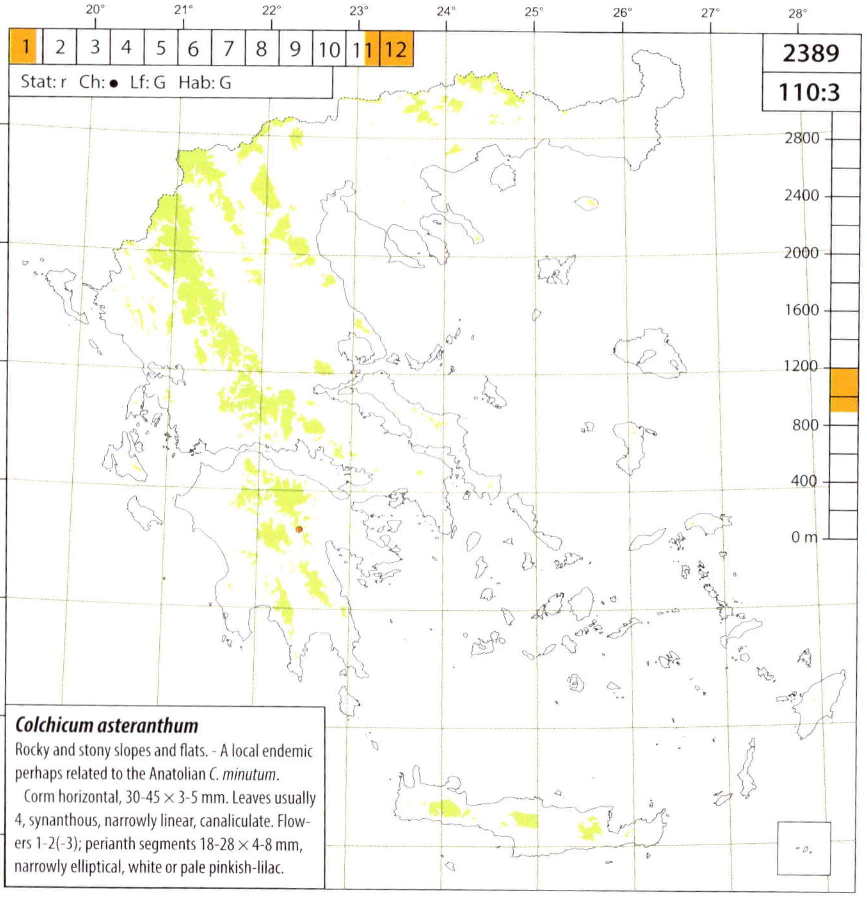

Colchicum asteranthum

Rocky and stony slopes and flats. - A local endemic perhaps related to the Anatolian *C. minutum*.

Corm horizontal, 30-45 × 3-5 mm. Leaves usually 4, synanthous, narrowly linear, canaliculate. Flowers 1-2(-3); perianth segments 18-28 × 4-8 mm, narrowly elliptical, white or pale pinkish-lilac.

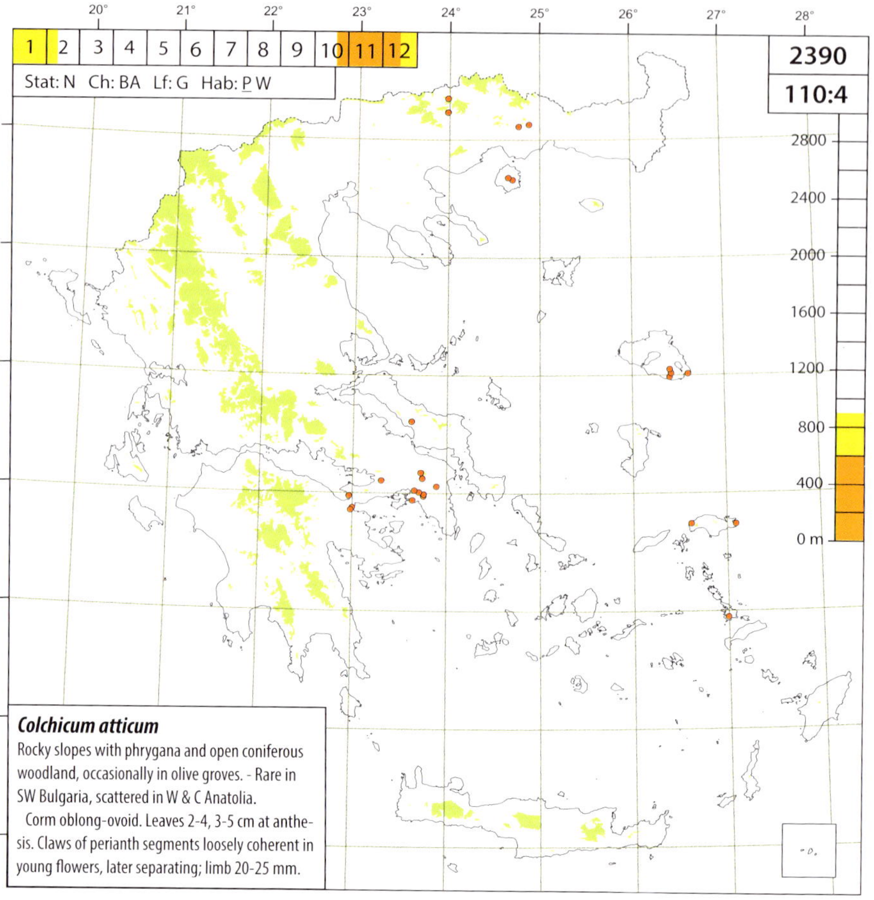

Colchicum atticum

Rocky slopes with phrygana and open coniferous woodland, occasionally in olive groves. - Rare in SW Bulgaria, scattered in W & C Anatolia.

Corm oblong-ovoid. Leaves 2-4, 3-5 cm at anthesis. Claws of perianth segments loosely coherent in young flowers, later separating; limb 20-25 mm.

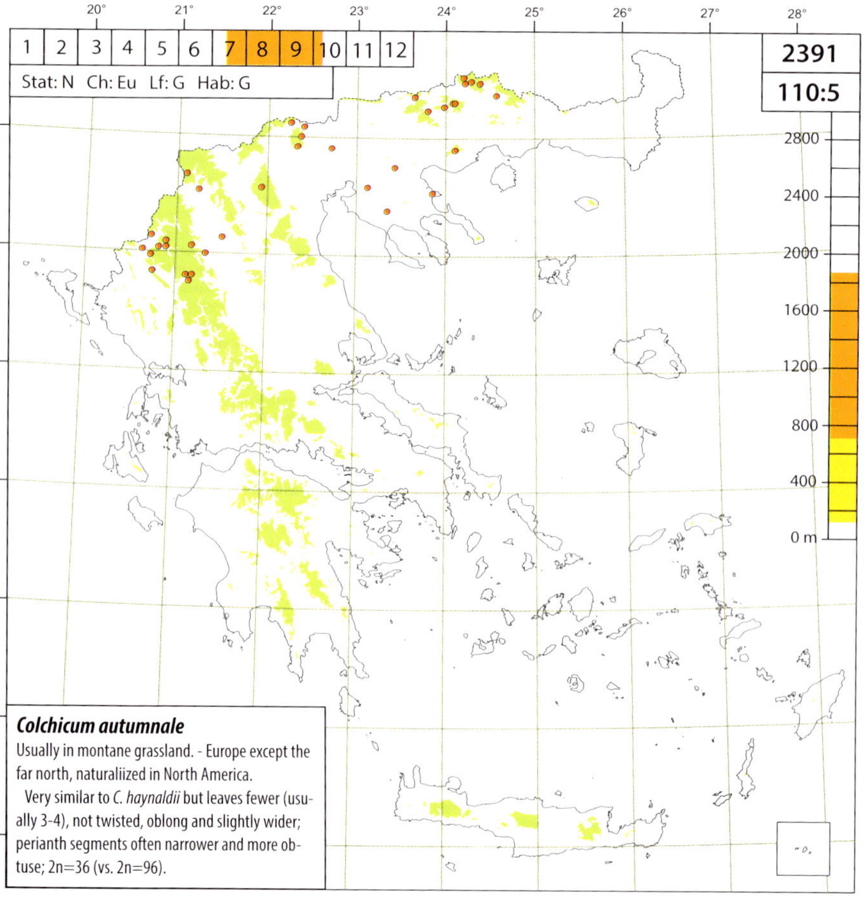

Colchicum autumnale

Usually in montane grassland. - Europe except the far north, naturaliized in North America.

Very similar to *C. haynaldii* but leaves fewer (usually 3-4), not twisted, oblong and slightly wider; perianth segments often narrower and more obtuse; 2n=36 (vs. 2n=96).

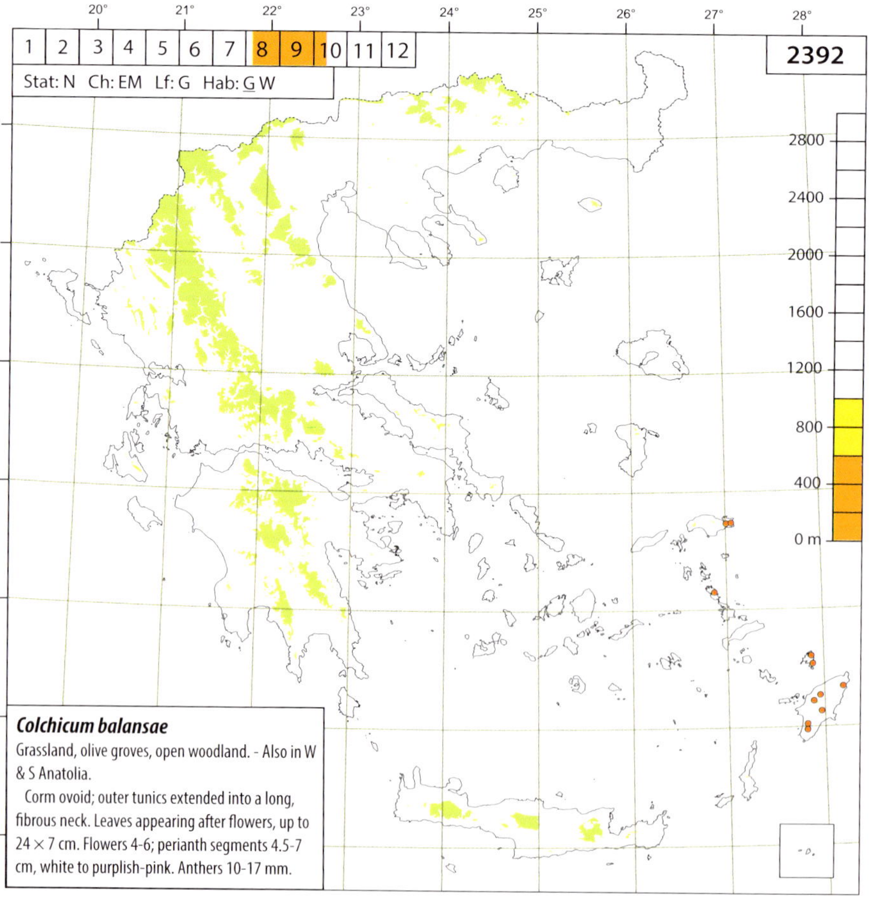

Colchicum balansae

Grassland, olive groves, open woodland. - Also in W & S Anatolia.

Corm ovoid; outer tunics extended into a long, fibrous neck. Leaves appearing after flowers, up to 24 × 7 cm. Flowers 4-6; perianth segments 4.5-7 cm, white to purplish-pink. Anthers 10-17 mm.

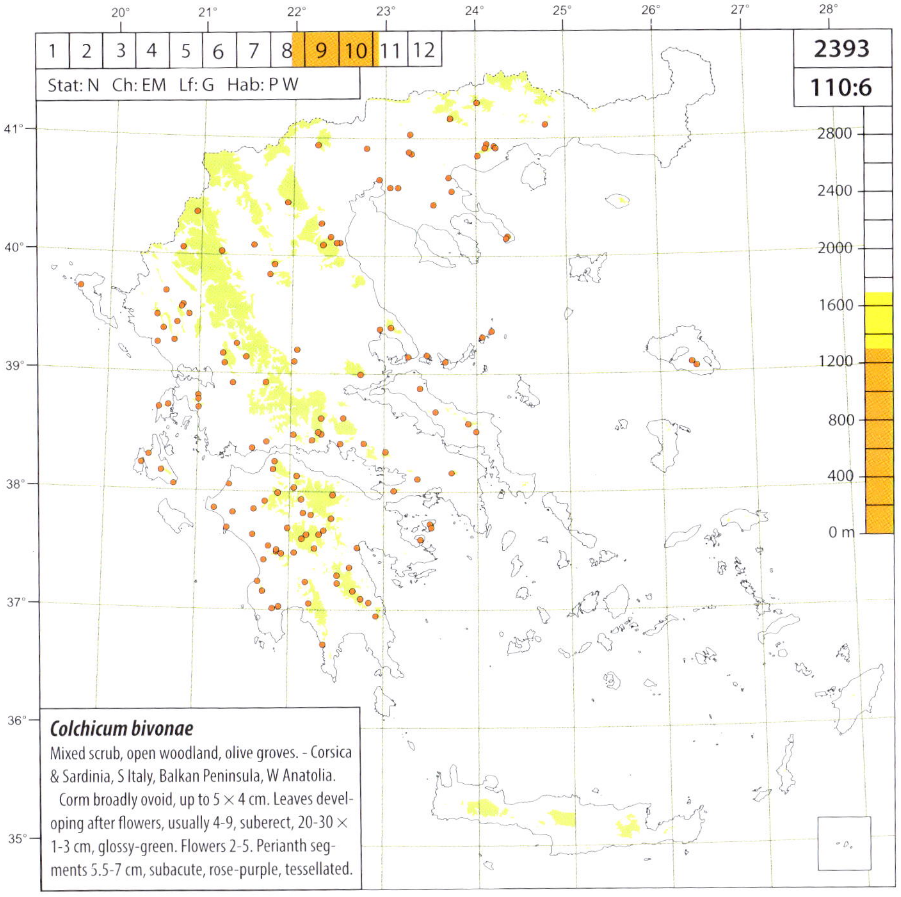

Colchicum bivonae
Mixed scrub, open woodland, olive groves. - Corsica & Sardinia, S Italy, Balkan Peninsula, W Anatolia.
Corm broadly ovoid, up to 5 × 4 cm. Leaves developing after flowers, usually 4-9, suberect, 20-30 × 1-3 cm, glossy-green. Flowers 2-5. Perianth segments 5.5-7 cm, subacute, rose-purple, tessellated.

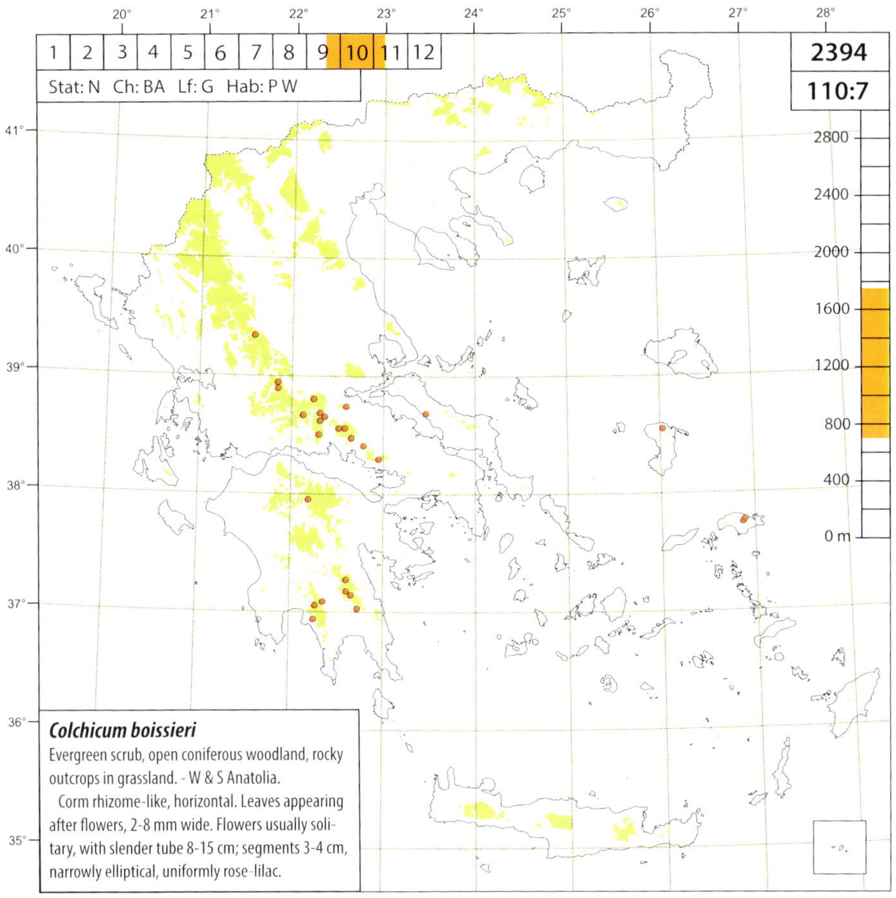

Colchicum boissieri
Evergreen scrub, open coniferous woodland, rocky outcrops in grassland. - W & S Anatolia.
Corm rhizome-like, horizontal. Leaves appearing after flowers, 2-8 mm wide. Flowers usually solitary, with slender tube 8-15 cm; segments 3-4 cm, narrowly elliptical, uniformly rose-lilac.

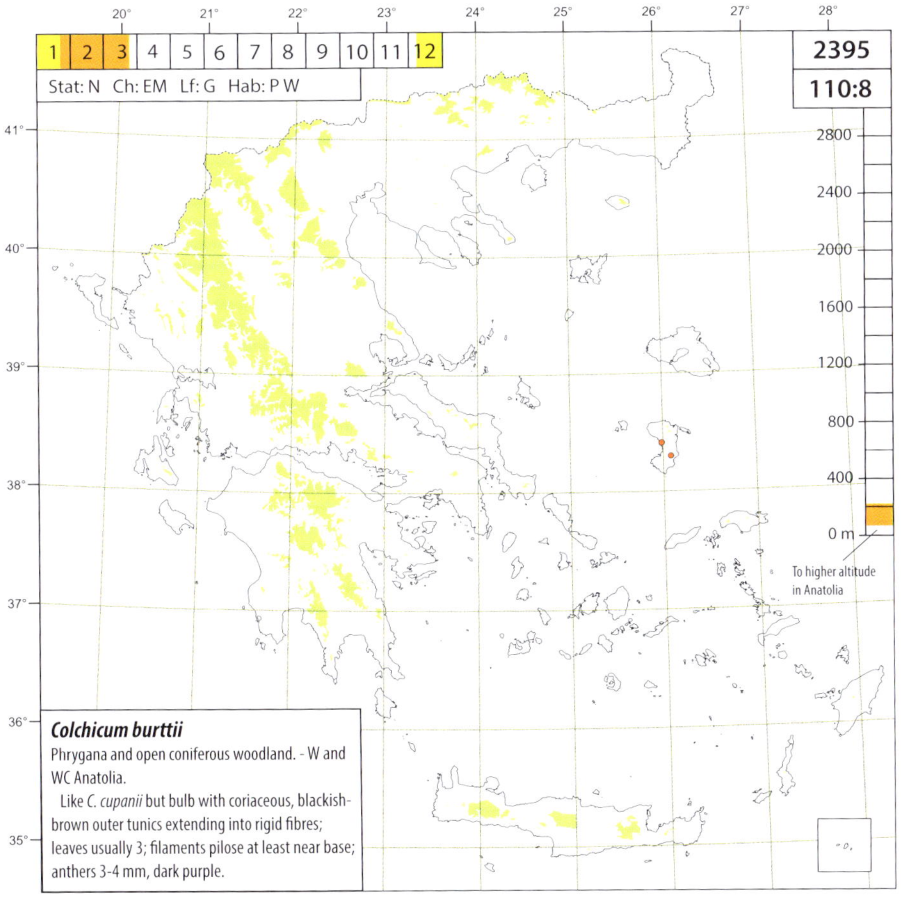

Colchicum burttii
Phrygana and open coniferous woodland. - W and WC Anatolia.
Like *C. cupanii* but bulb with coriaceous, blackish-brown outer tunics extending into rigid fibres; leaves usually 3; filaments pilose at least near base; anthers 3-4 mm, dark purple.

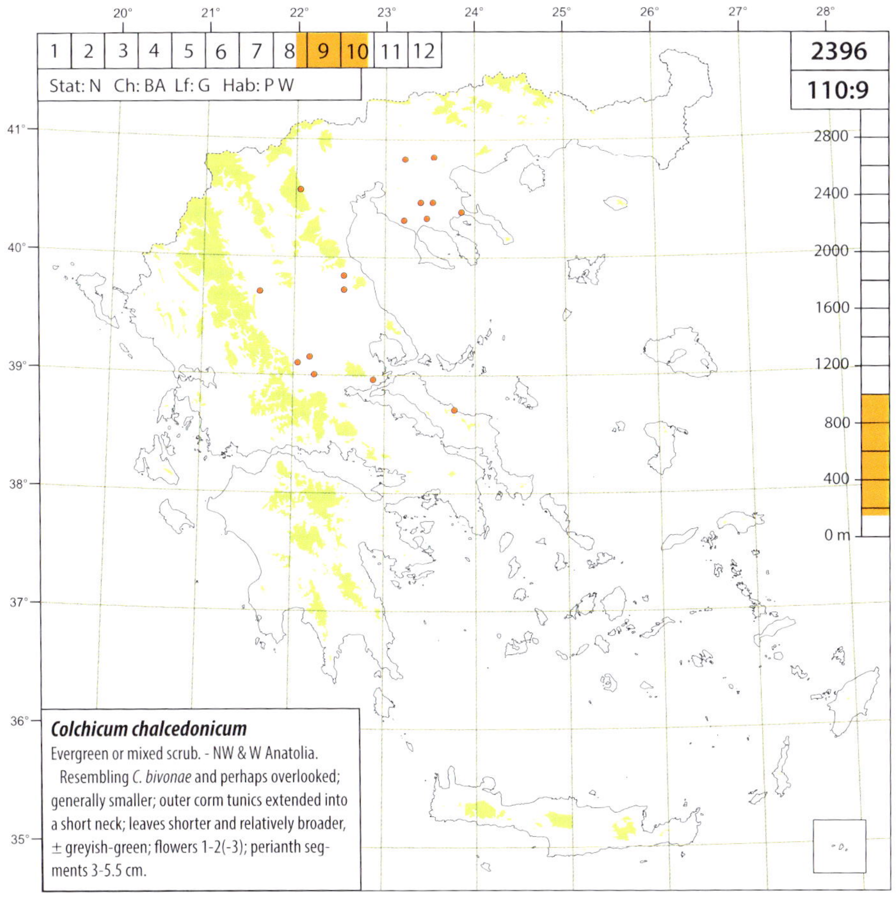

Colchicum chalcedonicum
Evergreen or mixed scrub. - NW & W Anatolia.
Resembling *C. bivonae* and perhaps overlooked; generally smaller; outer corm tunics extended into a short neck; leaves shorter and relatively broader, ± greyish-green; flowers 1-2(-3); perianth segments 3-5.5 cm.

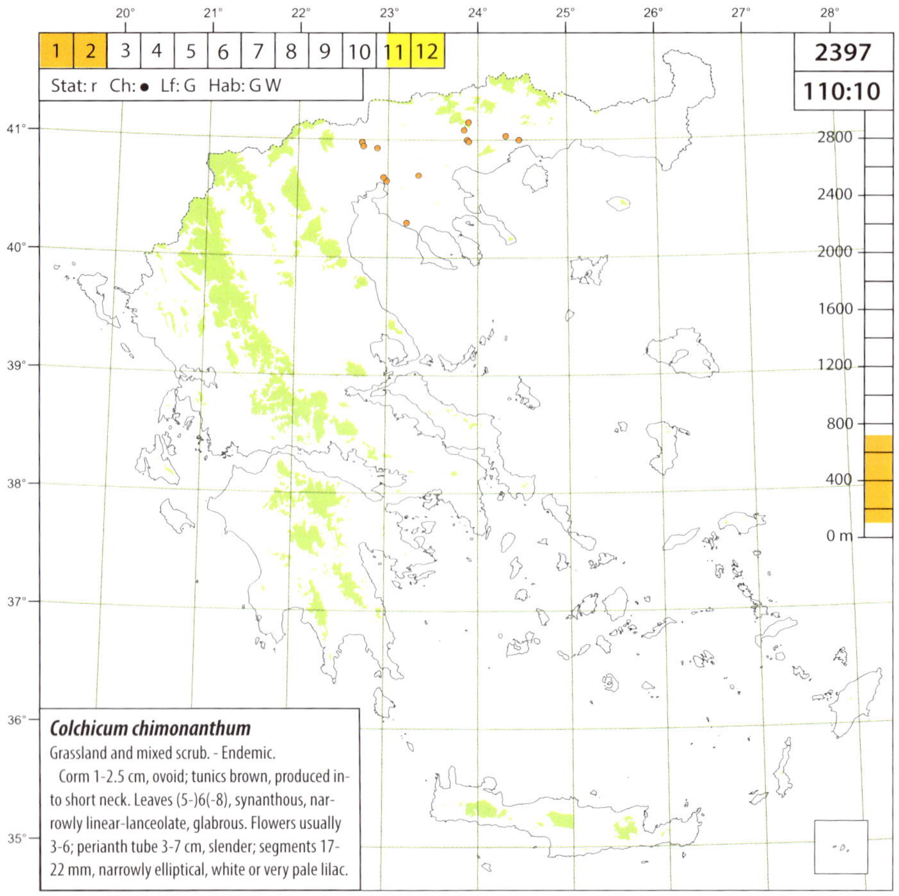

Colchicum chimonanthum

Grassland and mixed scrub. - Endemic.

Corm 1-2.5 cm, ovoid; tunics brown, produced into short neck. Leaves (5-)6(-8), synanthous, narrowly linear-lanceolate, glabrous. Flowers usually 3-6; perianth tube 3-7 cm, slender; segments 17-22 mm, narrowly elliptical, white or very pale lilac.

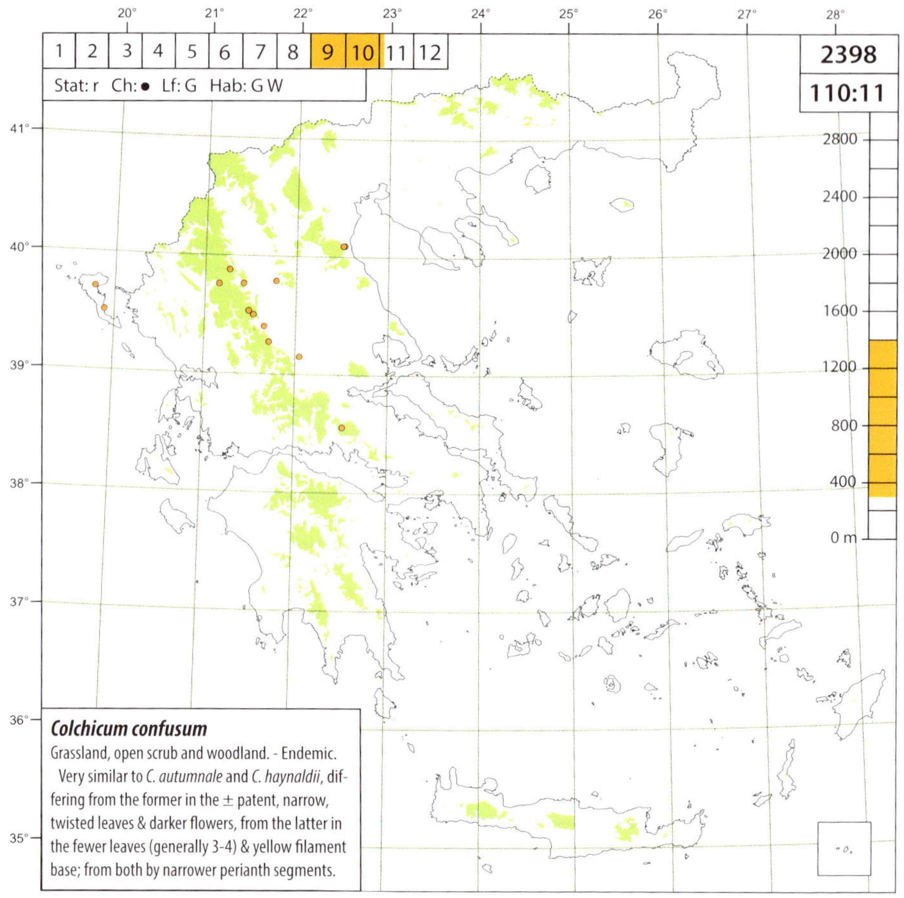

Colchicum confusum

Grassland, open scrub and woodland. - Endemic.

Very similar to *C. autumnale* and *C. haynaldii*, differing from the former in the ± patent, narrow, twisted leaves & darker flowers, from the latter in the fewer leaves (generally 3-4) & yellow filament base; from both by narrower perianth segments.

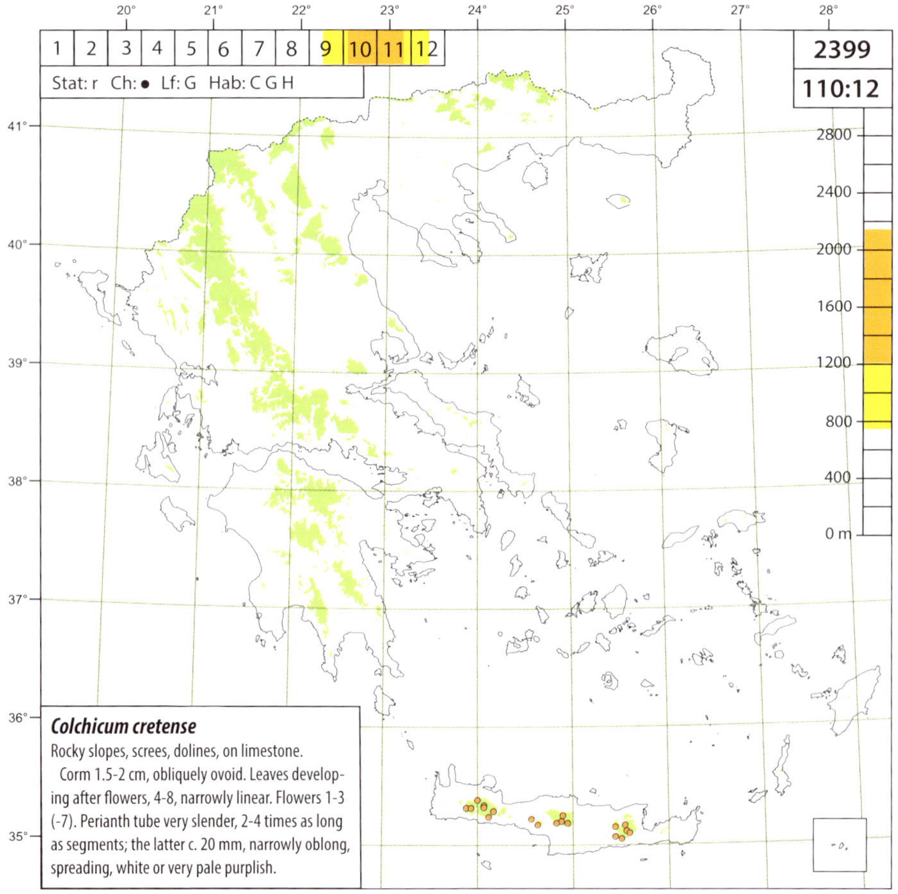

Colchicum cretense

Rocky slopes, screes, dolines, on limestone.

Corm 1.5-2 cm, obliquely ovoid. Leaves developing after flowers, 4-8, narrowly linear. Flowers 1-3 (-7). Perianth tube very slender, 2-4 times as long as segments; the latter c. 20 mm, narrowly oblong, spreading, white or very pale purplish.

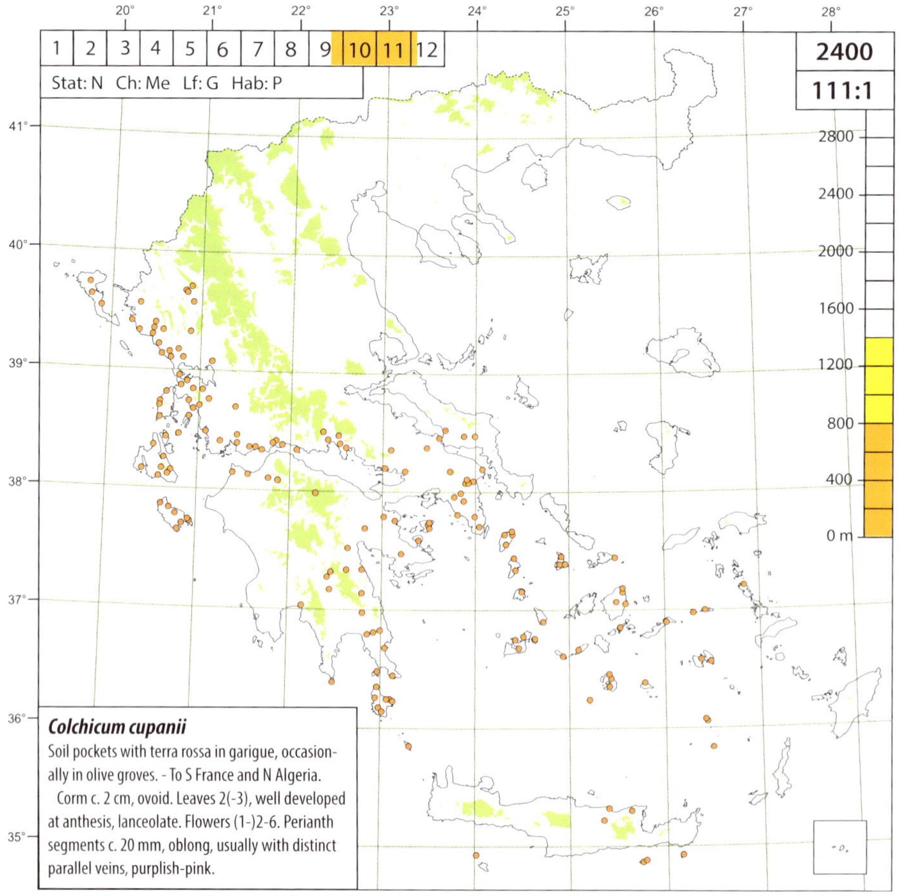

Colchicum cupanii

Soil pockets with terra rossa in garigue, occasionally in olive groves. - To S France and N Algeria.

Corm c. 2 cm, ovoid. Leaves 2(-3), well developed at anthesis, lanceolate. Flowers (1-)2-6. Perianth segments c. 20 mm, oblong, usually with distinct parallel veins, purplish-pink.

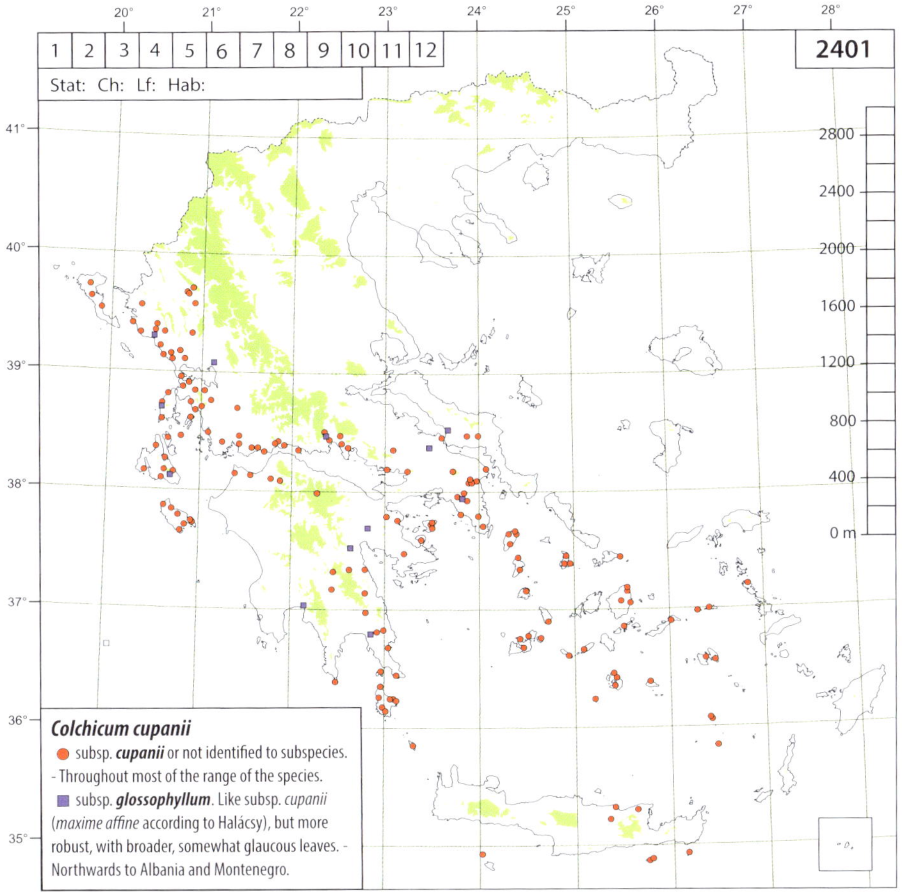

Colchicum cupanii
● subsp. ***cupanii*** or not identified to subspecies. - Throughout most of the range of the species.
■ subsp. ***glossophyllum***. Like subsp. *cupanii* (*maxime affine* according to Halácsy), but more robust, with broader, somewhat glaucous leaves. - Northwards to Albania and Montenegro.

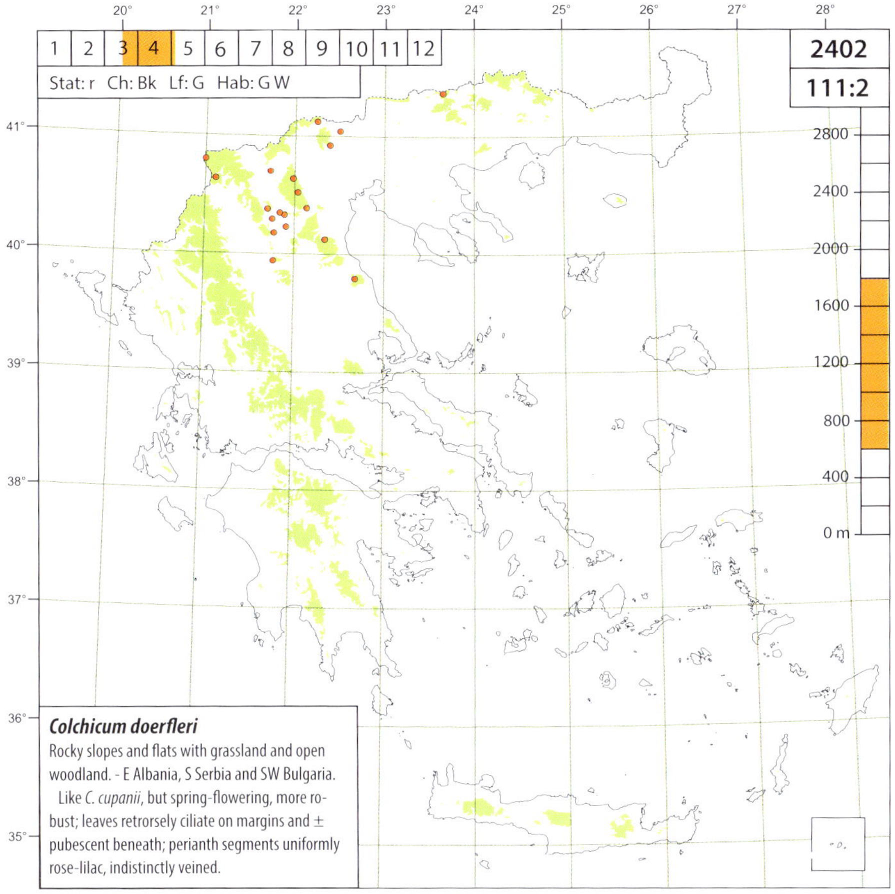

Colchicum doerfleri
Rocky slopes and flats with grassland and open woodland. - E Albania, S Serbia and SW Bulgaria.
Like *C. cupanii*, but spring-flowering, more robust; leaves retrorsely ciliate on margins and ± pubescent beneath; perianth segments uniformly rose-lilac, indistinctly veined.

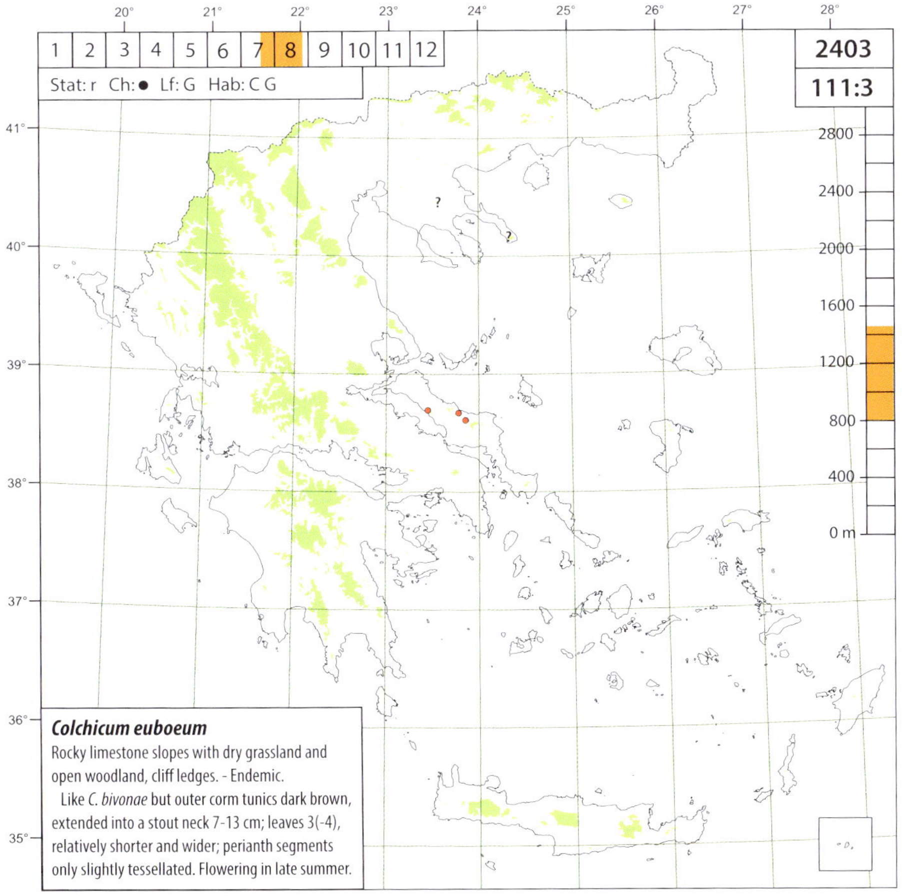

Colchicum euboeum
Rocky limestone slopes with dry grassland and open woodland, cliff ledges. - Endemic.
Like *C. bivonae* but outer corm tunics dark brown, extended into a stout neck 7-13 cm; leaves 3(-4), relatively shorter and wider; perianth segments only slightly tessellated. Flowering in late summer.

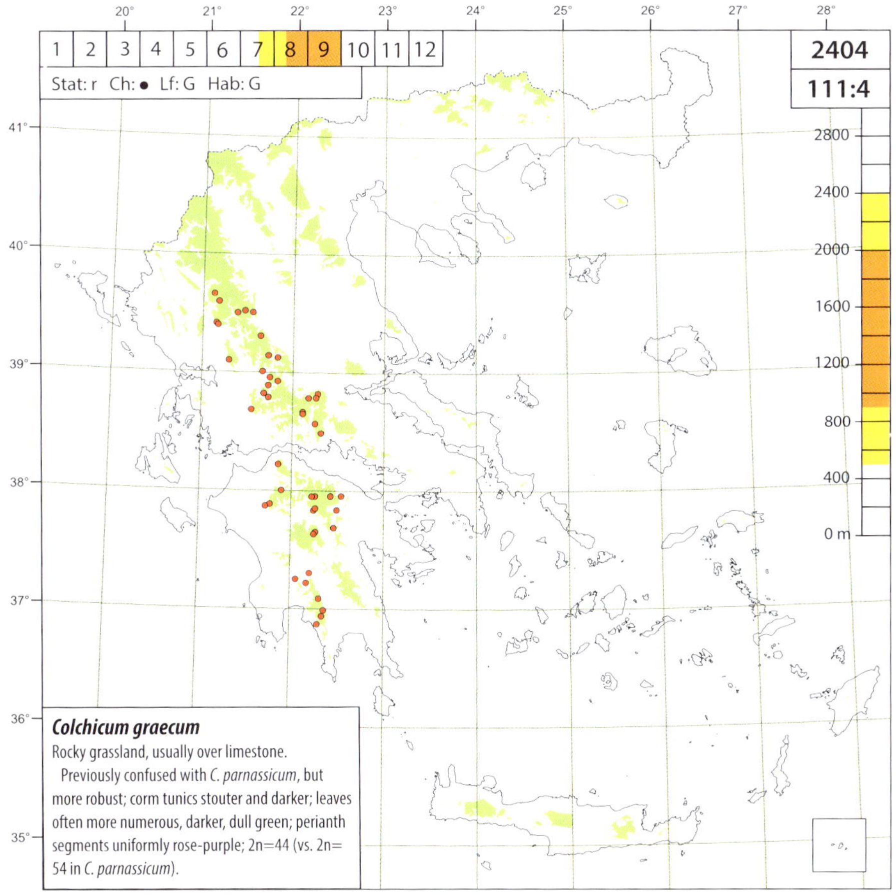

Colchicum graecum
Rocky grassland, usually over limestone.
Previously confused with *C. parnassicum*, but more robust; corm tunics stouter and darker; leaves often more numerous, darker, dull green; perianth segments uniformly rose-purple; 2n=44 (vs. 2n=54 in *C. parnassicum*).

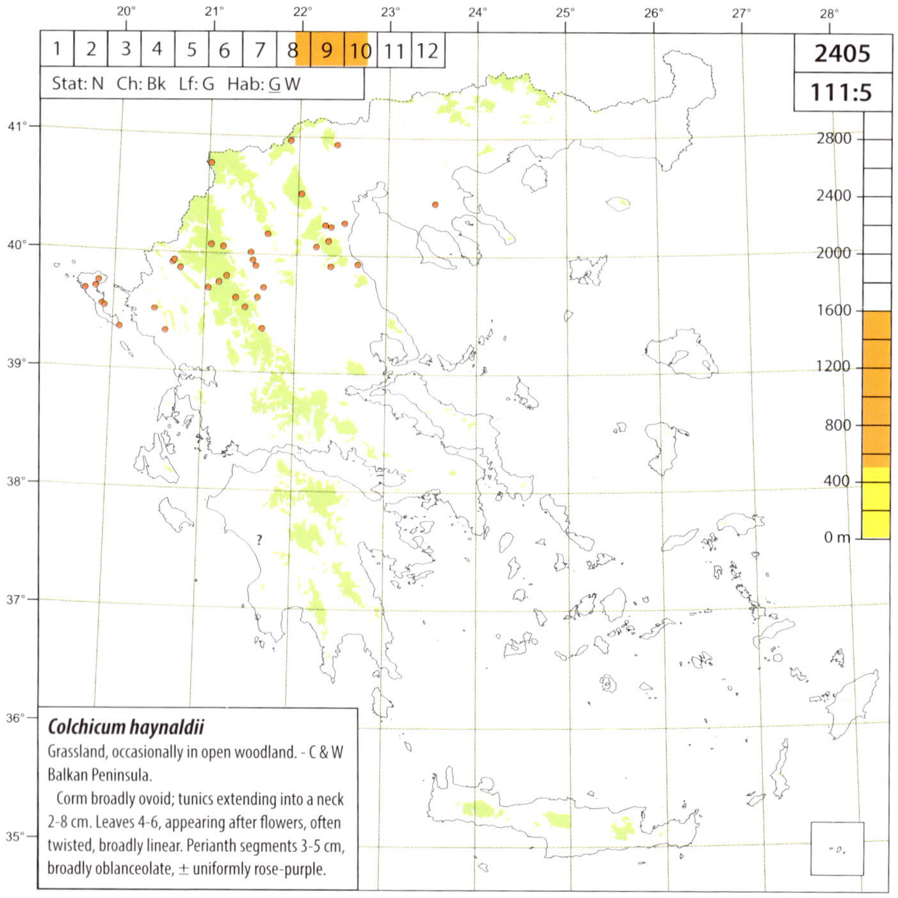

Colchicum haynaldii
Grassland, occasionally in open woodland. - C & W Balkan Peninsula.
Corm broadly ovoid; tunics extending into a neck 2-8 cm. Leaves 4-6, appearing after flowers, often twisted, broadly linear. Perianth segments 3-5 cm, broadly oblanceolate, ± uniformly rose-purple.

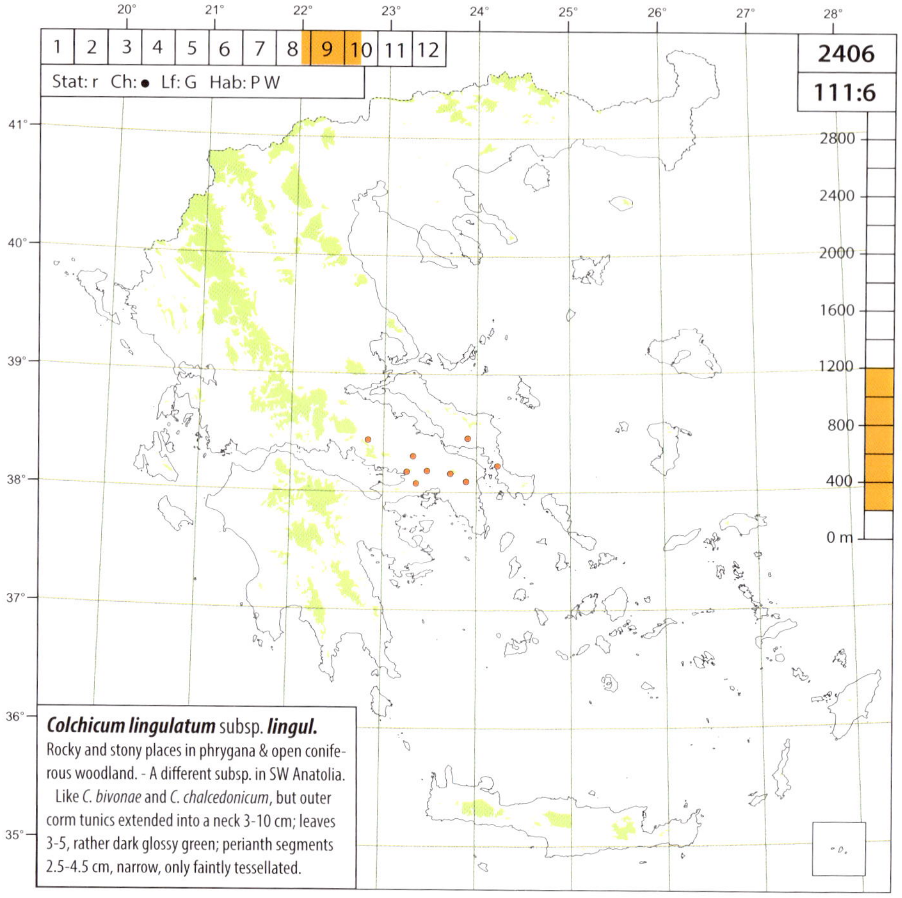

Colchicum lingulatum subsp. ***lingul.***
Rocky and stony places in phrygana & open coniferous woodland. - A different subsp. in SW Anatolia.
Like *C. bivonae* and *C. chalcedonicum*, but outer corm tunics extended into a neck 3-10 cm; leaves 3-5, rather dark glossy green; perianth segments 2.5-4.5 cm, narrow, only faintly tessellated.

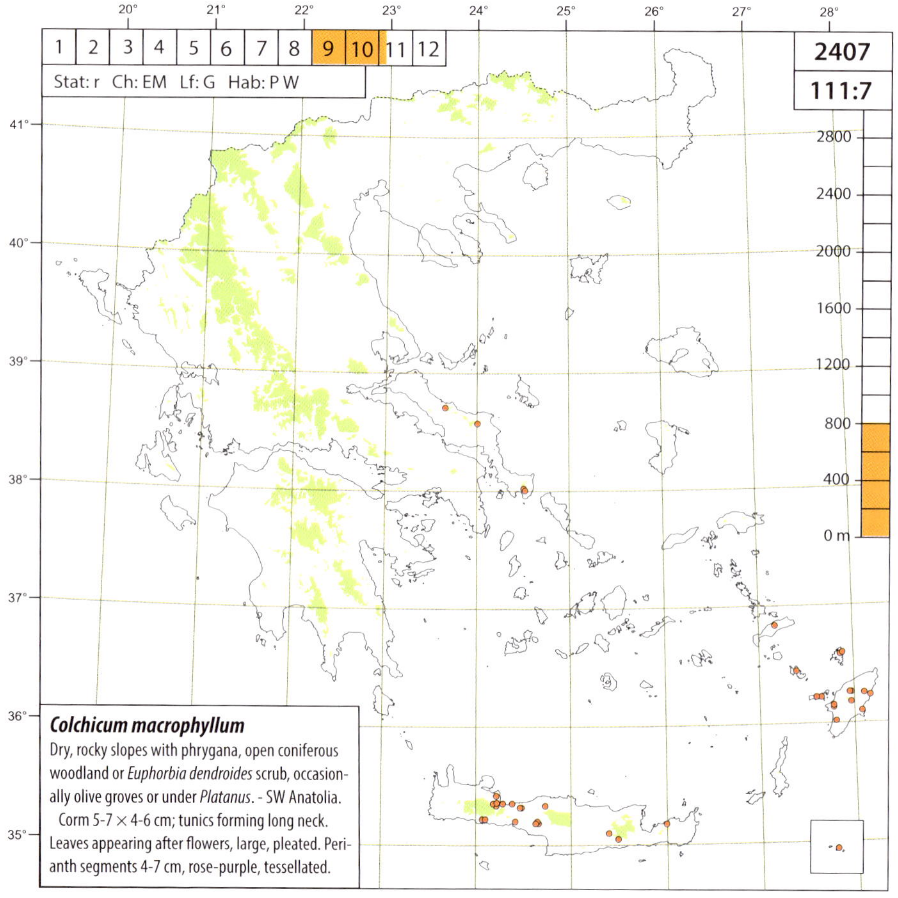

Colchicum macrophyllum
Dry, rocky slopes with phrygana, open coniferous woodland or *Euphorbia dendroides* scrub, occasionally olive groves or under *Platanus*. - SW Anatolia.
Corm 5-7 × 4-6 cm; tunics forming long neck. Leaves appearing after flowers, large, pleated. Perianth segments 4-7 cm, rose-purple, tessellated.

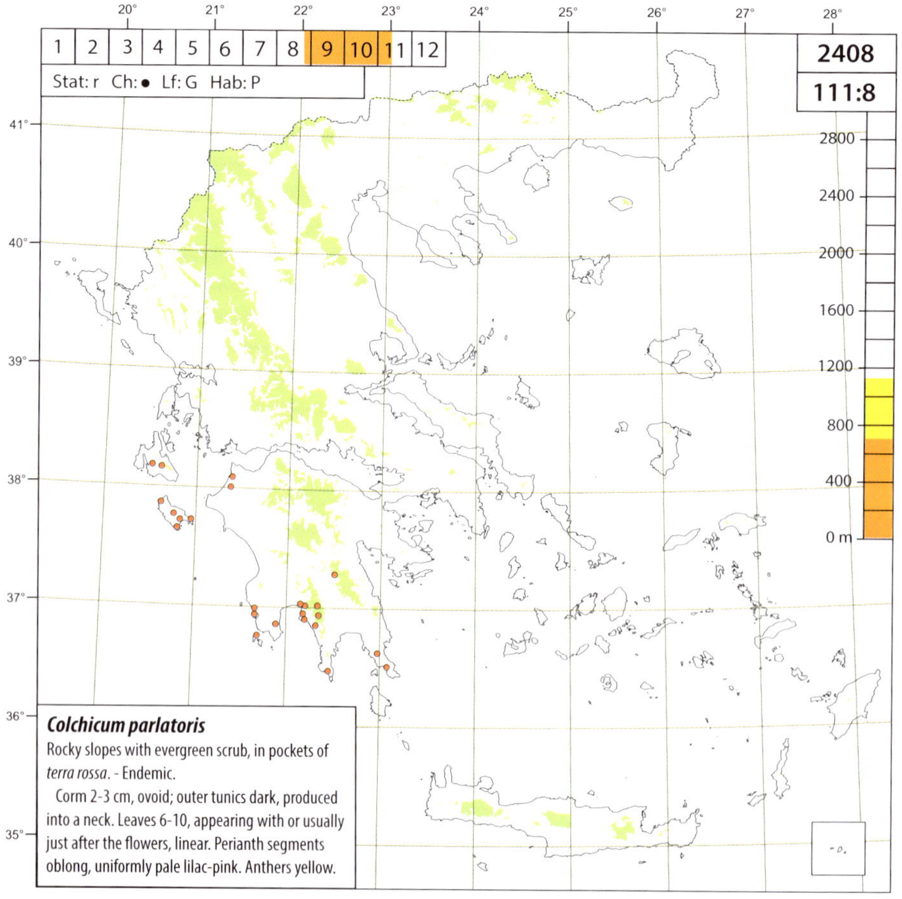

Colchicum parlatoris
Rocky slopes with evergreen scrub, in pockets of *terra rossa*. - Endemic.
Corm 2-3 cm, ovoid; outer tunics dark, produced into a neck. Leaves 6-10, appearing with or usually just after the flowers, linear. Perianth segments oblong, uniformly pale lilac-pink. Anthers yellow.

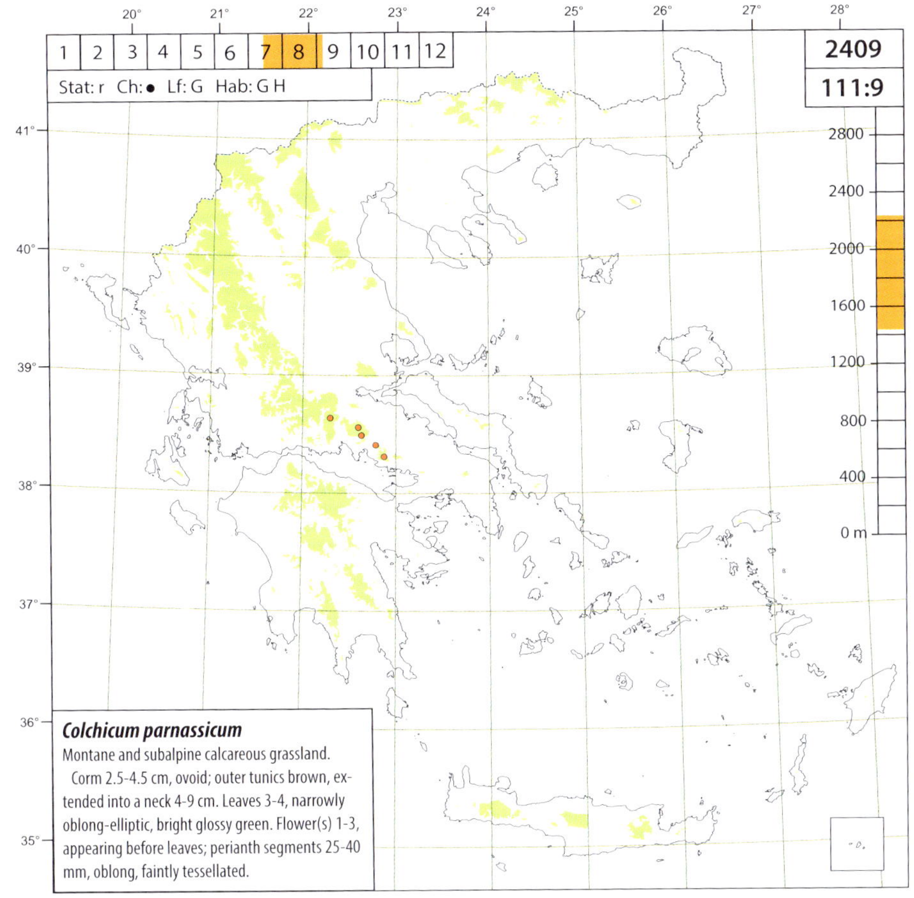

Colchicum parnassicum

Montane and subalpine calcareous grassland.

Corm 2.5-4.5 cm, ovoid; outer tunics brown, extended into a neck 4-9 cm. Leaves 3-4, narrowly oblong-elliptic, bright glossy green. Flower(s) 1-3, appearing before leaves; perianth segments 25-40 mm, oblong, faintly tessellated.

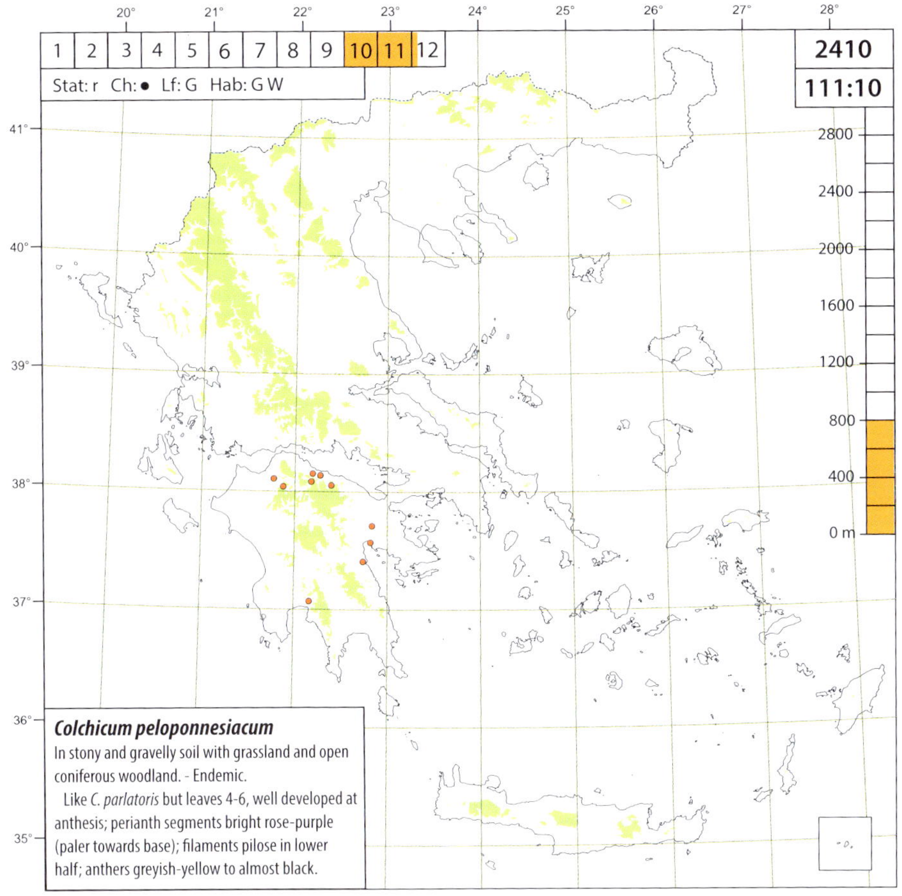

Colchicum peloponnesiacum

In stony and gravelly soil with grassland and open coniferous woodland. - Endemic.

Like *C. parlatoris* but leaves 4-6, well developed at anthesis; perianth segments bright rose-purple (paler towards base); filaments pilose in lower half; anthers greyish-yellow to almost black.

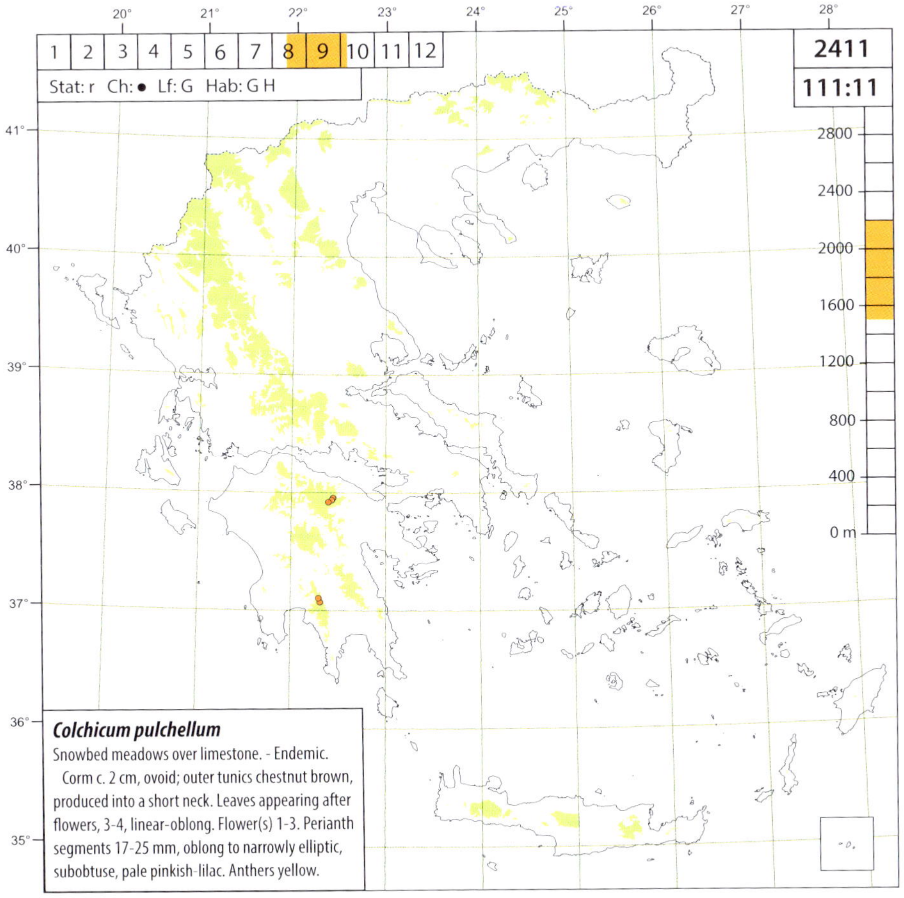

Colchicum pulchellum

Snowbed meadows over limestone. - Endemic.

Corm c. 2 cm, ovoid; outer tunics chestnut brown, produced into a short neck. Leaves appearing after flowers, 3-4, linear-oblong. Flower(s) 1-3. Perianth segments 17-25 mm, oblong to narrowly elliptic, subobtuse, pale pinkish-lilac. Anthers yellow.

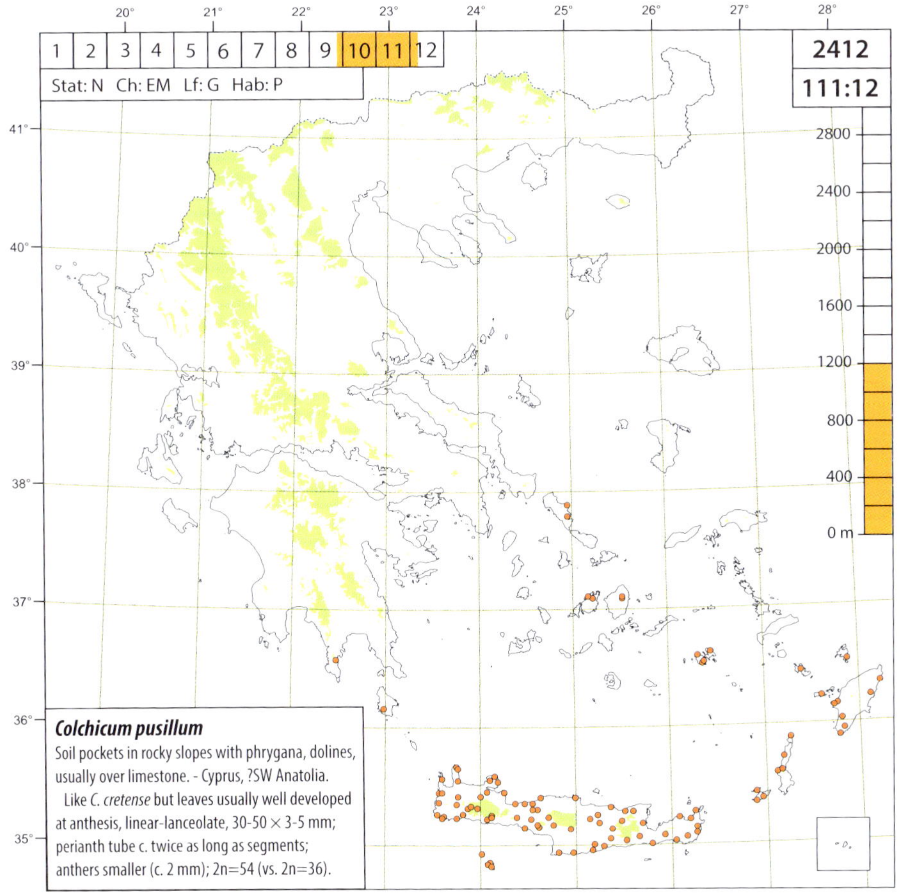

Colchicum pusillum

Soil pockets in rocky slopes with phrygana, dolines, usually over limestone. - Cyprus, ?SW Anatolia.

Like *C. cretense* but leaves usually well developed at anthesis, linear-lanceolate, 30-50 × 3-5 mm; perianth tube c. twice as long as segments; anthers smaller (c. 2 mm); 2n=54 (vs. 2n=36).

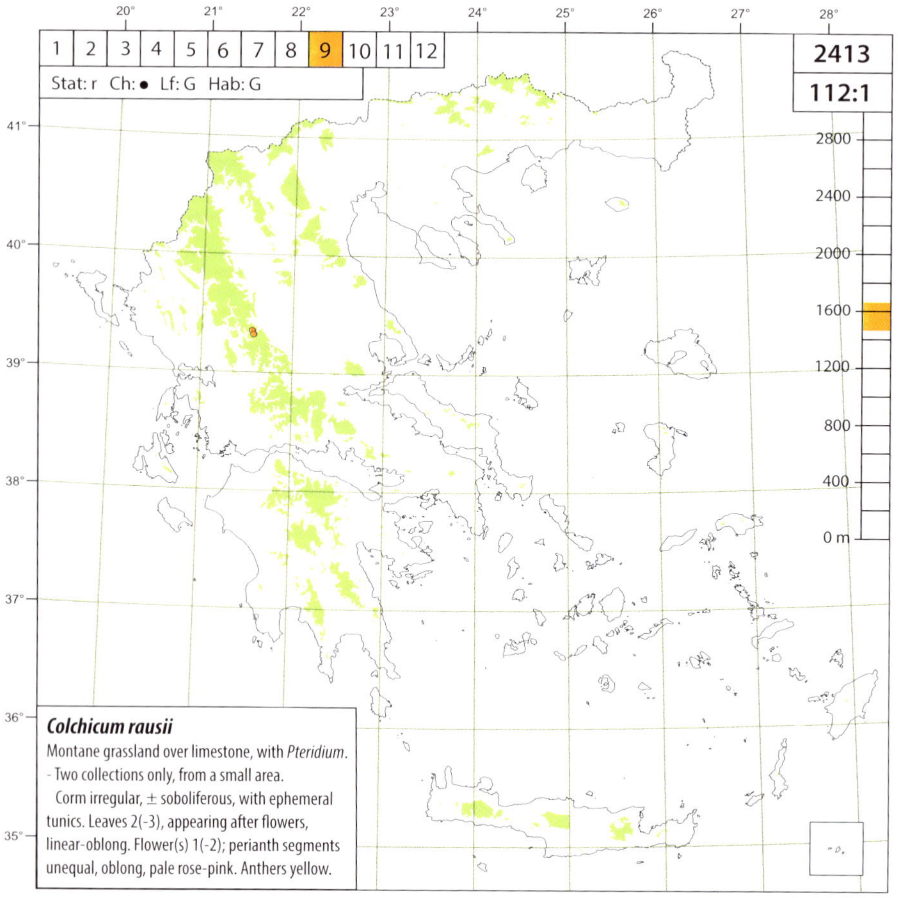

Colchicum rausii

Montane grassland over limestone, with *Pteridium*. - Two collections only, from a small area.

Corm irregular, ± soboliferous, with ephemeral tunics. Leaves 2(-3), appearing after flowers, linear-oblong. Flower(s) 1(-2); perianth segments unequal, oblong, pale rose-pink. Anthers yellow.

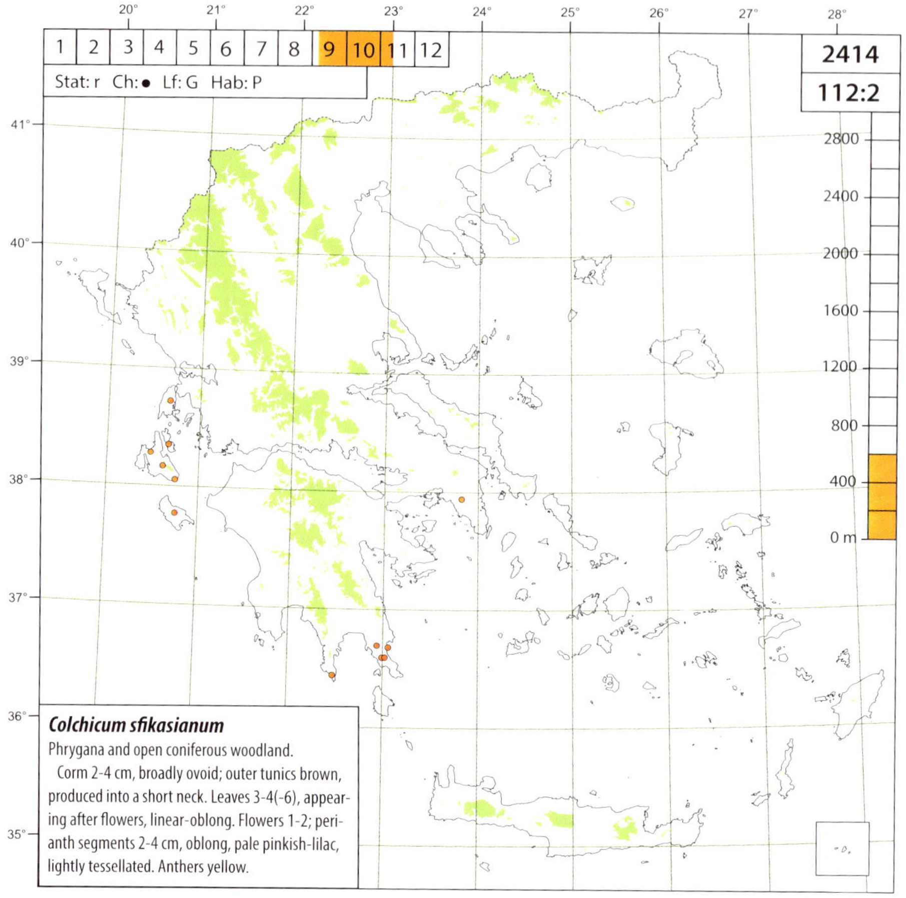

Colchicum sfikasianum

Phrygana and open coniferous woodland.

Corm 2-4 cm, broadly ovoid; outer tunics brown, produced into a short neck. Leaves 3-4(-6), appearing after flowers, linear-oblong. Flowers 1-2; perianth segments 2-4 cm, oblong, pale pinkish-lilac, lightly tessellated. Anthers yellow.

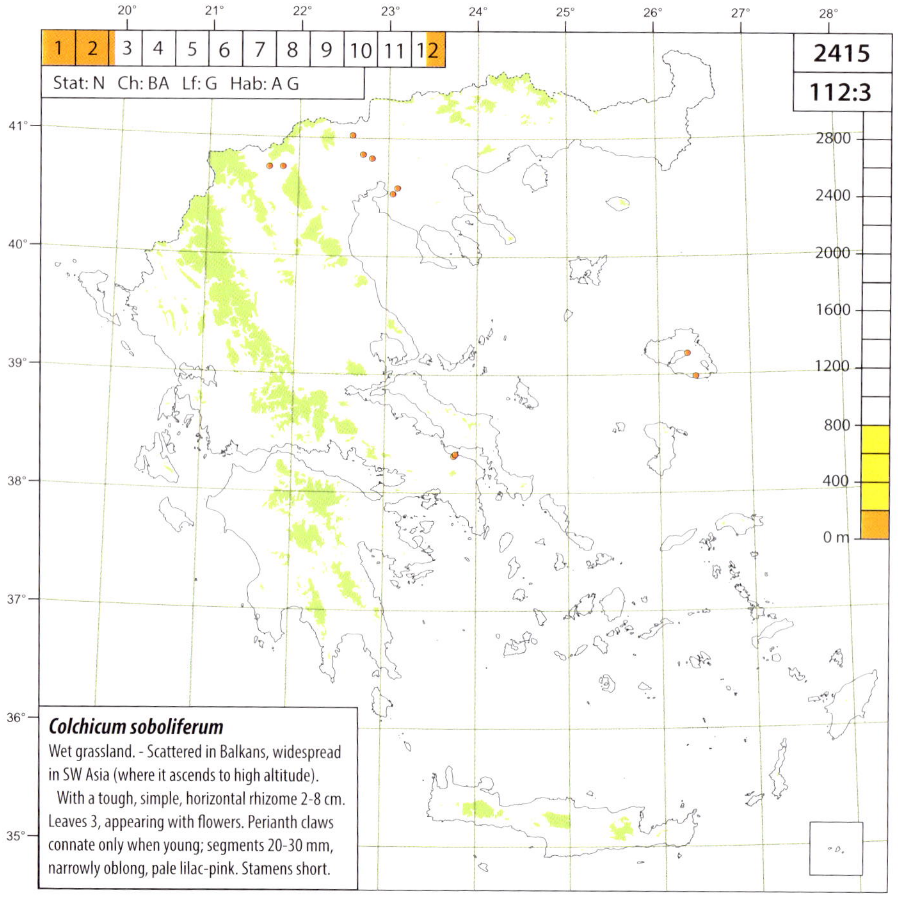

Colchicum soboliferum

Wet grassland. - Scattered in Balkans, widespread in SW Asia (where it ascends to high altitude).

With a tough, simple, horizontal rhizome 2-8 cm. Leaves 3, appearing with flowers. Perianth claws connate only when young; segments 20-30 mm, narrowly oblong, pale lilac-pink. Stamens short.

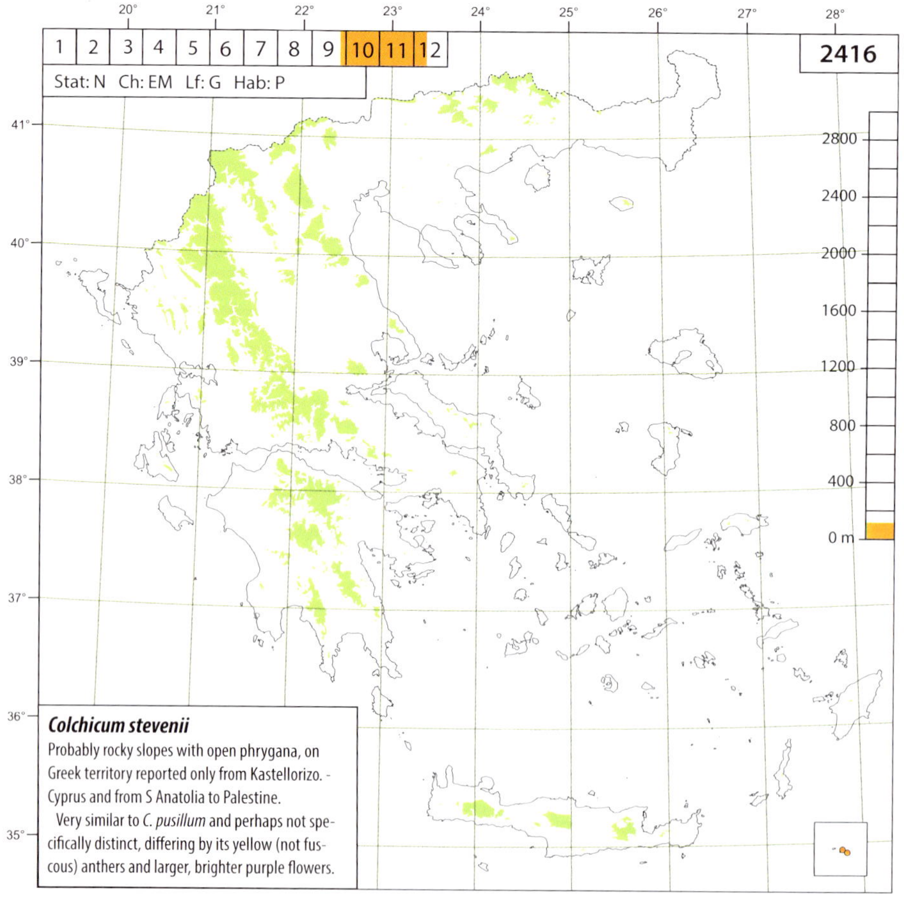

Colchicum stevenii

Probably rocky slopes with open phrygana, on Greek territory reported only from Kastellorizo. - Cyprus and from S Anatolia to Palestine.

Very similar to *C. pusillum* and perhaps not specifically distinct, differing by its yellow (not fuscous) anthers and larger, brighter purple flowers.

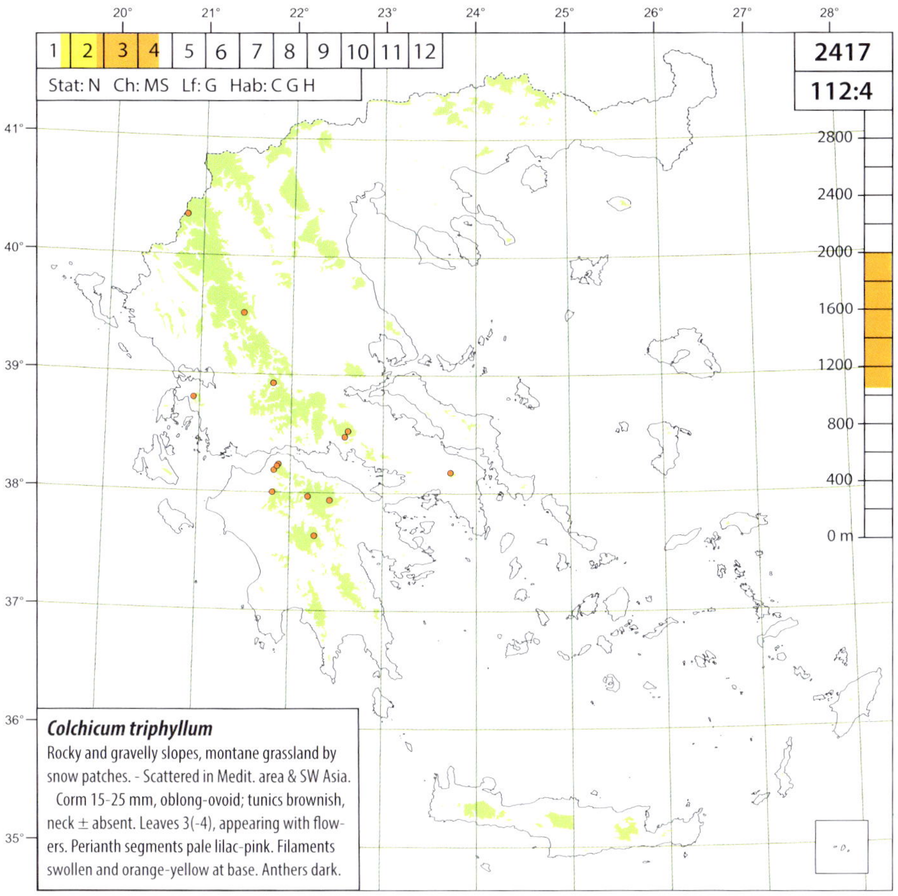

Colchicum triphyllum

Rocky and gravelly slopes, montane grassland by snow patches. - Scattered in Medit. area & SW Asia.

Corm 15-25 mm, oblong-ovoid; tunics brownish, neck ± absent. Leaves 3(-4), appearing with flowers. Perianth segments pale lilac-pink. Filaments swollen and orange-yellow at base. Anthers dark.

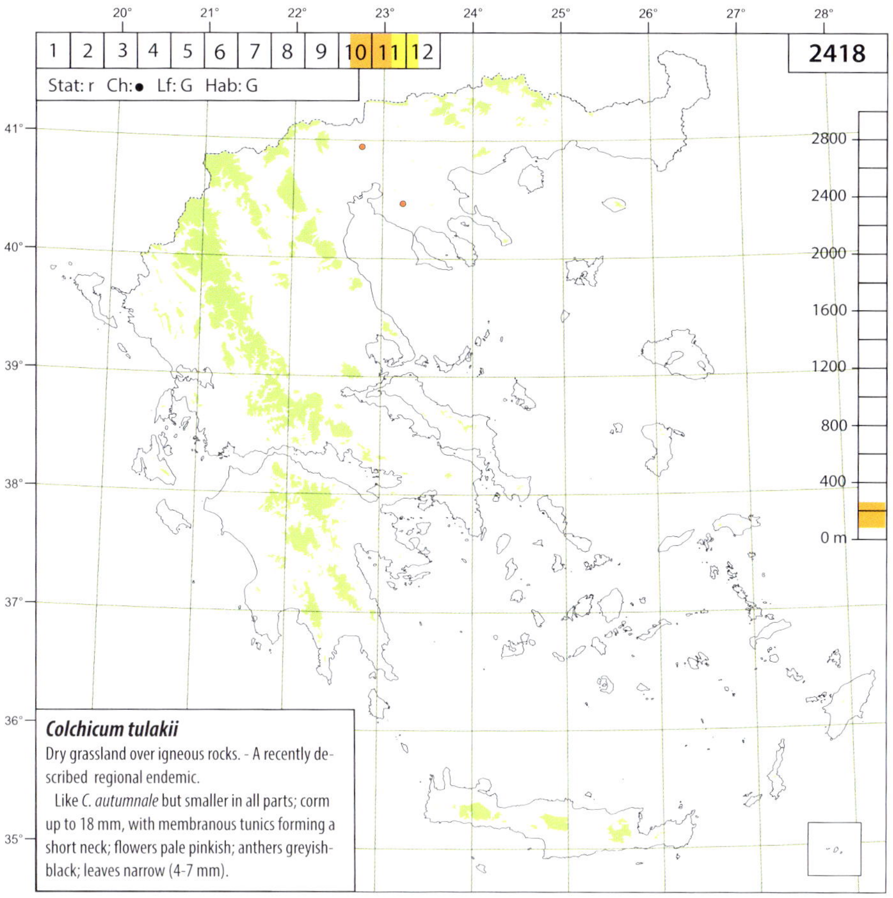

Colchicum tulakii

Dry grassland over igneous rocks. - A recently described regional endemic.

Like *C. autumnale* but smaller in all parts; corm up to 18 mm, with membranous tunics forming a short neck; flowers pale pinkish; anthers greyish-black; leaves narrow (4-7 mm).

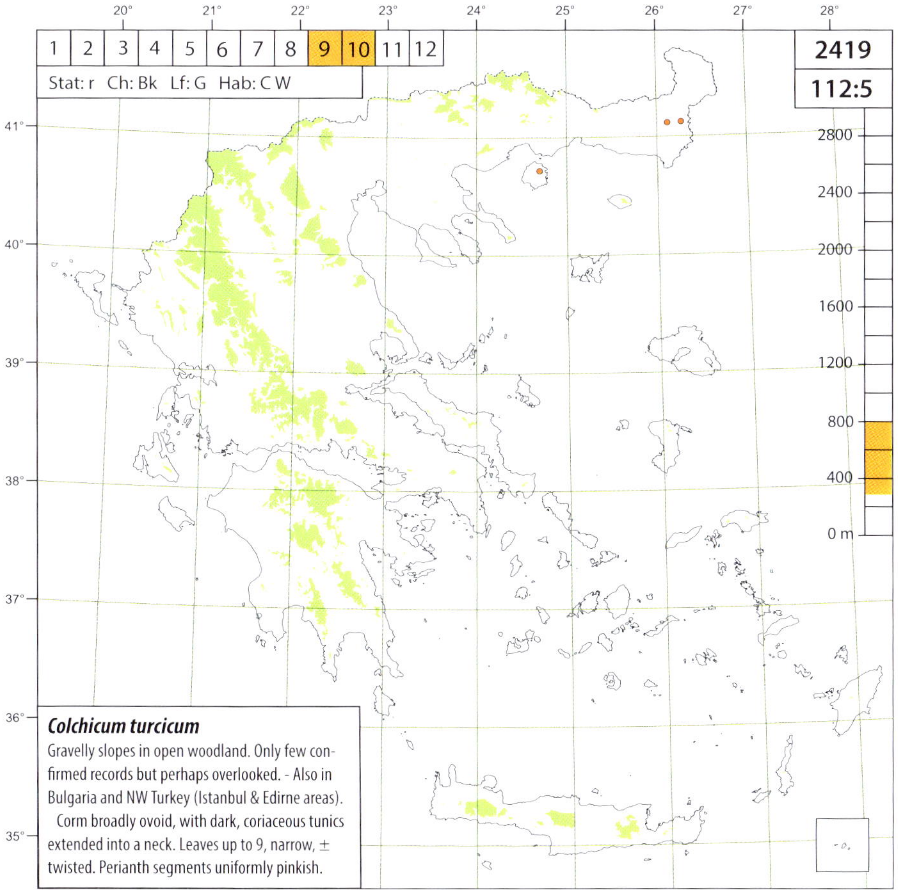

Colchicum turcicum

Gravelly slopes in open woodland. Only few confirmed records but perhaps overlooked. - Also in Bulgaria and NW Turkey (Istanbul & Edirne areas).

Corm broadly ovoid, with dark, coriaceous tunics extended into a neck. Leaves up to 9, narrow, ± twisted. Perianth segments uniformly pinkish.

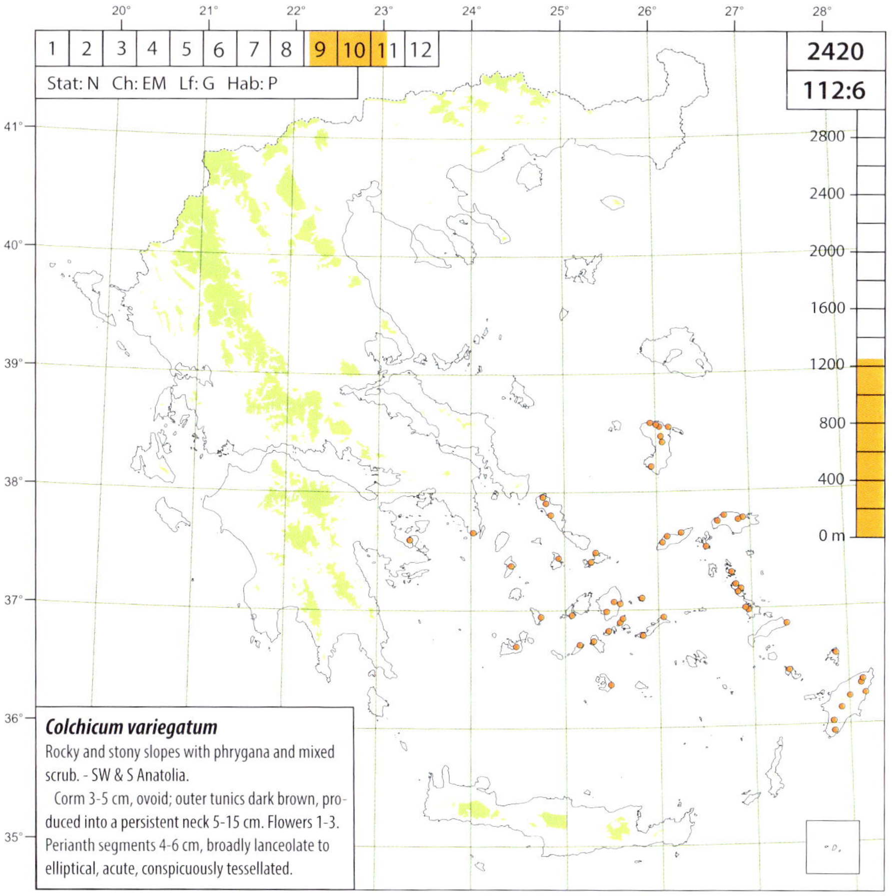

Colchicum variegatum

Rocky and stony slopes with phrygana and mixed scrub. - SW & S Anatolia.

Corm 3-5 cm, ovoid; outer tunics dark brown, produced into a persistent neck 5-15 cm. Flowers 1-3. Perianth segments 4-6 cm, broadly lanceolate to elliptical, acute, conspicuously tessellated.

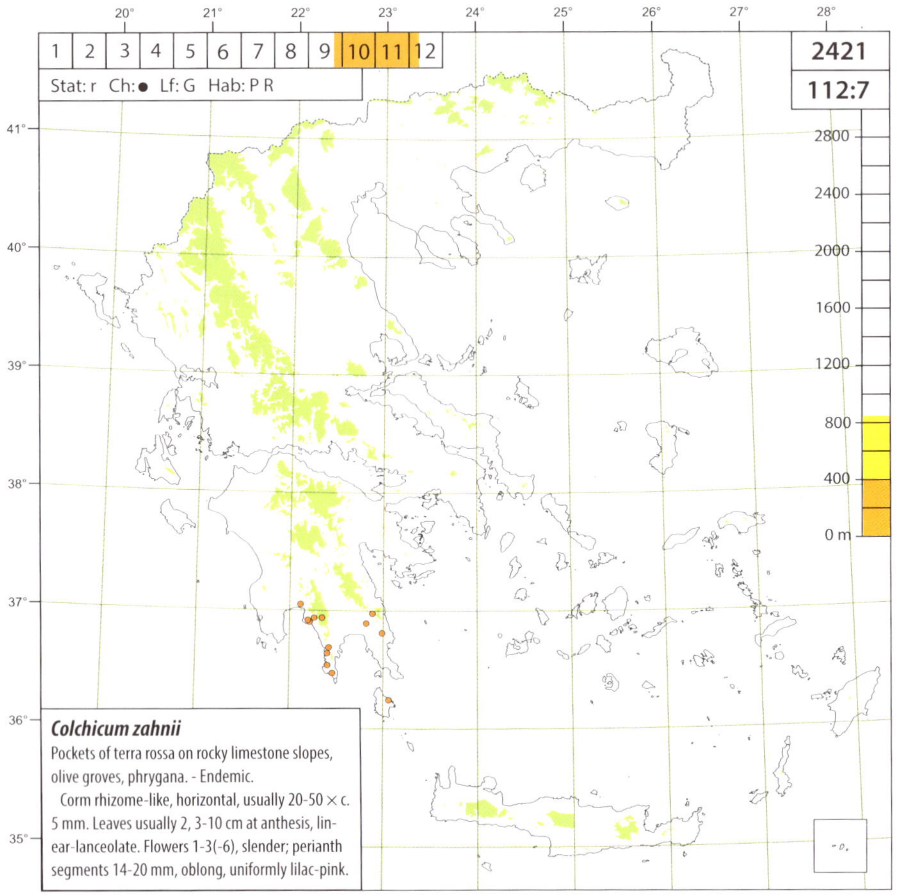

Colchicum zahnii
Pockets of terra rossa on rocky limestone slopes, olive groves, phrygana. - Endemic.
Corm rhizome-like, horizontal, usually 20-50 × c. 5 mm. Leaves usually 2, 3-10 cm at anthesis, linear-lanceolate. Flowers 1-3(-6), slender; perianth segments 14-20 mm, oblong, uniformly lilac-pink.

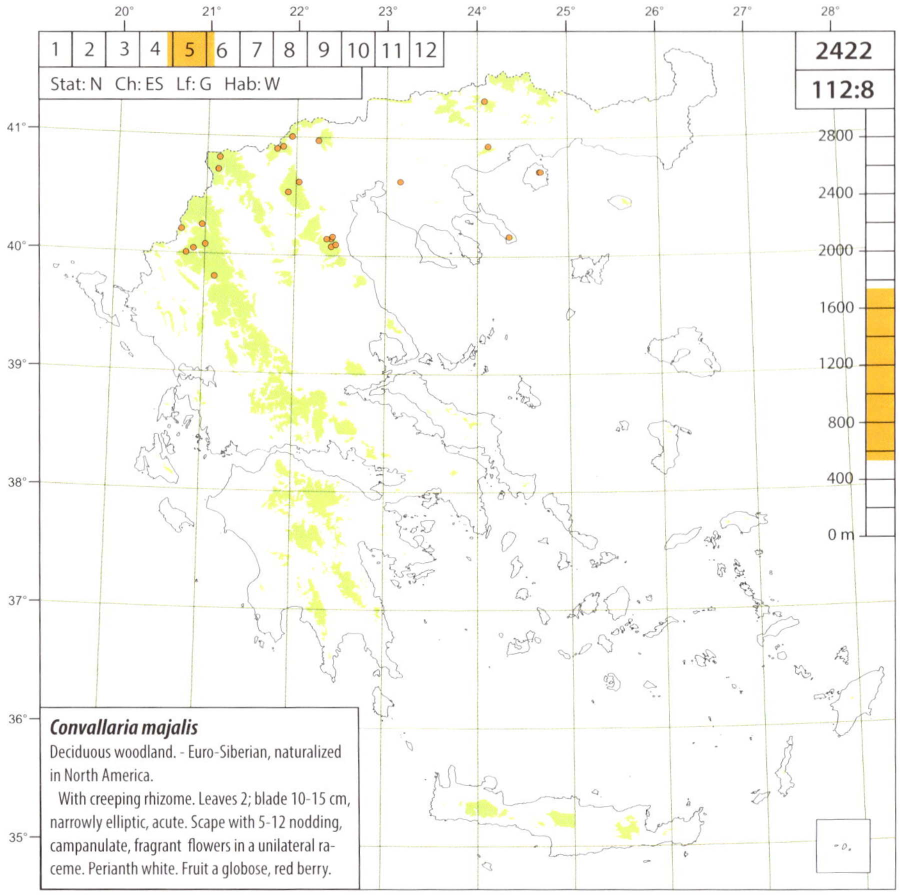

Convallaria majalis
Deciduous woodland. - Euro-Siberian, naturalized in North America.
With creeping rhizome. Leaves 2; blade 10-15 cm, narrowly elliptic, acute. Scape with 5-12 nodding, campanulate, fragrant flowers in a unilateral raceme. Perianth white. Fruit a globose, red berry.

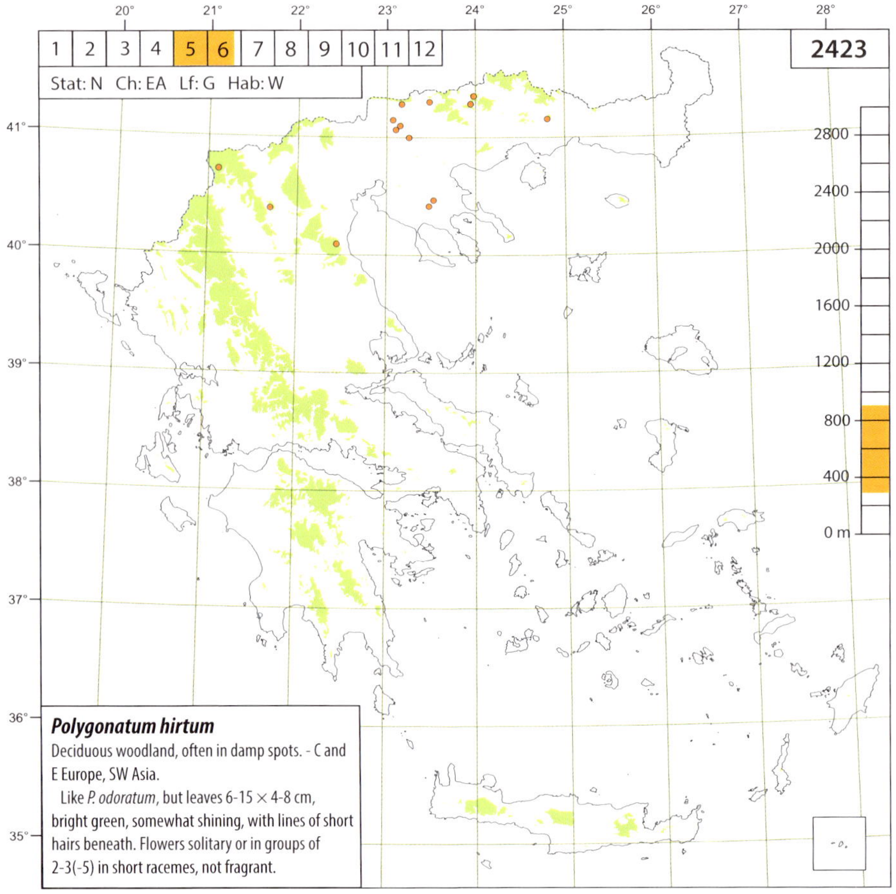

Polygonatum hirtum
Deciduous woodland, often in damp spots. - C and E Europe, SW Asia.
Like *P. odoratum*, but leaves 6-15 × 4-8 cm, bright green, somewhat shining, with lines of short hairs beneath. Flowers solitary or in groups of 2-3(-5) in short racemes, not fragrant.

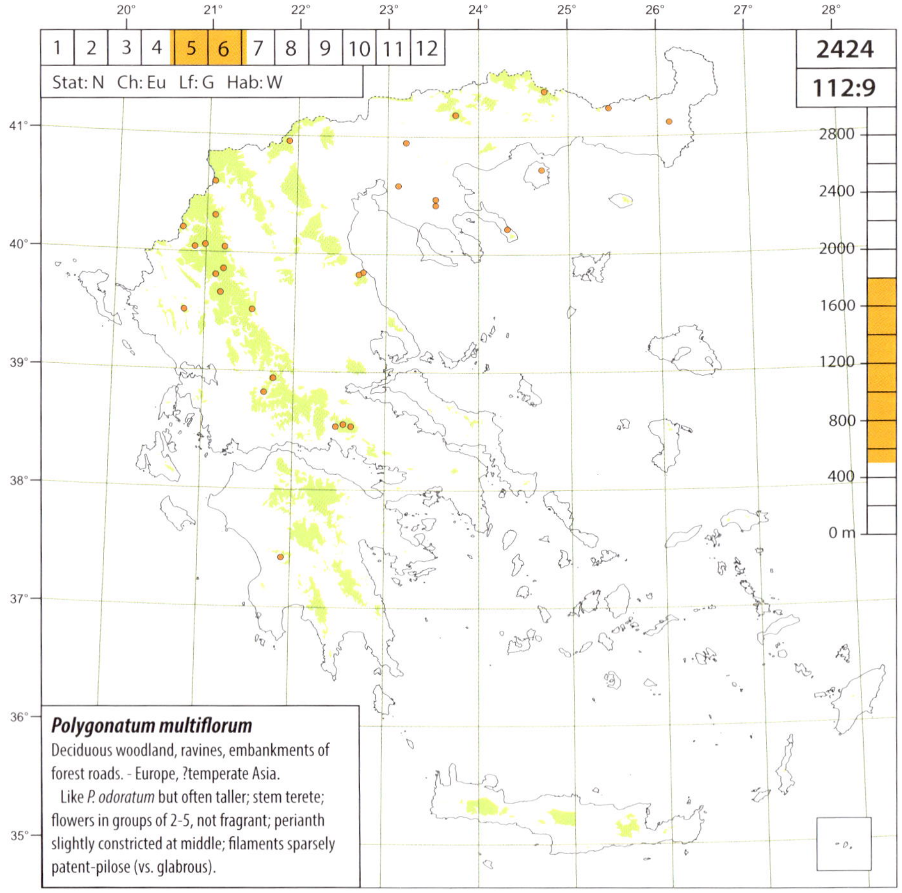

Polygonatum multiflorum
Deciduous woodland, ravines, embankments of forest roads. - Europe, ?temperate Asia.
Like *P. odoratum* but often taller; stem terete; flowers in groups of 2-5, not fragrant; perianth slightly constricted at middle; filaments sparsely patent-pilose (vs. glabrous).

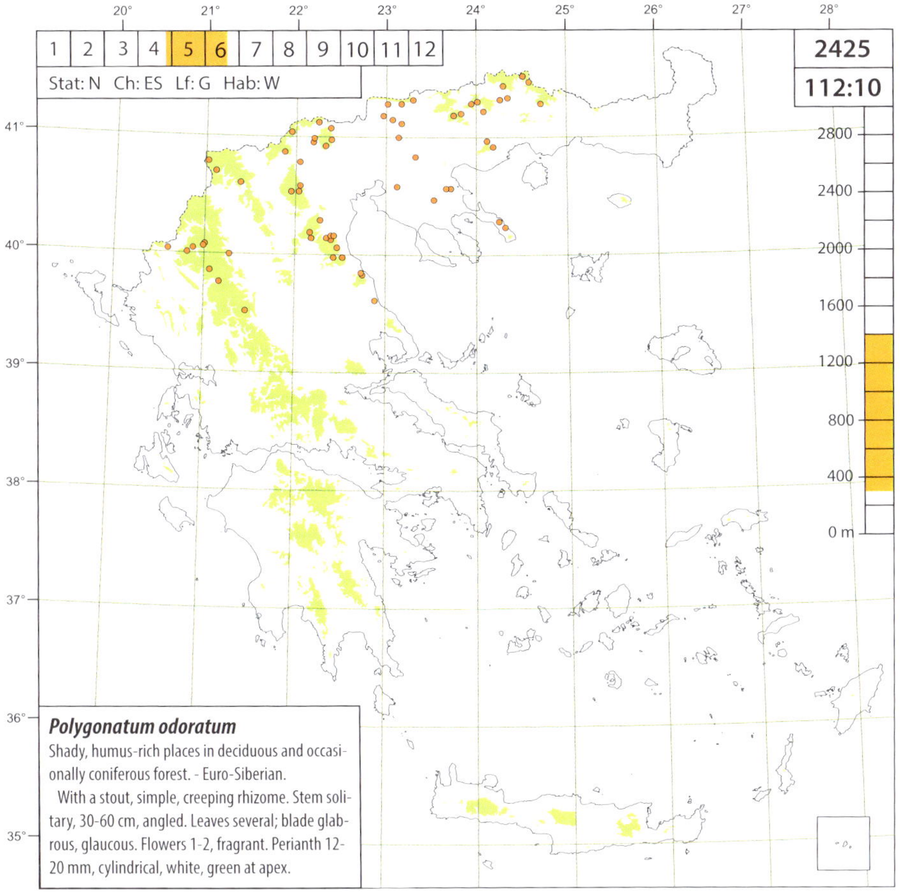

Polygonatum odoratum
Shady, humus-rich places in deciduous and occasionally coniferous forest. - Euro-Siberian.
With a stout, simple, creeping rhizome. Stem solitary, 30-60 cm, angled. Leaves several; blade glabrous, glaucous. Flowers 1-2, fragrant. Perianth 12-20 mm, cylindrical, white, green at apex.

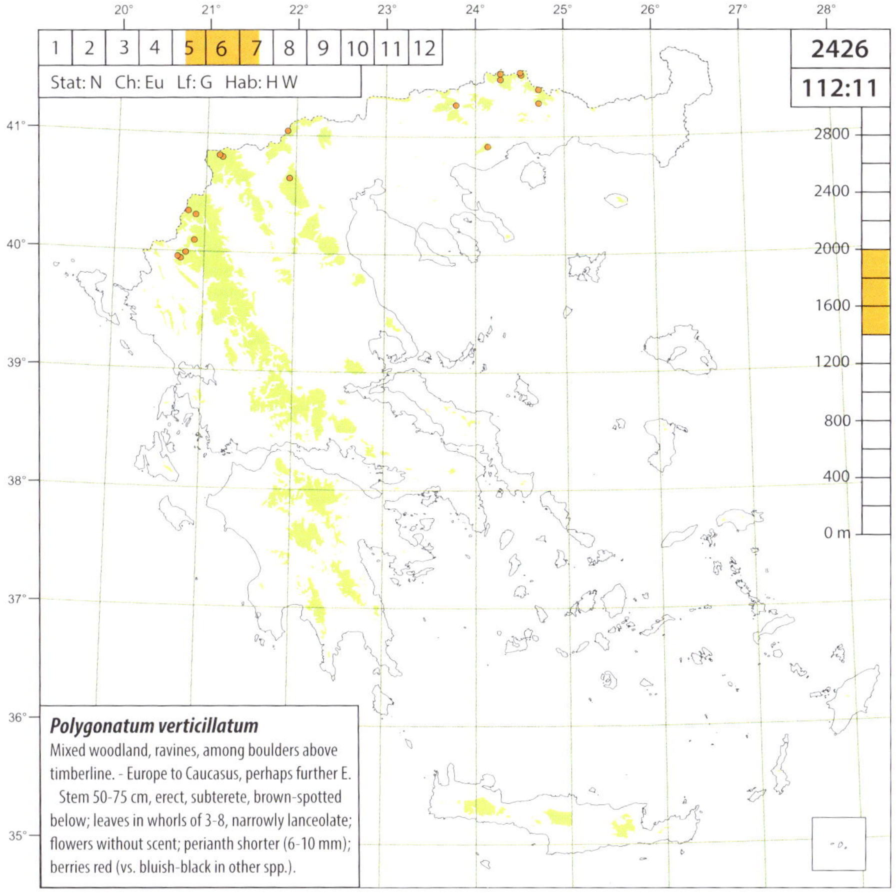

Polygonatum verticillatum
Mixed woodland, ravines, among boulders above timberline. - Europe to Caucasus, perhaps further E.
Stem 50-75 cm, erect, subterete, brown-spotted below; leaves in whorls of 3-8, narrowly lanceolate; flowers without scent; perianth shorter (6-10 mm); berries red (vs. bluish-black in other spp.).

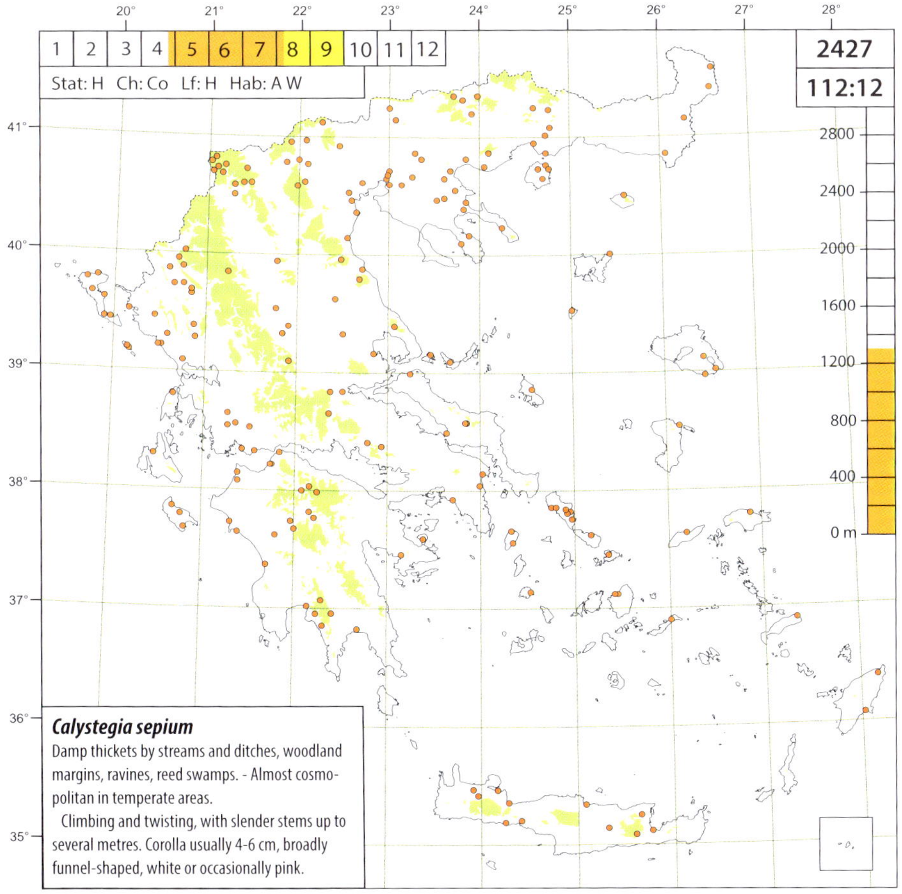

Calystegia sepium
Damp thickets by streams and ditches, woodland margins, ravines, reed swamps. - Almost cosmopolitan in temperate areas.
Climbing and twisting, with slender stems up to several metres. Corolla usually 4-6 cm, broadly funnel-shaped, white or occasionally pink.

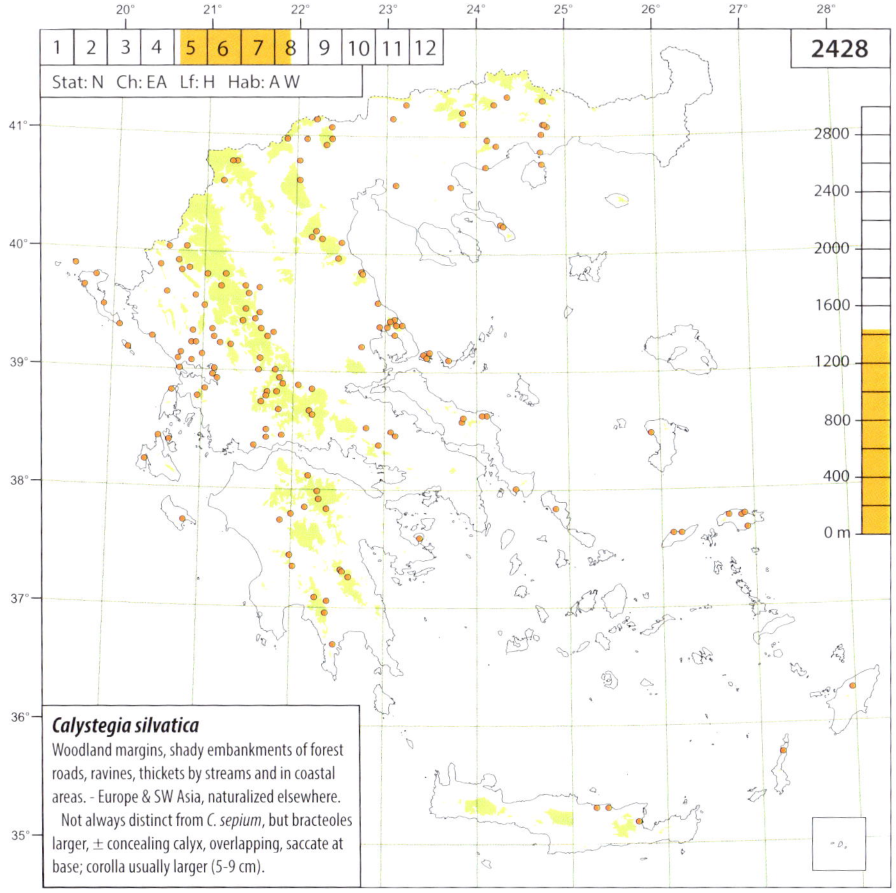

Calystegia silvatica
Woodland margins, shady embankments of forest roads, ravines, thickets by streams and in coastal areas. - Europe & SW Asia, naturalized elsewhere.
Not always distinct from *C. sepium*, but bracteoles larger, ± concealing calyx, overlapping, saccate at base; corolla usually larger (5-9 cm).

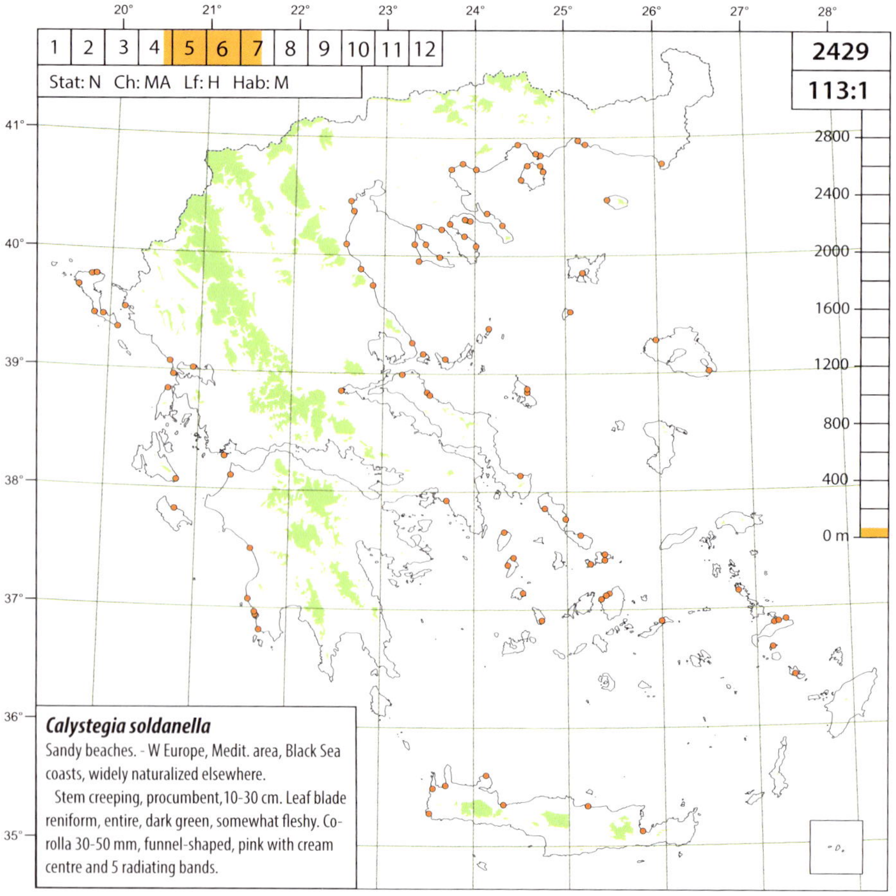

Calystegia soldanella

Sandy beaches. - W Europe, Medit. area, Black Sea coasts, widely naturalized elsewhere.

Stem creeping, procumbent, 10-30 cm. Leaf blade reniform, entire, dark green, somewhat fleshy. Corolla 30-50 mm, funnel-shaped, pink with cream centre and 5 radiating bands.

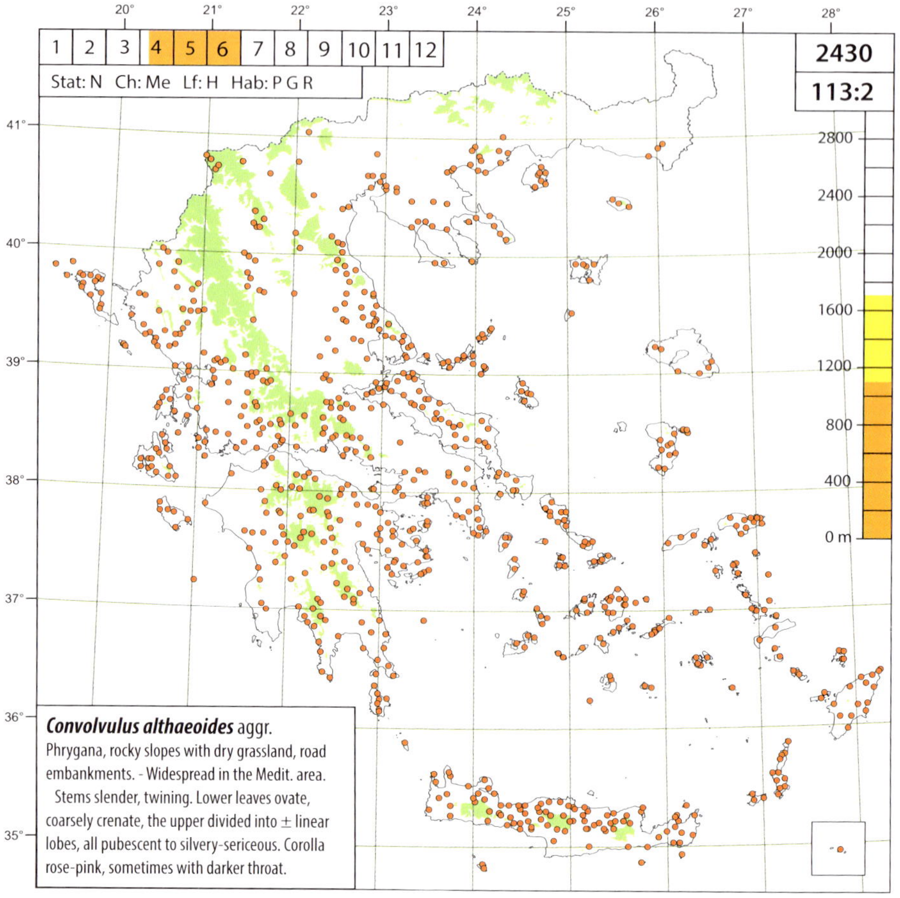

Convolvulus althaeoides **aggr.**

Phrygana, rocky slopes with dry grassland, road embankments. - Widespread in the Medit. area.

Stems slender, twining. Lower leaves ovate, coarsely crenate, the upper divided into ± linear lobes, all pubescent to silvery-sericeous. Corolla rose-pink, sometimes with darker throat.

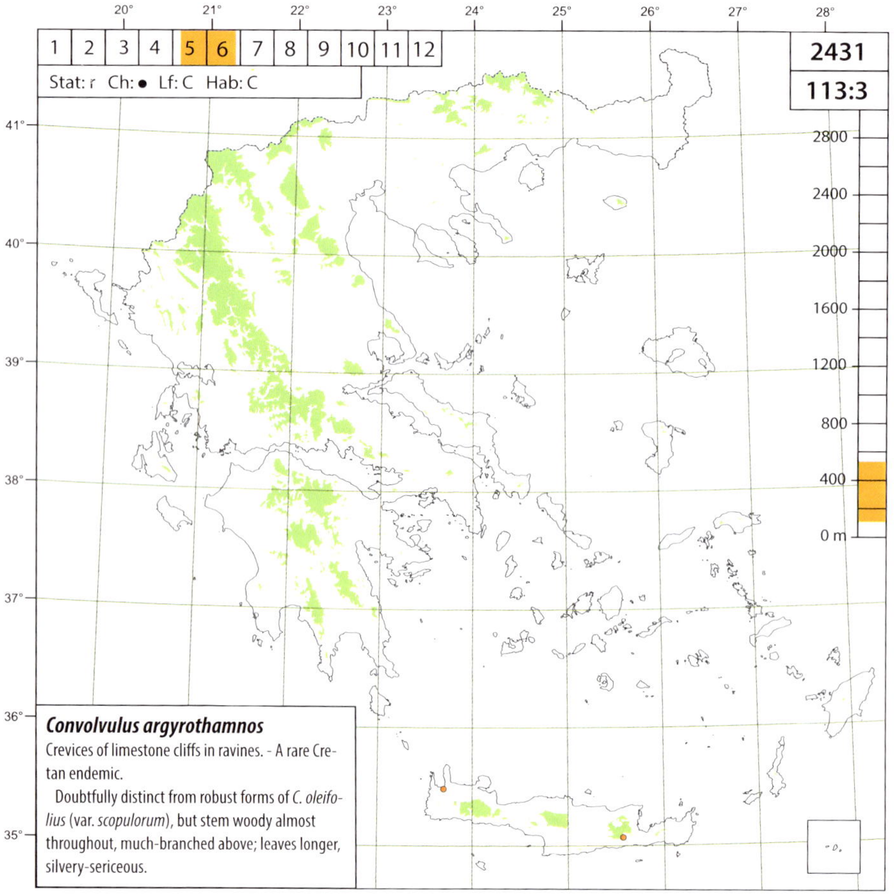

Convolvulus argyrothamnos

Crevices of limestone cliffs in ravines. - A rare Cretan endemic.

Doubtfully distinct from robust forms of *C. oleifolius* (var. *scopulorum*), but stem woody almost throughout, much-branched above; leaves longer, silvery-sericeous.

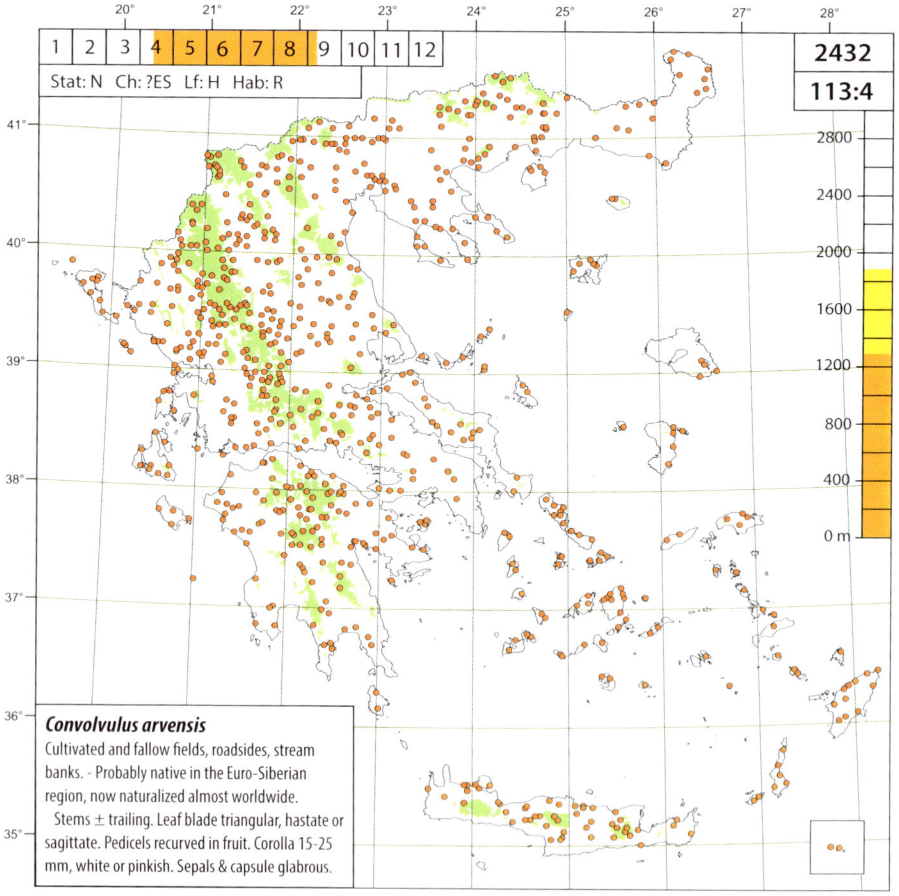

Convolvulus arvensis

Cultivated and fallow fields, roadsides, stream banks. - Probably native in the Euro-Siberian region, now naturalized almost worldwide.

Stems ± trailing. Leaf blade triangular, hastate or sagittate. Pedicels recurved in fruit. Corolla 15-25 mm, white or pinkish. Sepals & capsule glabrous.

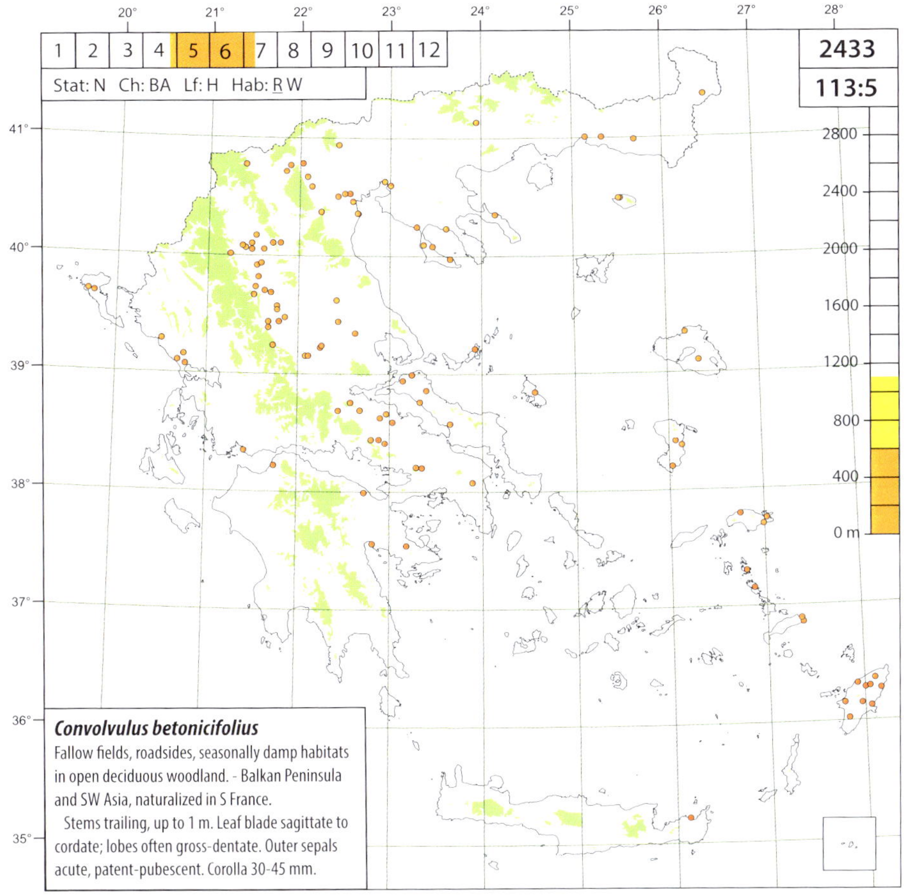

Convolvulus betonicifolius
Fallow fields, roadsides, seasonally damp habitats in open deciduous woodland. - Balkan Peninsula and SW Asia, naturalized in S France.
Stems trailing, up to 1 m. Leaf blade sagittate to cordate; lobes often gross-dentate. Outer sepals acute, patent-pubescent. Corolla 30-45 mm.

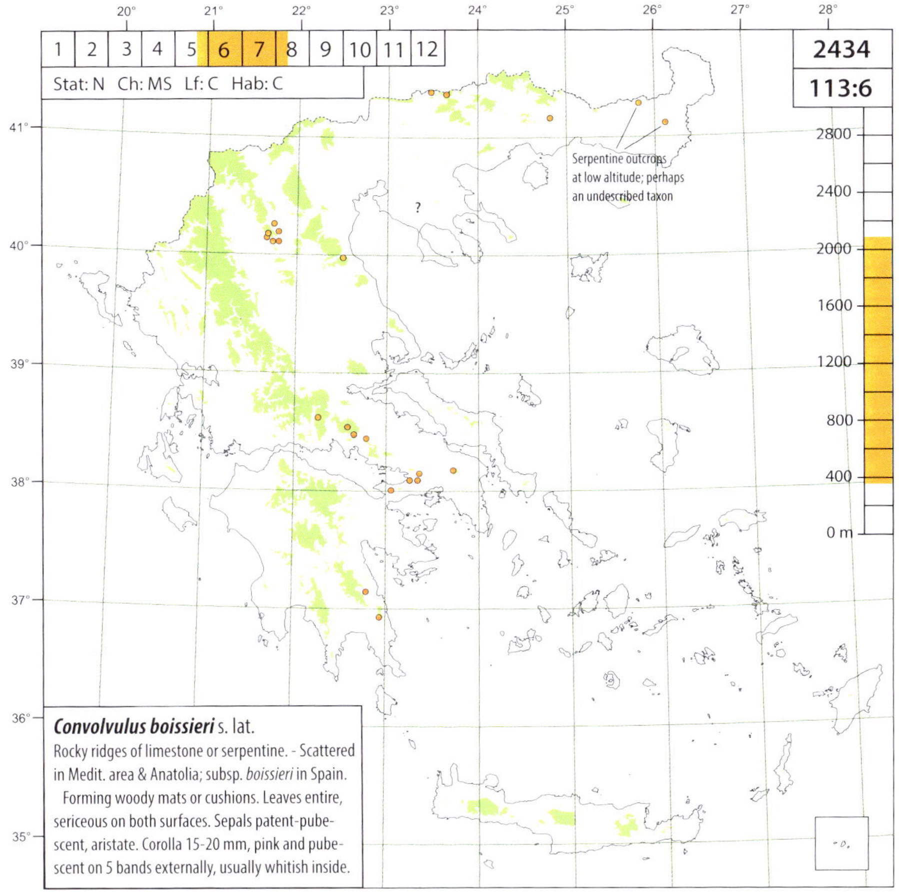

Convolvulus boissieri **s. lat.**
Rocky ridges of limestone or serpentine. - Scattered in Medit. area & Anatolia; subsp. *boissieri* in Spain.
Forming woody mats or cushions. Leaves entire, sericeous on both surfaces. Sepals patent-pubescent, aristate. Corolla 15-20 mm, pink and pubescent on 5 bands externally, usually whitish inside.

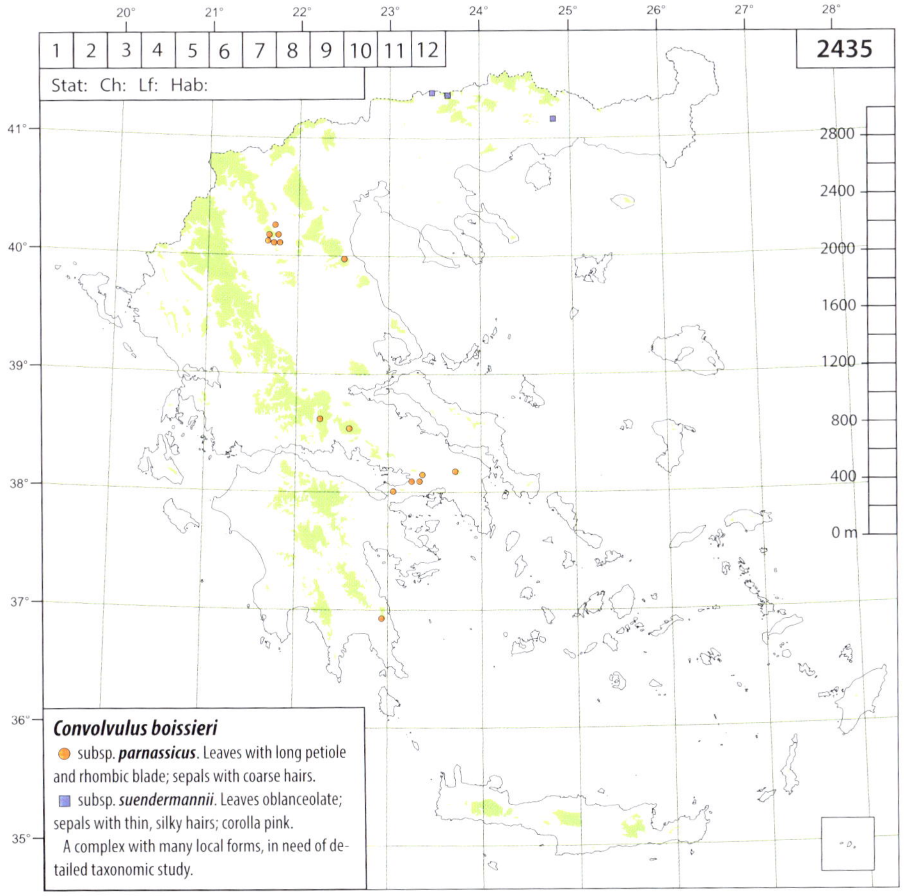

Convolvulus boissieri
● subsp. ***parnassicus***. Leaves with long petiole and rhombic blade; sepals with coarse hairs.
■ subsp. ***suendermannii***. Leaves oblanceolate; sepals with thin, silky hairs; corolla pink.
A complex with many local forms, in need of detailed taxonomic study.

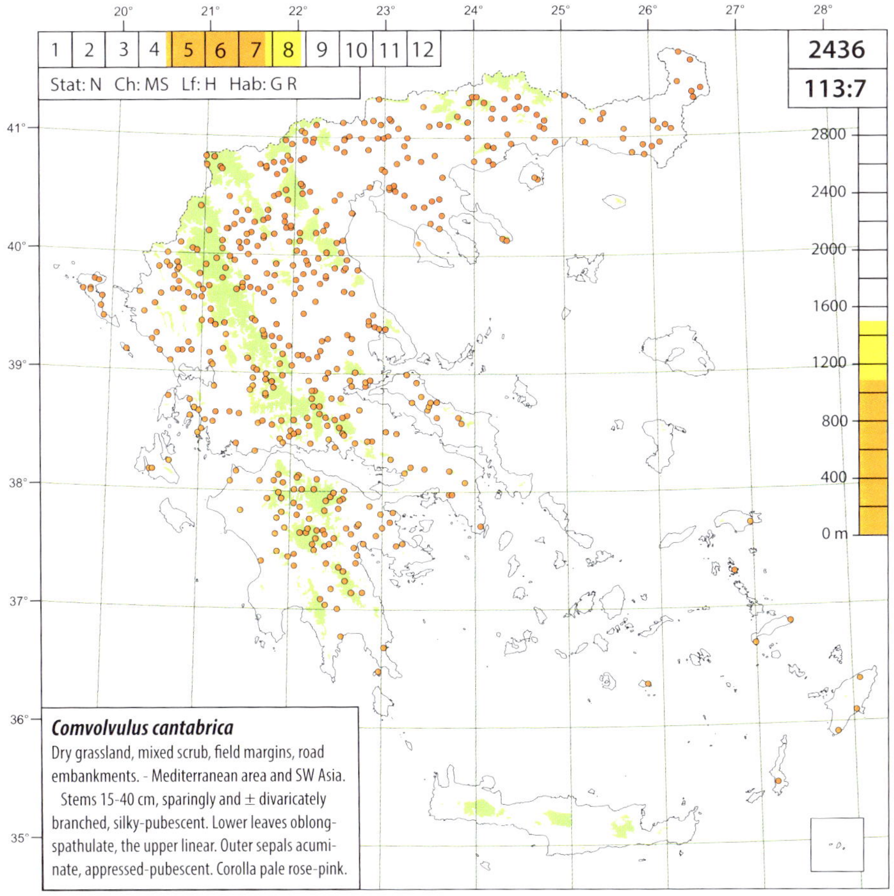

Comvolvulus cantabrica
Dry grassland, mixed scrub, field margins, road embankments. - Mediterranean area and SW Asia.
Stems 15-40 cm, sparingly and ± divaricately branched, silky-pubescent. Lower leaves oblong-spathulate, the upper linear. Outer sepals acuminate, appressed-pubescent. Corolla pale rose-pink.

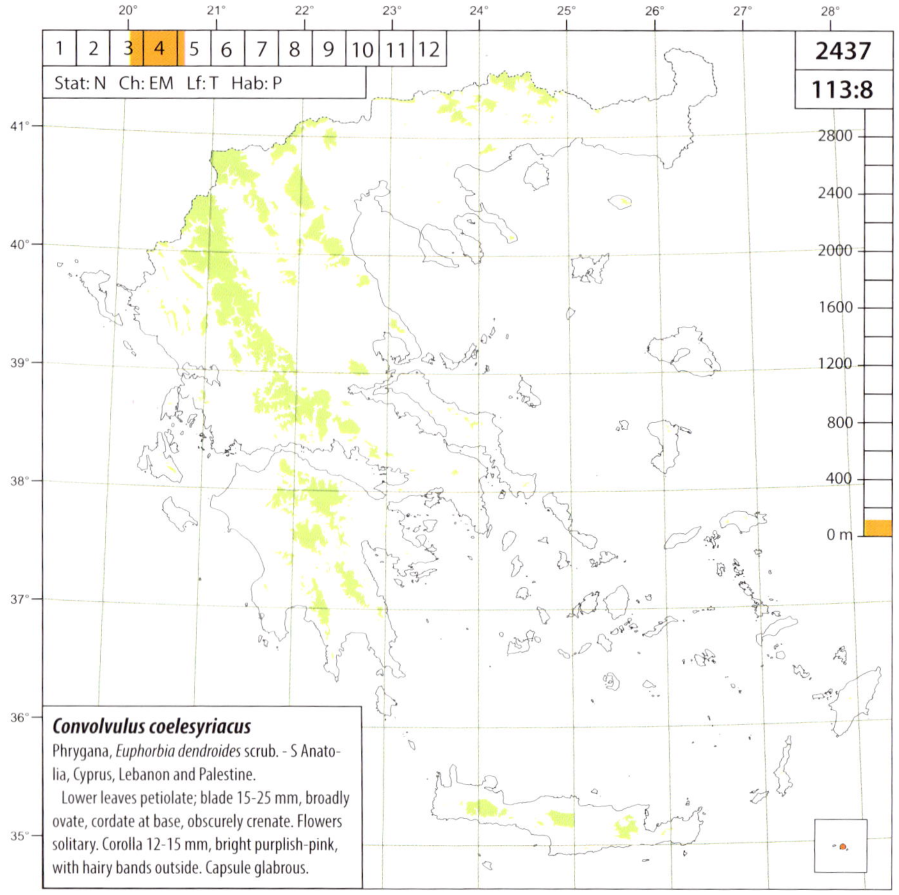

Convolvulus coelesyriacus

Phrygana, *Euphorbia dendroides* scrub. - S Anatolia, Cyprus, Lebanon and Palestine.

Lower leaves petiolate; blade 15-25 mm, broadly ovate, cordate at base, obscurely crenate. Flowers solitary. Corolla 12-15 mm, bright purplish-pink, with hairy bands outside. Capsule glabrous.

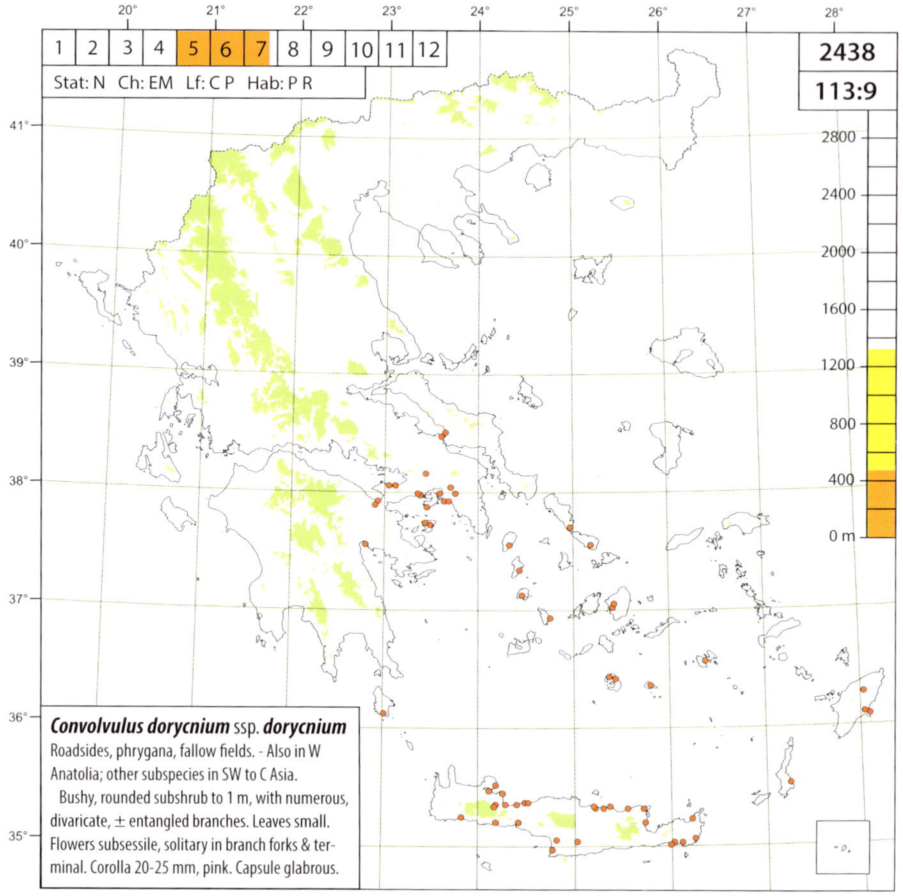

Convolvulus dorycnium ssp. ***dorycnium***

Roadsides, phrygana, fallow fields. - Also in W Anatolia; other subspecies in SW to C Asia.

Bushy, rounded subshrub to 1 m, with numerous, divaricate, ± entangled branches. Leaves small. Flowers subsessile, solitary in branch forks & terminal. Corolla 20-25 mm, pink. Capsule glabrous.

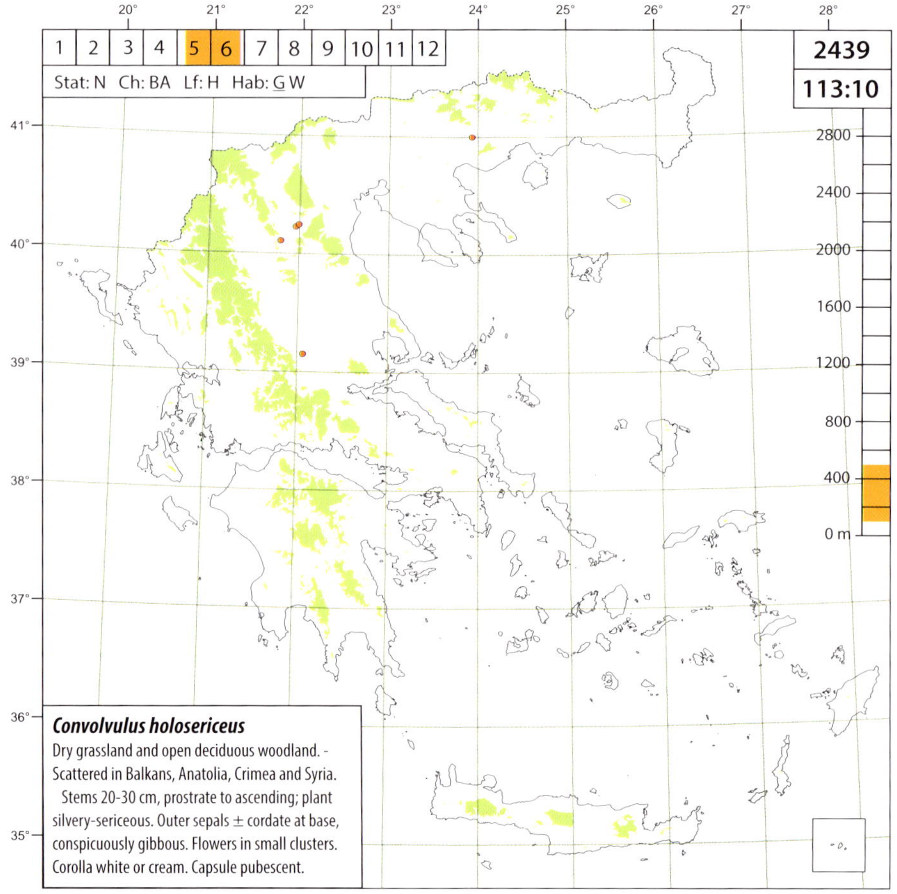

Convolvulus holosericeus

Dry grassland and open deciduous woodland. - Scattered in Balkans, Anatolia, Crimea and Syria.

Stems 20-30 cm, prostrate to ascending; plant silvery-sericeous. Outer sepals ± cordate at base, conspicuously gibbous. Flowers in small clusters. Corolla white or cream. Capsule pubescent.

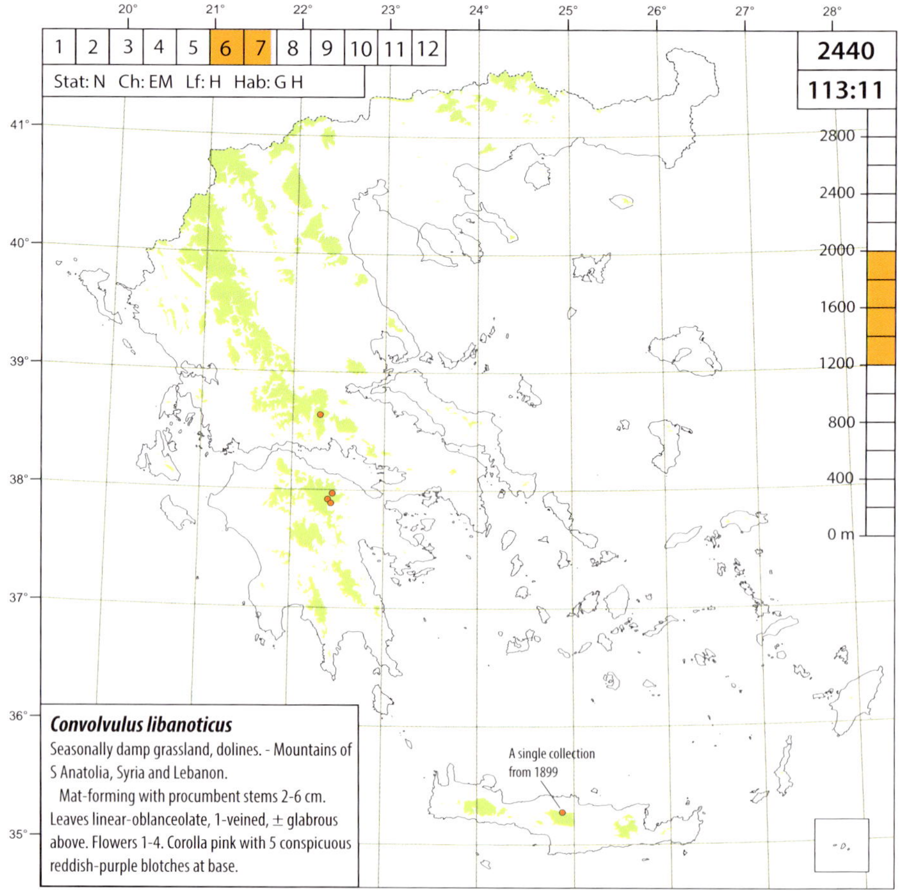

Convolvulus libanoticus

Seasonally damp grassland, dolines. - Mountains of S Anatolia, Syria and Lebanon.

Mat-forming with procumbent stems 2-6 cm. Leaves linear-oblanceolate, 1-veined, ± glabrous above. Flowers 1-4. Corolla pink with 5 conspicuous reddish-purple blotches at base.

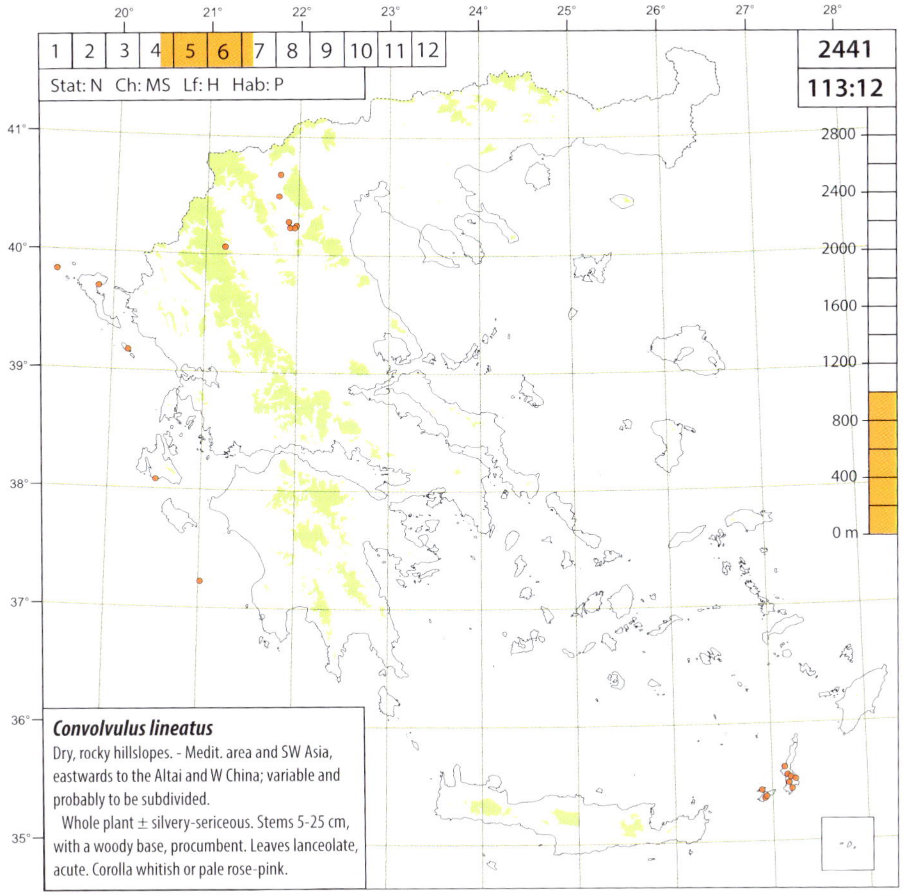

Convolvulus lineatus

Dry, rocky hillslopes. - Medit. area and SW Asia, eastwards to the Altai and W China; variable and probably to be subdivided.

Whole plant ± silvery-sericeous. Stems 5-25 cm, with a woody base, procumbent. Leaves lanceolate, acute. Corolla whitish or pale rose-pink.

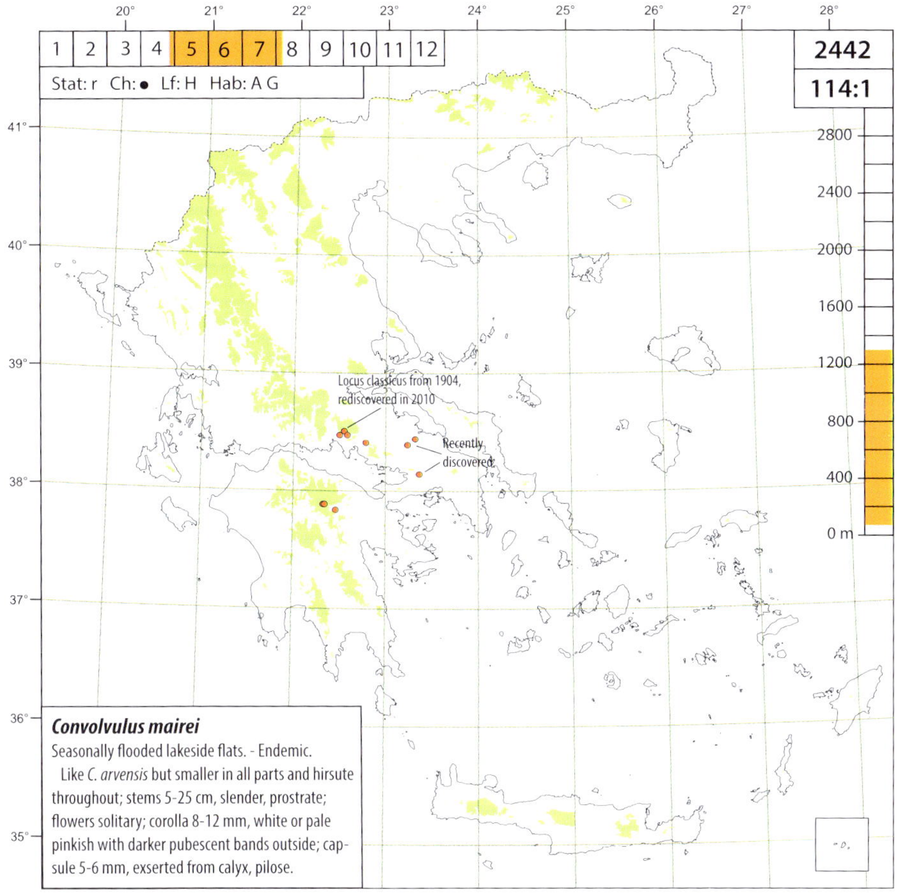

Convolvulus mairei

Seasonally flooded lakeside flats. - Endemic.

Like *C. arvensis* but smaller in all parts and hirsute throughout; stems 5-25 cm, slender, prostrate; flowers solitary; corolla 8-12 mm, white or pale pinkish with darker pubescent bands outside; capsule 5-6 mm, exserted from calyx, pilose.

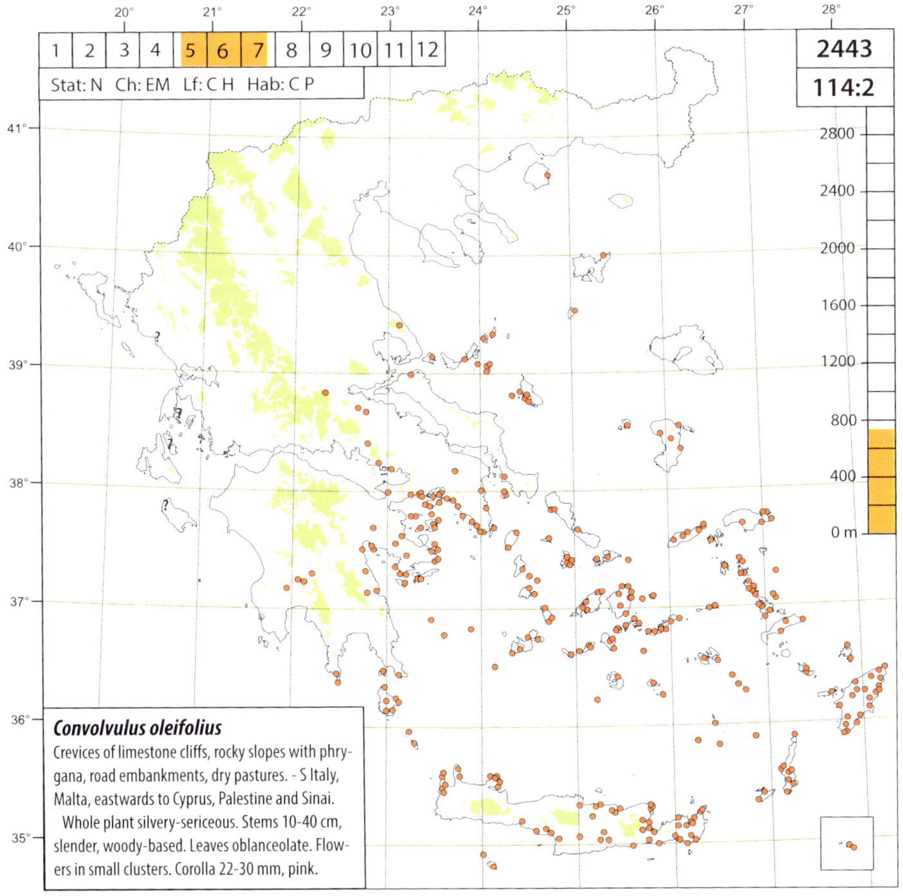

Convolvulus oleifolius

Crevices of limestone cliffs, rocky slopes with phrygana, road embankments, dry pastures. - S Italy, Malta, eastwards to Cyprus, Palestine and Sinai.

Whole plant silvery-sericeous. Stems 10-40 cm, slender, woody-based. Leaves oblanceolate. Flowers in small clusters. Corolla 22-30 mm, pink.

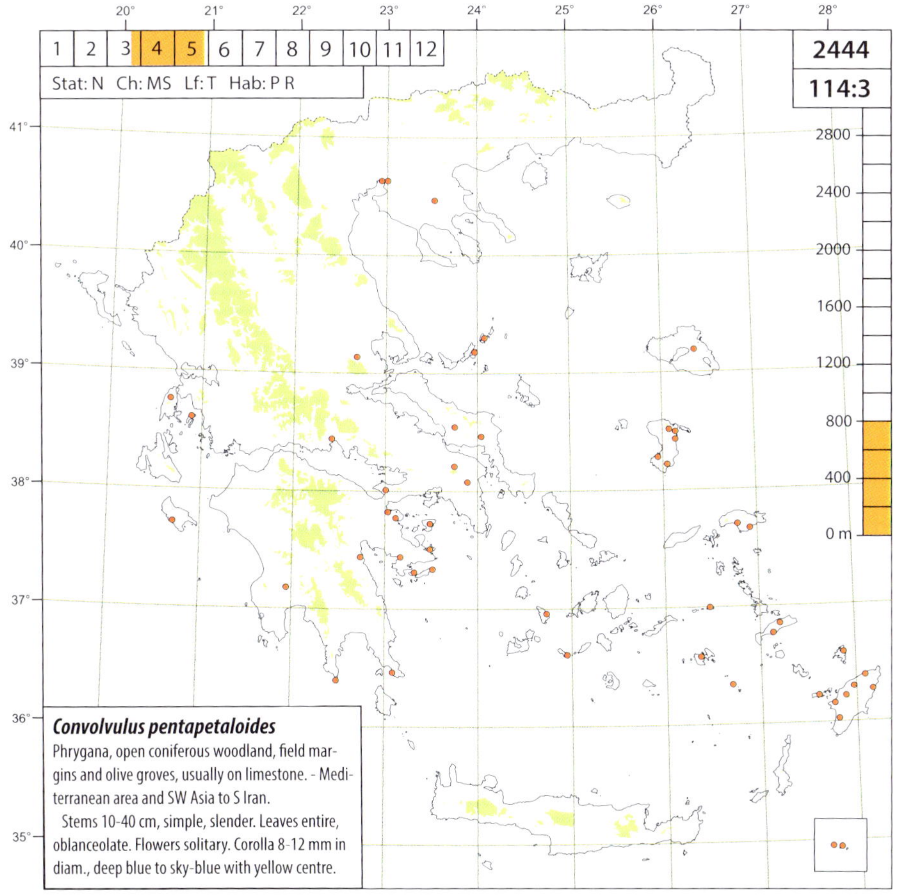

Convolvulus pentapetaloides

Phrygana, open coniferous woodland, field margins and olive groves, usually on limestone. - Mediterranean area and SW Asia to S Iran.

Stems 10-40 cm, simple, slender. Leaves entire, oblanceolate. Flowers solitary. Corolla 8-12 mm in diam., deep blue to sky-blue with yellow centre.

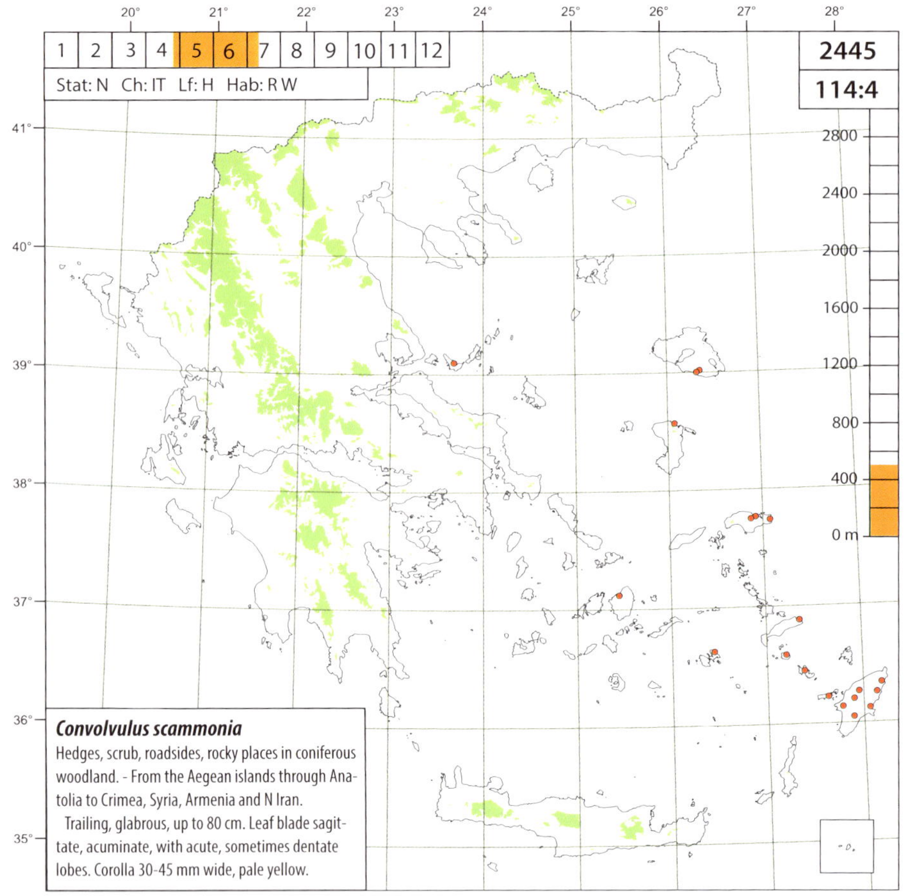

Convolvulus scammonia
Hedges, scrub, roadsides, rocky places in coniferous woodland. - From the Aegean islands through Anatolia to Crimea, Syria, Armenia and N Iran.
Trailing, glabrous, up to 80 cm. Leaf blade sagittate, acuminate, with acute, sometimes dentate lobes. Corolla 30-45 mm wide, pale yellow.

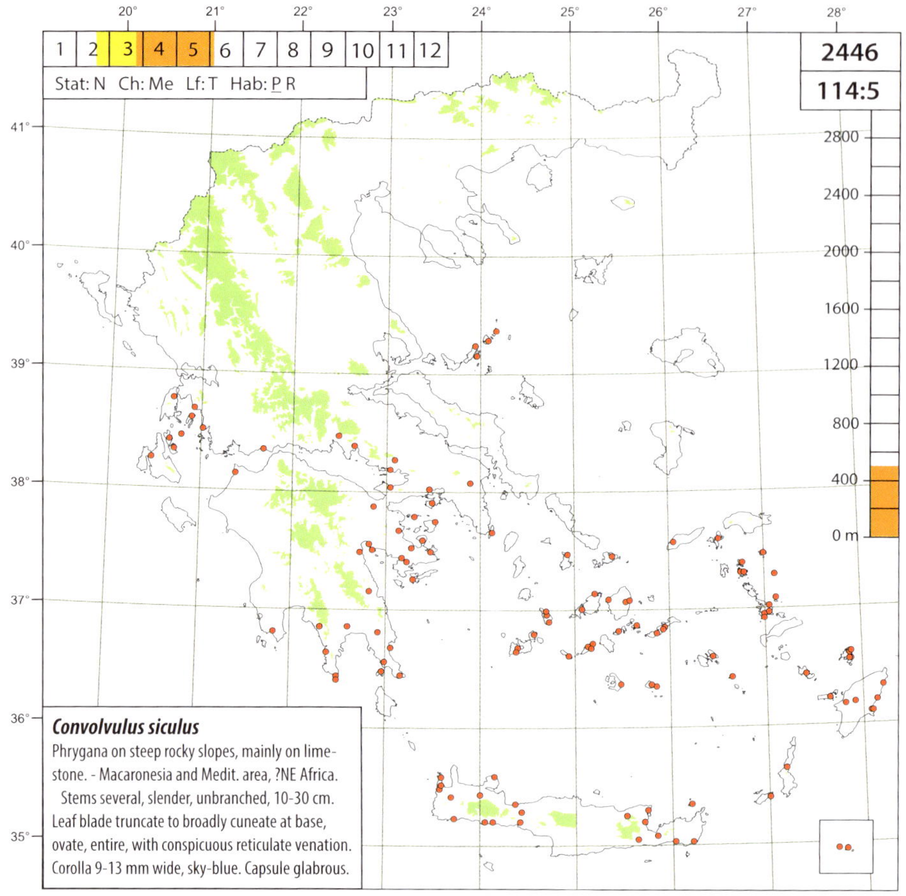

Convolvulus siculus
Phrygana on steep rocky slopes, mainly on limestone. - Macaronesia and Medit. area, ?NE Africa.
Stems several, slender, unbranched, 10-30 cm. Leaf blade truncate to broadly cuneate at base, ovate, entire, with conspicuous reticulate venation. Corolla 9-13 mm wide, sky-blue. Capsule glabrous.

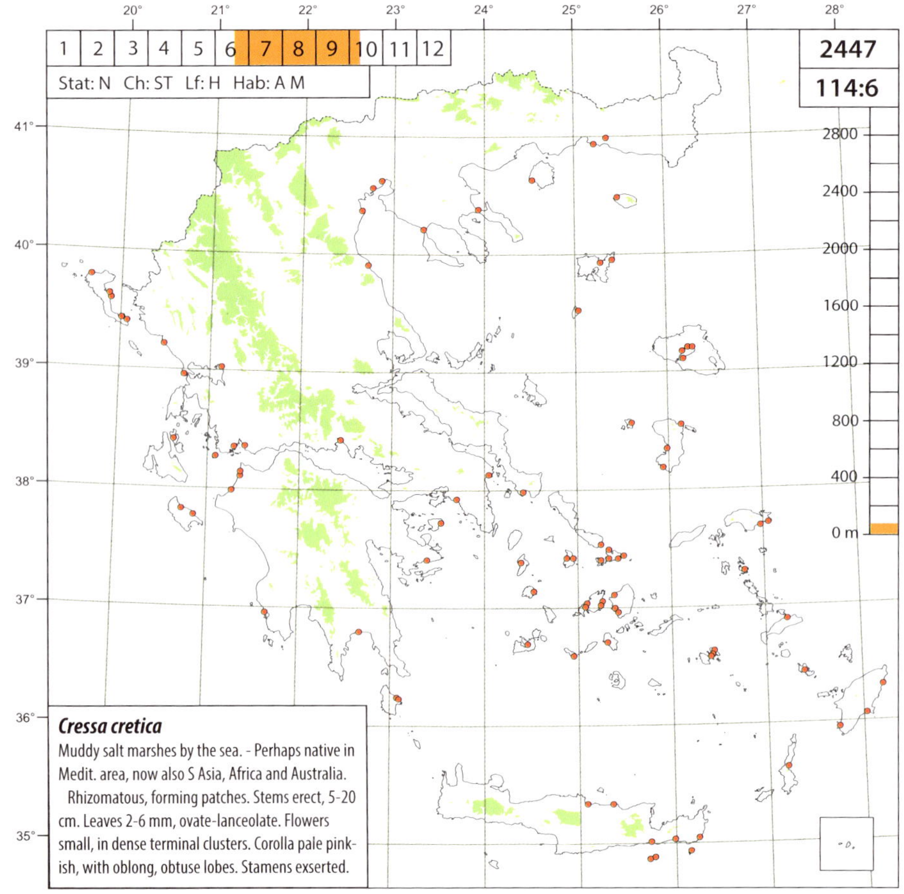

Cressa cretica
Muddy salt marshes by the sea. - Perhaps native in Medit. area, now also S Asia, Africa and Australia.
Rhizomatous, forming patches. Stems erect, 5-20 cm. Leaves 2-6 mm, ovate-lanceolate. Flowers small, in dense terminal clusters. Corolla pale pinkish, with oblong, obtuse lobes. Stamens exserted.

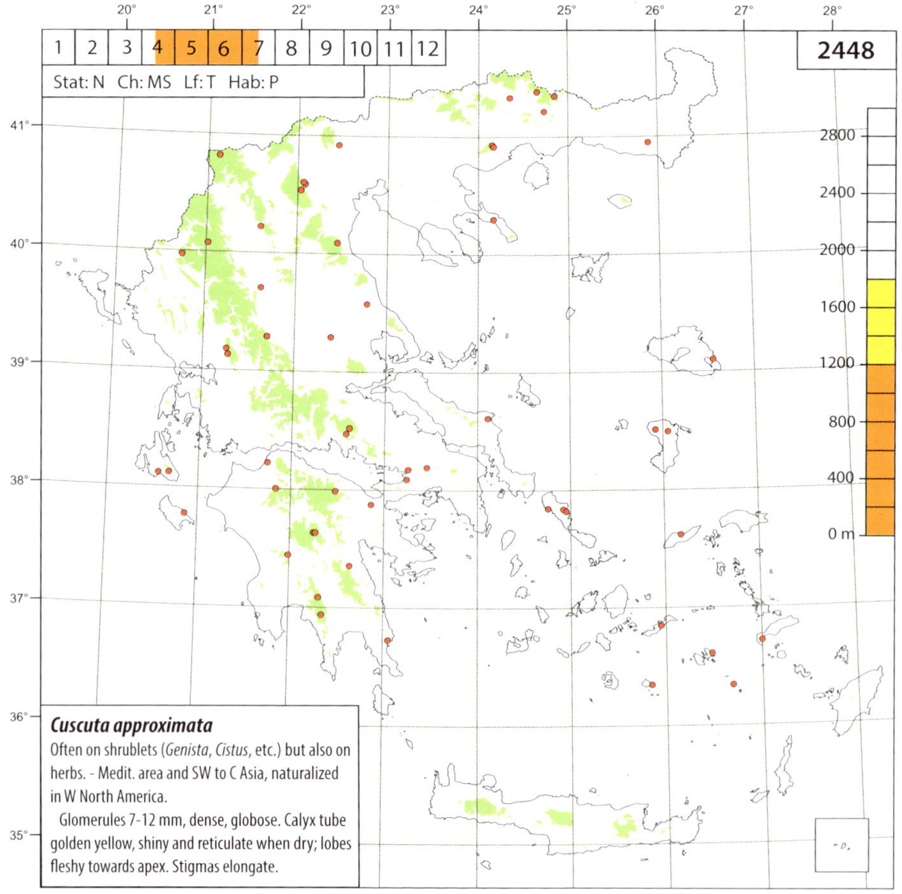

Cuscuta approximata
Often on shrublets (*Genista*, *Cistus*, etc.) but also on herbs. - Medit. area and SW to C Asia, naturalized in W North America.
Glomerules 7-12 mm, dense, globose. Calyx tube golden yellow, shiny and reticulate when dry; lobes fleshy towards apex. Stigmas elongate.

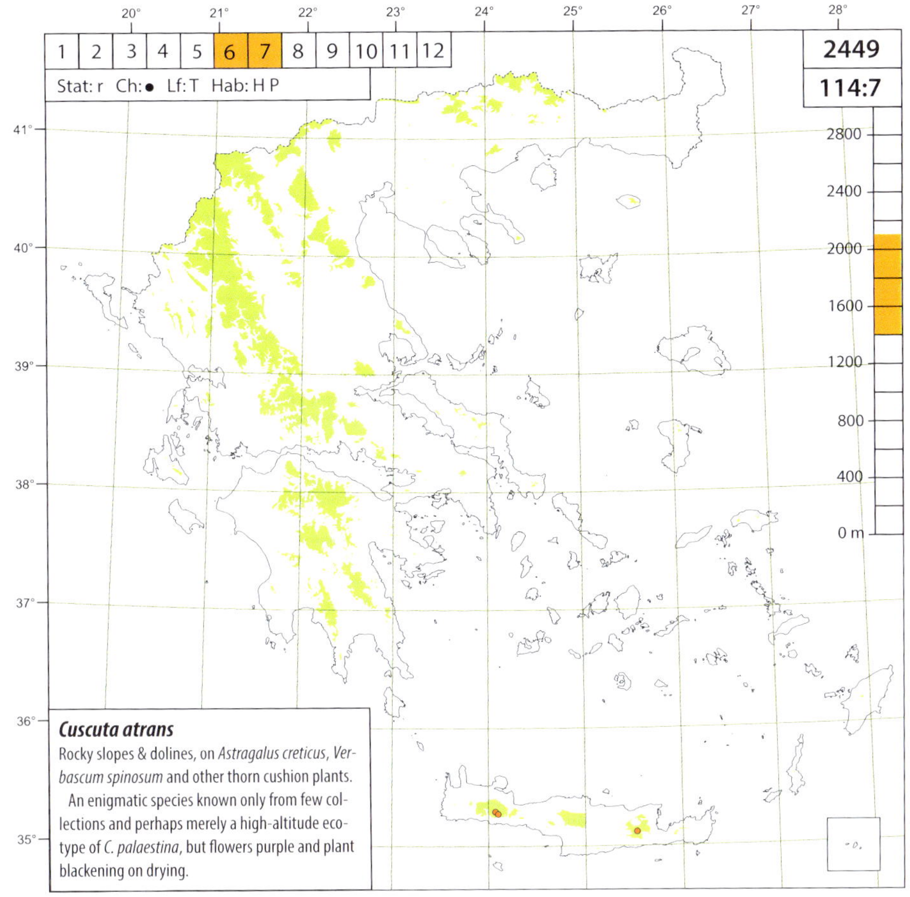

Cuscuta atrans
Rocky slopes & dolines, on *Astragalus creticus*, *Verbascum spinosum* and other thorn cushion plants.
An enigmatic species known only from few collections and perhaps merely a high-altitude ecotype of *C. palaestina*, but flowers purple and plant blackening on drying.

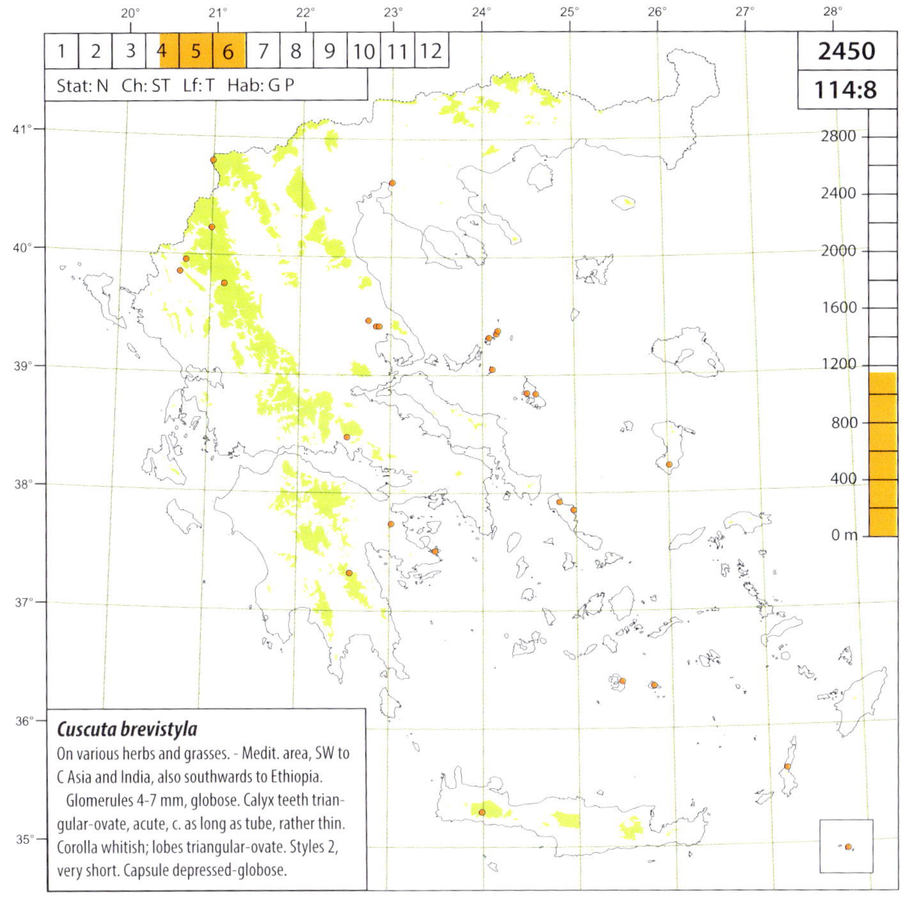

Cuscuta brevistyla
On various herbs and grasses. - Medit. area, SW to C Asia and India, also southwards to Ethiopia.
Glomerules 4-7 mm, globose. Calyx teeth triangular-ovate, acute, c. as long as tube, rather thin. Corolla whitish; lobes triangular-ovate. Styles 2, very short. Capsule depressed-globose.

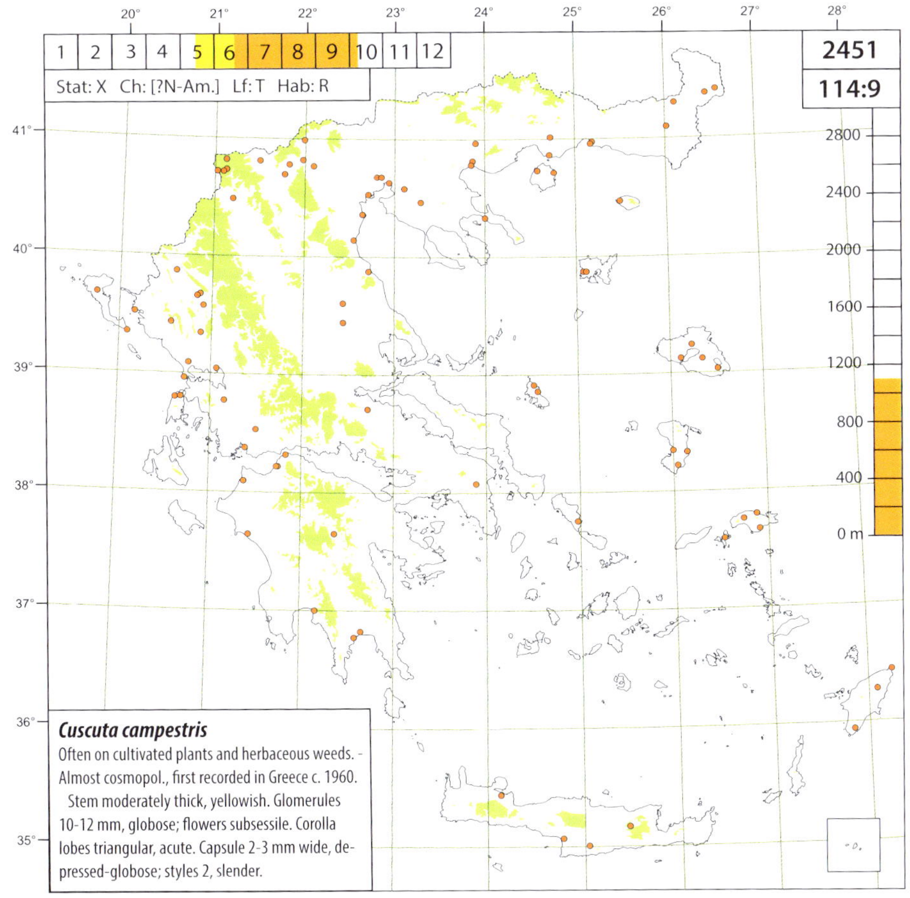

Cuscuta campestris
Often on cultivated plants and herbaceous weeds. - Almost cosmopol., first recorded in Greece c. 1960.
Stem moderately thick, yellowish. Glomerules 10-12 mm, globose; flowers subsessile. Corolla lobes triangular, acute. Capsule 2-3 mm wide, depressed-globose; styles 2, slender.

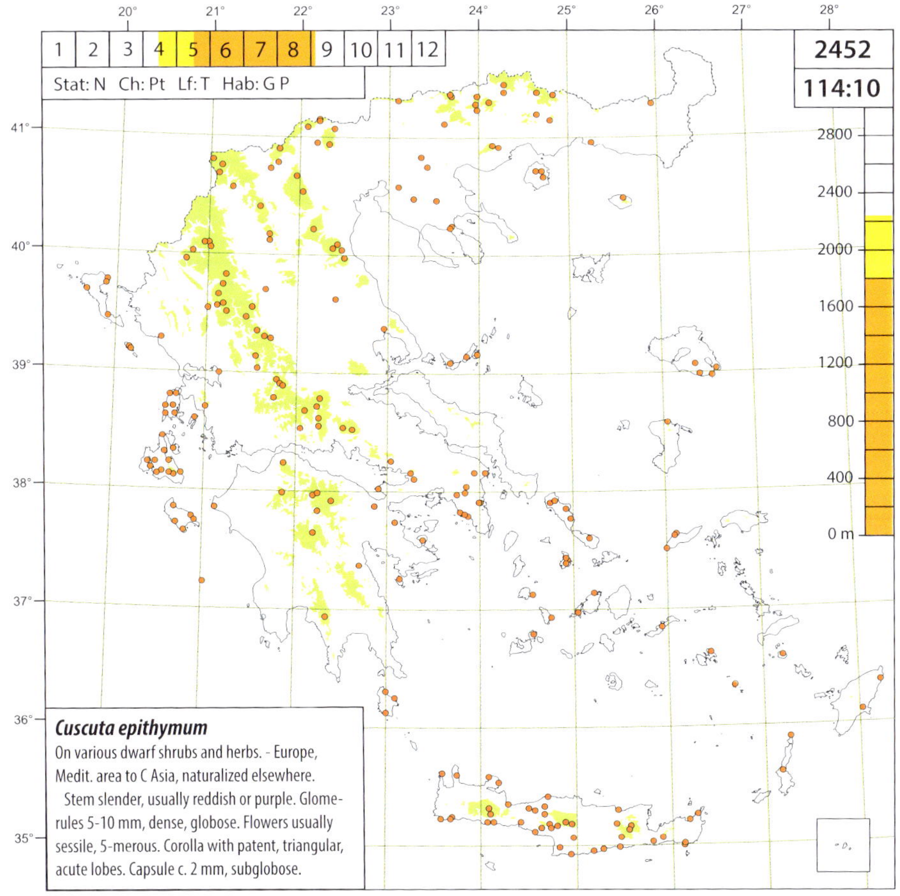

Cuscuta epithymum
On various dwarf shrubs and herbs. - Europe, Medit. area to C Asia, naturalized elsewhere.
Stem slender, usually reddish or purple. Glomerules 5-10 mm, dense, globose. Flowers usually sessile, 5-merous. Corolla with patent, triangular, acute lobes. Capsule c. 2 mm, subglobose.

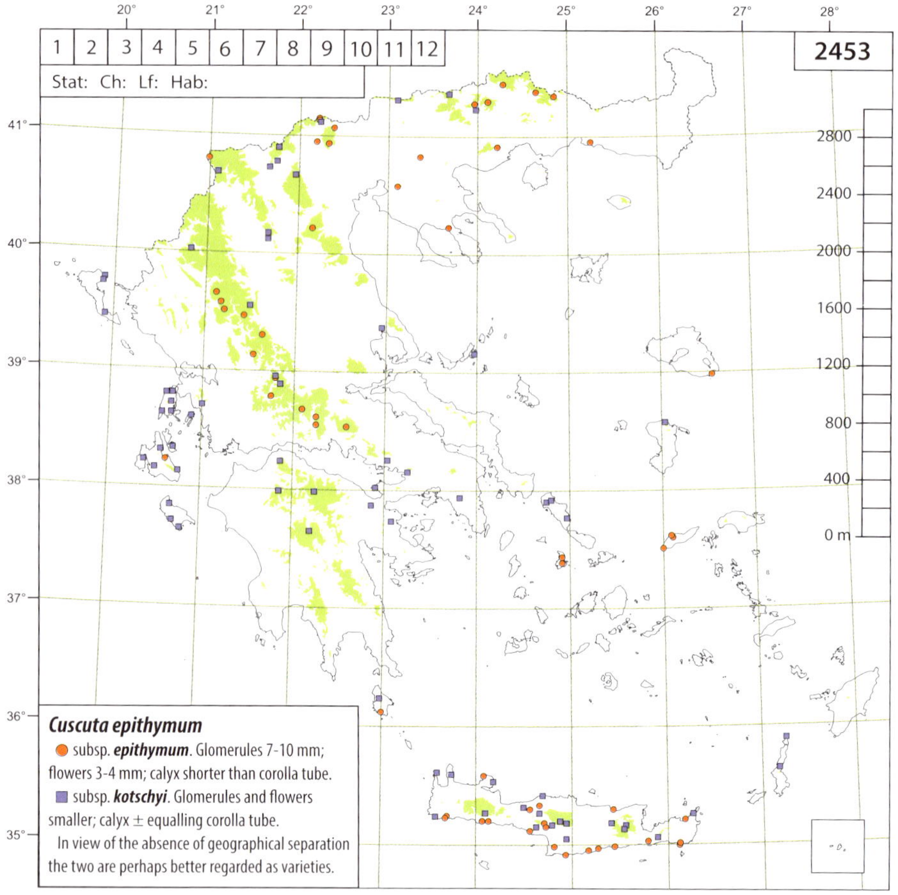

Cuscuta epithymum

● subsp. ***epithymum***. Glomerules 7-10 mm; flowers 3-4 mm; calyx shorter than corolla tube.
■ subsp. ***kotschyi***. Glomerules and flowers smaller; calyx ± equalling corolla tube.

In view of the absence of geographical separation the two are perhaps better regarded as varieties.

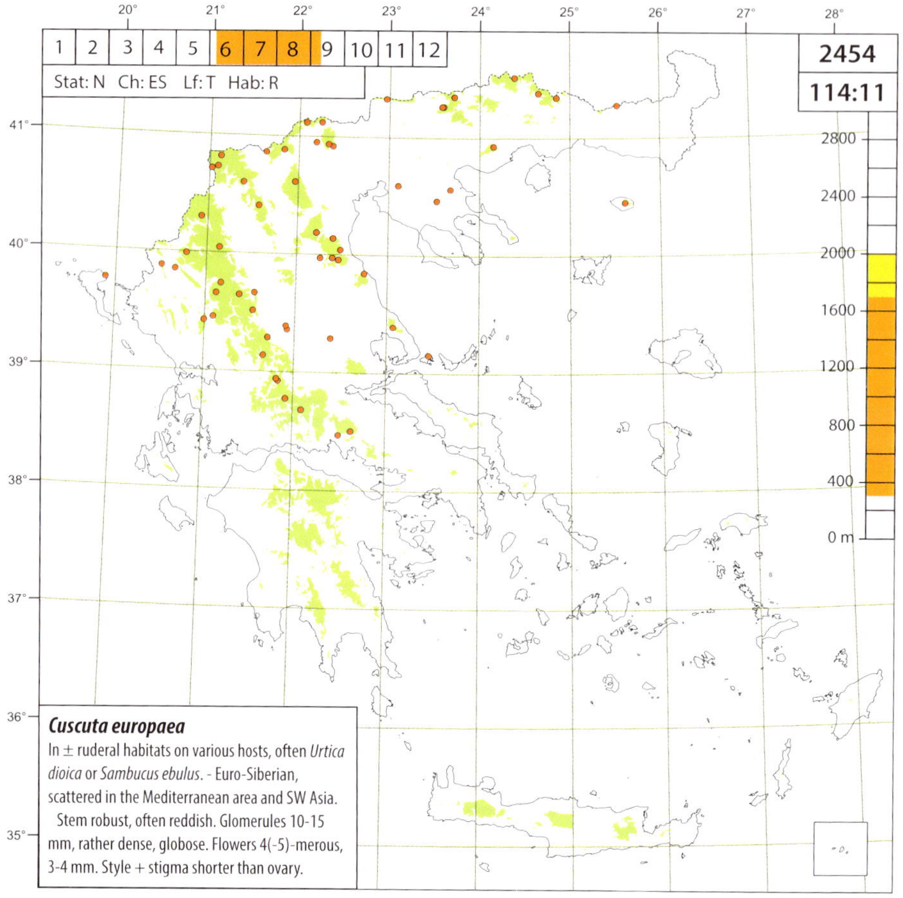

Cuscuta europaea

In ± ruderal habitats on various hosts, often *Urtica dioica* or *Sambucus ebulus*. - Euro-Siberian, scattered in the Mediterranean area and SW Asia.

Stem robust, often reddish. Glomerules 10-15 mm, rather dense, globose. Flowers 4(-5)-merous, 3-4 mm. Style + stigma shorter than ovary.

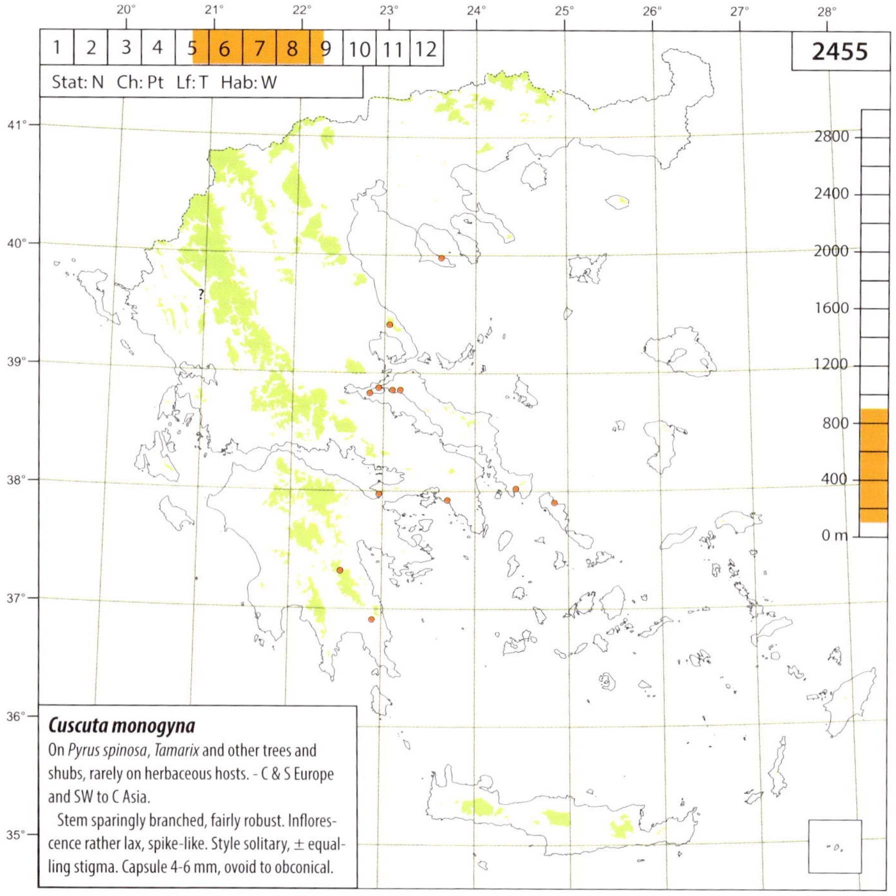

Cuscuta monogyna

On *Pyrus spinosa*, *Tamarix* and other trees and shubs, rarely on herbaceous hosts. - C & S Europe and SW to C Asia.

Stem sparingly branched, fairly robust. Inflorescence rather lax, spike-like. Style solitary, ± equalling stigma. Capsule 4-6 mm, ovoid to obconical.

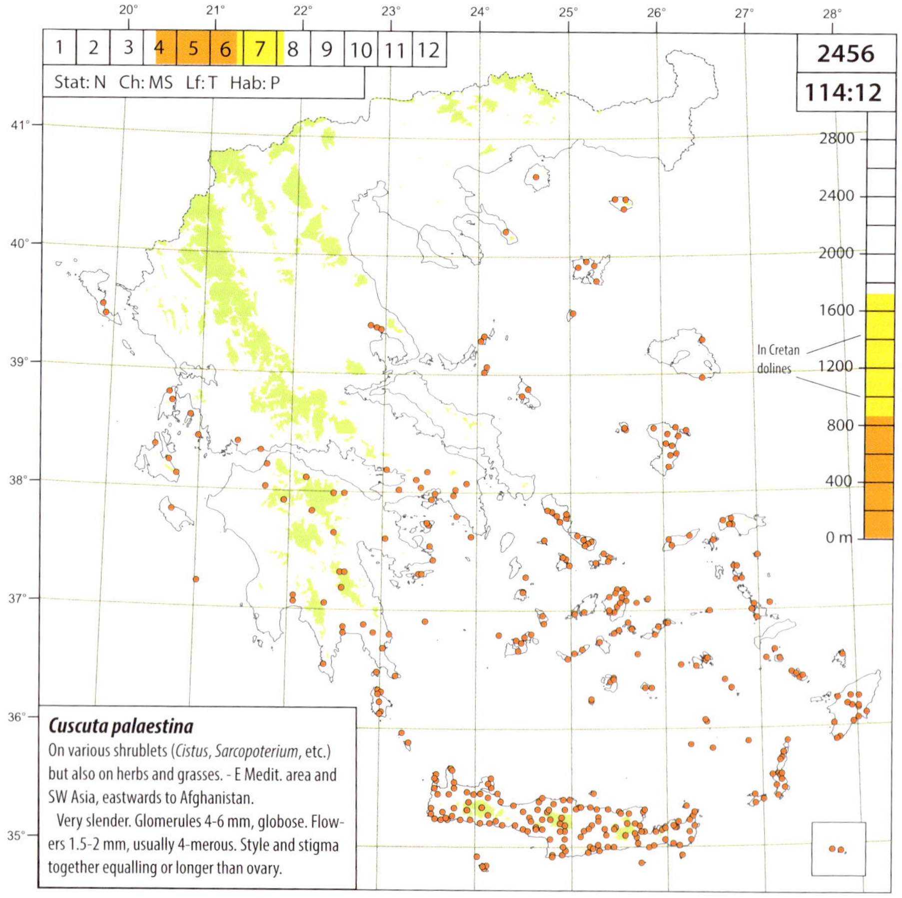

Cuscuta palaestina

On various shrublets (*Cistus*, *Sarcopoterium*, etc.) but also on herbs and grasses. - E Medit. area and SW Asia, eastwards to Afghanistan.

Very slender. Glomerules 4-6 mm, globose. Flowers 1.5-2 mm, usually 4-merous. Style and stigma together equalling or longer than ovary.

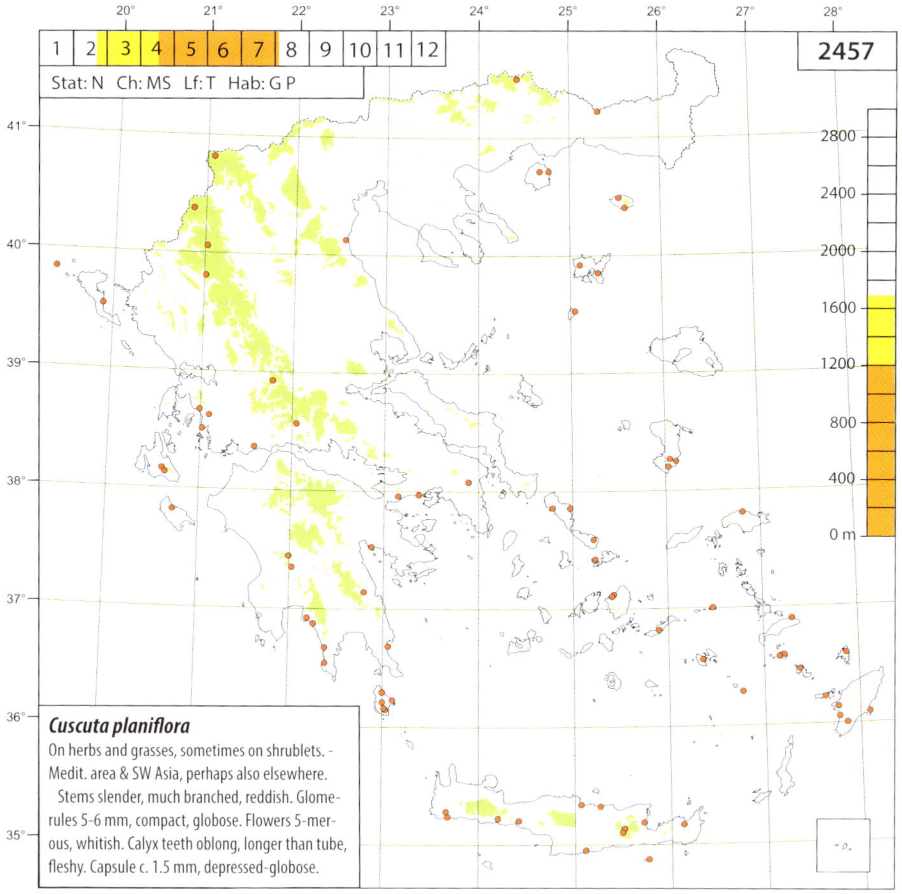

Cuscuta planiflora

On herbs and grasses, sometimes on shrublets. - Medit. area & SW Asia, perhaps also elsewhere.

Stems slender, much branched, reddish. Glomerules 5-6 mm, compact, globose. Flowers 5-merous, whitish. Calyx teeth oblong, longer than tube, fleshy. Capsule c. 1.5 mm, depressed-globose.

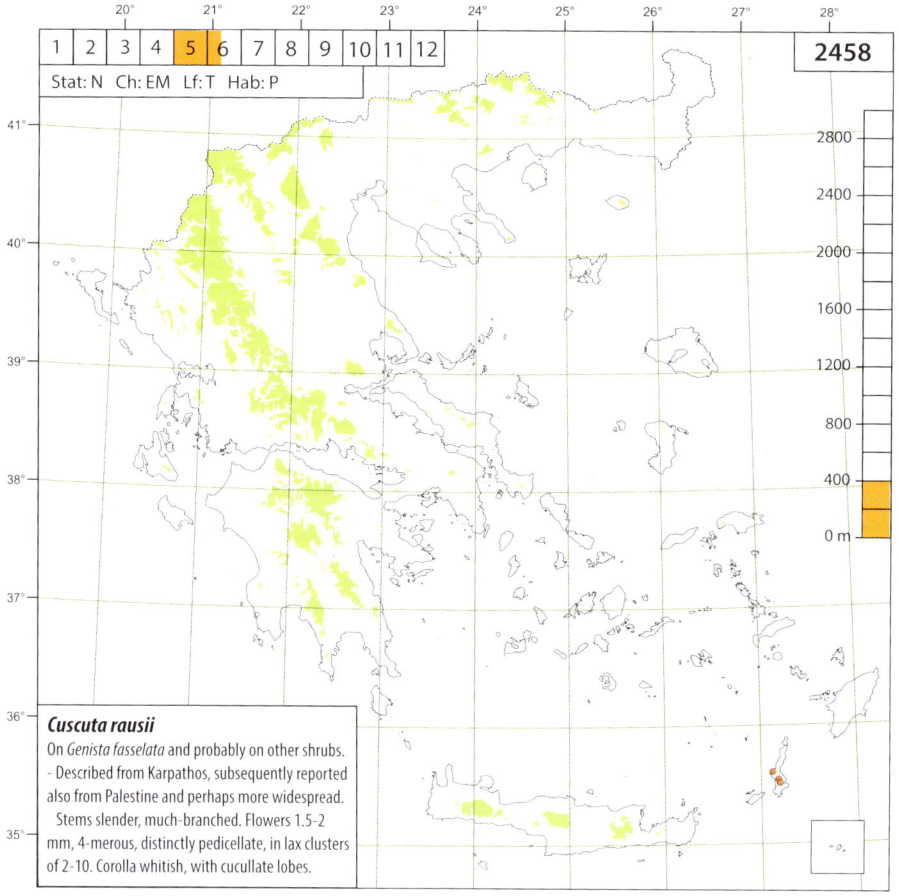

Cuscuta rausii

On *Genista fasselata* and probably on other shrubs. - Described from Karpathos, subsequently reported also from Palestine and perhaps more widespread.

Stems slender, much-branched. Flowers 1.5-2 mm, 4-merous, distinctly pedicellate, in lax clusters of 2-10. Corolla whitish, with cucullate lobes.

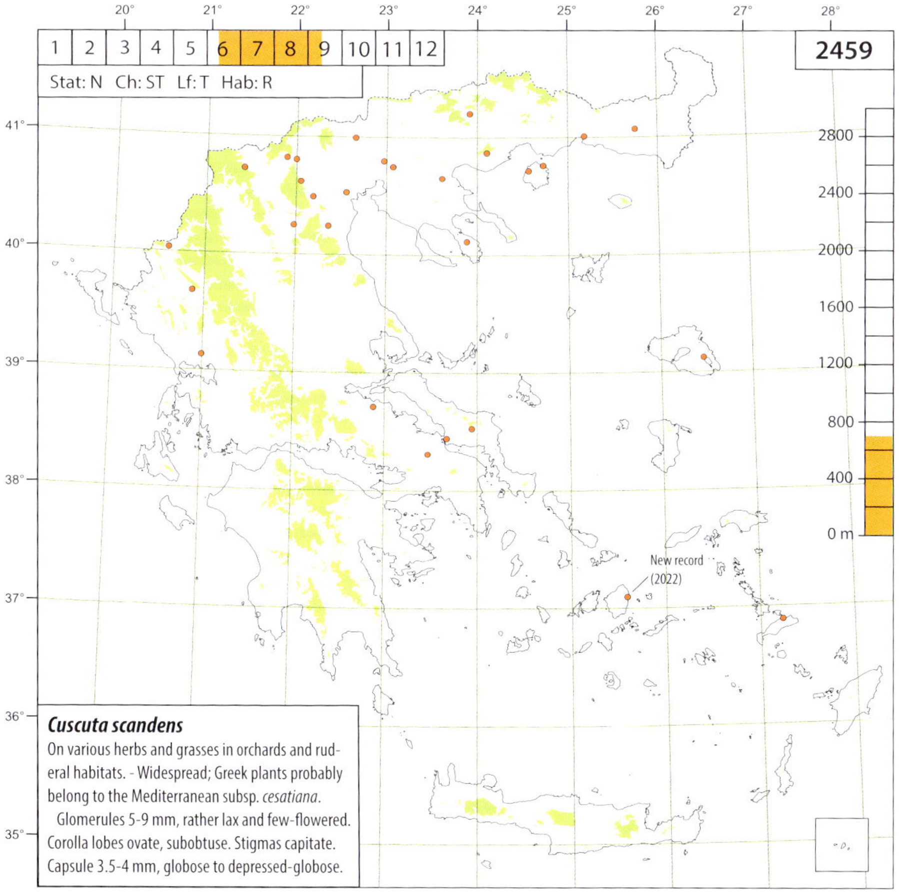

Cuscuta scandens

On various herbs and grasses in orchards and ruderal habitats. - Widespread; Greek plants probably belong to the Mediterranean subsp. *cesatiana*.

Glomerules 5-9 mm, rather lax and few-flowered. Corolla lobes ovate, subobtuse. Stigmas capitate. Capsule 3.5-4 mm, globose to depressed-globose.

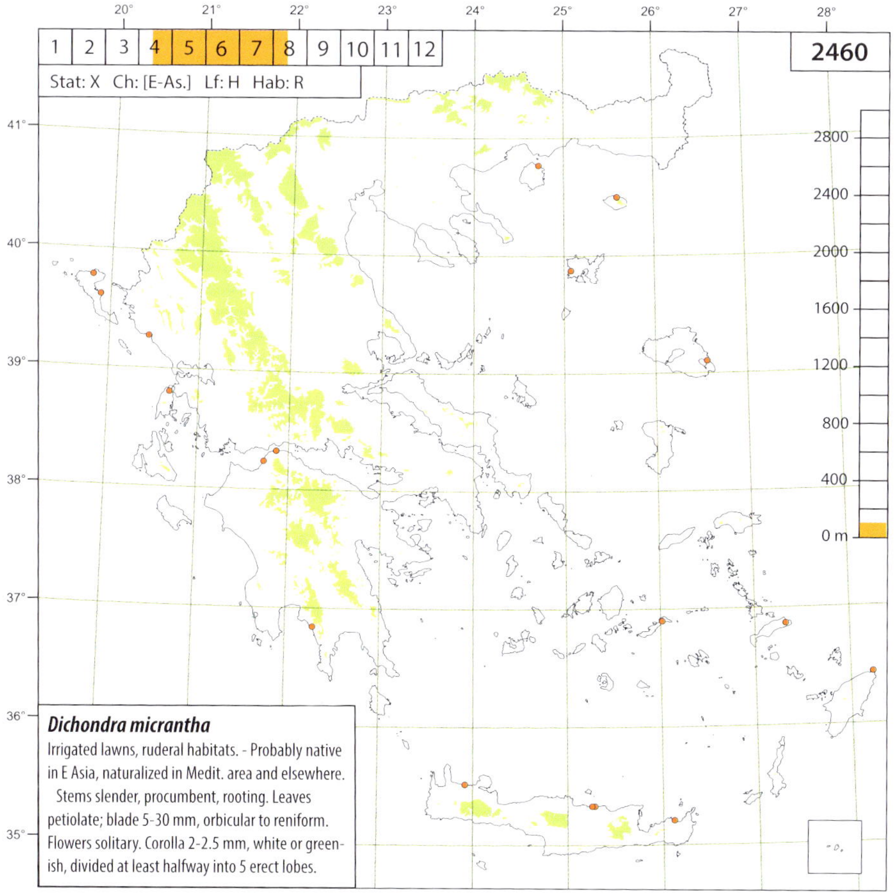

Dichondra micrantha

Irrigated lawns, ruderal habitats. - Probably native in E Asia, naturalized in Medit. area and elsewhere.

Stems slender, procumbent, rooting. Leaves petiolate; blade 5-30 mm, orbicular to reniform. Flowers solitary. Corolla 2-2.5 mm, white or greenish, divided at least halfway into 5 erect lobes.

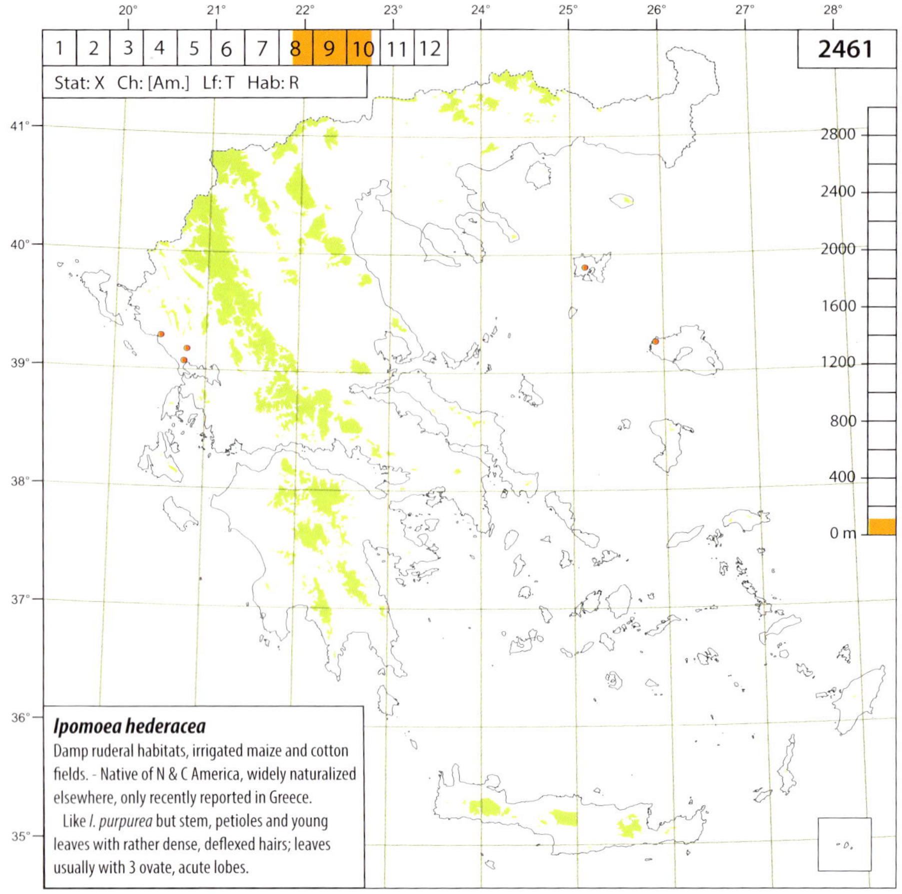

Ipomoea hederacea

Damp ruderal habitats, irrigated maize and cotton fields. - Native of N & C America, widely naturalized elsewhere, only recently reported in Greece.

Like *I. purpurea* but stem, petioles and young leaves with rather dense, deflexed hairs; leaves usually with 3 ovate, acute lobes.

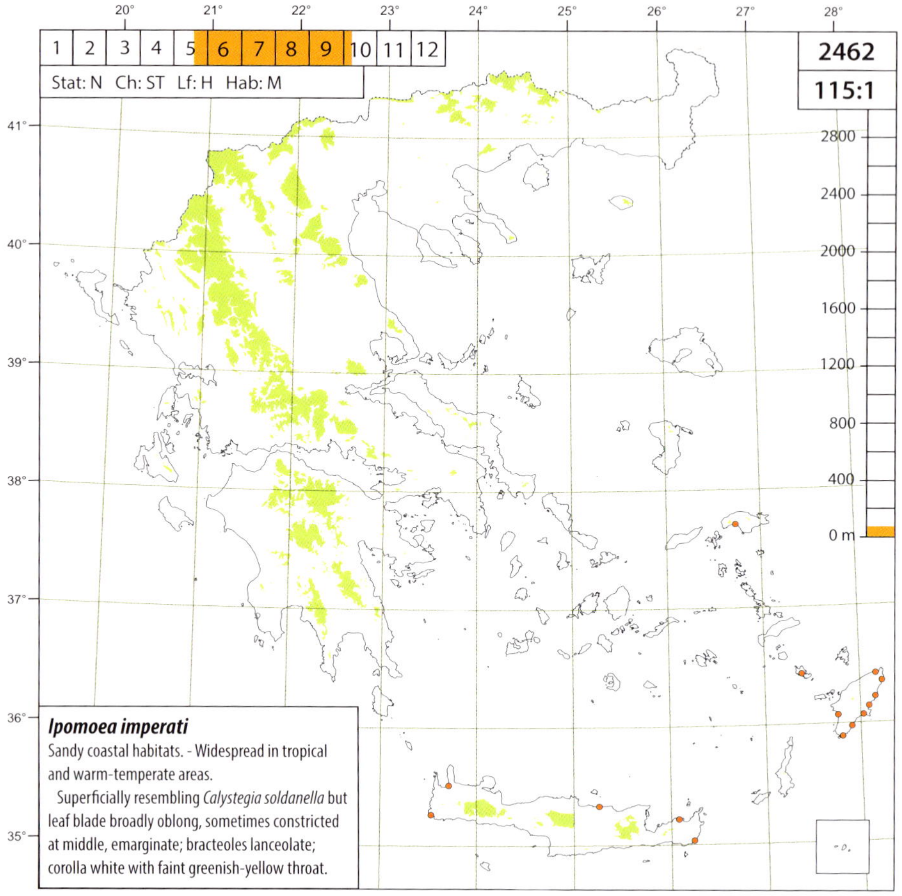

Ipomoea imperati

Sandy coastal habitats. - Widespread in tropical and warm-temperate areas.

Superficially resembling *Calystegia soldanella* but leaf blade broadly oblong, sometimes constricted at middle, emarginate; bracteoles lanceolate; corolla white with faint greenish-yellow throat.

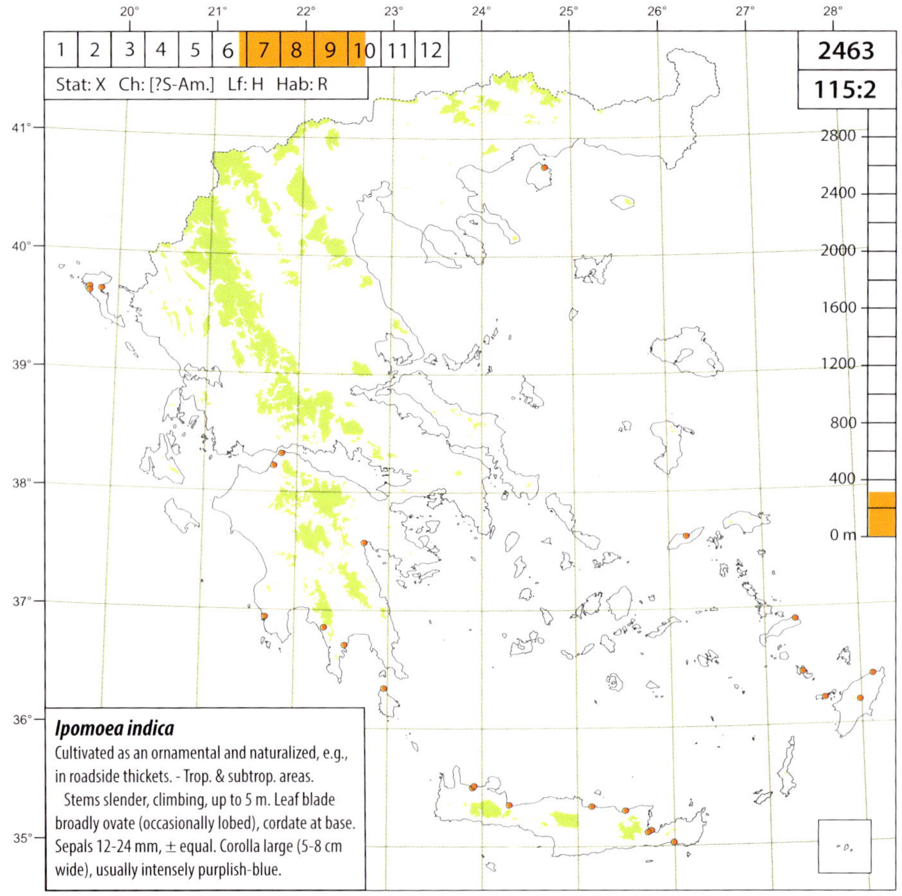

Ipomoea indica

Cultivated as an ornamental and naturalized, e.g., in roadside thickets. - Trop. & subtrop. areas.

Stems slender, climbing, up to 5 m. Leaf blade broadly ovate (occasionally lobed), cordate at base. Sepals 12-24 mm, ± equal. Corolla large (5-8 cm wide), usually intensely purplish-blue.

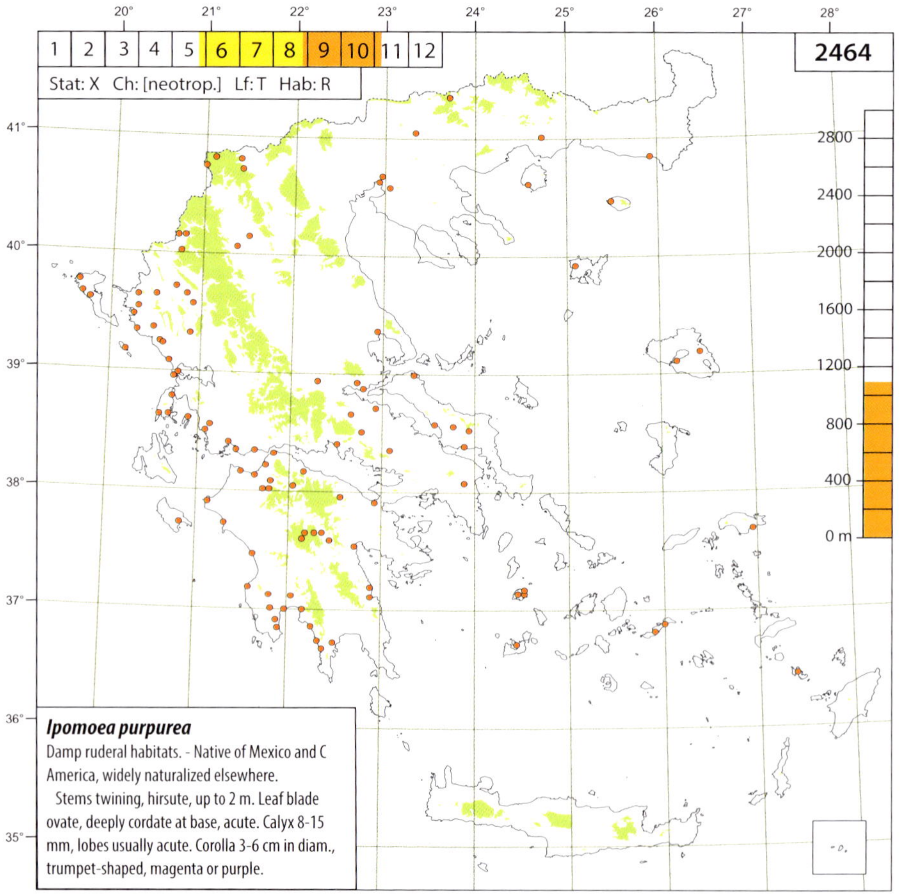

Ipomoea purpurea

Damp ruderal habitats. - Native of Mexico and C America, widely naturalized elsewhere.

Stems twining, hirsute, up to 2 m. Leaf blade ovate, deeply cordate at base, acute. Calyx 8-15 mm, lobes usually acute. Corolla 3-6 cm in diam., trumpet-shaped, magenta or purple.

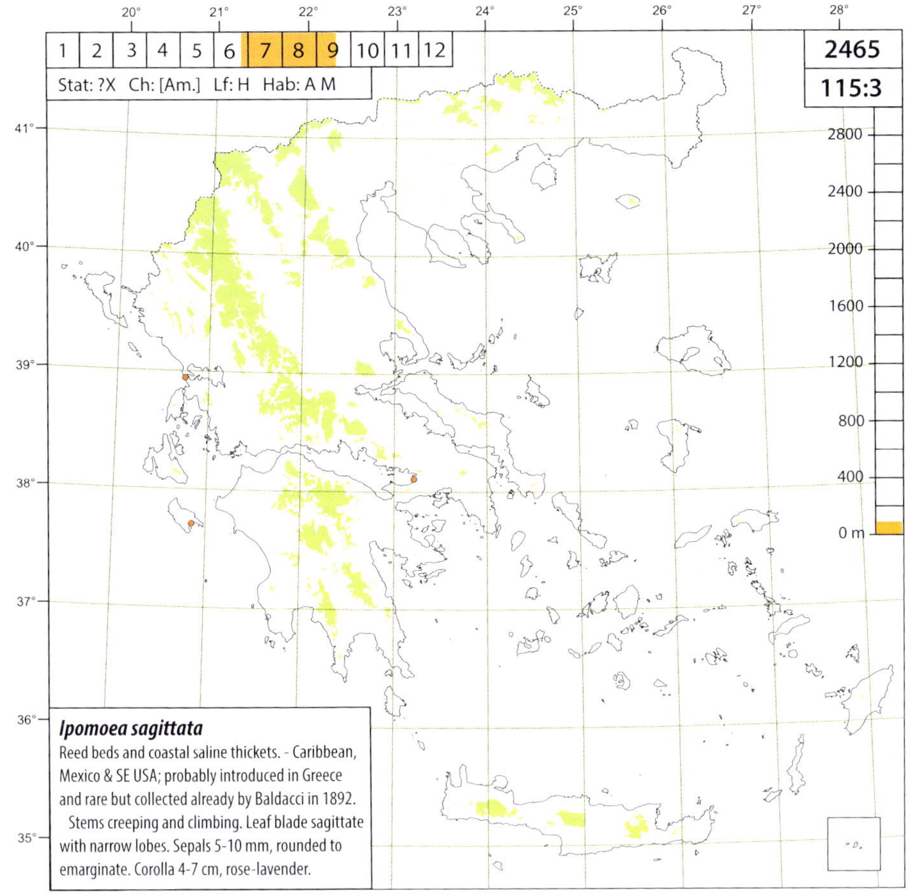

Ipomoea sagittata

Reed beds and coastal saline thickets. - Caribbean, Mexico & SE USA; probably introduced in Greece and rare but collected already by Baldacci in 1892.

Stems creeping and climbing. Leaf blade sagittate with narrow lobes. Sepals 5-10 mm, rounded to emarginate. Corolla 4-7 cm, rose-lavender.

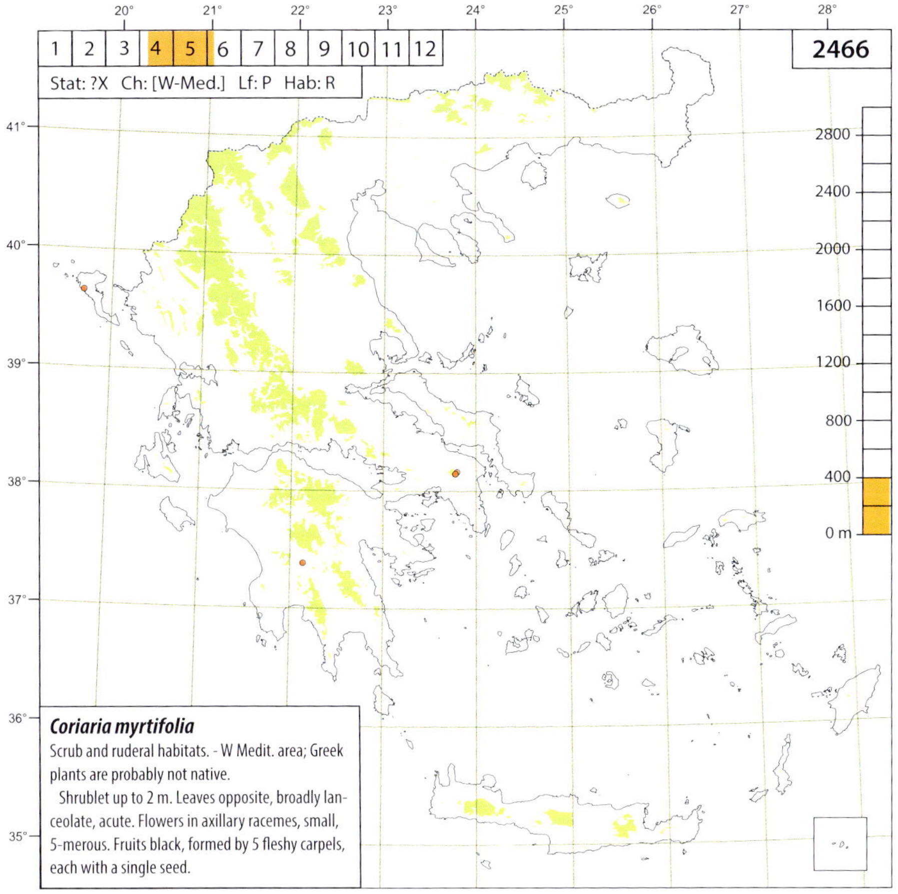

Coriaria myrtifolia

Scrub and ruderal habitats. - W Medit. area; Greek plants are probably not native.

Shrublet up to 2 m. Leaves opposite, broadly lanceolate, acute. Flowers in axillary racemes, small, 5-merous. Fruits black, formed by 5 fleshy carpels, each with a single seed.

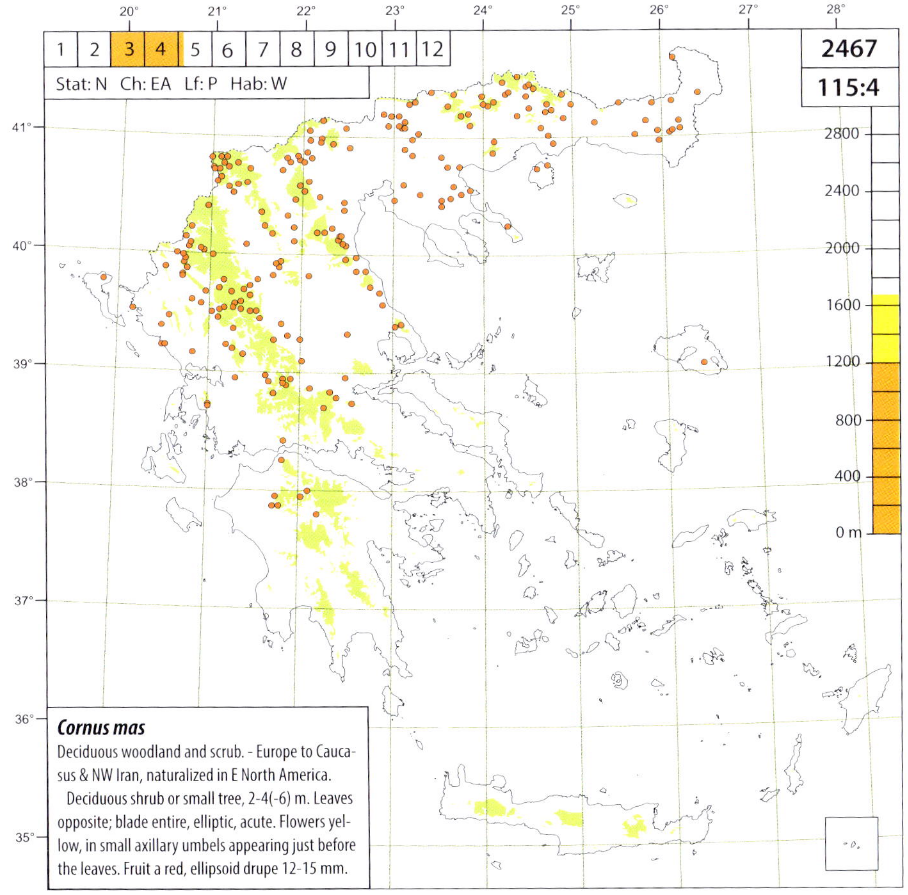

Cornus mas

Deciduous woodland and scrub. - Europe to Caucasus & NW Iran, naturalized in E North America.

Deciduous shrub or small tree, 2-4(-6) m. Leaves opposite; blade entire, elliptic, acute. Flowers yellow, in small axillary umbels appearing just before the leaves. Fruit a red, ellipsoid drupe 12-15 mm.

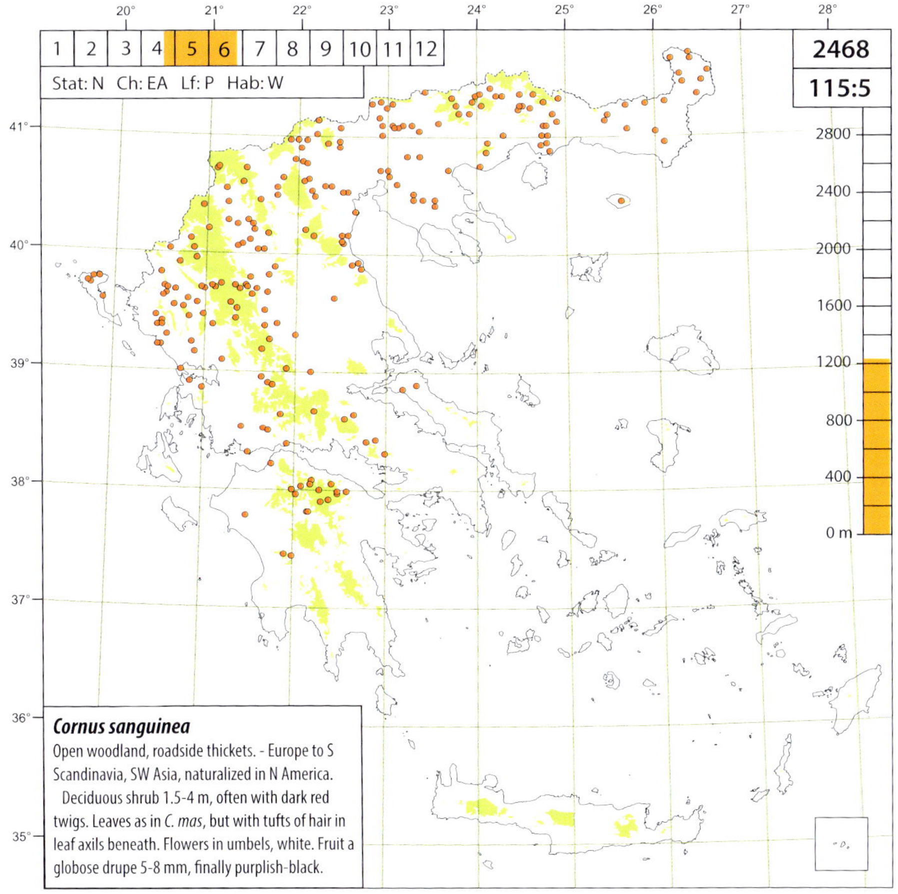

Cornus sanguinea

Open woodland, roadside thickets. - Europe to S Scandinavia, SW Asia, naturalized in N America.

Deciduous shrub 1.5-4 m, often with dark red twigs. Leaves as in *C. mas*, but with tufts of hair in leaf axils beneath. Flowers in umbels, white. Fruit a globose drupe 5-8 mm, finally purplish-black.

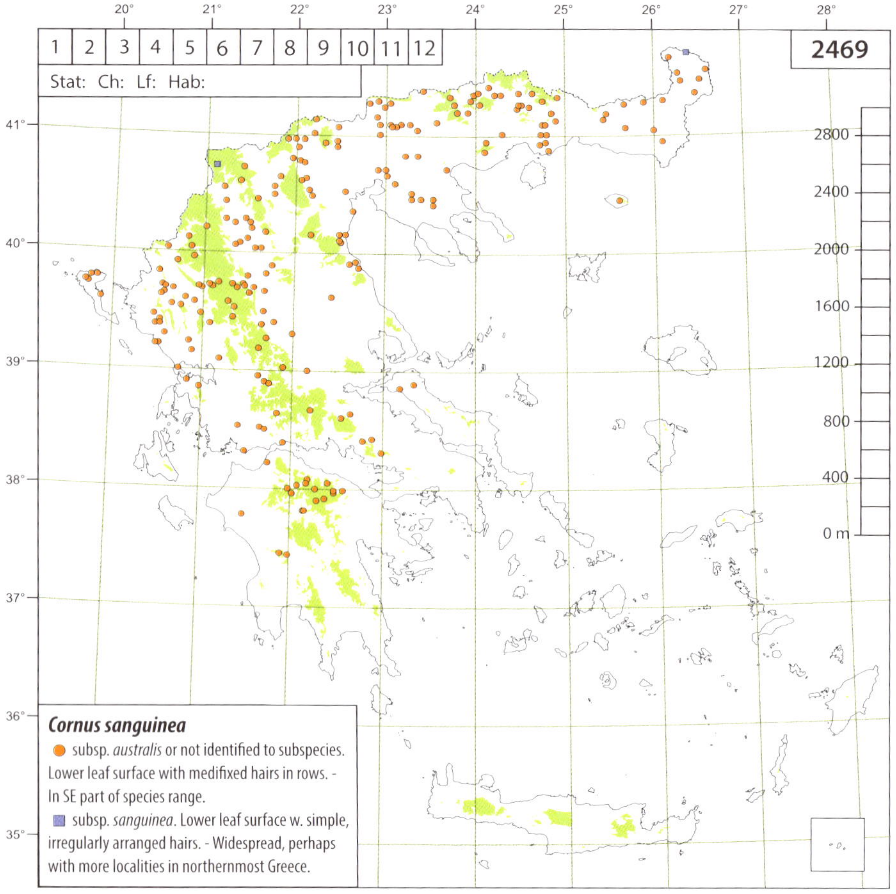

Cornus sanguinea

● subsp. *australis* or not identified to subspecies. Lower leaf surface with medifixed hairs in rows. - In SE part of species range.

■ subsp. *sanguinea*. Lower leaf surface w. simple, irregularly arranged hairs. - Widespread, perhaps with more localities in northernmost Greece.

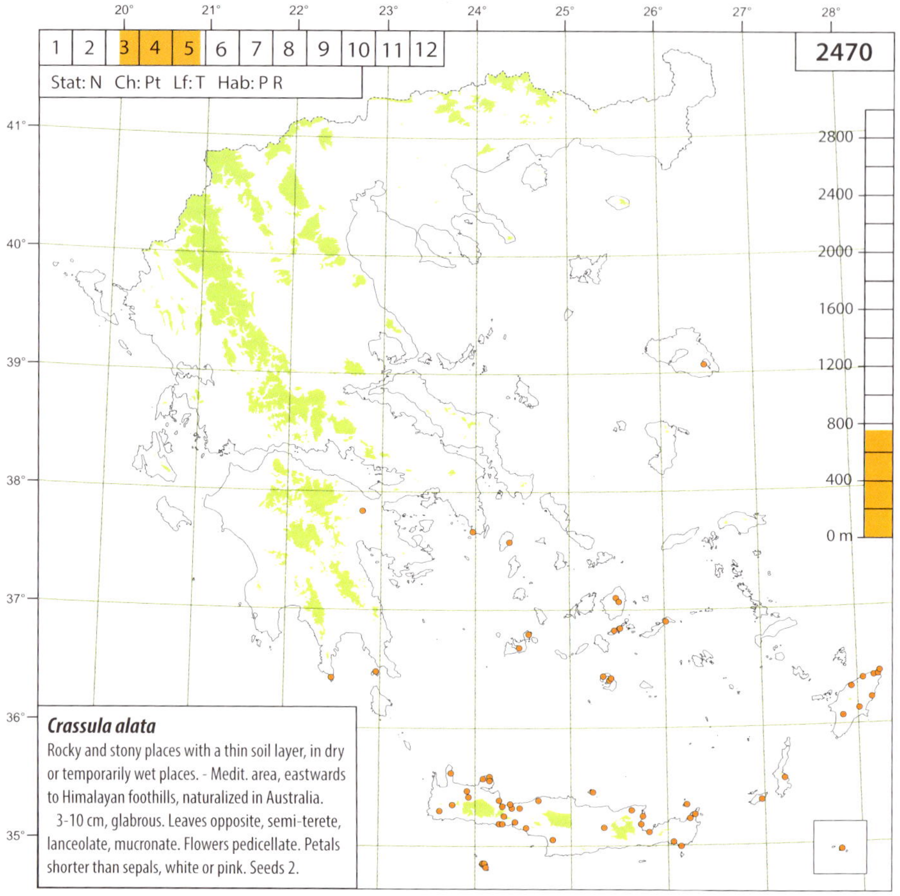

Crassula alata

Rocky and stony places with a thin soil layer, in dry or temporarily wet places. - Medit. area, eastwards to Himalayan foothills, naturalized in Australia.

3-10 cm, glabrous. Leaves opposite, semi-terete, lanceolate, mucronate. Flowers pedicellate. Petals shorter than sepals, white or pink. Seeds 2.

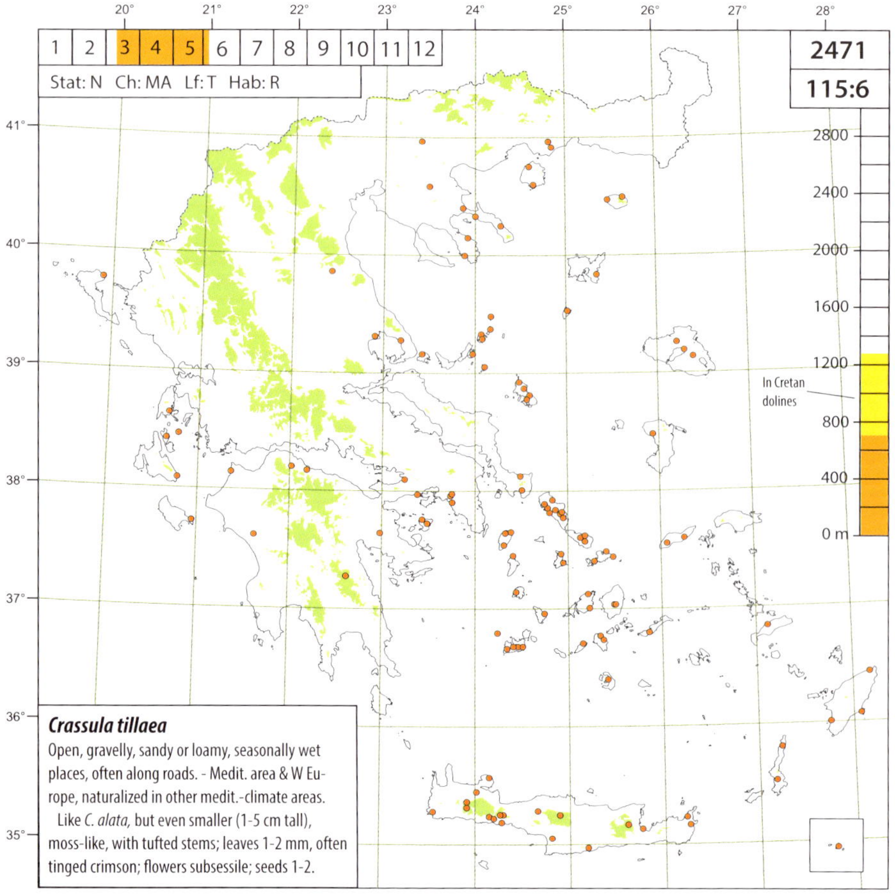

Crassula tillaea

Open, gravelly, sandy or loamy, seasonally wet places, often along roads. - Medit. area & W Europe, naturalized in other medit.-climate areas.

Like *C. alata*, but even smaller (1-5 cm tall), moss-like, with tufted stems; leaves 1-2 mm, often tinged crimson; flowers subsessile; seeds 1-2.

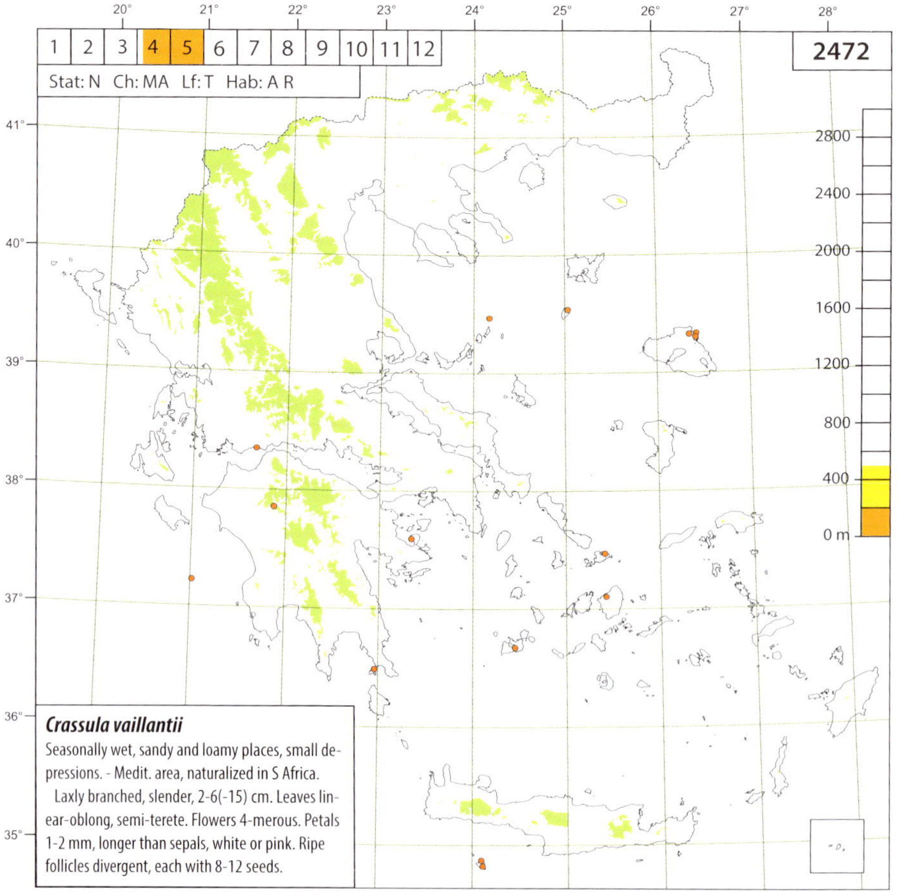

Crassula vaillantii

Seasonally wet, sandy and loamy places, small depressions. - Medit. area, naturalized in S Africa.

Laxly branched, slender, 2-6(-15) cm. Leaves linear-oblong, semi-terete. Flowers 4-merous. Petals 1-2 mm, longer than sepals, white or pink. Ripe follicles divergent, each with 8-12 seeds.

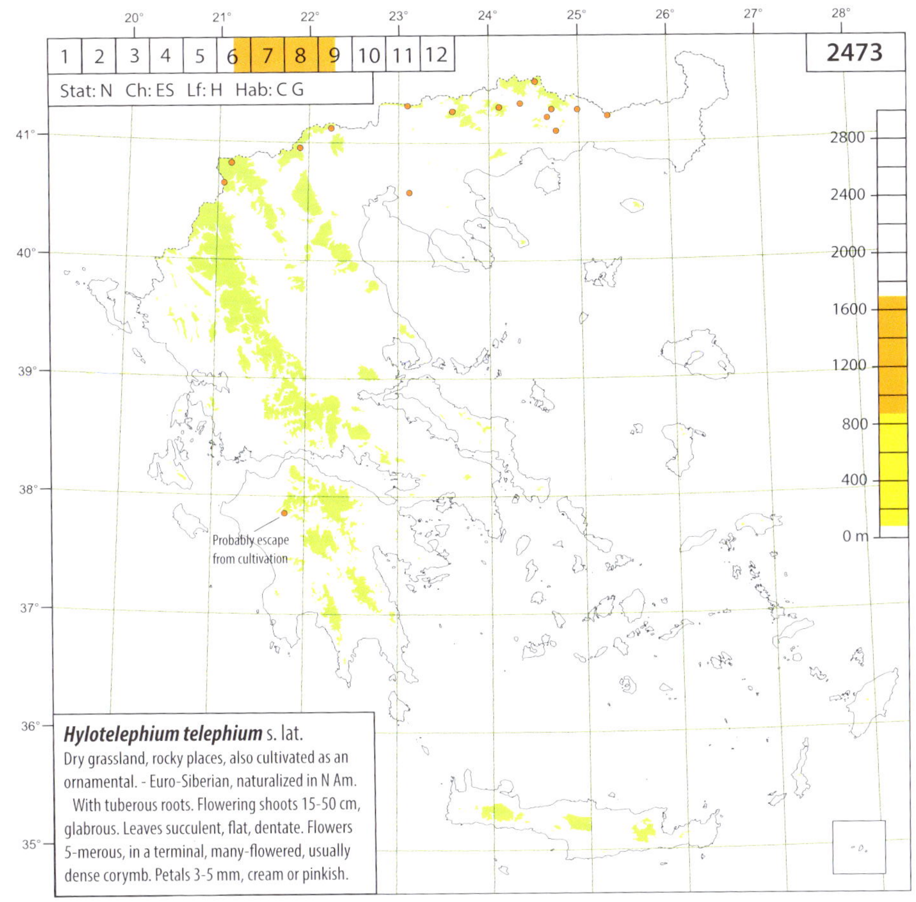

Hylotelephium telephium s. lat.
Dry grassland, rocky places, also cultivated as an ornamental. - Euro-Siberian, naturalized in N Am.
With tuberous roots. Flowering shoots 15-50 cm, glabrous. Leaves succulent, flat, dentate. Flowers 5-merous, in a terminal, many-flowered, usually dense corymb. Petals 3-5 mm, cream or pinkish.

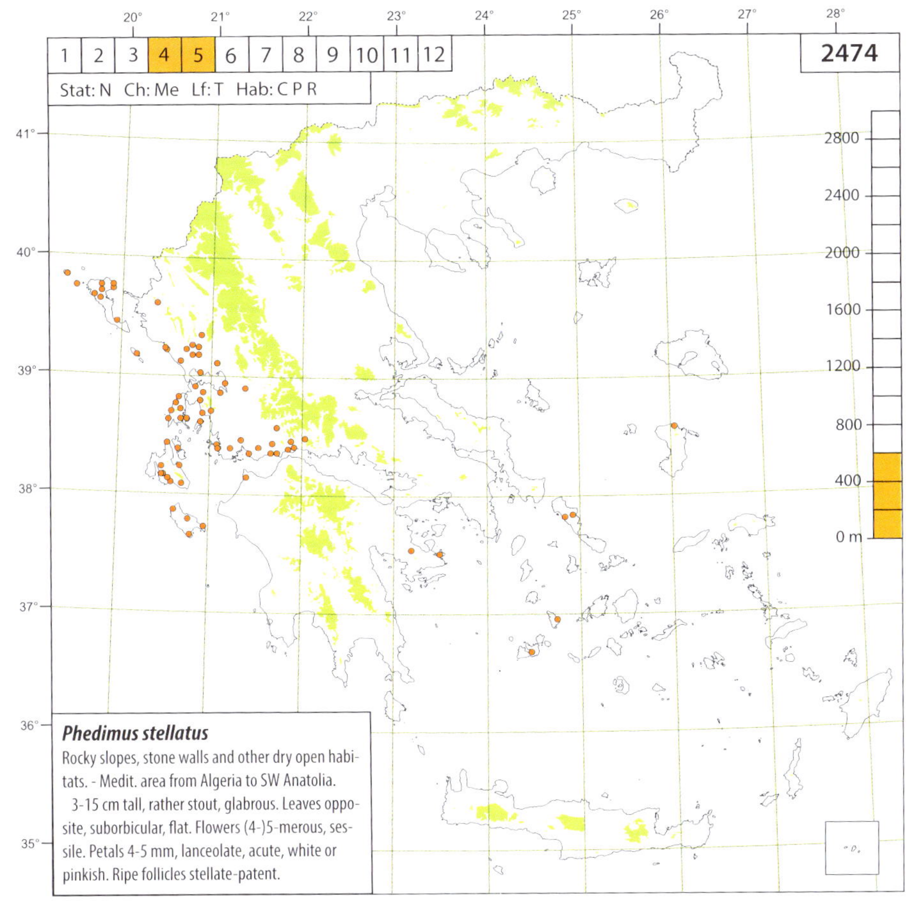

Phedimus stellatus
Rocky slopes, stone walls and other dry open habitats. - Medit. area from Algeria to SW Anatolia.
3-15 cm tall, rather stout, glabrous. Leaves opposite, suborbicular, flat. Flowers (4-)5-merous, sessile. Petals 4-5 mm, lanceolate, acute, white or pinkish. Ripe follicles stellate-patent.

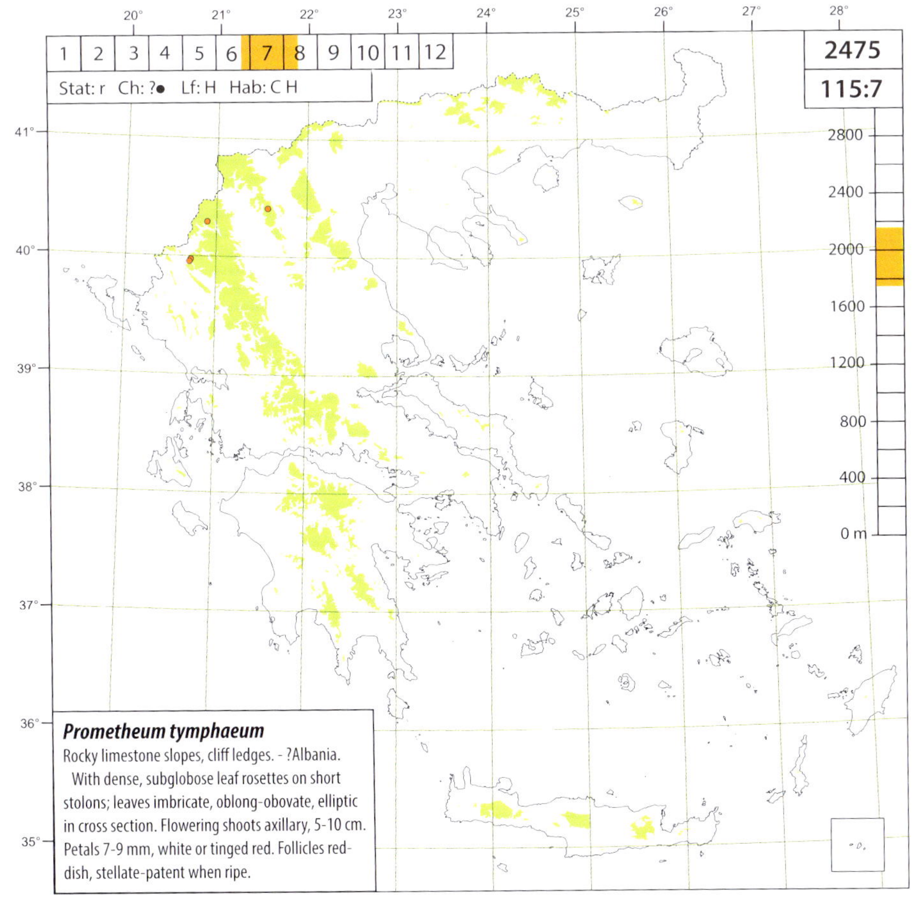

Prometheum tymphaeum
Rocky limestone slopes, cliff ledges. - ?Albania.
With dense, subglobose leaf rosettes on short stolons; leaves imbricate, oblong-obovate, elliptic in cross section. Flowering shoots axillary, 5-10 cm. Petals 7-9 mm, white or tinged red. Follicles reddish, stellate-patent when ripe.

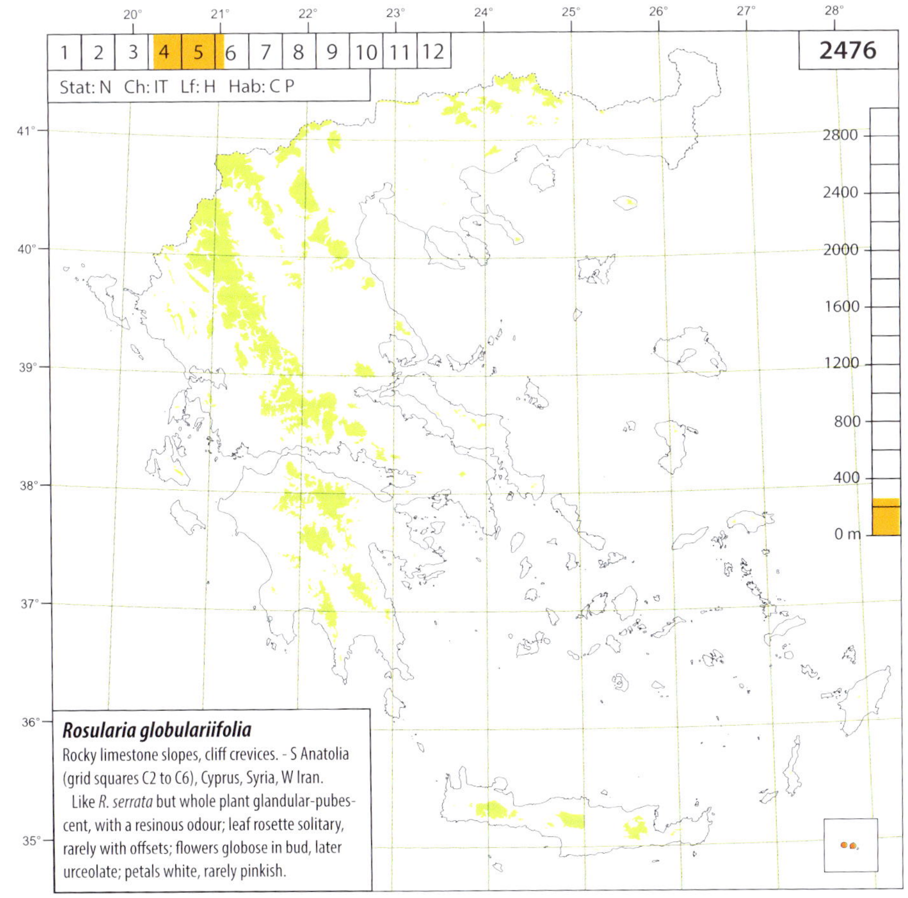

Rosularia globulariifolia
Rocky limestone slopes, cliff crevices. - S Anatolia (grid squares C2 to C6), Cyprus, Syria, W Iran.
Like *R. serrata* but whole plant glandular-pubescent, with a resinous odour; leaf rosette solitary, rarely with offsets; flowers globose in bud, later urceolate; petals white, rarely pinkish.

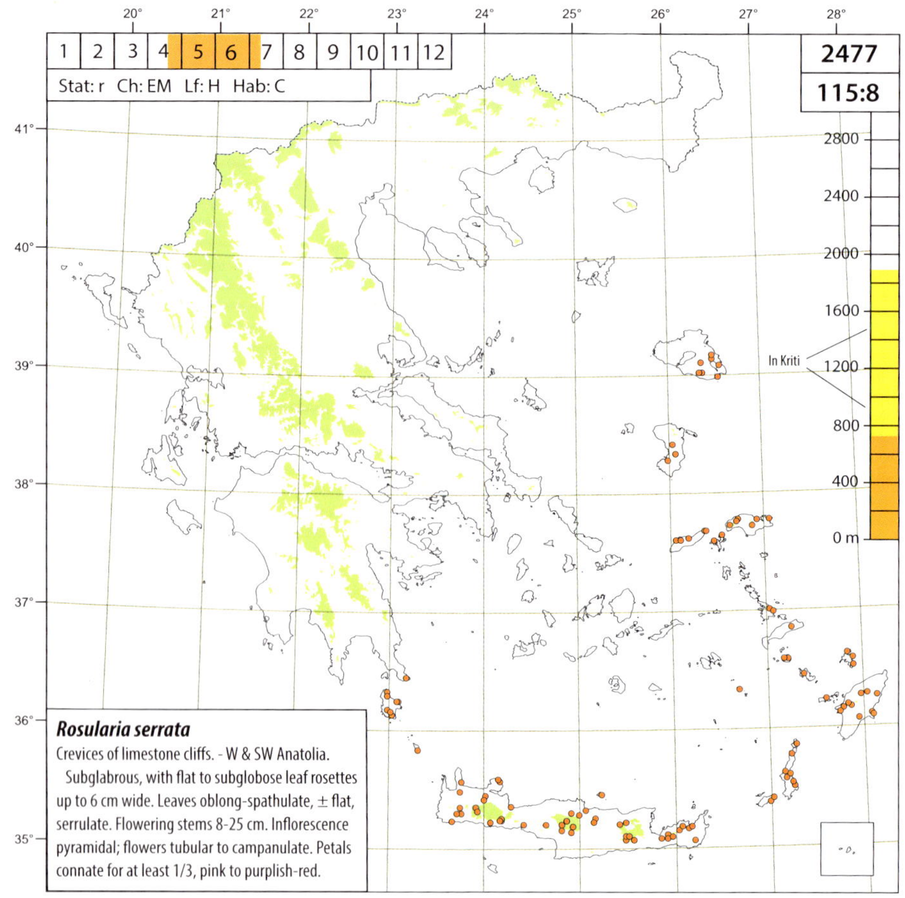

Rosularia serrata

Crevices of limestone cliffs. - W & SW Anatolia.

Subglabrous, with flat to subglobose leaf rosettes up to 6 cm wide. Leaves oblong-spathulate, ± flat, serrulate. Flowering stems 8-25 cm. Inflorescence pyramidal; flowers tubular to campanulate. Petals connate for at least 1/3, pink to purplish-red.

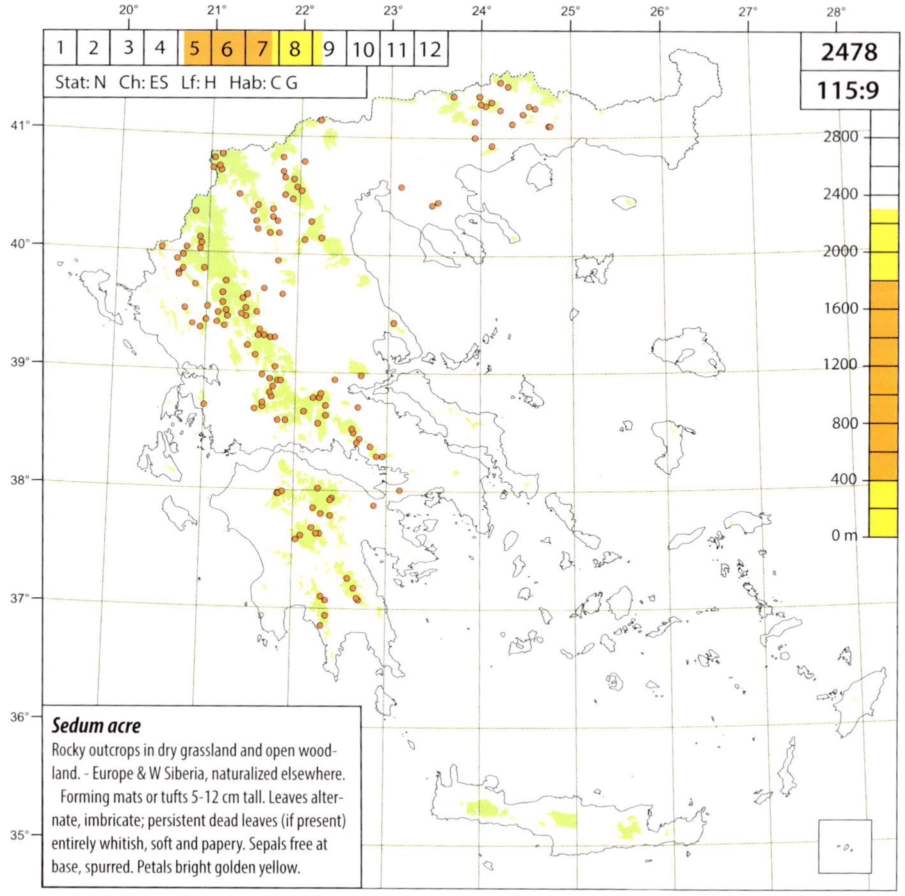

Sedum acre

Rocky outcrops in dry grassland and open woodland. - Europe & W Siberia, naturalized elsewhere.

Forming mats or tufts 5-12 cm tall. Leaves alternate, imbricate; persistent dead leaves (if present) entirely whitish, soft and papery. Sepals free at base, spurred. Petals bright golden yellow.

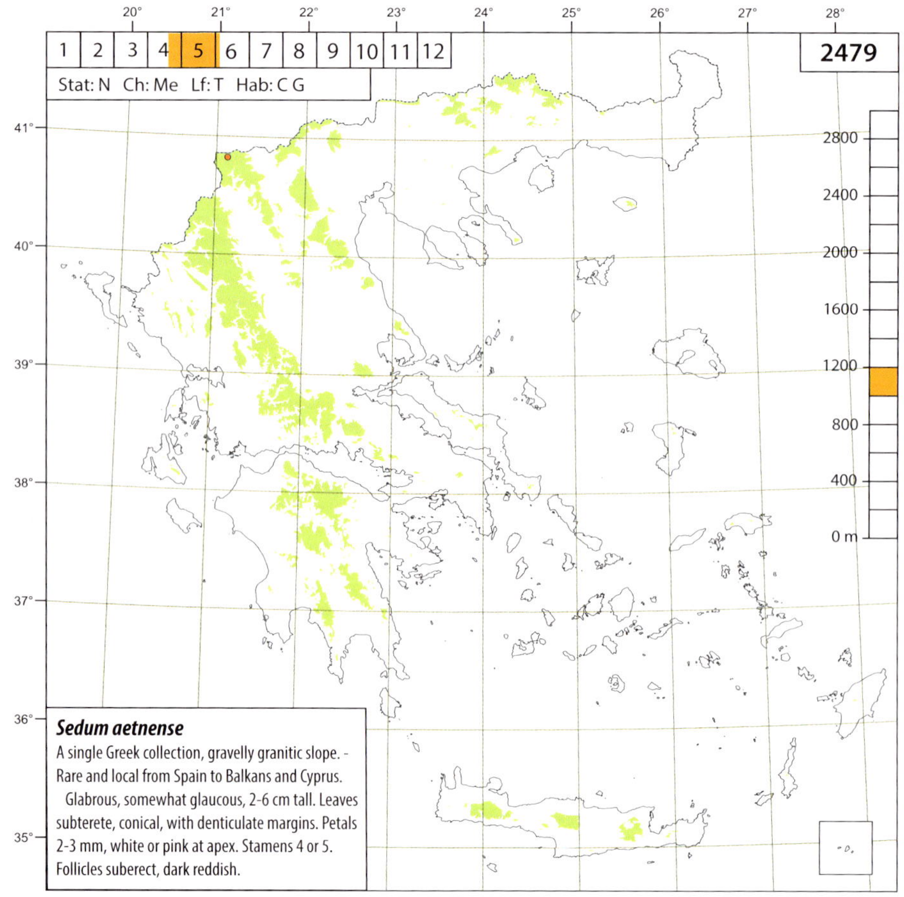

Sedum aetnense

A single Greek collection, gravelly granitic slope. - Rare and local from Spain to Balkans and Cyprus.

Glabrous, somewhat glaucous, 2-6 cm tall. Leaves subterete, conical, with denticulate margins. Petals 2-3 mm, white or pink at apex. Stamens 4 or 5. Follicles suberect, dark reddish.

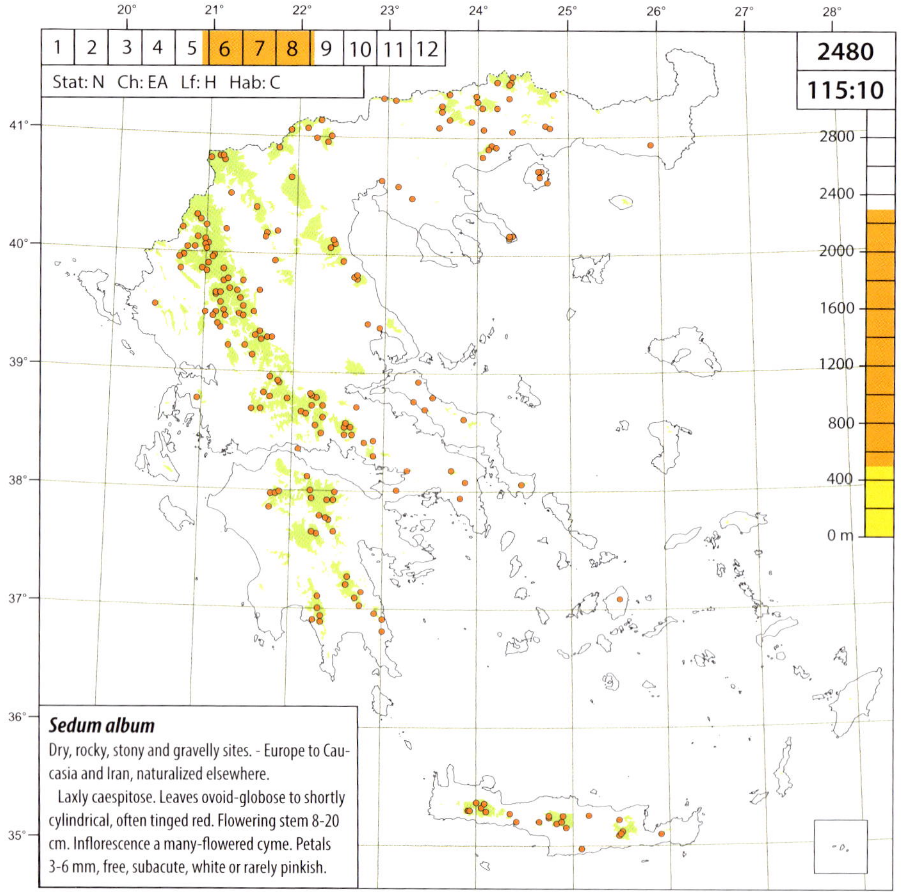

Sedum album

Dry, rocky, stony and gravelly sites. - Europe to Caucasia and Iran, naturalized elsewhere.

Laxly caespitose. Leaves ovoid-globose to shortly cylindrical, often tinged red. Flowering stem 8-20 cm. Inflorescence a many-flowered cyme. Petals 3-6 mm, free, subacute, white or rarely pinkish.

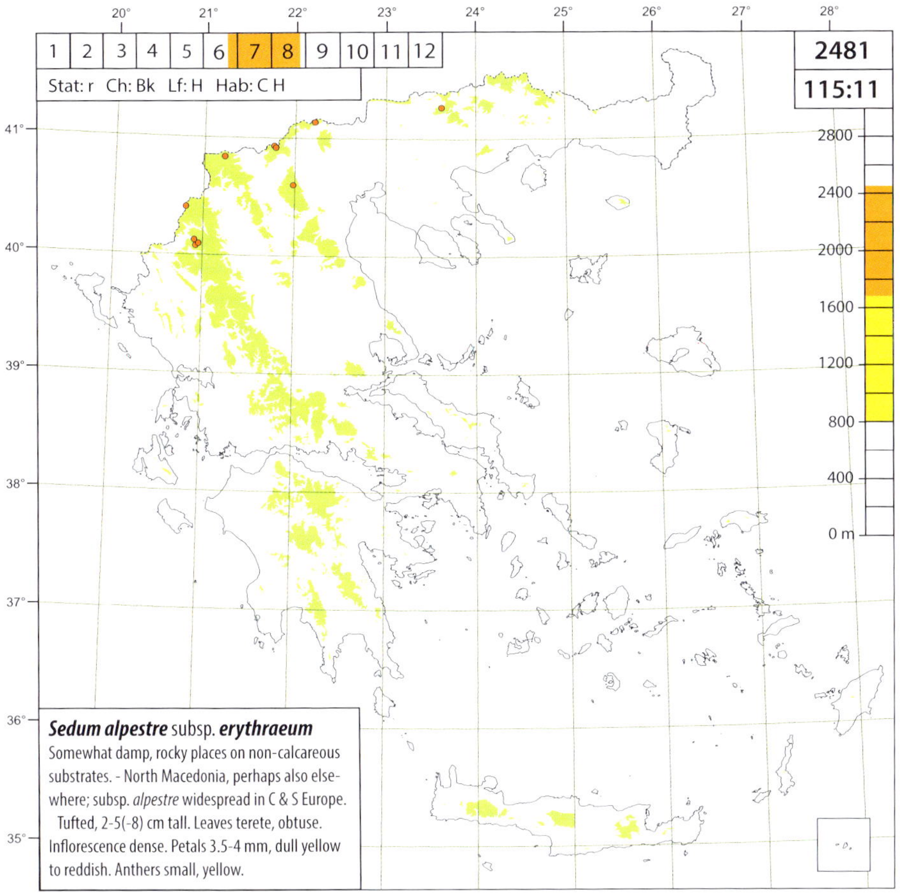

Sedum alpestre subsp. **erythraeum**
Somewhat damp, rocky places on non-calcareous substrates. - North Macedonia, perhaps also elsewhere; subsp. *alpestre* widespread in C & S Europe.
Tufted, 2-5(-8) cm tall. Leaves terete, obtuse. Inflorescence dense. Petals 3.5-4 mm, dull yellow to reddish. Anthers small, yellow.

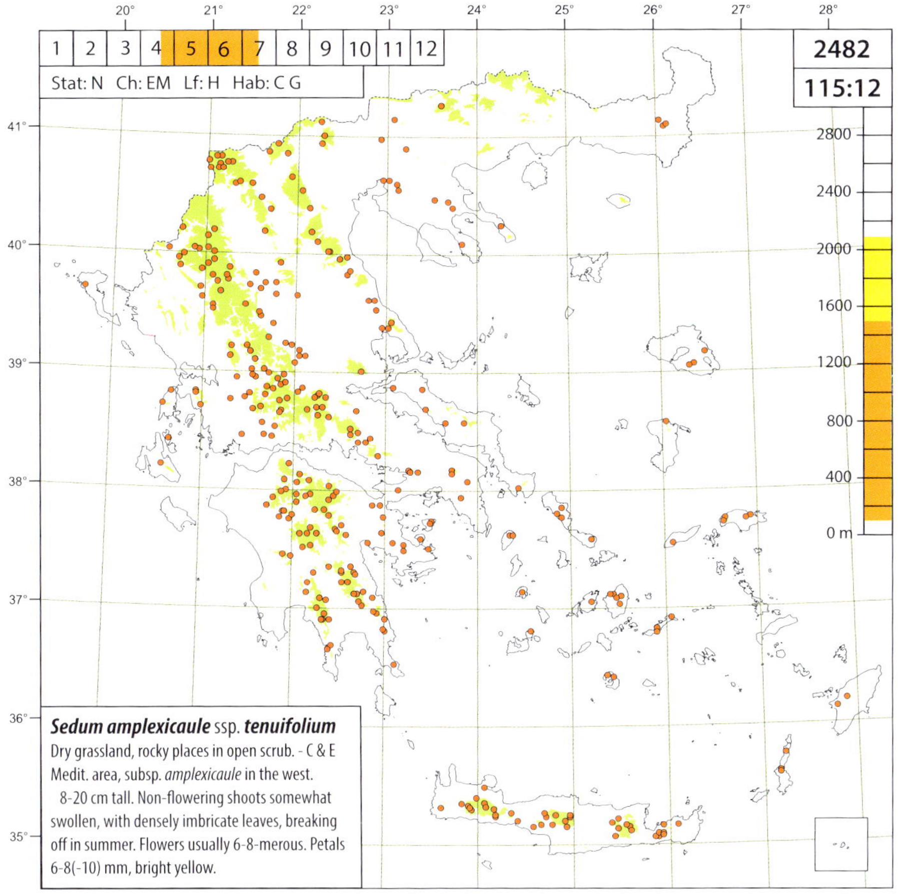

Sedum amplexicaule ssp. **tenuifolium**
Dry grassland, rocky places in open scrub. - C & E Medit. area, subsp. *amplexicaule* in the west.
8-20 cm tall. Non-flowering shoots somewhat swollen, with densely imbricate leaves, breaking off in summer. Flowers usually 6-8-merous. Petals 6-8(-10) mm, bright yellow.

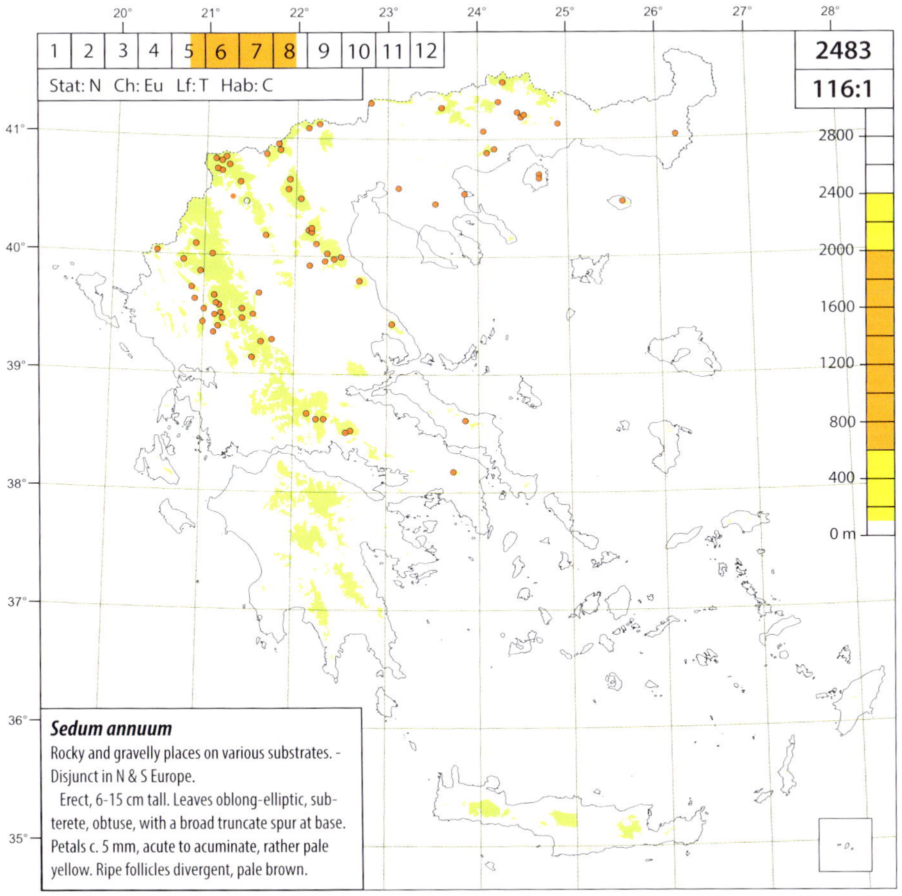

Sedum annuum
Rocky and gravelly places on various substrates. - Disjunct in N & S Europe.
Erect, 6-15 cm tall. Leaves oblong-elliptic, subterete, obtuse, with a broad truncate spur at base. Petals c. 5 mm, acute to acuminate, rather pale yellow. Ripe follicles divergent, pale brown.

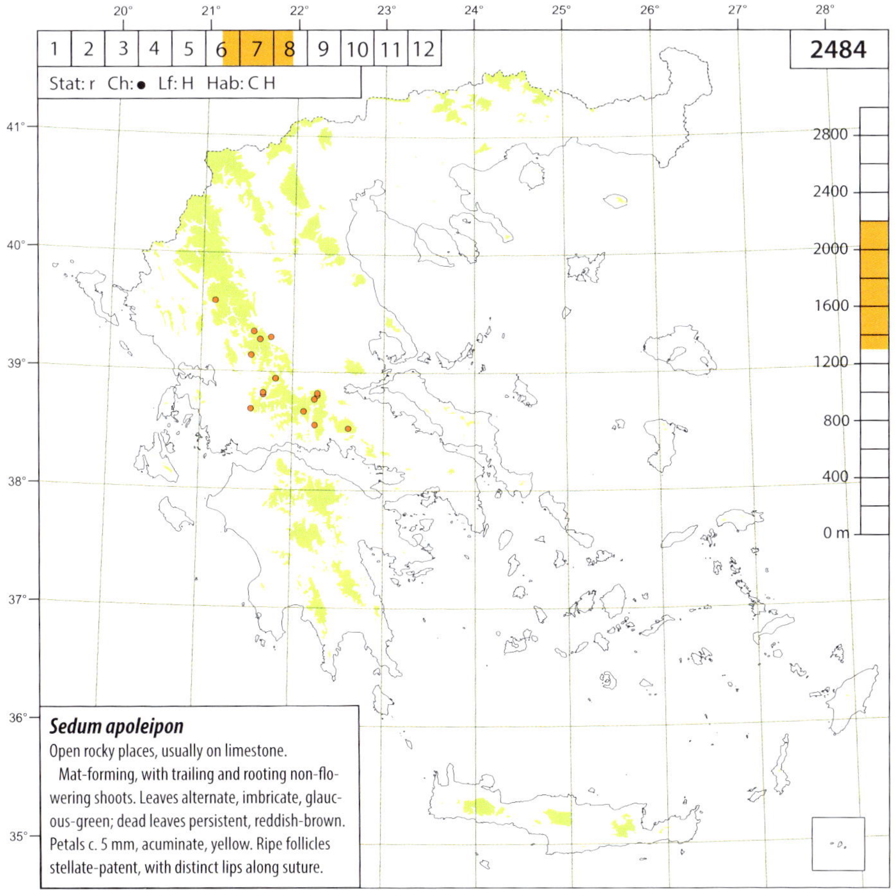

Sedum apoleipon
Open rocky places, usually on limestone.
Mat-forming, with trailing and rooting non-flowering shoots. Leaves alternate, imbricate, glaucous-green; dead leaves persistent, reddish-brown. Petals c. 5 mm, acuminate, yellow. Ripe follicles stellate-patent, with distinct lips along suture.

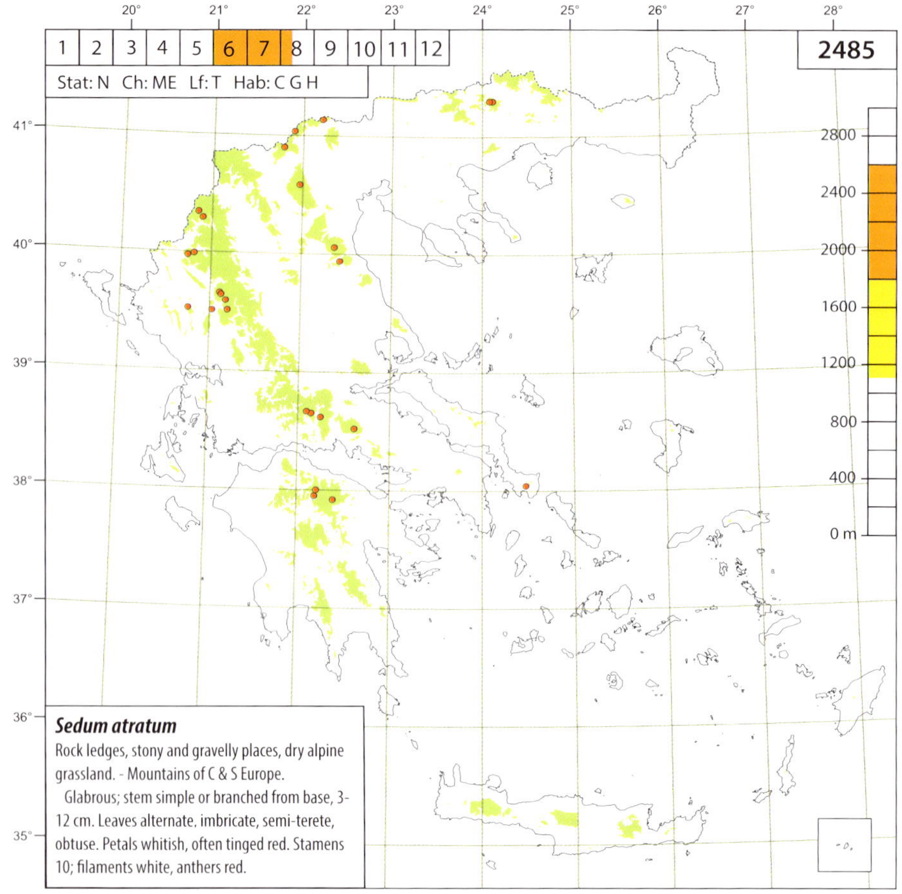

Sedum atratum

Rock ledges, stony and gravelly places, dry alpine grassland. - Mountains of C & S Europe.

Glabrous; stem simple or branched from base, 3-12 cm. Leaves alternate. imbricate, semi-terete, obtuse. Petals whitish, often tinged red. Stamens 10; filaments white, anthers red.

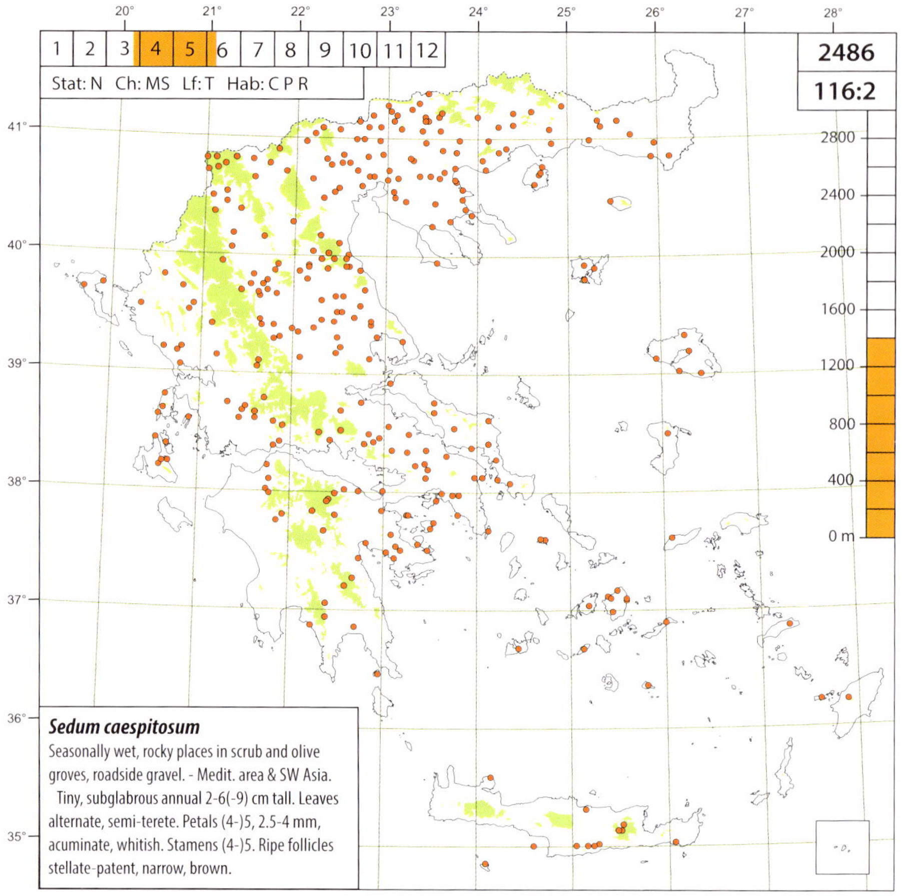

Sedum caespitosum

Seasonally wet, rocky places in scrub and olive groves, roadside gravel. - Medit. area & SW Asia.

Tiny, subglabrous annual 2-6(-9) cm tall. Leaves alternate, semi-terete. Petals (4-)5, 2.5-4 mm, acuminate, whitish. Stamens (4-)5. Ripe follicles stellate-patent, narrow, brown.

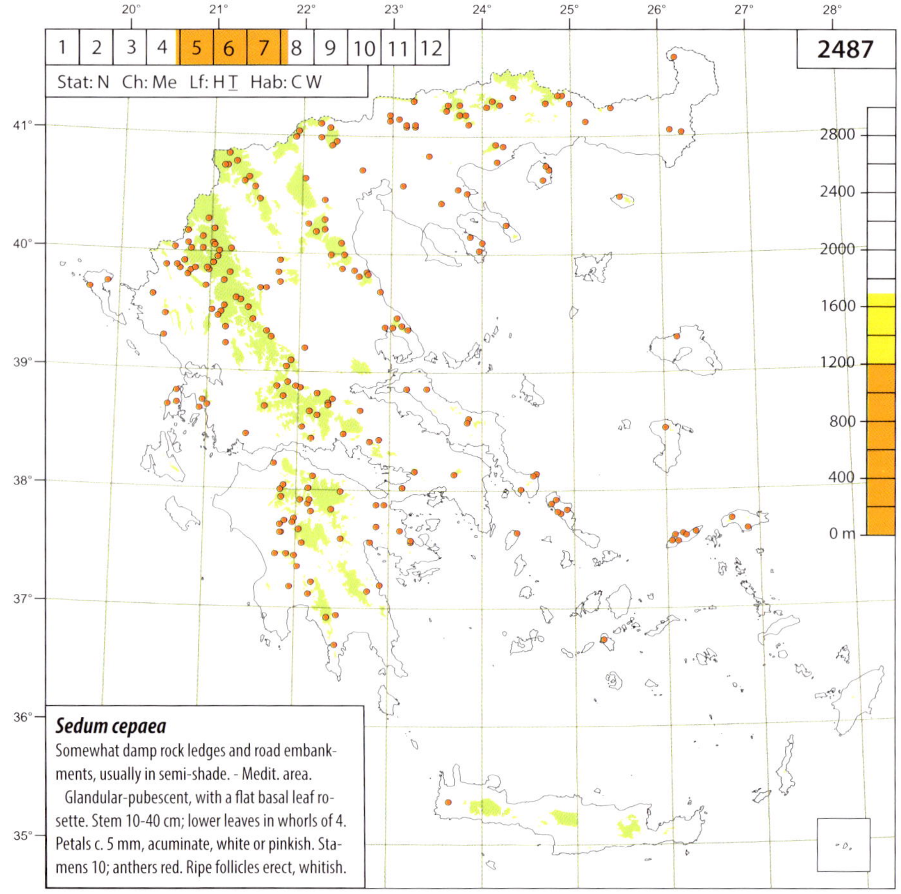

Sedum cepaea

Somewhat damp rock ledges and road embankments, usually in semi-shade. - Medit. area.

Glandular-pubescent, with a flat basal leaf rosette. Stem 10-40 cm; lower leaves in whorls of 4. Petals c. 5 mm, acuminate, white or pinkish. Stamens 10; anthers red. Ripe follicles erect, whitish.

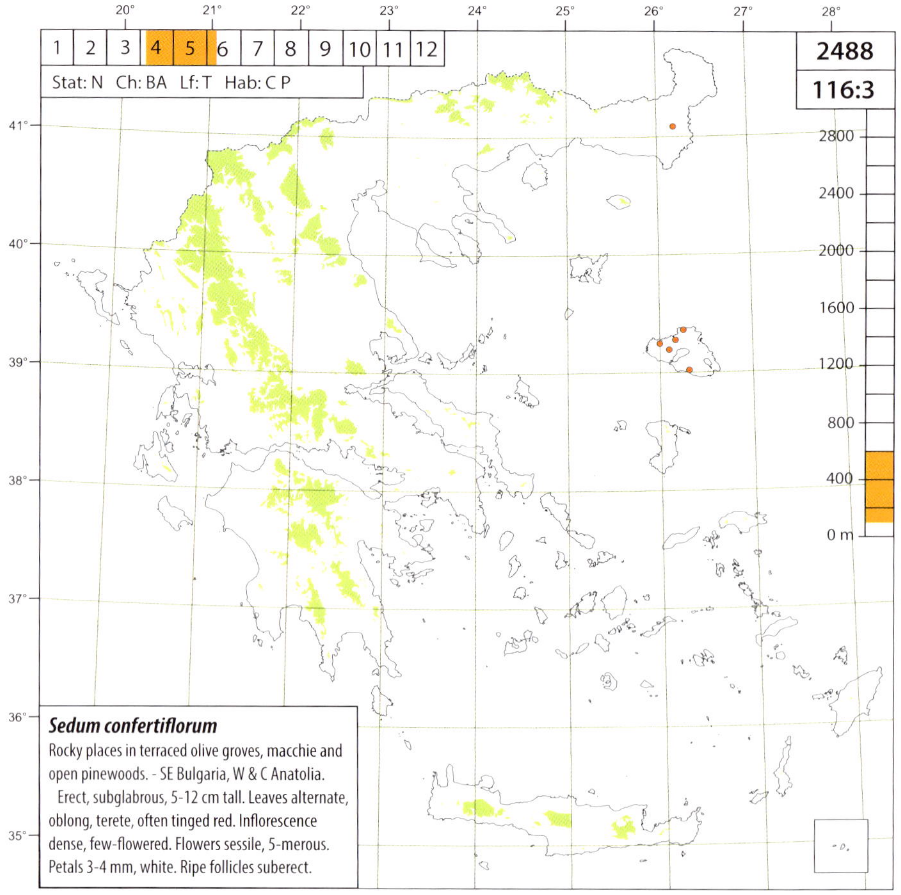

Sedum confertiflorum

Rocky places in terraced olive groves, macchie and open pinewoods. - SE Bulgaria, W & C Anatolia.

Erect, subglabrous, 5-12 cm tall. Leaves alternate, oblong, terete, often tinged red. Inflorescence dense, few-flowered. Flowers sessile, 5-merous. Petals 3-4 mm, white. Ripe follicles suberect.

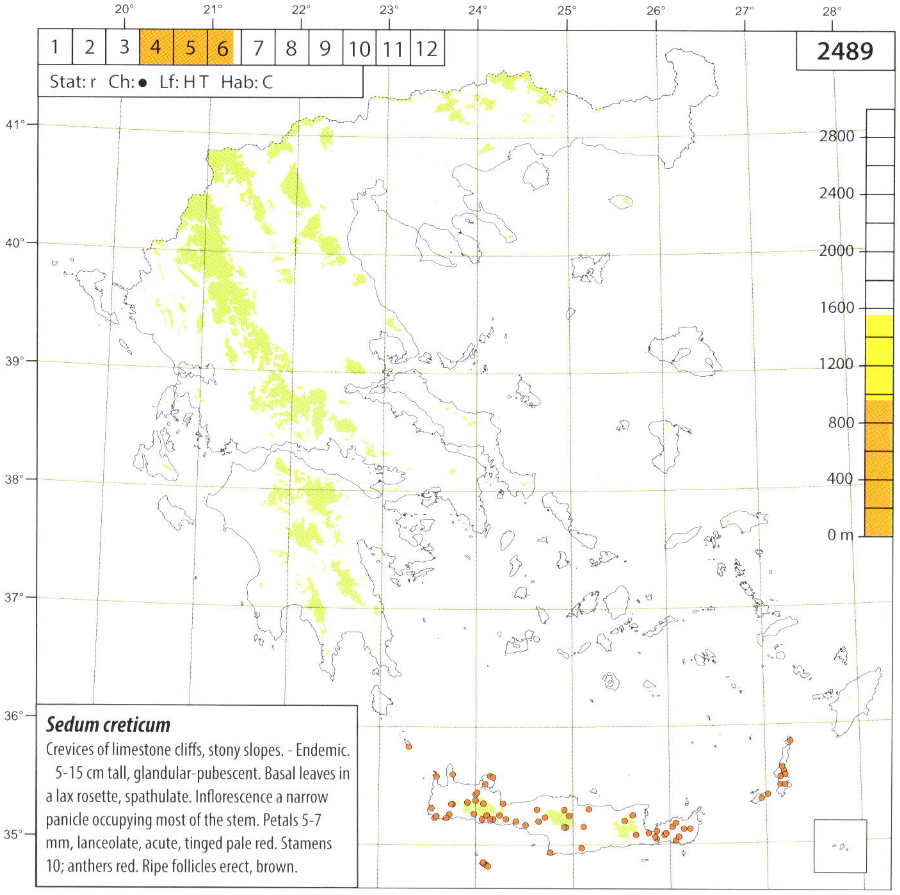

Sedum creticum

Crevices of limestone cliffs, stony slopes. - Endemic.

5-15 cm tall, glandular-pubescent. Basal leaves in a lax rosette, spathulate. Inflorescence a narrow panicle occupying most of the stem. Petals 5-7 mm, lanceolate, acute, tinged pale red. Stamens 10; anthers red. Ripe follicles erect, brown.

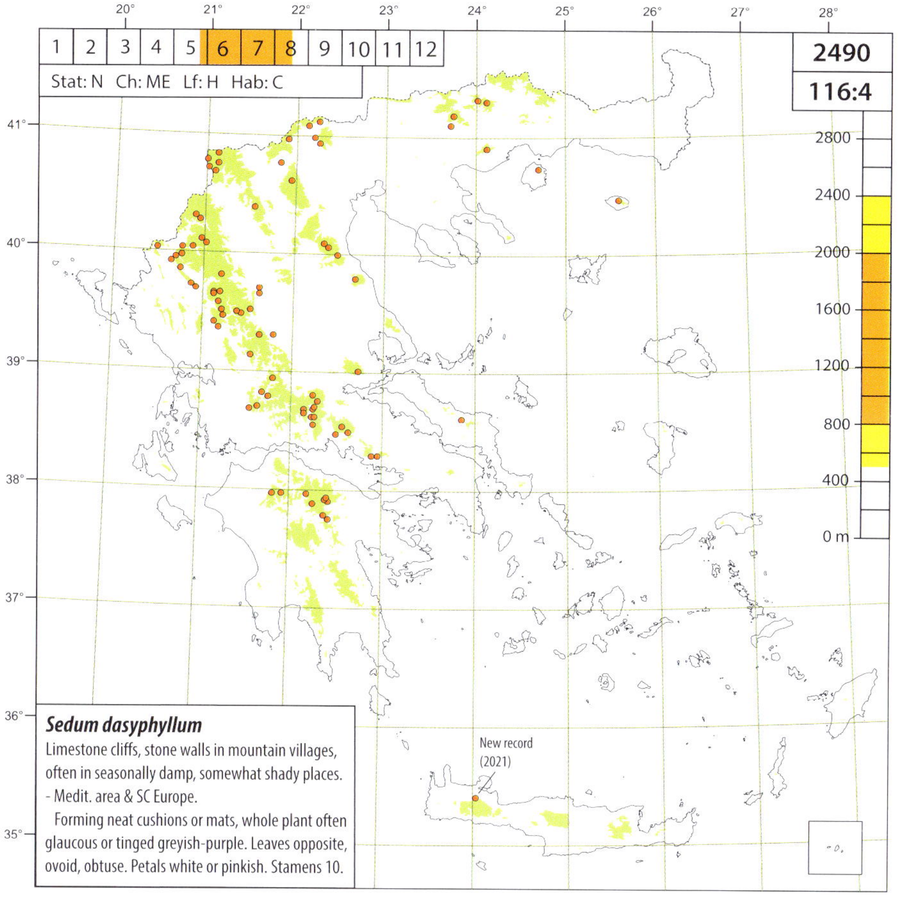

Sedum dasyphyllum

Limestone cliffs, stone walls in mountain villages, often in seasonally damp, somewhat shady places. - Medit. area & SC Europe.

Forming neat cushions or mats, whole plant often glaucous or tinged greyish-purple. Leaves opposite, ovoid, obtuse. Petals white or pinkish. Stamens 10.

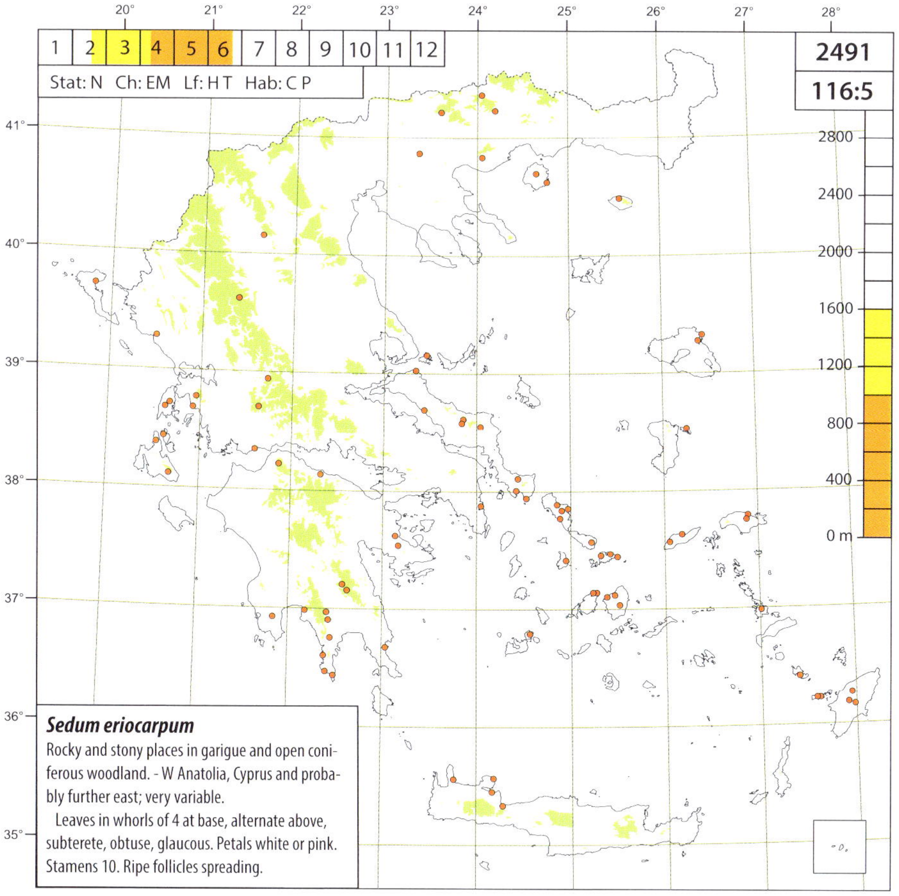

Sedum eriocarpum

Rocky and stony places in garigue and open coniferous woodland. - W Anatolia, Cyprus and probably further east; very variable.

Leaves in whorls of 4 at base, alternate above, subterete, obtuse, glaucous. Petals white or pink. Stamens 10. Ripe follicles spreading.

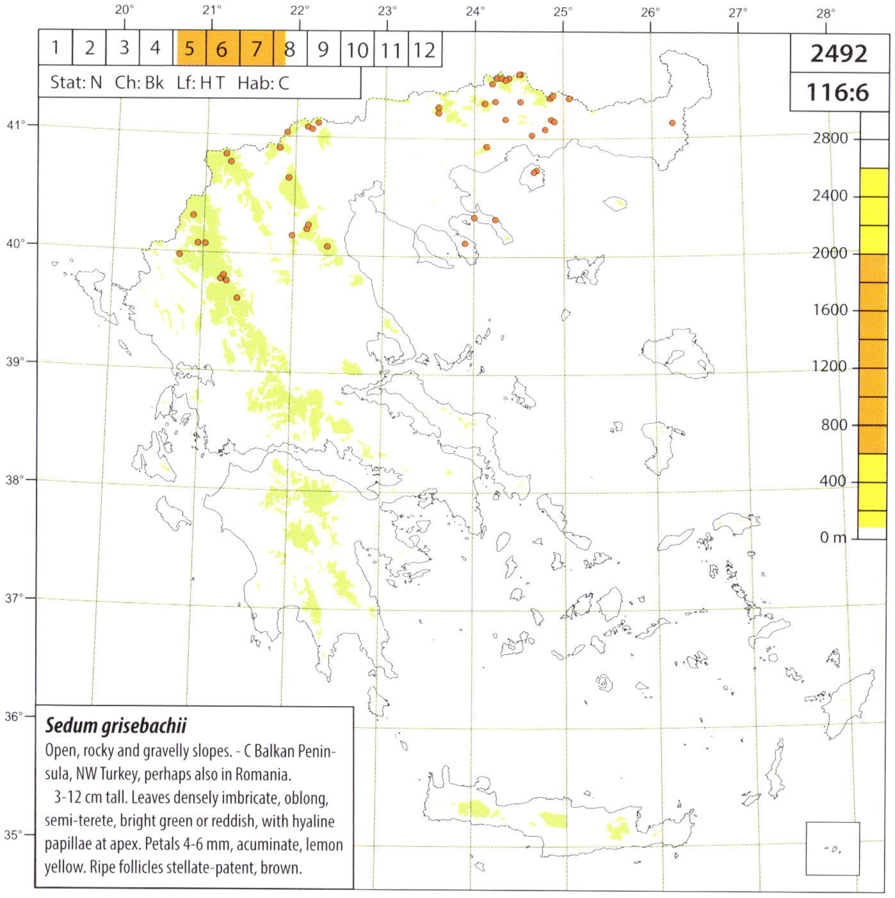

Sedum grisebachii

Open, rocky and gravelly slopes. - C Balkan Peninsula, NW Turkey, perhaps also in Romania.

3-12 cm tall. Leaves densely imbricate, oblong, semi-terete, bright green or reddish, with hyaline papillae at apex. Petals 4-6 mm, acuminate, lemon yellow. Ripe follicles stellate-patent, brown.

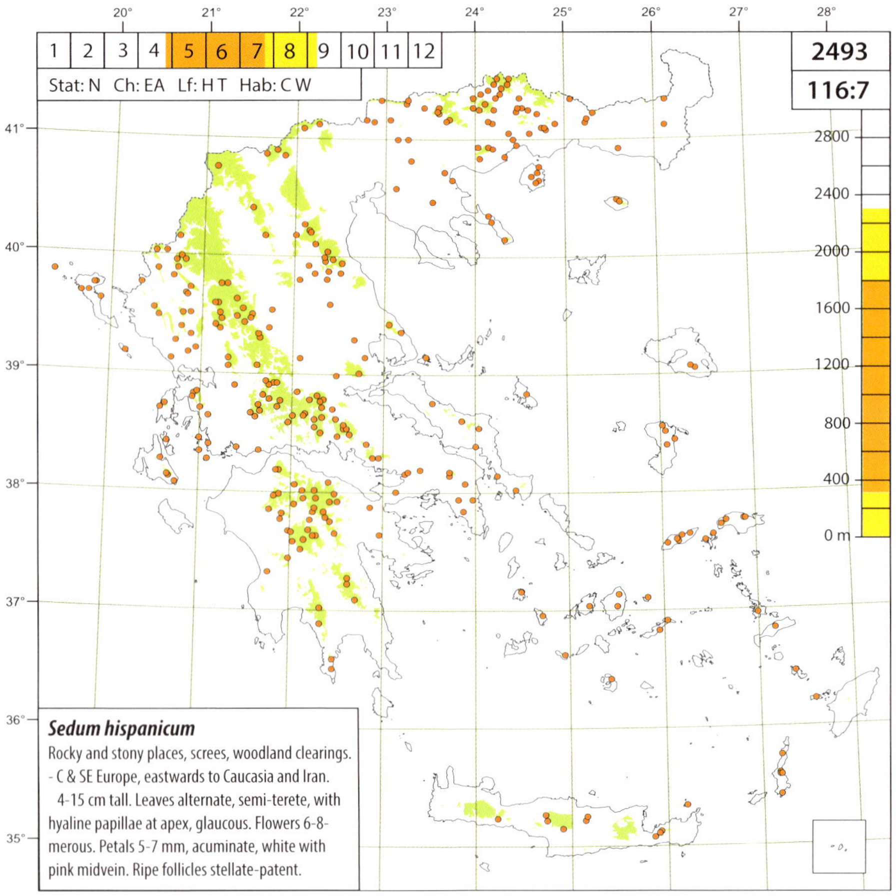

Sedum hispanicum
Rocky and stony places, screes, woodland clearings. - C & SE Europe, eastwards to Caucasia and Iran.
4-15 cm tall. Leaves alternate, semi-terete, with hyaline papillae at apex, glaucous. Flowers 6-8-merous. Petals 5-7 mm, acuminate, white with pink midvein. Ripe follicles stellate-patent.

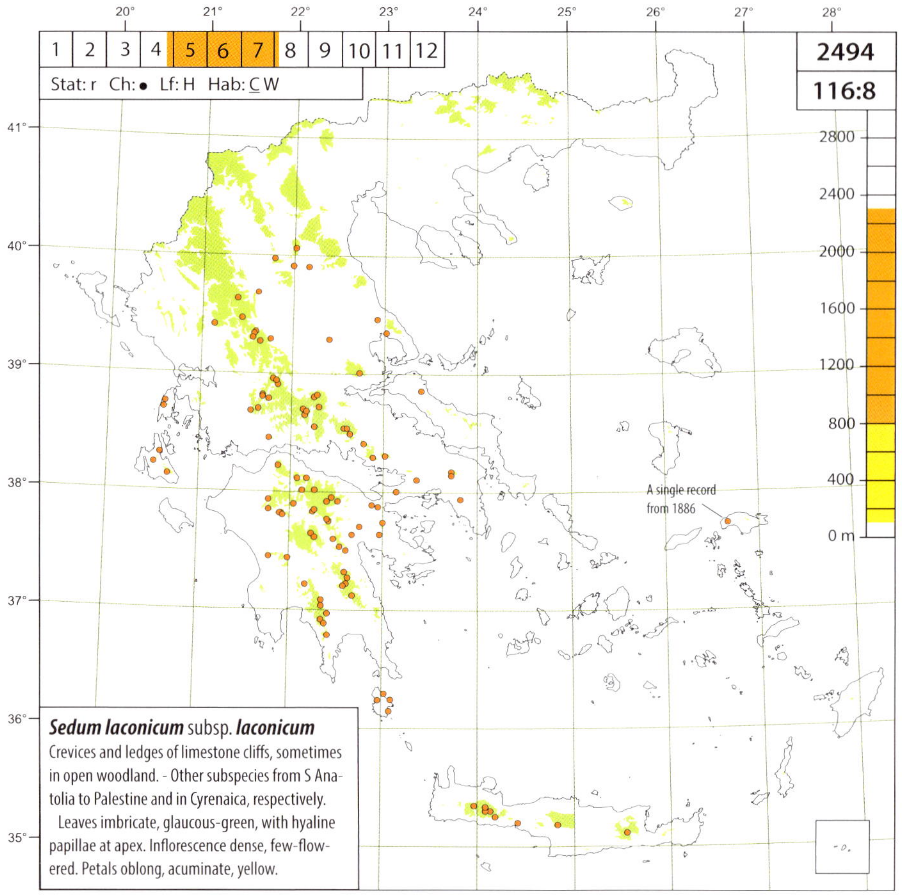

Sedum laconicum subsp. ***laconicum***
Crevices and ledges of limestone cliffs, sometimes in open woodland. - Other subspecies from S Anatolia to Palestine and in Cyrenaica, respectively.
Leaves imbricate, glaucous-green, with hyaline papillae at apex. Inflorescence dense, few-flowered. Petals oblong, acuminate, yellow.

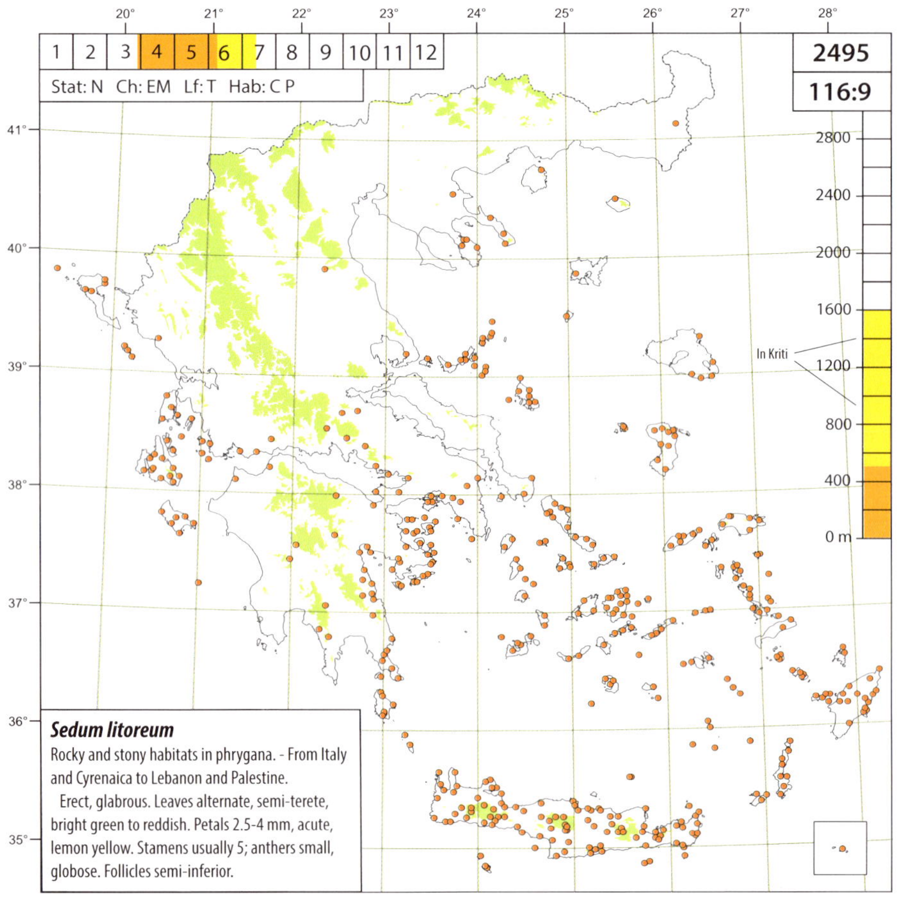

Sedum litoreum
Rocky and stony habitats in phrygana. - From Italy and Cyrenaica to Lebanon and Palestine.
Erect, glabrous. Leaves alternate, semi-terete, bright green to reddish. Petals 2.5-4 mm, acute, lemon yellow. Stamens usually 5; anthers small, globose. Follicles semi-inferior.

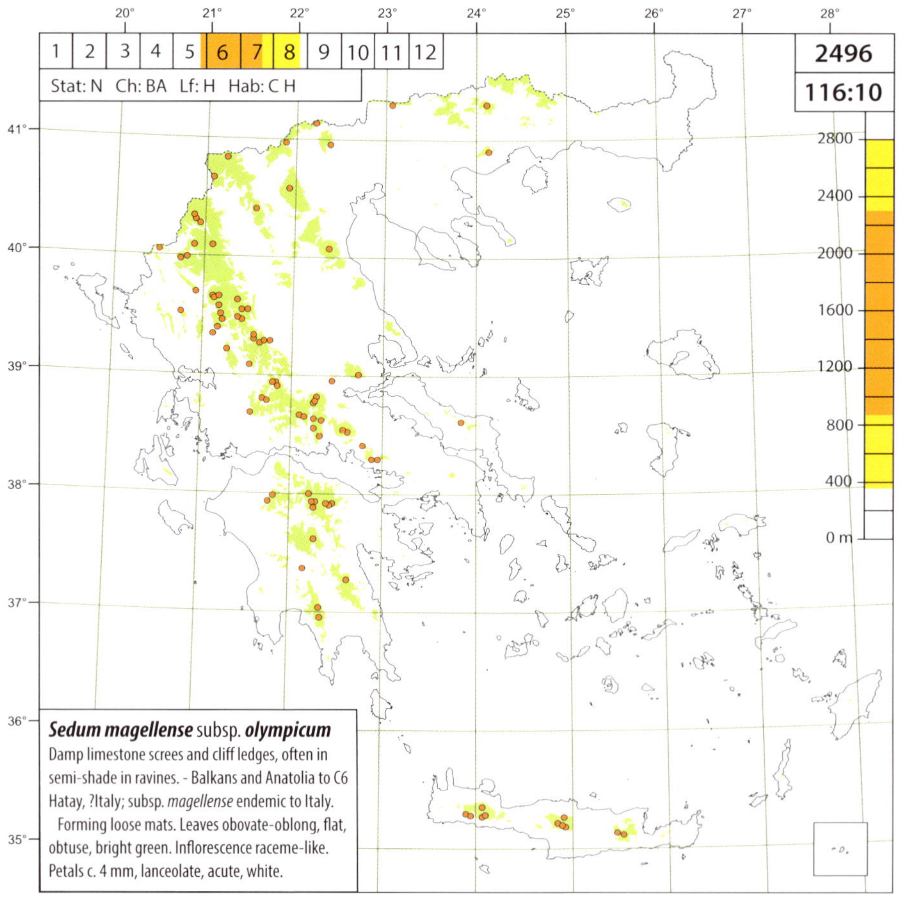

Sedum magellense subsp. ***olympicum***
Damp limestone screes and cliff ledges, often in semi-shade in ravines. - Balkans and Anatolia to C6 Hatay, ?Italy; subsp. *magellense* endemic to Italy.
Forming loose mats. Leaves obovate-oblong, flat, obtuse, bright green. Inflorescence raceme-like. Petals c. 4 mm, lanceolate, acute, white.

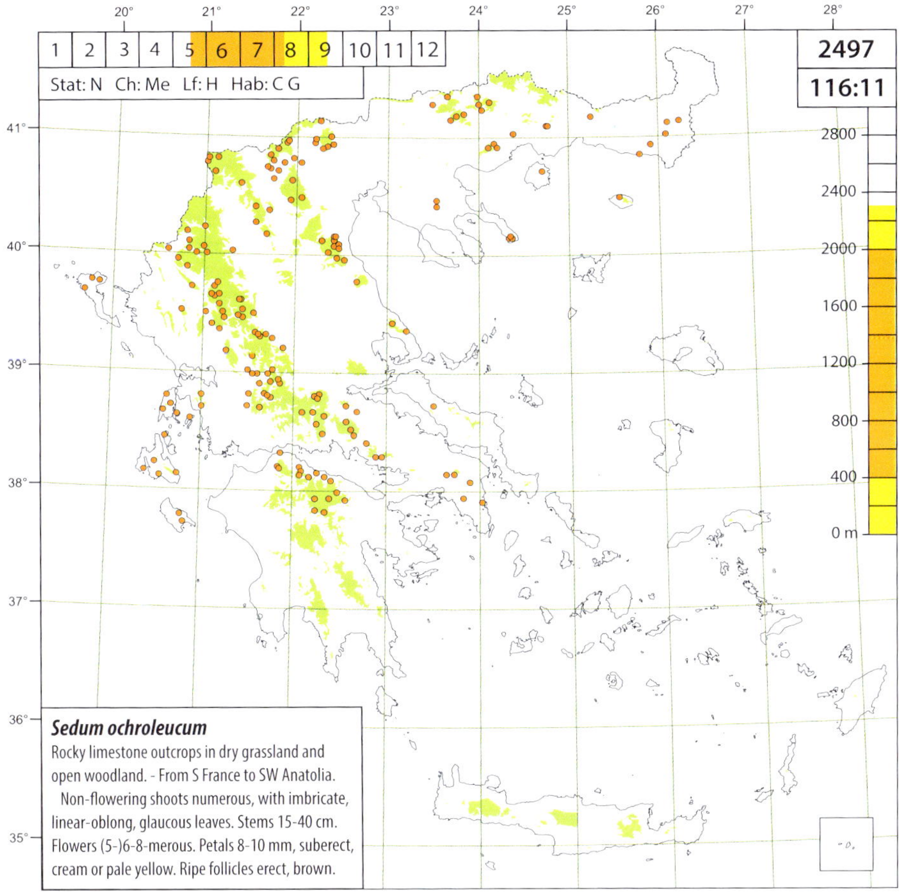

Sedum ochroleucum

Rocky limestone outcrops in dry grassland and open woodland. - From S France to SW Anatolia.

Non-flowering shoots numerous, with imbricate, linear-oblong, glaucous leaves. Stems 15-40 cm. Flowers (5-)6-8-merous. Petals 8-10 mm, suberect, cream or pale yellow. Ripe follicles erect, brown.

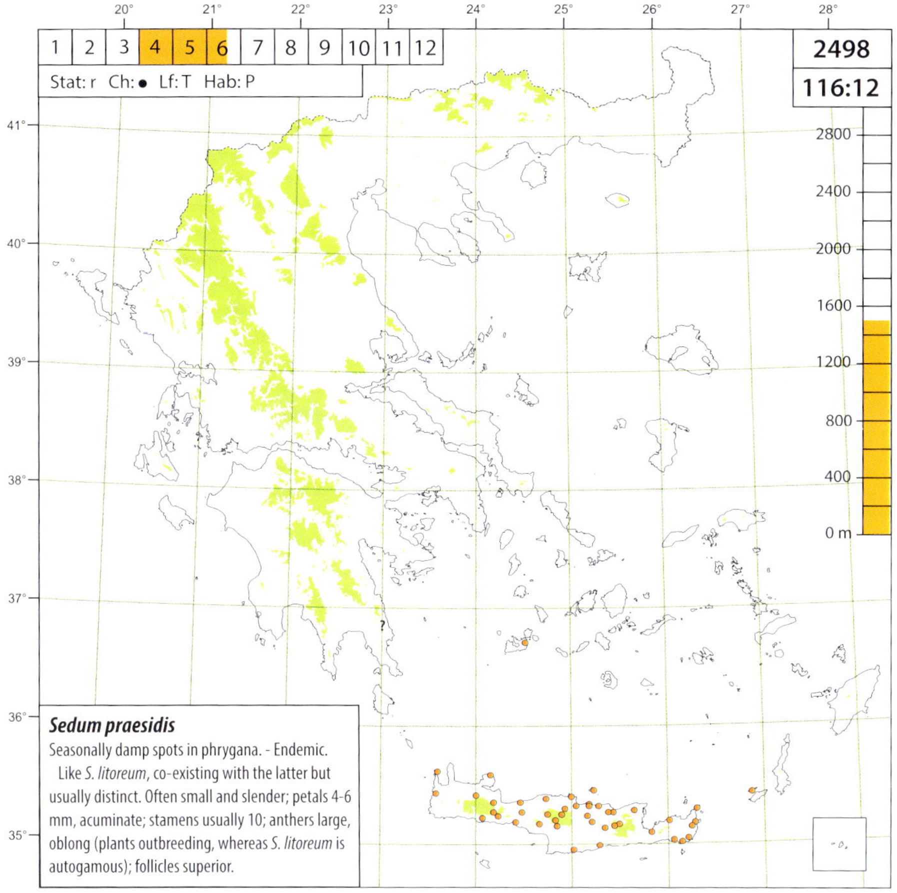

Sedum praesidis

Seasonally damp spots in phrygana. - Endemic.

Like *S. litoreum*, co-existing with the latter but usually distinct. Often small and slender; petals 4-6 mm, acuminate; stamens usually 10; anthers large, oblong (plants outbreeding, whereas *S. litoreum* is autogamous); follicles superior.

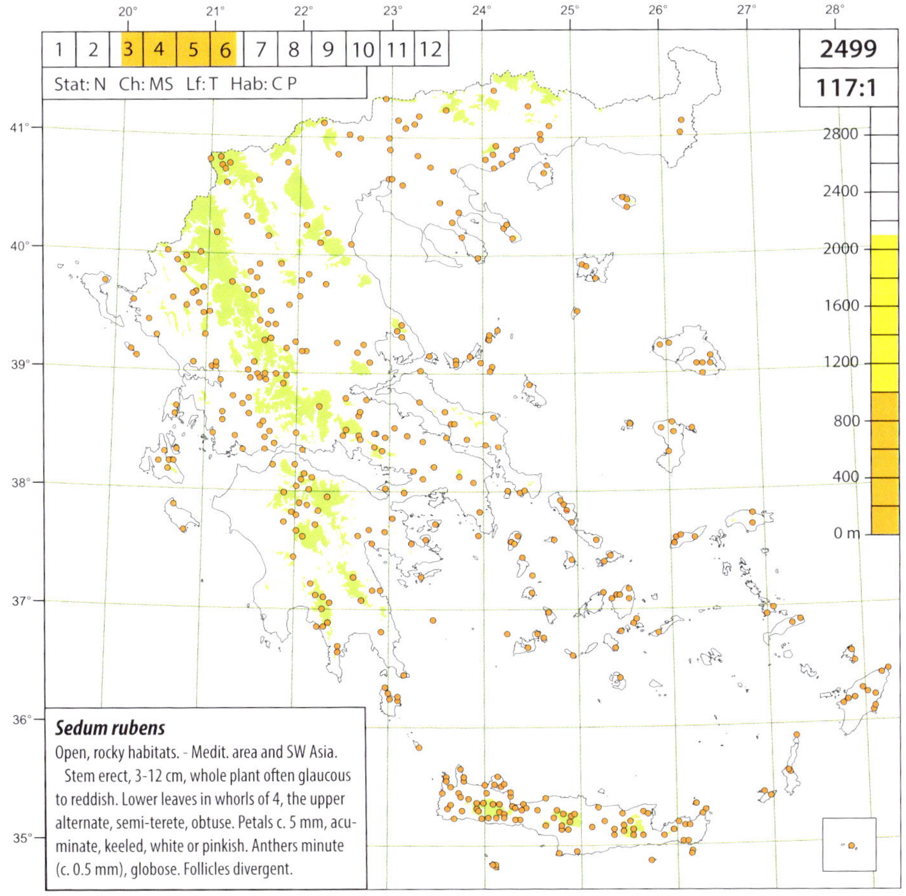

Sedum rubens

Open, rocky habitats. - Medit. area and SW Asia.

Stem erect, 3-12 cm, whole plant often glaucous to reddish. Lower leaves in whorls of 4, the upper alternate, semi-terete, obtuse. Petals c. 5 mm, acuminate, keeled, white or pinkish. Anthers minute (c. 0.5 mm), globose. Follicles divergent.

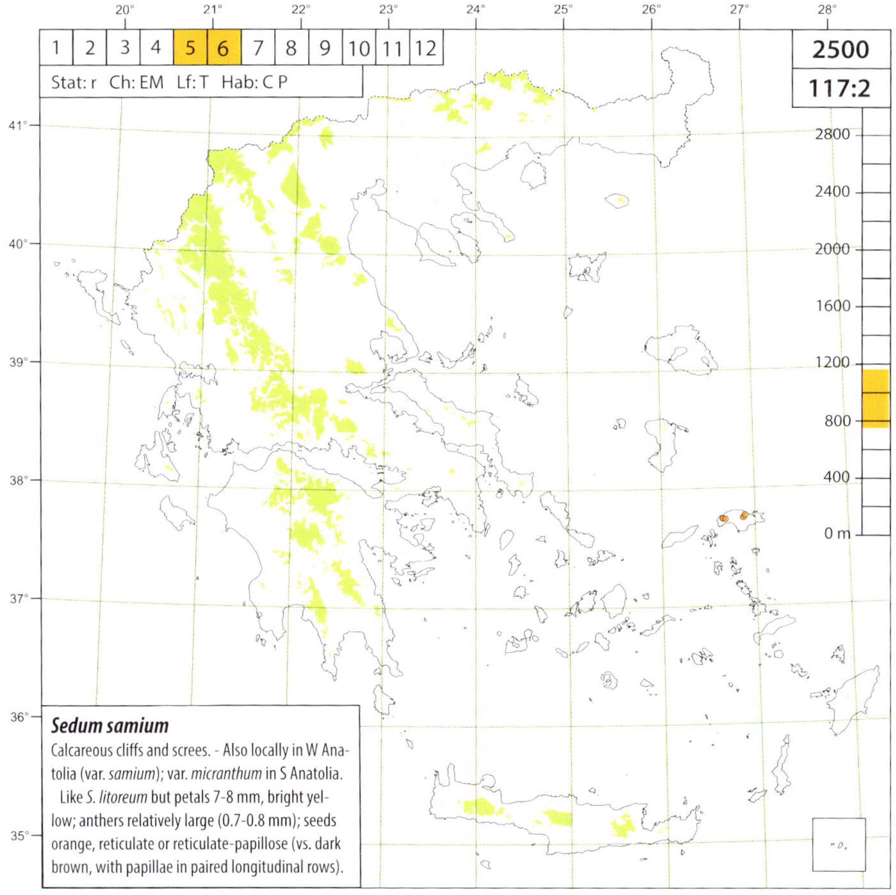

Sedum samium

Calcareous cliffs and screes. - Also locally in W Anatolia (var. *samium*); var. *micranthum* in S Anatolia.

Like *S. litoreum* but petals 7-8 mm, bright yellow; anthers relatively large (0.7-0.8 mm); seeds orange, reticulate or reticulate-papillose (vs. dark brown, with papillae in paired longitudinal rows).

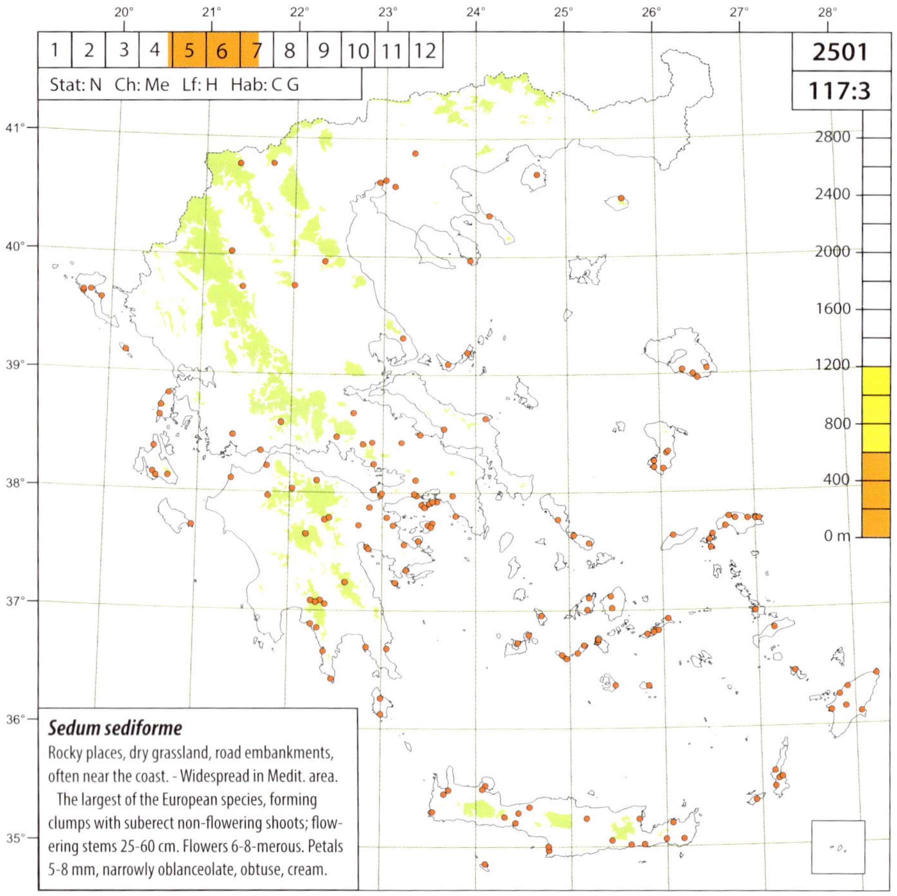

Sedum sediforme

Rocky places, dry grassland, road embankments, often near the coast. - Widespread in Medit. area.

The largest of the European species, forming clumps with suberect non-flowering shoots; flowering stems 25-60 cm. Flowers 6-8-merous. Petals 5-8 mm, narrowly oblanceolate, obtuse, cream.

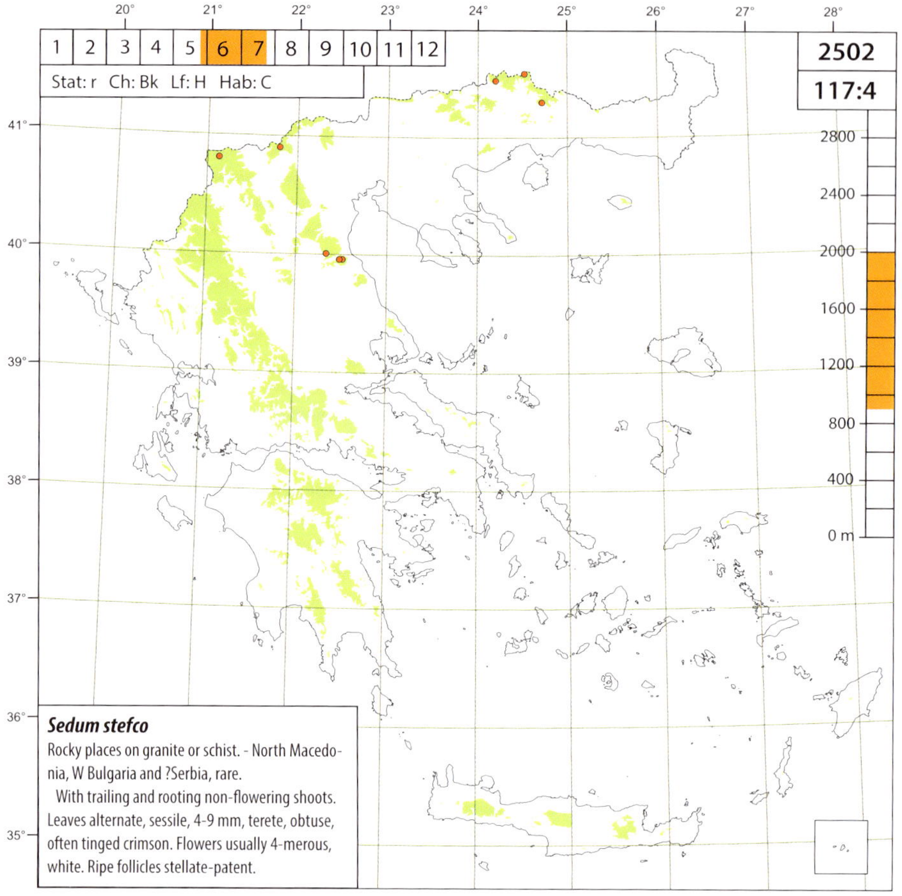

Sedum stefco

Rocky places on granite or schist. - North Macedonia, W Bulgaria and ?Serbia, rare.

With trailing and rooting non-flowering shoots. Leaves alternate, sessile, 4-9 mm, terete, obtuse, often tinged crimson. Flowers usually 4-merous, white. Ripe follicles stellate-patent.

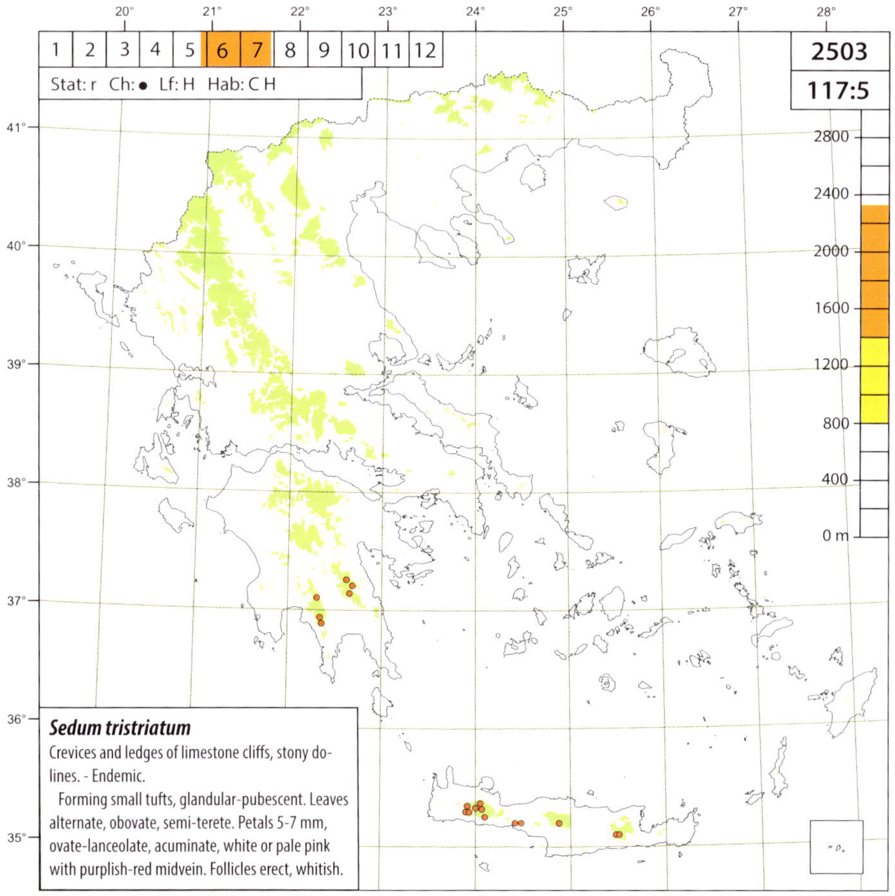

Sedum tristriatum

Crevices and ledges of limestone cliffs, stony dolines. - Endemic.

Forming small tufts, glandular-pubescent. Leaves alternate, obovate, semi-terete. Petals 5-7 mm, ovate-lanceolate, acuminate, white or pale pink with purplish-red midvein. Follicles erect, whitish.

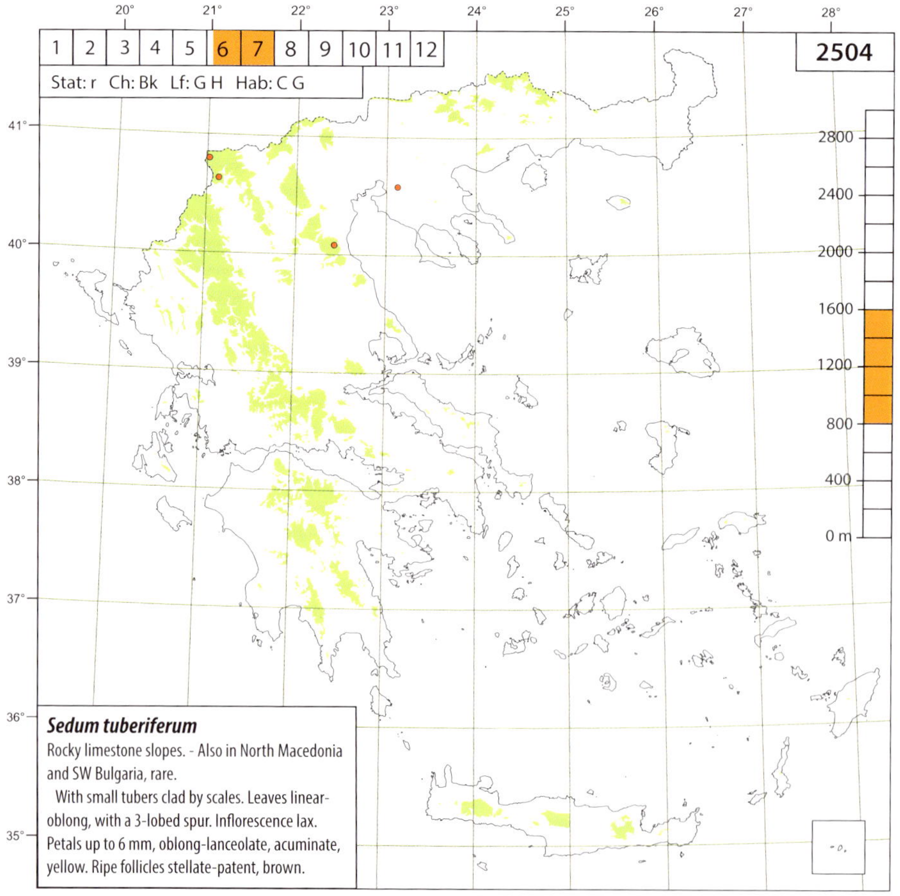

Sedum tuberiferum

Rocky limestone slopes. - Also in North Macedonia and SW Bulgaria, rare.

With small tubers clad by scales. Leaves linear-oblong, with a 3-lobed spur. Inflorescence lax. Petals up to 6 mm, oblong-lanceolate, acuminate, yellow. Ripe follicles stellate-patent, brown.

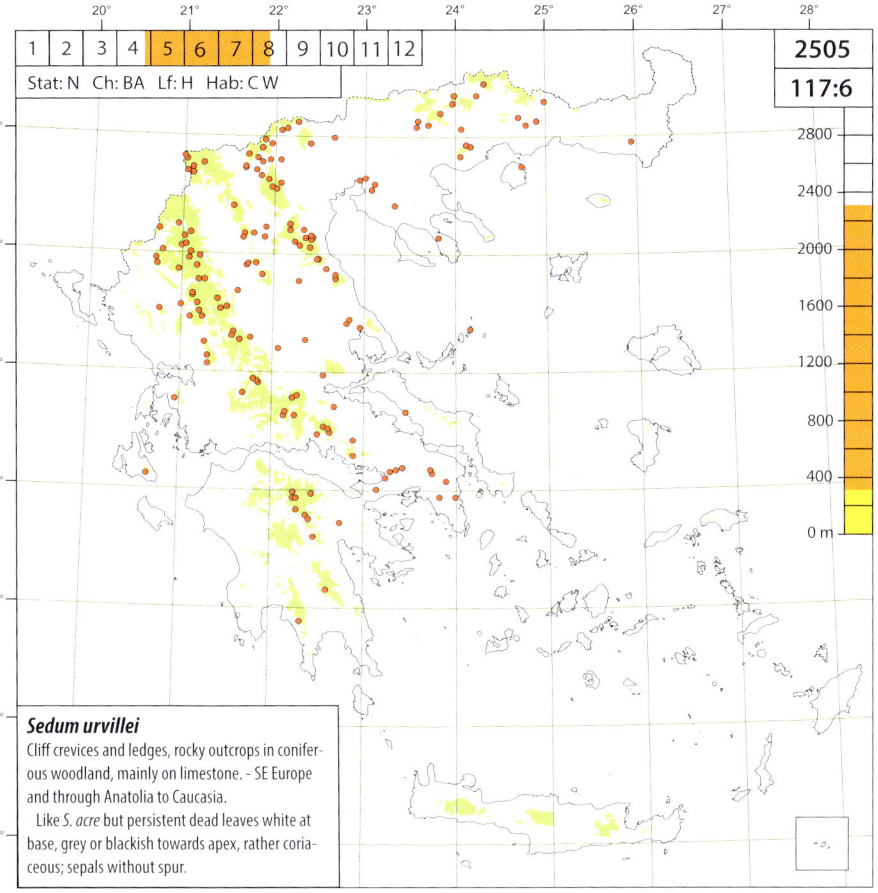

Sedum urvillei
Cliff crevices and ledges, rocky outcrops in coniferous woodland, mainly on limestone. - SE Europe and through Anatolia to Caucasia.
Like *S. acre* but persistent dead leaves white at base, grey or blackish towards apex, rather coriaceous; sepals without spur.

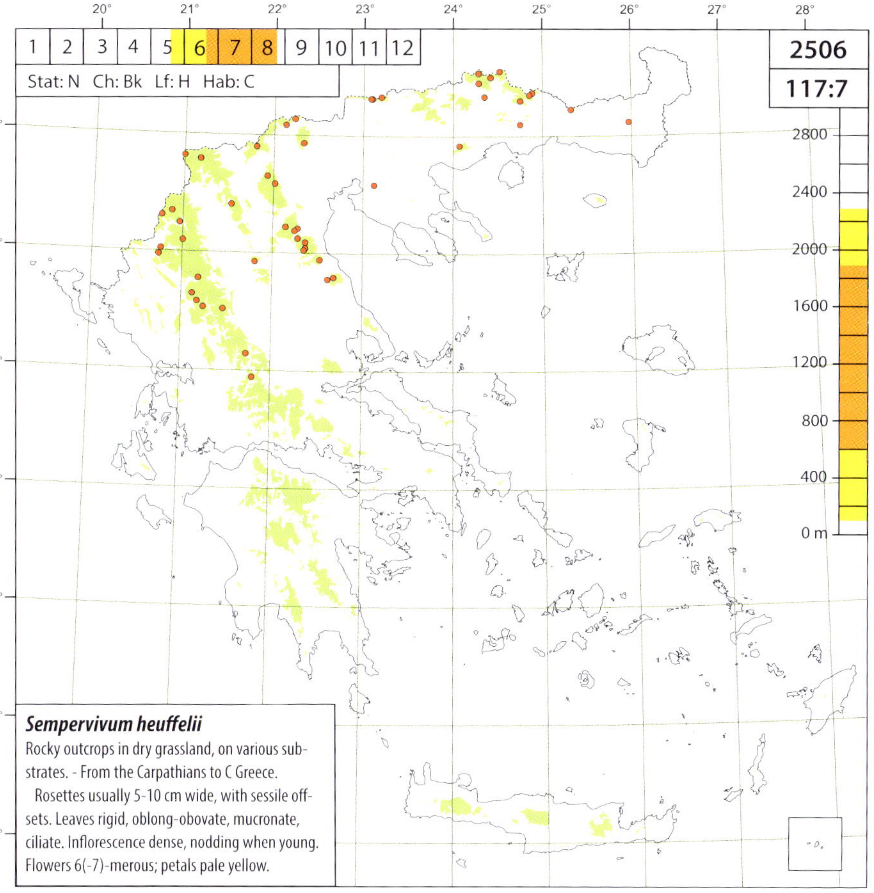

Sempervivum heuffelii
Rocky outcrops in dry grassland, on various substrates. - From the Carpathians to C Greece.
Rosettes usually 5-10 cm wide, with sessile offsets. Leaves rigid, oblong-obovate, mucronate, ciliate. Inflorescence dense, nodding when young. Flowers 6(-7)-merous; petals pale yellow.

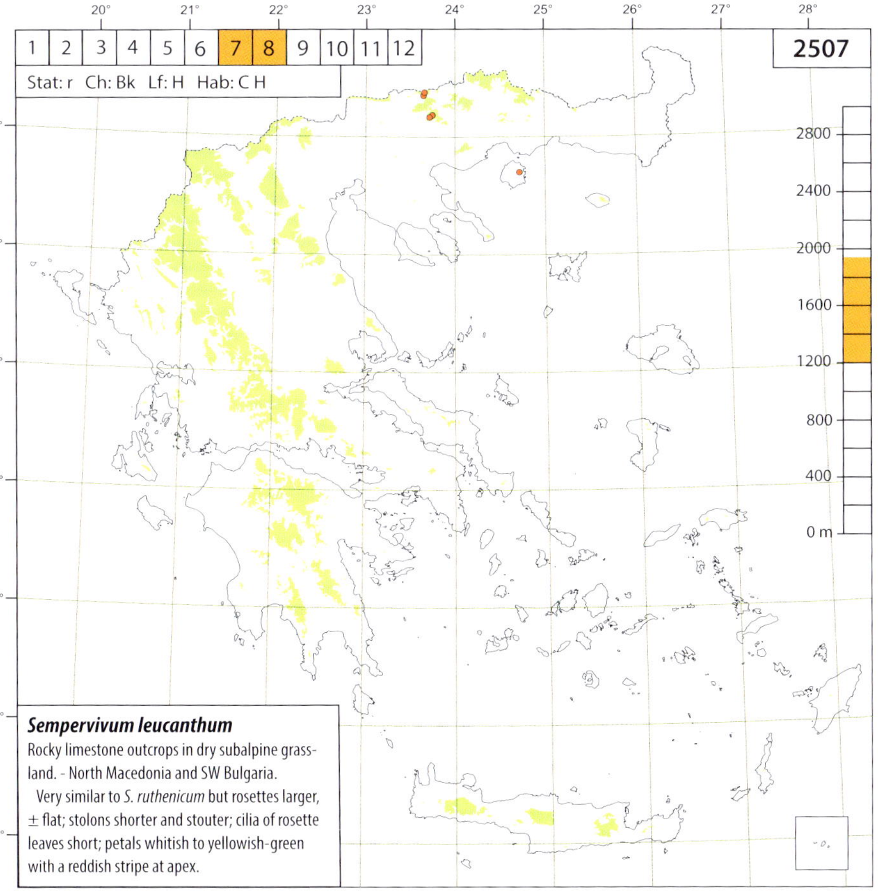

Sempervivum leucanthum
Rocky limestone outcrops in dry subalpine grassland. - North Macedonia and SW Bulgaria.
Very similar to *S. ruthenicum* but rosettes larger, ± flat; stolons shorter and stouter; cilia of rosette leaves short; petals whitish to yellowish-green with a reddish stripe at apex.

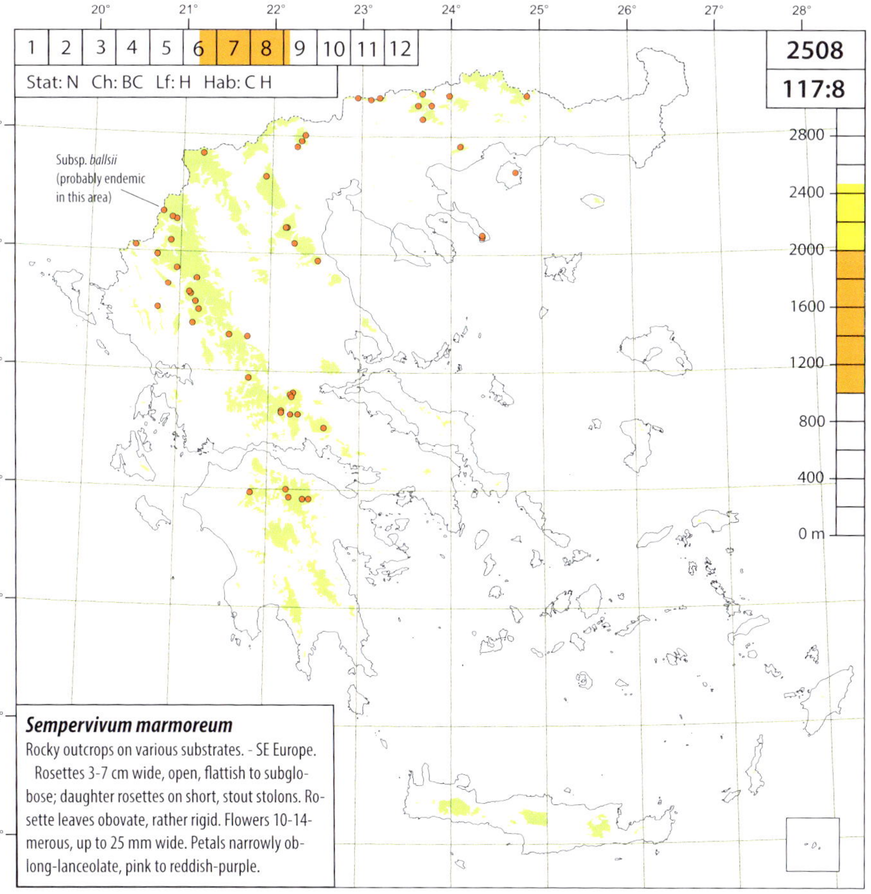

Sempervivum marmoreum
Rocky outcrops on various substrates. - SE Europe.
Rosettes 3-7 cm wide, open, flattish to subglobose; daughter rosettes on short, stout stolons. Rosette leaves obovate, rather rigid. Flowers 10-14-merous, up to 25 mm wide. Petals narrowly oblong-lanceolate, pink to reddish-purple.

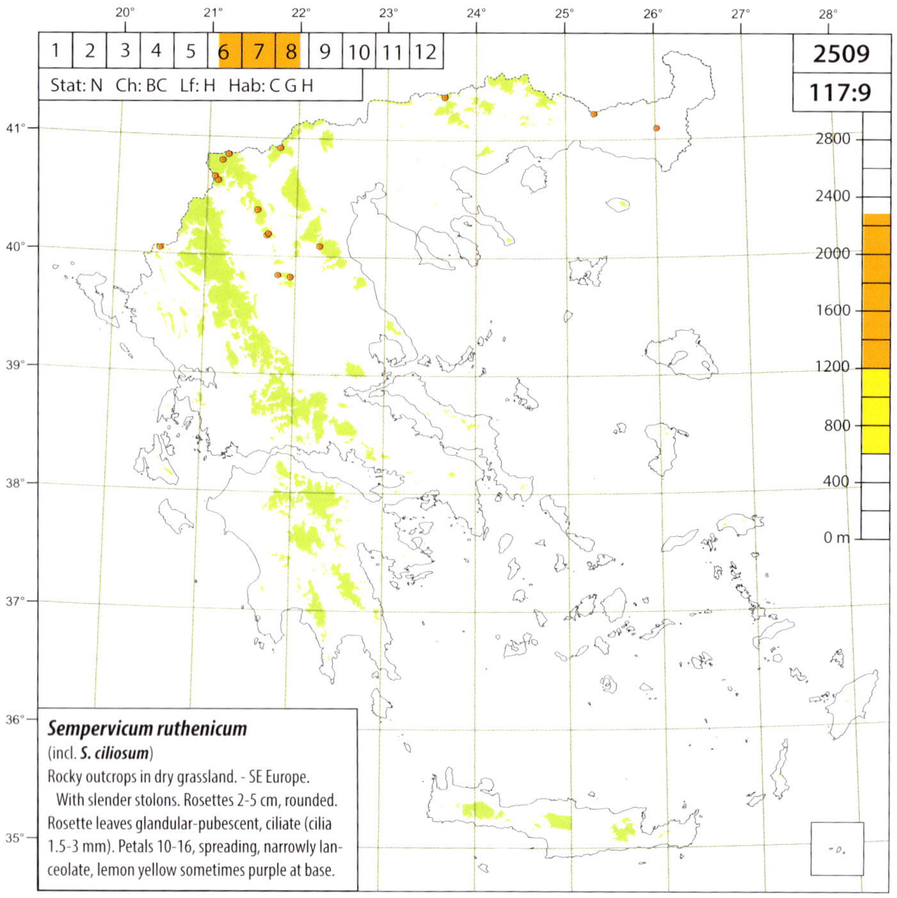

Sempervicum ruthenicum
(incl. ***S. ciliosum***)
Rocky outcrops in dry grassland. - SE Europe.
With slender stolons. Rosettes 2-5 cm, rounded. Rosette leaves glandular-pubescent, ciliate (cilia 1.5-3 mm). Petals 10-16, spreading, narrowly lanceolate, lemon yellow sometimes purple at base.

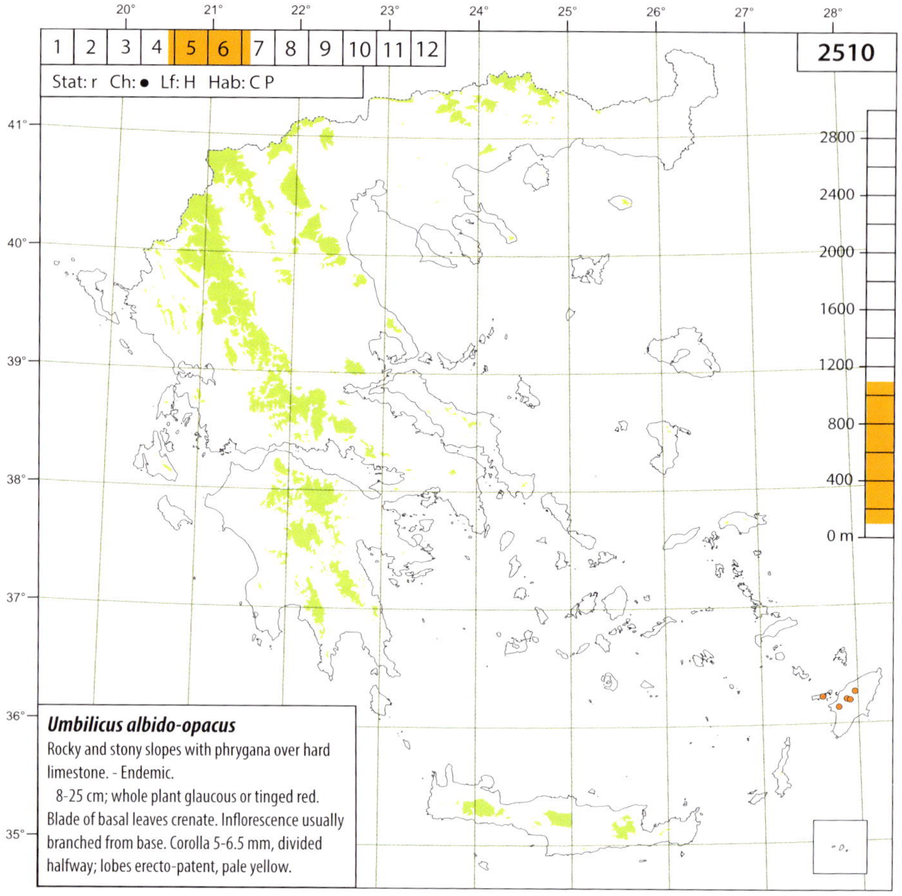

Umbilicus albido-opacus
Rocky and stony slopes with phrygana over hard limestone. - Endemic.
8-25 cm; whole plant glaucous or tinged red. Blade of basal leaves crenate. Inflorescence usually branched from base. Corolla 5-6.5 mm, divided halfway; lobes erecto-patent, pale yellow.

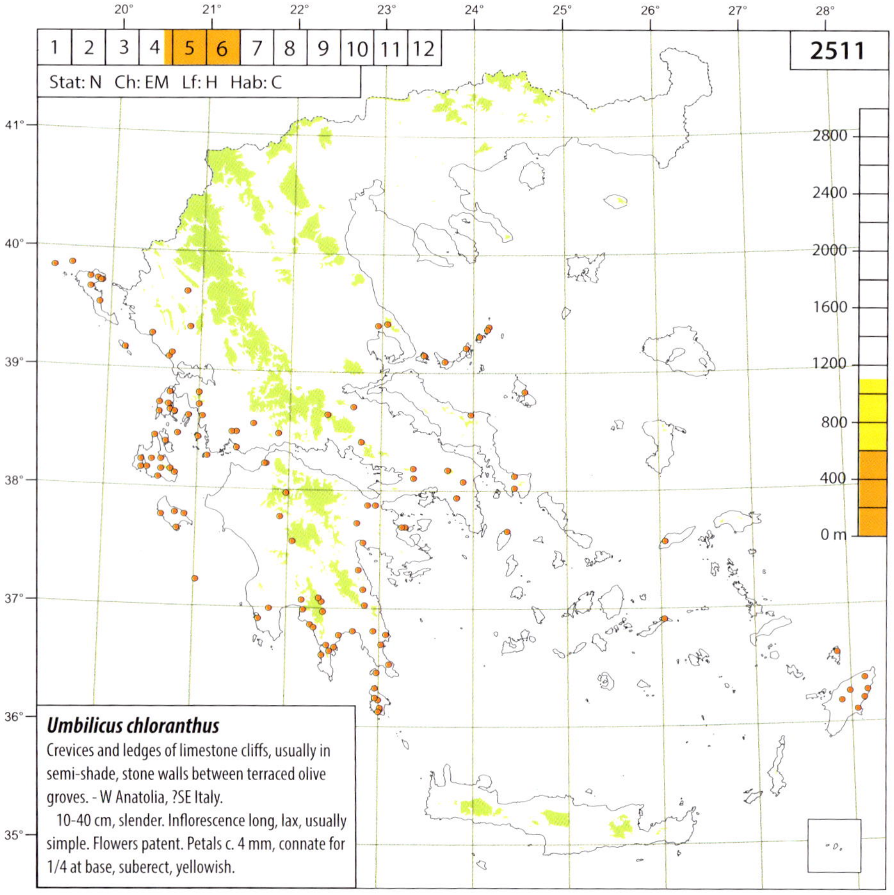

Umbilicus chloranthus
Crevices and ledges of limestone cliffs, usually in semi-shade, stone walls between terraced olive groves. - W Anatolia, ?SE Italy.
10-40 cm, slender. Inflorescence long, lax, usually simple. Flowers patent. Petals c. 4 mm, connate for 1/4 at base, suberect, yellowish.

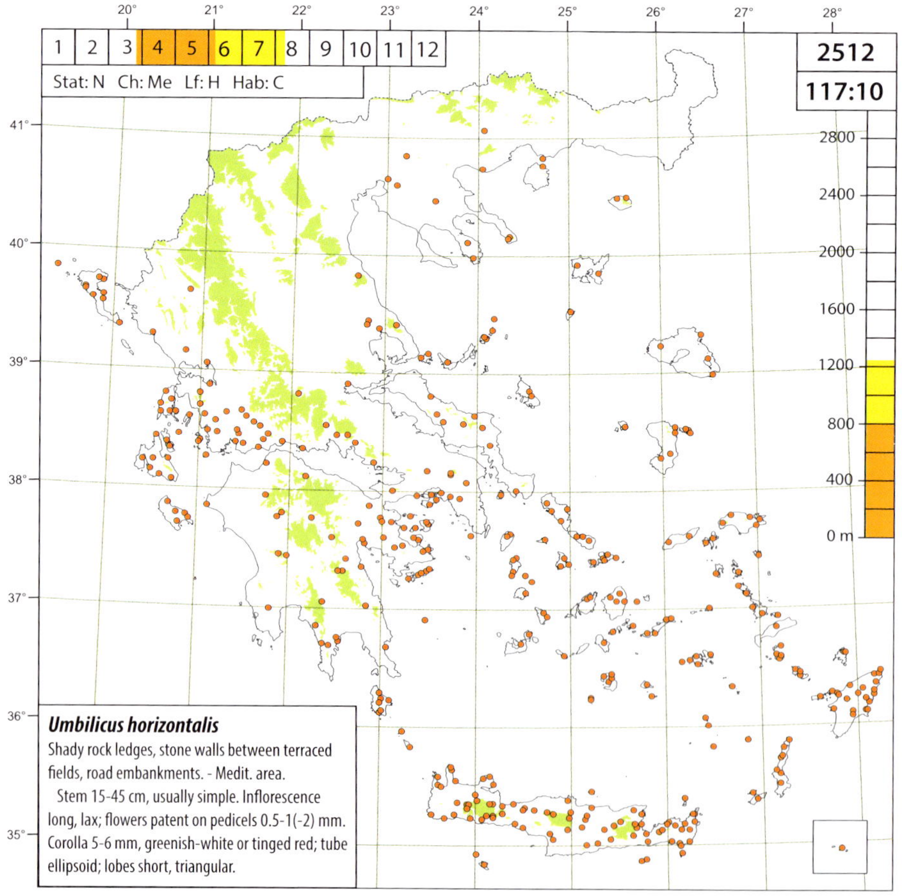

Umbilicus horizontalis
Shady rock ledges, stone walls between terraced fields, road embankments. - Medit. area.
Stem 15-45 cm, usually simple. Inflorescence long, lax; flowers patent on pedicels 0.5-1(-2) mm. Corolla 5-6 mm, greenish-white or tinged red; tube ellipsoid; lobes short, triangular.

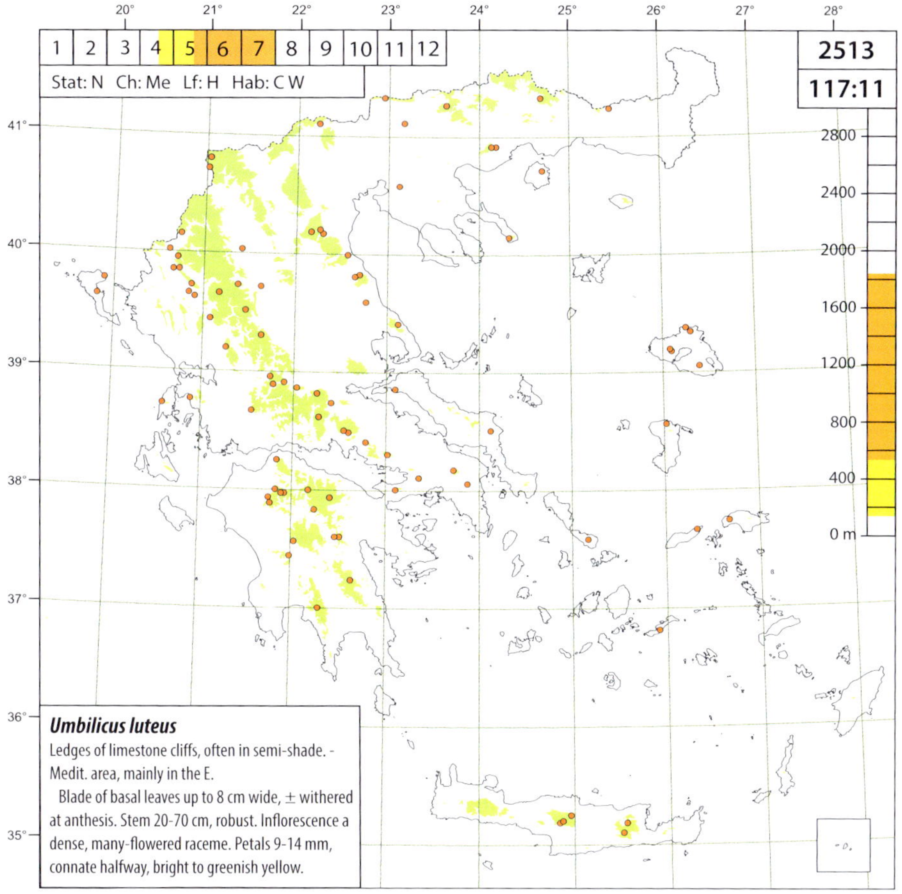

Umbilicus luteus

Ledges of limestone cliffs, often in semi-shade. - Medit. area, mainly in the E.

Blade of basal leaves up to 8 cm wide, ± withered at anthesis. Stem 20-70 cm, robust. Inflorescence a dense, many-flowered raceme. Petals 9-14 mm, connate halfway, bright to greenish yellow.

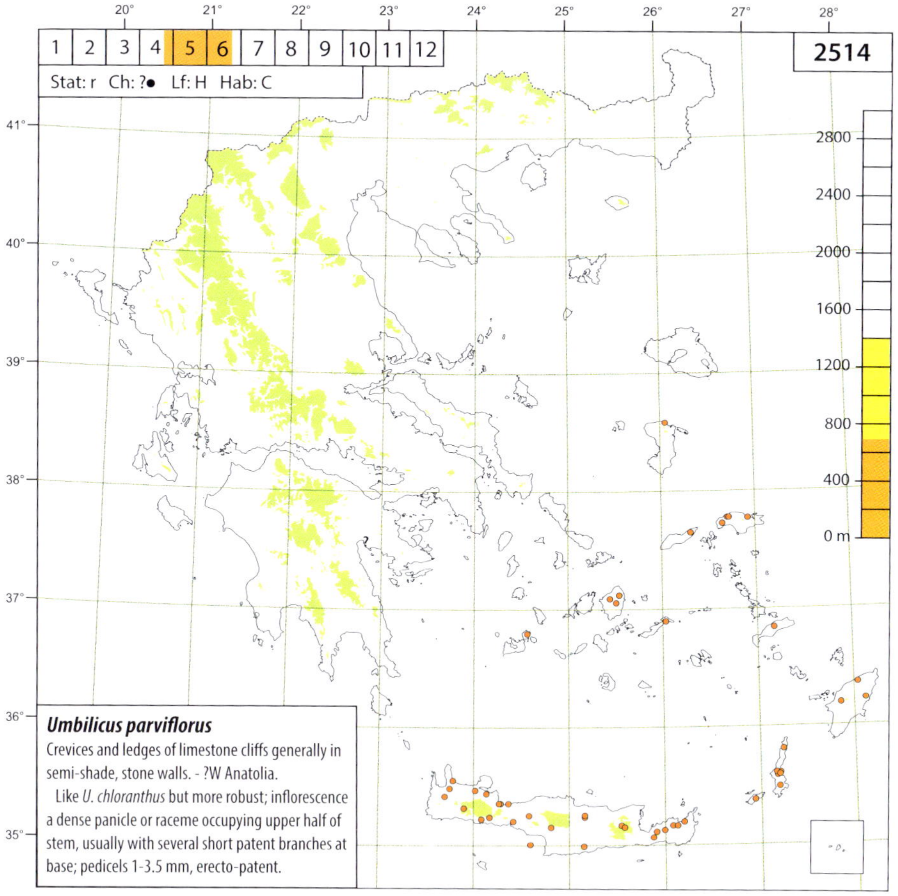

Umbilicus parviflorus

Crevices and ledges of limestone cliffs generally in semi-shade, stone walls. - ?W Anatolia.

Like *U. chloranthus* but more robust; inflorescence a dense panicle or raceme occupying upper half of stem, usually with several short patent branches at base; pedicels 1-3.5 mm, erecto-patent.

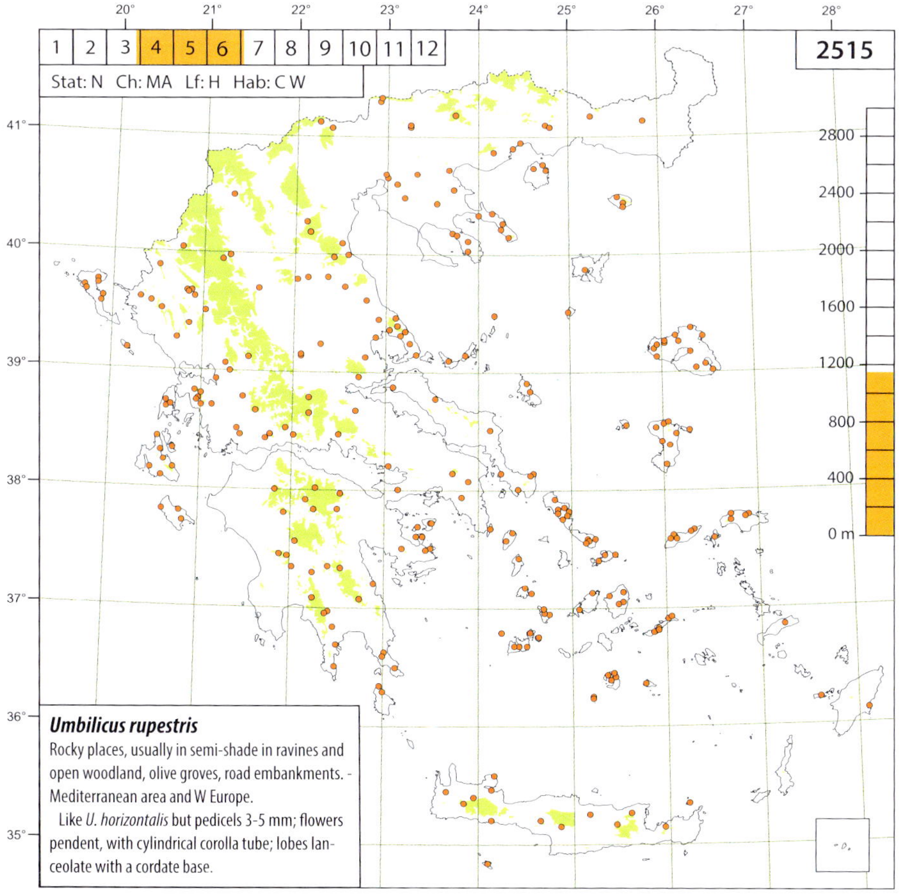

Umbilicus rupestris

Rocky places, usually in semi-shade in ravines and open woodland, olive groves, road embankments. - Mediterranean area and W Europe.

Like *U. horizontalis* but pedicels 3-5 mm; flowers pendent, with cylindrical corolla tube; lobes lanceolate with a cordate base.

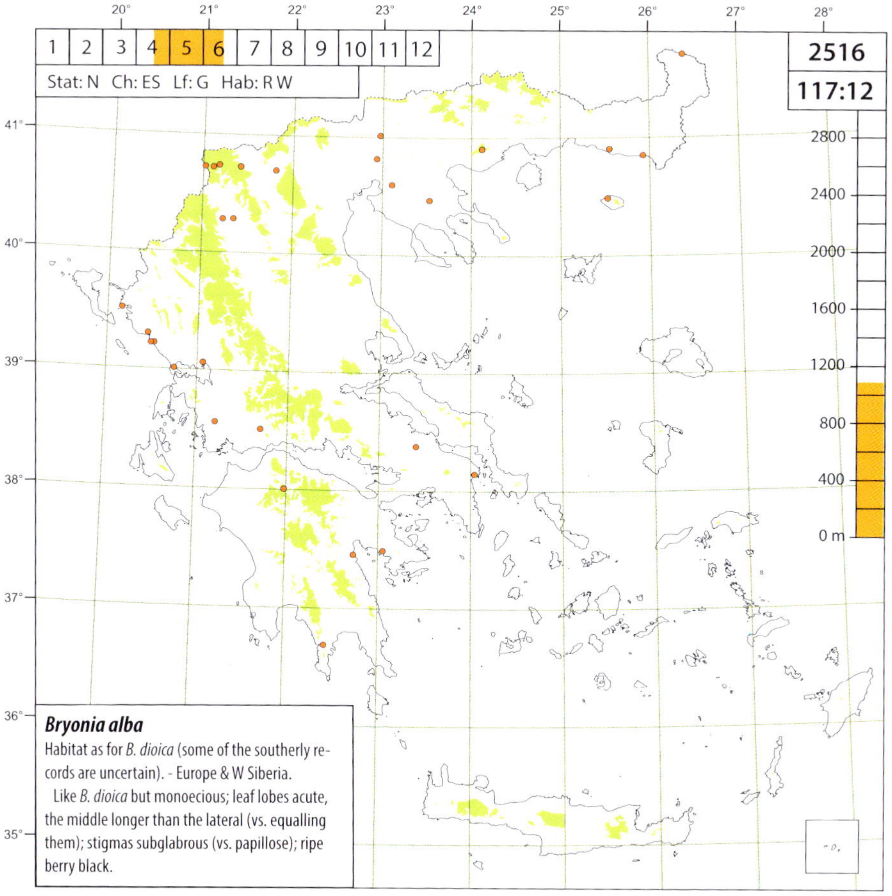

Bryonia alba

Habitat as for *B. dioica* (some of the southerly records are uncertain). - Europe & W Siberia.

Like *B. dioica* but monoecious; leaf lobes acute, the middle longer than the lateral (vs. equalling them); stigmas subglabrous (vs. papillose); ripe berry black.

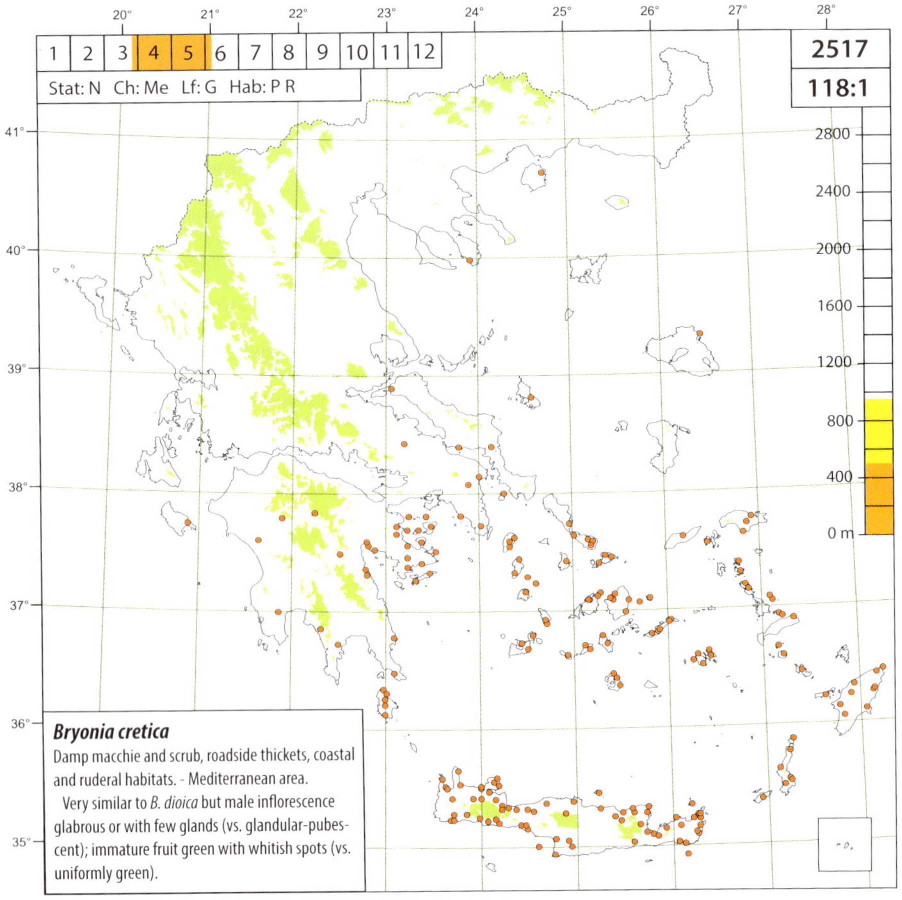

Bryonia cretica

Damp macchie and scrub, roadside thickets, coastal and ruderal habitats. - Mediterranean area.

Very similar to *B. dioica* but male inflorescence glabrous or with few glands (vs. glandular-pubescent); immature fruit green with whitish spots (vs. uniformly green).

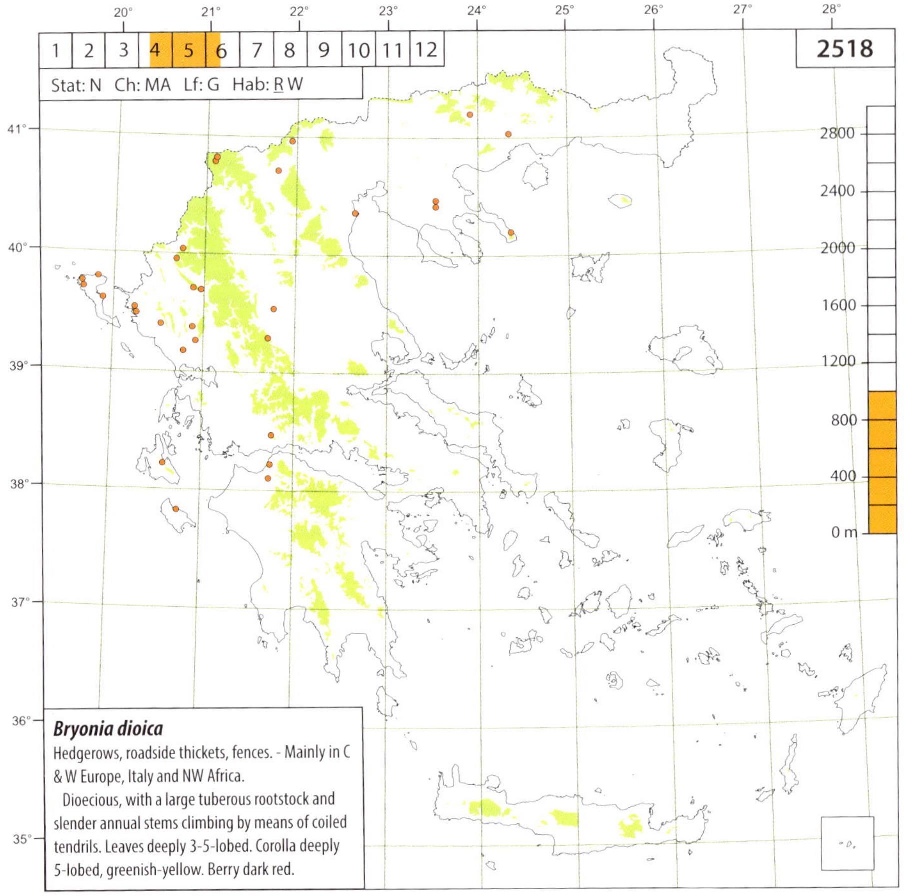

Bryonia dioica

Hedgerows, roadside thickets, fences. - Mainly in C & W Europe, Italy and NW Africa.

Dioecious, with a large tuberous rootstock and slender annual stems climbing by means of coiled tendrils. Leaves deeply 3-5-lobed. Corolla deeply 5-lobed, greenish-yellow. Berry dark red.

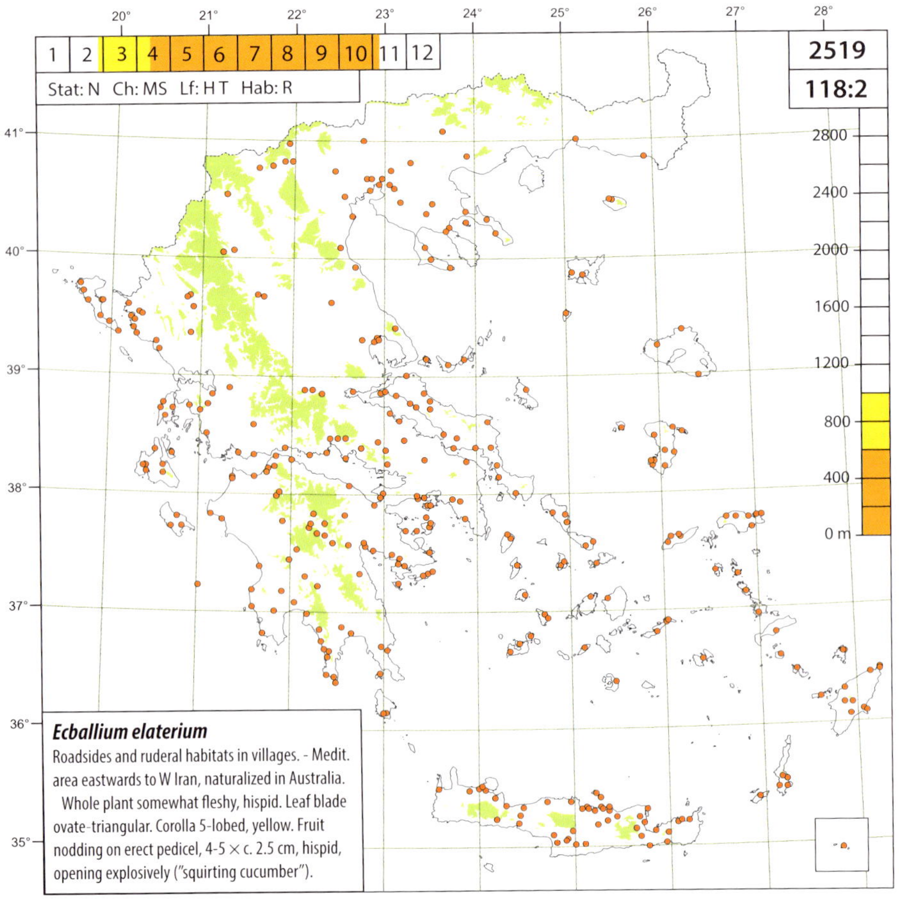

Ecballium elaterium

Roadsides and ruderal habitats in villages. - Medit. area eastwards to W Iran, naturalized in Australia.

Whole plant somewhat fleshy, hispid. Leaf blade ovate-triangular. Corolla 5-lobed, yellow. Fruit nodding on erect pedicel, 4-5 × c. 2.5 cm, hispid, opening explosively ("squirting cucumber").

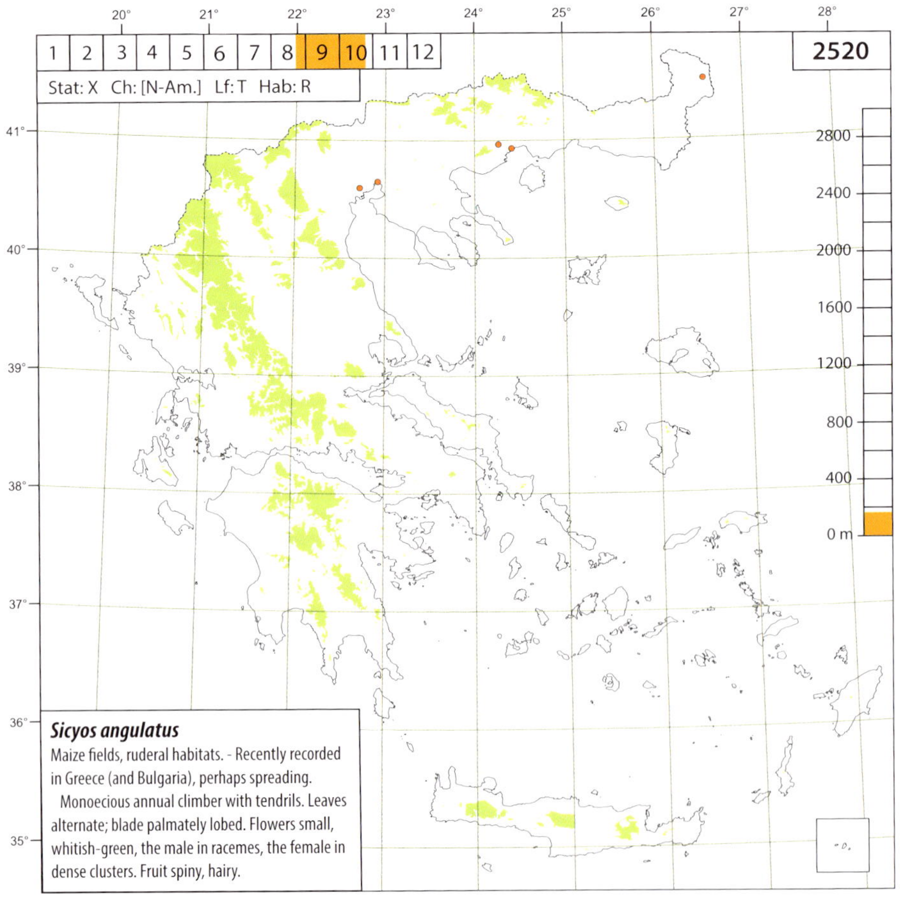

Sicyos angulatus

Maize fields, ruderal habitats. - Recently recorded in Greece (and Bulgaria), perhaps spreading.

Monoecious annual climber with tendrils. Leaves alternate; blade palmately lobed. Flowers small, whitish-green, the male in racemes, the female in dense clusters. Fruit spiny, hairy.

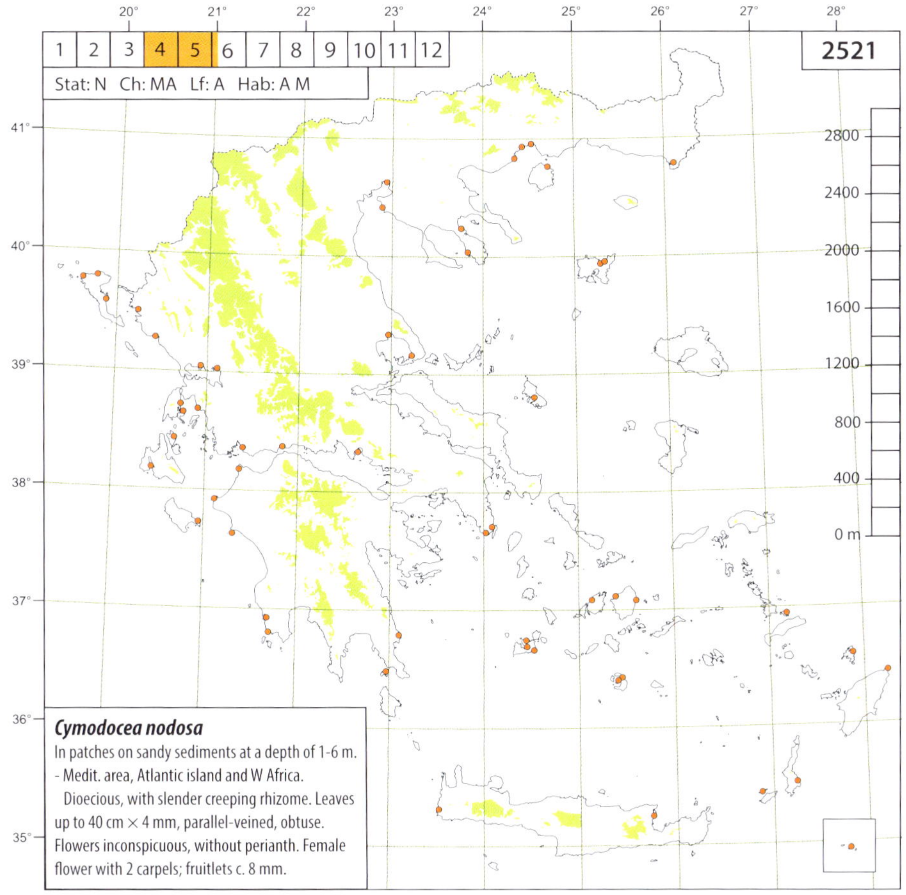

Cymodocea nodosa

In patches on sandy sediments at a depth of 1-6 m. - Medit. area, Atlantic island and W Africa.

Dioecious, with slender creeping rhizome. Leaves up to 40 cm × 4 mm, parallel-veined, obtuse. Flowers inconspicuous, without perianth. Female flower with 2 carpels; fruitlets c. 8 mm.

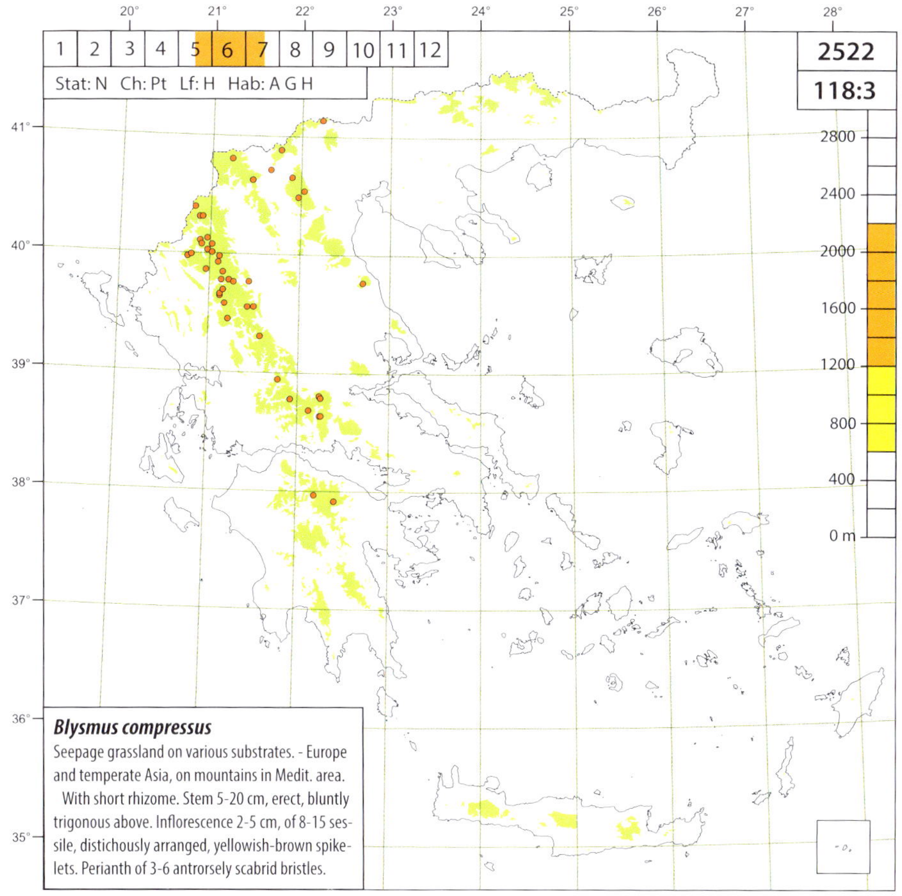

Blysmus compressus

Seepage grassland on various substrates. - Europe and temperate Asia, on mountains in Medit. area.

With short rhizome. Stem 5-20 cm, erect, bluntly trigonous above. Inflorescence 2-5 cm, of 8-15 sessile, distichously arranged, yellowish-brown spikelets. Perianth of 3-6 antrorsely scabrid bristles.

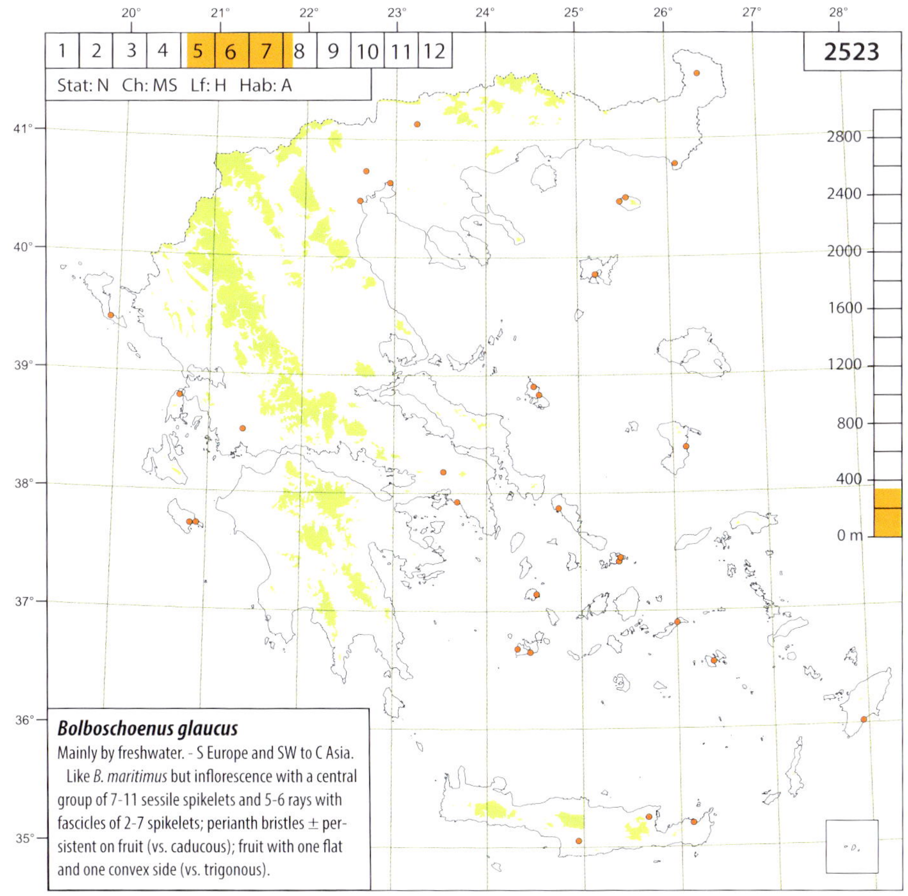

Bolboschoenus glaucus

Mainly by freshwater. - S Europe and SW to C Asia.

Like *B. maritimus* but inflorescence with a central group of 7-11 sessile spikelets and 5-6 rays with fascicles of 2-7 spikelets; perianth bristles ± persistent on fruit (vs. caducous); fruit with one flat and one convex side (vs. trigonous).

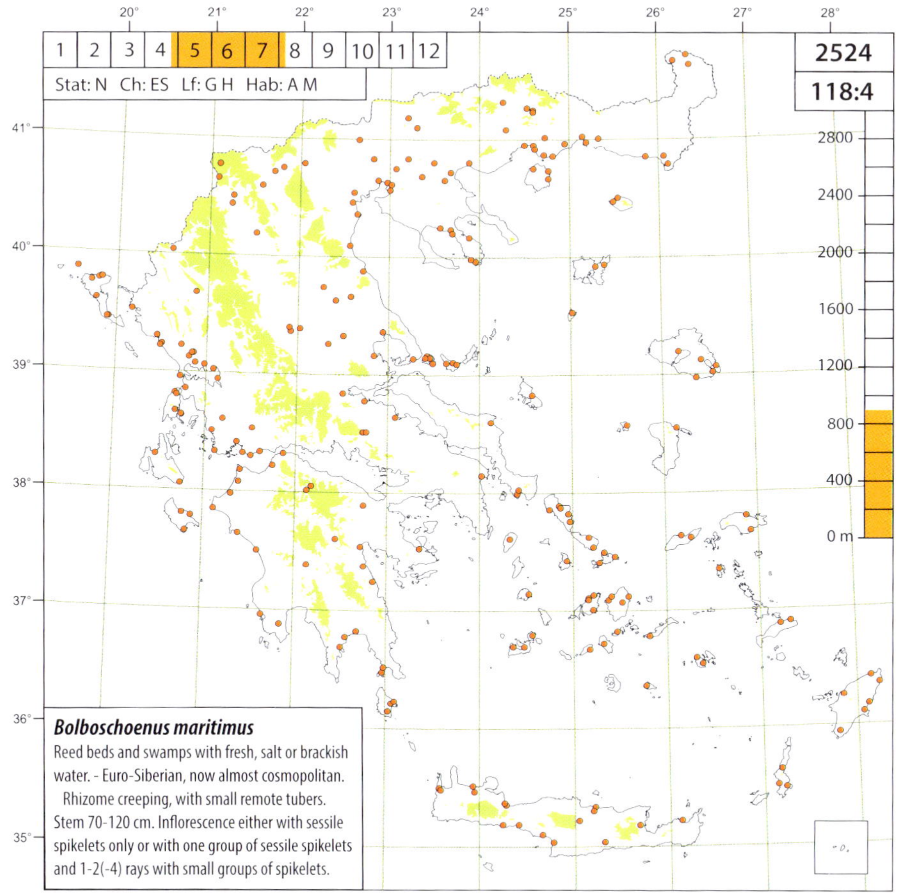

Bolboschoenus maritimus

Reed beds and swamps with fresh, salt or brackish water. - Euro-Siberian, now almost cosmopolitan.

Rhizome creeping, with small remote tubers. Stem 70-120 cm. Inflorescence either with sessile spikelets only or with one group of sessile spikelets and 1-2(-4) rays with small groups of spikelets.

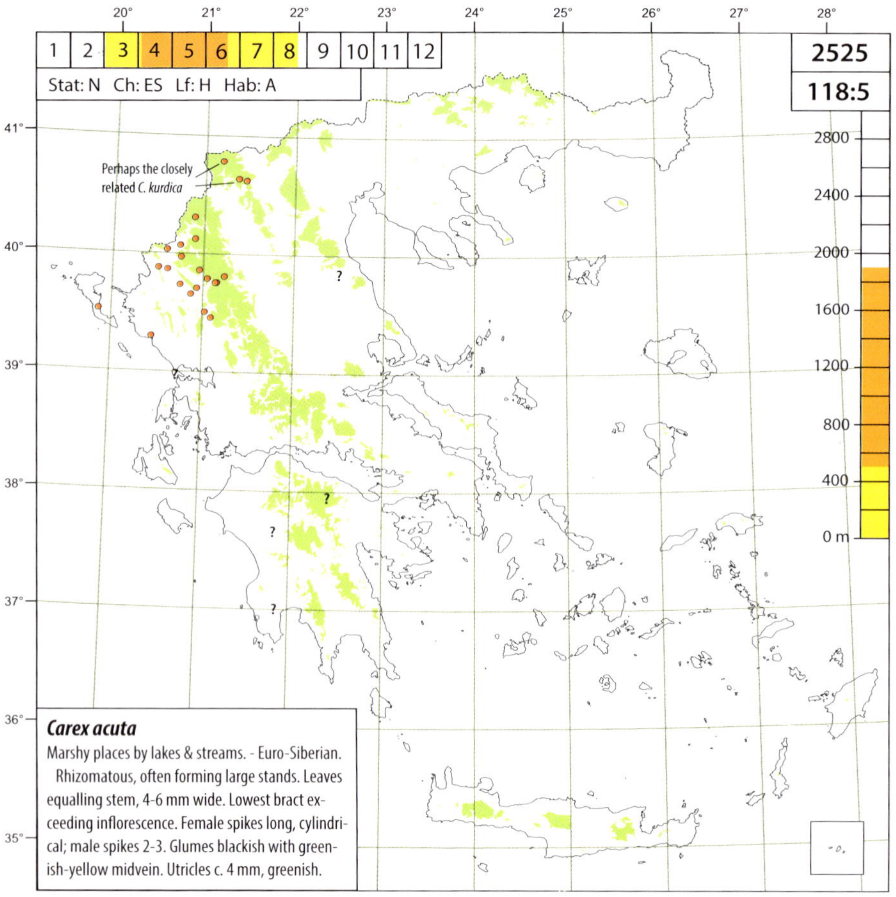

Carex acuta

Marshy places by lakes & streams. - Euro-Siberian.

Rhizomatous, often forming large stands. Leaves equalling stem, 4-6 mm wide. Lowest bract exceeding inflorescence. Female spikes long, cylindrical; male spikes 2-3. Glumes blackish with green-ish-yellow midvein. Utricles c. 4 mm, greenish.

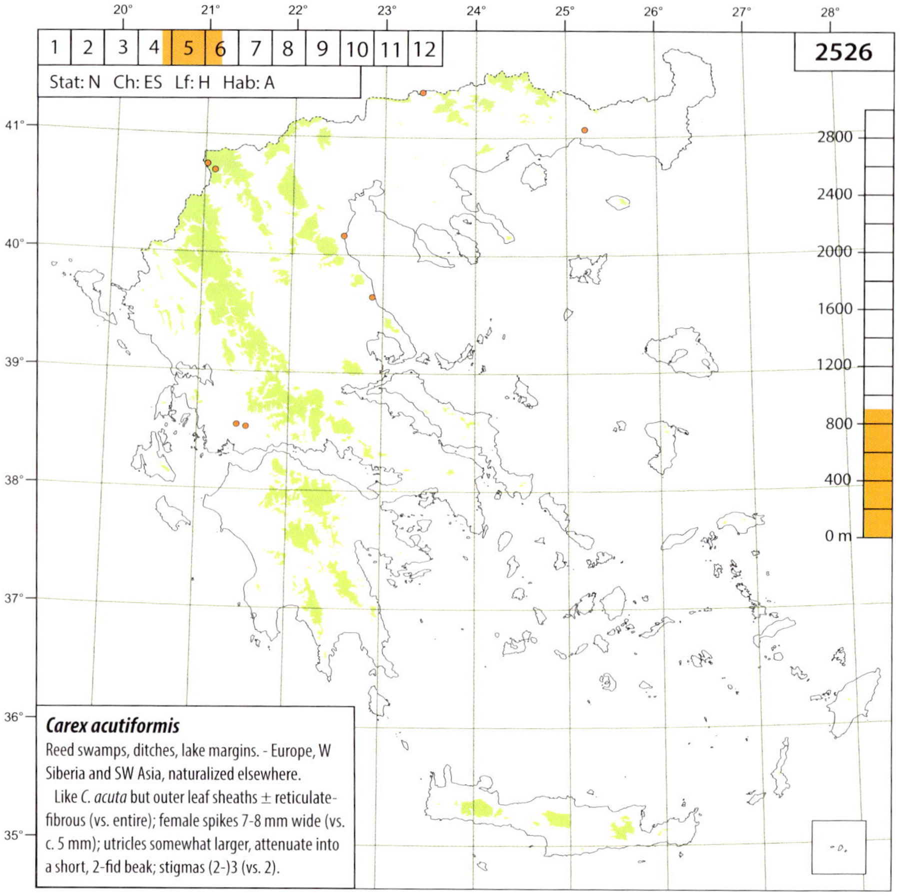

Carex acutiformis

Reed swamps, ditches, lake margins. - Europe, W Siberia and SW Asia, naturalized elsewhere.

Like *C. acuta* but outer leaf sheaths ± reticulate-fibrous (vs. entire); female spikes 7-8 mm wide (vs. c. 5 mm); utricles somewhat larger, attenuate into a short, 2-fid beak; stigmas (2-)3 (vs. 2).

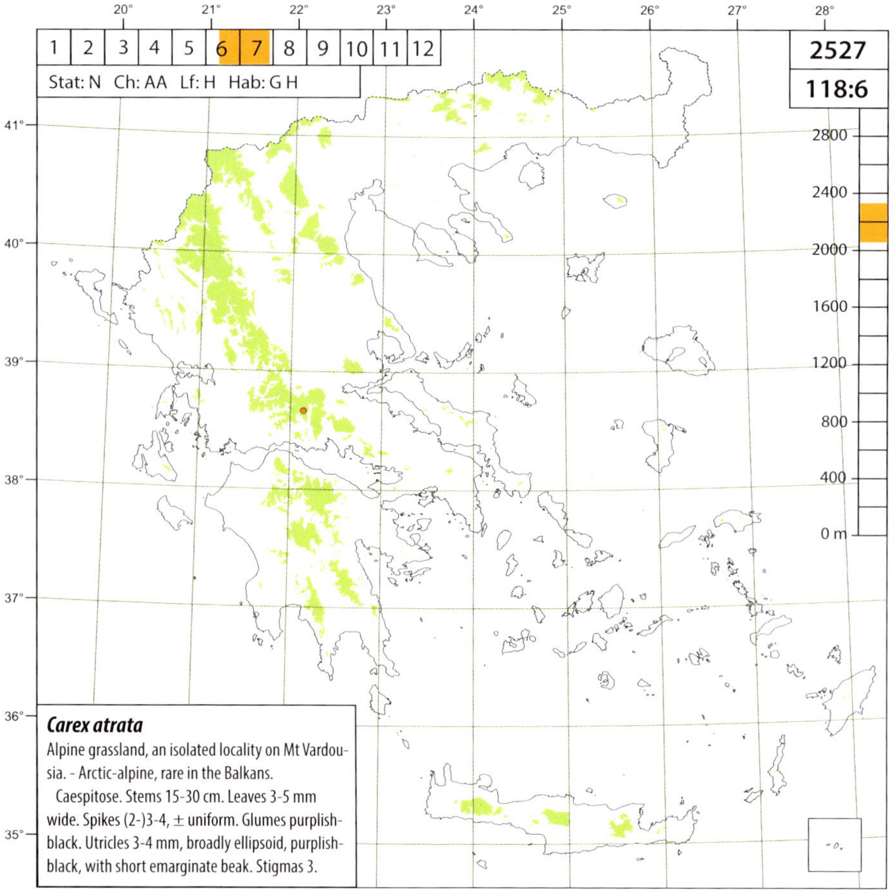

Carex atrata

Alpine grassland, an isolated locality on Mt Vardousia. - Arctic-alpine, rare in the Balkans.

Caespitose. Stems 15-30 cm. Leaves 3-5 mm wide. Spikes (2-)3-4, ± uniform. Glumes purplish-black. Utricles 3-4 mm, broadly ellipsoid, purplish-black, with short emarginate beak. Stigmas 3.

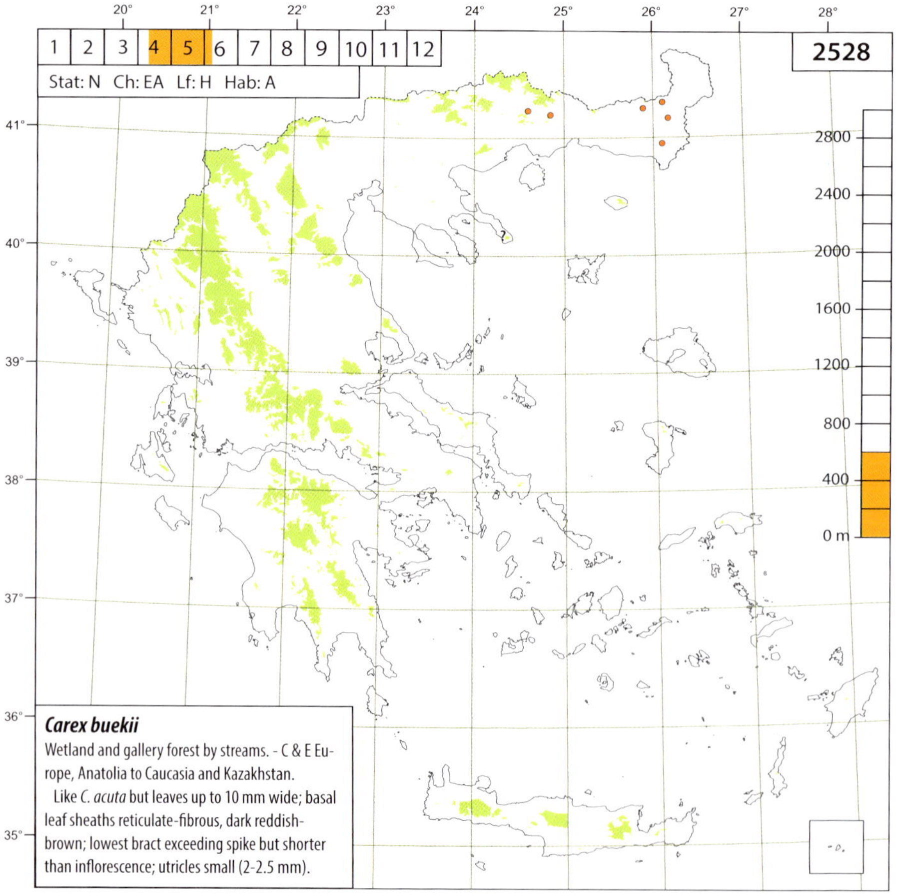

Carex buekii

Wetland and gallery forest by streams. - C & E Europe, Anatolia to Caucasia and Kazakhstan.

Like *C. acuta* but leaves up to 10 mm wide; basal leaf sheaths reticulate-fibrous, dark reddish-brown; lowest bract exceeding spike but shorter than inflorescence; utricles small (2-2.5 mm).

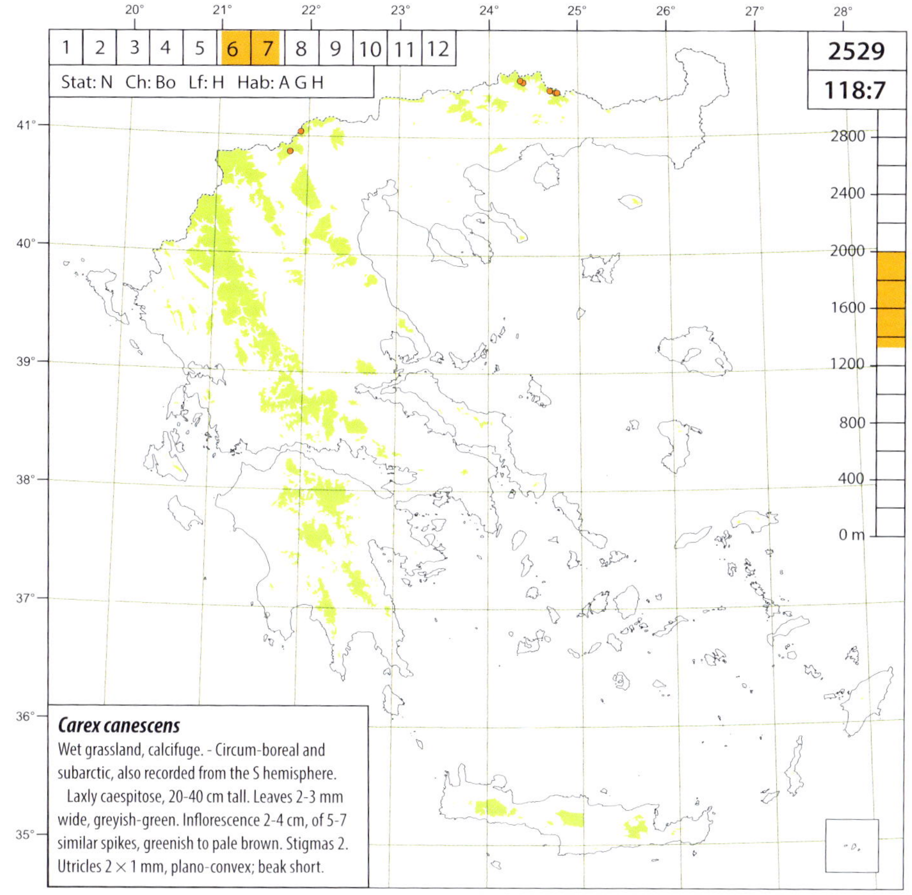

Carex canescens
Wet grassland, calcifuge. - Circum-boreal and subarctic, also recorded from the S hemisphere.
Laxly caespitose, 20-40 cm tall. Leaves 2-3 mm wide, greyish-green. Inflorescence 2-4 cm, of 5-7 similar spikes, greenish to pale brown. Stigmas 2. Utricles 2 × 1 mm, plano-convex; beak short.

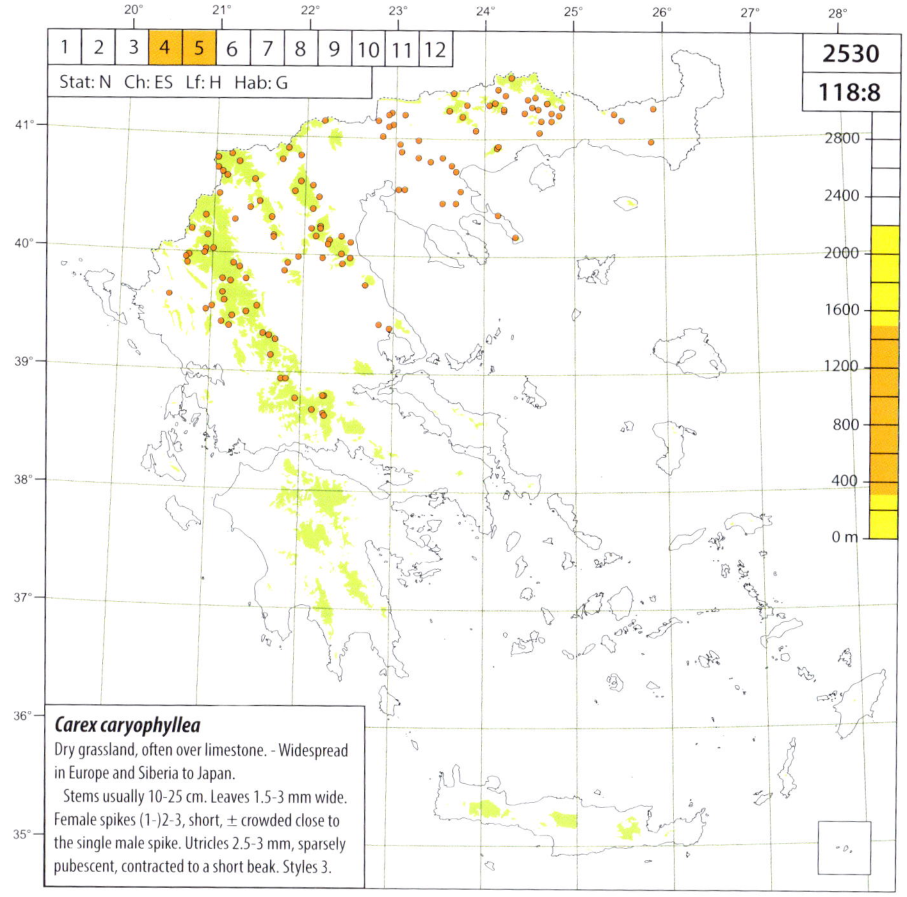

Carex caryophyllea
Dry grassland, often over limestone. - Widespread in Europe and Siberia to Japan.
Stems usually 10-25 cm. Leaves 1.5-3 mm wide. Female spikes (1-)2-3, short, ± crowded close to the single male spike. Utricles 2.5-3 mm, sparsely pubescent, contracted to a short beak. Styles 3.

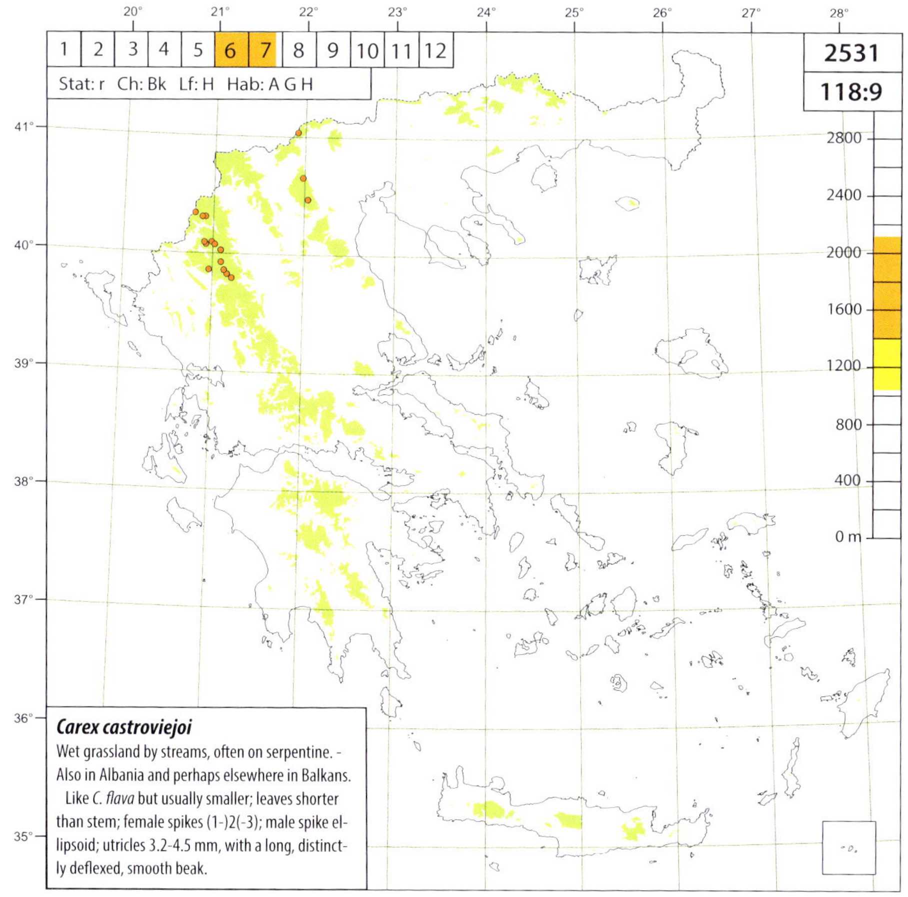

Carex castroviejoi
Wet grassland by streams, often on serpentine. - Also in Albania and perhaps elsewhere in Balkans.
Like *C. flava* but usually smaller; leaves shorter than stem; female spikes (1-)2(-3); male spike ellipsoid; utricles 3.2-4.5 mm, with a long, distinctly deflexed, smooth beak.

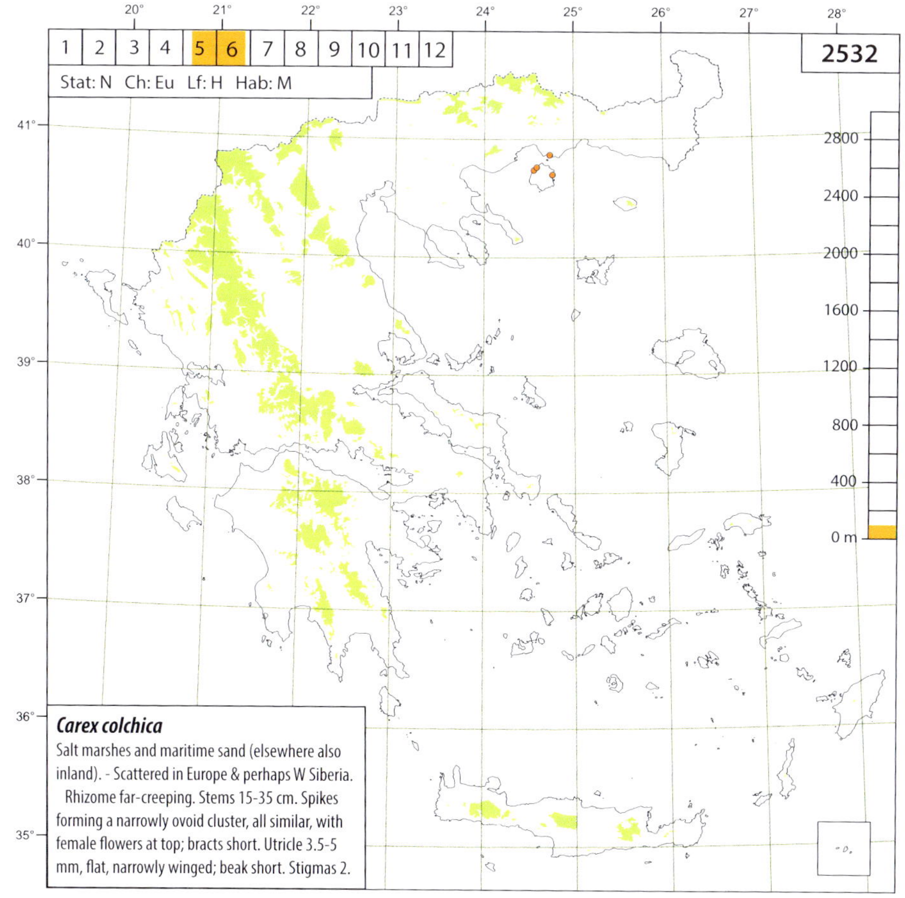

Carex colchica
Salt marshes and maritime sand (elsewhere also inland). - Scattered in Europe & perhaps W Siberia.
Rhizome far-creeping. Stems 15-35 cm. Spikes forming a narrowly ovoid cluster, all similar, with female flowers at top; bracts short. Utricle 3.5-5 mm, flat, narrowly winged; beak short. Stigmas 2.

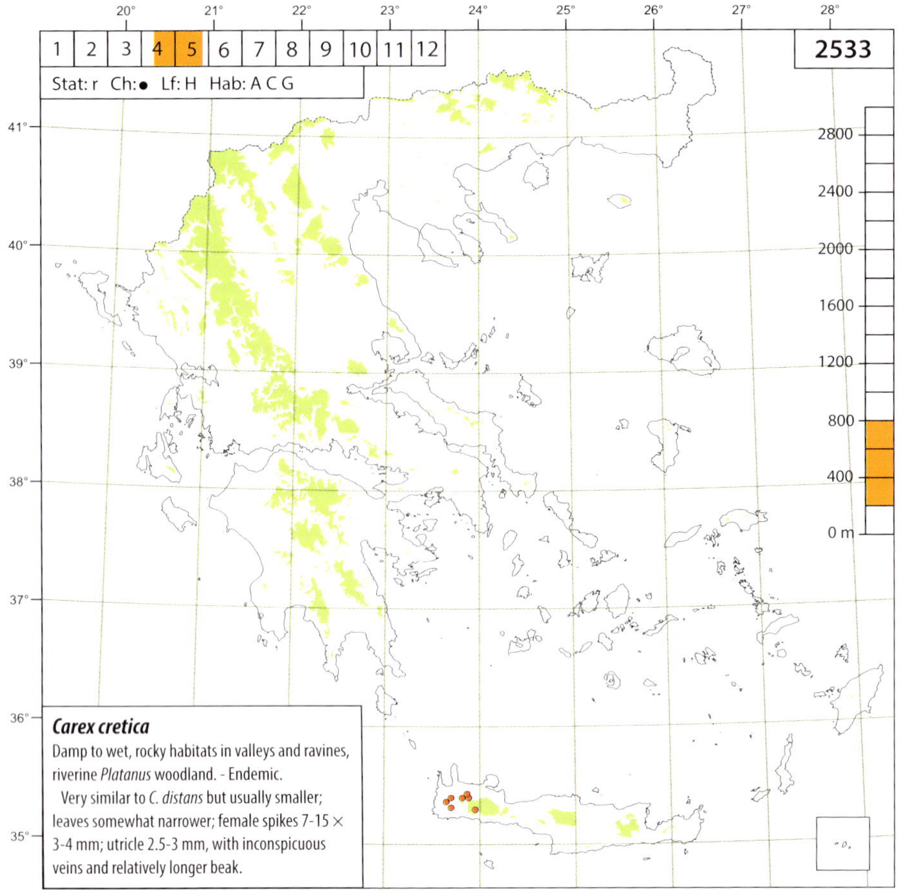

Carex cretica
Damp to wet, rocky habitats in valleys and ravines, riverine *Platanus* woodland. - Endemic.
Very similar to *C. distans* but usually smaller; leaves somewhat narrower; female spikes 7-15 × 3-4 mm; utricle 2.5-3 mm, with inconspicuous veins and relatively longer beak.

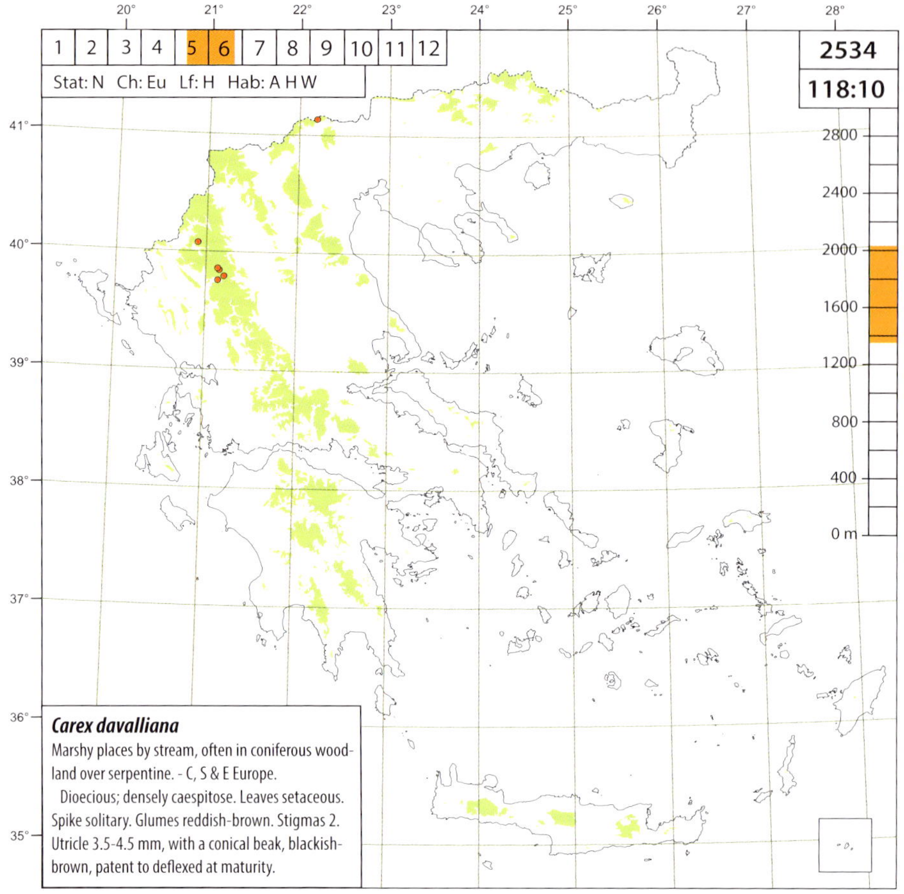

Carex davalliana
Marshy places by stream, often in coniferous woodland over serpentine. - C, S & E Europe.
Dioecious; densely caespitose. Leaves setaceous. Spike solitary. Glumes reddish-brown. Stigmas 2. Utricle 3.5-4.5 mm, with a conical beak, blackish-brown, patent to deflexed at maturity.

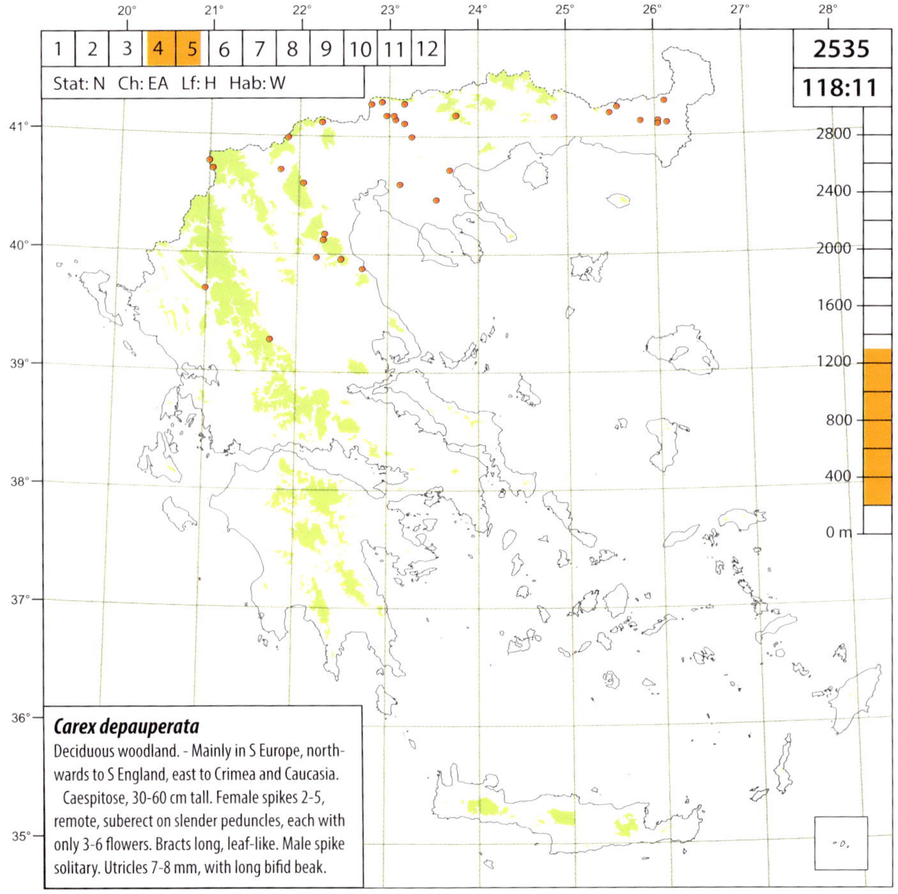

Carex depauperata
Deciduous woodland. - Mainly in S Europe, northwards to S England, east to Crimea and Caucasia.
Caespitose, 30-60 cm tall. Female spikes 2-5, remote, suberect on slender peduncles, each with only 3-6 flowers. Bracts long, leaf-like. Male spike solitary. Utricles 7-8 mm, with long bifid beak.

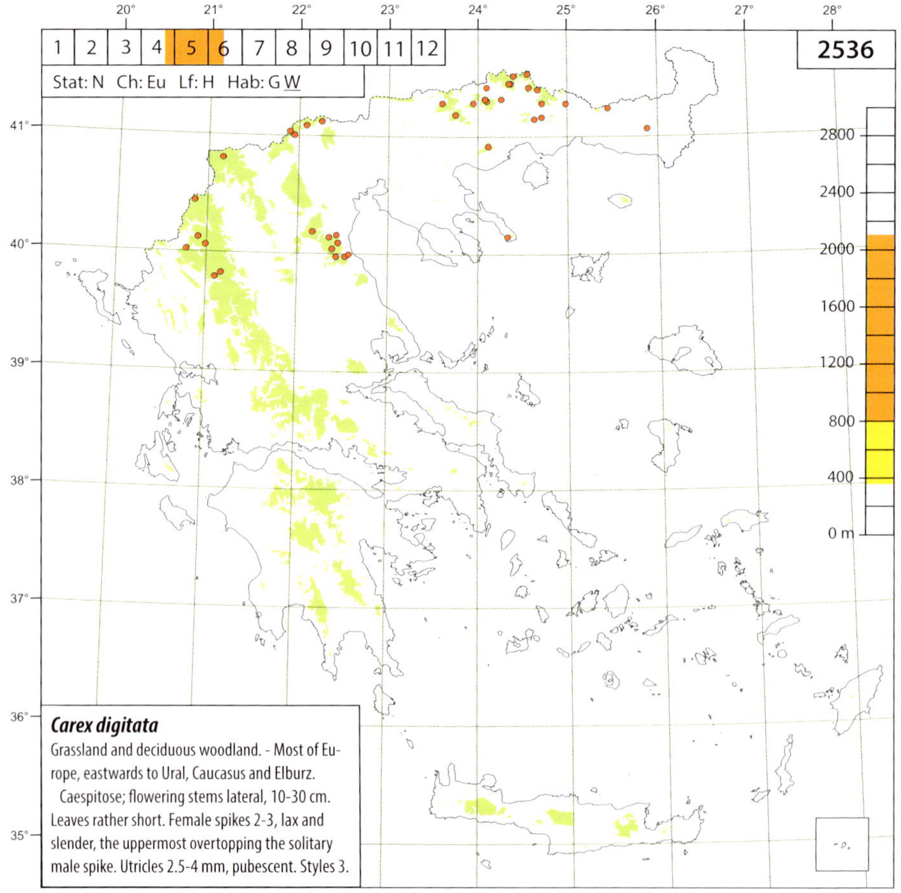

Carex digitata
Grassland and deciduous woodland. - Most of Europe, eastwards to Ural, Caucasus and Elburz.
Caespitose; flowering stems lateral, 10-30 cm. Leaves rather short. Female spikes 2-3, lax and slender, the uppermost overtopping the solitary male spike. Utricles 2.5-4 mm, pubescent. Styles 3.

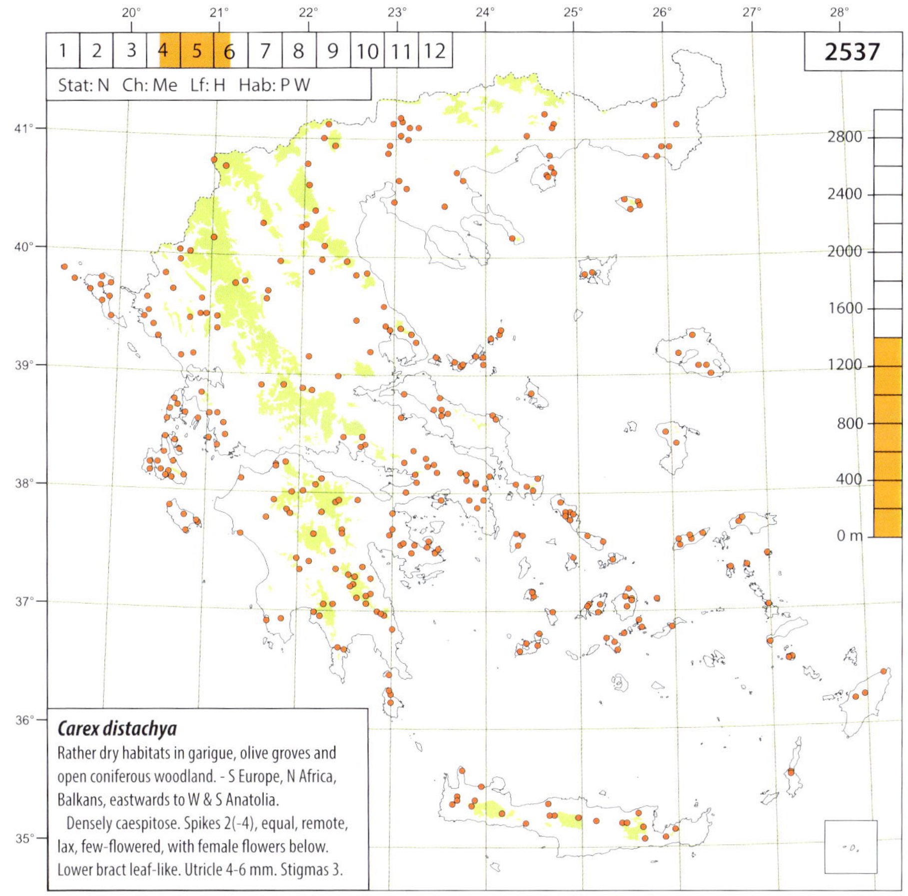

Carex distachya
Rather dry habitats in garigue, olive groves and open coniferous woodland. - S Europe, N Africa, Balkans, eastwards to W & S Anatolia.
Densely caespitose. Spikes 2(-4), equal, remote, lax, few-flowered, with female flowers below. Lower bract leaf-like. Utricle 4-6 mm. Stigmas 3.

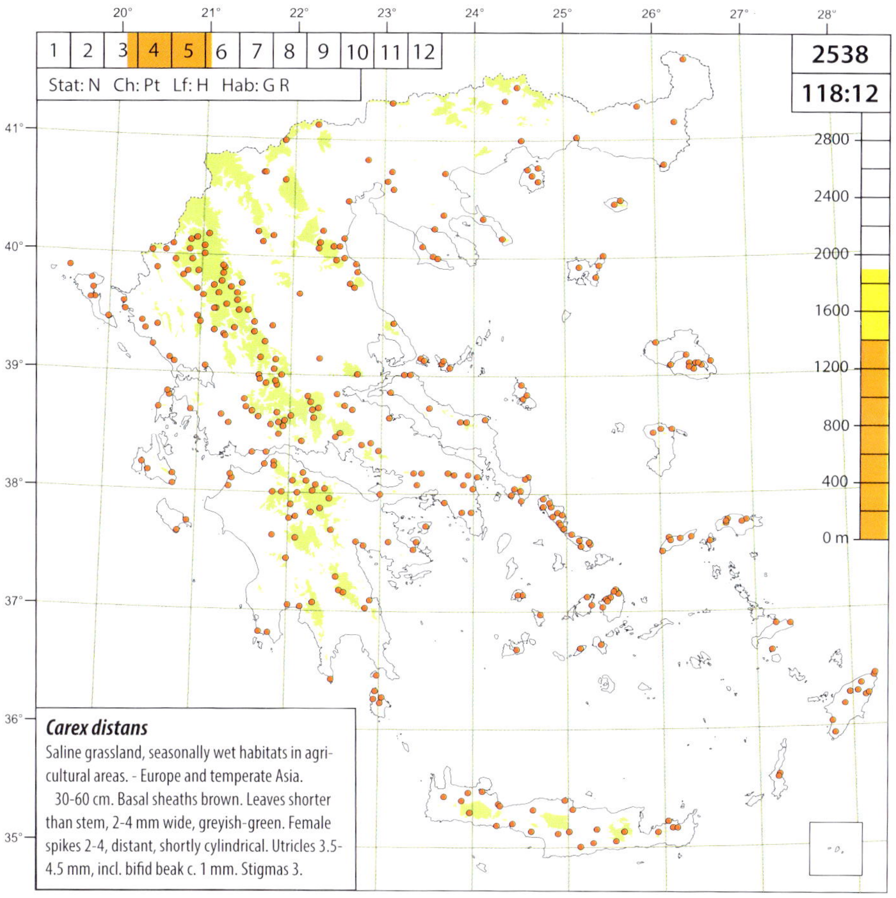

Carex distans
Saline grassland, seasonally wet habitats in agricultural areas. - Europe and temperate Asia.
30-60 cm. Basal sheaths brown. Leaves shorter than stem, 2-4 mm wide, greyish-green. Female spikes 2-4, distant, shortly cylindrical. Utricles 3.5-4.5 mm, incl. bifid beak c. 1 mm. Stigmas 3.

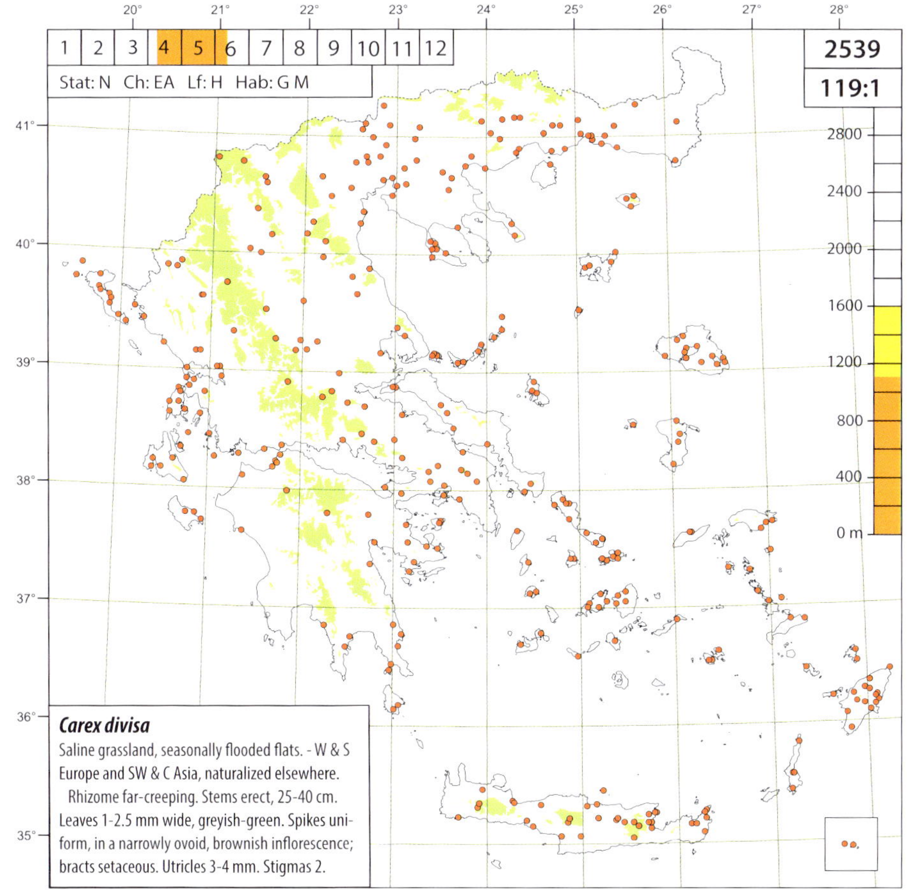

Carex divisa
Saline grassland, seasonally flooded flats. - W & S Europe and SW & C Asia, naturalized elsewhere.
Rhizome far-creeping. Stems erect, 25-40 cm. Leaves 1-2.5 mm wide, greyish-green. Spikes uniform, in a narrowly ovoid, brownish inflorescence; bracts setaceous. Utricles 3-4 mm. Stigmas 2.

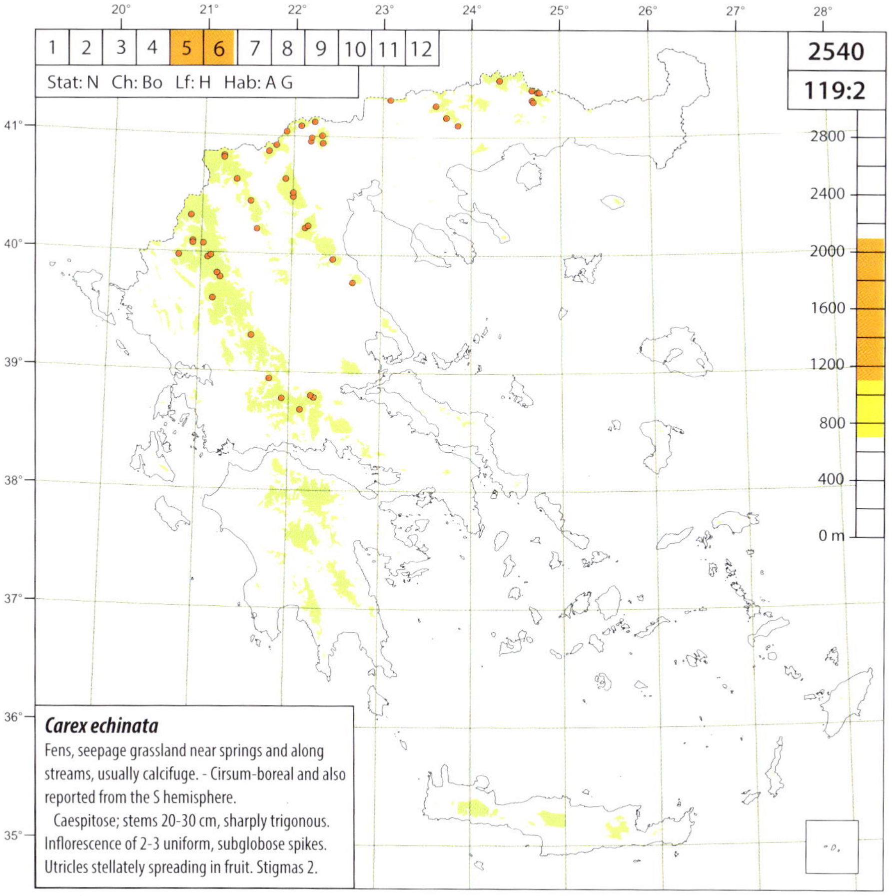

Carex echinata
Fens, seepage grassland near springs and along streams, usually calcifuge. - Circum-boreal and also reported from the S hemisphere.
Caespitose; stems 20-30 cm, sharply trigonous. Inflorescence of 2-3 uniform, subglobose spikes. Utricles stellately spreading in fruit. Stigmas 2.

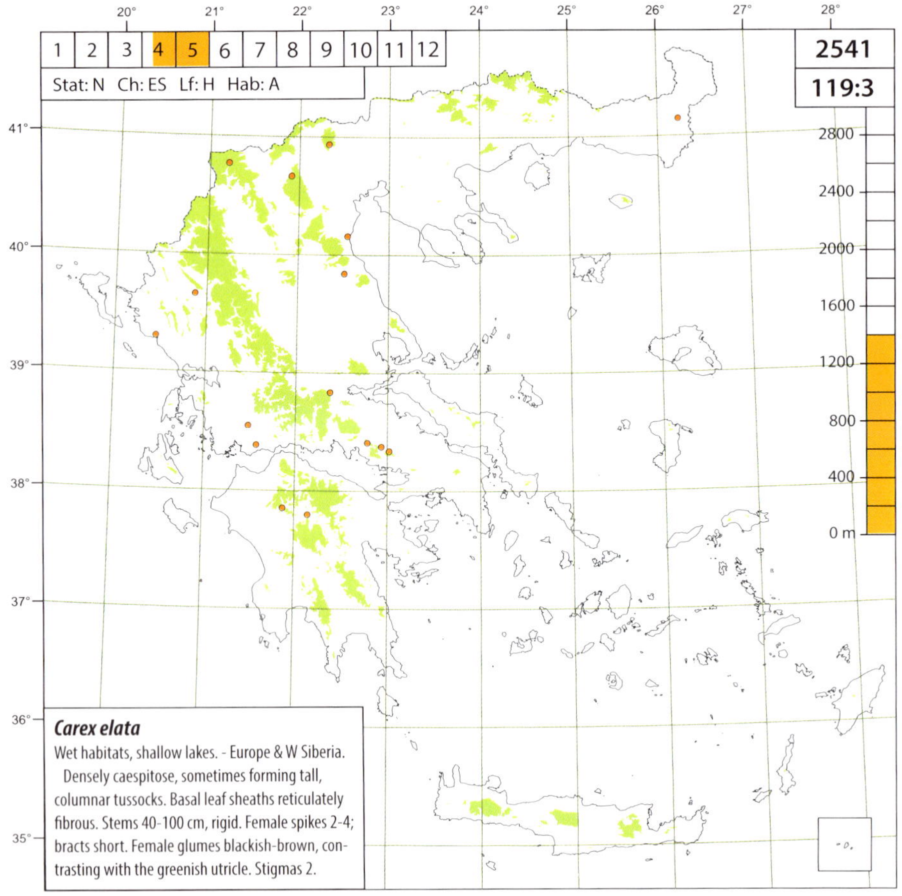

Carex elata

Wet habitats, shallow lakes. - Europe & W Siberia.

Densely caespitose, sometimes forming tall, columnar tussocks. Basal leaf sheaths reticulately fibrous. Stems 40-100 cm, rigid. Female spikes 2-4; bracts short. Female glumes blackish-brown, contrasting with the greenish utricle. Stigmas 2.

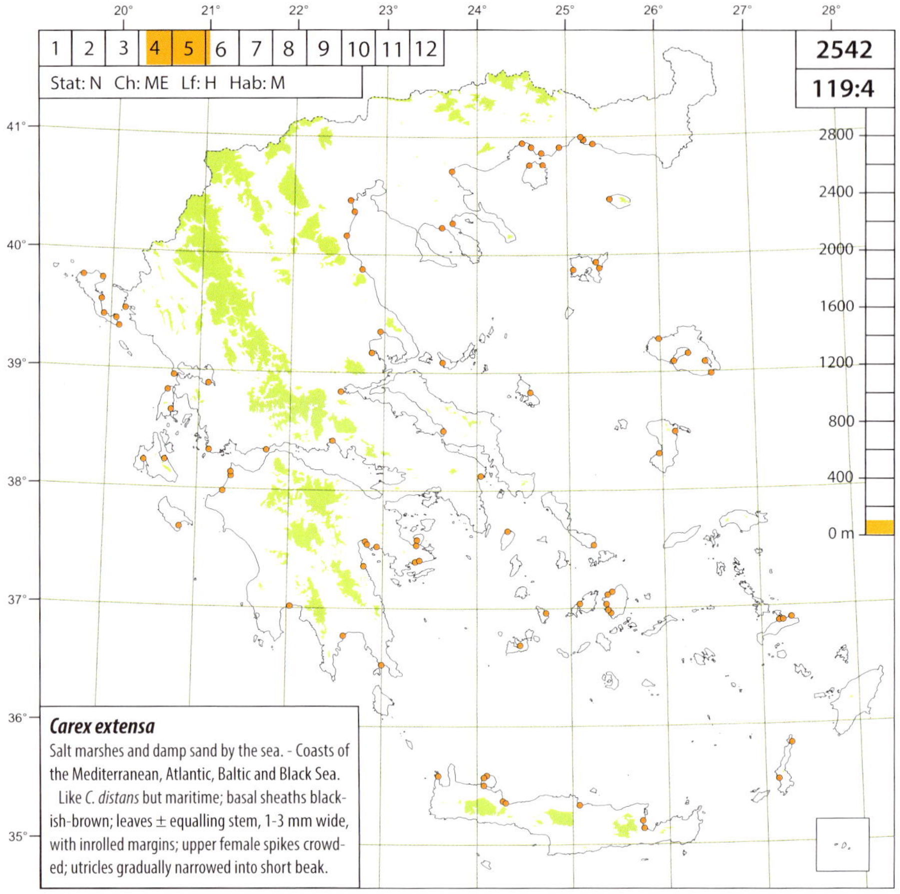

Carex extensa

Salt marshes and damp sand by the sea. - Coasts of the Mediterranean, Atlantic, Baltic and Black Sea.

Like *C. distans* but maritime; basal sheaths blackish-brown; leaves ± equalling stem, 1-3 mm wide, with inrolled margins; upper female spikes crowded; utricles gradually narrowed into short beak.

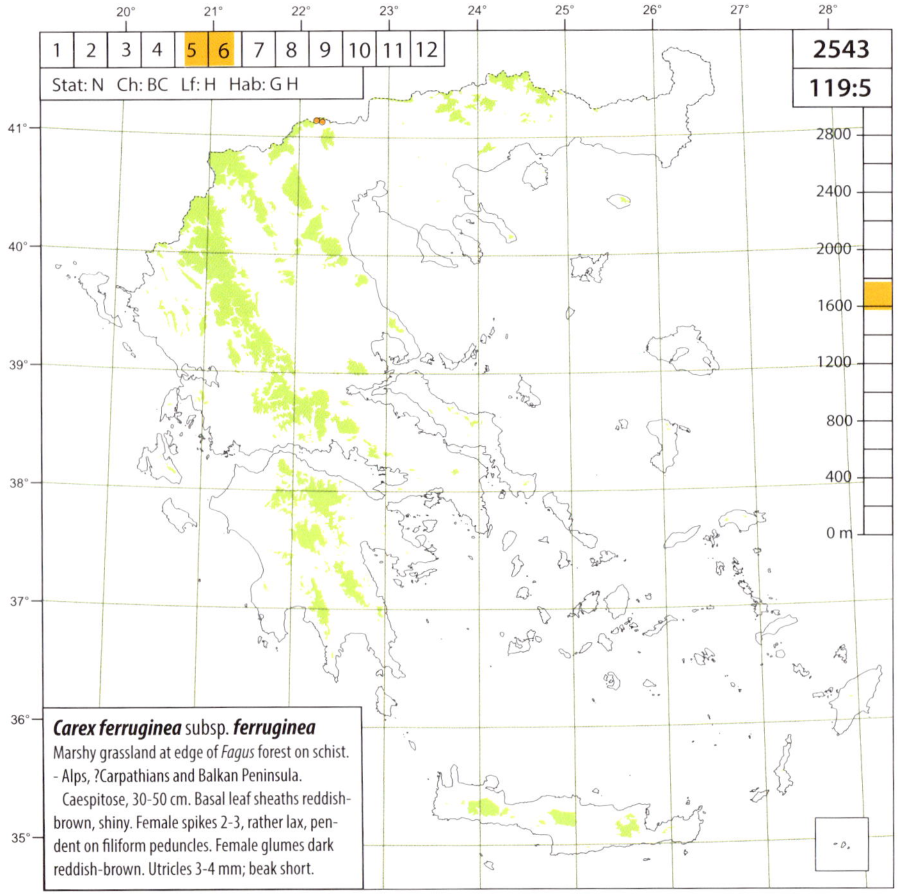

Carex ferruginea subsp. ***ferruginea***

Marshy grassland at edge of *Fagus* forest on schist. - Alps, ?Carpathians and Balkan Peninsula.

Caespitose, 30-50 cm. Basal leaf sheaths reddish-brown, shiny. Female spikes 2-3, rather lax, pendent on filiform peduncles. Female glumes dark reddish-brown. Utricles 3-4 mm; beak short.

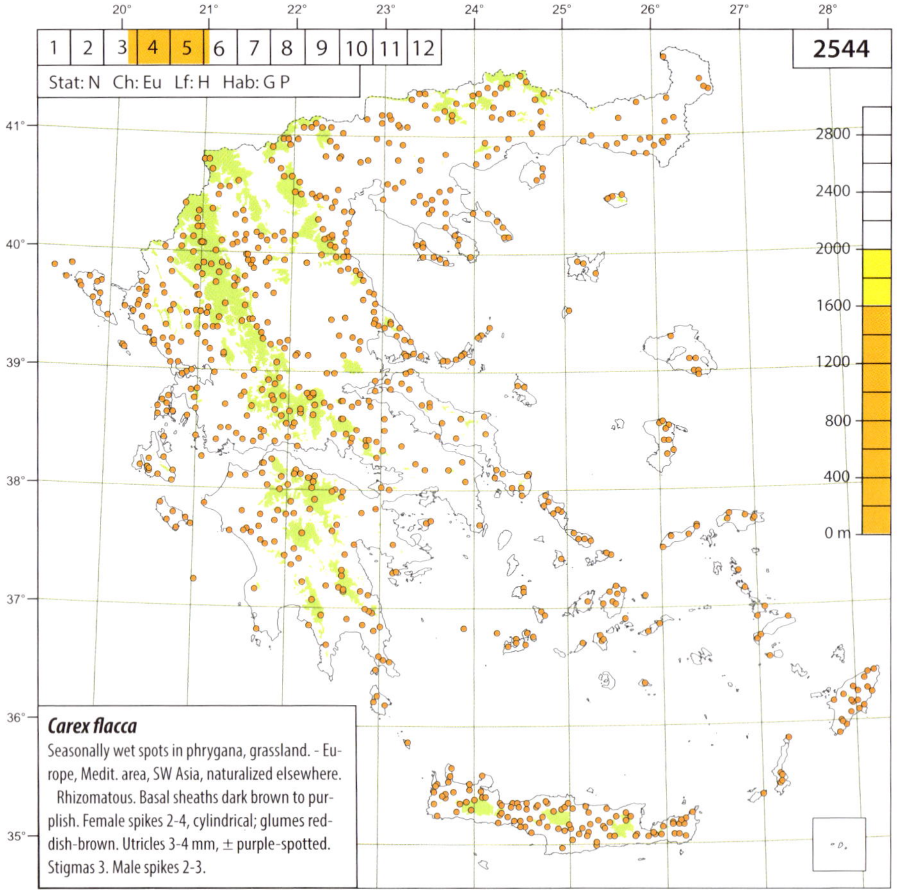

Carex flacca

Seasonally wet spots in phrygana, grassland. - Europe, Medit. area, SW Asia, naturalized elsewhere.

Rhizomatous. Basal sheaths dark brown to purplish. Female spikes 2-4, cylindrical; glumes red-dish-brown. Utricles 3-4 mm, ± purple-spotted. Stigmas 3. Male spikes 2-3.

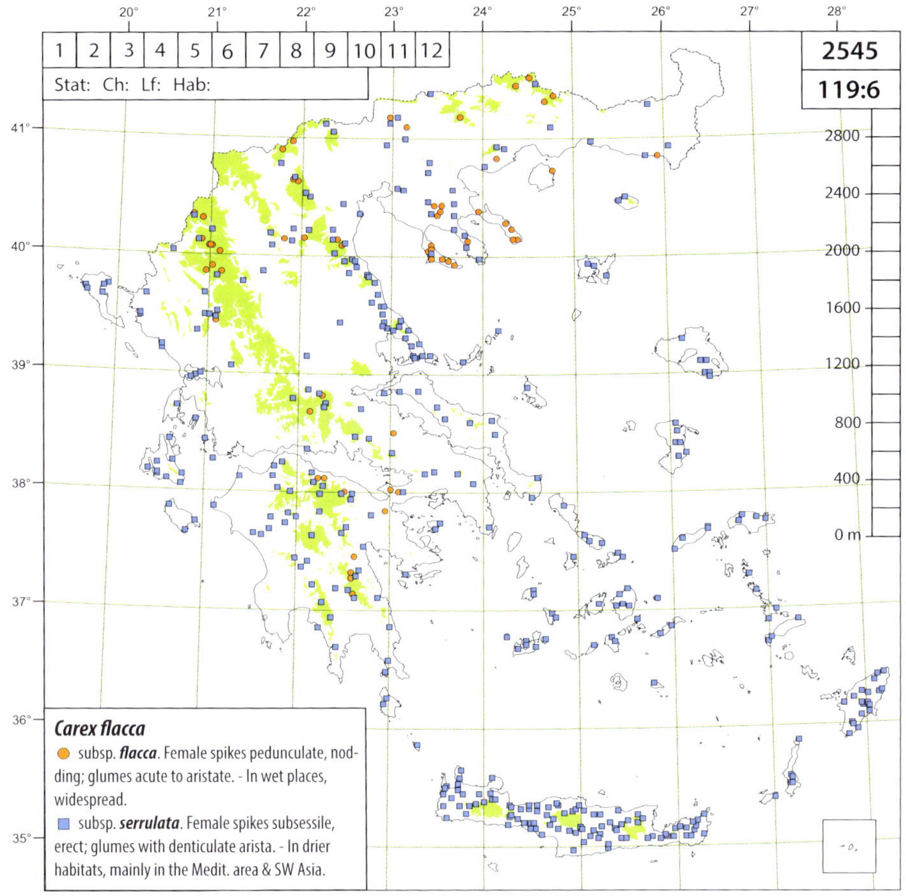

Carex flacca

● subsp. ***flacca***. Female spikes pedunculate, nodding; glumes acute to aristate. - In wet places, widespread.

■ subsp. ***serrulata***. Female spikes subsessile, erect; glumes with denticulate arista. - In drier habitats, mainly in the Medit. area & SW Asia.

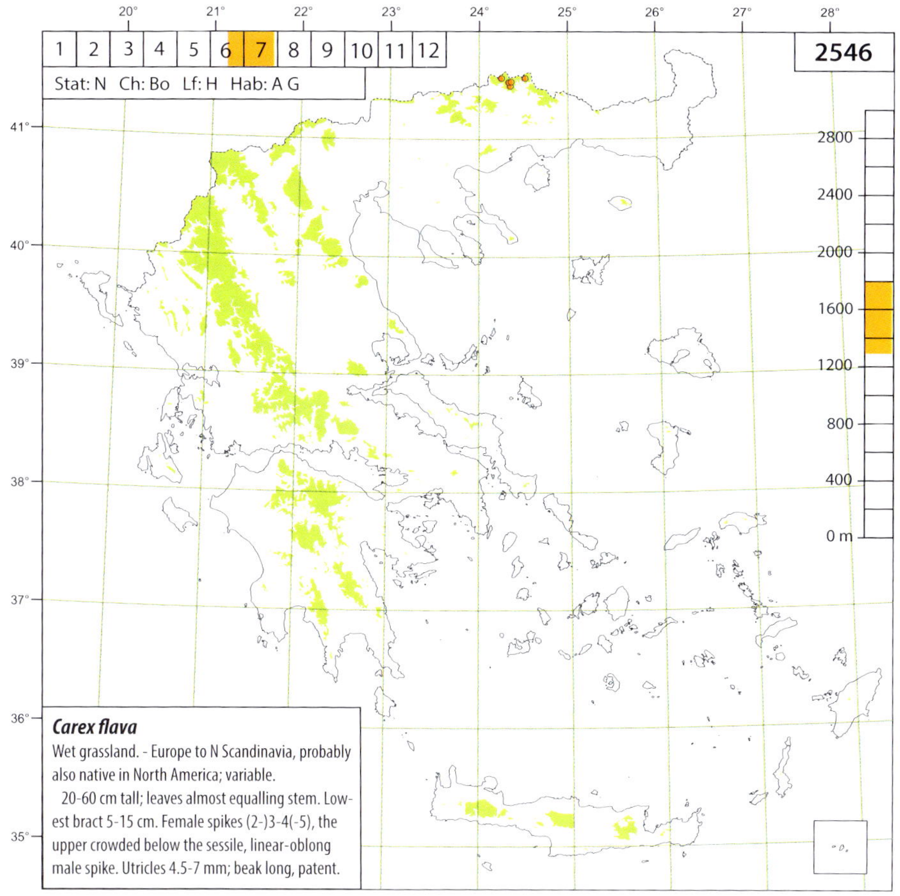

Carex flava

Wet grassland. - Europe to N Scandinavia, probably also native in North America; variable.

20-60 cm tall; leaves almost equalling stem. Lowest bract 5-15 cm. Female spikes (2-)3-4(-5), the upper crowded below the sessile, linear-oblong male spike. Utricles 4.5-7 mm; beak long, patent.

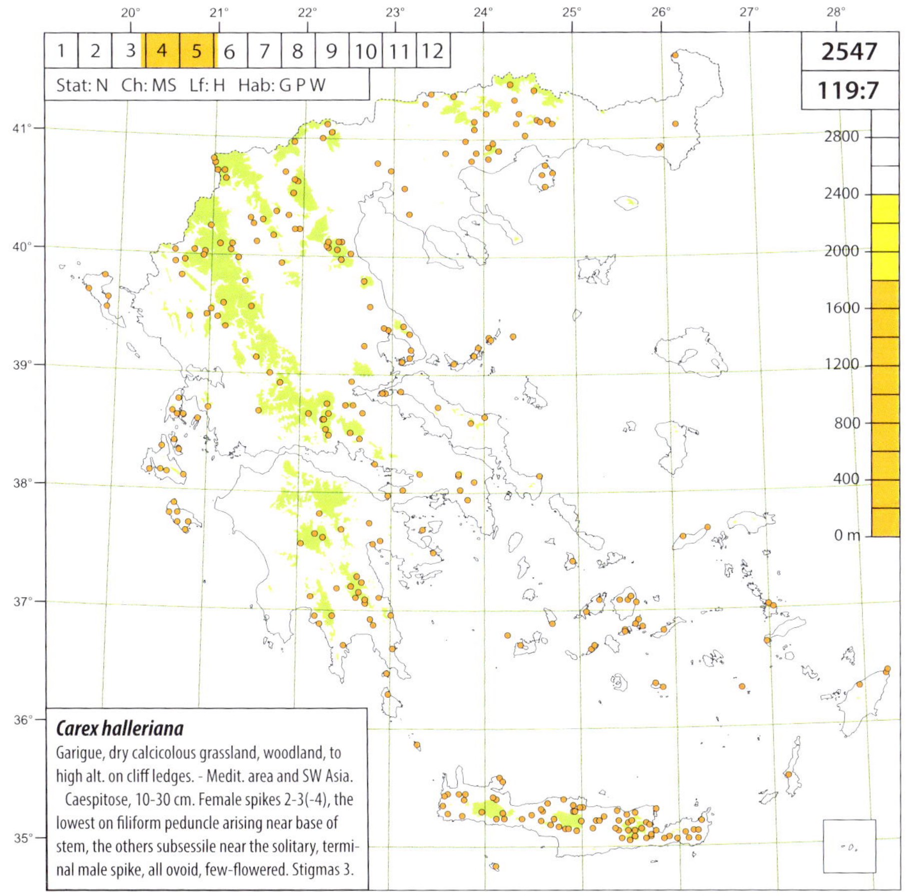

Carex halleriana

Garigue, dry calcicolous grassland, woodland, to high alt. on cliff ledges. - Medit. area and SW Asia.

Caespitose, 10-30 cm. Female spikes 2-3(-4), the lowest on filiform peduncle arising near base of stem, the others subsessile near the solitary, terminal male spike, all ovoid, few-flowered. Stigmas 3.

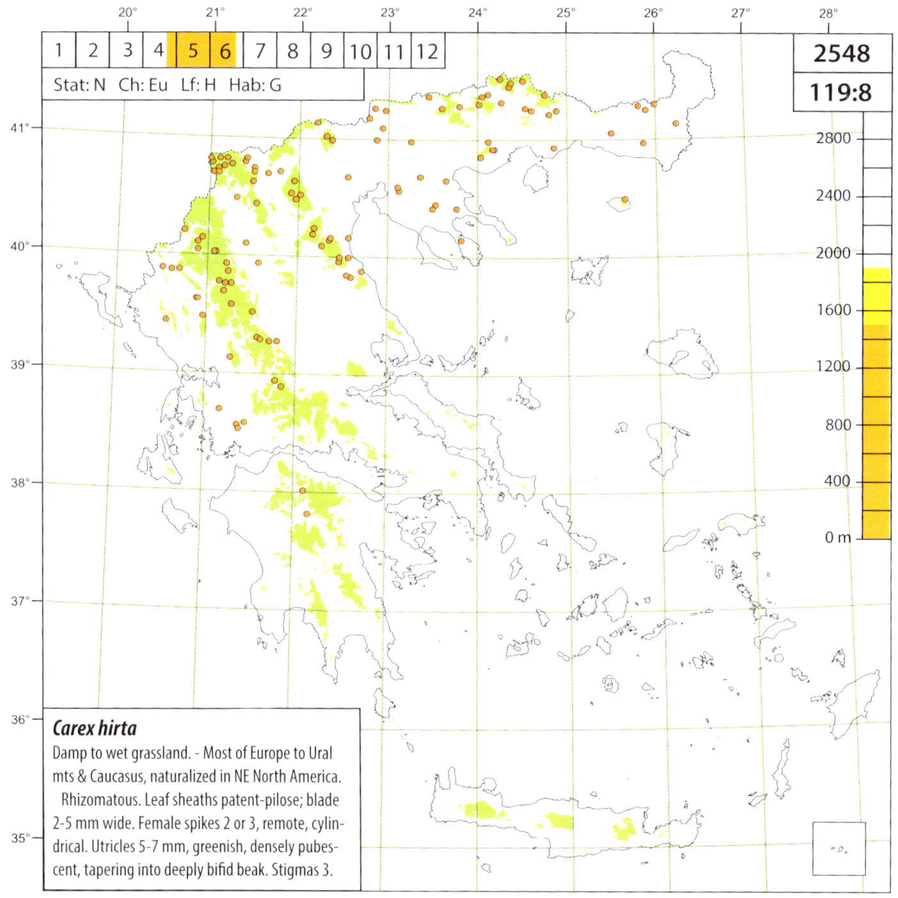

Carex hirta

Damp to wet grassland. - Most of Europe to Ural mts & Caucasus, naturalized in NE North America.

Rhizomatous. Leaf sheaths patent-pilose; blade 2-5 mm wide. Female spikes 2 or 3, remote, cylindrical. Utricles 5-7 mm, greenish, densely pubescent, tapering into deeply bifid beak. Stigmas 3.

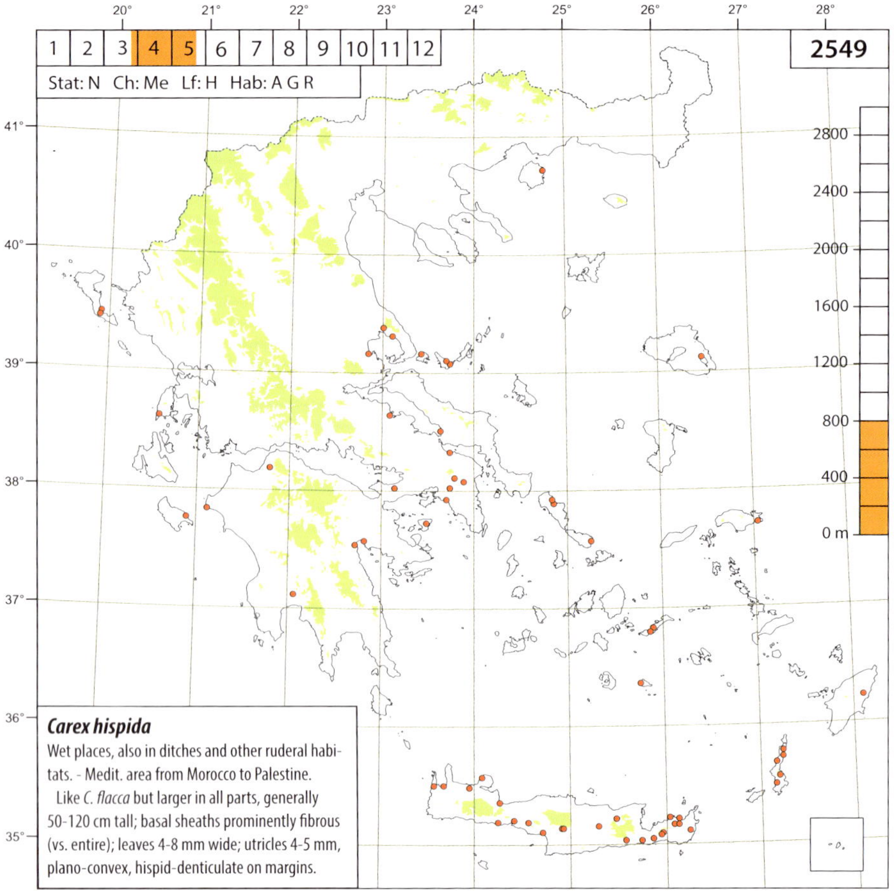

Carex hispida

Wet places, also in ditches and other ruderal habitats. - Medit. area from Morocco to Palestine.

Like *C. flacca* but larger in all parts, generally 50-120 cm tall; basal sheaths prominently fibrous (vs. entire); leaves 4-8 mm wide; utricles 4-5 mm, plano-convex, hispid-denticulate on margins.

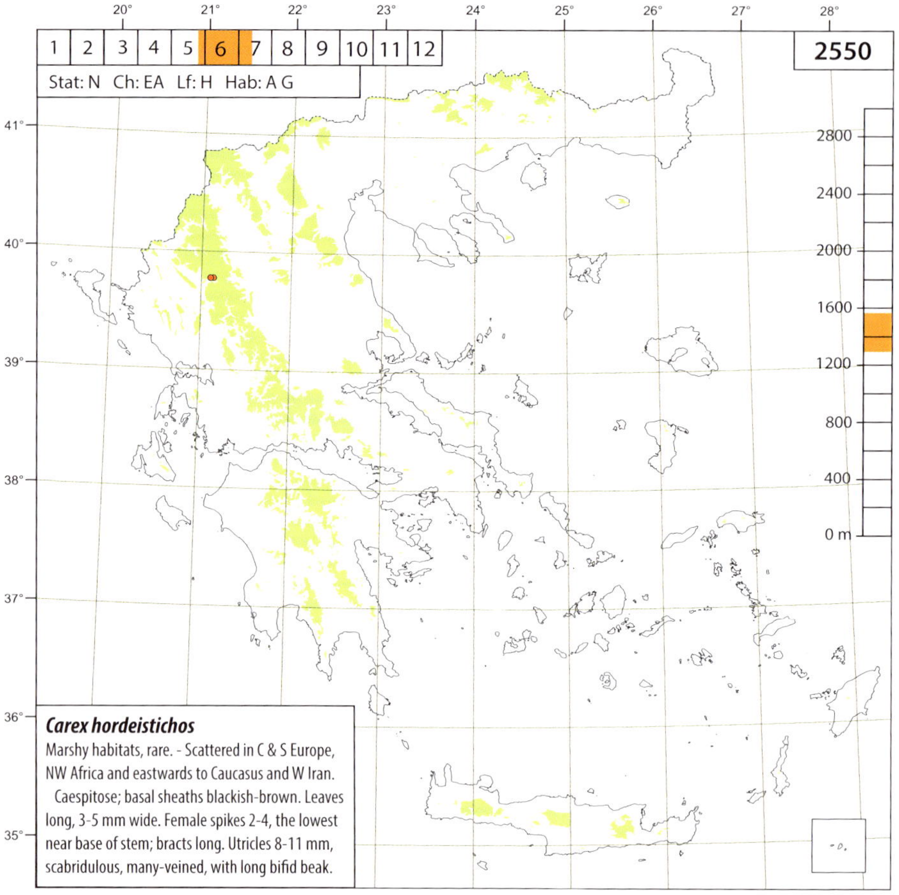

Carex hordeistichos

Marshy habitats, rare. - Scattered in C & S Europe, NW Africa and eastwards to Caucasus and W Iran.

Caespitose; basal sheaths blackish-brown. Leaves long, 3-5 mm wide. Female spikes 2-4, the lowest near base of stem; bracts long. Utricles 8-11 mm, scabridulous, many-veined, with long bifid beak.

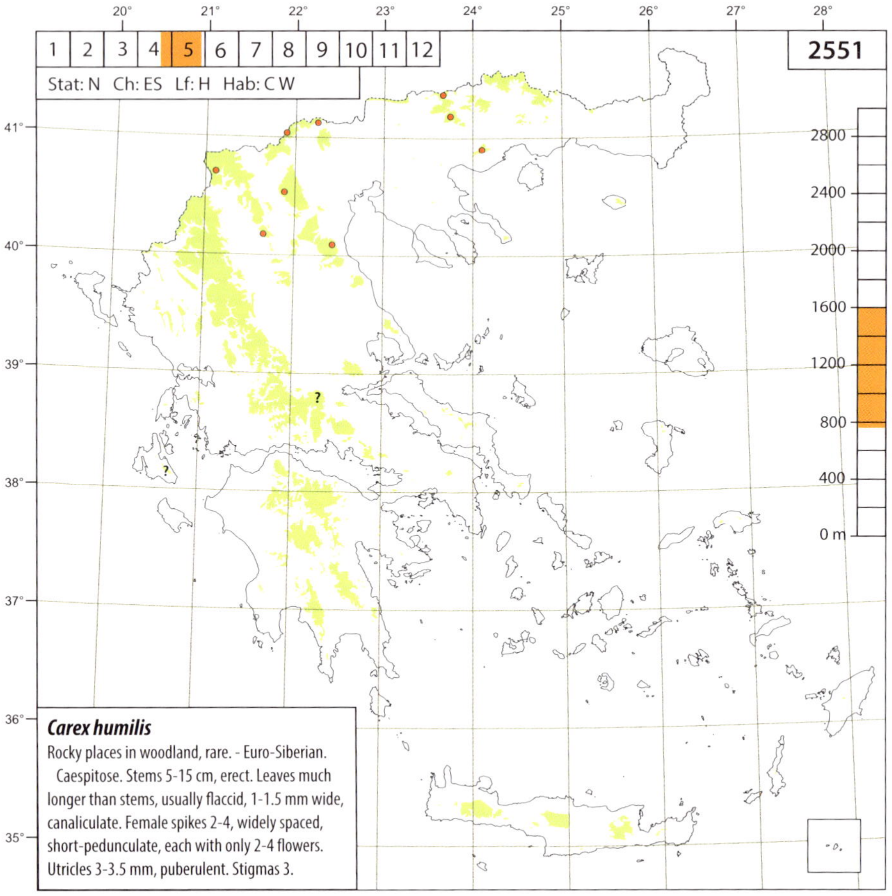

Carex humilis

Rocky places in woodland, rare. - Euro-Siberian.

Caespitose. Stems 5-15 cm, erect. Leaves much longer than stems, usually flaccid, 1-1.5 mm wide, canaliculate. Female spikes 2-4, widely spaced, short-pedunculate, each with only 2-4 flowers. Utricles 3-3.5 mm, puberulent. Stigmas 3.

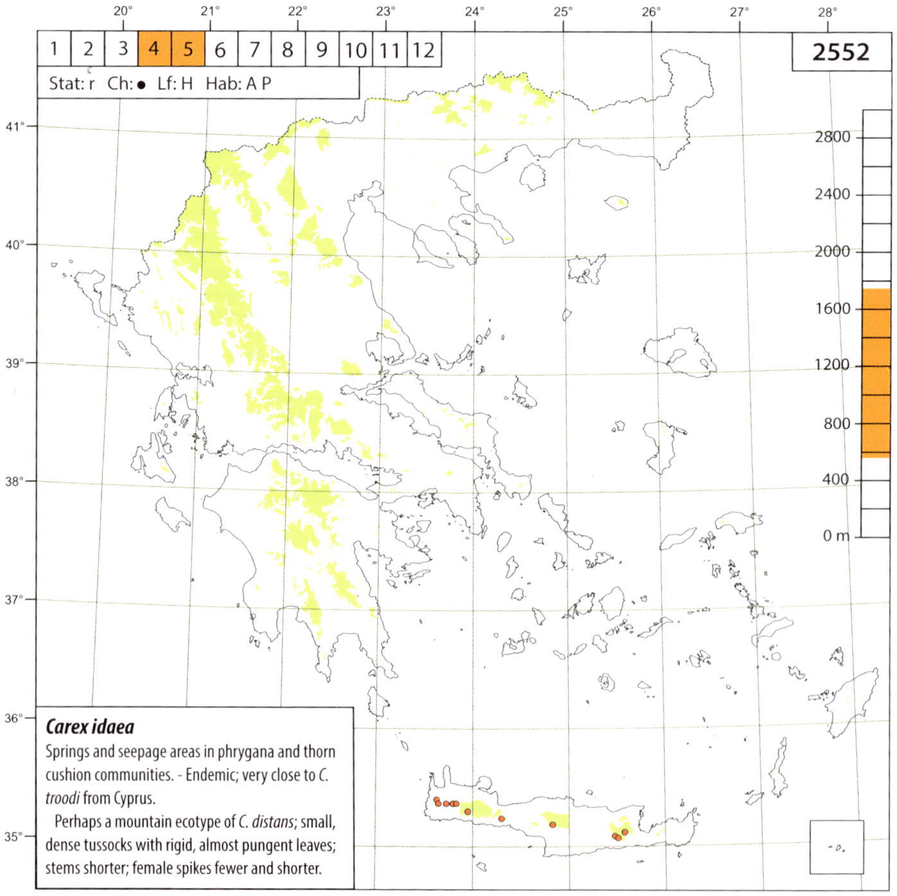

Carex idaea

Springs and seepage areas in phrygana and thorn cushion communities. - Endemic; very close to *C. troodi* from Cyprus.

Perhaps a mountain ecotype of *C. distans*; small, dense tussocks with rigid, almost pungent leaves; stems shorter; female spikes fewer and shorter.

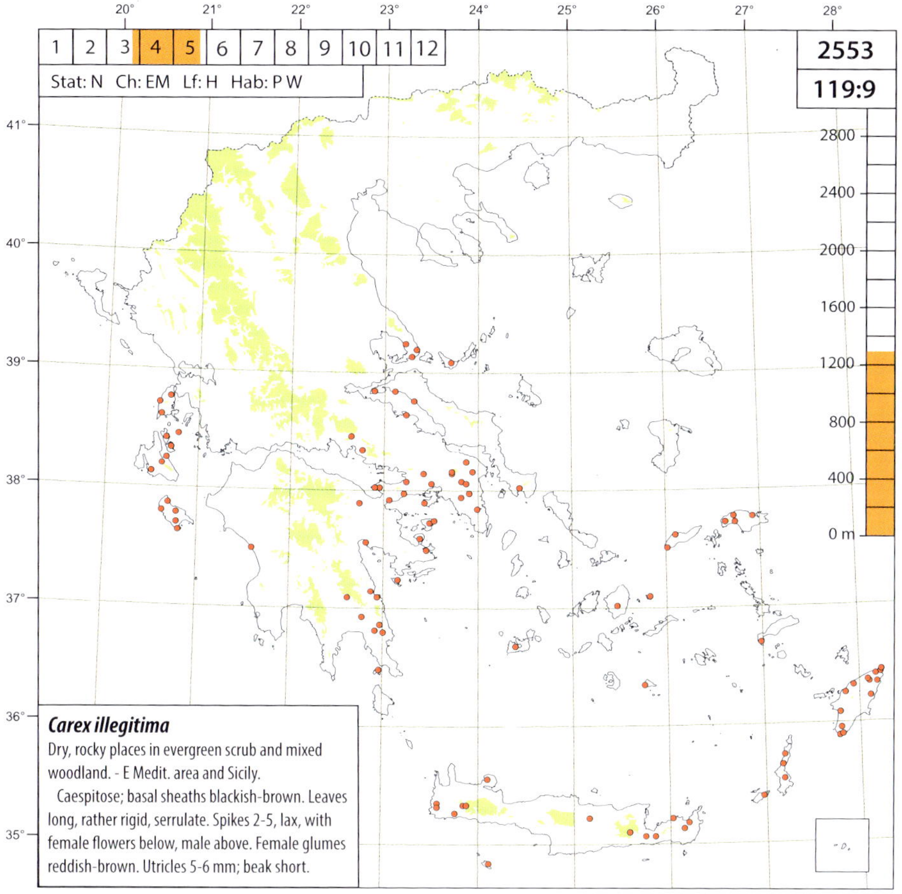

Carex illegitima

Dry, rocky places in evergreen scrub and mixed woodland. - E Medit. area and Sicily.

Caespitose; basal sheaths blackish-brown. Leaves long, rather rigid, serrulate. Spikes 2-5, lax, with female flowers below, male above. Female glumes reddish-brown. Utricles 5-6 mm; beak short.

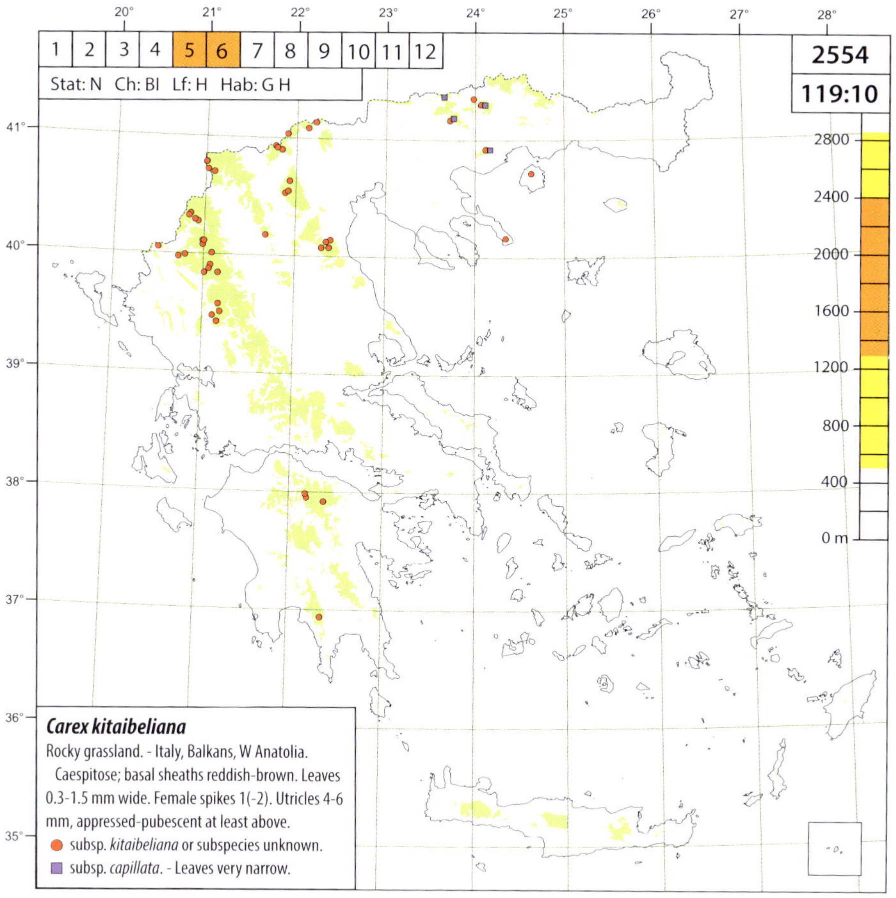

Carex kitaibeliana

Rocky grassland. - Italy, Balkans, W Anatolia.

Caespitose; basal sheaths reddish-brown. Leaves 0.3-1.5 mm wide. Female spikes 1(-2). Utricles 4-6 mm, appressed-pubescent at least above.

- subsp. *kitaibeliana* or subspecies unknown.
- subsp. *capillata*. - Leaves very narrow.

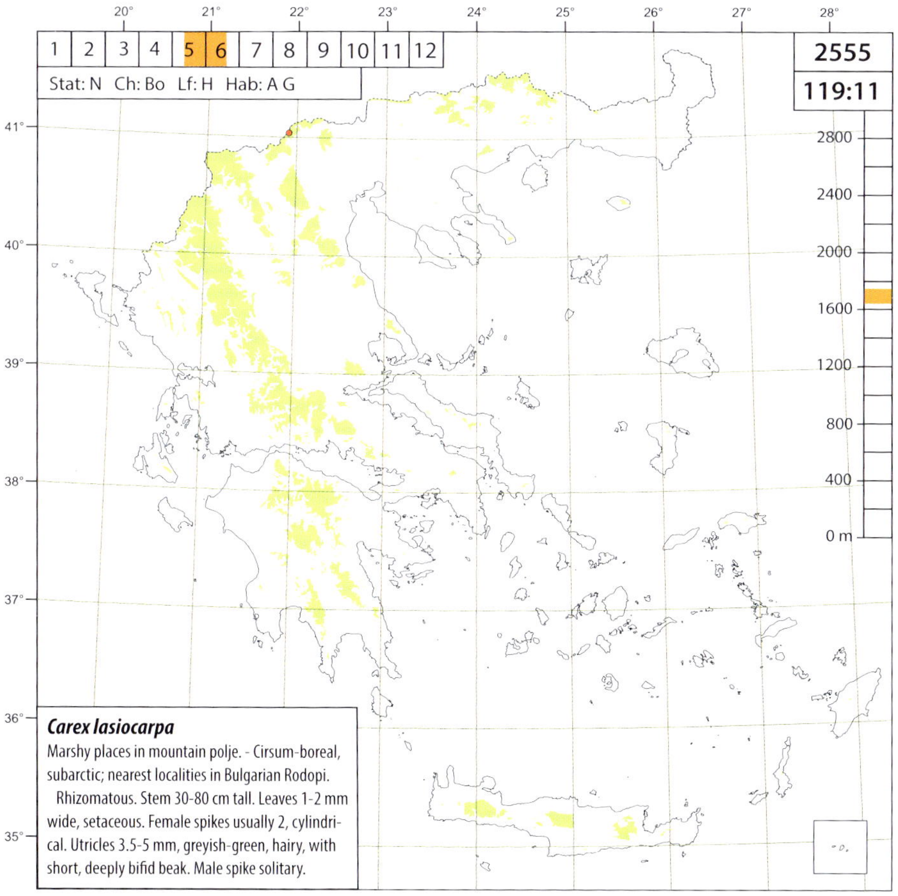

Carex lasiocarpa

Marshy places in mountain polje. - Cirsum-boreal, subarctic; nearest localities in Bulgarian Rodopi.

Rhizomatous. Stem 30-80 cm tall. Leaves 1-2 mm wide, setaceous. Female spikes usually 2, cylindrical. Utricles 3.5-5 mm, greyish-green, hairy, with short, deeply bifid beak. Male spike solitary.

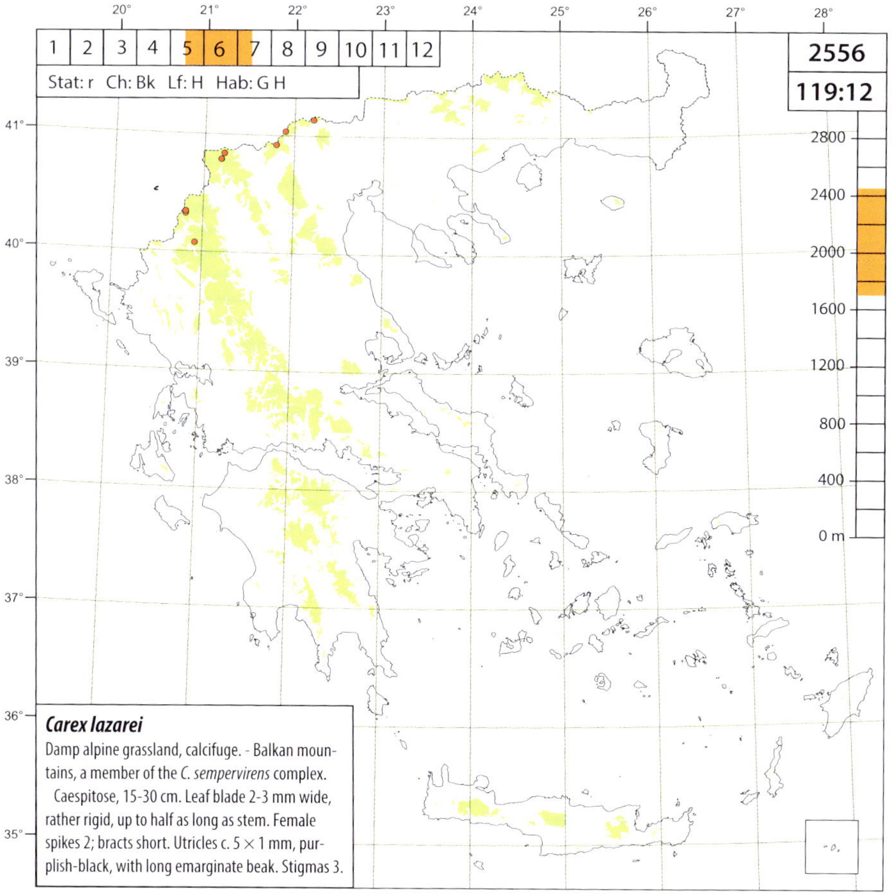

Carex lazarei

Damp alpine grassland, calcifuge. - Balkan mountains, a member of the *C. sempervirens* complex.

Caespitose, 15-30 cm. Leaf blade 2-3 mm wide, rather rigid, up to half as long as stem. Female spikes 2; bracts short. Utricles c. 5 × 1 mm, purplish-black, with long emarginate beak. Stigmas 3.

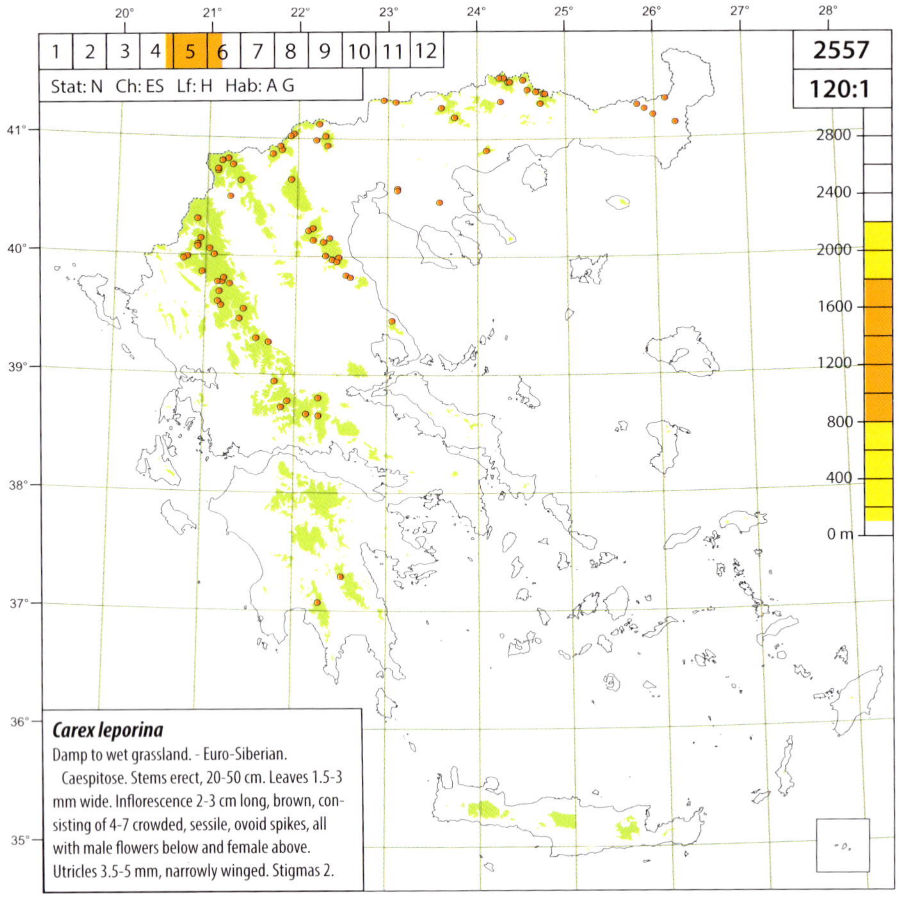

Carex leporina

Damp to wet grassland. - Euro-Siberian.

Caespitose. Stems erect, 20-50 cm. Leaves 1.5-3 mm wide. Inflorescence 2-3 cm long, brown, consisting of 4-7 crowded, sessile, ovoid spikes, all with male flowers below and female above. Utricles 3.5-5 mm, narrowly winged. Stigmas 2.

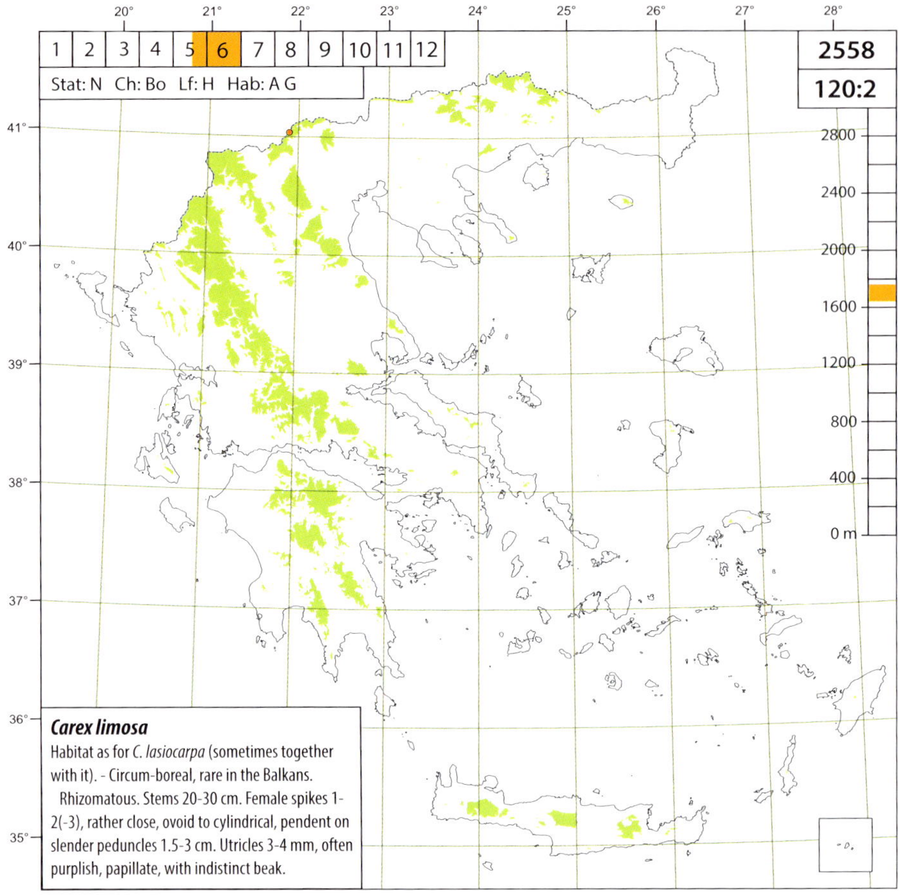

Carex limosa

Habitat as for *C. lasiocarpa* (sometimes together with it). - Circum-boreal, rare in the Balkans.

Rhizomatous. Stems 20-30 cm. Female spikes 1-2(-3), rather close, ovoid to cylindrical, pendent on slender peduncles 1.5-3 cm. Utricles 3-4 mm, often purplish, papillate, with indistinct beak.

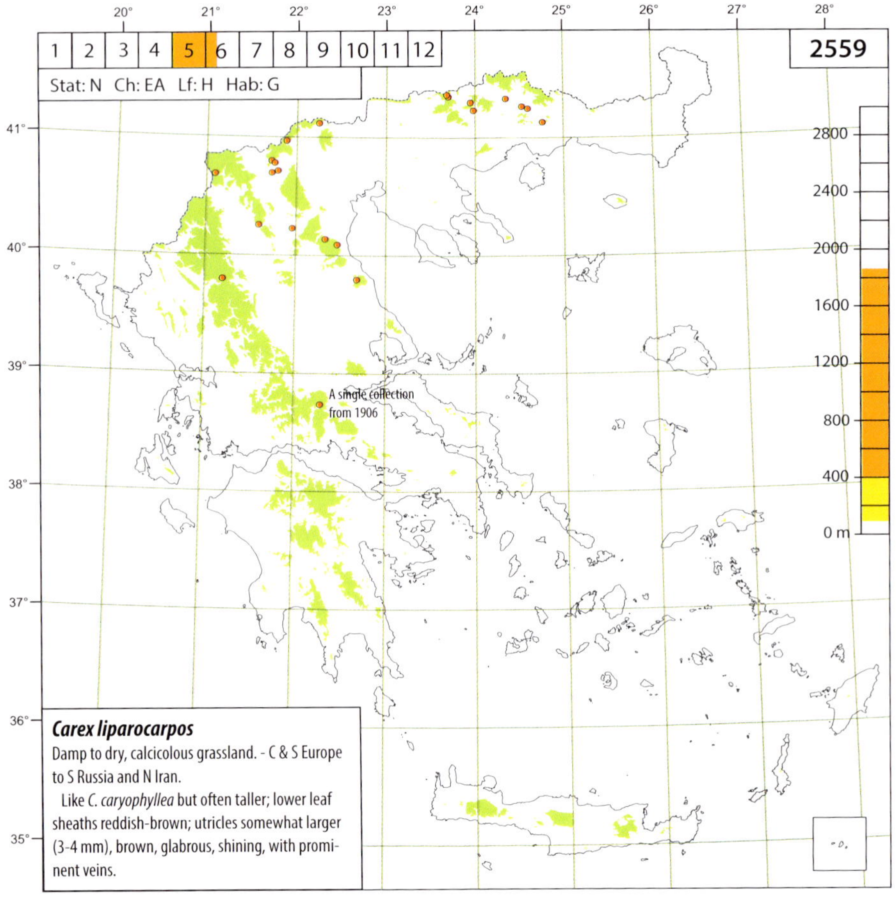

Carex liparocarpos

Damp to dry, calcicolous grassland. - C & S Europe to S Russia and N Iran.

Like *C. caryophyllea* but often taller; lower leaf sheaths reddish-brown; utricles somewhat larger (3-4 mm), brown, glabrous, shining, with prominent veins.

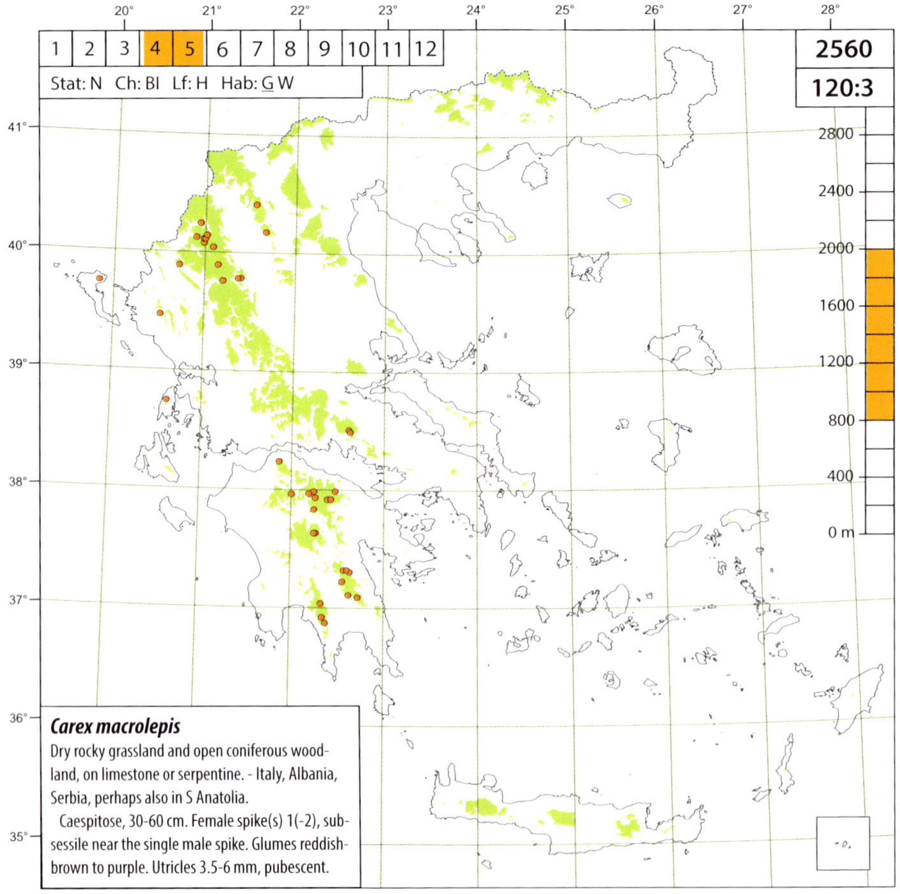

Carex macrolepis

Dry rocky grassland and open coniferous woodland, on limestone or serpentine. - Italy, Albania, Serbia, perhaps also in S Anatolia.

Caespitose, 30-60 cm. Female spike(s) 1(-2), subsessile near the single male spike. Glumes reddish-brown to purple. Utricles 3.5-6 mm, pubescent.

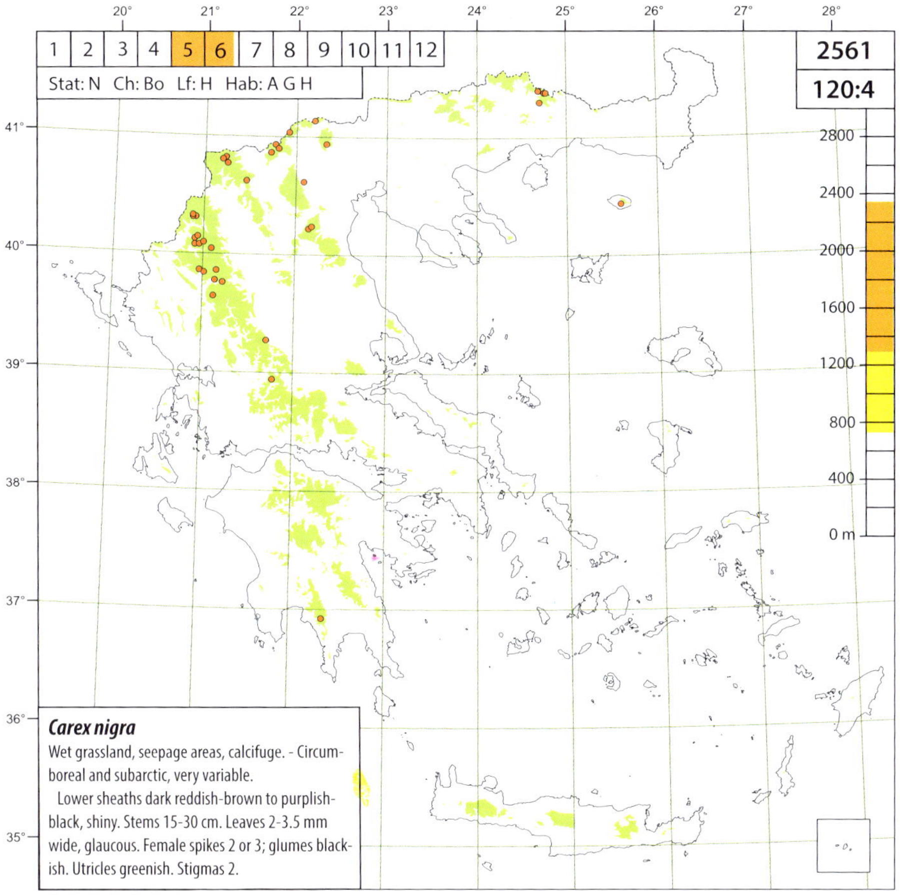

Carex nigra

Wet grassland, seepage areas, calcifuge. - Circumboreal and subarctic, very variable.

Lower sheaths dark reddish-brown to purplish-black, shiny. Stems 15-30 cm. Leaves 2-3.5 mm wide, glaucous. Female spikes 2 or 3; glumes blackish. Utricles greenish. Stigmas 2.

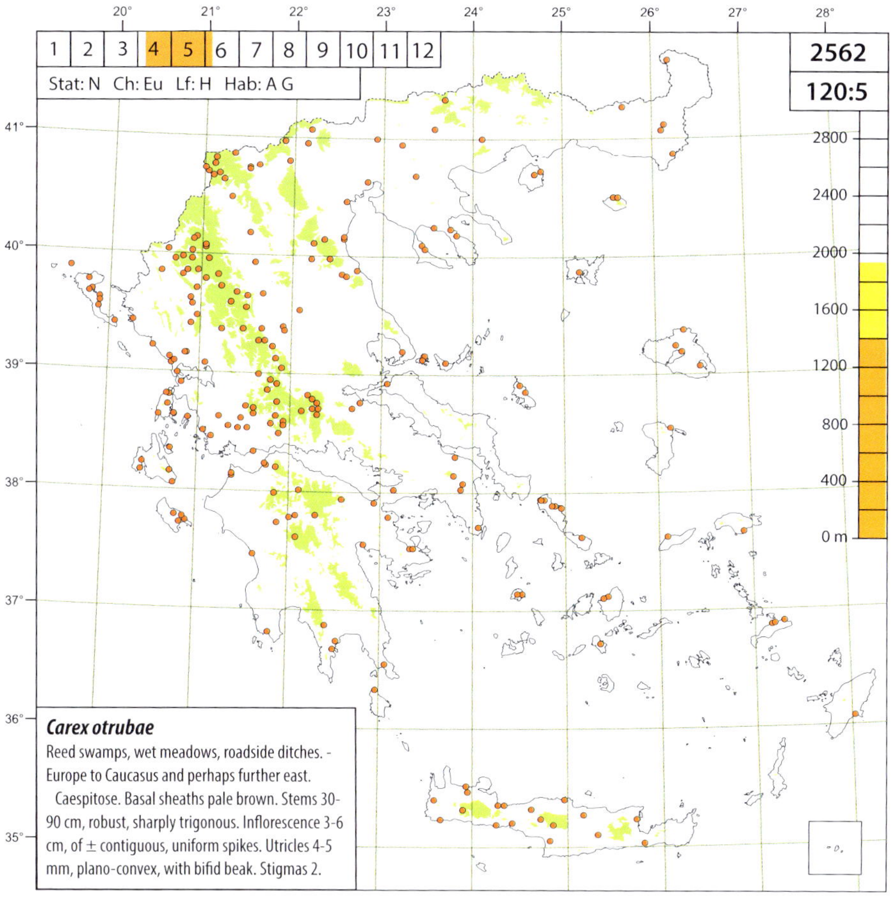

Carex otrubae

Reed swamps, wet meadows, roadside ditches. - Europe to Caucasus and perhaps further east.

Caespitose. Basal sheaths pale brown. Stems 30-90 cm, robust, sharply trigonous. Inflorescence 3-6 cm, of ± contiguous, uniform spikes. Utricles 4-5 mm, plano-convex, with bifid beak. Stigmas 2.

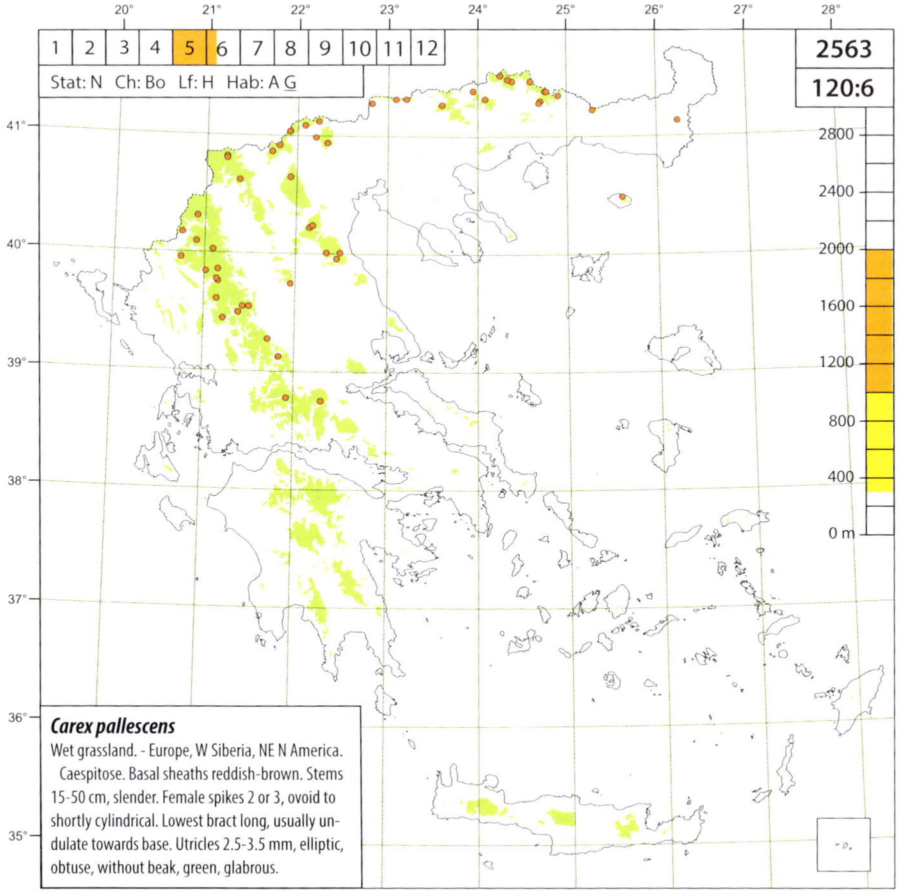

Carex pallescens

Wet grassland. - Europe, W Siberia, NE N America.

Caespitose. Basal sheaths reddish-brown. Stems 15-50 cm, slender. Female spikes 2 or 3, ovoid to shortly cylindrical. Lowest bract long, usually undulate towards base. Utricles 2.5-3.5 mm, elliptic, obtuse, without beak, green, glabrous.

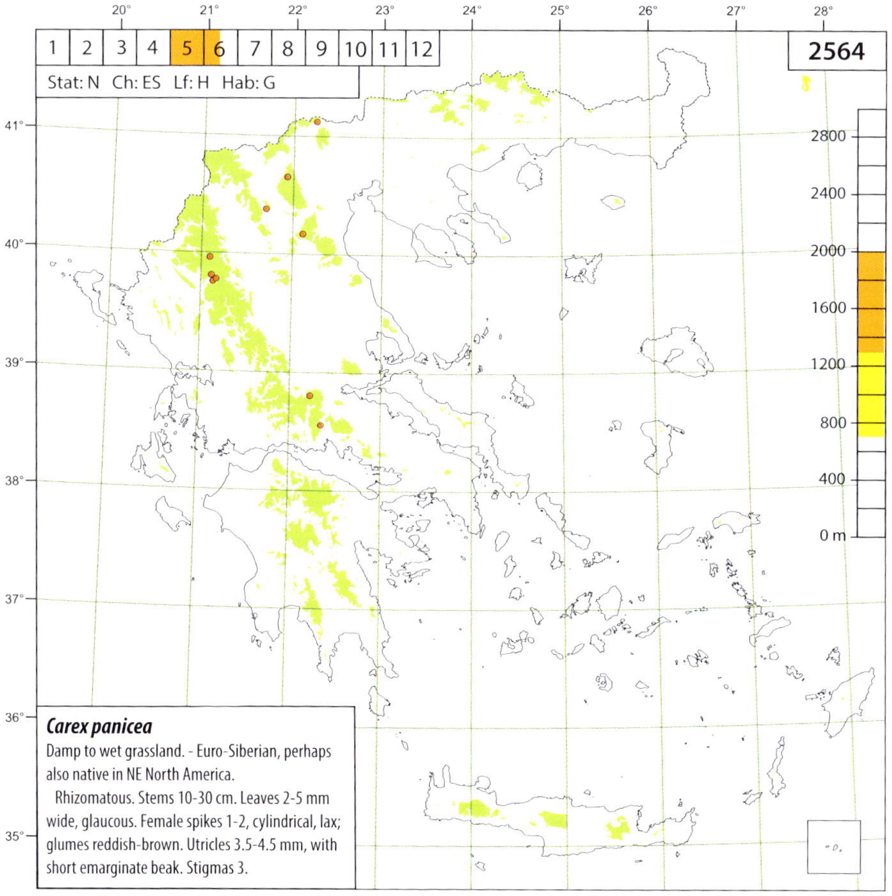

Carex panicea

Damp to wet grassland. - Euro-Siberian, perhaps also native in NE North America.

Rhizomatous. Stems 10-30 cm. Leaves 2-5 mm wide, glaucous. Female spikes 1-2, cylindrical, lax; glumes reddish-brown. Utricles 3.5-4.5 mm, with short emarginate beak. Stigmas 3.

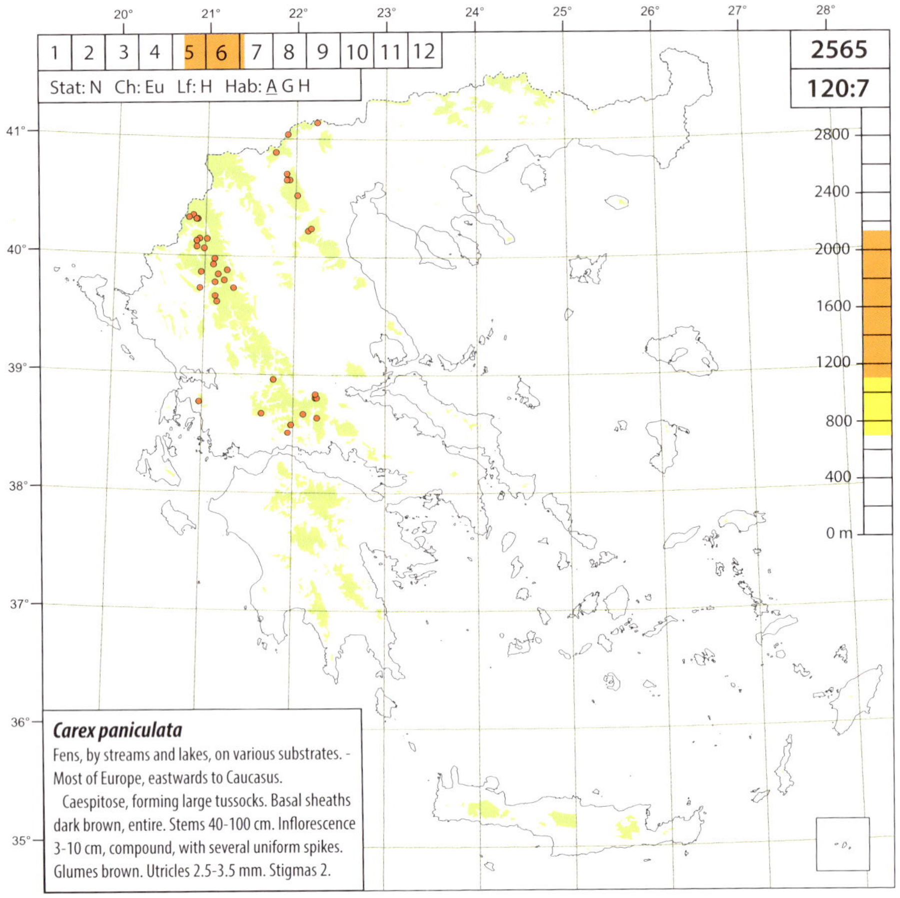

Carex paniculata
Fens, by streams and lakes, on various substrates. - Most of Europe, eastwards to Caucasus.
Caespitose, forming large tussocks. Basal sheaths dark brown, entire. Stems 40-100 cm. Inflorescence 3-10 cm, compound, with several uniform spikes. Glumes brown. Utricles 2.5-3.5 mm. Stigmas 2.

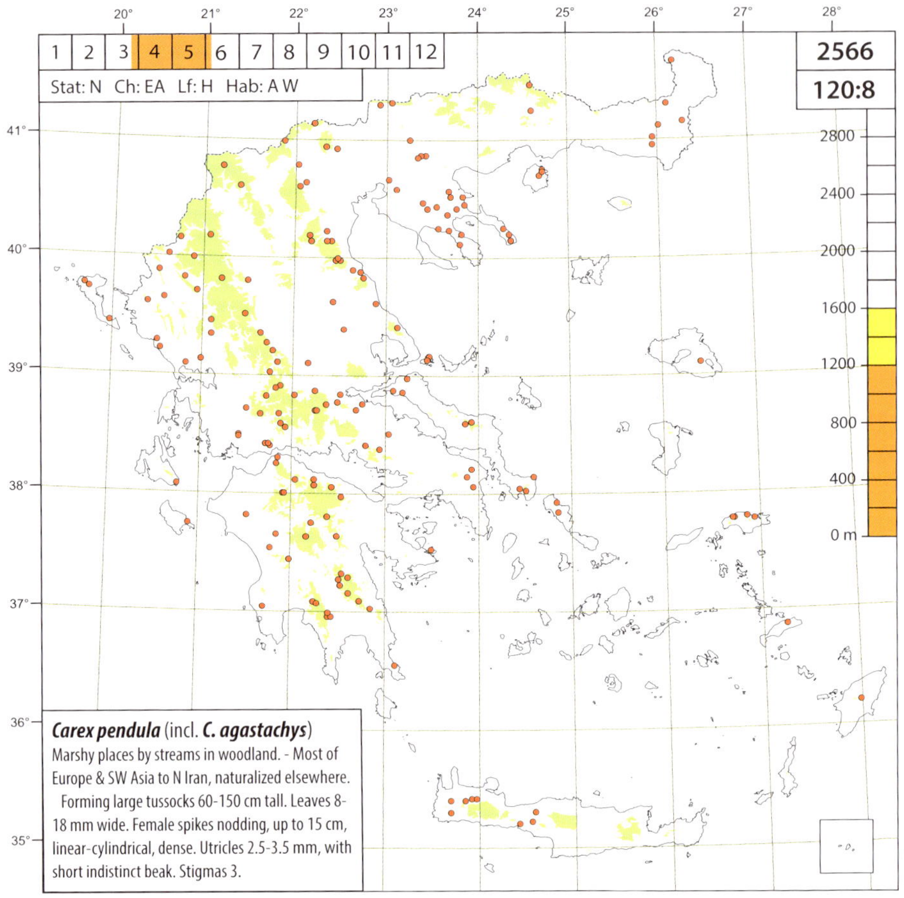

Carex pendula (incl. ***C. agastachys***)
Marshy places by streams in woodland. - Most of Europe & SW Asia to N Iran, naturalized elsewhere.
Forming large tussocks 60-150 cm tall. Leaves 8-18 mm wide. Female spikes nodding, up to 15 cm, linear-cylindrical, dense. Utricles 2.5-3.5 mm, with short indistinct beak. Stigmas 3.

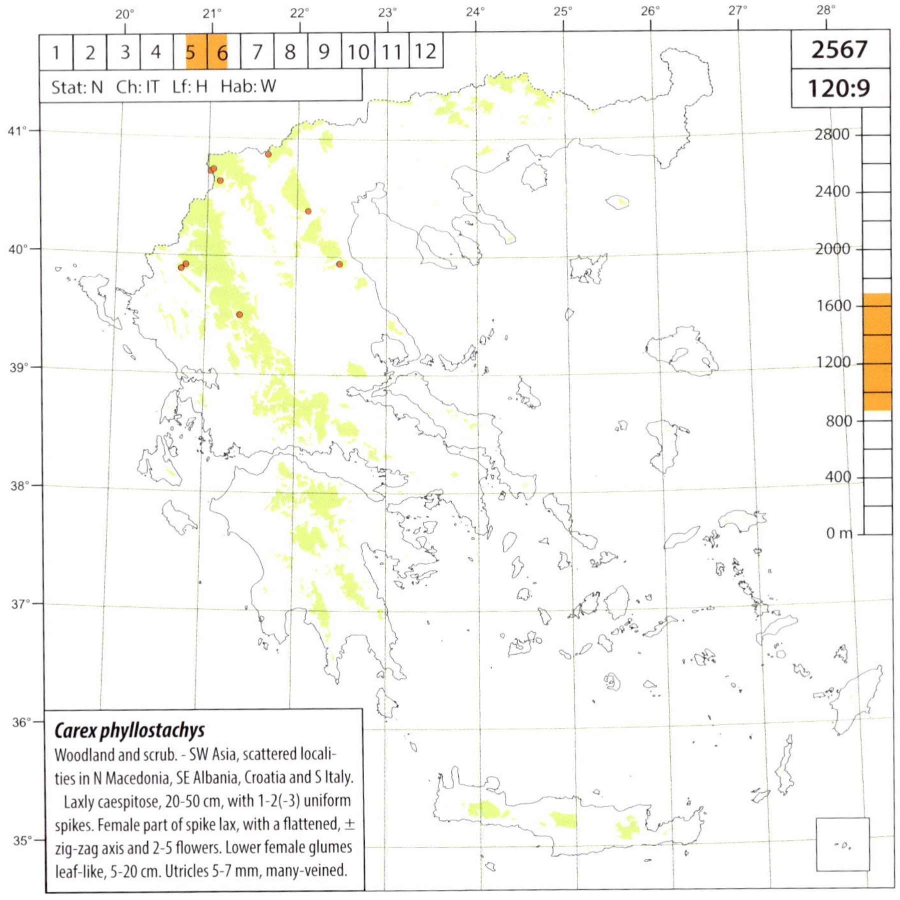

Carex phyllostachys
Woodland and scrub. - SW Asia, scattered localities in N Macedonia, SE Albania, Croatia and S Italy.
Laxly caespitose, 20-50 cm, with 1-2(-3) uniform spikes. Female part of spike lax, with a flattened, ± zig-zag axis and 2-5 flowers. Lower female glumes leaf-like, 5-20 cm. Utricles 5-7 mm, many-veined.

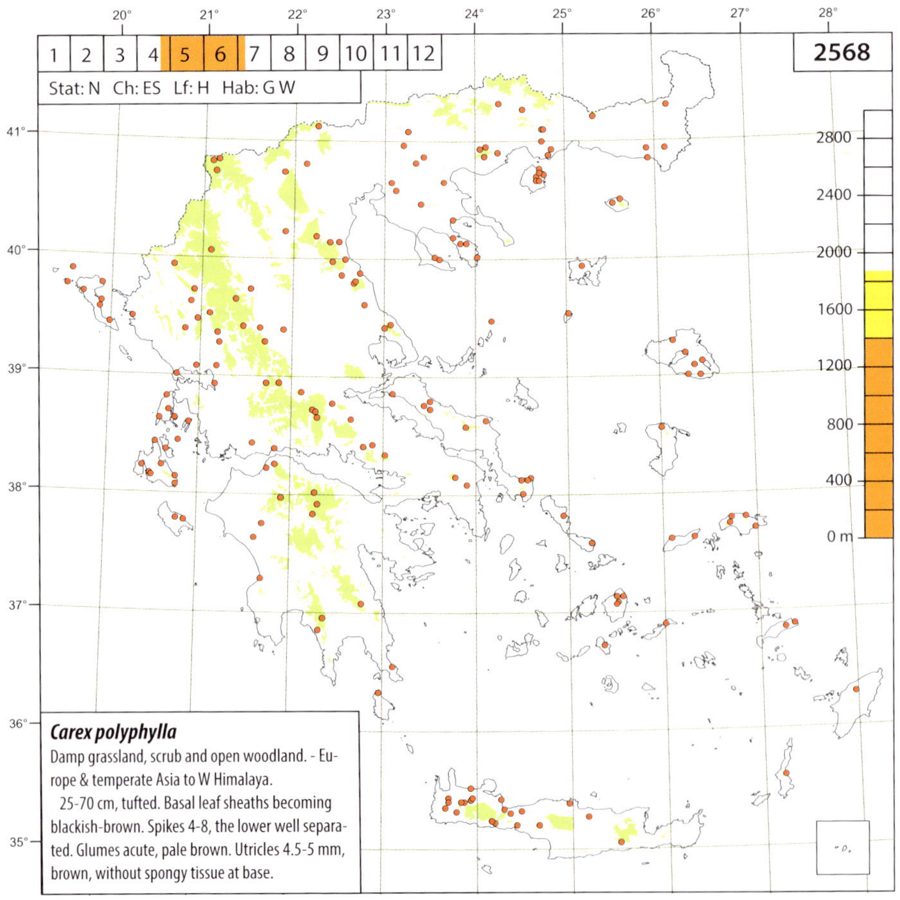

Carex polyphylla
Damp grassland, scrub and open woodland. - Europe & temperate Asia to W Himalaya.
25-70 cm, tufted. Basal leaf sheaths becoming blackish-brown. Spikes 4-8, the lower well separated. Glumes acute, pale brown. Utricles 4.5-5 mm, brown, without spongy tissue at base.

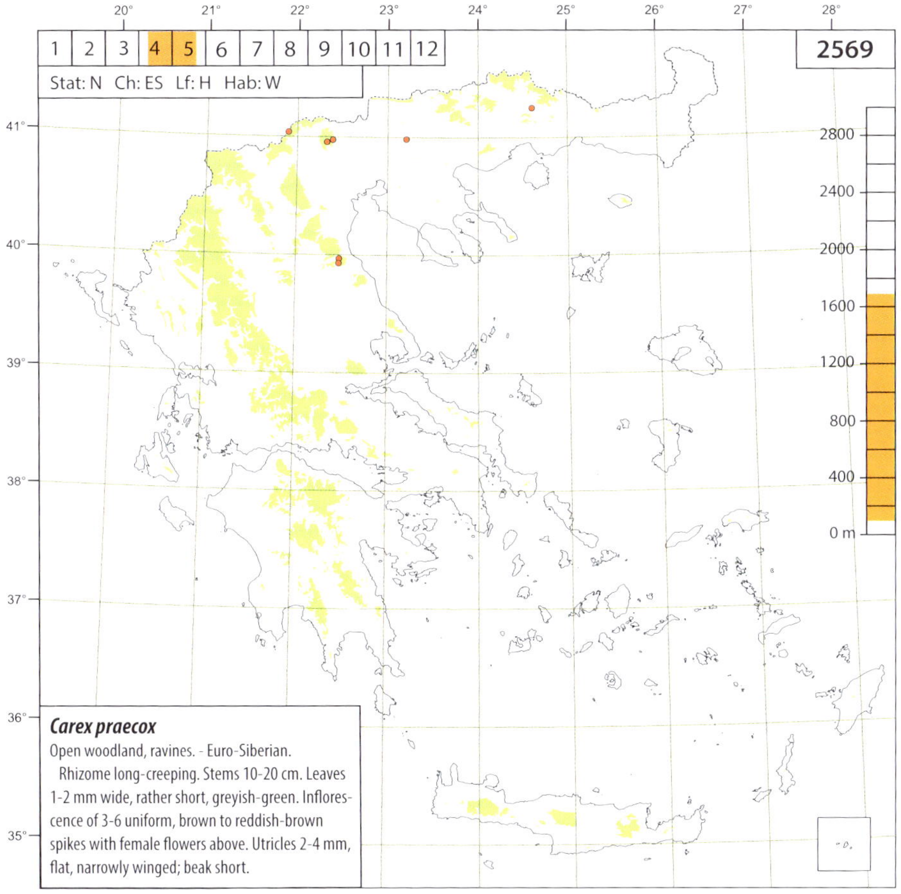

Carex praecox
Open woodland, ravines. - Euro-Siberian.
Rhizome long-creeping. Stems 10-20 cm. Leaves 1-2 mm wide, rather short, greyish-green. Inflorescence of 3-6 uniform, brown to reddish-brown spikes with female flowers above. Utricles 2-4 mm, flat, narrowly winged; beak short.

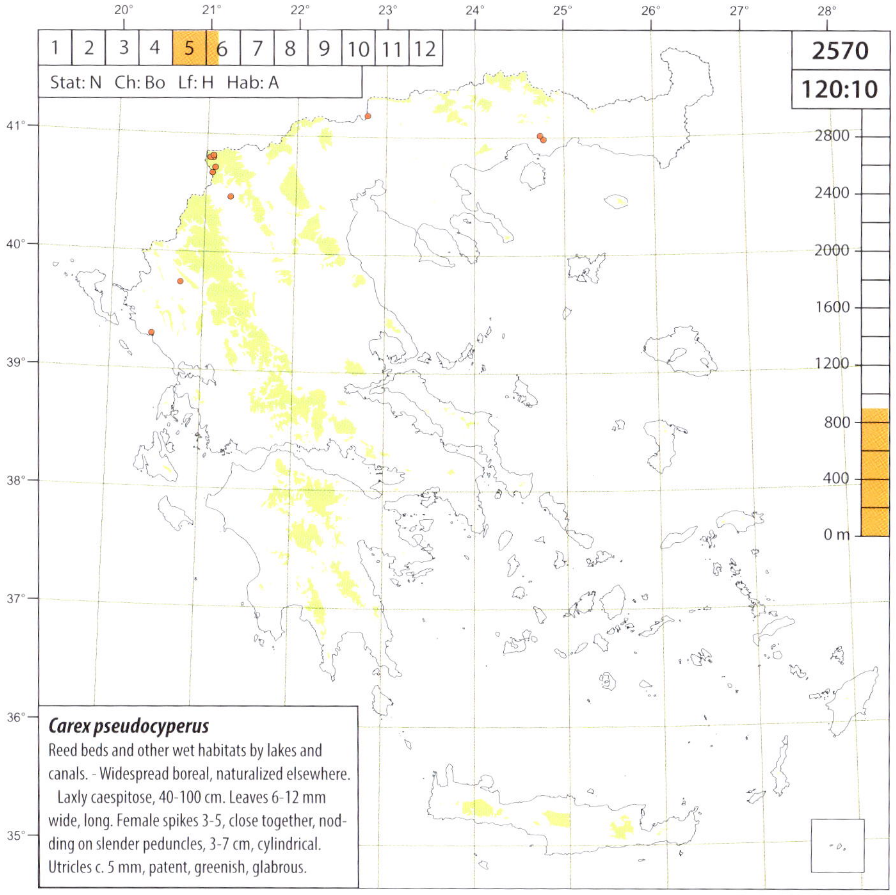

Carex pseudocyperus
Reed beds and other wet habitats by lakes and canals. - Widespread boreal, naturalized elsewhere.
Laxly caespitose, 40-100 cm. Leaves 6-12 mm wide, long. Female spikes 3-5, close together, nodding on slender peduncles, 3-7 cm, cylindrical. Utricles c. 5 mm, patent, greenish, glabrous.

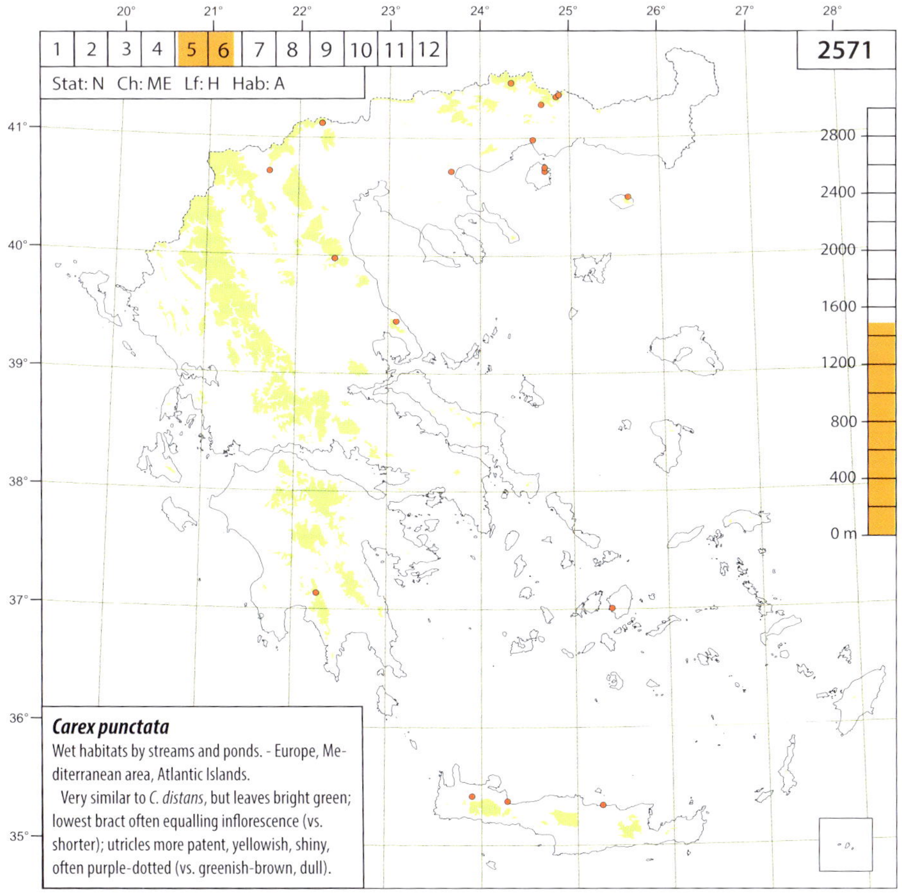

Carex punctata
Wet habitats by streams and ponds. - Europe, Mediterranean area, Atlantic Islands.
Very similar to *C. distans*, but leaves bright green; lowest bract often equalling inflorescence (vs. shorter); utricles more patent, yellowish, shiny, often purple-dotted (vs. greenish-brown, dull).

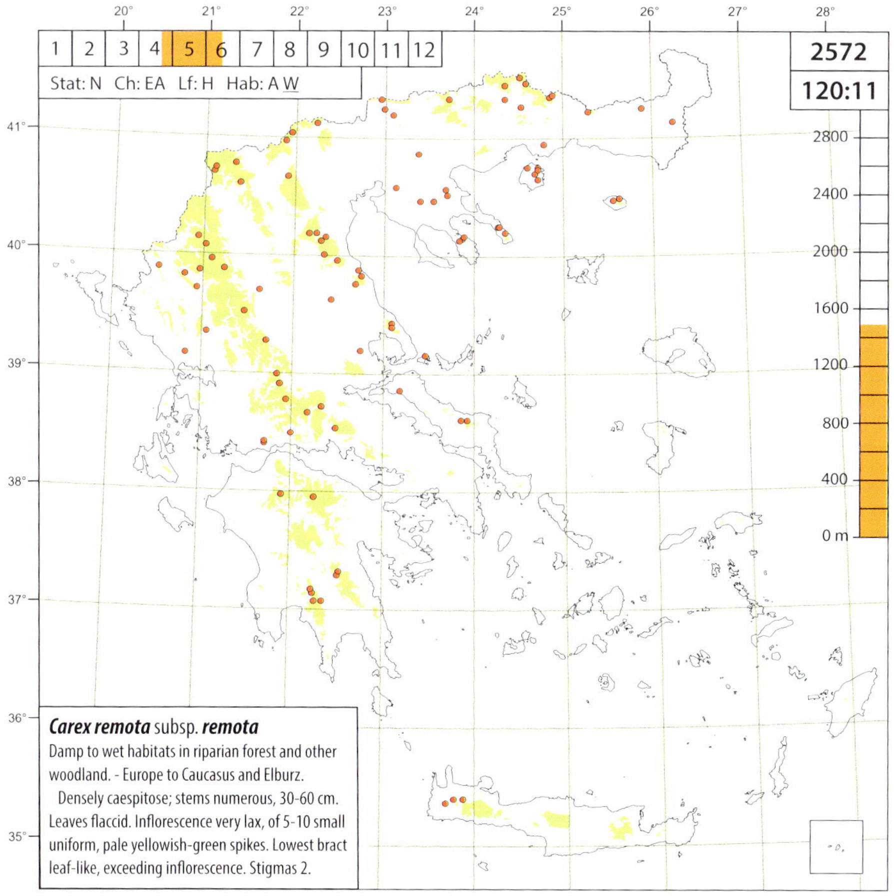

Carex remota subsp. ***remota***
Damp to wet habitats in riparian forest and other woodland. - Europe to Caucasus and Elburz.
Densely caespitose; stems numerous, 30-60 cm. Leaves flaccid. Inflorescence very lax, of 5-10 small uniform, pale yellowish-green spikes. Lowest bract leaf-like, exceeding inflorescence. Stigmas 2.

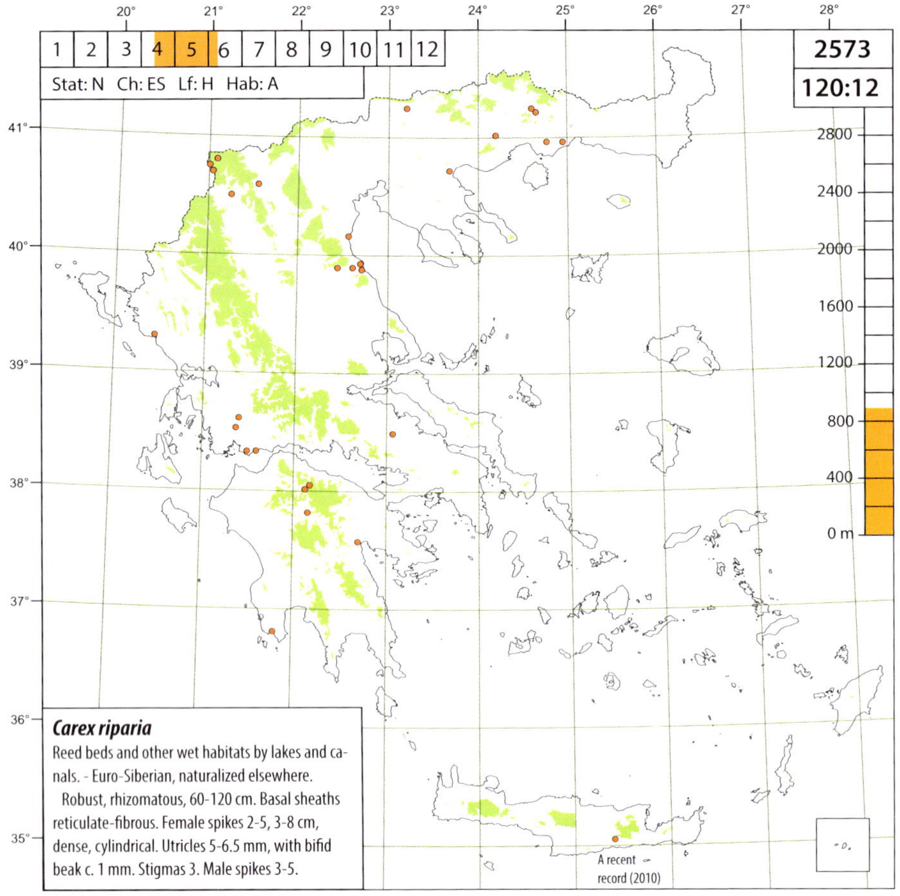

Carex riparia
Reed beds and other wet habitats by lakes and canals. - Euro-Siberian, naturalized elsewhere.
Robust, rhizomatous, 60-120 cm. Basal sheaths reticulate-fibrous. Female spikes 2-5, 3-8 cm, dense, cylindrical. Utricles 5-6.5 mm, with bifid beak c. 1 mm. Stigmas 3. Male spikes 3-5.

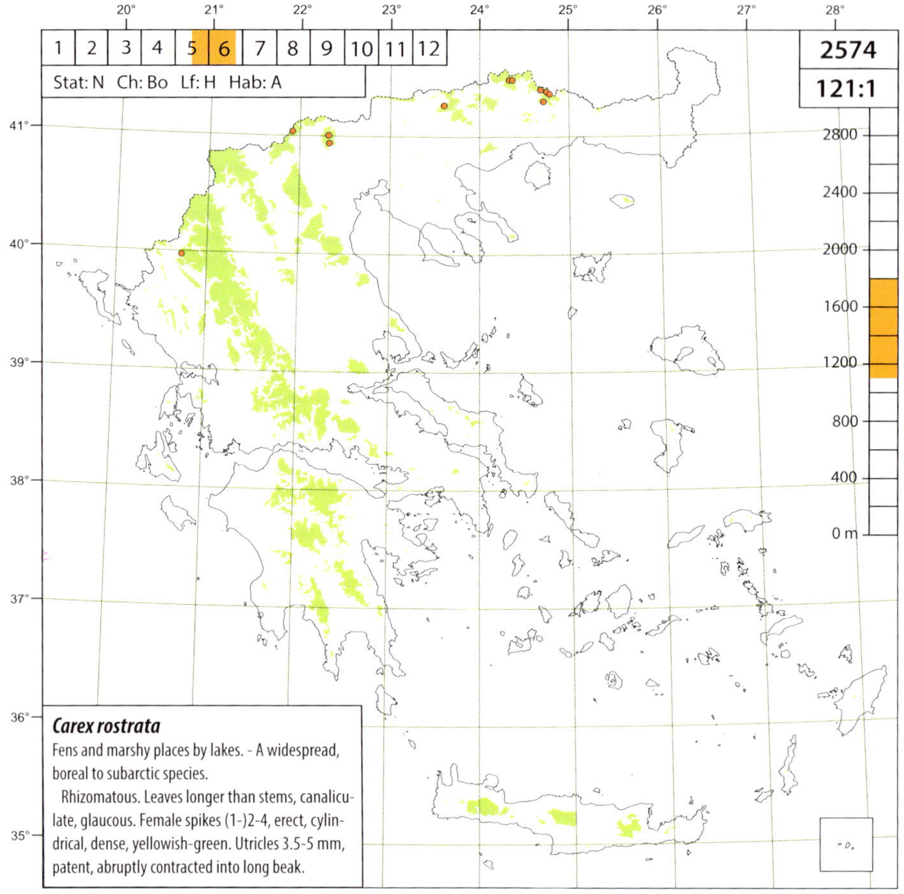

Carex rostrata
Fens and marshy places by lakes. - A widespread, boreal to subarctic species.
Rhizomatous. Leaves longer than stems, canaliculate, glaucous. Female spikes (1-)2-4, erect, cylindrical, dense, yellowish-green. Utricles 3.5-5 mm, patent, abruptly contracted into long beak.

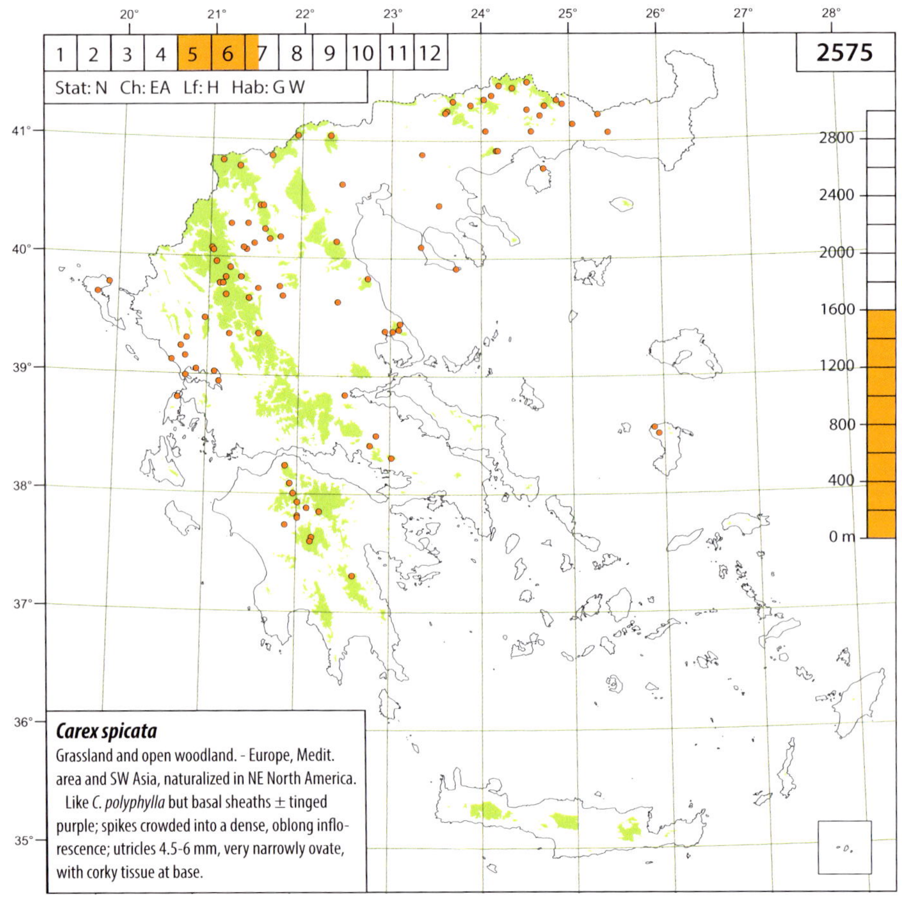

Carex spicata
Grassland and open woodland. - Europe, Medit. area and SW Asia, naturalized in NE North America.
Like *C. polyphylla* but basal sheaths ± tinged purple; spikes crowded into a dense, oblong inflorescence; utricles 4.5-6 mm, very narrowly ovate, with corky tissue at base.

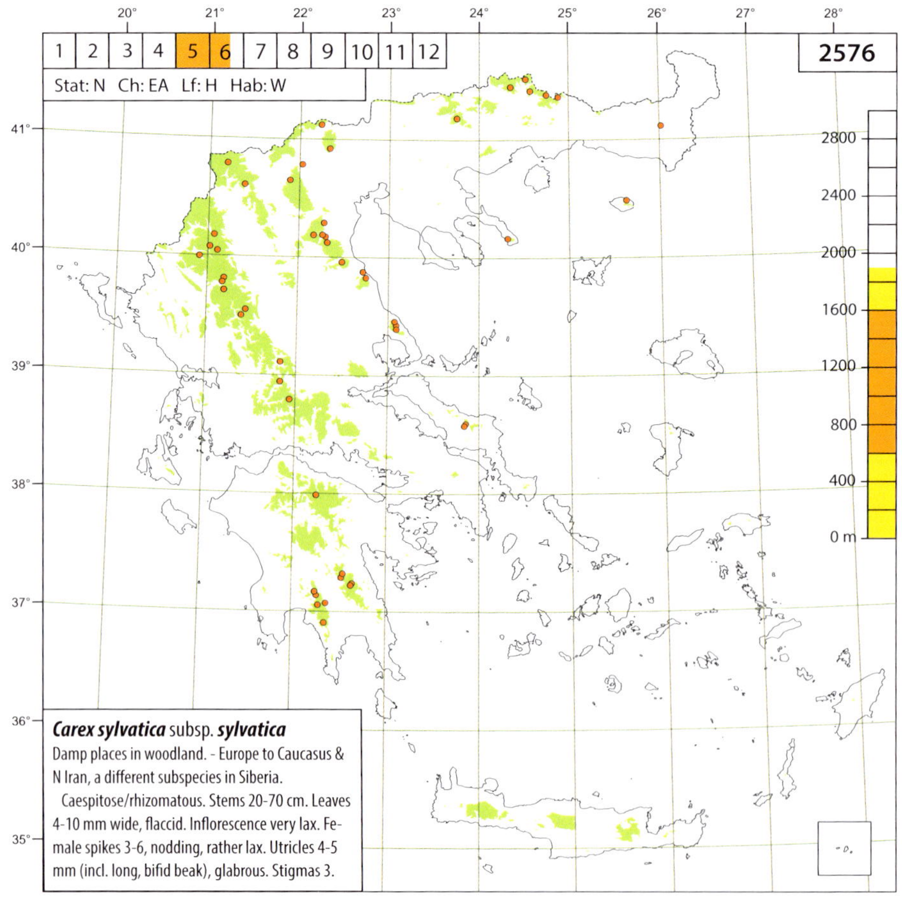

Carex sylvatica subsp. ***sylvatica***
Damp places in woodland. - Europe to Caucasus & N Iran, a different subspecies in Siberia.
Caespitose/rhizomatous. Stems 20-70 cm. Leaves 4-10 mm wide, flaccid. Inflorescence very lax. Female spikes 3-6, nodding, rather lax. Utricles 4-5 mm (incl. long, bifid beak), glabrous. Stigmas 3.

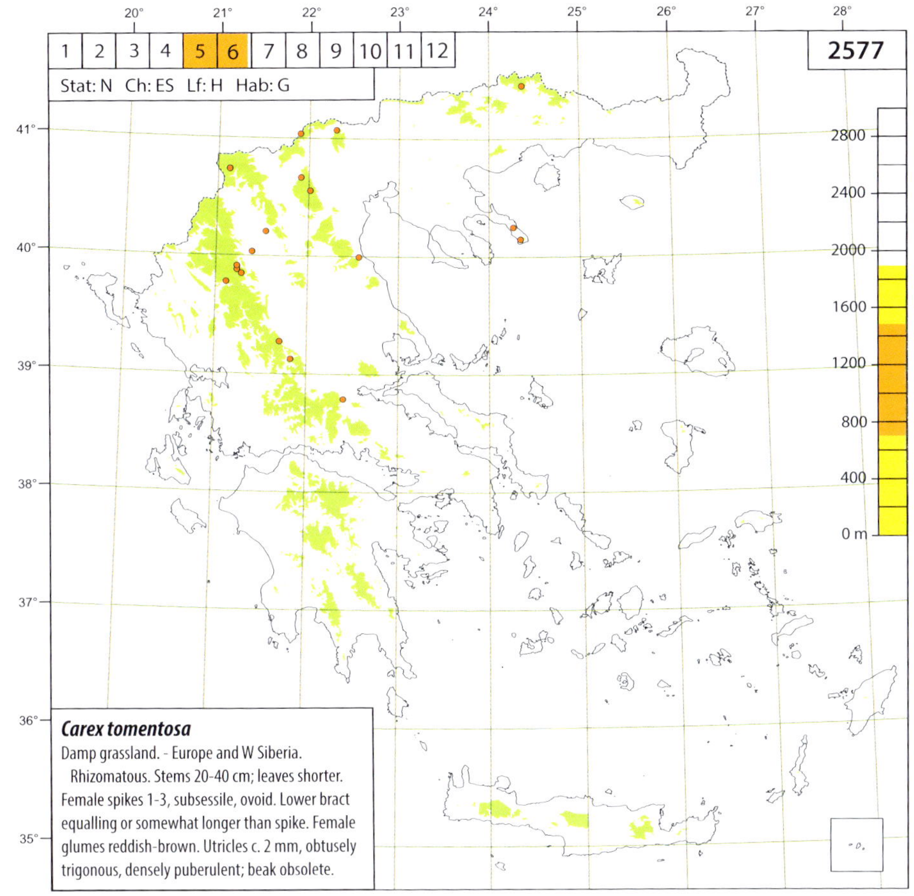

Carex tomentosa

Damp grassland. - Europe and W Siberia.

Rhizomatous. Stems 20-40 cm; leaves shorter. Female spikes 1-3, subsessile, ovoid. Lower bract equalling or somewhat longer than spike. Female glumes reddish-brown. Utricles c. 2 mm, obtusely trigonous, densely puberulent; beak obsolete.

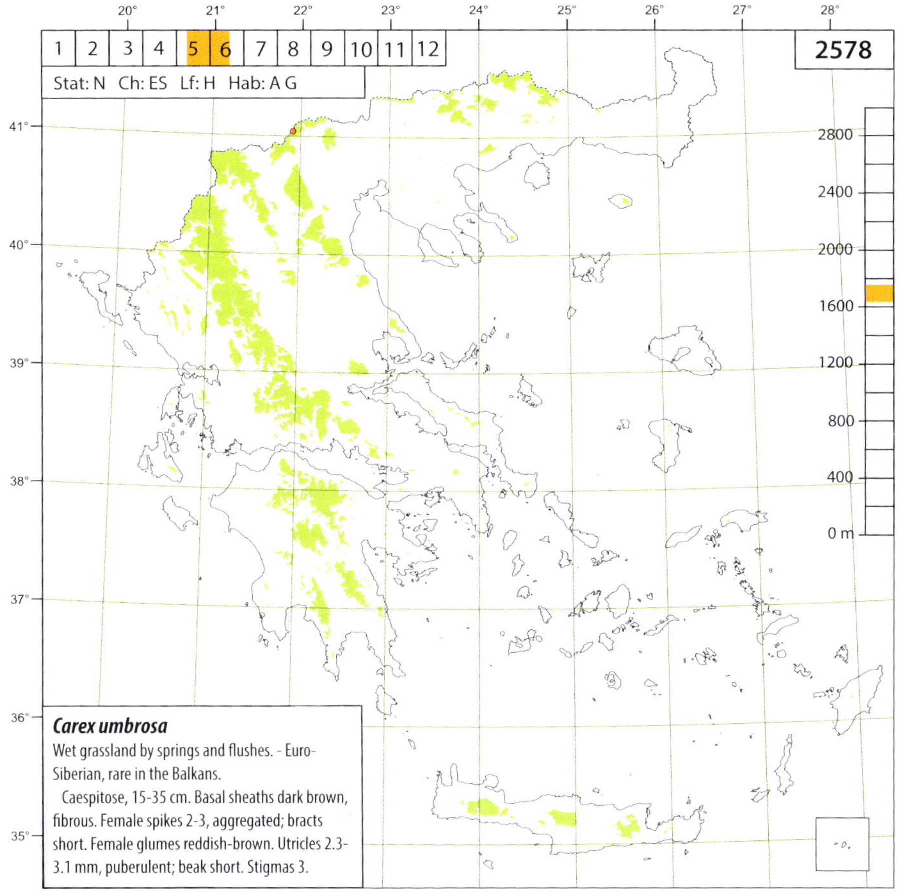

Carex umbrosa

Wet grassland by springs and flushes. - Euro-Siberian, rare in the Balkans.

Caespitose, 15-35 cm. Basal sheaths dark brown, fibrous. Female spikes 2-3, aggregated; bracts short. Female glumes reddish-brown. Utricles 2.3-3.1 mm, puberulent; beak short. Stigmas 3.

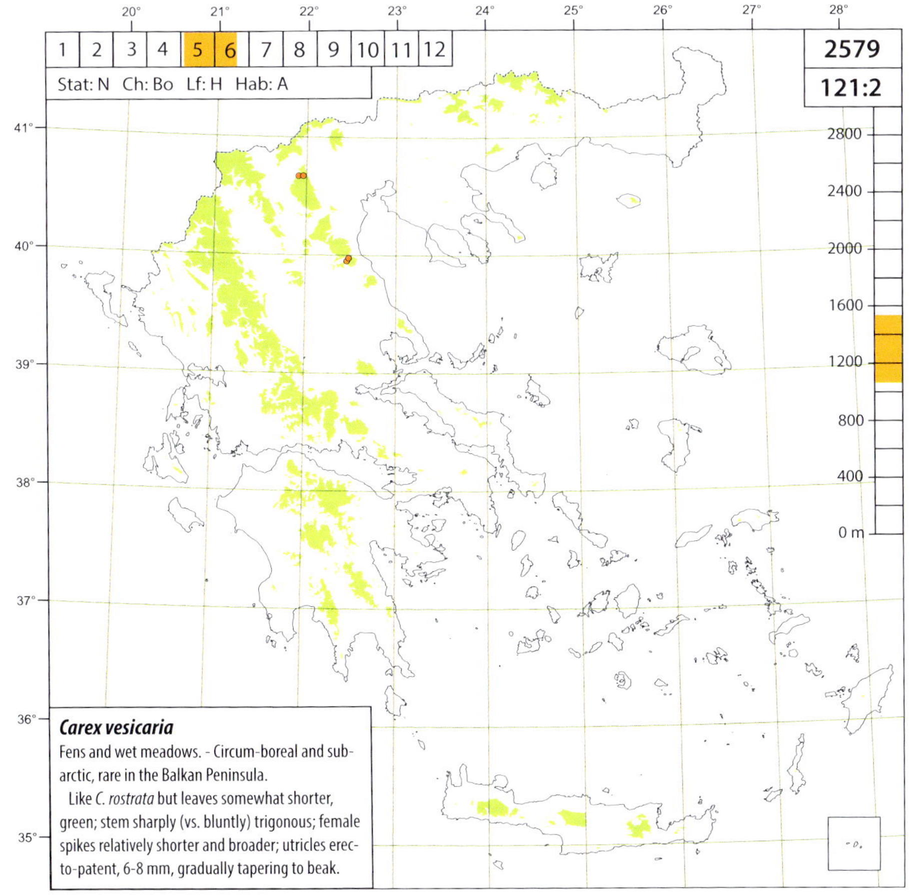

Carex vesicaria

Fens and wet meadows. - Circum-boreal and sub-arctic, rare in the Balkan Peninsula.

Like *C. rostrata* but leaves somewhat shorter, green; stem sharply (vs. bluntly) trigonous; female spikes relatively shorter and broader; utricles erec-to-patent, 6-8 mm, gradually tapering to beak.

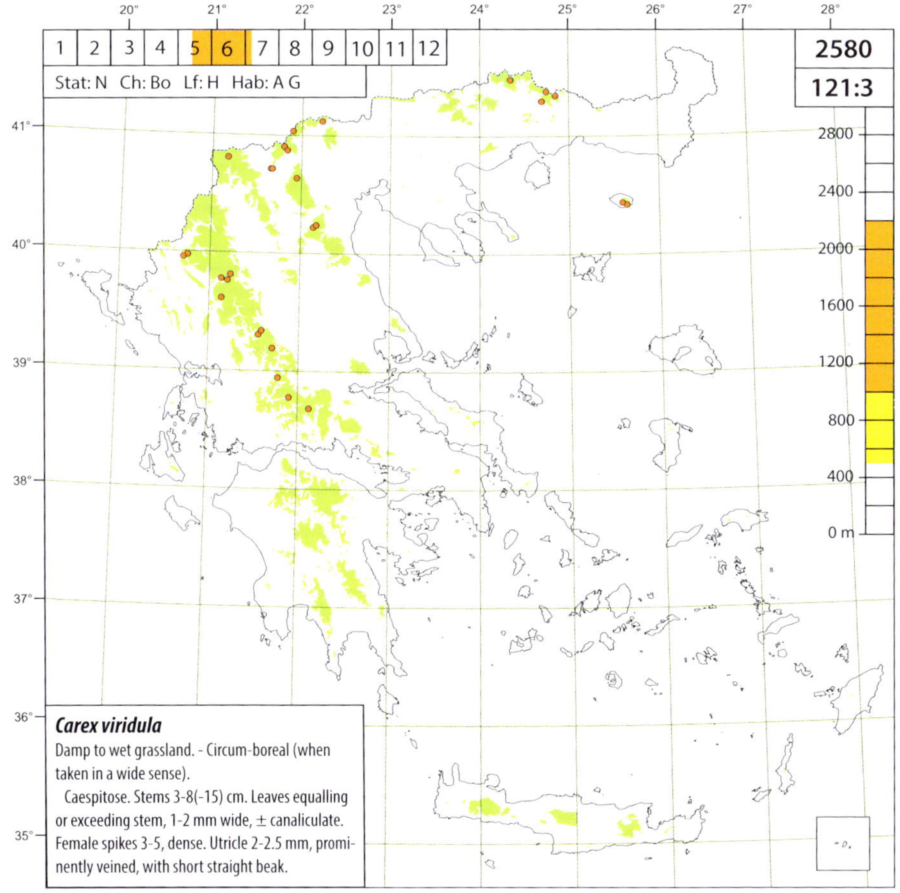

Carex viridula

Damp to wet grassland. - Circum-boreal (when taken in a wide sense).

Caespitose. Stems 3-8(-15) cm. Leaves equalling or exceeding stem, 1-2 mm wide, ± canaliculate. Female spikes 3-5, dense. Utricle 2-2.5 mm, prominently veined, with short straight beak.

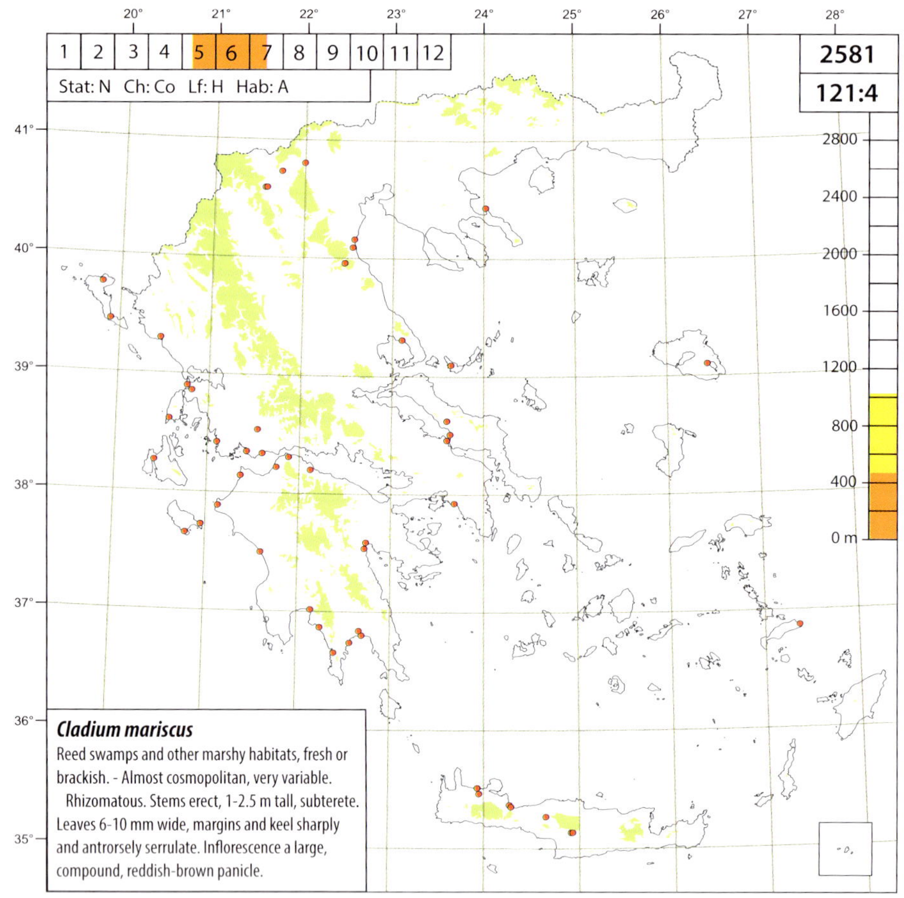

Cladium mariscus
Reed swamps and other marshy habitats, fresh or brackish. - Almost cosmopolitan, very variable.
Rhizomatous. Stems erect, 1-2.5 m tall, subterete. Leaves 6-10 mm wide, margins and keel sharply and antrorsely serrulate. Inflorescence a large, compound, reddish-brown panicle.

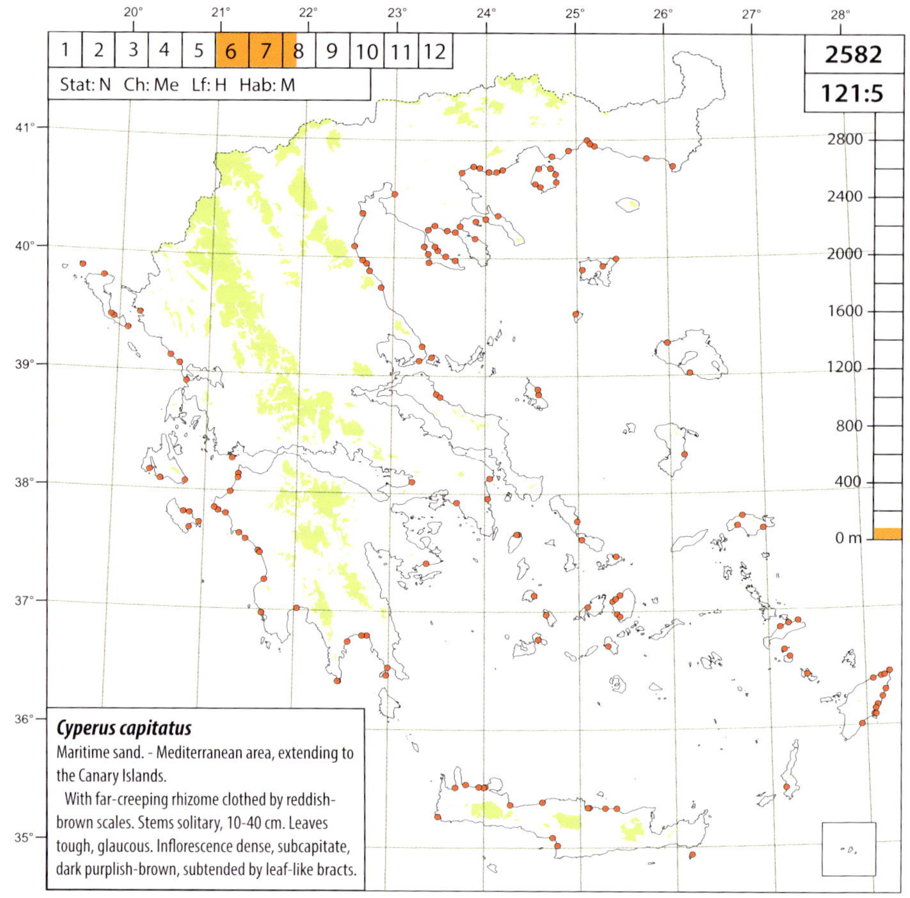

Cyperus capitatus
Maritime sand. - Mediterranean area, extending to the Canary Islands.
With far-creeping rhizome clothed by reddish-brown scales. Stems solitary, 10-40 cm. Leaves tough, glaucous. Inflorescence dense, subcapitate, dark purplish-brown, subtended by leaf-like bracts.

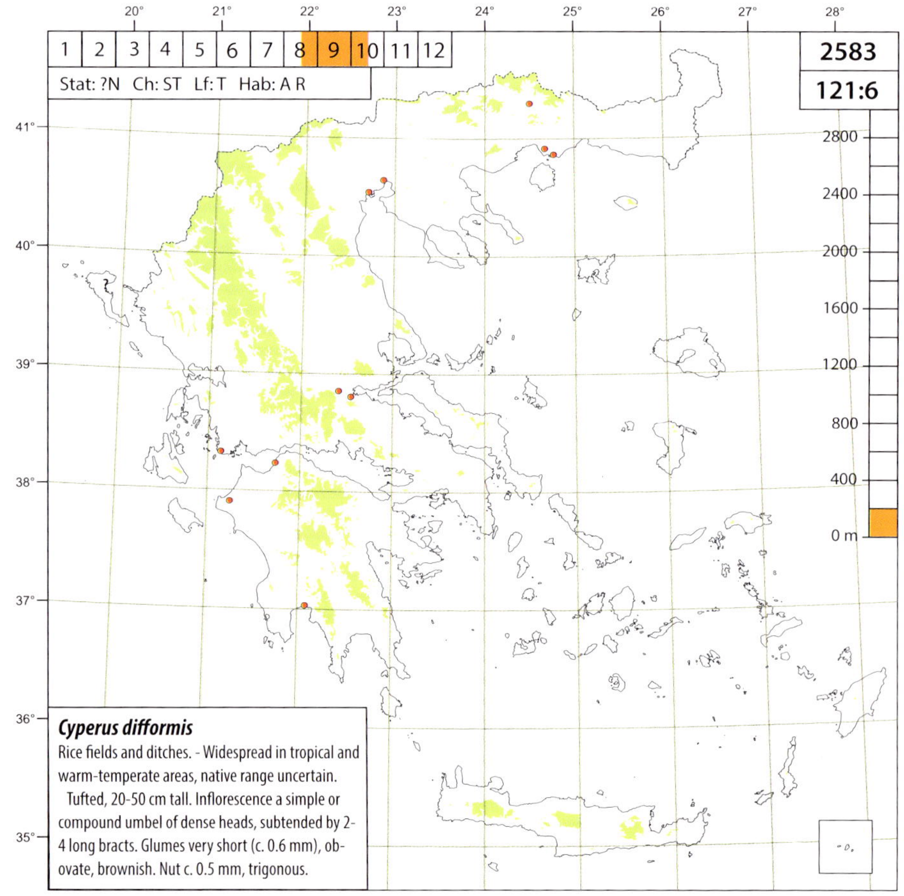

Cyperus difformis
Rice fields and ditches. - Widespread in tropical and warm-temperate areas, native range uncertain.
Tufted, 20-50 cm tall. Inflorescence a simple or compound umbel of dense heads, subtended by 2-4 long bracts. Glumes very short (c. 0.6 mm), ob-ovate, brownish. Nut c. 0.5 mm, trigonous.

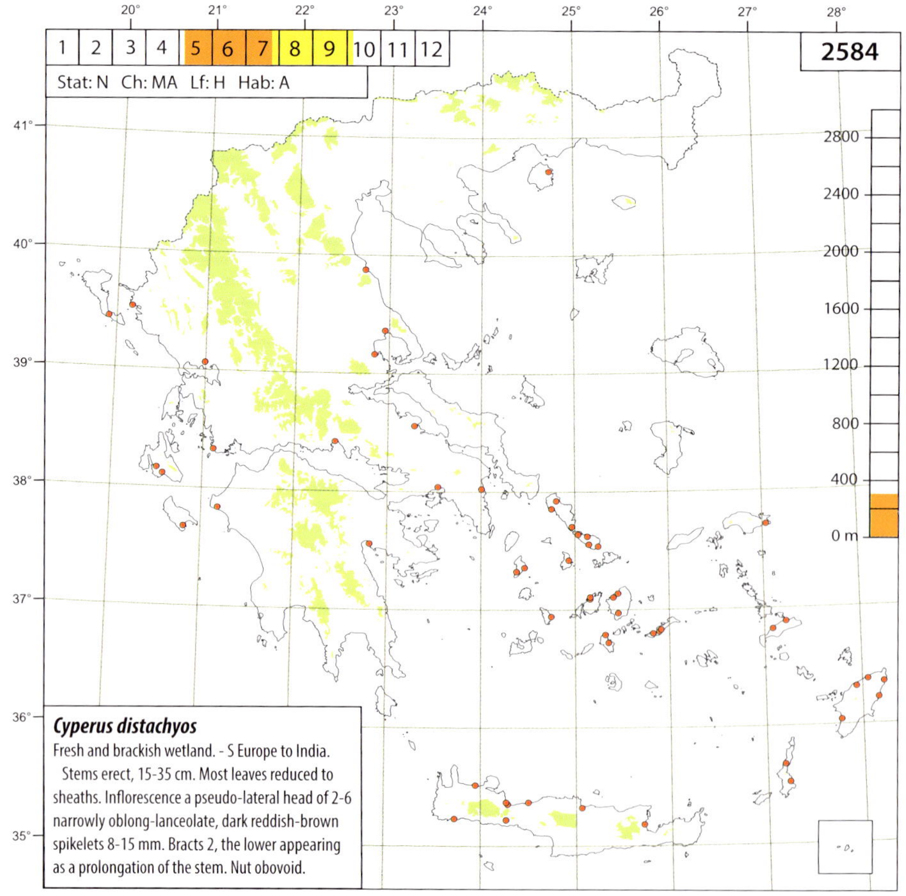

Cyperus distachyos
Fresh and brackish wetland. - S Europe to India.
Stems erect, 15-35 cm. Most leaves reduced to sheaths. Inflorescence a pseudo-lateral head of 2-6 narrowly oblong-lanceolate, dark reddish-brown spikelets 8-15 mm. Bracts 2, the lower appearing as a prolongation of the stem. Nut obovoid.

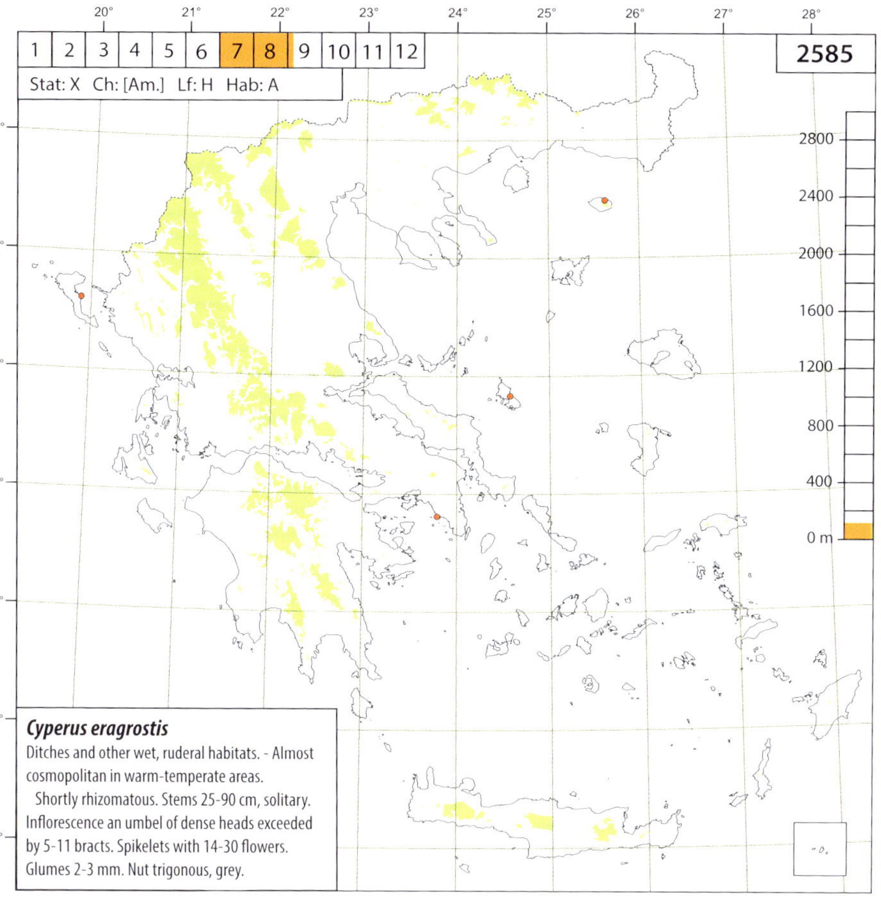

Cyperus eragrostis

Ditches and other wet, ruderal habitats. - Almost cosmopolitan in warm-temperate areas.

Shortly rhizomatous. Stems 25-90 cm, solitary. Inflorescence an umbel of dense heads exceeded by 5-11 bracts. Spikelets with 14-30 flowers. Glumes 2-3 mm. Nut trigonous, grey.

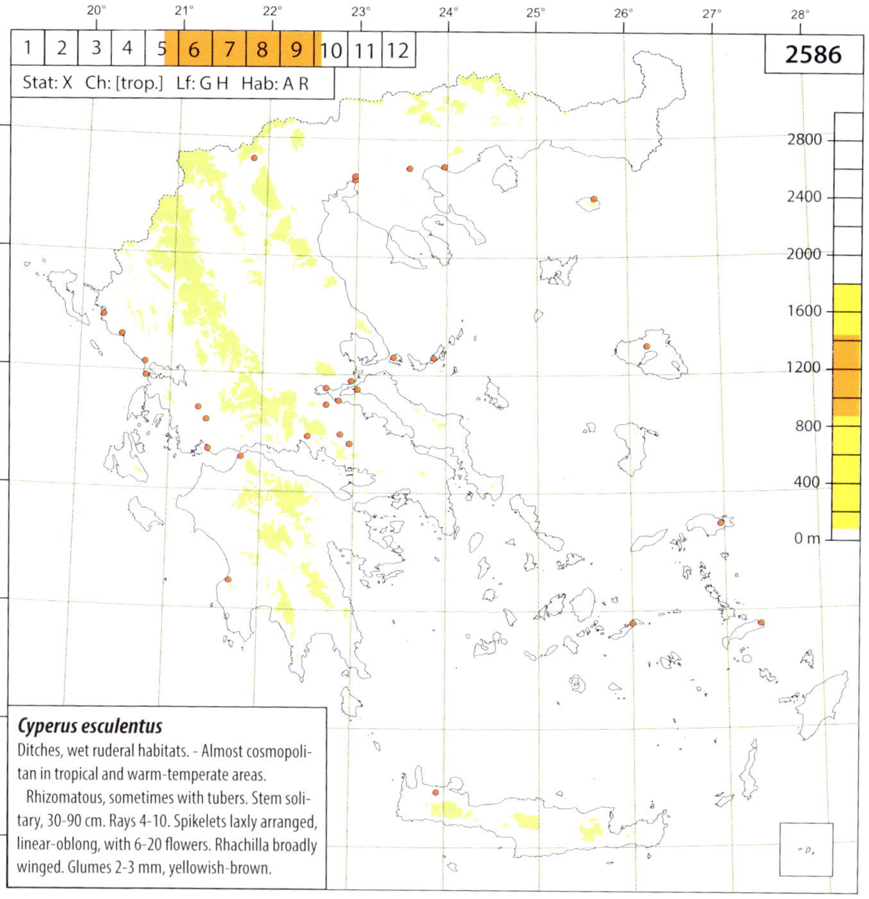

Cyperus esculentus

Ditches, wet ruderal habitats. - Almost cosmopolitan in tropical and warm-temperate areas.

Rhizomatous, sometimes with tubers. Stem solitary, 30-90 cm. Rays 4-10. Spikelets laxly arranged, linear-oblong, with 6-20 flowers. Rhachilla broadly winged. Glumes 2-3 mm, yellowish-brown.

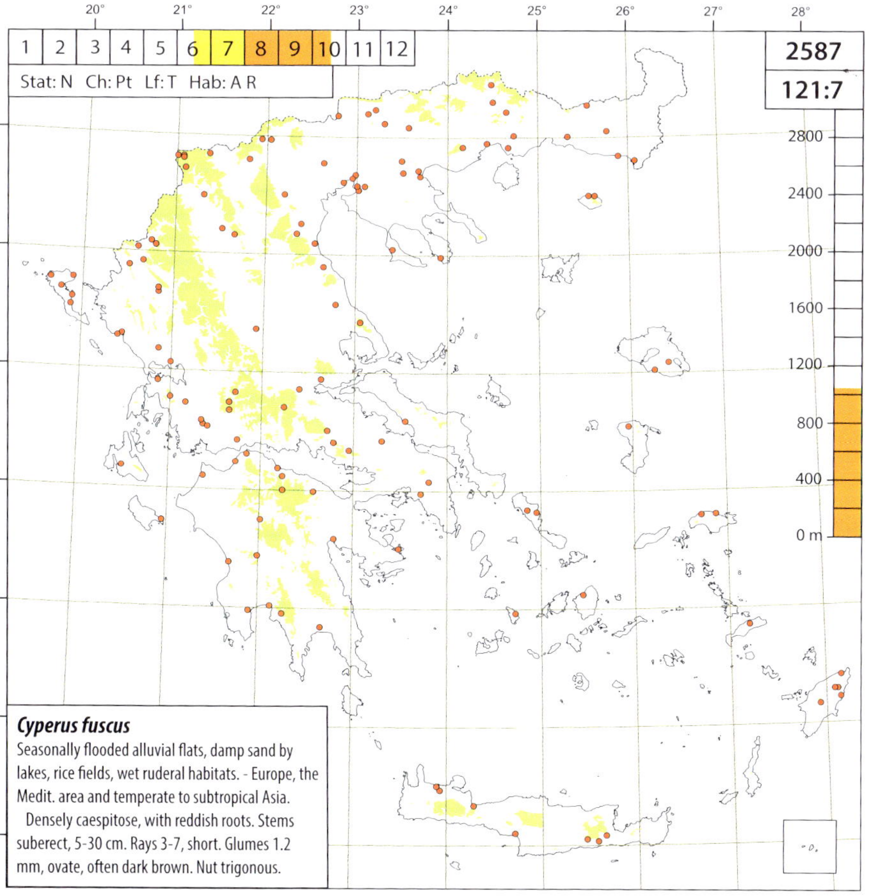

Cyperus fuscus

Seasonally flooded alluvial flats, damp sand by lakes, rice fields, wet ruderal habitats. - Europe, the Medit. area and temperate to subtropical Asia.

Densely caespitose, with reddish roots. Stems suberect, 5-30 cm. Rays 3-7, short. Glumes 1.2 mm, ovate, often dark brown. Nut trigonous.

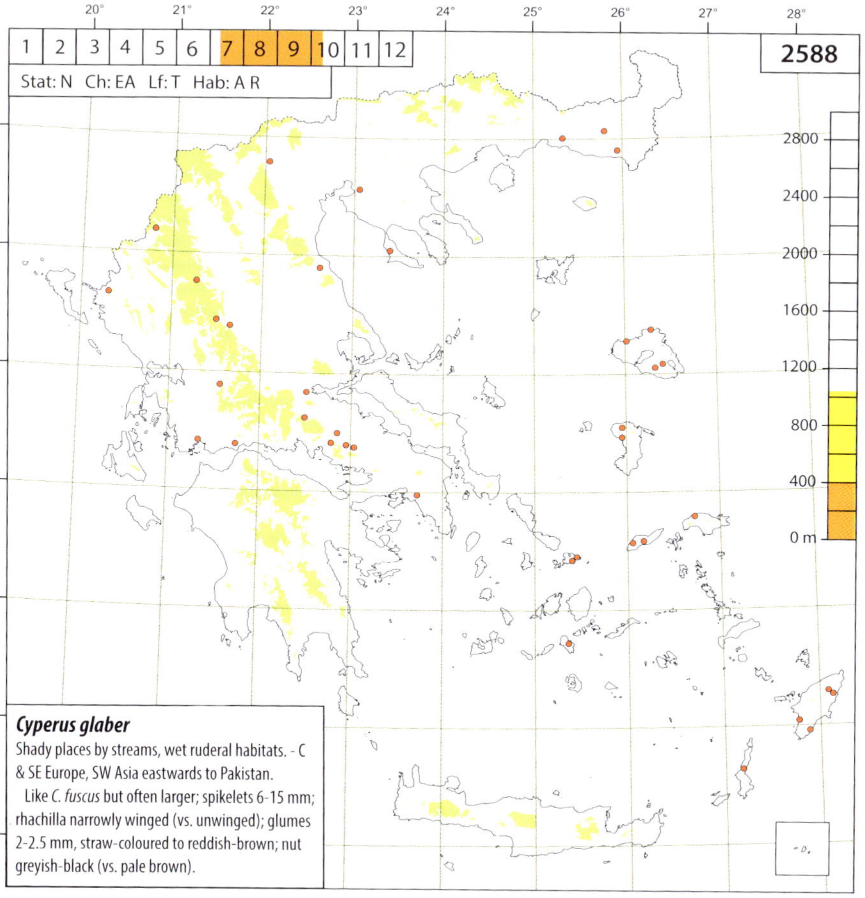

Cyperus glaber

Shady places by streams, wet ruderal habitats. - C & SE Europe, SW Asia eastwards to Pakistan.

Like *C. fuscus* but often larger; spikelets 6-15 mm; rhachilla narrowly winged (vs. unwinged); glumes 2-2.5 mm, straw-coloured to reddish-brown; nut greyish-black (vs. pale brown).

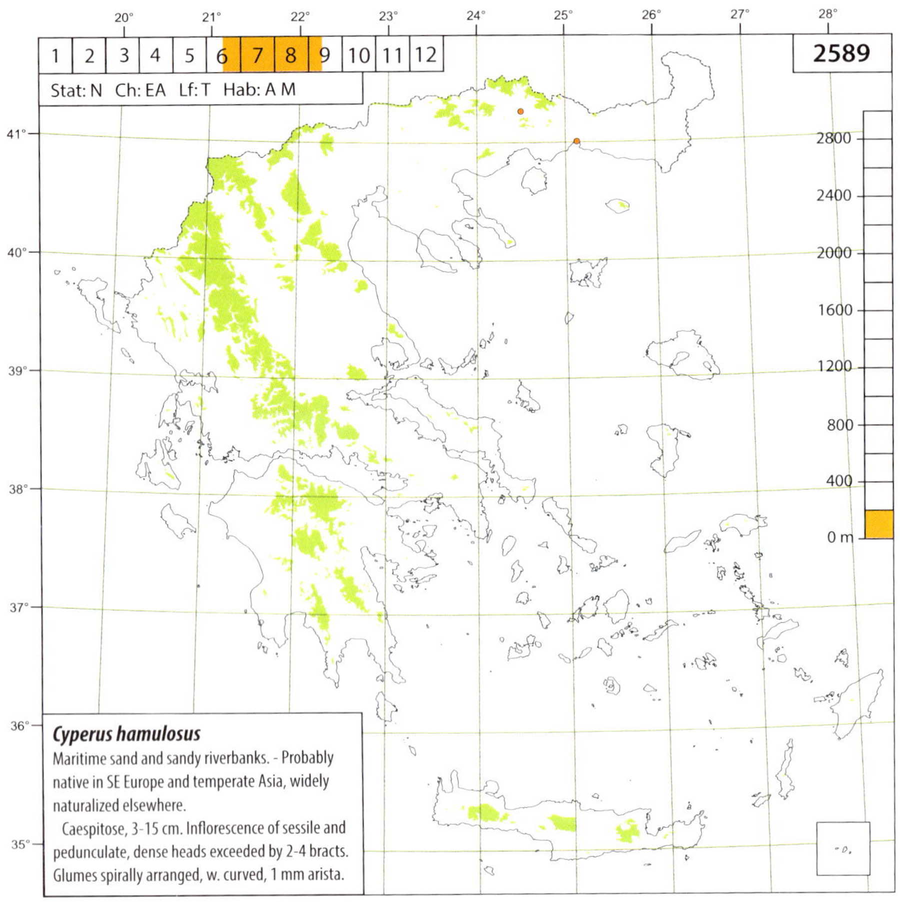

Cyperus hamulosus

Maritime sand and sandy riverbanks. - Probably native in SE Europe and temperate Asia, widely naturalized elsewhere.

Caespitose, 3-15 cm. Inflorescence of sessile and pedunculate, dense heads exceeded by 2-4 bracts. Glumes spirally arranged, w. curved, 1 mm arista.

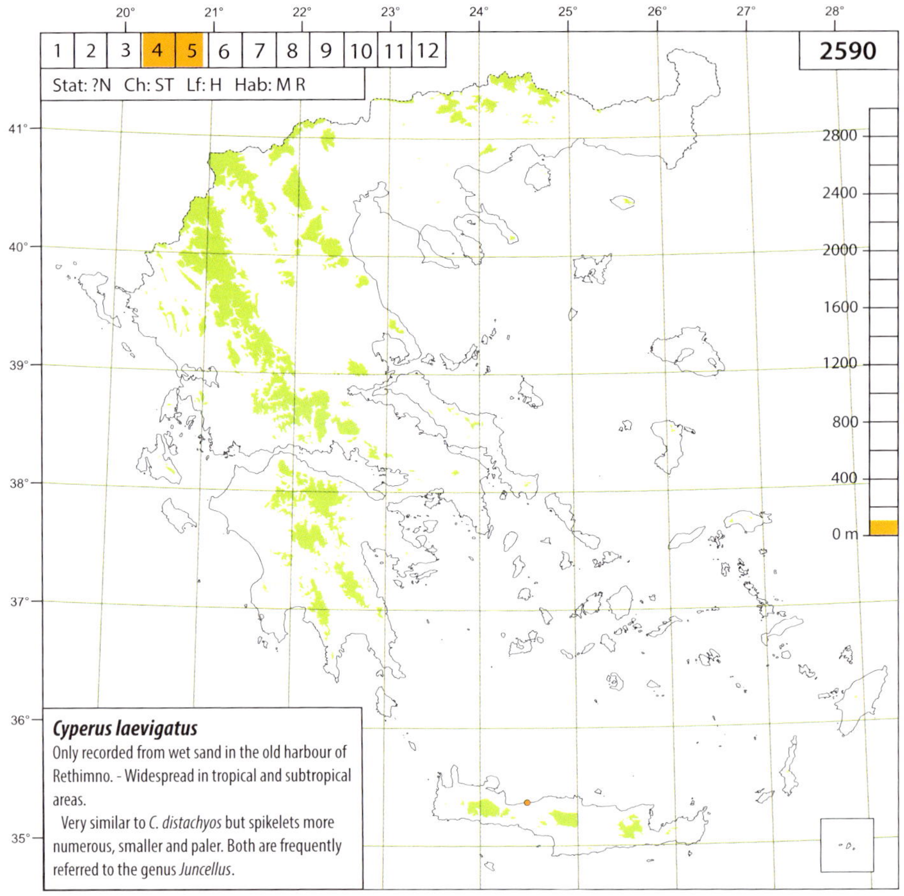

Cyperus laevigatus

Only recorded from wet sand in the old harbour of Rethimno. - Widespread in tropical and subtropical areas.

Very similar to *C. distachyos* but spikelets more numerous, smaller and paler. Both are frequently referred to the genus *Juncellus*.

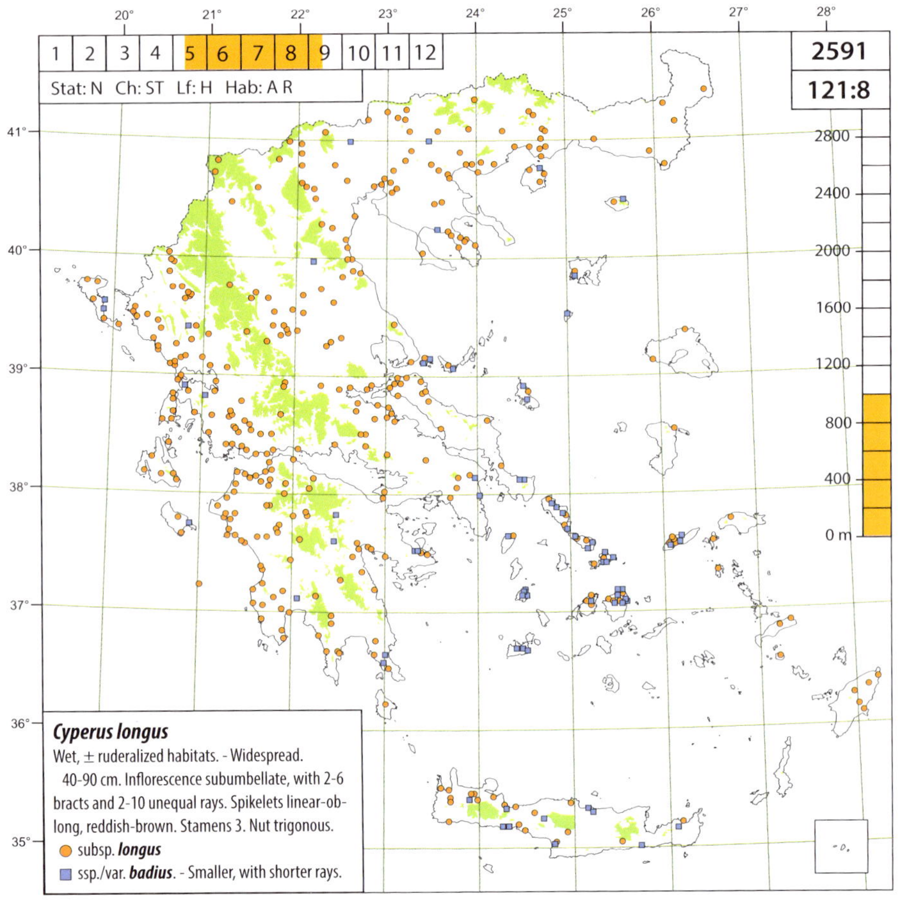

Cyperus longus

Wet, ± ruderalized habitats. - Widespread.

40-90 cm. Inflorescence subumbellate, with 2-6 bracts and 2-10 unequal rays. Spikelets linear-oblong, reddish-brown. Stamens 3. Nut trigonous.

● subsp. ***longus***

■ ssp./var. ***badius***. - Smaller, with shorter rays.

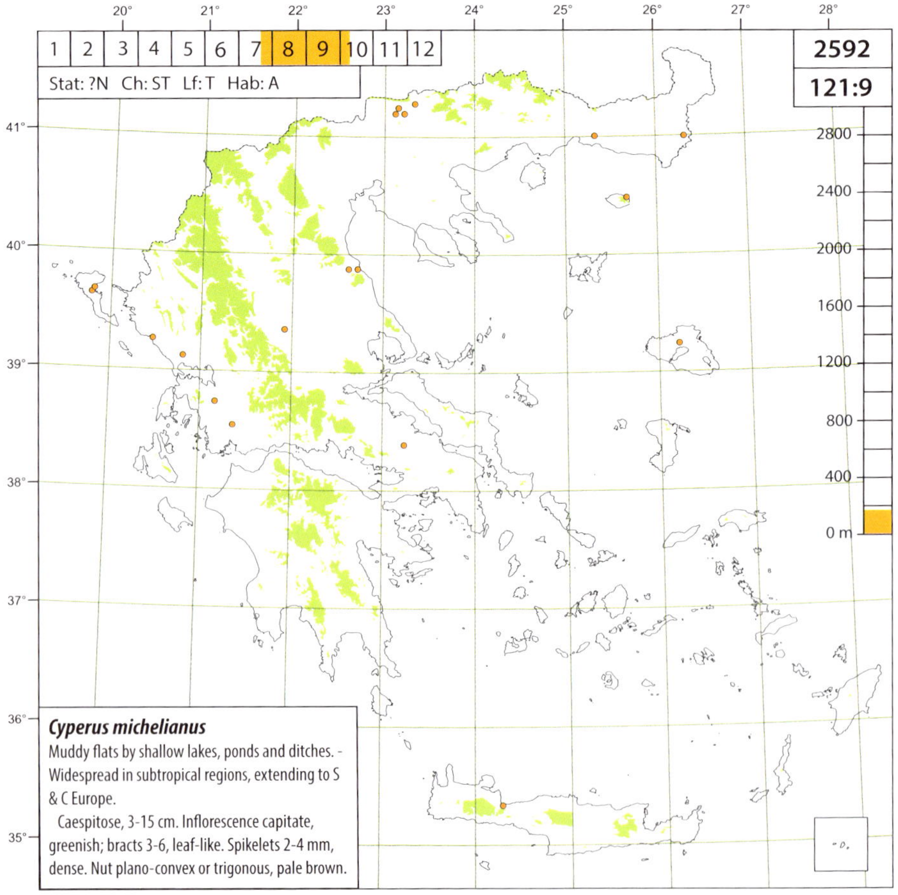

Cyperus michelianus

Muddy flats by shallow lakes, ponds and ditches. - Widespread in subtropical regions, extending to S & C Europe.

Caespitose, 3-15 cm. Inflorescence capitate, greenish; bracts 3-6, leaf-like. Spikelets 2-4 mm, dense. Nut plano-convex or trigonous, pale brown.

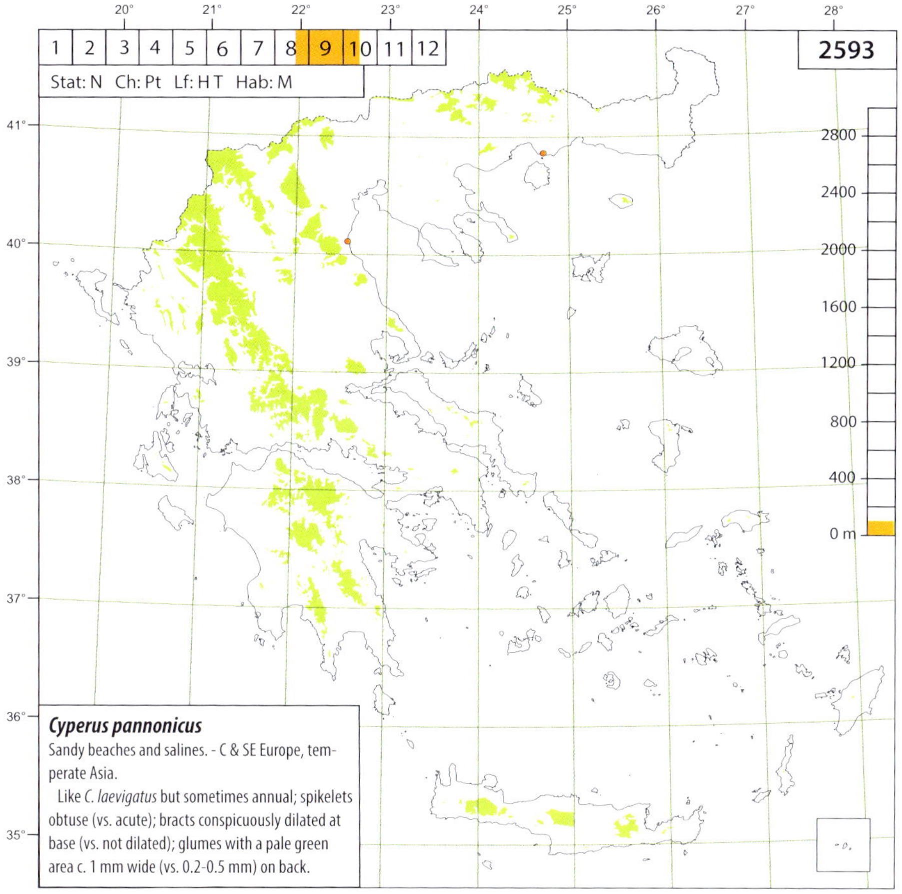

Cyperus pannonicus
Sandy beaches and salines. - C & SE Europe, temperate Asia.
Like *C. laevigatus* but sometimes annual; spikelets obtuse (vs. acute); bracts conspicuously dilated at base (vs. not dilated); glumes with a pale green area c. 1 mm wide (vs. 0.2-0.5 mm) on back.

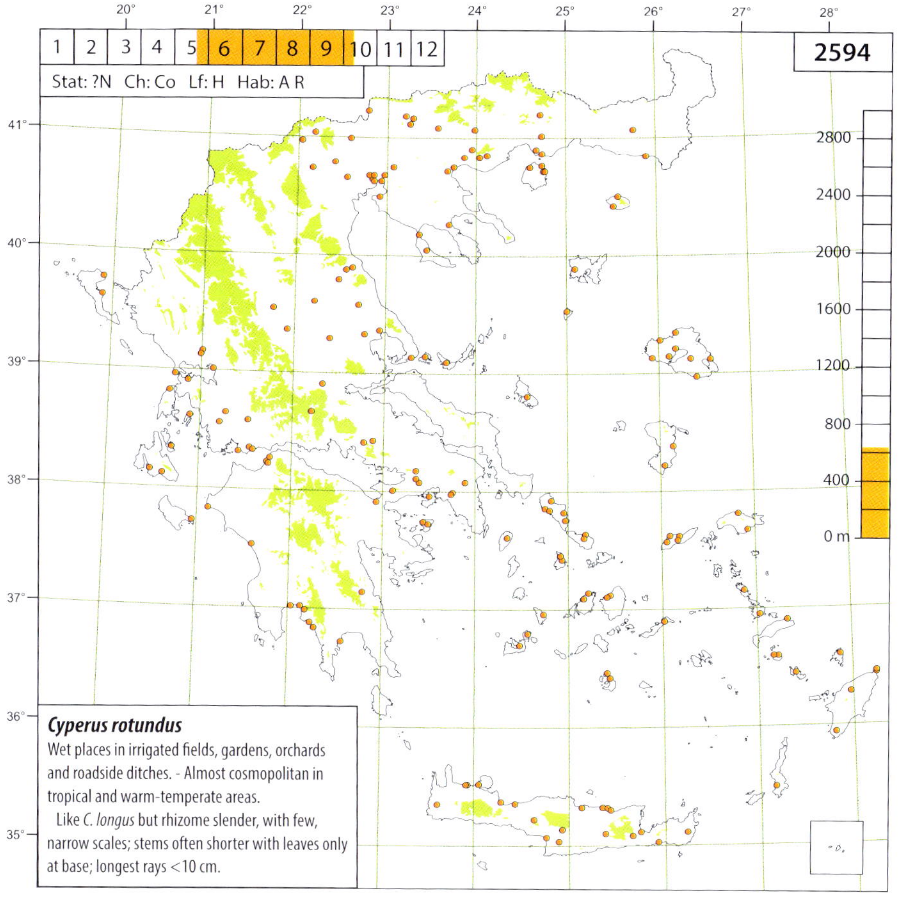

Cyperus rotundus
Wet places in irrigated fields, gardens, orchards and roadside ditches. - Almost cosmopolitan in tropical and warm-temperate areas.
Like *C. longus* but rhizome slender, with few, narrow scales; stems often shorter with leaves only at base; longest rays <10 cm.

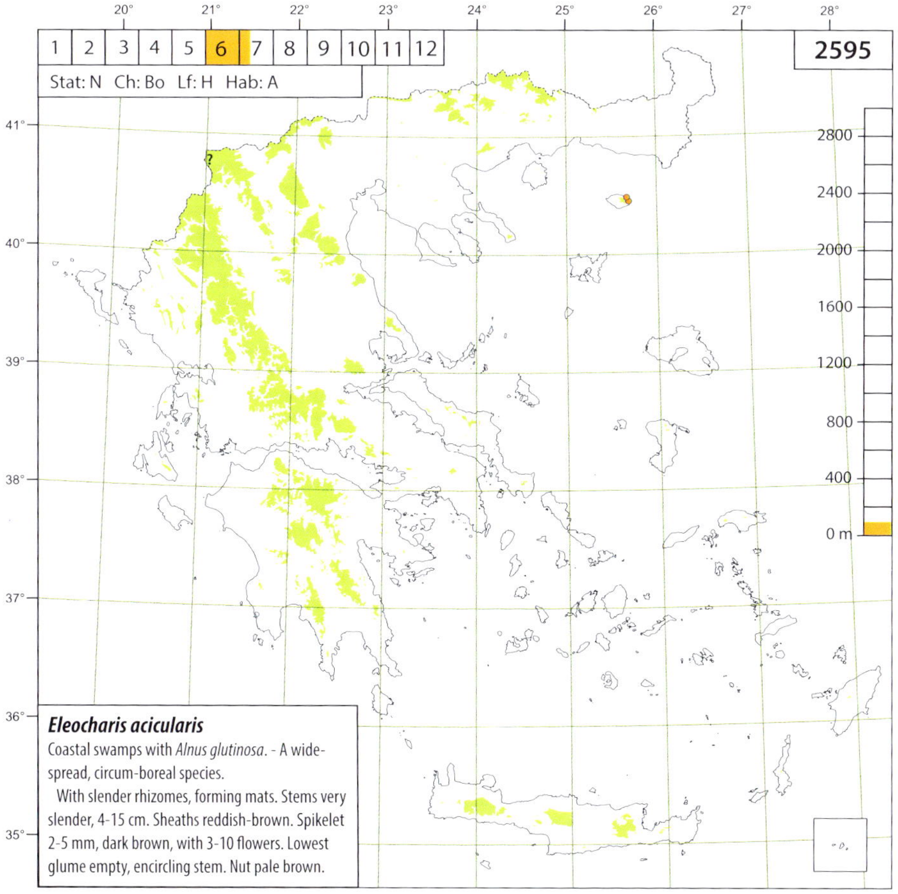

Eleocharis acicularis
Coastal swamps with *Alnus glutinosa*. - A widespread, circum-boreal species.
With slender rhizomes, forming mats. Stems very slender, 4-15 cm. Sheaths reddish-brown. Spikelet 2-5 mm, dark brown, with 3-10 flowers. Lowest glume empty, encircling stem. Nut pale brown.

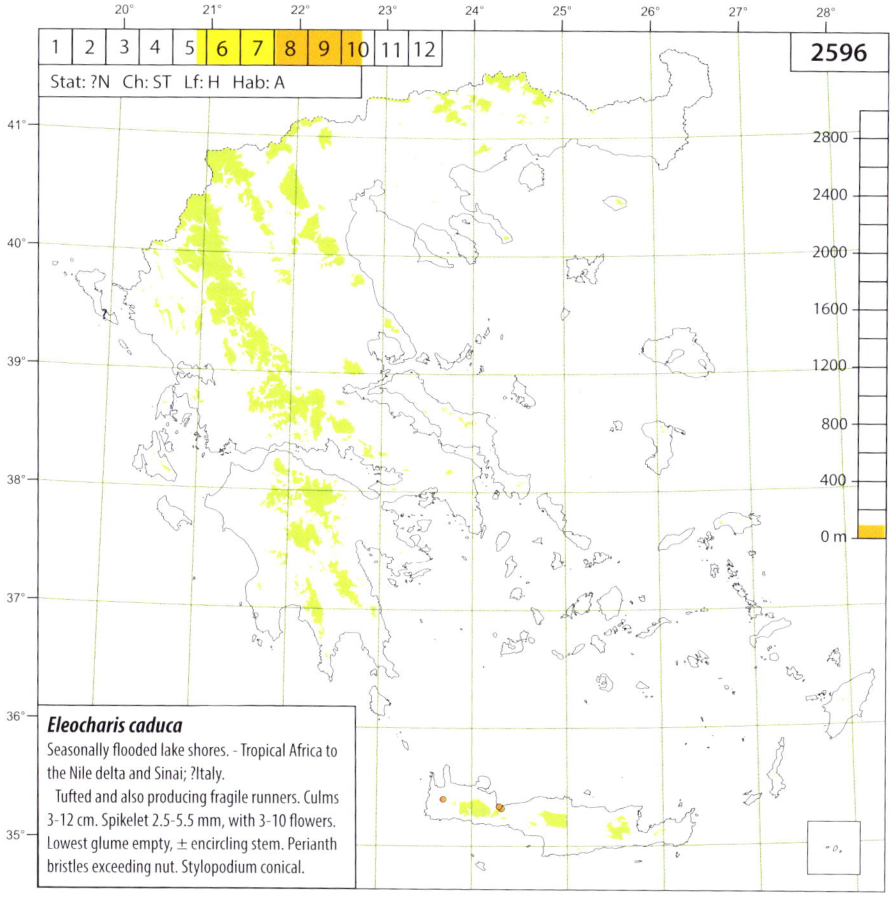

Eleocharis caduca
Seasonally flooded lake shores. - Tropical Africa to the Nile delta and Sinai; ?Italy.
Tufted and also producing fragile runners. Culms 3-12 cm. Spikelet 2.5-5.5 mm, with 3-10 flowers. Lowest glume empty, ± encircling stem. Perianth bristles exceeding nut. Stylopodium conical.

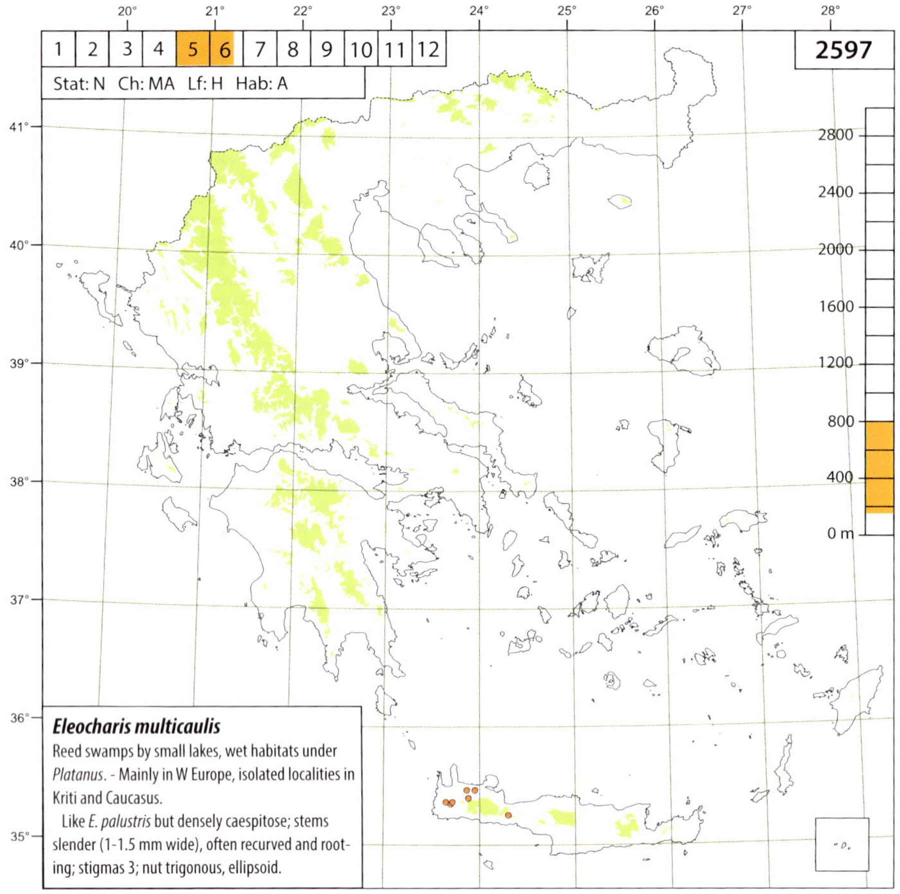

Eleocharis multicaulis
Reed swamps by small lakes, wet habitats under *Platanus*. - Mainly in W Europe, isolated localities in Kriti and Caucasus.
Like *E. palustris* but densely caespitose; stems slender (1-1.5 mm wide), often recurved and rooting; stigmas 3; nut trigonous, ellipsoid.

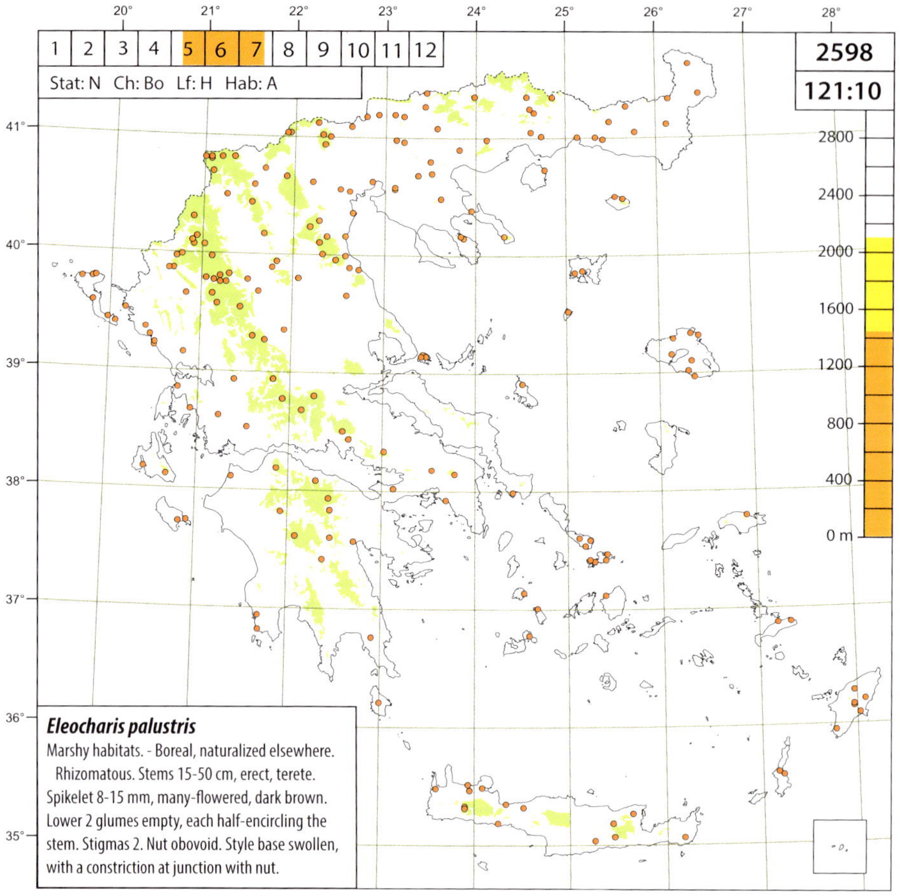

Eleocharis palustris
Marshy habitats. - Boreal, naturalized elsewhere.
Rhizomatous. Stems 15-50 cm, erect, terete. Spikelet 8-15 mm, many-flowered, dark brown. Lower 2 glumes empty, each half-encircling the stem. Stigmas 2. Nut obovoid. Style base swollen, with a constriction at junction with nut.

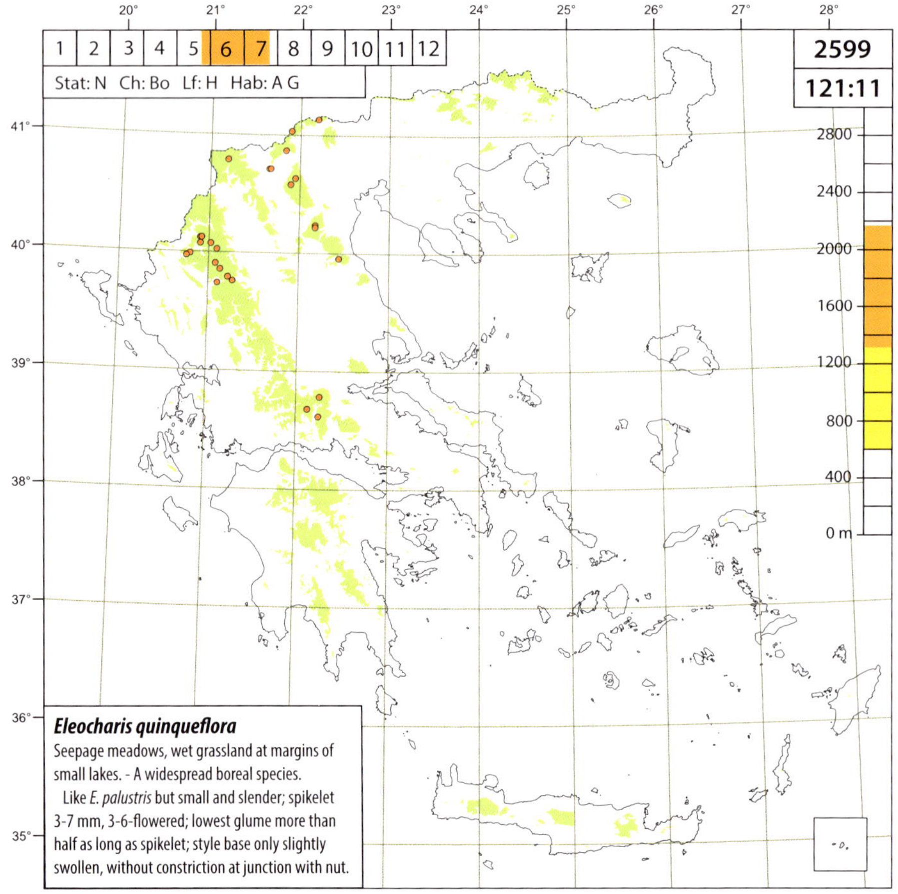

Eleocharis quinqueflora
Seepage meadows, wet grassland at margins of small lakes. - A widespread boreal species.
Like *E. palustris* but small and slender; spikelet 3-7 mm, 3-6-flowered; lowest glume more than half as long as spikelet; style base only slightly swollen, without constriction at junction with nut.

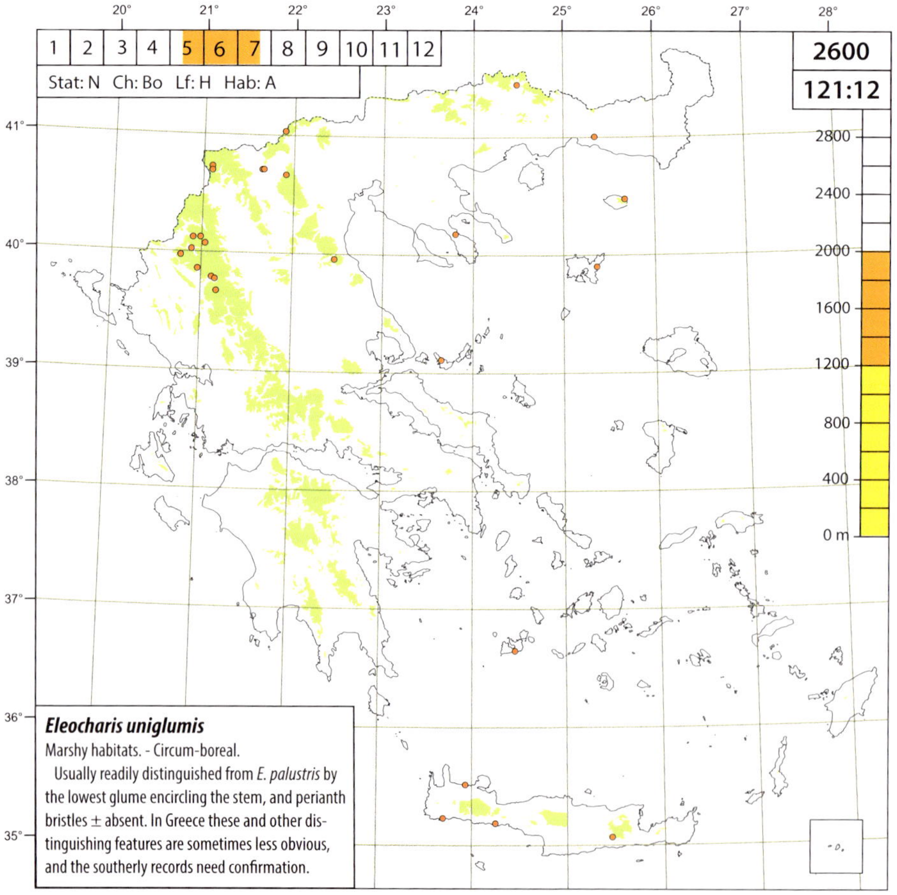

Eleocharis uniglumis
Marshy habitats. - Circum-boreal.
Usually readily distinguished from *E. palustris* by the lowest glume encircling the stem, and perianth bristles ± absent. In Greece these and other distinguishing features are sometimes less obvious, and the southerly records need confirmation.

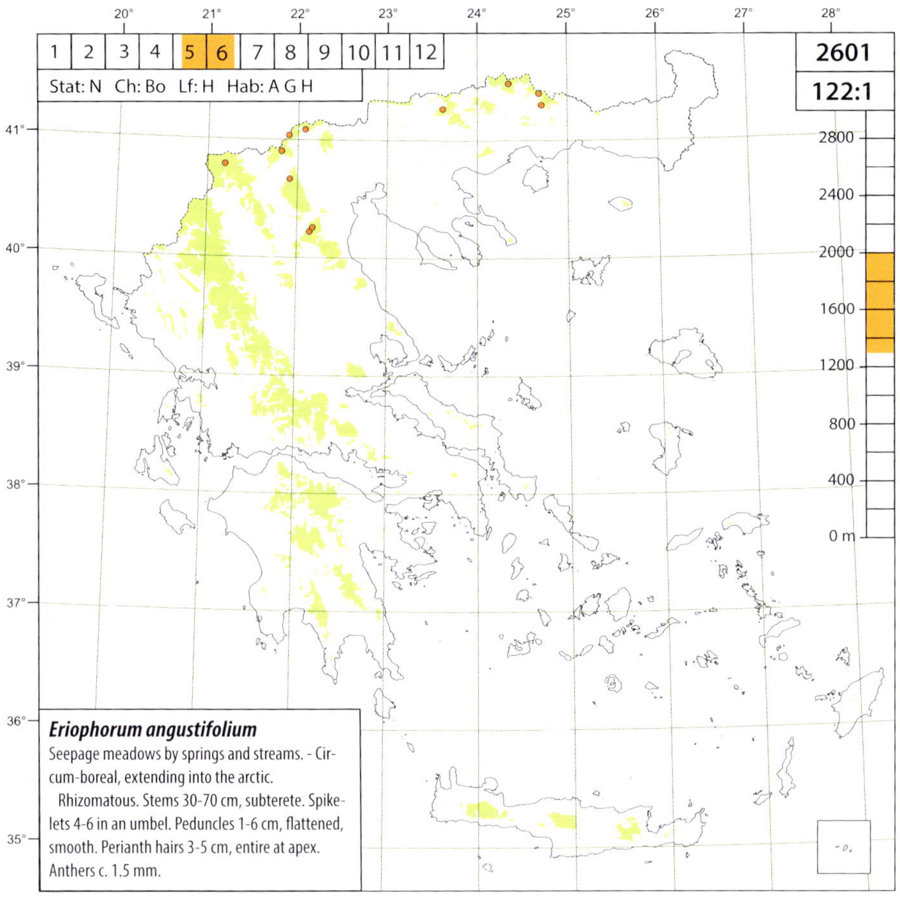

Eriophorum angustifolium

Seepage meadows by springs and streams. - Circum-boreal, extending into the arctic.

Rhizomatous. Stems 30-70 cm, subterete. Spikelets 4-6 in an umbel. Peduncles 1-6 cm, flattened, smooth. Perianth hairs 3-5 cm, entire at apex. Anthers c. 1.5 mm.

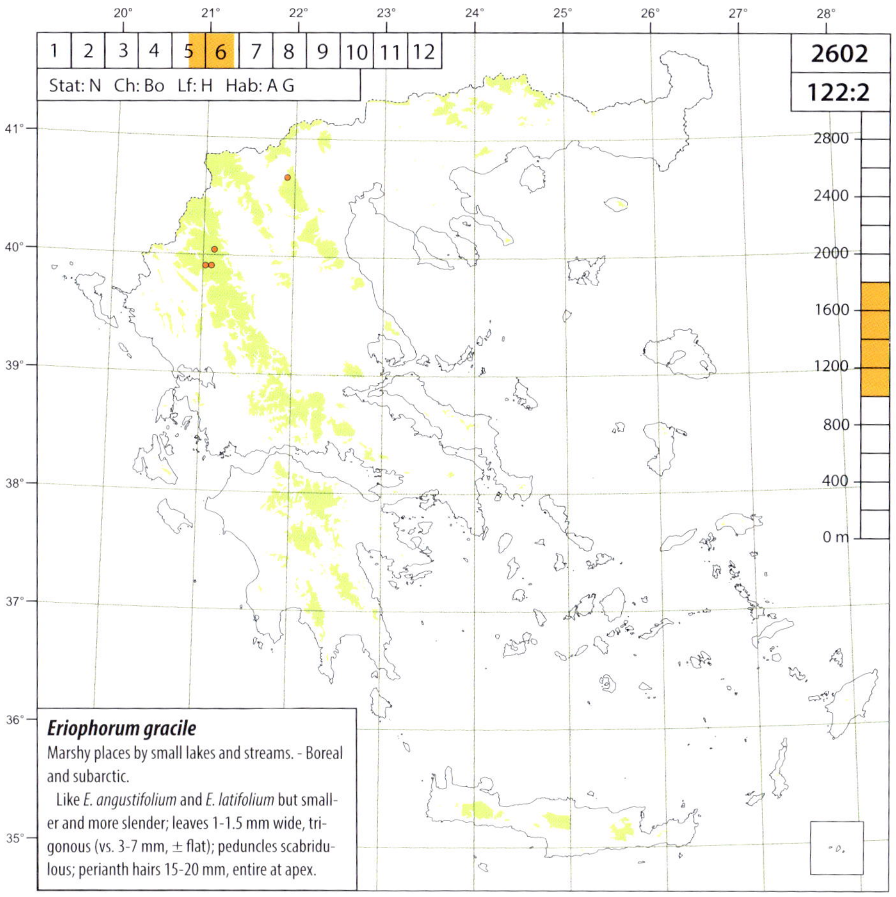

Eriophorum gracile

Marshy places by small lakes and streams. - Boreal and subarctic.

Like *E. angustifolium* and *E. latifolium* but smaller and more slender; leaves 1-1.5 mm wide, trigonous (vs. 3-7 mm, ± flat); peduncles scabridulous; perianth hairs 15-20 mm, entire at apex.

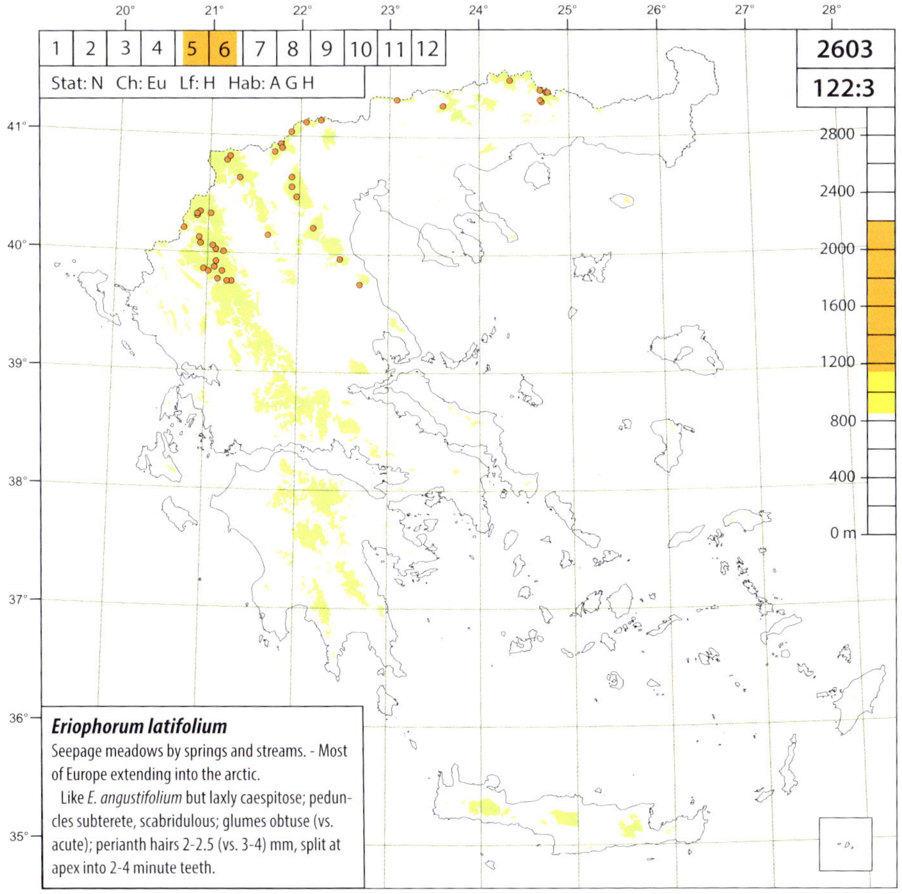

Eriophorum latifolium

Seepage meadows by springs and streams. - Most of Europe extending into the arctic.

Like *E. angustifolium* but laxly caespitose; peduncles subterete, scabridulous; glumes obtuse (vs. acute); perianth hairs 2-2.5 (vs. 3-4) mm, split at apex into 2-4 minute teeth.

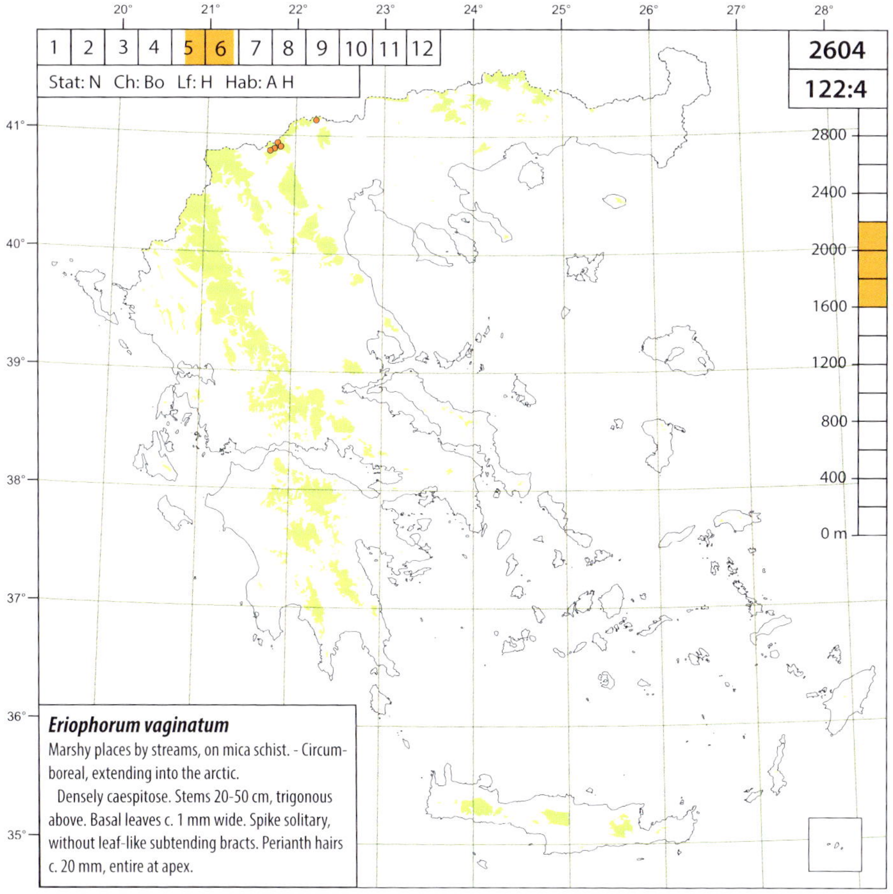

Eriophorum vaginatum

Marshy places by streams, on mica schist. - Circum-boreal, extending into the arctic.

Densely caespitose. Stems 20-50 cm, trigonous above. Basal leaves c. 1 mm wide. Spike solitary, without leaf-like subtending bracts. Perianth hairs c. 20 mm, entire at apex.

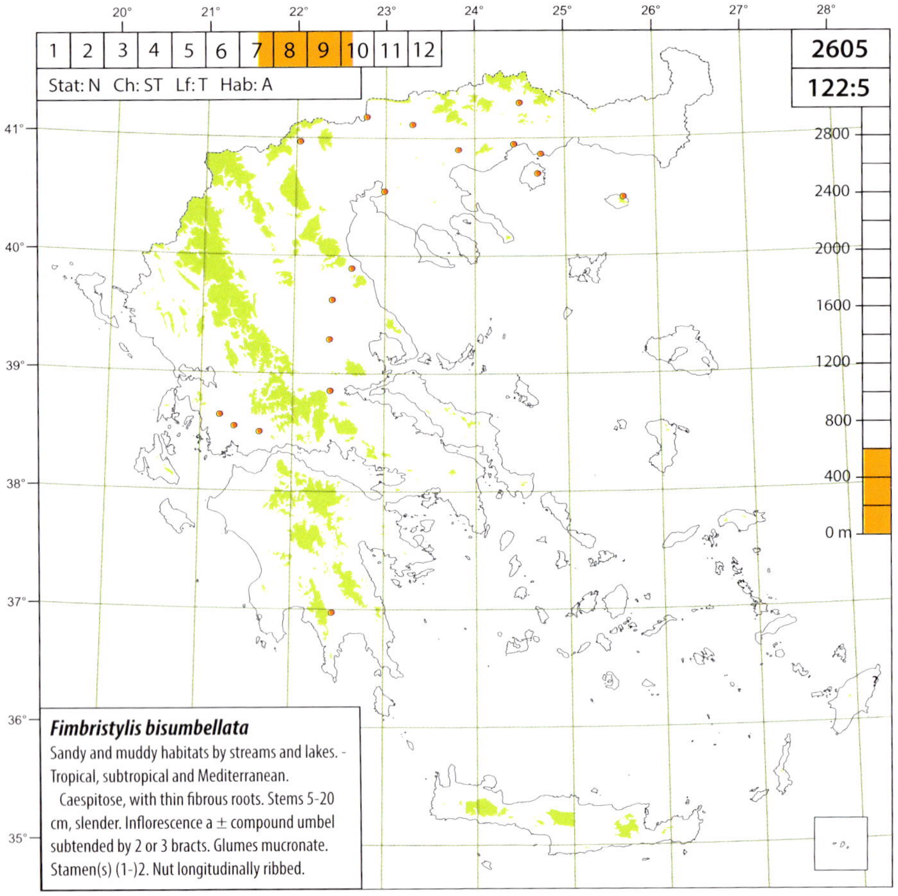

Fimbristylis bisumbellata

Sandy and muddy habitats by streams and lakes. - Tropical, subtropical and Mediterranean.

Caespitose, with thin fibrous roots. Stems 5-20 cm, slender. Inflorescence a ± compound umbel subtended by 2 or 3 bracts. Glumes mucronate. Stamen(s) (1-)2. Nut longitudinally ribbed.

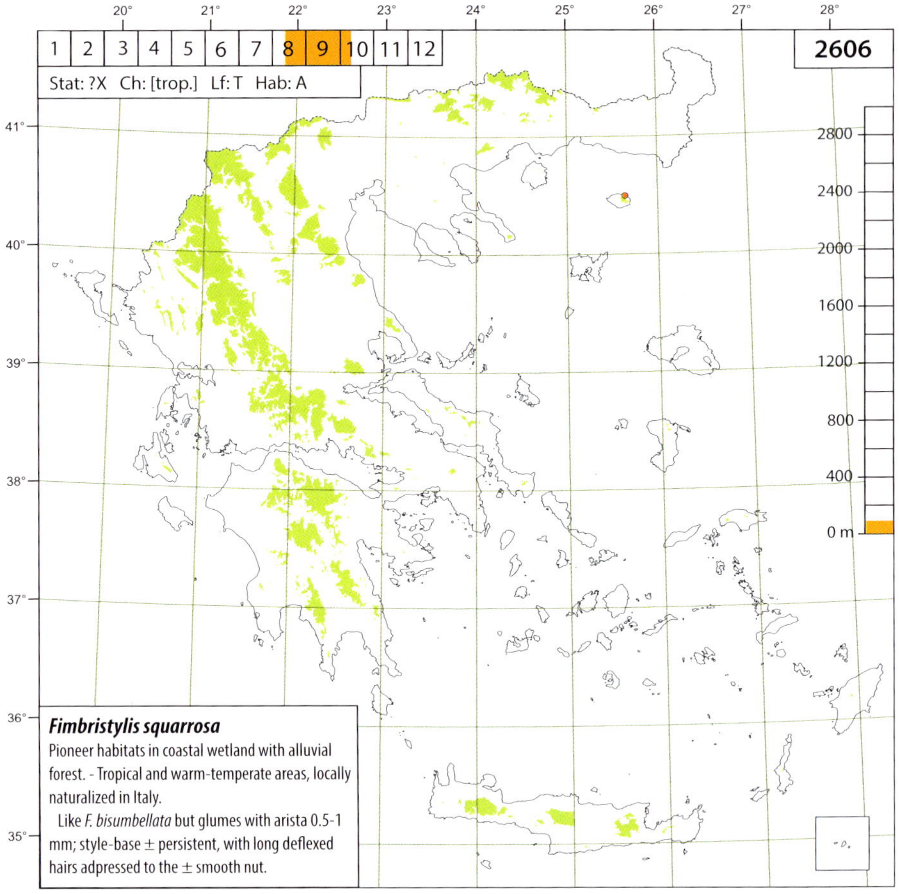

Fimbristylis squarrosa

Pioneer habitats in coastal wetland with alluvial forest. - Tropical and warm-temperate areas, locally naturalized in Italy.

Like *F. bisumbellata* but glumes with arista 0.5-1 mm; style-base ± persistent, with long deflexed hairs adpressed to the ± smooth nut.

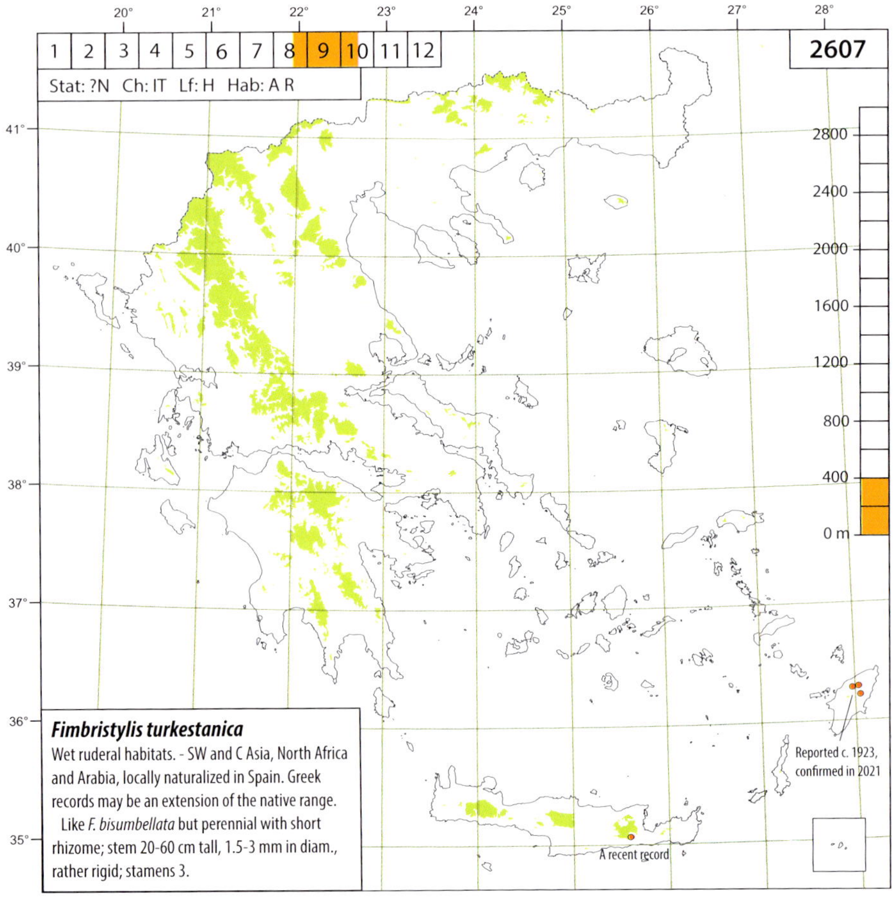

Fimbristylis turkestanica

Wet ruderal habitats. - SW and C Asia, North Africa and Arabia, locally naturalized in Spain. Greek records may be an extension of the native range.

Like *F. bisumbellata* but perennial with short rhizome; stem 20-60 cm tall, 1.5-3 mm in diam., rather rigid; stamens 3.

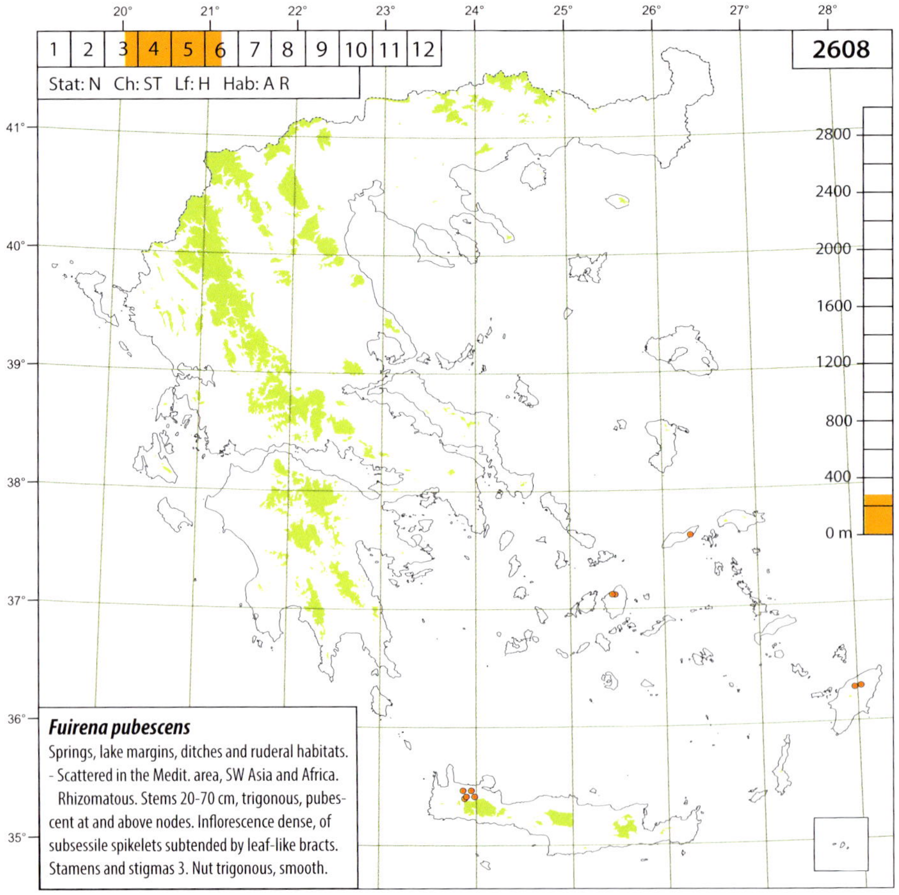

Fuirena pubescens

Springs, lake margins, ditches and ruderal habitats. - Scattered in the Medit. area, SW Asia and Africa.

Rhizomatous. Stems 20-70 cm, trigonous, pubescent at and above nodes. Inflorescence dense, of subsessile spikelets subtended by leaf-like bracts. Stamens and stigmas 3. Nut trigonous, smooth.

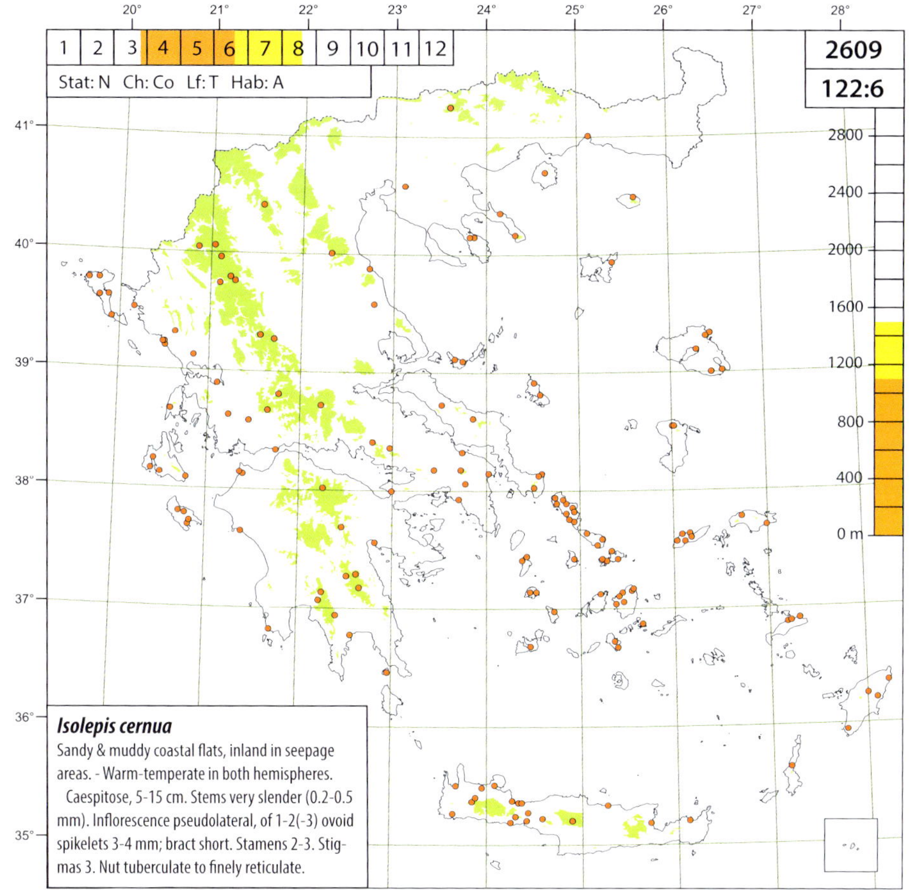

Isolepis cernua

Sandy & muddy coastal flats, inland in seepage areas. - Warm-temperate in both hemispheres.
Caespitose, 5-15 cm. Stems very slender (0.2-0.5 mm). Inflorescence pseudolateral, of 1-2(-3) ovoid spikelets 3-4 mm; bract short. Stamens 2-3. Stigmas 3. Nut tuberculate to finely reticulate.

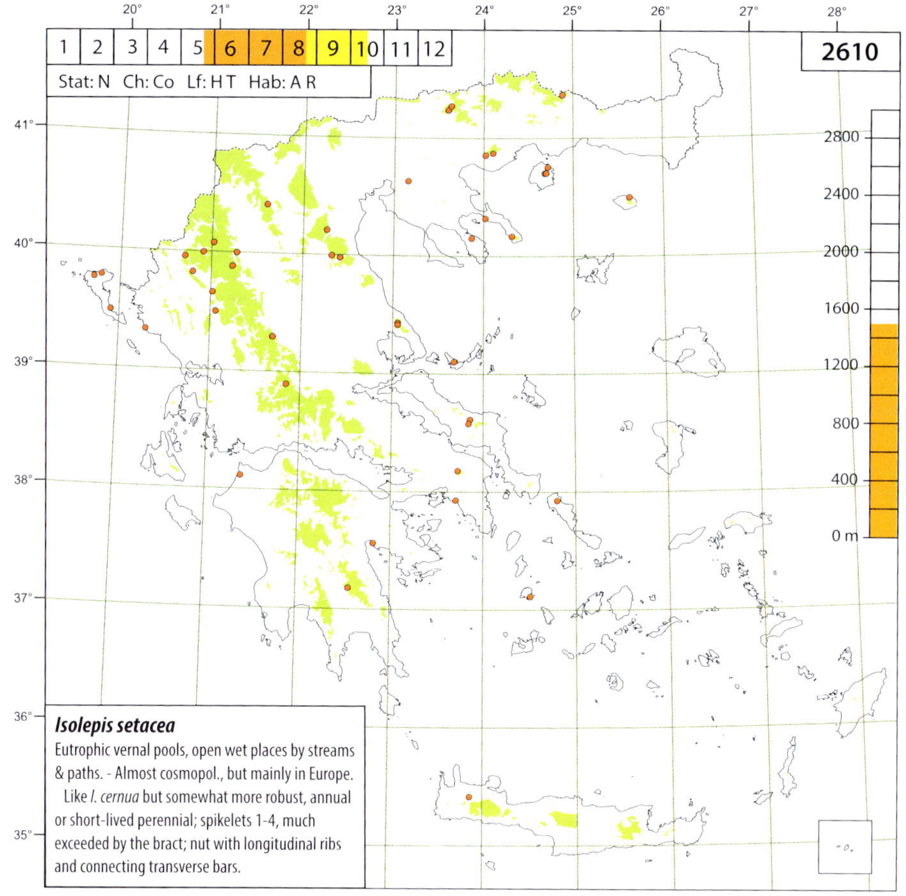

Isolepis setacea

Eutrophic vernal pools, open wet places by streams & paths. - Almost cosmopol., but mainly in Europe.
Like *I. cernua* but somewhat more robust, annual or short-lived perennial; spikelets 1-4, much exceeded by the bract; nut with longitudinal ribs and connecting transverse bars.

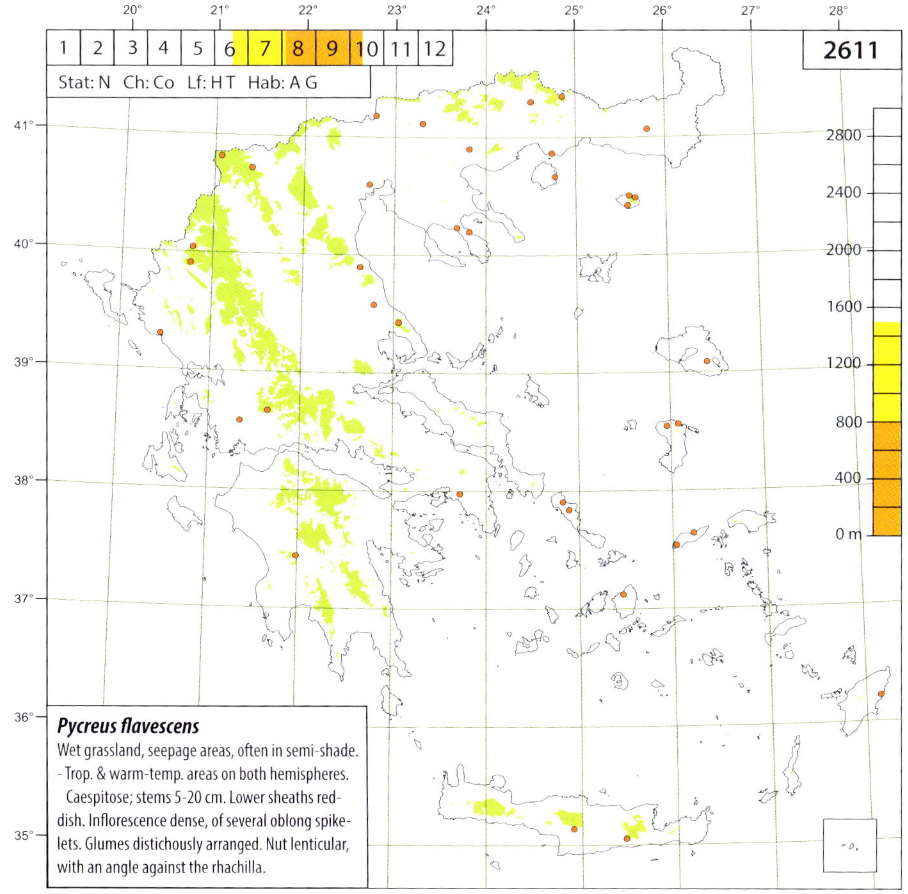

Pycreus flavescens

Wet grassland, seepage areas, often in semi-shade. - Trop. & warm-temp. areas on both hemispheres.
Caespitose; stems 5-20 cm. Lower sheaths reddish. Inflorescence dense, of several oblong spikelets. Glumes distichously arranged. Nut lenticular, with an angle against the rhachilla.

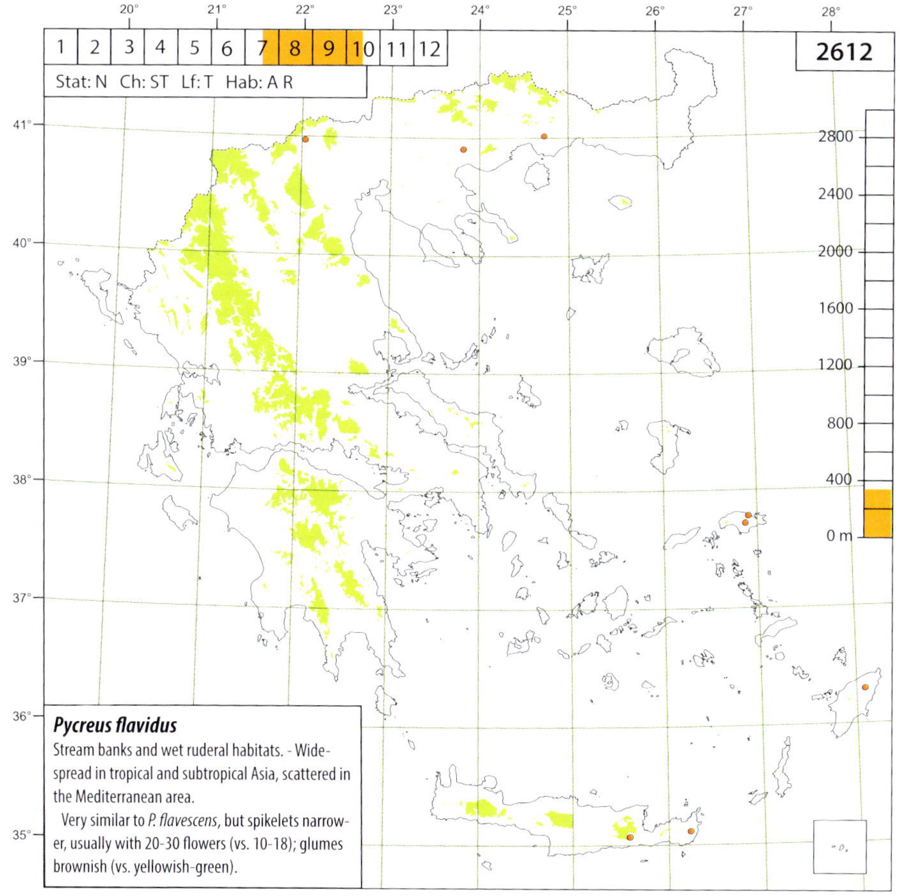

Pycreus flavidus

Stream banks and wet ruderal habitats. - Widespread in tropical and subtropical Asia, scattered in the Mediterranean area.
Very similar to *P. flavescens*, but spikelets narrower, usually with 20-30 flowers (vs. 10-18); glumes brownish (vs. yellowish-green).

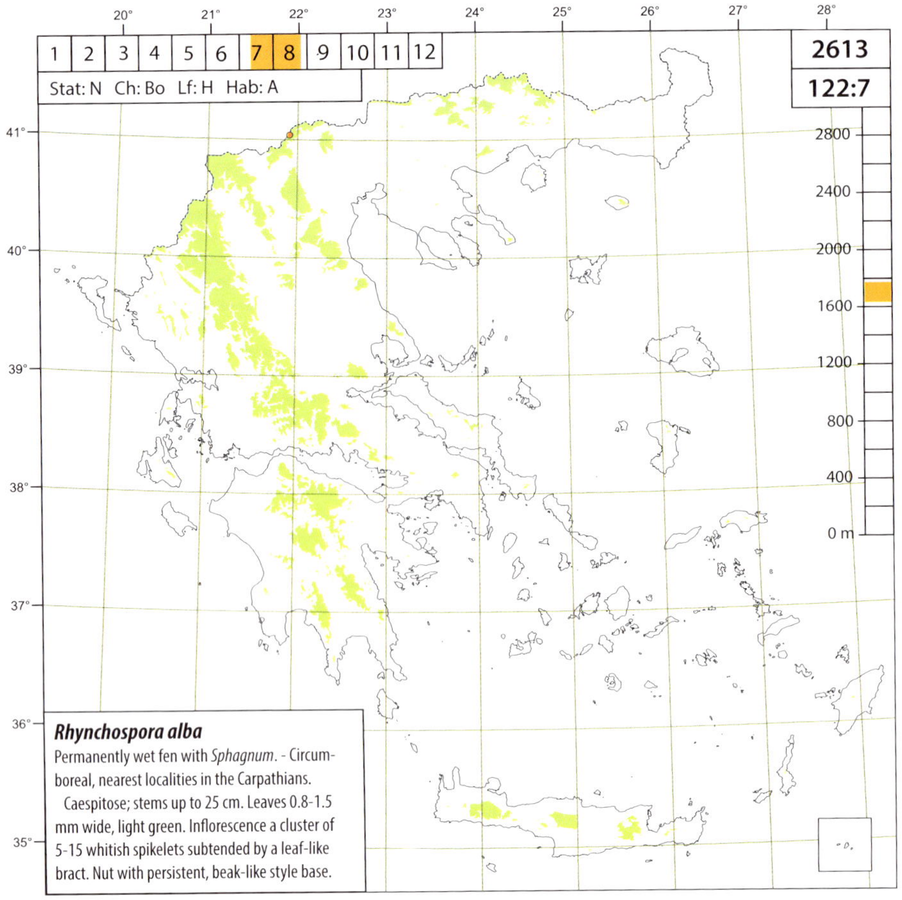

Rhynchospora alba
Permanently wet fen with *Sphagnum*. - Circumboreal, nearest localities in the Carpathians.
Caespitose; stems up to 25 cm. Leaves 0.8-1.5 mm wide, light green. Inflorescence a cluster of 5-15 whitish spikelets subtended by a leaf-like bract. Nut with persistent, beak-like style base.

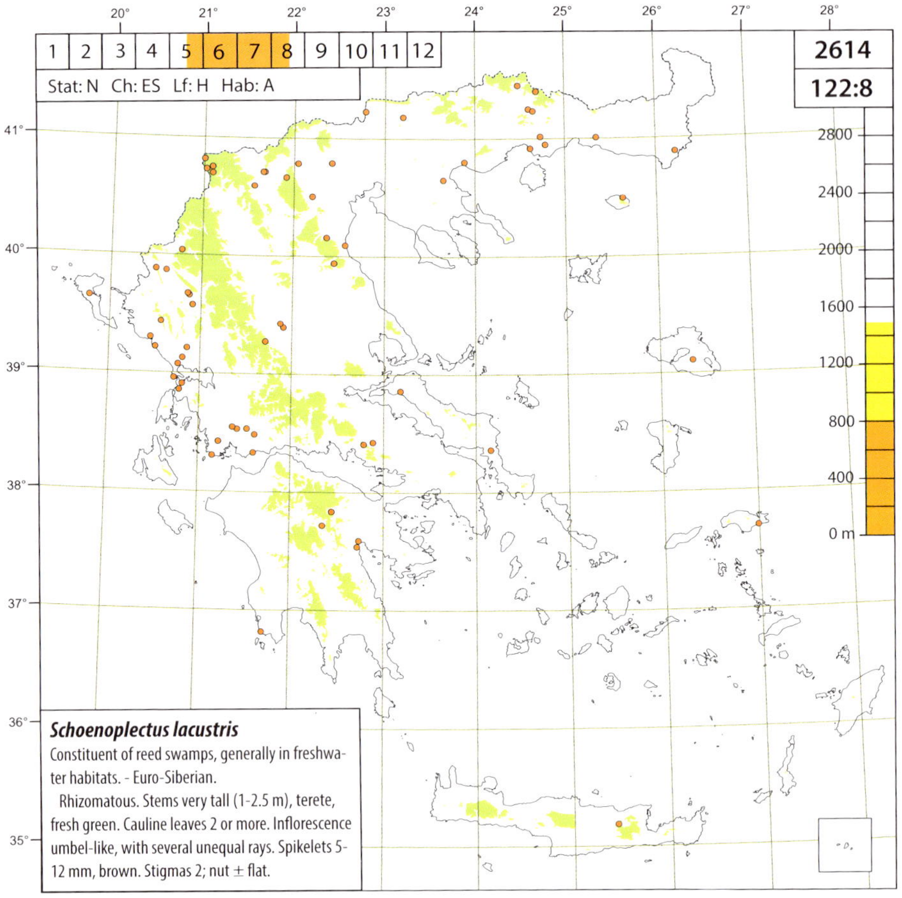

Schoenoplectus lacustris
Constituent of reed swamps, generally in freshwater habitats. - Euro-Siberian.
Rhizomatous. Stems very tall (1-2.5 m), terete, fresh green. Cauline leaves 2 or more. Inflorescence umbel-like, with several unequal rays. Spikelets 5-12 mm, brown. Stigmas 2; nut ± flat.

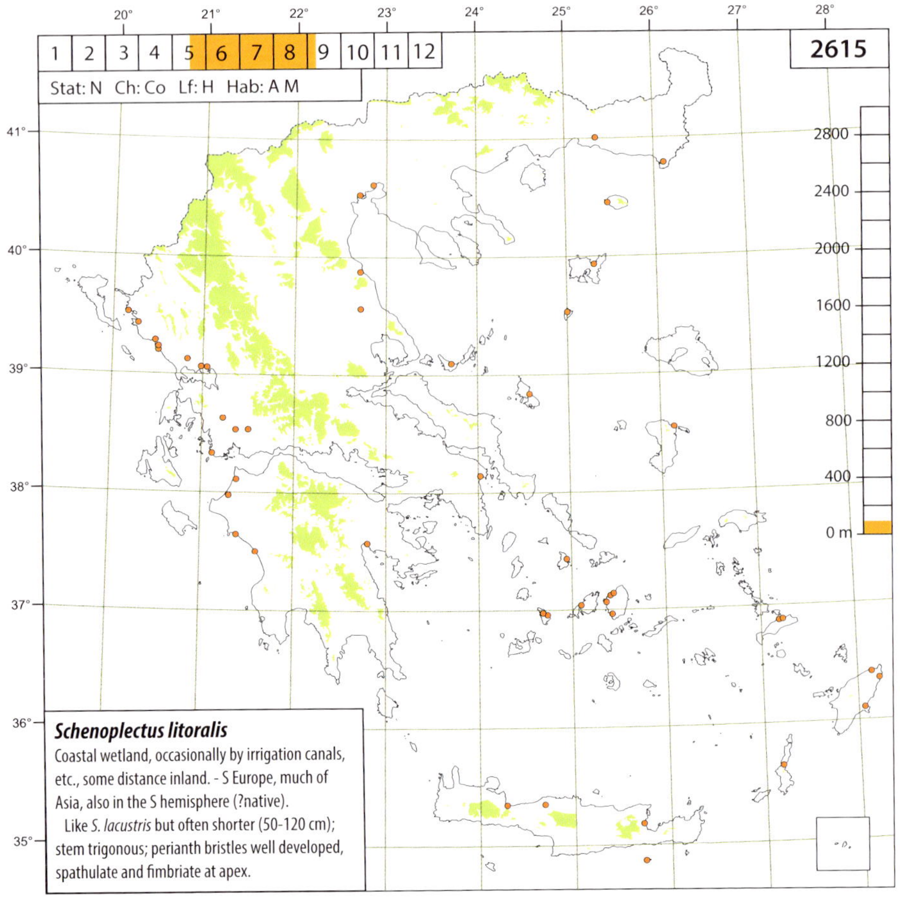

Schoenoplectus litoralis
Coastal wetland, occasionally by irrigation canals, etc., some distance inland. - S Europe, much of Asia, also in the S hemisphere (?native).
Like *S. lacustris* but often shorter (50-120 cm); stem trigonous; perianth bristles well developed, spathulate and fimbriate at apex.

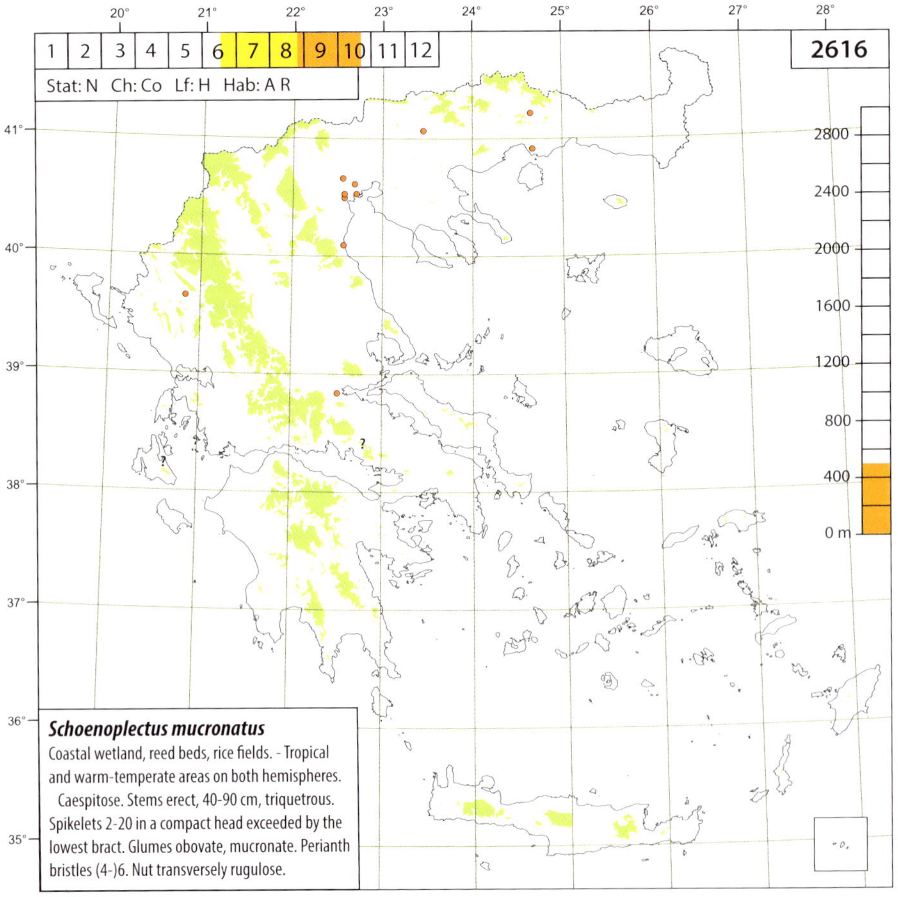

Schoenoplectus mucronatus
Coastal wetland, reed beds, rice fields. - Tropical and warm-temperate areas on both hemispheres.
Caespitose. Stems erect, 40-90 cm, triquetrous. Spikelets 2-20 in a compact head exceeded by the lowest bract. Glumes obovate, mucronate. Perianth bristles (4-)6. Nut transversely rugulose.

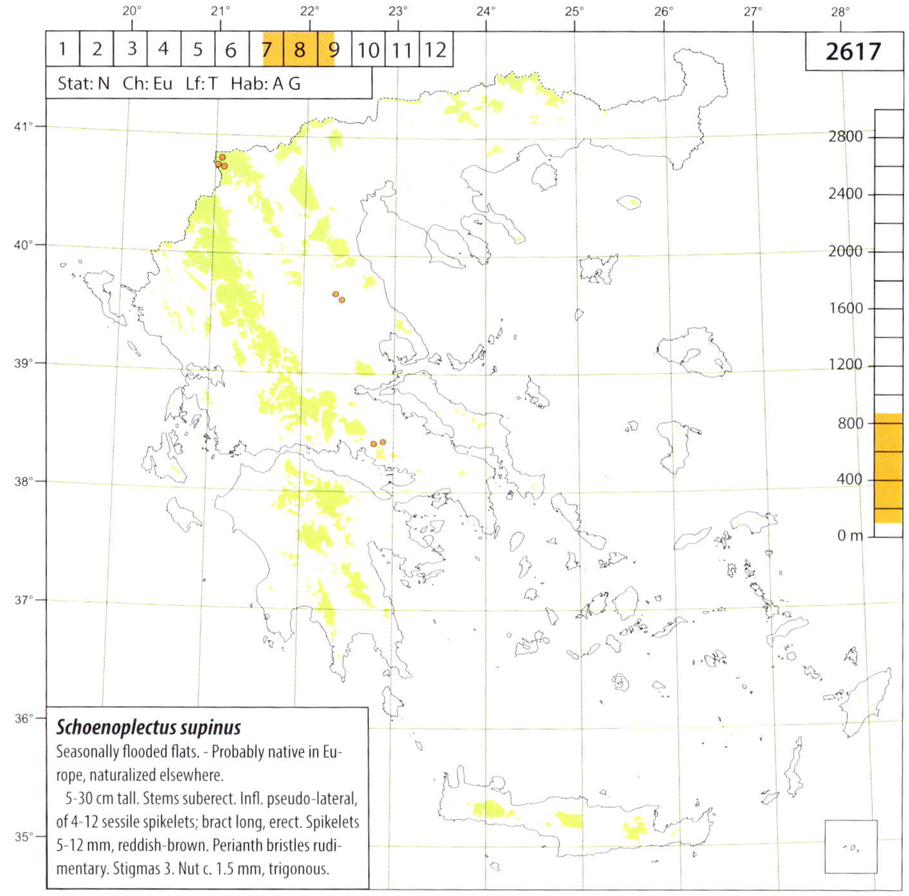

Schoenoplectus supinus
Seasonally flooded flats. - Probably native in Europe, naturalized elsewhere.
5-30 cm tall. Stems suberect. Infl. pseudo-lateral, of 4-12 sessile spikelets; bract long, erect. Spikelets 5-12 mm, reddish-brown. Perianth bristles rudimentary. Stigmas 3. Nut c. 1.5 mm, trigonous.

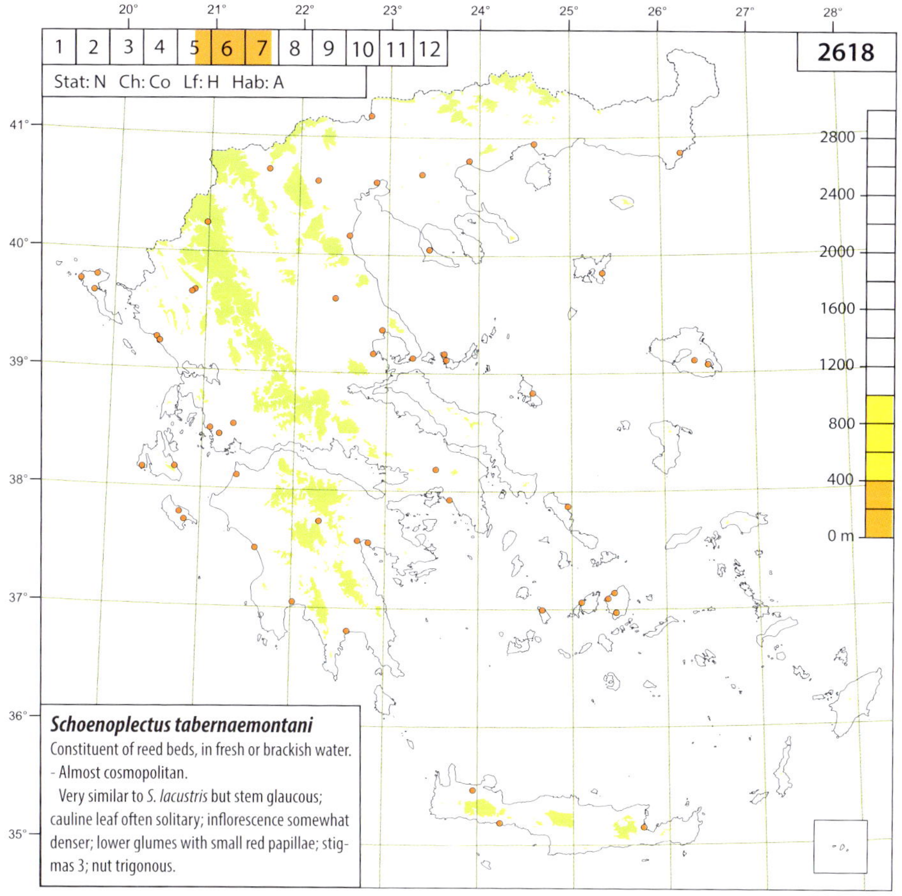

Schoenoplectus tabernaemontani
Constituent of reed beds, in fresh or brackish water. - Almost cosmopolitan.
Very similar to *S. lacustris* but stem glaucous; cauline leaf often solitary; inflorescence somewhat denser; lower glumes with small red papillae; stigmas 3; nut trigonous.

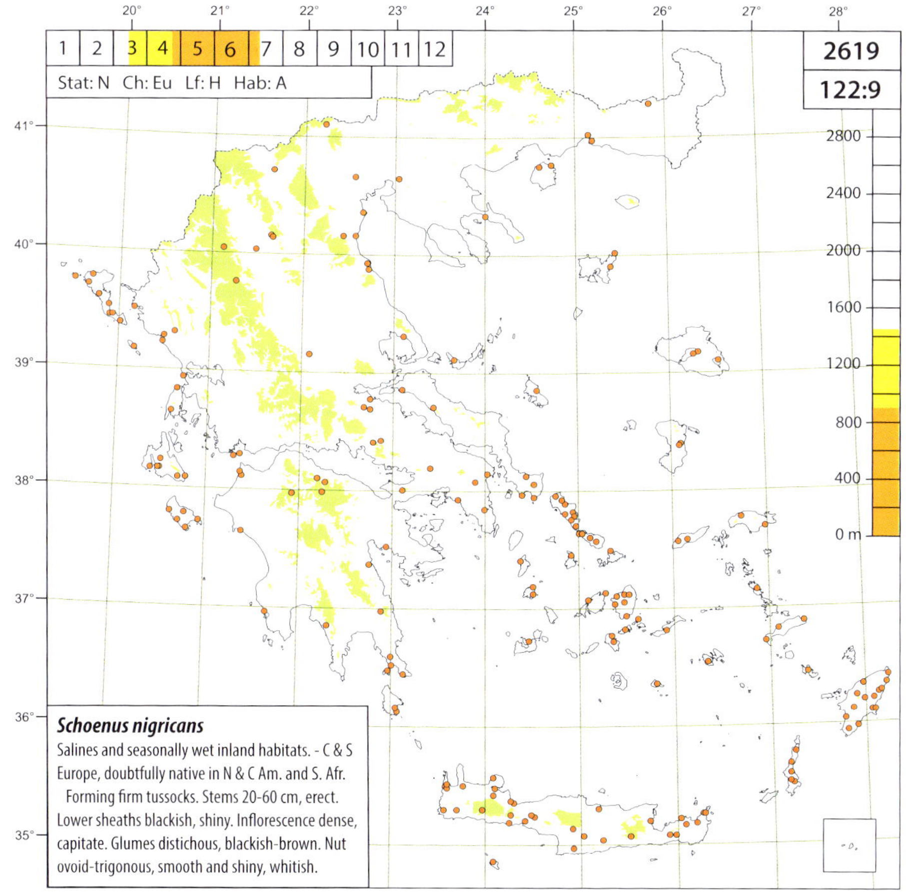

Schoenus nigricans
Salines and seasonally wet inland habitats. - C & S Europe, doubtfully native in N & C Am. and S. Afr.
Forming firm tussocks. Stems 20-60 cm, erect. Lower sheaths blackish, shiny. Inflorescence dense, capitate. Glumes distichous, blackish-brown. Nut ovoid-trigonous, smooth and shiny, whitish.

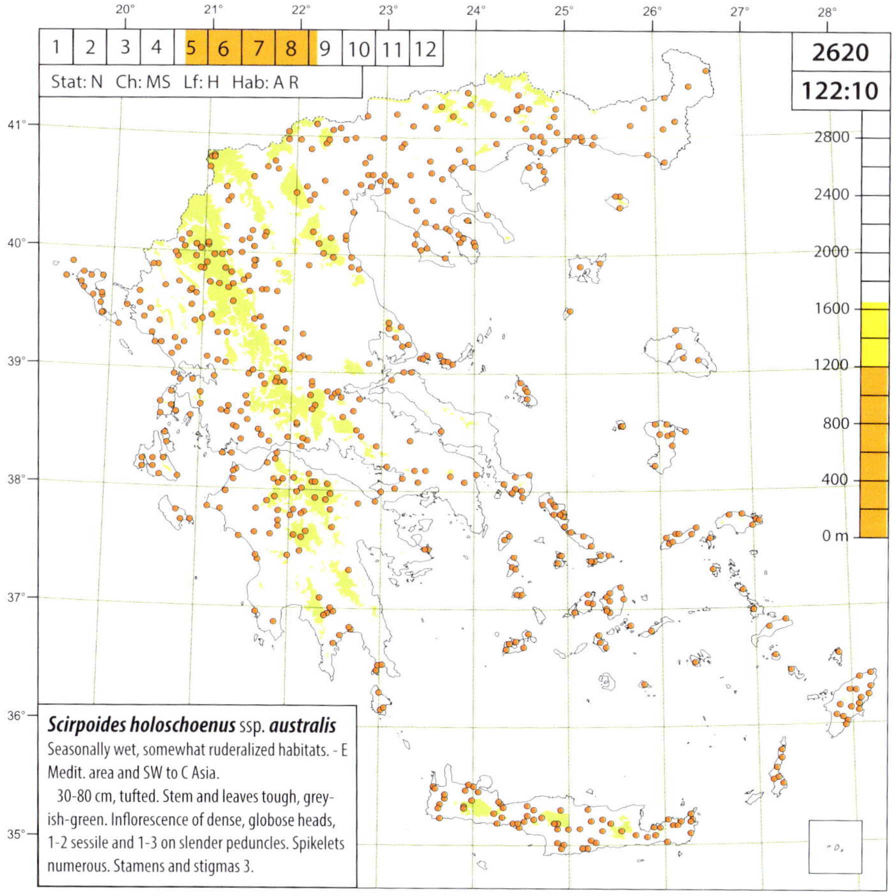

Scirpoides holoschoenus* ssp. *australis
Seasonally wet, somewhat ruderalized habitats. - E Medit. area and SW to C Asia.
30-80 cm, tufted. Stem and leaves tough, greyish-green. Inflorescence of dense, globose heads, 1-2 sessile and 1-3 on slender peduncles. Spikelets numerous. Stamens and stigmas 3.

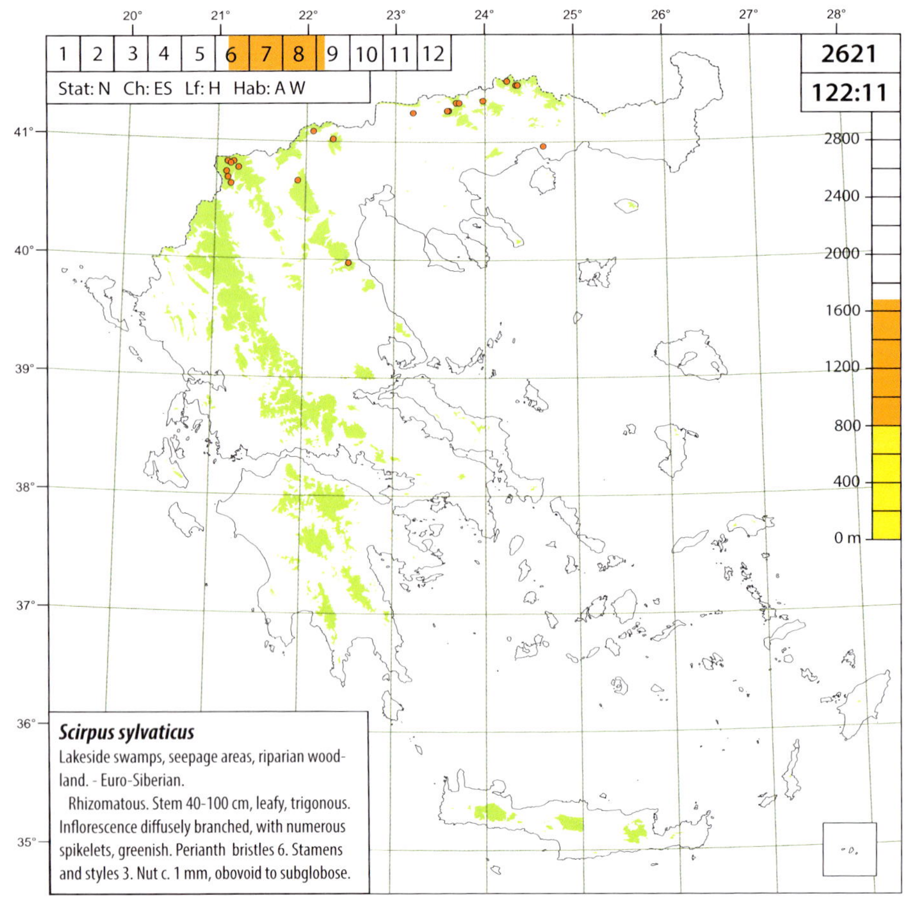

Scirpus sylvaticus
Lakeside swamps, seepage areas, riparian woodland. - Euro-Siberian.
Rhizomatous. Stem 40-100 cm, leafy, trigonous. Inflorescence diffusely branched, with numerous spikelets, greenish. Perianth bristles 6. Stamens and styles 3. Nut c. 1 mm, obovoid to subglobose.

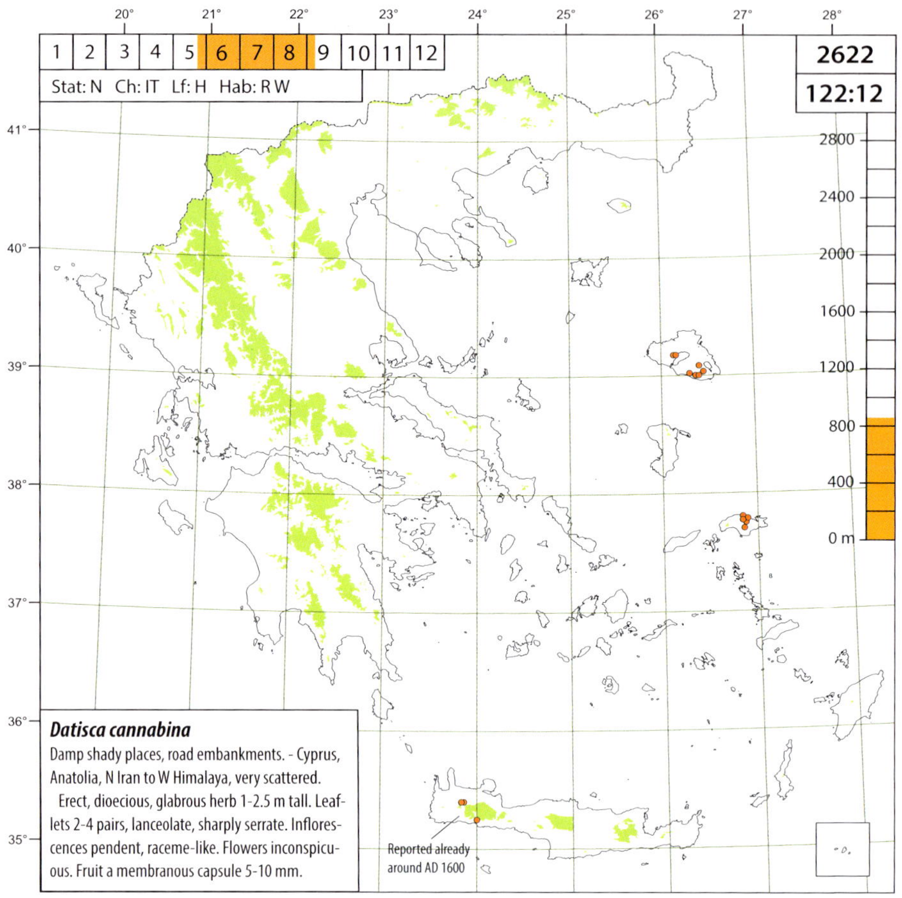

Datisca cannabina
Damp shady places, road embankments. - Cyprus, Anatolia, N Iran to W Himalaya, very scattered.
Erect, dioecious, glabrous herb 1-2.5 m tall. Leaflets 2-4 pairs, lanceolate, sharply serrate. Inflorescences pendent, raceme-like. Flowers inconspicuous. Fruit a membranous capsule 5-10 mm.

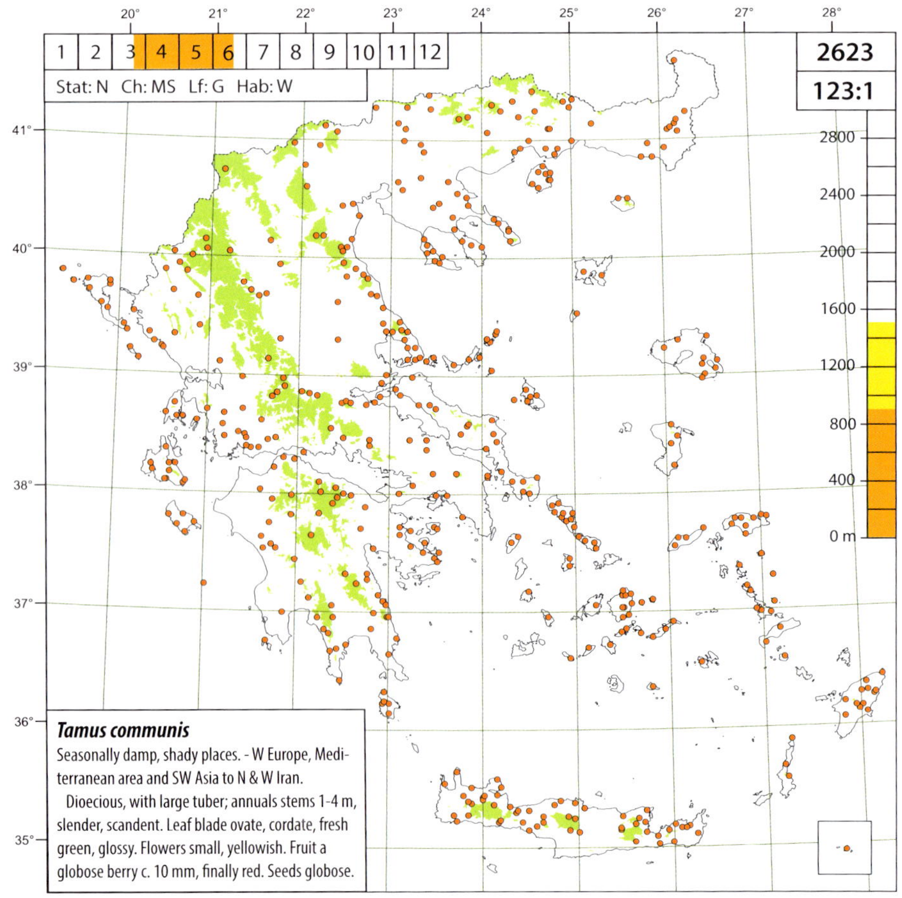

Tamus communis
Seasonally damp, shady places. - W Europe, Mediterranean area and SW Asia to N & W Iran.
Dioecious, with large tuber; annuals stems 1-4 m, slender, scandent. Leaf blade ovate, cordate, fresh green, glossy. Flowers small, yellowish. Fruit a globose berry c. 10 mm, finally red. Seeds globose.

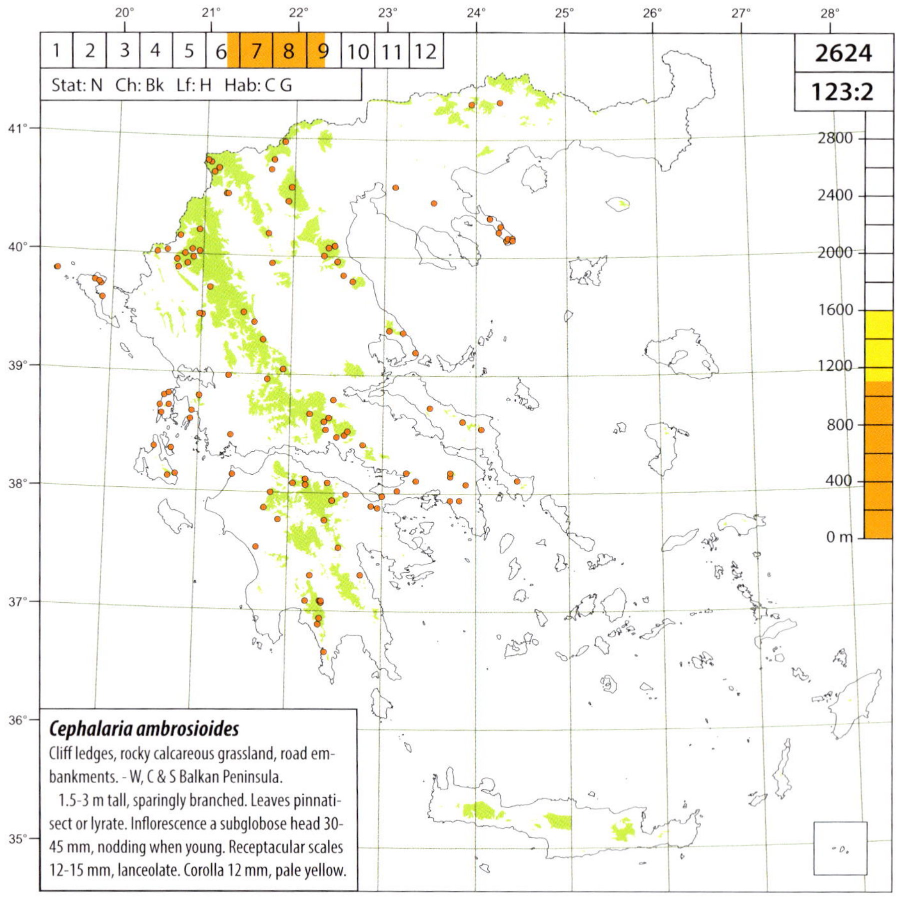

Cephalaria ambrosioides
Cliff ledges, rocky calcareous grassland, road embankments. - W, C & S Balkan Peninsula.
1.5-3 m tall, sparingly branched. Leaves pinnatisect or lyrate. Inflorescence a subglobose head 30-45 mm, nodding when young. Receptacular scales 12-15 mm, lanceolate. Corolla 12 mm, pale yellow.

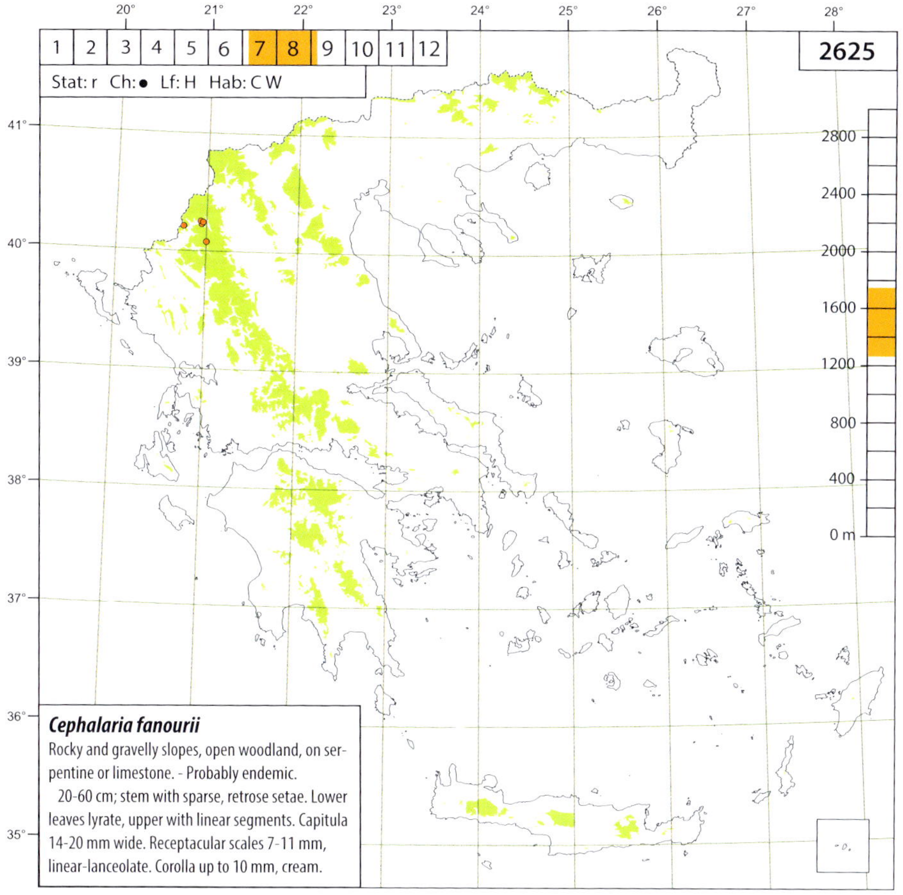

Cephalaria fanourii
Rocky and gravelly slopes, open woodland, on serpentine or limestone. - Probably endemic.
20-60 cm; stem with sparse, retrose setae. Lower leaves lyrate, upper with linear segments. Capitula 14-20 mm wide. Receptacular scales 7-11 mm, linear-lanceolate. Corolla up to 10 mm, cream.

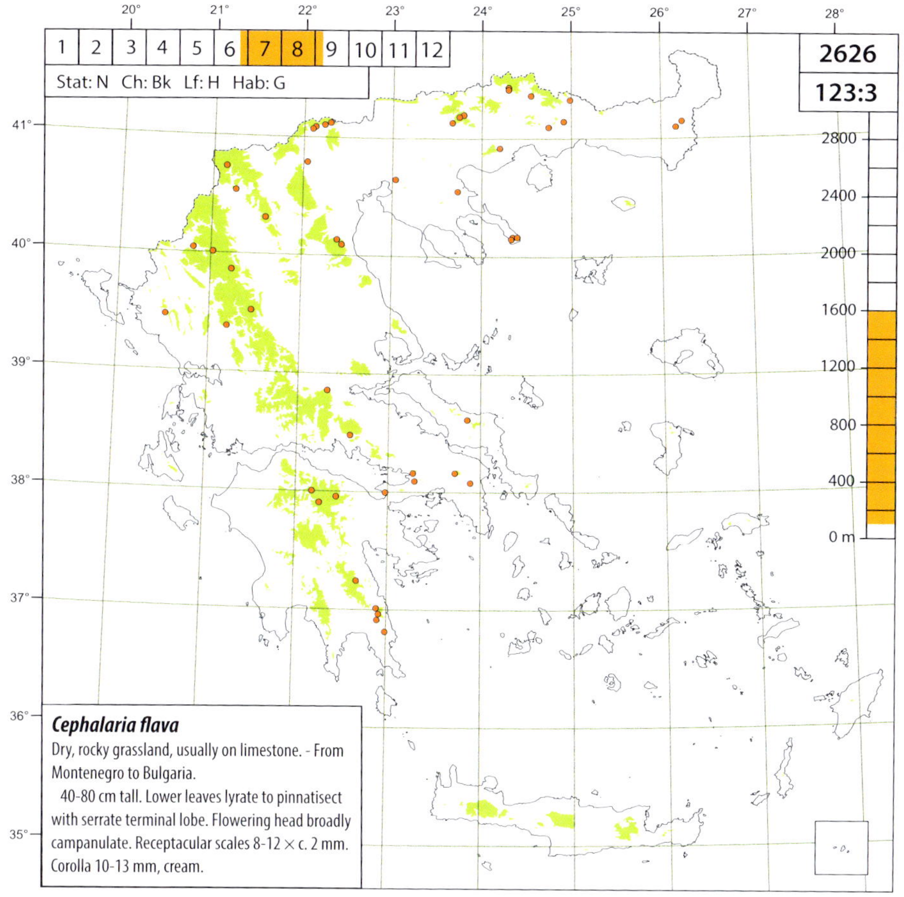

Cephalaria flava
Dry, rocky grassland, usually on limestone. - From Montenegro to Bulgaria.
40-80 cm tall. Lower leaves lyrate to pinnatisect with serrate terminal lobe. Flowering head broadly campanulate. Receptacular scales 8-12 × c. 2 mm. Corolla 10-13 mm, cream.

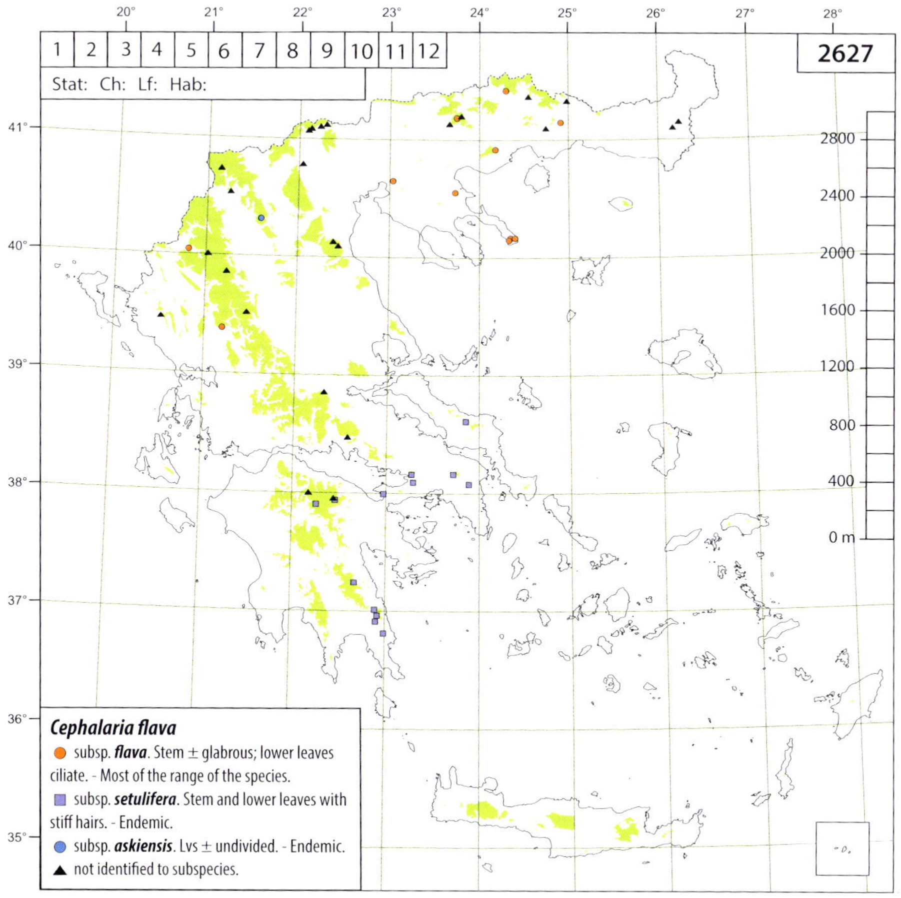

Cephalaria flava
● subsp. ***flava***. Stem ± glabrous; lower leaves ciliate. - Most of the range of the species.
■ subsp. ***setulifera***. Stem and lower leaves with stiff hairs. - Endemic.
● subsp. ***askiensis***. Lvs ± undivided. - Endemic.
▲ not identified to subspecies.

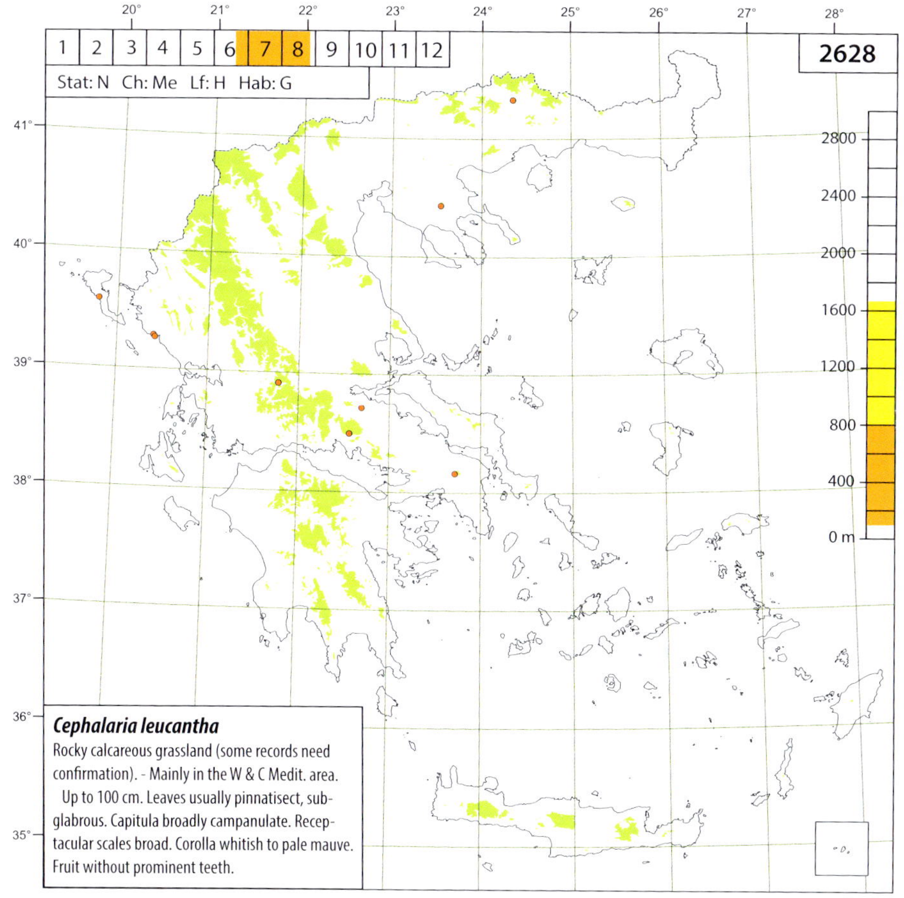

Cephalaria leucantha
Rocky calcareous grassland (some records need confirmation). - Mainly in the W & C Medit. area.
Up to 100 cm. Leaves usually pinnatisect, subglabrous. Capitula broadly campanulate. Receptacular scales broad. Corolla whitish to pale mauve. Fruit without prominent teeth.

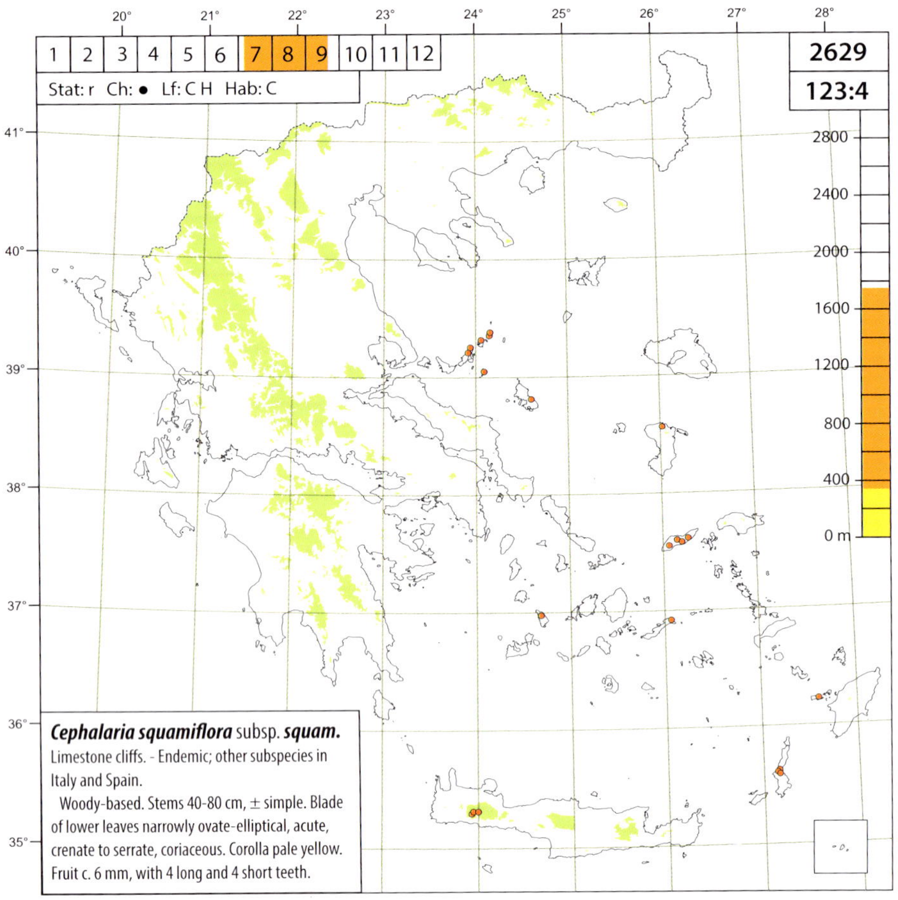

Cephalaria squamiflora* subsp. *squam.
Limestone cliffs. - Endemic; other subspecies in Italy and Spain.
Woody-based. Stems 40-80 cm, ± simple. Blade of lower leaves narrowly ovate-elliptical, acute, crenate to serrate, coriaceous. Corolla pale yellow. Fruit c. 6 mm, with 4 long and 4 short teeth.

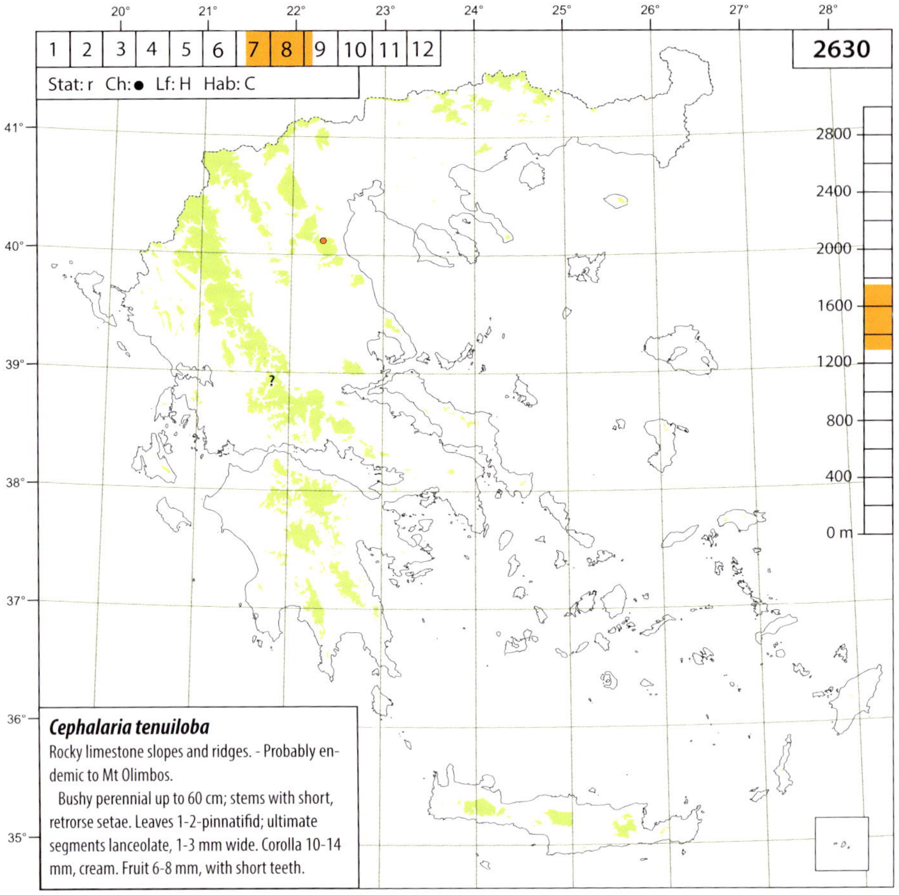

Cephalaria tenuiloba
Rocky limestone slopes and ridges. - Probably endemic to Mt Olimbos.
Bushy perennial up to 60 cm; stems with short, retrorse setae. Leaves 1-2-pinnatifid; ultimate segments lanceolate, 1-3 mm wide. Corolla 10-14 mm, cream. Fruit 6-8 mm, with short teeth.

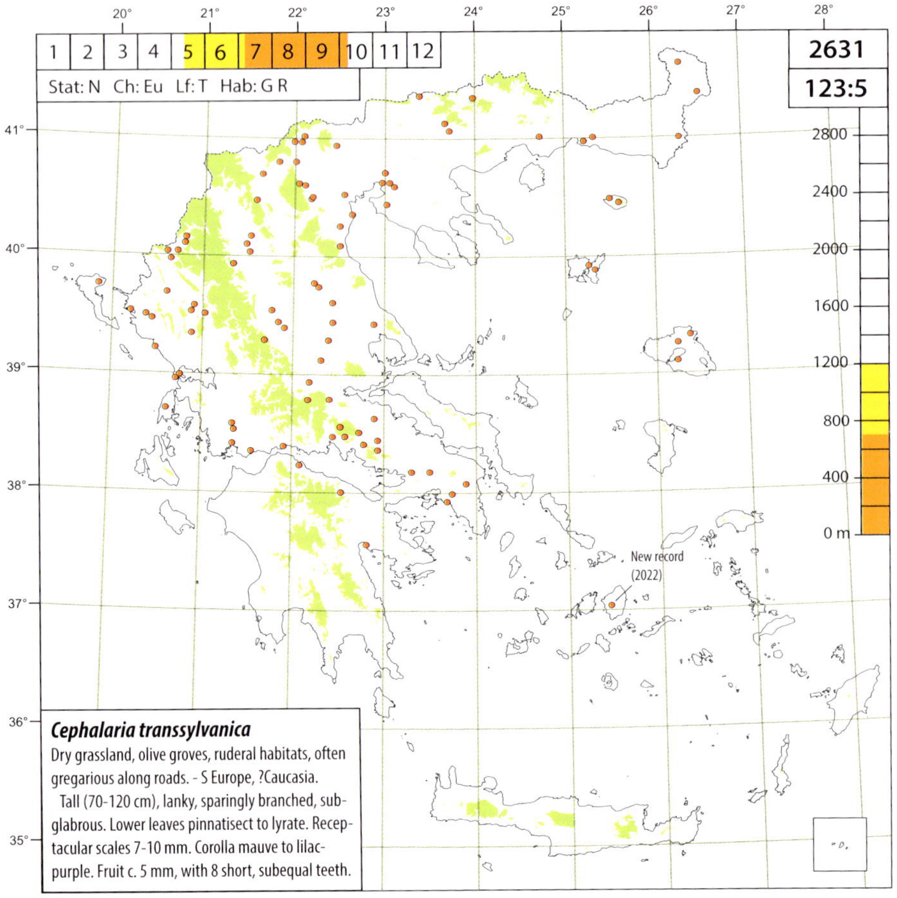

Cephalaria transsylvanica
Dry grassland, olive groves, ruderal habitats, often gregarious along roads. - S Europe, ?Caucasia.
Tall (70-120 cm), lanky, sparingly branched, subglabrous. Lower leaves pinnatisect to lyrate. Receptacular scales 7-10 mm. Corolla mauve to lilac-purple. Fruit c. 5 mm, with 8 short, subequal teeth.

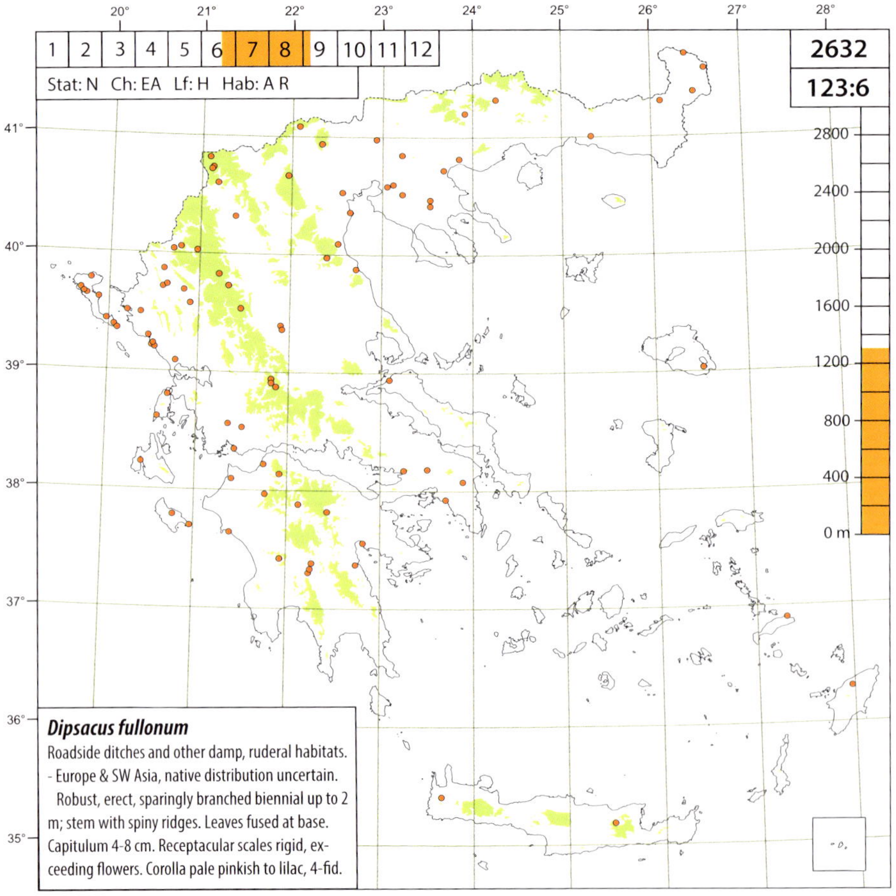

Dipsacus fullonum
Roadside ditches and other damp, ruderal habitats. - Europe & SW Asia, native distribution uncertain.
Robust, erect, sparingly branched biennial up to 2 m; stem with spiny ridges. Leaves fused at base. Capitulum 4-8 cm. Receptacular scales rigid, exceeding flowers. Corolla pale pinkish to lilac, 4-fid.

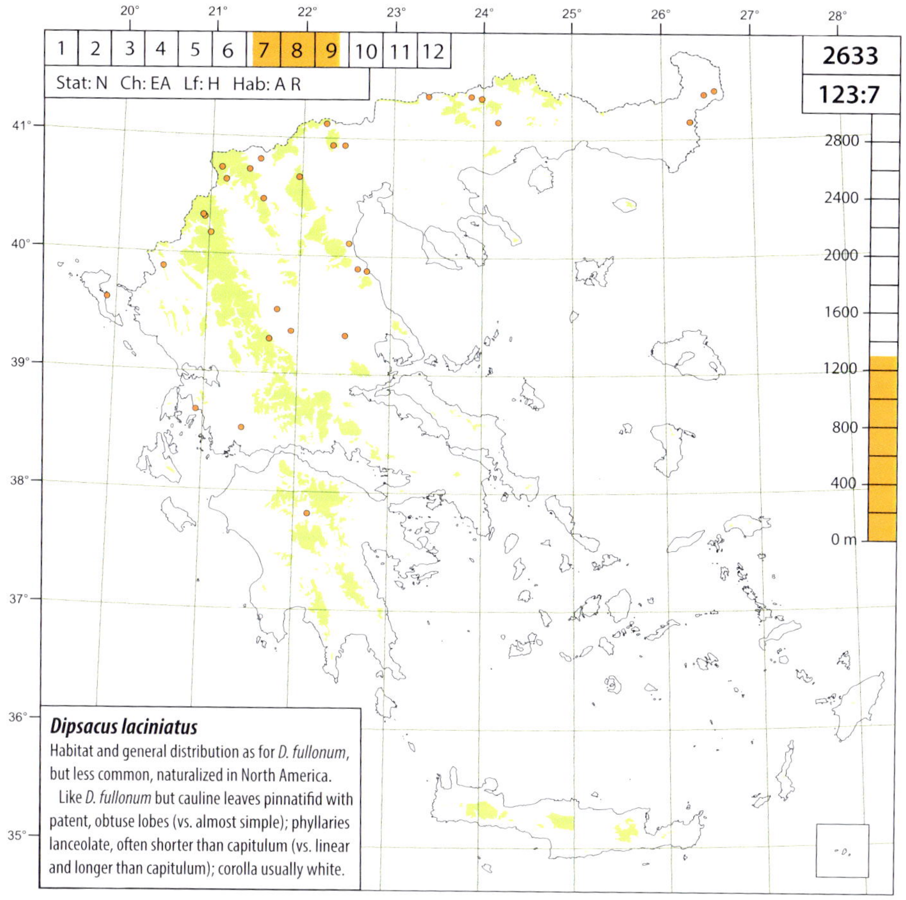

Dipsacus laciniatus

Habitat and general distribution as for *D. fullonum*, but less common, naturalized in North America.

Like *D. fullonum* but cauline leaves pinnatifid with patent, obtuse lobes (vs. almost simple); phyllaries lanceolate, often shorter than capitulum (vs. linear and longer than capitulum); corolla usually white.

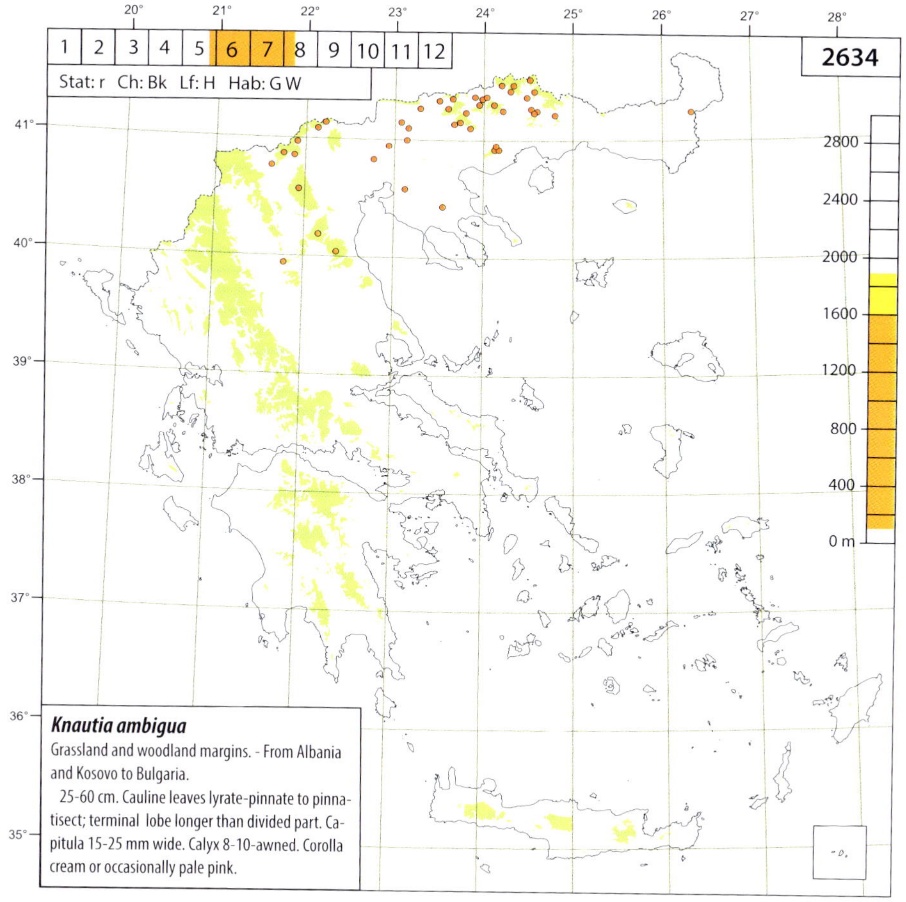

Knautia ambigua

Grassland and woodland margins. - From Albania and Kosovo to Bulgaria.

25-60 cm. Cauline leaves lyrate-pinnate to pinnatisect; terminal lobe longer than divided part. Capitula 15-25 mm wide. Calyx 8-10-awned. Corolla cream or occasionally pale pink.

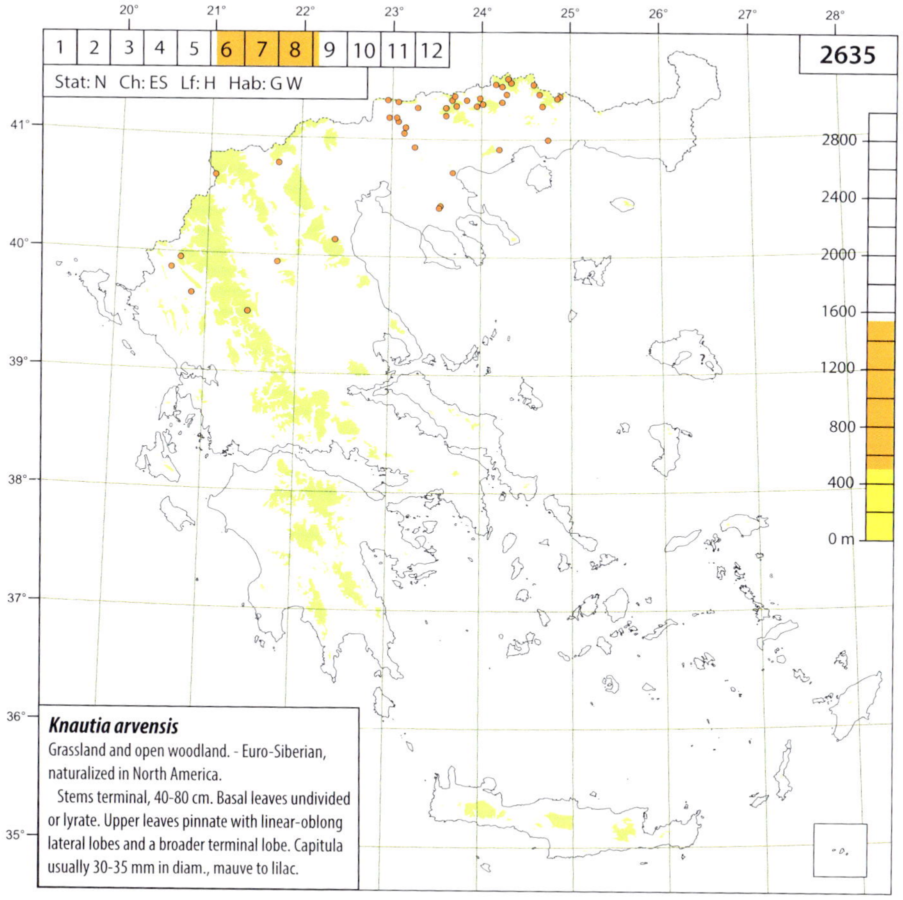

Knautia arvensis

Grassland and open woodland. - Euro-Siberian, naturalized in North America.

Stems terminal, 40-80 cm. Basal leaves undivided or lyrate. Upper leaves pinnate with linear-oblong lateral lobes and a broader terminal lobe. Capitula usually 30-35 mm in diam., mauve to lilac.

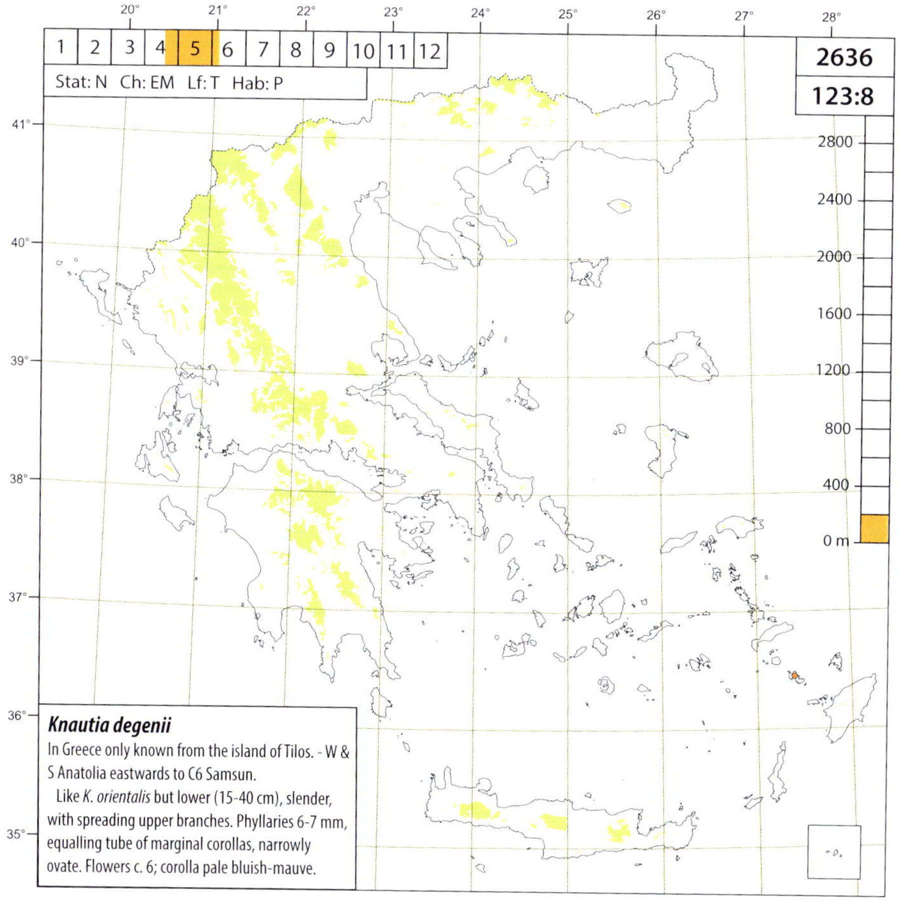

Knautia degenii

In Greece only known from the island of Tilos. - W & S Anatolia eastwards to C6 Samsun.

Like *K. orientalis* but lower (15-40 cm), slender, with spreading upper branches. Phyllaries 6-7 mm, equalling tube of marginal corollas, narrowly ovate. Flowers c. 6; corolla pale bluish-mauve.

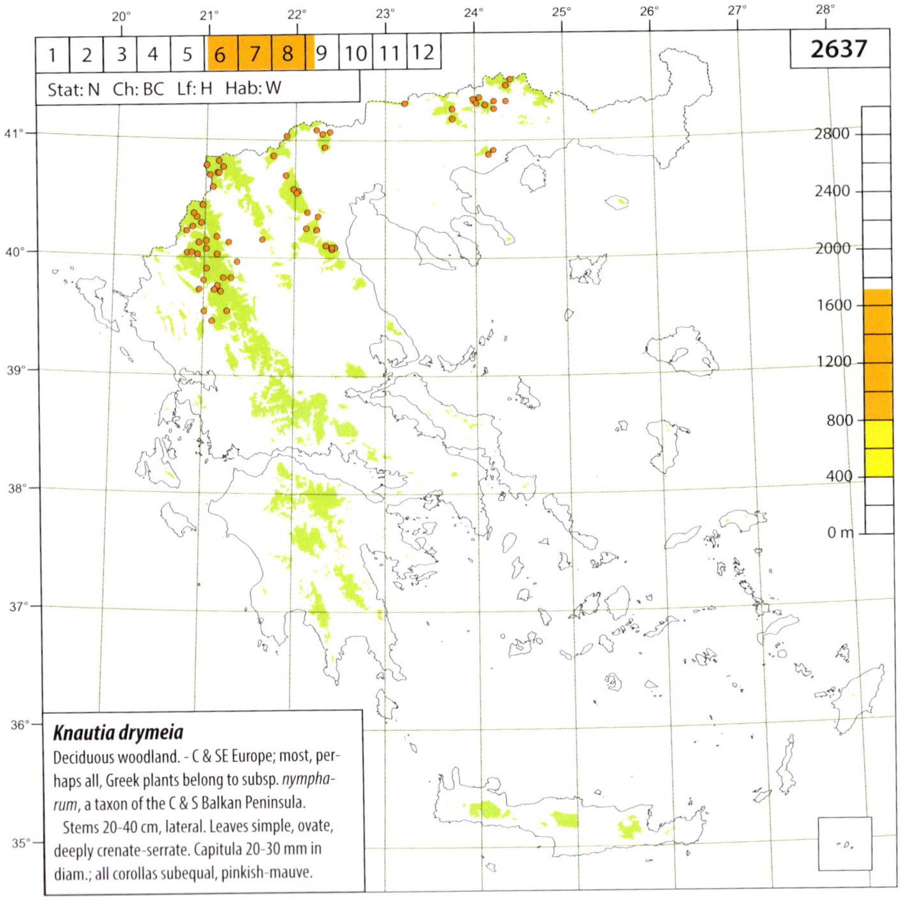

Knautia drymeia

Deciduous woodland. - C & SE Europe; most, perhaps all, Greek plants belong to subsp. *nympharum*, a taxon of the C & S Balkan Peninsula.

Stems 20-40 cm, lateral. Leaves simple, ovate, deeply crenate-serrate. Capitula 20-30 mm in diam.; all corollas subequal, pinkish-mauve.

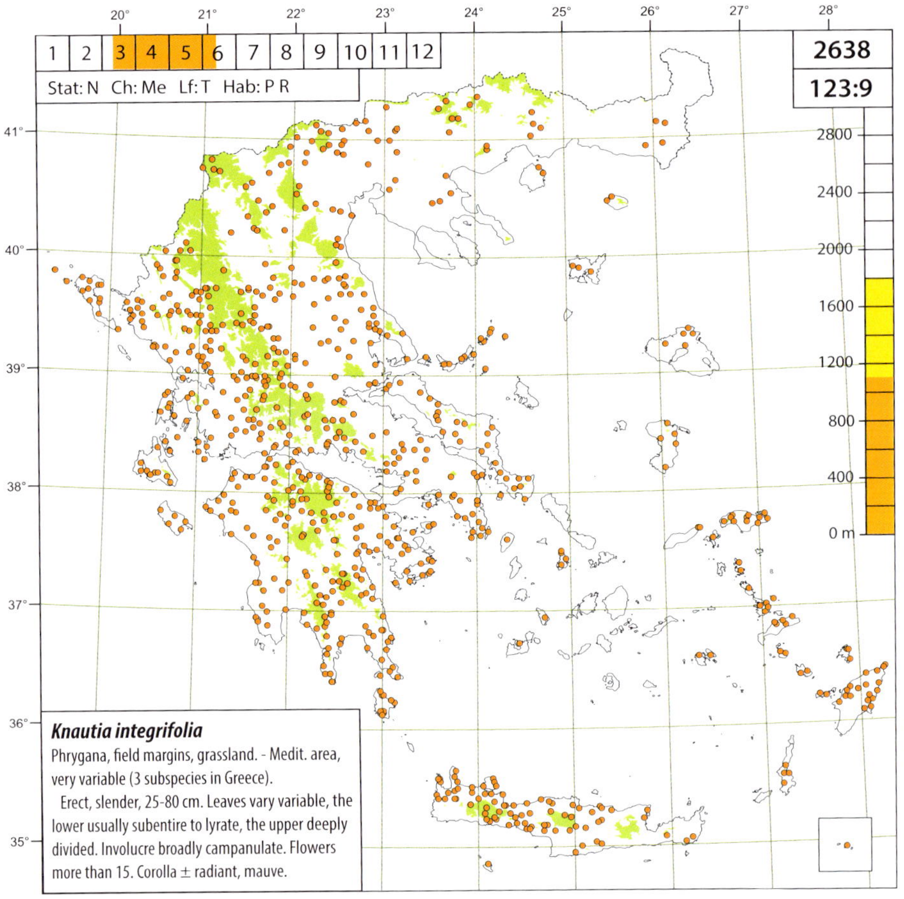

Knautia integrifolia

Phrygana, field margins, grassland. - Medit. area, very variable (3 subspecies in Greece).

Erect, slender, 25-80 cm. Leaves vary variable, the lower usually subentire to lyrate, the upper deeply divided. Involucre broadly campanulate. Flowers more than 15. Corolla ± radiant, mauve.

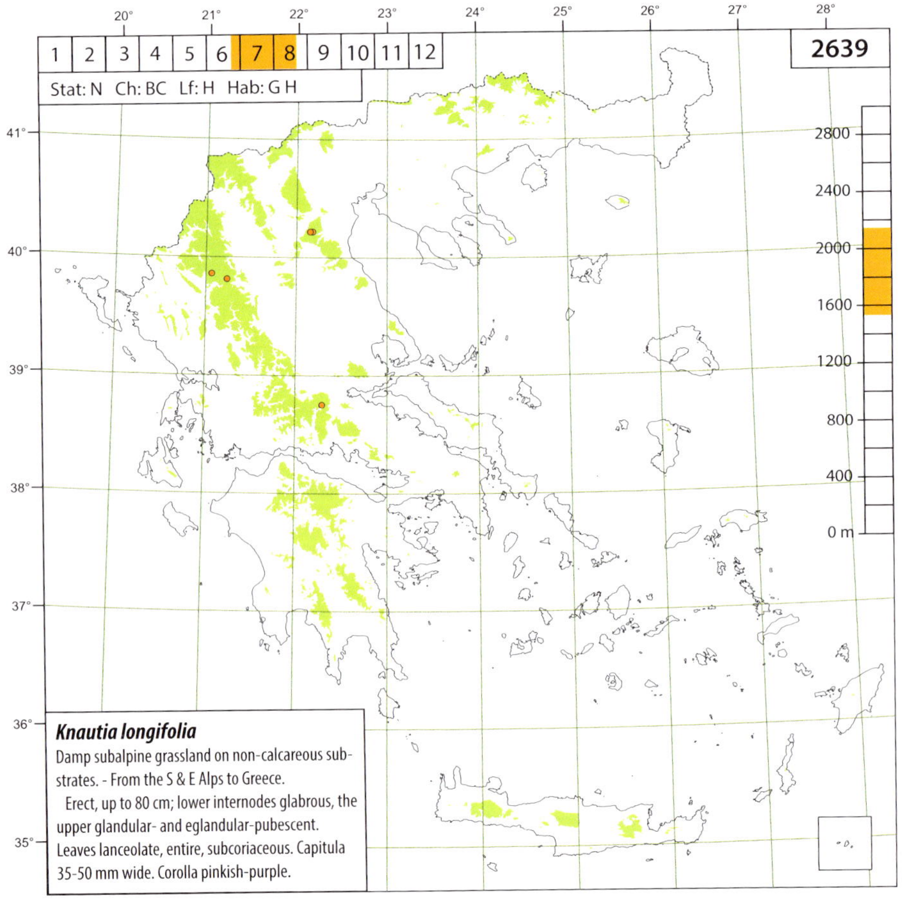

Knautia longifolia

Damp subalpine grassland on non-calcareous substrates. - From the S & E Alps to Greece.

Erect, up to 80 cm; lower internodes glabrous, the upper glandular- and eglandular-pubescent. Leaves lanceolate, entire, subcoriaceous. Capitula 35-50 mm wide. Corolla pinkish-purple.

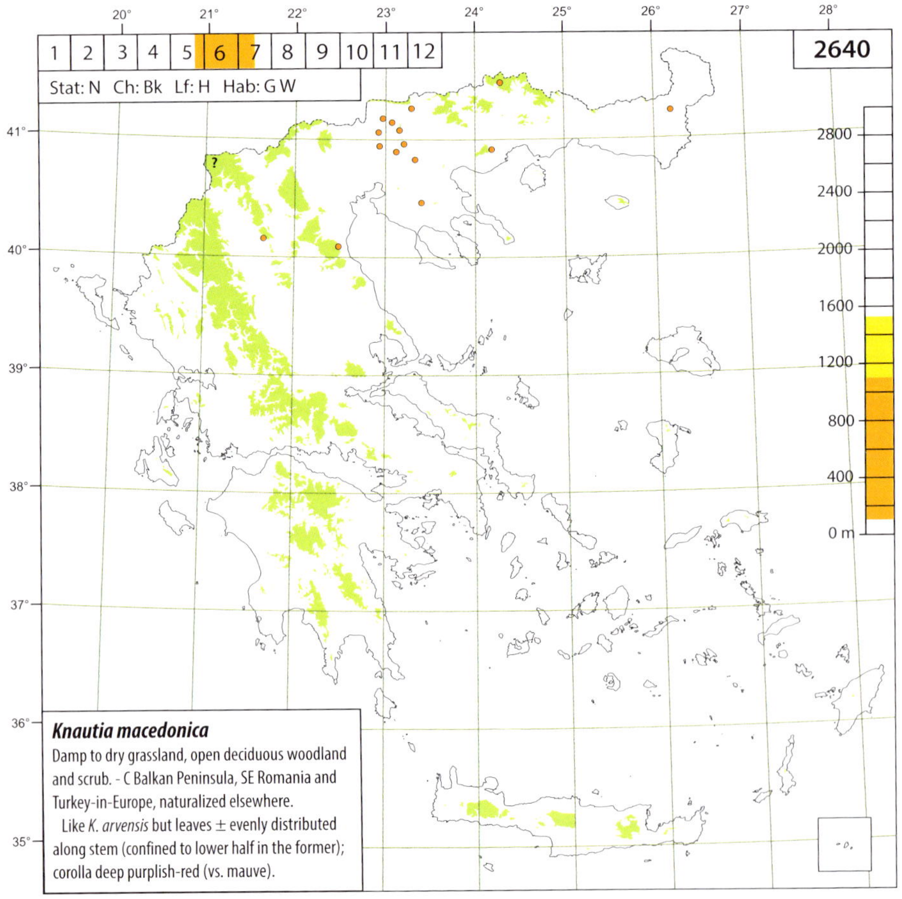

Knautia macedonica

Damp to dry grassland, open deciduous woodland and scrub. - C Balkan Peninsula, SE Romania and Turkey-in-Europe, naturalized elsewhere.

Like *K. arvensis* but leaves ± evenly distributed along stem (confined to lower half in the former); corolla deep purplish-red (vs. mauve).

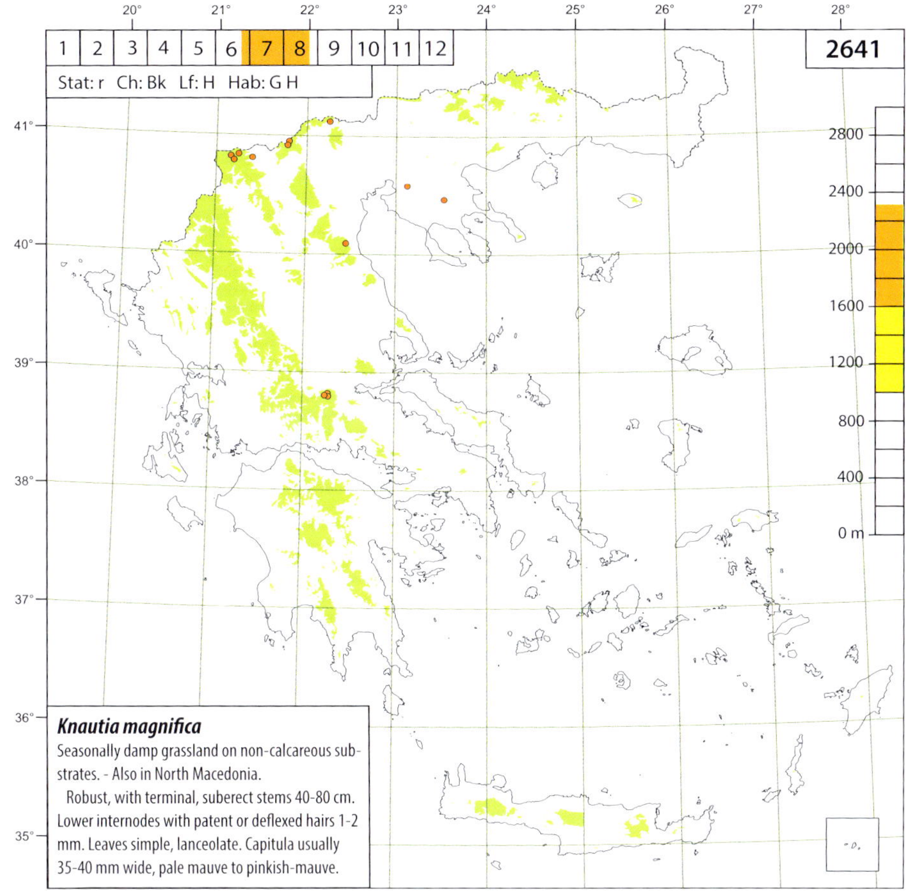

Knautia magnifica
Seasonally damp grassland on non-calcareous substrates. - Also in North Macedonia.
Robust, with terminal, suberect stems 40-80 cm. Lower internodes with patent or deflexed hairs 1-2 mm. Leaves simple, lanceolate. Capitula usually 35-40 mm wide, pale mauve to pinkish-mauve.

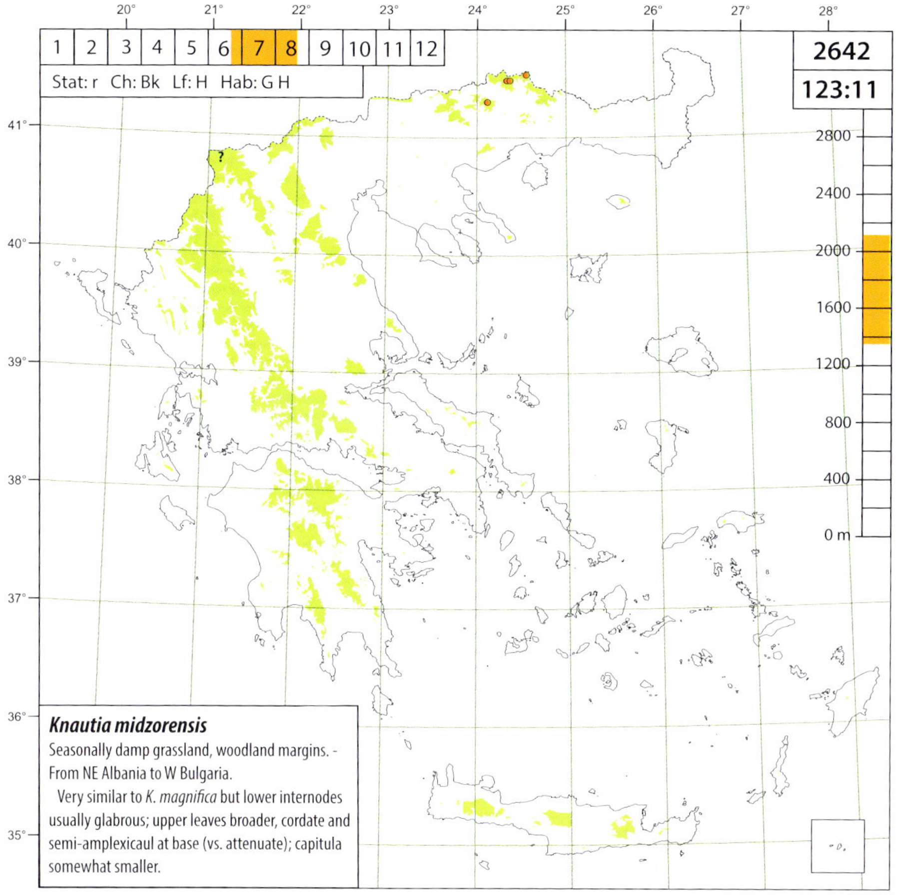

Knautia midzorensis
Seasonally damp grassland, woodland margins. - From NE Albania to W Bulgaria.
Very similar to *K. magnifica* but lower internodes usually glabrous; upper leaves broader, cordate and semi-amplexicaul at base (vs. attenuate); capitula somewhat smaller.

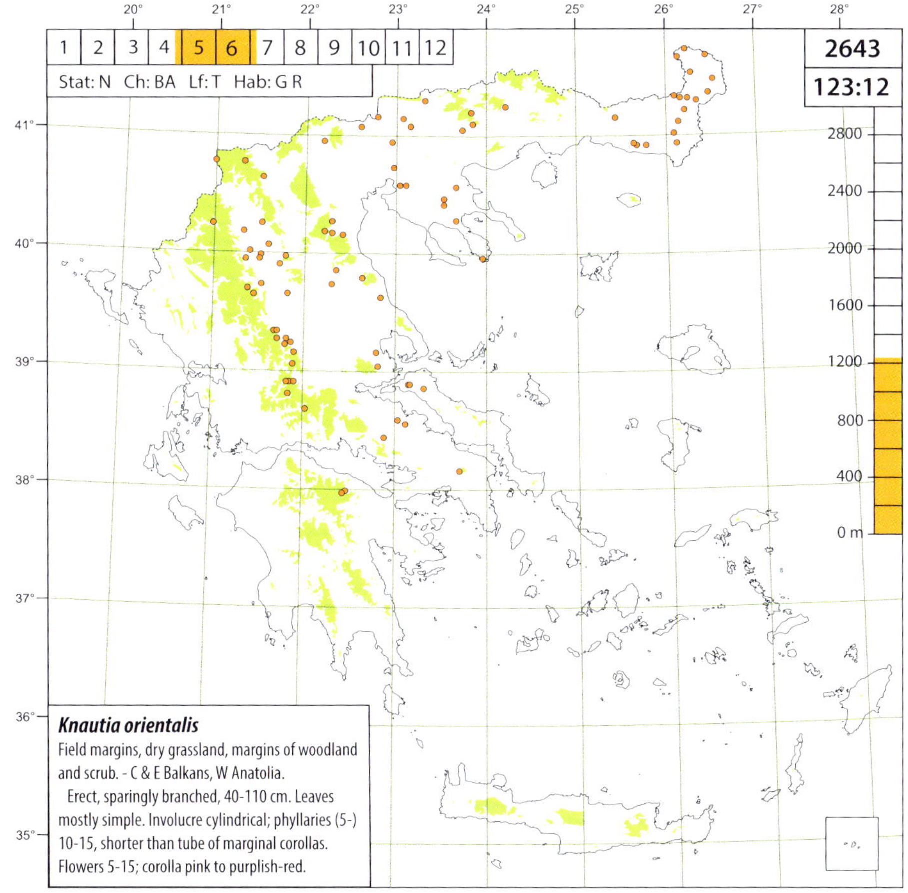

Knautia orientalis
Field margins, dry grassland, margins of woodland and scrub. - C & E Balkans, W Anatolia.
Erect, sparingly branched, 40-110 cm. Leaves mostly simple. Involucre cylindrical; phyllaries (5-) 10-15, shorter than tube of marginal corollas. Flowers 5-15; corolla pink to purplish-red.

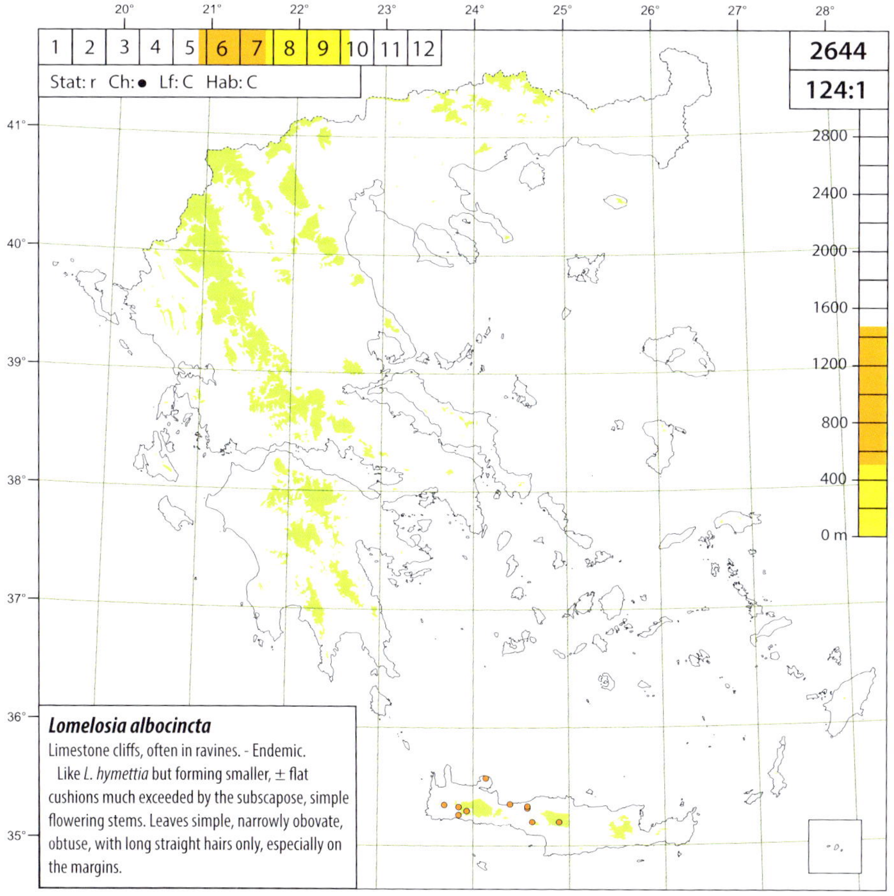

Lomelosia albocincta
Limestone cliffs, often in ravines. - Endemic.
Like *L. hymettia* but forming smaller, ± flat cushions much exceeded by the subscapose, simple flowering stems. Leaves simple, narrowly obovate, obtuse, with long straight hairs only, especially on the margins.

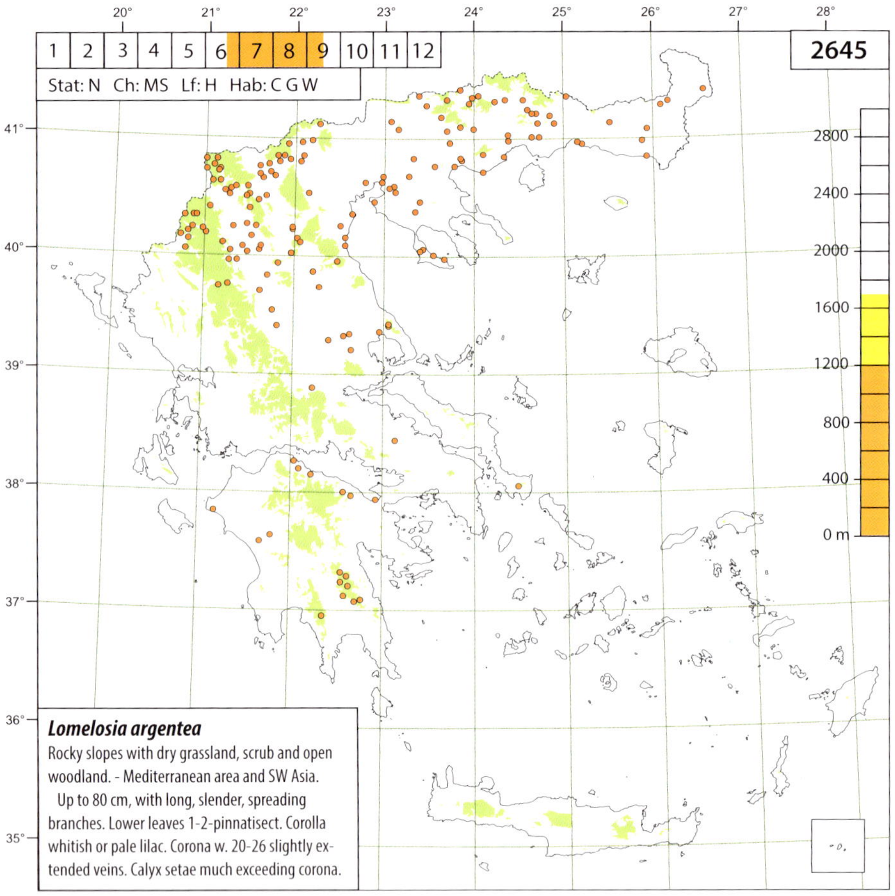

Lomelosia argentea
Rocky slopes with dry grassland, scrub and open woodland. - Mediterranean area and SW Asia.
Up to 80 cm, with long, slender, spreading branches. Lower leaves 1-2-pinnatisect. Corolla whitish or pale lilac. Corona w. 20-26 slightly extended veins. Calyx setae much exceeding corona.

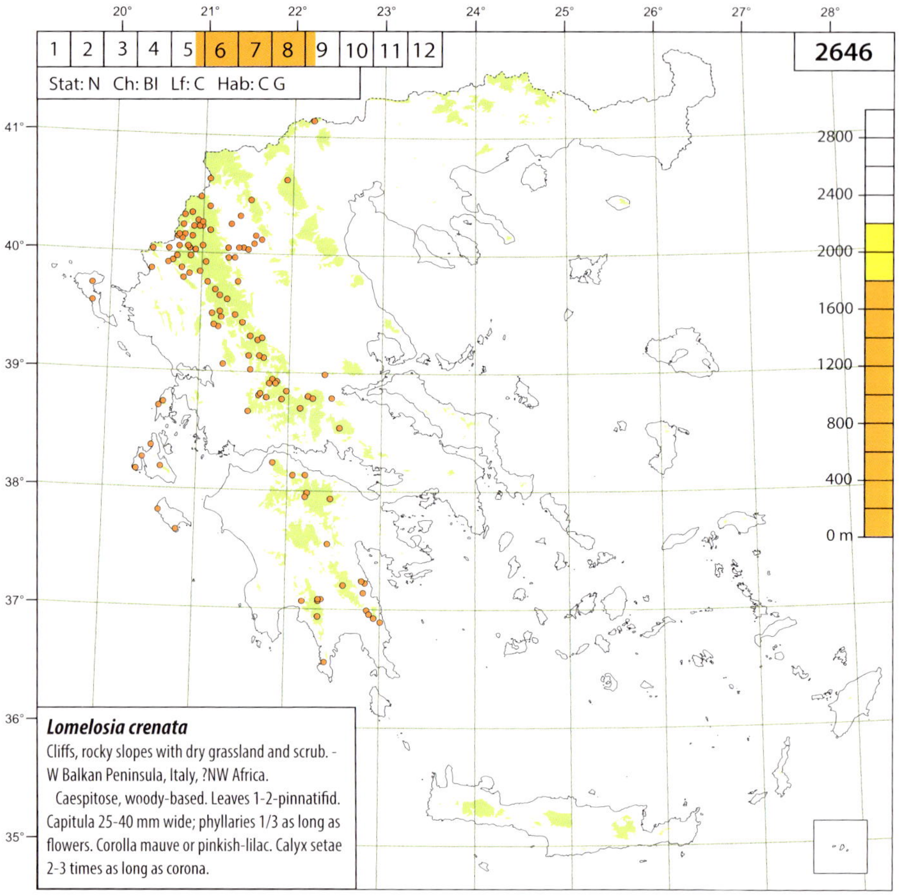

Lomelosia crenata
Cliffs, rocky slopes with dry grassland and scrub. - W Balkan Peninsula, Italy, ?NW Africa.
Caespitose, woody-based. Leaves 1-2-pinnatifid. Capitula 25-40 mm wide; phyllaries 1/3 as long as flowers. Corolla mauve or pinkish-lilac. Calyx setae 2-3 times as long as corona.

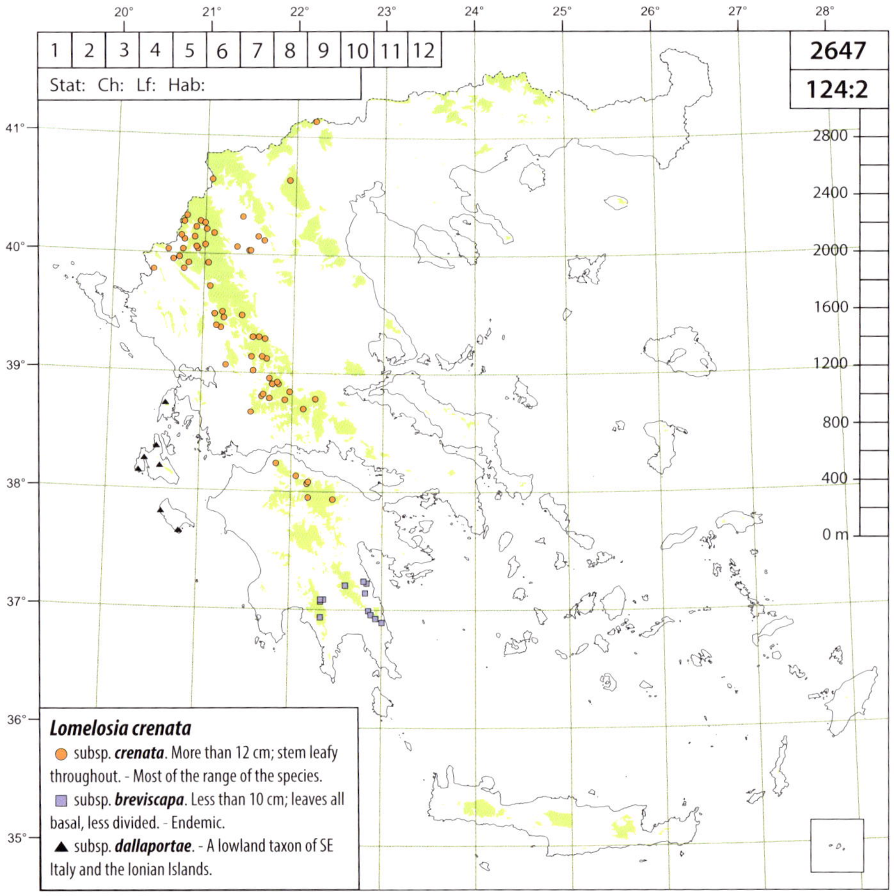

Lomelosia crenata
● subsp. ***crenata***. More than 12 cm; stem leafy throughout. - Most of the range of the species.
■ subsp. ***breviscapa***. Less than 10 cm; leaves all basal, less divided. - Endemic.
▲ subsp. ***dallaportae***. - A lowland taxon of SE Italy and the Ionian Islands.

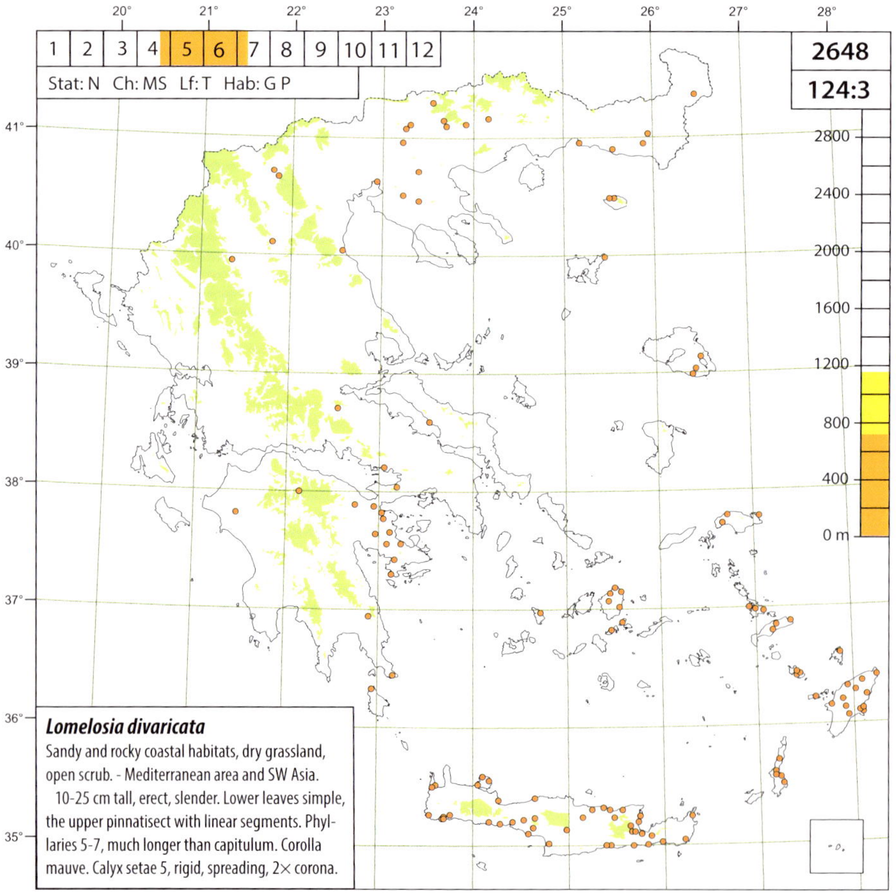

Lomelosia divaricata
Sandy and rocky coastal habitats, dry grassland, open scrub. - Mediterranean area and SW Asia.
10-25 cm tall, erect, slender. Lower leaves simple, the upper pinnatisect with linear segments. Phyllaries 5-7, much longer than capitulum. Corolla mauve. Calyx setae 5, rigid, spreading, 2× corona.

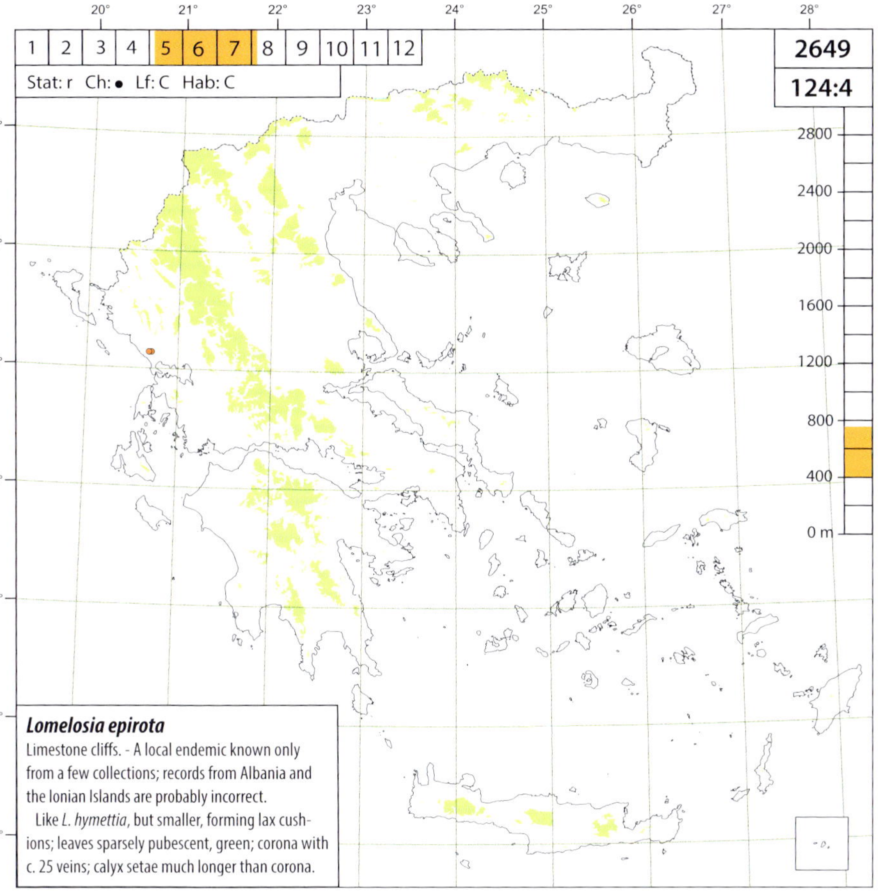

Lomelosia epirota

Limestone cliffs. - A local endemic known only from a few collections; records from Albania and the Ionian Islands are probably incorrect.

Like *L. hymettia*, but smaller, forming lax cushions; leaves sparsely pubescent, green; corona with c. 25 veins; calyx setae much longer than corona.

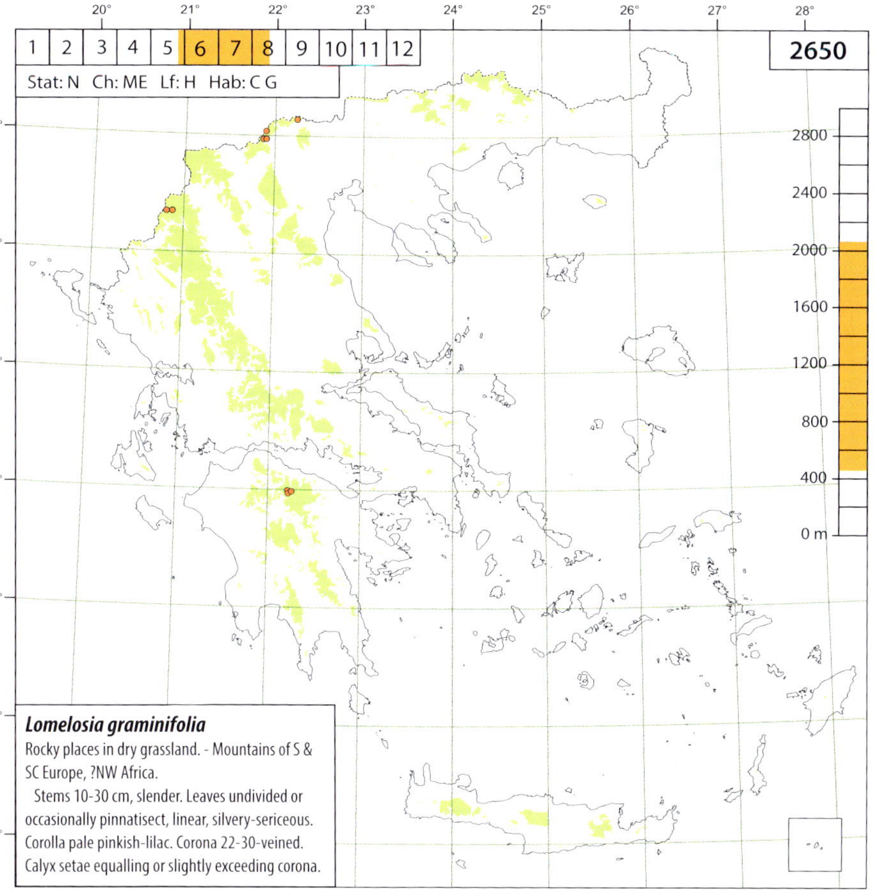

Lomelosia graminifolia

Rocky places in dry grassland. - Mountains of S & SC Europe, ?NW Africa.

Stems 10-30 cm, slender. Leaves undivided or occasionally pinnatisect, linear, silvery-sericeous. Corolla pale pinkish-lilac. Corona 22-30-veined. Calyx setae equalling or slightly exceeding corona.

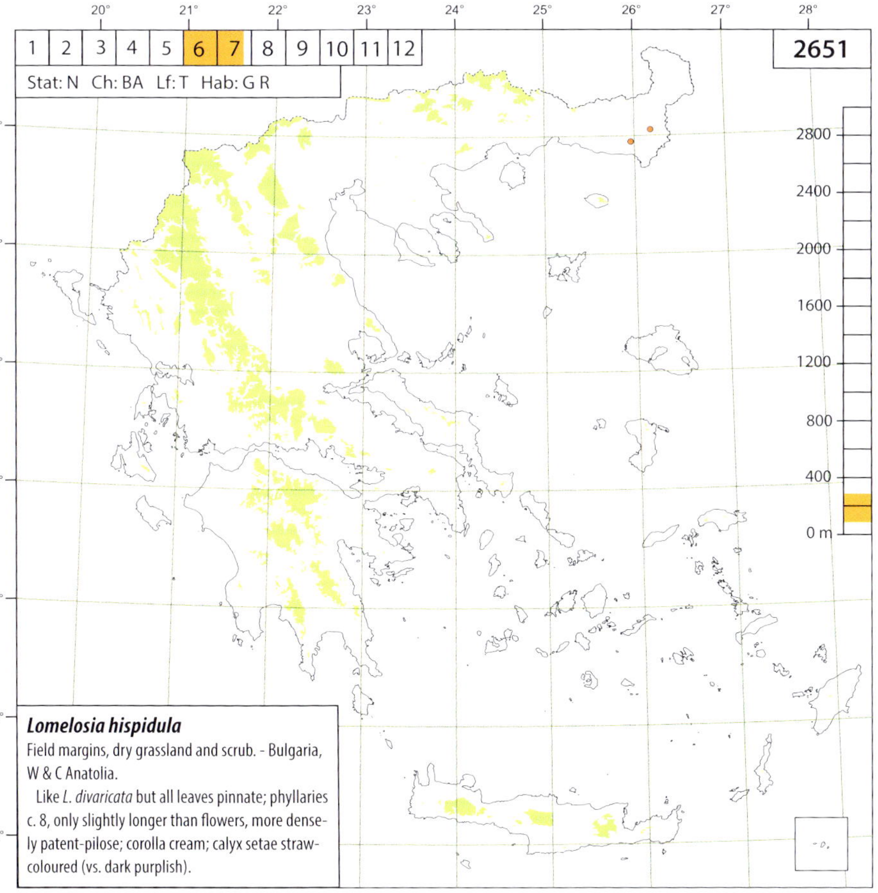

Lomelosia hispidula

Field margins, dry grassland and scrub. - Bulgaria, W & C Anatolia.

Like *L. divaricata* but all leaves pinnate; phyllaries c. 8, only slightly longer than flowers, more densely patent-pilose; corolla cream; calyx setae straw-coloured (vs. dark purplish).

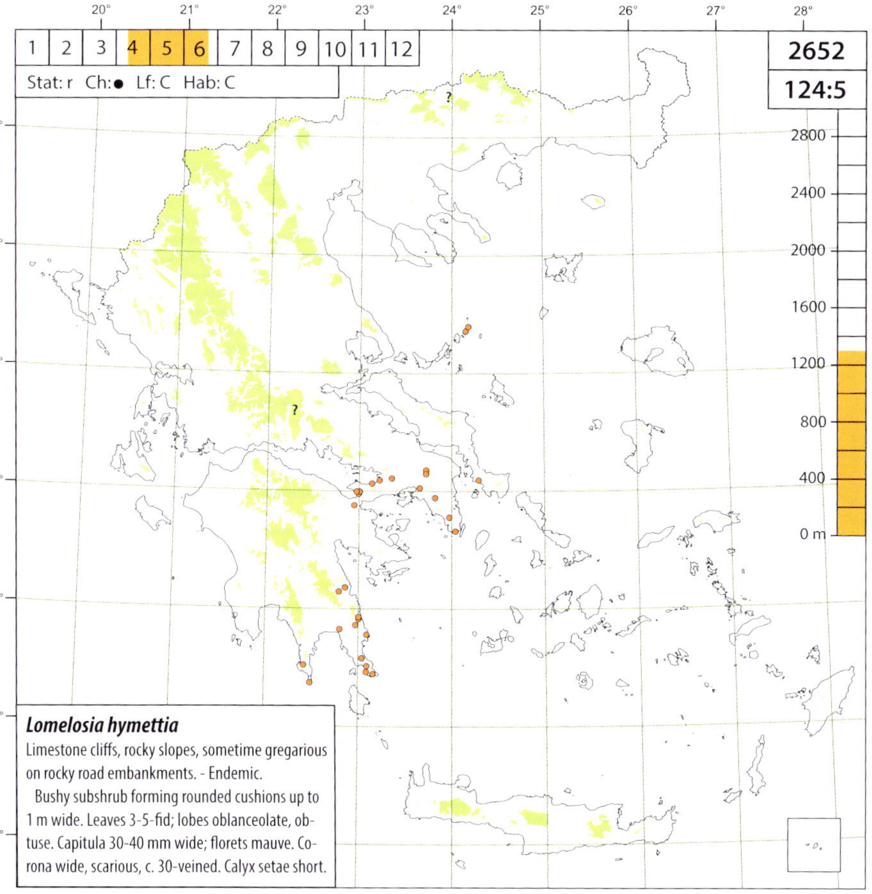

Lomelosia hymettia

Limestone cliffs, rocky slopes, sometime gregarious on rocky road embankments. - Endemic.

Bushy subshrub forming rounded cushions up to 1 m wide. Leaves 3-5-fid; lobes oblanceolate, obtuse. Capitula 30-40 mm wide; florets mauve. Corona wide, scarious, c. 30-veined. Calyx setae short.

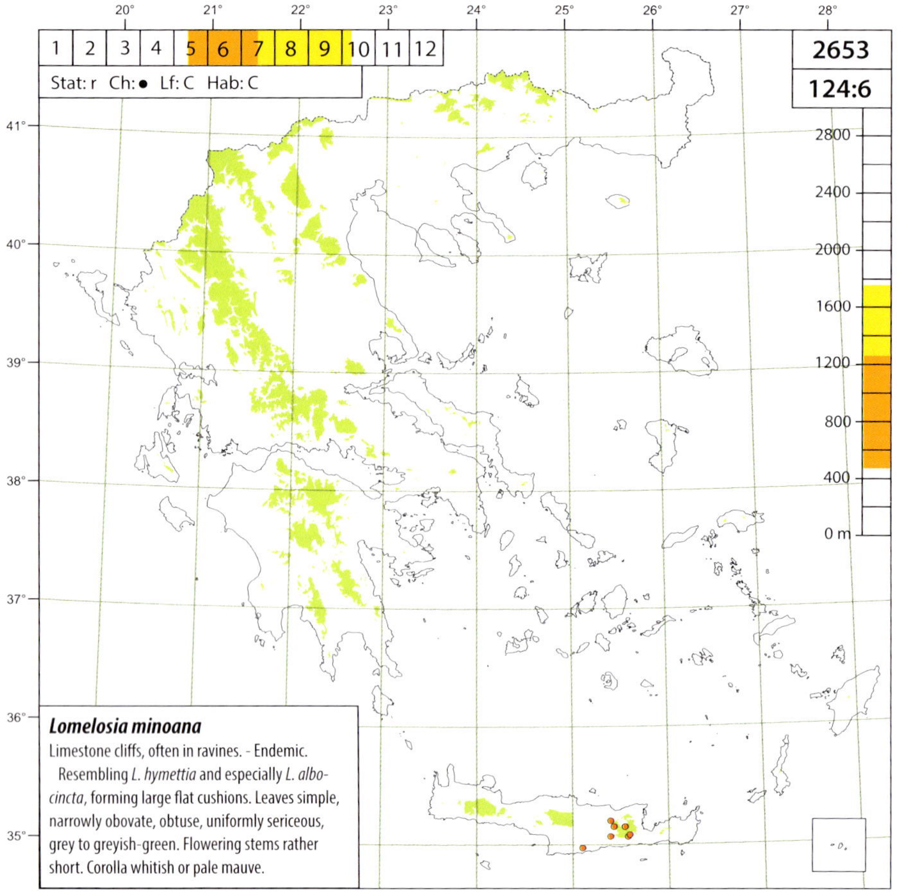

Lomelosia minoana

Limestone cliffs, often in ravines. - Endemic.

Resembling *L. hymettia* and especially *L. albocincta*, forming large flat cushions. Leaves simple, narrowly obovate, obtuse, uniformly sericeous, grey to greyish-green. Flowering stems rather short. Corolla whitish or pale mauve.

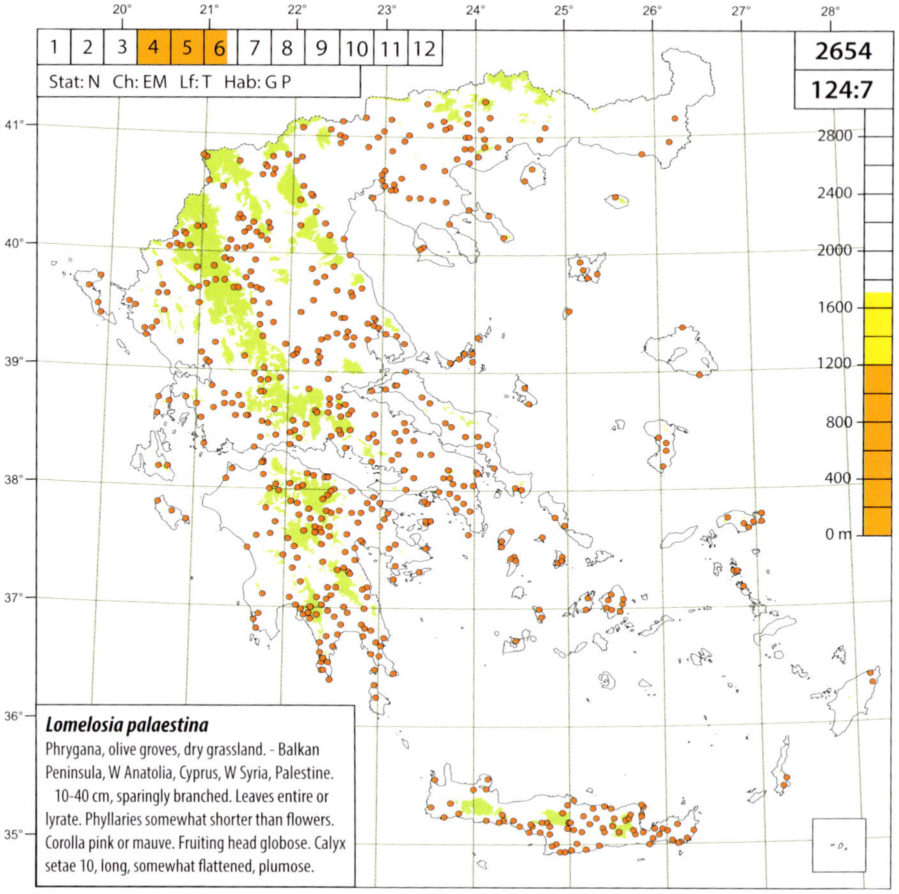

Lomelosia palaestina

Phrygana, olive groves, dry grassland. - Balkan Peninsula, W Anatolia, Cyprus, W Syria, Palestine.

10-40 cm, sparingly branched. Leaves entire or lyrate. Phyllaries somewhat shorter than flowers. Corolla pink or mauve. Fruiting head globose. Calyx setae 10, long, somewhat flattened, plumose.

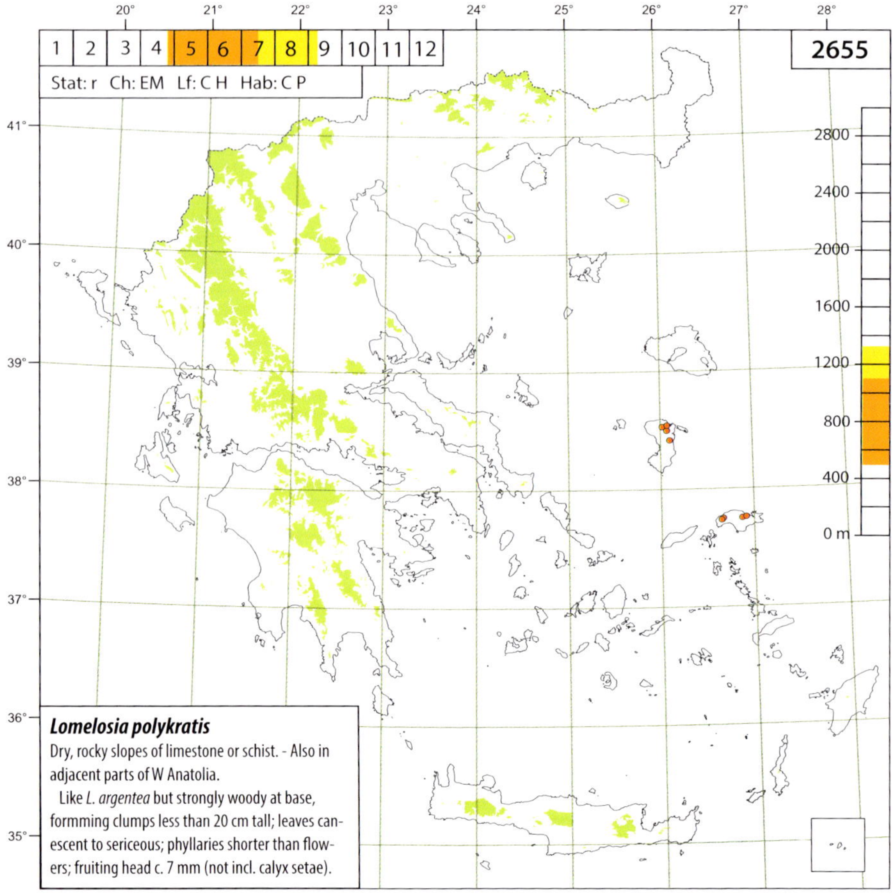

Lomelosia polykratis

Dry, rocky slopes of limestone or schist. - Also in adjacent parts of W Anatolia.

Like *L. argentea* but strongly woody at base, formming clumps less than 20 cm tall; leaves canescent to sericeous; phyllaries shorter than flowers; fruiting head c. 7 mm (not incl. calyx setae).

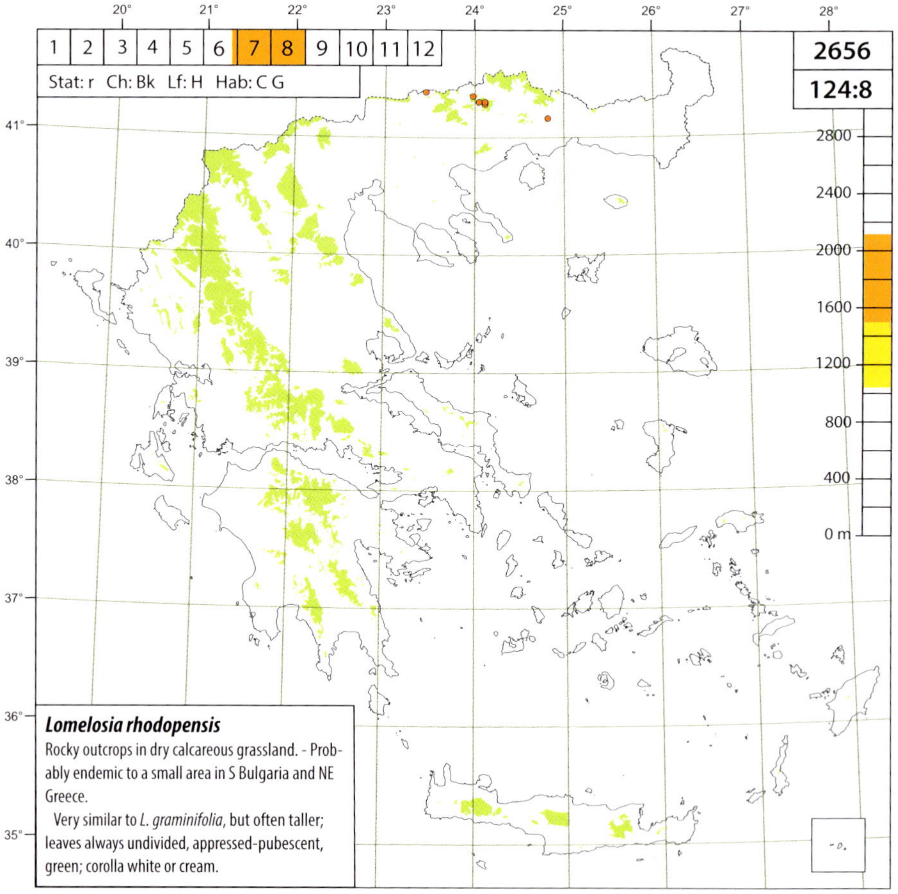

Lomelosia rhodopensis

Rocky outcrops in dry calcareous grassland. - Probably endemic to a small area in S Bulgaria and NE Greece.

Very similar to *L. graminifolia*, but often taller; leaves always undivided, appressed-pubescent, green; corolla white or cream.

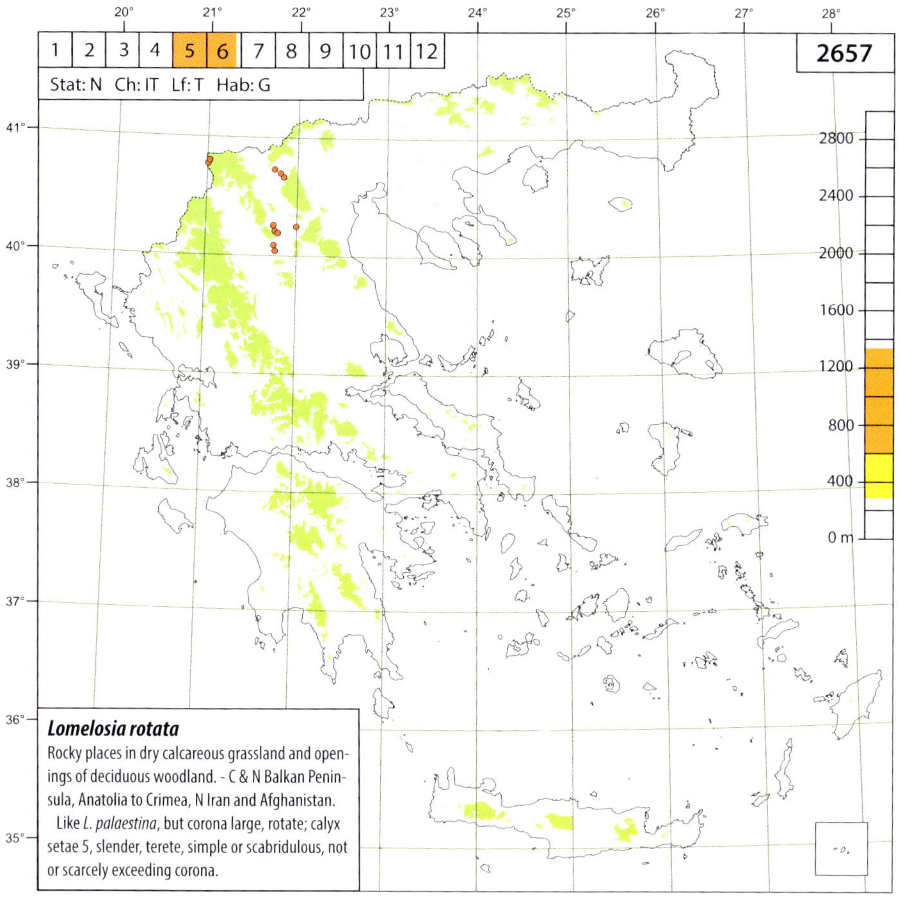

Lomelosia rotata

Rocky places in dry calcareous grassland and openings of deciduous woodland. - C & N Balkan Peninsula, Anatolia to Crimea, N Iran and Afghanistan.

Like *L. palaestina*, but corona large, rotate; calyx setae 5, slender, terete, simple or scabridulous, not or scarcely exceeding corona.

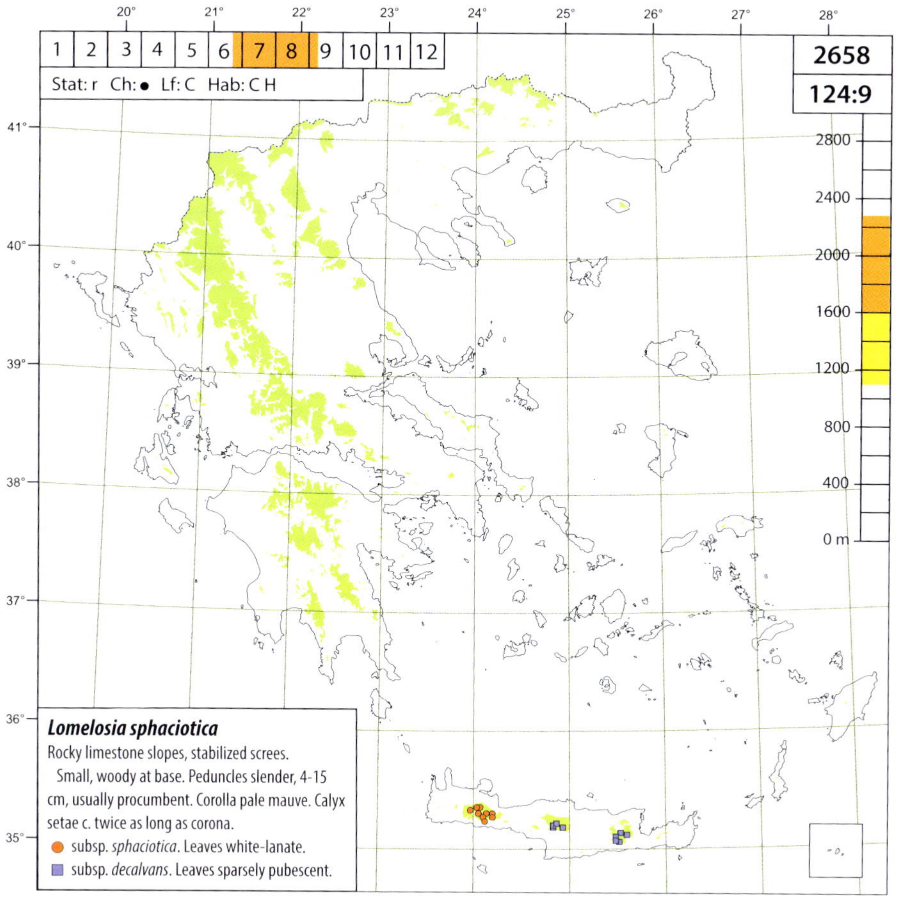

Lomelosia sphaciotica

Rocky limestone slopes, stabilized screes.

Small, woody at base. Peduncles slender, 4-15 cm, usually procumbent. Corolla pale mauve. Calyx setae c. twice as long as corona.

● subsp. *sphaciotica*. Leaves white-lanate.

■ subsp. *decalvans*. Leaves sparsely pubescent.

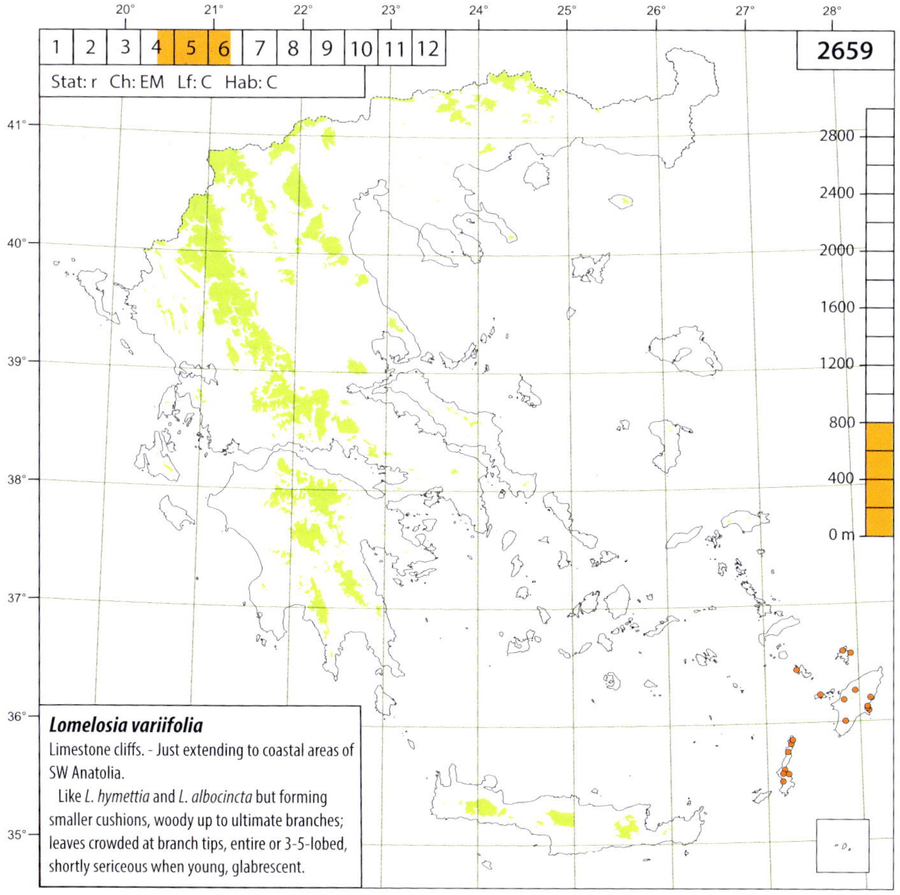

Lomelosia variifolia

Limestone cliffs. - Just extending to coastal areas of SW Anatolia.

Like *L. hymettia* and *L. albocincta* but forming smaller cushions, woody up to ultimate branches; leaves crowded at branch tips, entire or 3-5-lobed, shortly sericeous when young, glabrescent.

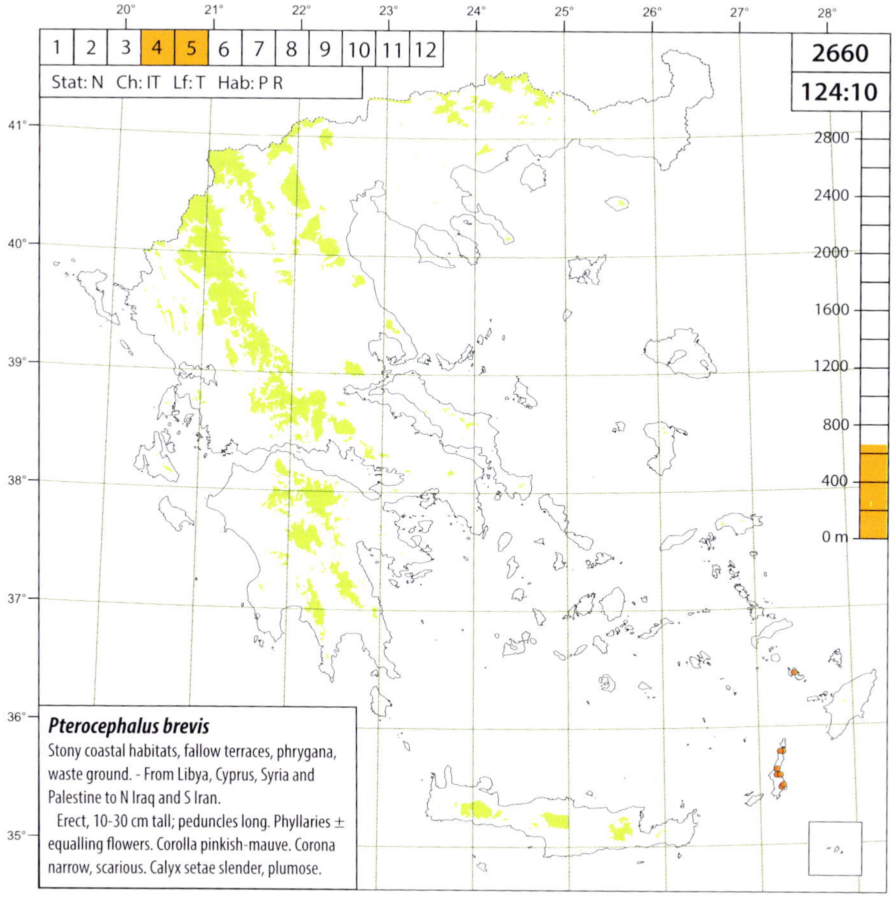

Pterocephalus brevis

Stony coastal habitats, fallow terraces, phrygana, waste ground. - From Libya, Cyprus, Syria and Palestine to N Iraq and S Iran.

Erect, 10-30 cm tall; peduncles long. Phyllaries ± equalling flowers. Corolla pinkish-mauve. Corona narrow, scarious. Calyx setae slender, plumose.

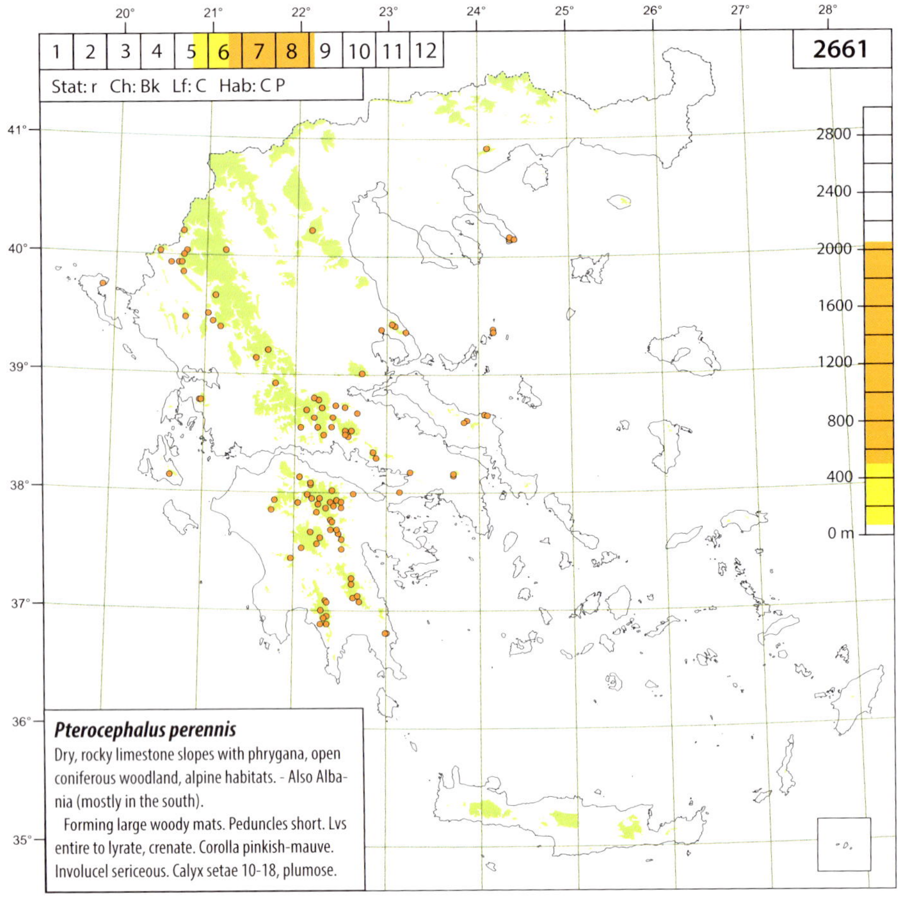

Pterocephalus perennis
Dry, rocky limestone slopes with phrygana, open coniferous woodland, alpine habitats. - Also Albania (mostly in the south).
Forming large woody mats. Peduncles short. Lvs entire to lyrate, crenate. Corolla pinkish-mauve. Involucel sericeous. Calyx setae 10-18, plumose.

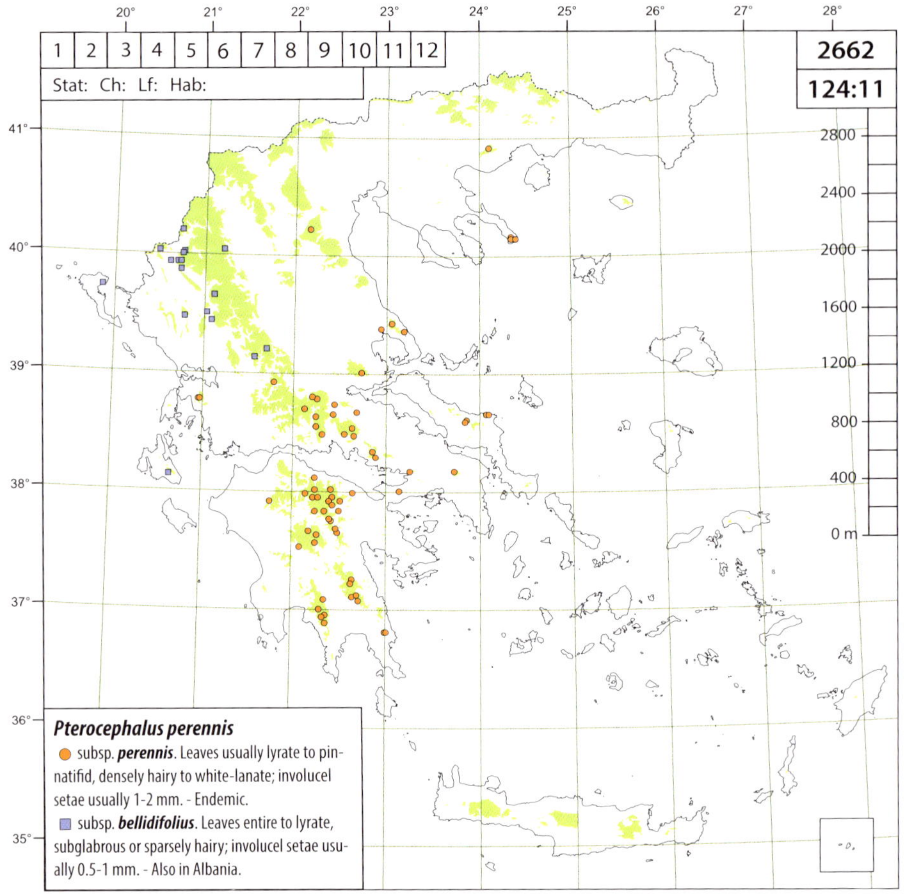

Pterocephalus perennis
● subsp. ***perennis***. Leaves usually lyrate to pinnatifid, densely hairy to white-lanate; involucel setae usually 1-2 mm. - Endemic.
■ subsp. ***bellidifolius***. Leaves entire to lyrate, subglabrous or sparsely hairy; involucel setae usually 0.5-1 mm. - Also in Albania.

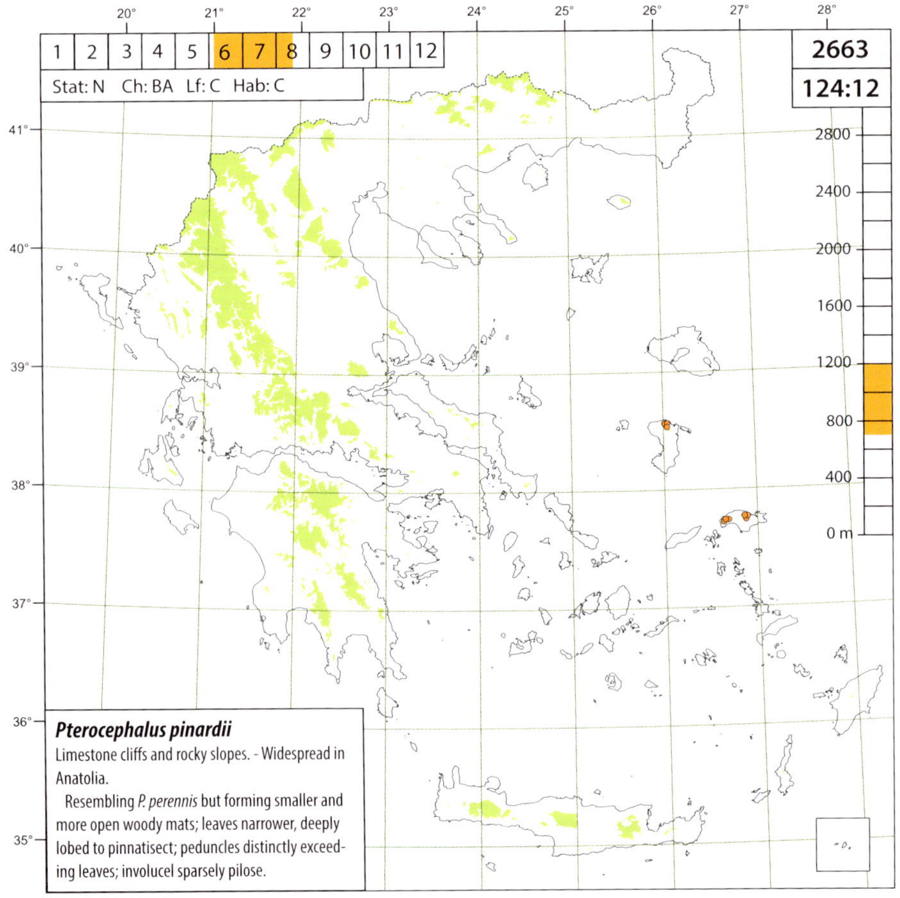

Pterocephalus pinardii
Limestone cliffs and rocky slopes. - Widespread in Anatolia.
Resembling *P. perennis* but forming smaller and more open woody mats; leaves narrower, deeply lobed to pinnatisect; peduncles distinctly exceeding leaves; involucel sparsely pilose.

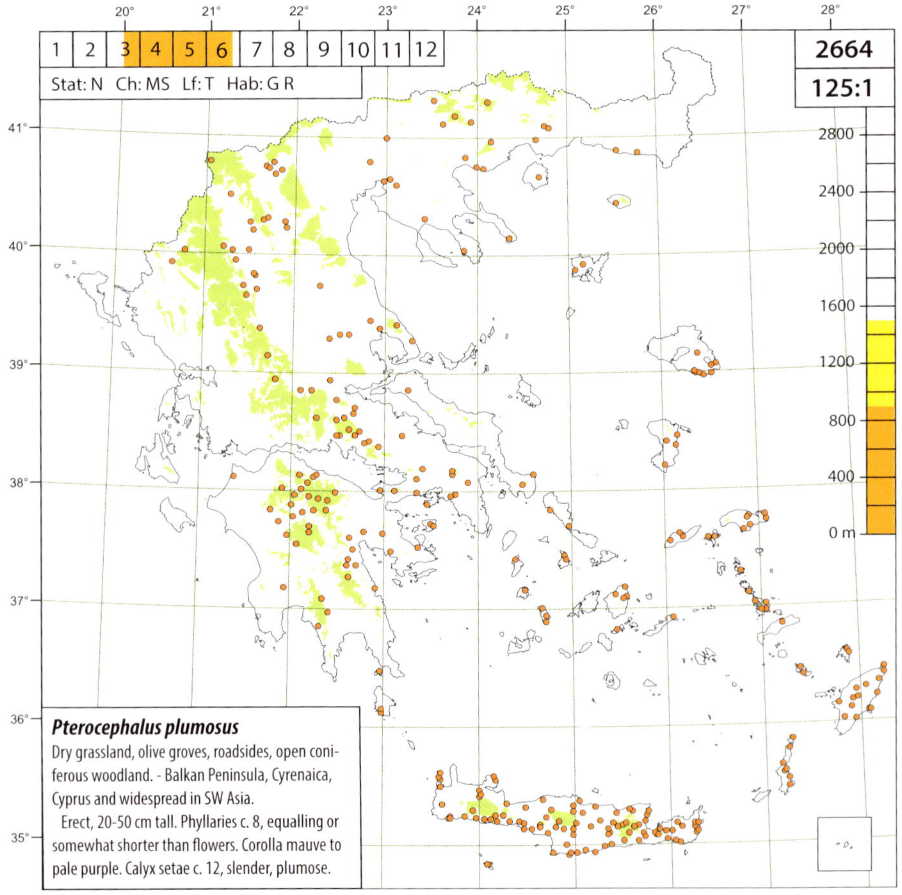

Pterocephalus plumosus
Dry grassland, olive groves, roadsides, open coniferous woodland. - Balkan Peninsula, Cyrenaica, Cyprus and widespread in SW Asia.
Erect, 20-50 cm tall. Phyllaries c. 8, equalling or somewhat shorter than flowers. Corolla mauve to pale purple. Calyx setae c. 12, slender, plumose.

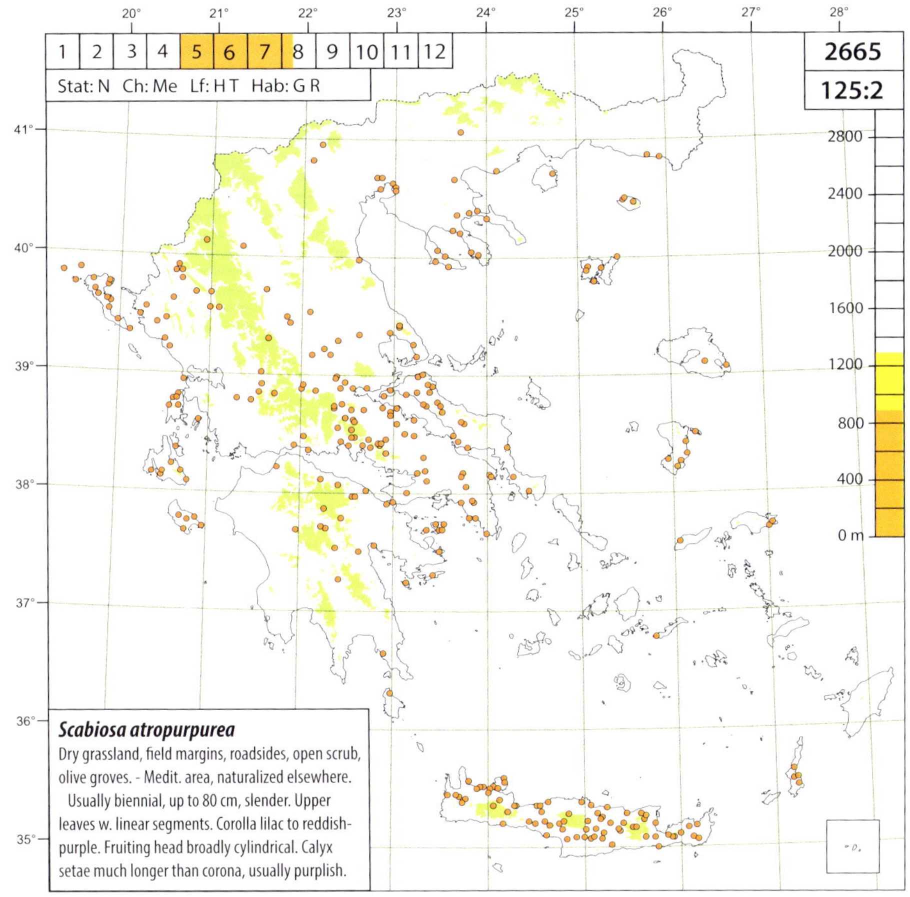

Scabiosa atropurpurea

Dry grassland, field margins, roadsides, open scrub, olive groves. - Medit. area, naturalized elsewhere.

Usually biennial, up to 80 cm, slender. Upper leaves w. linear segments. Corolla lilac to reddish-purple. Fruiting head broadly cylindrical. Calyx setae much longer than corona, usually purplish.

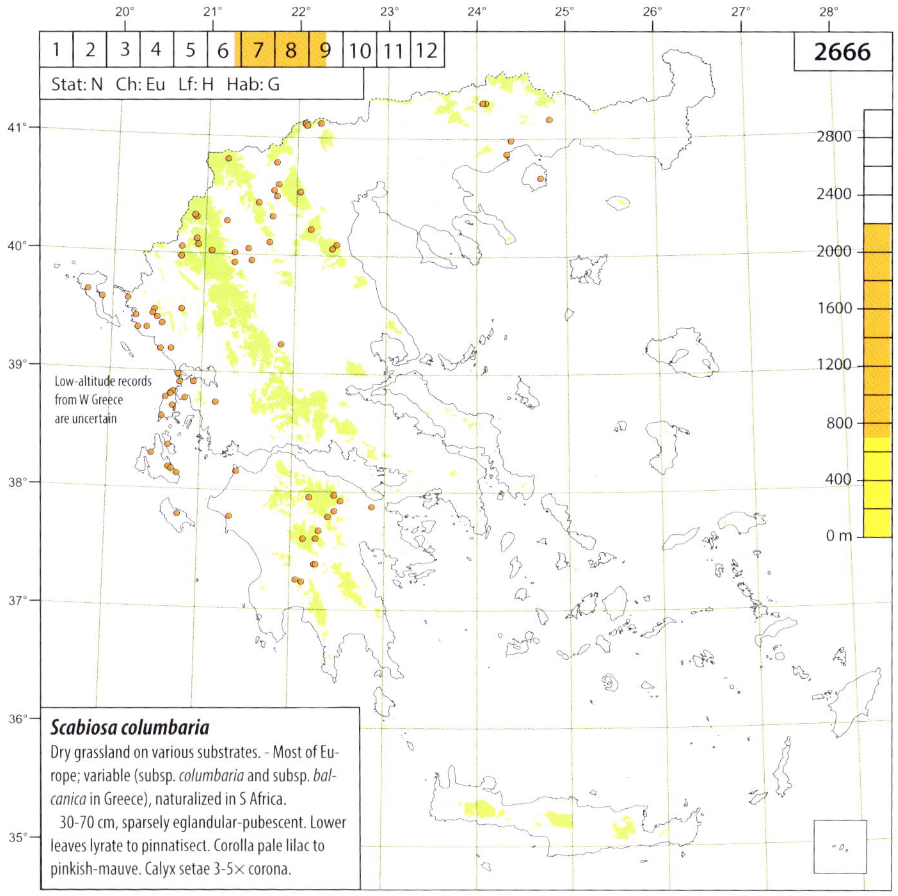

Scabiosa columbaria

Dry grassland on various substrates. - Most of Europe; variable (subsp. *columbaria* and subsp. *balcanica* in Greece), naturalized in S Africa.

30-70 cm, sparsely eglandular-pubescent. Lower leaves lyrate to pinnatisect. Corolla pale lilac to pinkish-mauve. Calyx setae 3-5× corona.

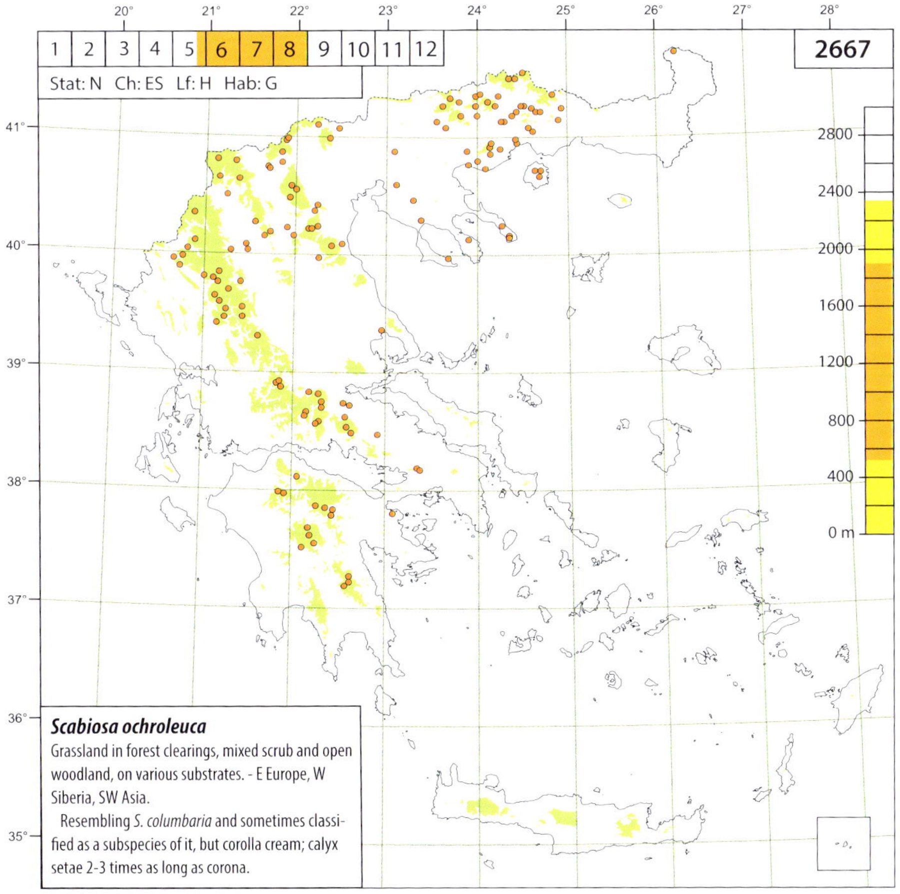

Scabiosa ochroleuca

Grassland in forest clearings, mixed scrub and open woodland, on various substrates. - E Europe, W Siberia, SW Asia.

Resembling *S. columbaria* and sometimes classified as a subspecies of it, but corolla cream; calyx setae 2-3 times as long as corona.

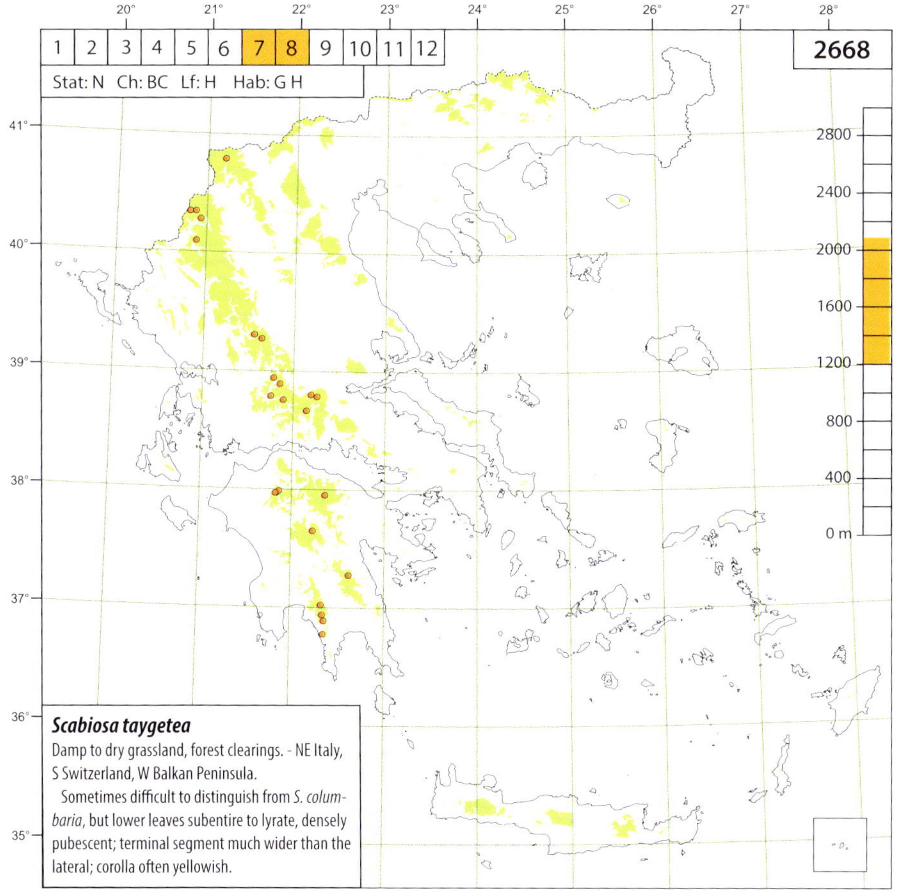

Scabiosa taygetea

Damp to dry grassland, forest clearings. - NE Italy, S Switzerland, W Balkan Peninsula.

Sometimes difficult to distinguish from *S. columbaria*, but lower leaves subentire to lyrate, densely pubescent; terminal segment much wider than the lateral; corolla often yellowish.

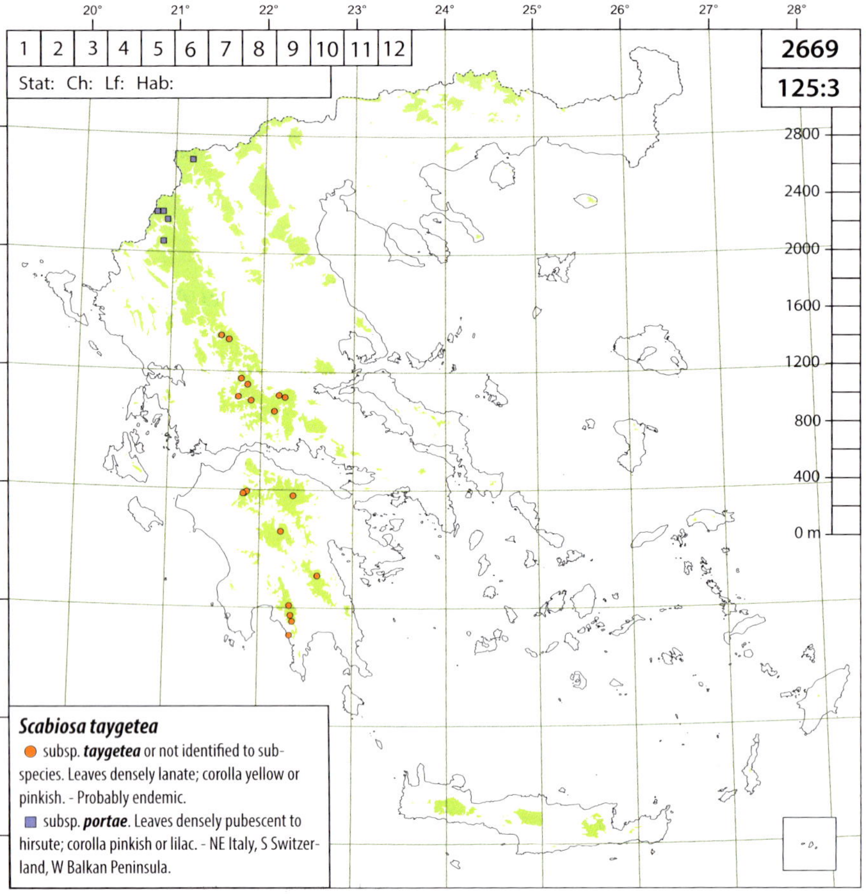

Scabiosa taygetea

● subsp. ***taygetea*** or not identified to subspecies. Leaves densely lanate; corolla yellow or pinkish. - Probably endemic.

■ subsp. ***portae***. Leaves densely pubescent to hirsute; corolla pinkish or lilac. - NE Italy, S Switzerland, W Balkan Peninsula.

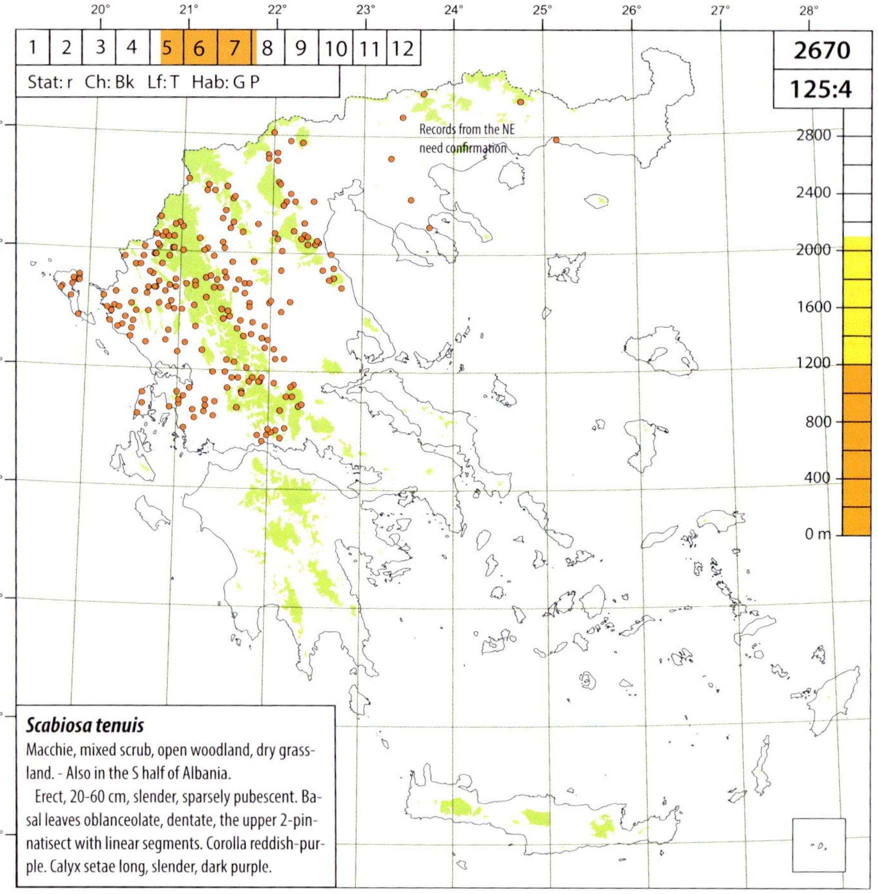

Scabiosa tenuis

Macchie, mixed scrub, open woodland, dry grassland. - Also in the S half of Albania.

Erect, 20-60 cm, slender, sparsely pubescent. Basal leaves oblanceolate, dentate, the upper 2-pinnatisect with linear segments. Corolla reddish-purple. Calyx setae long, slender, dark purple.

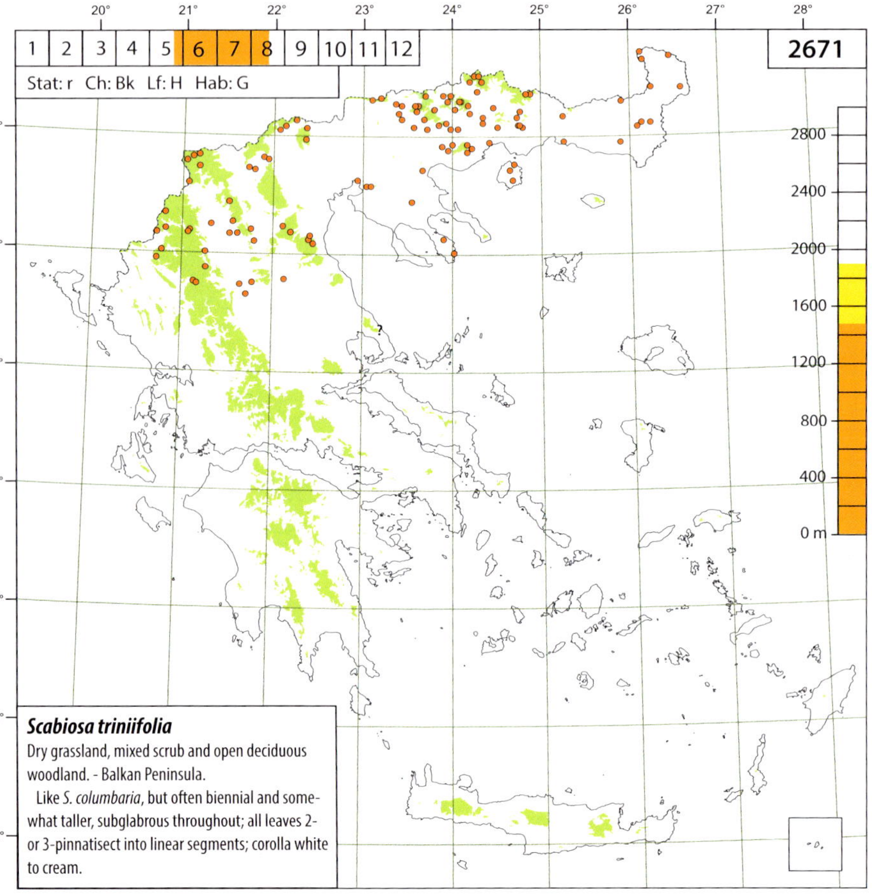

Scabiosa triniifolia

Dry grassland, mixed scrub and open deciduous woodland. - Balkan Peninsula.

Like *S. columbaria*, but often biennial and somewhat taller, subglabrous throughout; all leaves 2- or 3-pinnatisect into linear segments; corolla white to cream.

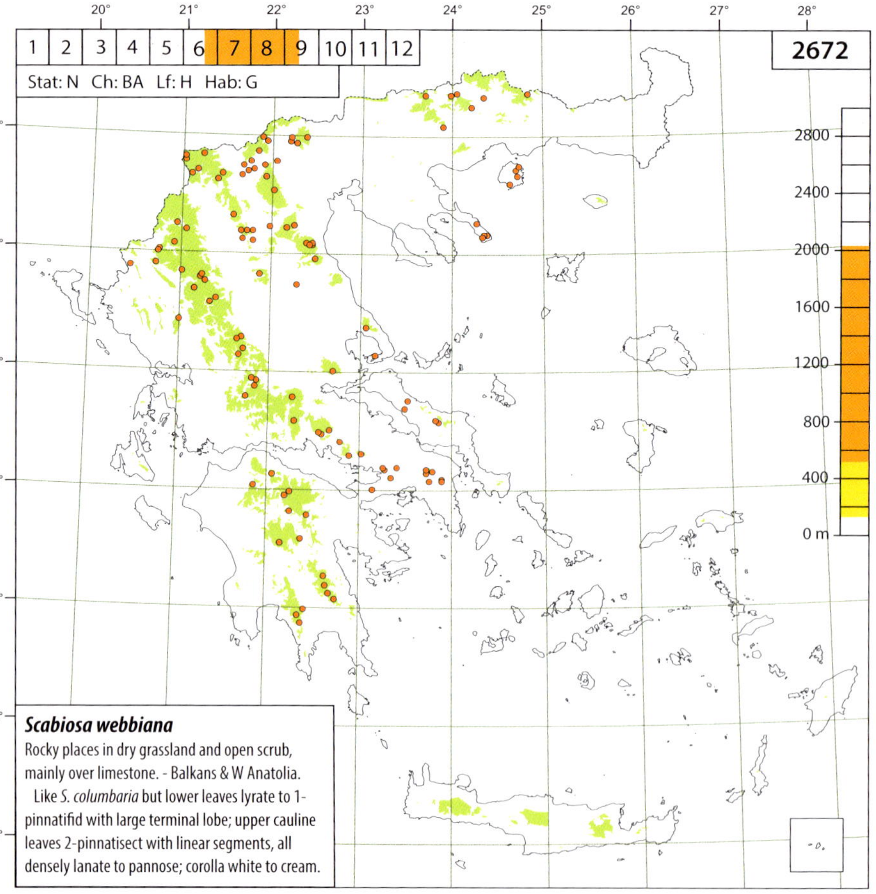

Scabiosa webbiana

Rocky places in dry grassland and open scrub, mainly over limestone. - Balkans & W Anatolia.

Like *S. columbaria* but lower leaves lyrate to 1-pinnatifid with large terminal lobe; upper cauline leaves 2-pinnatisect with linear segments, all densely lanate to pannose; corolla white to cream.

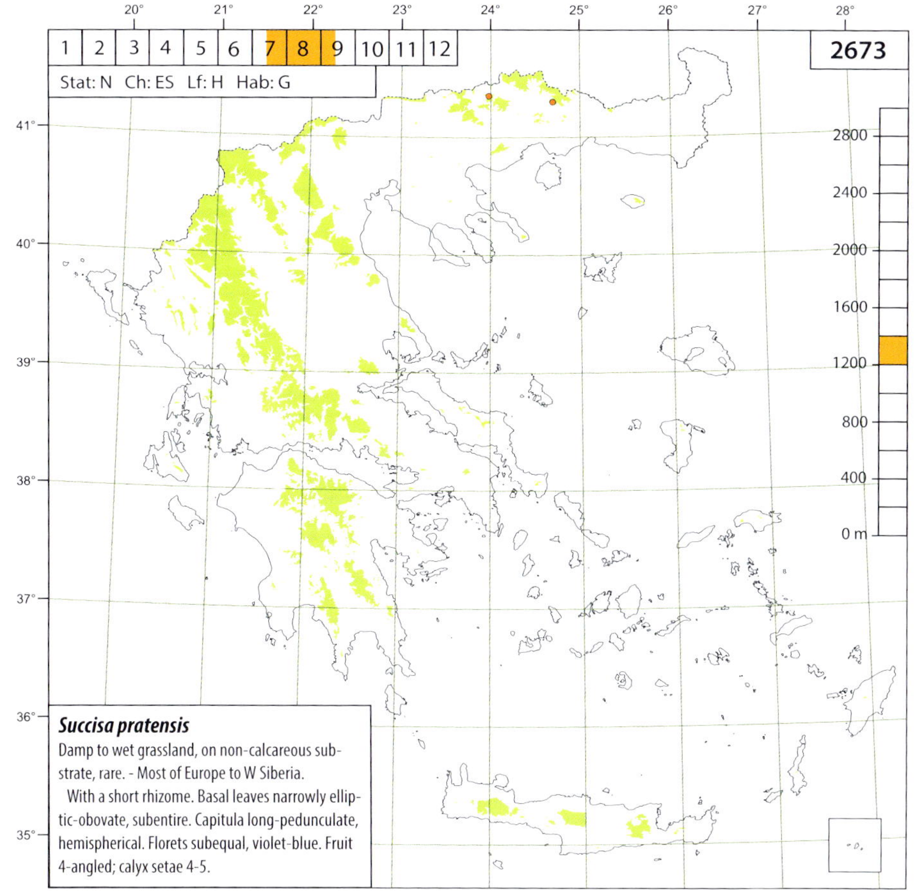

Succisa pratensis

Damp to wet grassland, on non-calcareous substrate, rare. - Most of Europe to W Siberia.

With a short rhizome. Basal leaves narrowly elliptic-obovate, subentire. Capitula long-pedunculate, hemispherical. Florets subequal, violet-blue. Fruit 4-angled; calyx setae 4-5.

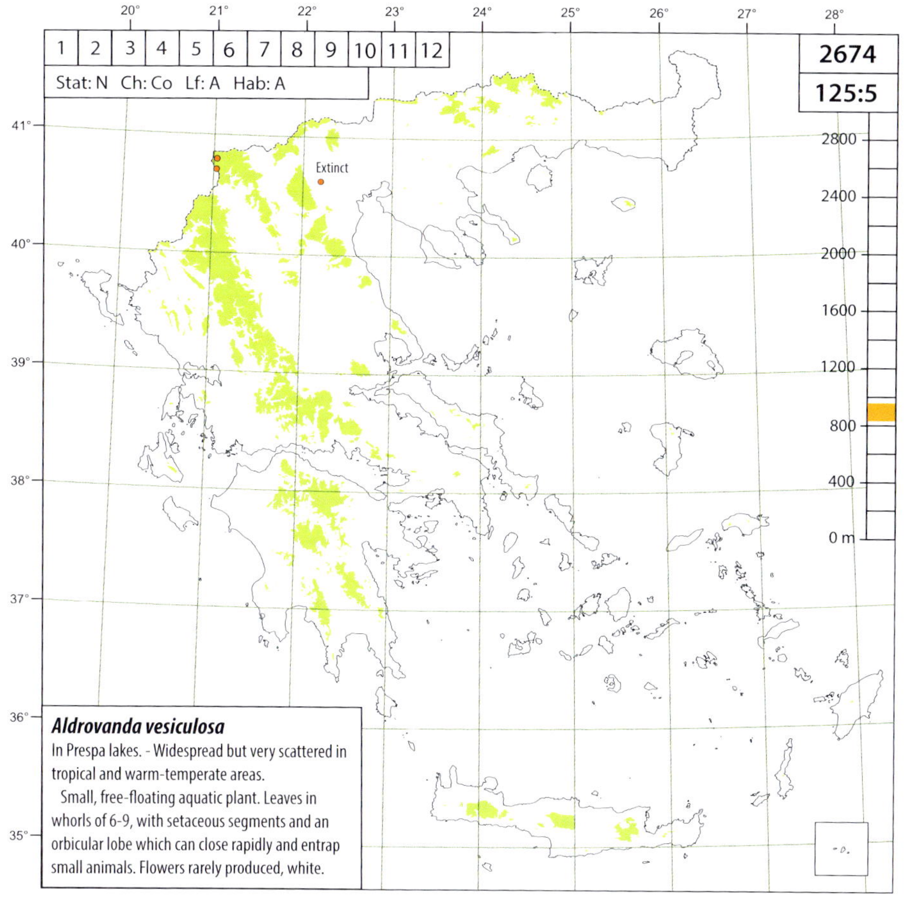

Aldrovanda vesiculosa

In Prespa lakes. - Widespread but very scattered in tropical and warm-temperate areas.

Small, free-floating aquatic plant. Leaves in whorls of 6-9, with setaceous segments and an orbicular lobe which can close rapidly and entrap small animals. Flowers rarely produced, white.

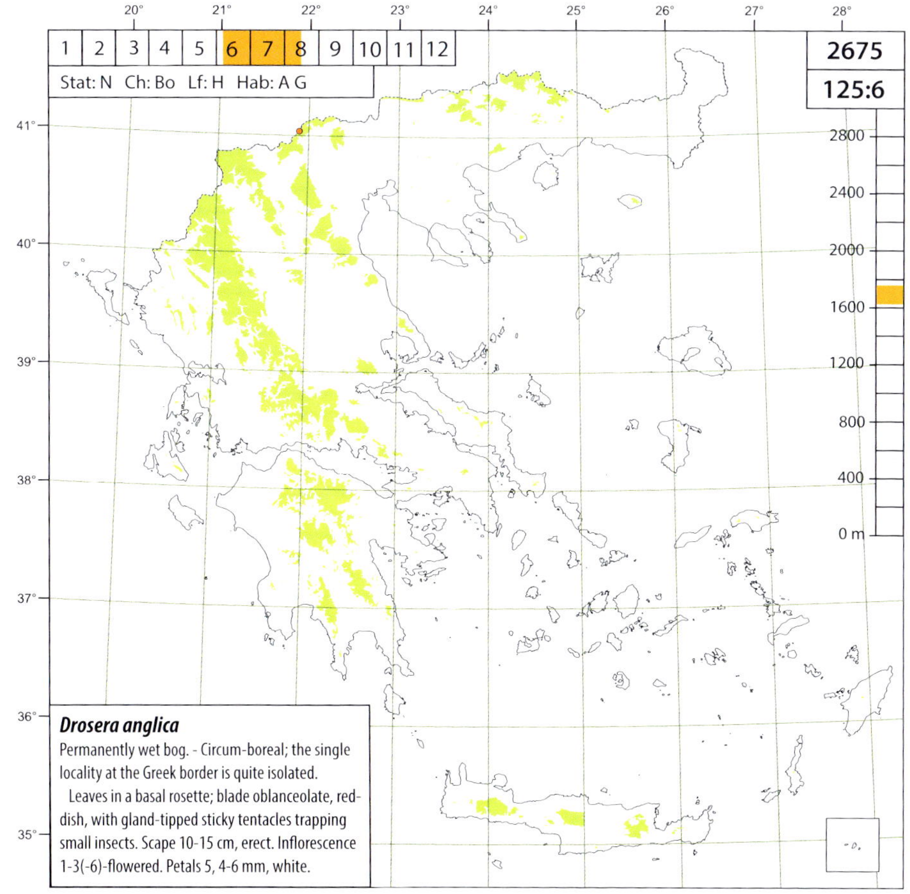

Drosera anglica

Permanently wet bog. - Circum-boreal; the single locality at the Greek border is quite isolated.

Leaves in a basal rosette; blade oblanceolate, reddish, with gland-tipped sticky tentacles trapping small insects. Scape 10-15 cm, erect. Inflorescence 1-3(-6)-flowered. Petals 5, 4-6 mm, white.

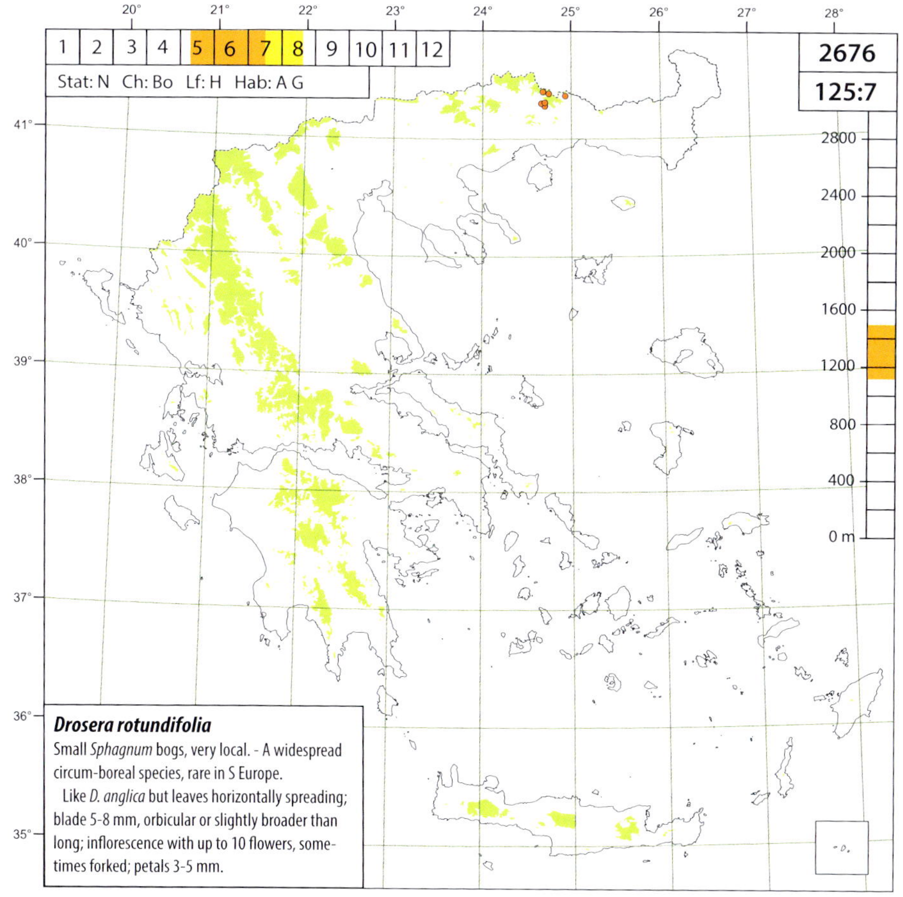

Drosera rotundifolia

Small _Sphagnum_ bogs, very local. - A widespread circum-boreal species, rare in S Europe.

Like _D. anglica_ but leaves horizontally spreading; blade 5-8 mm, orbicular or slightly broader than long; inflorescence with up to 10 flowers, sometimes forked; petals 3-5 mm.

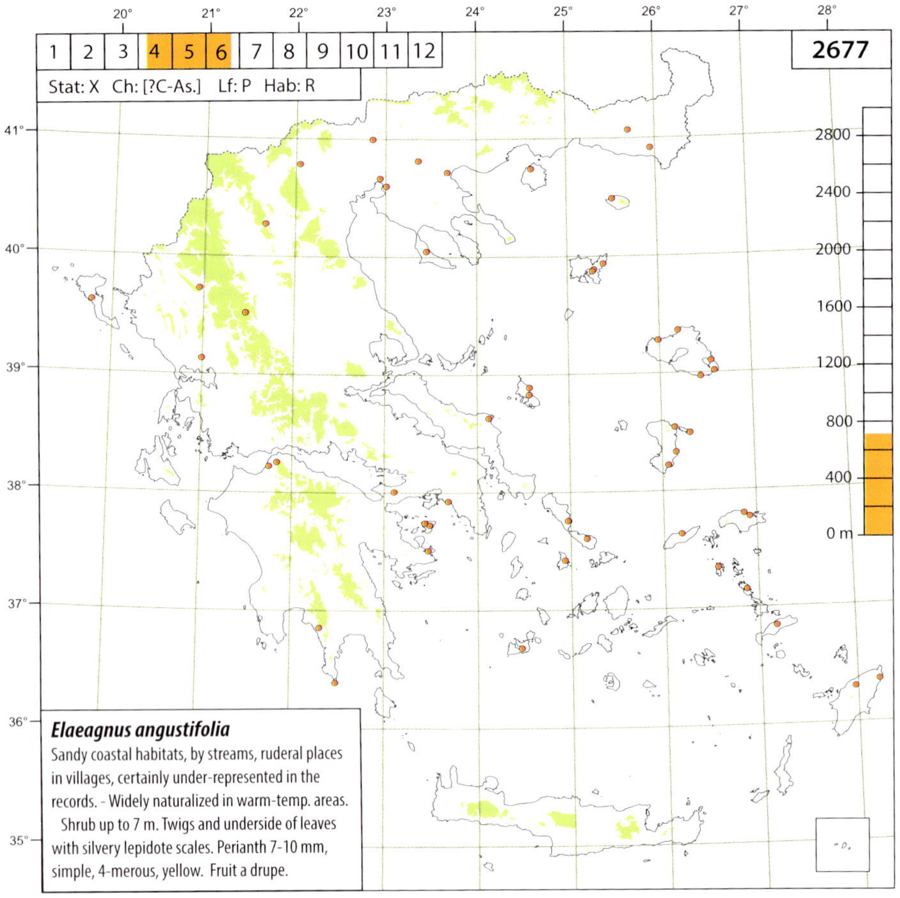

Elaeagnus angustifolia
Sandy coastal habitats, by streams, ruderal places in villages, certainly under-represented in the records. - Widely naturalized in warm-temp. areas.
Shrub up to 7 m. Twigs and underside of leaves with silvery lepidote scales. Perianth 7-10 mm, simple, 4-merous, yellow. Fruit a drupe.

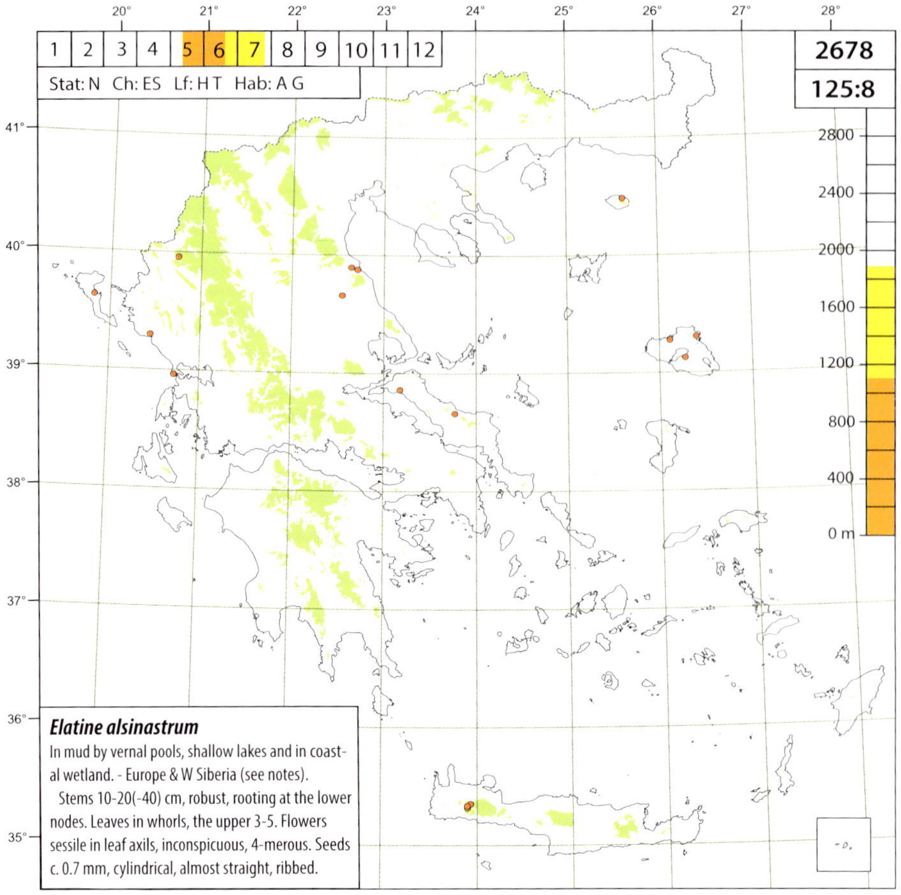

Elatine alsinastrum
In mud by vernal pools, shallow lakes and in coastal wetland. - Europe & W Siberia (see notes).
Stems 10-20(-40) cm, robust, rooting at the lower nodes. Leaves in whorls, the upper 3-5. Flowers sessile in leaf axils, inconspicuous, 4-merous. Seeds c. 0.7 mm, cylindrical, almost straight, ribbed.

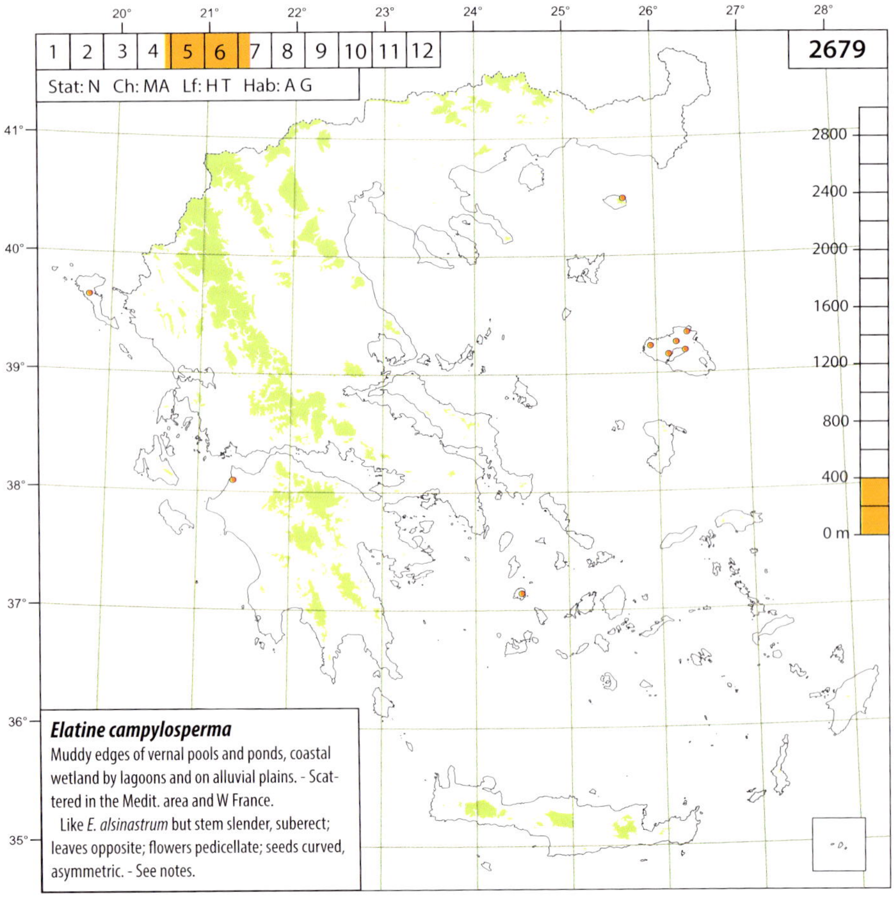

Elatine campylosperma
Muddy edges of vernal pools and ponds, coastal wetland by lagoons and on alluvial plains. - Scattered in the Medit. area and W France.
Like *E. alsinastrum* but stem slender, suberect; leaves opposite; flowers pedicellate; seeds curved, asymmetric. - See notes.

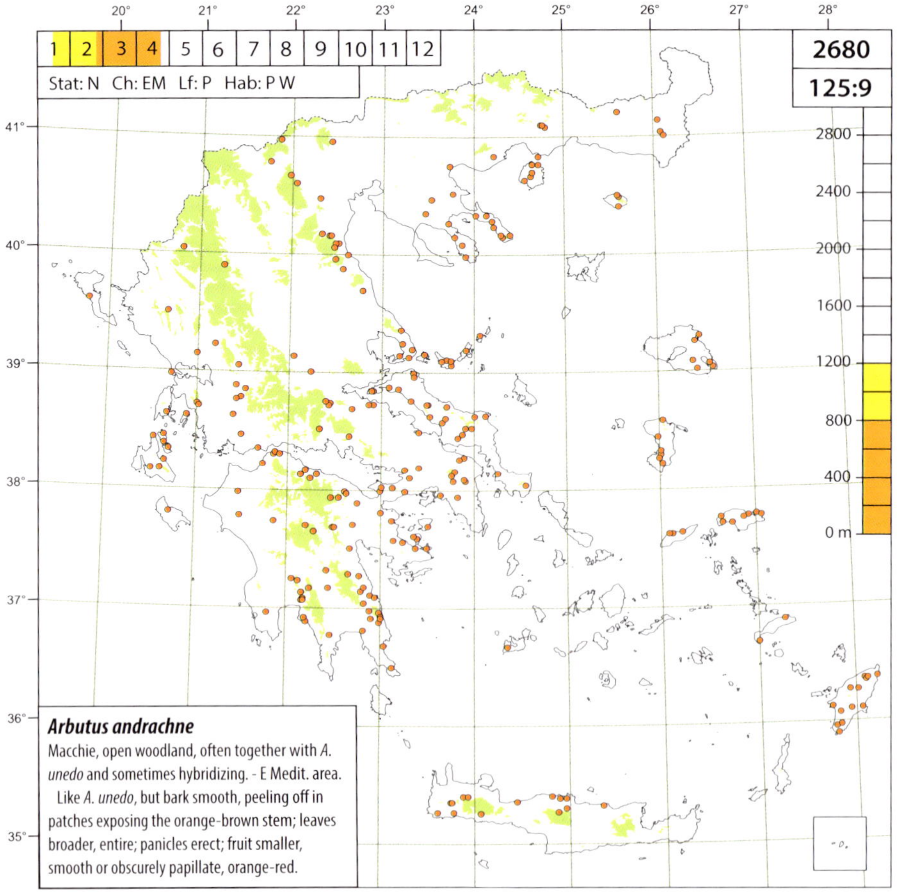

Arbutus andrachne
Macchie, open woodland, often together with *A. unedo* and sometimes hybridizing. - E Medit. area.
Like *A. unedo*, but bark smooth, peeling off in patches exposing the orange-brown stem; leaves broader, entire; panicles erect; fruit smaller, smooth or obscurely papillate, orange-red.

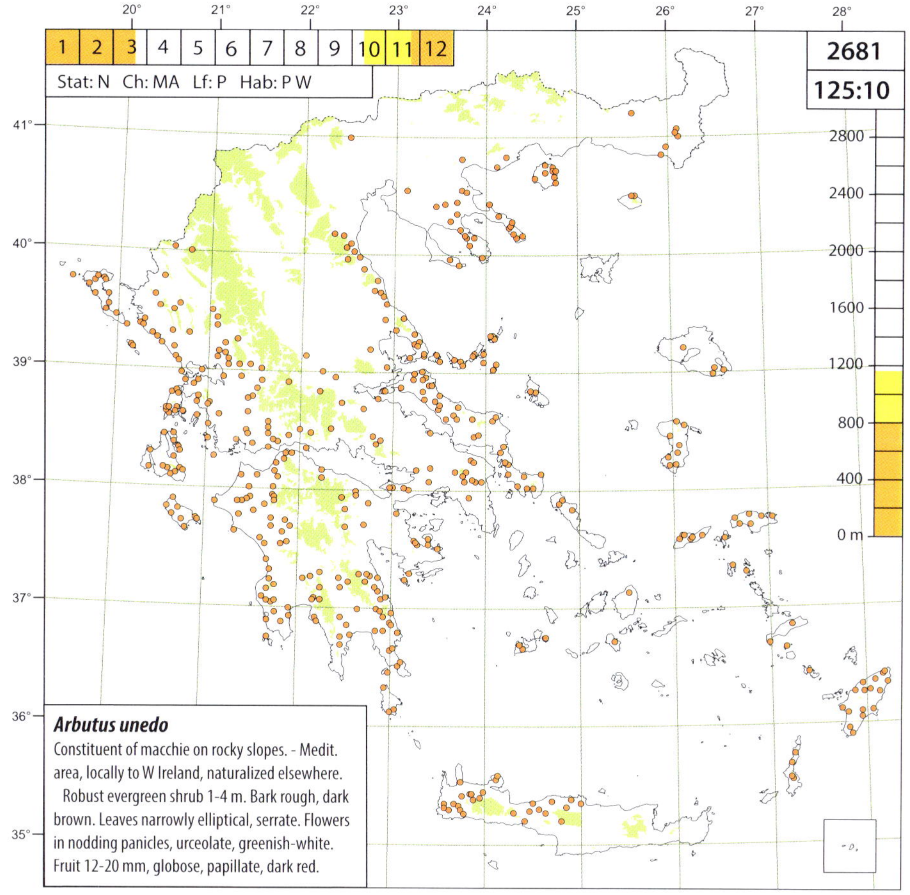

Arbutus unedo
Constituent of macchie on rocky slopes. - Medit. area, locally to W Ireland, naturalized elsewhere.
Robust evergreen shrub 1-4 m. Bark rough, dark brown. Leaves narrowly elliptical, serrate. Flowers in nodding panicles, urceolate, greenish-white. Fruit 12-20 mm, globose, papillate, dark red.

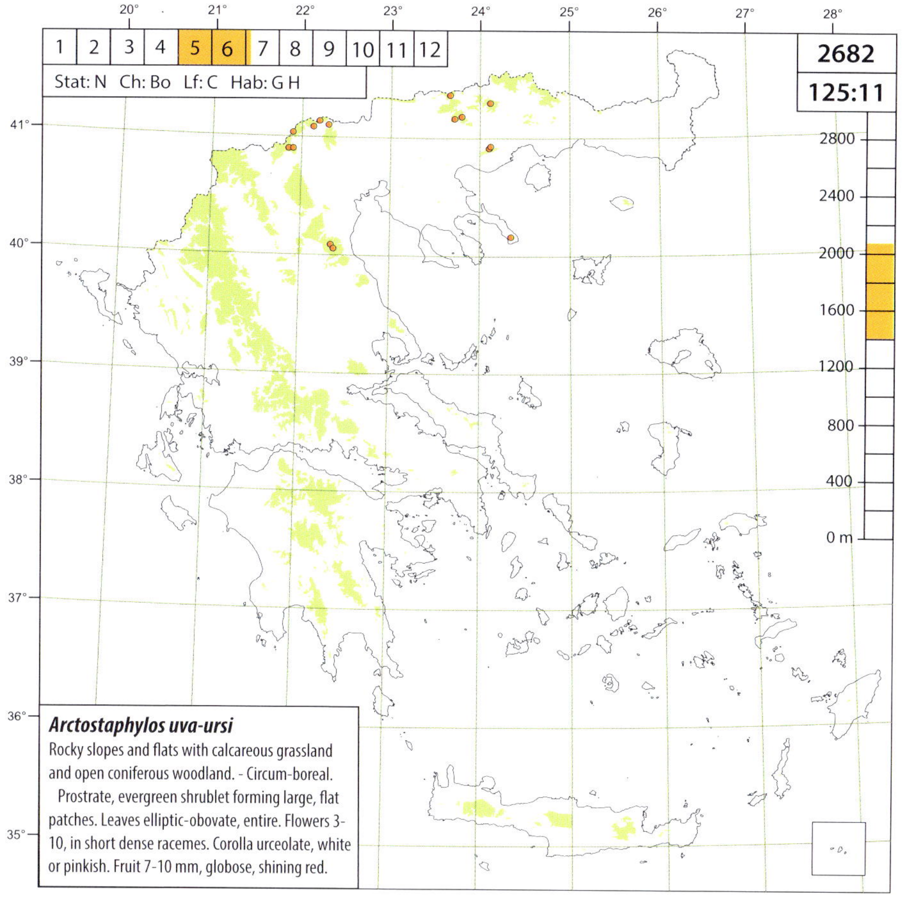

Arctostaphylos uva-ursi
Rocky slopes and flats with calcareous grassland and open coniferous woodland. - Circum-boreal.
Prostrate, evergreen shrublet forming large, flat patches. Leaves elliptic-obovate, entire. Flowers 3-10, in short dense racemes. Corolla urceolate, white or pinkish. Fruit 7-10 mm, globose, shining red.

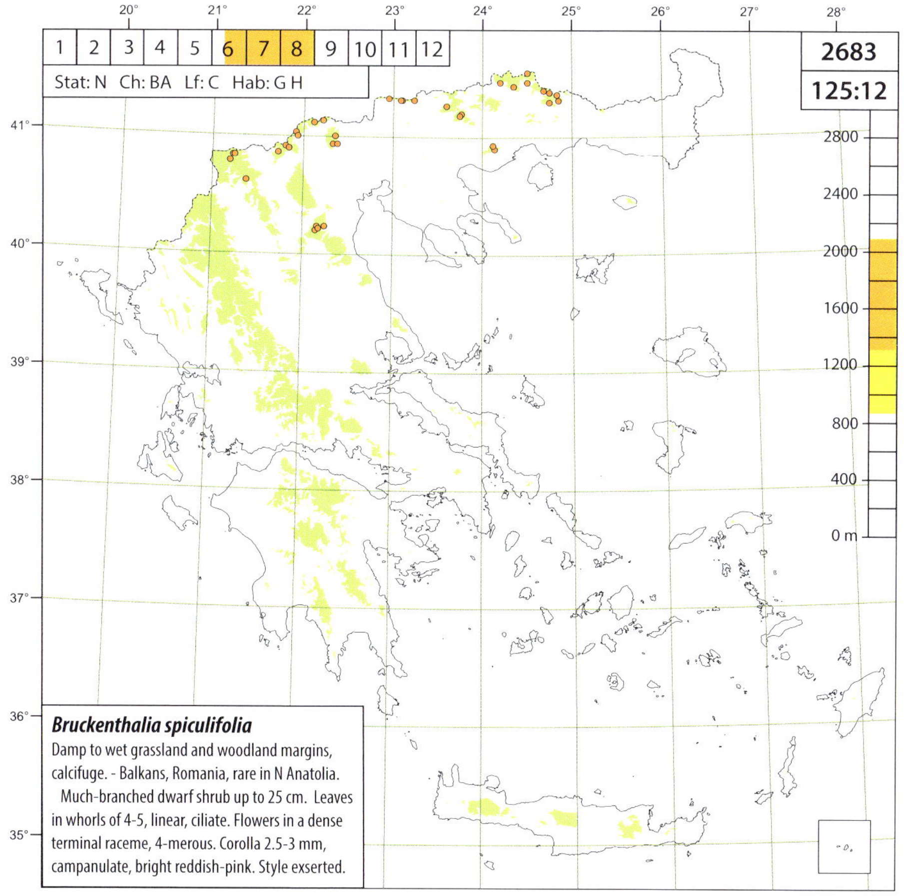

Bruckenthalia spiculifolia
Damp to wet grassland and woodland margins, calcifuge. - Balkans, Romania, rare in N Anatolia.
Much-branched dwarf shrub up to 25 cm. Leaves in whorls of 4-5, linear, ciliate. Flowers in a dense terminal raceme, 4-merous. Corolla 2.5-3 mm, campanulate, bright reddish-pink. Style exserted.

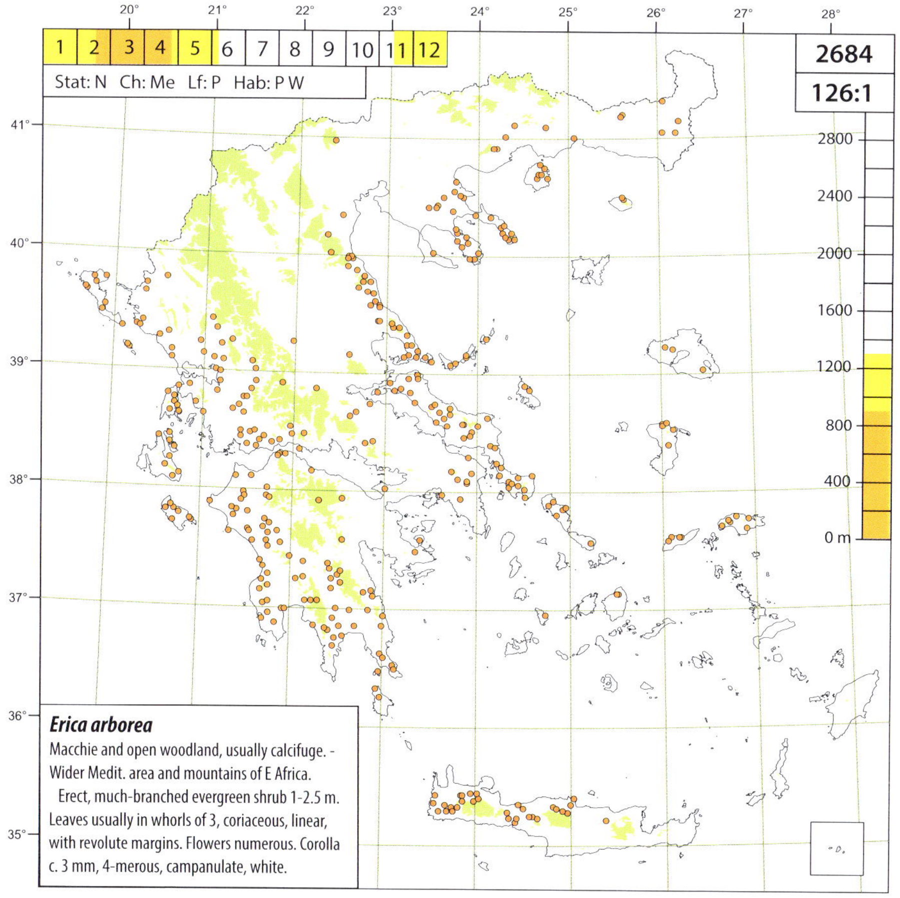

Erica arborea
Macchie and open woodland, usually calcifuge. - Wider Medit. area and mountains of E Africa.
Erect, much-branched evergreen shrub 1-2.5 m. Leaves usually in whorls of 3, coriaceous, linear, with revolute margins. Flowers numerous. Corolla c. 3 mm, 4-merous, campanulate, white.

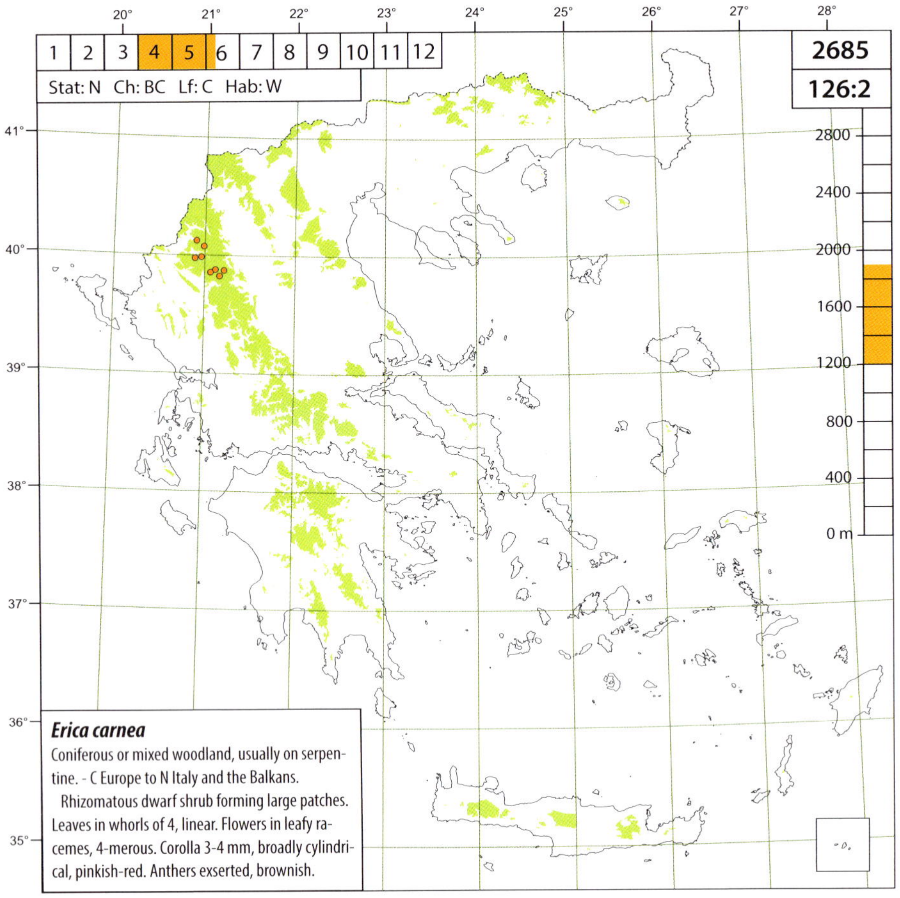

Erica carnea

Coniferous or mixed woodland, usually on serpentine. - C Europe to N Italy and the Balkans.

Rhizomatous dwarf shrub forming large patches. Leaves in whorls of 4, linear. Flowers in leafy racemes, 4-merous. Corolla 3-4 mm, broadly cylindrical, pinkish-red. Anthers exserted, brownish.

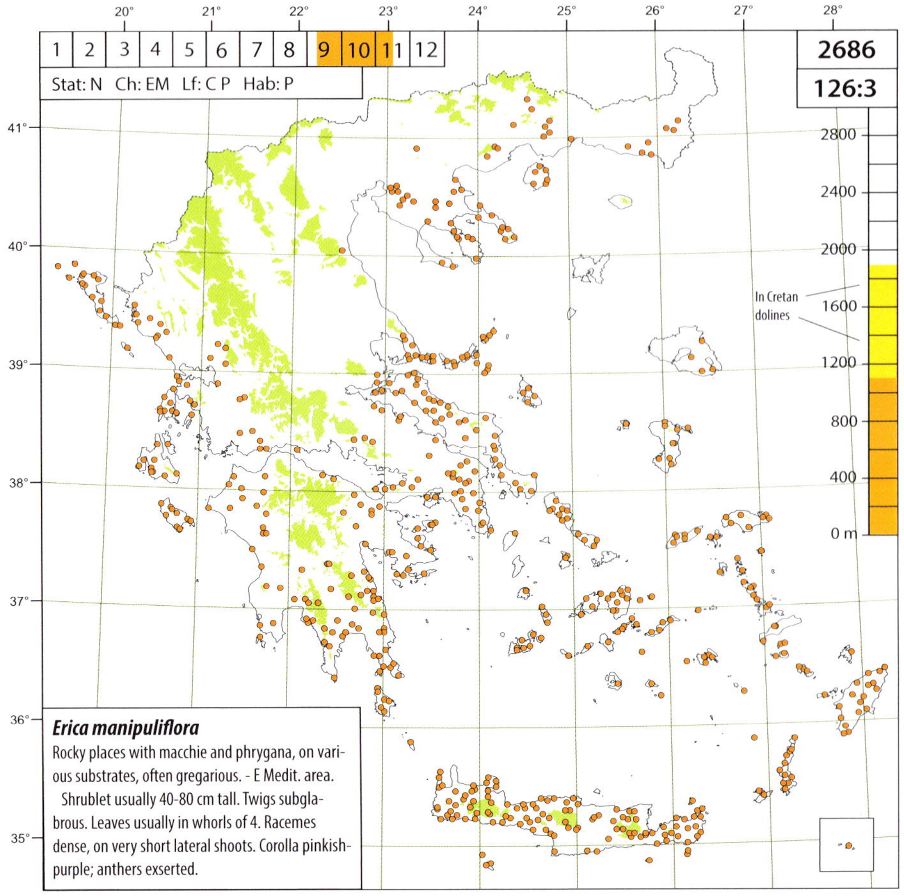

Erica manipuliflora

Rocky places with macchie and phrygana, on various substrates, often gregarious. - E Medit. area.

Shrublet usually 40-80 cm tall. Twigs subglabrous. Leaves usually in whorls of 4. Racemes dense, on very short lateral shoots. Corolla pinkish-purple; anthers exserted.

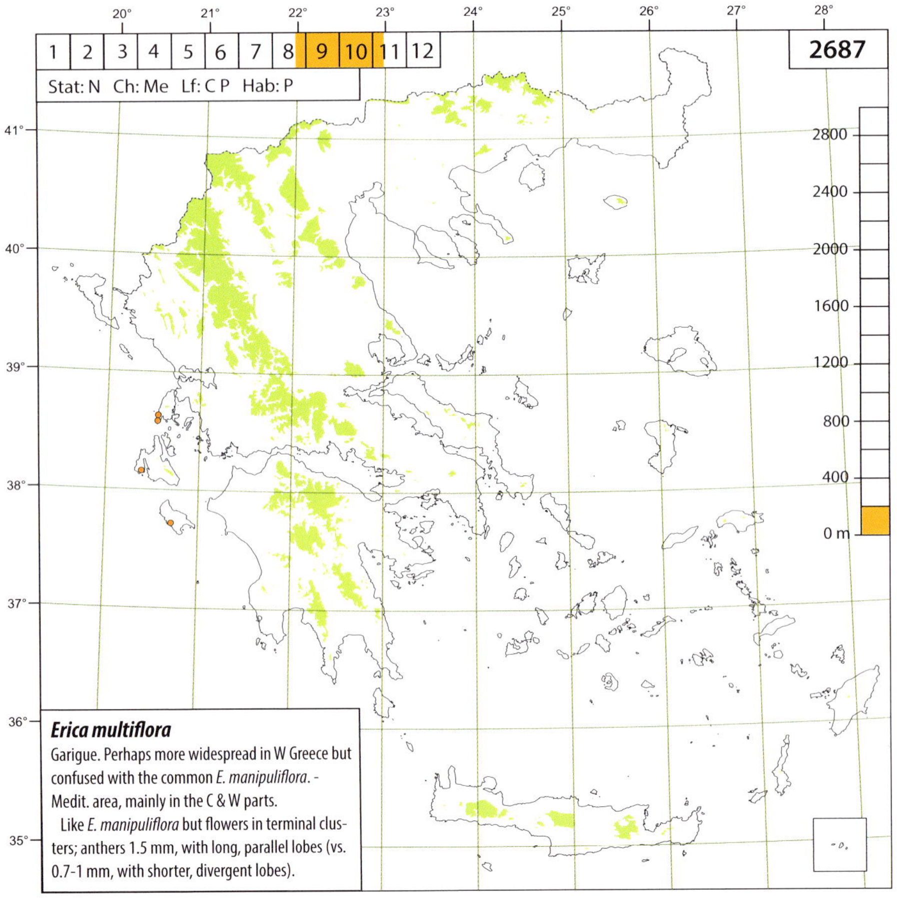

Erica multiflora

Garigue. Perhaps more widespread in W Greece but confused with the common *E. manipuliflora*. - Medit. area, mainly in the C & W parts.

Like *E. manipuliflora* but flowers in terminal clusters; anthers 1.5 mm, with long, parallel lobes (vs. 0.7-1 mm, with shorter, divergent lobes).

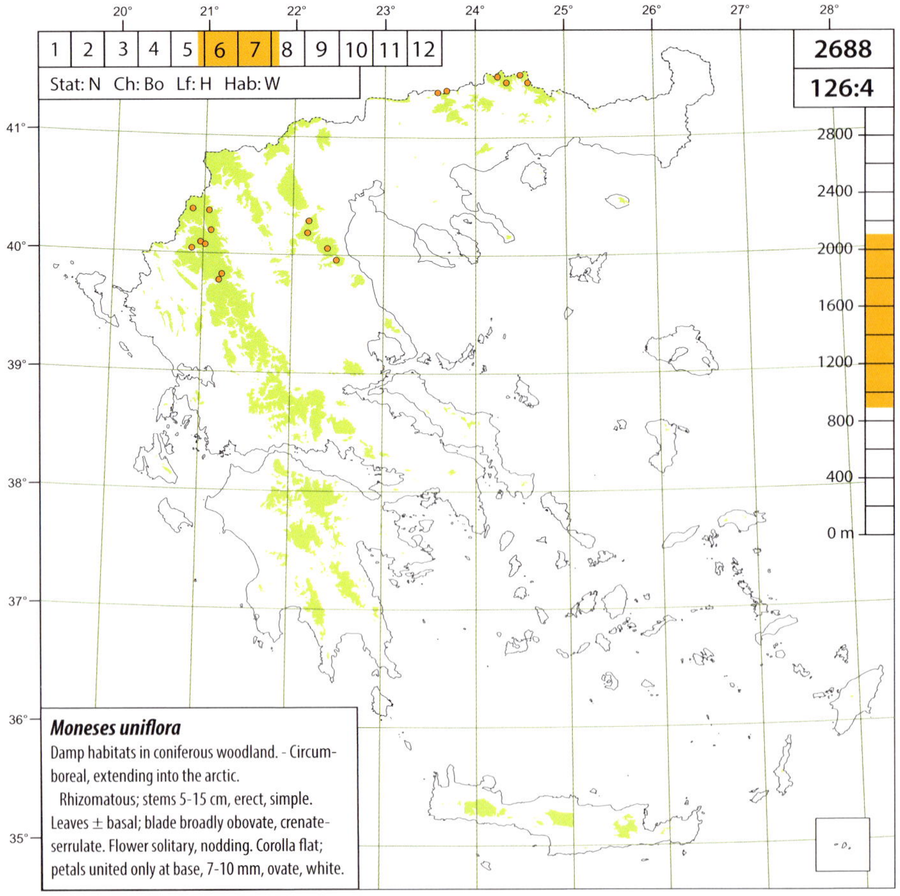

Moneses uniflora

Damp habitats in coniferous woodland. - Circumboreal, extending into the arctic.

Rhizomatous; stems 5-15 cm, erect, simple. Leaves ± basal; blade broadly obovate, crenate-serrulate. Flower solitary, nodding. Corolla flat; petals united only at base, 7-10 mm, ovate, white.

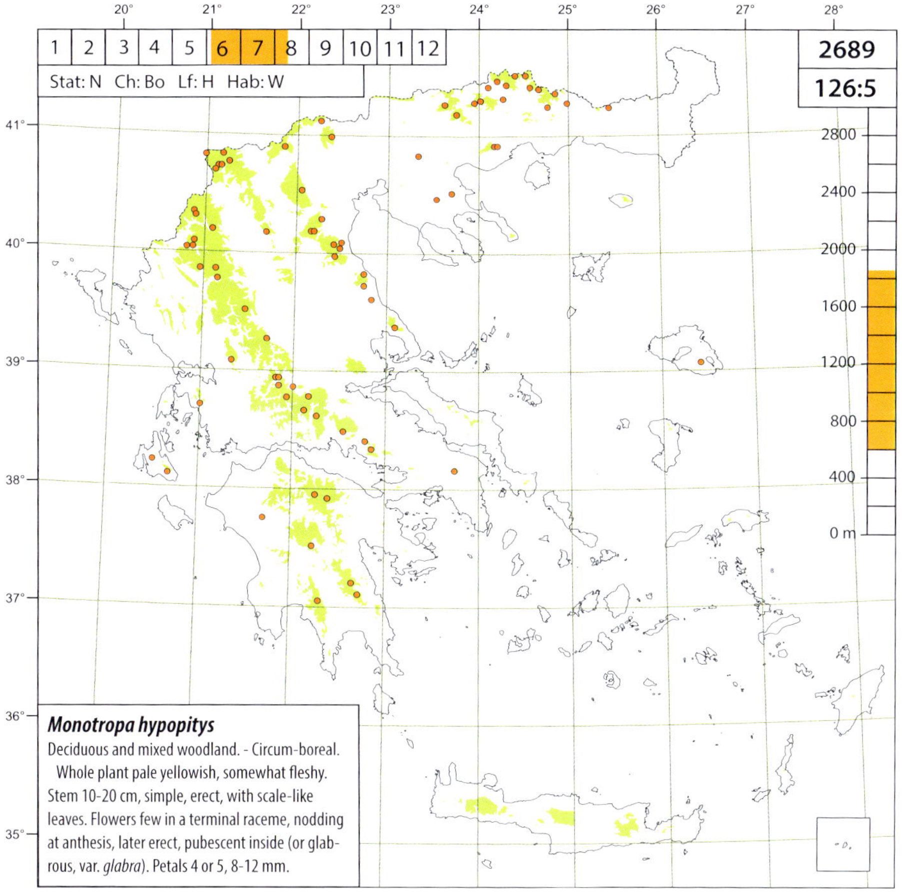

Monotropa hypopitys
Deciduous and mixed woodland. - Circum-boreal.
Whole plant pale yellowish, somewhat fleshy. Stem 10-20 cm, simple, erect, with scale-like leaves. Flowers few in a terminal raceme, nodding at anthesis, later erect, pubescent inside (or glabrous, var. *glabra*). Petals 4 or 5, 8-12 mm.

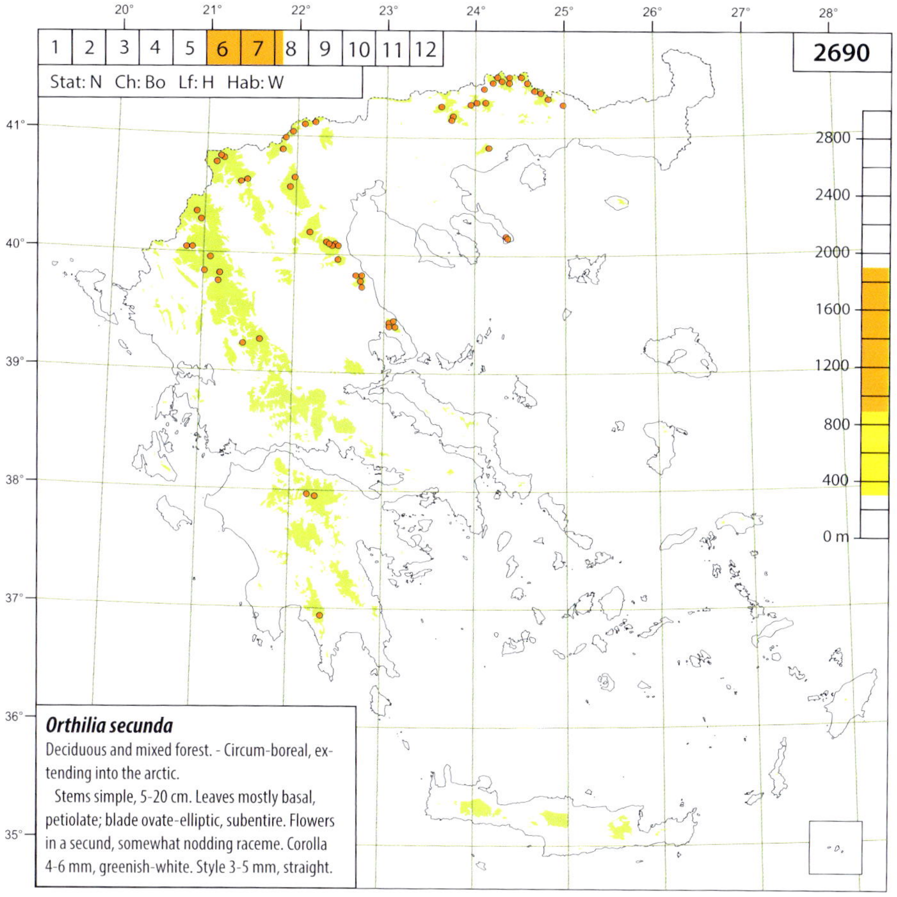

Orthilia secunda
Deciduous and mixed forest. - Circum-boreal, extending into the arctic.
Stems simple, 5-20 cm. Leaves mostly basal, petiolate; blade ovate-elliptic, subentire. Flowers in a secund, somewhat nodding raceme. Corolla 4-6 mm, greenish-white. Style 3-5 mm, straight.

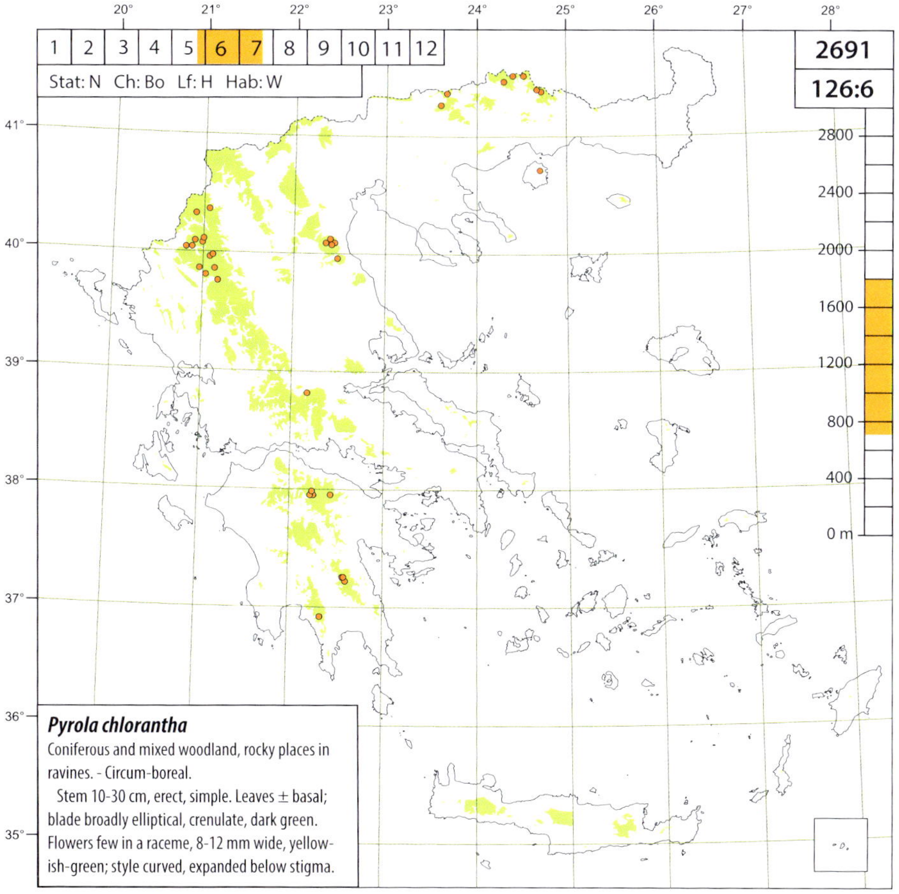

Pyrola chlorantha
Coniferous and mixed woodland, rocky places in ravines. - Circum-boreal.
Stem 10-30 cm, erect, simple. Leaves ± basal; blade broadly elliptical, crenulate, dark green. Flowers few in a raceme, 8-12 mm wide, yellowish-green; style curved, expanded below stigma.

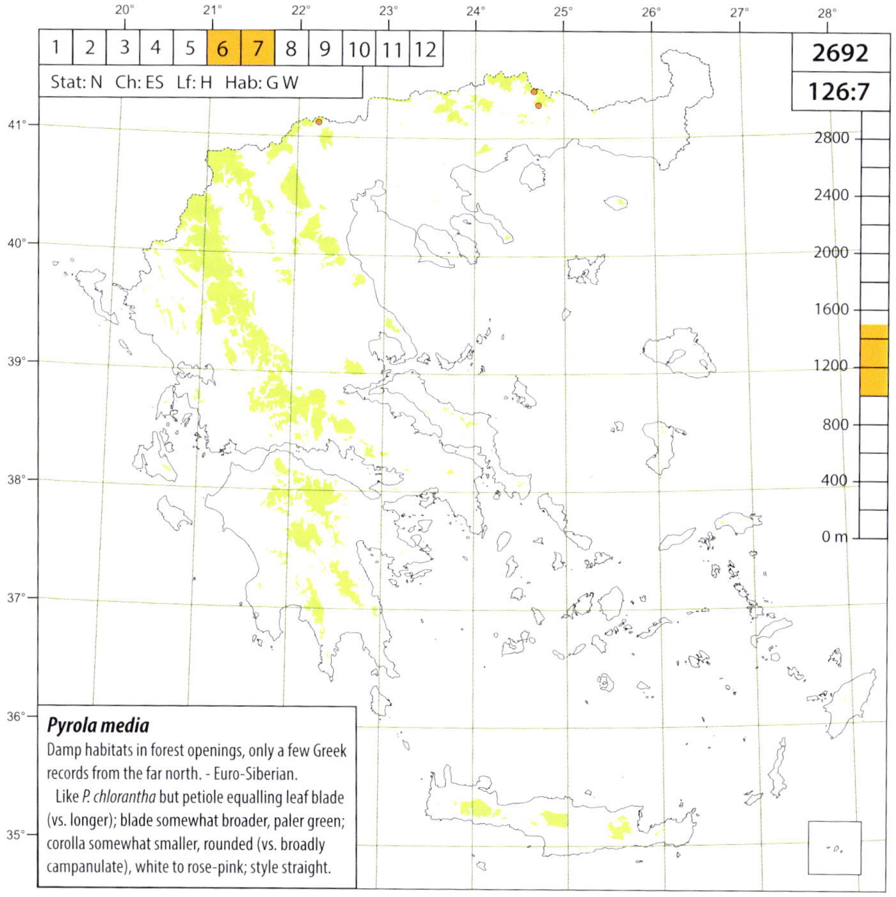

Pyrola media
Damp habitats in forest openings, only a few Greek records from the far north. - Euro-Siberian.
Like *P. chlorantha* but petiole equalling leaf blade (vs. longer); blade somewhat broader, paler green; corolla somewhat smaller, rounded (vs. broadly campanulate), white to rose-pink; style straight.

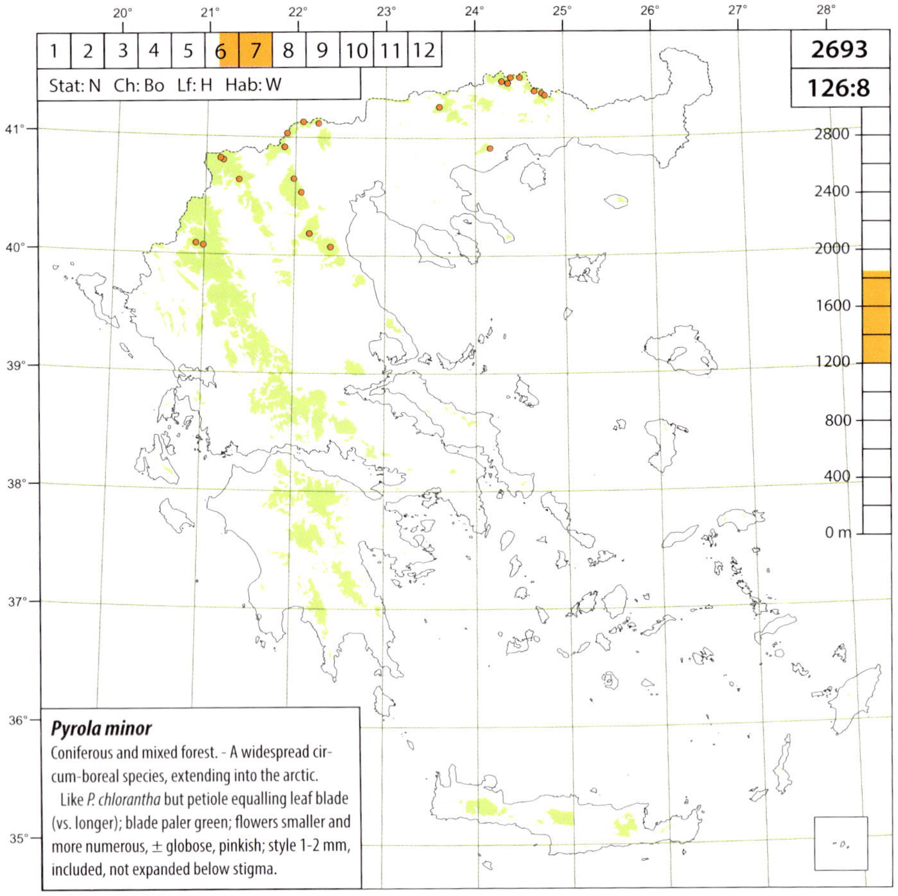

Pyrola minor
Coniferous and mixed forest. - A widespread circum-boreal species, extending into the arctic.
Like *P. chlorantha* but petiole equalling leaf blade (vs. longer); blade paler green; flowers smaller and more numerous, ± globose, pinkish; style 1-2 mm, included, not expanded below stigma.

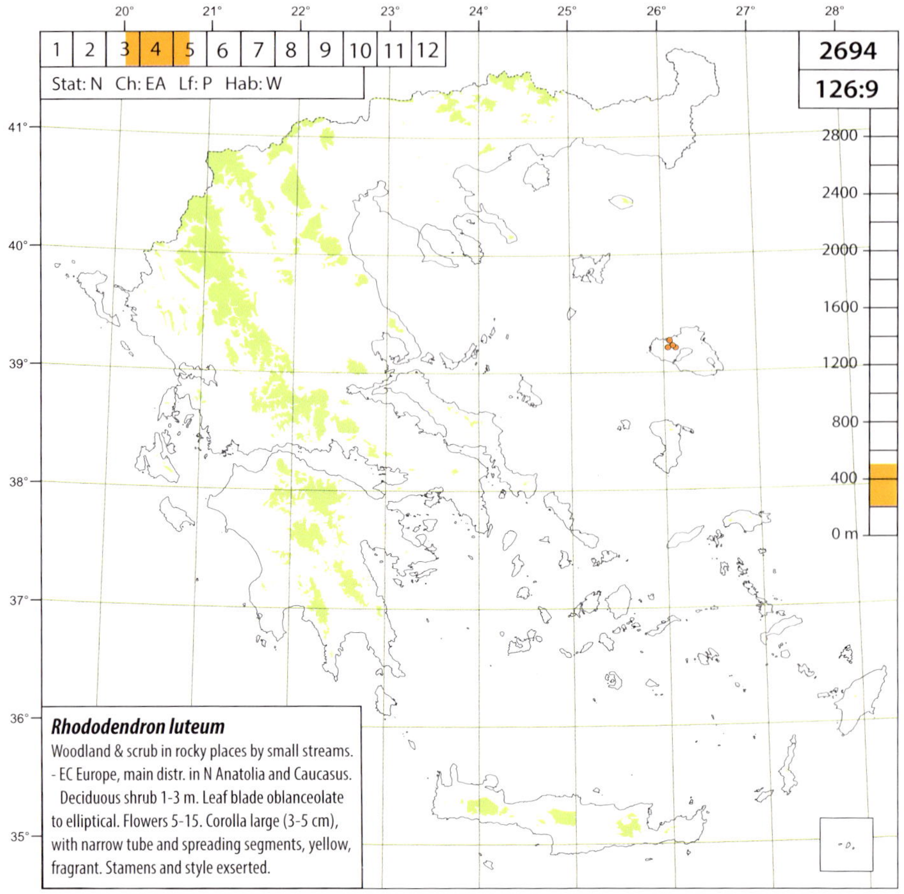

Rhododendron luteum
Woodland & scrub in rocky places by small streams. - EC Europe, main distr. in N Anatolia and Caucasus.
Deciduous shrub 1-3 m. Leaf blade oblanceolate to elliptical. Flowers 5-15. Corolla large (3-5 cm), with narrow tube and spreading segments, yellow, fragrant. Stamens and style exserted.

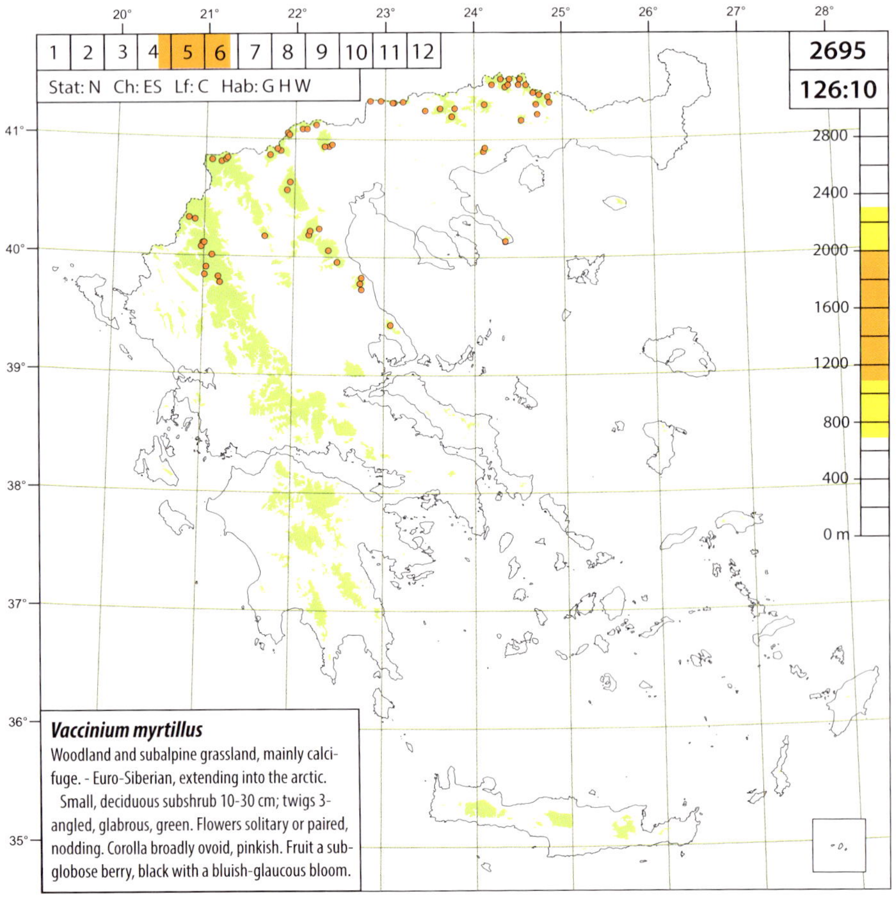

Vaccinium myrtillus
Woodland and subalpine grassland, mainly calcifuge. - Euro-Siberian, extending into the arctic.
Small, deciduous subshrub 10-30 cm; twigs 3-angled, glabrous, green. Flowers solitary or paired, nodding. Corolla broadly ovoid, pinkish. Fruit a subglobose berry, black with a bluish-glaucous bloom.

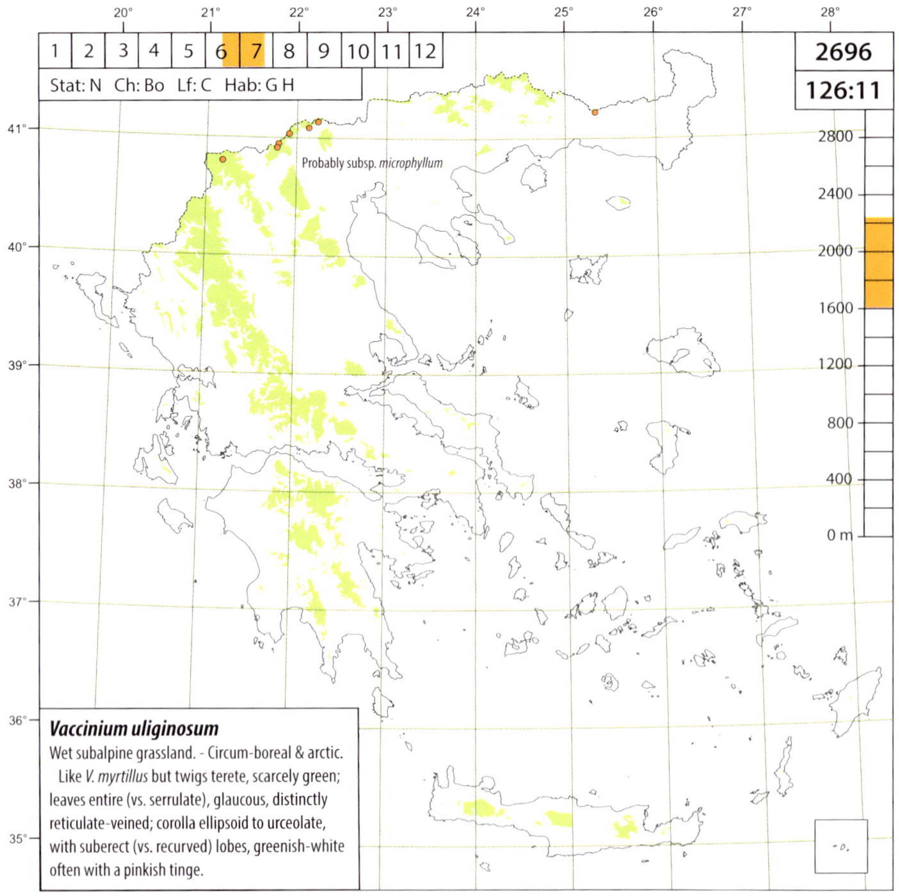

Vaccinium uliginosum
Wet subalpine grassland. - Circum-boreal & arctic.
Like *V. myrtillus* but twigs terete, scarcely green; leaves entire (vs. serrulate), glaucous, distinctly reticulate-veined; corolla ellipsoid to urceolate, with suberect (vs. recurved) lobes, greenish-white often with a pinkish tinge.

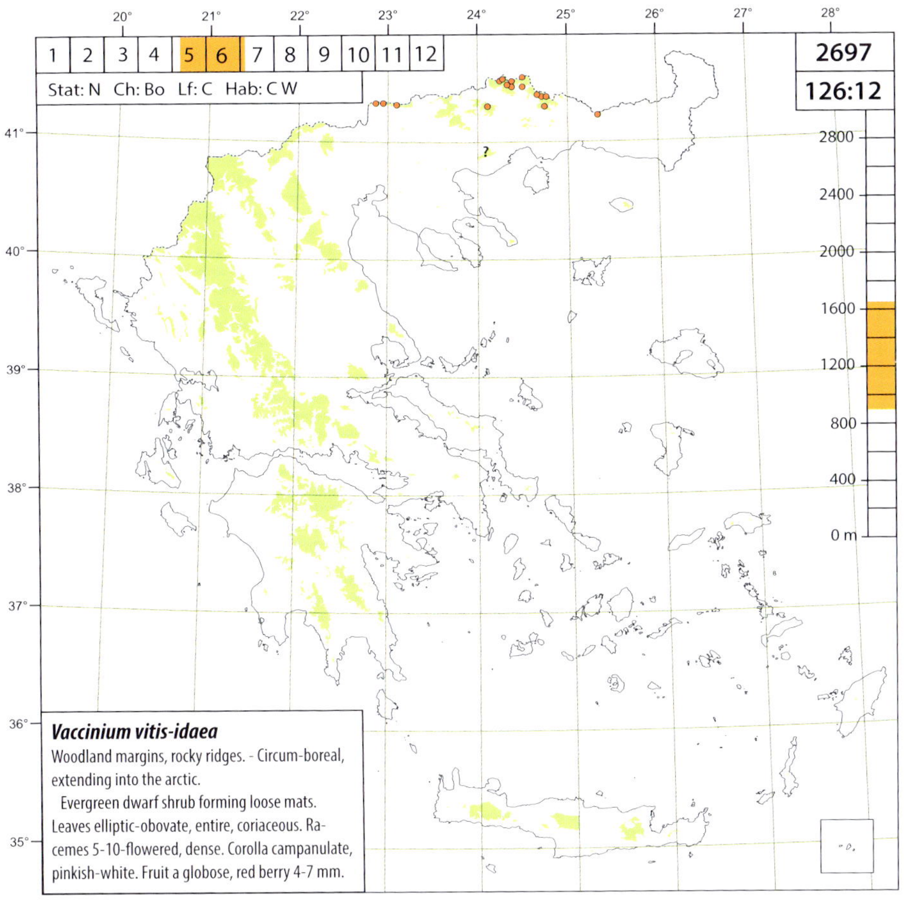

Vaccinium vitis-idaea
Woodland margins, rocky ridges. - Circum-boreal, extending into the arctic.
Evergreen dwarf shrub forming loose mats. Leaves elliptic-obovate, entire, coriaceous. Racemes 5-10-flowered, dense. Corolla campanulate, pinkish-white. Fruit a globose, red berry 4-7 mm.

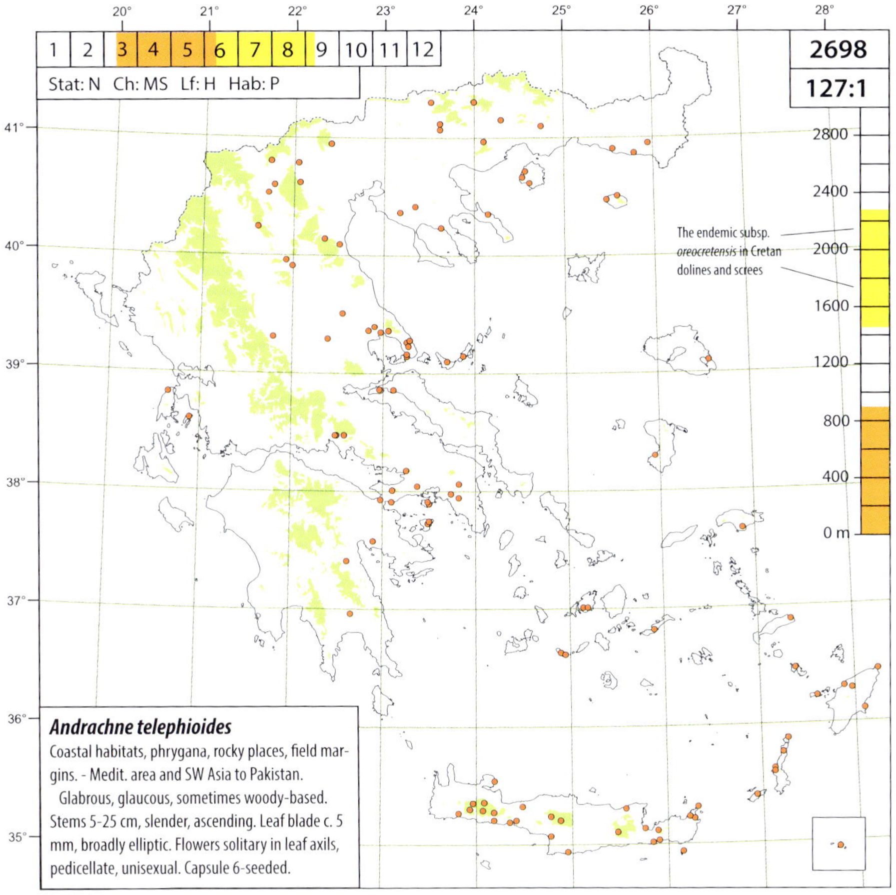

Andrachne telephioides
Coastal habitats, phrygana, rocky places, field margins. - Medit. area and SW Asia to Pakistan.
Glabrous, glaucous, sometimes woody-based. Stems 5-25 cm, slender, ascending. Leaf blade c. 5 mm, broadly elliptic. Flowers solitary in leaf axils, pedicellate, unisexual. Capsule 6-seeded.

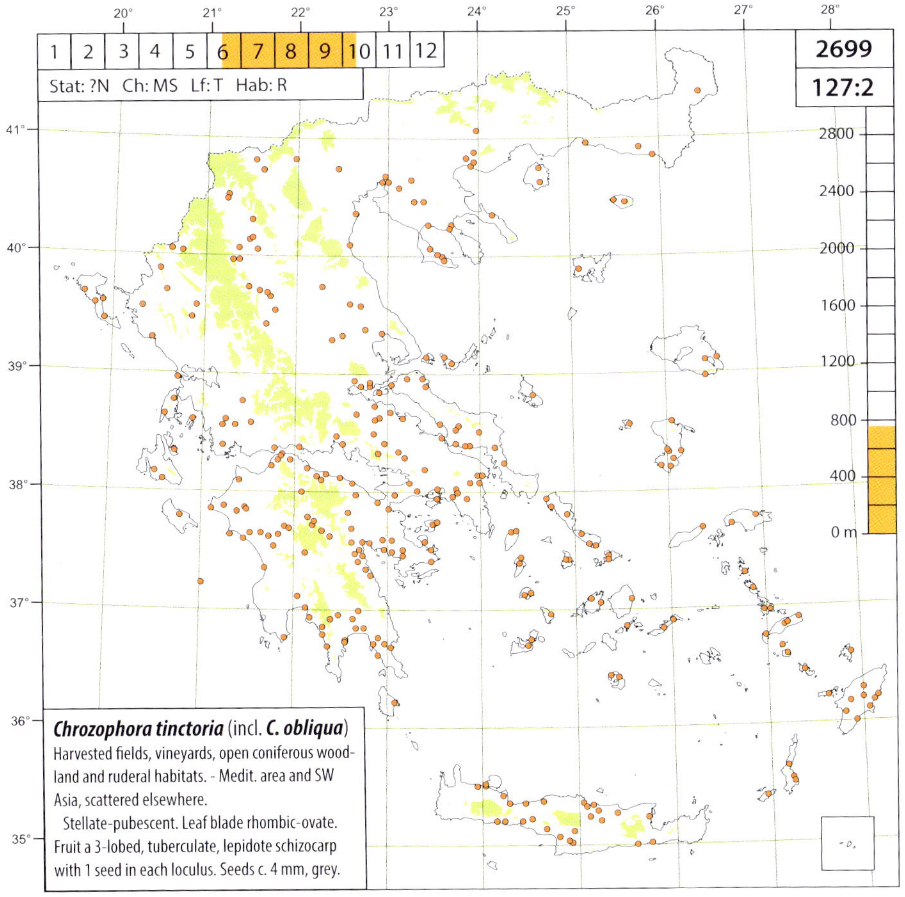

Chrozophora tinctoria (incl. ***C. obliqua***)
Harvested fields, vineyards, open coniferous woodland and ruderal habitats. - Medit. area and SW Asia, scattered elsewhere.
Stellate-pubescent. Leaf blade rhombic-ovate. Fruit a 3-lobed, tuberculate, lepidote schizocarp with 1 seed in each loculus. Seeds c. 4 mm, grey.

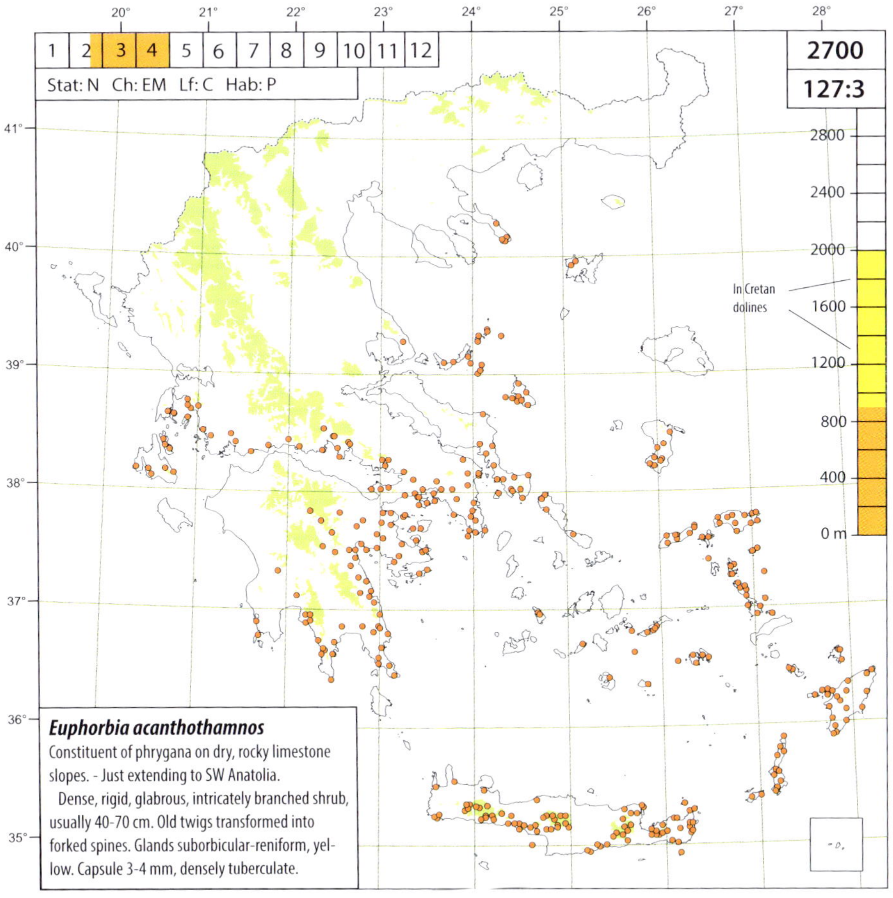

Euphorbia acanthothamnos
Constituent of phrygana on dry, rocky limestone slopes. - Just extending to SW Anatolia.
Dense, rigid, glabrous, intricately branched shrub, usually 40-70 cm. Old twigs transformed into forked spines. Glands suborbicular-reniform, yellow. Capsule 3-4 mm, densely tuberculate.

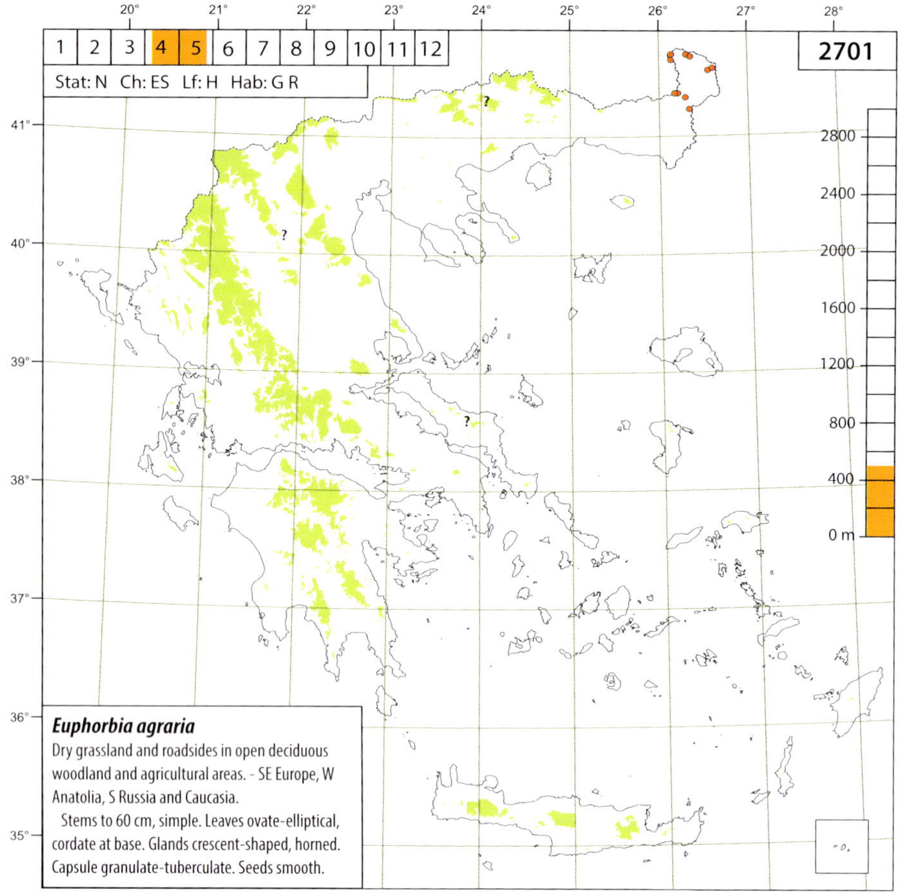

Euphorbia agraria
Dry grassland and roadsides in open deciduous woodland and agricultural areas. - SE Europe, W Anatolia, S Russia and Caucasia.
Stems to 60 cm, simple. Leaves ovate-elliptical, cordate at base. Glands crescent-shaped, horned. Capsule granulate-tuberculate. Seeds smooth.

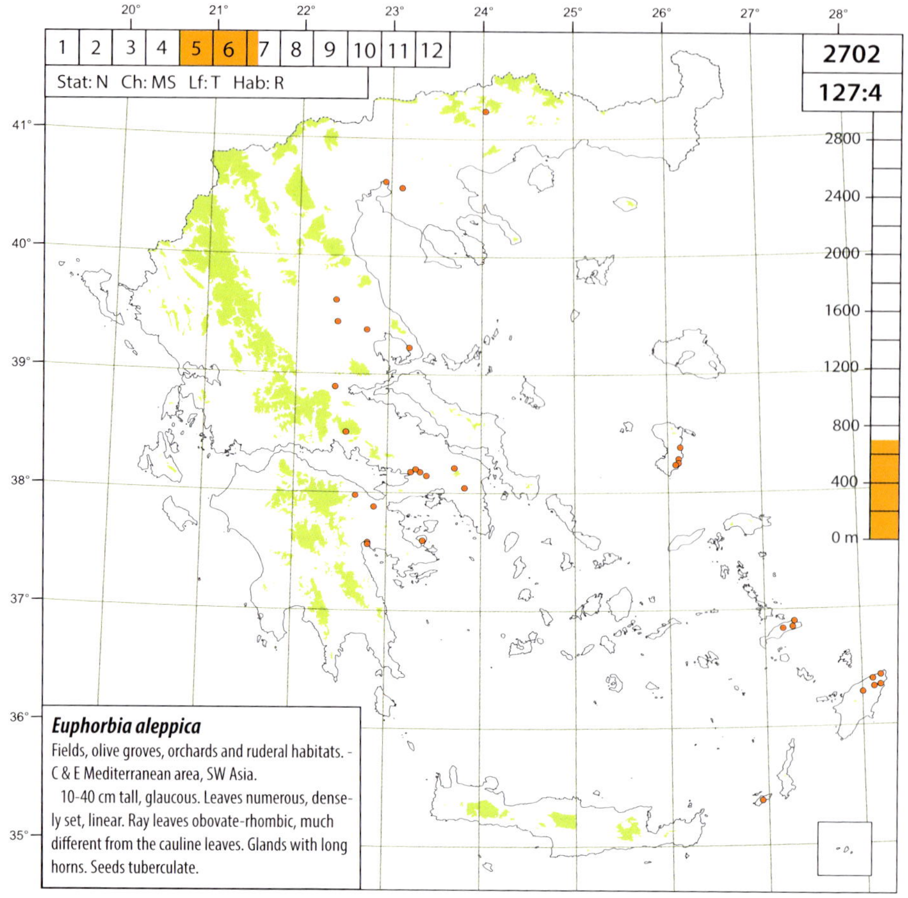

Euphorbia aleppica
Fields, olive groves, orchards and ruderal habitats. - C & E Mediterranean area, SW Asia.
10-40 cm tall, glaucous. Leaves numerous, densely set, linear. Ray leaves obovate-rhombic, much different from the cauline leaves. Glands with long horns. Seeds tuberculate.

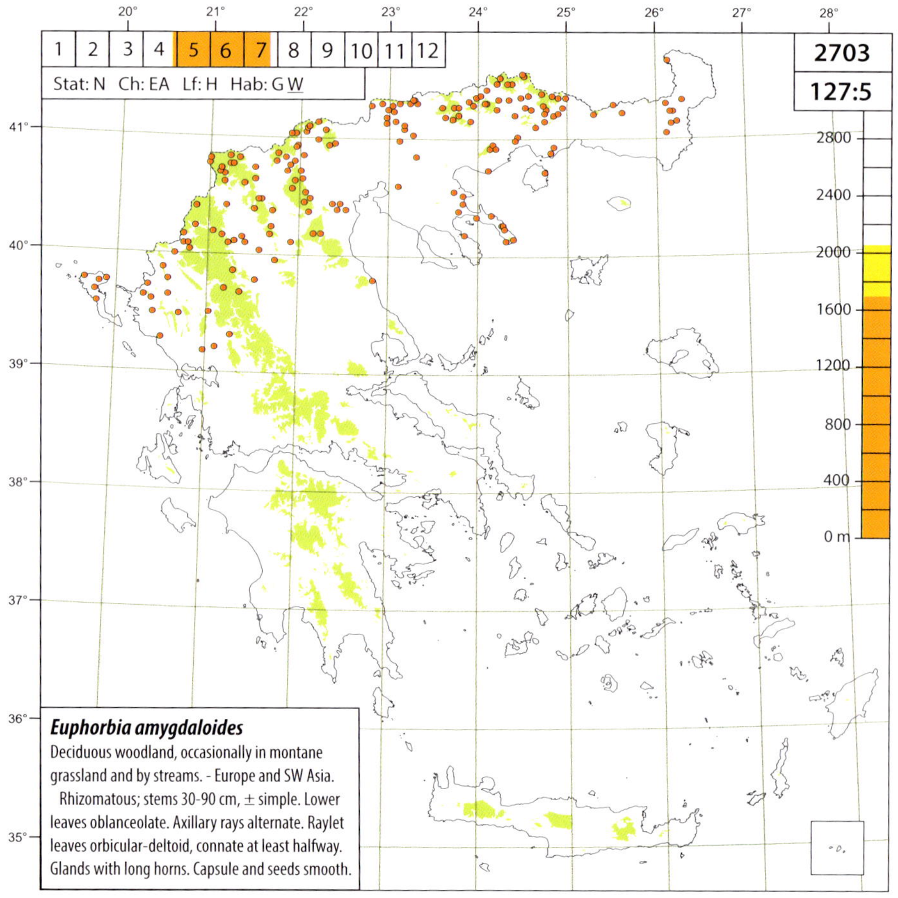

Euphorbia amygdaloides
Deciduous woodland, occasionally in montane grassland and by streams. - Europe and SW Asia.
Rhizomatous; stems 30-90 cm, ± simple. Lower leaves oblanceolate. Axillary rays alternate. Raylet leaves orbicular-deltoid, connate at least halfway. Glands with long horns. Capsule and seeds smooth.

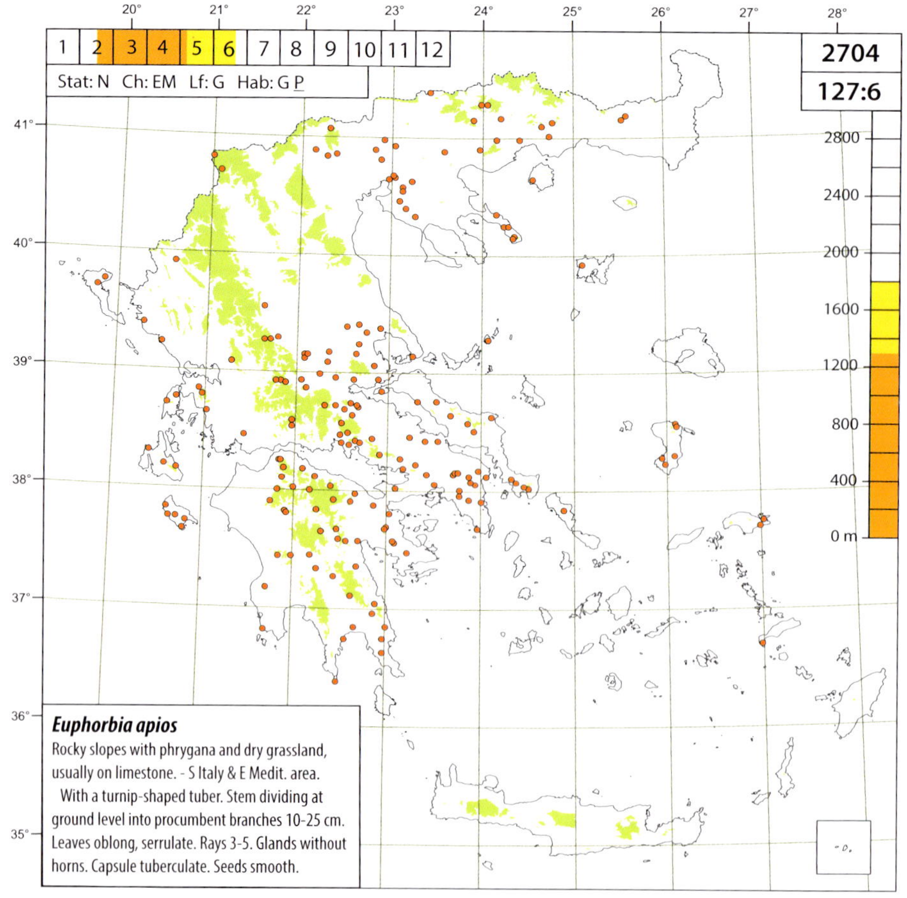

Euphorbia apios
Rocky slopes with phrygana and dry grassland, usually on limestone. - S Italy & E Medit. area.
With a turnip-shaped tuber. Stem dividing at ground level into procumbent branches 10-25 cm. Leaves oblong, serrulate. Rays 3-5. Glands without horns. Capsule tuberculate. Seeds smooth.

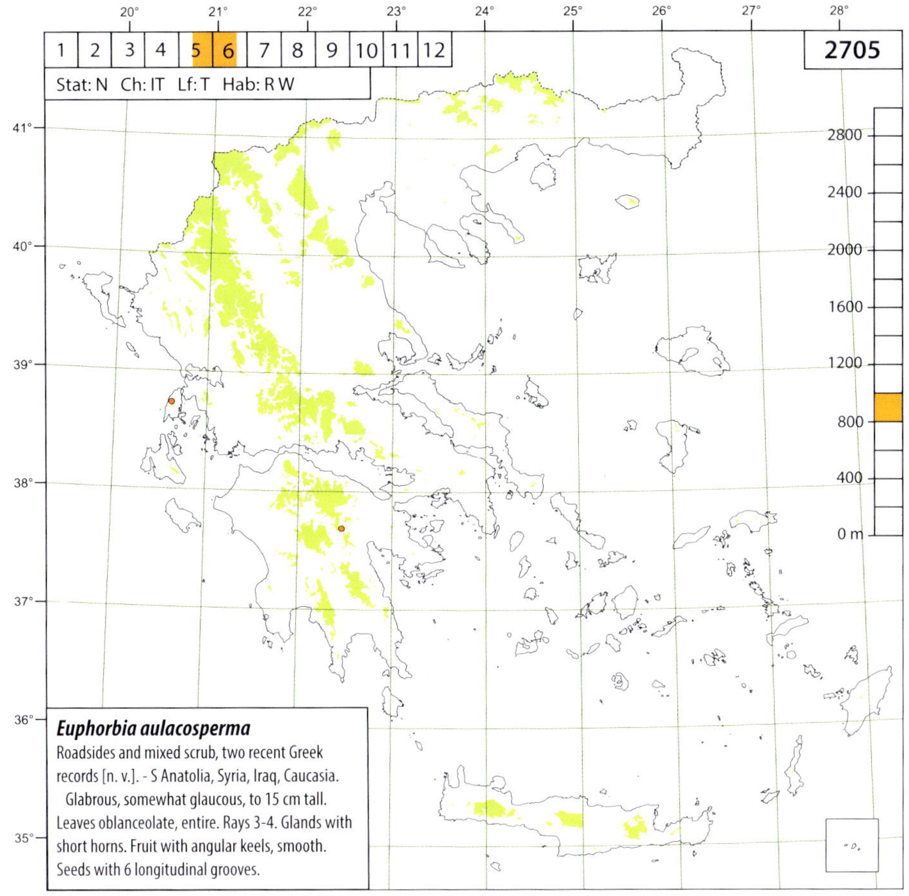

Euphorbia aulacosperma
Roadsides and mixed scrub, two recent Greek records [n. v.]. - S Anatolia, Syria, Iraq, Caucasia.
Glabrous, somewhat glaucous, to 15 cm tall. Leaves oblanceolate, entire. Rays 3-4. Glands with short horns. Fruit with angular keels, smooth. Seeds with 6 longitudinal grooves.

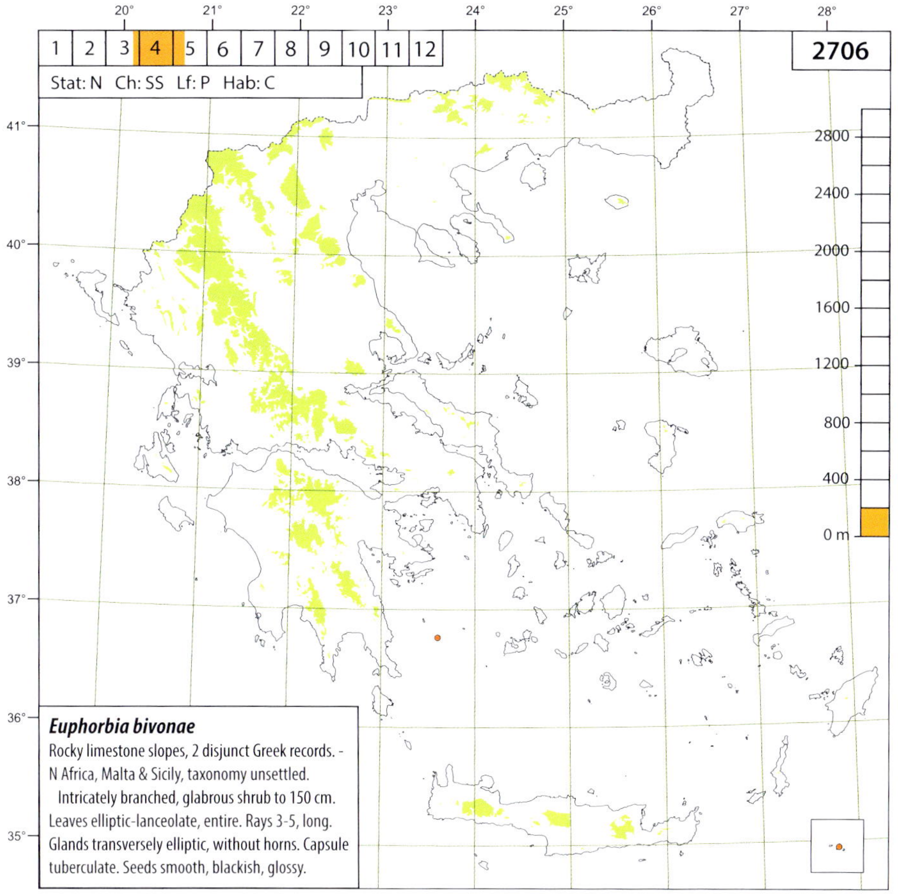

Euphorbia bivonae
Rocky limestone slopes, 2 disjunct Greek records. - N Africa, Malta & Sicily, taxonomy unsettled.
Intricately branched, glabrous shrub to 150 cm. Leaves elliptic-lanceolate, entire. Rays 3-5, long. Glands transversely elliptic, without horns. Capsule tuberculate. Seeds smooth, blackish, glossy.

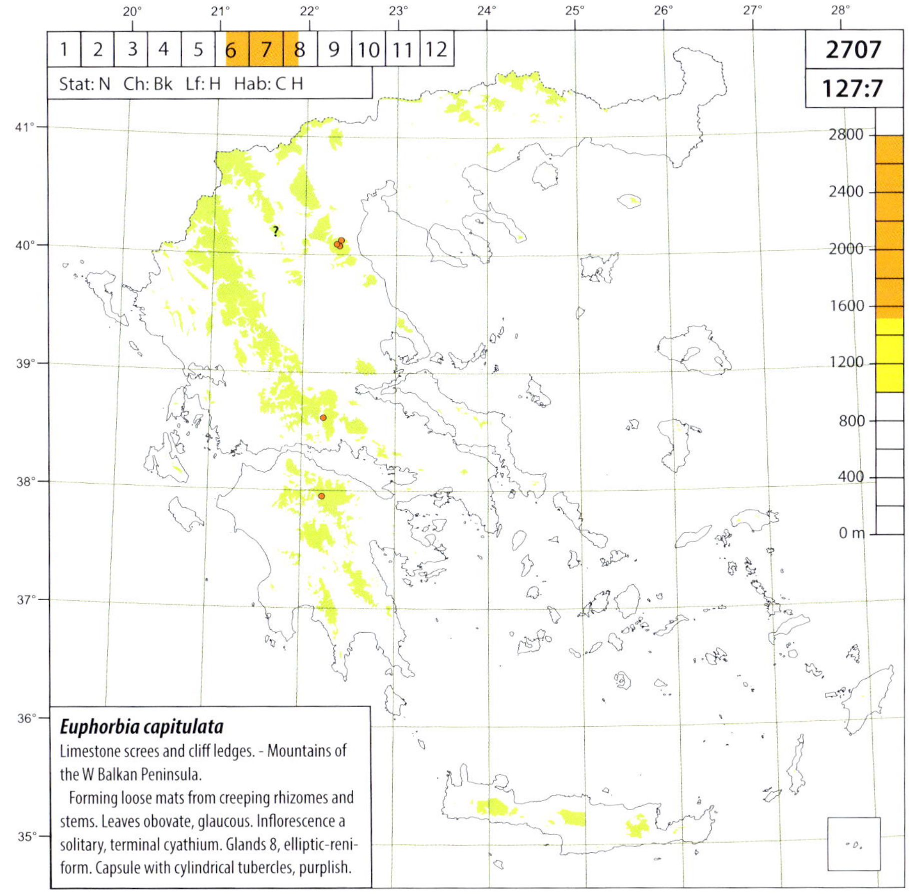

Euphorbia capitulata
Limestone screes and cliff ledges. - Mountains of the W Balkan Peninsula.
Forming loose mats from creeping rhizomes and stems. Leaves obovate, glaucous. Inflorescence a solitary, terminal cyathium. Glands 8, elliptic-reniform. Capsule with cylindrical tubercles, purplish.

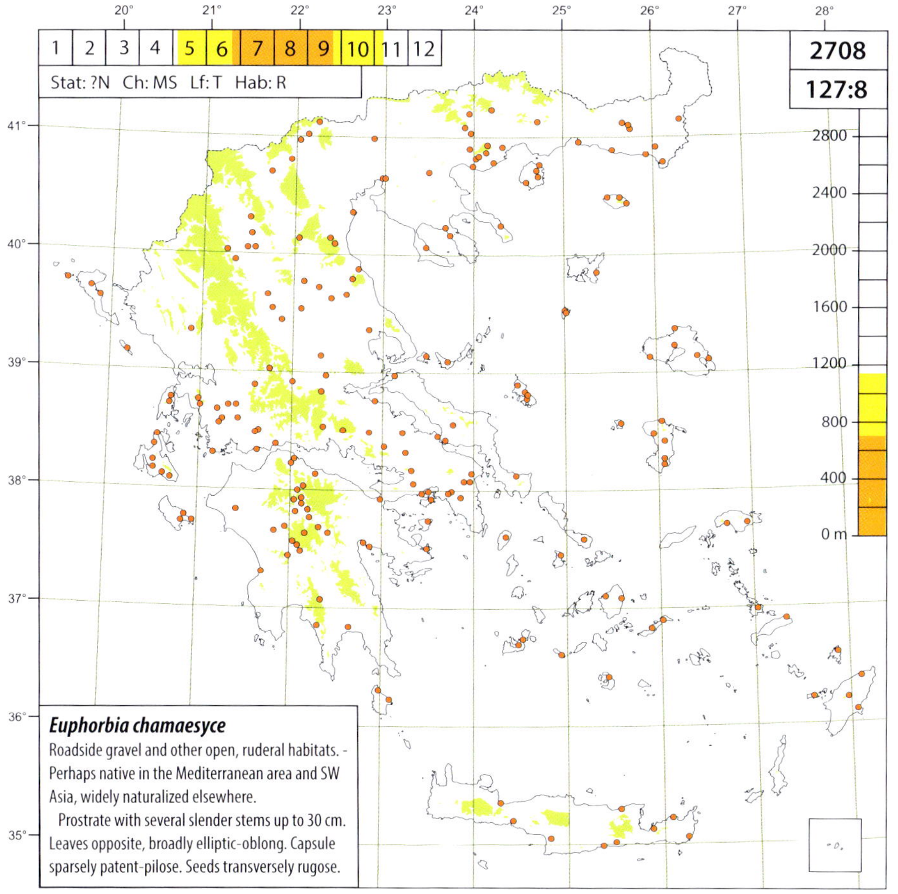

Euphorbia chamaesyce
Roadside gravel and other open, ruderal habitats. - Perhaps native in the Mediterranean area and SW Asia, widely naturalized elsewhere.
Prostrate with several slender stems up to 30 cm. Leaves opposite, broadly elliptic-oblong. Capsule sparsely patent-pilose. Seeds transversely rugose.

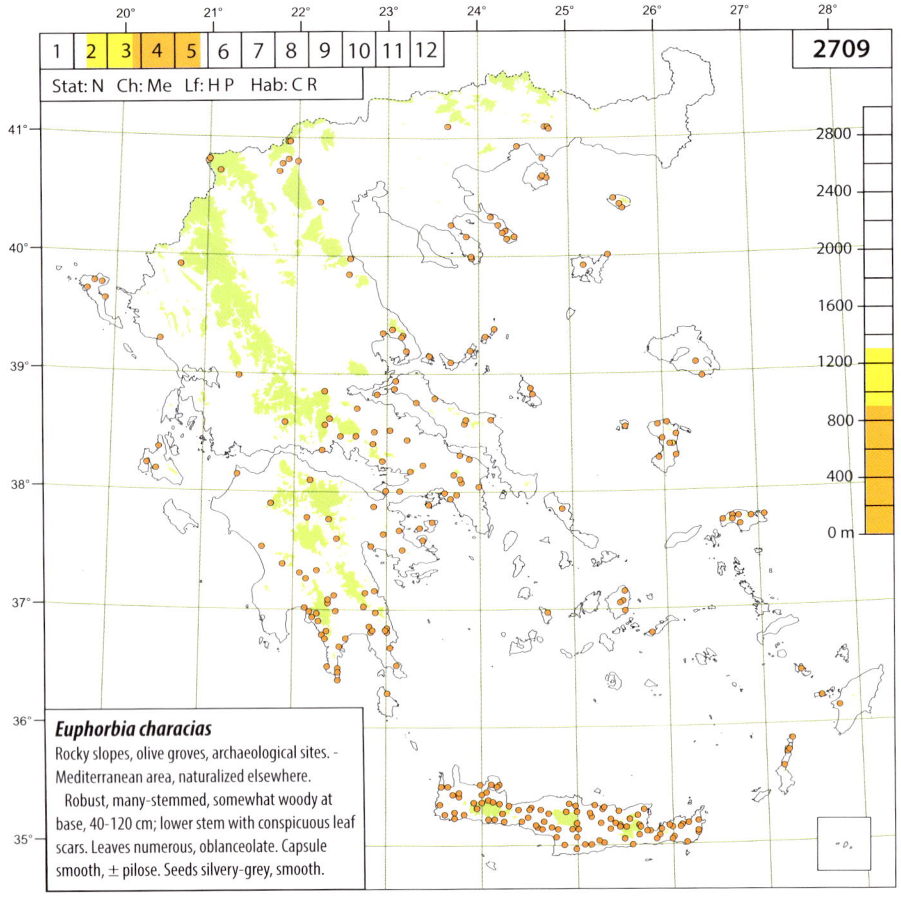
1 2 3 4 5 6 7 8 9 10 11 12
2709
Stat: N Ch: Me Lf: H P Hab: C R
Euphorbia characias
Rocky slopes, olive groves, archaeological sites. - Mediterranean area, naturalized elsewhere.
Robust, many-stemmed, somewhat woody at base, 40-120 cm; lower stem with conspicuous leaf scars. Leaves numerous, oblanceolate. Capsule smooth, ± pilose. Seeds silvery-grey, smooth.

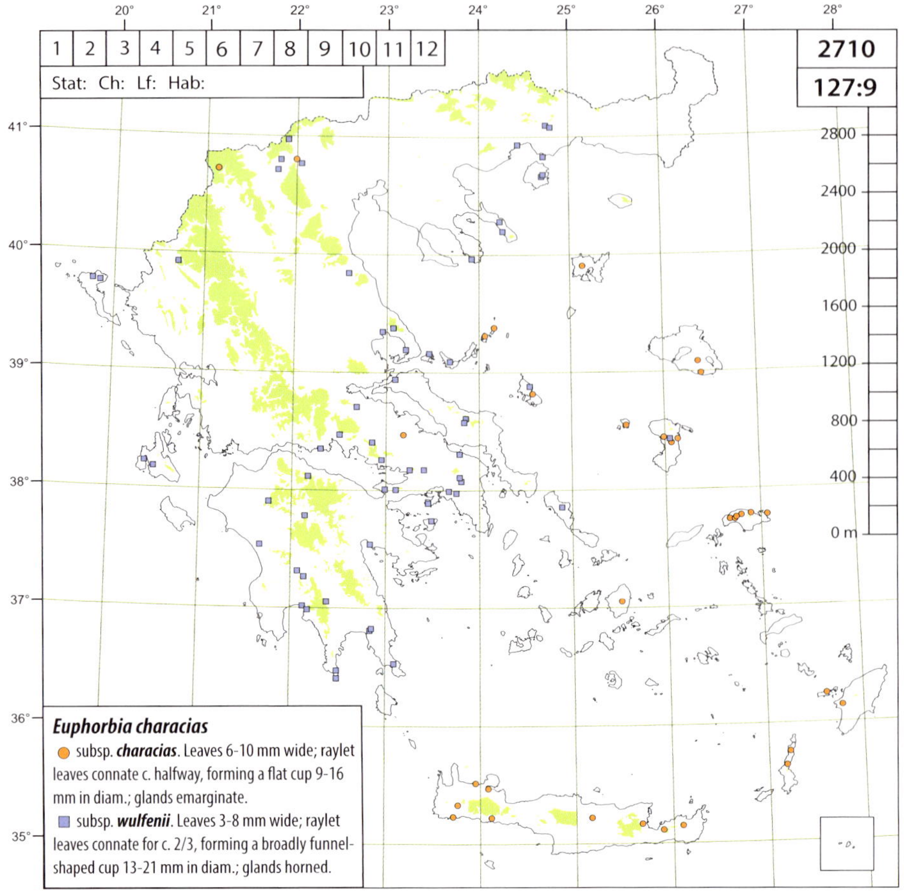
1 2 3 4 5 6 7 8 9 10 11 12
2710
127:9
Stat: Ch: Lf: Hab:
Euphorbia characias
subsp. characias. Leaves 6-10 mm wide; raylet leaves connate c. halfway, forming a flat cup 9-16 mm in diam.; glands emarginate.
subsp. wulfenii. Leaves 3-8 mm wide; raylet leaves connate for c. 2/3, forming a broadly funnel-shaped cup 13-21 mm in diam.; glands horned.

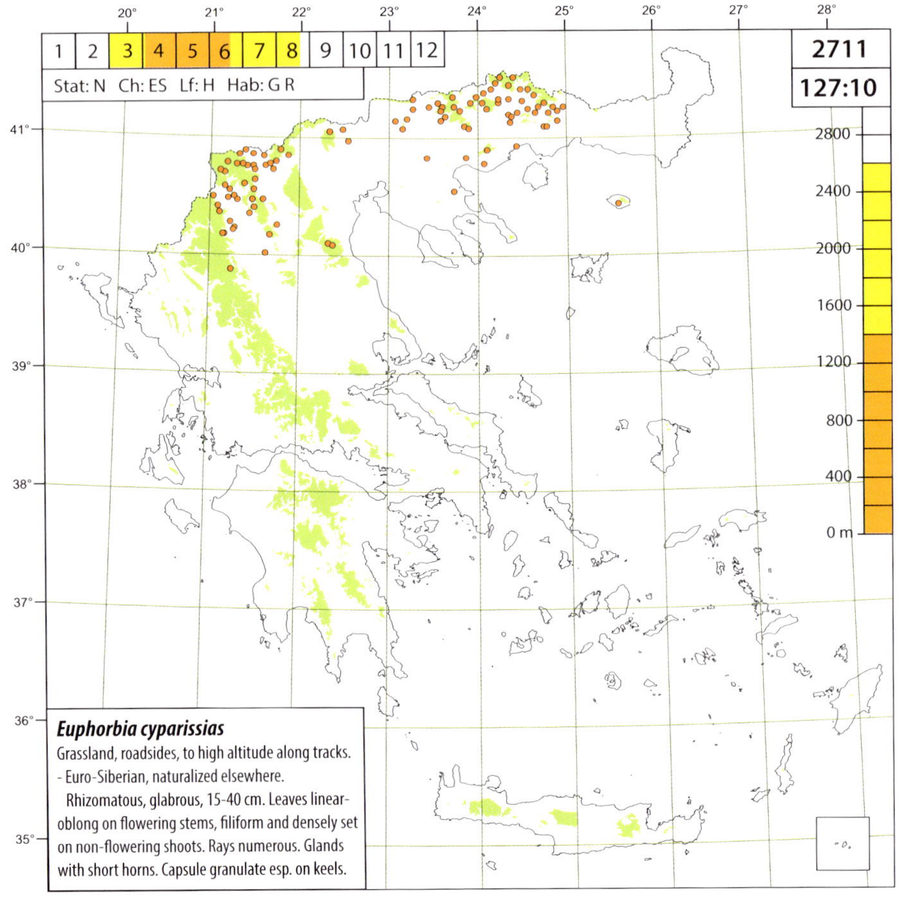
1 2 3 4 5 6 7 8 9 10 11 12
2711
127:10
Stat: N Ch: ES Lf: H Hab: G R
Euphorbia cyparissias
Grassland, roadsides, to high altitude along tracks. - Euro-Siberian, naturalized elsewhere.
Rhizomatous, glabrous, 15-40 cm. Leaves linear-oblong on flowering stems, filiform and densely set on non-flowering shoots. Rays numerous. Glands with short horns. Capsule granulate esp. on keels.

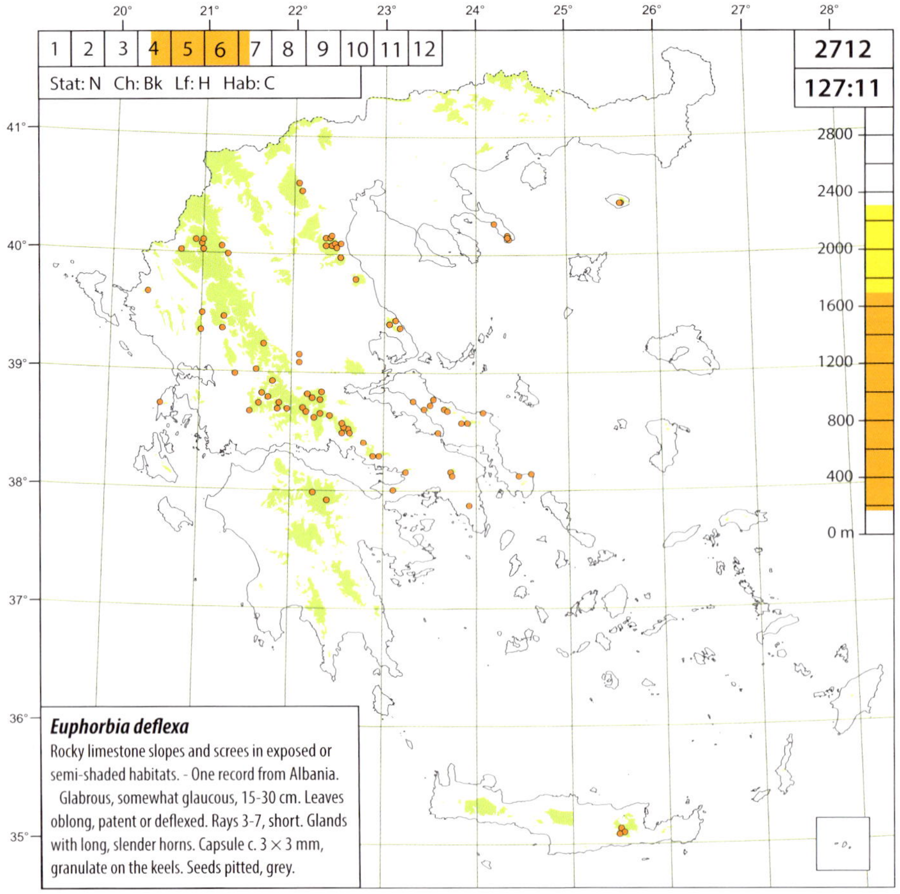
1 2 3 4 5 6 7 8 9 10 11 12
2712
127:11
Stat: N Ch: Bk Lf: H Hab: C
Euphorbia deflexa
Rocky limestone slopes and screes in exposed or semi-shaded habitats. - One record from Albania.
Glabrous, somewhat glaucous, 15-30 cm. Leaves oblong, patent or deflexed. Rays 3-7, short. Glands with long, slender horns. Capsule c. 3 × 3 mm, granulate on the keels. Seeds pitted, grey.

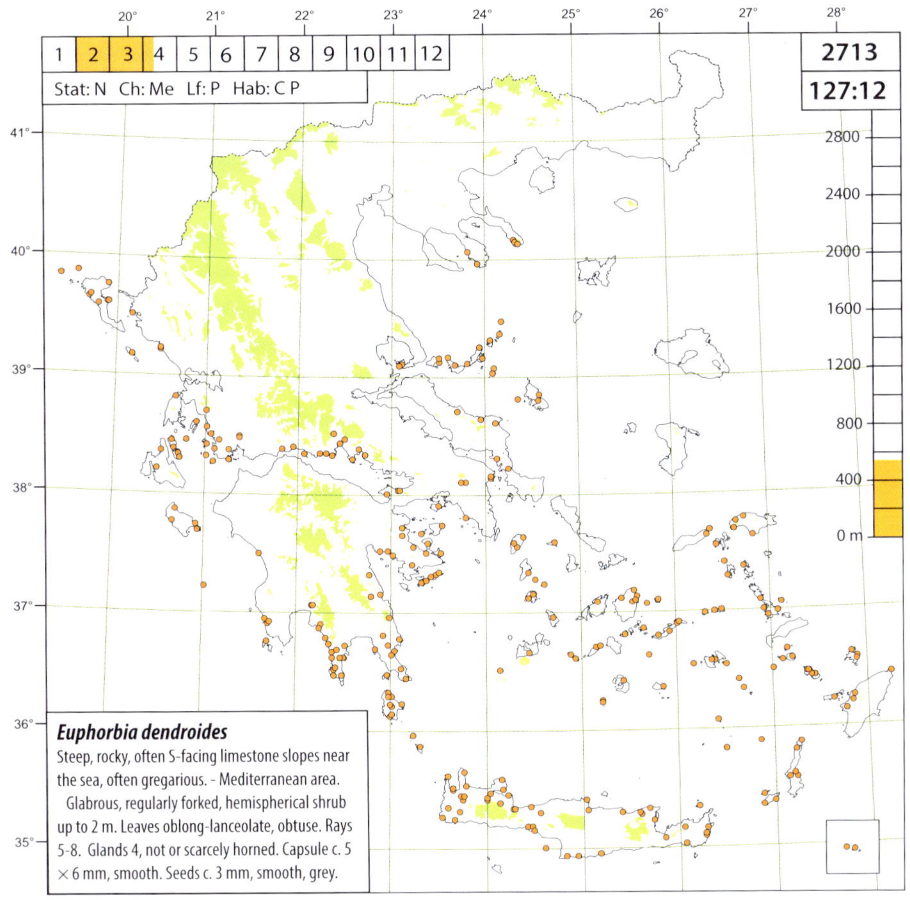

Euphorbia dendroides

Steep, rocky, often S-facing limestone slopes near the sea, often gregarious. - Mediterranean area.

Glabrous, regularly forked, hemispherical shrub up to 2 m. Leaves oblong-lanceolate, obtuse. Rays 5-8. Glands 4, not or scarcely horned. Capsule c. 5 × 6 mm, smooth. Seeds c. 3 mm, smooth, grey.

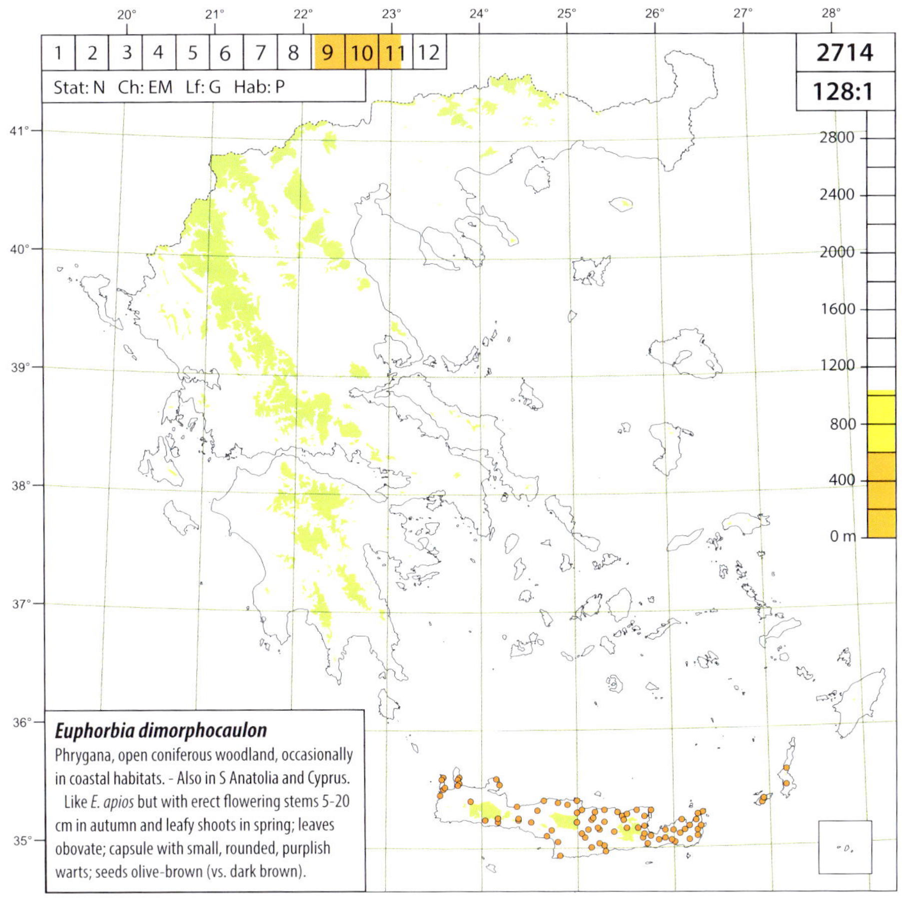

Euphorbia dimorphocaulon

Phrygana, open coniferous woodland, occasionally in coastal habitats. - Also in S Anatolia and Cyprus.

Like *E. apios* but with erect flowering stems 5-20 cm in autumn and leafy shoots in spring; leaves obovate; capsule with small, rounded, purplish warts; seeds olive-brown (vs. dark brown).

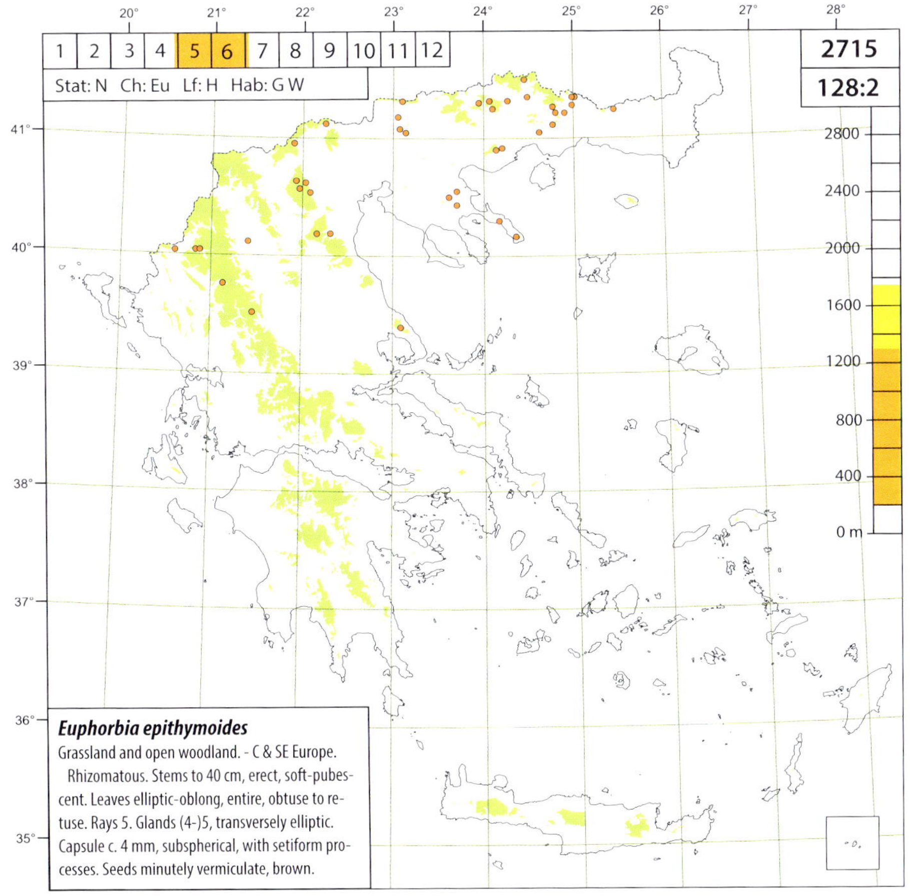

Euphorbia epithymoides

Grassland and open woodland. - C & SE Europe.

Rhizomatous. Stems to 40 cm, erect, soft-pubescent. Leaves elliptic-oblong, entire, obtuse to retuse. Rays 5. Glands (4-)5, transversely elliptic. Capsule c. 4 mm, subspherical, with setiform processes. Seeds minutely vermiculate, brown.

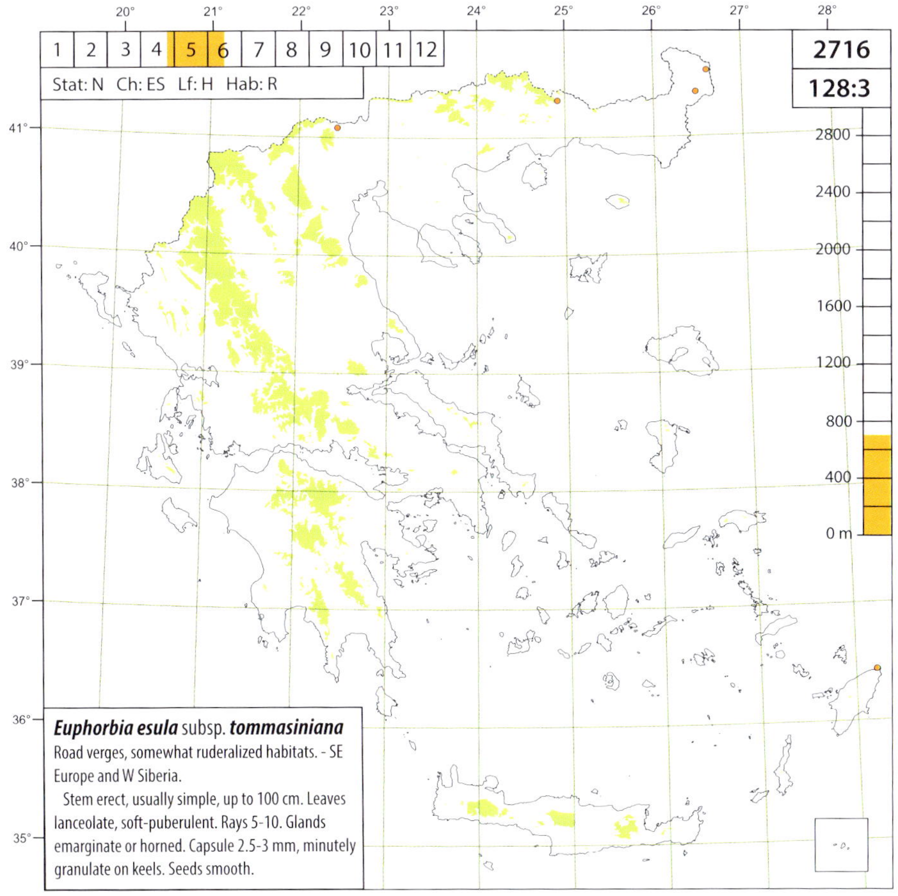

Euphorbia esula subsp. *tommasiniana*

Road verges, somewhat ruderalized habitats. - SE Europe and W Siberia.

Stem erect, usually simple, up to 100 cm. Leaves lanceolate, soft-puberulent. Rays 5-10. Glands emarginate or horned. Capsule 2.5-3 mm, minutely granulate on keels. Seeds smooth.

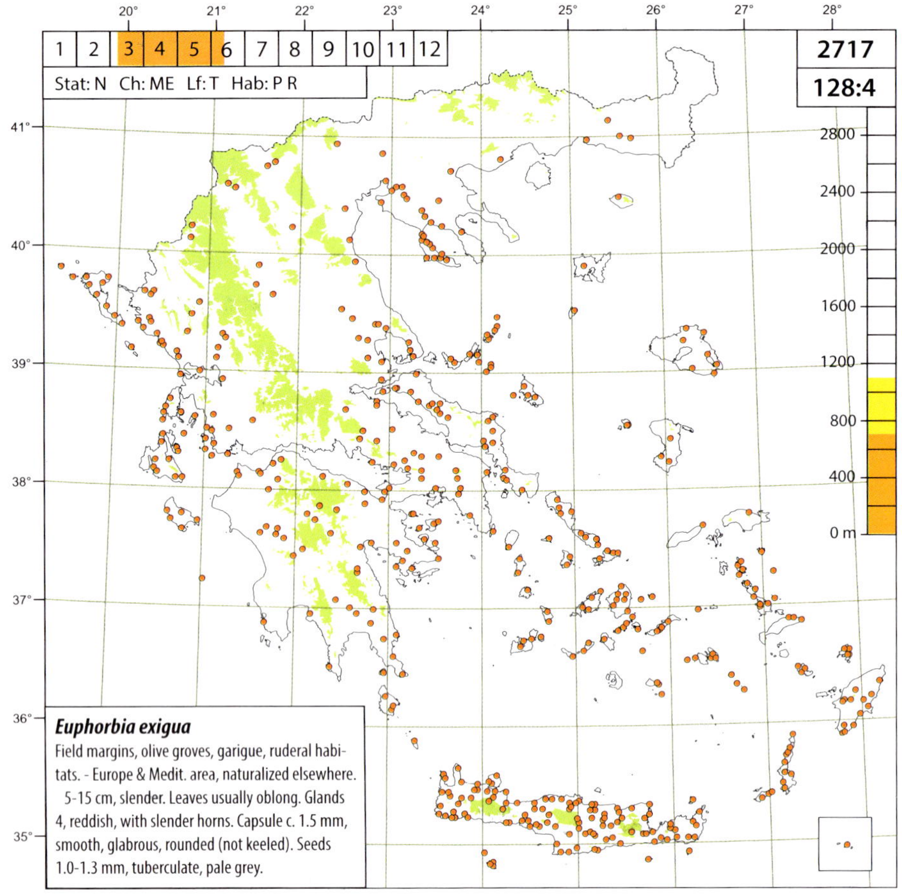

Euphorbia exigua
Field margins, olive groves, garigue, ruderal habitats. - Europe & Medit. area, naturalized elsewhere.
5-15 cm, slender. Leaves usually oblong. Glands 4, reddish, with slender horns. Capsule c. 1.5 mm, smooth, glabrous, rounded (not keeled). Seeds 1.0-1.3 mm, tuberculate, pale grey.

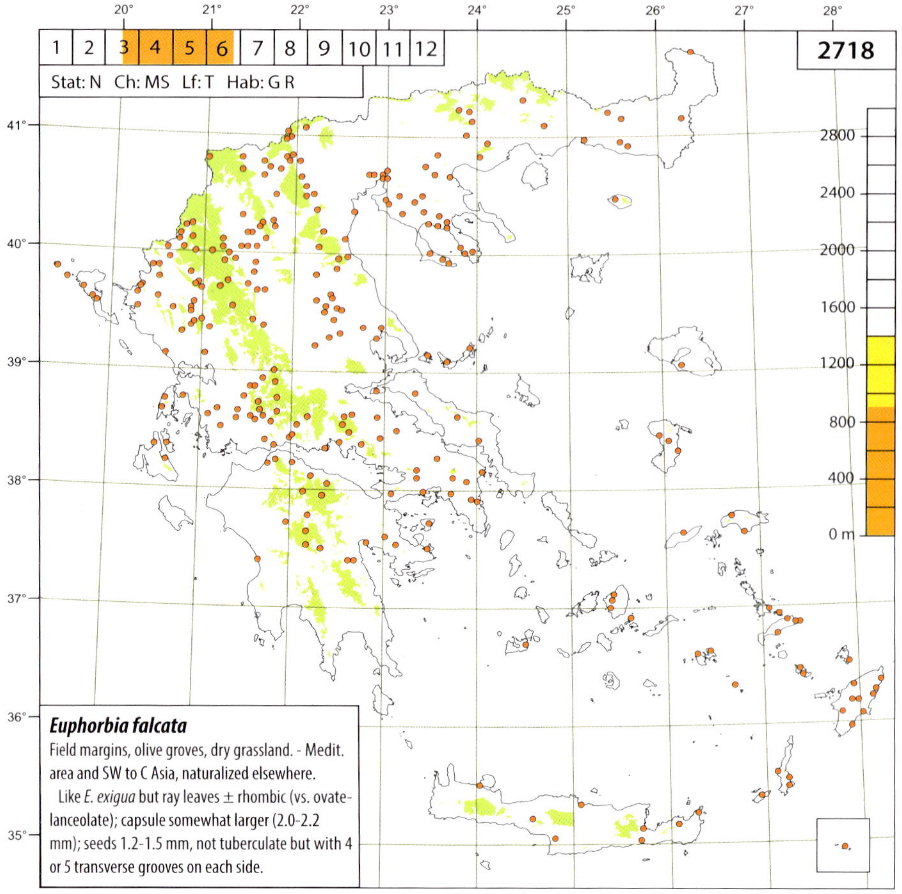

Euphorbia falcata
Field margins, olive groves, dry grassland. - Medit. area and SW to C Asia, naturalized elsewhere.
Like *E. exigua* but ray leaves ± rhombic (vs. ovate-lanceolate); capsule somewhat larger (2.0-2.2 mm); seeds 1.2-1.5 mm, not tuberculate but with 4 or 5 transverse grooves on each side.

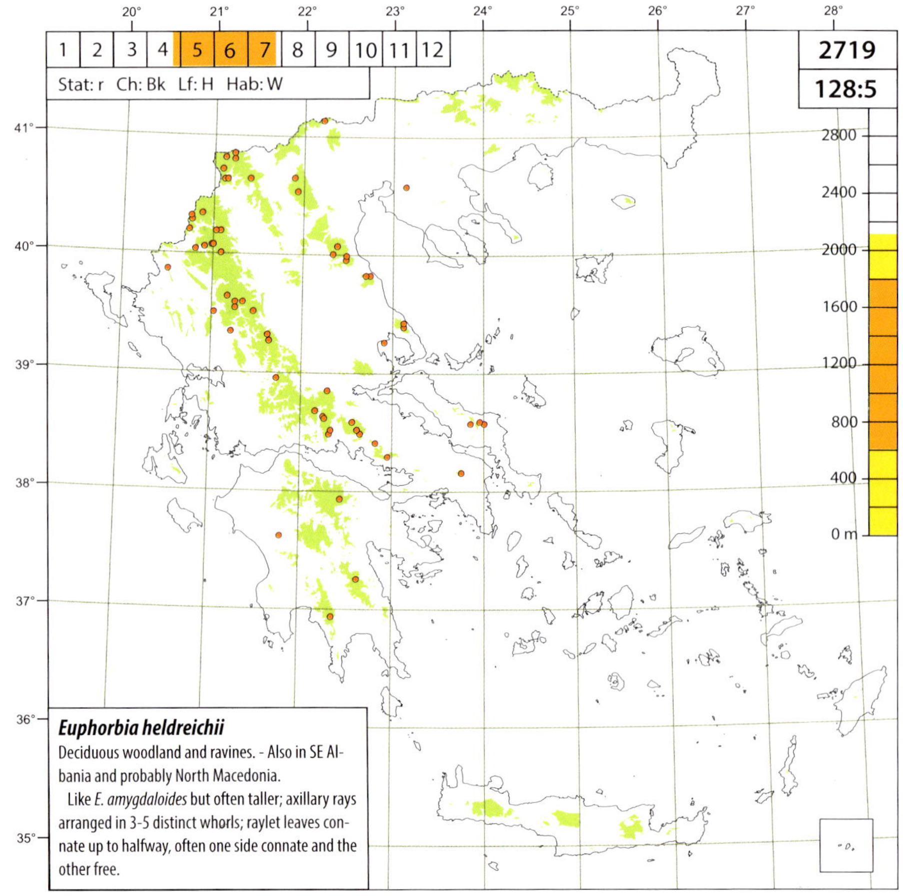

Euphorbia heldreichii
Deciduous woodland and ravines. - Also in SE Albania and probably North Macedonia.
Like *E. amygdaloides* but often taller; axillary rays arranged in 3-5 distinct whorls; raylet leaves connate up to halfway, often one side connate and the other free.

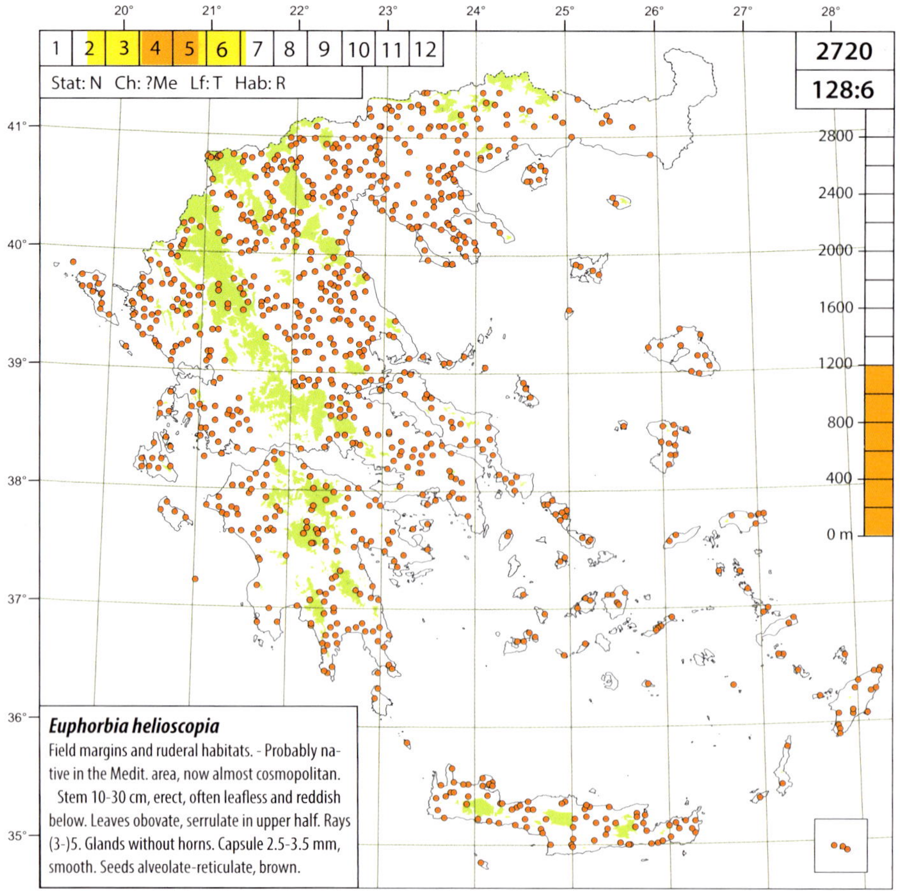

Euphorbia helioscopia
Field margins and ruderal habitats. - Probably native in the Medit. area, now almost cosmopolitan.
Stem 10-30 cm, erect, often leafless and reddish below. Leaves obovate, serrulate in upper half. Rays (3-)5. Glands without horns. Capsule 2.5-3.5 mm, smooth. Seeds alveolate-reticulate, brown.

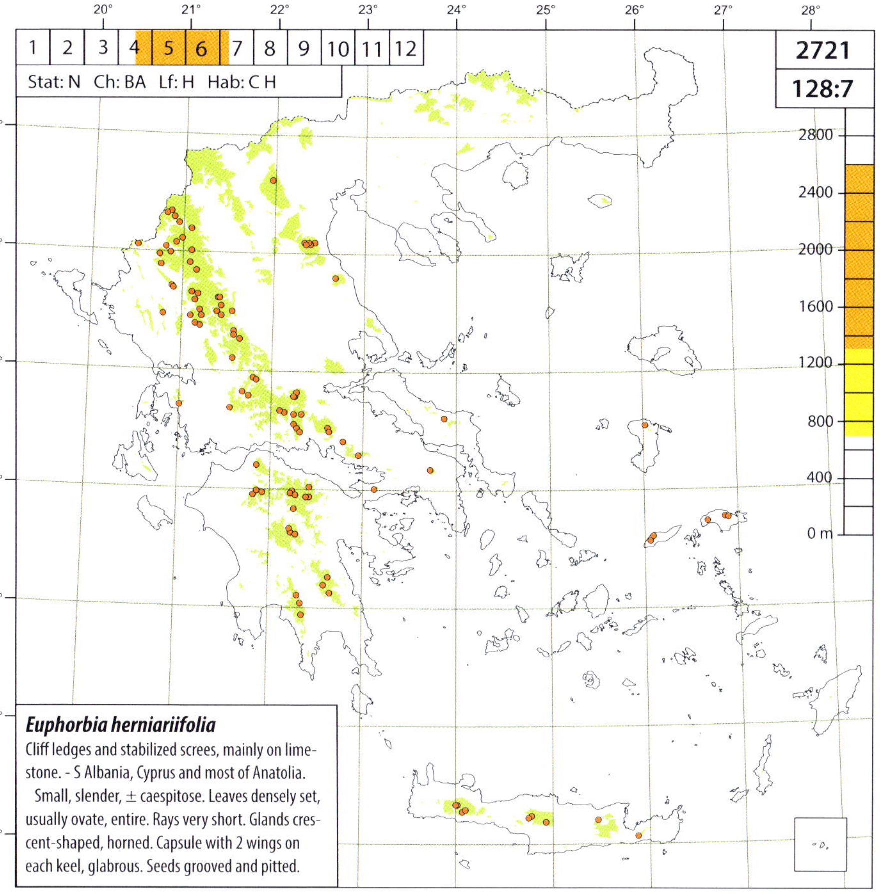

Euphorbia herniariifolia

Cliff ledges and stabilized screes, mainly on limestone. - S Albania, Cyprus and most of Anatolia.

Small, slender, ± caespitose. Leaves densely set, usually ovate, entire. Rays very short. Glands crescent-shaped, horned. Capsule with 2 wings on each keel, glabrous. Seeds grooved and pitted.

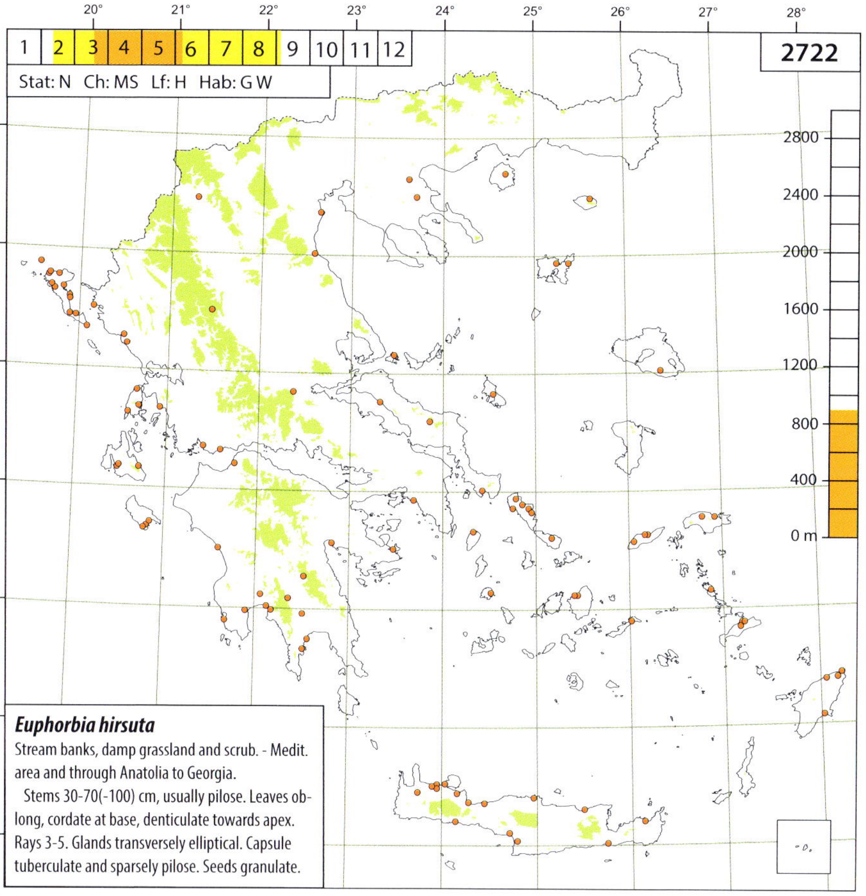

Euphorbia hirsuta

Stream banks, damp grassland and scrub. - Medit. area and through Anatolia to Georgia.

Stems 30-70(-100) cm, usually pilose. Leaves oblong, cordate at base, denticulate towards apex. Rays 3-5. Glands transversely elliptical. Capsule tuberculate and sparsely pilose. Seeds granulate.

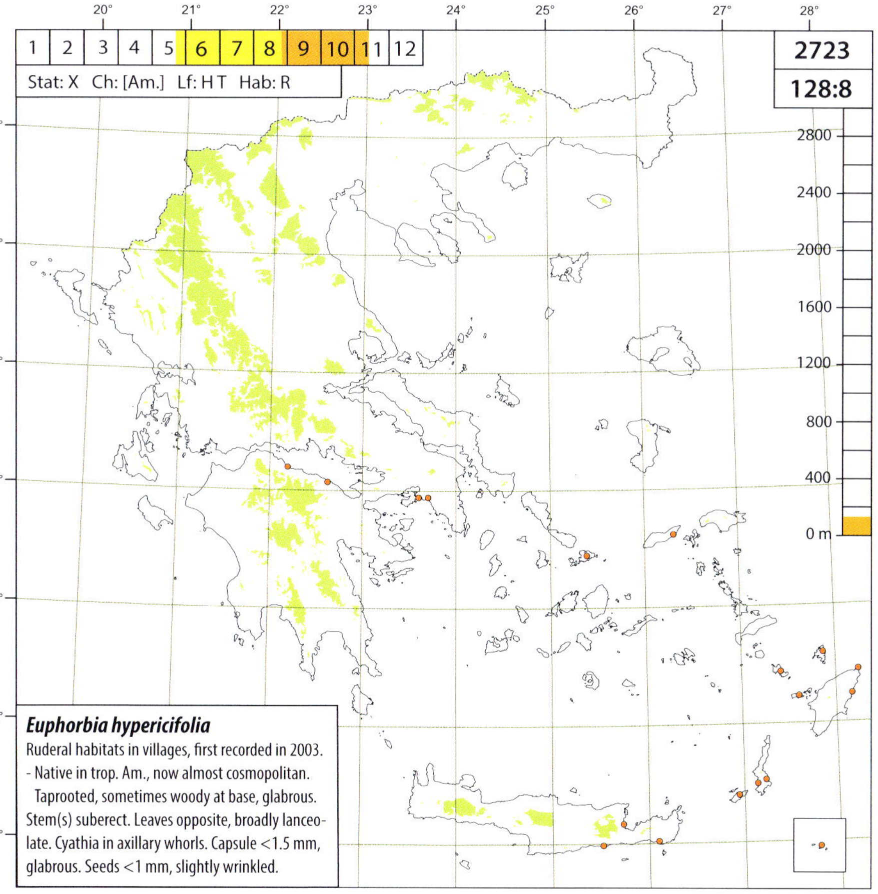

Euphorbia hypericifolia

Ruderal habitats in villages, first recorded in 2003. - Native in trop. Am., now almost cosmopolitan.

Taprooted, sometimes woody at base, glabrous. Stem(s) suberect. Leaves opposite, broadly lanceolate. Cyathia in axillary whorls. Capsule <1.5 mm, glabrous. Seeds <1 mm, slightly wrinkled.

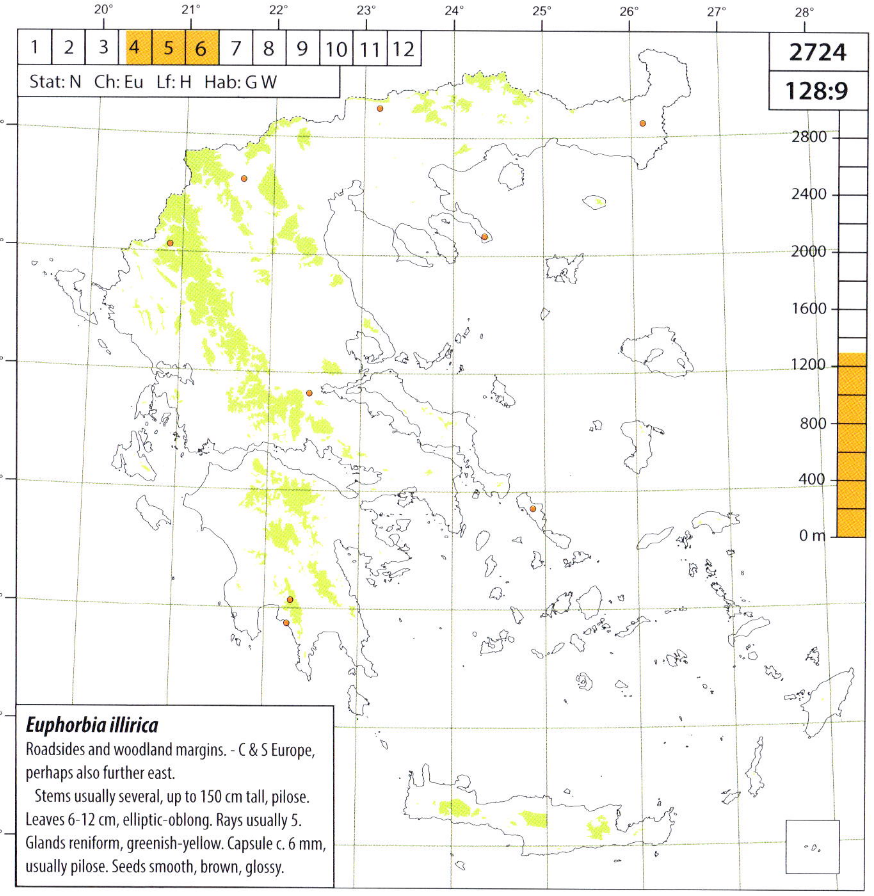

Euphorbia illirica

Roadsides and woodland margins. - C & S Europe, perhaps also further east.

Stems usually several, up to 150 cm tall, pilose. Leaves 6-12 cm, elliptic-oblong. Rays usually 5. Glands reniform, greenish-yellow. Capsule c. 6 mm, usually pilose. Seeds smooth, brown, glossy.

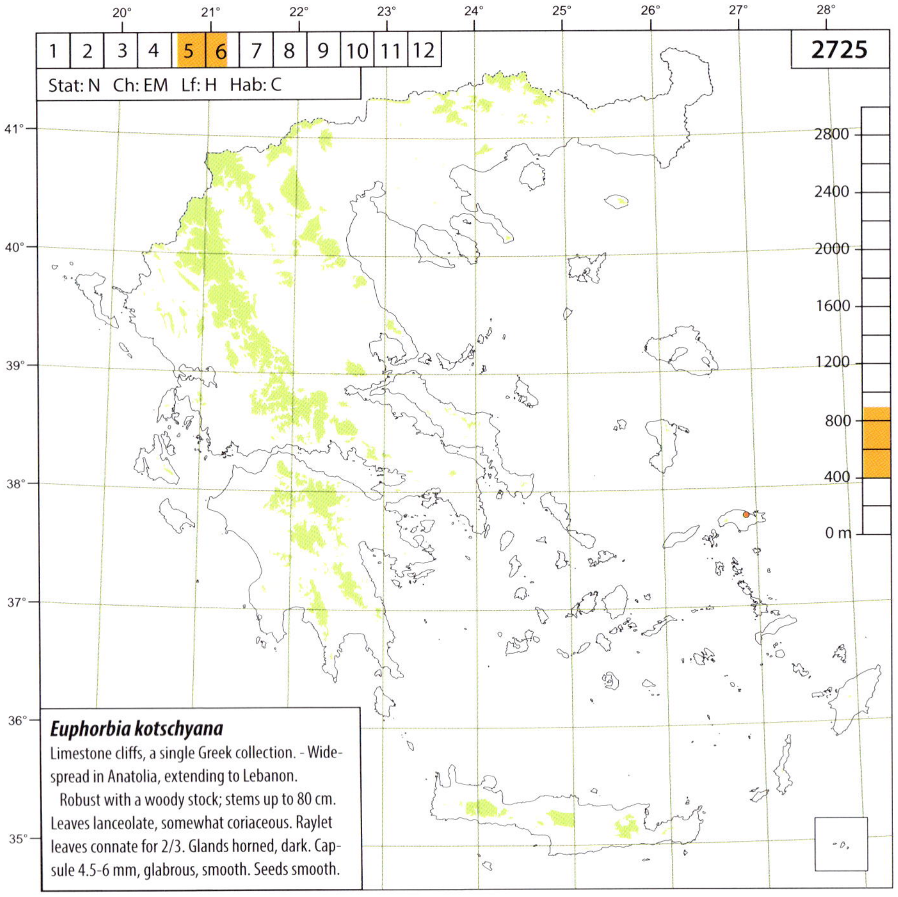

Euphorbia kotschyana
Limestone cliffs, a single Greek collection. - Widespread in Anatolia, extending to Lebanon.
Robust with a woody stock; stems up to 80 cm. Leaves lanceolate, somewhat coriaceous. Raylet leaves connate for 2/3. Glands horned, dark. Capsule 4.5-6 mm, glabrous, smooth. Seeds smooth.

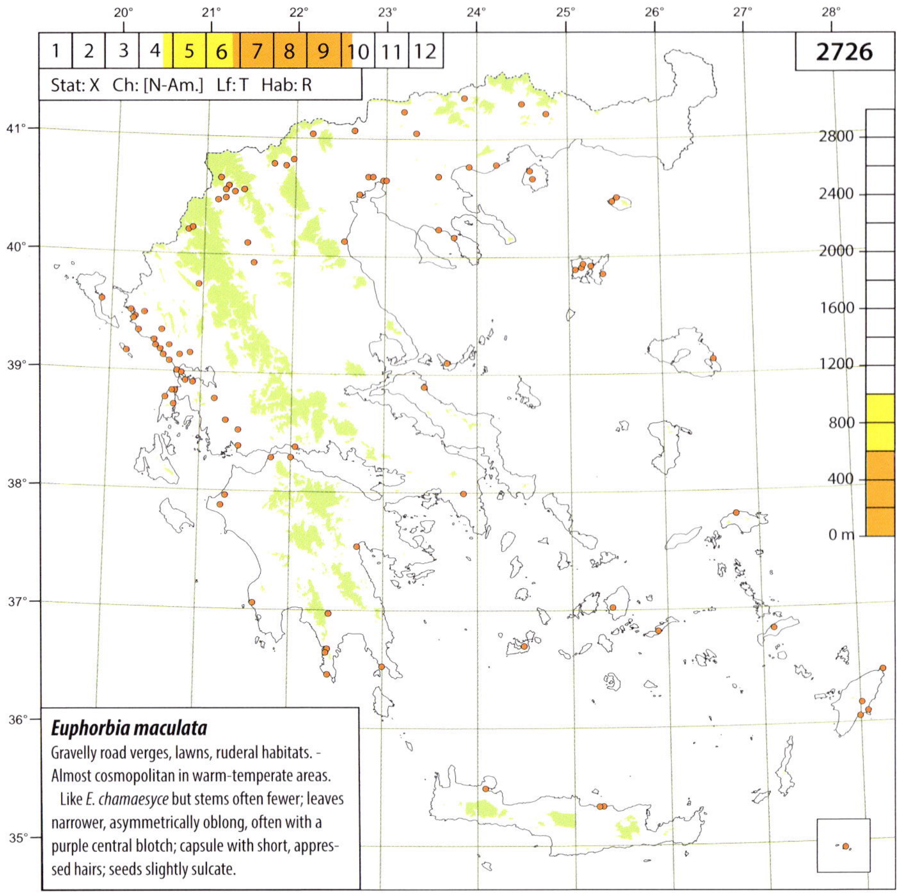

Euphorbia maculata
Gravelly road verges, lawns, ruderal habitats. - Almost cosmopolitan in warm-temperate areas.
Like *E. chamaesyce* but stems often fewer; leaves narrower, asymmetrically oblong, often with a purple central blotch; capsule with short, appressed hairs; seeds slightly sulcate.

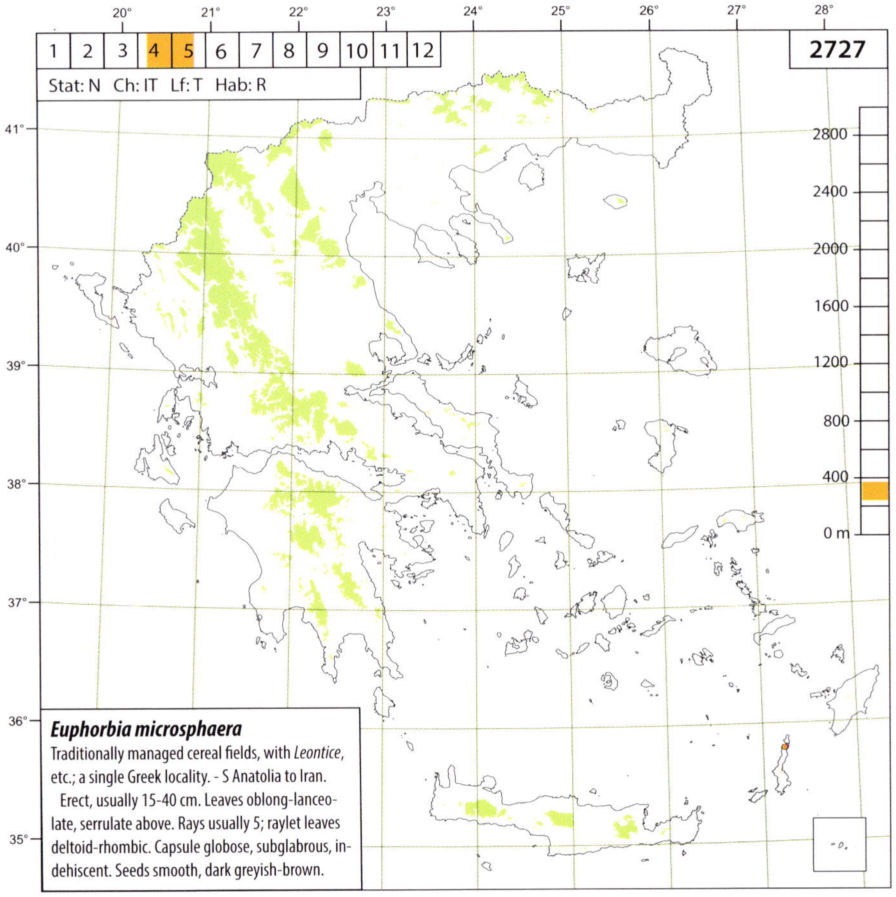

Euphorbia microsphaera
Traditionally managed cereal fields, with *Leontice*, etc.; a single Greek locality. - S Anatolia to Iran.
Erect, usually 15-40 cm. Leaves oblong-lanceolate, serrulate above. Rays usually 5; raylet leaves deltoid-rhombic. Capsule globose, subglabrous, indehiscent. Seeds smooth, dark greyish-brown.

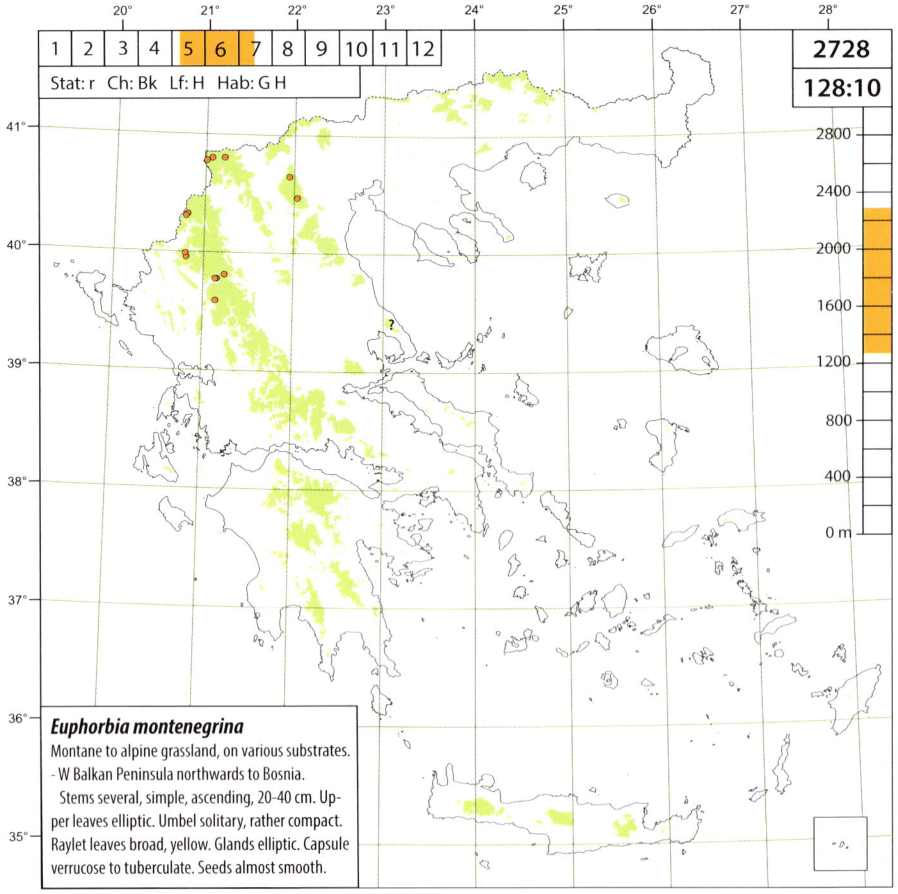

Euphorbia montenegrina
Montane to alpine grassland, on various substrates. - W Balkan Peninsula northwards to Bosnia.
Stems several, simple, ascending, 20-40 cm. Upper leaves elliptic. Umbel solitary, rather compact. Raylet leaves broad, yellow. Glands elliptic. Capsule verrucose to tuberculate. Seeds almost smooth.

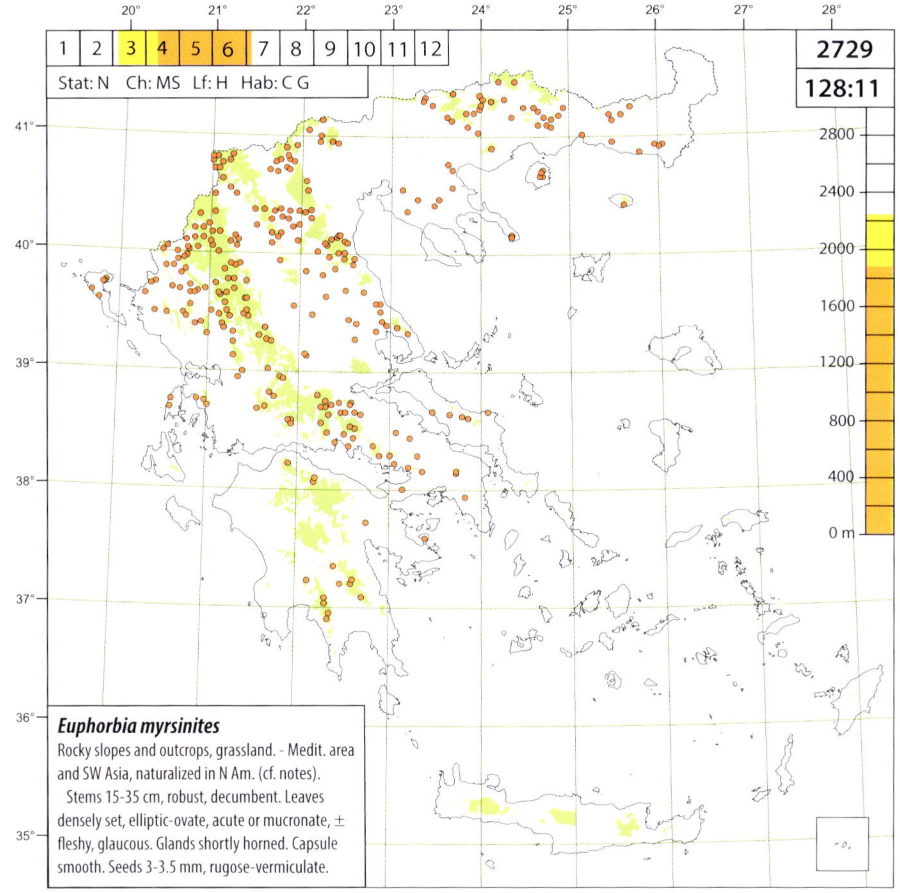

Euphorbia myrsinites

Rocky slopes and outcrops, grassland. - Medit. area and SW Asia, naturalized in N Am. (cf. notes).

Stems 15-35 cm, robust, decumbent. Leaves densely set, elliptic-ovate, acute or mucronate, ± fleshy, glaucous. Glands shortly horned. Capsule smooth. Seeds 3-3.5 mm, rugose-vermiculate.

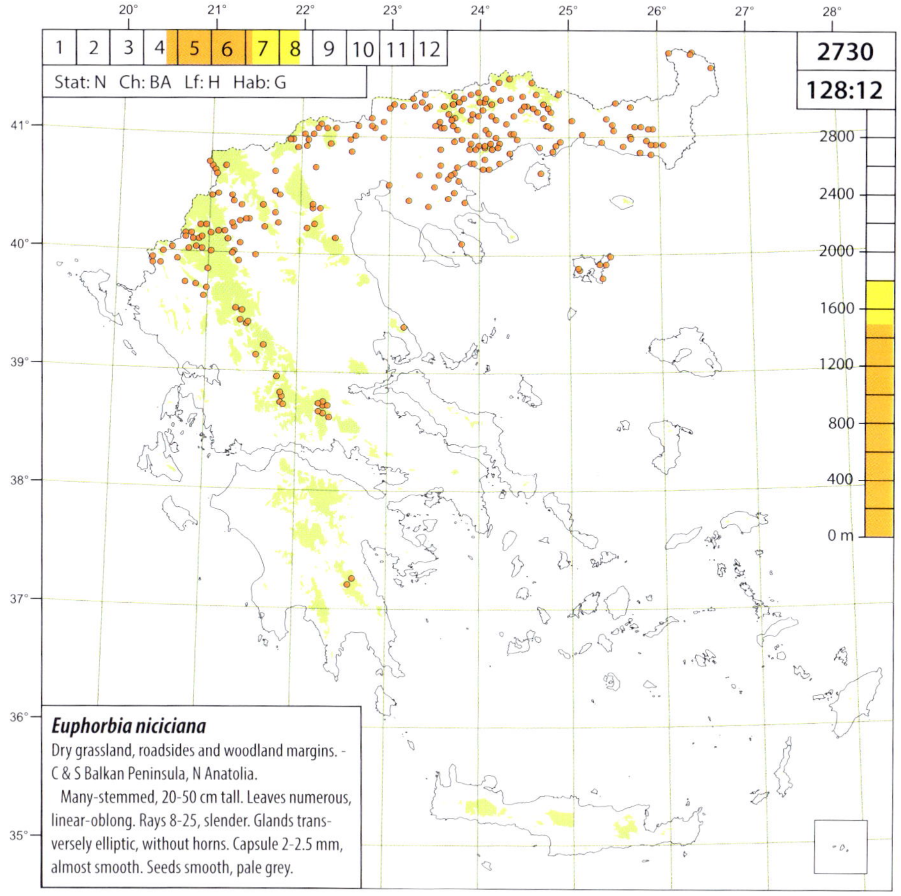

Euphorbia niciciana

Dry grassland, roadsides and woodland margins. - C & S Balkan Peninsula, N Anatolia.

Many-stemmed, 20-50 cm tall. Leaves numerous, linear-oblong. Rays 8-25, slender. Glands transversely elliptic, without horns. Capsule 2-2.5 mm, almost smooth. Seeds smooth, pale grey.

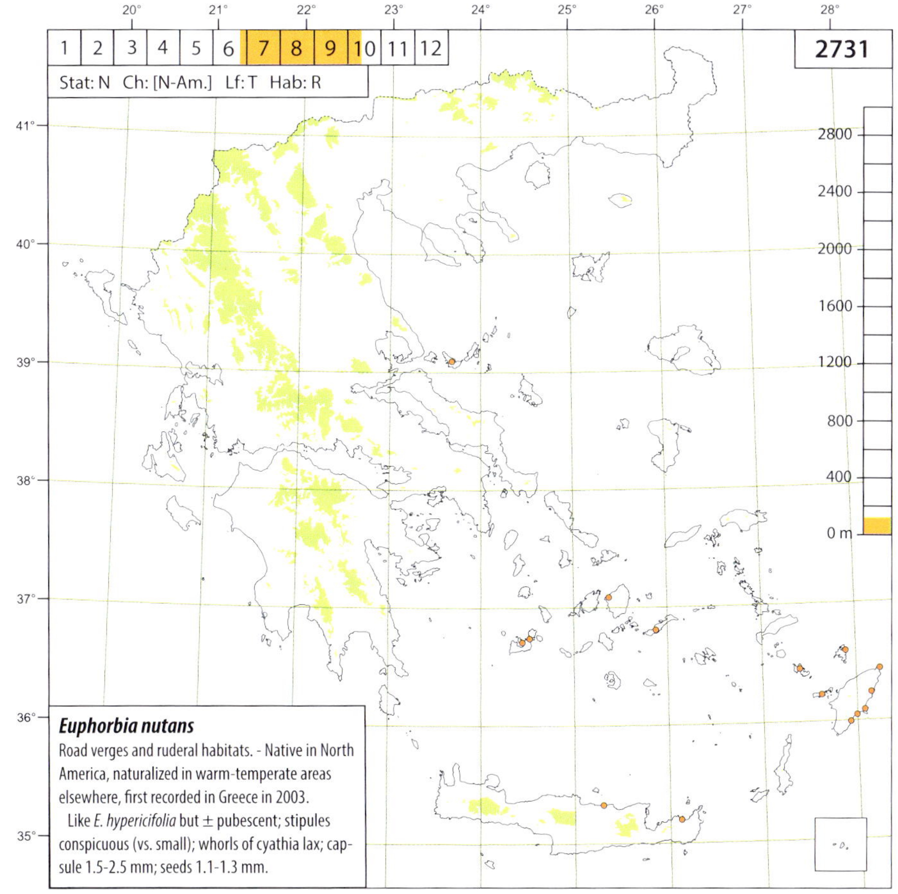

Euphorbia nutans

Road verges and ruderal habitats. - Native in North America, naturalized in warm-temperate areas elsewhere, first recorded in Greece in 2003.

Like *E. hypericifolia* but ± pubescent; stipules conspicuous (vs. small); whorls of cyathia lax; capsule 1.5-2.5 mm; seeds 1.1-1.3 mm.

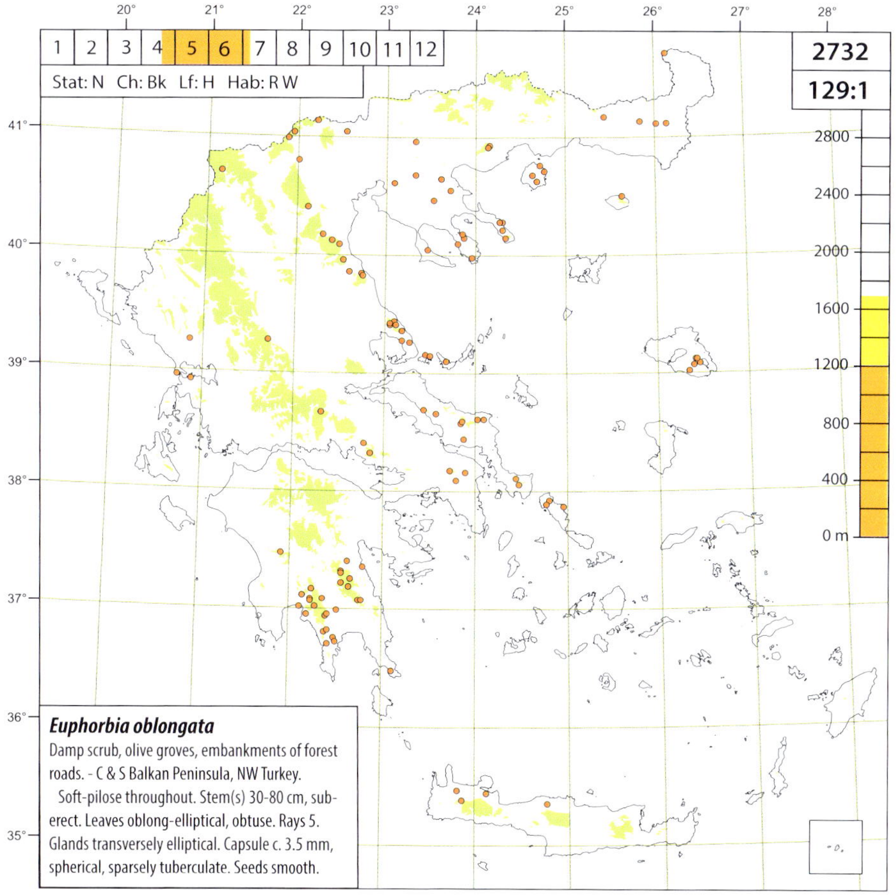

Euphorbia oblongata

Damp scrub, olive groves, embankments of forest roads. - C & S Balkan Peninsula, NW Turkey.

Soft-pilose throughout. Stem(s) 30-80 cm, suberect. Leaves oblong-elliptical, obtuse. Rays 5. Glands transversely elliptical. Capsule c. 3.5 mm, spherical, sparsely tuberculate. Seeds smooth.

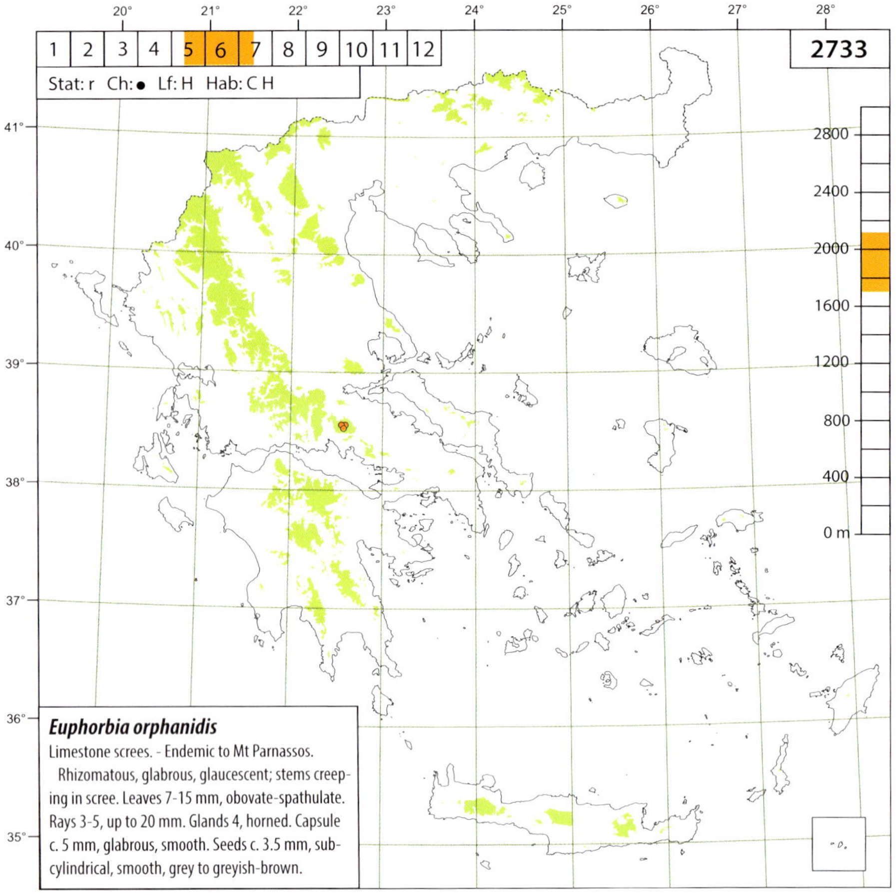

Euphorbia orphanidis
Limestone screes. - Endemic to Mt Parnassos.
Rhizomatous, glabrous, glaucescent; stems creeping in scree. Leaves 7-15 mm, obovate-spathulate. Rays 3-5, up to 20 mm. Glands 4, horned. Capsule c. 5 mm, glabrous, smooth. Seeds c. 3.5 mm, subcylindrical, smooth, grey to greyish-brown.

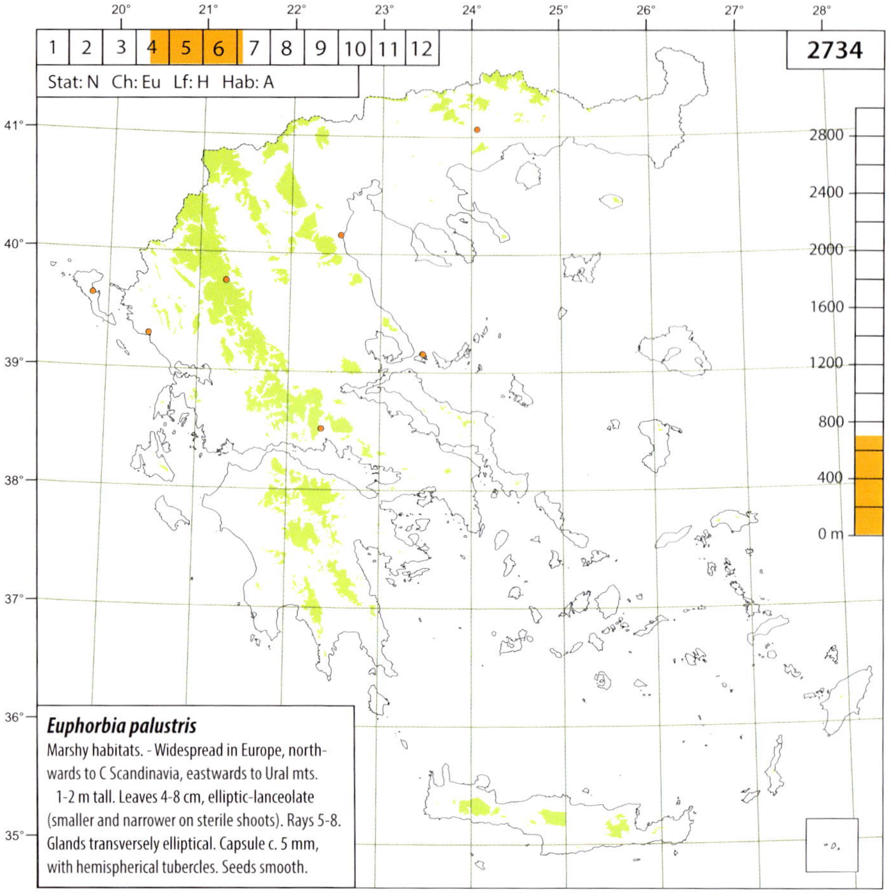

Euphorbia palustris
Marshy habitats. - Widespread in Europe, northwards to C Scandinavia, eastwards to Ural mts.
1-2 m tall. Leaves 4-8 cm, elliptic-lanceolate (smaller and narrower on sterile shoots). Rays 5-8. Glands transversely elliptical. Capsule c. 5 mm, with hemispherical tubercles. Seeds smooth.

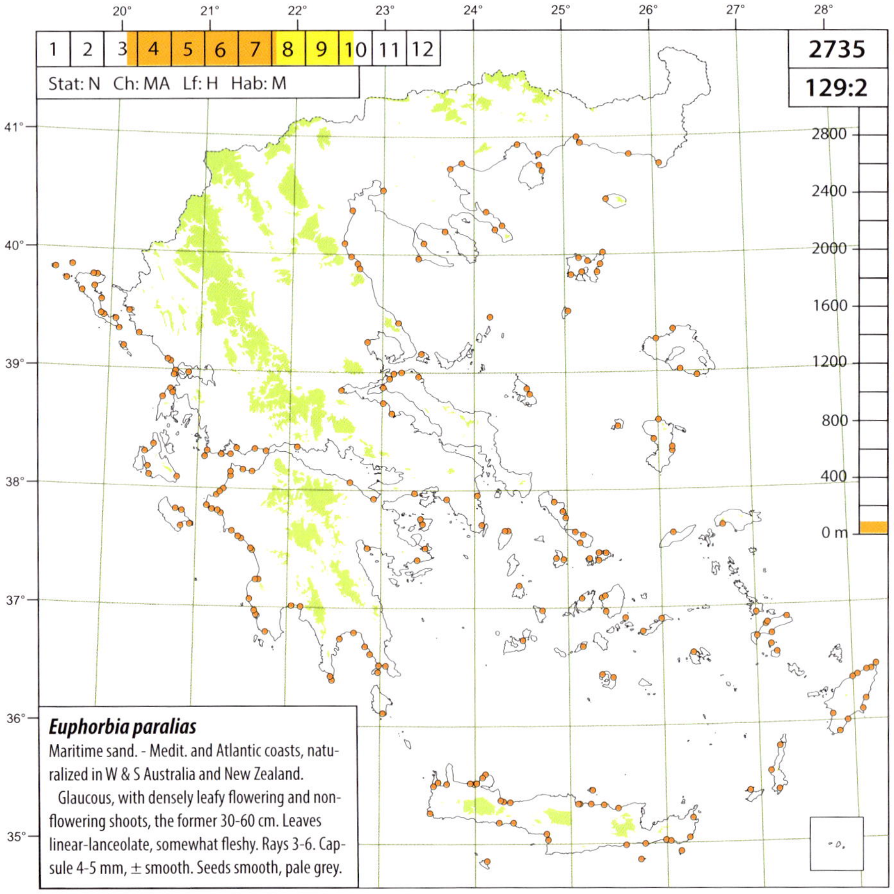

Euphorbia paralias
Maritime sand. - Medit. and Atlantic coasts, naturalized in W & S Australia and New Zealand.
Glaucous, with densely leafy flowering and non-flowering shoots, the former 30-60 cm. Leaves linear-lanceolate, somewhat fleshy. Rays 3-6. Capsule 4-5 mm, ± smooth. Seeds smooth, pale grey.

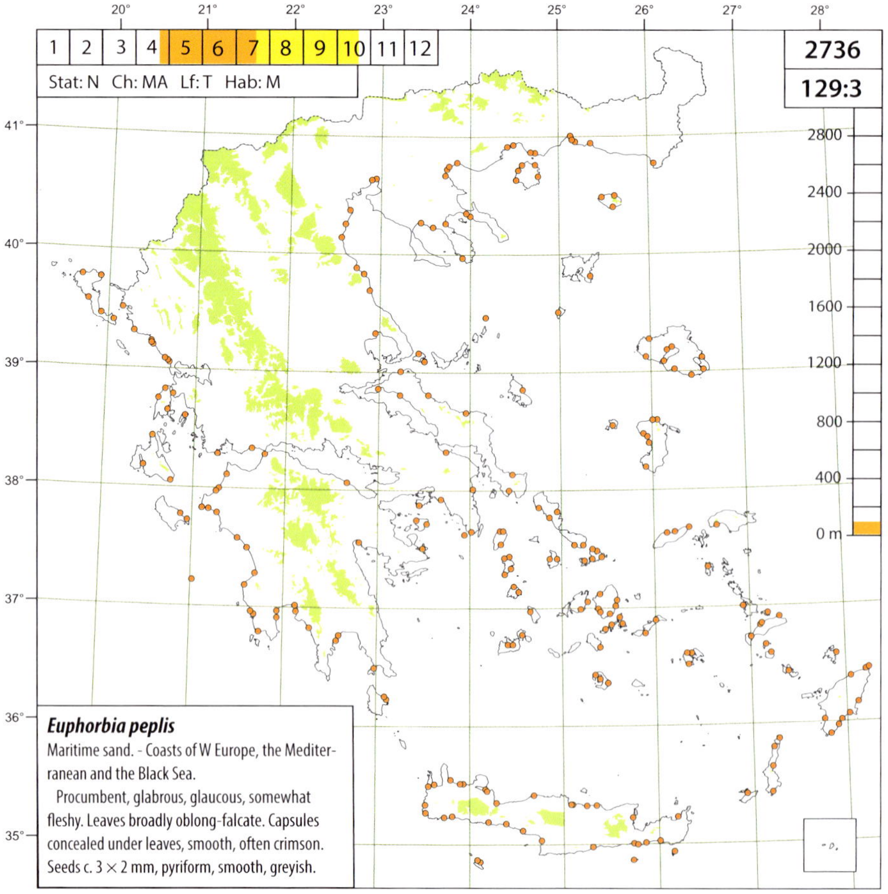

Euphorbia peplis
Maritime sand. - Coasts of W Europe, the Mediterranean and the Black Sea.
Procumbent, glabrous, glaucous, somewhat fleshy. Leaves broadly oblong-falcate. Capsules concealed under leaves, smooth, often crimson. Seeds c. 3 × 2 mm, pyriform, smooth, greyish.

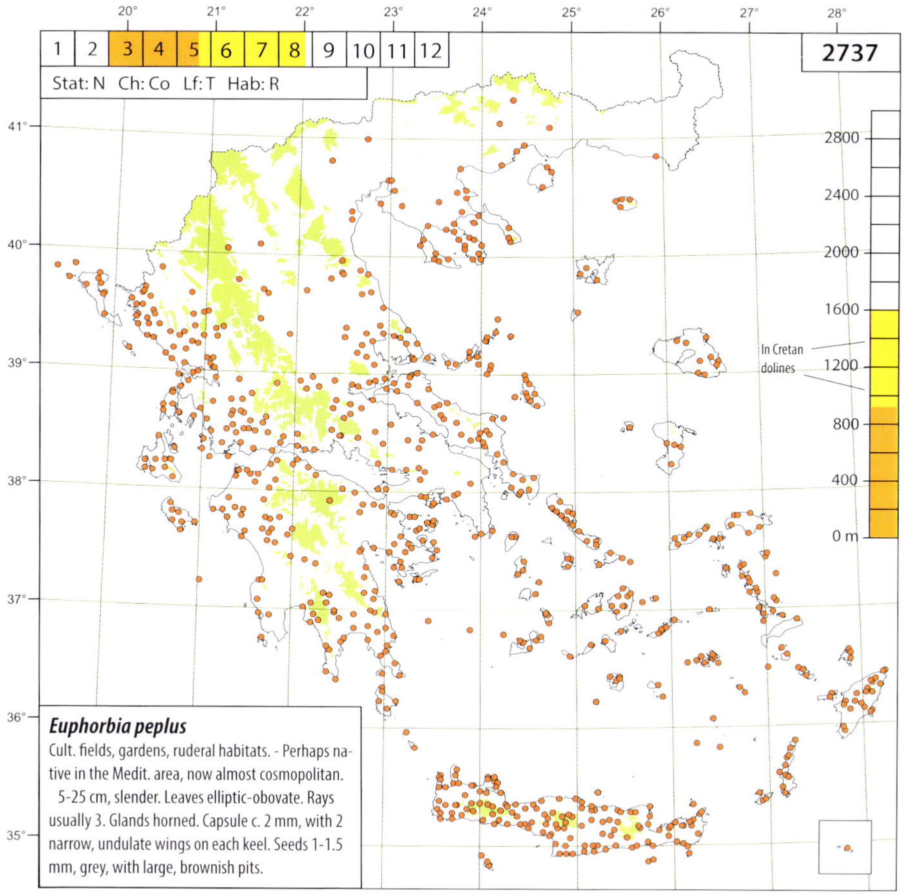

Euphorbia peplus

Cult. fields, gardens, ruderal habitats. - Perhaps native in the Medit. area, now almost cosmopolitan.

5-25 cm, slender. Leaves elliptic-obovate. Rays usually 3. Glands horned. Capsule c. 2 mm, with 2 narrow, undulate wings on each keel. Seeds 1-1.5 mm, grey, with large, brownish pits.

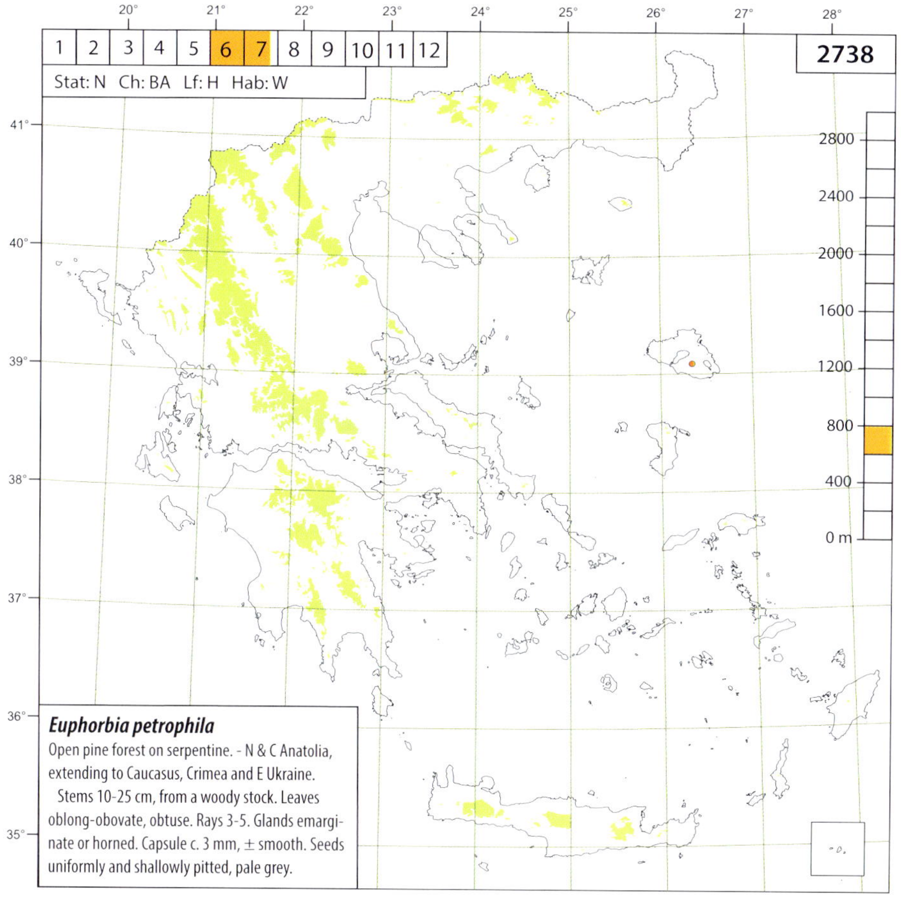

Euphorbia petrophila

Open pine forest on serpentine. - N & C Anatolia, extending to Caucasus, Crimea and E Ukraine.

Stems 10-25 cm, from a woody stock. Leaves oblong-obovate, obtuse. Rays 3-5. Glands emarginate or horned. Capsule c. 3 mm, ± smooth. Seeds uniformly and shallowly pitted, pale grey.

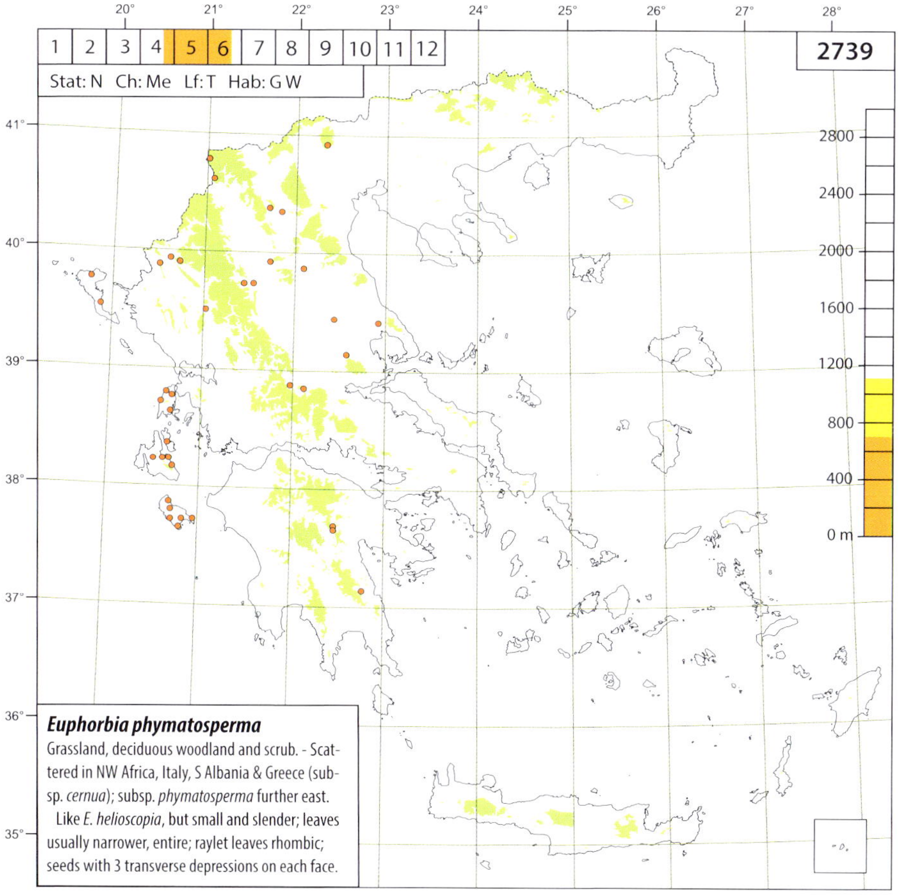

Euphorbia phymatosperma

Grassland, deciduous woodland and scrub. - Scattered in NW Africa, Italy, S Albania & Greece (subsp. *cernua*); subsp. *phymatosperma* further east.

Like *E. helioscopia*, but small and slender; leaves usually narrower, entire; raylet leaves rhombic; seeds with 3 transverse depressions on each face.

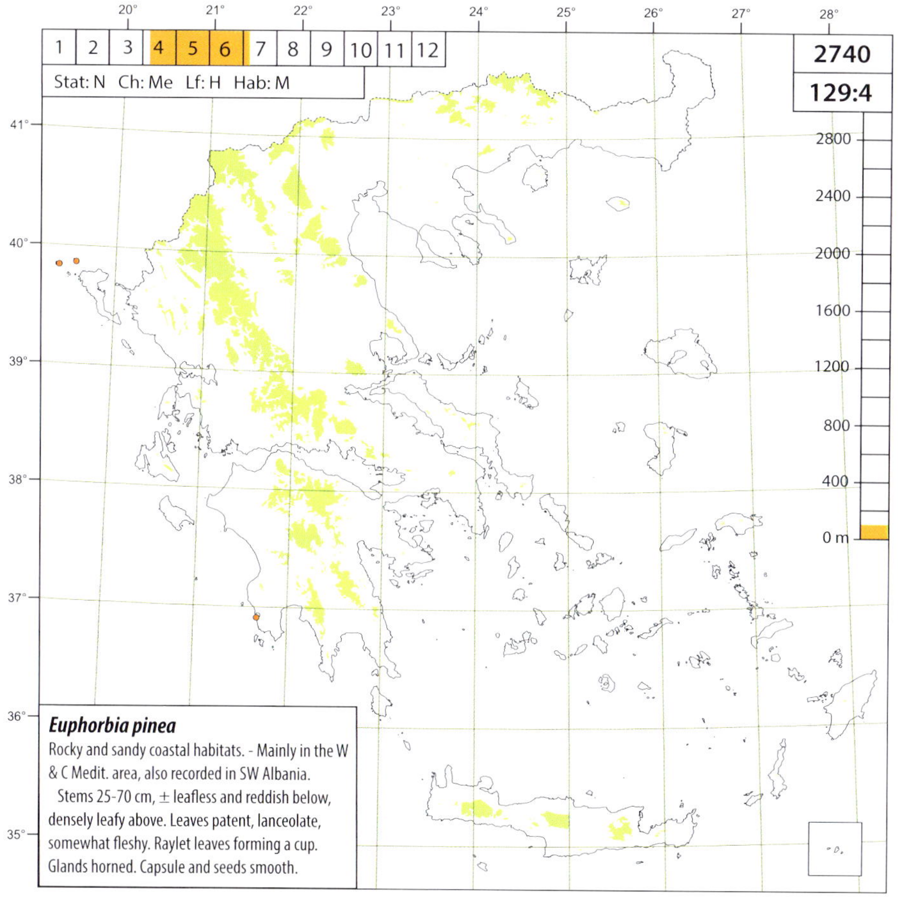

Euphorbia pinea

Rocky and sandy coastal habitats. - Mainly in the W & C Medit. area, also recorded in SW Albania.

Stems 25-70 cm, ± leafless and reddish below, densely leafy above. Leaves patent, lanceolate, somewhat fleshy. Raylet leaves forming a cup. Glands horned. Capsule and seeds smooth.

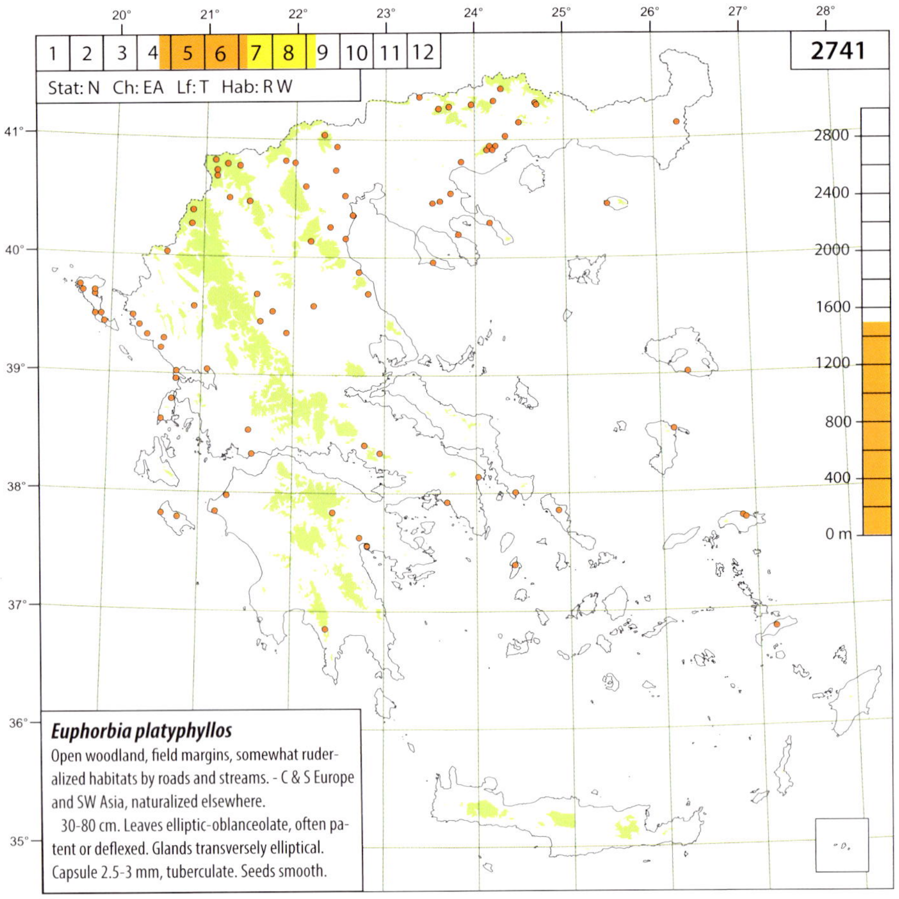

Euphorbia platyphyllos
Open woodland, field margins, somewhat ruderalized habitats by roads and streams. - C & S Europe and SW Asia, naturalized elsewhere.
30-80 cm. Leaves elliptic-oblanceolate, often patent or deflexed. Glands transversely elliptical. Capsule 2.5-3 mm, tuberculate. Seeds smooth.

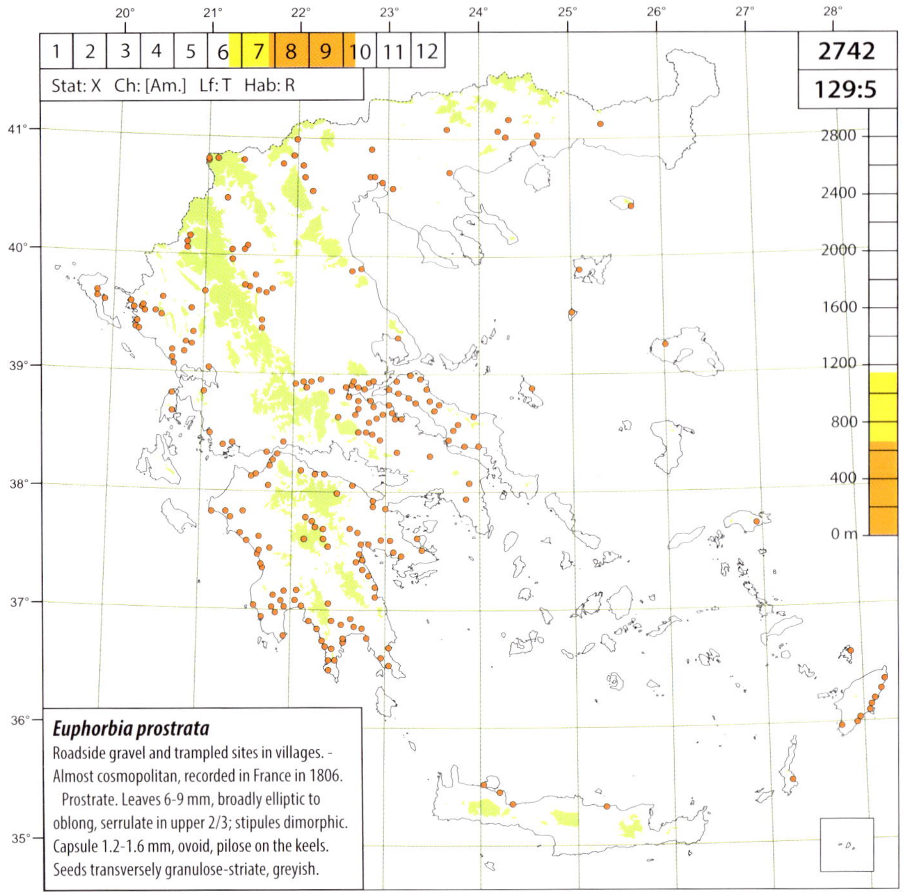

Euphorbia prostrata
Roadside gravel and trampled sites in villages. - Almost cosmopolitan, recorded in France in 1806.
Prostrate. Leaves 6-9 mm, broadly elliptic to oblong, serrulate in upper 2/3; stipules dimorphic. Capsule 1.2-1.6 mm, ovoid, pilose on the keels. Seeds transversely granulose-striate, greyish.

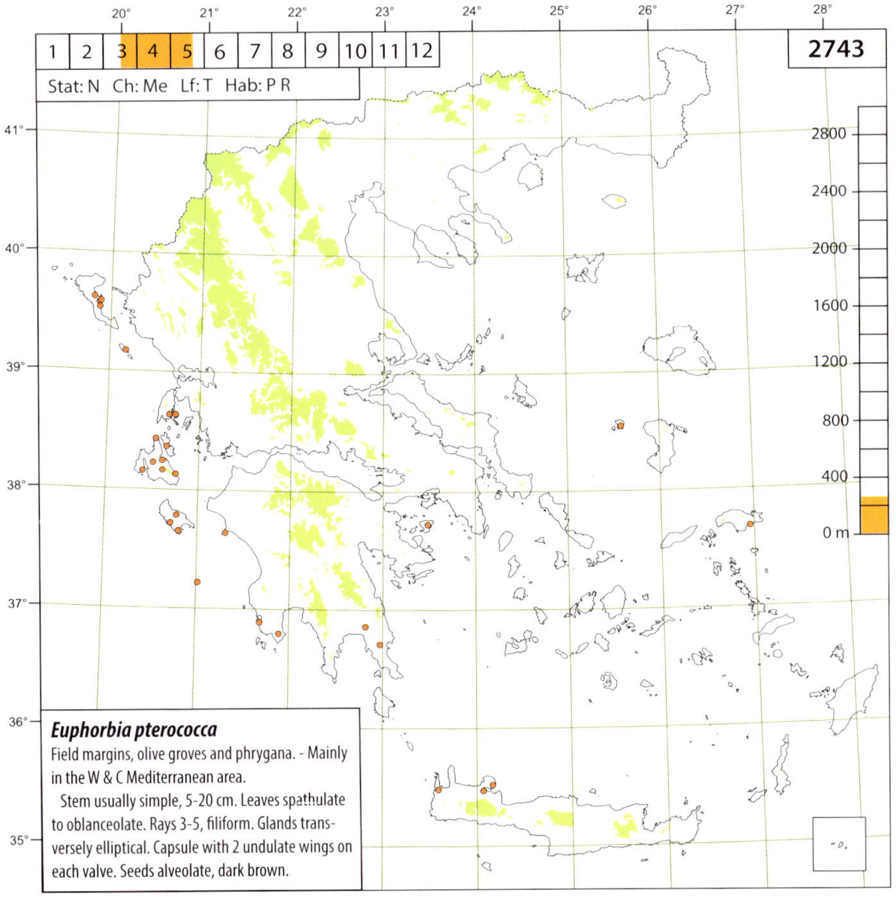

Euphorbia pterococca
Field margins, olive groves and phrygana. - Mainly in the W & C Mediterranean area.
Stem usually simple, 5-20 cm. Leaves spathulate to oblanceolate. Rays 3-5, filiform. Glands transversely elliptical. Capsule with 2 undulate wings on each valve. Seeds alveolate, dark brown.

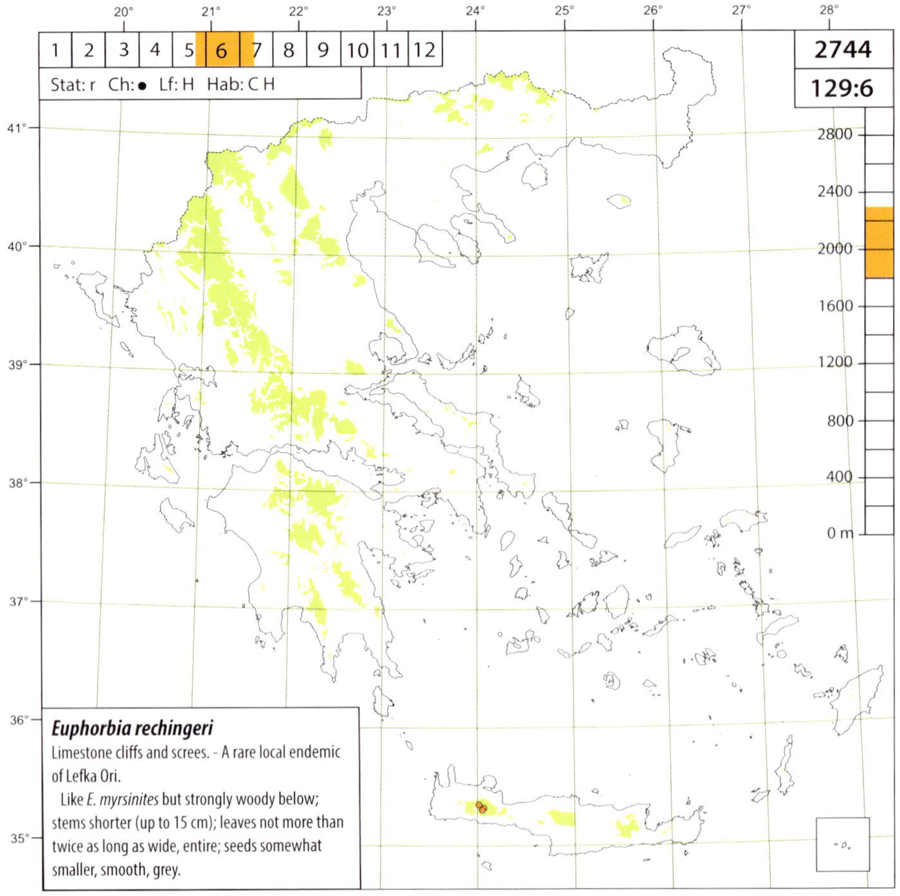

Euphorbia rechingeri
Limestone cliffs and screes. - A rare local endemic of Lefka Ori.
Like *E. myrsinites* but strongly woody below; stems shorter (up to 15 cm); leaves not more than twice as long as wide, entire; seeds somewhat smaller, smooth, grey.

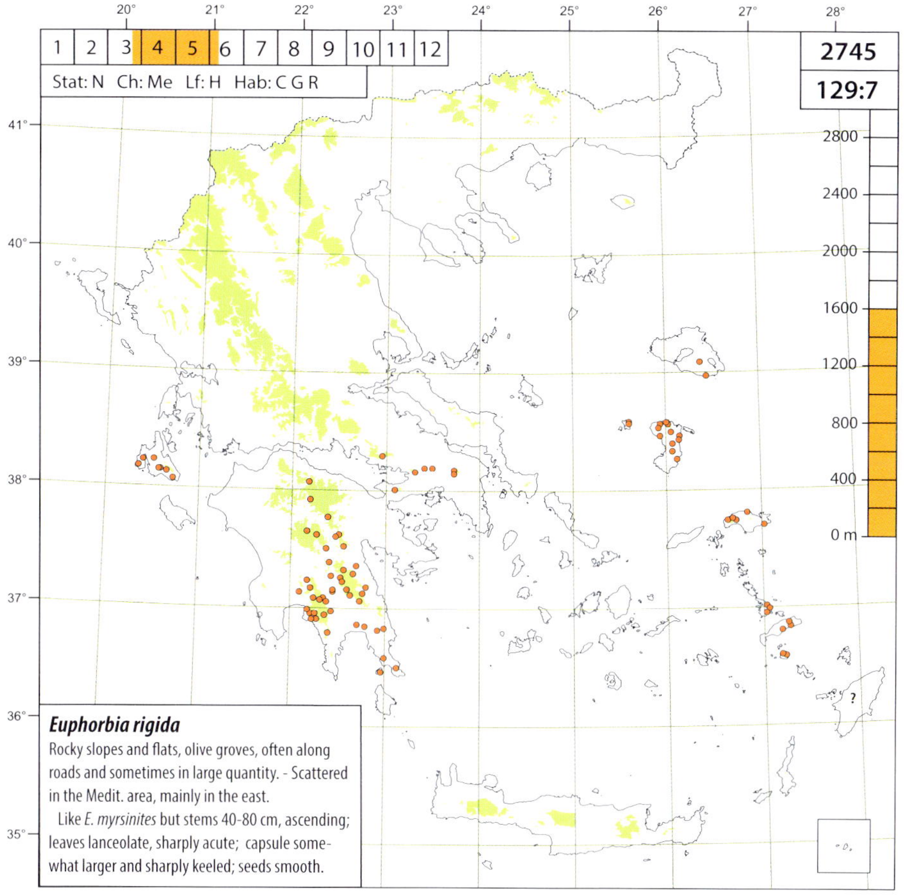

Euphorbia rigida

Rocky slopes and flats, olive groves, often along roads and sometimes in large quantity. - Scattered in the Medit. area, mainly in the east.

Like *E. myrsinites* but stems 40-80 cm, ascending; leaves lanceolate, sharply acute; capsule somewhat larger and sharply keeled; seeds smooth.

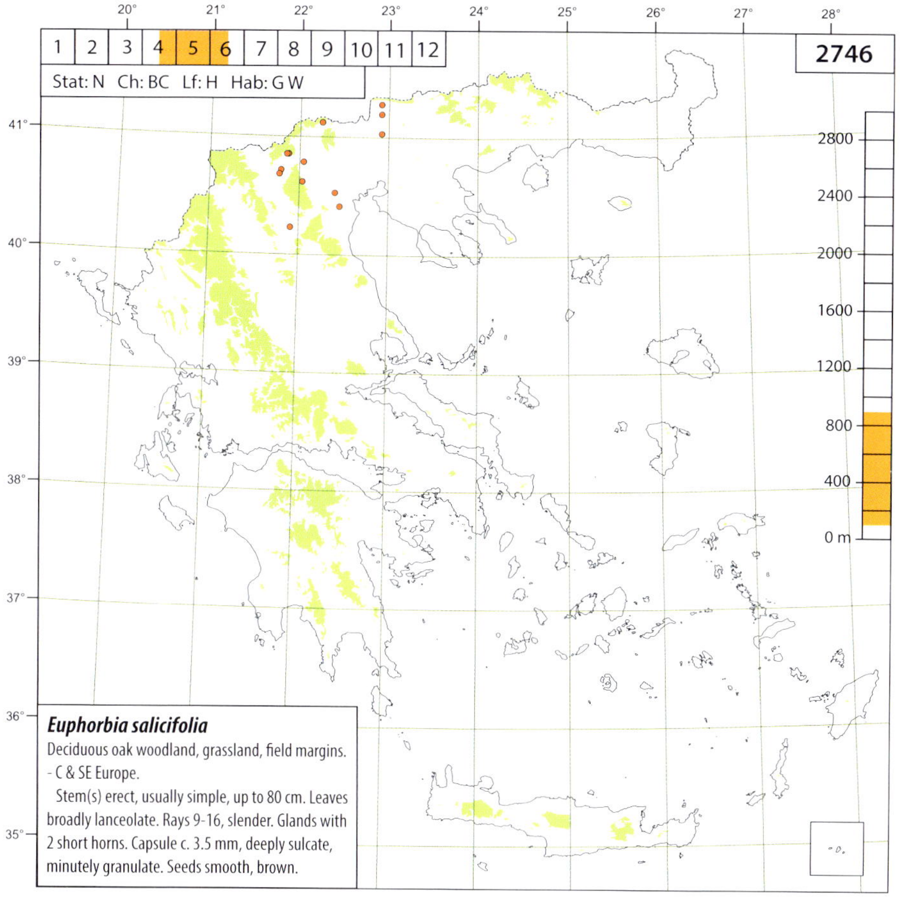

Euphorbia salicifolia

Deciduous oak woodland, grassland, field margins. - C & SE Europe.

Stem(s) erect, usually simple, up to 80 cm. Leaves broadly lanceolate. Rays 9-16, slender. Glands with 2 short horns. Capsule c. 3.5 mm, deeply sulcate, minutely granulate. Seeds smooth, brown.

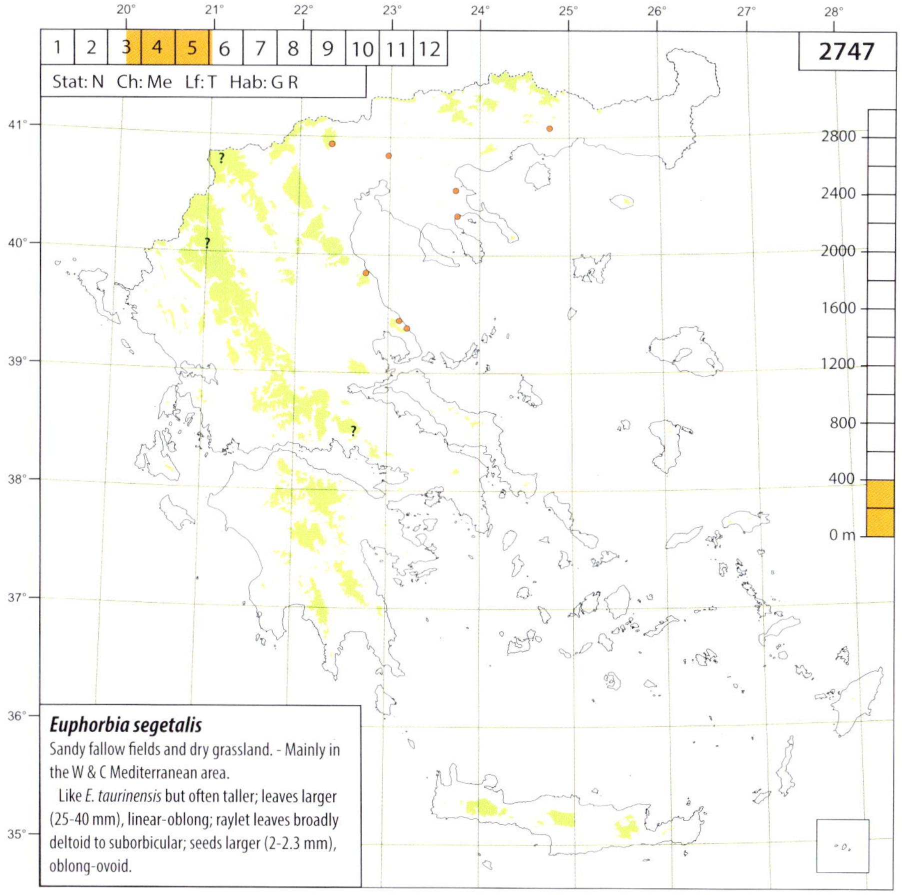

Euphorbia segetalis

Sandy fallow fields and dry grassland. - Mainly in the W & C Mediterranean area.

Like *E. taurinensis* but often taller; leaves larger (25-40 mm), linear-oblong; raylet leaves broadly deltoid to suborbicular; seeds larger (2-2.3 mm), oblong-ovoid.

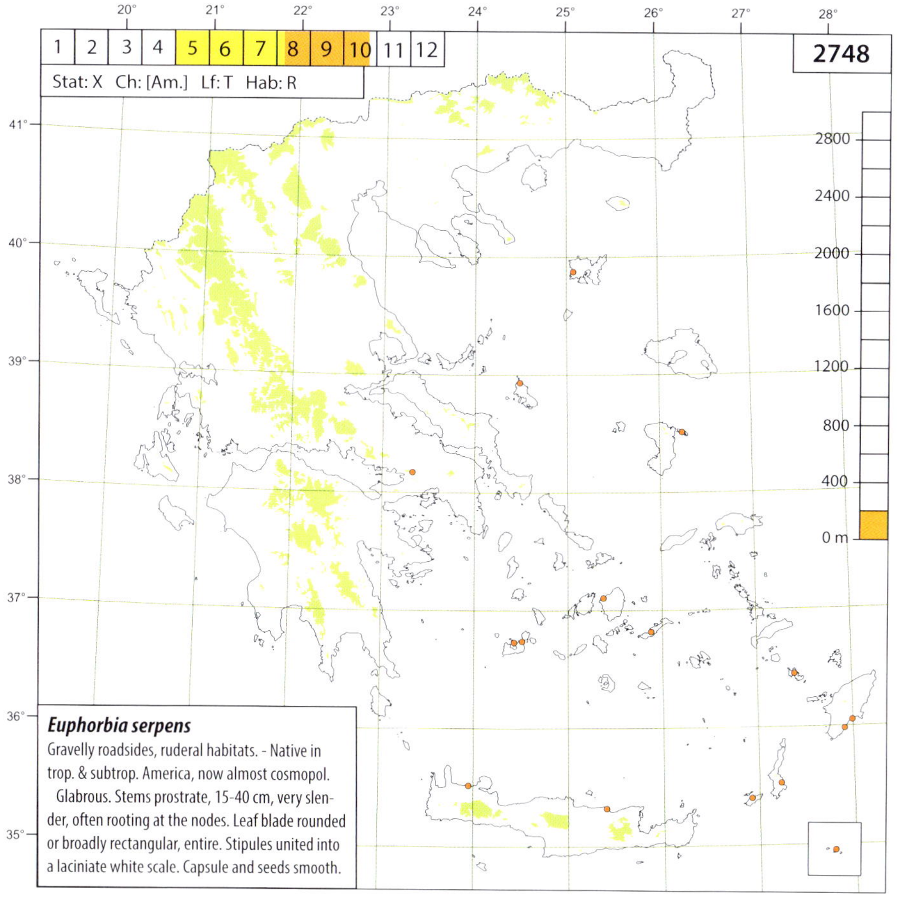

Euphorbia serpens

Gravelly roadsides, ruderal habitats. - Native in trop. & subtrop. America, now almost cosmopol.

Glabrous. Stems prostrate, 15-40 cm, very slender, often rooting at the nodes. Leaf blade rounded or broadly rectangular, entire. Stipules united into a laciniate white scale. Capsule and seeds smooth.

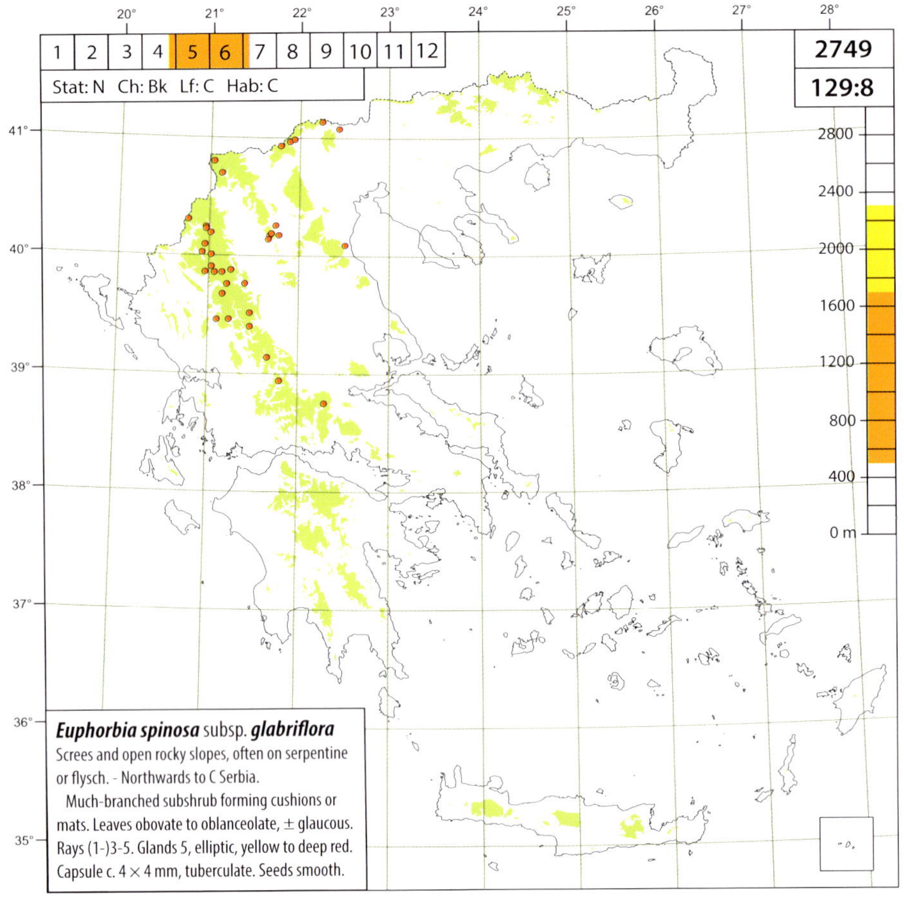

Euphorbia spinosa subsp. ***glabriflora***
Screes and open rocky slopes, often on serpentine or flysch. - Northwards to C Serbia.
Much-branched subshrub forming cushions or mats. Leaves obovate to oblanceolate, ± glaucous. Rays (1-)3-5. Glands 5, elliptic, yellow to deep red. Capsule c. 4 × 4 mm, tuberculate. Seeds smooth.

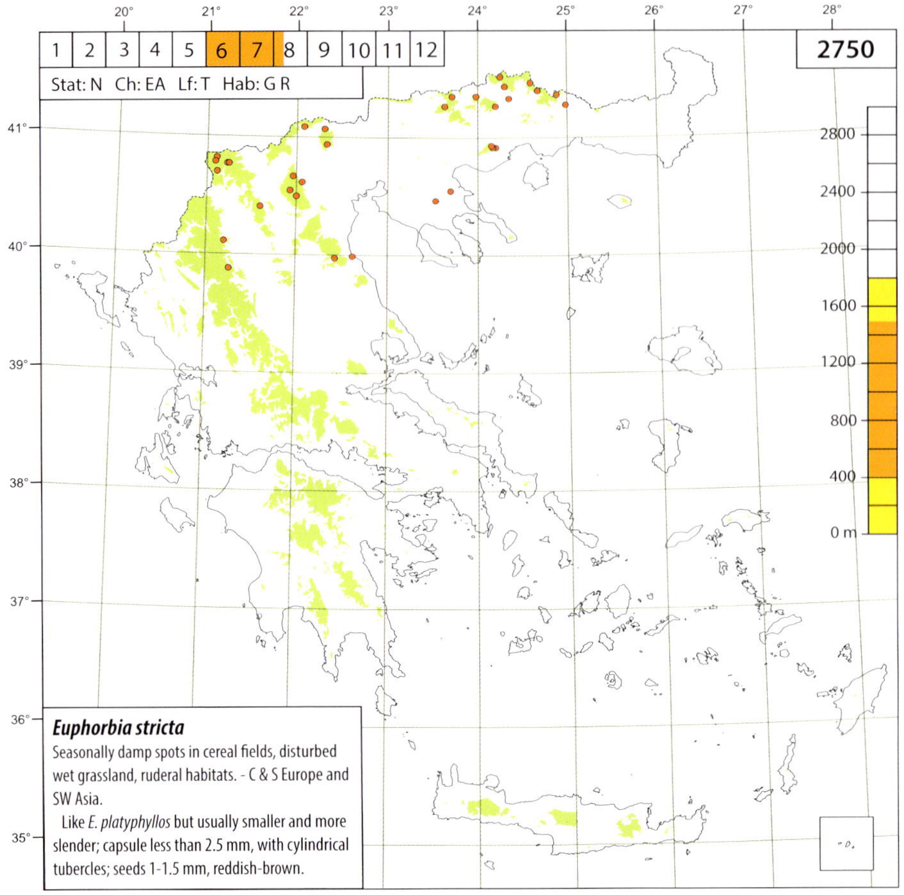

Euphorbia stricta
Seasonally damp spots in cereal fields, disturbed wet grassland, ruderal habitats. - C & S Europe and SW Asia.
Like *E. platyphyllos* but usually smaller and more slender; capsule less than 2.5 mm, with cylindrical tubercles; seeds 1-1.5 mm, reddish-brown.

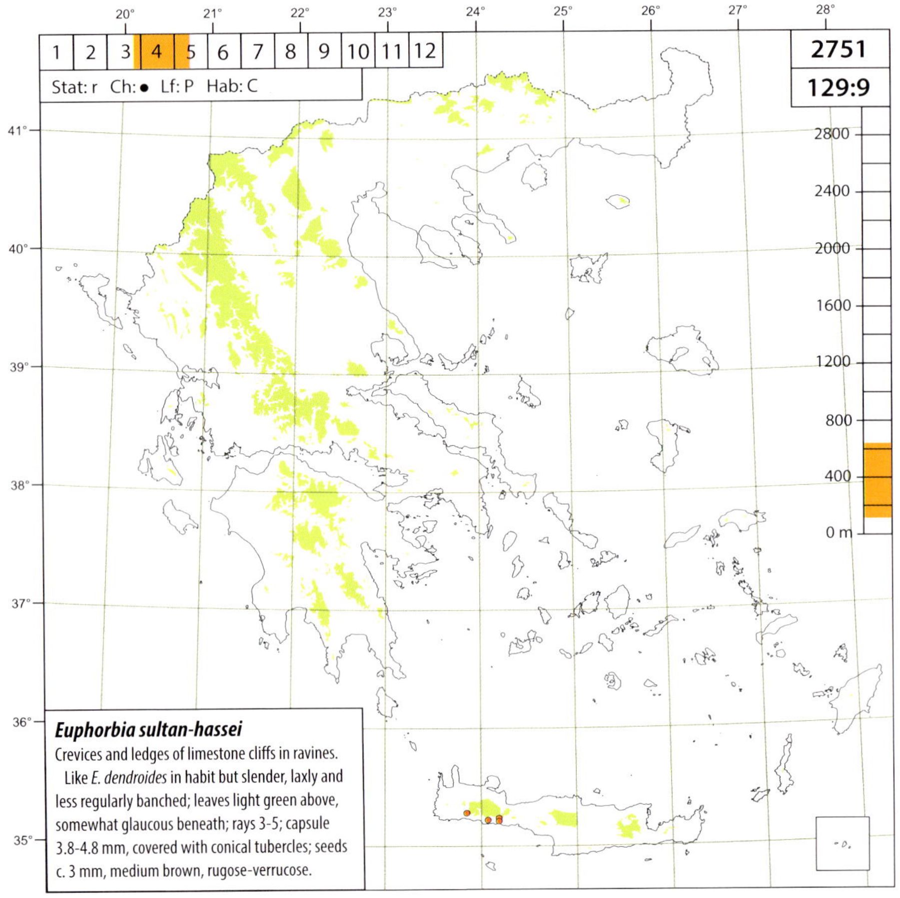

Euphorbia sultan-hassei
Crevices and ledges of limestone cliffs in ravines.
Like *E. dendroides* in habit but slender, laxly and less regularly banched; leaves light green above, somewhat glaucous beneath; rays 3-5; capsule 3.8-4.8 mm, covered with conical tubercles; seeds c. 3 mm, medium brown, rugose-verrucose.

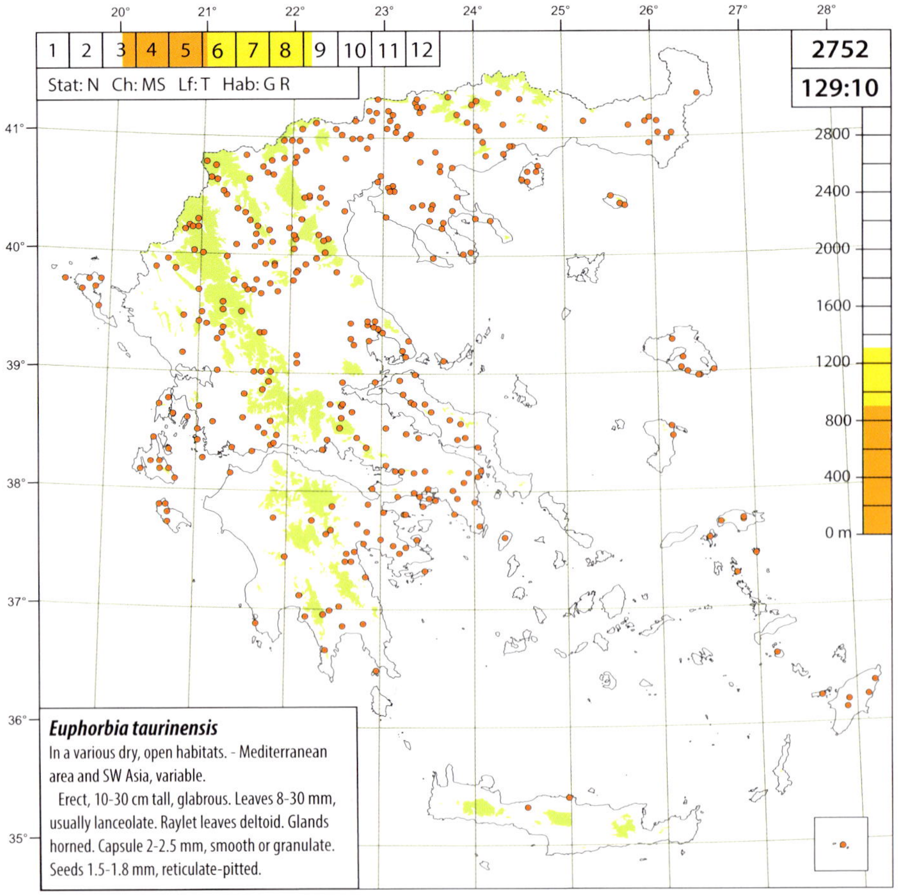

Euphorbia taurinensis
In a various dry, open habitats. - Mediterranean area and SW Asia, variable.
Erect, 10-30 cm tall, glabrous. Leaves 8-30 mm, usually lanceolate. Raylet leaves deltoid. Glands horned. Capsule 2-2.5 mm, smooth or granulate. Seeds 1.5-1.8 mm, reticulate-pitted.

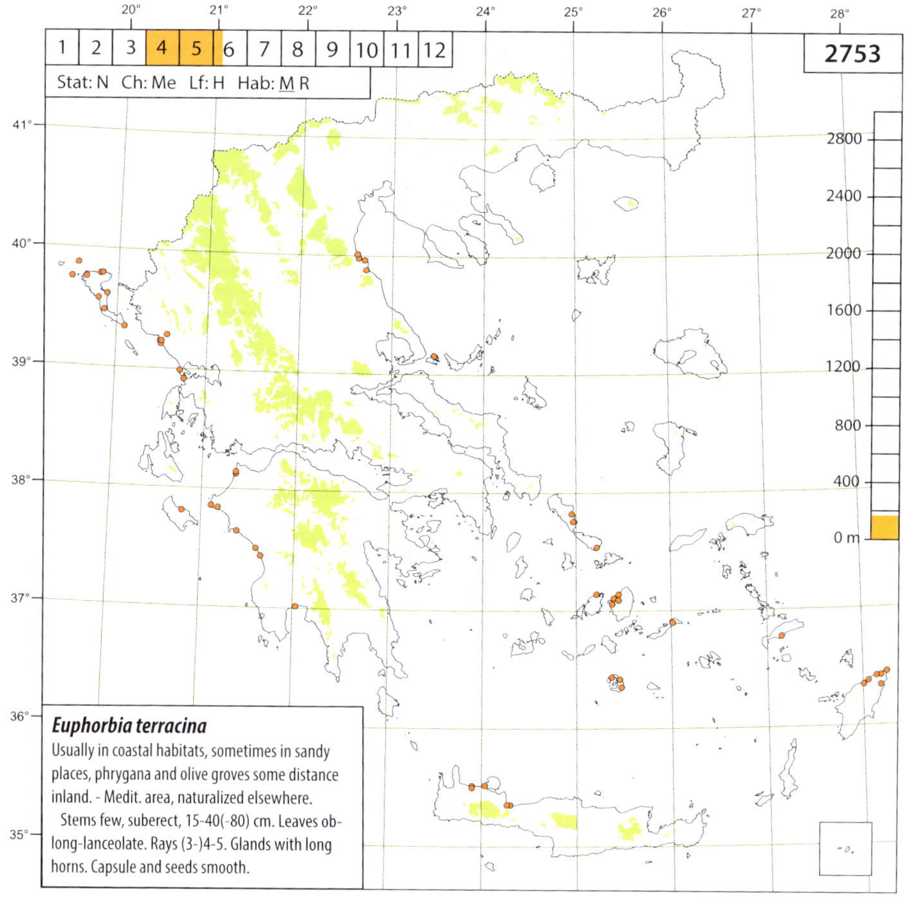

Euphorbia terracina
Usually in coastal habitats, sometimes in sandy places, phrygana and olive groves some distance inland. - Medit. area, naturalized elsewhere.
Stems few, suberect, 15-40(-80) cm. Leaves oblong-lanceolate. Rays (3-)4-5. Glands with long horns. Capsule and seeds smooth.

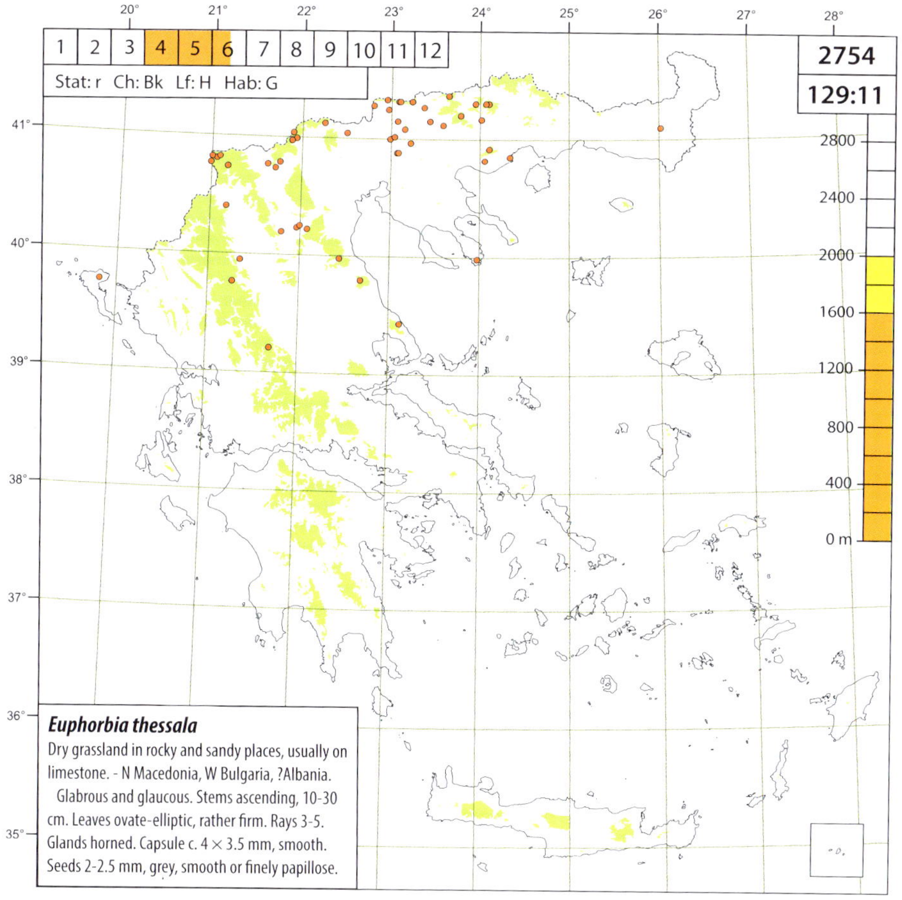

Euphorbia thessala
Dry grassland in rocky and sandy places, usually on limestone. - N Macedonia, W Bulgaria, ?Albania.
Glabrous and glaucous. Stems ascending, 10-30 cm. Leaves ovate-elliptic, rather firm. Rays 3-5. Glands horned. Capsule c. 4 × 3.5 mm, smooth. Seeds 2-2.5 mm, grey, smooth or finely papillose.

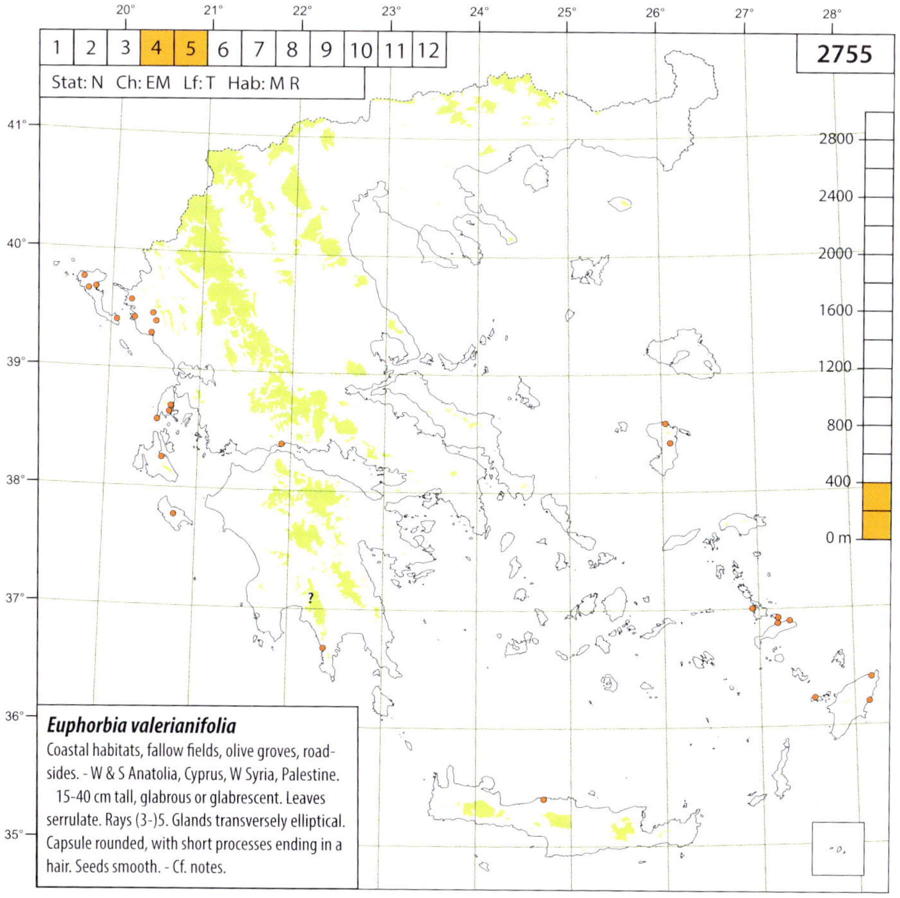

Euphorbia valerianifolia
Coastal habitats, fallow fields, olive groves, roadsides. - W & S Anatolia, Cyprus, W Syria, Palestine.
15-40 cm tall, glabrous or glabrescent. Leaves serrulate. Rays (3-)5. Glands transversely elliptical. Capsule rounded, with short processes ending in a hair. Seeds smooth. - Cf. notes.

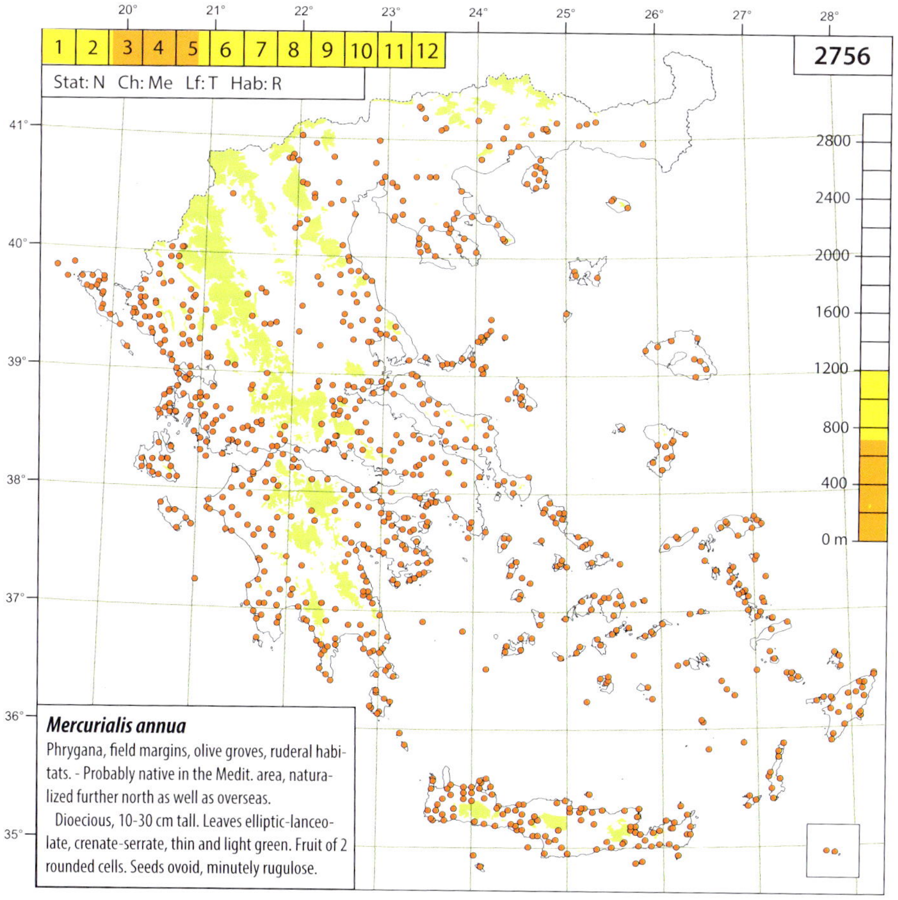

Mercurialis annua
Phrygana, field margins, olive groves, ruderal habitats. - Probably native in the Medit. area, naturalized further north as well as overseas.
Dioecious, 10-30 cm tall. Leaves elliptic-lanceolate, crenate-serrate, thin and light green. Fruit of 2 rounded cells. Seeds ovoid, minutely rugulose.

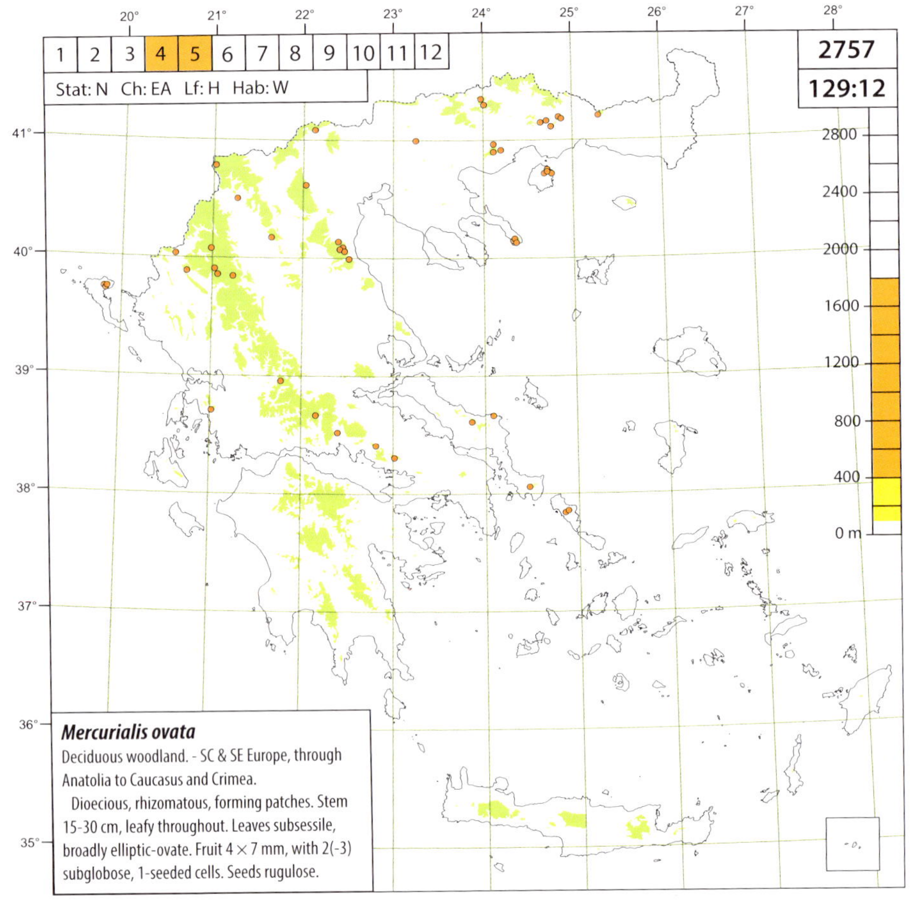

Mercurialis ovata
Deciduous woodland. - SC & SE Europe, through Anatolia to Caucasus and Crimea.
Dioecious, rhizomatous, forming patches. Stem 15-30 cm, leafy throughout. Leaves subsessile, broadly elliptic-ovate. Fruit 4 × 7 mm, with 2(-3) subglobose, 1-seeded cells. Seeds rugulose.

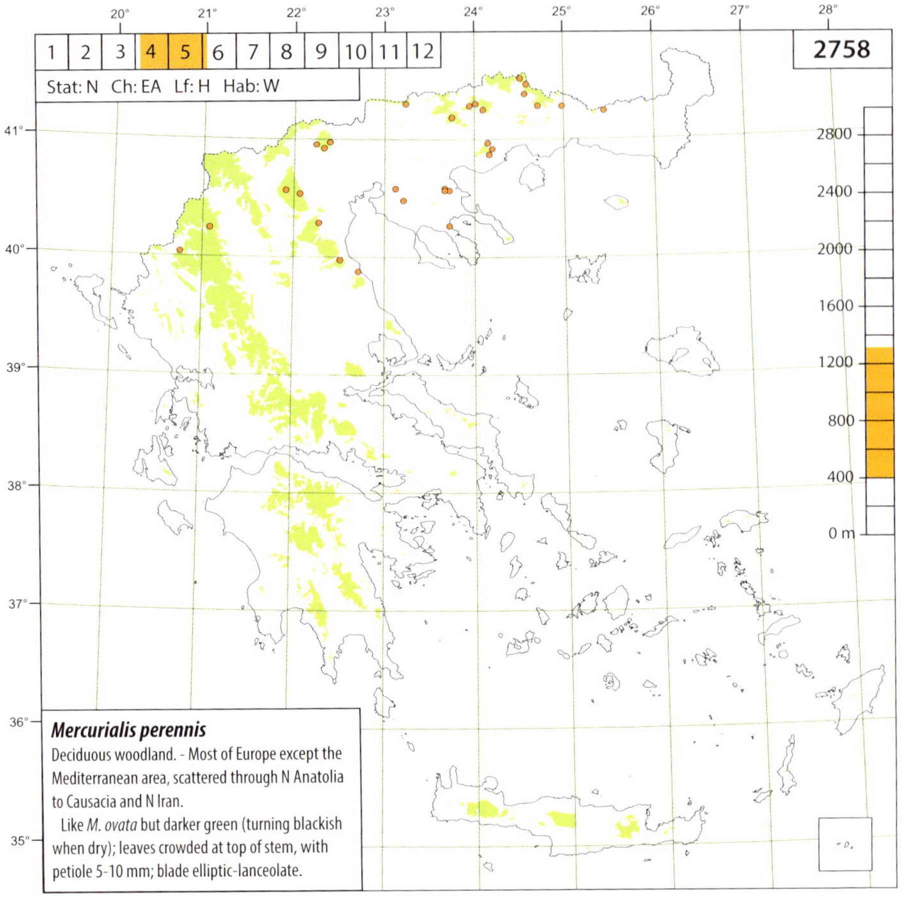

Mercurialis perennis
Deciduous woodland. - Most of Europe except the Mediterranean area, scattered through N Anatolia to Causacia and N Iran.
Like *M. ovata* but darker green (turning blackish when dry); leaves crowded at top of stem, with petiole 5-10 mm; blade elliptic-lanceolate.

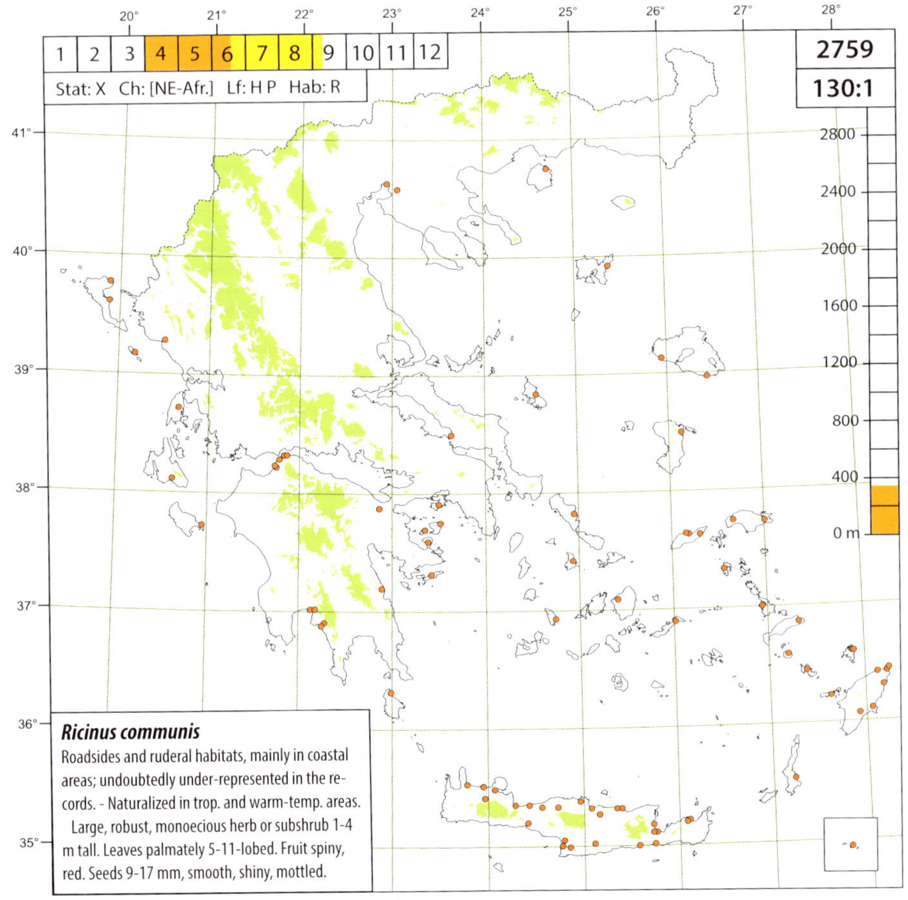

Ricinus communis
Roadsides and ruderal habitats, mainly in coastal areas; undoubtedly under-represented in the records. - Naturalized in trop. and warm-temp. areas.
Large, robust, monoecious herb or subshrub 1-4 m tall. Leaves palmately 5-11-lobed. Fruit spiny, red. Seeds 9-17 mm, smooth, shiny, mottled.

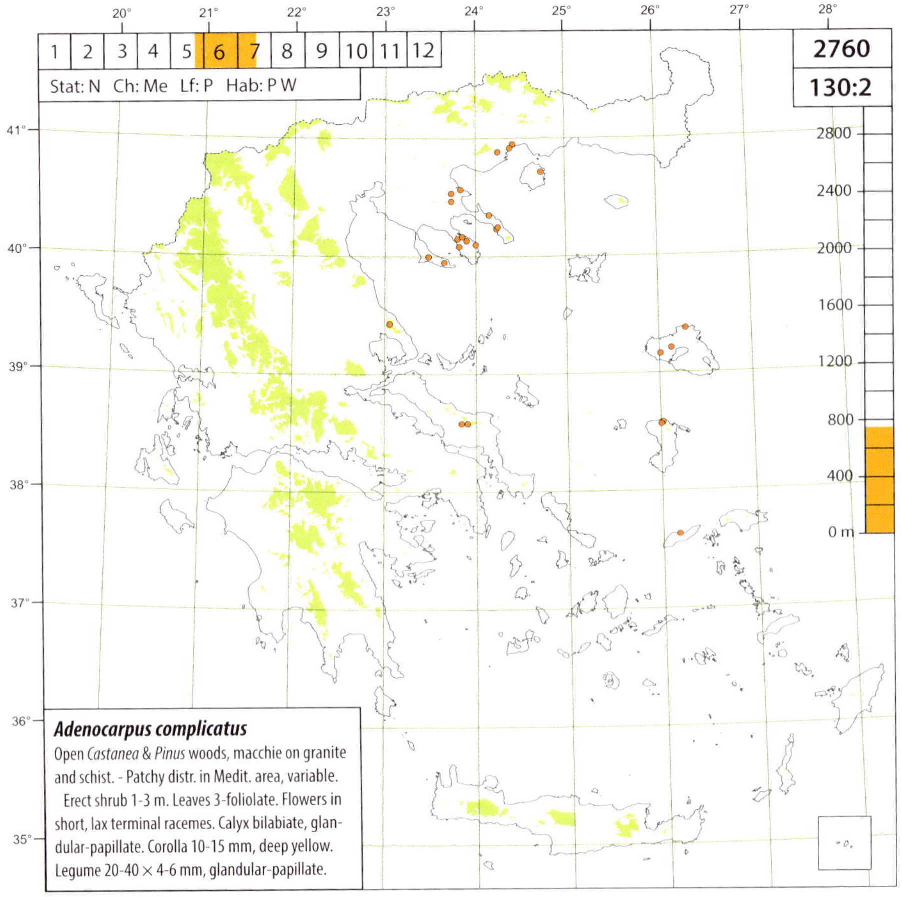

Adenocarpus complicatus
Open *Castanea* & *Pinus* woods, macchie on granite and schist. - Patchy distr. in Medit. area, variable.
Erect shrub 1-3 m. Leaves 3-foliolate. Flowers in short, lax terminal racemes. Calyx bilabiate, glandular-papillate. Corolla 10-15 mm, deep yellow. Legume 20-40 × 4-6 mm, glandular-papillate.

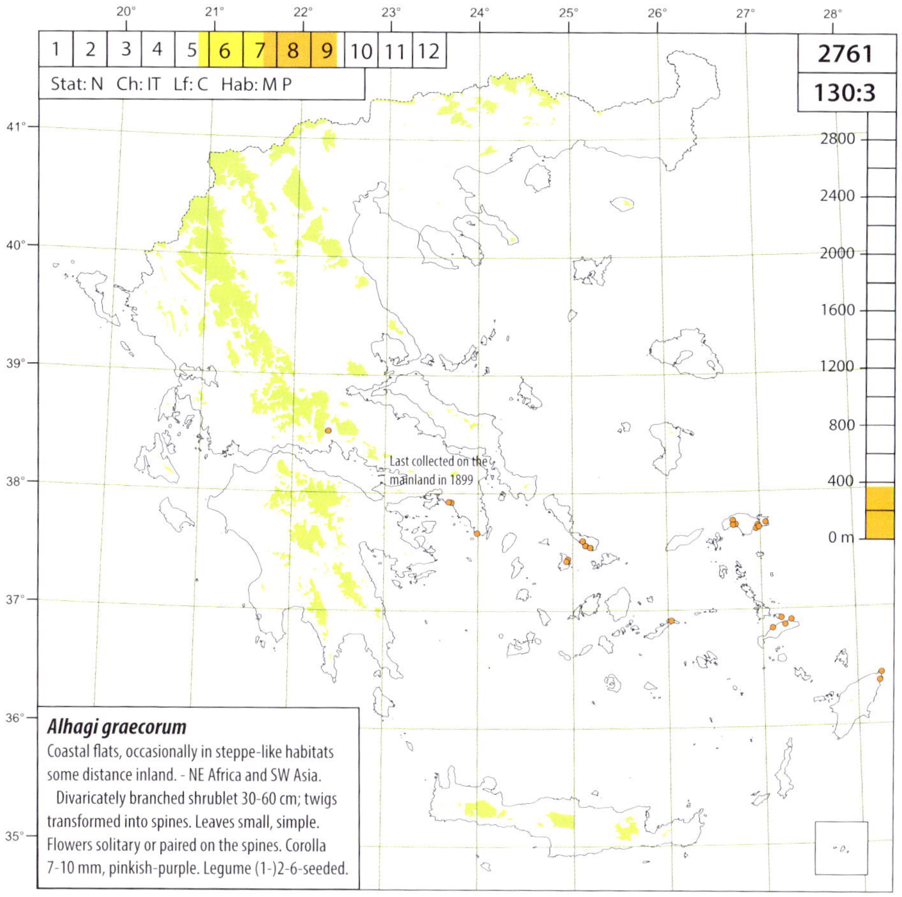

Alhagi graecorum

Coastal flats, occasionally in steppe-like habitats some distance inland. - NE Africa and SW Asia.

Divaricately branched shrublet 30-60 cm; twigs transformed into spines. Leaves small, simple. Flowers solitary or paired on the spines. Corolla 7-10 mm, pinkish-purple. Legume (1-)2-6-seeded.

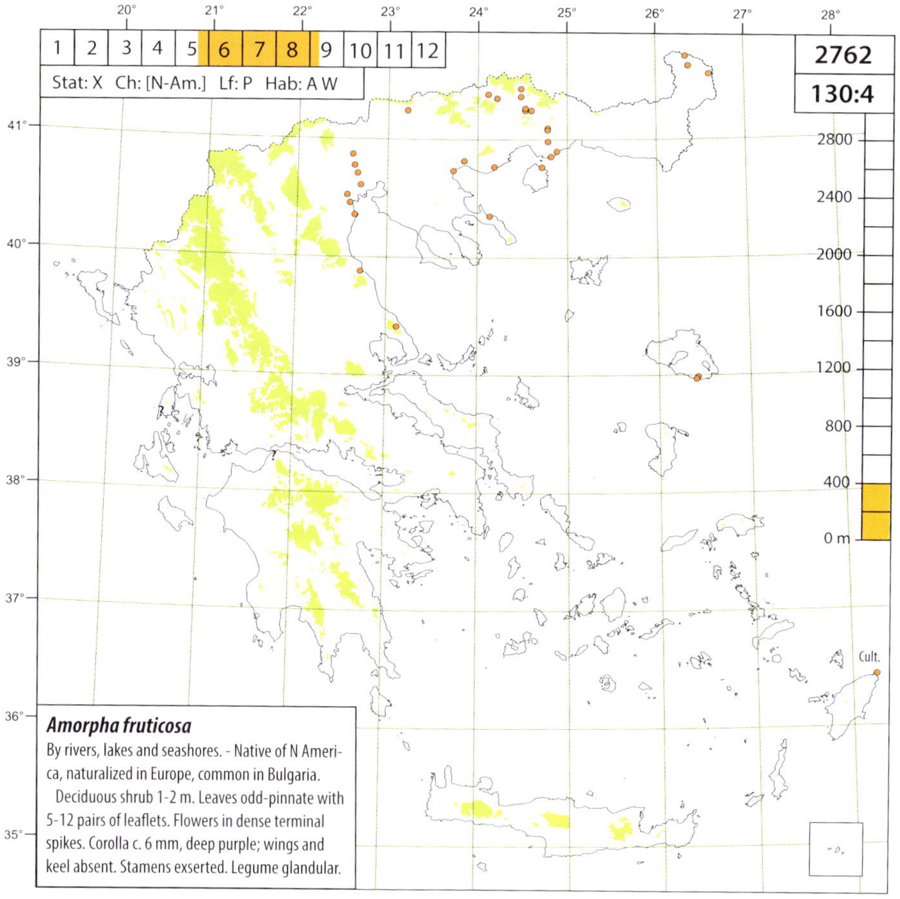

Amorpha fruticosa

By rivers, lakes and seashores. - Native of N America, naturalized in Europe, common in Bulgaria.

Deciduous shrub 1-2 m. Leaves odd-pinnate with 5-12 pairs of leaflets. Flowers in dense terminal spikes. Corolla c. 6 mm, deep purple; wings and keel absent. Stamens exserted. Legume glandular.

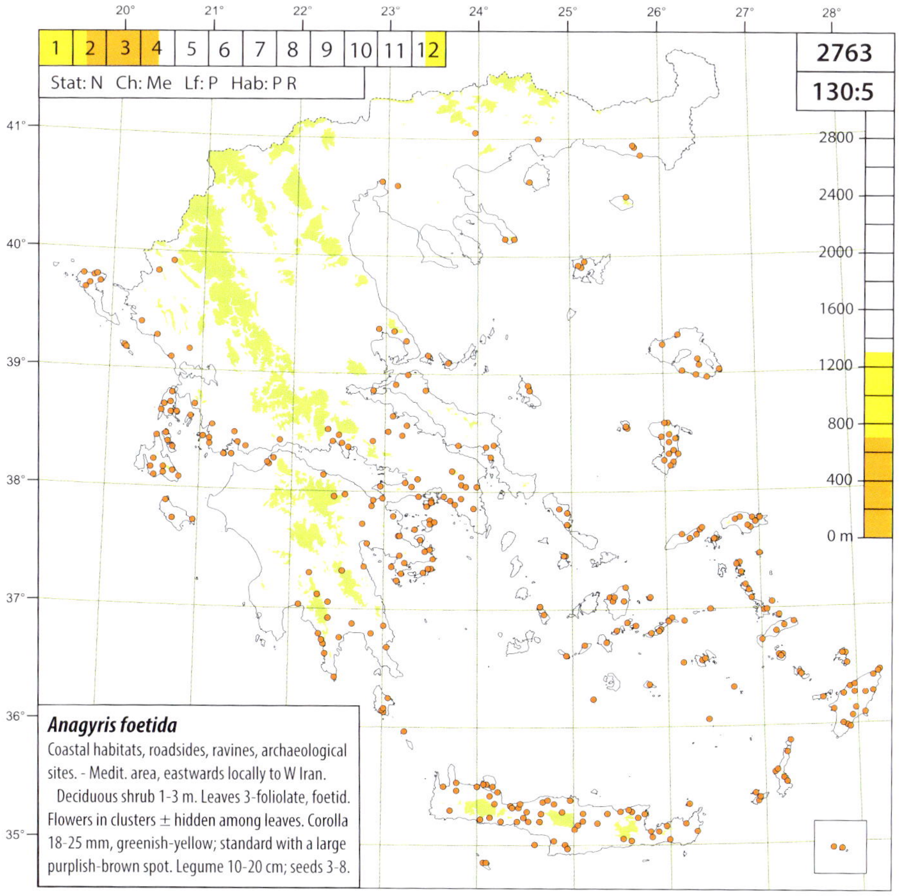

Anagyris foetida

Coastal habitats, roadsides, ravines, archaeological sites. - Medit. area, eastwards locally to W Iran.

Deciduous shrub 1-3 m. Leaves 3-foliolate, foetid. Flowers in clusters ± hidden among leaves. Corolla 18-25 mm, greenish-yellow; standard with a large purplish-brown spot. Legume 10-20 cm; seeds 3-8.

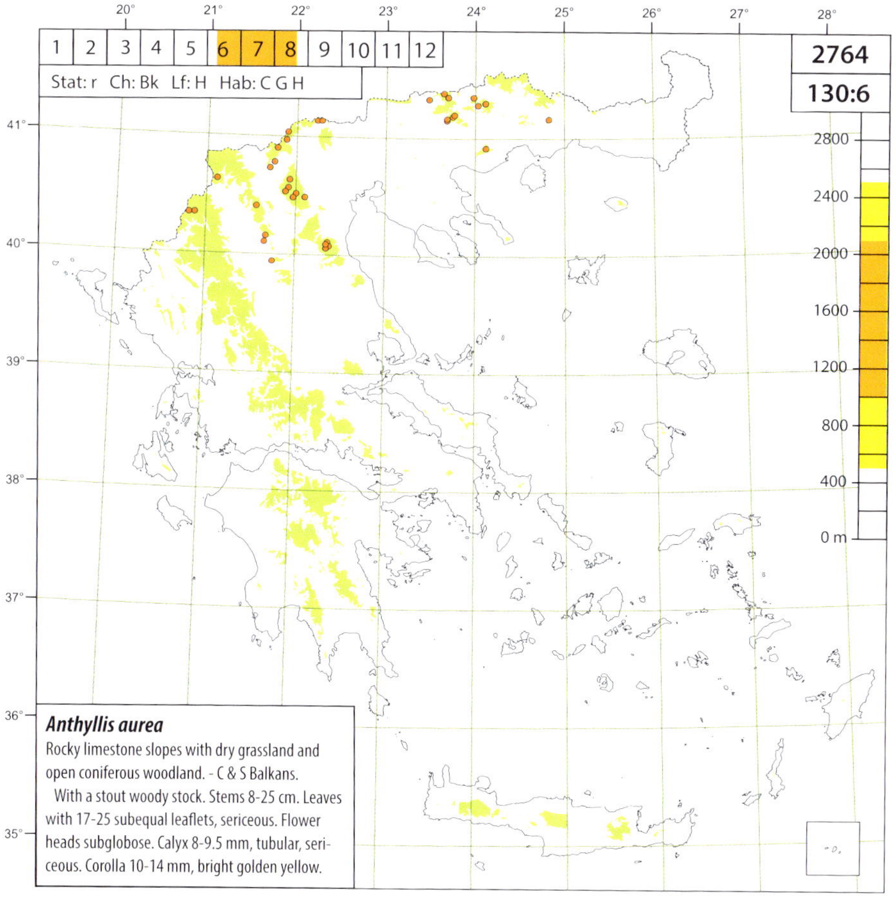

Anthyllis aurea

Rocky limestone slopes with dry grassland and open coniferous woodland. - C & S Balkans.

With a stout woody stock. Stems 8-25 cm. Leaves with 17-25 subequal leaflets, sericeous. Flower heads subglobose. Calyx 8-9.5 mm, tubular, sericeous. Corolla 10-14 mm, bright golden yellow.

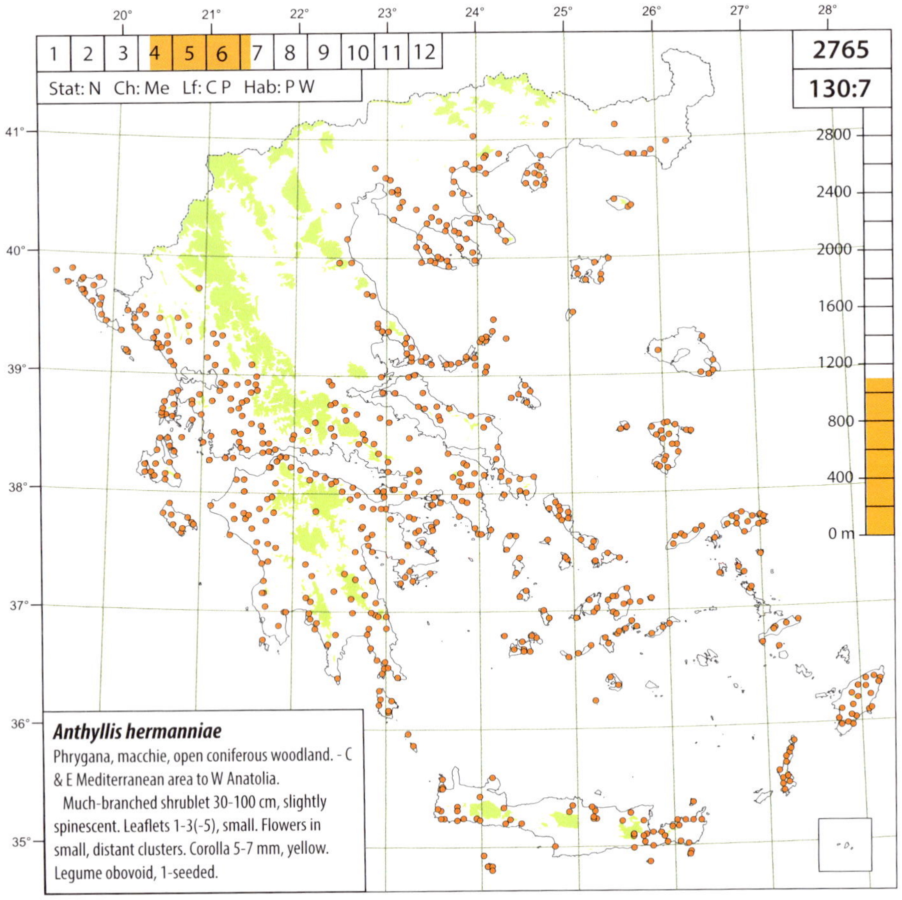

Anthyllis hermanniae
Phrygana, macchie, open coniferous woodland. - C & E Mediterranean area to W Anatolia.
Much-branched shrublet 30-100 cm, slightly spinescent. Leaflets 1-3(-5), small. Flowers in small, distant clusters. Corolla 5-7 mm, yellow. Legume obovoid, 1-seeded.

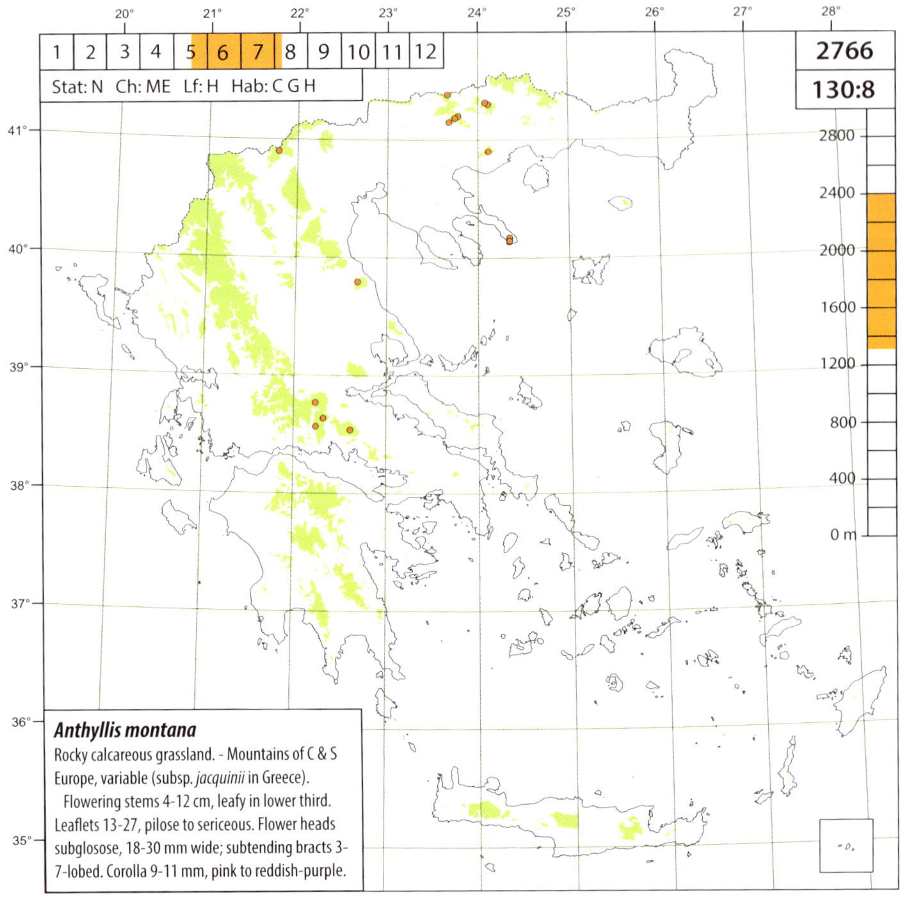

Anthyllis montana
Rocky calcareous grassland. - Mountains of C & S Europe, variable (subsp. *jacquinii* in Greece).
Flowering stems 4-12 cm, leafy in lower third. Leaflets 13-27, pilose to sericeous. Flower heads subglosose, 18-30 mm wide; subtending bracts 3-7-lobed. Corolla 9-11 mm, pink to reddish-purple.

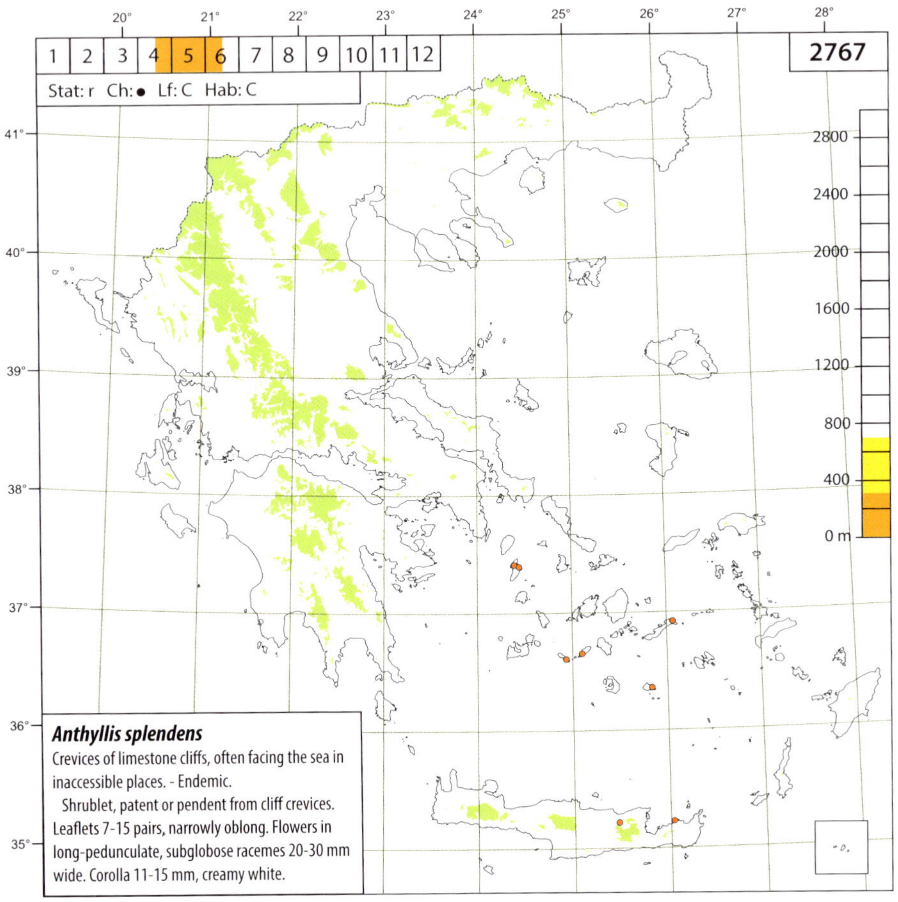

Anthyllis splendens
Crevices of limestone cliffs, often facing the sea in inaccessible places. - Endemic.
Shrublet, patent or pendent from cliff crevices. Leaflets 7-15 pairs, narrowly oblong. Flowers in long-pedunculate, subglobose racemes 20-30 mm wide. Corolla 11-15 mm, creamy white.

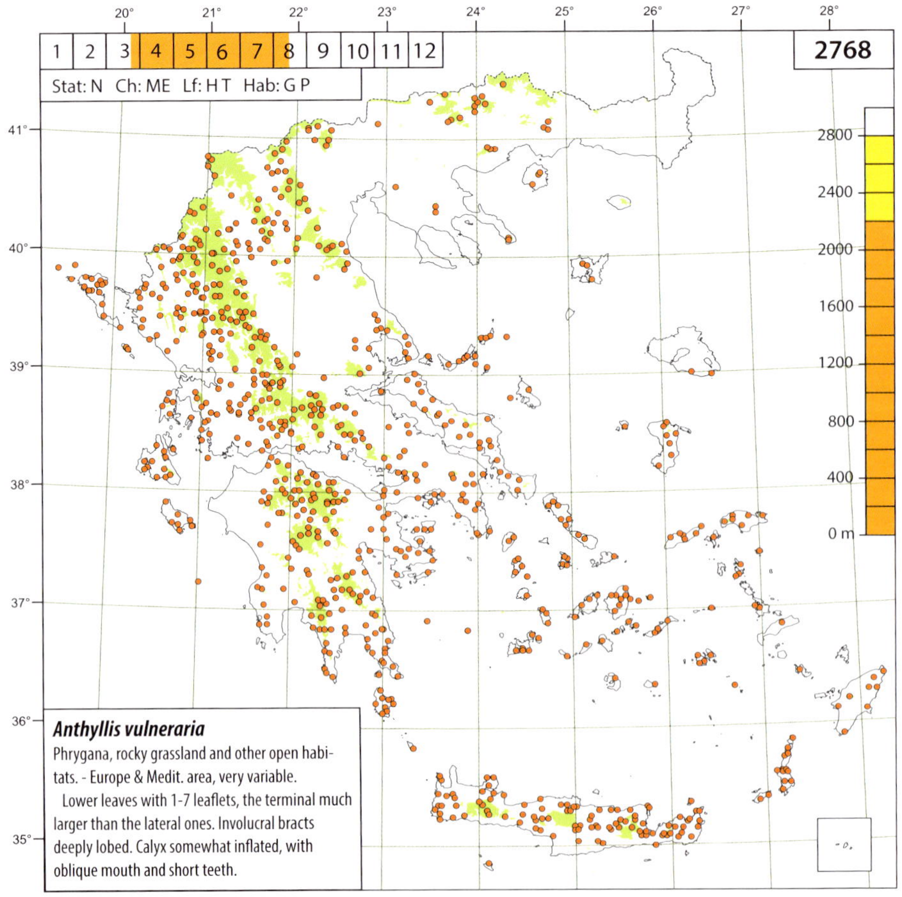

Anthyllis vulneraria
Phrygana, rocky grassland and other open habitats. - Europe & Medit. area, very variable.
Lower leaves with 1-7 leaflets, the terminal much larger than the lateral ones. Involucral bracts deeply lobed. Calyx somewhat inflated, with oblique mouth and short teeth.

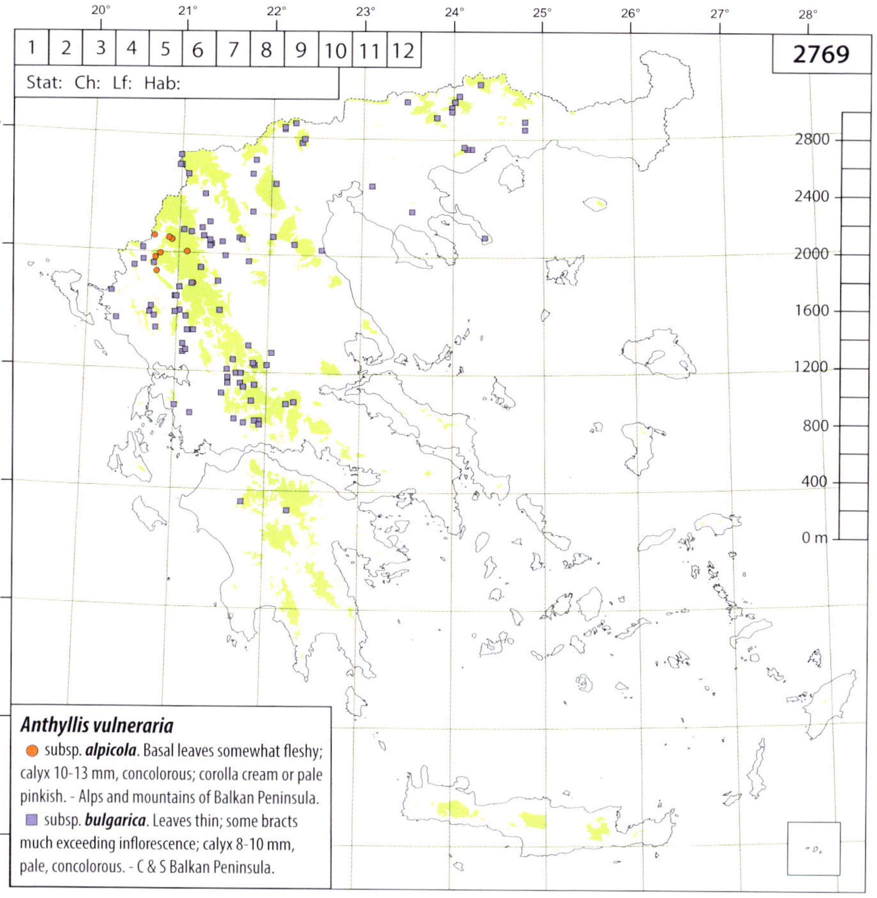

Anthyllis vulneraria
● subsp. ***alpicola***. Basal leaves somewhat fleshy; calyx 10-13 mm, concolorous; corolla cream or pale pinkish. - Alps and mountains of Balkan Peninsula.
■ subsp. ***bulgarica***. Leaves thin; some bracts much exceeding inflorescence; calyx 8-10 mm, pale, concolorous. - C & S Balkan Peninsula.

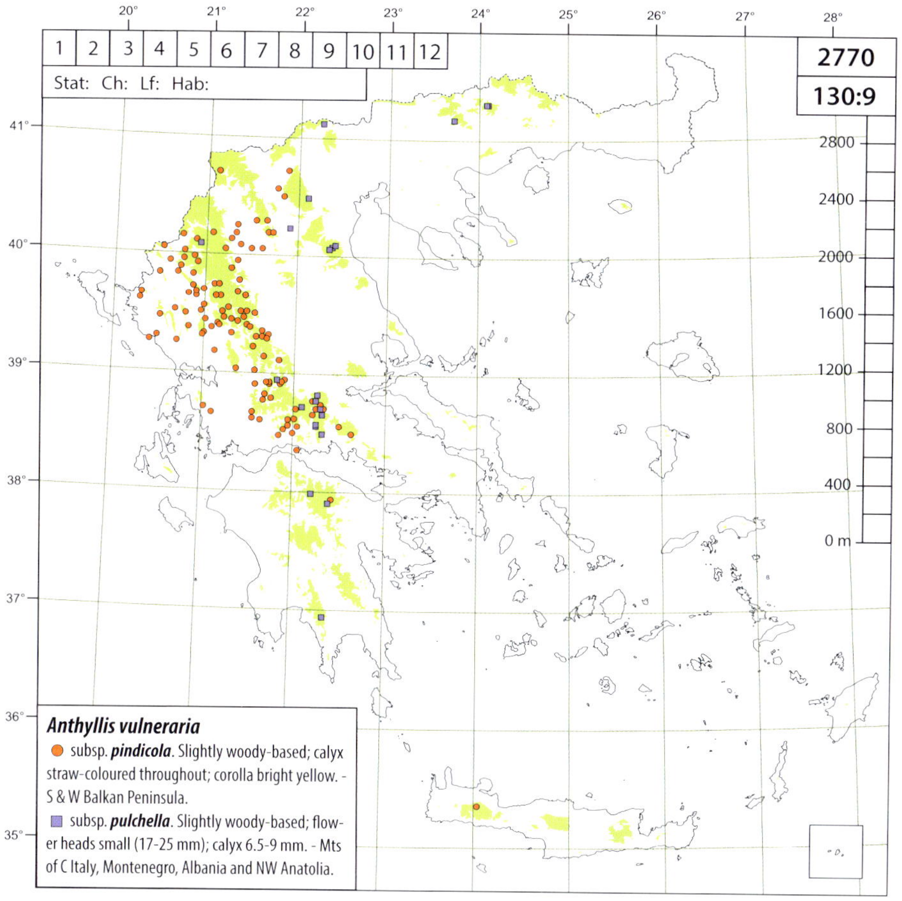

Anthyllis vulneraria
● subsp. ***pindicola***. Slightly woody-based; calyx straw-coloured throughout; corolla bright yellow. - S & W Balkan Peninsula.
■ subsp. ***pulchella***. Slightly woody-based; flower heads small (17-25 mm); calyx 6.5-9 mm. - Mts of C Italy, Montenegro, Albania and NW Anatolia.

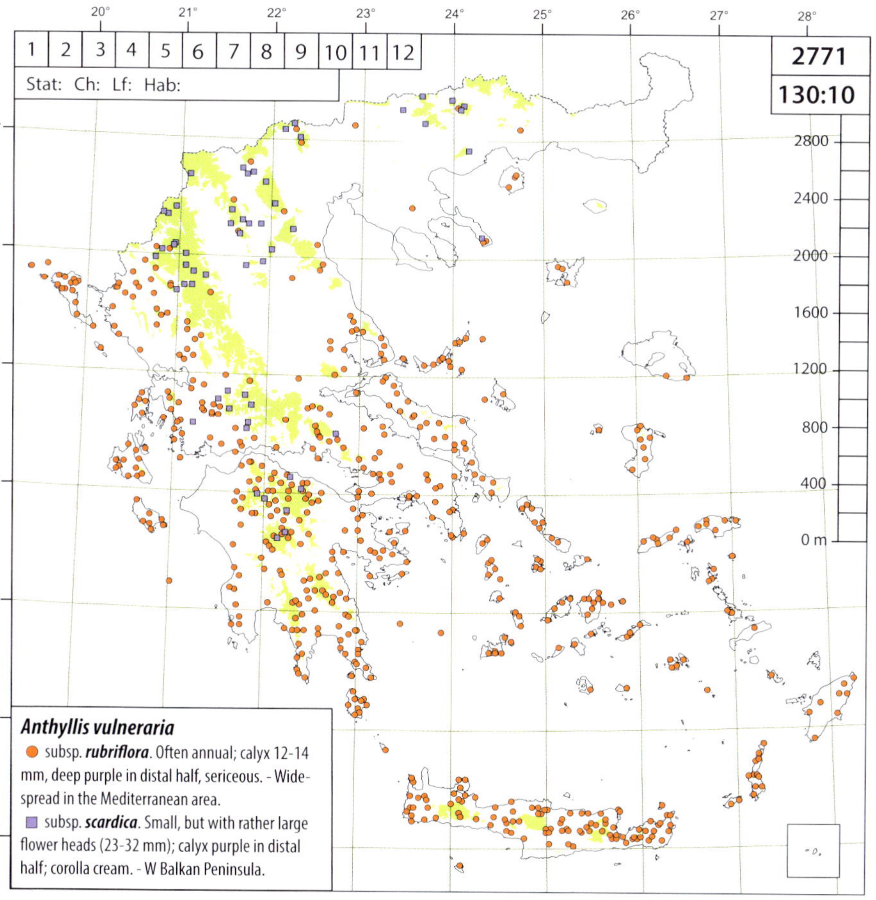

Anthyllis vulneraria
● subsp. ***rubriflora***. Often annual; calyx 12-14 mm, deep purple in distal half, sericeous. - Widespread in the Mediterranean area.
■ subsp. ***scardica***. Small, but with rather large flower heads (23-32 mm); calyx purple in distal half; corolla cream. - W Balkan Peninsula.

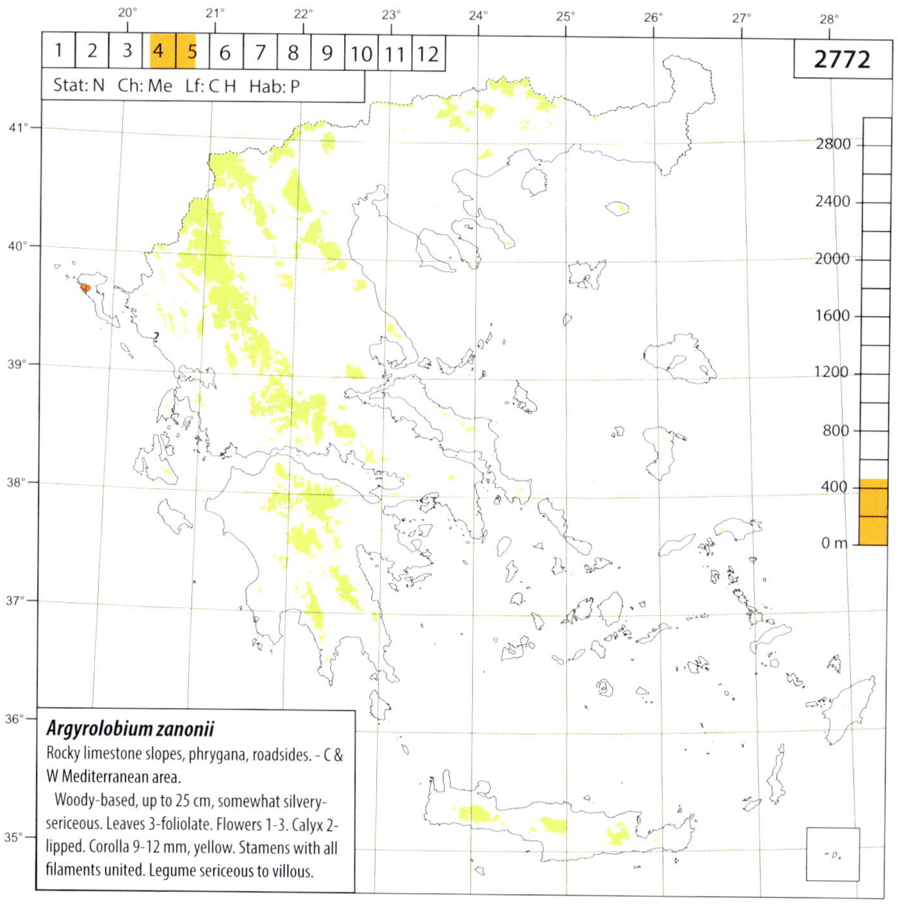

Argyrolobium zanonii
Rocky limestone slopes, phrygana, roadsides. - C & W Mediterranean area.
Woody-based, up to 25 cm, somewhat silvery-sericeous. Leaves 3-foliolate. Flowers 1-3. Calyx 2-lipped. Corolla 9-12 mm, yellow. Stamens with all filaments united. Legume sericeous to villous.

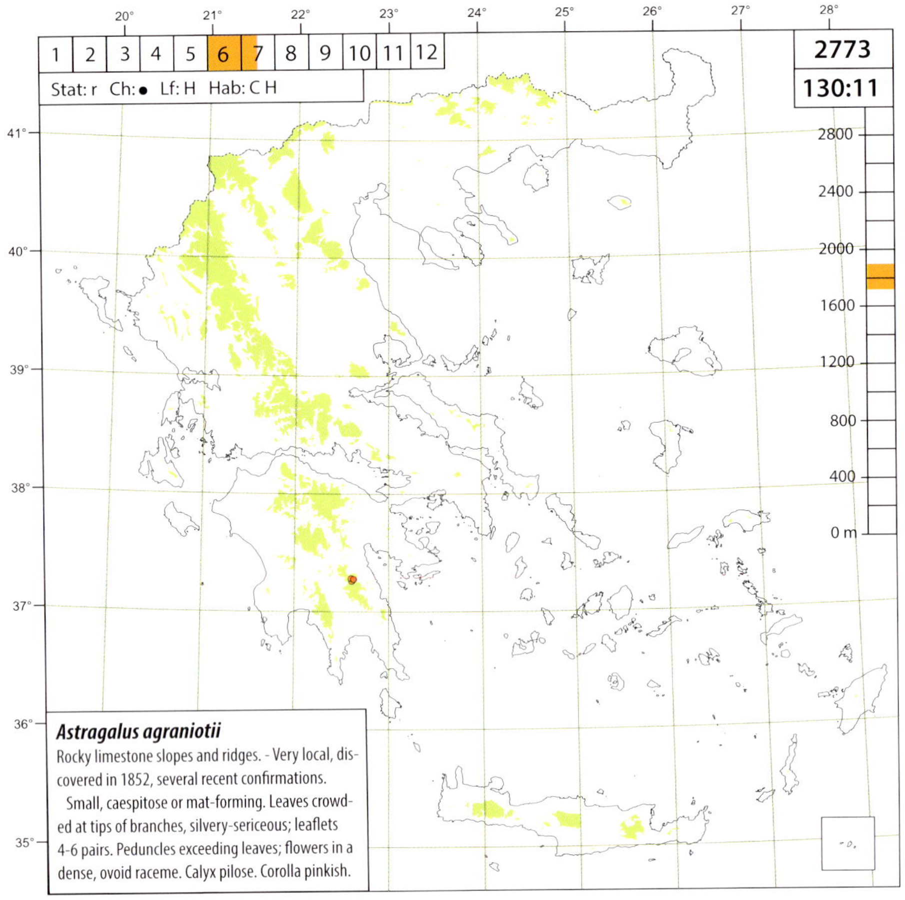

Astragalus agraniotii

Rocky limestone slopes and ridges. - Very local, discovered in 1852, several recent confirmations.

Small, caespitose or mat-forming. Leaves crowded at tips of branches, silvery-sericeous; leaflets 4-6 pairs. Peduncles exceeding leaves; flowers in a dense, ovoid raceme. Calyx pilose. Corolla pinkish.

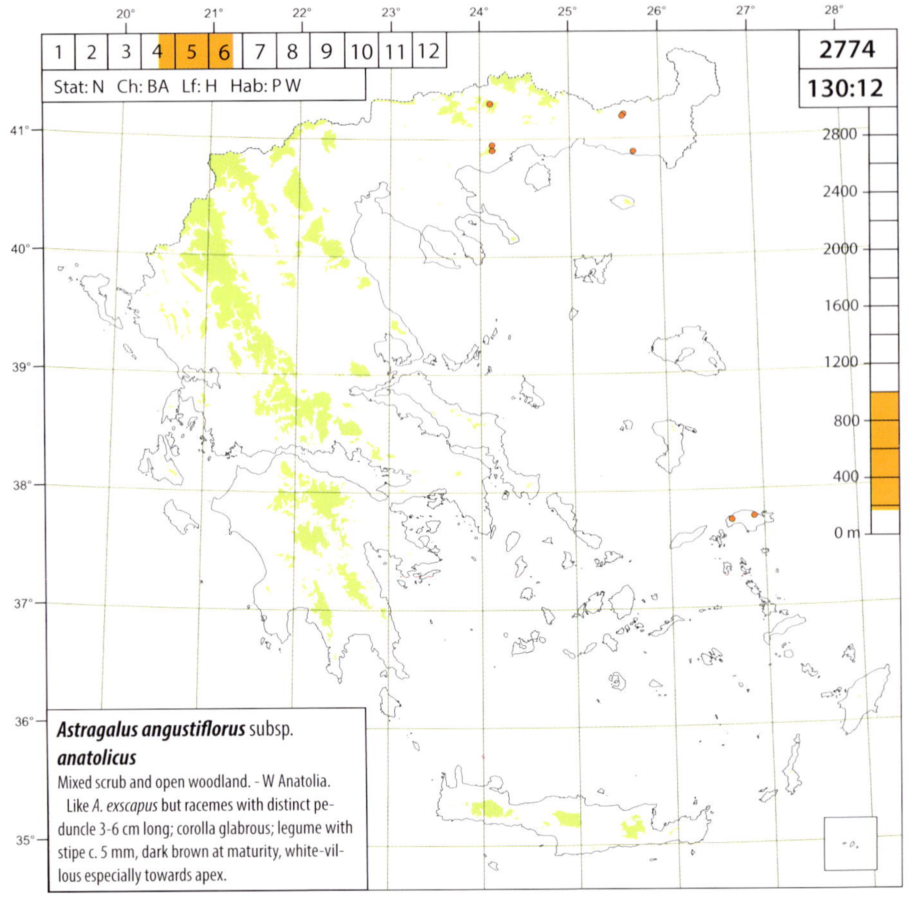

Astragalus angustiflorus subsp. ***anatolicus***

Mixed scrub and open woodland. - W Anatolia.

Like *A. exscapus* but racemes with distinct peduncle 3-6 cm long; corolla glabrous; legume with stipe c. 5 mm, dark brown at maturity, white-villous especially towards apex.

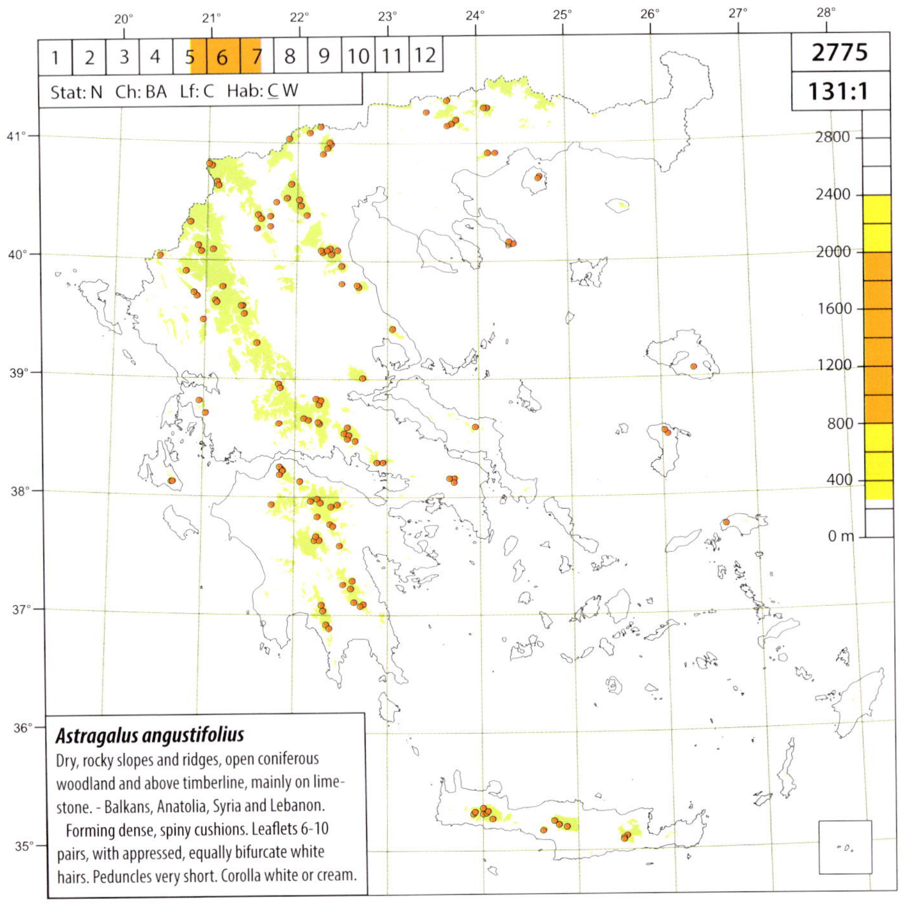

Astragalus angustifolius

Dry, rocky slopes and ridges, open coniferous woodland and above timberline, mainly on limestone. - Balkans, Anatolia, Syria and Lebanon.

Forming dense, spiny cushions. Leaflets 6-10 pairs, with appressed, equally bifurcate white hairs. Peduncles very short. Corolla white or cream.

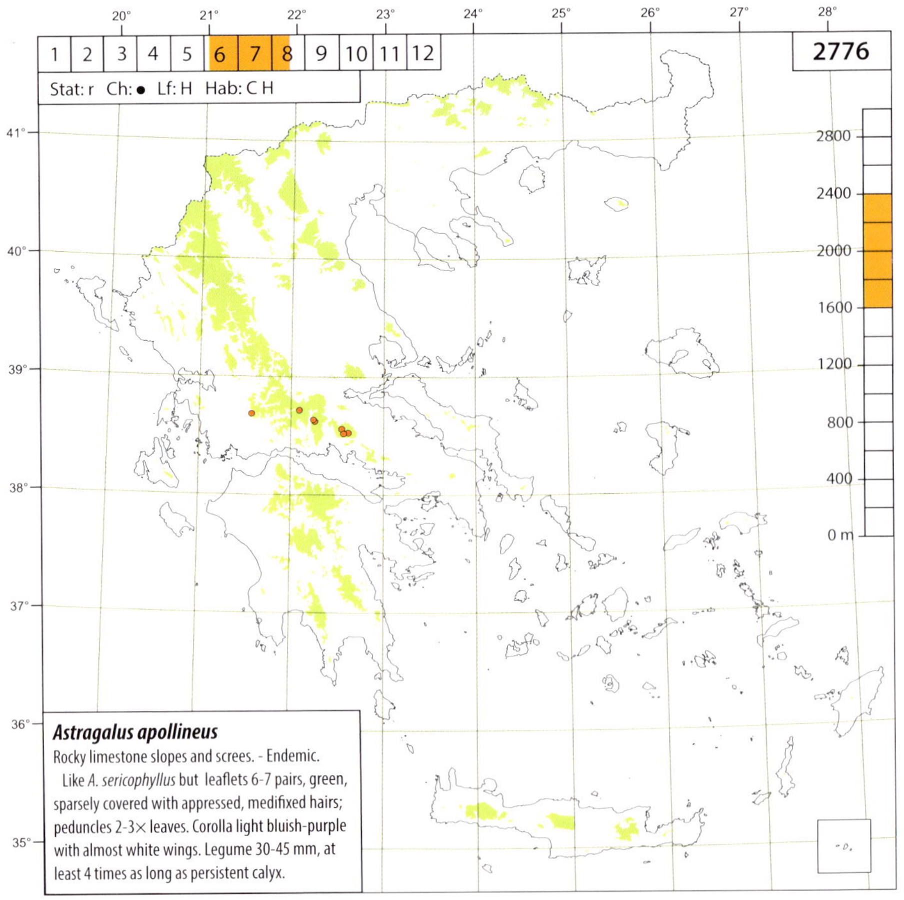

Astragalus apollineus

Rocky limestone slopes and screes. - Endemic.

Like *A. sericophyllus* but leaflets 6-7 pairs, green, sparsely covered with appressed, medifixed hairs; peduncles 2-3× leaves. Corolla light bluish-purple with almost white wings. Legume 30-45 mm, at least 4 times as long as persistent calyx.

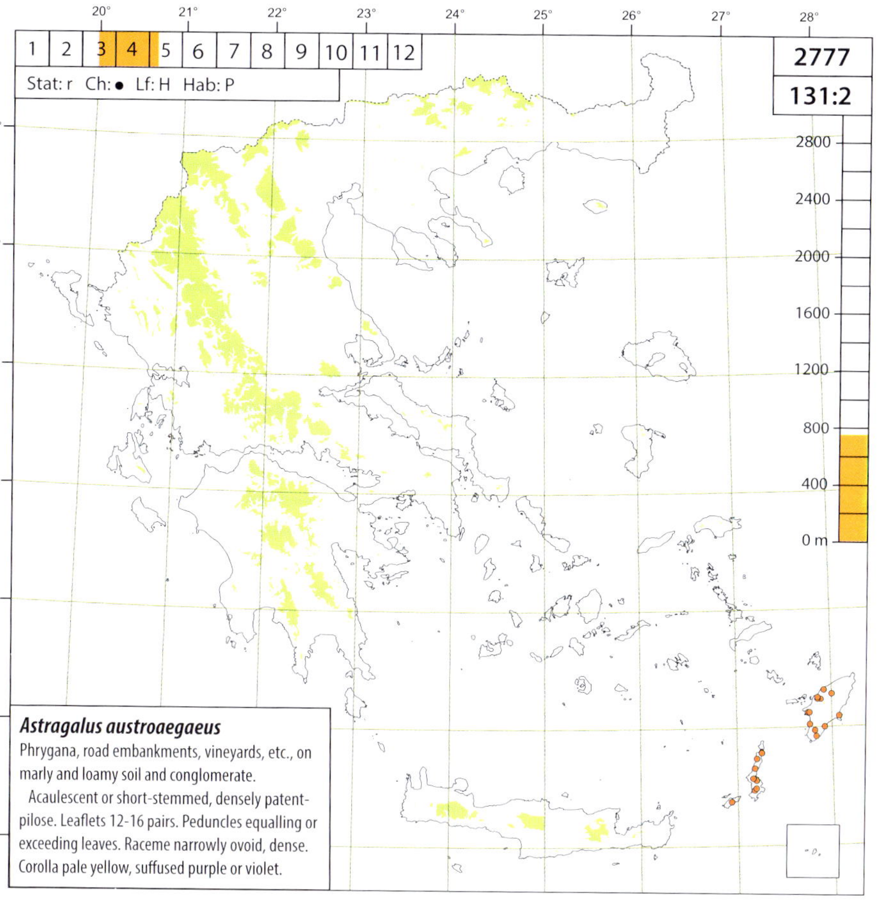

Astragalus austroaegaeus

Phrygana, road embankments, vineyards, etc., on marly and loamy soil and conglomerate.

Acaulescent or short-stemmed, densely patent-pilose. Leaflets 12-16 pairs. Peduncles equalling or exceeding leaves. Raceme narrowly ovoid, dense. Corolla pale yellow, suffused purple or violet.

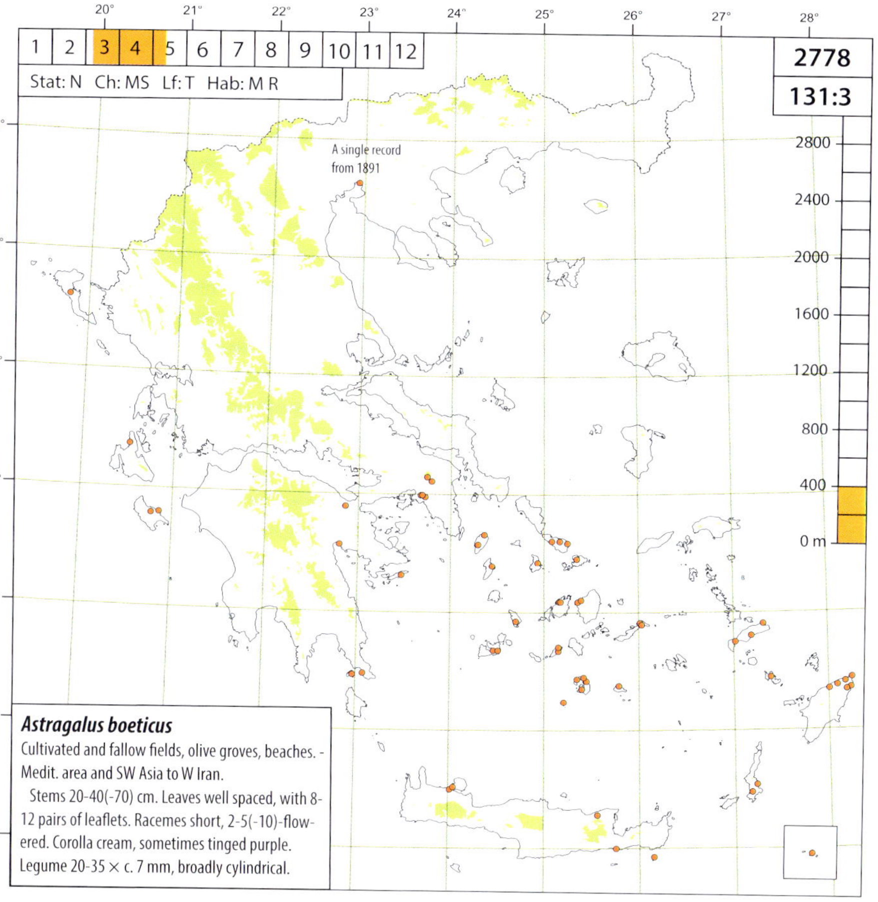

Astragalus boeticus

Cultivated and fallow fields, olive groves, beaches. - Medit. area and SW Asia to W Iran.

Stems 20-40(-70) cm. Leaves well spaced, with 8-12 pairs of leaflets. Racemes short, 2-5(-10)-flowered. Corolla cream, sometimes tinged purple. Legume 20-35 × c. 7 mm, broadly cylindrical.

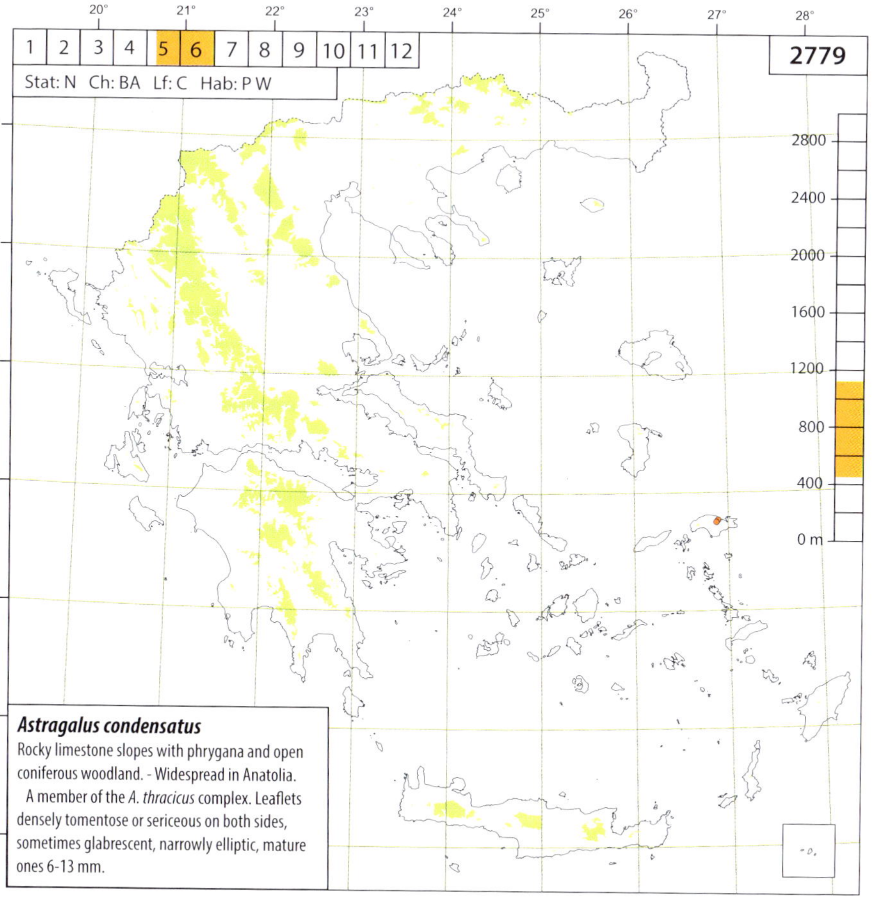

Astragalus condensatus

Rocky limestone slopes with phrygana and open coniferous woodland. - Widespread in Anatolia.

A member of the *A. thracicus* complex. Leaflets densely tomentose or sericeous on both sides, sometimes glabrescent, narrowly elliptic, mature ones 6-13 mm.

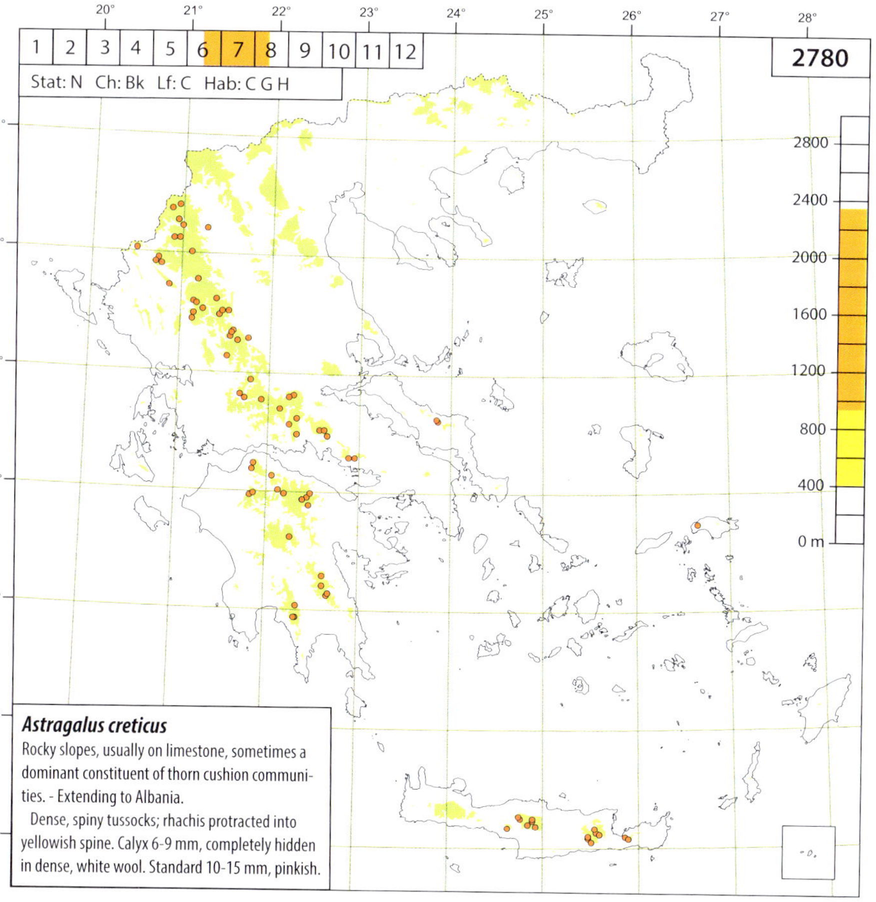

Astragalus creticus

Rocky slopes, usually on limestone, sometimes a dominant constituent of thorn cushion communities. - Extending to Albania.

Dense, spiny tussocks; rhachis protracted into yellowish spine. Calyx 6-9 mm, completely hidden in dense, white wool. Standard 10-15 mm, pinkish.

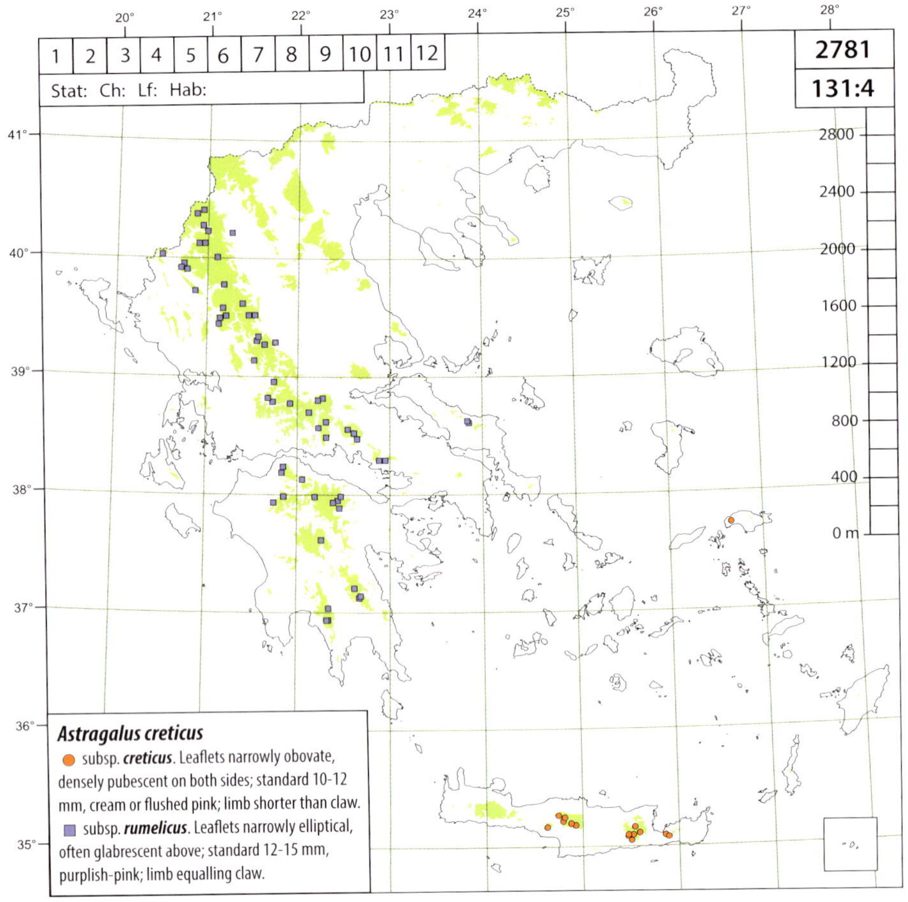

Astragalus creticus

● subsp. ***creticus***. Leaflets narrowly obovate, densely pubescent on both sides; standard 10-12 mm, cream or flushed pink; limb shorter than claw.

■ subsp. ***rumelicus***. Leaflets narrowly elliptical, often glabrescent above; standard 12-15 mm, purplish-pink; limb equalling claw.

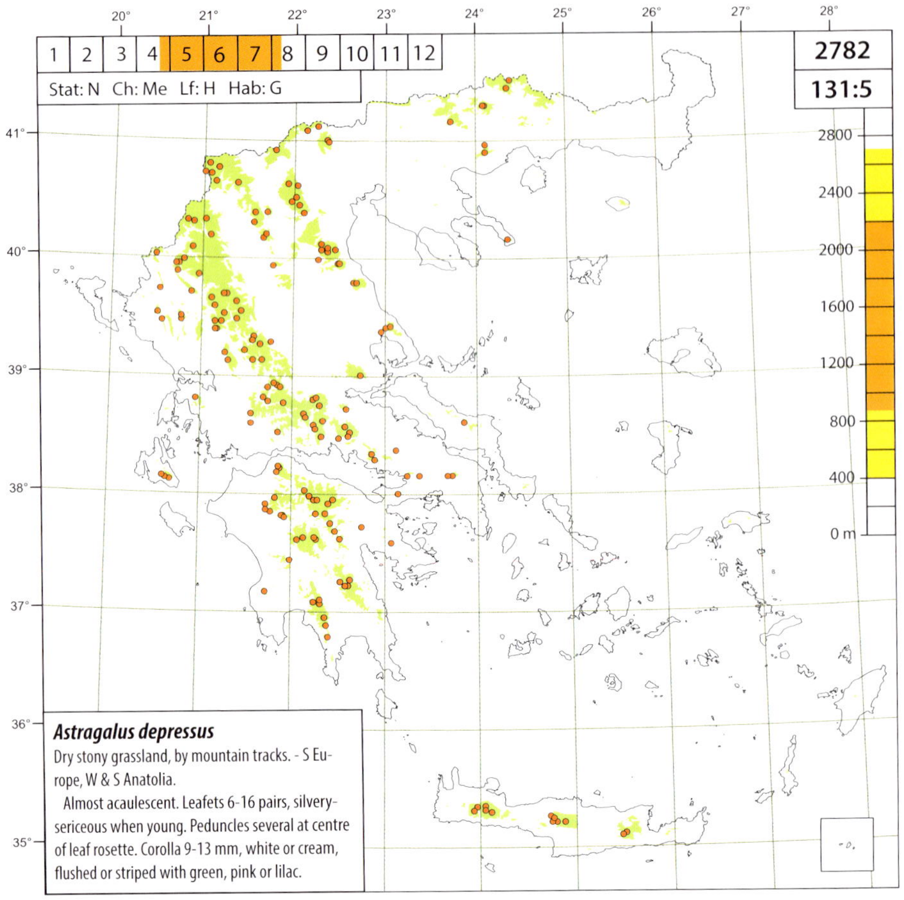

Astragalus depressus

Dry stony grassland, by mountain tracks. - S Europe, W & S Anatolia.

Almost acaulescent. Leafets 6-16 pairs, silvery-sericeous when young. Peduncles several at centre of leaf rosette. Corolla 9-13 mm, white or cream, flushed or striped with green, pink or lilac.

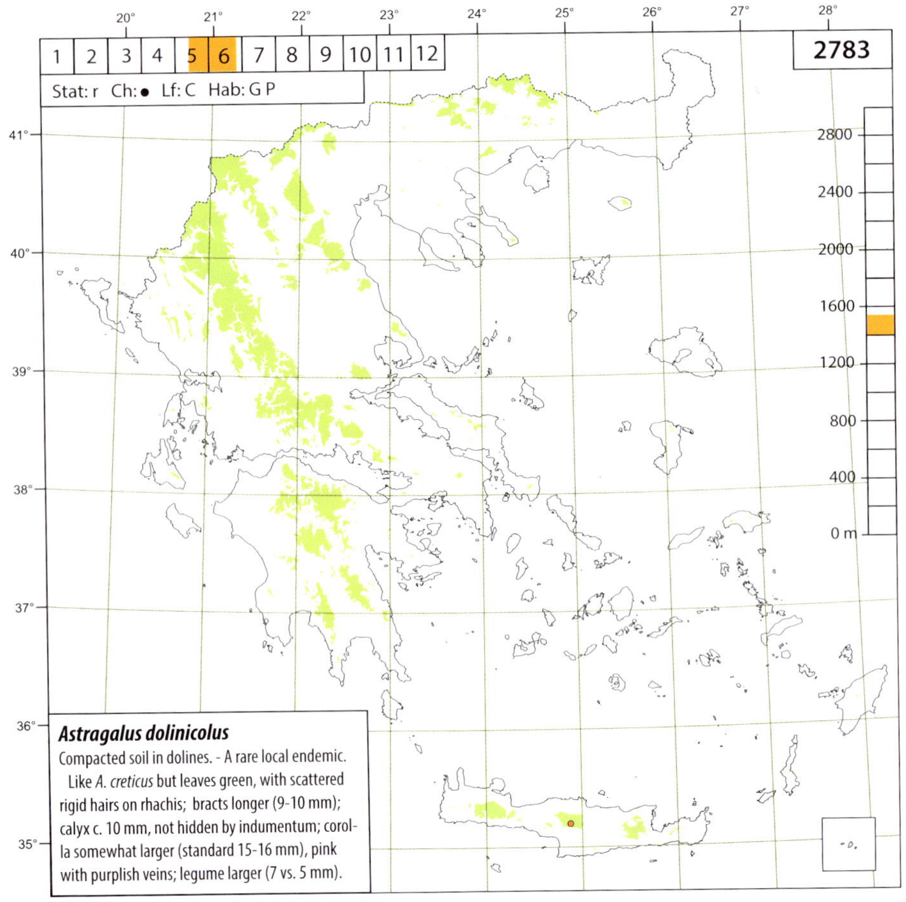

Astragalus dolinicolus

Compacted soil in dolines. - A rare local endemic.

Like *A. creticus* but leaves green, with scattered rigid hairs on rhachis; bracts longer (9-10 mm); calyx c. 10 mm, not hidden by indumentum; corolla somewhat larger (standard 15-16 mm), pink with purplish veins; legume larger (7 vs. 5 mm).

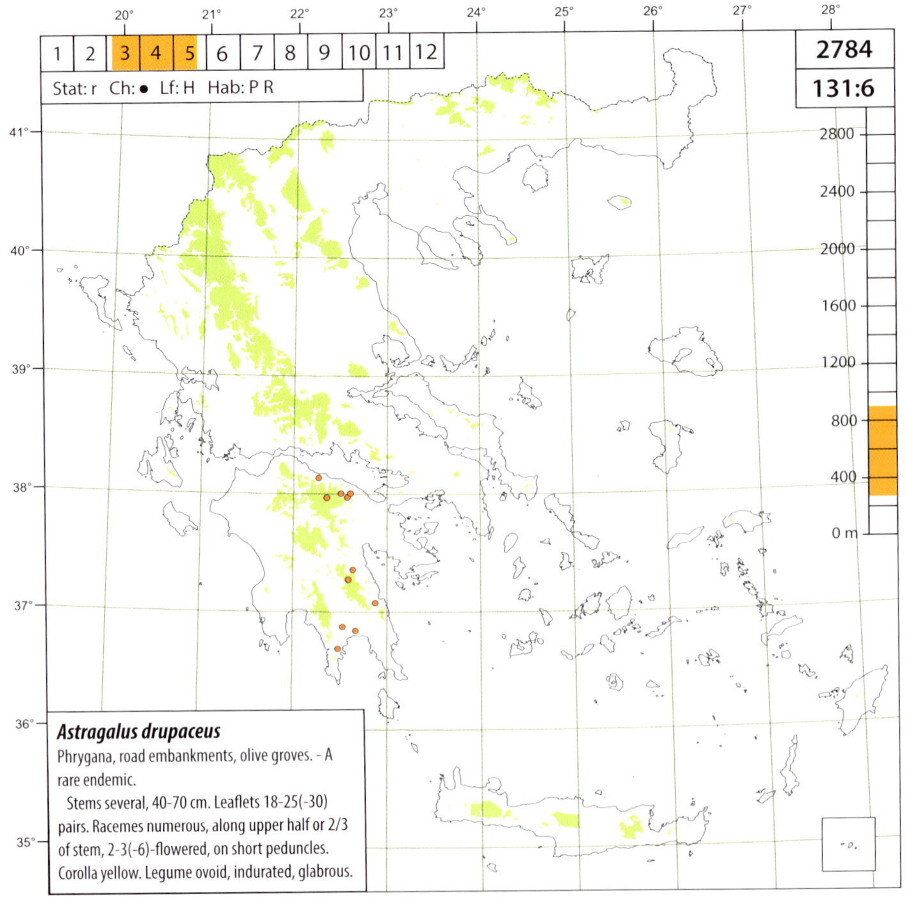

Astragalus drupaceus

Phrygana, road embankments, olive groves. - A rare endemic.

Stems several, 40-70 cm. Leaflets 18-25(-30) pairs. Racemes numerous, along upper half or 2/3 of stem, 2-3(-6)-flowered, on short peduncles. Corolla yellow. Legume ovoid, indurated, glabrous.

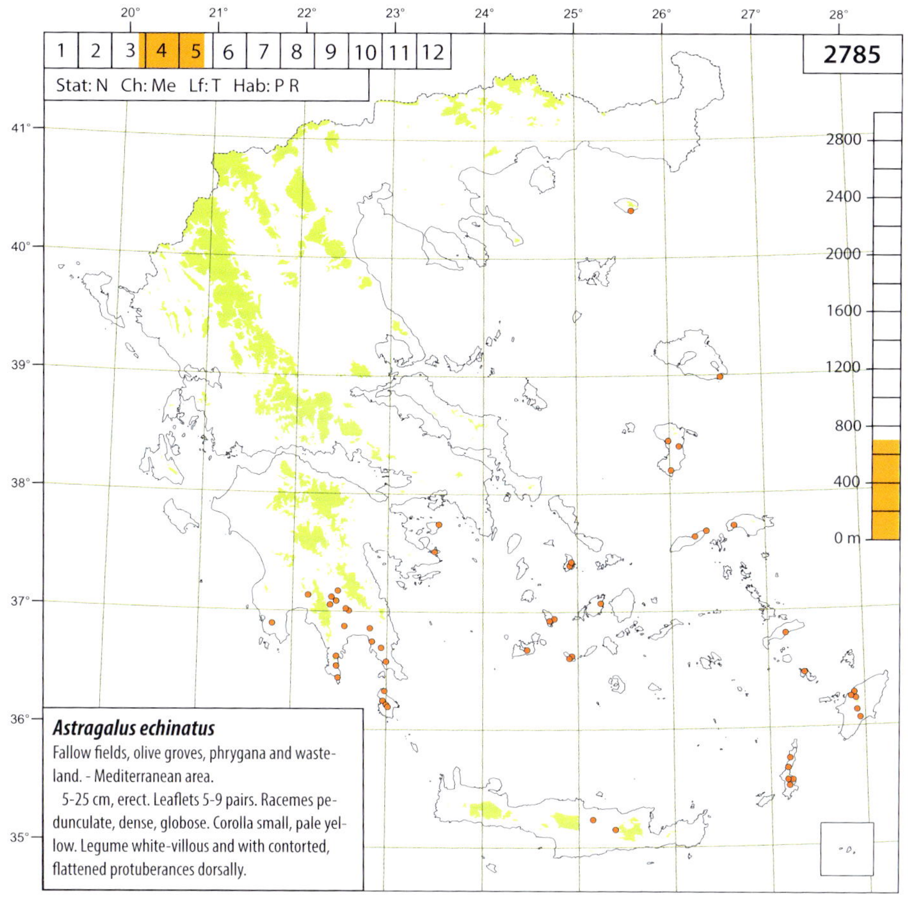

Astragalus echinatus
Fallow fields, olive groves, phrygana and wasteland. - Mediterranean area.
5-25 cm, erect. Leaflets 5-9 pairs. Racemes pedunculate, dense, globose. Corolla small, pale yellow. Legume white-villous and with contorted, flattened protuberances dorsally.

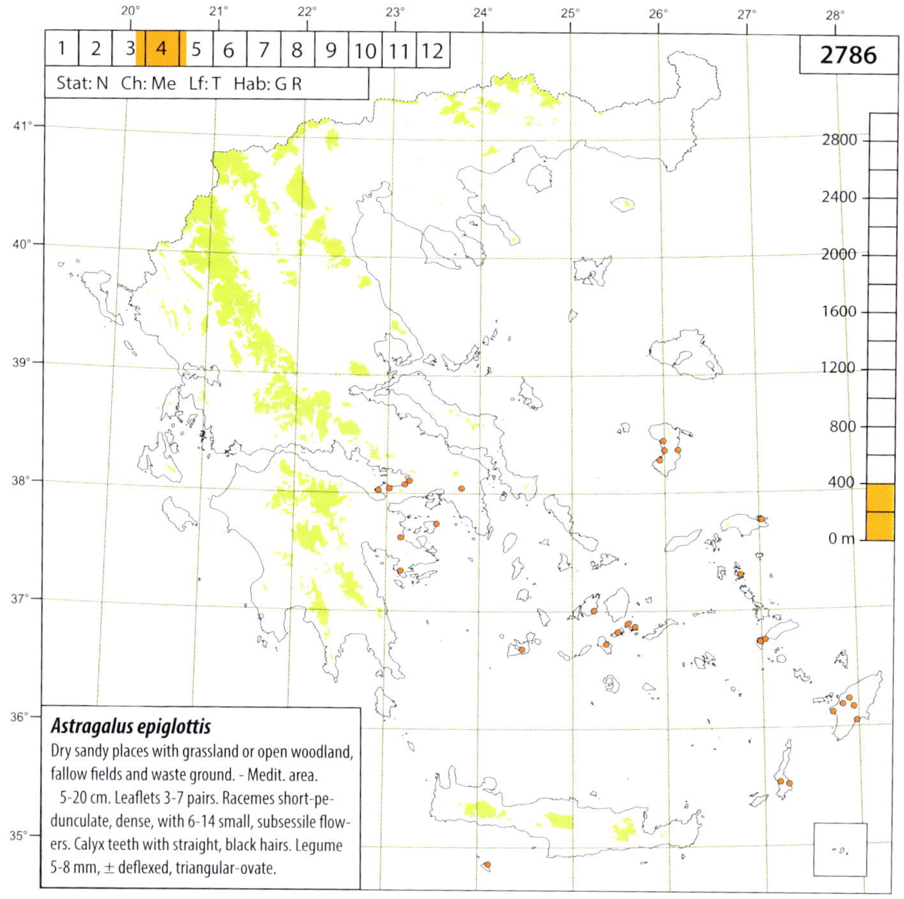

Astragalus epiglottis
Dry sandy places with grassland or open woodland, fallow fields and waste ground. - Medit. area.
5-20 cm. Leaflets 3-7 pairs. Racemes short-pedunculate, dense, with 6-14 small, subsessile flowers. Calyx teeth with straight, black hairs. Legume 5-8 mm, ± deflexed, triangular-ovate.

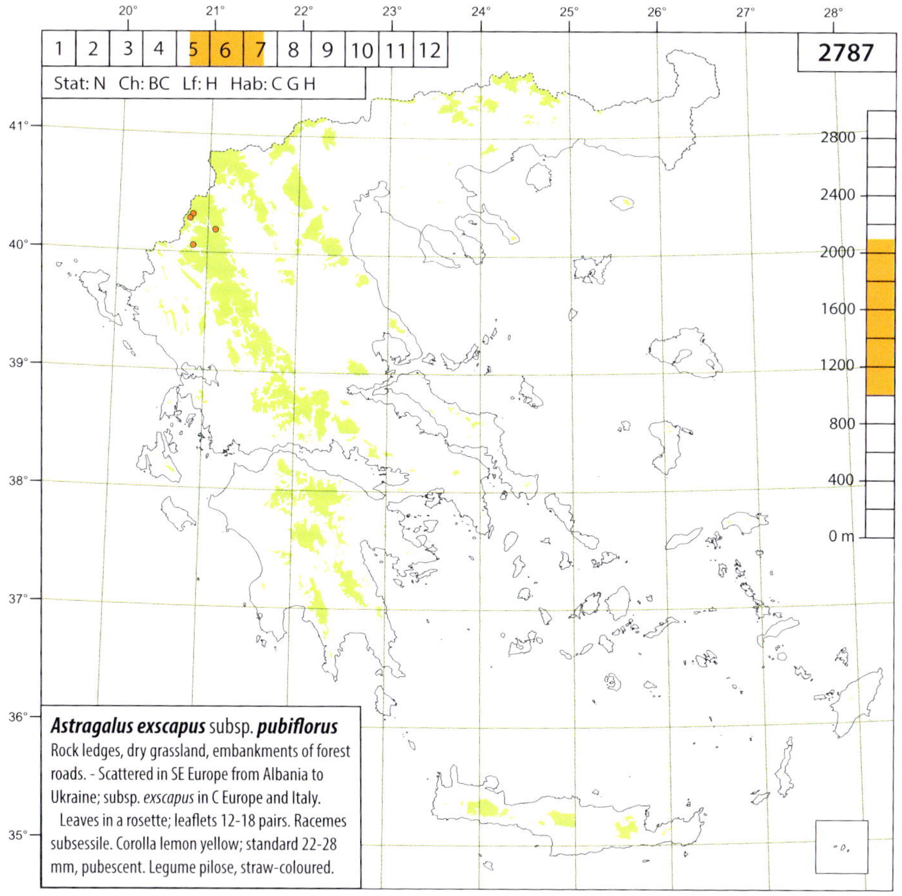

Astragalus exscapus subsp. ***pubiflorus***
Rock ledges, dry grassland, embankments of forest roads. - Scattered in SE Europe from Albania to Ukraine; subsp. *exscapus* in C Europe and Italy.
Leaves in a rosette; leaflets 12-18 pairs. Racemes subsessile. Corolla lemon yellow; standard 22-28 mm, pubescent. Legume pilose, straw-coloured.

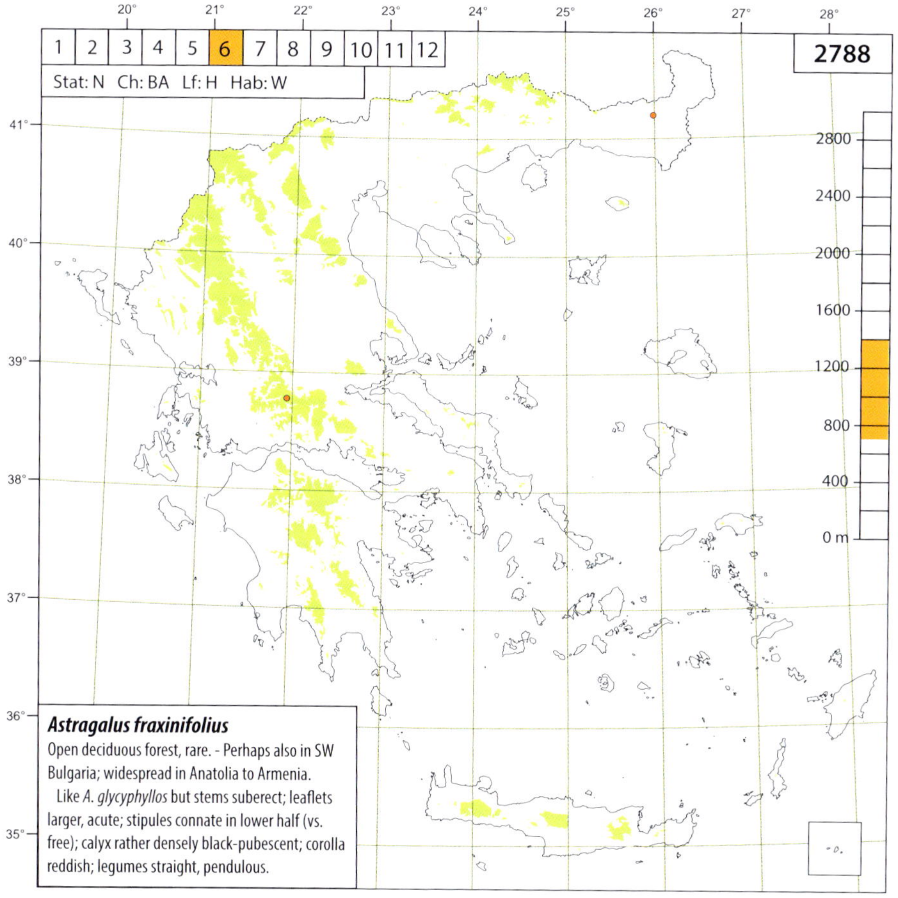

Astragalus fraxinifolius
Open deciduous forest, rare. - Perhaps also in SW Bulgaria; widespread in Anatolia to Armenia.
Like *A. glycyphyllos* but stems suberect; leaflets larger, acute; stipules connate in lower half (vs. free); calyx rather densely black-pubescent; corolla reddish; legumes straight, pendulous.

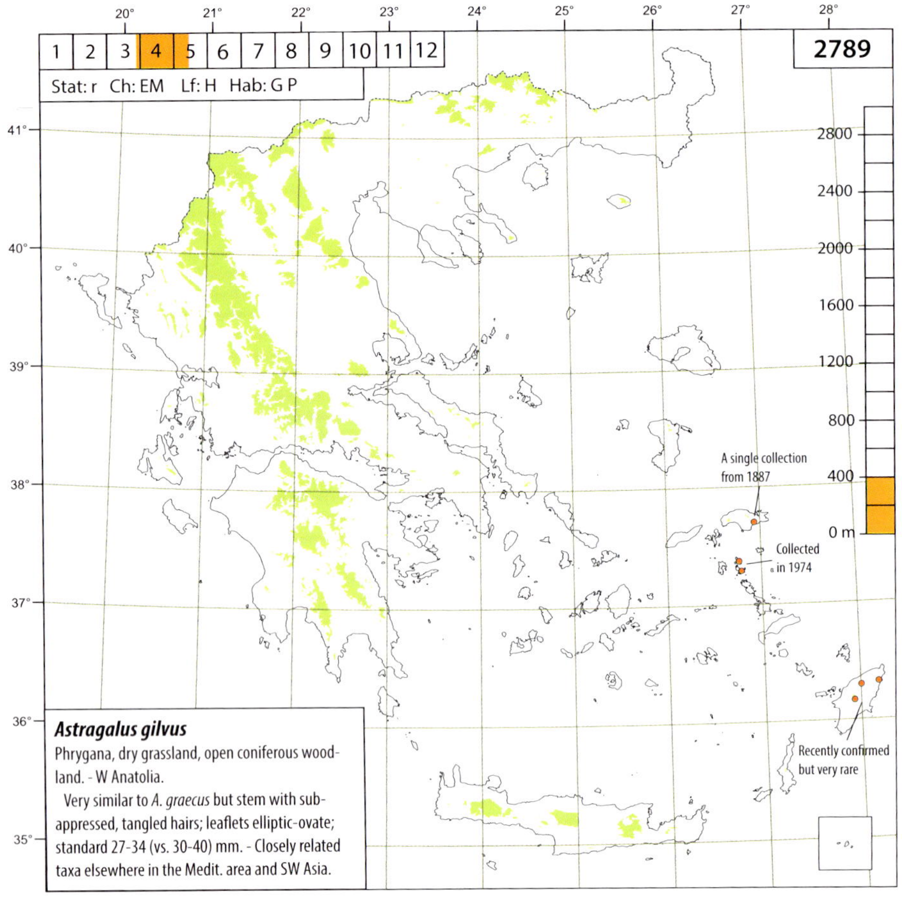

Astragalus gilvus
Phrygana, dry grassland, open coniferous woodland. - W Anatolia.
Very similar to *A. graecus* but stem with sub-appressed, tangled hairs; leaflets elliptic-ovate; standard 27-34 (vs. 30-40) mm. - Closely related taxa elsewhere in the Medit. area and SW Asia.

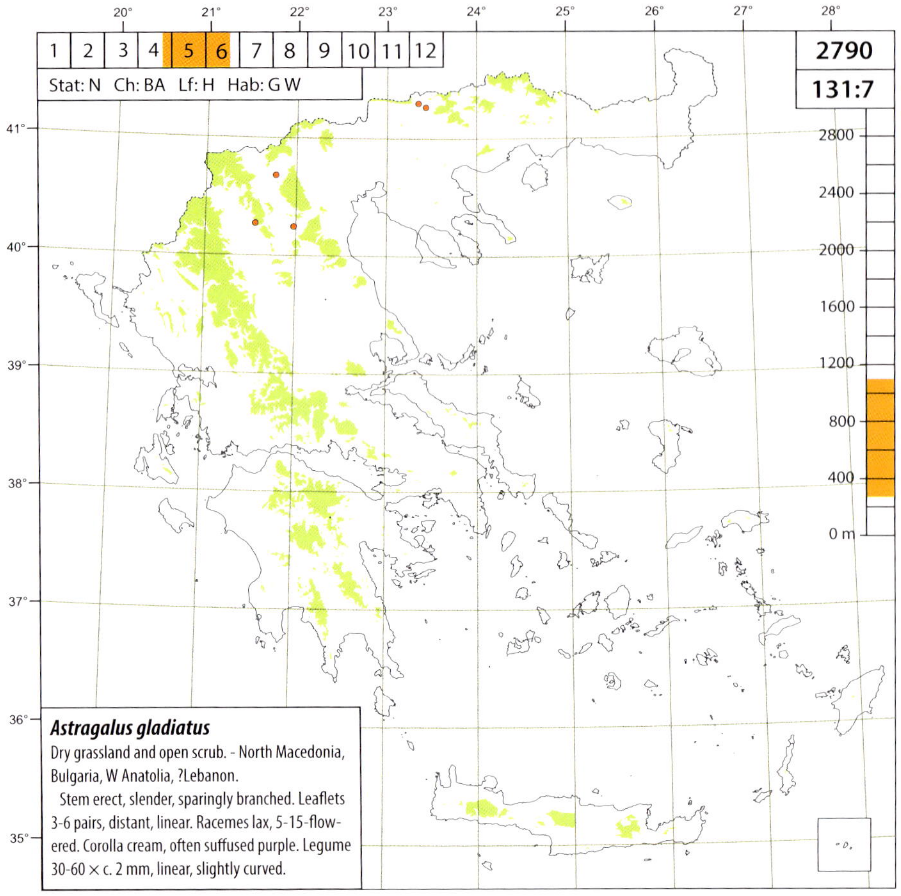

Astragalus gladiatus
Dry grassland and open scrub. - North Macedonia, Bulgaria, W Anatolia, ?Lebanon.
Stem erect, slender, sparingly branched. Leaflets 3-6 pairs, distant, linear. Racemes lax, 5-15-flowered. Corolla cream, often suffused purple. Legume 30-60 × c. 2 mm, linear, slightly curved.

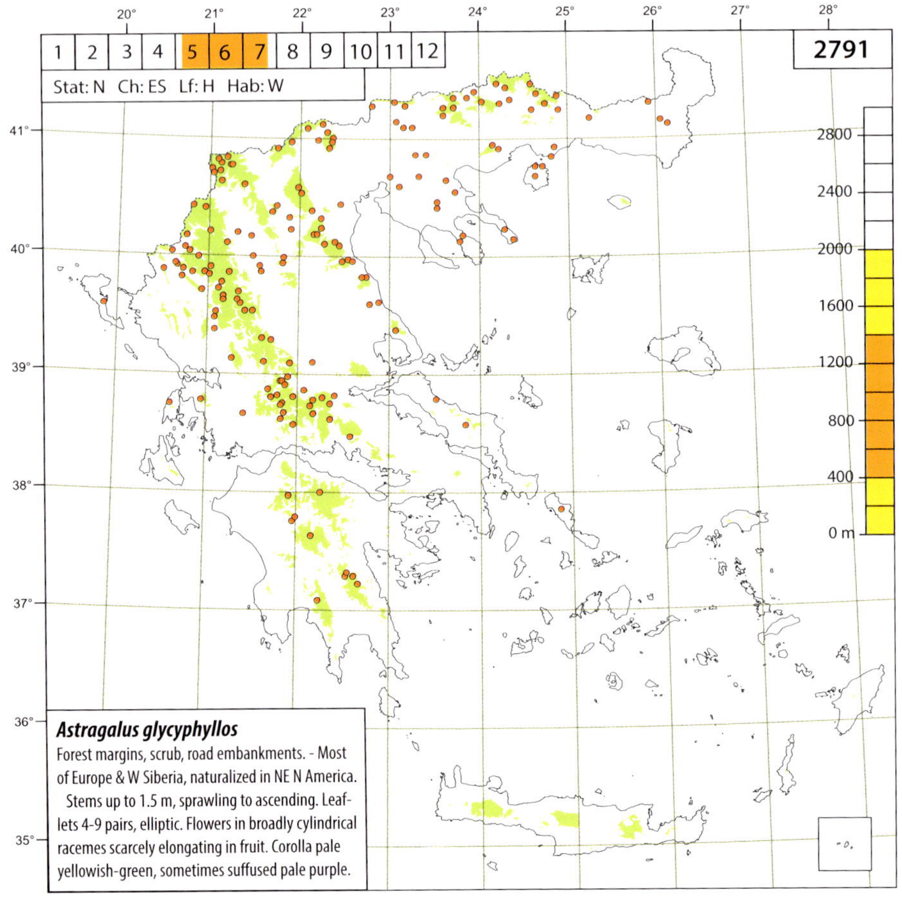

Astragalus glycyphyllos
Forest margins, scrub, road embankments. - Most of Europe & W Siberia, naturalized in NE N America.
Stems up to 1.5 m, sprawling to ascending. Leaflets 4-9 pairs, elliptic. Flowers in broadly cylindrical racemes scarcely elongating in fruit. Corolla pale yellowish-green, sometimes suffused pale purple.

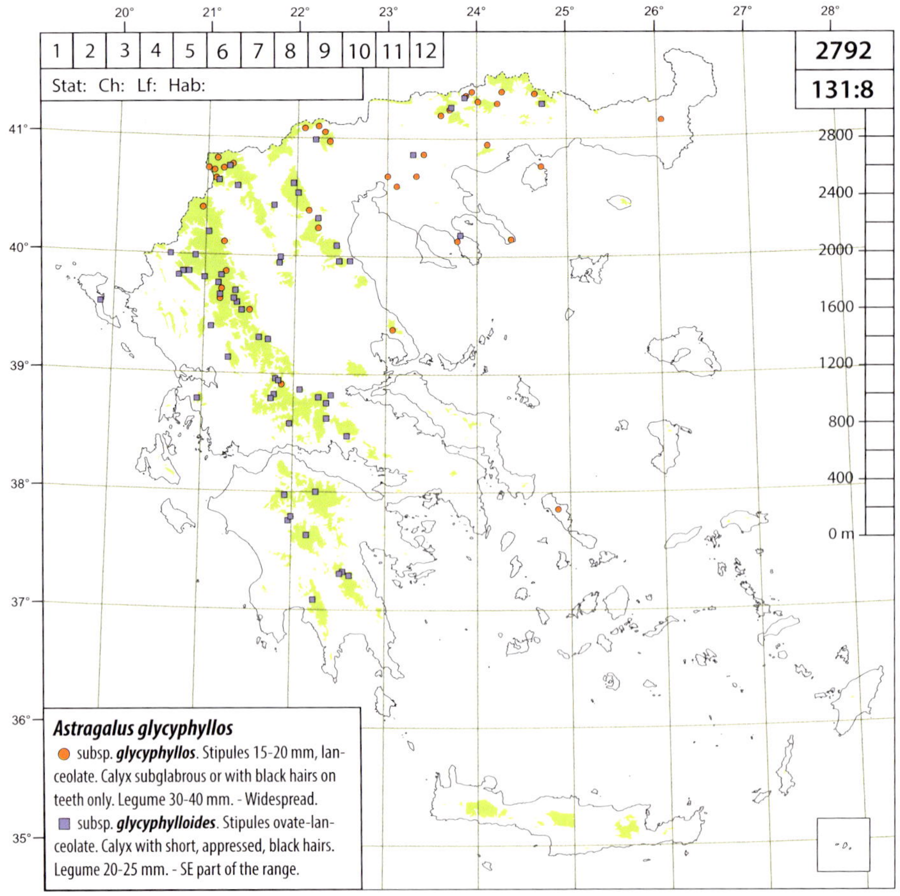

Astragalus glycyphyllos
● subsp. ***glycyphyllos***. Stipules 15-20 mm, lanceolate. Calyx subglabrous or with black hairs on teeth only. Legume 30-40 mm. - Widespread.
■ subsp. ***glycyphylloides***. Stipules ovate-lanceolate. Calyx with short, appressed, black hairs. Legume 20-25 mm. - SE part of the range.

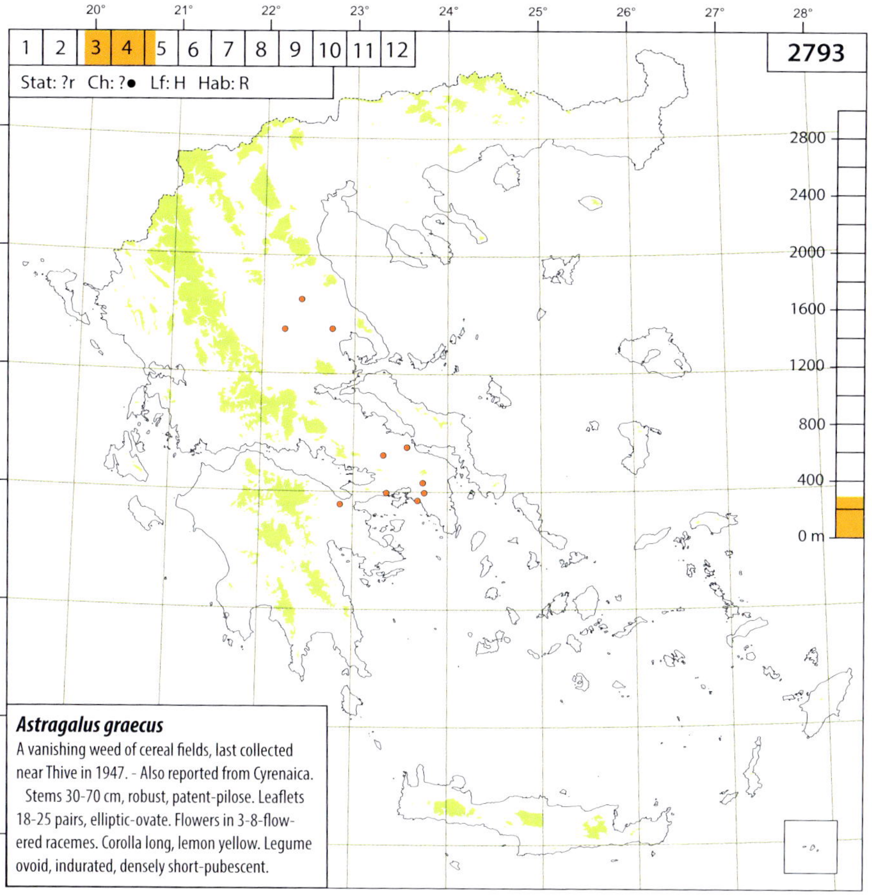

Astragalus graecus
A vanishing weed of cereal fields, last collected near Thive in 1947. - Also reported from Cyrenaica.
Stems 30-70 cm, robust, patent-pilose. Leaflets 18-25 pairs, elliptic-ovate. Flowers in 3-8-flowered racemes. Corolla long, lemon yellow. Legume ovoid, indurated, densely short-pubescent.

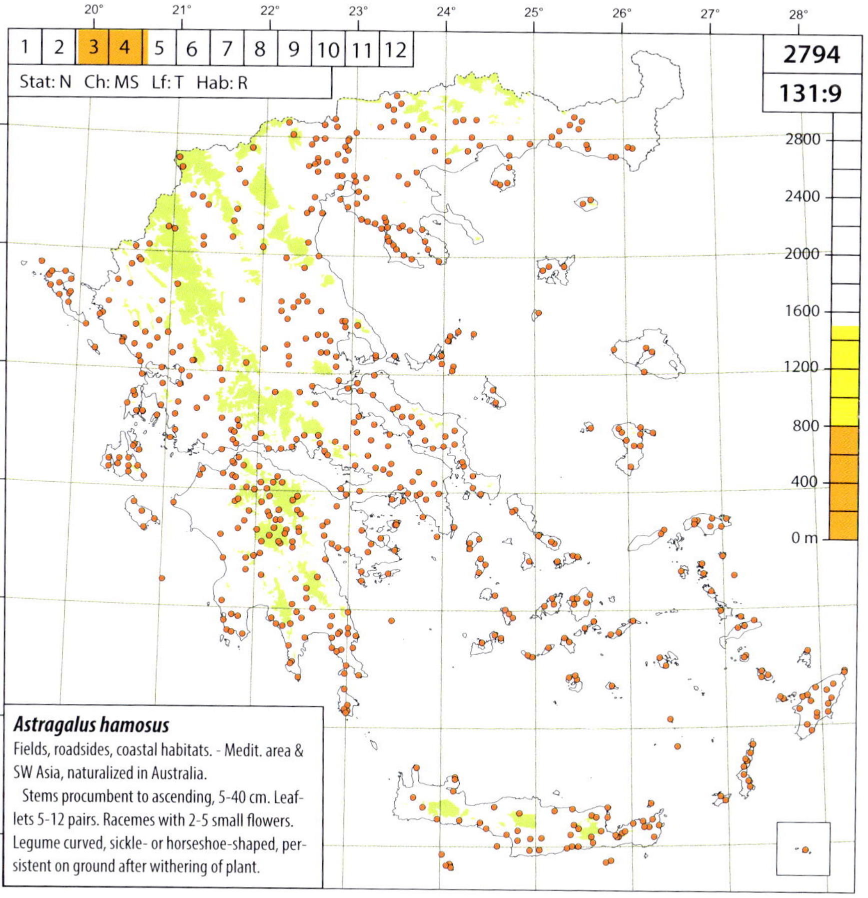

Astragalus hamosus
Fields, roadsides, coastal habitats. - Medit. area & SW Asia, naturalized in Australia.
Stems procumbent to ascending, 5-40 cm. Leaflets 5-12 pairs. Racemes with 2-5 small flowers. Legume curved, sickle- or horseshoe-shaped, persistent on ground after withering of plant.

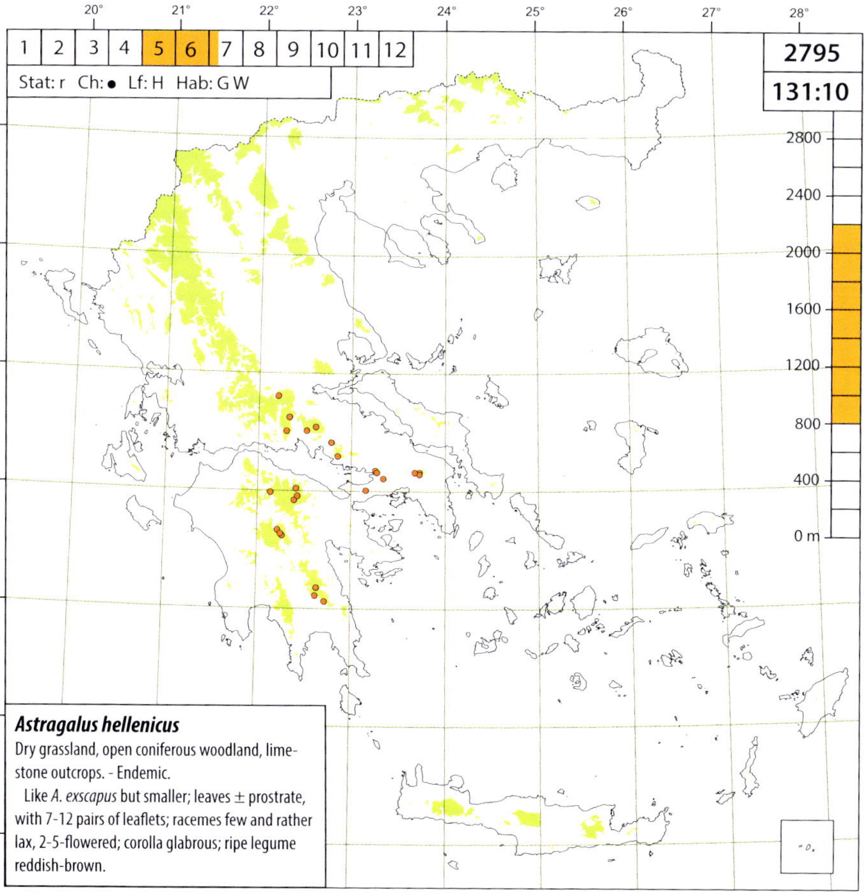

Astragalus hellenicus
Dry grassland, open coniferous woodland, limestone outcrops. - Endemic.
Like *A. exscapus* but smaller; leaves ± prostrate, with 7-12 pairs of leaflets; racemes few and rather lax, 2-5-flowered; corolla glabrous; ripe legume reddish-brown.

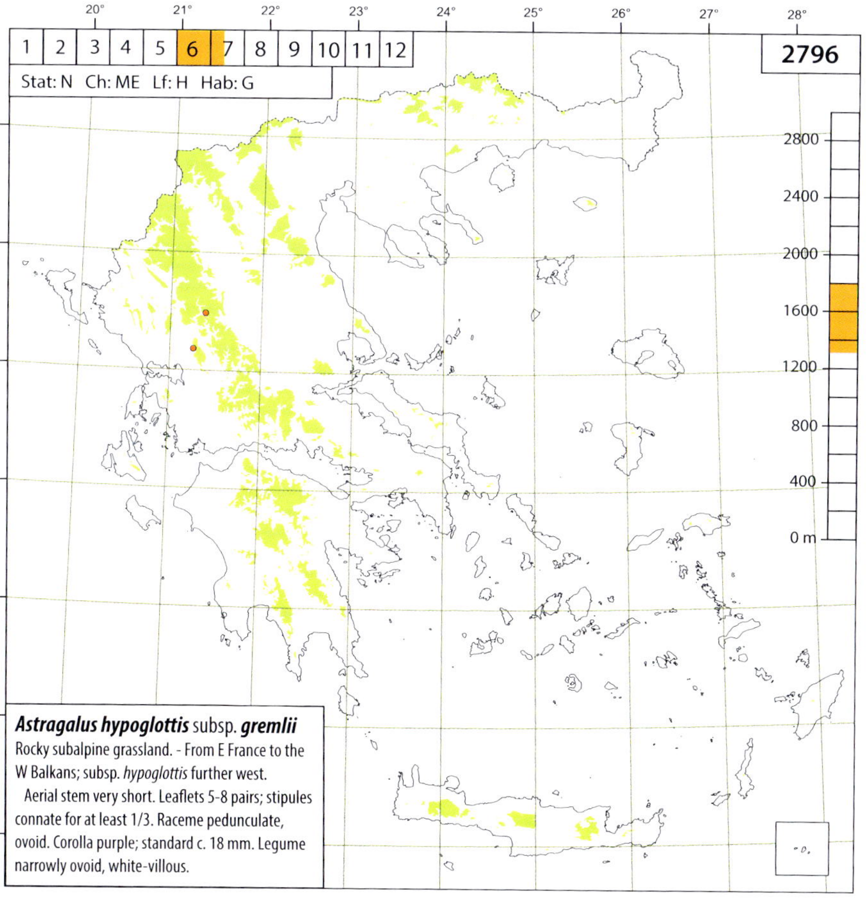

Astragalus hypoglottis subsp. **_gremlii_**
Rocky subalpine grassland. - From E France to the W Balkans; subsp. *hypoglottis* further west.
Aerial stem very short. Leaflets 5-8 pairs; stipules connate for at least 1/3. Raceme pedunculate, ovoid. Corolla purple; standard c. 18 mm. Legume narrowly ovoid, white-villous.

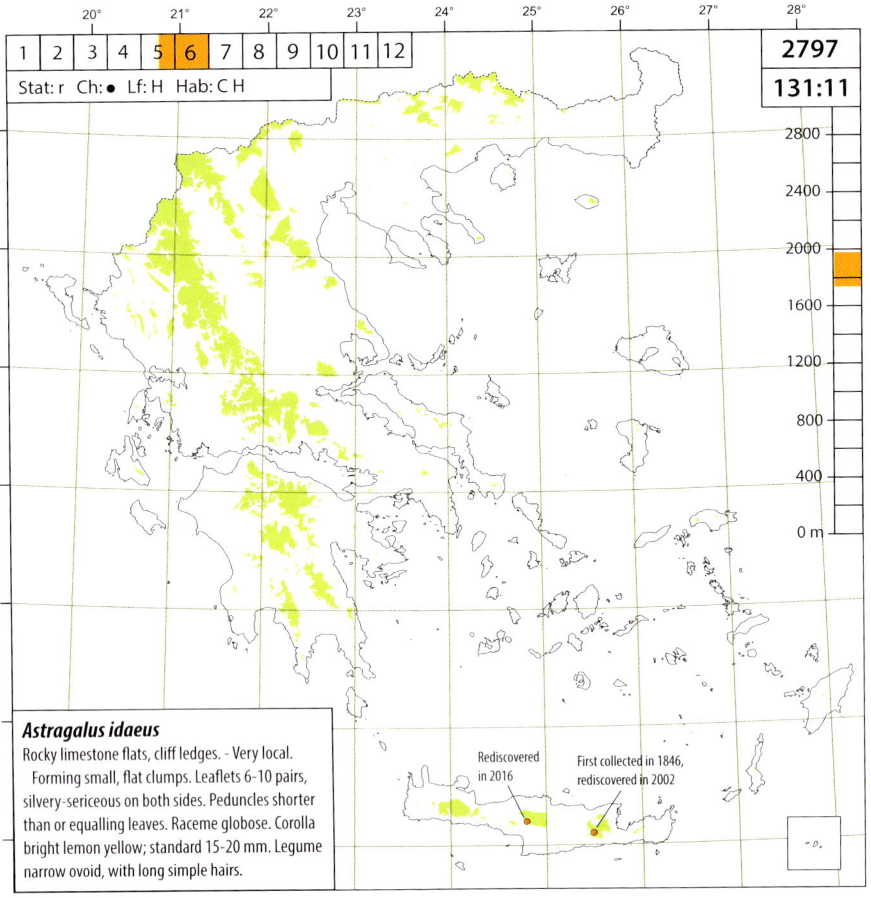

Astragalus idaeus

Rocky limestone flats, cliff ledges. - Very local.

Forming small, flat clumps. Leaflets 6-10 pairs, silvery-sericeous on both sides. Peduncles shorter than or equalling leaves. Raceme globose. Corolla bright lemon yellow; standard 15-20 mm. Legume narrow ovoid, with long simple hairs.

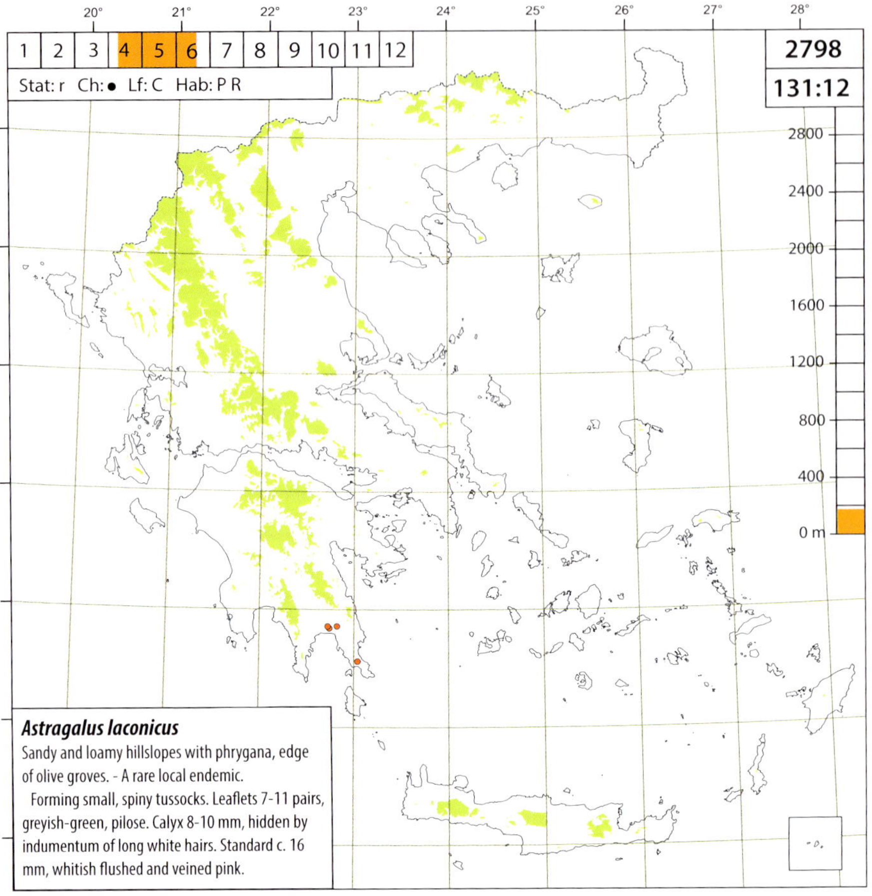

Astragalus laconicus

Sandy and loamy hillslopes with phrygana, edge of olive groves. - A rare local endemic.

Forming small, spiny tussocks. Leaflets 7-11 pairs, greyish-green, pilose. Calyx 8-10 mm, hidden by indumentum of long white hairs. Standard c. 16 mm, whitish flushed and veined pink.

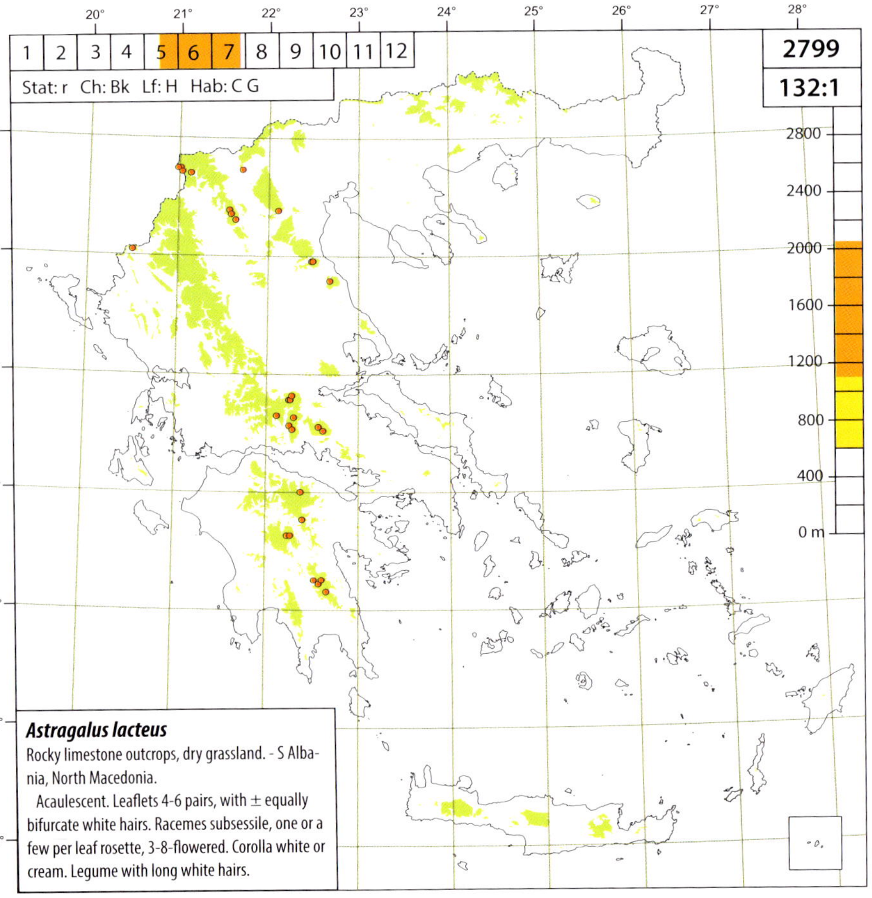

Astragalus lacteus

Rocky limestone outcrops, dry grassland. - S Albania, North Macedonia.

Acaulescent. Leaflets 4-6 pairs, with ± equally bifurcate white hairs. Racemes subsessile, one or a few per leaf rosette, 3-8-flowered. Corolla white or cream. Legume with long white hairs.

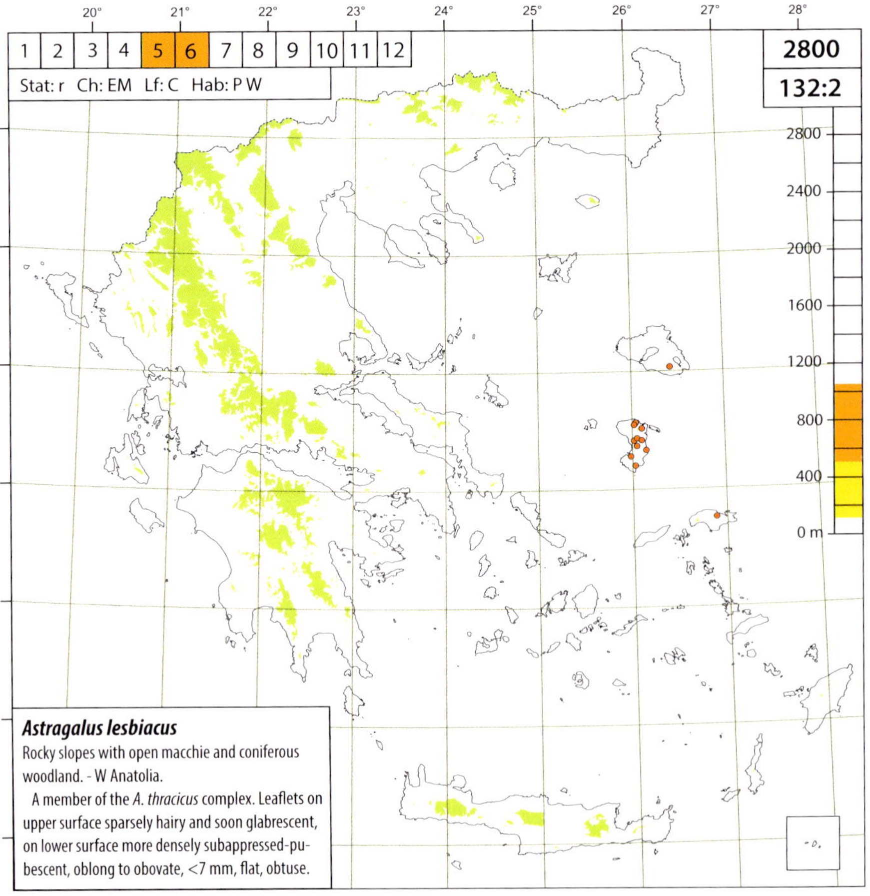

Astragalus lesbiacus

Rocky slopes with open macchie and coniferous woodland. - W Anatolia.

A member of the *A. thracicus* complex. Leaflets on upper surface sparsely hairy and soon glabrescent, on lower surface more densely subappressed-pubescent, oblong to obovate, <7 mm, flat, obtuse.

Atlas of the Hellenic Flora

II

Arne Strid

Atlas of the Hellenic Flora

II

Published with support from the
Aage V. Jensen Charity Foundation

Broken Hill Publishers Ltd

ISBN: 978-9925-35-026-1

16 Zenas Kanther Princess De Tyras Str.
1065 Nicosia, CYPRUS
e-mail: info@brokenhill.com.cy
website: http://www.brokenhill.com.cy

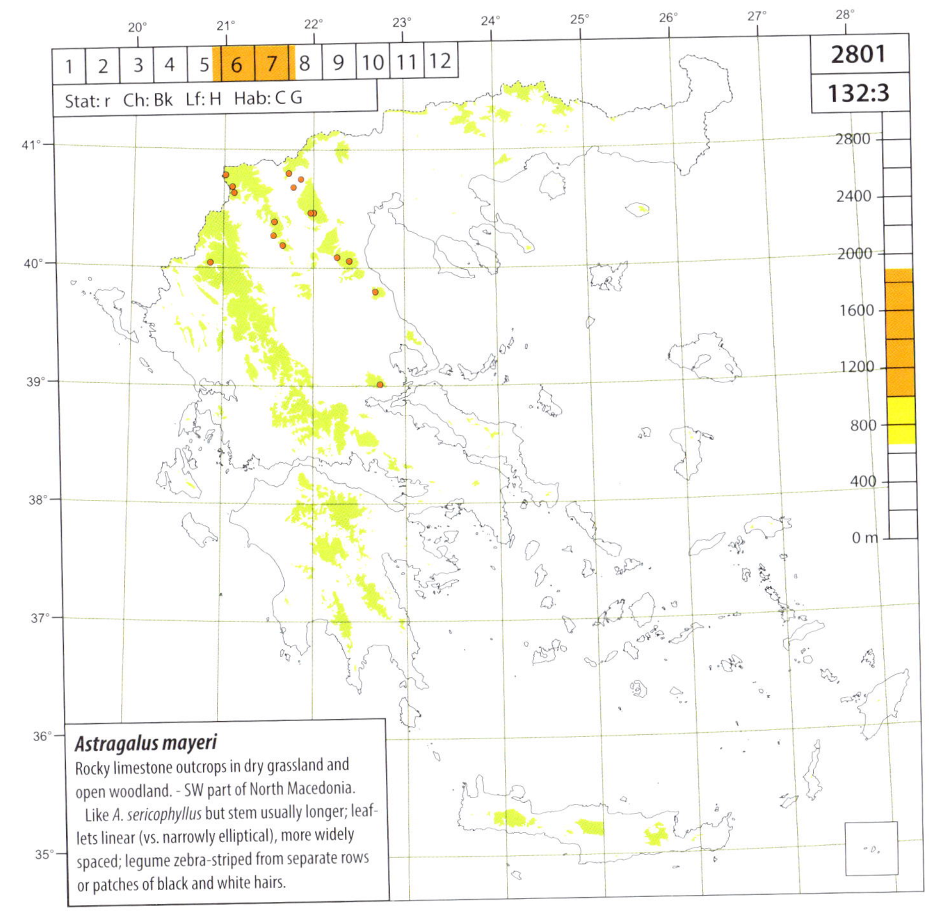

Astragalus mayeri
Rocky limestone outcrops in dry grassland and open woodland. - SW part of North Macedonia.
Like *A. sericophyllus* but stem usually longer; leaflets linear (vs. narrowly elliptical), more widely spaced; legume zebra-striped from separate rows or patches of black and white hairs.

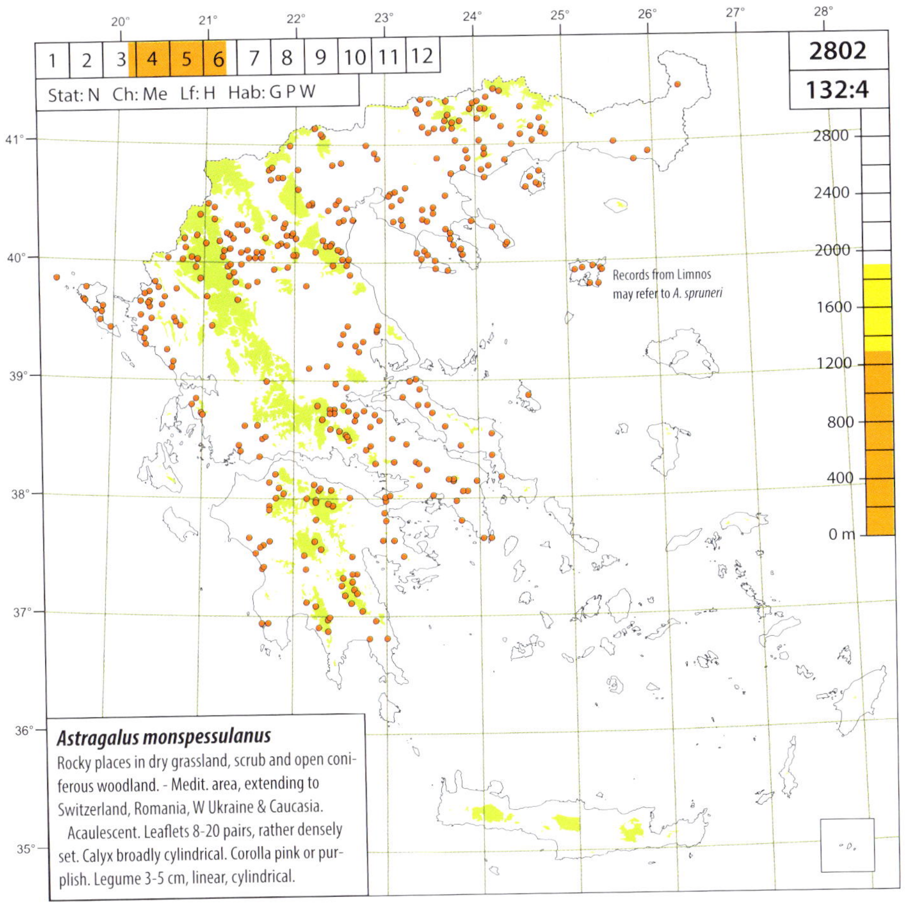

Astragalus monspessulanus
Rocky places in dry grassland, scrub and open coniferous woodland. - Medit. area, extending to Switzerland, Romania, W Ukraine & Caucasia.
Acaulescent. Leaflets 8-20 pairs, rather densely set. Calyx broadly cylindrical. Corolla pink or purplish. Legume 3-5 cm, linear, cylindrical.

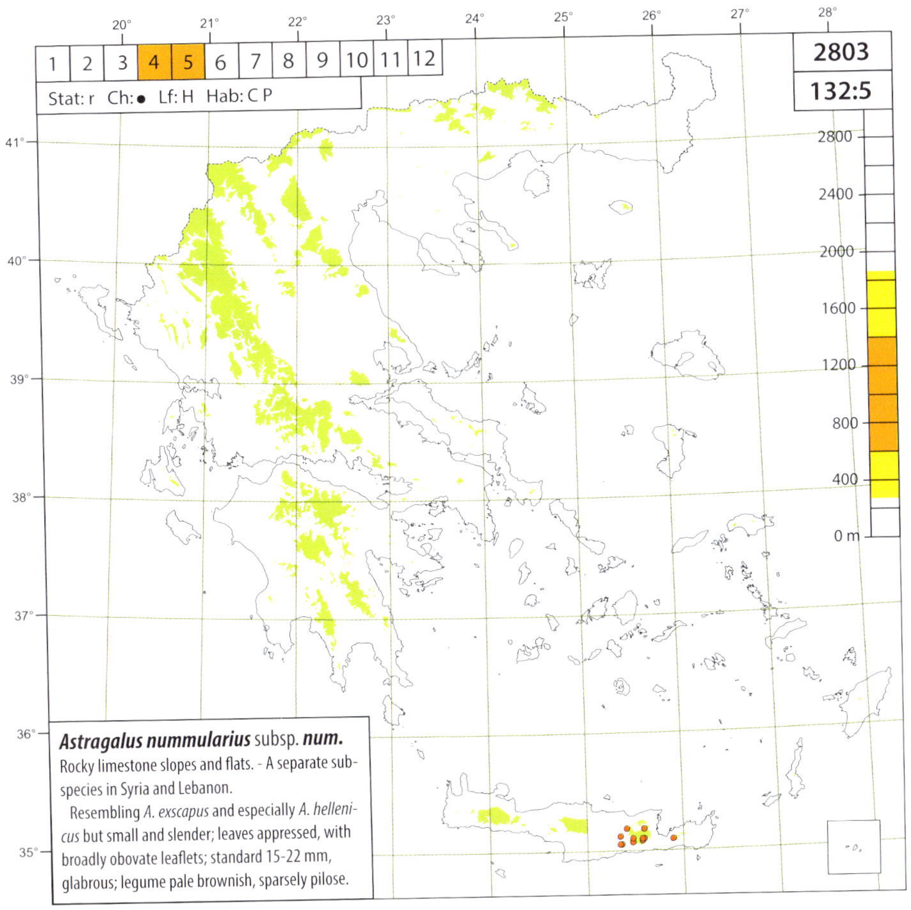

Astragalus nummularius subsp. ***num.***
Rocky limestone slopes and flats. - A separate subspecies in Syria and Lebanon.
Resembling *A. exscapus* and especially *A. hellenicus* but small and slender; leaves appressed, with broadly obovate leaflets; standard 15-22 mm, glabrous; legume pale brownish, sparsely pilose.

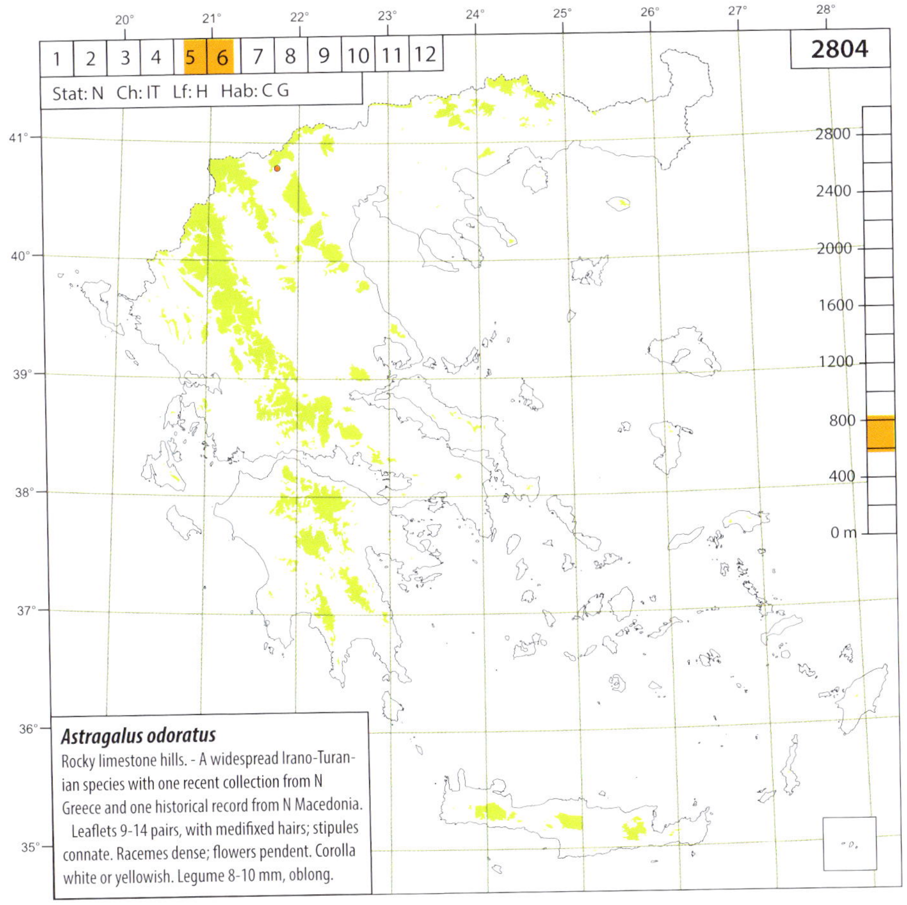

Astragalus odoratus
Rocky limestone hills. - A widespread Irano-Turanian species with one recent collection from N Greece and one historical record from N Macedonia.
Leaflets 9-14 pairs, with medifixed hairs; stipules connate. Racemes dense; flowers pendent. Corolla white or yellowish. Legume 8-10 mm, oblong.

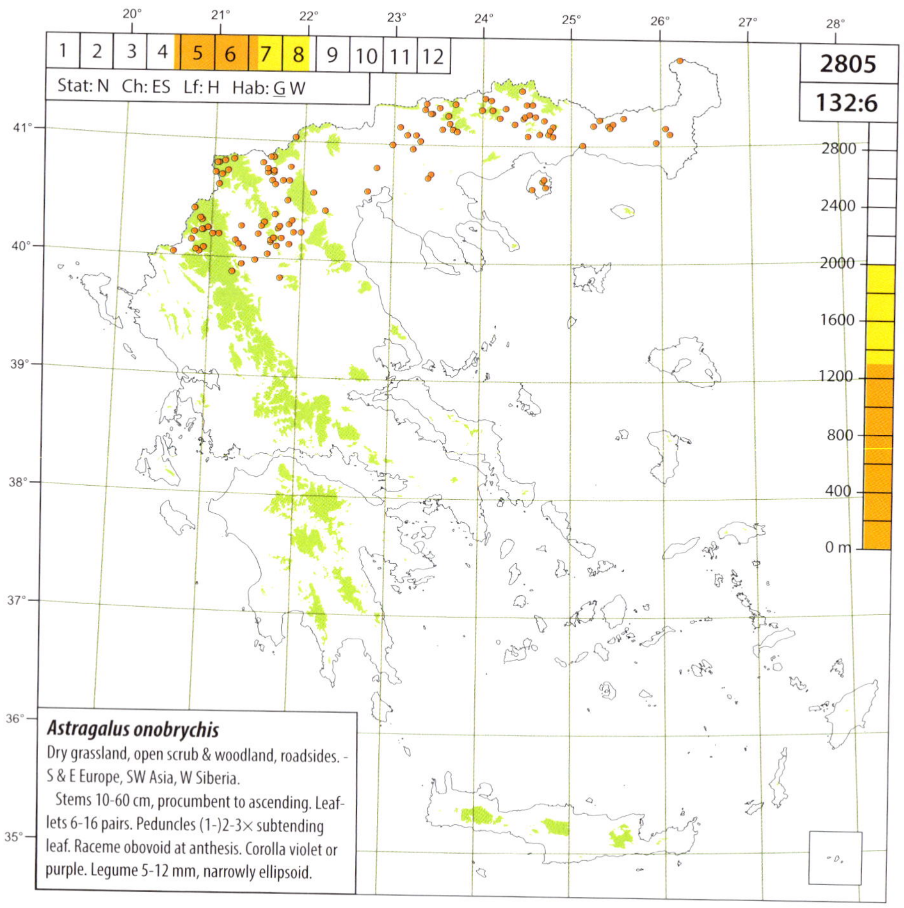

Astragalus onobrychis
Dry grassland, open scrub & woodland, roadsides. - S & E Europe, SW Asia, W Siberia.
Stems 10-60 cm, procumbent to ascending. Leaflets 6-16 pairs. Peduncles (1-)2-3× subtending leaf. Raceme obovoid at anthesis. Corolla violet or purple. Legume 5-12 mm, narrowly ellipsoid.

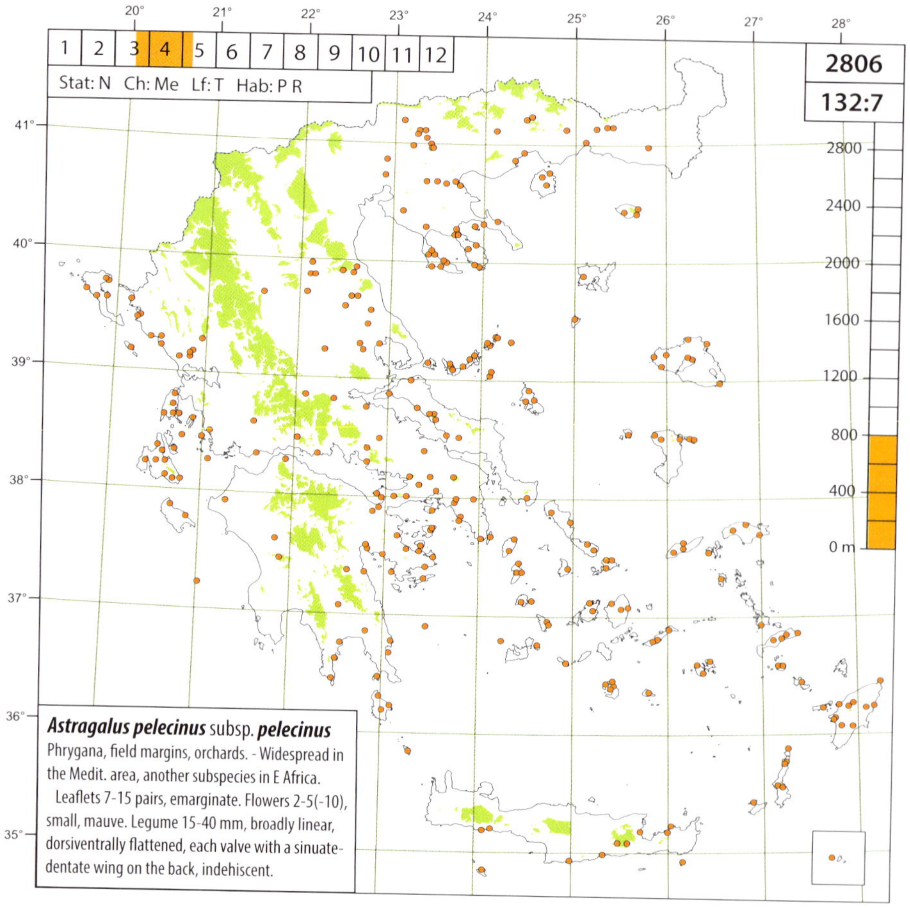

Astragalus pelecinus subsp. **pelecinus**
Phrygana, field margins, orchards. - Widespread in the Medit. area, another subspecies in E Africa.
Leaflets 7-15 pairs, emarginate. Flowers 2-5(-10), small, mauve. Legume 15-40 mm, broadly linear, dorsiventrally flattened, each valve with a sinuate-dentate wing on the back, indehiscent.

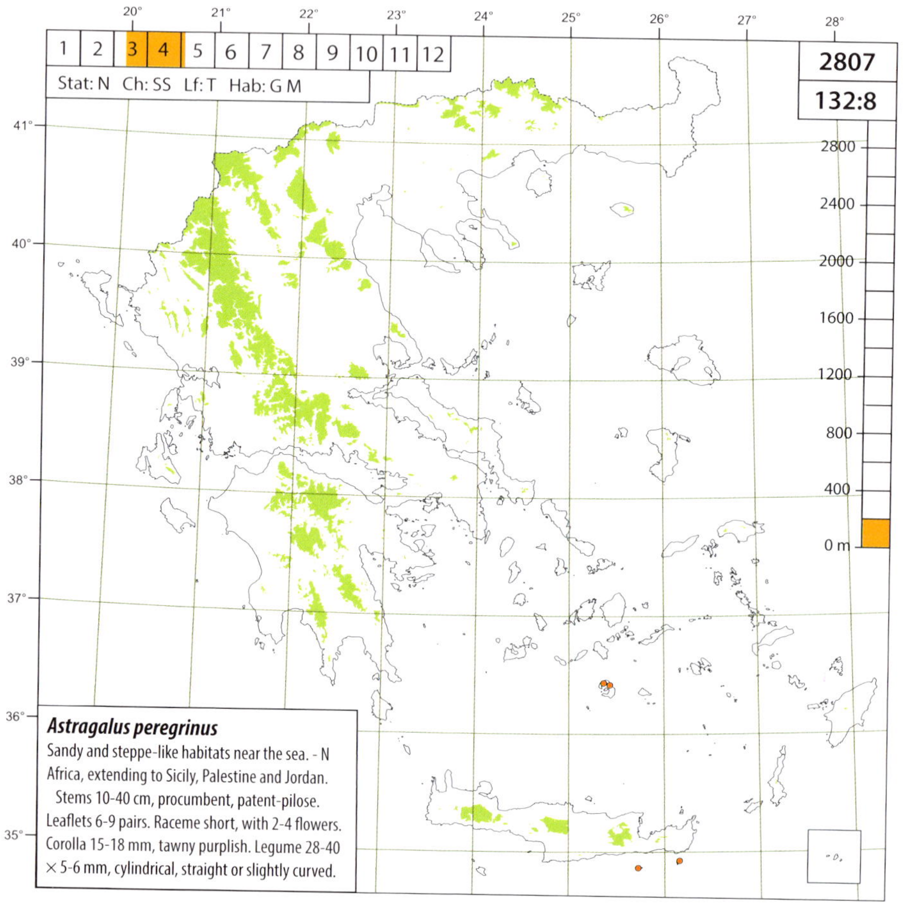

Astragalus peregrinus
Sandy and steppe-like habitats near the sea. - N Africa, extending to Sicily, Palestine and Jordan.
Stems 10-40 cm, procumbent, patent-pilose. Leaflets 6-9 pairs. Raceme short, with 2-4 flowers. Corolla 15-18 mm, tawny purplish. Legume 28-40 × 5-6 mm, cylindrical, straight or slightly curved.

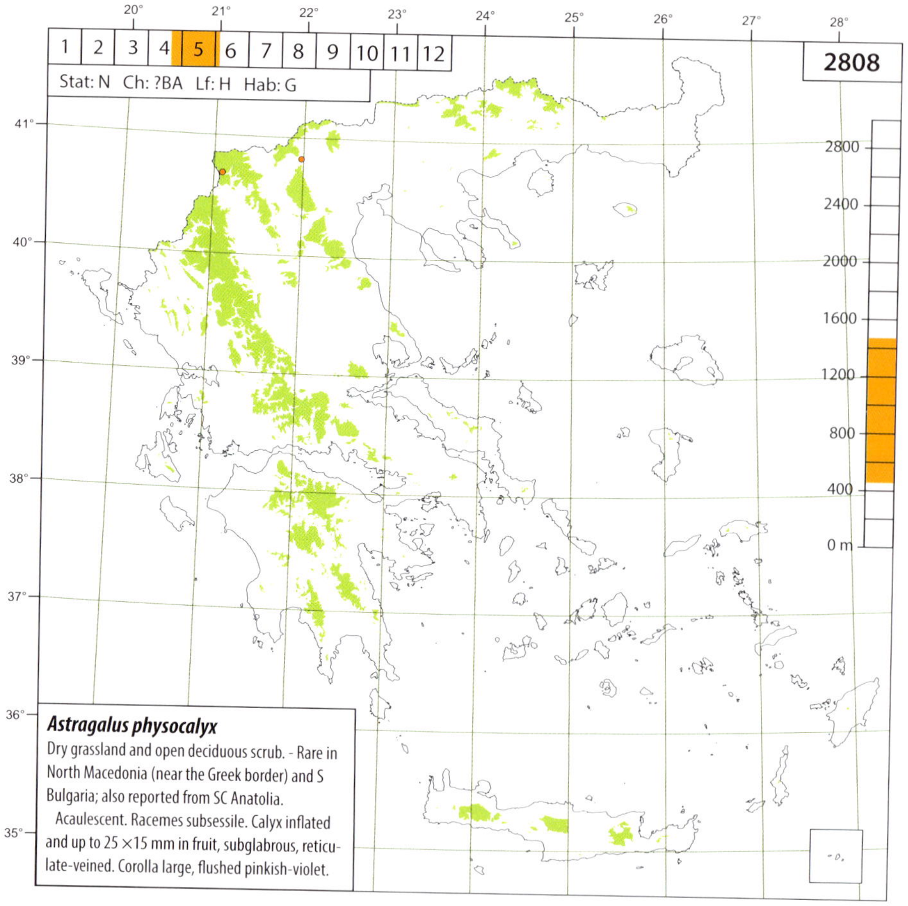

Astragalus physocalyx
Dry grassland and open deciduous scrub. - Rare in North Macedonia (near the Greek border) and S Bulgaria; also reported from SC Anatolia.
Acaulescent. Racemes subsessile. Calyx inflated and up to 25 ×15 mm in fruit, subglabrous, reticulate-veined. Corolla large, flushed pinkish-violet.

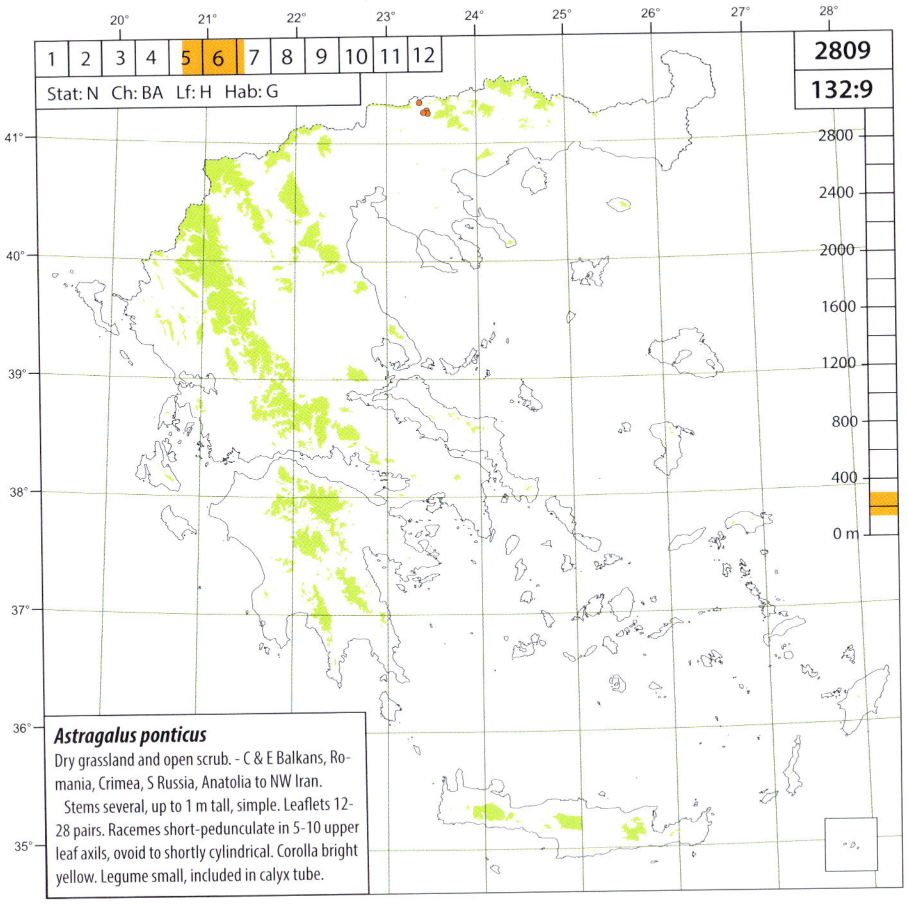

Astragalus ponticus

Dry grassland and open scrub. - C & E Balkans, Romania, Crimea, S Russia, Anatolia to NW Iran.

Stems several, up to 1 m tall, simple. Leaflets 12-28 pairs. Racemes short-pedunculate in 5-10 upper leaf axils, ovoid to shortly cylindrical. Corolla bright yellow. Legume small, included in calyx tube.

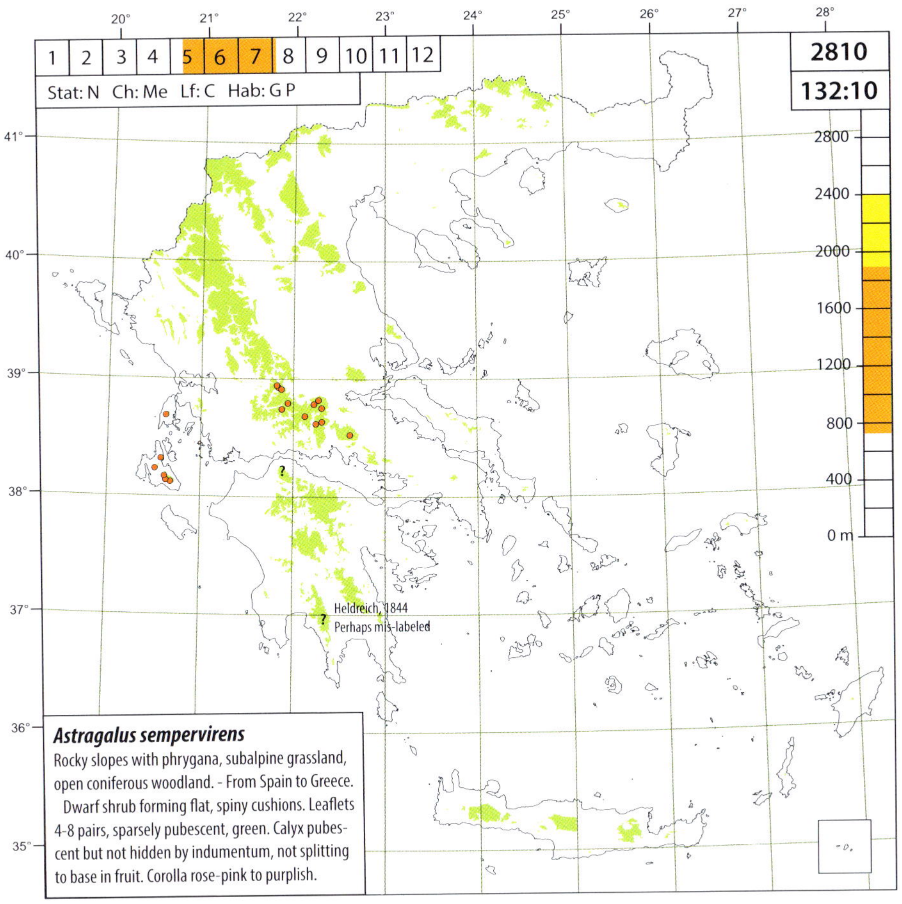

Astragalus sempervirens

Rocky slopes with phrygana, subalpine grassland, open coniferous woodland. - From Spain to Greece.

Dwarf shrub forming flat, spiny cushions. Leaflets 4-8 pairs, sparsely pubescent, green. Calyx pubescent but not hidden by indumentum, not splitting to base in fruit. Corolla rose-pink to purplish.

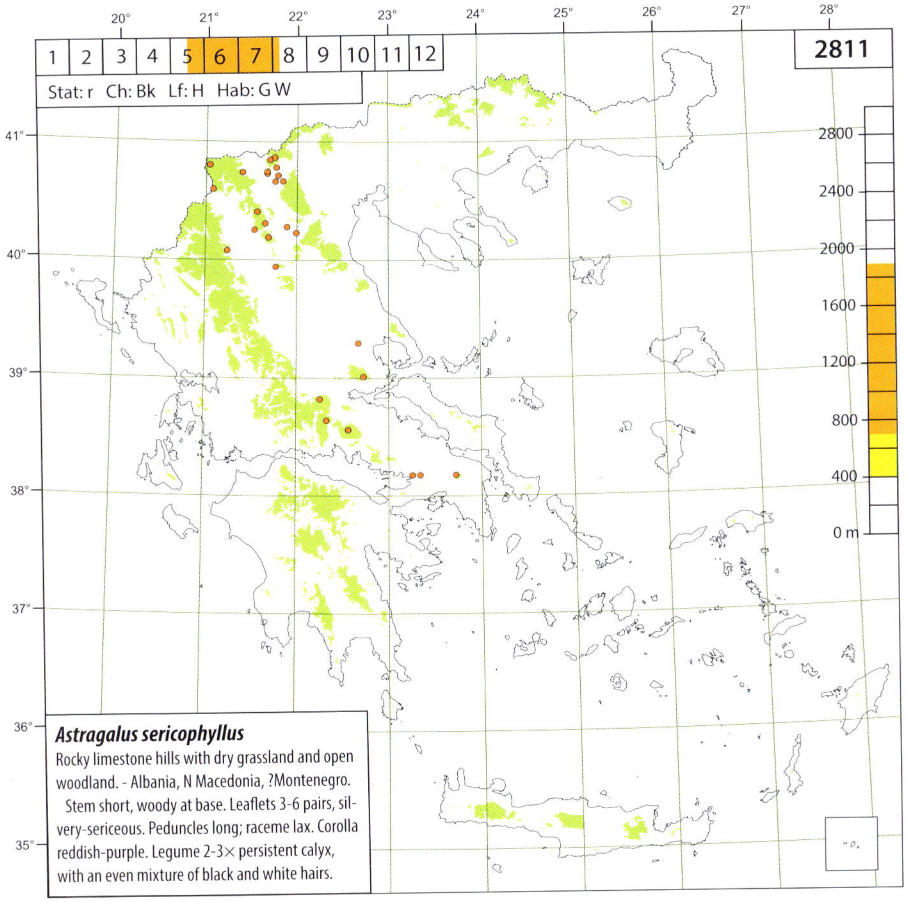

Astragalus sericophyllus

Rocky limestone hills with dry grassland and open woodland. - Albania, N Macedonia, ?Montenegro.

Stem short, woody at base. Leaflets 3-6 pairs, silvery-sericeous. Peduncles long; raceme lax. Corolla reddish-purple. Legume 2-3× persistent calyx, with an even mixture of black and white hairs.

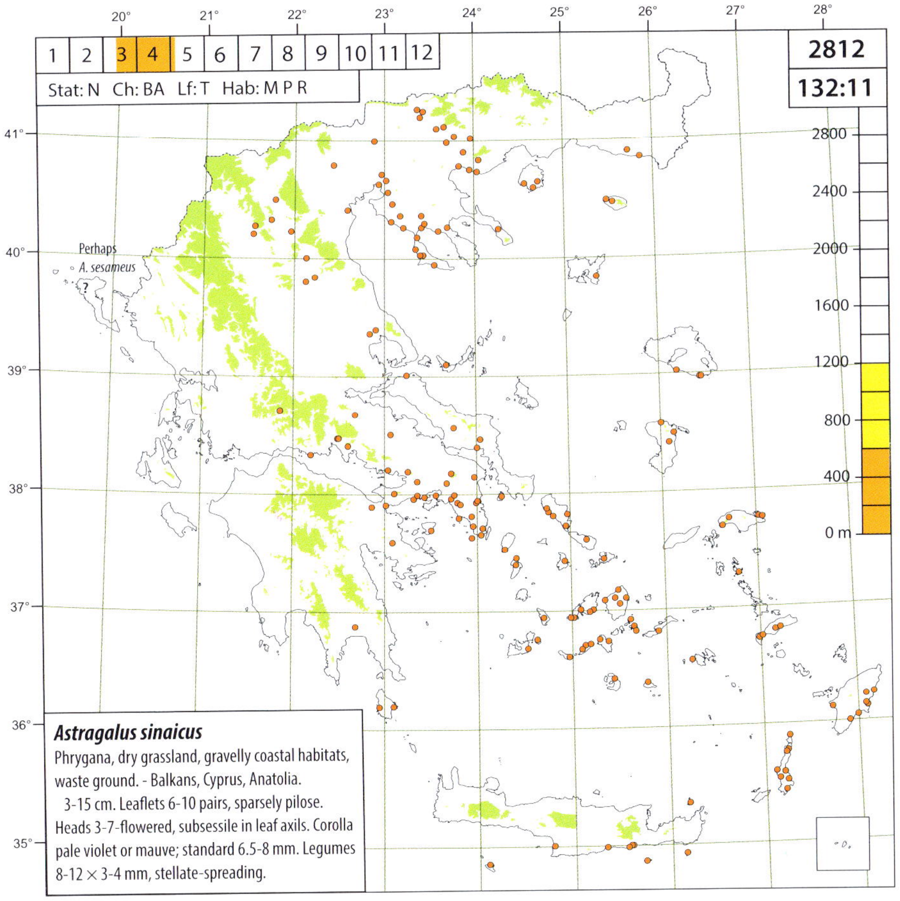

Astragalus sinaicus

Phrygana, dry grassland, gravelly coastal habitats, waste ground. - Balkans, Cyprus, Anatolia.

3-15 cm. Leaflets 6-10 pairs, sparsely pilose. Heads 3-7-flowered, subsessile in leaf axils. Corolla pale violet or mauve; standard 6.5-8 mm. Legumes 8-12 × 3-4 mm, stellate-spreading.

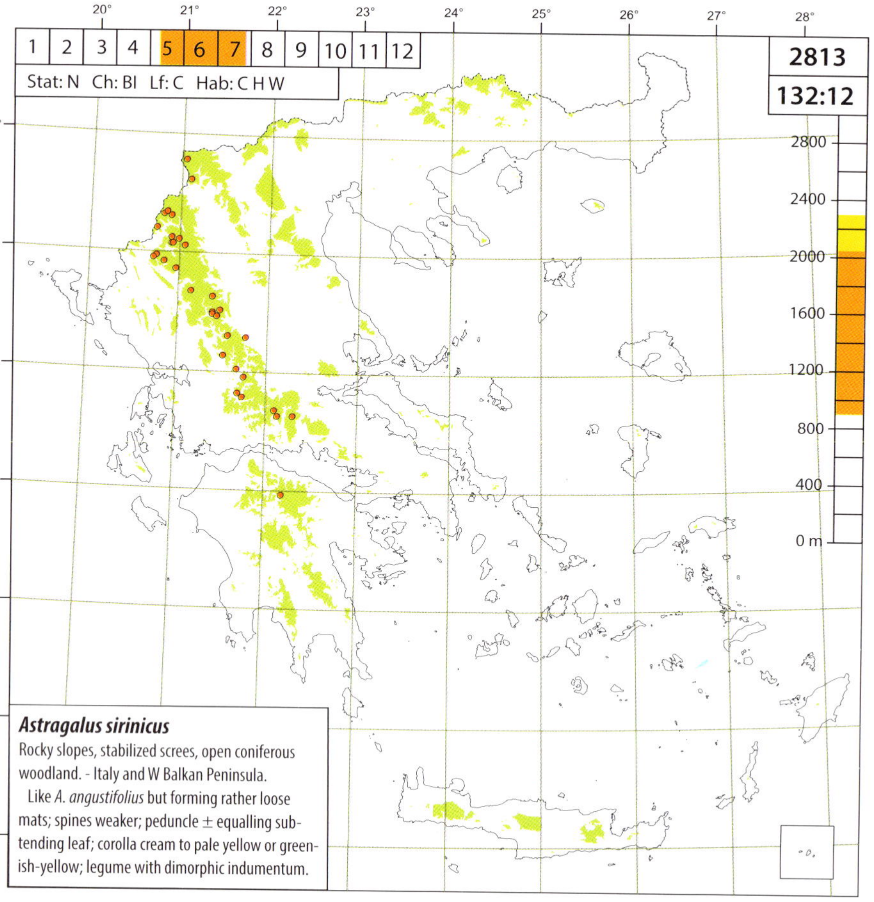

Astragalus sirinicus
Rocky slopes, stabilized screes, open coniferous woodland. - Italy and W Balkan Peninsula.
Like *A. angustifolius* but forming rather loose mats; spines weaker; peduncle ± equalling subtending leaf; corolla cream to pale yellow or greenish-yellow; legume with dimorphic indumentum.

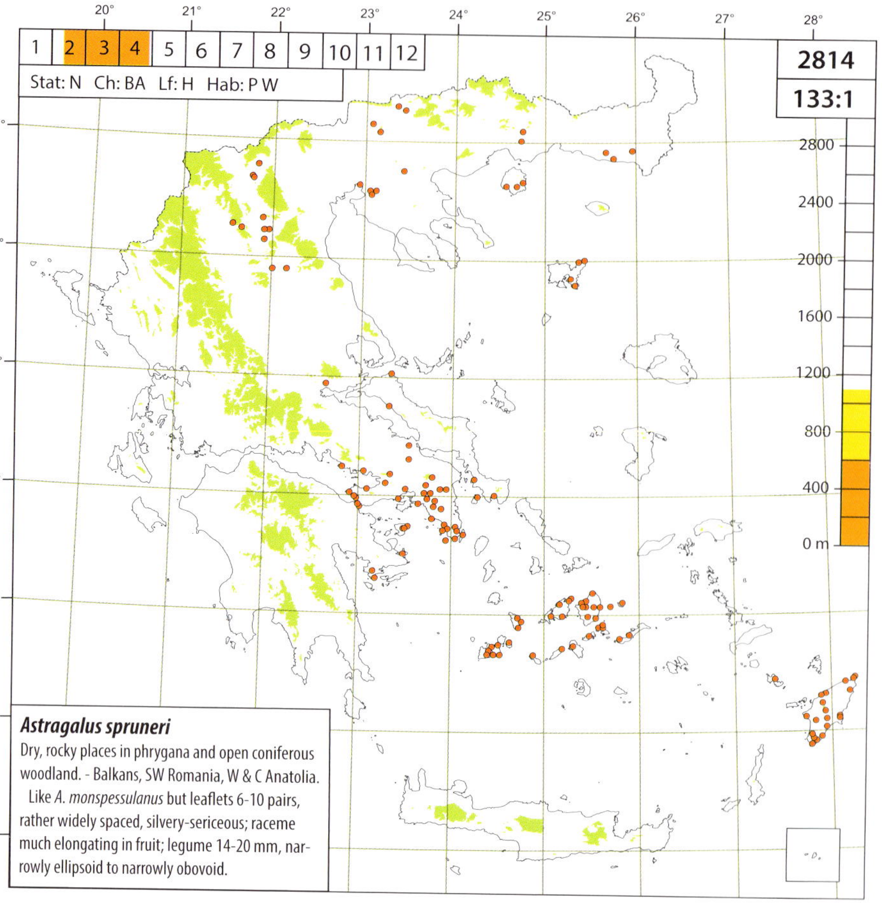

Astragalus spruneri
Dry, rocky places in phrygana and open coniferous woodland. - Balkans, SW Romania, W & C Anatolia.
Like *A. monspessulanus* but leaflets 6-10 pairs, rather widely spaced, silvery-sericeous; raceme much elongating in fruit; legume 14-20 mm, narrowly ellipsoid to narrowly obovoid.

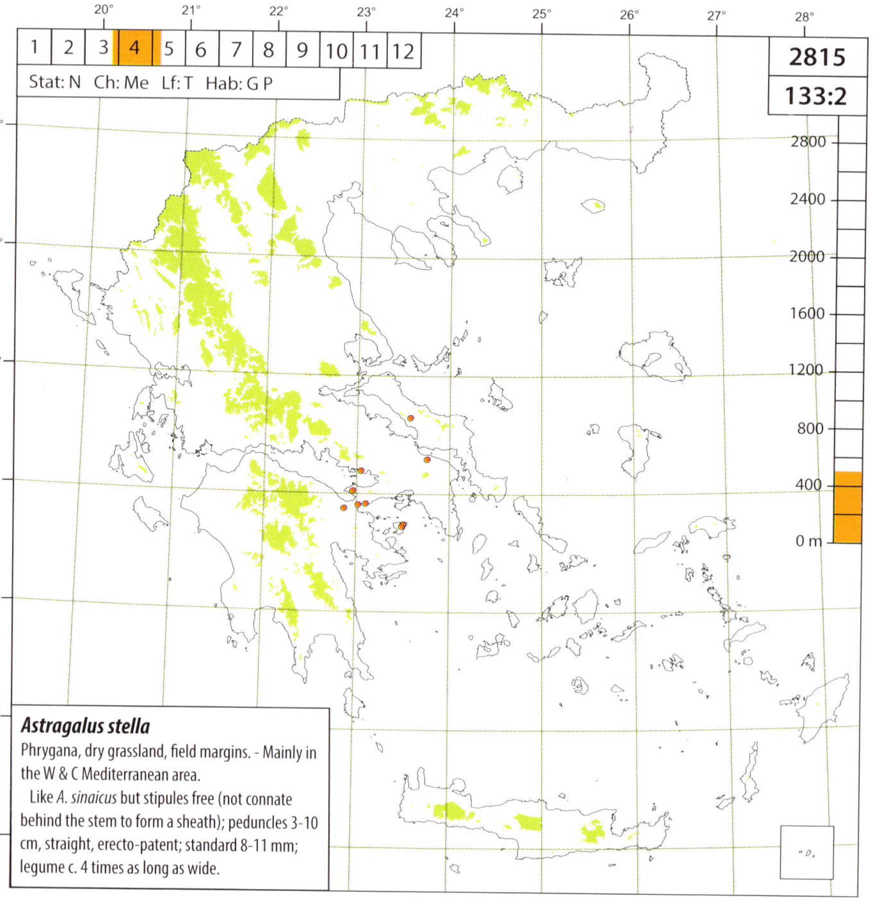

Astragalus stella
Phrygana, dry grassland, field margins. - Mainly in the W & C Mediterranean area.
Like *A. sinaicus* but stipules free (not connate behind the stem to form a sheath); peduncles 3-10 cm, straight, erecto-patent; standard 8-11 mm; legume c. 4 times as long as wide.

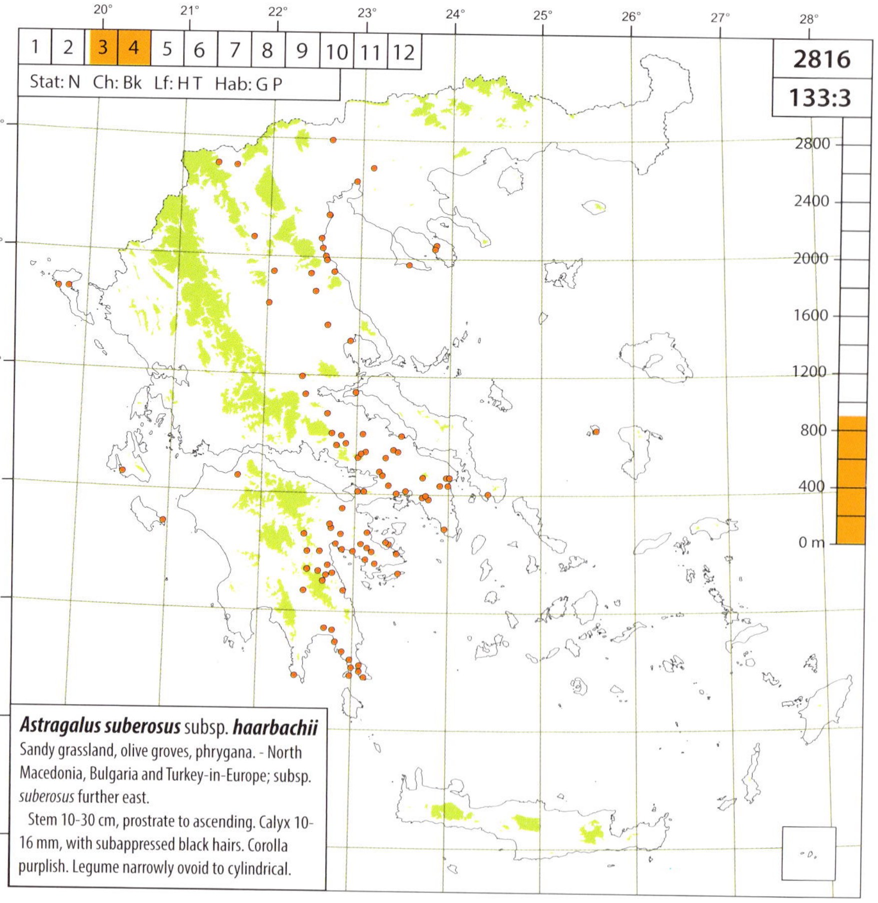

Astragalus suberosus subsp. haarbachii
Sandy grassland, olive groves, phrygana. - North Macedonia, Bulgaria and Turkey-in-Europe; subsp. *suberosus* further east.
Stem 10-30 cm, prostrate to ascending. Calyx 10-16 mm, with subappressed black hairs. Corolla purplish. Legume narrowly ovoid to cylindrical.

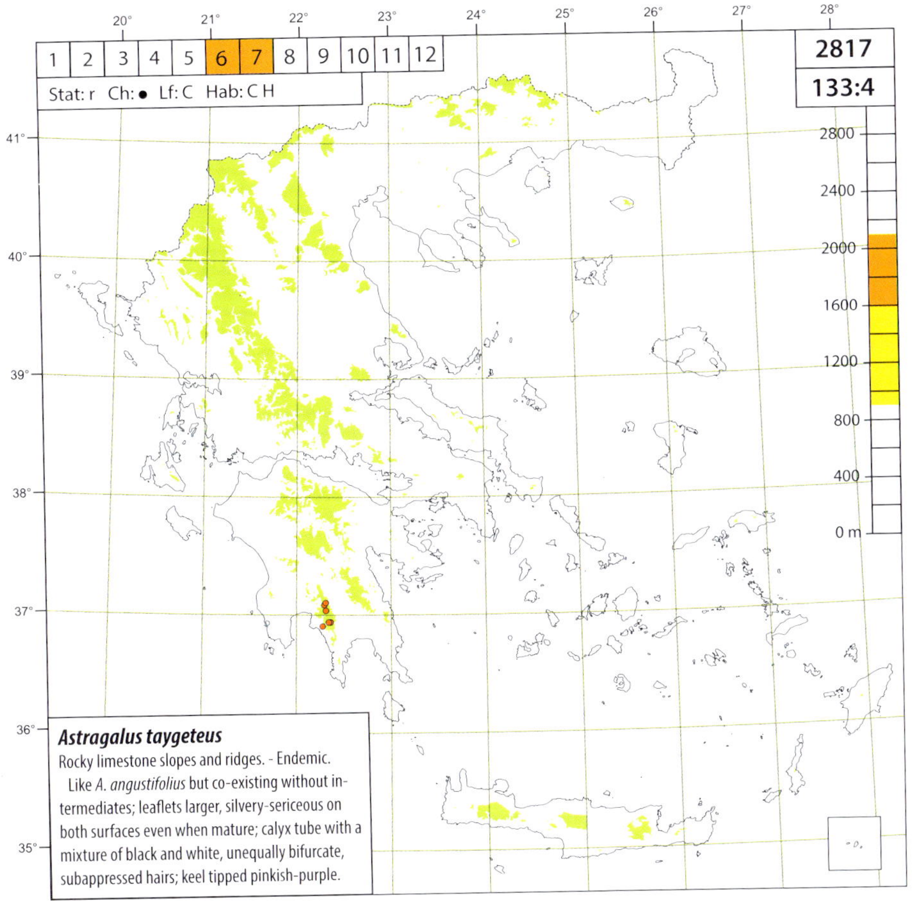

Astragalus taygeteus
Rocky limestone slopes and ridges. - Endemic.
Like *A. angustifolius* but co-existing without intermediates; leaflets larger, silvery-sericeous on both surfaces even when mature; calyx tube with a mixture of black and white, unequally bifurcate, subappressed hairs; keel tipped pinkish-purple.

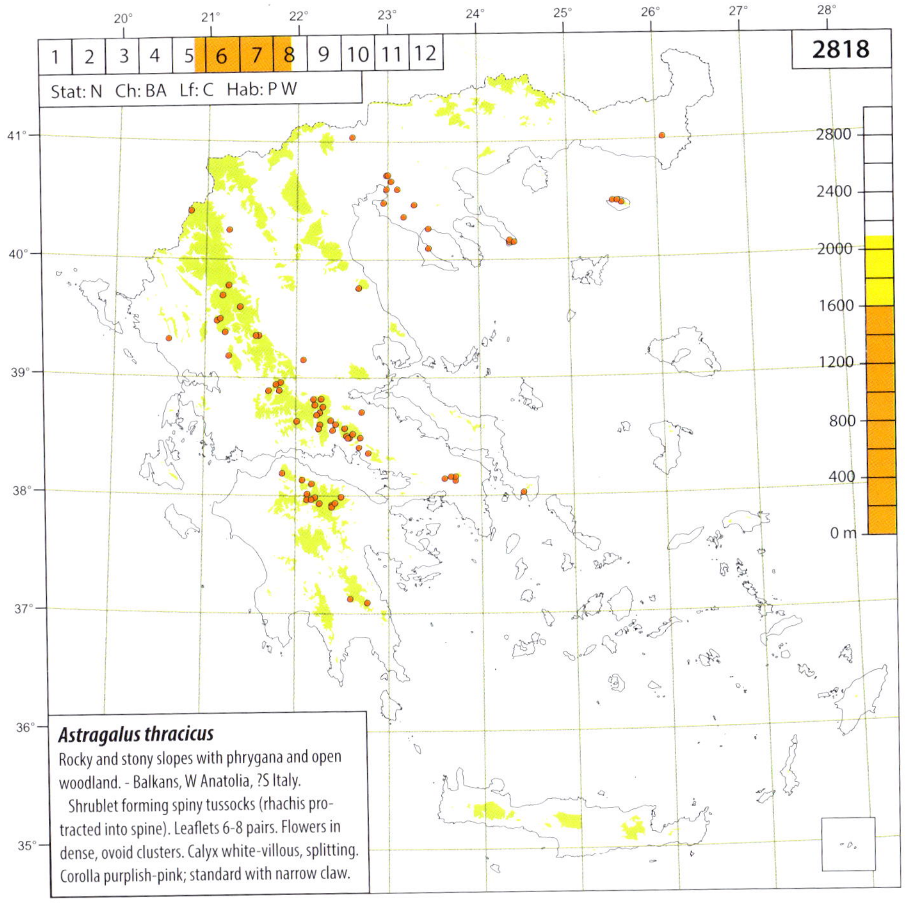

Astragalus thracicus
Rocky and stony slopes with phrygana and open woodland. - Balkans, W Anatolia, ?S Italy.
Shrublet forming spiny tussocks (rhachis protracted into spine). Leaflets 6-8 pairs. Flowers in dense, ovoid clusters. Calyx white-villous, splitting. Corolla purplish-pink; standard with narrow claw.

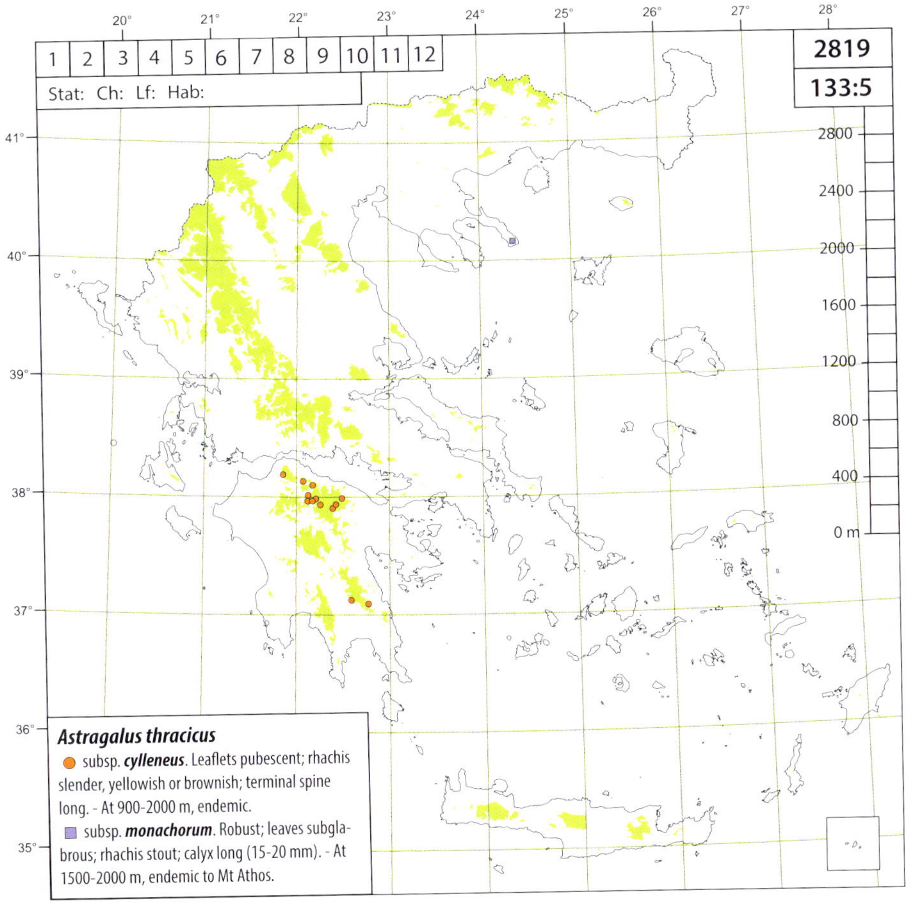

Astragalus thracicus
● subsp. ***cylleneus***. Leaflets pubescent; rhachis slender, yellowish or brownish; terminal spine long. - At 900-2000 m, endemic.
■ subsp. ***monachorum***. Robust; leaves subglabrous; rhachis stout; calyx long (15-20 mm). - At 1500-2000 m, endemic to Mt Athos.

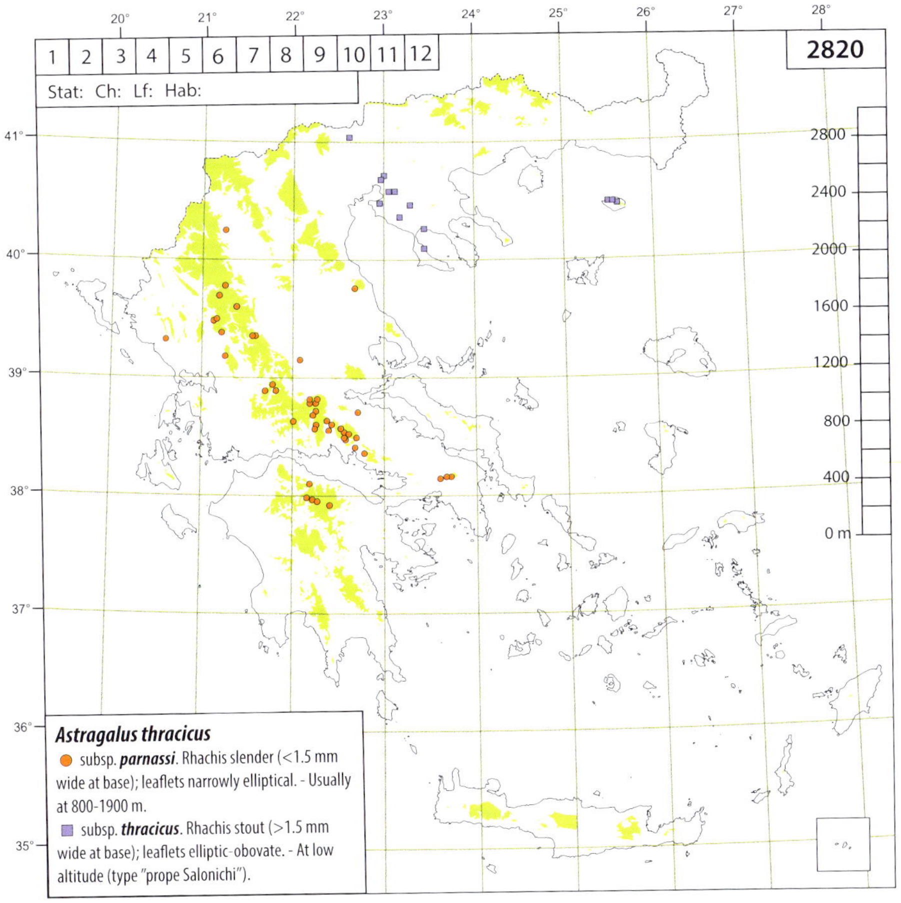

Astragalus thracicus
● subsp. ***parnassi***. Rhachis slender (<1.5 mm wide at base); leaflets narrowly elliptical. - Usually at 800-1900 m.
■ subsp. ***thracicus***. Rhachis stout (>1.5 mm wide at base); leaflets elliptic-obovate. - At low altitude (type "prope Salonichi").

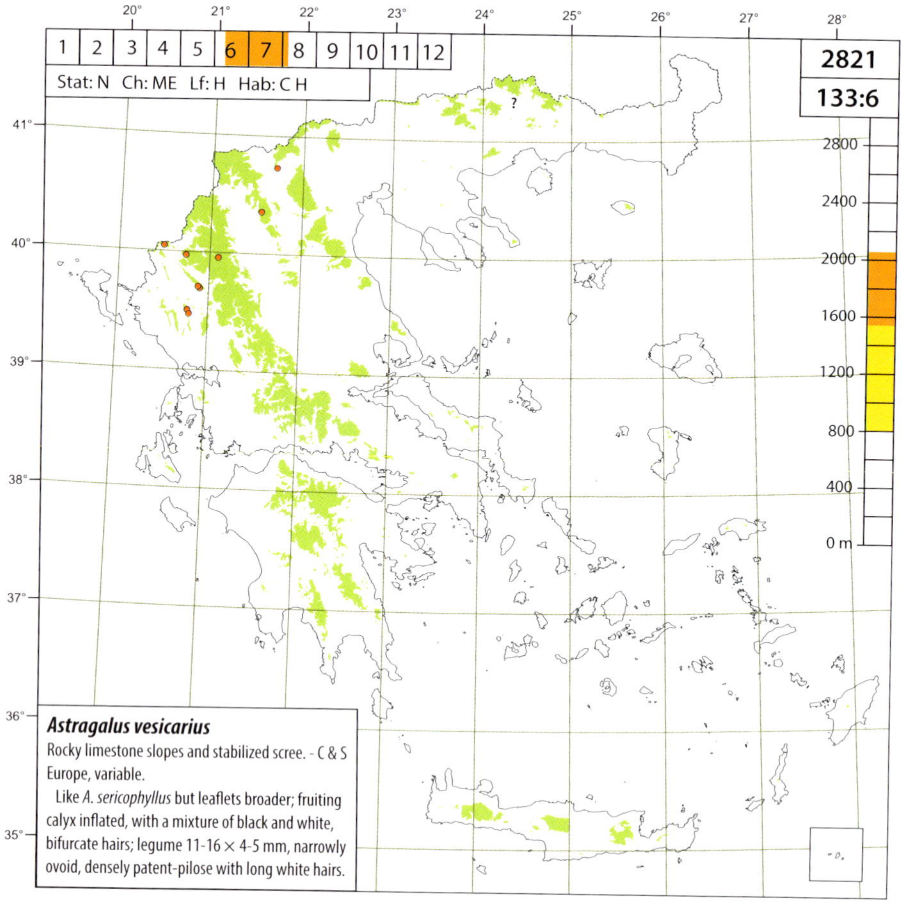

Astragalus vesicarius

Rocky limestone slopes and stabilized scree. - C & S Europe, variable.

Like *A. sericophyllus* but leaflets broader; fruiting calyx inflated, with a mixture of black and white, bifurcate hairs; legume 11-16 × 4-5 mm, narrowly ovoid, densely patent-pilose with long white hairs.

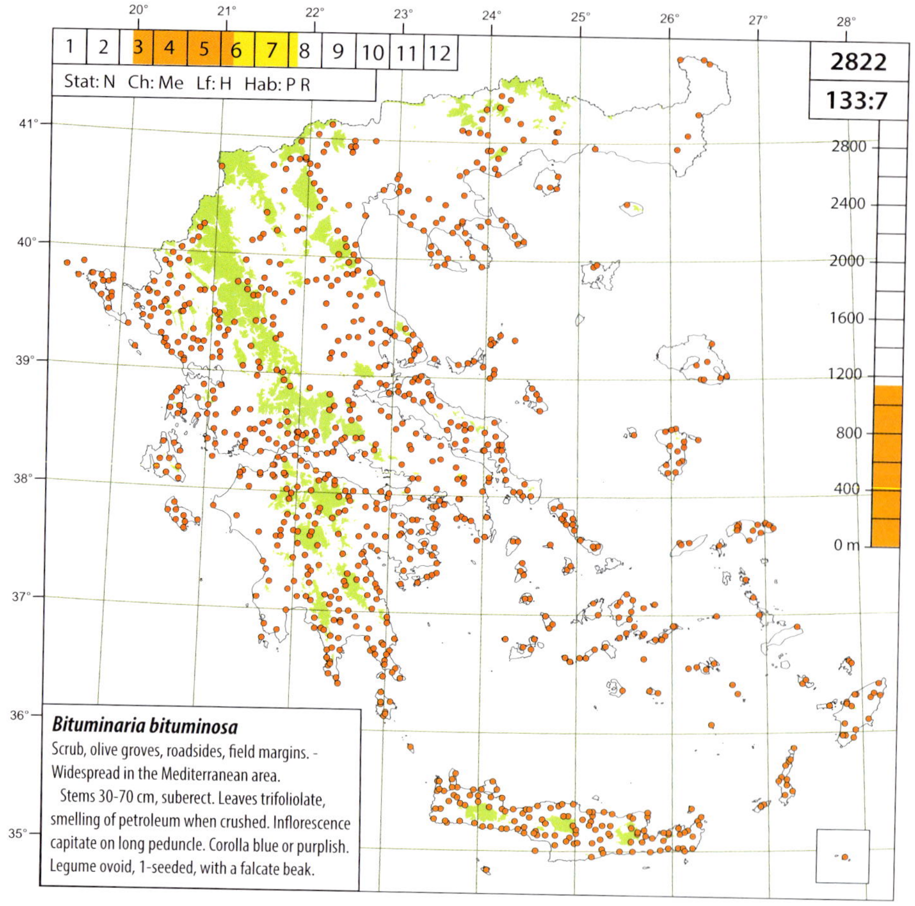

Bituminaria bituminosa

Scrub, olive groves, roadsides, field margins. - Widespread in the Mediterranean area.

Stems 30-70 cm, suberect. Leaves trifoliolate, smelling of petroleum when crushed. Inflorescence capitate on long peduncle. Corolla blue or purplish. Legume ovoid, 1-seeded, with a falcate beak.

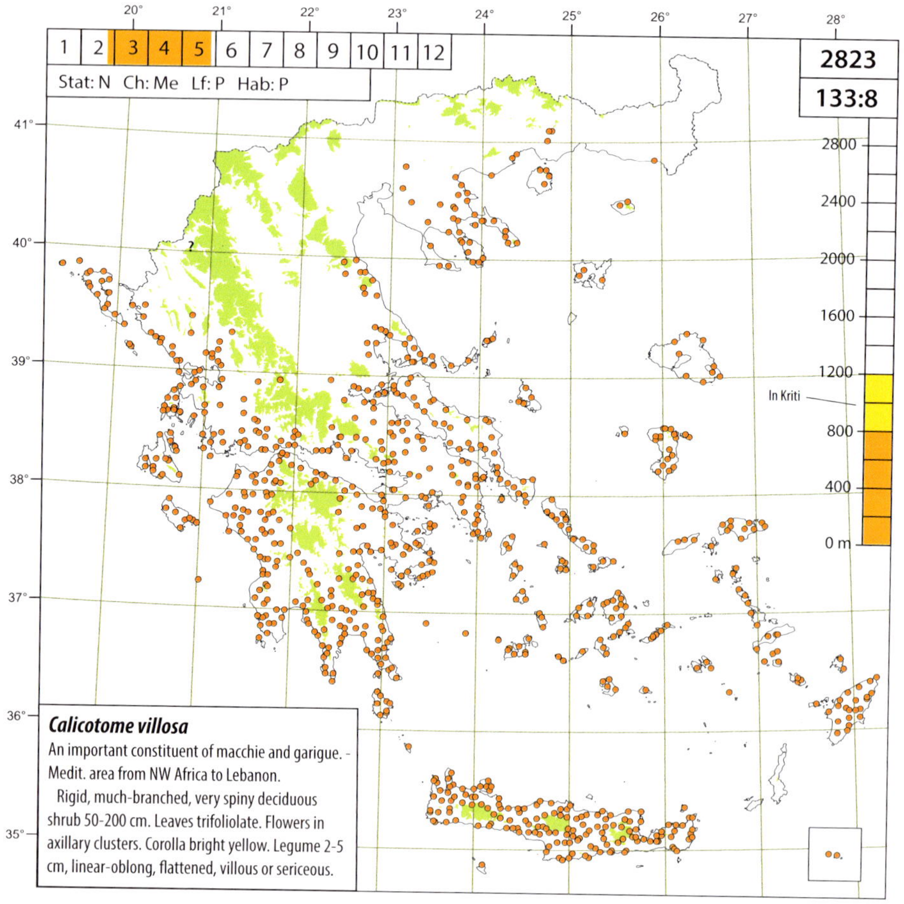

Calicotome villosa

An important constituent of macchie and garigue. - Medit. area from NW Africa to Lebanon.

Rigid, much-branched, very spiny deciduous shrub 50-200 cm. Leaves trifoliolate. Flowers in axillary clusters. Corolla bright yellow. Legume 2-5 cm, linear-oblong, flattened, villous or sericeous.

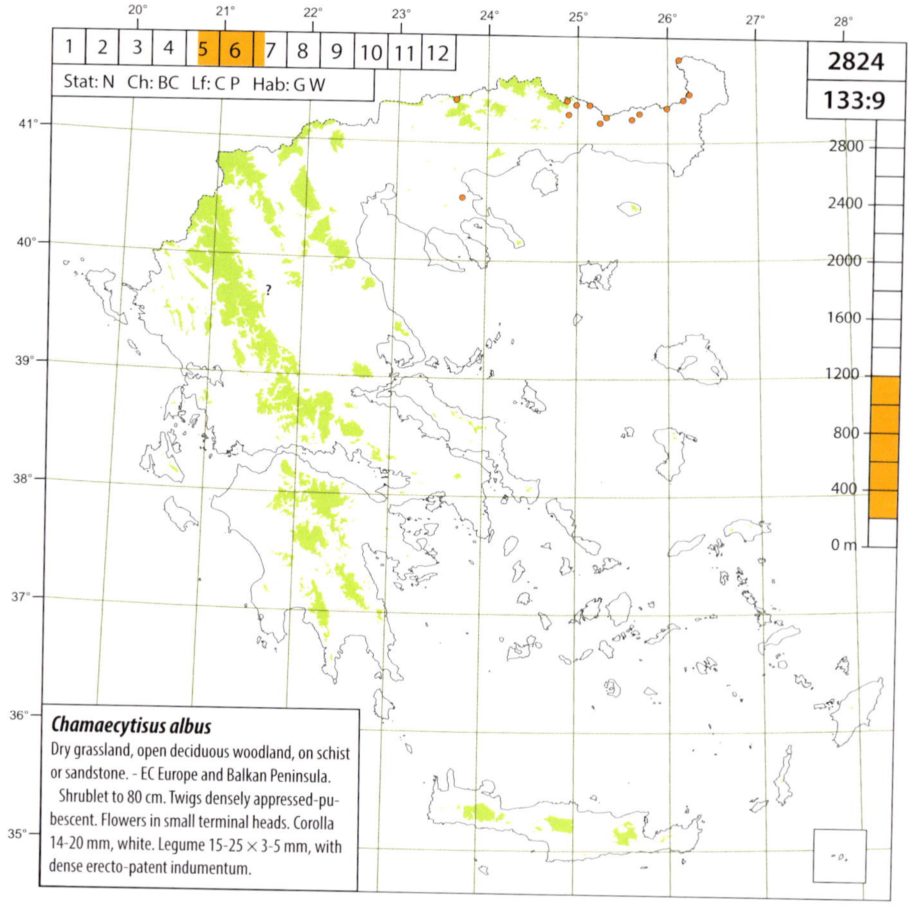

Chamaecytisus albus

Dry grassland, open deciduous woodland, on schist or sandstone. - EC Europe and Balkan Peninsula.

Shrublet to 80 cm. Twigs densely appressed-pubescent. Flowers in small terminal heads. Corolla 14-20 mm, white. Legume 15-25 × 3-5 mm, with dense erecto-patent indumentum.

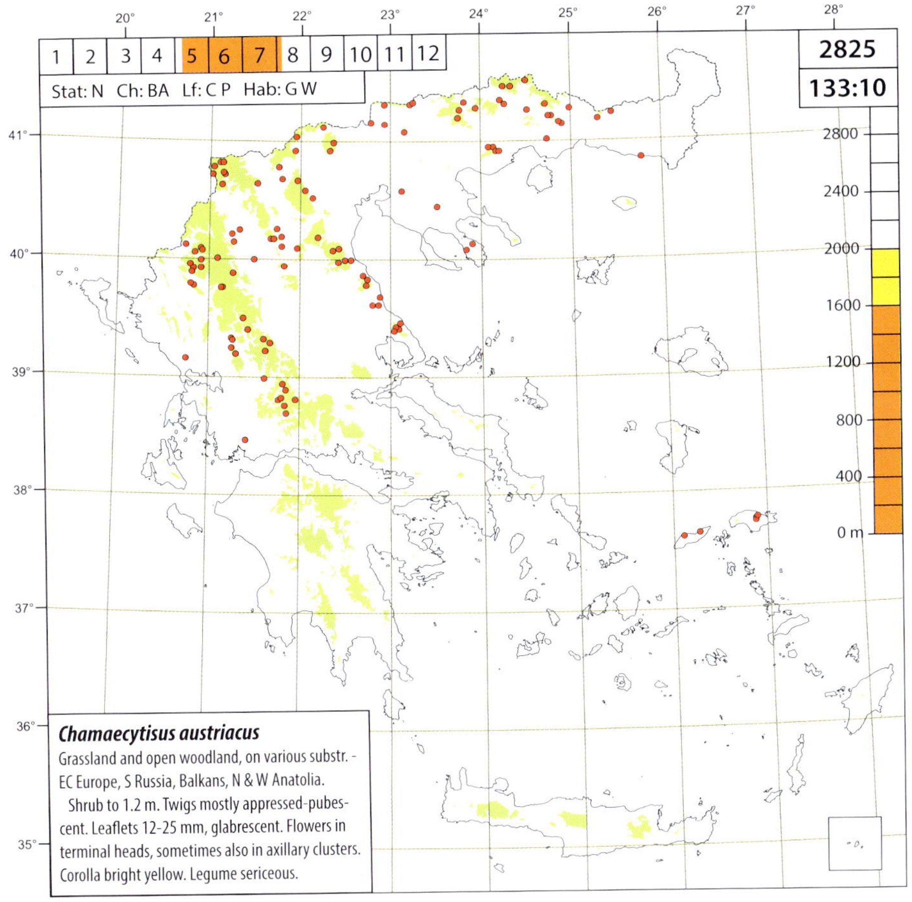

Chamaecytisus austriacus
Grassland and open woodland, on various substr. - EC Europe, S Russia, Balkans, N & W Anatolia.
Shrub to 1.2 m. Twigs mostly appressed-pubescent. Leaflets 12-25 mm, glabrescent. Flowers in terminal heads, sometimes also in axillary clusters. Corolla bright yellow. Legume sericeous.

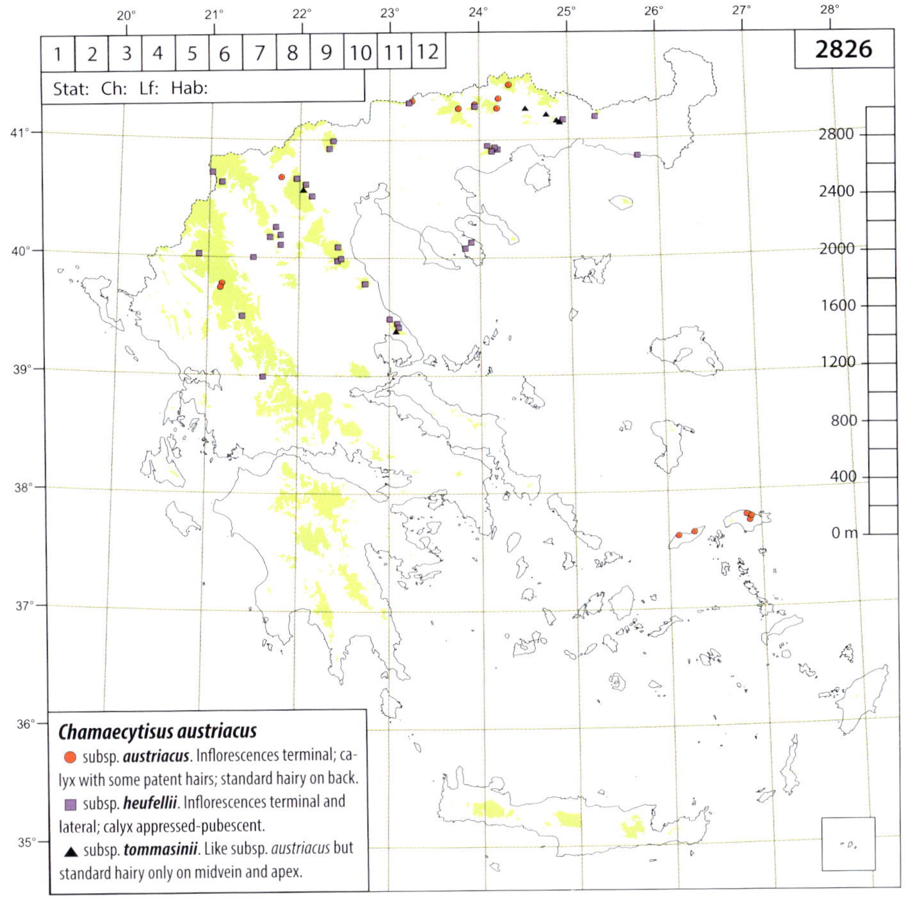

Chamaecytisus austriacus
● subsp. ***austriacus***. Inflorescences terminal; calyx with some patent hairs; standard hairy on back.
■ subsp. ***heufellii***. Inflorescences terminal and lateral; calyx appressed-pubescent.
▲ subsp. ***tommasinii***. Like subsp. *austriacus* but standard hairy only on midvein and apex.

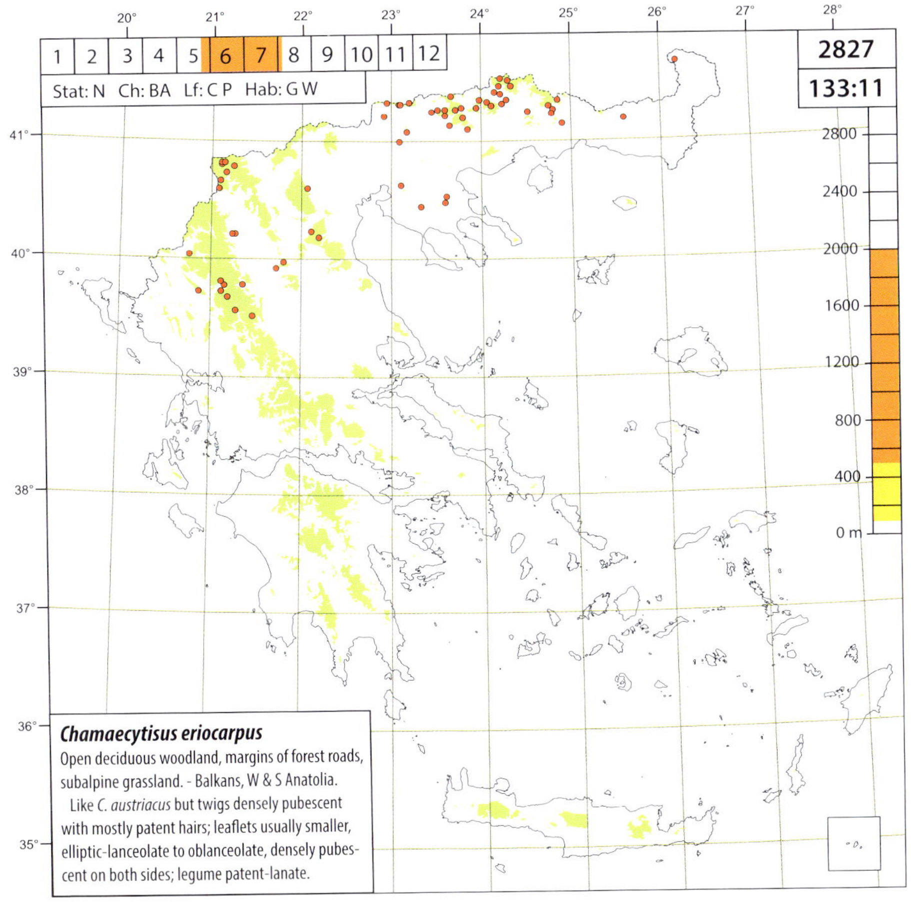

Chamaecytisus eriocarpus
Open deciduous woodland, margins of forest roads, subalpine grassland. - Balkans, W & S Anatolia.
Like *C. austriacus* but twigs densely pubescent with mostly patent hairs; leaflets usually smaller, elliptic-lanceolate to oblanceolate, densely pubescent on both sides; legume patent-lanate.

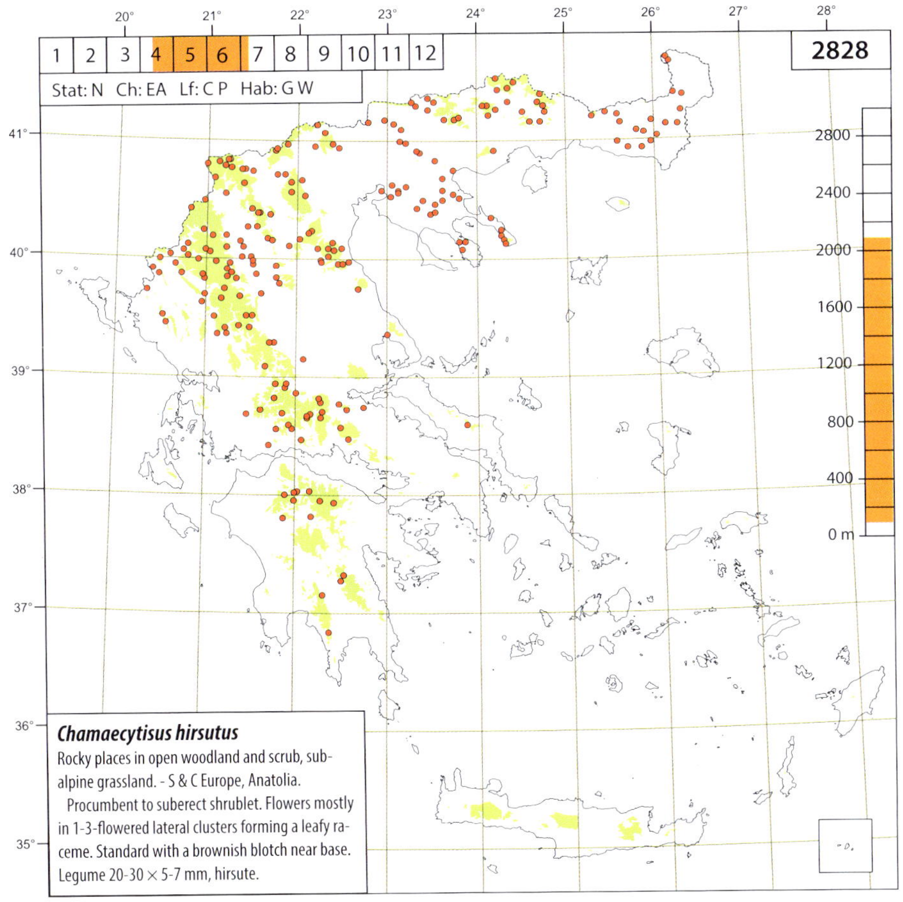

Chamaecytisus hirsutus
Rocky places in open woodland and scrub, subalpine grassland. - S & C Europe, Anatolia.
Procumbent to suberect shrublet. Flowers mostly in 1-3-flowered lateral clusters forming a leafy raceme. Standard with a brownish blotch near base. Legume 20-30 × 5-7 mm, hirsute.

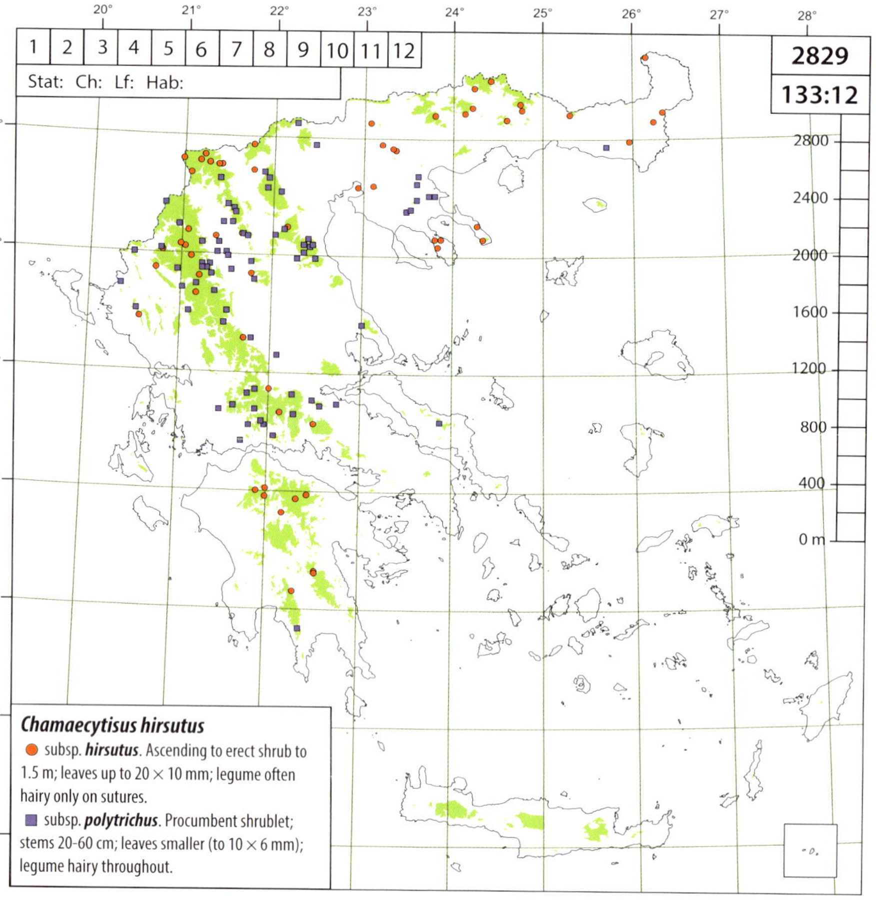

Chamaecytisus hirsutus

● subsp. ***hirsutus***. Ascending to erect shrub to 1.5 m; leaves up to 20 × 10 mm; legume often hairy only on sutures.

■ subsp. ***polytrichus***. Procumbent shrublet; stems 20-60 cm; leaves smaller (to 10 × 6 mm); legume hairy throughout.

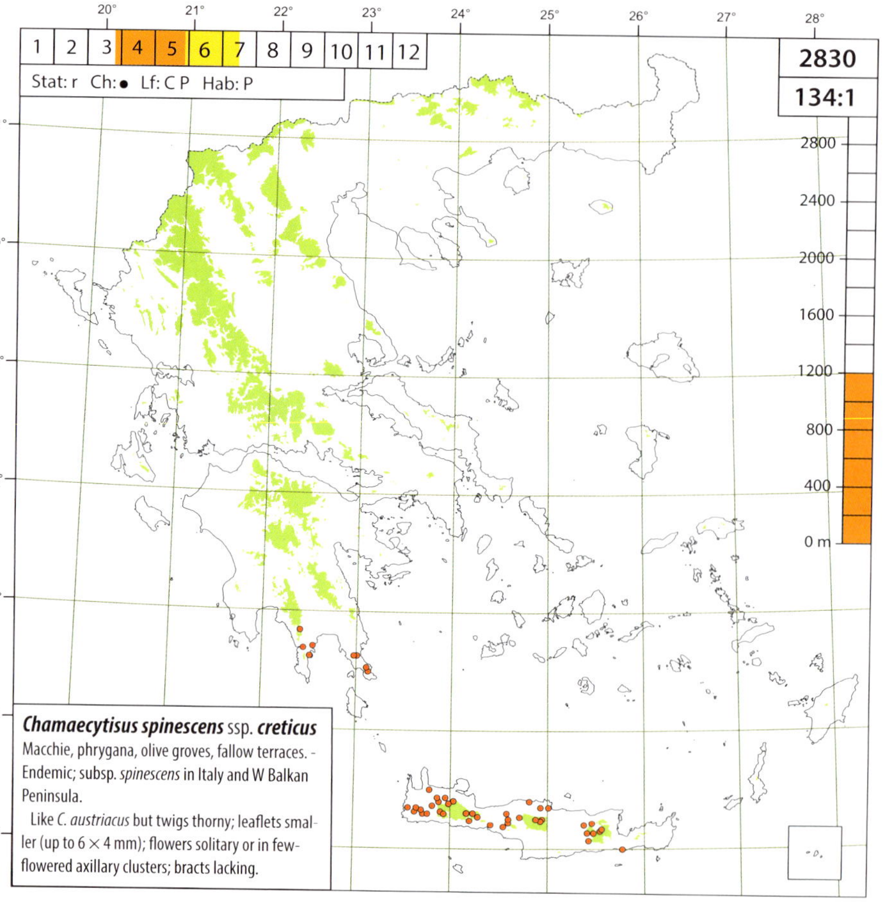

Chamaecytisus spinescens* ssp. *creticus

Macchie, phrygana, olive groves, fallow terraces. - Endemic; subsp. *spinescens* in Italy and W Balkan Peninsula.

Like *C. austriacus* but twigs thorny; leaflets smaller (up to 6 × 4 mm); flowers solitary or in few-flowered axillary clusters; bracts lacking.

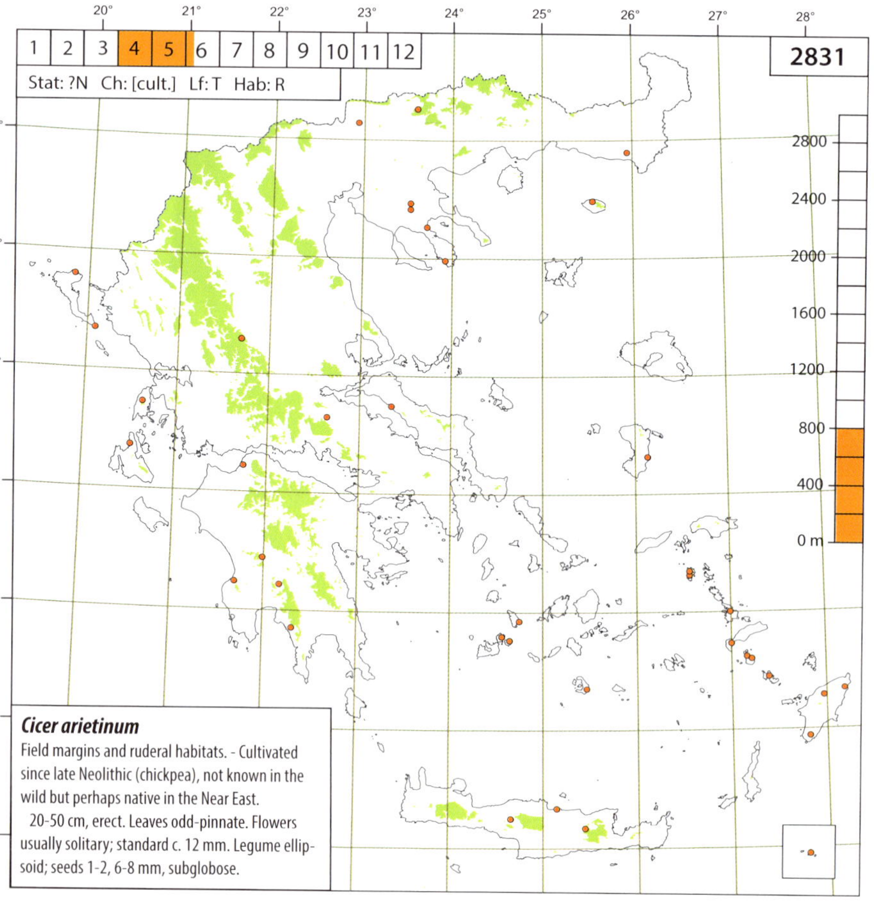

Cicer arietinum

Field margins and ruderal habitats. - Cultivated since late Neolithic (chickpea), not known in the wild but perhaps native in the Near East.

20-50 cm, erect. Leaves odd-pinnate. Flowers usually solitary; standard c. 12 mm. Legume ellipsoid; seeds 1-2, 6-8 mm, subglobose.

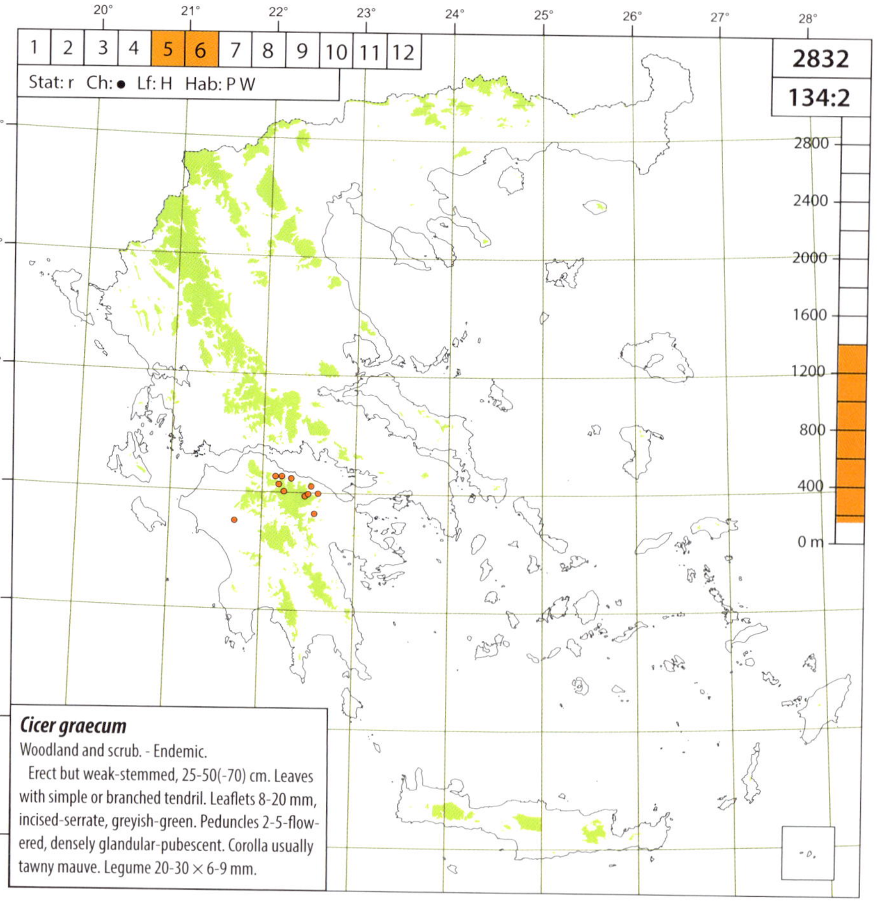

Cicer graecum

Woodland and scrub. - Endemic.

Erect but weak-stemmed, 25-50(-70) cm. Leaves with simple or branched tendril. Leaflets 8-20 mm, incised-serrate, greyish-green. Peduncles 2-5-flowered, densely glandular-pubescent. Corolla usually tawny mauve. Legume 20-30 × 6-9 mm.

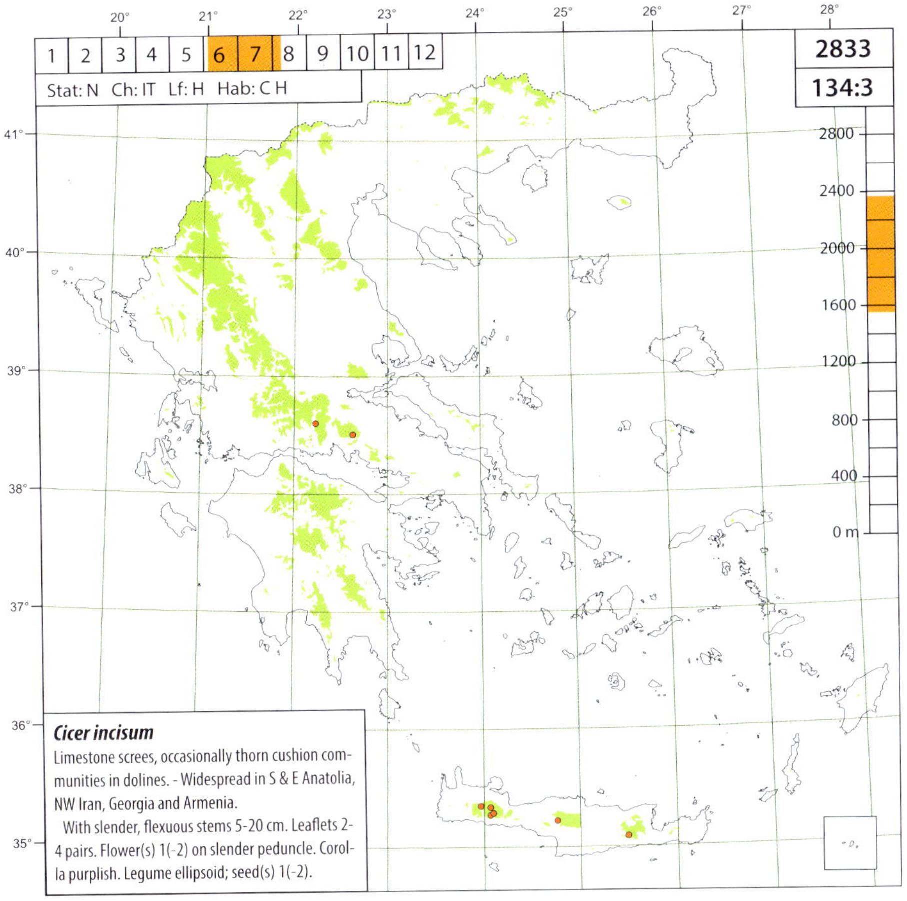

Cicer incisum
Limestone screes, occasionally thorn cushion communities in dolines. - Widespread in S & E Anatolia, NW Iran, Georgia and Armenia.
With slender, flexuous stems 5-20 cm. Leaflets 2-4 pairs. Flower(s) 1(-2) on slender peduncle. Corolla purplish. Legume ellipsoid; seed(s) 1(-2).

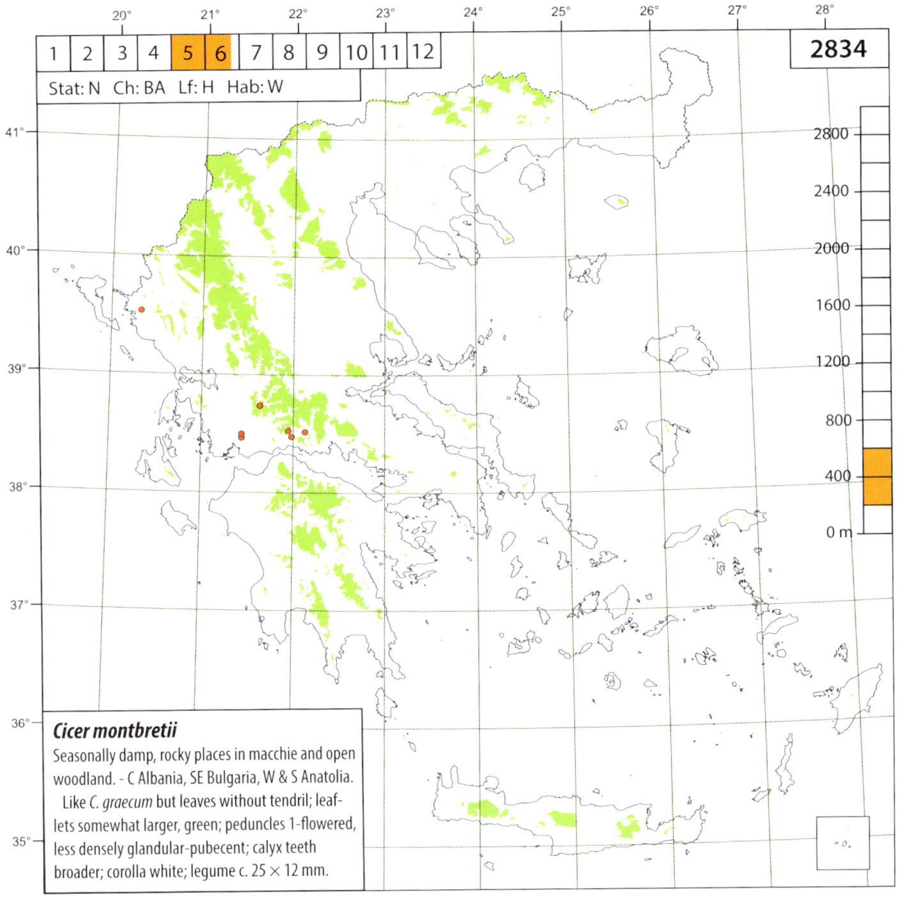

Cicer montbretii
Seasonally damp, rocky places in macchie and open woodland. - C Albania, SE Bulgaria, W & S Anatolia.
Like *C. graecum* but leaves without tendril; leaflets somewhat larger, green; peduncles 1-flowered, less densely glandular-pubecent; calyx teeth broader; corolla white; legume c. 25 × 12 mm.

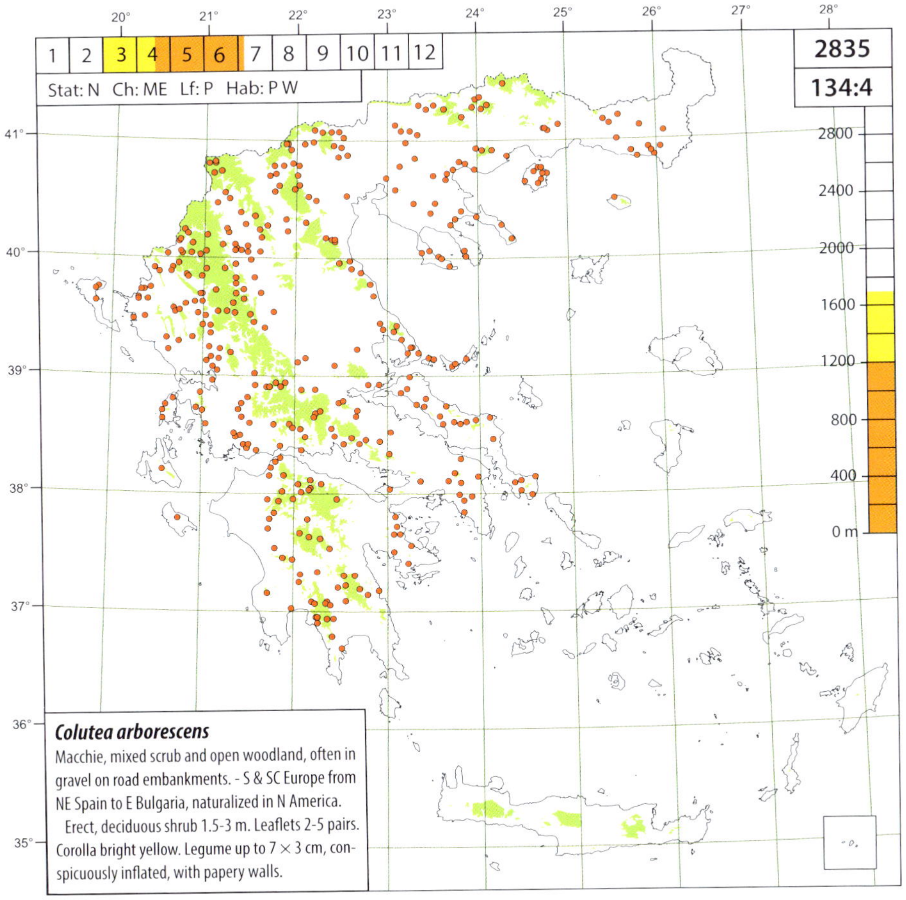

Colutea arborescens
Macchie, mixed scrub and open woodland, often in gravel on road embankments. - S & SC Europe from NE Spain to E Bulgaria, naturalized in N America.
Erect, deciduous shrub 1.5-3 m. Leaflets 2-5 pairs. Corolla bright yellow. Legume up to 7 × 3 cm, conspicuously inflated, with papery walls.

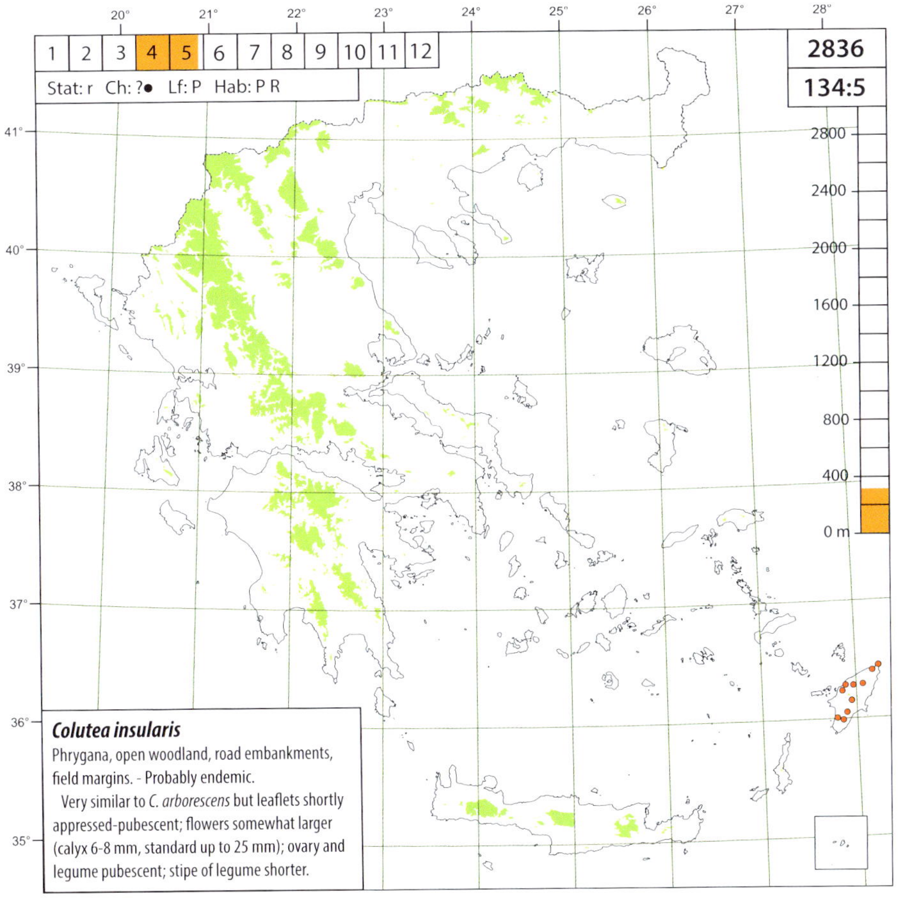

Colutea insularis
Phrygana, open woodland, road embankments, field margins. - Probably endemic.
Very similar to *C. arborescens* but leaflets shortly appressed-pubescent; flowers somewhat larger (calyx 6-8 mm, standard up to 25 mm); ovary and legume pubescent; stipe of legume shorter.

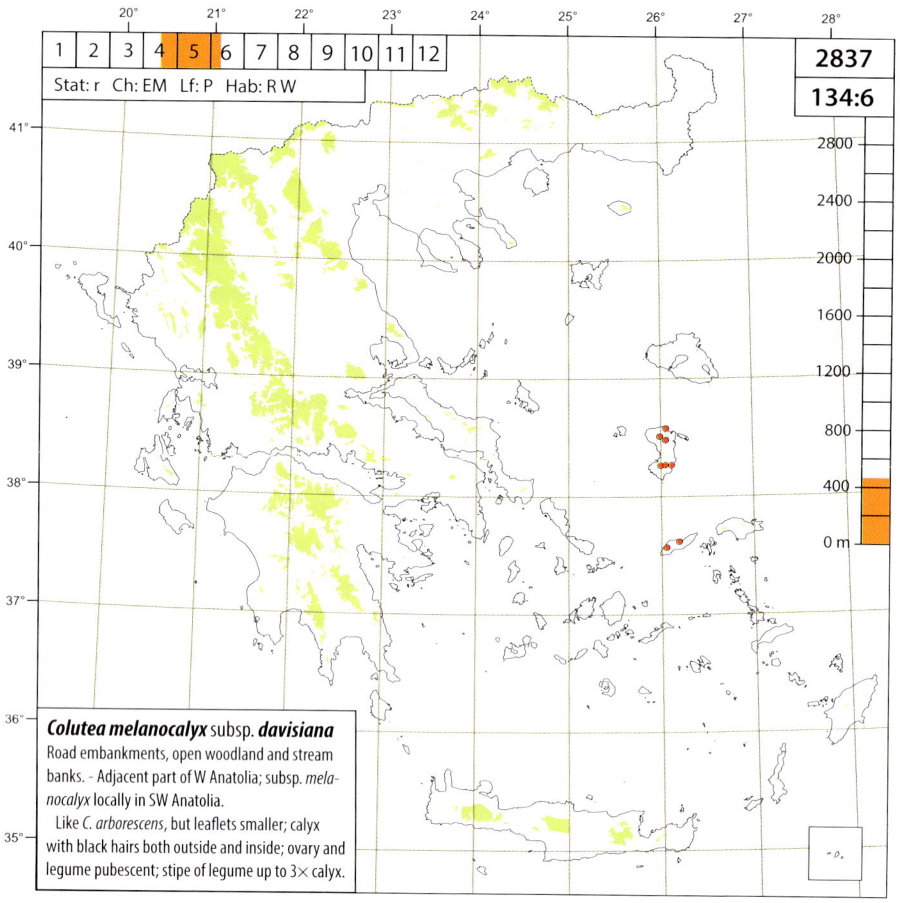

Colutea melanocalyx subsp. _davisiana_
Road embankments, open woodland and stream banks. - Adjacent part of W Anatolia; subsp. *melanocalyx* locally in SW Anatolia.
Like *C. arborescens*, but leaflets smaller; calyx with black hairs both outside and inside; ovary and legume pubescent; stipe of legume up to 3× calyx.

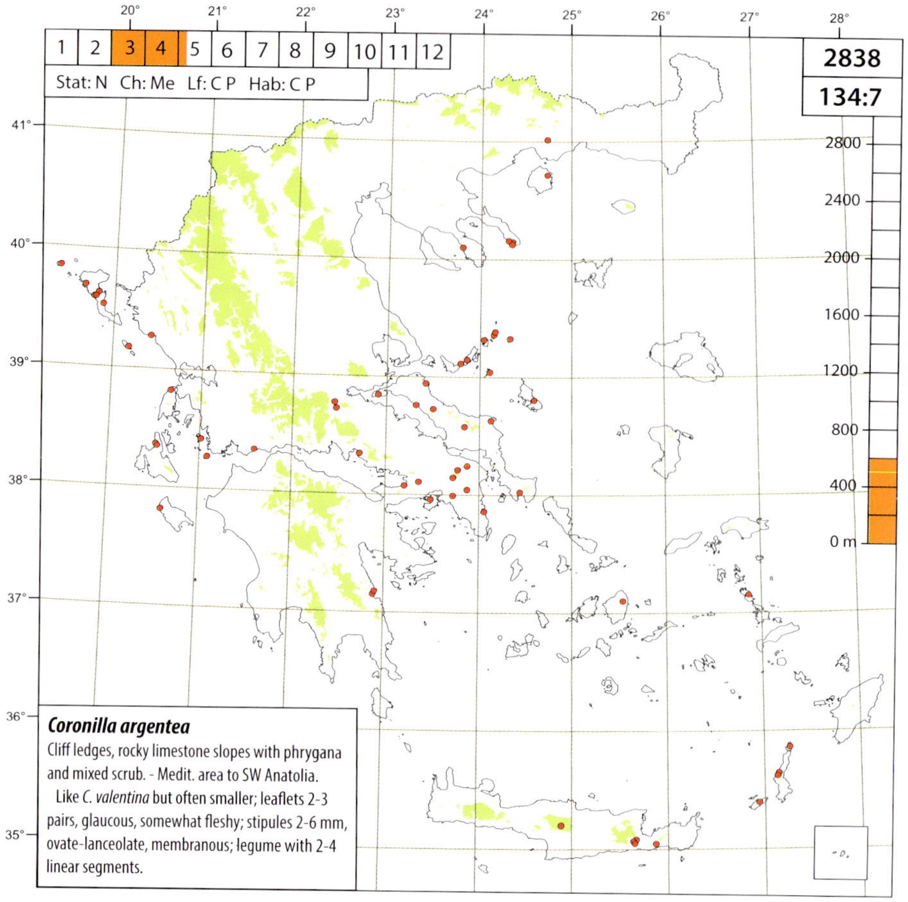

Coronilla argentea
Cliff ledges, rocky limestone slopes with phrygana and mixed scrub. - Medit. area to SW Anatolia.
Like *C. valentina* but often smaller; leaflets 2-3 pairs, glaucous, somewhat fleshy; stipules 2-6 mm, ovate-lanceolate, membranous; legume with 2-4 linear segments.

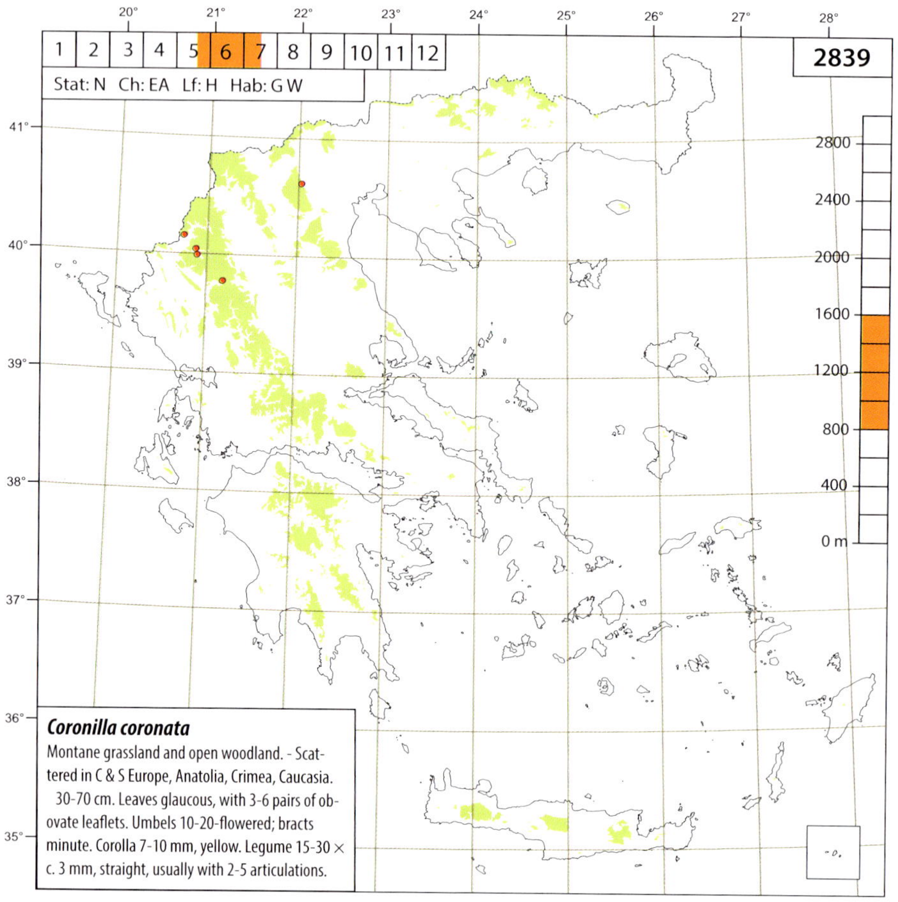

Coronilla coronata
Montane grassland and open woodland. - Scattered in C & S Europe, Anatolia, Crimea, Caucasia.
30-70 cm. Leaves glaucous, with 3-6 pairs of obovate leaflets. Umbels 10-20-flowered; bracts minute. Corolla 7-10 mm, yellow. Legume 15-30 × c. 3 mm, straight, usually with 2-5 articulations.

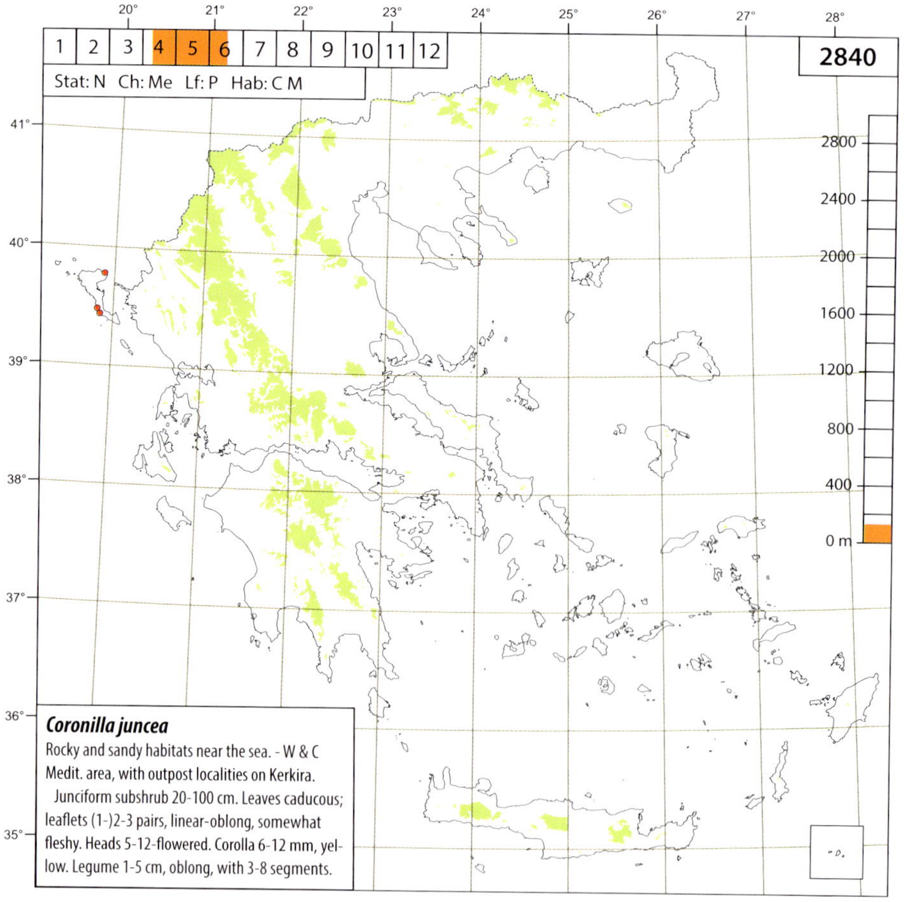

Coronilla juncea
Rocky and sandy habitats near the sea. - W & C Medit. area, with outpost localities on Kerkira.
Junciform subshrub 20-100 cm. Leaves caducous; leaflets (1-)2-3 pairs, linear-oblong, somewhat fleshy. Heads 5-12-flowered. Corolla 6-12 mm, yellow. Legume 1-5 cm, oblong, with 3-8 segments.

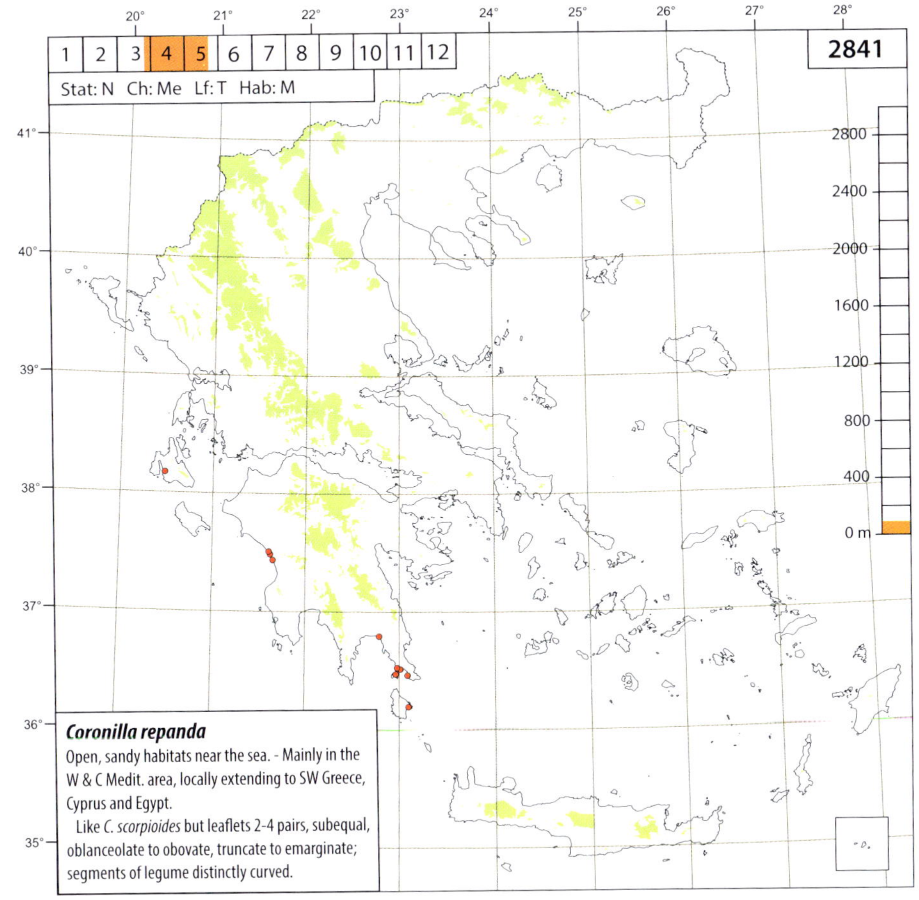

Coronilla repanda

Open, sandy habitats near the sea. - Mainly in the W & C Medit. area, locally extending to SW Greece, Cyprus and Egypt.

Like *C. scorpioides* but leaflets 2-4 pairs, subequal, oblanceolate to obovate, truncate to emarginate; segments of legume distinctly curved.

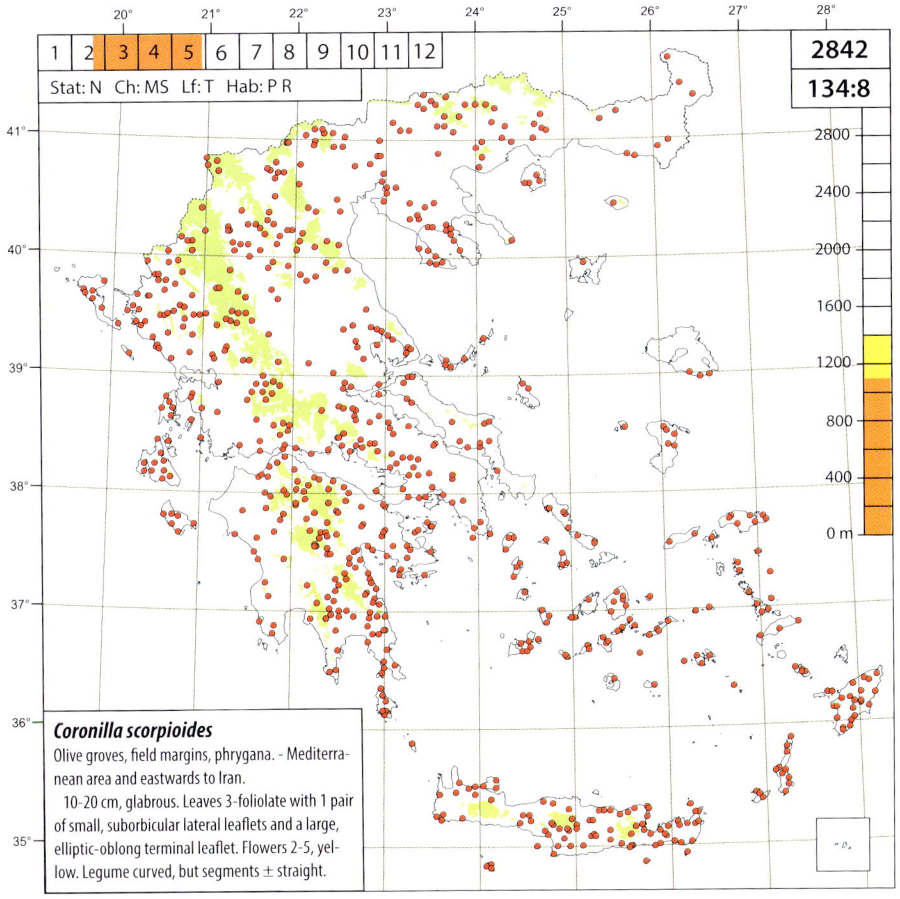

Coronilla scorpioides

Olive groves, field margins, phrygana. - Mediterranean area and eastwards to Iran.

10-20 cm, glabrous. Leaves 3-foliolate with 1 pair of small, suborbicular lateral leaflets and a large, elliptic-oblong terminal leaflet. Flowers 2-5, yellow. Legume curved, but segments ± straight.

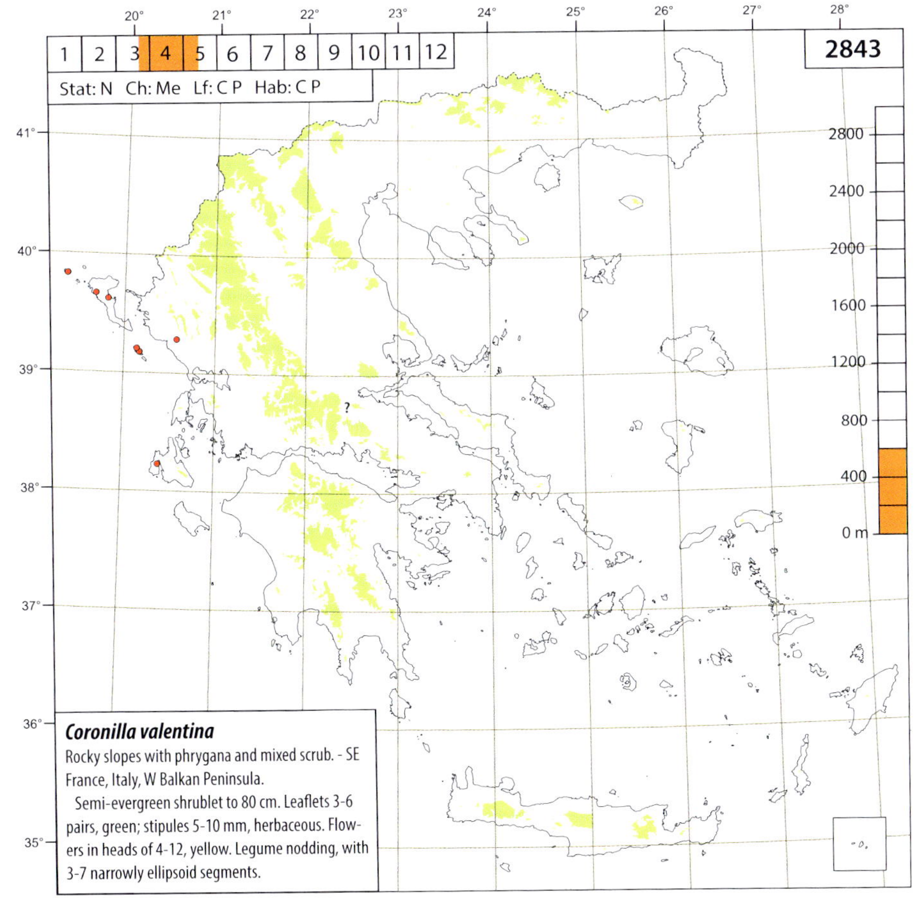

Coronilla valentina

Rocky slopes with phrygana and mixed scrub. - SE France, Italy, W Balkan Peninsula.

Semi-evergreen shrublet to 80 cm. Leaflets 3-6 pairs, green; stipules 5-10 mm, herbaceous. Flowers in heads of 4-12, yellow. Legume nodding, with 3-7 narrowly ellipsoid segments.

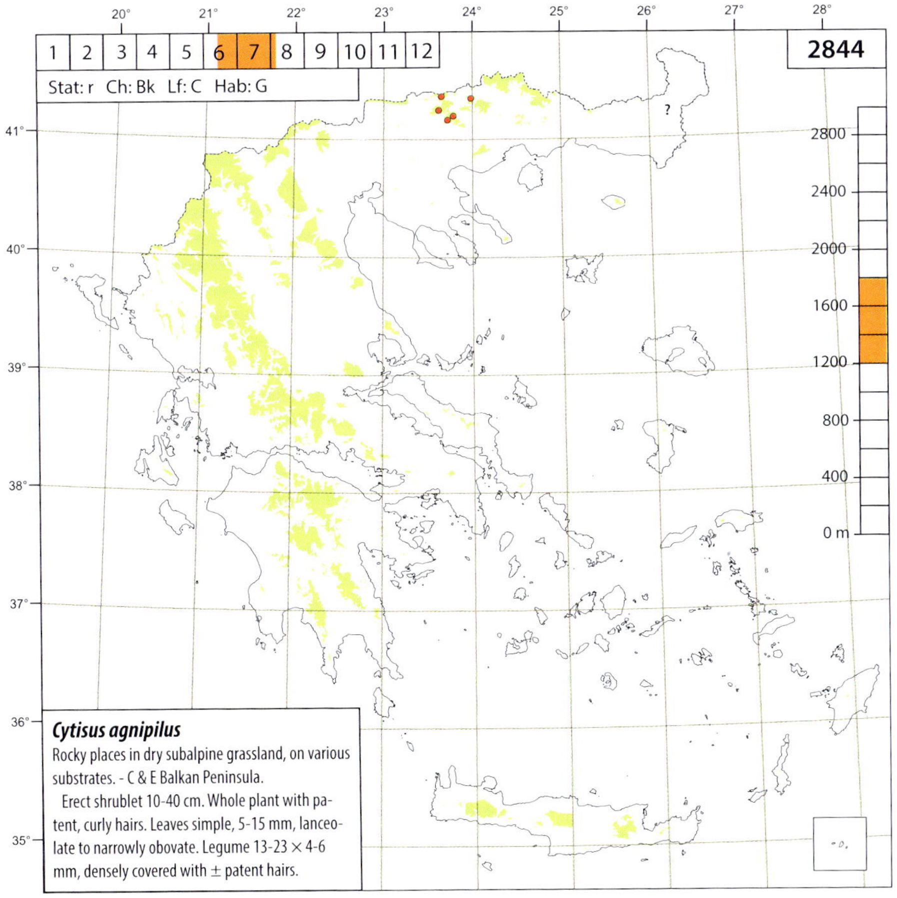

Cytisus agnipilus

Rocky places in dry subalpine grassland, on various substrates. - C & E Balkan Peninsula.

Erect shrublet 10-40 cm. Whole plant with patent, curly hairs. Leaves simple, 5-15 mm, lanceolate to narrowly obovate. Legume 13-23 × 4-6 mm, densely covered with ± patent hairs.

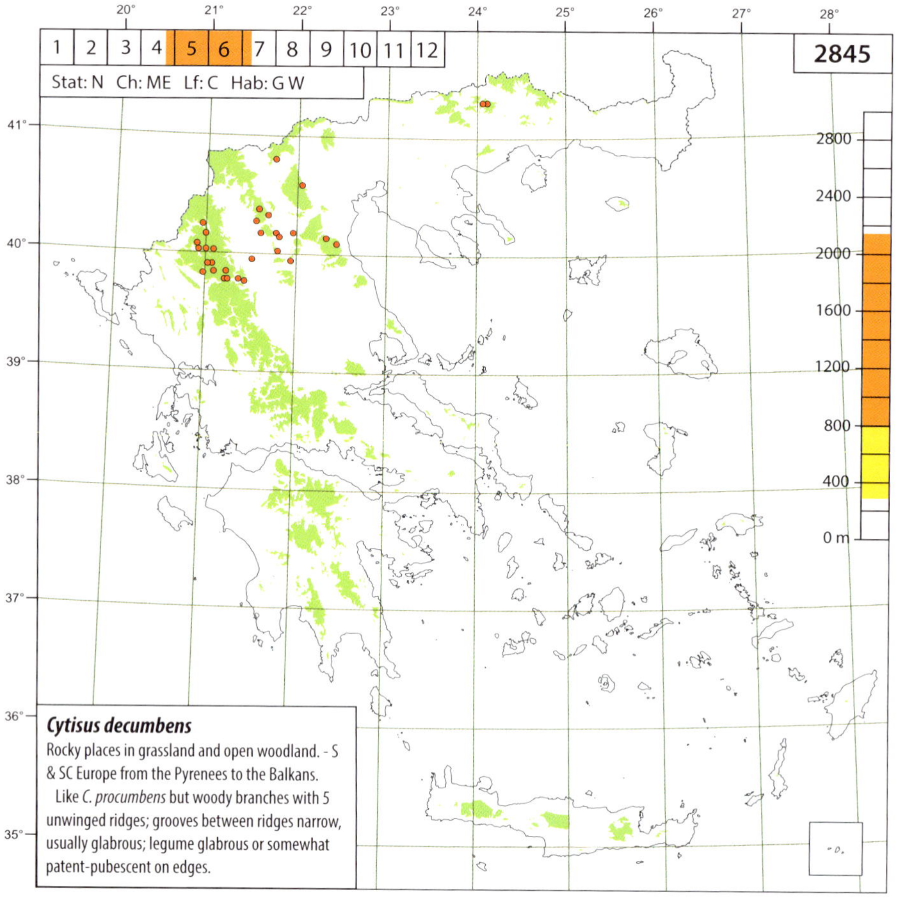

Cytisus decumbens

Rocky places in grassland and open woodland. - S & SC Europe from the Pyrenees to the Balkans.

Like *C. procumbens* but woody branches with 5 unwinged ridges; grooves between ridges narrow, usually glabrous; legume glabrous or somewhat patent-pubescent on edges.

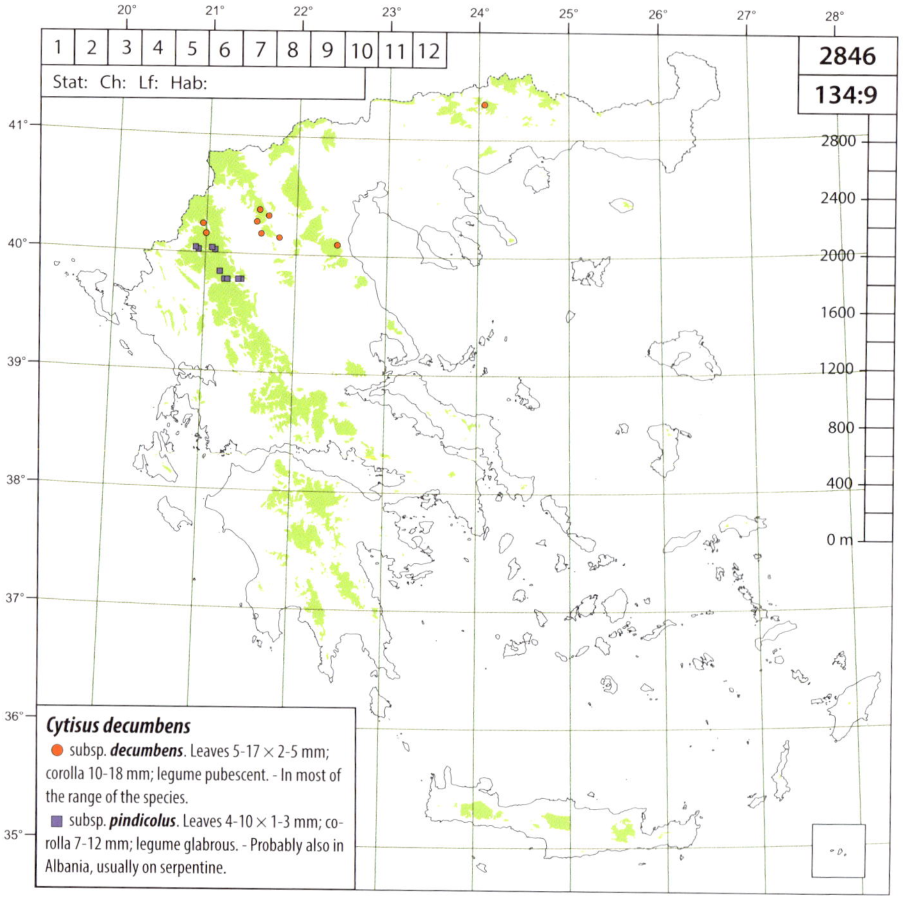

Cytisus decumbens

● subsp. ***decumbens***. Leaves 5-17 × 2-5 mm; corolla 10-18 mm; legume pubescent. - In most of the range of the species.

■ subsp. ***pindicolus***. Leaves 4-10 × 1-3 mm; corolla 7-12 mm; legume glabrous. - Probably also in Albania, usually on serpentine.

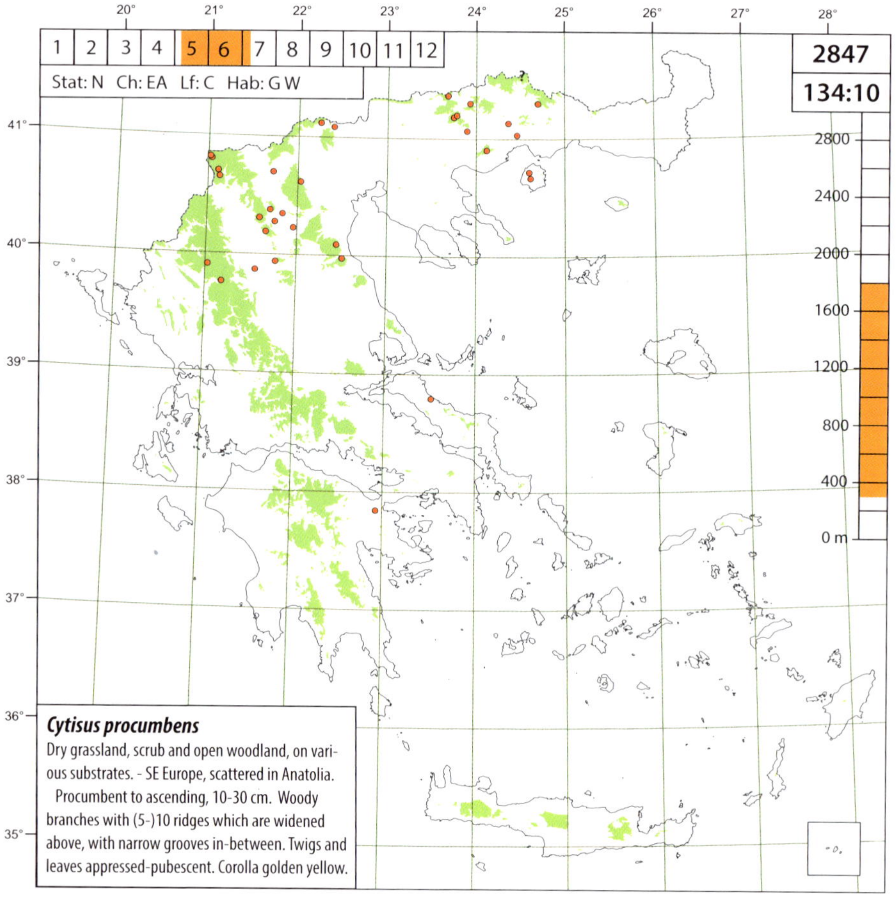

Cytisus procumbens

Dry grassland, scrub and open woodland, on various substrates. - SE Europe, scattered in Anatolia.

Procumbent to ascending, 10-30 cm. Woody branches with (5-)10 ridges which are widened above, with narrow grooves in-between. Twigs and leaves appressed-pubescent. Corolla golden yellow.

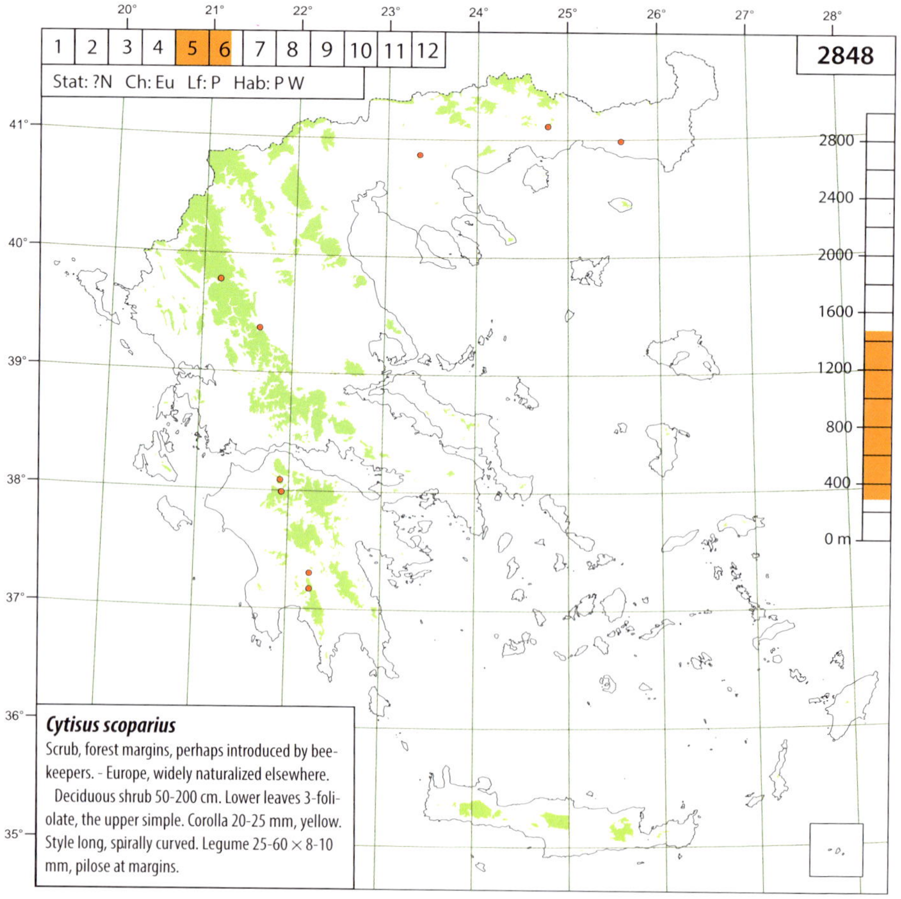

Cytisus scoparius

Scrub, forest margins, perhaps introduced by bee-keepers. - Europe, widely naturalized elsewhere.

Deciduous shrub 50-200 cm. Lower leaves 3-foliolate, the upper simple. Corolla 20-25 mm, yellow. Style long, spirally curved. Legume 25-60 × 8-10 mm, pilose at margins.

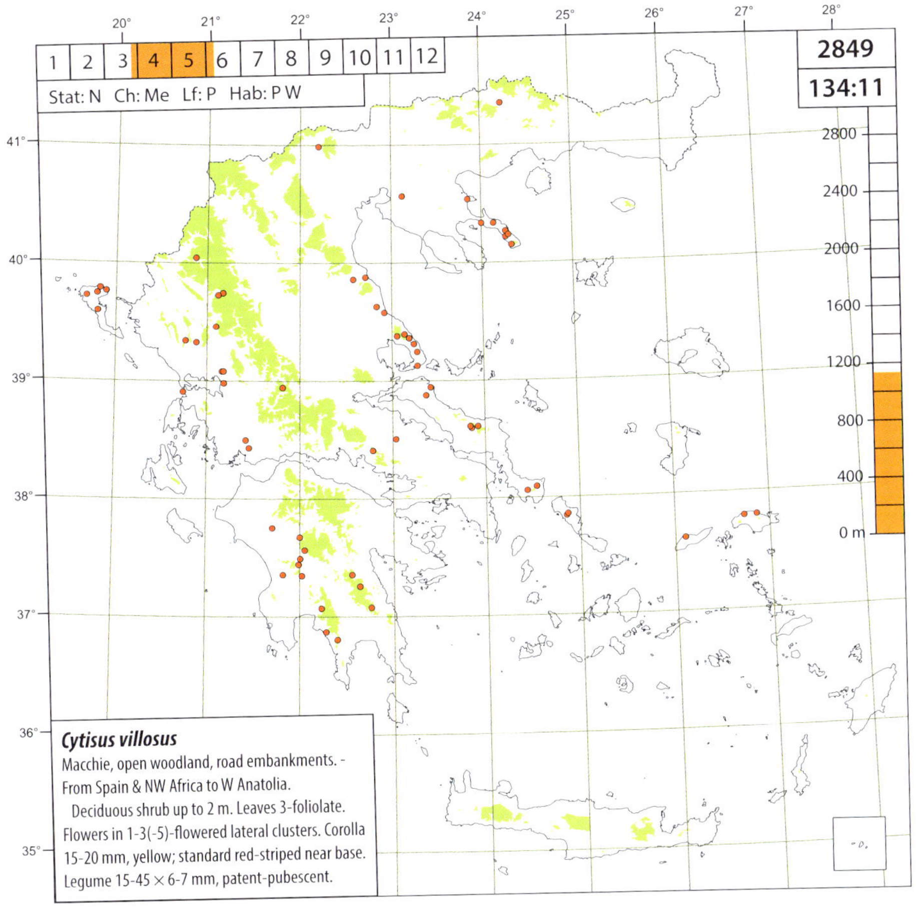

Cytisus villosus

Macchie, open woodland, road embankments. - From Spain & NW Africa to W Anatolia.

Deciduous shrub up to 2 m. Leaves 3-foliolate. Flowers in 1-3(-5)-flowered lateral clusters. Corolla 15-20 mm, yellow; standard red-striped near base. Legume 15-45 × 6-7 mm, patent-pubescent.

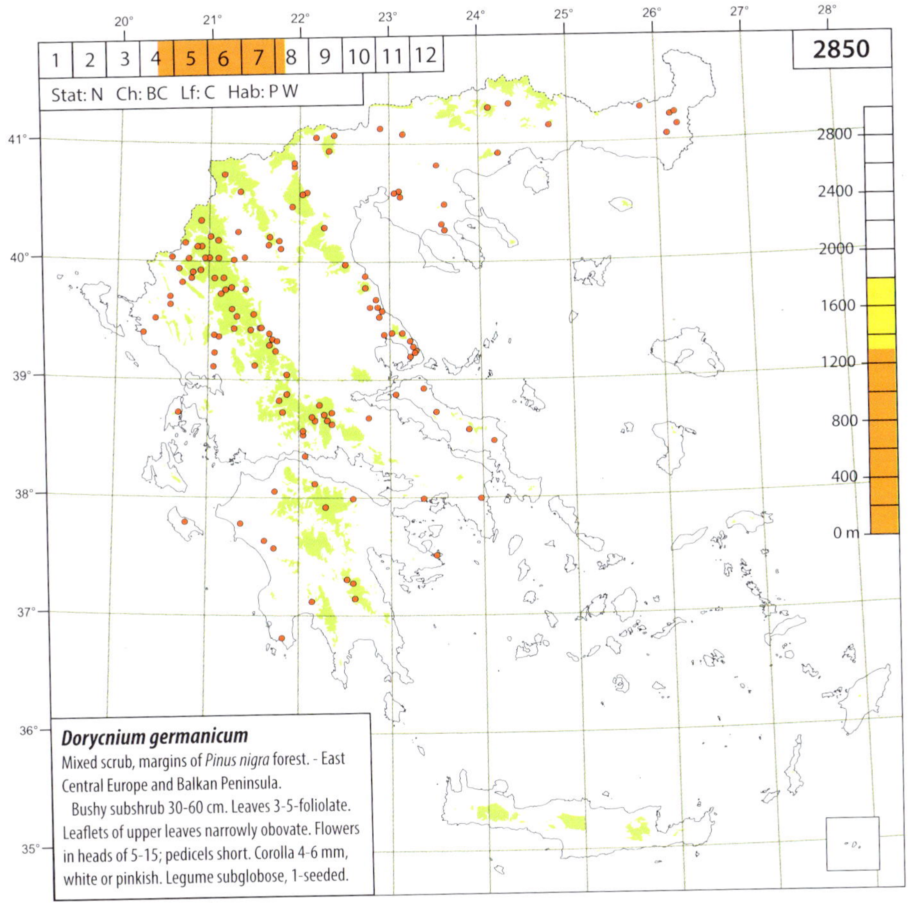

Dorycnium germanicum

Mixed scrub, margins of *Pinus nigra* forest. - East Central Europe and Balkan Peninsula.

Bushy subshrub 30-60 cm. Leaves 3-5-foliolate. Leaflets of upper leaves narrowly obovate. Flowers in heads of 5-15; pedicels short. Corolla 4-6 mm, white or pinkish. Legume subglobose, 1-seeded.

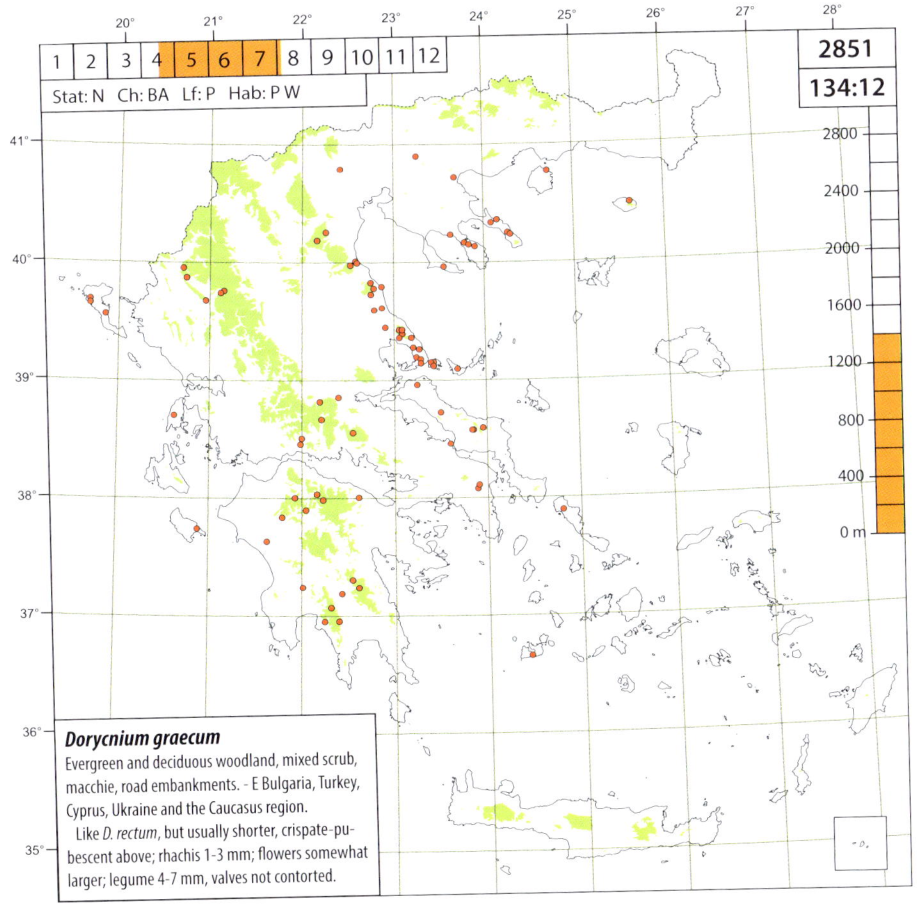

Dorycnium graecum

Evergreen and deciduous woodland, mixed scrub, macchie, road embankments. - E Bulgaria, Turkey, Cyprus, Ukraine and the Caucasus region.

Like *D. rectum*, but usually shorter, crispate-pubescent above; rhachis 1-3 mm; flowers somewhat larger; legume 4-7 mm, valves not contorted.

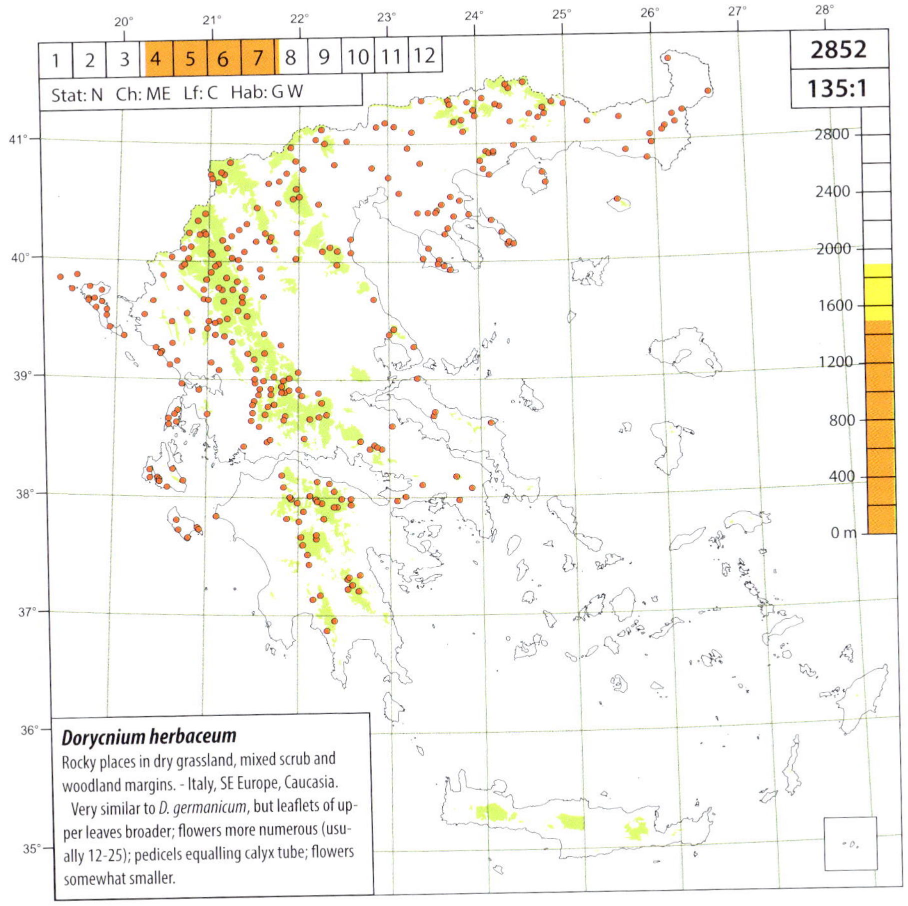

Dorycnium herbaceum

Rocky places in dry grassland, mixed scrub and woodland margins. - Italy, SE Europe, Caucasia.

Very similar to *D. germanicum*, but leaflets of upper leaves broader; flowers more numerous (usually 12-25); pedicels equalling calyx tube; flowers somewhat smaller.

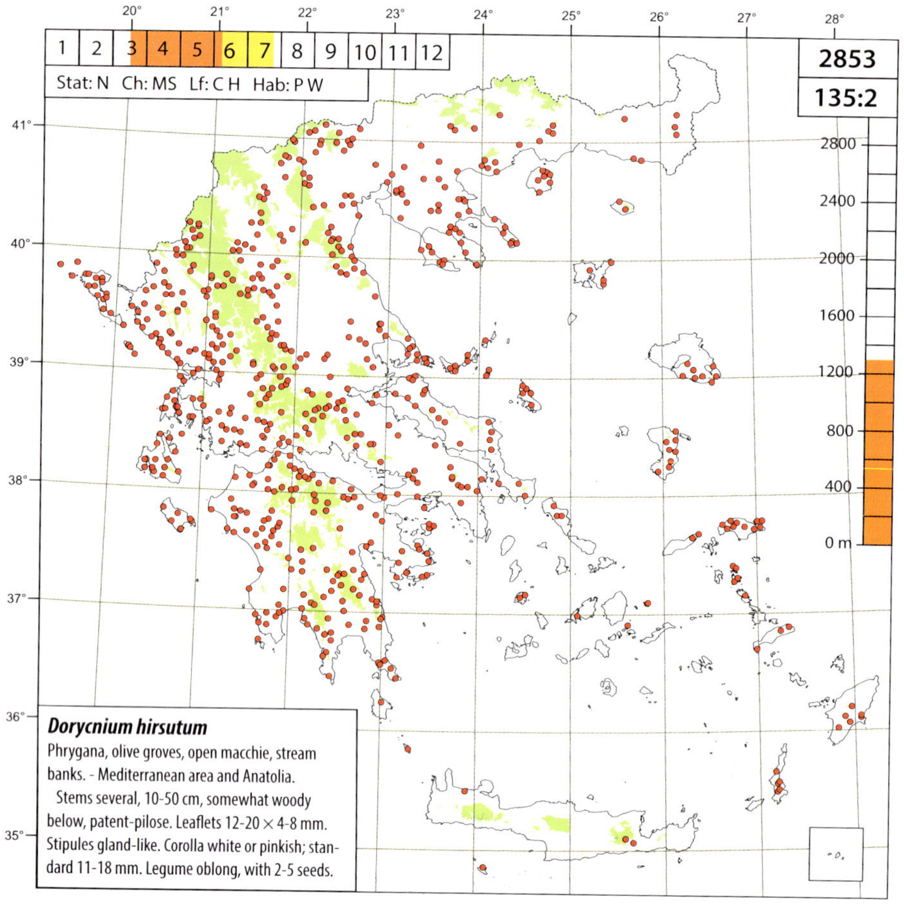

Dorycnium hirsutum
Phrygana, olive groves, open macchie, stream banks. - Mediterranean area and Anatolia.
Stems several, 10-50 cm, somewhat woody below, patent-pilose. Leaflets 12-20 × 4-8 mm. Stipules gland-like. Corolla white or pinkish; standard 11-18 mm. Legume oblong, with 2-5 seeds.

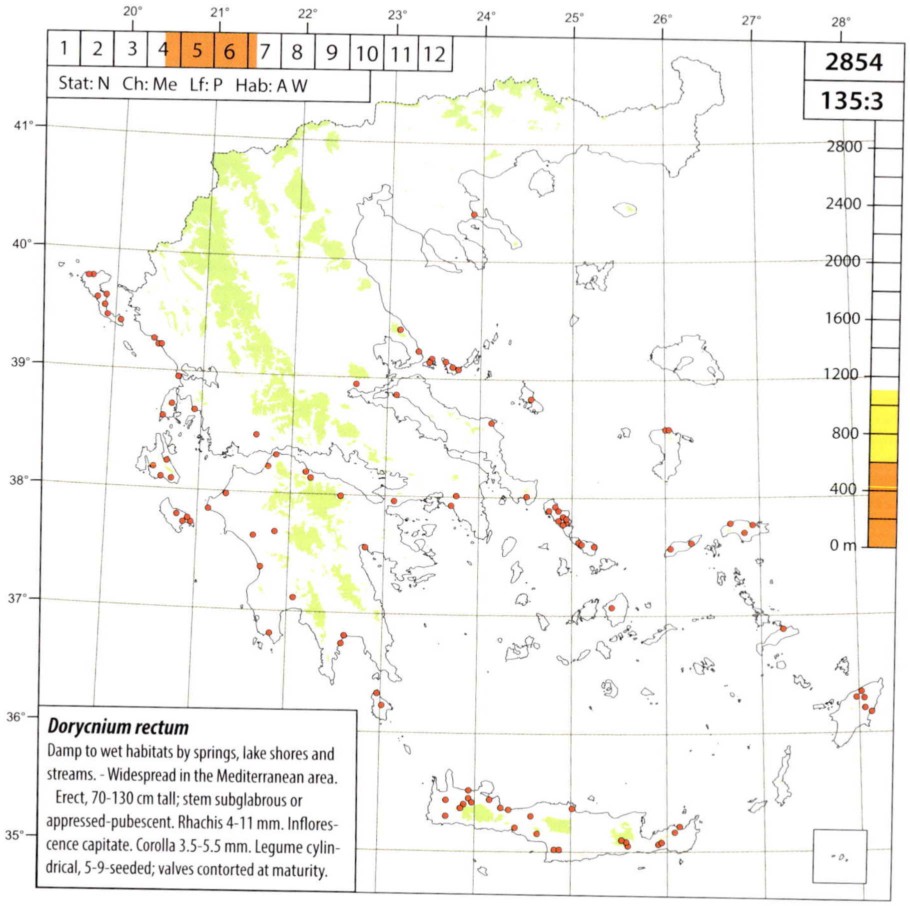

Dorycnium rectum
Damp to wet habitats by springs, lake shores and streams. - Widespread in the Mediterranean area.
Erect, 70-130 cm tall; stem subglabrous or appressed-pubescent. Rhachis 4-11 mm. Inflorescence capitate. Corolla 3.5-5.5 mm. Legume cylindrical, 5-9-seeded; valves contorted at maturity.

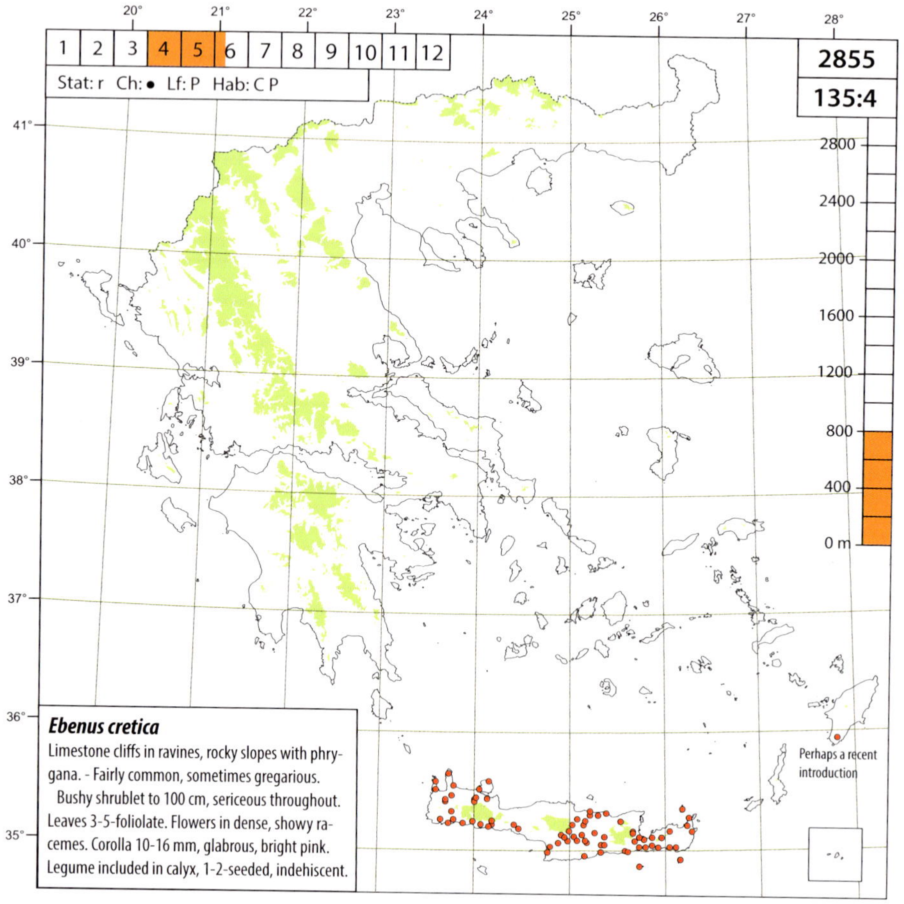

Ebenus cretica
Limestone cliffs in ravines, rocky slopes with phrygana. - Fairly common, sometimes gregarious.
Bushy shrublet to 100 cm, sericeous throughout. Leaves 3-5-foliolate. Flowers in dense, showy racemes. Corolla 10-16 mm, glabrous, bright pink. Legume included in calyx, 1-2-seeded, indehiscent.

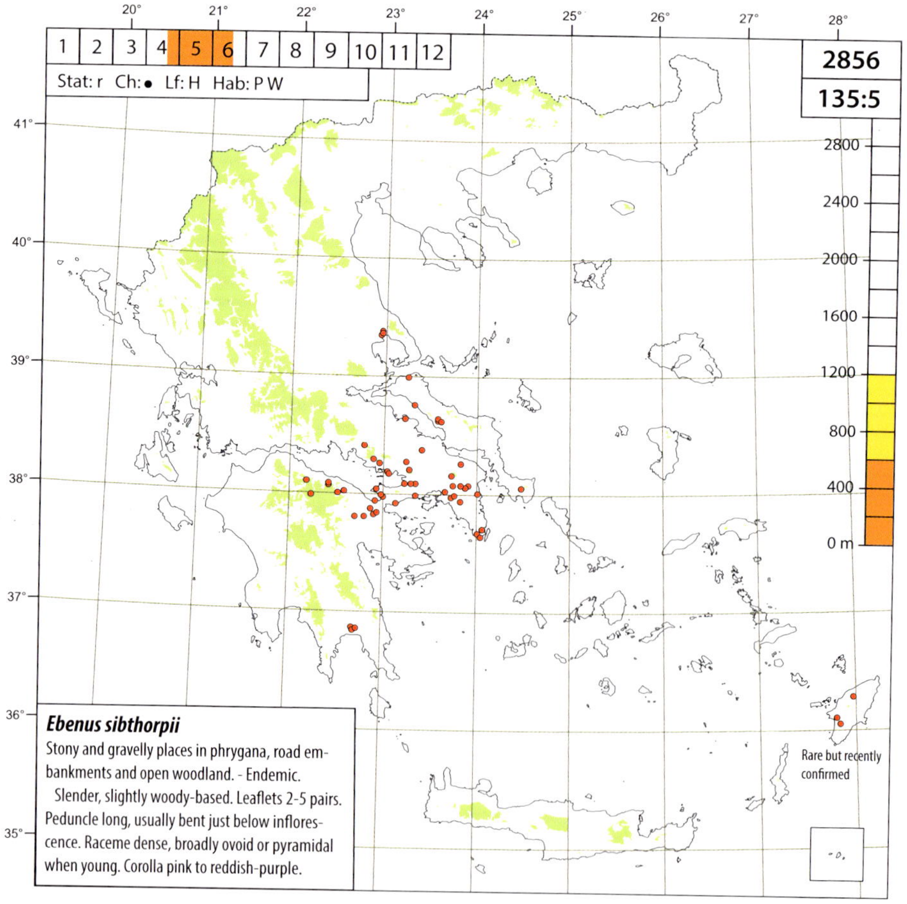

Ebenus sibthorpii
Stony and gravelly places in phrygana, road embankments and open woodland. - Endemic.
Slender, slightly woody-based. Leaflets 2-5 pairs. Peduncle long, usually bent just below inflorescence. Raceme dense, broadly ovoid or pyramidal when young. Corolla pink to reddish-purple.

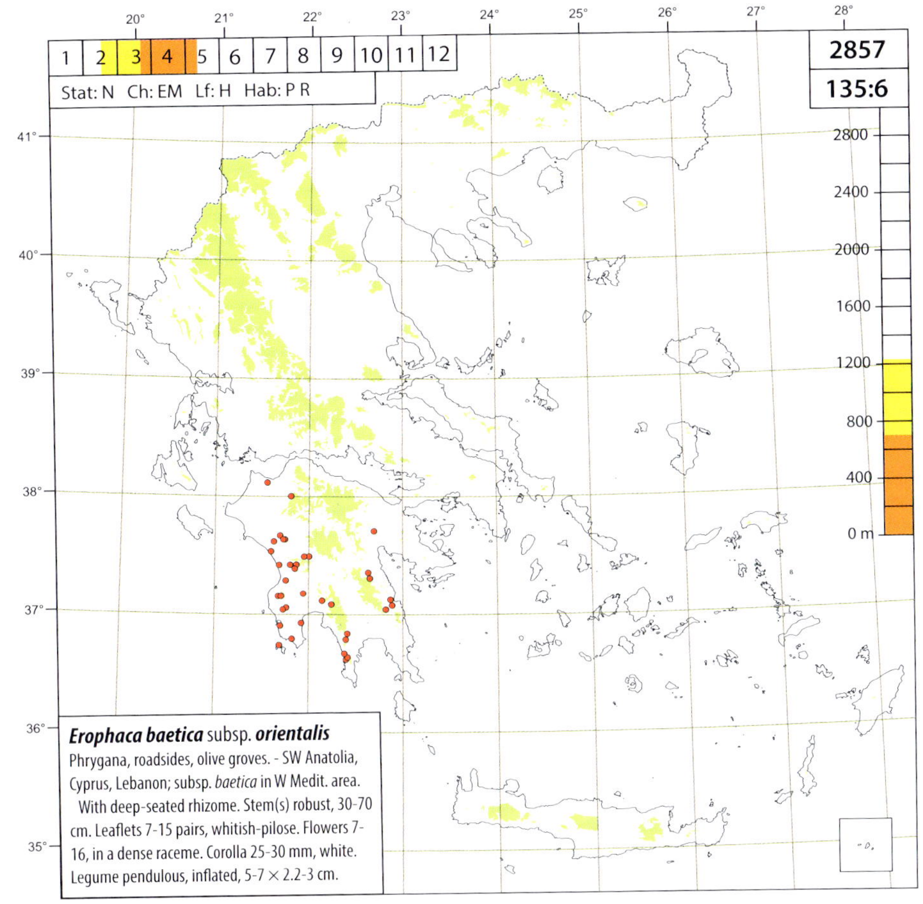

Erophaca baetica subsp. ***orientalis***
Phrygana, roadsides, olive groves. - SW Anatolia, Cyprus, Lebanon; subsp. *baetica* in W Medit. area.
With deep-seated rhizome. Stem(s) robust, 30-70 cm. Leaflets 7-15 pairs, whitish-pilose. Flowers 7-16, in a dense raceme. Corolla 25-30 mm, white. Legume pendulous, inflated, 5-7 × 2.2-3 cm.

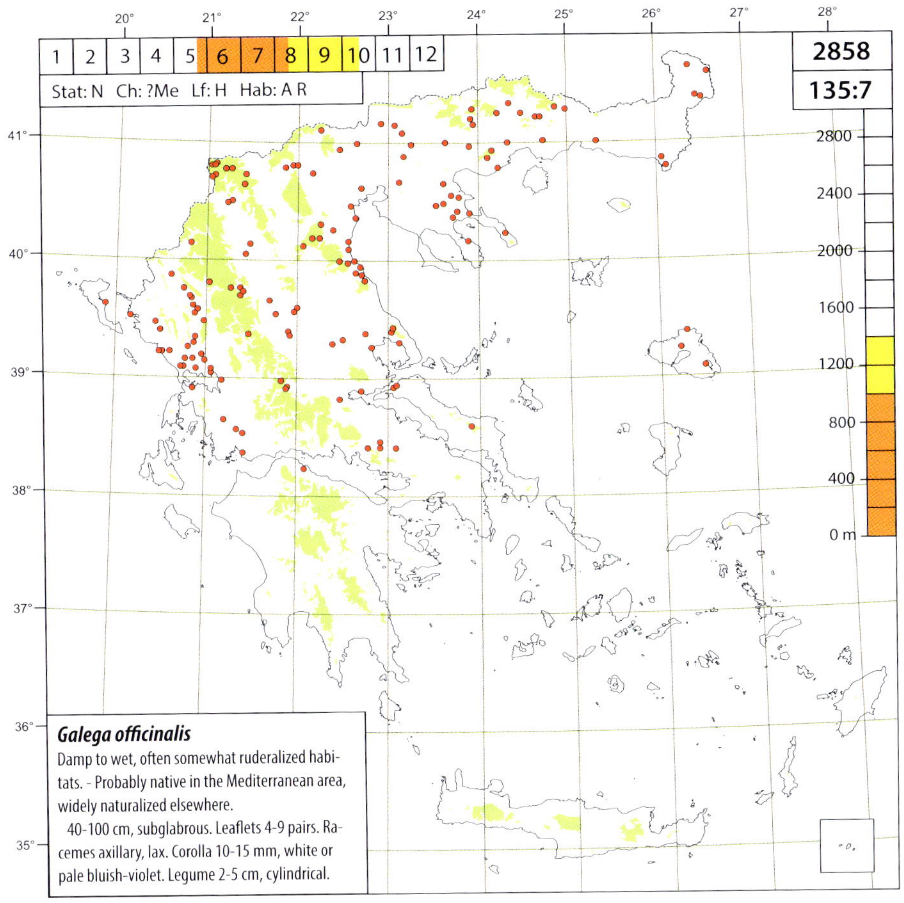

Galega officinalis
Damp to wet, often somewhat ruderalized habitats. - Probably native in the Mediterranean area, widely naturalized elsewhere.
40-100 cm, subglabrous. Leaflets 4-9 pairs. Racemes axillary, lax. Corolla 10-15 mm, white or pale bluish-violet. Legume 2-5 cm, cylindrical.

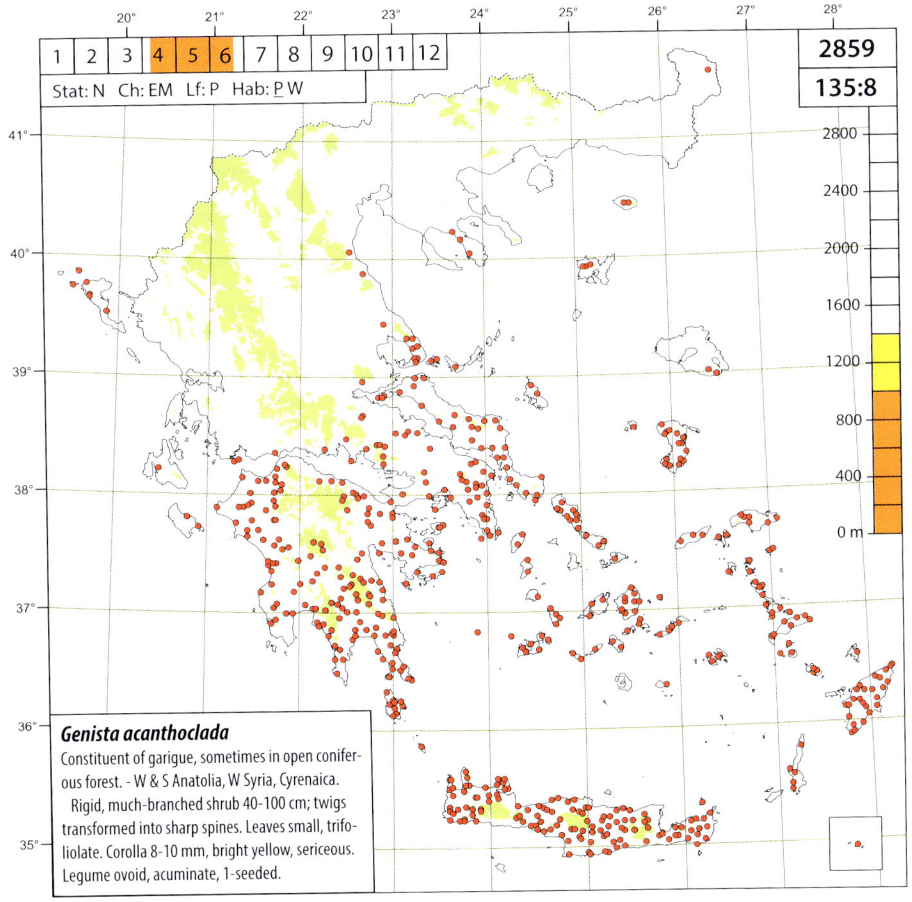

Genista acanthoclada
Constituent of garigue, sometimes in open coniferous forest. - W & S Anatolia, W Syria, Cyrenaica.
Rigid, much-branched shrub 40-100 cm; twigs transformed into sharp spines. Leaves small, trifoliolate. Corolla 8-10 mm, bright yellow, sericeous. Legume ovoid, acuminate, 1-seeded.

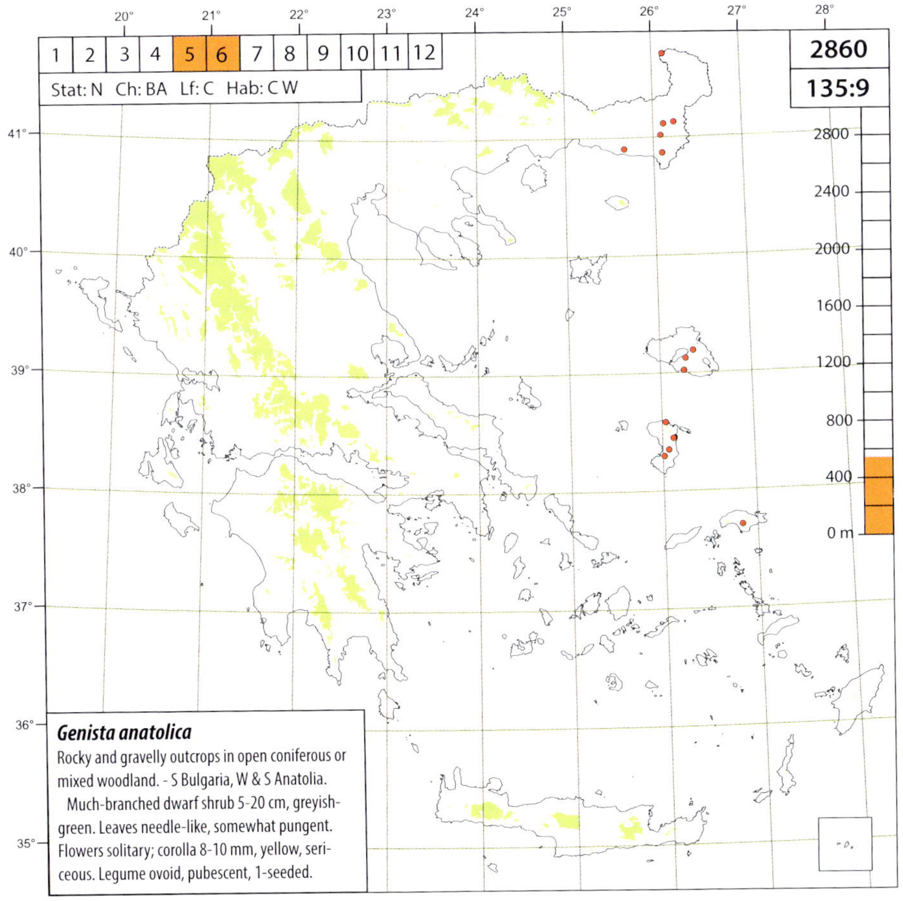

Genista anatolica
Rocky and gravelly outcrops in open coniferous or mixed woodland. - S Bulgaria, W & S Anatolia.
Much-branched dwarf shrub 5-20 cm, greyish-green. Leaves needle-like, somewhat pungent. Flowers solitary; corolla 8-10 mm, yellow, sericeous. Legume ovoid, pubescent, 1-seeded.

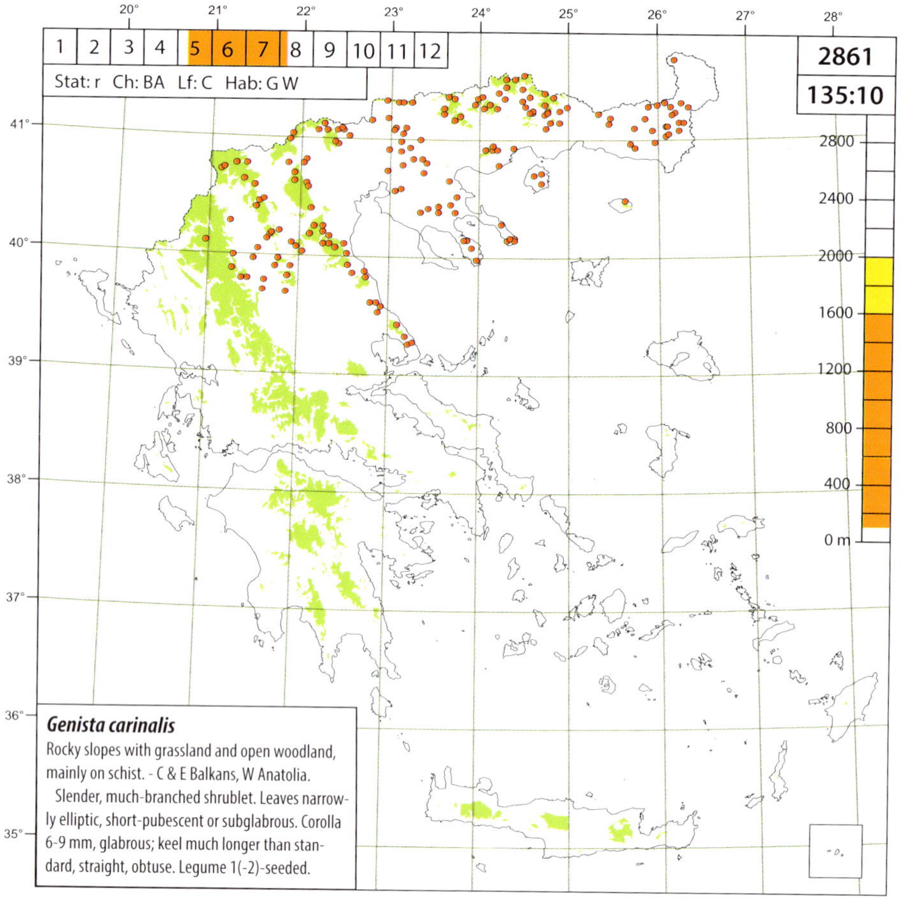

Genista carinalis
Rocky slopes with grassland and open woodland, mainly on schist. - C & E Balkans, W Anatolia.
Slender, much-branched shrublet. Leaves narrowly elliptic, short-pubescent or subglabrous. Corolla 6-9 mm, glabrous; keel much longer than standard, straight, obtuse. Legume 1(-2)-seeded.

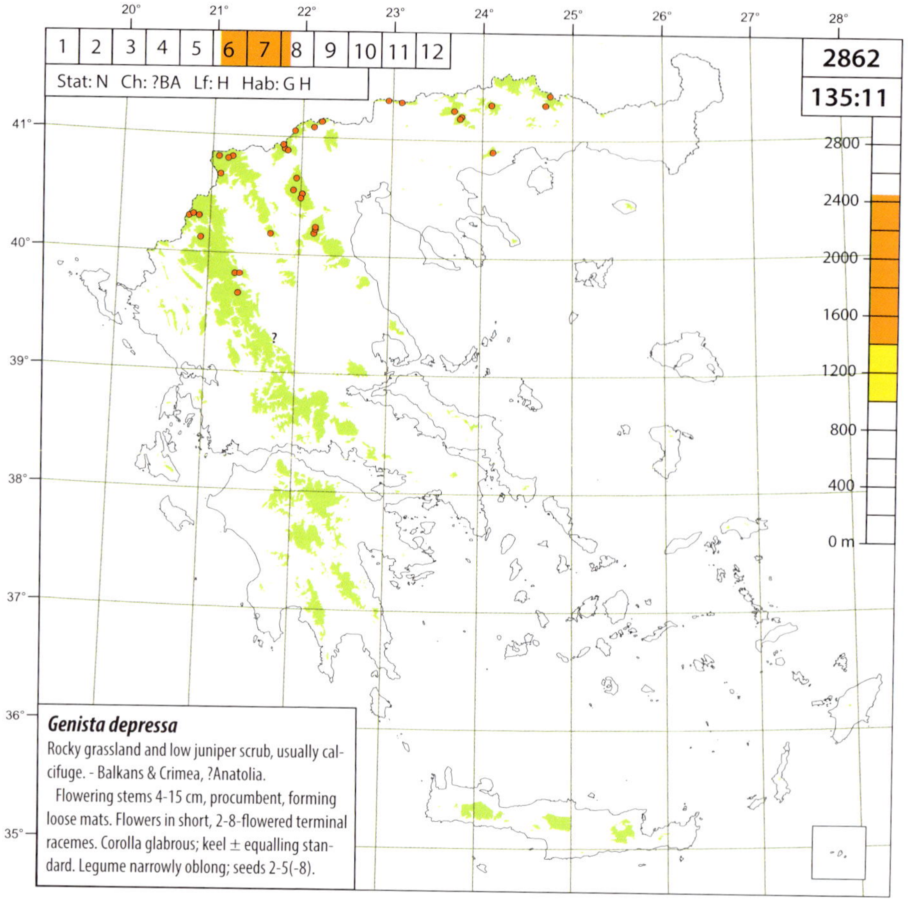

Genista depressa
Rocky grassland and low juniper scrub, usually calcifuge. - Balkans & Crimea, ?Anatolia.
Flowering stems 4-15 cm, procumbent, forming loose mats. Flowers in short, 2-8-flowered terminal racemes. Corolla glabrous; keel ± equalling standard. Legume narrowly oblong; seeds 2-5(-8).

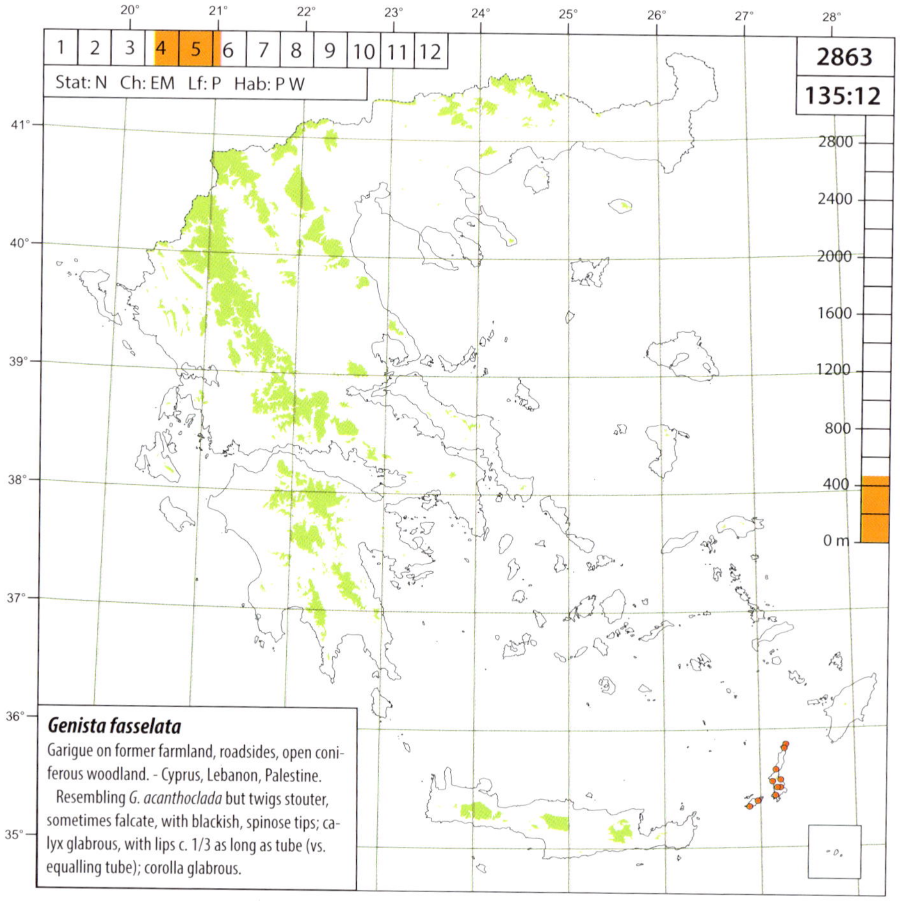

Genista fasselata
Garigue on former farmland, roadsides, open coniferous woodland. - Cyprus, Lebanon, Palestine.
Resembling *G. acanthoclada* but twigs stouter, sometimes falcate, with blackish, spinose tips; calyx glabrous, with lips c. 1/3 as long as tube (vs. equalling tube); corolla glabrous.

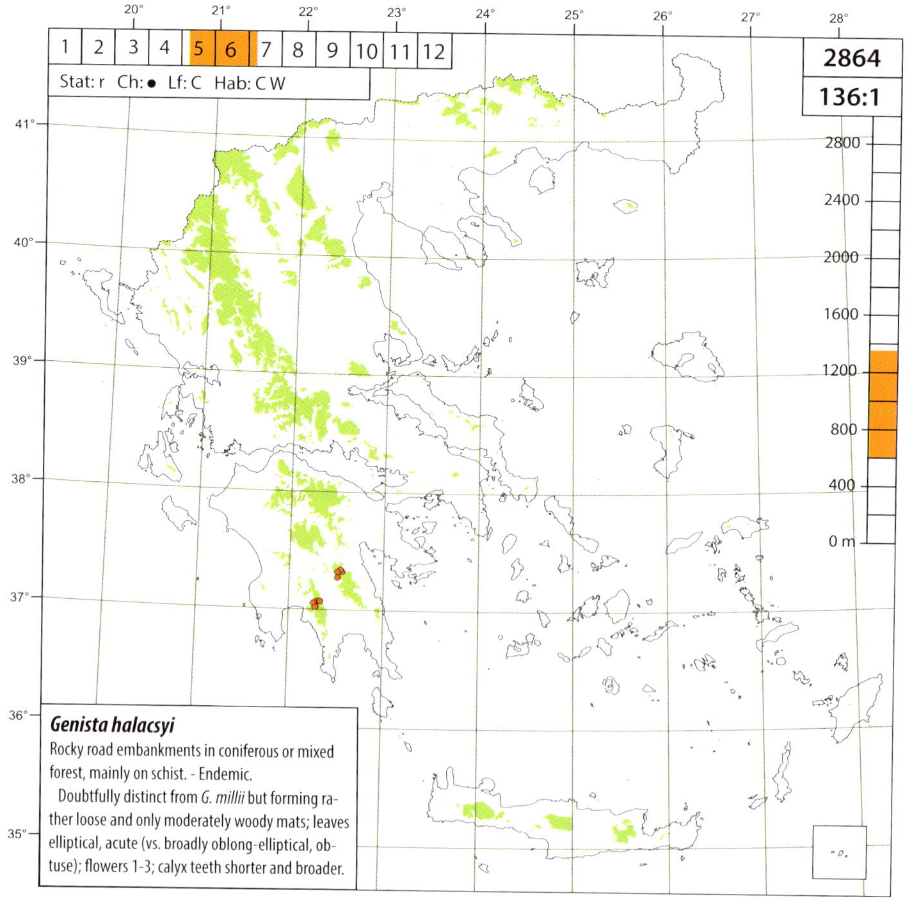

Genista halacsyi
Rocky road embankments in coniferous or mixed forest, mainly on schist. - Endemic.
Doubtfully distinct from *G. millii* but forming rather loose and only moderately woody mats; leaves elliptical, acute (vs. broadly oblong-elliptical, obtuse); flowers 1-3; calyx teeth shorter and broader.

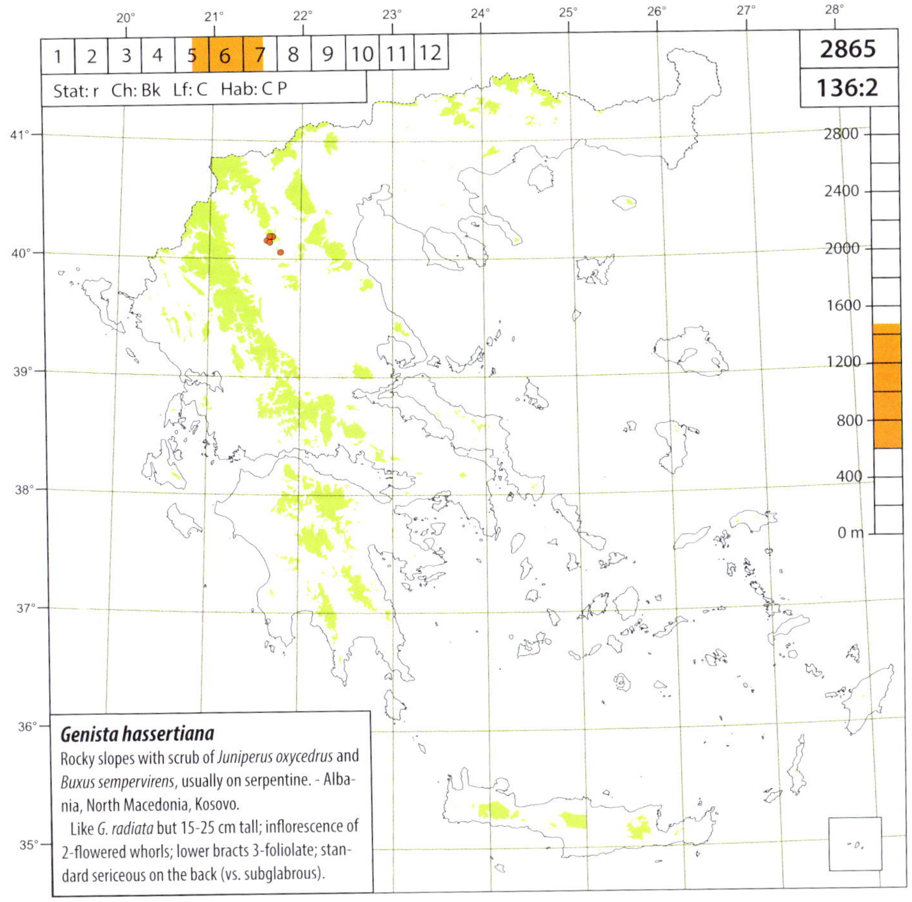

Genista hassertiana

Rocky slopes with scrub of *Juniperus oxycedrus* and *Buxus sempervirens*, usually on serpentine. - Albania, North Macedonia, Kosovo.

Like *G. radiata* but 15-25 cm tall; inflorescence of 2-flowered whorls; lower bracts 3-foliolate; standard sericeous on the back (vs. subglabrous).

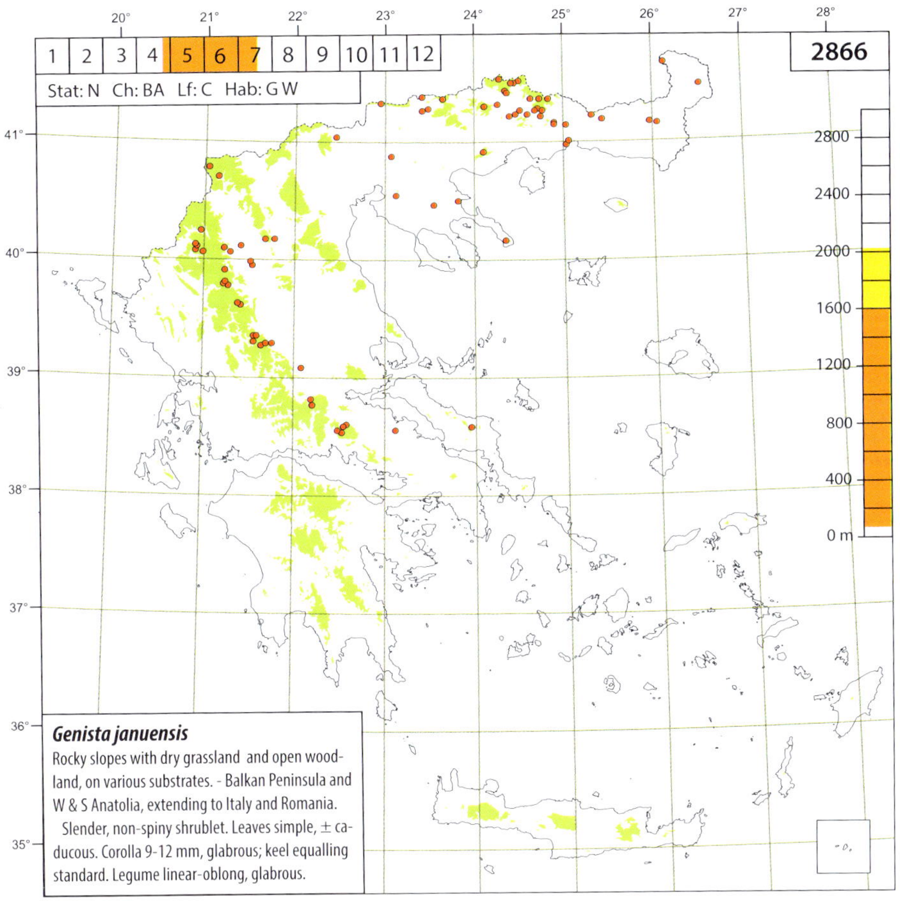

Genista januensis

Rocky slopes with dry grassland and open woodland, on various substrates. - Balkan Peninsula and W & S Anatolia, extending to Italy and Romania.

Slender, non-spiny shrublet. Leaves simple, ± caducous. Corolla 9-12 mm, glabrous; keel equalling standard. Legume linear-oblong, glabrous.

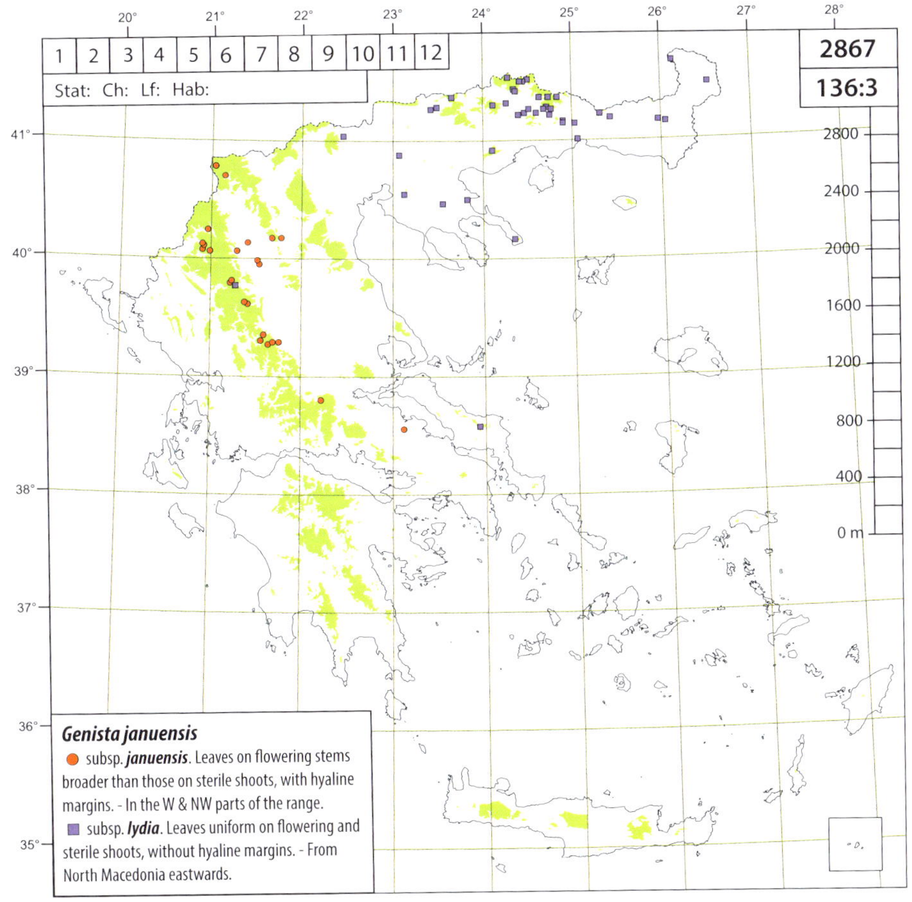

Genista januensis

● subsp. ***januensis***. Leaves on flowering stems broader than those on sterile shoots, with hyaline margins. - In the W & NW parts of the range.

■ subsp. ***lydia***. Leaves uniform on flowering and sterile shoots, without hyaline margins. - From North Macedonia eastwards.

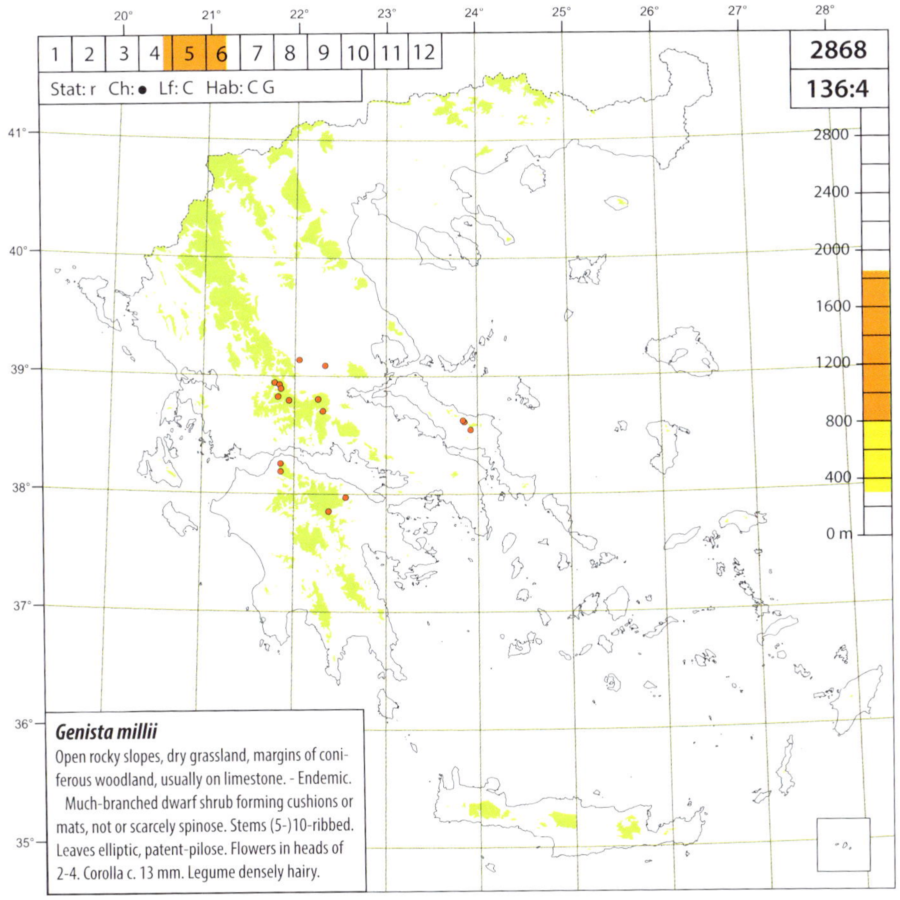

Genista millii

Open rocky slopes, dry grassland, margins of coniferous woodland, usually on limestone. - Endemic.

Much-branched dwarf shrub forming cushions or mats, not or scarcely spinose. Stems (5-)10-ribbed. Leaves elliptic, patent-pilose. Flowers in heads of 2-4. Corolla c. 13 mm. Legume densely hairy.

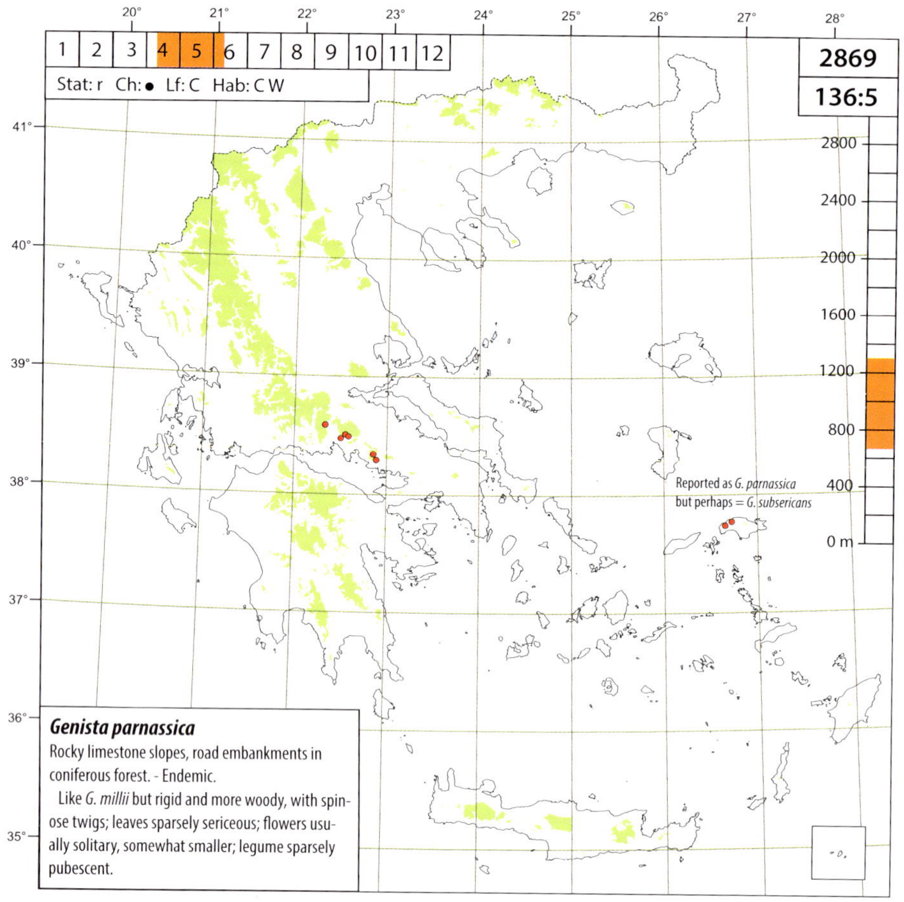

Genista parnassica
Rocky limestone slopes, road embankments in coniferous forest. - Endemic.
Like *G. millii* but rigid and more woody, with spinose twigs; leaves sparsely sericeous; flowers usually solitary, somewhat smaller; legume sparsely pubescent.

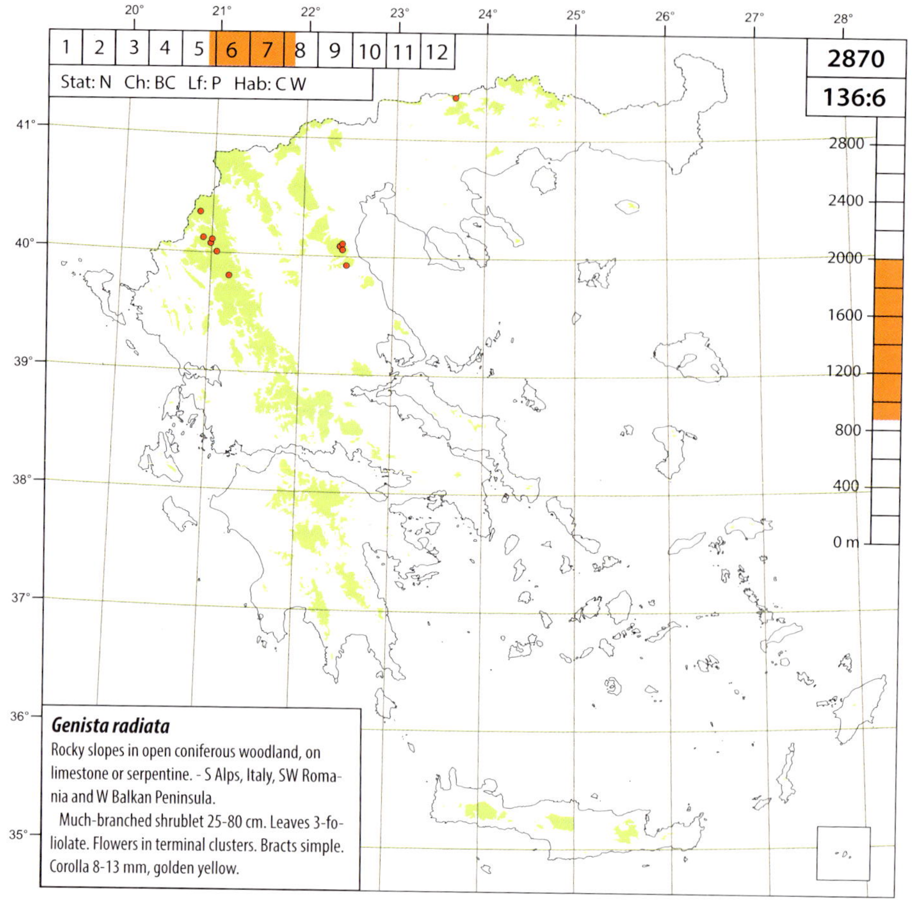

Genista radiata
Rocky slopes in open coniferous woodland, on limestone or serpentine. - S Alps, Italy, SW Romania and W Balkan Peninsula.
Much-branched shrublet 25-80 cm. Leaves 3-foliolate. Flowers in terminal clusters. Bracts simple. Corolla 8-13 mm, golden yellow.

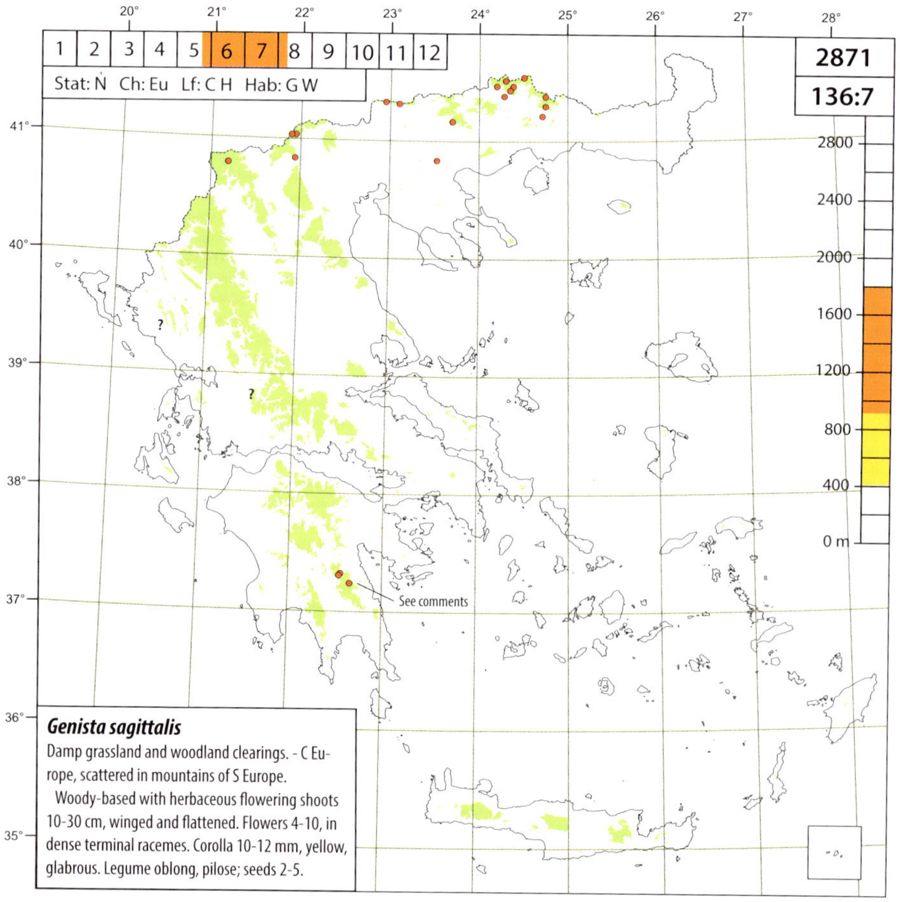

Genista sagittalis
Damp grassland and woodland clearings. - C Europe, scattered in mountains of S Europe.
Woody-based with herbaceous flowering shoots 10-30 cm, winged and flattened. Flowers 4-10, in dense terminal racemes. Corolla 10-12 mm, yellow, glabrous. Legume oblong, pilose; seeds 2-5.

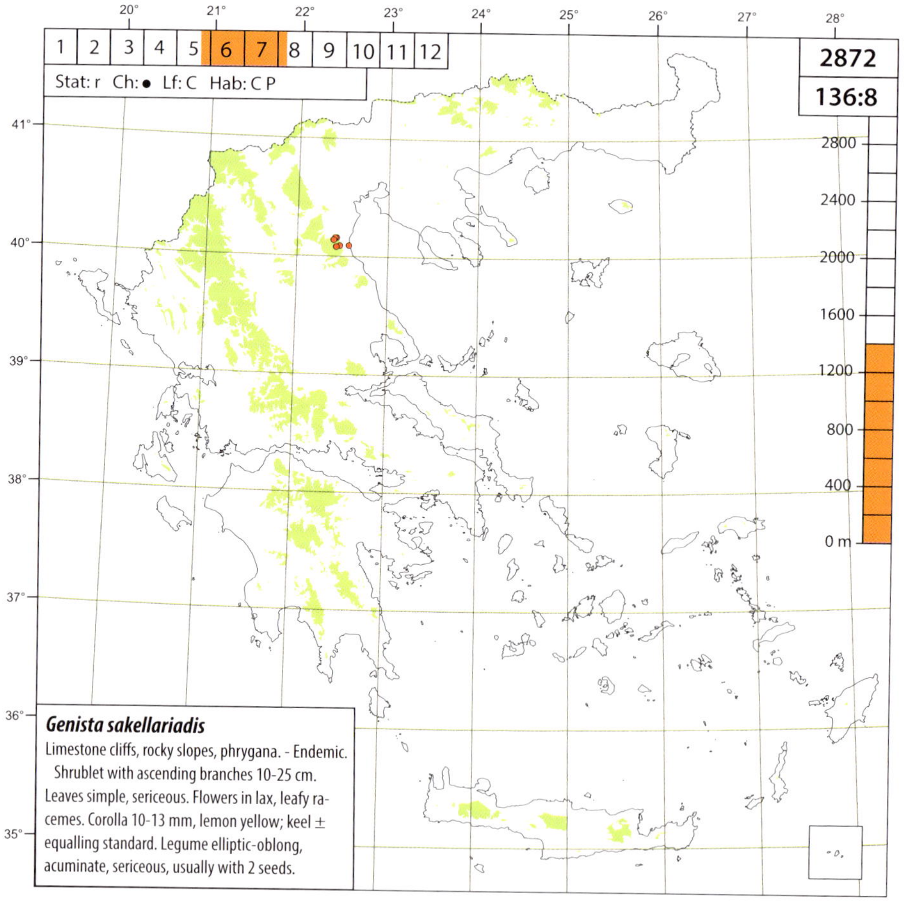

Genista sakellariadis
Limestone cliffs, rocky slopes, phrygana. - Endemic.
Shrublet with ascending branches 10-25 cm. Leaves simple, sericeous. Flowers in lax, leafy racemes. Corolla 10-13 mm, lemon yellow; keel ± equalling standard. Legume elliptic-oblong, acuminate, sericeous, usually with 2 seeds.

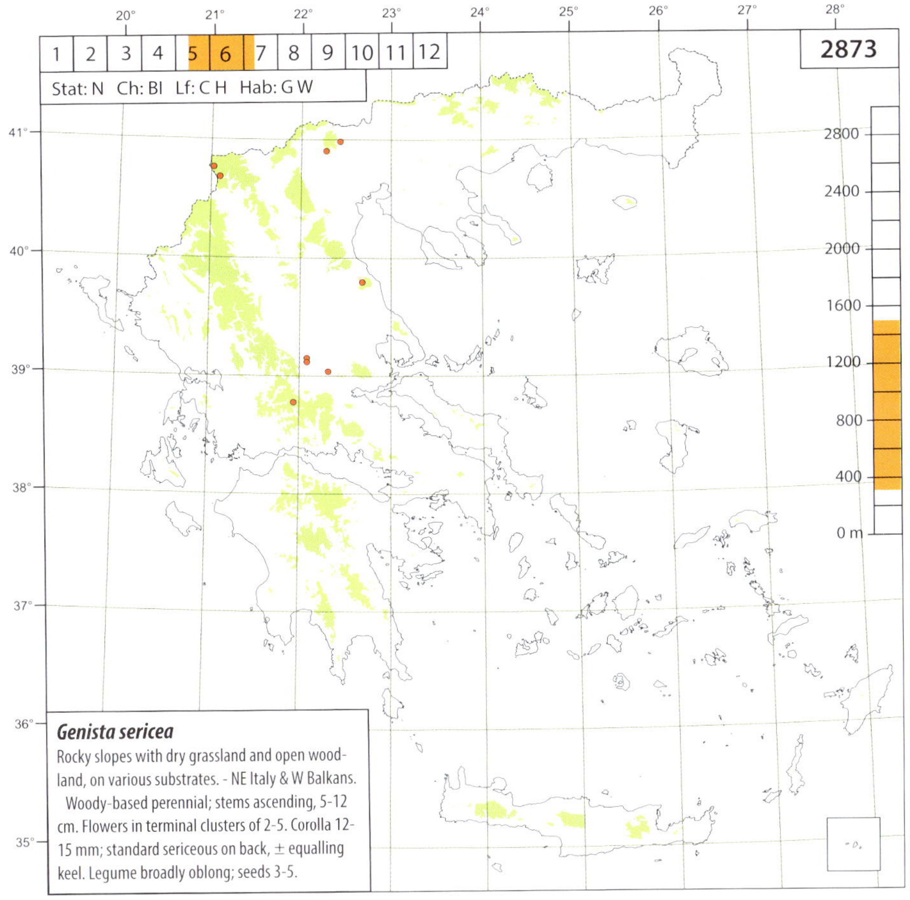

Genista sericea

Rocky slopes with dry grassland and open woodland, on various substrates. - NE Italy & W Balkans.

Woody-based perennial; stems ascending, 5-12 cm. Flowers in terminal clusters of 2-5. Corolla 12-15 mm; standard sericeous on back, ± equalling keel. Legume broadly oblong; seeds 3-5.

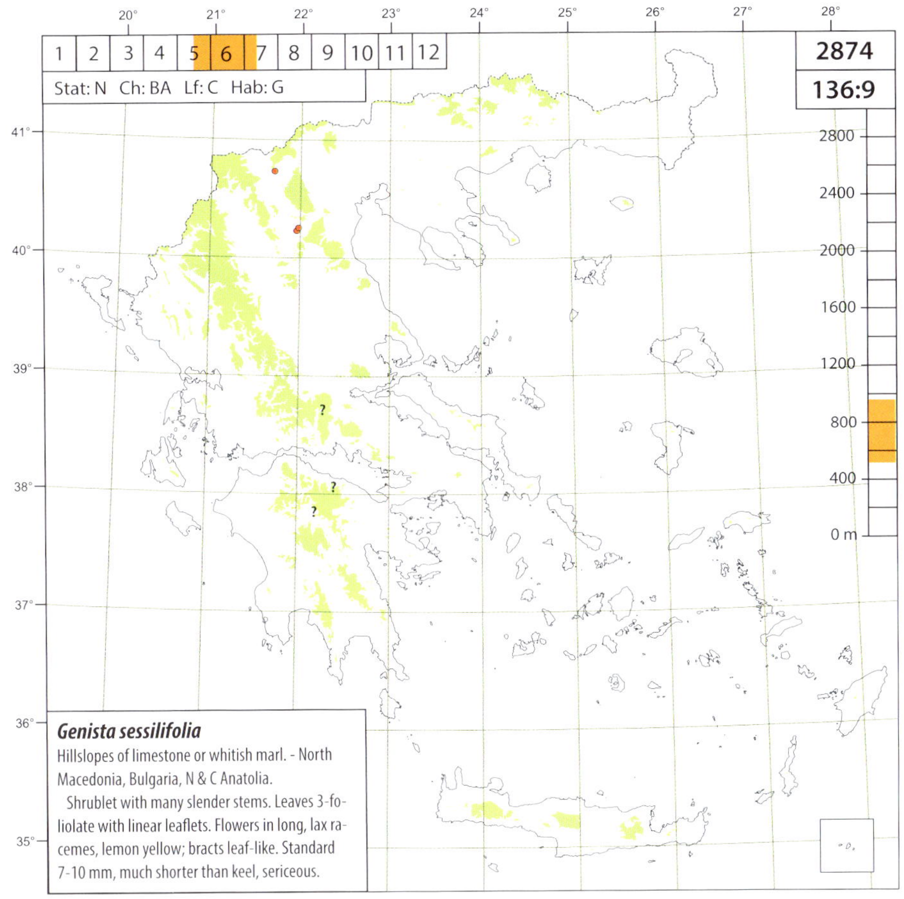

Genista sessilifolia

Hillslopes of limestone or whitish marl. - North Macedonia, Bulgaria, N & C Anatolia.

Shrublet with many slender stems. Leaves 3-foliolate with linear leaflets. Flowers in long, lax racemes, lemon yellow; bracts leaf-like. Standard 7-10 mm, much shorter than keel, sericeous.

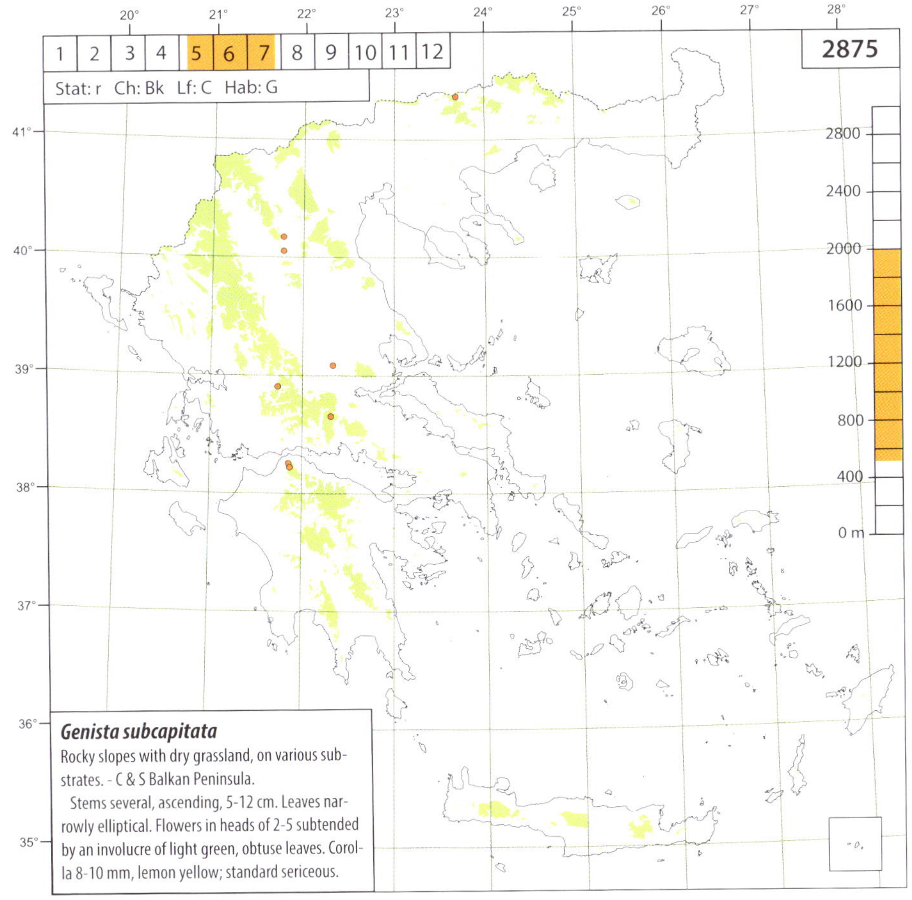

Genista subcapitata

Rocky slopes with dry grassland, on various substrates. - C & S Balkan Peninsula.

Stems several, ascending, 5-12 cm. Leaves narrowly elliptical. Flowers in heads of 2-5 subtended by an involucre of light green, obtuse leaves. Corolla 8-10 mm, lemon yellow; standard sericeous.

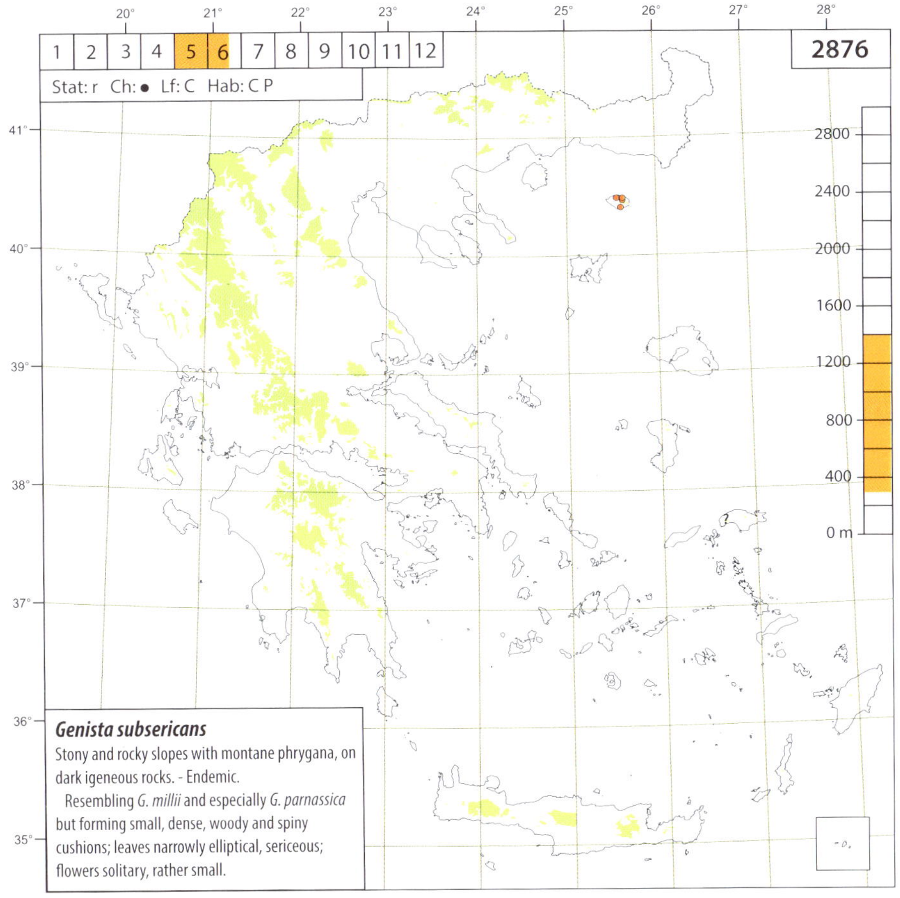

Genista subsericans

Stony and rocky slopes with montane phrygana, on dark igeneous rocks. - Endemic.

Resembling *G. millii* and especially *G. parnassica* but forming small, dense, woody and spiny cushions; leaves narrowly elliptical, sericeous; flowers solitary, rather small.

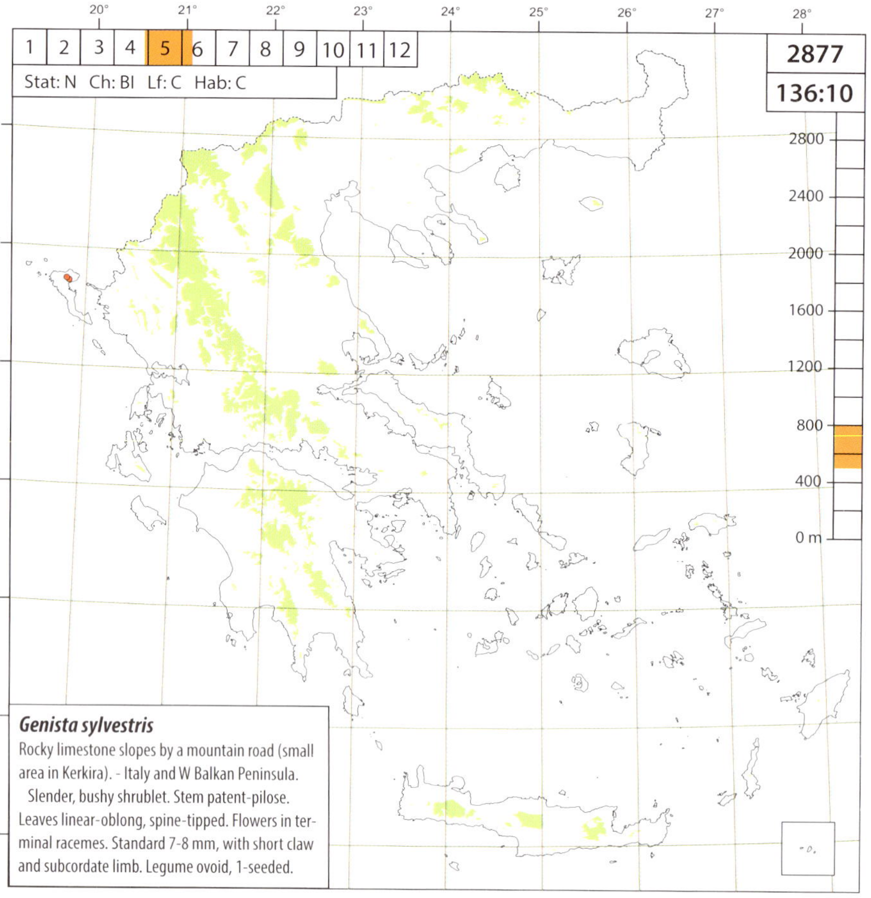

Genista sylvestris
Rocky limestone slopes by a mountain road (small area in Kerkira). - Italy and W Balkan Peninsula.
Slender, bushy shrublet. Stem patent-pilose. Leaves linear-oblong, spine-tipped. Flowers in terminal racemes. Standard 7-8 mm, with short claw and subcordate limb. Legume ovoid, 1-seeded.

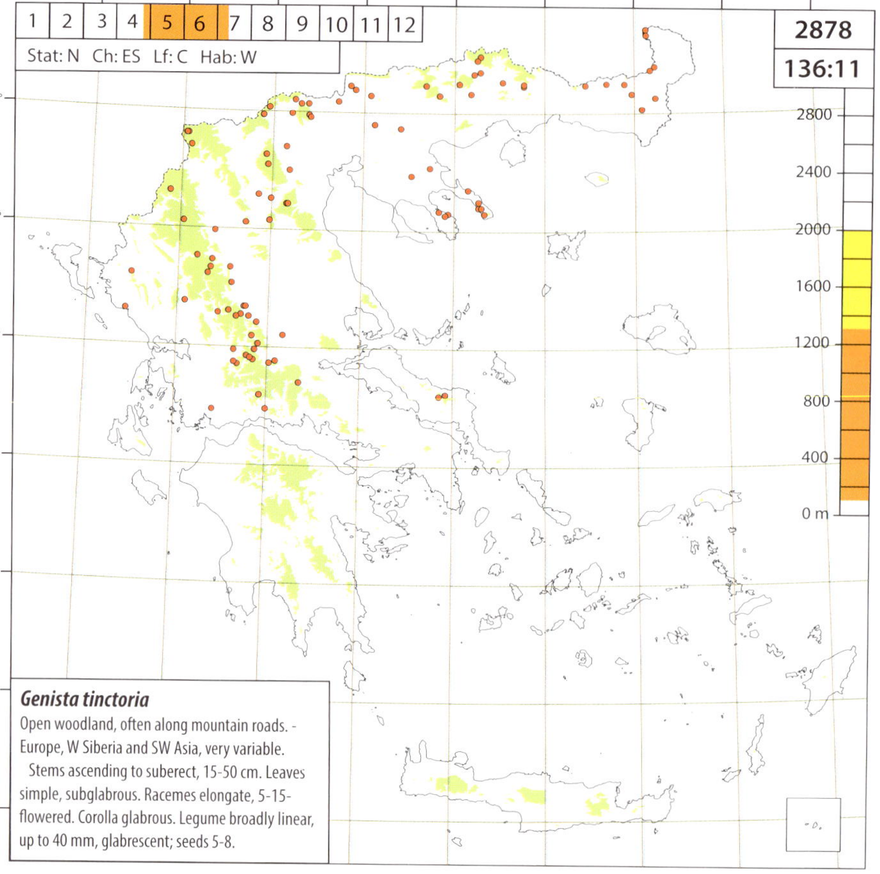

Genista tinctoria
Open woodland, often along mountain roads. - Europe, W Siberia and SW Asia, very variable.
Stems ascending to suberect, 15-50 cm. Leaves simple, subglabrous. Racemes elongate, 5-15-flowered. Corolla glabrous. Legume broadly linear, up to 40 mm, glabrescent; seeds 5-8.

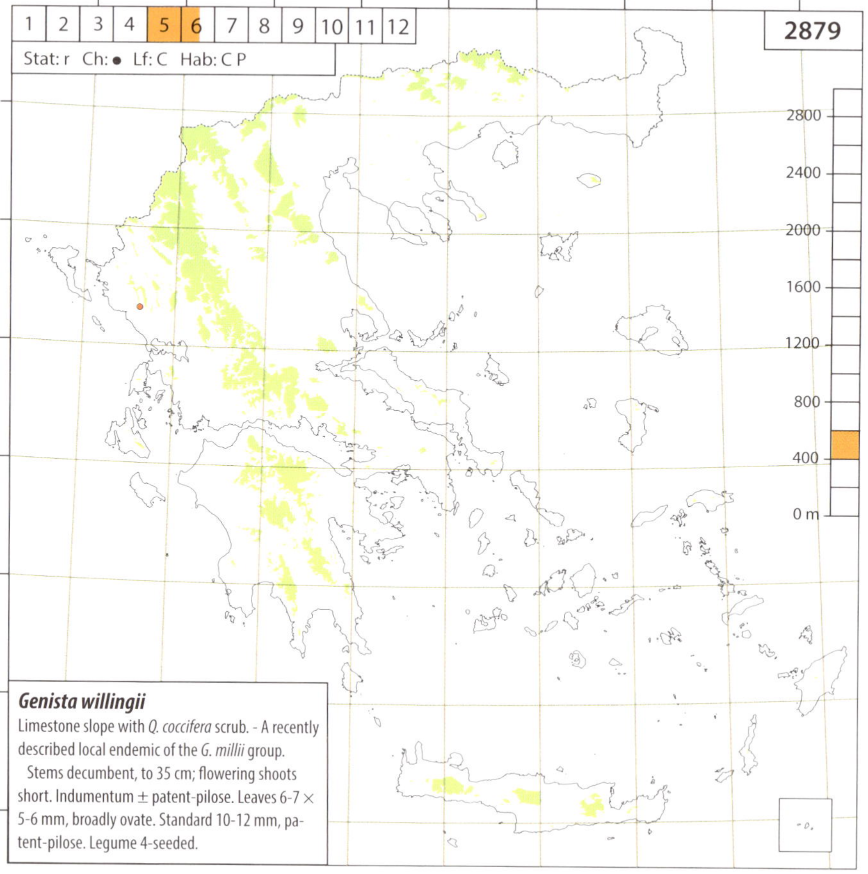

Genista willingii
Limestone slope with *Q. coccifera* scrub. - A recently described local endemic of the *G. millii* group.
Stems decumbent, to 35 cm; flowering shoots short. Indumentum ± patent-pilose. Leaves 6-7 × 5-6 mm, broadly ovate. Standard 10-12 mm, patent-pilose. Legume 4-seeded.

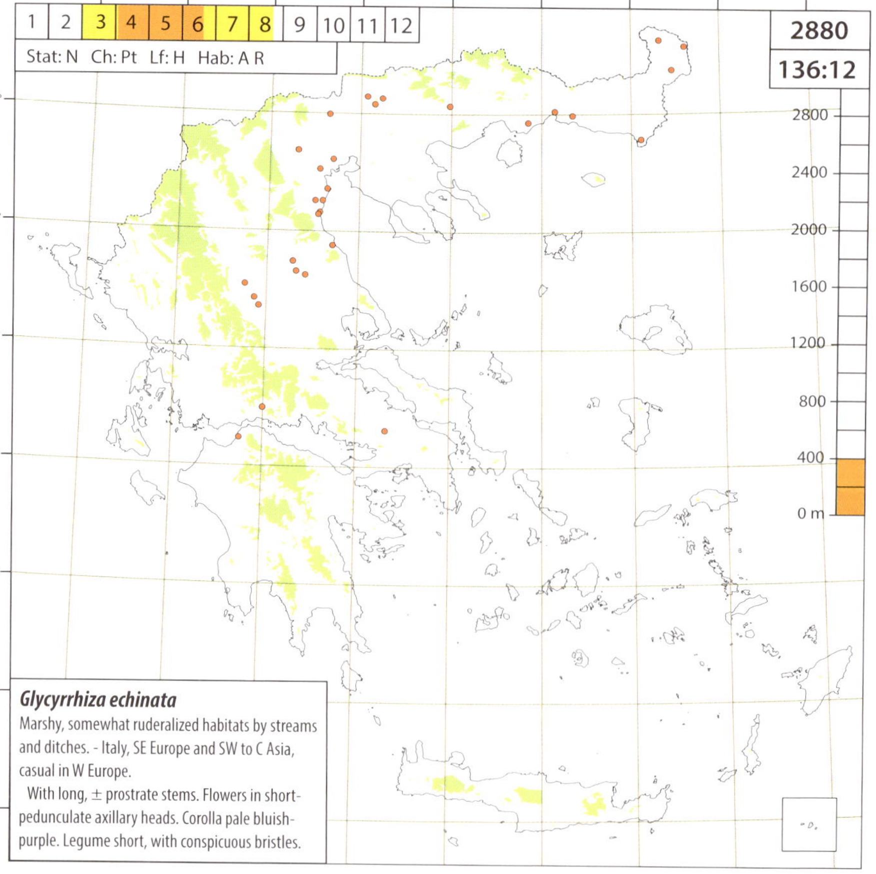

Glycyrrhiza echinata
Marshy, somewhat ruderalized habitats by streams and ditches. - Italy, SE Europe and SW to C Asia, casual in W Europe.
With long, ± prostrate stems. Flowers in short-pedunculate axillary heads. Corolla pale bluish-purple. Legume short, with conspicuous bristles.

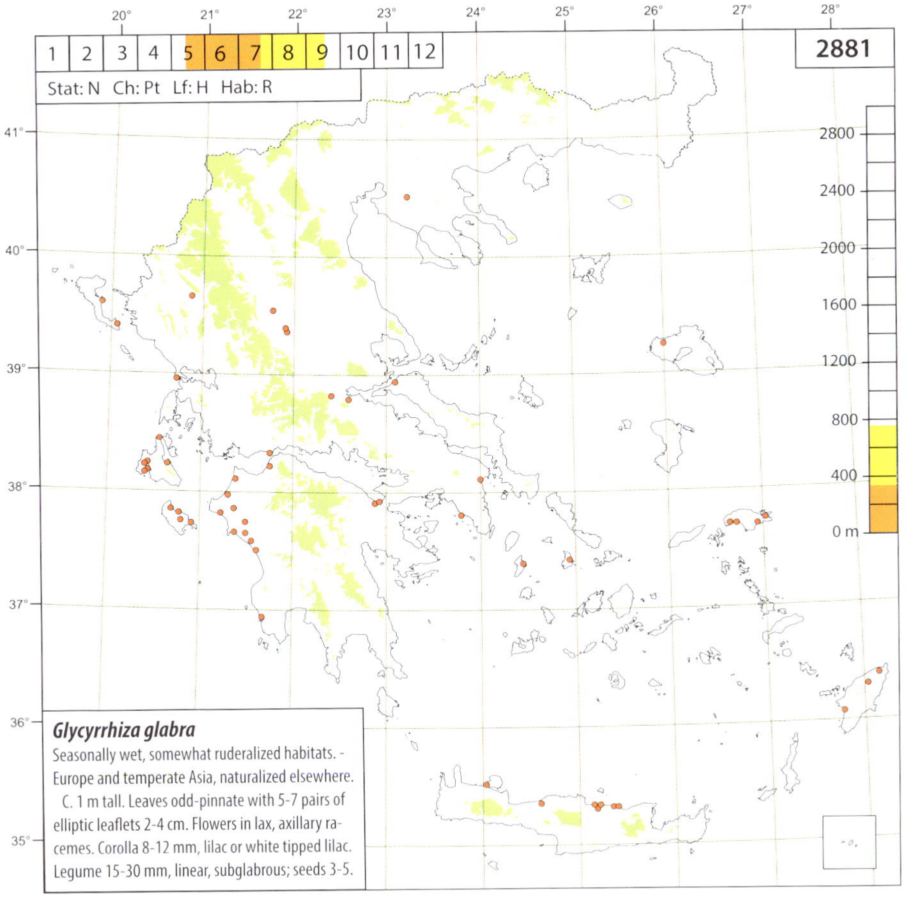

Glycyrrhiza glabra
Seasonally wet, somewhat ruderalized habitats. - Europe and temperate Asia, naturalized elsewhere.
C. 1 m tall. Leaves odd-pinnate with 5-7 pairs of elliptic leaflets 2-4 cm. Flowers in lax, axillary racemes. Corolla 8-12 mm, lilac or white tipped lilac. Legume 15-30 mm, linear, subglabrous; seeds 3-5.

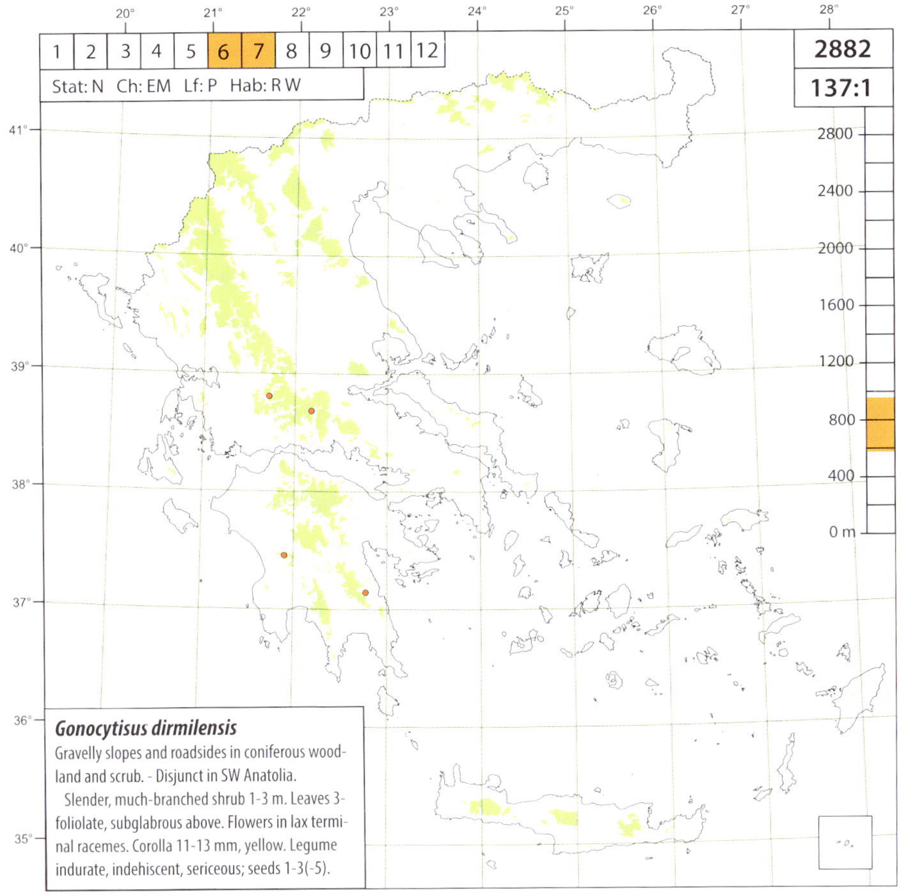

Gonocytisus dirmilensis
Gravelly slopes and roadsides in coniferous woodland and scrub. - Disjunct in SW Anatolia.
Slender, much-branched shrub 1-3 m. Leaves 3-foliolate, subglabrous above. Flowers in lax terminal racemes. Corolla 11-13 mm, yellow. Legume indurate, indehiscent, sericeous; seeds 1-3(-5).

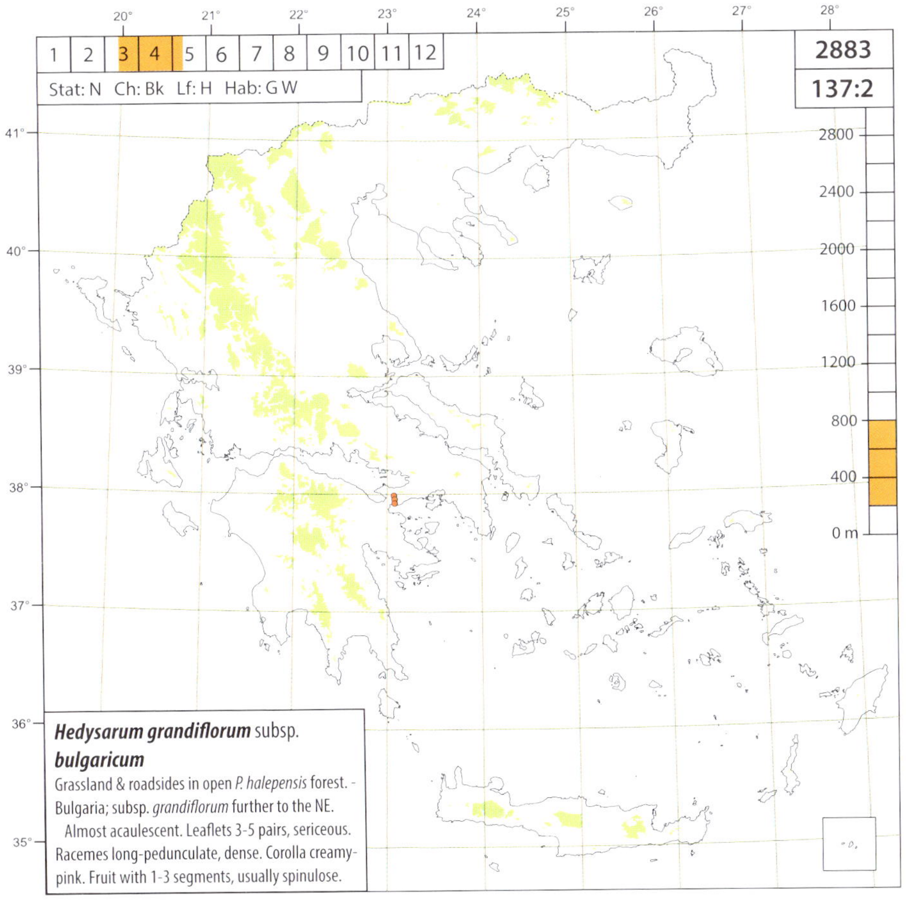

Hedysarum grandiflorum subsp. ***bulgaricum***
Grassland & roadsides in open *P. halepensis* forest. - Bulgaria; subsp. *grandiflorum* further to the NE.
Almost acaulescent. Leaflets 3-5 pairs, sericeous. Racemes long-pedunculate, dense. Corolla creamy-pink. Fruit with 1-3 segments, usually spinulose.

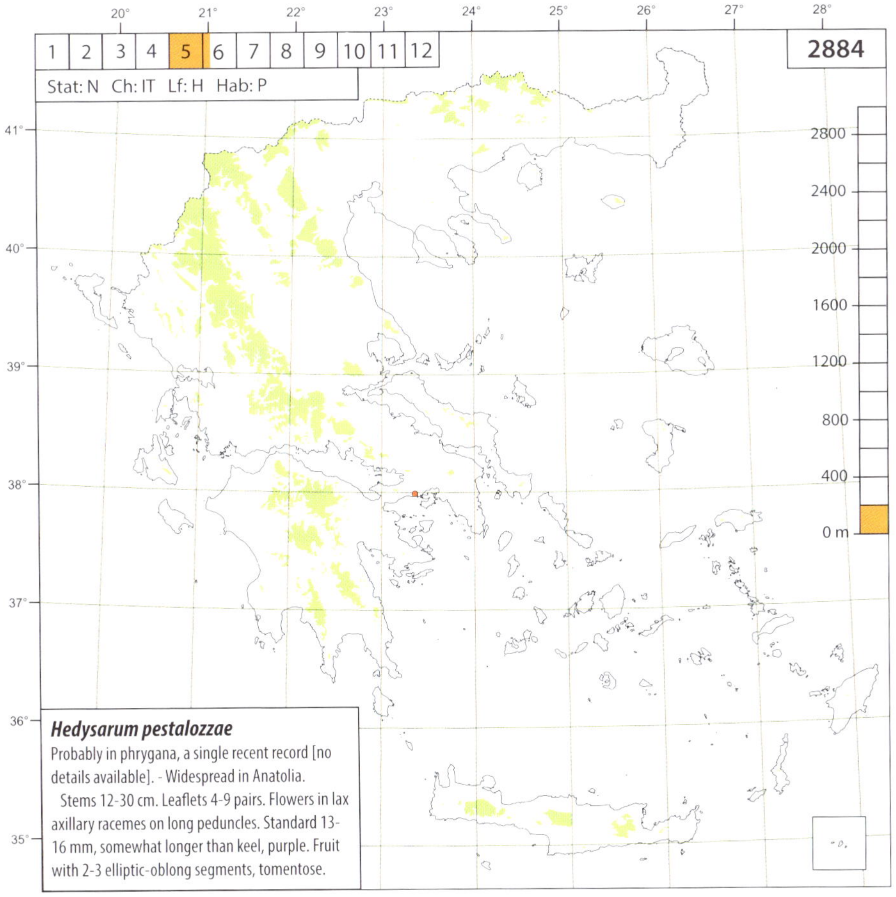

Hedysarum pestalozzae
Probably in phrygana, a single recent record [no details available]. - Widespread in Anatolia.
Stems 12-30 cm. Leaflets 4-9 pairs. Flowers in lax axillary racemes on long peduncles. Standard 13-16 mm, somewhat longer than keel, purple. Fruit with 2-3 elliptic-oblong segments, tomentose.

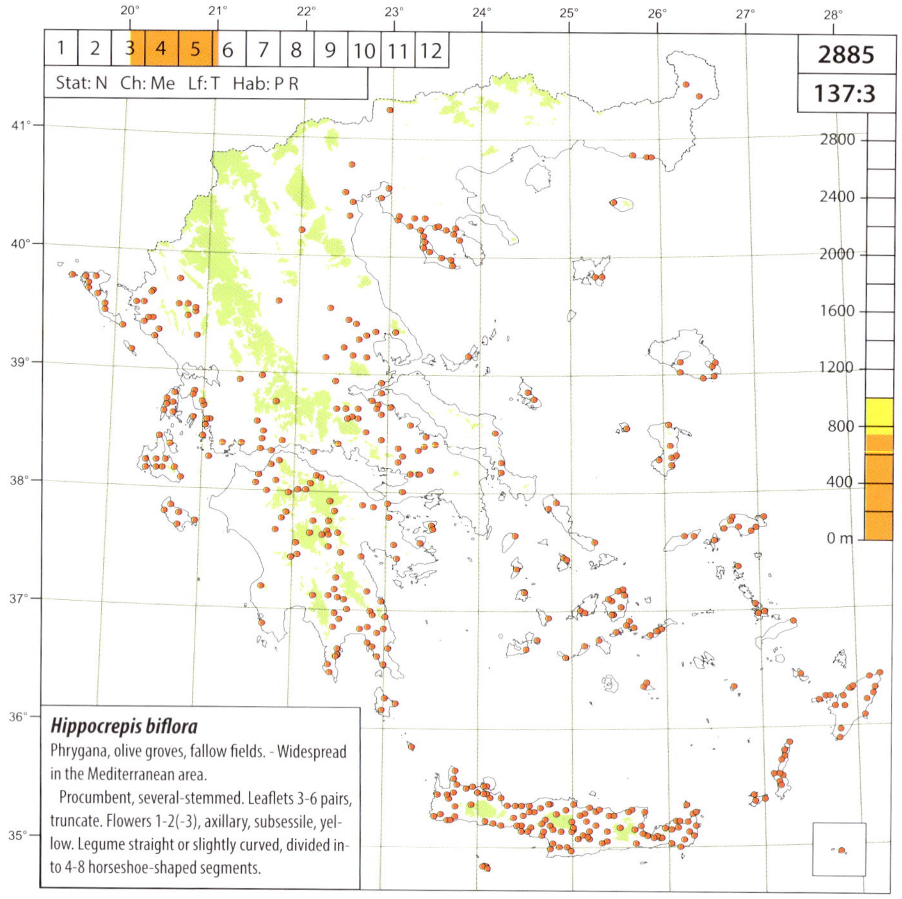

Hippocrepis biflora

Phrygana, olive groves, fallow fields. - Widespread in the Mediterranean area.

Procumbent, several-stemmed. Leaflets 3-6 pairs, truncate. Flowers 1-2(-3), axillary, subsessile, yellow. Legume straight or slightly curved, divided into 4-8 horseshoe-shaped segments.

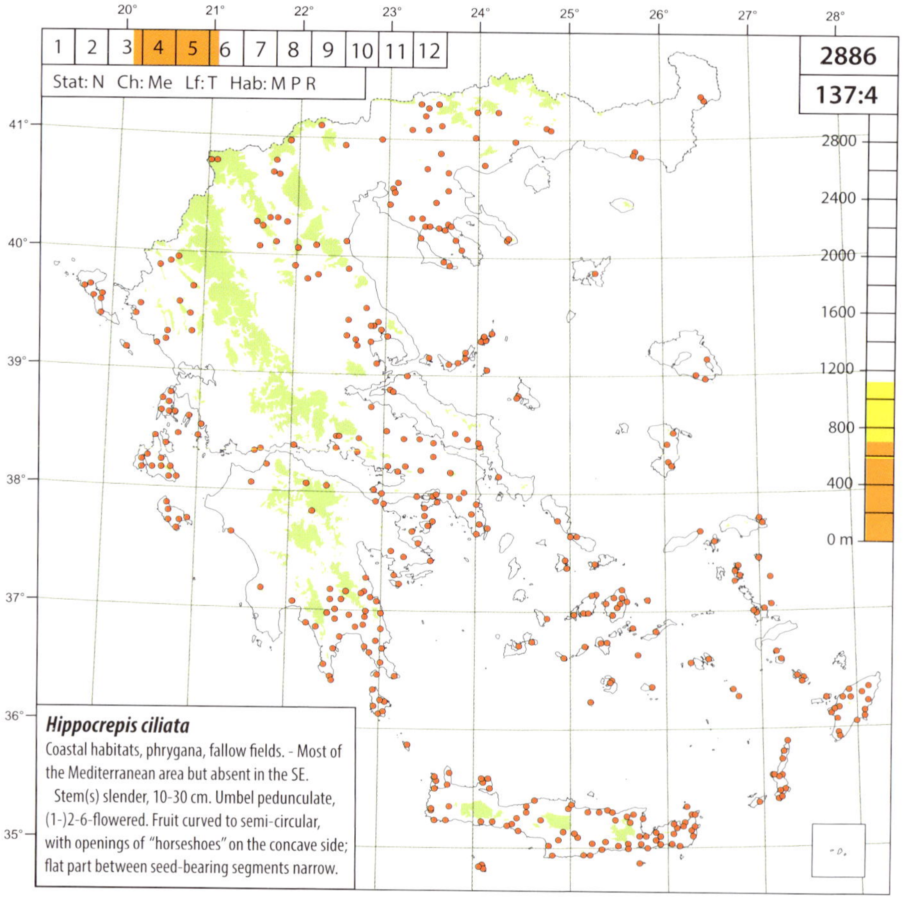

Hippocrepis ciliata

Coastal habitats, phrygana, fallow fields. - Most of the Mediterranean area but absent in the SE.

Stem(s) slender, 10-30 cm. Umbel pedunculate, (1-)2-6-flowered. Fruit curved to semi-circular, with openings of "horseshoes" on the concave side; flat part between seed-bearing segments narrow.

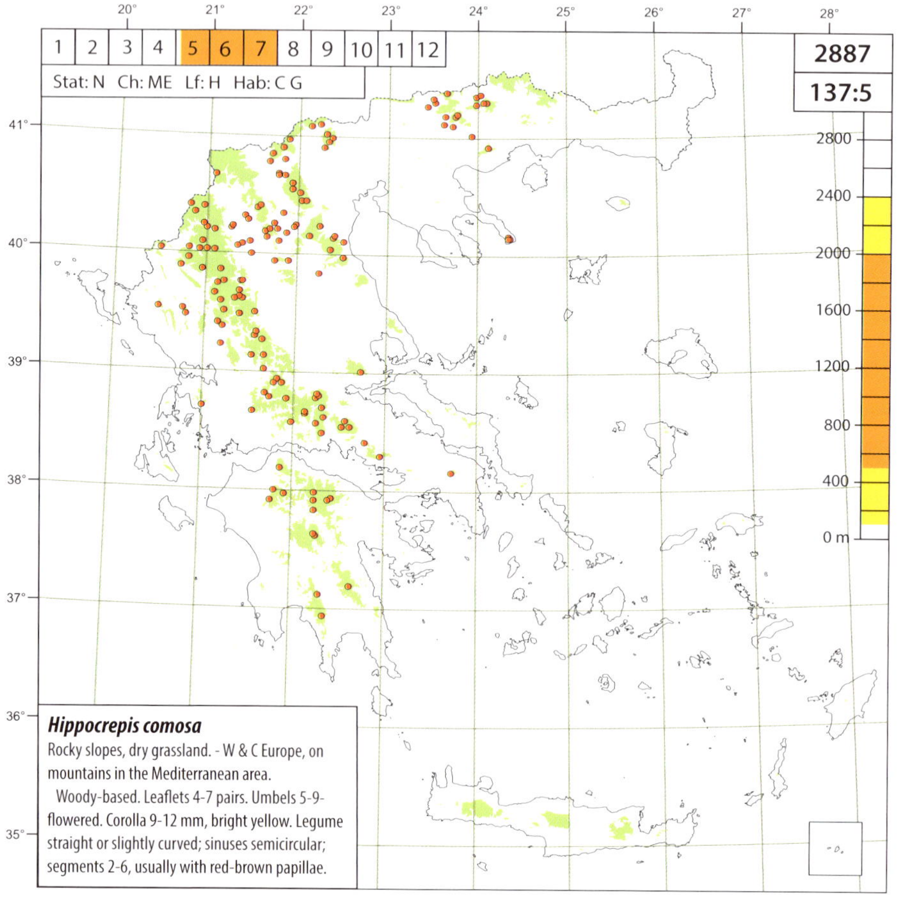

Hippocrepis comosa

Rocky slopes, dry grassland. - W & C Europe, on mountains in the Mediterranean area.

Woody-based. Leaflets 4-7 pairs. Umbels 5-9-flowered. Corolla 9-12 mm, bright yellow. Legume straight or slightly curved; sinuses semicircular; segments 2-6, usually with red-brown papillae.

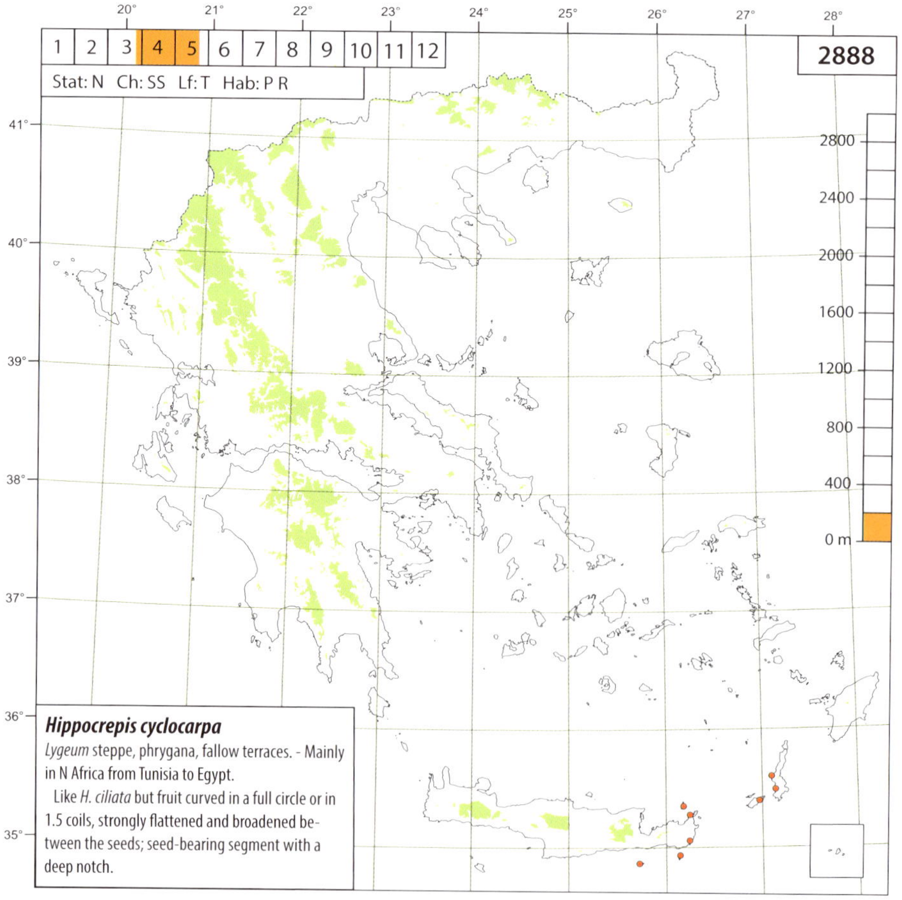

Hippocrepis cyclocarpa

Lygeum steppe, phrygana, fallow terraces. - Mainly in N Africa from Tunisia to Egypt.

Like *H. ciliata* but fruit curved in a full circle or in 1.5 coils, strongly flattened and broadened between the seeds; seed-bearing segment with a deep notch.

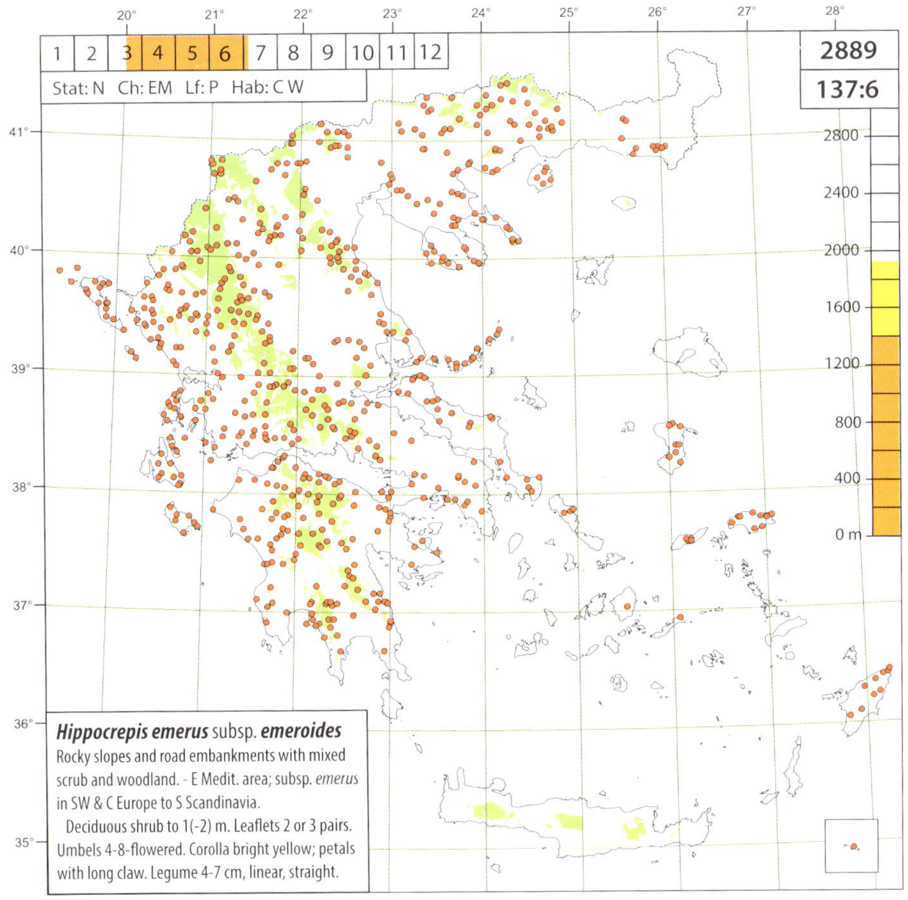

Hippocrepis emerus subsp. **emeroides**
Rocky slopes and road embankments with mixed scrub and woodland. - E Medit. area; subsp. *emerus* in SW & C Europe to S Scandinavia.
Deciduous shrub to 1(-2) m. Leaflets 2 or 3 pairs. Umbels 4-8-flowered. Corolla bright yellow; petals with long claw. Legume 4-7 cm, linear, straight.

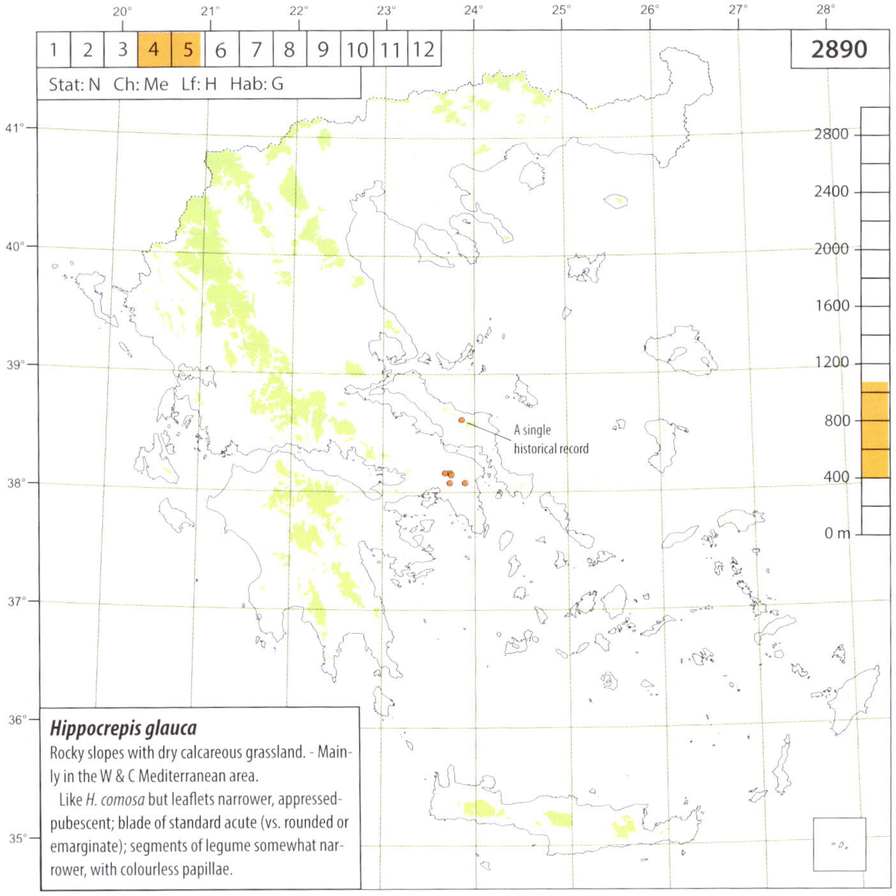

Hippocrepis glauca
Rocky slopes with dry calcareous grassland. - Mainly in the W & C Mediterranean area.
Like *H. comosa* but leaflets narrower, appressed-pubescent; blade of standard acute (vs. rounded or emarginate); segments of legume somewhat narrower, with colourless papillae.

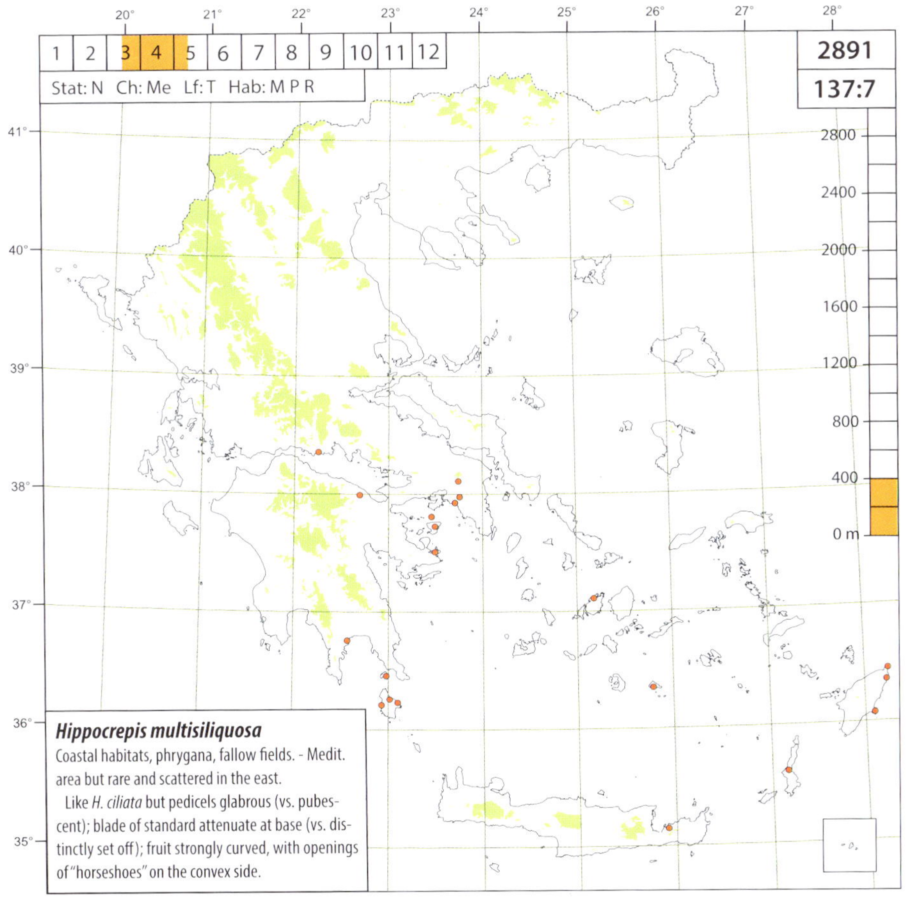

Hippocrepis multisiliquosa
Coastal habitats, phrygana, fallow fields. - Medit. area but rare and scattered in the east.
Like *H. ciliata* but pedicels glabrous (vs. pubescent); blade of standard attenuate at base (vs. distinctly set off); fruit strongly curved, with openings of "horseshoes" on the convex side.

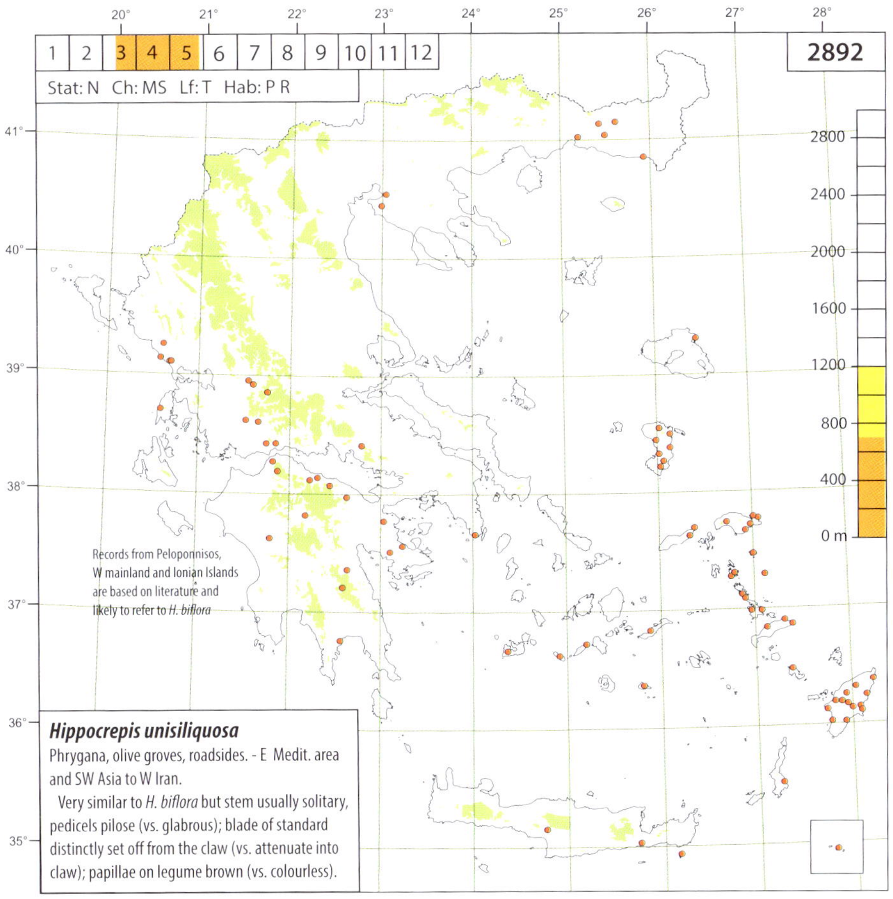

Hippocrepis unisiliquosa
Phrygana, olive groves, roadsides. - E Medit. area and SW Asia to W Iran.
Very similar to *H. biflora* but stem usually solitary, pedicels pilose (vs. glabrous); blade of standard distinctly set off from the claw (vs. attenuate into claw); papillae on legume brown (vs. colourless).

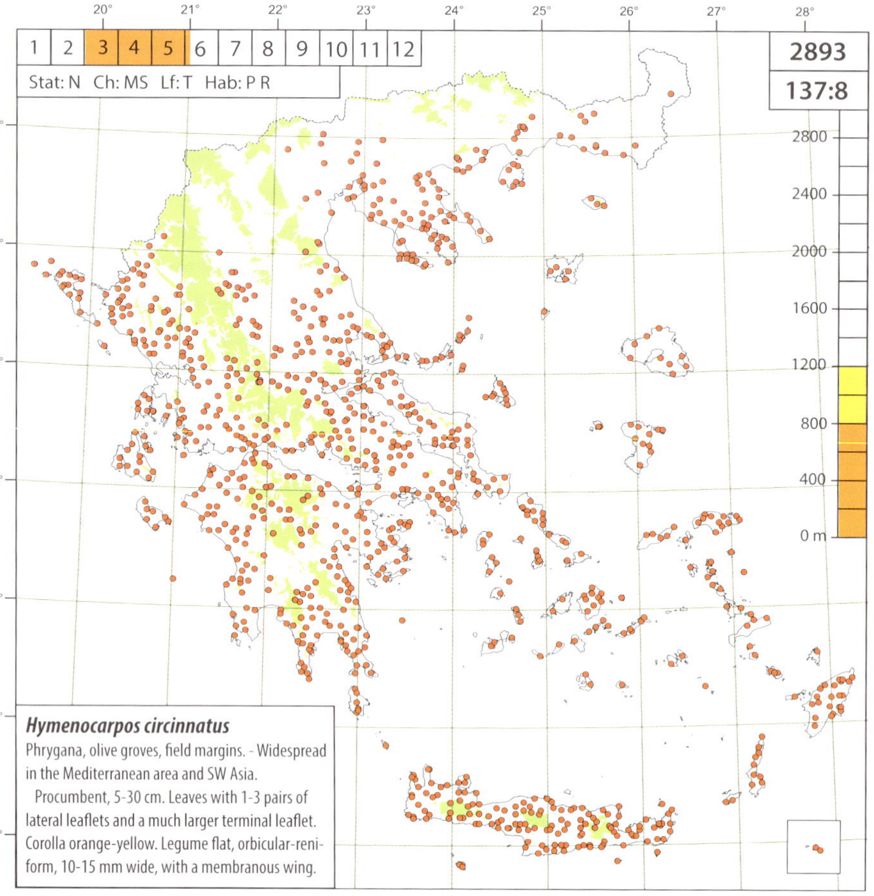

Hymenocarpos circinnatus

Phrygana, olive groves, field margins. - Widespread in the Mediterranean area and SW Asia.

Procumbent, 5-30 cm. Leaves with 1-3 pairs of lateral leaflets and a much larger terminal leaflet. Corolla orange-yellow. Legume flat, orbicular-reniform, 10-15 mm wide, with a membranous wing.

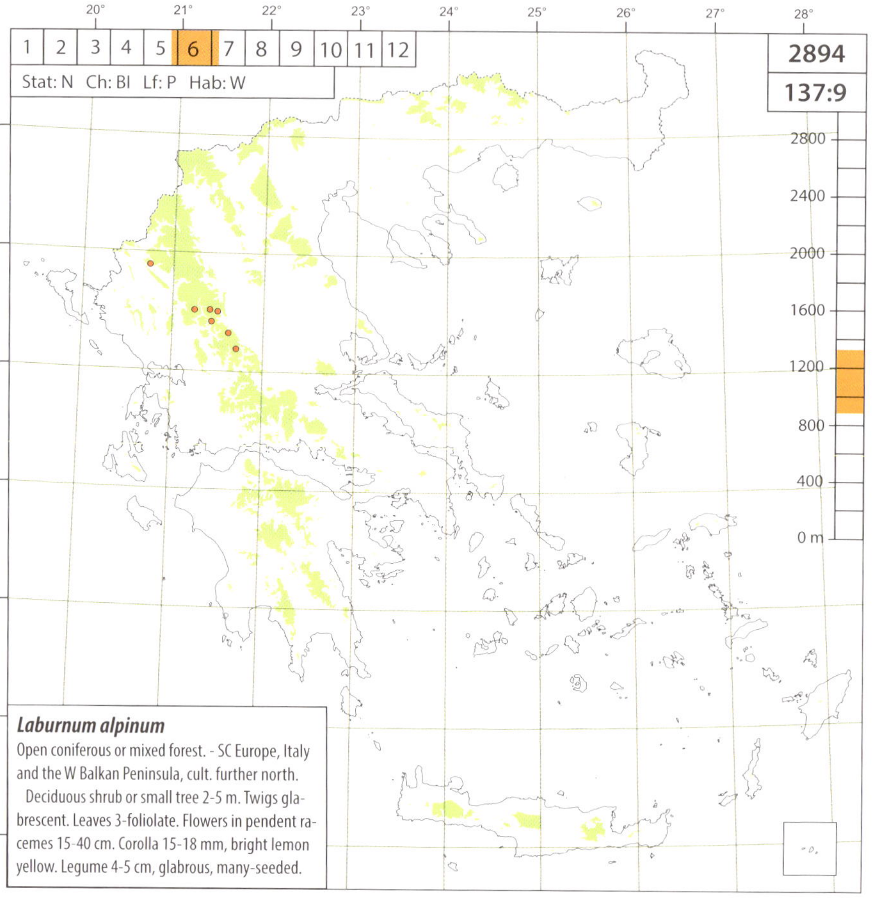

Laburnum alpinum

Open coniferous or mixed forest. - SC Europe, Italy and the W Balkan Peninsula, cult. further north.

Deciduous shrub or small tree 2-5 m. Twigs glabrescent. Leaves 3-foliolate. Flowers in pendent racemes 15-40 cm. Corolla 15-18 mm, bright lemon yellow. Legume 4-5 cm, glabrous, many-seeded.

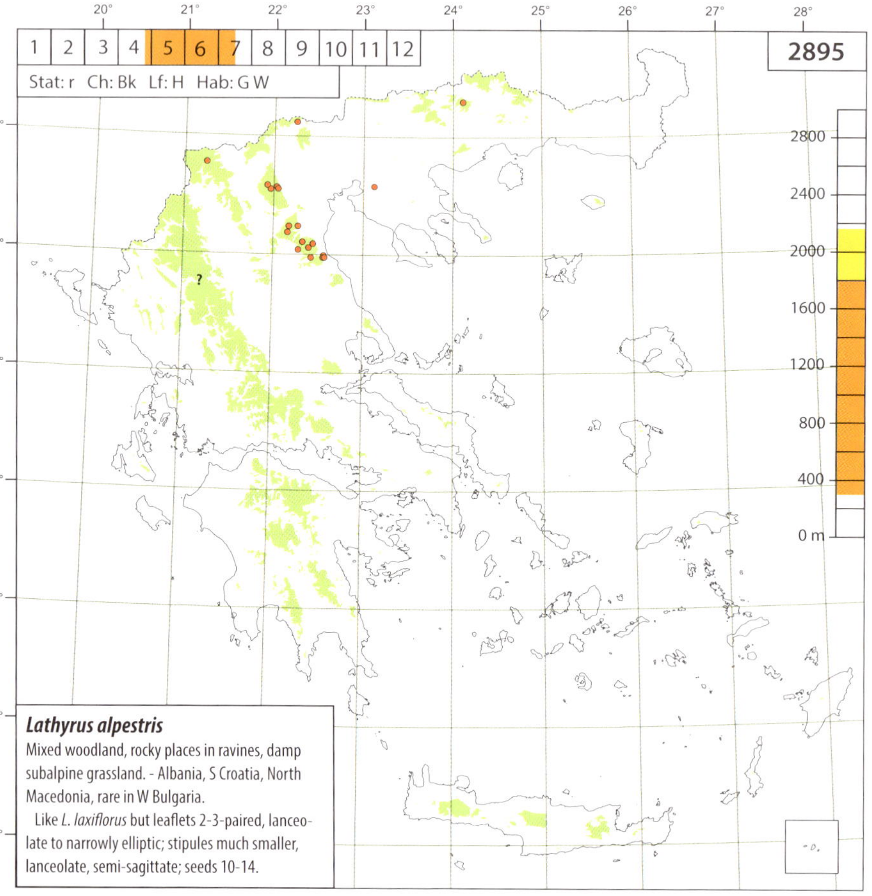

Lathyrus alpestris

Mixed woodland, rocky places in ravines, damp subalpine grassland. - Albania, S Croatia, North Macedonia, rare in W Bulgaria.

Like *L. laxiflorus* but leaflets 2-3-paired, lanceolate to narrowly elliptic; stipules much smaller, lanceolate, semi-sagittate; seeds 10-14.

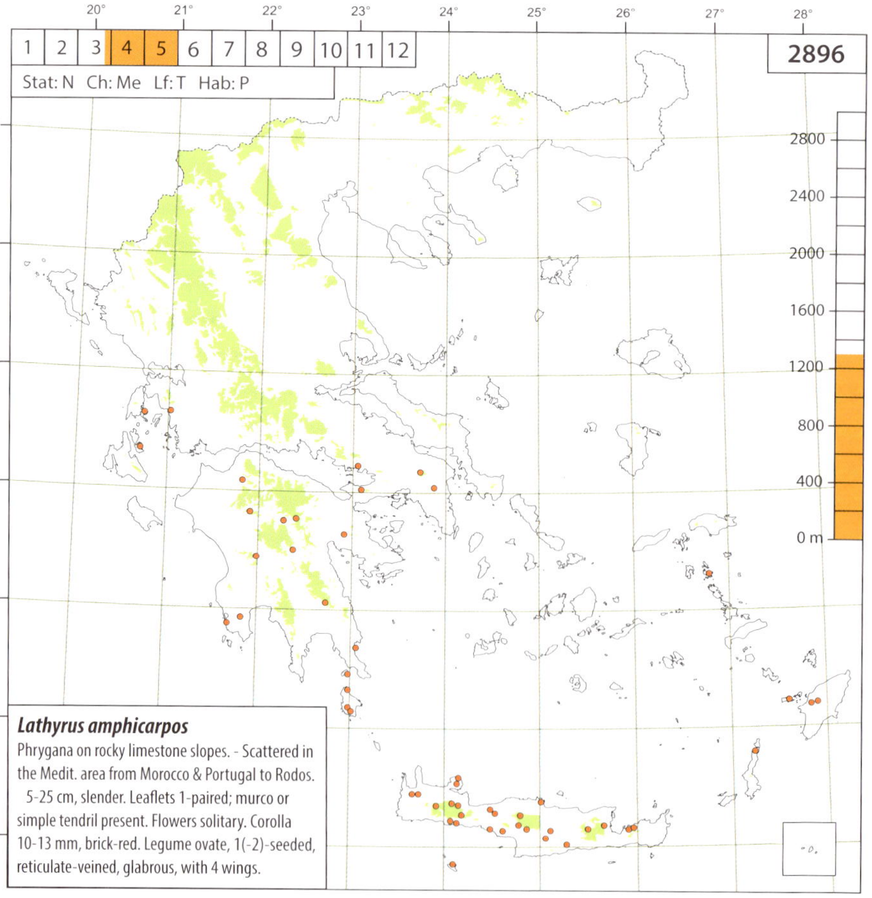

Lathyrus amphicarpos

Phrygana on rocky limestone slopes. - Scattered in the Medit. area from Morocco & Portugal to Rodos.

5-25 cm, slender. Leaflets 1-paired; murco or simple tendril present. Flowers solitary. Corolla 10-13 mm, brick-red. Legume ovate, 1(-2)-seeded, reticulate-veined, glabrous, with 4 wings.

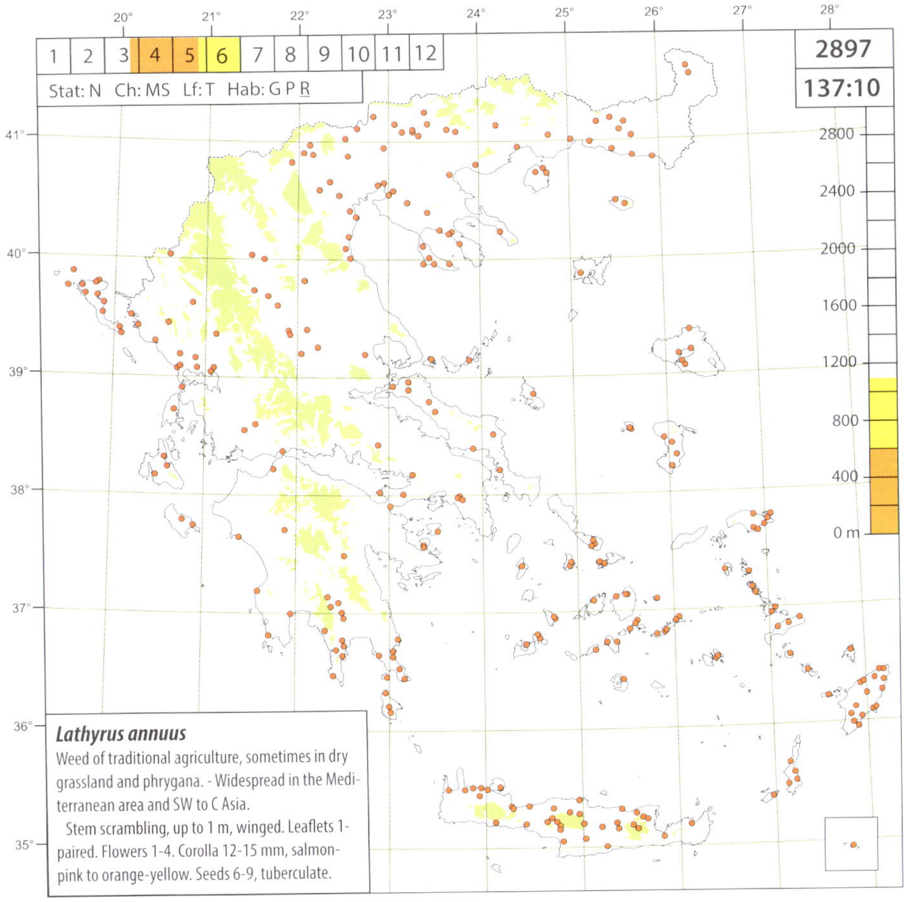

Lathyrus annuus

Weed of traditional agriculture, sometimes in dry grassland and phrygana. - Widespread in the Mediterranean area and SW to C Asia.

Stem scrambling, up to 1 m, winged. Leaflets 1-paired. Flowers 1-4. Corolla 12-15 mm, salmon-pink to orange-yellow. Seeds 6-9, tuberculate.

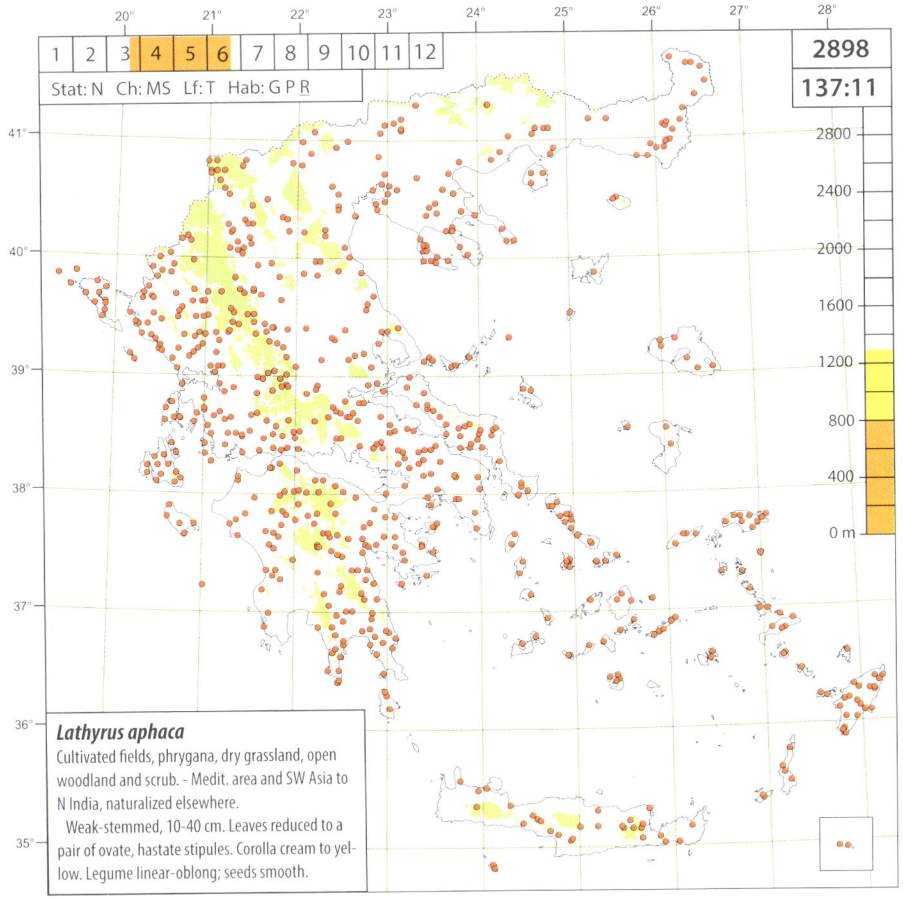

Lathyrus aphaca

Cultivated fields, phrygana, dry grassland, open woodland and scrub. - Medit. area and SW Asia to N India, naturalized elsewhere.

Weak-stemmed, 10-40 cm. Leaves reduced to a pair of ovate, hastate stipules. Corolla cream to yellow. Legume linear-oblong; seeds smooth.

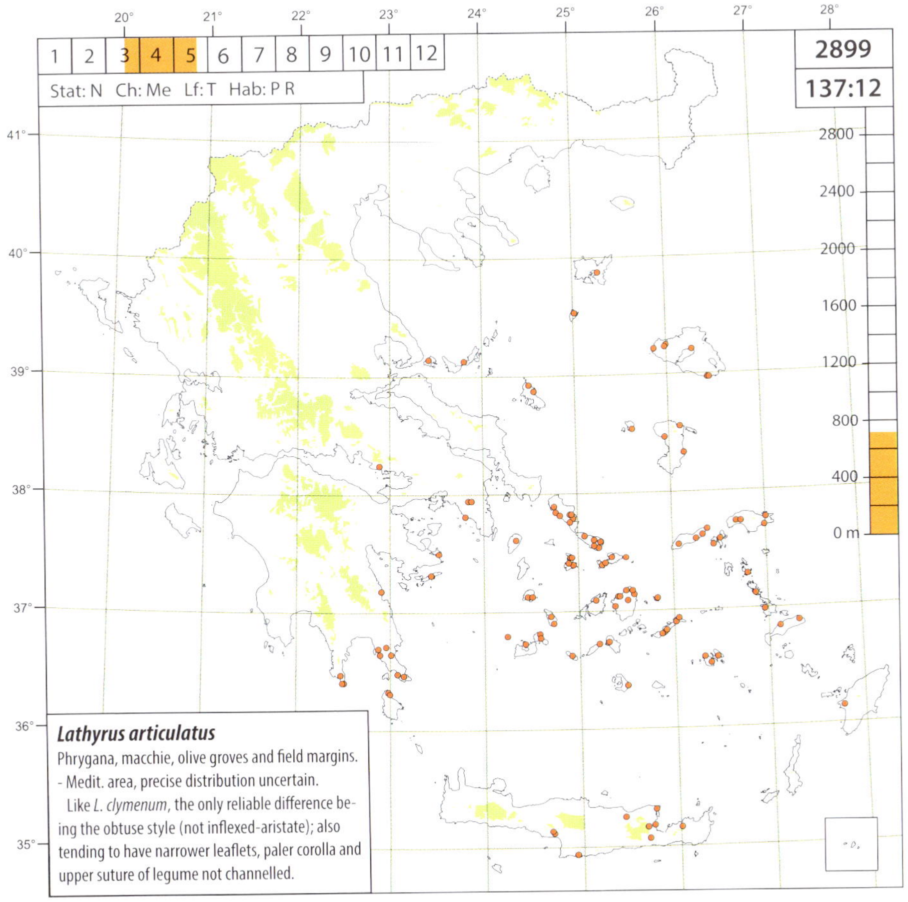

Lathyrus articulatus

Phrygana, macchie, olive groves and field margins. - Medit. area, precise distribution uncertain.

Like *L. clymenum*, the only reliable difference being the obtuse style (not inflexed-aristate); also tending to have narrower leaflets, paler corolla and upper suture of legume not channelled.

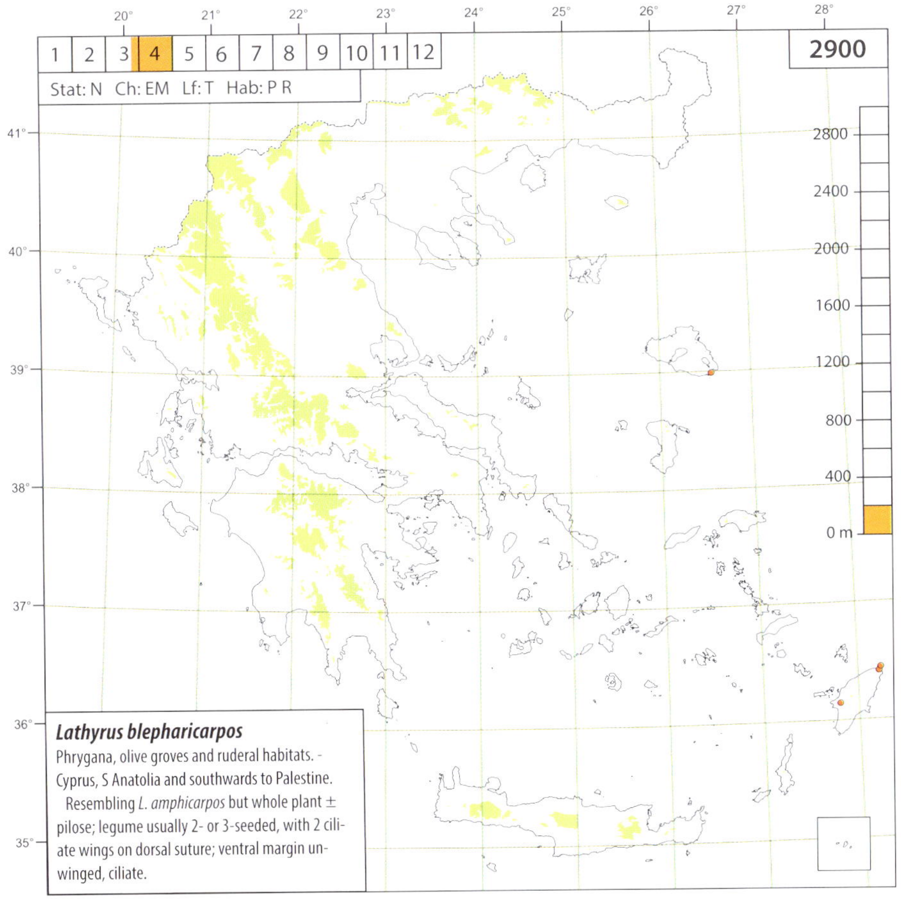

Lathyrus blepharicarpos

Phrygana, olive groves and ruderal habitats. - Cyprus, S Anatolia and southwards to Palestine.

Resembling *L. amphicarpos* but whole plant ± pilose; legume usually 2- or 3-seeded, with 2 ciliate wings on dorsal suture; ventral margin unwinged, ciliate.

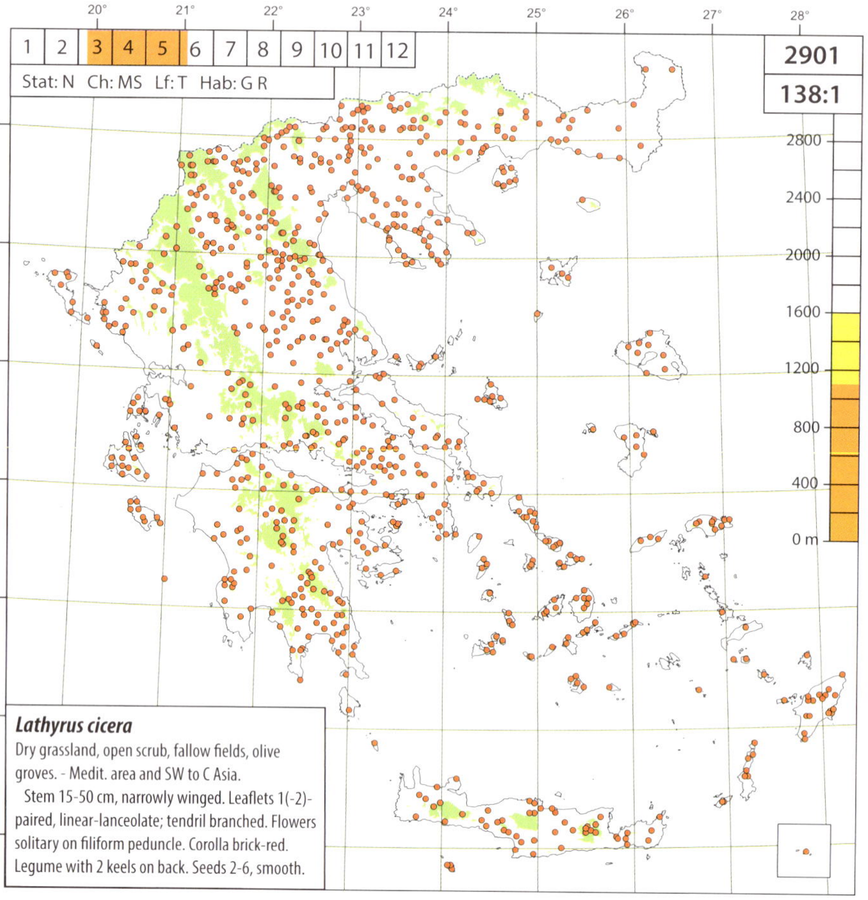

Lathyrus cicera
Dry grassland, open scrub, fallow fields, olive groves. - Medit. area and SW to C Asia.
Stem 15-50 cm, narrowly winged. Leaflets 1(-2)-paired, linear-lanceolate; tendril branched. Flowers solitary on filiform peduncle. Corolla brick-red. Legume with 2 keels on back. Seeds 2-6, smooth.

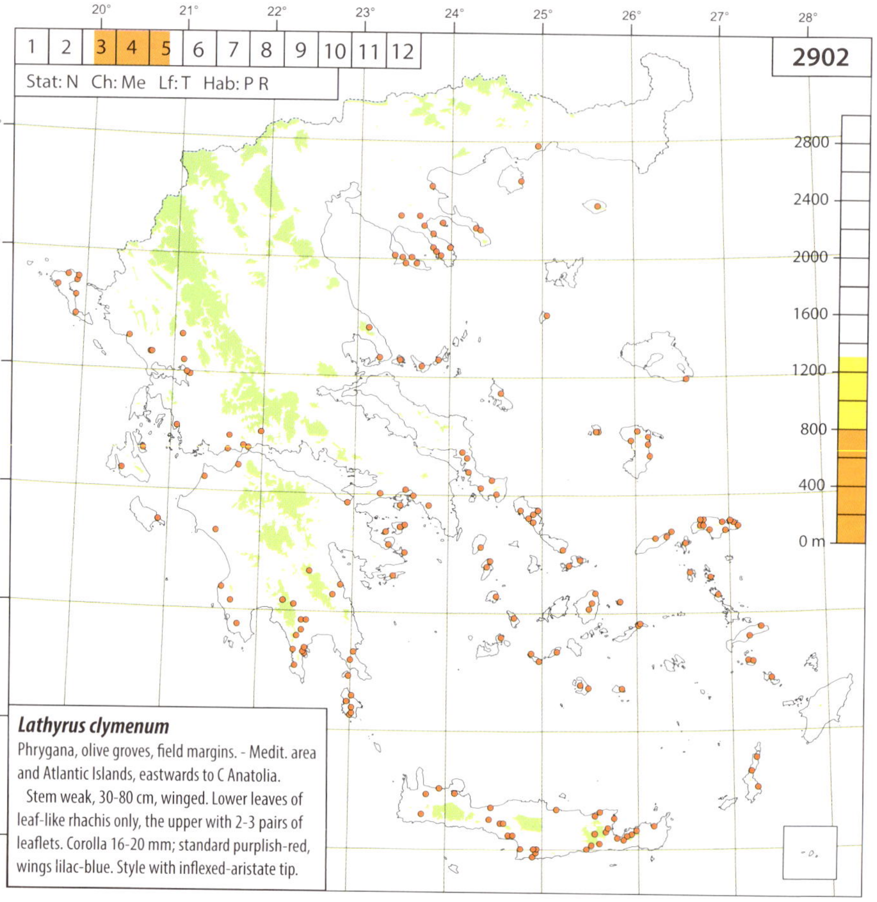

Lathyrus clymenum
Phrygana, olive groves, field margins. - Medit. area and Atlantic Islands, eastwards to C Anatolia.
Stem weak, 30-80 cm, winged. Lower leaves of leaf-like rhachis only, the upper with 2-3 pairs of leaflets. Corolla 16-20 mm; standard purplish-red, wings lilac-blue. Style with inflexed-aristate tip.

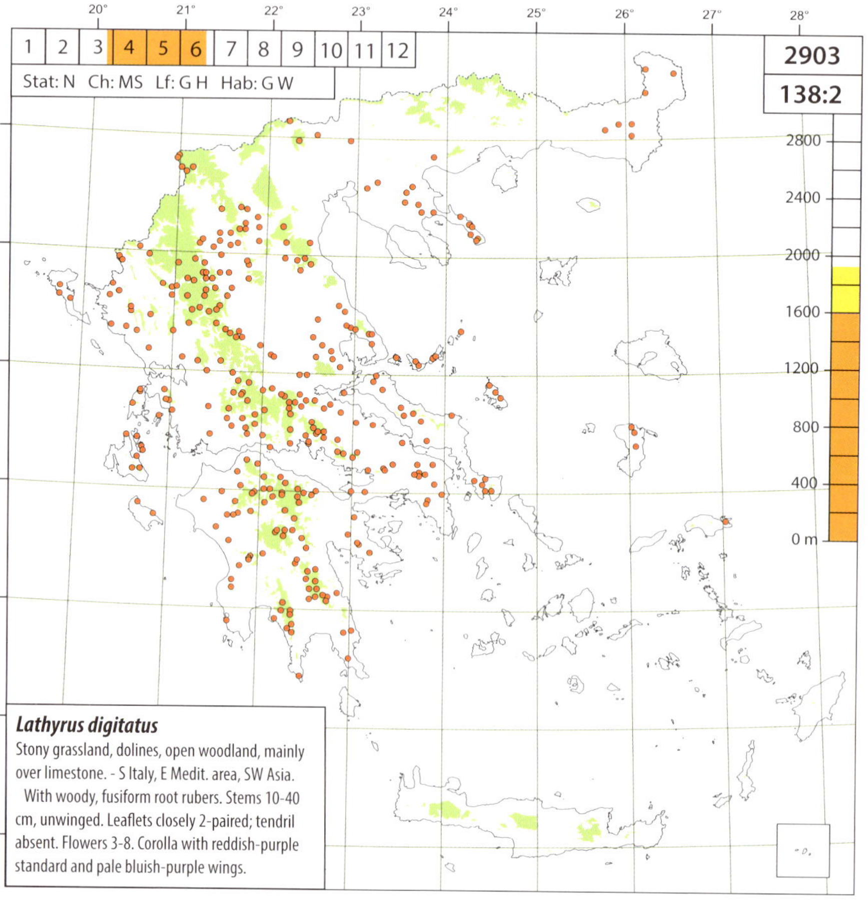

Lathyrus digitatus
Stony grassland, dolines, open woodland, mainly over limestone. - S Italy, E Medit. area, SW Asia.
With woody, fusiform root rubers. Stems 10-40 cm, unwinged. Leaflets closely 2-paired; tendril absent. Flowers 3-8. Corolla with reddish-purple standard and pale bluish-purple wings.

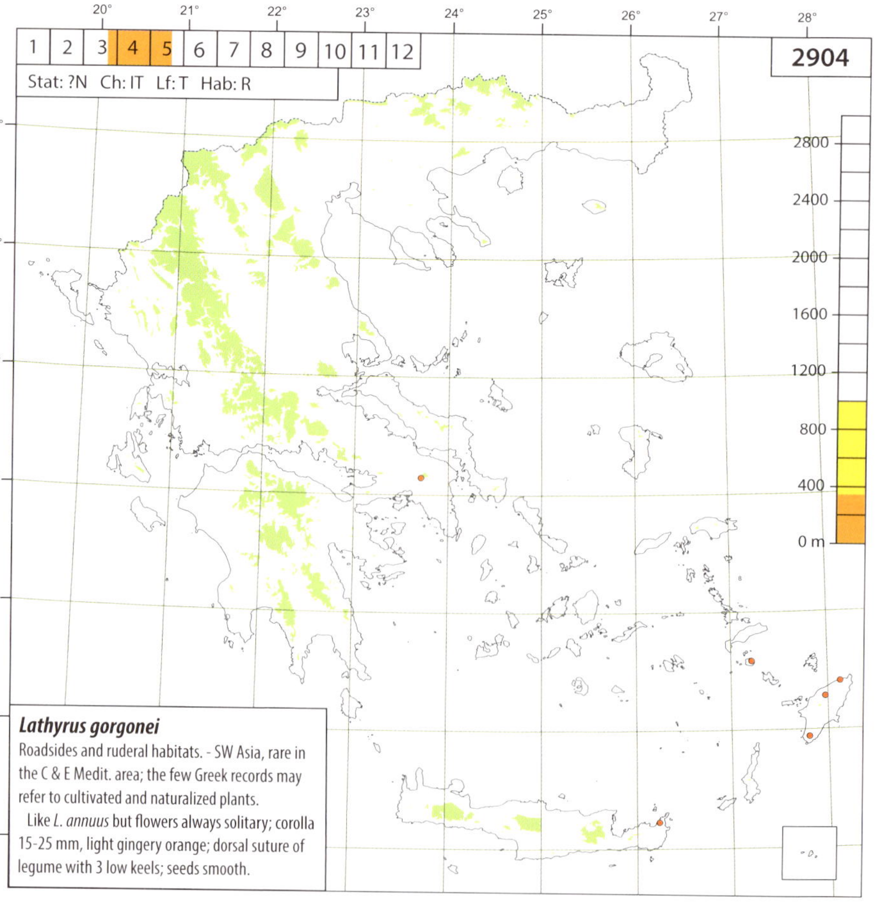

Lathyrus gorgonei
Roadsides and ruderal habitats. - SW Asia, rare in the C & E Medit. area; the few Greek records may refer to cultivated and naturalized plants.
Like *L. annuus* but flowers always solitary; corolla 15-25 mm, light gingery orange; dorsal suture of legume with 3 low keels; seeds smooth.

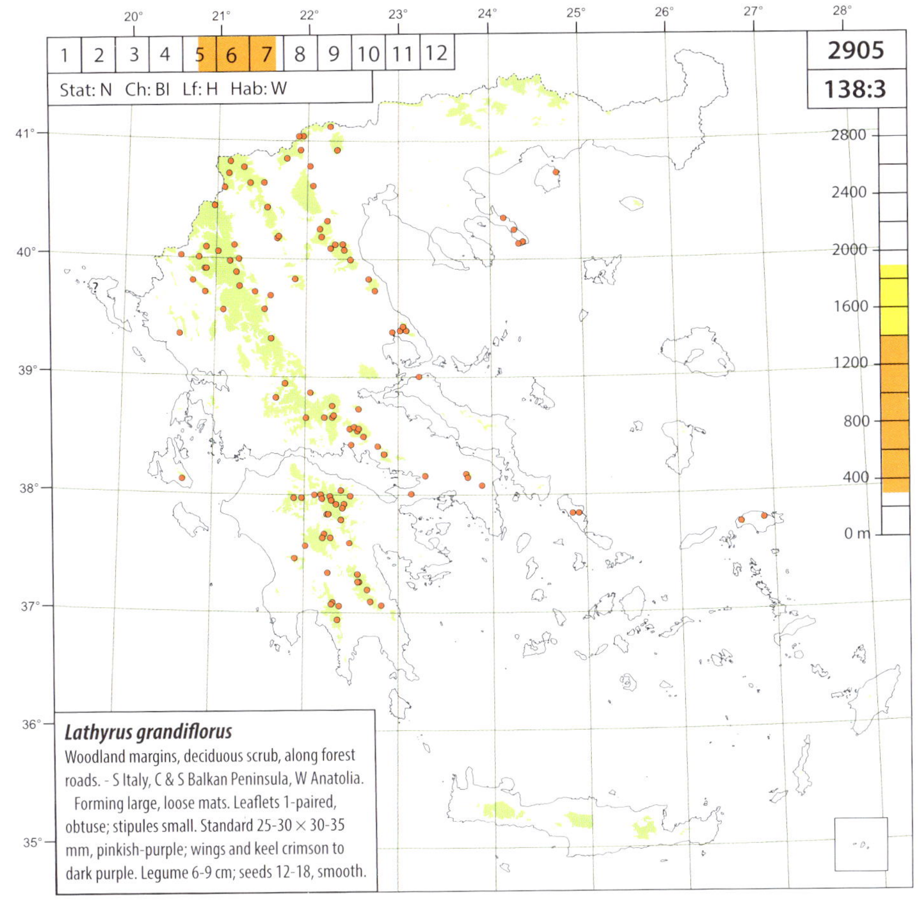

Lathyrus grandiflorus

Woodland margins, deciduous scrub, along forest roads. - S Italy, C & S Balkan Peninsula, W Anatolia.

Forming large, loose mats. Leaflets 1-paired, obtuse; stipules small. Standard 25-30 × 30-35 mm, pinkish-purple; wings and keel crimson to dark purple. Legume 6-9 cm; seeds 12-18, smooth.

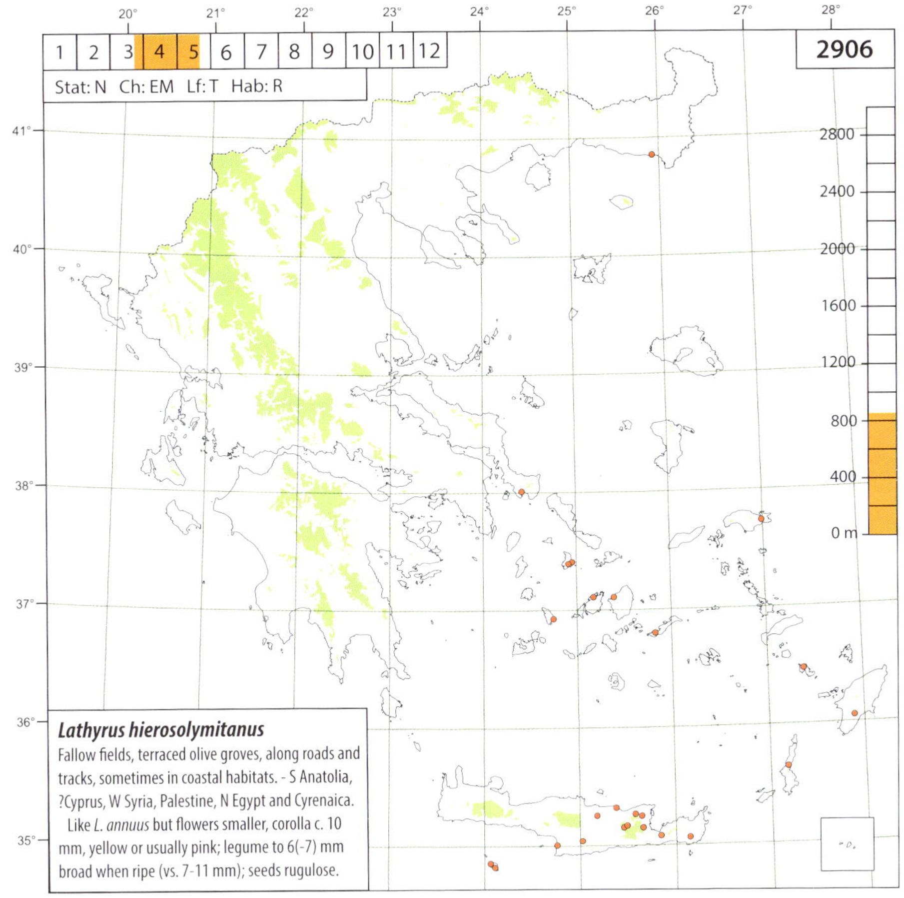

Lathyrus hierosolymitanus

Fallow fields, terraced olive groves, along roads and tracks, sometimes in coastal habitats. - S Anatolia, ?Cyprus, W Syria, Palestine, N Egypt and Cyrenaica.

Like *L. annuus* but flowers smaller, corolla c. 10 mm, yellow or usually pink; legume to 6(-7) mm broad when ripe (vs. 7-11 mm); seeds rugulose.

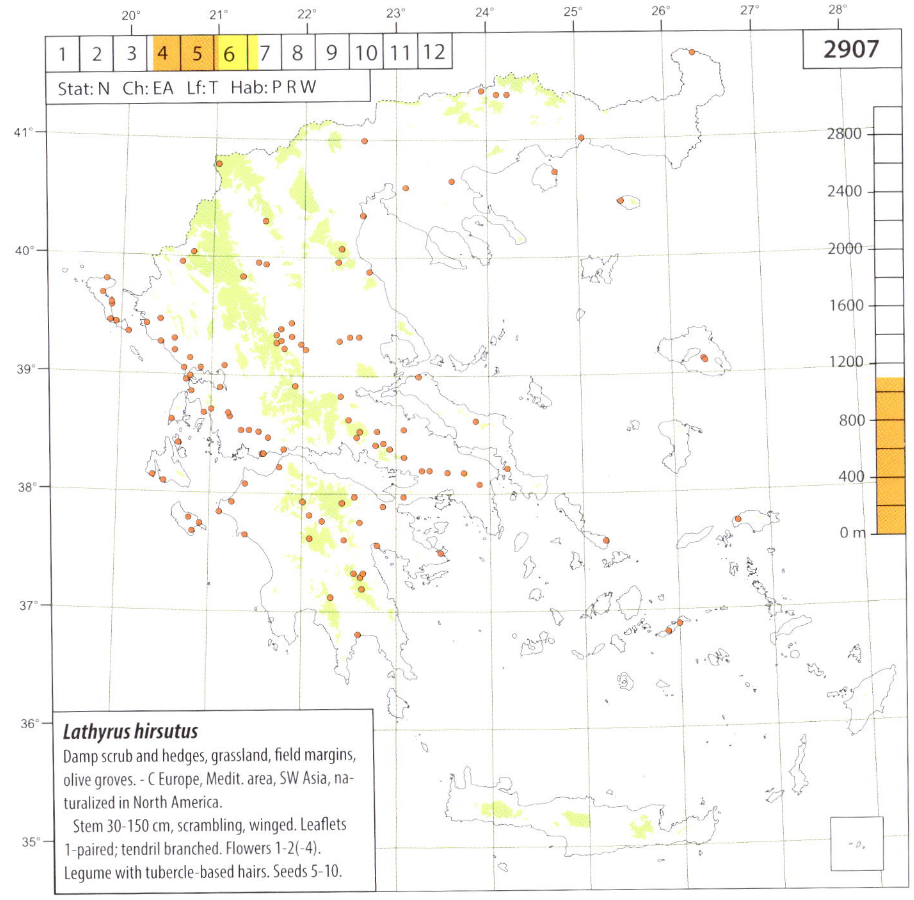

Lathyrus hirsutus

Damp scrub and hedges, grassland, field margins, olive groves. - C Europe, Medit. area, SW Asia, naturalized in North America.

Stem 30-150 cm, scrambling, winged. Leaflets 1-paired; tendril branched. Flowers 1-2(-4). Legume with tubercle-based hairs. Seeds 5-10.

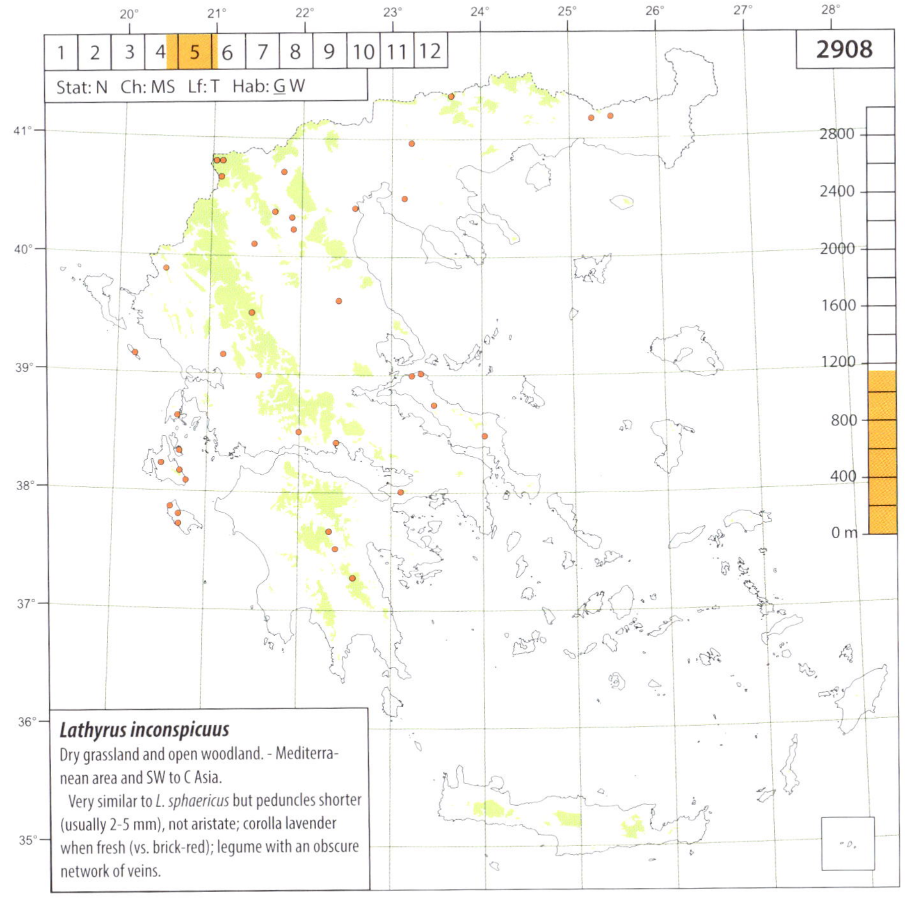

Lathyrus inconspicuus

Dry grassland and open woodland. - Mediterranean area and SW to C Asia.

Very similar to *L. sphaericus* but peduncles shorter (usually 2-5 mm), not aristate; corolla lavender when fresh (vs. brick-red); legume with an obscure network of veins.

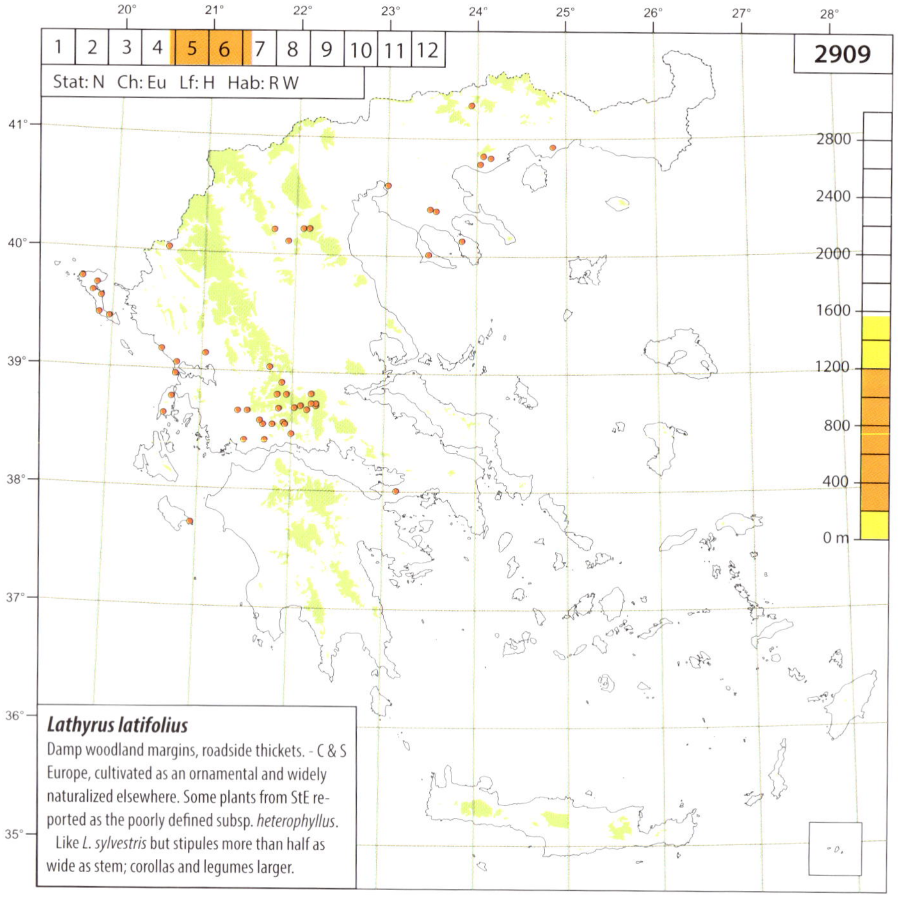

Lathyrus latifolius
Damp woodland margins, roadside thickets. - C & S Europe, cultivated as an ornamental and widely naturalized elsewhere. Some plants from StE reported as the poorly defined subsp. *heterophyllus*.
Like *L. sylvestris* but stipules more than half as wide as stem; corollas and legumes larger.

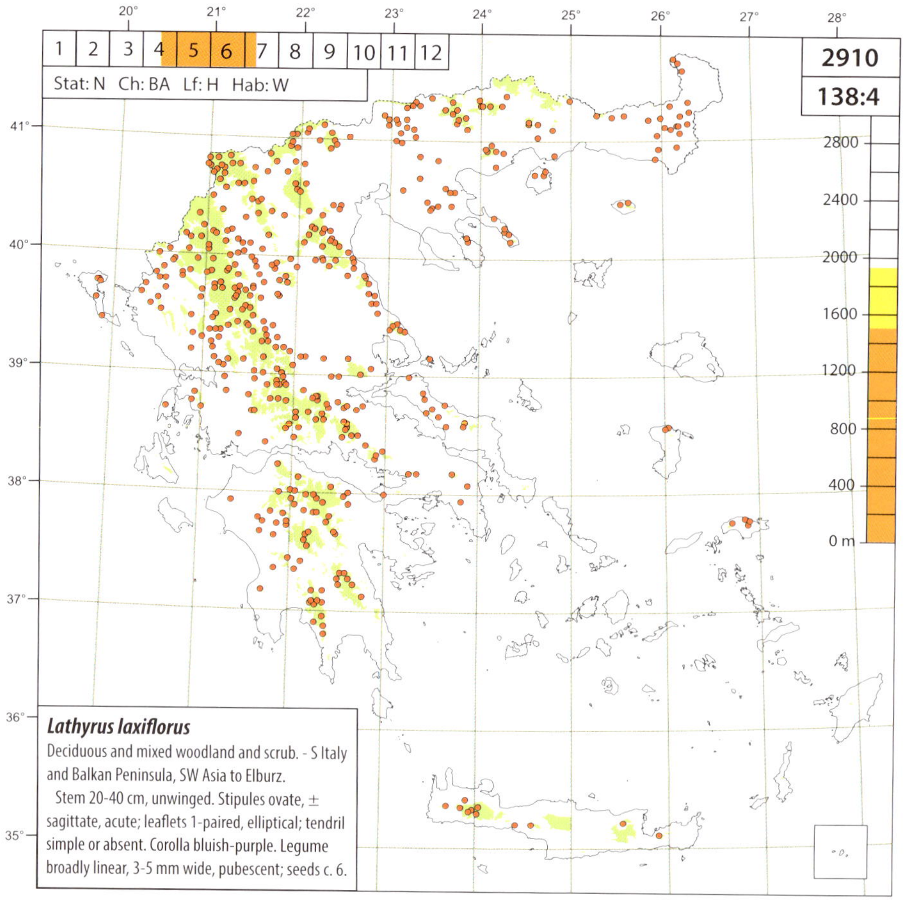

Lathyrus laxiflorus
Deciduous and mixed woodland and scrub. - S Italy and Balkan Peninsula, SW Asia to Elburz.
Stem 20-40 cm, unwinged. Stipules ovate, ± sagittate, acute; leaflets 1-paired, elliptical; tendril simple or absent. Corolla bluish-purple. Legume broadly linear, 3-5 mm wide, pubescent; seeds c. 6.

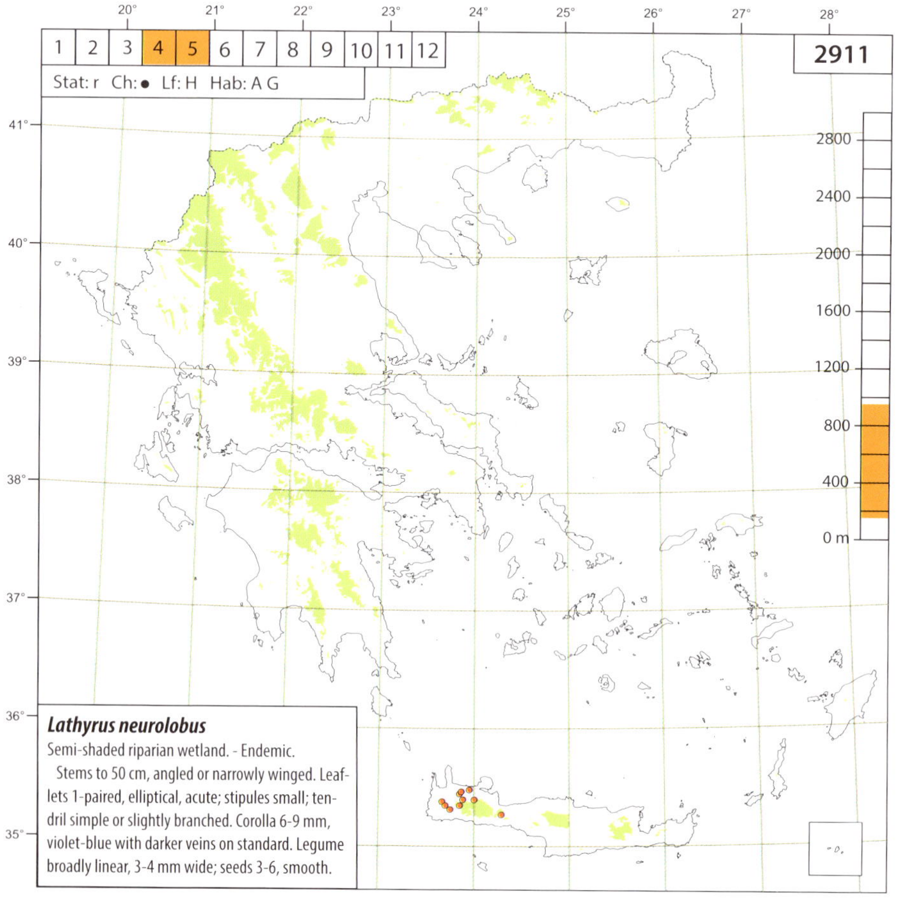

Lathyrus neurolobus
Semi-shaded riparian wetland. - Endemic.
Stems to 50 cm, angled or narrowly winged. Leaflets 1-paired, elliptical, acute; stipules small; tendril simple or slightly branched. Corolla 6-9 mm, violet-blue with darker veins on standard. Legume broadly linear, 3-4 mm wide; seeds 3-6, smooth.

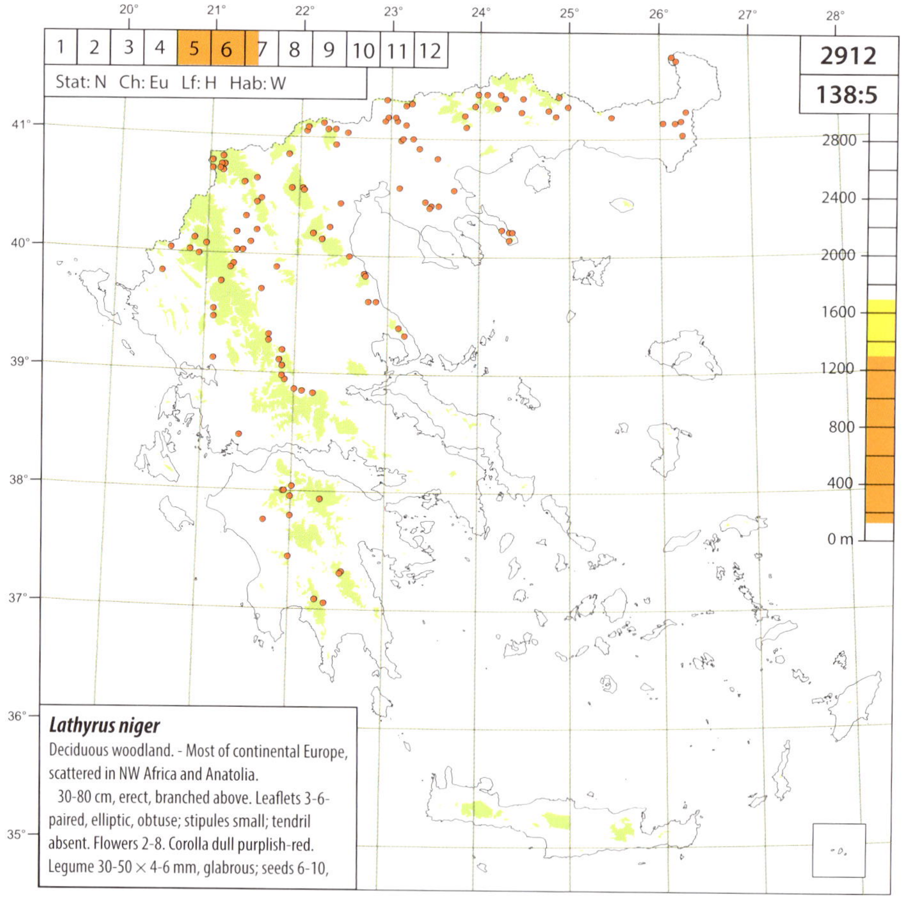

Lathyrus niger
Deciduous woodland. - Most of continental Europe, scattered in NW Africa and Anatolia.
30-80 cm, erect, branched above. Leaflets 3-6-paired, elliptic, obtuse; stipules small; tendril absent. Flowers 2-8. Corolla dull purplish-red. Legume 30-50 × 4-6 mm, glabrous; seeds 6-10,

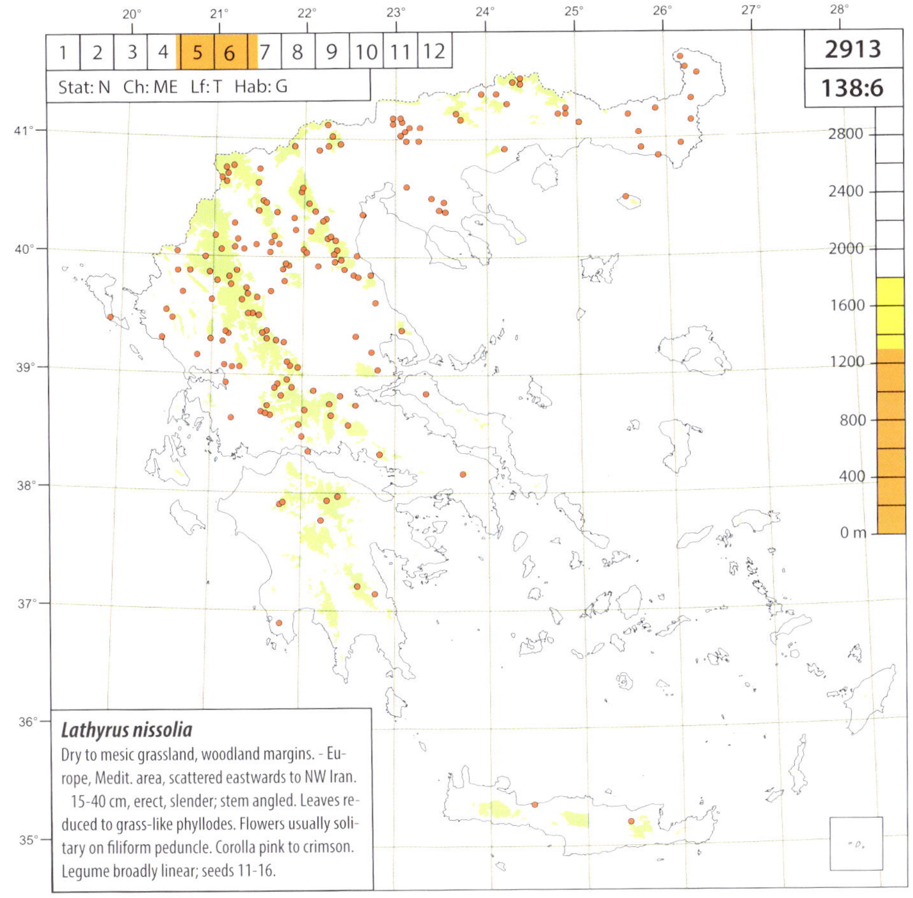

Lathyrus nissolia

Dry to mesic grassland, woodland margins. - Europe, Medit. area, scattered eastwards to NW Iran.

15-40 cm, erect, slender; stem angled. Leaves reduced to grass-like phyllodes. Flowers usually solitary on filiform peduncle. Corolla pink to crimson. Legume broadly linear; seeds 11-16.

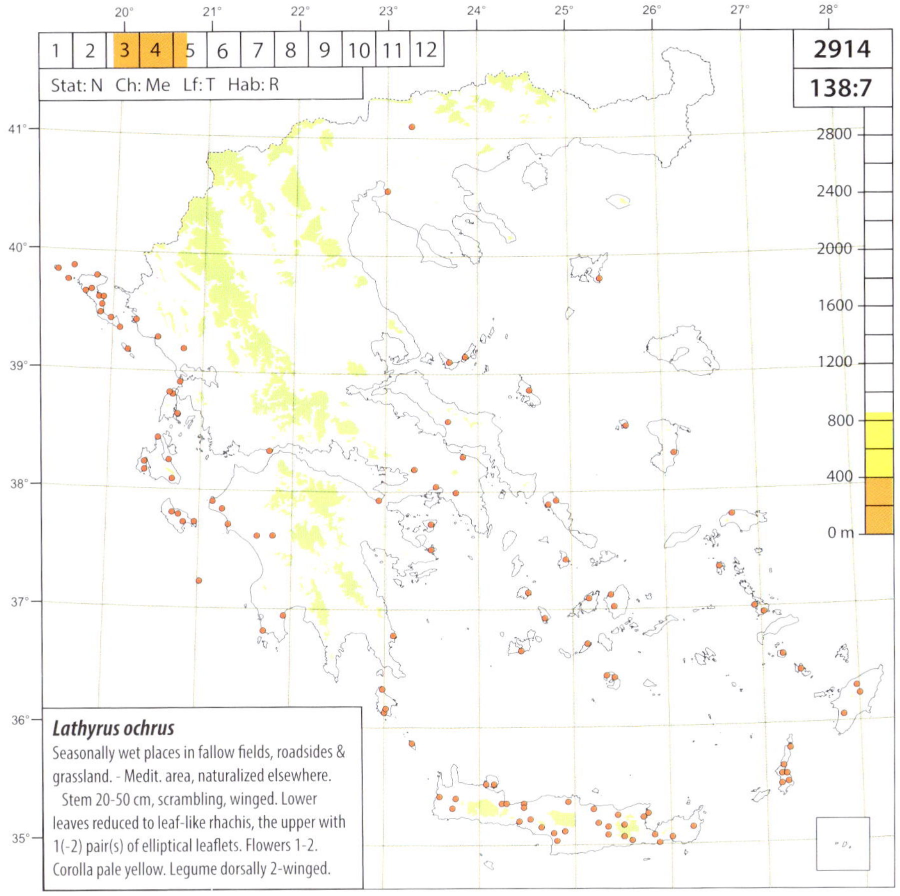

Lathyrus ochrus

Seasonally wet places in fallow fields, roadsides & grassland. - Medit. area, naturalized elsewhere.

Stem 20-50 cm, scrambling, winged. Lower leaves reduced to leaf-like rhachis, the upper with 1(-2) pair(s) of elliptical leaflets. Flowers 1-2. Corolla pale yellow. Legume dorsally 2-winged.

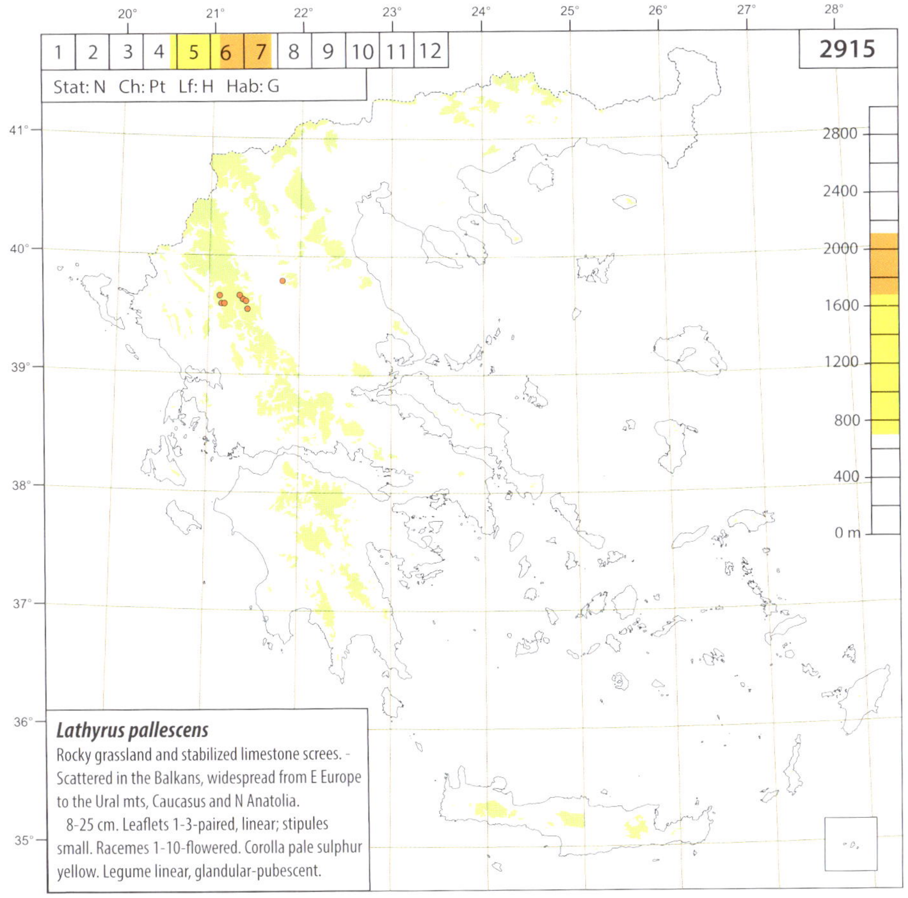

Lathyrus pallescens

Rocky grassland and stabilized limestone screes. - Scattered in the Balkans, widespread from E Europe to the Ural mts, Caucasus and N Anatolia.

8-25 cm. Leaflets 1-3-paired, linear; stipules small. Racemes 1-10-flowered. Corolla pale sulphur yellow. Legume linear, glandular-pubescent.

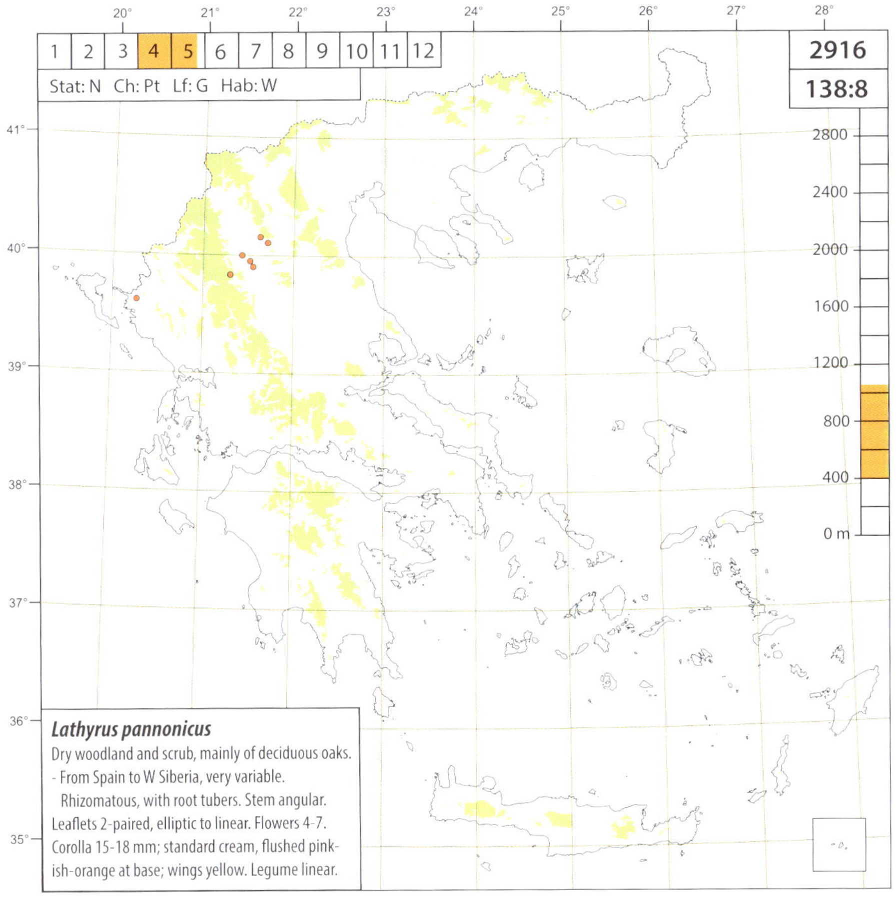

Lathyrus pannonicus

Dry woodland and scrub, mainly of deciduous oaks. - From Spain to W Siberia, very variable.

Rhizomatous, with root tubers. Stem angular. Leaflets 2-paired, elliptic to linear. Flowers 4-7. Corolla 15-18 mm; standard cream, flushed pinkish-orange at base; wings yellow. Legume linear.

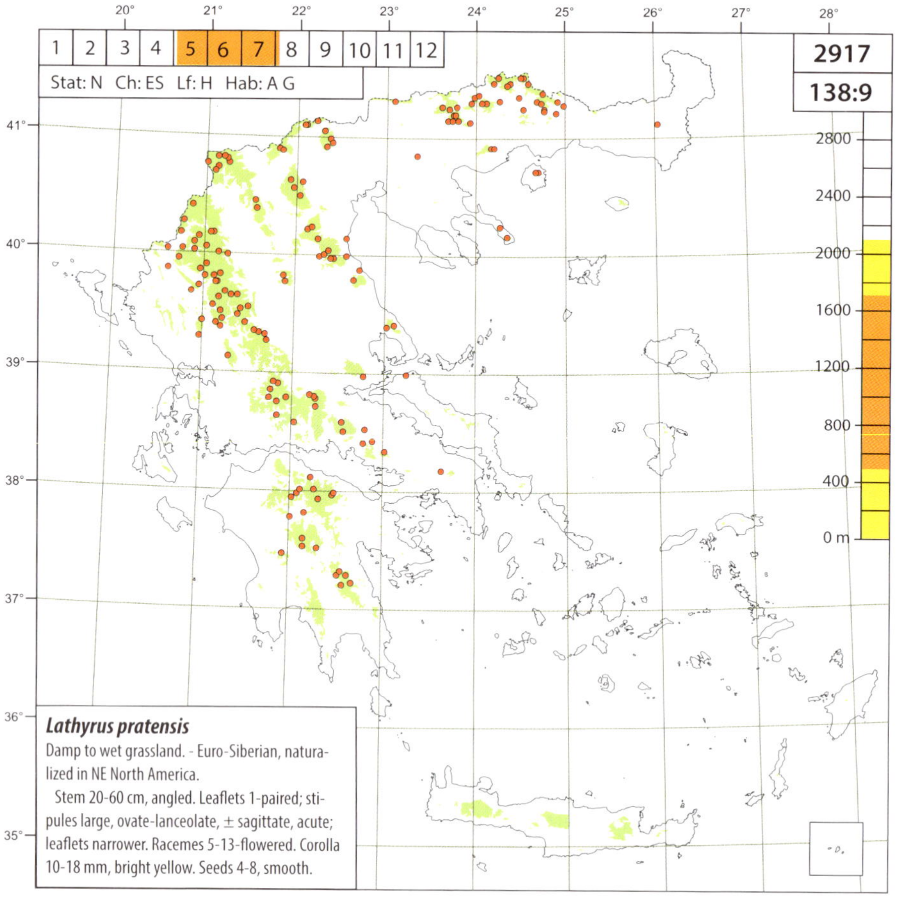

Lathyrus pratensis

Damp to wet grassland. - Euro-Siberian, naturalized in NE North America.

Stem 20-60 cm, angled. Leaflets 1-paired; stipules large, ovate-lanceolate, ± sagittate, acute; leaflets narrower. Racemes 5-13-flowered. Corolla 10-18 mm, bright yellow. Seeds 4-8, smooth.

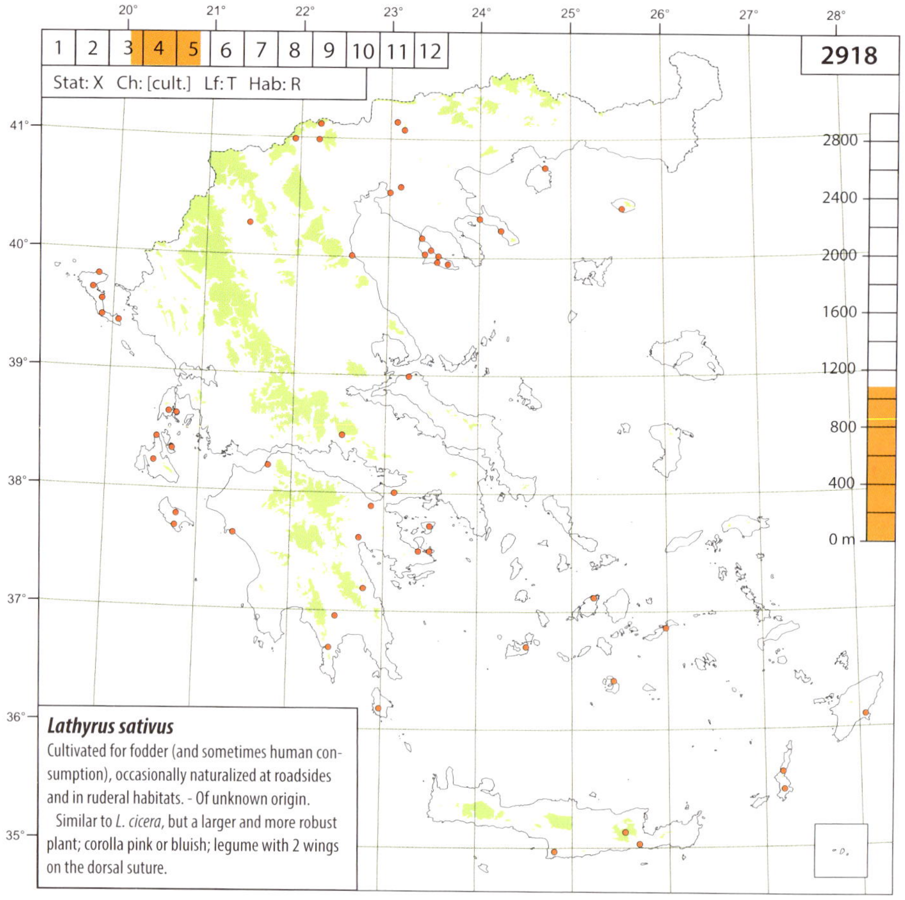

Lathyrus sativus

Cultivated for fodder (and sometimes human consumption), occasionally naturalized at roadsides and in ruderal habitats. - Of unknown origin.

Similar to *L. cicera*, but a larger and more robust plant; corolla pink or bluish; legume with 2 wings on the dorsal suture.

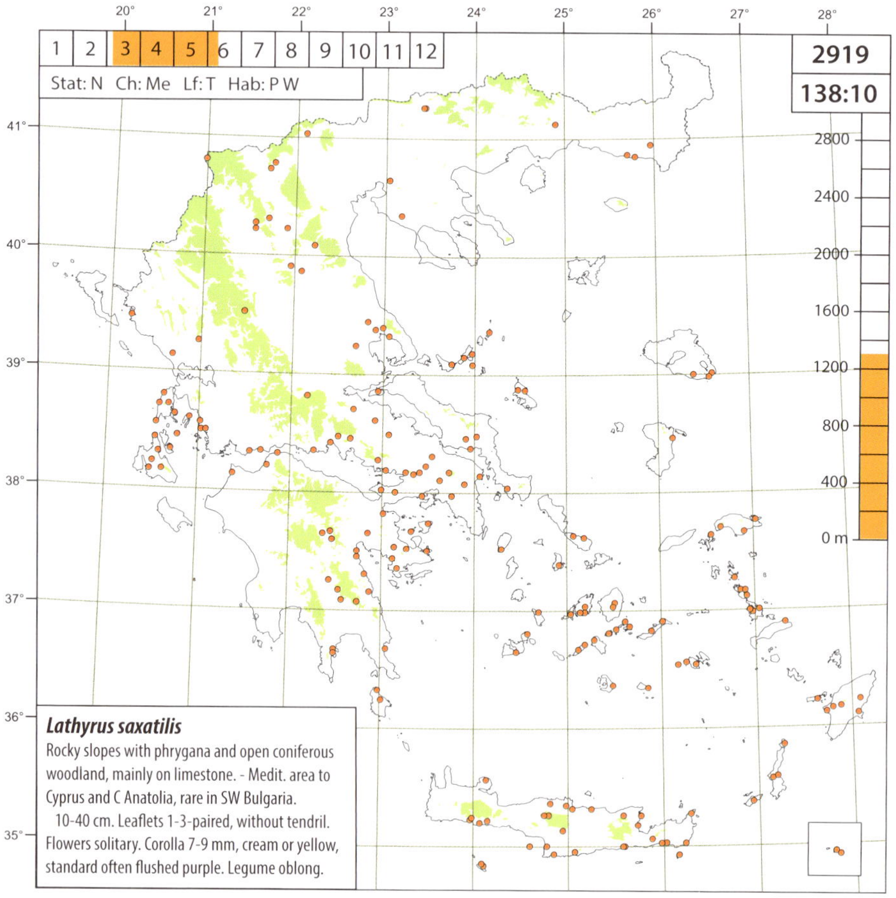

Lathyrus saxatilis

Rocky slopes with phrygana and open coniferous woodland, mainly on limestone. - Medit. area to Cyprus and C Anatolia, rare in SW Bulgaria.

10-40 cm. Leaflets 1-3-paired, without tendril. Flowers solitary. Corolla 7-9 mm, cream or yellow, standard often flushed purple. Legume oblong.

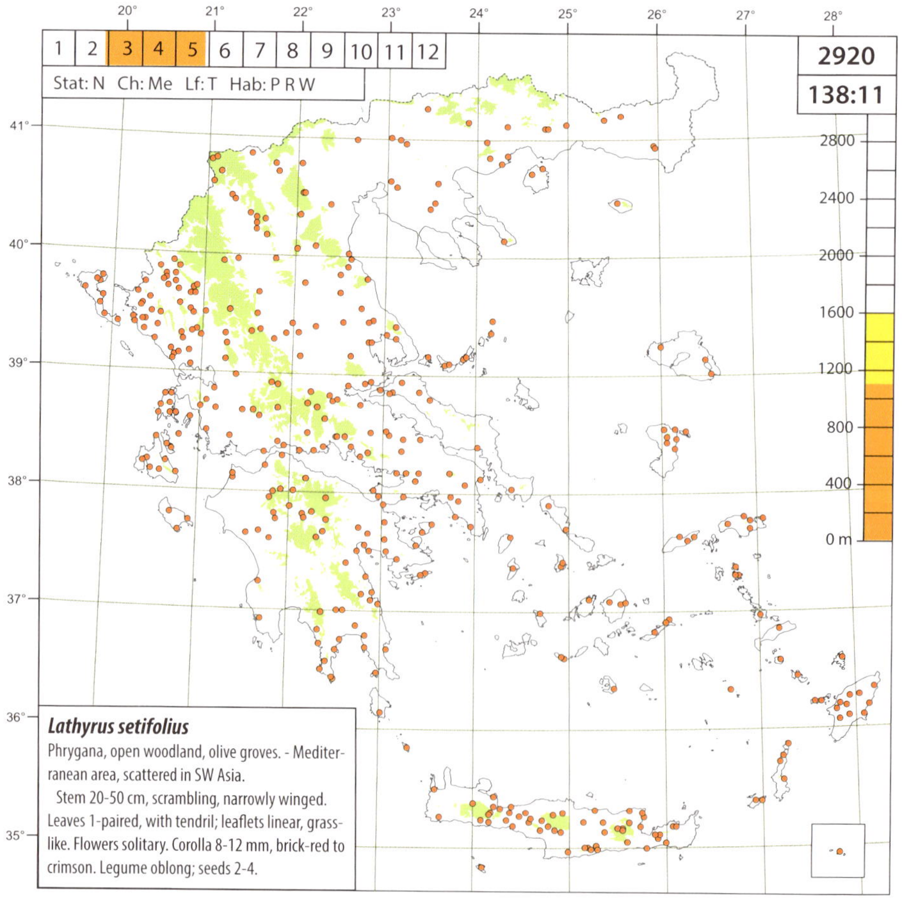

Lathyrus setifolius

Phrygana, open woodland, olive groves. - Mediterranean area, scattered in SW Asia.

Stem 20-50 cm, scrambling, narrowly winged. Leaves 1-paired, with tendril; leaflets linear, grass-like. Flowers solitary. Corolla 8-12 mm, brick-red to crimson. Legume oblong; seeds 2-4.

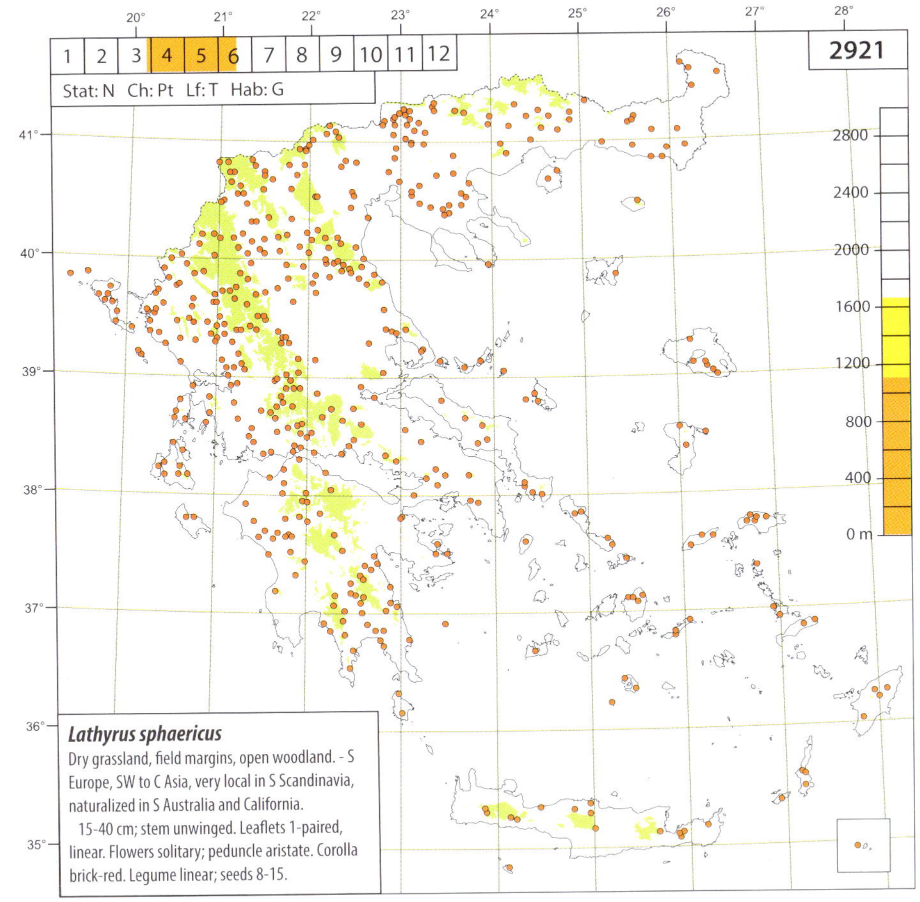

Lathyrus sphaericus

Dry grassland, field margins, open woodland. - S Europe, SW to C Asia, very local in S Scandinavia, naturalized in S Australia and California.

15-40 cm; stem unwinged. Leaflets 1-paired, linear. Flowers solitary; peduncle aristate. Corolla brick-red. Legume linear; seeds 8-15.

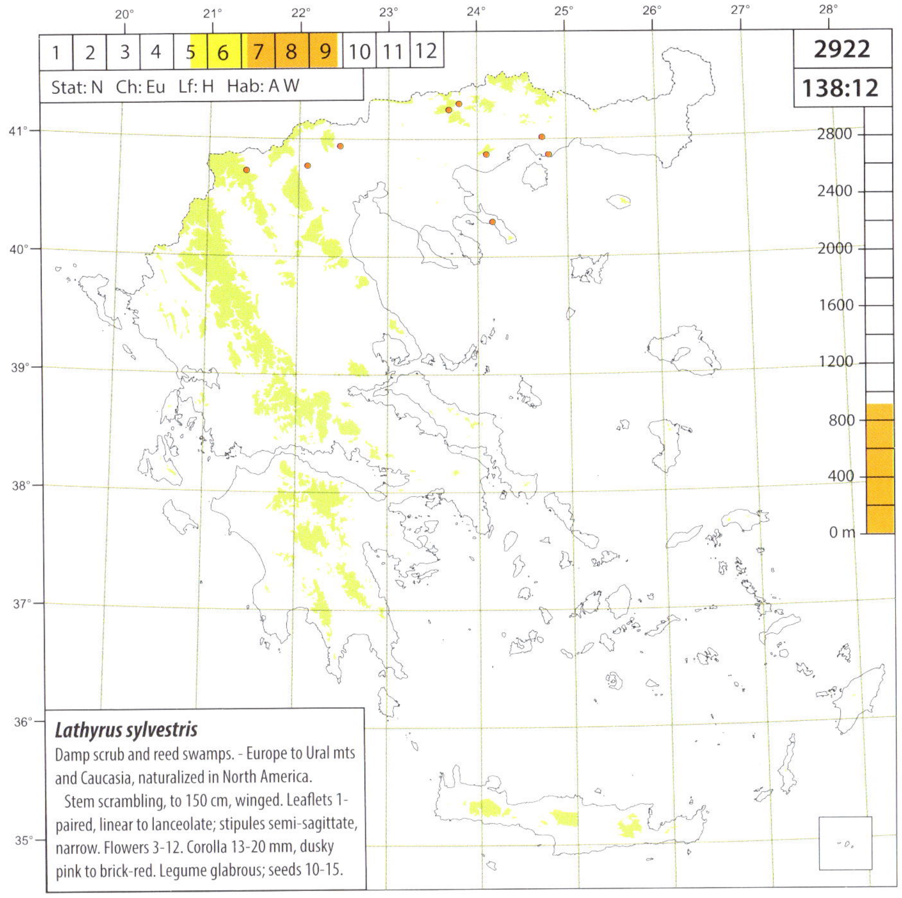

Lathyrus sylvestris

Damp scrub and reed swamps. - Europe to Ural mts and Caucasia, naturalized in North America.

Stem scrambling, to 150 cm, winged. Leaflets 1-paired, linear to lanceolate; stipules semi-sagittate, narrow. Flowers 3-12. Corolla 13-20 mm, dusky pink to brick-red. Legume glabrous; seeds 10-15.

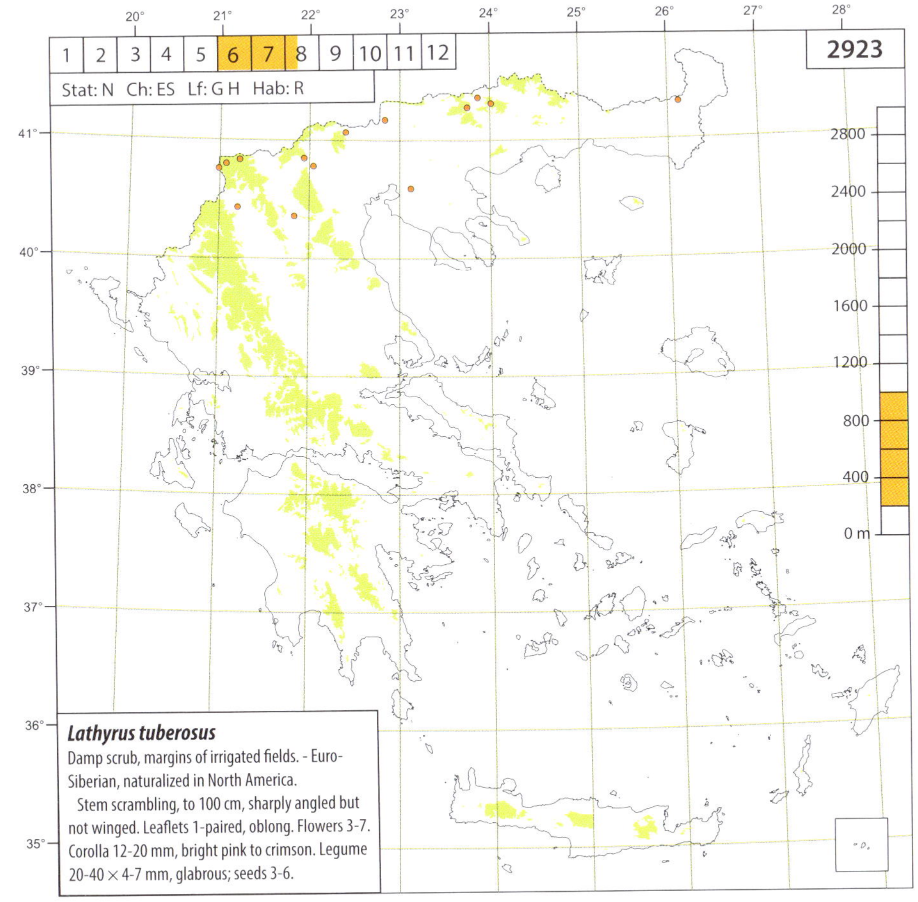

Lathyrus tuberosus

Damp scrub, margins of irrigated fields. - Euro-Siberian, naturalized in North America.

Stem scrambling, to 100 cm, sharply angled but not winged. Leaflets 1-paired, oblong. Flowers 3-7. Corolla 12-20 mm, bright pink to crimson. Legume 20-40 × 4-7 mm, glabrous; seeds 3-6.

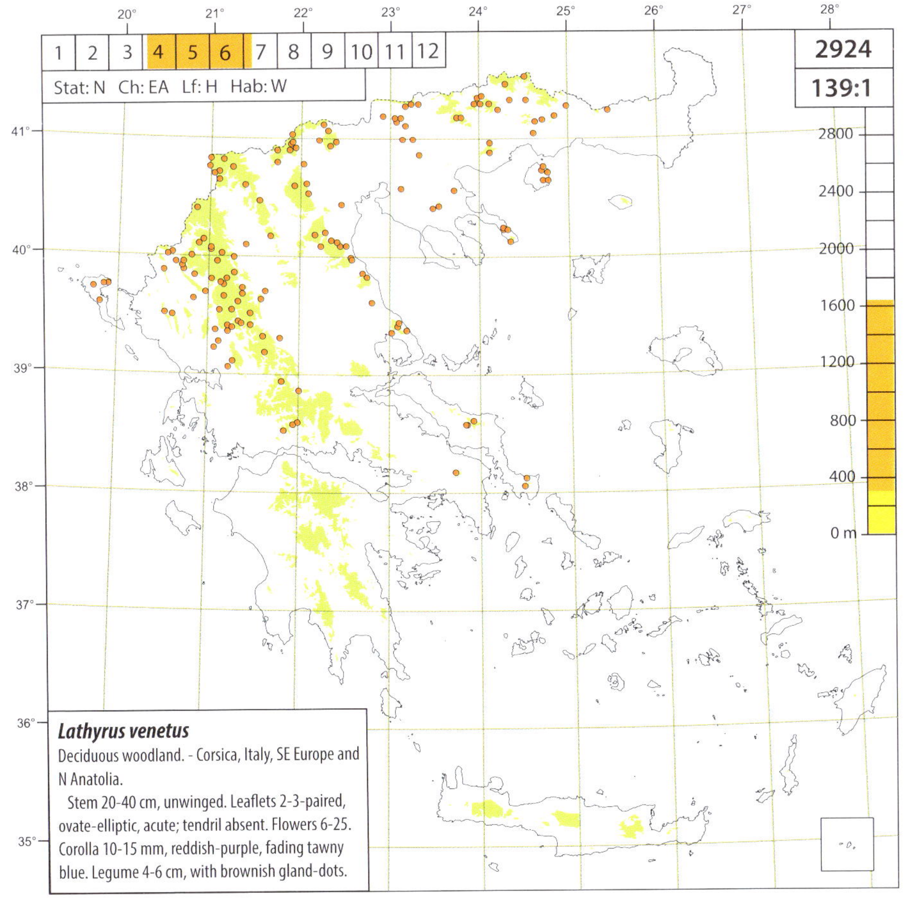

Lathyrus venetus

Deciduous woodland. - Corsica, Italy, SE Europe and N Anatolia.

Stem 20-40 cm, unwinged. Leaflets 2-3-paired, ovate-elliptic, acute; tendril absent. Flowers 6-25. Corolla 10-15 mm, reddish-purple, fading tawny blue. Legume 4-6 cm, with brownish gland-dots.

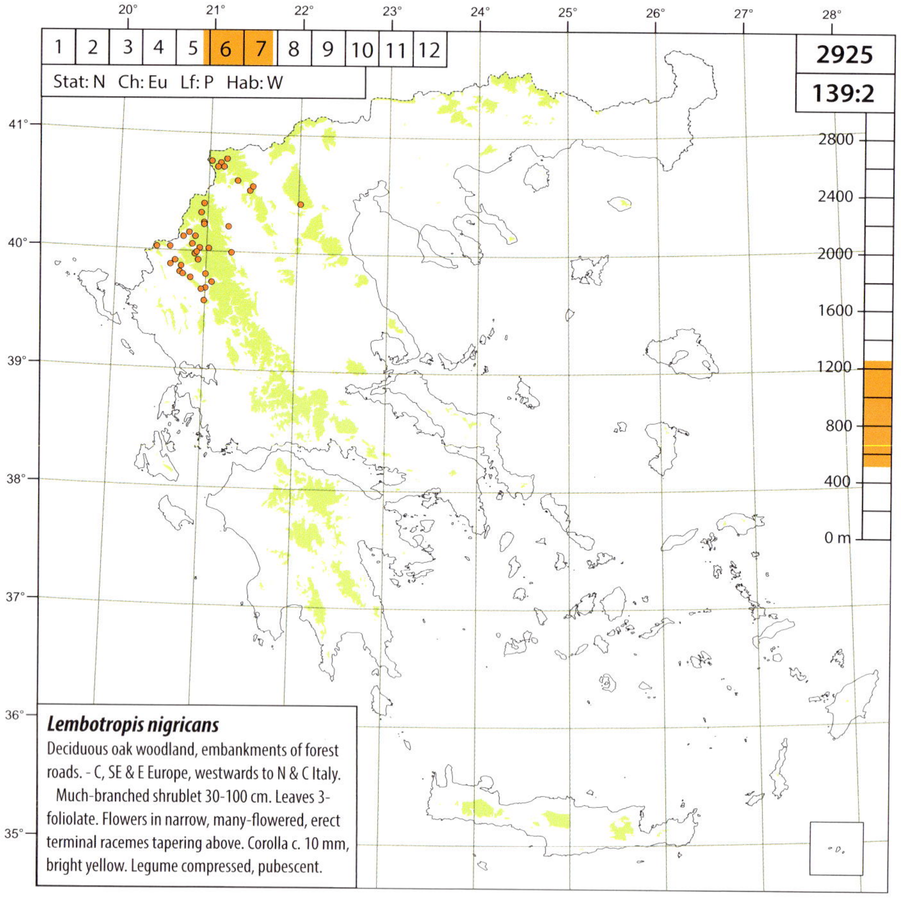

Lembotropis nigricans

Deciduous oak woodland, embankments of forest roads. - C, SE & E Europe, westwards to N & C Italy.

Much-branched shrublet 30-100 cm. Leaves 3-foliolate. Flowers in narrow, many-flowered, erect terminal racemes tapering above. Corolla c. 10 mm, bright yellow. Legume compressed, pubescent.

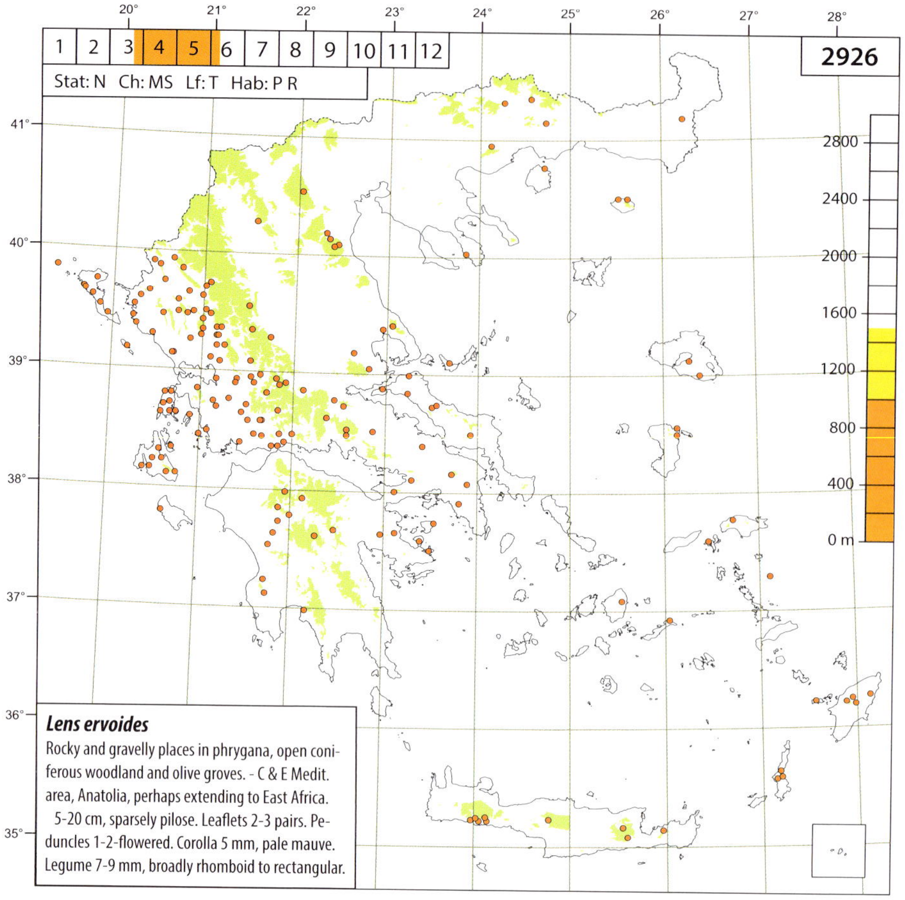

Lens ervoides

Rocky and gravelly places in phrygana, open coniferous woodland and olive groves. - C & E Medit. area, Anatolia, perhaps extending to East Africa.

5-20 cm, sparsely pilose. Leaflets 2-3 pairs. Peduncles 1-2-flowered. Corolla 5 mm, pale mauve. Legume 7-9 mm, broadly rhomboid to rectangular.

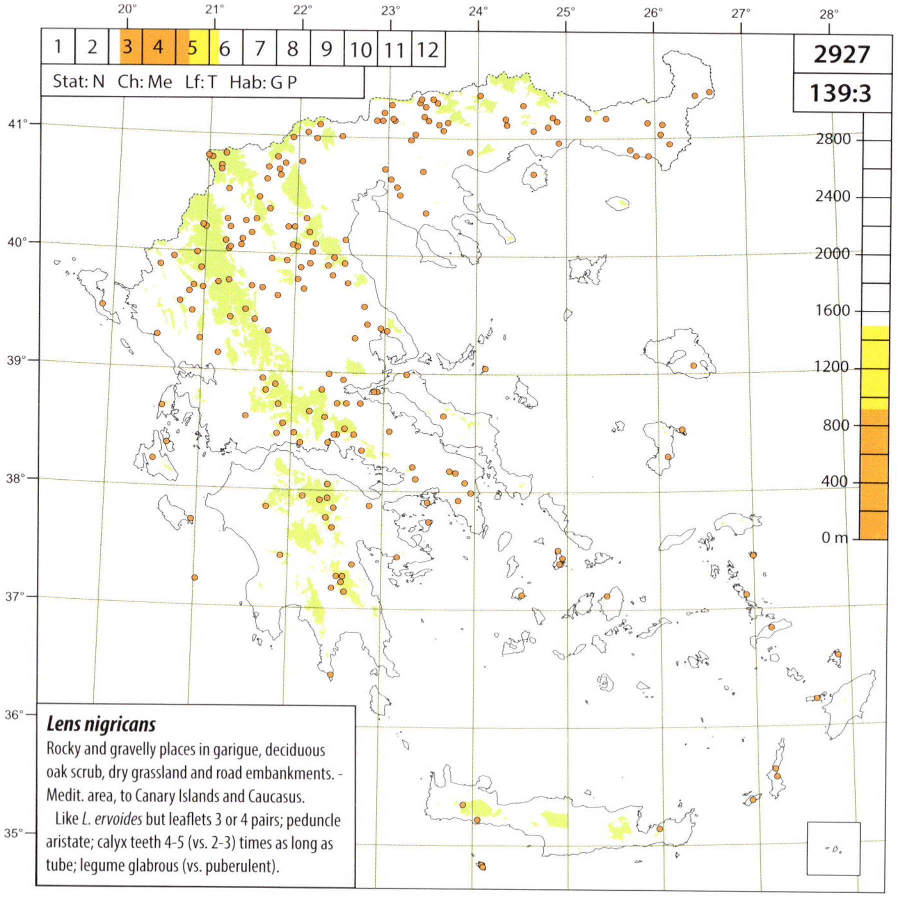

Lens nigricans

Rocky and gravelly places in garigue, deciduous oak scrub, dry grassland and road embankments. - Medit. area, to Canary Islands and Caucasus.

Like *L. ervoides* but leaflets 3 or 4 pairs; peduncle aristate; calyx teeth 4-5 (vs. 2-3) times as long as tube; legume glabrous (vs. puberulent).

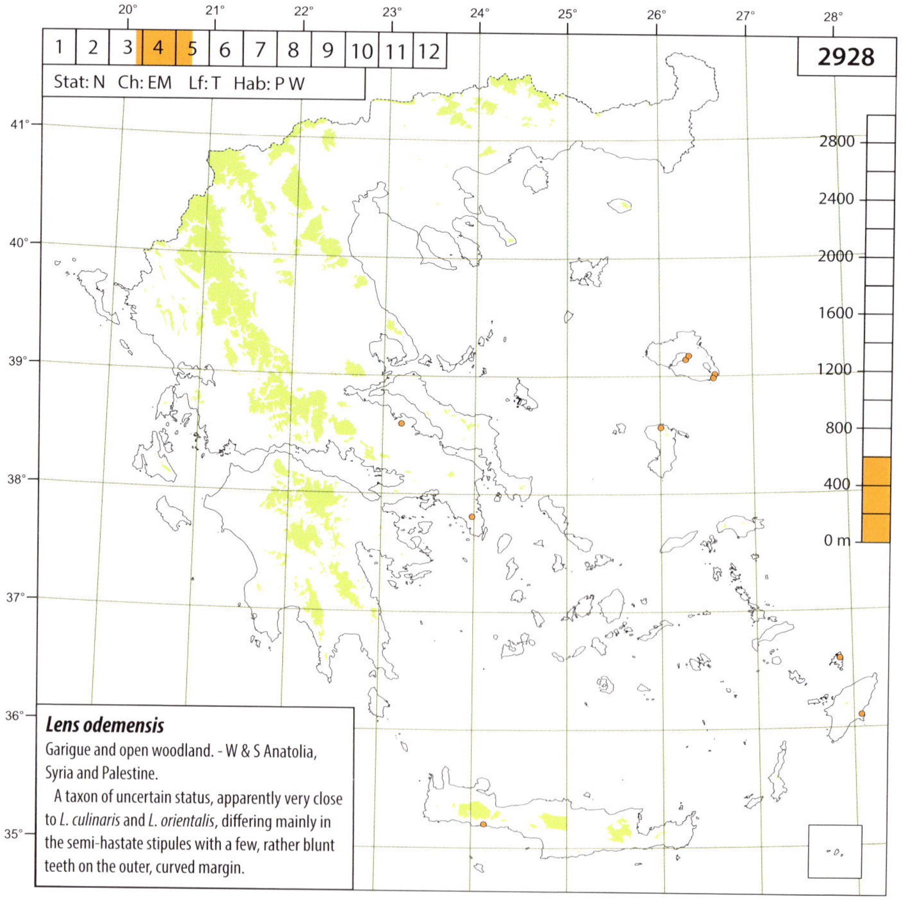

Lens odemensis

Garigue and open woodland. - W & S Anatolia, Syria and Palestine.

A taxon of uncertain status, apparently very close to *L. culinaris* and *L. orientalis*, differing mainly in the semi-hastate stipules with a few, rather blunt teeth on the outer, curved margin.

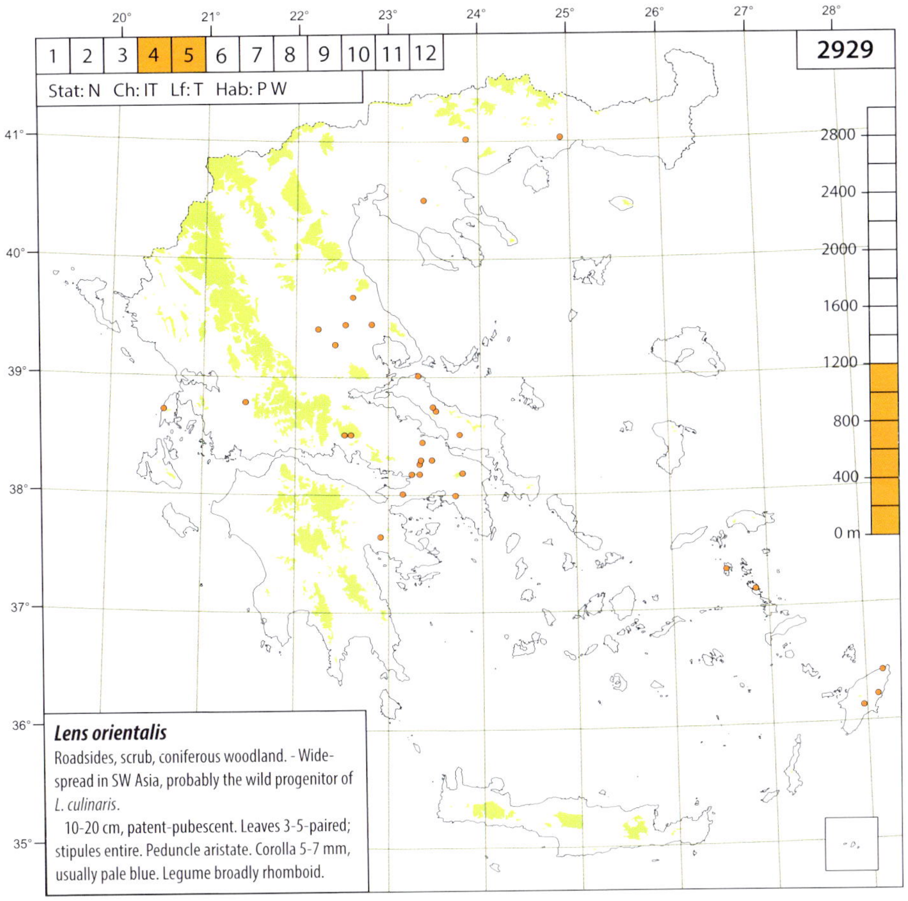

Lens orientalis

Roadsides, scrub, coniferous woodland. - Widespread in SW Asia, probably the wild progenitor of *L. culinaris*.

10-20 cm, patent-pubescent. Leaves 3-5-paired; stipules entire. Peduncle aristate. Corolla 5-7 mm, usually pale blue. Legume broadly rhomboid.

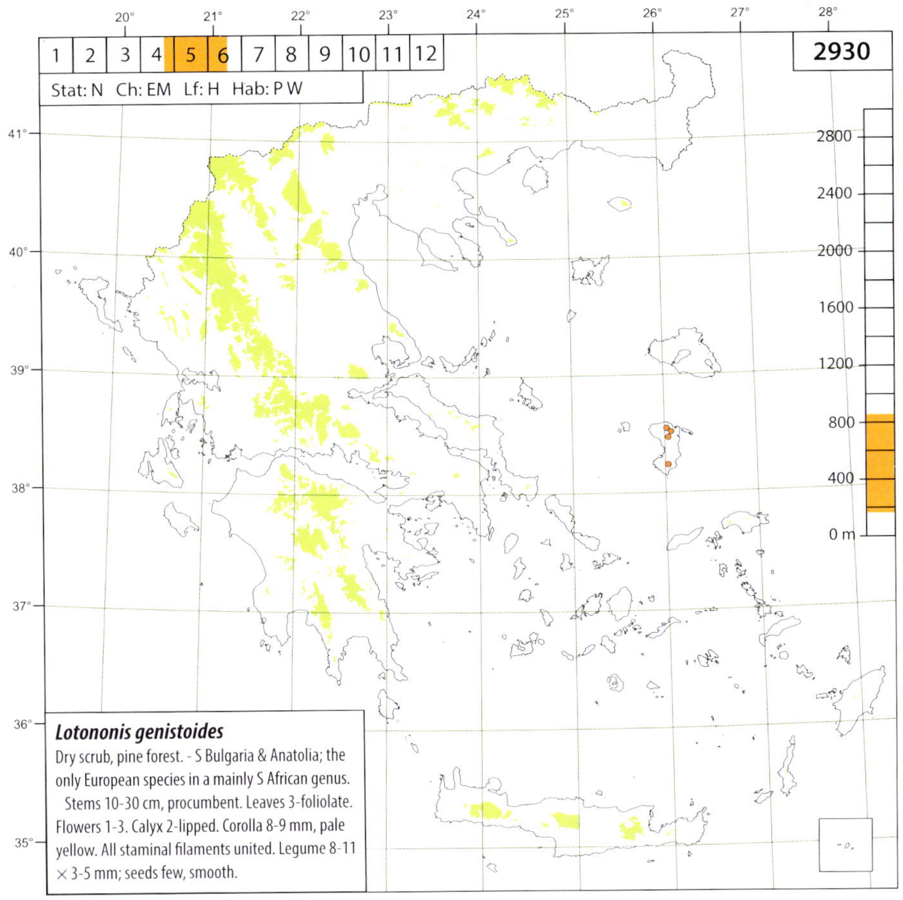

Lotononis genistoides

Dry scrub, pine forest. - S Bulgaria & Anatolia; the only European species in a mainly S African genus.

Stems 10-30 cm, procumbent. Leaves 3-foliolate. Flowers 1-3. Calyx 2-lipped. Corolla 8-9 mm, pale yellow. All staminal filaments united. Legume 8-11 × 3-5 mm; seeds few, smooth.

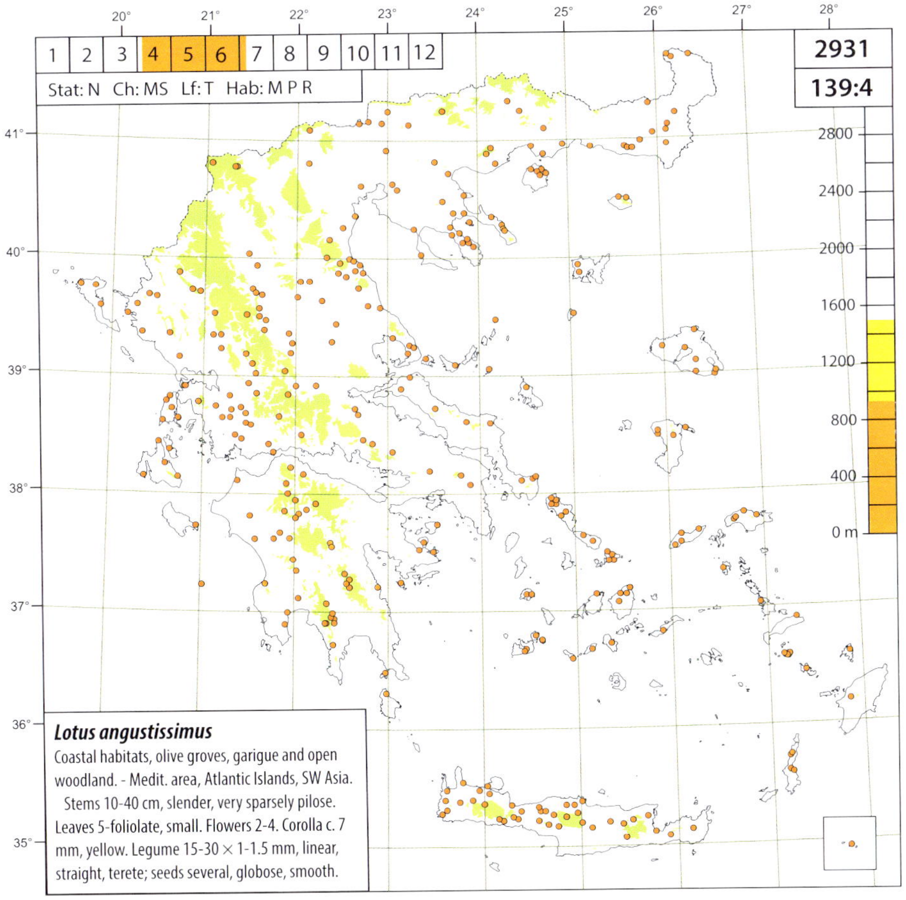

Lotus angustissimus

Coastal habitats, olive groves, garigue and open woodland. - Medit. area, Atlantic Islands, SW Asia.

Stems 10-40 cm, slender, very sparsely pilose. Leaves 5-foliolate, small. Flowers 2-4. Corolla c. 7 mm, yellow. Legume 15-30 × 1-1.5 mm, linear, straight, terete; seeds several, globose, smooth.

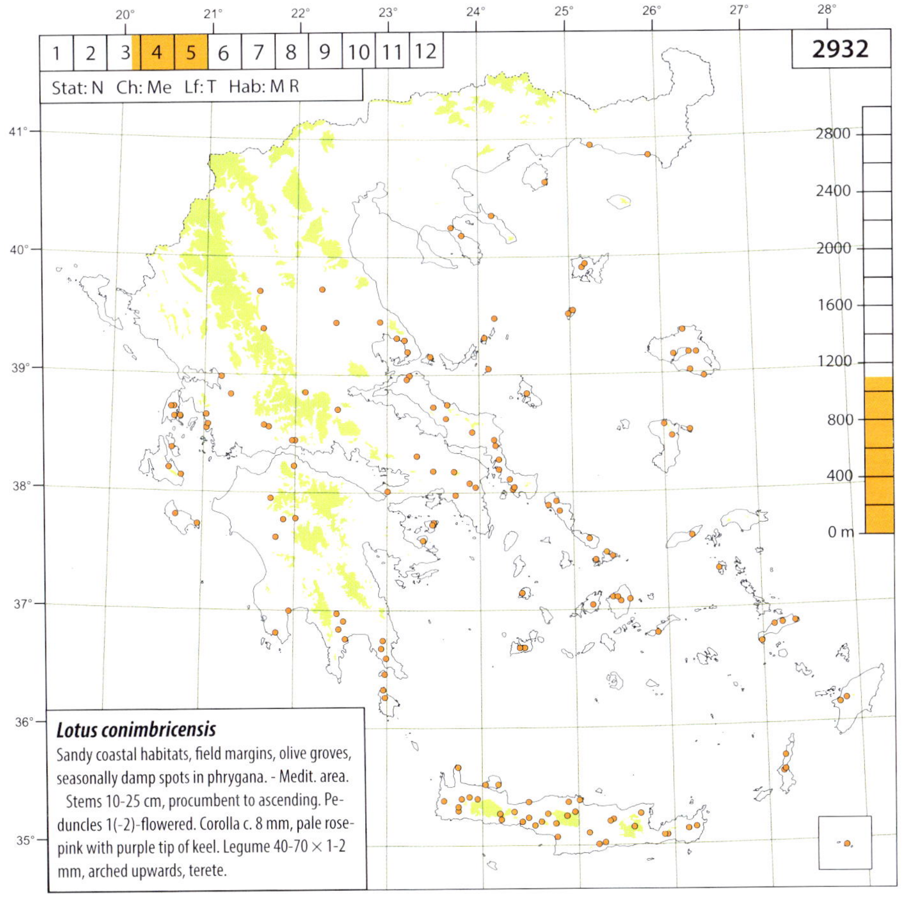

Lotus conimbricensis

Sandy coastal habitats, field margins, olive groves, seasonally damp spots in phrygana. - Medit. area.

Stems 10-25 cm, procumbent to ascending. Peduncles 1(-2)-flowered. Corolla c. 8 mm, pale rose-pink with purple tip of keel. Legume 40-70 × 1-2 mm, arched upwards, terete.

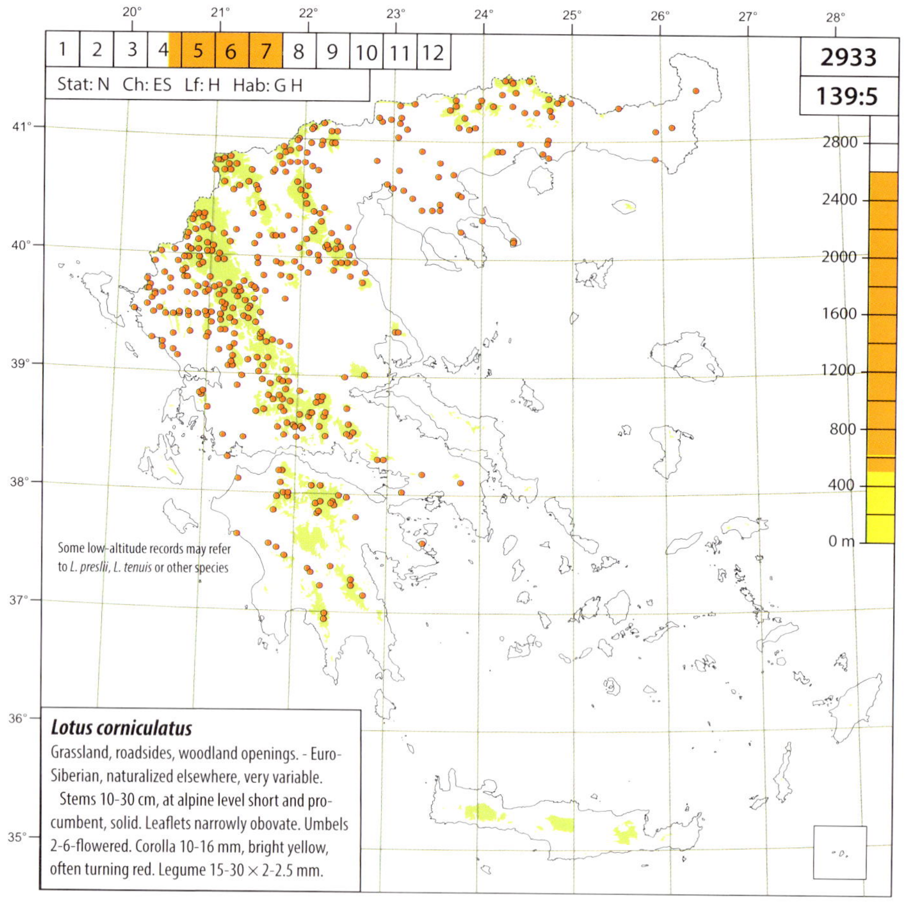

Lotus corniculatus

Grassland, roadsides, woodland openings. - Euro-Siberian, naturalized elsewhere, very variable.

Stems 10-30 cm, at alpine level short and procumbent, solid. Leaflets narrowly obovate. Umbels 2-6-flowered. Corolla 10-16 mm, bright yellow, often turning red. Legume 15-30 × 2-2.5 mm.

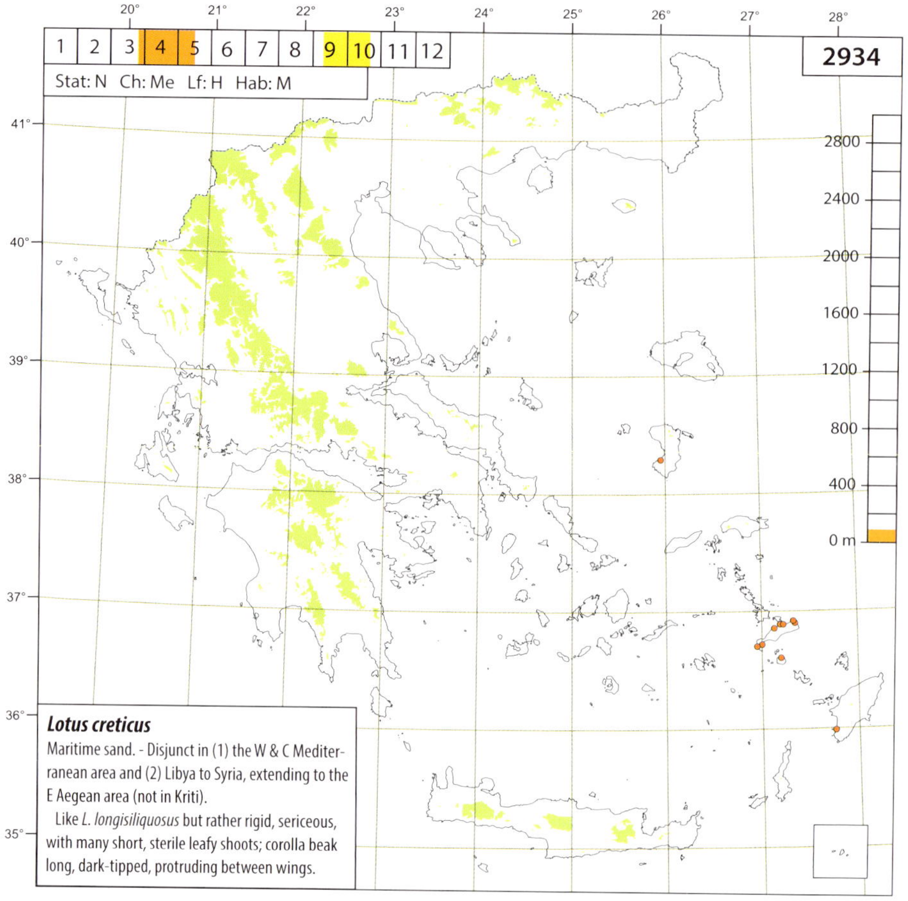

Lotus creticus

Maritime sand. - Disjunct in (1) the W & C Mediterranean area and (2) Libya to Syria, extending to the E Aegean area (not in Kriti).

Like *L. longisiliquosus* but rather rigid, sericeous, with many short, sterile leafy shoots; corolla beak long, dark-tipped, protruding between wings.

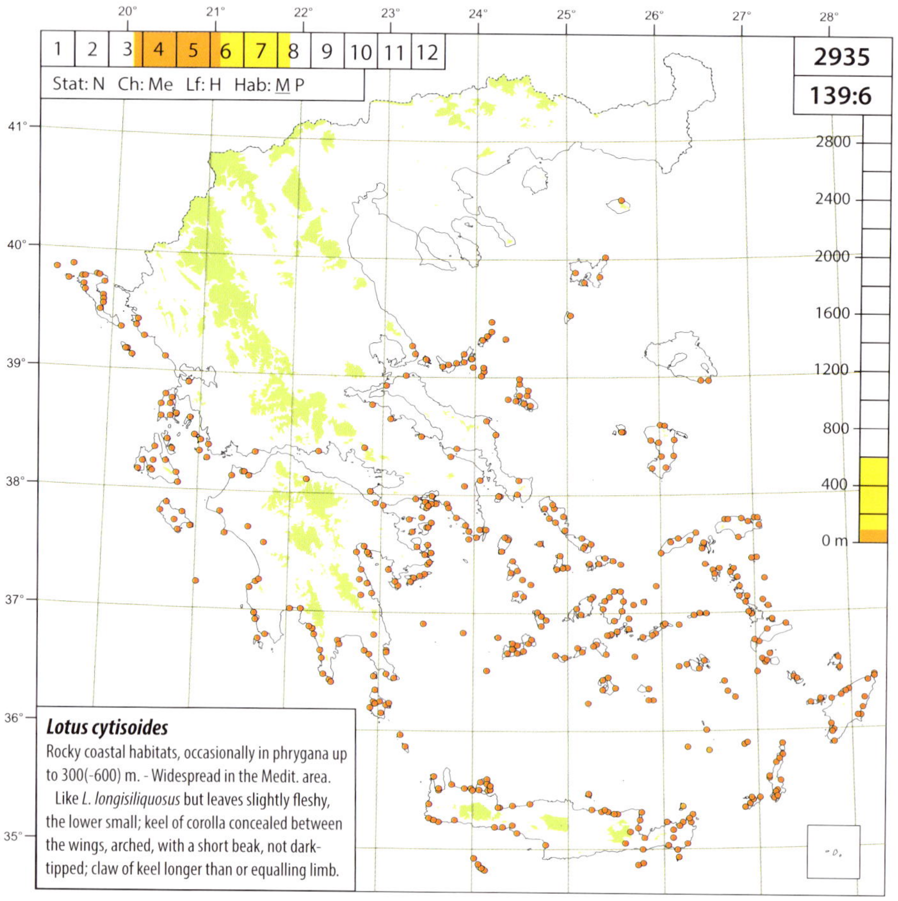

Lotus cytisoides

Rocky coastal habitats, occasionally in phrygana up to 300(-600) m. - Widespread in the Medit. area.

Like *L. longisiliquosus* but leaves slightly fleshy, the lower small; keel of corolla concealed between the wings, arched, with a short beak, not dark-tipped; claw of keel longer than or equalling limb.

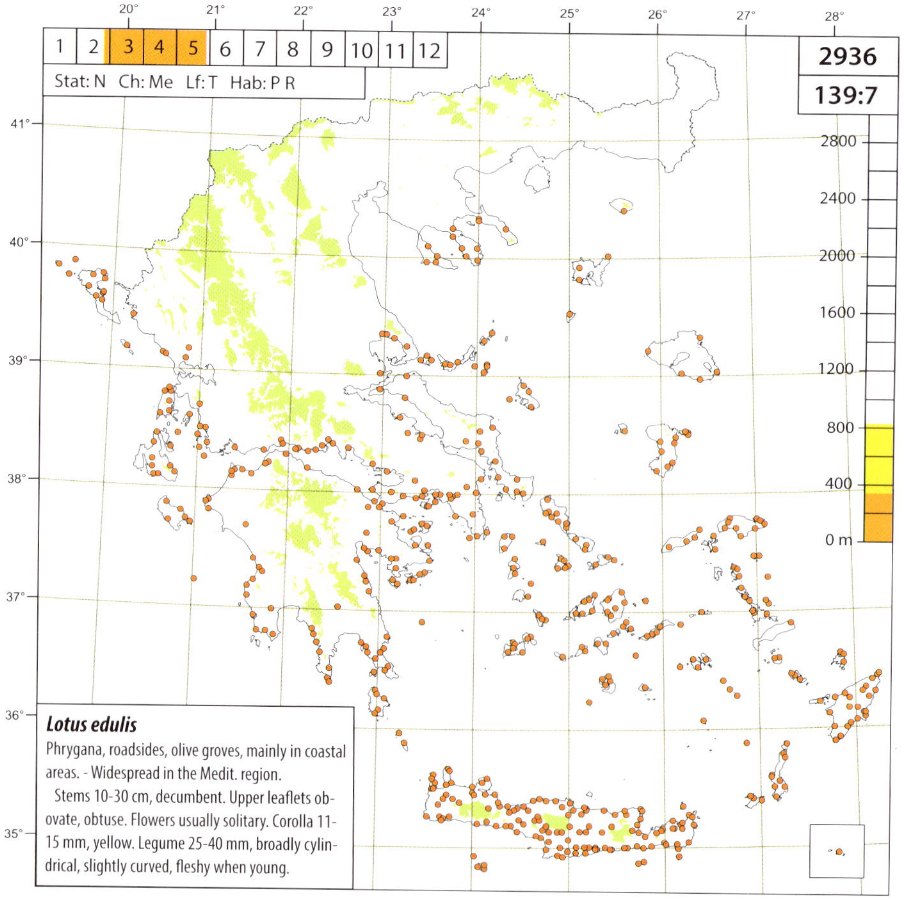

Lotus edulis

Phrygana, roadsides, olive groves, mainly in coastal areas. - Widespread in the Medit. region.

Stems 10-30 cm, decumbent. Upper leaflets obovate, obtuse. Flowers usually solitary. Corolla 11-15 mm, yellow. Legume 25-40 mm, broadly cylindrical, slightly curved, fleshy when young.

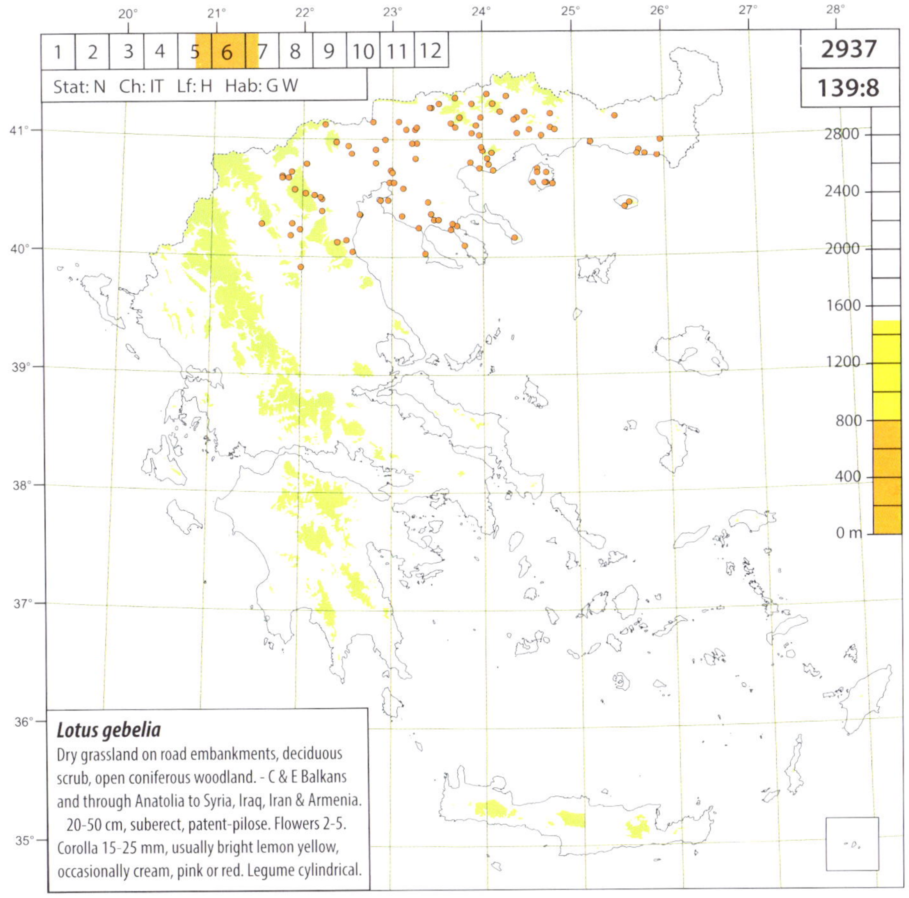

Lotus gebelia
Dry grassland on road embankments, deciduous scrub, open coniferous woodland. - C & E Balkans and through Anatolia to Syria, Iraq, Iran & Armenia.
20-50 cm, suberect, patent-pilose. Flowers 2-5. Corolla 15-25 mm, usually bright lemon yellow, occasionally cream, pink or red. Legume cylindrical.

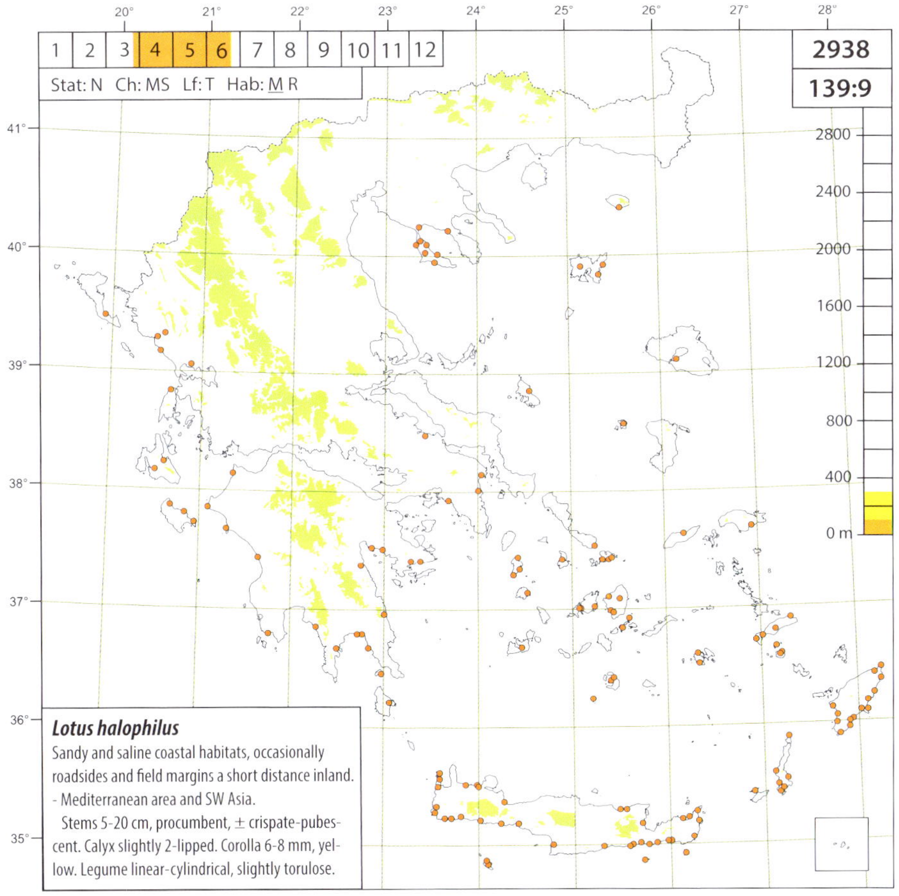

Lotus halophilus
Sandy and saline coastal habitats, occasionally roadsides and field margins a short distance inland. - Mediterranean area and SW Asia.
Stems 5-20 cm, procumbent, ± crispate-pubescent. Calyx slightly 2-lipped. Corolla 6-8 mm, yellow. Legume linear-cylindrical, slightly torulose.

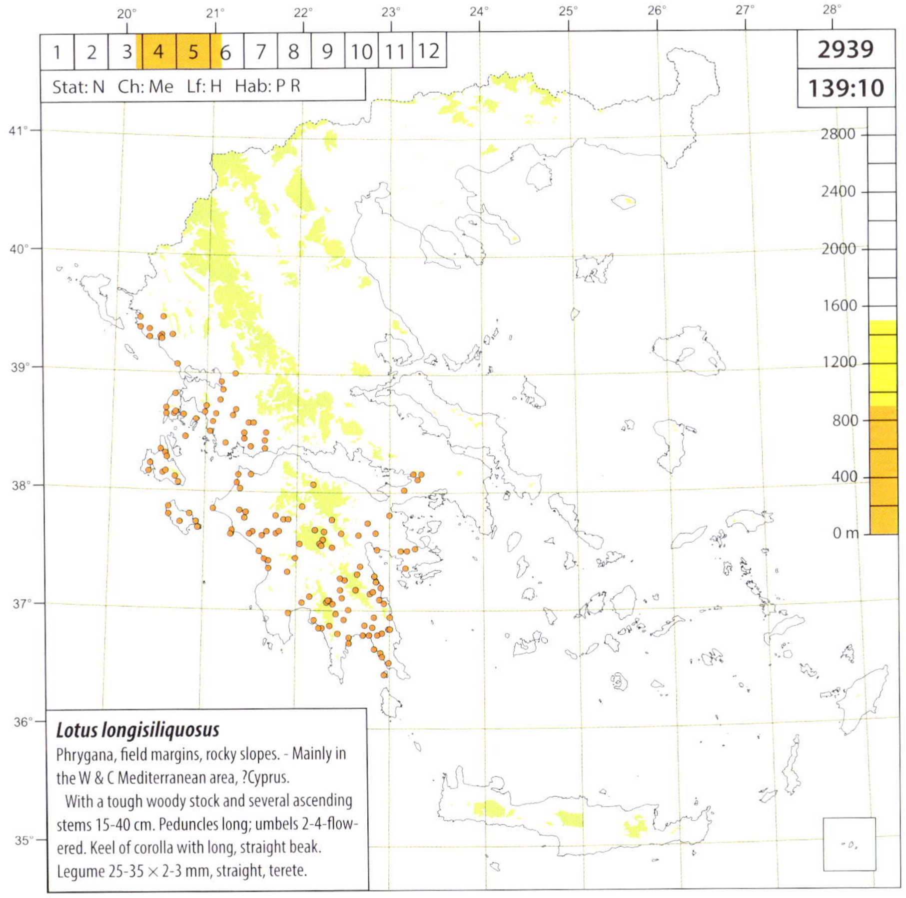

Lotus longisiliquosus
Phrygana, field margins, rocky slopes. - Mainly in the W & C Mediterranean area, ?Cyprus.
With a tough woody stock and several ascending stems 15-40 cm. Peduncles long; umbels 2-4-flowered. Keel of corolla with long, straight beak. Legume 25-35 × 2-3 mm, straight, terete.

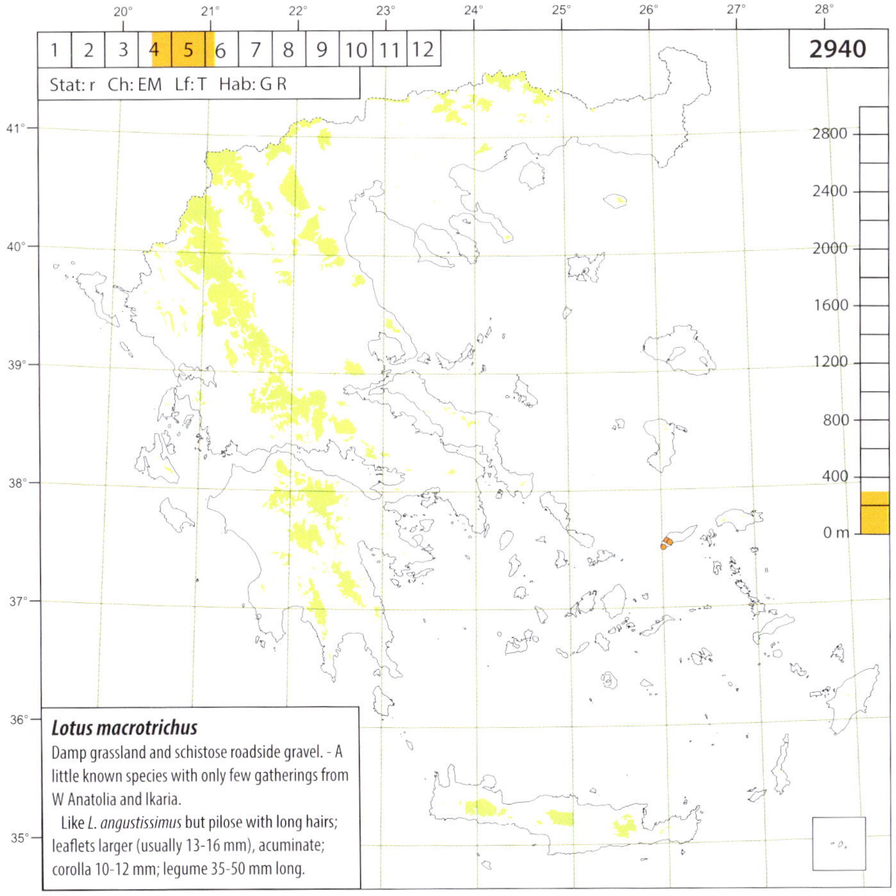

Lotus macrotrichus
Damp grassland and schistose roadside gravel. - A little known species with only few gatherings from W Anatolia and Ikaria.
Like *L. angustissimus* but pilose with long hairs; leaflets larger (usually 13-16 mm), acuminate; corolla 10-12 mm; legume 35-50 mm long.

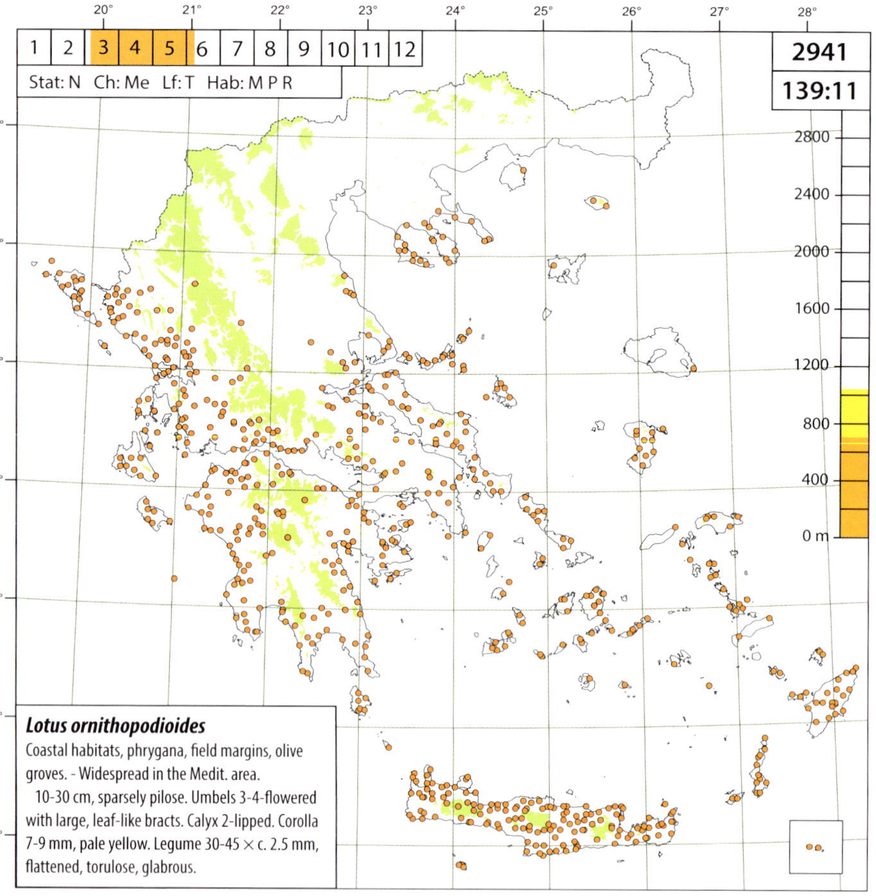

Lotus ornithopodioides
Coastal habitats, phrygana, field margins, olive groves. - Widespread in the Medit. area.
10-30 cm, sparsely pilose. Umbels 3-4-flowered with large, leaf-like bracts. Calyx 2-lipped. Corolla 7-9 mm, pale yellow. Legume 30-45 × c. 2.5 mm, flattened, torulose, glabrous.

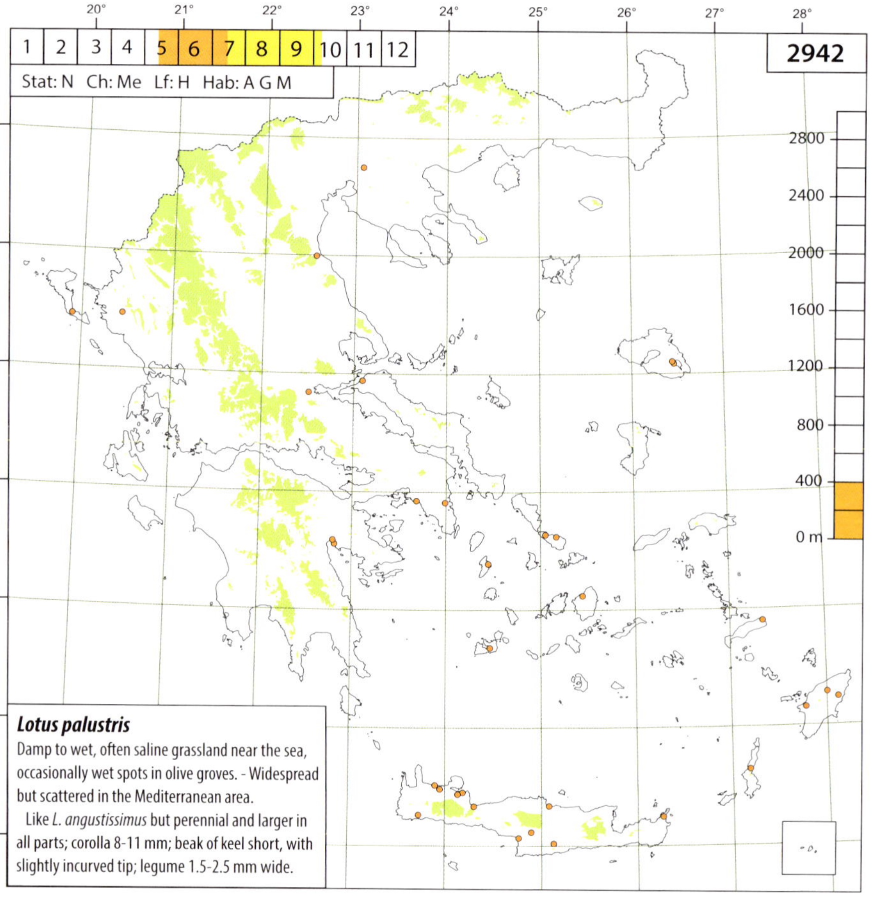

Lotus palustris
Damp to wet, often saline grassland near the sea, occasionally wet spots in olive groves. - Widespread but scattered in the Mediterranean area.
Like *L. angustissimus* but perennial and larger in all parts; corolla 8-11 mm; beak of keel short, with slightly incurved tip; legume 1.5-2.5 mm wide.

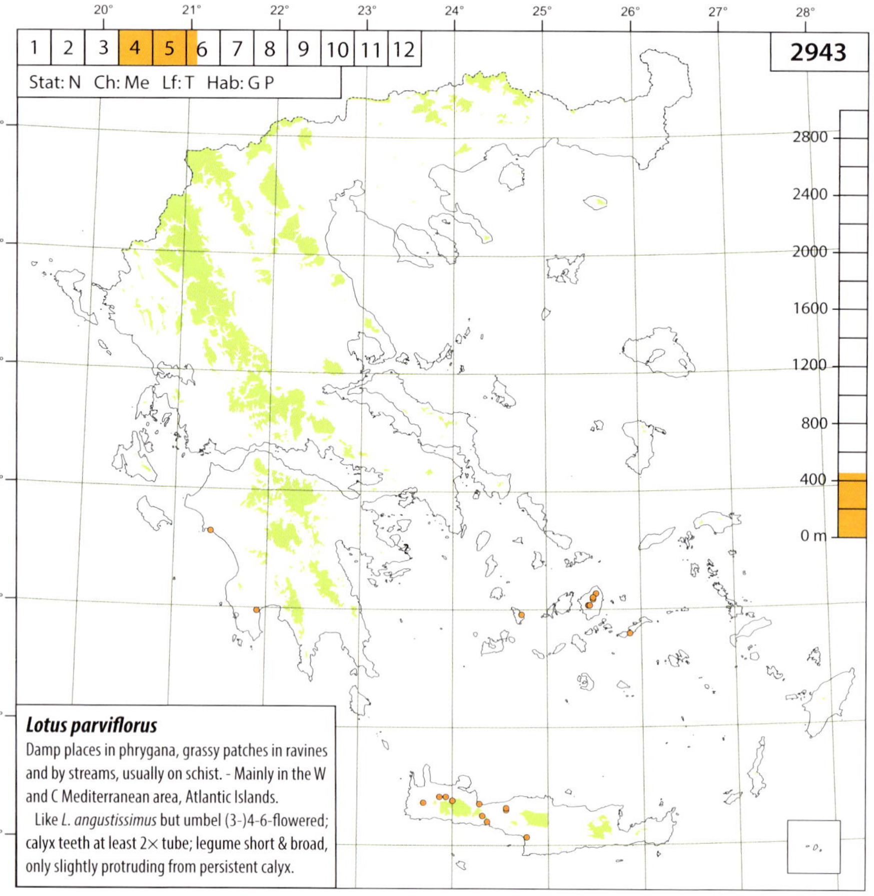

Lotus parviflorus
Damp places in phrygana, grassy patches in ravines and by streams, usually on schist. - Mainly in the W and C Mediterranean area, Atlantic Islands.
Like *L. angustissimus* but umbel (3-)4-6-flowered; calyx teeth at least 2× tube; legume short & broad, only slightly protruding from persistent calyx.

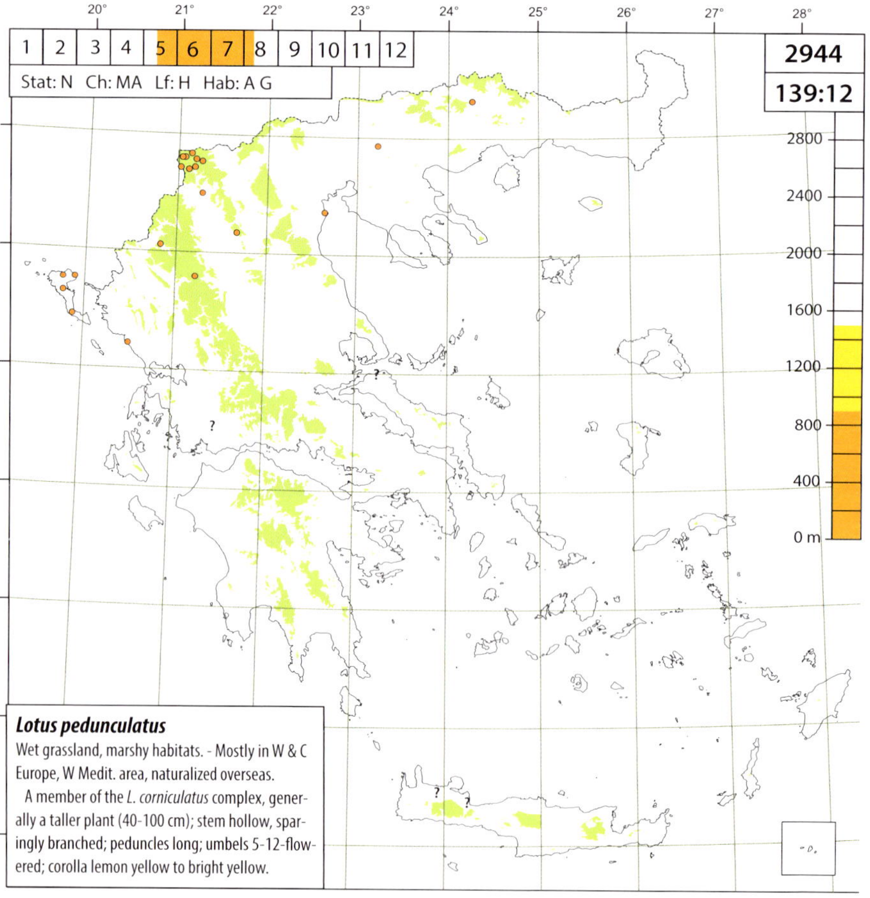

Lotus pedunculatus
Wet grassland, marshy habitats. - Mostly in W & C Europe, W Medit. area, naturalized overseas.
A member of the *L. corniculatus* complex, generally a taller plant (40-100 cm); stem hollow, sparingly branched; peduncles long; umbels 5-12-flowered; corolla lemon yellow to bright yellow.

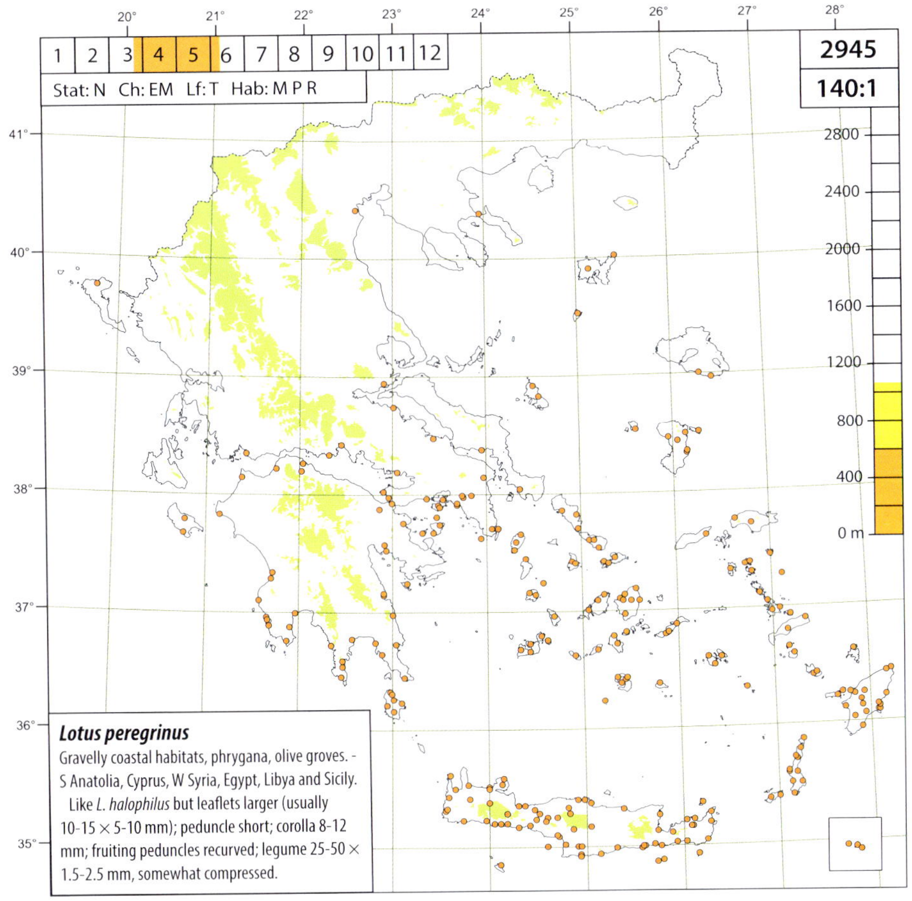

Lotus peregrinus
Gravelly coastal habitats, phrygana, olive groves. - S Anatolia, Cyprus, W Syria, Egypt, Libya and Sicily.
Like *L. halophilus* but leaflets larger (usually 10-15 × 5-10 mm); peduncle short; corolla 8-12 mm; fruiting peduncles recurved; legume 25-50 × 1.5-2.5 mm, somewhat compressed.

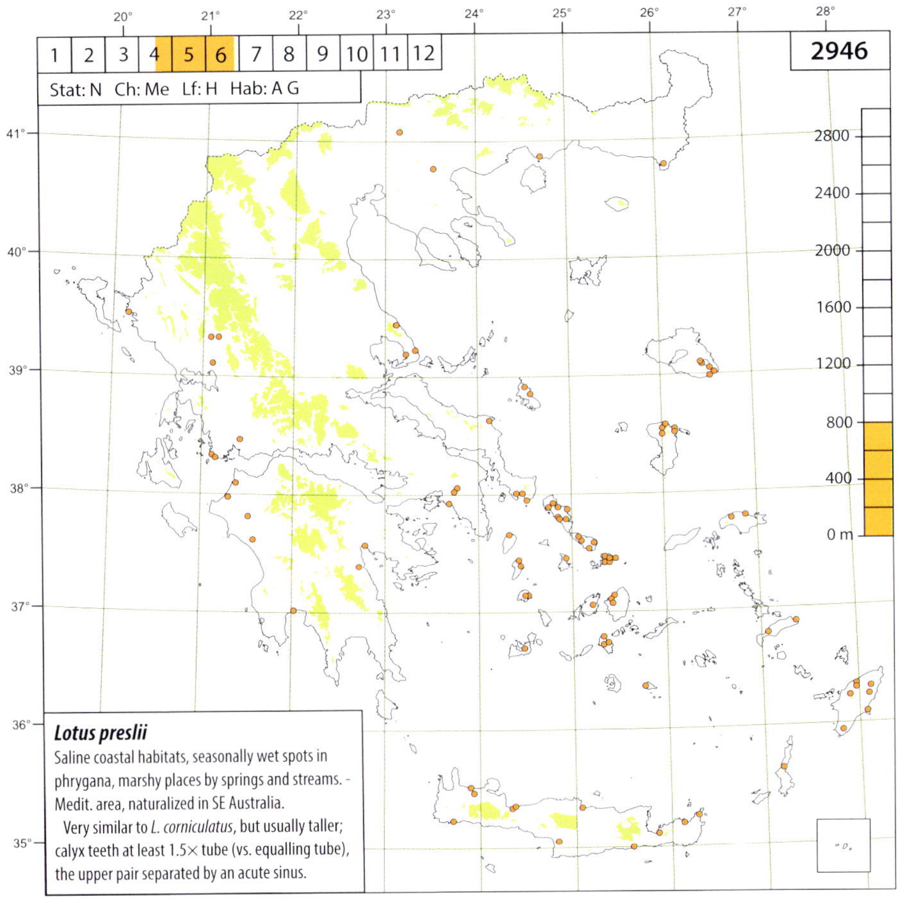

Lotus preslii
Saline coastal habitats, seasonally wet spots in phrygana, marshy places by springs and streams. - Medit. area, naturalized in SE Australia.
Very similar to *L. corniculatus*, but usually taller; calyx teeth at least 1.5× tube (vs. equalling tube), the upper pair separated by an acute sinus.

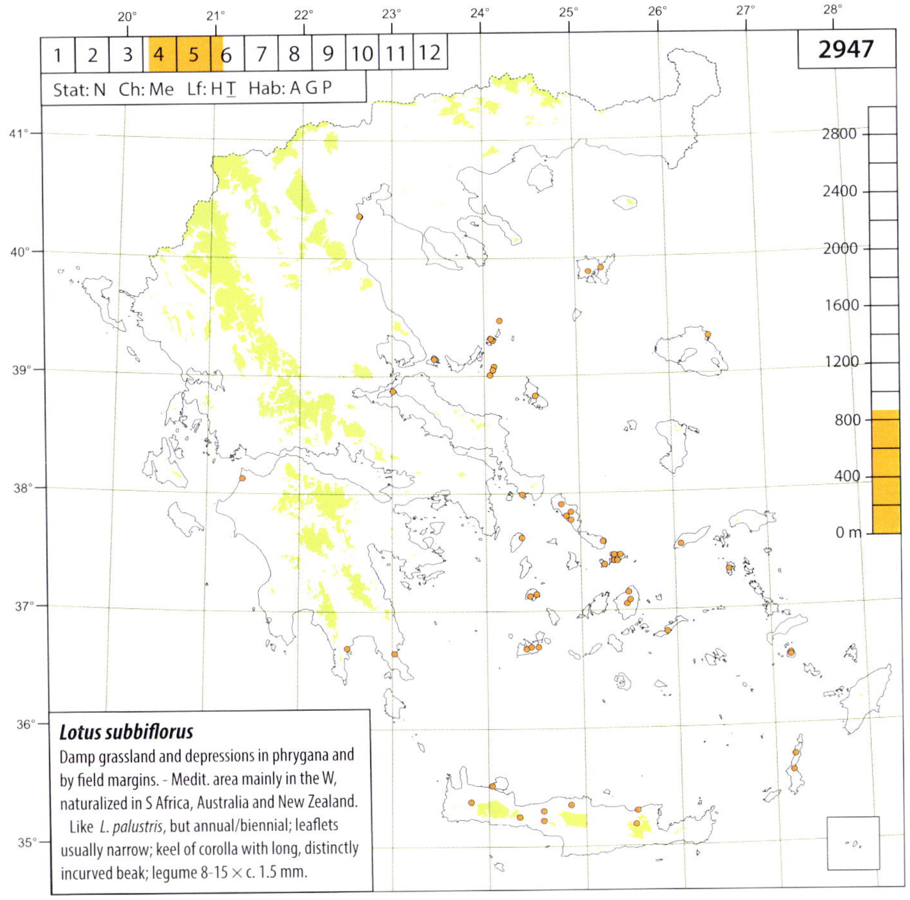

Lotus subbiflorus
Damp grassland and depressions in phrygana and by field margins. - Medit. area mainly in the W, naturalized in S Africa, Australia and New Zealand.
Like *L. palustris*, but annual/biennial; leaflets usually narrow; keel of corolla with long, distinctly incurved beak; legume 8-15 × c. 1.5 mm.

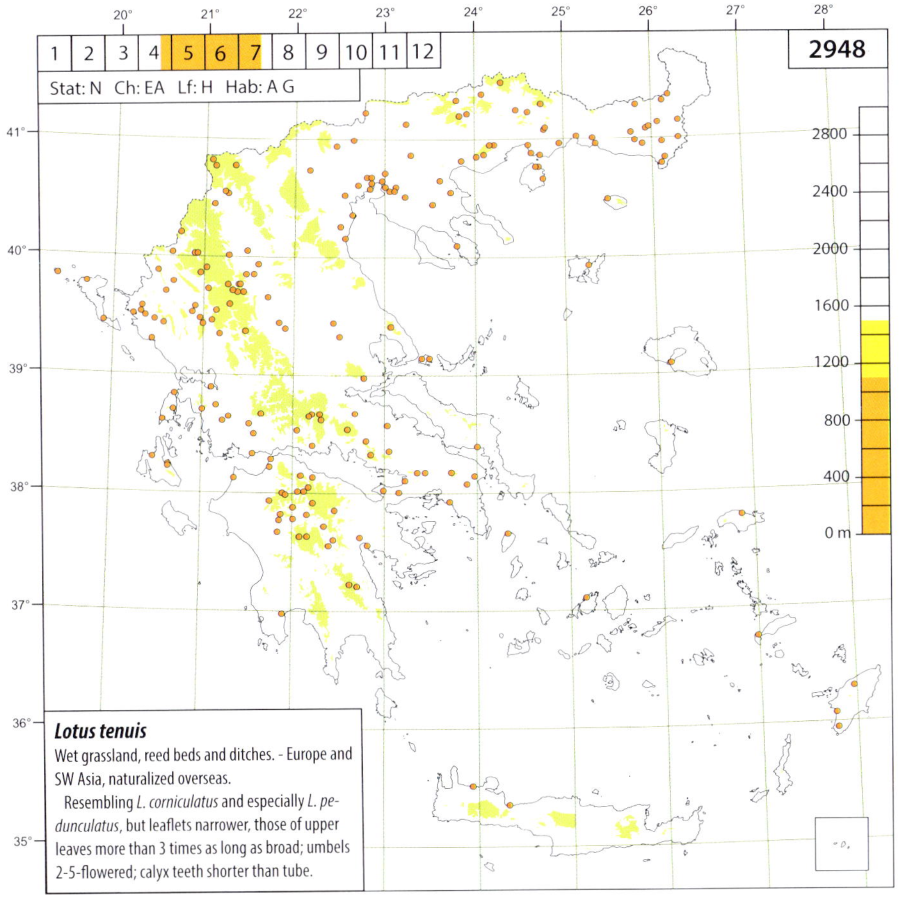

Lotus tenuis
Wet grassland, reed beds and ditches. - Europe and SW Asia, naturalized overseas.
Resembling *L. corniculatus* and especially *L. pedunculatus*, but leaflets narrower, those of upper leaves more than 3 times as long as broad; umbels 2-5-flowered; calyx teeth shorter than tube.

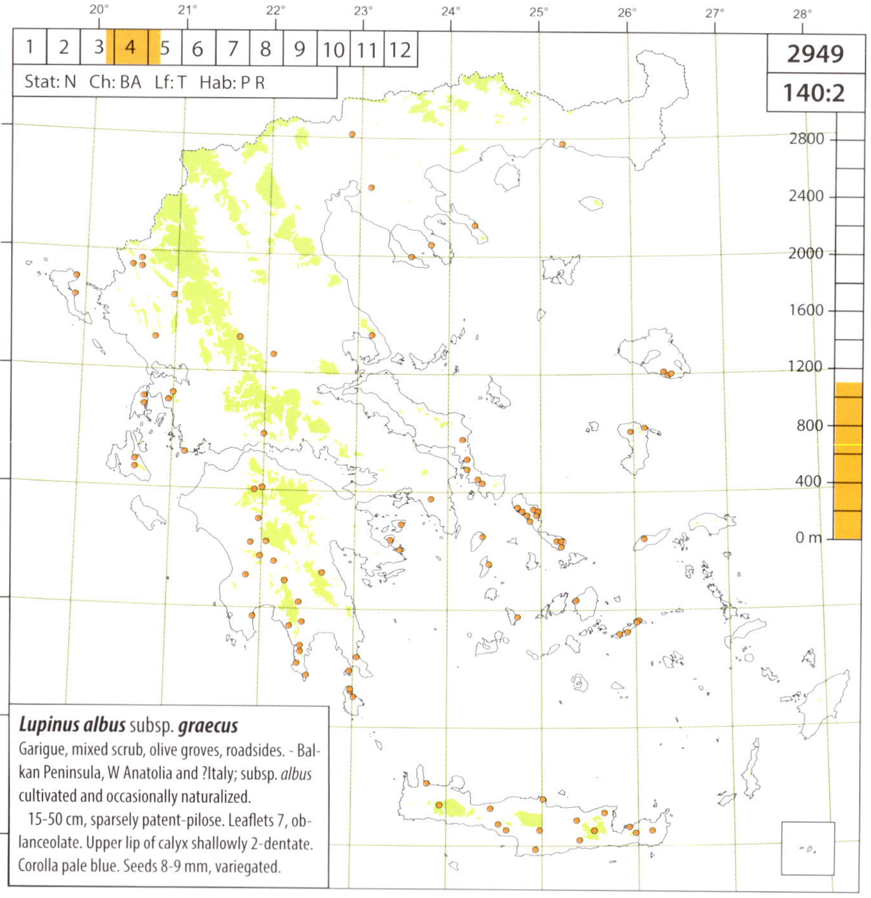

Lupinus albus subsp. ***graecus***
Garigue, mixed scrub, olive groves, roadsides. - Balkan Peninsula, W Anatolia and ?Italy; subsp. *albus* cultivated and occasionally naturalized.
15-50 cm, sparsely patent-pilose. Leaflets 7, oblanceolate. Upper lip of calyx shallowly 2-dentate. Corolla pale blue. Seeds 8-9 mm, variegated.

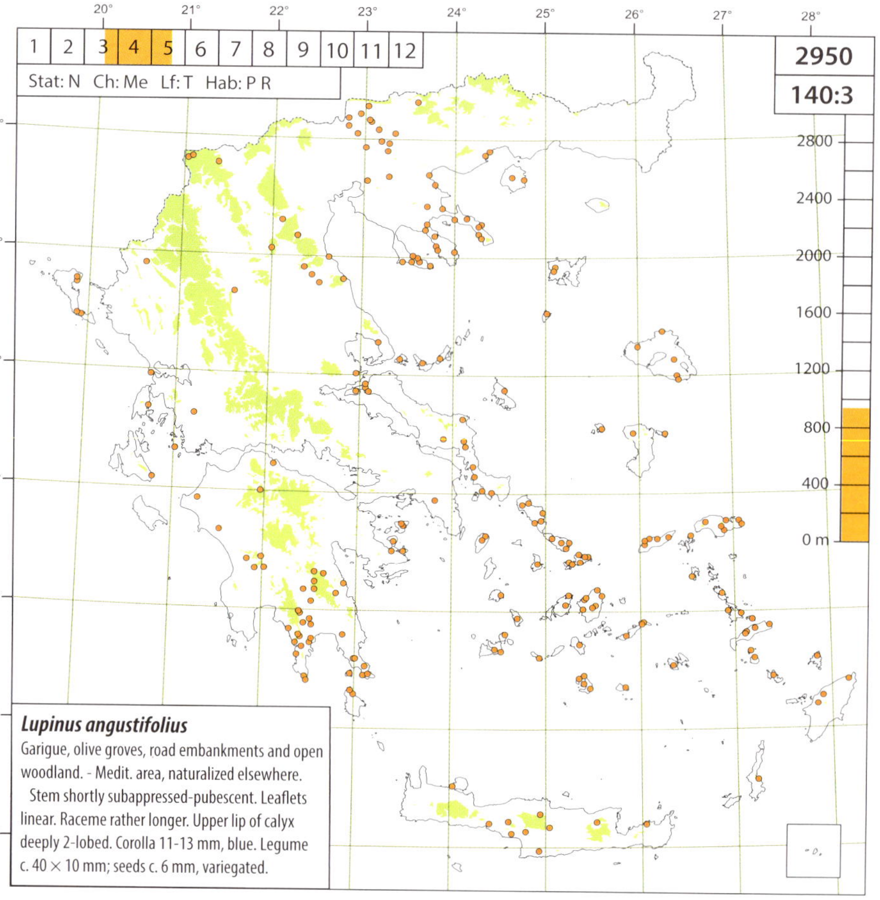

Lupinus angustifolius
Garigue, olive groves, road embankments and open woodland. - Medit. area, naturalized elsewhere.
Stem shortly subappressed-pubescent. Leaflets linear. Raceme rather longer. Upper lip of calyx deeply 2-lobed. Corolla 11-13 mm, blue. Legume c. 40 × 10 mm; seeds c. 6 mm, variegated.

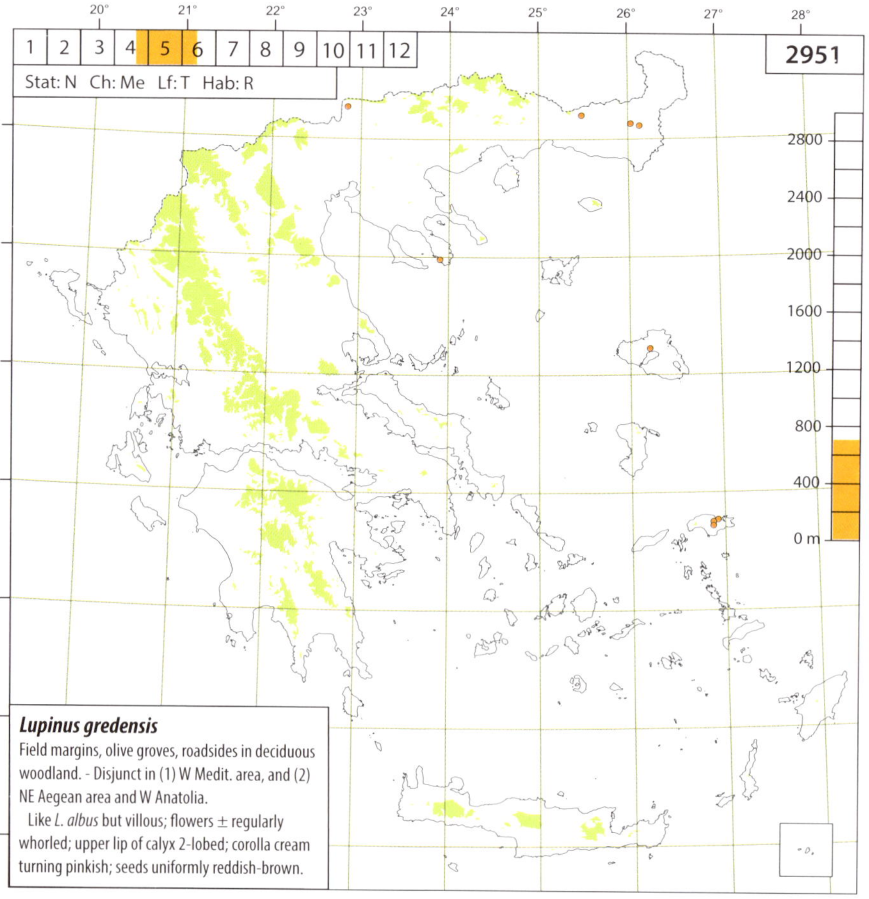

Lupinus gredensis
Field margins, olive groves, roadsides in deciduous woodland. - Disjunct in (1) W Medit. area, and (2) NE Aegean area and W Anatolia.
Like *L. albus* but villous; flowers ± regularly whorled; upper lip of calyx 2-lobed; corolla cream turning pinkish; seeds uniformly reddish-brown.

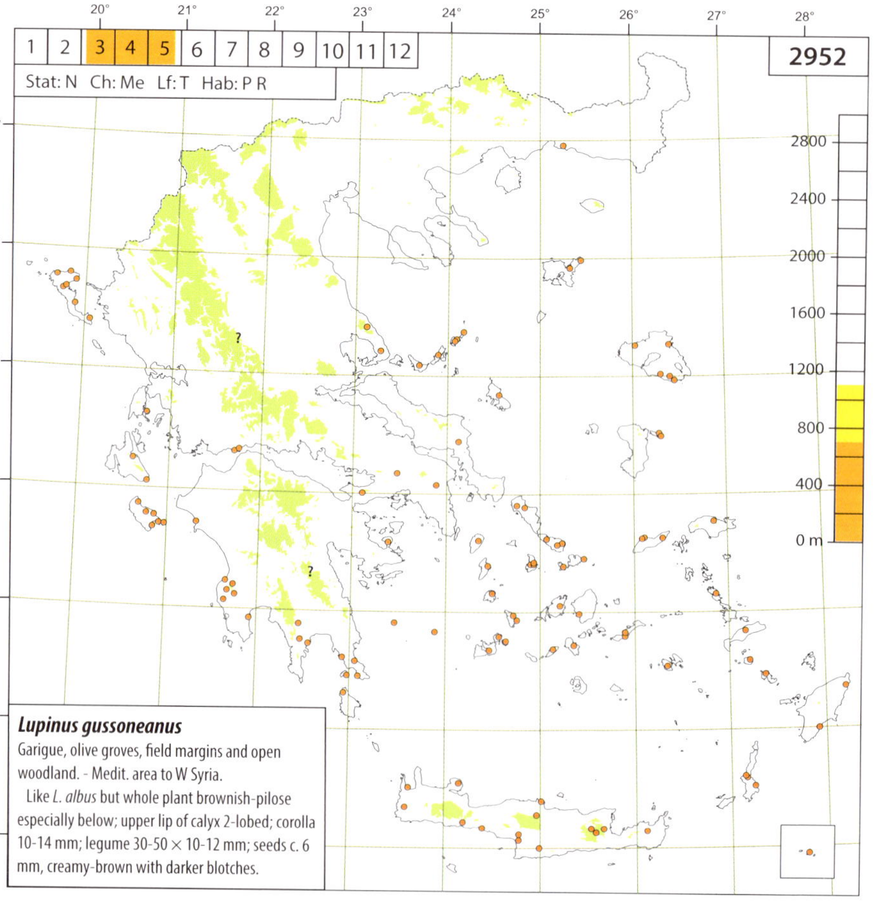

Lupinus gussoneanus
Garigue, olive groves, field margins and open woodland. - Medit. area to W Syria.
Like *L. albus* but whole plant brownish-pilose especially below; upper lip of calyx 2-lobed; corolla 10-14 mm; legume 30-50 × 10-12 mm; seeds c. 6 mm, creamy-brown with darker blotches.

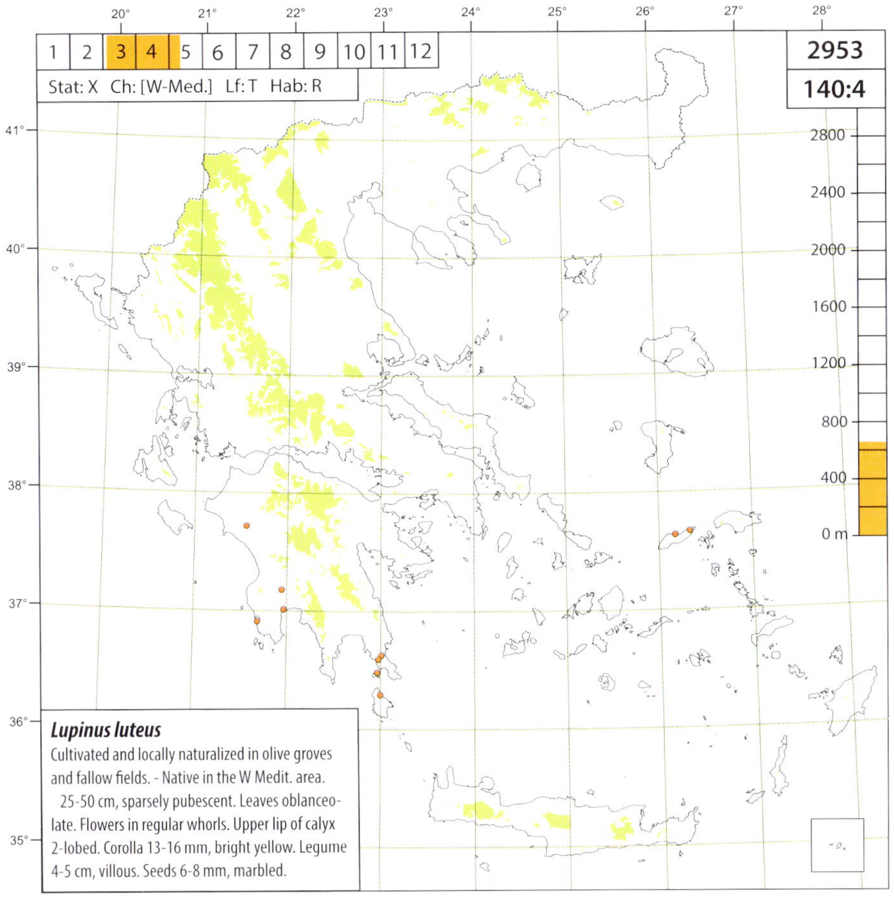

Lupinus luteus

Cultivated and locally naturalized in olive groves and fallow fields. - Native in the W Medit. area.

25-50 cm, sparsely pubescent. Leaves oblanceolate. Flowers in regular whorls. Upper lip of calyx 2-lobed. Corolla 13-16 mm, bright yellow. Legume 4-5 cm, villous. Seeds 6-8 mm, marbled.

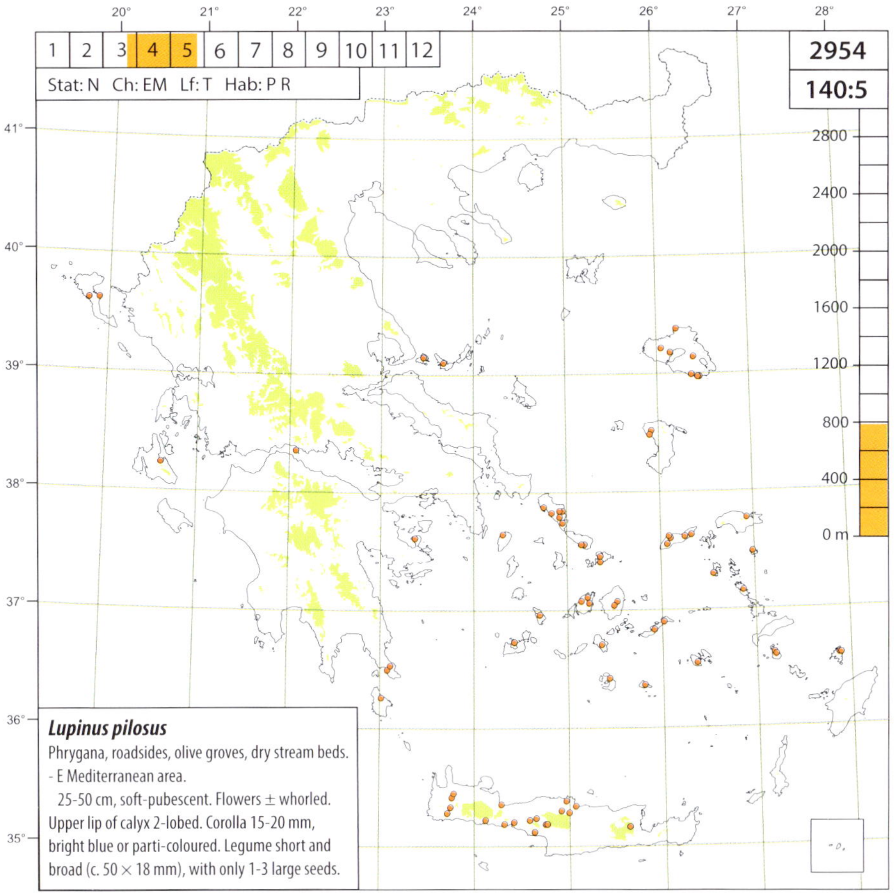

Lupinus pilosus

Phrygana, roadsides, olive groves, dry stream beds. - E Mediterranean area.

25-50 cm, soft-pubescent. Flowers ± whorled. Upper lip of calyx 2-lobed. Corolla 15-20 mm, bright blue or parti-coloured. Legume short and broad (c. 50 × 18 mm), with only 1-3 large seeds.

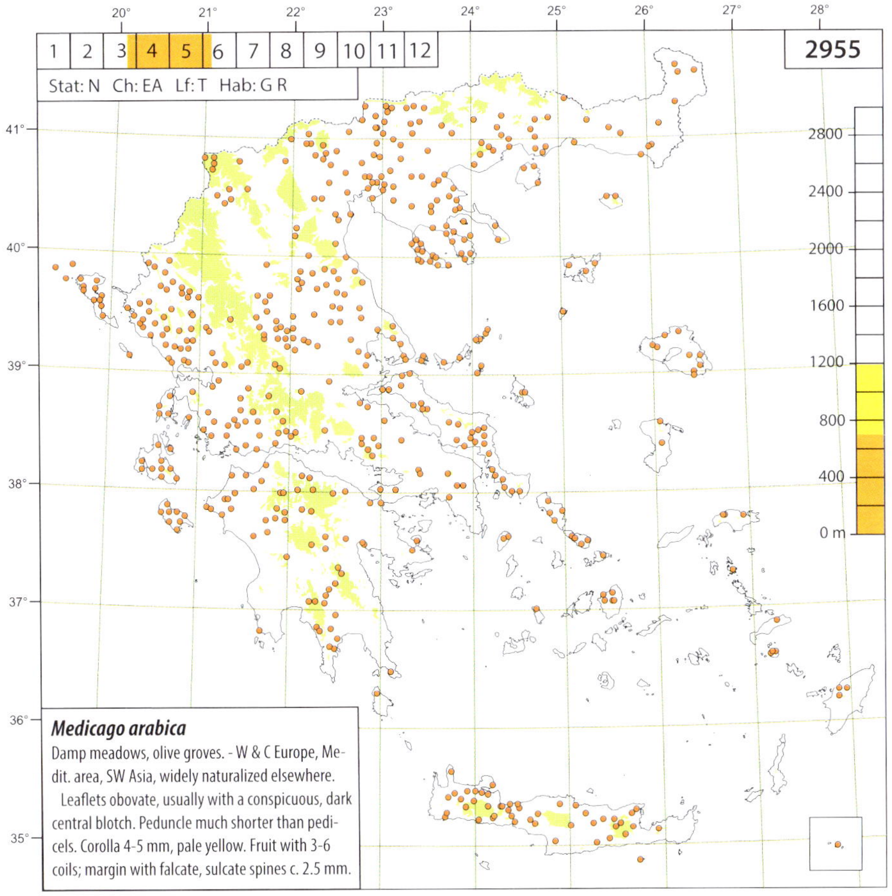

Medicago arabica

Damp meadows, olive groves. - W & C Europe, Medit. area, SW Asia, widely naturalized elsewhere.

Leaflets obovate, usually with a conspicuous, dark central blotch. Peduncle much shorter than pedicels. Corolla 4-5 mm, pale yellow. Fruit with 3-6 coils; margin with falcate, sulcate spines c. 2.5 mm.

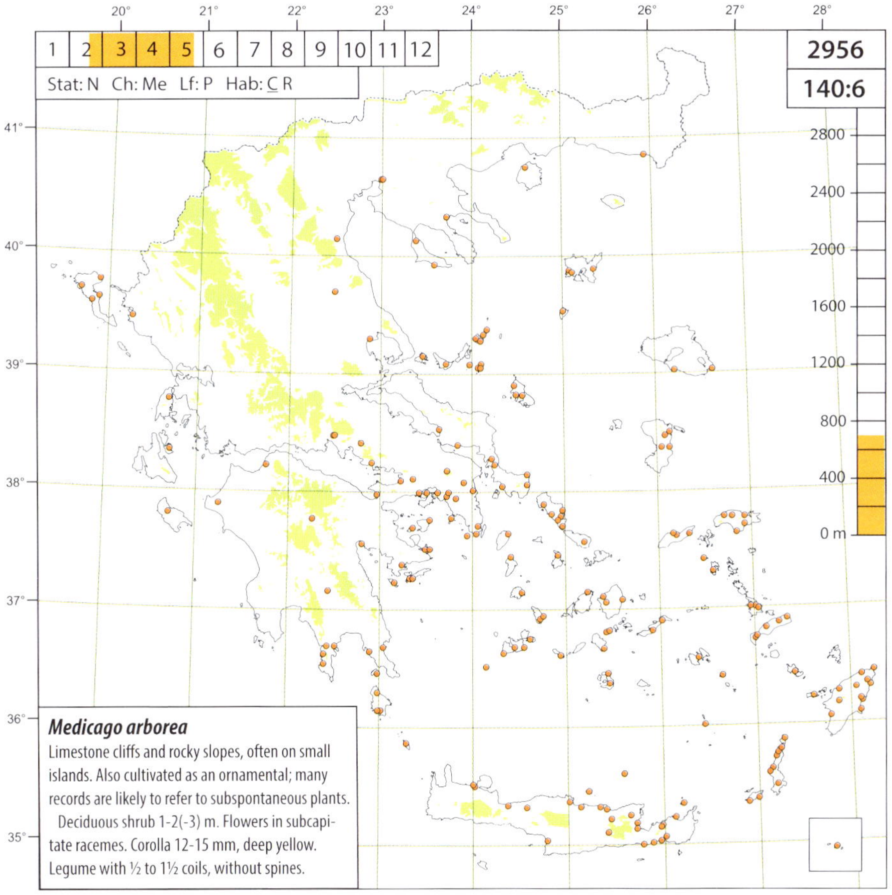

Medicago arborea

Limestone cliffs and rocky slopes, often on small islands. Also cultivated as an ornamental; many records are likely to refer to subspontaneous plants.

Deciduous shrub 1-2(-3) m. Flowers in subcapitate racemes. Corolla 12-15 mm, deep yellow. Legume with ½ to 1½ coils, without spines.

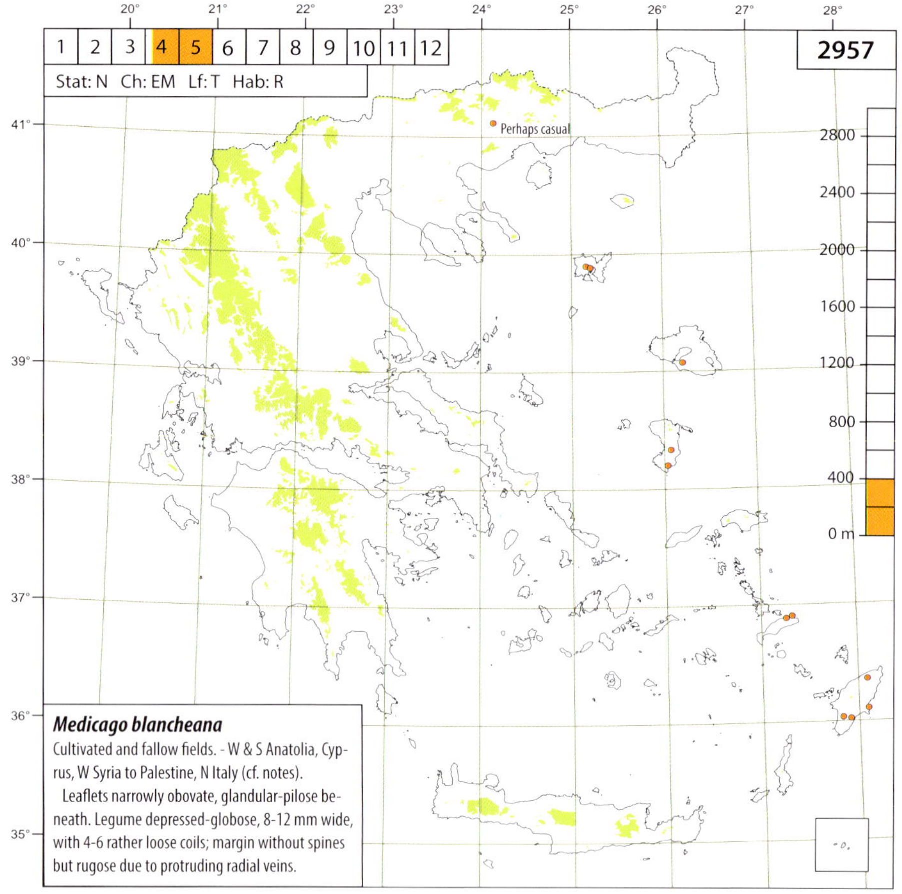

Medicago blancheana
Cultivated and fallow fields. - W & S Anatolia, Cyprus, W Syria to Palestine, N Italy (cf. notes).
Leaflets narrowly obovate, glandular-pilose beneath. Legume depressed-globose, 8-12 mm wide, with 4-6 rather loose coils; margin without spines but rugose due to protruding radial veins.

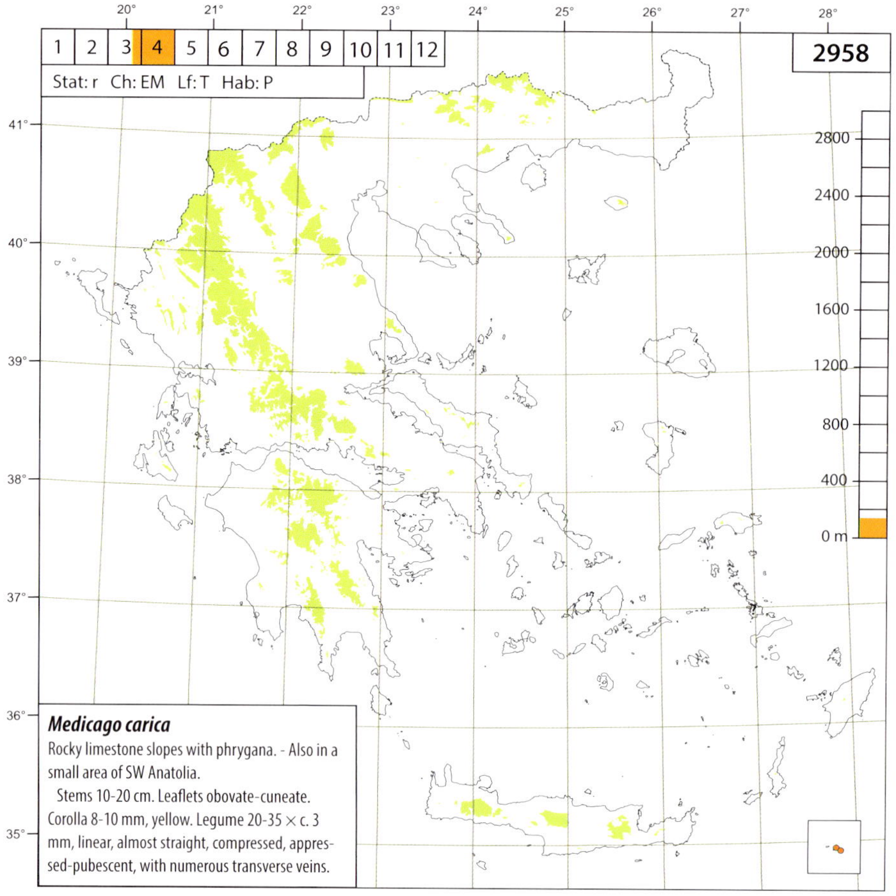

Medicago carica
Rocky limestone slopes with phrygana. - Also in a small area of SW Anatolia.
Stems 10-20 cm. Leaflets obovate-cuneate. Corolla 8-10 mm, yellow. Legume 20-35 × c. 3 mm, linear, almost straight, compressed, appressed-pubescent, with numerous transverse veins.

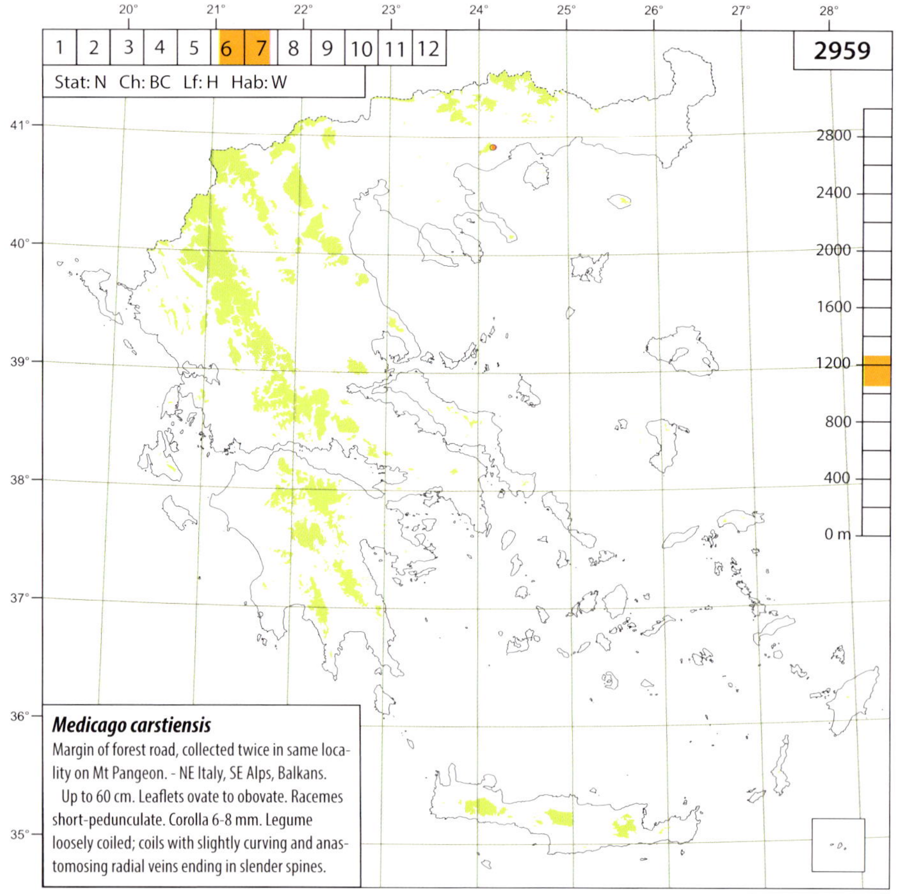

Medicago carstiensis
Margin of forest road, collected twice in same locality on Mt Pangeon. - NE Italy, SE Alps, Balkans.
Up to 60 cm. Leaflets ovate to obovate. Racemes short-pedunculate. Corolla 6-8 mm. Legume loosely coiled; coils with slightly curving and anastomosing radial veins ending in slender spines.

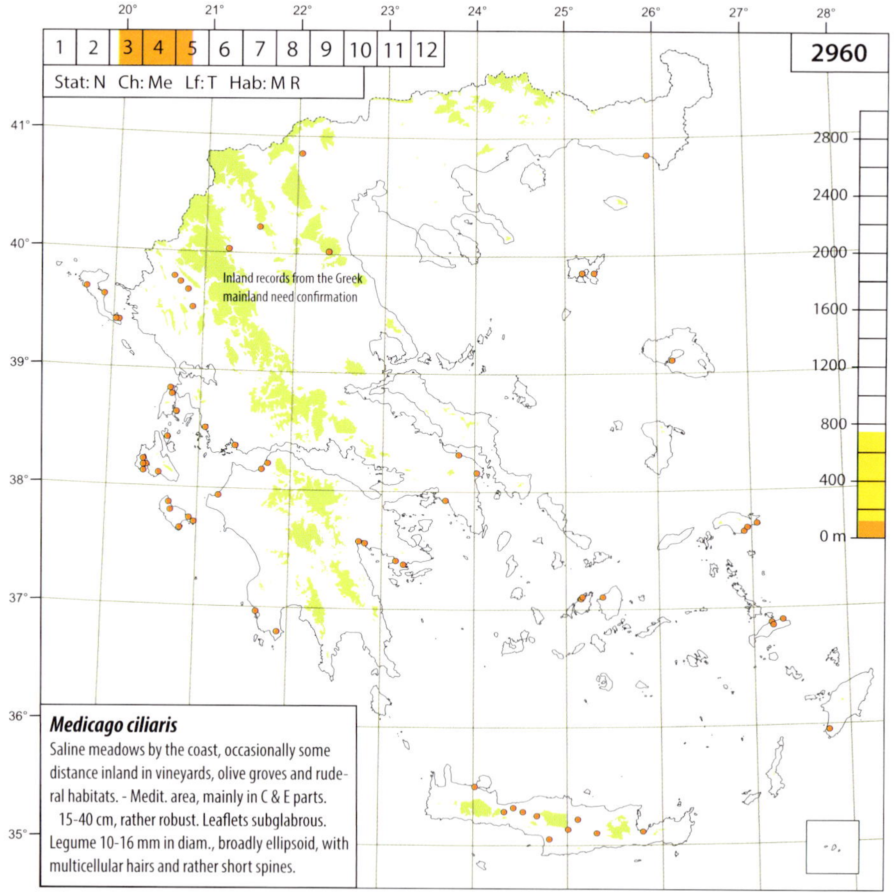

Medicago ciliaris
Saline meadows by the coast, occasionally some distance inland in vineyards, olive groves and ruderal habitats. - Medit. area, mainly in C & E parts.
15-40 cm, rather robust. Leaflets subglabrous. Legume 10-16 mm in diam., broadly ellipsoid, with multicellular hairs and rather short spines.

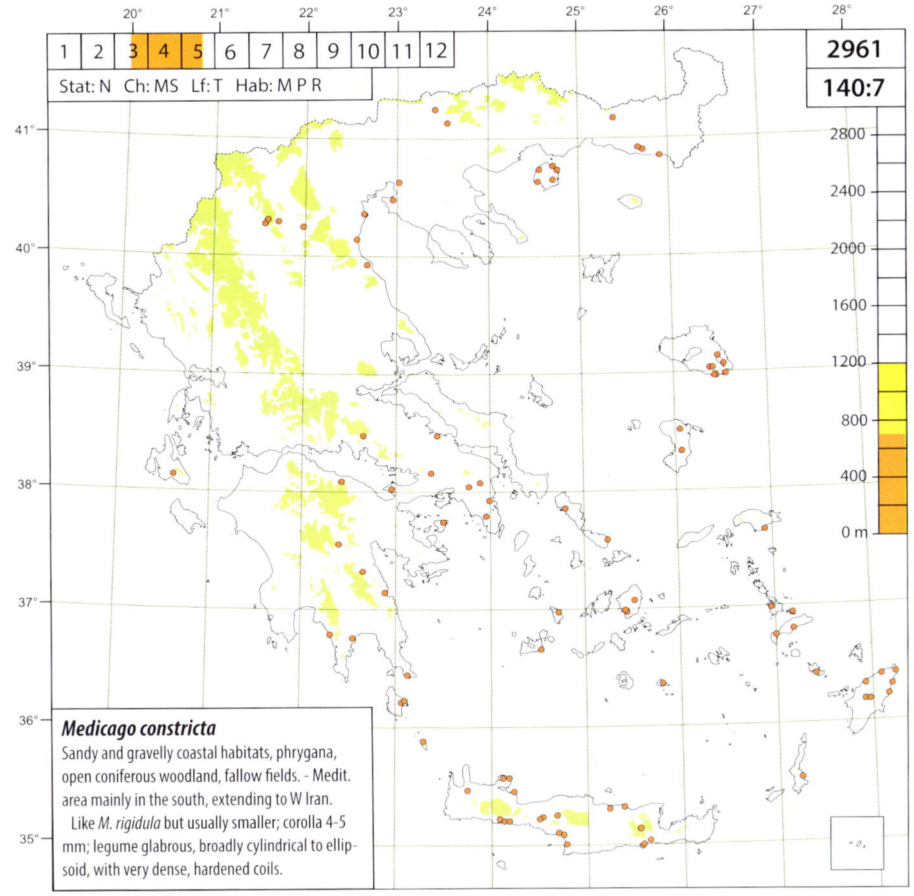

Medicago constricta
Sandy and gravelly coastal habitats, phrygana, open coniferous woodland, fallow fields. - Medit. area mainly in the south, extending to W Iran.
Like *M. rigidula* but usually smaller; corolla 4-5 mm; legume glabrous, broadly cylindrical to ellipsoid, with very dense, hardened coils.

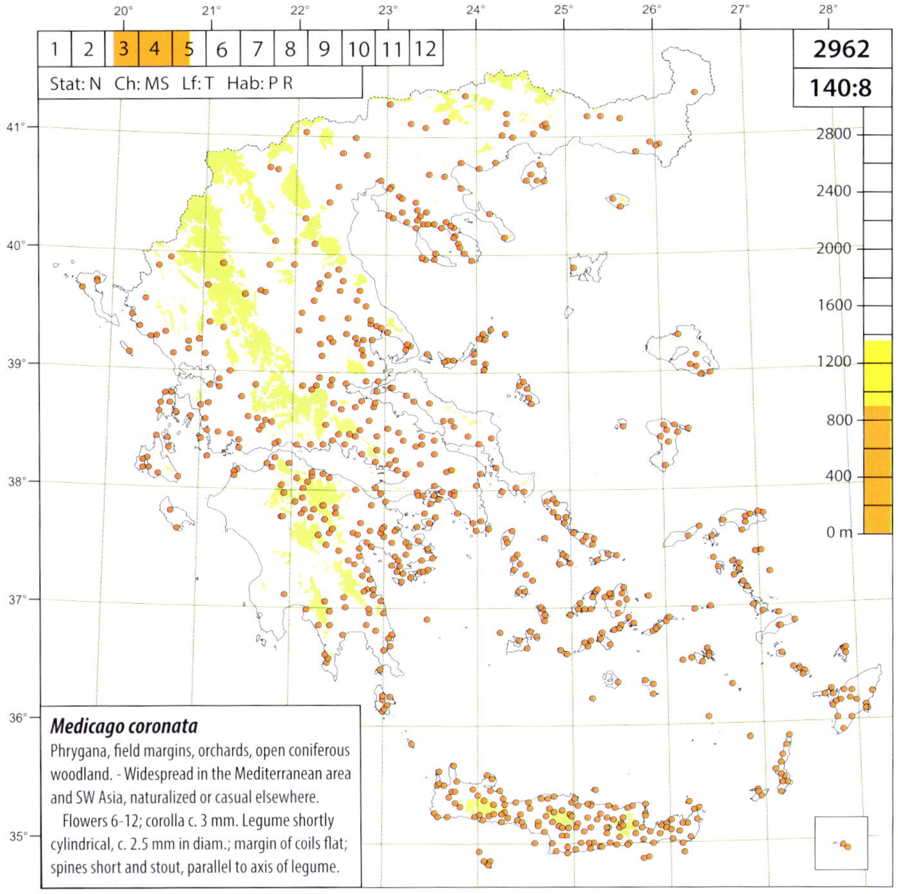

Medicago coronata
Phrygana, field margins, orchards, open coniferous woodland. - Widespread in the Mediterranean area and SW Asia, naturalized or casual elsewhere.
Flowers 6-12; corolla c. 3 mm. Legume shortly cylindrical, c. 2.5 mm in diam.; margin of coils flat; spines short and stout, parallel to axis of legume.

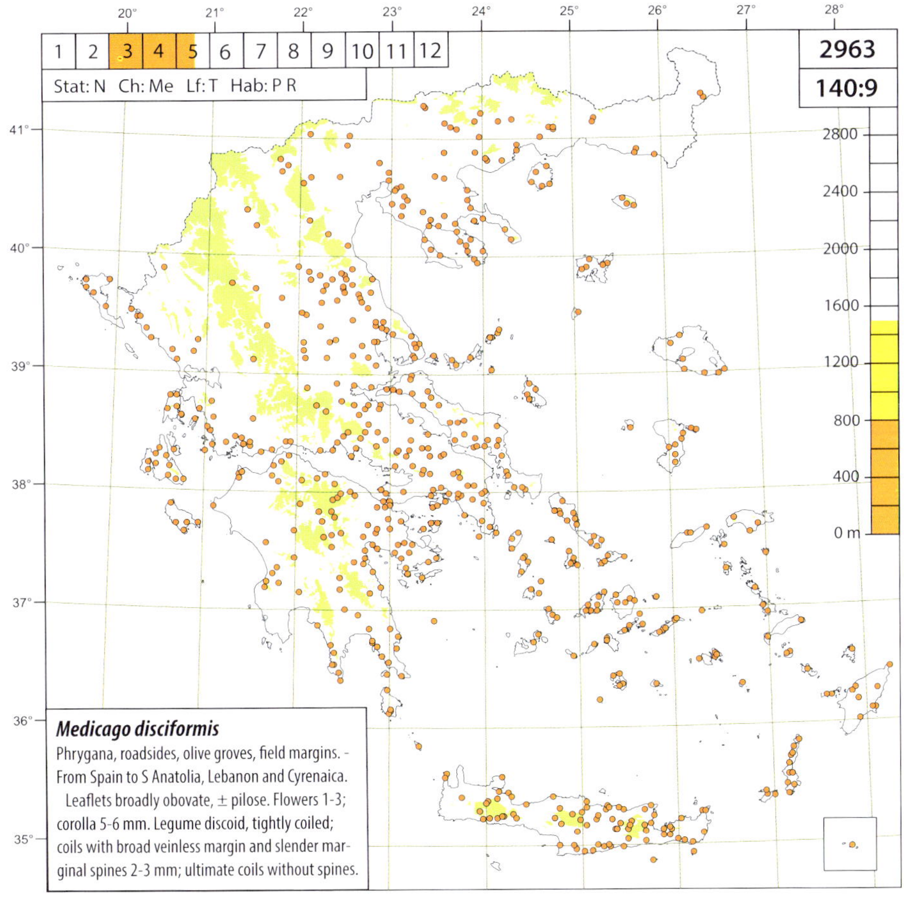

Medicago disciformis
Phrygana, roadsides, olive groves, field margins. - From Spain to S Anatolia, Lebanon and Cyrenaica.
Leaflets broadly obovate, ± pilose. Flowers 1-3; corolla 5-6 mm. Legume discoid, tightly coiled; coils with broad veinless margin and slender marginal spines 2-3 mm; ultimate coils without spines.

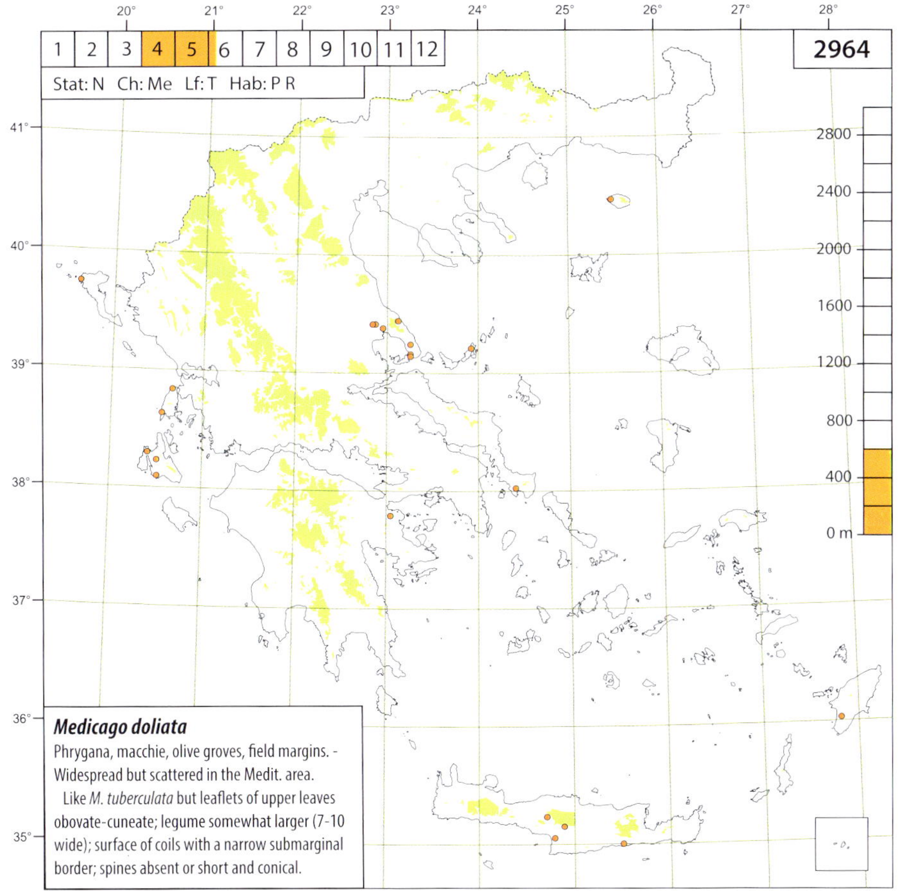

Medicago doliata
Phrygana, macchie, olive groves, field margins. - Widespread but scattered in the Medit. area.
Like *M. tuberculata* but leaflets of upper leaves obovate-cuneate; legume somewhat larger (7-10 wide); surface of coils with a narrow submarginal border; spines absent or short and conical.

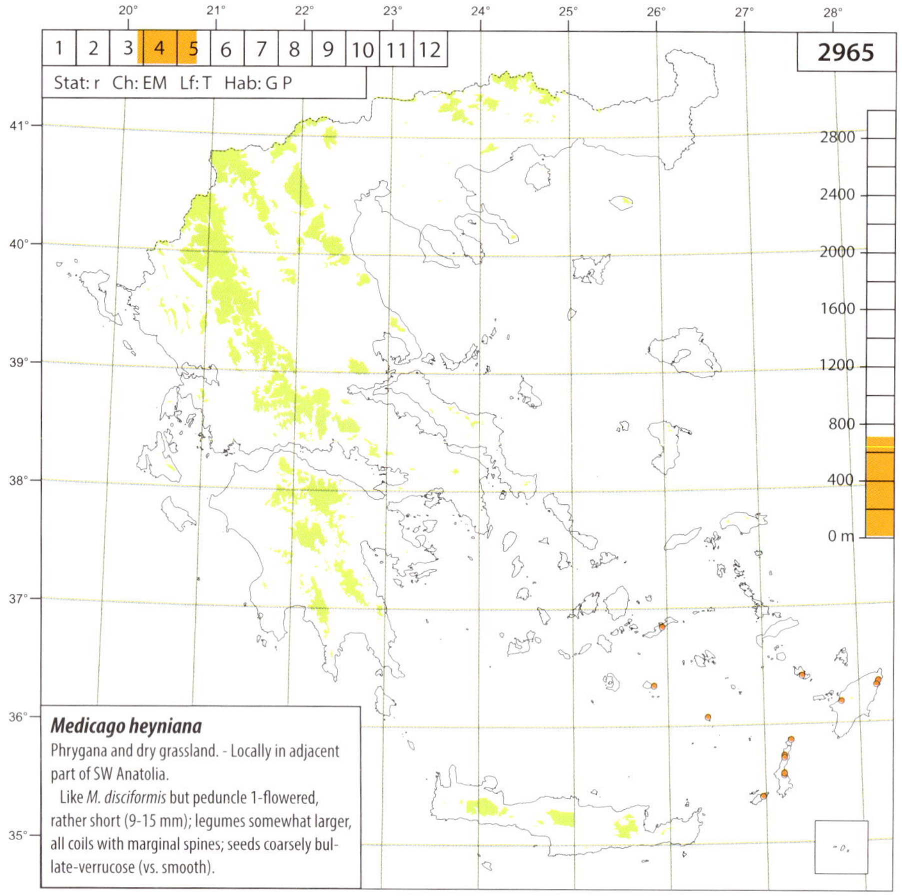

Medicago heyniana
Phrygana and dry grassland. - Locally in adjacent part of SW Anatolia.
Like *M. disciformis* but peduncle 1-flowered, rather short (9-15 mm); legumes somewhat larger, all coils with marginal spines; seeds coarsely bullate-verrucose (vs. smooth).

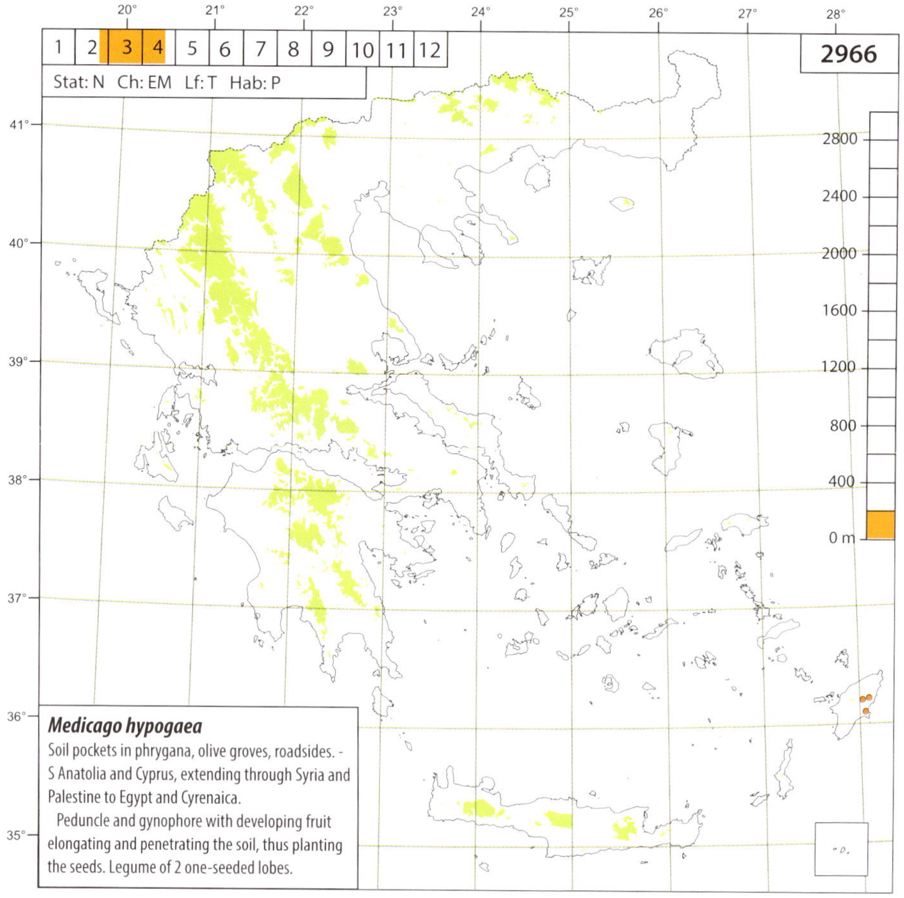

Medicago hypogaea
Soil pockets in phrygana, olive groves, roadsides. - S Anatolia and Cyprus, extending through Syria and Palestine to Egypt and Cyrenaica.
Peduncle and gynophore with developing fruit elongating and penetrating the soil, thus planting the seeds. Legume of 2 one-seeded lobes.

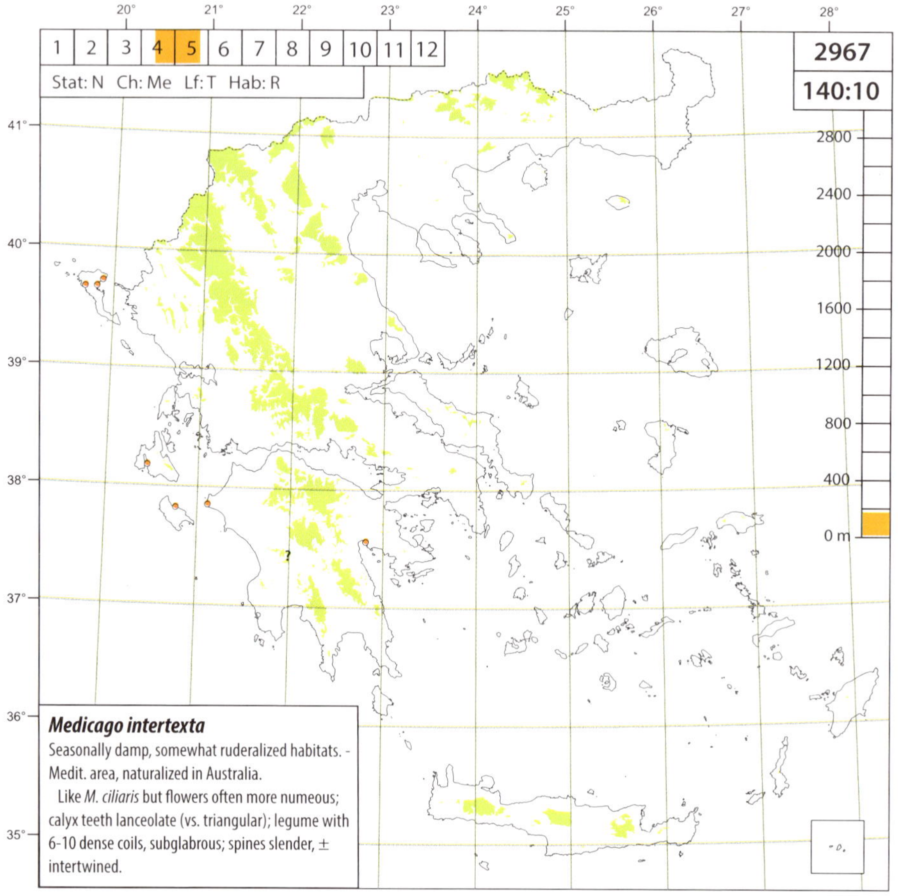

Medicago intertexta
Seasonally damp, somewhat ruderalized habitats. - Medit. area, naturalized in Australia.
Like *M. ciliaris* but flowers often more numeous; calyx teeth lanceolate (vs. triangular); legume with 6-10 dense coils, subglabrous; spines slender, ± intertwined.

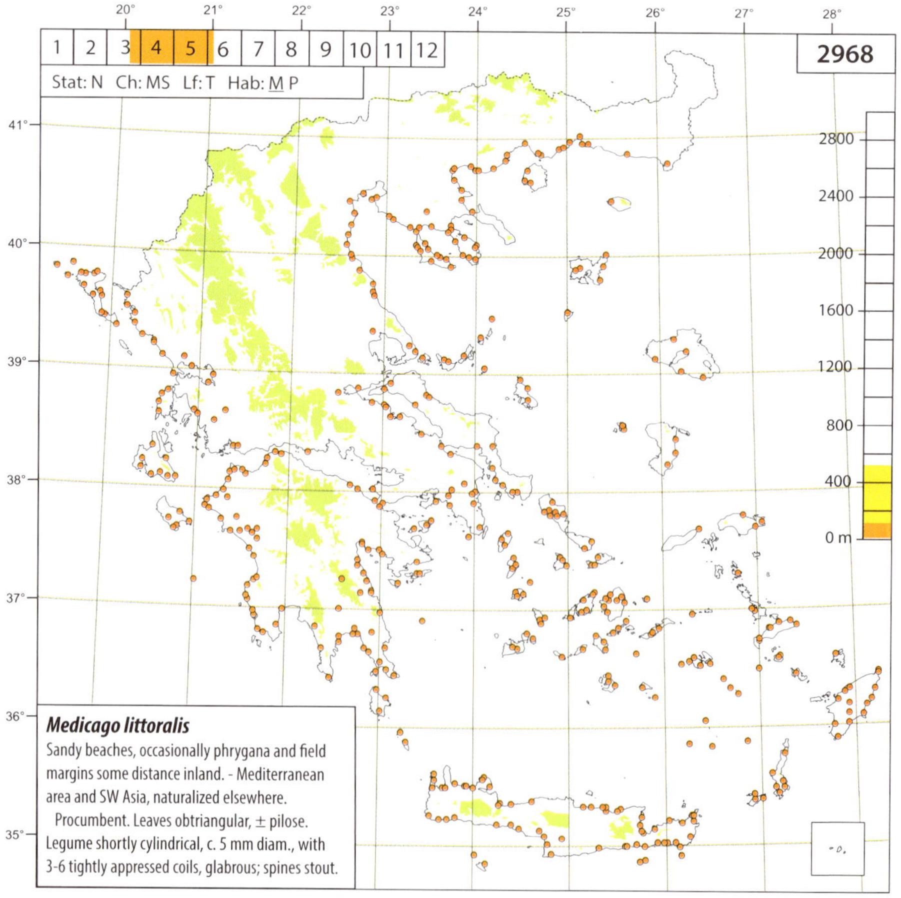

Medicago littoralis
Sandy beaches, occasionally phrygana and field margins some distance inland. - Mediterranean area and SW Asia, naturalized elsewhere.
Procumbent. Leaves obtriangular, ± pilose. Legume shortly cylindrical, c. 5 mm diam., with 3-6 tightly appressed coils, glabrous; spines stout.

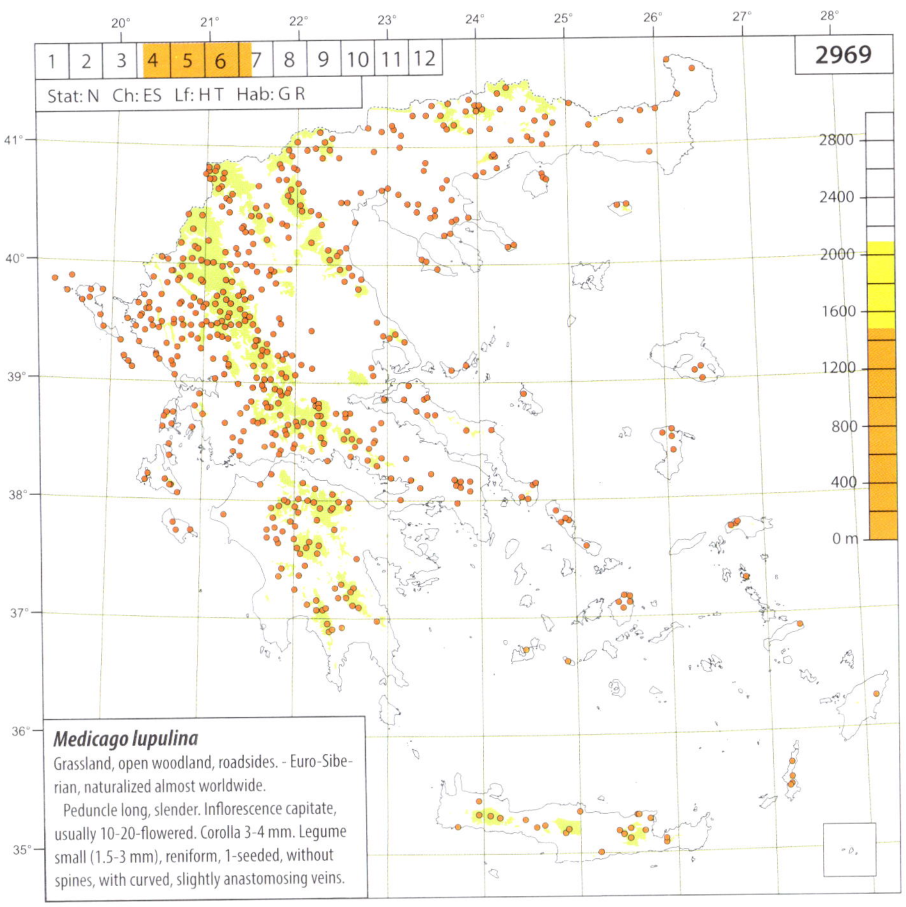

Medicago lupulina

Grassland, open woodland, roadsides. - Euro-Siberian, naturalized almost worldwide.

Peduncle long, slender. Inflorescence capitate, usually 10-20-flowered. Corolla 3-4 mm. Legume small (1.5-3 mm), reniform, 1-seeded, without spines, with curved, slightly anastomosing veins.

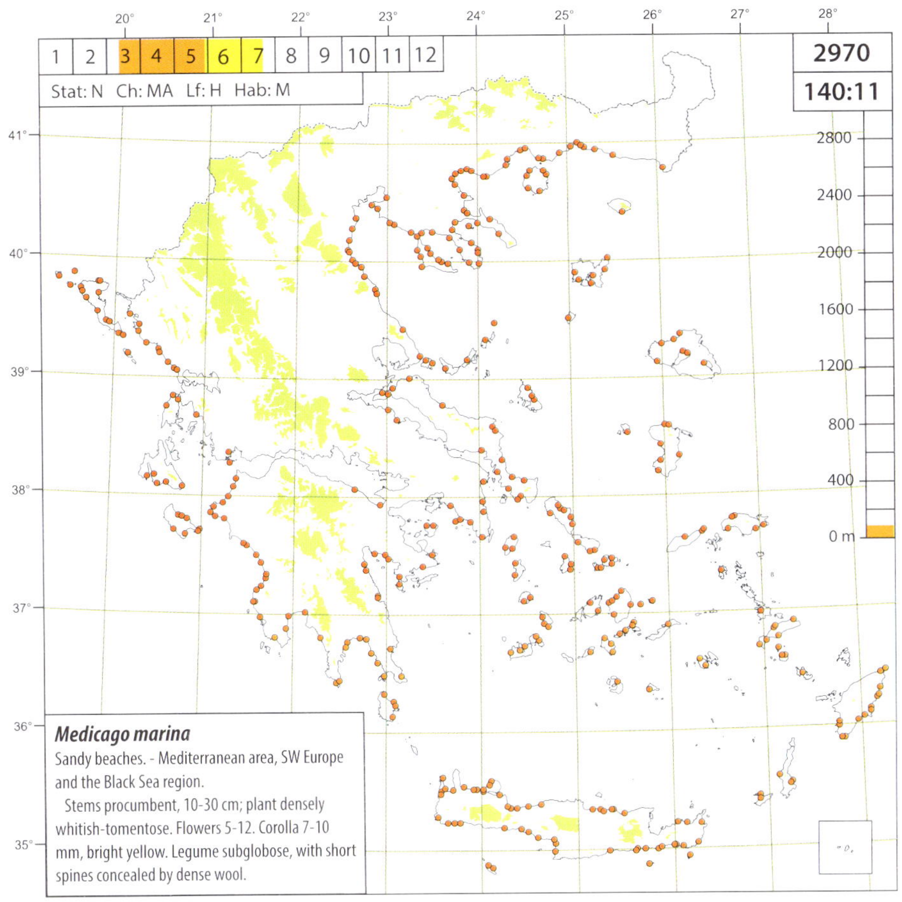

Medicago marina

Sandy beaches. - Mediterranean area, SW Europe and the Black Sea region.

Stems procumbent, 10-30 cm; plant densely whitish-tomentose. Flowers 5-12. Corolla 7-10 mm, bright yellow. Legume subglobose, with short spines concealed by dense wool.

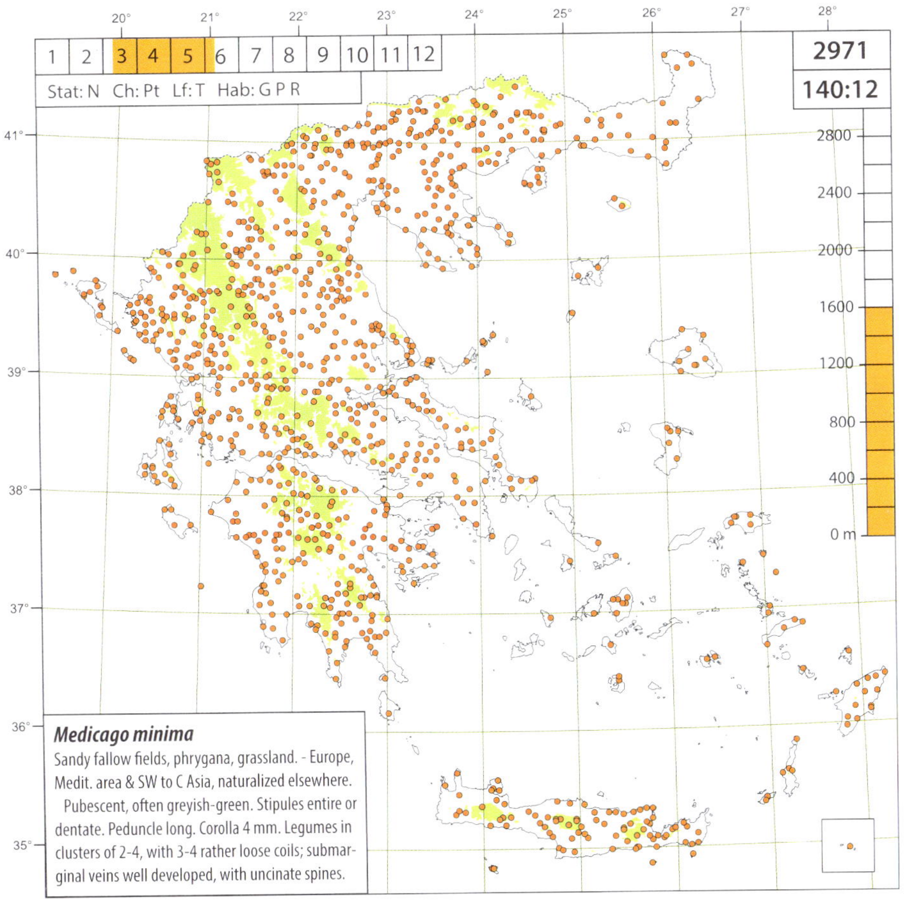

Medicago minima

Sandy fallow fields, phrygana, grassland. - Europe, Medit. area & SW to C Asia, naturalized elsewhere.

Pubescent, often greyish-green. Stipules entire or dentate. Peduncle long. Corolla 4 mm. Legumes in clusters of 2-4, with 3-4 rather loose coils; submarginal veins well developed, with uncinate spines.

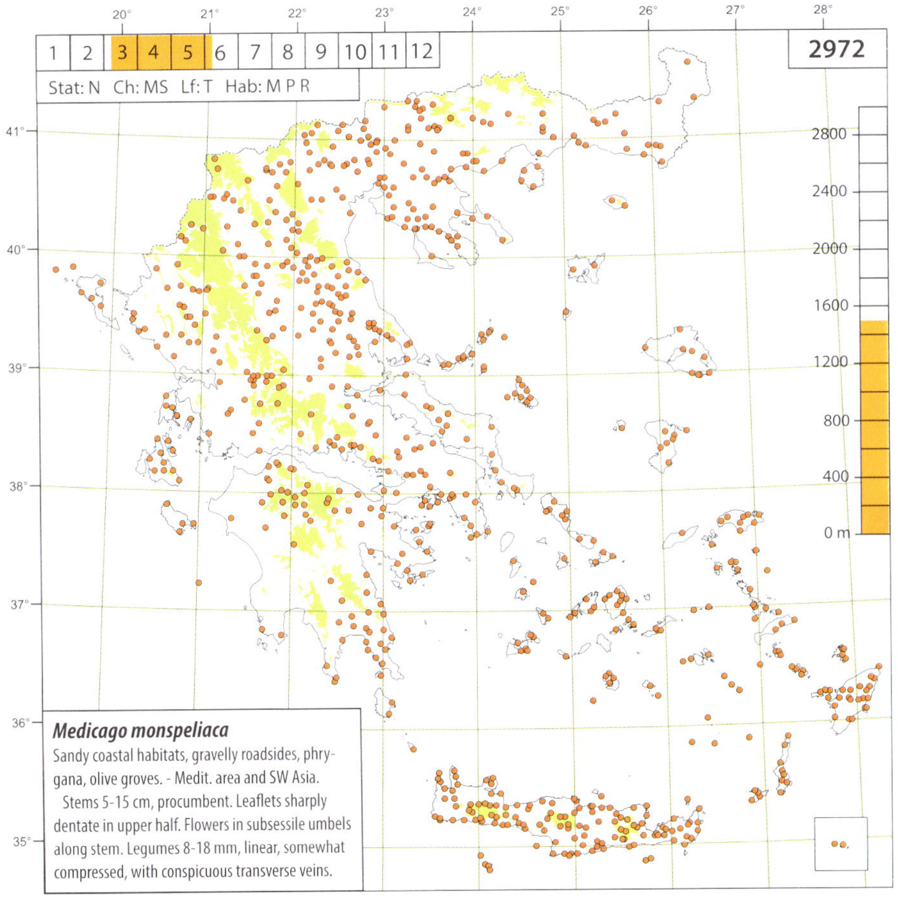

Medicago monspeliaca

Sandy coastal habitats, gravelly roadsides, phrygana, olive groves. - Medit. area and SW Asia.

Stems 5-15 cm, procumbent. Leaflets sharply dentate in upper half. Flowers in subsessile umbels along stem. Legumes 8-18 mm, linear, somewhat compressed, with conspicuous transverse veins.

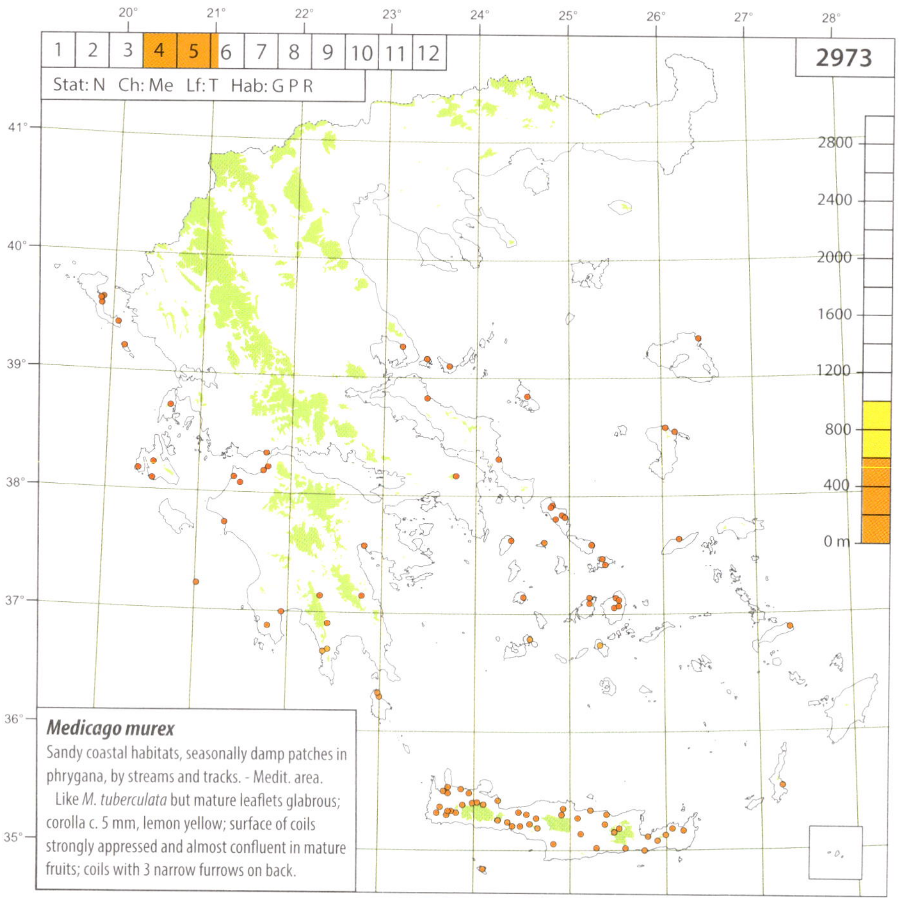

Medicago murex

Sandy coastal habitats, seasonally damp patches in phrygana, by streams and tracks. - Medit. area.

Like *M. tuberculata* but mature leaflets glabrous; corolla c. 5 mm, lemon yellow; surface of coils strongly appressed and almost confluent in mature fruits; coils with 3 narrow furrows on back.

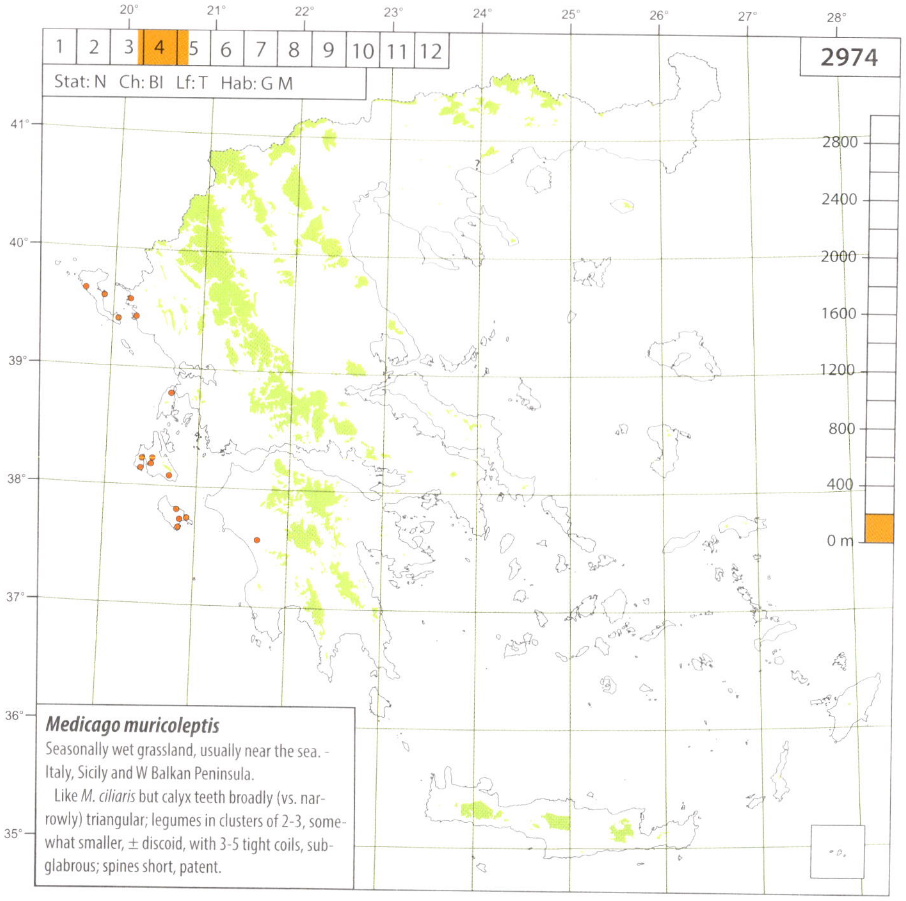

Medicago muricoleptis

Seasonally wet grassland, usually near the sea. - Italy, Sicily and W Balkan Peninsula.

Like *M. ciliaris* but calyx teeth broadly (vs. narrowly) triangular; legumes in clusters of 2-3, somewhat smaller, ± discoid, with 3-5 tight coils, subglabrous; spines short, patent.

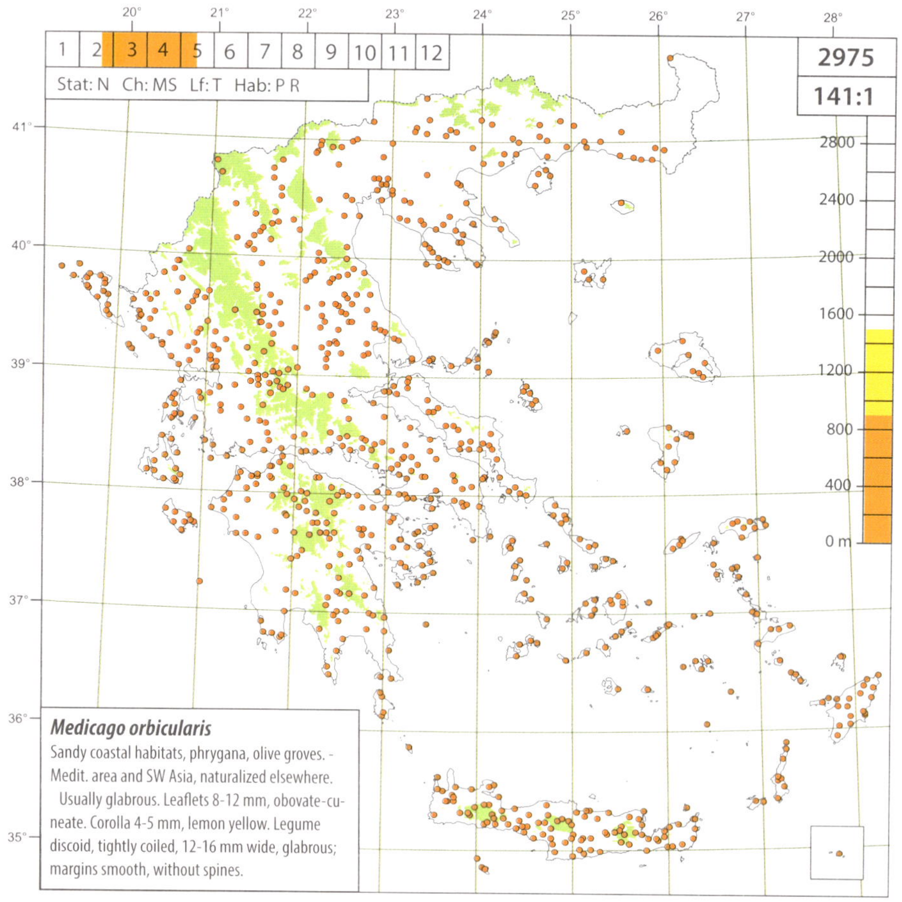

Medicago orbicularis

Sandy coastal habitats, phrygana, olive groves. - Medit. area and SW Asia, naturalized elsewhere.

Usually glabrous. Leaflets 8-12 mm, obovate-cuneate. Corolla 4-5 mm, lemon yellow. Legume discoid, tightly coiled, 12-16 mm wide, glabrous; margins smooth, without spines.

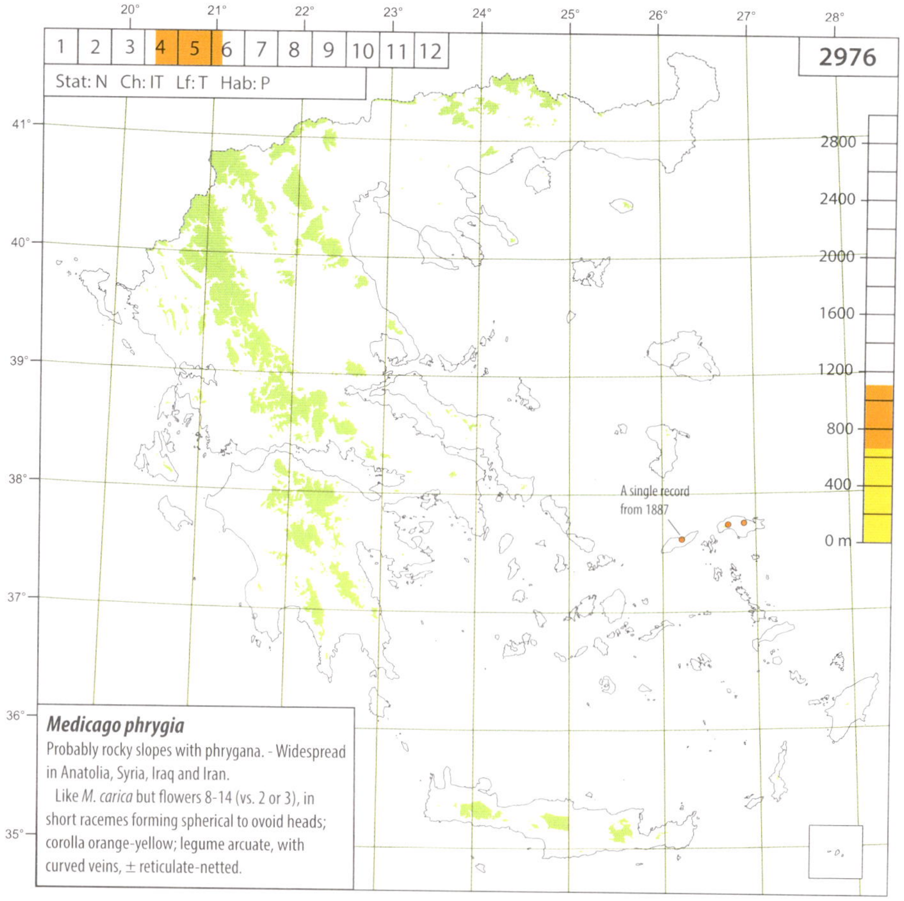

Medicago phrygia

Probably rocky slopes with phrygana. - Widespread in Anatolia, Syria, Iraq and Iran.

Like *M. carica* but flowers 8-14 (vs. 2 or 3), in short racemes forming spherical to ovoid heads; corolla orange-yellow; legume arcuate, with curved veins, ± reticulate-netted.

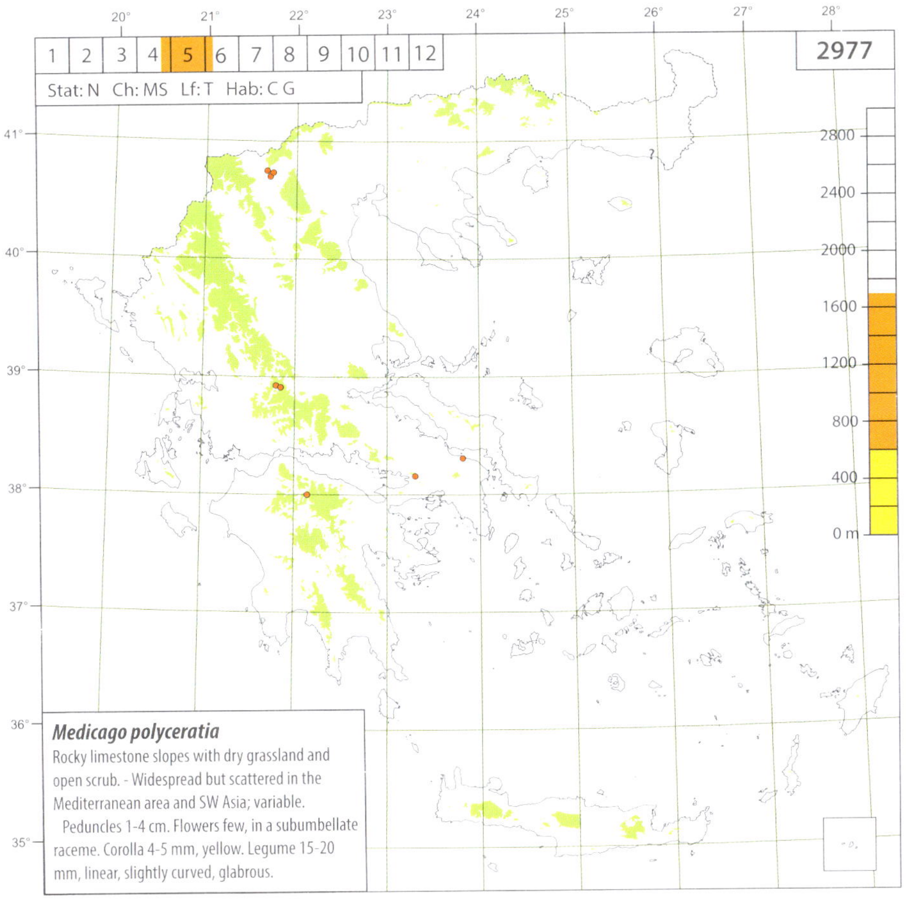

Medicago polyceratia
Rocky limestone slopes with dry grassland and open scrub. - Widespread but scattered in the Mediterranean area and SW Asia; variable.
Peduncles 1-4 cm. Flowers few, in a subumbellate raceme. Corolla 4-5 mm, yellow. Legume 15-20 mm, linear, slightly curved, glabrous.

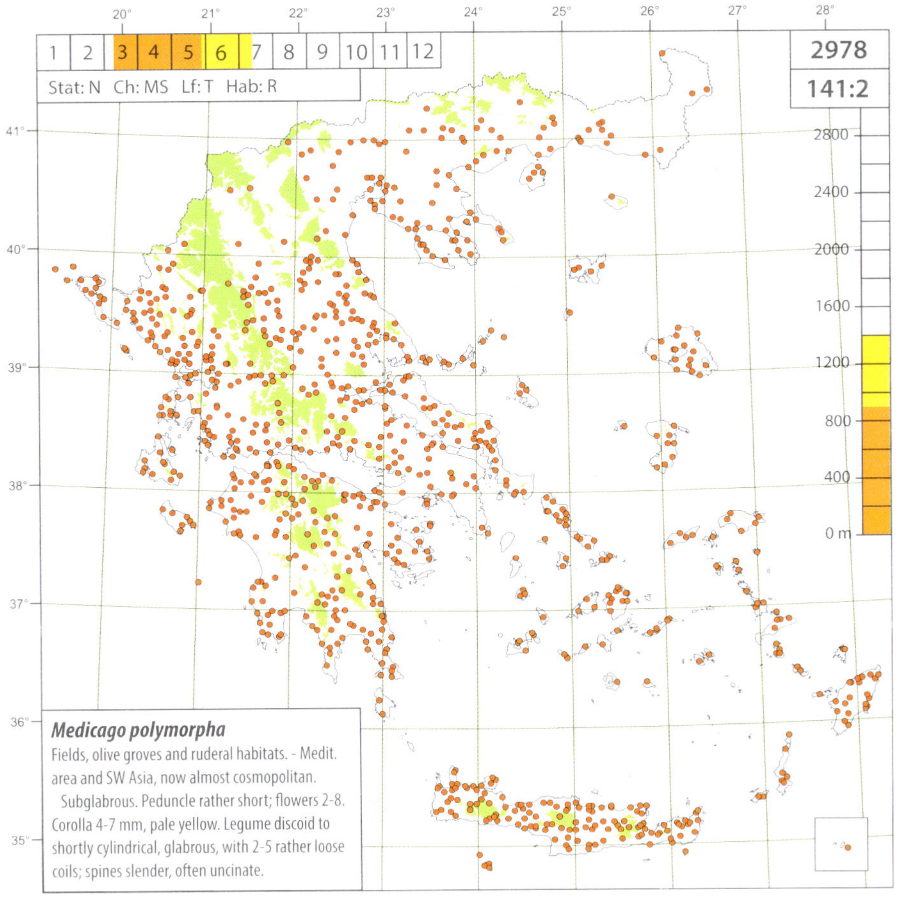

Medicago polymorpha
Fields, olive groves and ruderal habitats. - Medit. area and SW Asia, now almost cosmopolitan.
Subglabrous. Peduncle rather short; flowers 2-8. Corolla 4-7 mm, pale yellow. Legume discoid to shortly cylindrical, glabrous, with 2-5 rather loose coils; spines slender, often uncinate.

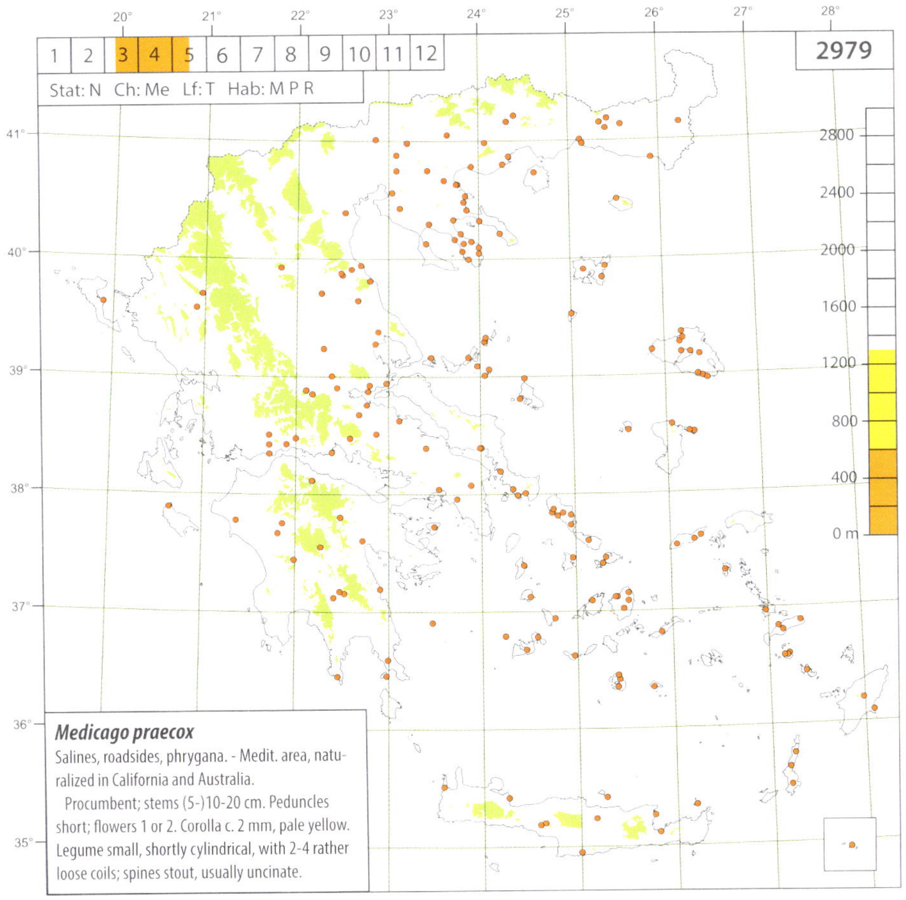

Medicago praecox
Salines, roadsides, phrygana. - Medit. area, naturalized in California and Australia.
Procumbent; stems (5-)10-20 cm. Peduncles short; flowers 1 or 2. Corolla c. 2 mm, pale yellow. Legume small, shortly cylindrical, with 2-4 rather loose coils; spines stout, usually uncinate.

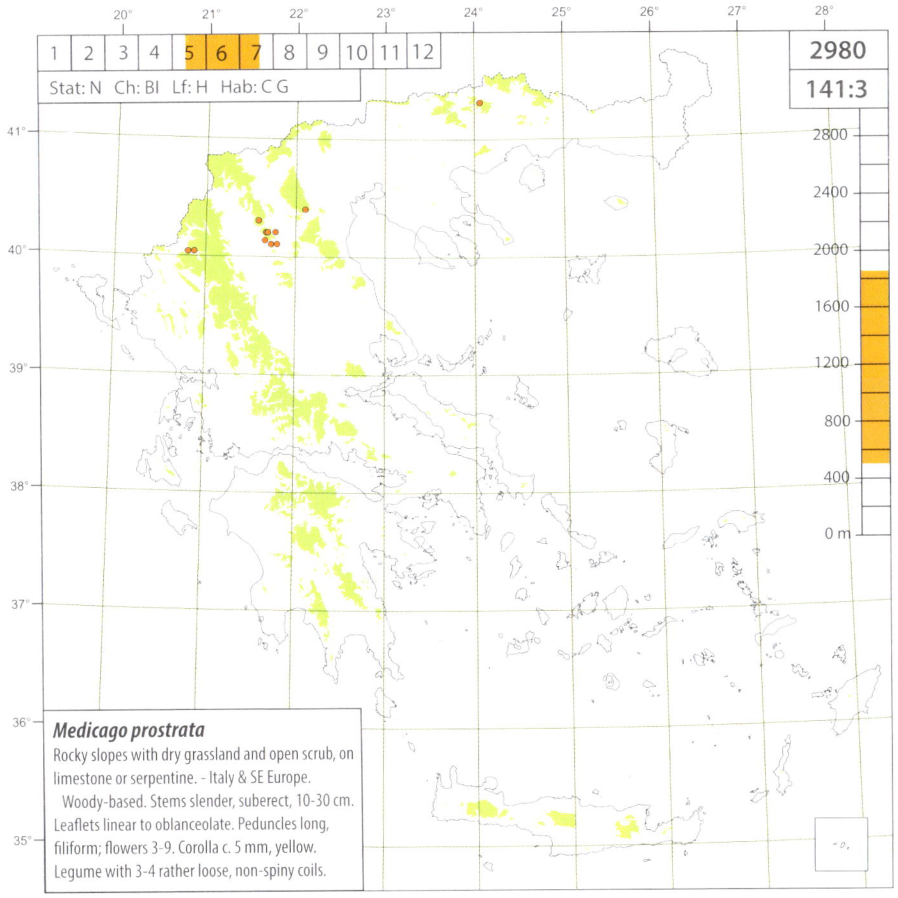

Medicago prostrata
Rocky slopes with dry grassland and open scrub, on limestone or serpentine. - Italy & SE Europe.
Woody-based. Stems slender, suberect, 10-30 cm. Leaflets linear to oblanceolate. Peduncles long, filiform; flowers 3-9. Corolla c. 5 mm, yellow. Legume with 3-4 rather loose, non-spiny coils.

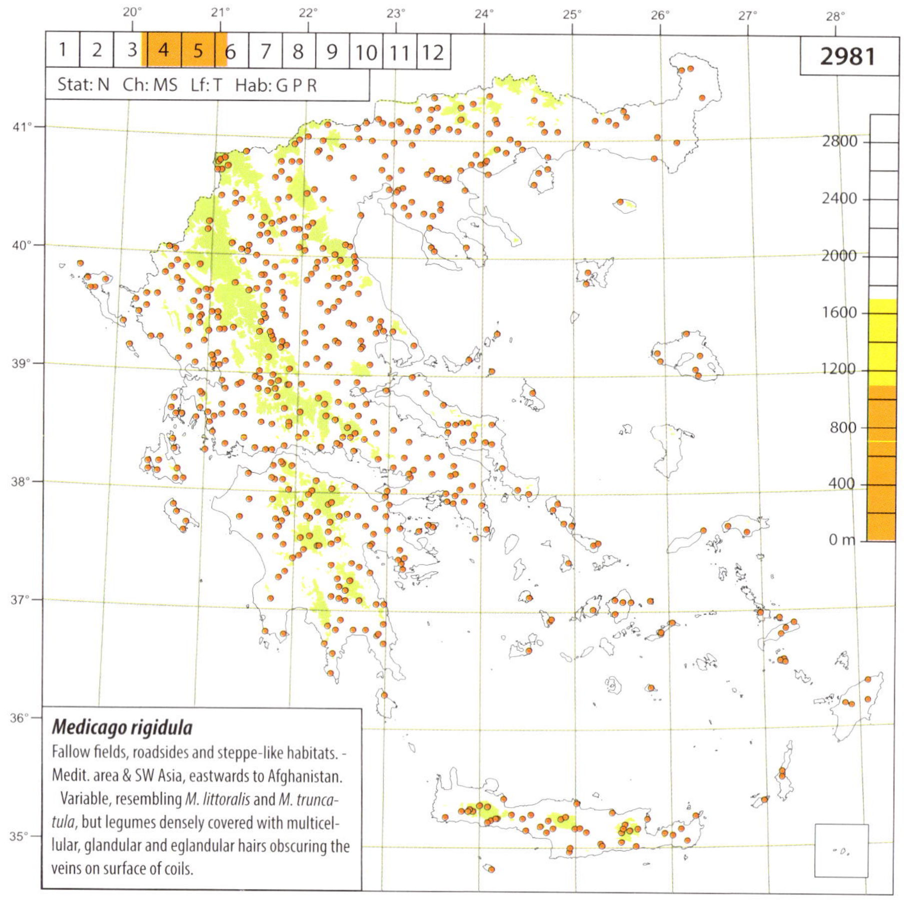

Medicago rigidula

Fallow fields, roadsides and steppe-like habitats. - Medit. area & SW Asia, eastwards to Afghanistan.

Variable, resembling *M. littoralis* and *M. truncatula*, but legumes densely covered with multicellular, glandular and eglandular hairs obscuring the veins on surface of coils.

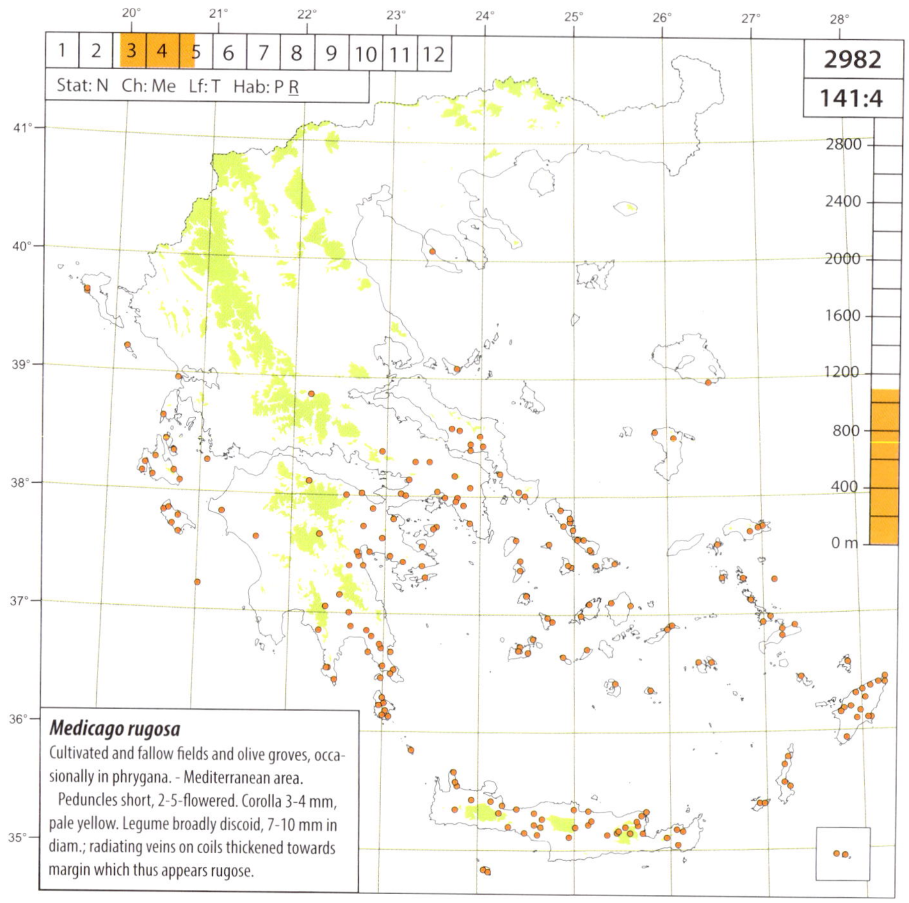

Medicago rugosa

Cultivated and fallow fields and olive groves, occasionally in phrygana. - Mediterranean area.

Peduncles short, 2-5-flowered. Corolla 3-4 mm, pale yellow. Legume broadly discoid, 7-10 mm in diam.; radiating veins on coils thickened towards margin which thus appears rugose.

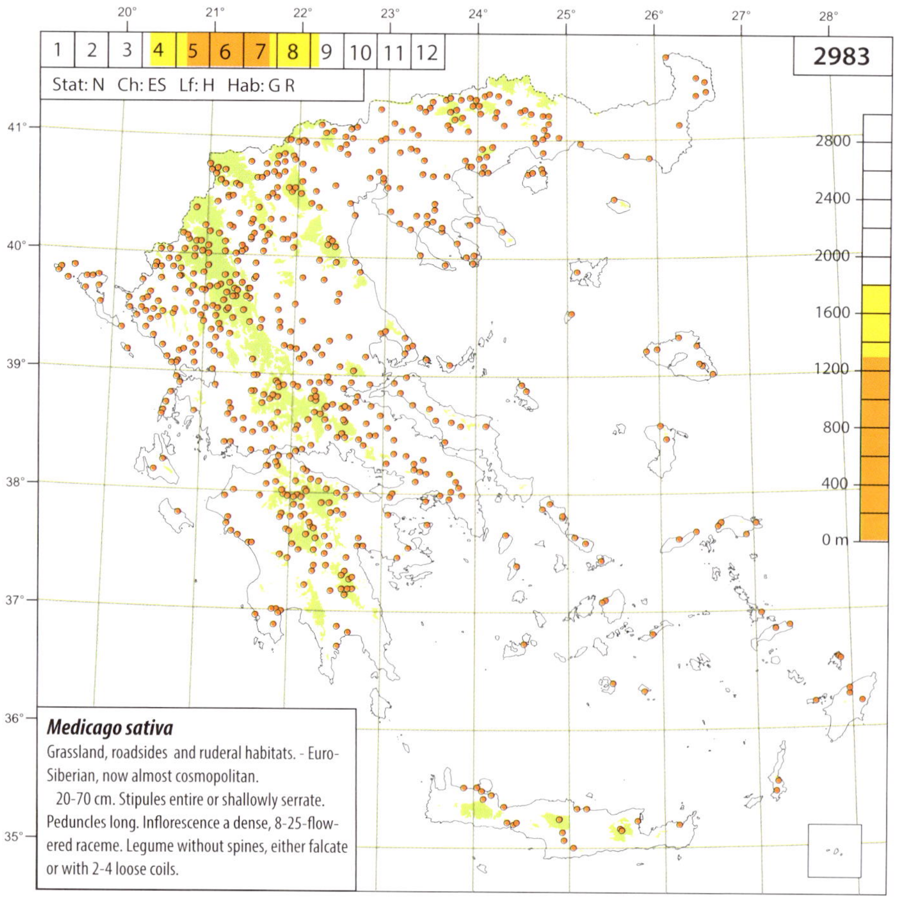

Medicago sativa

Grassland, roadsides and ruderal habitats. - Euro-Siberian, now almost cosmopolitan.

20-70 cm. Stipules entire or shallowly serrate. Peduncles long. Inflorescence a dense, 8-25-flowered raceme. Legume without spines, either falcate or with 2-4 loose coils.

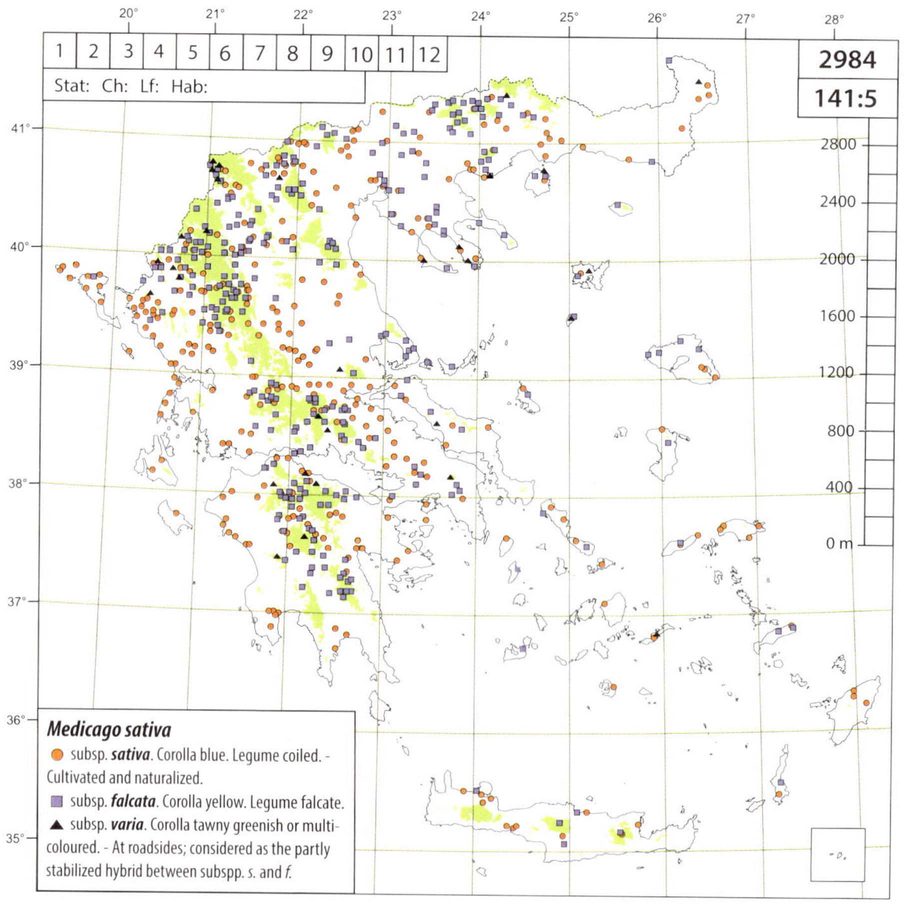

Medicago sativa

● subsp. ***sativa***. Corolla blue. Legume coiled. - Cultivated and naturalized.

■ subsp. ***falcata***. Corolla yellow. Legume falcate.

▲ subsp. ***varia***. Corolla tawny greenish or multi-coloured. - At roadsides; considered as the partly stabilized hybrid between subspp. *s.* and *f.*

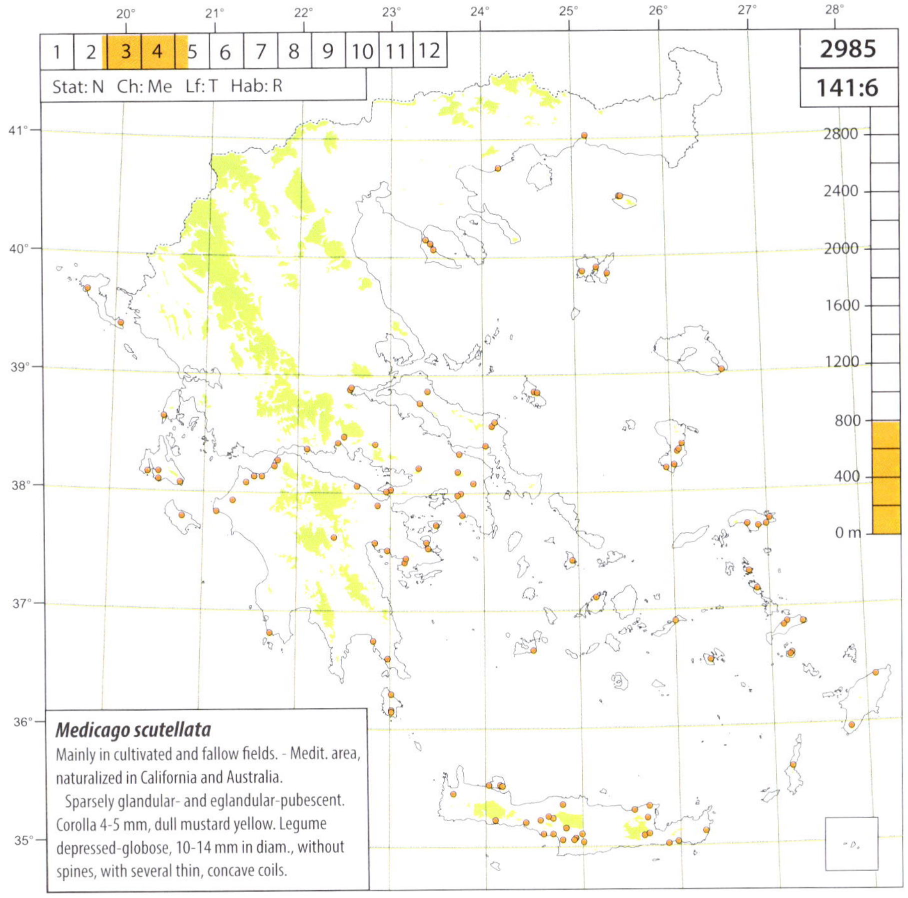

Medicago scutellata
Mainly in cultivated and fallow fields. - Medit. area, naturalized in California and Australia.
Sparsely glandular- and eglandular-pubescent. Corolla 4-5 mm, dull mustard yellow. Legume depressed-globose, 10-14 mm in diam., without spines, with several thin, concave coils.

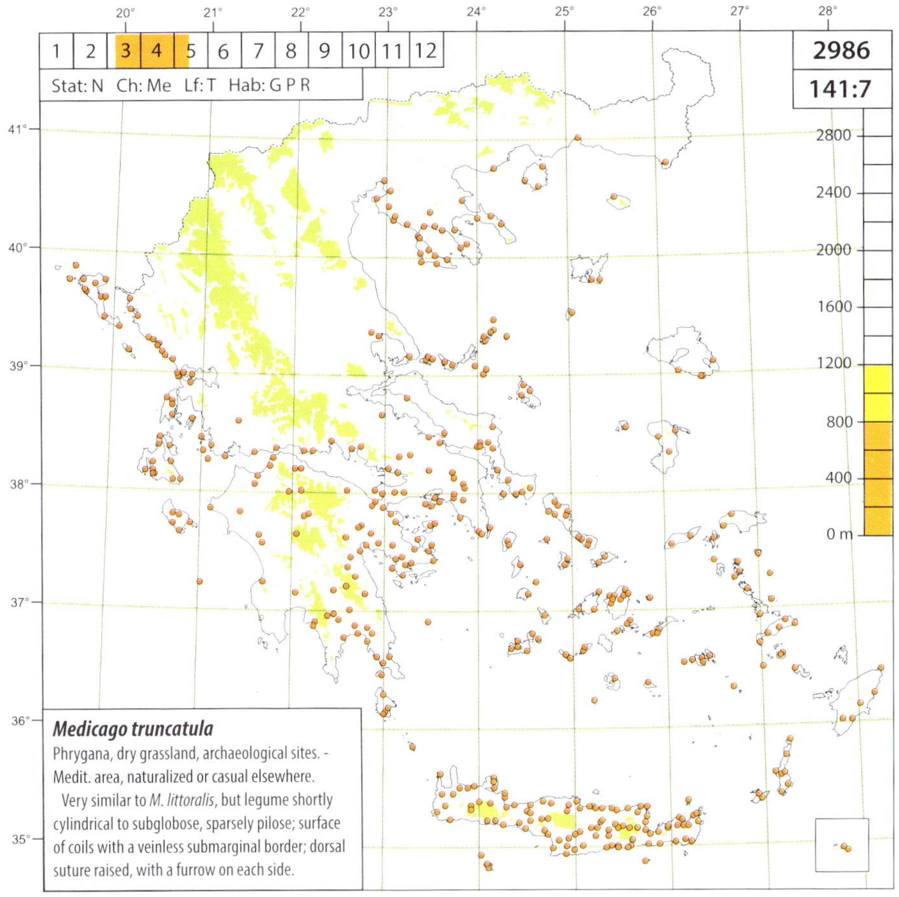

Medicago truncatula
Phrygana, dry grassland, archaeological sites. - Medit. area, naturalized or casual elsewhere.
Very similar to *M. littoralis*, but legume shortly cylindrical to subglobose, sparsely pilose; surface of coils with a veinless submarginal border; dorsal suture raised, with a furrow on each side.

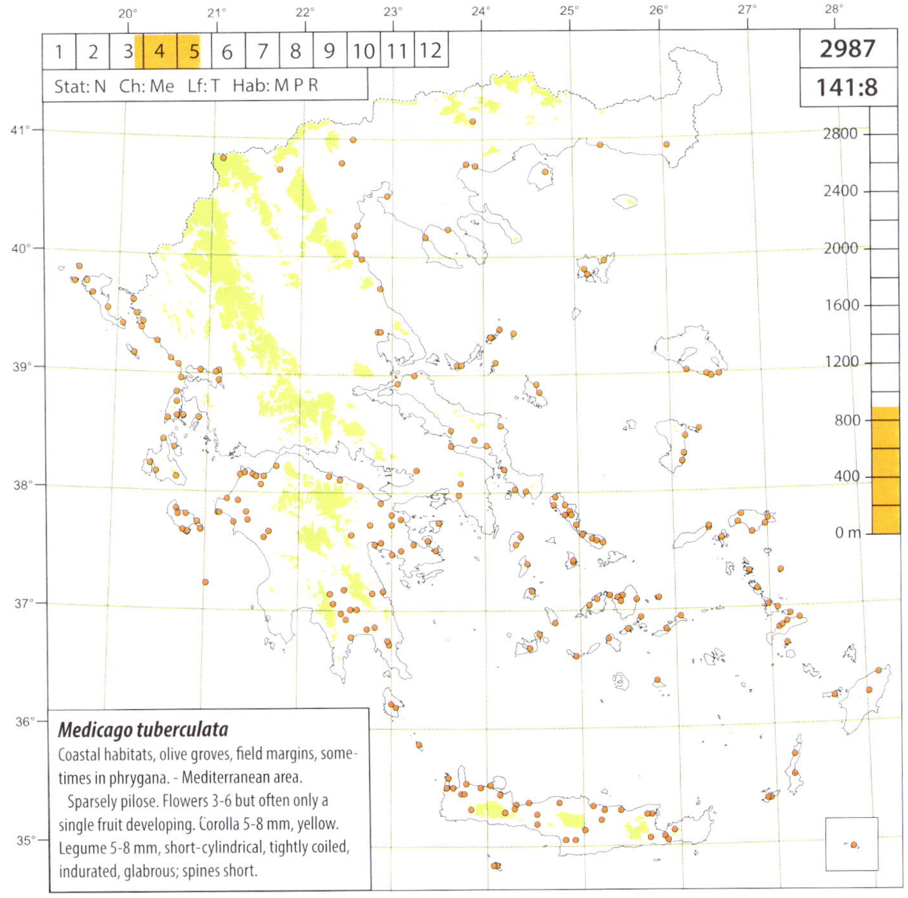

Medicago tuberculata
Coastal habitats, olive groves, field margins, sometimes in phrygana. - Mediterranean area.
Sparsely pilose. Flowers 3-6 but often only a single fruit developing. Corolla 5-8 mm, yellow. Legume 5-8 mm, short-cylindrical, tightly coiled, indurated, glabrous; spines short.

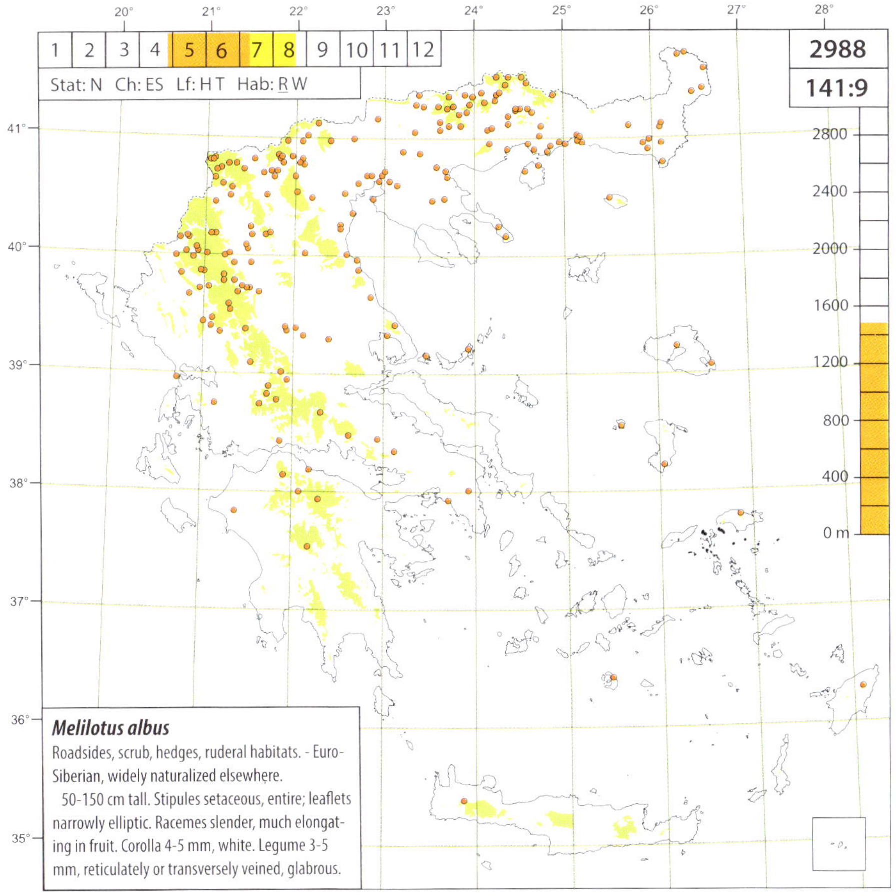

Melilotus albus
Roadsides, scrub, hedges, ruderal habitats. - Euro-Siberian, widely naturalized elsewhere.
50-150 cm tall. Stipules setaceous, entire; leaflets narrowly elliptic. Racemes slender, much elongating in fruit. Corolla 4-5 mm, white. Legume 3-5 mm, reticulately or transversely veined, glabrous.

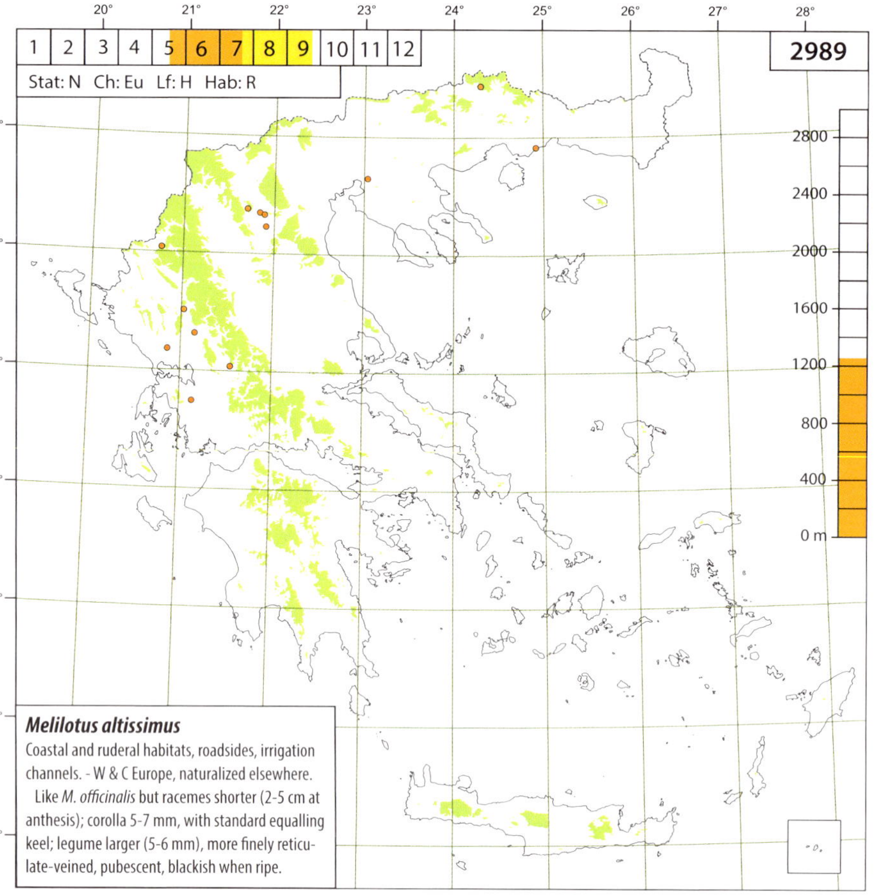

Melilotus altissimus

Coastal and ruderal habitats, roadsides, irrigation channels. - W & C Europe, naturalized elsewhere.

Like *M. officinalis* but racemes shorter (2-5 cm at anthesis); corolla 5-7 mm, with standard equalling keel; legume larger (5-6 mm), more finely reticulate-veined, pubescent, blackish when ripe.

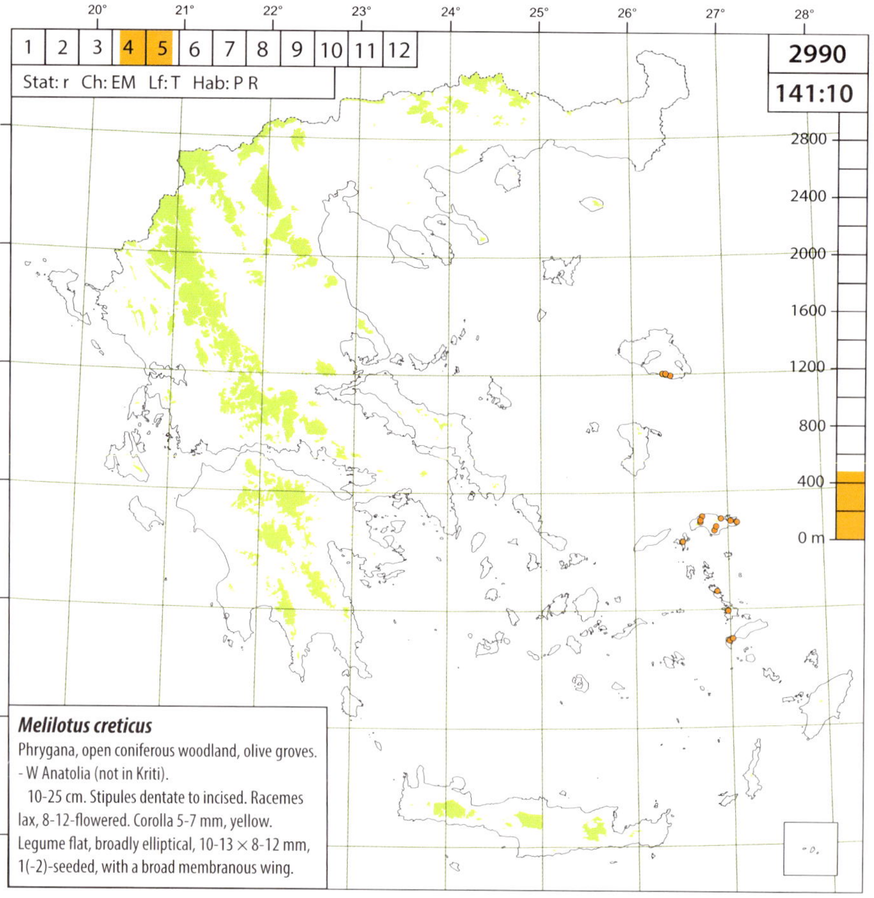

Melilotus creticus

Phrygana, open coniferous woodland, olive groves. - W Anatolia (not in Kriti).

10-25 cm. Stipules dentate to incised. Racemes lax, 8-12-flowered. Corolla 5-7 mm, yellow. Legume flat, broadly elliptical, 10-13 × 8-12 mm, 1(-2)-seeded, with a broad membranous wing.

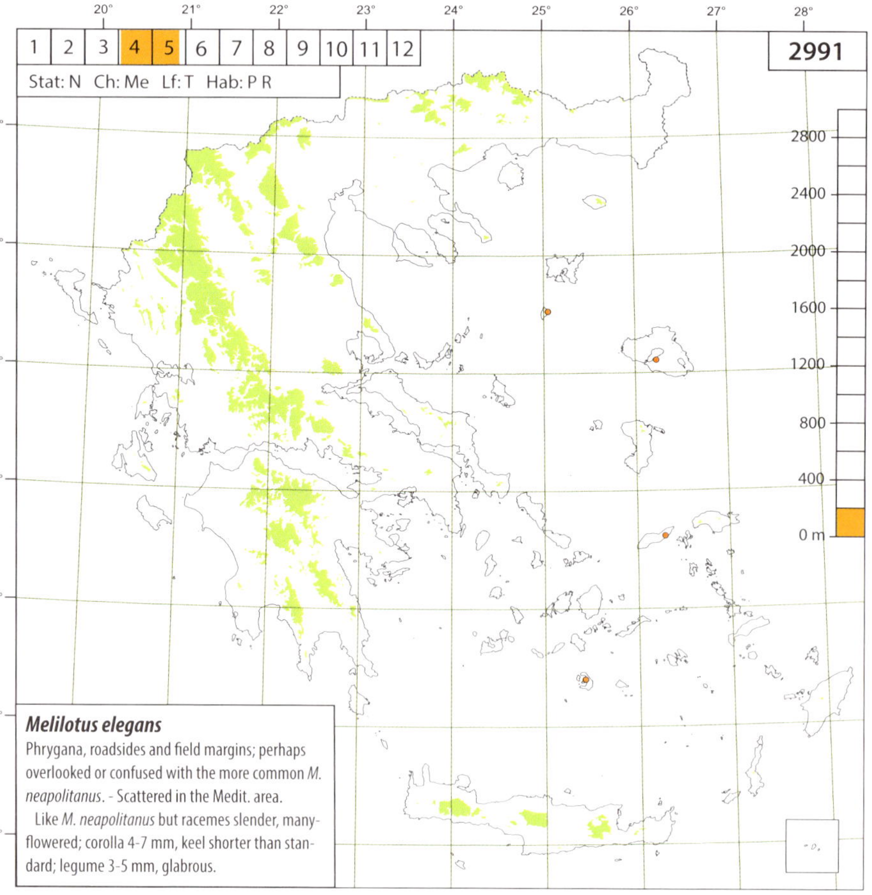

Melilotus elegans

Phrygana, roadsides and field margins; perhaps overlooked or confused with the more common *M. neapolitanus*. - Scattered in the Medit. area.

Like *M. neapolitanus* but racemes slender, many-flowered; corolla 4-7 mm, keel shorter than standard; legume 3-5 mm, glabrous.

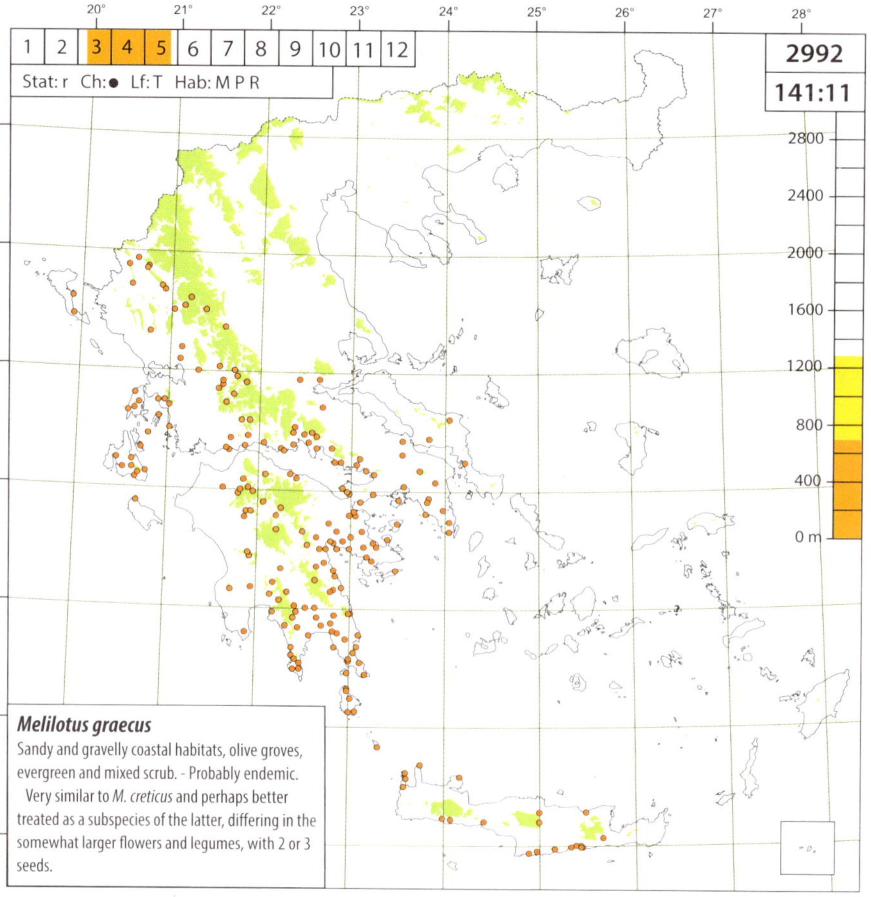

Melilotus graecus

Sandy and gravelly coastal habitats, olive groves, evergreen and mixed scrub. - Probably endemic.

Very similar to *M. creticus* and perhaps better treated as a subspecies of the latter, differing in the somewhat larger flowers and legumes, with 2 or 3 seeds.

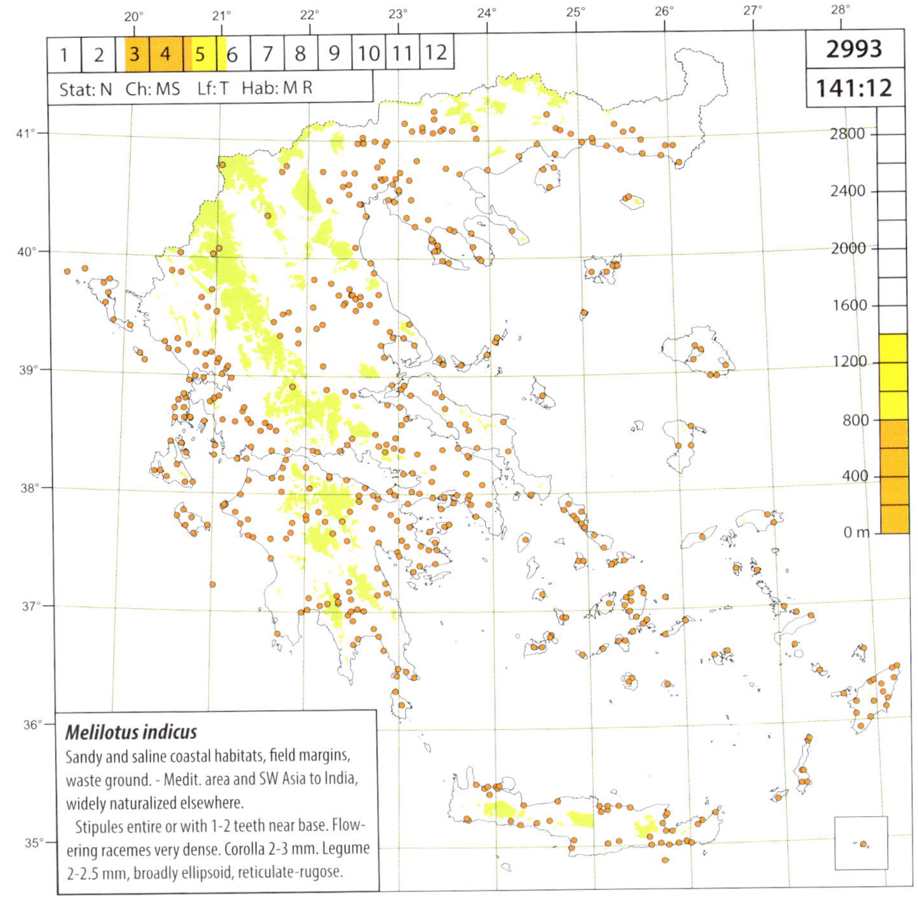

Melilotus indicus

Sandy and saline coastal habitats, field margins, waste ground. - Medit. area and SW Asia to India, widely naturalized elsewhere.

Stipules entire or with 1-2 teeth near base. Flowering racemes very dense. Corolla 2-3 mm. Legume 2-2.5 mm, broadly ellipsoid, reticulate-rugose.

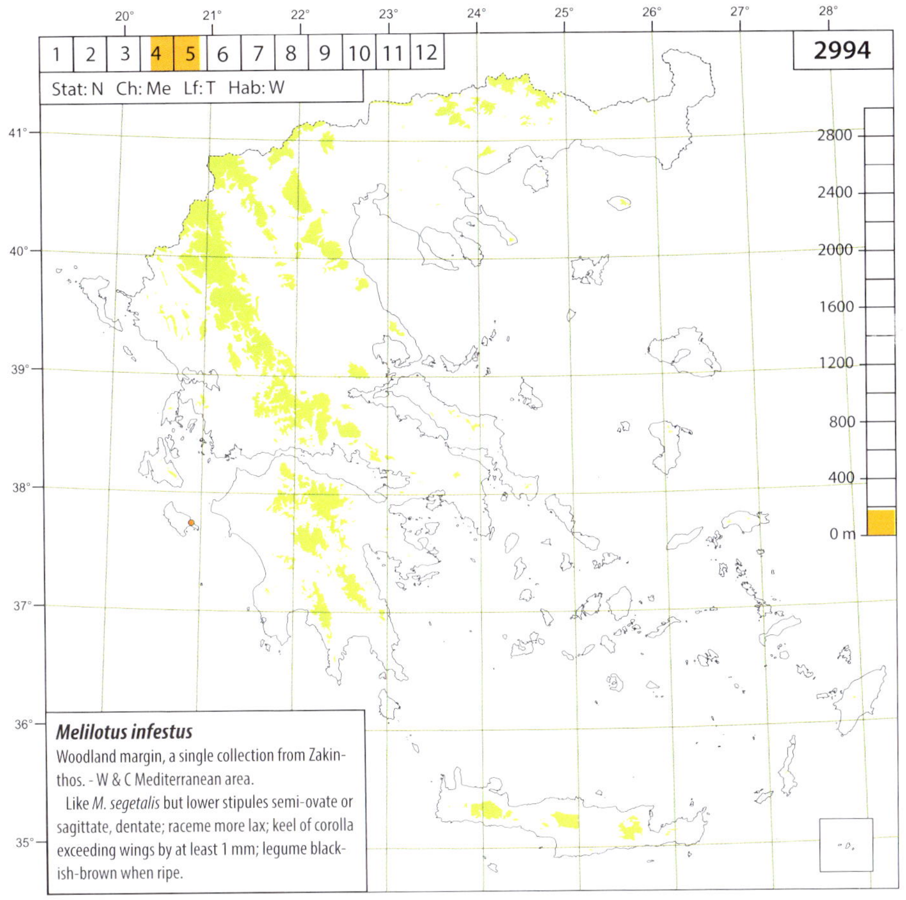

Melilotus infestus

Woodland margin, a single collection from Zakinthos. - W & C Mediterranean area.

Like *M. segetalis* but lower stipules semi-ovate or sagittate, dentate; raceme more lax; keel of corolla exceeding wings by at least 1 mm; legume blackish-brown when ripe.

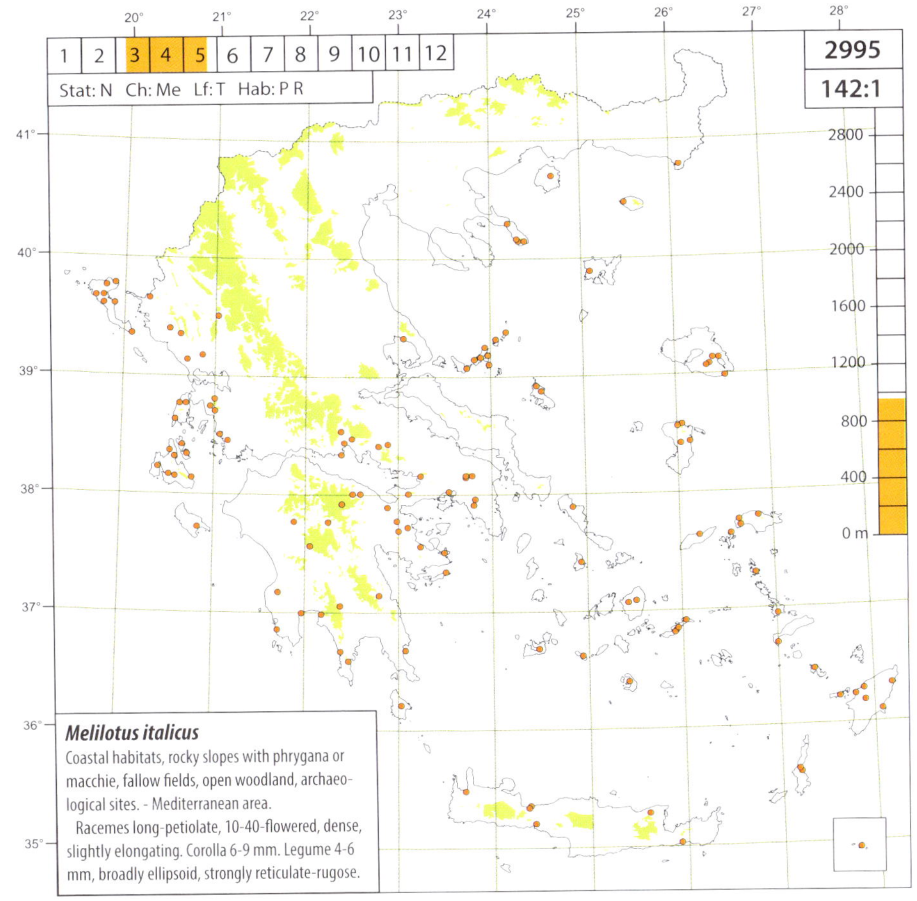

Melilotus italicus

Coastal habitats, rocky slopes with phrygana or macchie, fallow fields, open woodland, archaeological sites. - Mediterranean area.

Racemes long-petiolate, 10-40-flowered, dense, slightly elongating. Corolla 6-9 mm. Legume 4-6 mm, broadly ellipsoid, strongly reticulate-rugose.

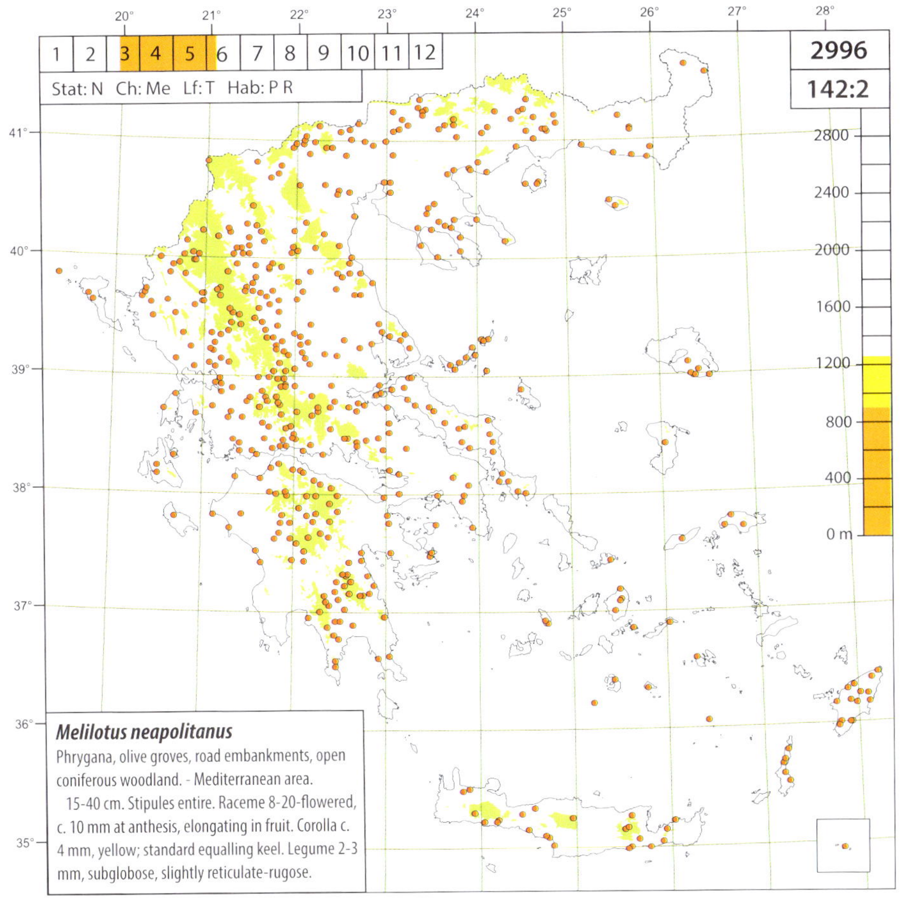

Melilotus neapolitanus

Phrygana, olive groves, road embankments, open coniferous woodland. - Mediterranean area.

15-40 cm. Stipules entire. Raceme 8-20-flowered, c. 10 mm at anthesis, elongating in fruit. Corolla c. 4 mm, yellow; standard equalling keel. Legume 2-3 mm, subglobose, slightly reticulate-rugose.

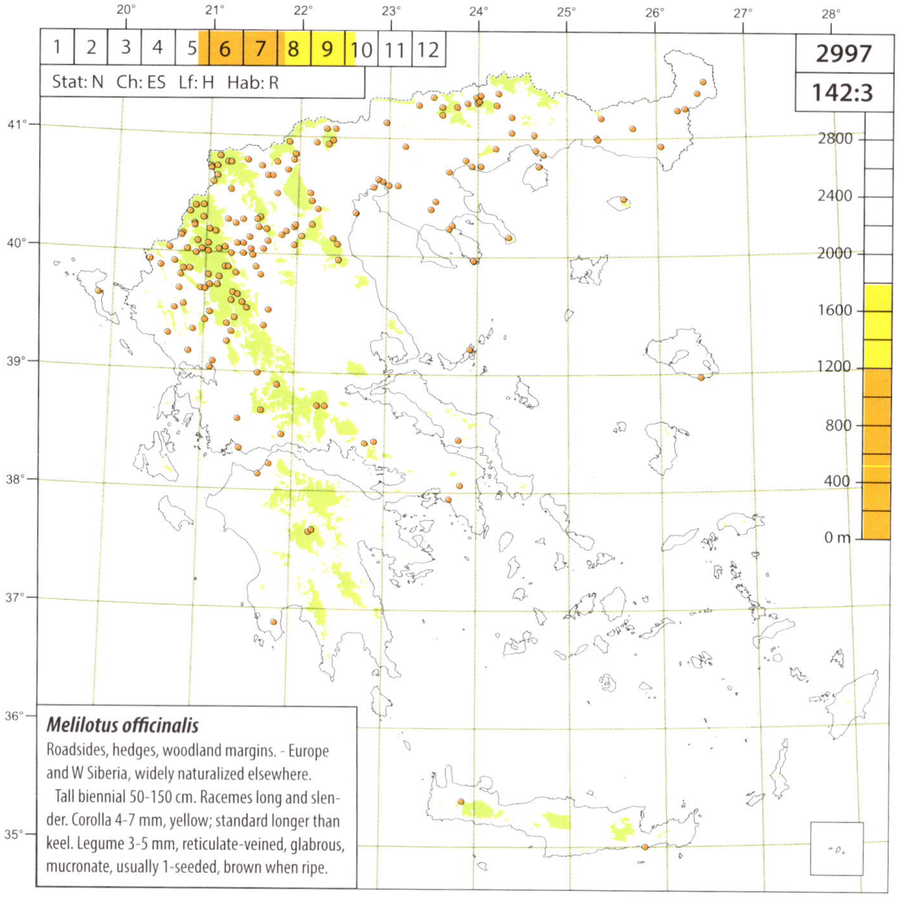

Melilotus officinalis

Roadsides, hedges, woodland margins. - Europe and W Siberia, widely naturalized elsewhere.

Tall biennial 50-150 cm. Racemes long and slender. Corolla 4-7 mm, yellow; standard longer than keel. Legume 3-5 mm, reticulate-veined, glabrous, mucronate, usually 1-seeded, brown when ripe.

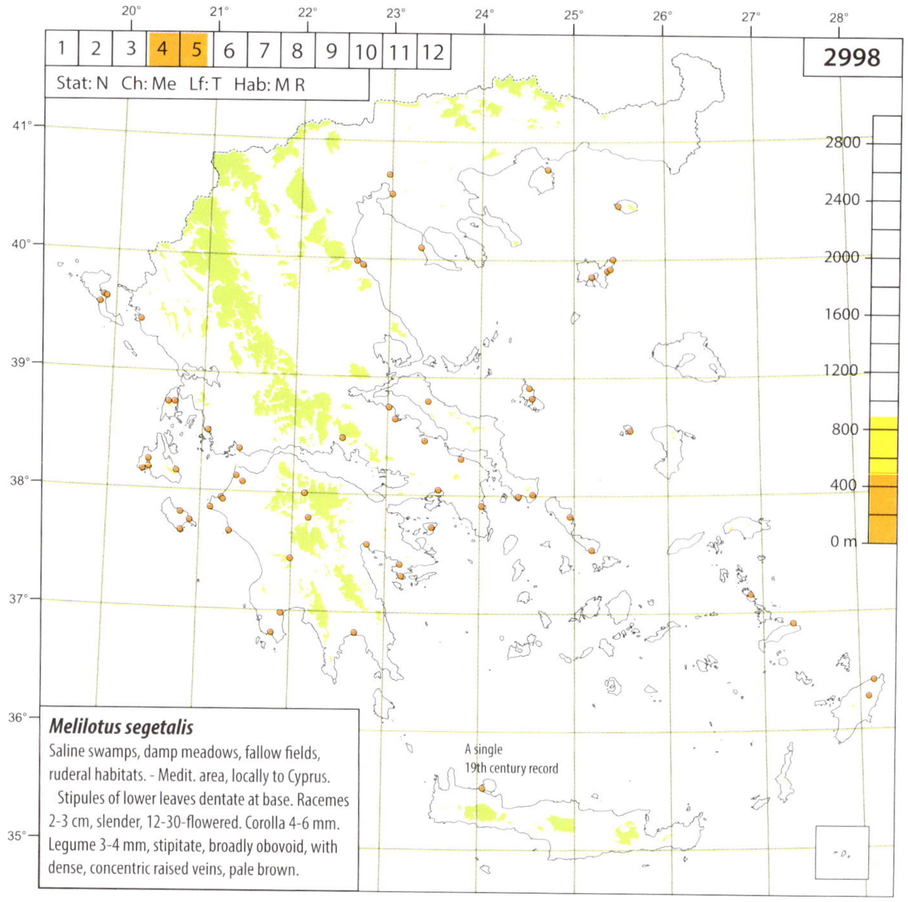

Melilotus segetalis

Saline swamps, damp meadows, fallow fields, ruderal habitats. - Medit. area, locally to Cyprus.

Stipules of lower leaves dentate at base. Racemes 2-3 cm, slender, 12-30-flowered. Corolla 4-6 mm. Legume 3-4 mm, stipitate, broadly obovoid, with dense, concentric raised veins, pale brown.

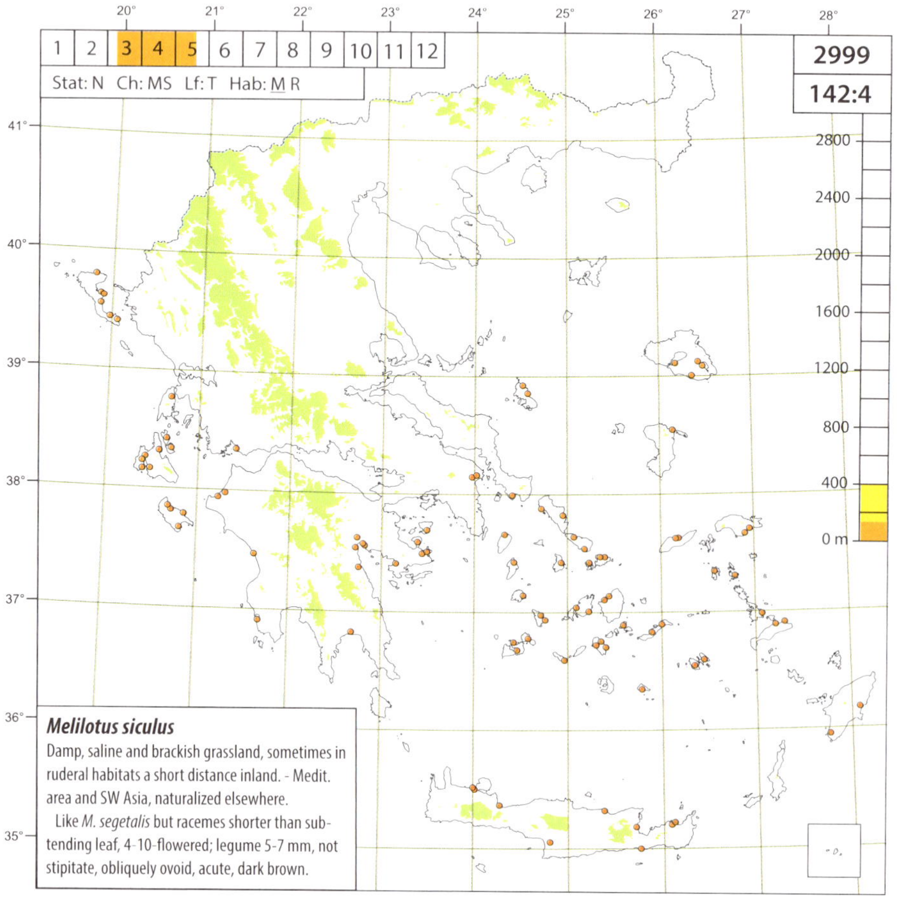

Melilotus siculus

Damp, saline and brackish grassland, sometimes in ruderal habitats a short distance inland. - Medit. area and SW Asia, naturalized elsewhere.

Like *M. segetalis* but racemes shorter than subtending leaf, 4-10-flowered; legume 5-7 mm, not stipitate, obliquely ovoid, acute, dark brown.

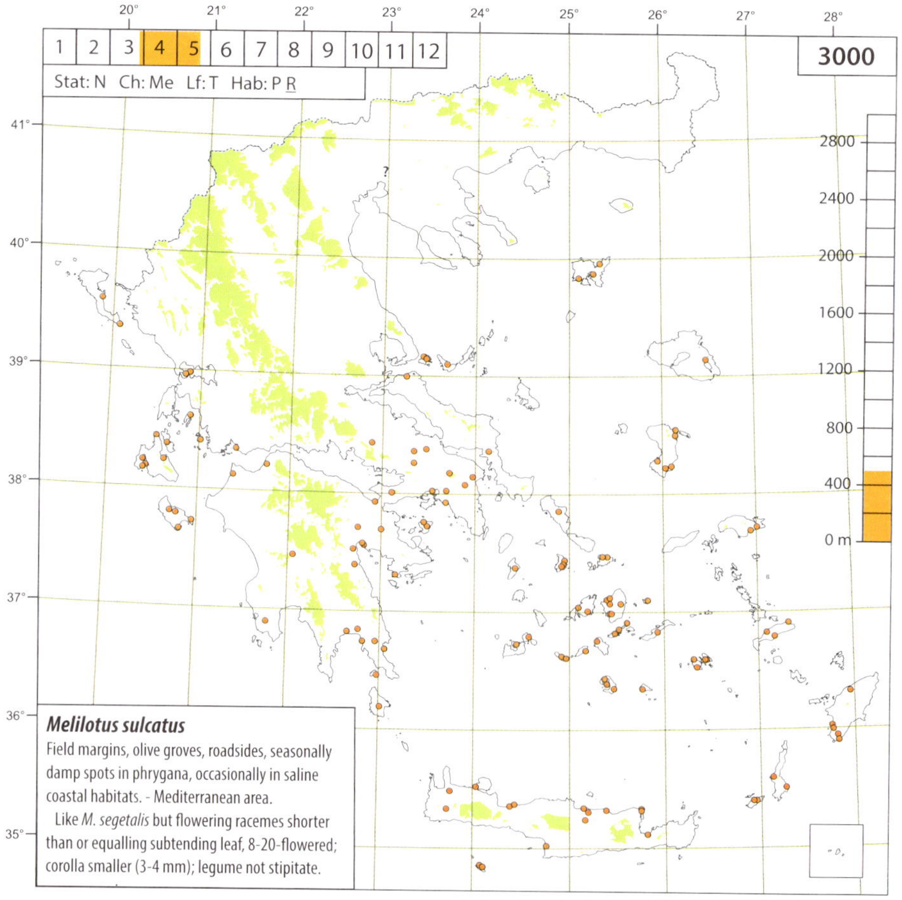

Melilotus sulcatus

Field margins, olive groves, roadsides, seasonally damp spots in phrygana, occasionally in saline coastal habitats. - Mediterranean area.

Like *M. segetalis* but flowering racemes shorter than or equalling subtending leaf, 8-20-flowered; corolla smaller (3-4 mm); legume not stipitate.

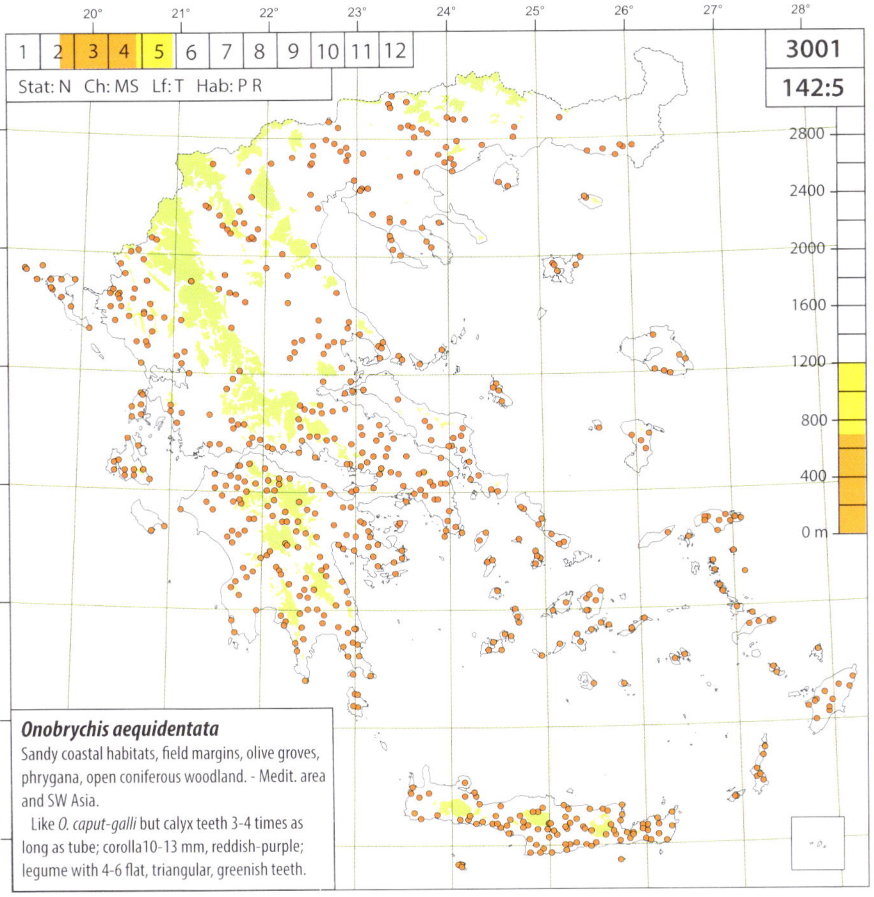

Onobrychis aequidentata
Sandy coastal habitats, field margins, olive groves, phrygana, open coniferous woodland. - Medit. area and SW Asia.
Like *O. caput-galli* but calyx teeth 3-4 times as long as tube; corolla10-13 mm, reddish-purple; legume with 4-6 flat, triangular, greenish teeth.

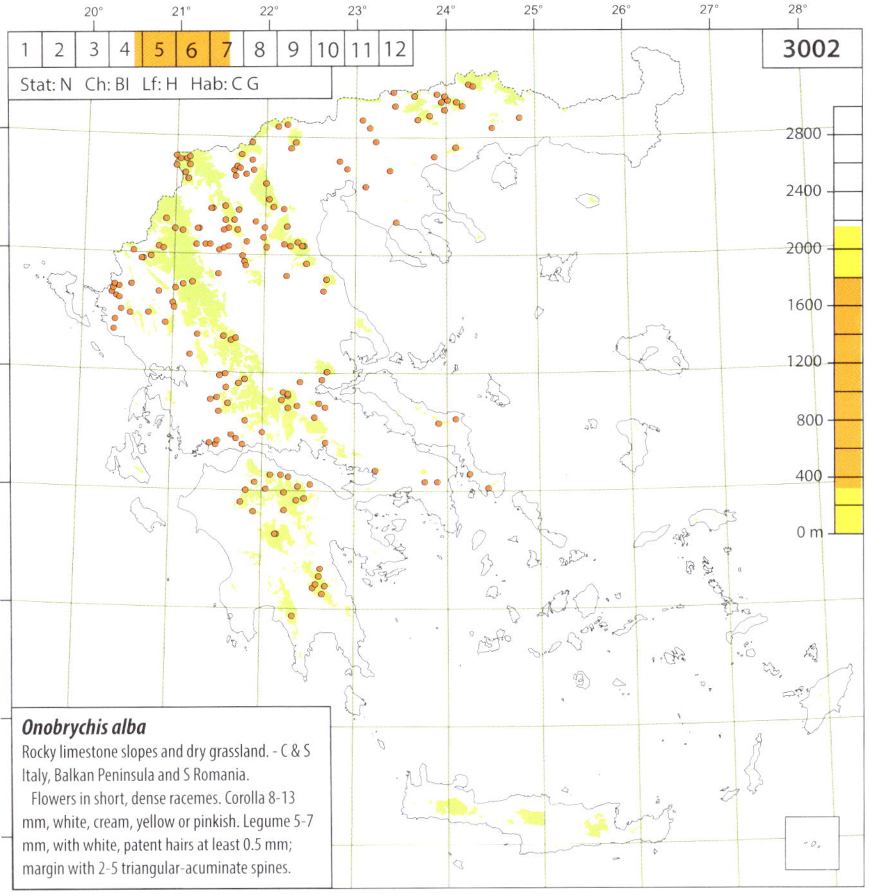

Onobrychis alba
Rocky limestone slopes and dry grassland. - C & S Italy, Balkan Peninsula and S Romania.
Flowers in short, dense racemes. Corolla 8-13 mm, white, cream, yellow or pinkish. Legume 5-7 mm, with white, patent hairs at least 0.5 mm; margin with 2-5 triangular-acuminate spines.

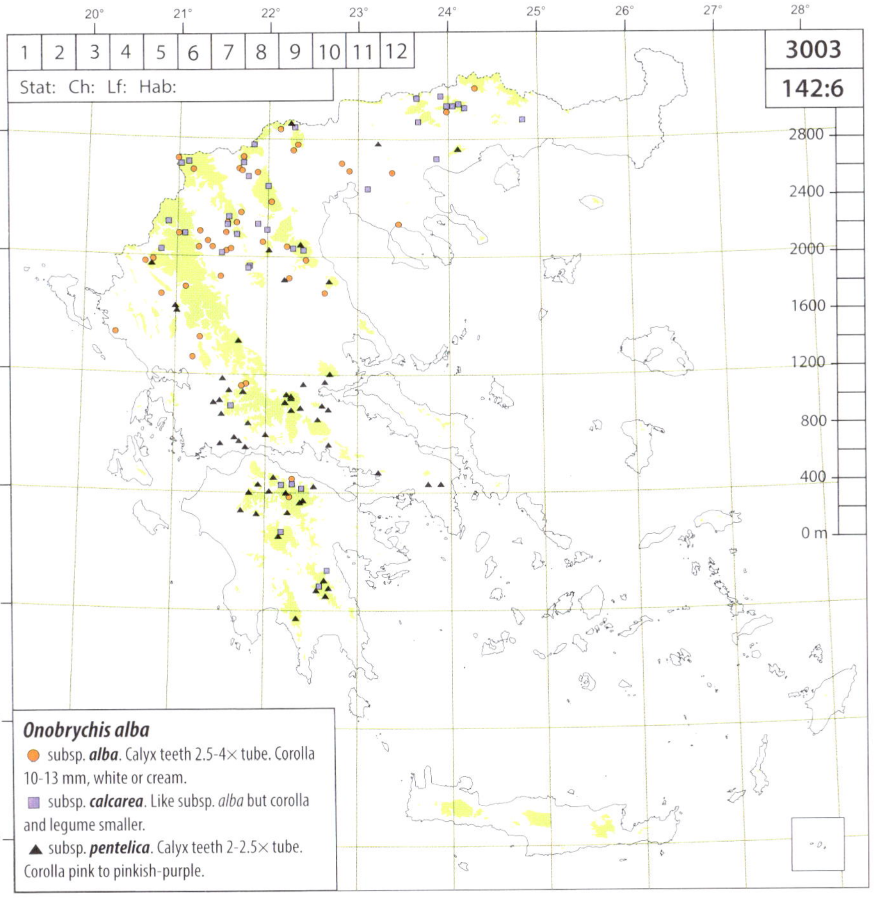

Onobrychis alba
● subsp. ***alba***. Calyx teeth 2.5-4× tube. Corolla 10-13 mm, white or cream.
■ subsp. ***calcarea***. Like subsp. *alba* but corolla and legume smaller.
▲ subsp. ***pentelica***. Calyx teeth 2-2.5× tube. Corolla pink to pinkish-purple.

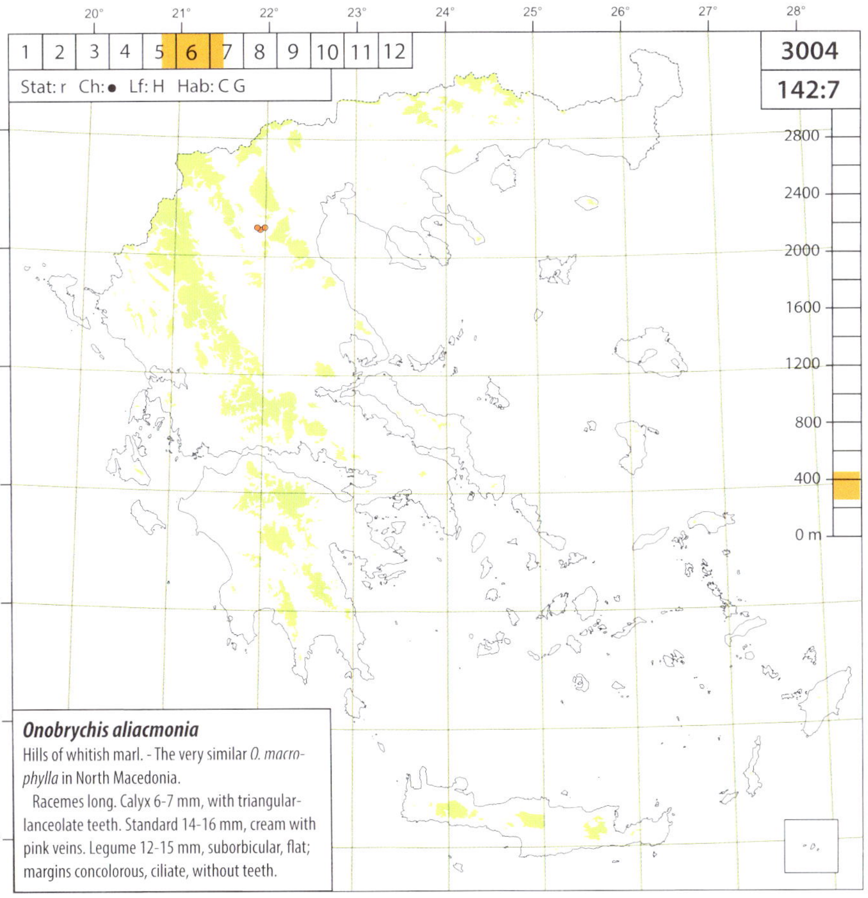

Onobrychis aliacmonia
Hills of whitish marl. - The very similar *O. macrophylla* in North Macedonia.
Racemes long. Calyx 6-7 mm, with triangular-lanceolate teeth. Standard 14-16 mm, cream with pink veins. Legume 12-15 mm, suborbicular, flat; margins concolorous, ciliate, without teeth.

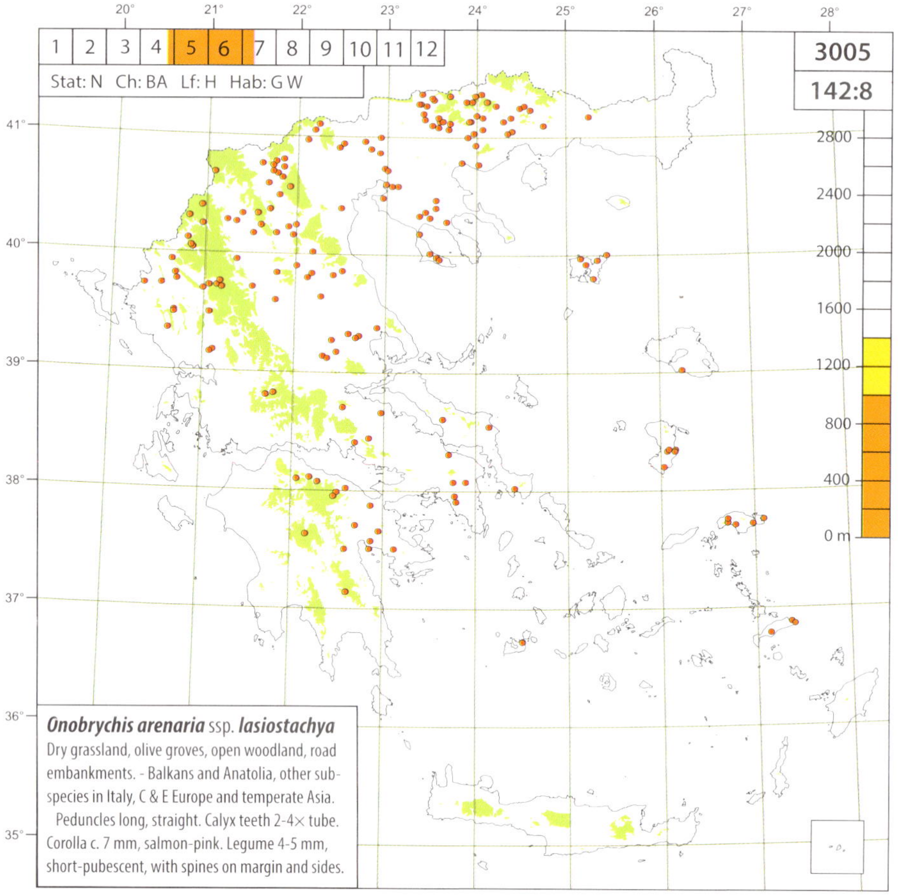

Onobrychis arenaria* ssp. *lasiostachya
Dry grassland, olive groves, open woodland, road embankments. - Balkans and Anatolia, other subspecies in Italy, C & E Europe and temperate Asia.
Peduncles long, straight. Calyx teeth 2-4× tube. Corolla c. 7 mm, salmon-pink. Legume 4-5 mm, short-pubescent, with spines on margin and sides.

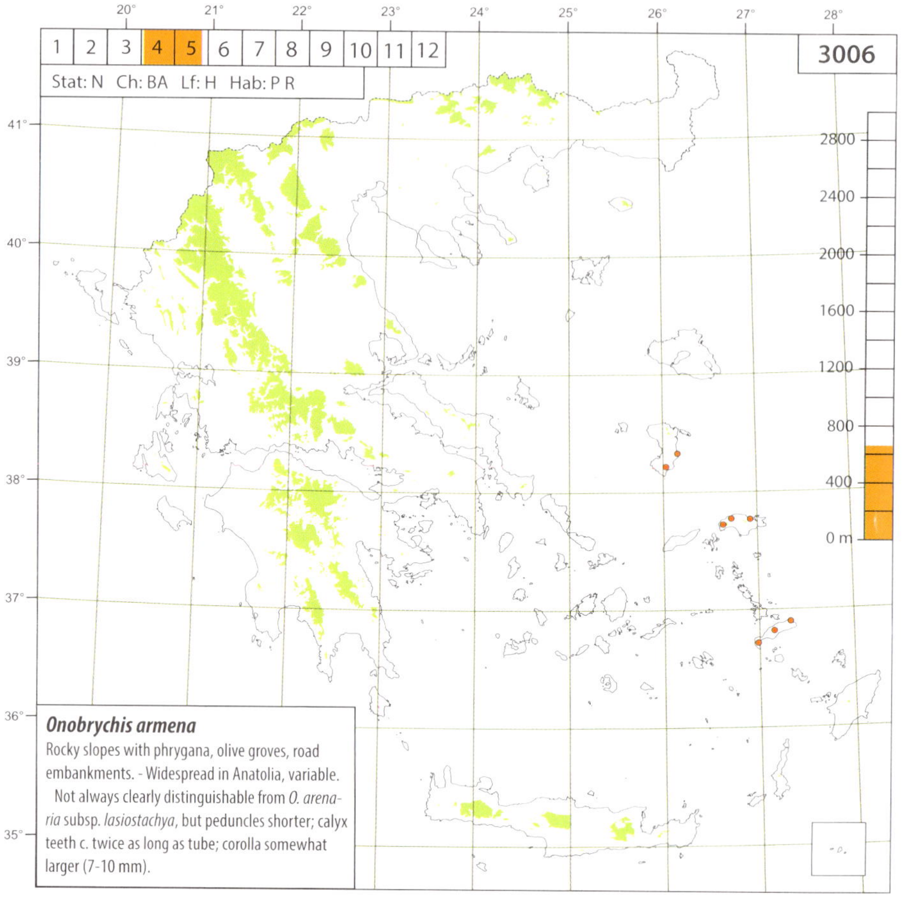

Onobrychis armena
Rocky slopes with phrygana, olive groves, road embankments. - Widespread in Anatolia, variable.
Not always clearly distinguishable from *O. arenaria* subsp. *lasiostachya*, but peduncles shorter; calyx teeth c. twice as long as tube; corolla somewhat larger (7-10 mm).

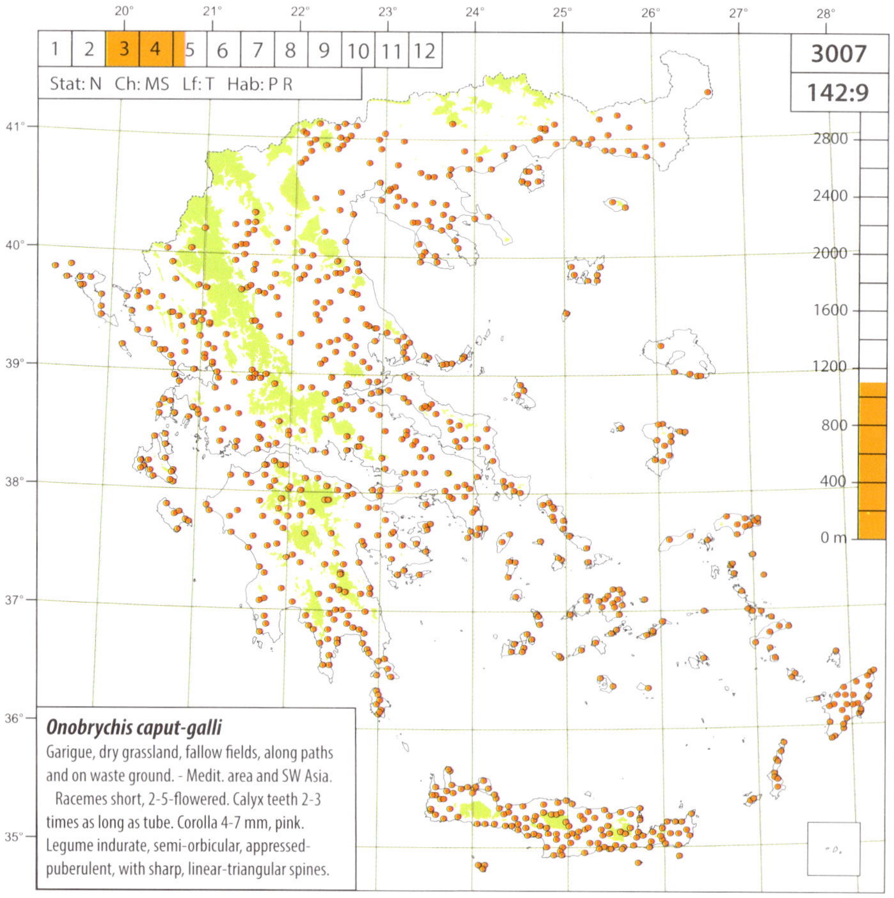

Onobrychis caput-galli
Garigue, dry grassland, fallow fields, along paths and on waste ground. - Medit. area and SW Asia.
Racemes short, 2-5-flowered. Calyx teeth 2-3 times as long as tube. Corolla 4-7 mm, pink. Legume indurate, semi-orbicular, appressed-puberulent, with sharp, linear-triangular spines.

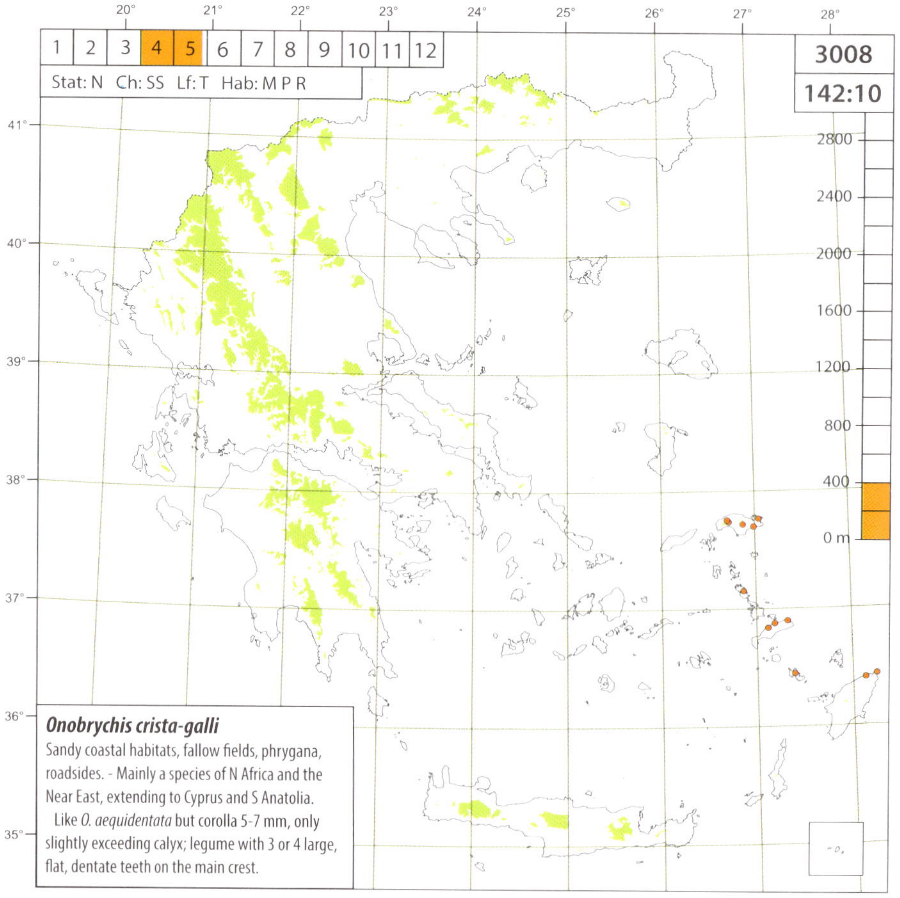

Onobrychis crista-galli
Sandy coastal habitats, fallow fields, phrygana, roadsides. - Mainly a species of N Africa and the Near East, extending to Cyprus and S Anatolia.
Like *O. aequidentata* but corolla 5-7 mm, only slightly exceeding calyx; legume with 3 or 4 large, flat, dentate teeth on the main crest.

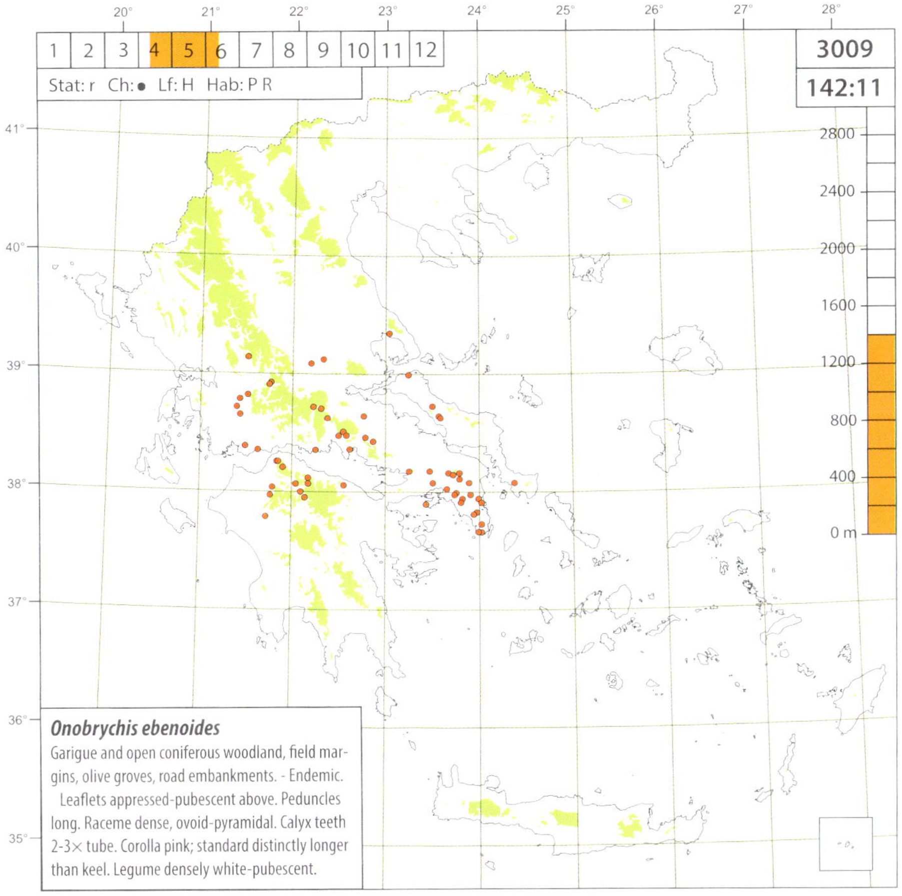

Onobrychis ebenoides

Garigue and open coniferous woodland, field margins, olive groves, road embankments. - Endemic.

Leaflets appressed-pubescent above. Peduncles long. Raceme dense, ovoid-pyramidal. Calyx teeth 2-3× tube. Corolla pink; standard distinctly longer than keel. Legume densely white-pubescent.

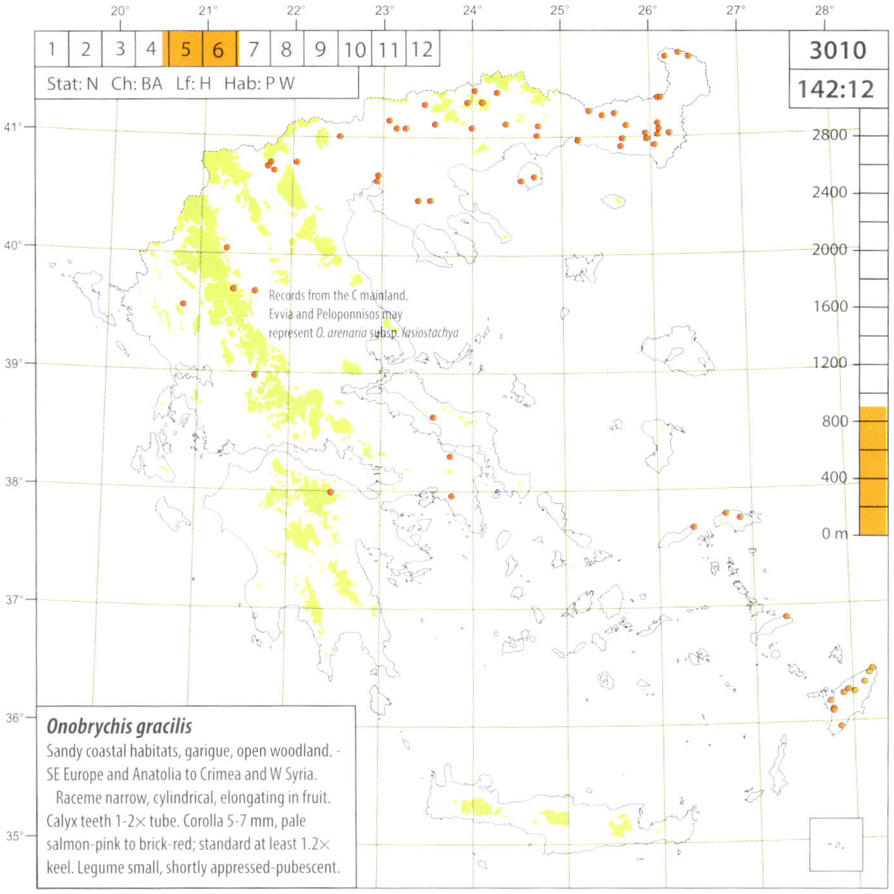

Onobrychis gracilis

Sandy coastal habitats, garigue, open woodland. - SE Europe and Anatolia to Crimea and W Syria.

Raceme narrow, cylindrical, elongating in fruit. Calyx teeth 1-2× tube. Corolla 5-7 mm, pale salmon-pink to brick-red; standard at least 1.2× keel. Legume small, shortly appressed-pubescent.

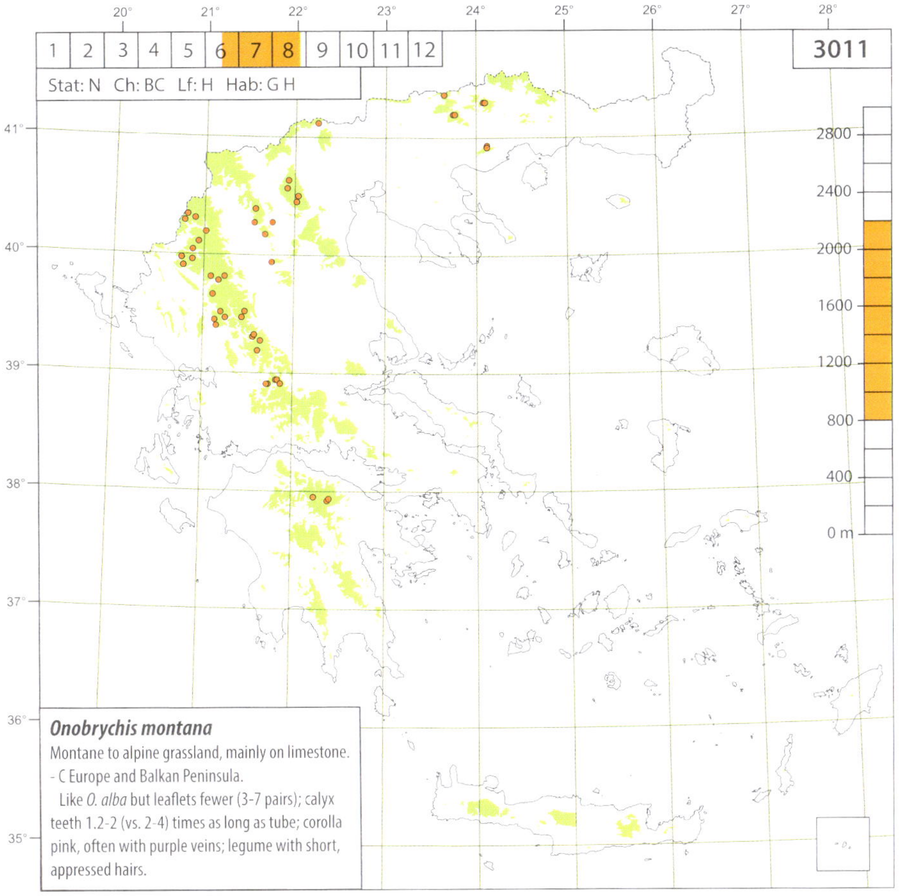

Onobrychis montana

Montane to alpine grassland, mainly on limestone. - C Europe and Balkan Peninsula.

Like *O. alba* but leaflets fewer (3-7 pairs); calyx teeth 1.2-2 (vs. 2-4) times as long as tube; corolla pink, often with purple veins; legume with short, appressed hairs.

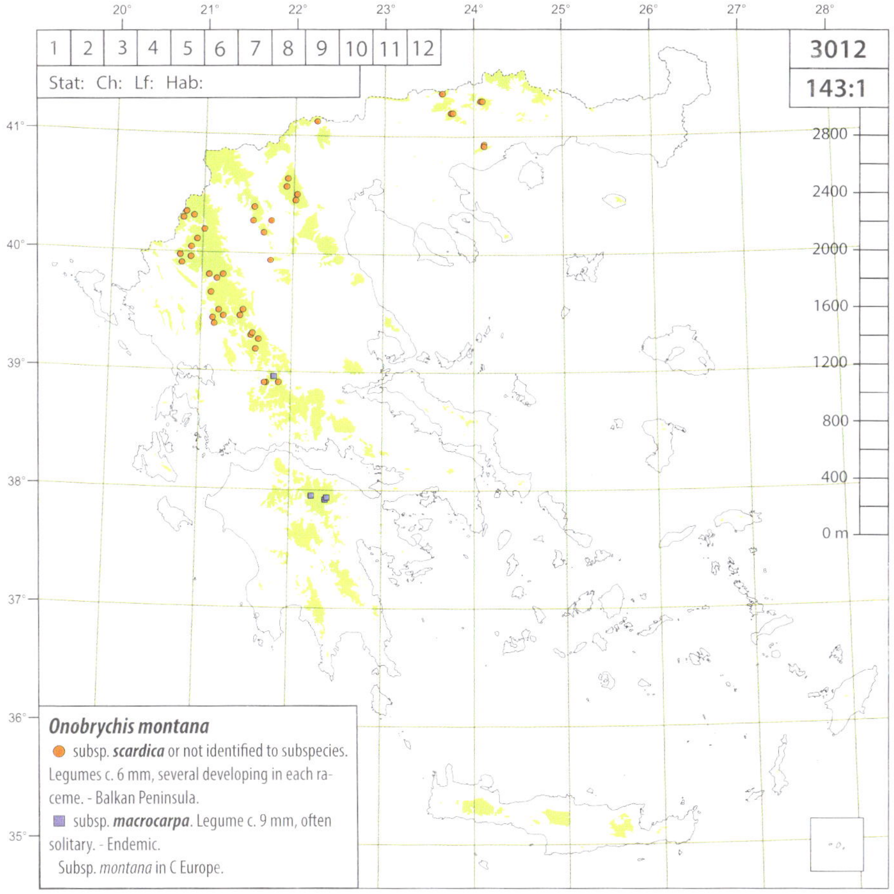

Onobrychis montana

● subsp. ***scardica*** or not identified to subspecies. Legumes c. 6 mm, several developing in each raceme. - Balkan Peninsula.

■ subsp. ***macrocarpa***. Legume c. 9 mm, often solitary. - Endemic.

Subsp. *montana* in C Europe.

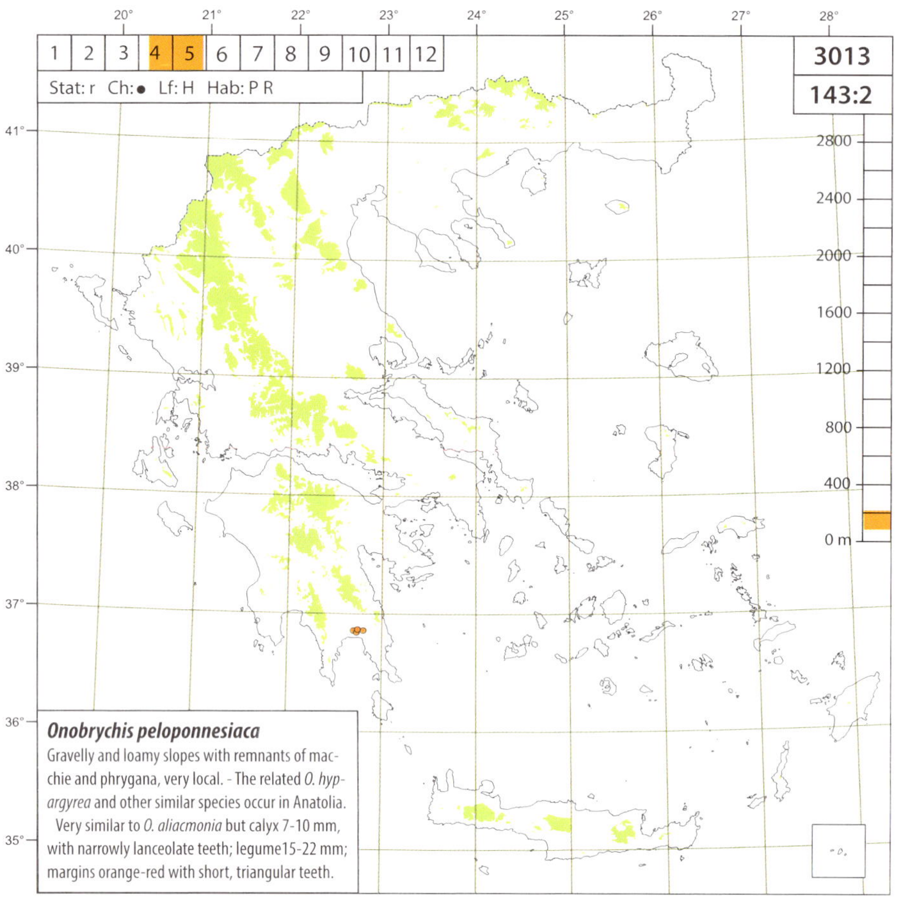

Onobrychis peloponnesiaca
Gravelly and loamy slopes with remnants of macchie and phrygana, very local. - The related *O. hypargyrea* and other similar species occur in Anatolia.
Very similar to *O. aliacmonia* but calyx 7-10 mm, with narrowly lanceolate teeth; legume15-22 mm; margins orange-red with short, triangular teeth.

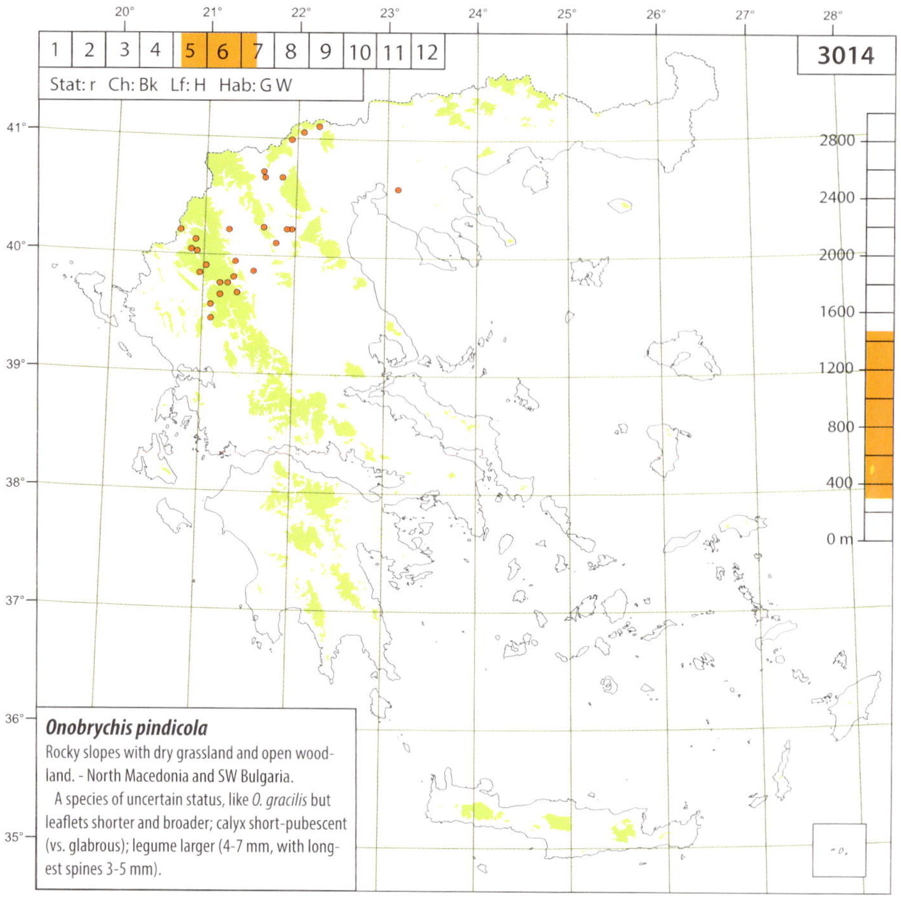

Onobrychis pindicola
Rocky slopes with dry grassland and open woodland. - North Macedonia and SW Bulgaria.
A species of uncertain status, like *O. gracilis* but leaflets shorter and broader; calyx short-pubescent (vs. glabrous); legume larger (4-7 mm, with longest spines 3-5 mm).

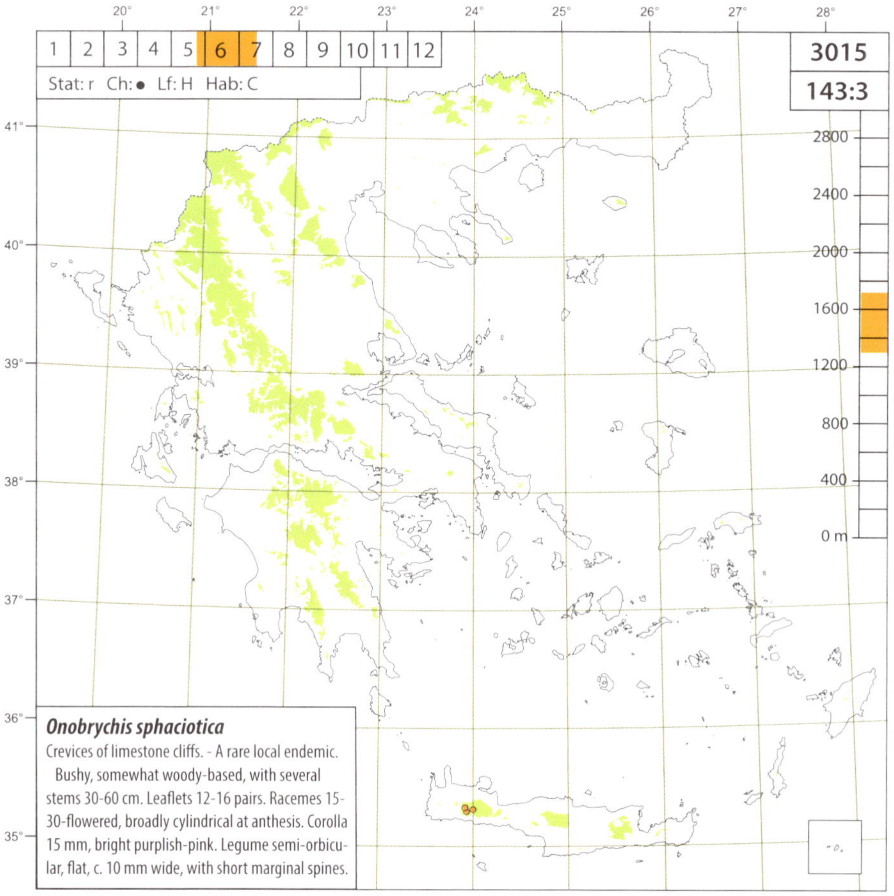

Onobrychis sphaciotica
Crevices of limestone cliffs. - A rare local endemic.
Bushy, somewhat woody-based, with several stems 30-60 cm. Leaflets 12-16 pairs. Racemes 15-30-flowered, broadly cylindrical at anthesis. Corolla 15 mm, bright purplish-pink. Legume semi-orbicular, flat, c. 10 mm wide, with short marginal spines.

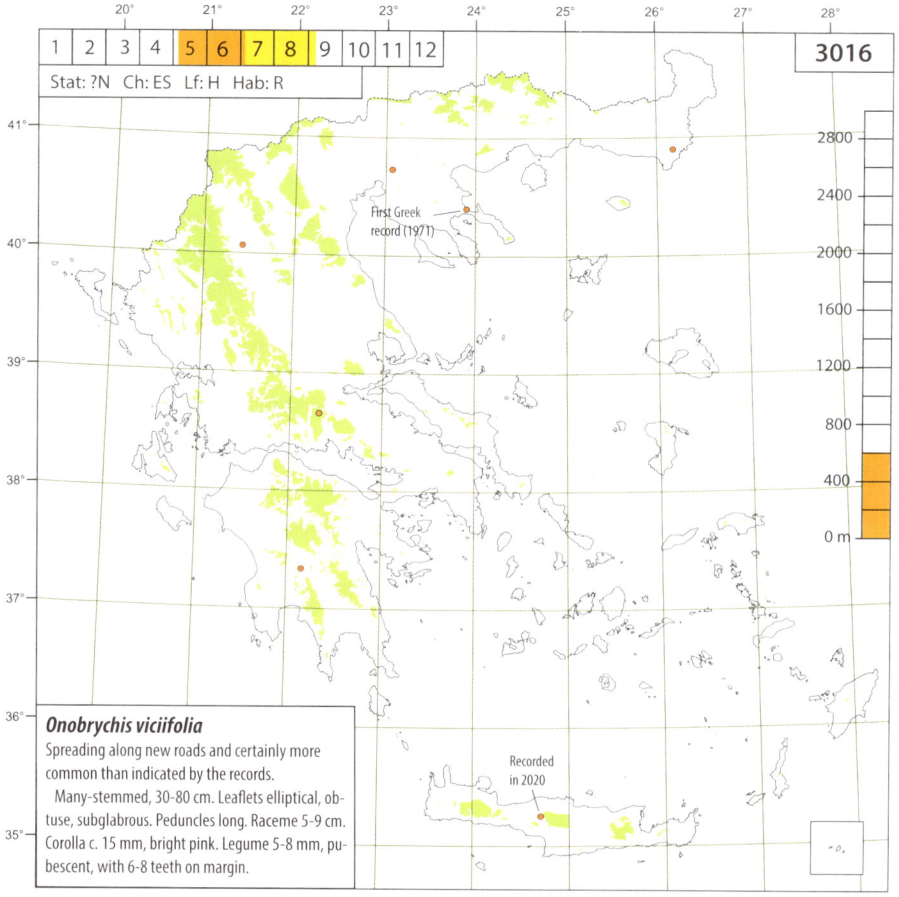

Onobrychis viciifolia
Spreading along new roads and certainly more common than indicated by the records.
Many-stemmed, 30-80 cm. Leaflets elliptical, obtuse, subglabrous. Peduncles long. Raceme 5-9 cm. Corolla c. 15 mm, bright pink. Legume 5-8 mm, pubescent, with 6-8 teeth on margin.

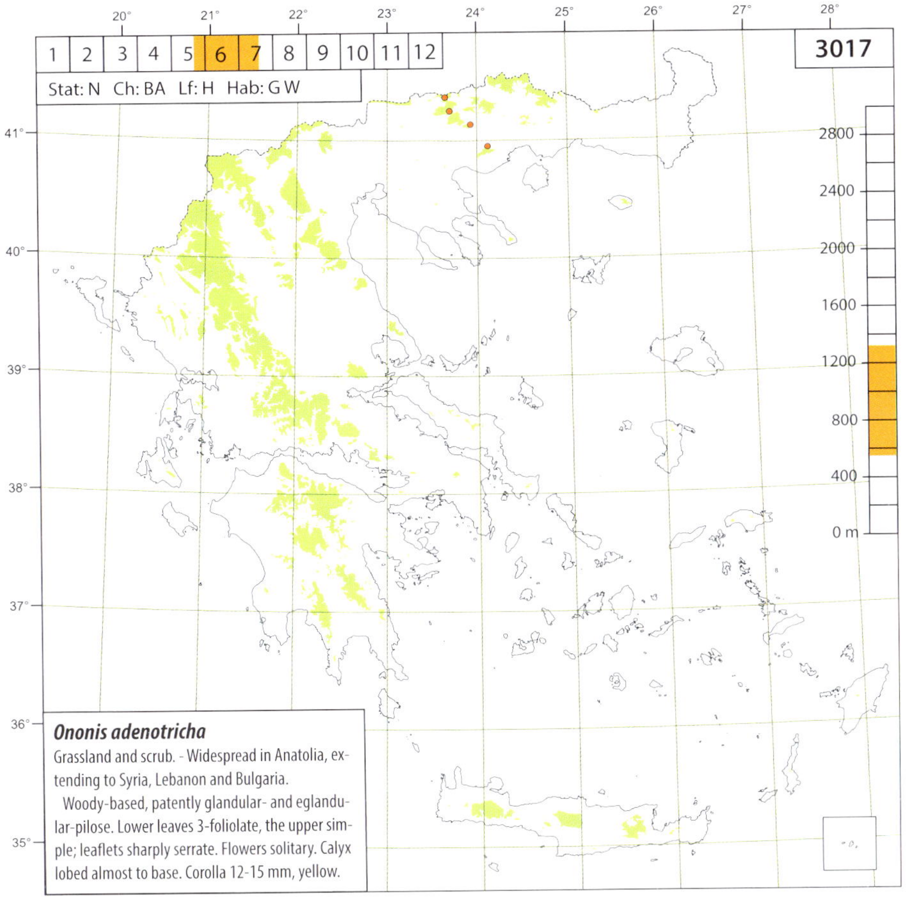

Ononis adenotricha

Grassland and scrub. - Widespread in Anatolia, extending to Syria, Lebanon and Bulgaria.

Woody-based, patently glandular- and eglandular-pilose. Lower leaves 3-foliolate, the upper simple; leaflets sharply serrate. Flowers solitary. Calyx lobed almost to base. Corolla 12-15 mm, yellow.

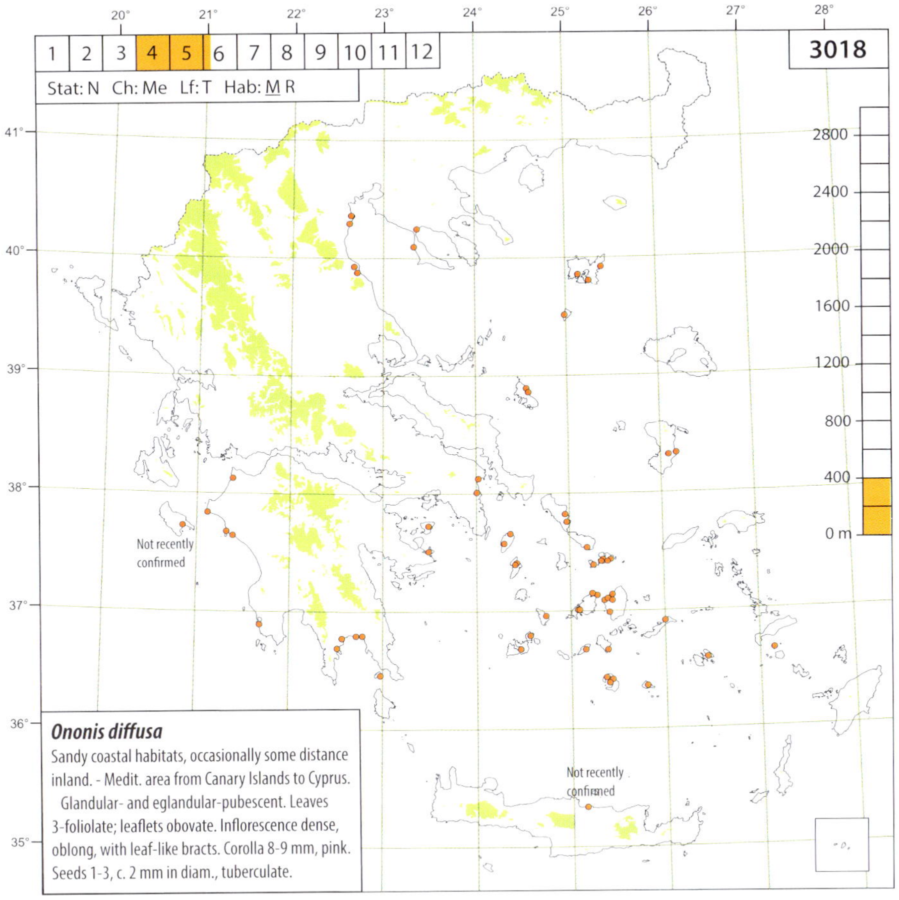

Ononis diffusa

Sandy coastal habitats, occasionally some distance inland. - Medit. area from Canary Islands to Cyprus.

Glandular- and eglandular-pubescent. Leaves 3-foliolate; leaflets obovate. Inflorescence dense, oblong, with leaf-like bracts. Corolla 8-9 mm, pink. Seeds 1-3, c. 2 mm in diam., tuberculate.

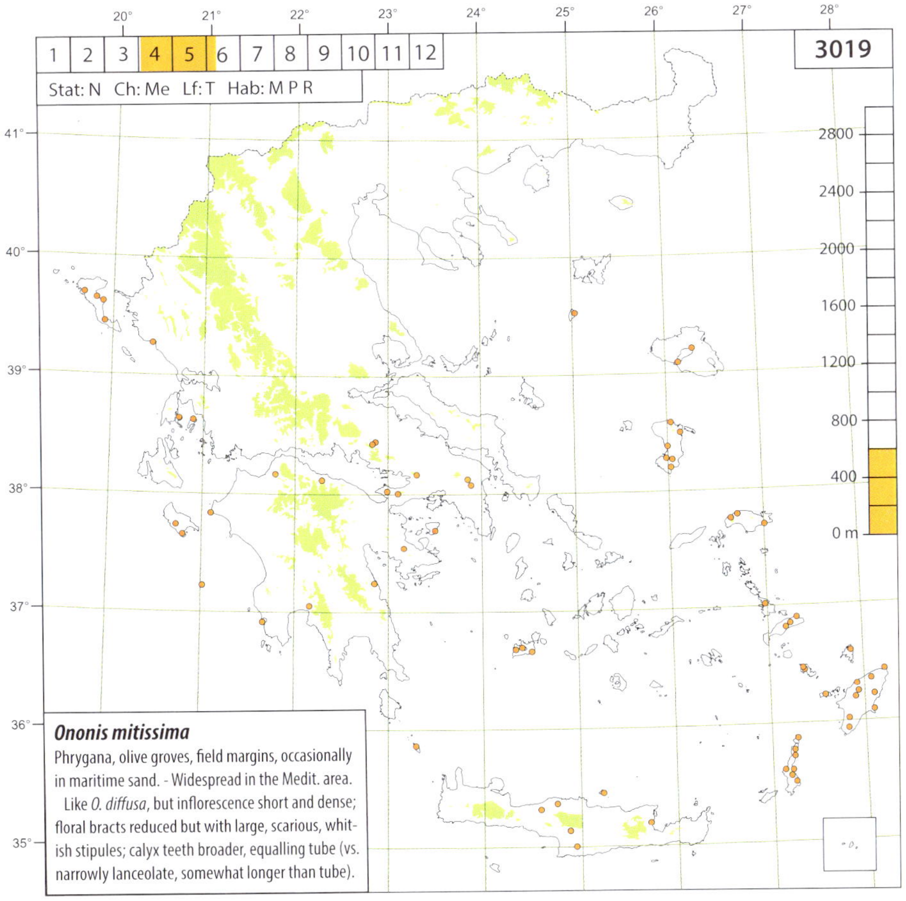

Ononis mitissima

Phrygana, olive groves, field margins, occasionally in maritime sand. - Widespread in the Medit. area.

Like *O. diffusa*, but inflorescence short and dense; floral bracts reduced but with large, scarious, whitish stipules; calyx teeth broader, equalling tube (vs. narrowly lanceolate, somewhat longer than tube).

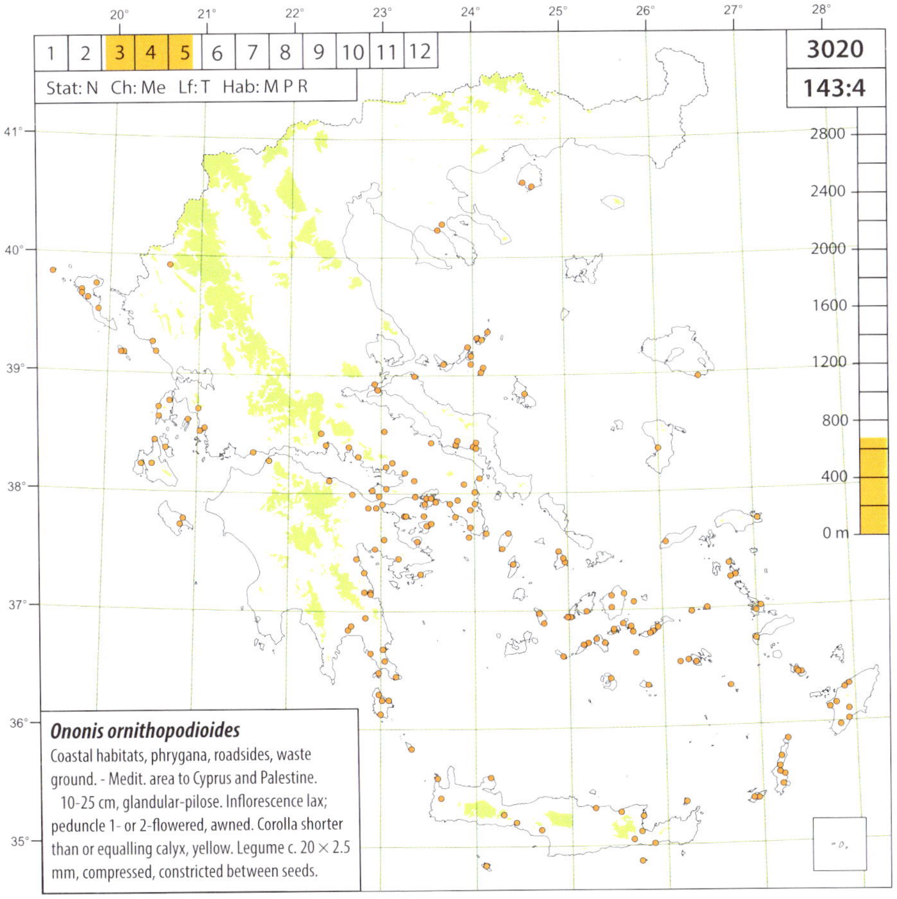

Ononis ornithopodioides

Coastal habitats, phrygana, roadsides, waste ground. - Medit. area to Cyprus and Palestine.

10-25 cm, glandular-pilose. Inflorescence lax; peduncle 1- or 2-flowered, awned. Corolla shorter than or equalling calyx, yellow. Legume c. 20 × 2.5 mm, compressed, constricted between seeds.

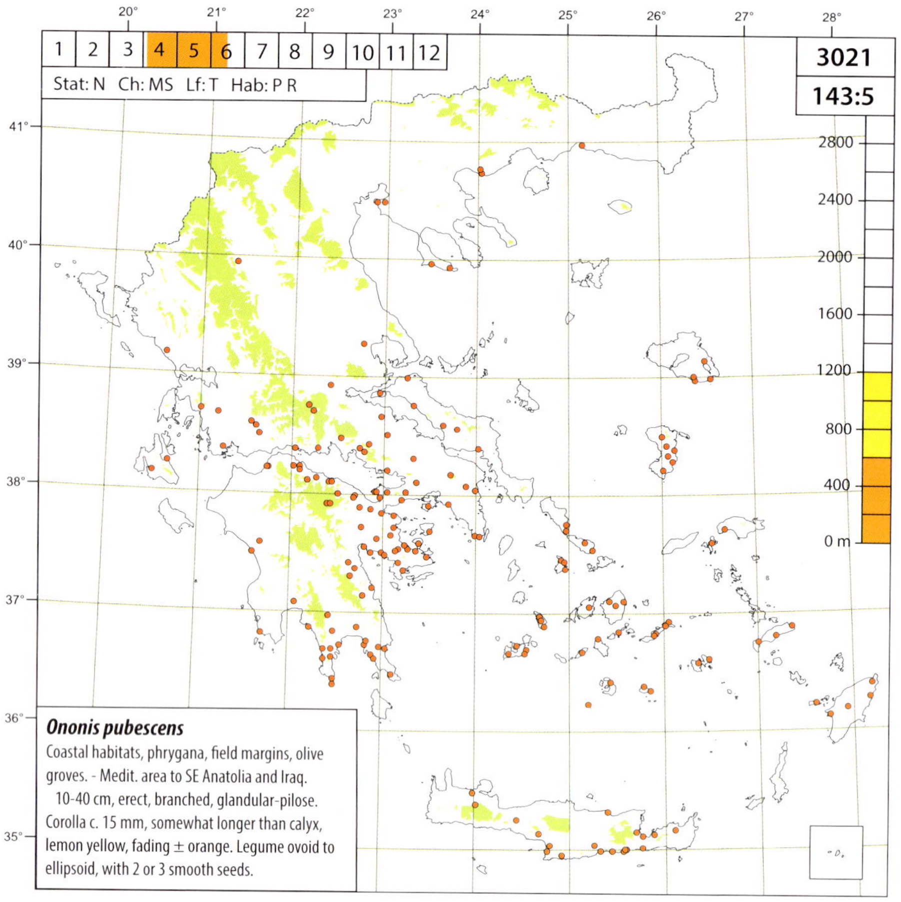

Ononis pubescens
Coastal habitats, phrygana, field margins, olive groves. - Medit. area to SE Anatolia and Iraq.
10-40 cm, erect, branched, glandular-pilose. Corolla c. 15 mm, somewhat longer than calyx, lemon yellow, fading ± orange. Legume ovoid to ellipsoid, with 2 or 3 smooth seeds.

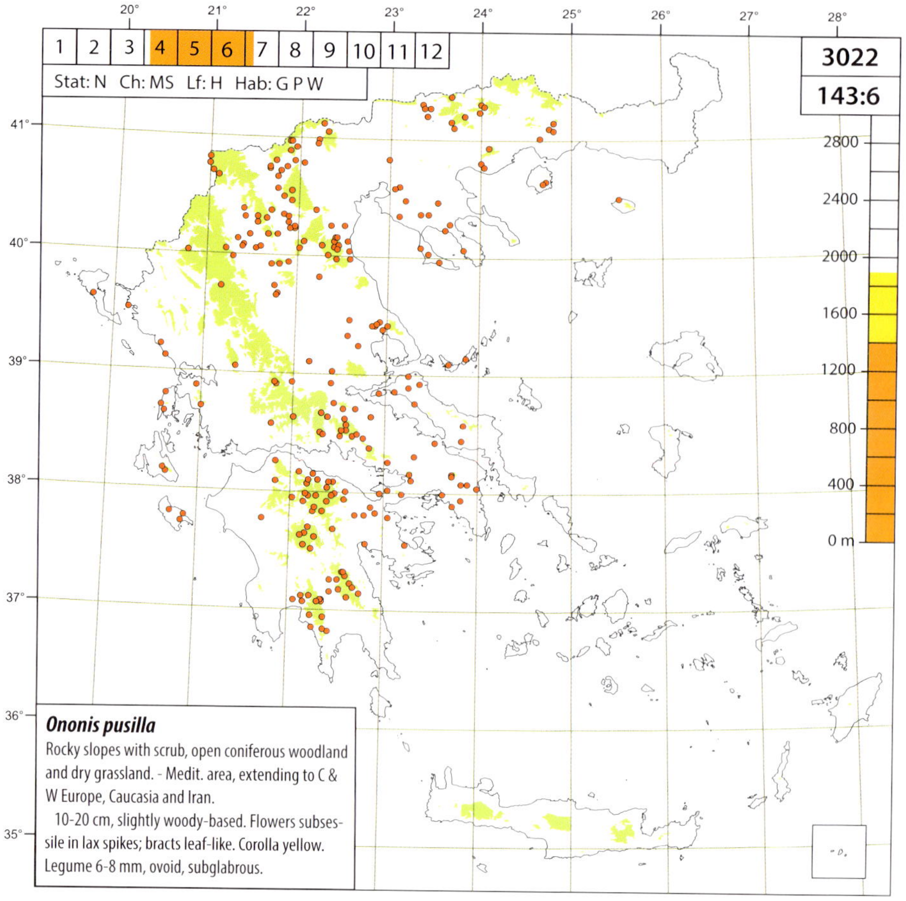

Ononis pusilla
Rocky slopes with scrub, open coniferous woodland and dry grassland. - Medit. area, extending to C & W Europe, Caucasia and Iran.
10-20 cm, slightly woody-based. Flowers subsessile in lax spikes; bracts leaf-like. Corolla yellow. Legume 6-8 mm, ovoid, subglabrous.

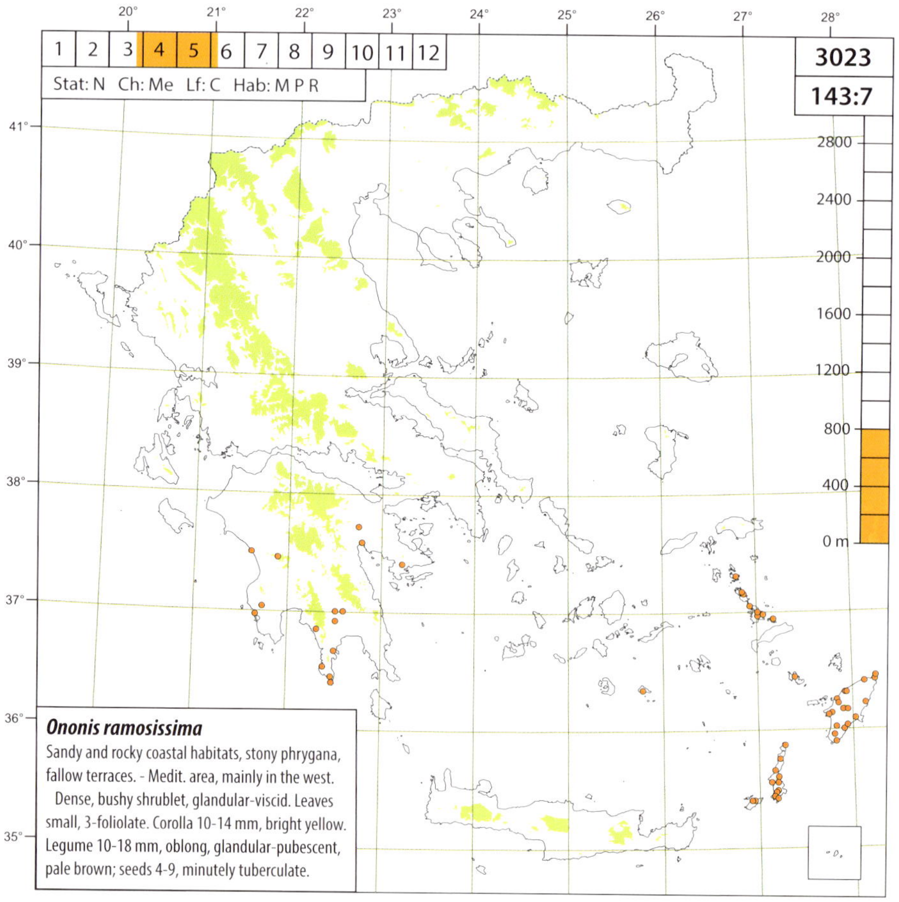

Ononis ramosissima
Sandy and rocky coastal habitats, stony phrygana, fallow terraces. - Medit. area, mainly in the west.
Dense, bushy shrublet, glandular-viscid. Leaves small, 3-foliolate. Corolla 10-14 mm, bright yellow. Legume 10-18 mm, oblong, glandular-pubescent, pale brown; seeds 4-9, minutely tuberculate.

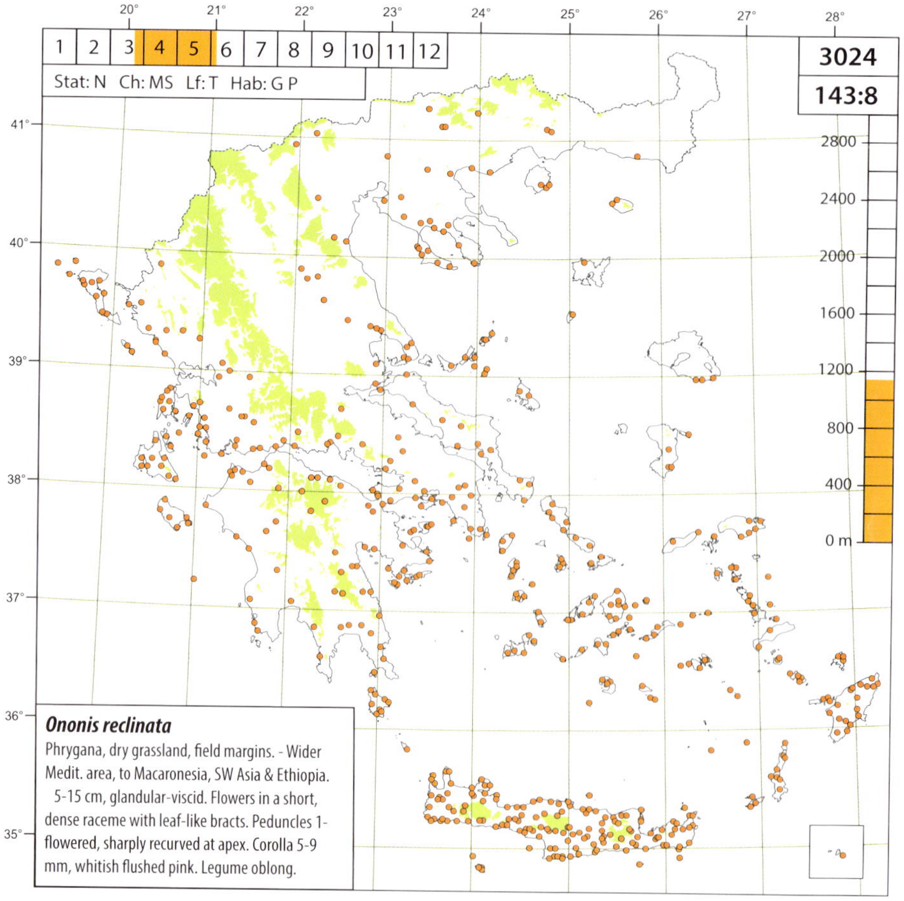

Ononis reclinata
Phrygana, dry grassland, field margins. - Wider Medit. area, to Macaronesia, SW Asia & Ethiopia.
5-15 cm, glandular-viscid. Flowers in a short, dense raceme with leaf-like bracts. Peduncles 1-flowered, sharply recurved at apex. Corolla 5-9 mm, whitish flushed pink. Legume oblong.

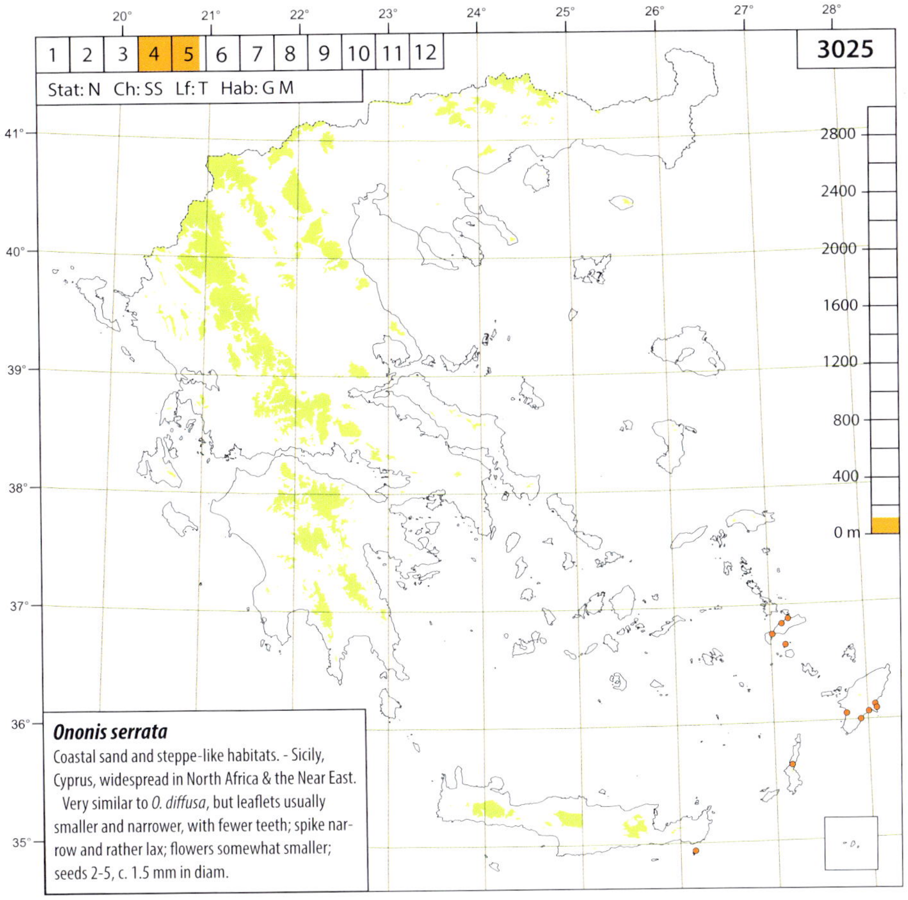

Ononis serrata

Coastal sand and steppe-like habitats. - Sicily, Cyprus, widespread in North Africa & the Near East.

Very similar to *O. diffusa*, but leaflets usually smaller and narrower, with fewer teeth; spike narrow and rather lax; flowers somewhat smaller; seeds 2-5, c. 1.5 mm in diam.

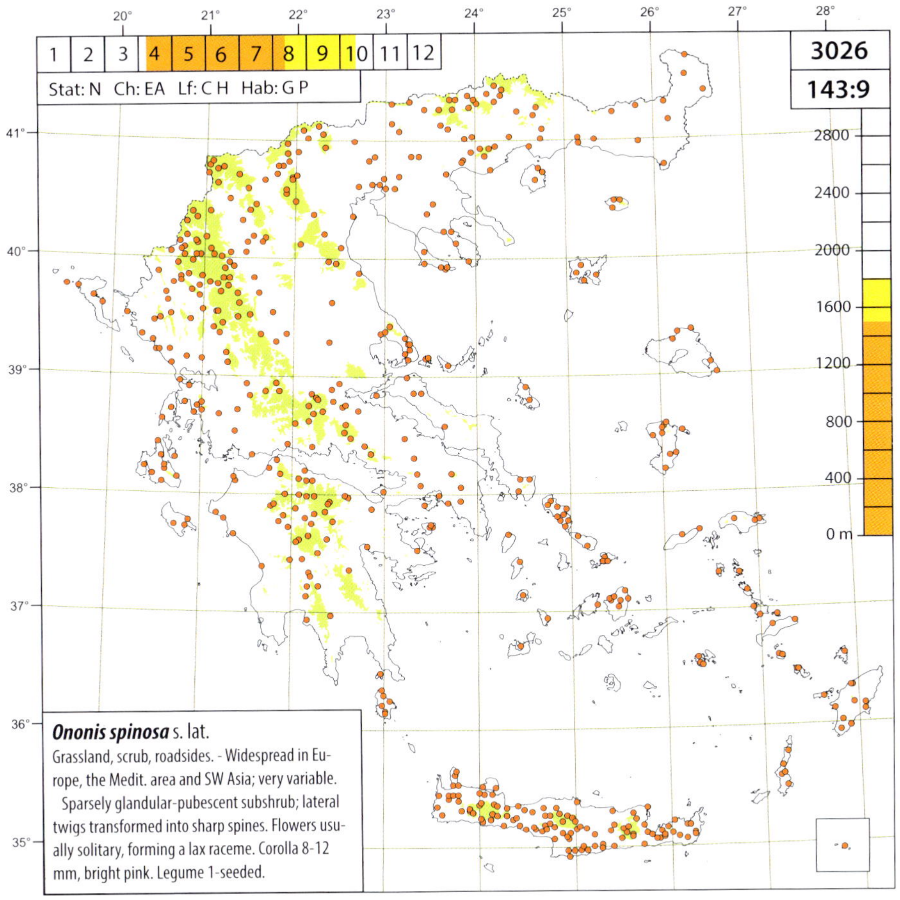

Ononis spinosa s. lat.

Grassland, scrub, roadsides. - Widespread in Europe, the Medit. area and SW Asia; very variable.

Sparsely glandular-pubescent subshrub; lateral twigs transformed into sharp spines. Flowers usually solitary, forming a lax raceme. Corolla 8-12 mm, bright pink. Legume 1-seeded.

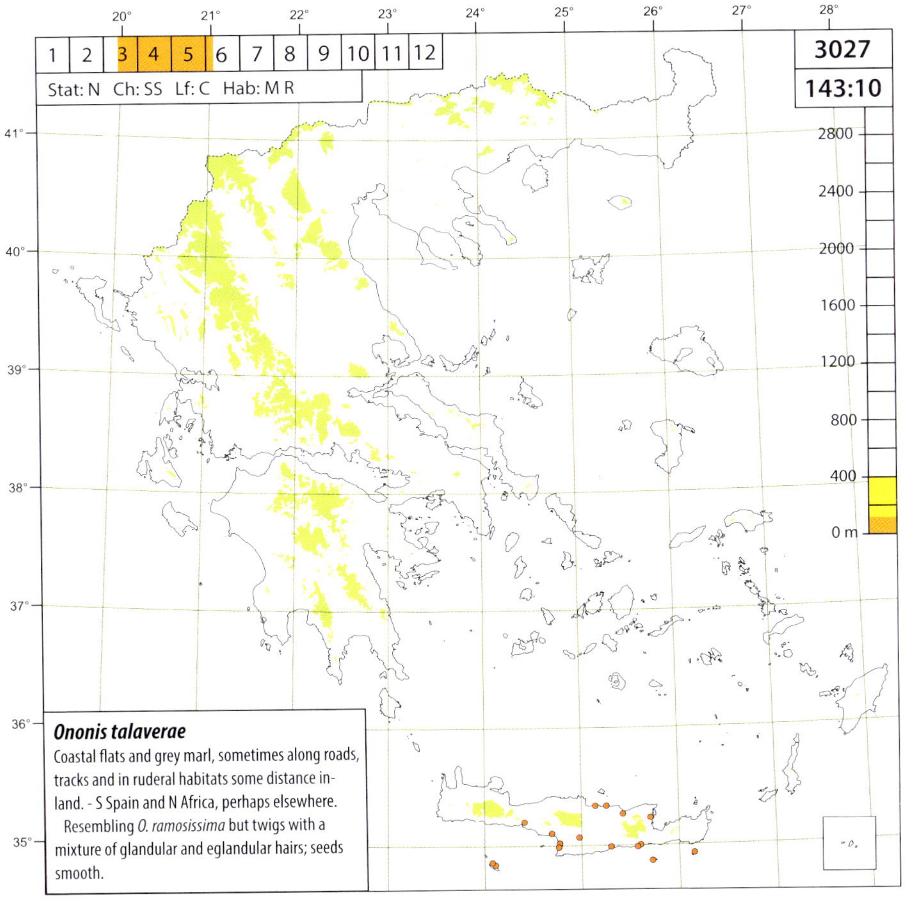

Ononis talaverae

Coastal flats and grey marl, sometimes along roads, tracks and in ruderal habitats some distance inland. - S Spain and N Africa, perhaps elsewhere.

Resembling *O. ramosissima* but twigs with a mixture of glandular and eglandular hairs; seeds smooth.

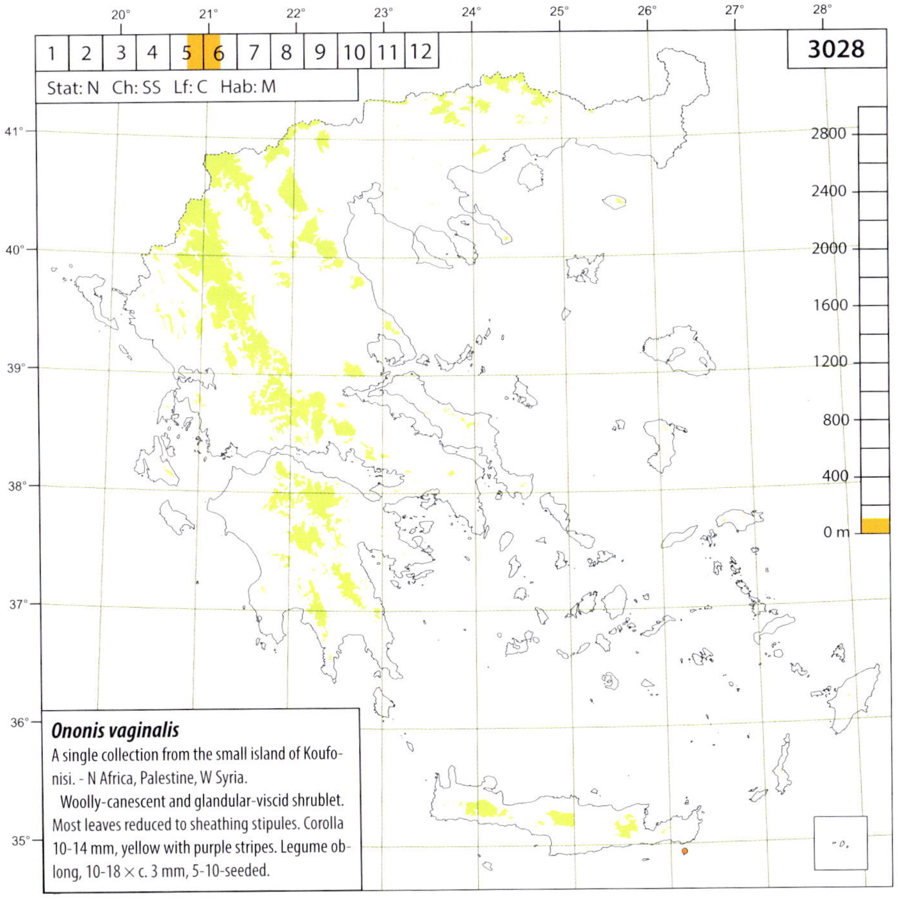

Ononis vaginalis

A single collection from the small island of Koufonisi. - N Africa, Palestine, W Syria.

Woolly-canescent and glandular-viscid shrublet. Most leaves reduced to sheathing stipules. Corolla 10-14 mm, yellow with purple stripes. Legume oblong, 10-18 × c. 3 mm, 5-10-seeded.

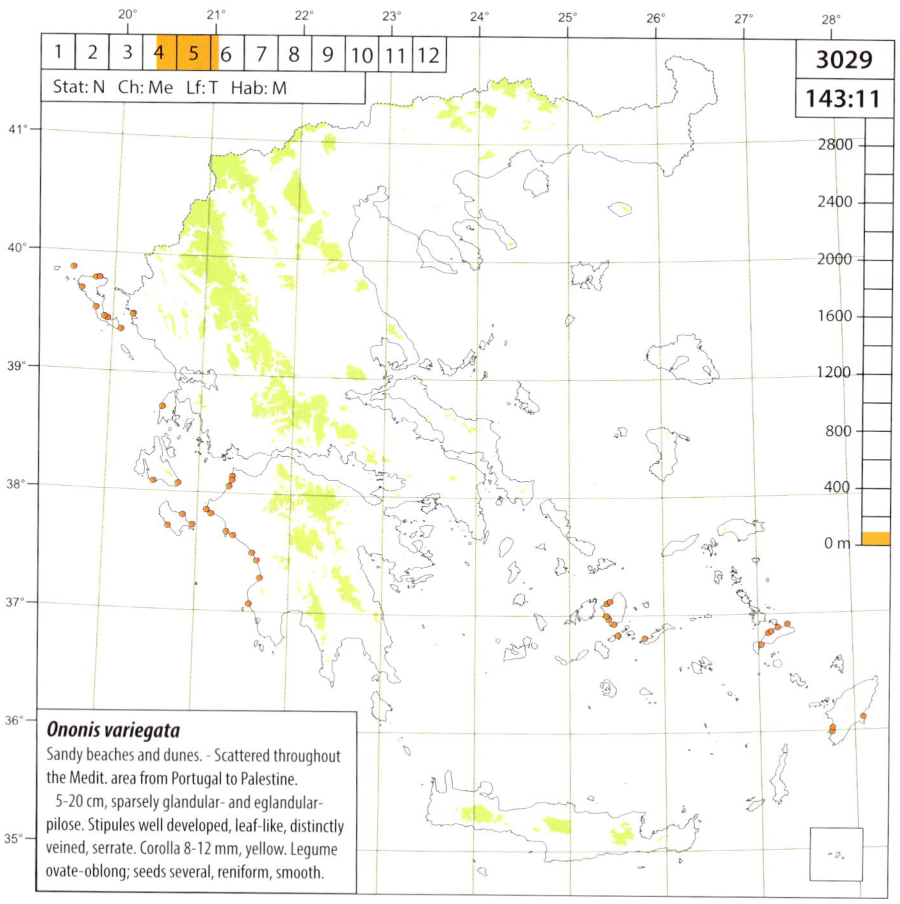

Ononis variegata

Sandy beaches and dunes. - Scattered throughout the Medit. area from Portugal to Palestine.

5-20 cm, sparsely glandular- and eglandular-pilose. Stipules well developed, leaf-like, distinctly veined, serrate. Corolla 8-12 mm, yellow. Legume ovate-oblong; seeds several, reniform, smooth.

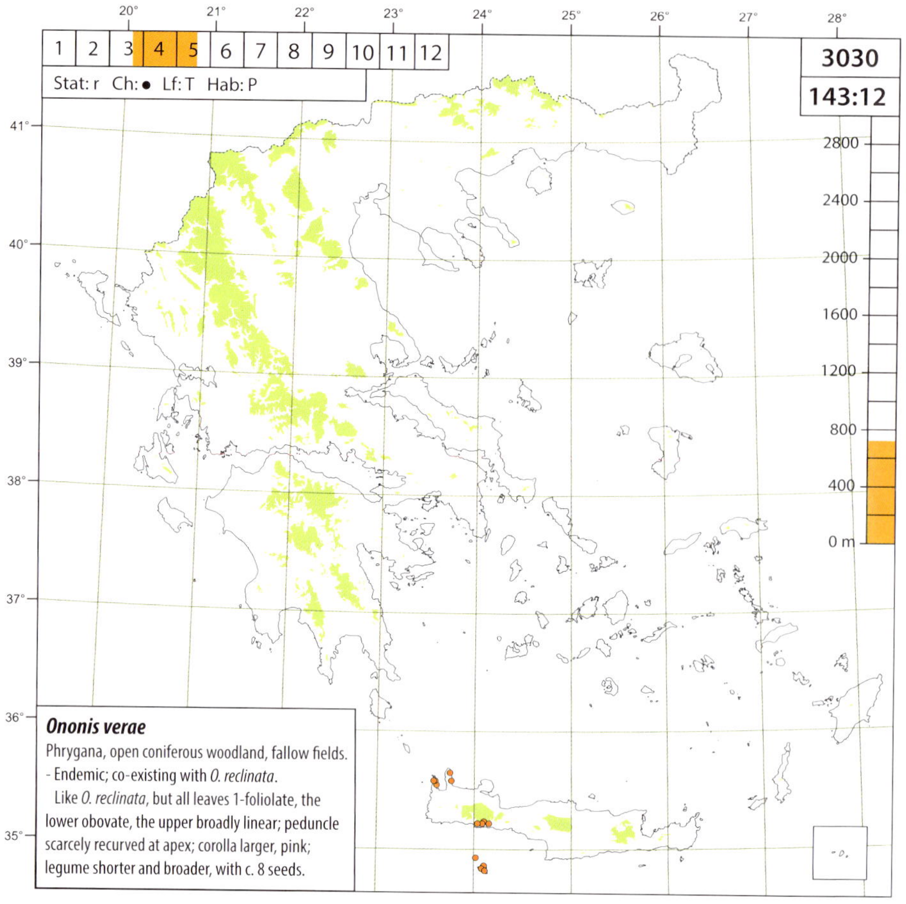

Ononis verae

Phrygana, open coniferous woodland, fallow fields. - Endemic; co-existing with *O. reclinata*.

Like *O. reclinata*, but all leaves 1-foliolate, the lower obovate, the upper broadly linear; peduncle scarcely recurved at apex; corolla larger, pink; legume shorter and broader, with c. 8 seeds.

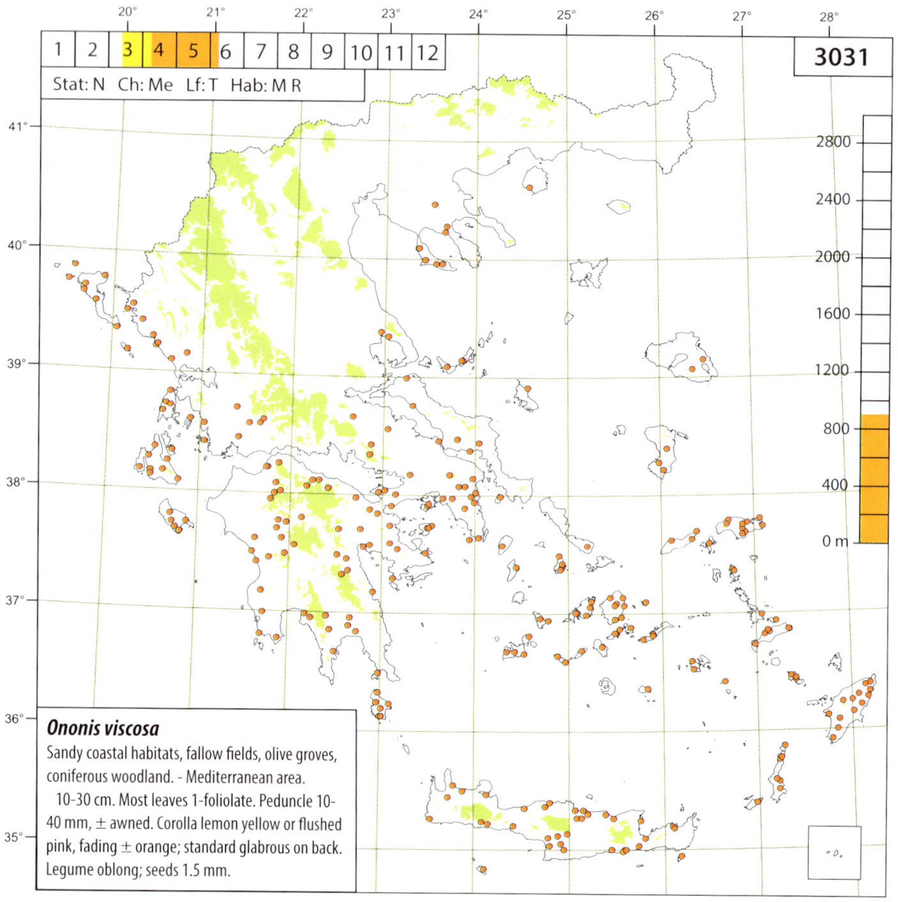

Ononis viscosa

Sandy coastal habitats, fallow fields, olive groves, coniferous woodland. - Mediterranean area.

10-30 cm. Most leaves 1-foliolate. Peduncle 10-40 mm, ± awned. Corolla lemon yellow or flushed pink, fading ± orange; standard glabrous on back. Legume oblong; seeds 1.5 mm.

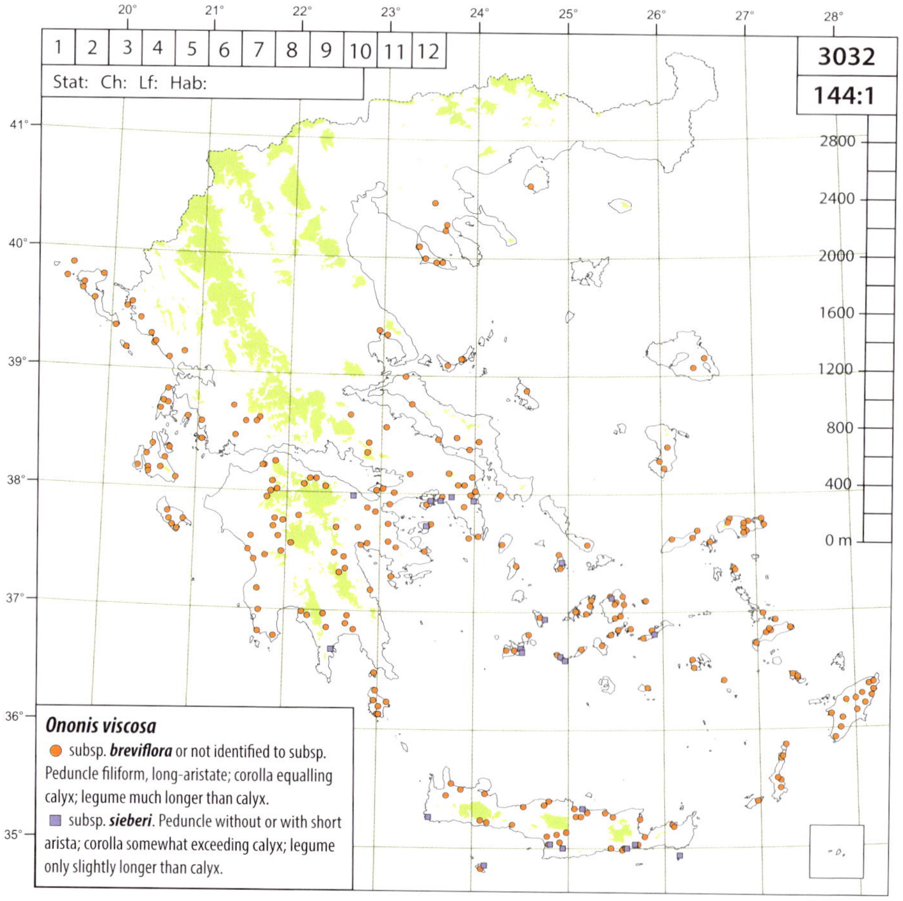

Ononis viscosa

● subsp. ***breviflora*** or not identified to subsp. Peduncle filiform, long-aristate; corolla equalling calyx; legume much longer than calyx.

■ subsp. ***sieberi***. Peduncle without or with short arista; corolla somewhat exceeding calyx; legume only slightly longer than calyx.

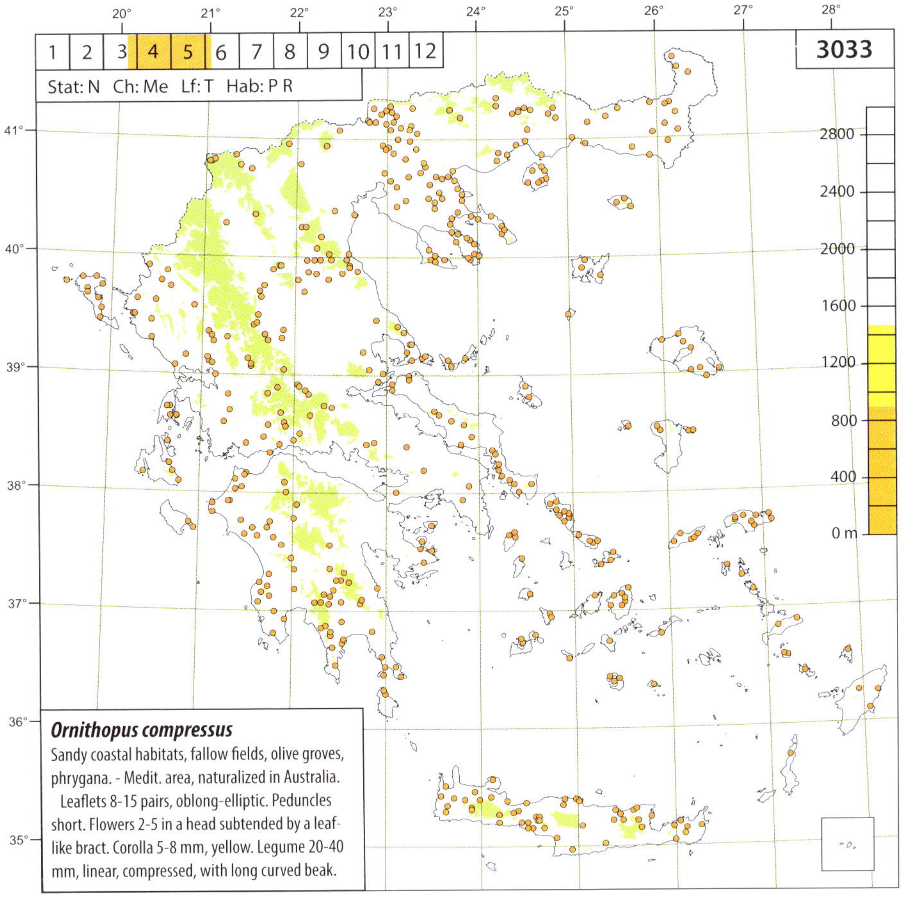

Ornithopus compressus

Sandy coastal habitats, fallow fields, olive groves, phrygana. - Medit. area, naturalized in Australia.

Leaflets 8-15 pairs, oblong-elliptic. Peduncles short. Flowers 2-5 in a head subtended by a leaf-like bract. Corolla 5-8 mm, yellow. Legume 20-40 mm, linear, compressed, with long curved beak.

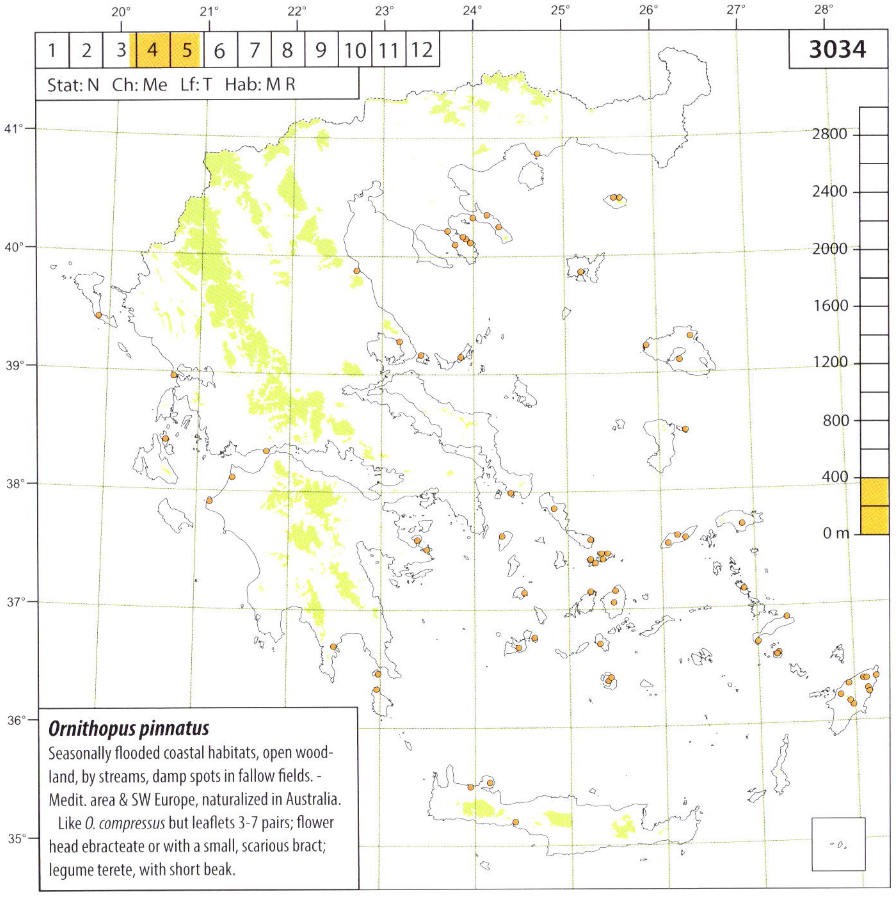

Ornithopus pinnatus

Seasonally flooded coastal habitats, open woodland, by streams, damp spots in fallow fields. - Medit. area & SW Europe, naturalized in Australia.

Like *O. compressus* but leaflets 3-7 pairs; flower head ebracteate or with a small, scarious bract; legume terete, with short beak.

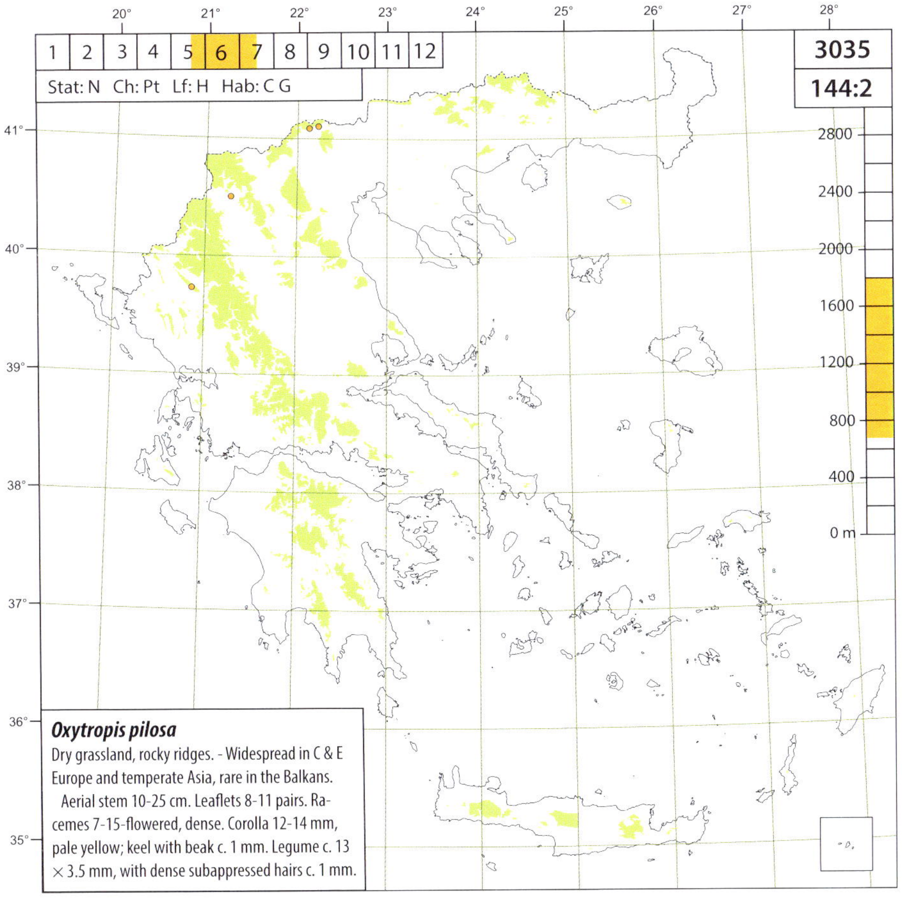

Oxytropis pilosa

Dry grassland, rocky ridges. - Widespread in C & E Europe and temperate Asia, rare in the Balkans.

Aerial stem 10-25 cm. Leaflets 8-11 pairs. Racemes 7-15-flowered, dense. Corolla 12-14 mm, pale yellow; keel with beak c. 1 mm. Legume c. 13 × 3.5 mm, with dense subappressed hairs c. 1 mm.

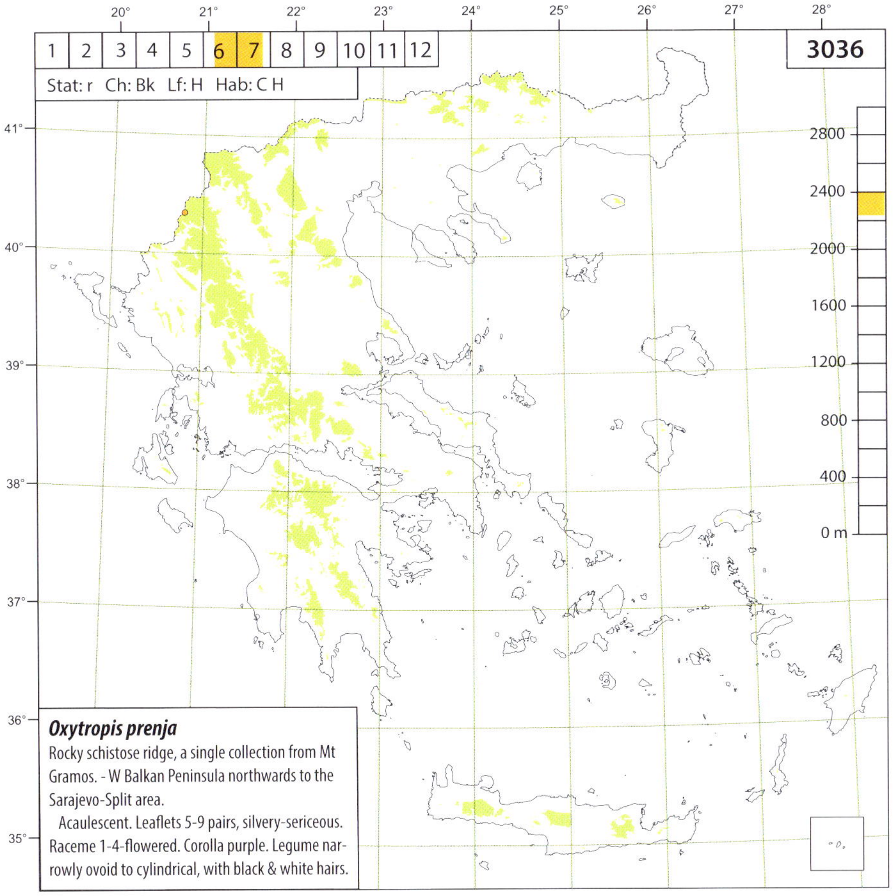

Oxytropis prenja

Rocky schistose ridge, a single collection from Mt Gramos. - W Balkan Peninsula northwards to the Sarajevo-Split area.

Acaulescent. Leaflets 5-9 pairs, silvery-sericeous. Raceme 1-4-flowered. Corolla purple. Legume narrowly ovoid to cylindrical, with black & white hairs.

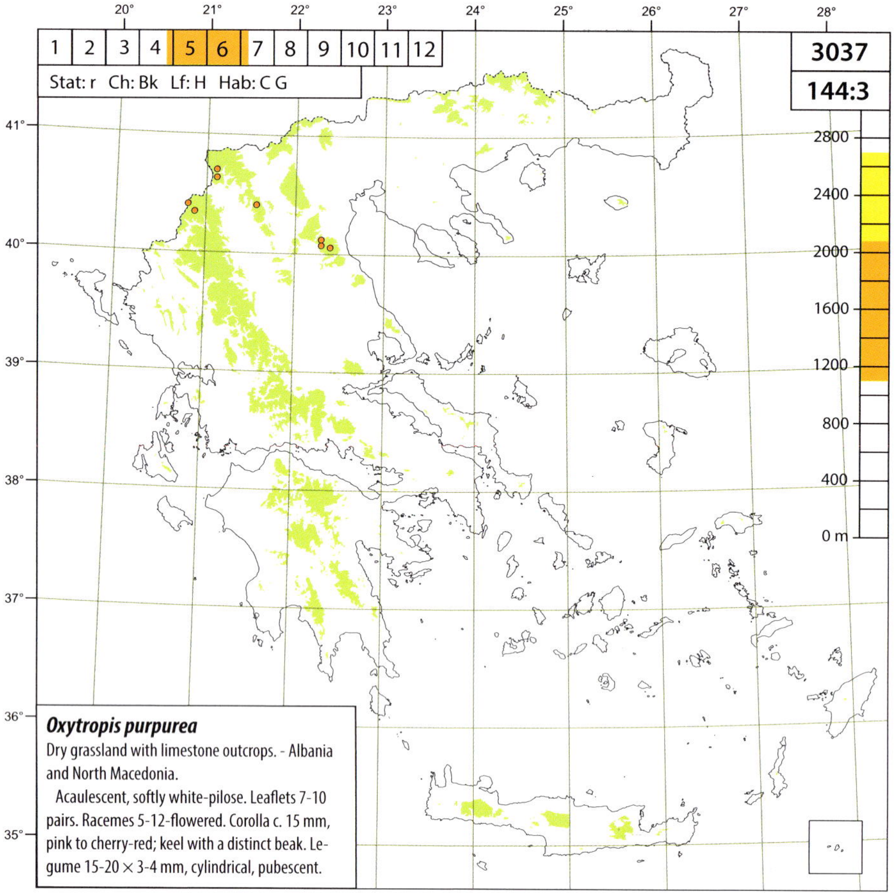

Oxytropis purpurea

Dry grassland with limestone outcrops. - Albania and North Macedonia.

Acaulescent, softly white-pilose. Leaflets 7-10 pairs. Racemes 5-12-flowered. Corolla c. 15 mm, pink to cherry-red; keel with a distinct beak. Legume 15-20 × 3-4 mm, cylindrical, pubescent.

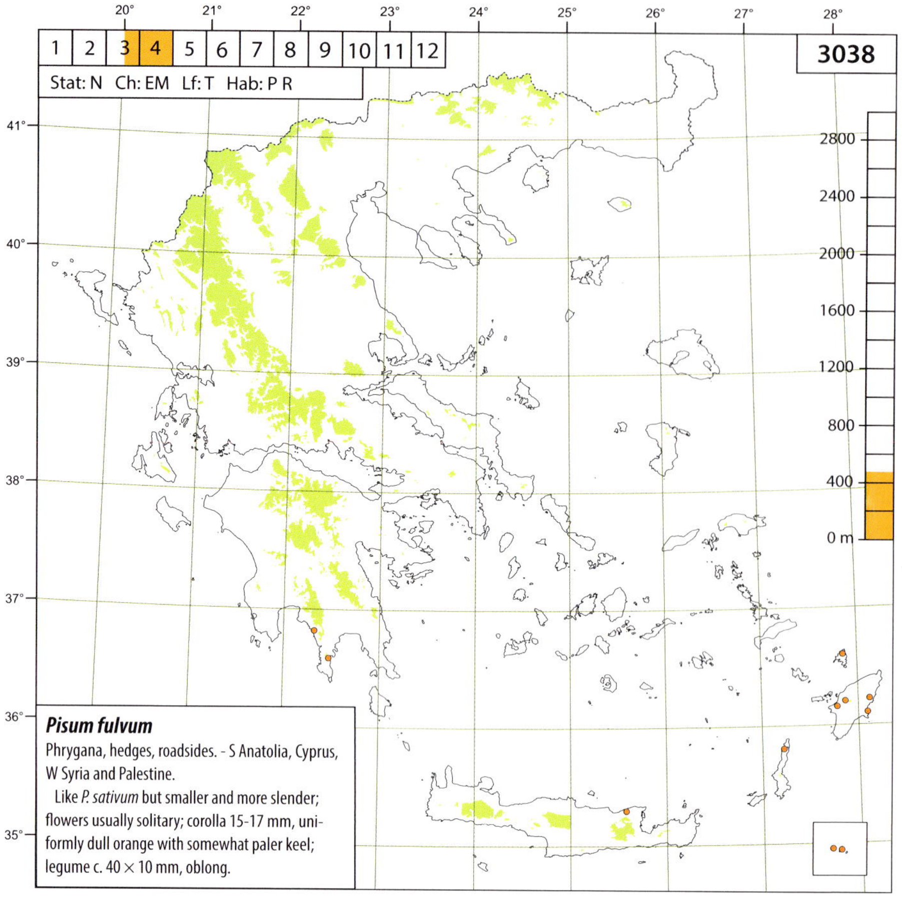

Pisum fulvum

Phrygana, hedges, roadsides. - S Anatolia, Cyprus, W Syria and Palestine.

Like *P. sativum* but smaller and more slender; flowers usually solitary; corolla 15-17 mm, uniformly dull orange with somewhat paler keel; legume c. 40 × 10 mm, oblong.

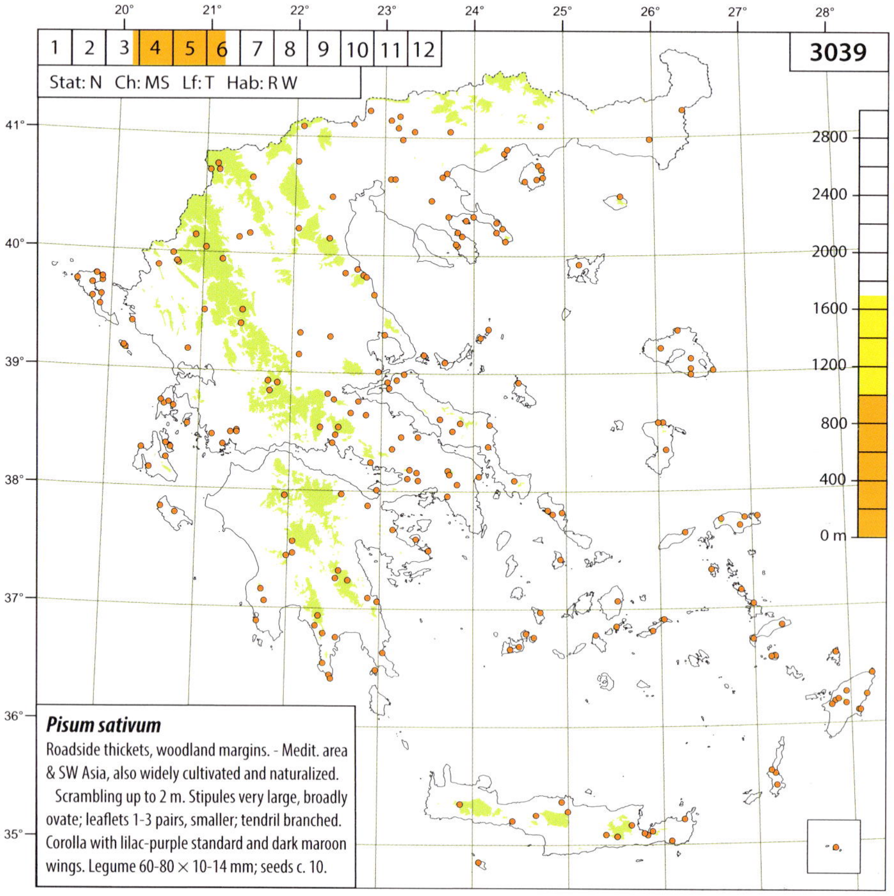

Pisum sativum

Roadside thickets, woodland margins. - Medit. area & SW Asia, also widely cultivated and naturalized.

Scrambling up to 2 m. Stipules very large, broadly ovate; leaflets 1-3 pairs, smaller; tendril branched. Corolla with lilac-purple standard and dark maroon wings. Legume 60-80 × 10-14 mm; seeds c. 10.

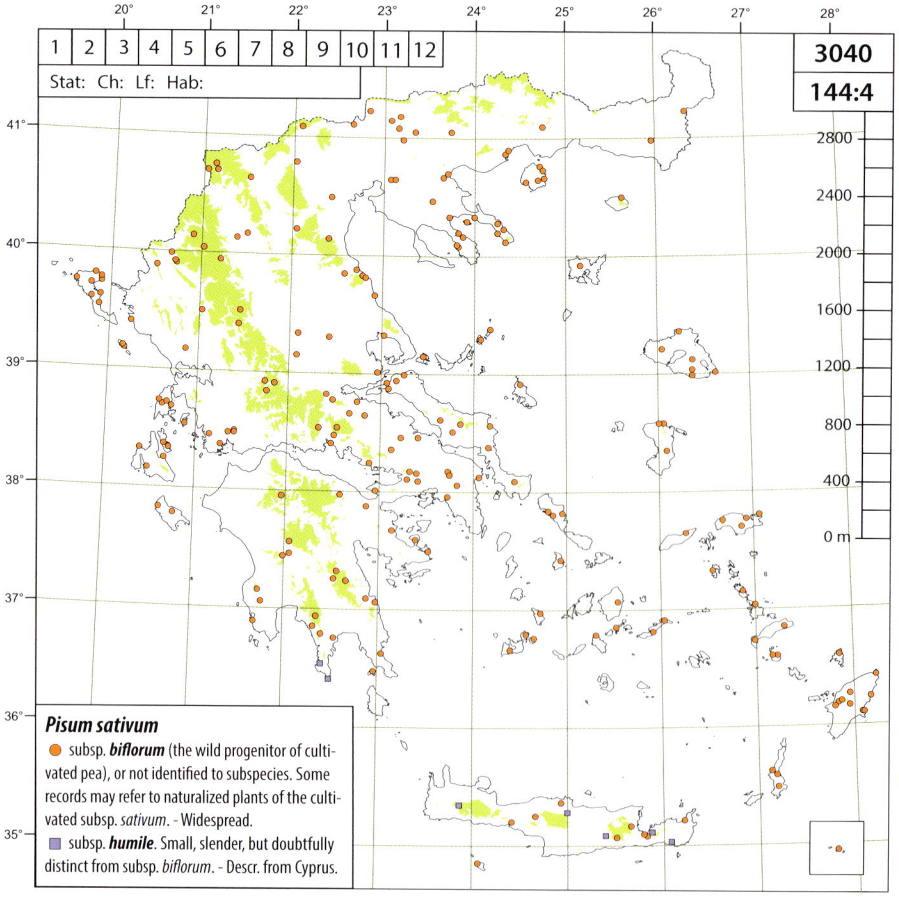

Pisum sativum

● subsp. ***biflorum*** (the wild progenitor of cultivated pea), or not identified to subspecies. Some records may refer to naturalized plants of the cultivated subsp. *sativum*. - Widespread.

■ subsp. ***humile***. Small, slender, but doubtfully distinct from subsp. *biflorum*. - Descr. from Cyprus.

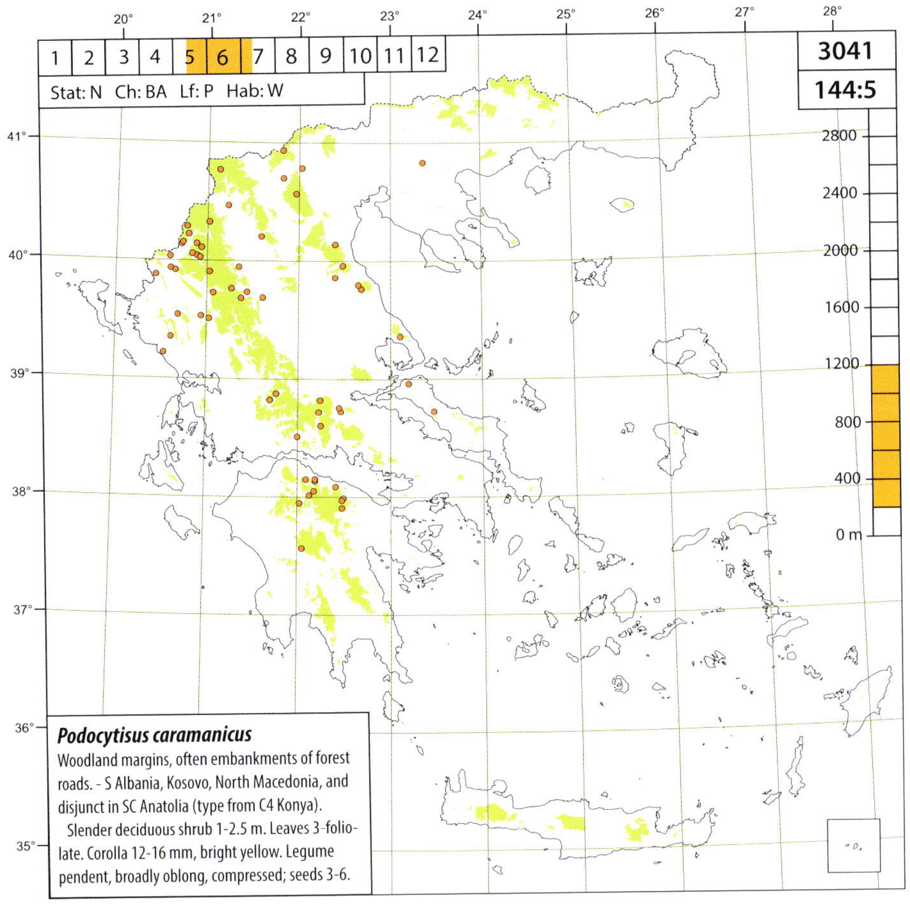

Podocytisus caramanicus
Woodland margins, often embankments of forest roads. - S Albania, Kosovo, North Macedonia, and disjunct in SC Anatolia (type from C4 Konya).
Slender deciduous shrub 1-2.5 m. Leaves 3-foliolate. Corolla 12-16 mm, bright yellow. Legume pendent, broadly oblong, compressed; seeds 3-6.

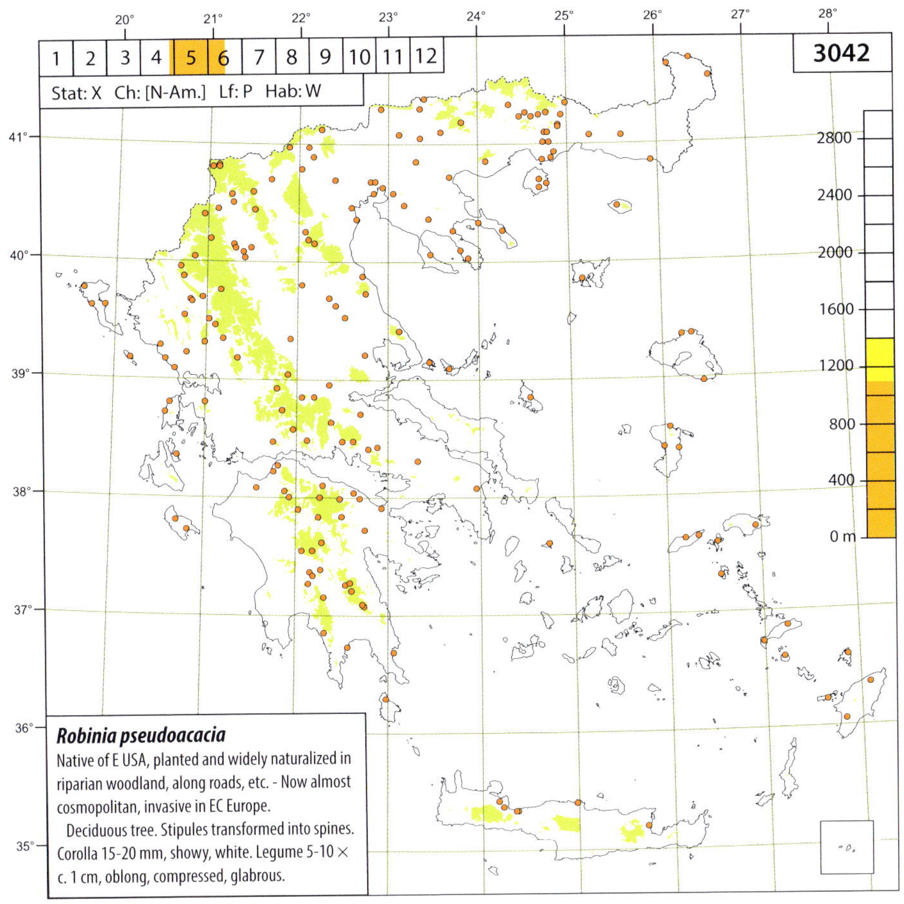

Robinia pseudoacacia
Native of E USA, planted and widely naturalized in riparian woodland, along roads, etc. - Now almost cosmopolitan, invasive in EC Europe.
Deciduous tree. Stipules transformed into spines. Corolla 15-20 mm, showy, white. Legume 5-10 × c. 1 cm, oblong, compressed, glabrous.

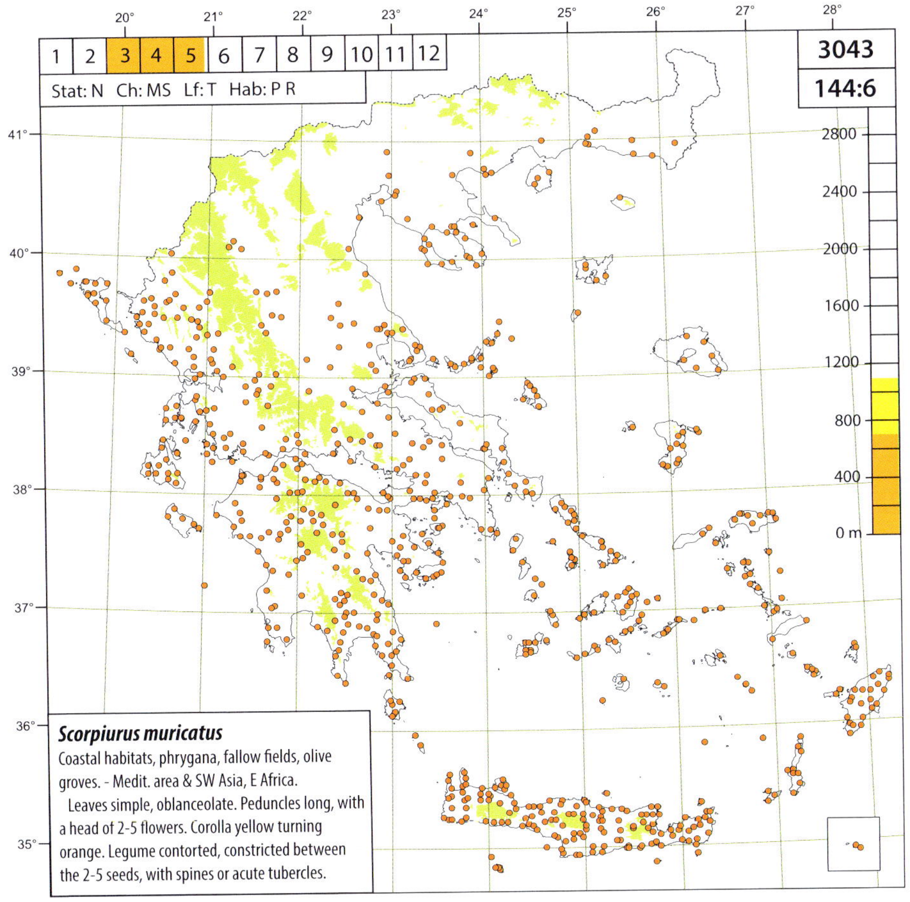

Scorpiurus muricatus
Coastal habitats, phrygana, fallow fields, olive groves. - Medit. area & SW Asia, E Africa.
Leaves simple, oblanceolate. Peduncles long, with a head of 2-5 flowers. Corolla yellow turning orange. Legume contorted, constricted between the 2-5 seeds, with spines or acute tubercles.

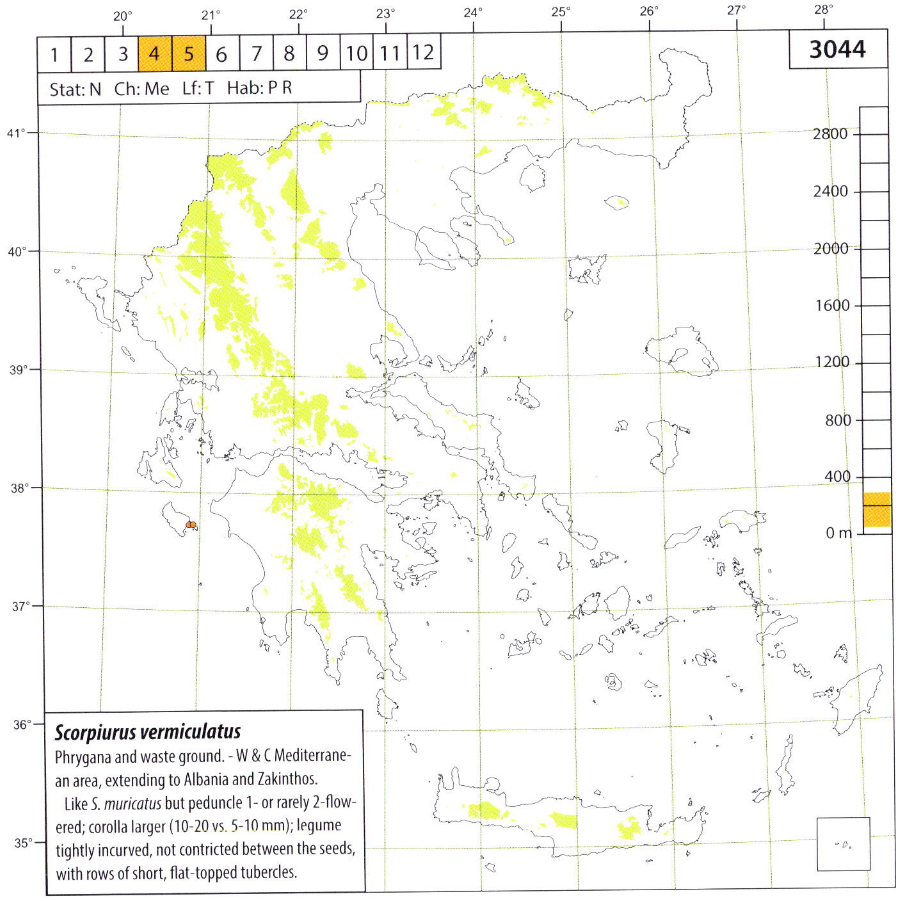

Scorpiurus vermiculatus
Phrygana and waste ground. - W & C Mediterranean area, extending to Albania and Zakinthos.
Like *S. muricatus* but peduncle 1- or rarely 2-flowered; corolla larger (10-20 vs. 5-10 mm); legume tightly incurved, not contricted between the seeds, with rows of short, flat-topped tubercles.

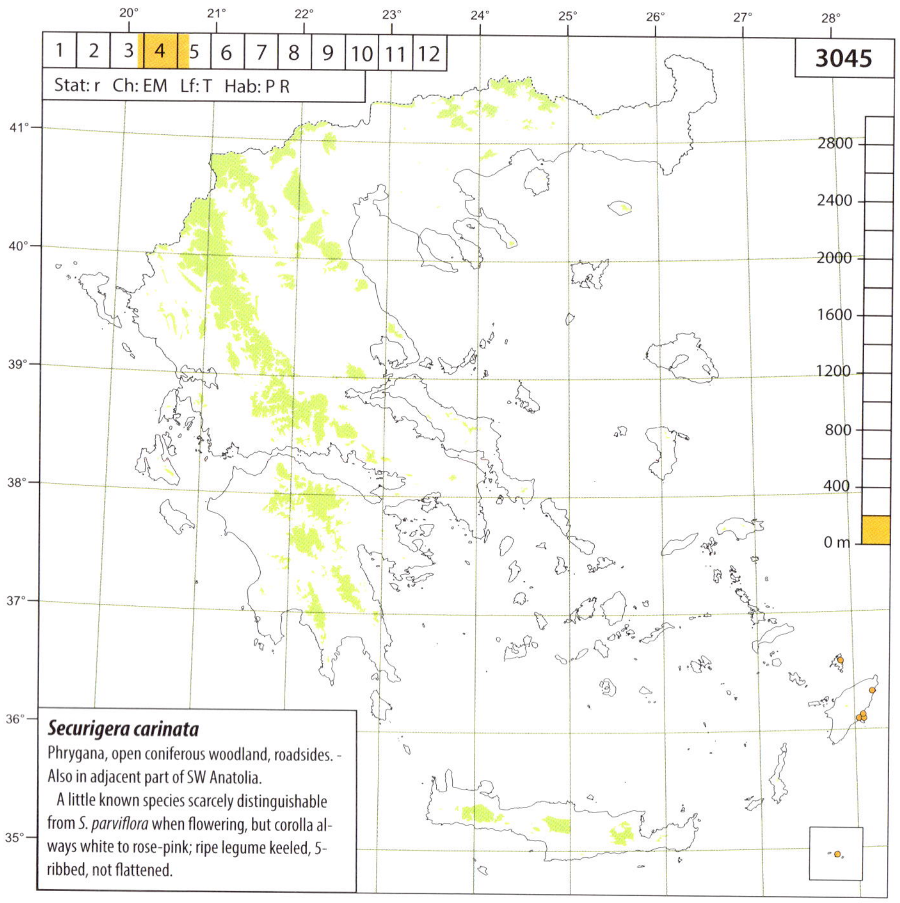

Securigera carinata

Phrygana, open coniferous woodland, roadsides. - Also in adjacent part of SW Anatolia.

A little known species scarcely distinguishable from *S. parviflora* when flowering, but corolla always white to rose-pink; ripe legume keeled, 5-ribbed, not flattened.

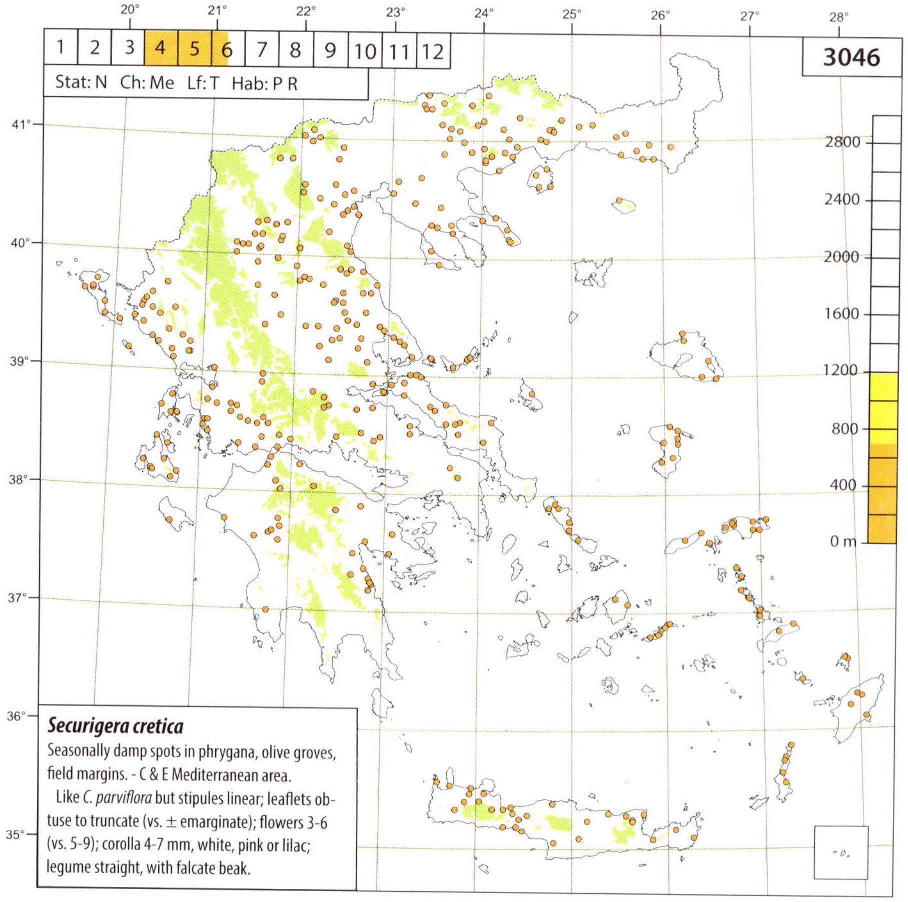

Securigera cretica

Seasonally damp spots in phrygana, olive groves, field margins. - C & E Mediterranean area.

Like *C. parviflora* but stipules linear; leaflets obtuse to truncate (vs. ± emarginate); flowers 3-6 (vs. 5-9); corolla 4-7 mm, white, pink or lilac; legume straight, with falcate beak.

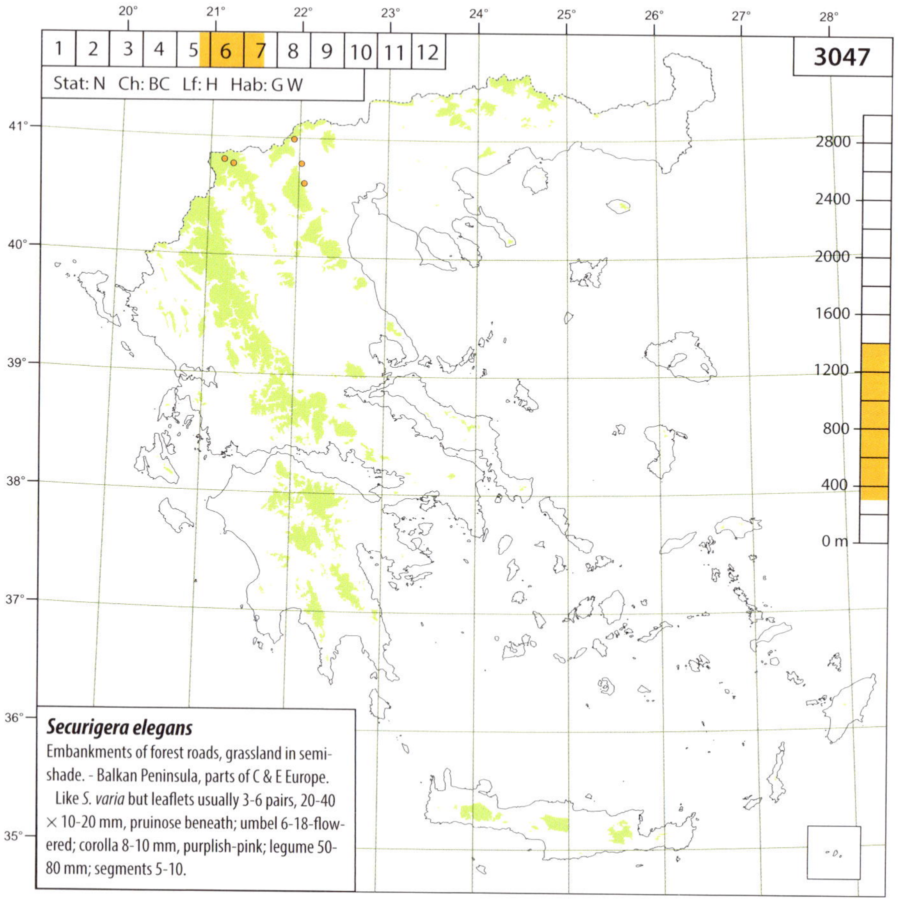

Securigera elegans

Embankments of forest roads, grassland in semi-shade. - Balkan Peninsula, parts of C & E Europe.

Like *S. varia* but leaflets usually 3-6 pairs, 20-40 × 10-20 mm, pruinose beneath; umbel 6-18-flowered; corolla 8-10 mm, purplish-pink; legume 50-80 mm; segments 5-10.

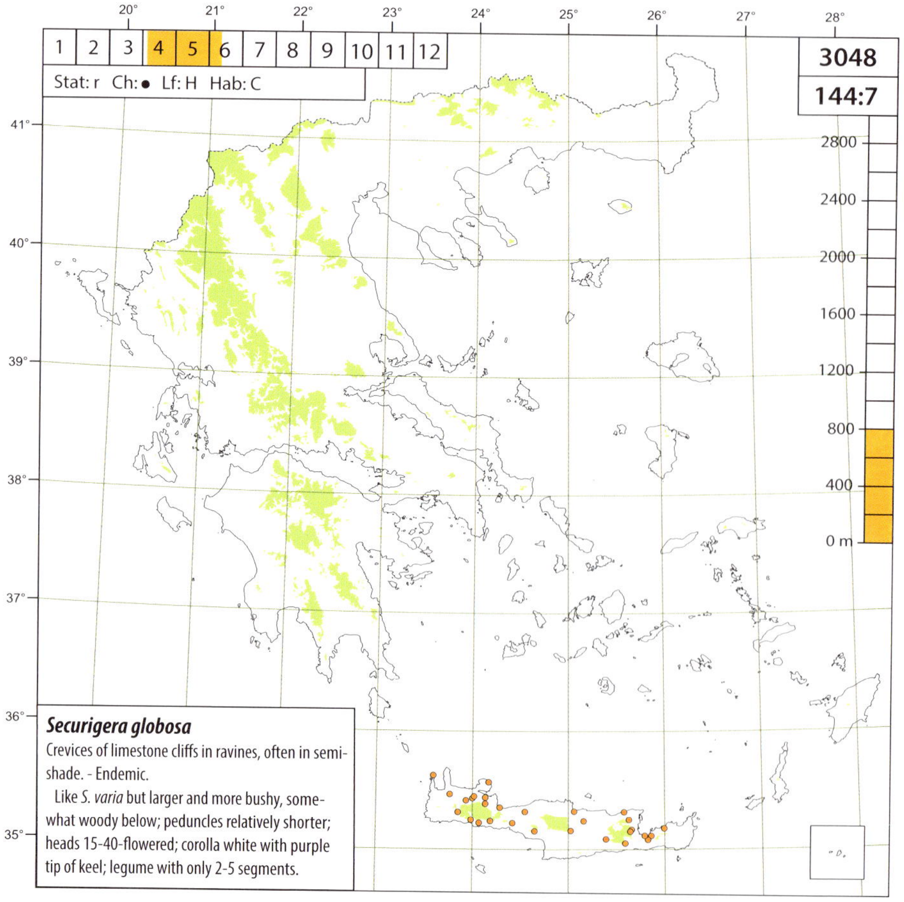

Securigera globosa

Crevices of limestone cliffs in ravines, often in semi-shade. - Endemic.

Like *S. varia* but larger and more bushy, somewhat woody below; peduncles relatively shorter; heads 15-40-flowered; corolla white with purple tip of keel; legume with only 2-5 segments.

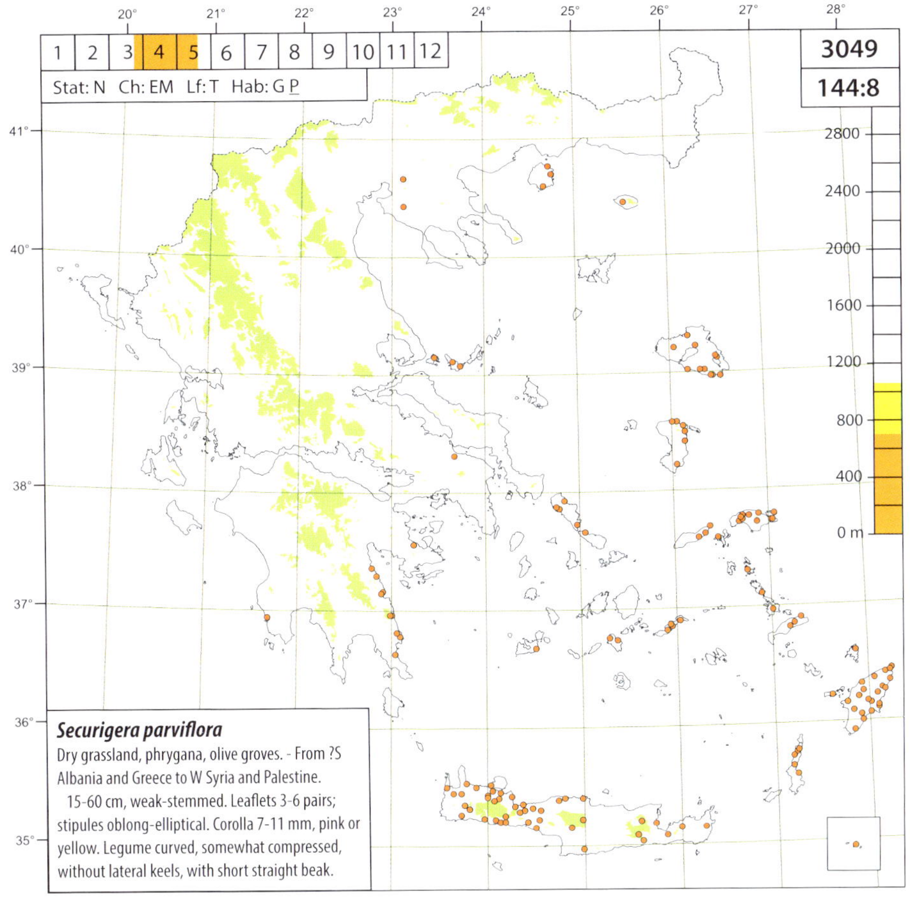

Securigera parviflora

Dry grassland, phrygana, olive groves. - From ?S Albania and Greece to W Syria and Palestine.

15-60 cm, weak-stemmed. Leaflets 3-6 pairs; stipules oblong-elliptical. Corolla 7-11 mm, pink or yellow. Legume curved, somewhat compressed, without lateral keels, with short straight beak.

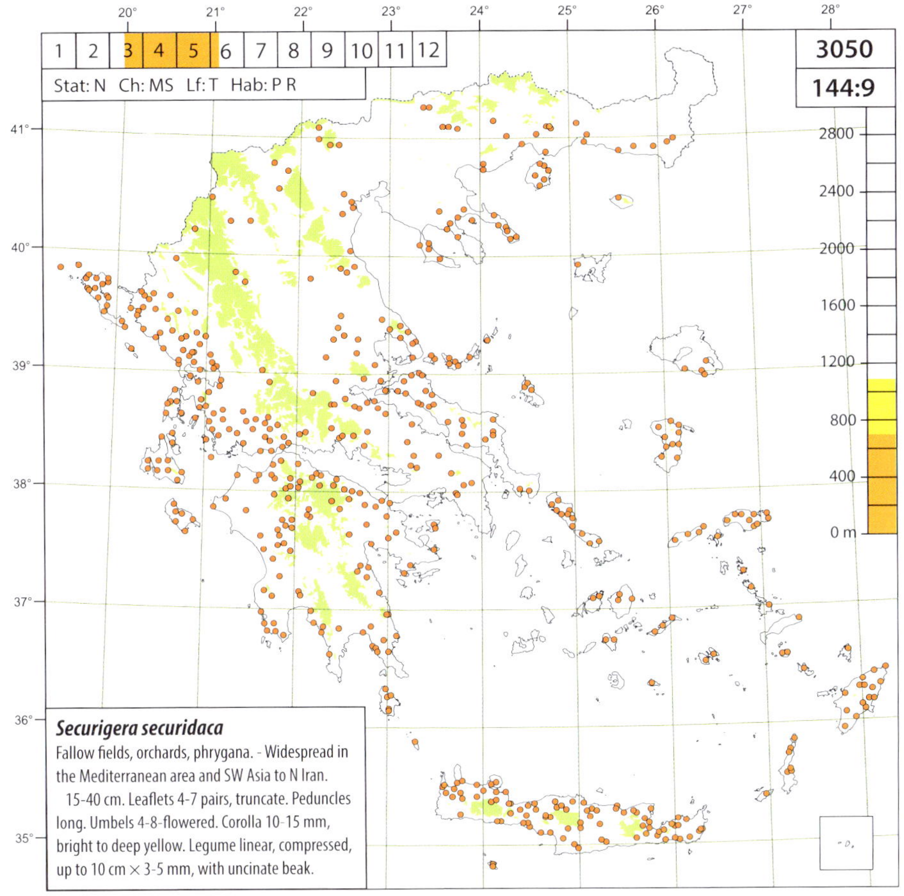

Securigera securidaca

Fallow fields, orchards, phrygana. - Widespread in the Mediterranean area and SW Asia to N Iran.

15-40 cm. Leaflets 4-7 pairs, truncate. Peduncles long. Umbels 4-8-flowered. Corolla 10-15 mm, bright to deep yellow. Legume linear, compressed, up to 10 cm × 3-5 mm, with uncinate beak.

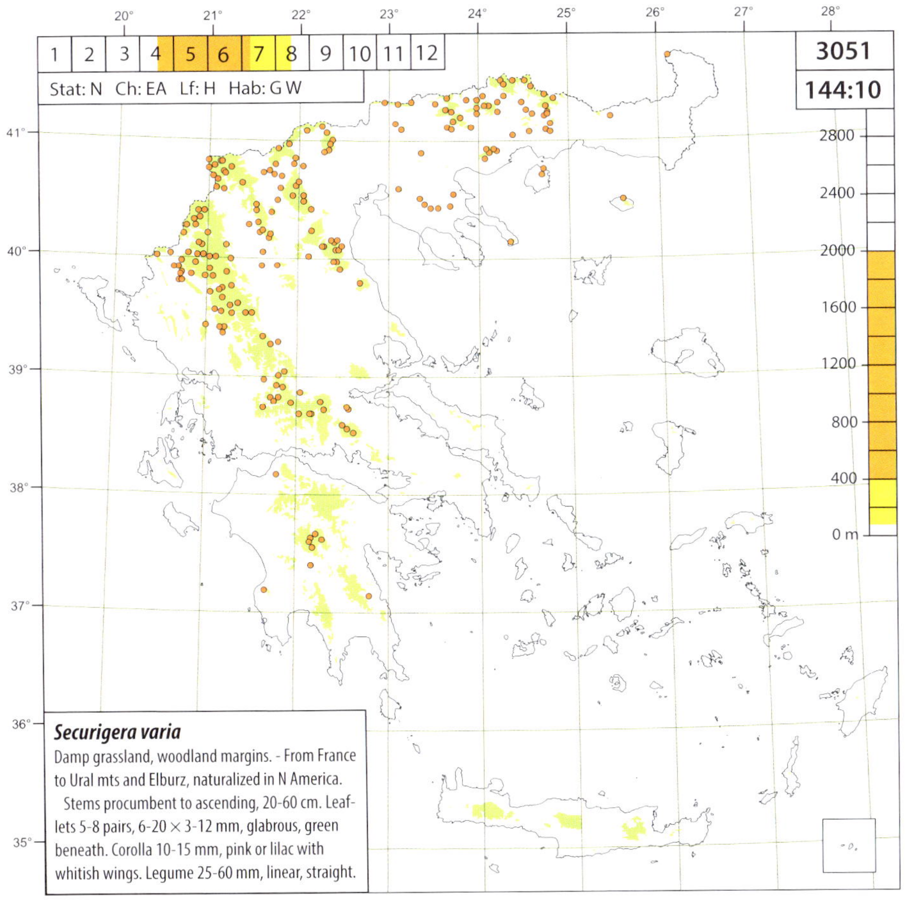

Securigera varia

Damp grassland, woodland margins. - From France to Ural mts and Elburz, naturalized in N America.

Stems procumbent to ascending, 20-60 cm. Leaflets 5-8 pairs, 6-20 × 3-12 mm, glabrous, green beneath. Corolla 10-15 mm, pink or lilac with whitish wings. Legume 25-60 mm, linear, straight.

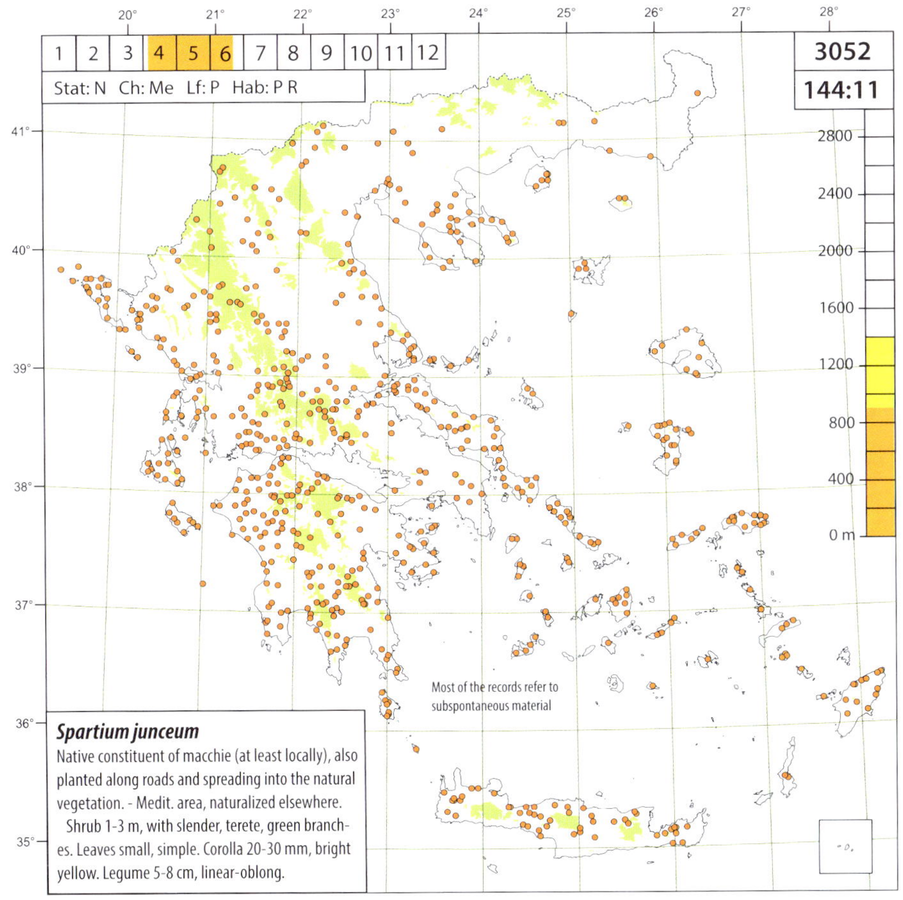

Spartium junceum

Native constituent of macchie (at least locally), also planted along roads and spreading into the natural vegetation. - Medit. area, naturalized elsewhere.

Shrub 1-3 m, with slender, terete, green branches. Leaves small, simple. Corolla 20-30 mm, bright yellow. Legume 5-8 cm, linear-oblong.

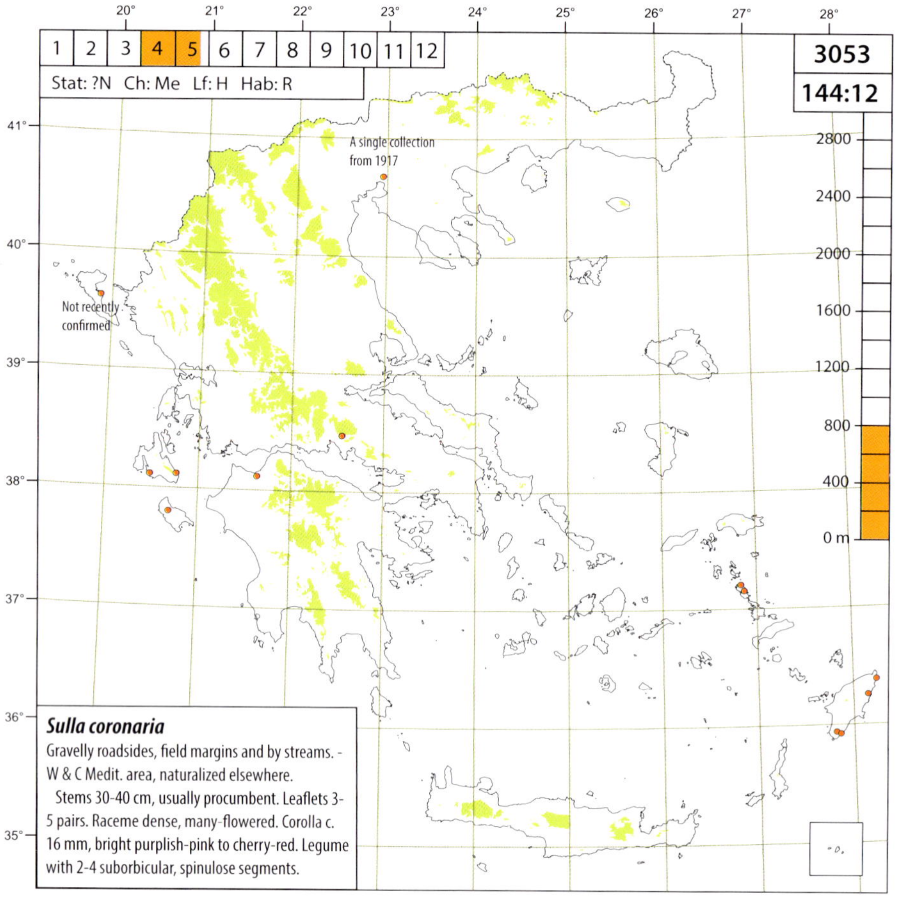

Sulla coronaria

Gravelly roadsides, field margins and by streams. - W & C Medit. area, naturalized elsewhere.

Stems 30-40 cm, usually procumbent. Leaflets 3-5 pairs. Raceme dense, many-flowered. Corolla c. 16 mm, bright purplish-pink to cherry-red. Legume with 2-4 suborbicular, spinulose segments.

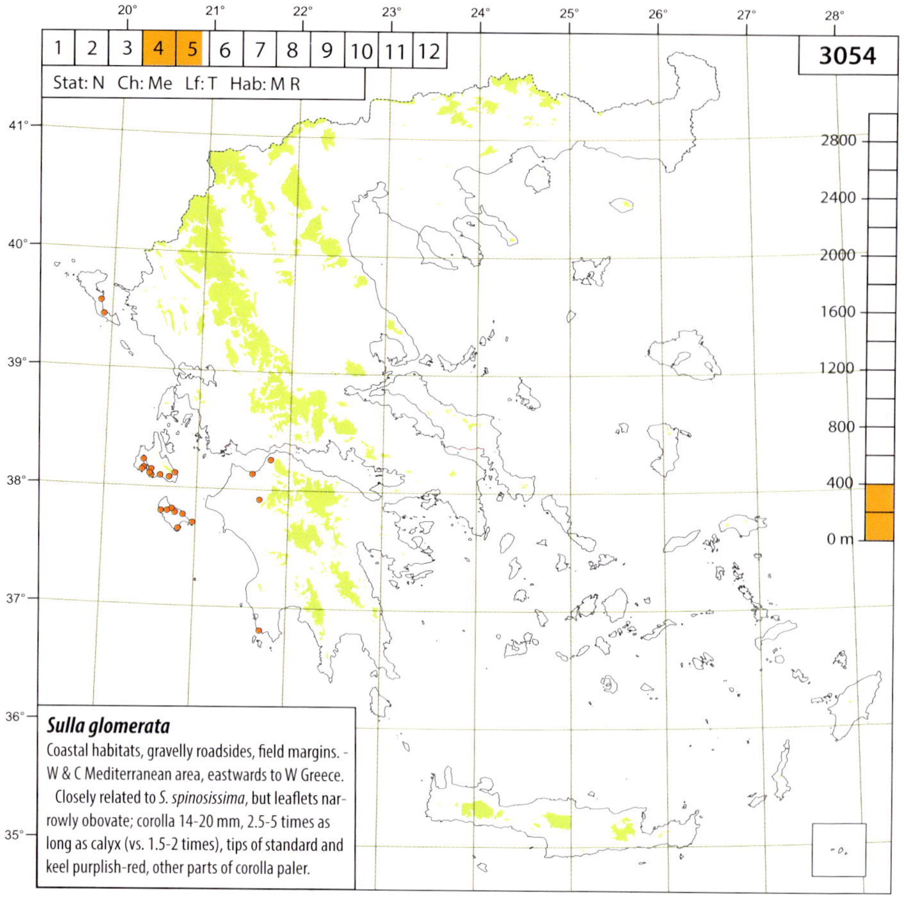

Sulla glomerata

Coastal habitats, gravelly roadsides, field margins. - W & C Mediterranean area, eastwards to W Greece.

Closely related to *S. spinosissima*, but leaflets narrowly obovate; corolla 14-20 mm, 2.5-5 times as long as calyx (vs. 1.5-2 times), tips of standard and keel purplish-red, other parts of corolla paler.

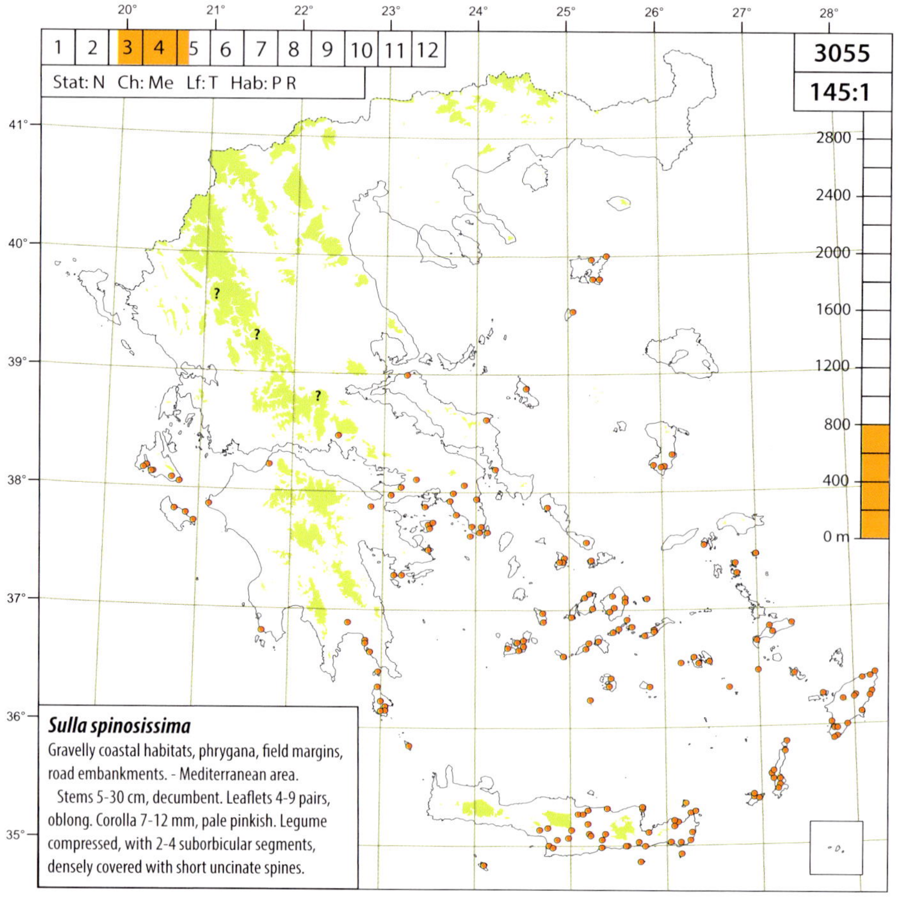

Sulla spinosissima

Gravelly coastal habitats, phrygana, field margins, road embankments. - Mediterranean area.

Stems 5-30 cm, decumbent. Leaflets 4-9 pairs, oblong. Corolla 7-12 mm, pale pinkish. Legume compressed, with 2-4 suborbicular segments, densely covered with short uncinate spines.

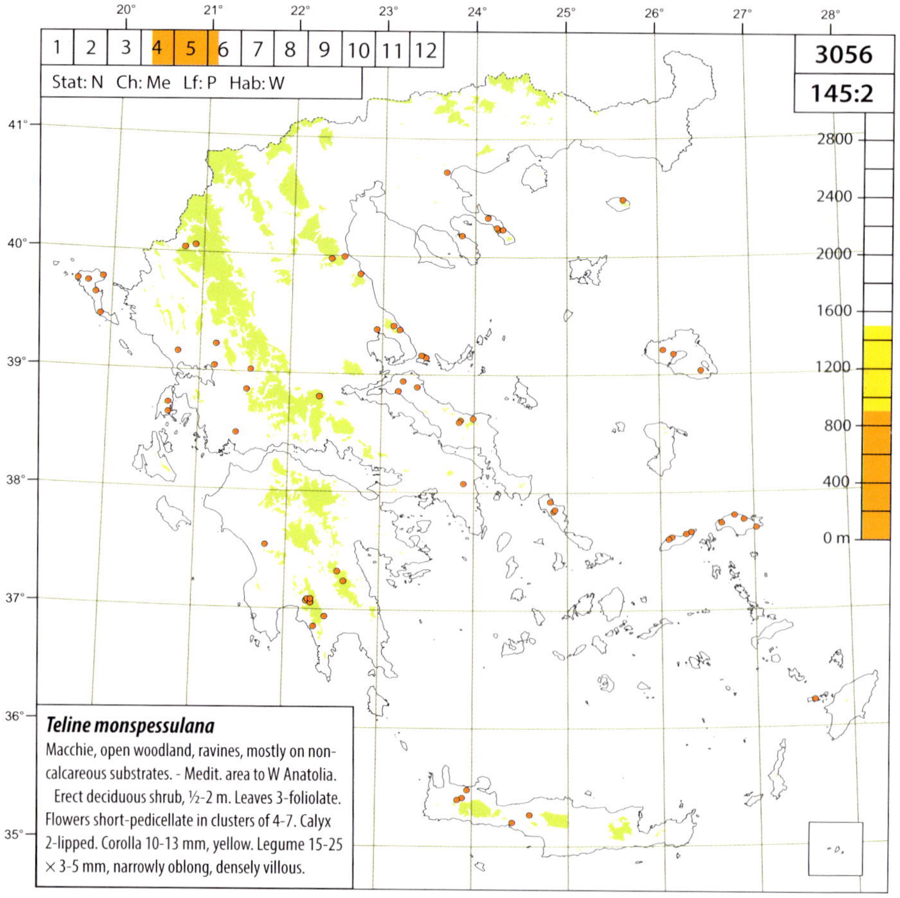

Teline monspessulana

Macchie, open woodland, ravines, mostly on non-calcareous substrates. - Medit. area to W Anatolia.

Erect deciduous shrub, ½-2 m. Leaves 3-foliolate. Flowers short-pedicellate in clusters of 4-7. Calyx 2-lipped. Corolla 10-13 mm, yellow. Legume 15-25 × 3-5 mm, narrowly oblong, densely villous.

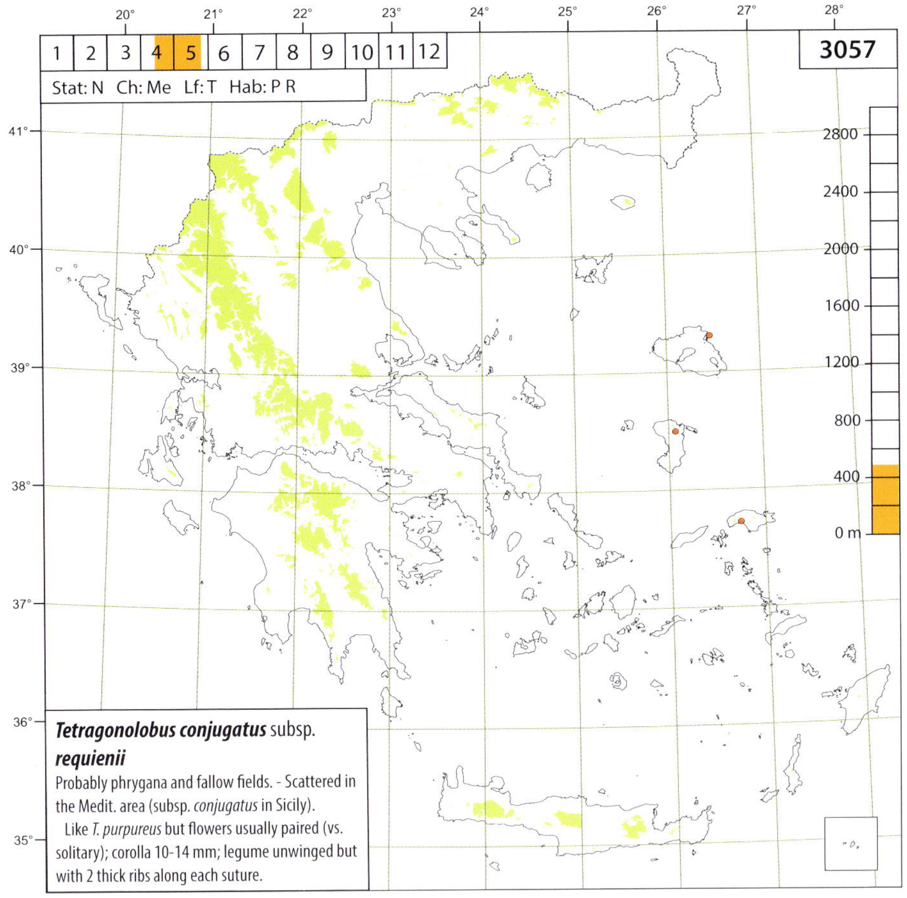

Tetragonolobus conjugatus subsp. ***requienii***
Probably phrygana and fallow fields. - Scattered in the Medit. area (subsp. *conjugatus* in Sicily).
Like *T. purpureus* but flowers usually paired (vs. solitary); corolla 10-14 mm; legume unwinged but with 2 thick ribs along each suture.

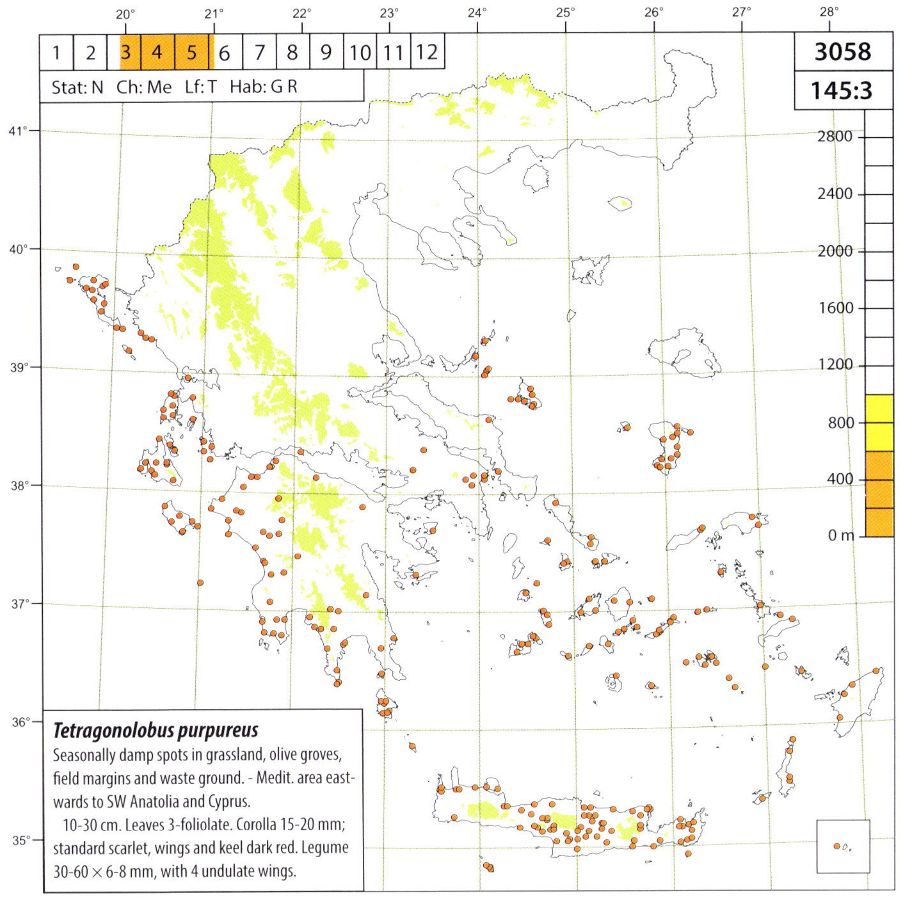

Tetragonolobus purpureus
Seasonally damp spots in grassland, olive groves, field margins and waste ground. - Medit. area eastwards to SW Anatolia and Cyprus.
10-30 cm. Leaves 3-foliolate. Corolla 15-20 mm; standard scarlet, wings and keel dark red. Legume 30-60 × 6-8 mm, with 4 undulate wings.

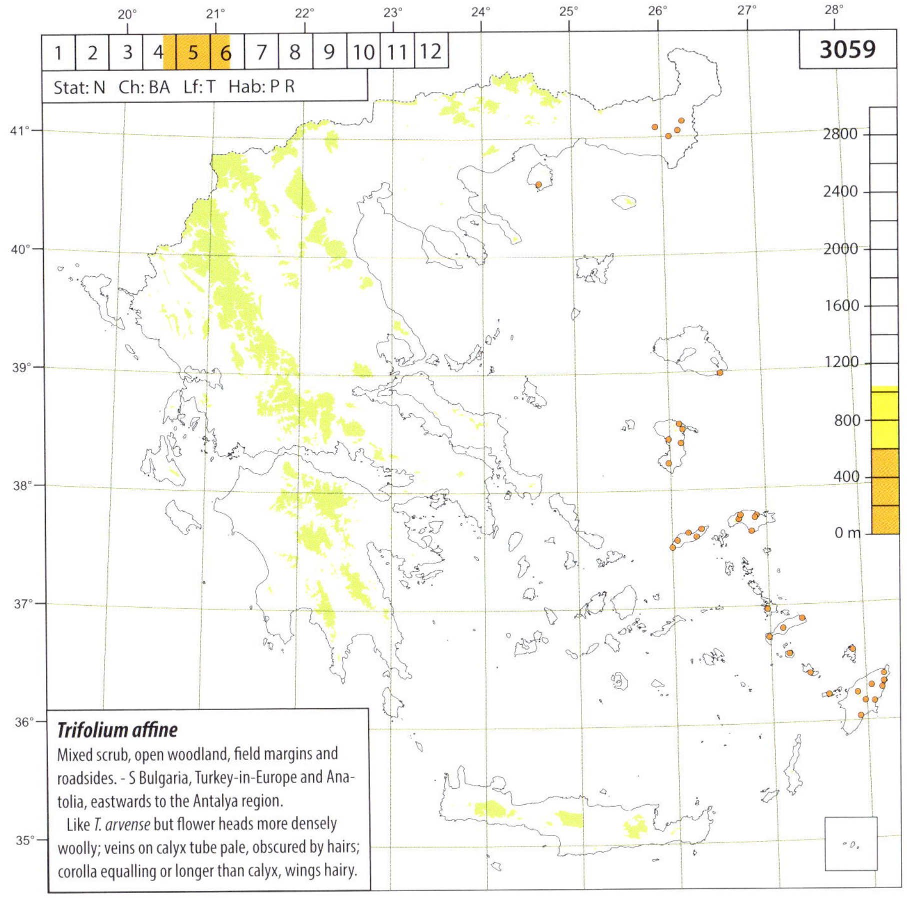

Trifolium affine
Mixed scrub, open woodland, field margins and roadsides. - S Bulgaria, Turkey-in-Europe and Anatolia, eastwards to the Antalya region.
Like *T. arvense* but flower heads more densely woolly; veins on calyx tube pale, obscured by hairs; corolla equalling or longer than calyx, wings hairy.

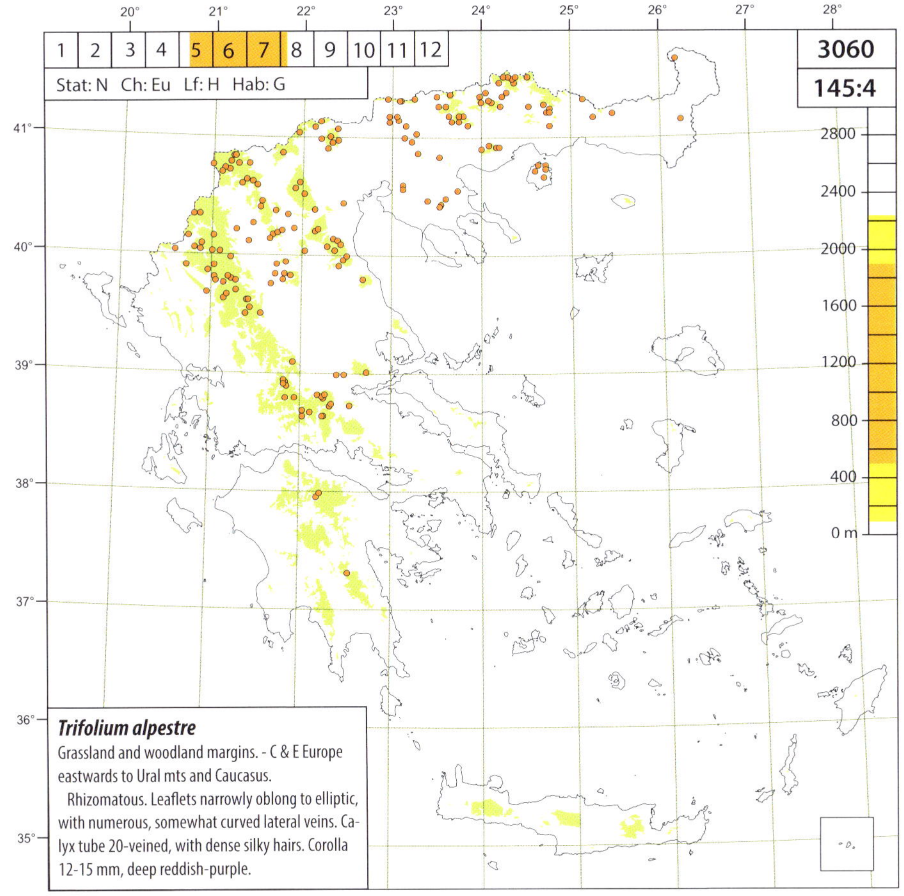

Trifolium alpestre
Grassland and woodland margins. - C & E Europe eastwards to Ural mts and Caucasus.
Rhizomatous. Leaflets narrowly oblong to elliptic, with numerous, somewhat curved lateral veins. Calyx tube 20-veined, with dense silky hairs. Corolla 12-15 mm, deep reddish-purple.

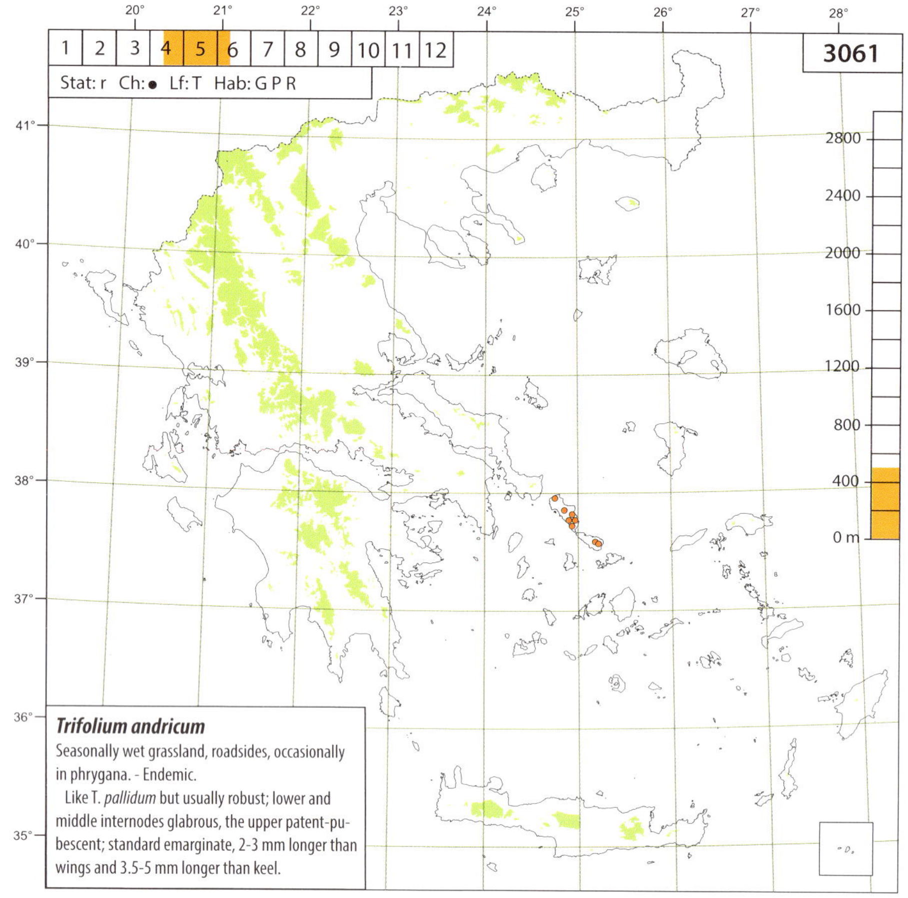

Trifolium andricum
Seasonally wet grassland, roadsides, occasionally in phrygana. - Endemic.
Like T. *pallidum* but usually robust; lower and middle internodes glabrous, the upper patent-pubescent; standard emarginate, 2-3 mm longer than wings and 3.5-5 mm longer than keel.

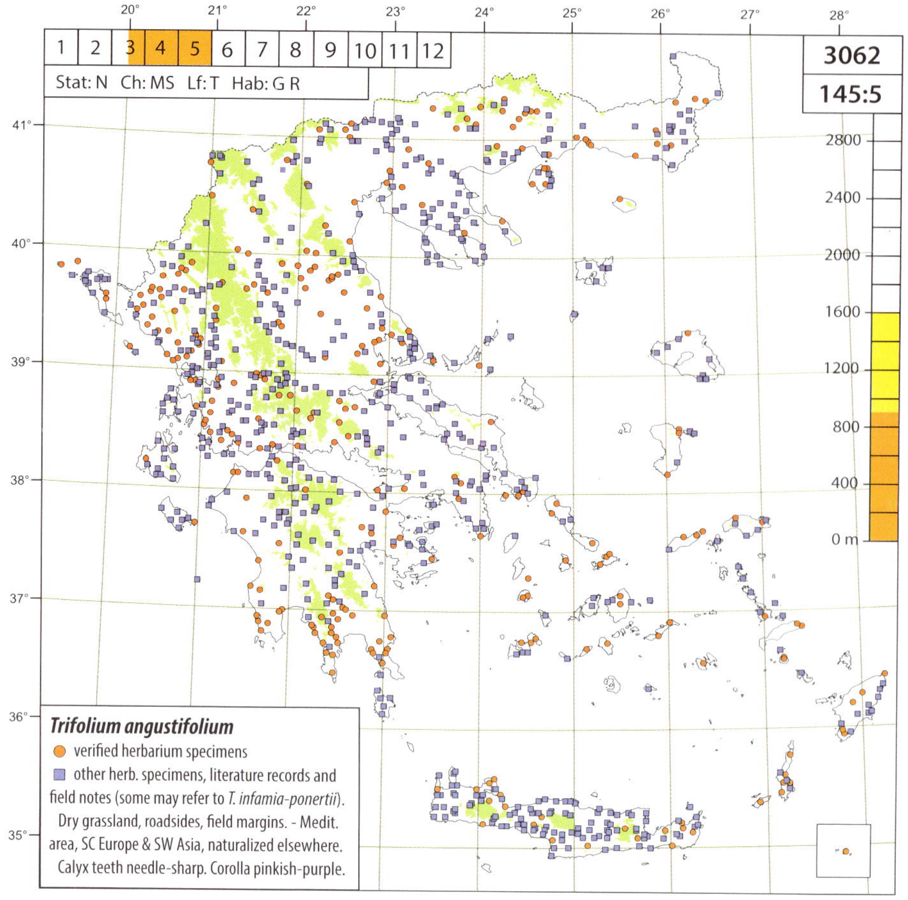

Trifolium angustifolium
● verified herbarium specimens
■ other herb. specimens, literature records and field notes (some may refer to *T. infamia-ponertii*).
Dry grassland, roadsides, field margins. - Medit. area, SC Europe & SW Asia, naturalized elsewhere.
Calyx teeth needle-sharp. Corolla pinkish-purple.

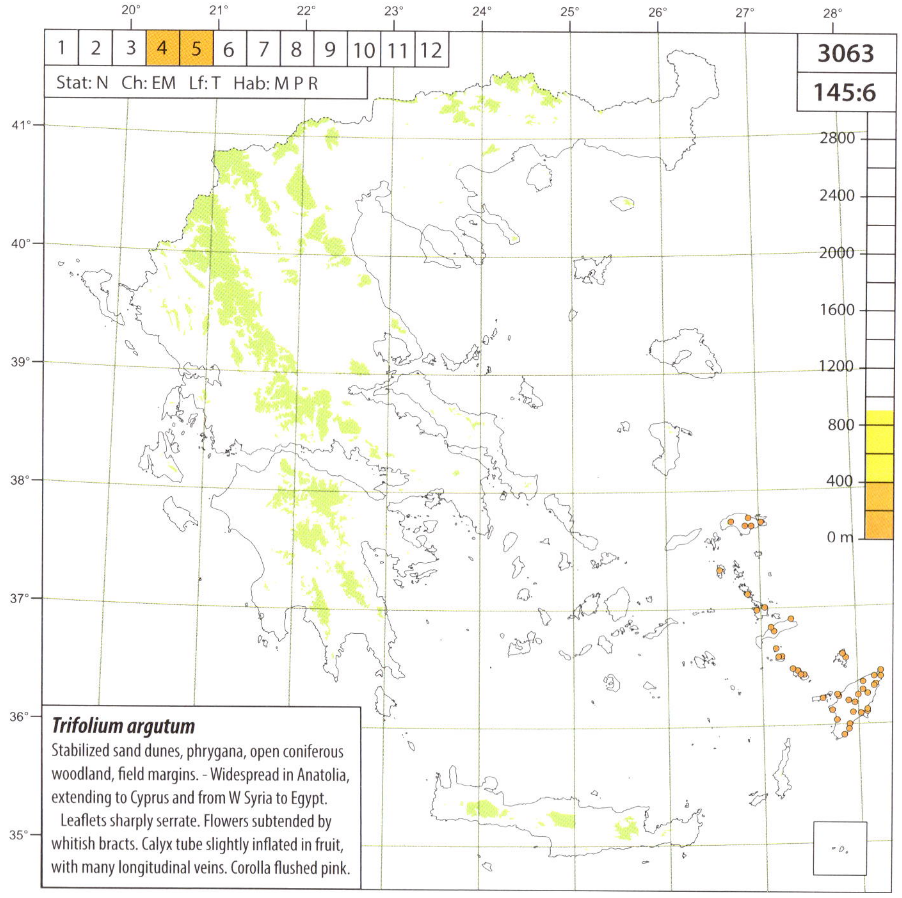

Trifolium argutum
Stabilized sand dunes, phrygana, open coniferous woodland, field margins. - Widespread in Anatolia, extending to Cyprus and from W Syria to Egypt.
Leaflets sharply serrate. Flowers subtended by whitish bracts. Calyx tube slightly inflated in fruit, with many longitudinal veins. Corolla flushed pink.

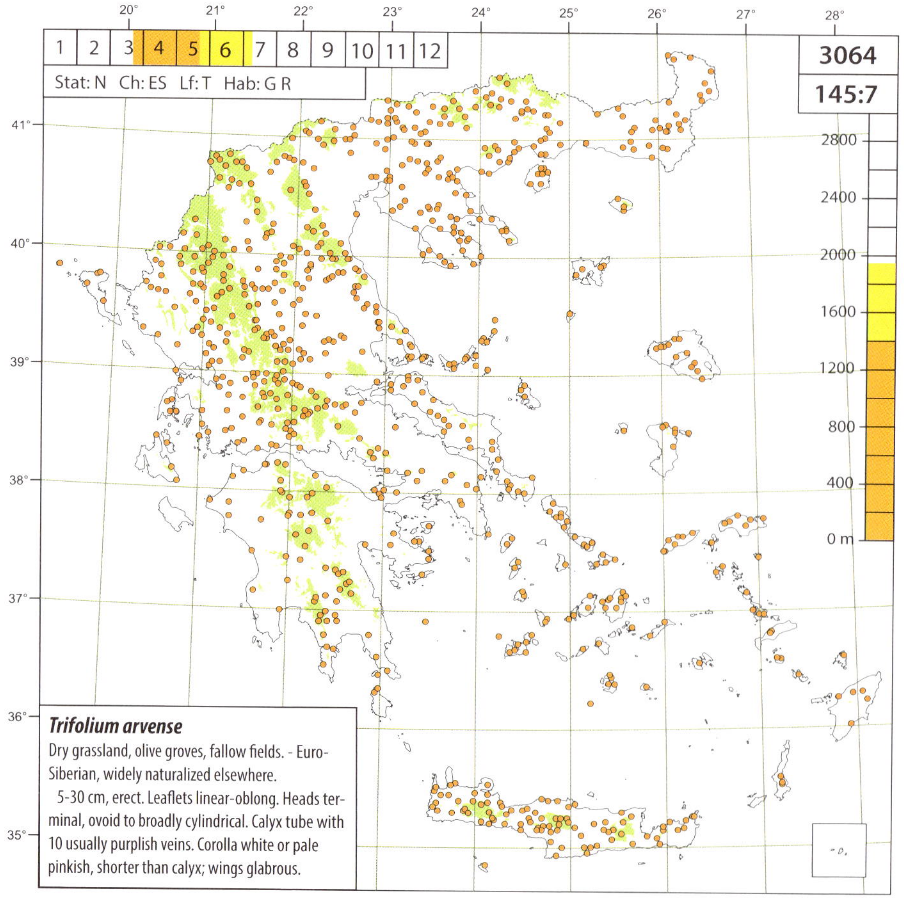

Trifolium arvense
Dry grassland, olive groves, fallow fields. - Euro-Siberian, widely naturalized elsewhere.
5-30 cm, erect. Leaflets linear-oblong. Heads terminal, ovoid to broadly cylindrical. Calyx tube with 10 usually purplish veins. Corolla white or pale pinkish, shorter than calyx; wings glabrous.

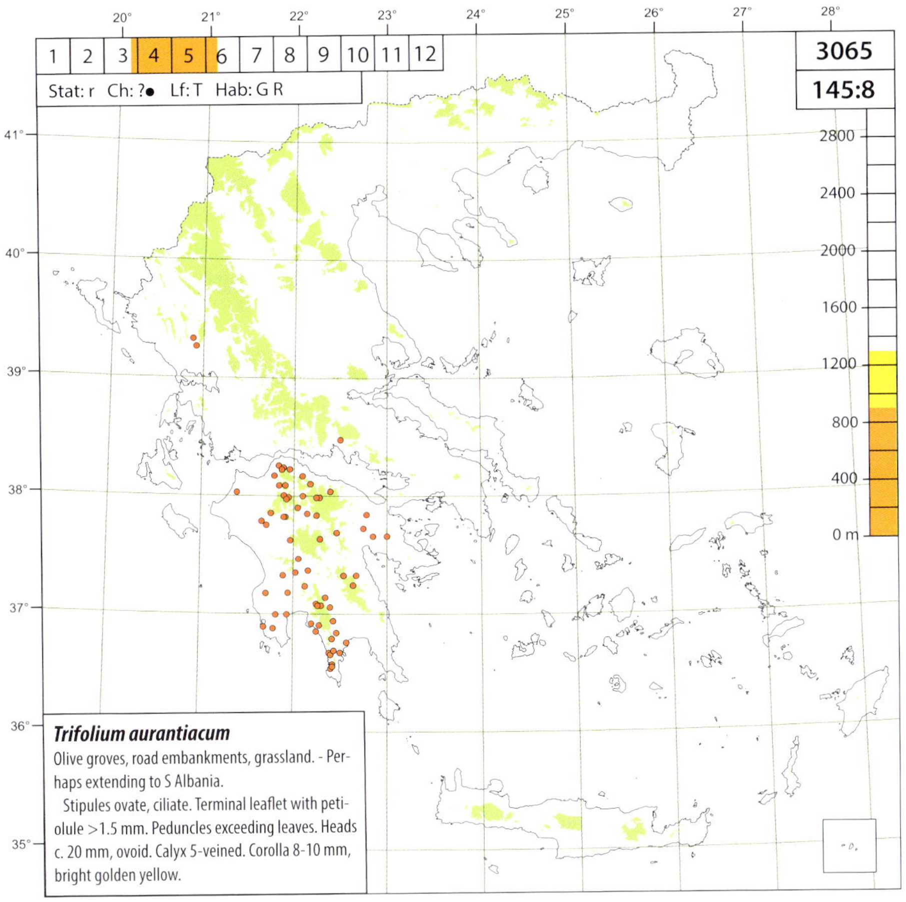

Trifolium aurantiacum

Olive groves, road embankments, grassland. - Perhaps extending to S Albania.

Stipules ovate, ciliate. Terminal leaflet with petiolule >1.5 mm. Peduncles exceeding leaves. Heads c. 20 mm, ovoid. Calyx 5-veined. Corolla 8-10 mm, bright golden yellow.

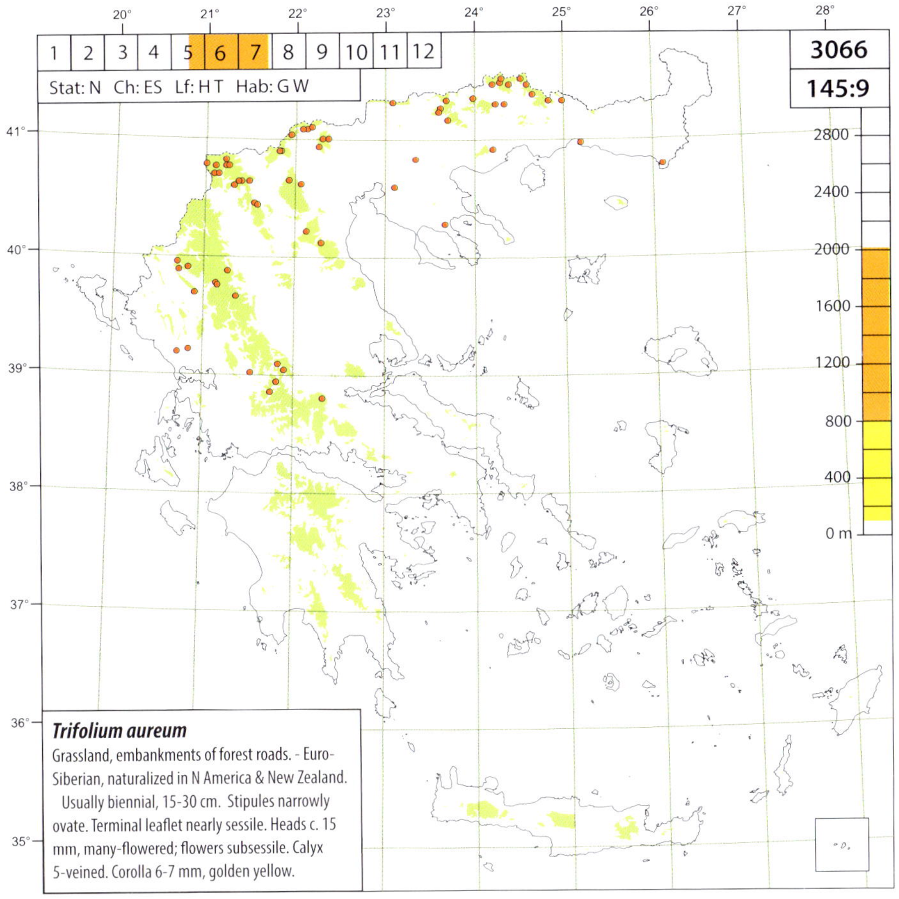

Trifolium aureum

Grassland, embankments of forest roads. - Euro-Siberian, naturalized in N America & New Zealand.

Usually biennial, 15-30 cm. Stipules narrowly ovate. Terminal leaflet nearly sessile. Heads c. 15 mm, many-flowered; flowers subsessile. Calyx 5-veined. Corolla 6-7 mm, golden yellow.

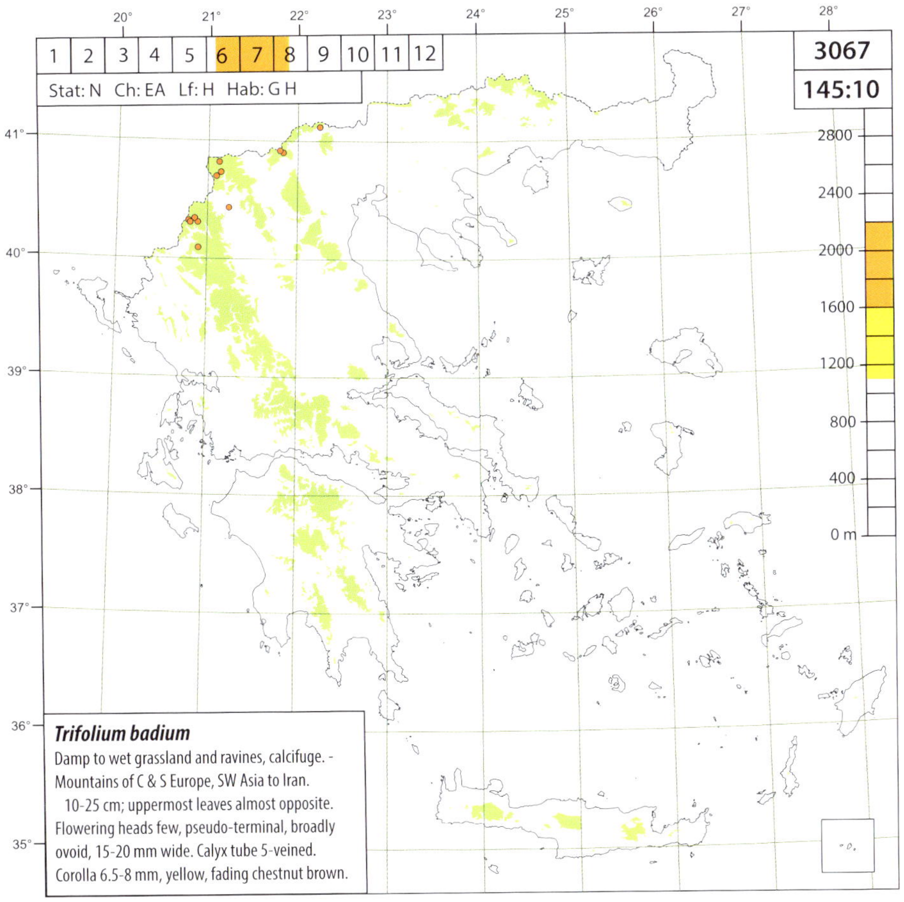

Trifolium badium

Damp to wet grassland and ravines, calcifuge. - Mountains of C & S Europe, SW Asia to Iran.

10-25 cm; uppermost leaves almost opposite. Flowering heads few, pseudo-terminal, broadly ovoid, 15-20 mm wide. Calyx tube 5-veined. Corolla 6.5-8 mm, yellow, fading chestnut brown.

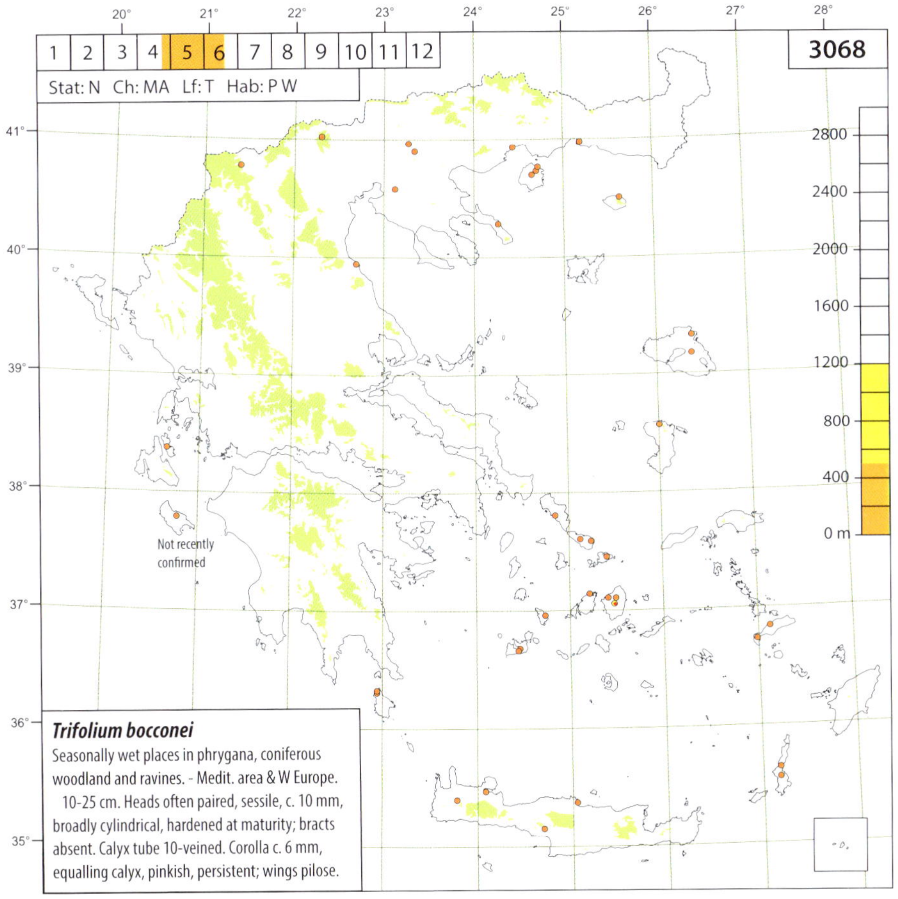

Trifolium bocconei

Seasonally wet places in phrygana, coniferous woodland and ravines. - Medit. area & W Europe.

10-25 cm. Heads often paired, sessile, c. 10 mm, broadly cylindrical, hardened at maturity; bracts absent. Calyx tube 10-veined. Corolla c. 6 mm, equalling calyx, pinkish, persistent; wings pilose.

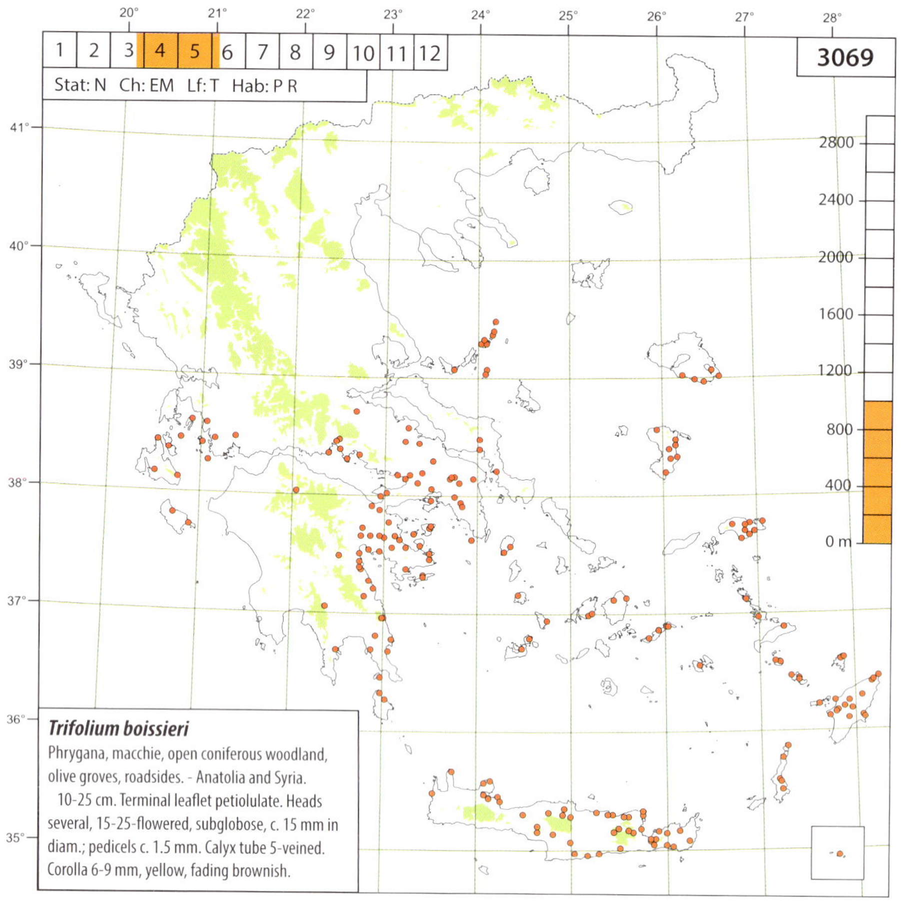

Trifolium boissieri

Phrygana, macchie, open coniferous woodland, olive groves, roadsides. - Anatolia and Syria.

10-25 cm. Terminal leaflet petiolulate. Heads several, 15-25-flowered, subglobose, c. 15 mm in diam.; pedicels c. 1.5 mm. Calyx tube 5-veined. Corolla 6-9 mm, yellow, fading brownish.

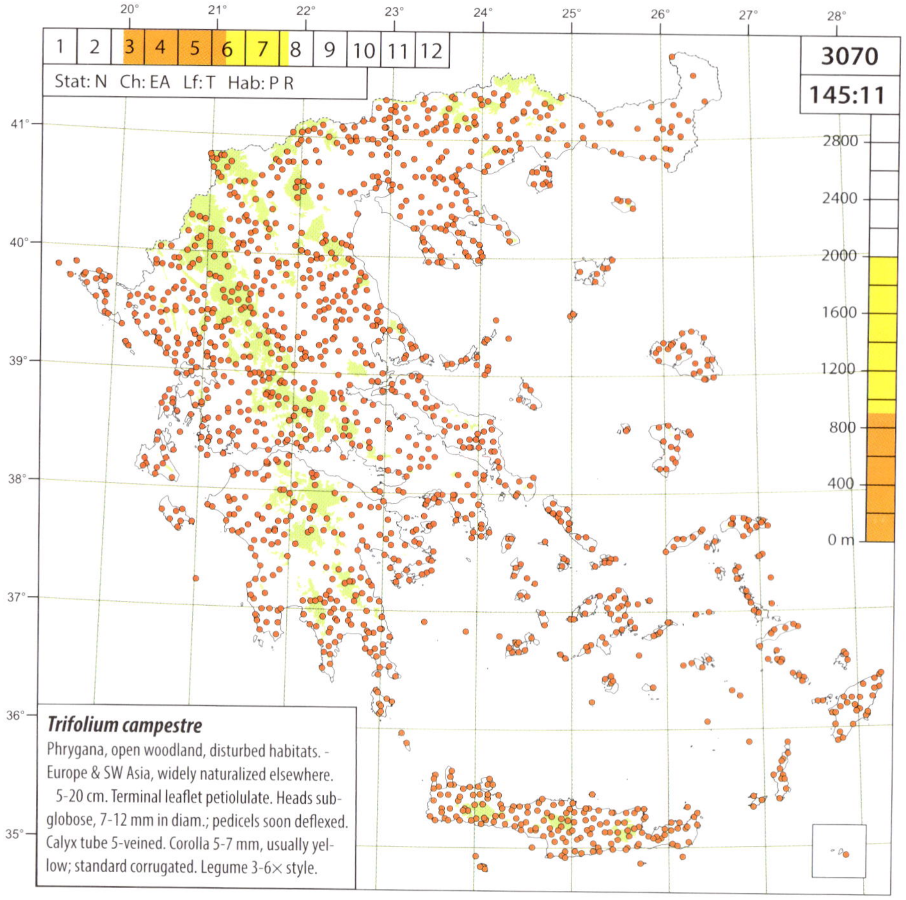

Trifolium campestre

Phrygana, open woodland, disturbed habitats. - Europe & SW Asia, widely naturalized elsewhere.

5-20 cm. Terminal leaflet petiolulate. Heads subglobose, 7-12 mm in diam.; pedicels soon deflexed. Calyx tube 5-veined. Corolla 5-7 mm, usually yellow; standard corrugated. Legume 3-6× style.

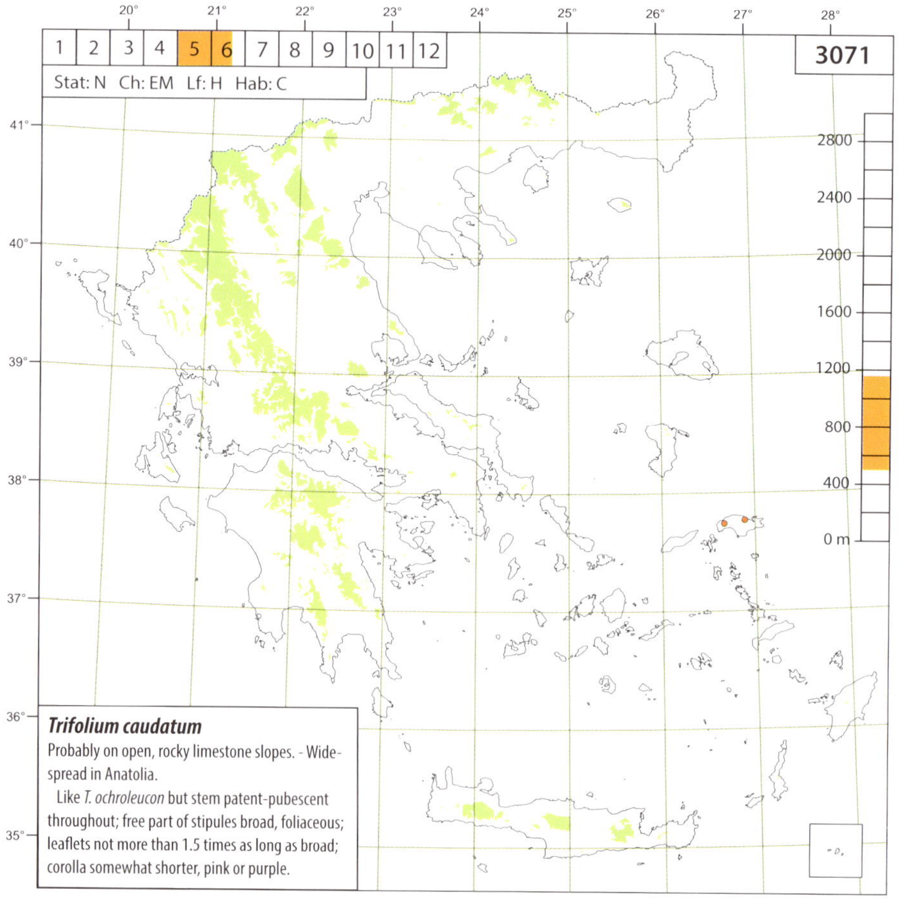

Trifolium caudatum

Probably on open, rocky limestone slopes. - Widespread in Anatolia.

Like *T. ochroleucon* but stem patent-pubescent throughout; free part of stipules broad, foliaceous; leaflets not more than 1.5 times as long as broad; corolla somewhat shorter, pink or purple.

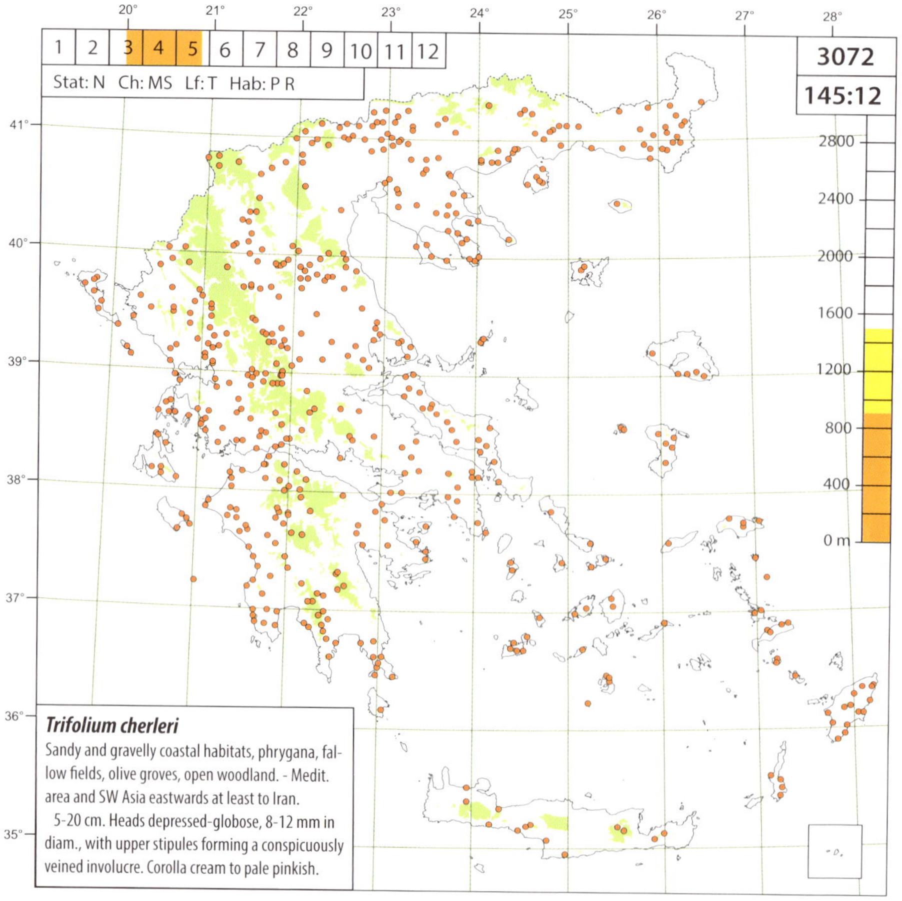

Trifolium cherleri

Sandy and gravelly coastal habitats, phrygana, fallow fields, olive groves, open woodland. - Medit. area and SW Asia eastwards at least to Iran.

5-20 cm. Heads depressed-globose, 8-12 mm in diam., with upper stipules forming a conspicuously veined involucre. Corolla cream to pale pinkish.

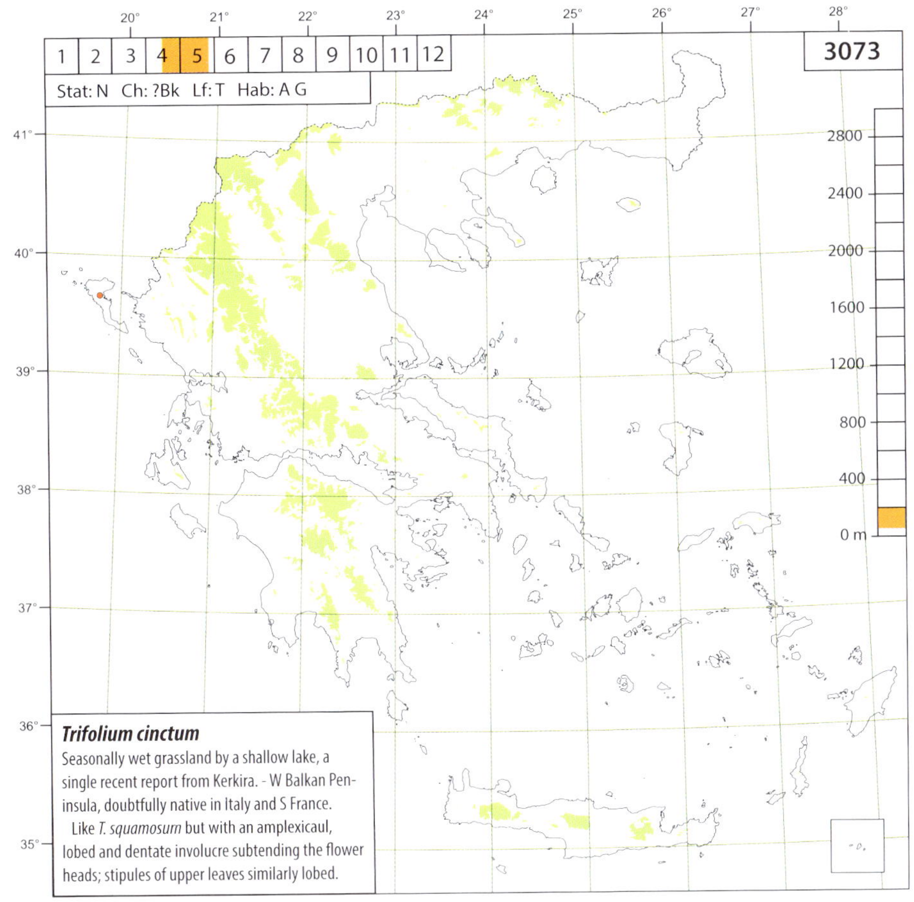

Trifolium cinctum

Seasonally wet grassland by a shallow lake, a single recent report from Kerkira. - W Balkan Peninsula, doubtfully native in Italy and S France.

Like *T. squamosum* but with an amplexicaul, lobed and dentate involucre subtending the flower heads; stipules of upper leaves similarly lobed.

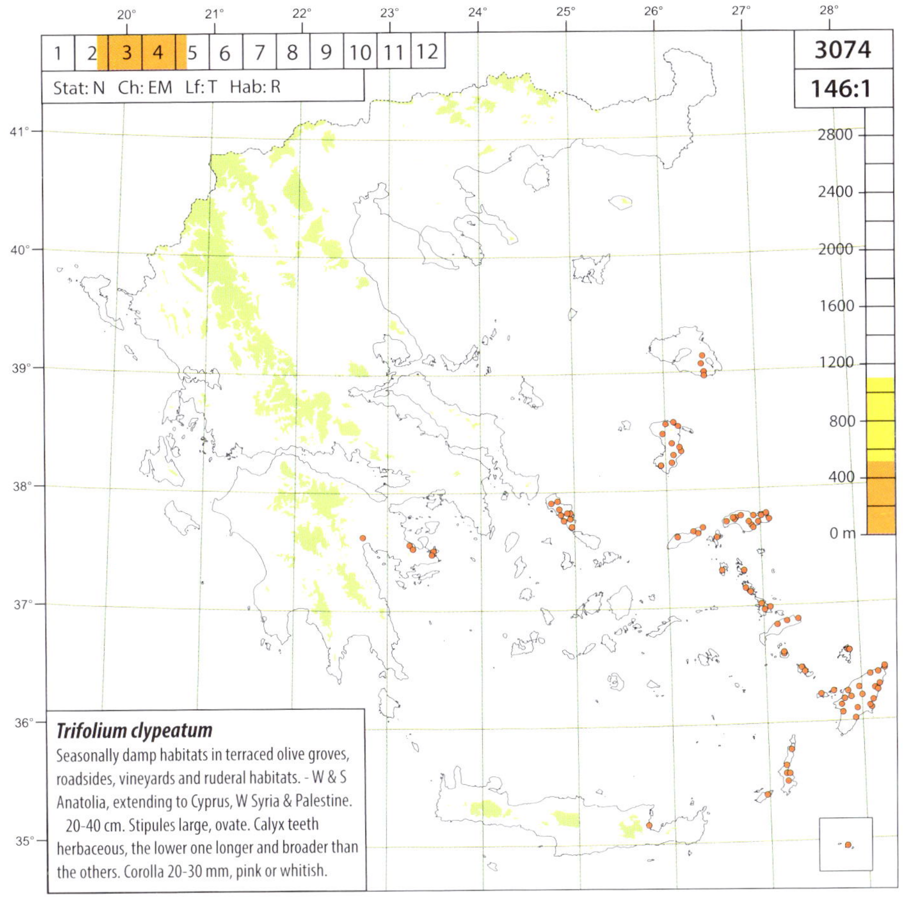

Trifolium clypeatum

Seasonally damp habitats in terraced olive groves, roadsides, vineyards and ruderal habitats. - W & S Anatolia, extending to Cyprus, W Syria & Palestine.

20-40 cm. Stipules large, ovate. Calyx teeth herbaceous, the lower one longer and broader than the others. Corolla 20-30 mm, pink or whitish.

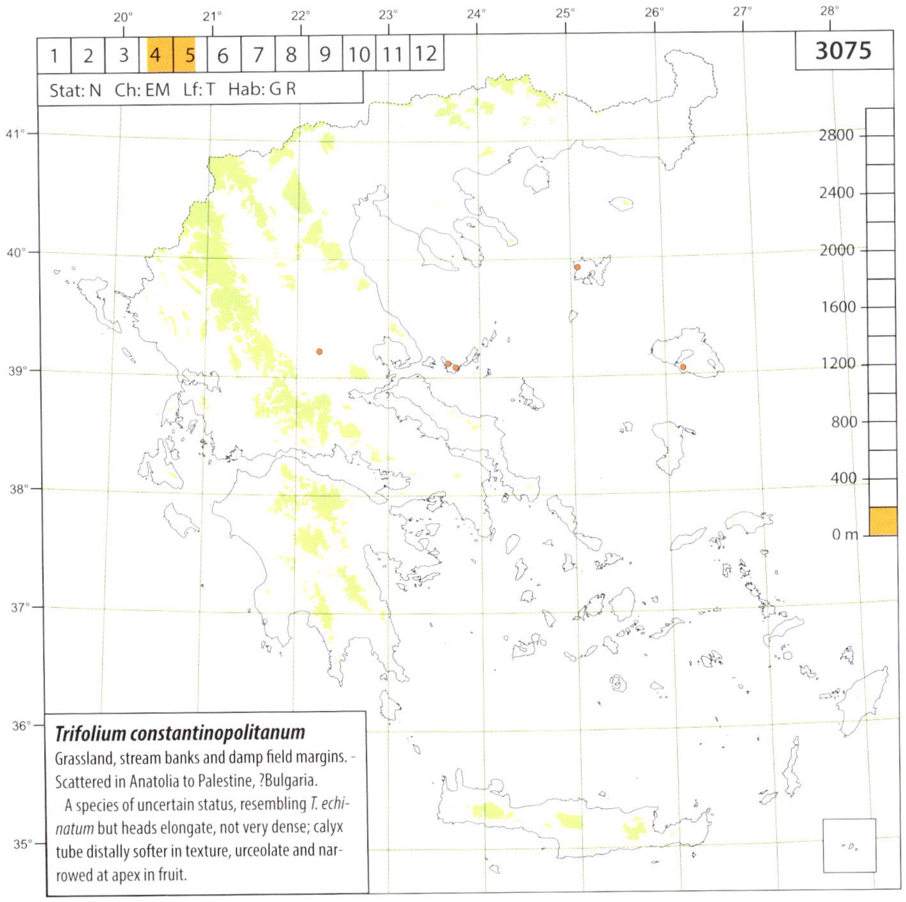

Trifolium constantinopolitanum

Grassland, stream banks and damp field margins. - Scattered in Anatolia to Palestine, ?Bulgaria.

A species of uncertain status, resembling *T. echinatum* but heads elongate, not very dense; calyx tube distally softer in texture, urceolate and narrowed at apex in fruit.

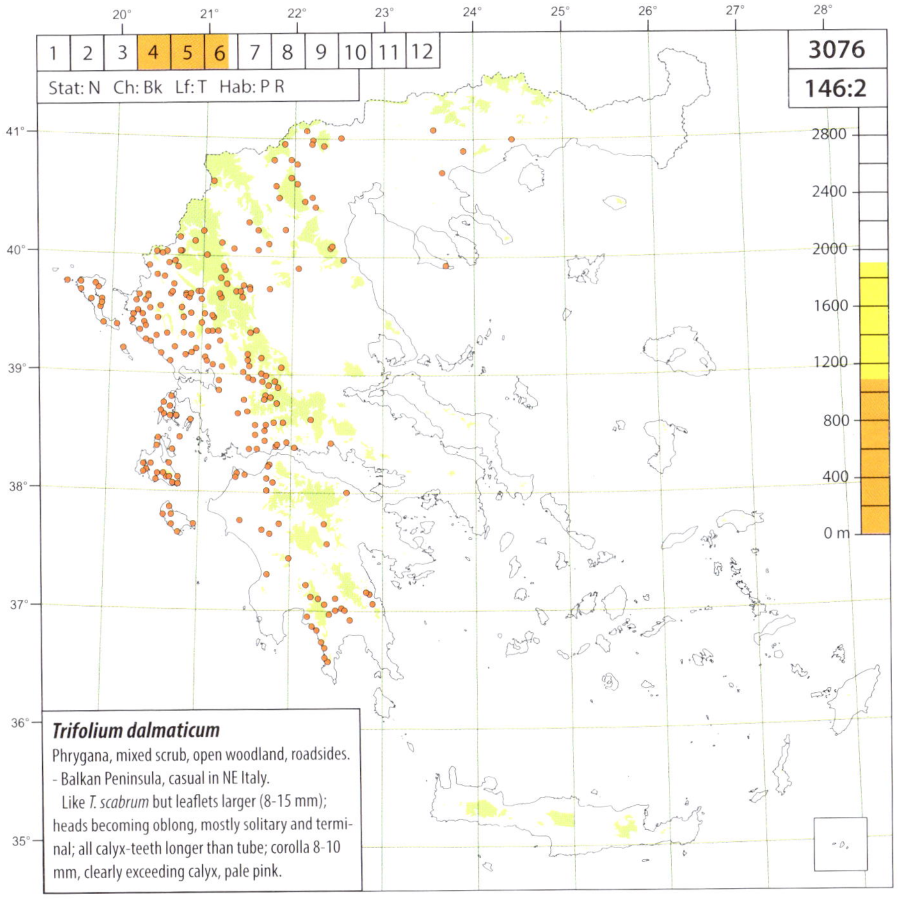

Trifolium dalmaticum

Phrygana, mixed scrub, open woodland, roadsides. - Balkan Peninsula, casual in NE Italy.

Like *T. scabrum* but leaflets larger (8-15 mm); heads becoming oblong, mostly solitary and terminal; all calyx-teeth longer than tube; corolla 8-10 mm, clearly exceeding calyx, pale pink.

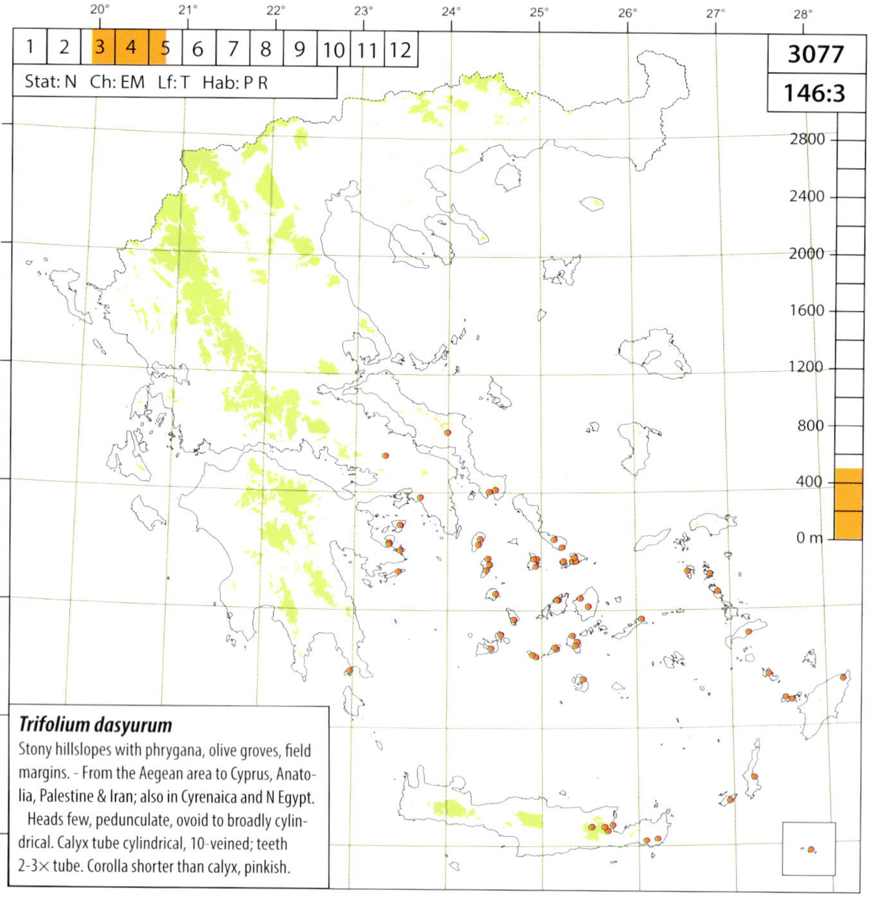

Trifolium dasyurum

Stony hillslopes with phrygana, olive groves, field margins. - From the Aegean area to Cyprus, Anatolia, Palestine & Iran; also in Cyrenaica and N Egypt.

Heads few, pedunculate, ovoid to broadly cylindrical. Calyx tube cylindrical, 10-veined; teeth 2-3× tube. Corolla shorter than calyx, pinkish.

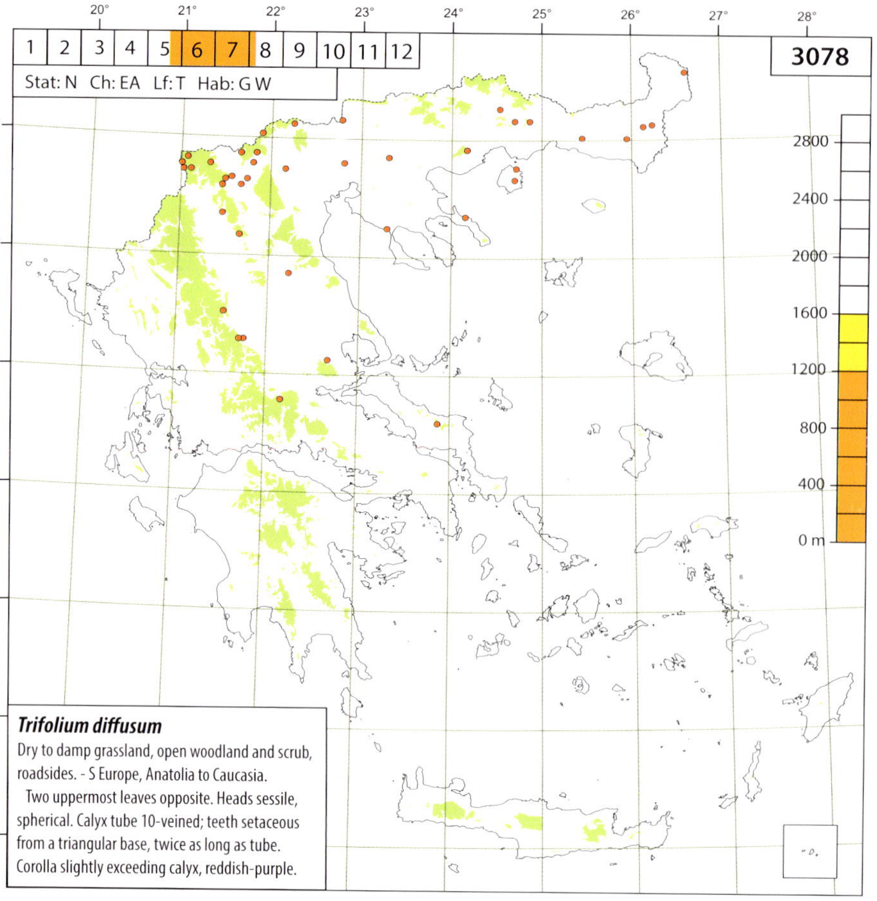

Trifolium diffusum

Dry to damp grassland, open woodland and scrub, roadsides. - S Europe, Anatolia to Caucasia.

Two uppermost leaves opposite. Heads sessile, spherical. Calyx tube 10-veined; teeth setaceous from a triangular base, twice as long as tube. Corolla slightly exceeding calyx, reddish-purple.

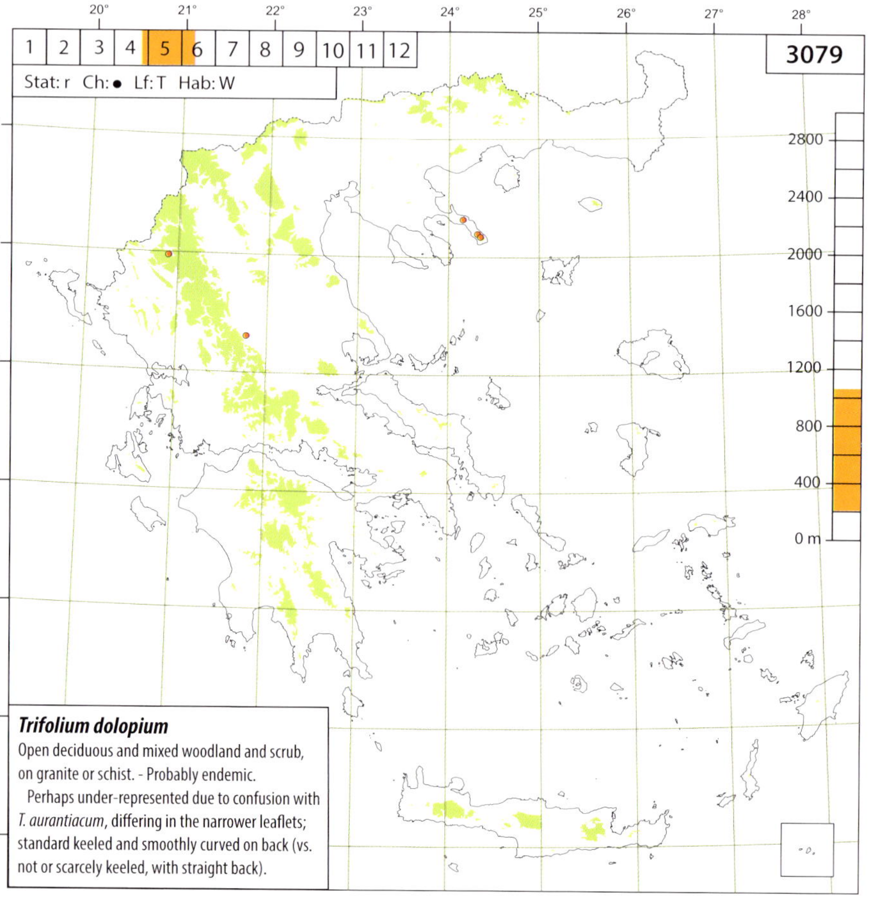

Trifolium dolopium

Open deciduous and mixed woodland and scrub, on granite or schist. - Probably endemic.

Perhaps under-represented due to confusion with *T. aurantiacum*, differing in the narrower leaflets; standard keeled and smoothly curved on back (vs. not or scarcely keeled, with straight back).

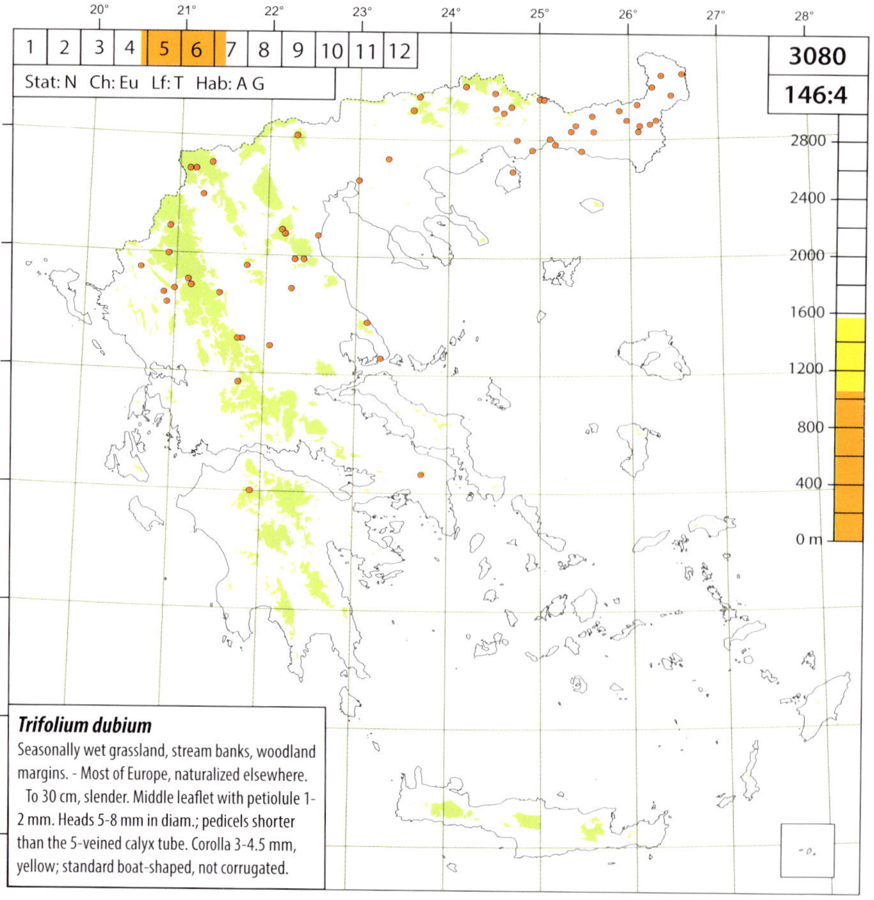

Trifolium dubium

Seasonally wet grassland, stream banks, woodland margins. - Most of Europe, naturalized elsewhere.

To 30 cm, slender. Middle leaflet with petiolule 1-2 mm. Heads 5-8 mm in diam.; pedicels shorter than the 5-veined calyx tube. Corolla 3-4.5 mm, yellow; standard boat-shaped, not corrugated.

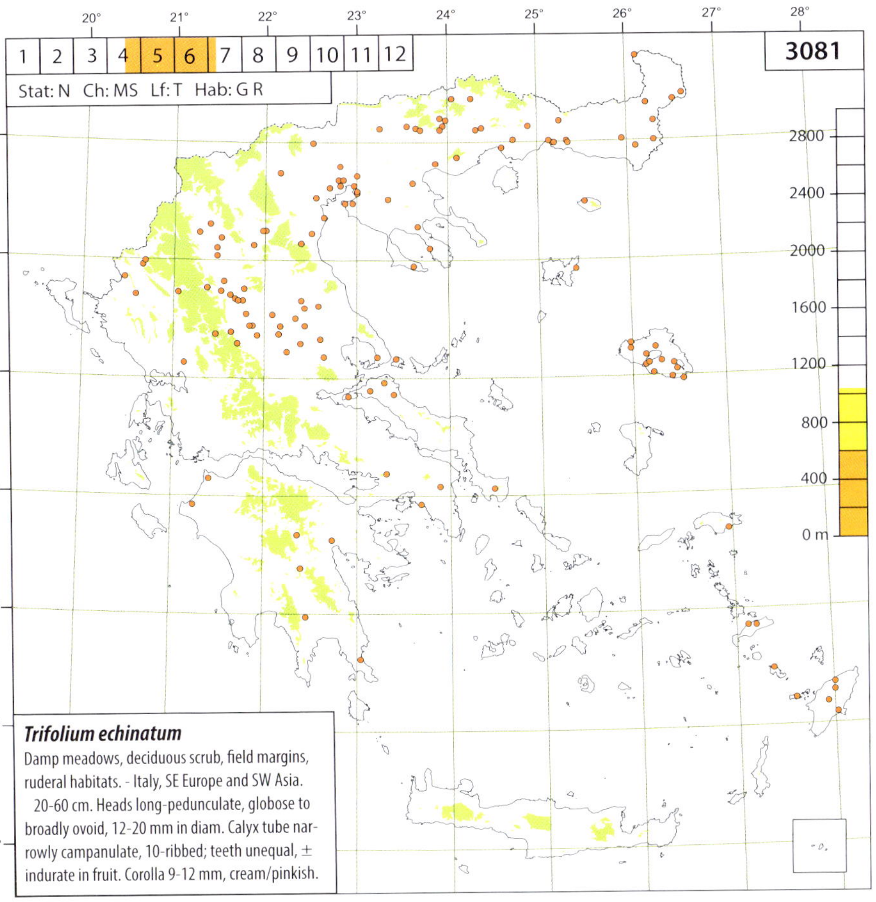

Trifolium echinatum

Damp meadows, deciduous scrub, field margins, ruderal habitats. - Italy, SE Europe and SW Asia.

20-60 cm. Heads long-pedunculate, globose to broadly ovoid, 12-20 mm in diam. Calyx tube narrowly campanulate, 10-ribbed; teeth unequal, ± indurate in fruit. Corolla 9-12 mm, cream/pinkish.

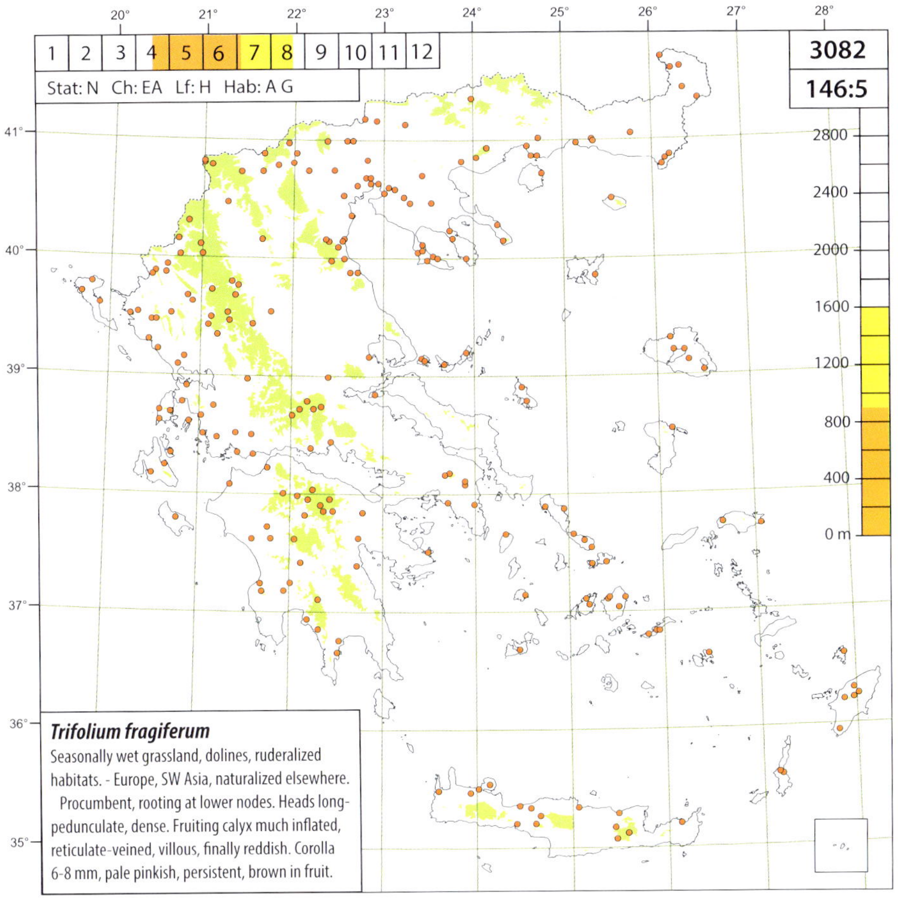

Trifolium fragiferum

Seasonally wet grassland, dolines, ruderalized habitats. - Europe, SW Asia, naturalized elsewhere.

Procumbent, rooting at lower nodes. Heads long-pedunculate, dense. Fruiting calyx much inflated, reticulate-veined, villous, finally reddish. Corolla 6-8 mm, pale pinkish, persistent, brown in fruit.

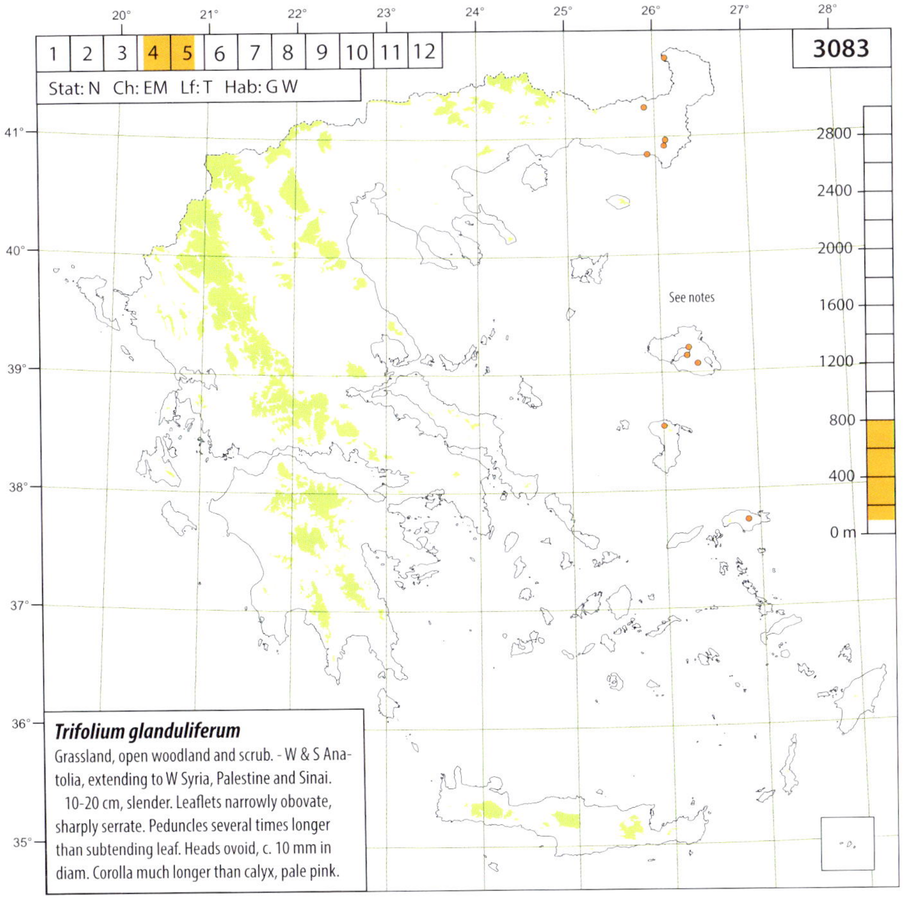

Trifolium glanduliferum

Grassland, open woodland and scrub. - W & S Anatolia, extending to W Syria, Palestine and Sinai.

10-20 cm, slender. Leaflets narrowly obovate, sharply serrate. Peduncles several times longer than subtending leaf. Heads ovoid, c. 10 mm in diam. Corolla much longer than calyx, pale pink.

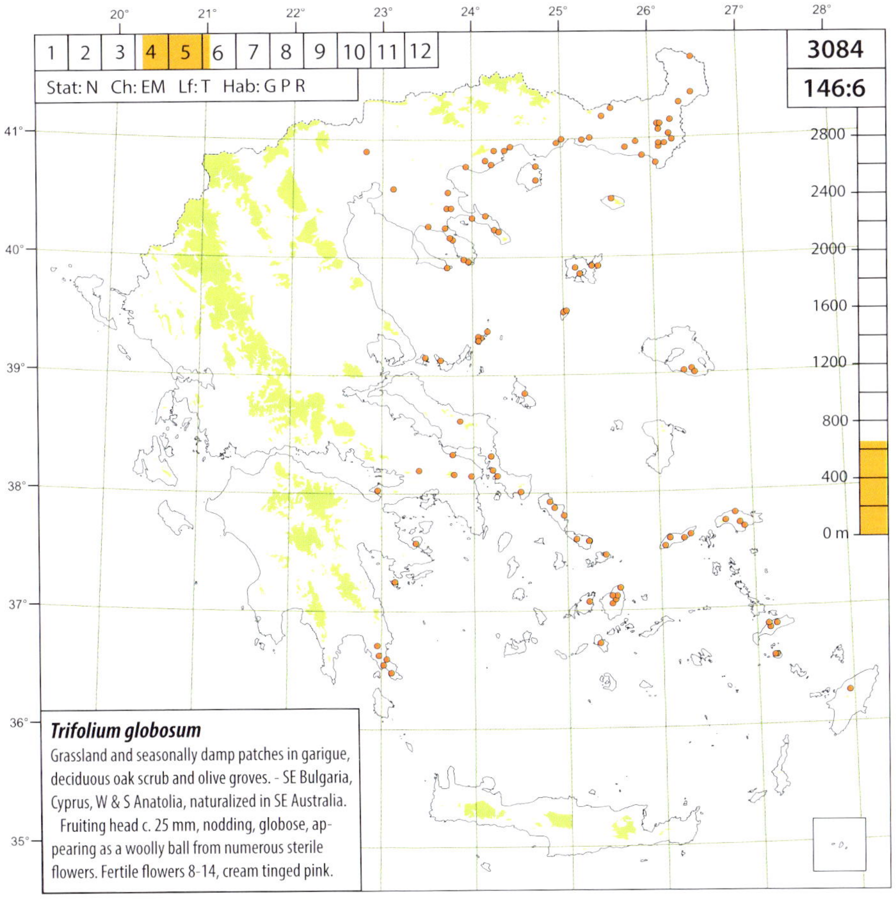

Trifolium globosum

Grassland and seasonally damp patches in garigue, deciduous oak scrub and olive groves. - SE Bulgaria, Cyprus, W & S Anatolia, naturalized in SE Australia.

Fruiting head c. 25 mm, nodding, globose, appearing as a woolly ball from numerous sterile flowers. Fertile flowers 8-14, cream tinged pink.

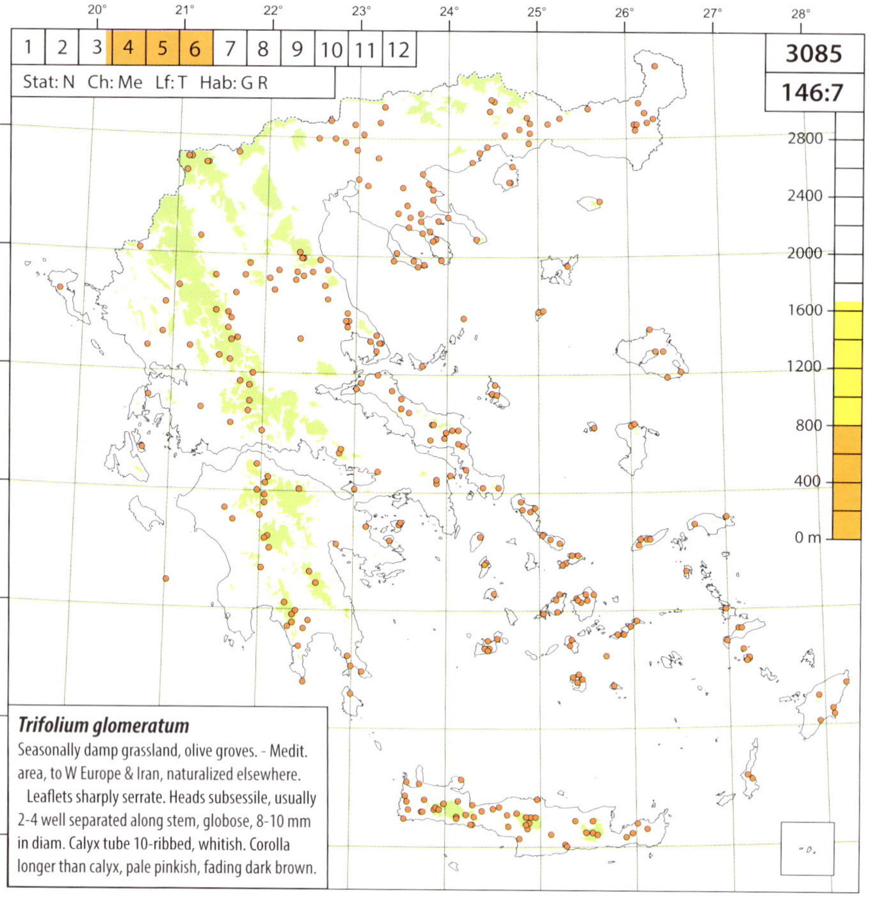

Trifolium glomeratum

Seasonally damp grassland, olive groves. - Medit. area, to W Europe & Iran, naturalized elsewhere.

Leaflets sharply serrate. Heads subsessile, usually 2-4 well separated along stem, globose, 8-10 mm in diam. Calyx tube 10-ribbed, whitish. Corolla longer than calyx, pale pinkish, fading dark brown.

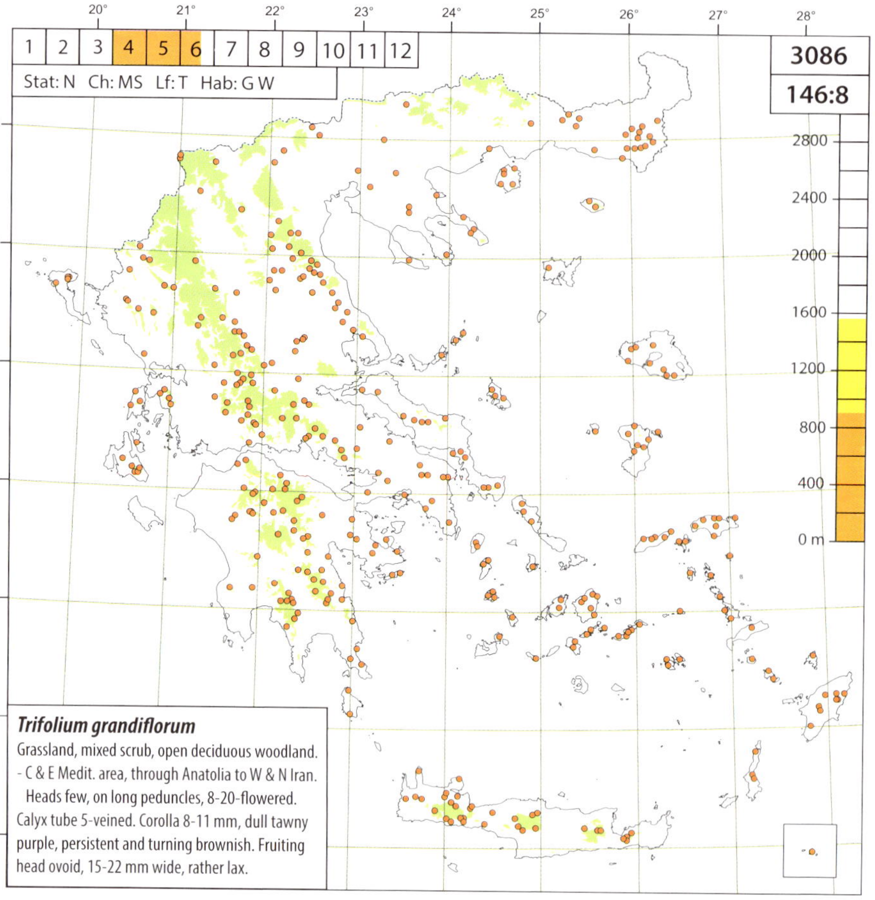

Trifolium grandiflorum

Grassland, mixed scrub, open deciduous woodland. - C & E Medit. area, through Anatolia to W & N Iran.

Heads few, on long peduncles, 8-20-flowered. Calyx tube 5-veined. Corolla 8-11 mm, dull tawny purple, persistent and turning brownish. Fruiting head ovoid, 15-22 mm wide, rather lax.

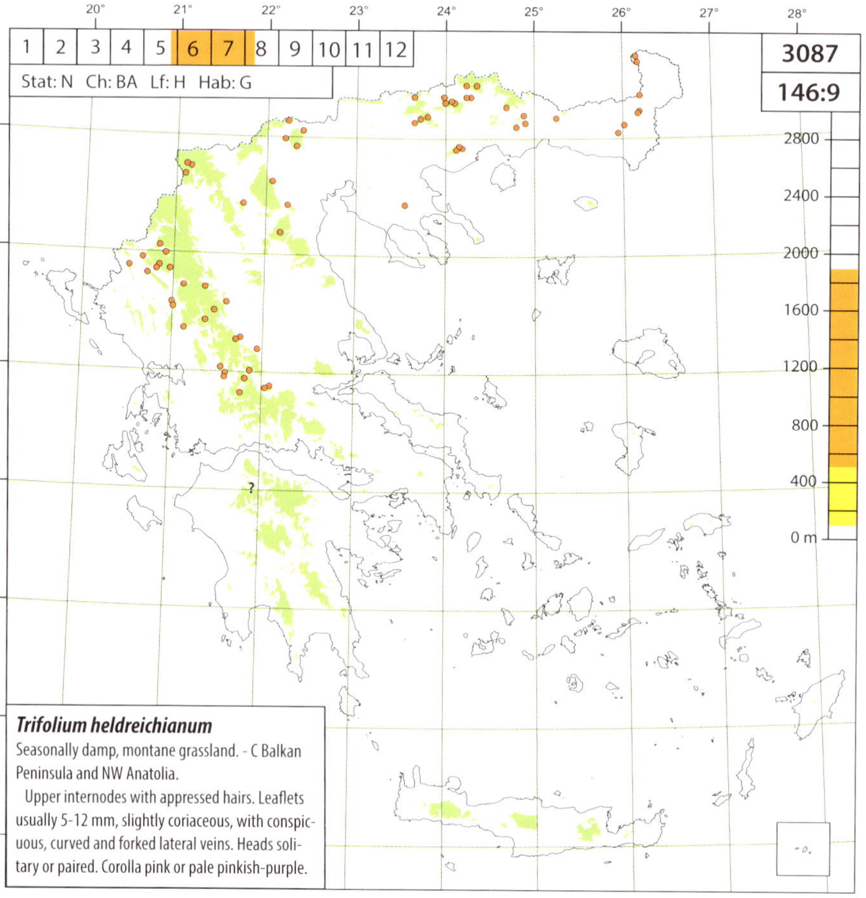

Trifolium heldreichianum

Seasonally damp, montane grassland. - C Balkan Peninsula and NW Anatolia.

Upper internodes with appressed hairs. Leaflets usually 5-12 mm, slightly coriaceous, with conspicuous, curved and forked lateral veins. Heads solitary or paired. Corolla pink or pale pinkish-purple.

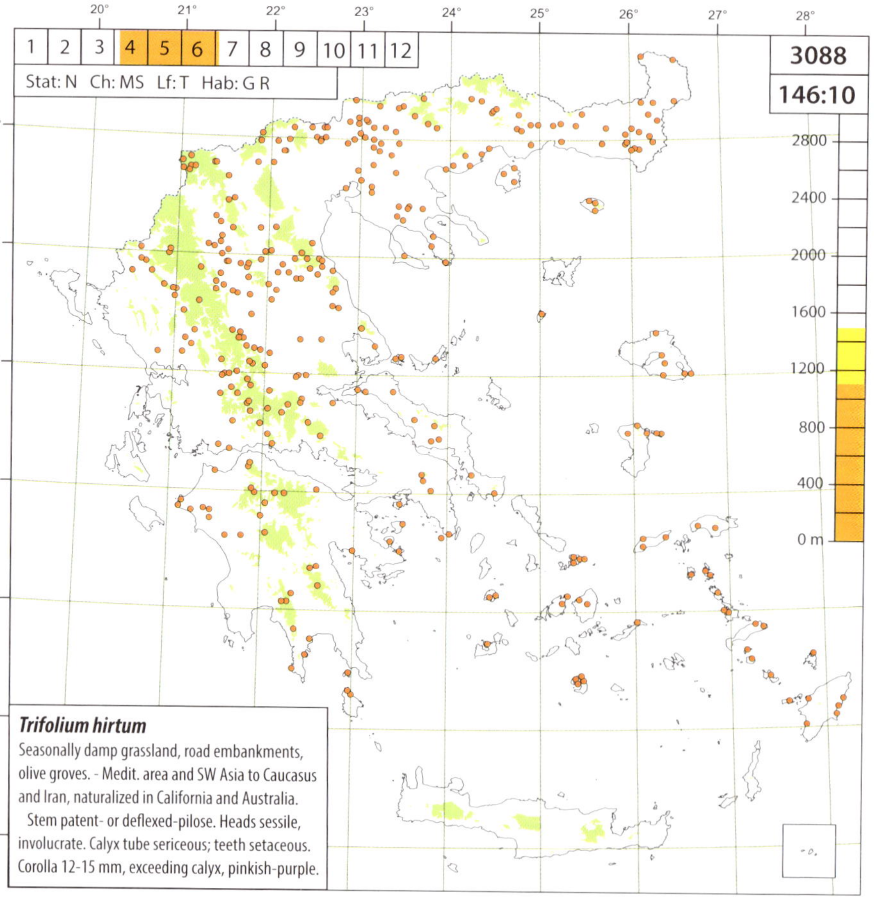

Trifolium hirtum

Seasonally damp grassland, road embankments, olive groves. - Medit. area and SW Asia to Caucasus and Iran, naturalized in California and Australia.

Stem patent- or deflexed-pilose. Heads sessile, involucrate. Calyx tube sericeous; teeth setaceous. Corolla 12-15 mm, exceeding calyx, pinkish-purple.

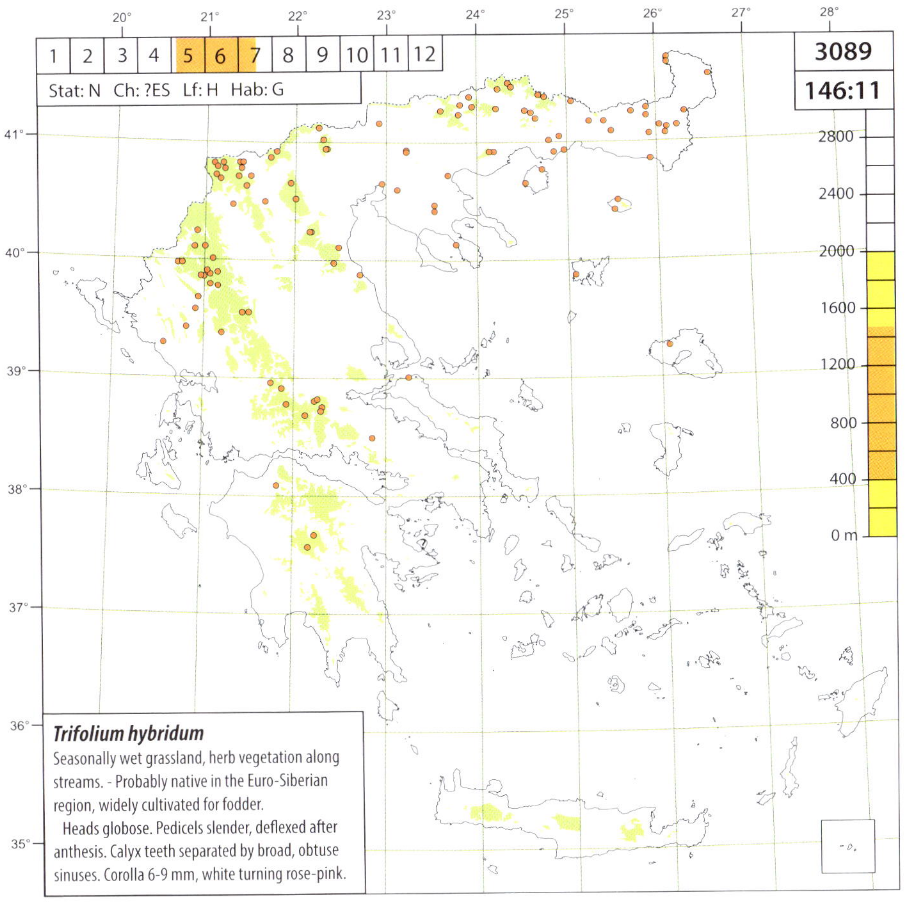

Trifolium hybridum

Seasonally wet grassland, herb vegetation along streams. - Probably native in the Euro-Siberian region, widely cultivated for fodder.

Heads globose. Pedicels slender, deflexed after anthesis. Calyx teeth separated by broad, obtuse sinuses. Corolla 6-9 mm, white turning rose-pink.

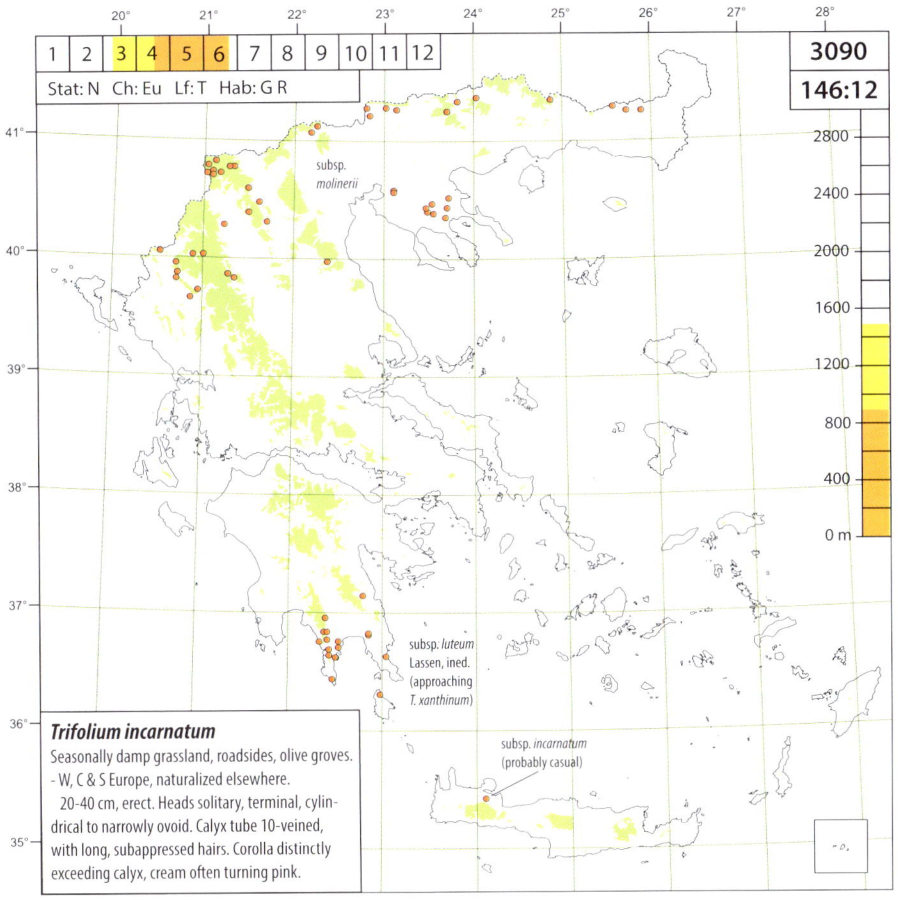

Trifolium incarnatum

Seasonally damp grassland, roadsides, olive groves. - W, C & S Europe, naturalized elsewhere.

20-40 cm, erect. Heads solitary, terminal, cylindrical to narrowly ovoid. Calyx tube 10-veined, with long, subappressed hairs. Corolla distinctly exceeding calyx, cream often turning pink.

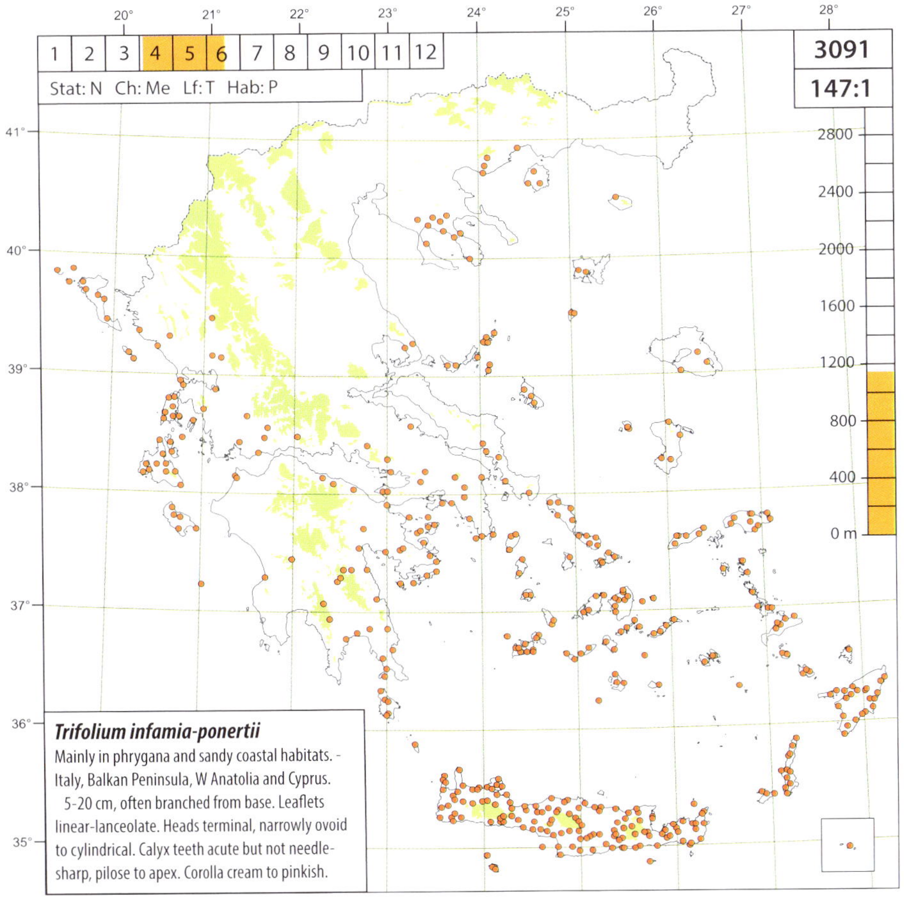

Trifolium infamia-ponertii

Mainly in phrygana and sandy coastal habitats. - Italy, Balkan Peninsula, W Anatolia and Cyprus.

5-20 cm, often branched from base. Leaflets linear-lanceolate. Heads terminal, narrowly ovoid to cylindrical. Calyx teeth acute but not needle-sharp, pilose to apex. Corolla cream to pinkish.

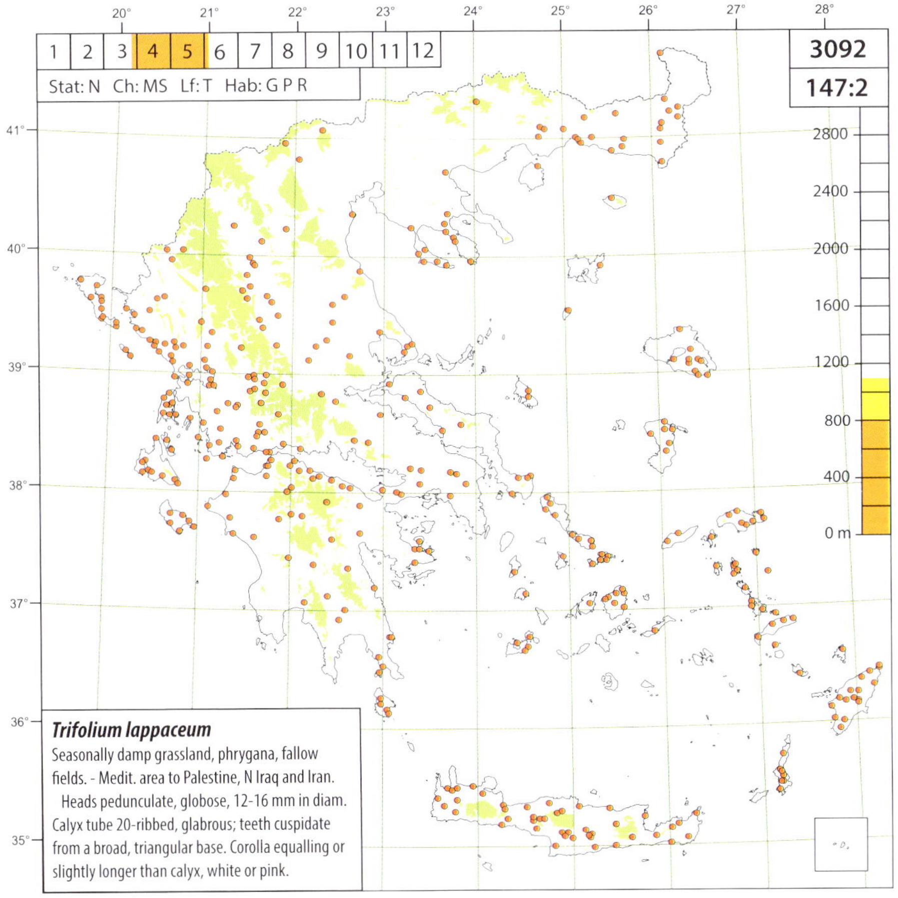

Trifolium lappaceum

Seasonally damp grassland, phrygana, fallow fields. - Medit. area to Palestine, N Iraq and Iran.

Heads pedunculate, globose, 12-16 mm in diam. Calyx tube 20-ribbed, glabrous; teeth cuspidate from a broad, triangular base. Corolla equalling or slightly longer than calyx, white or pink.

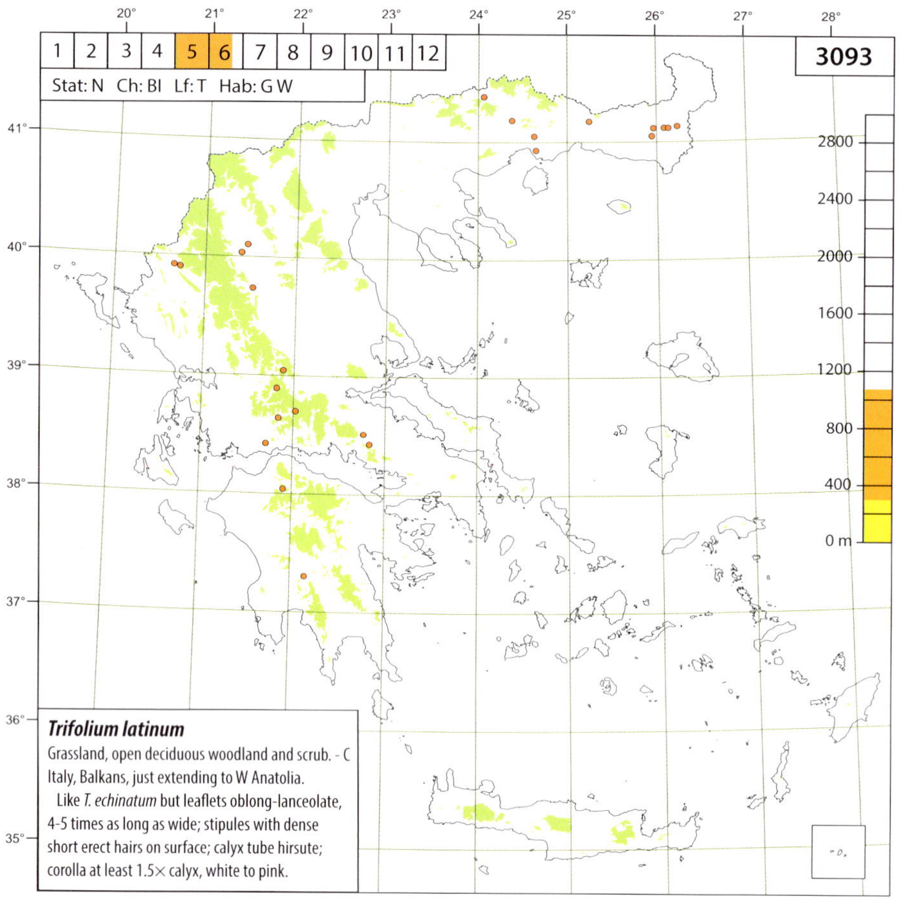

Trifolium latinum

Grassland, open deciduous woodland and scrub. - C Italy, Balkans, just extending to W Anatolia.

Like *T. echinatum* but leaflets oblong-lanceolate, 4-5 times as long as wide; stipules with dense short erect hairs on surface; calyx tube hirsute; corolla at least 1.5× calyx, white to pink.

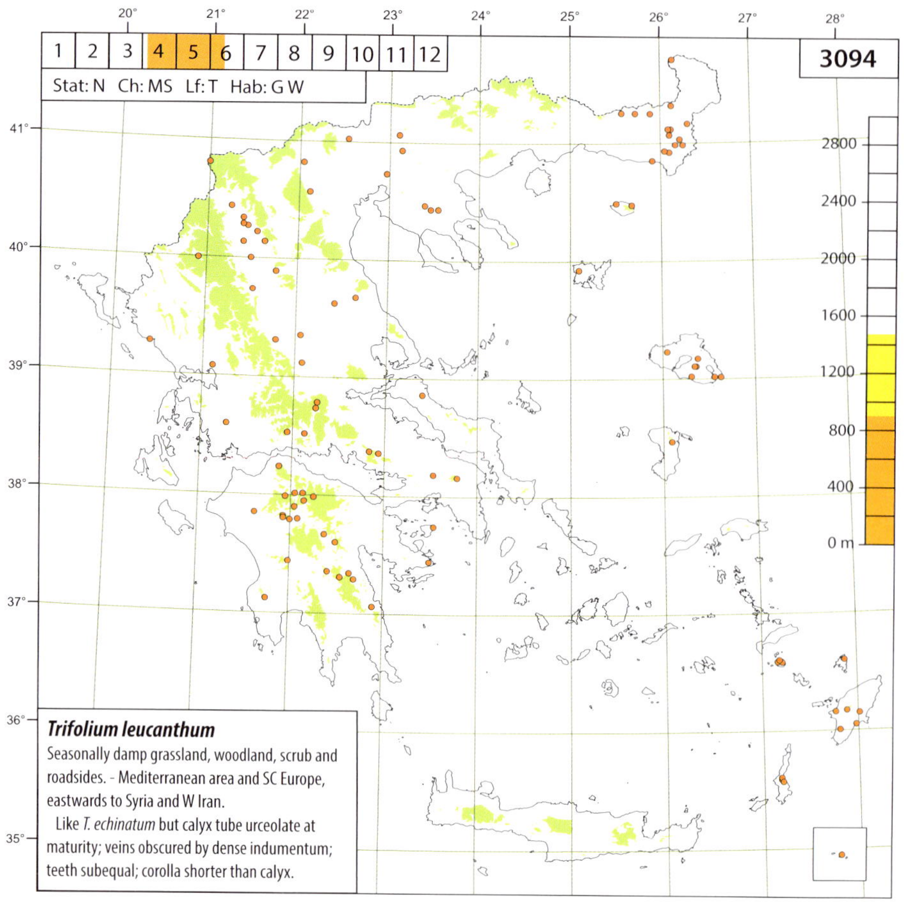

Trifolium leucanthum

Seasonally damp grassland, woodland, scrub and roadsides. - Mediterranean area and SC Europe, eastwards to Syria and W Iran.

Like *T. echinatum* but calyx tube urceolate at maturity; veins obscured by dense indumentum; teeth subequal; corolla shorter than calyx.

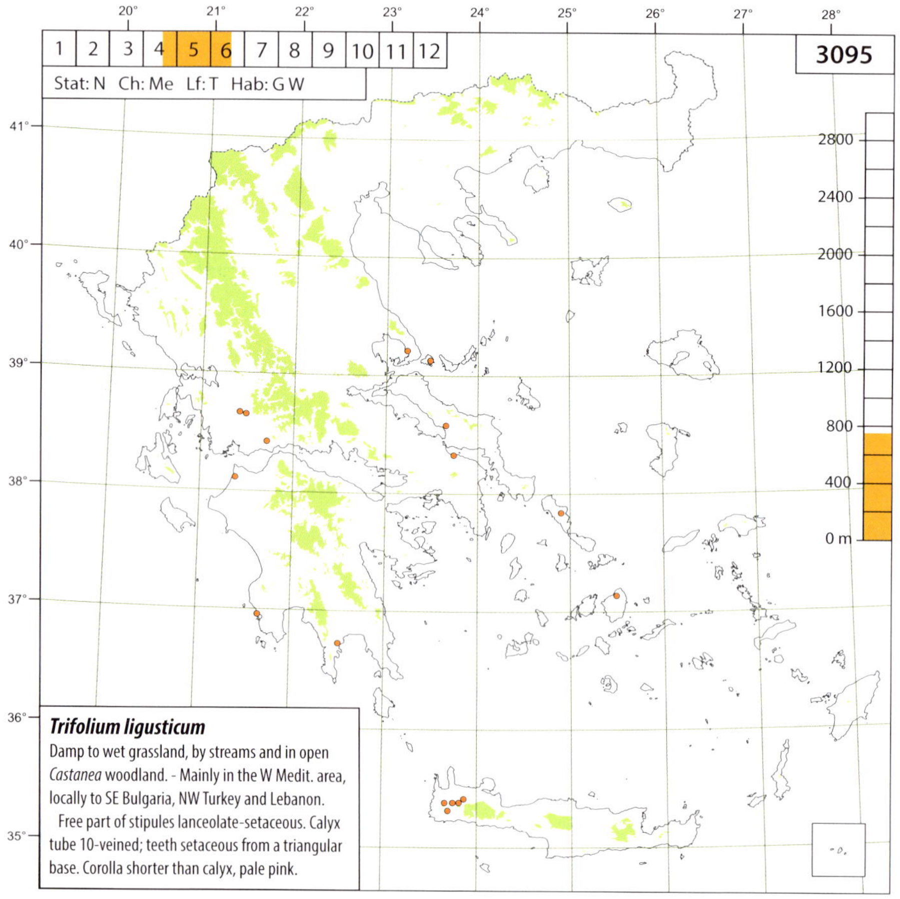

Trifolium ligusticum

Damp to wet grassland, by streams and in open *Castanea* woodland. - Mainly in the W Medit. area, locally to SE Bulgaria, NW Turkey and Lebanon.

Free part of stipules lanceolate-setaceous. Calyx tube 10-veined; teeth setaceous from a triangular base. Corolla shorter than calyx, pale pink.

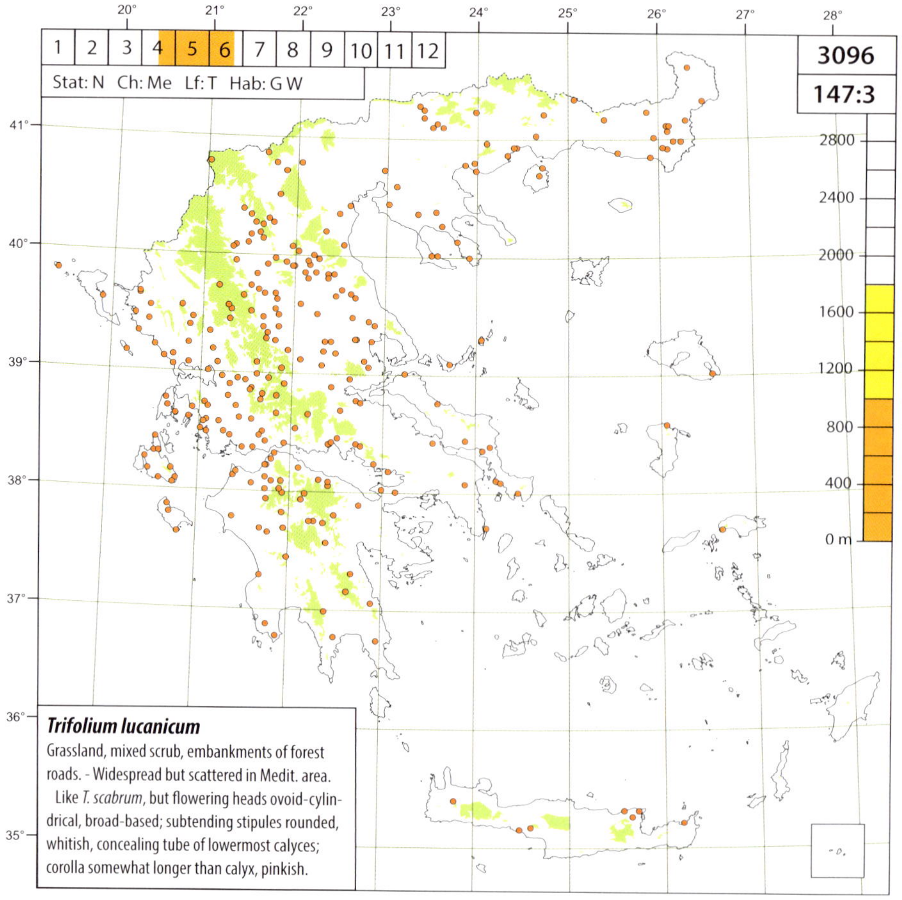

Trifolium lucanicum

Grassland, mixed scrub, embankments of forest roads. - Widespread but scattered in Medit. area.

Like *T. scabrum*, but flowering heads ovoid-cylindrical, broad-based; subtending stipules rounded, whitish, concealing tube of lowermost calyces; corolla somewhat longer than calyx, pinkish.

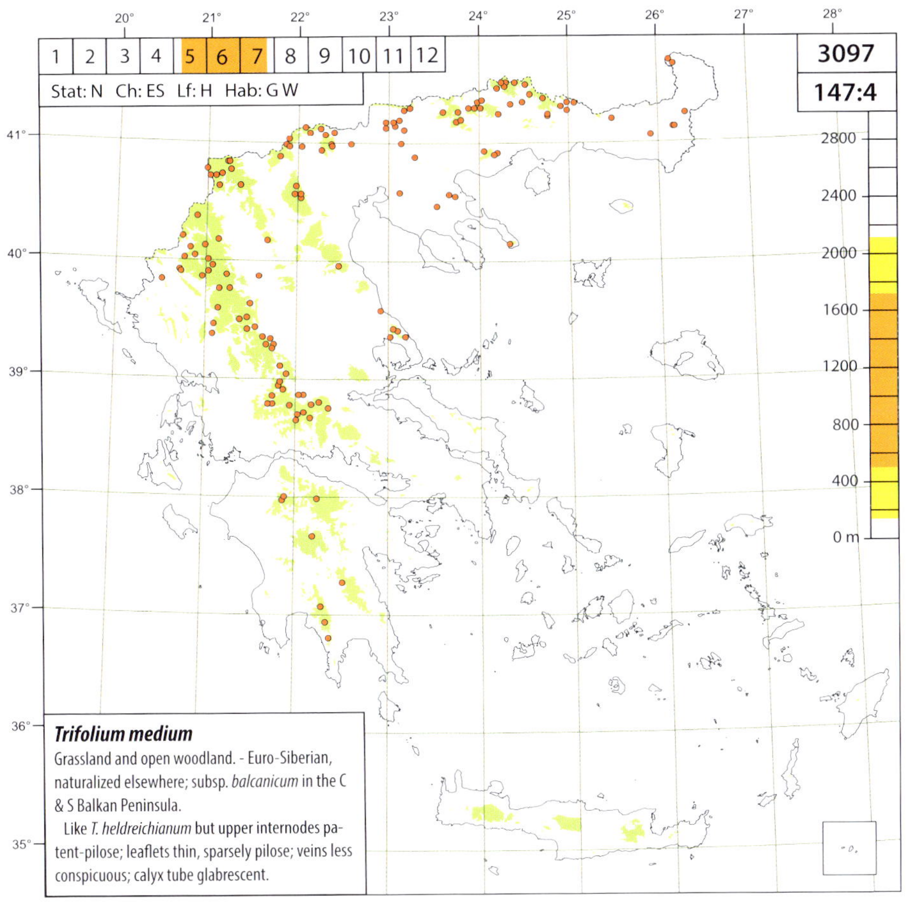

Trifolium medium

Grassland and open woodland. - Euro-Siberian, naturalized elsewhere; subsp. *balcanicum* in the C & S Balkan Peninsula.

Like *T. heldreichianum* but upper internodes patent-pilose; leaflets thin, sparsely pilose; veins less conspicuous; calyx tube glabrescent.

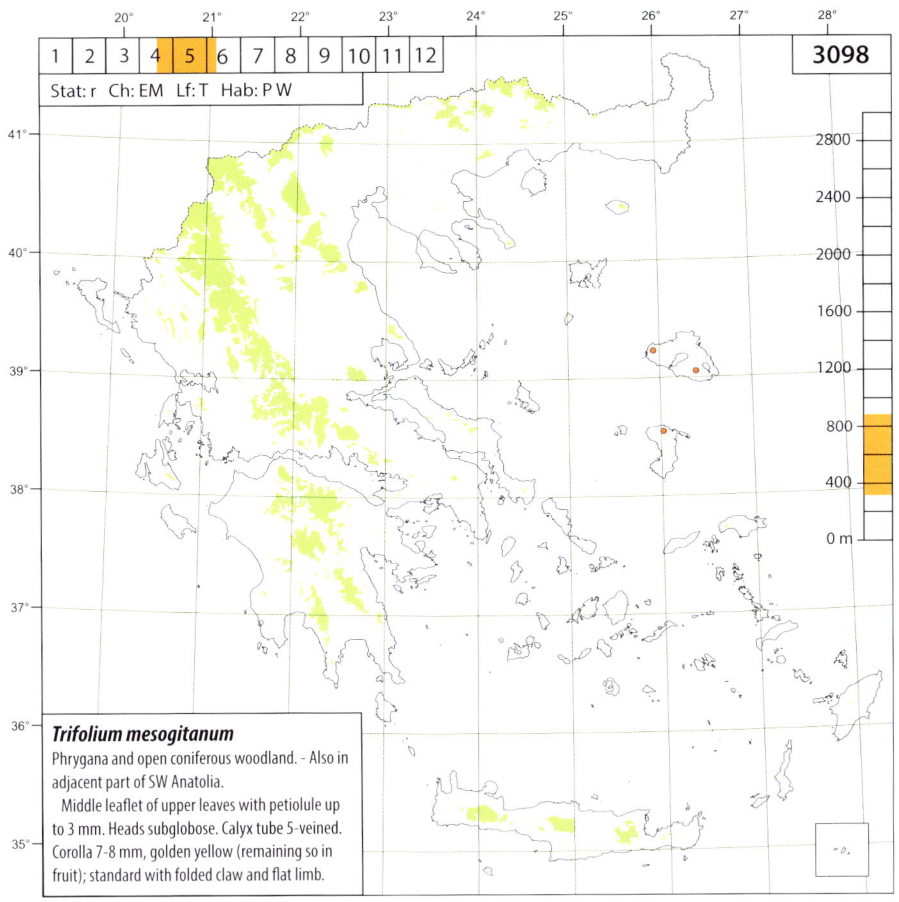

Trifolium mesogitanum

Phrygana and open coniferous woodland. - Also in adjacent part of SW Anatolia.

Middle leaflet of upper leaves with petiolule up to 3 mm. Heads subglobose. Calyx tube 5-veined. Corolla 7-8 mm, golden yellow (remaining so in fruit); standard with folded claw and flat limb.

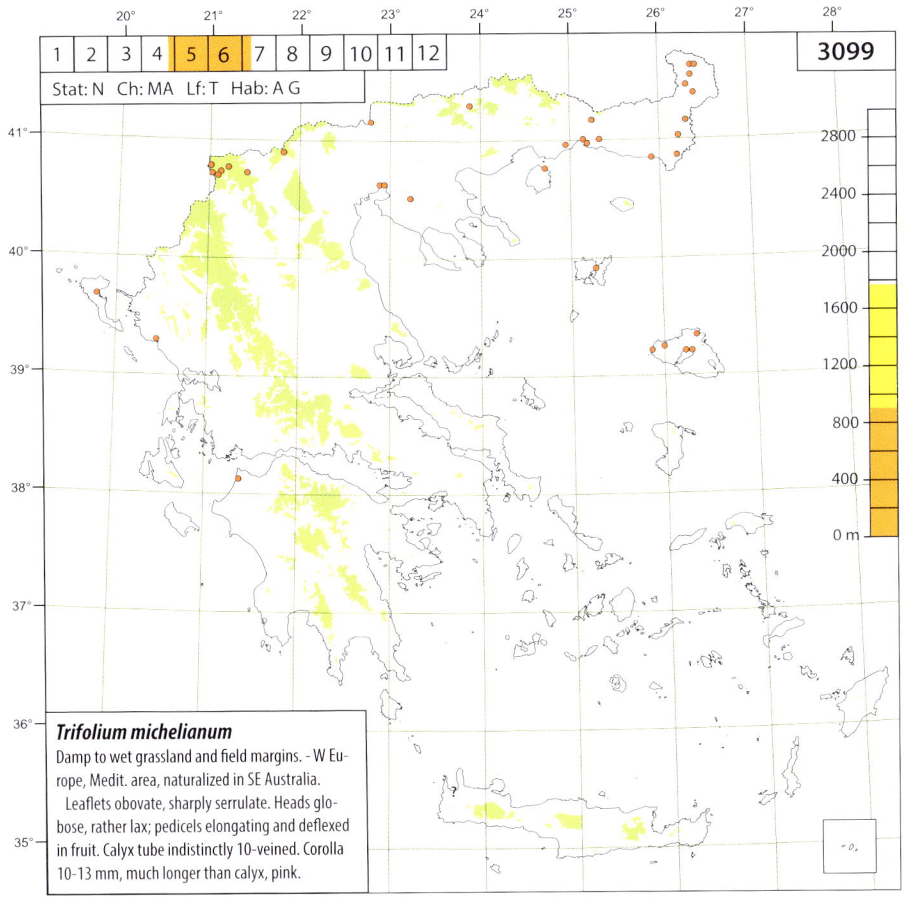

Trifolium michelianum

Damp to wet grassland and field margins. - W Europe, Medit. area, naturalized in SE Australia.

Leaflets obovate, sharply serrulate. Heads globose, rather lax; pedicels elongating and deflexed in fruit. Calyx tube indistinctly 10-veined. Corolla 10-13 mm, much longer than calyx, pink.

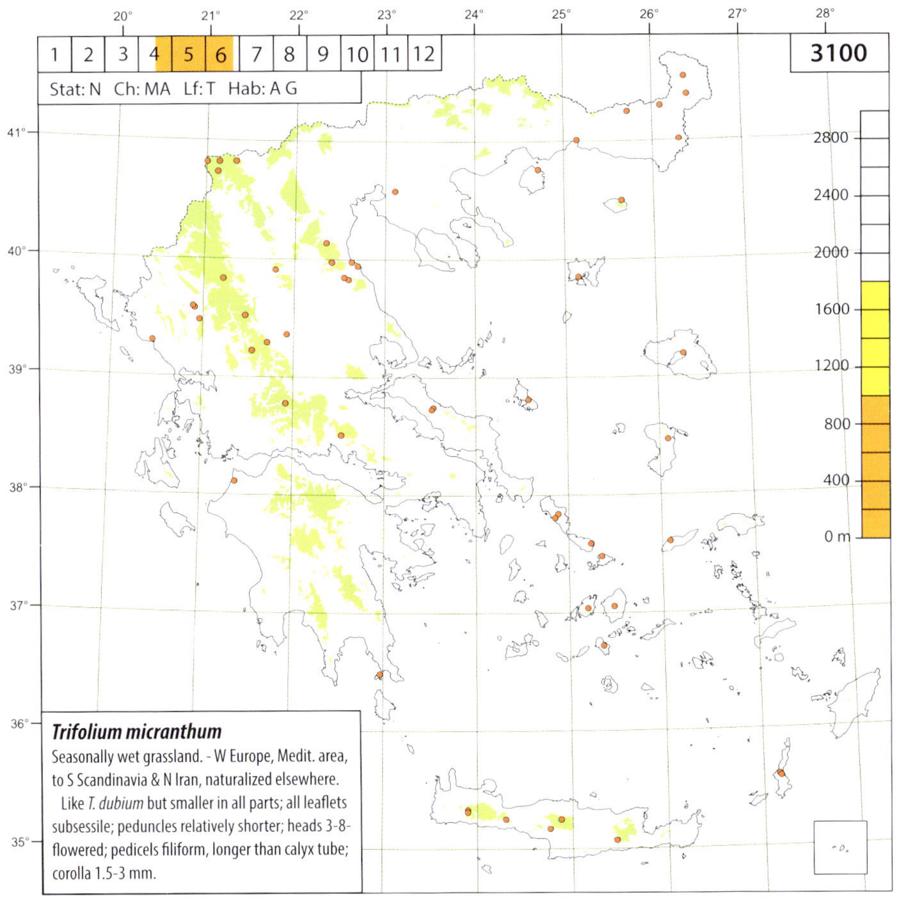

Trifolium micranthum

Seasonally wet grassland. - W Europe, Medit. area, to S Scandinavia & N Iran, naturalized elsewhere.

Like *T. dubium* but smaller in all parts; all leaflets subsessile; peduncles relatively shorter; heads 3-8-flowered; pedicels filiform, longer than calyx tube; corolla 1.5-3 mm.

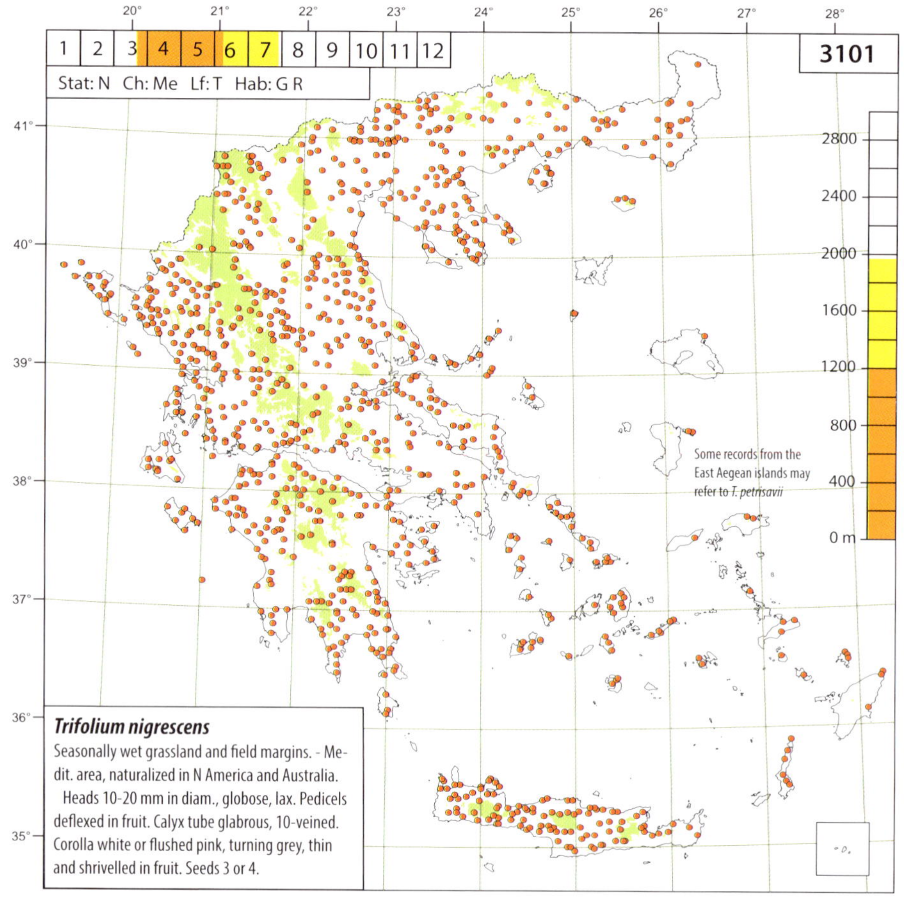

Trifolium nigrescens

Seasonally wet grassland and field margins. - Medit. area, naturalized in N America and Australia.

Heads 10-20 mm in diam., globose, lax. Pedicels deflexed in fruit. Calyx tube glabrous, 10-veined. Corolla white or flushed pink, turning grey, thin and shrivelled in fruit. Seeds 3 or 4.

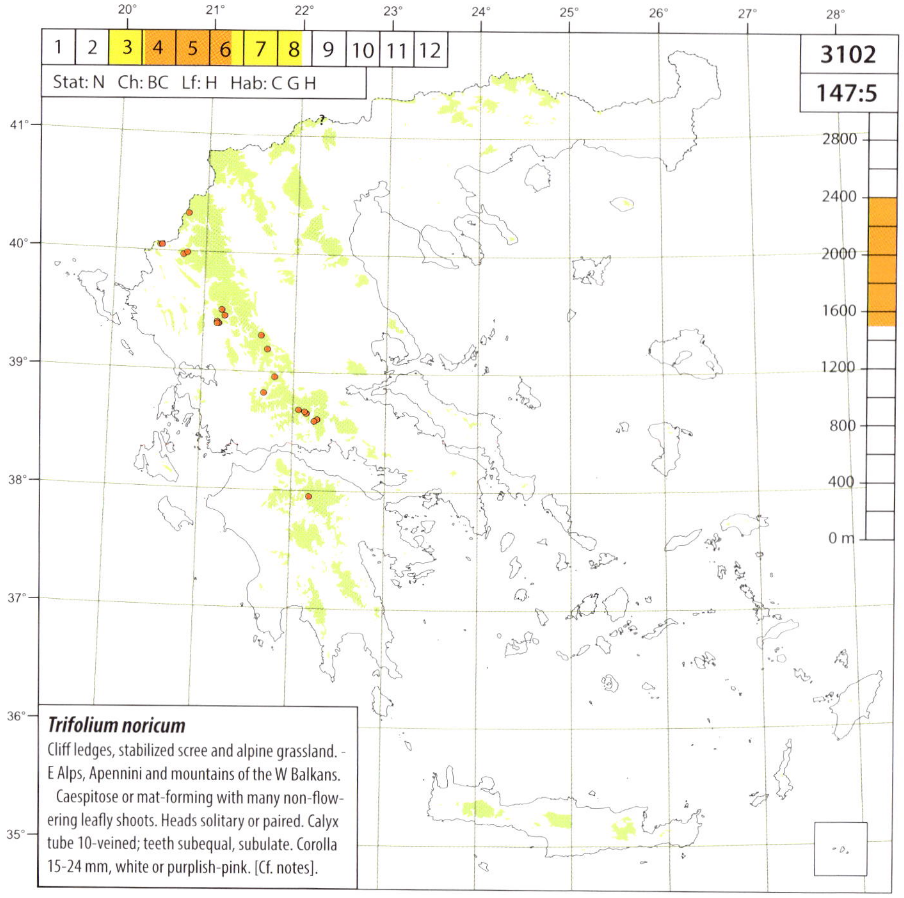

Trifolium noricum

Cliff ledges, stabilized scree and alpine grassland. - E Alps, Apennini and mountains of the W Balkans.

Caespitose or mat-forming with many non-flowering leafy shoots. Heads solitary or paired. Calyx tube 10-veined; teeth subequal, subulate. Corolla 15-24 mm, white or purplish-pink. [Cf. notes].

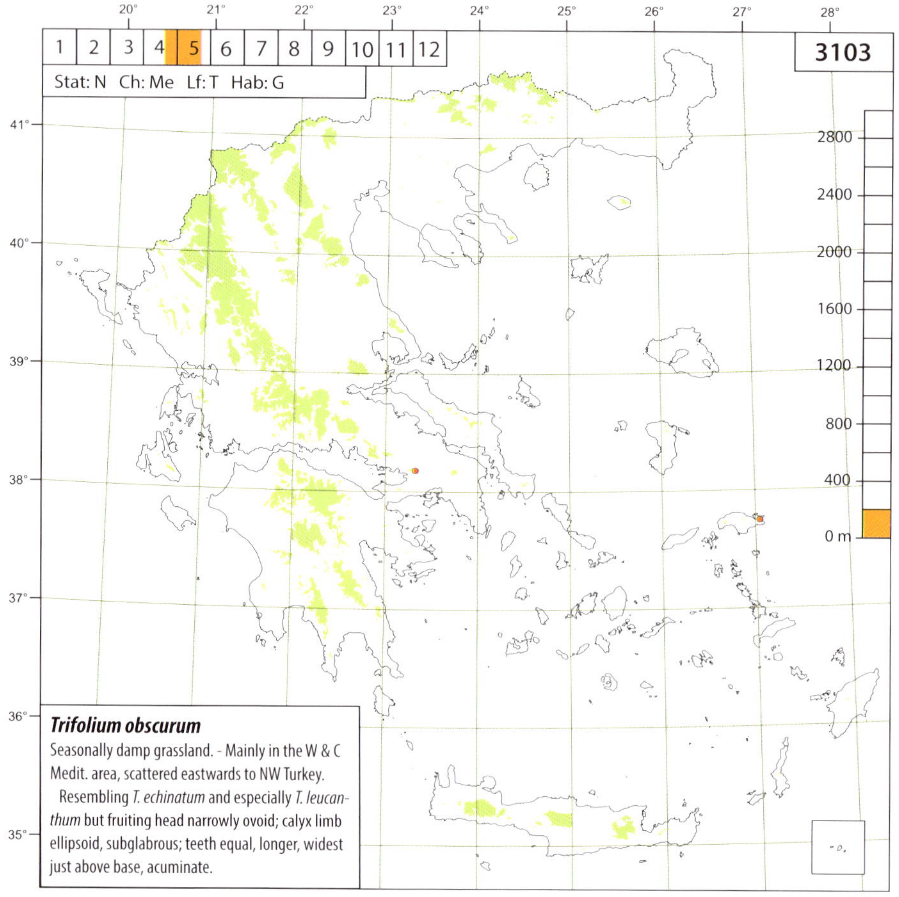

Trifolium obscurum

Seasonally damp grassland. - Mainly in the W & C Medit. area, scattered eastwards to NW Turkey.

Resembling *T. echinatum* and especially *T. leucanthum* but fruiting head narrowly ovoid; calyx limb ellipsoid, subglabrous; teeth equal, longer, widest just above base, acuminate.

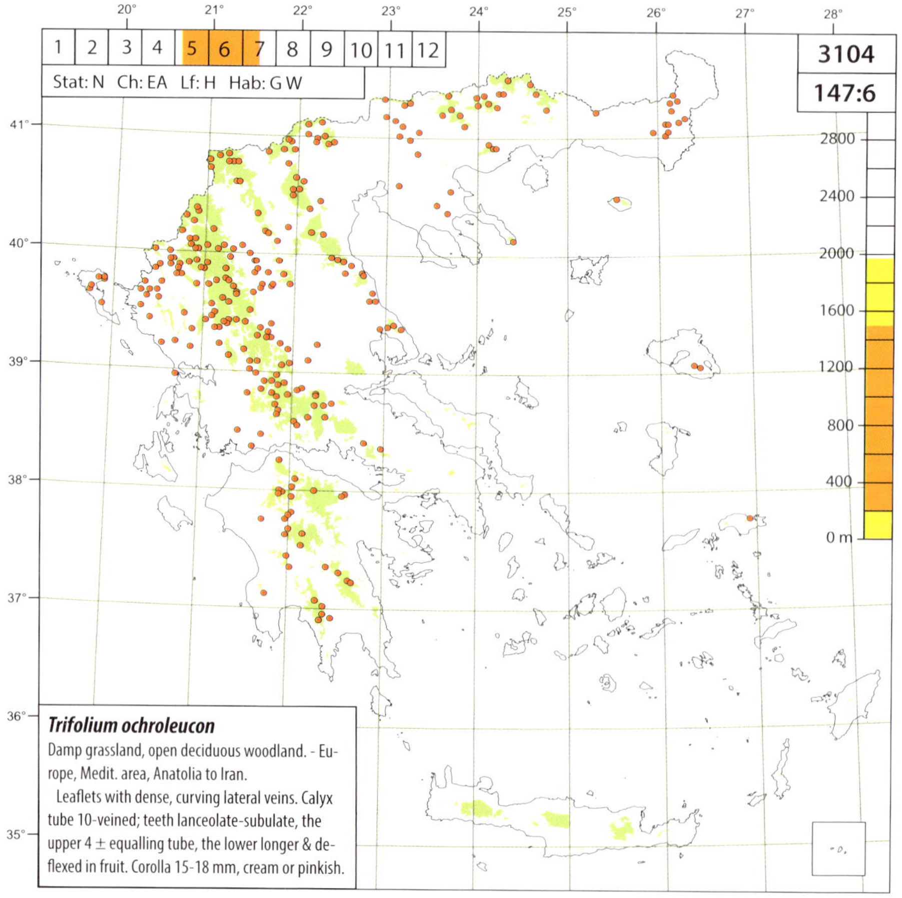

Trifolium ochroleucon

Damp grassland, open deciduous woodland. - Europe, Medit. area, Anatolia to Iran.

Leaflets with dense, curving lateral veins. Calyx tube 10-veined; teeth lanceolate-subulate, the upper 4 ± equalling tube, the lower longer & deflexed in fruit. Corolla 15-18 mm, cream or pinkish.

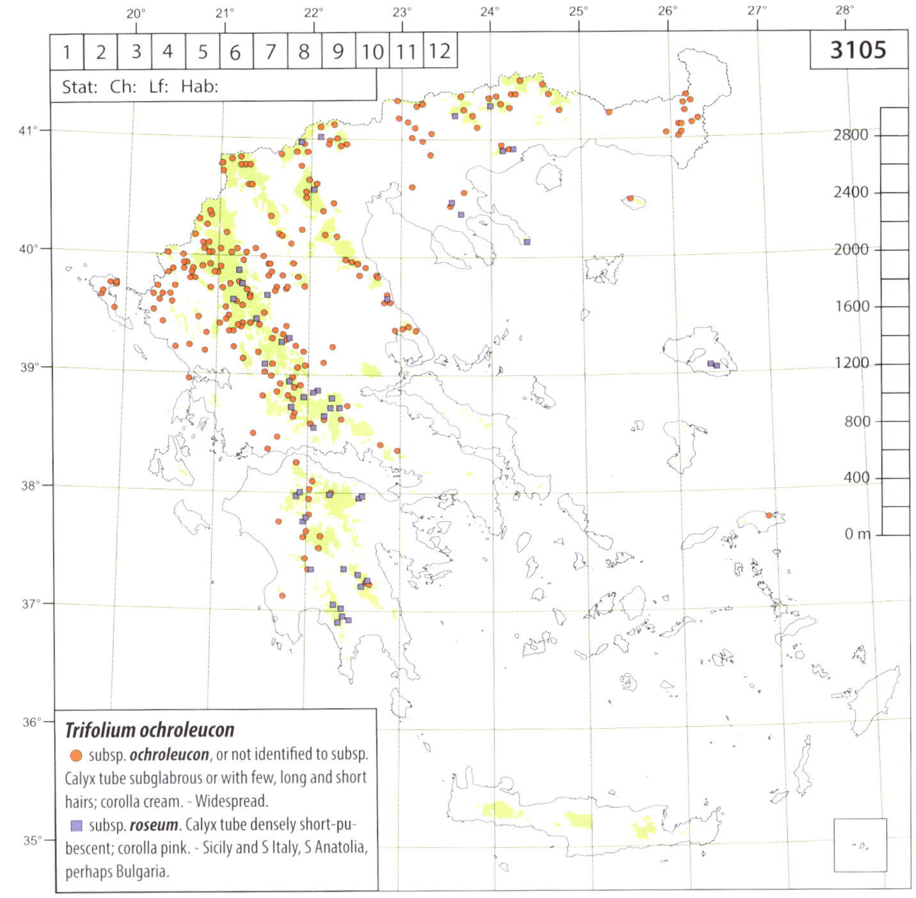

Trifolium ochroleucon

● subsp. ***ochroleucon***, or not identified to subsp. Calyx tube subglabrous or with few, long and short hairs; corolla cream. - Widespread.

■ subsp. ***roseum***. Calyx tube densely short-pubescent; corolla pink. - Sicily and S Italy, S Anatolia, perhaps Bulgaria.

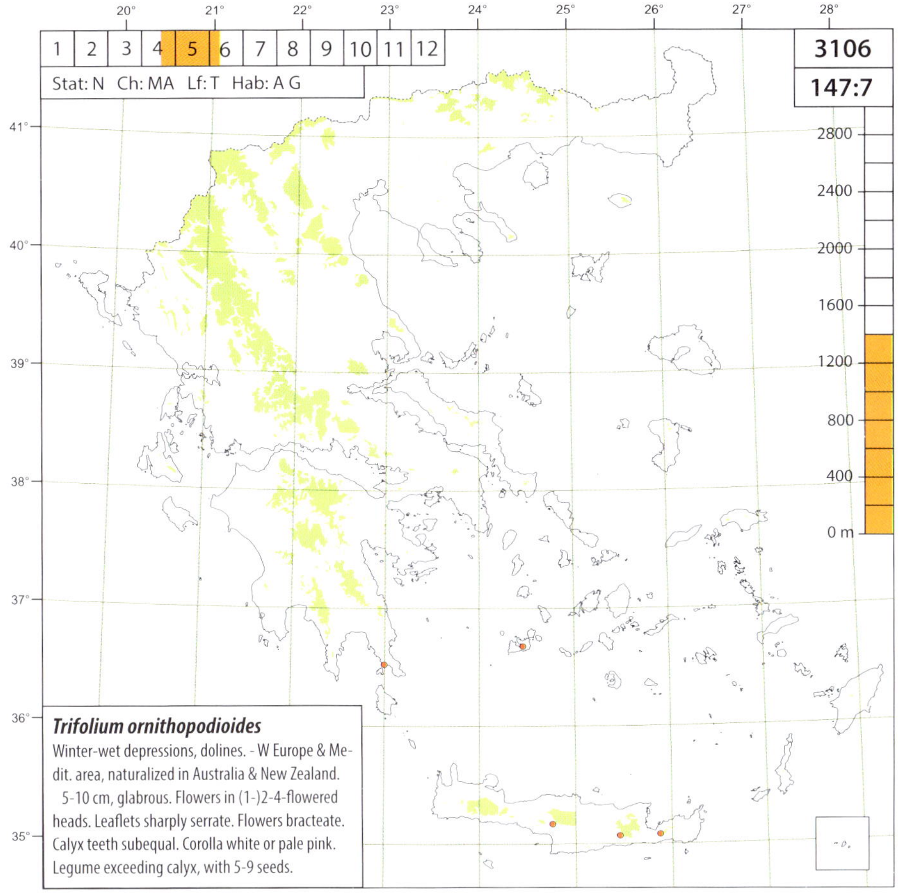

Trifolium ornithopodioides

Winter-wet depressions, dolines. - W Europe & Medit. area, naturalized in Australia & New Zealand.

5-10 cm, glabrous. Flowers in (1-)2-4-flowered heads. Leaflets sharply serrate. Flowers bracteate. Calyx teeth subequal. Corolla white or pale pink. Legume exceeding calyx, with 5-9 seeds.

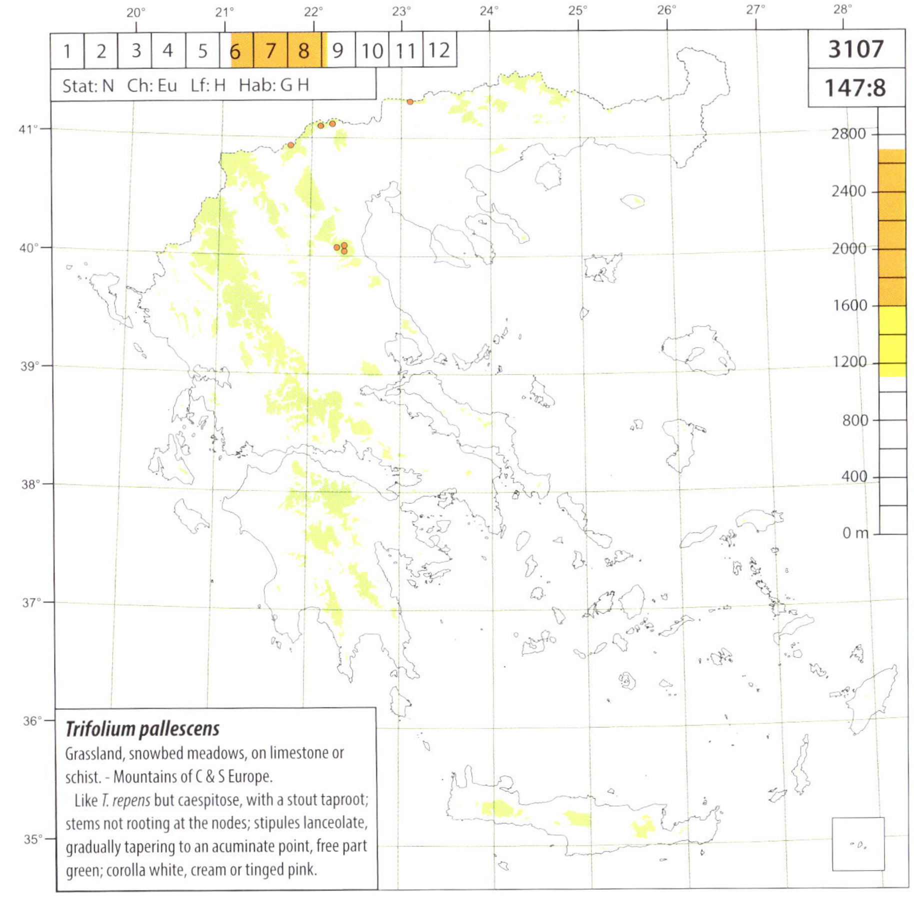

Trifolium pallescens

Grassland, snowbed meadows, on limestone or schist. - Mountains of C & S Europe.

Like *T. repens* but caespitose, with a stout taproot; stems not rooting at the nodes; stipules lanceolate, gradually tapering to an acuminate point, free part green; corolla white, cream or tinged pink.

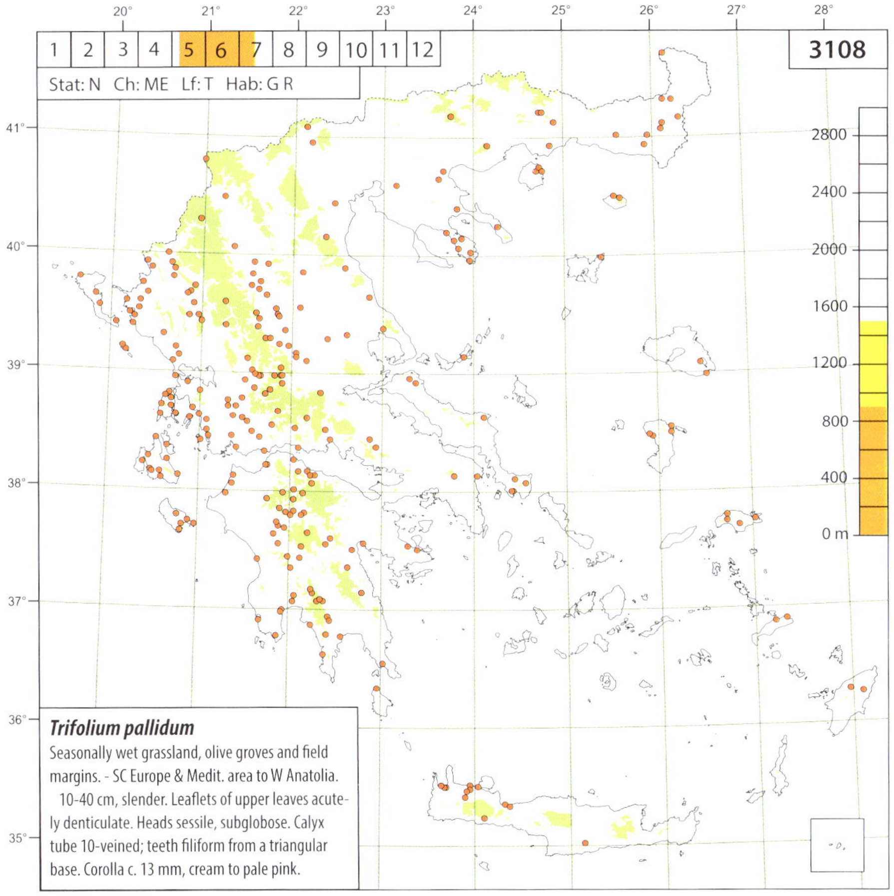

Trifolium pallidum

Seasonally wet grassland, olive groves and field margins. - SC Europe & Medit. area to W Anatolia.

10-40 cm, slender. Leaflets of upper leaves acutely denticulate. Heads sessile, subglobose. Calyx tube 10-veined; teeth filiform from a triangular base. Corolla c. 13 mm, cream to pale pink.

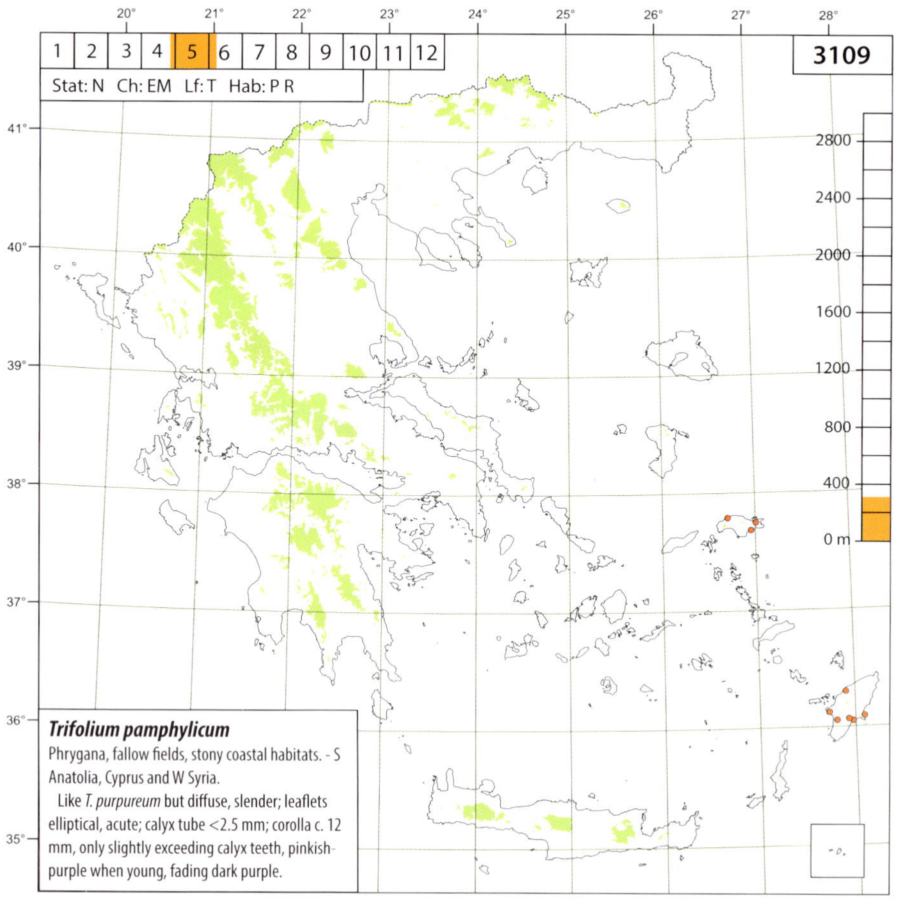

Trifolium pamphylicum

Phrygana, fallow fields, stony coastal habitats. - S Anatolia, Cyprus and W Syria.

Like *T. purpureum* but diffuse, slender; leaflets elliptical, acute; calyx tube <2.5 mm; corolla c. 12 mm, only slightly exceeding calyx teeth, pinkish-purple when young, fading dark purple.

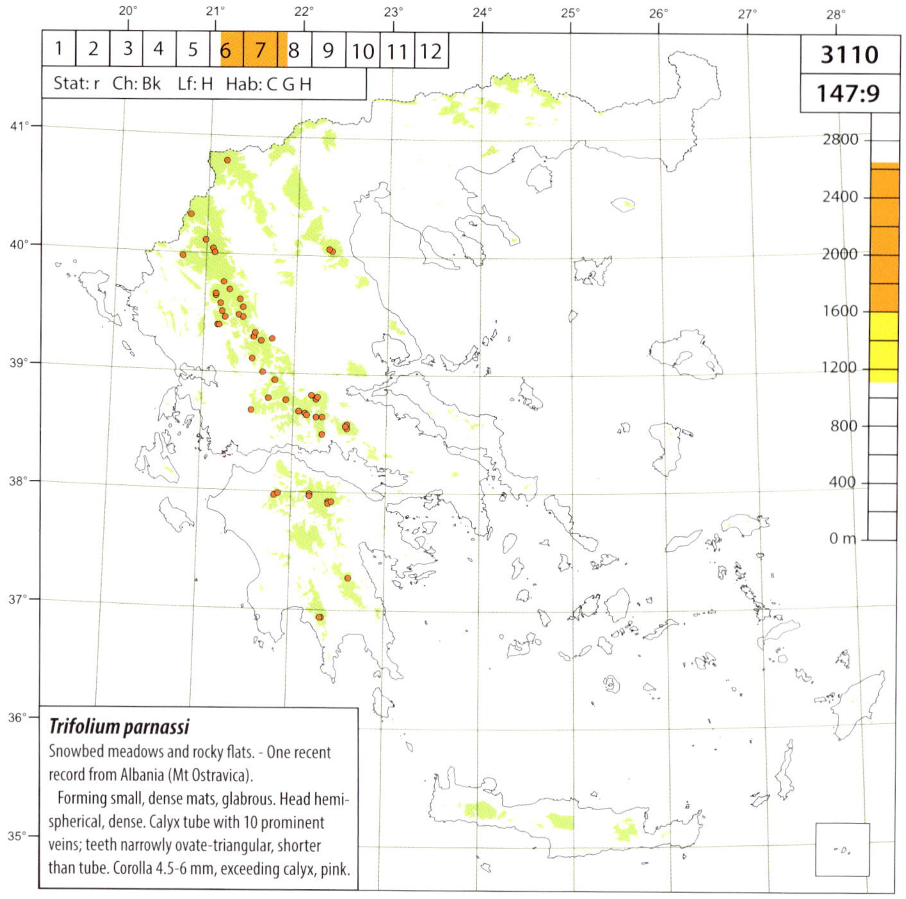

Trifolium parnassi

Snowbed meadows and rocky flats. - One recent record from Albania (Mt Ostravica).

Forming small, dense mats, glabrous. Head hemispherical, dense. Calyx tube with 10 prominent veins; teeth narrowly ovate-triangular, shorter than tube. Corolla 4.5-6 mm, exceeding calyx, pink.

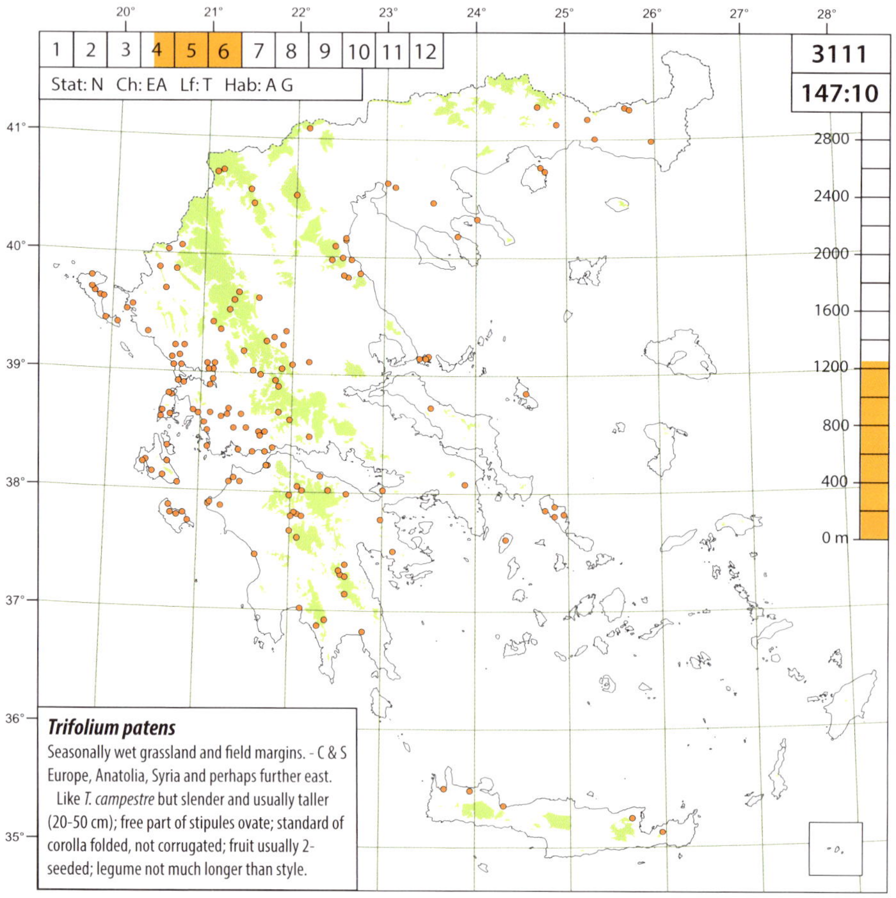

Trifolium patens

Seasonally wet grassland and field margins. - C & S Europe, Anatolia, Syria and perhaps further east.

Like *T. campestre* but slender and usually taller (20-50 cm); free part of stipules ovate; standard of corolla folded, not corrugated; fruit usually 2-seeded; legume not much longer than style.

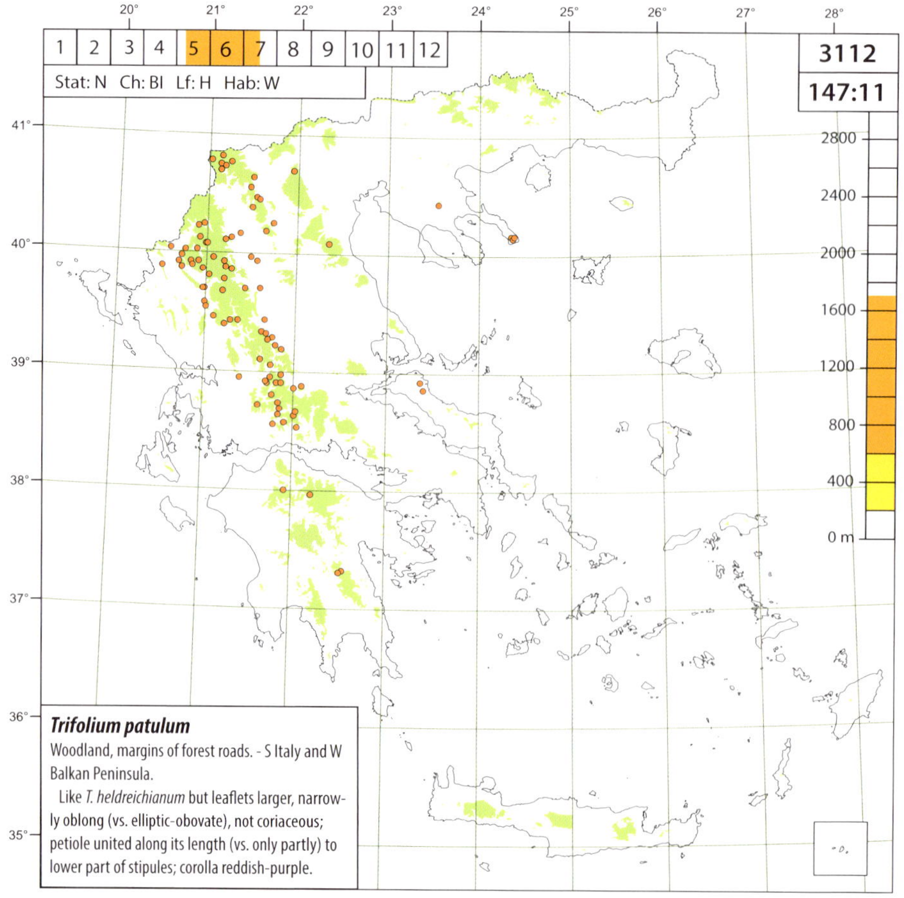

Trifolium patulum

Woodland, margins of forest roads. - S Italy and W Balkan Peninsula.

Like *T. heldreichianum* but leaflets larger, narrowly oblong (vs. elliptic-obovate), not coriaceous; petiole united along its length (vs. only partly) to lower part of stipules; corolla reddish-purple.

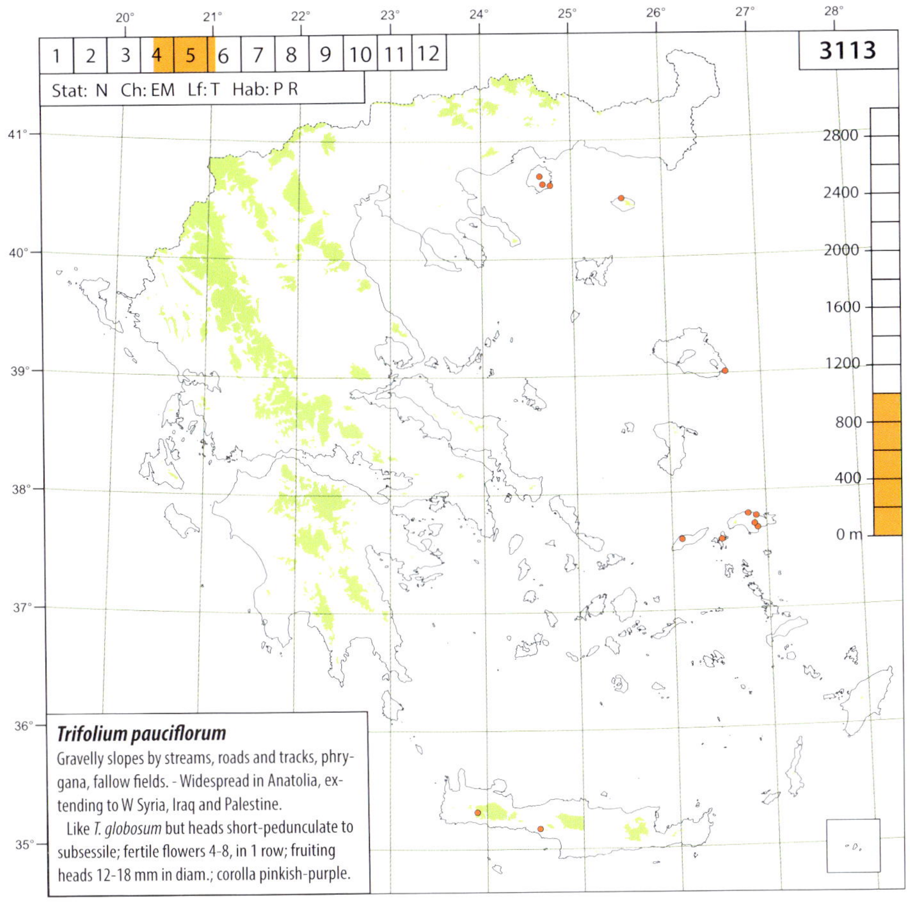

Trifolium pauciflorum

Gravelly slopes by streams, roads and tracks, phrygana, fallow fields. - Widespread in Anatolia, extending to W Syria, Iraq and Palestine.

Like *T. globosum* but heads short-pedunculate to subsessile; fertile flowers 4-8, in 1 row; fruiting heads 12-18 mm in diam.; corolla pinkish-purple.

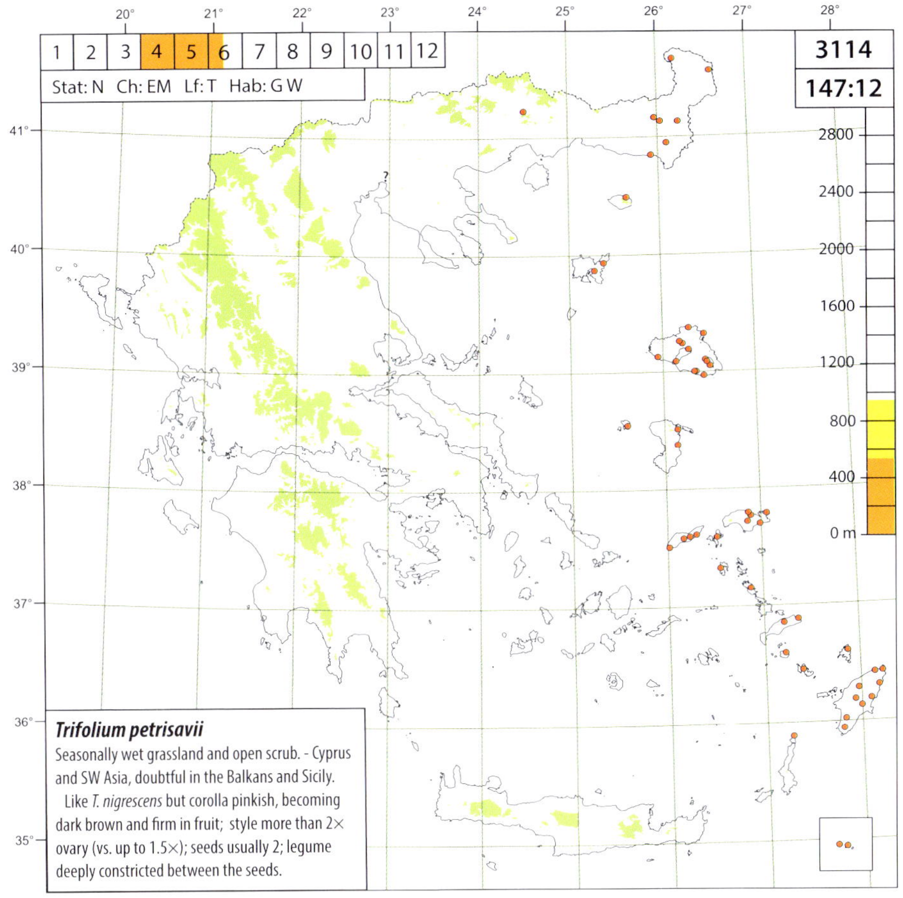

Trifolium petrisavii

Seasonally wet grassland and open scrub. - Cyprus and SW Asia, doubtful in the Balkans and Sicily.

Like *T. nigrescens* but corolla pinkish, becoming dark brown and firm in fruit; style more than 2× ovary (vs. up to 1.5×); seeds usually 2; legume deeply constricted between the seeds.

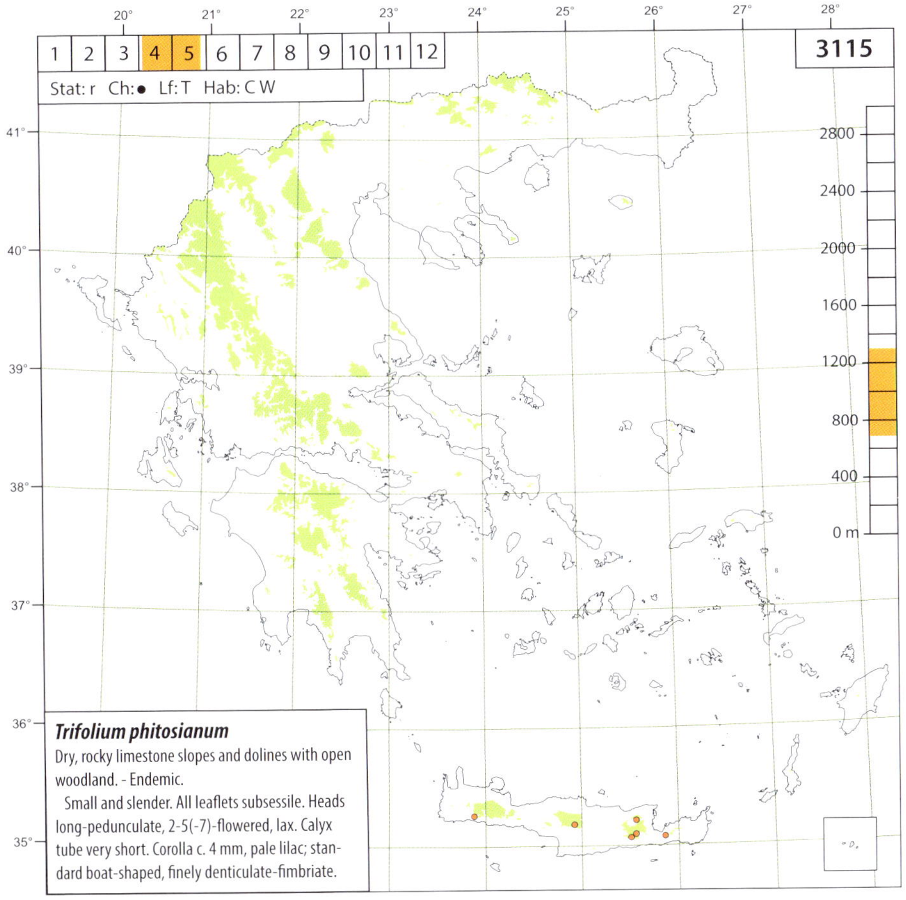

Trifolium phitosianum

Dry, rocky limestone slopes and dolines with open woodland. - Endemic.

Small and slender. All leaflets subsessile. Heads long-pedunculate, 2-5(-7)-flowered, lax. Calyx tube very short. Corolla c. 4 mm, pale lilac; standard boat-shaped, finely denticulate-fimbriate.

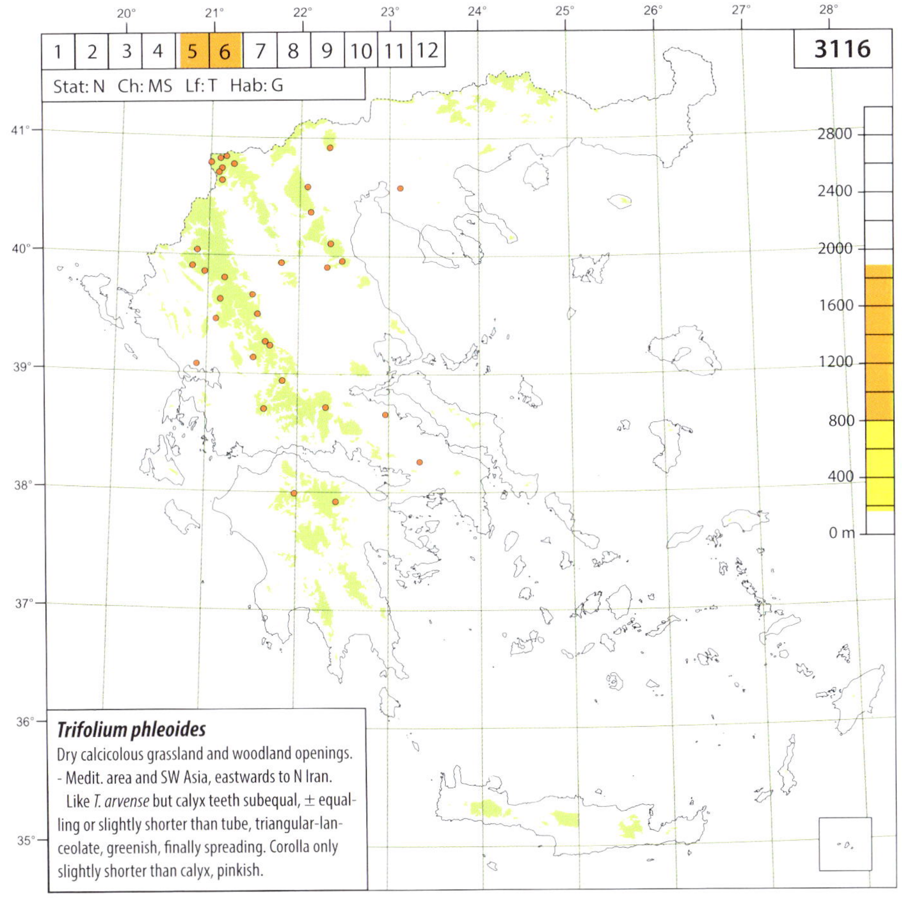

Trifolium phleoides

Dry calcicolous grassland and woodland openings. - Medit. area and SW Asia, eastwards to N Iran.

Like *T. arvense* but calyx teeth subequal, ± equalling or slightly shorter than tube, triangular-lanceolate, greenish, finally spreading. Corolla only slightly shorter than calyx, pinkish.

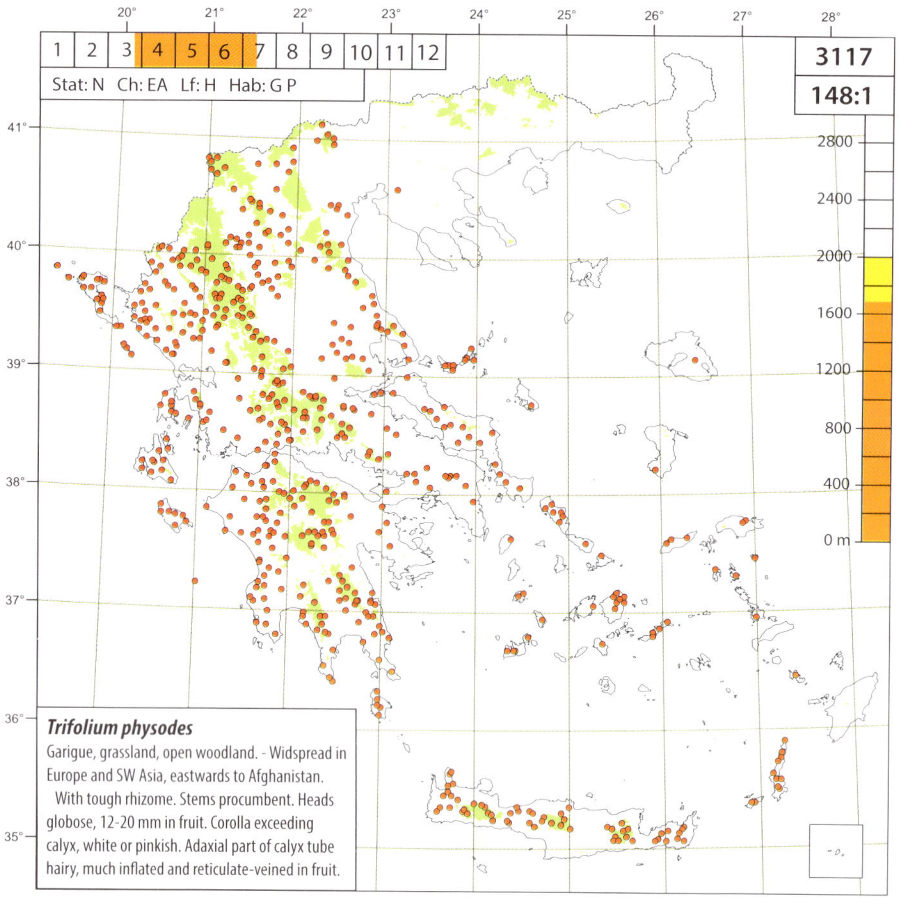

Trifolium physodes

Garigue, grassland, open woodland. - Widspread in Europe and SW Asia, eastwards to Afghanistan.

With tough rhizome. Stems procumbent. Heads globose, 12-20 mm in fruit. Corolla exceeding calyx, white or pinkish. Adaxial part of calyx tube hairy, much inflated and reticulate-veined in fruit.

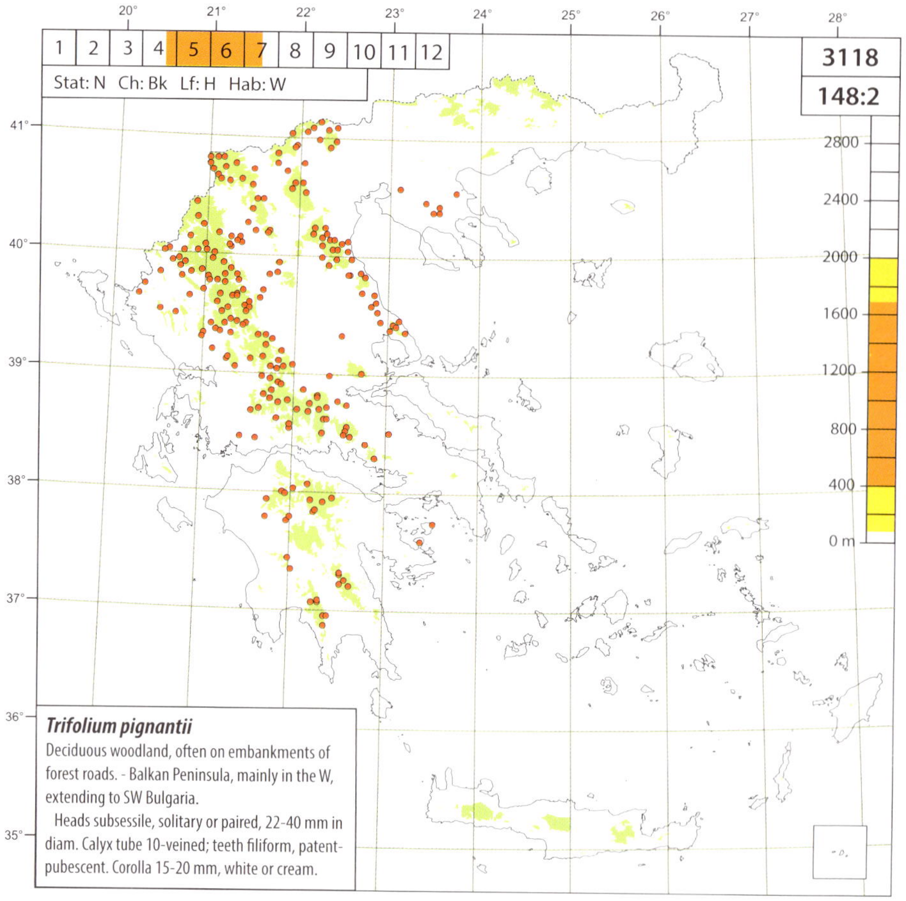

Trifolium pignantii

Deciduous woodland, often on embankments of forest roads. - Balkan Peninsula, mainly in the W, extending to SW Bulgaria.

Heads subsessile, solitary or paired, 22-40 mm in diam. Calyx tube 10-veined; teeth filiform, patent-pubescent. Corolla 15-20 mm, white or cream.

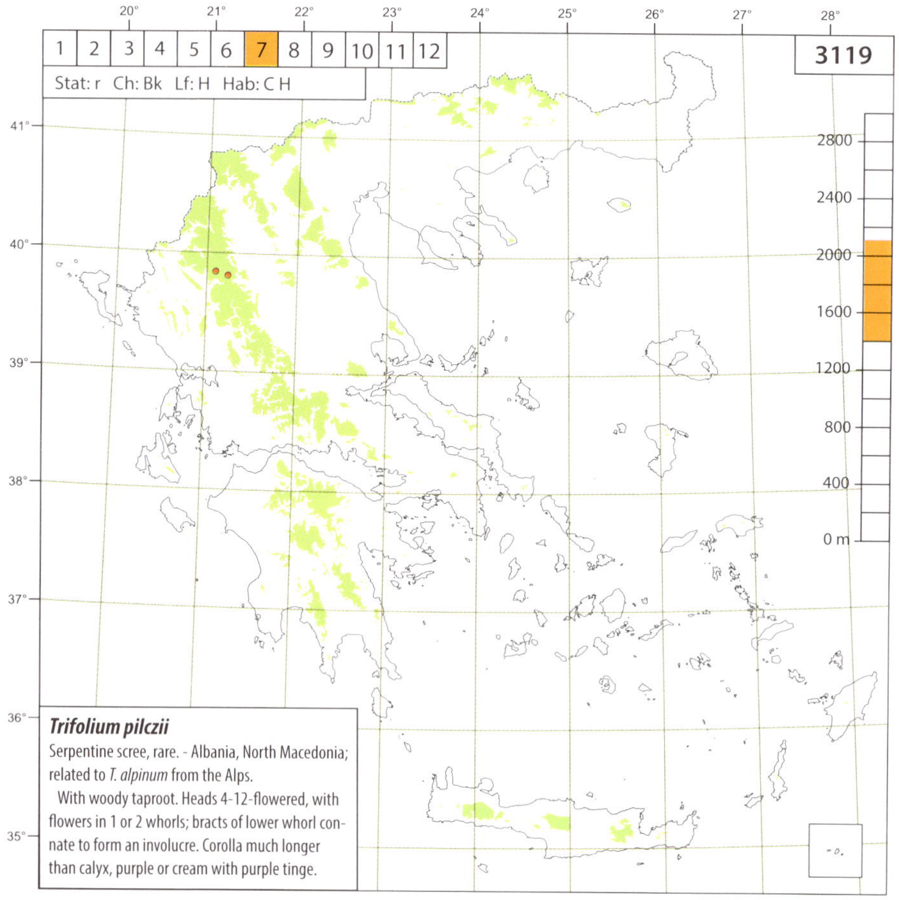

Trifolium pilczii

Serpentine scree, rare. - Albania, North Macedonia; related to *T. alpinum* from the Alps.

With woody taproot. Heads 4-12-flowered, with flowers in 1 or 2 whorls; bracts of lower whorl connate to form an involucre. Corolla much longer than calyx, purple or cream with purple tinge.

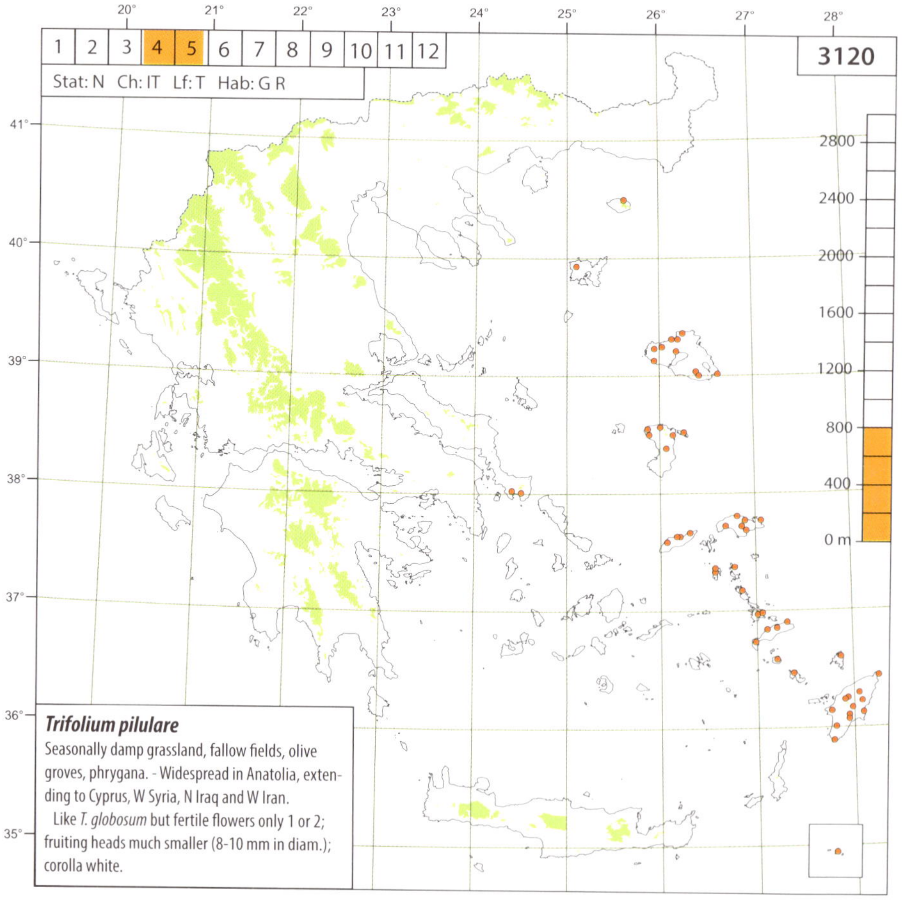

Trifolium pilulare

Seasonally damp grassland, fallow fields, olive groves, phrygana. - Widespread in Anatolia, extending to Cyprus, W Syria, N Iraq and W Iran.

Like *T. globosum* but fertile flowers only 1 or 2; fruiting heads much smaller (8-10 mm in diam.); corolla white.

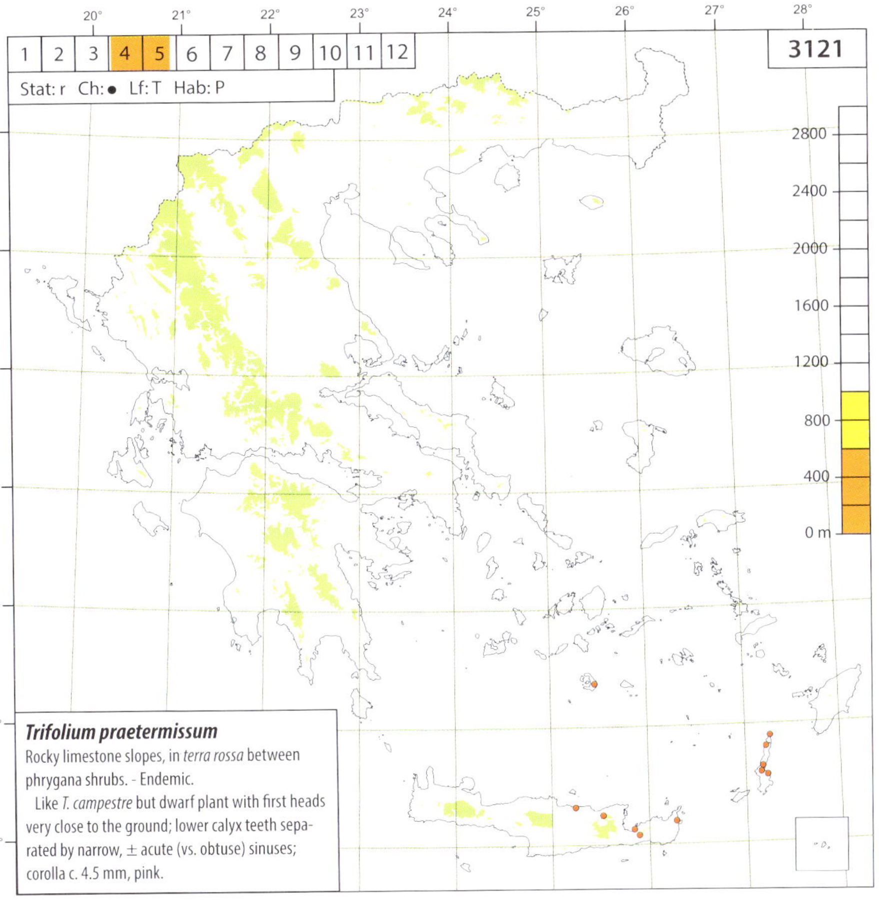

Trifolium praetermissum

Rocky limestone slopes, in *terra rossa* between phrygana shrubs. - Endemic.

Like *T. campestre* but dwarf plant with first heads very close to the ground; lower calyx teeth separated by narrow, ± acute (vs. obtuse) sinuses; corolla c. 4.5 mm, pink.

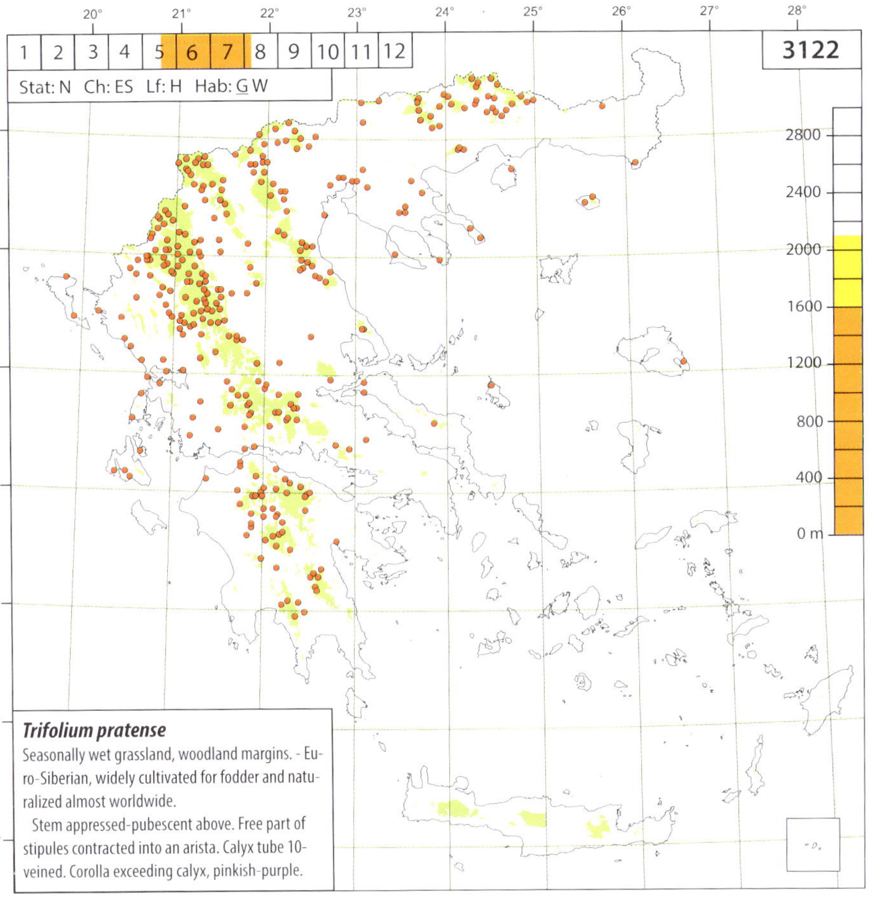

Trifolium pratense

Seasonally wet grassland, woodland margins. - Euro-Siberian, widely cultivated for fodder and naturalized almost worldwide.

Stem appressed-pubescent above. Free part of stipules contracted into an arista. Calyx tube 10-veined. Corolla exceeding calyx, pinkish-purple.

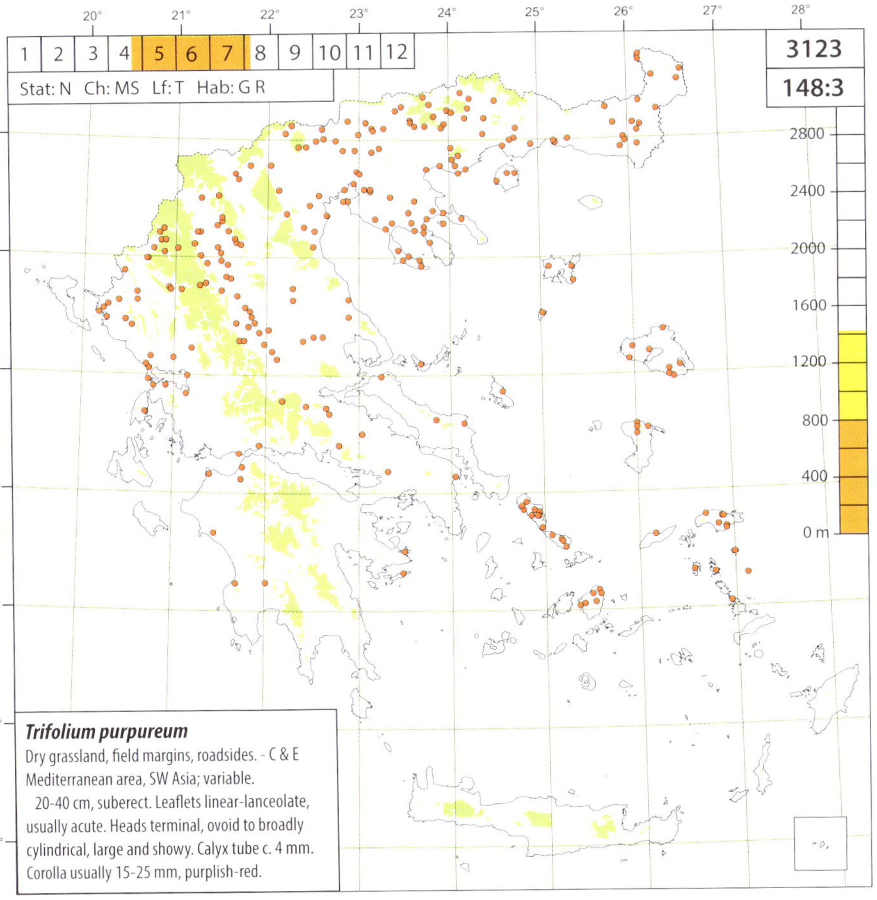

Trifolium purpureum

Dry grassland, field margins, roadsides. - C & E Mediterranean area, SW Asia; variable.

20-40 cm, suberect. Leaflets linear-lanceolate, usually acute. Heads terminal, ovoid to broadly cylindrical, large and showy. Calyx tube c. 4 mm. Corolla usually 15-25 mm, purplish-red.

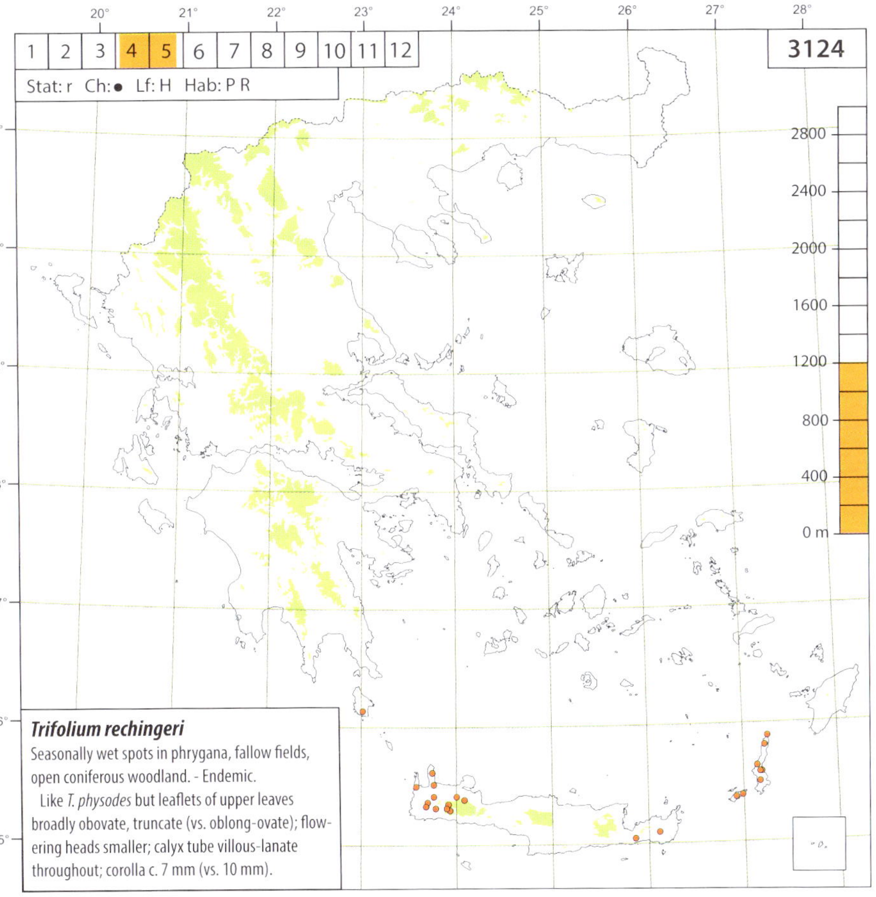

Trifolium rechingeri

Seasonally wet spots in phrygana, fallow fields, open coniferous woodland. - Endemic.

Like *T. physodes* but leaflets of upper leaves broadly obovate, truncate (vs. oblong-ovate); flowering heads smaller; calyx tube villous-lanate throughout; corolla c. 7 mm (vs. 10 mm).

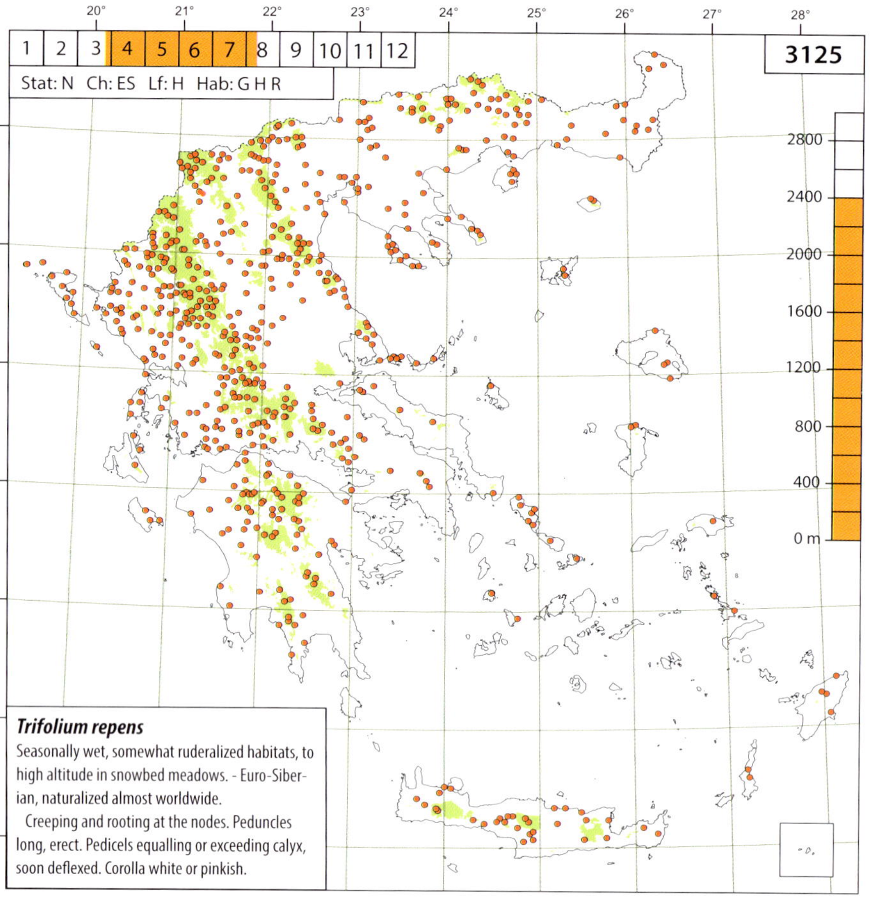

Trifolium repens

Seasonally wet, somewhat ruderalized habitats, to high altitude in snowbed meadows. - Euro-Siberian, naturalized almost worldwide.

Creeping and rooting at the nodes. Peduncles long, erect. Pedicels equalling or exceeding calyx, soon deflexed. Corolla white or pinkish.

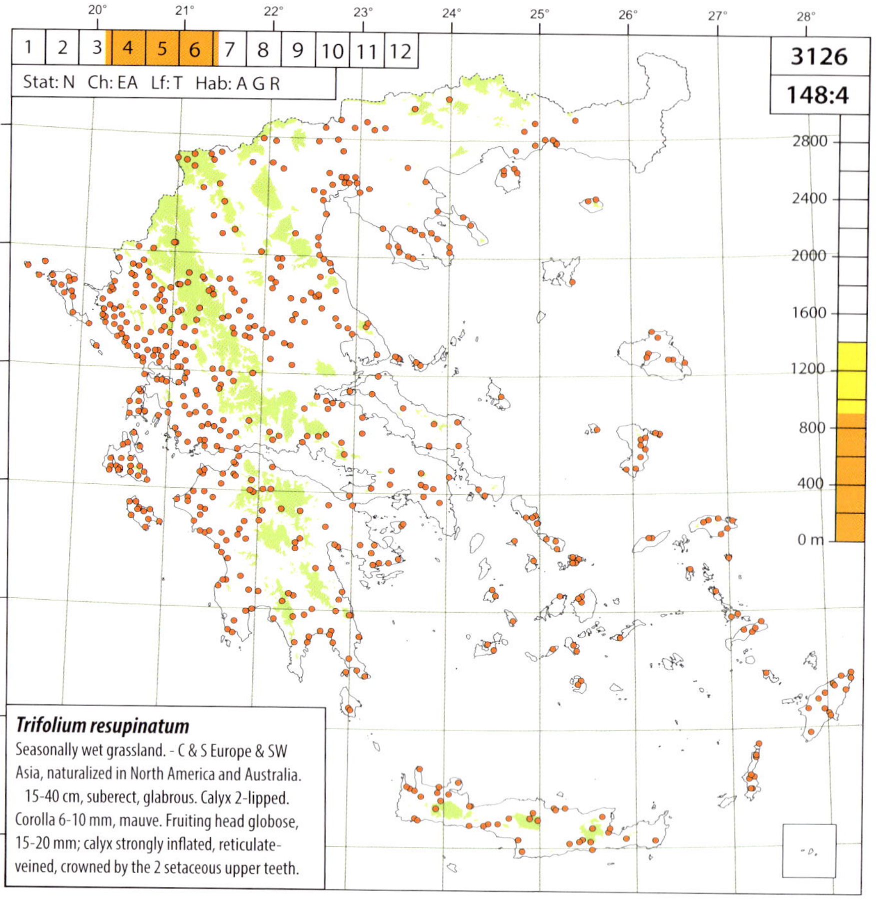

Trifolium resupinatum

Seasonally wet grassland. - C & S Europe & SW Asia, naturalized in North America and Australia.

15-40 cm, suberect, glabrous. Calyx 2-lipped. Corolla 6-10 mm, mauve. Fruiting head globose, 15-20 mm; calyx strongly inflated, reticulate-veined, crowned by the 2 setaceous upper teeth.

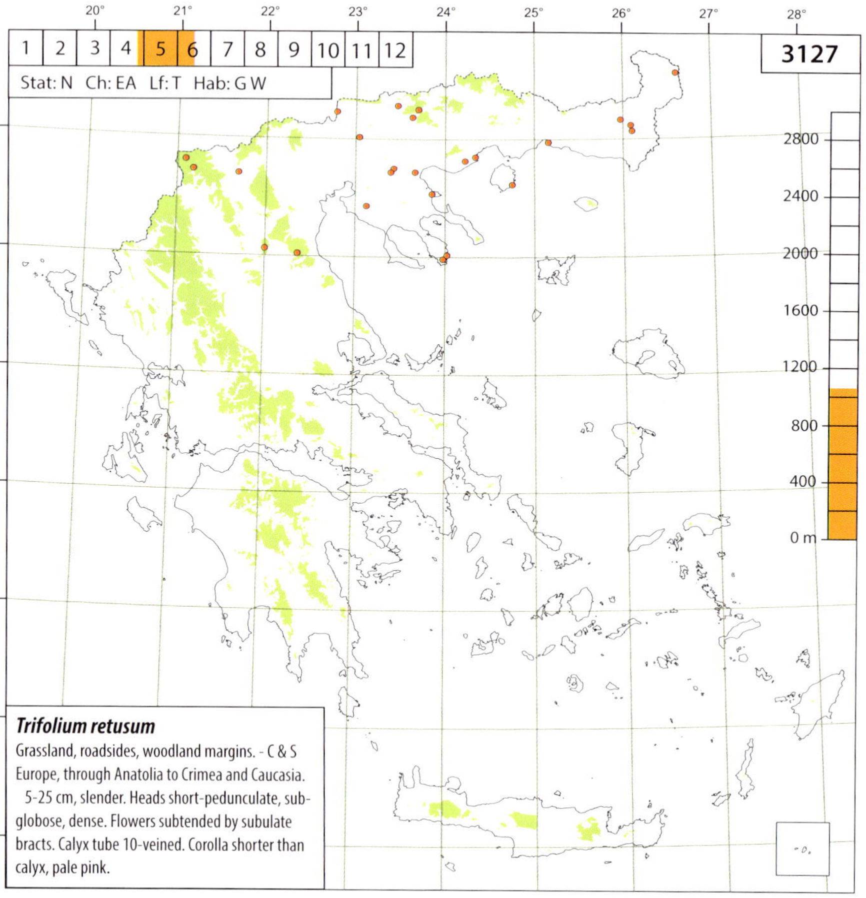

Trifolium retusum

Grassland, roadsides, woodland margins. - C & S Europe, through Anatolia to Crimea and Caucasia.

5-25 cm, slender. Heads short-pedunculate, subglobose, dense. Flowers subtended by subulate bracts. Calyx tube 10-veined. Corolla shorter than calyx, pale pink.

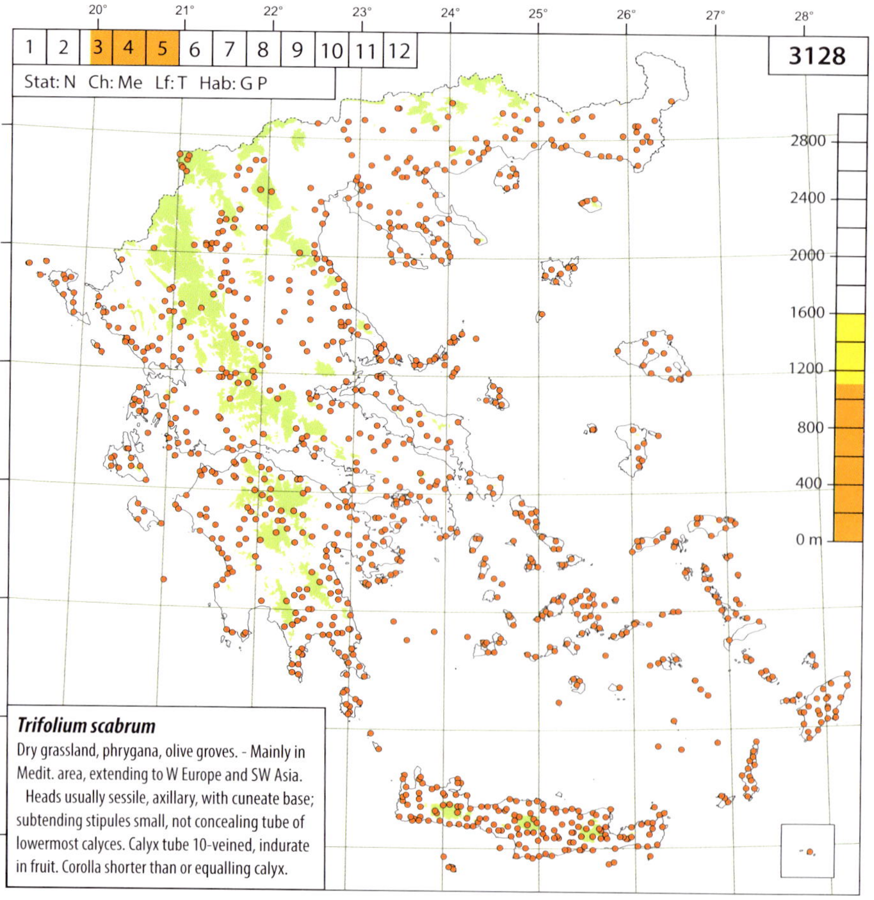

Trifolium scabrum

Dry grassland, phrygana, olive groves. - Mainly in Medit. area, extending to W Europe and SW Asia.

Heads usually sessile, axillary, with cuneate base; subtending stipules small, not concealing tube of lowermost calyces. Calyx tube 10-veined, indurate in fruit. Corolla shorter than or equalling calyx.

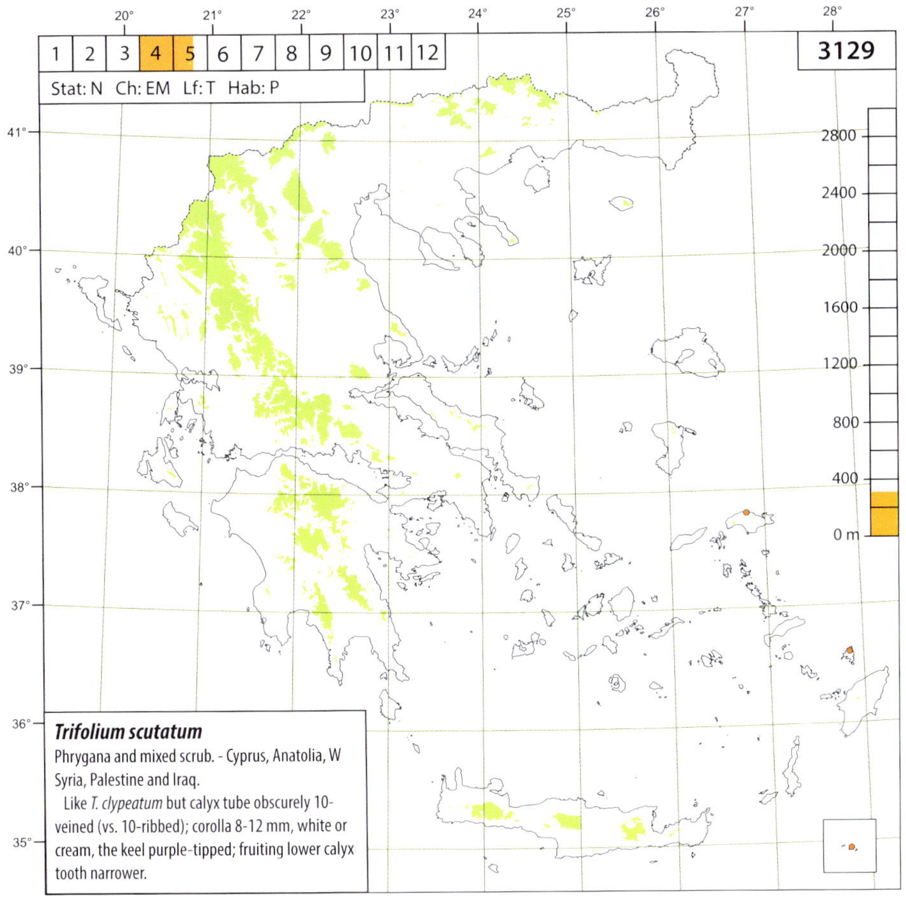

3129

Months: 1 2 3 4 5 6 7 8 9 10 11 12 (4–5 highlighted)

Stat: N Ch: EM Lf: T Hab: P

Trifolium scutatum

Phrygana and mixed scrub. - Cyprus, Anatolia, W Syria, Palestine and Iraq.

Like *T. clypeatum* but calyx tube obscurely 10-veined (vs. 10-ribbed); corolla 8-12 mm, white or cream, the keel purple-tipped; fruiting lower calyx tooth narrower.

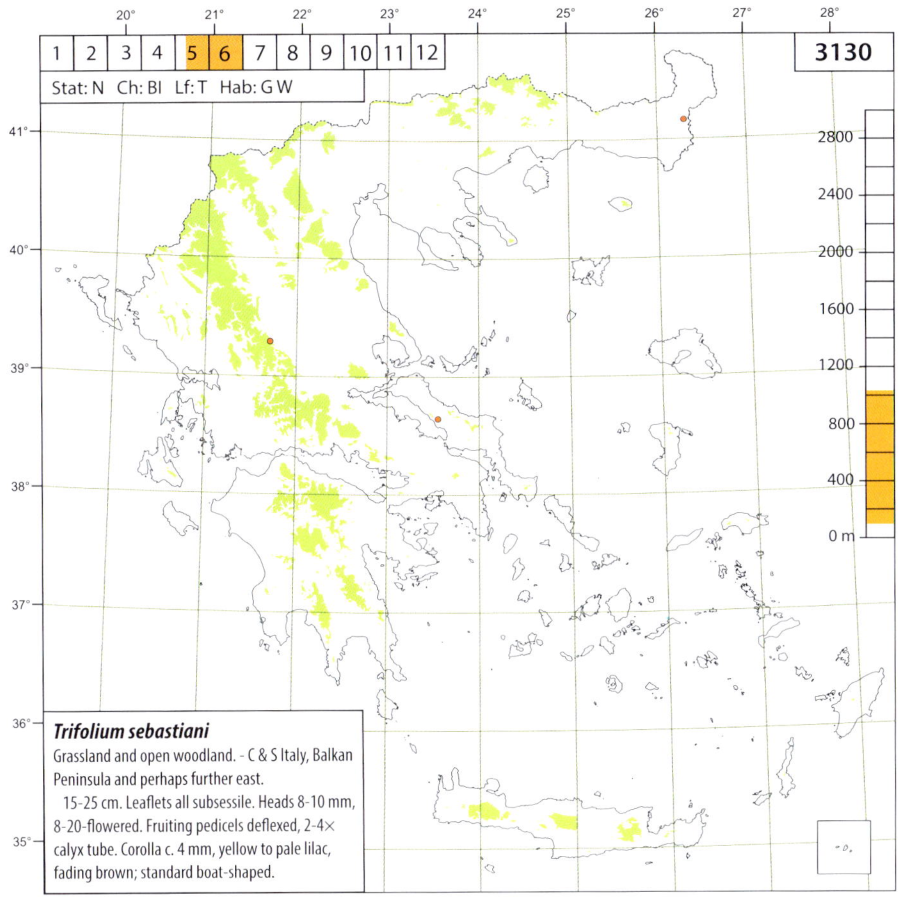

3130

Months: 1 2 3 4 5 6 7 8 9 10 11 12 (5–6 highlighted)

Stat: N Ch: Bl Lf: T Hab: G W

Trifolium sebastiani

Grassland and open woodland. - C & S Italy, Balkan Peninsula and perhaps further east.

15-25 cm. Leaflets all subsessile. Heads 8-10 mm, 8-20-flowered. Fruiting pedicels deflexed, 2-4× calyx tube. Corolla c. 4 mm, yellow to pale lilac, fading brown; standard boat-shaped.

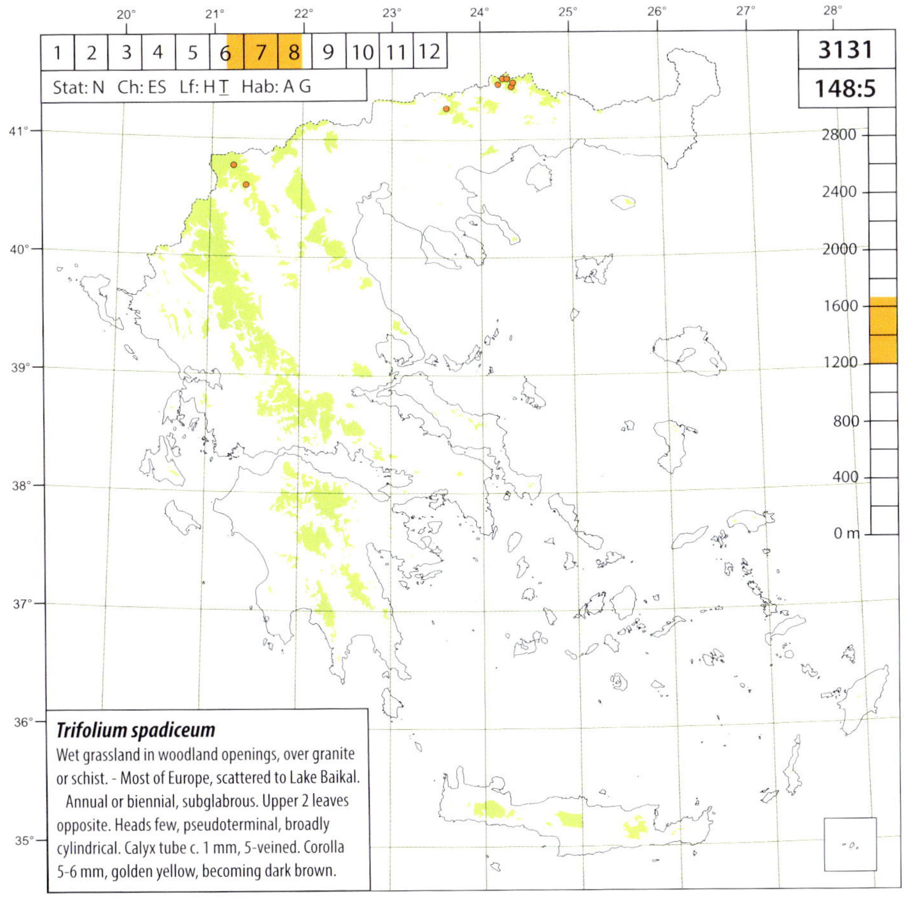

3131

148:5

Months: 1 2 3 4 5 6 7 8 9 10 11 12 (6–8 highlighted)

Stat: N Ch: ES Lf: H T Hab: A G

Trifolium spadiceum

Wet grassland in woodland openings, over granite or schist. - Most of Europe, scattered to Lake Baikal.

Annual or biennial, subglabrous. Upper 2 leaves opposite. Heads few, pseudoterminal, broadly cylindrical. Calyx tube c. 1 mm, 5-veined. Corolla 5-6 mm, golden yellow, becoming dark brown.

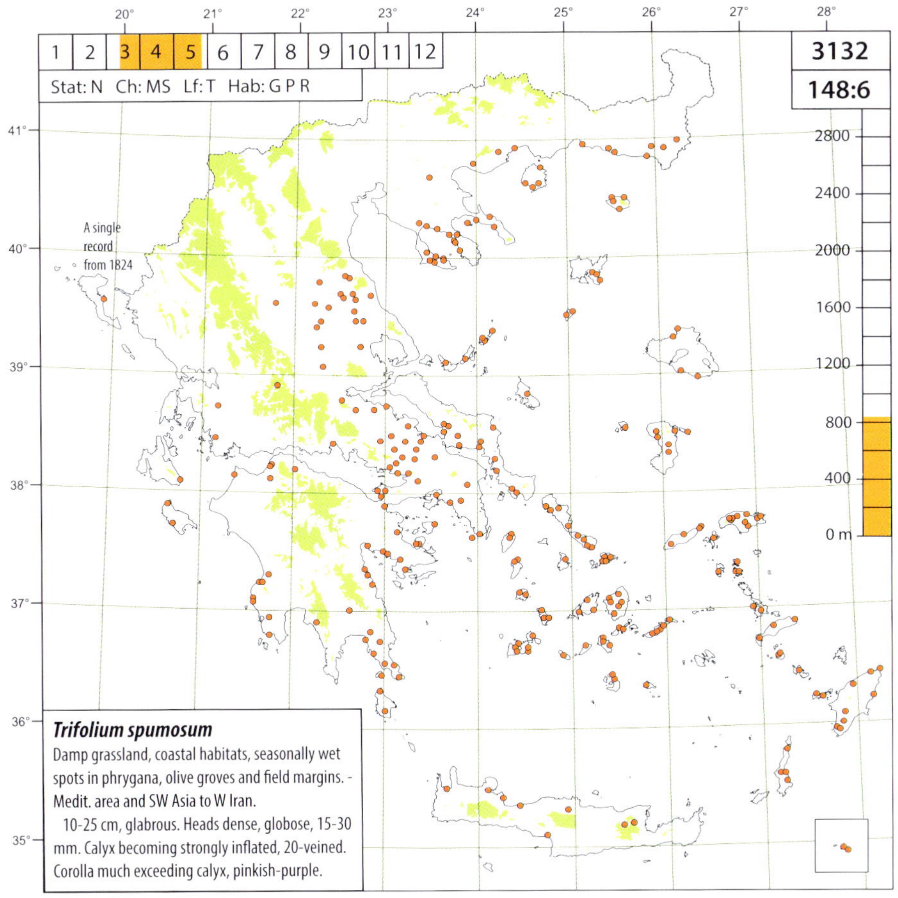

3132

148:6

Months: 1 2 3 4 5 6 7 8 9 10 11 12 (3–5 highlighted)

Stat: N Ch: MS Lf: T Hab: G P R

Trifolium spumosum

Damp grassland, coastal habitats, seasonally wet spots in phrygana, olive groves and field margins. - Medit. area and SW Asia to W Iran.

10-25 cm, glabrous. Heads dense, globose, 15-30 mm. Calyx becoming strongly inflated, 20-veined. Corolla much exceeding calyx, pinkish-purple.

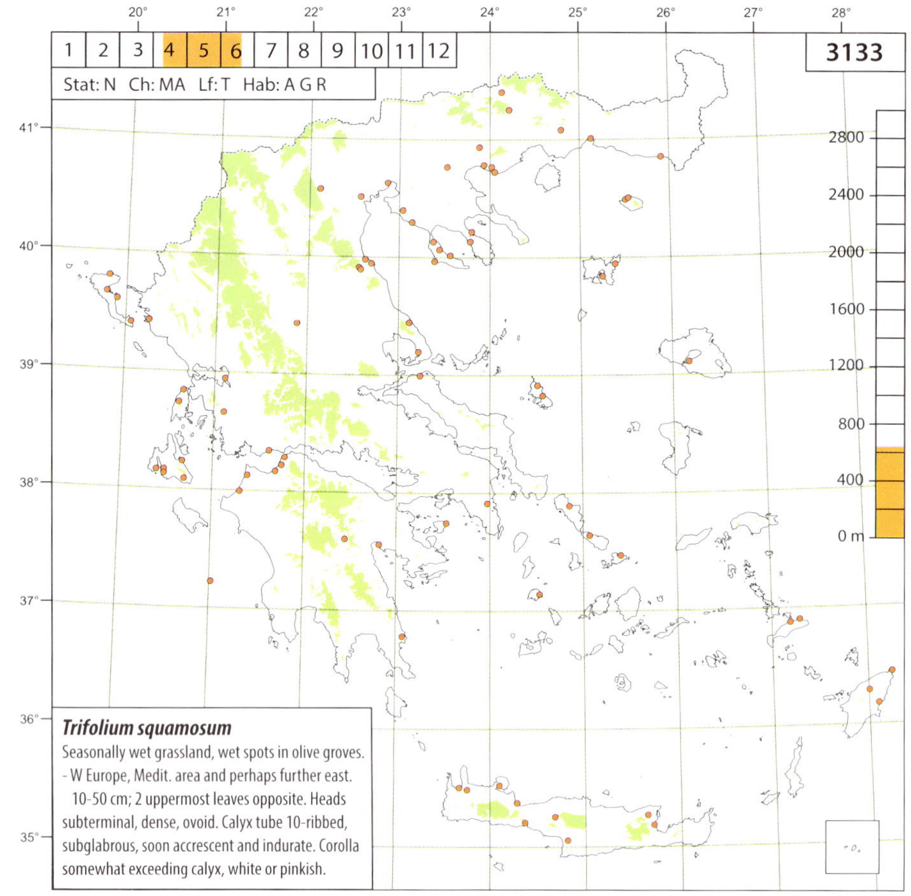

Trifolium squamosum

Seasonally wet grassland, wet spots in olive groves. - W Europe, Medit. area and perhaps further east.

10-50 cm; 2 uppermost leaves opposite. Heads subterminal, dense, ovoid. Calyx tube 10-ribbed, subglabrous, soon accrescent and indurate. Corolla somewhat exceeding calyx, white or pinkish.

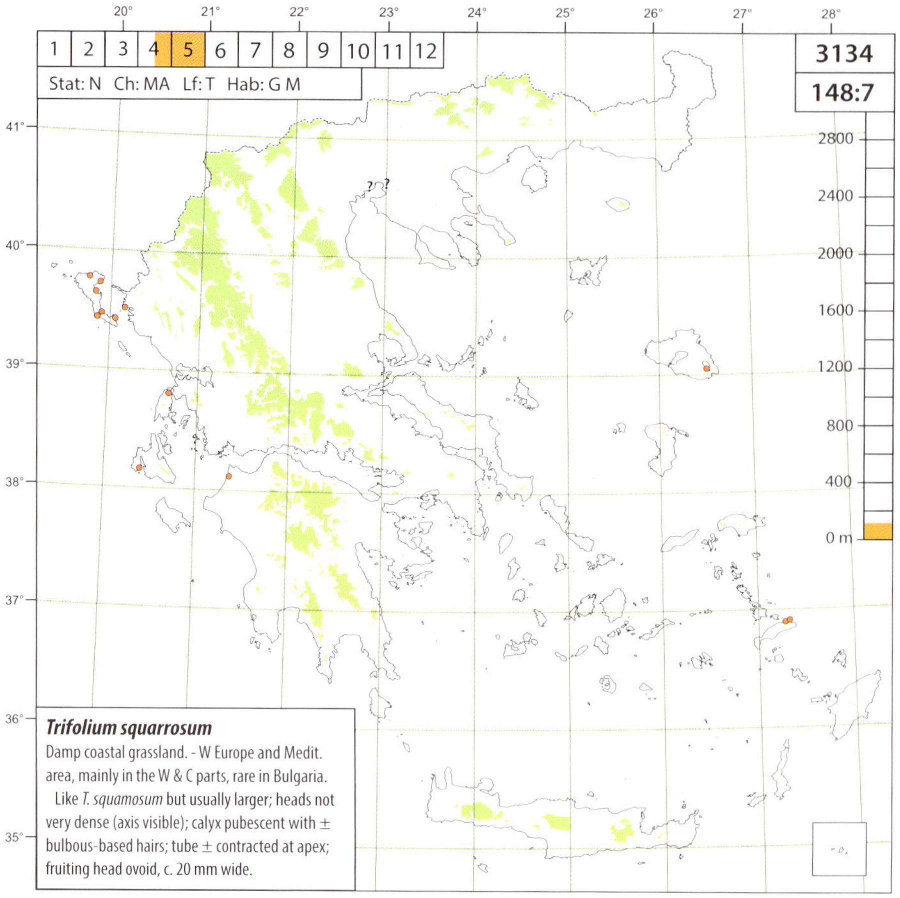

Trifolium squarrosum

Damp coastal grassland. - W Europe and Medit. area, mainly in the W & C parts, rare in Bulgaria.

Like *T. squamosum* but usually larger; heads not very dense (axis visible); calyx pubescent with ± bulbous-based hairs; tube ± contracted at apex; fruiting head ovoid, c. 20 mm wide.

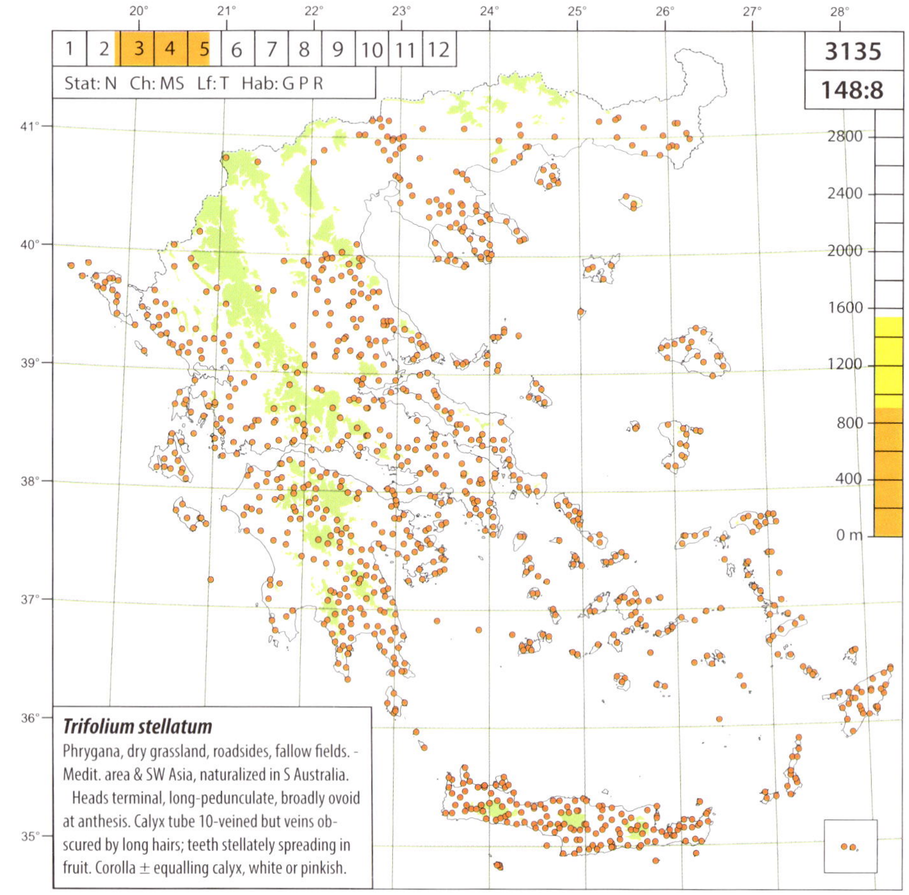

Trifolium stellatum

Phrygana, dry grassland, roadsides, fallow fields. - Medit. area & SW Asia, naturalized in S Australia.

Heads terminal, long-pedunculate, broadly ovoid at anthesis. Calyx tube 10-veined but veins obscured by long hairs; teeth stellately spreading in fruit. Corolla ± equalling calyx, white or pinkish.

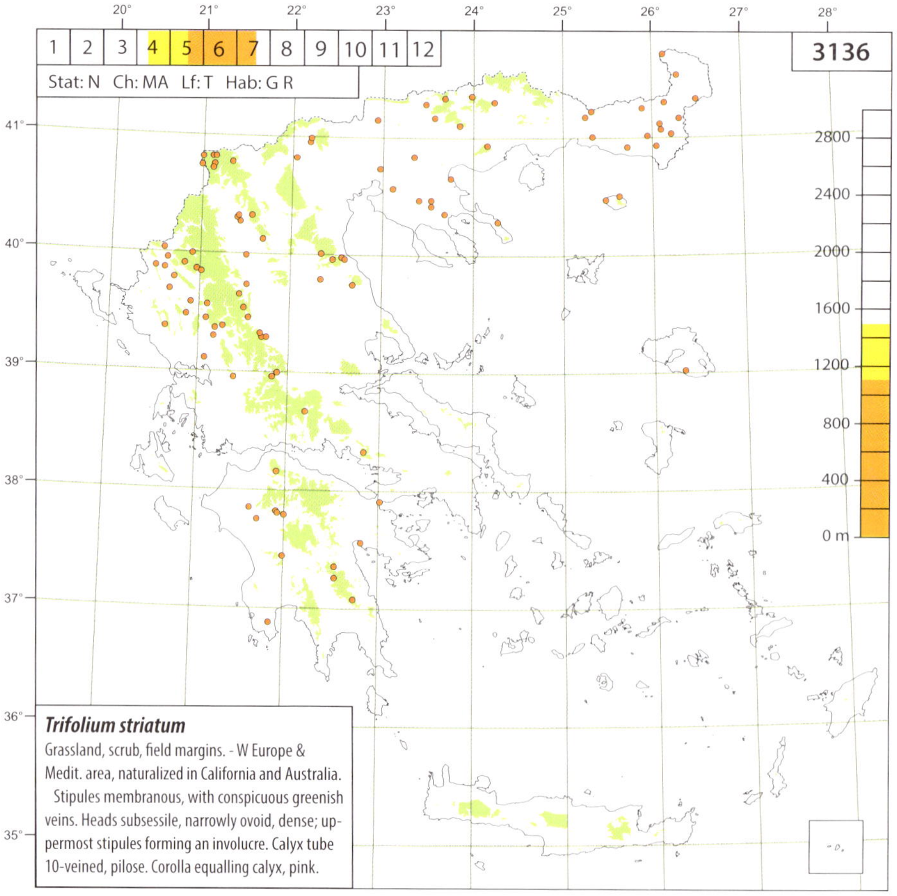

Trifolium striatum

Grassland, scrub, field margins. - W Europe & Medit. area, naturalized in California and Australia.

Stipules membranous, with conspicuous greenish veins. Heads subsessile, narrowly ovoid, dense; uppermost stipules forming an involucre. Calyx tube 10-veined, pilose. Corolla equalling calyx, pink.

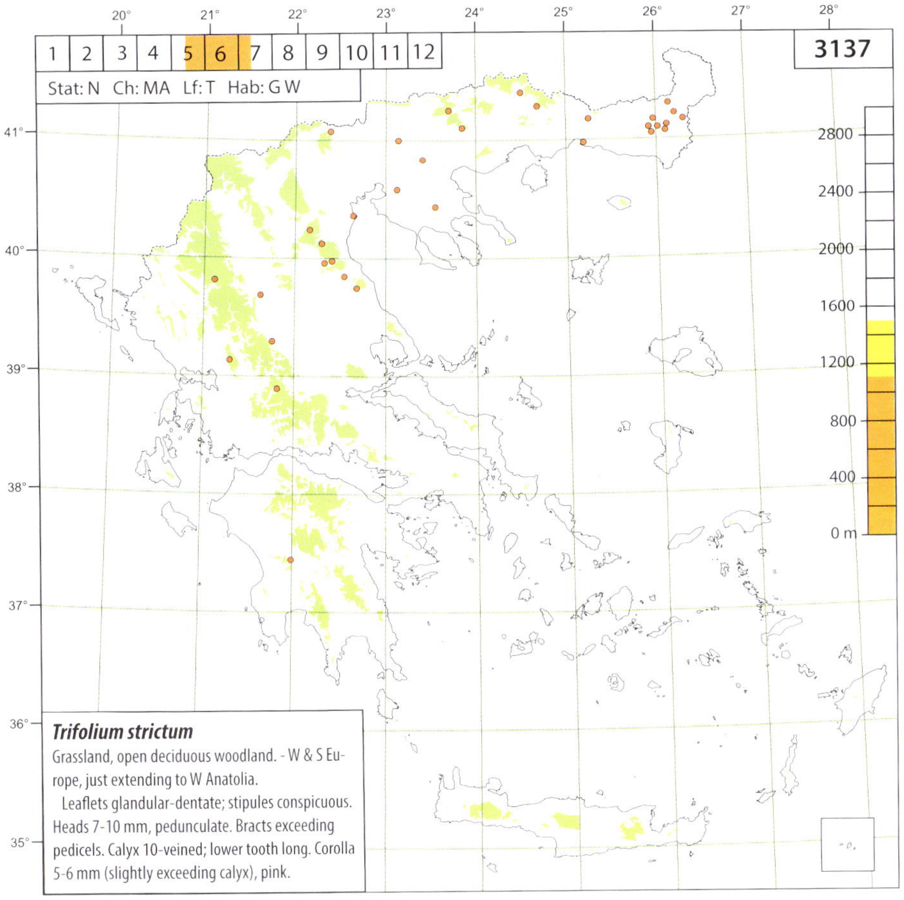

Trifolium strictum

Grassland, open deciduous woodland. - W & S Europe, just extending to W Anatolia.

Leaflets glandular-dentate; stipules conspicuous. Heads 7-10 mm, pedunculate. Bracts exceeding pedicels. Calyx 10-veined; lower tooth long. Corolla 5-6 mm (slightly exceeding calyx), pink.

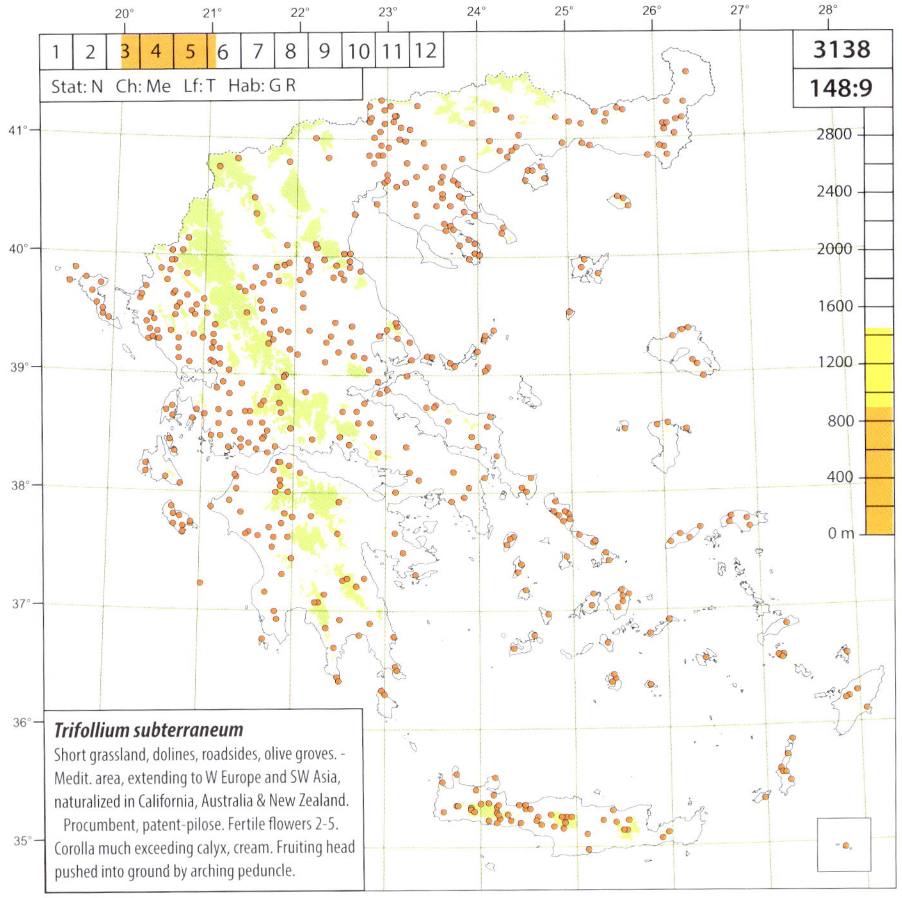

Trifollium subterraneum

Short grassland, dolines, roadsides, olive groves. - Medit. area, extending to W Europe and SW Asia, naturalized in California, Australia & New Zealand.

Procumbent, patent-pilose. Fertile flowers 2-5. Corolla much exceeding calyx, cream. Fruiting head pushed into ground by arching peduncle.

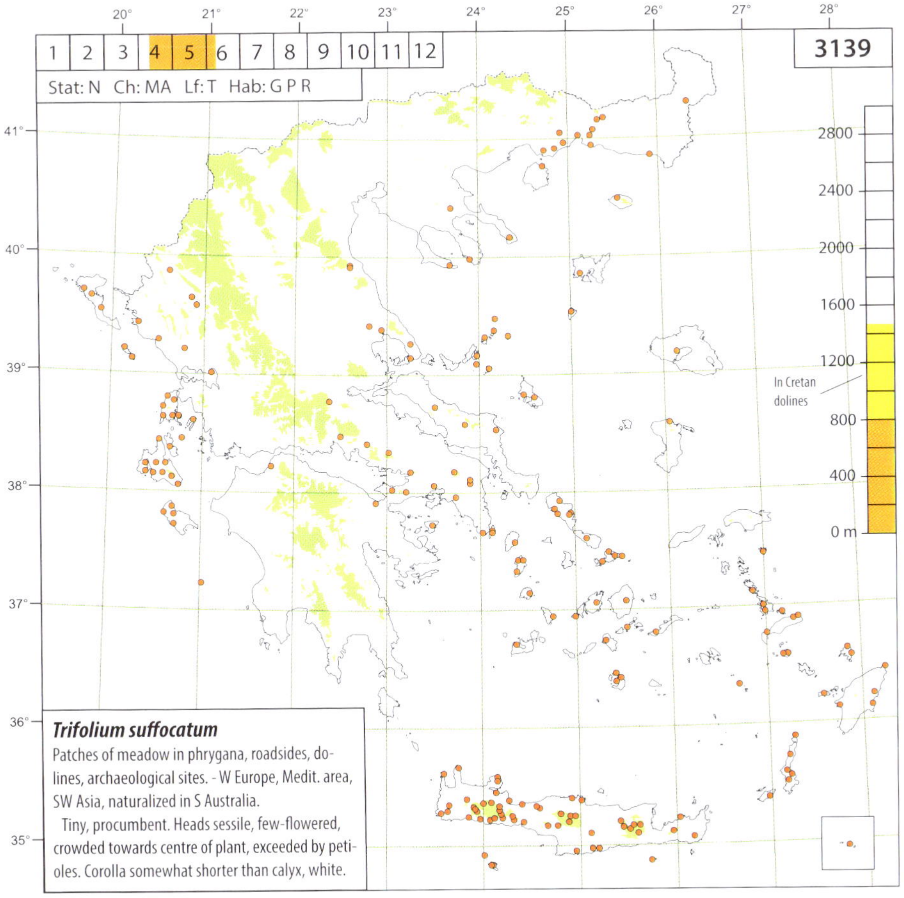

Trifolium suffocatum

Patches of meadow in phrygana, roadsides, dolines, archaeological sites. - W Europe, Medit. area, SW Asia, naturalized in S Australia.

Tiny, procumbent. Heads sessile, few-flowered, crowded towards centre of plant, exceeded by petioles. Corolla somewhat shorter than calyx, white.

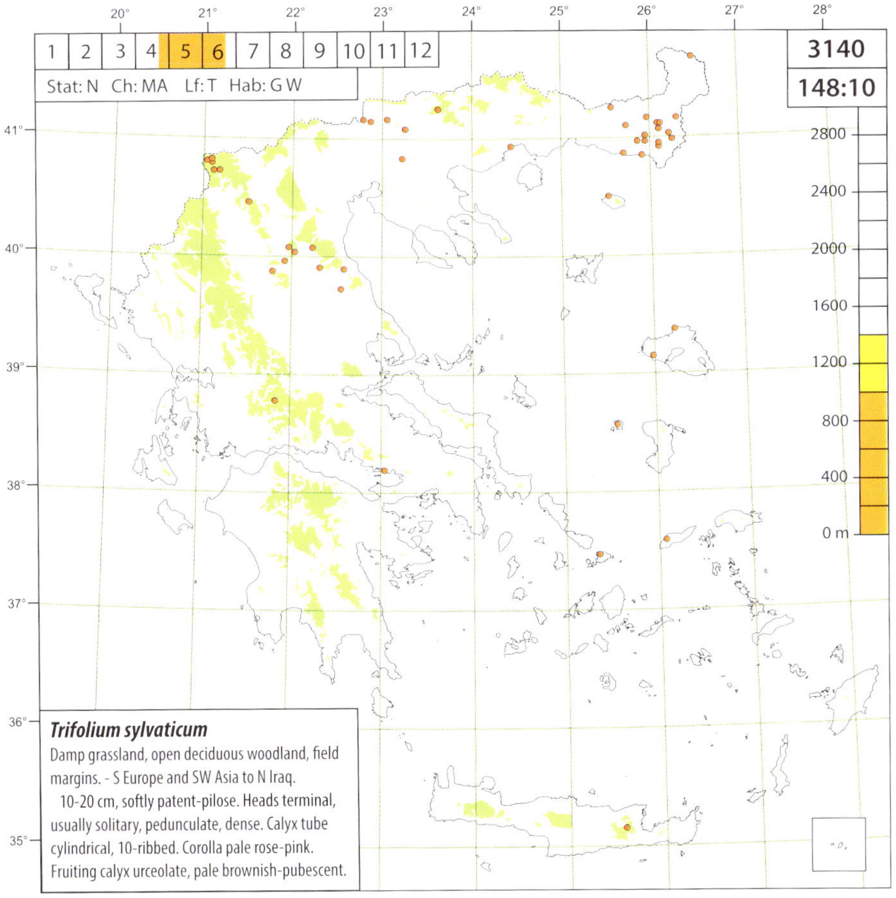

Trifolium sylvaticum

Damp grassland, open deciduous woodland, field margins. - S Europe and SW Asia to N Iraq.

10-20 cm, softly patent-pilose. Heads terminal, usually solitary, pedunculate, dense. Calyx tube cylindrical, 10-ribbed. Corolla pale rose-pink. Fruiting calyx urceolate, pale brownish-pubescent.

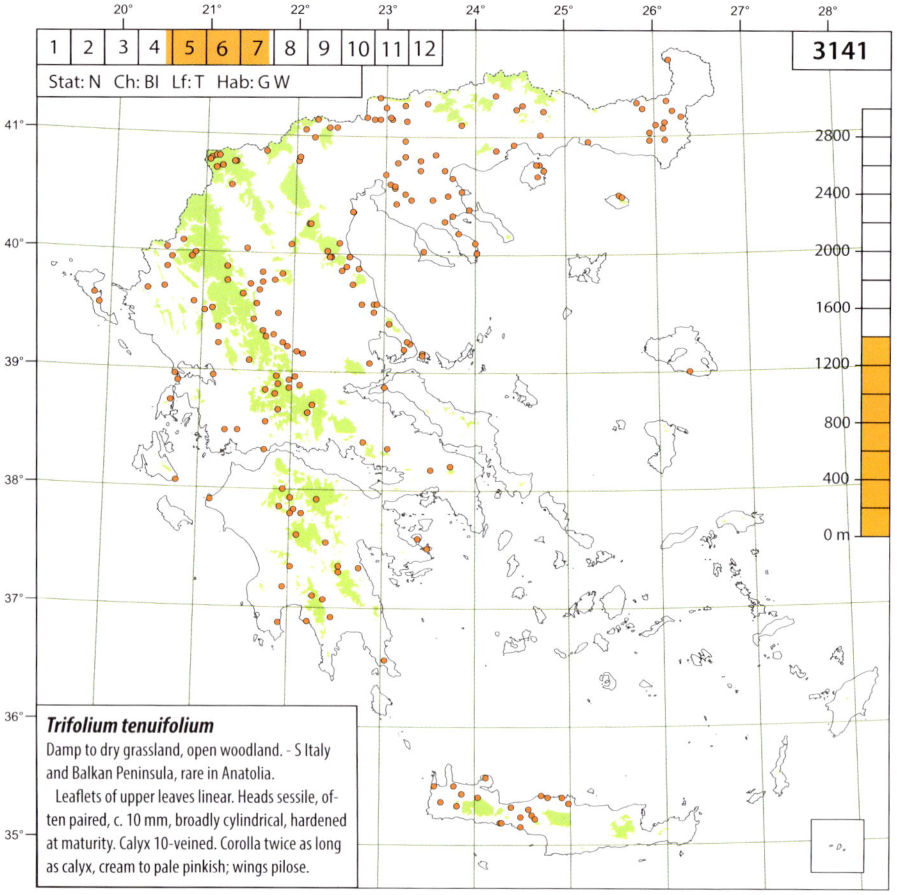

Trifolium tenuifolium

Damp to dry grassland, open woodland. - S Italy and Balkan Peninsula, rare in Anatolia.

Leaflets of upper leaves linear. Heads sessile, often paired, c. 10 mm, broadly cylindrical, hardened at maturity. Calyx 10-veined. Corolla twice as long as calyx, cream to pale pinkish; wings pilose.

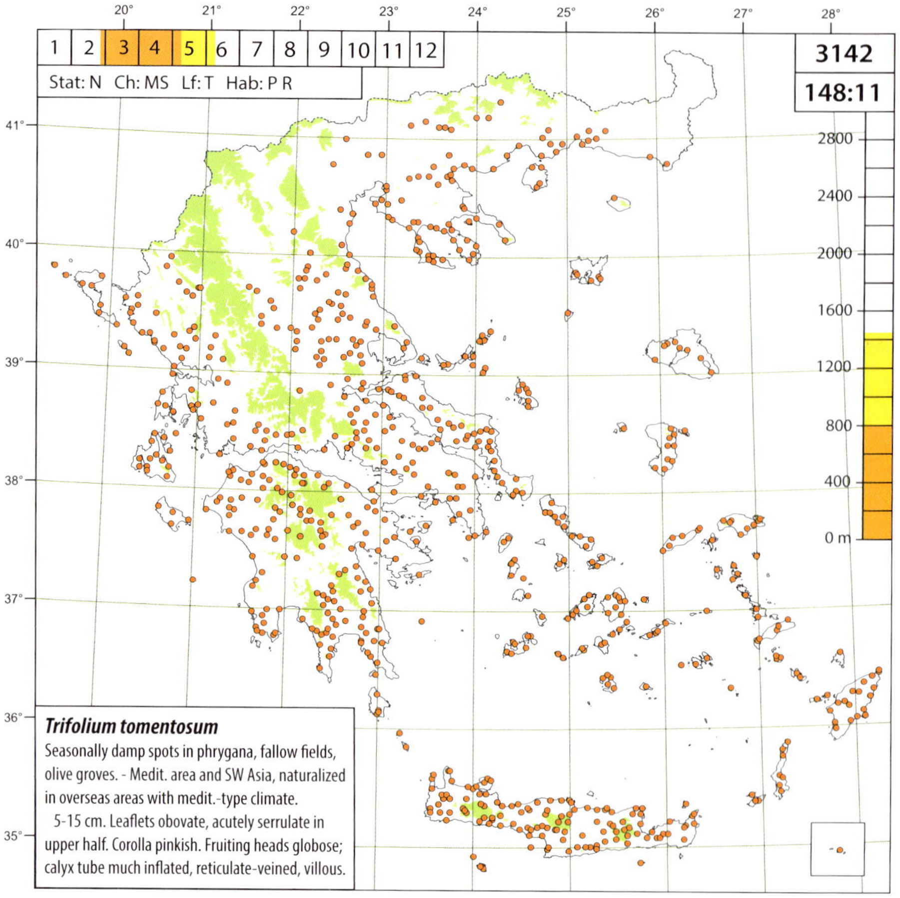

Trifolium tomentosum

Seasonally damp spots in phrygana, fallow fields, olive groves. - Medit. area and SW Asia, naturalized in overseas areas with medit.-type climate.

5-15 cm. Leaflets obovate, acutely serrulate in upper half. Corolla pinkish. Fruiting heads globose; calyx tube much inflated, reticulate-veined, villous.

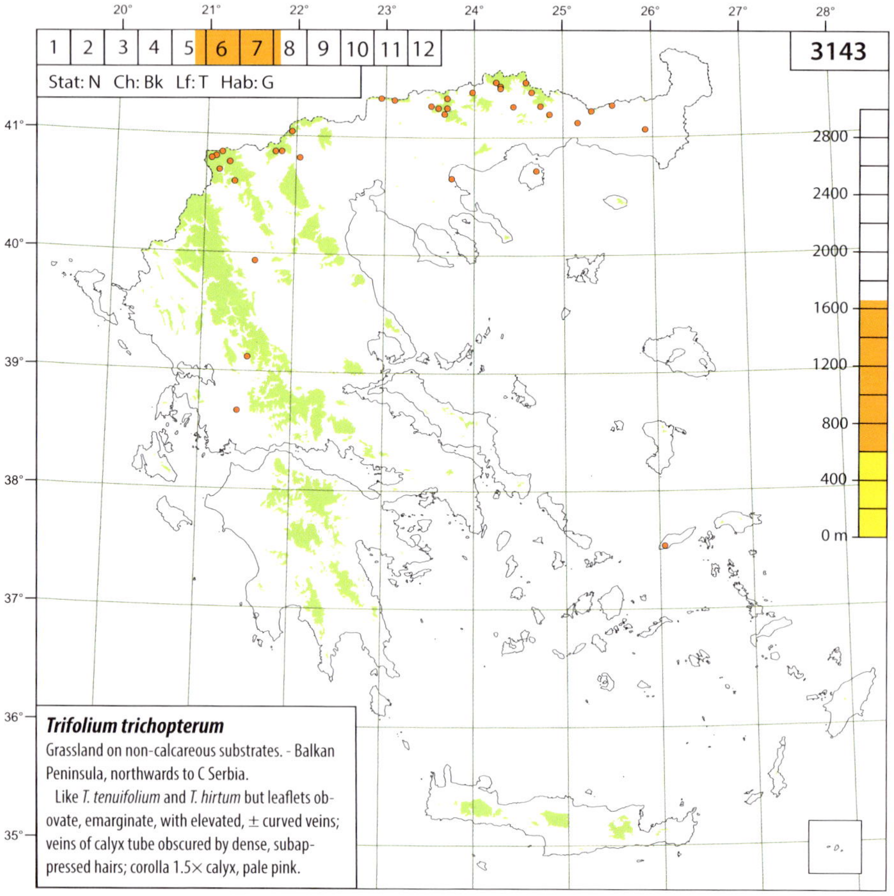

Trifolium trichopterum

Grassland on non-calcareous substrates. - Balkan Peninsula, northwards to C Serbia.

Like *T. tenuifolium* and *T. hirtum* but leaflets obovate, emarginate, with elevated, ± curved veins; veins of calyx tube obscured by dense, subappressed hairs; corolla 1.5× calyx, pale pink.

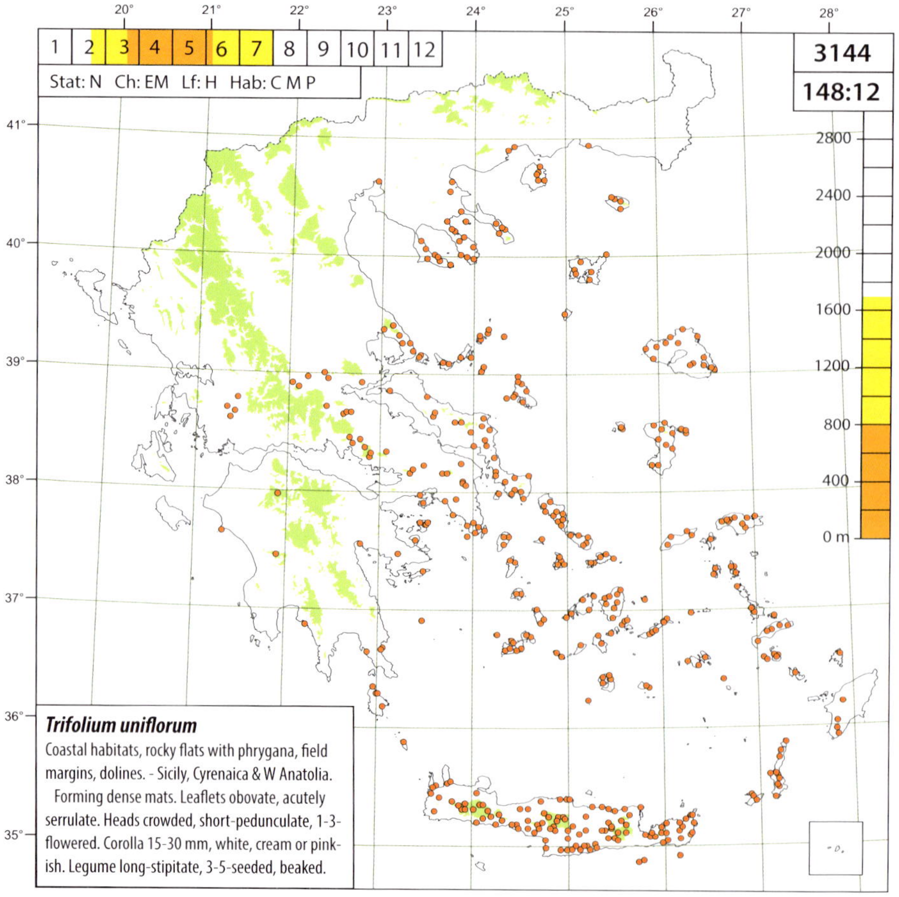

Trifolium uniflorum

Coastal habitats, rocky flats with phrygana, field margins, dolines. - Sicily, Cyrenaica & W Anatolia.

Forming dense mats. Leaflets obovate, acutely serrulate. Heads crowded, short-pedunculate, 1-3-flowered. Corolla 15-30 mm, white, cream or pinkish. Legume long-stipitate, 3-5-seeded, beaked.

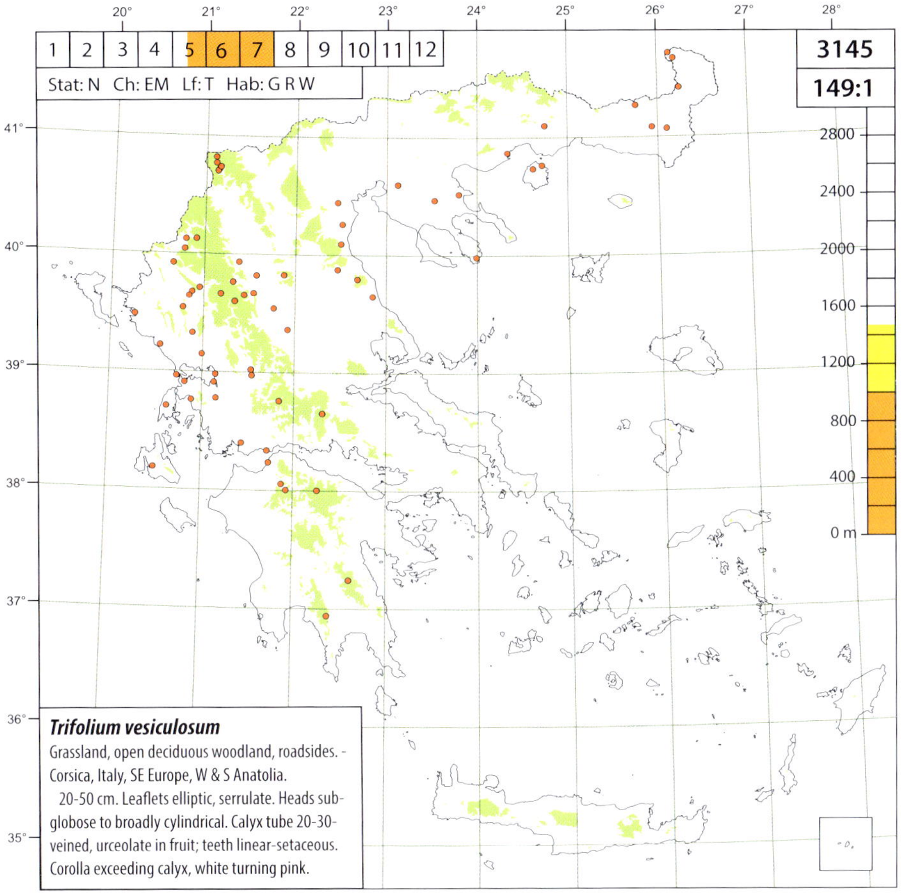

Trifolium vesiculosum

Grassland, open deciduous woodland, roadsides. - Corsica, Italy, SE Europe, W & S Anatolia.

20-50 cm. Leaflets elliptic, serrulate. Heads subglobose to broadly cylindrical. Calyx tube 20-30-veined, urceolate in fruit; teeth linear-setaceous. Corolla exceeding calyx, white turning pink.

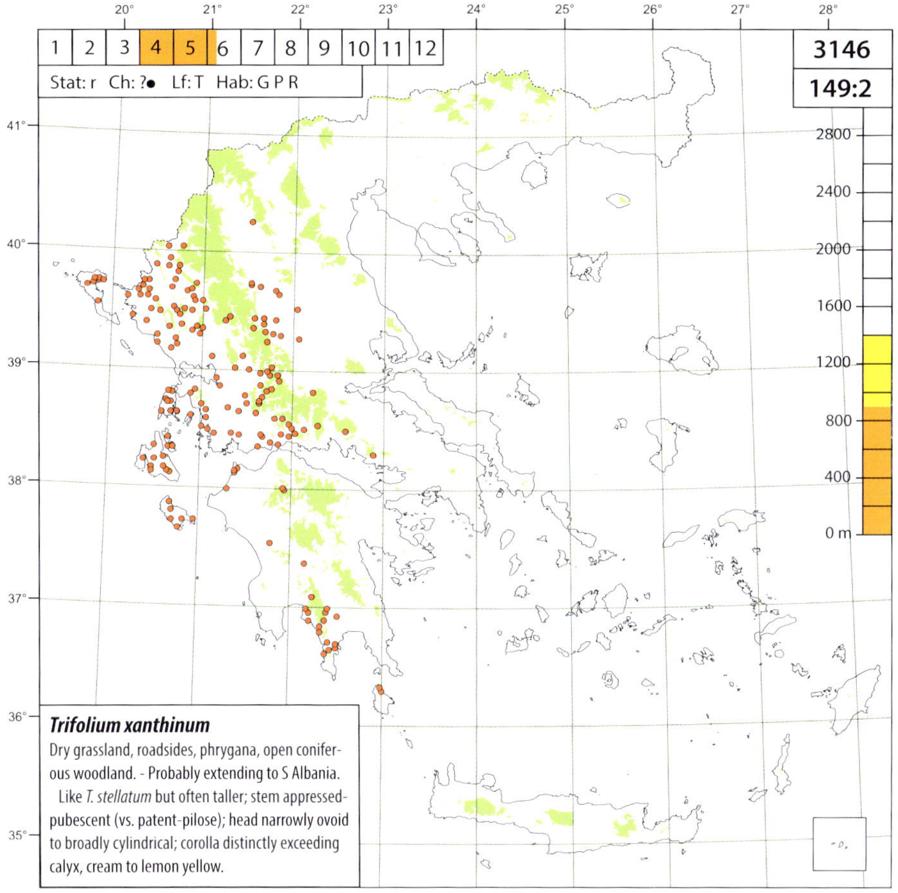

Trifolium xanthinum

Dry grassland, roadsides, phrygana, open coniferous woodland. - Probably extending to S Albania.

Like *T. stellatum* but often taller; stem appressed-pubescent (vs. patent-pilose); head narrowly ovoid to broadly cylindrical; corolla distinctly exceeding calyx, cream to lemon yellow.

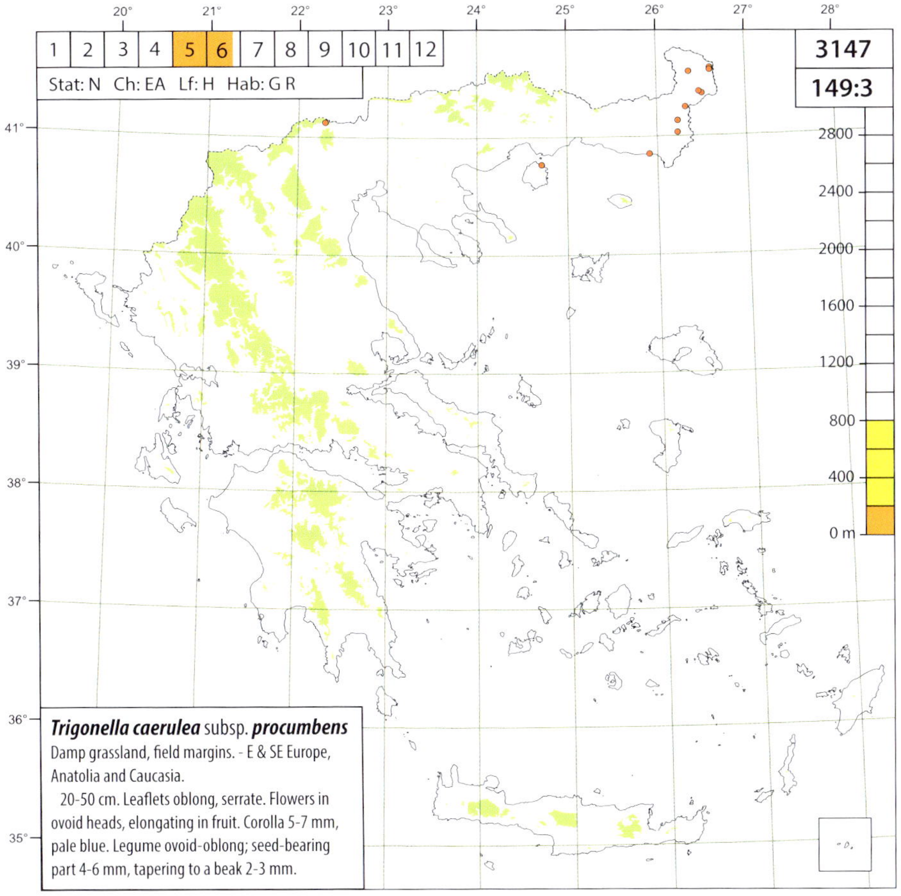

Trigonella caerulea* subsp. *procumbens

Damp grassland, field margins. - E & SE Europe, Anatolia and Caucasia.

20-50 cm. Leaflets oblong, serrate. Flowers in ovoid heads, elongating in fruit. Corolla 5-7 mm, pale blue. Legume ovoid-oblong; seed-bearing part 4-6 mm, tapering to a beak 2-3 mm.

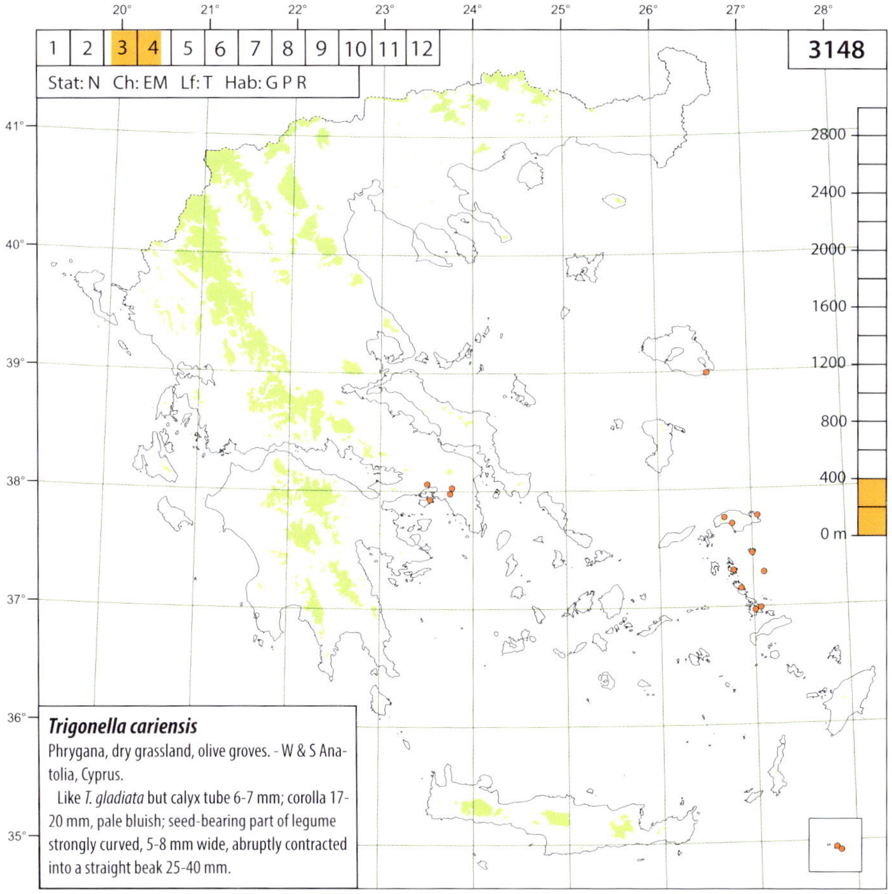

Trigonella cariensis

Phrygana, dry grassland, olive groves. - W & S Anatolia, Cyprus.

Like *T. gladiata* but calyx tube 6-7 mm; corolla 17-20 mm, pale bluish; seed-bearing part of legume strongly curved, 5-8 mm wide, abruptly contracted into a straight beak 25-40 mm.

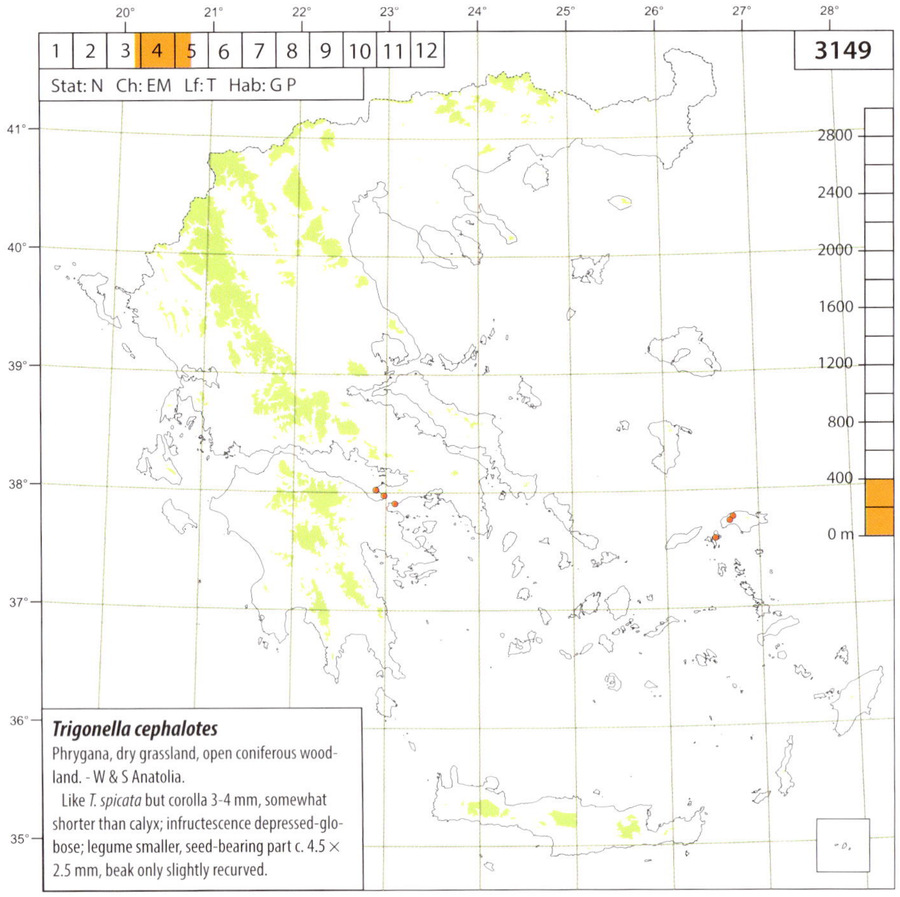

Trigonella cephalotes
Phrygana, dry grassland, open coniferous woodland. - W & S Anatolia.
Like *T. spicata* but corolla 3-4 mm, somewhat shorter than calyx; infructescence depressed-globose; legume smaller, seed-bearing part c. 4.5 × 2.5 mm, beak only slightly recurved.

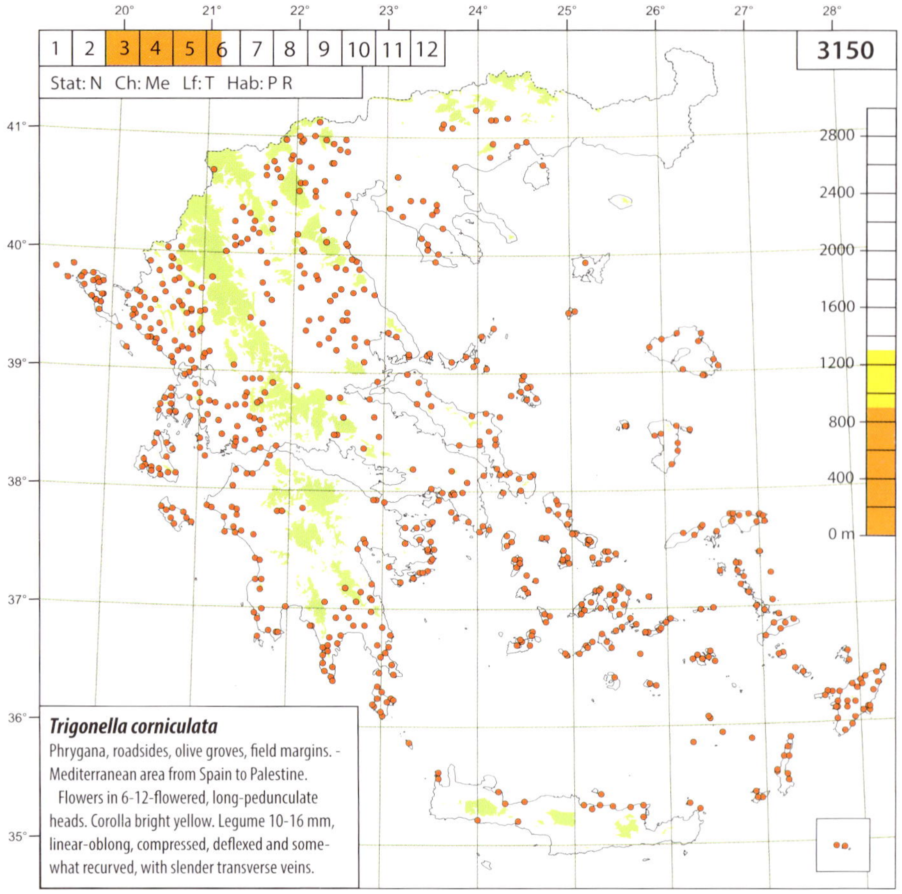

Trigonella corniculata
Phrygana, roadsides, olive groves, field margins. - Mediterranean area from Spain to Palestine.
Flowers in 6-12-flowered, long-pedunculate heads. Corolla bright yellow. Legume 10-16 mm, linear-oblong, compressed, deflexed and somewhat recurved, with slender transverse veins.

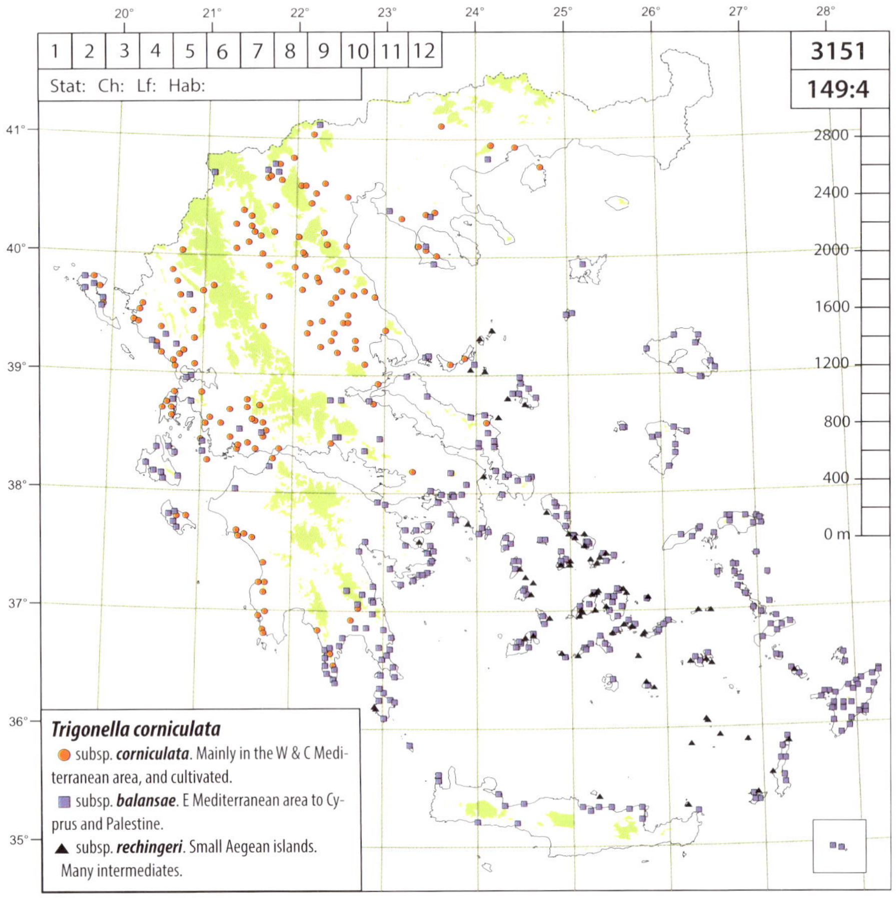

Trigonella corniculata
● subsp. ***corniculata***. Mainly in the W & C Mediterranean area, and cultivated.
■ subsp. ***balansae***. E Mediterranean area to Cyprus and Palestine.
▲ subsp. ***rechingeri***. Small Aegean islands.
Many intermediates.

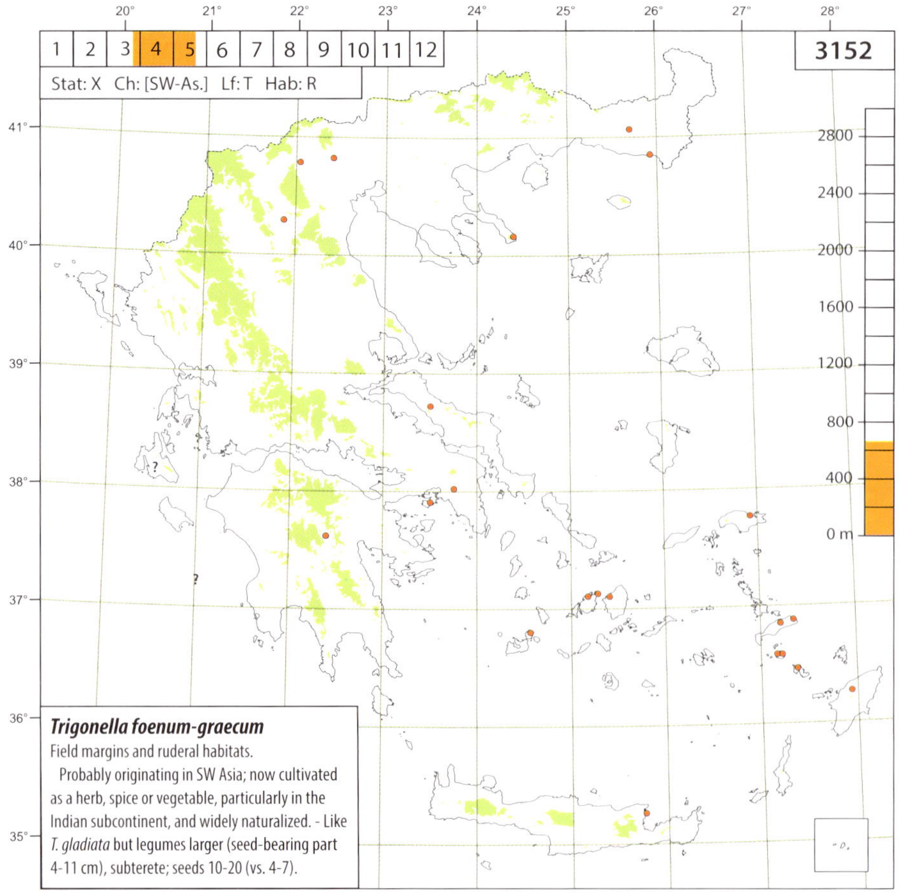

Trigonella foenum-graecum
Field margins and ruderal habitats.
Probably originating in SW Asia; now cultivated as a herb, spice or vegetable, particularly in the Indian subcontinent, and widely naturalized. - Like *T. gladiata* but legumes larger (seed-bearing part 4-11 cm), subterete; seeds 10-20 (vs. 4-7).

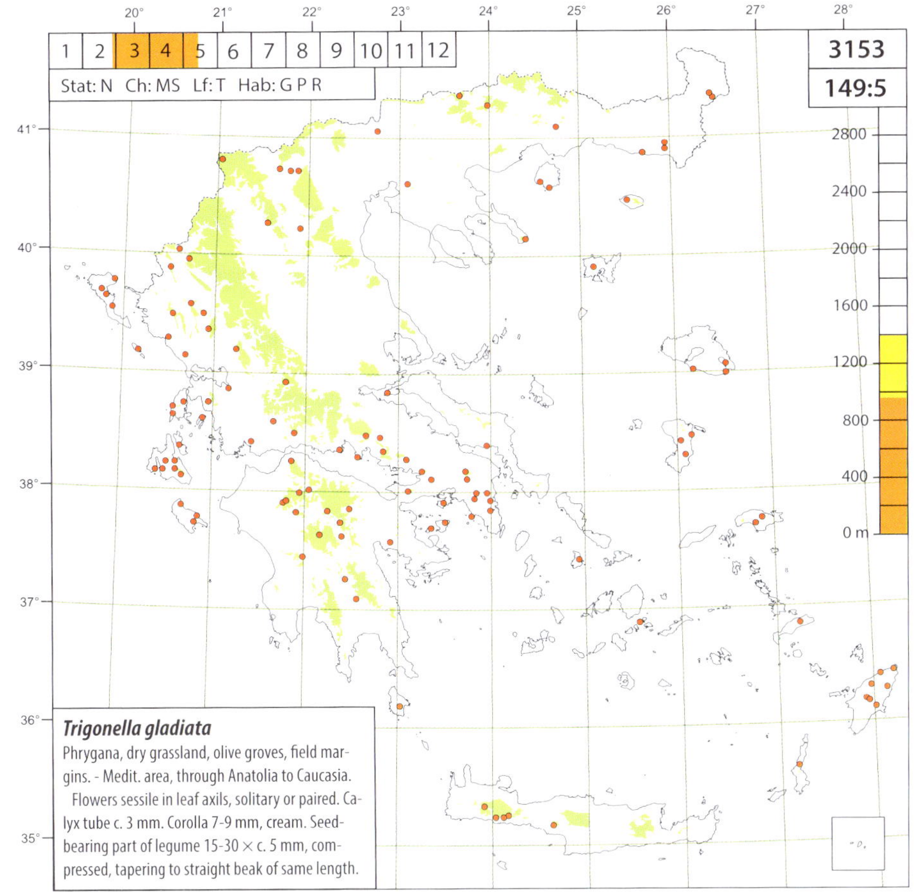

Trigonella gladiata
Phrygana, dry grassland, olive groves, field margins. - Medit. area, through Anatolia to Caucasia.
Flowers sessile in leaf axils, solitary or paired. Calyx tube c. 3 mm. Corolla 7-9 mm, cream. Seed-bearing part of legume 15-30 × c. 5 mm, compressed, tapering to straight beak of same length.

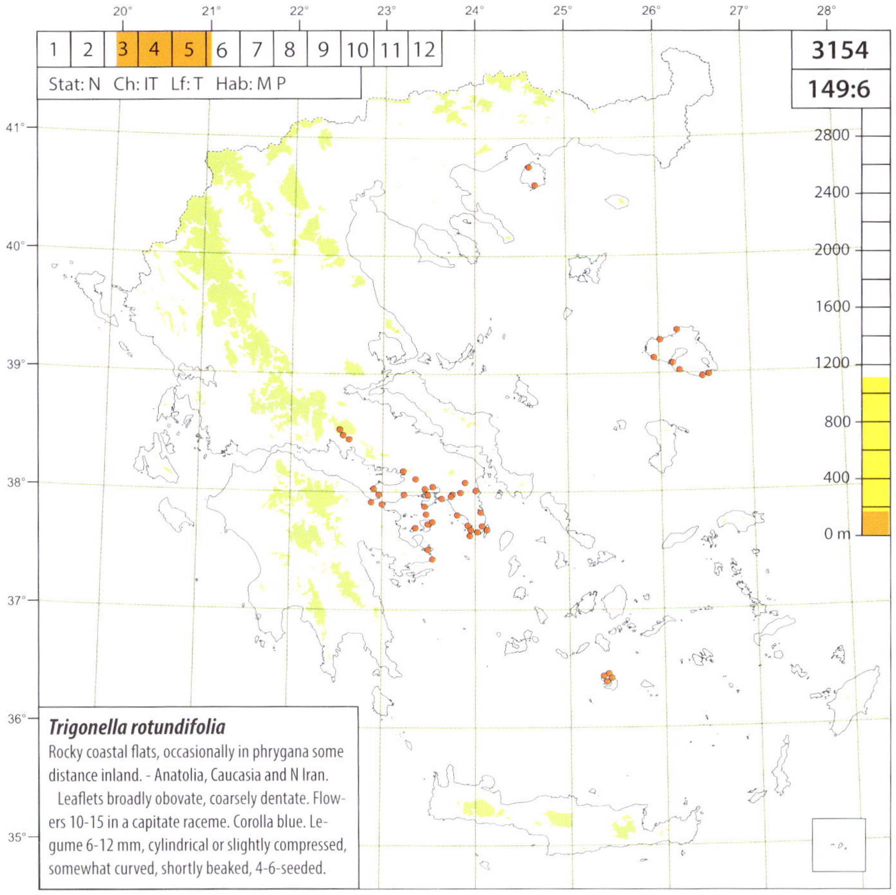

Trigonella rotundifolia
Rocky coastal flats, occasionally in phrygana some distance inland. - Anatolia, Caucasia and N Iran.
Leaflets broadly obovate, coarsely dentate. Flowers 10-15 in a capitate raceme. Corolla blue. Legume 6-12 mm, cylindrical or slightly compressed, somewhat curved, shortly beaked, 4-6-seeded.

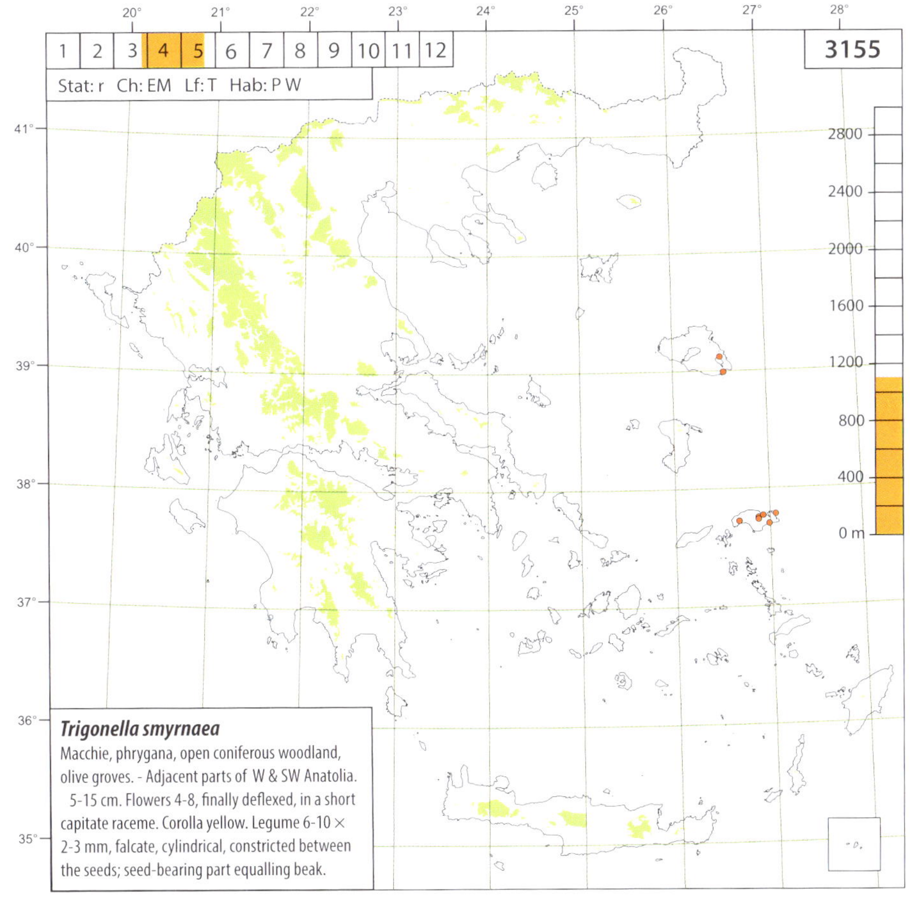

Trigonella smyrnaea
Macchie, phrygana, open coniferous woodland, olive groves. - Adjacent parts of W & SW Anatolia.
5-15 cm. Flowers 4-8, finally deflexed, in a short capitate raceme. Corolla yellow. Legume 6-10 × 2-3 mm, falcate, cylindrical, constricted between the seeds; seed-bearing part equalling beak.

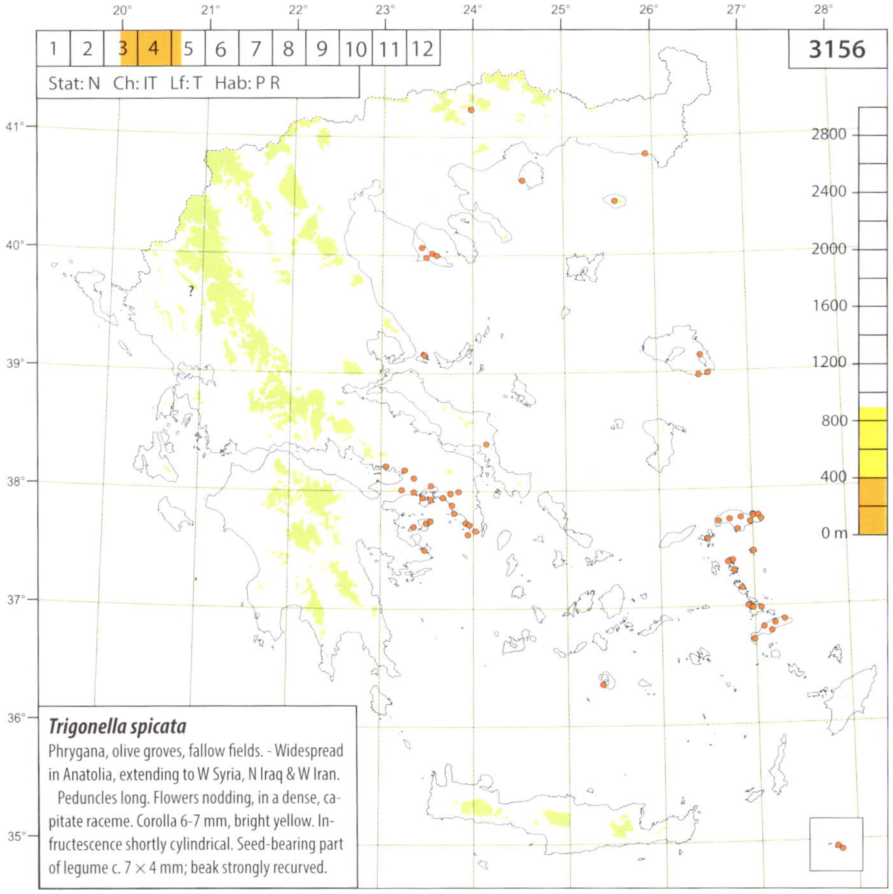

Trigonella spicata
Phrygana, olive groves, fallow fields. - Widespread in Anatolia, extending to W Syria, N Iraq & W Iran.
Peduncles long. Flowers nodding, in a dense, capitate raceme. Corolla 6-7 mm, bright yellow. Infructescence shortly cylindrical. Seed-bearing part of legume c. 7 × 4 mm; beak strongly recurved.

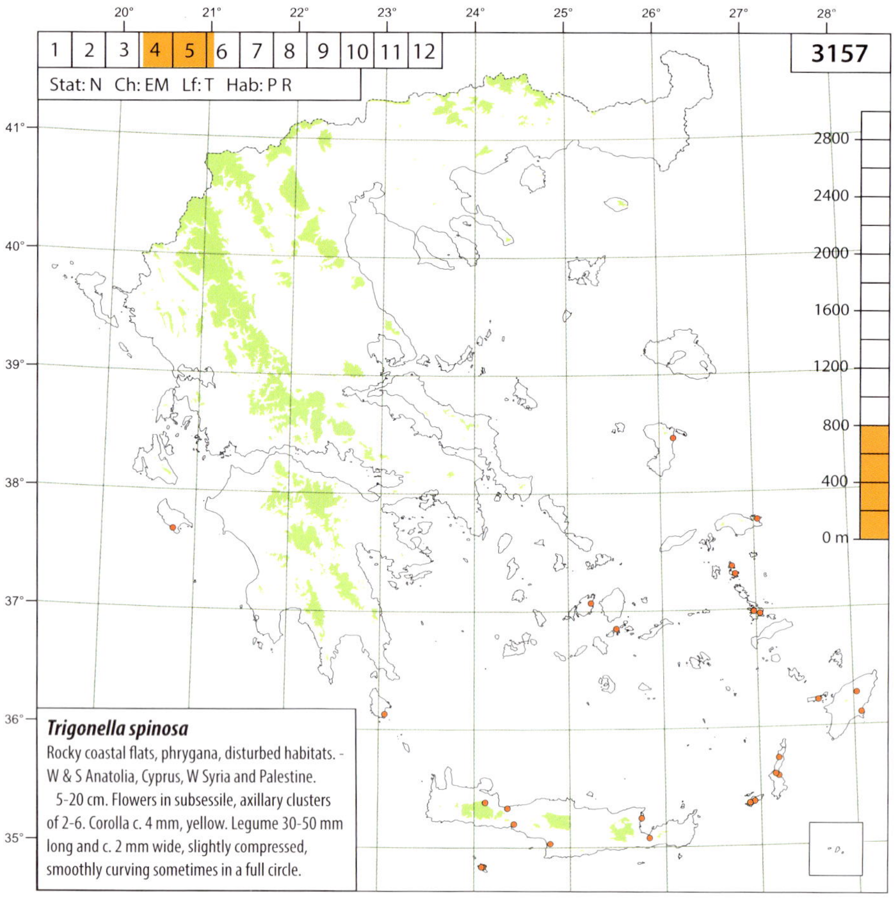

Trigonella spinosa
Rocky coastal flats, phrygana, disturbed habitats. - W & S Anatolia, Cyprus, W Syria and Palestine.
5-20 cm. Flowers in subsessile, axillary clusters of 2-6. Corolla c. 4 mm, yellow. Legume 30-50 mm long and c. 2 mm wide, slightly compressed, smoothly curving sometimes in a full circle.

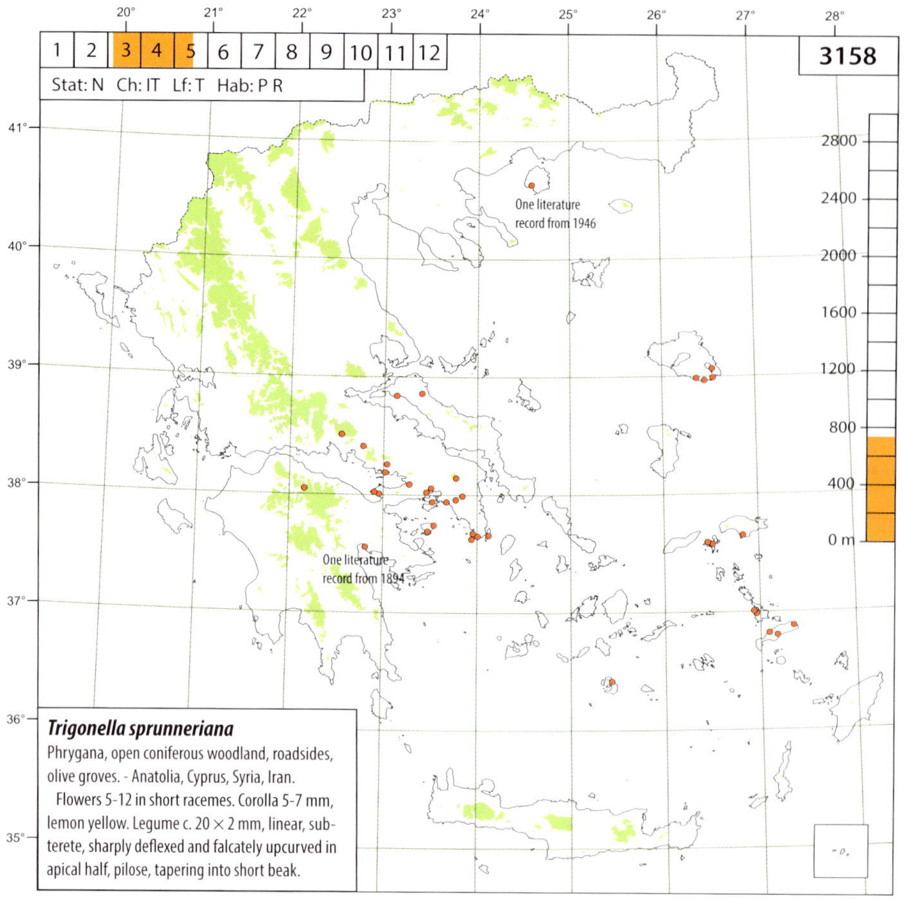

Trigonella sprunneriana
Phrygana, open coniferous woodland, roadsides, olive groves. - Anatolia, Cyprus, Syria, Iran.
Flowers 5-12 in short racemes. Corolla 5-7 mm, lemon yellow. Legume c. 20 × 2 mm, linear, subterete, sharply deflexed and falcately upcurved in apical half, pilose, tapering into short beak.

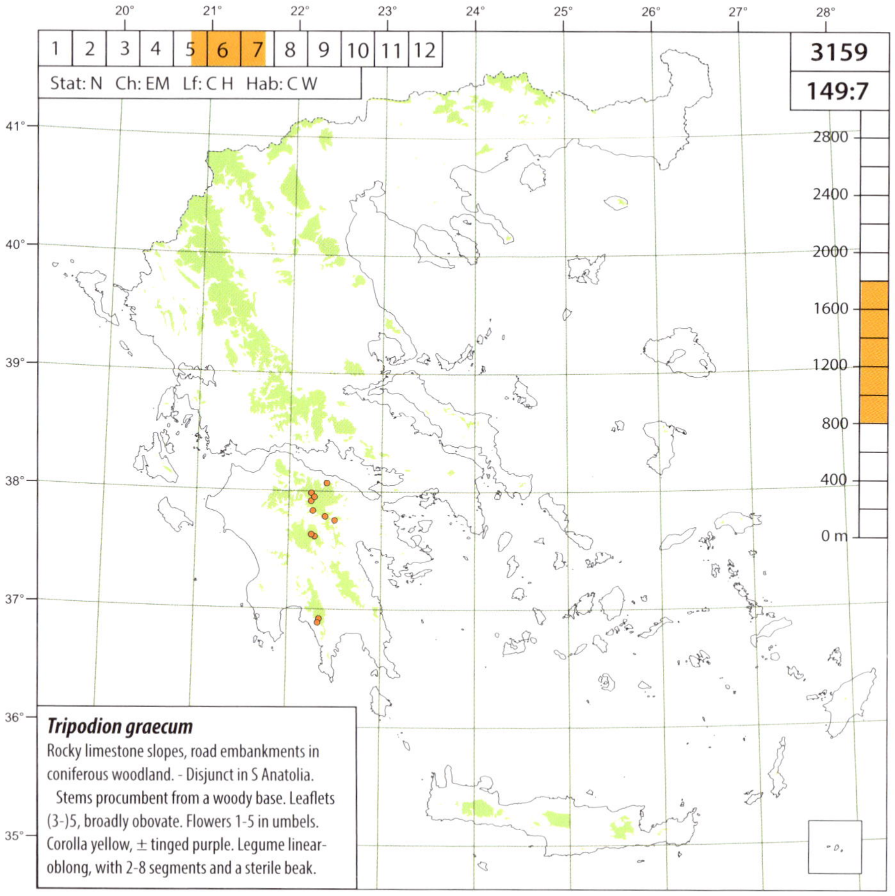

Tripodion graecum
Rocky limestone slopes, road embankments in coniferous woodland. - Disjunct in S Anatolia.
Stems procumbent from a woody base. Leaflets (3-)5, broadly obovate. Flowers 1-5 in umbels. Corolla yellow, ± tinged purple. Legume linear-oblong, with 2-8 segments and a sterile beak.

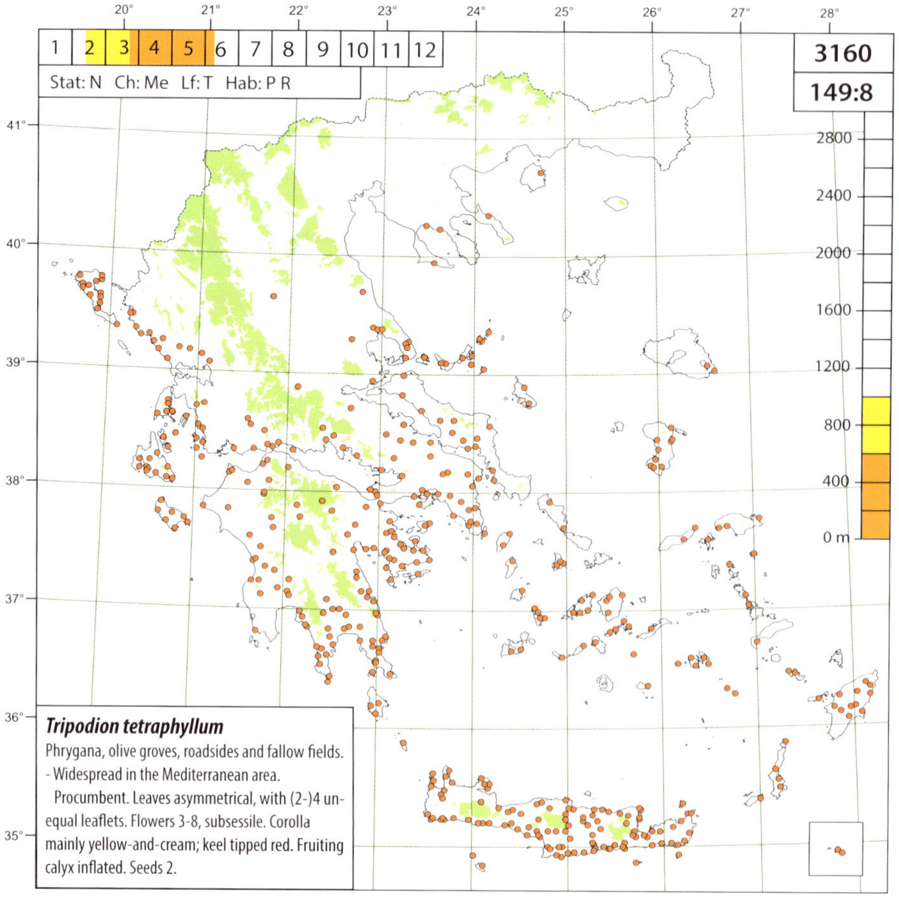

Tripodion tetraphyllum
Phrygana, olive groves, roadsides and fallow fields. - Widespread in the Mediterranean area.
Procumbent. Leaves asymmetrical, with (2-)4 unequal leaflets. Flowers 3-8, subsessile. Corolla mainly yellow-and-cream; keel tipped red. Fruiting calyx inflated. Seeds 2.

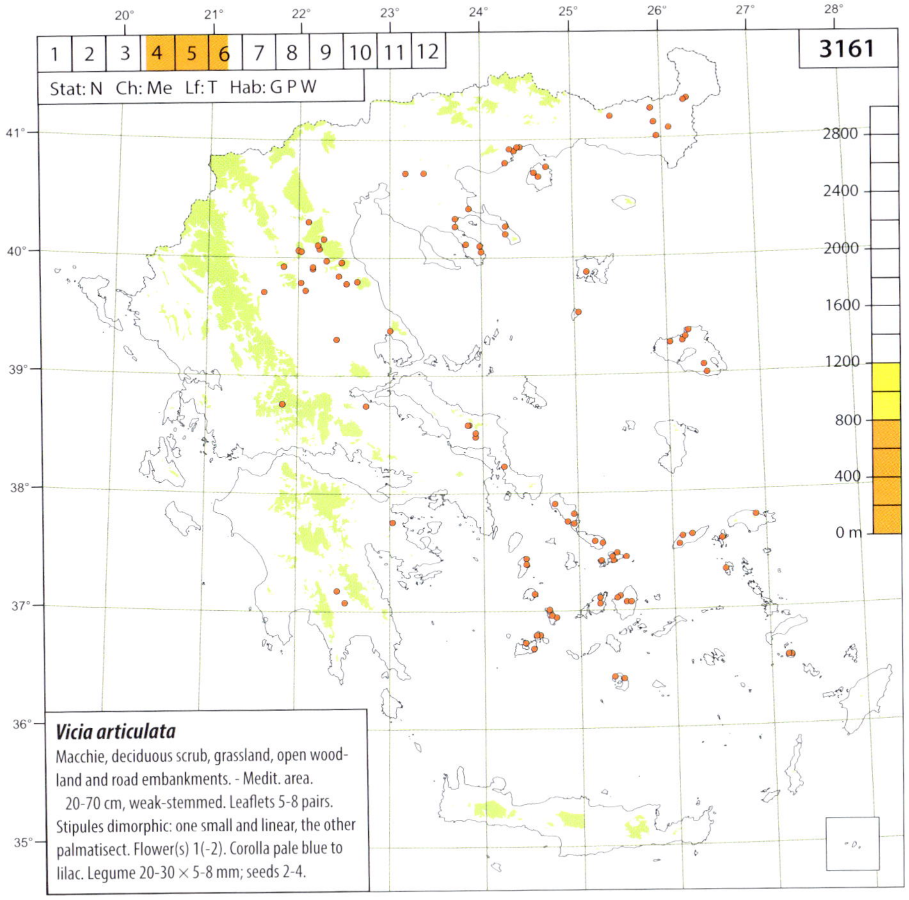

Vicia articulata
Macchie, deciduous scrub, grassland, open woodland and road embankments. - Medit. area.
20-70 cm, weak-stemmed. Leaflets 5-8 pairs. Stipules dimorphic: one small and linear, the other palmatisect. Flower(s) 1(-2). Corolla pale blue to lilac. Legume 20-30 × 5-8 mm; seeds 2-4.

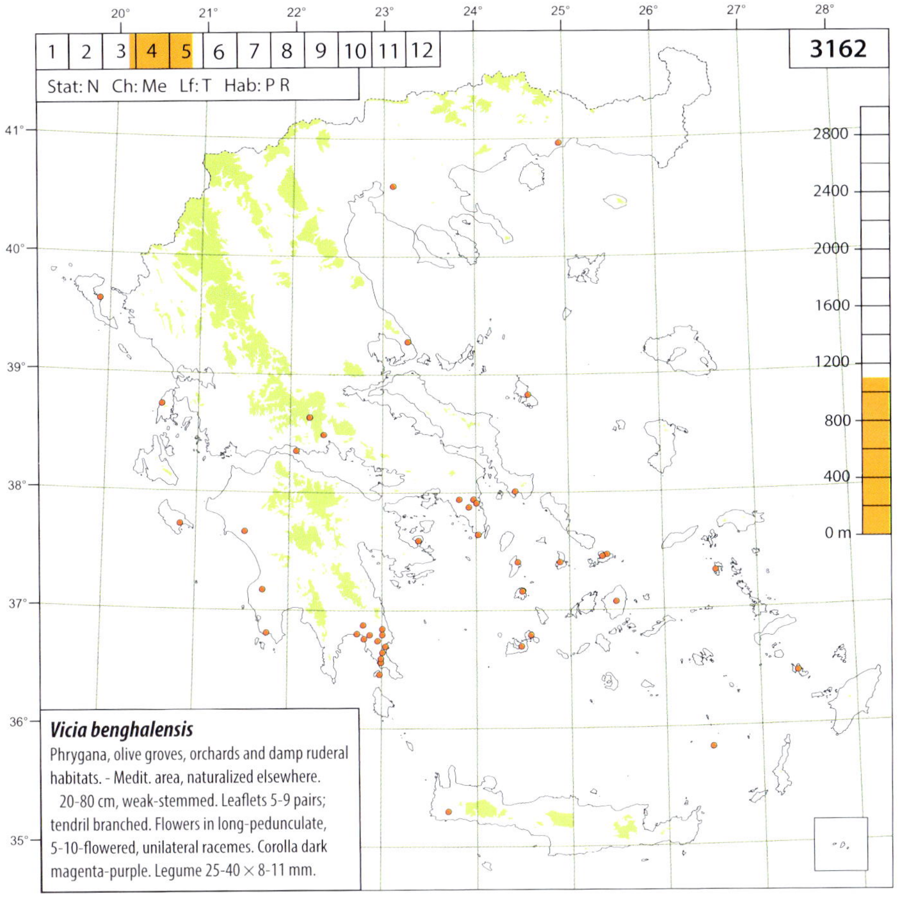

Vicia benghalensis
Phrygana, olive groves, orchards and damp ruderal habitats. - Medit. area, naturalized elsewhere.
20-80 cm, weak-stemmed. Leaflets 5-9 pairs; tendril branched. Flowers in long-pedunculate, 5-10-flowered, unilateral racemes. Corolla dark magenta-purple. Legume 25-40 × 8-11 mm.

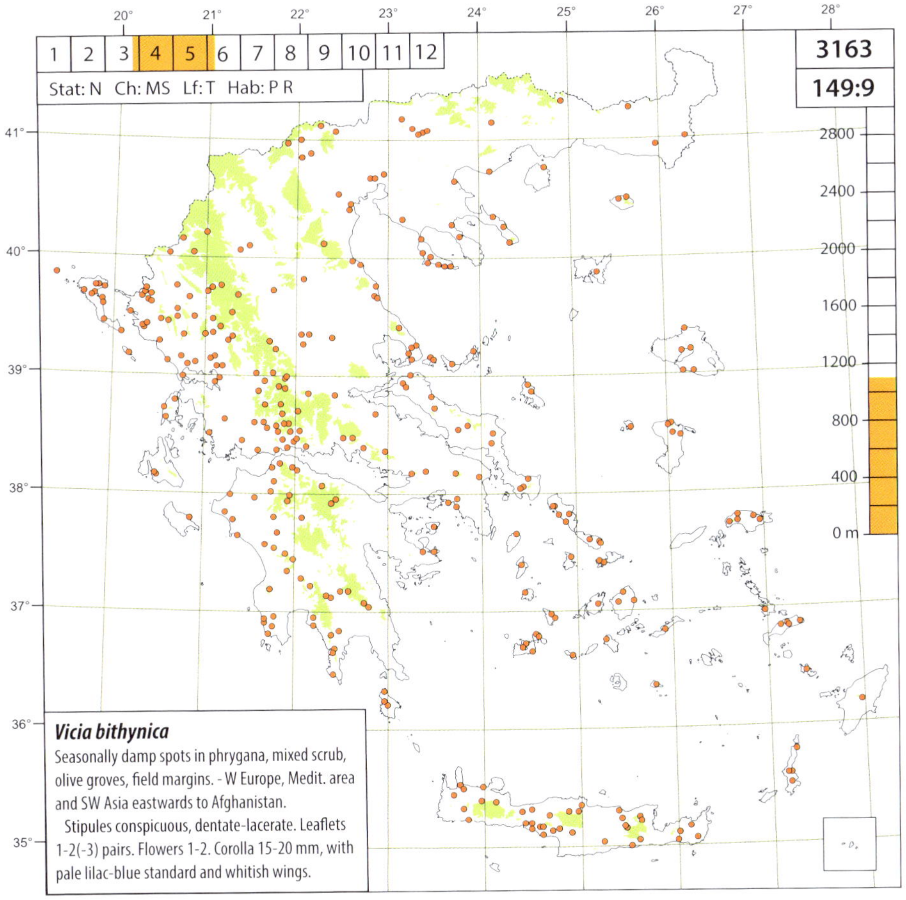

Vicia bithynica
Seasonally damp spots in phrygana, mixed scrub, olive groves, field margins. - W Europe, Medit. area and SW Asia eastwards to Afghanistan.
Stipules conspicuous, dentate-lacerate. Leaflets 1-2(-3) pairs. Flowers 1-2. Corolla 15-20 mm, with pale lilac-blue standard and whitish wings.

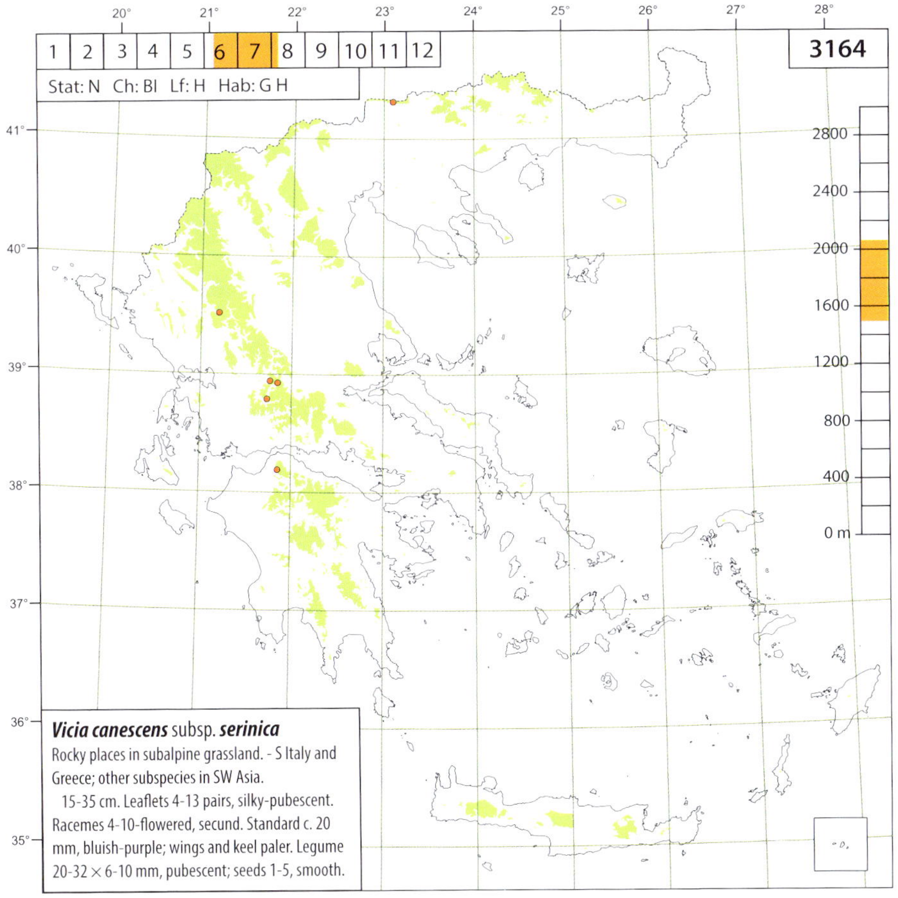

Vicia canescens subsp. ***serinica***
Rocky places in subalpine grassland. - S Italy and Greece; other subspecies in SW Asia.
15-35 cm. Leaflets 4-13 pairs, silky-pubescent. Racemes 4-10-flowered, secund. Standard c. 20 mm, bluish-purple; wings and keel paler. Legume 20-32 × 6-10 mm, pubescent; seeds 1-5, smooth.

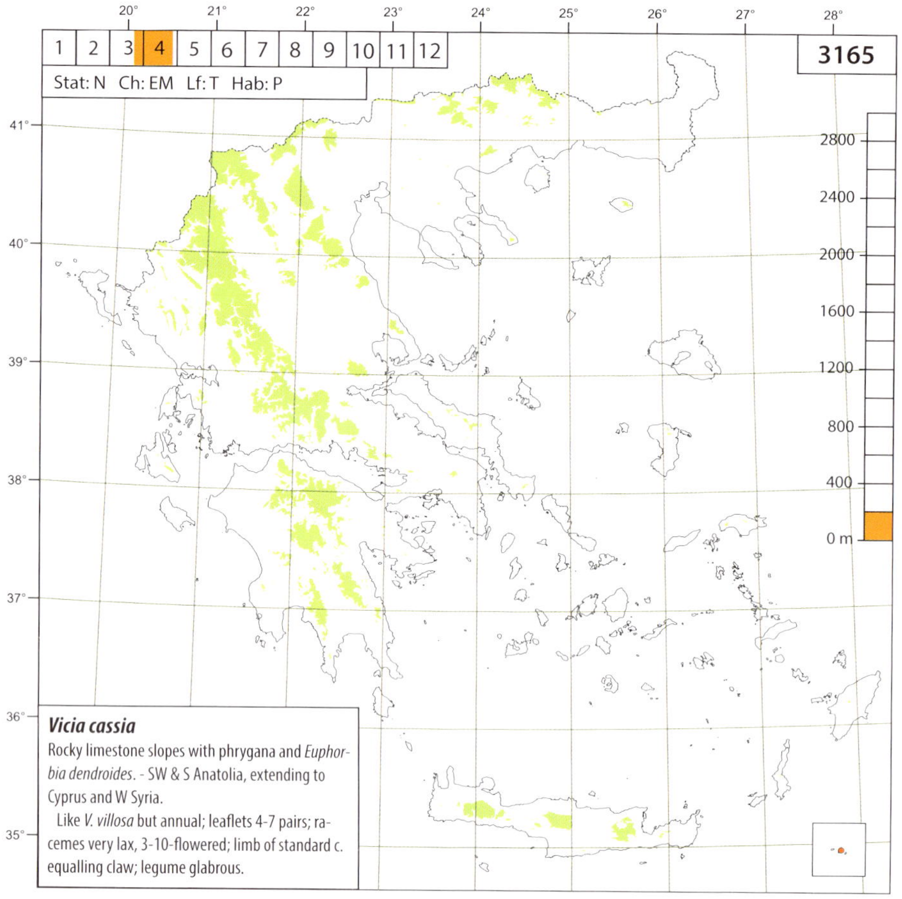

Vicia cassia

Rocky limestone slopes with phrygana and *Euphorbia dendroides*. - SW & S Anatolia, extending to Cyprus and W Syria.

Like *V. villosa* but annual; leaflets 4-7 pairs; racemes very lax, 3-10-flowered; limb of standard c. equalling claw; legume glabrous.

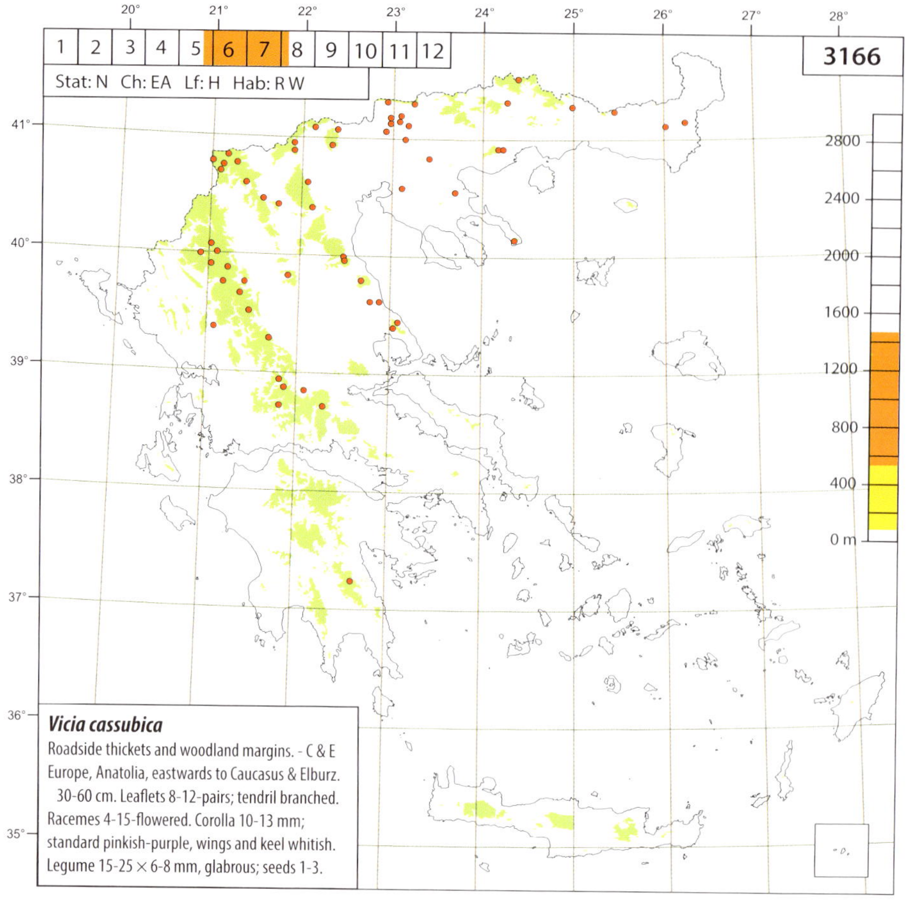

Vicia cassubica

Roadside thickets and woodland margins. - C & E Europe, Anatolia, eastwards to Caucasus & Elburz.

30-60 cm. Leaflets 8-12-pairs; tendril branched. Racemes 4-15-flowered. Corolla 10-13 mm; standard pinkish-purple, wings and keel whitish. Legume 15-25 × 6-8 mm, glabrous; seeds 1-3.

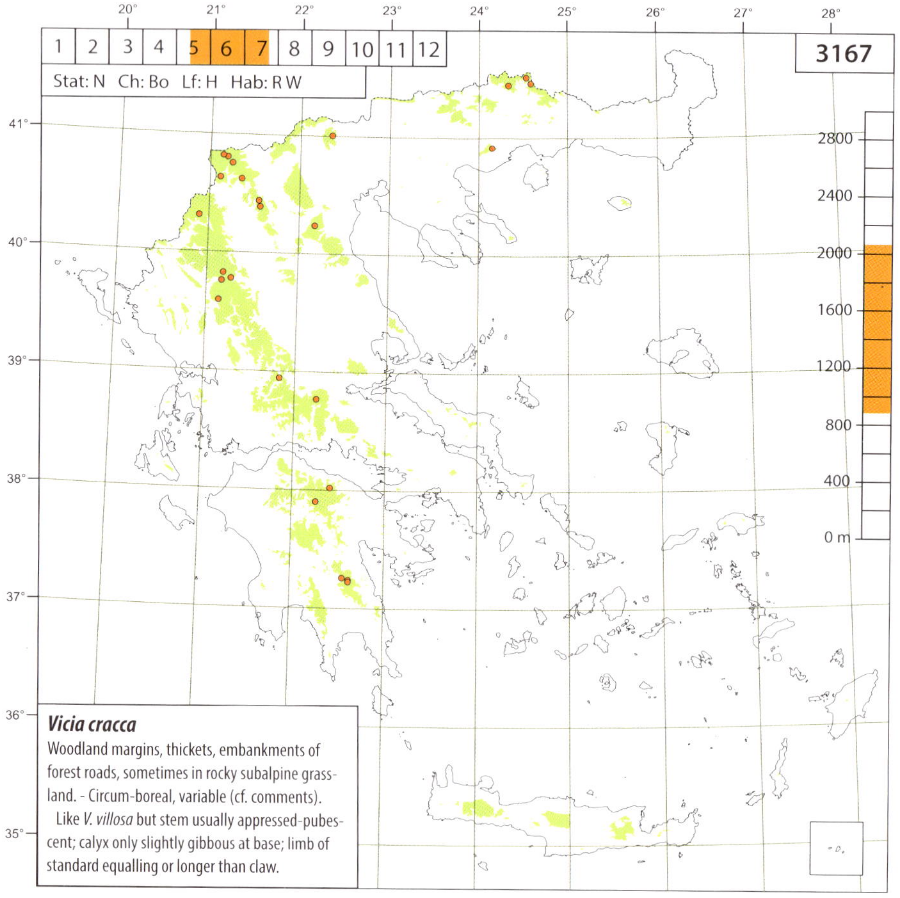

Vicia cracca

Woodland margins, thickets, embankments of forest roads, sometimes in rocky subalpine grassland. - Circum-boreal, variable (cf. comments).

Like *V. villosa* but stem usually appressed-pubescent; calyx only slightly gibbous at base; limb of standard equalling or longer than claw.

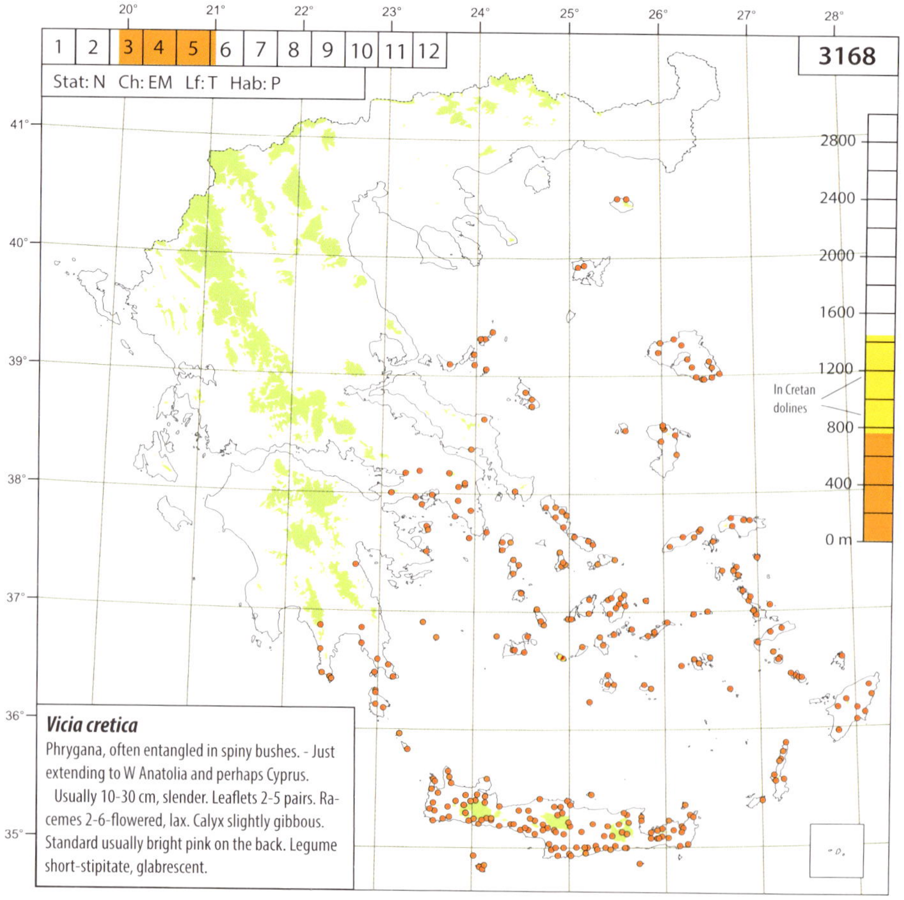

Vicia cretica

Phrygana, often entangled in spiny bushes. - Just extending to W Anatolia and perhaps Cyprus.

Usually 10-30 cm, slender. Leaflets 2-5 pairs. Racemes 2-6-flowered, lax. Calyx slightly gibbous. Standard usually bright pink on the back. Legume short-stipitate, glabrescent.

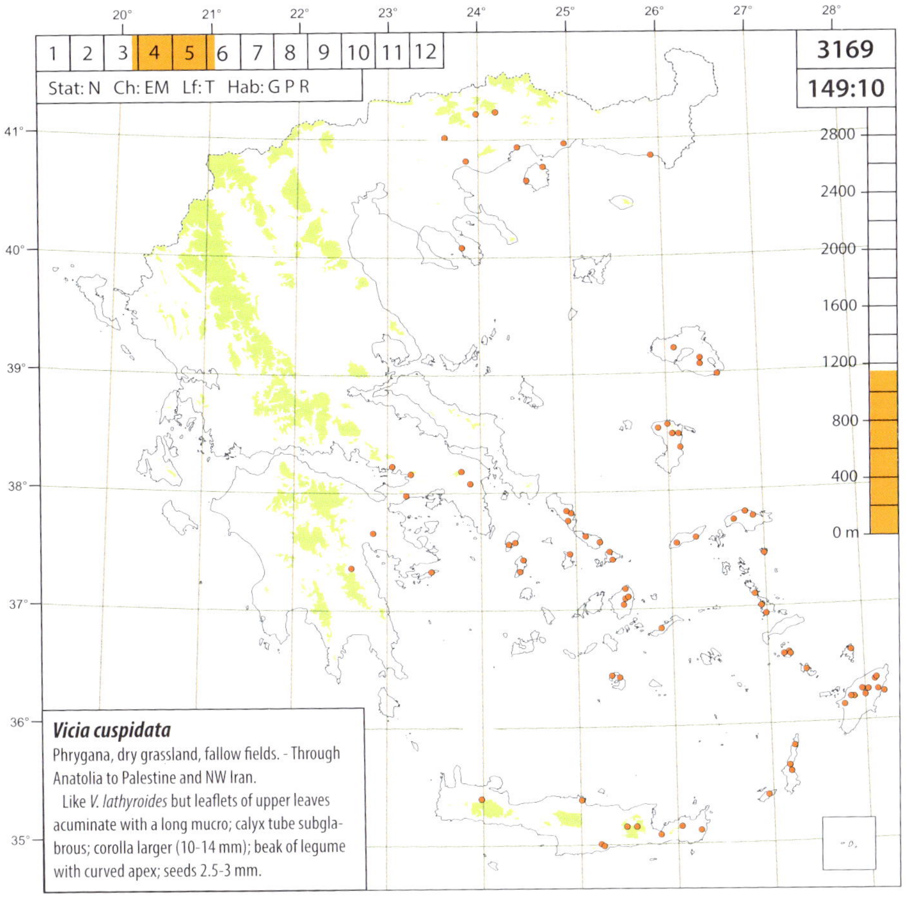

Vicia cuspidata

Phrygana, dry grassland, fallow fields. - Through Anatolia to Palestine and NW Iran.

Like *V. lathyroides* but leaflets of upper leaves acuminate with a long mucro; calyx tube subglabrous; corolla larger (10-14 mm); beak of legume with curved apex; seeds 2.5-3 mm.

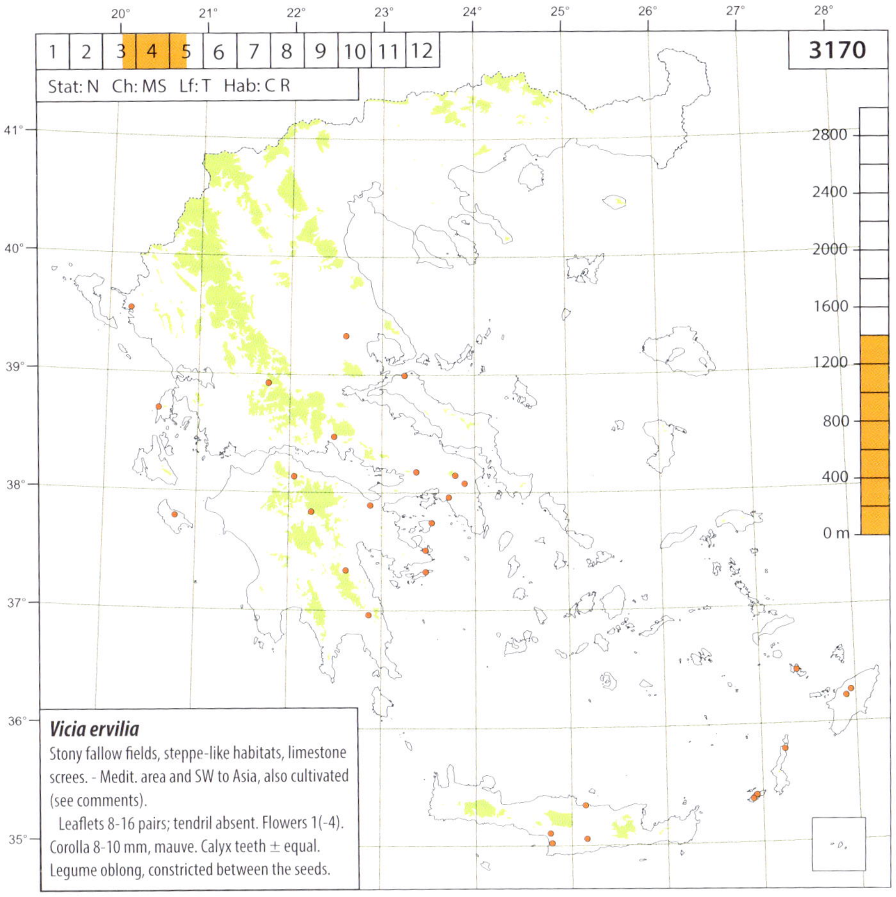

Vicia ervilia

Stony fallow fields, steppe-like habitats, limestone screes. - Medit. area and SW to Asia, also cultivated (see comments).

Leaflets 8-16 pairs; tendril absent. Flowers 1(-4). Corolla 8-10 mm, mauve. Calyx teeth ± equal. Legume oblong, constricted between the seeds.

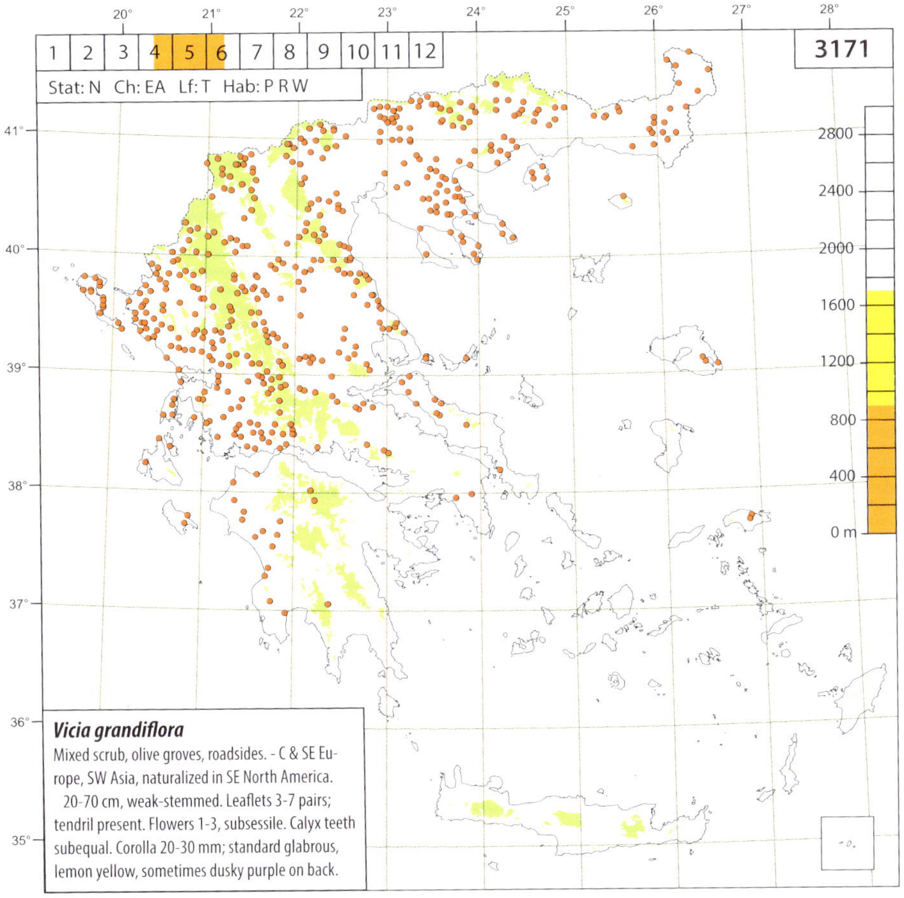

Vicia grandiflora

Mixed scrub, olive groves, roadsides. - C & SE Europe, SW Asia, naturalized in SE North America.

20-70 cm, weak-stemmed. Leaflets 3-7 pairs; tendril present. Flowers 1-3, subsessile. Calyx teeth subequal. Corolla 20-30 mm; standard glabrous, lemon yellow, sometimes dusky purple on back.

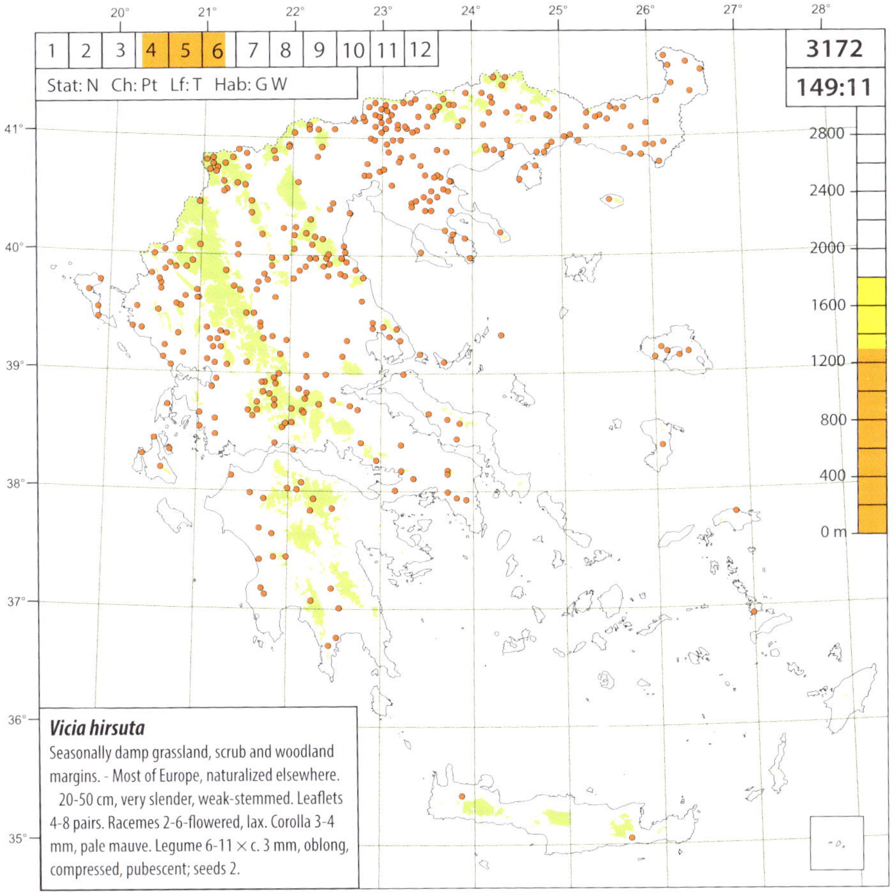

Vicia hirsuta

Seasonally damp grassland, scrub and woodland margins. - Most of Europe, naturalized elsewhere.

20-50 cm, very slender, weak-stemmed. Leaflets 4-8 pairs. Racemes 2-6-flowered, lax. Corolla 3-4 mm, pale mauve. Legume 6-11 × c. 3 mm, oblong, compressed, pubescent; seeds 2.

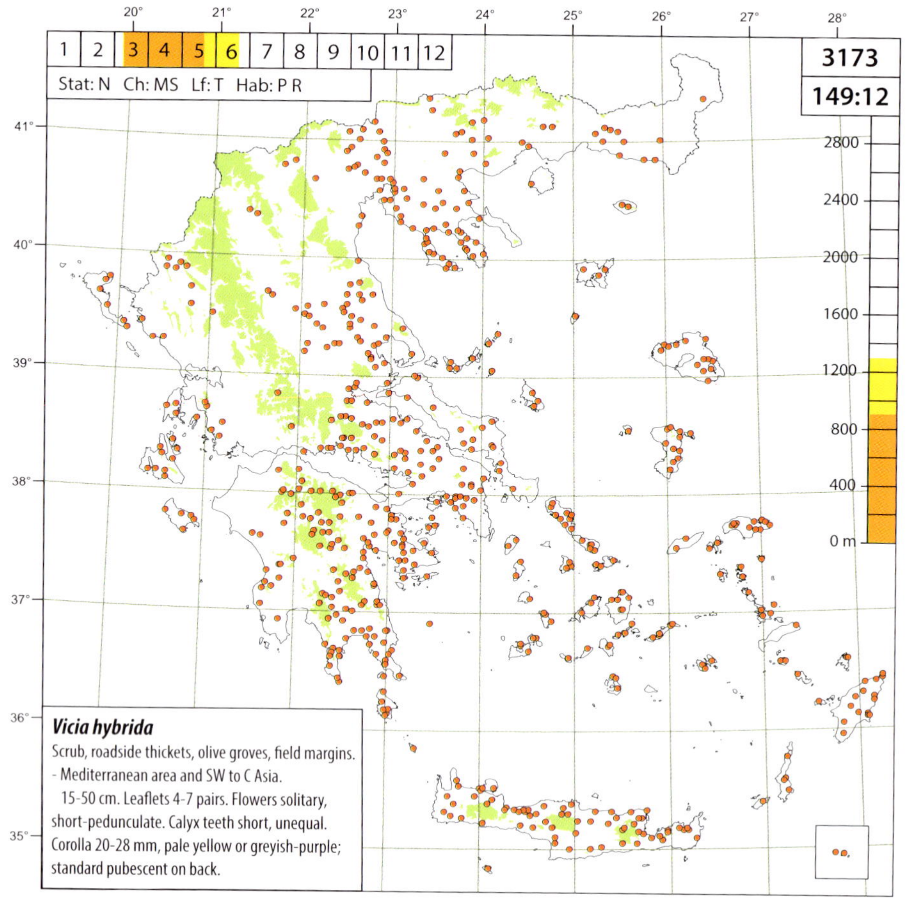

Vicia hybrida
Scrub, roadside thickets, olive groves, field margins. - Mediterranean area and SW to C Asia.
15-50 cm. Leaflets 4-7 pairs. Flowers solitary, short-pedunculate. Calyx teeth short, unequal. Corolla 20-28 mm, pale yellow or greyish-purple; standard pubescent on back.

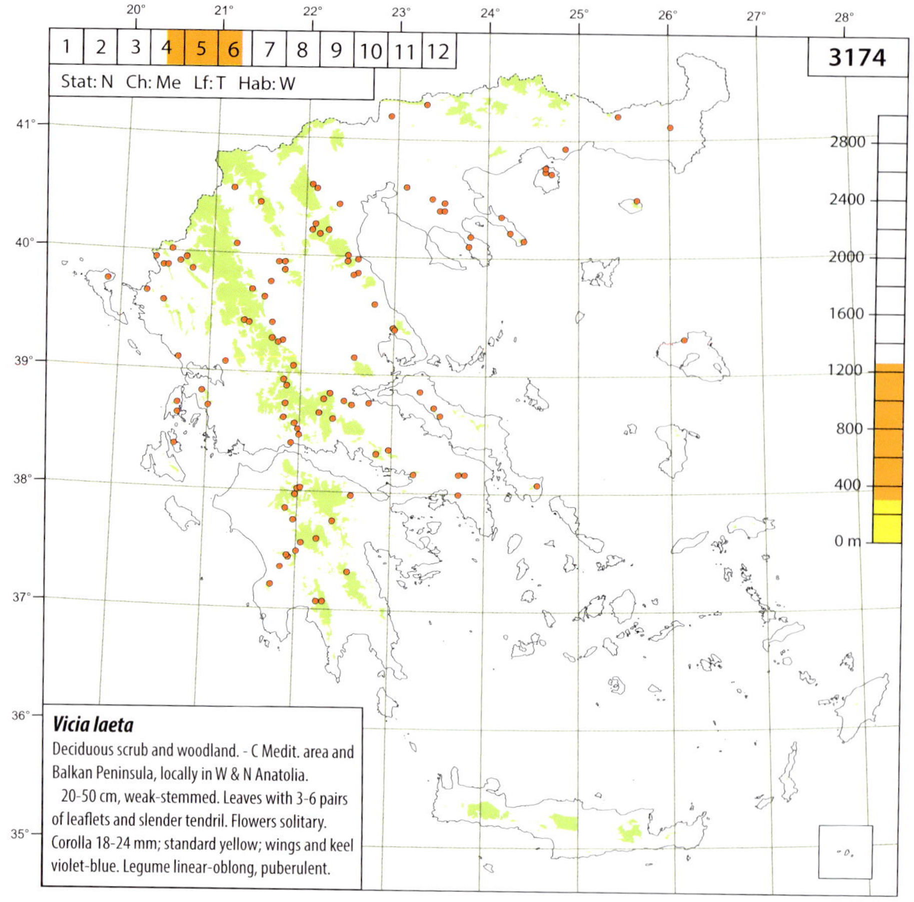

Vicia laeta
Deciduous scrub and woodland. - C Medit. area and Balkan Peninsula, locally in W & N Anatolia.
20-50 cm, weak-stemmed. Leaves with 3-6 pairs of leaflets and slender tendril. Flowers solitary. Corolla 18-24 mm; standard yellow; wings and keel violet-blue. Legume linear-oblong, puberulent.

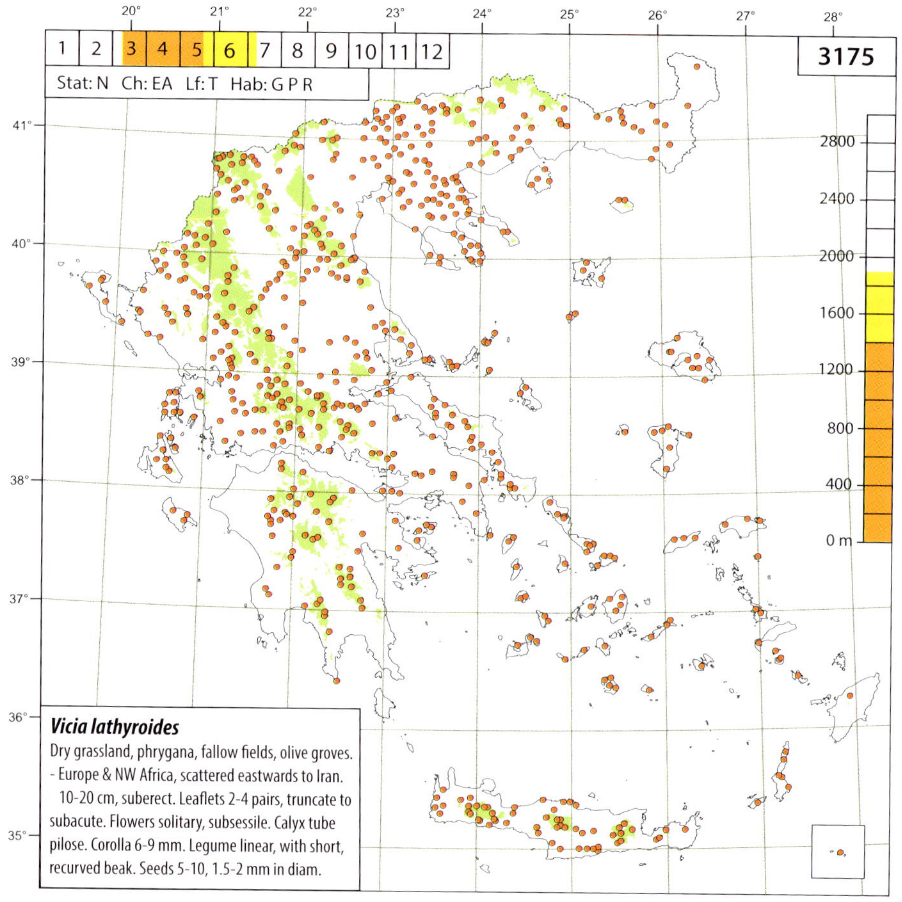

Vicia lathyroides
Dry grassland, phrygana, fallow fields, olive groves. - Europe & NW Africa, scattered eastwards to Iran.
10-20 cm, suberect. Leaflets 2-4 pairs, truncate to subacute. Flowers solitary, subsessile. Calyx tube pilose. Corolla 6-9 mm. Legume linear, with short, recurved beak. Seeds 5-10, 1.5-2 mm in diam.

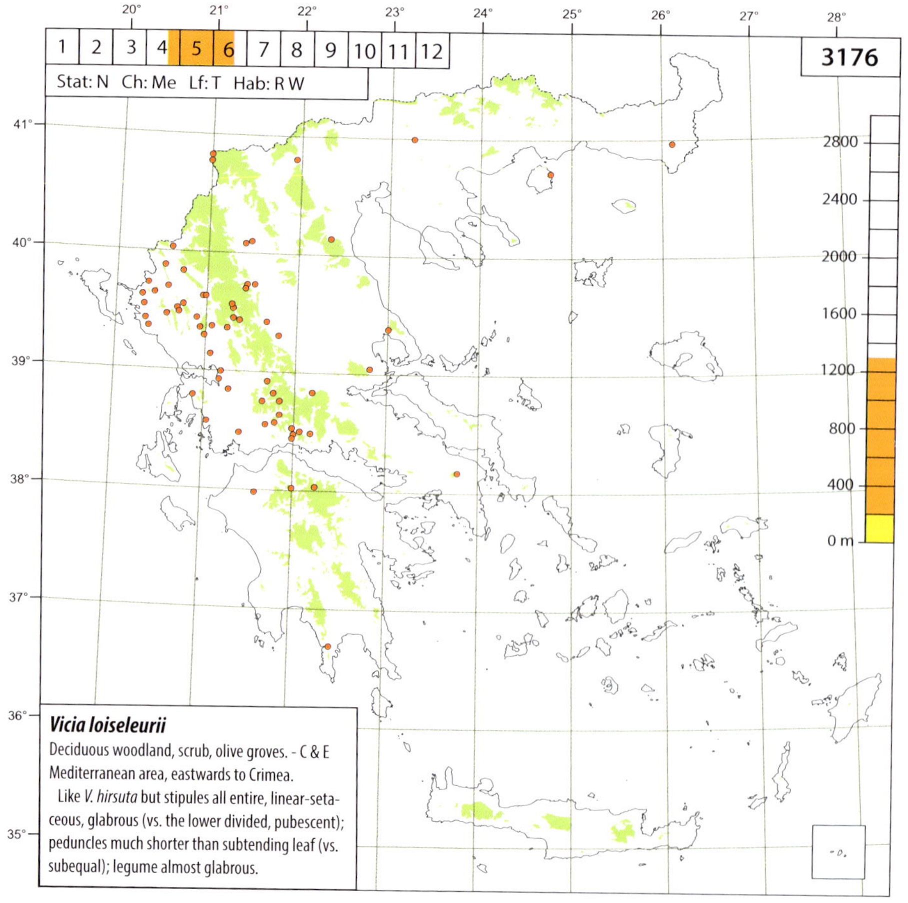

Vicia loiseleurii
Deciduous woodland, scrub, olive groves. - C & E Mediterranean area, eastwards to Crimea.
Like *V. hirsuta* but stipules all entire, linear-setaceous, glabrous (vs. the lower divided, pubescent); peduncles much shorter than subtending leaf (vs. subequal); legume almost glabrous.

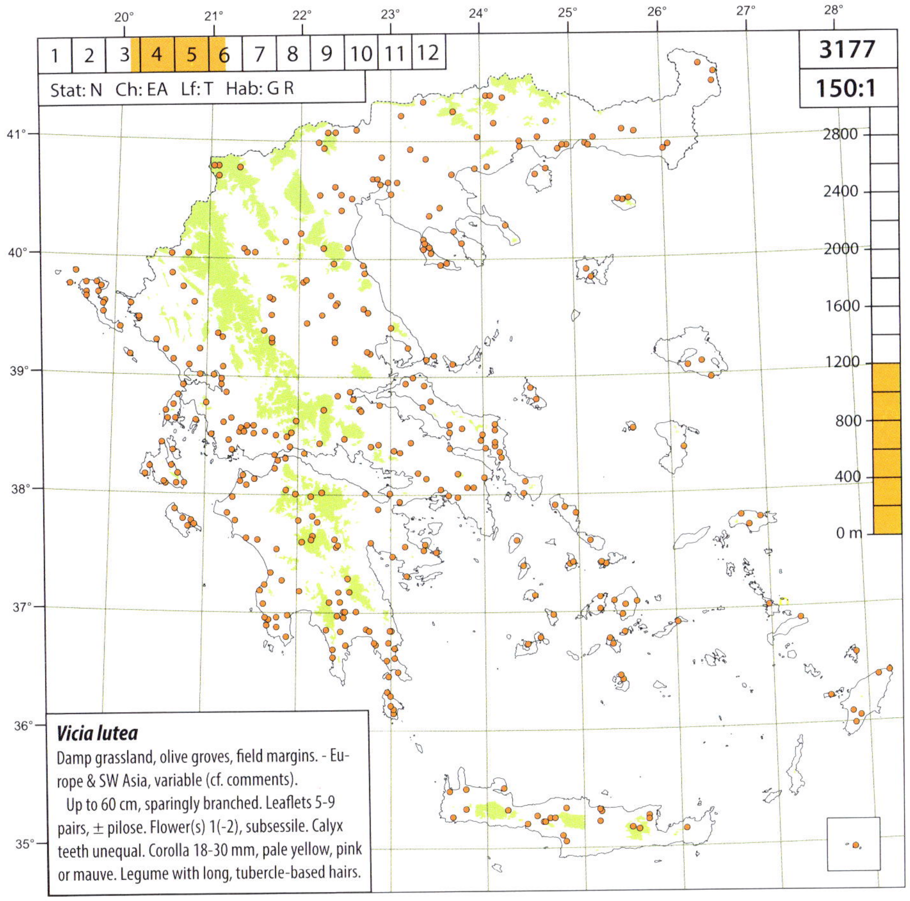

Vicia lutea
Damp grassland, olive groves, field margins. - Europe & SW Asia, variable (cf. comments).
Up to 60 cm, sparingly branched. Leaflets 5-9 pairs, ± pilose. Flower(s) 1(-2), subsessile. Calyx teeth unequal. Corolla 18-30 mm, pale yellow, pink or mauve. Legume with long, tubercle-based hairs.

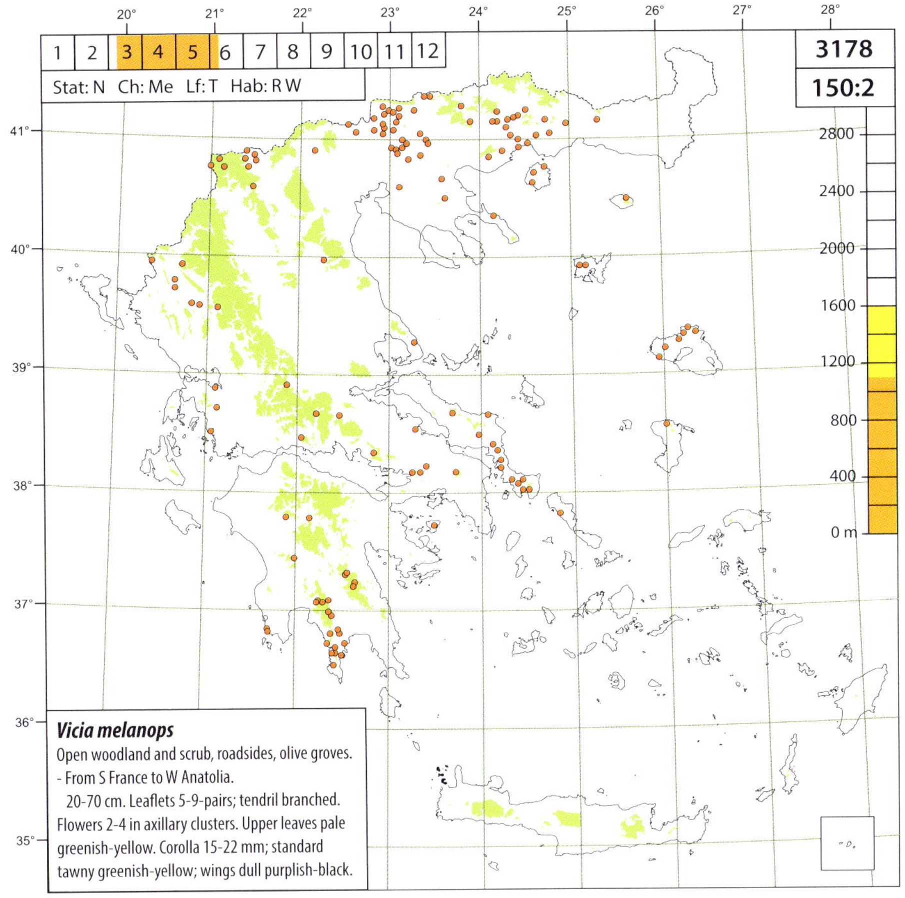

Vicia melanops
Open woodland and scrub, roadsides, olive groves. - From S France to W Anatolia.
20-70 cm. Leaflets 5-9-pairs; tendril branched. Flowers 2-4 in axillary clusters. Upper leaves pale greenish-yellow. Corolla 15-22 mm; standard tawny greenish-yellow; wings dull purplish-black.

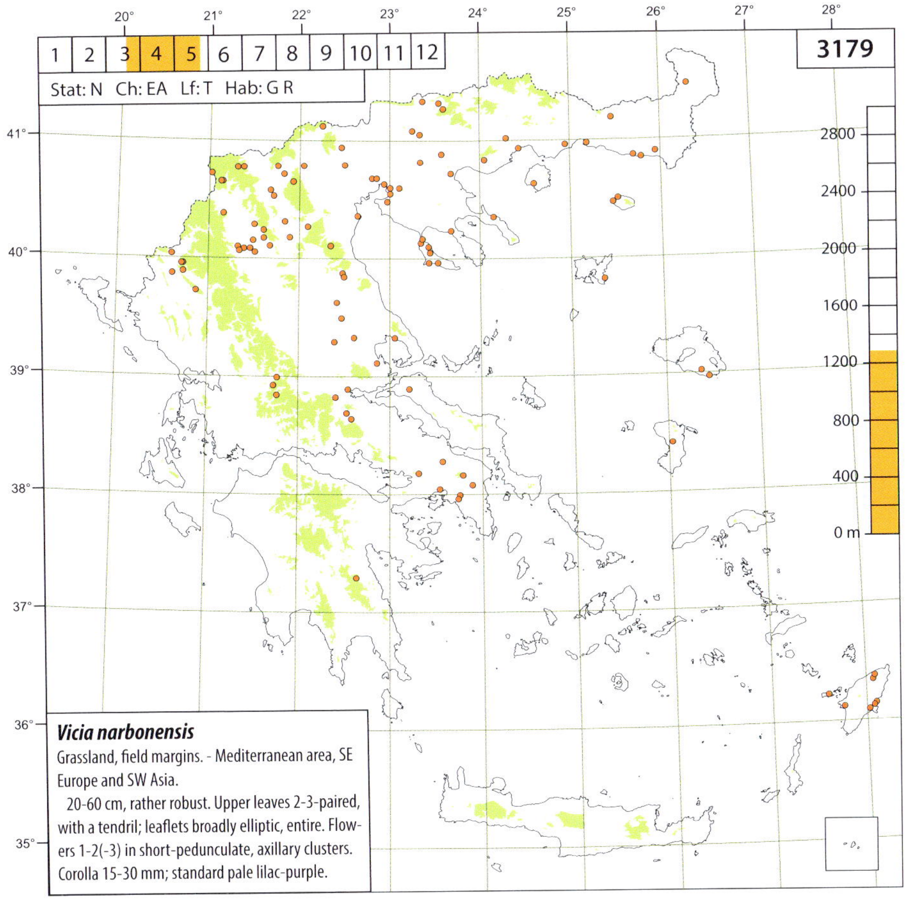

Vicia narbonensis
Grassland, field margins. - Mediterranean area, SE Europe and SW Asia.
20-60 cm, rather robust. Upper leaves 2-3-paired, with a tendril; leaflets broadly elliptic, entire. Flowers 1-2(-3) in short-pedunculate, axillary clusters. Corolla 15-30 mm; standard pale lilac-purple.

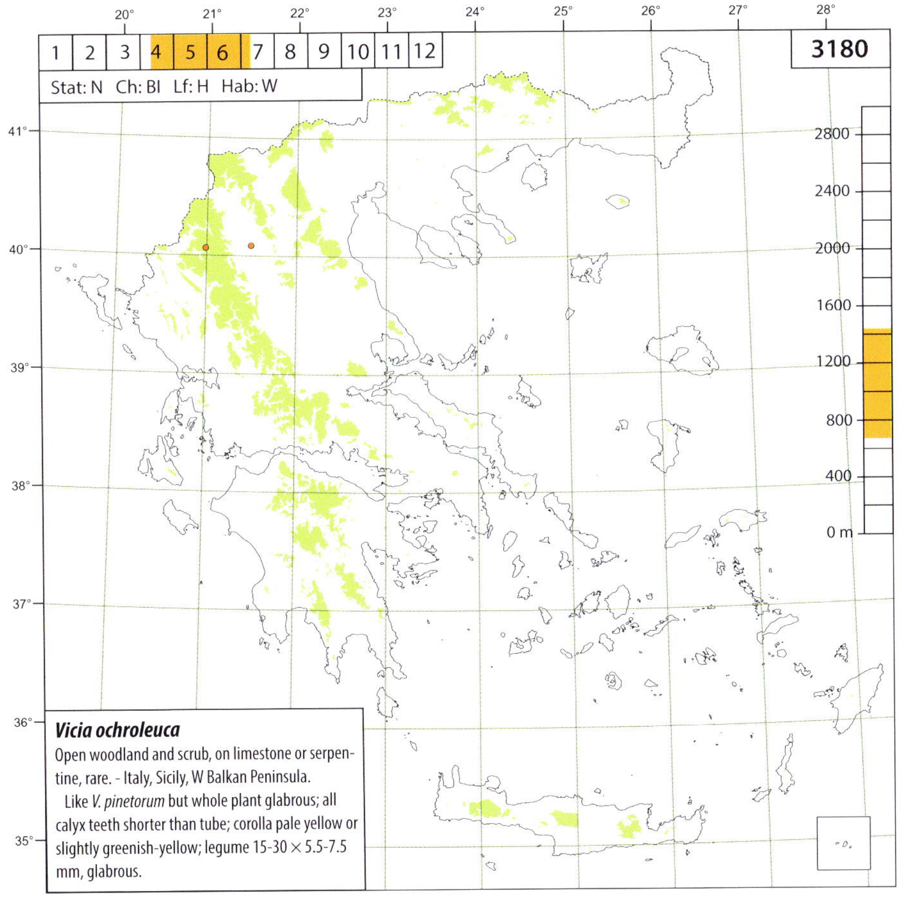

Vicia ochroleuca
Open woodland and scrub, on limestone or serpentine, rare. - Italy, Sicily, W Balkan Peninsula.
Like *V. pinetorum* but whole plant glabrous; all calyx teeth shorter than tube; corolla pale yellow or slightly greenish-yellow; legume 15-30 × 5.5-7.5 mm, glabrous.

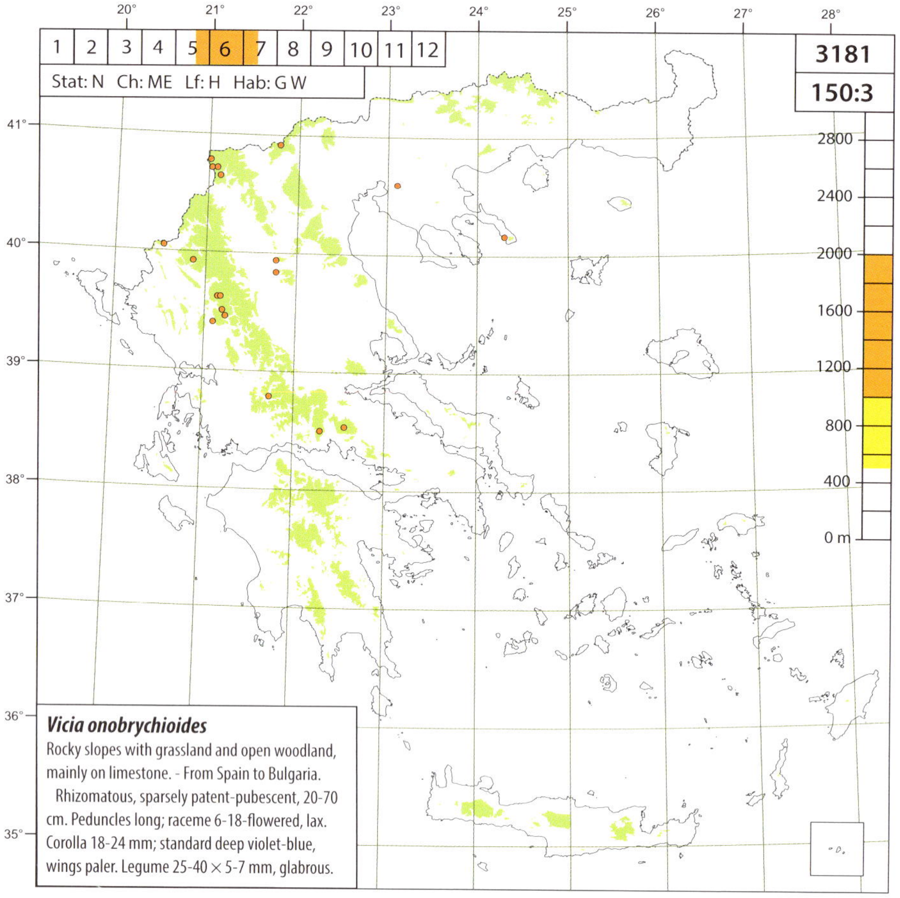

Vicia onobrychioides
Rocky slopes with grassland and open woodland, mainly on limestone. - From Spain to Bulgaria.
Rhizomatous, sparsely patent-pubescent, 20-70 cm. Peduncles long; raceme 6-18-flowered, lax. Corolla 18-24 mm; standard deep violet-blue, wings paler. Legume 25-40 × 5-7 mm, glabrous.

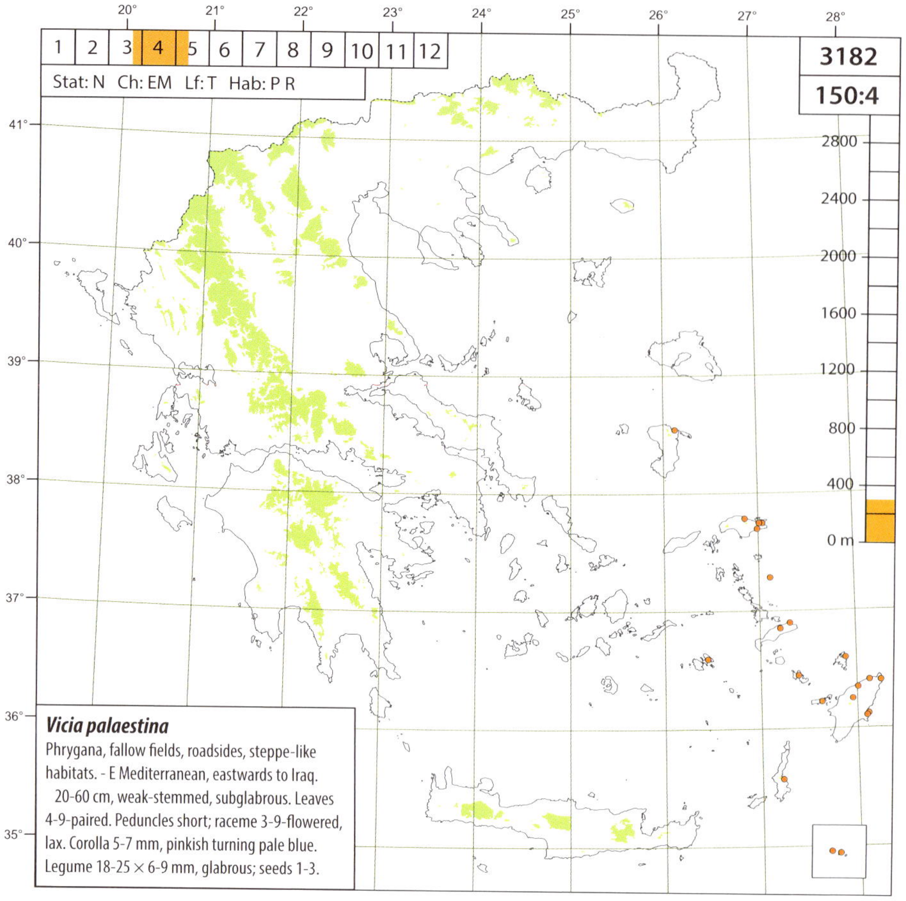

Vicia palaestina
Phrygana, fallow fields, roadsides, steppe-like habitats. - E Mediterranean, eastwards to Iraq.
20-60 cm, weak-stemmed, subglabrous. Leaves 4-9-paired. Peduncles short; raceme 3-9-flowered, lax. Corolla 5-7 mm, pinkish turning pale blue. Legume 18-25 × 6-9 mm, glabrous; seeds 1-3.

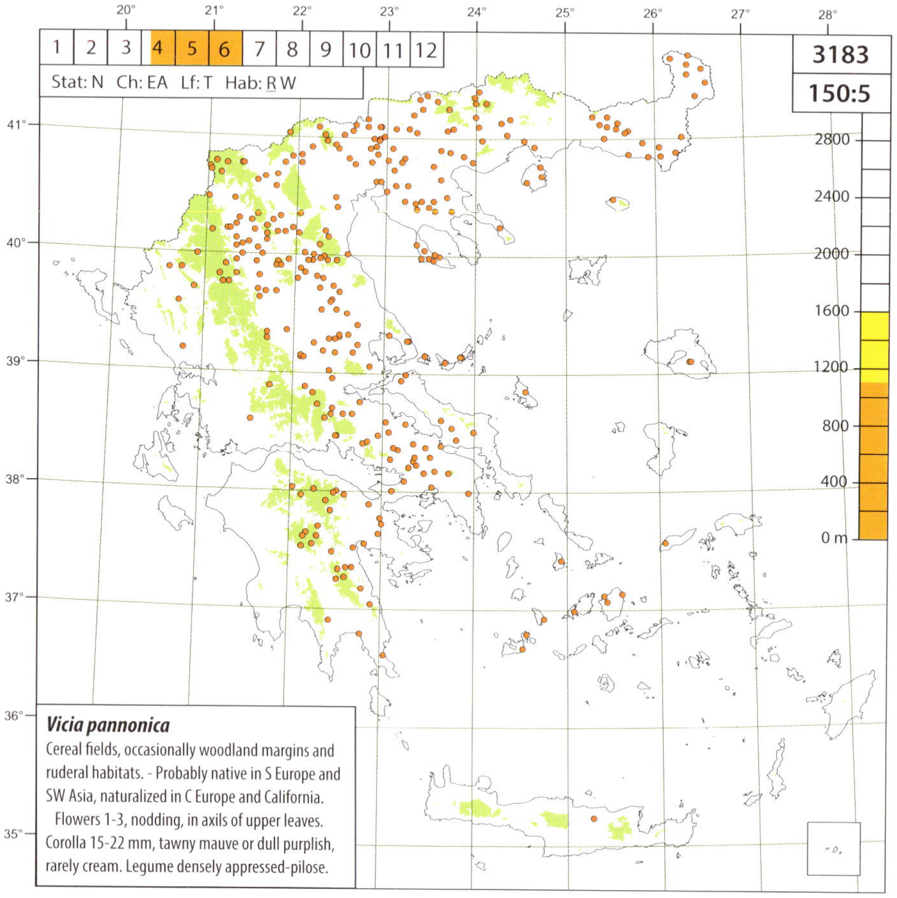

Vicia pannonica
Cereal fields, occasionally woodland margins and ruderal habitats. - Probably native in S Europe and SW Asia, naturalized in C Europe and California.
Flowers 1-3, nodding, in axils of upper leaves. Corolla 15-22 mm, tawny mauve or dull purplish, rarely cream. Legume densely appressed-pilose.

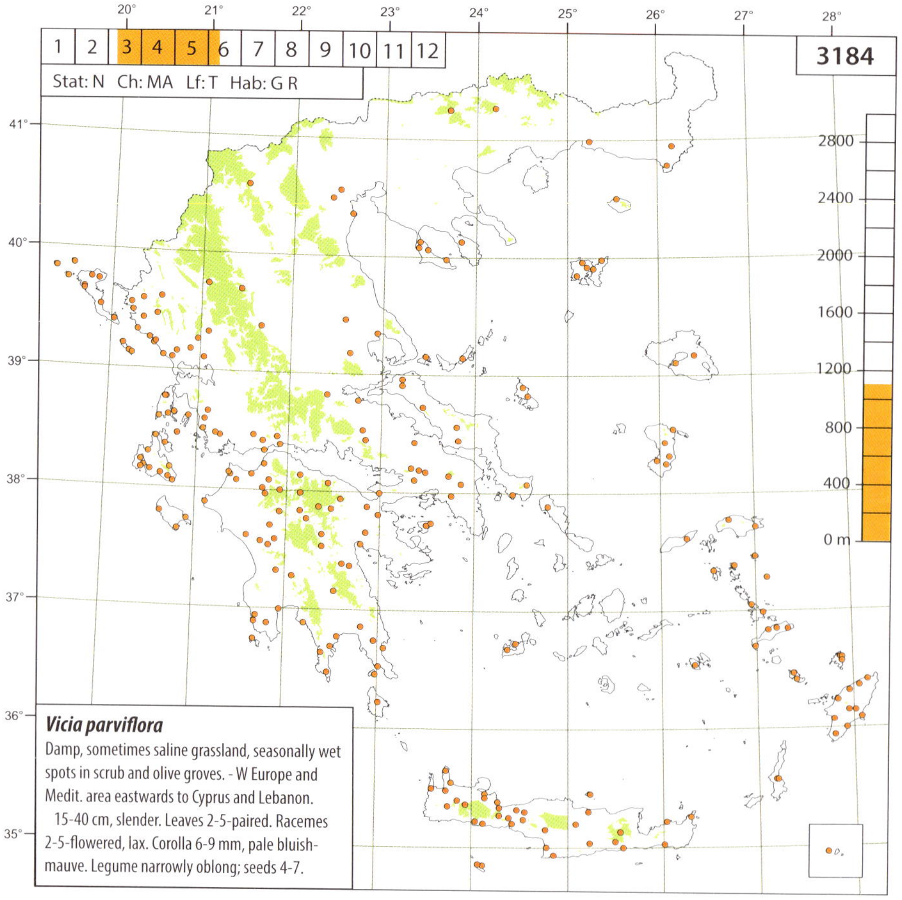

Vicia parviflora
Damp, sometimes saline grassland, seasonally wet spots in scrub and olive groves. - W Europe and Medit. area eastwards to Cyprus and Lebanon.
15-40 cm, slender. Leaves 2-5-paired. Racemes 2-5-flowered, lax. Corolla 6-9 mm, pale bluish-mauve. Legume narrowly oblong; seeds 4-7.

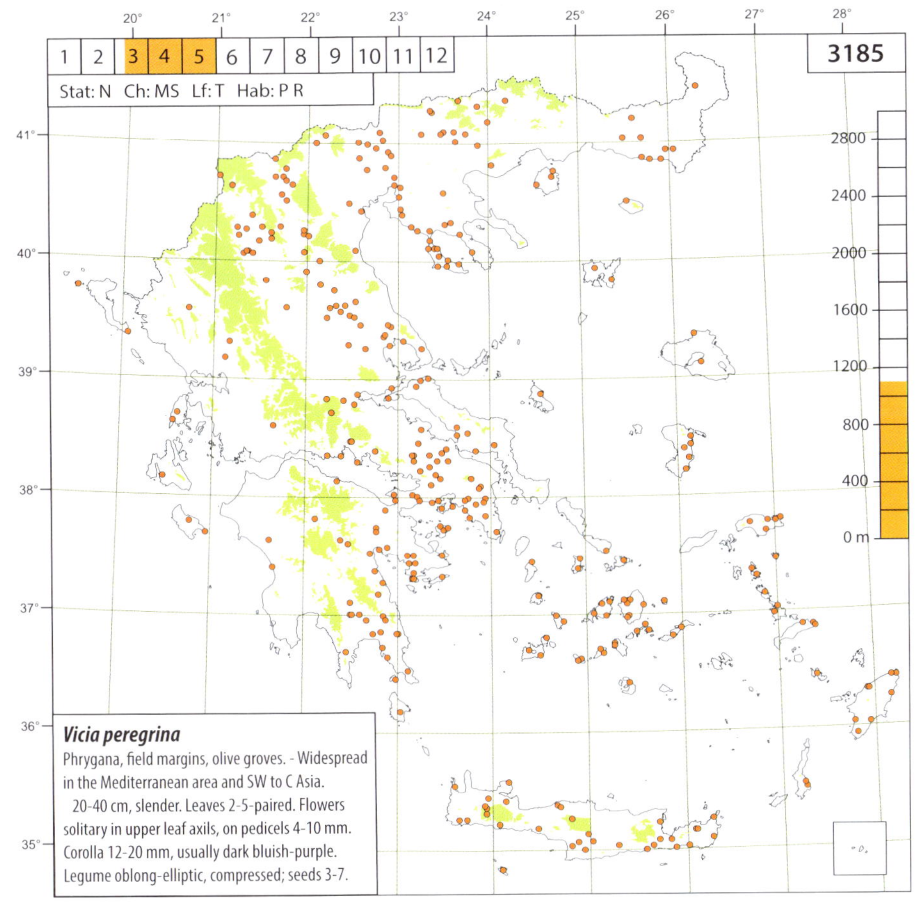

Vicia peregrina
Phrygana, field margins, olive groves. - Widespread in the Mediterranean area and SW to C Asia.
20-40 cm, slender. Leaves 2-5-paired. Flowers solitary in upper leaf axils, on pedicels 4-10 mm. Corolla 12-20 mm, usually dark bluish-purple. Legume oblong-elliptic, compressed; seeds 3-7.

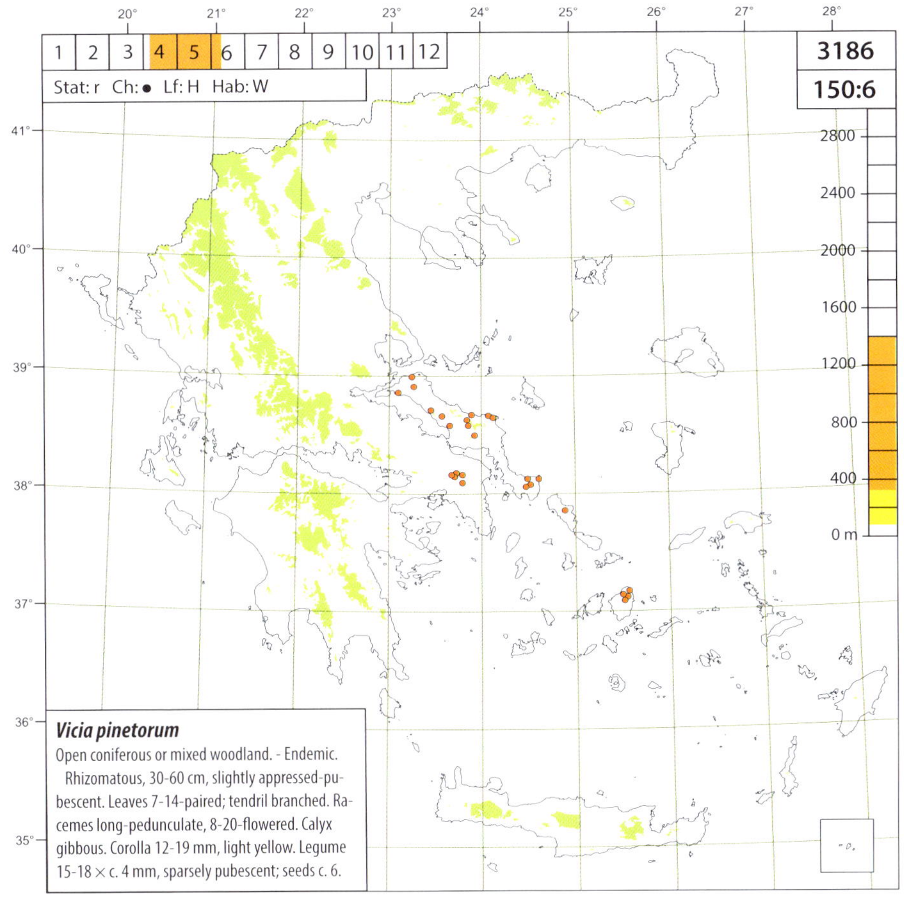

Vicia pinetorum
Open coniferous or mixed woodland. - Endemic.
Rhizomatous, 30-60 cm, slightly appressed-pubescent. Leaves 7-14-paired; tendril branched. Racemes long-pedunculate, 8-20-flowered. Calyx gibbous. Corolla 12-19 mm, light yellow. Legume 15-18 × c. 4 mm, sparsely pubescent; seeds c. 6.

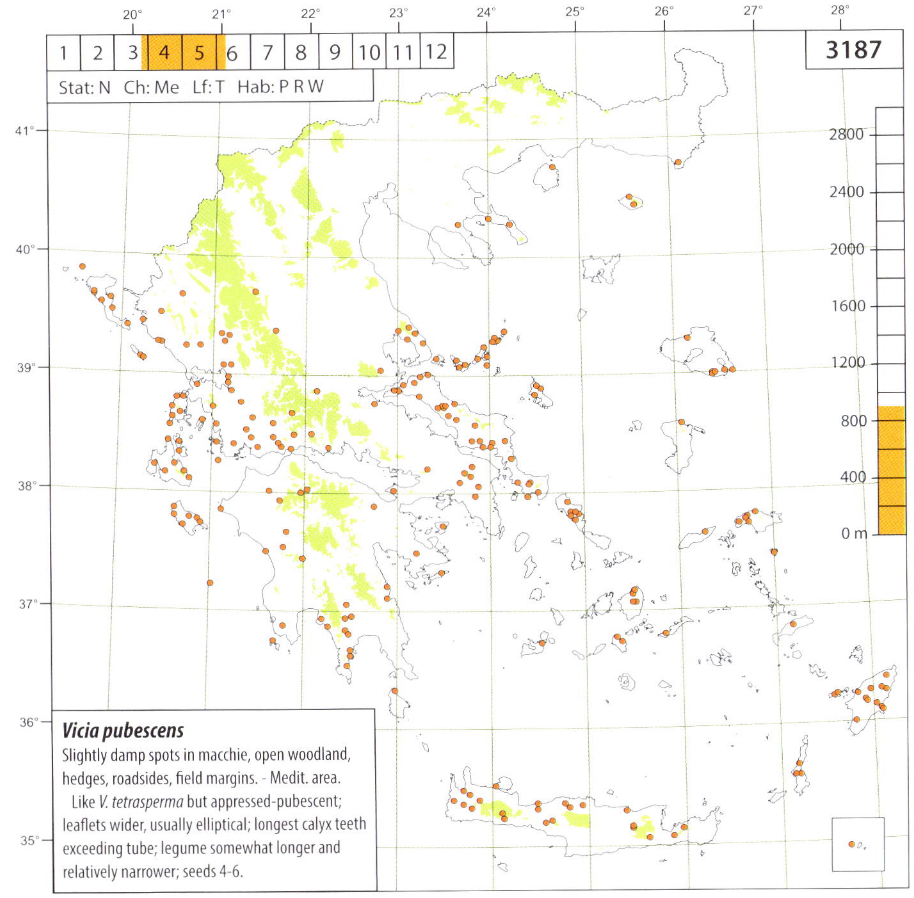

Vicia pubescens
Slightly damp spots in macchie, open woodland, hedges, roadsides, field margins. - Medit. area.
Like *V. tetrasperma* but appressed-pubescent; leaflets wider, usually elliptical; longest calyx teeth exceeding tube; legume somewhat longer and relatively narrower; seeds 4-6.

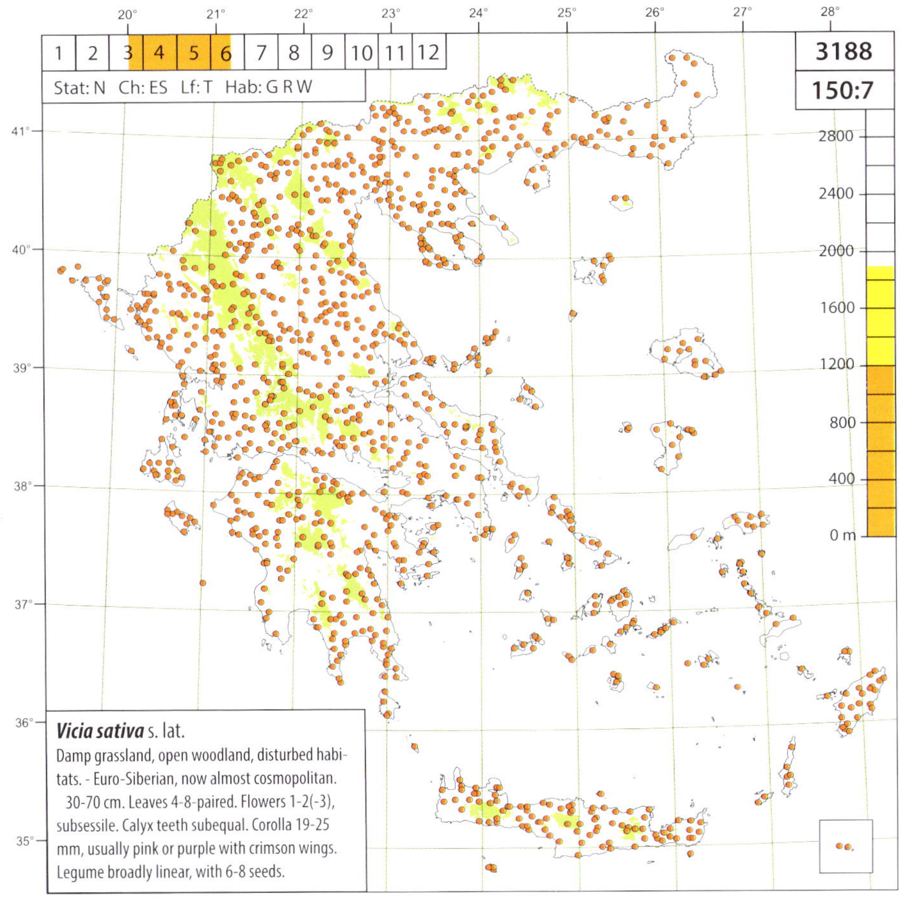

Vicia sativa s. lat.
Damp grassland, open woodland, disturbed habitats. - Euro-Siberian, now almost cosmopolitan.
30-70 cm. Leaves 4-8-paired. Flowers 1-2(-3), subsessile. Calyx teeth subequal. Corolla 19-25 mm, usually pink or purple with crimson wings. Legume broadly linear, with 6-8 seeds.

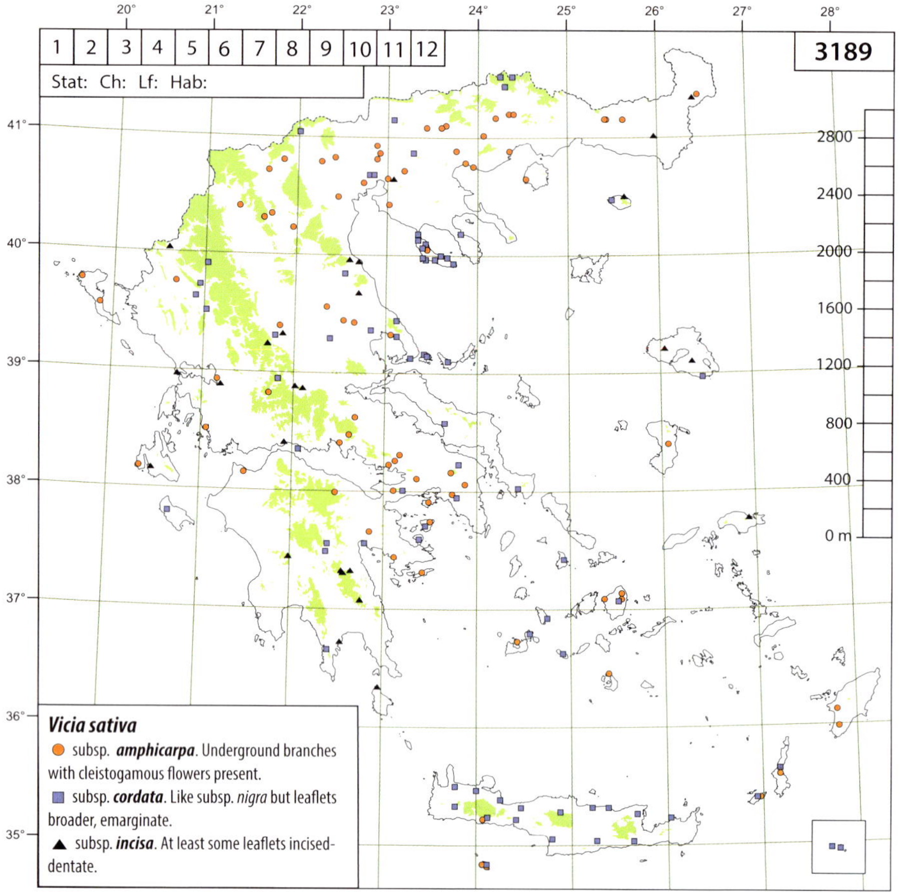

Vicia sativa

● subsp. ***amphicarpa***. Underground branches with cleistogamous flowers present.
■ subsp. ***cordata***. Like subsp. *nigra* but leaflets broader, emarginate.
▲ subsp. ***incisa***. At least some leaflets incised-dentate.

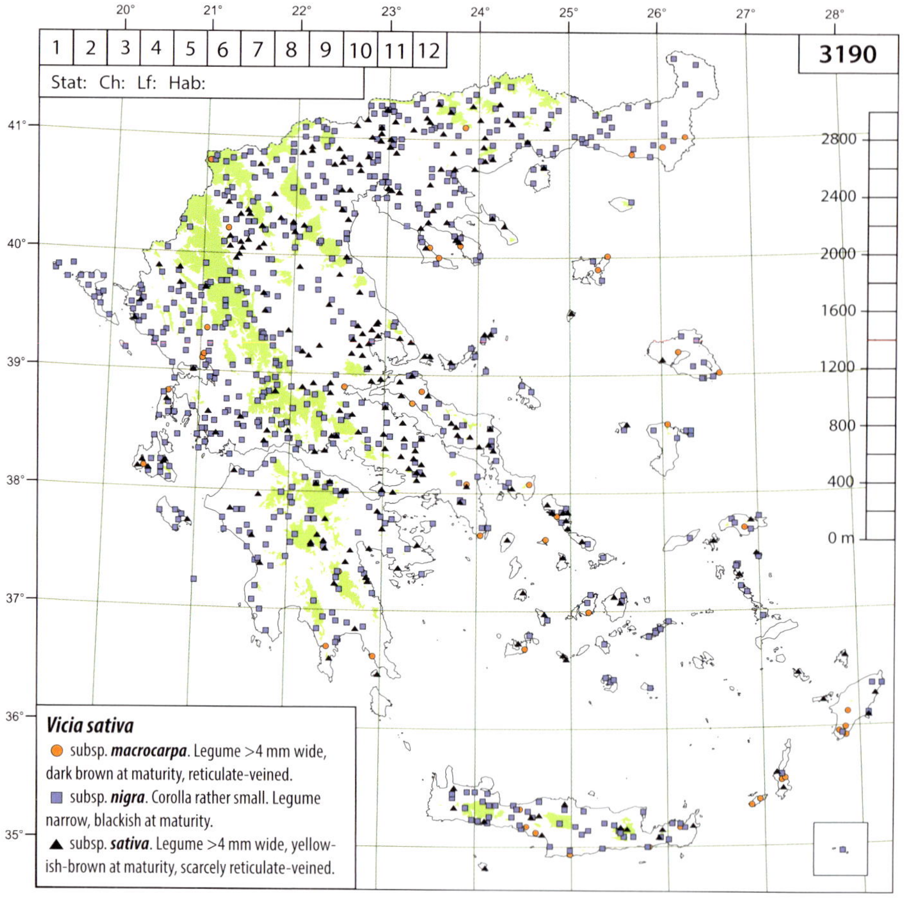

Vicia sativa

● subsp. ***macrocarpa***. Legume >4 mm wide, dark brown at maturity, reticulate-veined.
■ subsp. ***nigra***. Corolla rather small. Legume narrow, blackish at maturity.
▲ subsp. ***sativa***. Legume >4 mm wide, yellowish-brown at maturity, scarcely reticulate-veined.

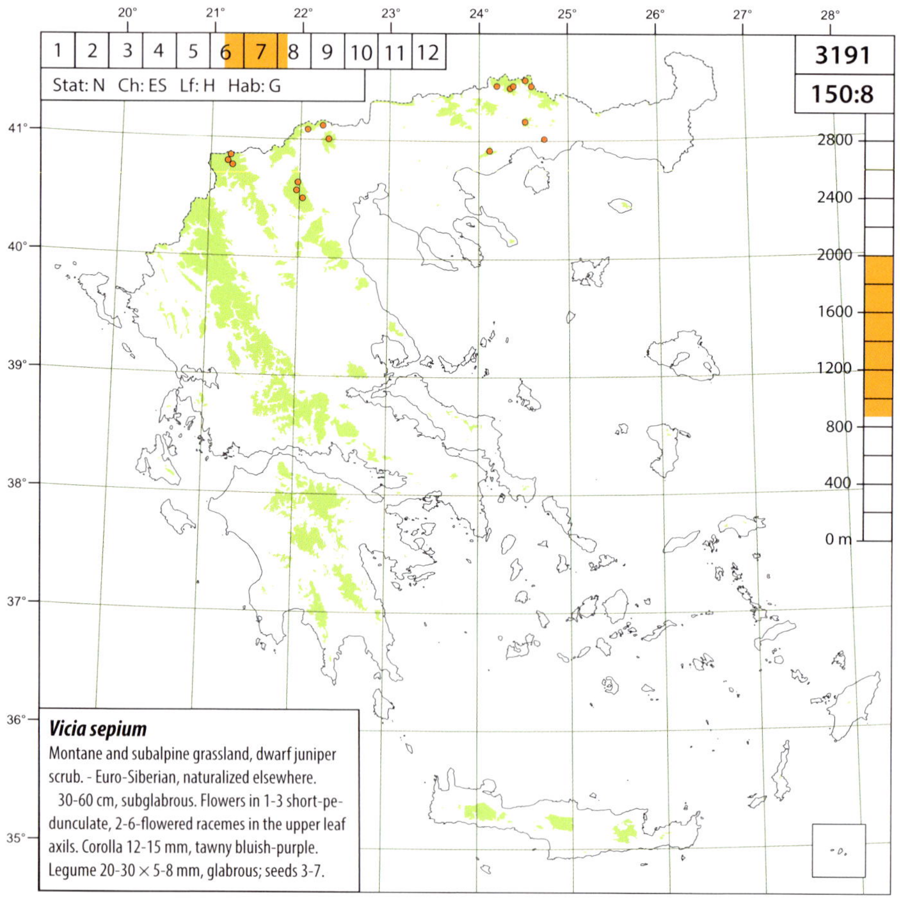

Vicia sepium

Montane and subalpine grassland, dwarf juniper scrub. - Euro-Siberian, naturalized elsewhere.

30-60 cm, subglabrous. Flowers in 1-3 short-pedunculate, 2-6-flowered racemes in the upper leaf axils. Corolla 12-15 mm, tawny bluish-purple. Legume 20-30 × 5-8 mm, glabrous; seeds 3-7.

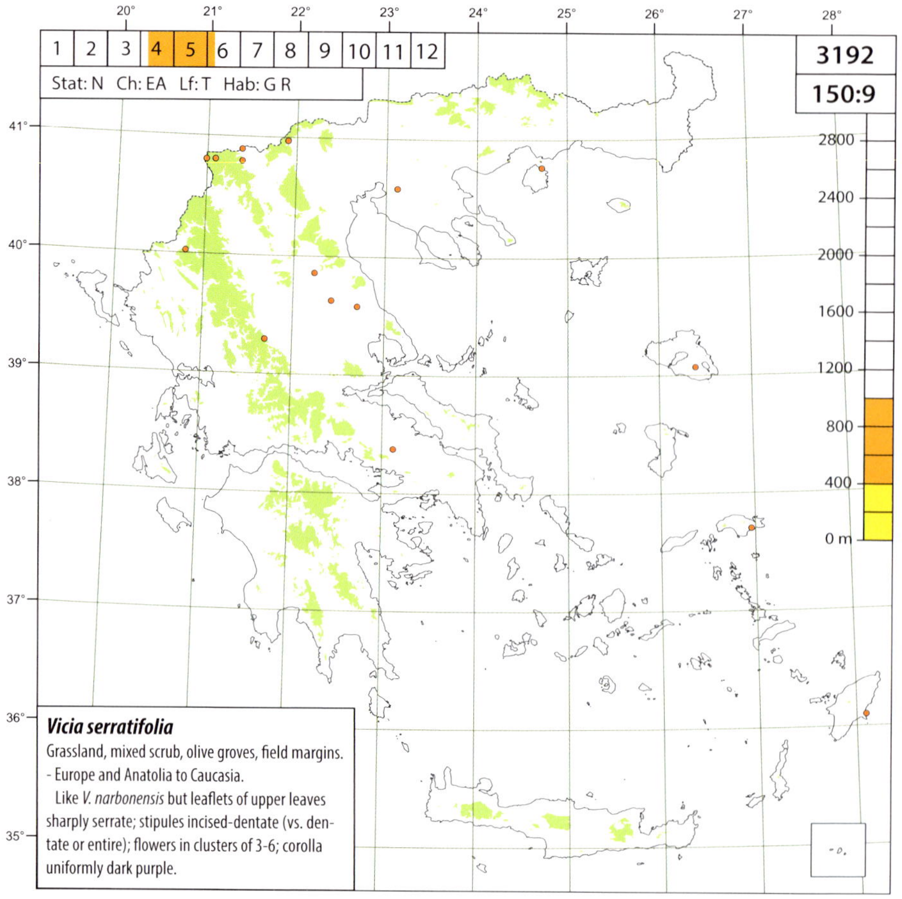

Vicia serratifolia

Grassland, mixed scrub, olive groves, field margins. - Europe and Anatolia to Caucasia.

Like *V. narbonensis* but leaflets of upper leaves sharply serrate; stipules incised-dentate (vs. dentate or entire); flowers in clusters of 3-6; corolla uniformly dark purple.

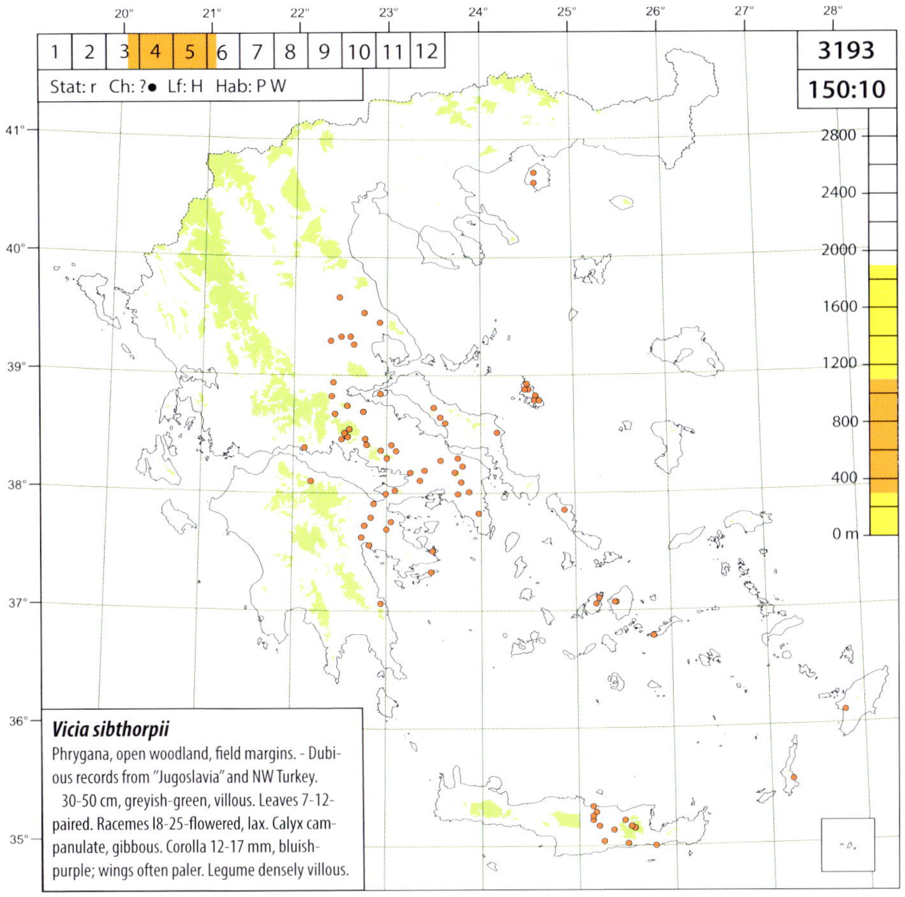

Vicia sibthorpii

Phrygana, open woodland, field margins. - Dubious records from "Jugoslavia" and NW Turkey.

30-50 cm, greyish-green, villous. Leaves 7-12-paired. Racemes 18-25-flowered, lax. Calyx campanulate, gibbous. Corolla 12-17 mm, bluish-purple; wings often paler. Legume densely villous.

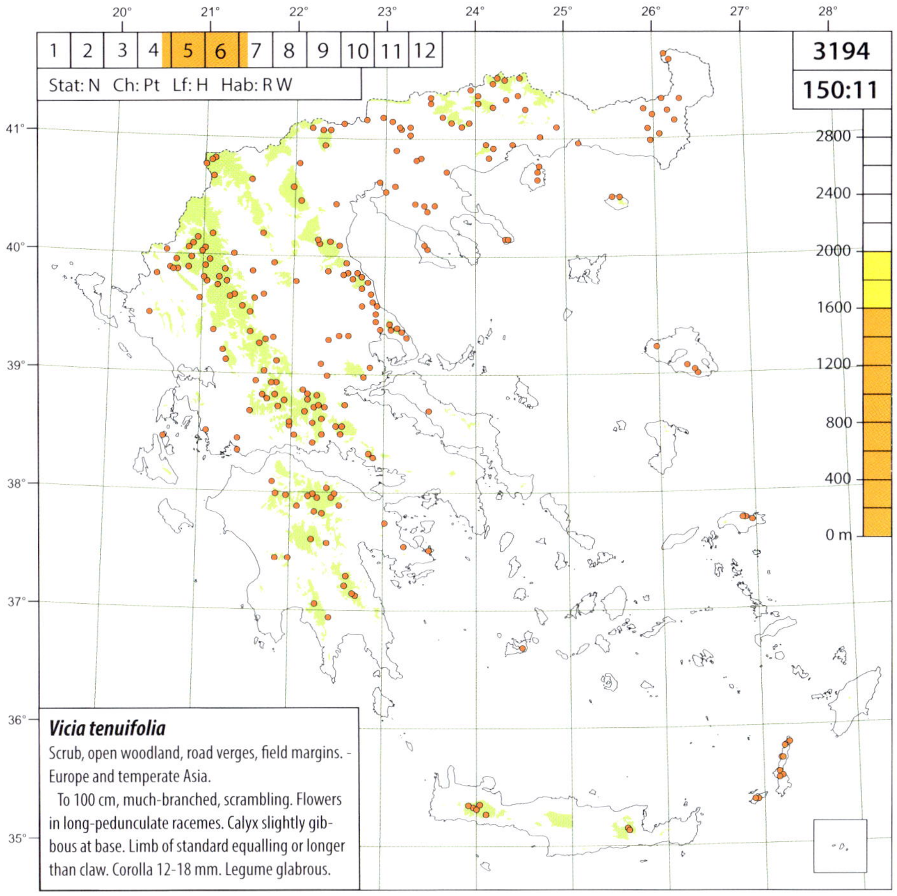

Vicia tenuifolia

Scrub, open woodland, road verges, field margins. - Europe and temperate Asia.

To 100 cm, much-branched, scrambling. Flowers in long-pedunculate racemes. Calyx slightly gibbous at base. Limb of standard equalling or longer than claw. Corolla 12-18 mm. Legume glabrous.

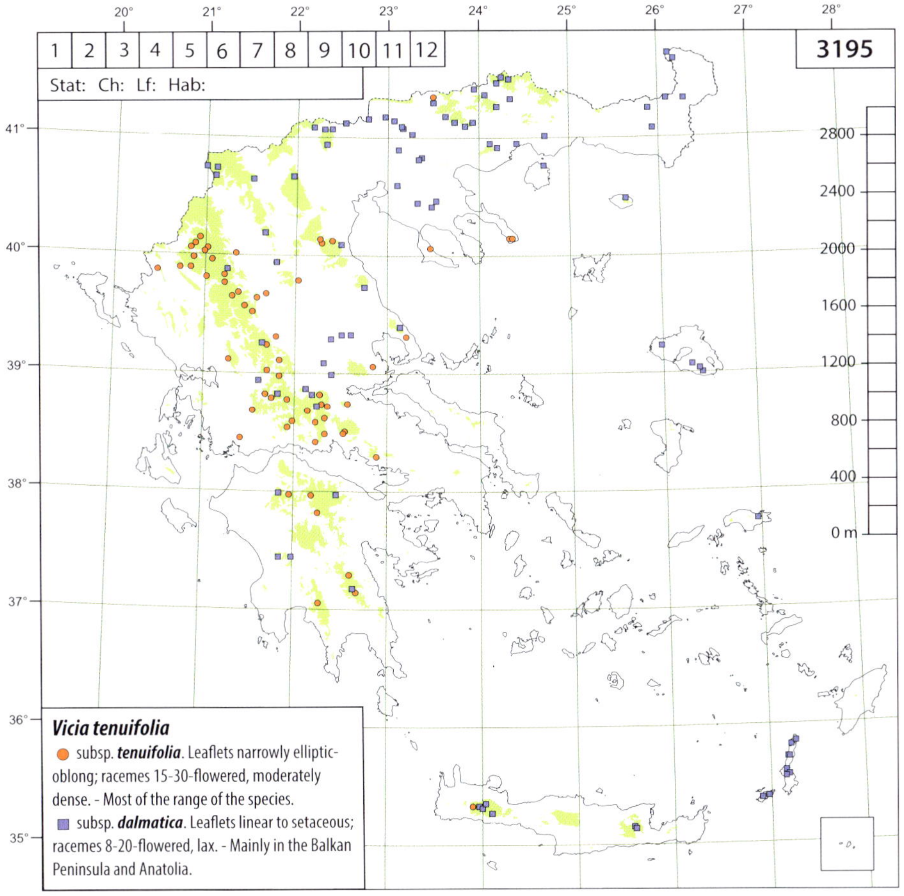

Vicia tenuifolia

● subsp. ***tenuifolia***. Leaflets narrowly elliptic-oblong; racemes 15-30-flowered, moderately dense. - Most of the range of the species.

■ subsp. ***dalmatica***. Leaflets linear to setaceous; racemes 8-20-flowered, lax. - Mainly in the Balkan Peninsula and Anatolia.

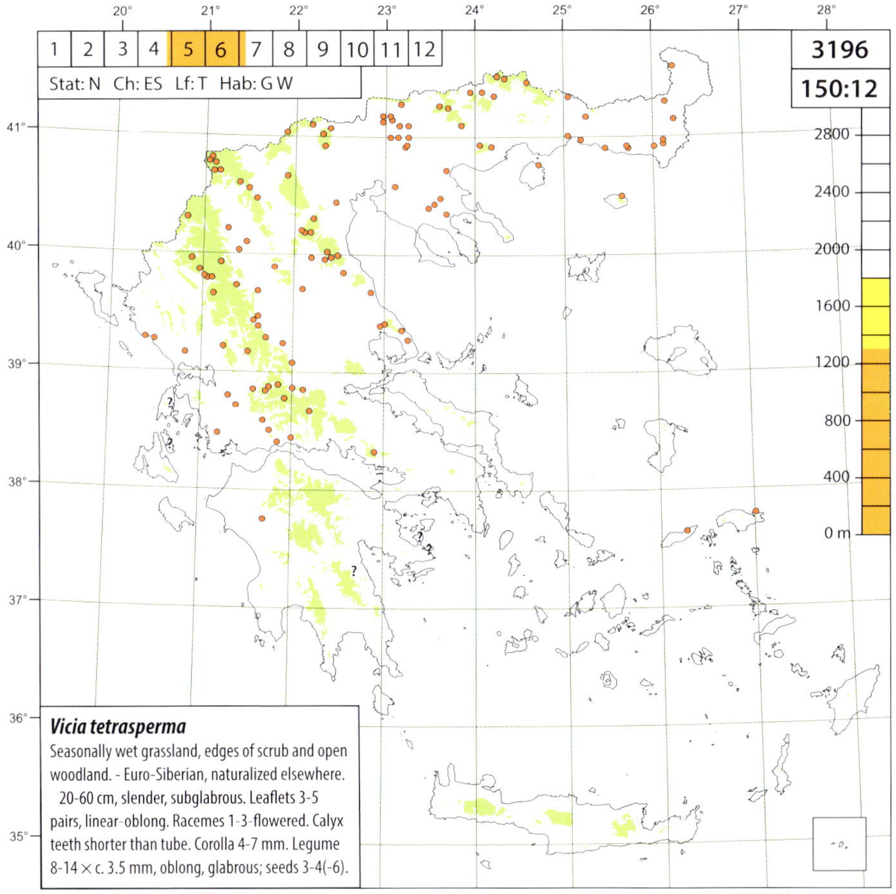

Vicia tetrasperma

Seasonally wet grassland, edges of scrub and open woodland. - Euro-Siberian, naturalized elsewhere.

20-60 cm, slender, subglabrous. Leaflets 3-5 pairs, linear-oblong. Racemes 1-3-flowered. Calyx teeth shorter than tube. Corolla 4-7 mm. Legume 8-14 × c. 3.5 mm, oblong, glabrous; seeds 3-4(-6).

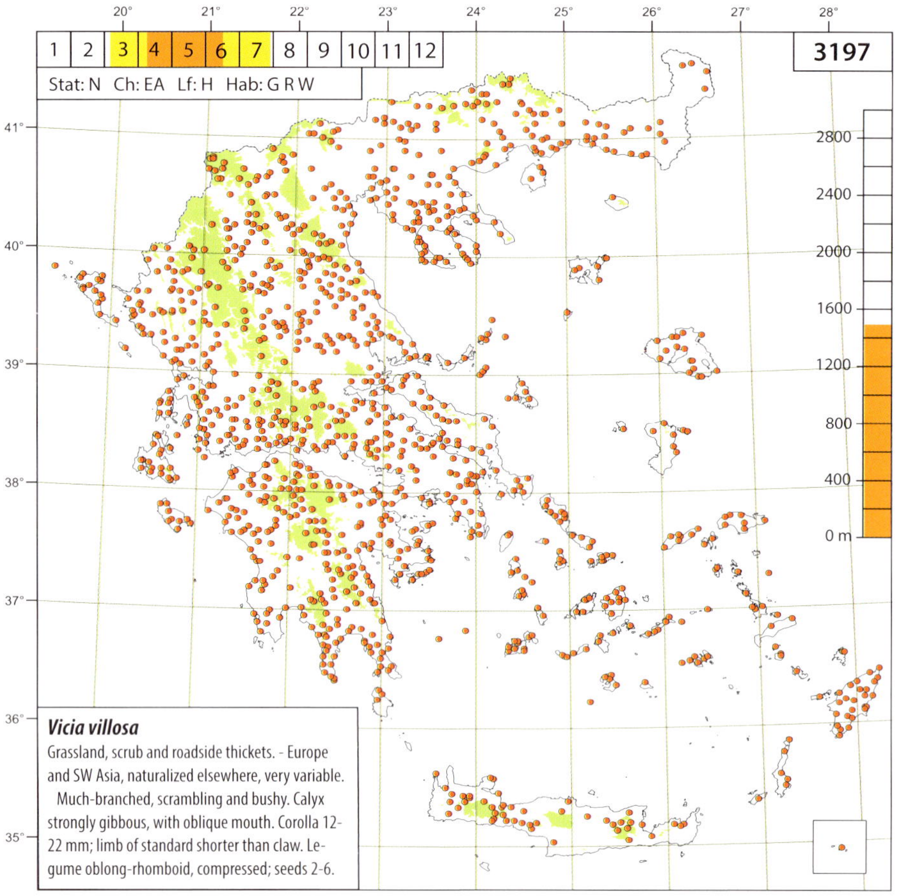

Vicia villosa
Grassland, scrub and roadside thickets. - Europe and SW Asia, naturalized elsewhere, very variable.
Much-branched, scrambling and bushy. Calyx strongly gibbous, with oblique mouth. Corolla 12-22 mm; limb of standard shorter than claw. Legume oblong-rhomboid, compressed; seeds 2-6.

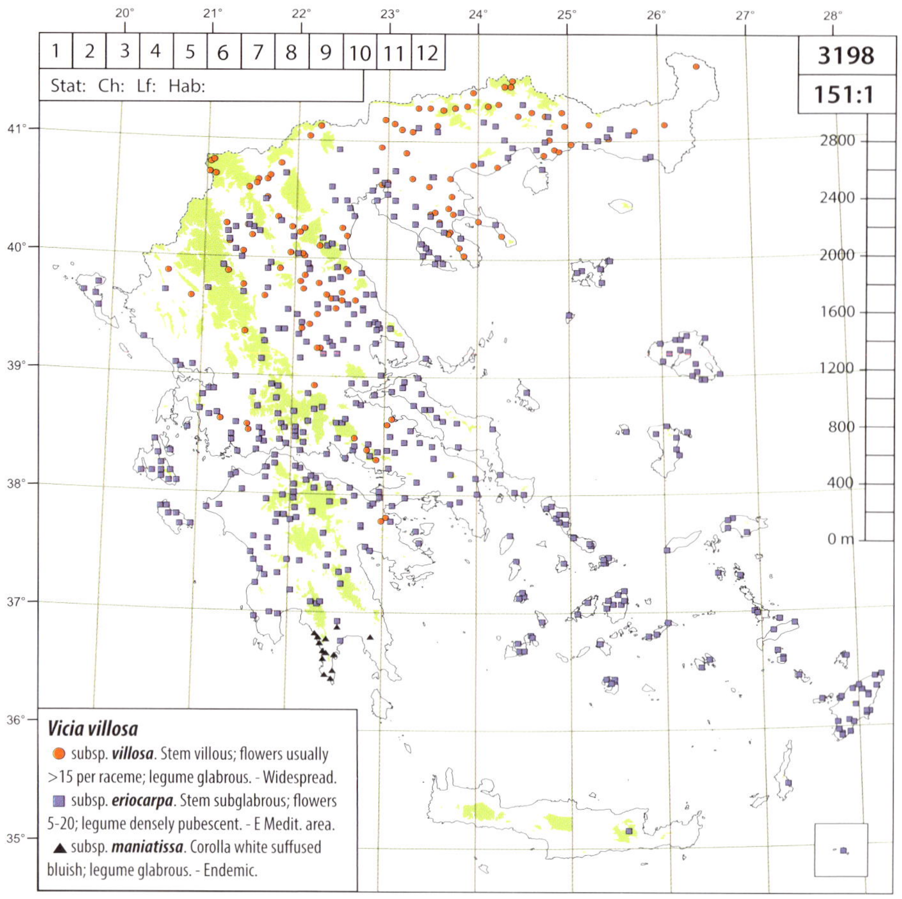

Vicia villosa
● subsp. ***villosa***. Stem villous; flowers usually >15 per raceme; legume glabrous. - Widespread.
■ subsp. ***eriocarpa***. Stem subglabrous; flowers 5-20; legume densely pubescent. - E Medit. area.
▲ subsp. ***maniatissa***. Corolla white suffused bluish; legume glabrous. - Endemic.

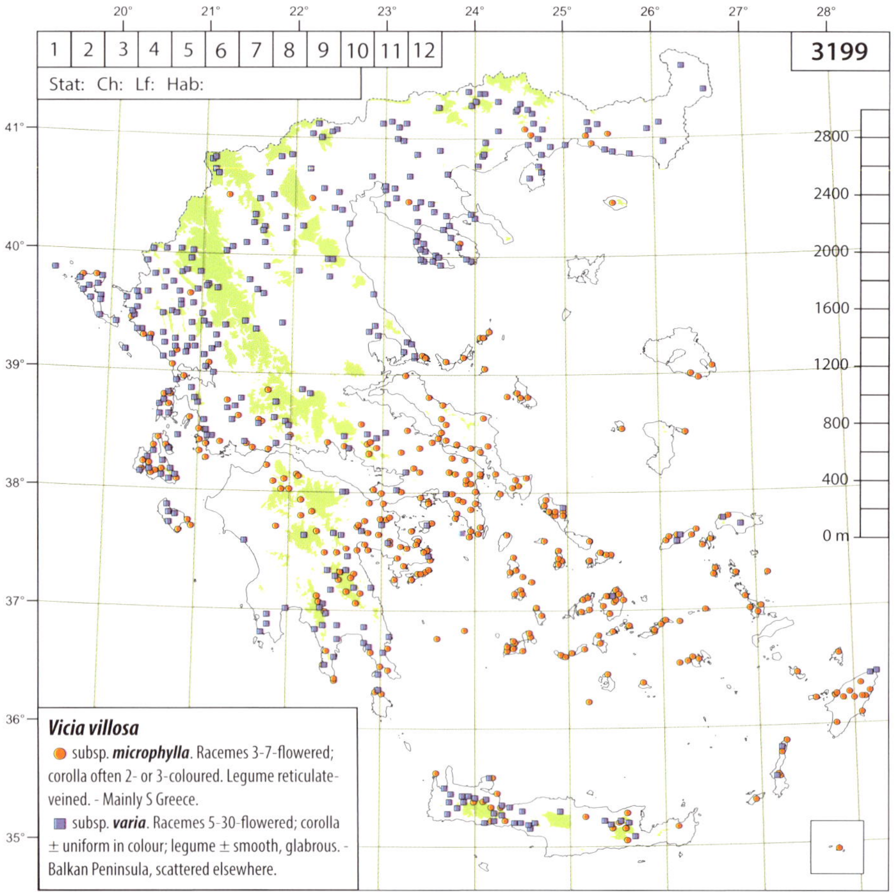

Vicia villosa
● subsp. ***microphylla***. Racemes 3-7-flowered; corolla often 2- or 3-coloured. Legume reticulate-veined. - Mainly S Greece.
■ subsp. ***varia***. Racemes 5-30-flowered; corolla ± uniform in colour; legume ± smooth, glabrous. - Balkan Peninsula, scattered elsewhere.

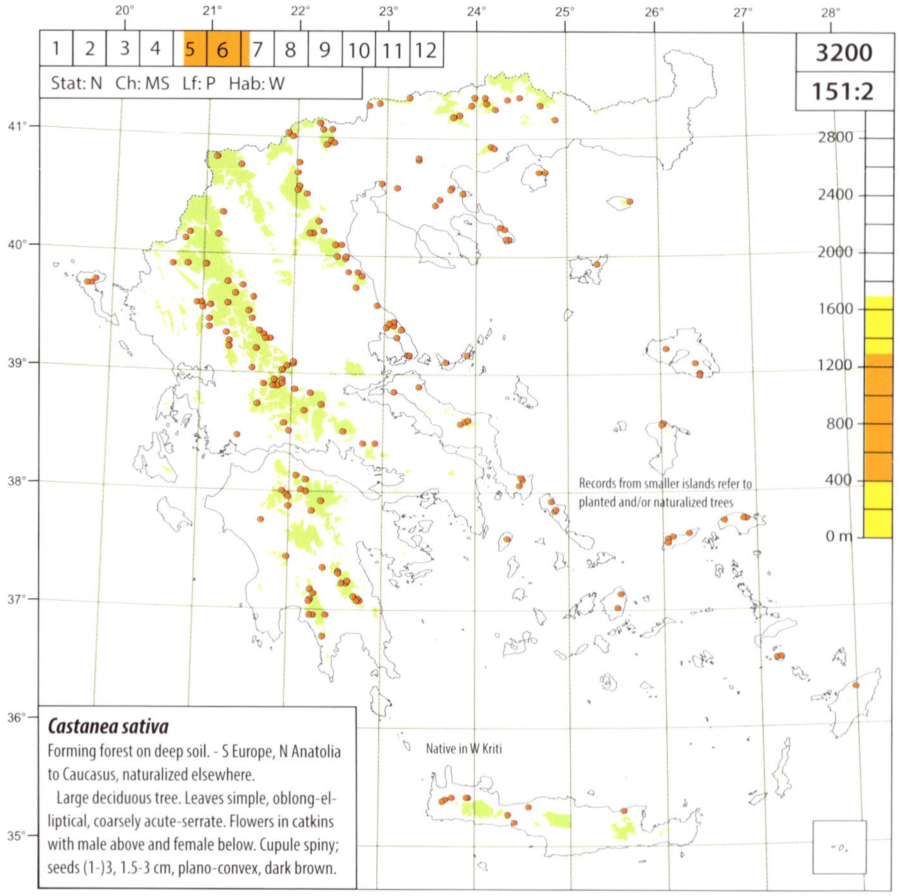

Castanea sativa
Forming forest on deep soil. - S Europe, N Anatolia to Caucasus, naturalized elsewhere.
Large deciduous tree. Leaves simple, oblong-elliptical, coarsely acute-serrate. Flowers in catkins with male above and female below. Cupule spiny; seeds (1-)3, 1.5-3 cm, plano-convex, dark brown.

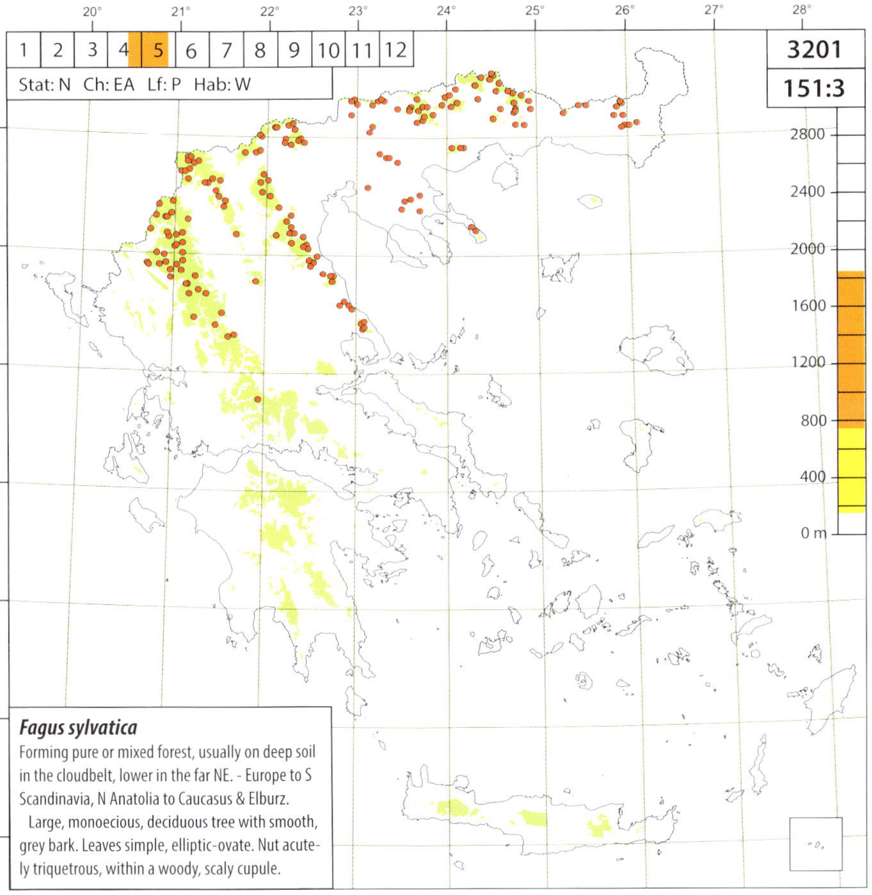

Fagus sylvatica
Forming pure or mixed forest, usually on deep soil in the cloudbelt, lower in the far NE. - Europe to S Scandinavia, N Anatolia to Caucasus & Elburz.
Large, monoecious, deciduous tree with smooth, grey bark. Leaves simple, elliptic-ovate. Nut acutely triquetrous, within a woody, scaly cupule.

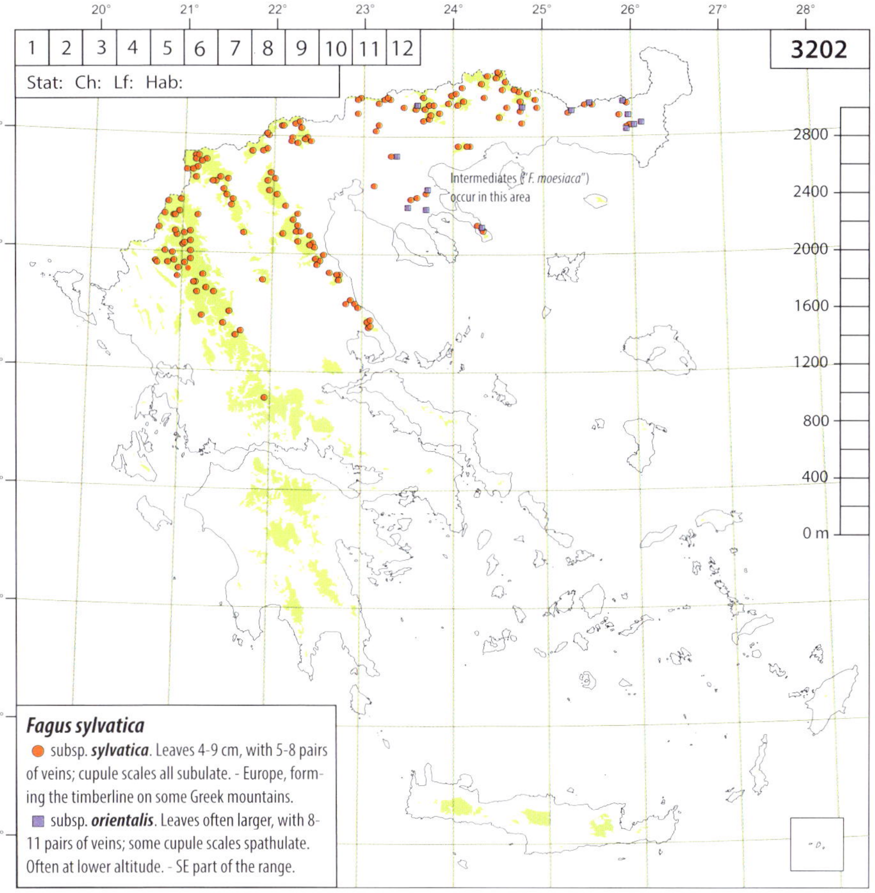

Fagus sylvatica
● subsp. ***sylvatica***. Leaves 4-9 cm, with 5-8 pairs of veins; cupule scales all subulate. - Europe, forming the timberline on some Greek mountains.
■ subsp. ***orientalis***. Leaves often larger, with 8-11 pairs of veins; some cupule scales spathulate. Often at lower altitude. - SE part of the range.

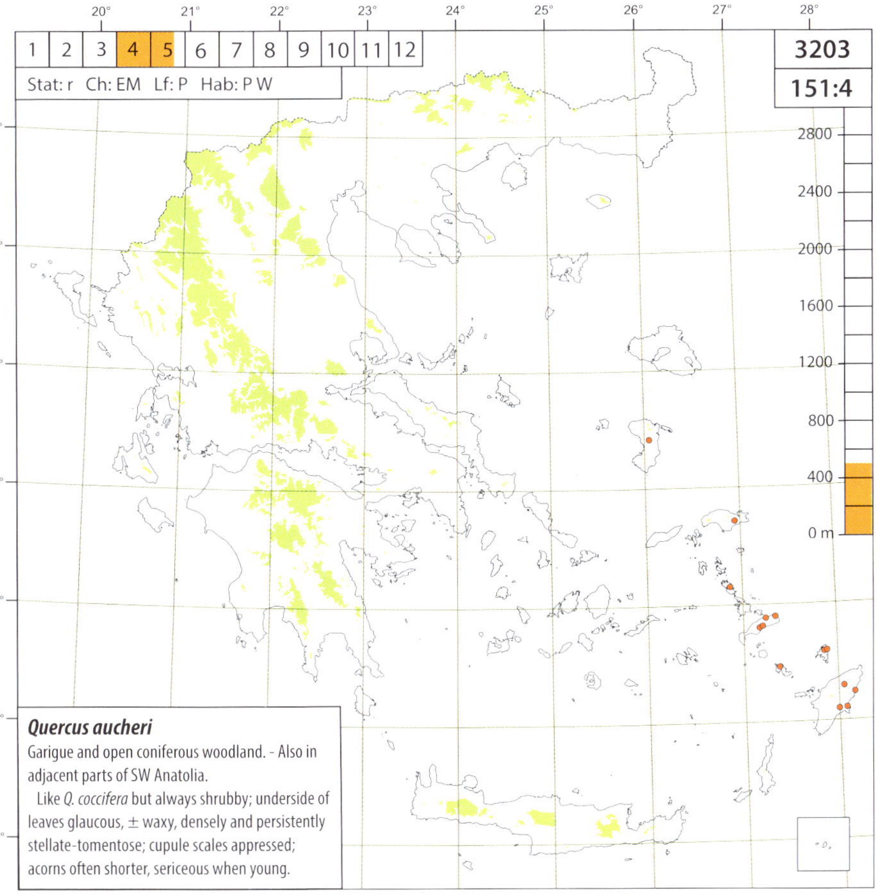

Quercus aucheri
Garigue and open coniferous woodland. - Also in adjacent parts of SW Anatolia.
Like *Q. coccifera* but always shrubby; underside of leaves glaucous, ± waxy, densely and persistently stellate-tomentose; cupule scales appressed; acorns often shorter, sericeous when young.

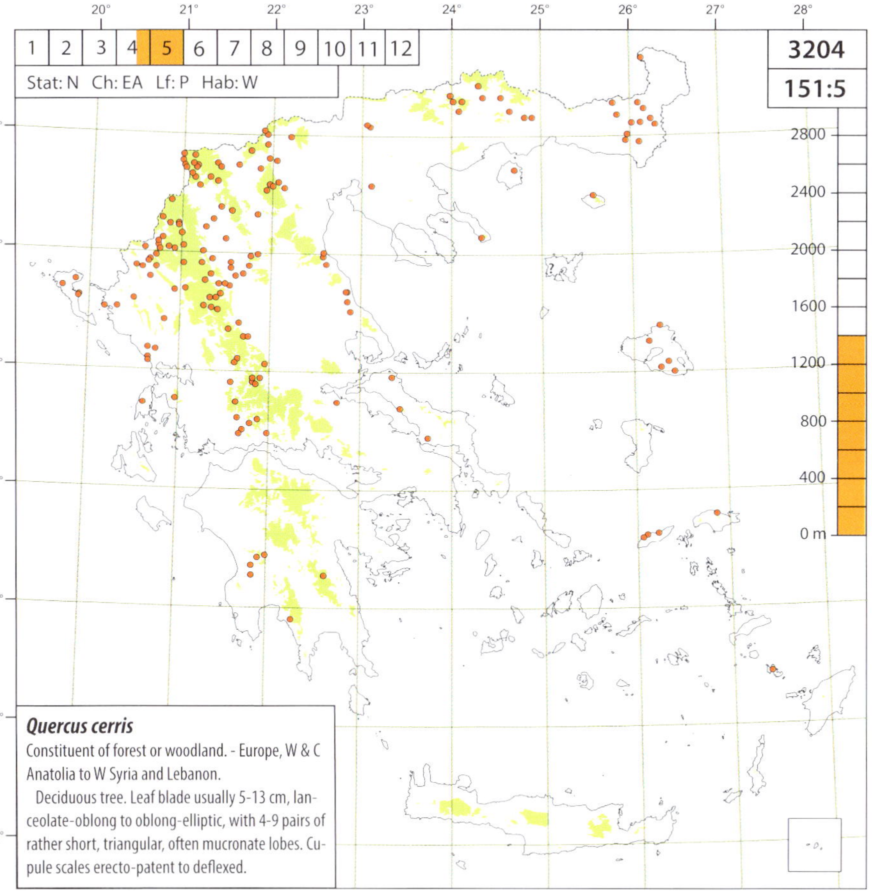

Quercus cerris
Constituent of forest or woodland. - Europe, W & C Anatolia to W Syria and Lebanon.
Deciduous tree. Leaf blade usually 5-13 cm, lanceolate-oblong to oblong-elliptic, with 4-9 pairs of rather short, triangular, often mucronate lobes. Cupule scales erecto-patent to deflexed.

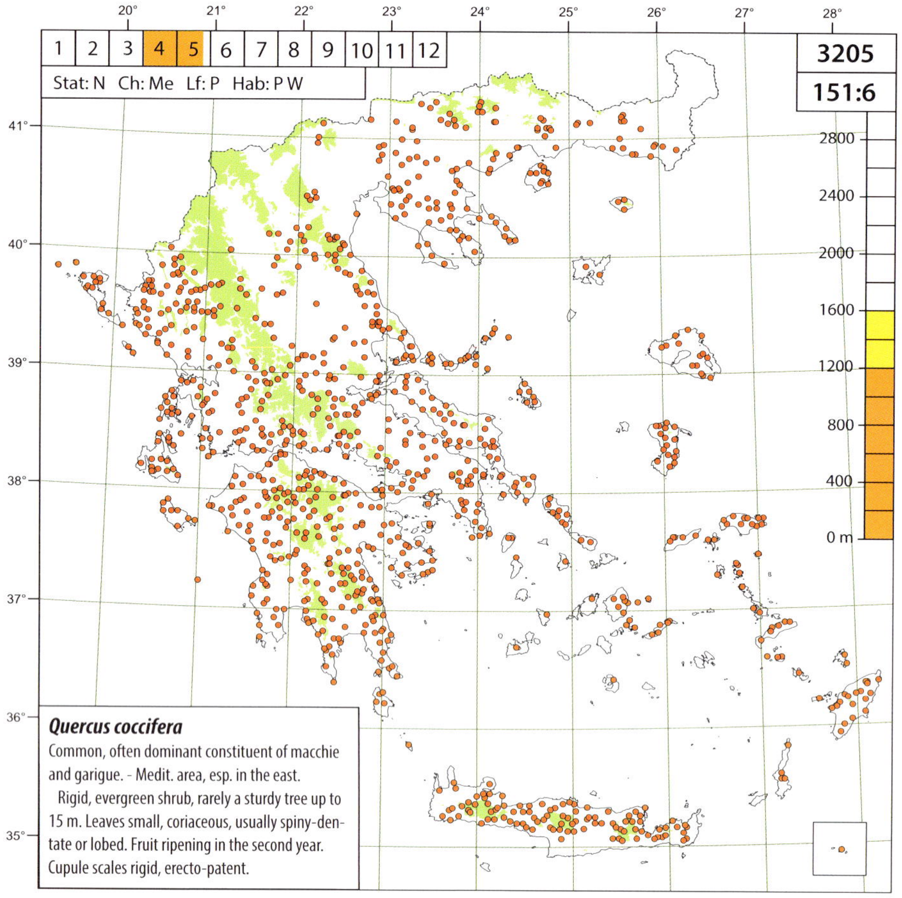

Quercus coccifera
Common, often dominant constituent of macchie and garigue. - Medit. area, esp. in the east.
Rigid, evergreen shrub, rarely a sturdy tree up to 15 m. Leaves small, coriaceous, usually spiny-dentate or lobed. Fruit ripening in the second year. Cupule scales rigid, erecto-patent.

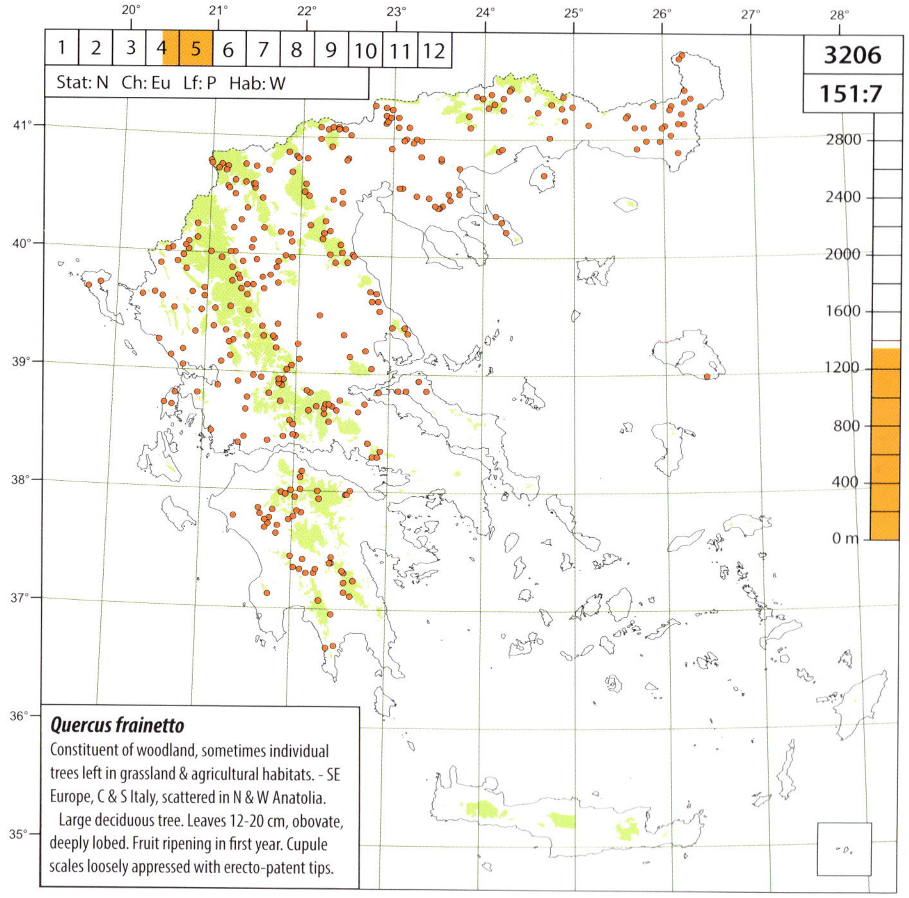

Quercus frainetto
Constituent of woodland, sometimes individual trees left in grassland & agricultural habitats. - SE Europe, C & S Italy, scattered in N & W Anatolia.
Large deciduous tree. Leaves 12-20 cm, obovate, deeply lobed. Fruit ripening in first year. Cupule scales loosely appressed with erecto-patent tips.

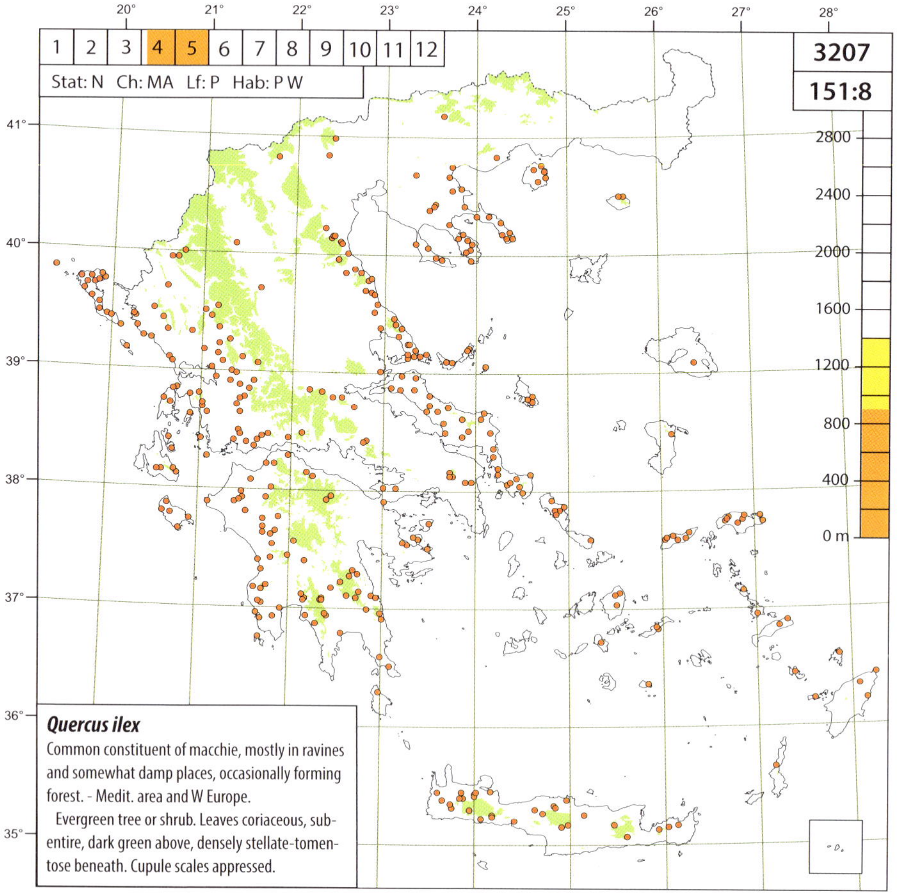

Quercus ilex
Common constituent of macchie, mostly in ravines and somewhat damp places, occasionally forming forest. - Medit. area and W Europe.
Evergreen tree or shrub. Leaves coriaceous, subentire, dark green above, densely stellate-tomentose beneath. Cupule scales appressed.

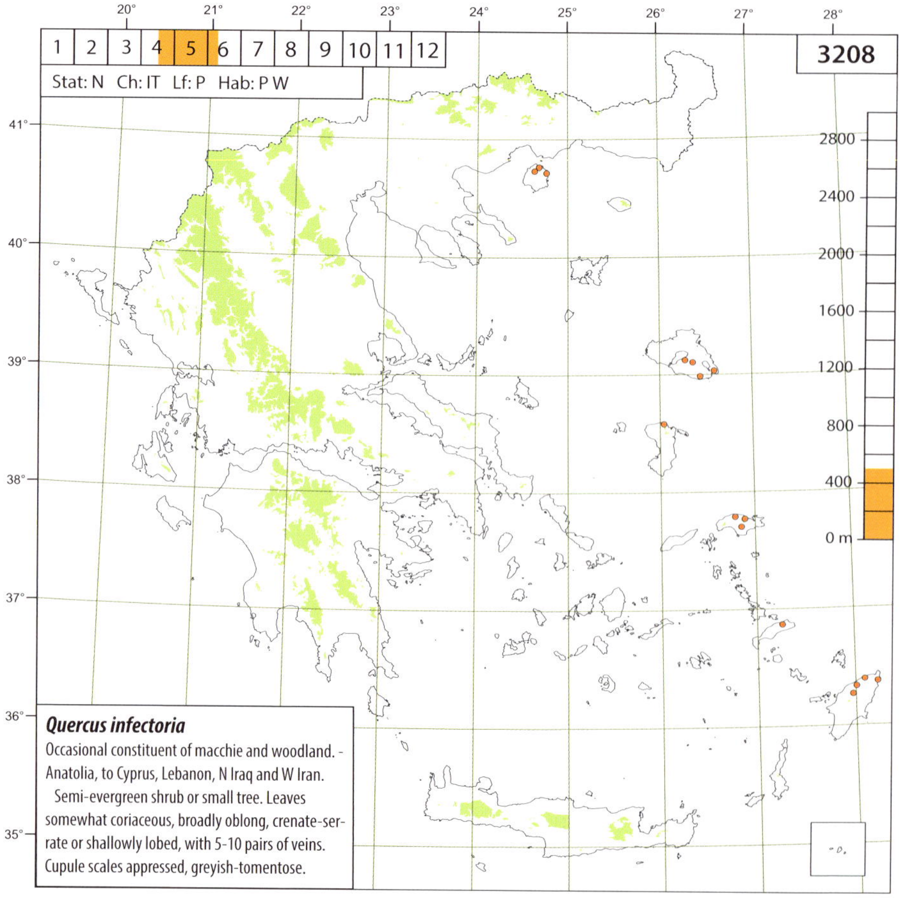

Quercus infectoria
Occasional constituent of macchie and woodland. - Anatolia, to Cyprus, Lebanon, N Iraq and W Iran.
Semi-evergreen shrub or small tree. Leaves somewhat coriaceous, broadly oblong, crenate-serrate or shallowly lobed, with 5-10 pairs of veins. Cupule scales appressed, greyish-tomentose.

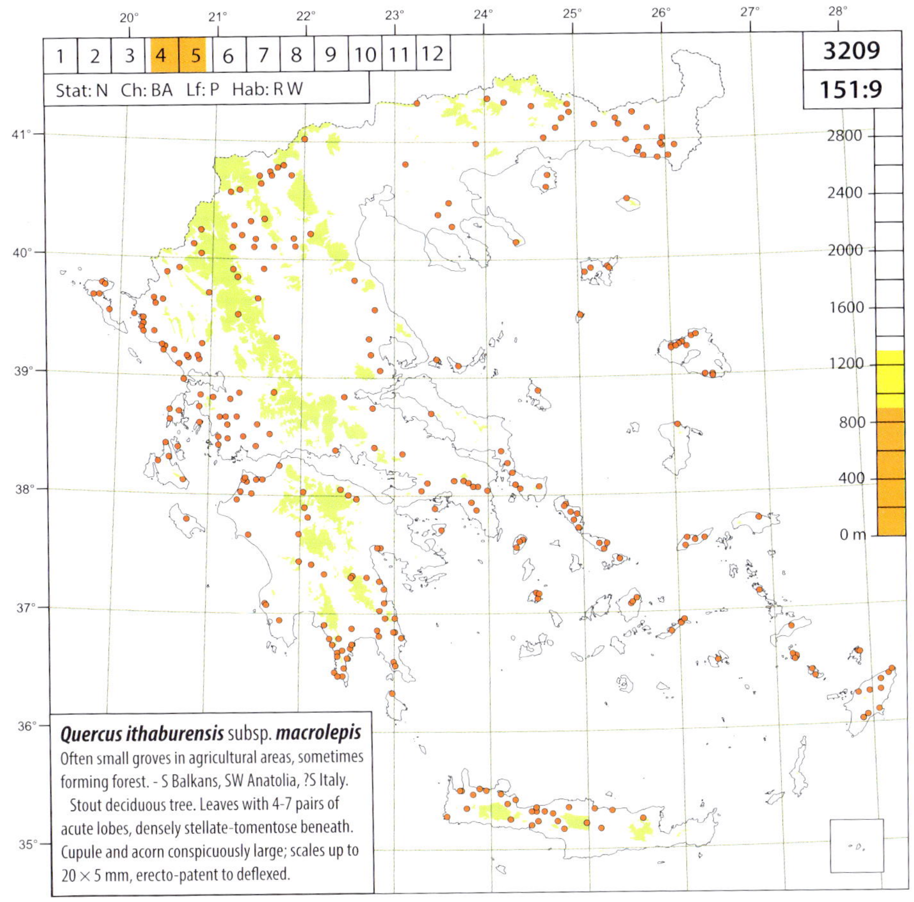

Quercus ithaburensis subsp. **macrolepis**
Often small groves in agricultural areas, sometimes forming forest. - S Balkans, SW Anatolia, ?S Italy.
Stout deciduous tree. Leaves with 4-7 pairs of acute lobes, densely stellate-tomentose beneath. Cupule and acorn conspicuously large; scales up to 20 × 5 mm, erecto-patent to deflexed.

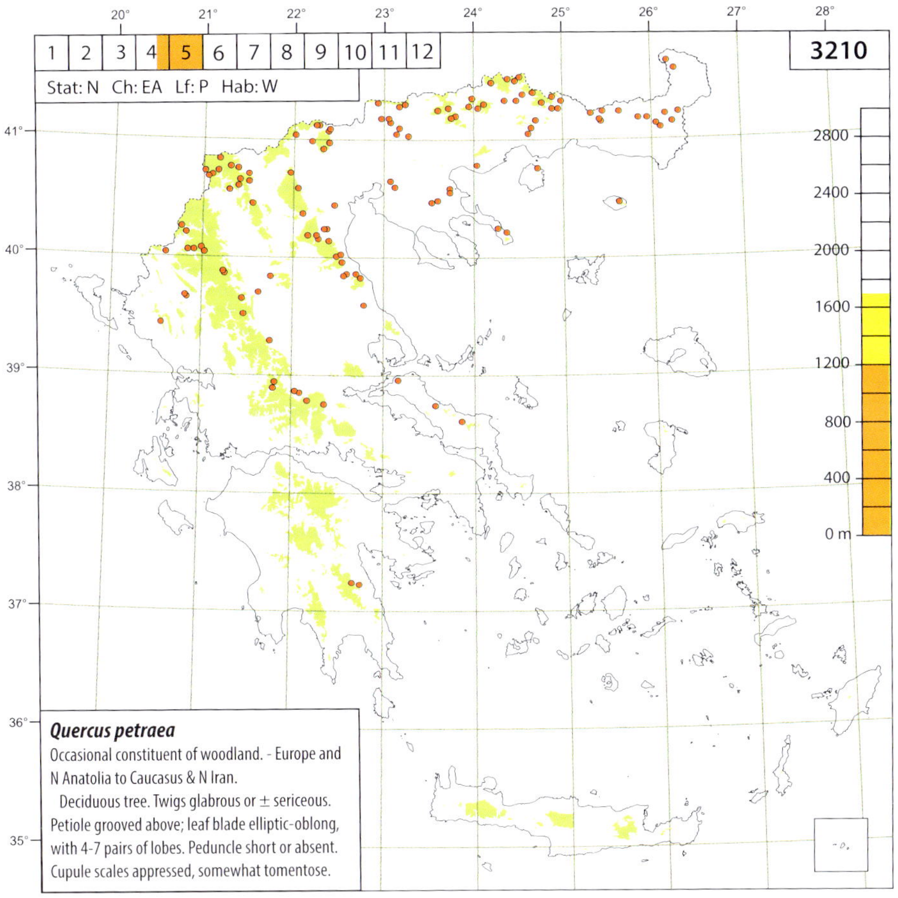

Quercus petraea
Occasional constituent of woodland. - Europe and N Anatolia to Caucasus & N Iran.
Deciduous tree. Twigs glabrous or ± sericeous. Petiole grooved above; leaf blade elliptic-oblong, with 4-7 pairs of lobes. Peduncle short or absent. Cupule scales appressed, somewhat tomentose.

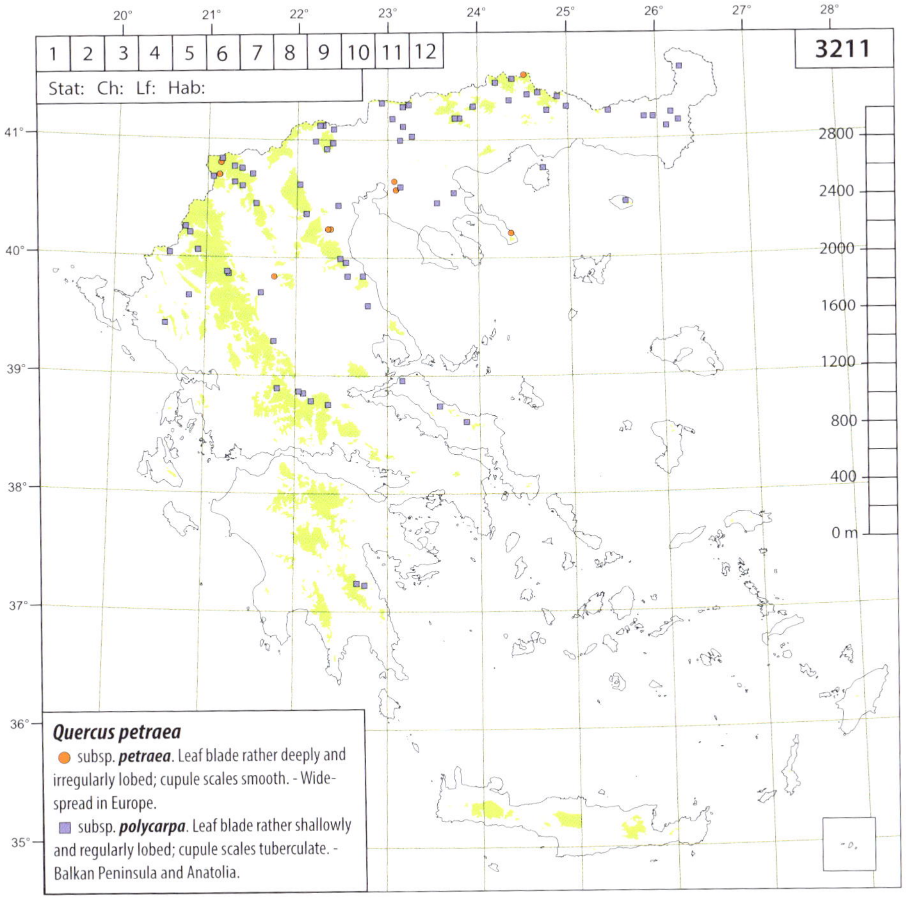

Quercus petraea
● subsp. ***petraea***. Leaf blade rather deeply and irregularly lobed; cupule scales smooth. - Widespread in Europe.
■ subsp. ***polycarpa***. Leaf blade rather shallowly and regularly lobed; cupule scales tuberculate. - Balkan Peninsula and Anatolia.

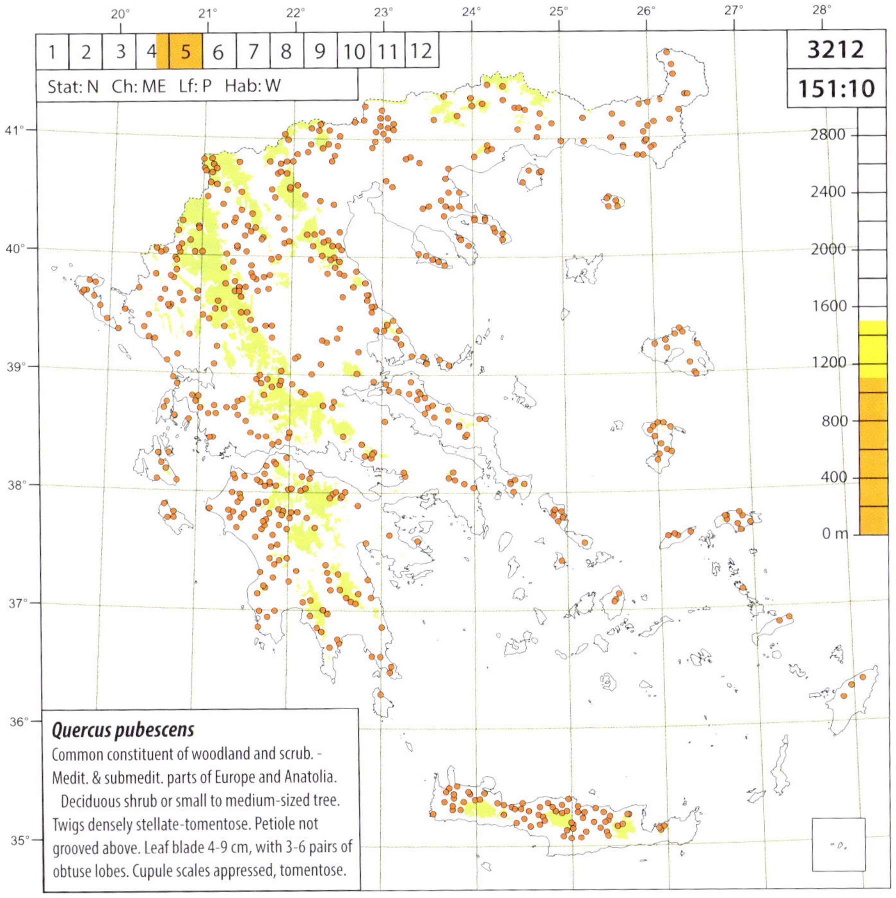

Quercus pubescens
Common constituent of woodland and scrub. - Medit. & submedit. parts of Europe and Anatolia.
Deciduous shrub or small to medium-sized tree. Twigs densely stellate-tomentose. Petiole not grooved above. Leaf blade 4-9 cm, with 3-6 pairs of obtuse lobes. Cupule scales appressed, tomentose.

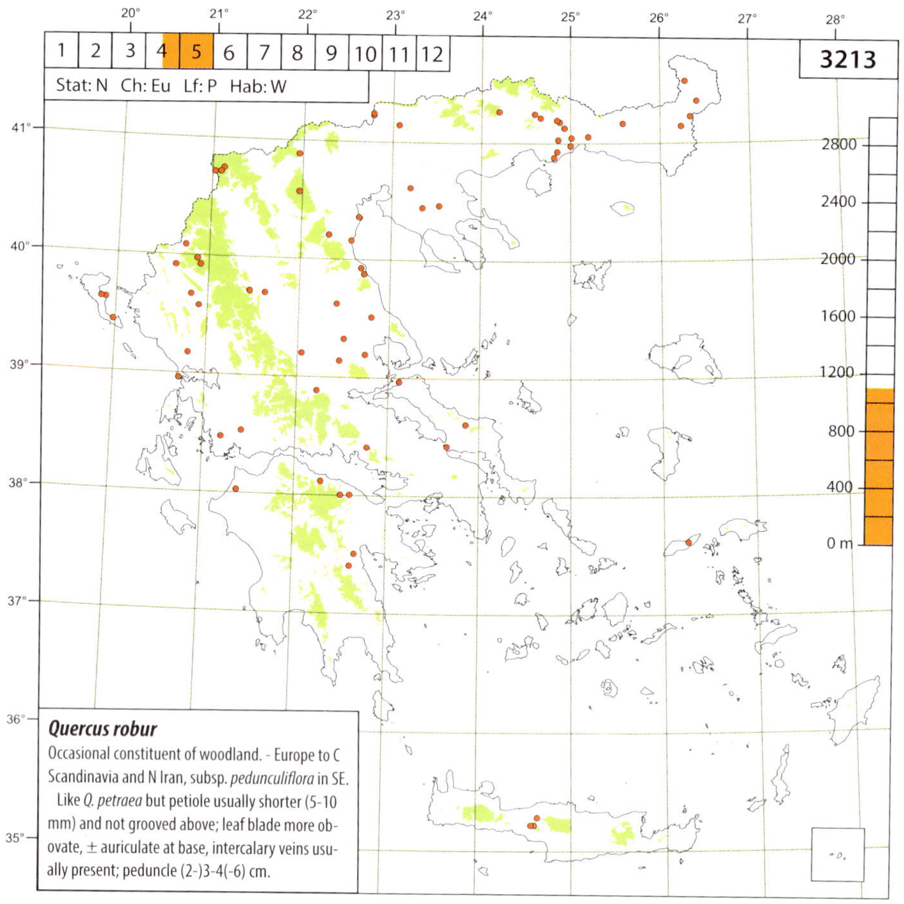

Quercus robur
Occasional constituent of woodland. - Europe to C Scandinavia and N Iran, subsp. *pedunculiflora* in SE.
Like *Q. petraea* but petiole usually shorter (5-10 mm) and not grooved above; leaf blade more obovate, ± auriculate at base, intercalary veins usually present; peduncle (2-)3-4(-6) cm.

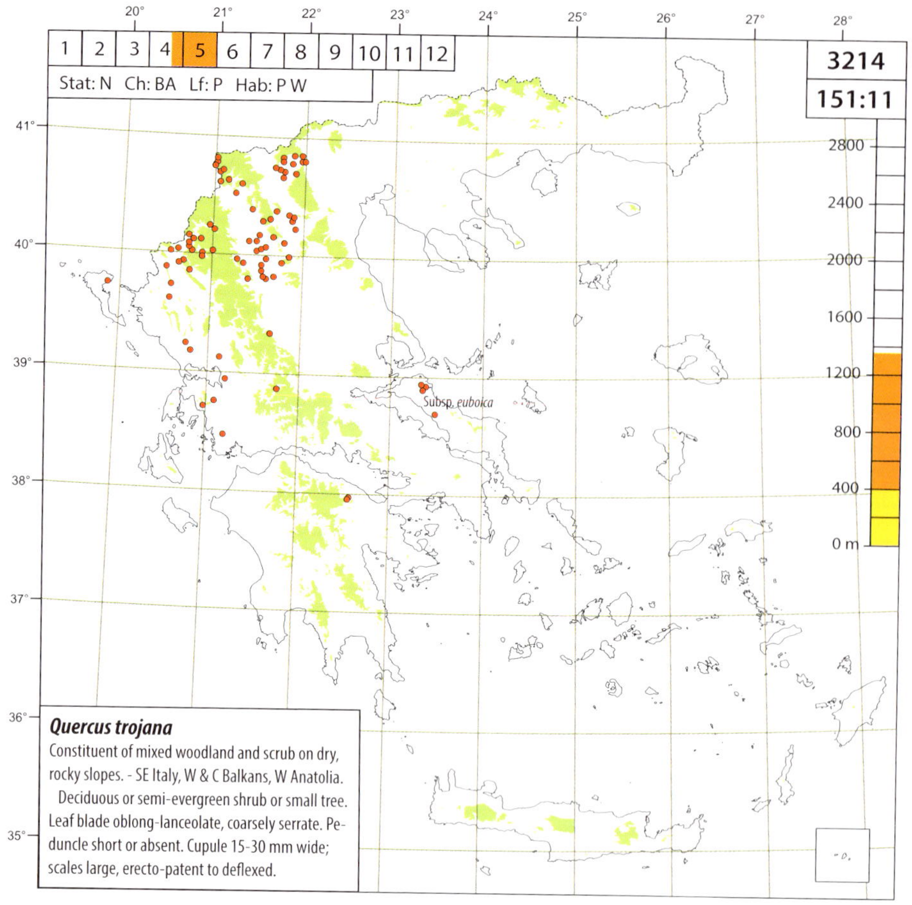

Quercus trojana
Constituent of mixed woodland and scrub on dry, rocky slopes. - SE Italy, W & C Balkans, W Anatolia.
Deciduous or semi-evergreen shrub or small tree. Leaf blade oblong-lanceolate, coarsely serrate. Peduncle short or absent. Cupule 15-30 mm wide; scales large, erecto-patent to deflexed.

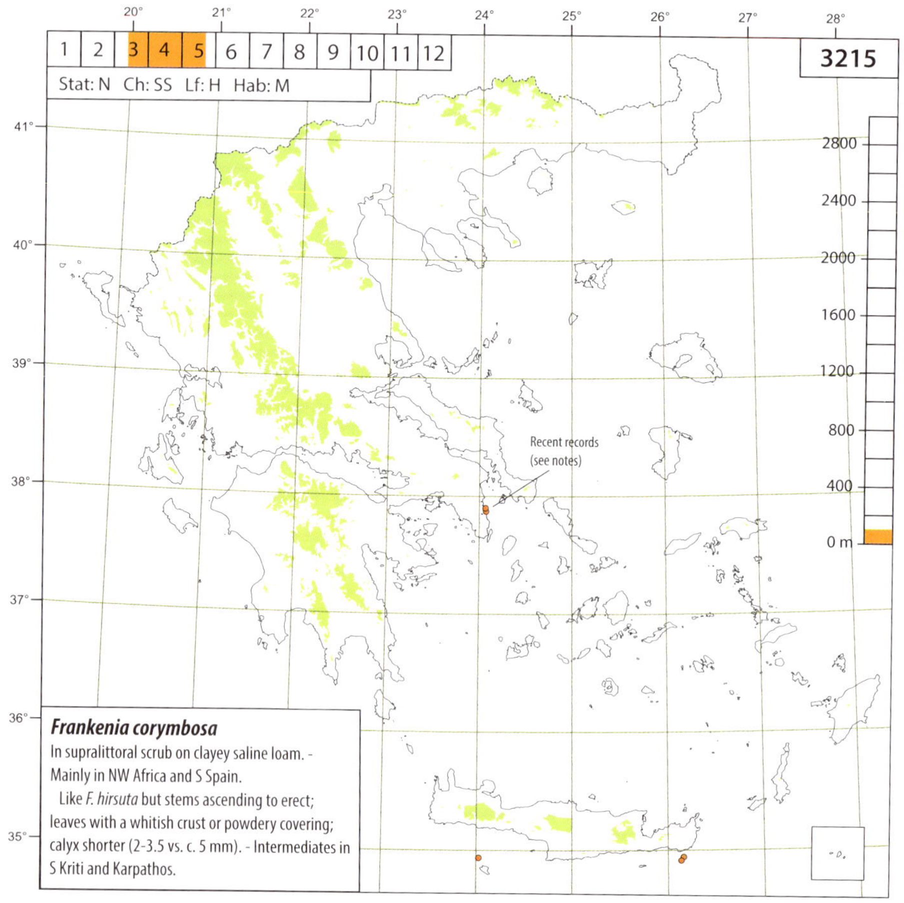

Frankenia corymbosa
In supralittoral scrub on clayey saline loam. - Mainly in NW Africa and S Spain.
Like *F. hirsuta* but stems ascending to erect; leaves with a whitish crust or powdery covering; calyx shorter (2-3.5 vs. c. 5 mm). - Intermediates in S Kriti and Karpathos.

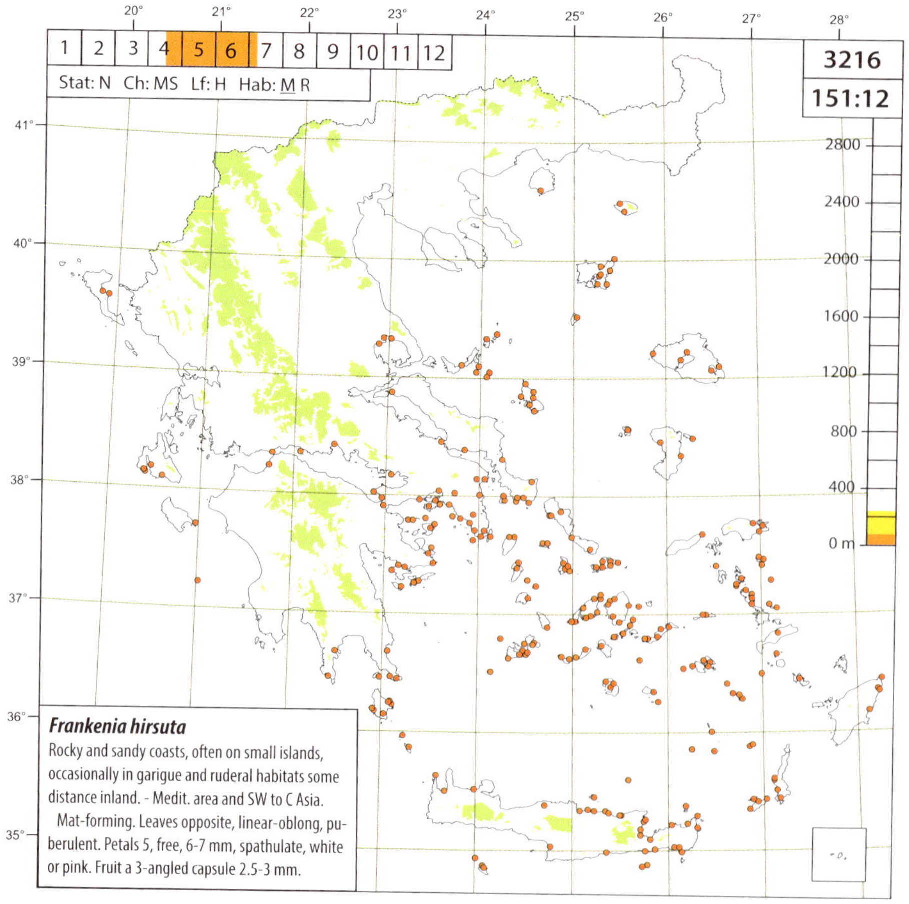

Frankenia hirsuta
Rocky and sandy coasts, often on small islands, occasionally in garigue and ruderal habitats some distance inland. - Medit. area and SW to C Asia.
Mat-forming. Leaves opposite, linear-oblong, puberulent. Petals 5, free, 6-7 mm, spathulate, white or pink. Fruit a 3-angled capsule 2.5-3 mm.

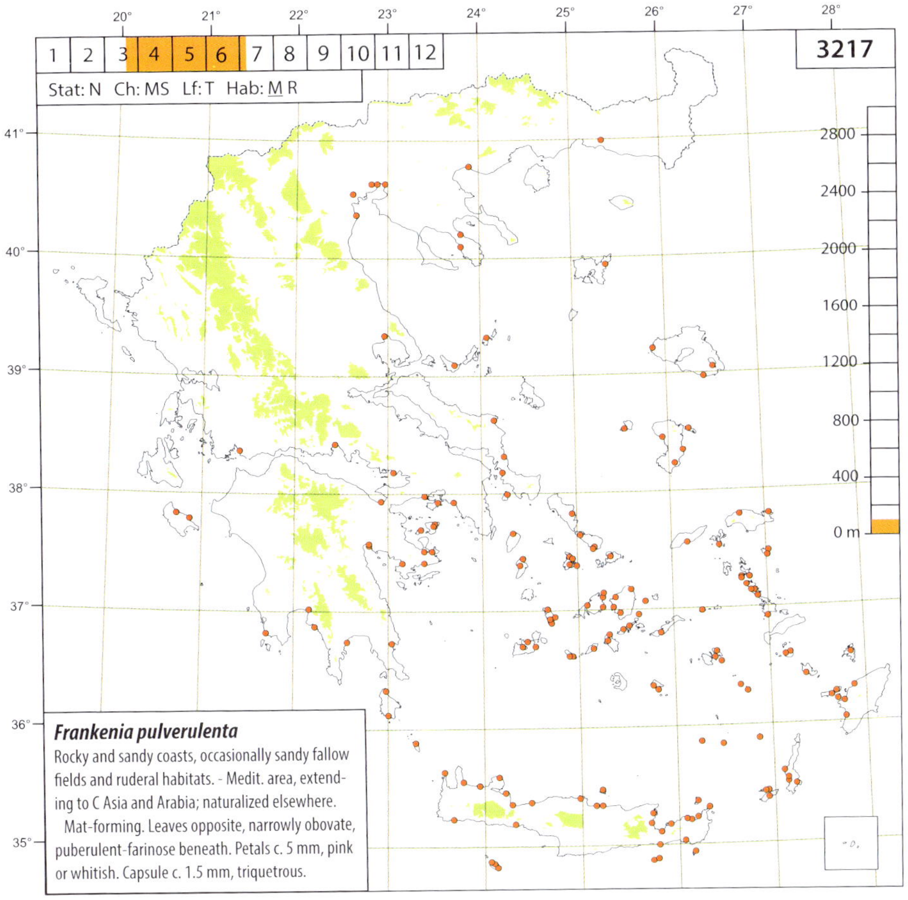

Frankenia pulverulenta
Rocky and sandy coasts, occasionally sandy fallow fields and ruderal habitats. - Medit. area, extending to C Asia and Arabia; naturalized elsewhere.
Mat-forming. Leaves opposite, narrowly obovate, puberulent-farinose beneath. Petals c. 5 mm, pink or whitish. Capsule c. 1.5 mm, triquetrous.

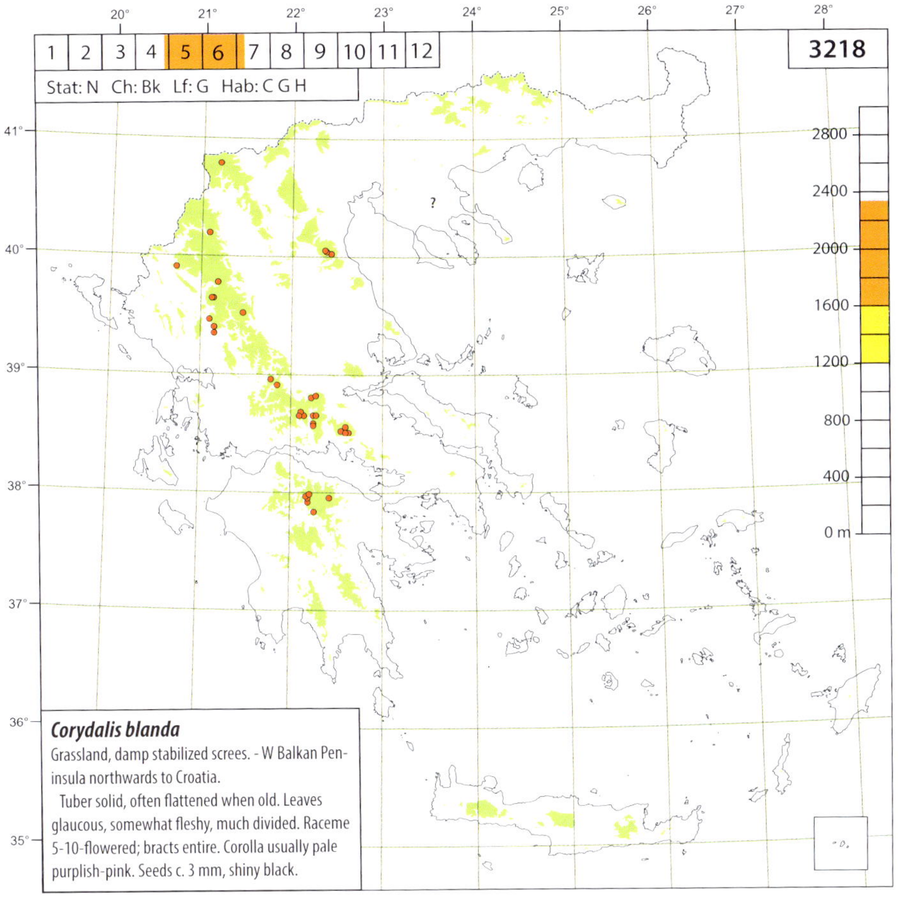

Corydalis blanda
Grassland, damp stabilized screes. - W Balkan Peninsula northwards to Croatia.
Tuber solid, often flattened when old. Leaves glaucous, somewhat fleshy, much divided. Raceme 5-10-flowered; bracts entire. Corolla usually pale purplish-pink. Seeds c. 3 mm, shiny black.

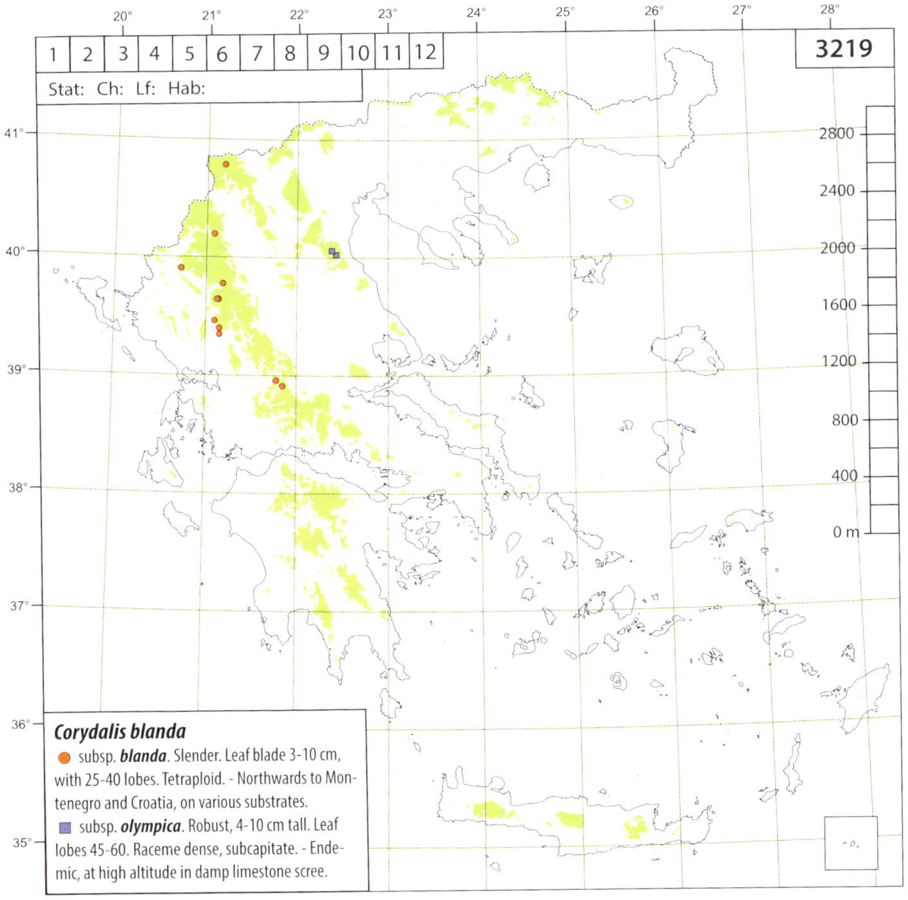

Corydalis blanda
● subsp. ***blanda***. Slender. Leaf blade 3-10 cm, with 25-40 lobes. Tetraploid. - Northwards to Montenegro and Croatia, on various substrates.
■ subsp. ***olympica***. Robust, 4-10 cm tall. Leaf lobes 45-60. Raceme dense, subcapitate. - Endemic, at high altitude in damp limestone scree.

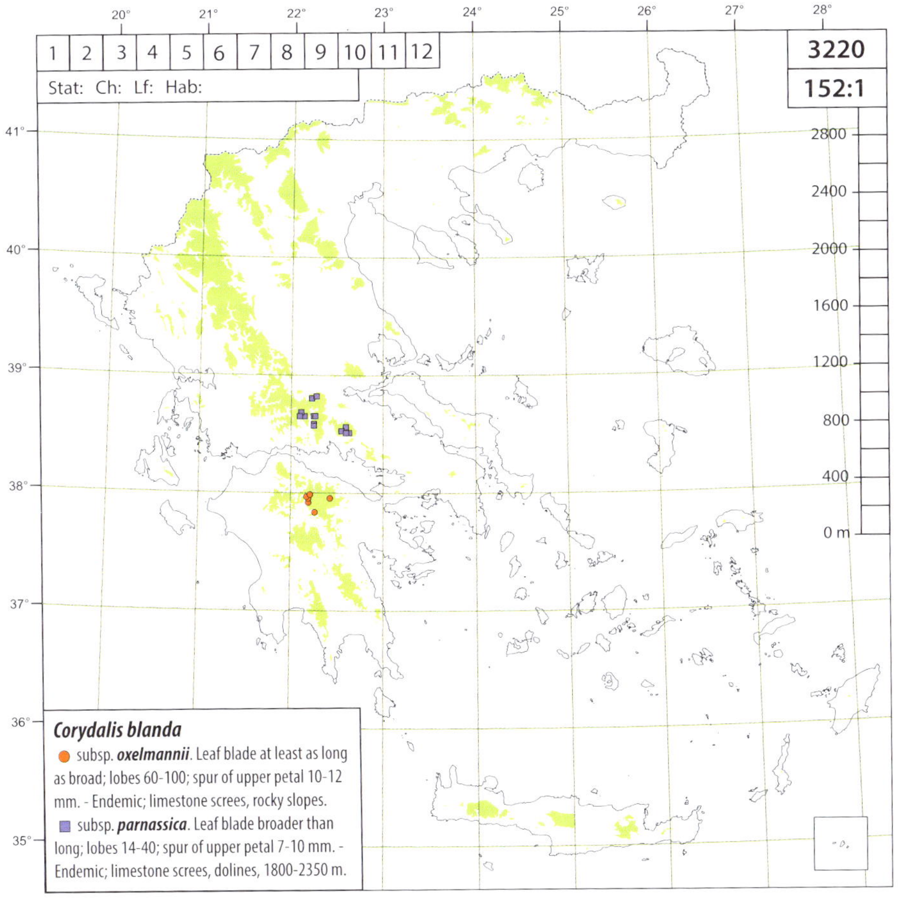

Corydalis blanda
● subsp. ***oxelmannii***. Leaf blade at least as long as broad; lobes 60-100; spur of upper petal 10-12 mm. - Endemic; limestone screes, rocky slopes.
■ subsp. ***parnassica***. Leaf blade broader than long; lobes 14-40; spur of upper petal 7-10 mm. - Endemic; limestone screes, dolines, 1800-2350 m.

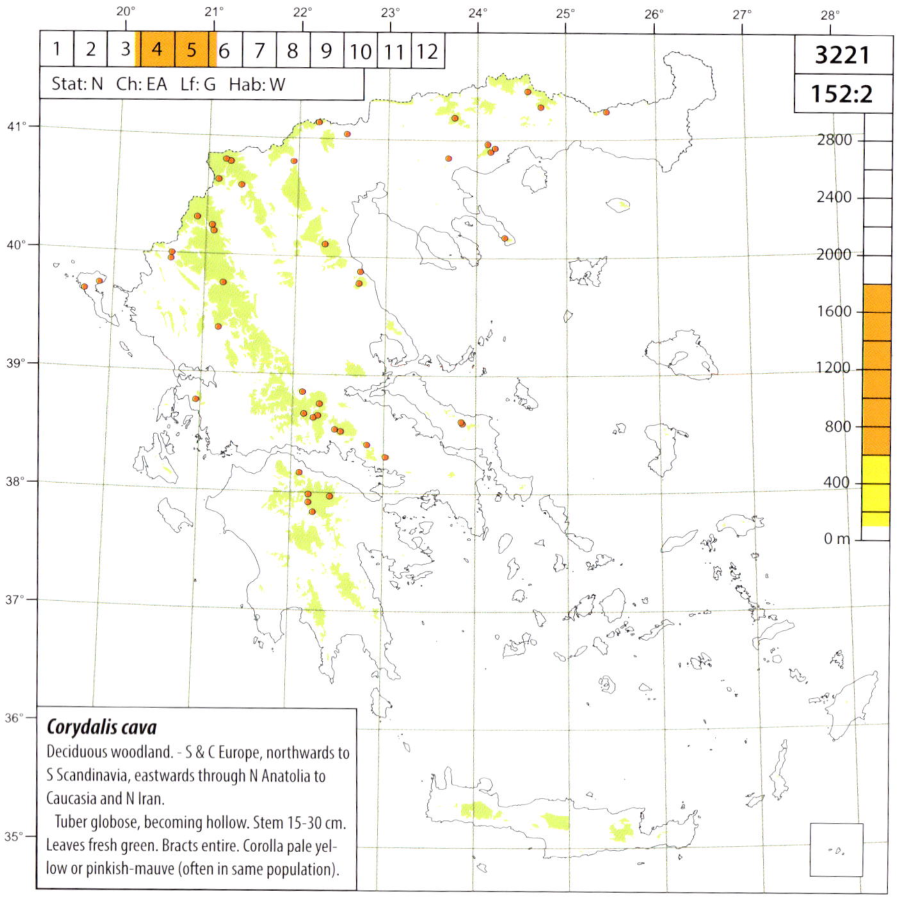

Corydalis cava

Deciduous woodland. - S & C Europe, northwards to S Scandinavia, eastwards through N Anatolia to Caucasia and N Iran.

Tuber globose, becoming hollow. Stem 15-30 cm. Leaves fresh green. Bracts entire. Corolla pale yellow or pinkish-mauve (often in same population).

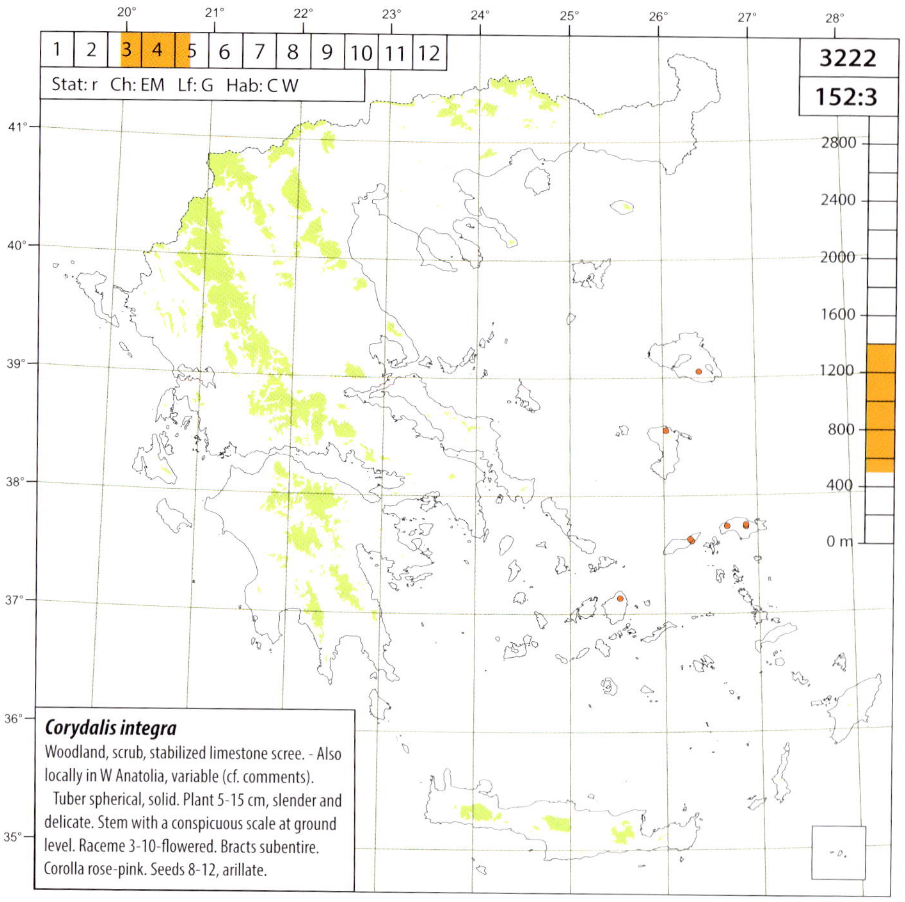

Corydalis integra

Woodland, scrub, stabilized limestone scree. - Also locally in W Anatolia, variable (cf. comments).

Tuber spherical, solid. Plant 5-15 cm, slender and delicate. Stem with a conspicuous scale at ground level. Raceme 3-10-flowered. Bracts subentire. Corolla rose-pink. Seeds 8-12, arillate.

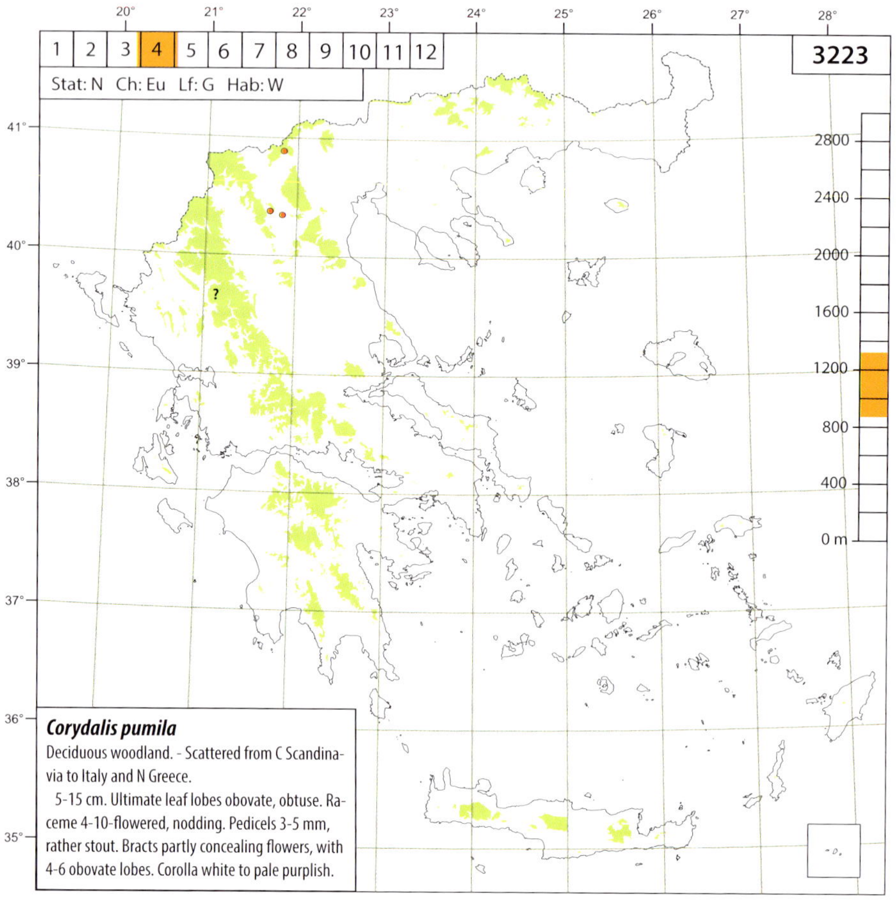

Corydalis pumila

Deciduous woodland. - Scattered from C Scandinavia to Italy and N Greece.

5-15 cm. Ultimate leaf lobes obovate, obtuse. Raceme 4-10-flowered, nodding. Pedicels 3-5 mm, rather stout. Bracts partly concealing flowers, with 4-6 obovate lobes. Corolla white to pale purplish.

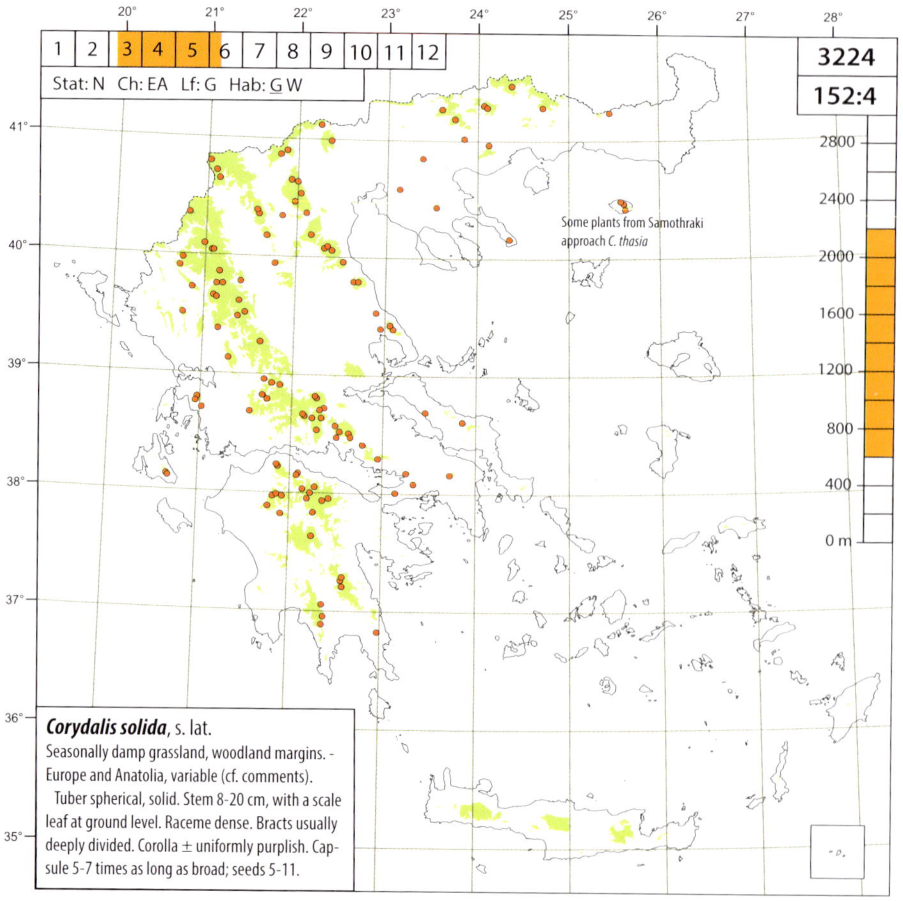

Corydalis solida, s. lat.

Seasonally damp grassland, woodland margins. - Europe and Anatolia, variable (cf. comments).

Tuber spherical, solid. Stem 8-20 cm, with a scale leaf at ground level. Raceme dense. Bracts usually deeply divided. Corolla ± uniformly purplish. Capsule 5-7 times as long as broad; seeds 5-11.

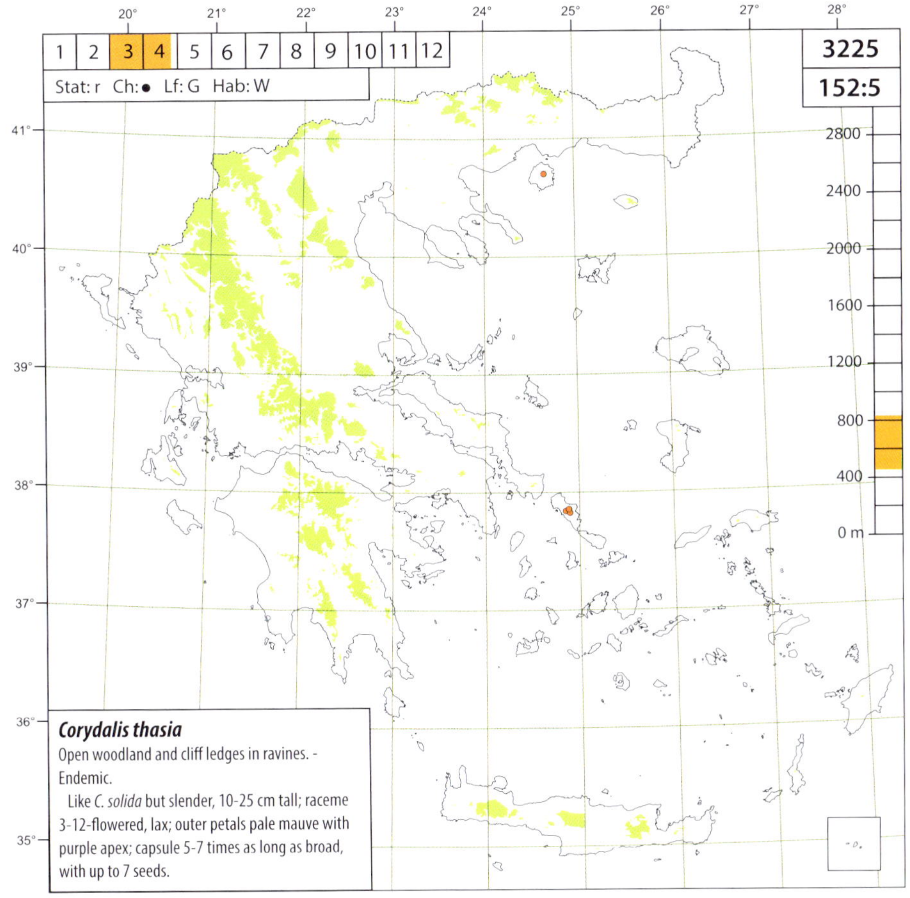

Corydalis thasia
Open woodland and cliff ledges in ravines. - Endemic.
Like *C. solida* but slender, 10-25 cm tall; raceme 3-12-flowered, lax; outer petals pale mauve with purple apex; capsule 5-7 times as long as broad, with up to 7 seeds.

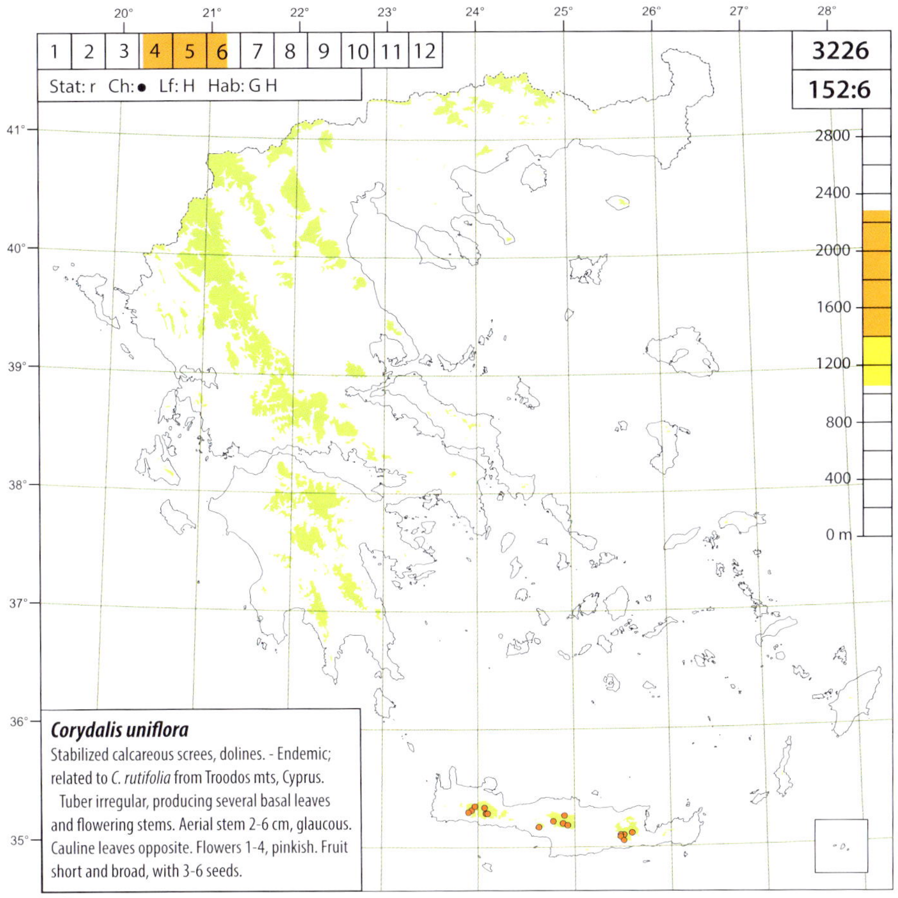

Corydalis uniflora
Stabilized calcareous screes, dolines. - Endemic; related to *C. rutifolia* from Troodos mts, Cyprus.
Tuber irregular, producing several basal leaves and flowering stems. Aerial stem 2-6 cm, glaucous. Cauline leaves opposite. Flowers 1-4, pinkish. Fruit short and broad, with 3-6 seeds.

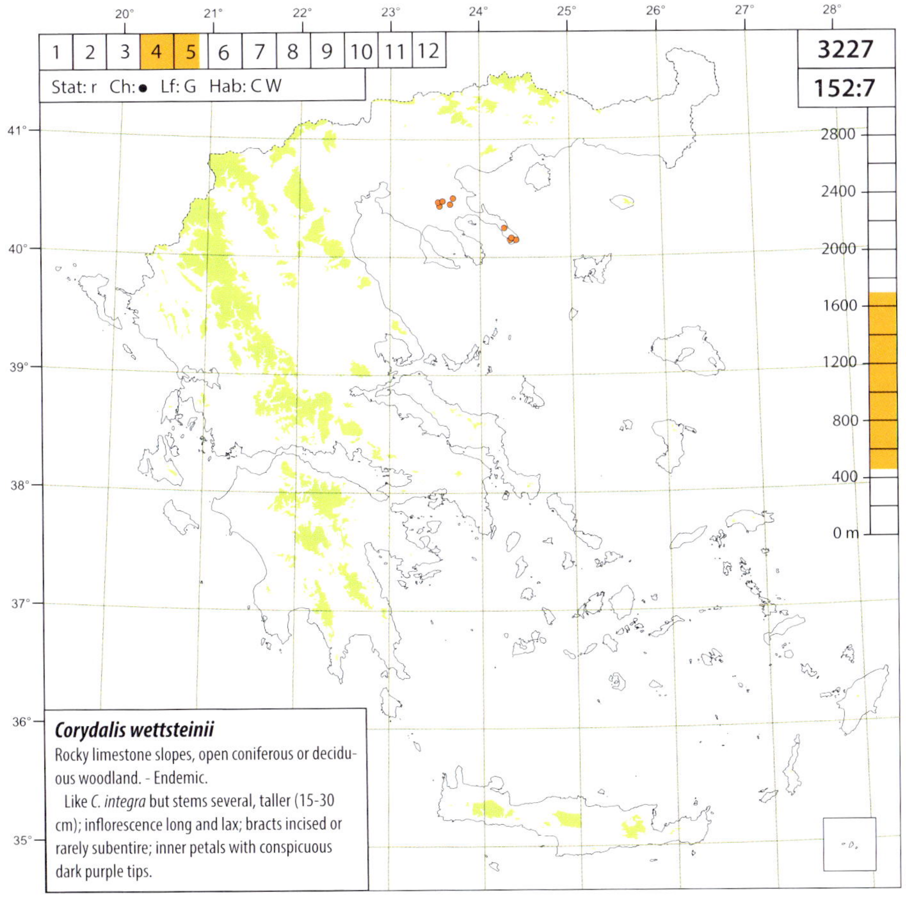

Corydalis wettsteinii
Rocky limestone slopes, open coniferous or deciduous woodland. - Endemic.
Like *C. integra* but stems several, taller (15-30 cm); inflorescence long and lax; bracts incised or rarely subentire; inner petals with conspicuous dark purple tips.

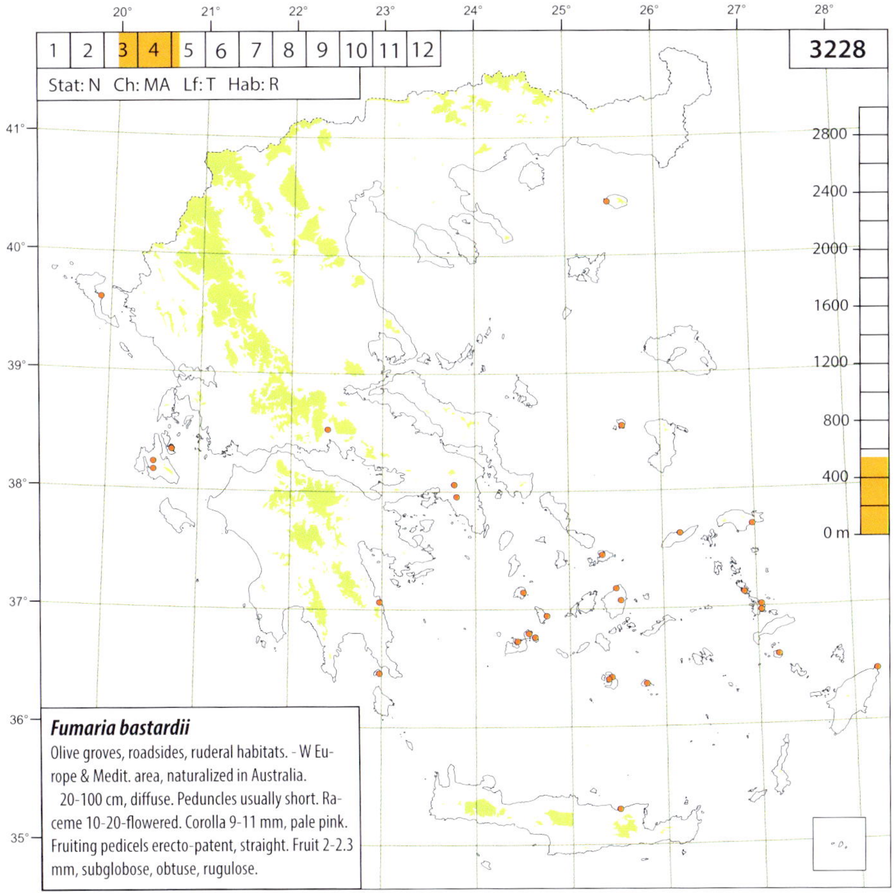

Fumaria bastardii
Olive groves, roadsides, ruderal habitats. - W Europe & Medit. area, naturalized in Australia.
20-100 cm, diffuse. Peduncles usually short. Raceme 10-20-flowered. Corolla 9-11 mm, pale pink. Fruiting pedicels erecto-patent, straight. Fruit 2-2.3 mm, subglobose, obtuse, rugulose.

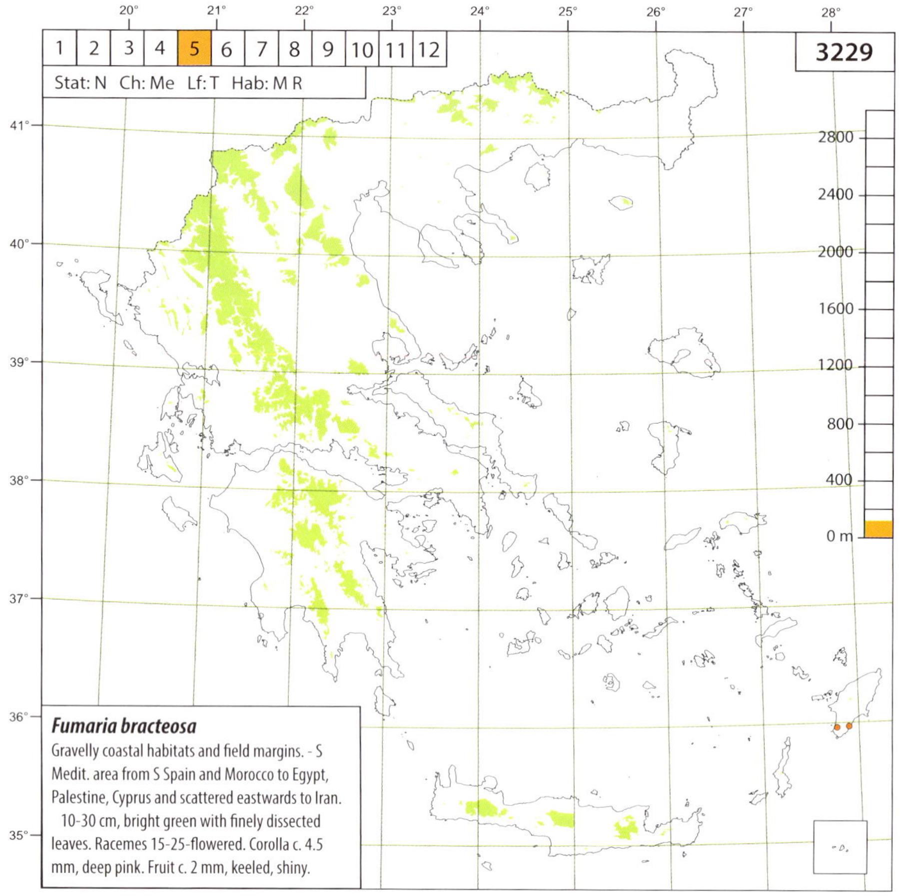

Fumaria bracteosa
Gravelly coastal habitats and field margins. - S Medit. area from S Spain and Morocco to Egypt, Palestine, Cyprus and scattered eastwards to Iran.
10-30 cm, bright green with finely dissected leaves. Racemes 15-25-flowered. Corolla c. 4.5 mm, deep pink. Fruit c. 2 mm, keeled, shiny.

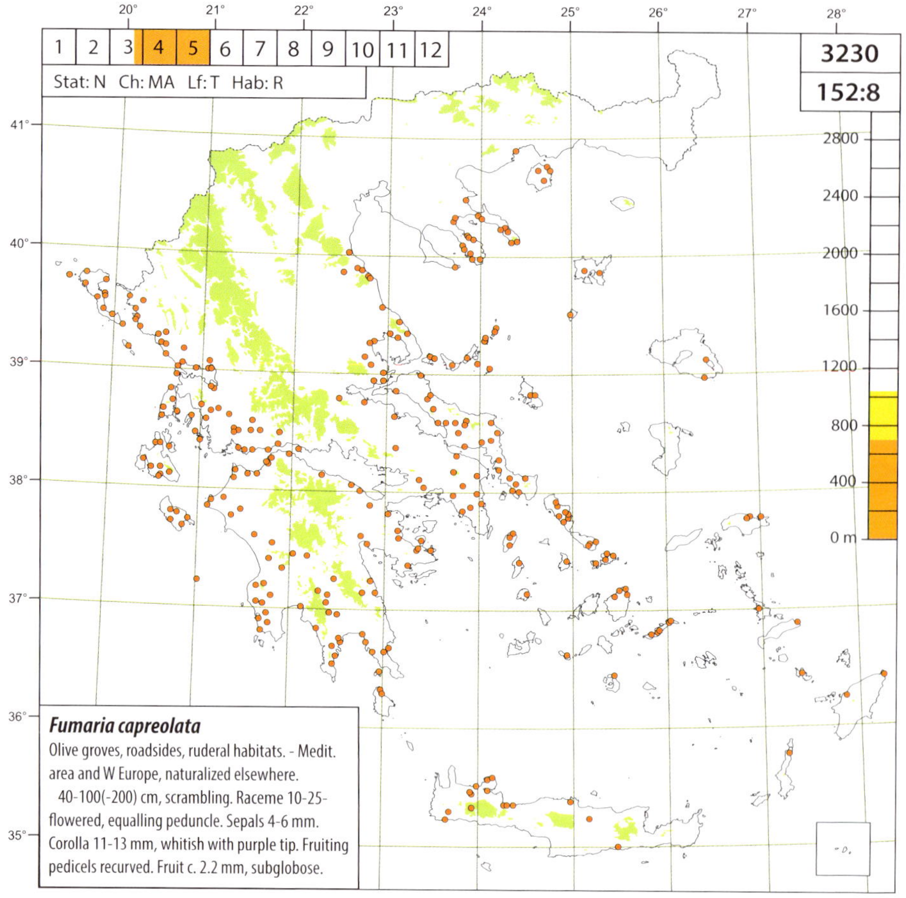

Fumaria capreolata
Olive groves, roadsides, ruderal habitats. - Medit. area and W Europe, naturalized elsewhere.
40-100(-200) cm, scrambling. Raceme 10-25-flowered, equalling peduncle. Sepals 4-6 mm. Corolla 11-13 mm, whitish with purple tip. Fruiting pedicels recurved. Fruit c. 2.2 mm, subglobose.

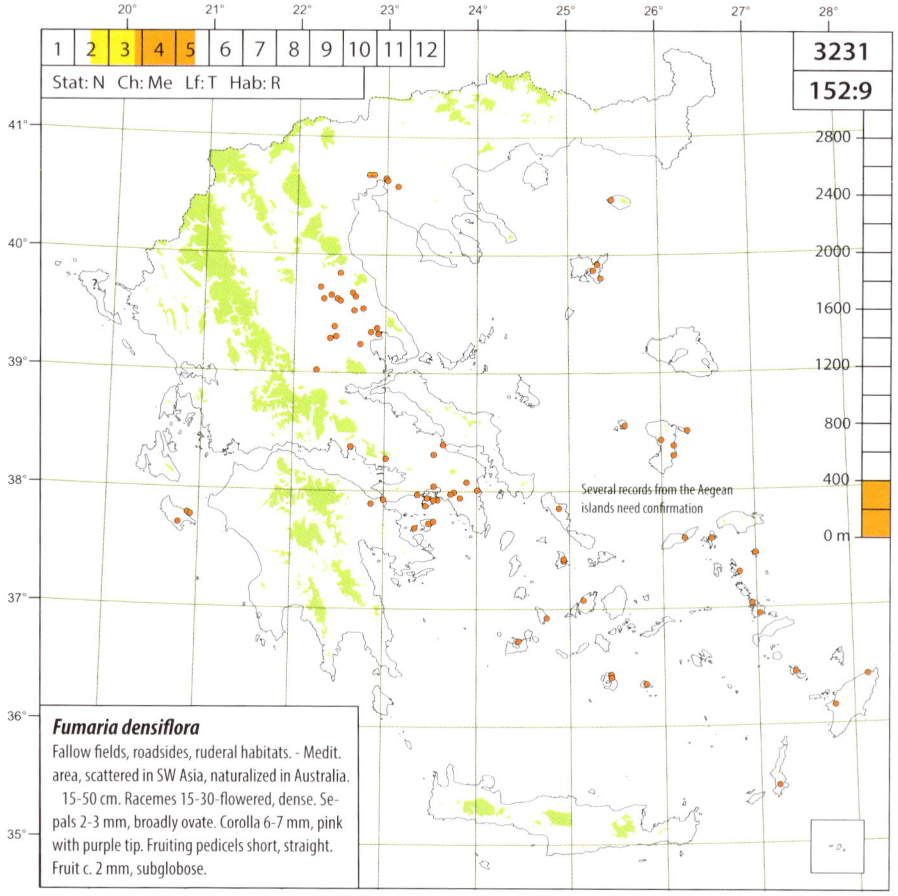

Fumaria densiflora
Fallow fields, roadsides, ruderal habitats. - Medit. area, scattered in SW Asia, naturalized in Australia.
15-50 cm. Racemes 15-30-flowered, dense. Sepals 2-3 mm, broadly ovate. Corolla 6-7 mm, pink with purple tip. Fruiting pedicels short, straight. Fruit c. 2 mm, subglobose.

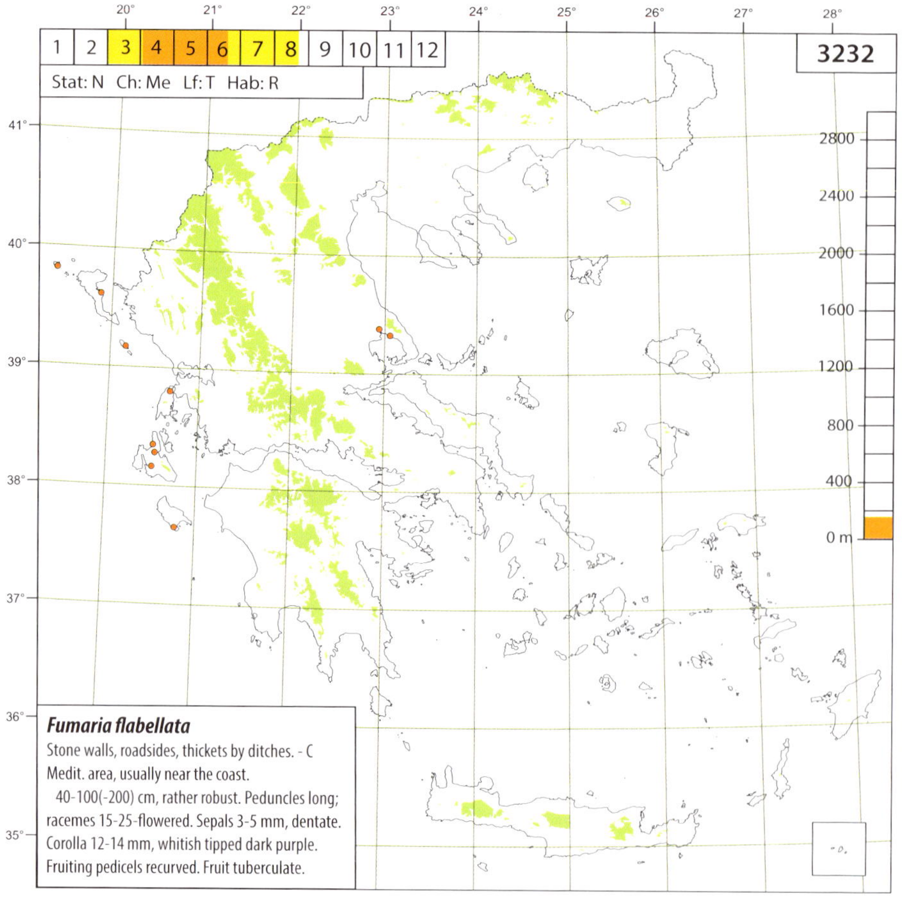

Fumaria flabellata
Stone walls, roadsides, thickets by ditches. - C Medit. area, usually near the coast.
40-100(-200) cm, rather robust. Peduncles long; racemes 15-25-flowered. Sepals 3-5 mm, dentate. Corolla 12-14 mm, whitish tipped dark purple. Fruiting pedicels recurved. Fruit tuberculate.

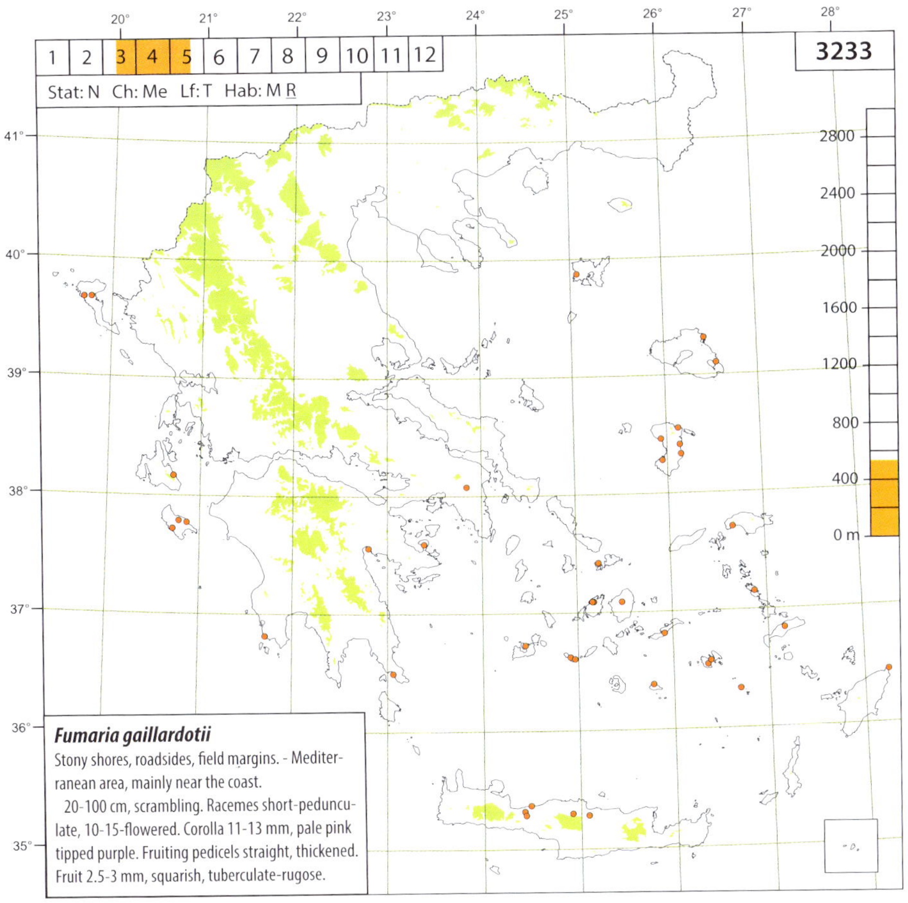

Fumaria gaillardotii

Stony shores, roadsides, field margins. - Mediterranean area, mainly near the coast.

20-100 cm, scrambling. Racemes short-pedunculate, 10-15-flowered. Corolla 11-13 mm, pale pink tipped purple. Fruiting pedicels straight, thickened. Fruit 2.5-3 mm, squarish, tuberculate-rugose.

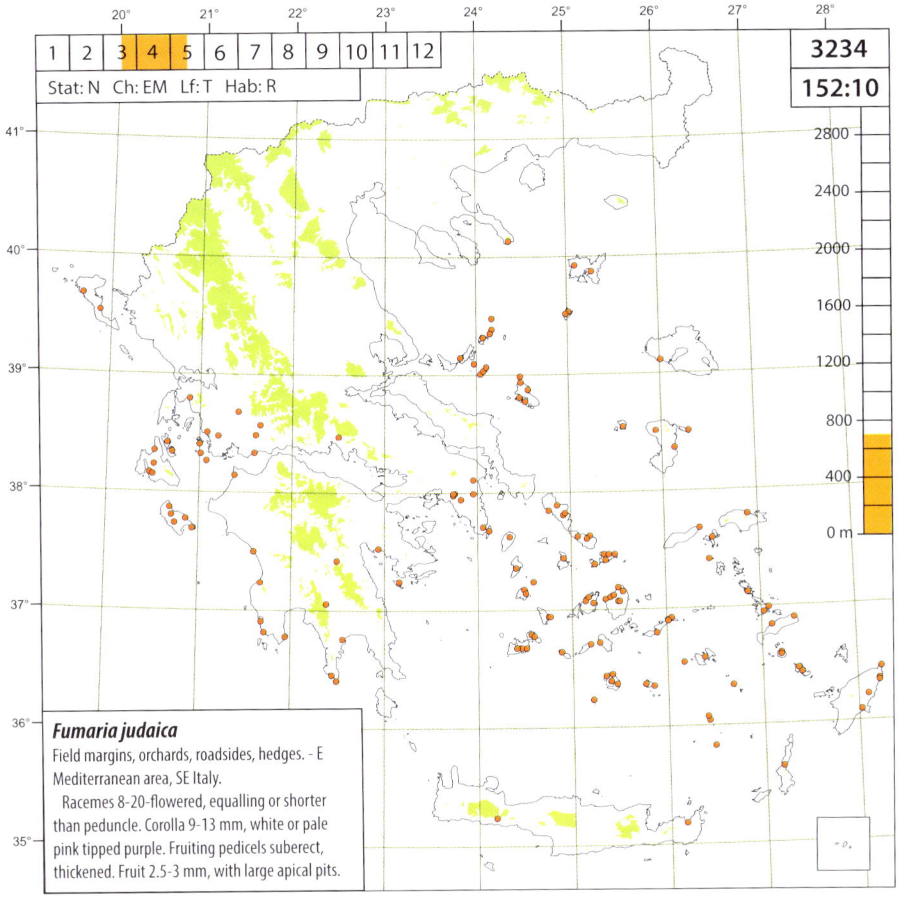

Fumaria judaica

Field margins, orchards, roadsides, hedges. - E Mediterranean area, SE Italy.

Racemes 8-20-flowered, equalling or shorter than peduncle. Corolla 9-13 mm, white or pale pink tipped purple. Fruiting pedicels suberect, thickened. Fruit 2.5-3 mm, with large apical pits.

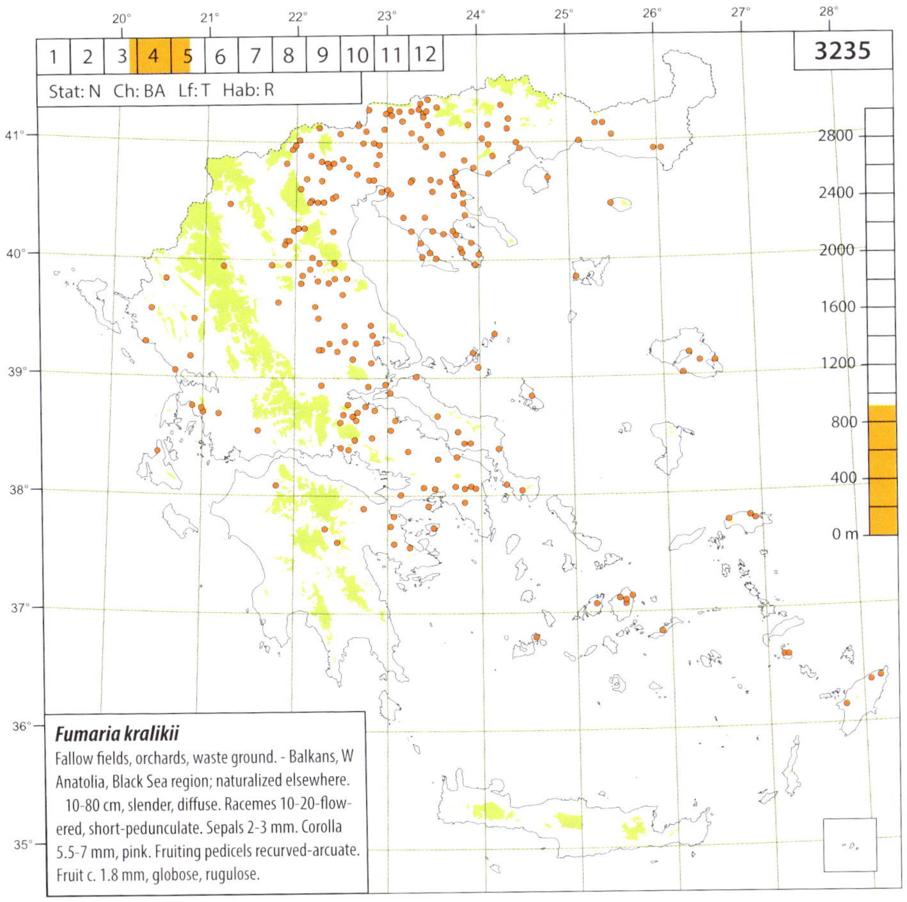

Fumaria kralikii

Fallow fields, orchards, waste ground. - Balkans, W Anatolia, Black Sea region; naturalized elsewhere.

10-80 cm, slender, diffuse. Racemes 10-20-flowered, short-pedunculate. Sepals 2-3 mm. Corolla 5.5-7 mm, pink. Fruiting pedicels recurved-arcuate. Fruit c. 1.8 mm, globose, rugulose.

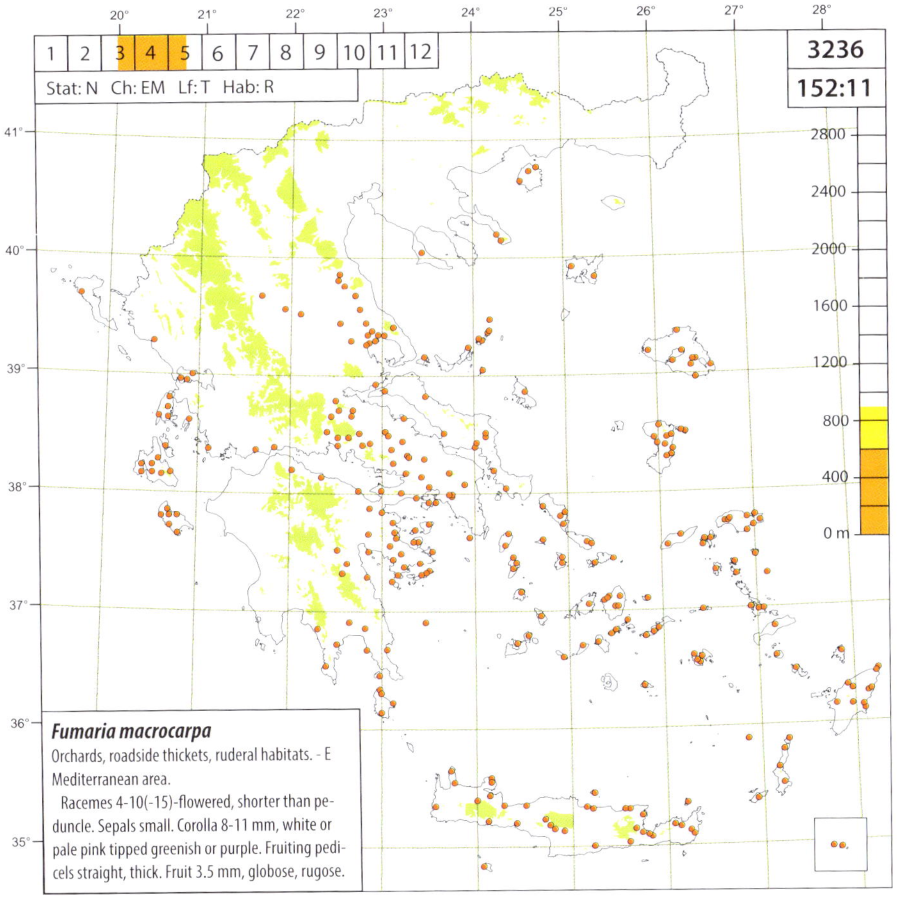

Fumaria macrocarpa

Orchards, roadside thickets, ruderal habitats. - E Mediterranean area.

Racemes 4-10(-15)-flowered, shorter than peduncle. Sepals small. Corolla 8-11 mm, white or pale pink tipped greenish or purple. Fruiting pedicels straight, thick. Fruit 3.5 mm, globose, rugose.

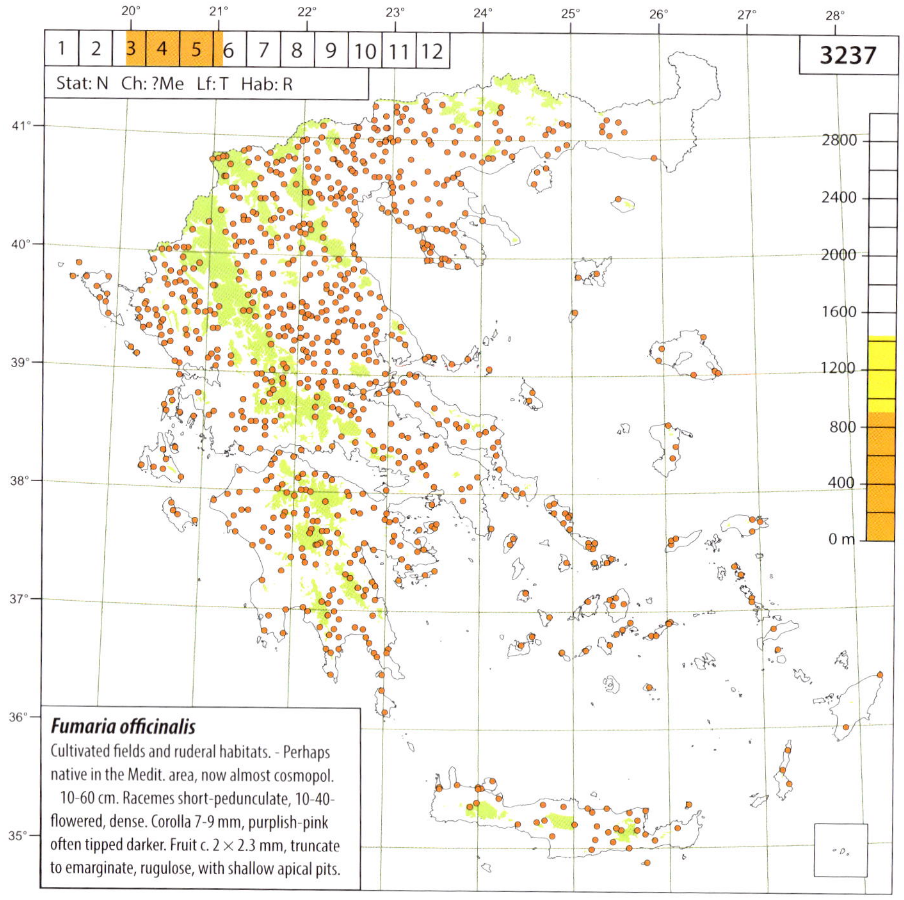

Fumaria officinalis

Cultivated fields and ruderal habitats. - Perhaps native in the Medit. area, now almost cosmopol.

10-60 cm. Racemes short-pedunculate, 10-40-flowered, dense. Corolla 7-9 mm, purplish-pink often tipped darker. Fruit c. 2 × 2.3 mm, truncate to emarginate, rugulose, with shallow apical pits.

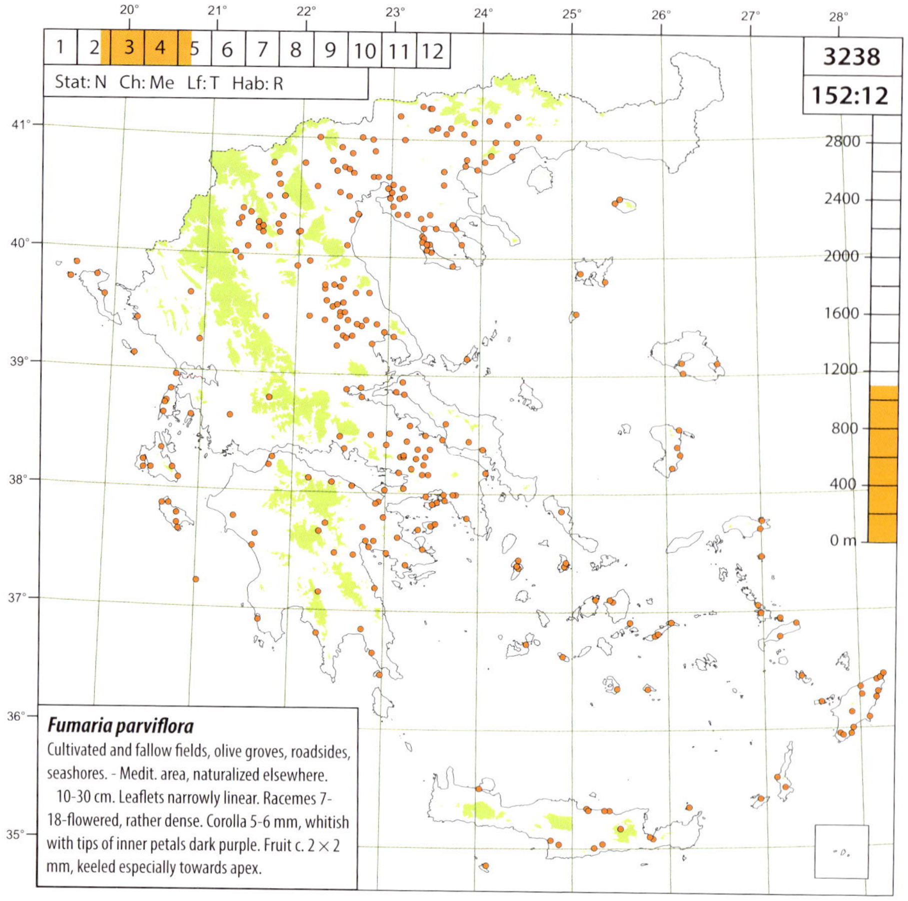

Fumaria parviflora

Cultivated and fallow fields, olive groves, roadsides, seashores. - Medit. area, naturalized elsewhere.

10-30 cm. Leaflets narrowly linear. Racemes 7-18-flowered, rather dense. Corolla 5-6 mm, whitish with tips of inner petals dark purple. Fruit c. 2 × 2 mm, keeled especially towards apex.

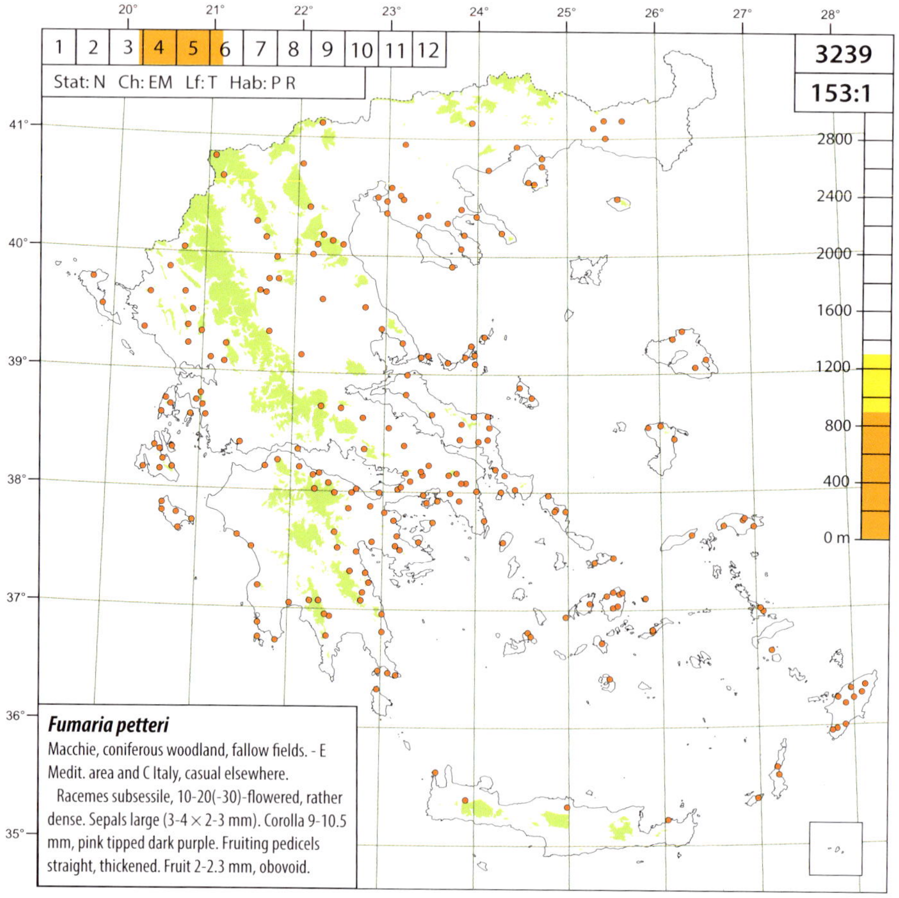

Fumaria petteri

Macchie, coniferous woodland, fallow fields. - E Medit. area and C Italy, casual elsewhere.

Racemes subsessile, 10-20(-30)-flowered, rather dense. Sepals large (3-4 × 2-3 mm). Corolla 9-10.5 mm, pink tipped dark purple. Fruiting pedicels straight, thickened. Fruit 2-2.3 mm, obovoid.

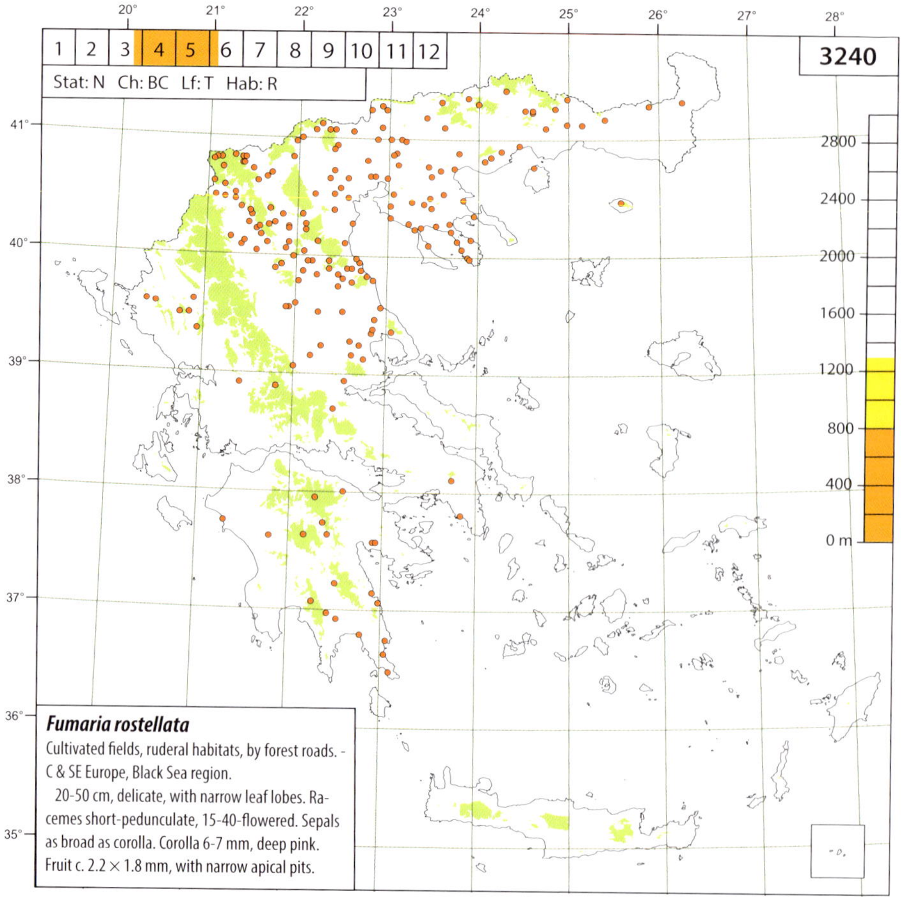

Fumaria rostellata

Cultivated fields, ruderal habitats, by forest roads. - C & SE Europe, Black Sea region.

20-50 cm, delicate, with narrow leaf lobes. Racemes short-pedunculate, 15-40-flowered. Sepals as broad as corolla. Corolla 6-7 mm, deep pink. Fruit c. 2.2 × 1.8 mm, with narrow apical pits.

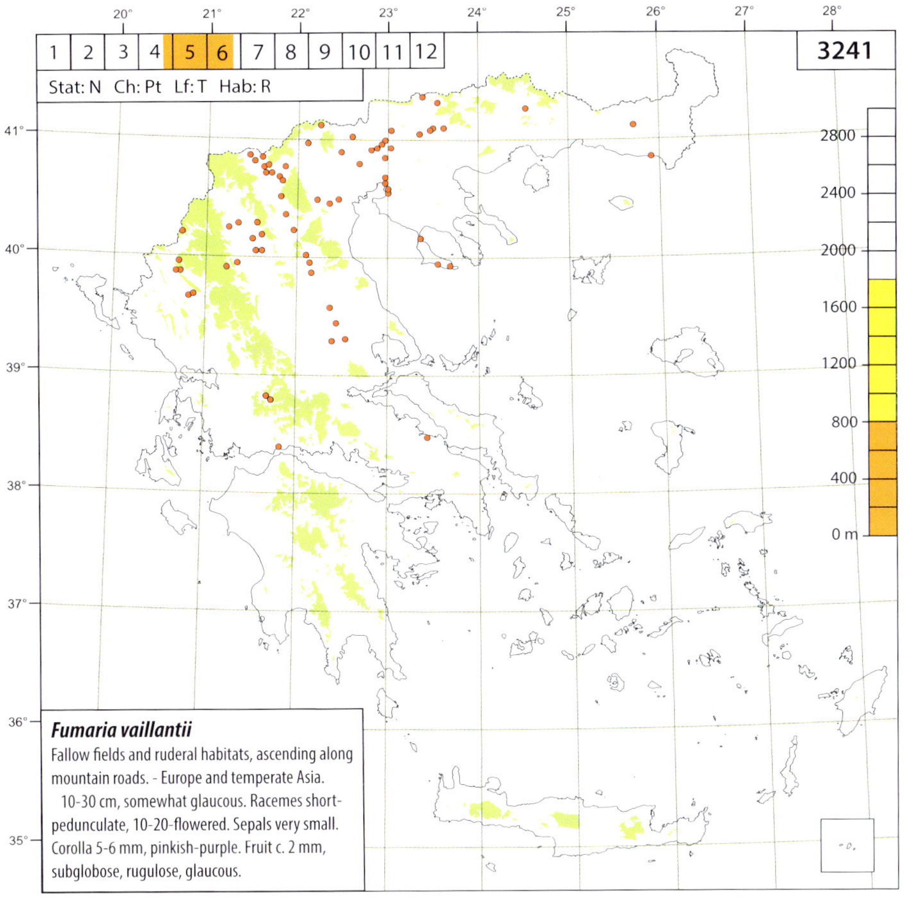

Fumaria vaillantii

Fallow fields and ruderal habitats, ascending along mountain roads. - Europe and temperate Asia.

10-30 cm, somewhat glaucous. Racemes short-pedunculate, 10-20-flowered. Sepals very small. Corolla 5-6 mm, pinkish-purple. Fruit c. 2 mm, subglobose, rugulose, glaucous.

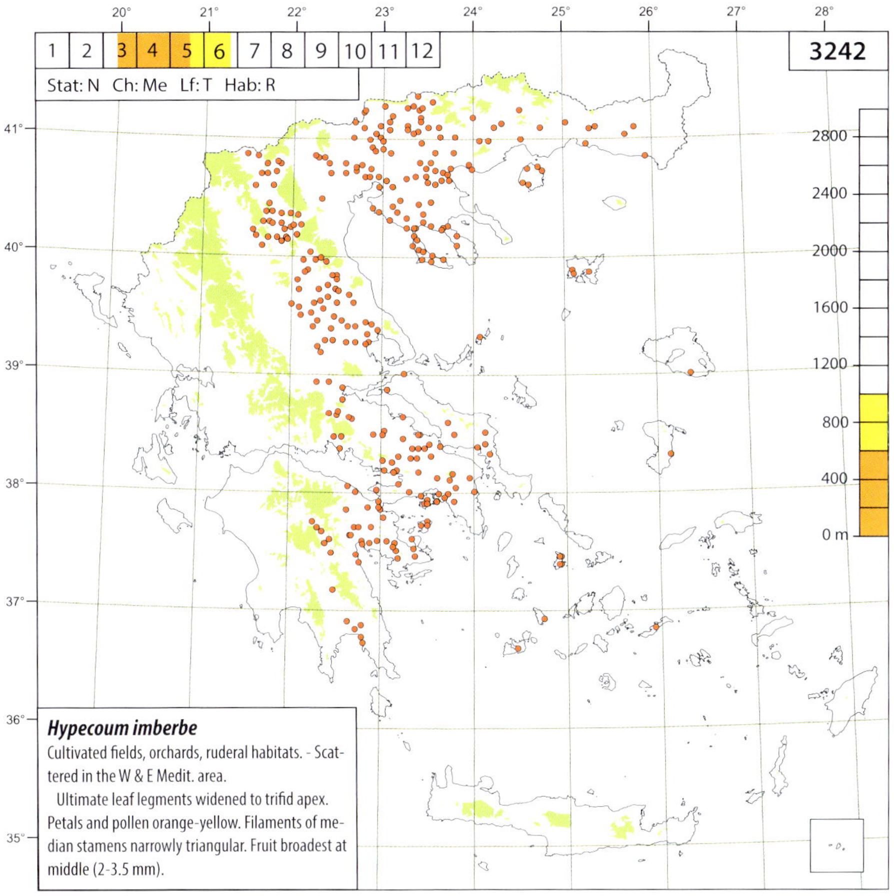

Hypecoum imberbe

Cultivated fields, orchards, ruderal habitats. - Scattered in the W & E Medit. area.

Ultimate leaf legments widened to trifid apex. Petals and pollen orange-yellow. Filaments of median stamens narrowly triangular. Fruit broadest at middle (2-3.5 mm).

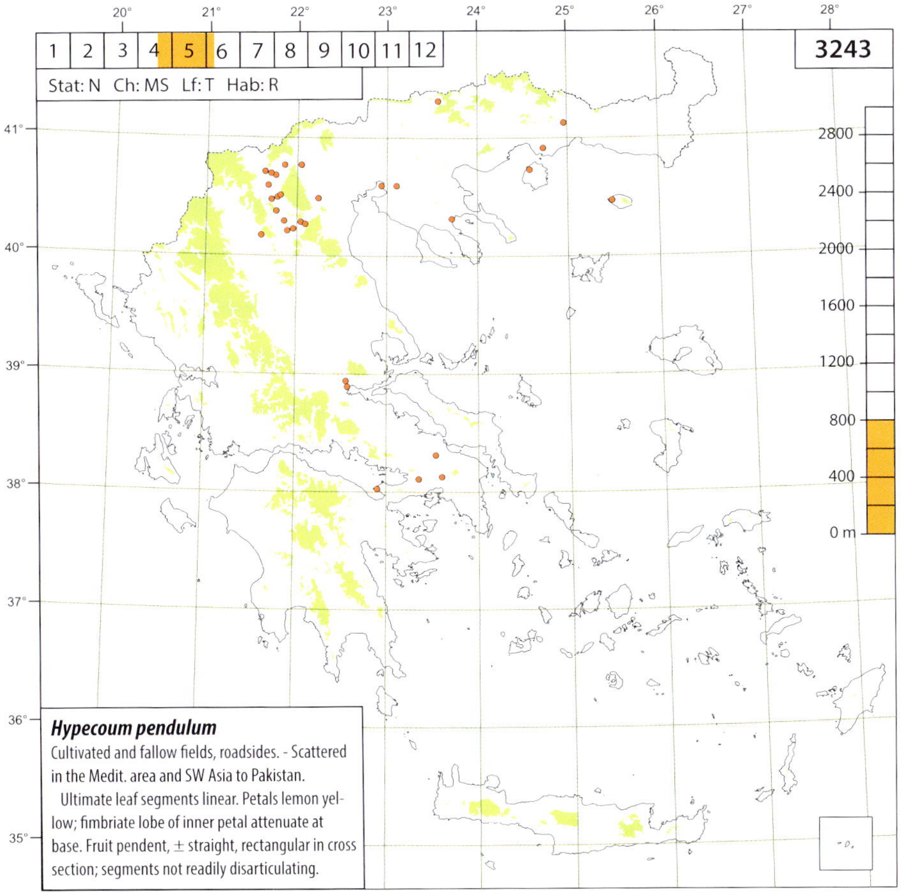

Hypecoum pendulum

Cultivated and fallow fields, roadsides. - Scattered in the Medit. area and SW Asia to Pakistan.

Ultimate leaf segments linear. Petals lemon yellow; fimbriate lobe of inner petal attenuate at base. Fruit pendent, ± straight, rectangular in cross section; segments not readily disarticulating.

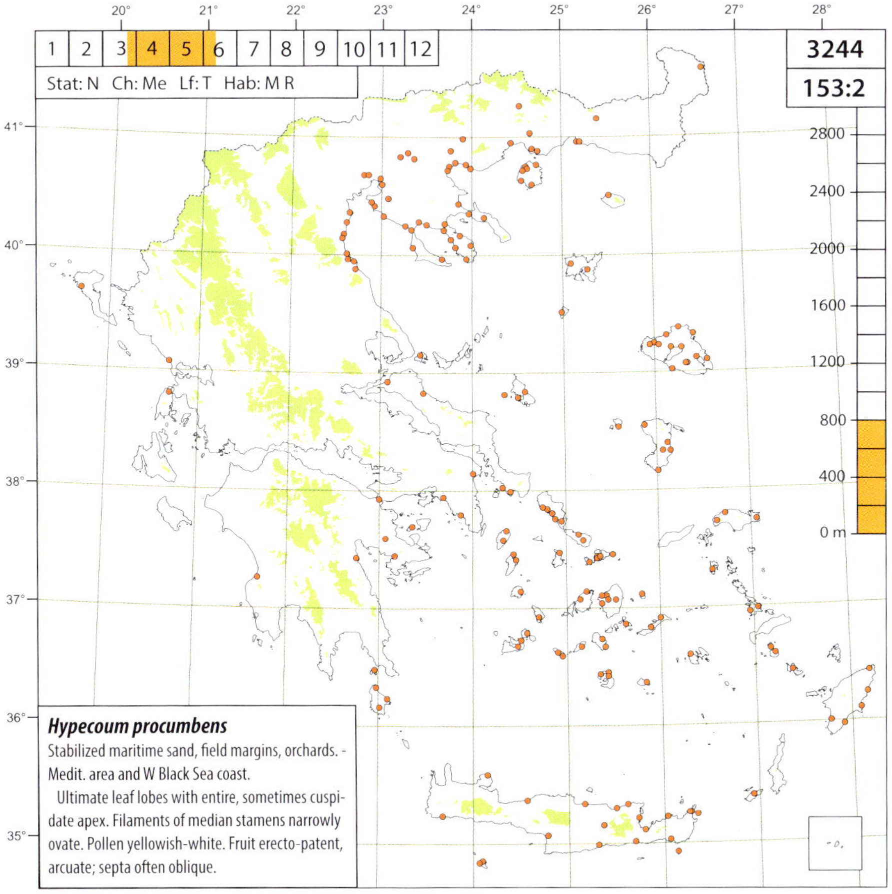

Hypecoum procumbens

Stabilized maritime sand, field margins, orchards. - Medit. area and W Black Sea coast.

Ultimate leaf lobes with entire, sometimes cuspidate apex. Filaments of median stamens narrowly ovate. Pollen yellowish-white. Fruit erecto-patent, arcuate; septa often oblique.

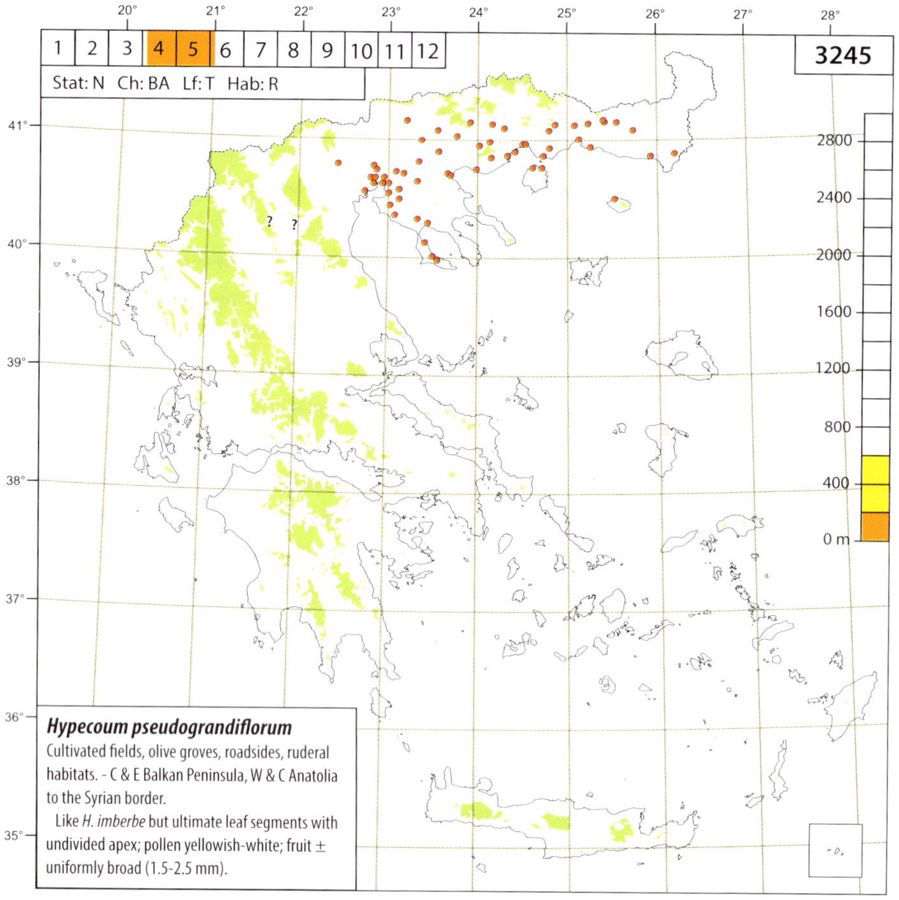

Hypecoum pseudograndiflorum
Cultivated fields, olive groves, roadsides, ruderal habitats. - C & E Balkan Peninsula, W & C Anatolia to the Syrian border.
Like *H. imberbe* but ultimate leaf segments with undivided apex; pollen yellowish-white; fruit ± uniformly broad (1.5-2.5 mm).

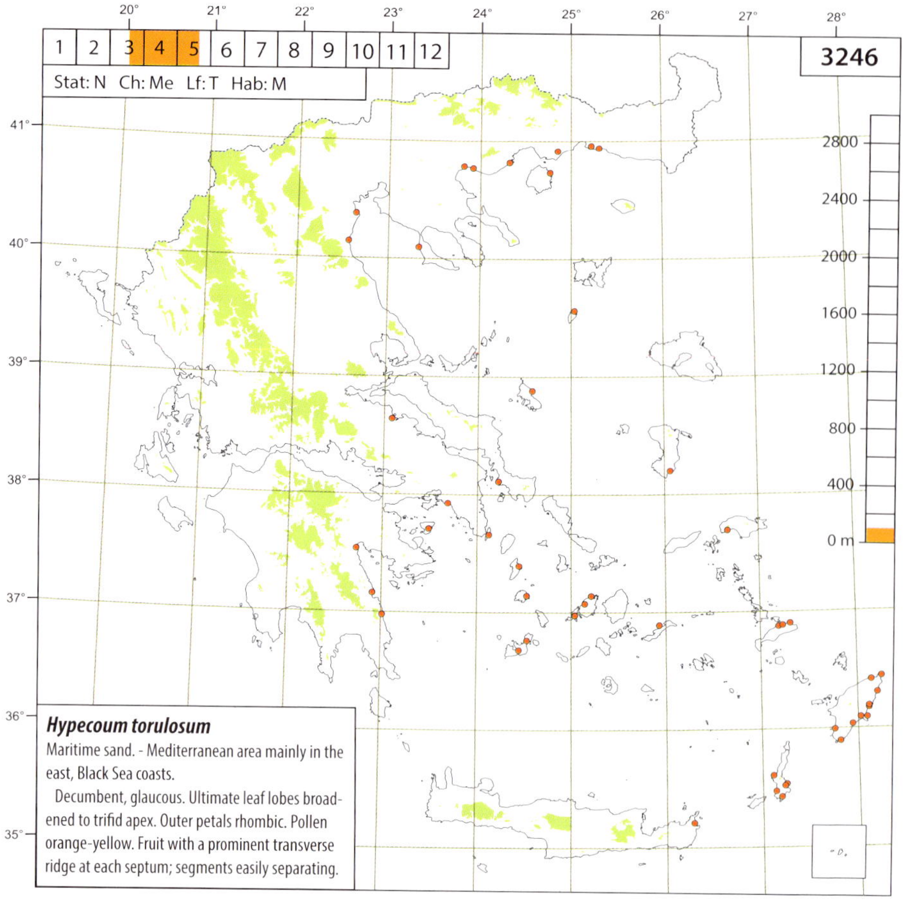

Hypecoum torulosum
Maritime sand. - Mediterranean area mainly in the east, Black Sea coasts.
Decumbent, glaucous. Ultimate leaf lobes broadened to trifid apex. Outer petals rhombic. Pollen orange-yellow. Fruit with a prominent transverse ridge at each septum; segments easily separating.

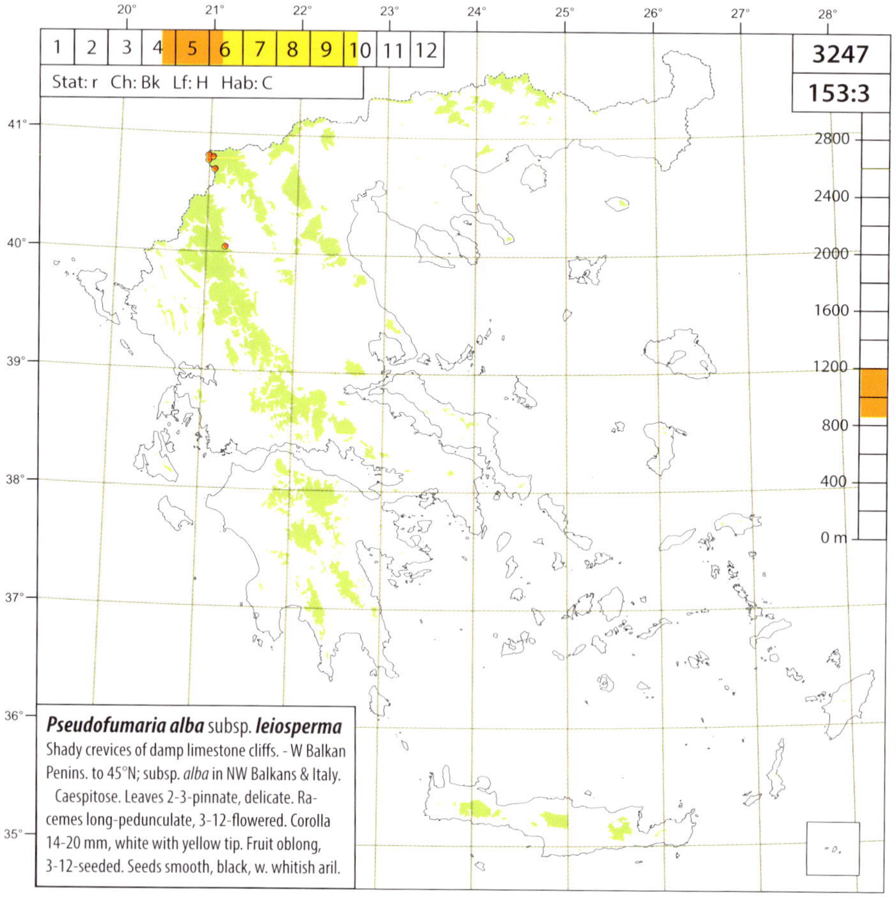

Pseudofumaria alba* subsp. *leiosperma
Shady crevices of damp limestone cliffs. - W Balkan Penins. to 45°N; subsp. *alba* in NW Balkans & Italy.
Caespitose. Leaves 2-3-pinnate, delicate. Racemes long-pedunculate, 3-12-flowered. Corolla 14-20 mm, white with yellow tip. Fruit oblong, 3-12-seeded. Seeds smooth, black, w. whitish aril.

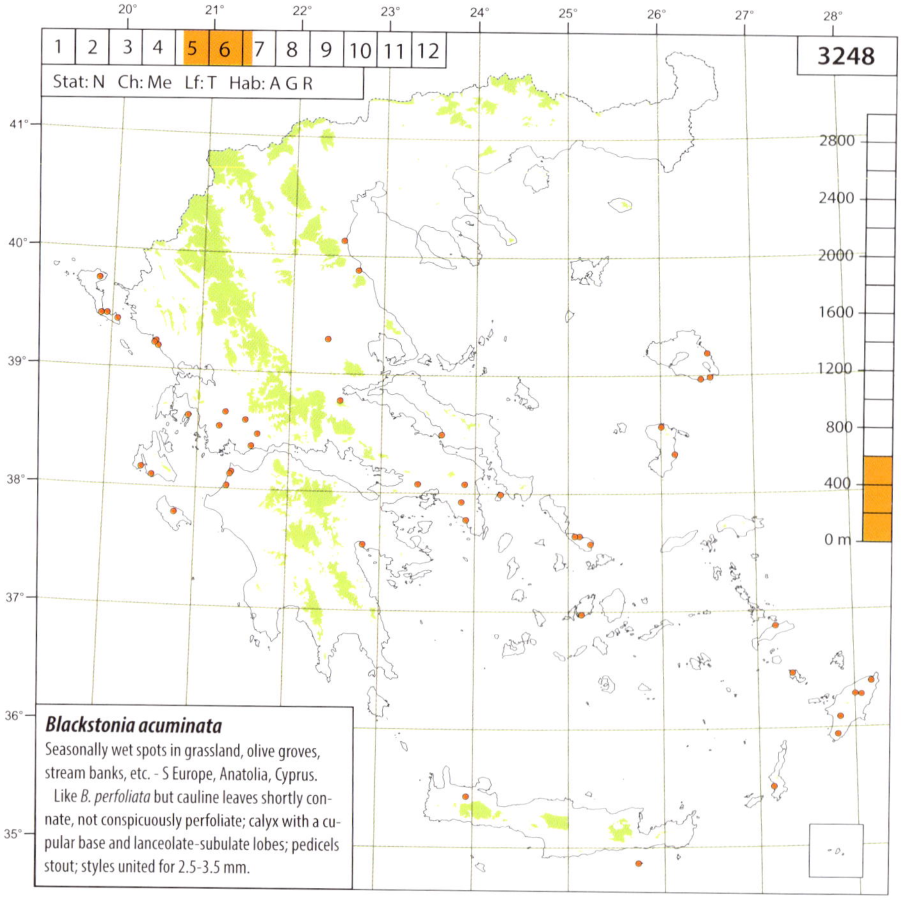

Blackstonia acuminata
Seasonally wet spots in grassland, olive groves, stream banks, etc. - S Europe, Anatolia, Cyprus.
Like *B. perfoliata* but cauline leaves shortly connate, not conspicuously perfoliate; calyx with a cupular base and lanceolate-subulate lobes; pedicels stout; styles united for 2.5-3.5 mm.

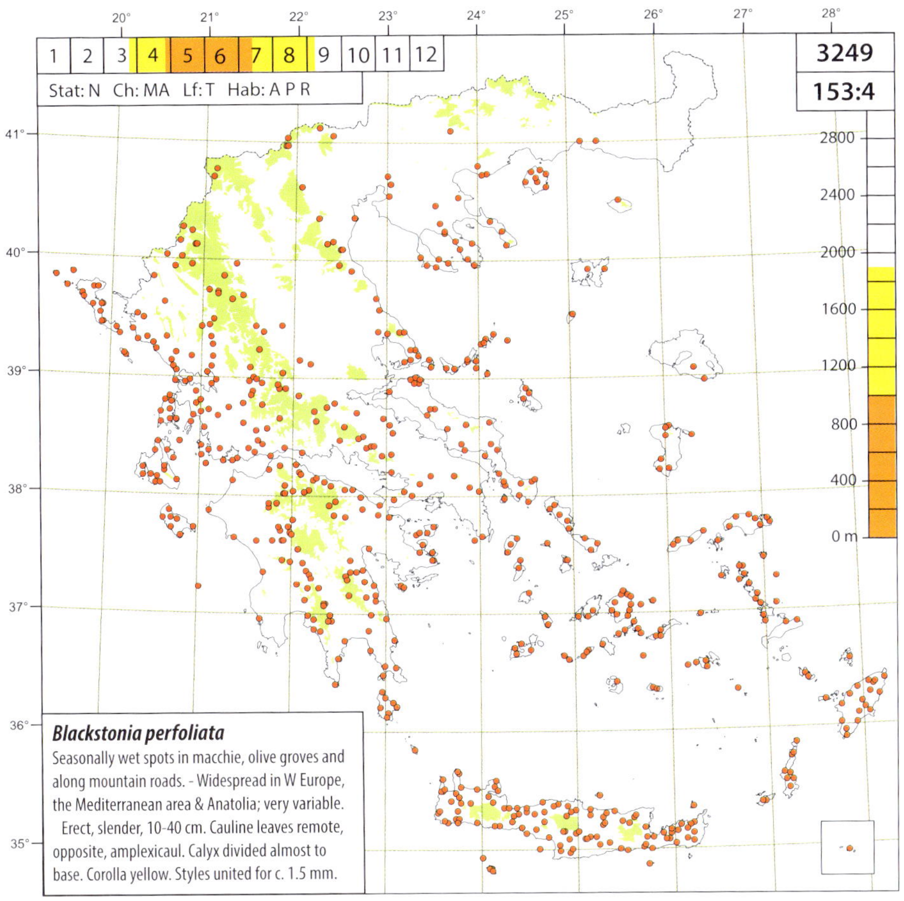

Blackstonia perfoliata

Seasonally wet spots in macchie, olive groves and along mountain roads. - Widespread in W Europe, the Mediterranean area & Anatolia; very variable.

Erect, slender, 10-40 cm. Cauline leaves remote, opposite, amplexicaul. Calyx divided almost to base. Corolla yellow. Styles united for c. 1.5 mm.

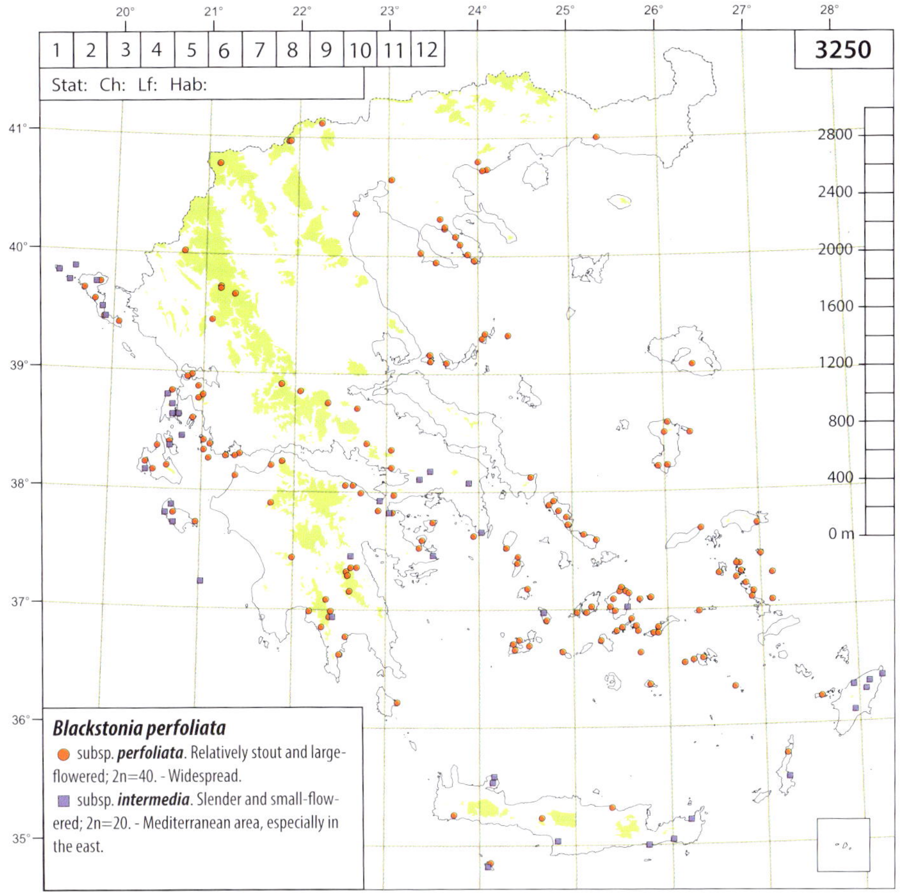

Blackstonia perfoliata

● subsp. ***perfoliata***. Relatively stout and large-flowered; 2n=40. - Widespread.

■ subsp. ***intermedia***. Slender and small-flowered; 2n=20. - Mediterranean area, especially in the east.

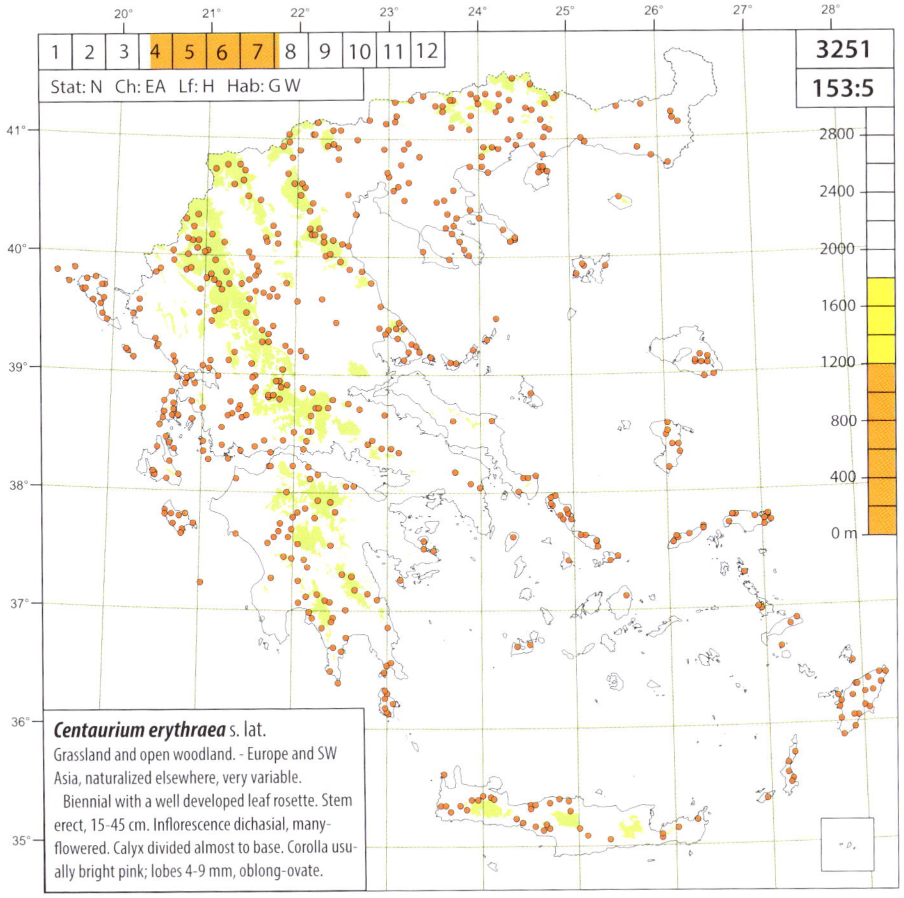

Centaurium erythraea s. lat.

Grassland and open woodland. - Europe and SW Asia, naturalized elsewhere, very variable.

Biennial with a well developed leaf rosette. Stem erect, 15-45 cm. Inflorescence dichasial, many-flowered. Calyx divided almost to base. Corolla usually bright pink; lobes 4-9 mm, oblong-ovate.

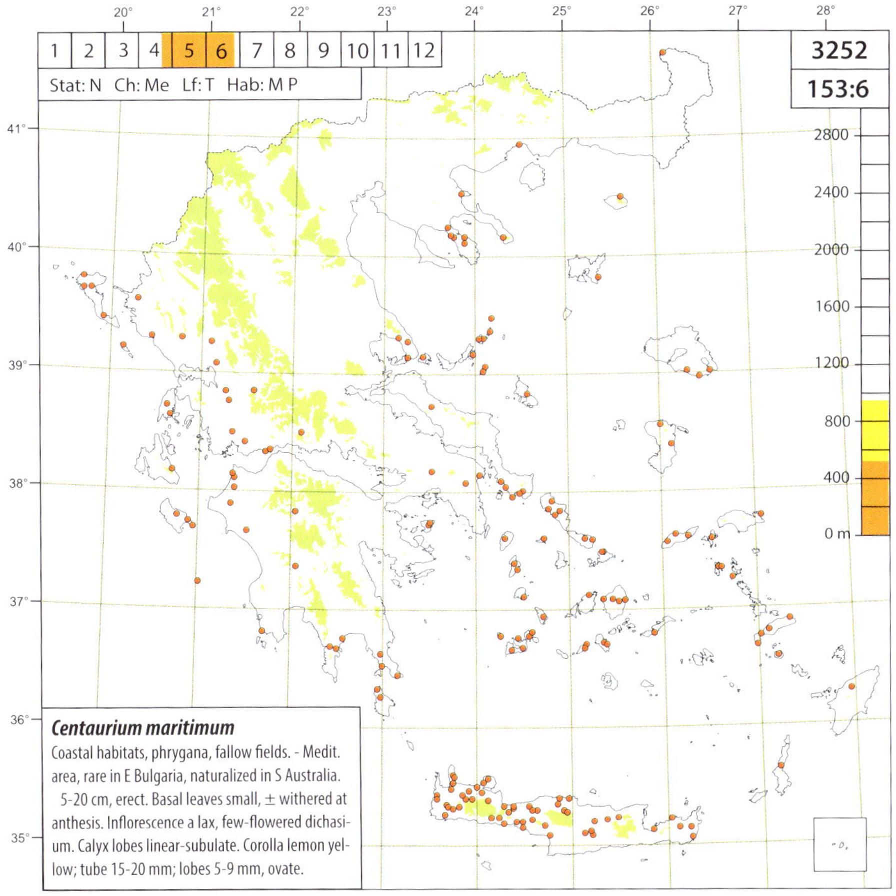

Centaurium maritimum

Coastal habitats, phrygana, fallow fields. - Medit. area, rare in E Bulgaria, naturalized in S Australia.

5-20 cm, erect. Basal leaves small, ± withered at anthesis. Inflorescence a lax, few-flowered dichasium. Calyx lobes linear-subulate. Corolla lemon yellow; tube 15-20 mm; lobes 5-9 mm, ovate.

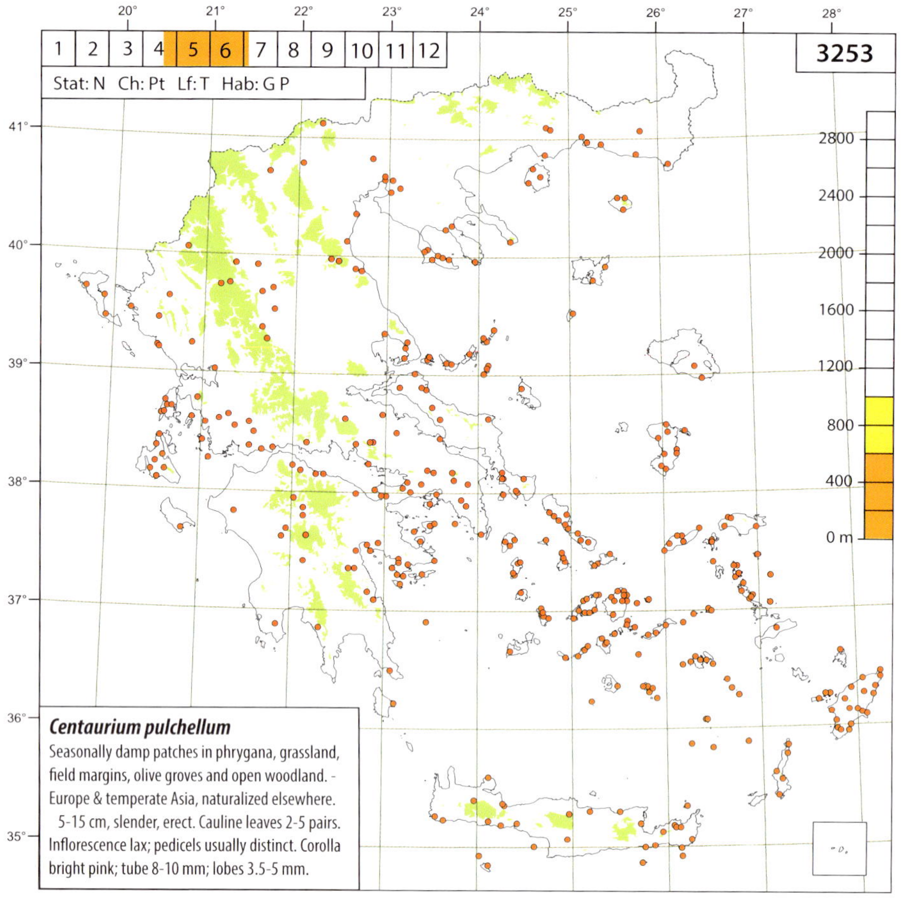

Centaurium pulchellum
Seasonally damp patches in phrygana, grassland, field margins, olive groves and open woodland. - Europe & temperate Asia, naturalized elsewhere.
5-15 cm, slender, erect. Cauline leaves 2-5 pairs. Inflorescence lax; pedicels usually distinct. Corolla bright pink; tube 8-10 mm; lobes 3.5-5 mm.

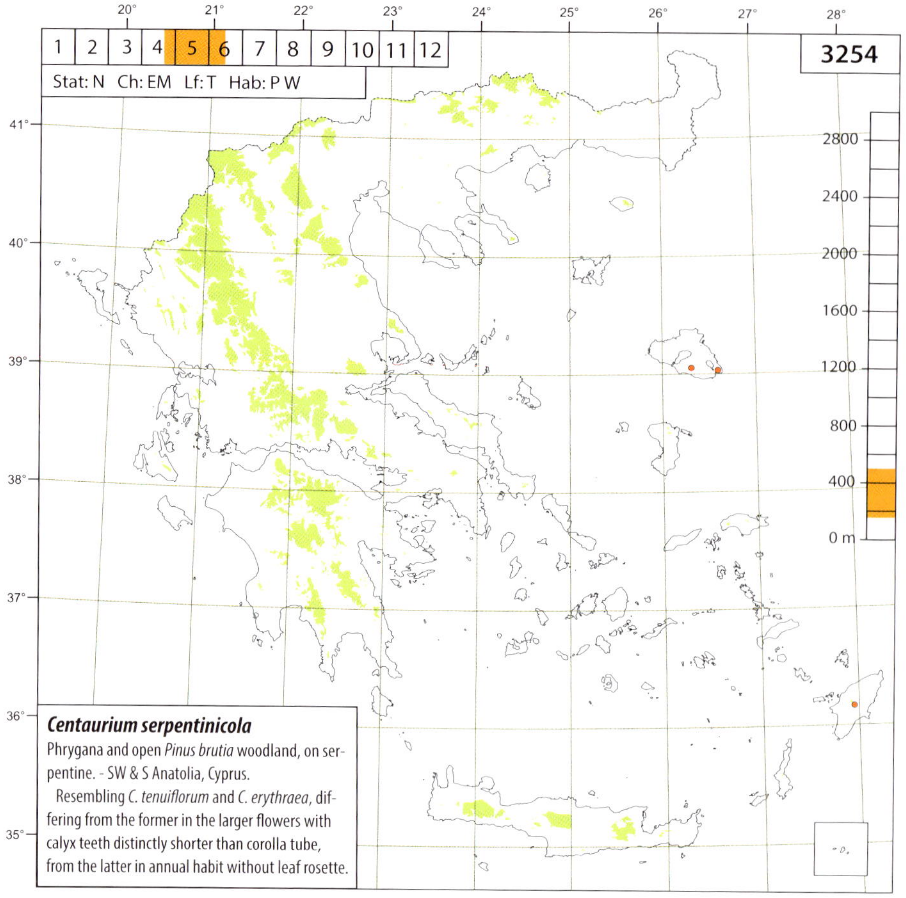

Centaurium serpentinicola
Phrygana and open *Pinus brutia* woodland, on serpentine. - SW & S Anatolia, Cyprus.
Resembling *C. tenuiflorum* and *C. erythraea*, differing from the former in the larger flowers with calyx teeth distinctly shorter than corolla tube, from the latter in annual habit without leaf rosette.

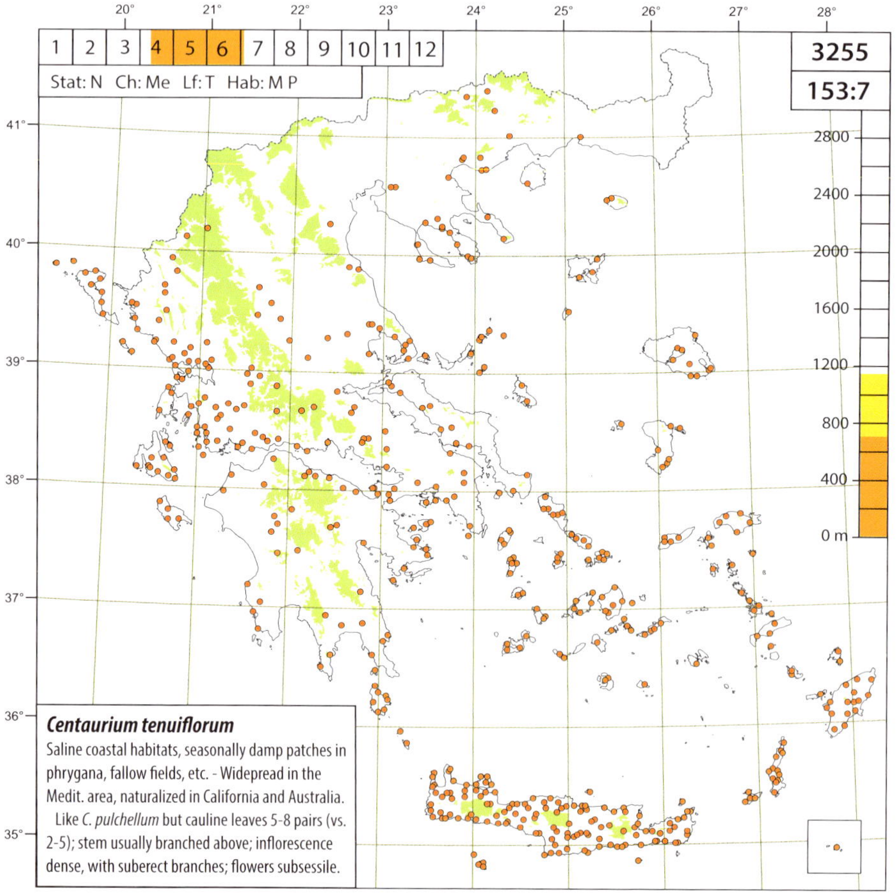

Centaurium tenuiflorum
Saline coastal habitats, seasonally damp patches in phrygana, fallow fields, etc. - Widespread in the Medit. area, naturalized in California and Australia.
Like *C. pulchellum* but cauline leaves 5-8 pairs (vs. 2-5); stem usually branched above; inflorescence dense, with suberect branches; flowers subsessile.

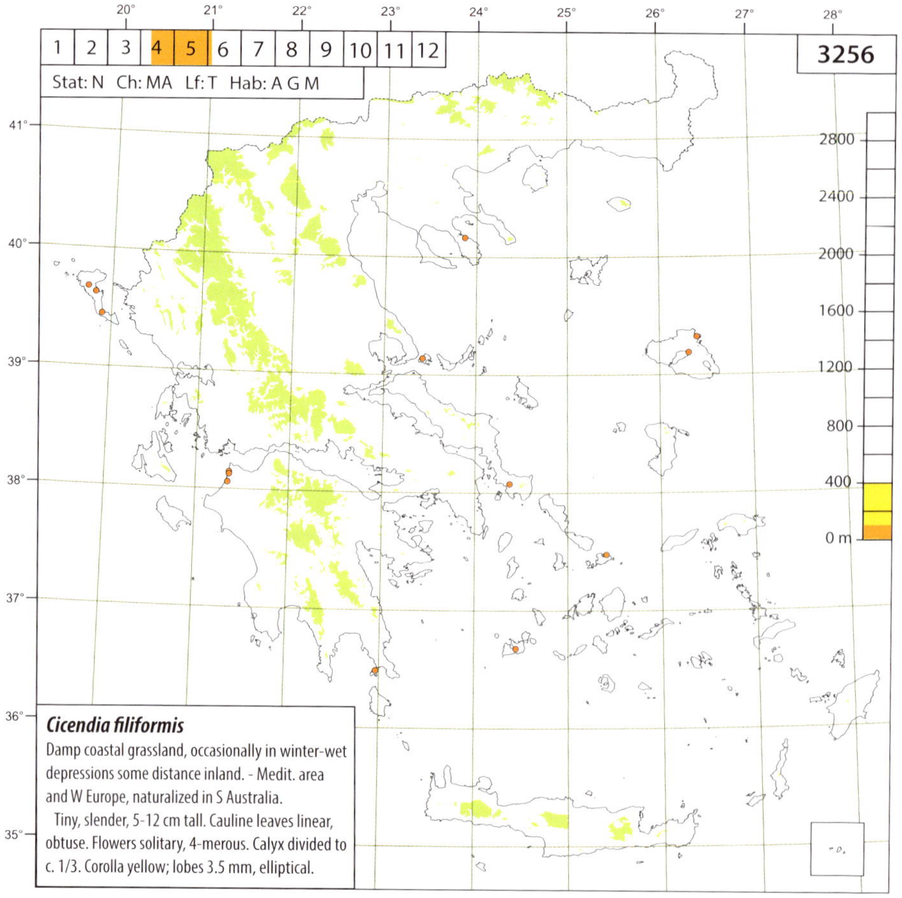

Cicendia filiformis
Damp coastal grassland, occasionally in winter-wet depressions some distance inland. - Medit. area and W Europe, naturalized in S Australia.
Tiny, slender, 5-12 cm tall. Cauline leaves linear, obtuse. Flowers solitary, 4-merous. Calyx divided to c. 1/3. Corolla yellow; lobes 3.5 mm, elliptical.

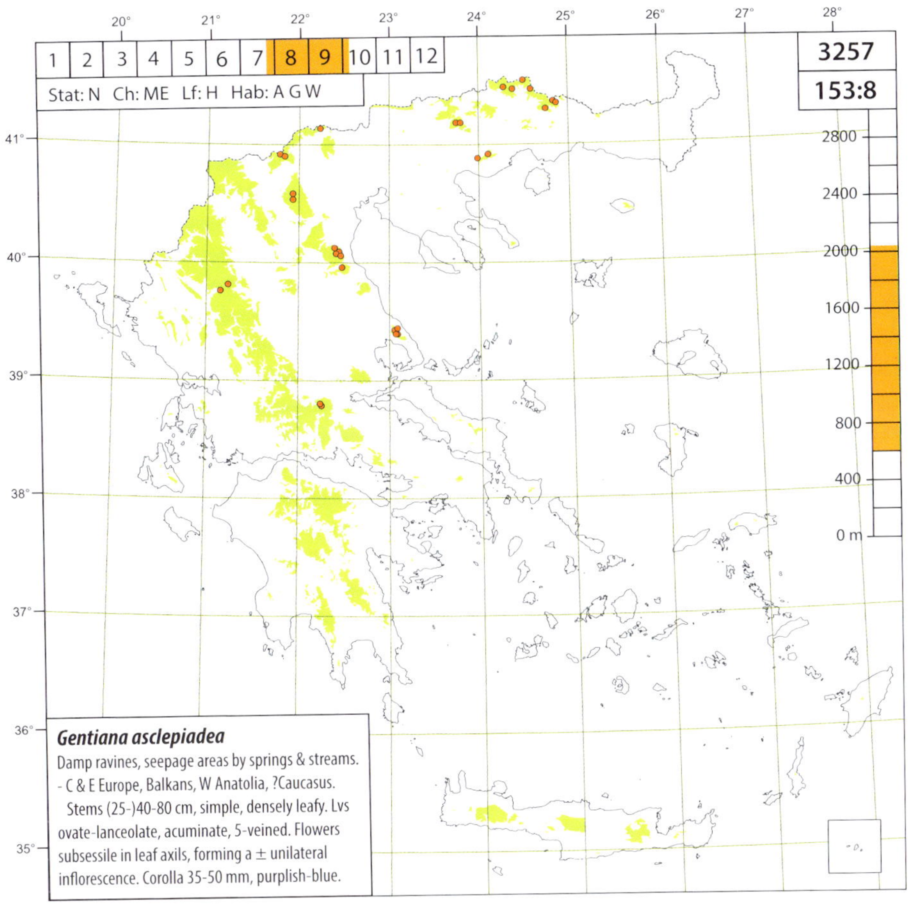

Gentiana asclepiadea
Damp ravines, seepage areas by springs & streams. - C & E Europe, Balkans, W Anatolia, ?Caucasus.
Stems (25-)40-80 cm, simple, densely leafy. Lvs ovate-lanceolate, acuminate, 5-veined. Flowers subsessile in leaf axils, forming a ± unilateral inflorescence. Corolla 35-50 mm, purplish-blue.

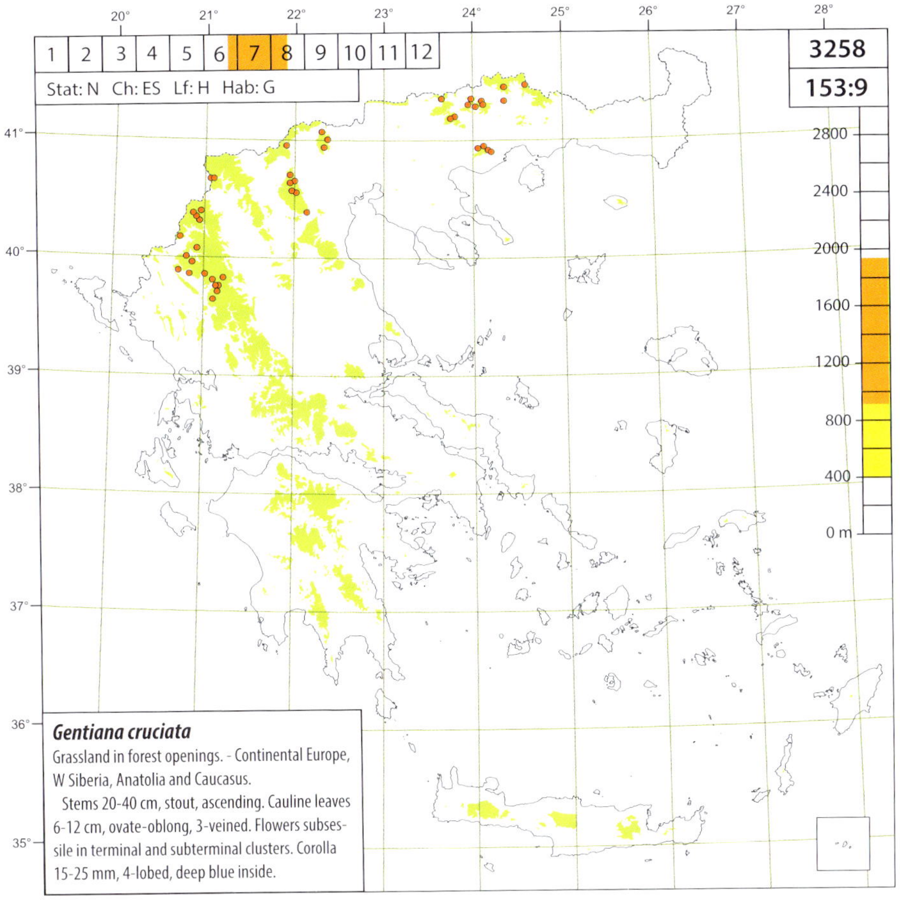

Gentiana cruciata
Grassland in forest openings. - Continental Europe, W Siberia, Anatolia and Caucasus.
Stems 20-40 cm, stout, ascending. Cauline leaves 6-12 cm, ovate-oblong, 3-veined. Flowers subsessile in terminal and subterminal clusters. Corolla 15-25 mm, 4-lobed, deep blue inside.

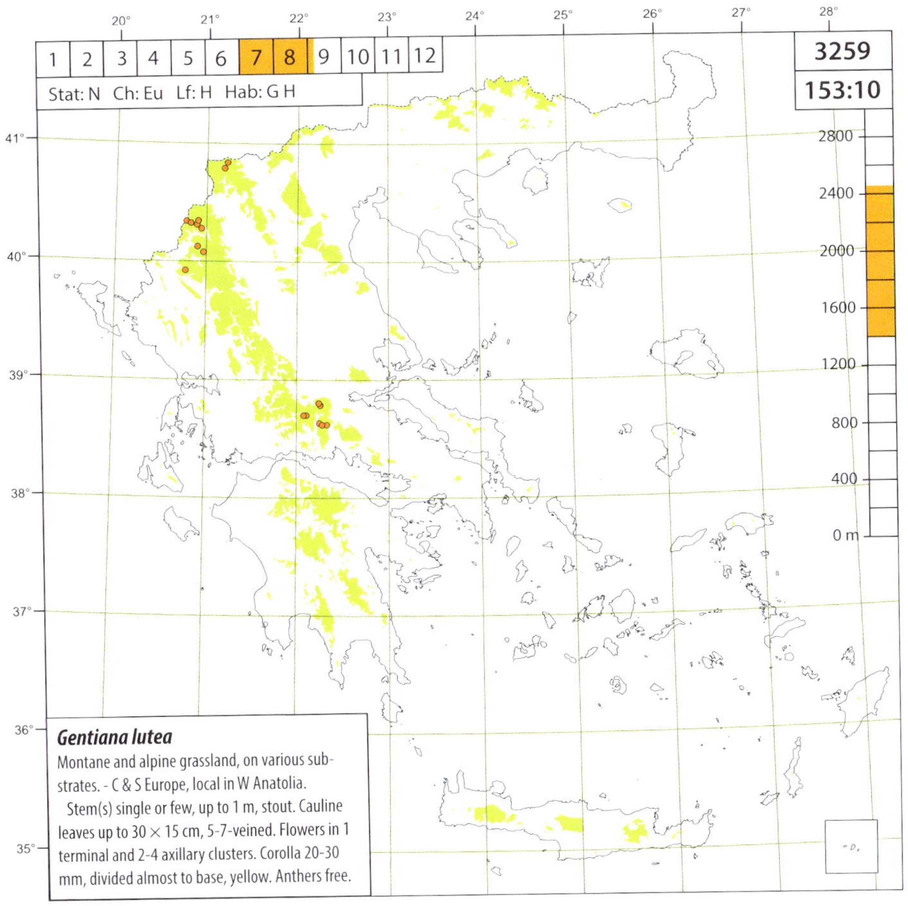

Gentiana lutea
Montane and alpine grassland, on various substrates. - C & S Europe, local in W Anatolia.
Stem(s) single or few, up to 1 m, stout. Cauline leaves up to 30 × 15 cm, 5-7-veined. Flowers in 1 terminal and 2-4 axillary clusters. Corolla 20-30 mm, divided almost to base, yellow. Anthers free.

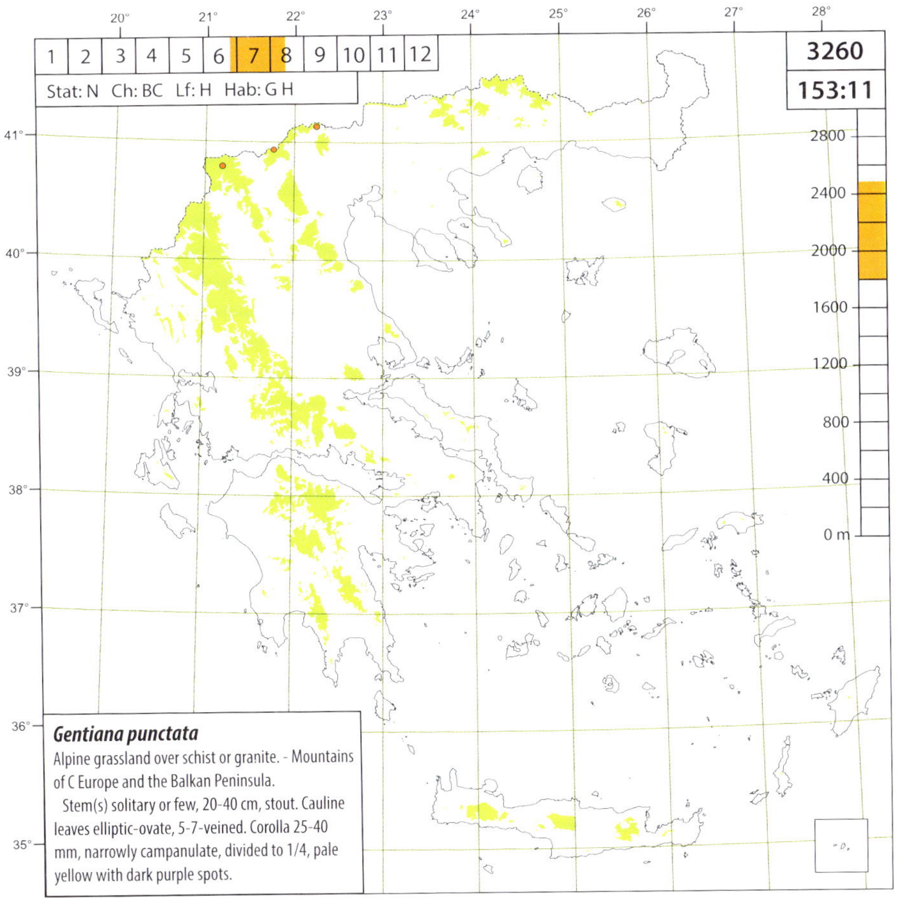

Gentiana punctata
Alpine grassland over schist or granite. - Mountains of C Europe and the Balkan Peninsula.
Stem(s) solitary or few, 20-40 cm, stout. Cauline leaves elliptic-ovate, 5-7-veined. Corolla 25-40 mm, narrowly campanulate, divided to 1/4, pale yellow with dark purple spots.

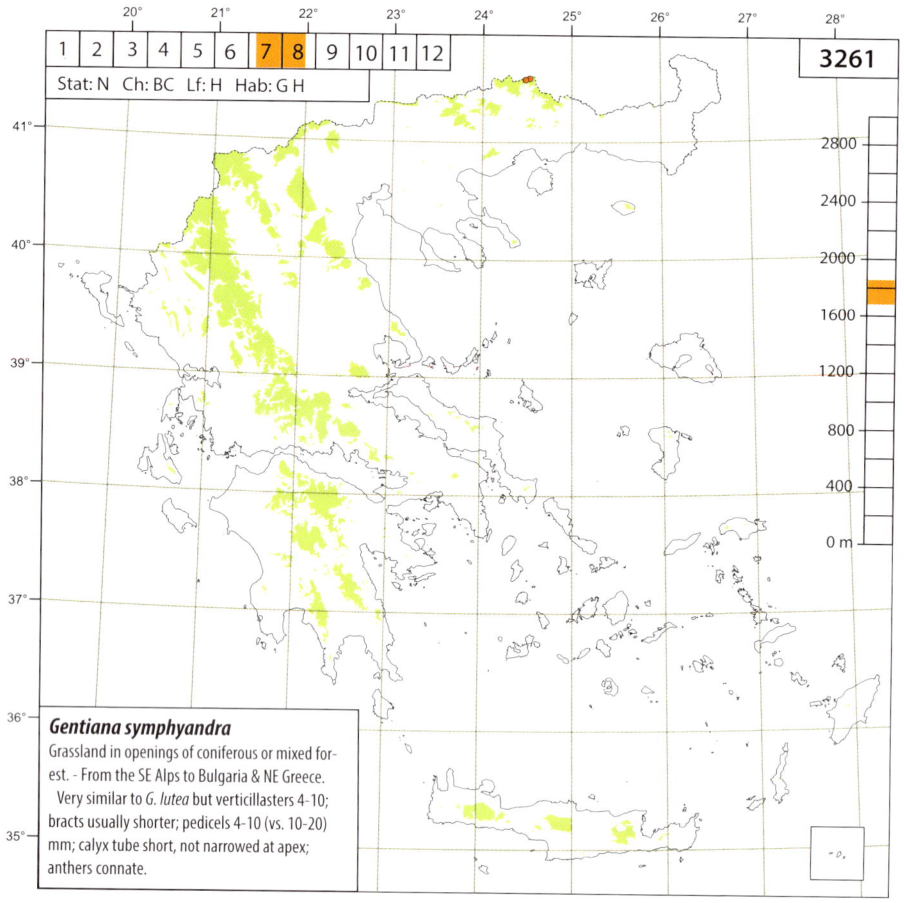

Gentiana symphyandra
Grassland in openings of coniferous or mixed forest. - From the SE Alps to Bulgaria & NE Greece.
Very similar to *G. lutea* but verticillasters 4-10; bracts usually shorter; pedicels 4-10 (vs. 10-20) mm; calyx tube short, not narrowed at apex; anthers connate.

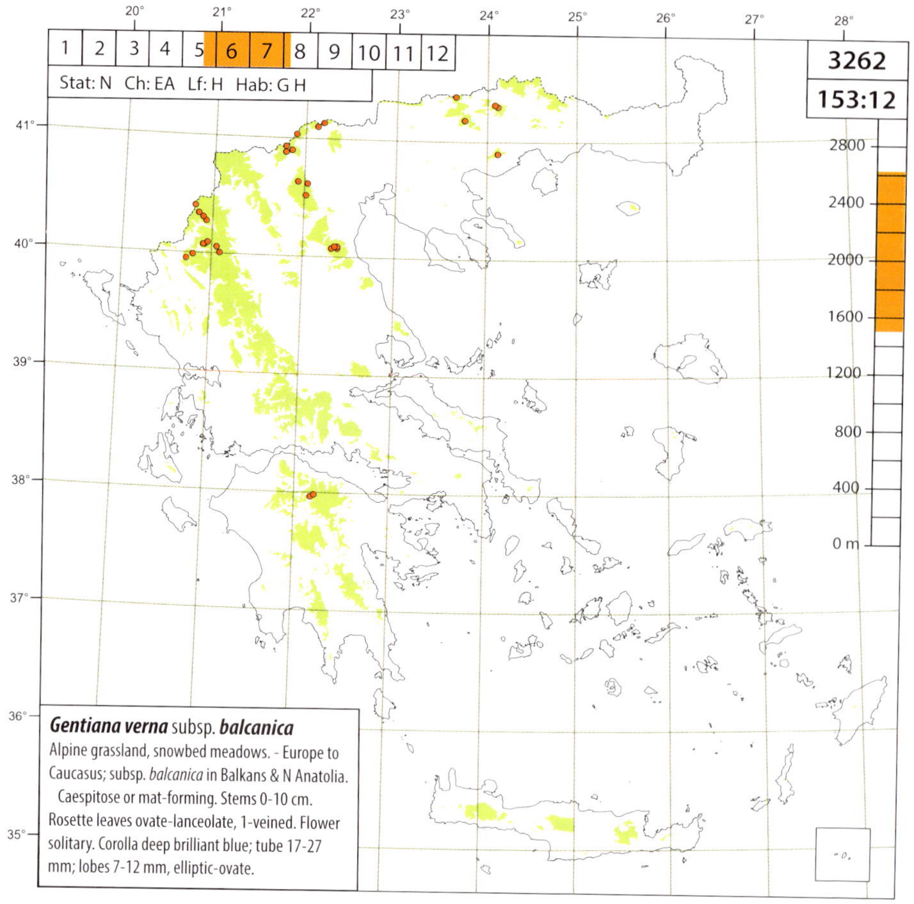

Gentiana verna* subsp. *balcanica
Alpine grassland, snowbed meadows. - Europe to Caucasus; subsp. *balcanica* in Balkans & N Anatolia.
Caespitose or mat-forming. Stems 0-10 cm. Rosette leaves ovate-lanceolate, 1-veined. Flower solitary. Corolla deep brilliant blue; tube 17-27 mm; lobes 7-12 mm, elliptic-ovate.

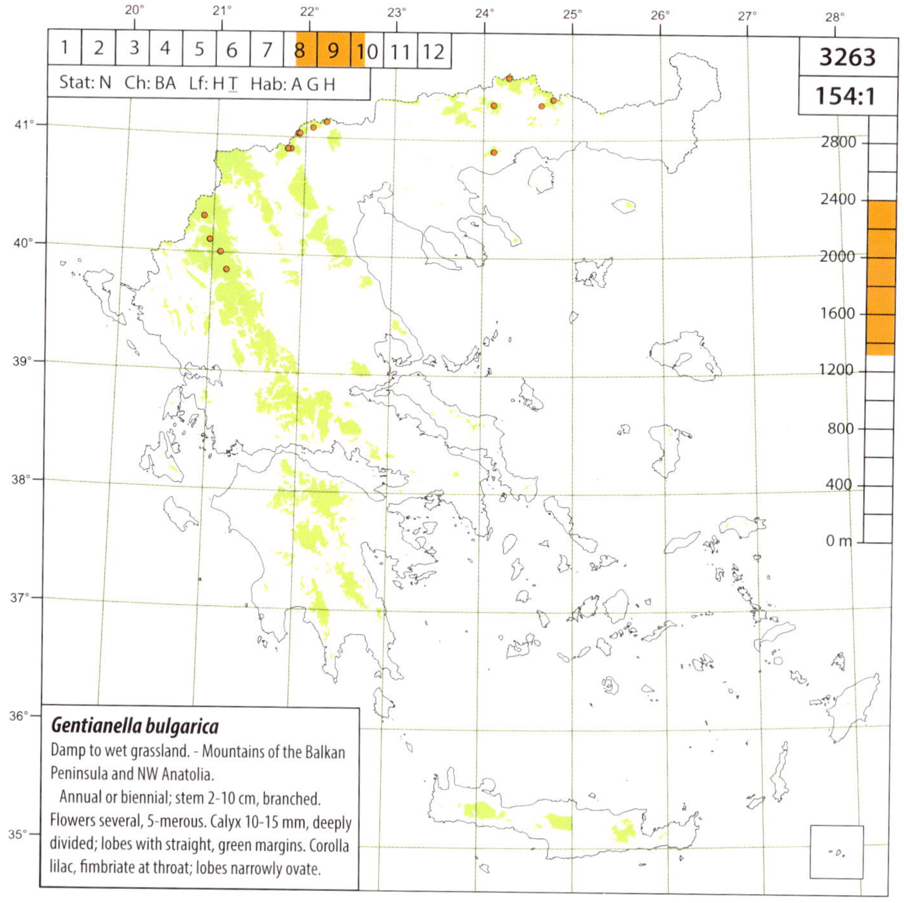

Gentianella bulgarica
Damp to wet grassland. - Mountains of the Balkan Peninsula and NW Anatolia.
Annual or biennial; stem 2-10 cm, branched. Flowers several, 5-merous. Calyx 10-15 mm, deeply divided; lobes with straight, green margins. Corolla lilac, fimbriate at throat; lobes narrowly ovate.

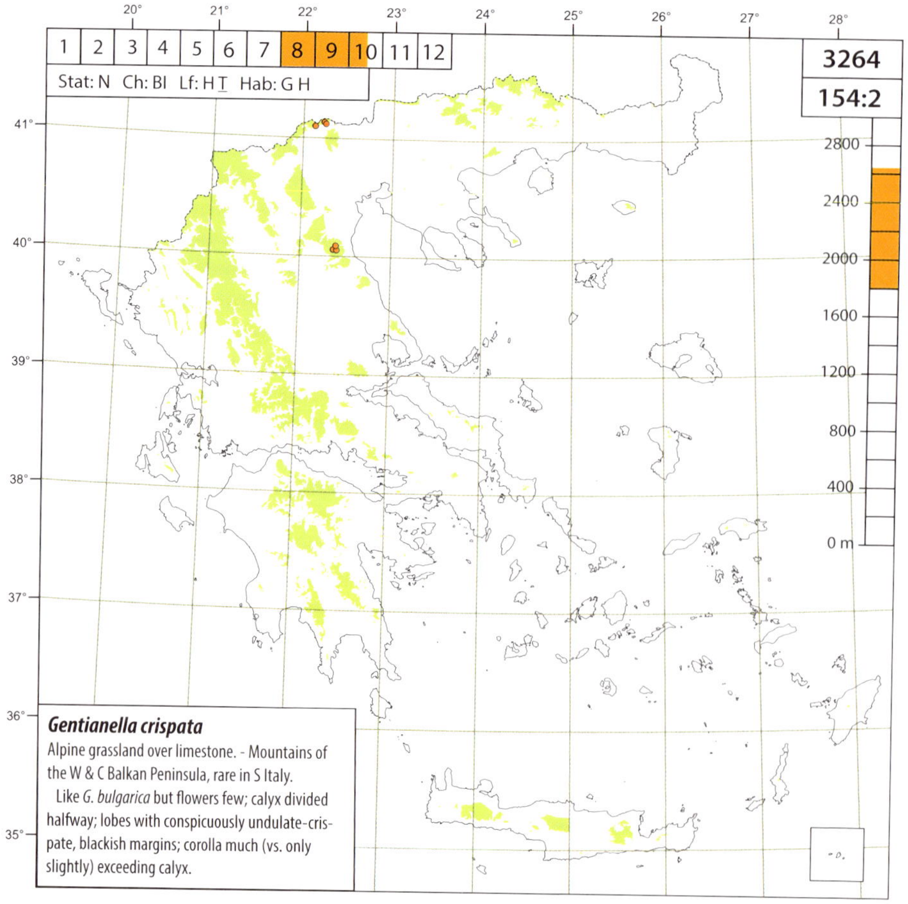

Gentianella crispata
Alpine grassland over limestone. - Mountains of the W & C Balkan Peninsula, rare in S Italy.
Like *G. bulgarica* but flowers few; calyx divided halfway; lobes with conspicuously undulate-crispate, blackish margins; corolla much (vs. only slightly) exceeding calyx.

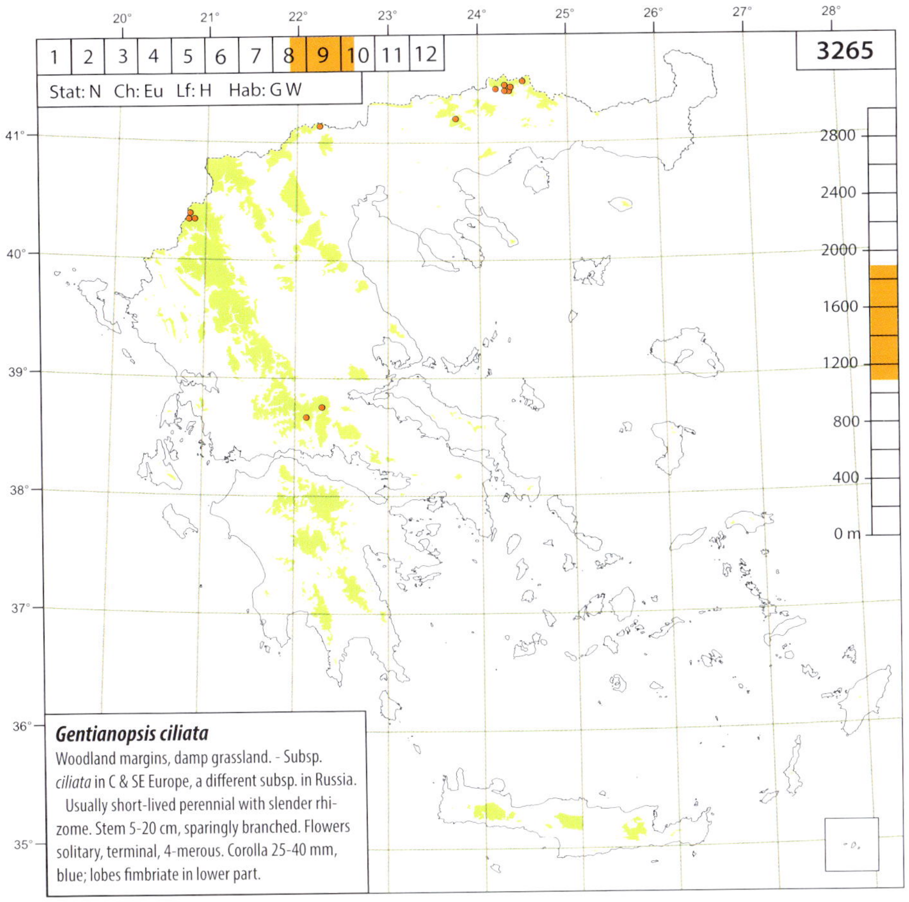

Gentianopsis ciliata

Woodland margins, damp grassland. - Subsp. *ciliata* in C & SE Europe, a different subsp. in Russia.

Usually short-lived perennial with slender rhizome. Stem 5-20 cm, sparingly branched. Flowers solitary, terminal, 4-merous. Corolla 25-40 mm, blue; lobes fimbriate in lower part.

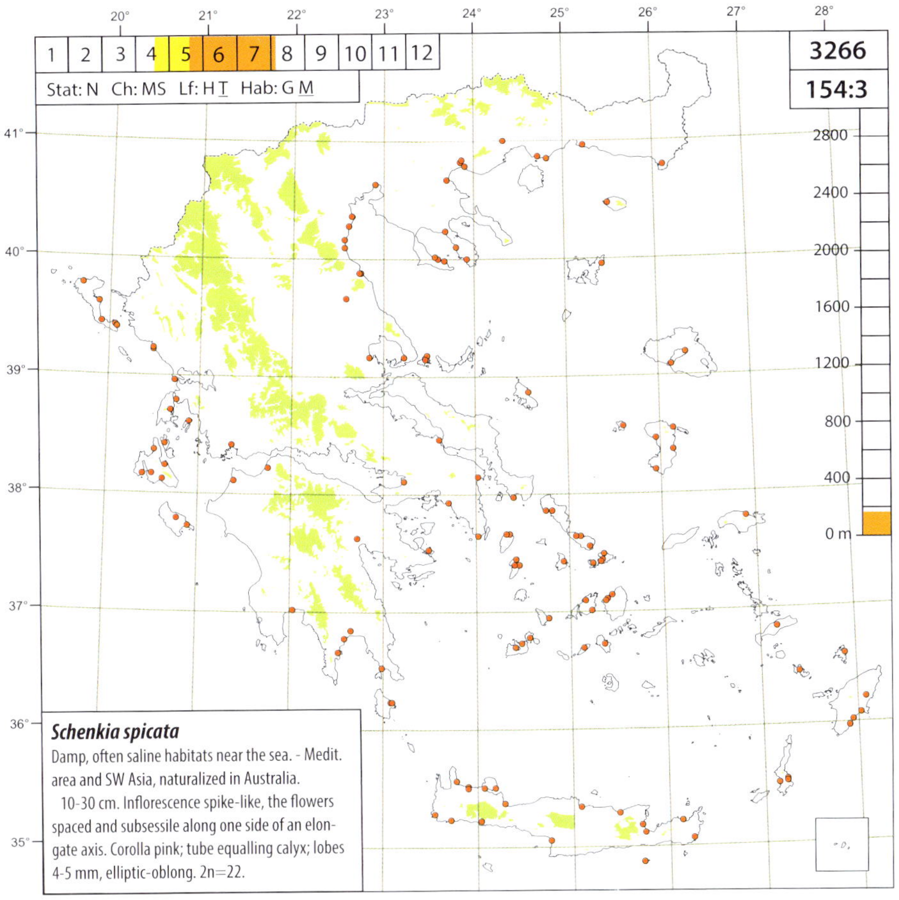

Schenkia spicata

Damp, often saline habitats near the sea. - Medit. area and SW Asia, naturalized in Australia.

10-30 cm. Inflorescence spike-like, the flowers spaced and subsessile along one side of an elongate axis. Corolla pink; tube equalling calyx; lobes 4-5 mm, elliptic-oblong. 2n=22.

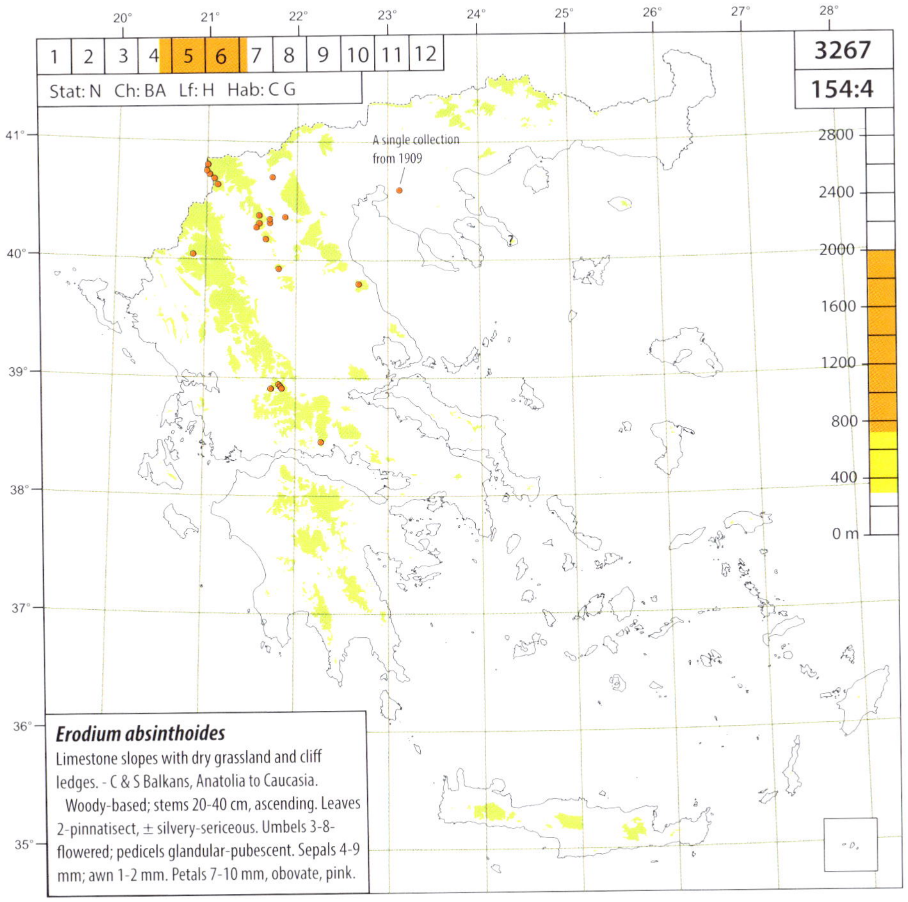

Erodium absinthoides

Limestone slopes with dry grassland and cliff ledges. - C & S Balkans, Anatolia to Caucasia.

Woody-based; stems 20-40 cm, ascending. Leaves 2-pinnatisect, ± silvery-sericeous. Umbels 3-8-flowered; pedicels glandular-pubescent. Sepals 4-9 mm; awn 1-2 mm. Petals 7-10 mm, obovate, pink.

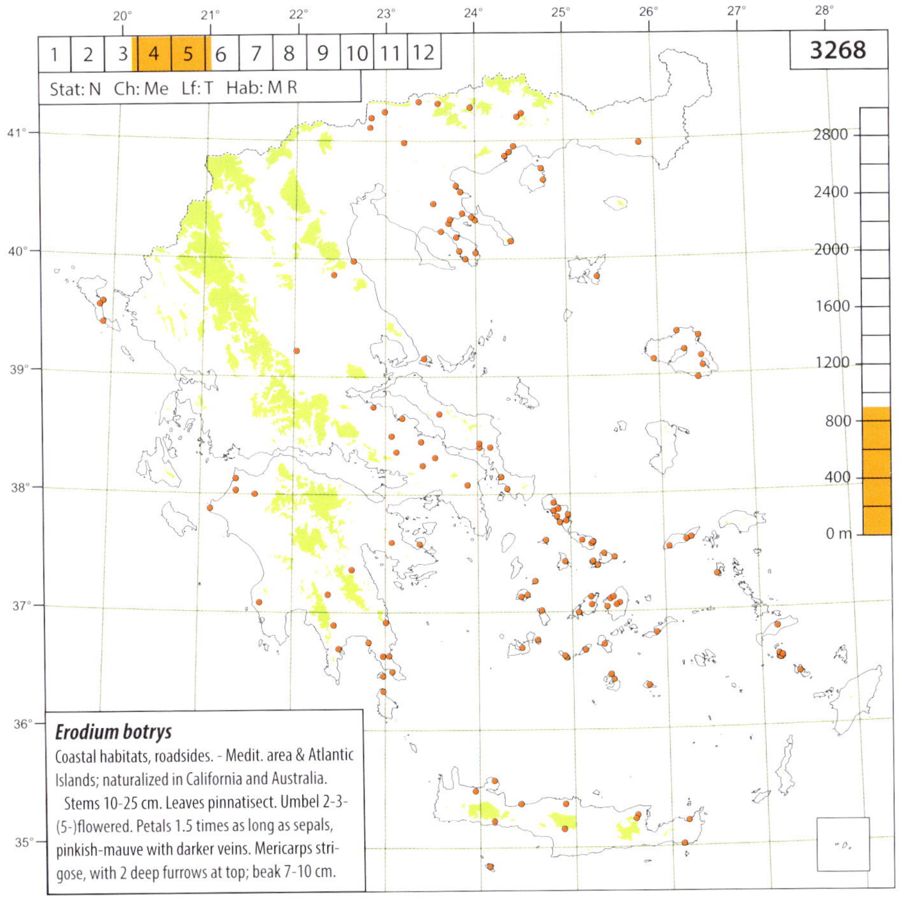

Erodium botrys

Coastal habitats, roadsides. - Medit. area & Atlantic Islands; naturalized in California and Australia.

Stems 10-25 cm. Leaves pinnatisect. Umbel 2-3-(5-)flowered. Petals 1.5 times as long as sepals, pinkish-mauve with darker veins. Mericarps strigose, with 2 deep furrows at top; beak 7-10 cm.

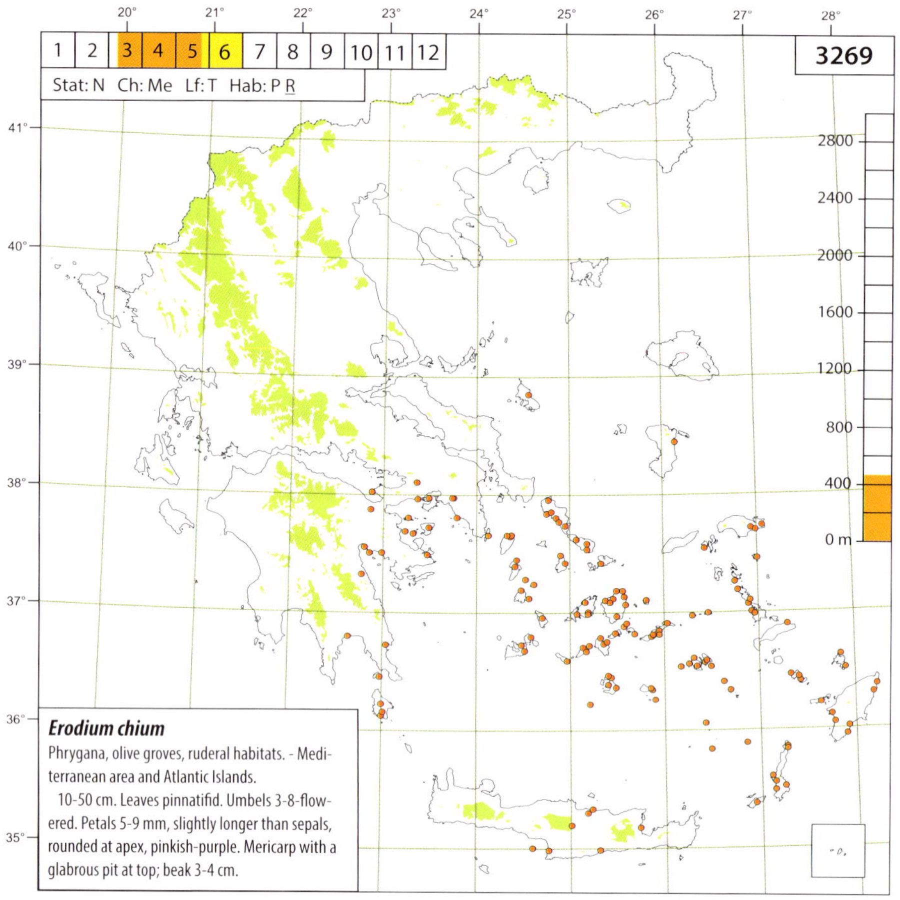

Erodium chium

Phrygana, olive groves, ruderal habitats. - Mediterranean area and Atlantic Islands.

10-50 cm. Leaves pinnatifid. Umbels 3-8-flowered. Petals 5-9 mm, slightly longer than sepals, rounded at apex, pinkish-purple. Mericarp with a glabrous pit at top; beak 3-4 cm.

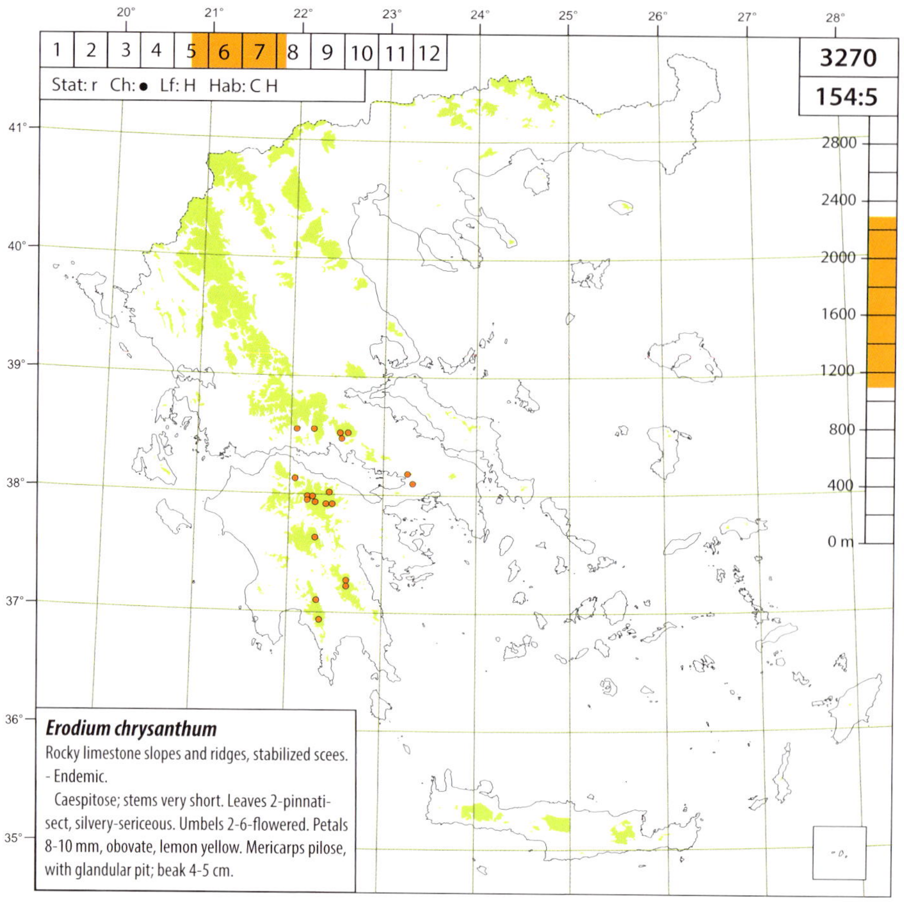

Erodium chrysanthum

Rocky limestone slopes and ridges, stabilized scees. - Endemic.

Caespitose; stems very short. Leaves 2-pinnatisect, silvery-sericeous. Umbels 2-6-flowered. Petals 8-10 mm, obovate, lemon yellow. Mericarps pilose, with glandular pit; beak 4-5 cm.

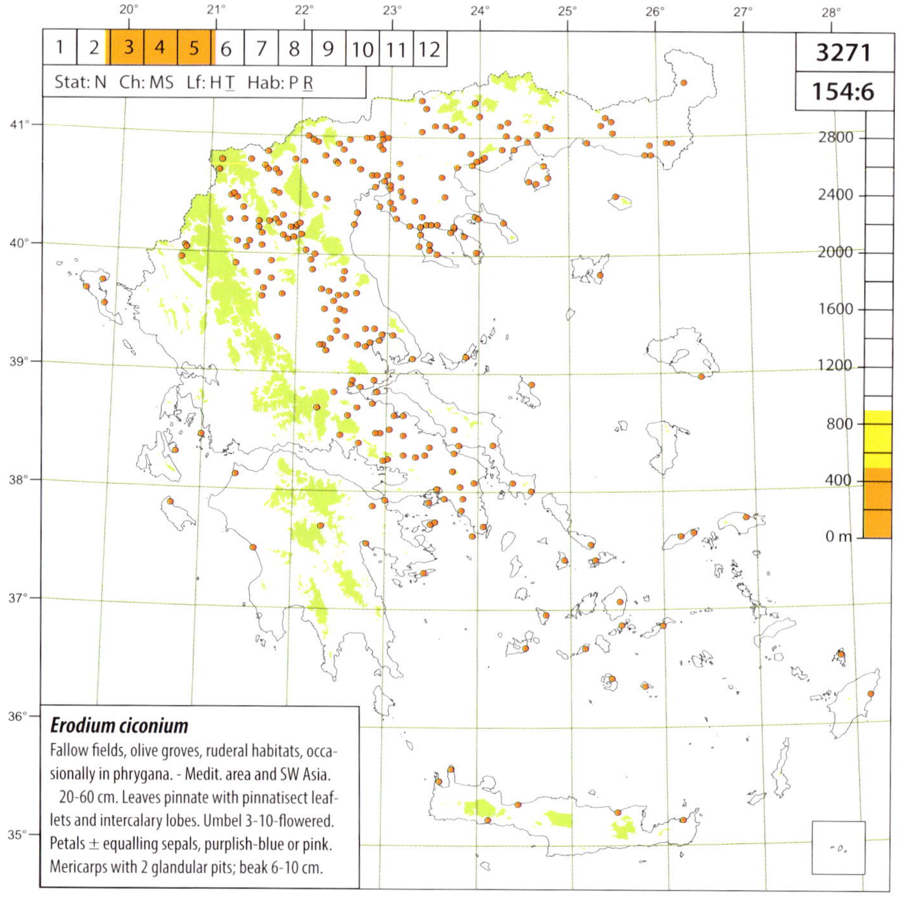

Erodium ciconium

Fallow fields, olive groves, ruderal habitats, occasionally in phrygana. - Medit. area and SW Asia.

20-60 cm. Leaves pinnate with pinnatisect leaflets and intercalary lobes. Umbel 3-10-flowered. Petals ± equalling sepals, purplish-blue or pink. Mericarps with 2 glandular pits; beak 6-10 cm.

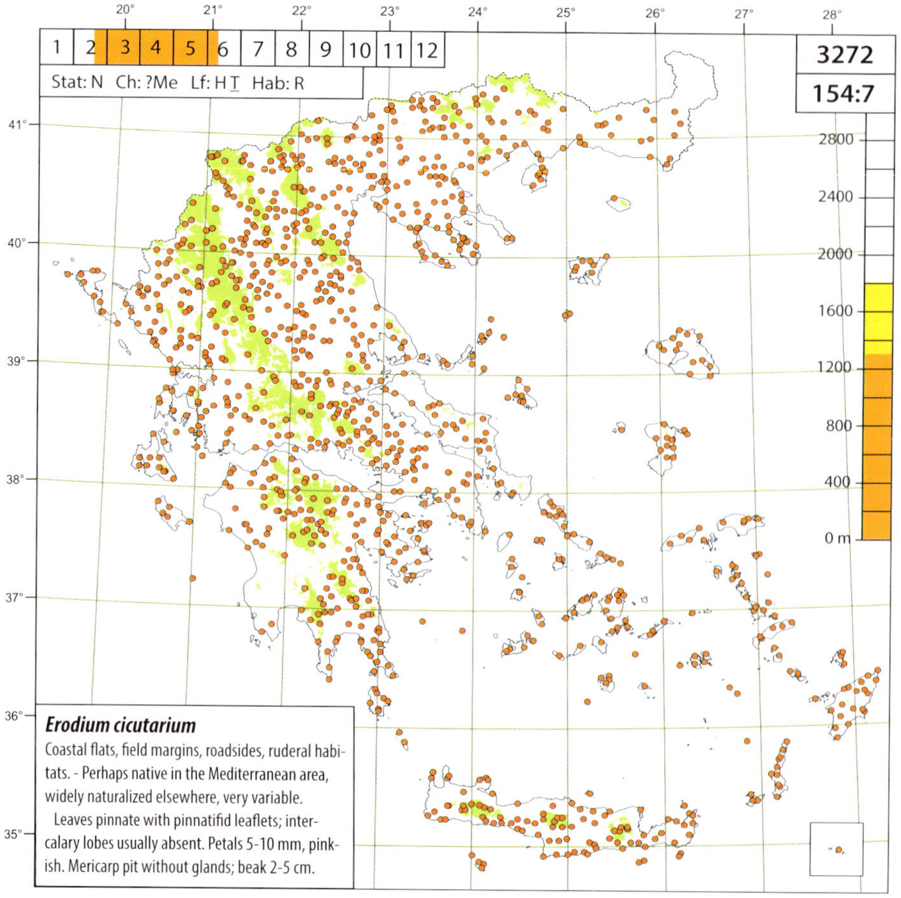

Erodium cicutarium

Coastal flats, field margins, roadsides, ruderal habitats. - Perhaps native in the Mediterranean area, widely naturalized elsewhere, very variable.

Leaves pinnate with pinnatifid leaflets; intercalary lobes usually absent. Petals 5-10 mm, pinkish. Mericarp pit without glands; beak 2-5 cm.

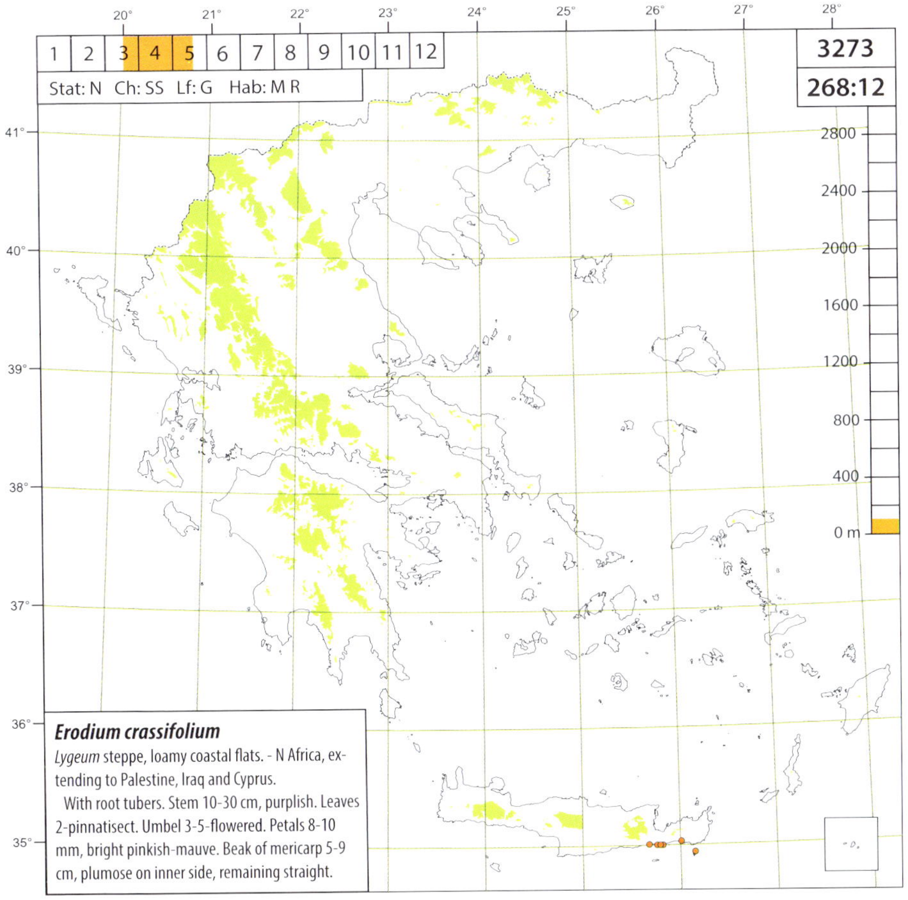

Erodium crassifolium

Lygeum steppe, loamy coastal flats. - N Africa, extending to Palestine, Iraq and Cyprus.

With root tubers. Stem 10-30 cm, purplish. Leaves 2-pinnatisect. Umbel 3-5-flowered. Petals 8-10 mm, bright pinkish-mauve. Beak of mericarp 5-9 cm, plumose on inner side, remaining straight.

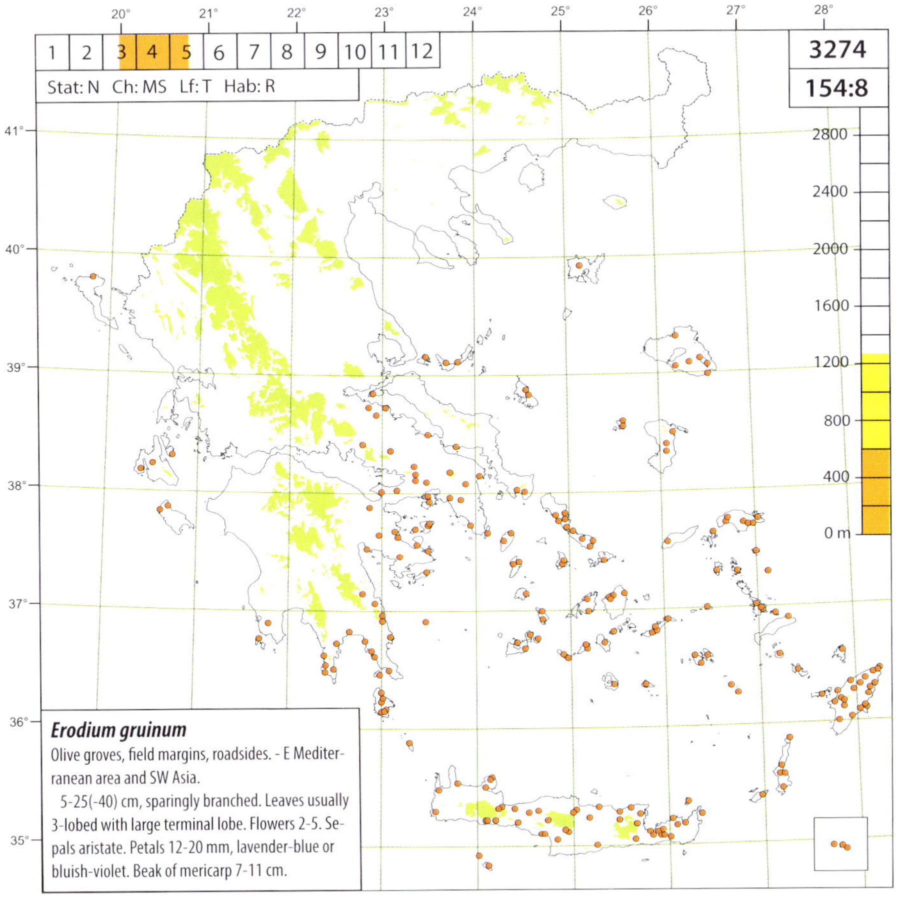

Erodium gruinum

Olive groves, field margins, roadsides. - E Mediterranean area and SW Asia.

5-25(-40) cm, sparingly branched. Leaves usually 3-lobed with large terminal lobe. Flowers 2-5. Sepals aristate. Petals 12-20 mm, lavender-blue or bluish-violet. Beak of mericarp 7-11 cm.

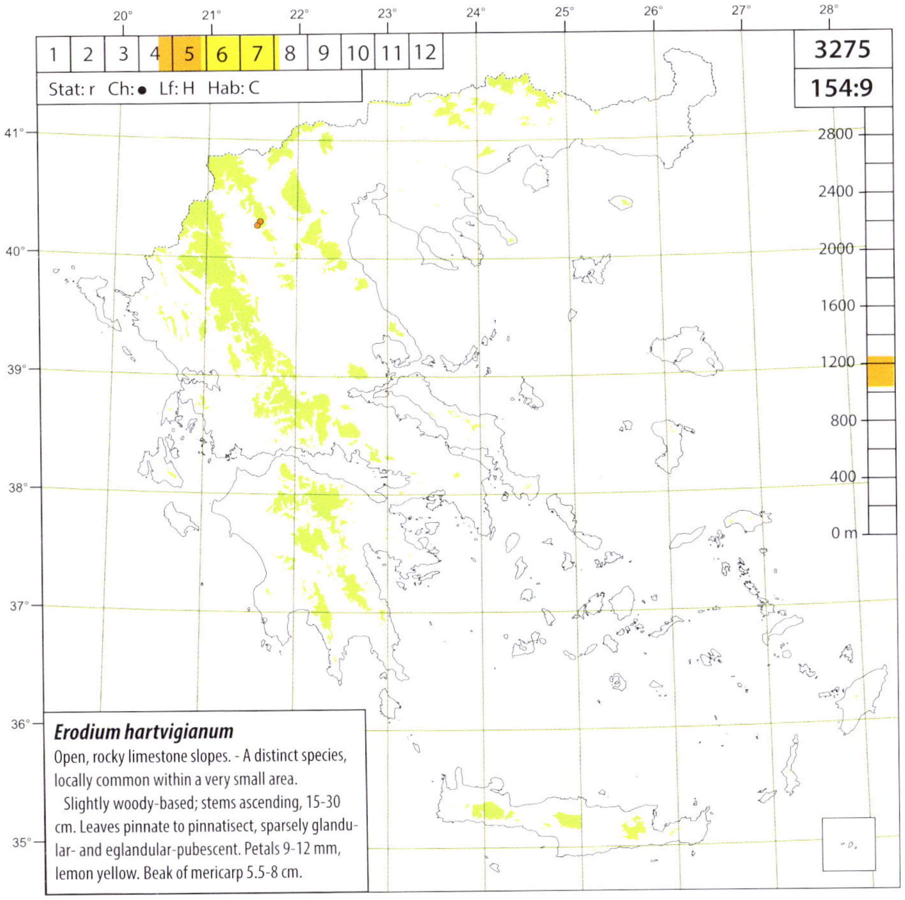

Erodium hartvigianum

Open, rocky limestone slopes. - A distinct species, locally common within a very small area.

Slightly woody-based; stems ascending, 15-30 cm. Leaves pinnate to pinnatisect, sparsely glandular- and eglandular-pubescent. Petals 9-12 mm, lemon yellow. Beak of mericarp 5.5-8 cm.

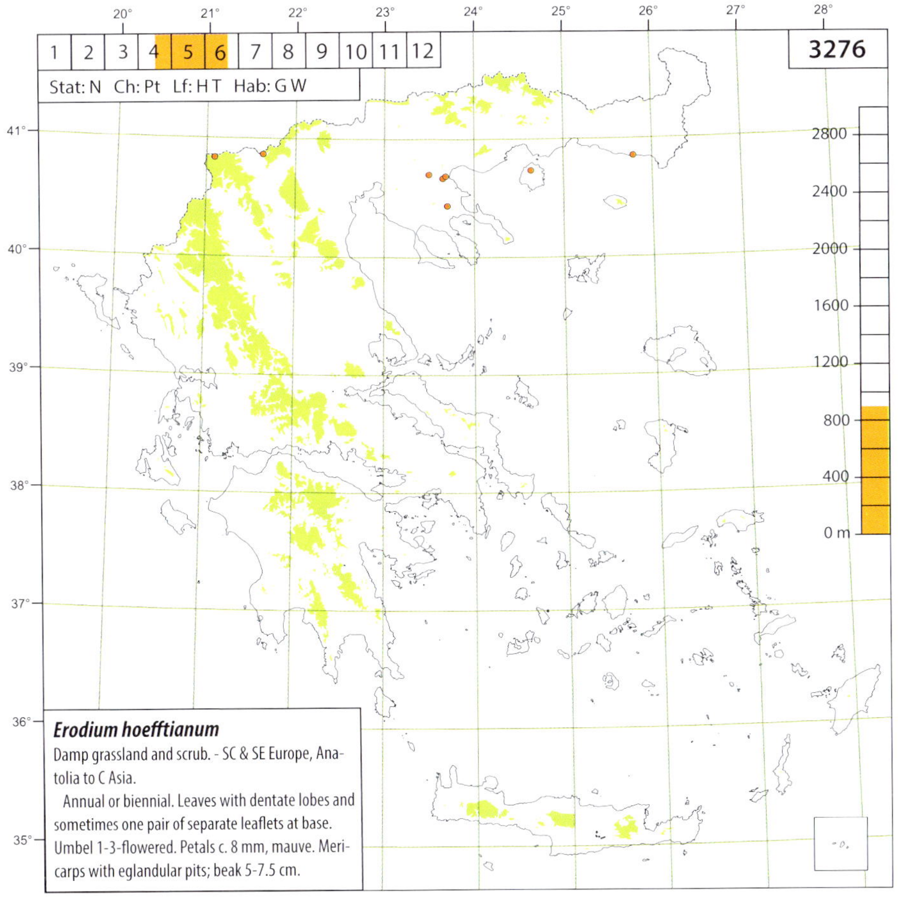

Erodium hoefftianum

Damp grassland and scrub. - SC & SE Europe, Anatolia to C Asia.

Annual or biennial. Leaves with dentate lobes and sometimes one pair of separate leaflets at base. Umbel 1-3-flowered. Petals c. 8 mm, mauve. Mericarps with eglandular pits; beak 5-7.5 cm.

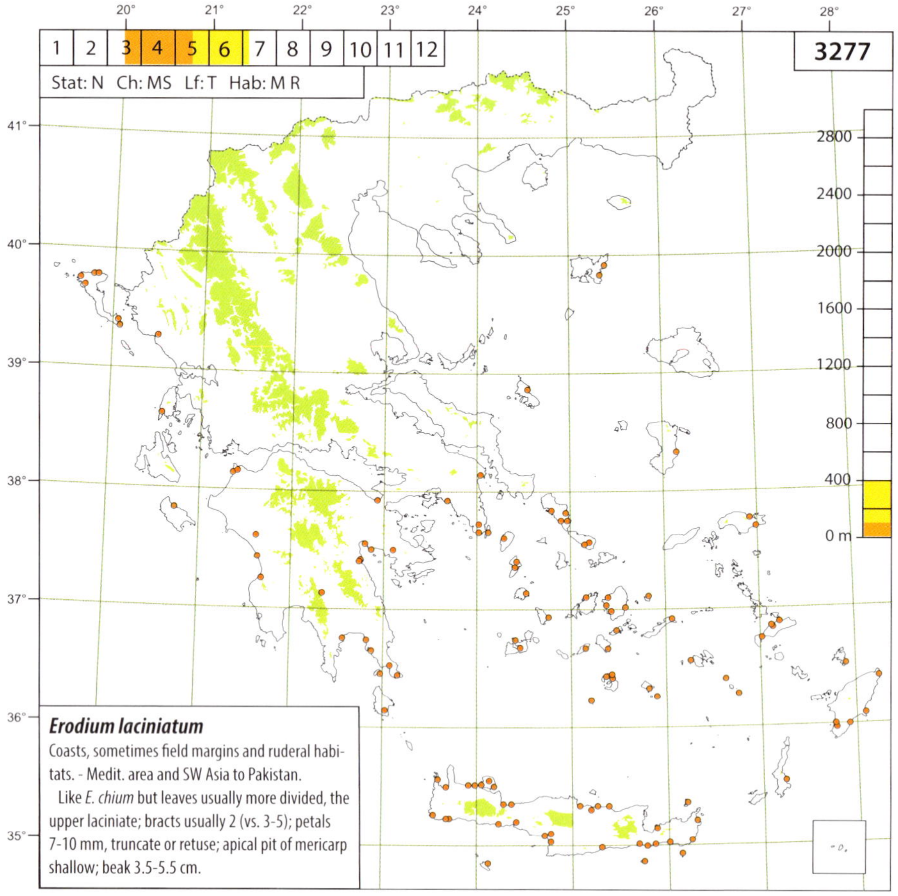

Erodium laciniatum

Coasts, sometimes field margins and ruderal habitats. - Medit. area and SW Asia to Pakistan.

Like *E. chium* but leaves usually more divided, the upper laciniate; bracts usually 2 (vs. 3-5); petals 7-10 mm, truncate or retuse; apical pit of mericarp shallow; beak 3.5-5.5 cm.

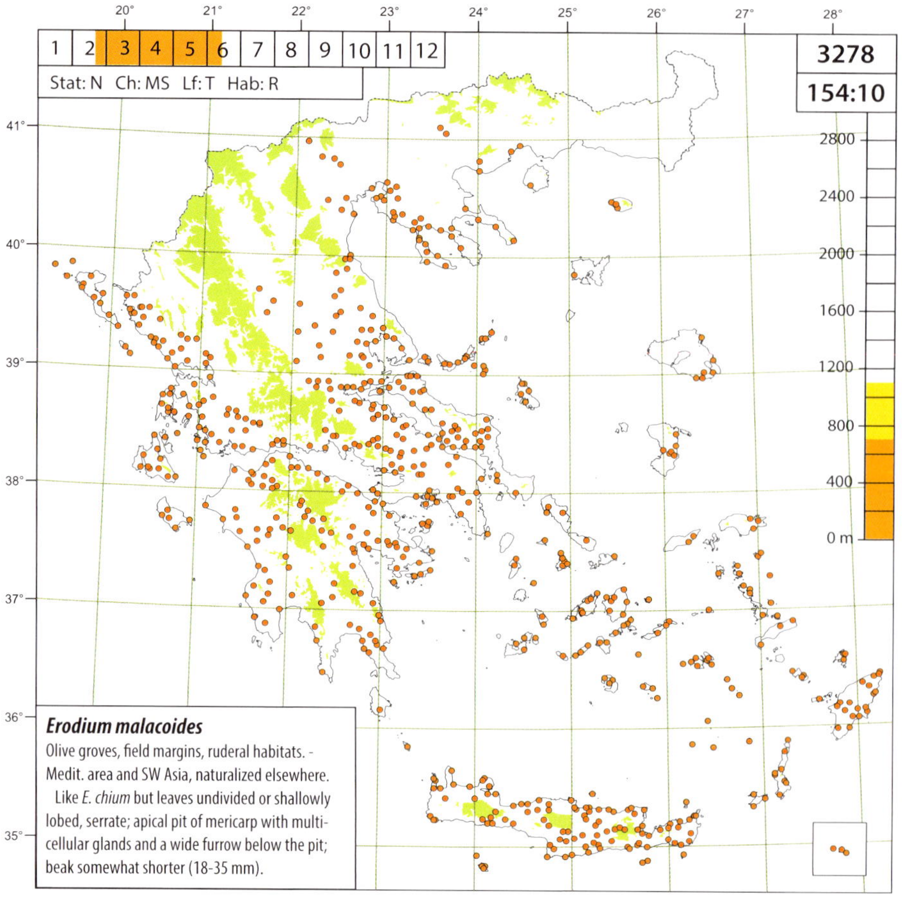

Erodium malacoides

Olive groves, field margins, ruderal habitats. - Medit. area and SW Asia, naturalized elsewhere.

Like *E. chium* but leaves undivided or shallowly lobed, serrate; apical pit of mericarp with multicellular glands and a wide furrow below the pit; beak somewhat shorter (18-35 mm).

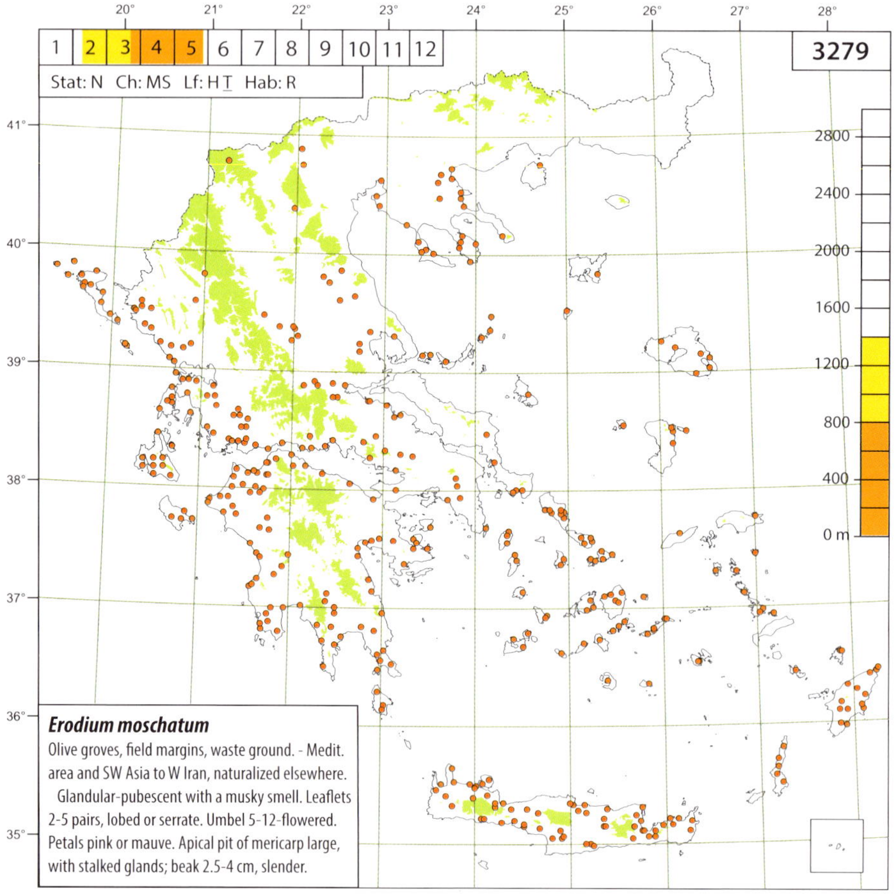

Erodium moschatum

Olive groves, field margins, waste ground. - Medit. area and SW Asia to W Iran, naturalized elsewhere.

Glandular-pubescent with a musky smell. Leaflets 2-5 pairs, lobed or serrate. Umbel 5-12-flowered. Petals pink or mauve. Apical pit of mericarp large, with stalked glands; beak 2.5-4 cm, slender.

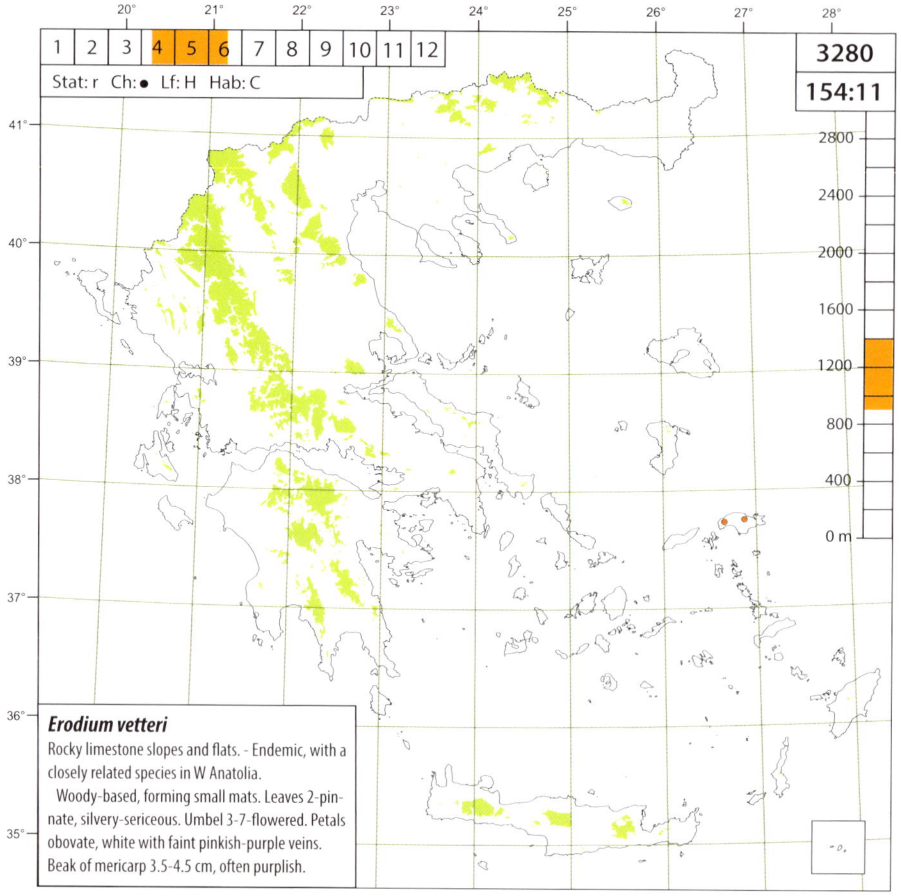

Erodium vetteri

Rocky limestone slopes and flats. - Endemic, with a closely related species in W Anatolia.

Woody-based, forming small mats. Leaves 2-pinnate, silvery-sericeous. Umbel 3-7-flowered. Petals obovate, white with faint pinkish-purple veins. Beak of mericarp 3.5-4.5 cm, often purplish.

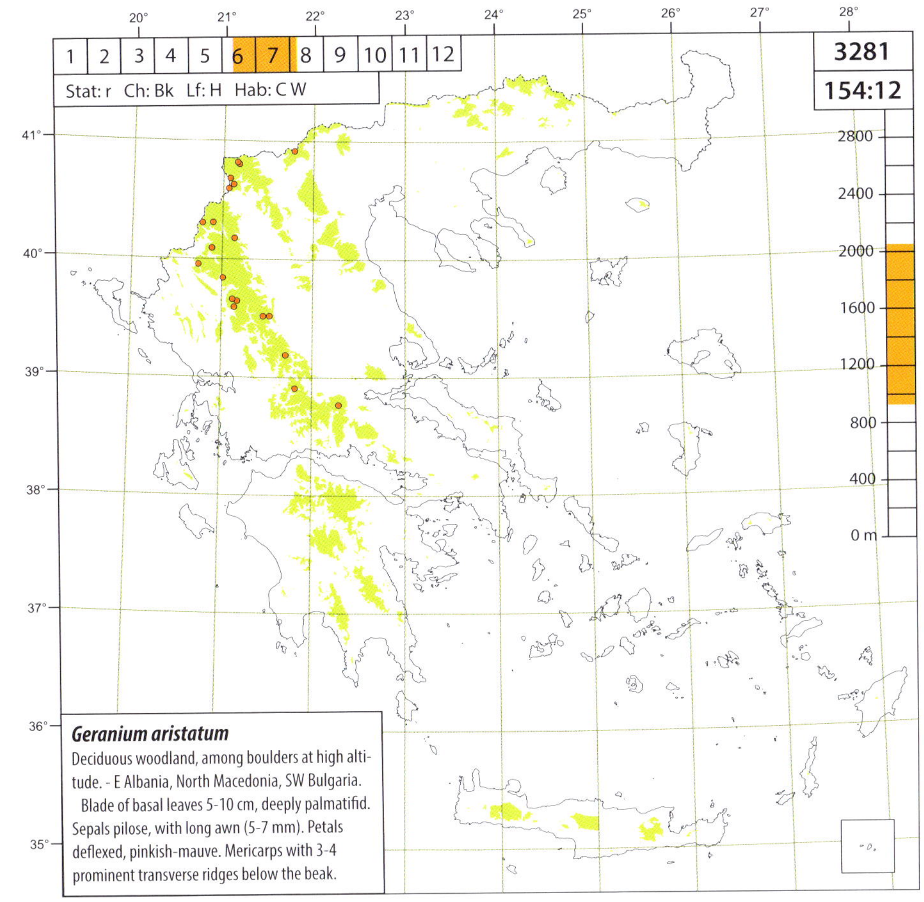

Geranium aristatum
Deciduous woodland, among boulders at high altitude. - E Albania, North Macedonia, SW Bulgaria.
Blade of basal leaves 5-10 cm, deeply palmatifid. Sepals pilose, with long awn (5-7 mm). Petals deflexed, pinkish-mauve. Mericarps with 3-4 prominent transverse ridges below the beak.

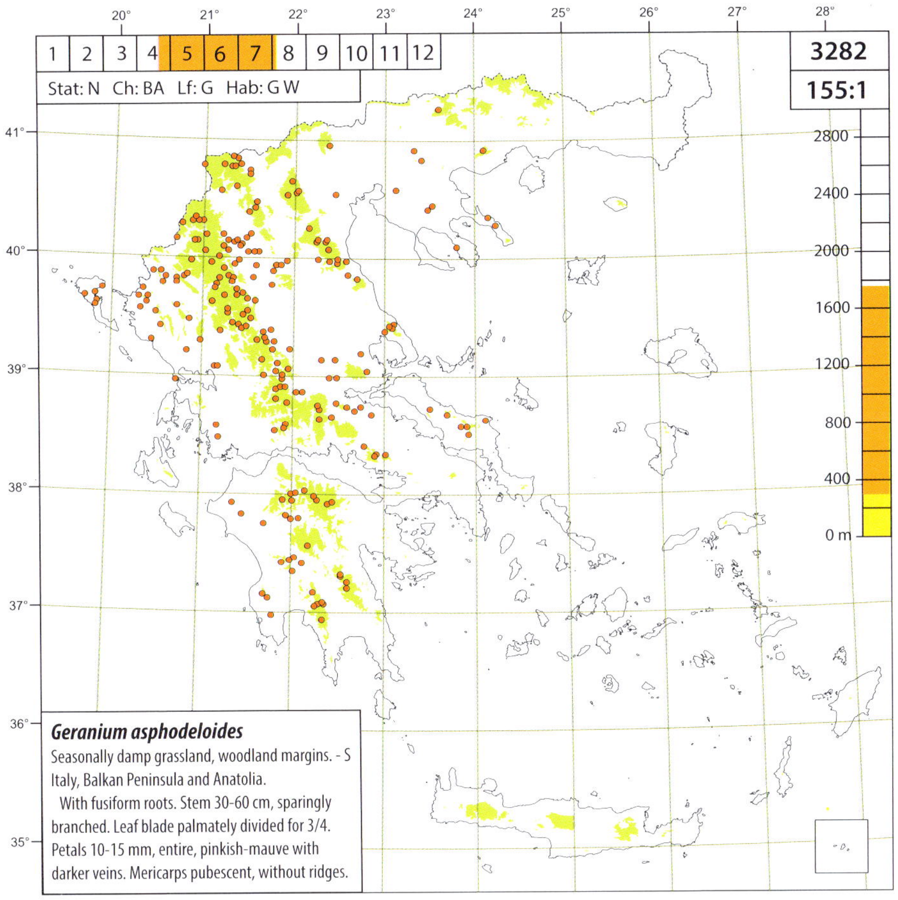

Geranium asphodeloides
Seasonally damp grassland, woodland margins. - S Italy, Balkan Peninsula and Anatolia.
With fusiform roots. Stem 30-60 cm, sparingly branched. Leaf blade palmately divided for 3/4. Petals 10-15 mm, entire, pinkish-mauve with darker veins. Mericarps pubescent, without ridges.

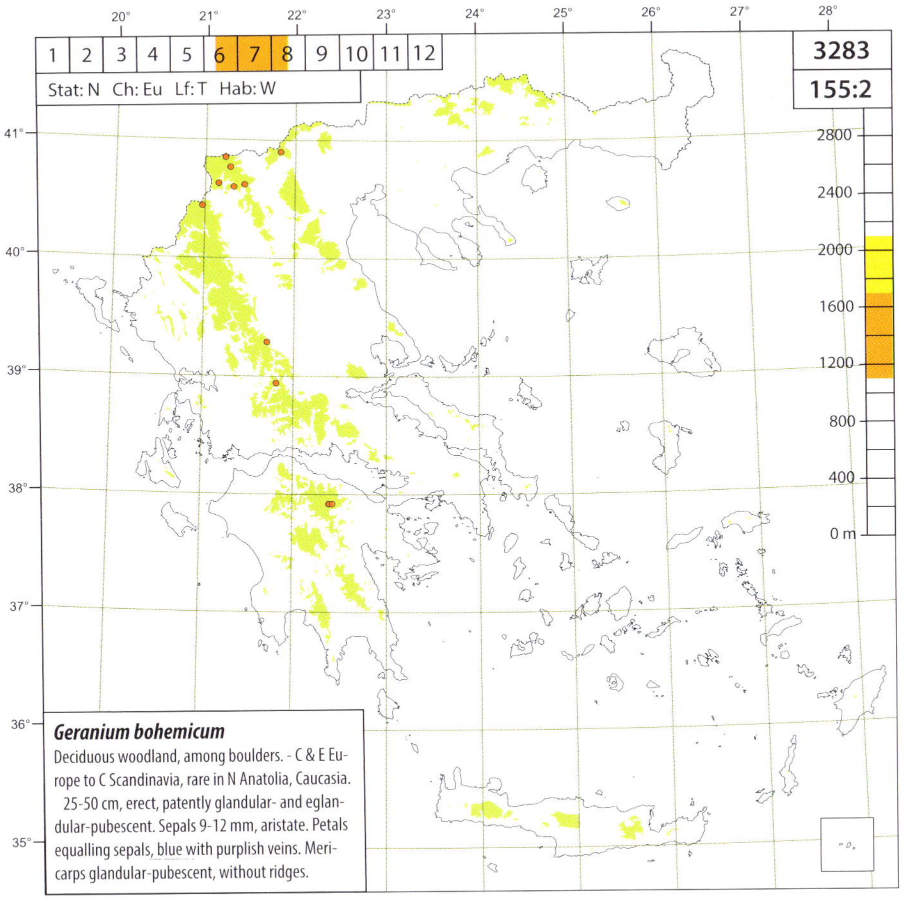

Geranium bohemicum
Deciduous woodland, among boulders. - C & E Europe to C Scandinavia, rare in N Anatolia, Caucasia.
25-50 cm, erect, patently glandular- and eglandular-pubescent. Sepals 9-12 mm, aristate. Petals equalling sepals, blue with purplish veins. Mericarps glandular-pubescent, without ridges.

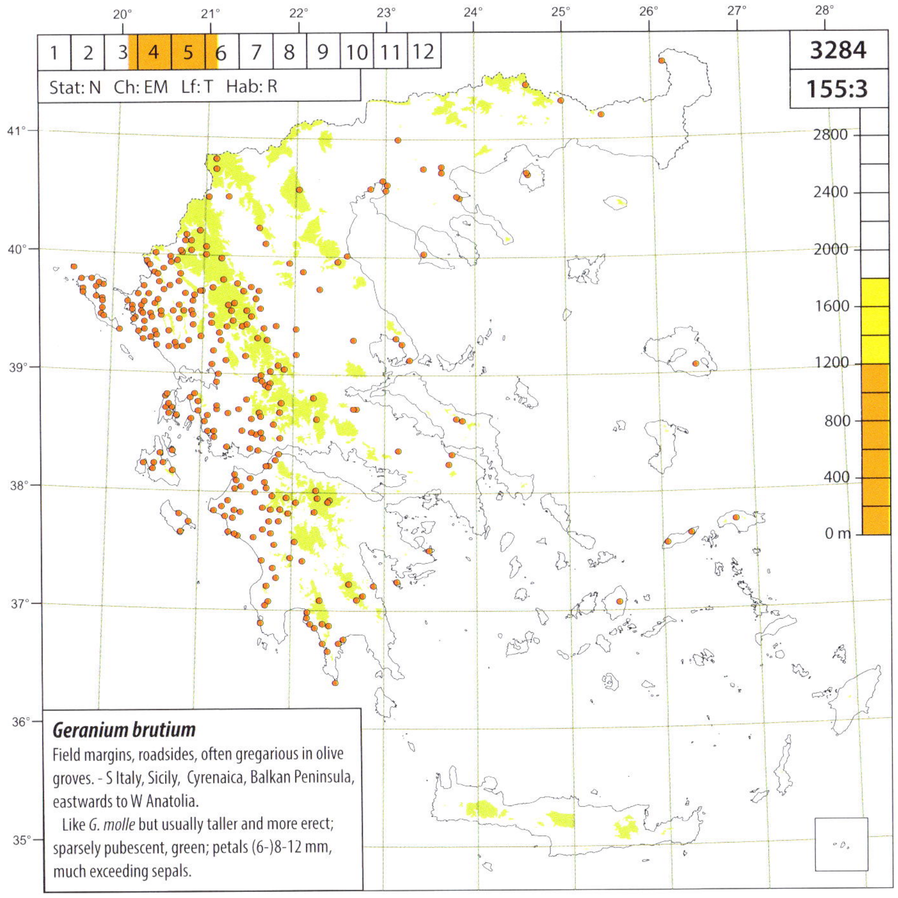

Geranium brutium
Field margins, roadsides, often gregarious in olive groves. - S Italy, Sicily, Cyrenaica, Balkan Peninsula, eastwards to W Anatolia.
Like *G. molle* but usually taller and more erect; sparsely pubescent, green; petals (6-)8-12 mm, much exceeding sepals.

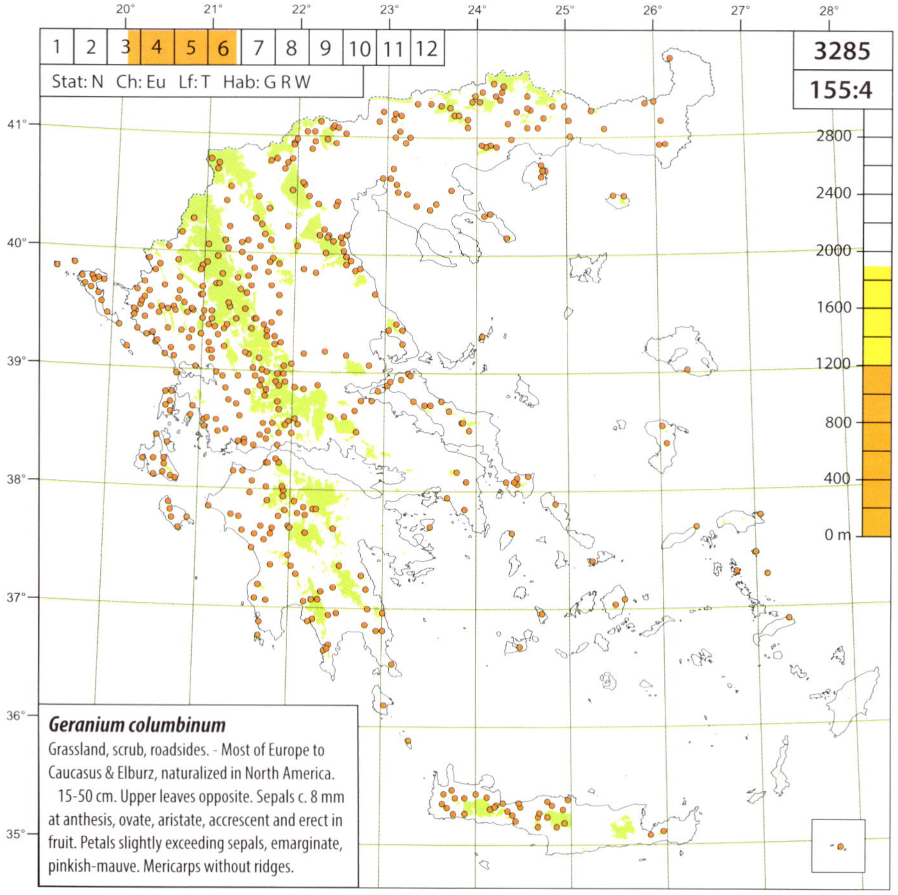

Geranium columbinum

Grassland, scrub, roadsides. - Most of Europe to Caucasus & Elburz, naturalized in North America.
15-50 cm. Upper leaves opposite. Sepals c. 8 mm at anthesis, ovate, aristate, accrescent and erect in fruit. Petals slightly exceeding sepals, emarginate, pinkish-mauve. Mericarps without ridges.

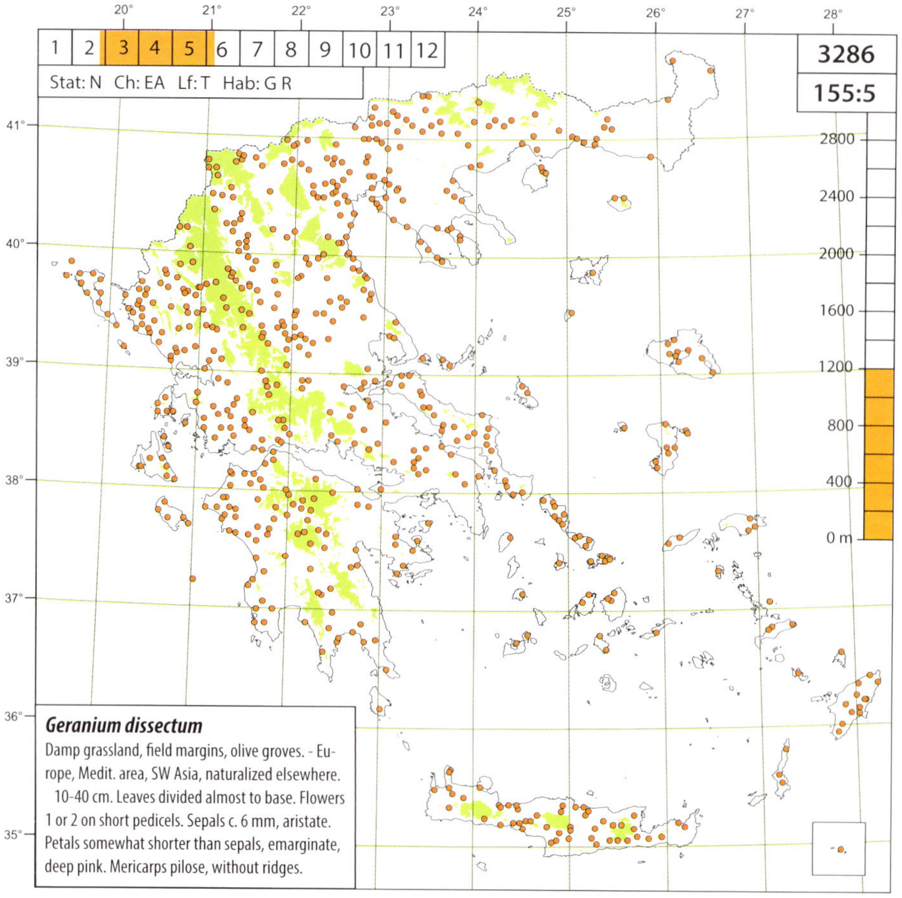

Geranium dissectum

Damp grassland, field margins, olive groves. - Europe, Medit. area, SW Asia, naturalized elsewhere.
10-40 cm. Leaves divided almost to base. Flowers 1 or 2 on short pedicels. Sepals c. 6 mm, aristate. Petals somewhat shorter than sepals, emarginate, deep pink. Mericarps pilose, without ridges.

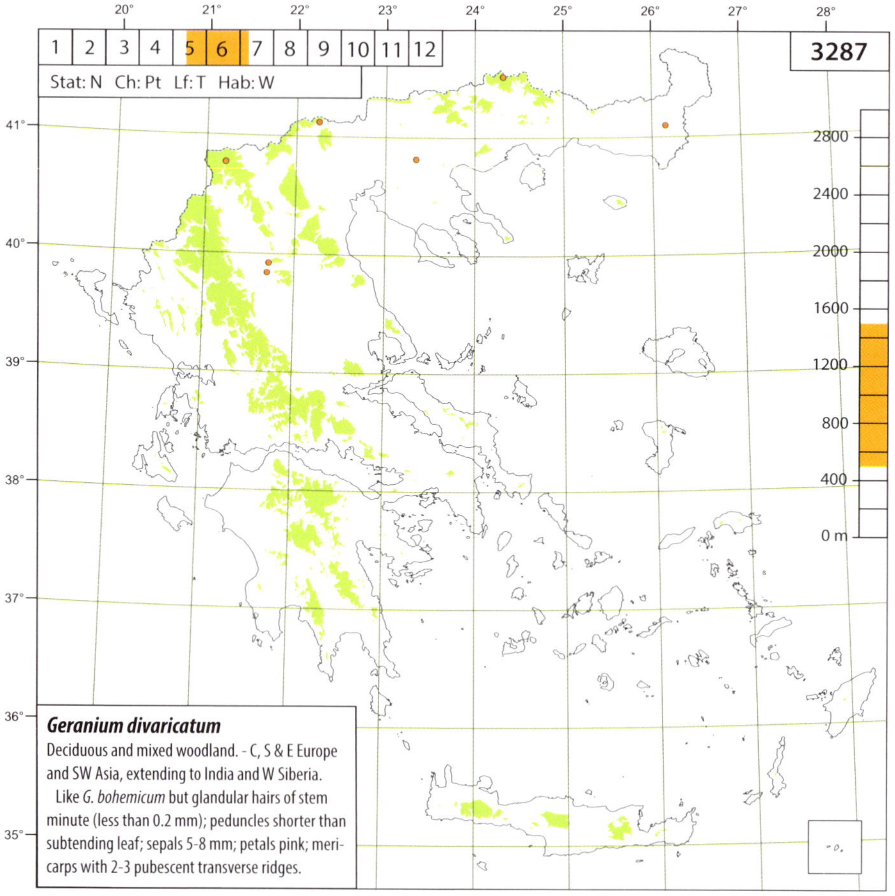

Geranium divaricatum

Deciduous and mixed woodland. - C, S & E Europe and SW Asia, extending to India and W Siberia.
Like *G. bohemicum* but glandular hairs of stem minute (less than 0.2 mm); peduncles shorter than subtending leaf; sepals 5-8 mm; petals pink; mericarps with 2-3 pubescent transverse ridges.

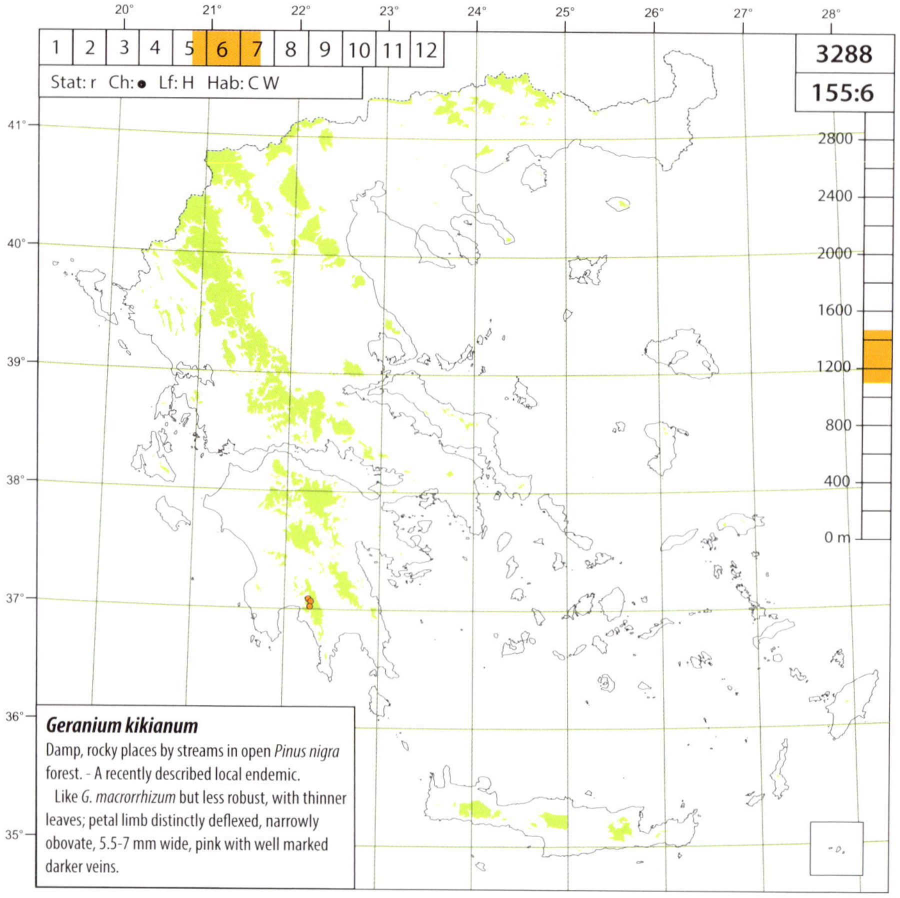

Geranium kikianum

Damp, rocky places by streams in open *Pinus nigra* forest. - A recently described local endemic.
Like *G. macrorrhizum* but less robust, with thinner leaves; petal limb distinctly deflexed, narrowly obovate, 5.5-7 mm wide, pink with well marked darker veins.

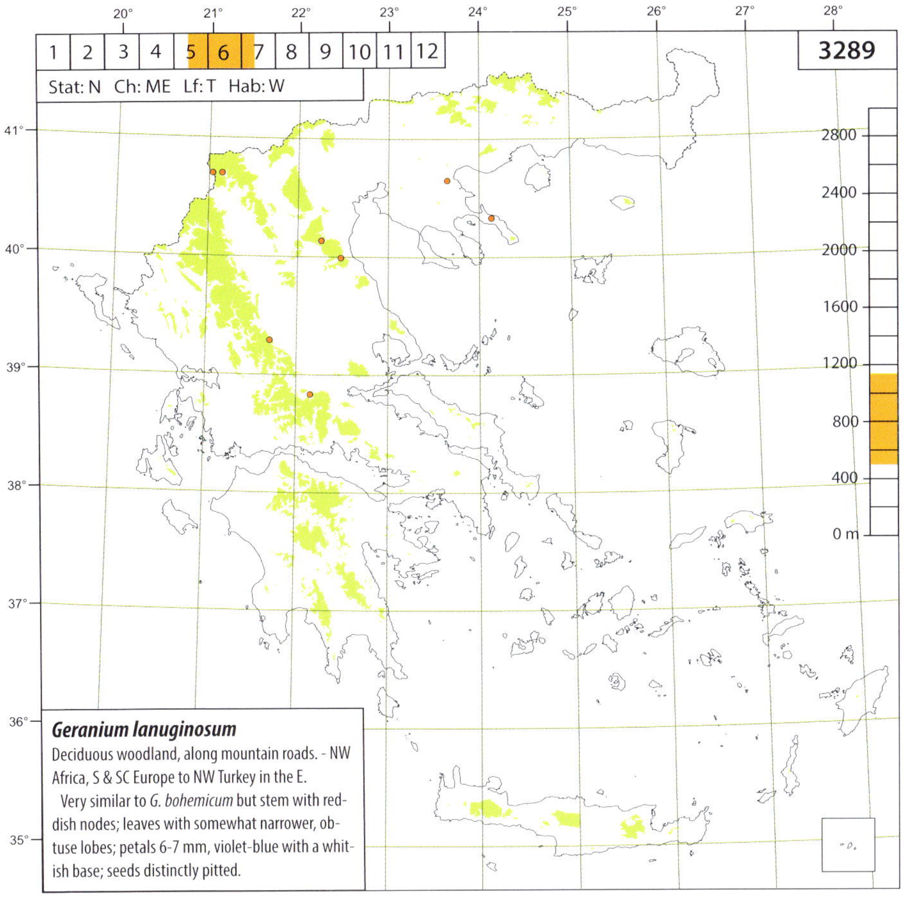

Geranium lanuginosum

Deciduous woodland, along mountain roads. - NW Africa, S & SC Europe to NW Turkey in the E.

Very similar to *G. bohemicum* but stem with reddish nodes; leaves with somewhat narrower, obtuse lobes; petals 6-7 mm, violet-blue with a whitish base; seeds distinctly pitted.

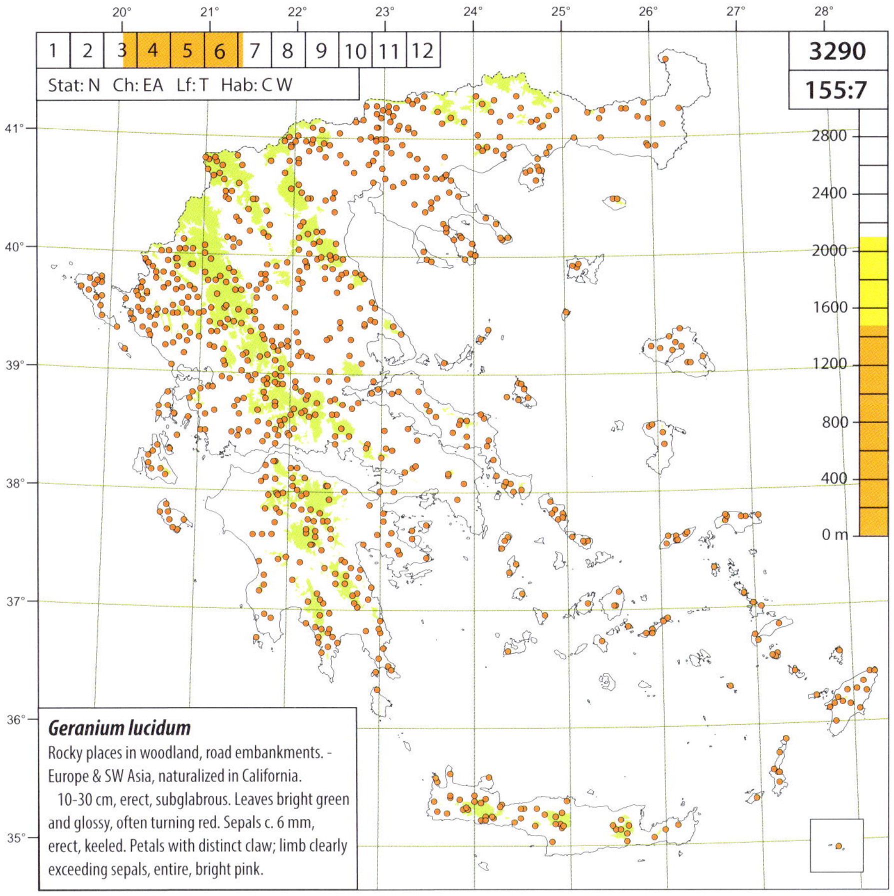

Geranium lucidum

Rocky places in woodland, road embankments. - Europe & SW Asia, naturalized in California.

10-30 cm, erect, subglabrous. Leaves bright green and glossy, often turning red. Sepals c. 6 mm, erect, keeled. Petals with distinct claw; limb clearly exceeding sepals, entire, bright pink.

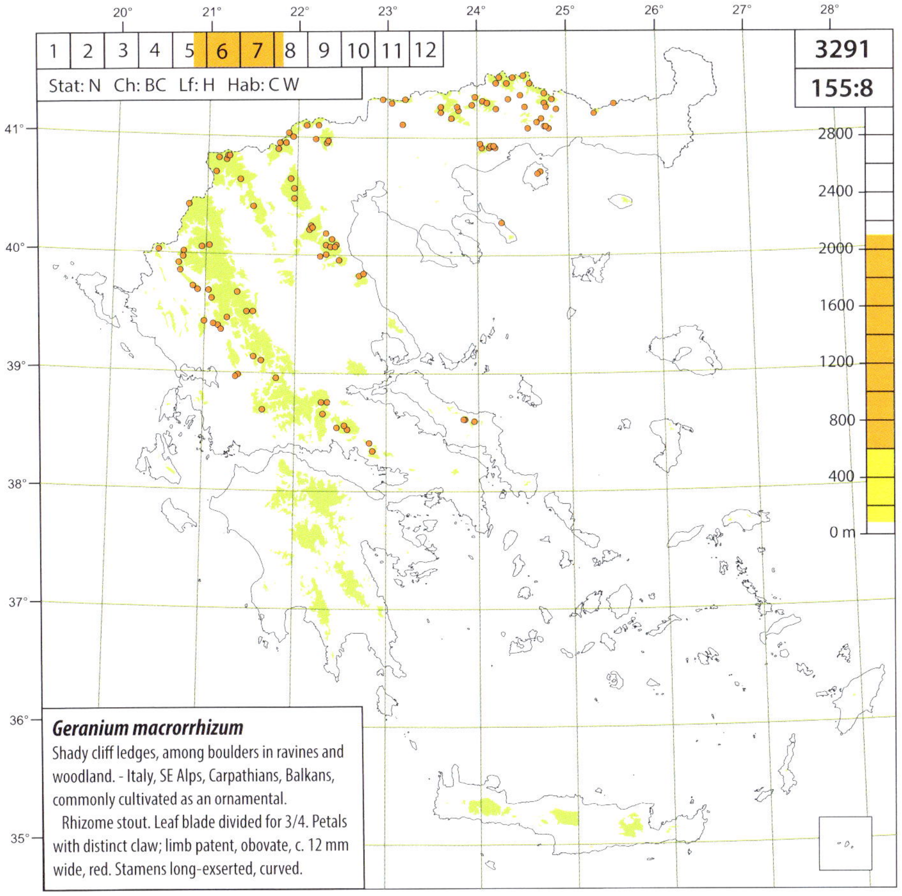

Geranium macrorrhizum

Shady cliff ledges, among boulders in ravines and woodland. - Italy, SE Alps, Carpathians, Balkans, commonly cultivated as an ornamental.

Rhizome stout. Leaf blade divided for 3/4. Petals with distinct claw; limb patent, obovate, c. 12 mm wide, red. Stamens long-exserted, curved.

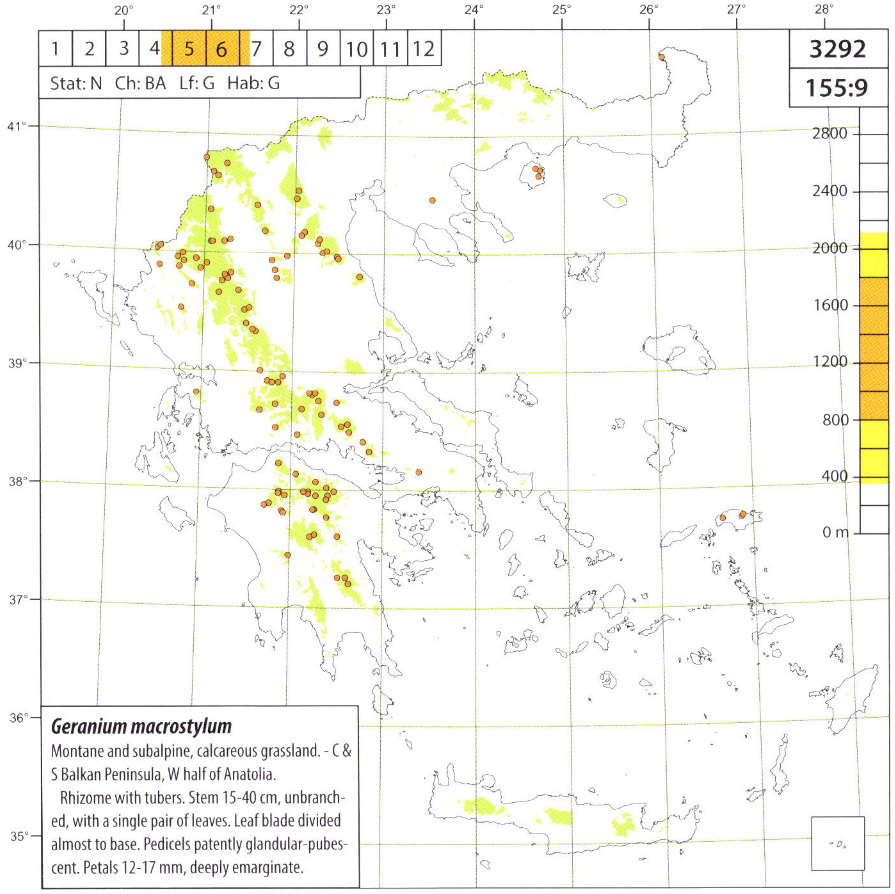

Geranium macrostylum

Montane and subalpine, calcareous grassland. - C & S Balkan Peninsula, W half of Anatolia.

Rhizome with tubers. Stem 15-40 cm, unbranched, with a single pair of leaves. Leaf blade divided almost to base. Pedicels patently glandular-pubescent. Petals 12-17 mm, deeply emarginate.

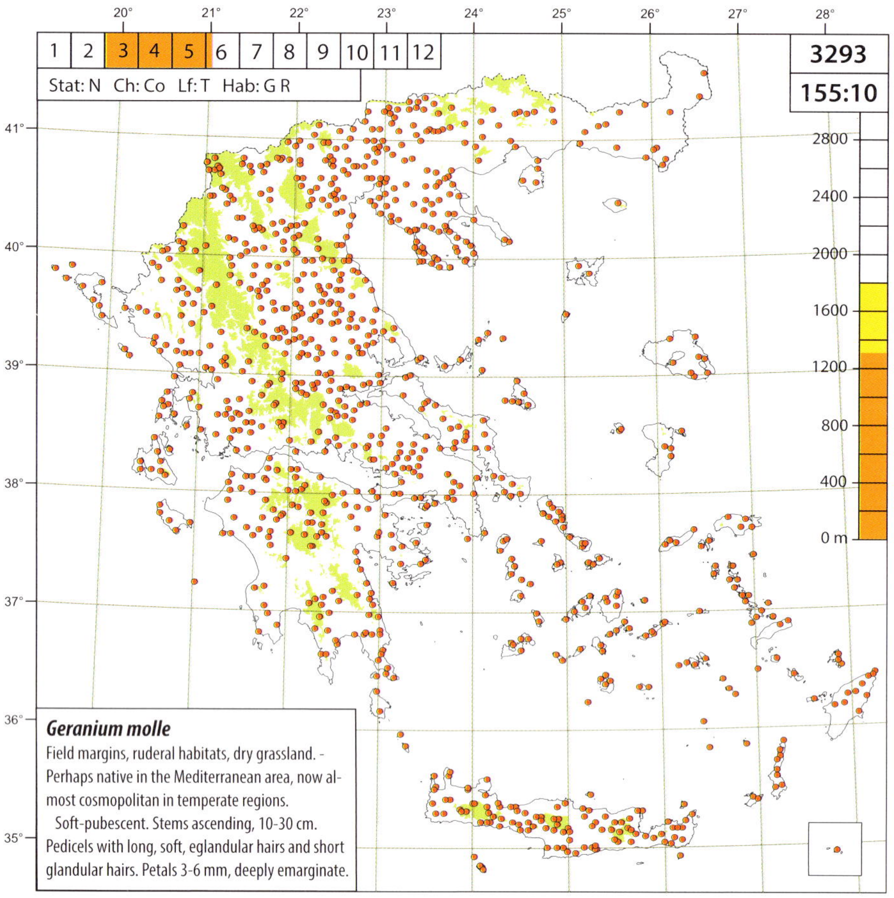

Geranium molle
Field margins, ruderal habitats, dry grassland. - Perhaps native in the Mediterranean area, now almost cosmopolitan in temperate regions.
Soft-pubescent. Stems ascending, 10-30 cm. Pedicels with long, soft, eglandular hairs and short glandular hairs. Petals 3-6 mm, deeply emarginate.

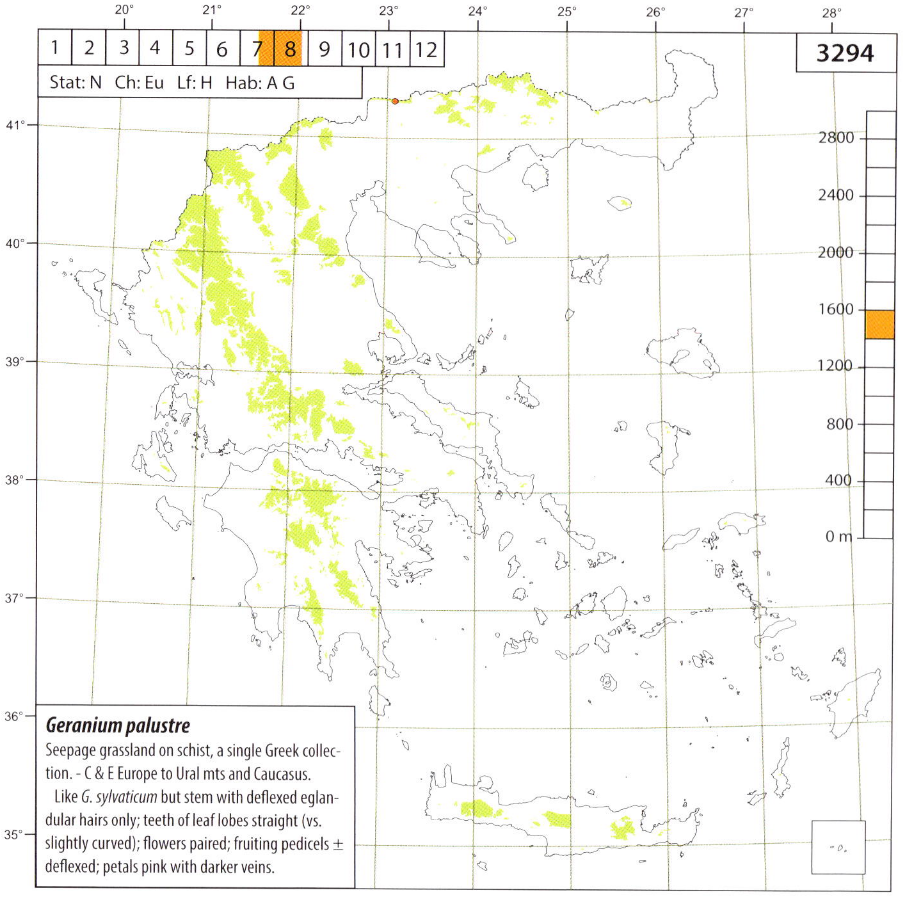

Geranium palustre
Seepage grassland on schist, a single Greek collection. - C & E Europe to Ural mts and Caucasus.
Like *G. sylvaticum* but stem with deflexed eglandular hairs only; teeth of leaf lobes straight (vs. slightly curved); flowers paired; fruiting pedicels ± deflexed; petals pink with darker veins.

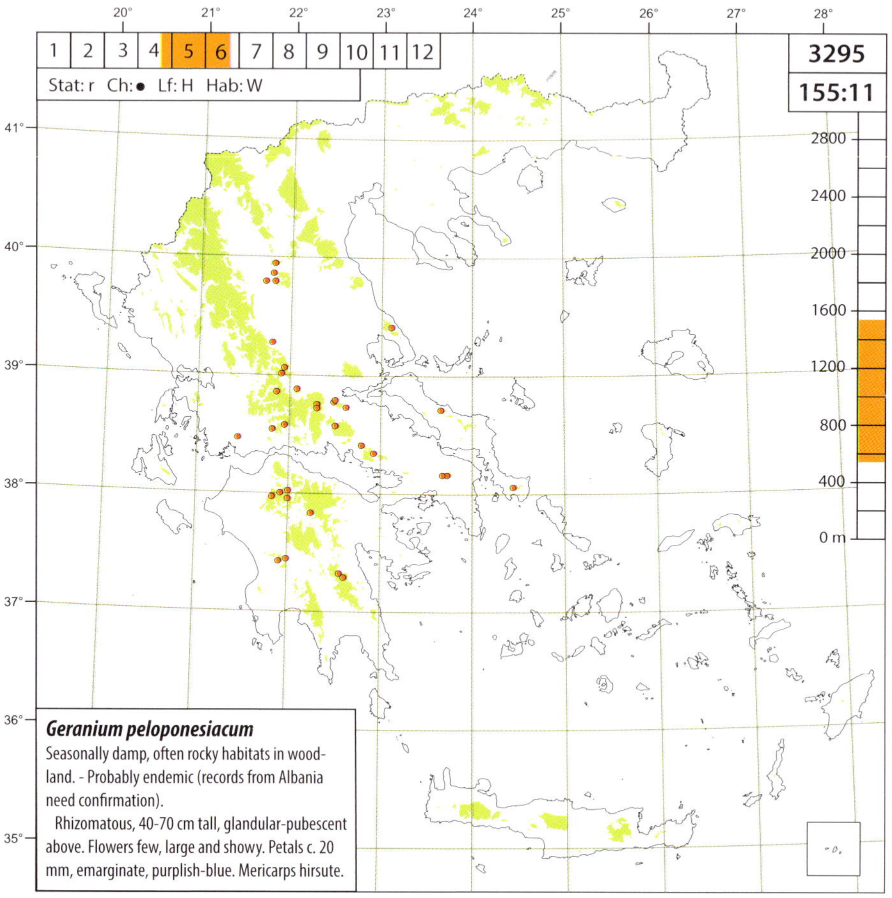

Geranium peloponesiacum
Seasonally damp, often rocky habitats in woodland. - Probably endemic (records from Albania need confirmation).
Rhizomatous, 40-70 cm tall, glandular-pubescent above. Flowers few, large and showy. Petals c. 20 mm, emarginate, purplish-blue. Mericarps hirsute.

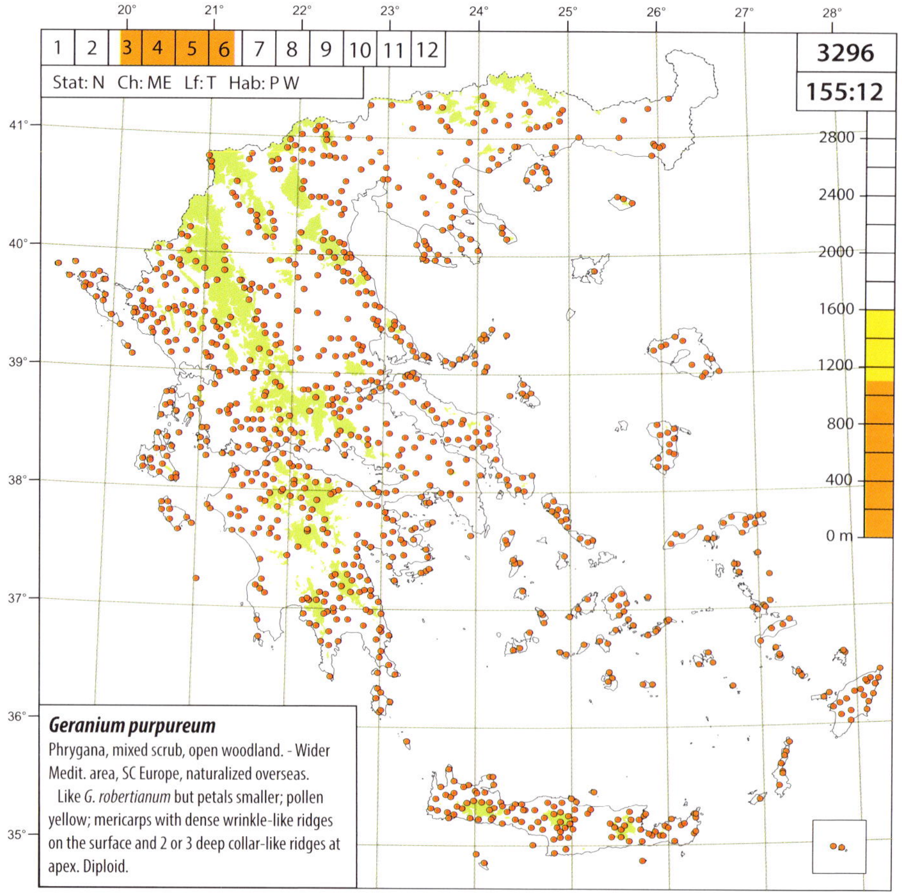

Geranium purpureum
Phrygana, mixed scrub, open woodland. - Wider Medit. area, SC Europe, naturalized overseas.
Like *G. robertianum* but petals smaller; pollen yellow; mericarps with dense wrinkle-like ridges on the surface and 2 or 3 deep collar-like ridges at apex. Diploid.

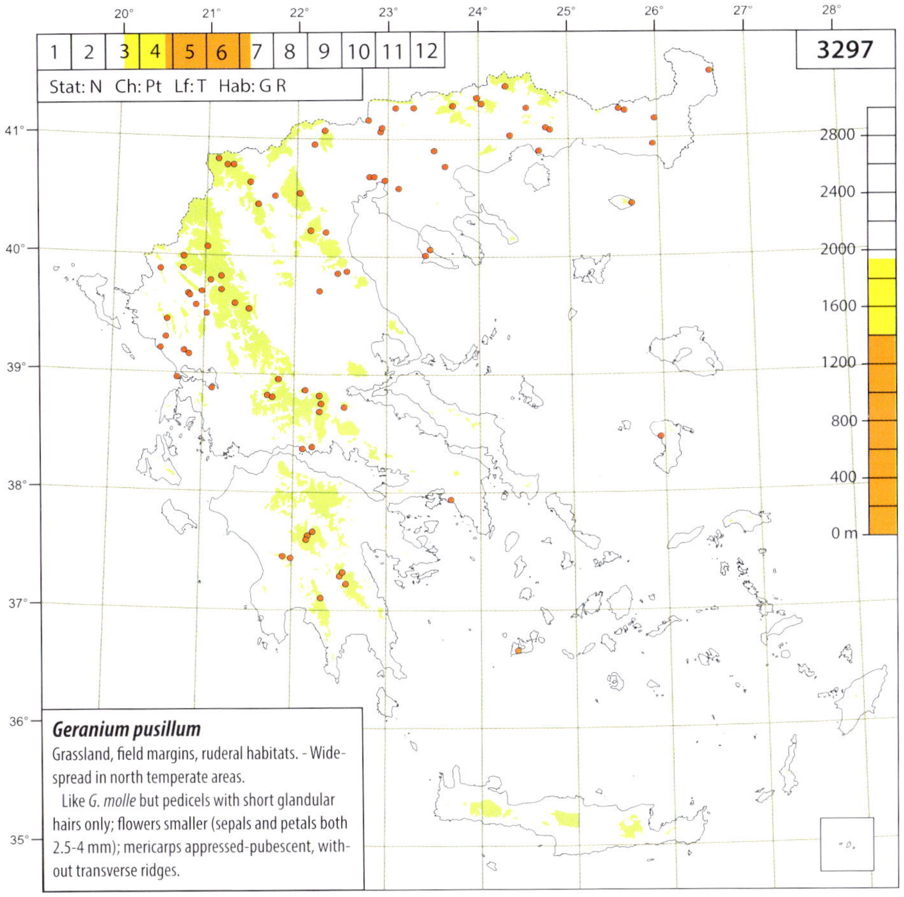

Geranium pusillum

Grassland, field margins, ruderal habitats. - Widespread in north temperate areas.

Like *G. molle* but pedicels with short glandular hairs only; flowers smaller (sepals and petals both 2.5-4 mm); mericarps appressed-pubescent, without transverse ridges.

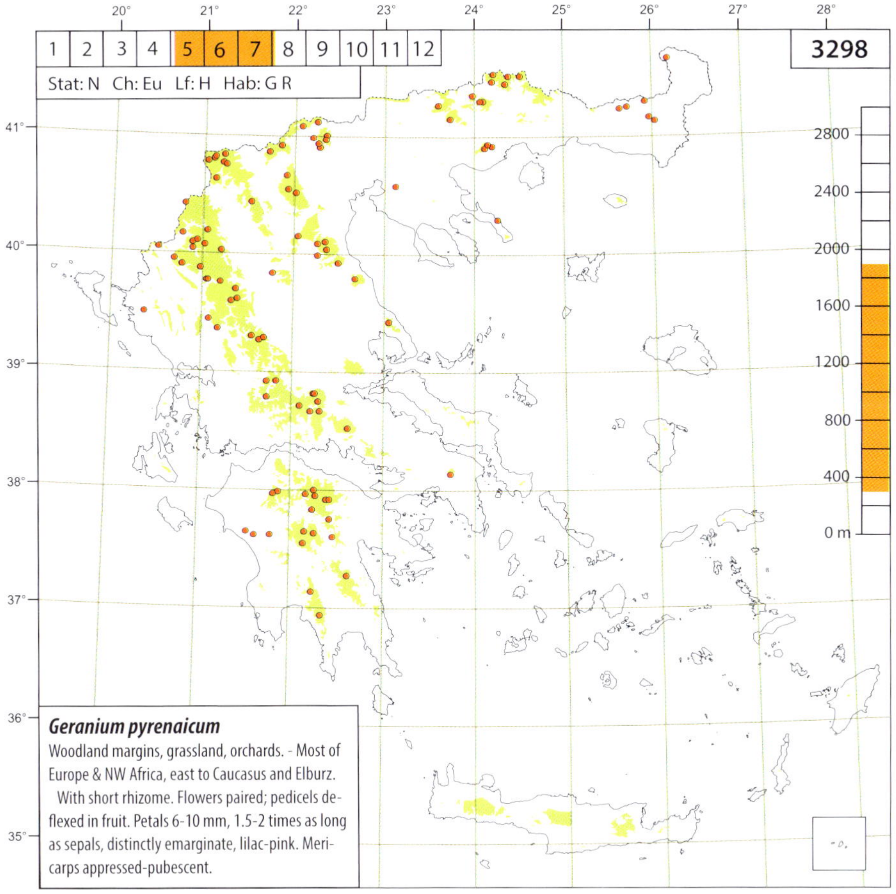

Geranium pyrenaicum

Woodland margins, grassland, orchards. - Most of Europe & NW Africa, east to Caucasus and Elburz.

With short rhizome. Flowers paired; pedicels deflexed in fruit. Petals 6-10 mm, 1.5-2 times as long as sepals, distinctly emarginate, lilac-pink. Mericarps appressed-pubescent.

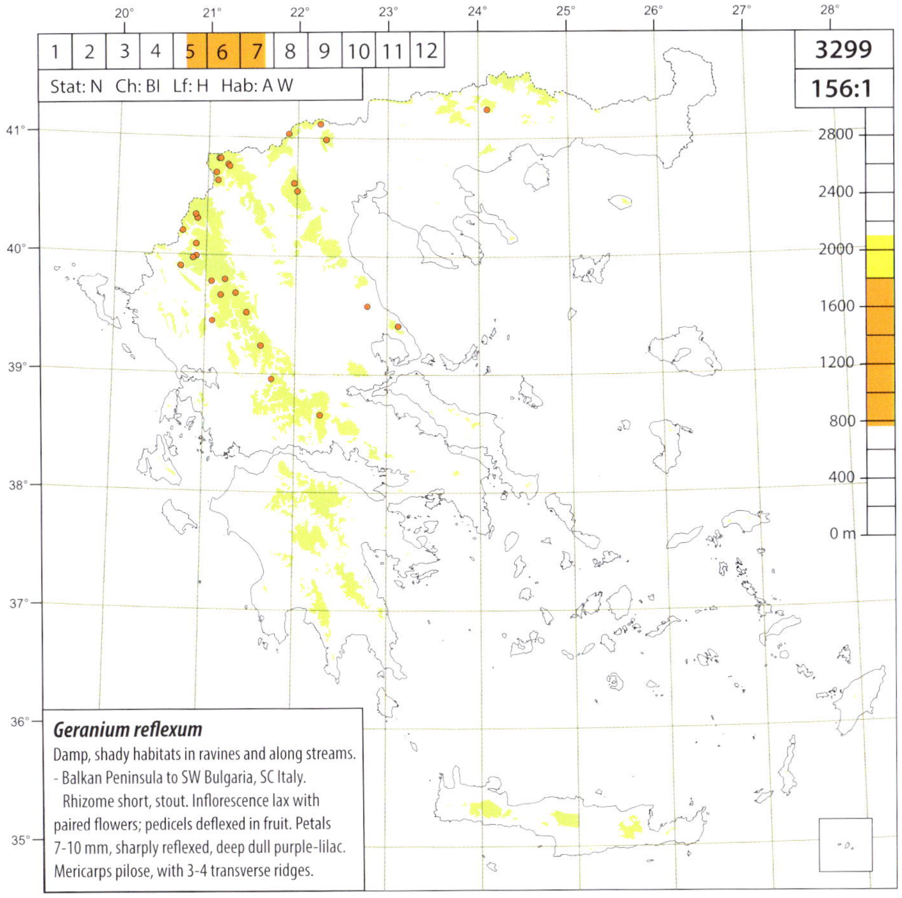

Geranium reflexum

Damp, shady habitats in ravines and along streams. - Balkan Peninsula to SW Bulgaria, SC Italy.

Rhizome short, stout. Inflorescence lax with paired flowers; pedicels deflexed in fruit. Petals 7-10 mm, sharply reflexed, deep dull purple-lilac. Mericarps pilose, with 3-4 transverse ridges.

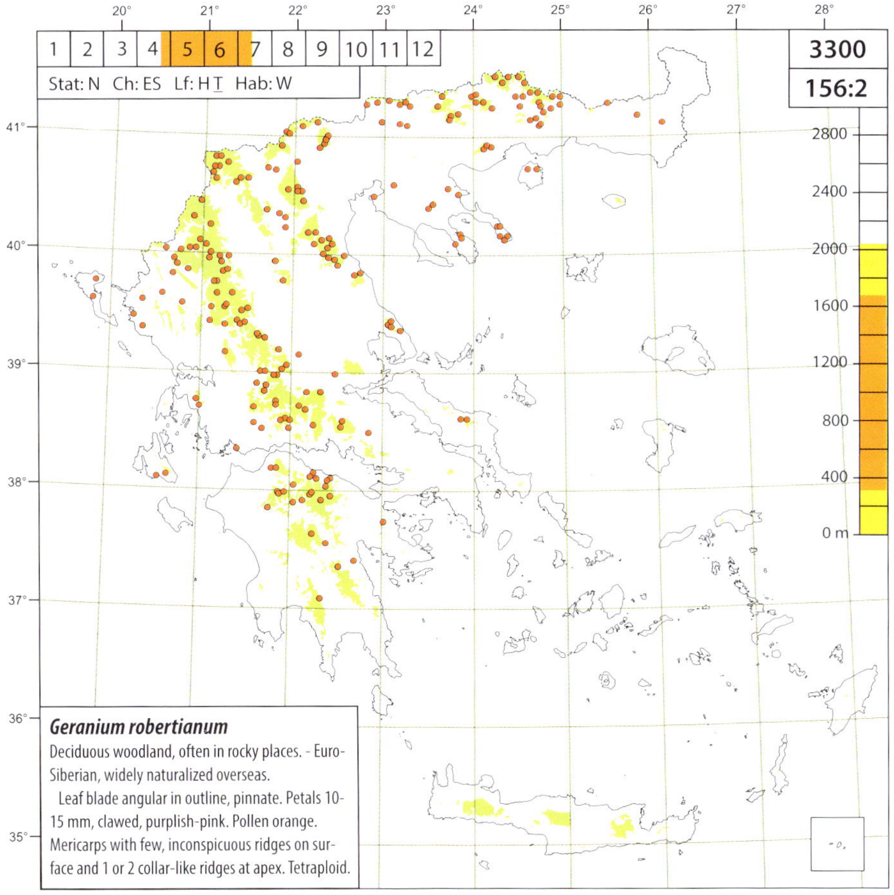

Geranium robertianum

Deciduous woodland, often in rocky places. - Euro-Siberian, widely naturalized overseas.

Leaf blade angular in outline, pinnate. Petals 10-15 mm, clawed, purplish-pink. Pollen orange. Mericarps with few, inconspicuous ridges on surface and 1 or 2 collar-like ridges at apex. Tetraploid.

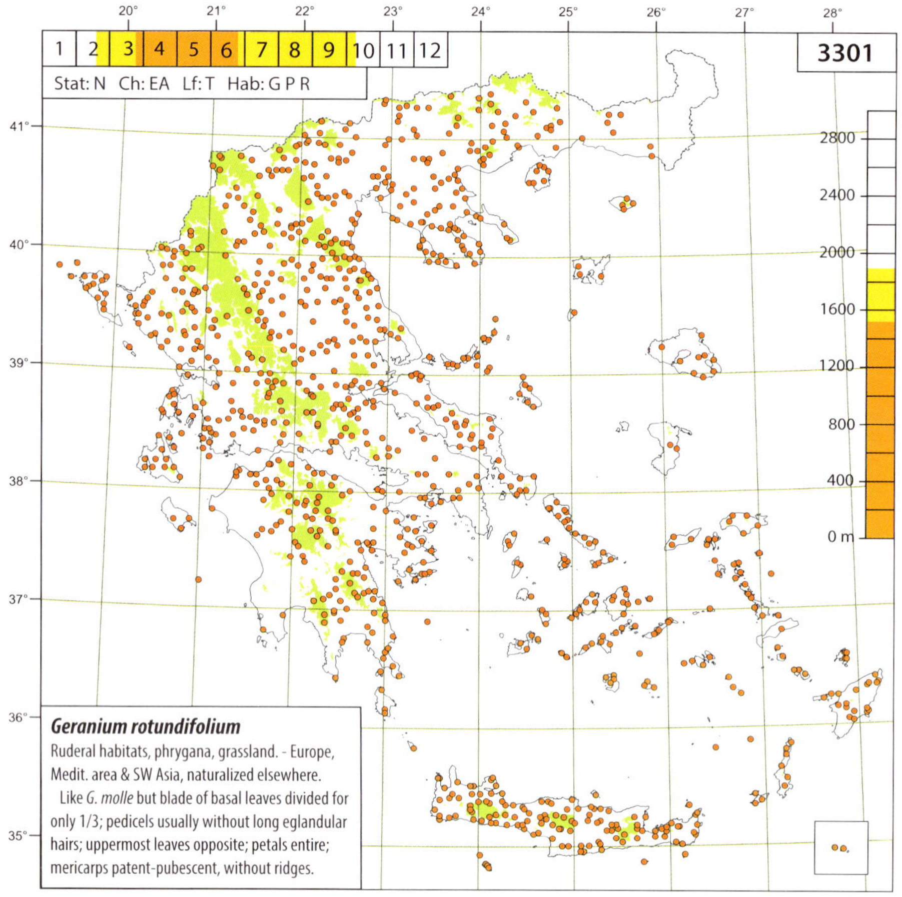

Geranium rotundifolium

Ruderal habitats, phrygana, grassland. - Europe, Medit. area & SW Asia, naturalized elsewhere.

Like *G. molle* but blade of basal leaves divided for only 1/3; pedicels usually without long eglandular hairs; uppermost leaves opposite; petals entire; mericarps patent-pubescent, without ridges.

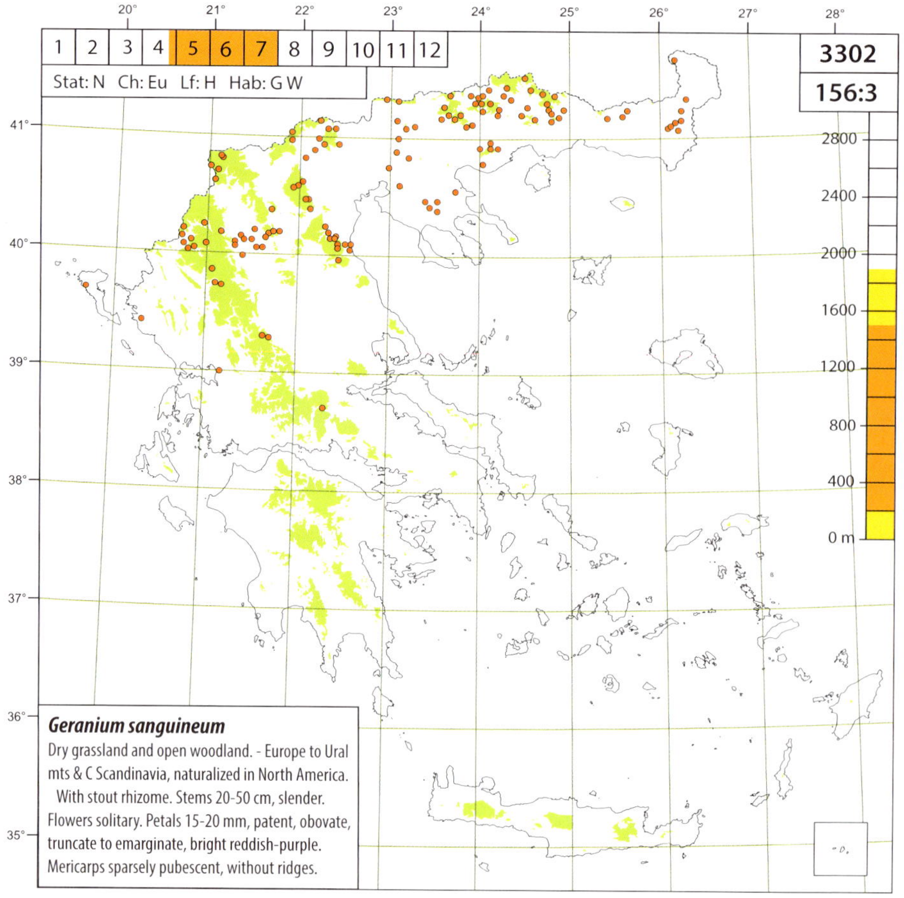

Geranium sanguineum

Dry grassland and open woodland. - Europe to Ural mts & C Scandinavia, naturalized in North America.

With stout rhizome. Stems 20-50 cm, slender. Flowers solitary. Petals 15-20 mm, patent, obovate, truncate to emarginate, bright reddish-purple. Mericarps sparsely pubescent, without ridges.

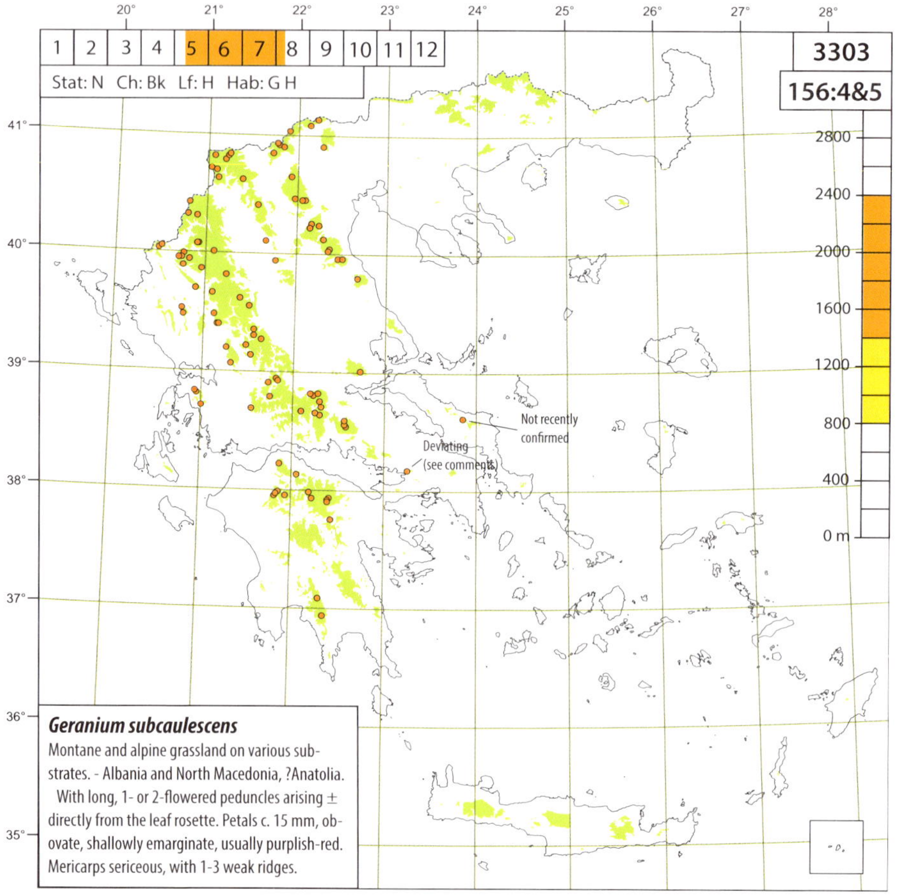

Geranium subcaulescens

Montane and alpine grassland on various substrates. - Albania and North Macedonia, ?Anatolia.

With long, 1- or 2-flowered peduncles arising ± directly from the leaf rosette. Petals c. 15 mm, obovate, shallowly emarginate, usually purplish-red. Mericarps sericeous, with 1-3 weak ridges.

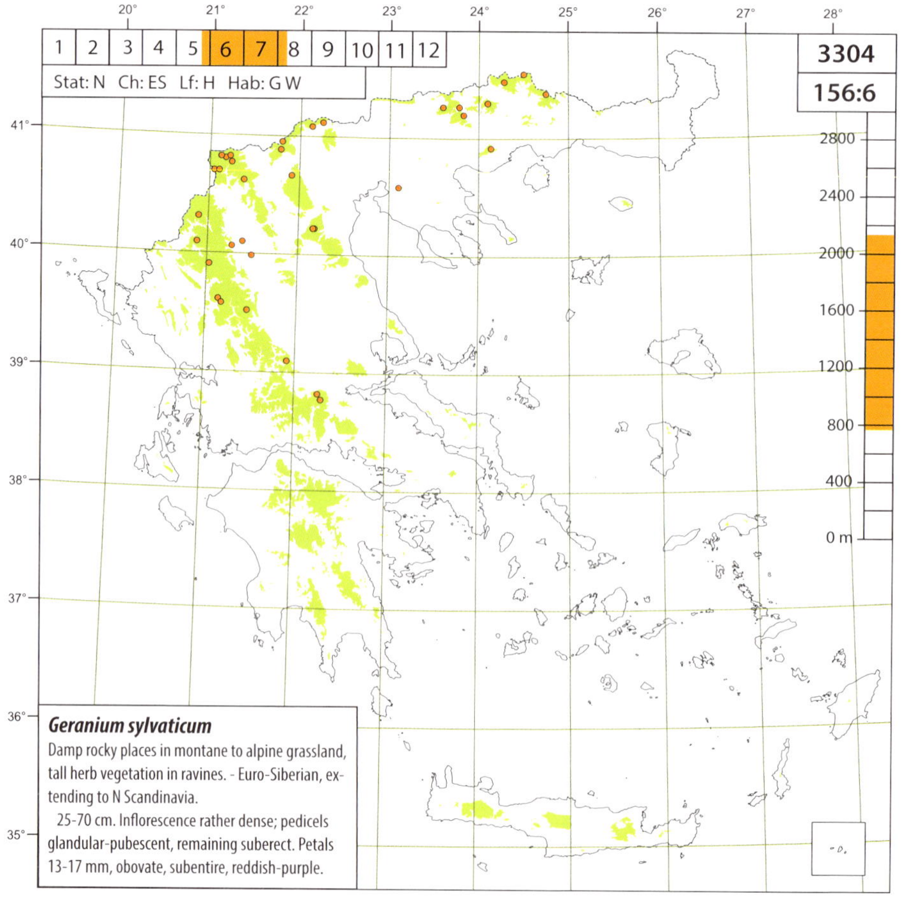

Geranium sylvaticum

Damp rocky places in montane to alpine grassland, tall herb vegetation in ravines. - Euro-Siberian, extending to N Scandinavia.

25-70 cm. Inflorescence rather dense; pedicels glandular-pubescent, remaining suberect. Petals 13-17 mm, obovate, subentire, reddish-purple.

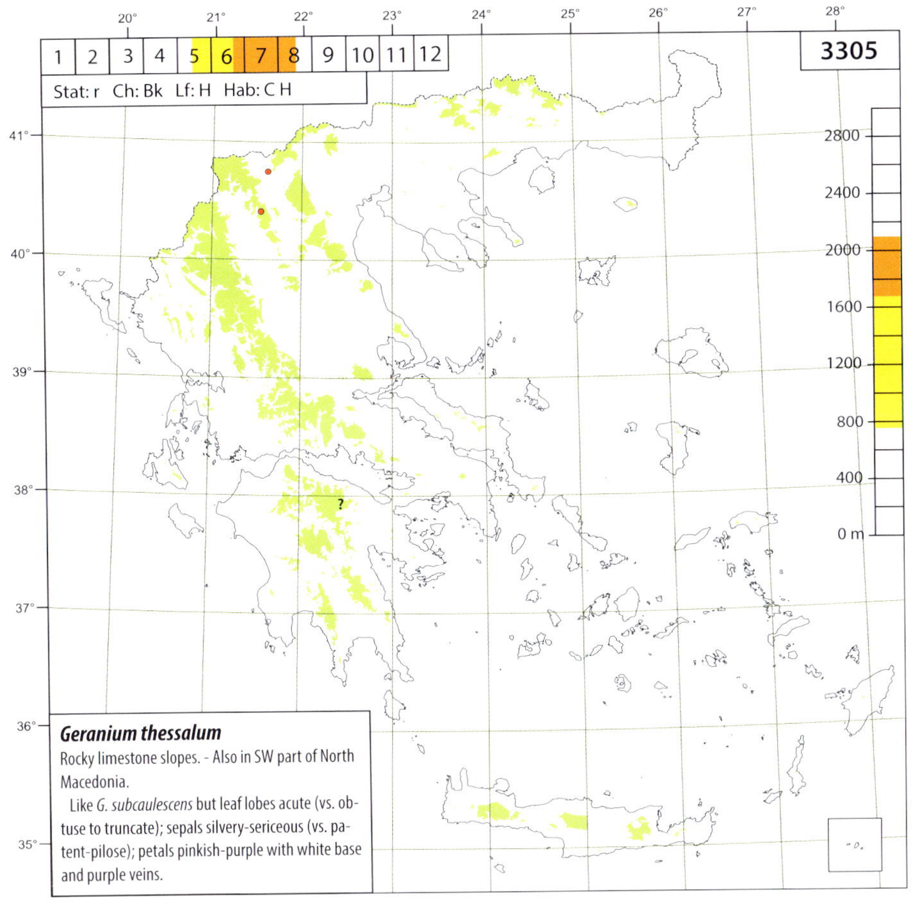

Geranium thessalum

Rocky limestone slopes. - Also in SW part of North Macedonia.

Like *G. subcaulescens* but leaf lobes acute (vs. obtuse to truncate); sepals silvery-sericeous (vs. patent-pilose); petals pinkish-purple with white base and purple veins.

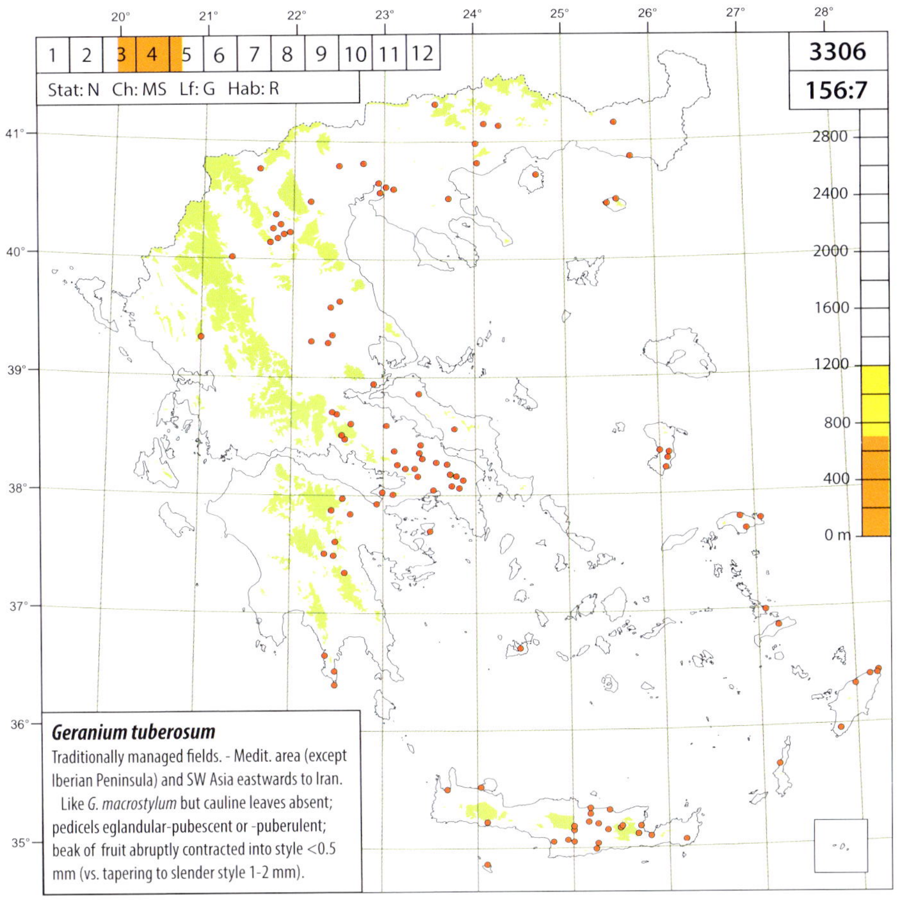

Geranium tuberosum

Traditionally managed fields. - Medit. area (except Iberian Peninsula) and SW Asia eastwards to Iran.

Like *G. macrostylum* but cauline leaves absent; pedicels eglandular-pubescent or -puberulent; beak of fruit abruptly contracted into style <0.5 mm (vs. tapering to slender style 1-2 mm).

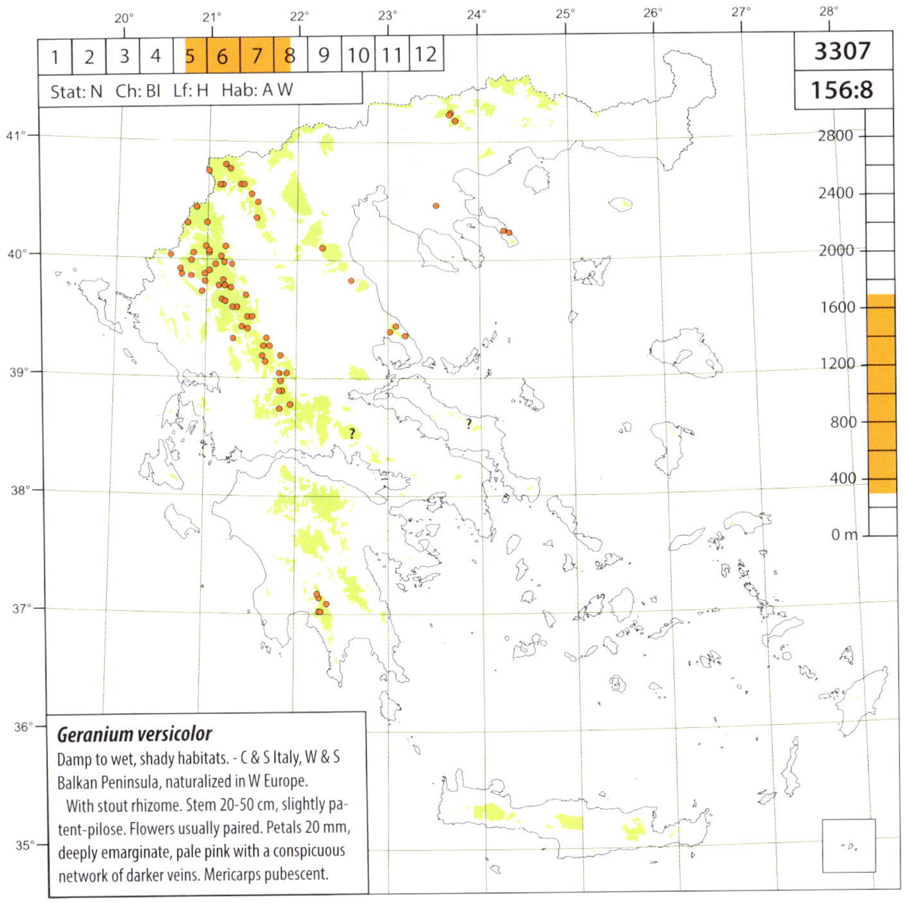

Geranium versicolor

Damp to wet, shady habitats. - C & S Italy, W & S Balkan Peninsula, naturalized in W Europe.

With stout rhizome. Stem 20-50 cm, slightly patent-pilose. Flowers usually paired. Petals 20 mm, deeply emarginate, pale pink with a conspicuous network of darker veins. Mericarps pubescent.

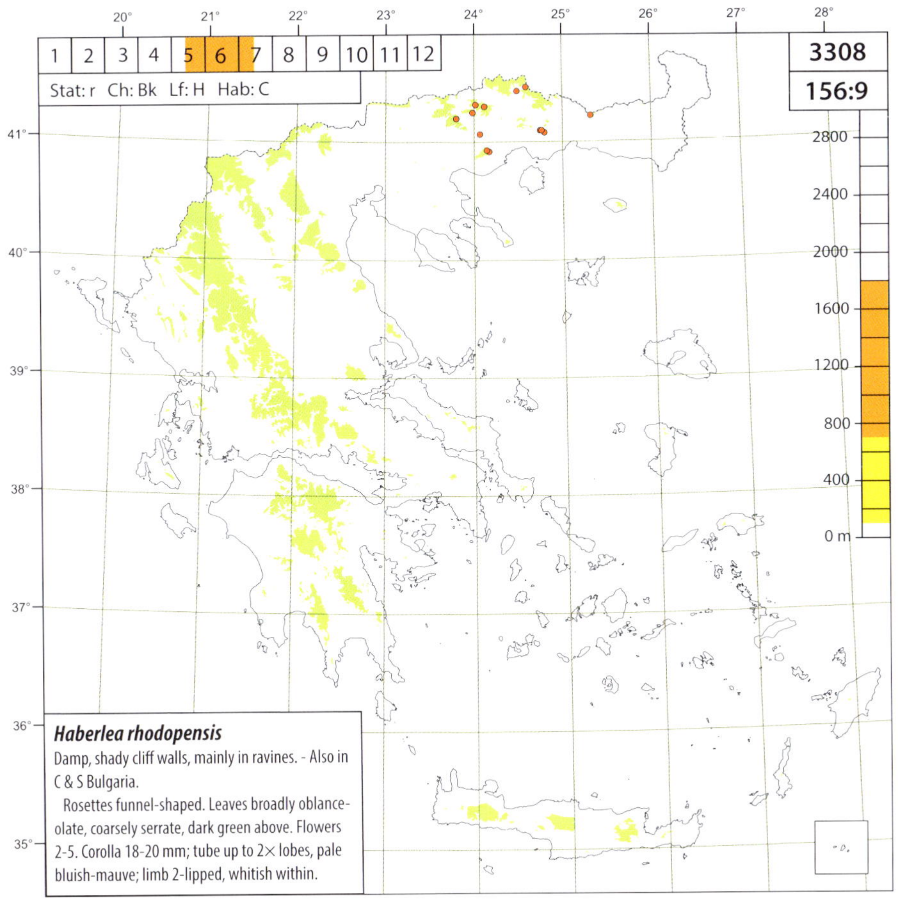

Haberlea rhodopensis

Damp, shady cliff walls, mainly in ravines. - Also in C & S Bulgaria.

Rosettes funnel-shaped. Leaves broadly oblanceolate, coarsely serrate, dark green above. Flowers 2-5. Corolla 18-20 mm; tube up to 2× lobes, pale bluish-mauve; limb 2-lipped, whitish within.

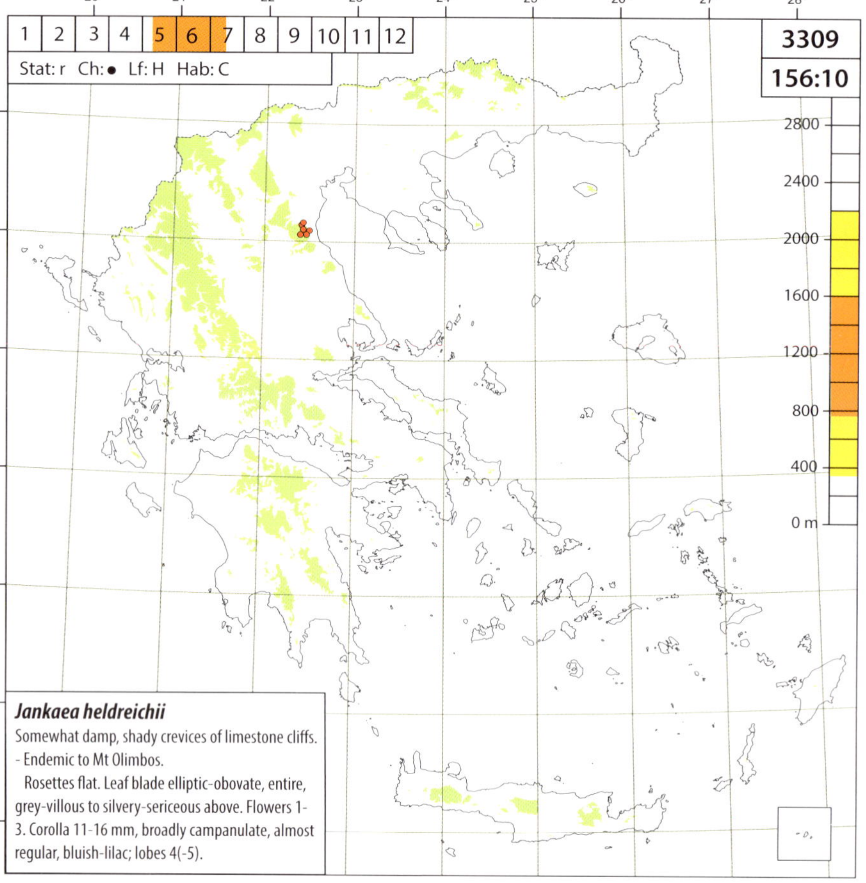

3309
156:10

Stat: r Ch: ● Lf: H Hab: C

Jankaea heldreichii

Somewhat damp, shady crevices of limestone cliffs. - Endemic to Mt Olimbos.

Rosettes flat. Leaf blade elliptic-obovate, entire, grey-villous to silvery-sericeous above. Flowers 1-3. Corolla 11-16 mm, broadly campanulate, almost regular, bluish-lilac; lobes 4(-5).

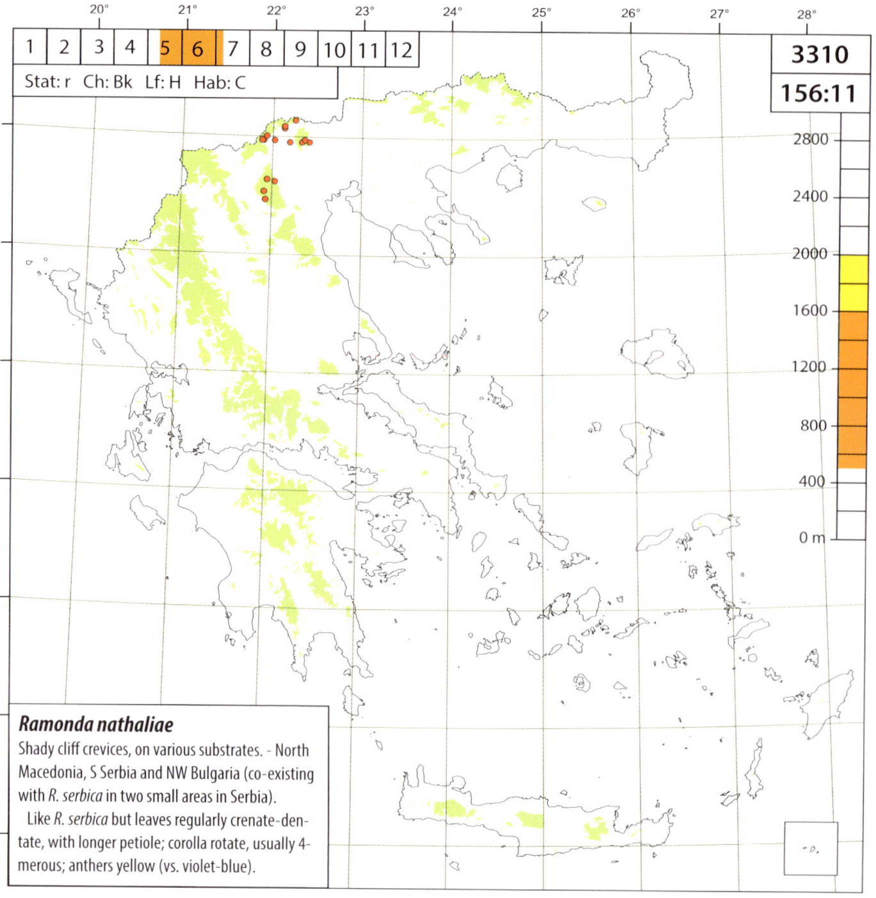

3310
156:11

Stat: r Ch: Bk Lf: H Hab: C

Ramonda nathaliae

Shady cliff crevices, on various substrates. - North Macedonia, S Serbia and NW Bulgaria (co-existing with *R. serbica* in two small areas in Serbia).

Like *R. serbica* but leaves regularly crenate-dentate, with longer petiole; corolla rotate, usually 4-merous; anthers yellow (vs. violet-blue).

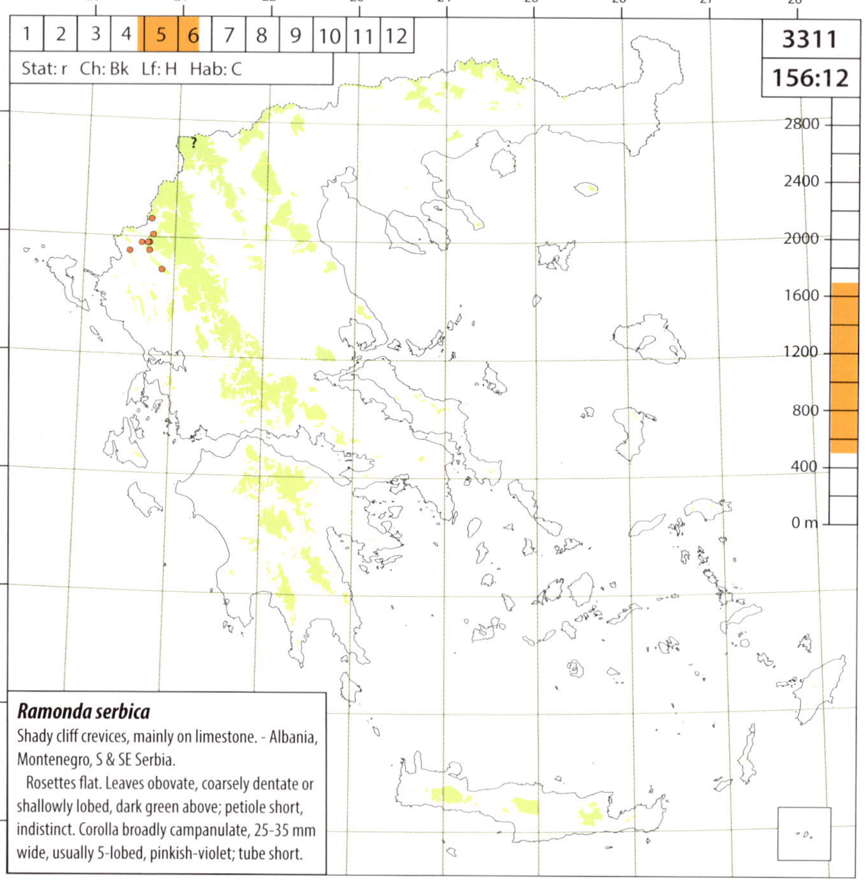

3311
156:12

Stat: r Ch: Bk Lf: H Hab: C

Ramonda serbica

Shady cliff crevices, mainly on limestone. - Albania, Montenegro, S & SE Serbia.

Rosettes flat. Leaves obovate, coarsely dentate or shallowly lobed, dark green above; petiole short, indistinct. Corolla broadly campanulate, 25-35 mm wide, usually 5-lobed, pinkish-violet; tube short.

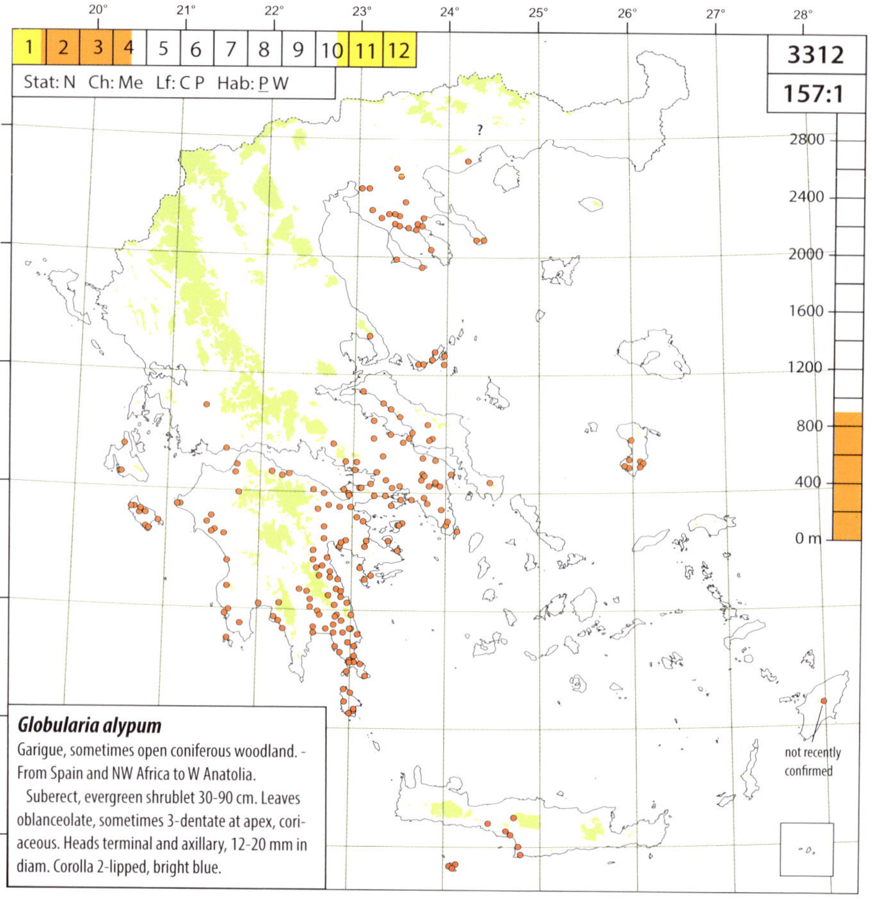

3312
157:1

Stat: N Ch: Me Lf: C P Hab: P W

Globularia alypum

Garigue, sometimes open coniferous woodland. - From Spain and NW Africa to W Anatolia.

Suberect, evergreen shrublet 30-90 cm. Leaves oblanceolate, sometimes 3-dentate at apex, coriaceous. Heads terminal and axillary, 12-20 mm in diam. Corolla 2-lipped, bright blue.

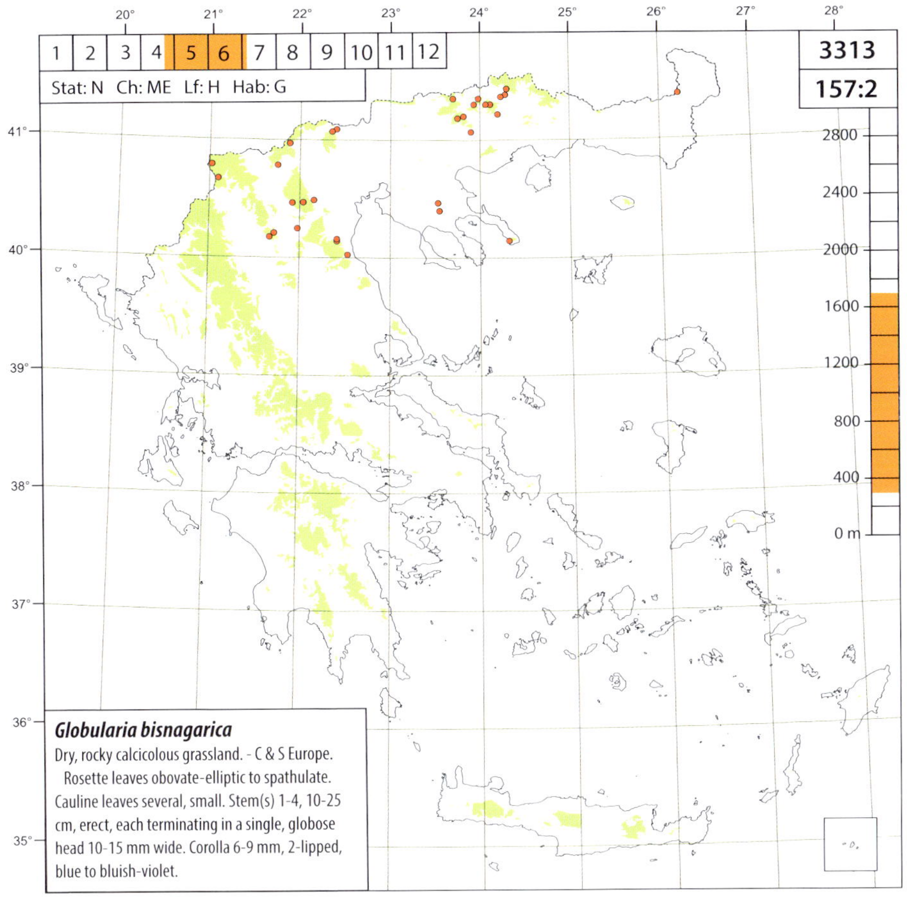

Globularia bisnagarica

Dry, rocky calcicolous grassland. - C & S Europe.

Rosette leaves obovate-elliptic to spathulate. Cauline leaves several, small. Stem(s) 1-4, 10-25 cm, erect, each terminating in a single, globose head 10-15 mm wide. Corolla 6-9 mm, 2-lipped, blue to bluish-violet.

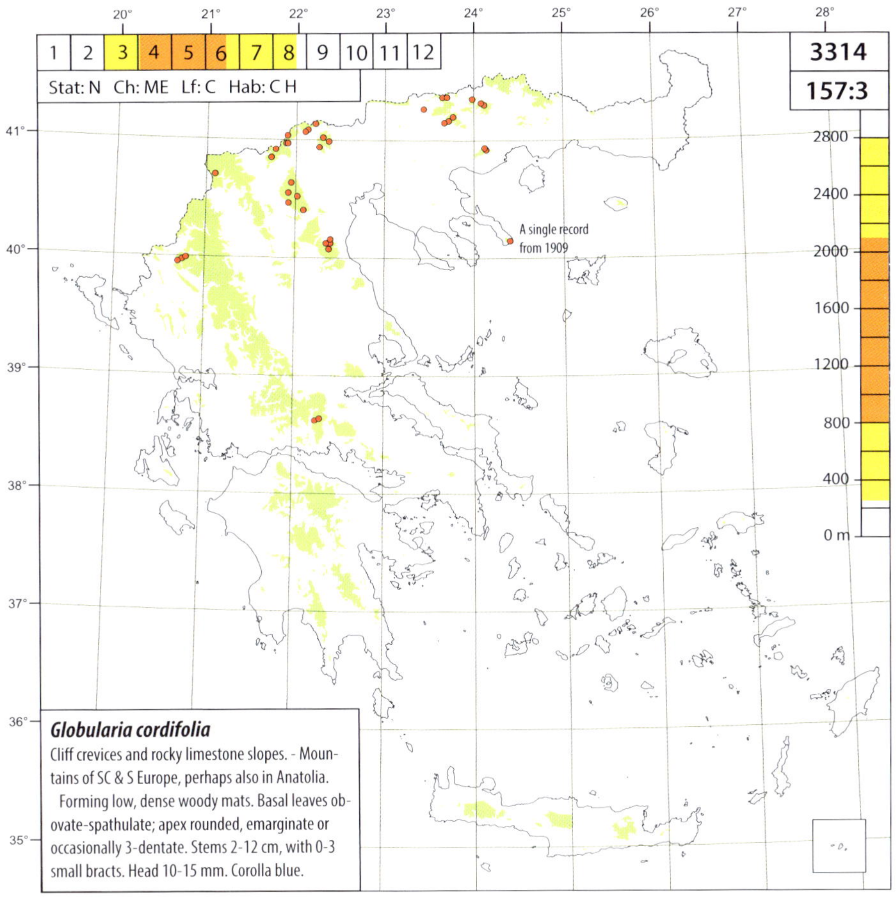

Globularia cordifolia

Cliff crevices and rocky limestone slopes. - Mountains of SC & S Europe, perhaps also in Anatolia.

Forming low, dense woody mats. Basal leaves obovate-spathulate; apex rounded, emarginate or occasionally 3-dentate. Stems 2-12 cm, with 0-3 small bracts. Head 10-15 mm. Corolla blue.

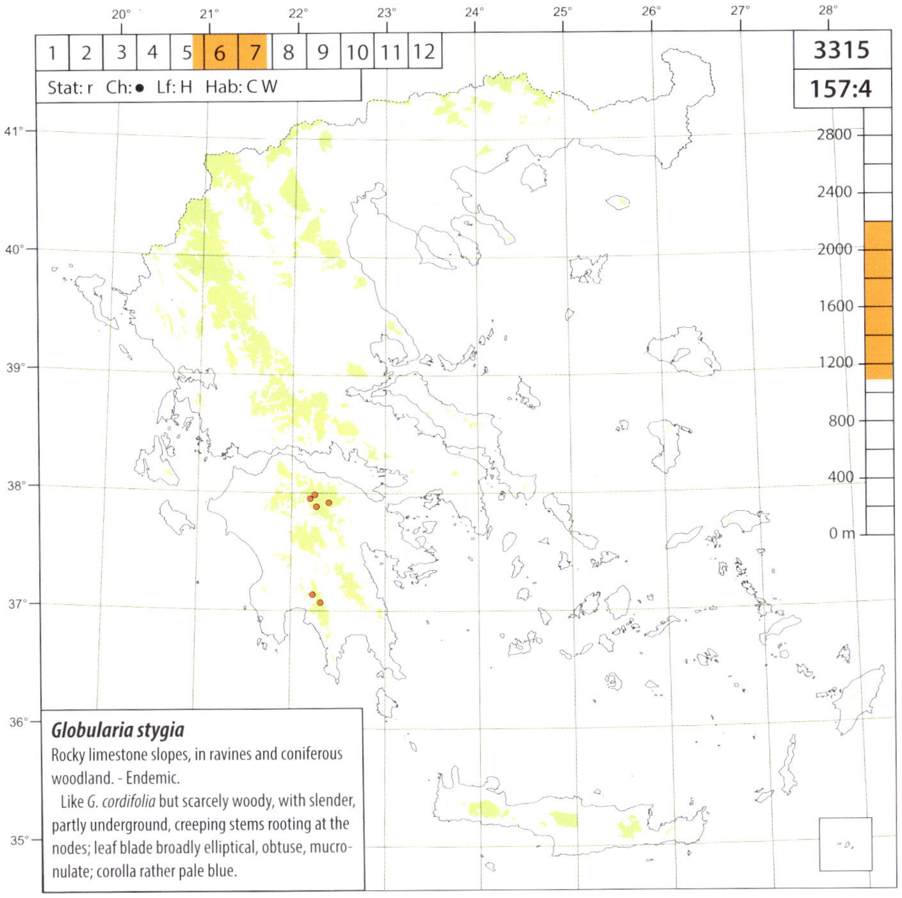

Globularia stygia

Rocky limestone slopes, in ravines and coniferous woodland. - Endemic.

Like *G. cordifolia* but scarcely woody, with slender, partly underground, creeping stems rooting at the nodes; leaf blade broadly elliptical, obtuse, mucronulate; corolla rather pale blue.

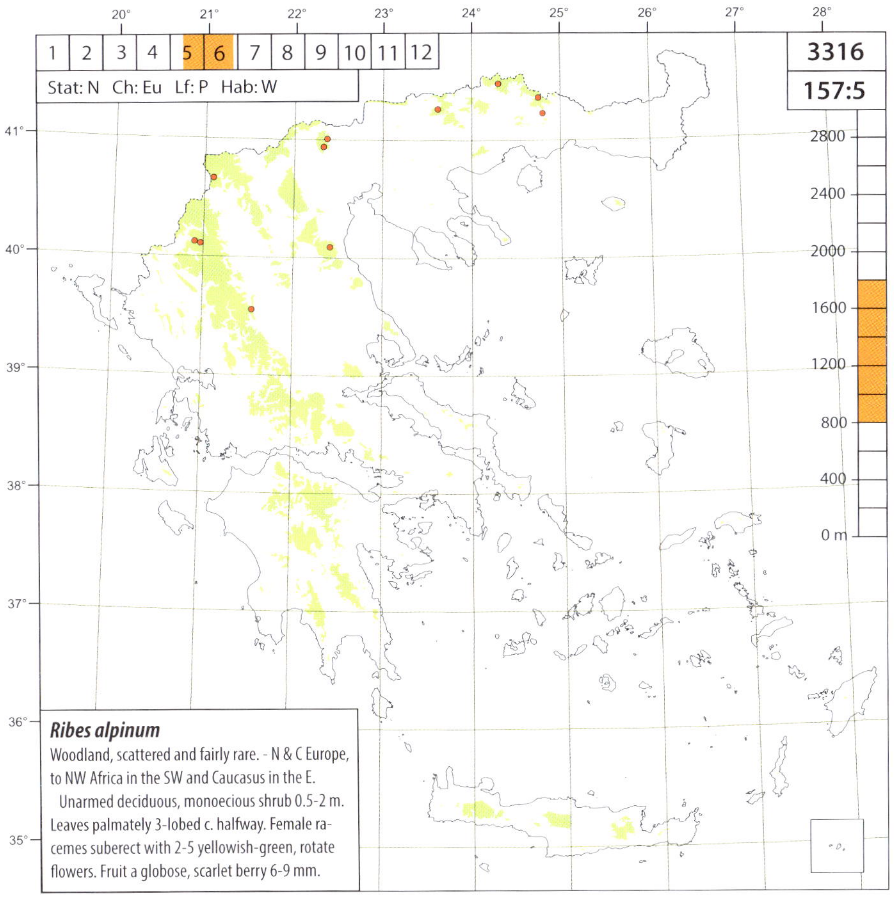

Ribes alpinum

Woodland, scattered and fairly rare. - N & C Europe, to NW Africa in the SW and Caucasus in the E.

Unarmed deciduous, monoecious shrub 0.5-2 m. Leaves palmately 3-lobed c. halfway. Female racemes suberect with 2-5 yellowish-green, rotate flowers. Fruit a globose, scarlet berry 6-9 mm.

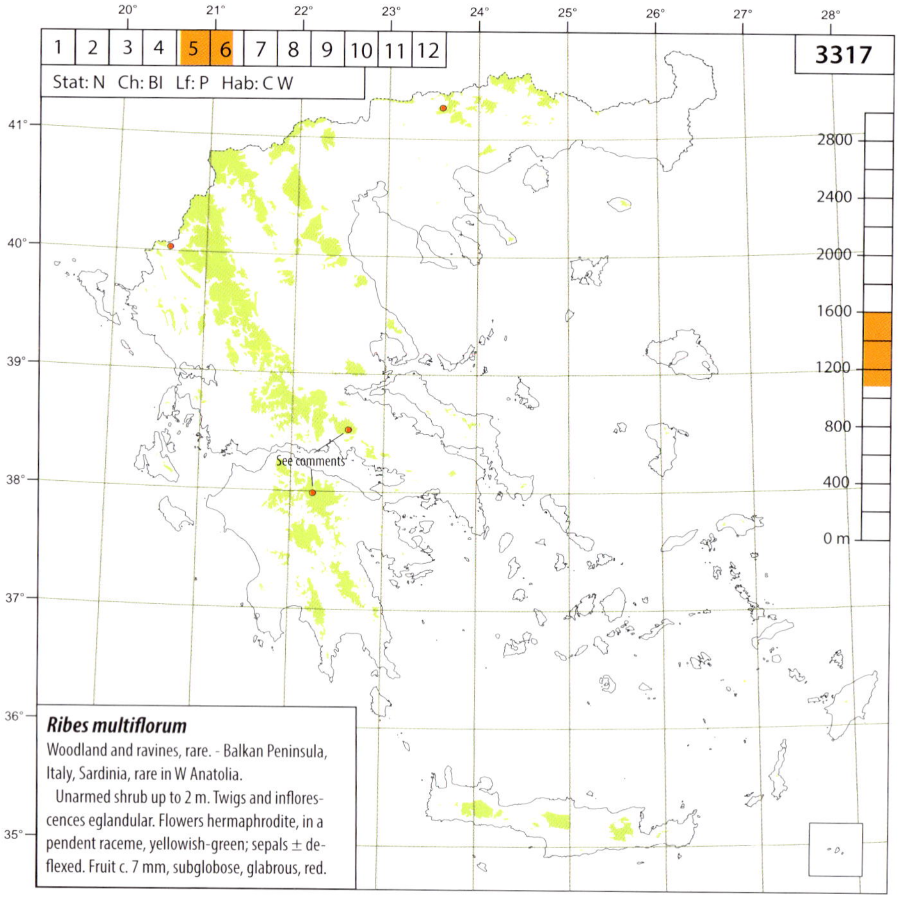

Ribes multiflorum

Woodland and ravines, rare. - Balkan Peninsula, Italy, Sardinia, rare in W Anatolia.

Unarmed shrub up to 2 m. Twigs and inflorescences eglandular. Flowers hermaphrodite, in a pendent raceme, yellowish-green; sepals ± deflexed. Fruit c. 7 mm, subglobose, glabrous, red.

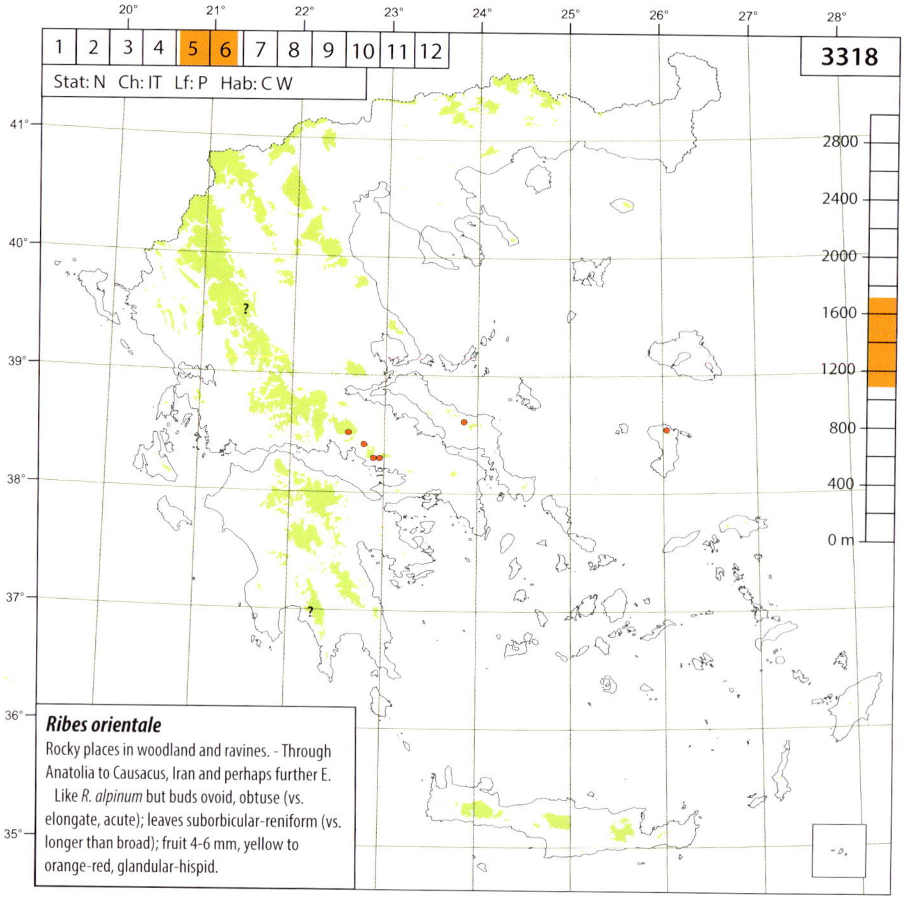

Ribes orientale

Rocky places in woodland and ravines. - Through Anatolia to Causacus, Iran and perhaps further E.

Like *R. alpinum* but buds ovoid, obtuse (vs. elongate, acute); leaves suborbicular-reniform (vs. longer than broad); fruit 4-6 mm, yellow to orange-red, glandular-hispid.

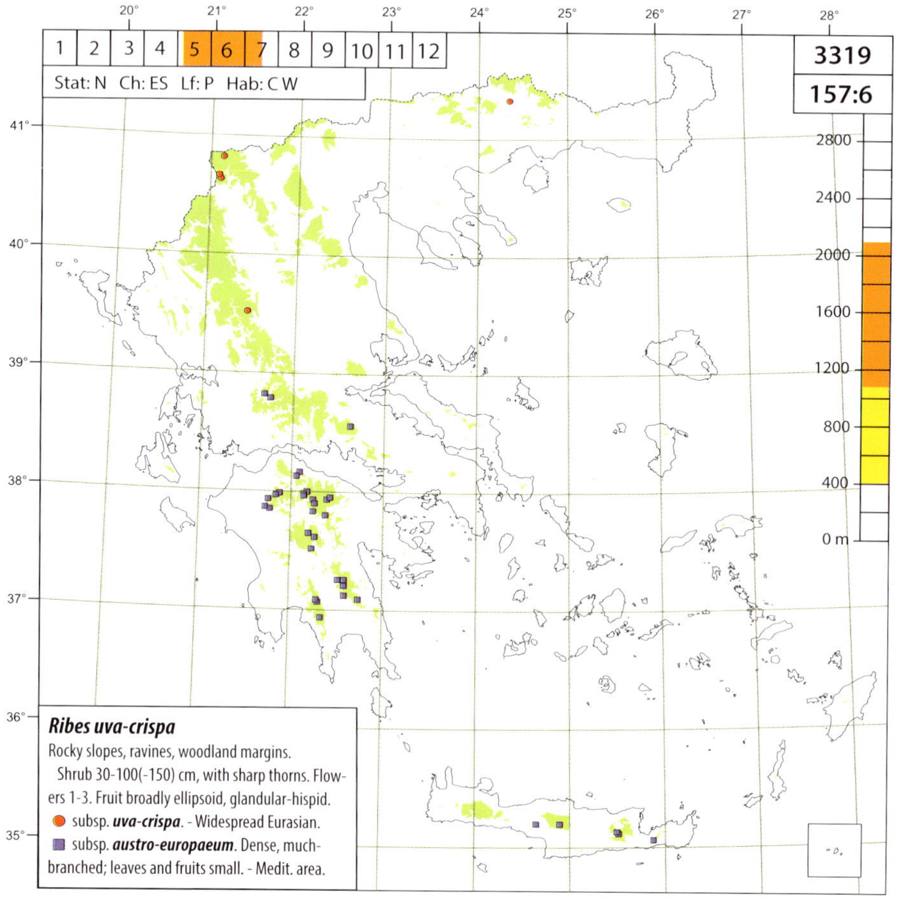

Ribes uva-crispa

Rocky slopes, ravines, woodland margins.

Shrub 30-100(-150) cm, with sharp thorns. Flowers 1-3. Fruit broadly ellipsoid, glandular-hispid.

● subsp. ***uva-crispa***. - Widespread Eurasian.

■ subsp. ***austro-europaeum***. Dense, much-branched; leaves and fruits small. - Medit. area.

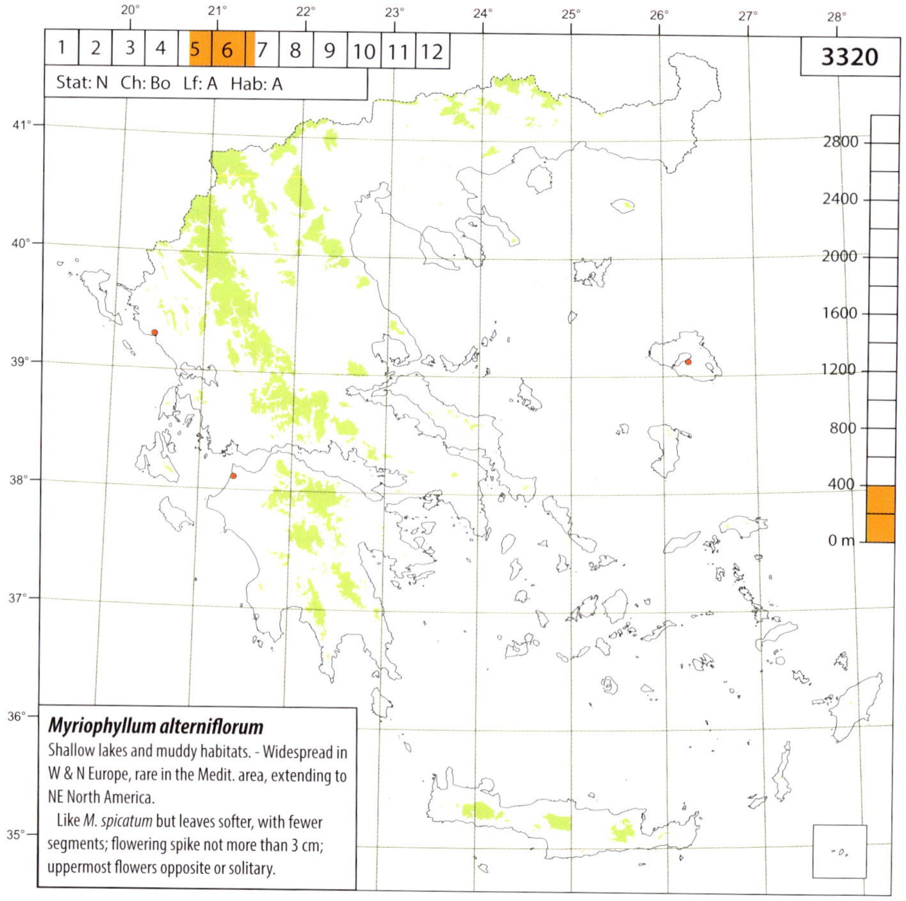

Myriophyllum alterniflorum

Shallow lakes and muddy habitats. - Widespread in W & N Europe, rare in the Medit. area, extending to NE North America.

Like *M. spicatum* but leaves softer, with fewer segments; flowering spike not more than 3 cm; uppermost flowers opposite or solitary.

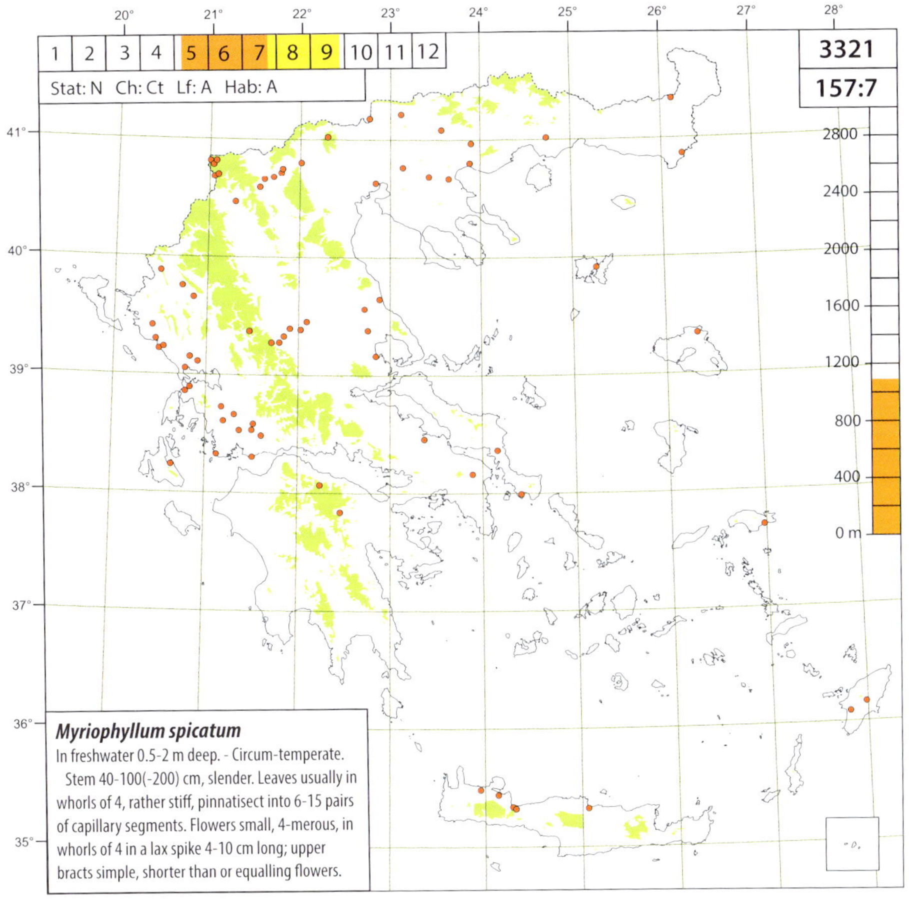

Myriophyllum spicatum

In freshwater 0.5-2 m deep. - Circum-temperate.

Stem 40-100(-200) cm, slender. Leaves usually in whorls of 4, rather stiff, pinnatisect into 6-15 pairs of capillary segments. Flowers small, 4-merous, in whorls of 4 in a lax spike 4-10 cm long; upper bracts simple, shorter than or equalling flowers.

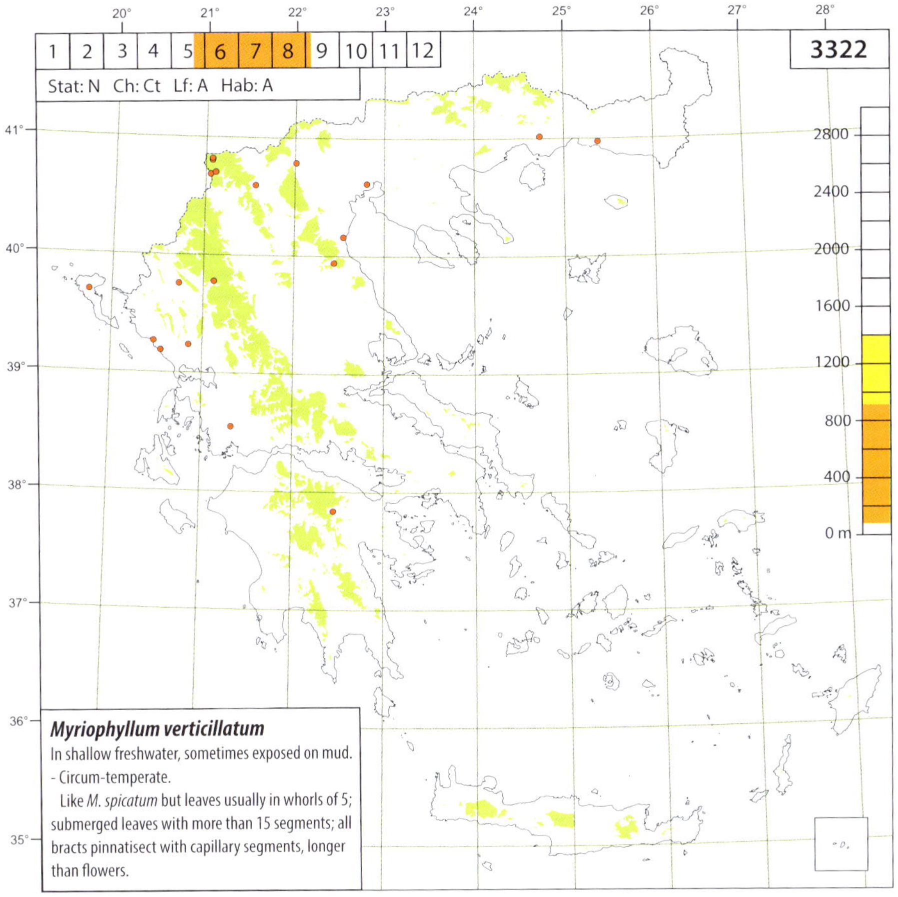

Myriophyllum verticillatum

In shallow freshwater, sometimes exposed on mud. - Circum-temperate.

Like *M. spicatum* but leaves usually in whorls of 5; submerged leaves with more than 15 segments; all bracts pinnatisect with capillary segments, longer than flowers.

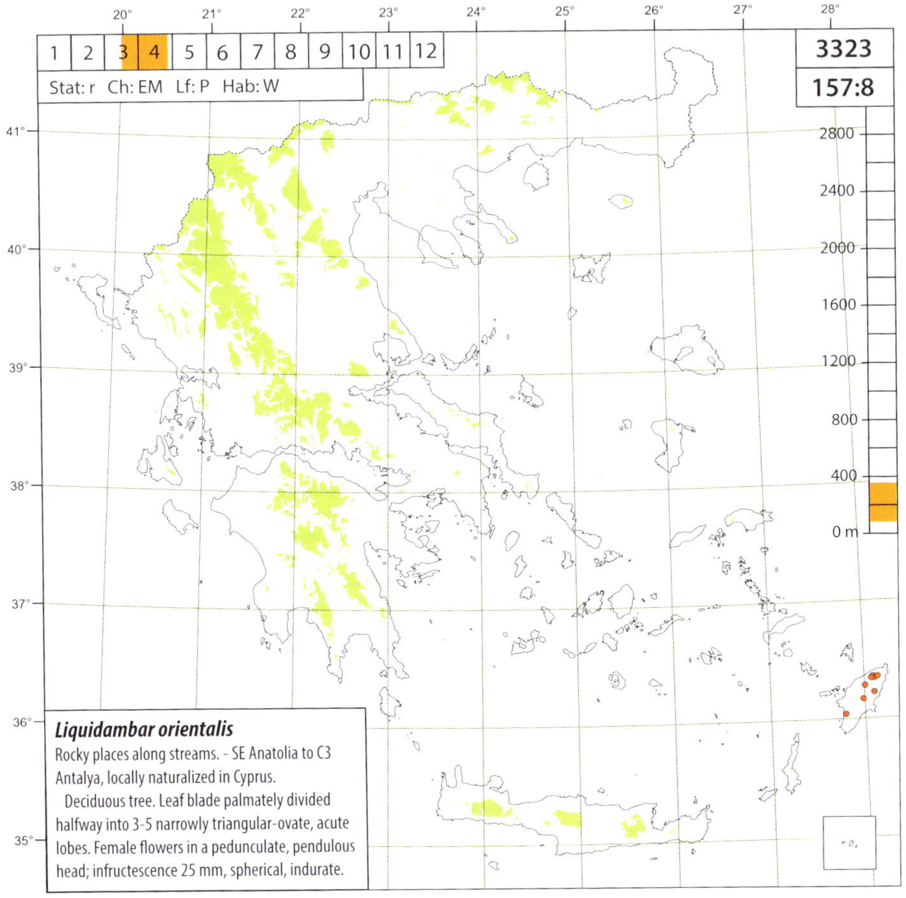

Liquidambar orientalis

Rocky places along streams. - SE Anatolia to C3 Antalya, locally naturalized in Cyprus.

Deciduous tree. Leaf blade palmately divided halfway into 3-5 narrowly triangular-ovate, acute lobes. Female flowers in a pedunculate, pendulous head; infructescence 25 mm, spherical, indurate.

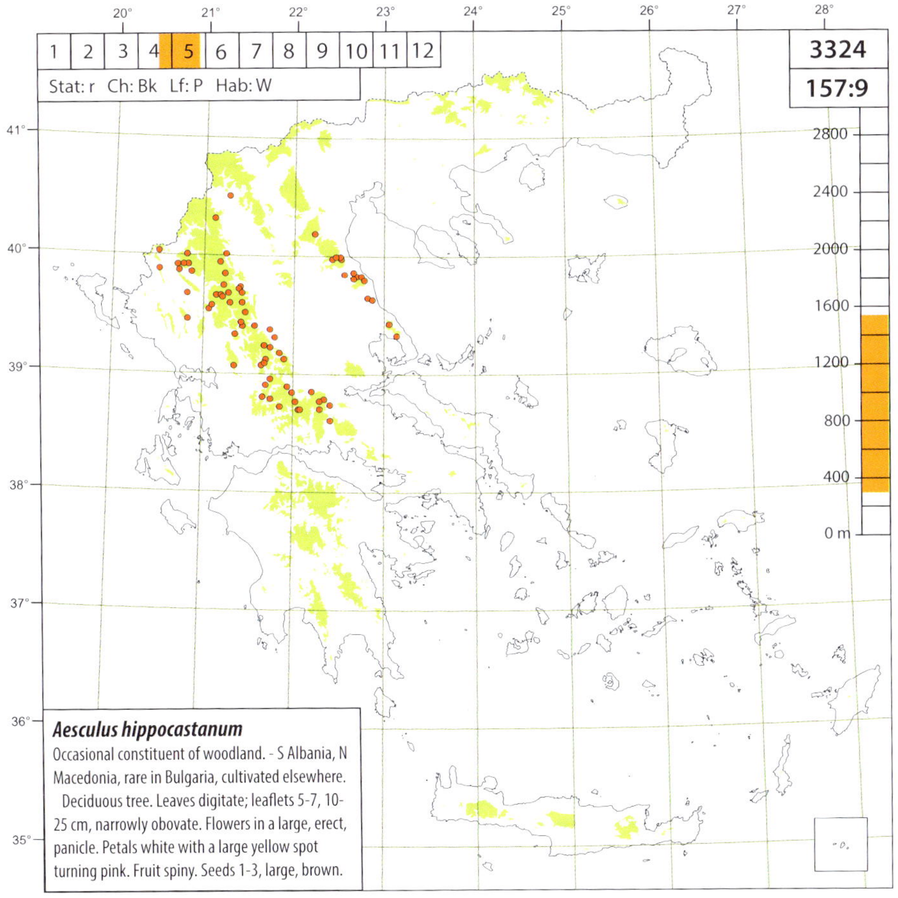

Aesculus hippocastanum

Occasional constituent of woodland. - S Albania, N Macedonia, rare in Bulgaria, cultivated elsewhere.

Deciduous tree. Leaves digitate; leaflets 5-7, 10-25 cm, narrowly obovate. Flowers in a large, erect, panicle. Petals white with a large yellow spot turning pink. Fruit spiny. Seeds 1-3, large, brown.

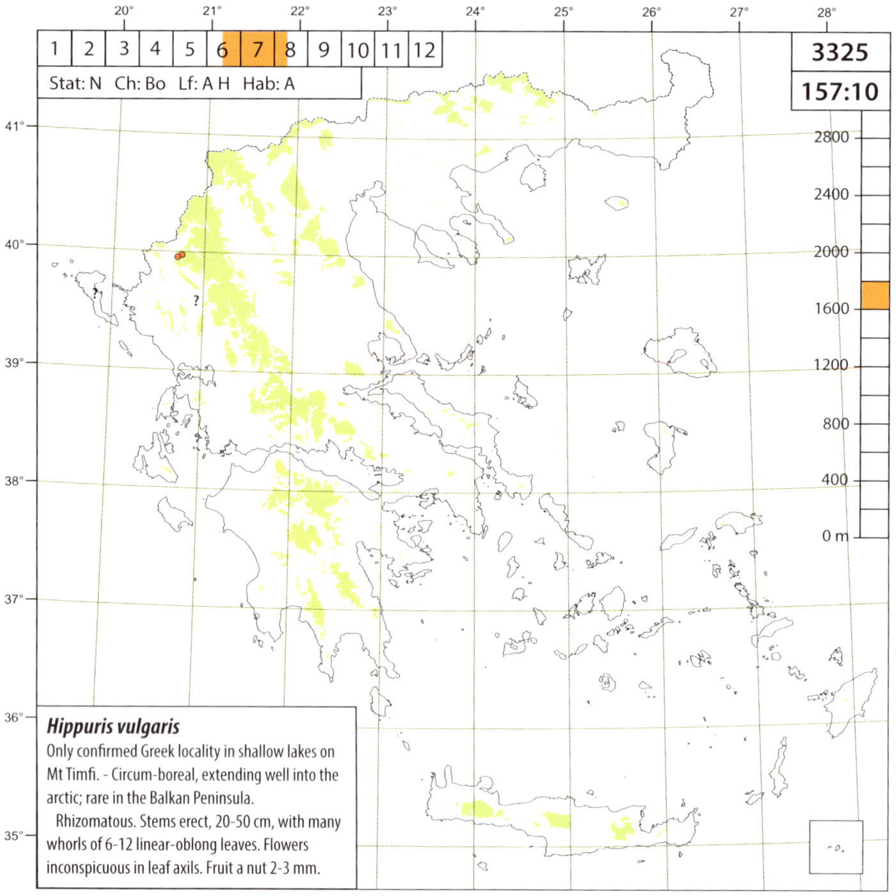

Hippuris vulgaris
Only confirmed Greek locality in shallow lakes on Mt Timfi. - Circum-boreal, extending well into the arctic; rare in the Balkan Peninsula.
Rhizomatous. Stems erect, 20-50 cm, with many whorls of 6-12 linear-oblong leaves. Flowers inconspicuous in leaf axils. Fruit a nut 2-3 mm.

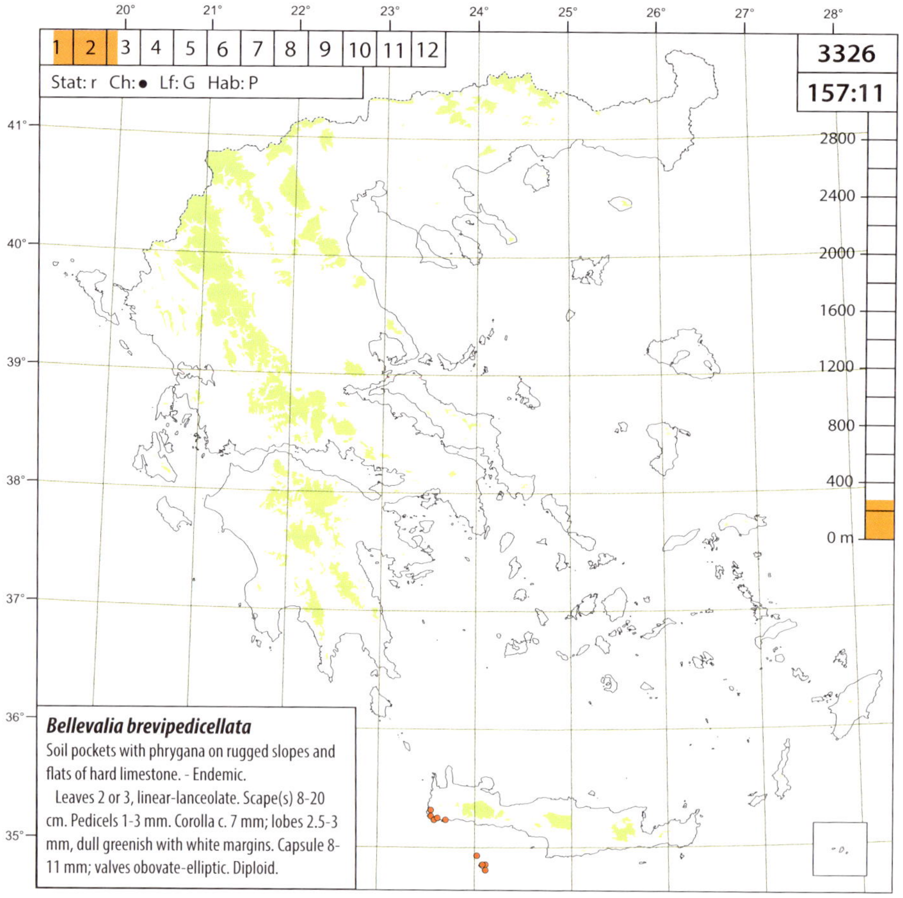

Bellevalia brevipedicellata
Soil pockets with phrygana on rugged slopes and flats of hard limestone. - Endemic.
Leaves 2 or 3, linear-lanceolate. Scape(s) 8-20 cm. Pedicels 1-3 mm. Corolla c. 7 mm; lobes 2.5-3 mm, dull greenish with white margins. Capsule 8-11 mm; valves obovate-elliptic. Diploid.

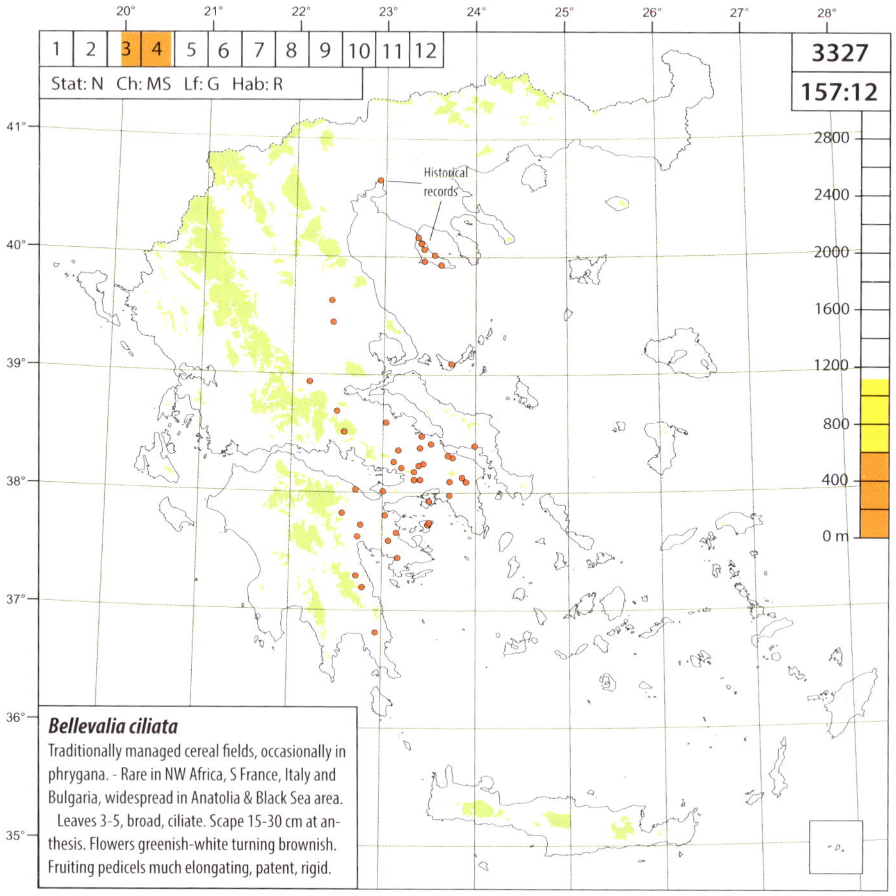

Bellevalia ciliata
Traditionally managed cereal fields, occasionally in phrygana. - Rare in NW Africa, S France, Italy and Bulgaria, widespread in Anatolia & Black Sea area.
Leaves 3-5, broad, ciliate. Scape 15-30 cm at anthesis. Flowers greenish-white turning brownish. Fruiting pedicels much elongating, patent, rigid.

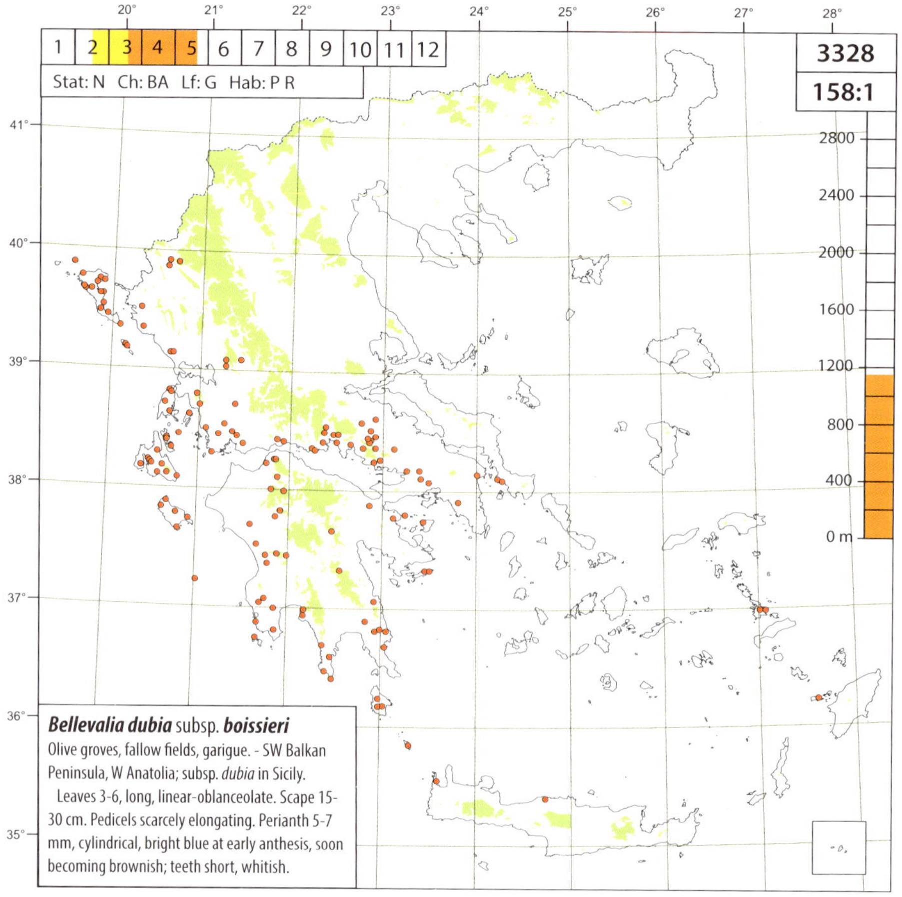

Bellevalia dubia subsp. ***boissieri***
Olive groves, fallow fields, garigue. - SW Balkan Peninsula, W Anatolia; subsp. *dubia* in Sicily.
Leaves 3-6, long, linear-oblanceolate. Scape 15-30 cm. Pedicels scarcely elongating. Perianth 5-7 mm, cylindrical, bright blue at early anthesis, soon becoming brownish; teeth short, whitish.

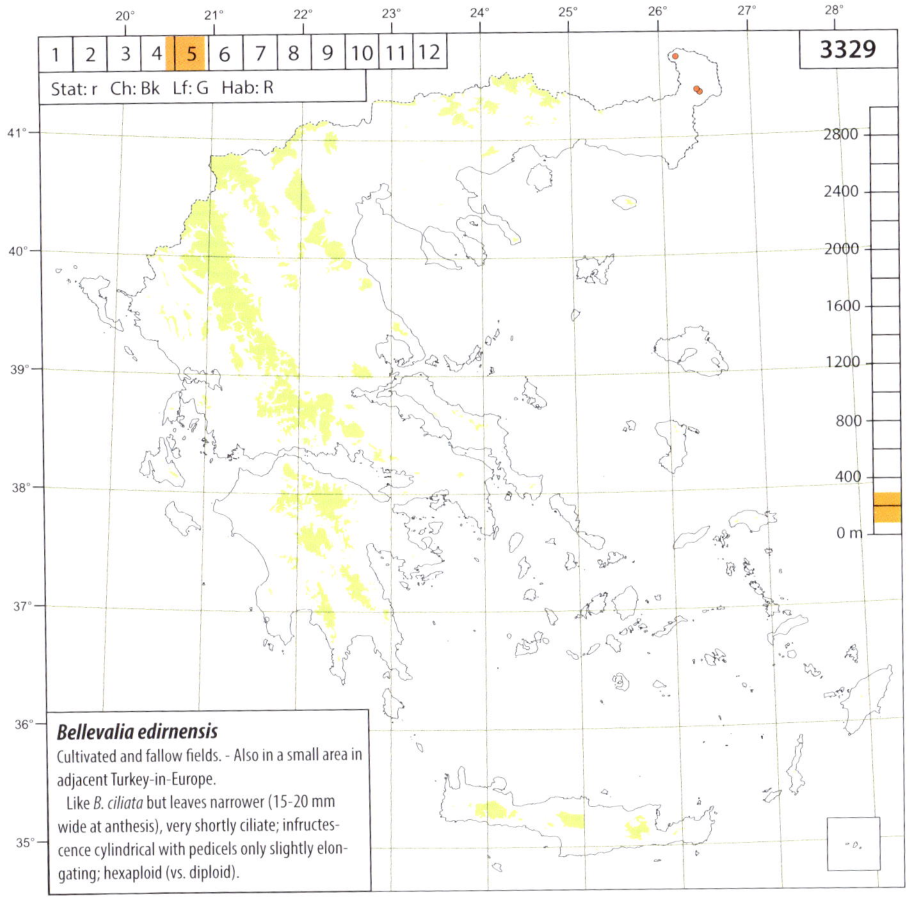

Bellevalia edirnensis

Cultivated and fallow fields. - Also in a small area in adjacent Turkey-in-Europe.

Like *B. ciliata* but leaves narrower (15-20 mm wide at anthesis), very shortly ciliate; infructescence cylindrical with pedicels only slightly elongating; hexaploid (vs. diploid).

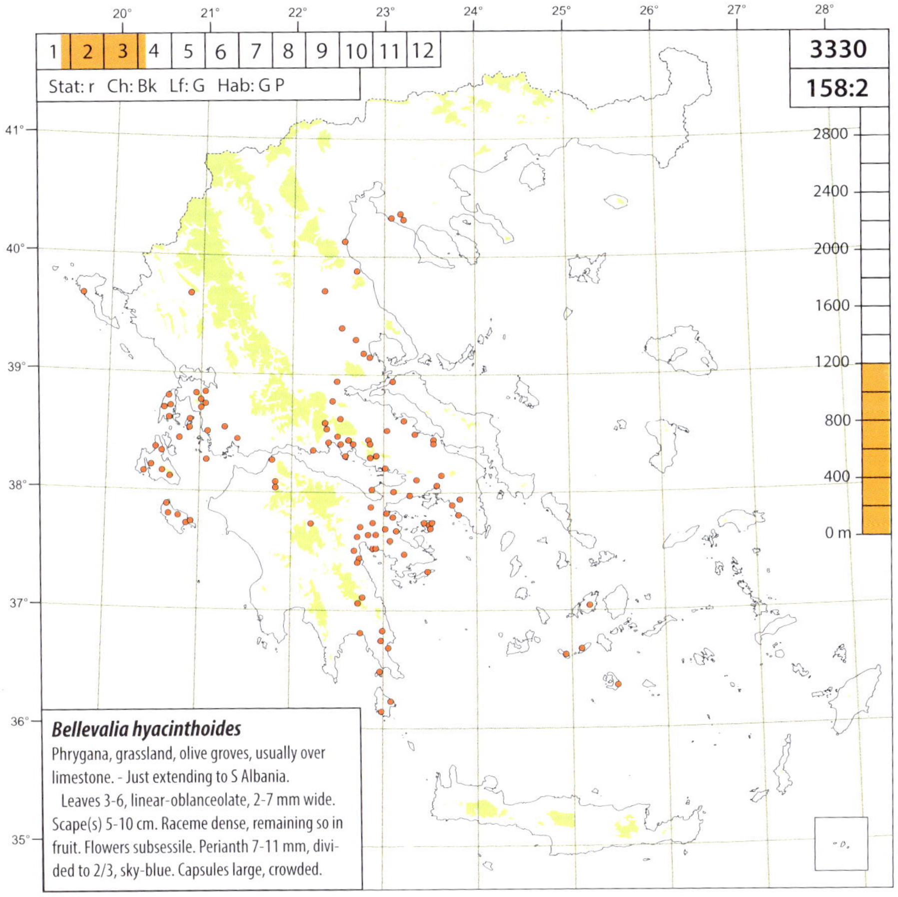

Bellevalia hyacinthoides

Phrygana, grassland, olive groves, usually over limestone. - Just extending to S Albania.

Leaves 3-6, linear-oblanceolate, 2-7 mm wide. Scape(s) 5-10 cm. Raceme dense, remaining so in fruit. Flowers subsessile. Perianth 7-11 mm, divided to 2/3, sky-blue. Capsules large, crowded.

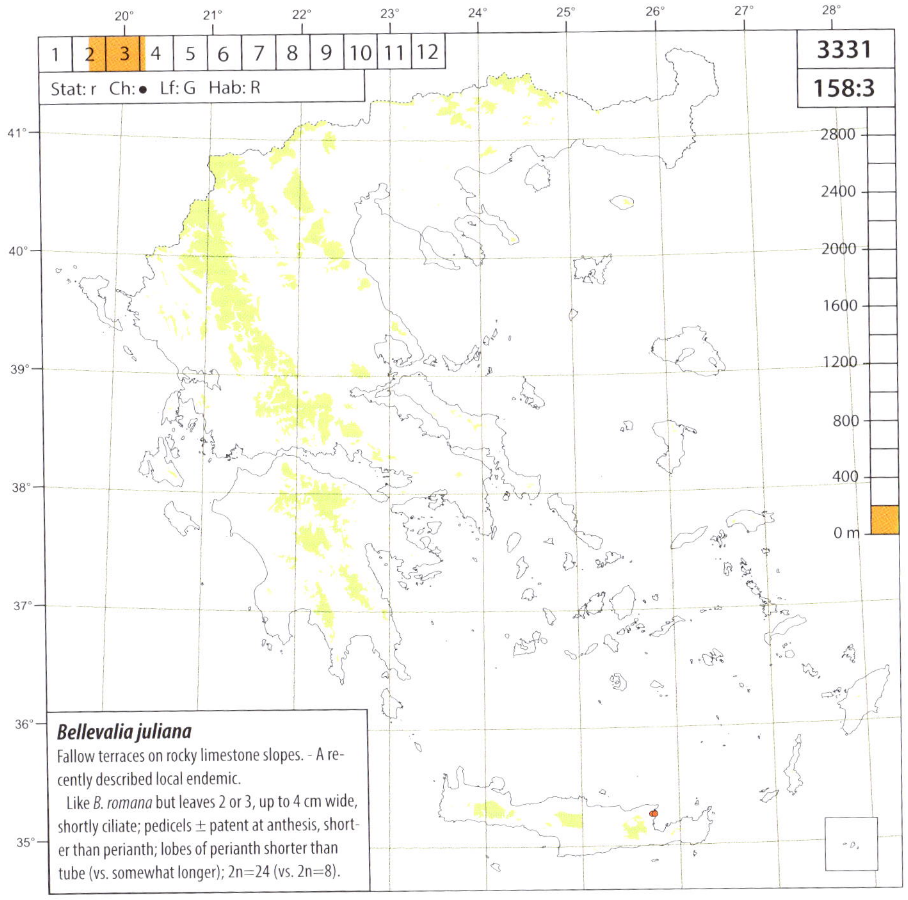

Bellevalia juliana

Fallow terraces on rocky limestone slopes. - A recently described local endemic.

Like *B. romana* but leaves 2 or 3, up to 4 cm wide, shortly ciliate; pedicels ± patent at anthesis, shorter than perianth; lobes of perianth shorter than tube (vs. somewhat longer); 2n=24 (vs. 2n=8).

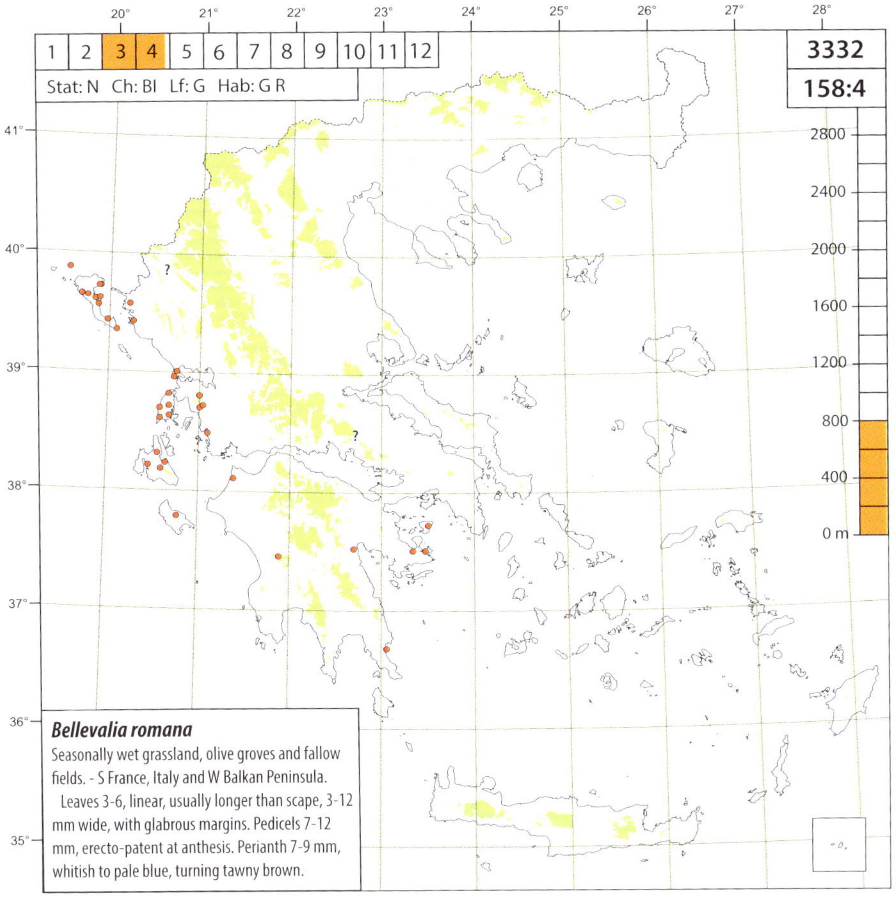

Bellevalia romana

Seasonally wet grassland, olive groves and fallow fields. - S France, Italy and W Balkan Peninsula.

Leaves 3-6, linear, usually longer than scape, 3-12 mm wide, with glabrous margins. Pedicels 7-12 mm, erecto-patent at anthesis. Perianth 7-9 mm, whitish to pale blue, turning tawny brown.

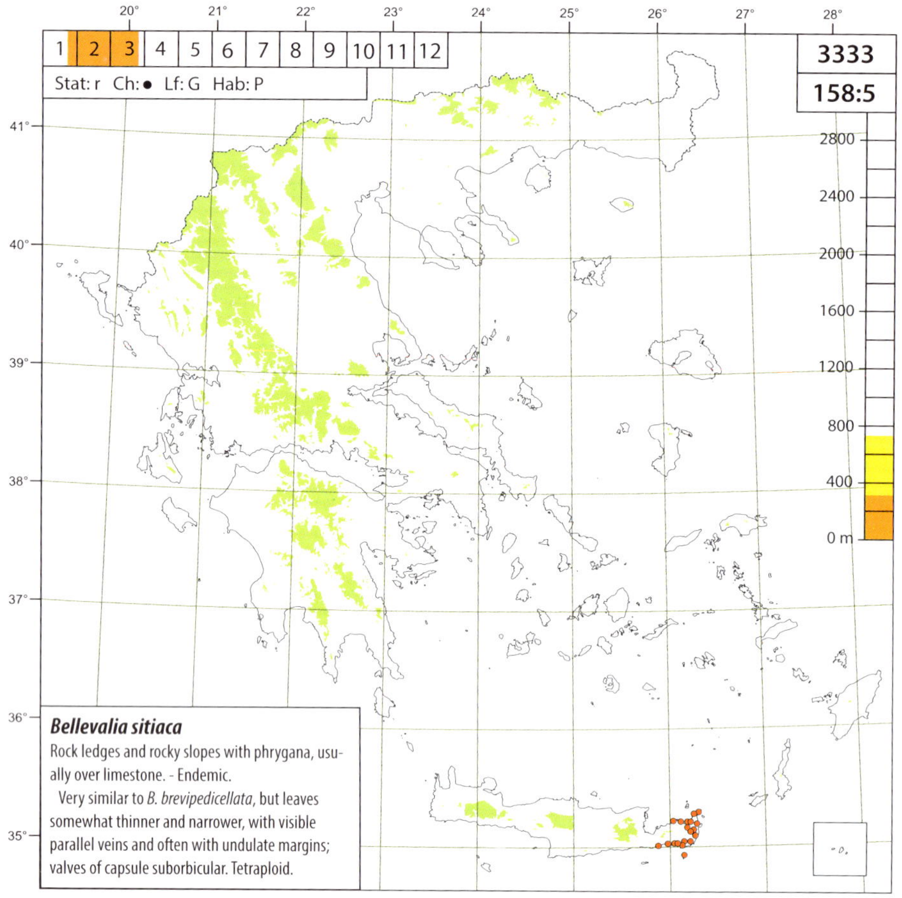

Bellevalia sitiaca
Rock ledges and rocky slopes with phrygana, usually over limestone. - Endemic.
Very similar to *B. brevipedicellata*, but leaves somewhat thinner and narrower, with visible parallel veins and often with undulate margins; valves of capsule suborbicular. Tetraploid.

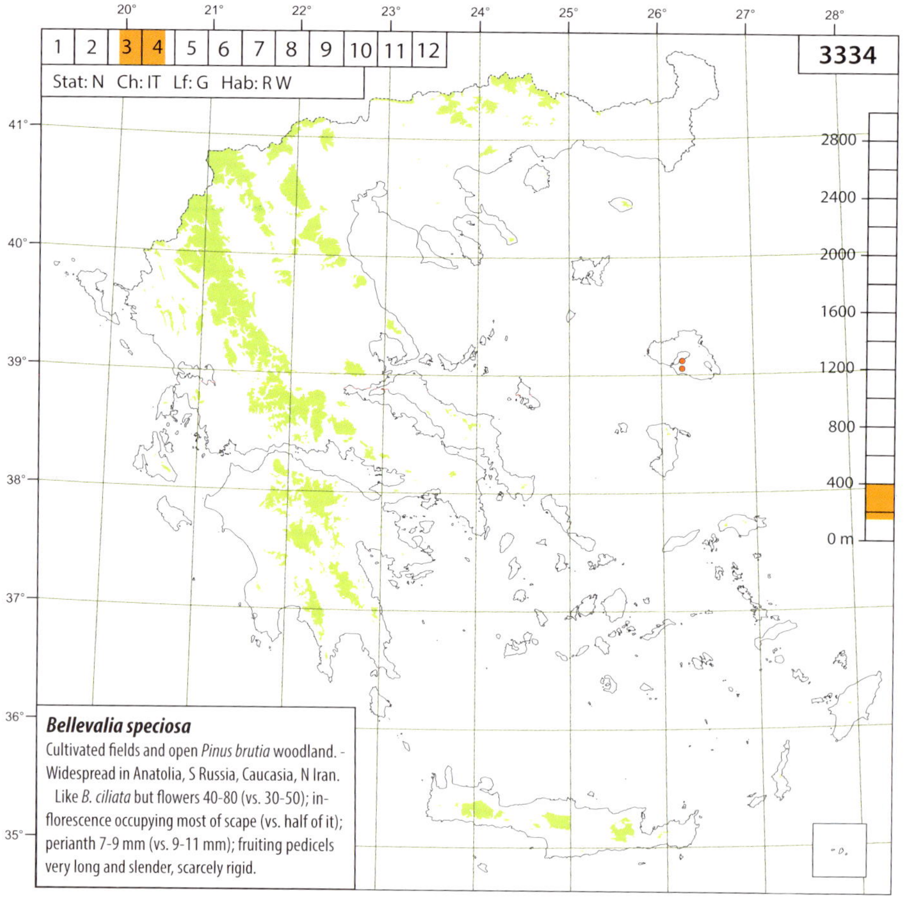

Bellevalia speciosa
Cultivated fields and open *Pinus brutia* woodland. - Widespread in Anatolia, S Russia, Caucasia, N Iran.
Like *B. ciliata* but flowers 40-80 (vs. 30-50); inflorescence occupying most of scape (vs. half of it); perianth 7-9 mm (vs. 9-11 mm); fruiting pedicels very long and slender, scarcely rigid.

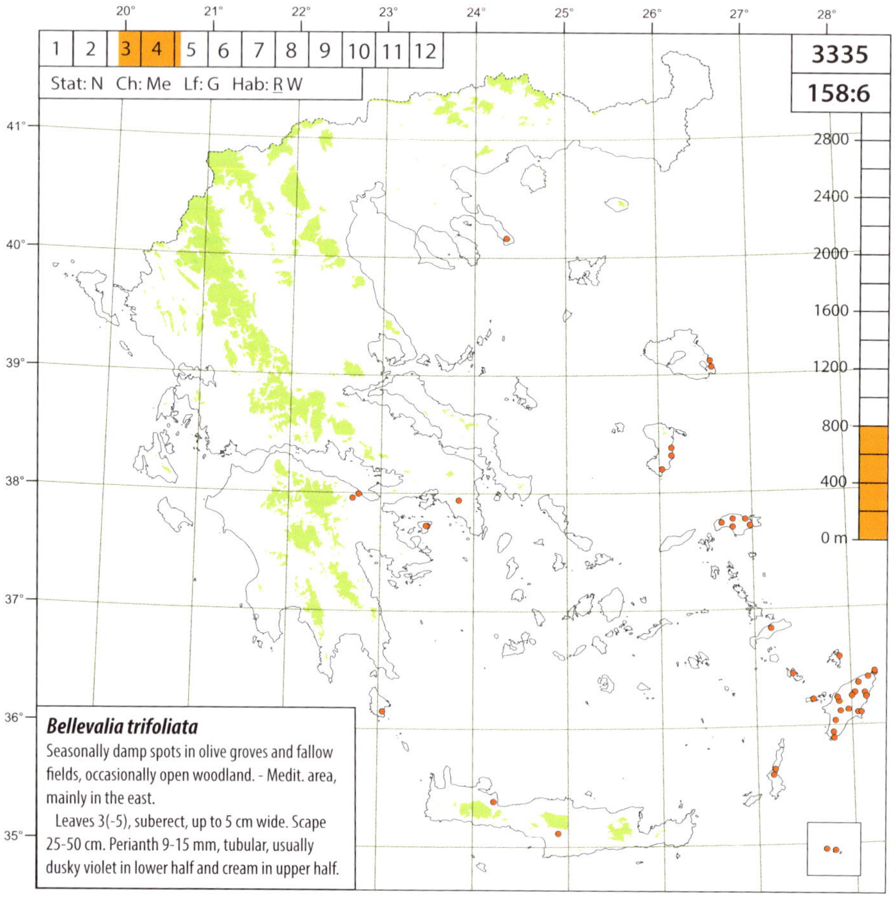

Bellevalia trifoliata
Seasonally damp spots in olive groves and fallow fields, occasionally open woodland. - Medit. area, mainly in the east.
Leaves 3(-5), suberect, up to 5 cm wide. Scape 25-50 cm. Perianth 9-15 mm, tubular, usually dusky violet in lower half and cream in upper half.

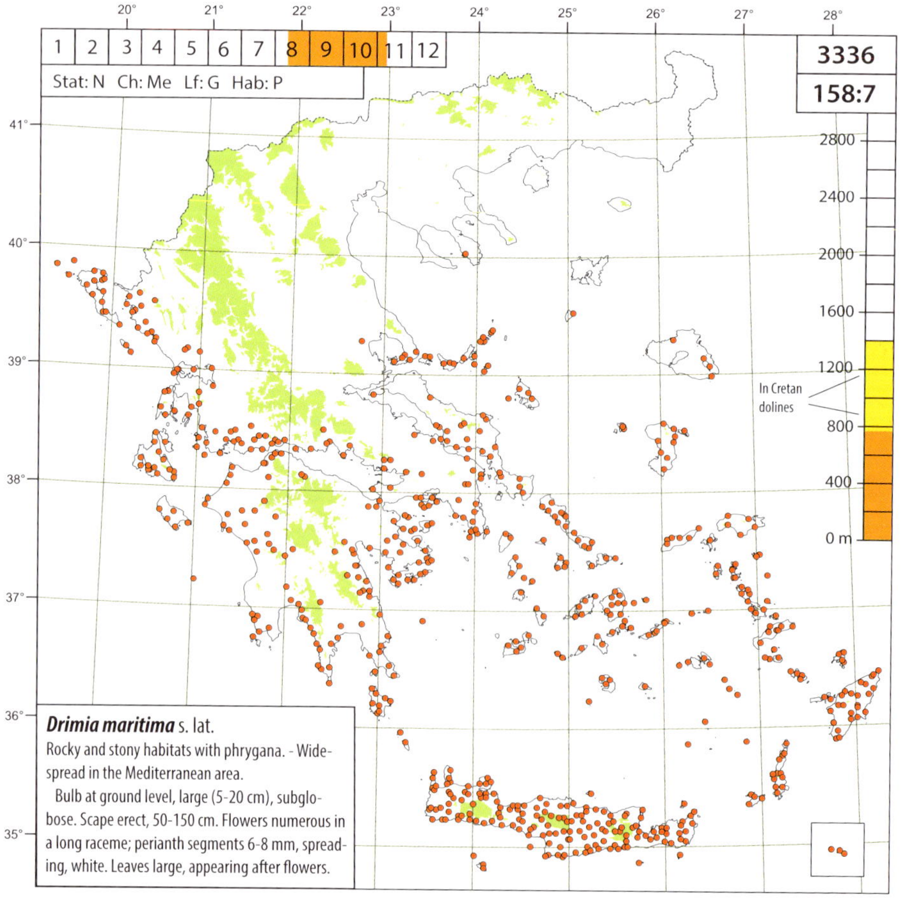

Drimia maritima s. lat.
Rocky and stony habitats with phrygana. - Widespread in the Mediterranean area.
Bulb at ground level, large (5-20 cm), subglobose. Scape erect, 50-150 cm. Flowers numerous in a long raceme; perianth segments 6-8 mm, spreading, white. Leaves large, appearing after flowers.

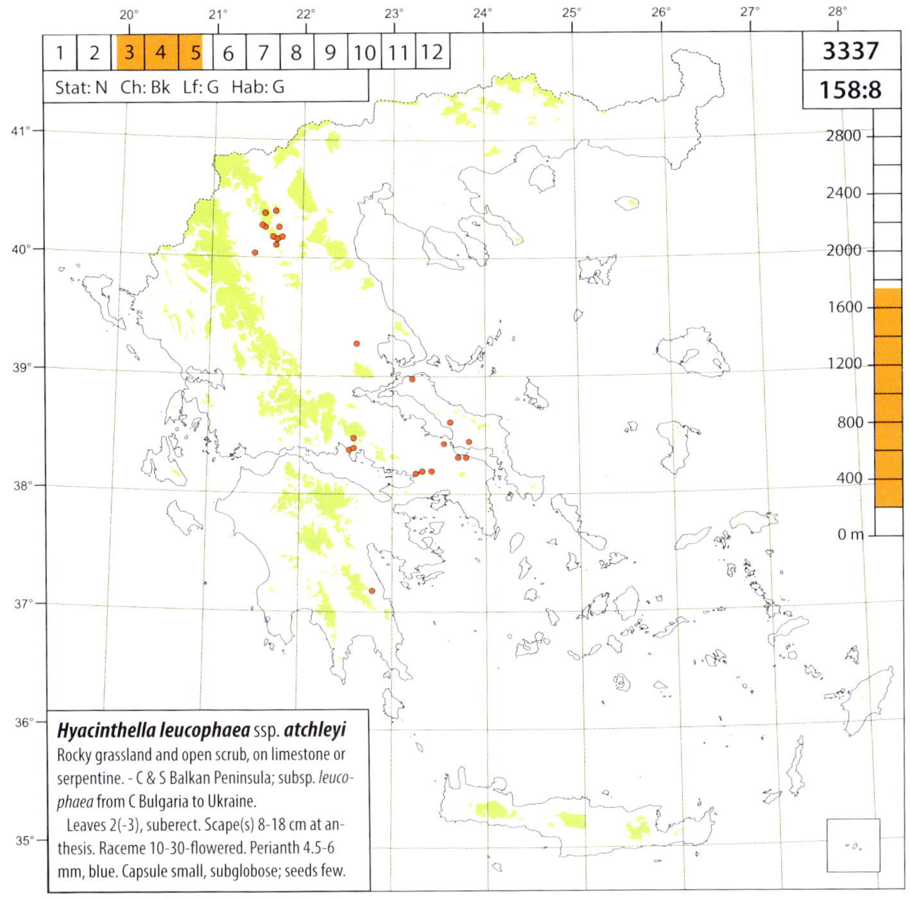

Hyacinthella leucophaea ssp. ***atchleyi***

Rocky grassland and open scrub, on limestone or serpentine. - C & S Balkan Peninsula; subsp. *leucophaea* from C Bulgaria to Ukraine.

Leaves 2(-3), suberect. Scape(s) 8-18 cm at anthesis. Raceme 10-30-flowered. Perianth 4.5-6 mm, blue. Capsule small, subglobose; seeds few.

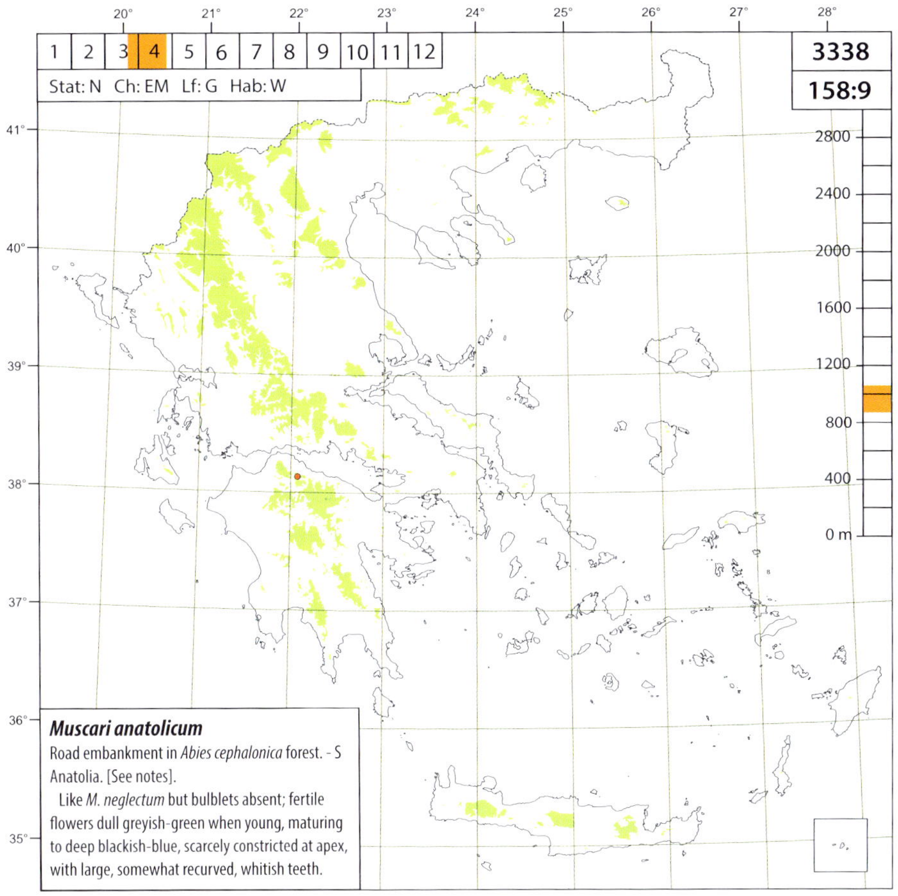

Muscari anatolicum

Road embankment in *Abies cephalonica* forest. - S Anatolia. [See notes].

Like *M. neglectum* but bulblets absent; fertile flowers dull greyish-green when young, maturing to deep blackish-blue, scarcely constricted at apex, with large, somewhat recurved, whitish teeth.

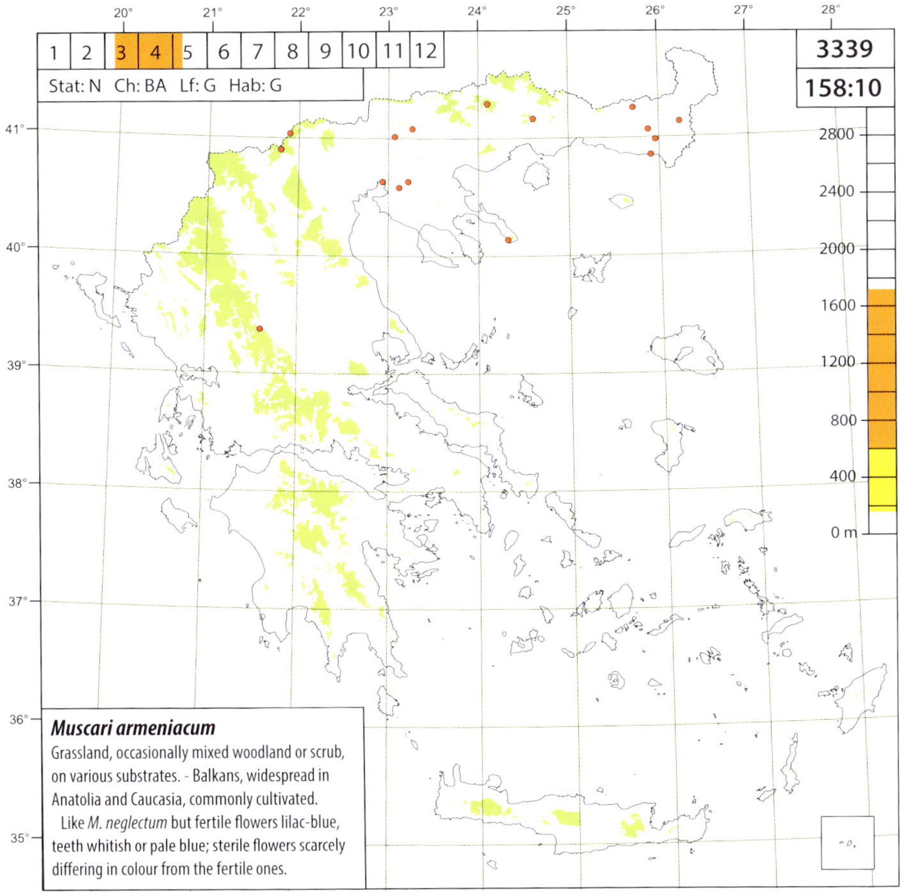

Muscari armeniacum

Grassland, occasionally mixed woodland or scrub, on various substrates. - Balkans, widespread in Anatolia and Caucasia, commonly cultivated.

Like *M. neglectum* but fertile flowers lilac-blue, teeth whitish or pale blue; sterile flowers scarcely differing in colour from the fertile ones.

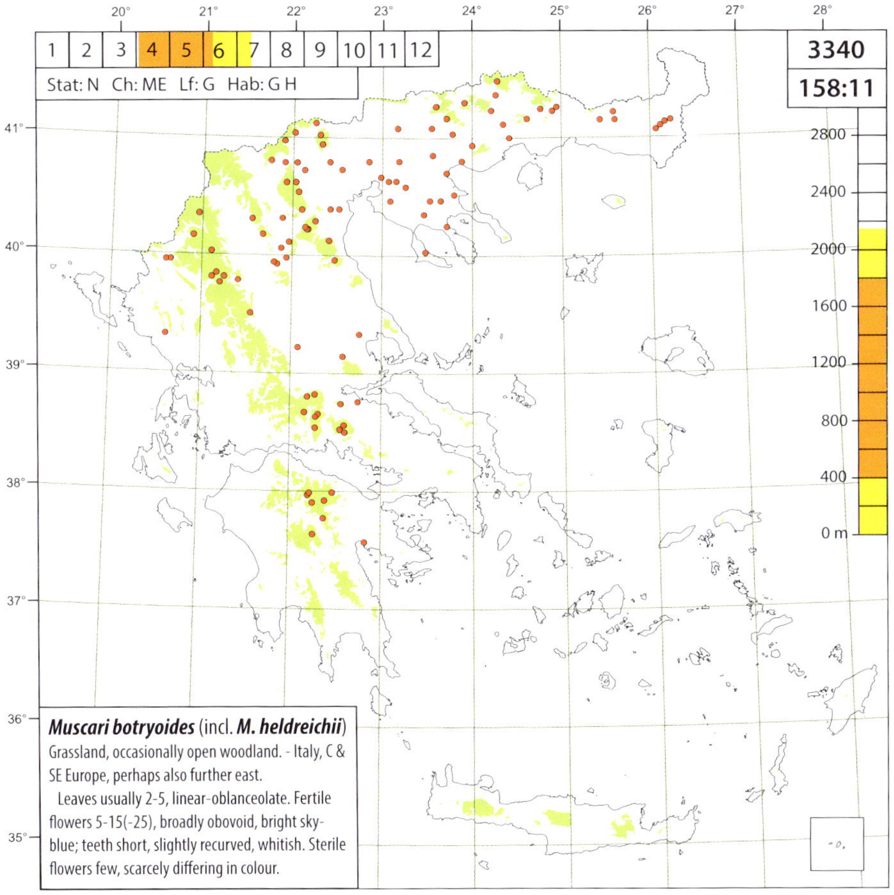

Muscari botryoides (incl. ***M. heldreichii***)

Grassland, occasionally open woodland. - Italy, C & SE Europe, perhaps also further east.

Leaves usually 2-5, linear-oblanceolate. Fertile flowers 5-15(-25), broadly obovoid, bright sky-blue; teeth short, slightly recurved, whitish. Sterile flowers few, scarcely differing in colour.

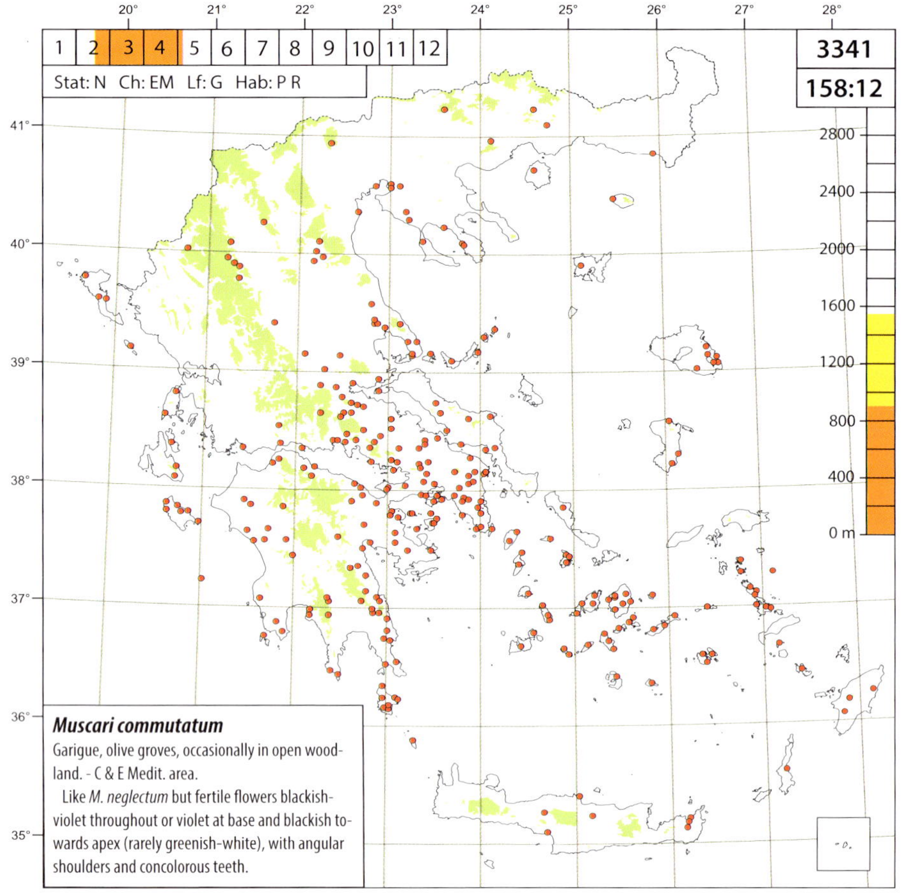

Muscari commutatum

Garigue, olive groves, occasionally in open woodland. - C & E Medit. area.

Like *M. neglectum* but fertile flowers blackish-violet throughout or violet at base and blackish towards apex (rarely greenish-white), with angular shoulders and concolorous teeth.

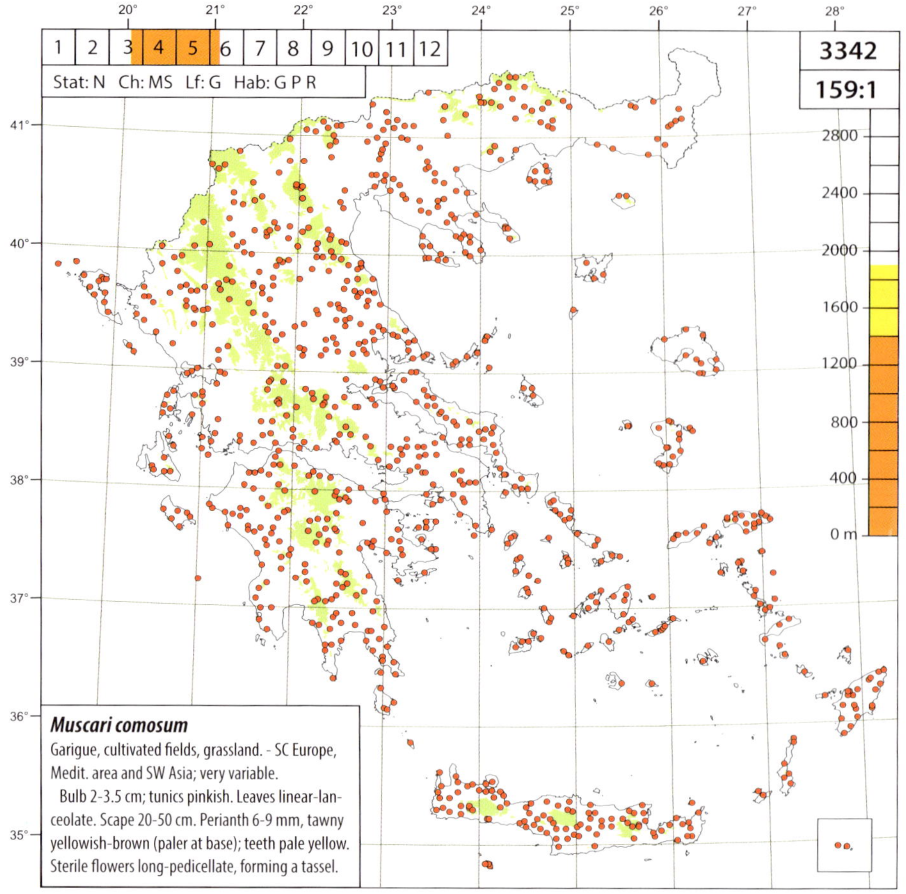

Muscari comosum

Garigue, cultivated fields, grassland. - SC Europe, Medit. area and SW Asia; very variable.

Bulb 2-3.5 cm; tunics pinkish. Leaves linear-lanceolate. Scape 20-50 cm. Perianth 6-9 mm, tawny yellowish-brown (paler at base); teeth pale yellow. Sterile flowers long-pedicellate, forming a tassel.

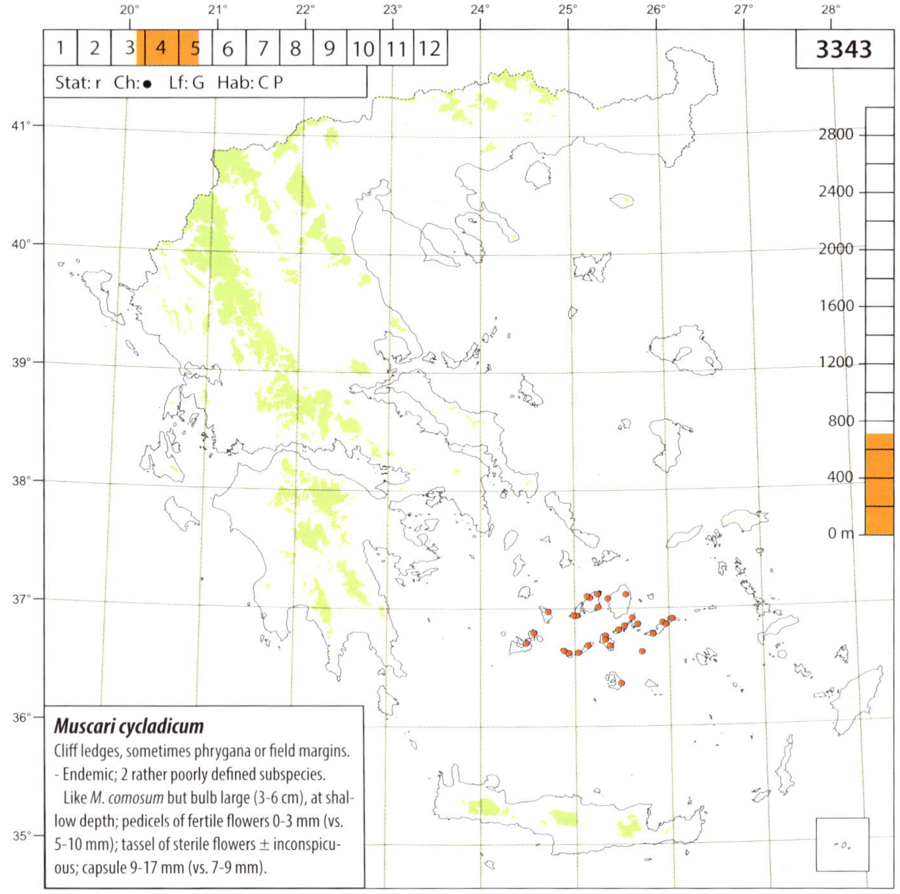

Muscari cycladicum

Cliff ledges, sometimes phrygana or field margins. - Endemic; 2 rather poorly defined subspecies.

Like *M. comosum* but bulb large (3-6 cm), at shallow depth; pedicels of fertile flowers 0-3 mm (vs. 5-10 mm); tassel of sterile flowers ± inconspicuous; capsule 9-17 mm (vs. 7-9 mm).

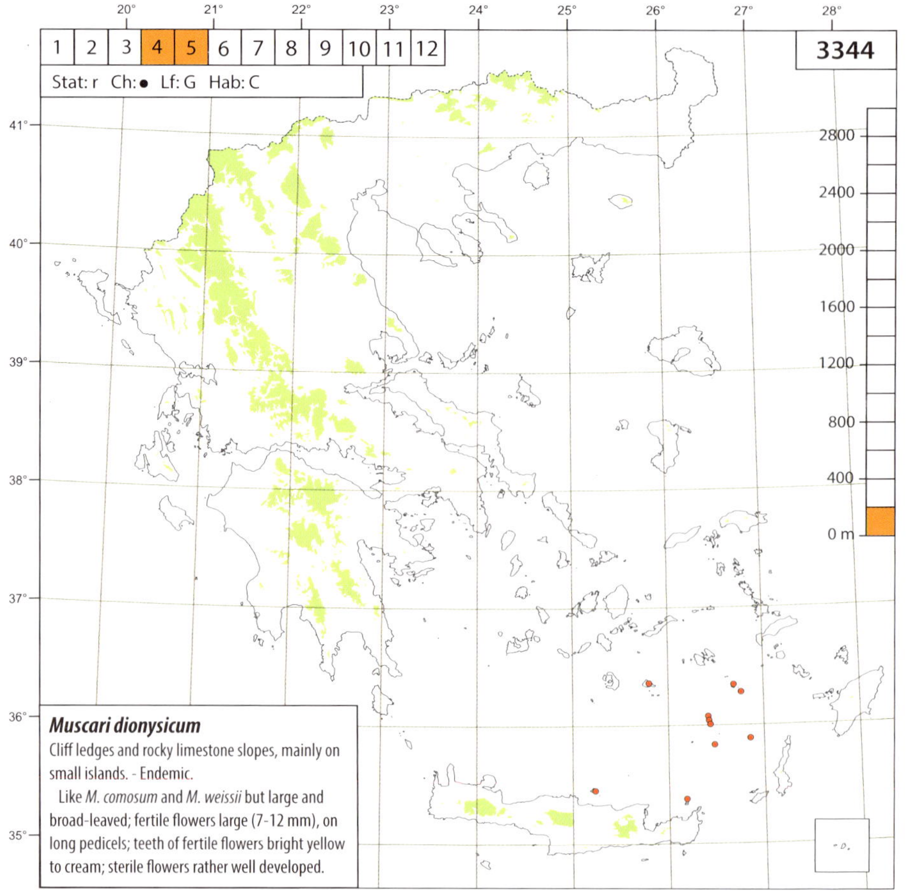

Muscari dionysicum

Cliff ledges and rocky limestone slopes, mainly on small islands. - Endemic.

Like *M. comosum* and *M. weissii* but large and broad-leaved; fertile flowers large (7-12 mm), on long pedicels; teeth of fertile flowers bright yellow to cream; sterile flowers rather well developed.

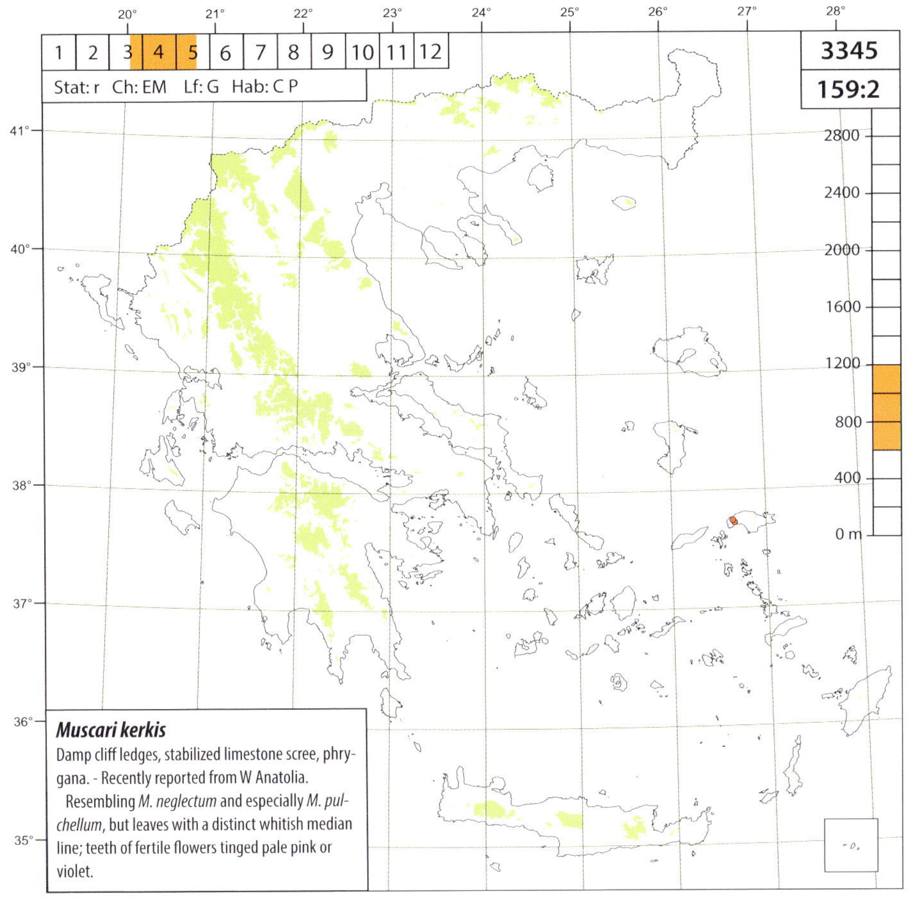

Muscari kerkis

Damp cliff ledges, stabilized limestone scree, phrygana. - Recently reported from W Anatolia.

Resembling *M. neglectum* and especially *M. pulchellum*, but leaves with a distinct whitish median line; teeth of fertile flowers tinged pale pink or violet.

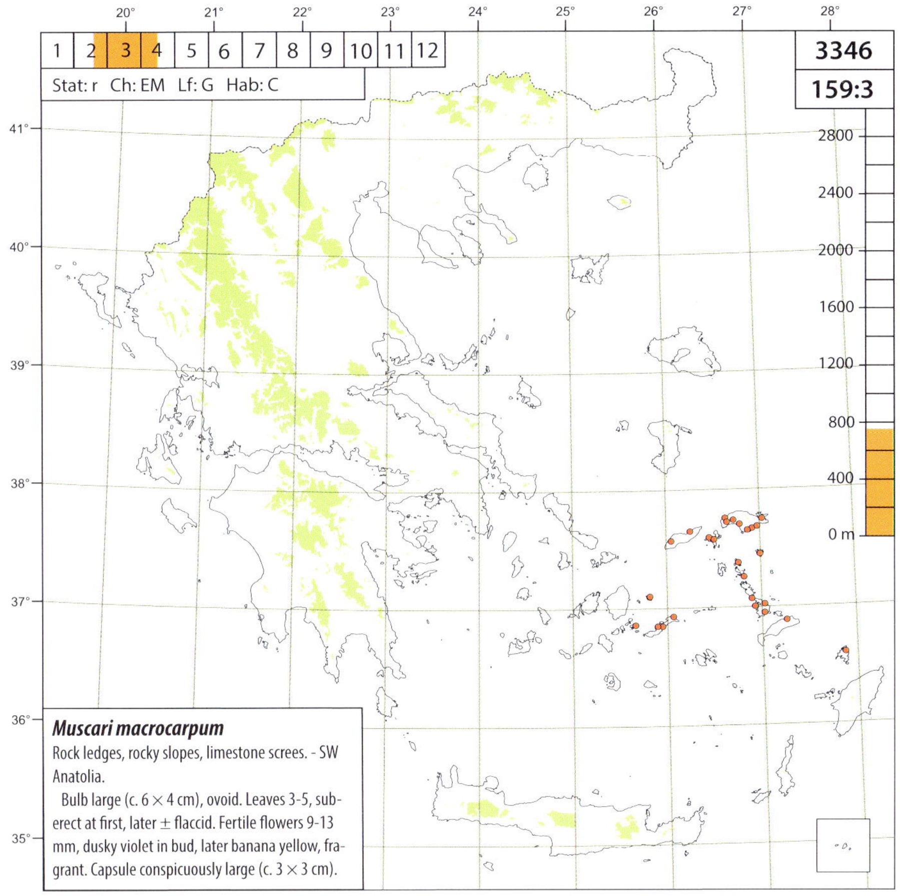

Muscari macrocarpum

Rock ledges, rocky slopes, limestone screes. - SW Anatolia.

Bulb large (c. 6 × 4 cm), ovoid. Leaves 3-5, suberect at first, later ± flaccid. Fertile flowers 9-13 mm, dusky violet in bud, later banana yellow, fragrant. Capsule conspicuously large (c. 3 × 3 cm).

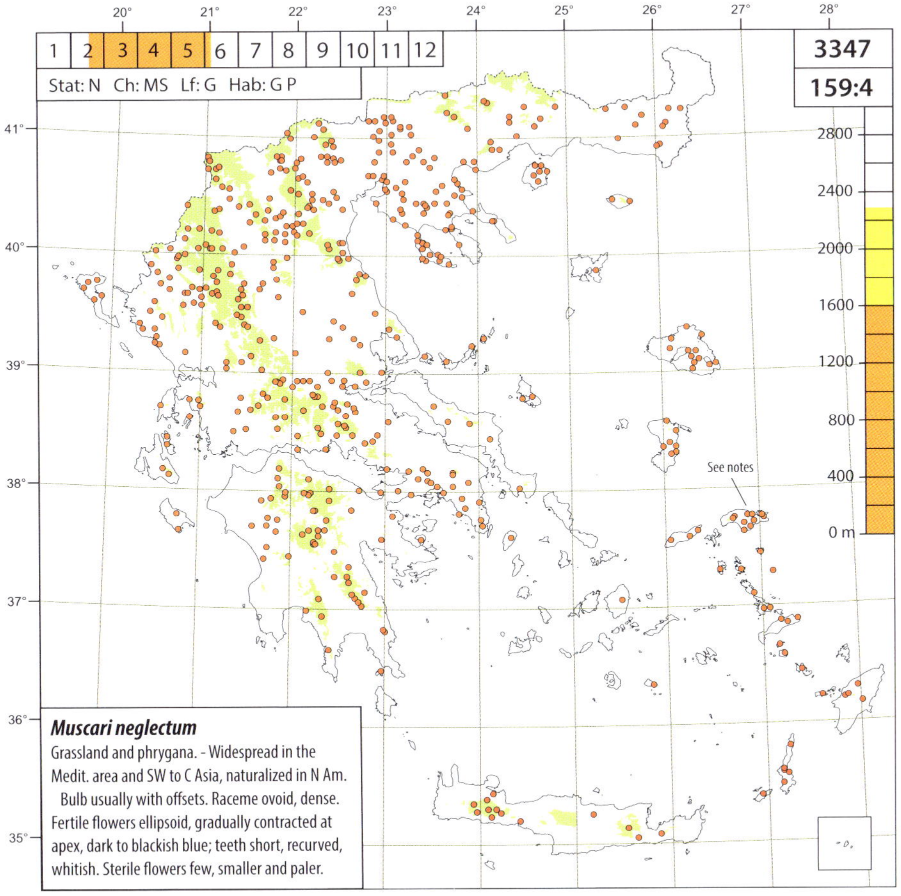

Muscari neglectum

Grassland and phrygana. - Widespread in the Medit. area and SW to C Asia, naturalized in N Am.

Bulb usually with offsets. Raceme ovoid, dense. Fertile flowers ellipsoid, gradually contracted at apex, dark to blackish blue; teeth short, recurved, whitish. Sterile flowers few, smaller and paler.

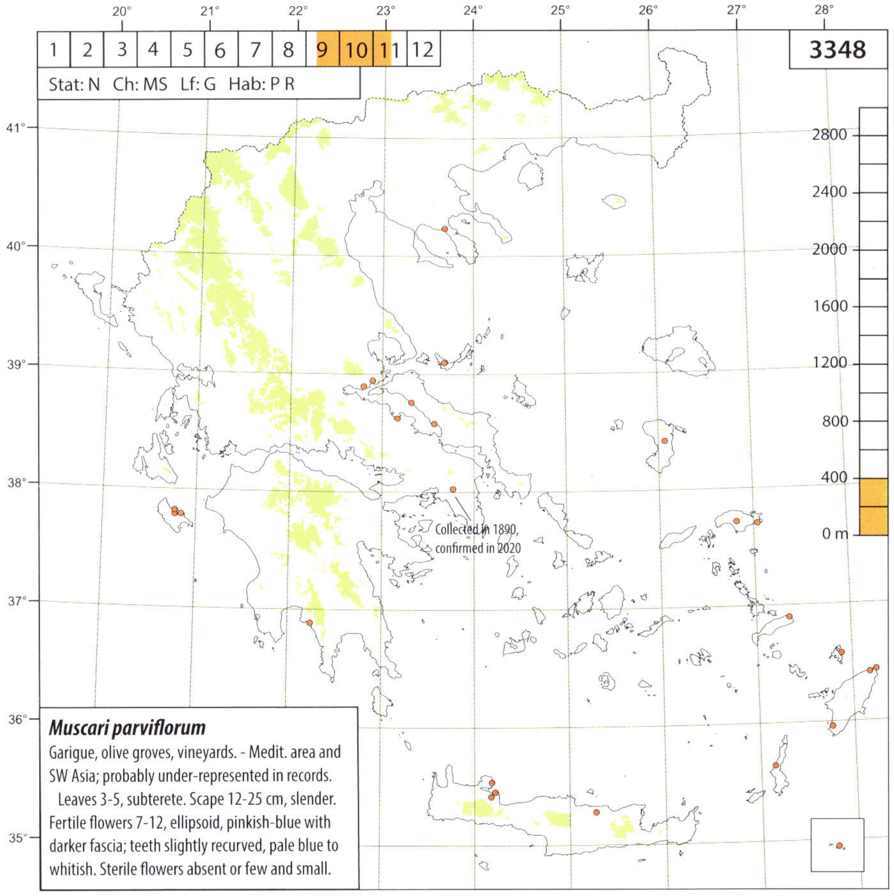

Muscari parviflorum

Garigue, olive groves, vineyards. - Medit. area and SW Asia; probably under-represented in records.

Leaves 3-5, subterete. Scape 12-25 cm, slender. Fertile flowers 7-12, ellipsoid, pinkish-blue with darker fascia; teeth slightly recurved, pale blue to whitish. Sterile flowers absent or few and small.

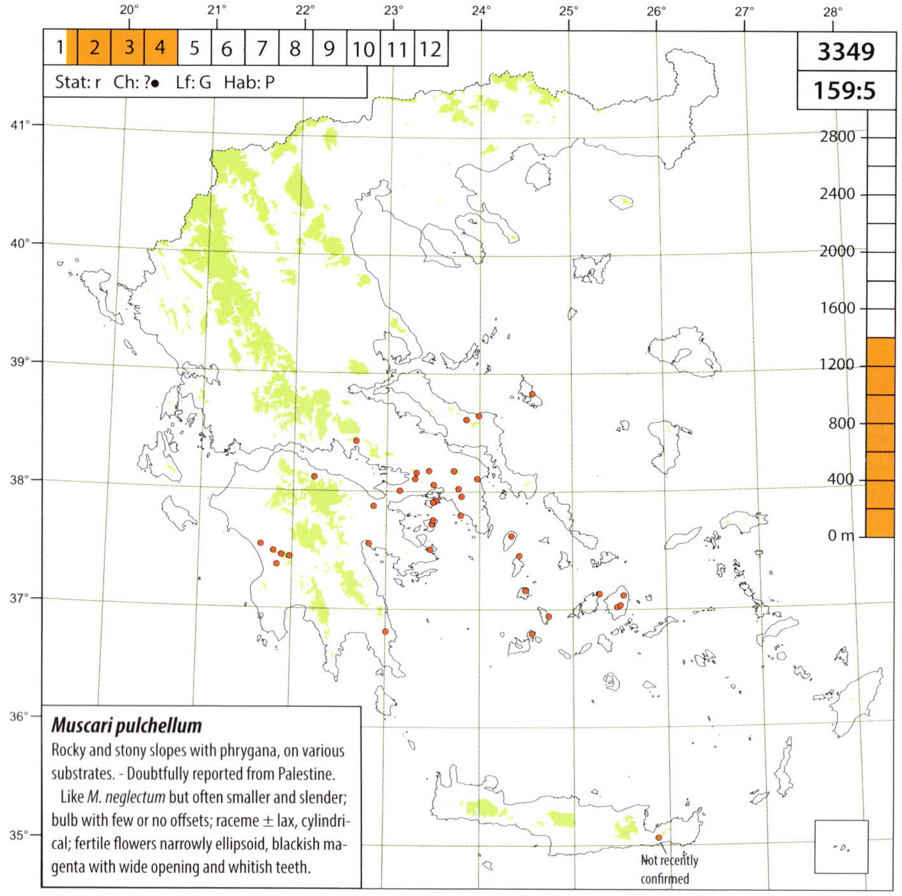

Muscari pulchellum

Rocky and stony slopes with phrygana, on various substrates. - Doubtfully reported from Palestine.

Like *M. neglectum* but often smaller and slender; bulb with few or no offsets; raceme ± lax, cylindrical; fertile flowers narrowly ellipsoid, blackish magenta with wide opening and whitish teeth.

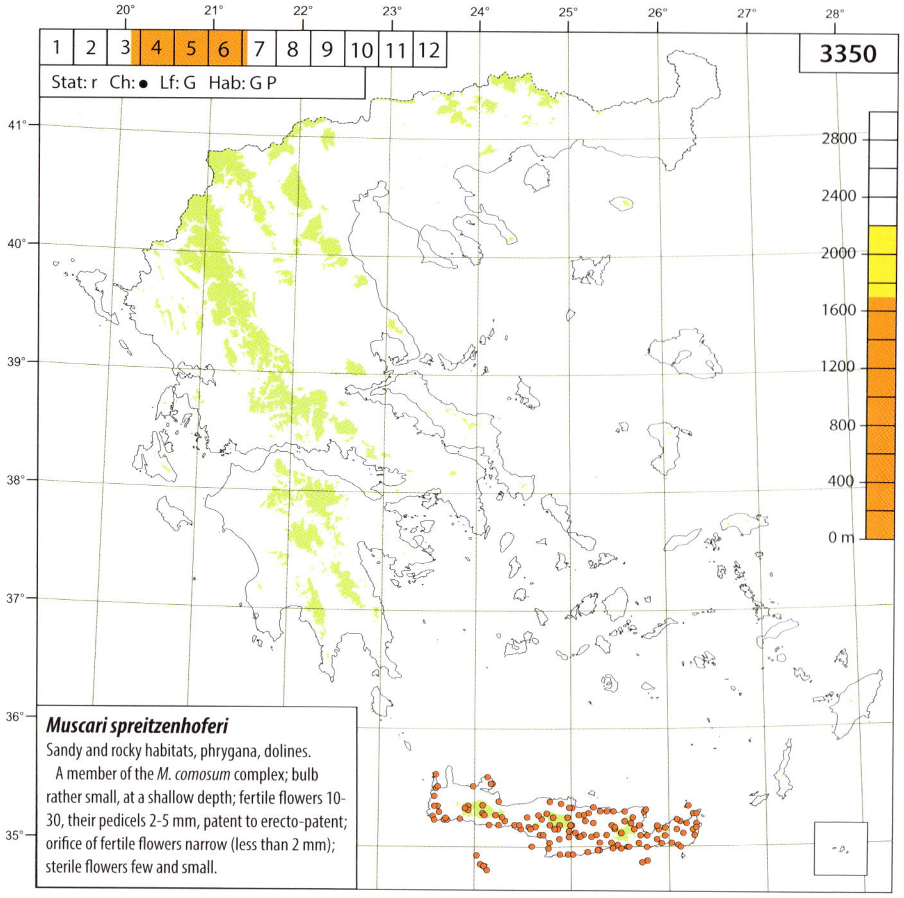

Muscari spreitzenhoferi

Sandy and rocky habitats, phrygana, dolines.

A member of the *M. comosum* complex; bulb rather small, at a shallow depth; fertile flowers 10-30, their pedicels 2-5 mm, patent to erecto-patent; orifice of fertile flowers narrow (less than 2 mm); sterile flowers few and small.

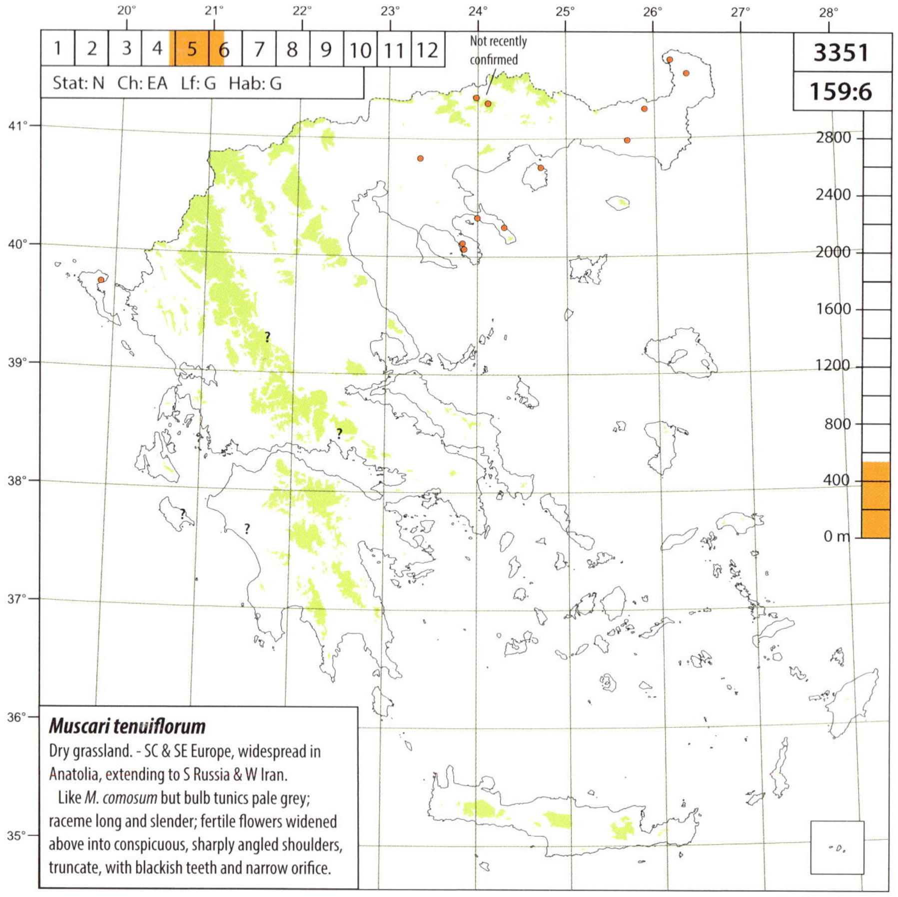

Muscari tenuiflorum

Dry grassland. - SC & SE Europe, widespread in Anatolia, extending to S Russia & W Iran.

Like *M. comosum* but bulb tunics pale grey; raceme long and slender; fertile flowers widened above into conspicuous, sharply angled shoulders, truncate, with blackish teeth and narrow orifice.

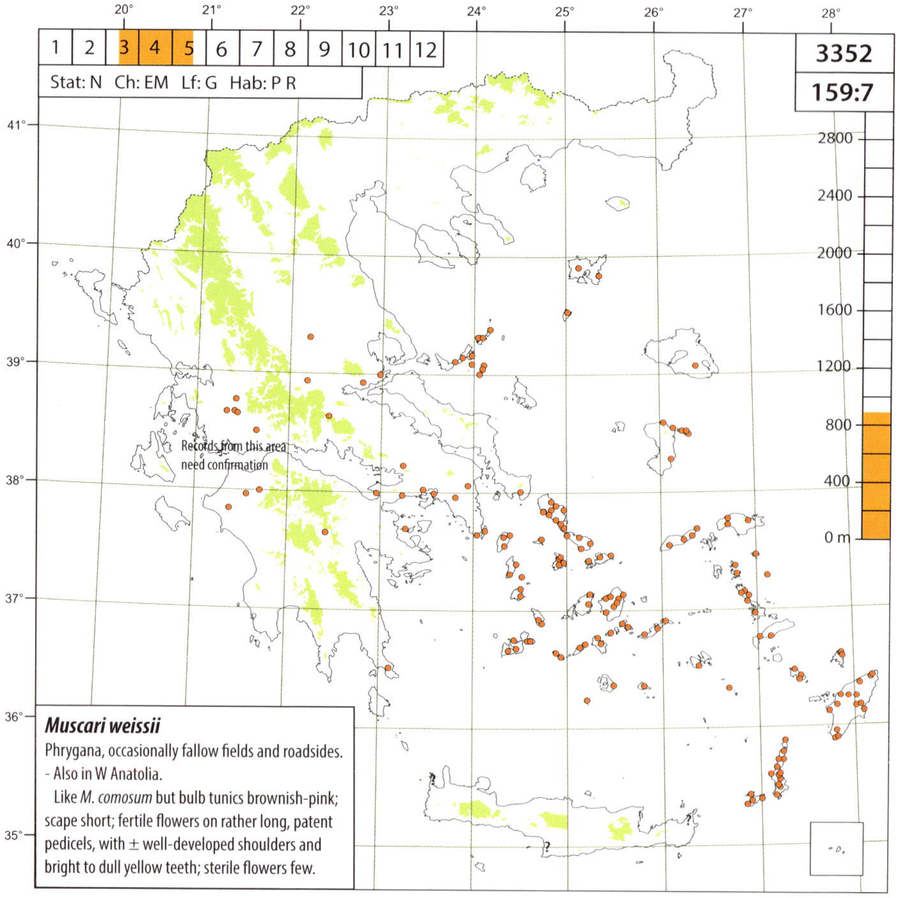

Muscari weissii

Phrygana, occasionally fallow fields and roadsides. - Also in W Anatolia.

Like *M. comosum* but bulb tunics brownish-pink; scape short; fertile flowers on rather long, patent pedicels, with ± well-developed shoulders and bright to dull yellow teeth; sterile flowers few.

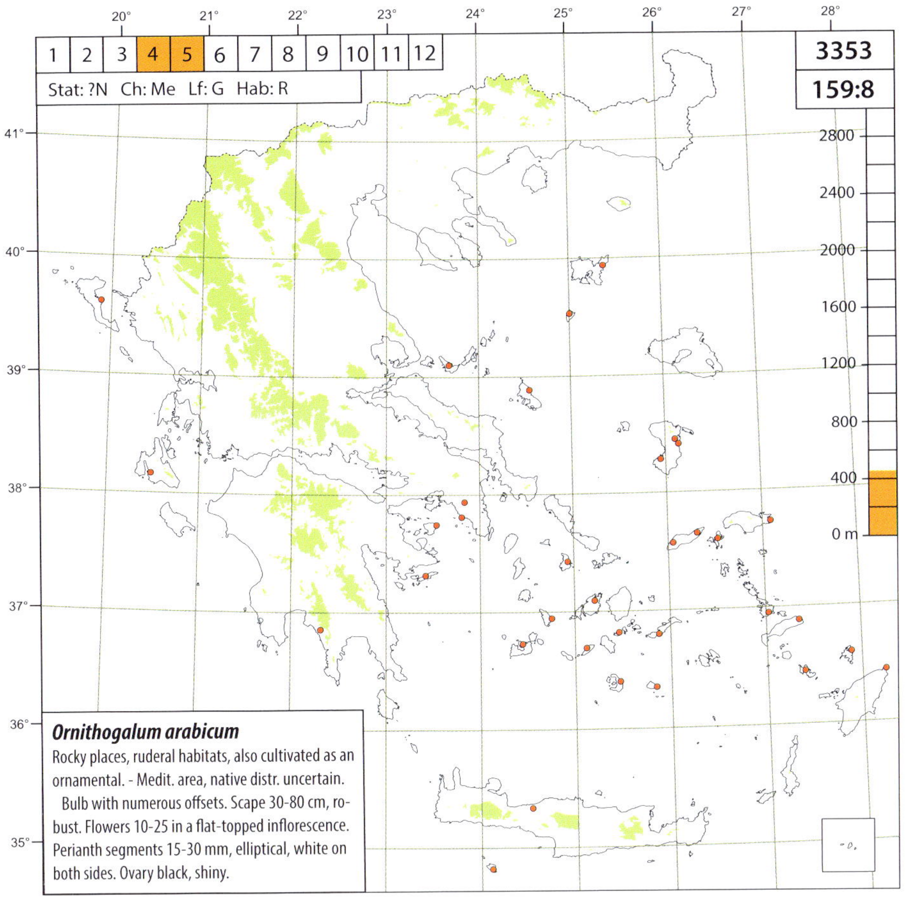

Ornithogalum arabicum
Rocky places, ruderal habitats, also cultivated as an ornamental. - Medit. area, native distr. uncertain.
Bulb with numerous offsets. Scape 30-80 cm, robust. Flowers 10-25 in a flat-topped inflorescence. Perianth segments 15-30 mm, elliptical, white on both sides. Ovary black, shiny.

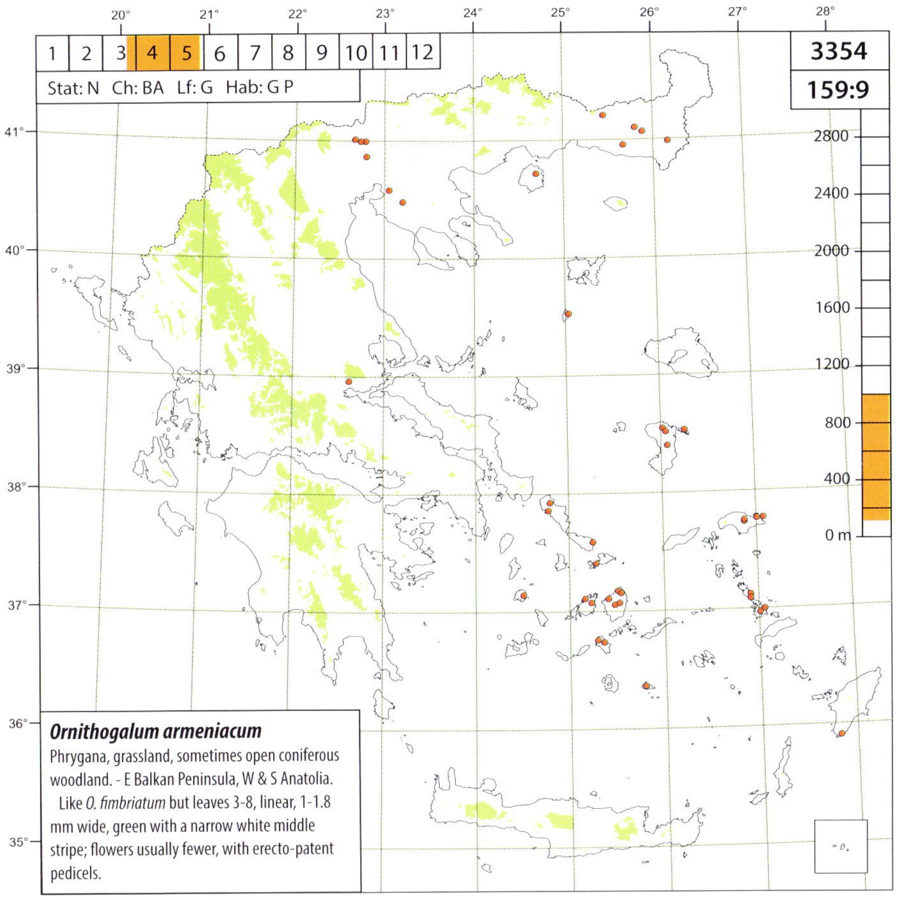

Ornithogalum armeniacum
Phrygana, grassland, sometimes open coniferous woodland. - E Balkan Peninsula, W & S Anatolia.
Like *O. fimbriatum* but leaves 3-8, linear, 1-1.8 mm wide, green with a narrow white middle stripe; flowers usually fewer, with erecto-patent pedicels.

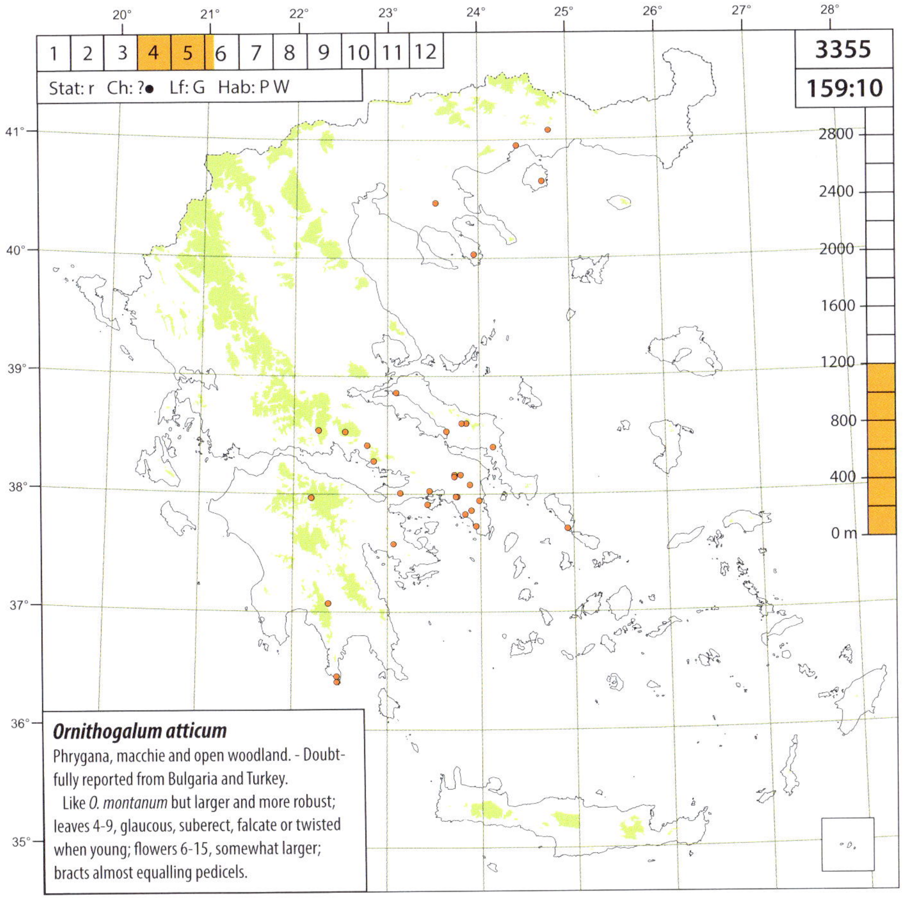

Ornithogalum atticum
Phrygana, macchie and open woodland. - Doubtfully reported from Bulgaria and Turkey.
Like *O. montanum* but larger and more robust; leaves 4-9, glaucous, suberect, falcate or twisted when young; flowers 6-15, somewhat larger; bracts almost equalling pedicels.

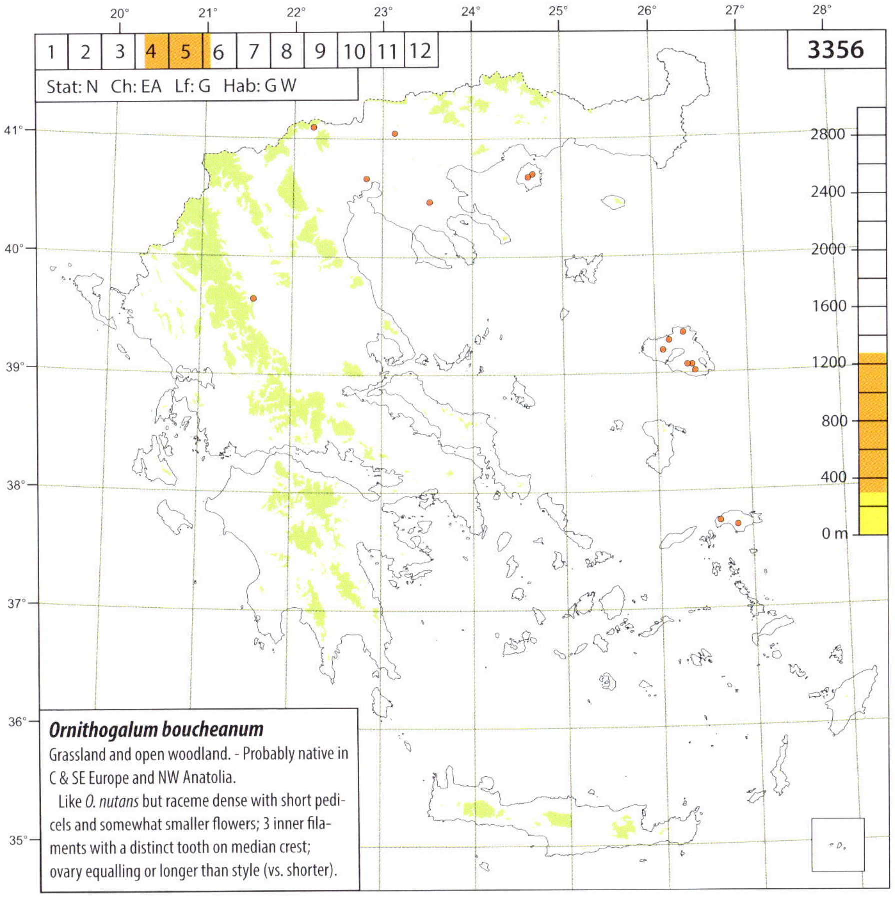

Ornithogalum boucheanum
Grassland and open woodland. - Probably native in C & SE Europe and NW Anatolia.
Like *O. nutans* but raceme dense with short pedicels and somewhat smaller flowers; 3 inner filaments with a distinct tooth on median crest; ovary equalling or longer than style (vs. shorter).

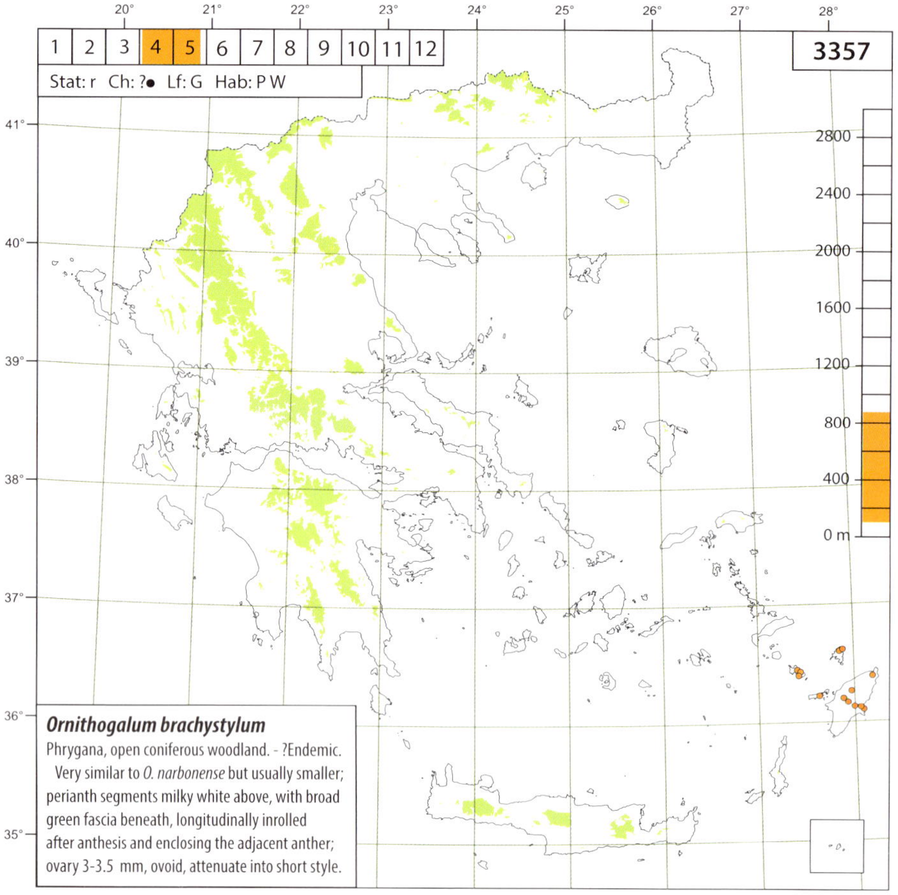

Ornithogalum brachystylum
Phrygana, open coniferous woodland. - ?Endemic.
Very similar to *O. narbonense* but usually smaller; perianth segments milky white above, with broad green fascia beneath, longitudinally inrolled after anthesis and enclosing the adjacent anther; ovary 3-3.5 mm, ovoid, attenuate into short style.

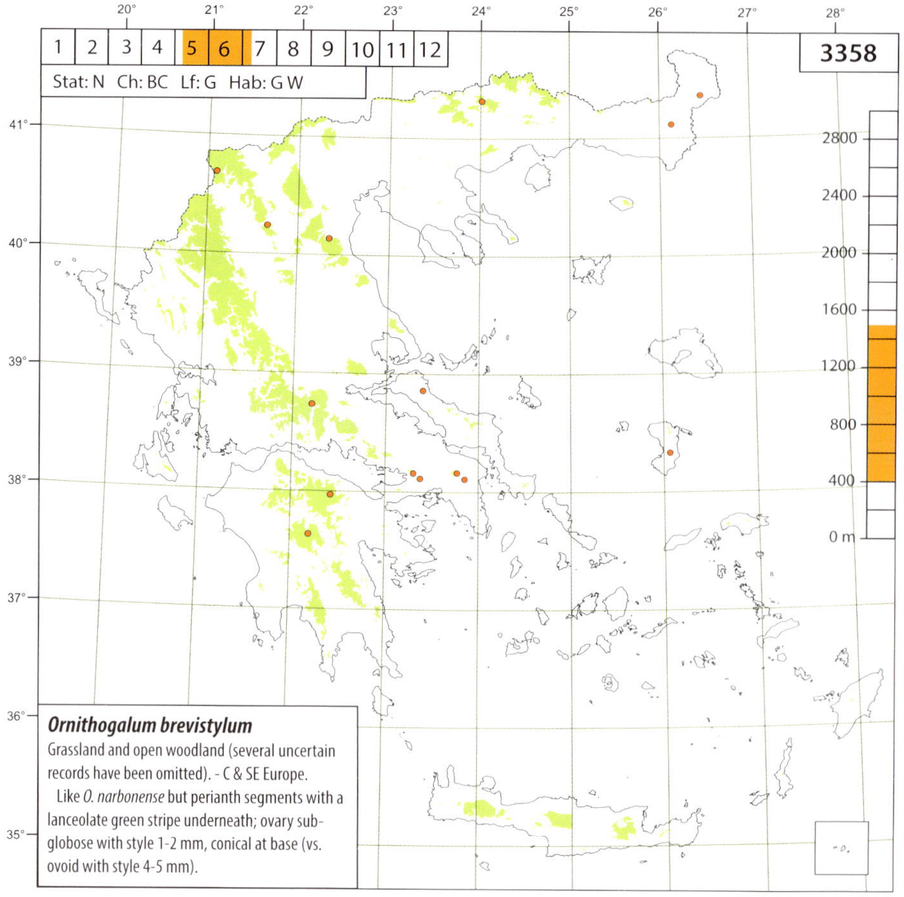

Ornithogalum brevistylum
Grassland and open woodland (several uncertain records have been omitted). - C & SE Europe.
Like *O. narbonense* but perianth segments with a lanceolate green stripe underneath; ovary subglobose with style 1-2 mm, conical at base (vs. ovoid with style 4-5 mm).

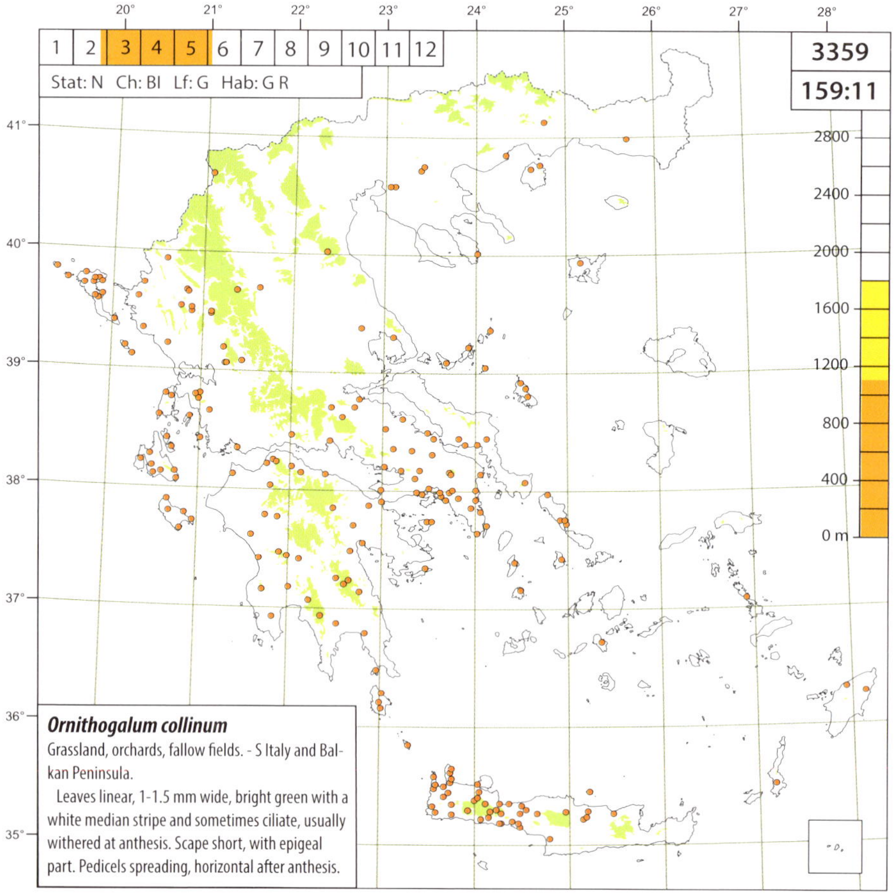

Ornithogalum collinum
Grassland, orchards, fallow fields. - S Italy and Balkan Peninsula.
Leaves linear, 1-1.5 mm wide, bright green with a white median stripe and sometimes ciliate, usually withered at anthesis. Scape short, with epigeal part. Pedicels spreading, horizontal after anthesis.

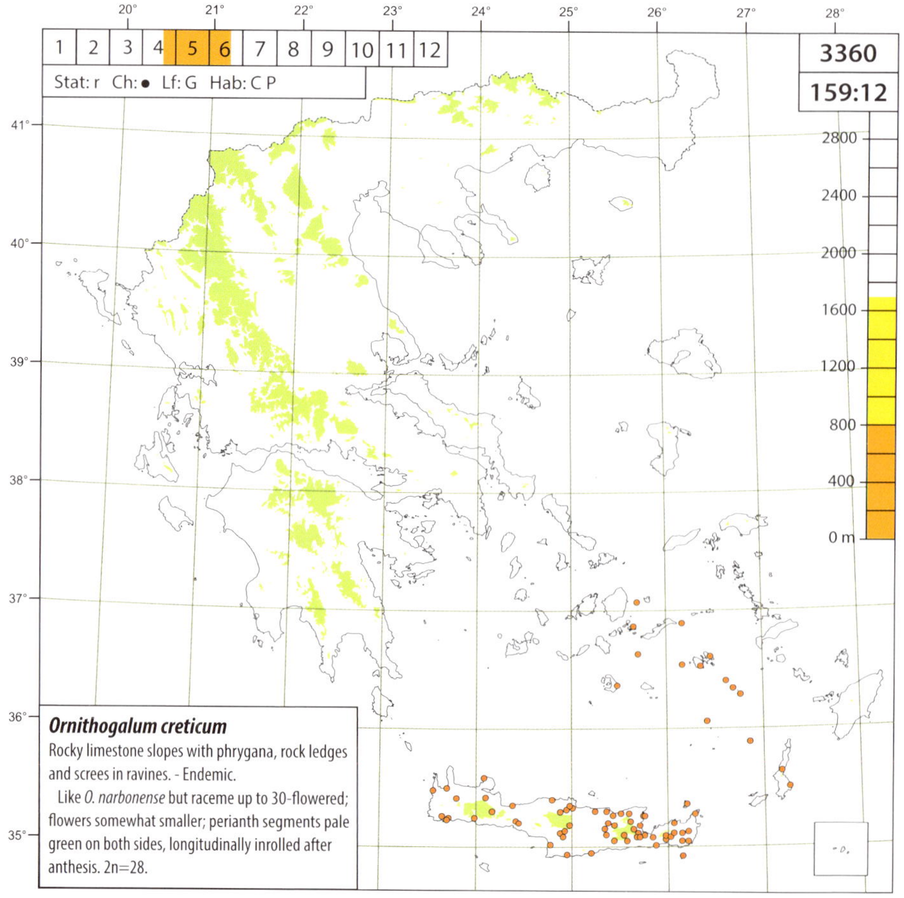

Ornithogalum creticum
Rocky limestone slopes with phrygana, rock ledges and screes in ravines. - Endemic.
Like *O. narbonense* but raceme up to 30-flowered; flowers somewhat smaller; perianth segments pale green on both sides, longitudinally inrolled after anthesis. 2n=28.

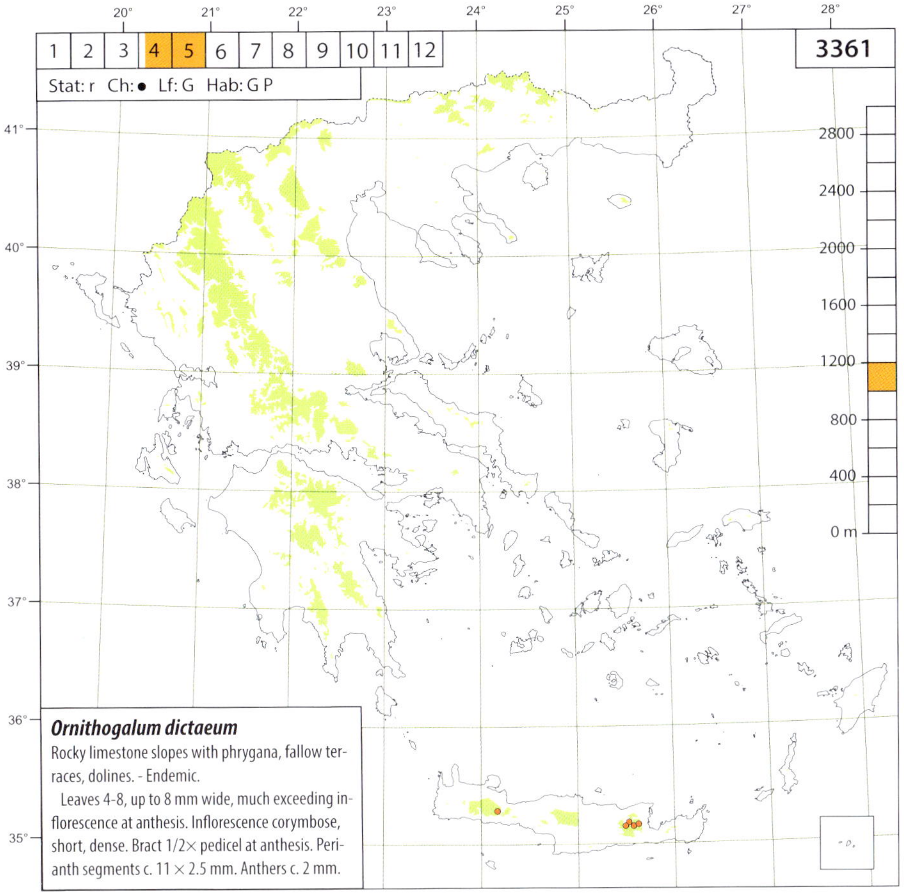

Ornithogalum dictaeum

Rocky limestone slopes with phrygana, fallow terraces, dolines. - Endemic.

Leaves 4-8, up to 8 mm wide, much exceeding inflorescence at anthesis. Inflorescence corymbose, short, dense. Bract 1/2× pedicel at anthesis. Perianth segments c. 11 × 2.5 mm. Anthers c. 2 mm.

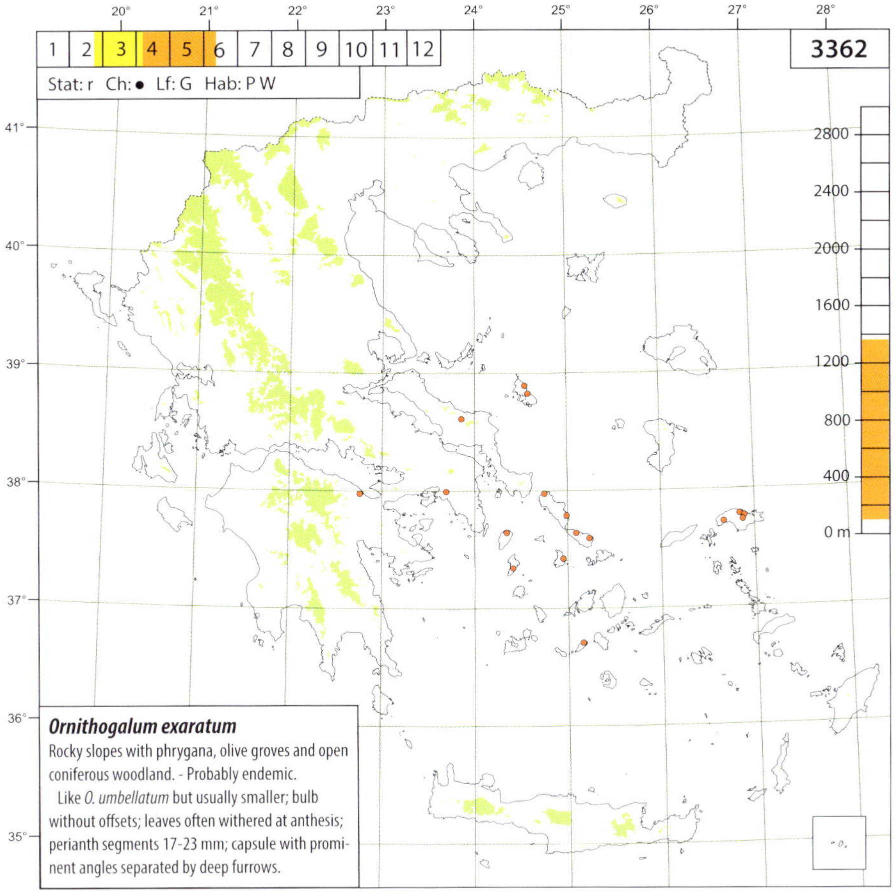

Ornithogalum exaratum

Rocky slopes with phrygana, olive groves and open coniferous woodland. - Probably endemic.

Like *O. umbellatum* but usually smaller; bulb without offsets; leaves often withered at anthesis; perianth segments 17-23 mm; capsule with prominent angles separated by deep furrows.

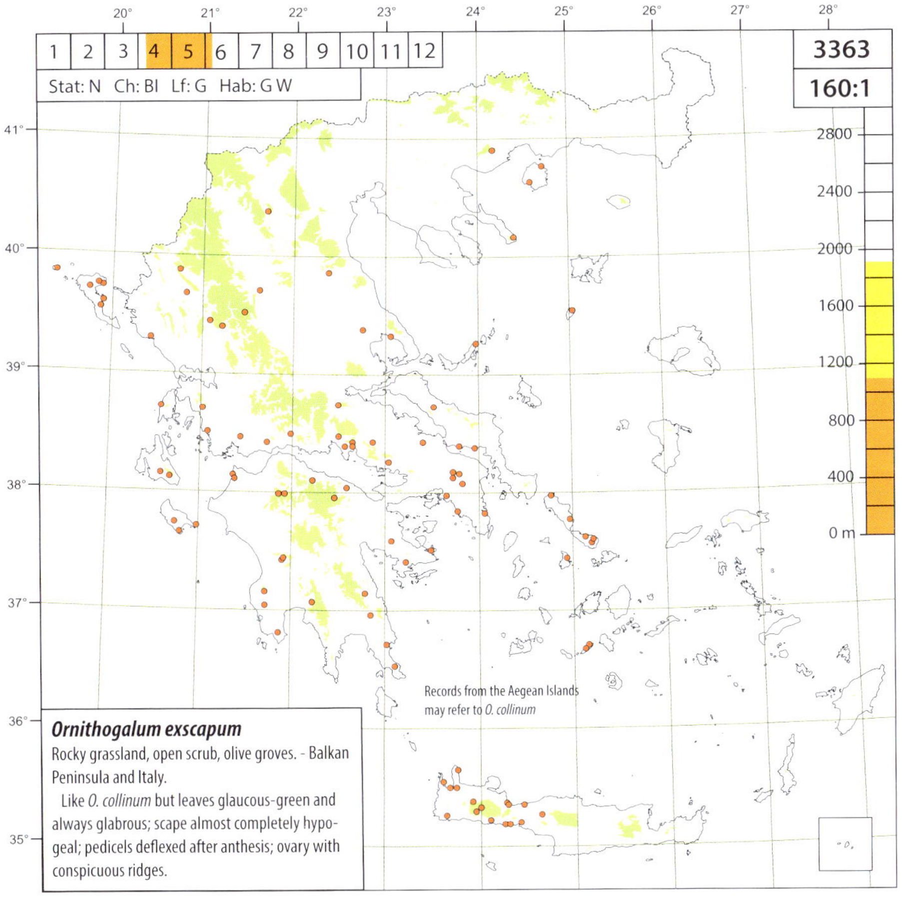

Ornithogalum exscapum

Rocky grassland, open scrub, olive groves. - Balkan Peninsula and Italy.

Like *O. collinum* but leaves glaucous-green and always glabrous; scape almost completely hypogeal; pedicels deflexed after anthesis; ovary with conspicuous ridges.

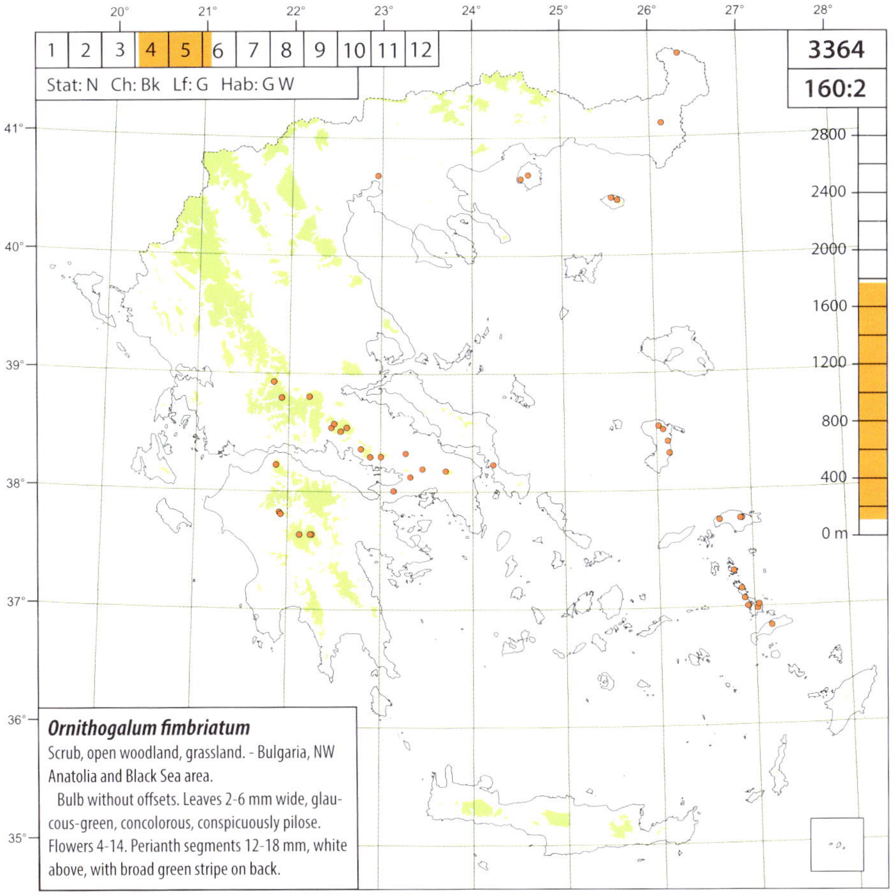

Ornithogalum fimbriatum

Scrub, open woodland, grassland. - Bulgaria, NW Anatolia and Black Sea area.

Bulb without offsets. Leaves 2-6 mm wide, glaucous-green, concolorous, conspicuously pilose. Flowers 4-14. Perianth segments 12-18 mm, white above, with broad green stripe on back.

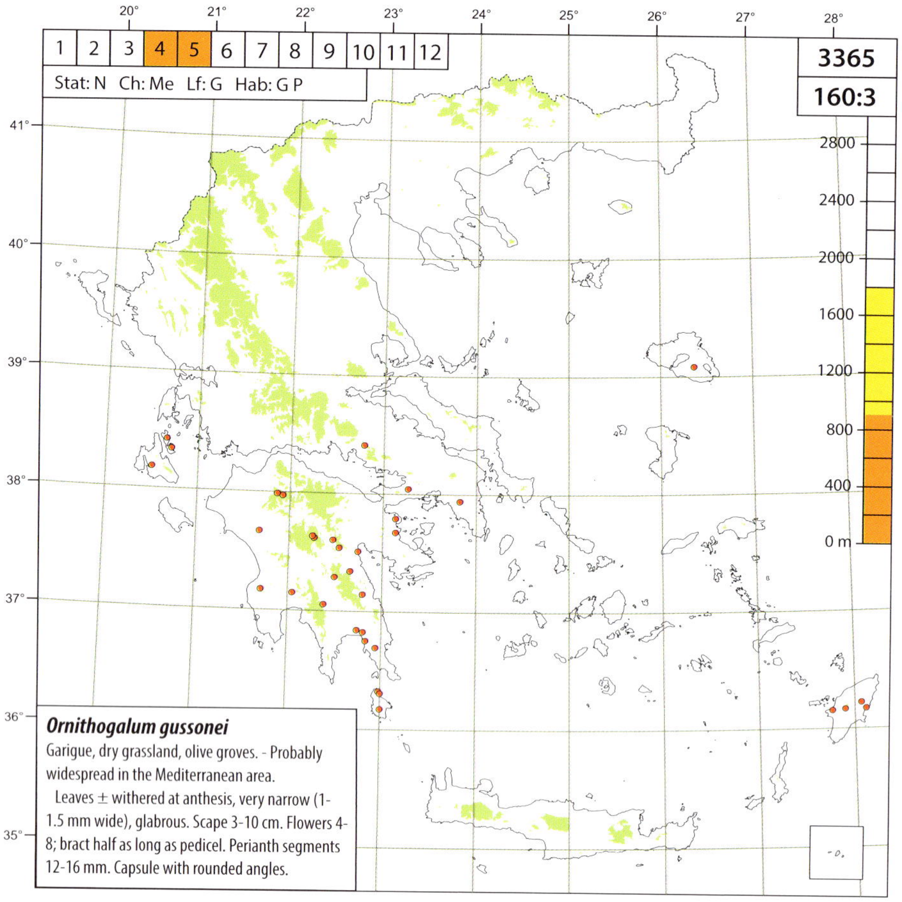

Ornithogalum gussonei

Garigue, dry grassland, olive groves. - Probably widespread in the Mediterranean area.

Leaves ± withered at anthesis, very narrow (1-1.5 mm wide), glabrous. Scape 3-10 cm. Flowers 4-8; bract half as long as pedicel. Perianth segments 12-16 mm. Capsule with rounded angles.

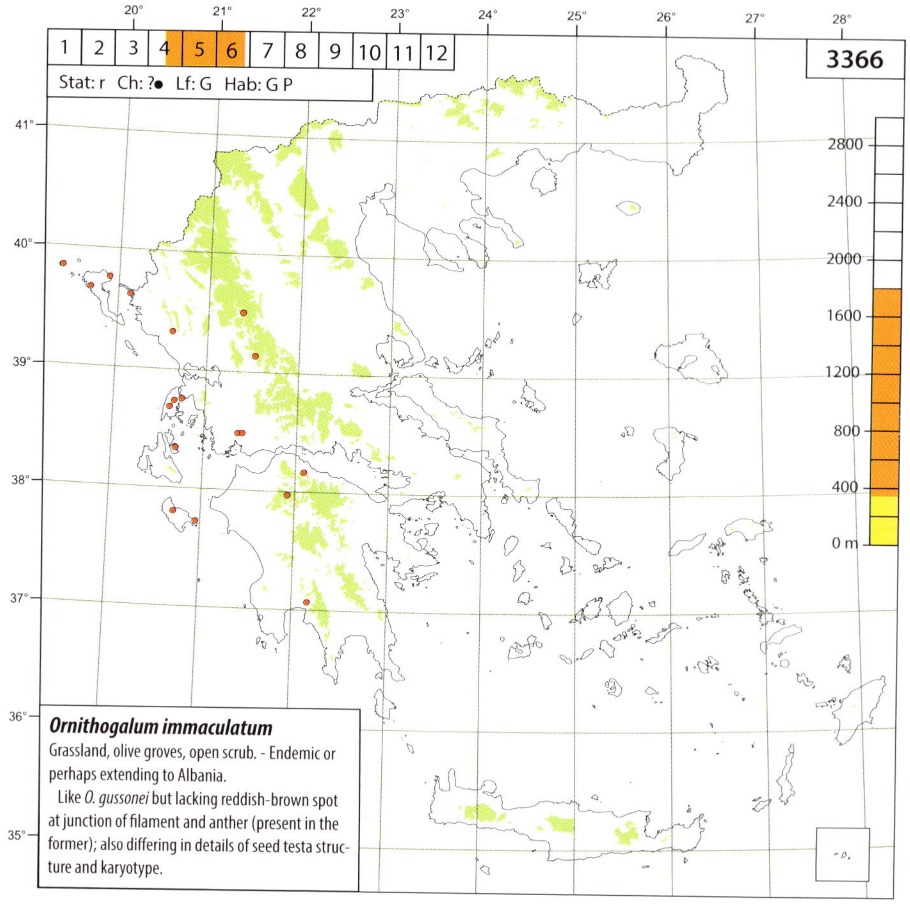

Ornithogalum immaculatum

Grassland, olive groves, open scrub. - Endemic or perhaps extending to Albania.

Like *O. gussonei* but lacking reddish-brown spot at junction of filament and anther (present in the former); also differing in details of seed testa structure and karyotype.

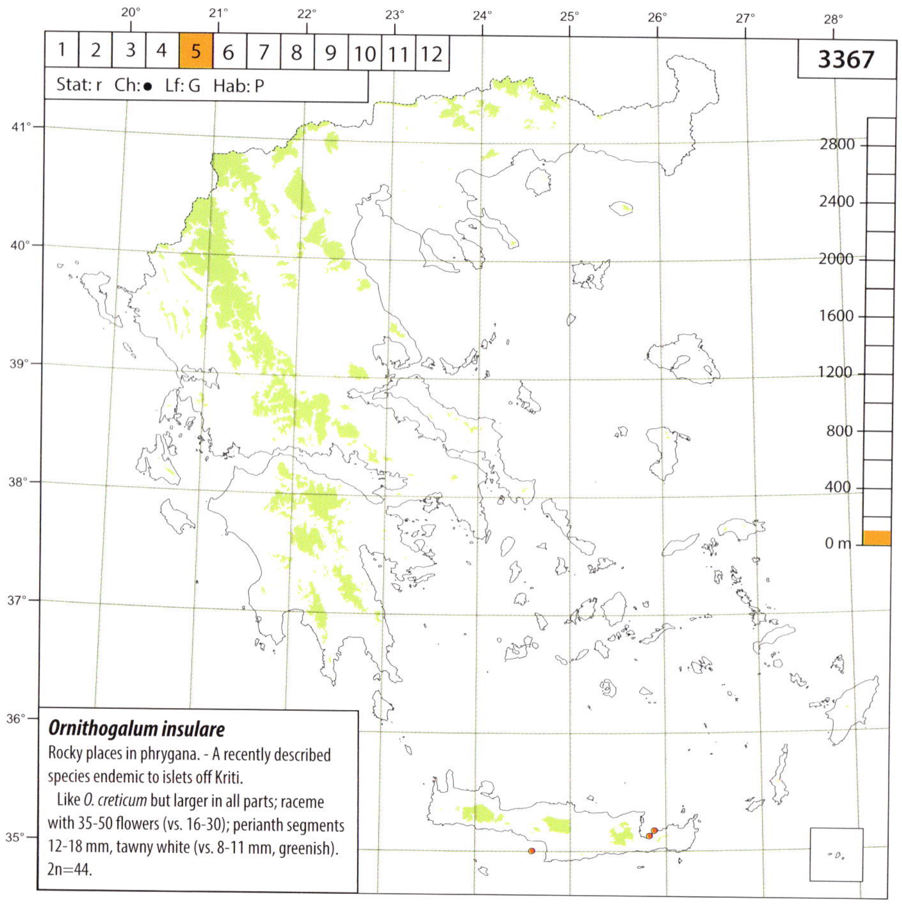

Ornithogalum insulare

Rocky places in phrygana. - A recently described species endemic to islets off Kriti.

Like *O. creticum* but larger in all parts; raceme with 35-50 flowers (vs. 16-30); perianth segments 12-18 mm, tawny white (vs. 8-11 mm, greenish). 2n=44.

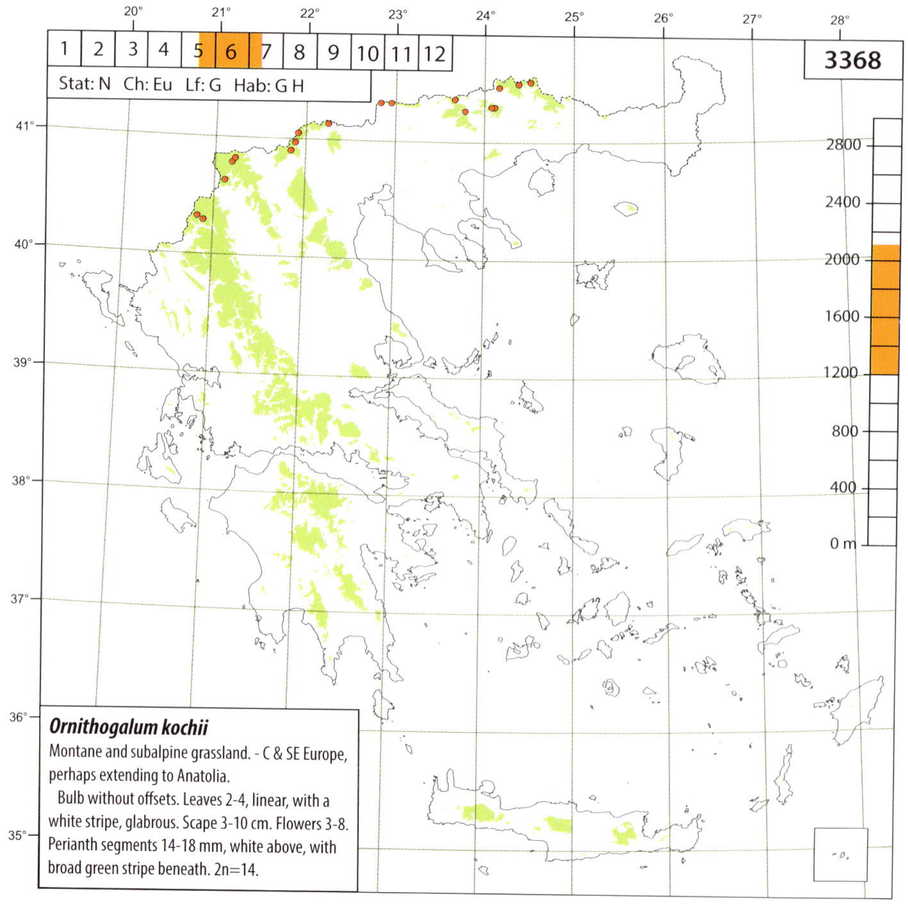

Ornithogalum kochii

Montane and subalpine grassland. - C & SE Europe, perhaps extending to Anatolia.

Bulb without offsets. Leaves 2-4, linear, with a white stripe, glabrous. Scape 3-10 cm. Flowers 3-8. Perianth segments 14-18 mm, white above, with broad green stripe beneath. 2n=14.

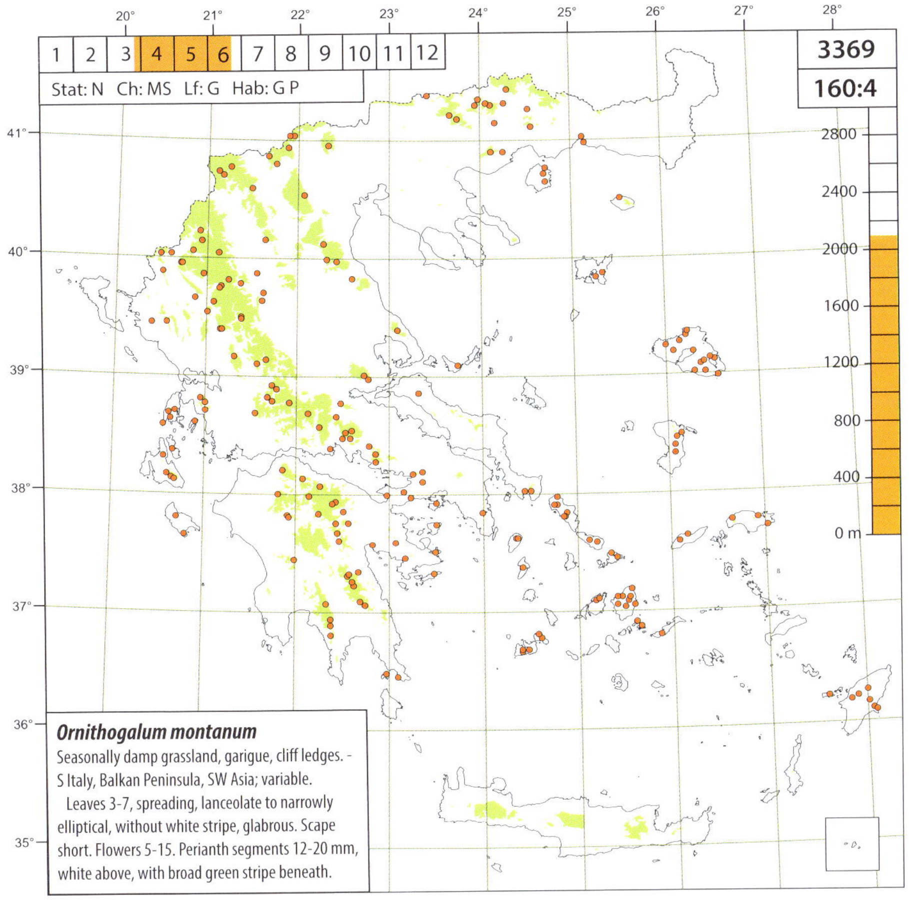

Ornithogalum montanum

Seasonally damp grassland, garigue, cliff ledges. - S Italy, Balkan Peninsula, SW Asia; variable.
Leaves 3-7, spreading, lanceolate to narrowly elliptical, without white stripe, glabrous. Scape short. Flowers 5-15. Perianth segments 12-20 mm, white above, with broad green stripe beneath.

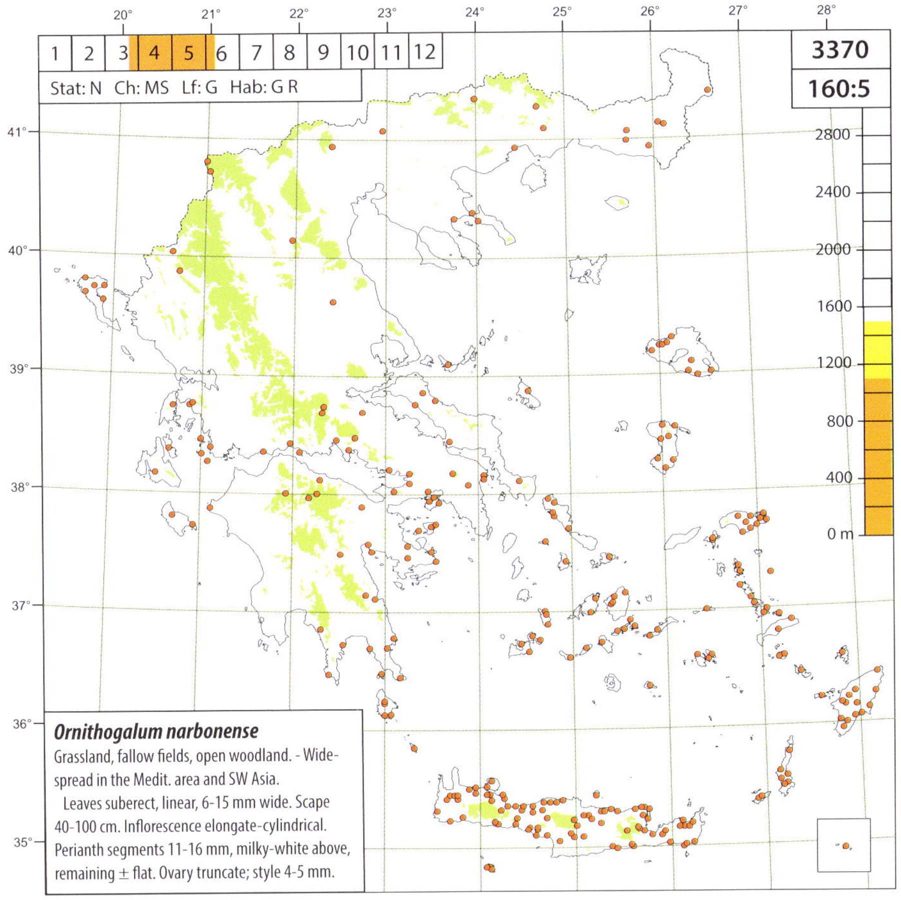

Ornithogalum narbonense

Grassland, fallow fields, open woodland. - Widespread in the Medit. area and SW Asia.
Leaves suberect, linear, 6-15 mm wide. Scape 40-100 cm. Inflorescence elongate-cylindrical. Perianth segments 11-16 mm, milky-white above, remaining ± flat. Ovary truncate; style 4-5 mm.

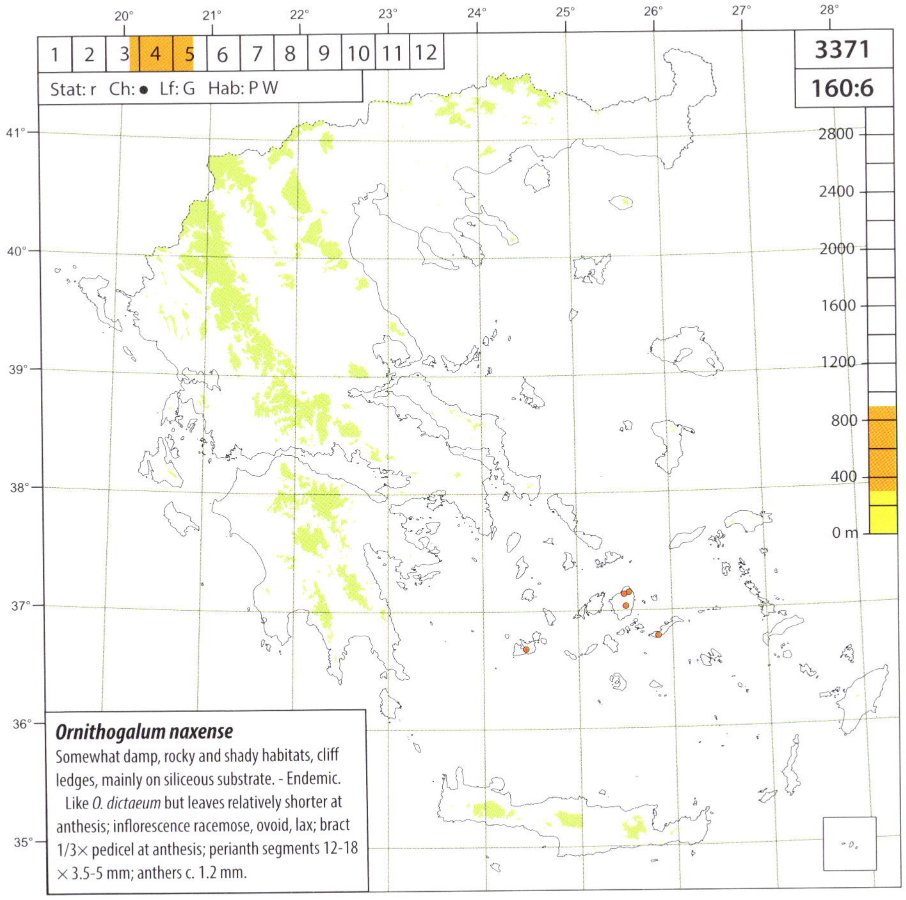

Ornithogalum naxense

Somewhat damp, rocky and shady habitats, cliff ledges, mainly on siliceous substrate. - Endemic.
Like *O. dictaeum* but leaves relatively shorter at anthesis; inflorescence racemose, ovoid, lax; bract 1/3× pedicel at anthesis; perianth segments 12-18 × 3.5-5 mm; anthers c. 1.2 mm.

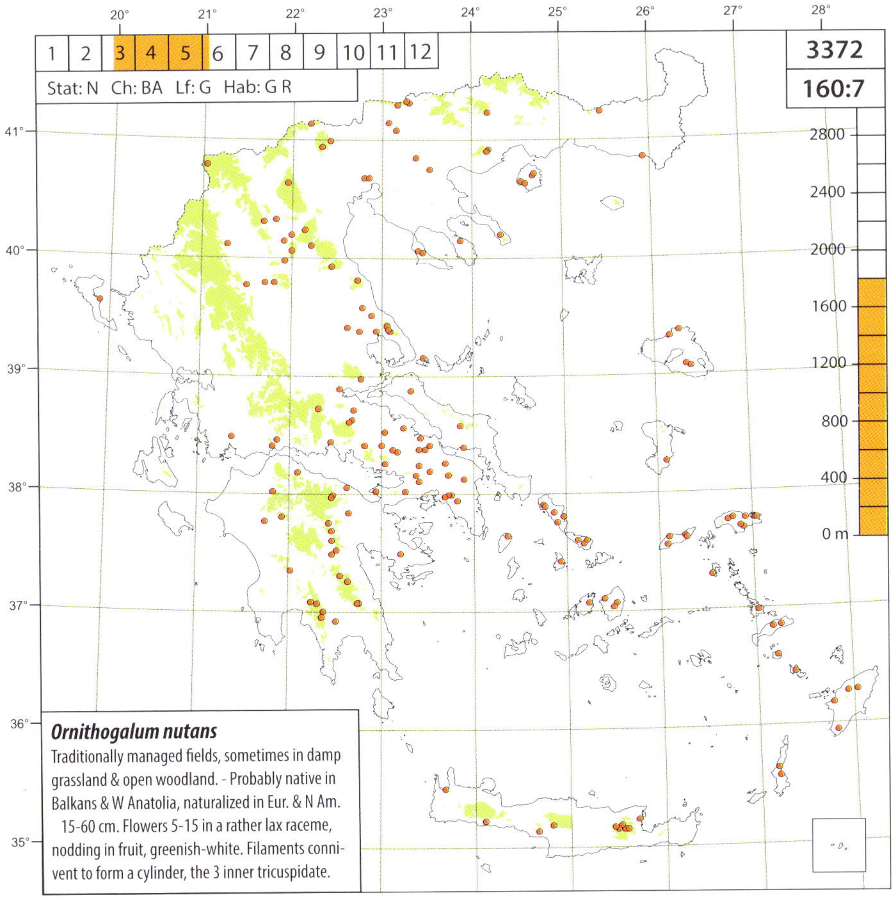

Ornithogalum nutans

Traditionally managed fields, sometimes in damp grassland & open woodland. - Probably native in Balkans & W Anatolia, naturalized in Eur. & N Am.
15-60 cm. Flowers 5-15 in a rather lax raceme, nodding in fruit, greenish-white. Filaments connivent to form a cylinder, the 3 inner tricuspidate.

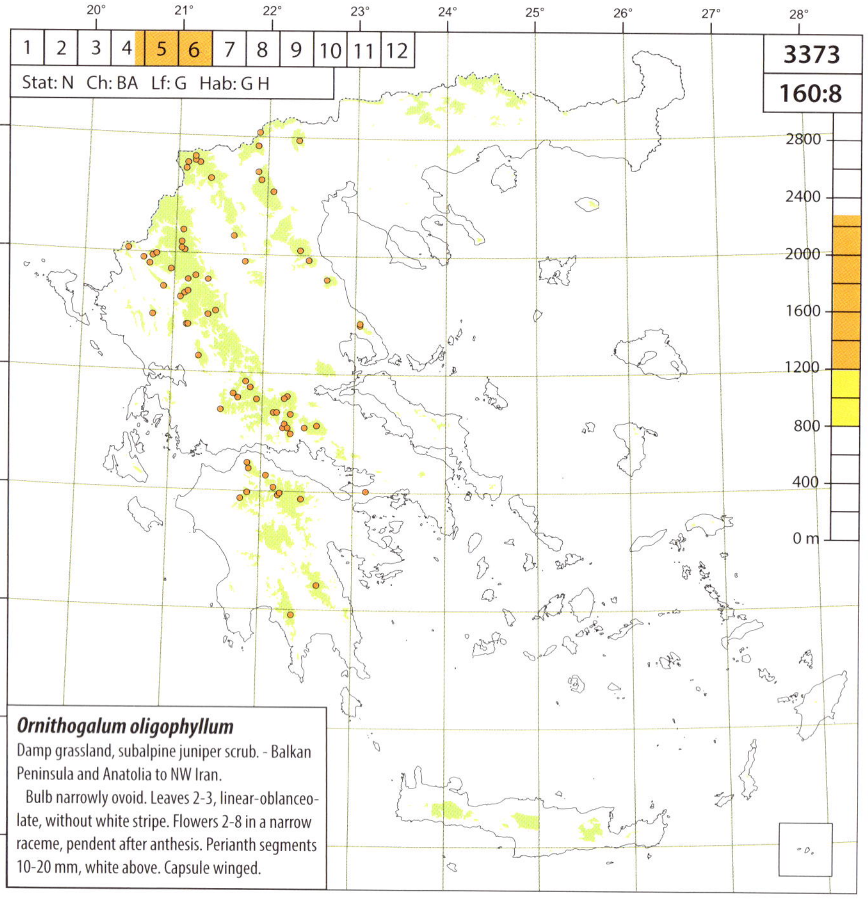

Ornithogalum oligophyllum
Damp grassland, subalpine juniper scrub. - Balkan Peninsula and Anatolia to NW Iran.
Bulb narrowly ovoid. Leaves 2-3, linear-oblanceolate, without white stripe. Flowers 2-8 in a narrow raceme, pendent after anthesis. Perianth segments 10-20 mm, white above. Capsule winged.

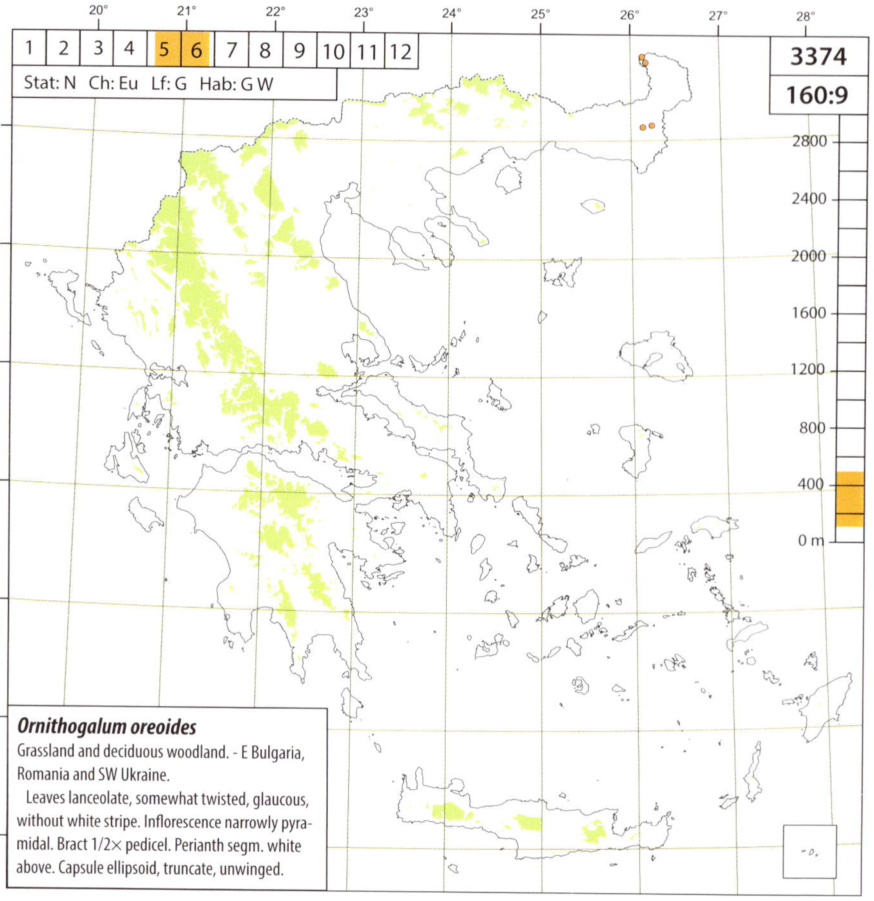

Ornithogalum oreoides
Grassland and deciduous woodland. - E Bulgaria, Romania and SW Ukraine.
Leaves lanceolate, somewhat twisted, glaucous, without white stripe. Inflorescence narrowly pyramidal. Bract 1/2× pedicel. Perianth segm. white above. Capsule ellipsoid, truncate, unwinged.

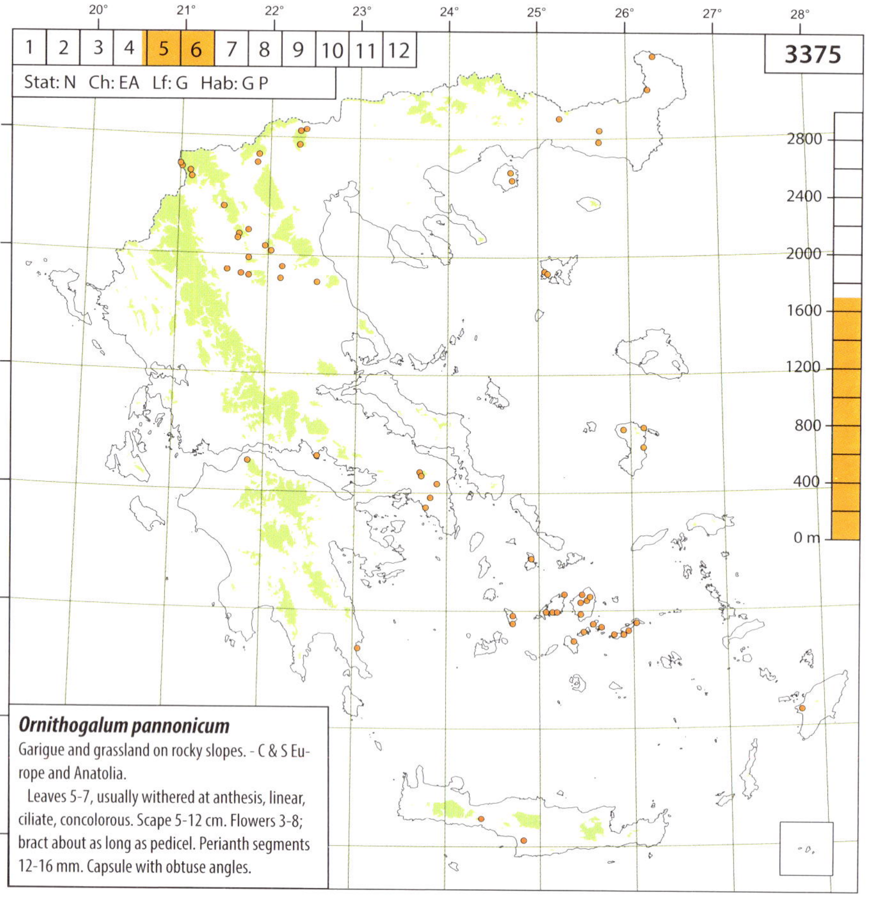

Ornithogalum pannonicum
Garigue and grassland on rocky slopes. - C & S Europe and Anatolia.
Leaves 5-7, usually withered at anthesis, linear, ciliate, concolorous. Scape 5-12 cm. Flowers 3-8; bract about as long as pedicel. Perianth segments 12-16 mm. Capsule with obtuse angles.

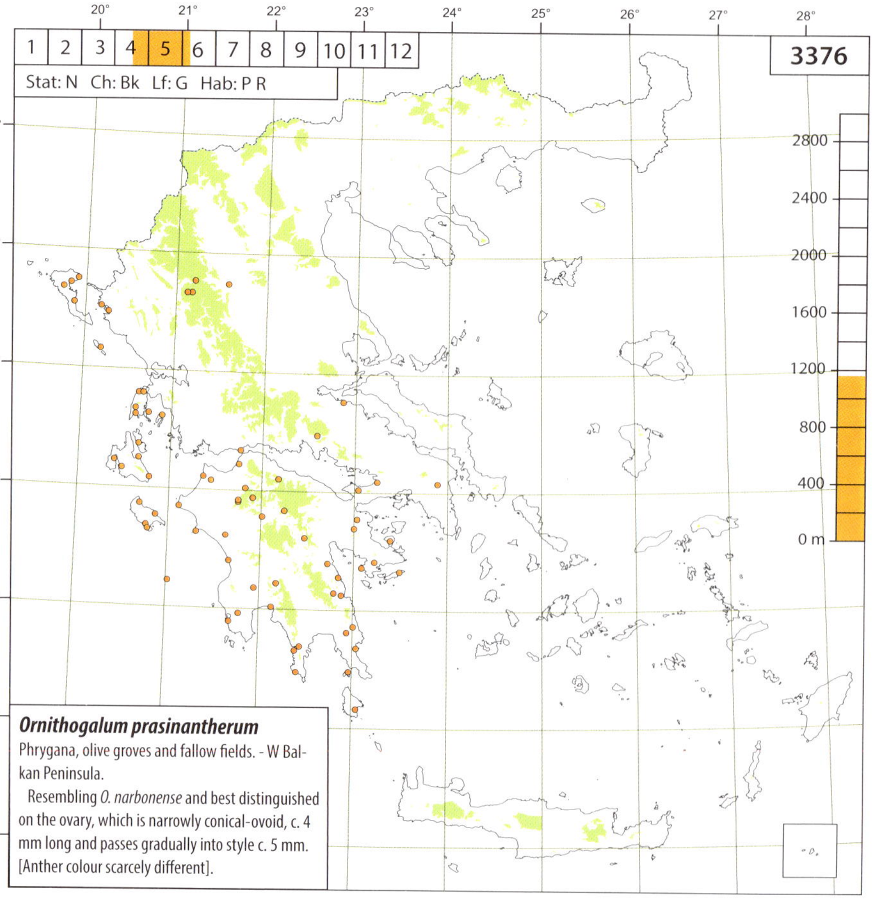

Ornithogalum prasinantherum
Phrygana, olive groves and fallow fields. - W Balkan Peninsula.
Resembling *O. narbonense* and best distinguished on the ovary, which is narrowly conical-ovoid, c. 4 mm long and passes gradually into style c. 5 mm. [Anther colour scarcely different].

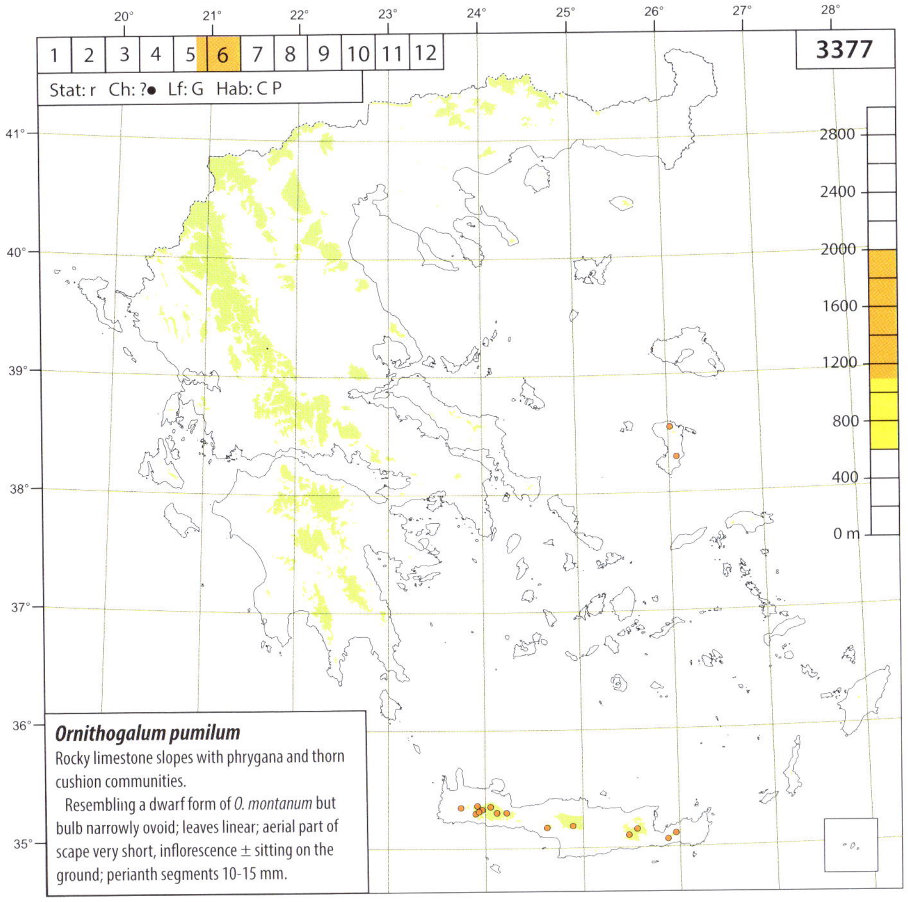

Ornithogalum pumilum

Rocky limestone slopes with phrygana and thorn cushion communities.

Resembling a dwarf form of *O. montanum* but bulb narrowly ovoid; leaves linear; aerial part of scape very short, inflorescence ± sitting on the ground; perianth segments 10-15 mm.

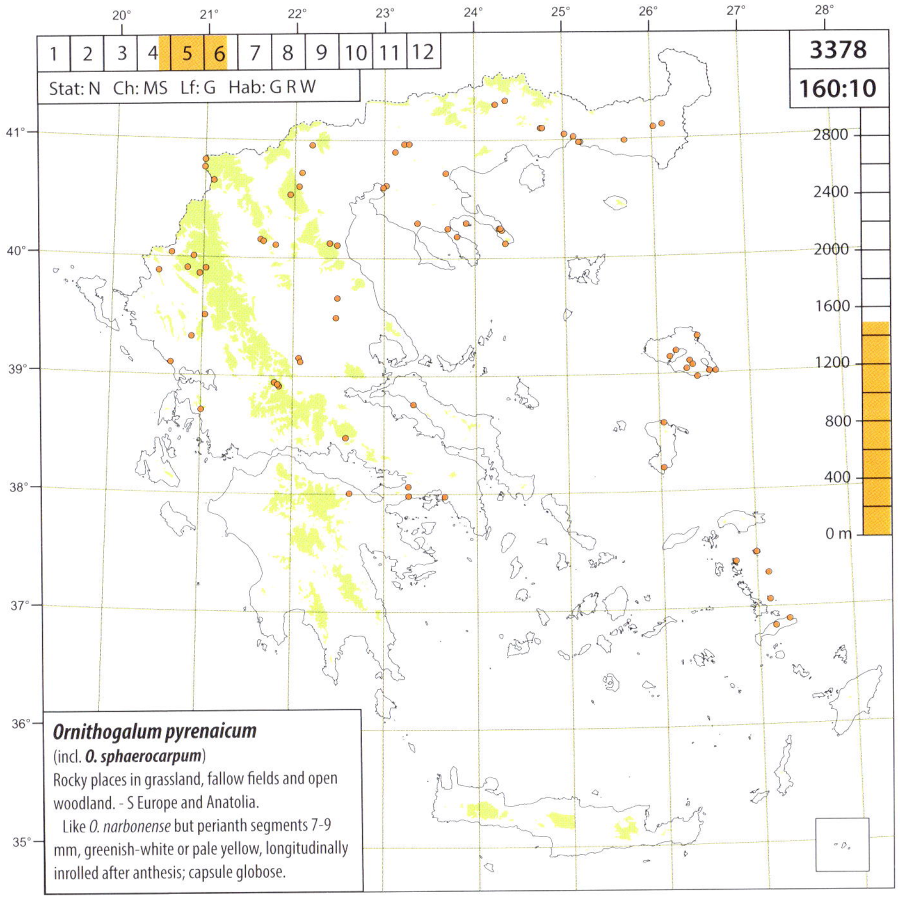

Ornithogalum pyrenaicum

(incl. ***O. sphaerocarpum***)

Rocky places in grassland, fallow fields and open woodland. - S Europe and Anatolia.

Like *O. narbonense* but perianth segments 7-9 mm, greenish-white or pale yellow, longitudinally inrolled after anthesis; capsule globose.

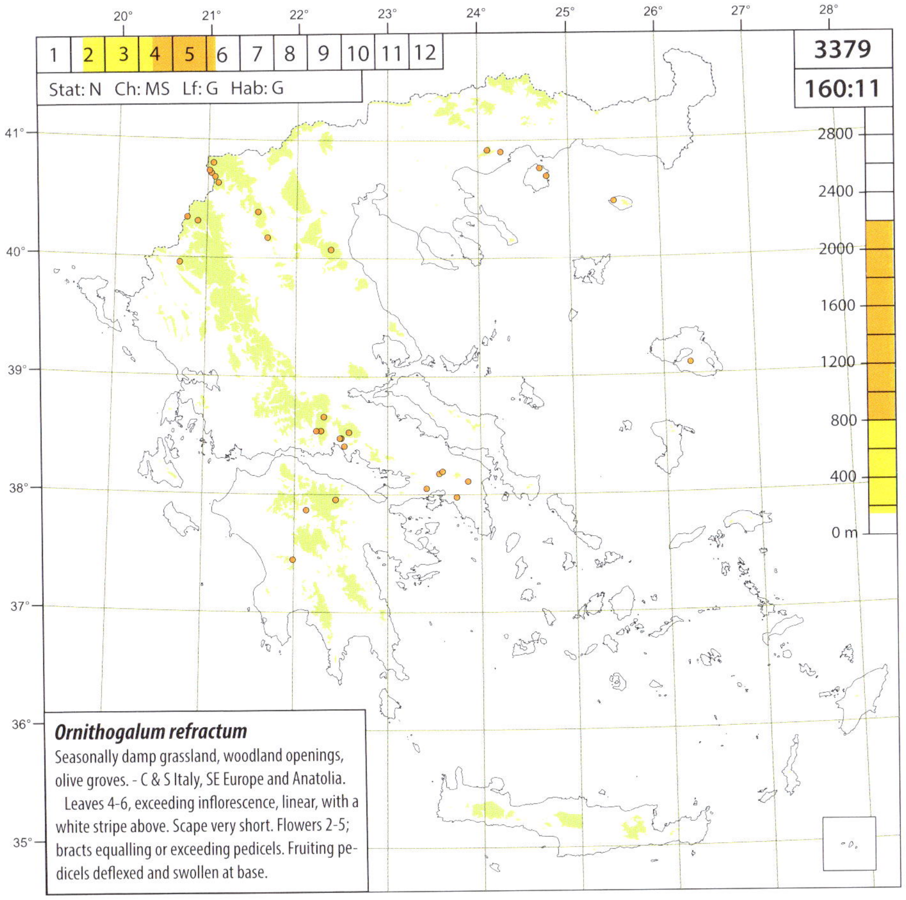

Ornithogalum refractum

Seasonally damp grassland, woodland openings, olive groves. - C & S Italy, SE Europe and Anatolia.

Leaves 4-6, exceeding inflorescence, linear, with a white stripe above. Scape very short. Flowers 2-5; bracts equalling or exceeding pedicels. Fruiting pedicels deflexed and swollen at base.

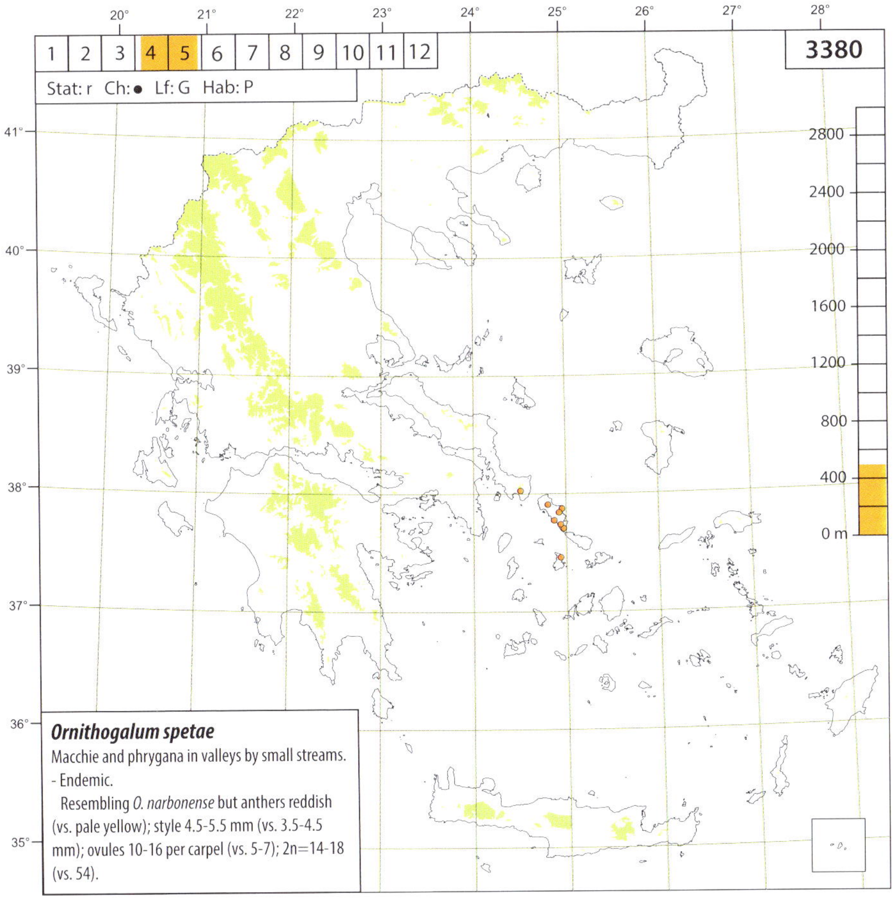

Ornithogalum spetae

Macchie and phrygana in valleys by small streams. - Endemic.

Resembling *O. narbonense* but anthers reddish (vs. pale yellow); style 4.5-5.5 mm (vs. 3.5-4.5 mm); ovules 10-16 per carpel (vs. 5-7); 2n=14-18 (vs. 54).

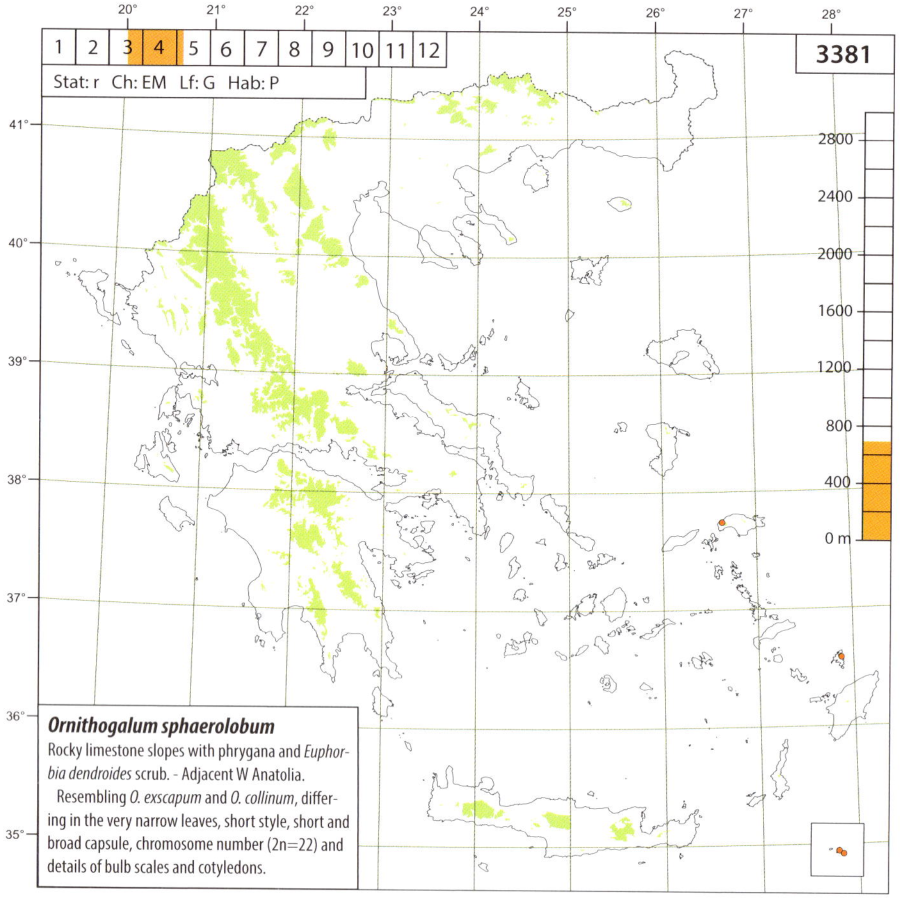

Ornithogalum sphaerolobum

Rocky limestone slopes with phrygana and *Euphorbia dendroides* scrub. - Adjacent W Anatolia.

Resembling *O. exscapum* and *O. collinum*, differing in the very narrow leaves, short style, short and broad capsule, chromosome number (2n=22) and details of bulb scales and cotyledons.

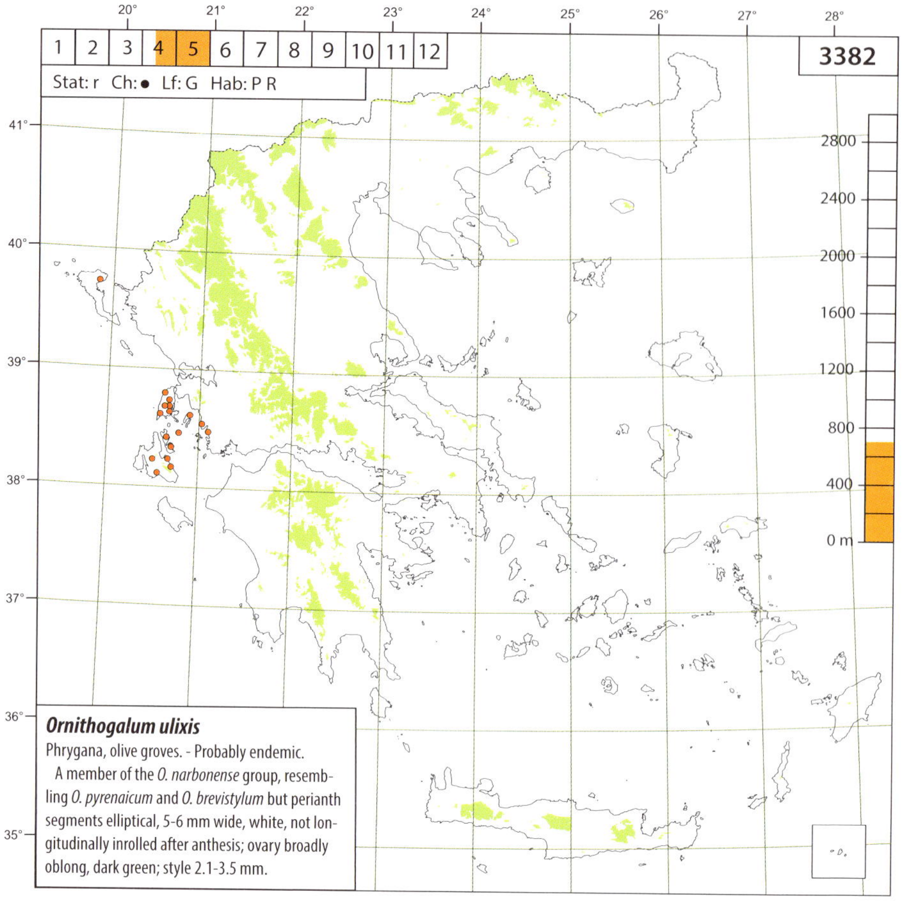

Ornithogalum ulixis

Phrygana, olive groves. - Probably endemic.

A member of the *O. narbonense* group, resembling *O. pyrenaicum* and *O. brevistylum* but perianth segments elliptical, 5-6 mm wide, white, not longitudinally inrolled after anthesis; ovary broadly oblong, dark green; style 2.1-3.5 mm.

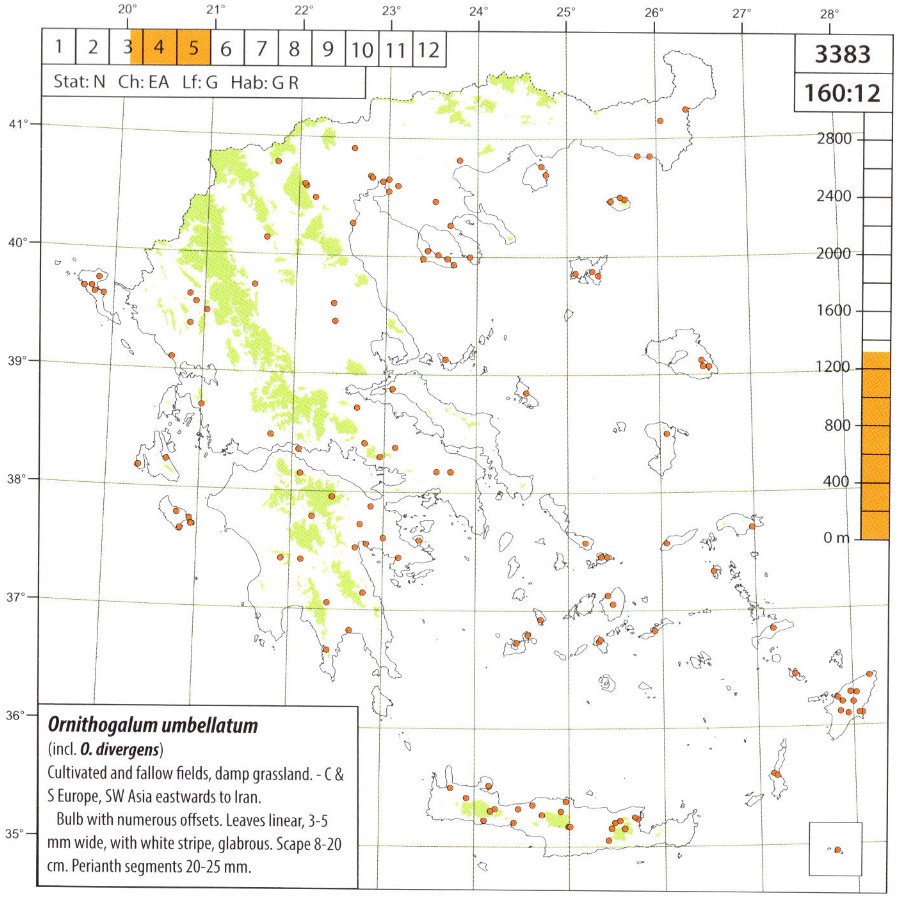

Ornithogalum umbellatum

(incl. ***O. divergens***)

Cultivated and fallow fields, damp grassland. - C & S Europe, SW Asia eastwards to Iran.

Bulb with numerous offsets. Leaves linear, 3-5 mm wide, with white stripe, glabrous. Scape 8-20 cm. Perianth segments 20-25 mm.

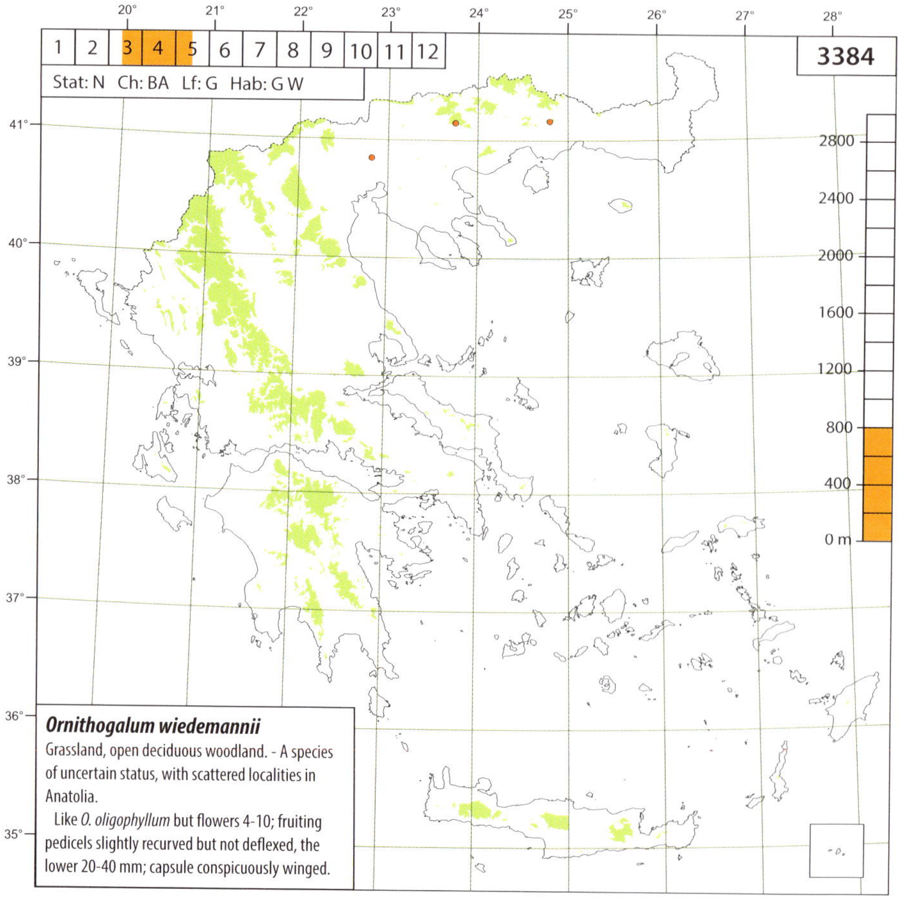

Ornithogalum wiedemannii

Grassland, open deciduous woodland. - A species of uncertain status, with scattered localities in Anatolia.

Like *O. oligophyllum* but flowers 4-10; fruiting pedicels slightly recurved but not deflexed, the lower 20-40 mm; capsule conspicuously winged.

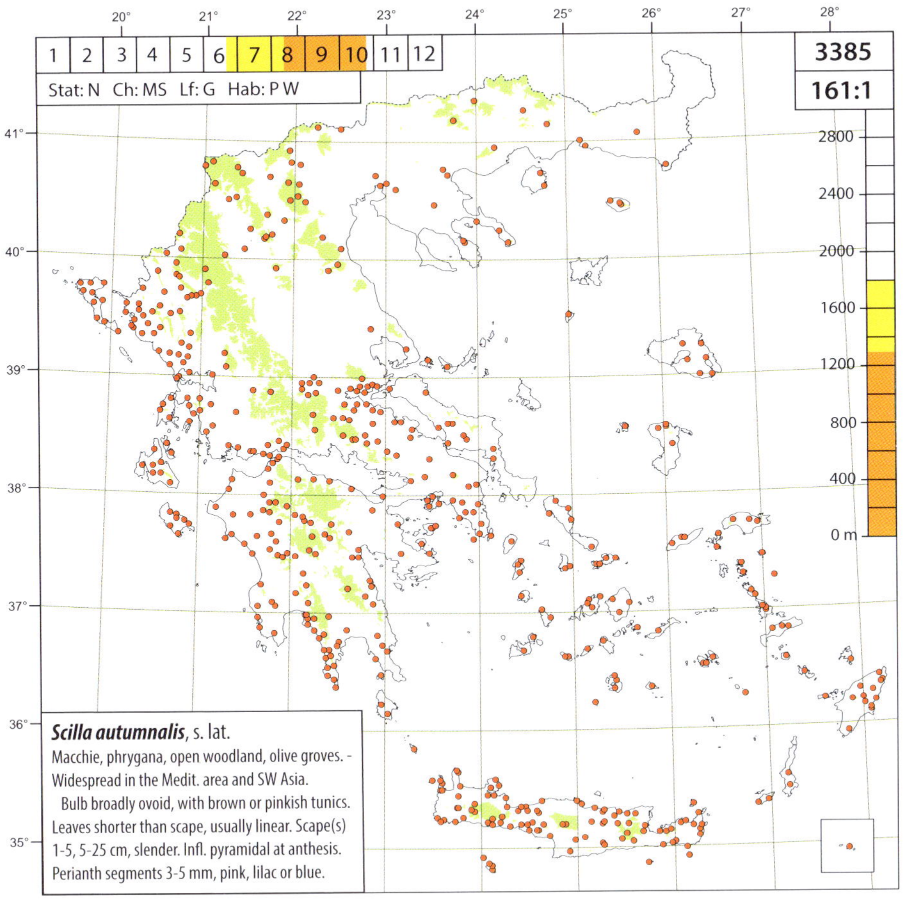

Scilla autumnalis, s. lat.
Macchie, phrygana, open woodland, olive groves. - Widespread in the Medit. area and SW Asia.
Bulb broadly ovoid, with brown or pinkish tunics. Leaves shorter than scape, usually linear. Scape(s) 1-5, 5-25 cm, slender. Infl. pyramidal at anthesis. Perianth segments 3-5 mm, pink, lilac or blue.

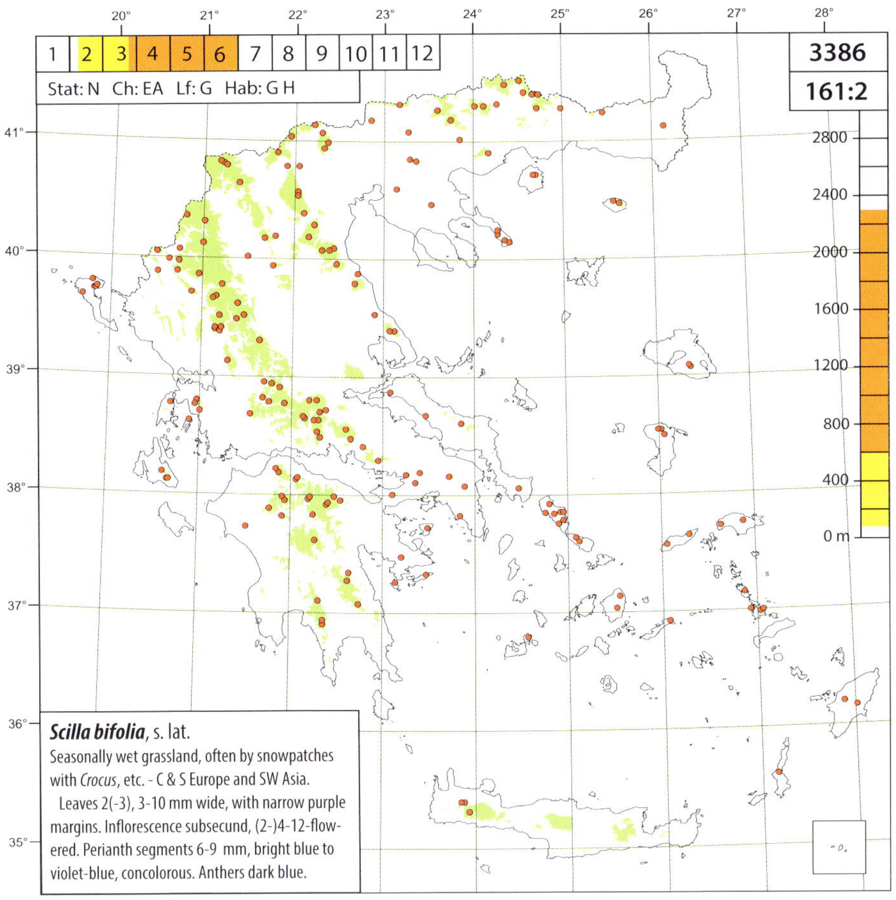

Scilla bifolia, s. lat.
Seasonally wet grassland, often by snowpatches with *Crocus*, etc. - C & S Europe and SW Asia.
Leaves 2(-3), 3-10 mm wide, with narrow purple margins. Inflorescence subsecund, (2-)4-12-flowered. Perianth segments 6-9 mm, bright blue to violet-blue, concolorous. Anthers dark blue.

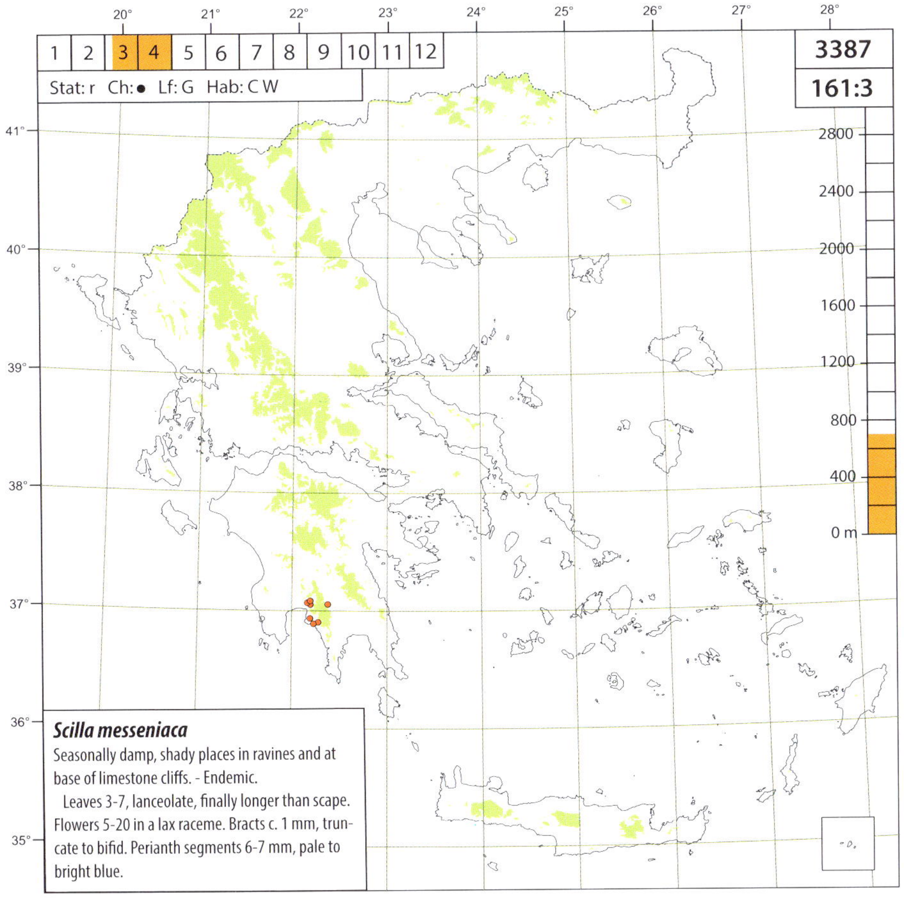

Scilla messeniaca
Seasonally damp, shady places in ravines and at base of limestone cliffs. - Endemic.
Leaves 3-7, lanceolate, finally longer than scape. Flowers 5-20 in a lax raceme. Bracts c. 1 mm, truncate to bifid. Perianth segments 6-7 mm, pale to bright blue.

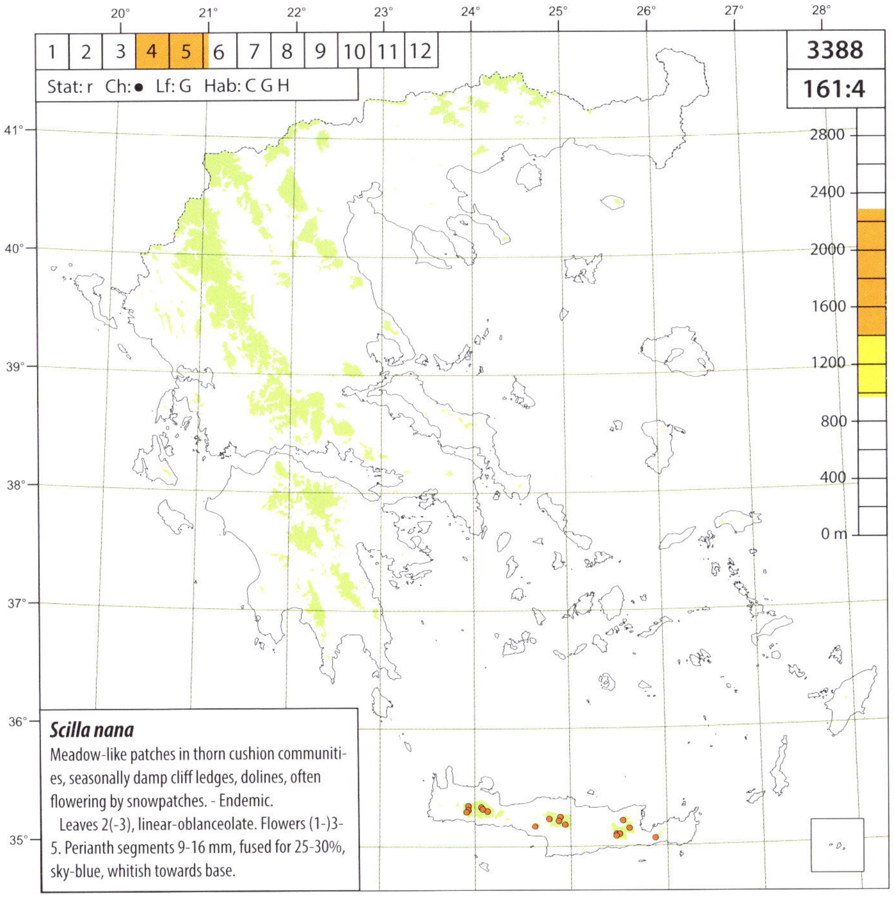

Scilla nana
Meadow-like patches in thorn cushion communities, seasonally damp cliff ledges, dolines, often flowering by snowpatches. - Endemic.
Leaves 2(-3), linear-oblanceolate. Flowers (1-)3-5. Perianth segments 9-16 mm, fused for 25-30%, sky-blue, whitish towards base.

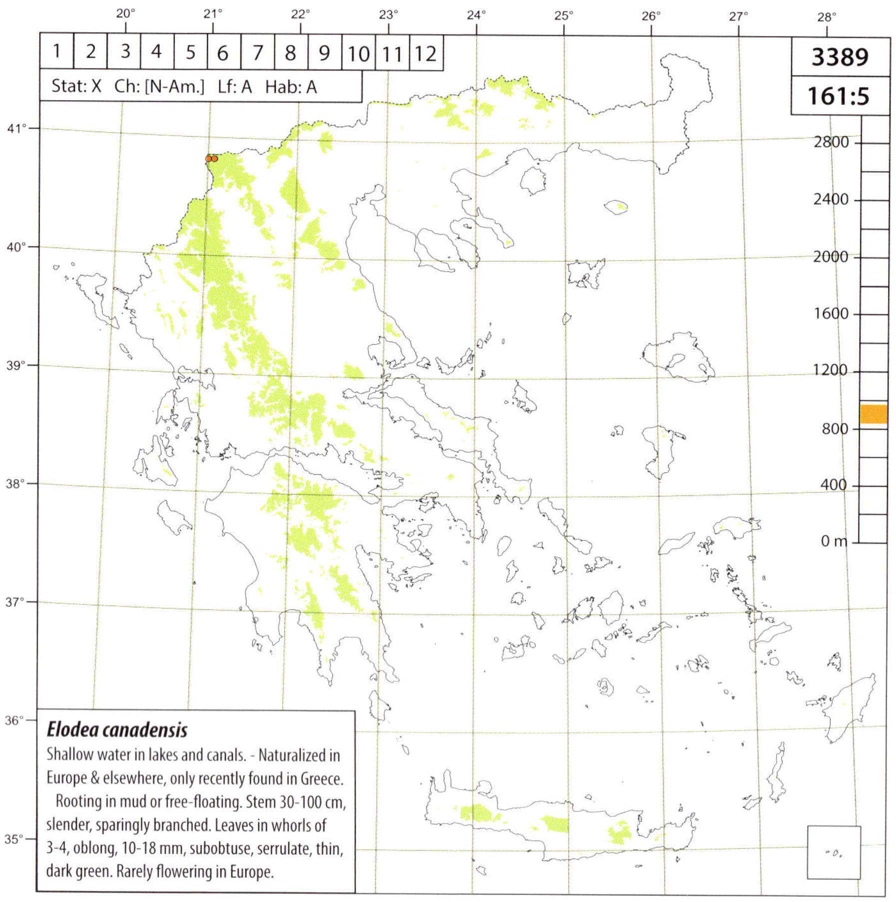

Elodea canadensis
Shallow water in lakes and canals. - Naturalized in Europe & elsewhere, only recently found in Greece.
Rooting in mud or free-floating. Stem 30-100 cm, slender, sparingly branched. Leaves in whorls of 3-4, oblong, 10-18 mm, subobtuse, serrulate, thin, dark green. Rarely flowering in Europe.

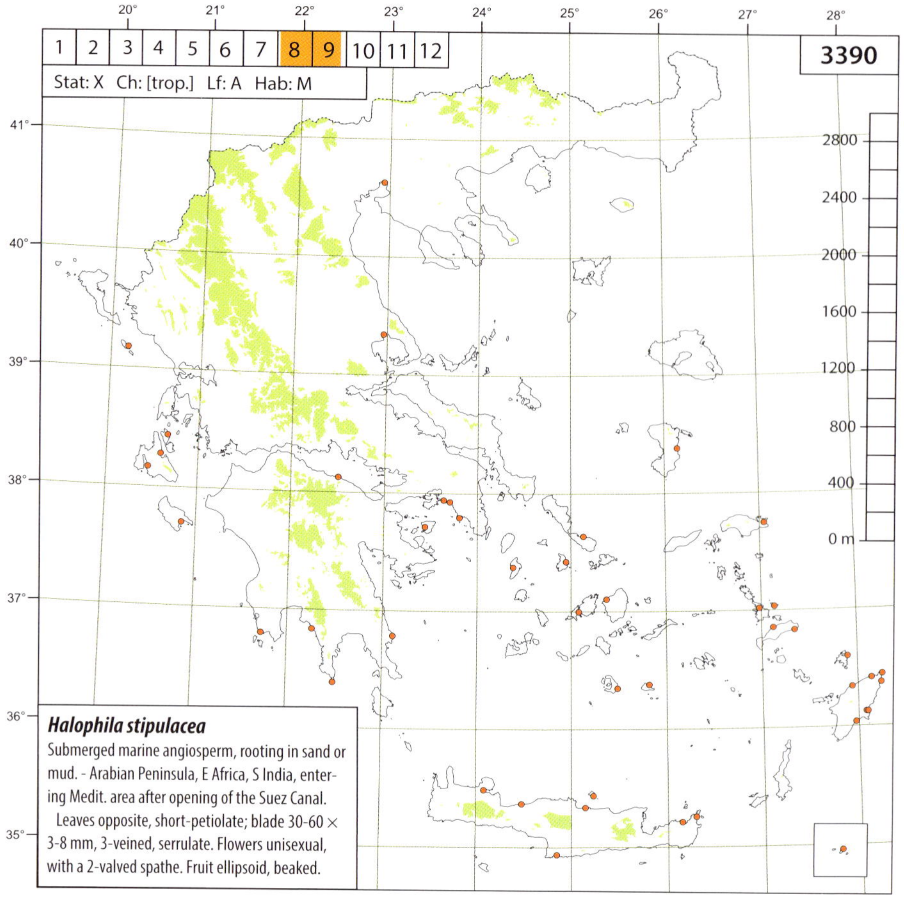

Halophila stipulacea
Submerged marine angiosperm, rooting in sand or mud. - Arabian Peninsula, E Africa, S India, entering Medit. area after opening of the Suez Canal.
Leaves opposite, short-petiolate; blade 30-60 × 3-8 mm, 3-veined, serrulate. Flowers unisexual, with a 2-valved spathe. Fruit ellipsoid, beaked.

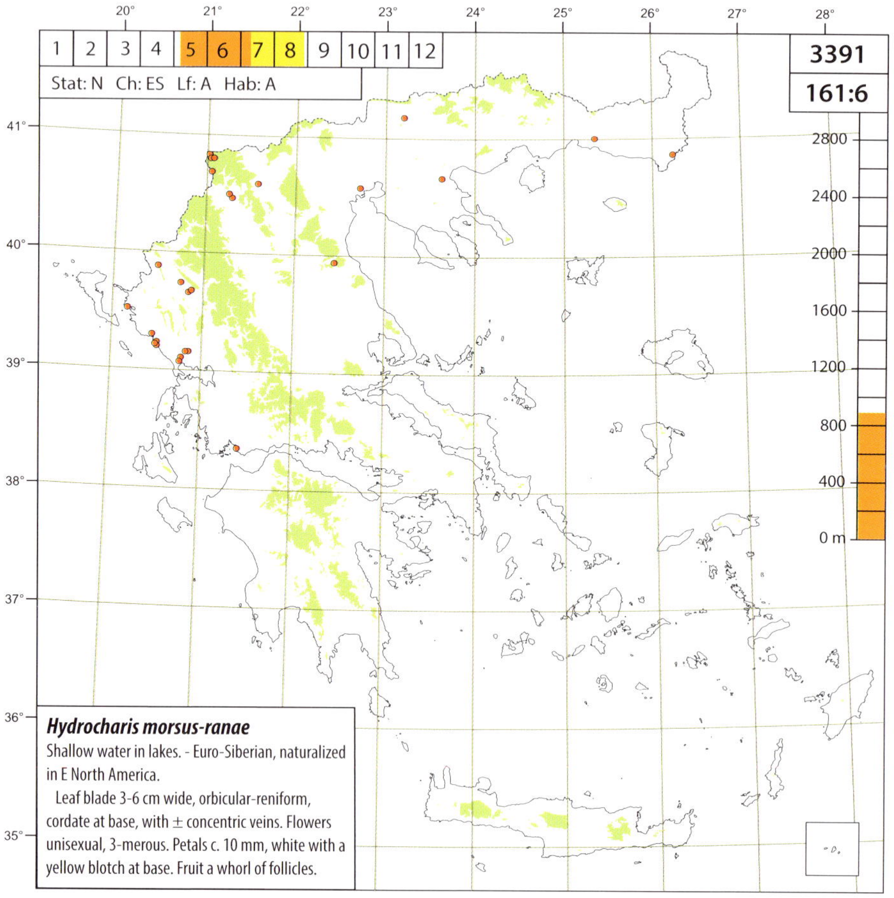

Hydrocharis morsus-ranae
Shallow water in lakes. - Euro-Siberian, naturalized in E North America.
Leaf blade 3-6 cm wide, orbicular-reniform, cordate at base, with ± concentric veins. Flowers unisexual, 3-merous. Petals c. 10 mm, white with a yellow blotch at base. Fruit a whorl of follicles.

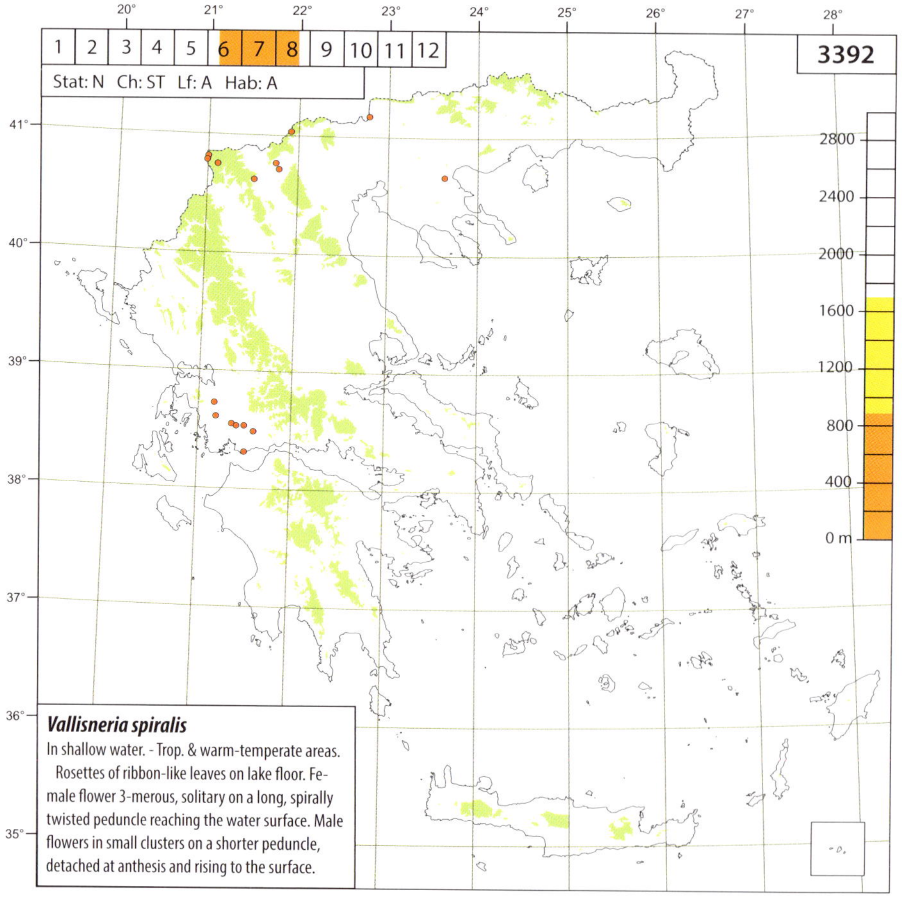

Vallisneria spiralis
In shallow water. - Trop. & warm-temperate areas.
Rosettes of ribbon-like leaves on lake floor. Female flower 3-merous, solitary on a long, spirally twisted peduncle reaching the water surface. Male flowers in small clusters on a shorter peduncle, detached at anthesis and rising to the surface.

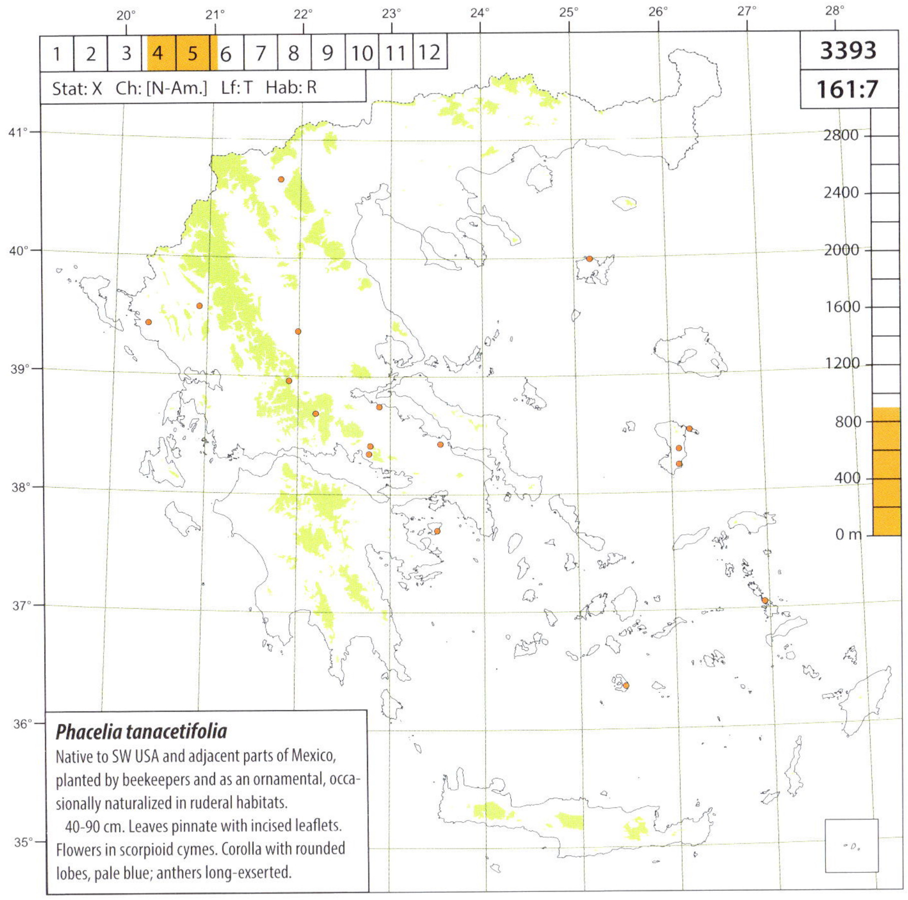

Phacelia tanacetifolia

Native to SW USA and adjacent parts of Mexico, planted by beekeepers and as an ornamental, occasionally naturalized in ruderal habitats.

40-90 cm. Leaves pinnate with incised leaflets. Flowers in scorpioid cymes. Corolla with rounded lobes, pale blue; anthers long-exserted.

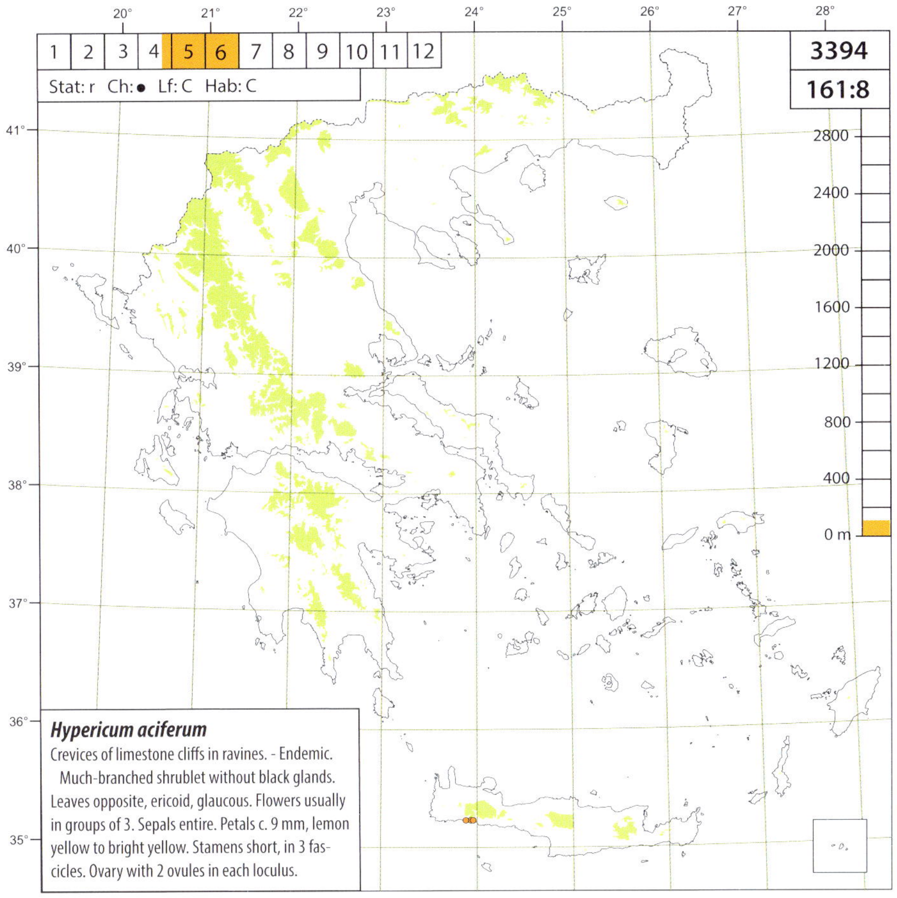

Hypericum aciferum

Crevices of limestone cliffs in ravines. - Endemic.

Much-branched shrublet without black glands. Leaves opposite, ericoid, glaucous. Flowers usually in groups of 3. Sepals entire. Petals c. 9 mm, lemon yellow to bright yellow. Stamens short, in 3 fascicles. Ovary with 2 ovules in each loculus.

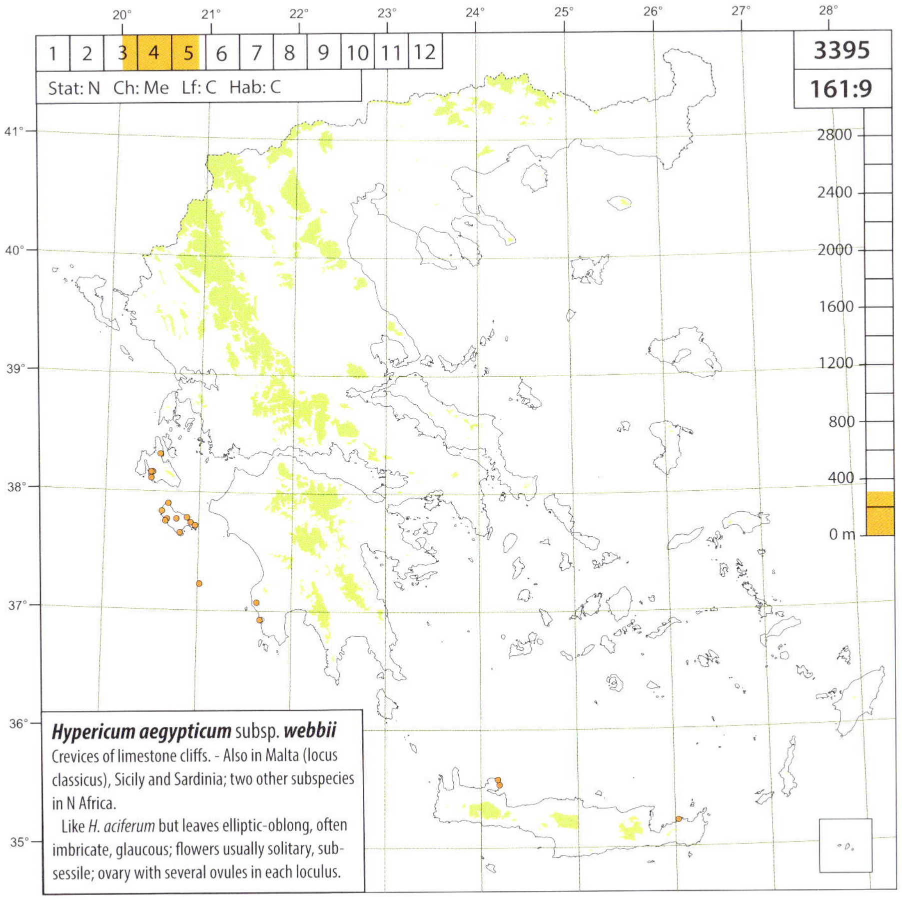

Hypericum aegypticum subsp. _webbii_

Crevices of limestone cliffs. - Also in Malta (locus classicus), Sicily and Sardinia; two other subspecies in N Africa.

Like *H. aciferum* but leaves elliptic-oblong, often imbricate, glaucous; flowers usually solitary, subsessile; ovary with several ovules in each loculus.

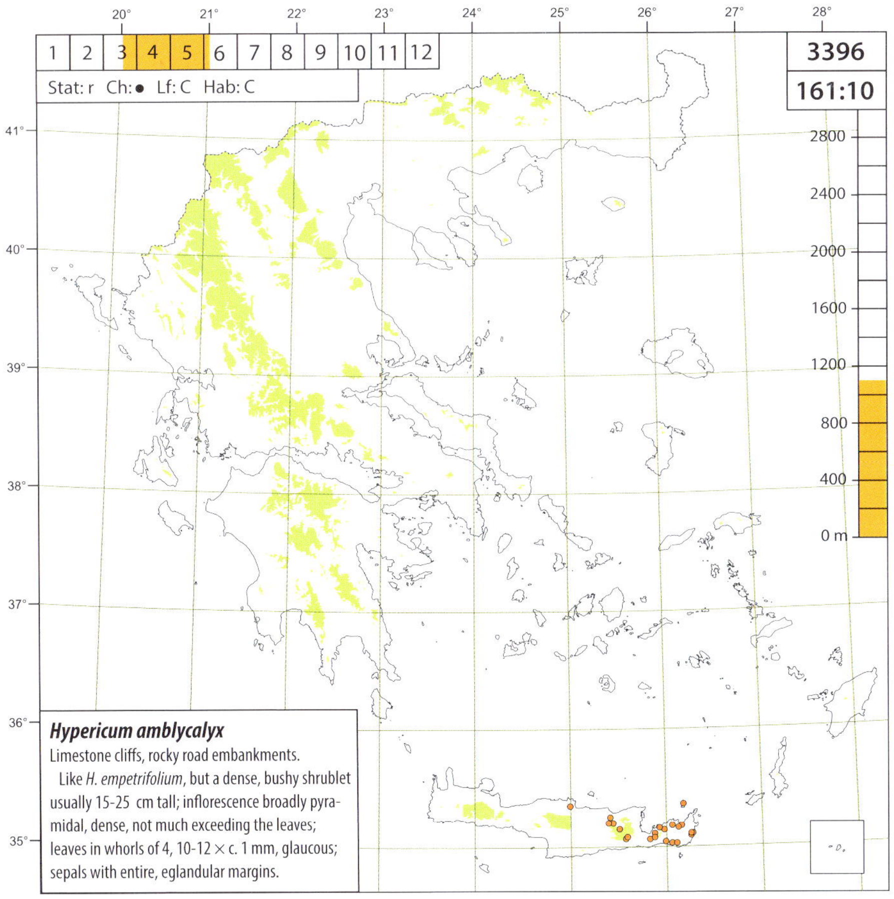

Hypericum amblycalyx

Limestone cliffs, rocky road embankments.

Like *H. empetrifolium*, but a dense, bushy shrublet usually 15-25 cm tall; inflorescence broadly pyramidal, dense, not much exceeding the leaves; leaves in whorls of 4, 10-12 × c. 1 mm, glaucous; sepals with entire, eglandular margins.

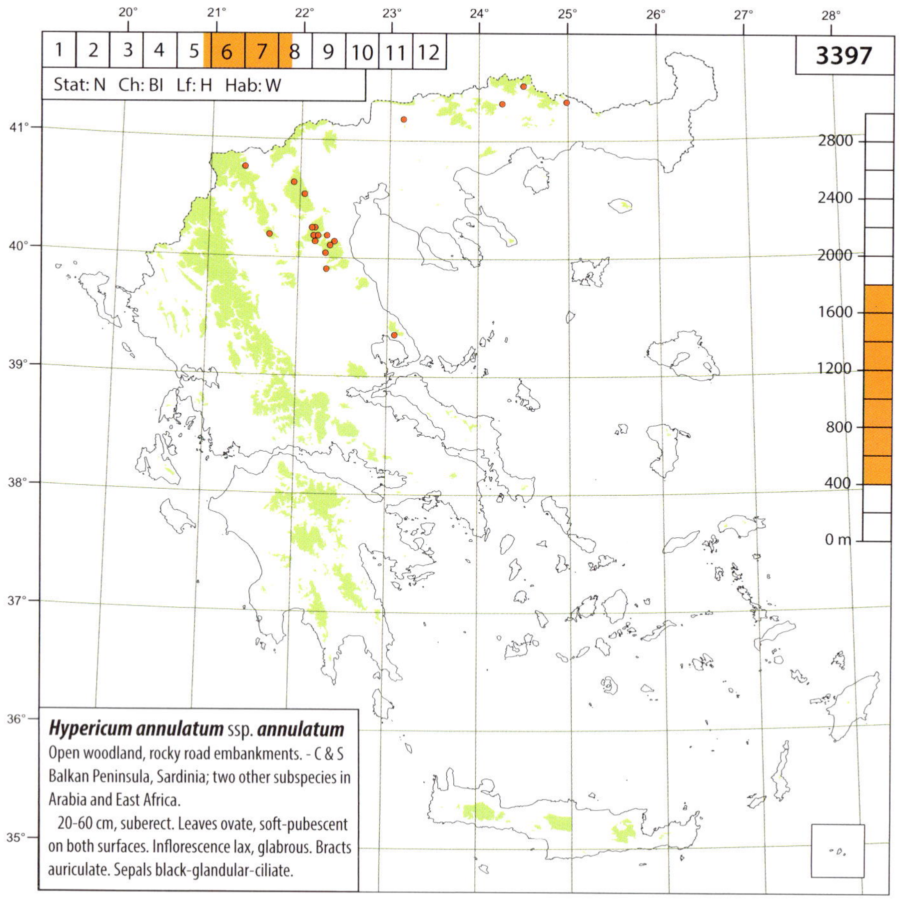

Hypericum annulatum ssp. **annulatum**
Open woodland, rocky road embankments. - C & S Balkan Peninsula, Sardinia; two other subspecies in Arabia and East Africa.
20-60 cm, suberect. Leaves ovate, soft-pubescent on both surfaces. Inflorescence lax, glabrous. Bracts auriculate. Sepals black-glandular-ciliate.

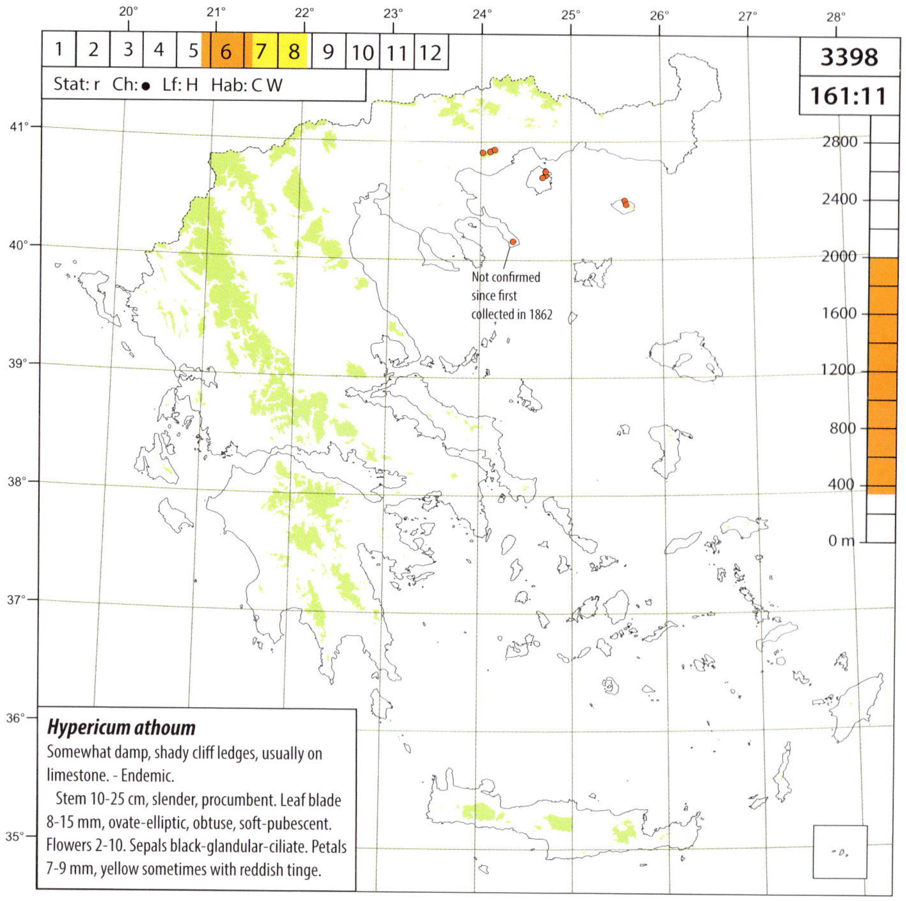

Hypericum athoum
Somewhat damp, shady cliff ledges, usually on limestone. - Endemic.
Stem 10-25 cm, slender, procumbent. Leaf blade 8-15 mm, ovate-elliptic, obtuse, soft-pubescent. Flowers 2-10. Sepals black-glandular-ciliate. Petals 7-9 mm, yellow sometimes with reddish tinge.

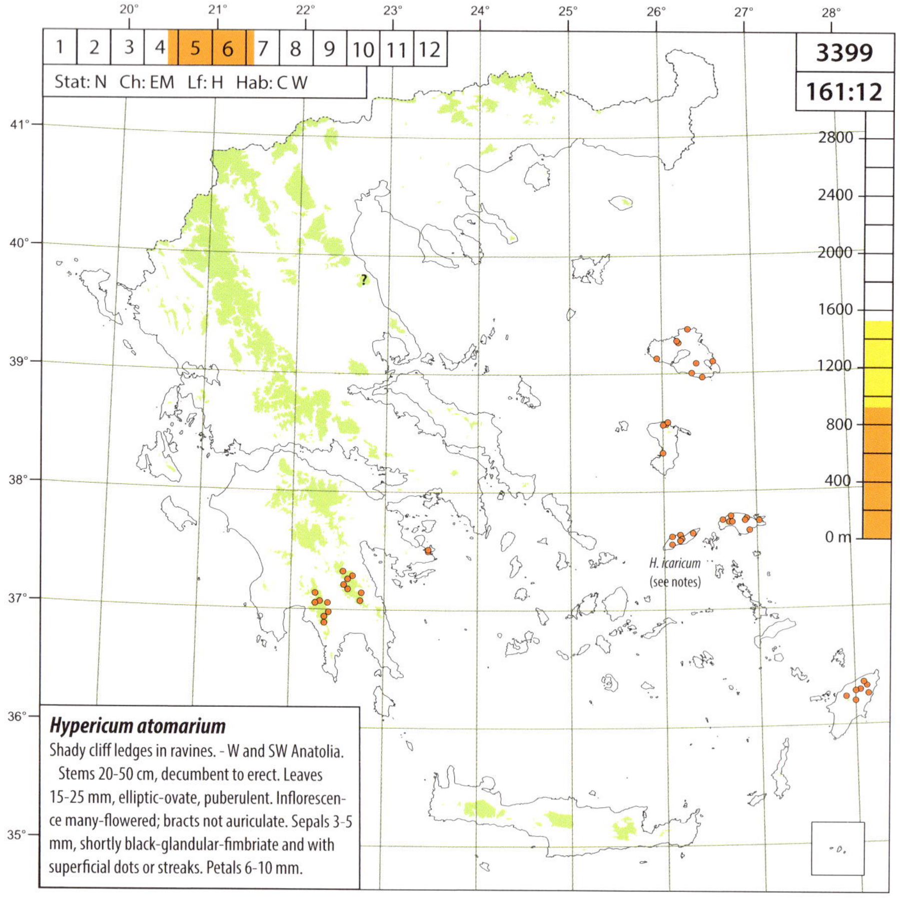

Hypericum atomarium
Shady cliff ledges in ravines. - W and SW Anatolia.
Stems 20-50 cm, decumbent to erect. Leaves 15-25 mm, elliptic-ovate, puberulent. Inflorescence many-flowered; bracts not auriculate. Sepals 3-5 mm, shortly black-glandular-fimbriate and with superficial dots or streaks. Petals 6-10 mm.

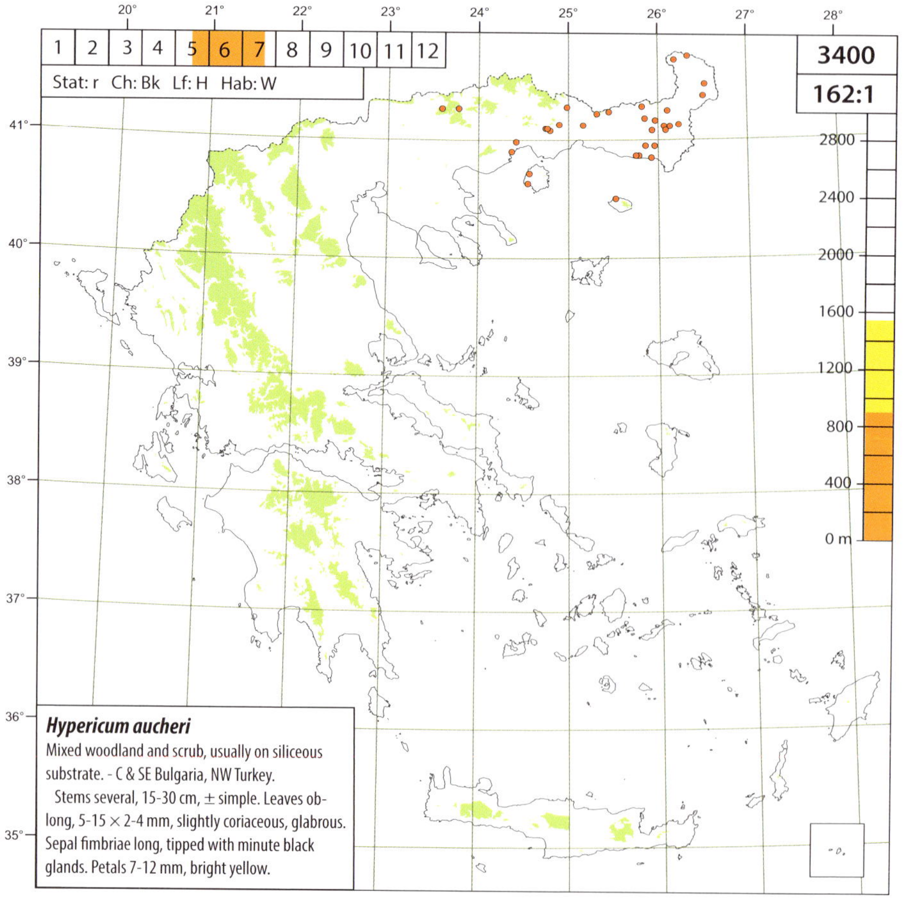

Hypericum aucheri
Mixed woodland and scrub, usually on siliceous substrate. - C & SE Bulgaria, NW Turkey.
Stems several, 15-30 cm, ± simple. Leaves oblong, 5-15 × 2-4 mm, slightly coriaceous, glabrous. Sepal fimbriae long, tipped with minute black glands. Petals 7-12 mm, bright yellow.

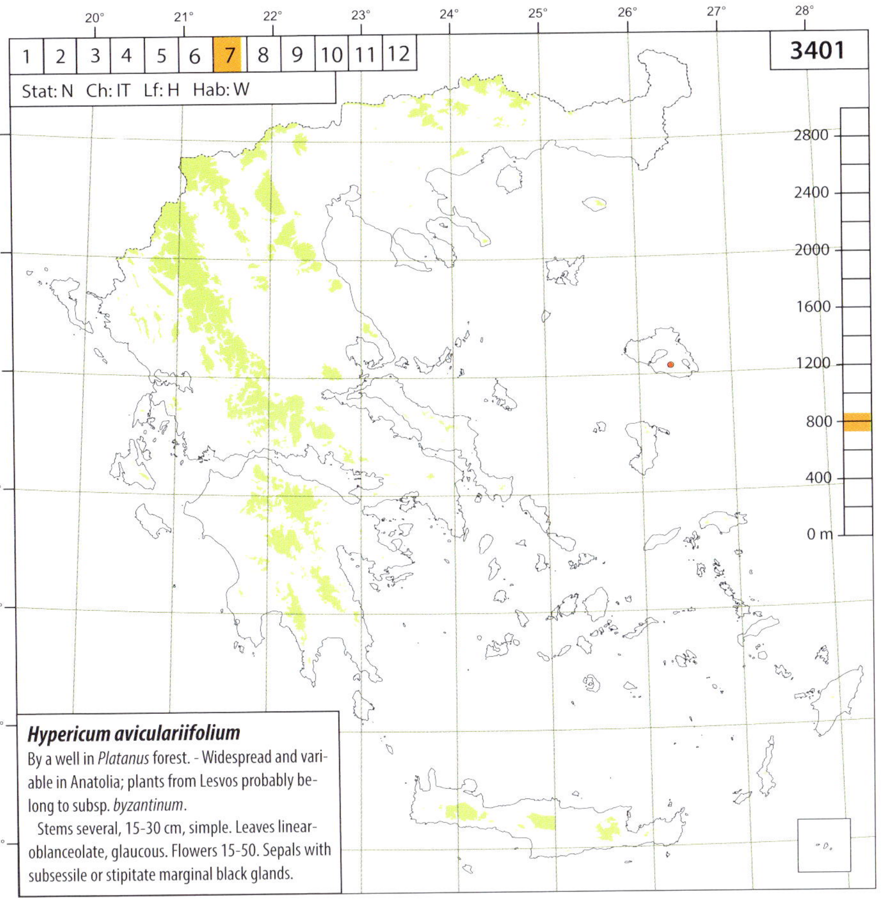

Hypericum aviculariifolium

By a well in *Platanus* forest. - Widespread and variable in Anatolia; plants from Lesvos probably belong to subsp. *byzantinum*.

Stems several, 15-30 cm, simple. Leaves linear-oblanceolate, glaucous. Flowers 15-50. Sepals with subsessile or stipitate marginal black glands.

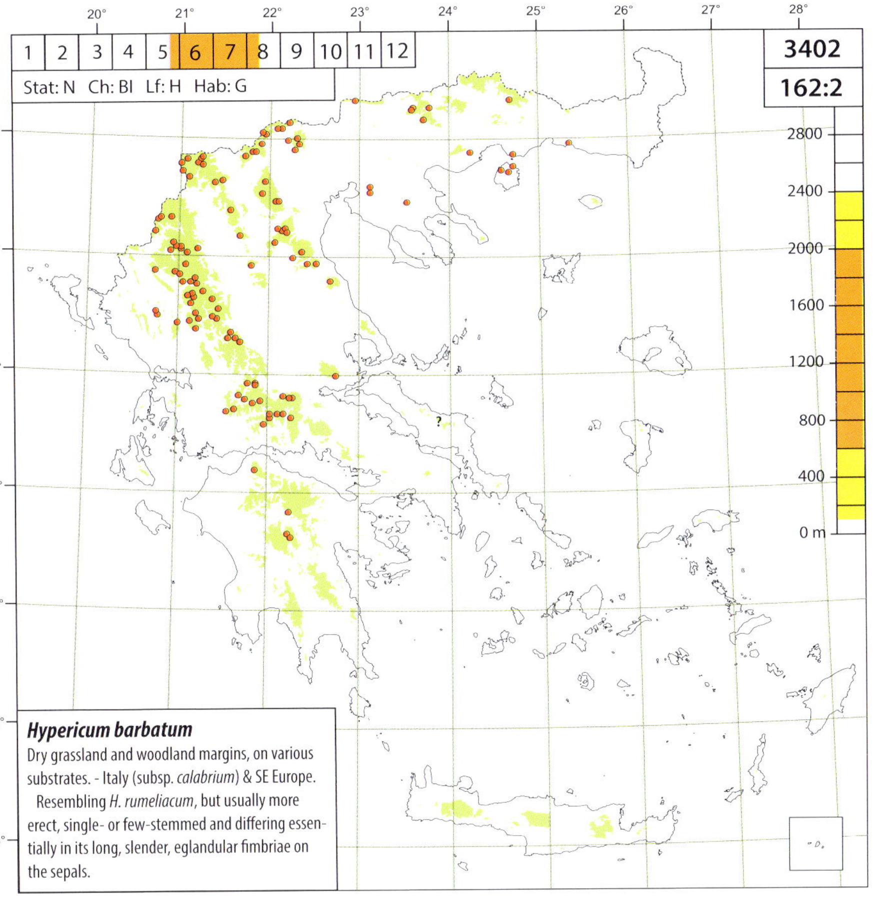

Hypericum barbatum

Dry grassland and woodland margins, on various substrates. - Italy (subsp. *calabrium*) & SE Europe.

Resembling *H. rumeliacum*, but usually more erect, single- or few-stemmed and differing essentially in its long, slender, eglandular fimbriae on the sepals.

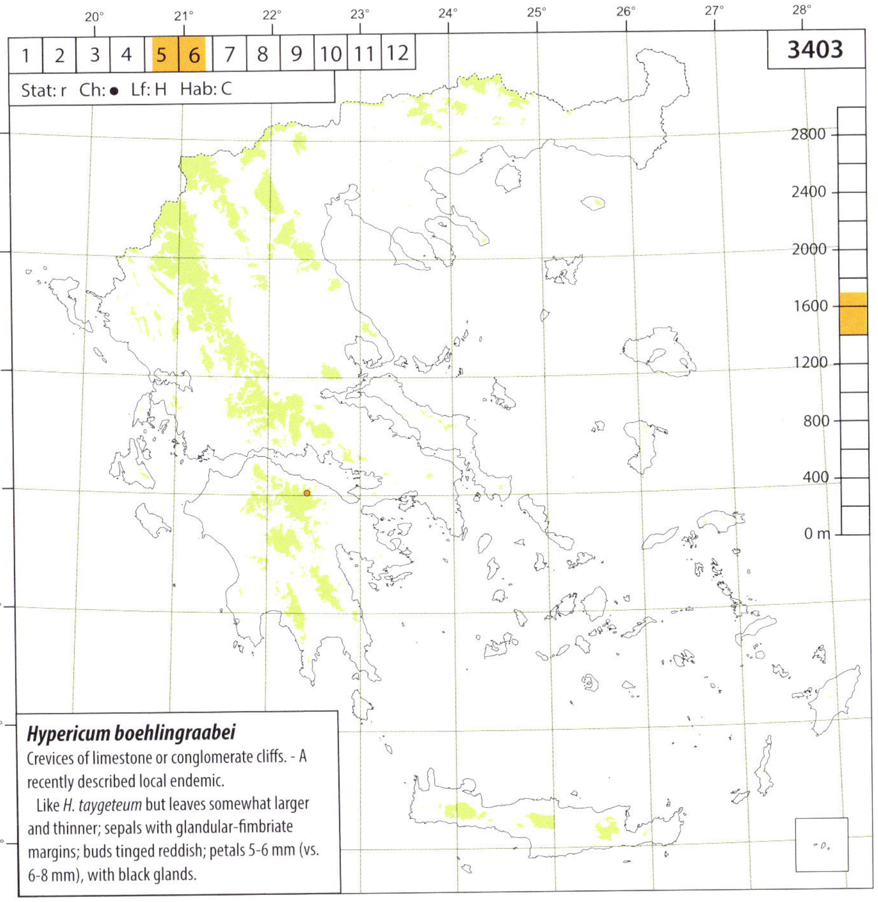

Hypericum boehlingraabei

Crevices of limestone or conglomerate cliffs. - A recently described local endemic.

Like *H. taygeteum* but leaves somewhat larger and thinner; sepals with glandular-fimbriate margins; buds tinged reddish; petals 5-6 mm (vs. 6-8 mm), with black glands.

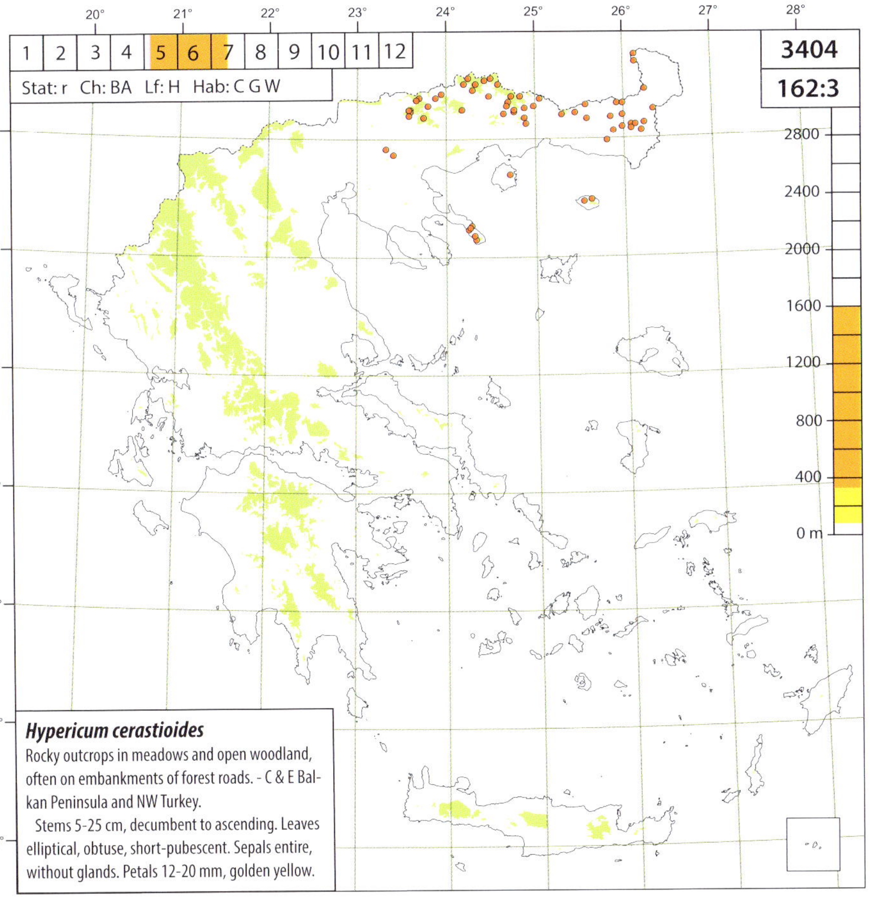

Hypericum cerastioides

Rocky outcrops in meadows and open woodland, often on embankments of forest roads. - C & E Balkan Peninsula and NW Turkey.

Stems 5-25 cm, decumbent to ascending. Leaves elliptical, obtuse, short-pubescent. Sepals entire, without glands. Petals 12-20 mm, golden yellow.

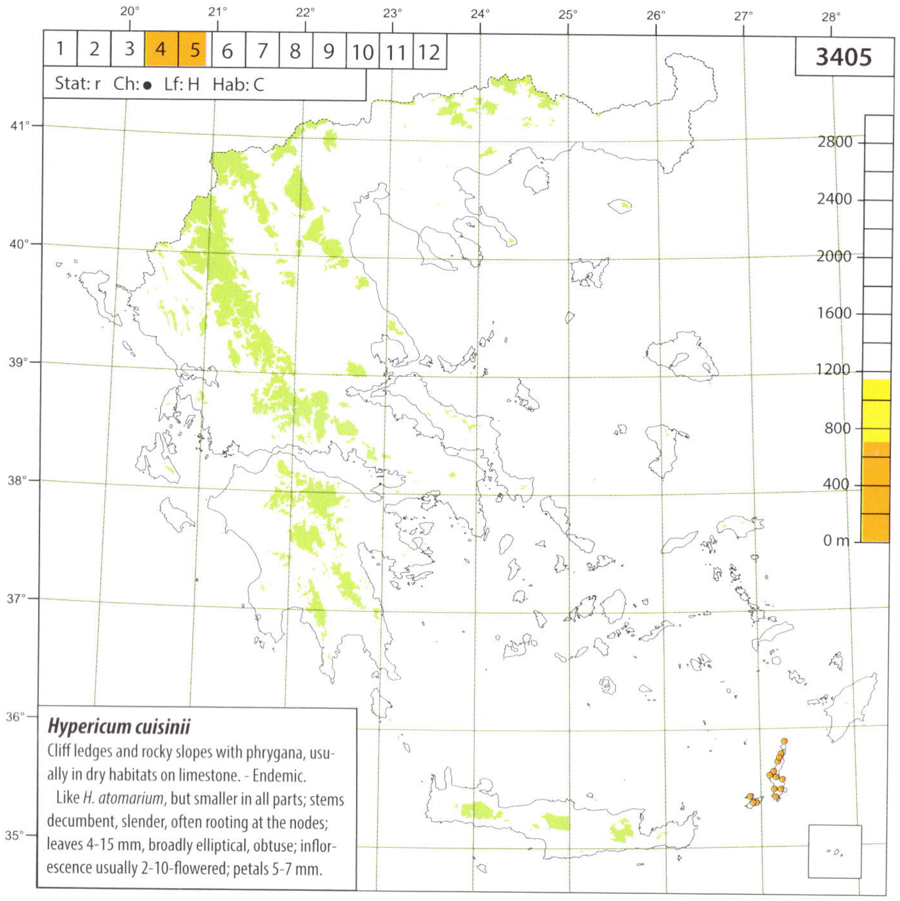

Hypericum cuisinii
Cliff ledges and rocky slopes with phrygana, usually in dry habitats on limestone. - Endemic.
Like *H. atomarium*, but smaller in all parts; stems decumbent, slender, often rooting at the nodes; leaves 4-15 mm, broadly elliptical, obtuse; inflorescence usually 2-10-flowered; petals 5-7 mm.

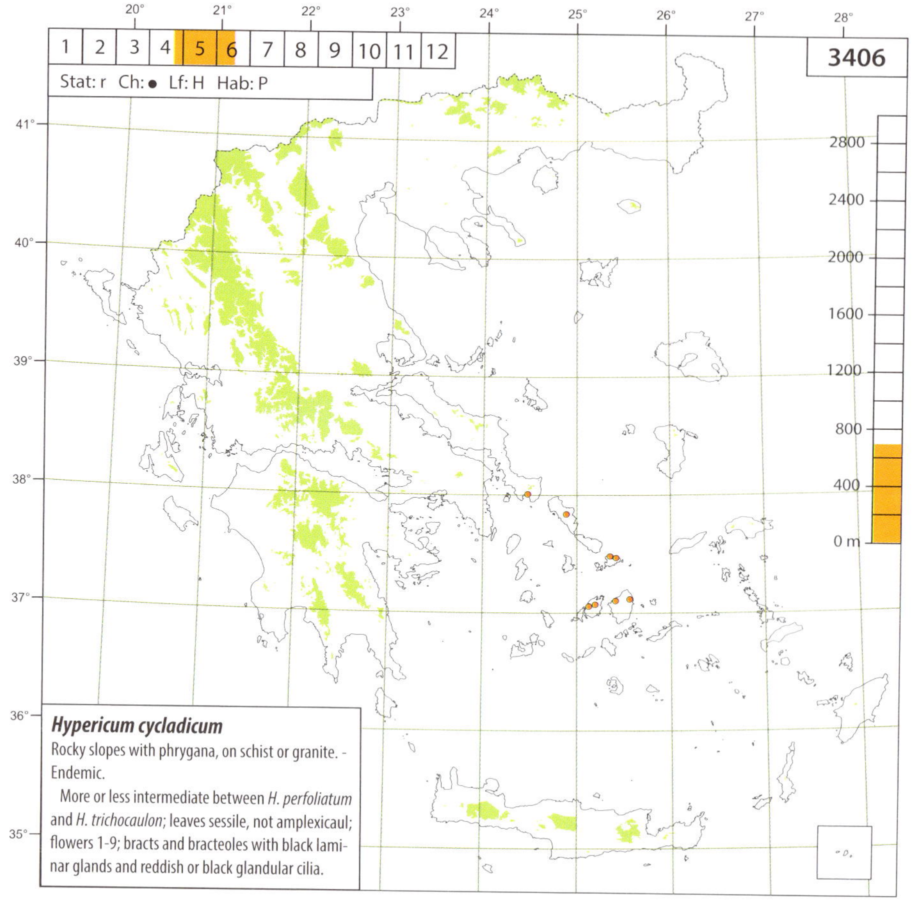

Hypericum cycladicum
Rocky slopes with phrygana, on schist or granite. - Endemic.
More or less intermediate between *H. perfoliatum* and *H. trichocaulon*; leaves sessile, not amplexicaul; flowers 1-9; bracts and bracteoles with black laminar glands and reddish or black glandular cilia.

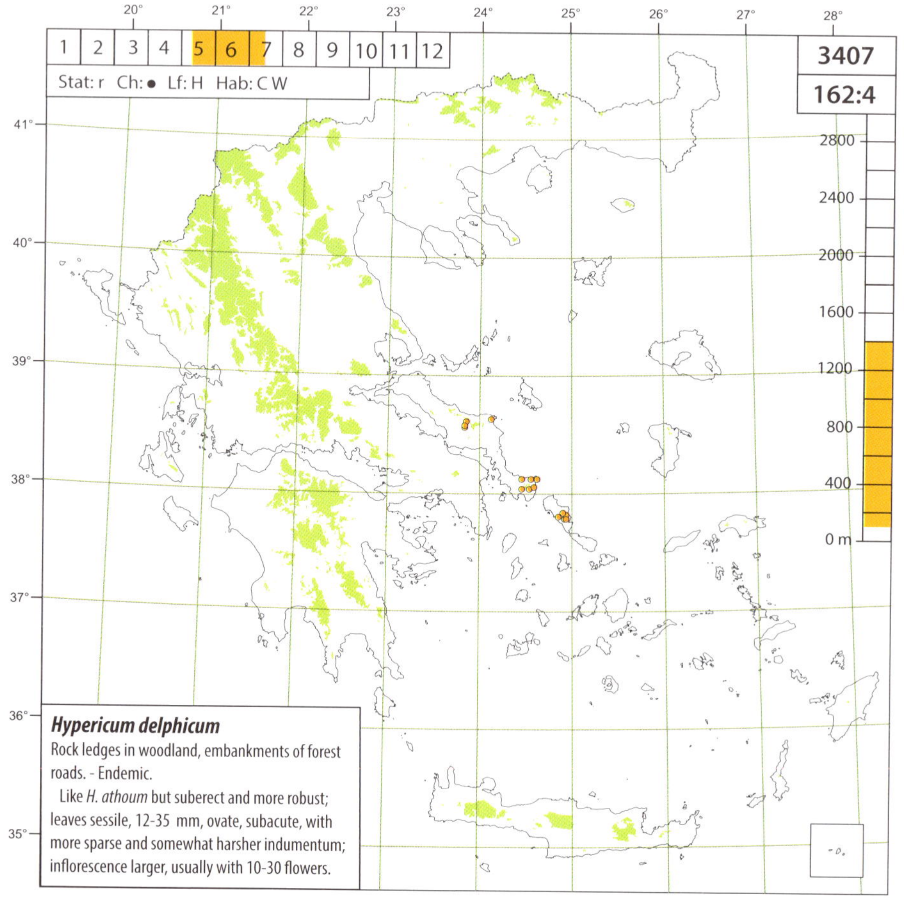

Hypericum delphicum
Rock ledges in woodland, embankments of forest roads. - Endemic.
Like *H. athoum* but suberect and more robust; leaves sessile, 12-35 mm, ovate, subacute, with more sparse and somewhat harsher indumentum; inflorescence larger, usually with 10-30 flowers.

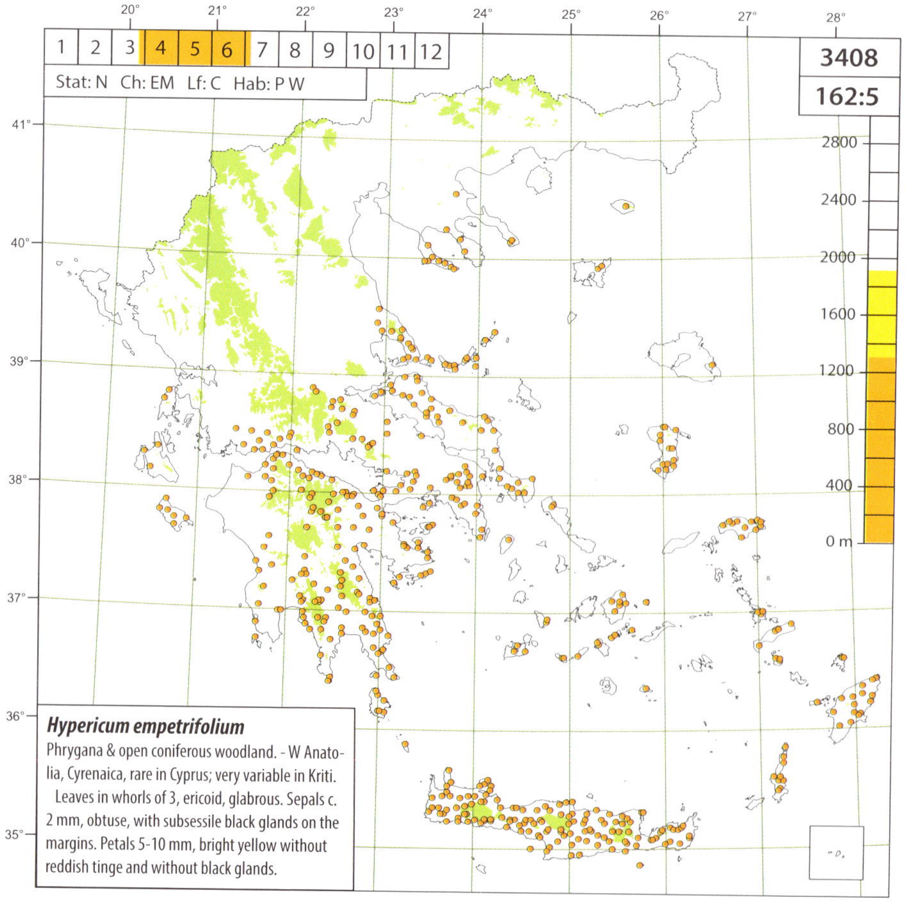

Hypericum empetrifolium
Phrygana & open coniferous woodland. - W Anatolia, Cyrenaica, rare in Cyprus; very variable in Kriti.
Leaves in whorls of 3, ericoid, glabrous. Sepals c. 2 mm, obtuse, with subsessile black glands on the margins. Petals 5-10 mm, bright yellow without reddish tinge and without black glands.

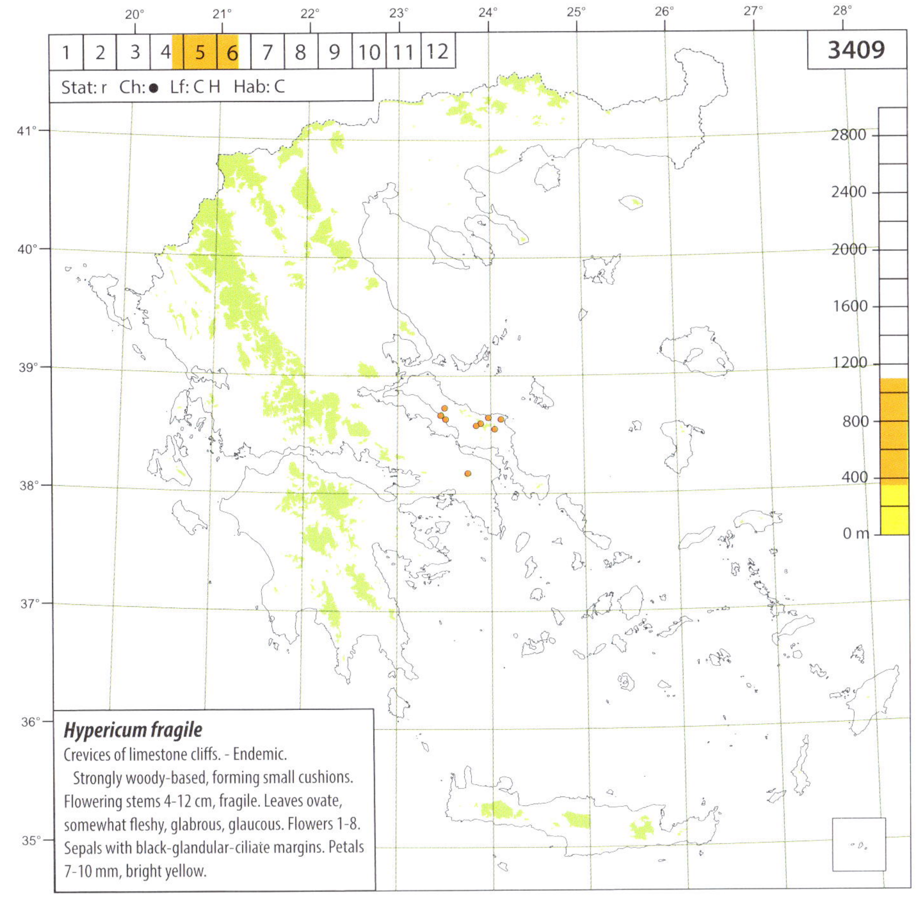

Hypericum fragile

Crevices of limestone cliffs. - Endemic.

Strongly woody-based, forming small cushions. Flowering stems 4-12 cm, fragile. Leaves ovate, somewhat fleshy, glabrous, glaucous. Flowers 1-8. Sepals with black-glandular-ciliate margins. Petals 7-10 mm, bright yellow.

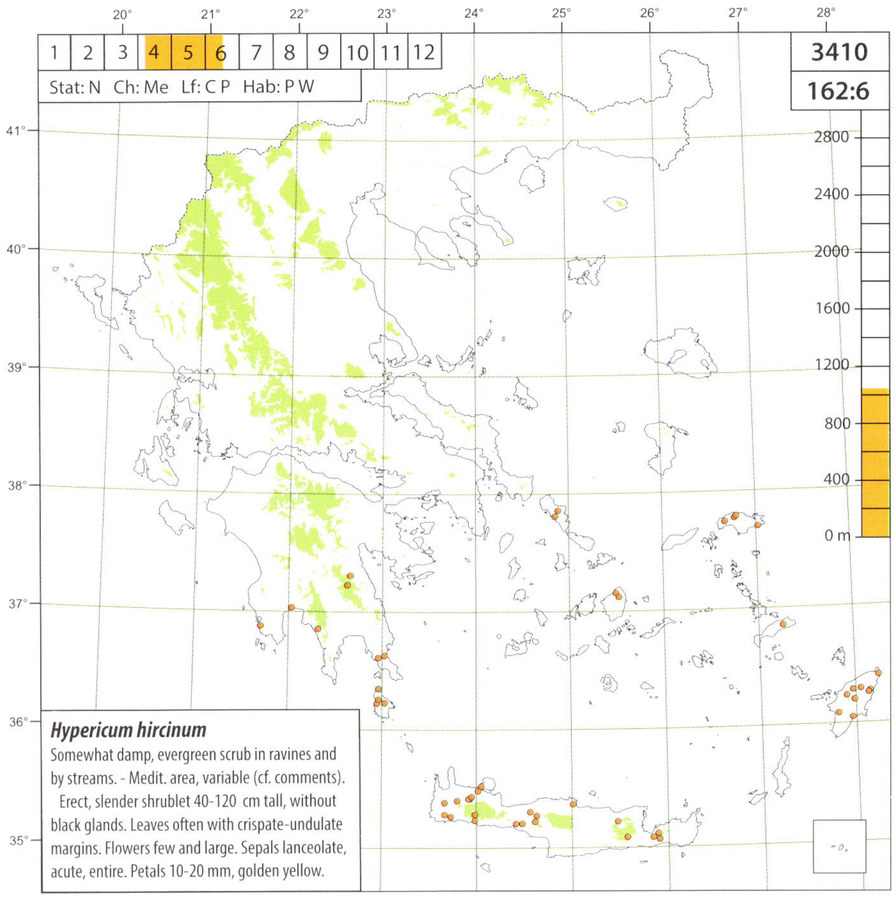

Hypericum hircinum

Somewhat damp, evergreen scrub in ravines and by streams. - Medit. area, variable (cf. comments).

Erect, slender shrublet 40-120 cm tall, without black glands. Leaves often with crispate-undulate margins. Flowers few and large. Sepals lanceolate, acute, entire. Petals 10-20 mm, golden yellow.

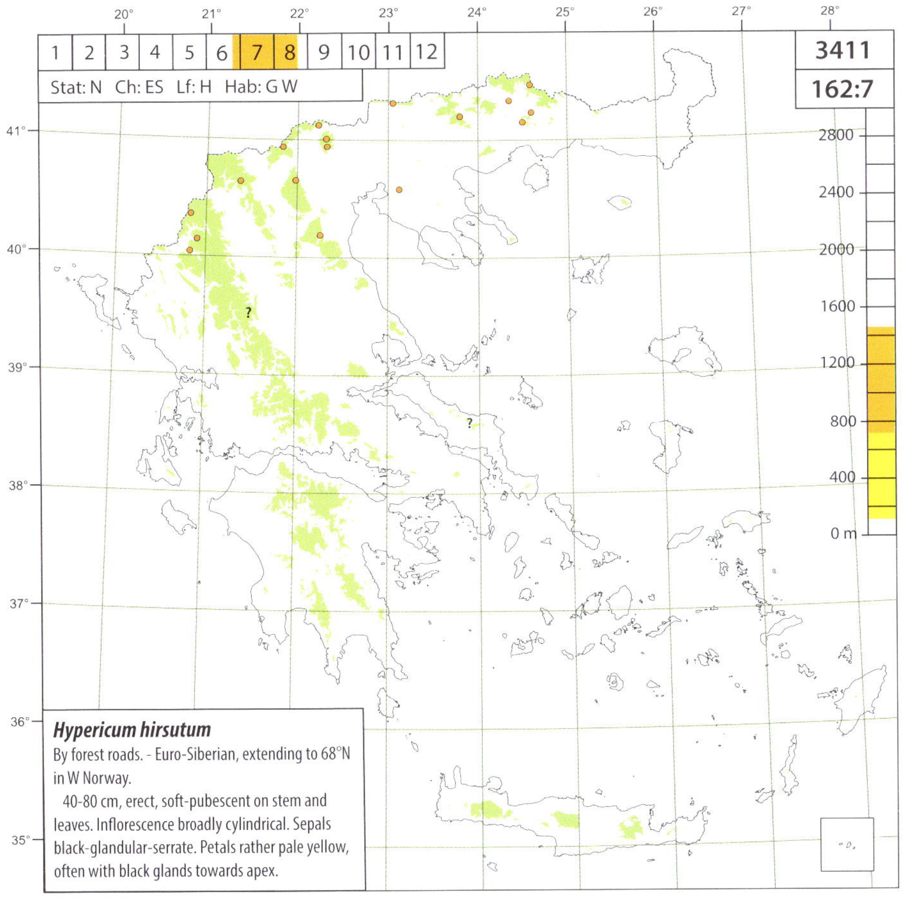

Hypericum hirsutum

By forest roads. - Euro-Siberian, extending to 68°N in W Norway.

40-80 cm, erect, soft-pubescent on stem and leaves. Inflorescence broadly cylindrical. Sepals black-glandular-serrate. Petals rather pale yellow, often with black glands towards apex.

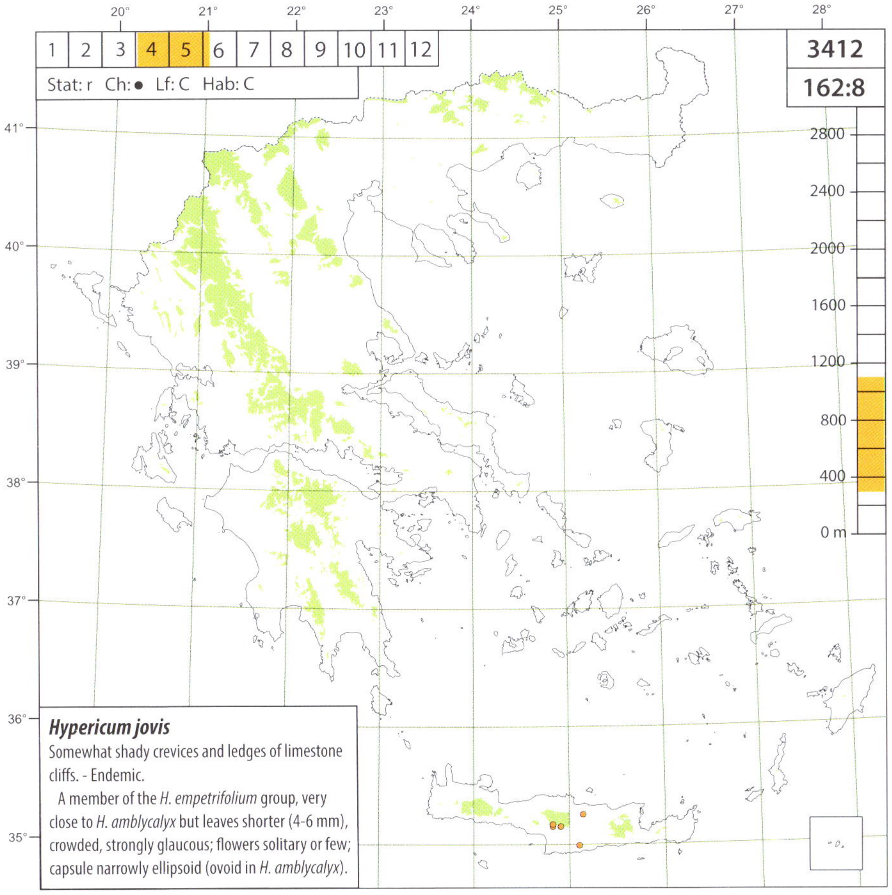

Hypericum jovis

Somewhat shady crevices and ledges of limestone cliffs. - Endemic.

A member of the *H. empetrifolium* group, very close to *H. amblycalyx* but leaves shorter (4-6 mm), crowded, strongly glaucous; flowers solitary or few; capsule narrowly ellipsoid (ovoid in *H. amblycalyx*).

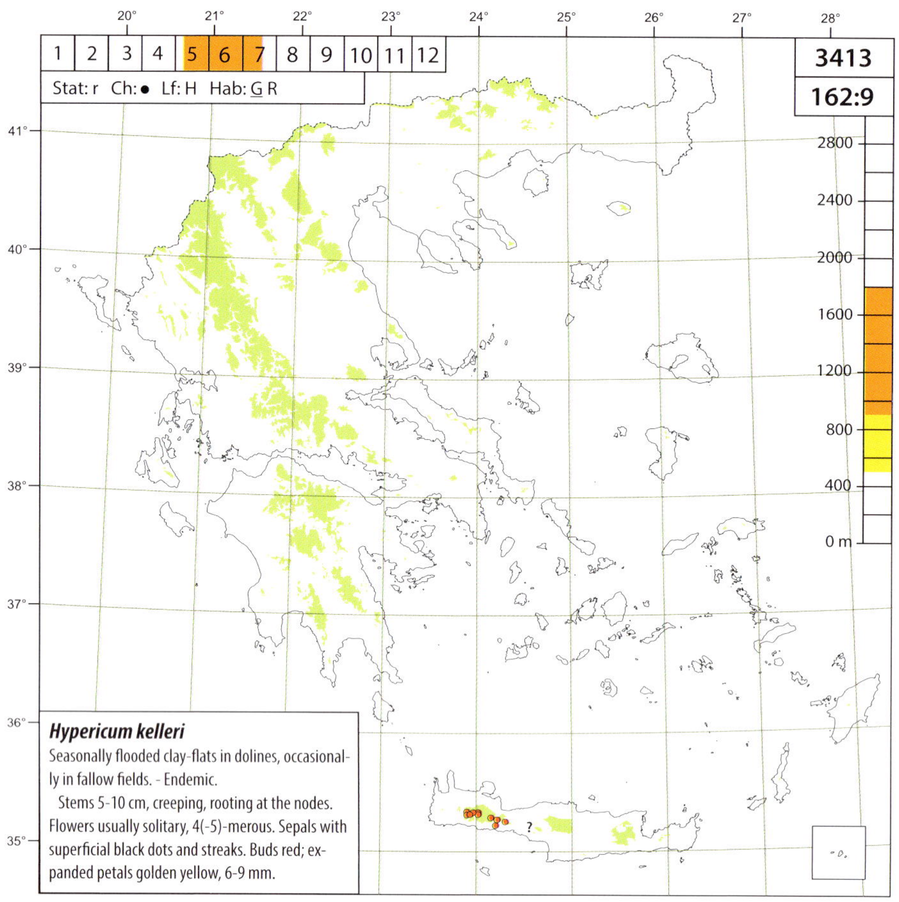

Hypericum kelleri
Seasonally flooded clay-flats in dolines, occasionally in fallow fields. - Endemic.
Stems 5-10 cm, creeping, rooting at the nodes. Flowers usually solitary, 4(-5)-merous. Sepals with superficial black dots and streaks. Buds red; expanded petals golden yellow, 6-9 mm.

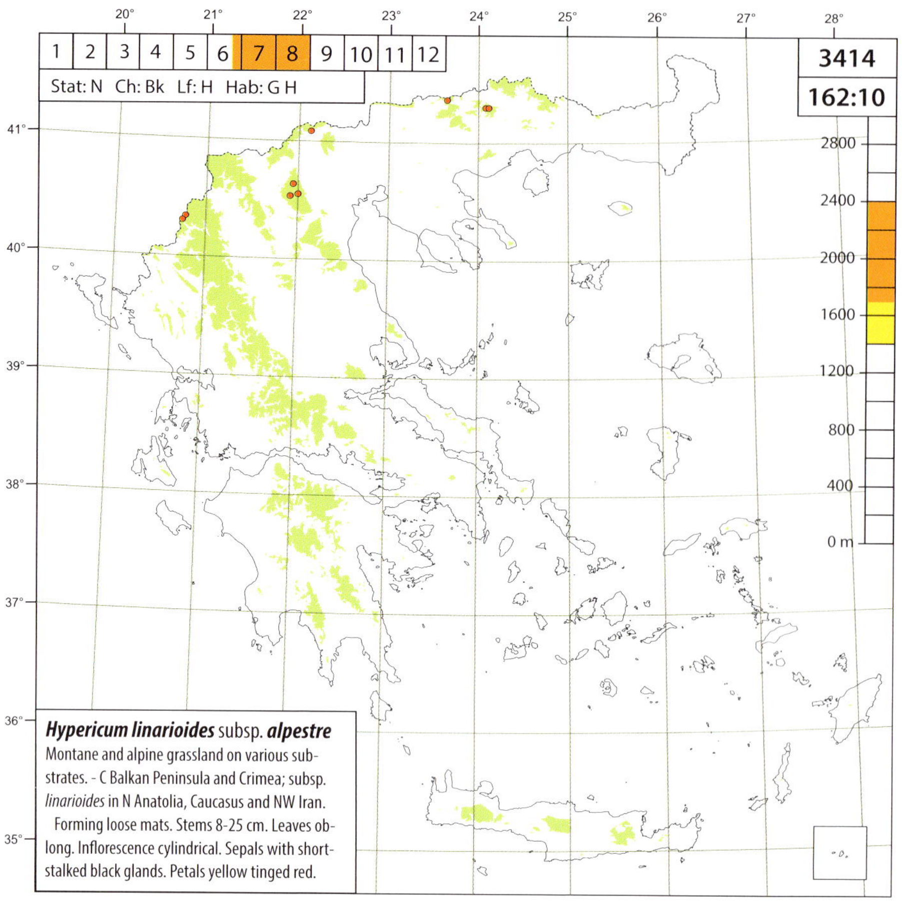

Hypericum linarioides subsp. ***alpestre***
Montane and alpine grassland on various substrates. - C Balkan Peninsula and Crimea; subsp. *linarioides* in N Anatolia, Caucasus and NW Iran.
Forming loose mats. Stems 8-25 cm. Leaves oblong. Inflorescence cylindrical. Sepals with short-stalked black glands. Petals yellow tinged red.

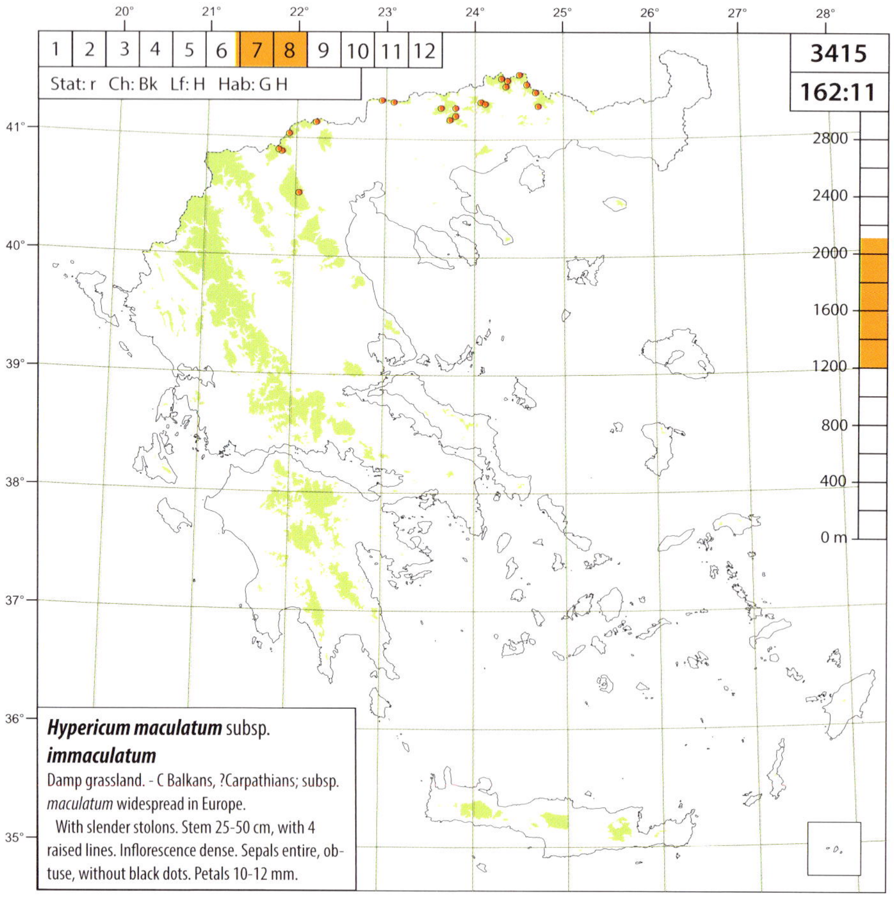

Hypericum maculatum subsp. ***immaculatum***
Damp grassland. - C Balkans, ?Carpathians; subsp. *maculatum* widespread in Europe.
With slender stolons. Stem 25-50 cm, with 4 raised lines. Inflorescence dense. Sepals entire, obtuse, without black dots. Petals 10-12 mm.

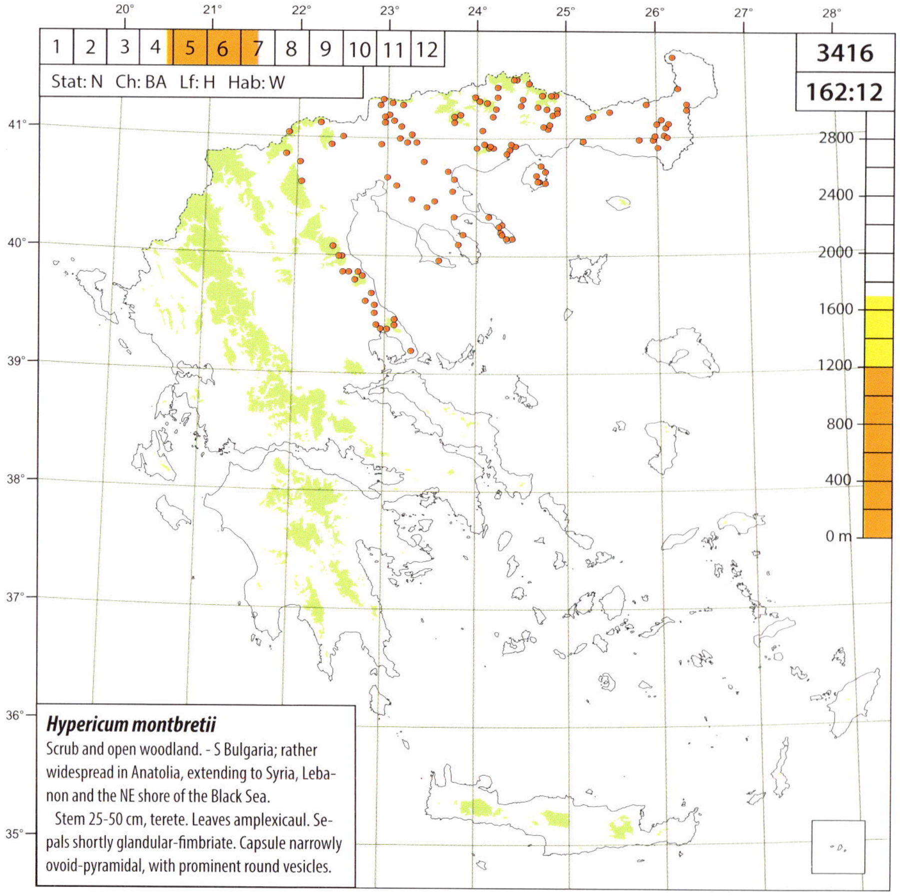

Hypericum montbretii
Scrub and open woodland. - S Bulgaria; rather widespread in Anatolia, extending to Syria, Lebanon and the NE shore of the Black Sea.
Stem 25-50 cm, terete. Leaves amplexicaul. Sepals shortly glandular-fimbriate. Capsule narrowly ovoid-pyramidal, with prominent round vesicles.

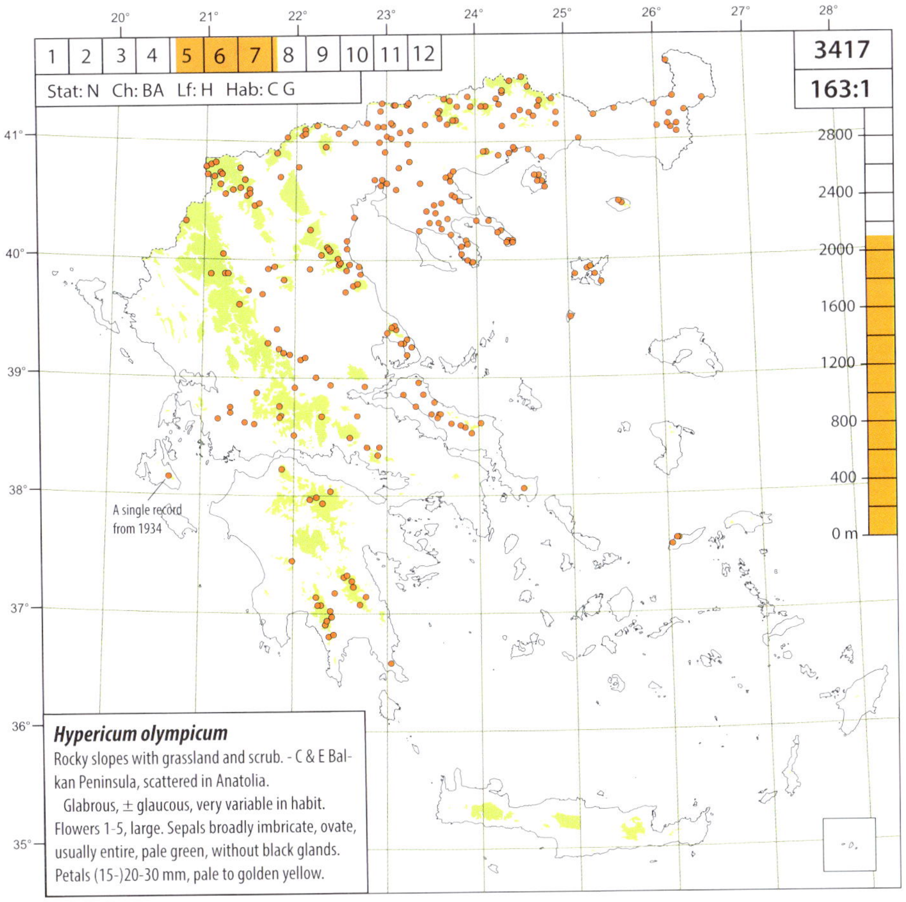

Hypericum olympicum

Rocky slopes with grassland and scrub. - C & E Balkan Peninsula, scattered in Anatolia.

Glabrous, ± glaucous, very variable in habit. Flowers 1-5, large. Sepals broadly imbricate, ovate, usually entire, pale green, without black glands. Petals (15-)20-30 mm, pale to golden yellow.

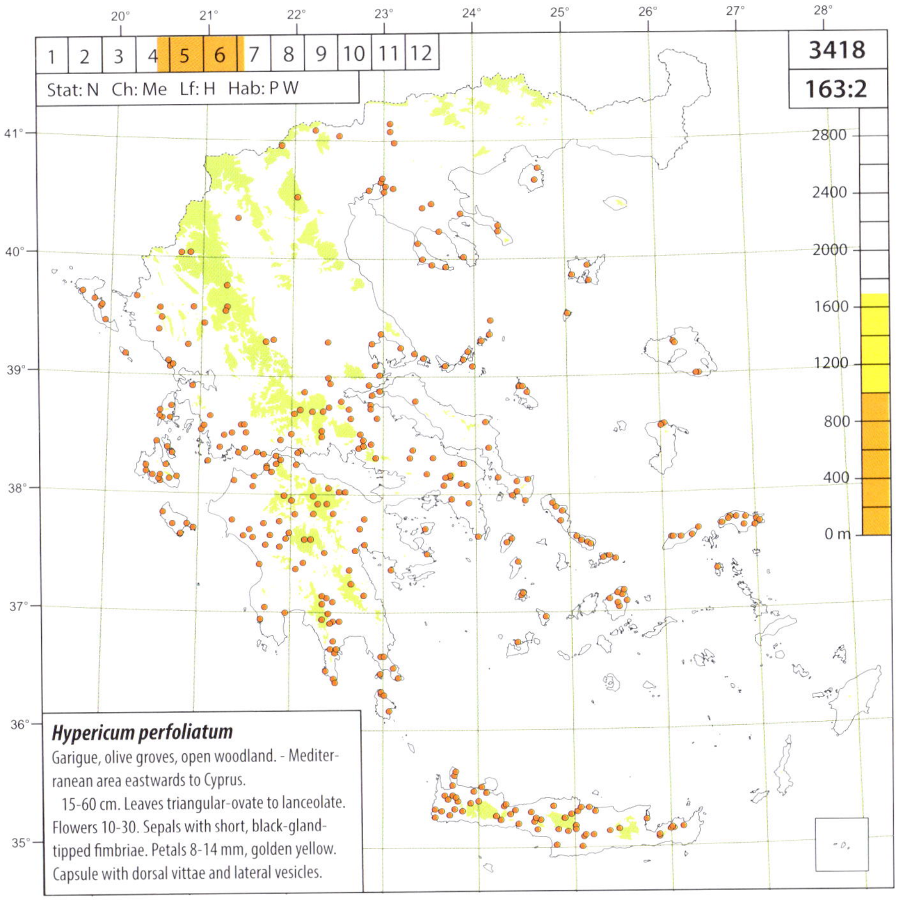

Hypericum perfoliatum

Garigue, olive groves, open woodland. - Mediterranean area eastwards to Cyprus.

15-60 cm. Leaves triangular-ovate to lanceolate. Flowers 10-30. Sepals with short, black-gland-tipped fimbriae. Petals 8-14 mm, golden yellow. Capsule with dorsal vittae and lateral vesicles.

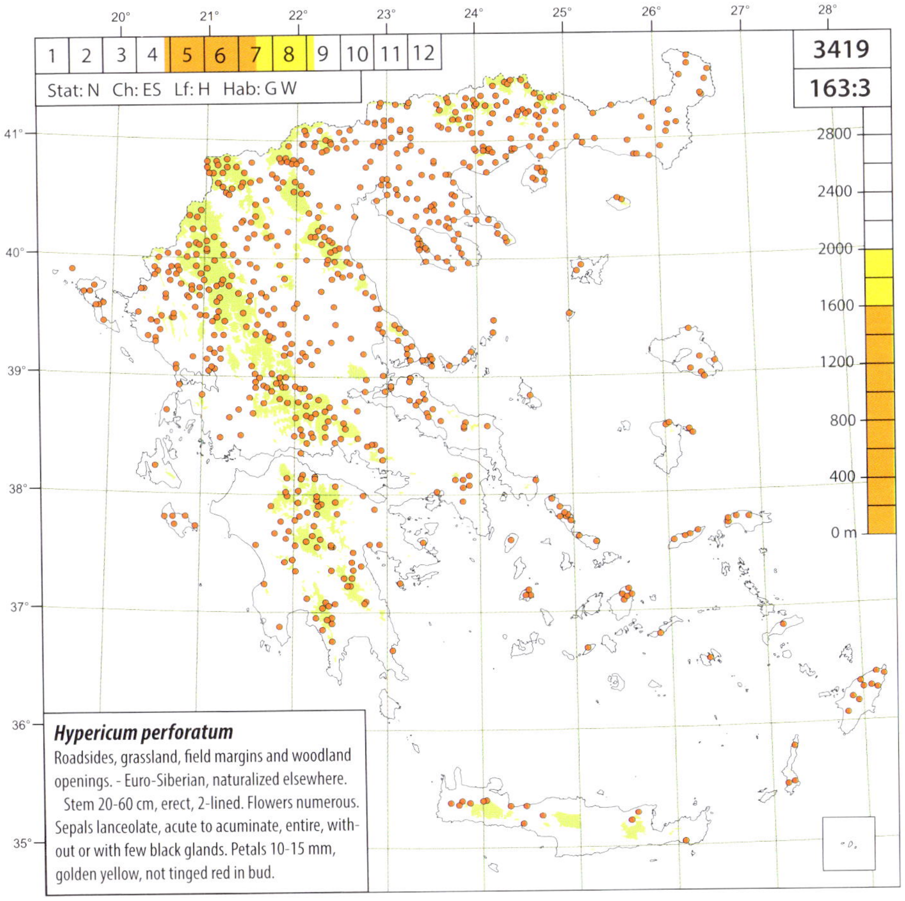

Hypericum perforatum

Roadsides, grassland, field margins and woodland openings. - Euro-Siberian, naturalized elsewhere.

Stem 20-60 cm, erect, 2-lined. Flowers numerous. Sepals lanceolate, acute to acuminate, entire, without or with few black glands. Petals 10-15 mm, golden yellow, not tinged red in bud.

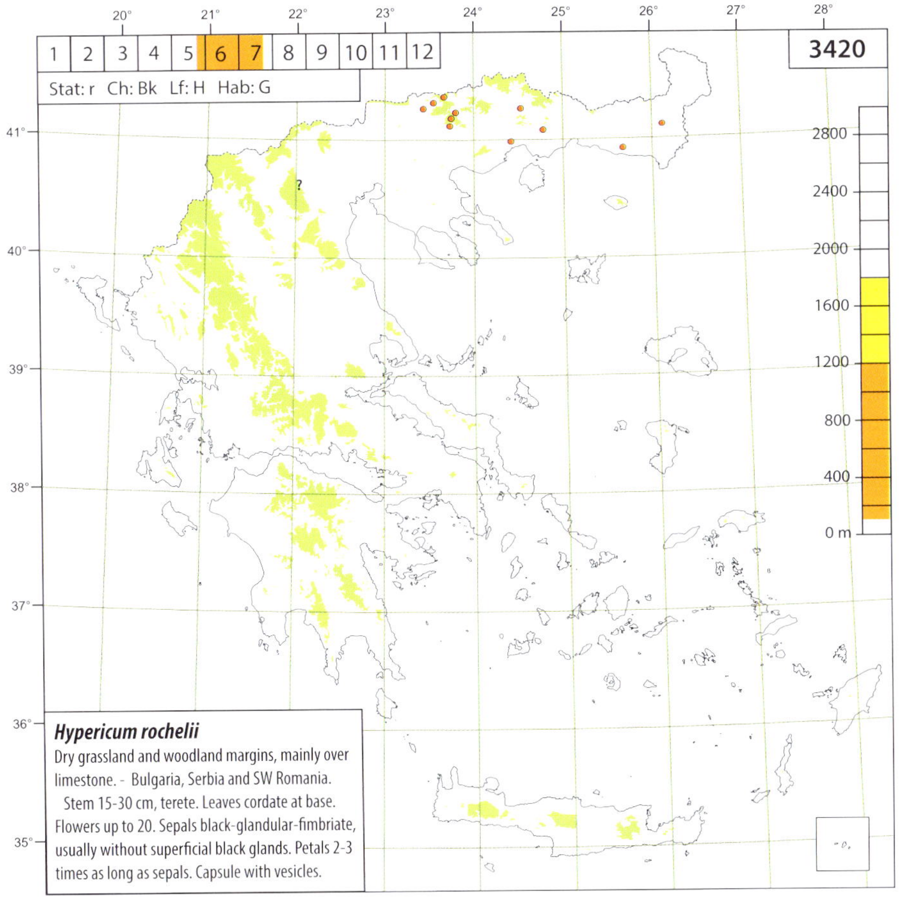

Hypericum rochelii

Dry grassland and woodland margins, mainly over limestone. - Bulgaria, Serbia and SW Romania.

Stem 15-30 cm, terete. Leaves cordate at base. Flowers up to 20. Sepals black-glandular-fimbriate, usually without superficial black glands. Petals 2-3 times as long as sepals. Capsule with vesicles.

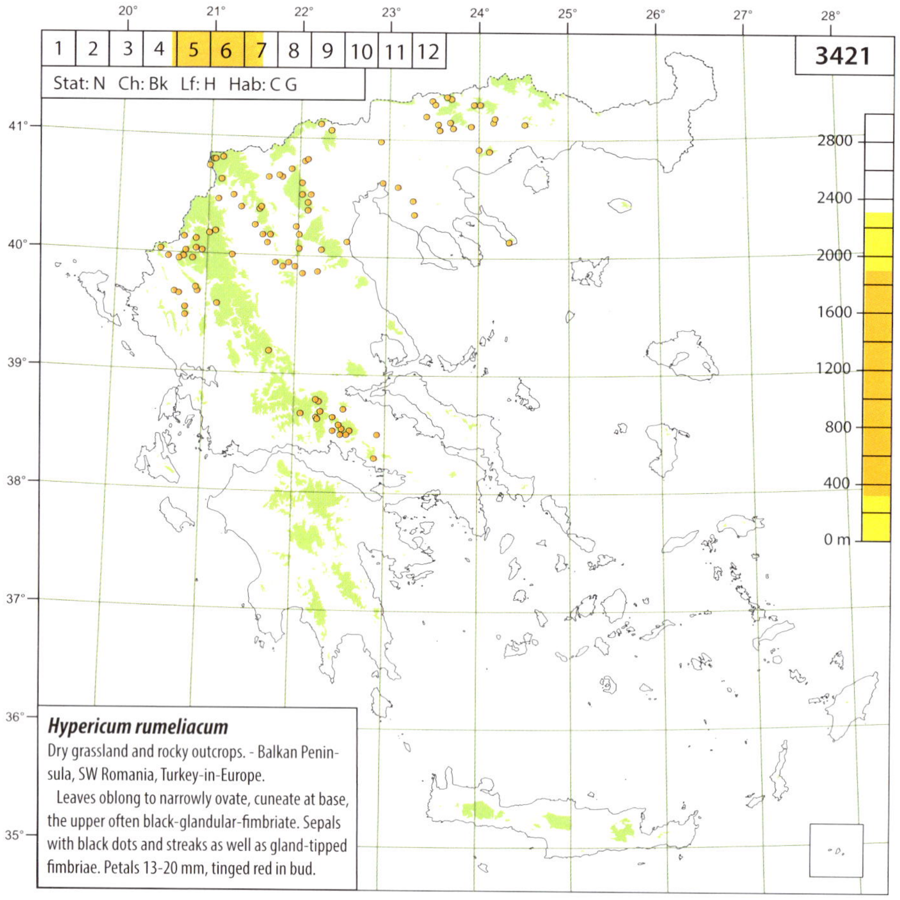

Hypericum rumeliacum

Dry grassland and rocky outcrops. - Balkan Peninsula, SW Romania, Turkey-in-Europe.

Leaves oblong to narrowly ovate, cuneate at base, the upper often black-glandular-fimbriate. Sepals with black dots and streaks as well as gland-tipped fimbriae. Petals 13-20 mm, tinged red in bud.

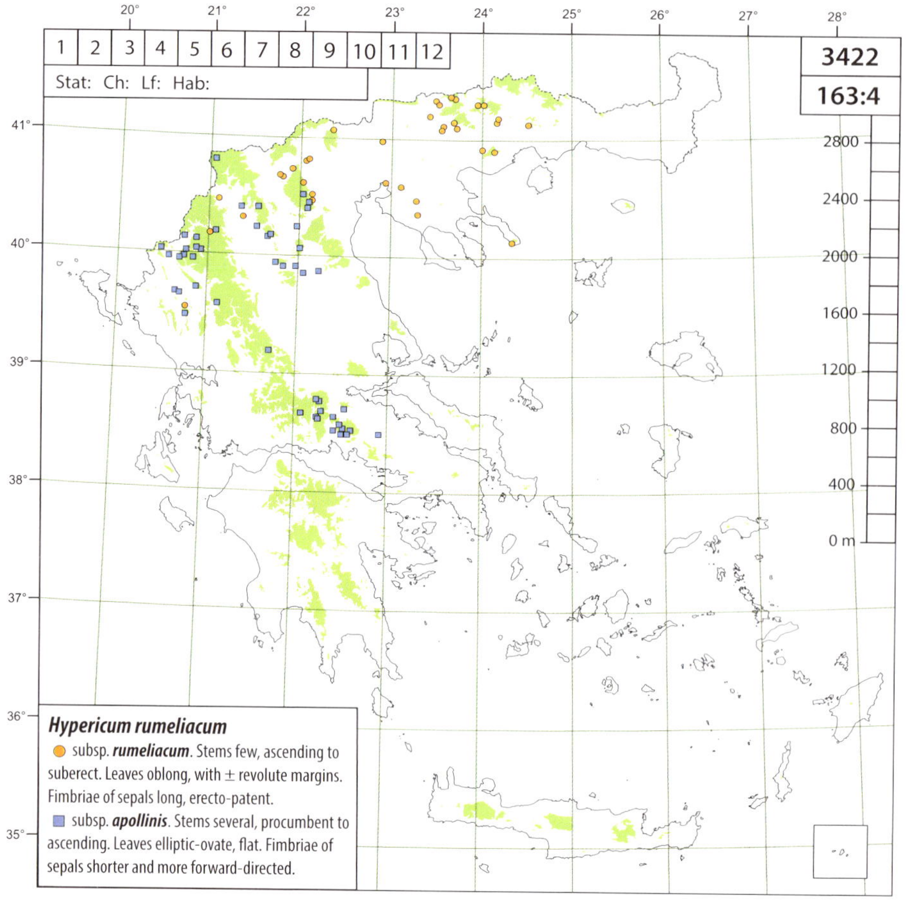

Hypericum rumeliacum

● subsp. ***rumeliacum***. Stems few, ascending to suberect. Leaves oblong, with ± revolute margins. Fimbriae of sepals long, erecto-patent.

■ subsp. ***apollinis***. Stems several, procumbent to ascending. Leaves elliptic-ovate, flat. Fimbriae of sepals shorter and more forward-directed.

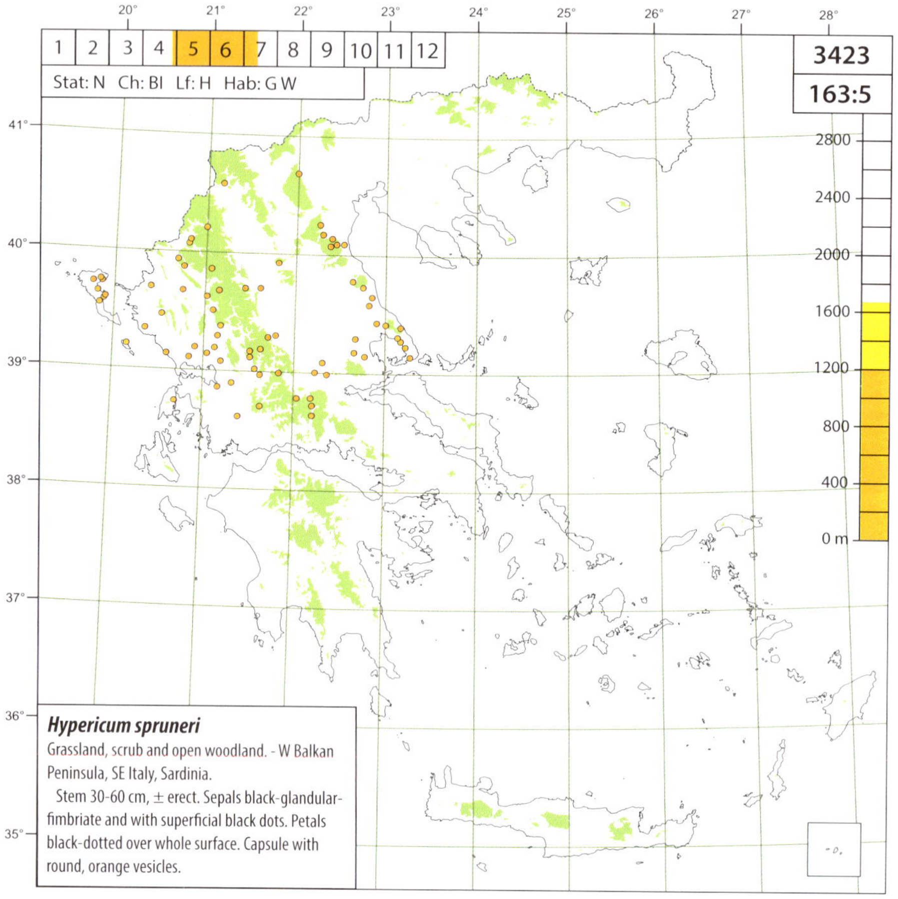

Hypericum spruneri

Grassland, scrub and open woodland. - W Balkan Peninsula, SE Italy, Sardinia.

Stem 30-60 cm, ± erect. Sepals black-glandular-fimbriate and with superficial black dots. Petals black-dotted over whole surface. Capsule with round, orange vesicles.

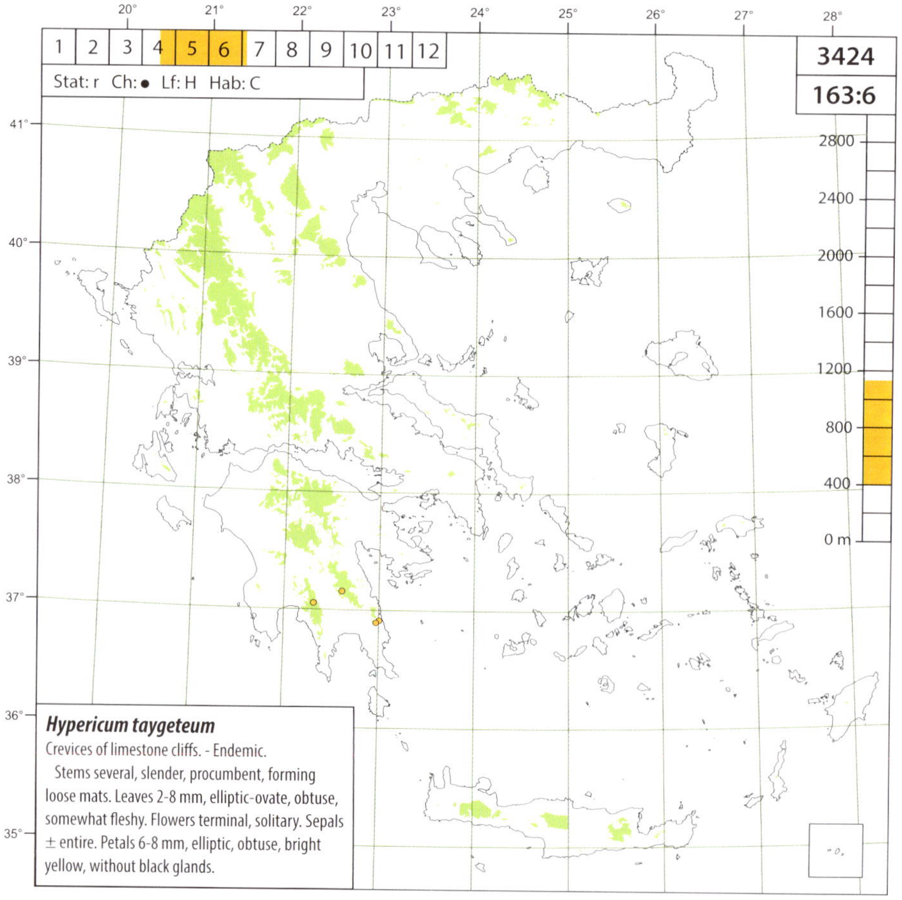

Hypericum taygeteum

Crevices of limestone cliffs. - Endemic.

Stems several, slender, procumbent, forming loose mats. Leaves 2-8 mm, elliptic-ovate, obtuse, somewhat fleshy. Flowers terminal, solitary. Sepals ± entire. Petals 6-8 mm, elliptic, obtuse, bright yellow, without black glands.

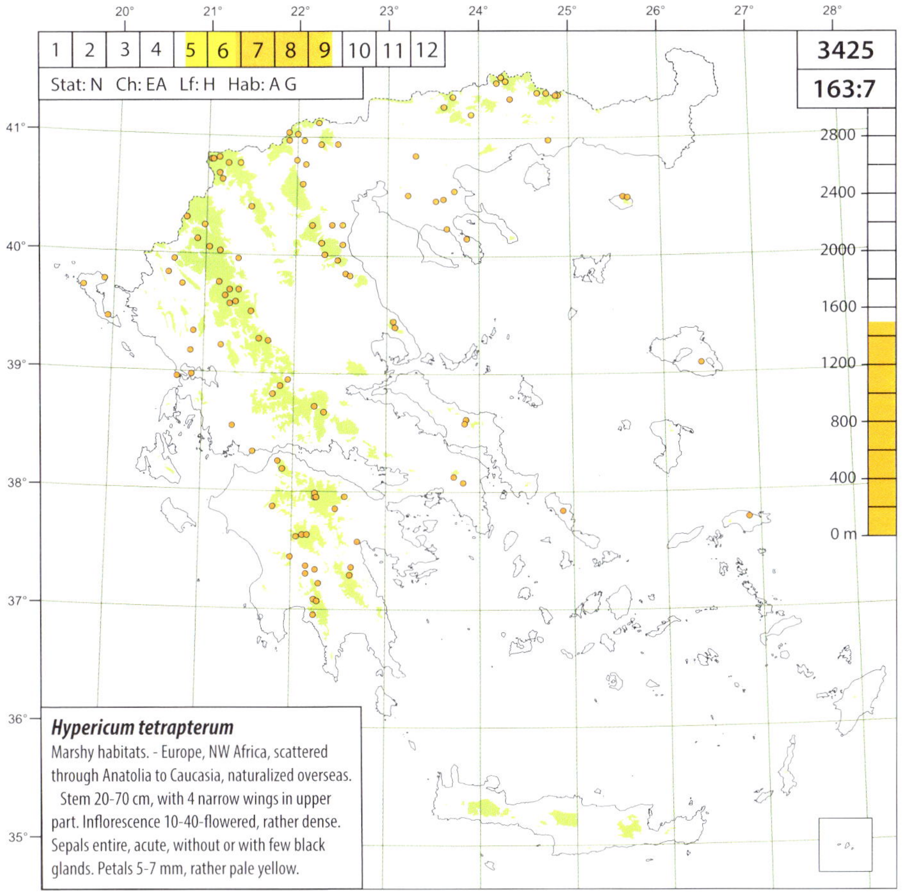

Hypericum tetrapterum

Marshy habitats. - Europe, NW Africa, scattered through Anatolia to Caucasia, naturalized overseas.

Stem 20-70 cm, with 4 narrow wings in upper part. Inflorescence 10-40-flowered, rather dense. Sepals entire, acute, without or with few black glands. Petals 5-7 mm, rather pale yellow.

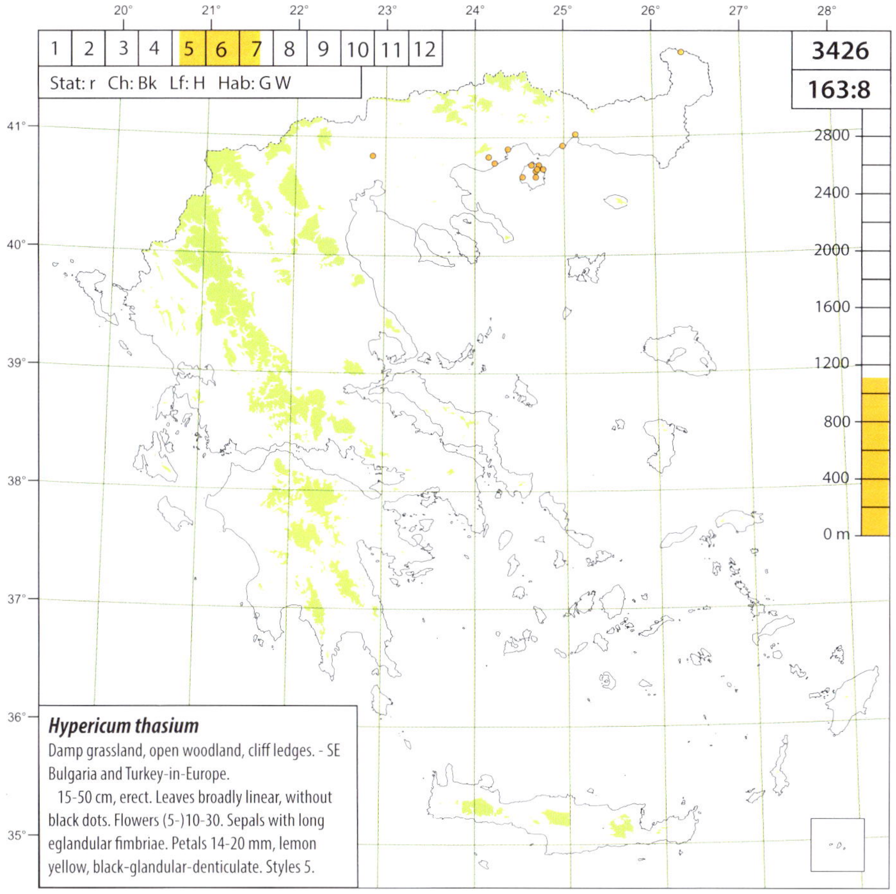

Hypericum thasium

Damp grassland, open woodland, cliff ledges. - SE Bulgaria and Turkey-in-Europe.

15-50 cm, erect. Leaves broadly linear, without black dots. Flowers (5-)10-30. Sepals with long eglandular fimbriae. Petals 14-20 mm, lemon yellow, black-glandular-denticulate. Styles 5.

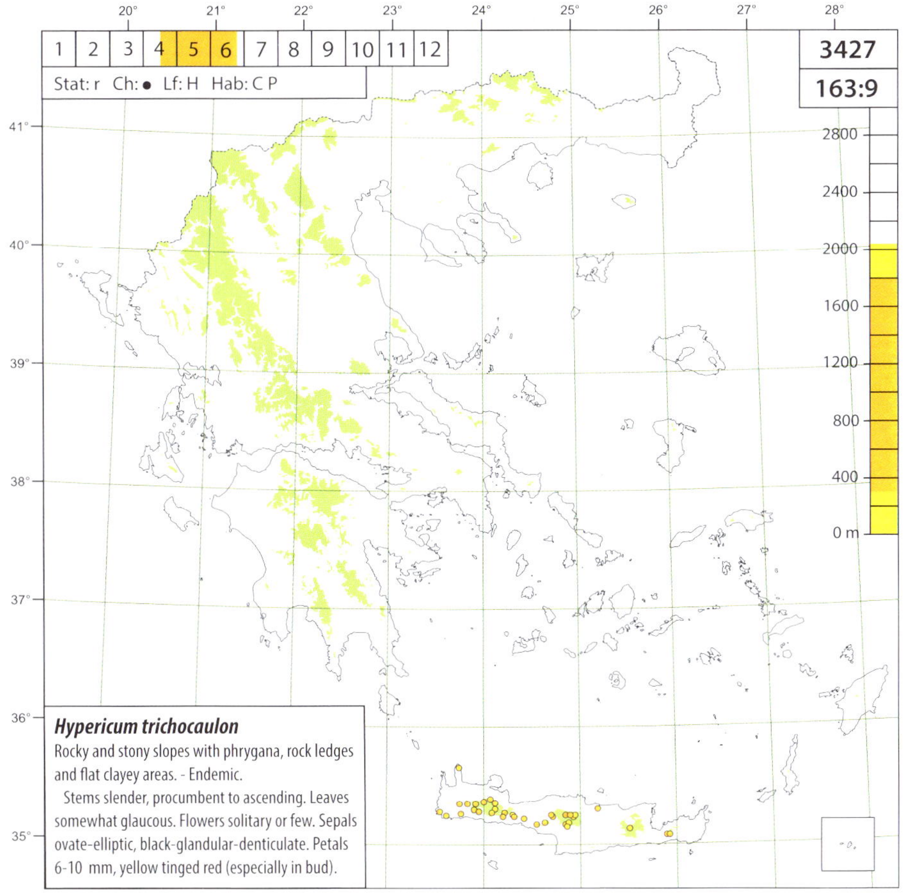

Hypericum trichocaulon

Rocky and stony slopes with phrygana, rock ledges and flat clayey areas. - Endemic.

Stems slender, procumbent to ascending. Leaves somewhat glaucous. Flowers solitary or few. Sepals ovate-elliptic, black-glandular-denticulate. Petals 6-10 mm, yellow tinged red (especially in bud).

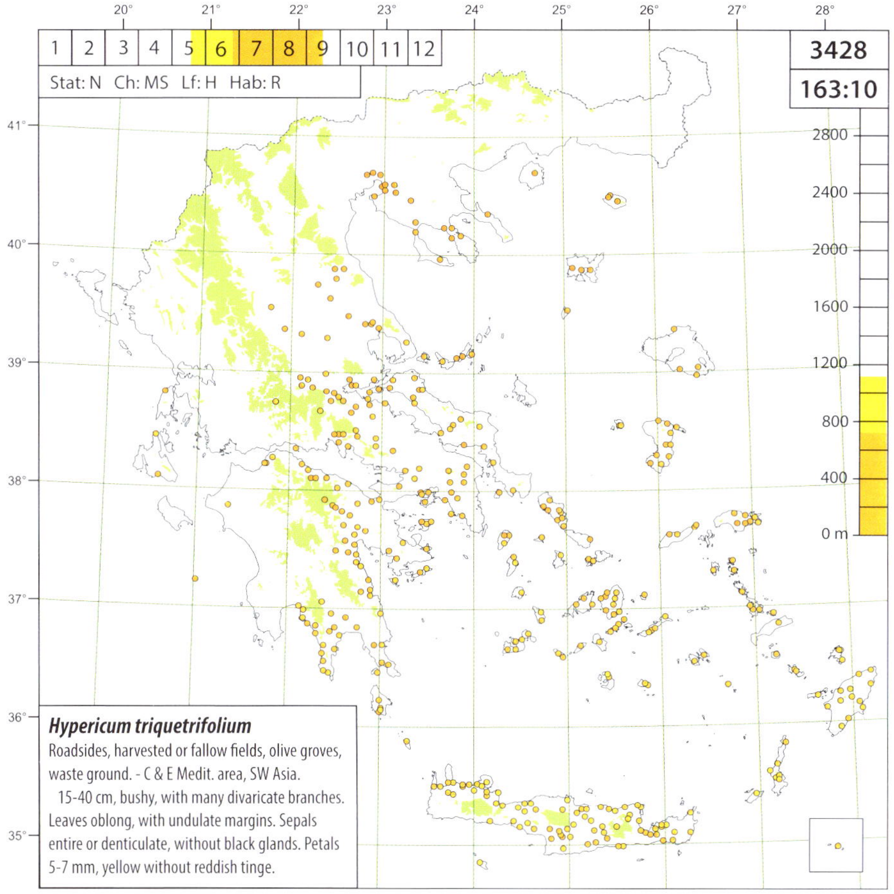

Hypericum triquetrifolium

Roadsides, harvested or fallow fields, olive groves, waste ground. - C & E Medit. area, SW Asia.

15-40 cm, bushy, with many divaricate branches. Leaves oblong, with undulate margins. Sepals entire or denticulate, without black glands. Petals 5-7 mm, yellow without reddish tinge.

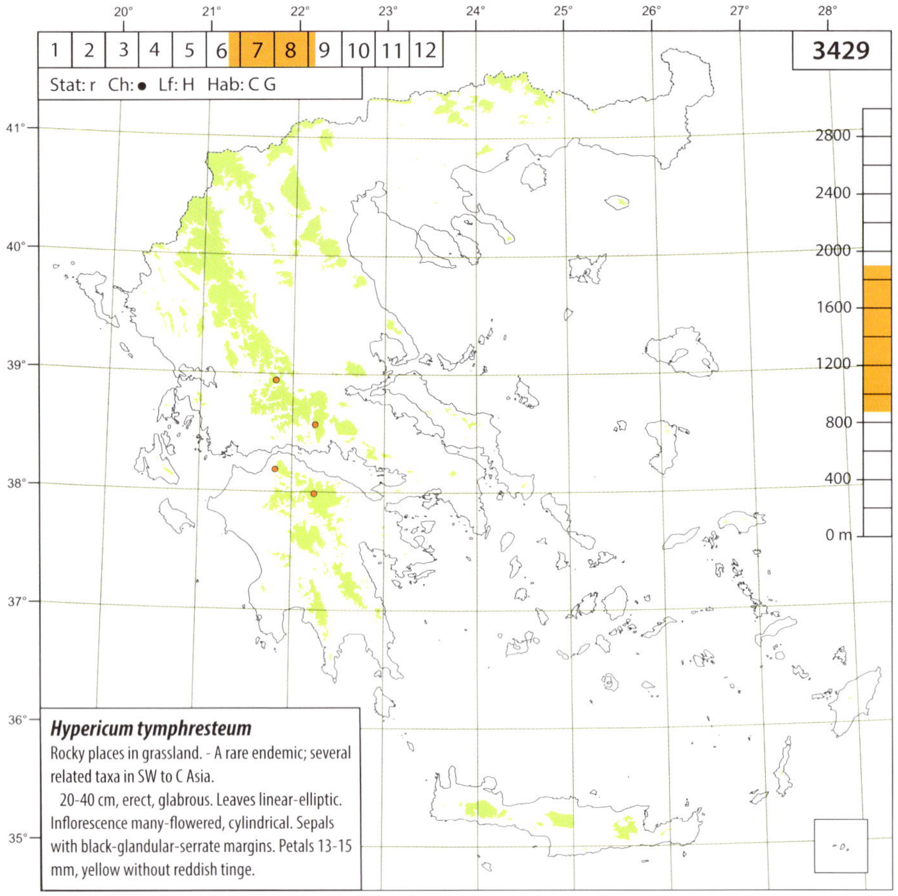

Hypericum tymphresteum

Rocky places in grassland. - A rare endemic; several related taxa in SW to C Asia.

20-40 cm, erect, glabrous. Leaves linear-elliptic. Inflorescence many-flowered, cylindrical. Sepals with black-glandular-serrate margins. Petals 13-15 mm, yellow without reddish tinge.

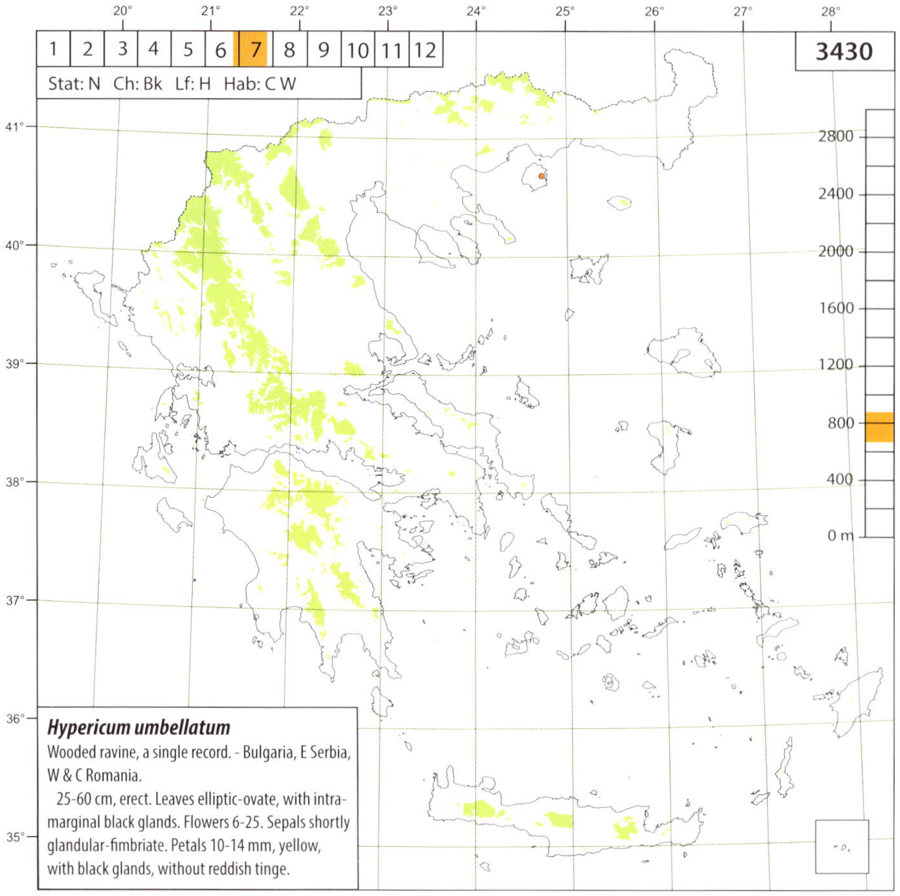

Hypericum umbellatum

Wooded ravine, a single record. - Bulgaria, E Serbia, W & C Romania.

25-60 cm, erect. Leaves elliptic-ovate, with intramarginal black glands. Flowers 6-25. Sepals shortly glandular-fimbriate. Petals 10-14 mm, yellow, with black glands, without reddish tinge.

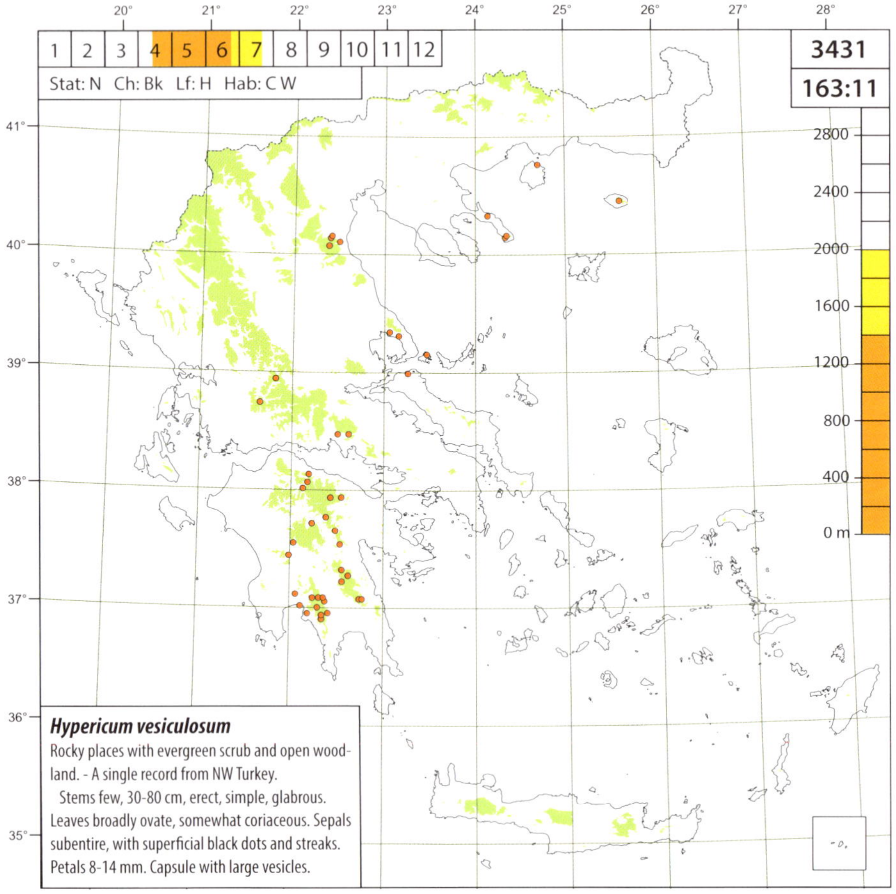

Hypericum vesiculosum

Rocky places with evergreen scrub and open woodland. - A single record from NW Turkey.

Stems few, 30-80 cm, erect, simple, glabrous. Leaves broadly ovate, somewhat coriaceous. Sepals subentire, with superficial black dots and streaks. Petals 8-14 mm. Capsule with large vesicles.

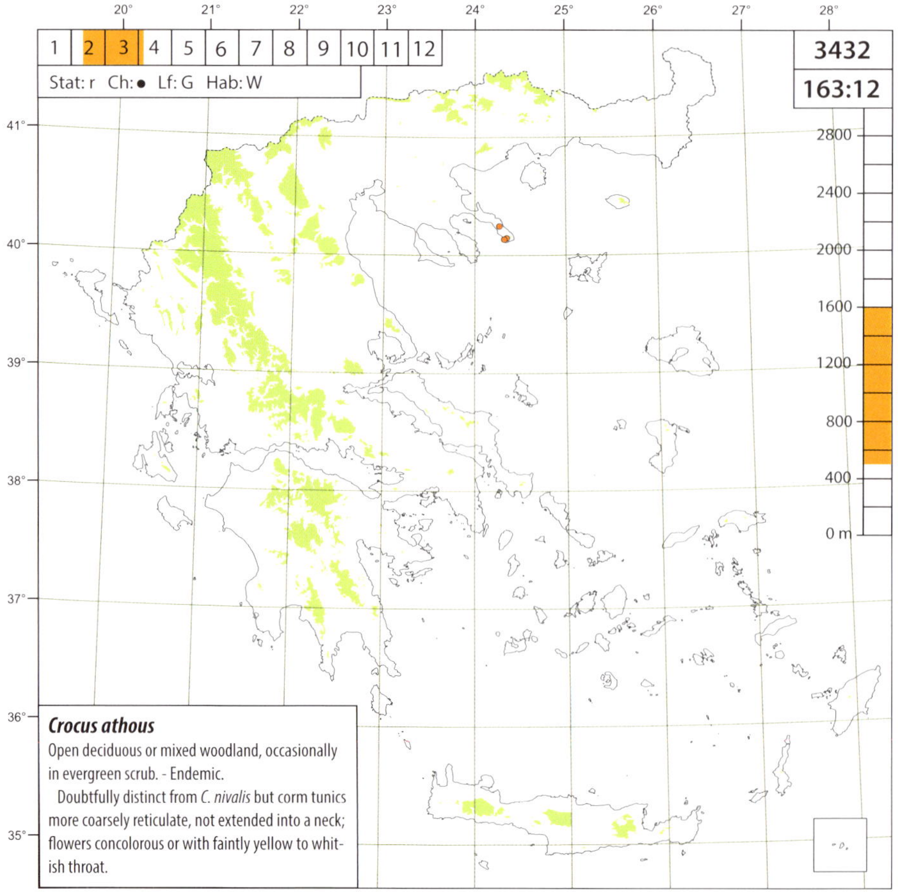

Crocus athous

Open deciduous or mixed woodland, occasionally in evergreen scrub. - Endemic.

Doubtfully distinct from *C. nivalis* but corm tunics more coarsely reticulate, not extended into a neck; flowers concolorous or with faintly yellow to whitish throat.

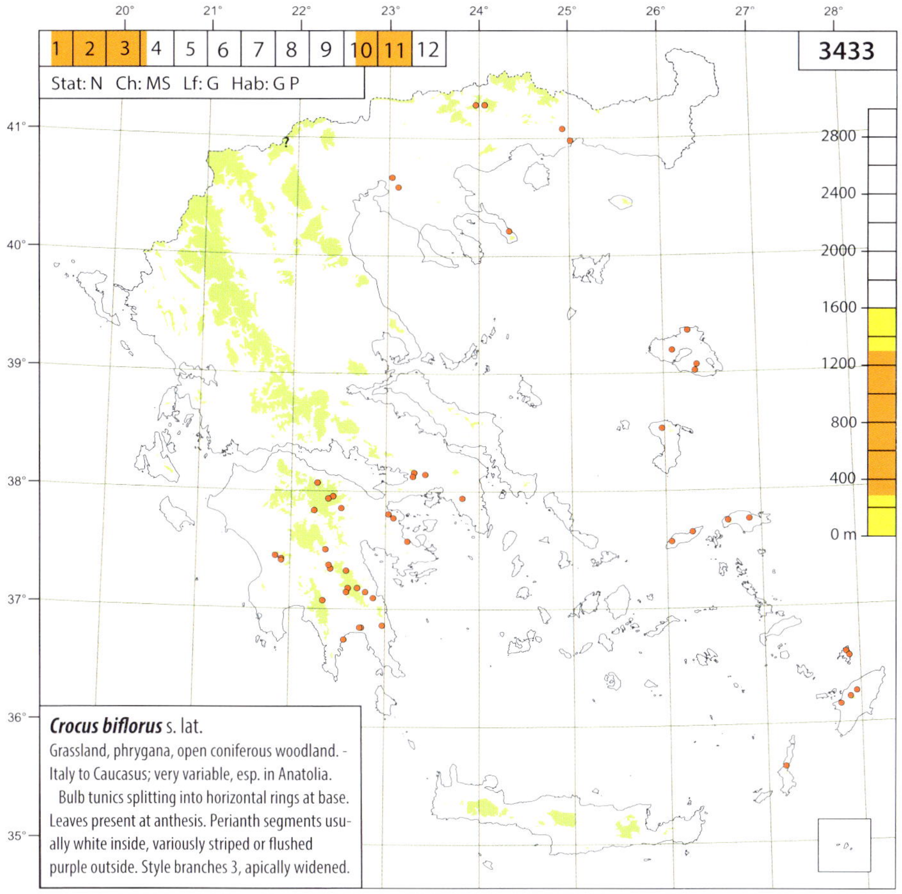

Crocus biflorus s. lat.
Grassland, phrygana, open coniferous woodland. - Italy to Caucasus; very variable, esp. in Anatolia.
Bulb tunics splitting into horizontal rings at base. Leaves present at anthesis. Perianth segments usually white inside, variously striped or flushed purple outside. Style branches 3, apically widened.

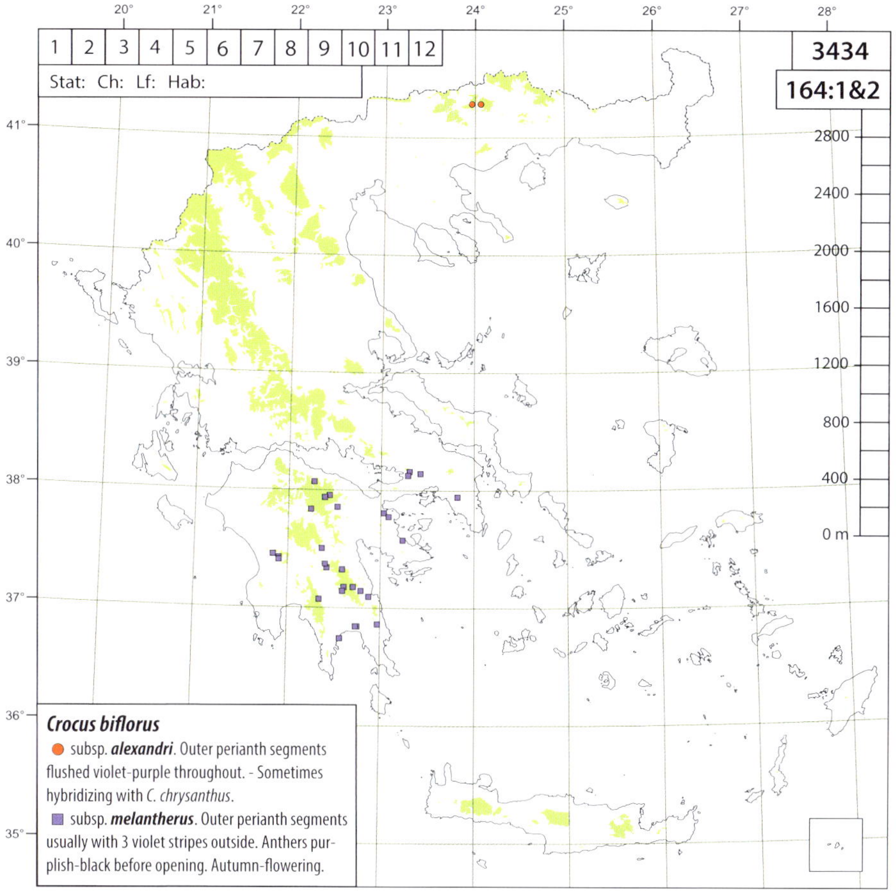

Crocus biflorus
● subsp. ***alexandri***. Outer perianth segments flushed violet-purple throughout. - Sometimes hybridizing with *C. chrysanthus*.
■ subsp. ***melantherus***. Outer perianth segments usually with 3 violet stripes outside. Anthers purplish-black before opening. Autumn-flowering.

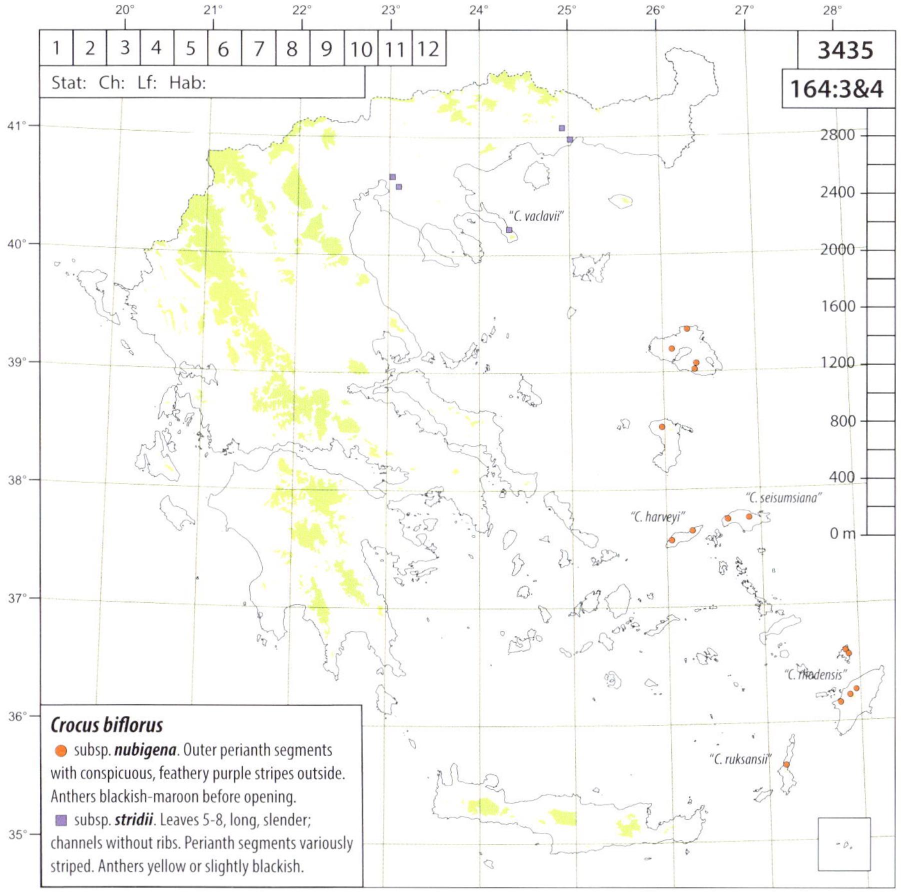

Crocus biflorus
● subsp. ***nubigena***. Outer perianth segments with conspicuous, feathery purple stripes outside. Anthers blackish-maroon before opening.
■ subsp. ***stridii***. Leaves 5-8, long, slender; channels without ribs. Perianth segments variously striped. Anthers yellow or slightly blackish.

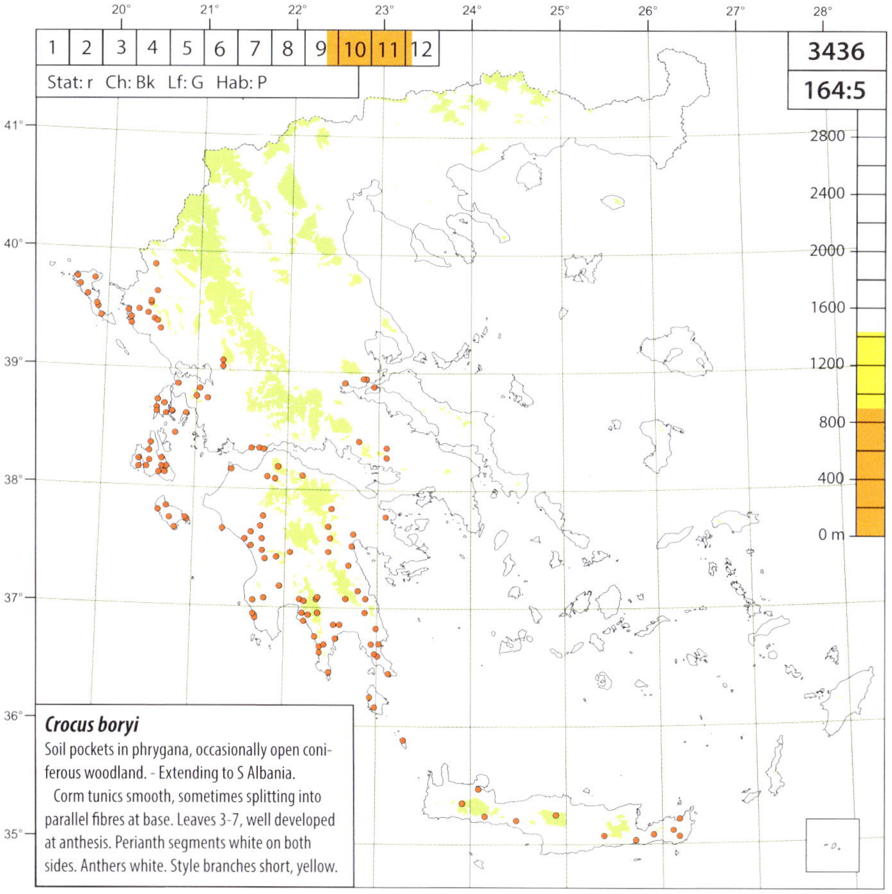

Crocus boryi
Soil pockets in phrygana, occasionally open coniferous woodland. - Extending to S Albania.
Corm tunics smooth, sometimes splitting into parallel fibres at base. Leaves 3-7, well developed at anthesis. Perianth segments white on both sides. Anthers white. Style branches short, yellow.

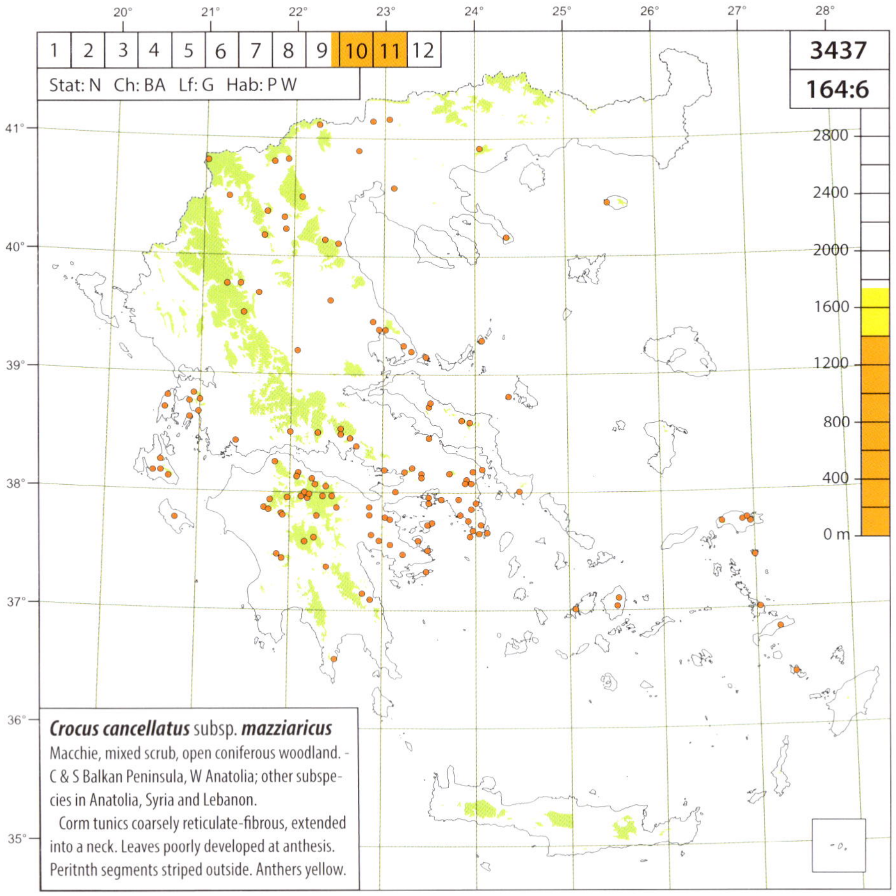

Crocus cancellatus subsp. ***mazziaricus***
Macchie, mixed scrub, open coniferous woodland. - C & S Balkan Peninsula, W Anatolia; other subspecies in Anatolia, Syria and Lebanon.
Corm tunics coarsely reticulate-fibrous, extended into a neck. Leaves poorly developed at anthesis. Peritnth segments striped outside. Anthers yellow.

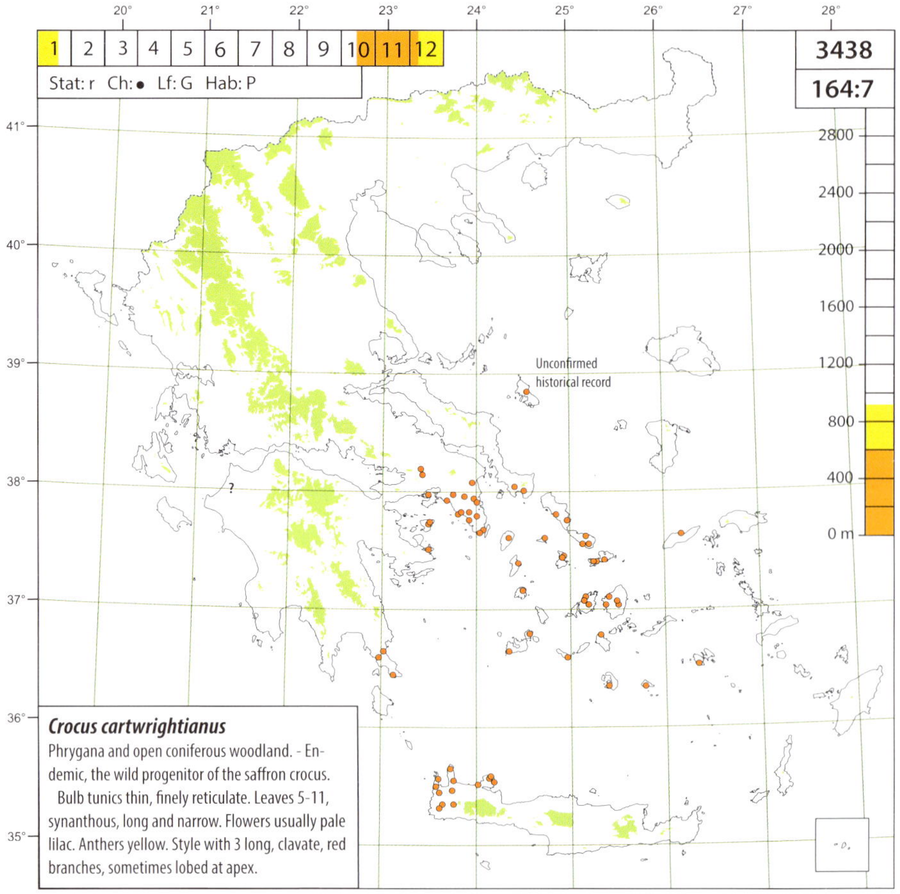

Crocus cartwrightianus
Phrygana and open coniferous woodland. - Endemic, the wild progenitor of the saffron crocus.
Bulb tunics thin, finely reticulate. Leaves 5-11, synanthous, long and narrow. Flowers usually pale lilac. Anthers yellow. Style with 3 long, clavate, red branches, sometimes lobed at apex.

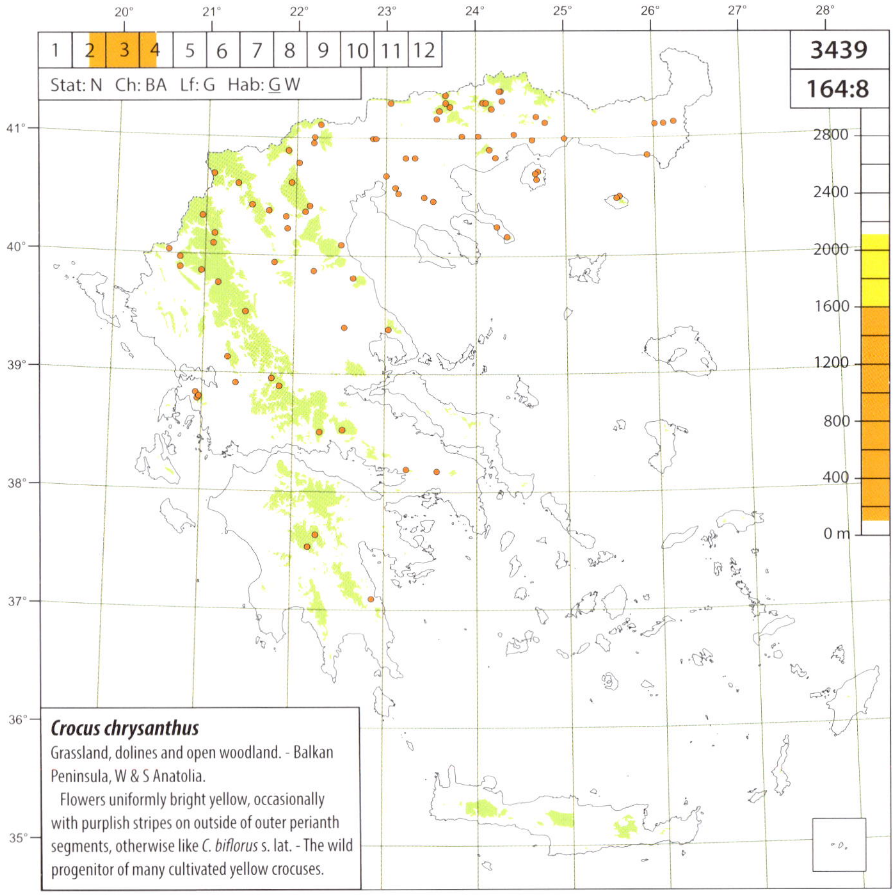

Crocus chrysanthus
Grassland, dolines and open woodland. - Balkan Peninsula, W & S Anatolia.
Flowers uniformly bright yellow, occasionally with purplish stripes on outside of outer perianth segments, otherwise like *C. biflorus* s. lat. - The wild progenitor of many cultivated yellow crocuses.

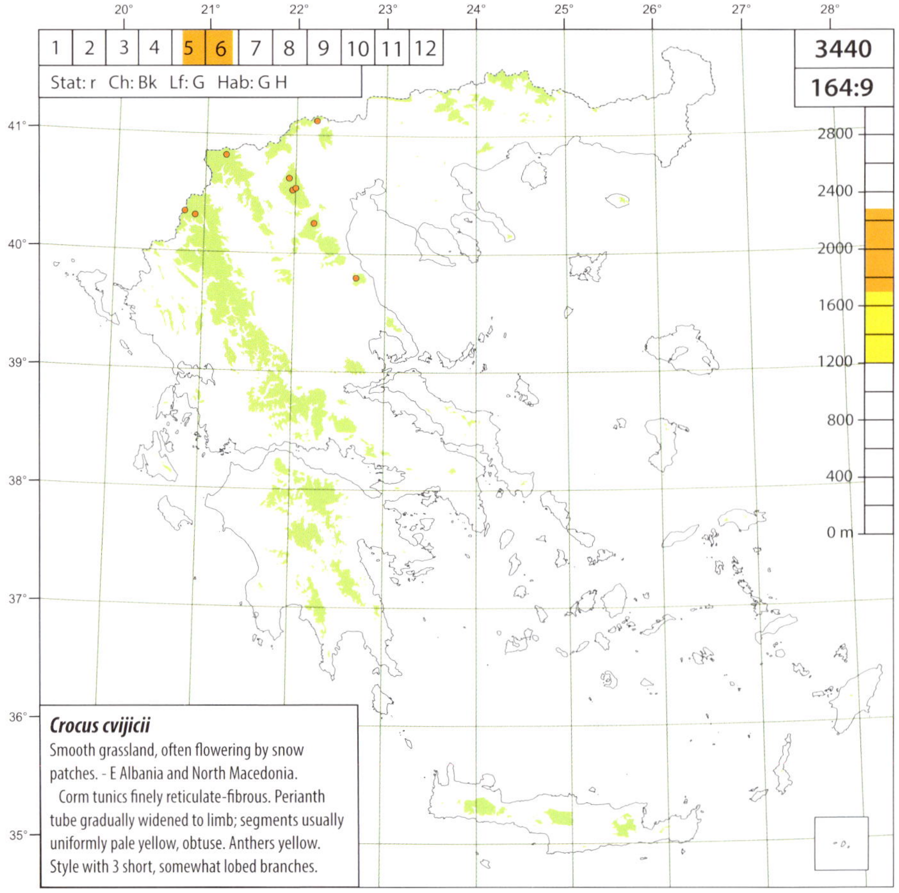

Crocus cvijicii
Smooth grassland, often flowering by snow patches. - E Albania and North Macedonia.
Corm tunics finely reticulate-fibrous. Perianth tube gradually widened to limb; segments usually uniformly pale yellow, obtuse. Anthers yellow. Style with 3 short, somewhat lobed branches.

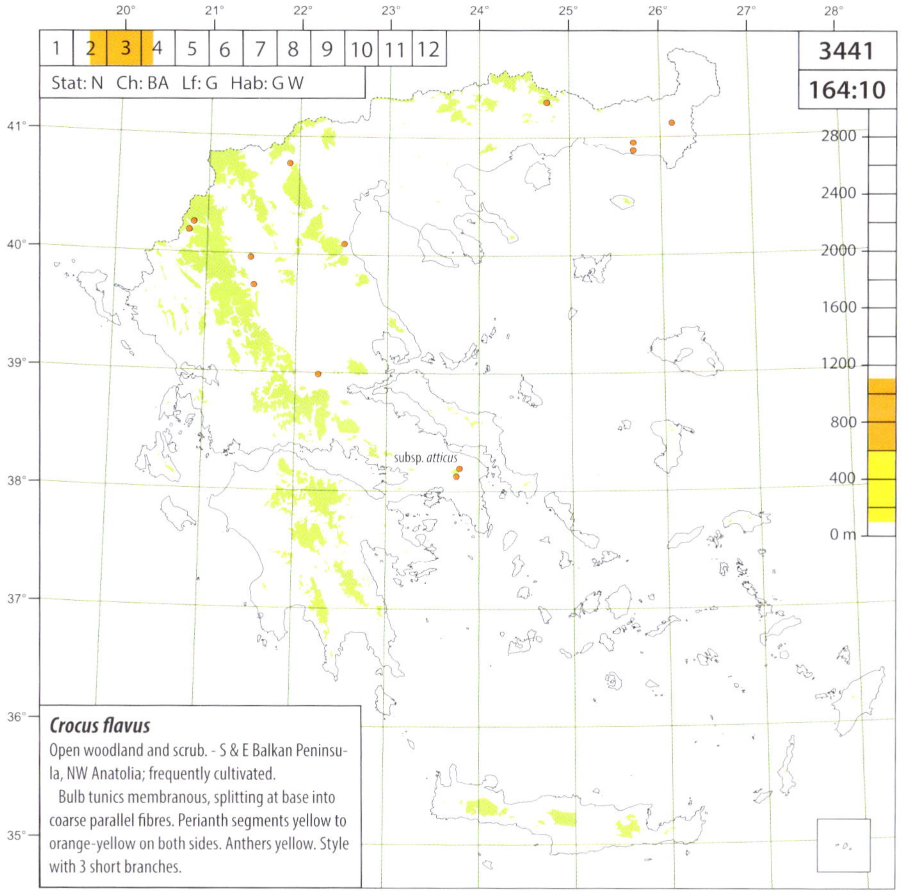

Crocus flavus
Open woodland and scrub. - S & E Balkan Peninsula, NW Anatolia; frequently cultivated.
Bulb tunics membranous, splitting at base into coarse parallel fibres. Perianth segments yellow to orange-yellow on both sides. Anthers yellow. Style with 3 short branches.

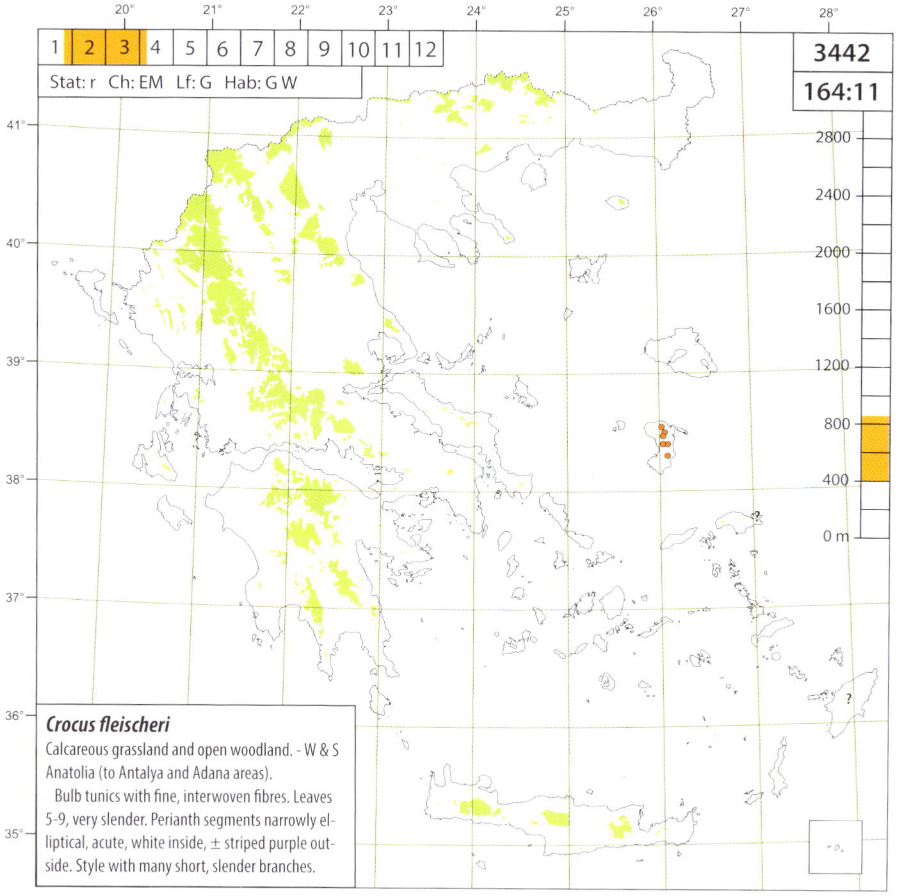

Crocus fleischeri
Calcareous grassland and open woodland. - W & S Anatolia (to Antalya and Adana areas).
Bulb tunics with fine, interwoven fibres. Leaves 5-9, very slender. Perianth segments narrowly elliptical, acute, white inside, ± striped purple outside. Style with many short, slender branches.

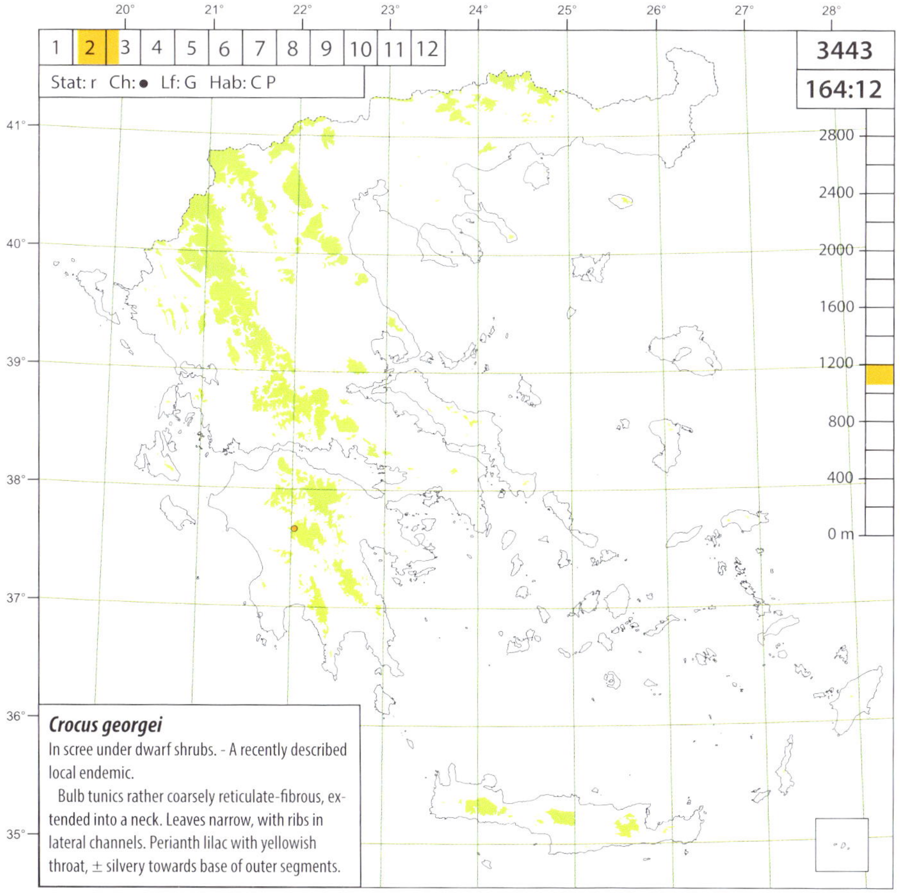

Crocus georgei
In scree under dwarf shrubs. - A recently described local endemic.
Bulb tunics rather coarsely reticulate-fibrous, extended into a neck. Leaves narrow, with ribs in lateral channels. Perianth lilac with yellowish throat, ± silvery towards base of outer segments.

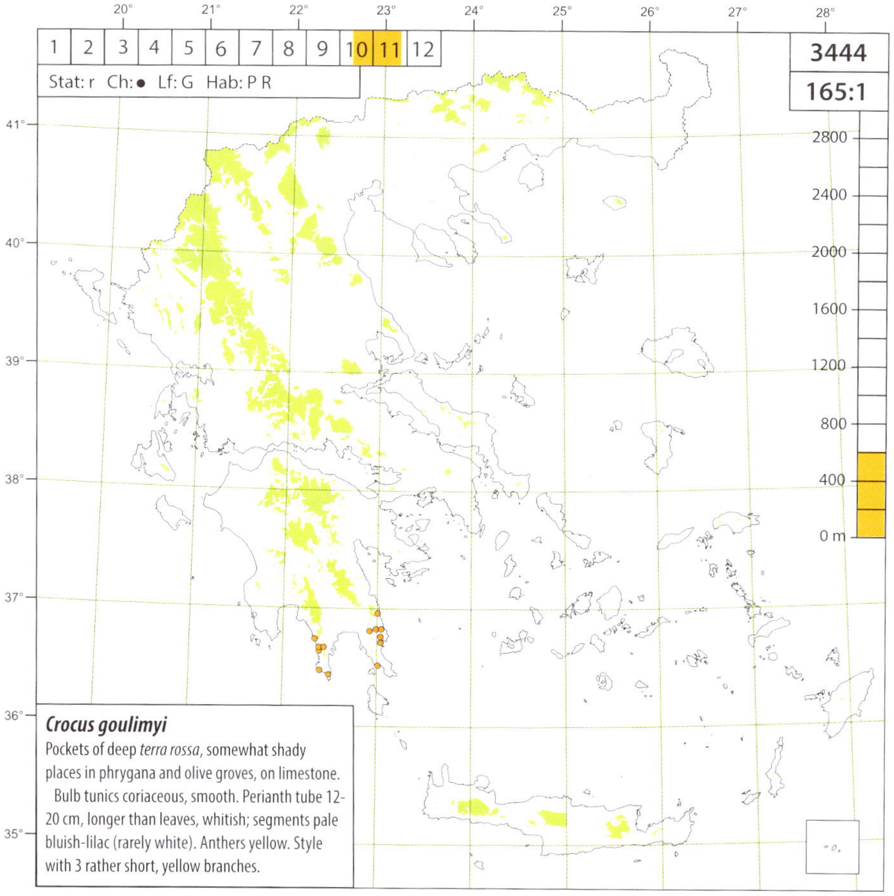

Crocus goulimyi
Pockets of deep *terra rossa*, somewhat shady places in phrygana and olive groves, on limestone.
Bulb tunics coriaceous, smooth. Perianth tube 12-20 cm, longer than leaves, whitish; segments pale bluish-lilac (rarely white). Anthers yellow. Style with 3 rather short, yellow branches.

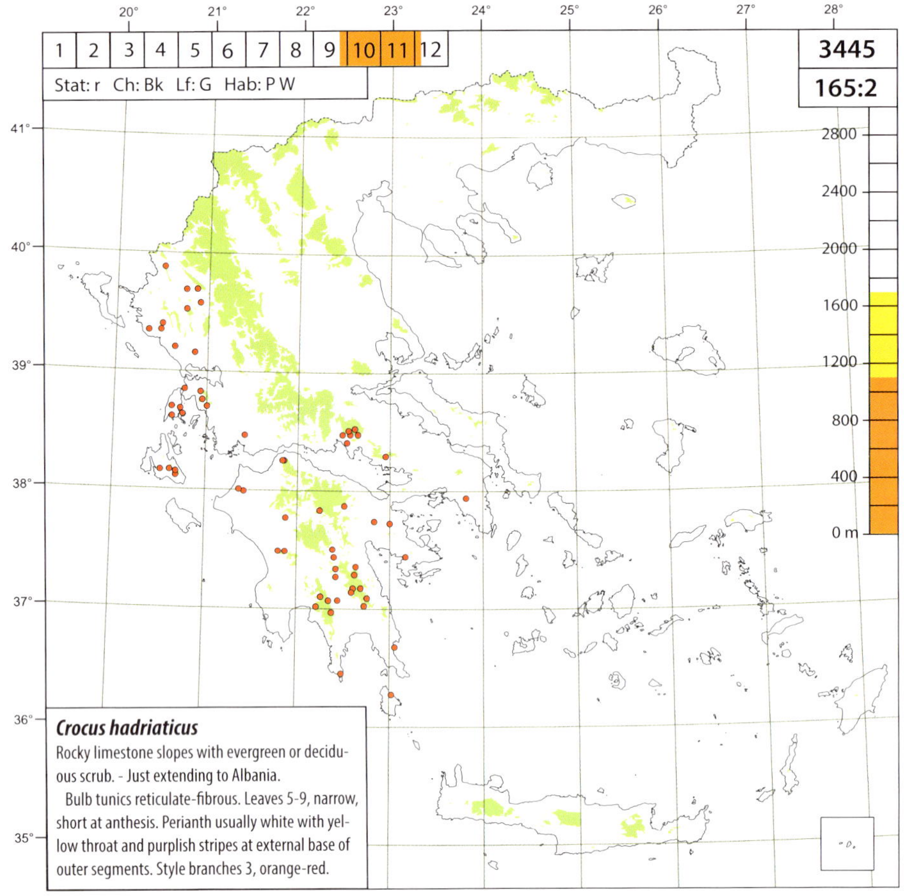

Crocus hadriaticus

Rocky limestone slopes with evergreen or deciduous scrub. - Just extending to Albania.

Bulb tunics reticulate-fibrous. Leaves 5-9, narrow, short at anthesis. Perianth usually white with yellow throat and purplish stripes at external base of outer segments. Style branches 3, orange-red.

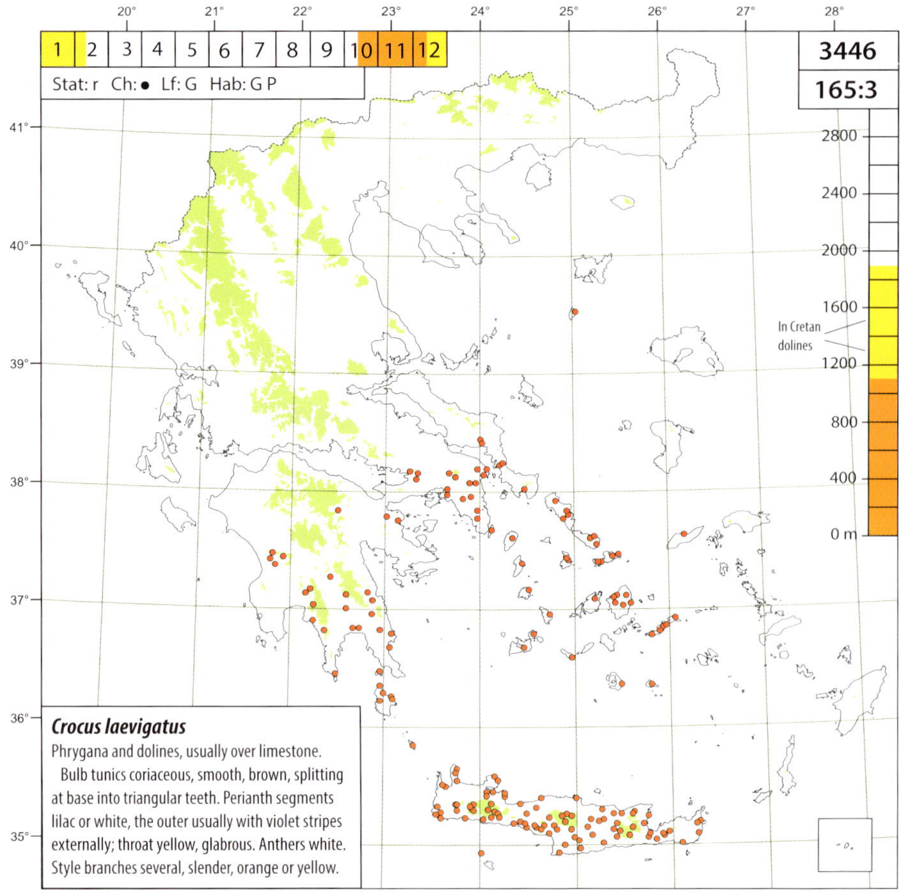

Crocus laevigatus

Phrygana and dolines, usually over limestone.

Bulb tunics coriaceous, smooth, brown, splitting at base into triangular teeth. Perianth segments lilac or white, the outer usually with violet stripes externally; throat yellow, glabrous. Anthers white. Style branches several, slender, orange or yellow.

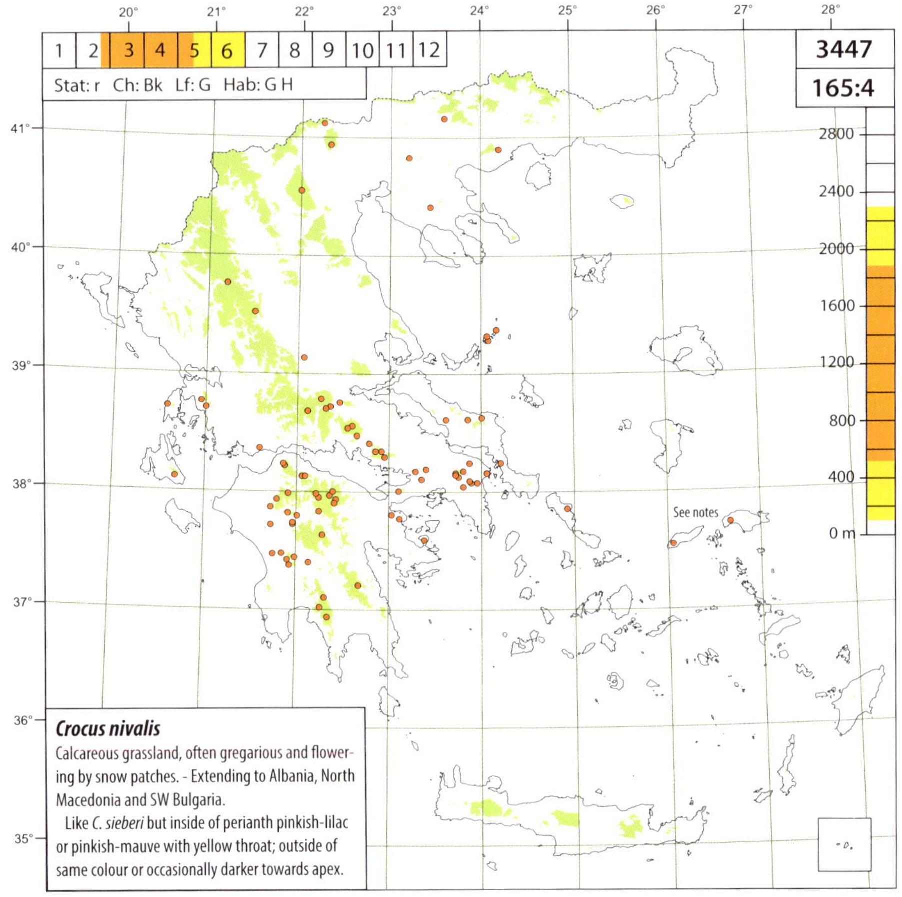

Crocus nivalis

Calcareous grassland, often gregarious and flowering by snow patches. - Extending to Albania, North Macedonia and SW Bulgaria.

Like *C. sieberi* but inside of perianth pinkish-lilac or pinkish-mauve with yellow throat; outside of same colour or occasionally darker towards apex.

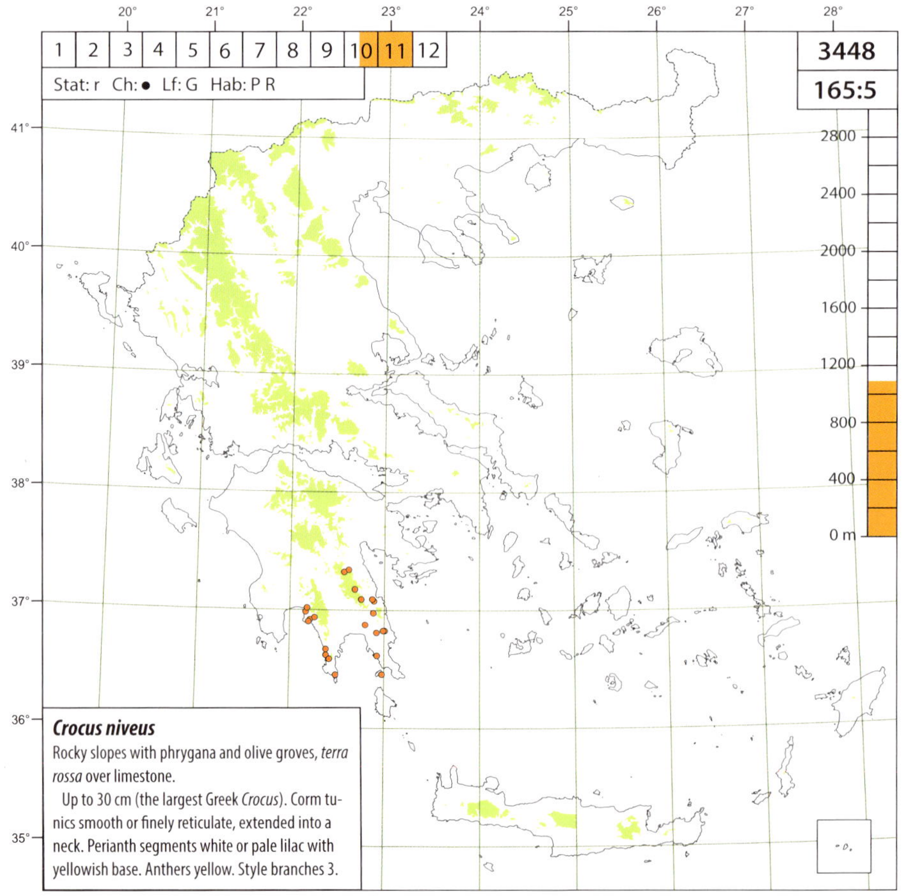

Crocus niveus

Rocky slopes with phrygana and olive groves, *terra rossa* over limestone.

Up to 30 cm (the largest Greek *Crocus*). Corm tunics smooth or finely reticulate, extended into a neck. Perianth segments white or pale lilac with yellowish base. Anthers yellow. Style branches 3.

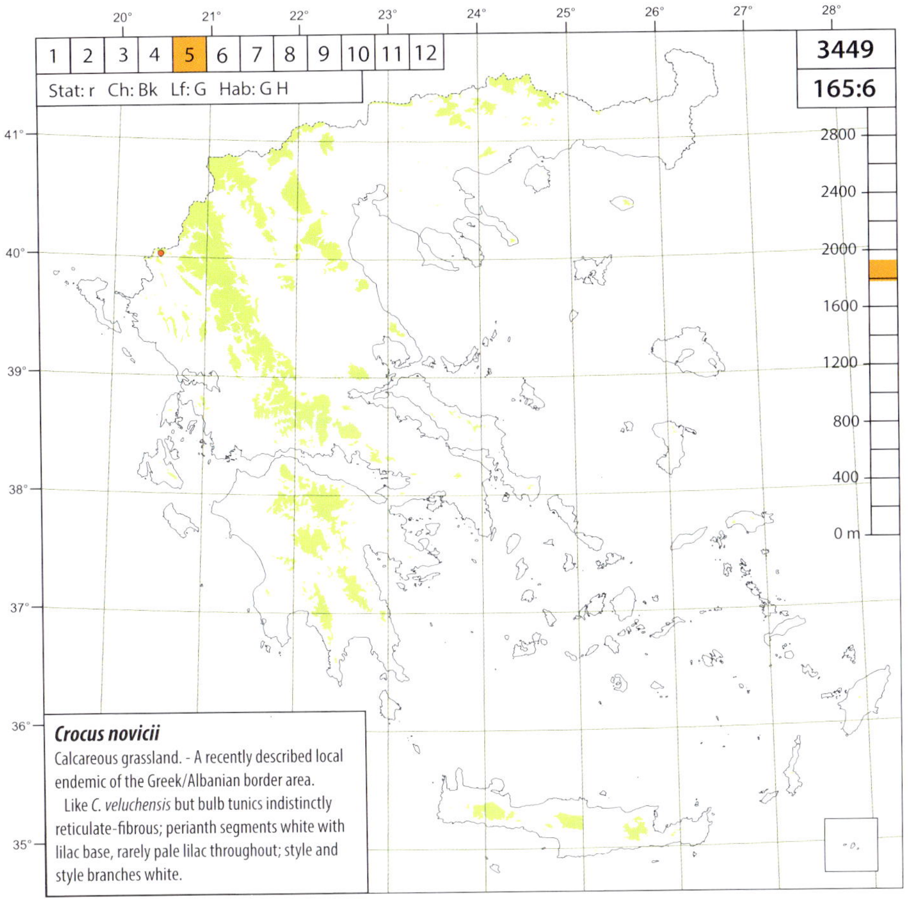

Crocus novicii
Calcareous grassland. - A recently described local endemic of the Greek/Albanian border area.
Like *C. veluchensis* but bulb tunics indistinctly reticulate-fibrous; perianth segments white with lilac base, rarely pale lilac throughout; style and style branches white.

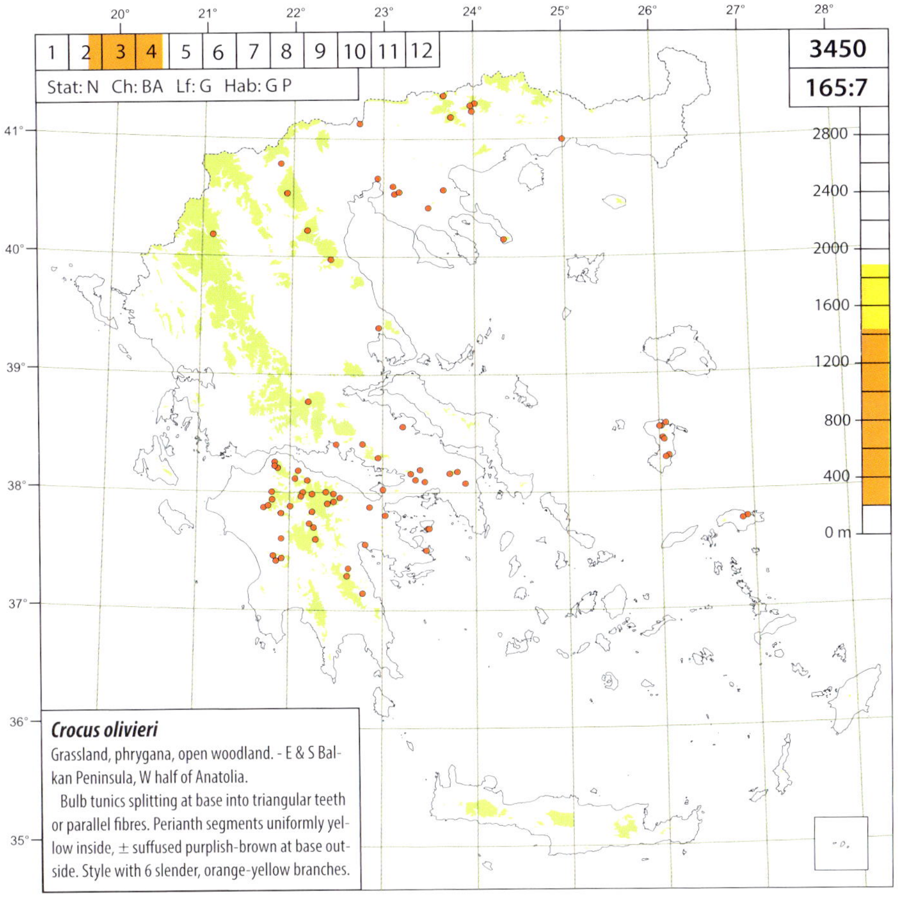

Crocus olivieri
Grassland, phrygana, open woodland. - E & S Balkan Peninsula, W half of Anatolia.
Bulb tunics splitting at base into triangular teeth or parallel fibres. Perianth segments uniformly yellow inside, ± suffused purplish-brown at base outside. Style with 6 slender, orange-yellow branches.

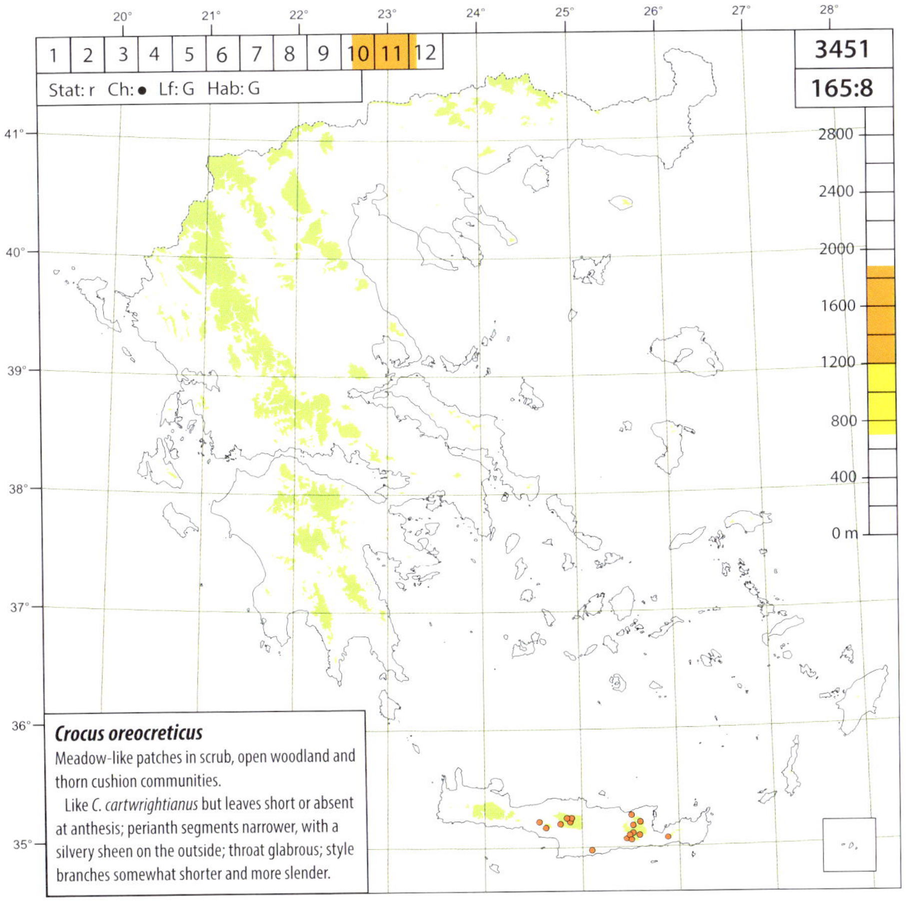

Crocus oreocreticus
Meadow-like patches in scrub, open woodland and thorn cushion communities.
Like *C. cartwrightianus* but leaves short or absent at anthesis; perianth segments narrower, with a silvery sheen on the outside; throat glabrous; style branches somewhat shorter and more slender.

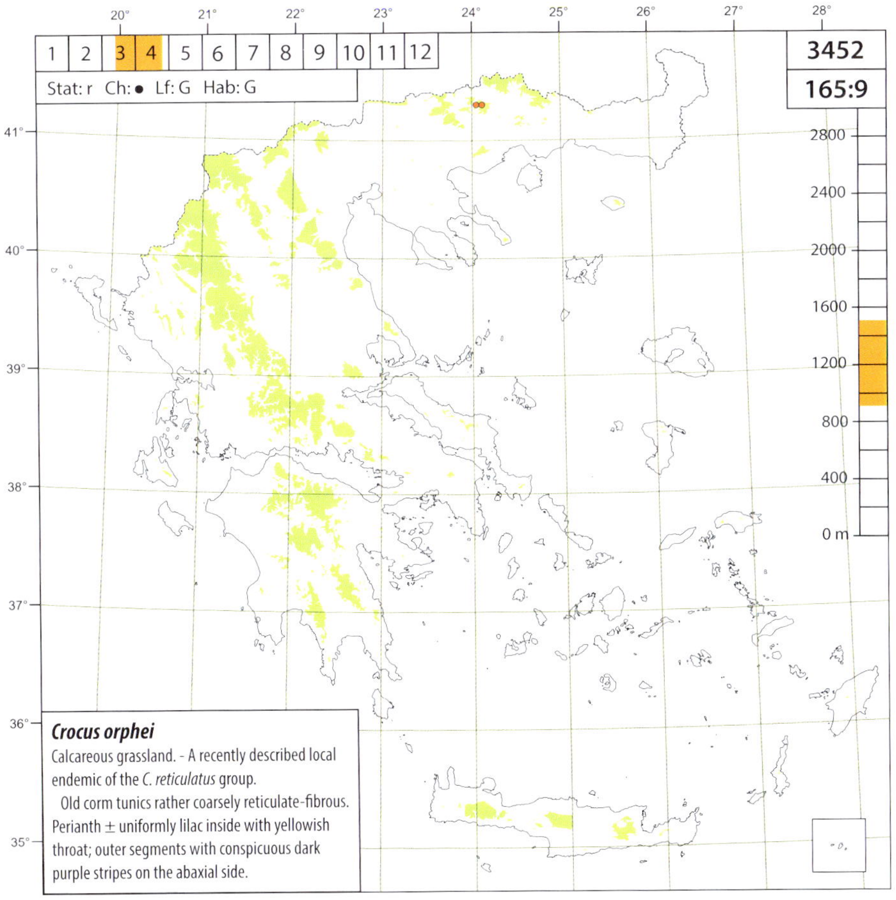

Crocus orphei
Calcareous grassland. - A recently described local endemic of the *C. reticulatus* group.
Old corm tunics rather coarsely reticulate-fibrous. Perianth ± uniformly lilac inside with yellowish throat; outer segments with conspicuous dark purple stripes on the abaxial side.

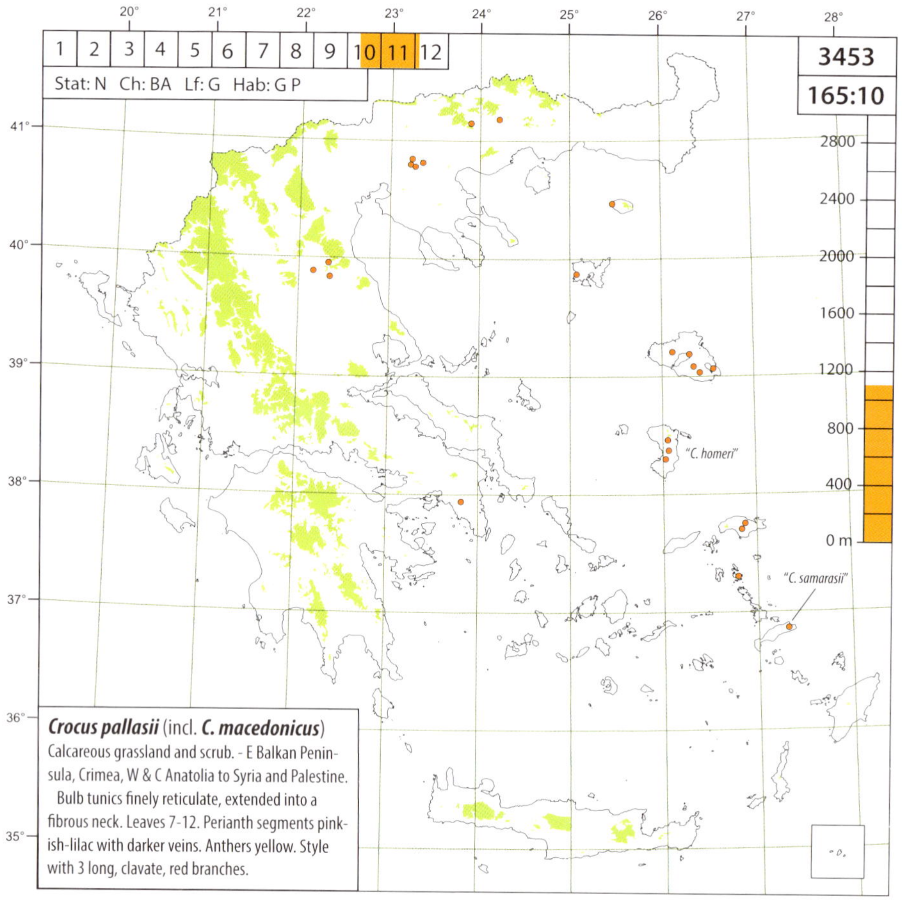

Crocus pallasii (incl. **C. macedonicus**)
Calcareous grassland and scrub. - E Balkan Peninsula, Crimea, W & C Anatolia to Syria and Palestine.
Bulb tunics finely reticulate, extended into a fibrous neck. Leaves 7-12. Perianth segments pinkish-lilac with darker veins. Anthers yellow. Style with 3 long, clavate, red branches.

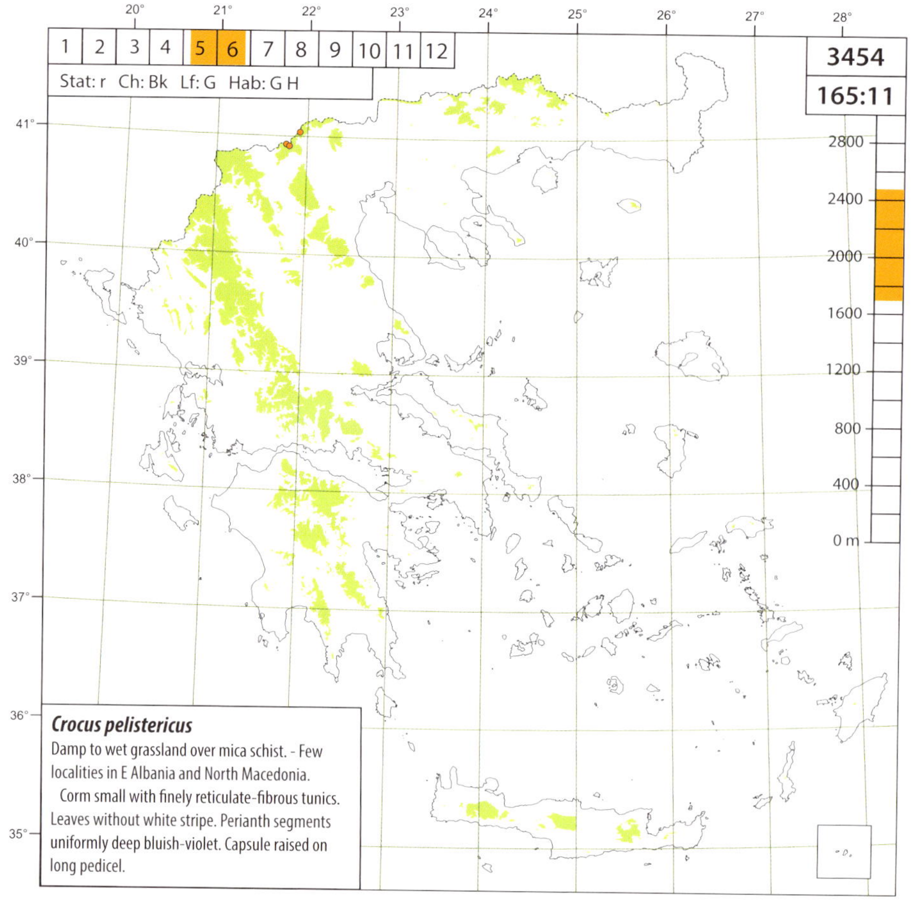

Crocus pelistericus
Damp to wet grassland over mica schist. - Few localities in E Albania and North Macedonia.
Corm small with finely reticulate-fibrous tunics. Leaves without white stripe. Perianth segments uniformly deep bluish-violet. Capsule raised on long pedicel.

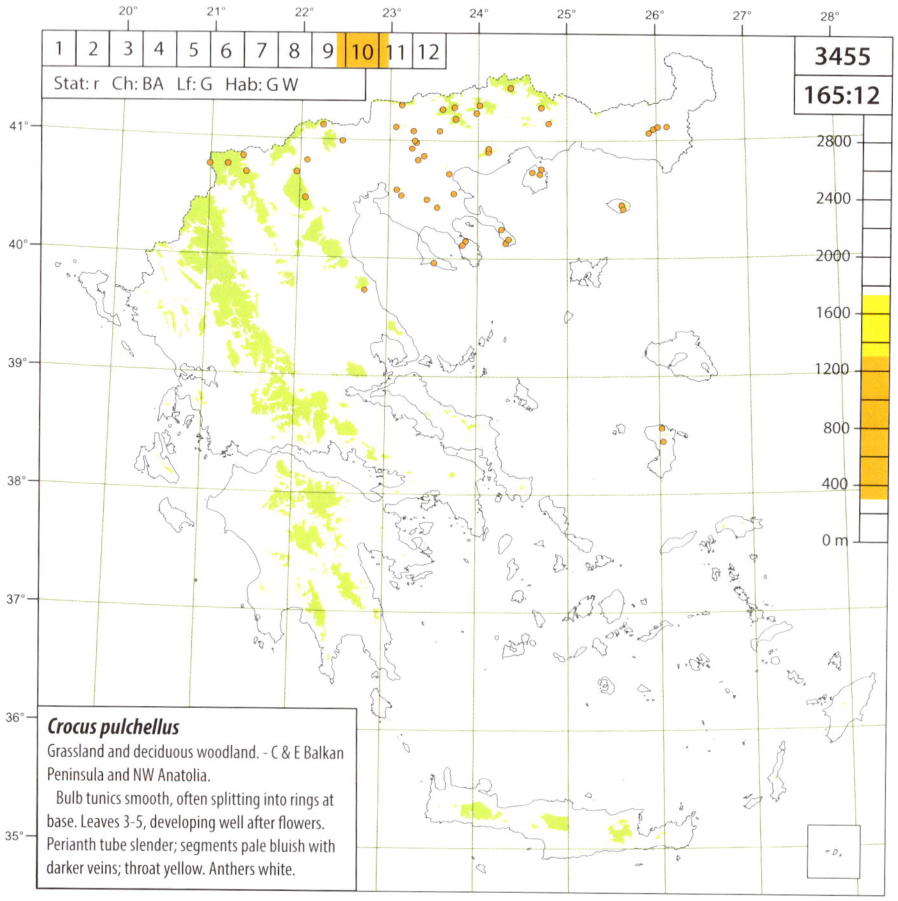

Crocus pulchellus
Grassland and deciduous woodland. - C & E Balkan Peninsula and NW Anatolia.
Bulb tunics smooth, often splitting into rings at base. Leaves 3-5, developing well after flowers. Perianth tube slender; segments pale bluish with darker veins; throat yellow. Anthers white.

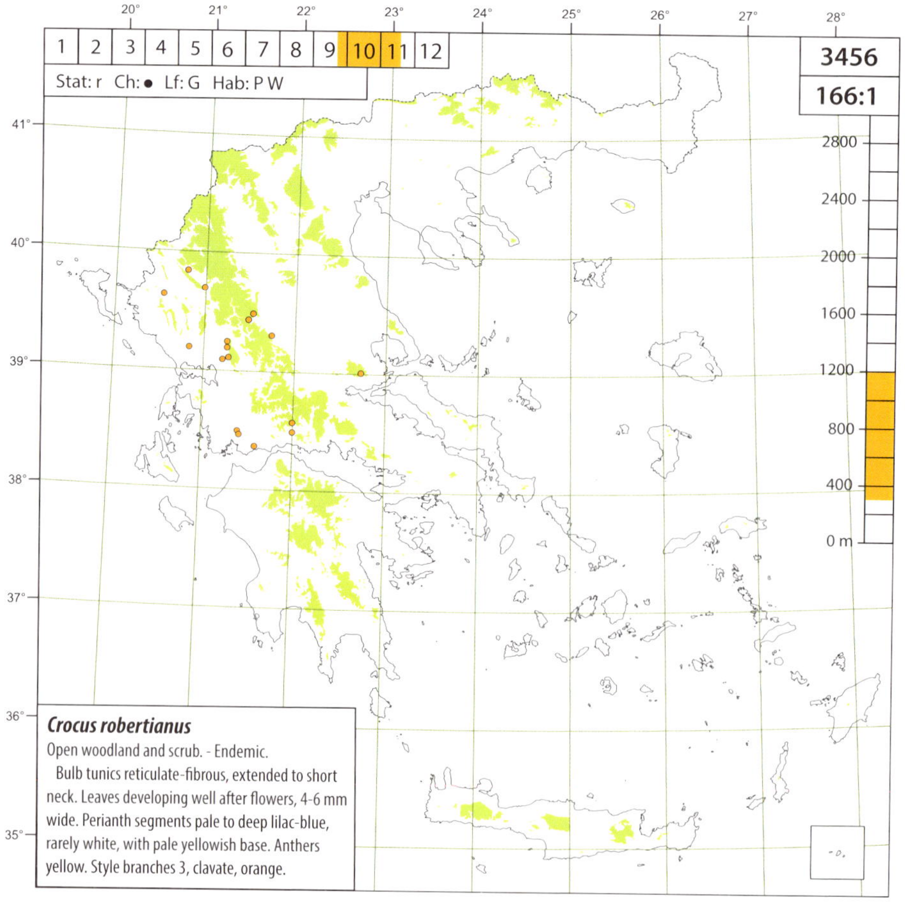

Crocus robertianus
Open woodland and scrub. - Endemic.
Bulb tunics reticulate-fibrous, extended to short neck. Leaves developing well after flowers, 4-6 mm wide. Perianth segments pale to deep lilac-blue, rarely white, with pale yellowish base. Anthers yellow. Style branches 3, clavate, orange.

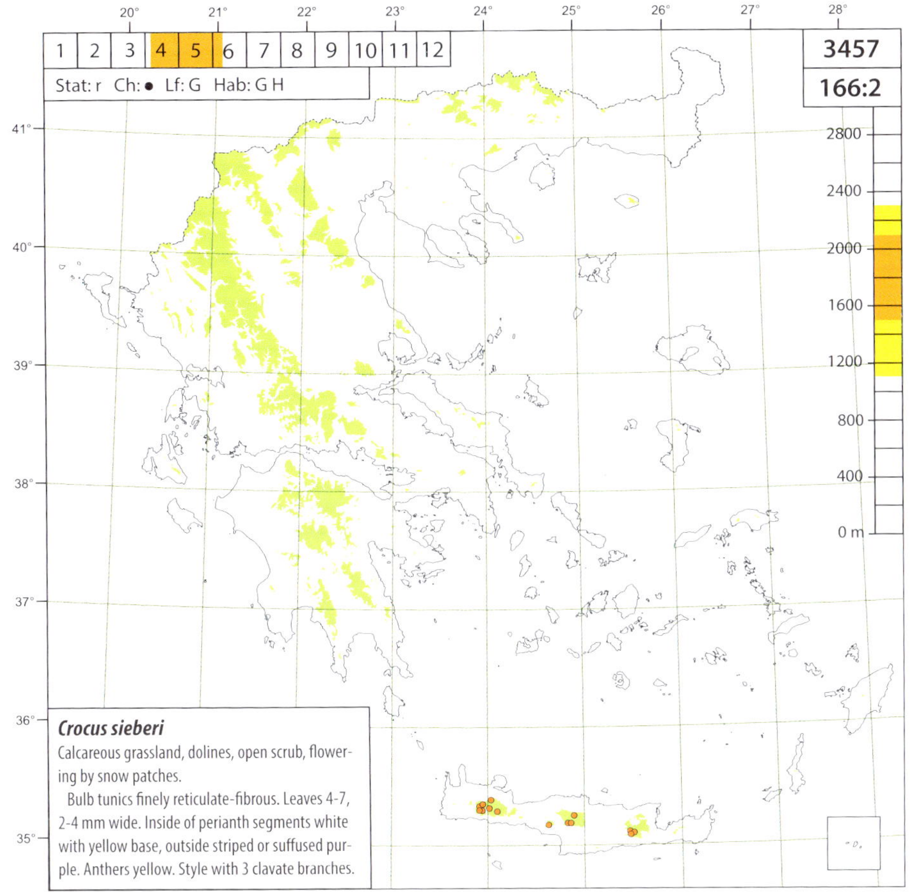

Crocus sieberi
Calcareous grassland, dolines, open scrub, flowering by snow patches.
Bulb tunics finely reticulate-fibrous. Leaves 4-7, 2-4 mm wide. Inside of perianth segments white with yellow base, outside striped or suffused purple. Anthers yellow. Style with 3 clavate branches.

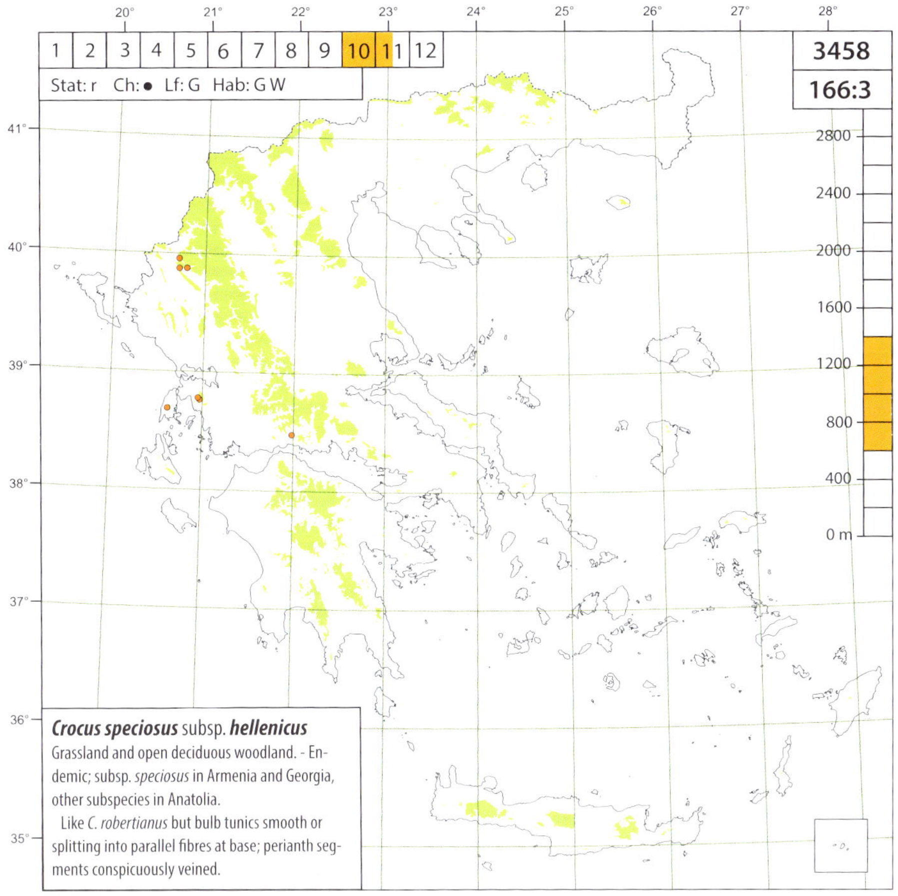

Crocus speciosus subsp. **hellenicus**
Grassland and open deciduous woodland. - Endemic; subsp. *speciosus* in Armenia and Georgia, other subspecies in Anatolia.
Like *C. robertianus* but bulb tunics smooth or splitting into parallel fibres at base; perianth segments conspicuously veined.

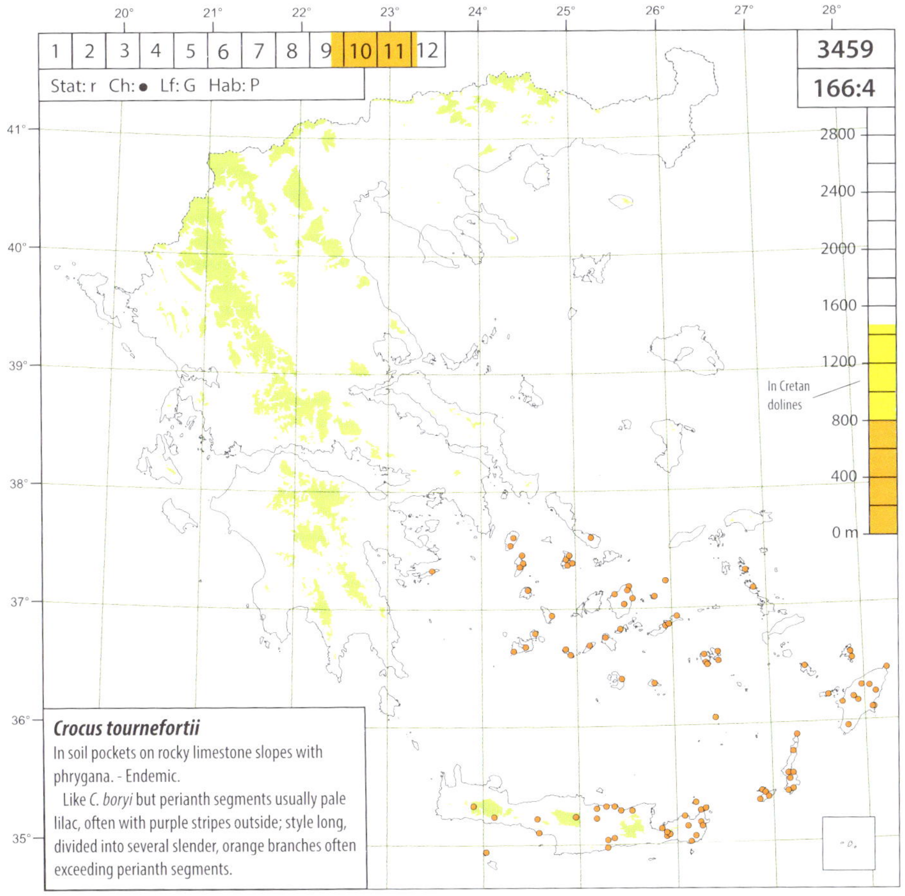

Crocus tournefortii
In soil pockets on rocky limestone slopes with phrygana. - Endemic.
Like *C. boryi* but perianth segments usually pale lilac, often with purple stripes outside; style long, divided into several slender, orange branches often exceeding perianth segments.

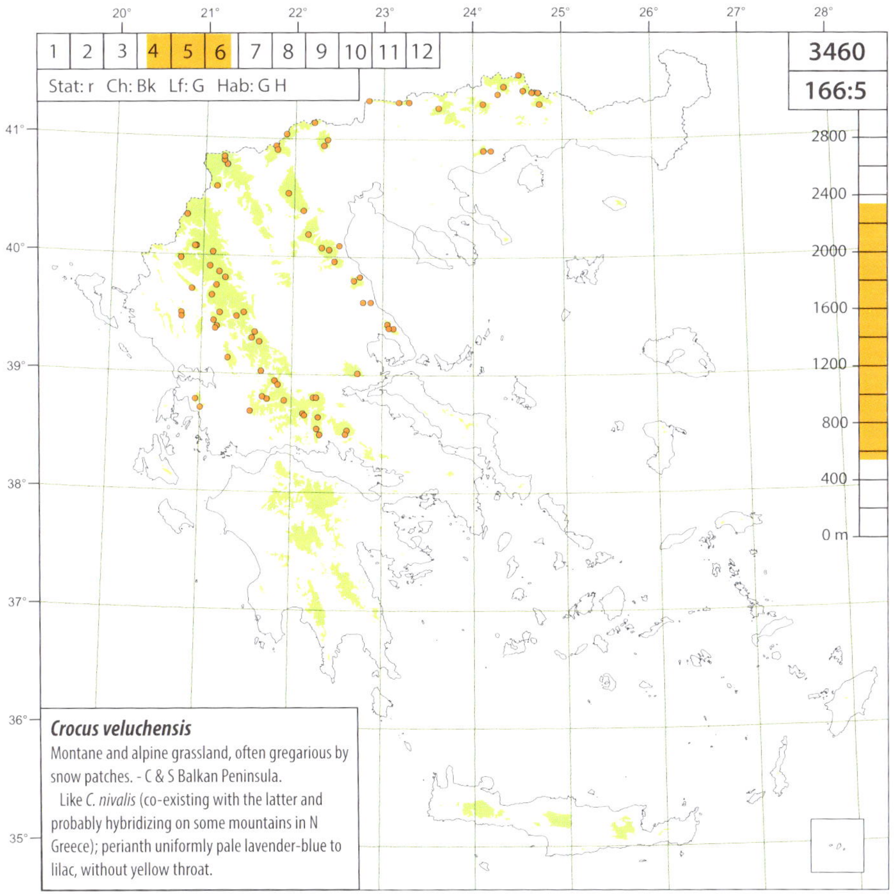

Crocus veluchensis
Montane and alpine grassland, often gregarious by snow patches. - C & S Balkan Peninsula.
Like *C. nivalis* (co-existing with the latter and probably hybridizing on some mountains in N Greece); perianth uniformly pale lavender-blue to lilac, without yellow throat.

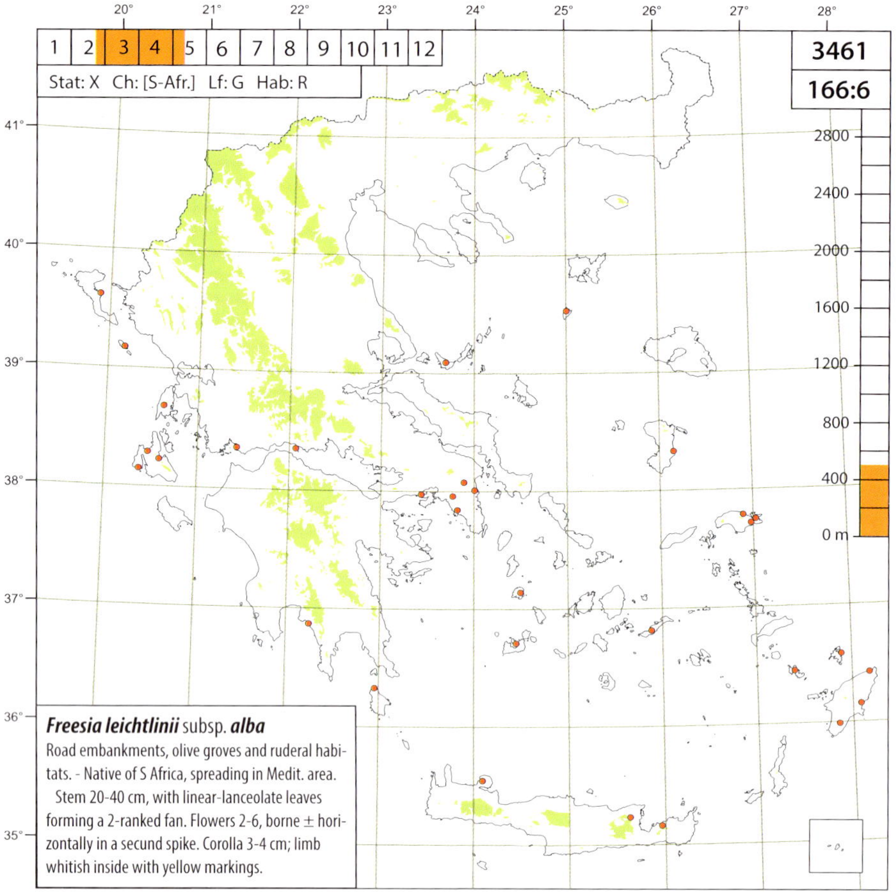

Freesia leichtlinii* subsp. *alba
Road embankments, olive groves and ruderal habitats. - Native of S Africa, spreading in Medit. area.
Stem 20-40 cm, with linear-lanceolate leaves forming a 2-ranked fan. Flowers 2-6, borne ± horizontally in a secund spike. Corolla 3-4 cm; limb whitish inside with yellow markings.

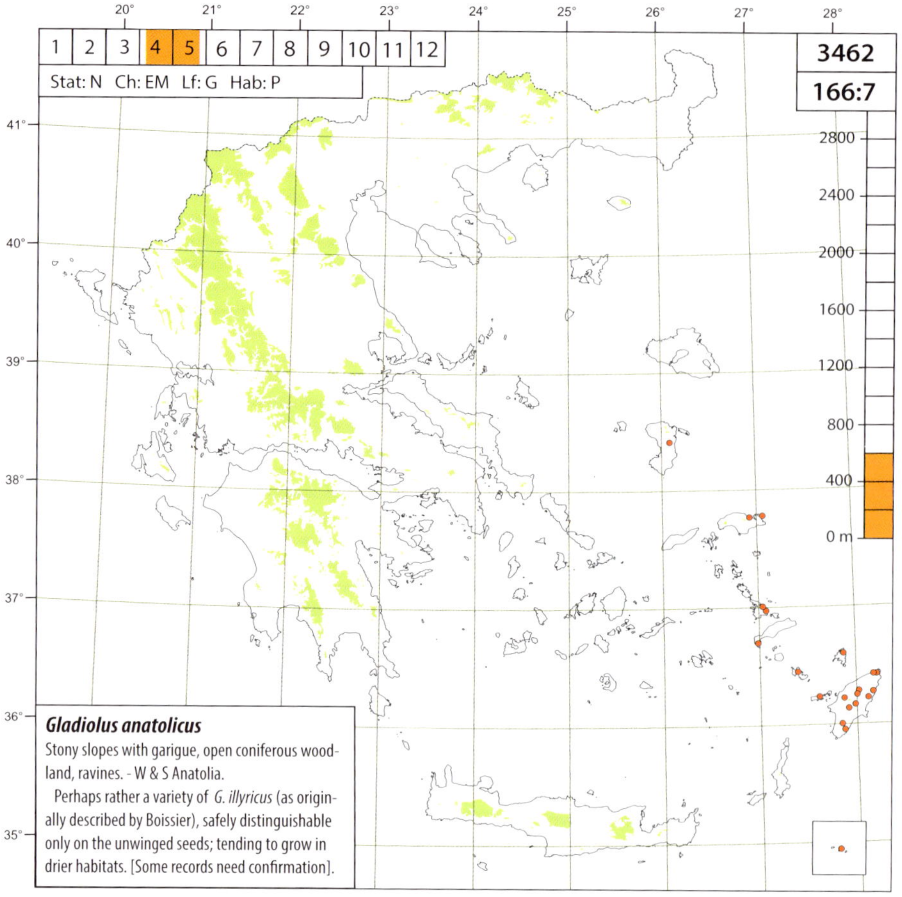

Gladiolus anatolicus
Stony slopes with garigue, open coniferous woodland, ravines. - W & S Anatolia.
Perhaps rather a variety of *G. illyricus* (as originally described by Boissier), safely distinguishable only on the unwinged seeds; tending to grow in drier habitats. [Some records need confirmation].

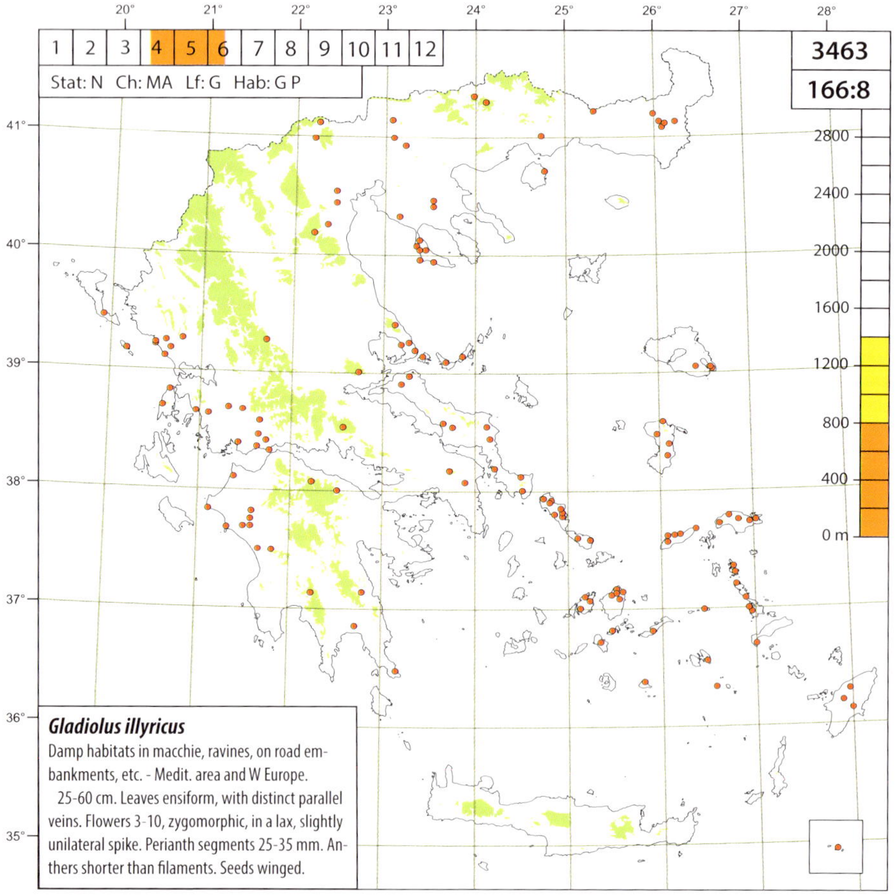

Gladiolus illyricus
Damp habitats in macchie, ravines, on road embankments, etc. - Medit. area and W Europe.
25-60 cm. Leaves ensiform, with distinct parallel veins. Flowers 3-10, zygomorphic, in a lax, slightly unilateral spike. Perianth segments 25-35 mm. Anthers shorter than filaments. Seeds winged.

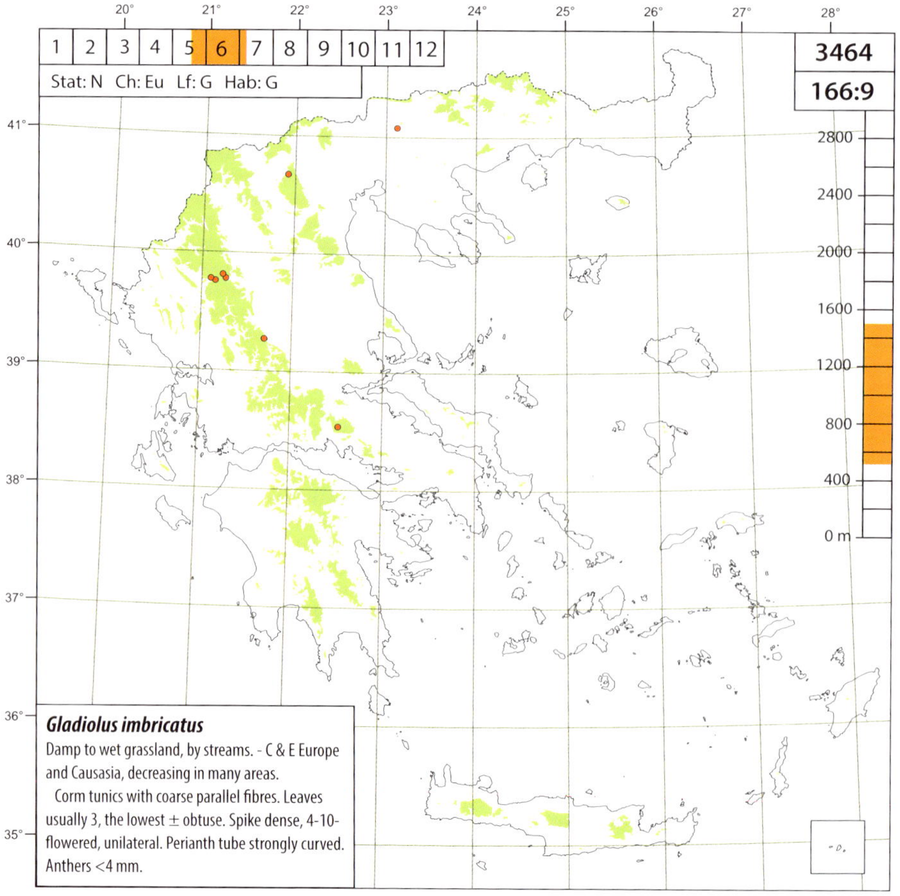

Gladiolus imbricatus
Damp to wet grassland, by streams. - C & E Europe and Causasia, decreasing in many areas.
Corm tunics with coarse parallel fibres. Leaves usually 3, the lowest ± obtuse. Spike dense, 4-10-flowered, unilateral. Perianth tube strongly curved. Anthers <4 mm.

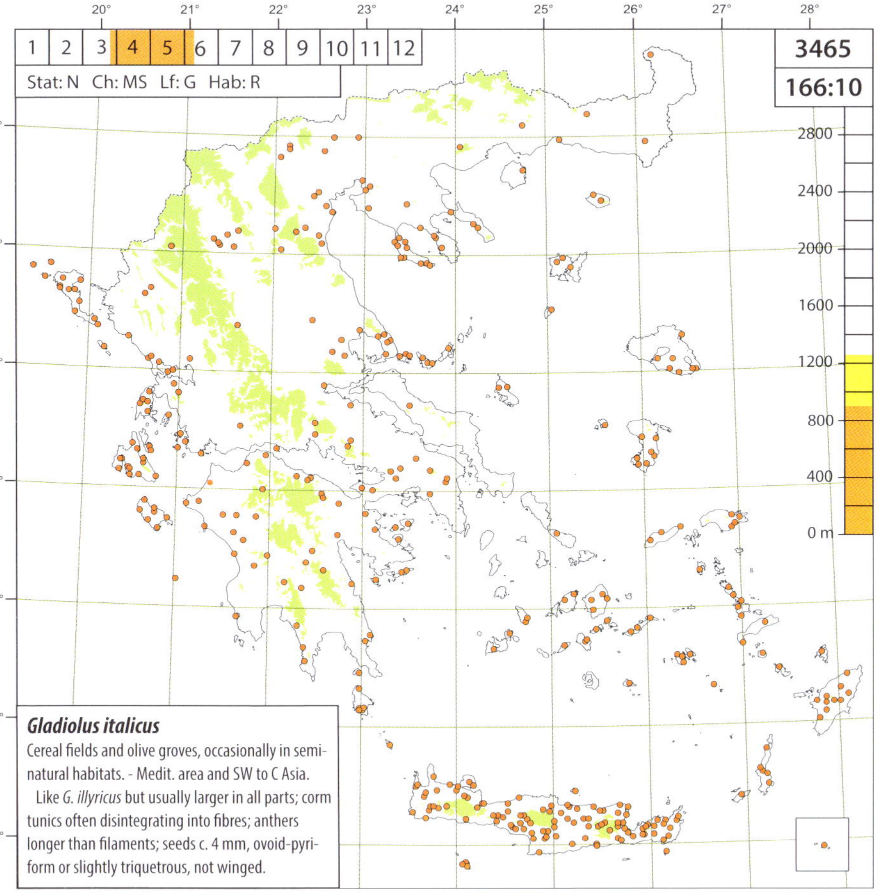

Gladiolus italicus

Cereal fields and olive groves, occasionally in semi-natural habitats. - Medit. area and SW to C Asia.

Like *G. illyricus* but usually larger in all parts; corm tunics often disintegrating into fibres; anthers longer than filaments; seeds c. 4 mm, ovoid-pyriform or slightly triquetrous, not winged.

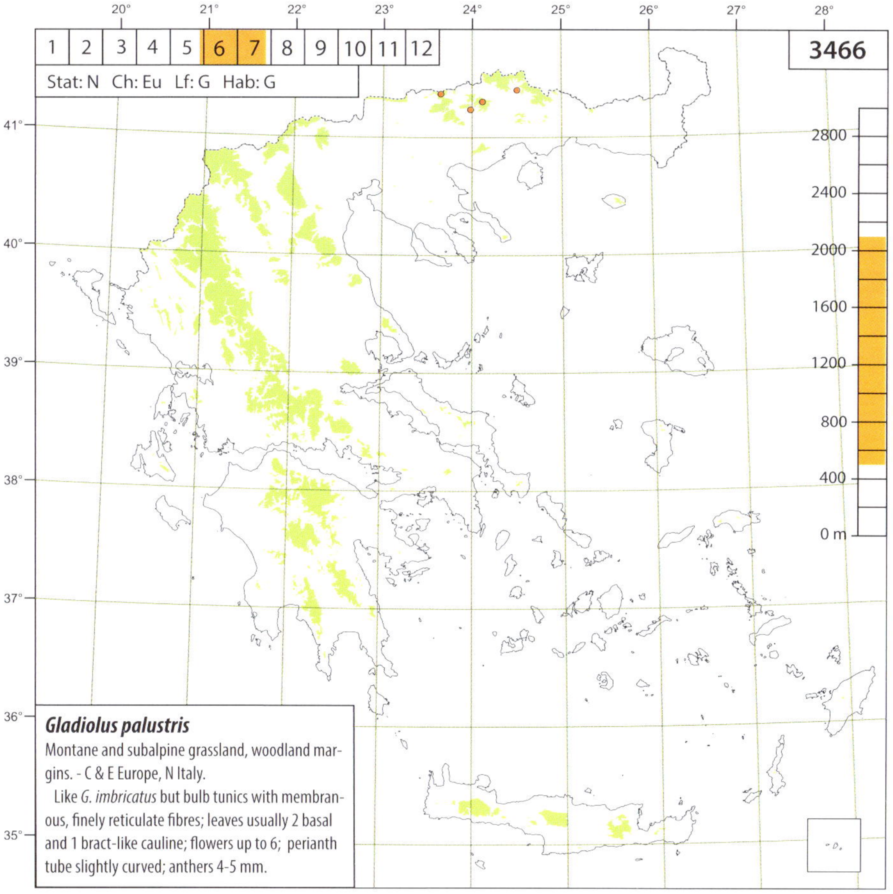

Gladiolus palustris

Montane and subalpine grassland, woodland margins. - C & E Europe, N Italy.

Like *G. imbricatus* but bulb tunics with membranous, finely reticulate fibres; leaves usually 2 basal and 1 bract-like cauline; flowers up to 6; perianth tube slightly curved; anthers 4-5 mm.

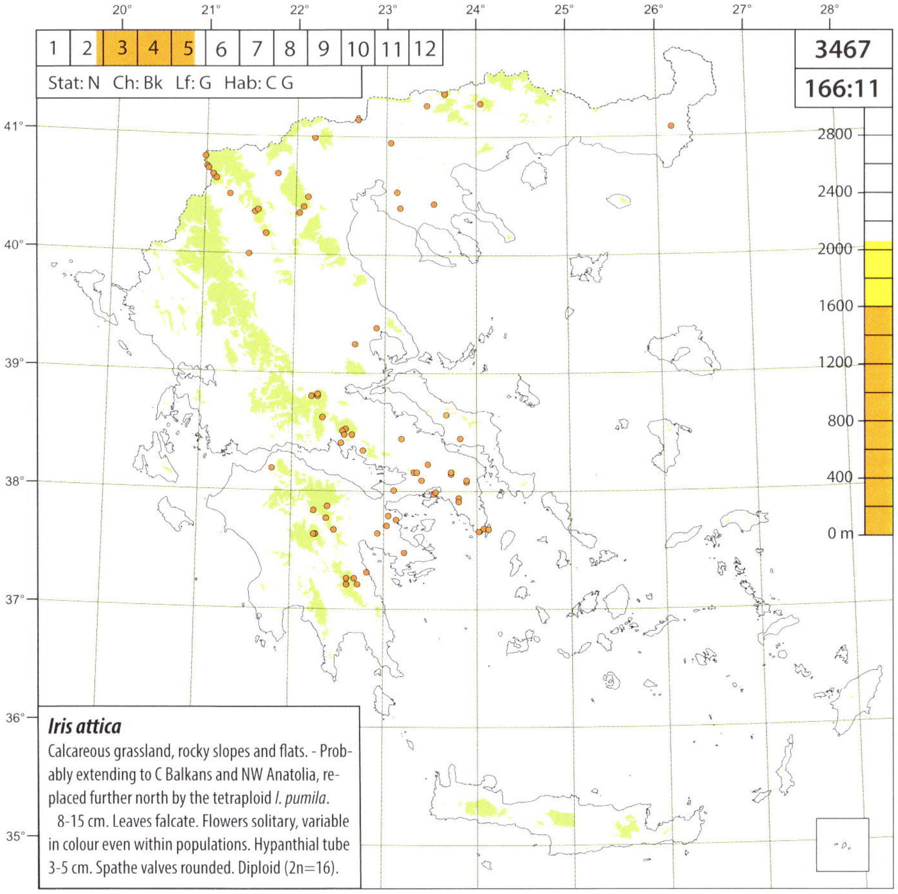

Iris attica

Calcareous grassland, rocky slopes and flats. - Probably extending to C Balkans and NW Anatolia, replaced further north by the tetraploid *I. pumila*.

8-15 cm. Leaves falcate. Flowers solitary, variable in colour even within populations. Hyphanthial tube 3-5 cm. Spathe valves rounded. Diploid (2n=16).

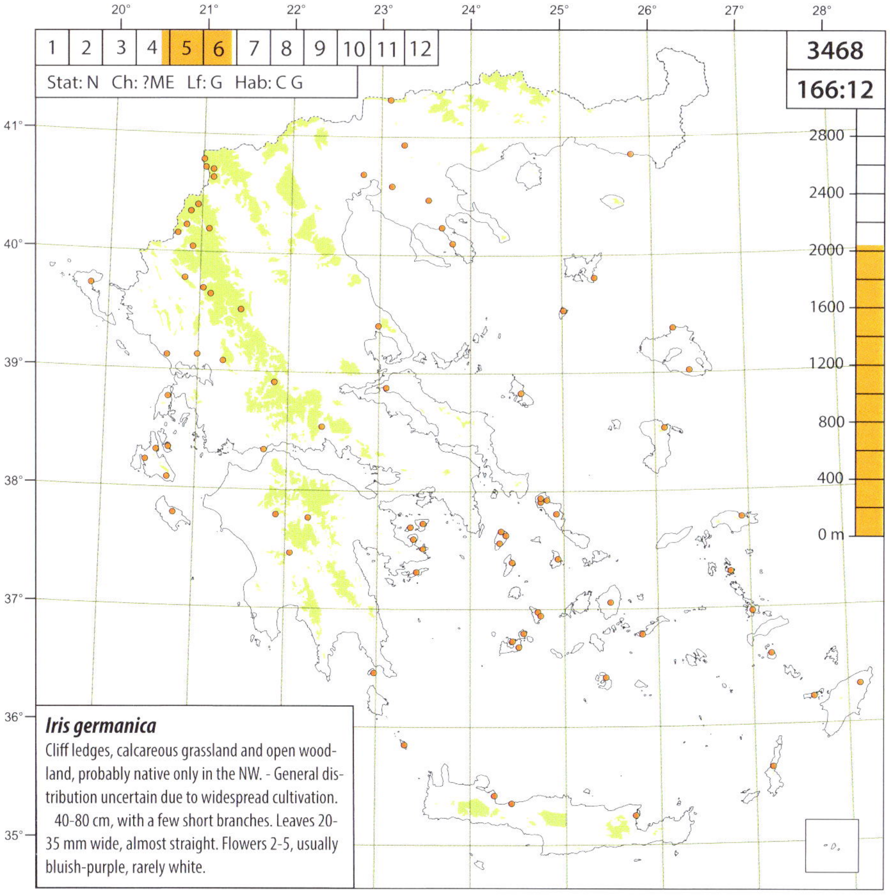

Iris germanica

Cliff ledges, calcareous grassland and open woodland, probably native only in the NW. - General distribution uncertain due to widespread cultivation.

40-80 cm, with a few short branches. Leaves 20-35 mm wide, almost straight. Flowers 2-5, usually bluish-purple, rarely white.

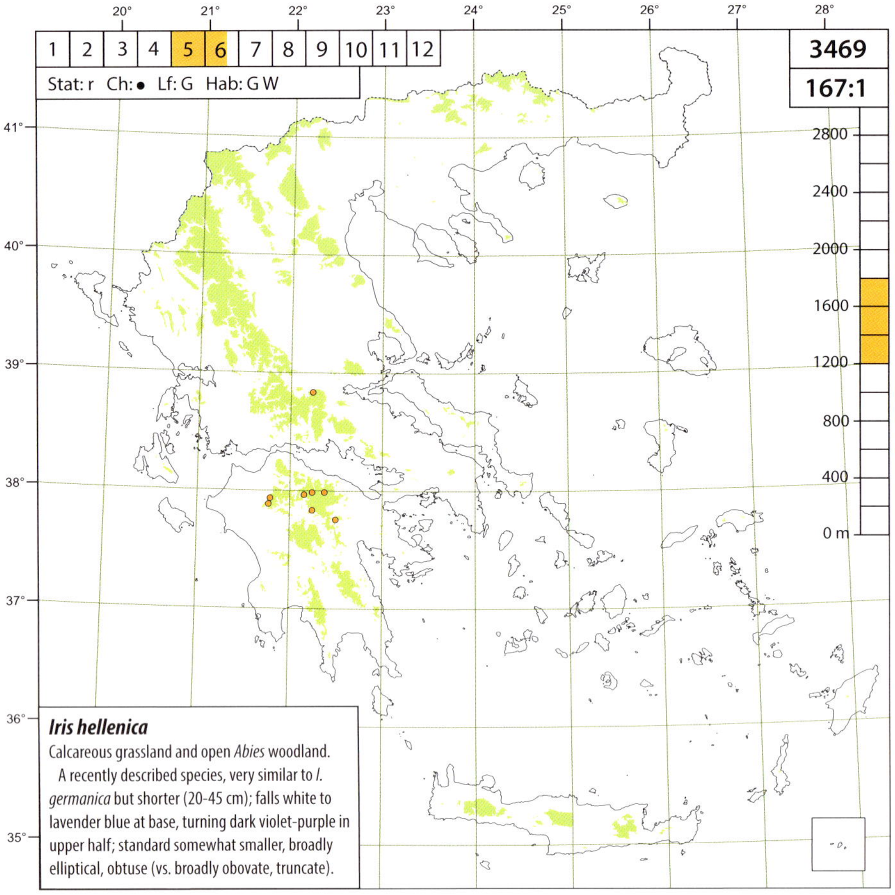

Iris hellenica

Calcareous grassland and open *Abies* woodland.

A recently described species, very similar to *I. germanica* but shorter (20-45 cm); falls white to lavender blue at base, turning dark violet-purple in upper half; standard somewhat smaller, broadly elliptical, obtuse (vs. broadly obovate, truncate).

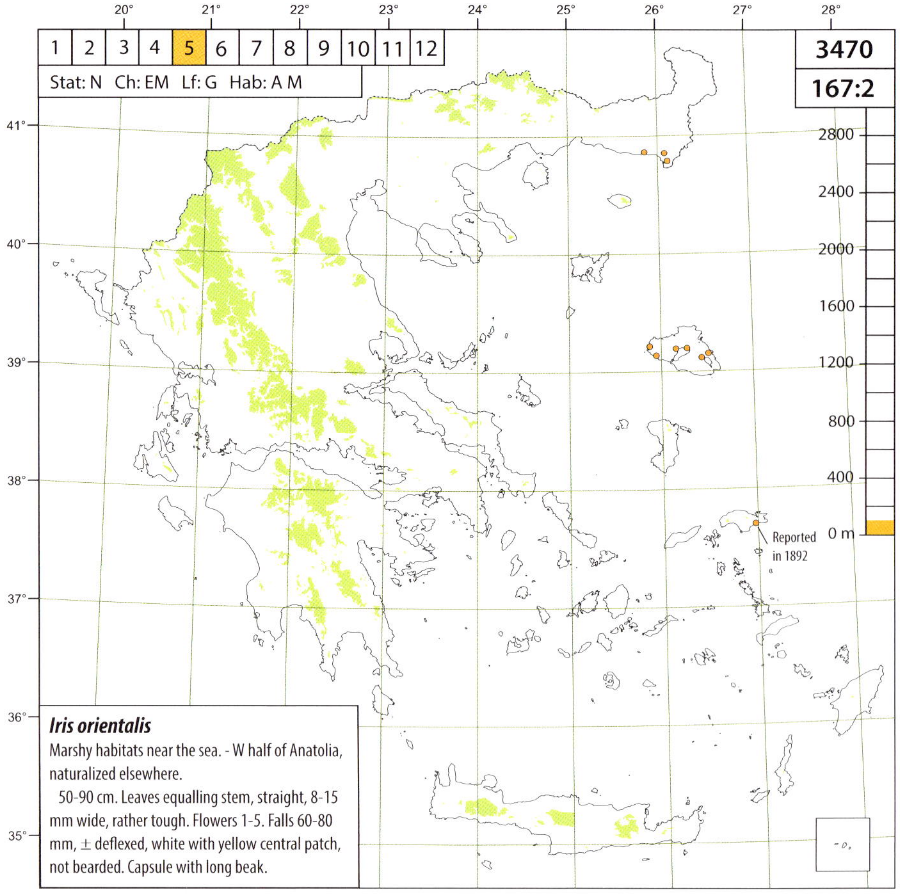

Iris orientalis

Marshy habitats near the sea. - W half of Anatolia, naturalized elsewhere.

50-90 cm. Leaves equalling stem, straight, 8-15 mm wide, rather tough. Flowers 1-5. Falls 60-80 mm, ± deflexed, white with yellow central patch, not bearded. Capsule with long beak.

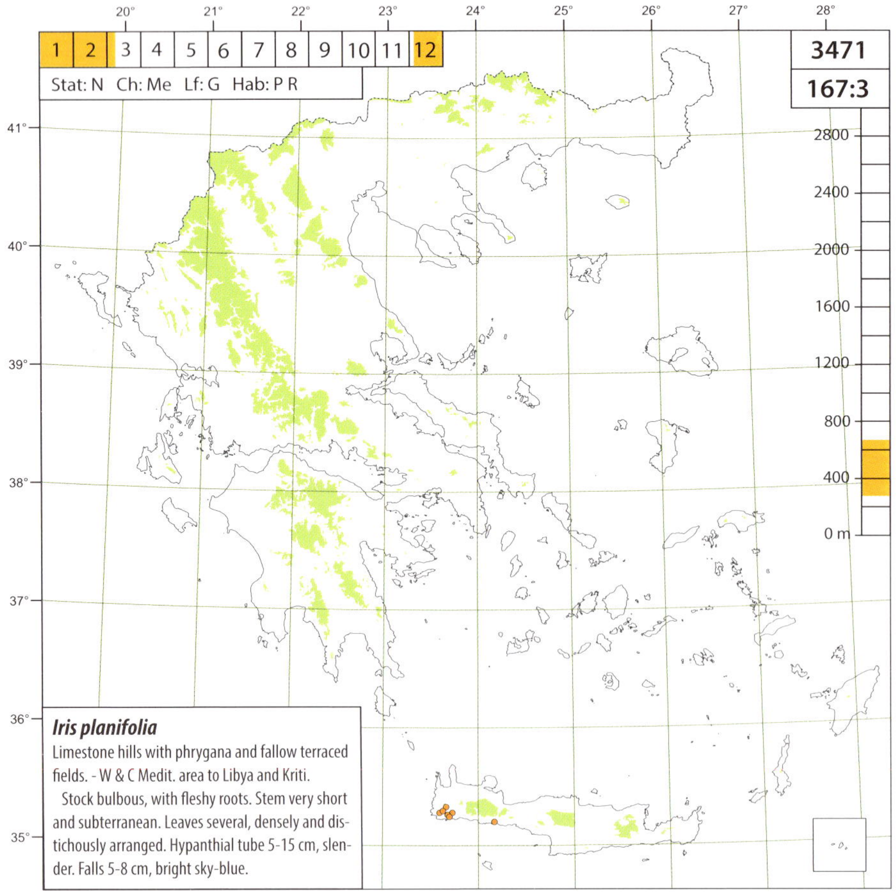

Iris planifolia

Limestone hills with phrygana and fallow terraced fields. - W & C Medit. area to Libya and Kriti.

Stock bulbous, with fleshy roots. Stem very short and subterranean. Leaves several, densely and distichously arranged. Hypanthial tube 5-15 cm, slender. Falls 5-8 cm, bright sky-blue.

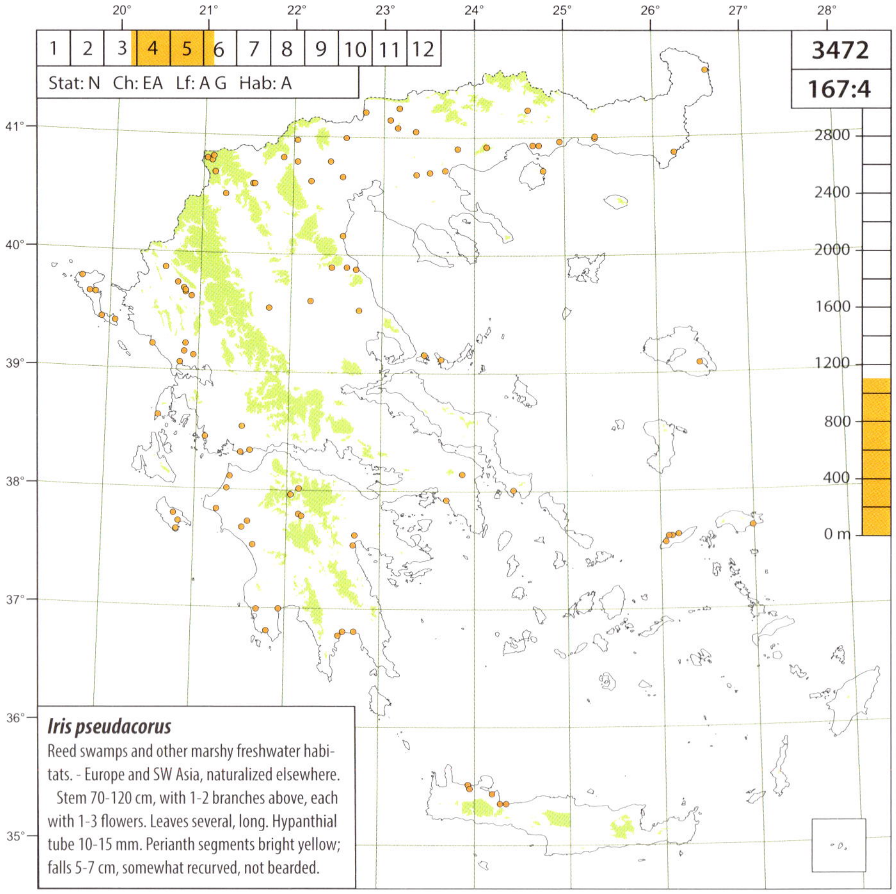

Iris pseudacorus

Reed swamps and other marshy freshwater habitats. - Europe and SW Asia, naturalized elsewhere.

Stem 70-120 cm, with 1-2 branches above, each with 1-3 flowers. Leaves several, long. Hypanthial tube 10-15 mm. Perianth segments bright yellow; falls 5-7 cm, somewhat recurved, not bearded.

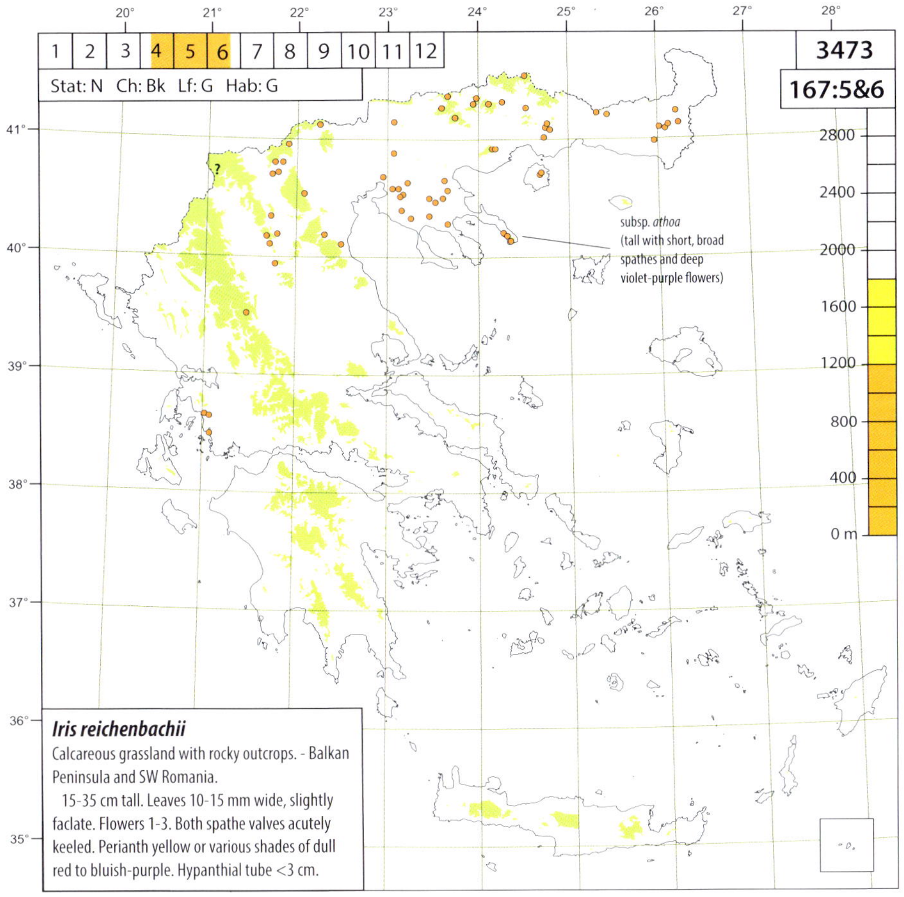

Iris reichenbachii

Calcareous grassland with rocky outcrops. - Balkan Peninsula and SW Romania.

15-35 cm tall. Leaves 10-15 mm wide, slightly faclate. Flowers 1-3. Both spathe valves acutely keeled. Perianth yellow or various shades of dull red to bluish-purple. Hypanthial tube <3 cm.

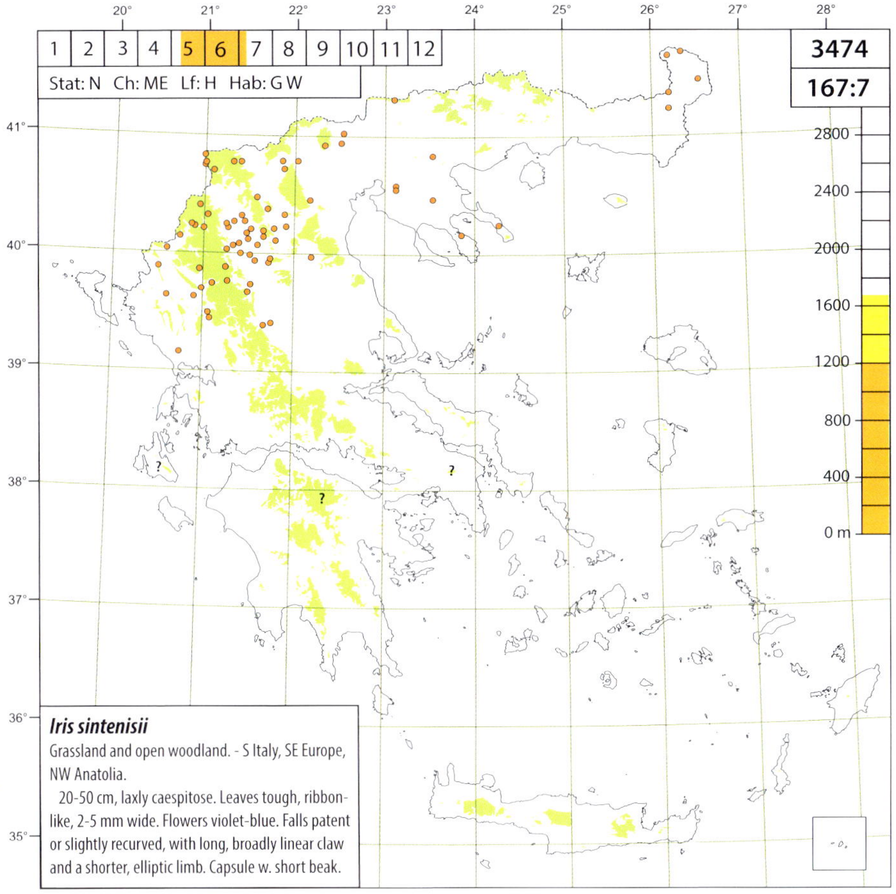

Iris sintenisii

Grassland and open woodland. - S Italy, SE Europe, NW Anatolia.

20-50 cm, laxly caespitose. Leaves tough, ribbon-like, 2-5 mm wide. Flowers violet-blue. Falls patent or slightly recurved, with long, broadly linear claw and a shorter, elliptic limb. Capsule w. short beak.

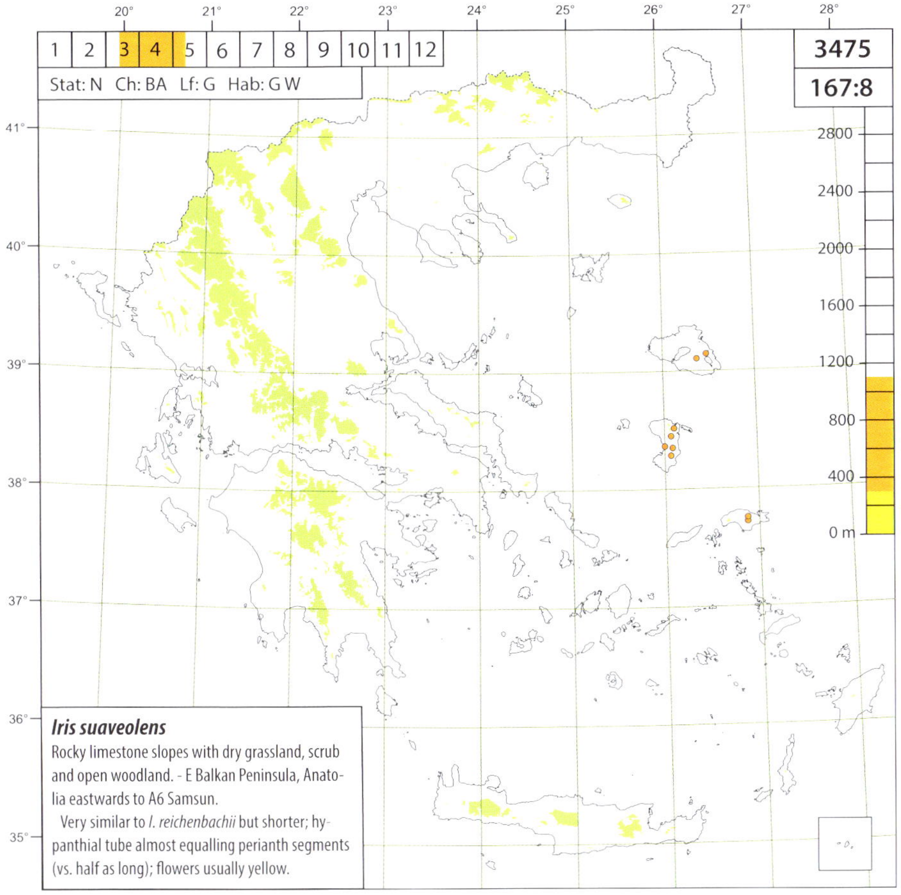

Iris suaveolens

Rocky limestone slopes with dry grassland, scrub and open woodland. - E Balkan Peninsula, Anatolia eastwards to A6 Samsun.

Very similar to *I. reichenbachii* but shorter; hypanthial tube almost equalling perianth segments (vs. half as long); flowers usually yellow.

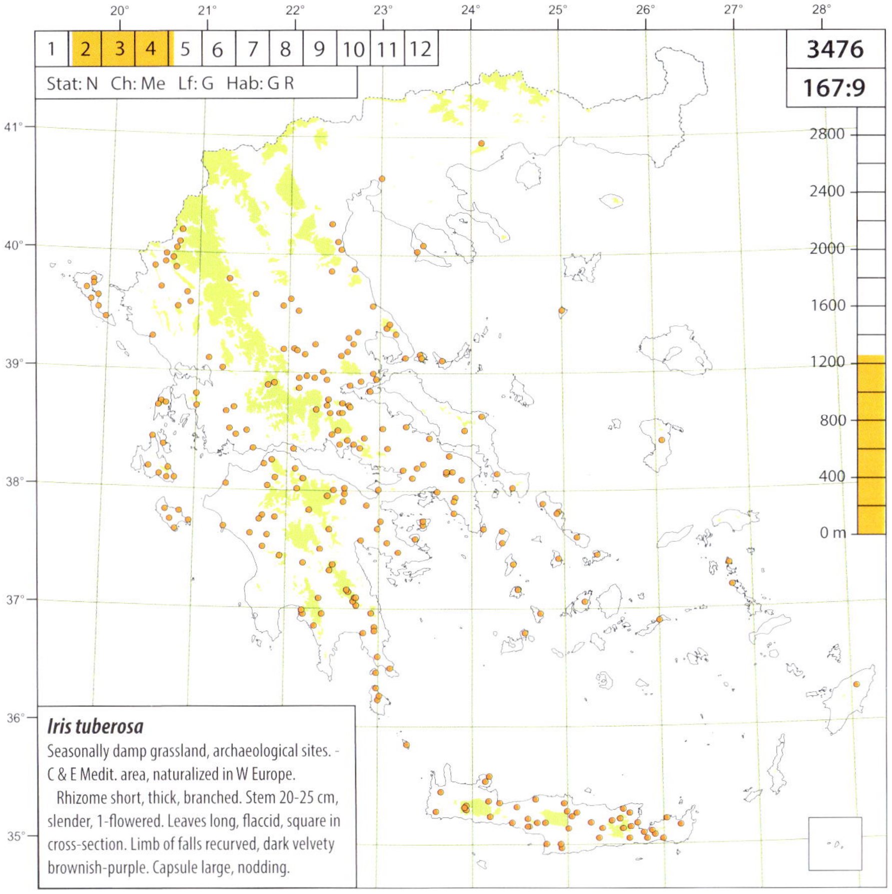

Iris tuberosa

Seasonally damp grassland, archaeological sites. - C & E Medit. area, naturalized in W Europe.

Rhizome short, thick, branched. Stem 20-25 cm, slender, 1-flowered. Leaves long, flaccid, square in cross-section. Limb of falls recurved, dark velvety brownish-purple. Capsule large, nodding.

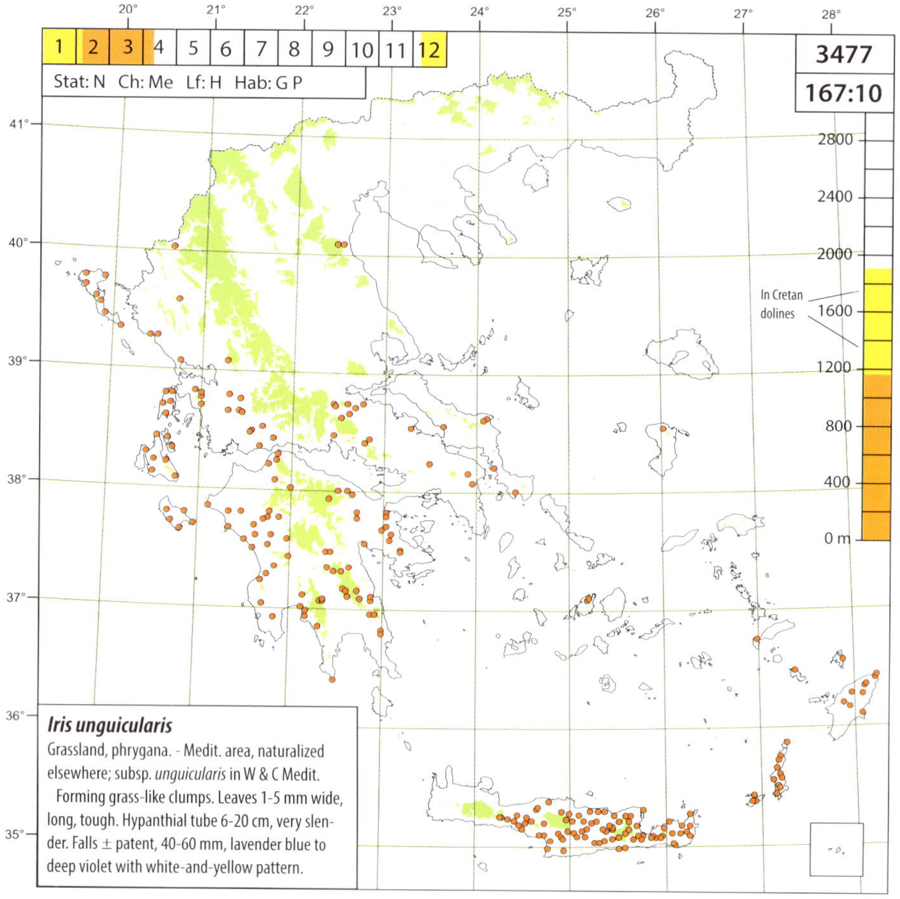

Iris unguicularis

Grassland, phrygana. - Medit. area, naturalized elsewhere; subsp. *unguicularis* in W & C Medit.

Forming grass-like clumps. Leaves 1-5 mm wide, long, tough. Hypanthial tube 6-20 cm, very slender. Falls ± patent, 40-60 mm, lavender blue to deep violet with white-and-yellow pattern.

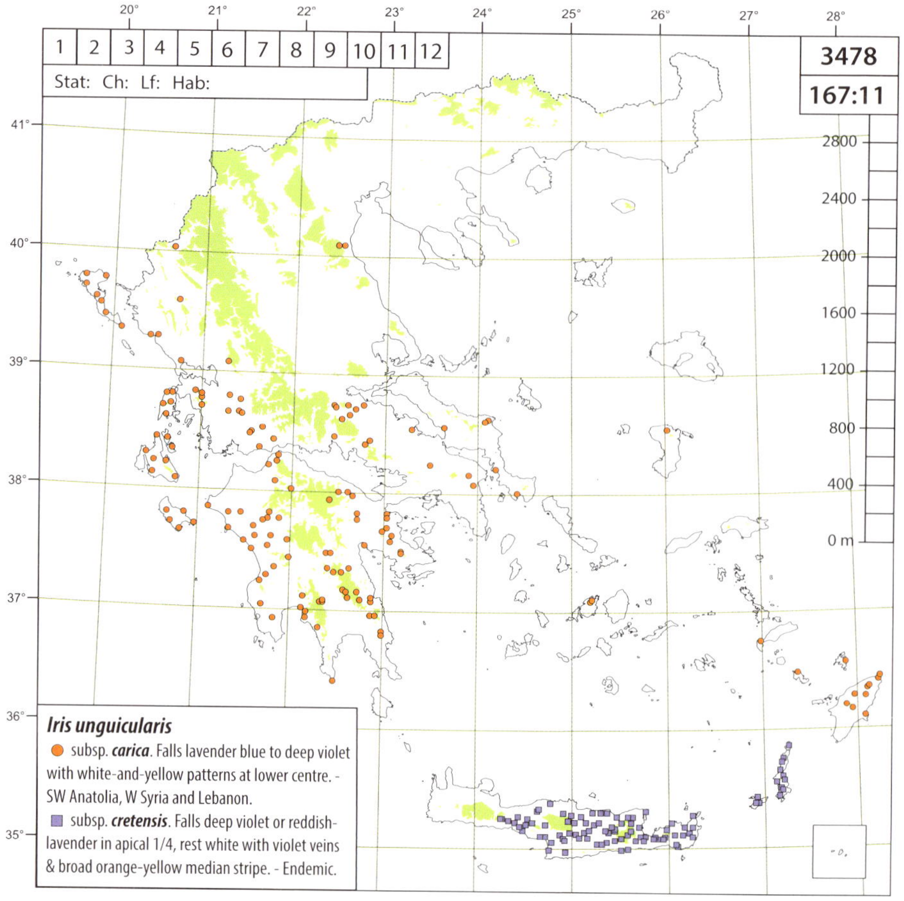

Iris unguicularis

● subsp. ***carica***. Falls lavender blue to deep violet with white-and-yellow patterns at lower centre. - SW Anatolia, W Syria and Lebanon.

■ subsp. ***cretensis***. Falls deep violet or reddish-lavender in apical 1/4, rest white with violet veins & broad orange-yellow median stripe. - Endemic.

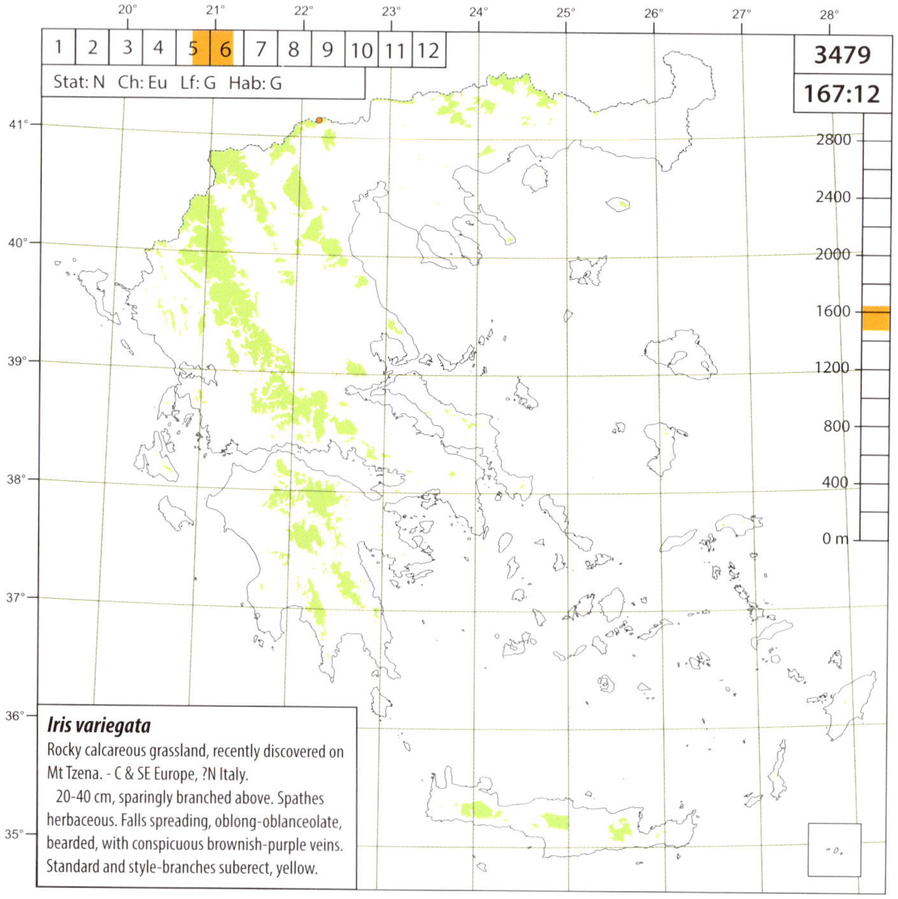

Iris variegata

Rocky calcareous grassland, recently discovered on Mt Tzena. - C & SE Europe, ?N Italy.

20-40 cm, sparingly branched above. Spathes herbaceous. Falls spreading, oblong-oblanceolate, bearded, with conspicuous brownish-purple veins. Standard and style-branches suberect, yellow.

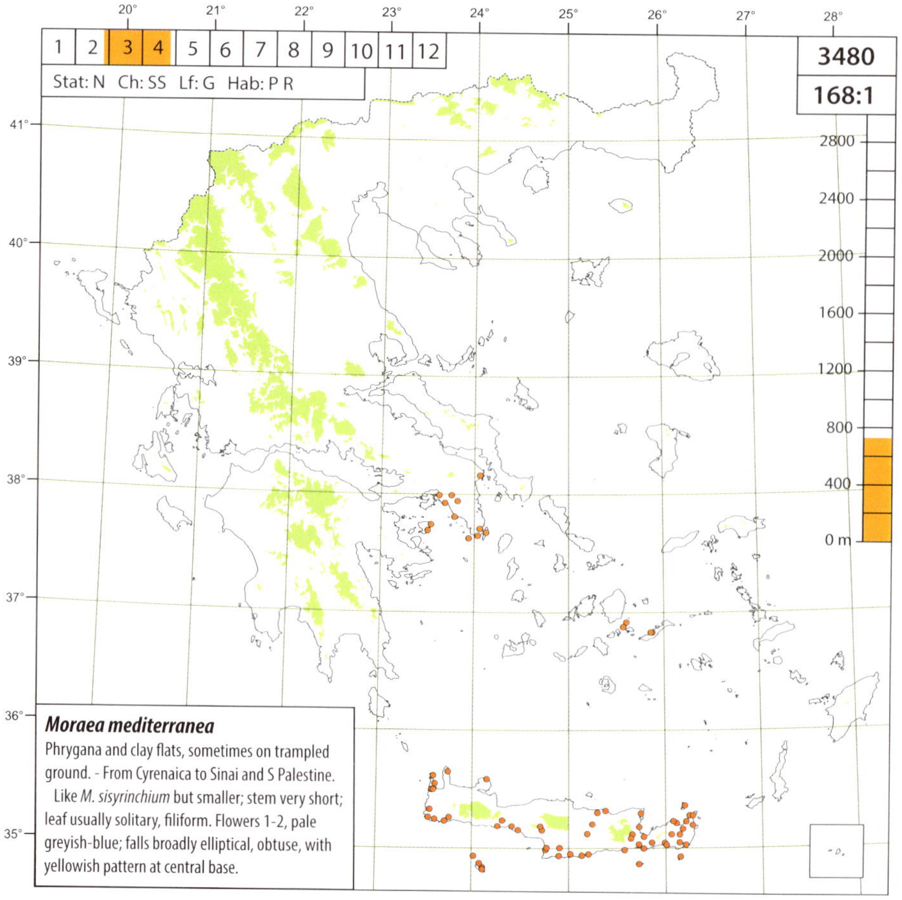

Moraea mediterranea

Phrygana and clay flats, sometimes on trampled ground. - From Cyrenaica to Sinai and S Palestine.

Like *M. sisyrinchium* but smaller; stem very short; leaf usually solitary, filiform. Flowers 1-2, pale greyish-blue; falls broadly elliptical, obtuse, with yellowish pattern at central base.

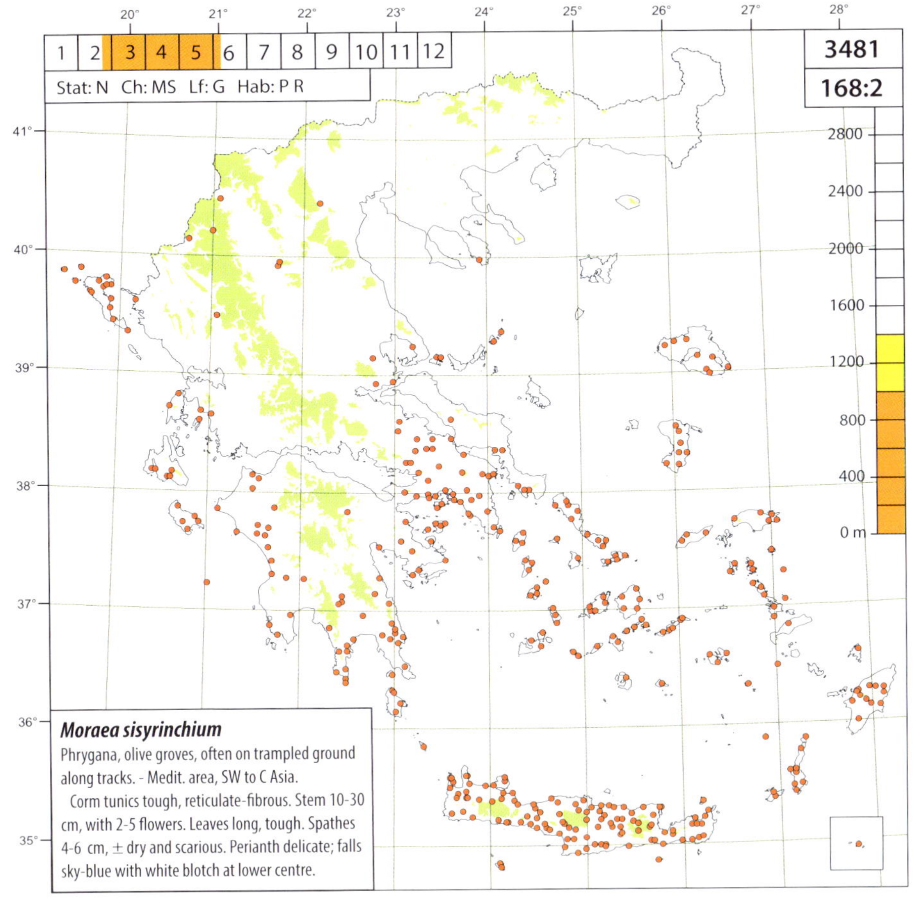

Moraea sisyrinchium
Phrygana, olive groves, often on trampled ground along tracks. - Medit. area, SW to C Asia.
Corm tunics tough, reticulate-fibrous. Stem 10-30 cm, with 2-5 flowers. Leaves long, tough. Spathes 4-6 cm, ± dry and scarious. Perianth delicate; falls sky-blue with white blotch at lower centre.

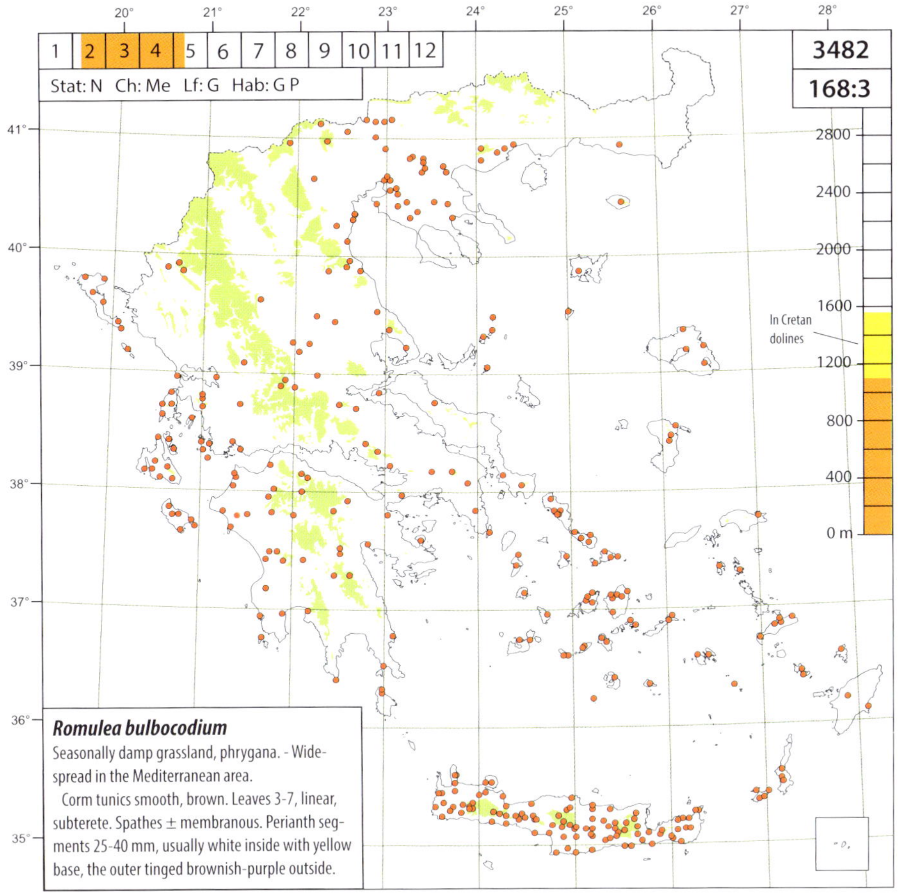

Romulea bulbocodium
Seasonally damp grassland, phrygana. - Widespread in the Mediterranean area.
Corm tunics smooth, brown. Leaves 3-7, linear, subterete. Spathes ± membranous. Perianth segments 25-40 mm, usually white inside with yellow base, the outer tinged brownish-purple outside.

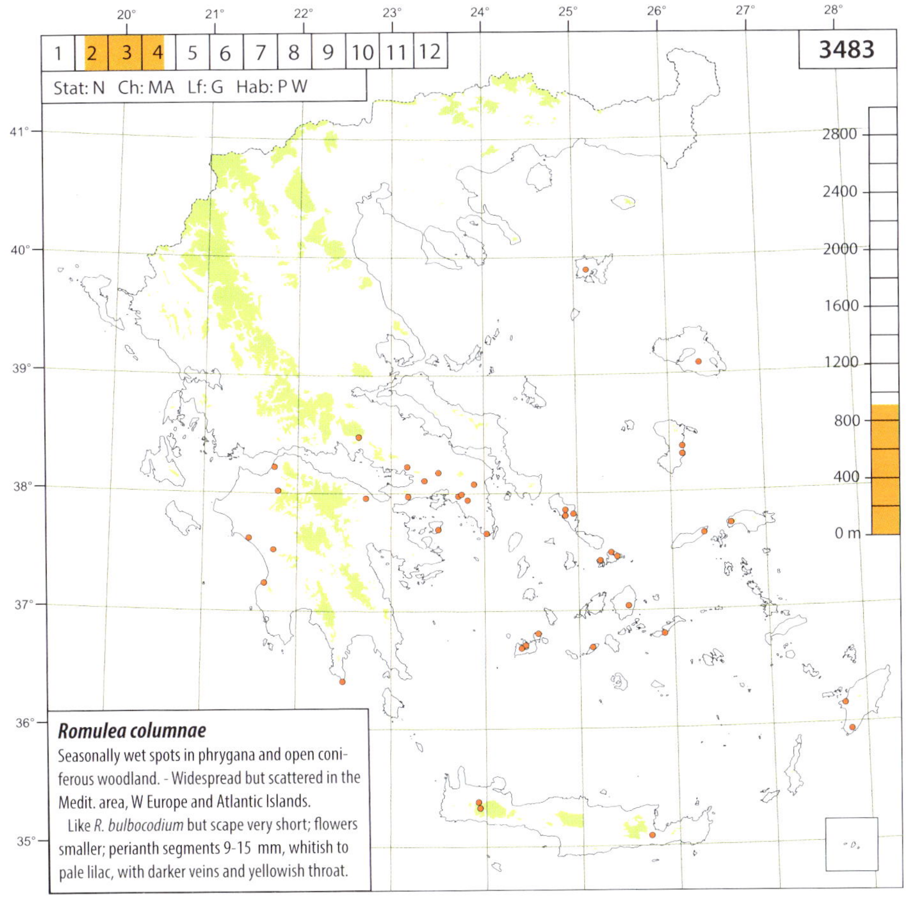

Romulea columnae
Seasonally wet spots in phrygana and open coniferous woodland. - Widespread but scattered in the Medit. area, W Europe and Atlantic Islands.
Like *R. bulbocodium* but scape very short; flowers smaller; perianth segments 9-15 mm, whitish to pale lilac, with darker veins and yellowish throat.

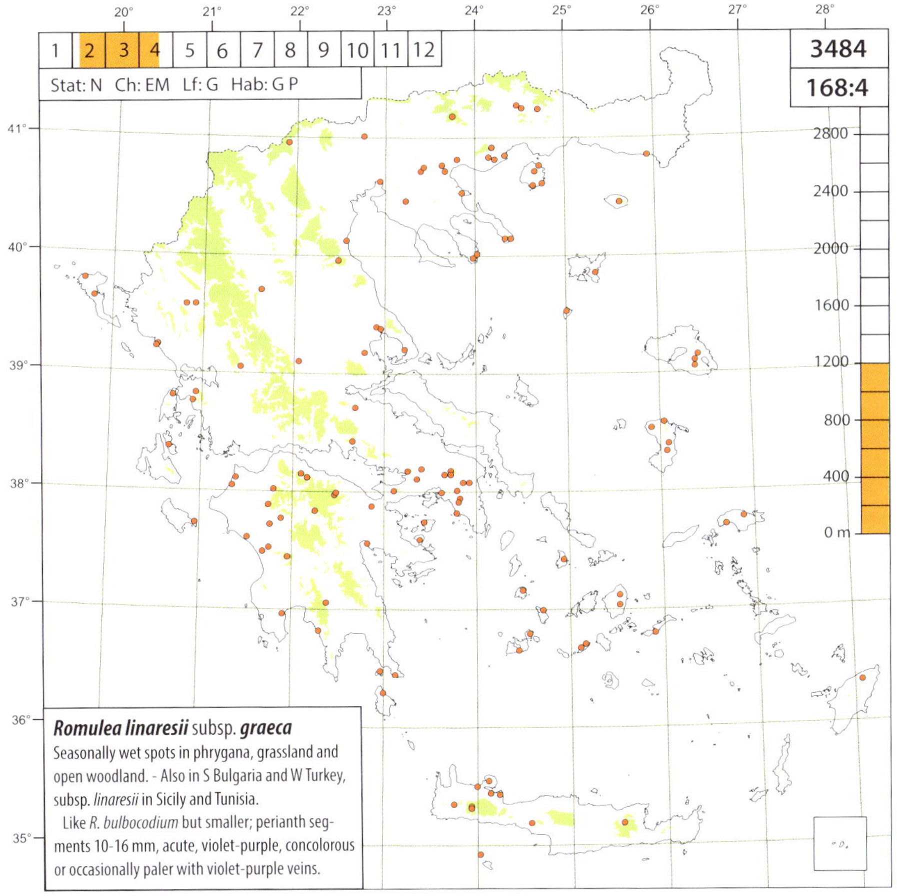

Romulea linaresii* subsp. *graeca
Seasonally wet spots in phrygana, grassland and open woodland. - Also in S Bulgaria and W Turkey, subsp. *linaresii* in Sicily and Tunisia.
Like *R. bulbocodium* but smaller; perianth segments 10-16 mm, acute, violet-purple, concolorous or occasionally paler with violet-purple veins.

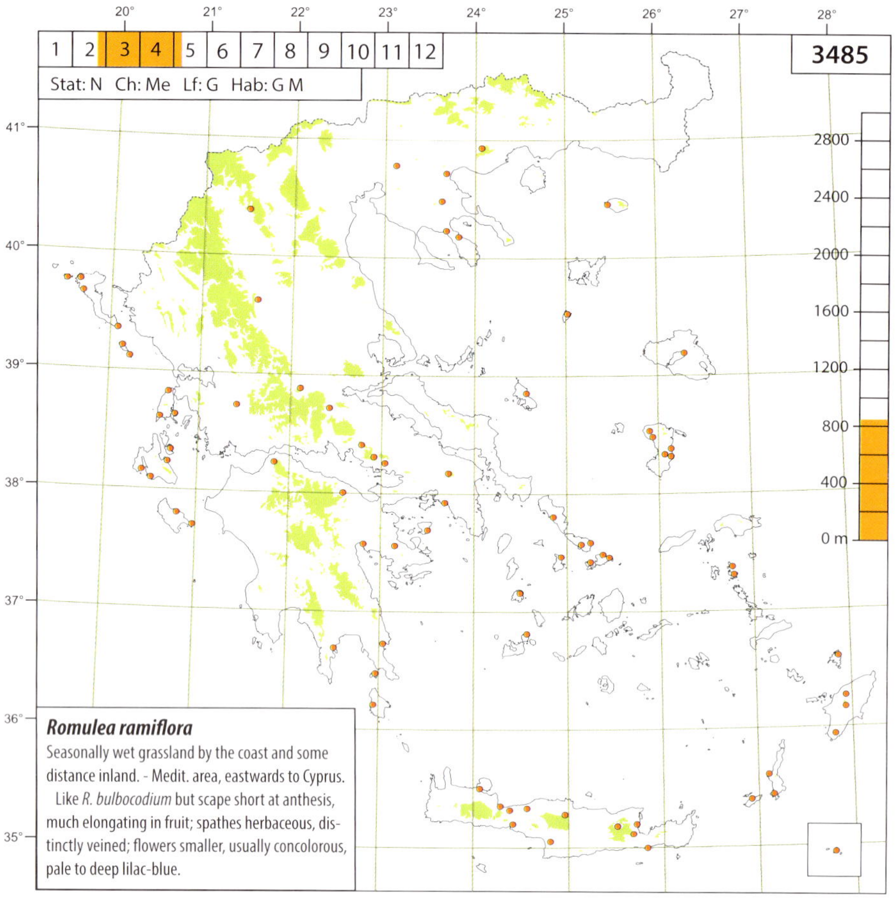

Romulea ramiflora
Seasonally wet grassland by the coast and some distance inland. - Medit. area, eastwards to Cyprus.
Like *R. bulbocodium* but scape short at anthesis, much elongating in fruit; spathes herbaceous, distinctly veined; flowers smaller, usually concolorous, pale to deep lilac-blue.

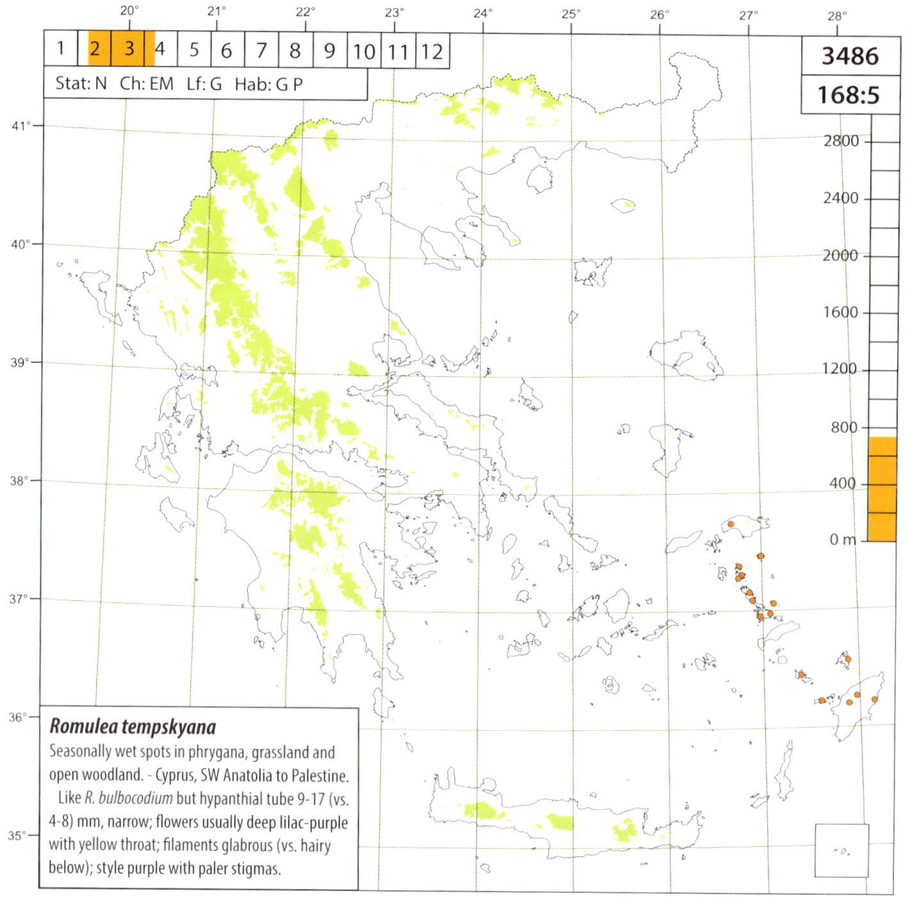

Romulea tempskyana
Seasonally wet spots in phrygana, grassland and open woodland. - Cyprus, SW Anatolia to Palestine.
Like *R. bulbocodium* but hypanthial tube 9-17 (vs. 4-8) mm, narrow; flowers usually deep lilac-purple with yellow throat; filaments glabrous (vs. hairy below); style purple with paler stigmas.

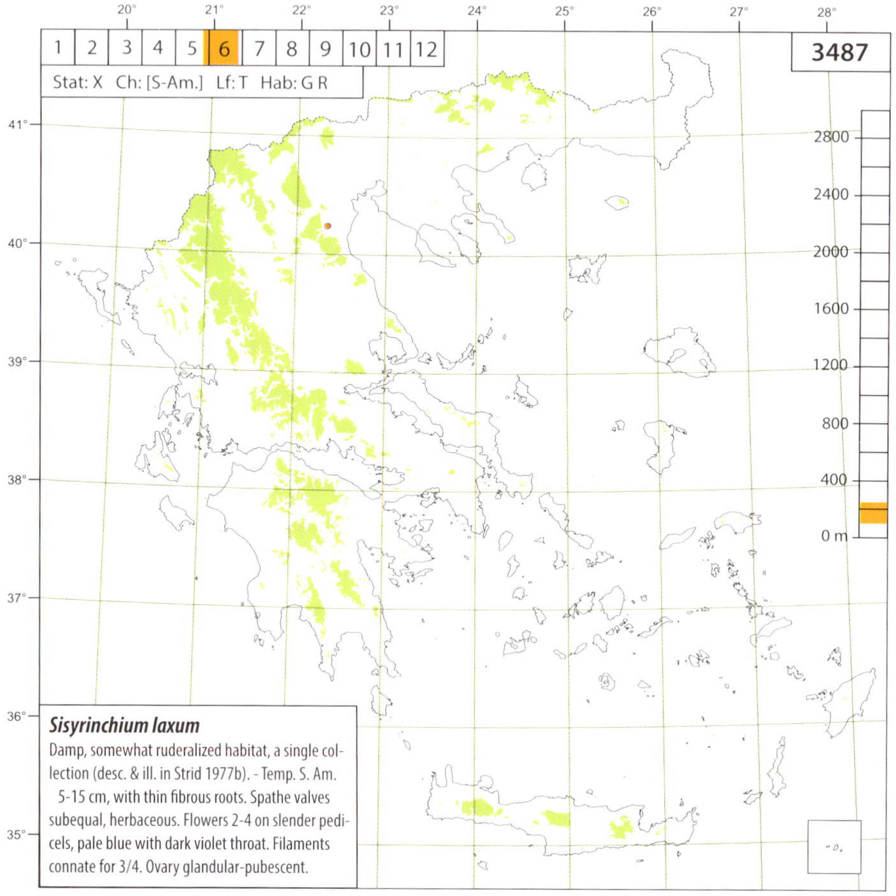

Sisyrinchium laxum
Damp, somewhat ruderalized habitat, a single collection (desc. & ill. in Strid 1977b). - Temp. S. Am.
5-15 cm, with thin fibrous roots. Spathe valves subequal, herbaceous. Flowers 2-4 on slender pedicels, pale blue with dark violet throat. Filaments connate for 3/4. Ovary glandular-pubescent.

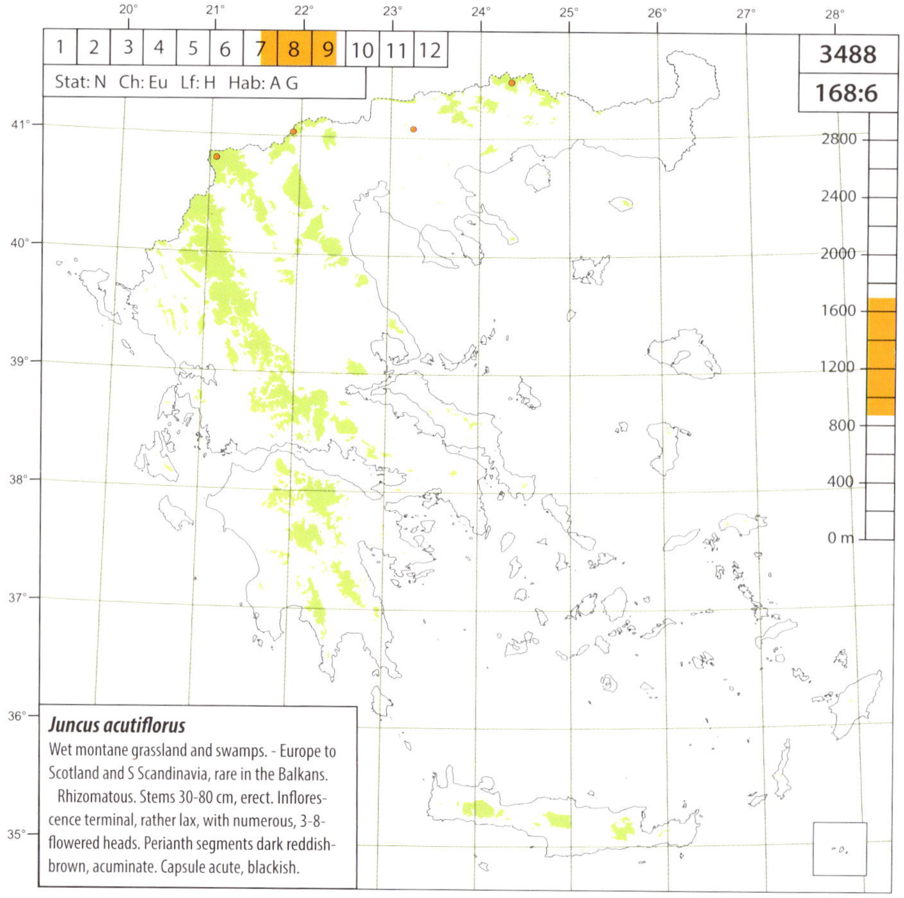

Juncus acutiflorus
Wet montane grassland and swamps. - Europe to Scotland and S Scandinavia, rare in the Balkans.
Rhizomatous. Stems 30-80 cm, erect. Inflorescence terminal, rather lax, with numerous, 3-8-flowered heads. Perianth segments dark reddish-brown, acuminate. Capsule acute, blackish.

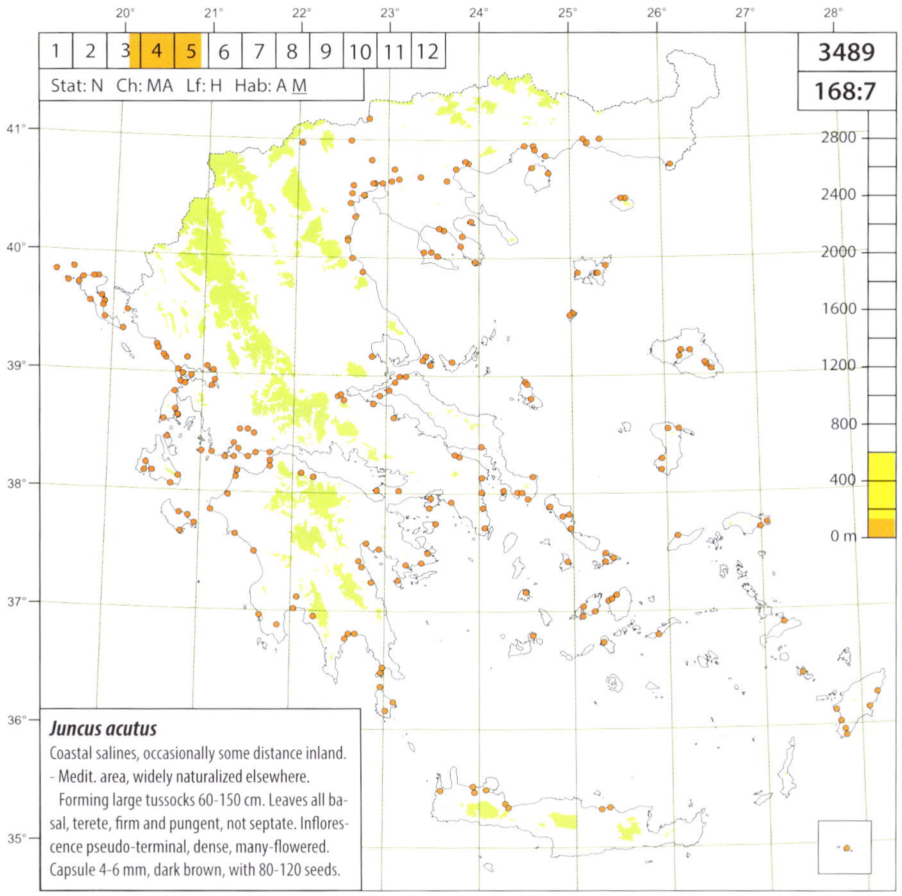

Juncus acutus

Coastal salines, occasionally some distance inland. - Medit. area, widely naturalized elsewhere.

Forming large tussocks 60-150 cm. Leaves all basal, terete, firm and pungent, not septate. Inflorescence pseudo-terminal, dense, many-flowered. Capsule 4-6 mm, dark brown, with 80-120 seeds.

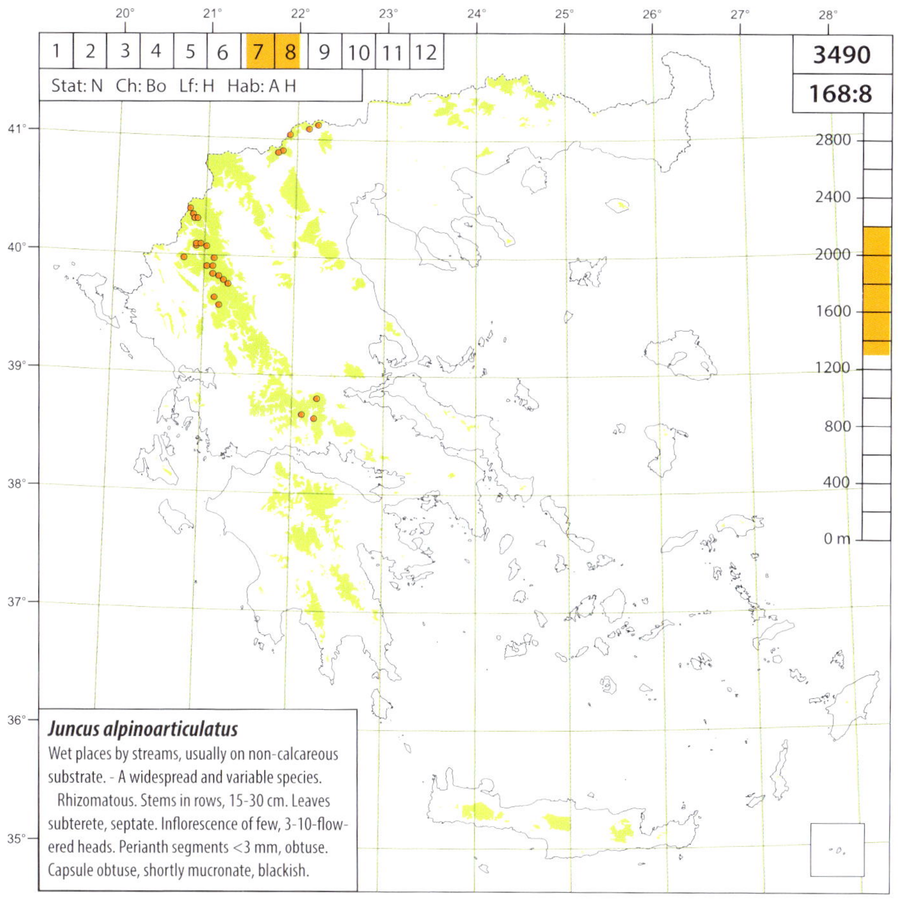

Juncus alpinoarticulatus

Wet places by streams, usually on non-calcareous substrate. - A widespread and variable species.

Rhizomatous. Stems in rows, 15-30 cm. Leaves subterete, septate. Inflorescence of few, 3-10-flowered heads. Perianth segments <3 mm, obtuse. Capsule obtuse, shortly mucronate, blackish.

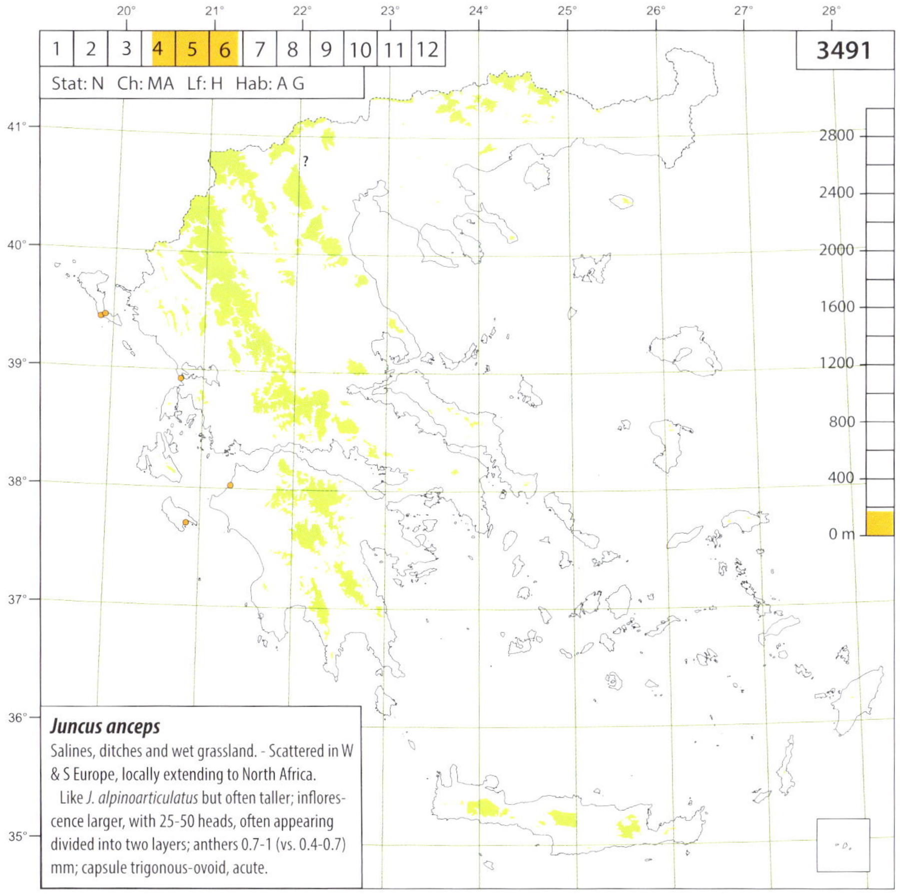

Juncus anceps

Salines, ditches and wet grassland. - Scattered in W & S Europe, locally extending to North Africa.

Like *J. alpinoarticulatus* but often taller; inflorescence larger, with 25-50 heads, often appearing divided into two layers; anthers 0.7-1 (vs. 0.4-0.7) mm; capsule trigonous-ovoid, acute.

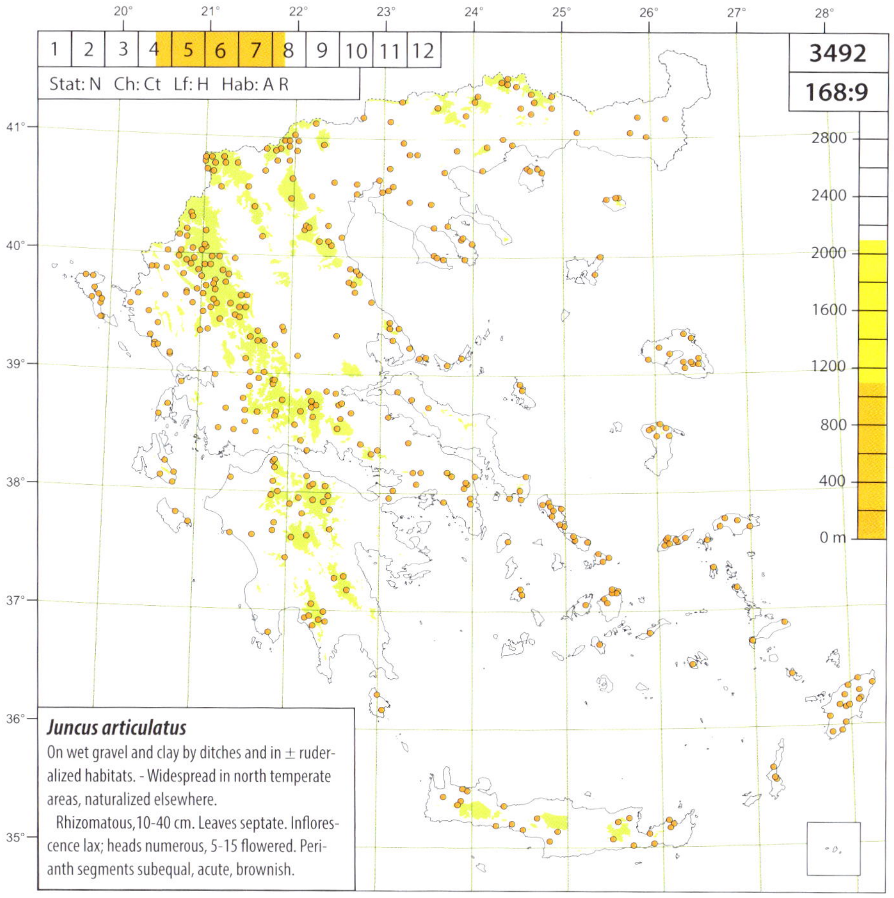

Juncus articulatus

On wet gravel and clay by ditches and in ± ruderalized habitats. - Widespread in north temperate areas, naturalized elsewhere.

Rhizomatous,10-40 cm. Leaves septate. Inflorescence lax; heads numerous, 5-15 flowered. Perianth segments subequal, acute, brownish.

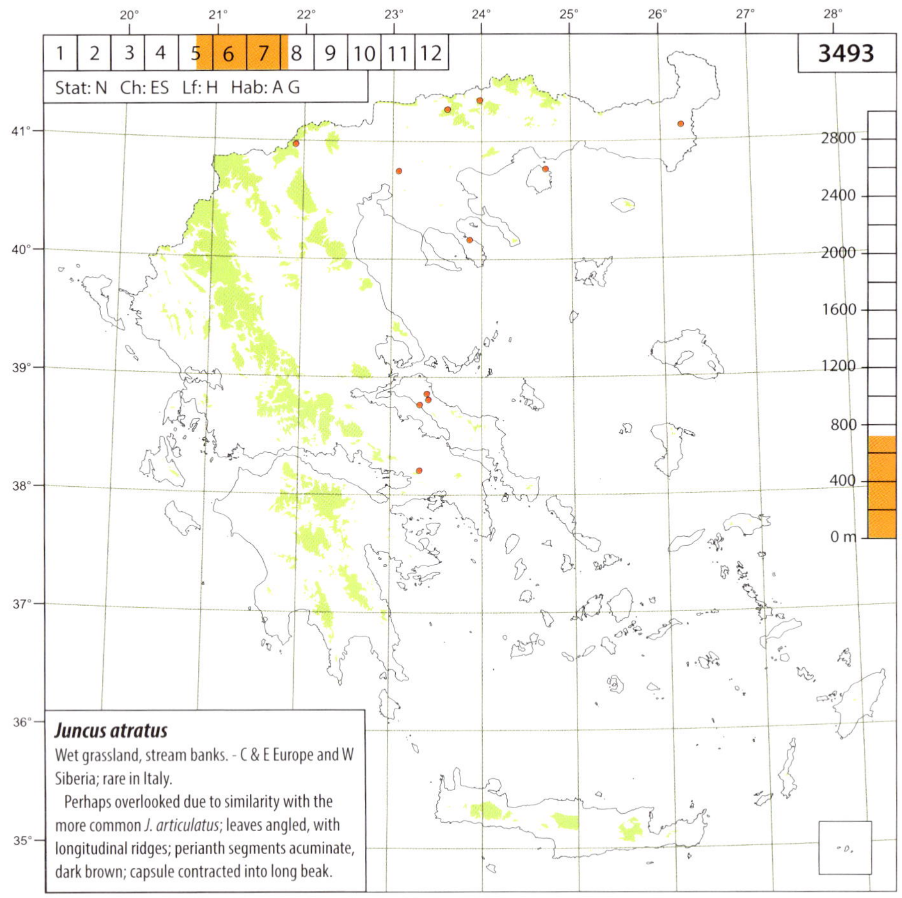

Juncus atratus

Wet grassland, stream banks. - C & E Europe and W Siberia; rare in Italy.

Perhaps overlooked due to similarity with the more common *J. articulatus*; leaves angled, with longitudinal ridges; perianth segments acuminate, dark brown; capsule contracted into long beak.

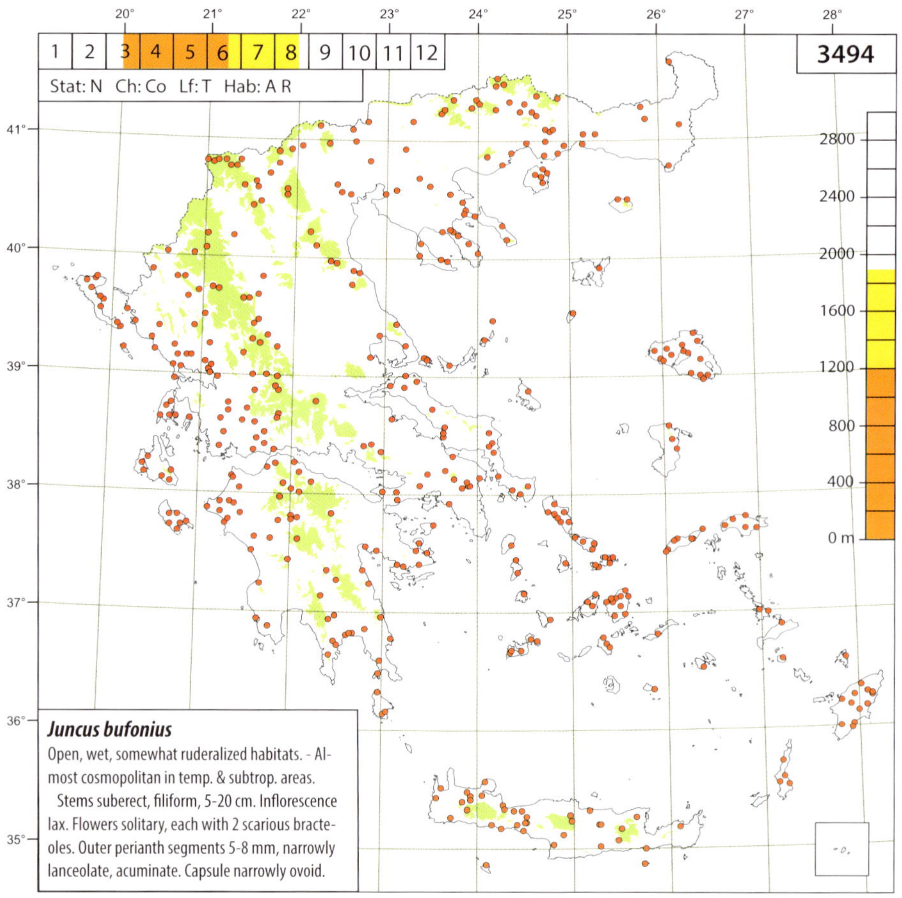

Juncus bufonius

Open, wet, somewhat ruderalized habitats. - Almost cosmopolitan in temp. & subtrop. areas.

Stems suberect, filiform, 5-20 cm. Inflorescence lax. Flowers solitary, each with 2 scarious bracteoles. Outer perianth segments 5-8 mm, narrowly lanceolate, acuminate. Capsule narrowly ovoid.

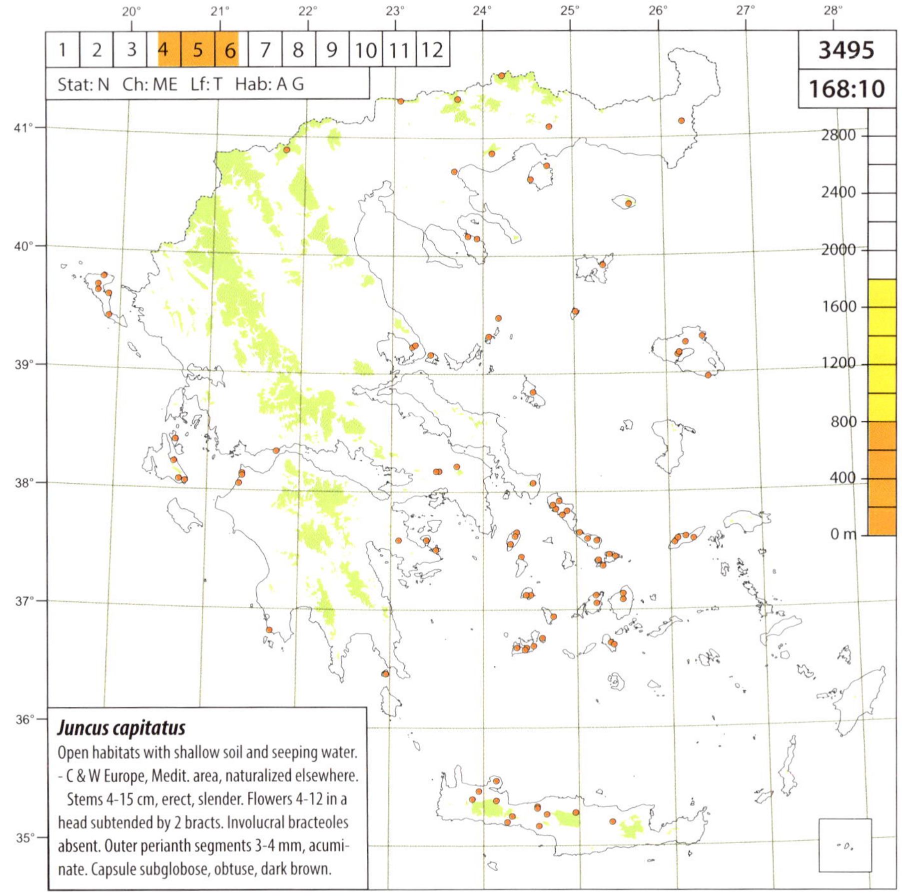

Juncus capitatus

Open habitats with shallow soil and seeping water. - C & W Europe, Medit. area, naturalized elsewhere.

Stems 4-15 cm, erect, slender. Flowers 4-12 in a head subtended by 2 bracts. Involucral bracteoles absent. Outer perianth segments 3-4 mm, acuminate. Capsule subglobose, obtuse, dark brown.

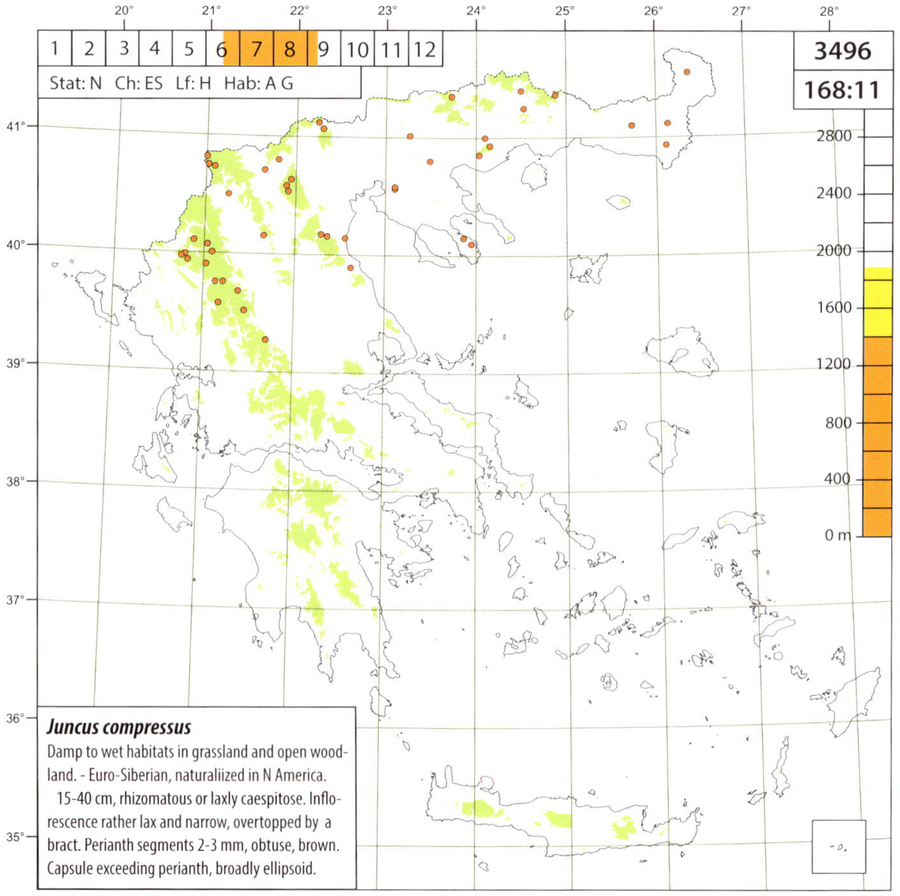

Juncus compressus

Damp to wet habitats in grassland and open woodland. - Euro-Siberian, naturalized in N America.

15-40 cm, rhizomatous or laxly caespitose. Inflorescence rather lax and narrow, overtopped by a bract. Perianth segments 2-3 mm, obtuse, brown. Capsule exceeding perianth, broadly ellipsoid.

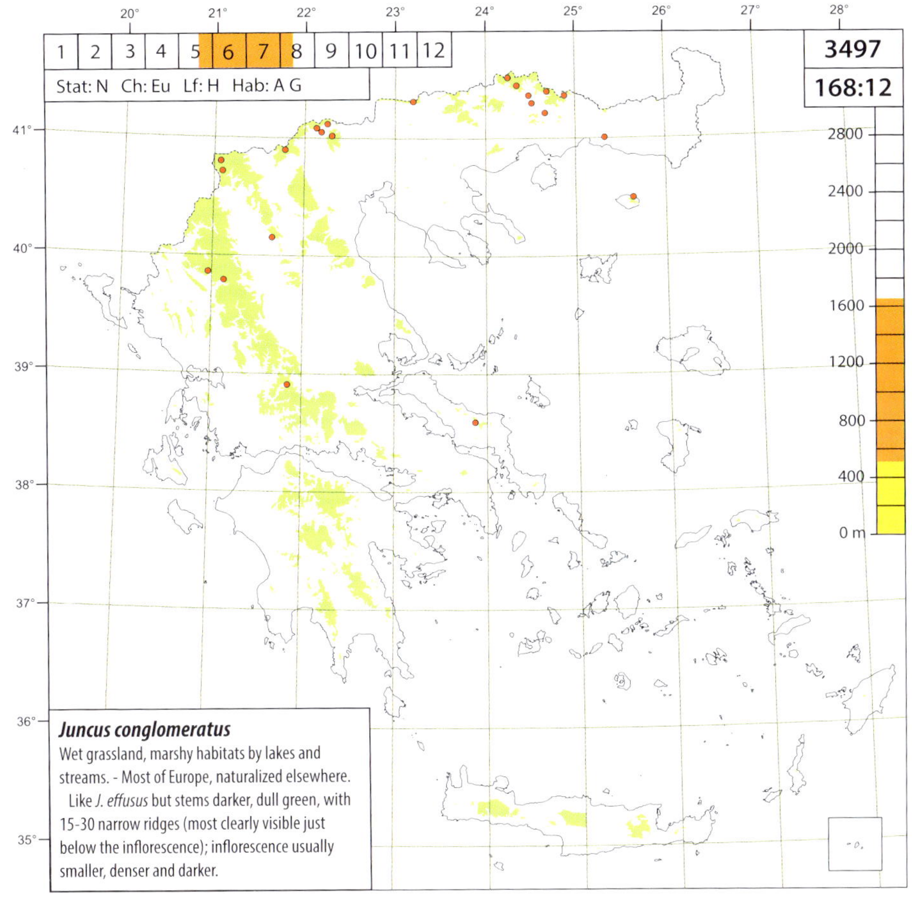

Juncus conglomeratus

Wet grassland, marshy habitats by lakes and streams. - Most of Europe, naturalized elsewhere.

Like *J. effusus* but stems darker, dull green, with 15-30 narrow ridges (most clearly visible just below the inflorescence); inflorescence usually smaller, denser and darker.

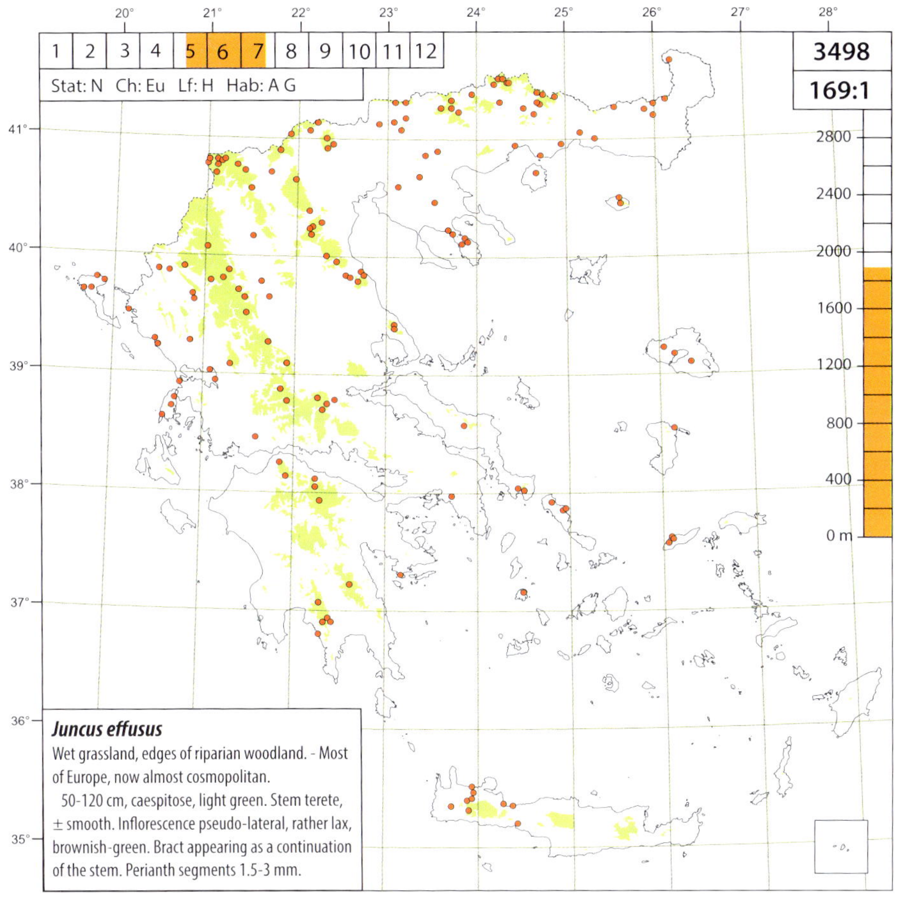

Juncus effusus

Wet grassland, edges of riparian woodland. - Most of Europe, now almost cosmopolitan.

50-120 cm, caespitose, light green. Stem terete, ± smooth. Inflorescence pseudo-lateral, rather lax, brownish-green. Bract appearing as a continuation of the stem. Perianth segments 1.5-3 mm.

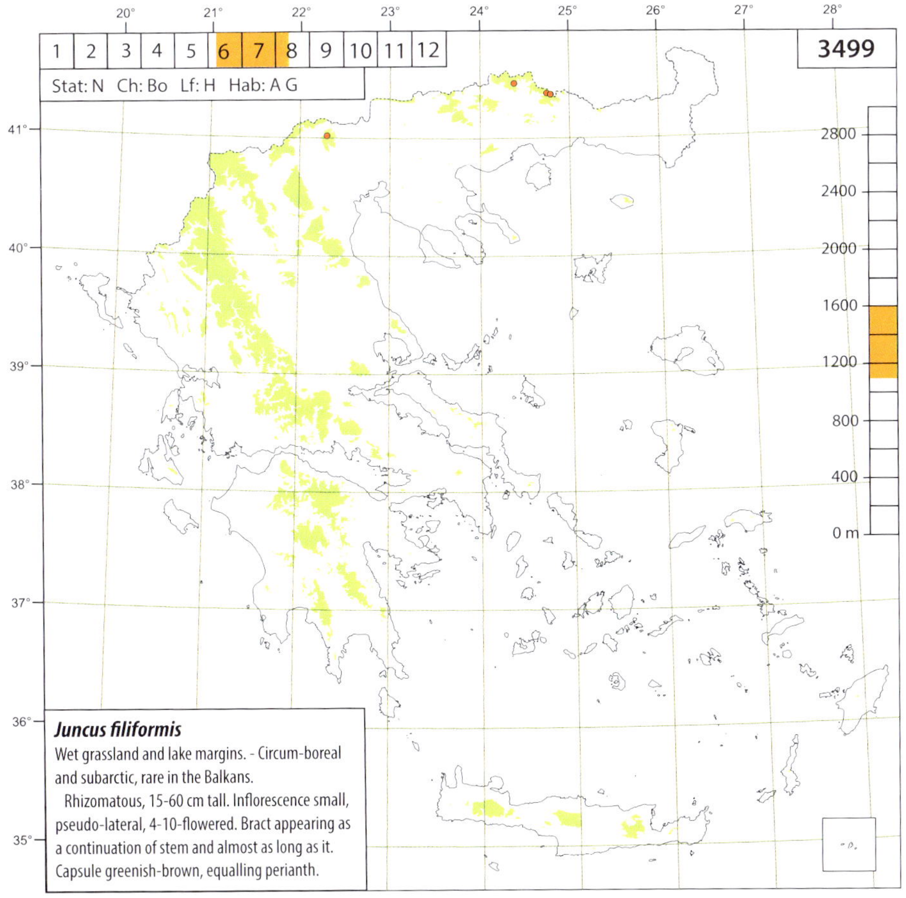

Juncus filiformis

Wet grassland and lake margins. - Circum-boreal and subarctic, rare in the Balkans.

Rhizomatous, 15-60 cm tall. Inflorescence small, pseudo-lateral, 4-10-flowered. Bract appearing as a continuation of stem and almost as long as it. Capsule greenish-brown, equalling perianth.

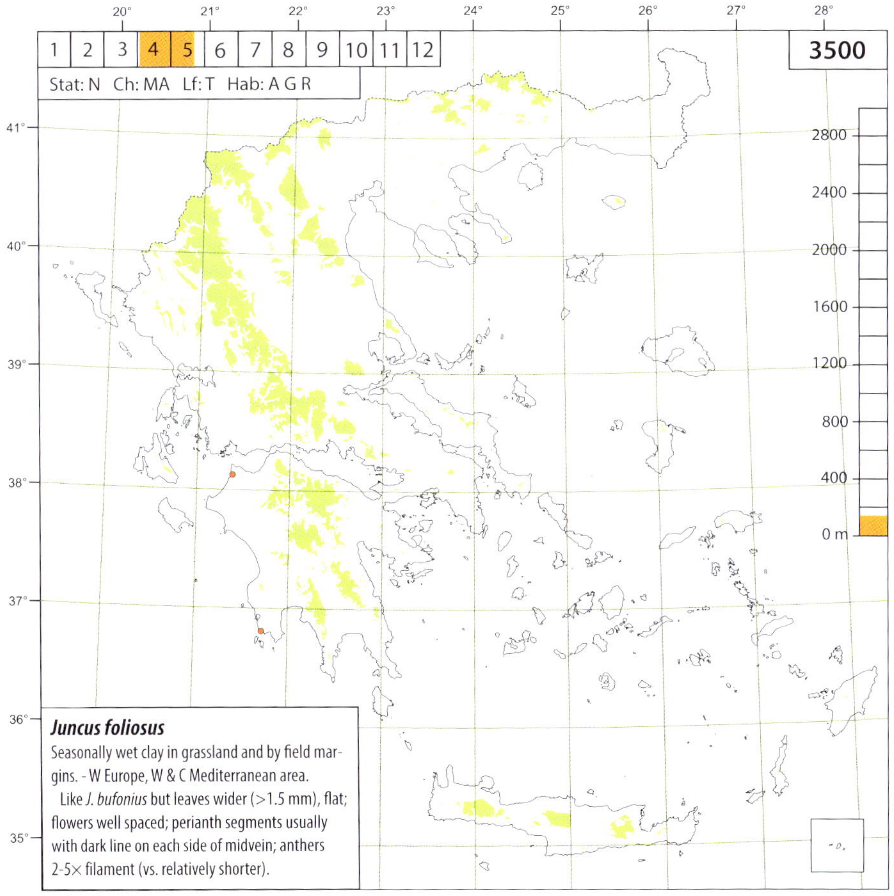

Juncus foliosus

Seasonally wet clay in grassland and by field margins. - W Europe, W & C Mediterranean area.

Like *J. bufonius* but leaves wider (>1.5 mm), flat; flowers well spaced; perianth segments usually with dark line on each side of midvein; anthers 2-5× filament (vs. relatively shorter).

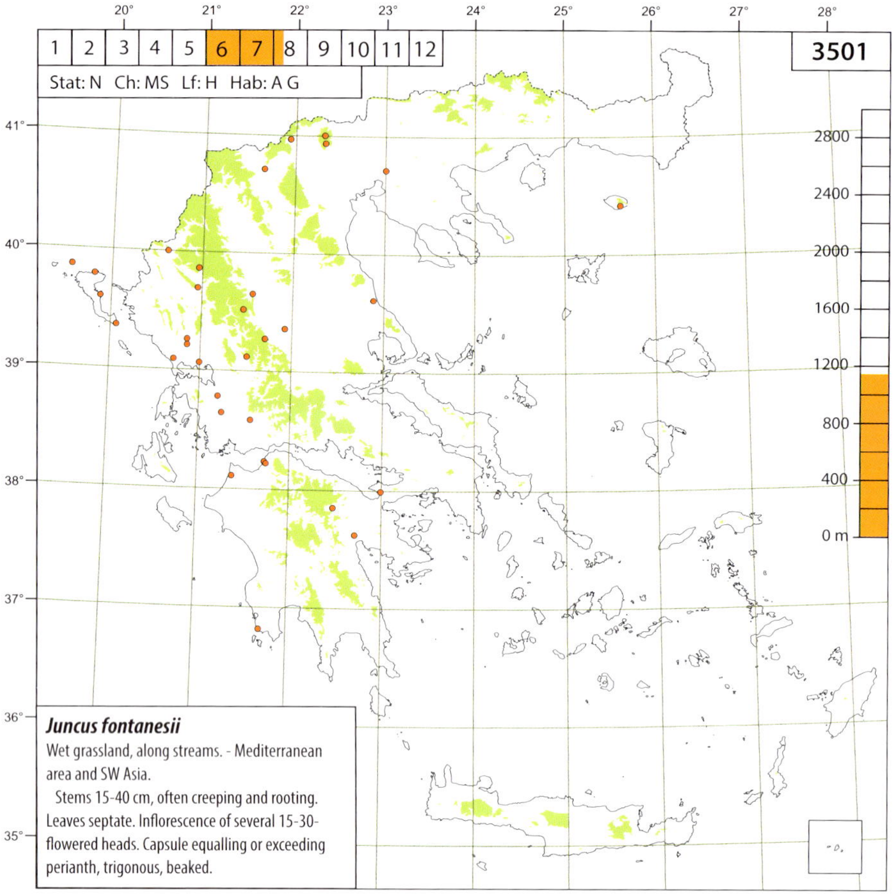

Juncus fontanesii

Wet grassland, along streams. - Mediterranean area and SW Asia.

Stems 15-40 cm, often creeping and rooting. Leaves septate. Inflorescence of several 15-30-flowered heads. Capsule equalling or exceeding perianth, trigonous, beaked.

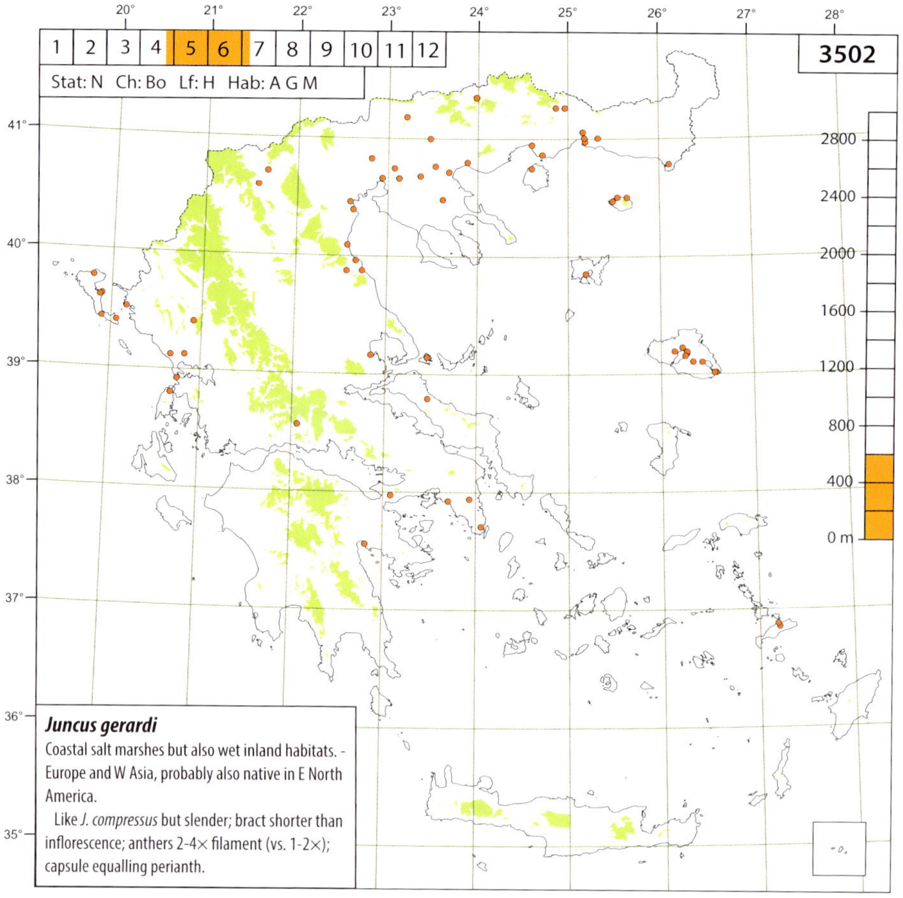

Juncus gerardi

Coastal salt marshes but also wet inland habitats. - Europe and W Asia, probably also native in E North America.

Like *J. compressus* but slender; bract shorter than inflorescence; anthers 2-4× filament (vs. 1-2×); capsule equalling perianth.

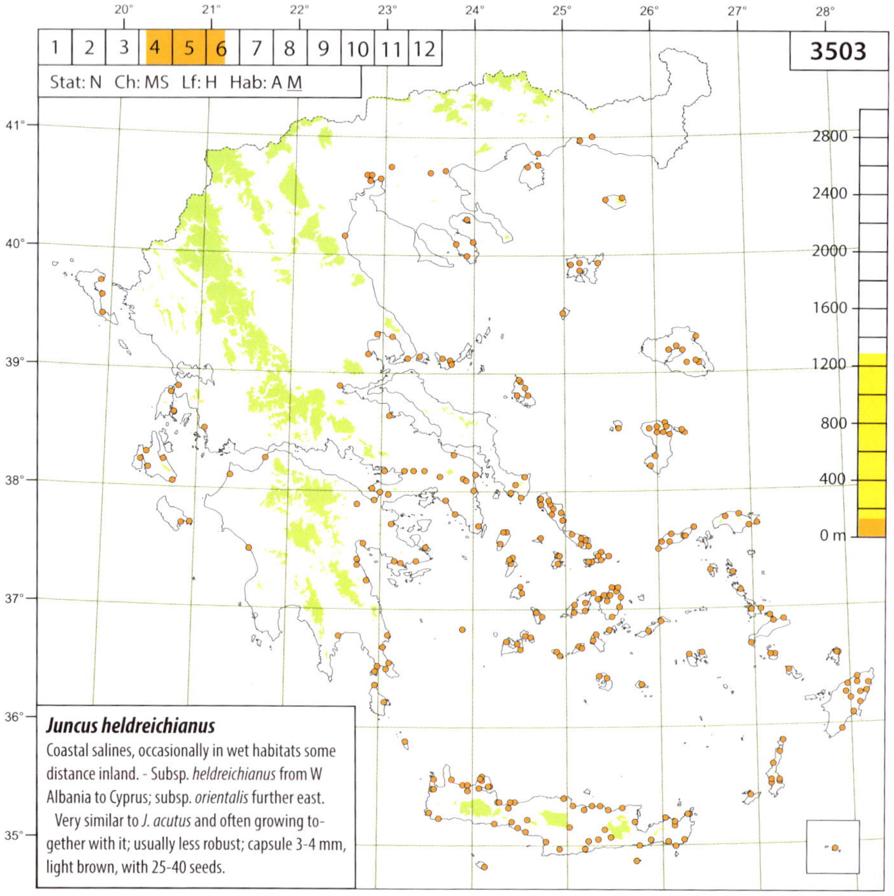

Juncus heldreichianus

Coastal salines, occasionally in wet habitats some distance inland. - Subsp. *heldreichianus* from W Albania to Cyprus; subsp. *orientalis* further east.

Very similar to *J. acutus* and often growing together with it; usually less robust; capsule 3-4 mm, light brown, with 25-40 seeds.

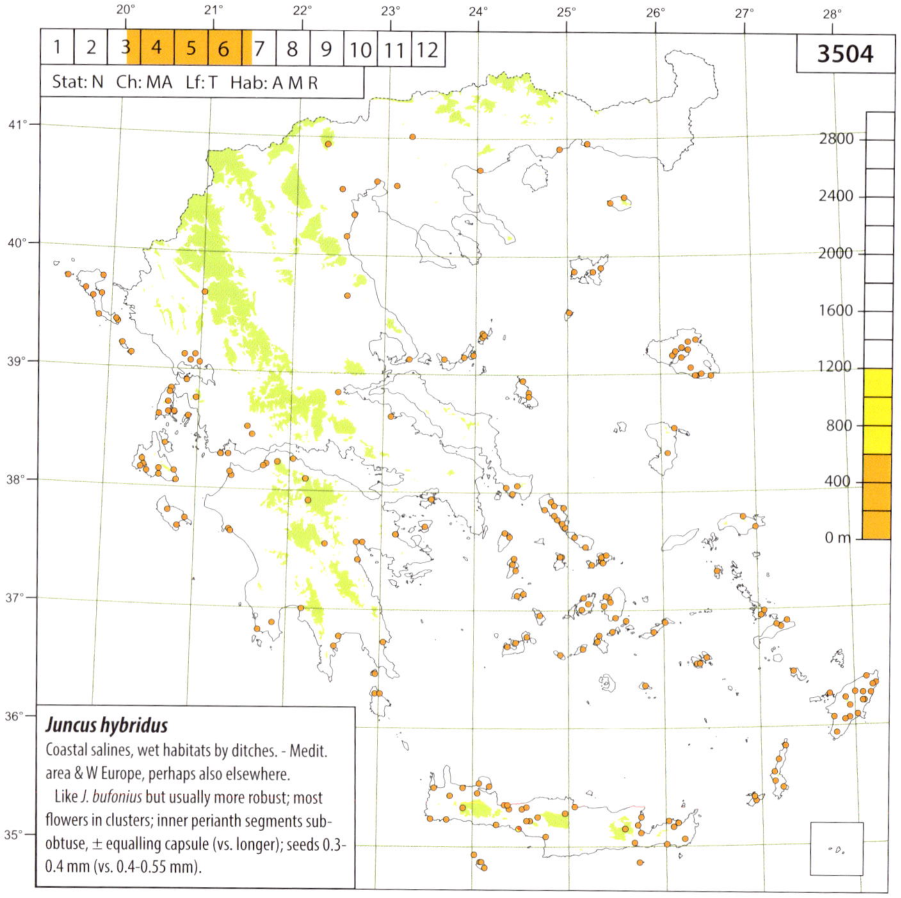

Juncus hybridus

Coastal salines, wet habitats by ditches. - Medit. area & W Europe, perhaps also elsewhere.

Like *J. bufonius* but usually more robust; most flowers in clusters; inner perianth segments sub-obtuse, ± equalling capsule (vs. longer); seeds 0.3-0.4 mm (vs. 0.4-0.55 mm).

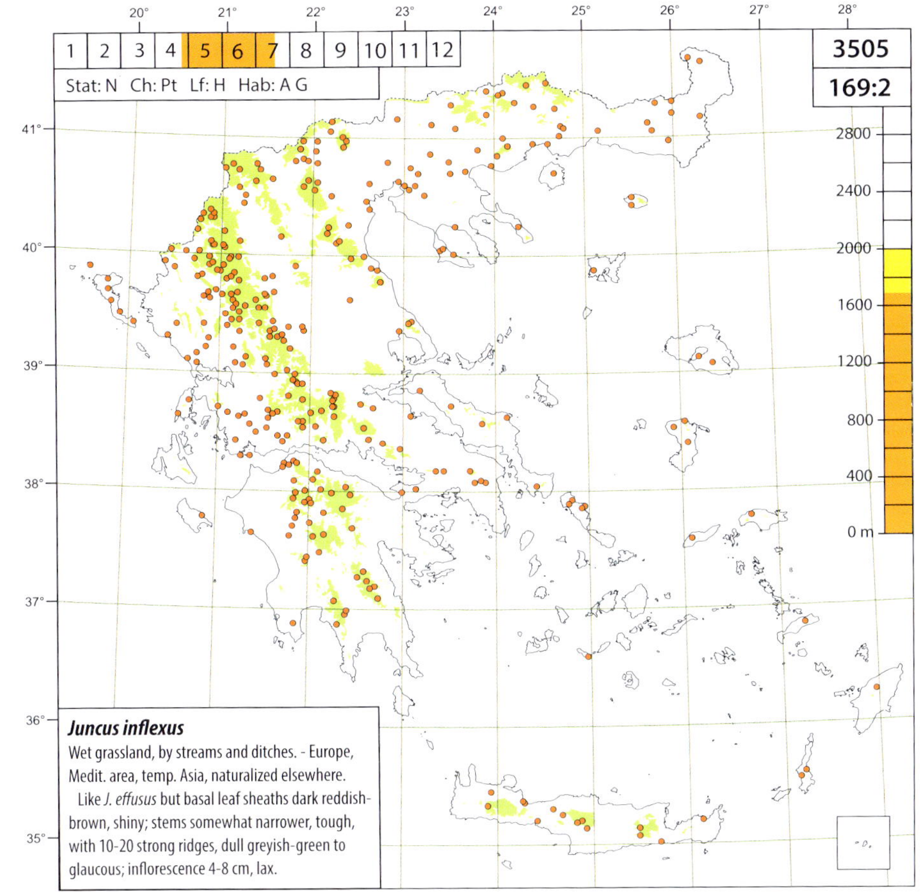

Juncus inflexus

Wet grassland, by streams and ditches. - Europe, Medit. area, temp. Asia, naturalized elsewhere.

Like *J. effusus* but basal leaf sheaths dark reddish-brown, shiny; stems somewhat narrower, tough, with 10-20 strong ridges, dull greyish-green to glaucous; inflorescence 4-8 cm, lax.

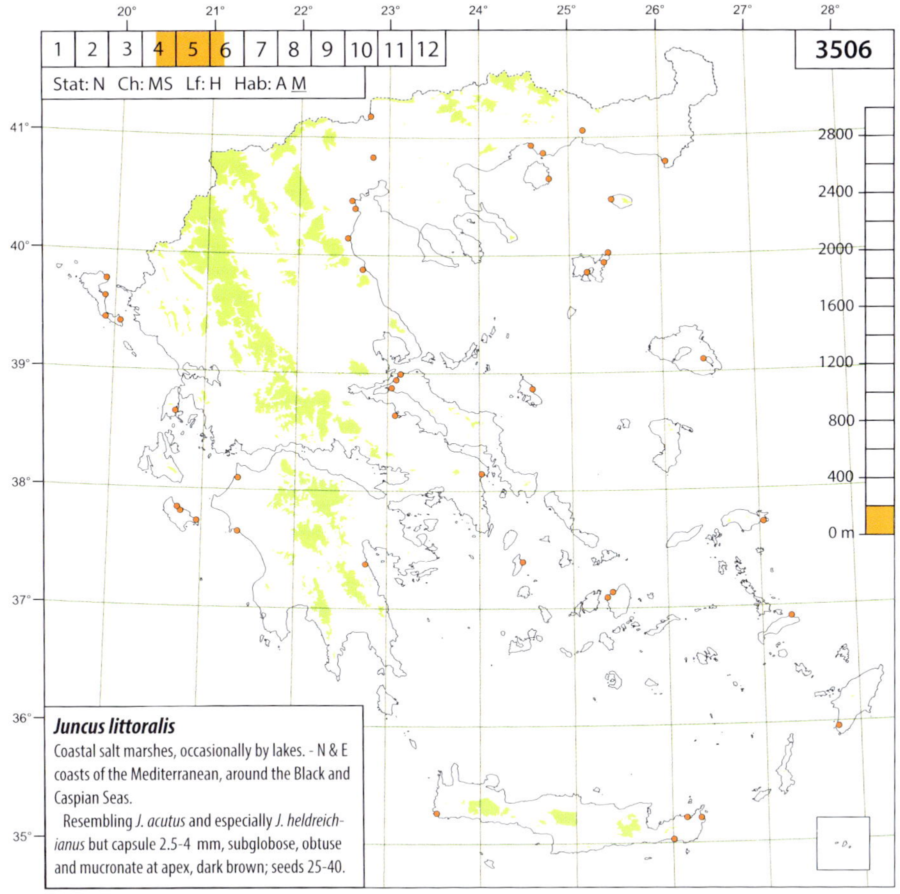

Juncus littoralis

Coastal salt marshes, occasionally by lakes. - N & E coasts of the Mediterranean, around the Black and Caspian Seas.

Resembling *J. acutus* and especially *J. heldreichianus* but capsule 2.5-4 mm, subglobose, obtuse and mucronate at apex, dark brown; seeds 25-40.

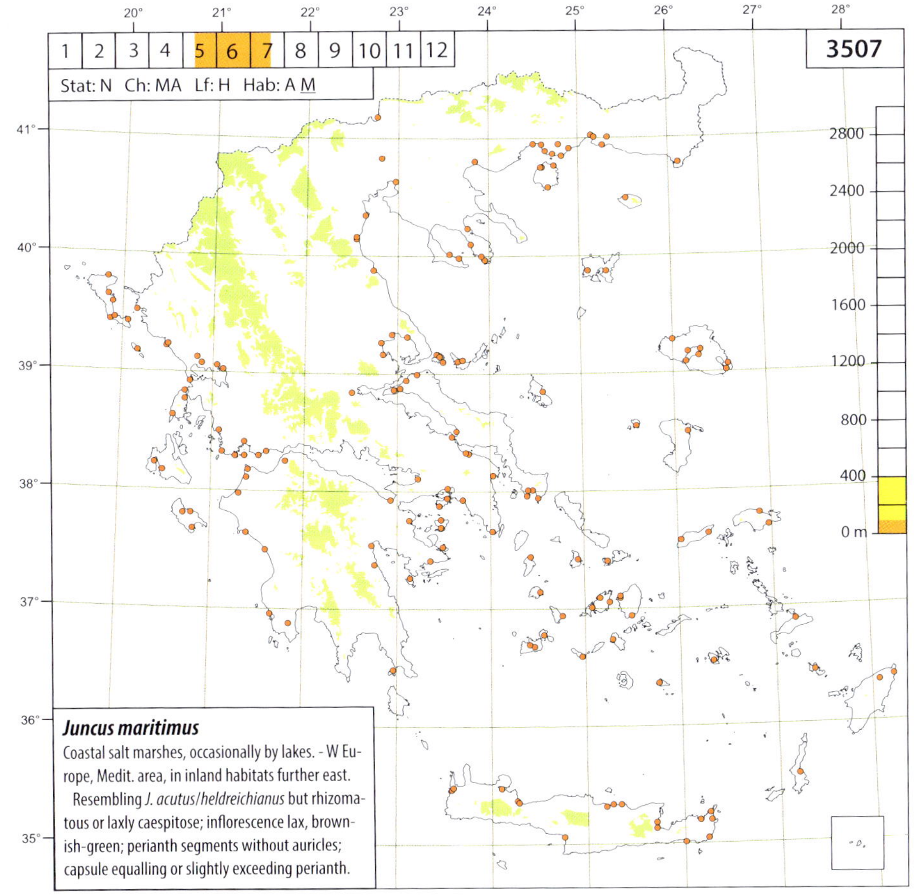

Juncus maritimus

Coastal salt marshes, occasionally by lakes. - W Europe, Medit. area, in inland habitats further east.

Resembling *J. acutus/heldreichianus* but rhizomatous or laxly caespitose; inflorescence lax, brownish-green; perianth segments without auricles; capsule equalling or slightly exceeding perianth.

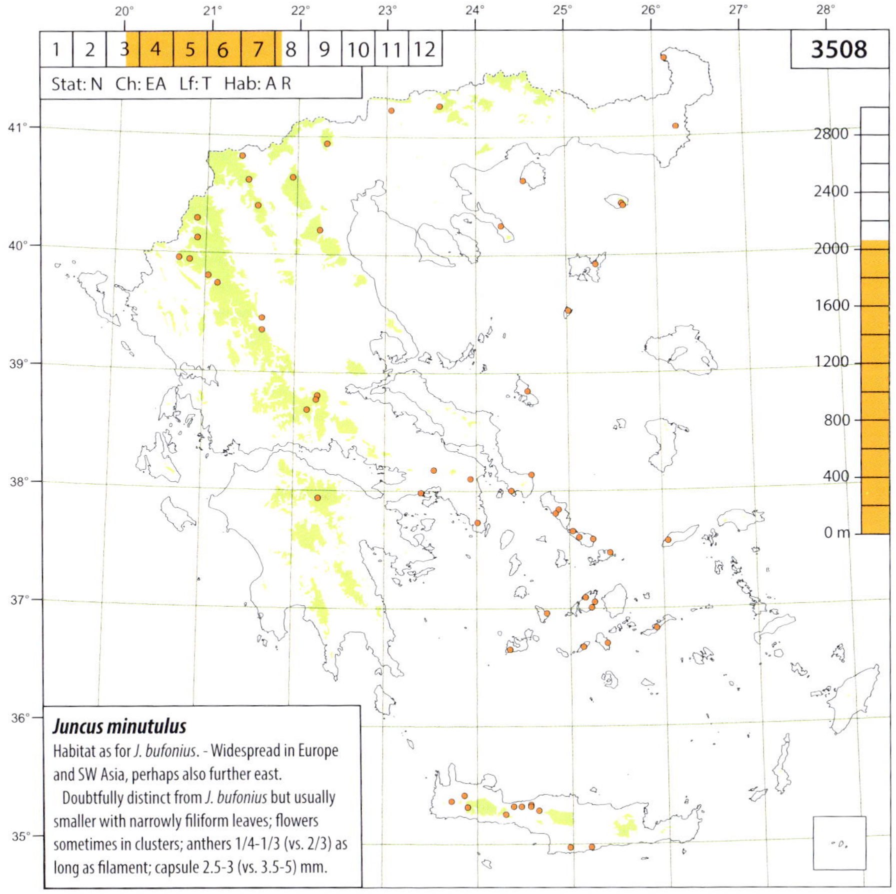

Juncus minutulus

Habitat as for *J. bufonius*. - Widespread in Europe and SW Asia, perhaps also further east.

Doubtfully distinct from *J. bufonius* but usually smaller with narrowly filiform leaves; flowers sometimes in clusters; anthers 1/4-1/3 (vs. 2/3) as long as filament; capsule 2.5-3 (vs. 3.5-5) mm.

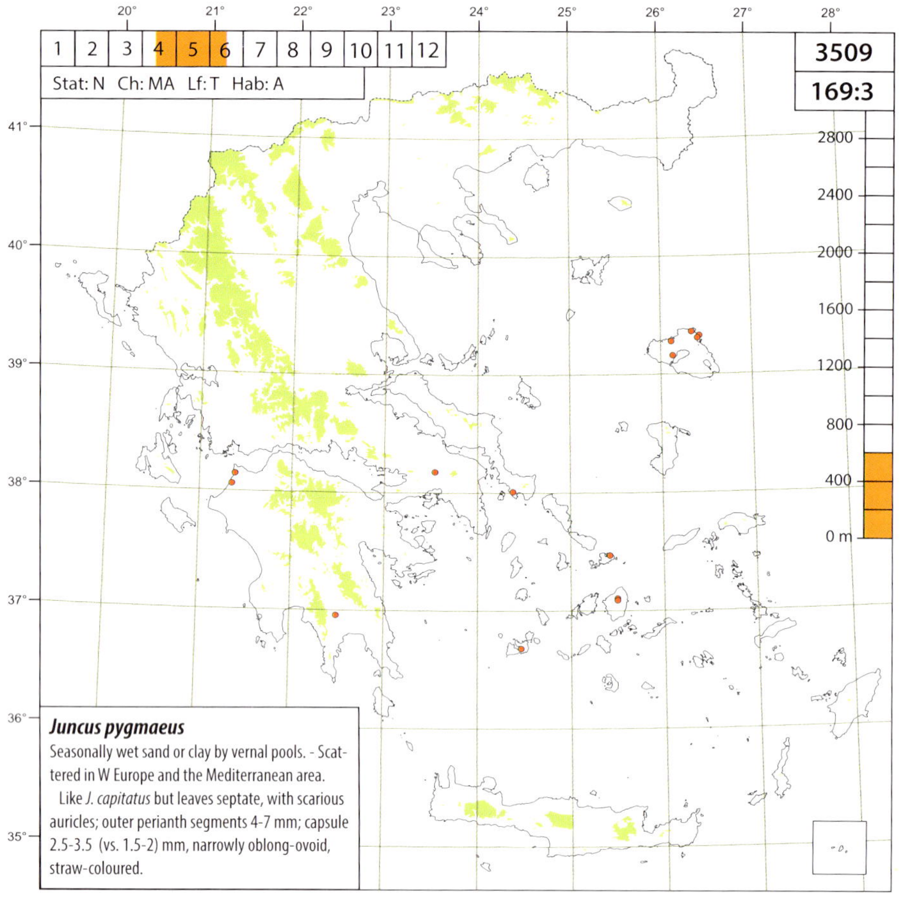

Juncus pygmaeus

Seasonally wet sand or clay by vernal pools. - Scattered in W Europe and the Mediterranean area.

Like *J. capitatus* but leaves septate, with scarious auricles; outer perianth segments 4-7 mm; capsule 2.5-3.5 (vs. 1.5-2) mm, narrowly oblong-ovoid, straw-coloured.

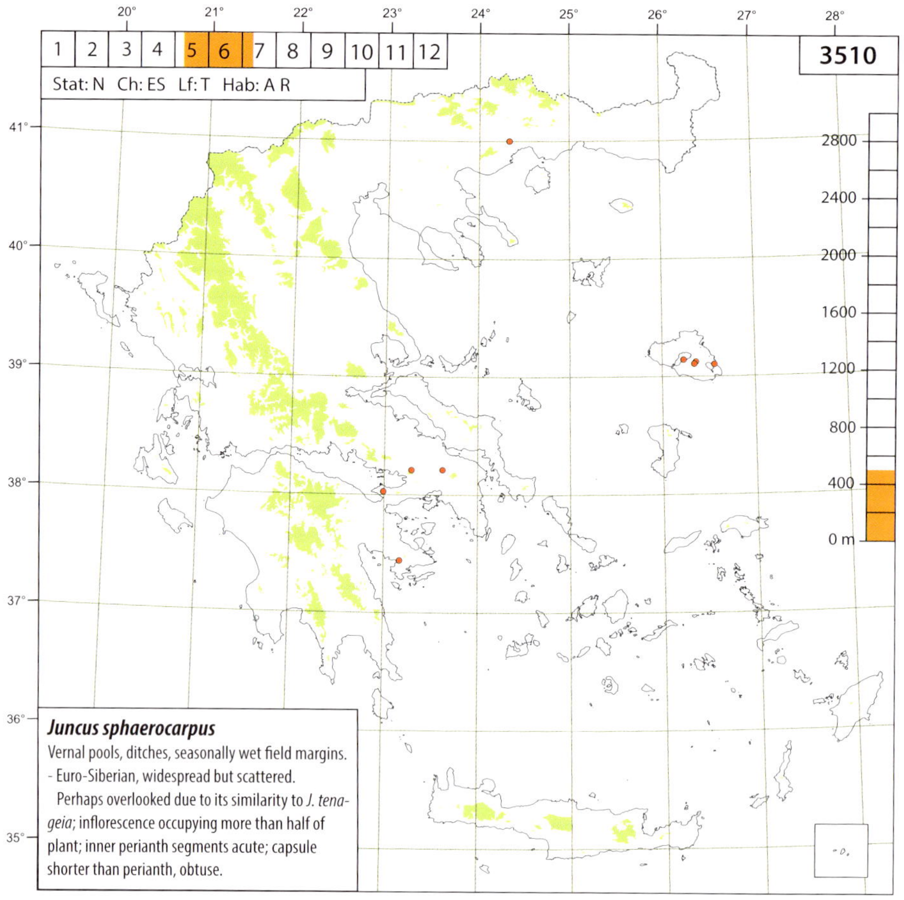

Juncus sphaerocarpus

Vernal pools, ditches, seasonally wet field margins. - Euro-Siberian, widespread but scattered.

Perhaps overlooked due to its similarity to *J. tenageia*; inflorescence occupying more than half of plant; inner perianth segments acute; capsule shorter than perianth, obtuse.

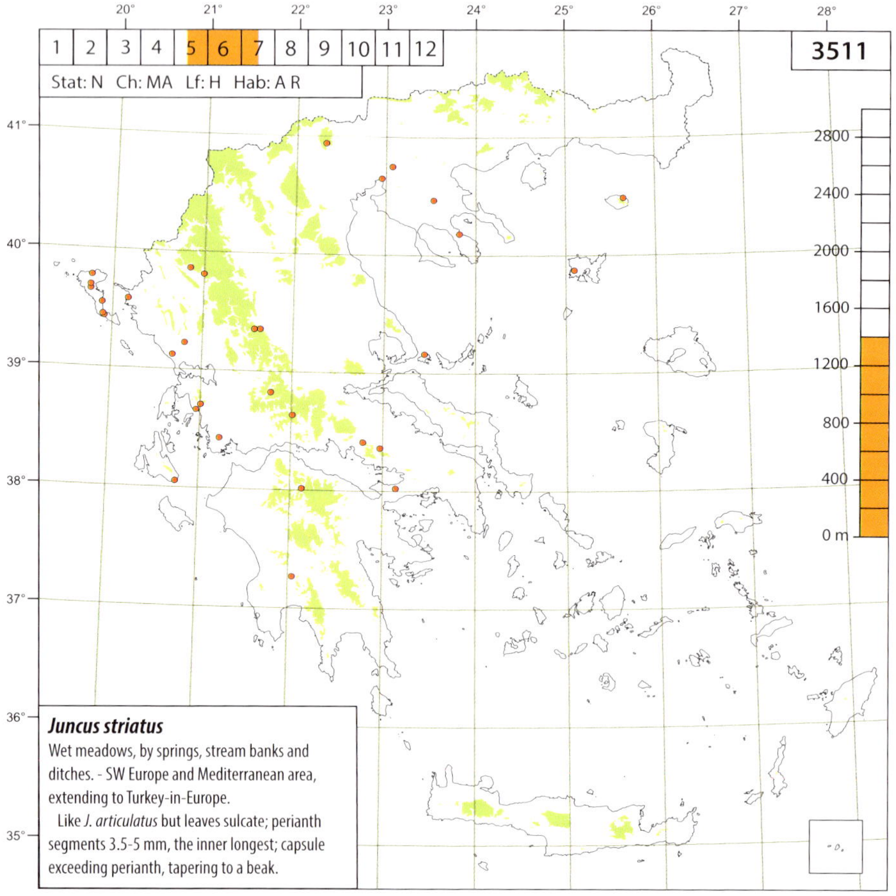

Juncus striatus

Wet meadows, by springs, stream banks and ditches. - SW Europe and Mediterranean area, extending to Turkey-in-Europe.

Like *J. articulatus* but leaves sulcate; perianth segments 3.5-5 mm, the inner longest; capsule exceeding perianth, tapering to a beak.

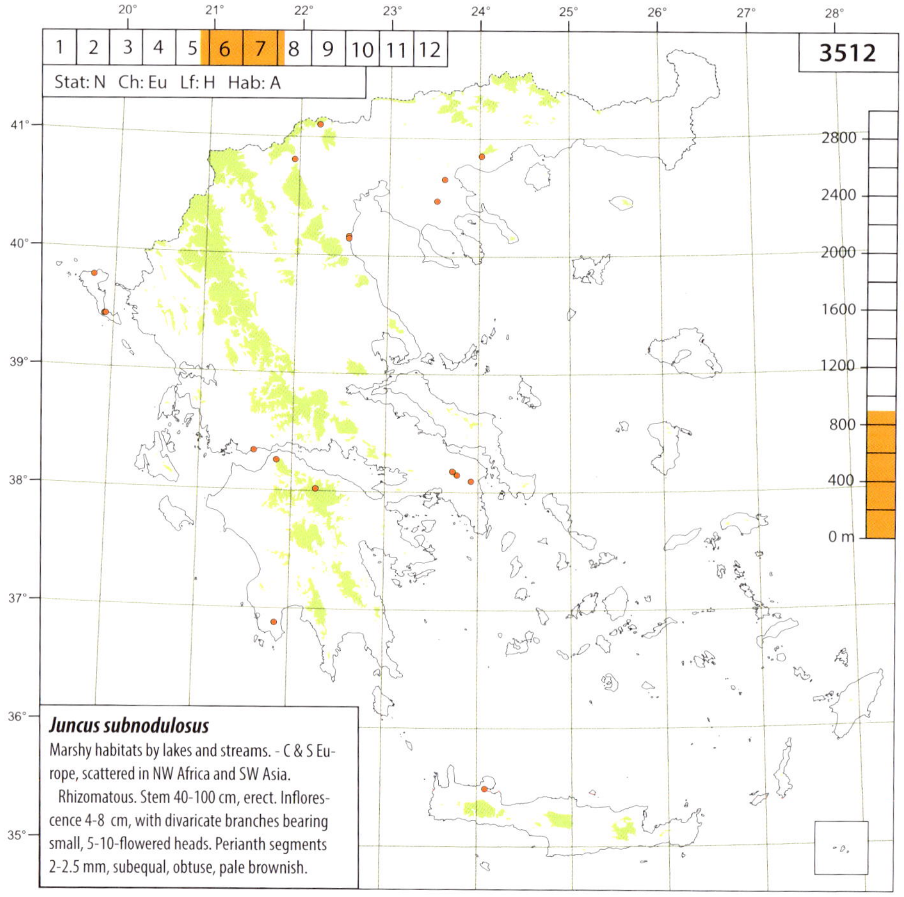

Juncus subnodulosus

Marshy habitats by lakes and streams. - C & S Europe, scattered in NW Africa and SW Asia.

Rhizomatous. Stem 40-100 cm, erect. Inflorescence 4-8 cm, with divaricate branches bearing small, 5-10-flowered heads. Perianth segments 2-2.5 mm, subequal, obtuse, pale brownish.

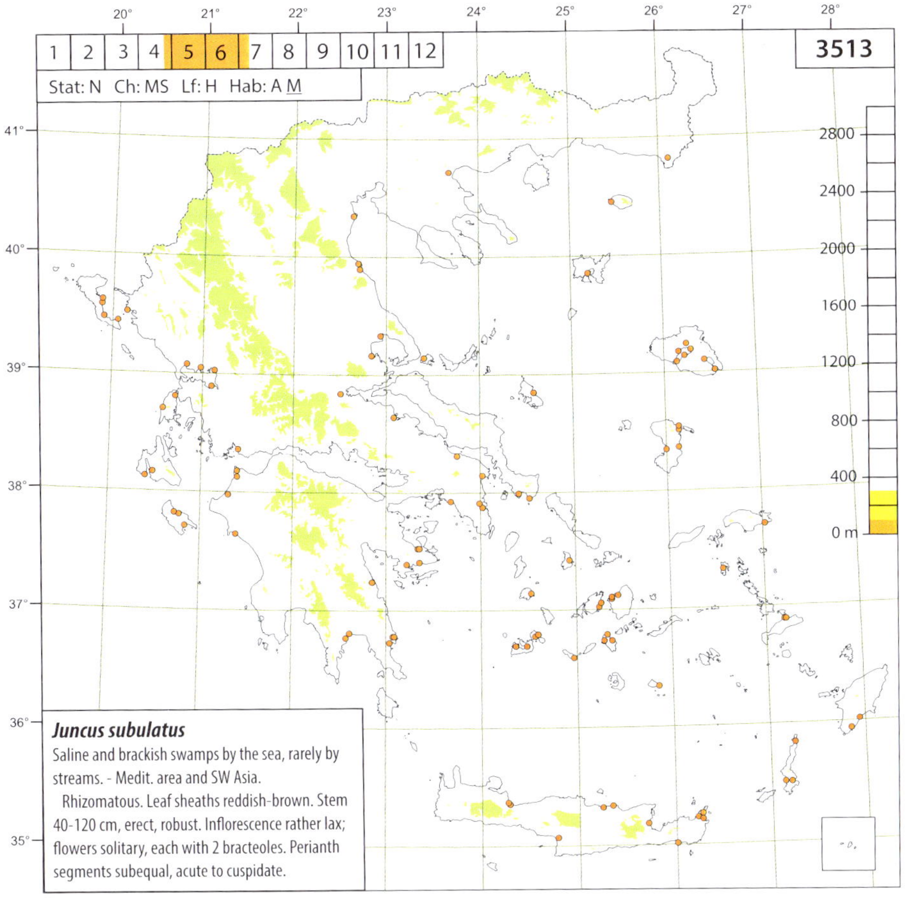

Juncus subulatus

Saline and brackish swamps by the sea, rarely by streams. - Medit. area and SW Asia.

Rhizomatous. Leaf sheaths reddish-brown. Stem 40-120 cm, erect, robust. Inflorescence rather lax; flowers solitary, each with 2 bracteoles. Perianth segments subequal, acute to cuspidate.

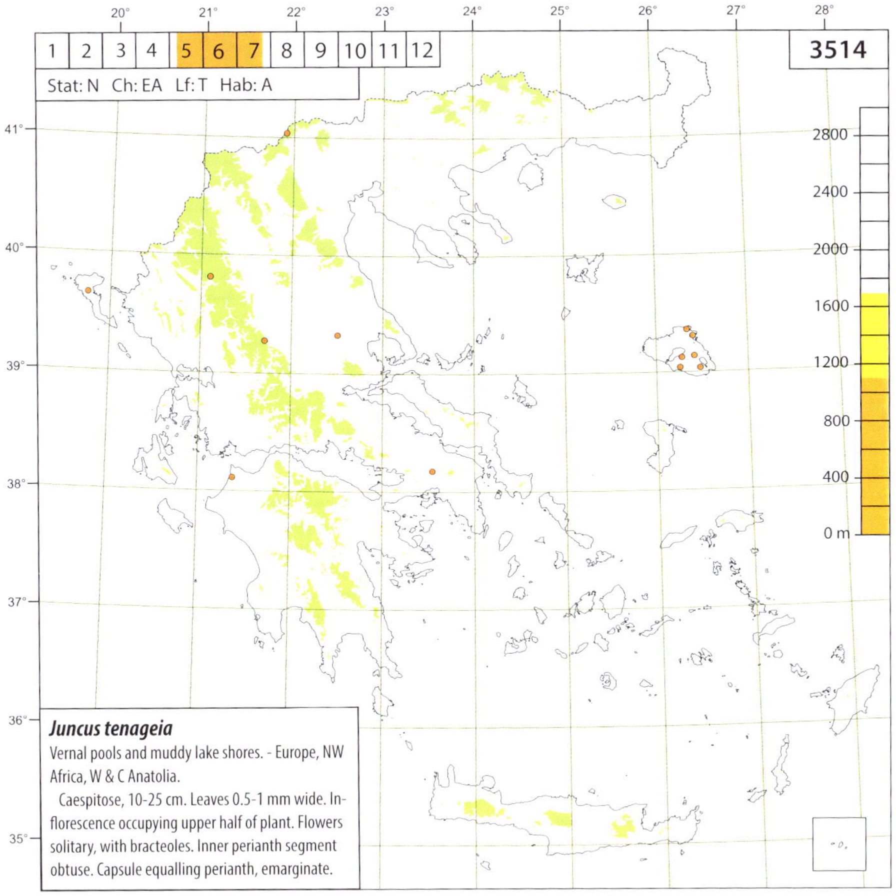

Juncus tenageia

Vernal pools and muddy lake shores. - Europe, NW Africa, W & C Anatolia.

Caespitose, 10-25 cm. Leaves 0.5-1 mm wide. Inflorescence occupying upper half of plant. Flowers solitary, with bracteoles. Inner perianth segment obtuse. Capsule equalling perianth, emarginate.

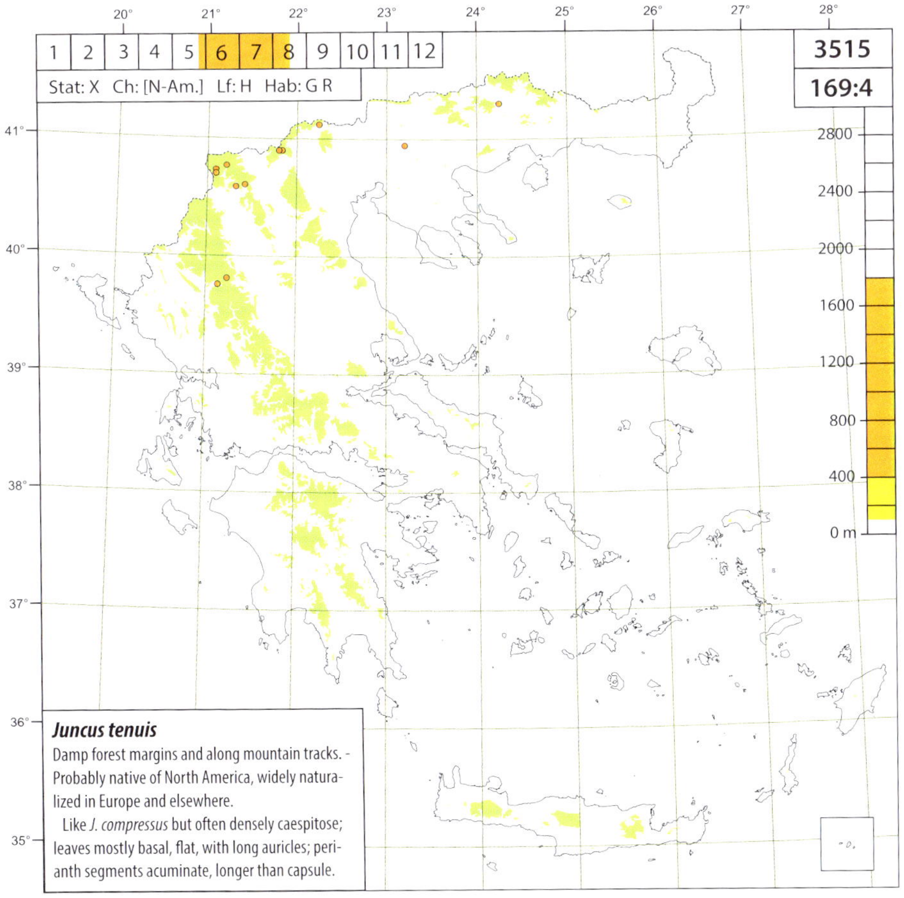

Juncus tenuis

Damp forest margins and along mountain tracks. - Probably native of North America, widely naturalized in Europe and elsewhere.

Like *J. compressus* but often densely caespitose; leaves mostly basal, flat, with long auricles; perianth segments acuminate, longer than capsule.

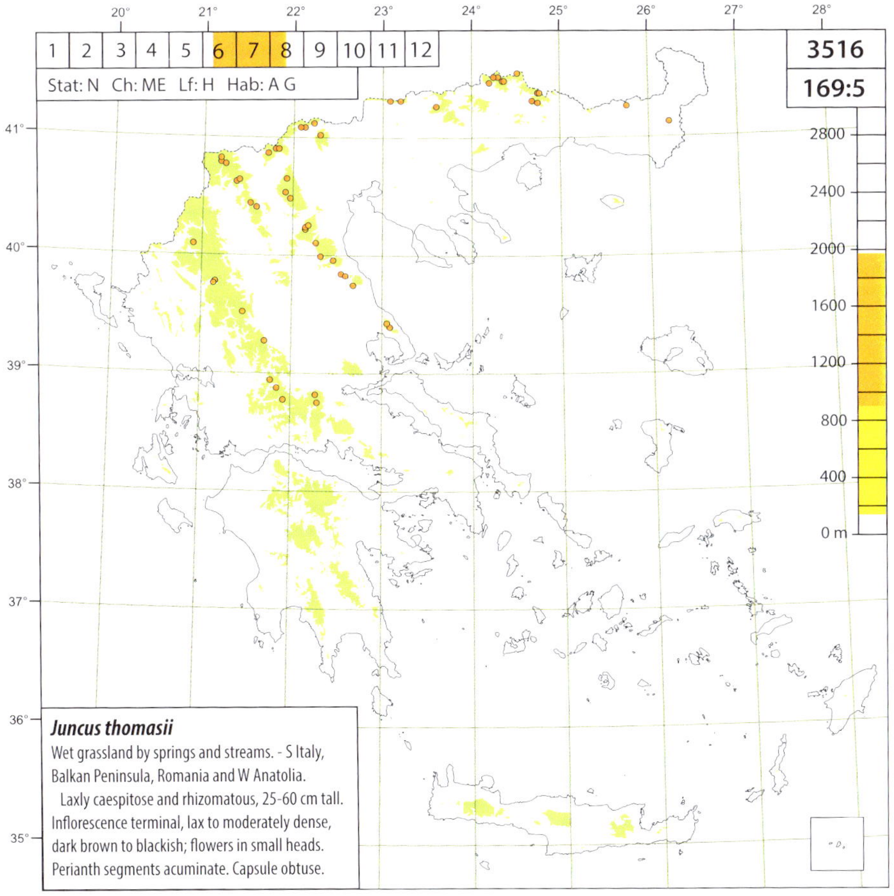

Juncus thomasii

Wet grassland by springs and streams. - S Italy, Balkan Peninsula, Romania and W Anatolia.

Laxly caespitose and rhizomatous, 25-60 cm tall. Inflorescence terminal, lax to moderately dense, dark brown to blackish; flowers in small heads. Perianth segments acuminate. Capsule obtuse.

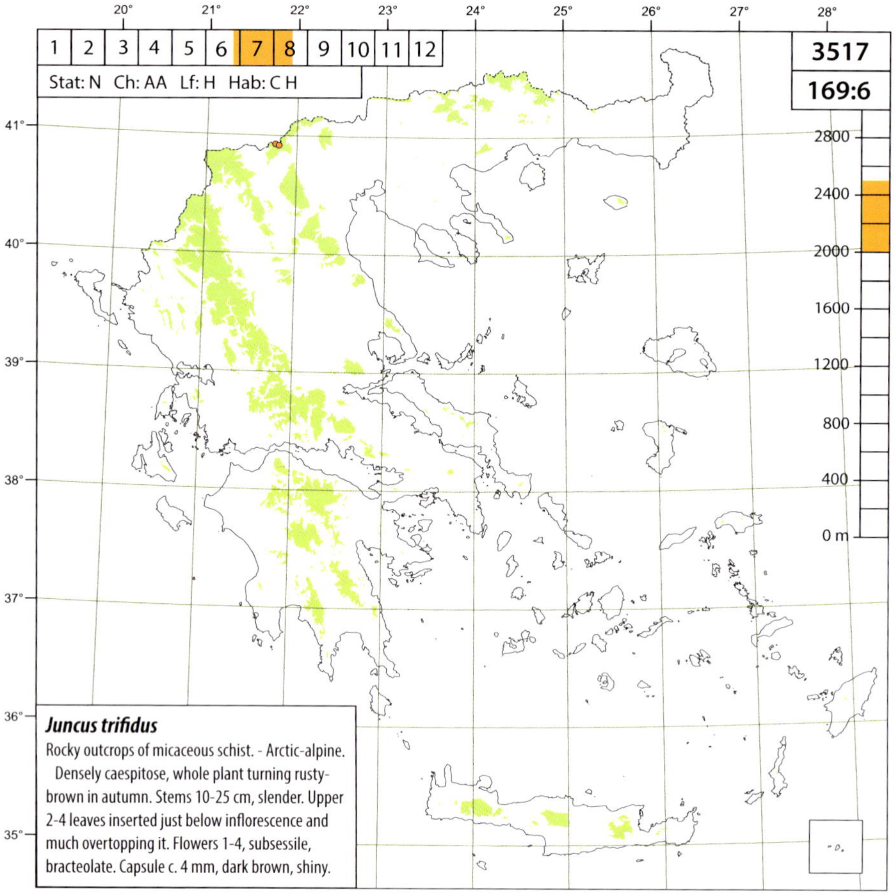

Juncus trifidus

Rocky outcrops of micaceous schist. - Arctic-alpine.

Densely caespitose, whole plant turning rusty-brown in autumn. Stems 10-25 cm, slender. Upper 2-4 leaves inserted just below inflorescence and much overtopping it. Flowers 1-4, subsessile, bracteolate. Capsule c. 4 mm, dark brown, shiny.

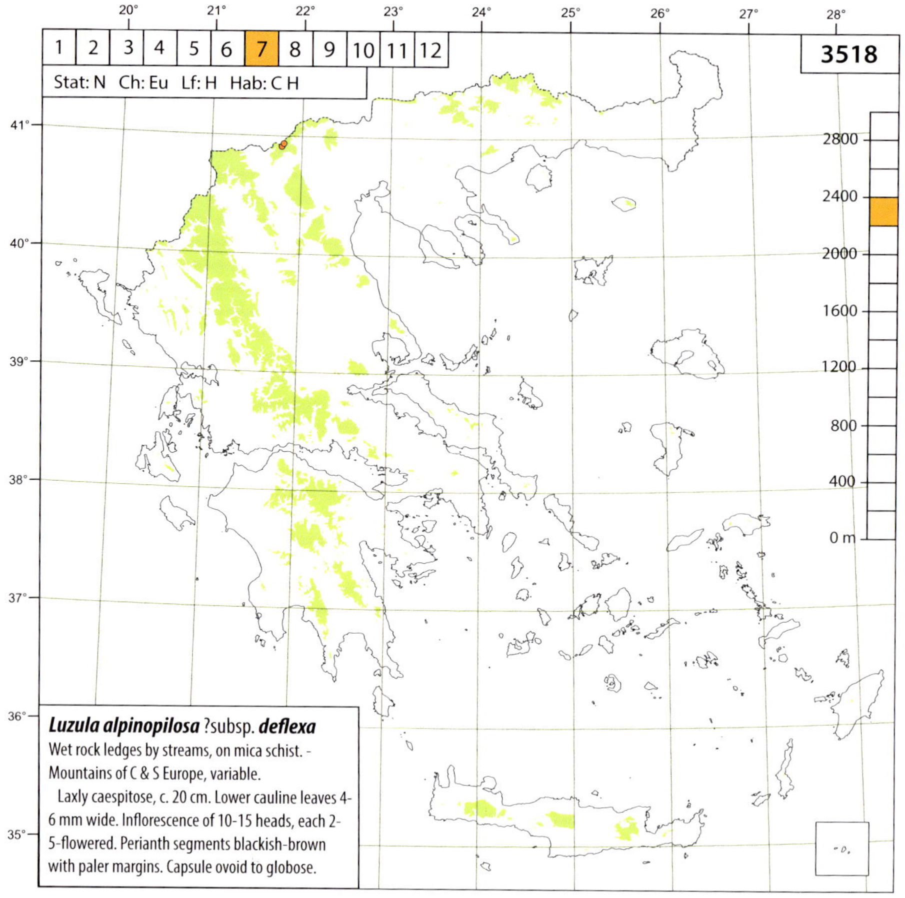

Luzula alpinopilosa ?subsp. ***deflexa***

Wet rock ledges by streams, on mica schist. - Mountains of C & S Europe, variable.

Laxly caespitose, c. 20 cm. Lower cauline leaves 4-6 mm wide. Inflorescence of 10-15 heads, each 2-5-flowered. Perianth segments blackish-brown with paler margins. Capsule ovoid to globose.

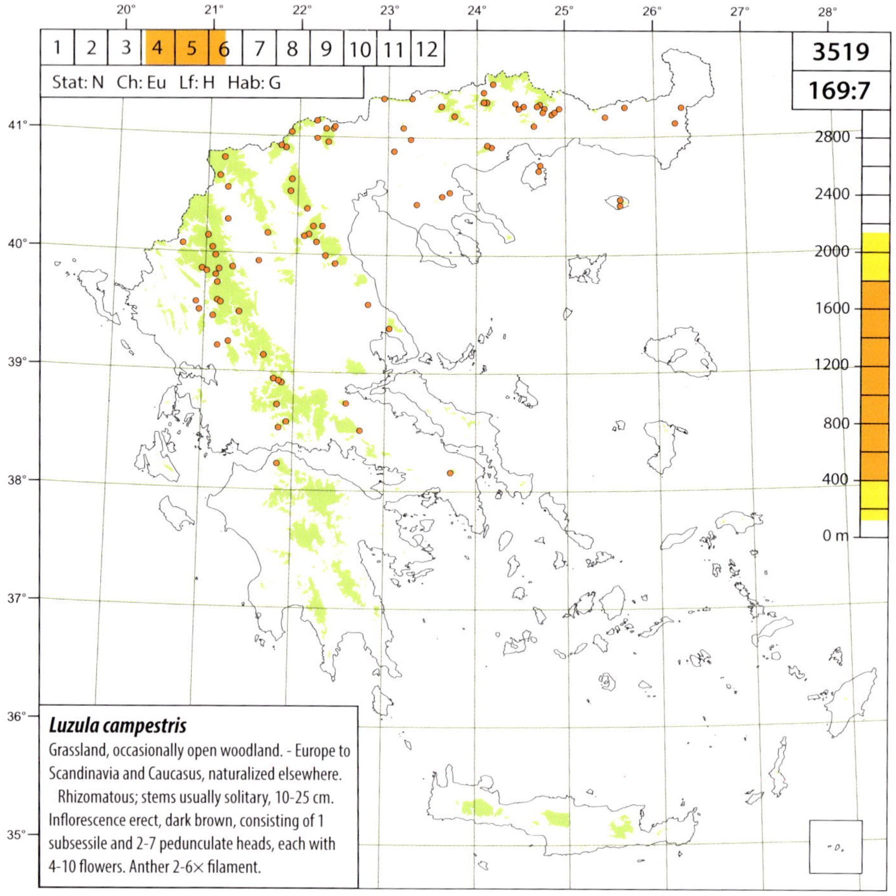

Luzula campestris

Grassland, occasionally open woodland. - Europe to Scandinavia and Caucasus, naturalized elsewhere.

Rhizomatous; stems usually solitary, 10-25 cm. Inflorescence erect, dark brown, consisting of 1 subsessile and 2-7 pedunculate heads, each with 4-10 flowers. Anther 2-6× filament.

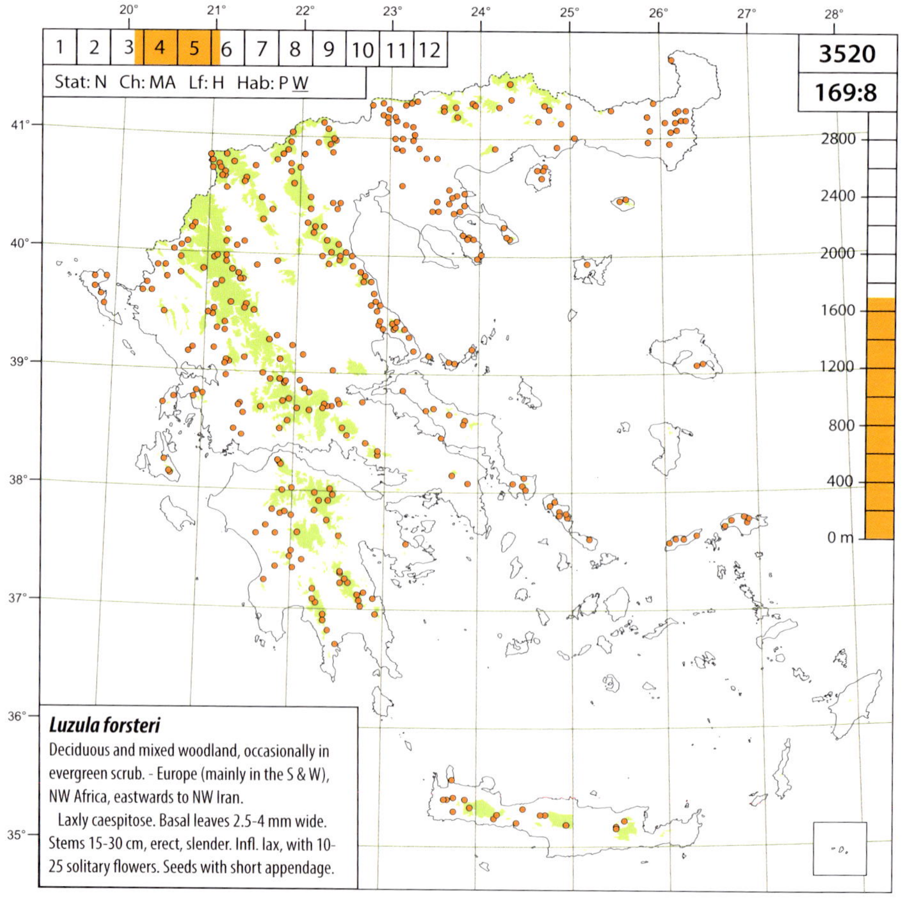

Luzula forsteri

Deciduous and mixed woodland, occasionally in evergreen scrub. - Europe (mainly in the S & W), NW Africa, eastwards to NW Iran.

Laxly caespitose. Basal leaves 2.5-4 mm wide. Stems 15-30 cm, erect, slender. Infl. lax, with 10-25 solitary flowers. Seeds with short appendage.

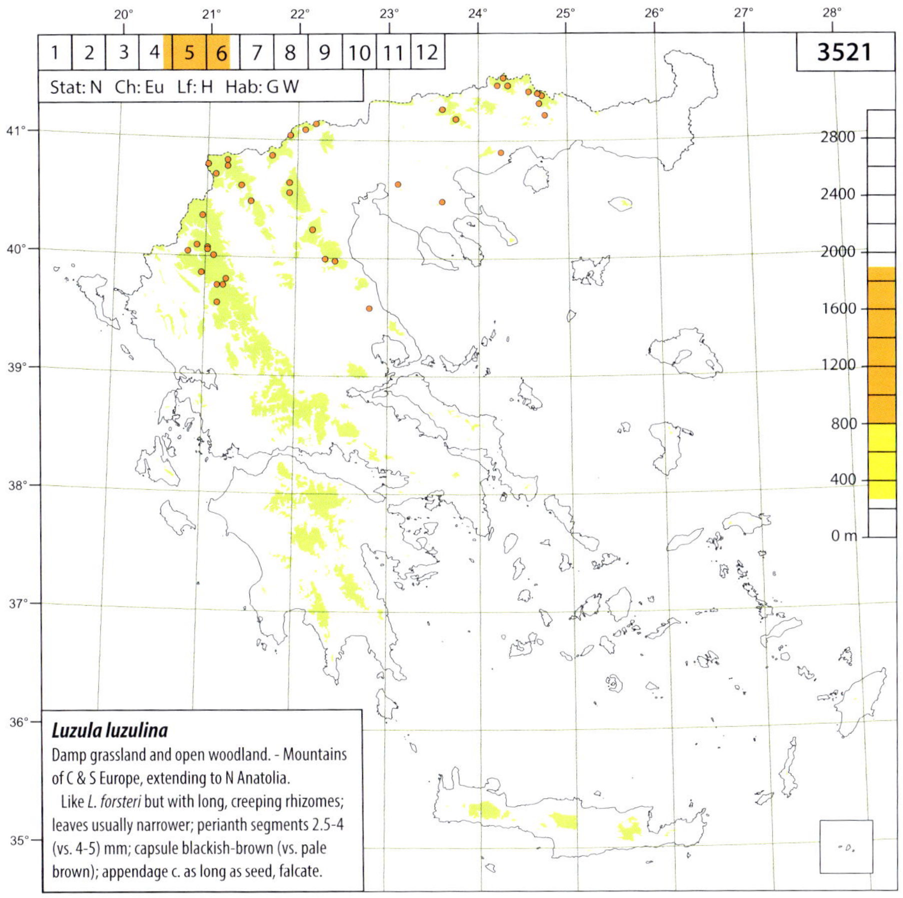

Luzula luzulina

Damp grassland and open woodland. - Mountains of C & S Europe, extending to N Anatolia.

Like *L. forsteri* but with long, creeping rhizomes; leaves usually narrower; perianth segments 2.5-4 (vs. 4-5) mm; capsule blackish-brown (vs. pale brown); appendage c. as long as seed, falcate.

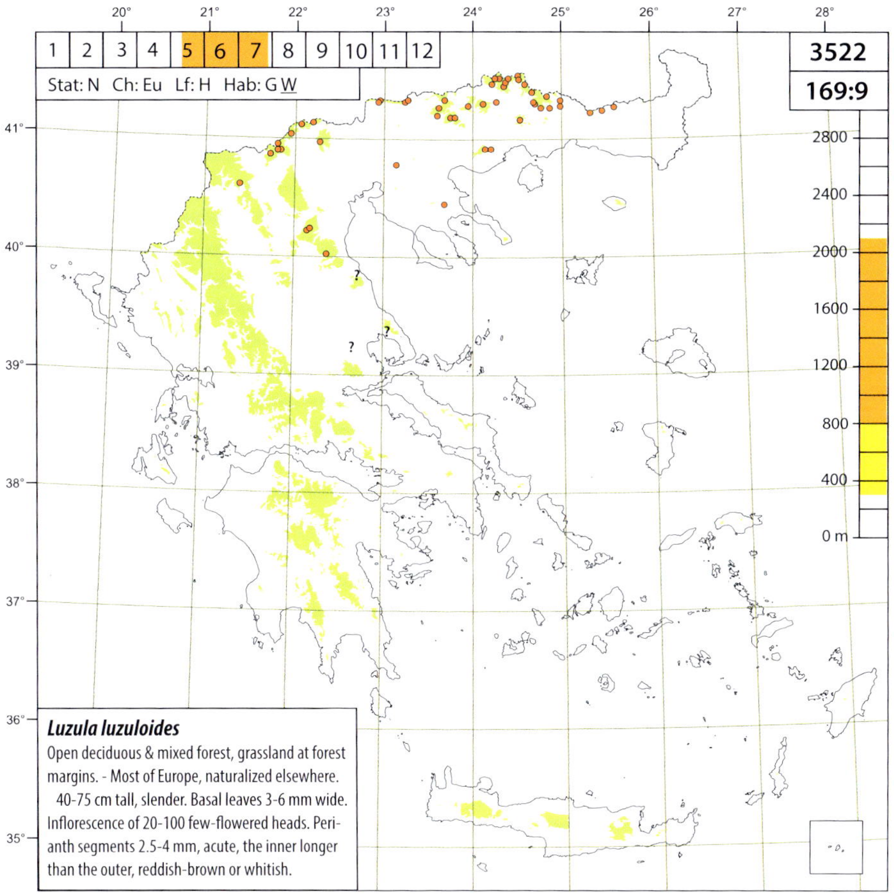

Luzula luzuloides

Open deciduous & mixed forest, grassland at forest margins. - Most of Europe, naturalized elsewhere.

40-75 cm tall, slender. Basal leaves 3-6 mm wide. Inflorescence of 20-100 few-flowered heads. Perianth segments 2.5-4 mm, acute, the inner longer than the outer, reddish-brown or whitish.

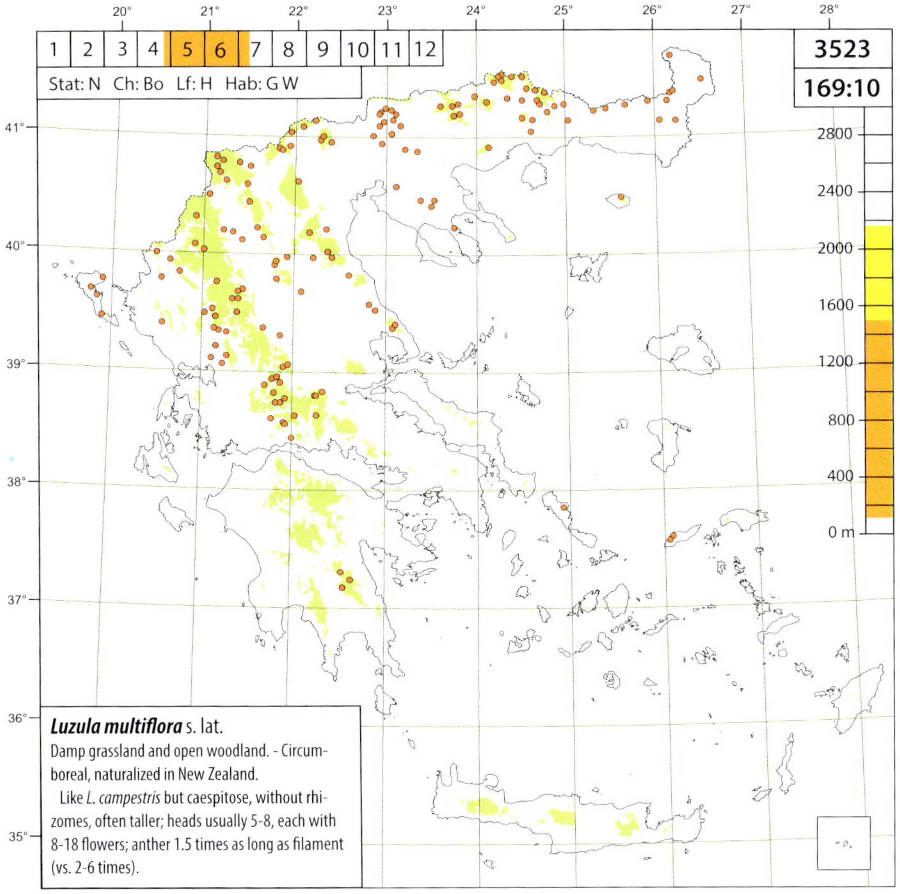

Luzula multiflora s. lat.

Damp grassland and open woodland. - Circumboreal, naturalized in New Zealand.

Like *L. campestris* but caespitose, without rhizomes, often taller; heads usually 5-8, each with 8-18 flowers; anther 1.5 times as long as filament (vs. 2-6 times).

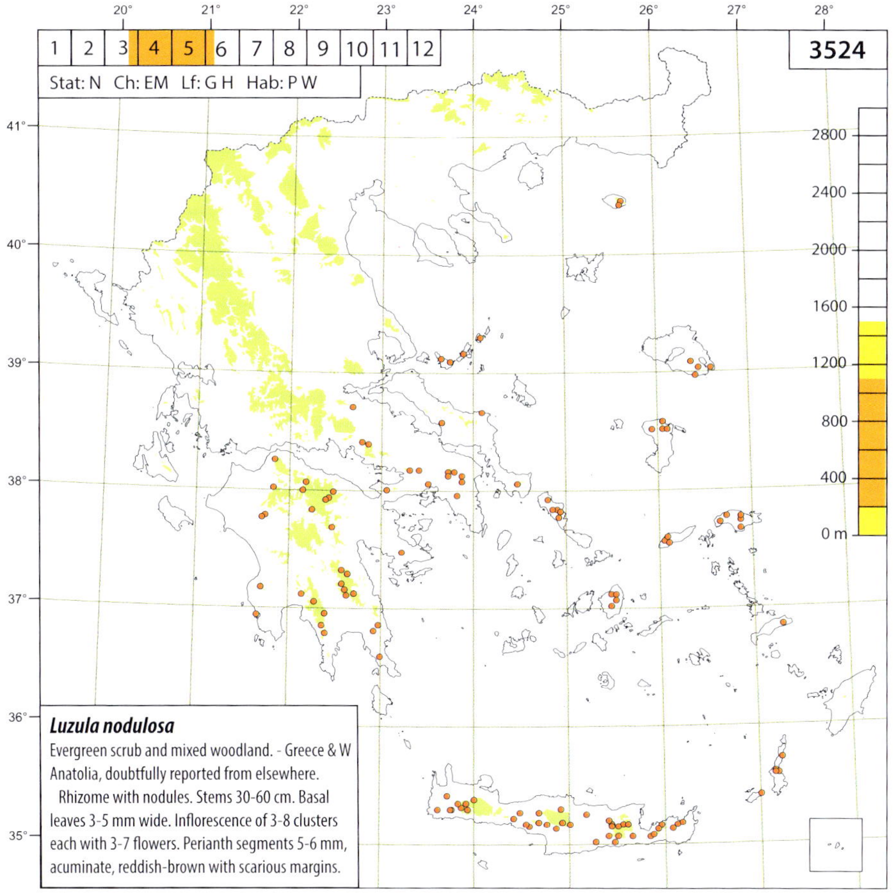

Luzula nodulosa

Evergreen scrub and mixed woodland. - Greece & W Anatolia, doubtfully reported from elsewhere.

Rhizome with nodules. Stems 30-60 cm. Basal leaves 3-5 mm wide. Inflorescence of 3-8 clusters each with 3-7 flowers. Perianth segments 5-6 mm, acuminate, reddish-brown with scarious margins.

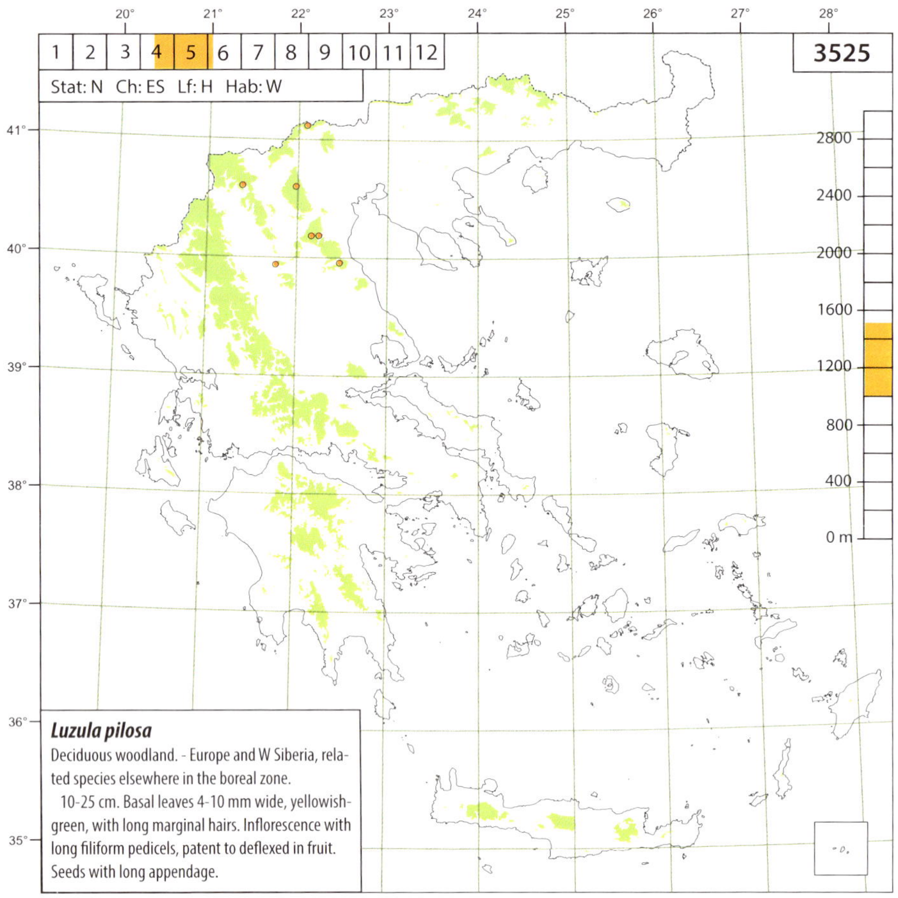

Luzula pilosa
Deciduous woodland. - Europe and W Siberia, related species elsewhere in the boreal zone.
10-25 cm. Basal leaves 4-10 mm wide, yellowish-green, with long marginal hairs. Inflorescence with long filiform pedicels, patent to deflexed in fruit. Seeds with long appendage.

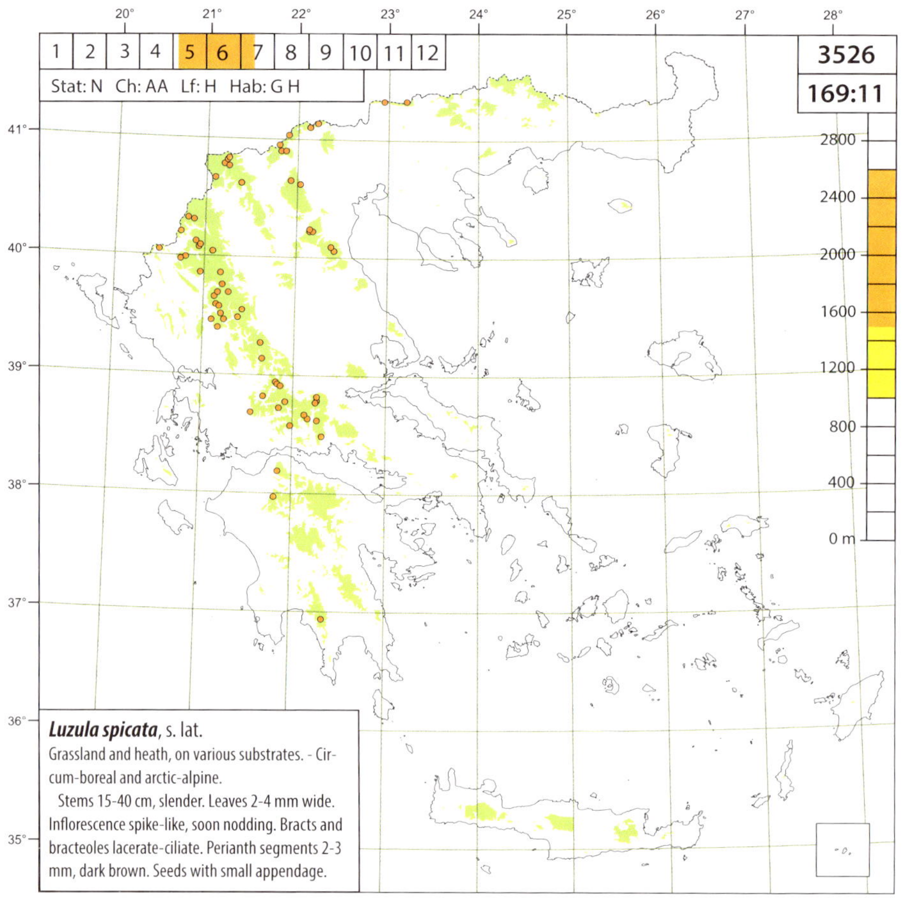

Luzula spicata, s. lat.
Grassland and heath, on various substrates. - Circum-boreal and arctic-alpine.
Stems 15-40 cm, slender. Leaves 2-4 mm wide. Inflorescence spike-like, soon nodding. Bracts and bracteoles lacerate-ciliate. Perianth segments 2-3 mm, dark brown. Seeds with small appendage.

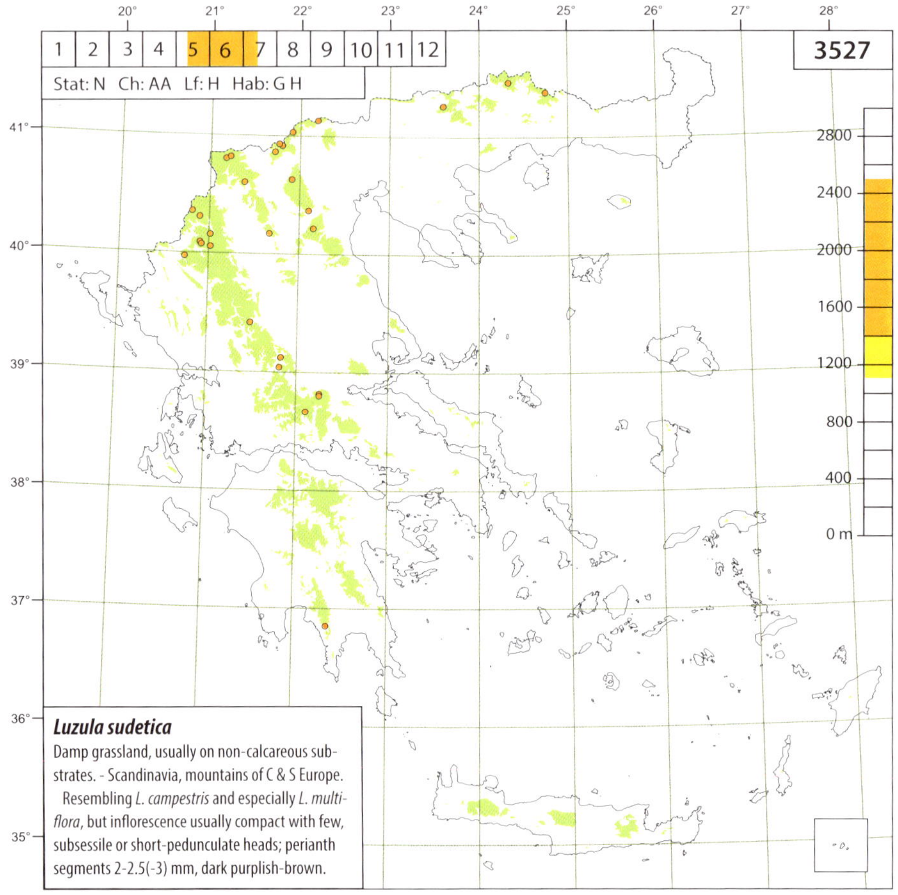

Luzula sudetica
Damp grassland, usually on non-calcareous substrates. - Scandinavia, mountains of C & S Europe.
Resembling *L. campestris* and especially *L. multiflora*, but inflorescence usually compact with few, subsessile or short-pedunculate heads; perianth segments 2-2.5(-3) mm, dark purplish-brown.

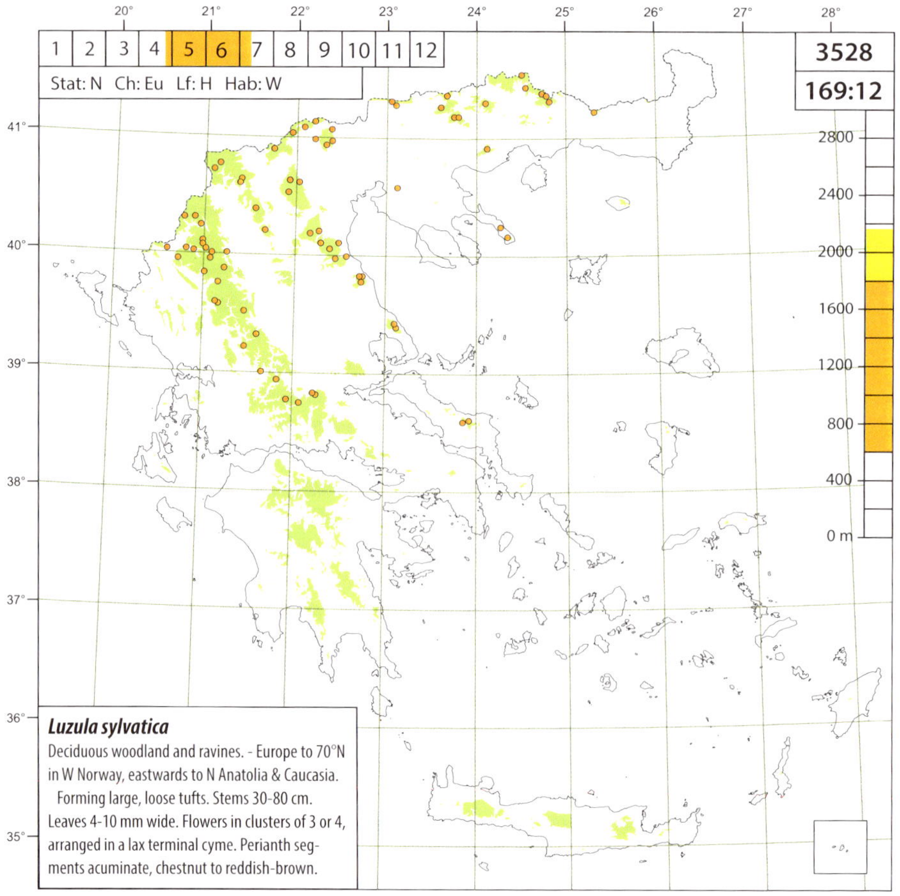

Luzula sylvatica
Deciduous woodland and ravines. - Europe to 70°N in W Norway, eastwards to N Anatolia & Caucasia.
Forming large, loose tufts. Stems 30-80 cm. Leaves 4-10 mm wide. Flowers in clusters of 3 or 4, arranged in a lax terminal cyme. Perianth segments acuminate, chestnut to reddish-brown.

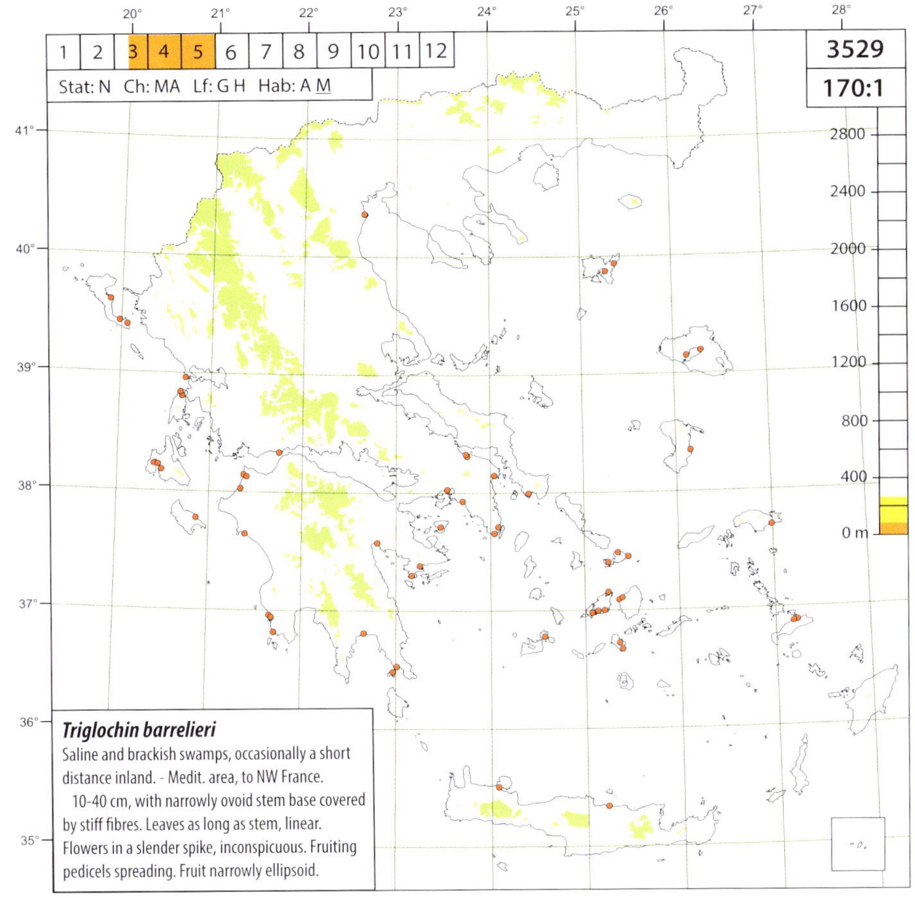

Triglochin barrelieri
Saline and brackish swamps, occasionally a short distance inland. - Medit. area, to NW France.
10-40 cm, with narrowly ovoid stem base covered by stiff fibres. Leaves as long as stem, linear. Flowers in a slender spike, inconspicuous. Fruiting pedicels spreading. Fruit narrowly ellipsoid.

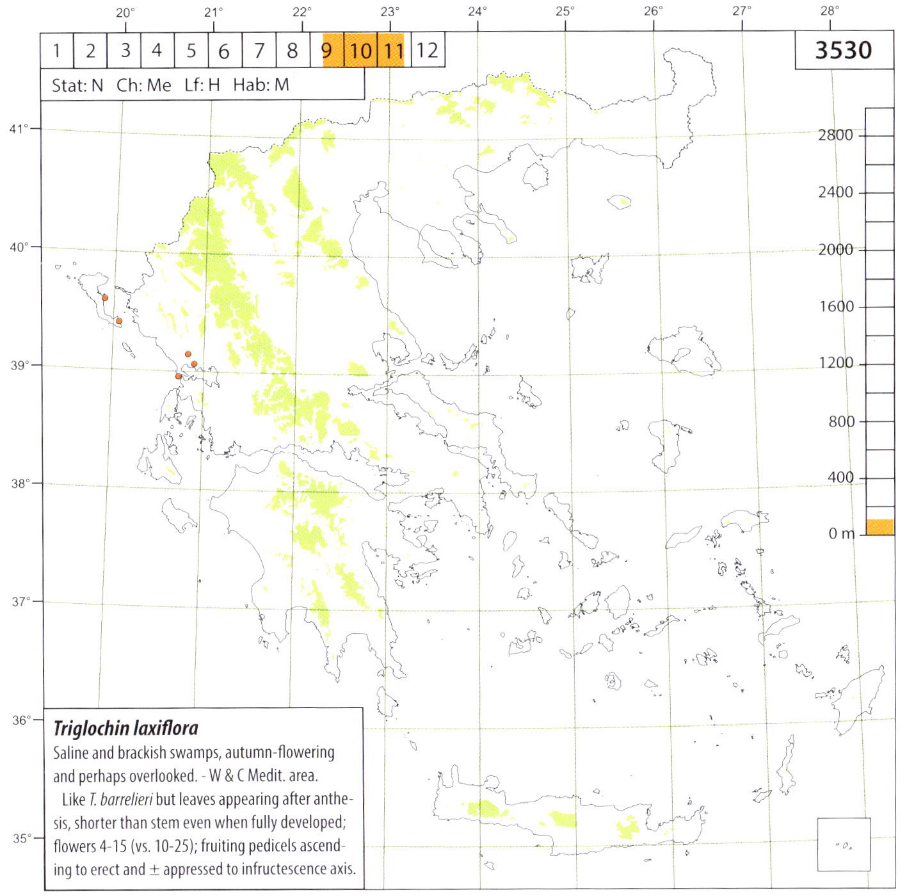

Triglochin laxiflora
Saline and brackish swamps, autumn-flowering and perhaps overlooked. - W & C Medit. area.
Like *T. barrelieri* but leaves appearing after anthesis, shorter than stem even when fully developed; flowers 4-15 (vs. 10-25); fruiting pedicels ascending to erect and ± appressed to infructescence axis.

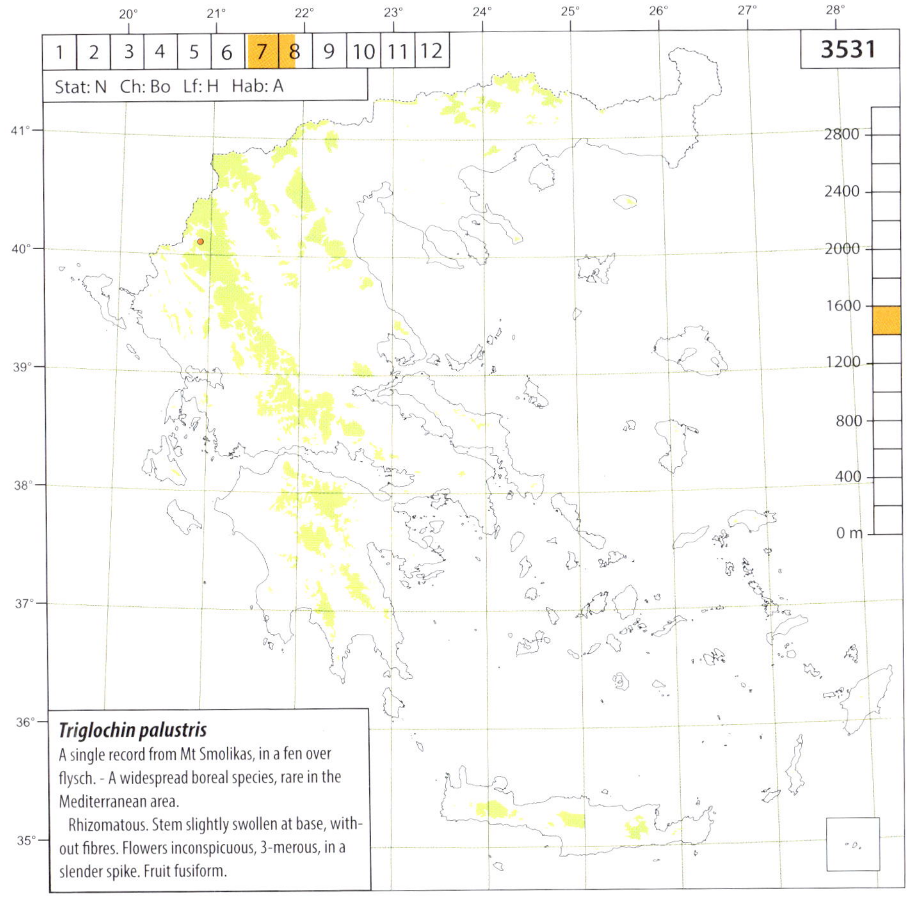

Triglochin palustris
A single record from Mt Smolikas, in a fen over flysch. - A widespread boreal species, rare in the Mediterranean area.
Rhizomatous. Stem slightly swollen at base, without fibres. Flowers inconspicuous, 3-merous, in a slender spike. Fruit fusiform.

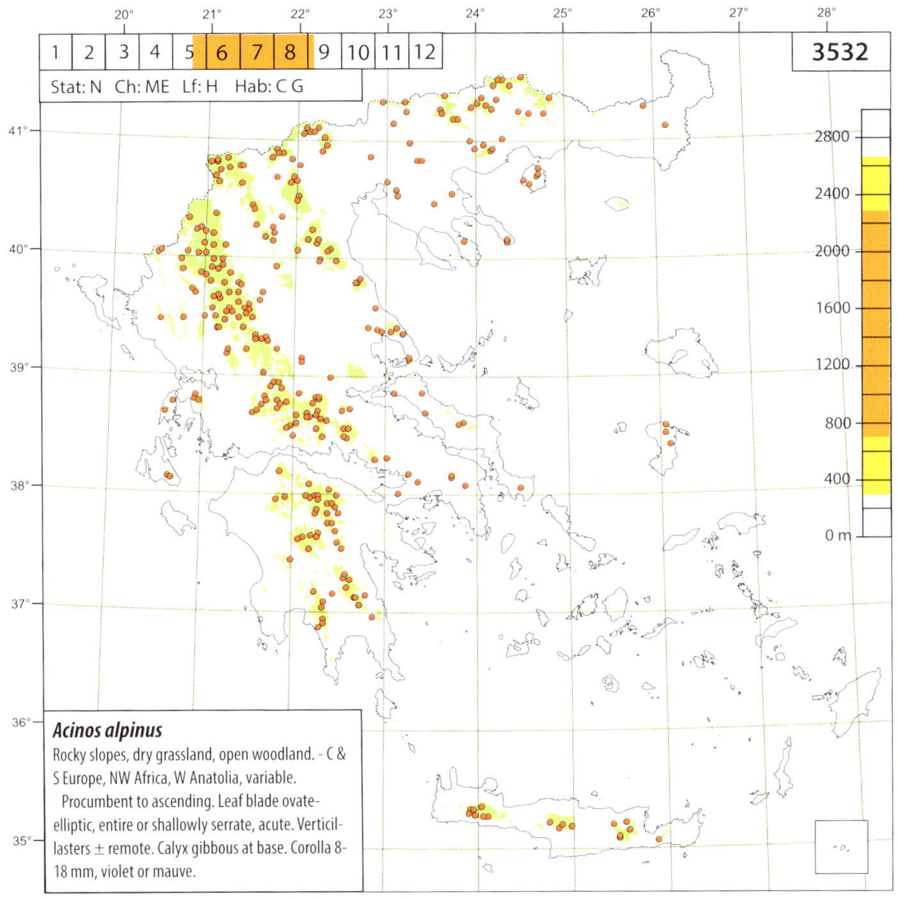

Acinos alpinus
Rocky slopes, dry grassland, open woodland. - C & S Europe, NW Africa, W Anatolia, variable.
Procumbent to ascending. Leaf blade ovate-elliptic, entire or shallowly serrate, acute. Verticillasters ± remote. Calyx gibbous at base. Corolla 8-18 mm, violet or mauve.

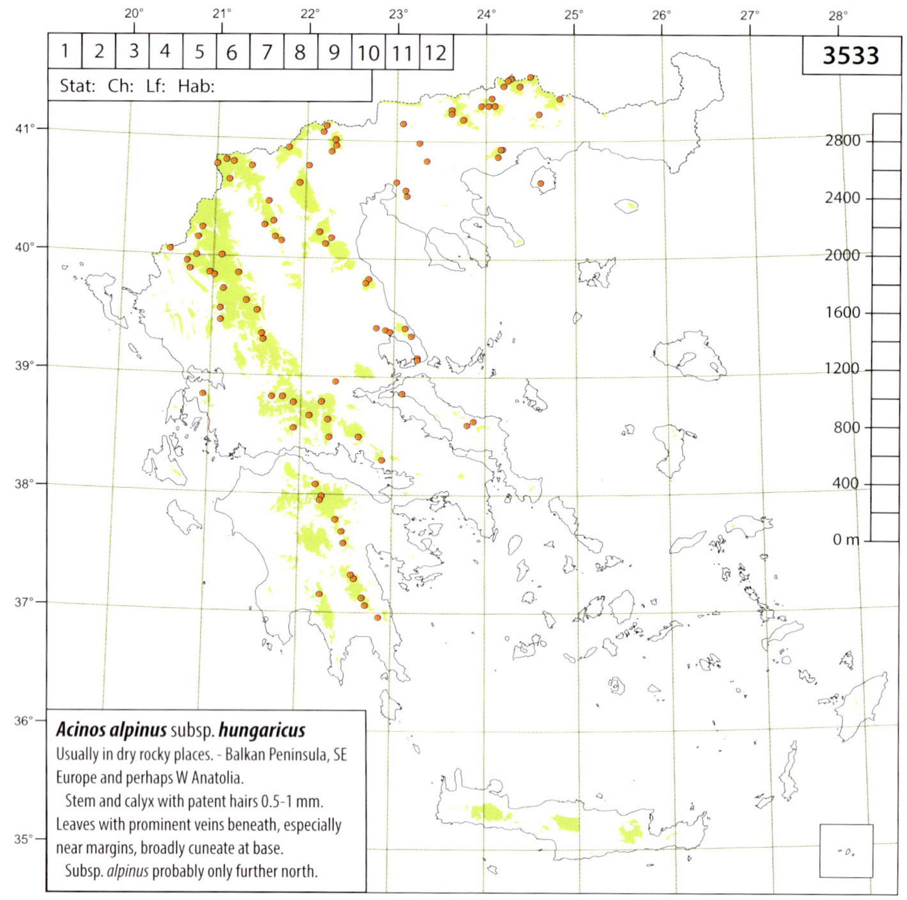

Acinos alpinus subsp. ***hungaricus***
Usually in dry rocky places. - Balkan Peninsula, SE Europe and perhaps W Anatolia.
Stem and calyx with patent hairs 0.5-1 mm. Leaves with prominent veins beneath, especially near margins, broadly cuneate at base.
Subsp. *alpinus* probably only further north.

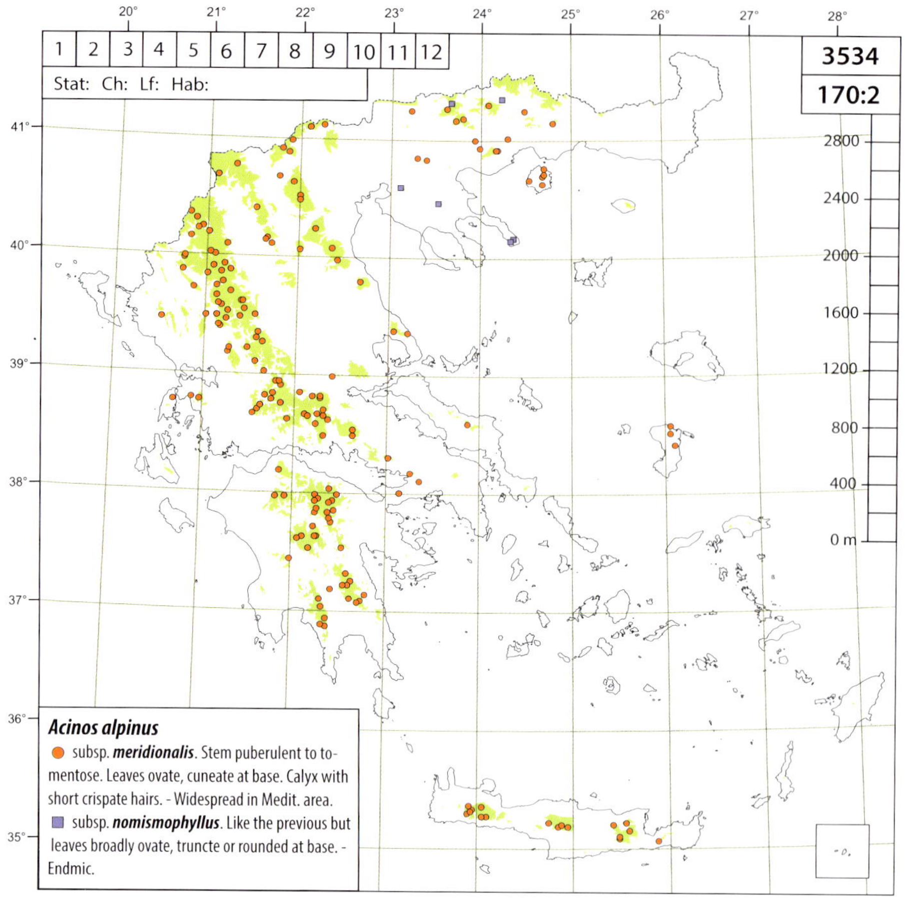

Acinos alpinus
● subsp. ***meridionalis***. Stem puberulent to tomentose. Leaves ovate, cuneate at base. Calyx with short crispate hairs. - Widespread in Medit. area.
■ subsp. ***nomismophyllus***. Like the previous but leaves broadly ovate, truncte or rounded at base. - Endmic.

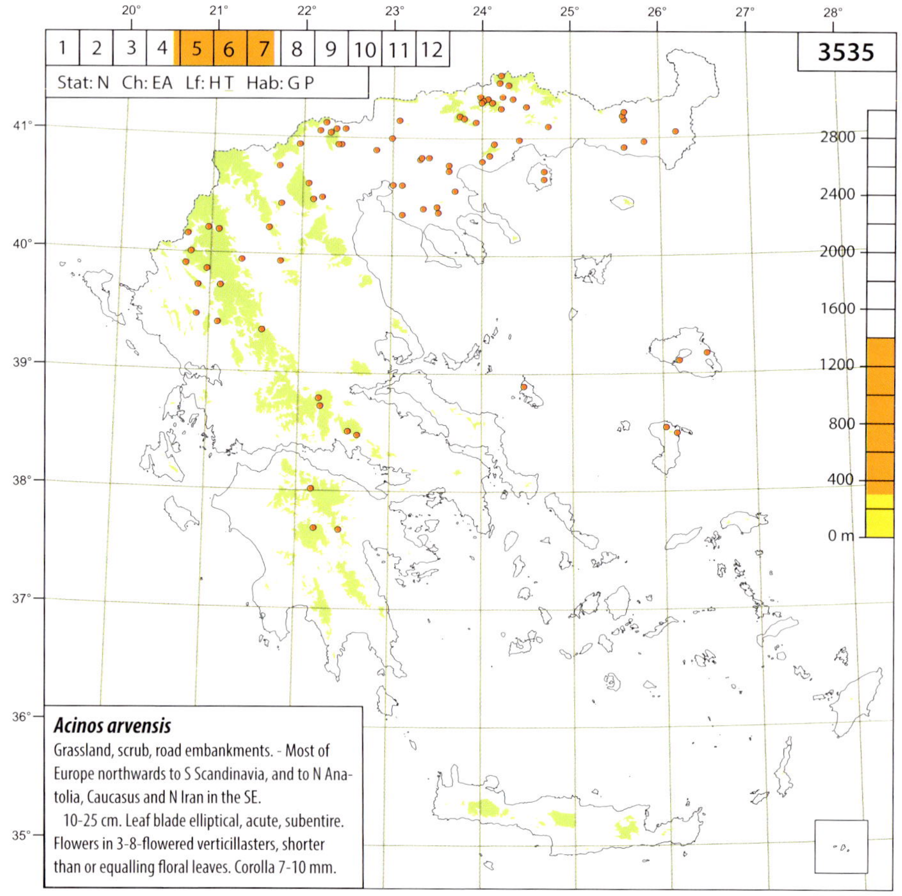

Acinos arvensis
Grassland, scrub, road embankments. - Most of Europe northwards to S Scandinavia, and to N Anatolia, Caucasus and N Iran in the SE.
10-25 cm. Leaf blade elliptical, acute, subentire. Flowers in 3-8-flowered verticillasters, shorter than or equalling floral leaves. Corolla 7-10 mm.

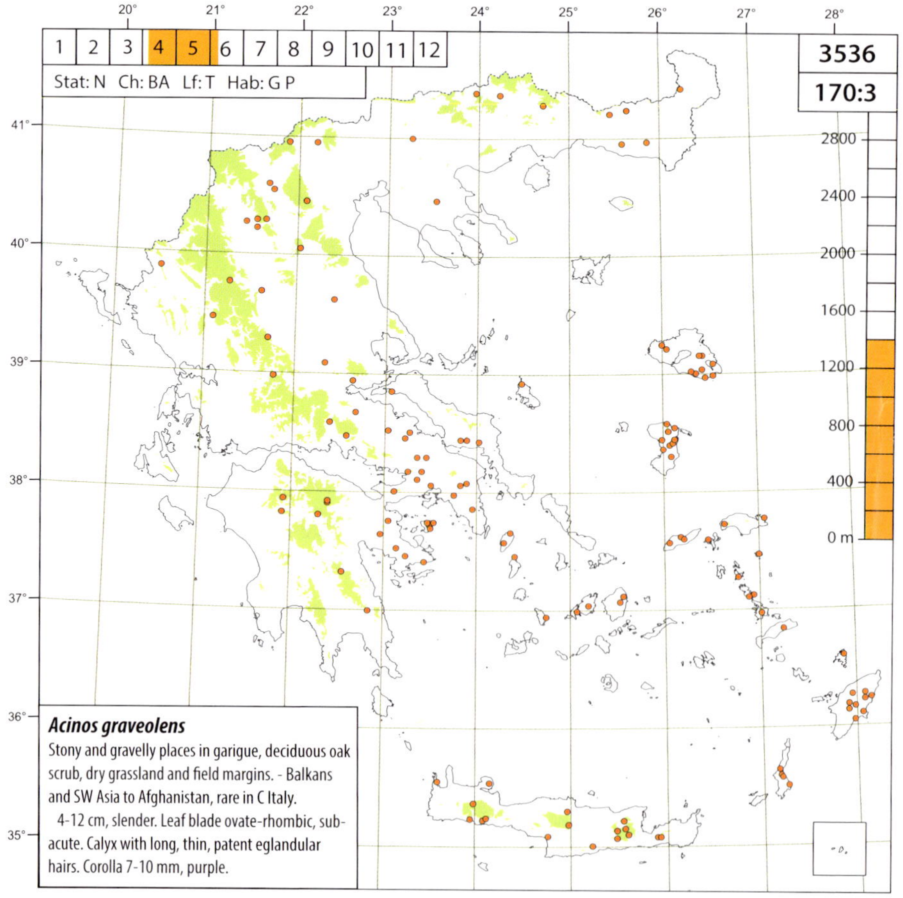

Acinos graveolens
Stony and gravelly places in garigue, deciduous oak scrub, dry grassland and field margins. - Balkans and SW Asia to Afghanistan, rare in C Italy.
4-12 cm, slender. Leaf blade ovate-rhombic, subacute. Calyx with long, thin, patent eglandular hairs. Corolla 7-10 mm, purple.

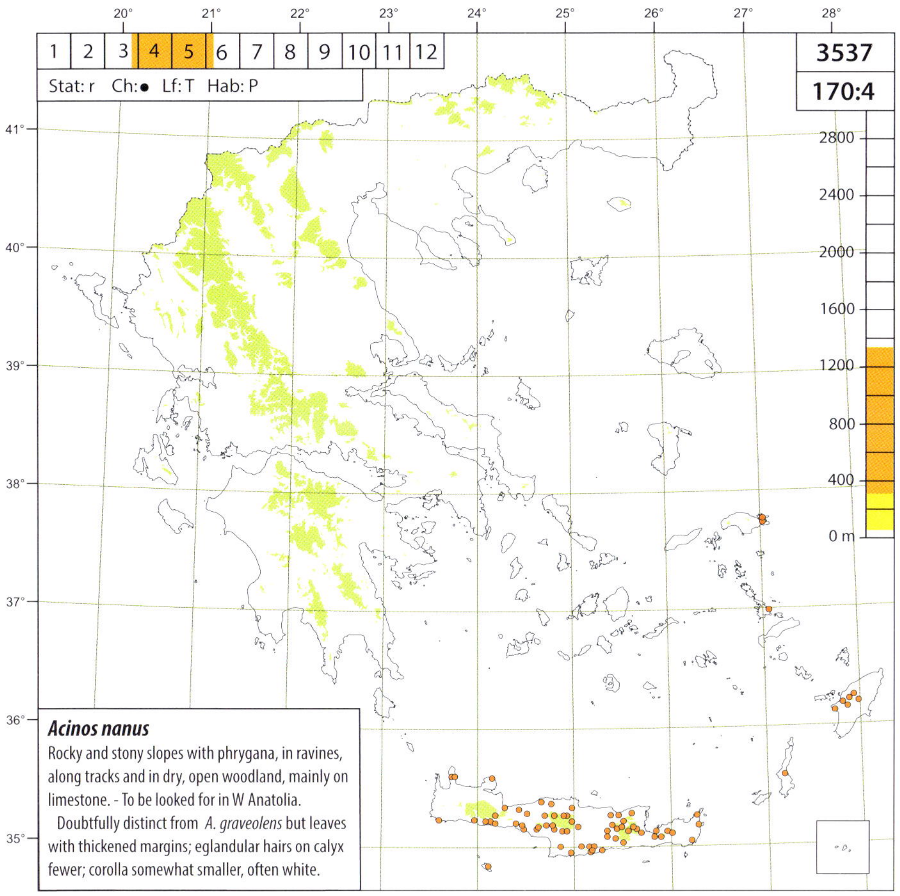

Acinos nanus
Rocky and stony slopes with phrygana, in ravines, along tracks and in dry, open woodland, mainly on limestone. - To be looked for in W Anatolia.
Doubtfully distinct from *A. graveolens* but leaves with thickened margins; eglandular hairs on calyx fewer; corolla somewhat smaller, often white.

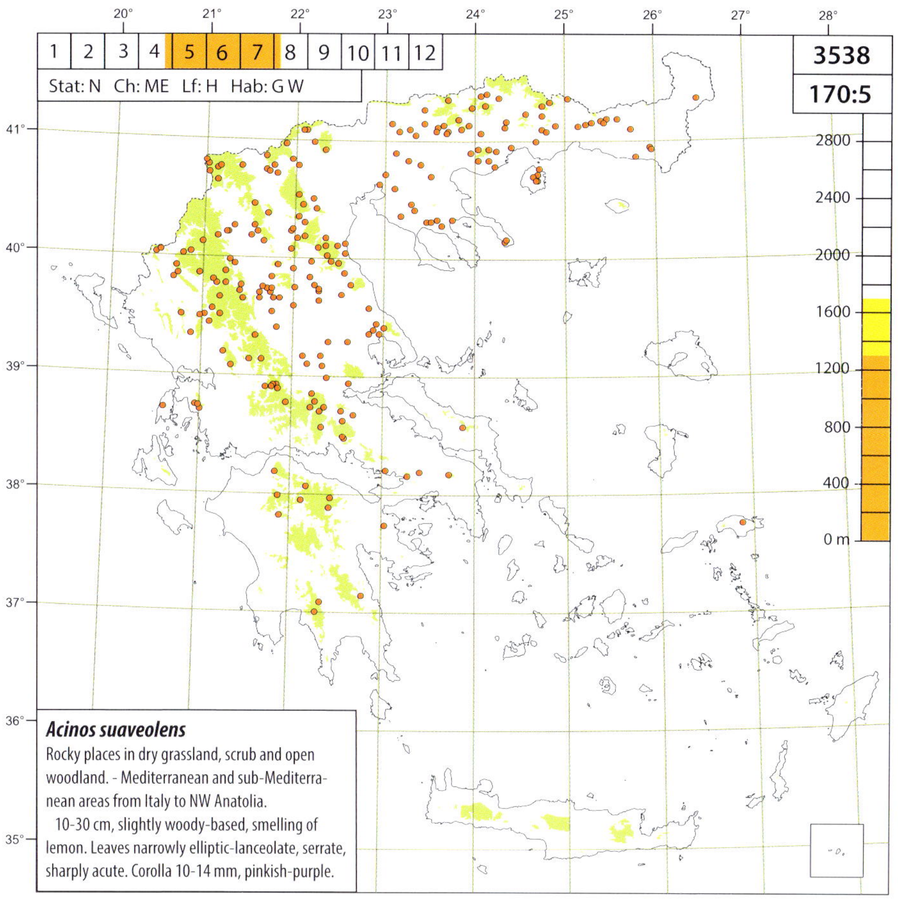

Acinos suaveolens
Rocky places in dry grassland, scrub and open woodland. - Mediterranean and sub-Mediterranean areas from Italy to NW Anatolia.
10-30 cm, slightly woody-based, smelling of lemon. Leaves narrowly elliptic-lanceolate, serrate, sharply acute. Corolla 10-14 mm, pinkish-purple.

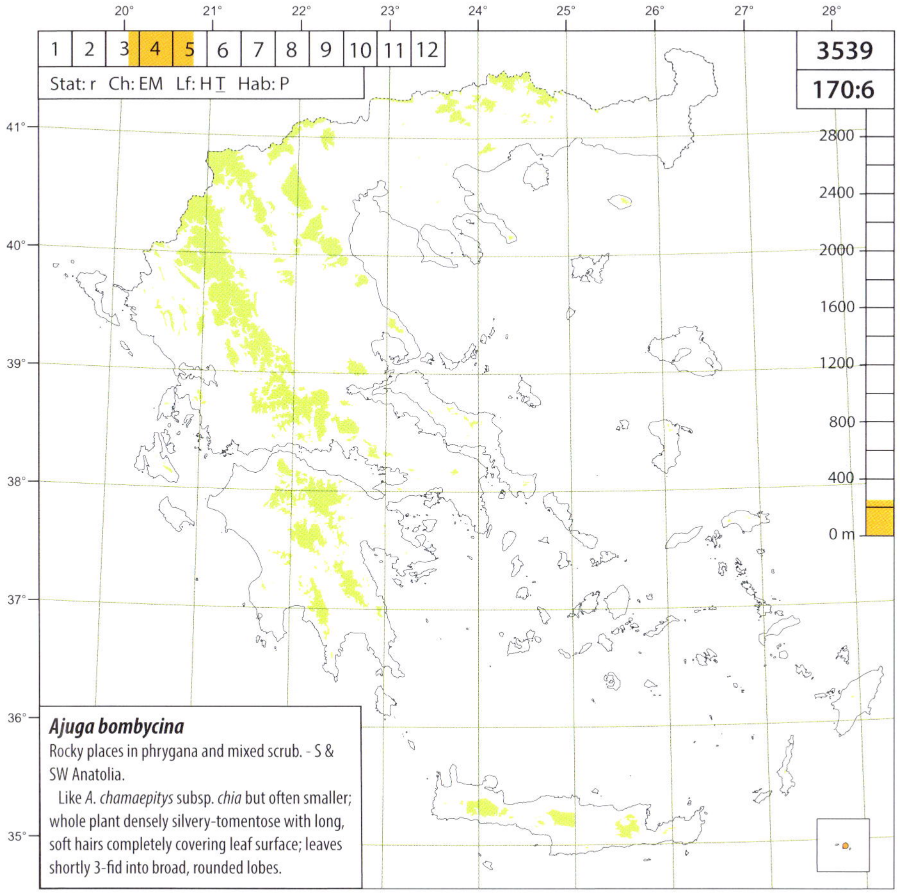

Ajuga bombycina
Rocky places in phrygana and mixed scrub. - S & SW Anatolia.
Like *A. chamaepitys* subsp. *chia* but often smaller; whole plant densely silvery-tomentose with long, soft hairs completely covering leaf surface; leaves shortly 3-fid into broad, rounded lobes.

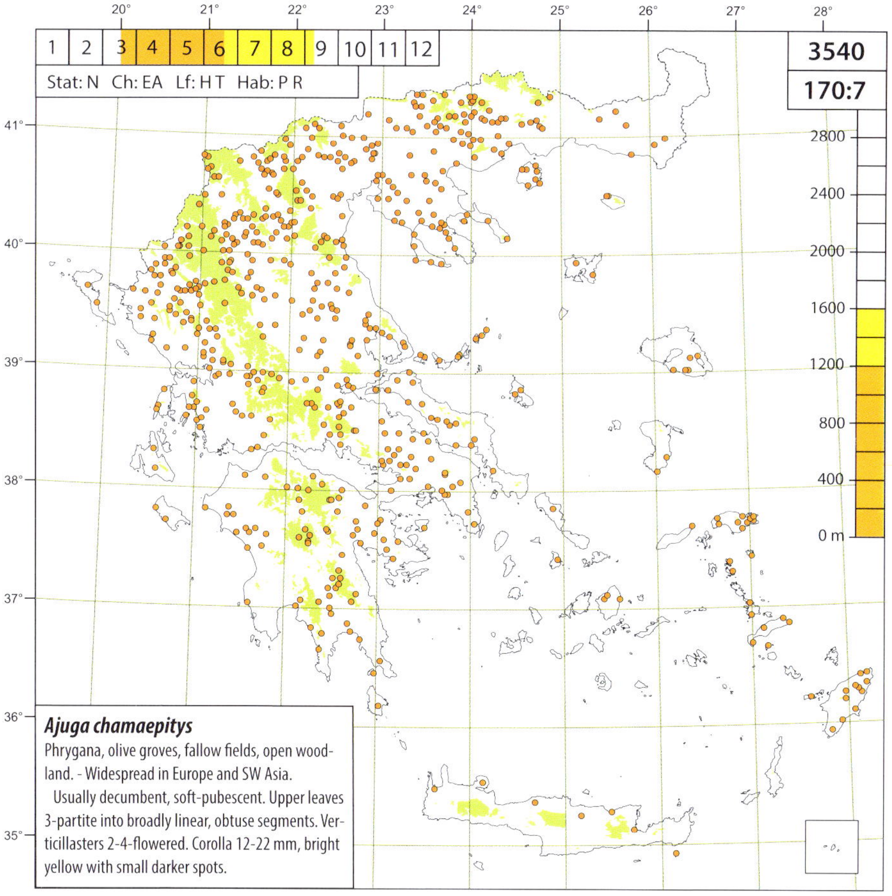

Ajuga chamaepitys
Phrygana, olive groves, fallow fields, open woodland. - Widespread in Europe and SW Asia.
Usually decumbent, soft-pubescent. Upper leaves 3-partite into broadly linear, obtuse segments. Verticillasters 2-4-flowered. Corolla 12-22 mm, bright yellow with small darker spots.

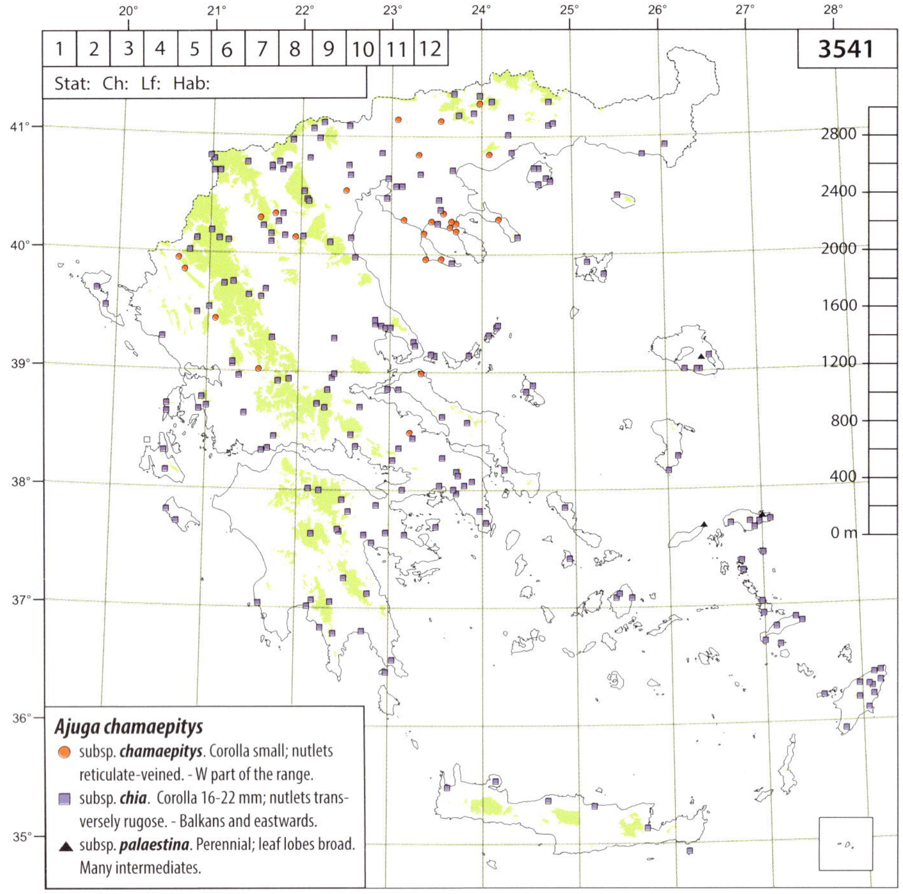

Ajuga chamaepitys

- subsp. ***chamaepitys***. Corolla small; nutlets reticulate-veined. - W part of the range.
- subsp. ***chia***. Corolla 16-22 mm; nutlets transversely rugose. - Balkans and eastwards.
- ▲ subsp. ***palaestina***. Perennial; leaf lobes broad.

Many intermediates.

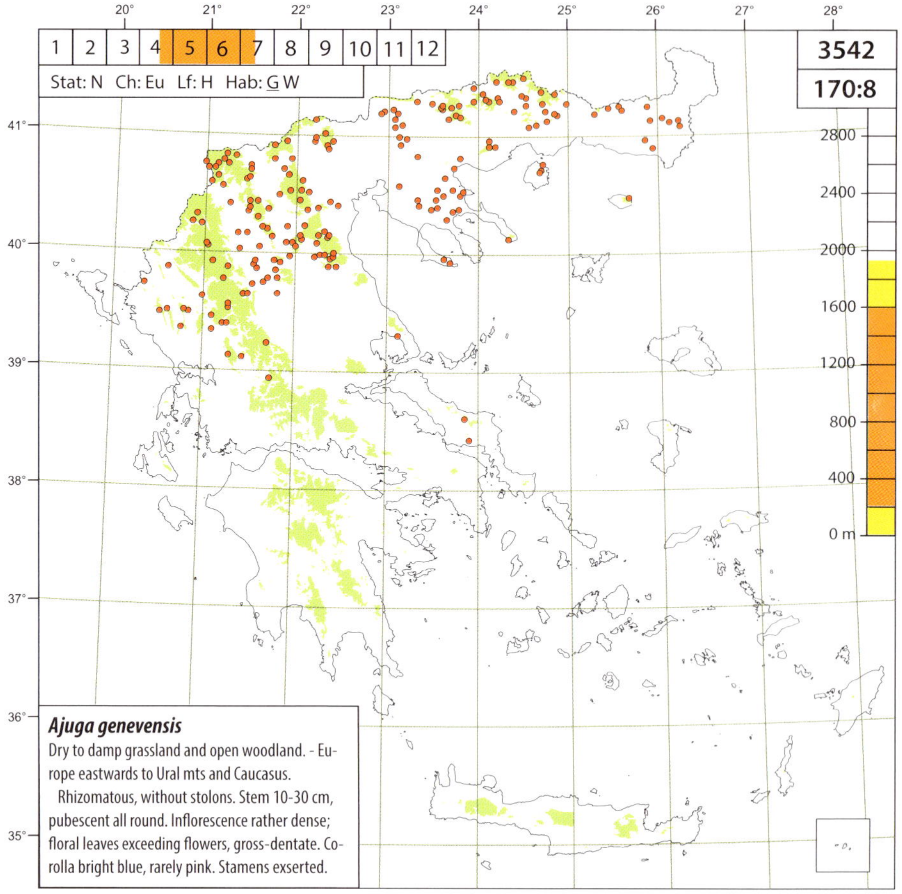

Ajuga genevensis

Dry to damp grassland and open woodland. - Europe eastwards to Ural mts and Caucasus.

Rhizomatous, without stolons. Stem 10-30 cm, pubescent all round. Inflorescence rather dense; floral leaves exceeding flowers, gross-dentate. Corolla bright blue, rarely pink. Stamens exserted.

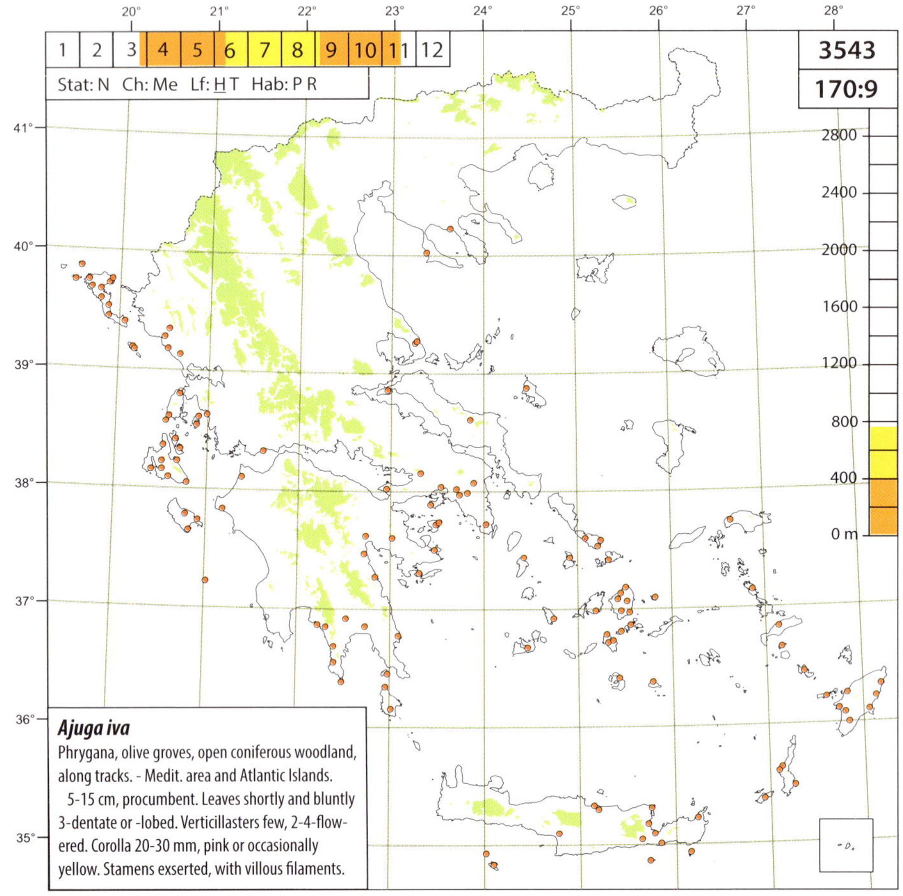

Ajuga iva

Phrygana, olive groves, open coniferous woodland, along tracks. - Medit. area and Atlantic Islands.

5-15 cm, procumbent. Leaves shortly and bluntly 3-dentate or -lobed. Verticillasters few, 2-4-flowered. Corolla 20-30 mm, pink or occasionally yellow. Stamens exserted, with villous filaments.

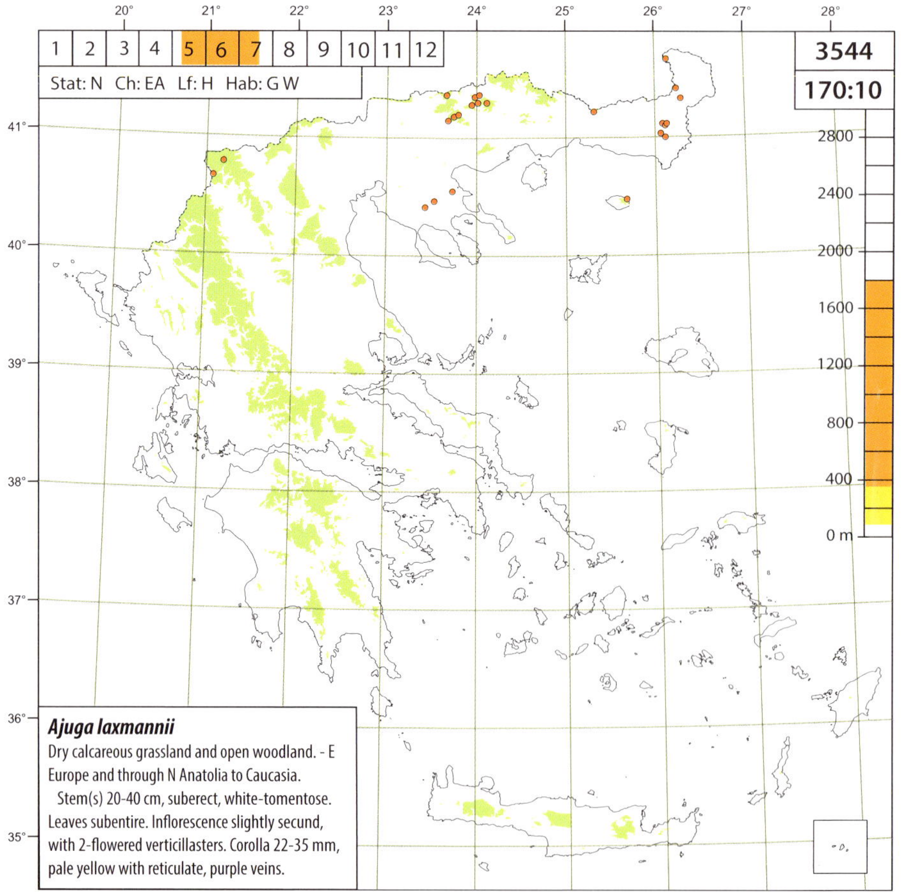

Ajuga laxmannii

Dry calcareous grassland and open woodland. - E Europe and through N Anatolia to Caucasia.

Stem(s) 20-40 cm, suberect, white-tomentose. Leaves subentire. Inflorescence slightly secund, with 2-flowered verticillasters. Corolla 22-35 mm, pale yellow with reticulate, purple veins.

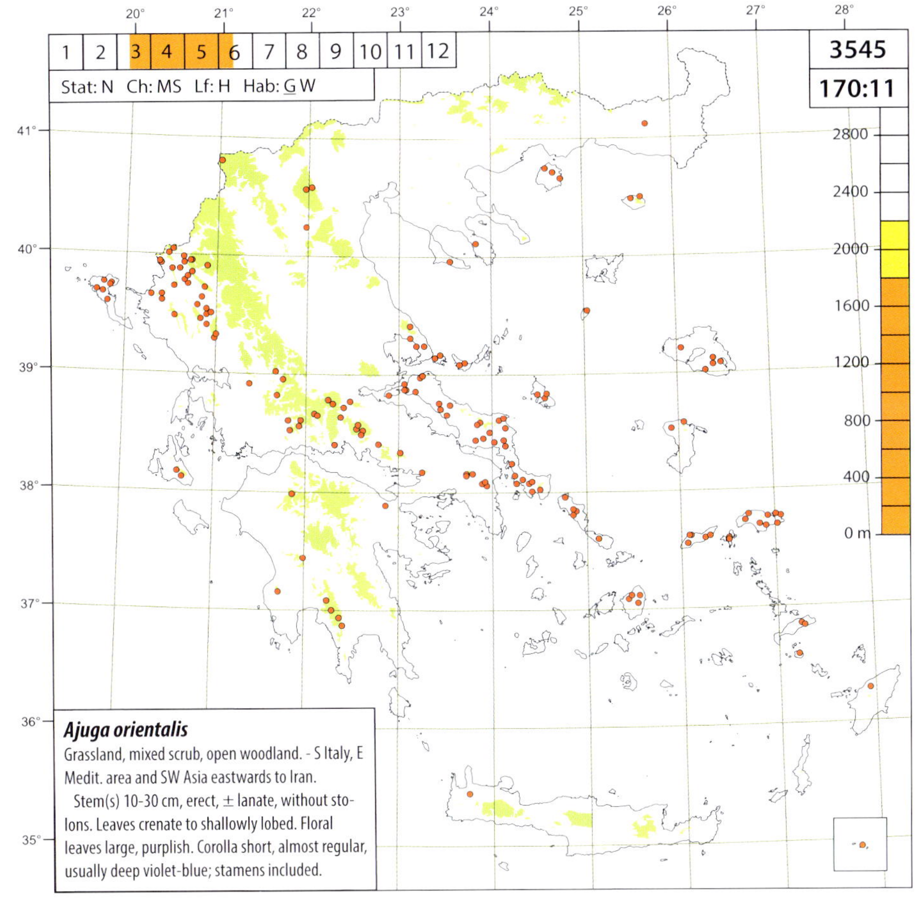

Ajuga orientalis

Grassland, mixed scrub, open woodland. - S Italy, E Medit. area and SW Asia eastwards to Iran.

Stem(s) 10-30 cm, erect, ± lanate, without stolons. Leaves crenate to shallowly lobed. Floral leaves large, purplish. Corolla short, almost regular, usually deep violet-blue; stamens included.

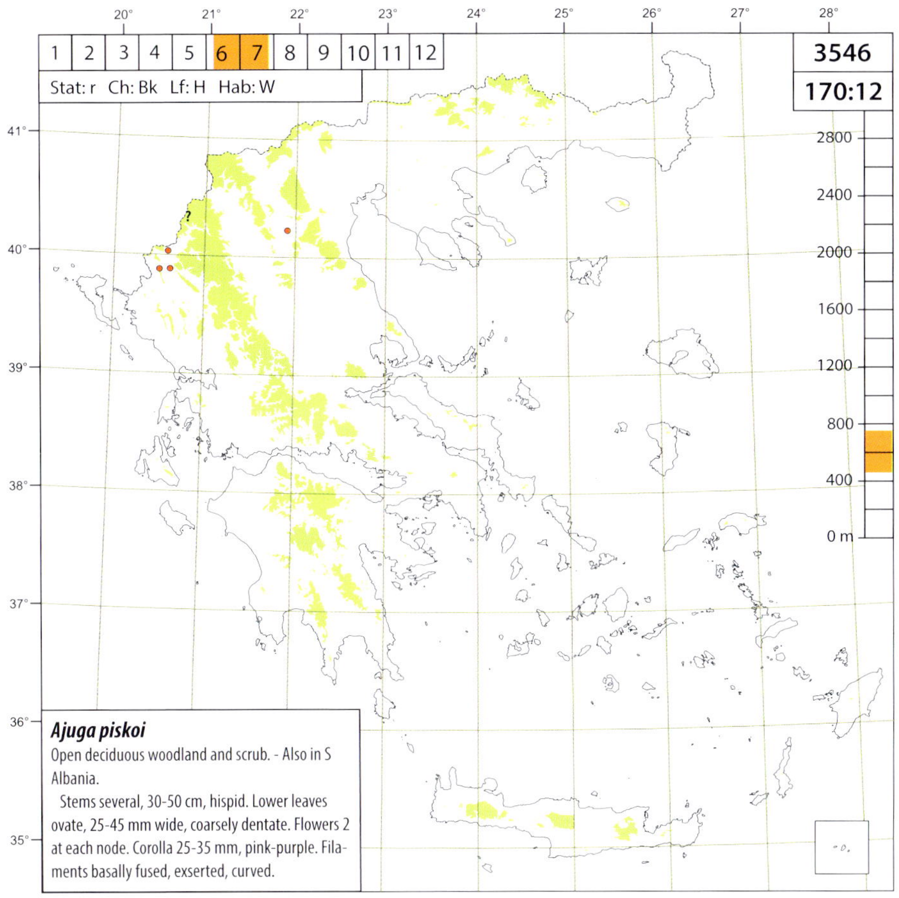

Ajuga piskoi

Open deciduous woodland and scrub. - Also in S Albania.

Stems several, 30-50 cm, hispid. Lower leaves ovate, 25-45 mm wide, coarsely dentate. Flowers 2 at each node. Corolla 25-35 mm, pink-purple. Filaments basally fused, exserted, curved.

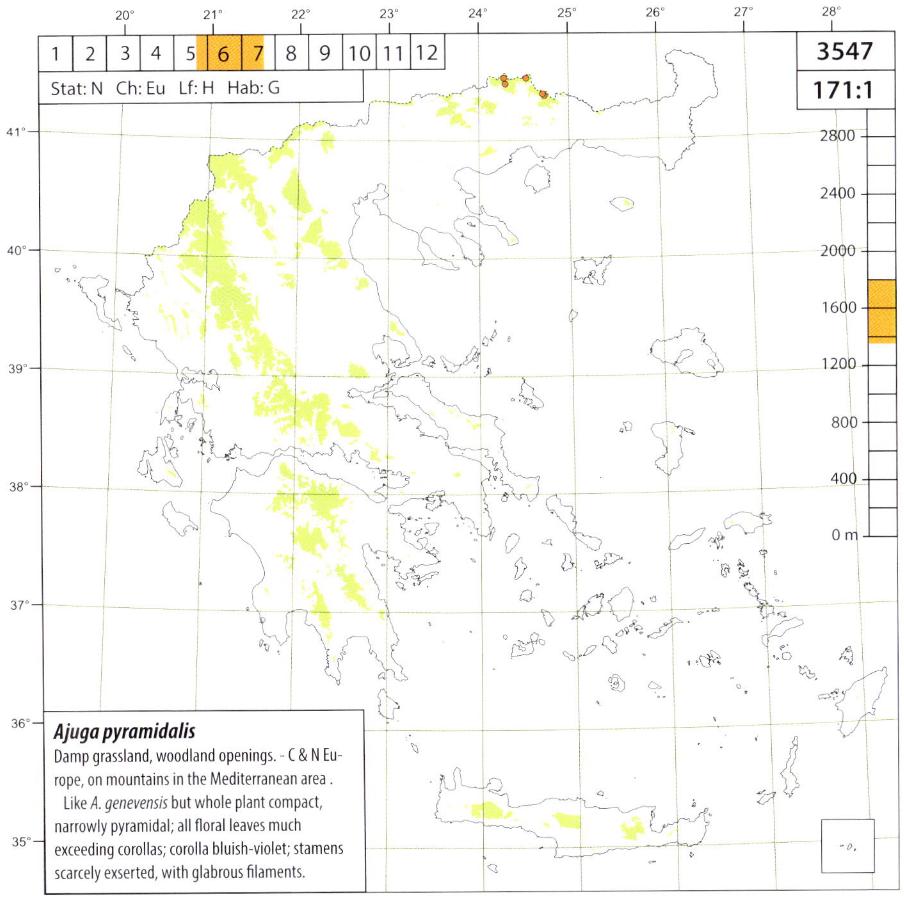

Ajuga pyramidalis

Damp grassland, woodland openings. - C & N Europe, on mountains in the Mediterranean area .

Like *A. genevensis* but whole plant compact, narrowly pyramidal; all floral leaves much exceeding corollas; corolla bluish-violet; stamens scarcely exserted, with glabrous filaments.

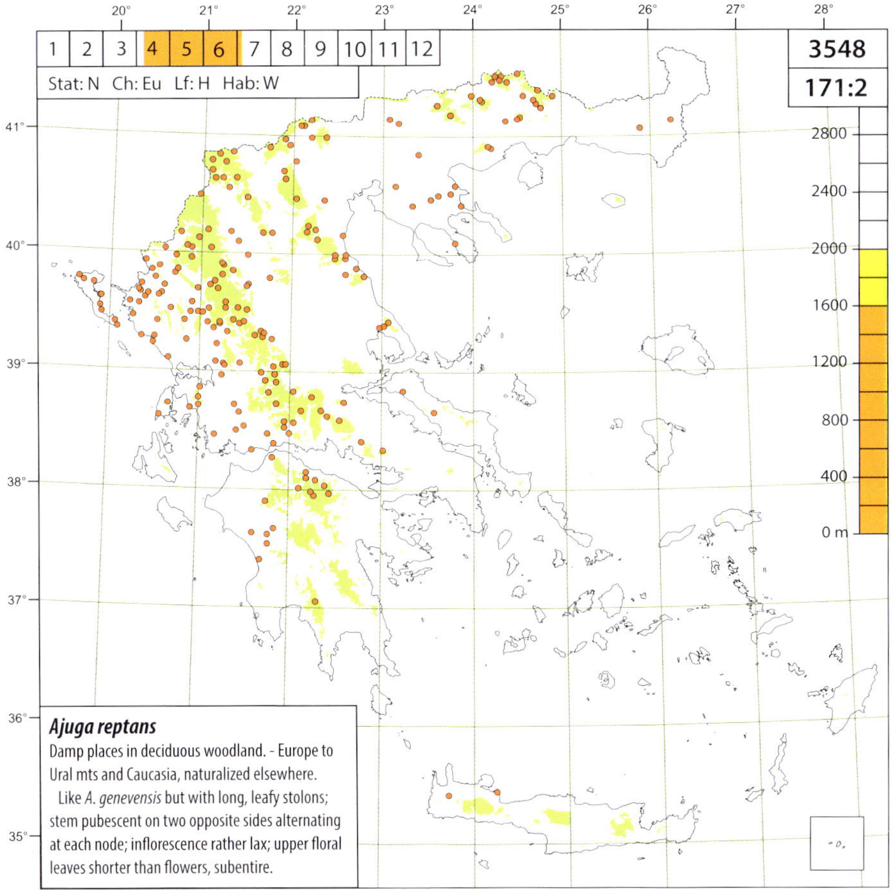

Ajuga reptans

Damp places in deciduous woodland. - Europe to Ural mts and Caucasia, naturalized elsewhere.

Like *A. genevensis* but with long, leafy stolons; stem pubescent on two opposite sides alternating at each node; inflorescence rather lax; upper floral leaves shorter than flowers, subentire.

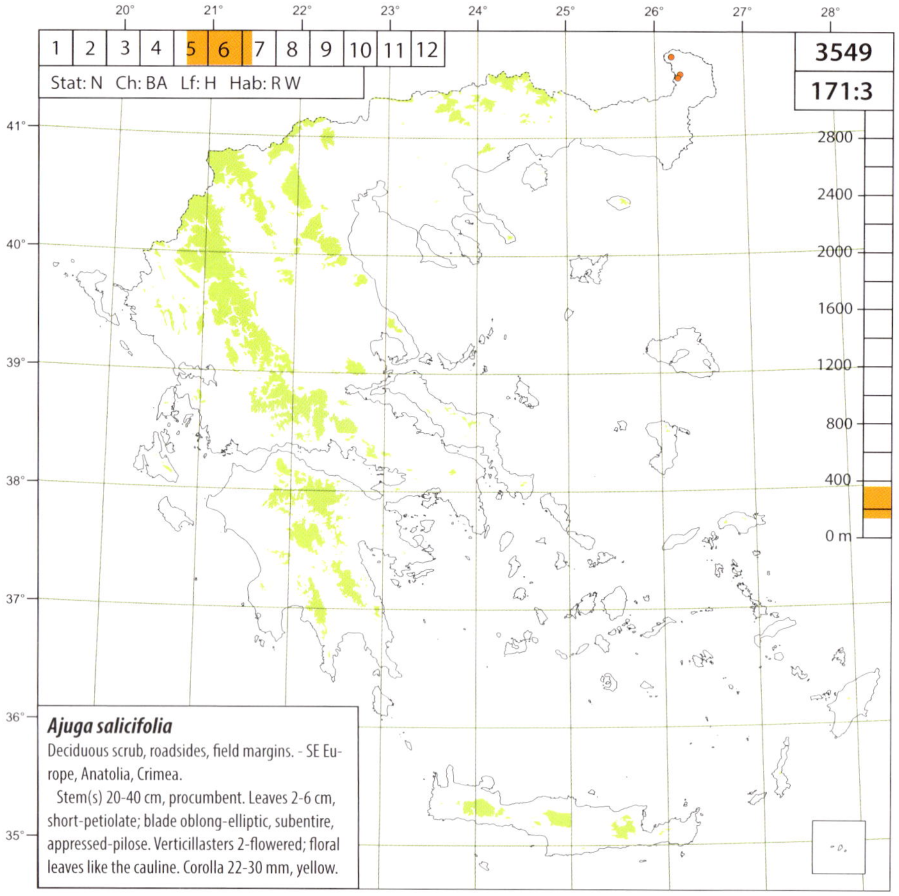

Ajuga salicifolia

Deciduous scrub, roadsides, field margins. - SE Europe, Anatolia, Crimea.

Stem(s) 20-40 cm, procumbent. Leaves 2-6 cm, short-petiolate; blade oblong-elliptic, subentire, appressed-pilose. Verticillasters 2-flowered; floral leaves like the cauline. Corolla 22-30 mm, yellow.

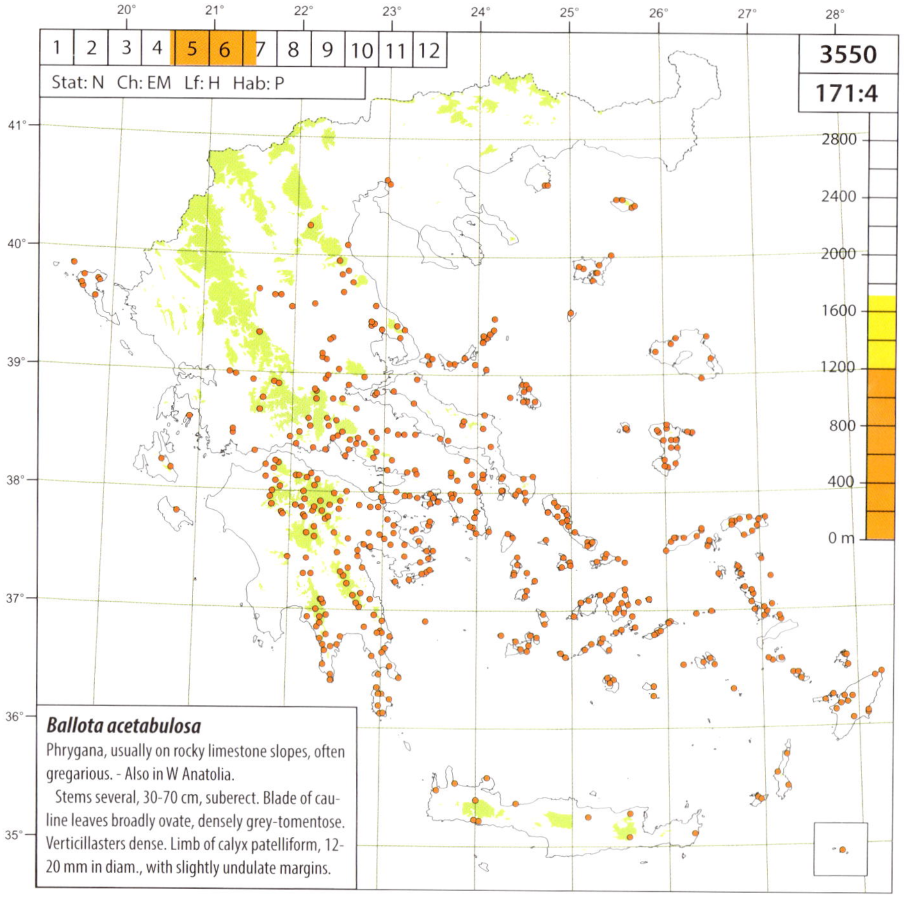

Ballota acetabulosa

Phrygana, usually on rocky limestone slopes, often gregarious. - Also in W Anatolia.

Stems several, 30-70 cm, suberect. Blade of cauline leaves broadly ovate, densely grey-tomentose. Verticillasters dense. Limb of calyx patelliform, 12-20 mm in diam., with slightly undulate margins.

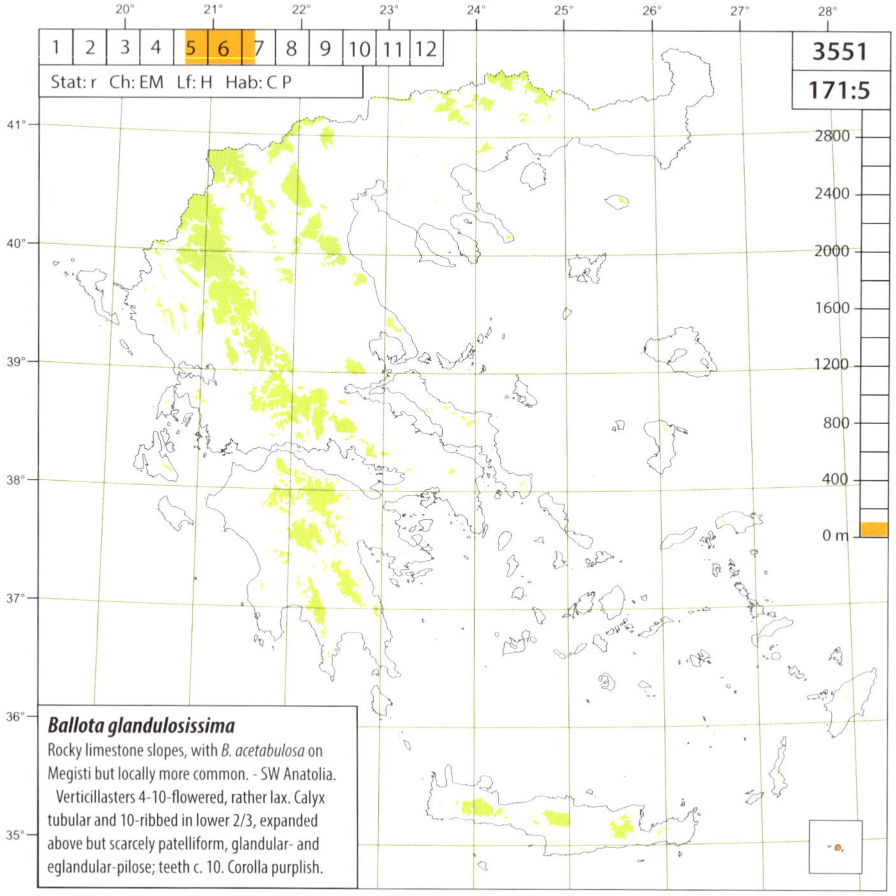

Ballota glandulosissima

Rocky limestone slopes, with *B. acetabulosa* on Megisti but locally more common. - SW Anatolia.

Verticillasters 4-10-flowered, rather lax. Calyx tubular and 10-ribbed in lower 2/3, expanded above but scarcely patelliform, glandular- and eglandular-pilose; teeth c. 10. Corolla purplish.

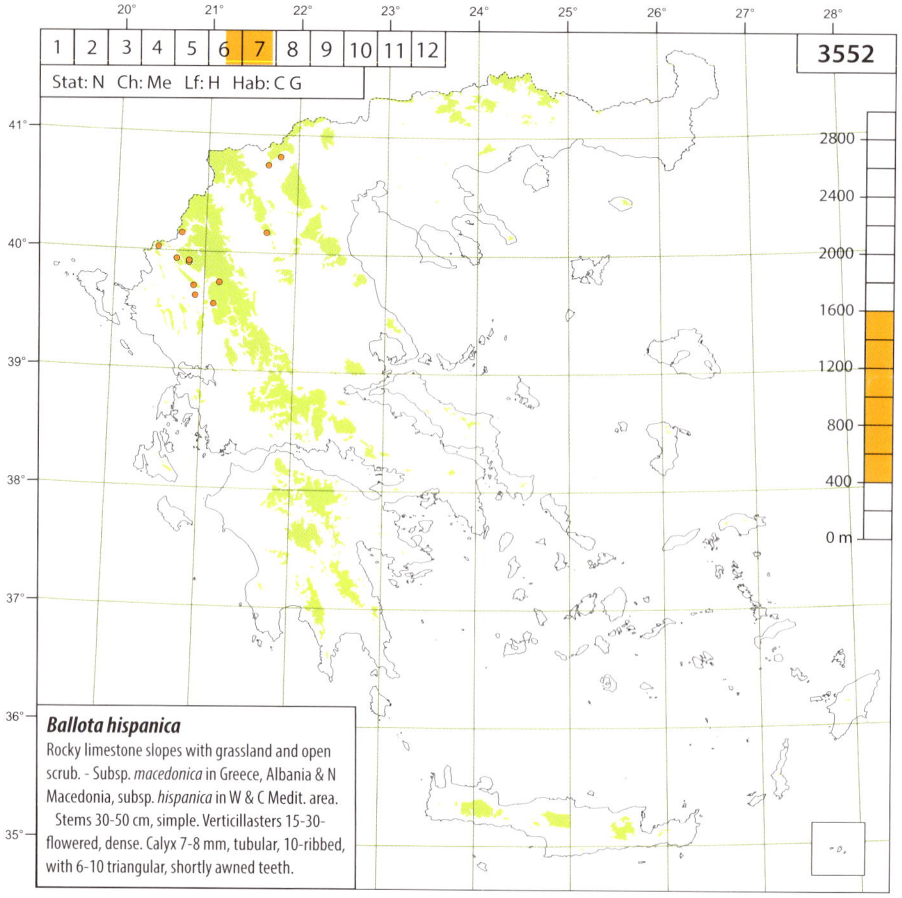

Ballota hispanica

Rocky limestone slopes with grassland and open scrub. - Subsp. *macedonica* in Greece, Albania & N Macedonia, subsp. *hispanica* in W & C Medit. area.

Stems 30-50 cm, simple. Verticillasters 15-30-flowered, dense. Calyx 7-8 mm, tubular, 10-ribbed, with 6-10 triangular, shortly awned teeth.

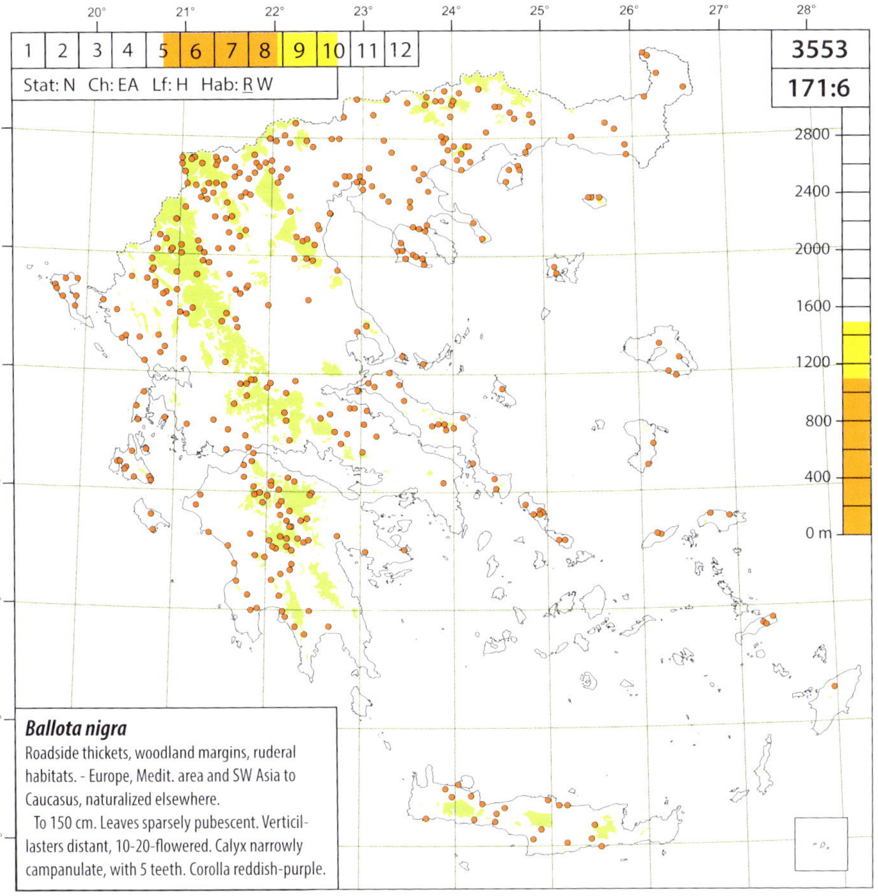

Ballota nigra
Roadside thickets, woodland margins, ruderal habitats. - Europe, Medit. area and SW Asia to Caucasus, naturalized elsewhere.
To 150 cm. Leaves sparsely pubescent. Verticillasters distant, 10-20-flowered. Calyx narrowly campanulate, with 5 teeth. Corolla reddish-purple.

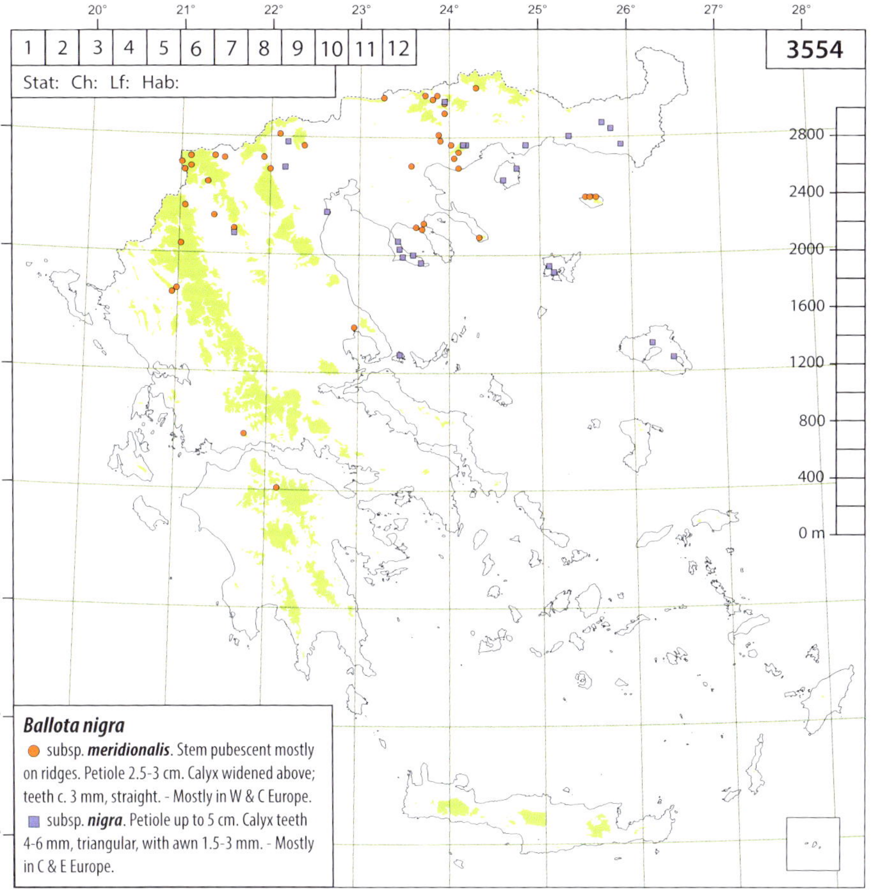

Ballota nigra
● subsp. ***meridionalis***. Stem pubescent mostly on ridges. Petiole 2.5-3 cm. Calyx widened above; teeth c. 3 mm, straight. - Mostly in W & C Europe.
■ subsp. ***nigra***. Petiole up to 5 cm. Calyx teeth 4-6 mm, triangular, with awn 1.5-3 mm. - Mostly in C & E Europe.

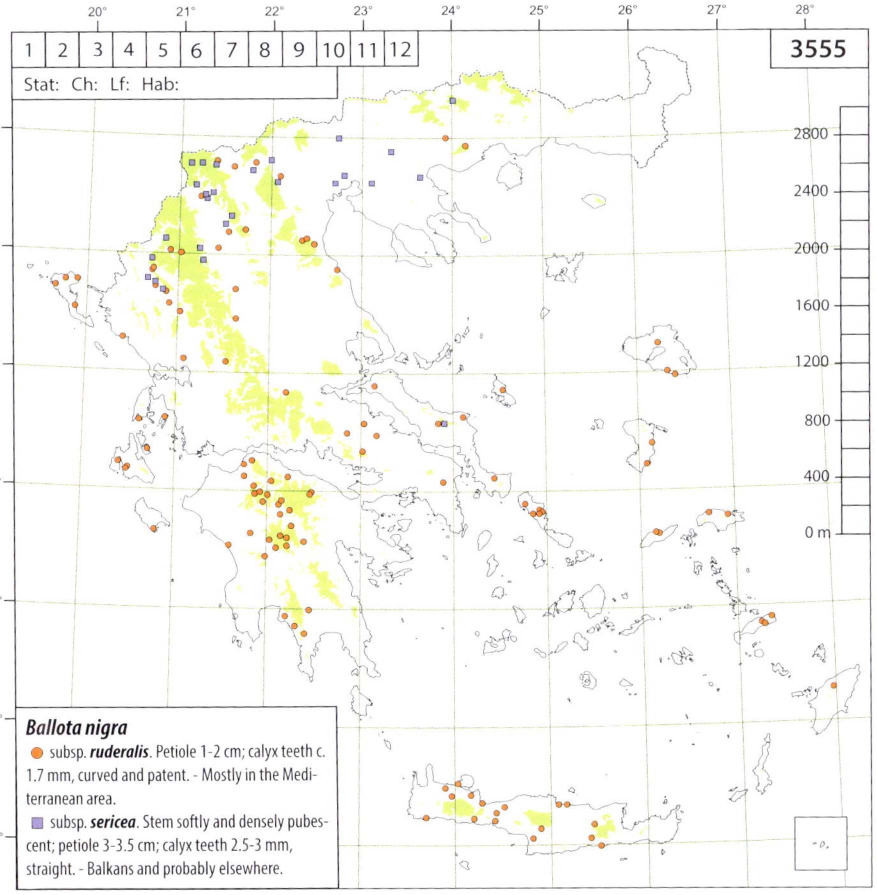

Ballota nigra
● subsp. ***ruderalis***. Petiole 1-2 cm; calyx teeth c. 1.7 mm, curved and patent. - Mostly in the Mediterranean area.
■ subsp. ***sericea***. Stem softly and densely pubescent; petiole 3-3.5 cm; calyx teeth 2.5-3 mm, straight. - Balkans and probably elsewhere.

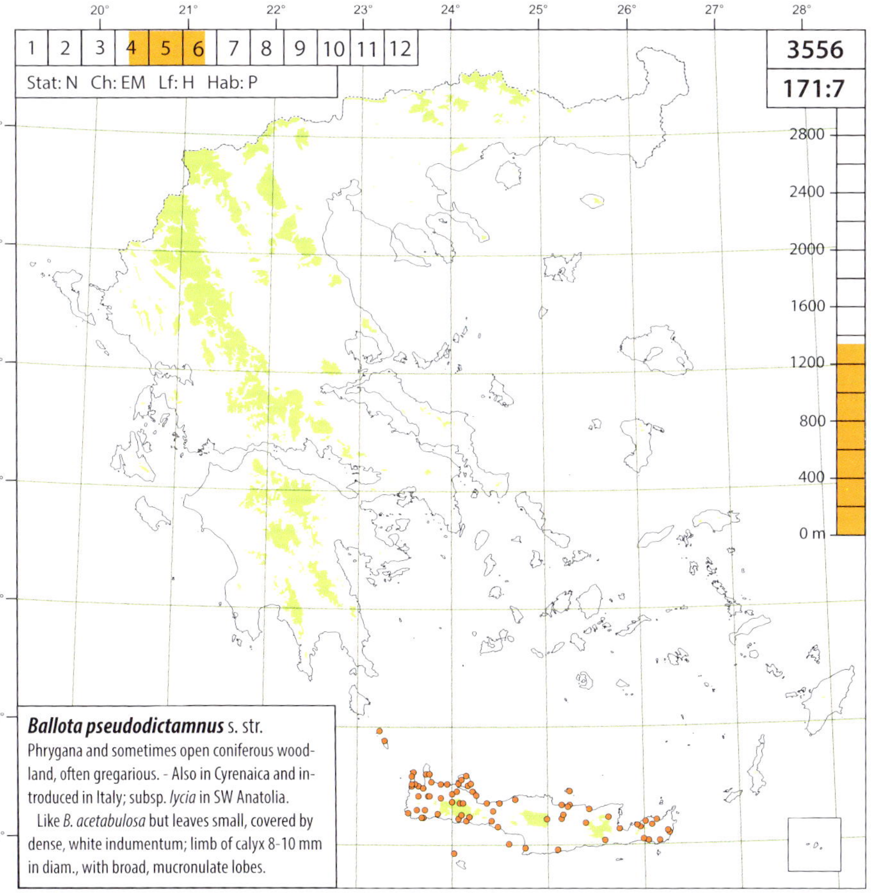

Ballota pseudodictamnus s. str.
Phrygana and sometimes open coniferous woodland, often gregarious. - Also in Cyrenaica and introduced in Italy; subsp. *lycia* in SW Anatolia.
Like *B. acetabulosa* but leaves small, covered by dense, white indumentum; limb of calyx 8-10 mm in diam., with broad, mucronulate lobes.

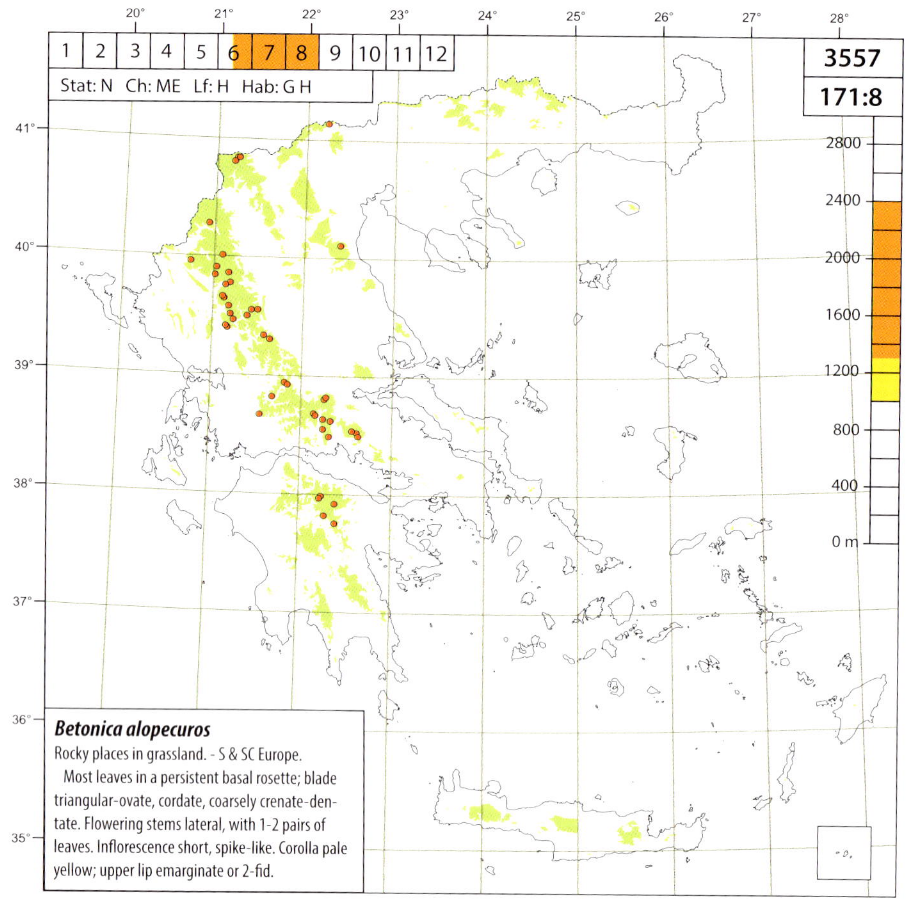

Betonica alopecuros
Rocky places in grassland. - S & SC Europe.
Most leaves in a persistent basal rosette; blade triangular-ovate, cordate, coarsely crenate-dentate. Flowering stems lateral, with 1-2 pairs of leaves. Inflorescence short, spike-like. Corolla pale yellow; upper lip emarginate or 2-fid.

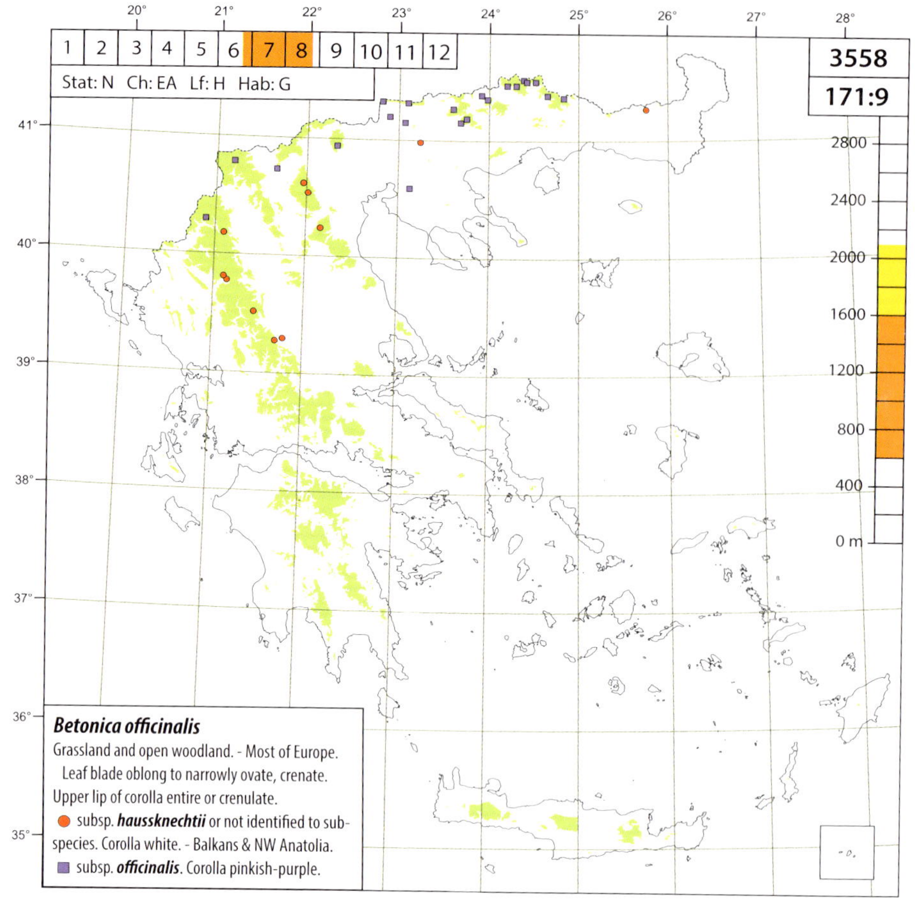

Betonica officinalis
Grassland and open woodland. - Most of Europe.
Leaf blade oblong to narrowly ovate, crenate. Upper lip of corolla entire or crenulate.
● subsp. ***haussknechtii*** or not identified to subspecies. Corolla white. - Balkans & NW Anatolia.
■ subsp. ***officinalis***. Corolla pinkish-purple.

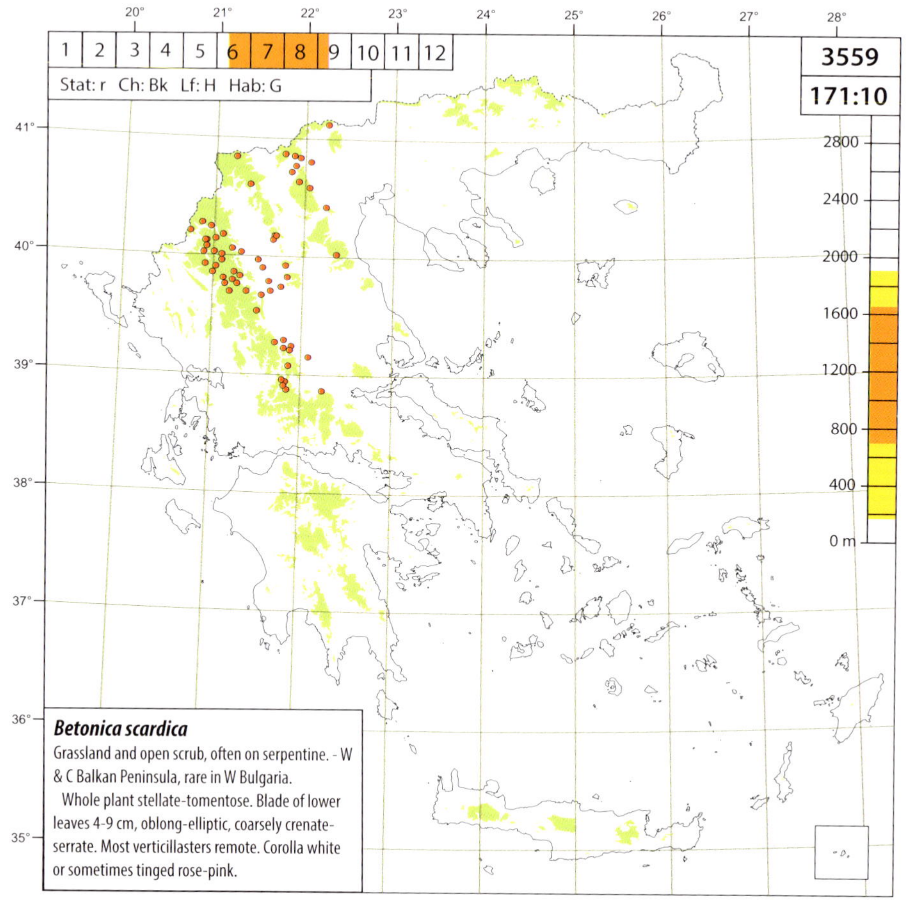

Betonica scardica
Grassland and open scrub, often on serpentine. - W & C Balkan Peninsula, rare in W Bulgaria.
Whole plant stellate-tomentose. Blade of lower leaves 4-9 cm, oblong-elliptic, coarsely crenate-serrate. Most verticillasters remote. Corolla white or sometimes tinged rose-pink.

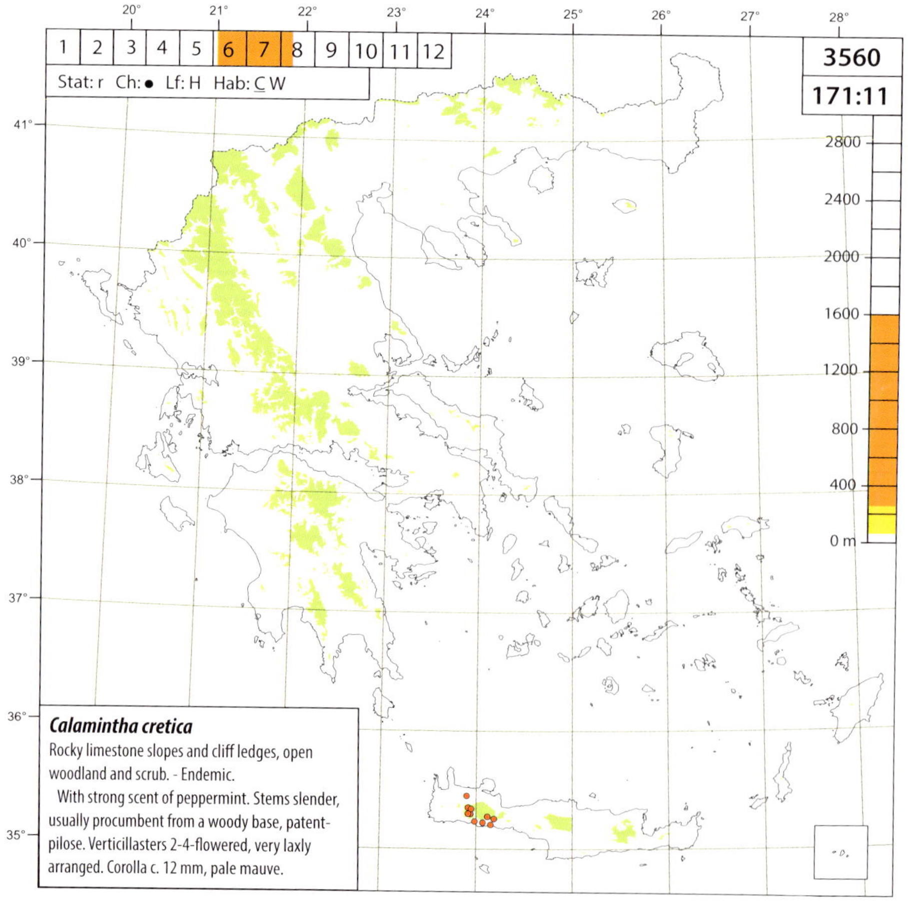

Calamintha cretica
Rocky limestone slopes and cliff ledges, open woodland and scrub. - Endemic.
With strong scent of peppermint. Stems slender, usually procumbent from a woody base, patent-pilose. Verticillasters 2-4-flowered, very laxly arranged. Corolla c. 12 mm, pale mauve.

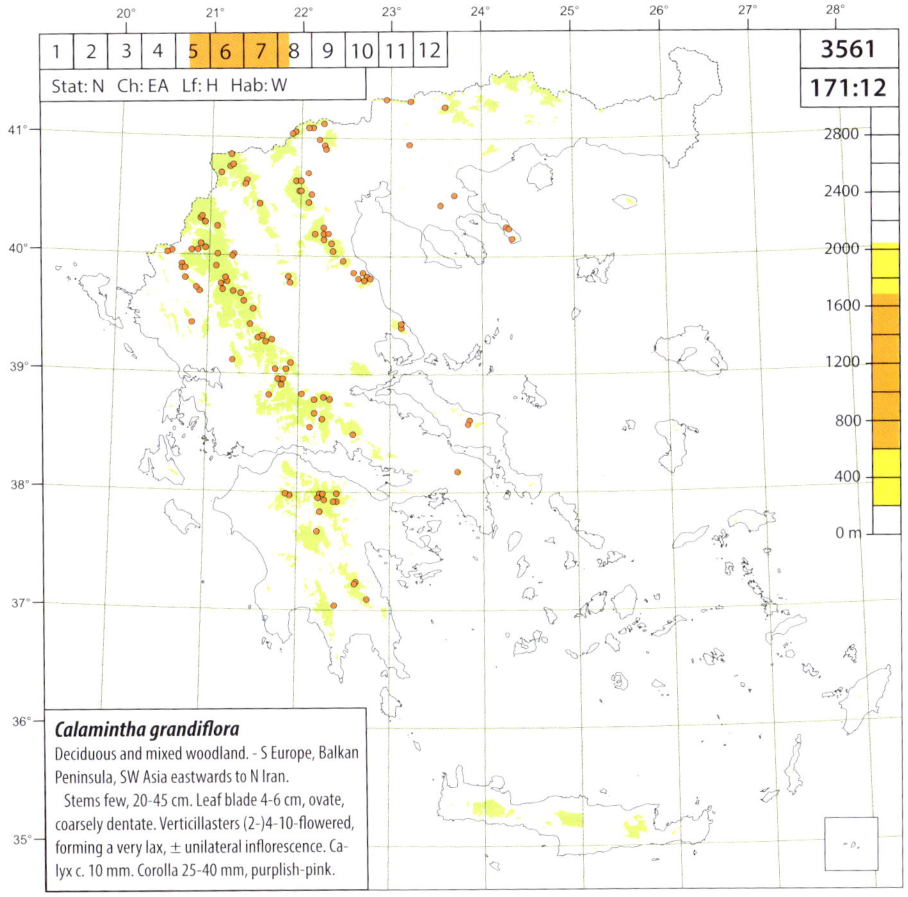

Calamintha grandiflora

Deciduous and mixed woodland. - S Europe, Balkan Peninsula, SW Asia eastwards to N Iran.

Stems few, 20-45 cm. Leaf blade 4-6 cm, ovate, coarsely dentate. Verticillasters (2-)4-10-flowered, forming a very lax, ± unilateral inflorescence. Calyx c. 10 mm. Corolla 25-40 mm, purplish-pink.

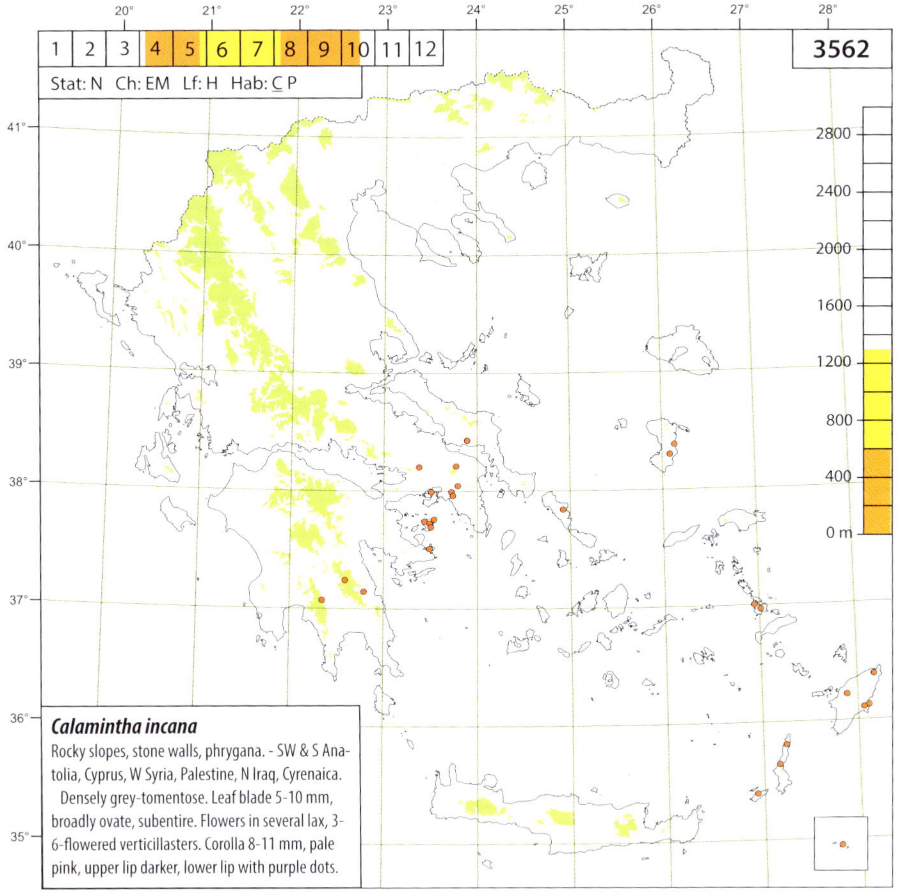

Calamintha incana

Rocky slopes, stone walls, phrygana. - SW & S Anatolia, Cyprus, W Syria, Palestine, N Iraq, Cyrenaica.

Densely grey-tomentose. Leaf blade 5-10 mm, broadly ovate, subentire. Flowers in several lax, 3-6-flowered verticillasters. Corolla 8-11 mm, pale pink, upper lip darker, lower lip with purple dots.

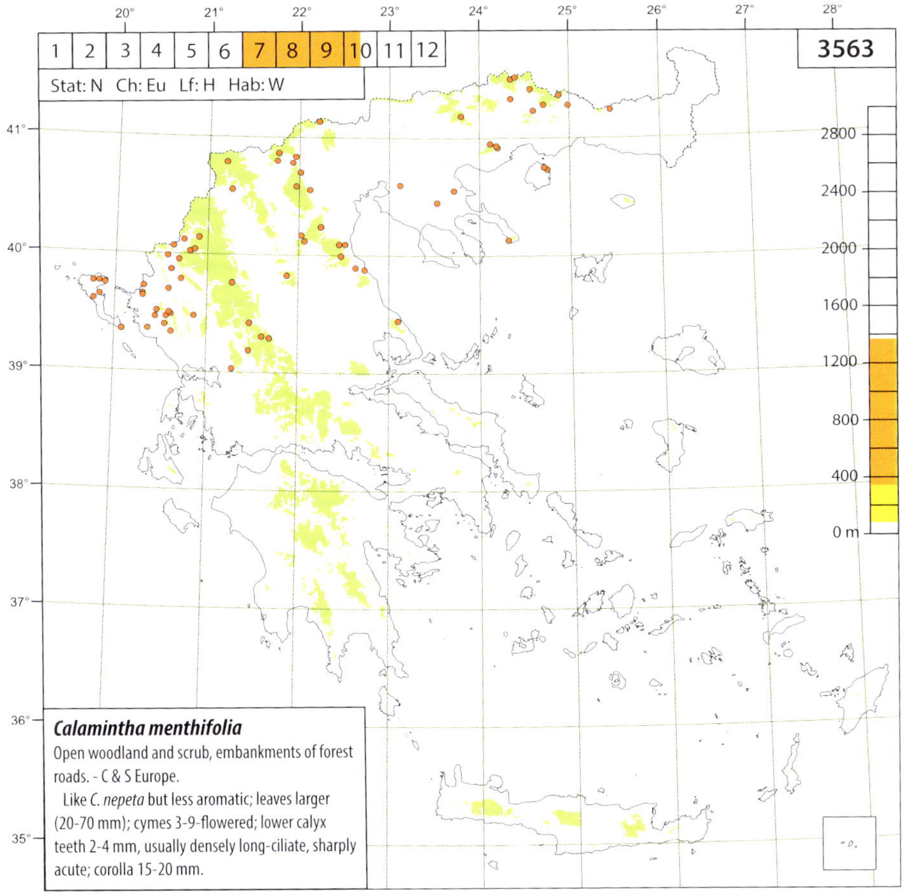

Calamintha menthifolia

Open woodland and scrub, embankments of forest roads. - C & S Europe.

Like *C. nepeta* but less aromatic; leaves larger (20-70 mm); cymes 3-9-flowered; lower calyx teeth 2-4 mm, usually densely long-ciliate, sharply acute; corolla 15-20 mm.

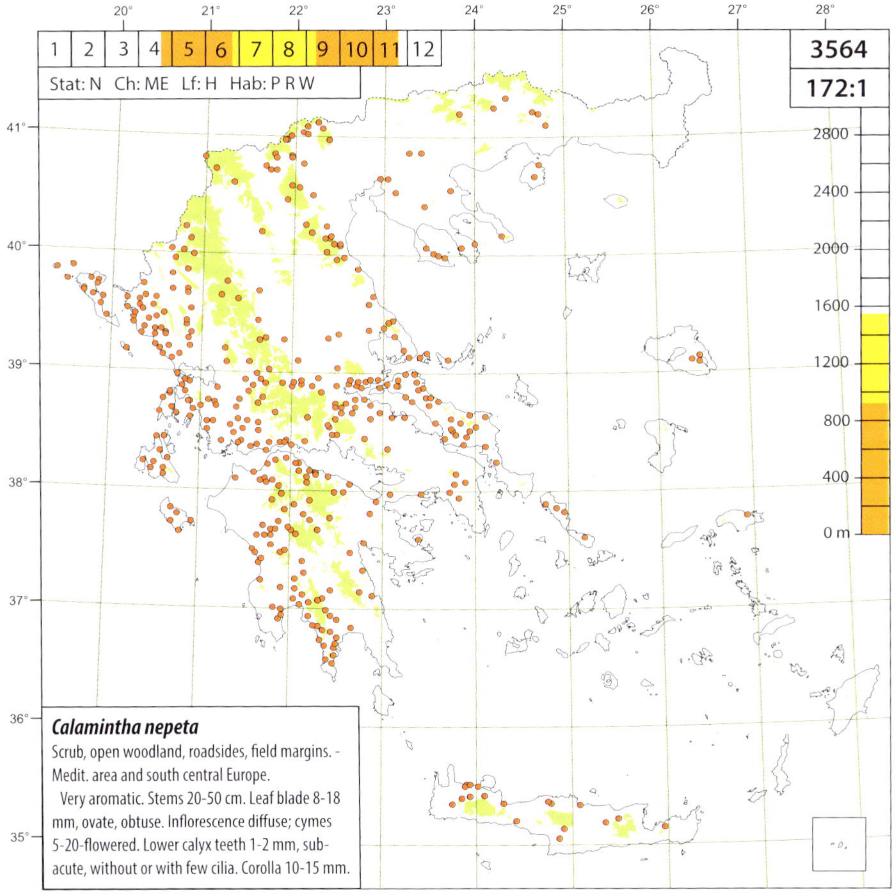

Calamintha nepeta

Scrub, open woodland, roadsides, field margins. - Medit. area and south central Europe.

Very aromatic. Stems 20-50 cm. Leaf blade 8-18 mm, ovate, obtuse. Inflorescence diffuse; cymes 5-20-flowered. Lower calyx teeth 1-2 mm, subacute, without or with few cilia. Corolla 10-15 mm.

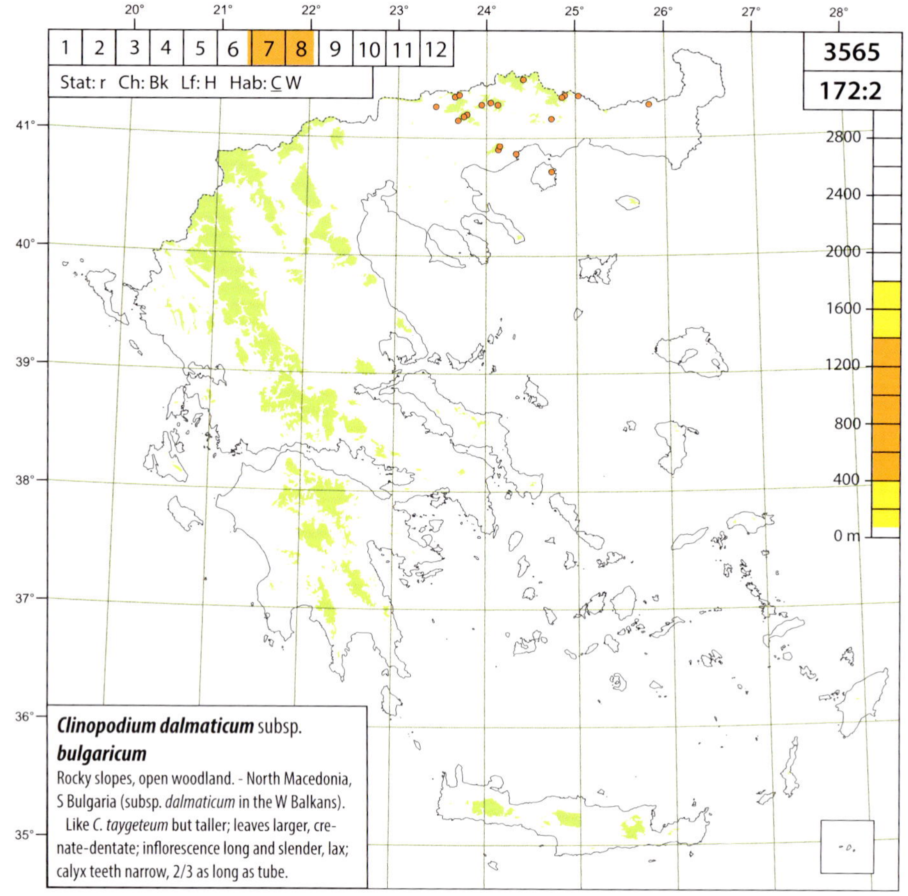

Clinopodium dalmaticum subsp. ***bulgaricum***

Rocky slopes, open woodland. - North Macedonia, S Bulgaria (subsp. *dalmaticum* in the W Balkans).

Like *C. taygeteum* but taller; leaves larger, crenate-dentate; inflorescence long and slender, lax; calyx teeth narrow, 2/3 as long as tube.

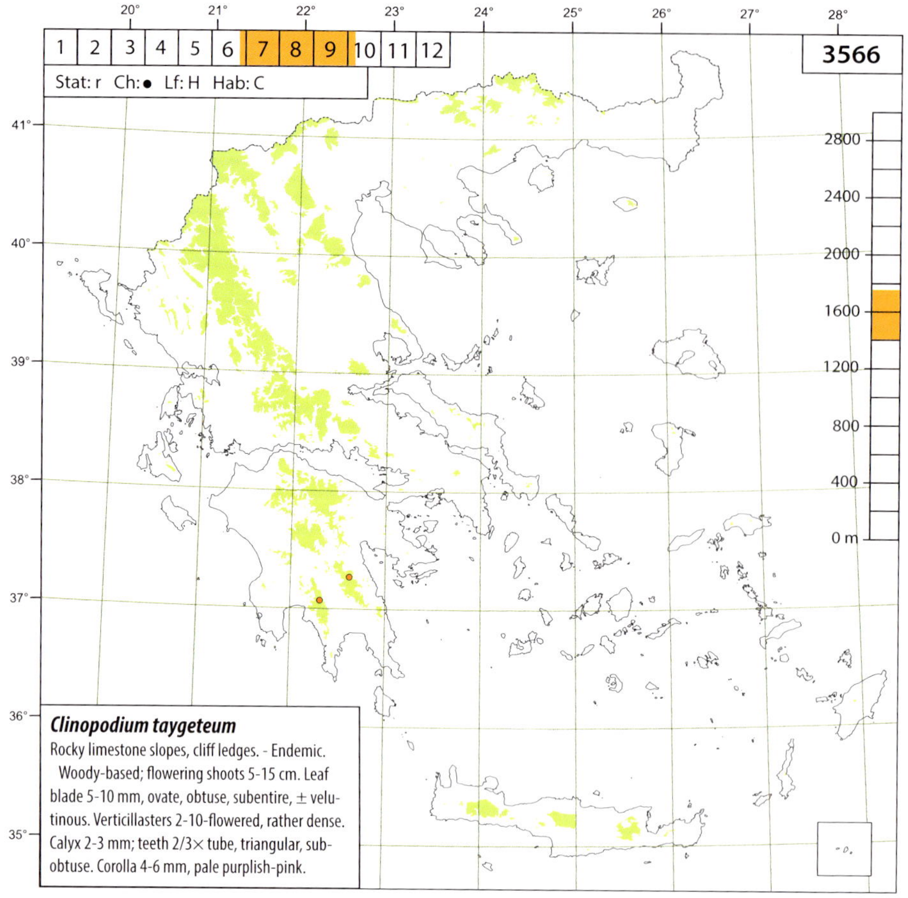

Clinopodium taygeteum

Rocky limestone slopes, cliff ledges. - Endemic.

Woody-based; flowering shoots 5-15 cm. Leaf blade 5-10 mm, ovate, obtuse, subentire, ± velutinous. Verticillasters 2-10-flowered, rather dense. Calyx 2-3 mm; teeth 2/3× tube, triangular, subobtuse. Corolla 4-6 mm, pale purplish-pink.

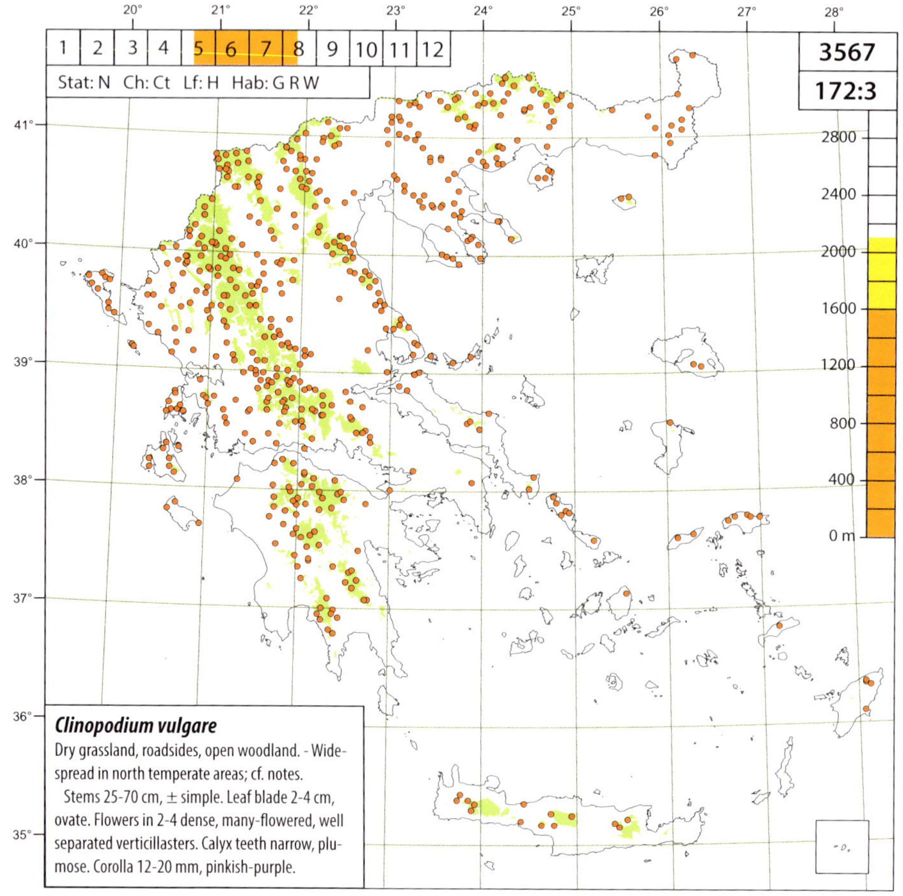

Clinopodium vulgare

Dry grassland, roadsides, open woodland. - Widespread in north temperate areas; cf. notes.

Stems 25-70 cm, ± simple. Leaf blade 2-4 cm, ovate. Flowers in 2-4 dense, many-flowered, well separated verticillasters. Calyx teeth narrow, plumose. Corolla 12-20 mm, pinkish-purple.

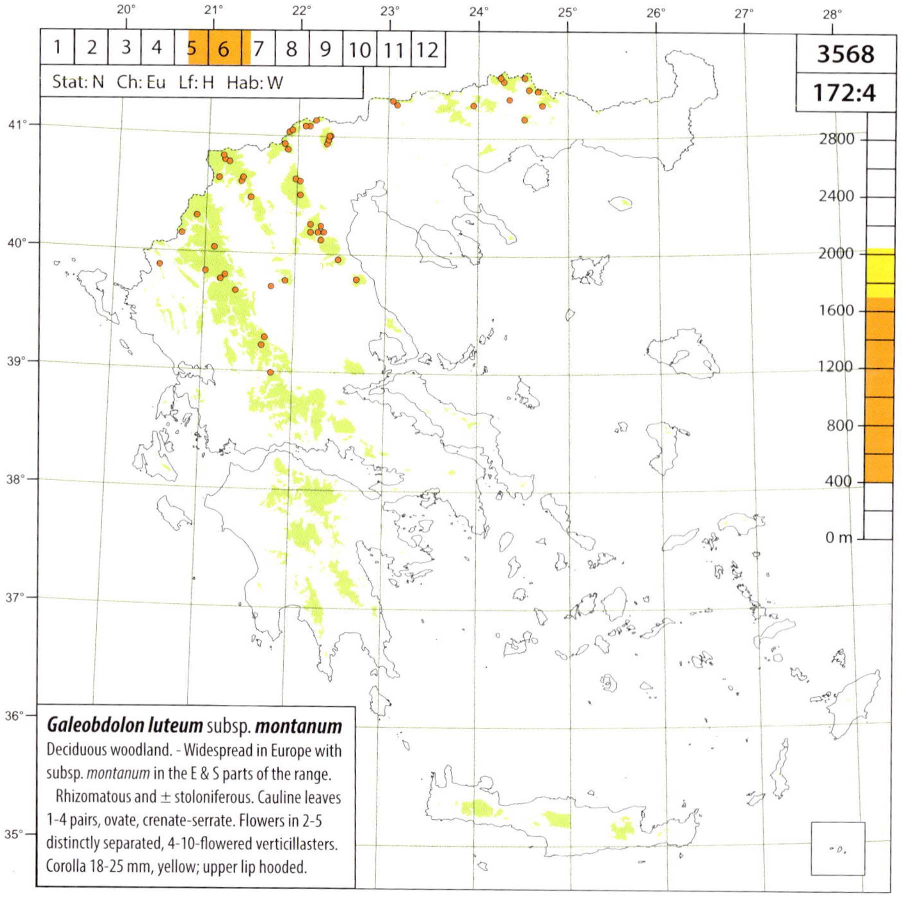

Galeobdolon luteum subsp. ***montanum***

Deciduous woodland. - Widespread in Europe with subsp. *montanum* in the E & S parts of the range.

Rhizomatous and ± stoloniferous. Cauline leaves 1-4 pairs, ovate, crenate-serrate. Flowers in 2-5 distinctly separated, 4-10-flowered verticillasters. Corolla 18-25 mm, yellow; upper lip hooded.

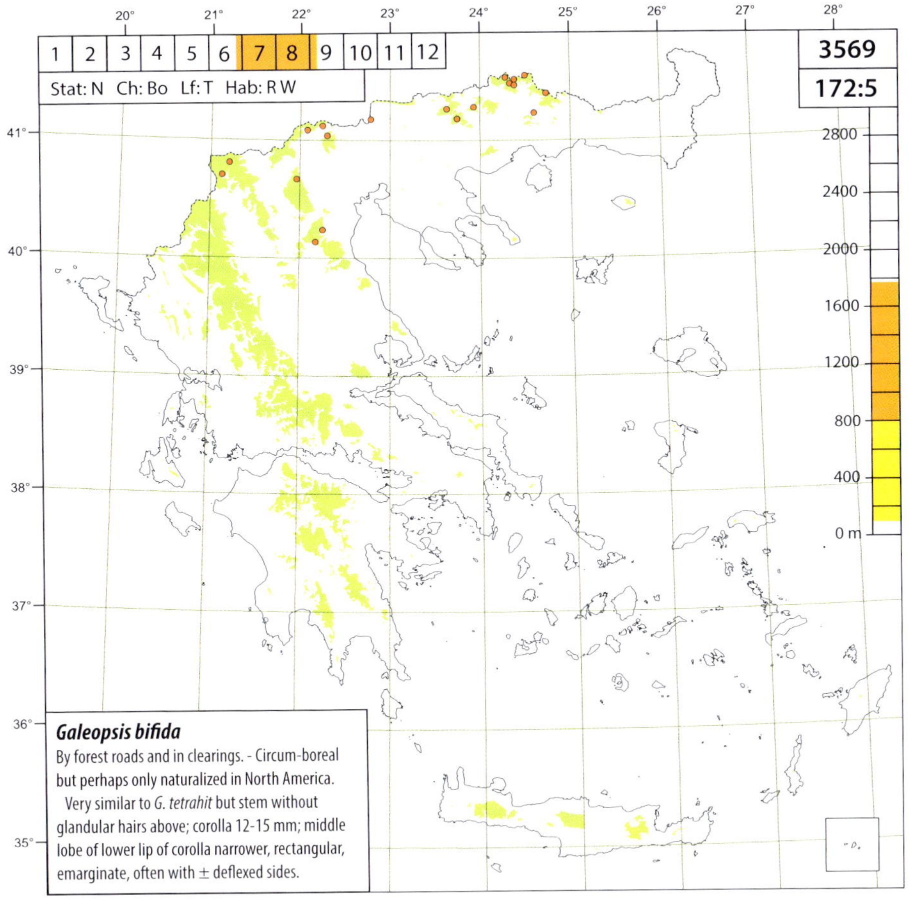

Galeopsis bifida

By forest roads and in clearings. - Circum-boreal but perhaps only naturalized in North America.

Very similar to *G. tetrahit* but stem without glandular hairs above; corolla 12-15 mm; middle lobe of lower lip of corolla narrower, rectangular, emarginate, often with ± deflexed sides.

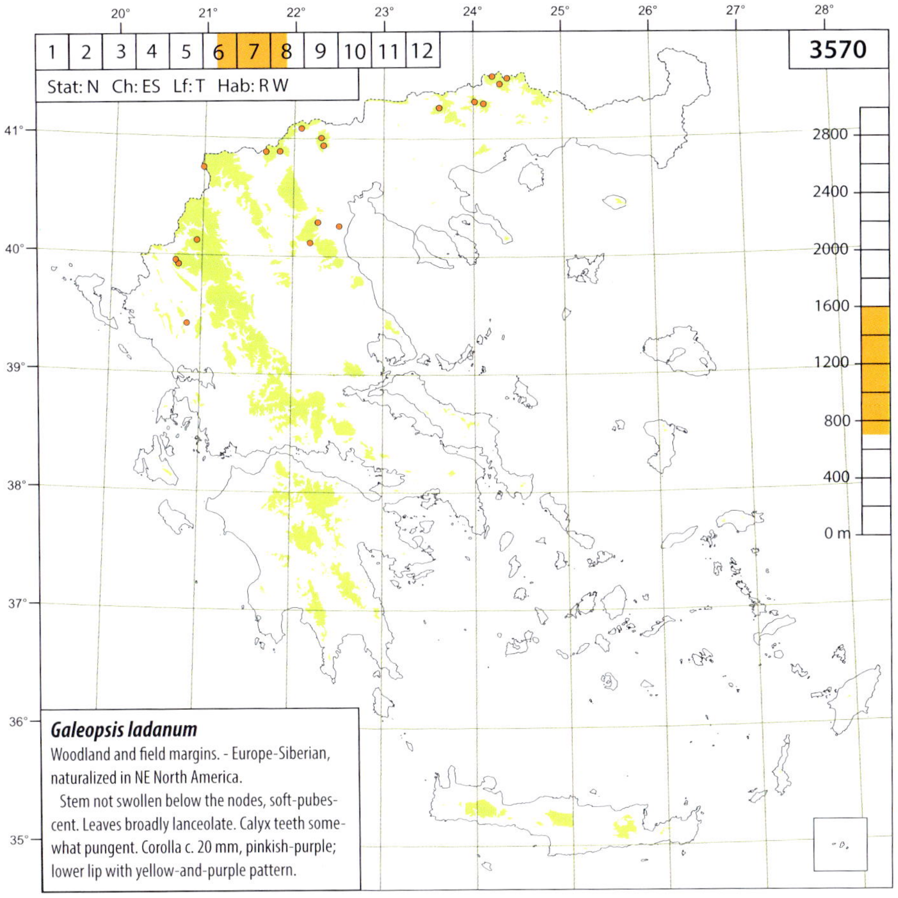

Galeopsis ladanum

Woodland and field margins. - Europe-Siberian, naturalized in NE North America.

Stem not swollen below the nodes, soft-pubescent. Leaves broadly lanceolate. Calyx teeth somewhat pungent. Corolla c. 20 mm, pinkish-purple; lower lip with yellow-and-purple pattern.

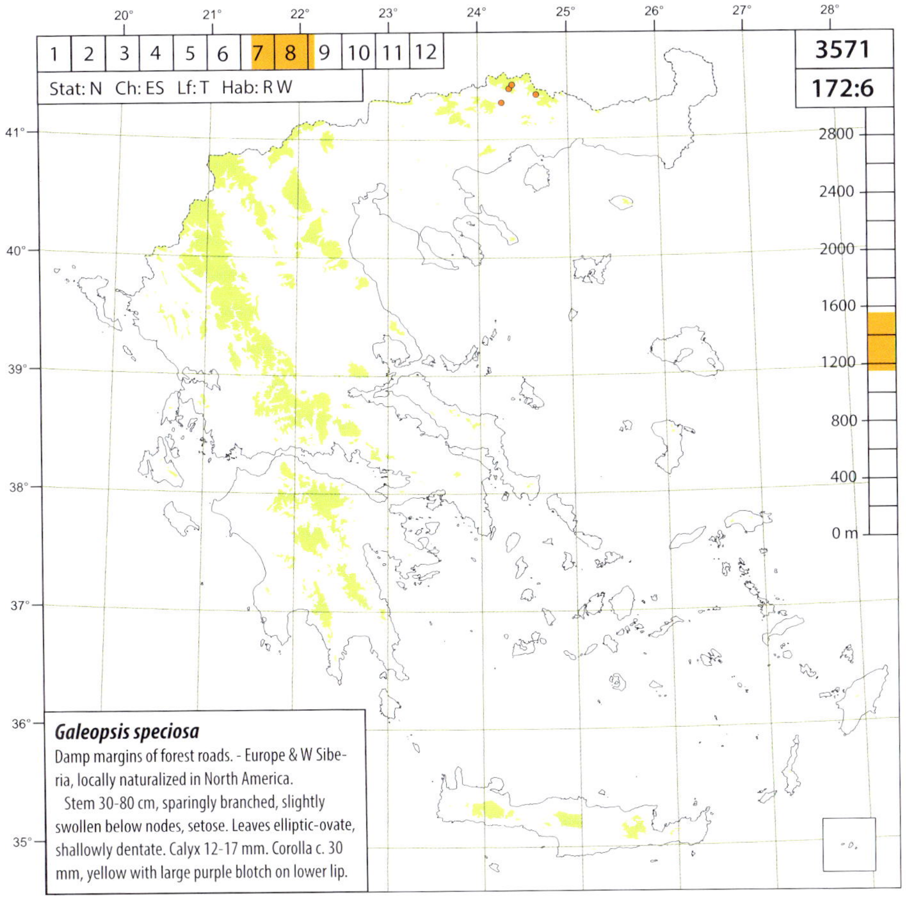

Galeopsis speciosa

Damp margins of forest roads. - Europe & W Siberia, locally naturalized in North America.

Stem 30-80 cm, sparingly branched, slightly swollen below nodes, setose. Leaves elliptic-ovate, shallowly dentate. Calyx 12-17 mm. Corolla c. 30 mm, yellow with large purple blotch on lower lip.

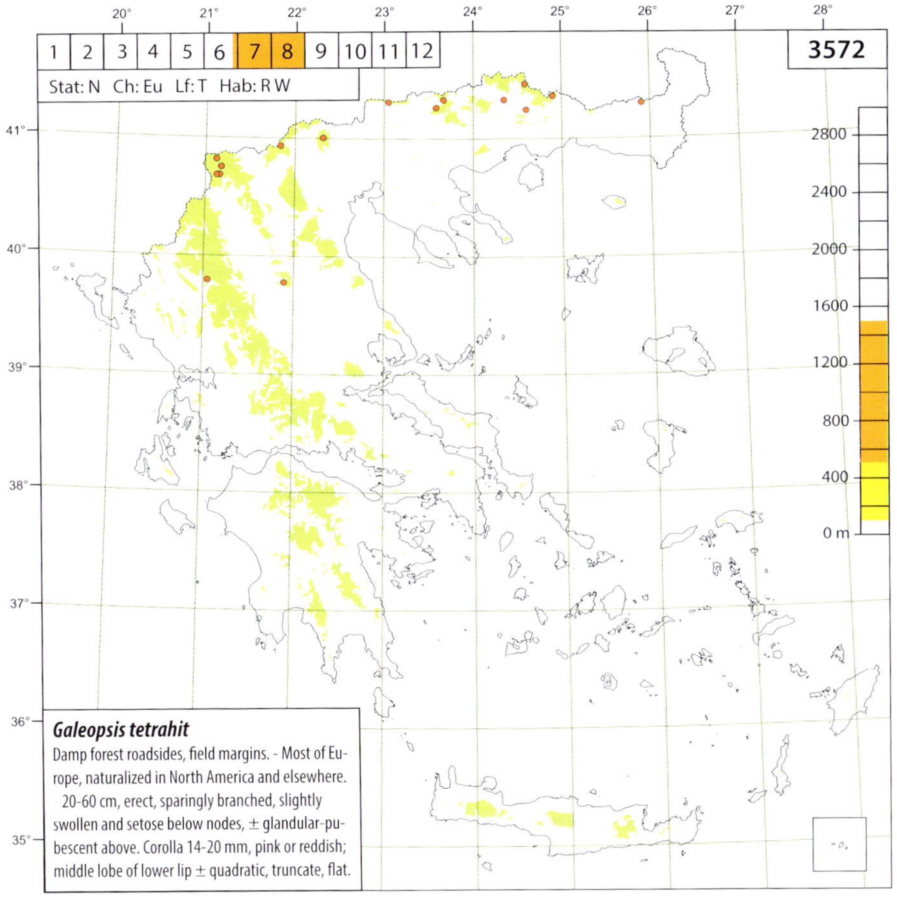

Galeopsis tetrahit

Damp forest roadsides, field margins. - Most of Europe, naturalized in North America and elsewhere.

20-60 cm, erect, sparingly branched, slightly swollen and setose below nodes, ± glandular-pubescent above. Corolla 14-20 mm, pink or reddish; middle lobe of lower lip ± quadratic, truncate, flat.

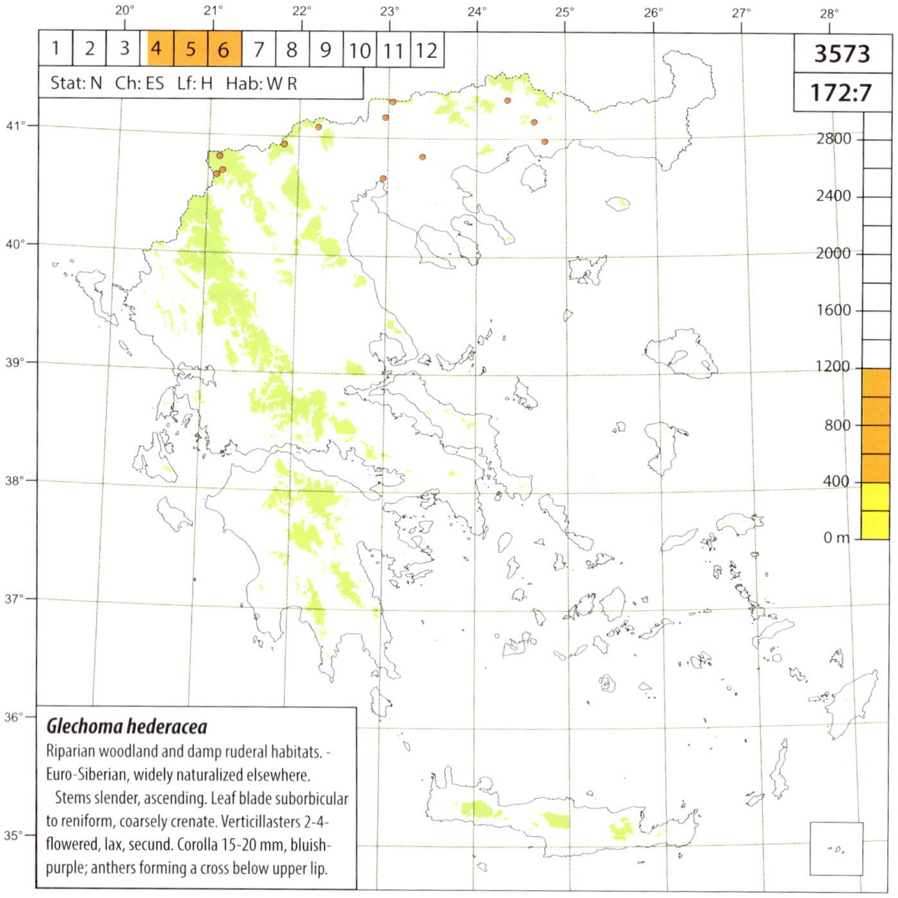

Glechoma hederacea
Riparian woodland and damp ruderal habitats. - Euro-Siberian, widely naturalized elsewhere.
Stems slender, ascending. Leaf blade suborbicular to reniform, coarsely crenate. Verticillasters 2-4-flowered, lax, secund. Corolla 15-20 mm, bluish-purple; anthers forming a cross below upper lip.

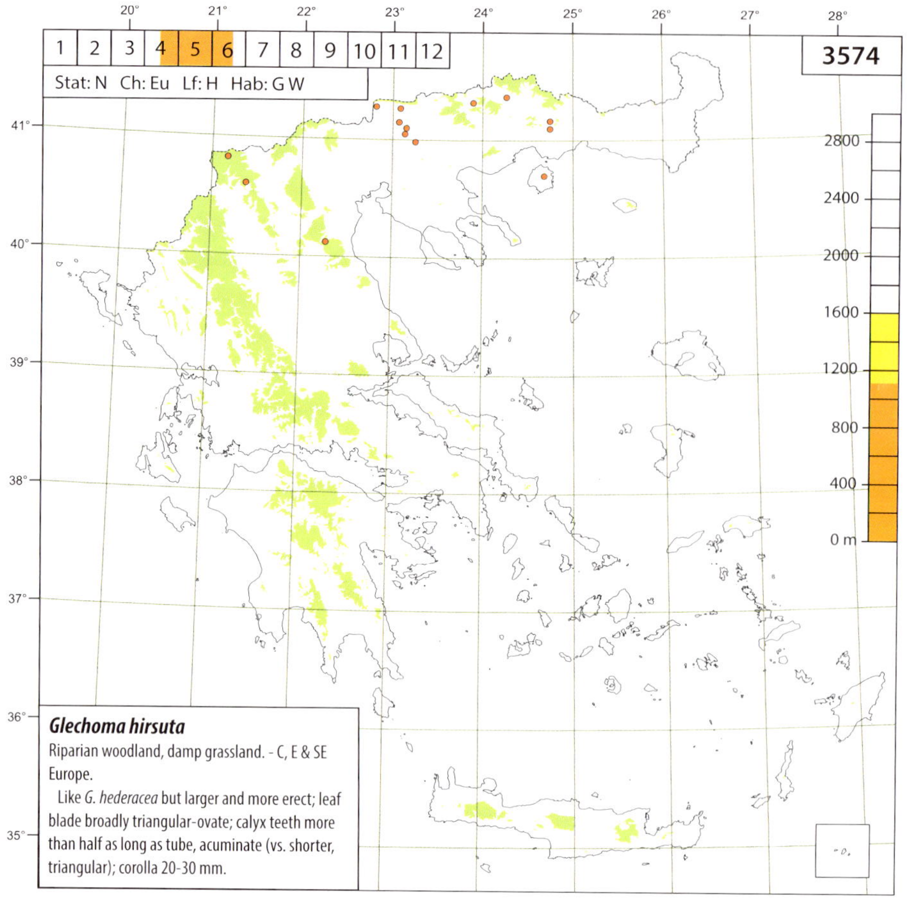

Glechoma hirsuta
Riparian woodland, damp grassland. - C, E & SE Europe.
Like *G. hederacea* but larger and more erect; leaf blade broadly triangular-ovate; calyx teeth more than half as long as tube, acuminate (vs. shorter, triangular); corolla 20-30 mm.

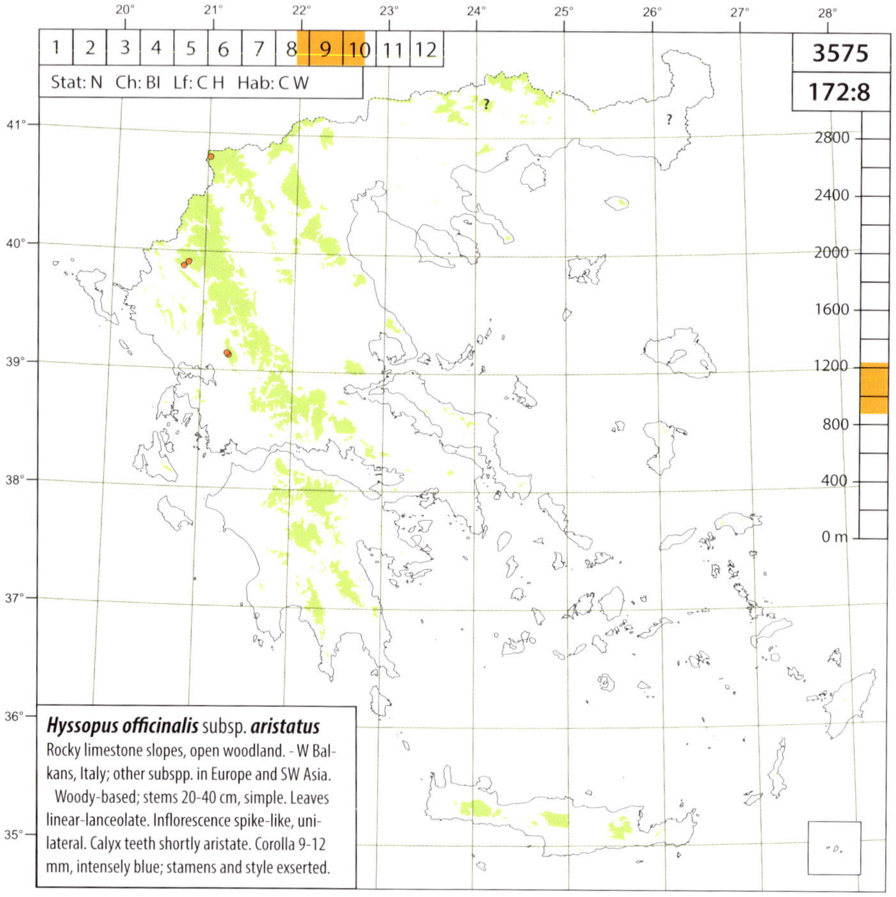

Hyssopus officinalis subsp. ***aristatus***
Rocky limestone slopes, open woodland. - W Balkans, Italy; other subspp. in Europe and SW Asia.
Woody-based; stems 20-40 cm, simple. Leaves linear-lanceolate. Inflorescence spike-like, unilateral. Calyx teeth shortly aristate. Corolla 9-12 mm, intensely blue; stamens and style exserted.

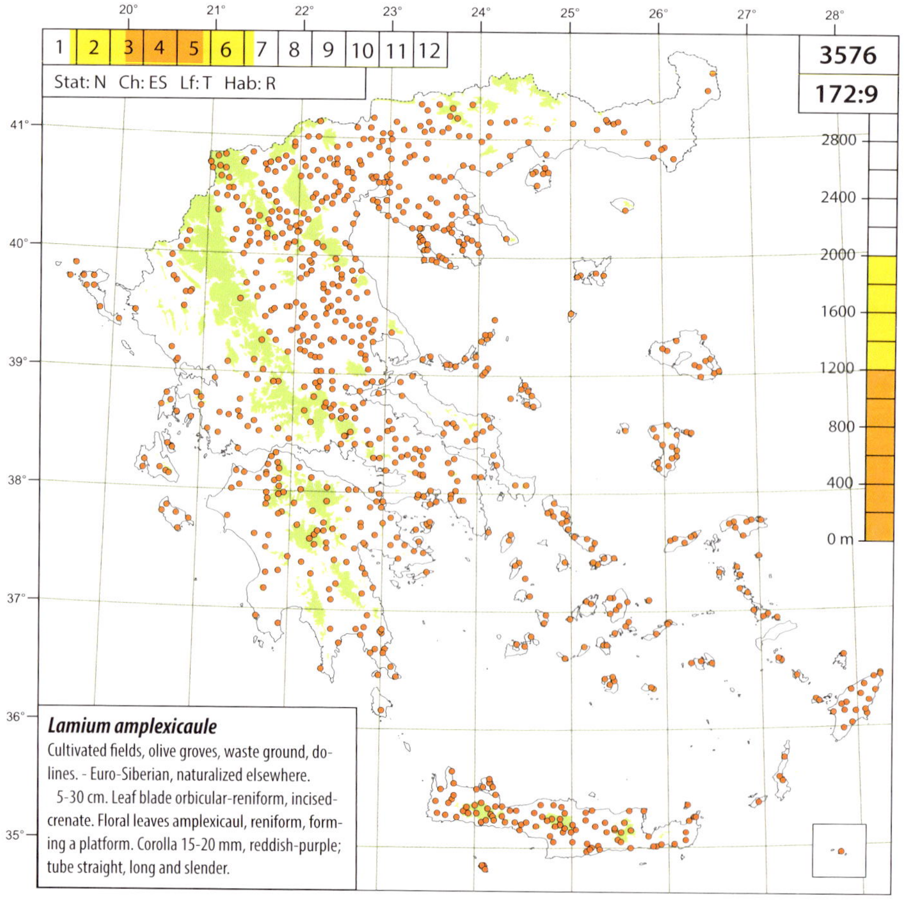

Lamium amplexicaule
Cultivated fields, olive groves, waste ground, dolines. - Euro-Siberian, naturalized elsewhere.
5-30 cm. Leaf blade orbicular-reniform, incised-crenate. Floral leaves amplexicaul, reniform, forming a platform. Corolla 15-20 mm, reddish-purple; tube straight, long and slender.

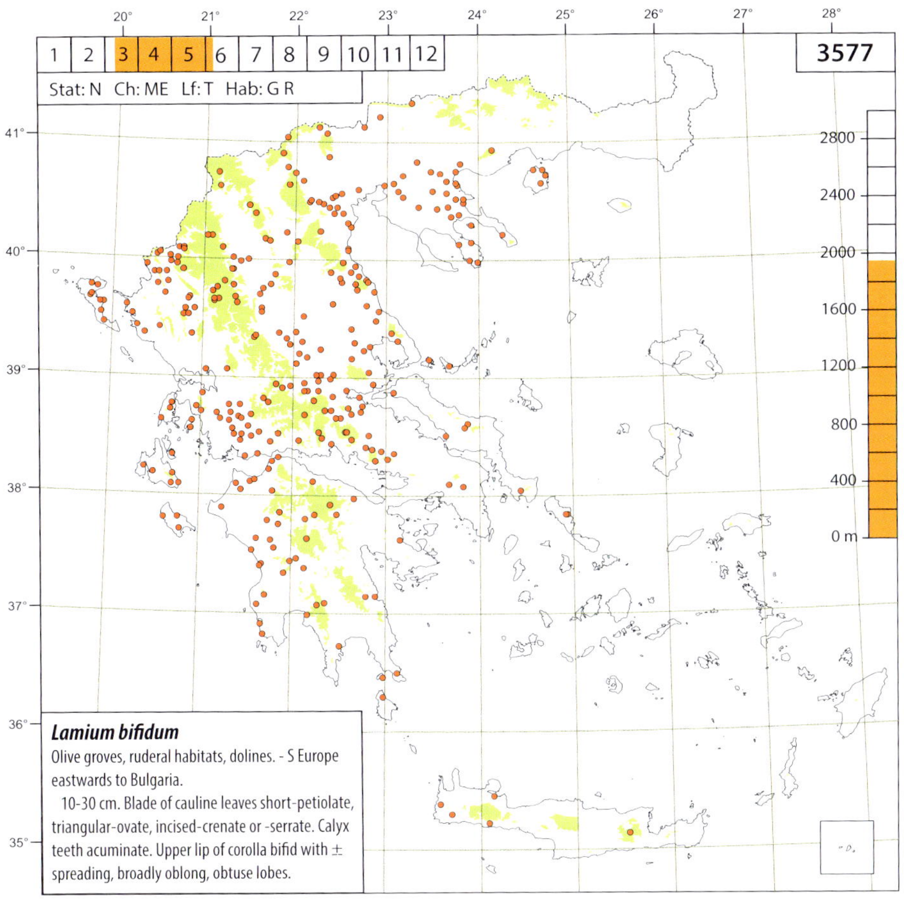

Lamium bifidum

Olive groves, ruderal habitats, dolines. - S Europe eastwards to Bulgaria.

10-30 cm. Blade of cauline leaves short-petiolate, triangular-ovate, incised-crenate or -serrate. Calyx teeth acuminate. Upper lip of corolla bifid with ± spreading, broadly oblong, obtuse lobes.

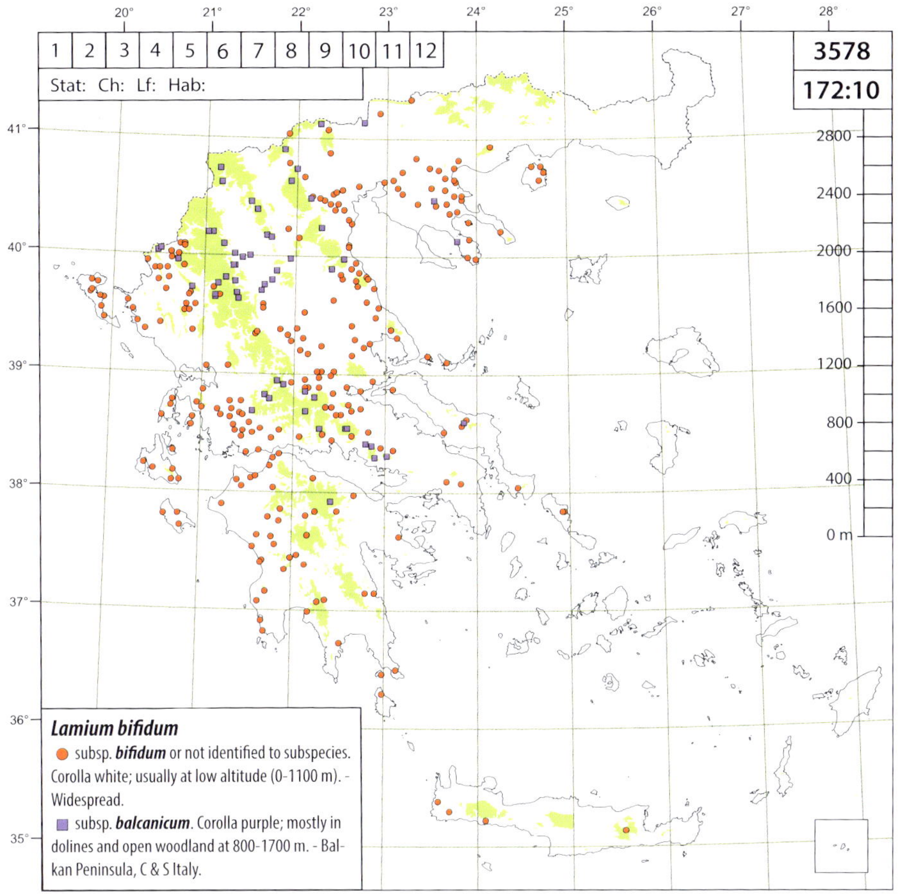

Lamium bifidum

● subsp. ***bifidum*** or not identified to subspecies. Corolla white; usually at low altitude (0-1100 m). - Widespread.

■ subsp. ***balcanicum***. Corolla purple; mostly in dolines and open woodland at 800-1700 m. - Balkan Peninsula, C & S Italy.

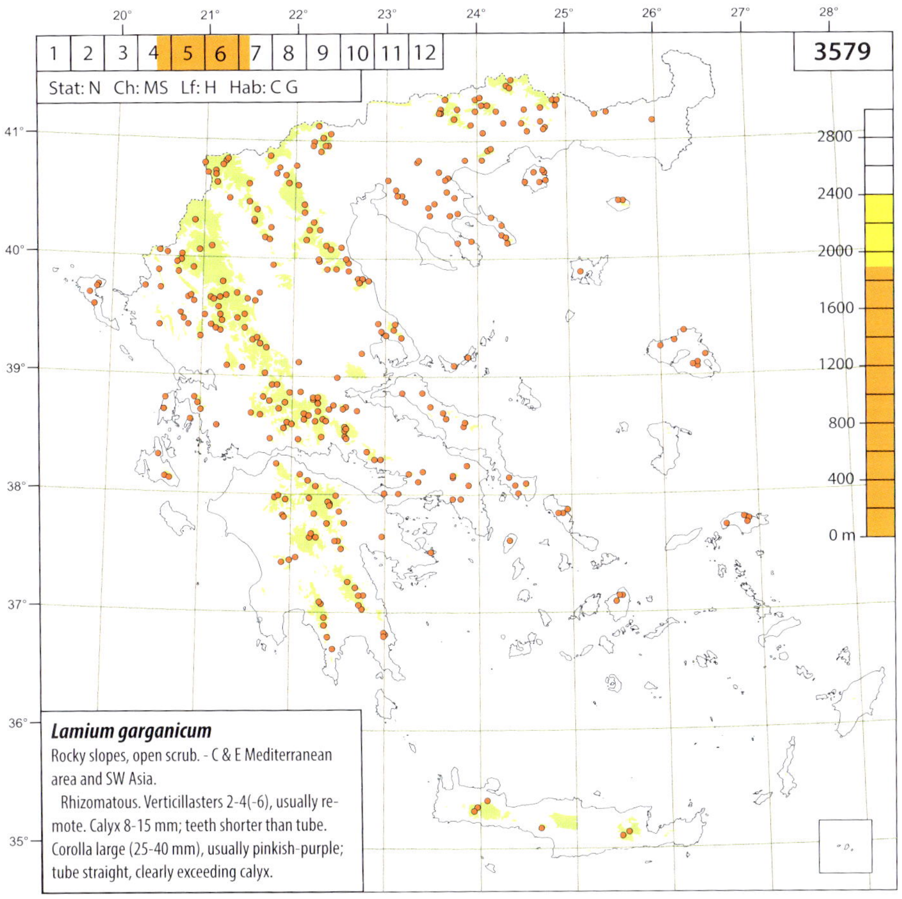

Lamium garganicum

Rocky slopes, open scrub. - C & E Mediterranean area and SW Asia.

Rhizomatous. Verticillasters 2-4(-6), usually remote. Calyx 8-15 mm; teeth shorter than tube. Corolla large (25-40 mm), usually pinkish-purple; tube straight, clearly exceeding calyx.

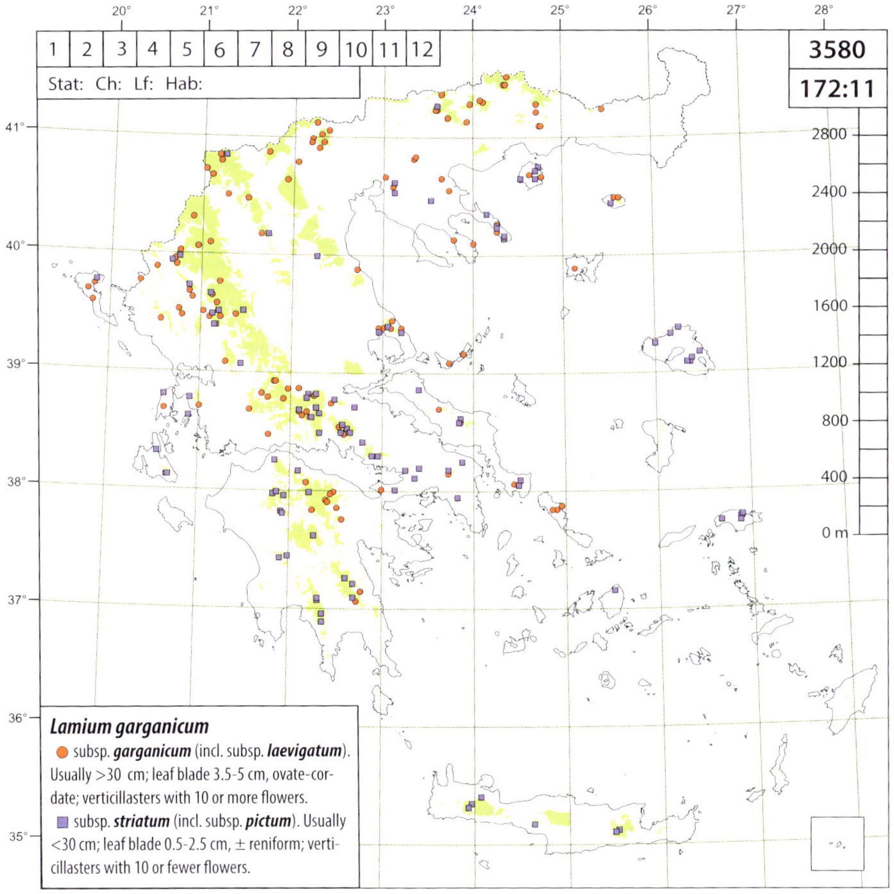

Lamium garganicum

● subsp. ***garganicum*** (incl. subsp. ***laevigatum***). Usually >30 cm; leaf blade 3.5-5 cm, ovate-cordate; verticillasters with 10 or more flowers.

■ subsp. ***striatum*** (incl. subsp. ***pictum***). Usually <30 cm; leaf blade 0.5-2.5 cm, ± reniform; verticillasters with 10 or fewer flowers.

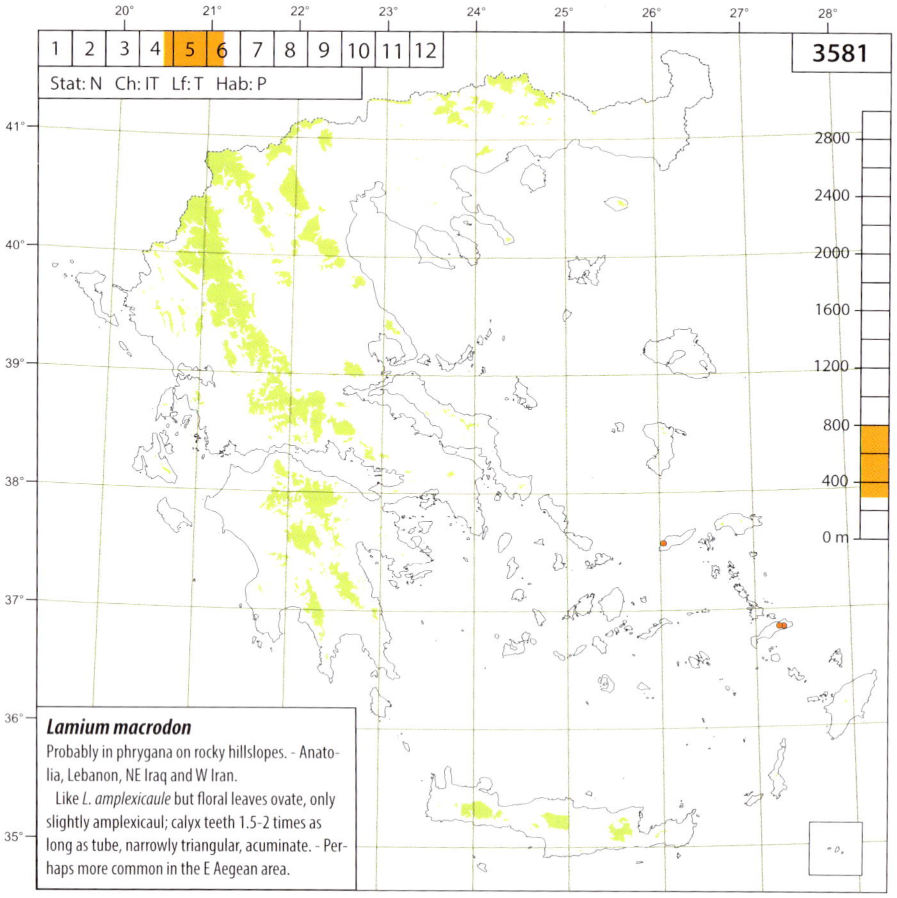

Lamium macrodon

Probably in phrygana on rocky hillslopes. - Anatolia, Lebanon, NE Iraq and W Iran.

Like *L. amplexicaule* but floral leaves ovate, only slightly amplexicaul; calyx teeth 1.5-2 times as long as tube, narrowly triangular, acuminate. - Perhaps more common in the E Aegean area.

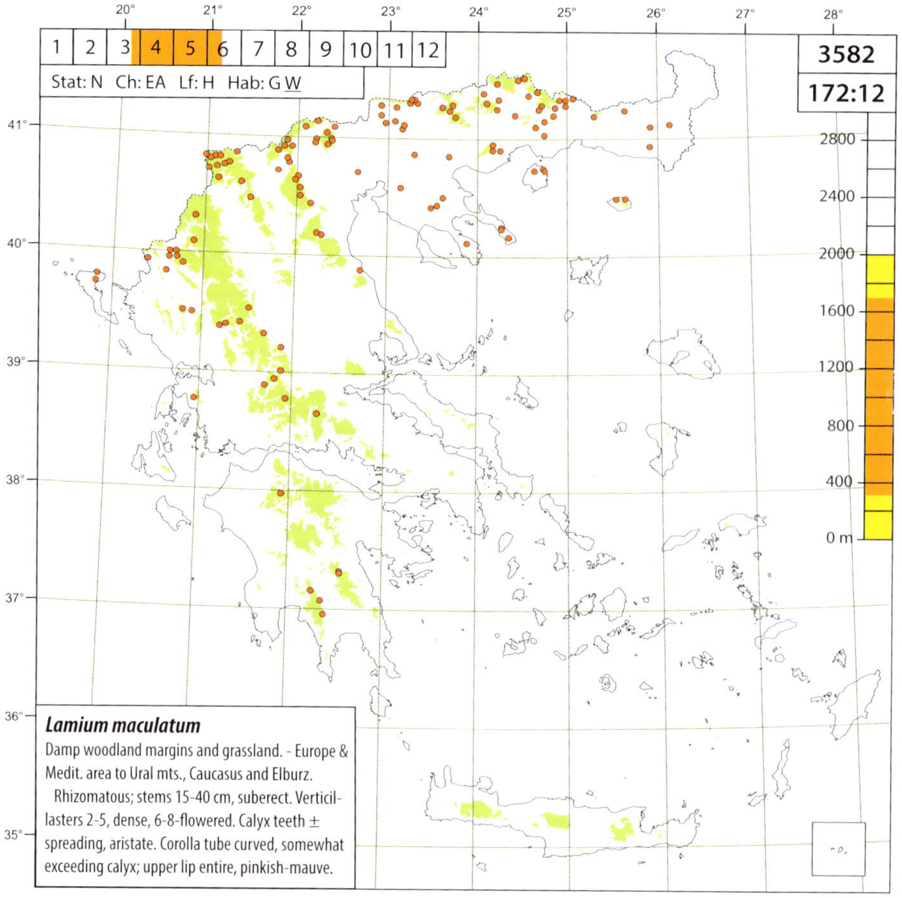

Lamium maculatum

Damp woodland margins and grassland. - Europe & Medit. area to Ural mts., Caucasus and Elburz.

Rhizomatous; stems 15-40 cm, suberect. Verticillasters 2-5, dense, 6-8-flowered. Calyx teeth ± spreading, aristate. Corolla tube curved, somewhat exceeding calyx; upper lip entire, pinkish-mauve.

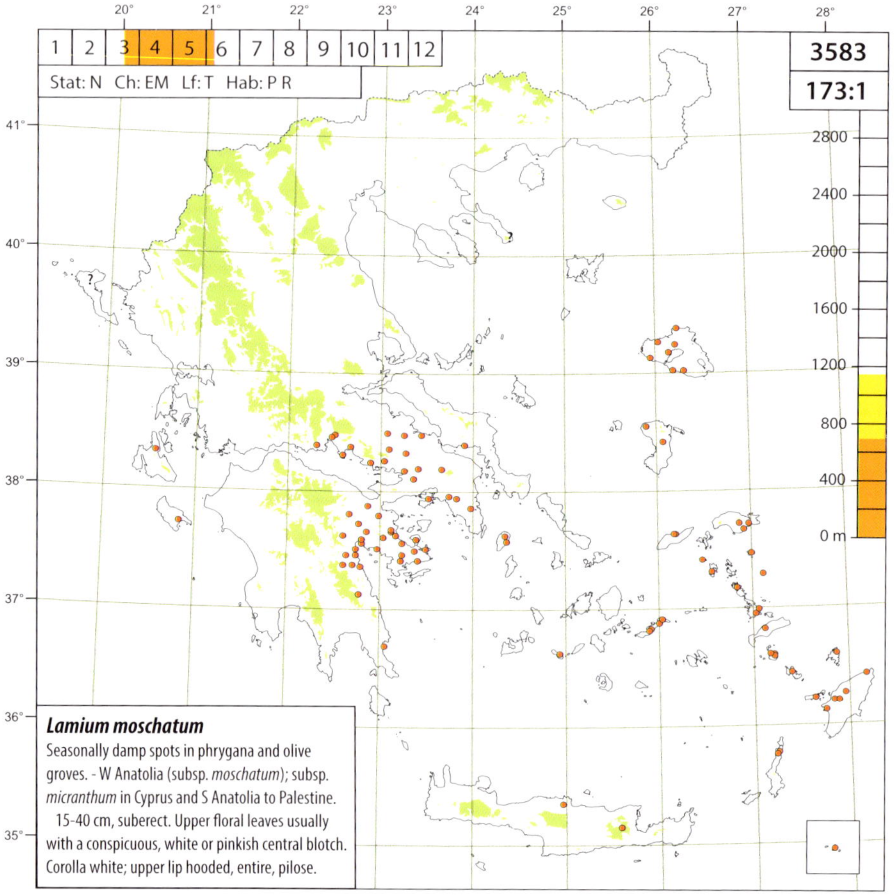

Lamium moschatum

Seasonally damp spots in phrygana and olive groves. - W Anatolia (subsp. *moschatum*); subsp. *micranthum* in Cyprus and S Anatolia to Palestine.

15-40 cm, suberect. Upper floral leaves usually with a conspicuous, white or pinkish central blotch. Corolla white; upper lip hooded, entire, pilose.

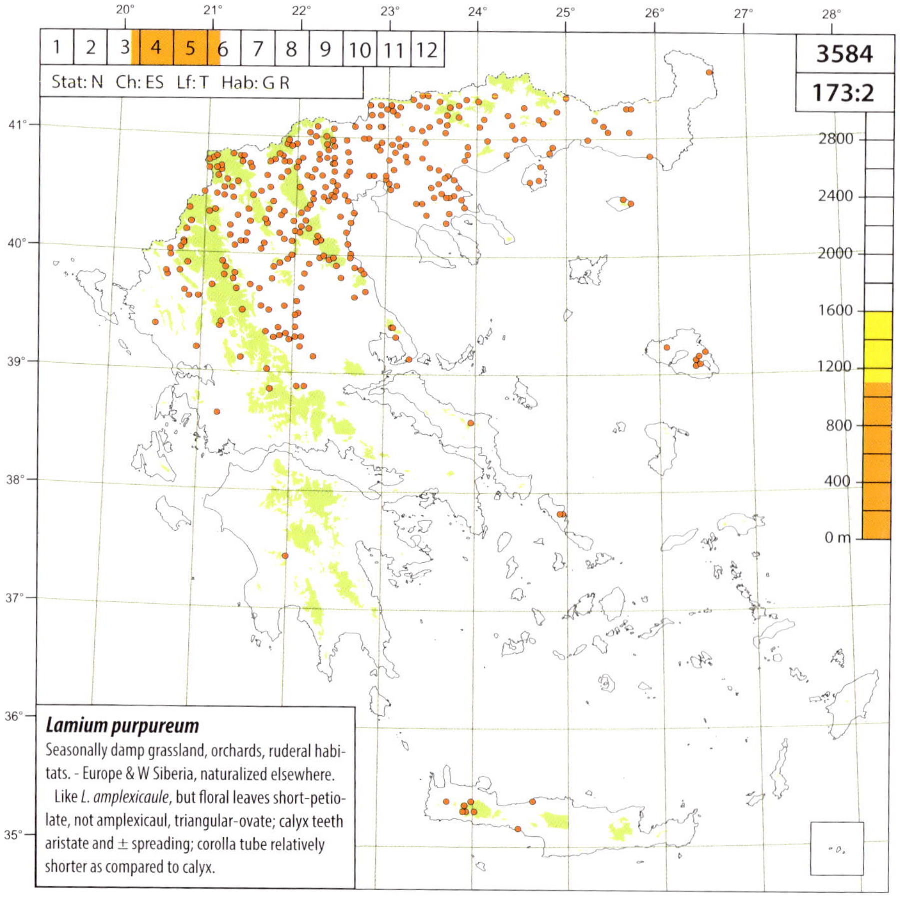

Lamium purpureum

Seasonally damp grassland, orchards, ruderal habitats. - Europe & W Siberia, naturalized elsewhere.

Like *L. amplexicaule*, but floral leaves short-petiolate, not amplexicaul, triangular-ovate; calyx teeth aristate and ± spreading; corolla tube relatively shorter as compared to calyx.

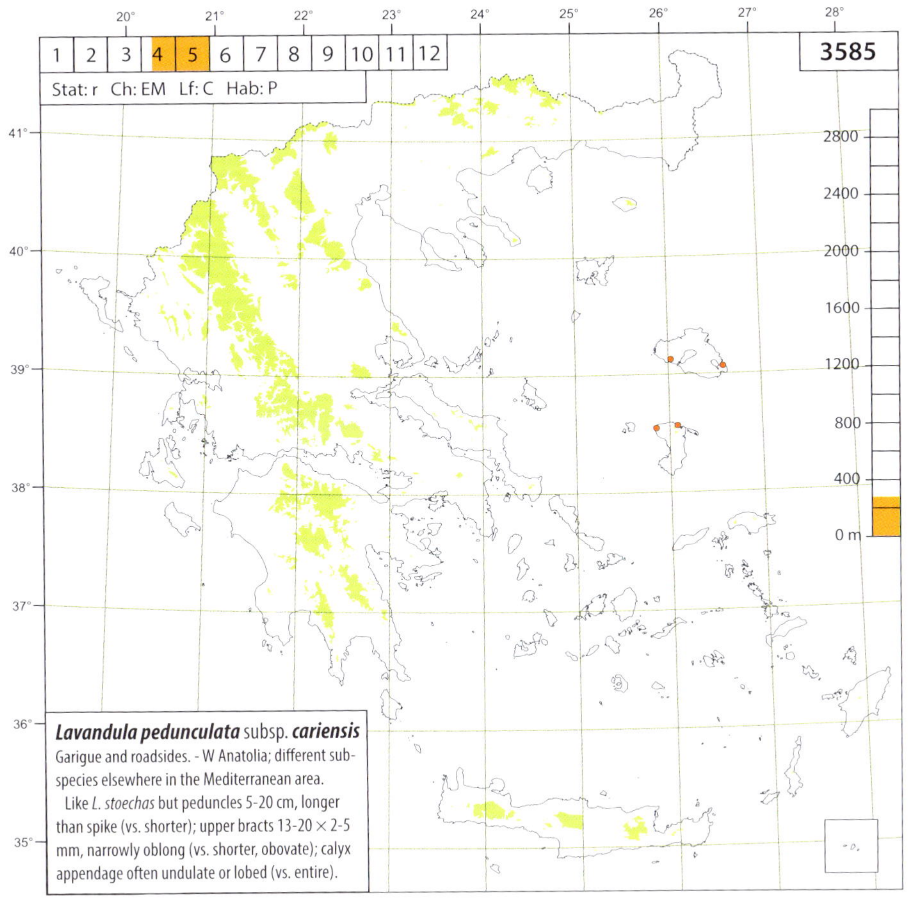

Lavandula pedunculata **subsp. *cariensis***
Garigue and roadsides. - W Anatolia; different subspecies elsewhere in the Mediterranean area.
Like *L. stoechas* but peduncles 5-20 cm, longer than spike (vs. shorter); upper bracts 13-20 × 2-5 mm, narrowly oblong (vs. shorter, obovate); calyx appendage often undulate or lobed (vs. entire).

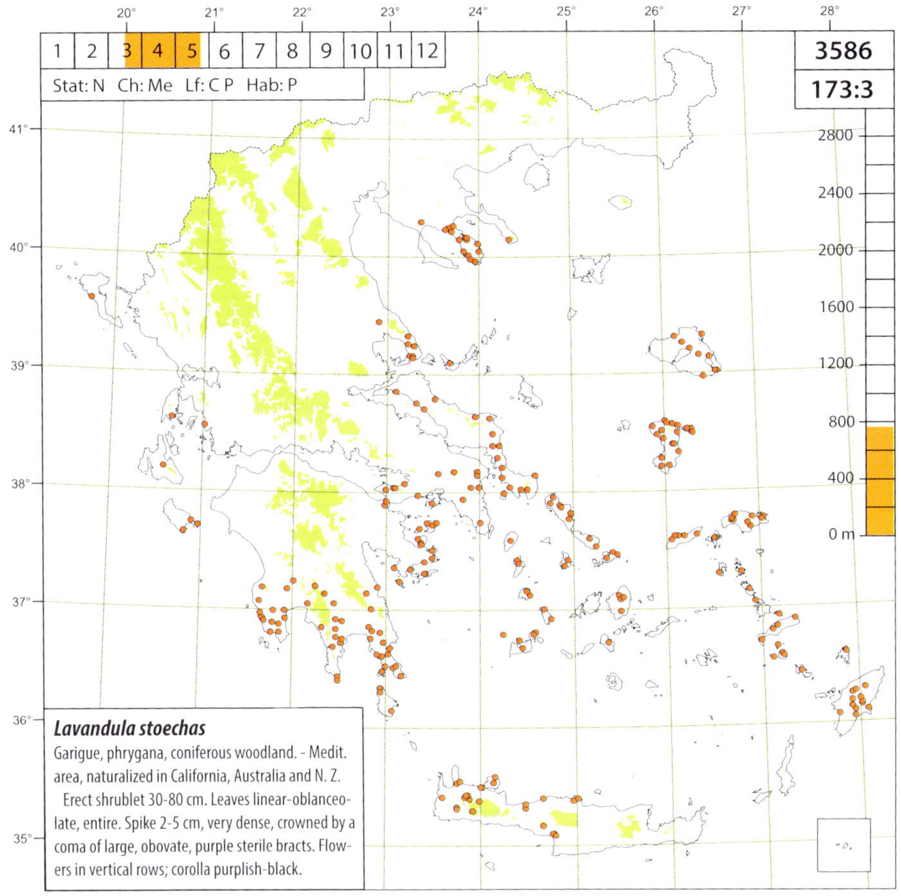

Lavandula stoechas
Garigue, phrygana, coniferous woodland. - Medit. area, naturalized in California, Australia and N. Z.
Erect shrublet 30-80 cm. Leaves linear-oblanceolate, entire. Spike 2-5 cm, very dense, crowned by a coma of large, obovate, purple sterile bracts. Flowers in vertical rows; corolla purplish-black.

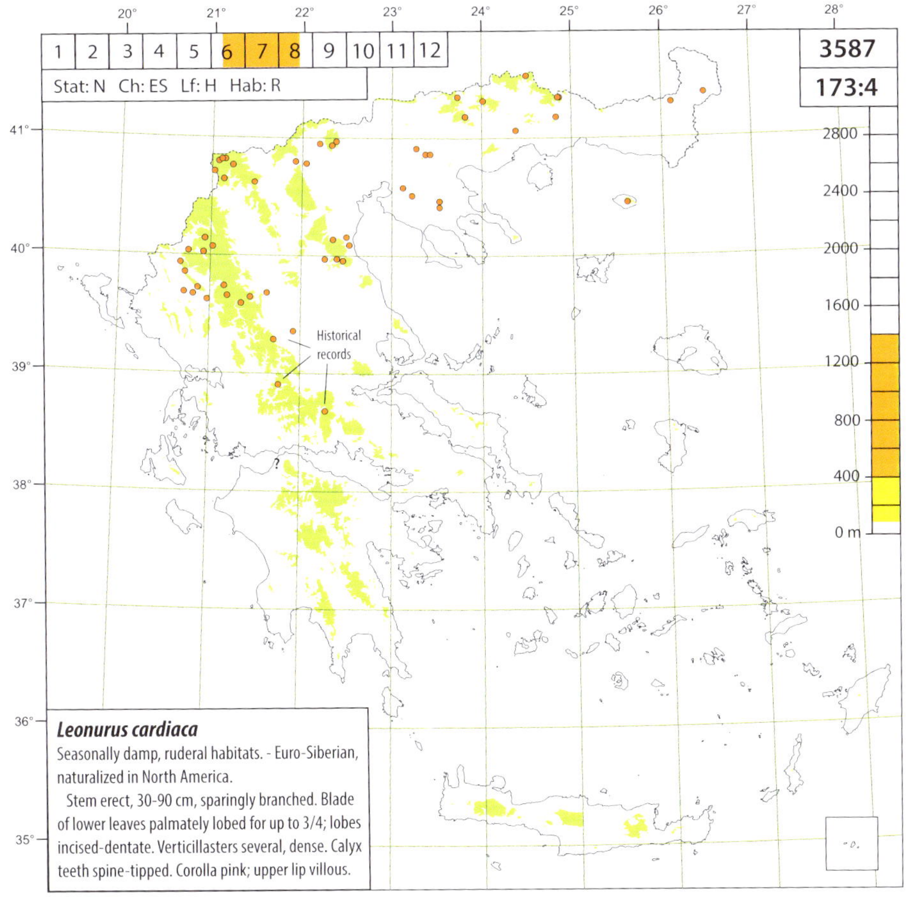

Leonurus cardiaca
Seasonally damp, ruderal habitats. - Euro-Siberian, naturalized in North America.
Stem erect, 30-90 cm, sparingly branched. Blade of lower leaves palmately lobed for up to 3/4; lobes incised-dentate. Verticillasters several, dense. Calyx teeth spine-tipped. Corolla pink; upper lip villous.

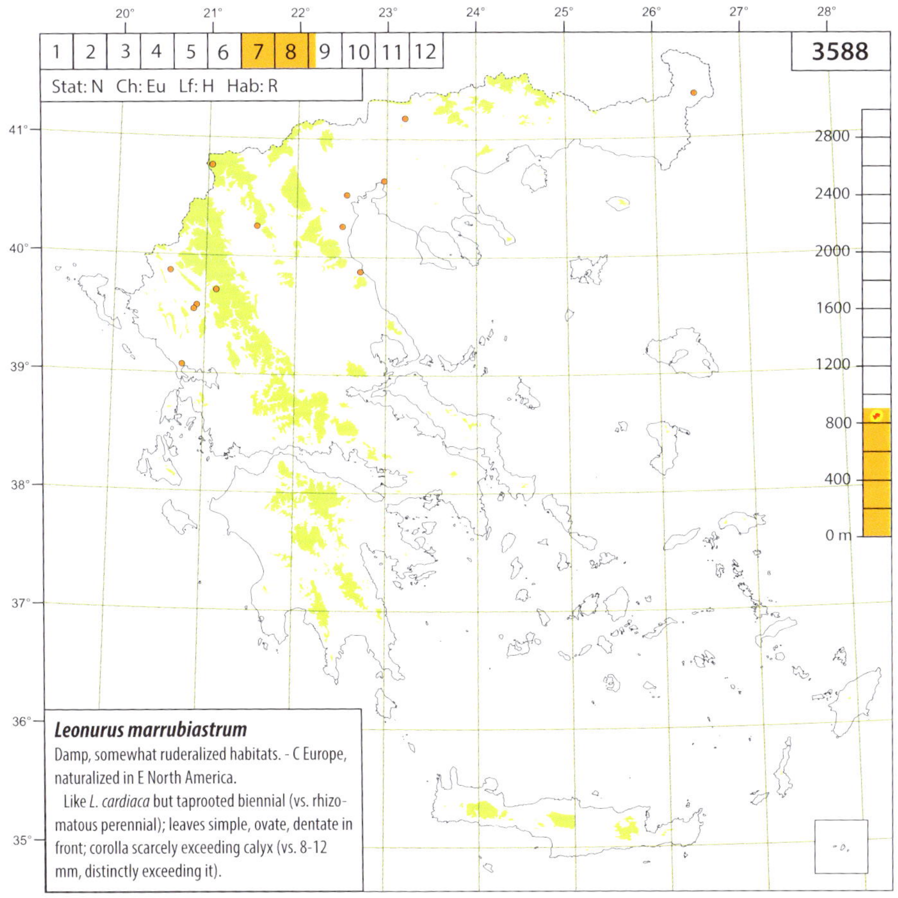

Leonurus marrubiastrum
Damp, somewhat ruderalized habitats. - C Europe, naturalized in E North America.
Like *L. cardiaca* but taprooted biennial (vs. rhizomatous perennial); leaves simple, ovate, dentate in front; corolla scarcely exceeding calyx (vs. 8-12 mm, distinctly exceeding it).

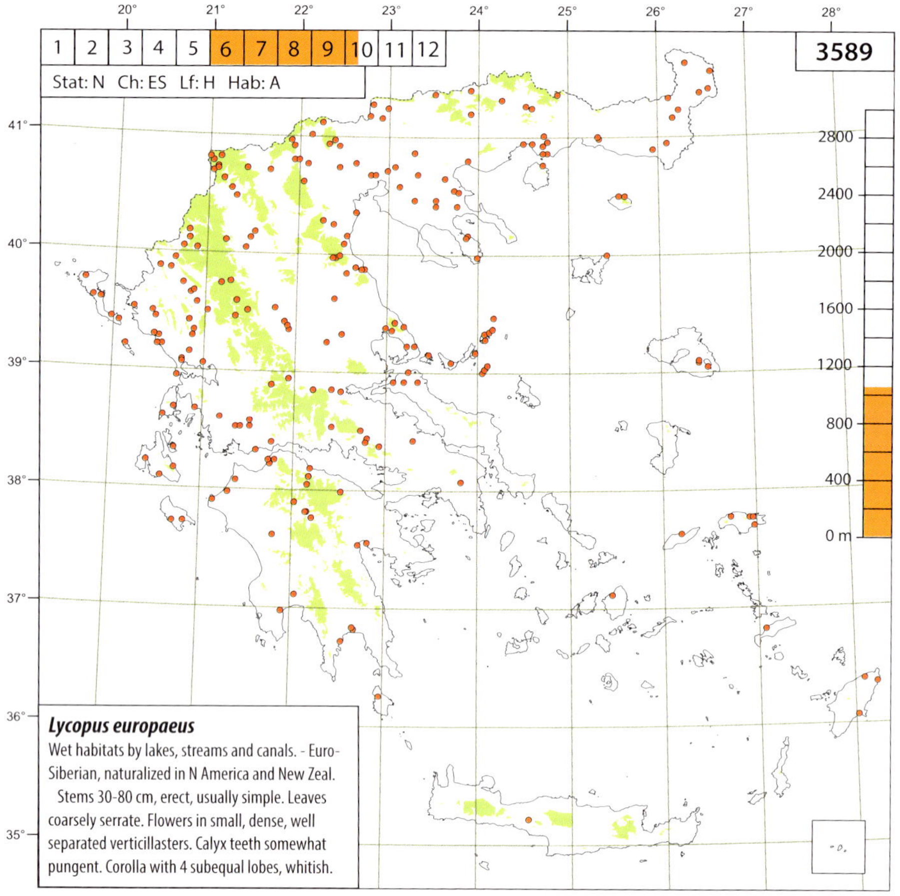

Lycopus europaeus

Wet habitats by lakes, streams and canals. - Euro-Siberian, naturalized in N America and New Zeal.

Stems 30-80 cm, erect, usually simple. Leaves coarsely serrate. Flowers in small, dense, well separated verticillasters. Calyx teeth somewhat pungent. Corolla with 4 subequal lobes, whitish.

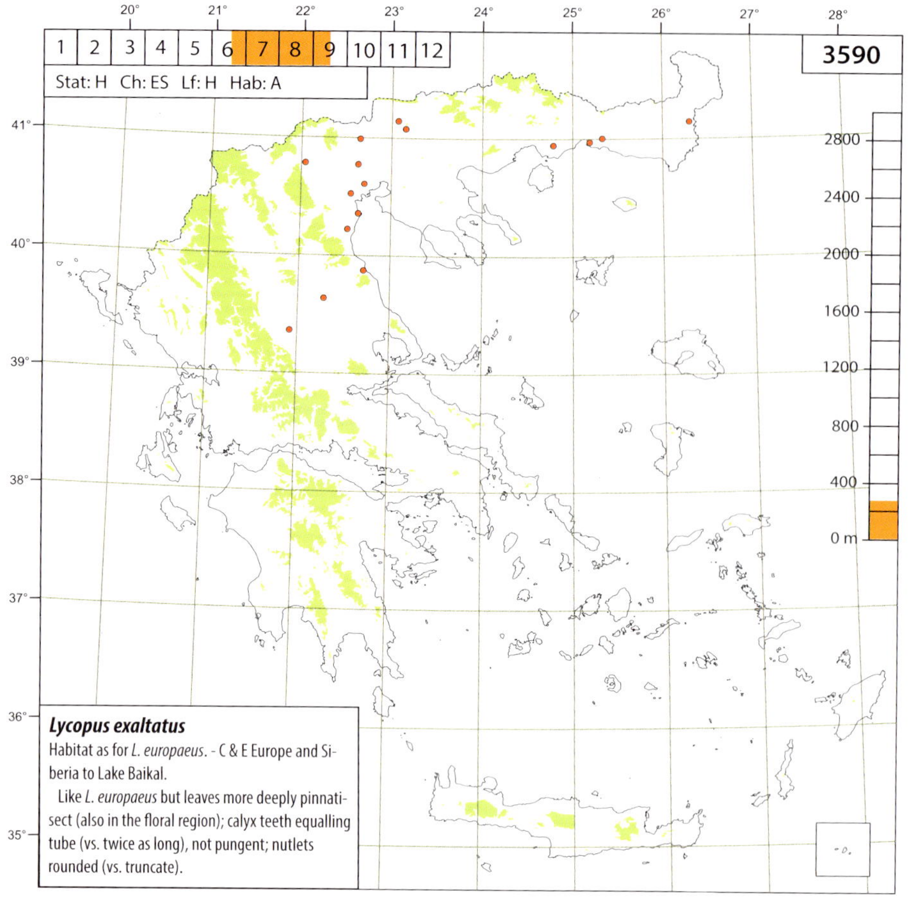

Lycopus exaltatus

Habitat as for *L. europaeus*. - C & E Europe and Siberia to Lake Baikal.

Like *L. europaeus* but leaves more deeply pinnatisect (also in the floral region); calyx teeth equalling tube (vs. twice as long), not pungent; nutlets rounded (vs. truncate).

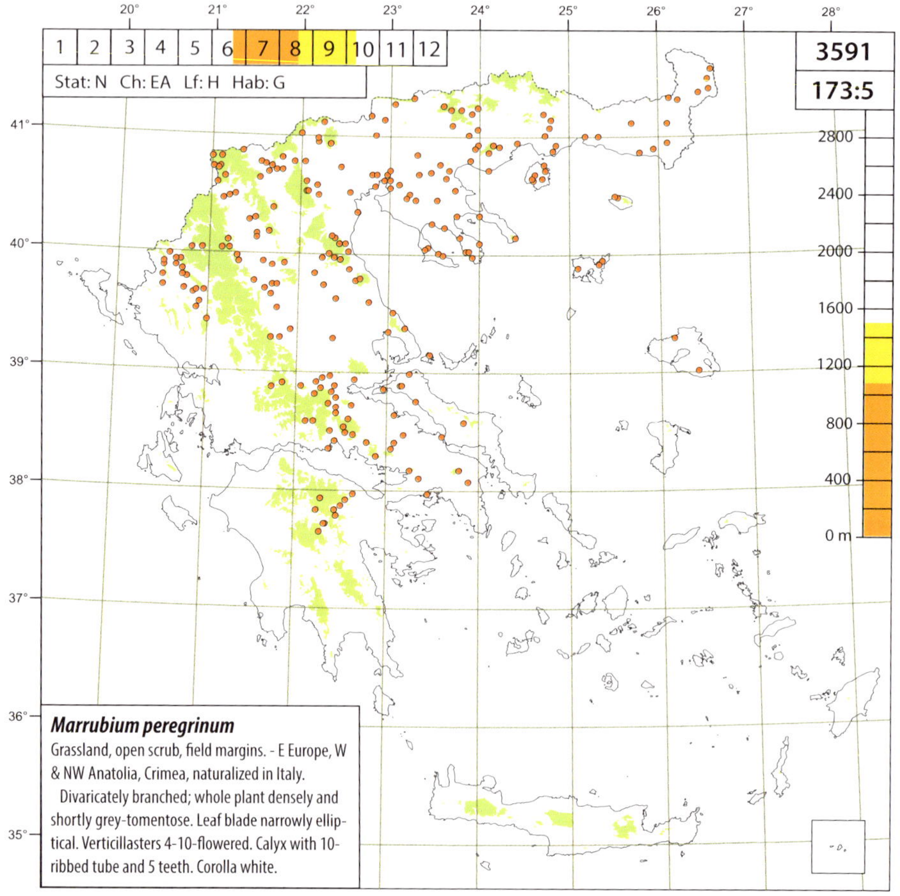

Marrubium peregrinum

Grassland, open scrub, field margins. - E Europe, W & NW Anatolia, Crimea, naturalized in Italy.

Divaricately branched; whole plant densely and shortly grey-tomentose. Leaf blade narrowly elliptical. Verticillasters 4-10-flowered. Calyx with 10-ribbed tube and 5 teeth. Corolla white.

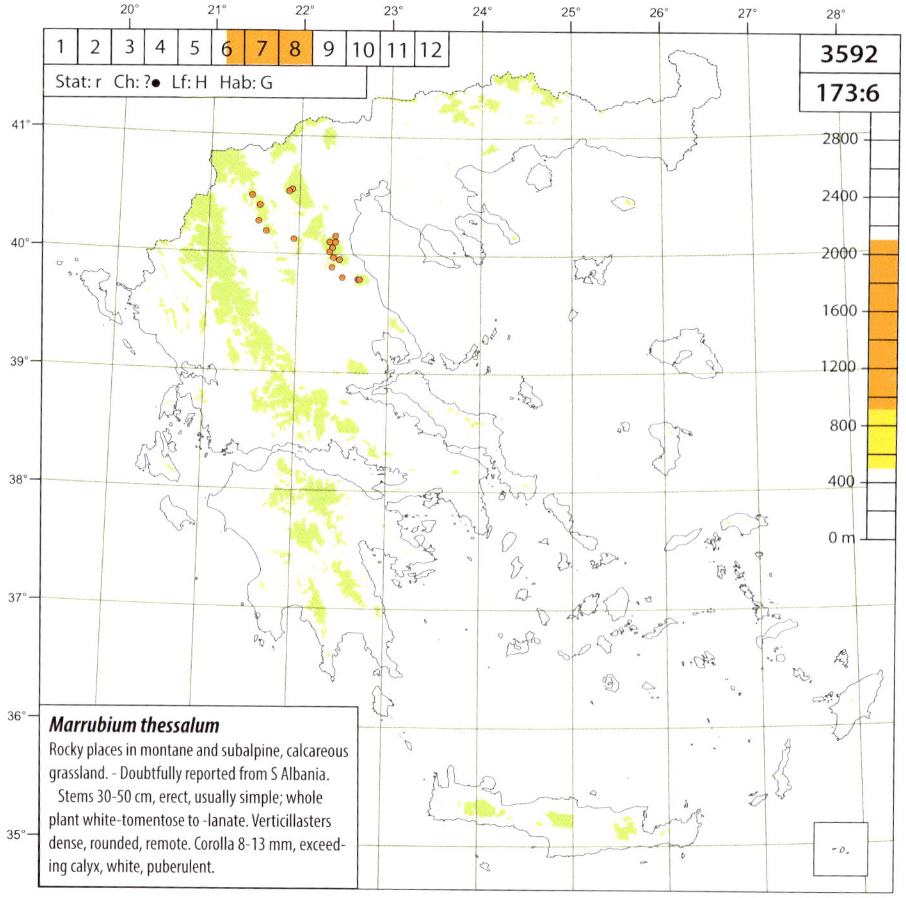

Marrubium thessalum

Rocky places in montane and subalpine, calcareous grassland. - Doubtfully reported from S Albania.

Stems 30-50 cm, erect, usually simple; whole plant white-tomentose to -lanate. Verticillasters dense, rounded, remote. Corolla 8-13 mm, exceeding calyx, white, puberulent.

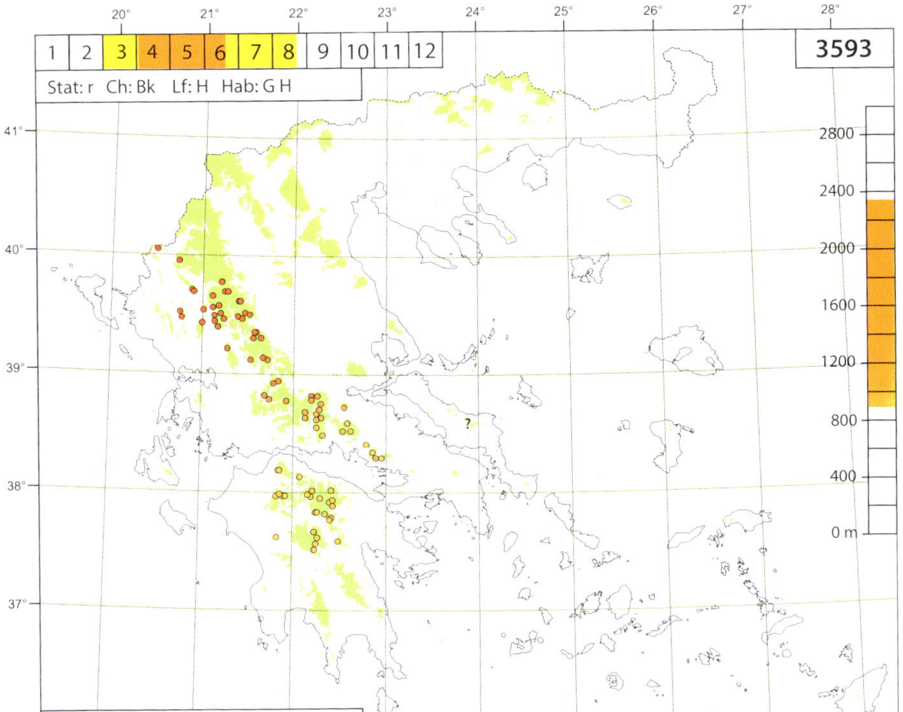

Marrubium velutinum
Rocky places in grassland, occasionally in open coniferous woodland, on various substrates but mostly limestone. - Also in S Albania.
Like *M. thessalum* but whole plant yellowish-tomentose; calyx ± equalling corolla, the latter pale yellow to lemon yellow.

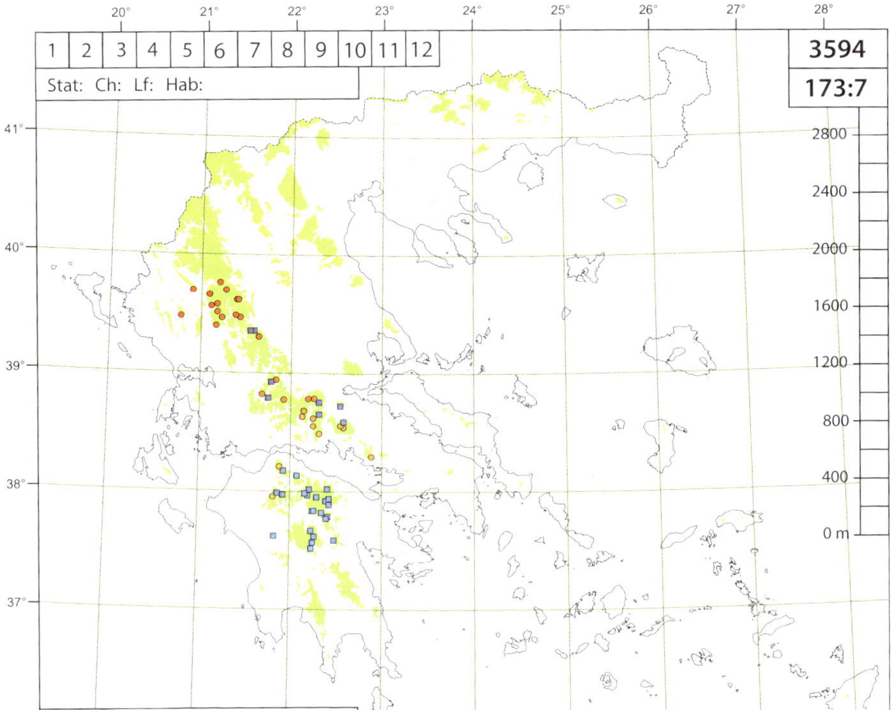

Marrubium velutinum
● subsp. ***velutinum***. Bracteoles 9-13 mm. Calyx teeth equalling tube; corolla shorter than or equalling calyx. - Probably extending to S Albania.
■ subsp. ***cylleneum***. Bracteoles less than 7 mm. Calyx teeth less than half as long as tube. Corolla exceeding calyx. - Endemic.

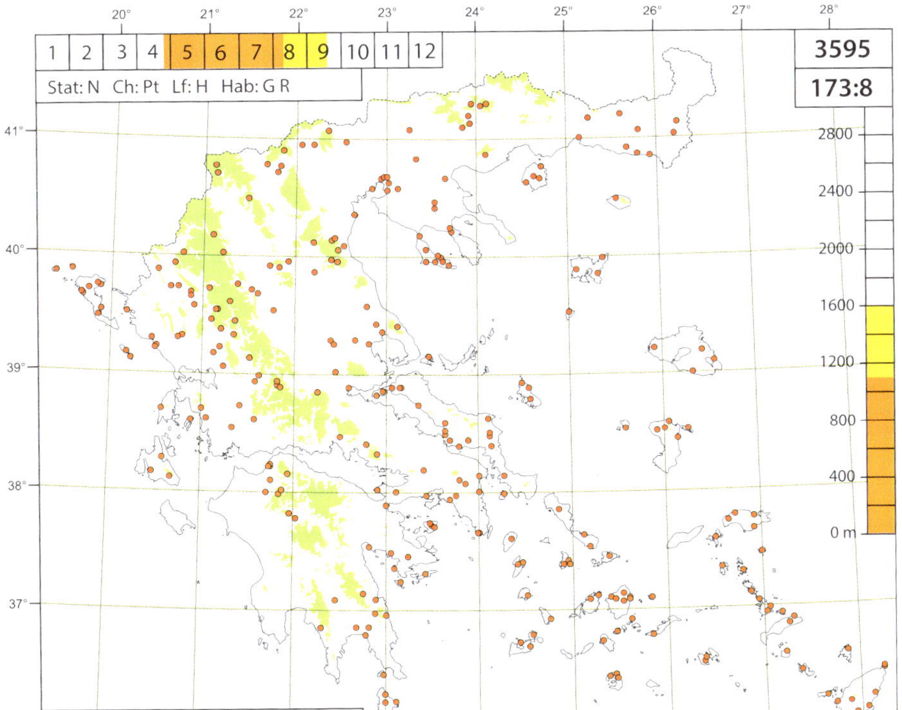

Marrubium vulgare
Dry grassland, scrub, steppe-like habitats. - Europe & SW to C Asia, widely naturalized elsewhere.
Stems 20-60 cm, erect, ± simple. Leaf blade obovate-suborbicular, crenate, reticulate-veined. Verticillasters dense, subglobose. Calyx with 10 narrow, uncinate teeth. Corolla white or cream.

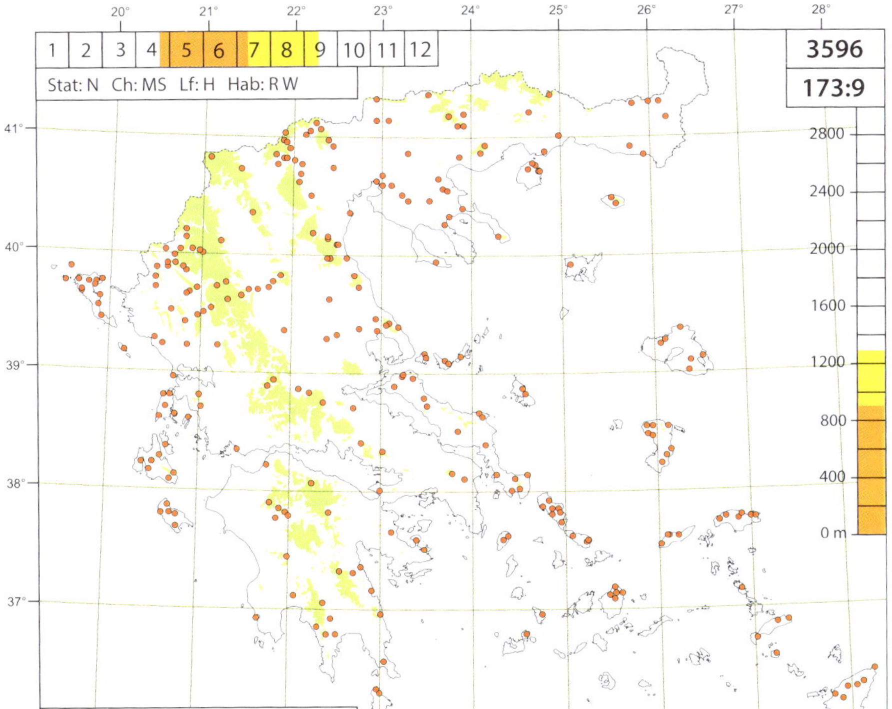

Melissa officinalis
Damp scrub, stream banks, ravines, olive groves and ruderal habitats. - Medit. area and SW Asia, widely cultivated and naturalized elsewhere.
30-120 cm, whole plant lemon-scented. Leaf blade ovate, crenate. Verticillasters remote, lax. Corolla 8-15 mm, 2-lipped, cream or tinged pink.

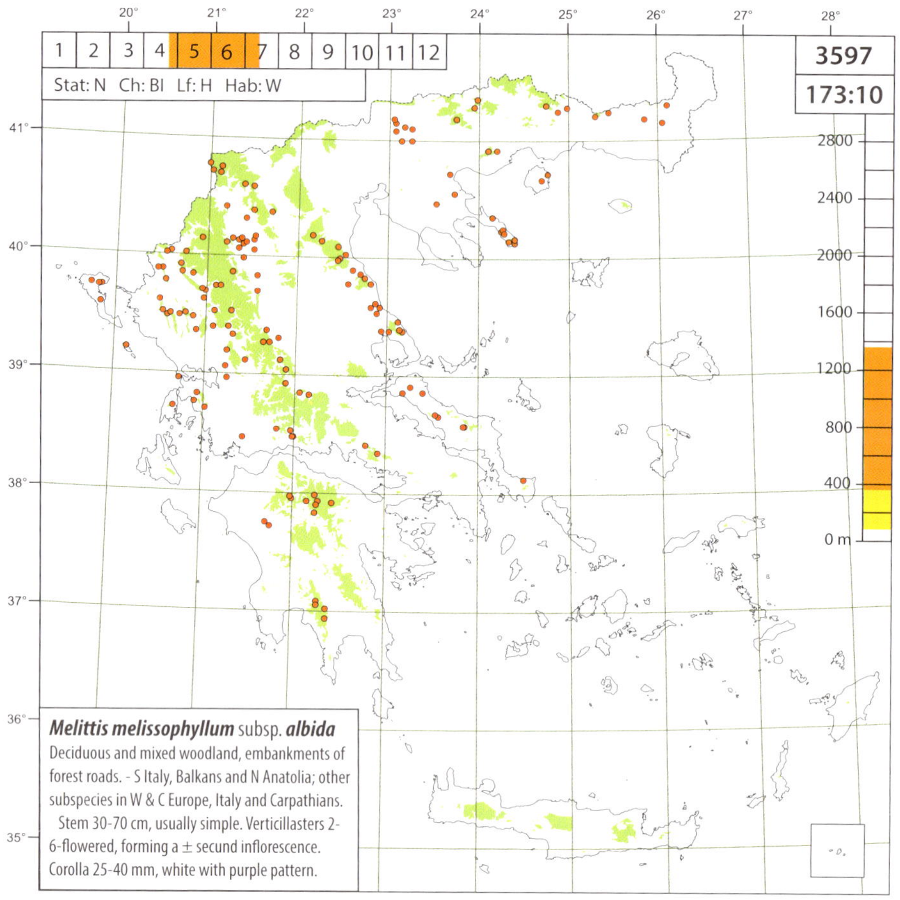

Melittis melissophyllum subsp. ***albida***
Deciduous and mixed woodland, embankments of forest roads. - S Italy, Balkans and N Anatolia; other subspecies in W & C Europe, Italy and Carpathians.
Stem 30-70 cm, usually simple. Verticillasters 2-6-flowered, forming a ± secund inflorescence. Corolla 25-40 mm, white with purple pattern.

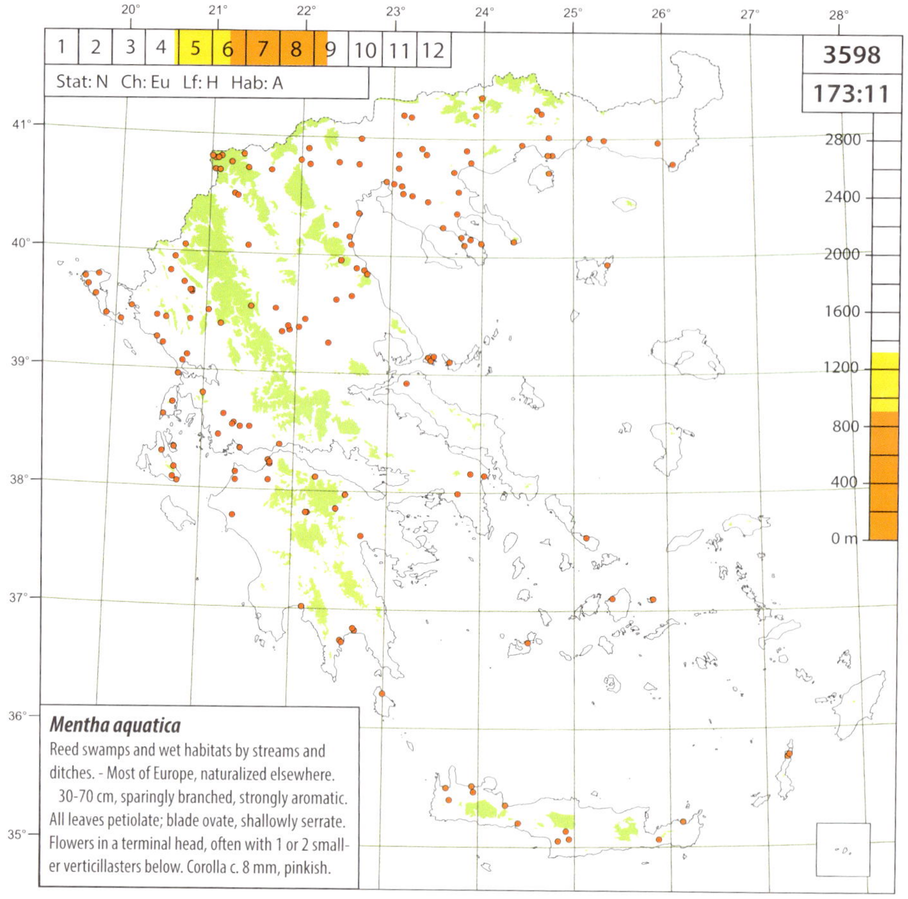

Mentha aquatica
Reed swamps and wet habitats by streams and ditches. - Most of Europe, naturalized elsewhere.
30-70 cm, sparingly branched, strongly aromatic. All leaves petiolate; blade ovate, shallowly serrate. Flowers in a terminal head, often with 1 or 2 smaller verticillasters below. Corolla c. 8 mm, pinkish.

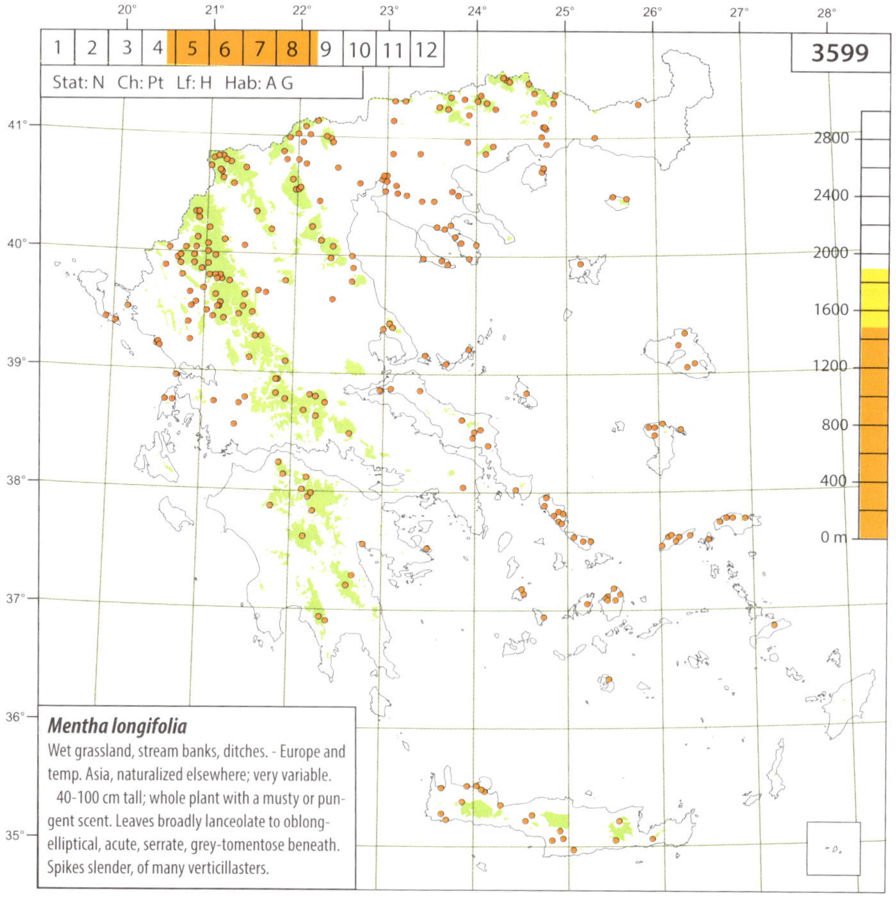

Mentha longifolia
Wet grassland, stream banks, ditches. - Europe and temp. Asia, naturalized elsewhere; very variable.
40-100 cm tall; whole plant with a musty or pungent scent. Leaves broadly lanceolate to oblong-elliptical, acute, serrate, grey-tomentose beneath. Spikes slender, of many verticillasters.

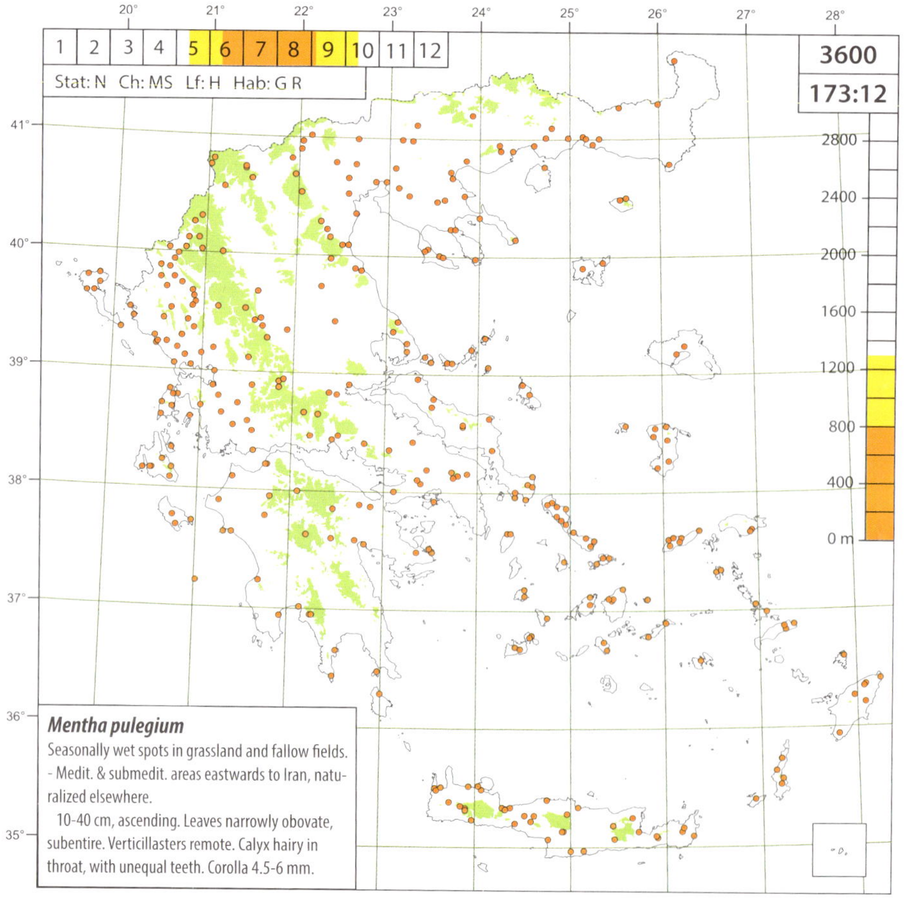

Mentha pulegium
Seasonally wet spots in grassland and fallow fields. - Medit. & submedit. areas eastwards to Iran, naturalized elsewhere.
10-40 cm, ascending. Leaves narrowly obovate, subentire. Verticillasters remote. Calyx hairy in throat, with unequal teeth. Corolla 4.5-6 mm.

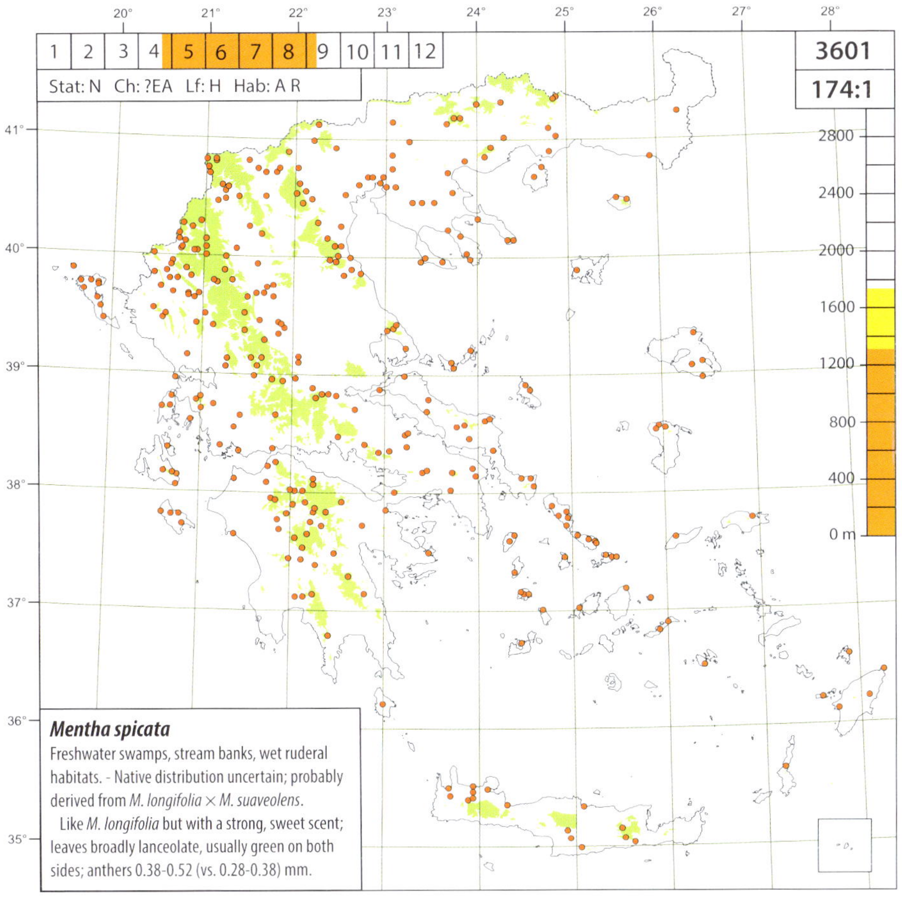

Mentha spicata

Freshwater swamps, stream banks, wet ruderal habitats. - Native distribution uncertain; probably derived from *M. longifolia* × *M. suaveolens*.

Like *M. longifolia* but with a strong, sweet scent; leaves broadly lanceolate, usually green on both sides; anthers 0.38-0.52 (vs. 0.28-0.38) mm.

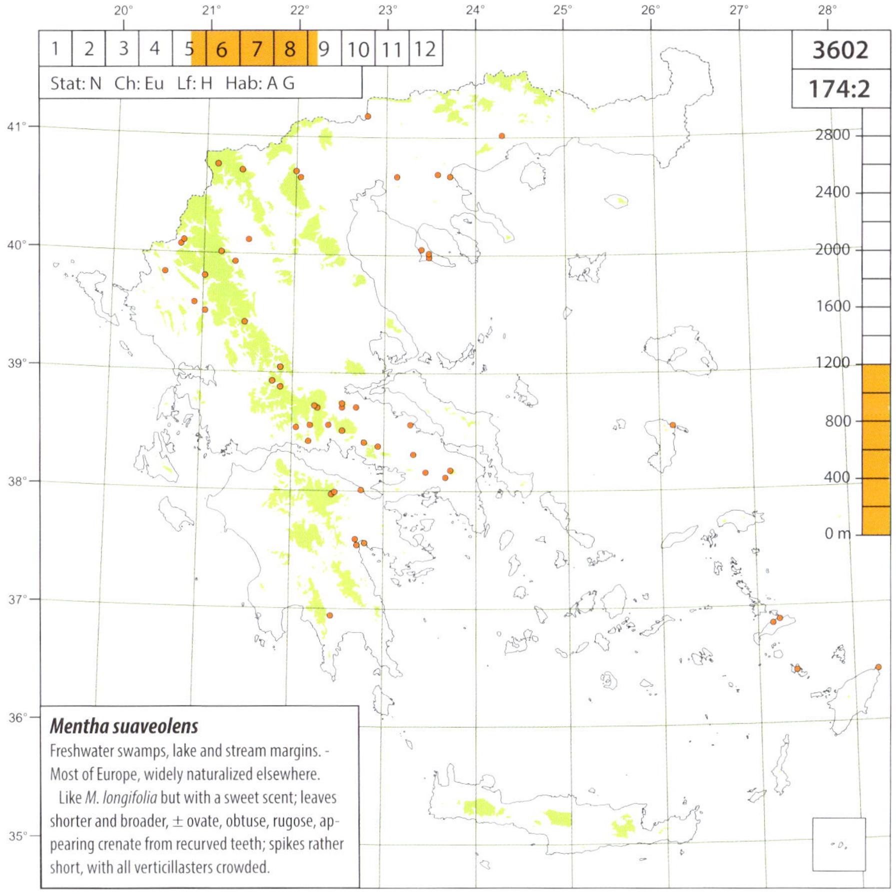

Mentha suaveolens

Freshwater swamps, lake and stream margins. - Most of Europe, widely naturalized elsewhere.

Like *M. longifolia* but with a sweet scent; leaves shorter and broader, ± ovate, obtuse, rugose, appearing crenate from recurved teeth; spikes rather short, with all verticillasters crowded.

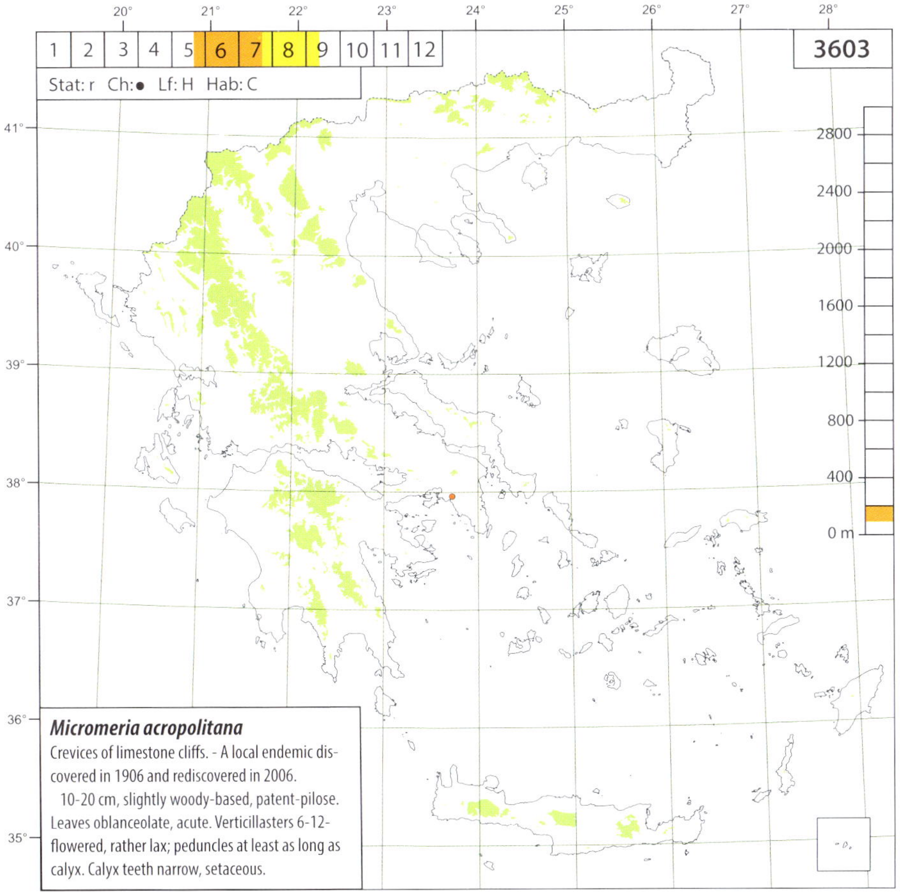

Micromeria acropolitana

Crevices of limestone cliffs. - A local endemic discovered in 1906 and rediscovered in 2006.

10-20 cm, slightly woody-based, patent-pilose. Leaves oblanceolate, acute. Verticillasters 6-12-flowered, rather lax; peduncles at least as long as calyx. Calyx teeth narrow, setaceous.

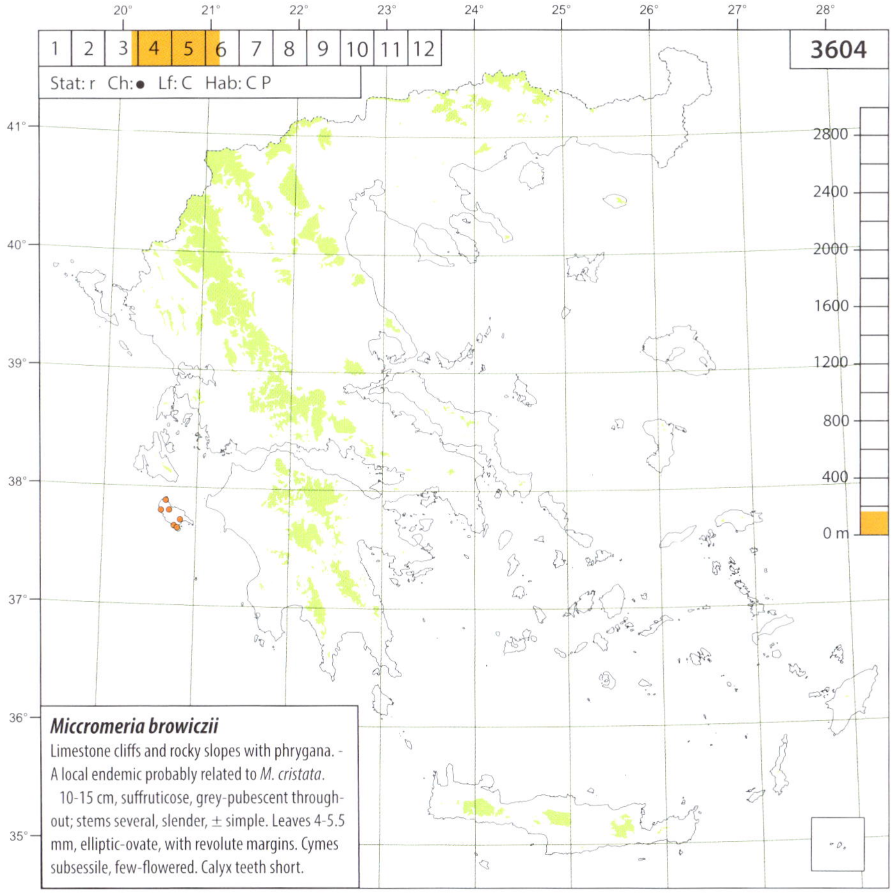

Micromeria browiczii

Limestone cliffs and rocky slopes with phrygana. - A local endemic probably related to *M. cristata*.

10-15 cm, suffruticose, grey-pubescent throughout; stems several, slender, ± simple. Leaves 4-5.5 mm, elliptic-ovate, with revolute margins. Cymes subsessile, few-flowered. Calyx teeth short.

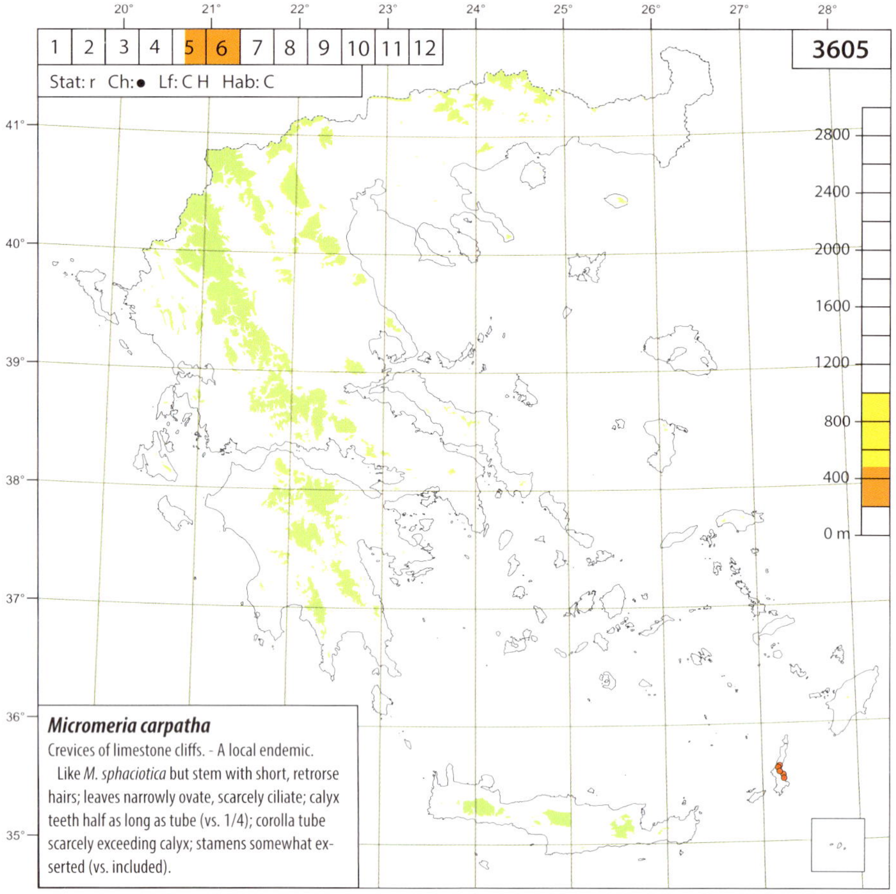

Micromeria carpatha
Crevices of limestone cliffs. - A local endemic.
Like *M. sphaciotica* but stem with short, retrorse hairs; leaves narrowly ovate, scarcely ciliate; calyx teeth half as long as tube (vs. 1/4); corolla tube scarcely exceeding calyx; stamens somewhat exserted (vs. included).

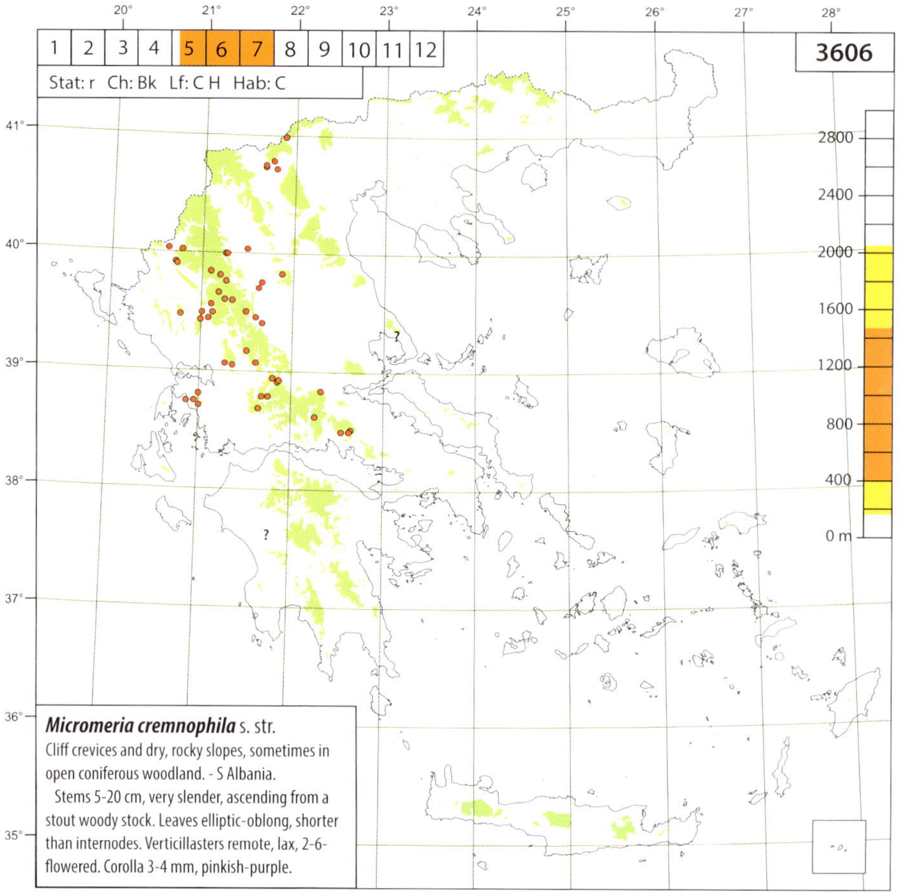

Micromeria cremnophila s. str.
Cliff crevices and dry, rocky slopes, sometimes in open coniferous woodland. - S Albania.
Stems 5-20 cm, very slender, ascending from a stout woody stock. Leaves elliptic-oblong, shorter than internodes. Verticillasters remote, lax, 2-6-flowered. Corolla 3-4 mm, pinkish-purple.

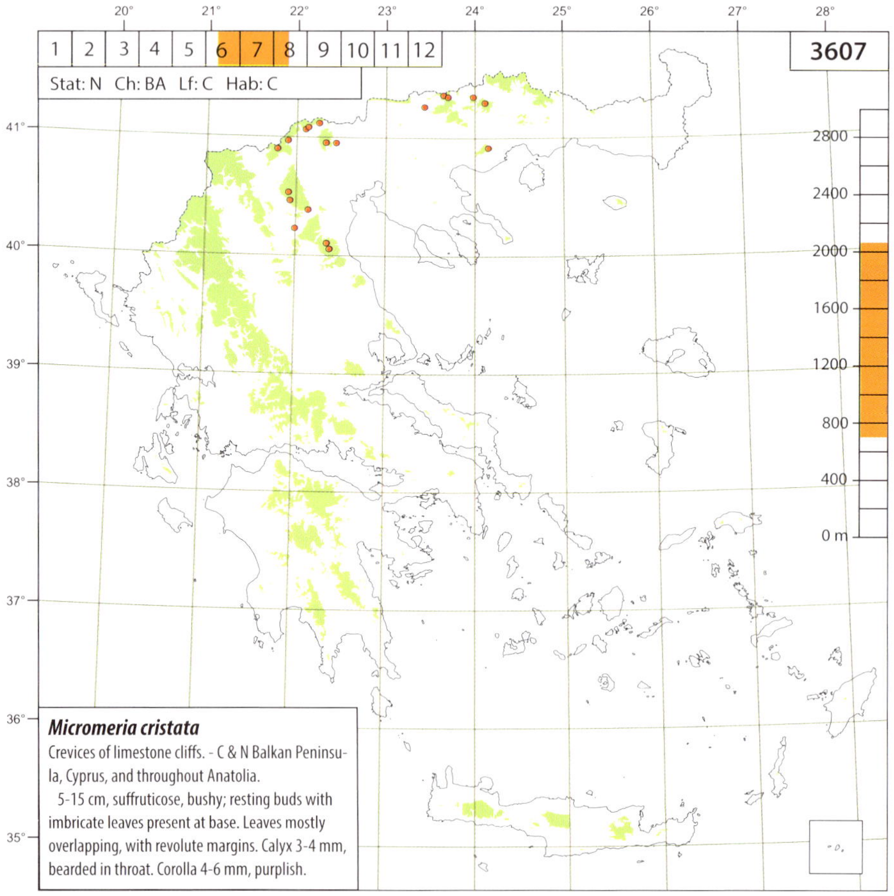

Micromeria cristata
Crevices of limestone cliffs. - C & N Balkan Peninsula, Cyprus, and throughout Anatolia.
5-15 cm, suffruticose, bushy; resting buds with imbricate leaves present at base. Leaves mostly overlapping, with revolute margins. Calyx 3-4 mm, bearded in throat. Corolla 4-6 mm, purplish.

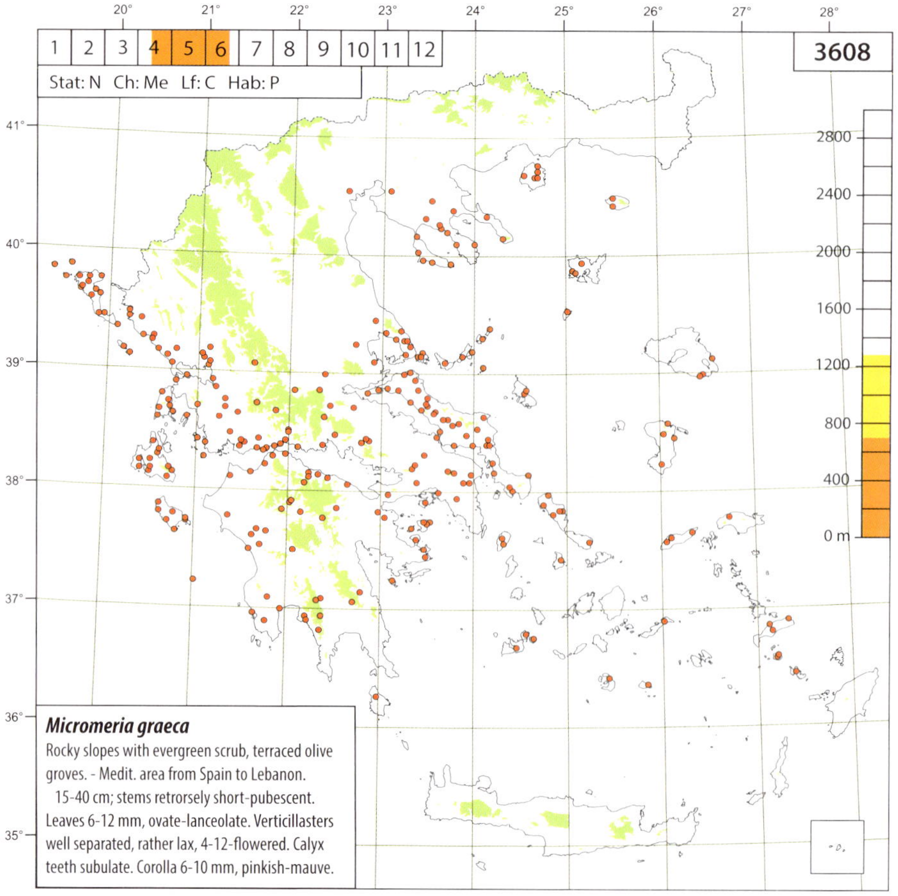

Micromeria graeca
Rocky slopes with evergreen scrub, terraced olive groves. - Medit. area from Spain to Lebanon.
15-40 cm; stems retrorsely short-pubescent. Leaves 6-12 mm, ovate-lanceolate. Verticillasters well separated, rather lax, 4-12-flowered. Calyx teeth subulate. Corolla 6-10 mm, pinkish-mauve.

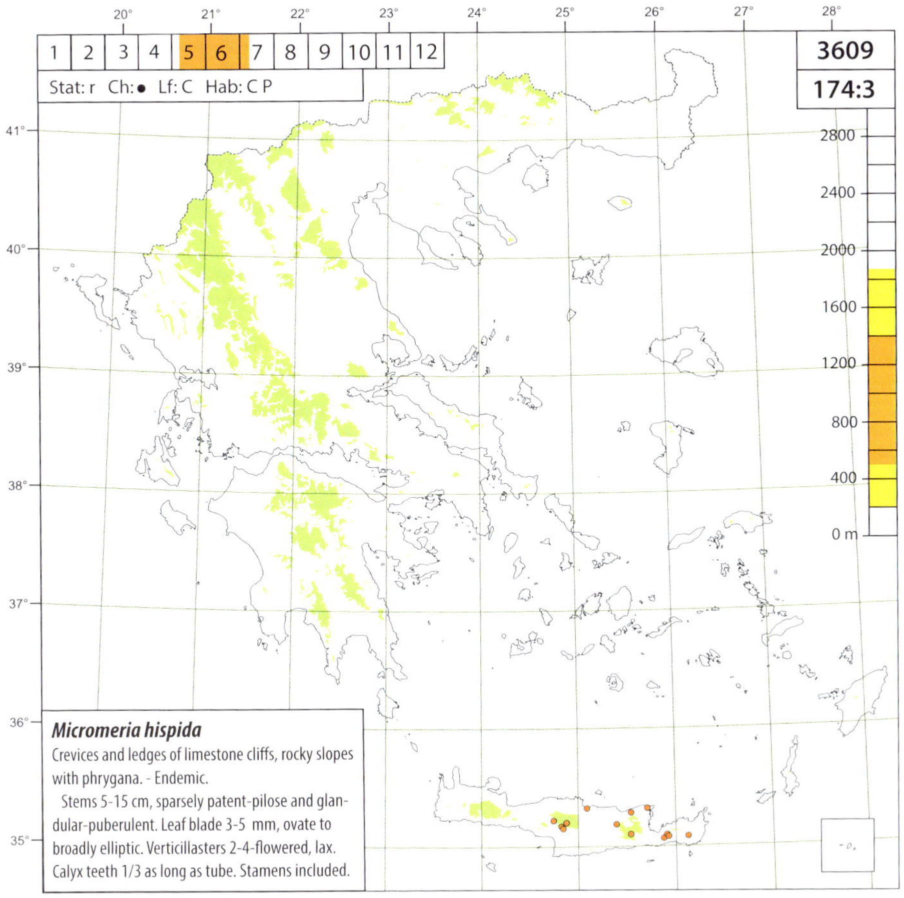

Micromeria hispida
Crevices and ledges of limestone cliffs, rocky slopes with phrygana. - Endemic.
Stems 5-15 cm, sparsely patent-pilose and glandular-puberulent. Leaf blade 3-5 mm, ovate to broadly elliptic. Verticillasters 2-4-flowered, lax. Calyx teeth 1/3 as long as tube. Stamens included.

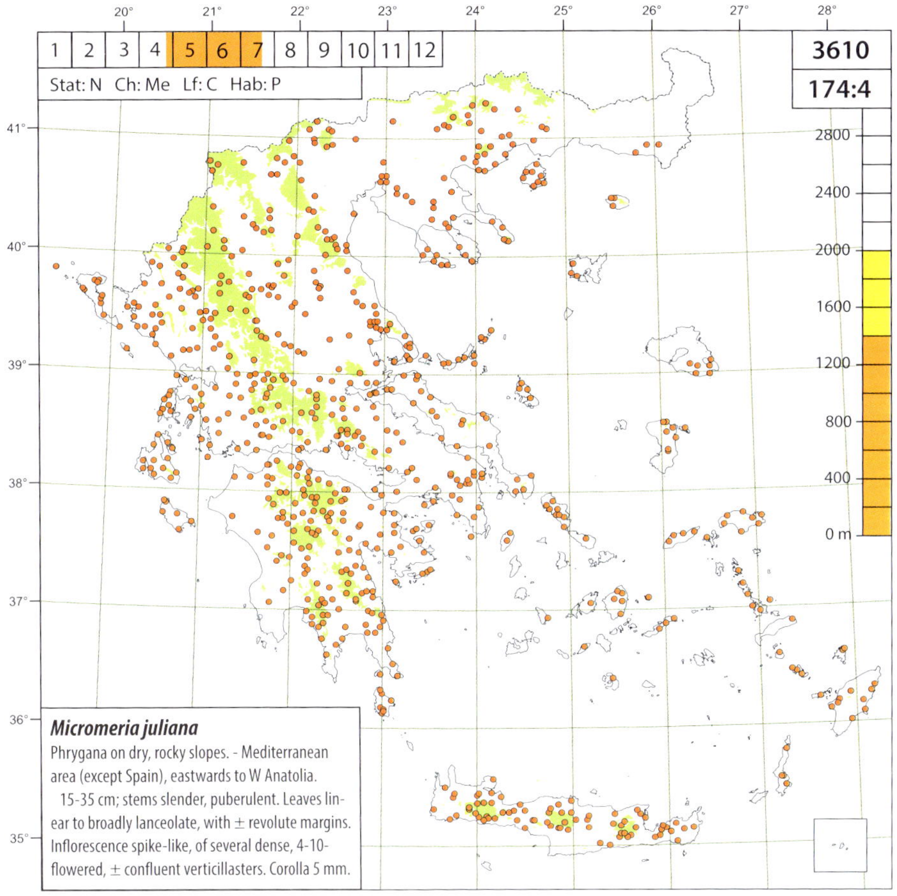

Micromeria juliana
Phrygana on dry, rocky slopes. - Mediterranean area (except Spain), eastwards to W Anatolia.
15-35 cm; stems slender, puberulent. Leaves linear to broadly lanceolate, with ± revolute margins. Inflorescence spike-like, of several dense, 4-10-flowered, ± confluent verticillasters. Corolla 5 mm.

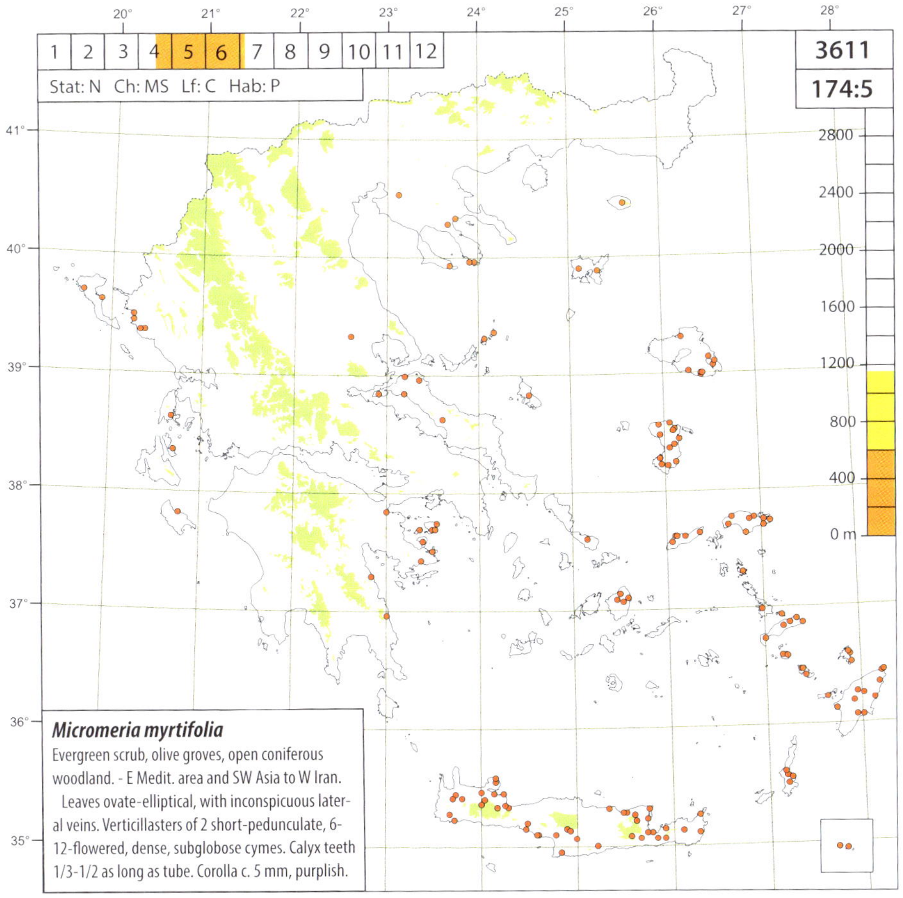

Micromeria myrtifolia
Evergreen scrub, olive groves, open coniferous woodland. - E Medit. area and SW Asia to W Iran.
Leaves ovate-elliptical, with inconspicuous lateral veins. Verticillasters of 2 short-pedunculate, 6-12-flowered, dense, subglobose cymes. Calyx teeth 1/3-1/2 as long as tube. Corolla c. 5 mm, purplish.

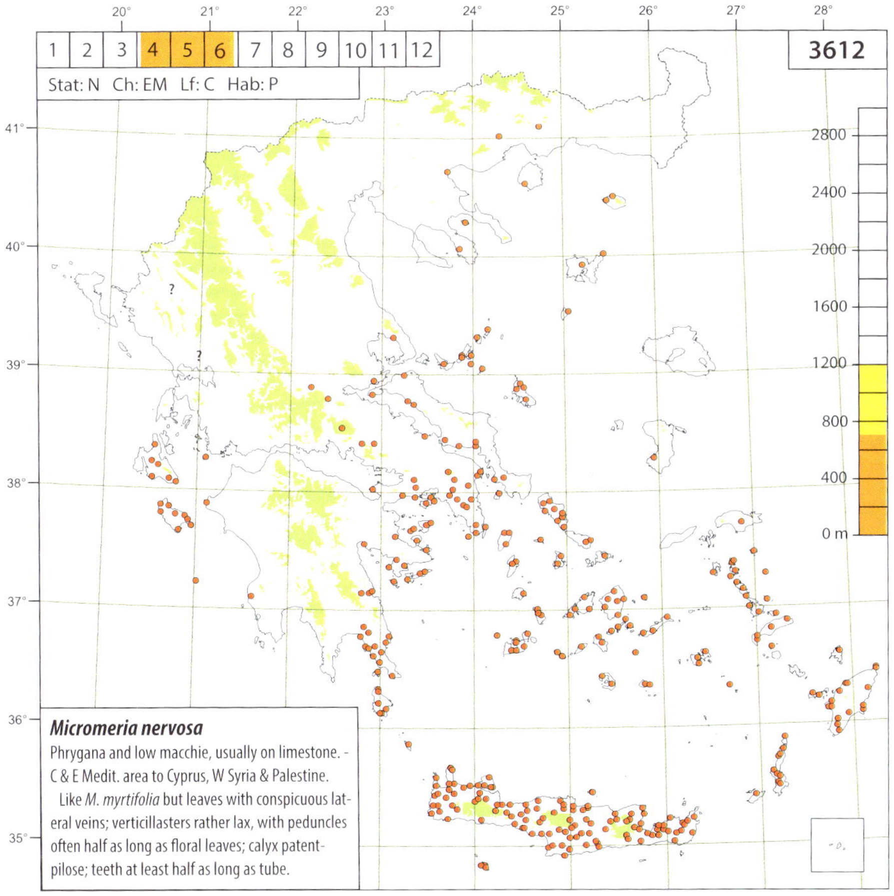

Micromeria nervosa
Phrygana and low macchie, usually on limestone. - C & E Medit. area to Cyprus, W Syria & Palestine.
Like *M. myrtifolia* but leaves with conspicuous lateral veins; verticillasters rather lax, with peduncles often half as long as floral leaves; calyx patent-pilose; teeth at least half as long as tube.

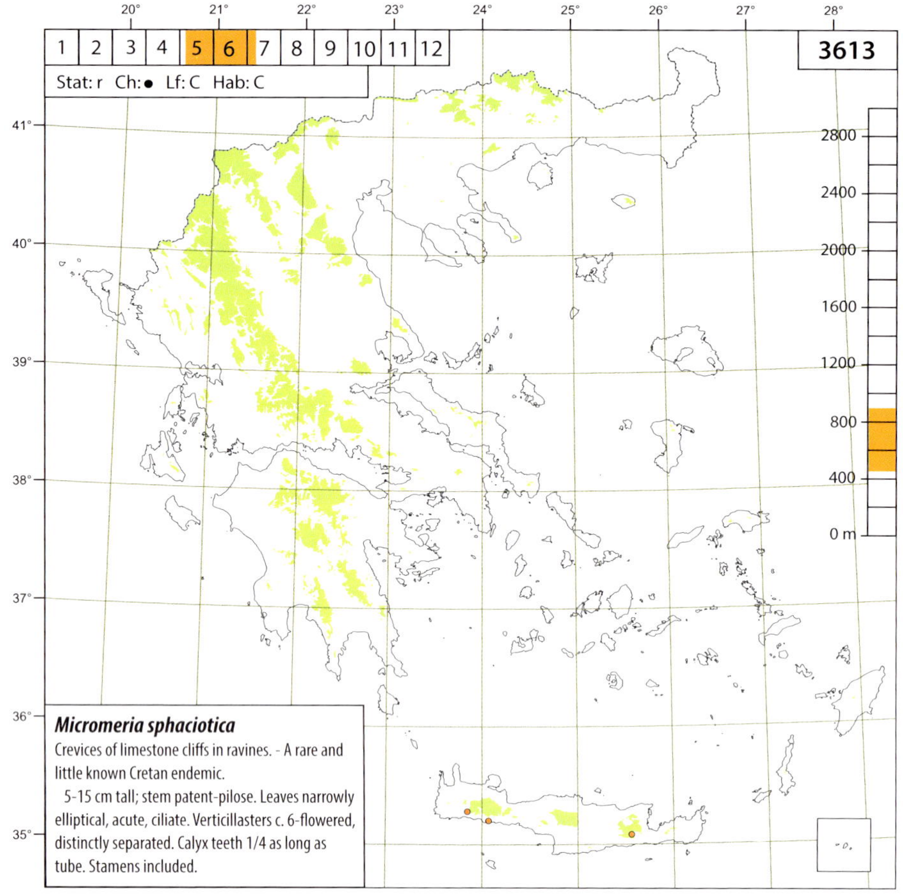

Micromeria sphaciotica
Crevices of limestone cliffs in ravines. - A rare and little known Cretan endemic.
5-15 cm tall; stem patent-pilose. Leaves narrowly elliptical, acute, ciliate. Verticillasters c. 6-flowered, distinctly separated. Calyx teeth 1/4 as long as tube. Stamens included.

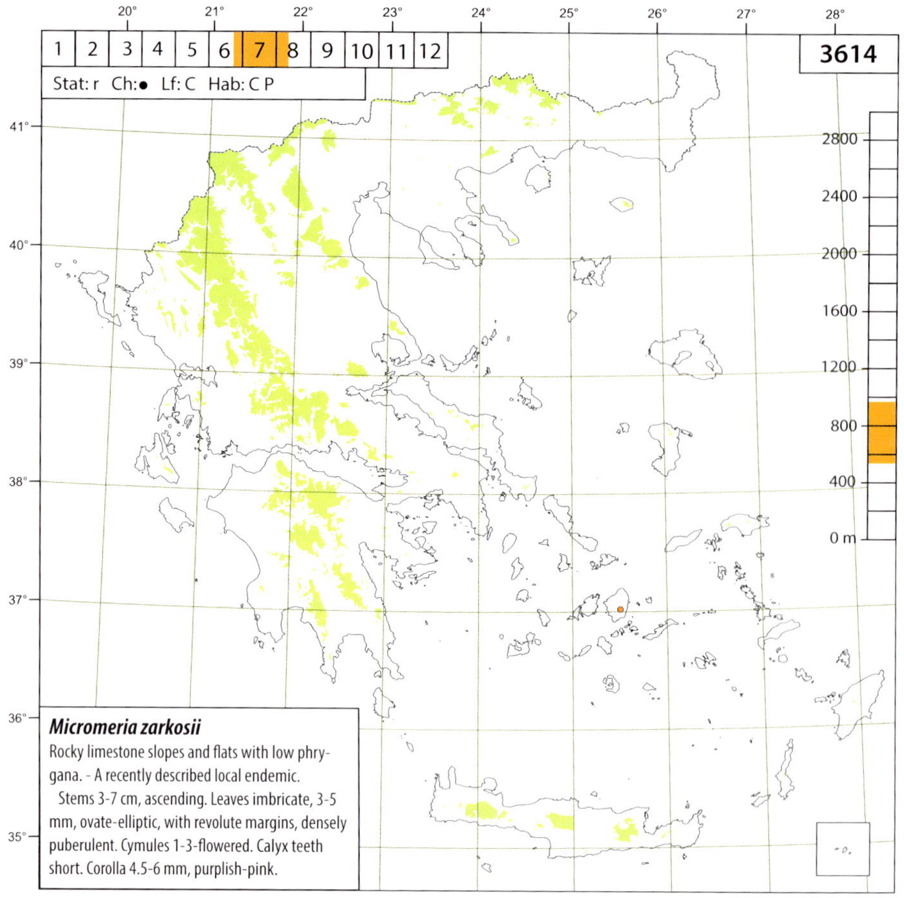

Micromeria zarkosii
Rocky limestone slopes and flats with low phrygana. - A recently described local endemic.
Stems 3-7 cm, ascending. Leaves imbricate, 3-5 mm, ovate-elliptic, with revolute margins, densely puberulent. Cymules 1-3-flowered. Calyx teeth short. Corolla 4.5-6 mm, purplish-pink.

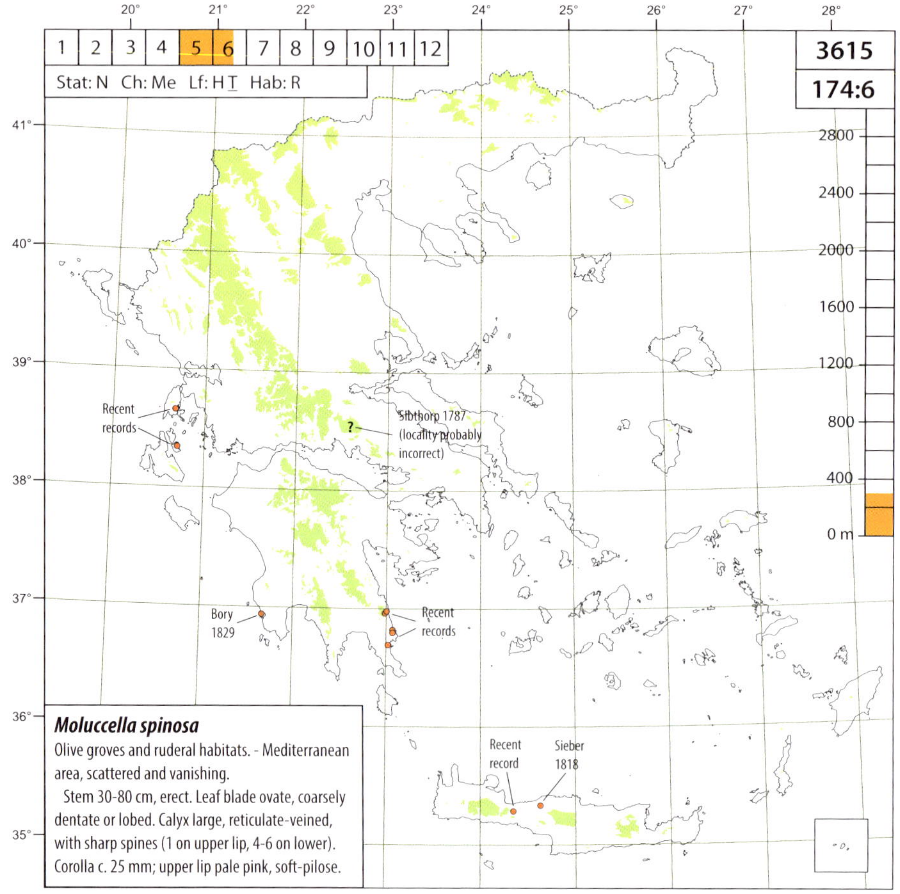

Moluccella spinosa
Olive groves and ruderal habitats. - Mediterranean area, scattered and vanishing.
Stem 30-80 cm, erect. Leaf blade ovate, coarsely dentate or lobed. Calyx large, reticulate-veined, with sharp spines (1 on upper lip, 4-6 on lower). Corolla c. 25 mm; upper lip pale pink, soft-pilose.

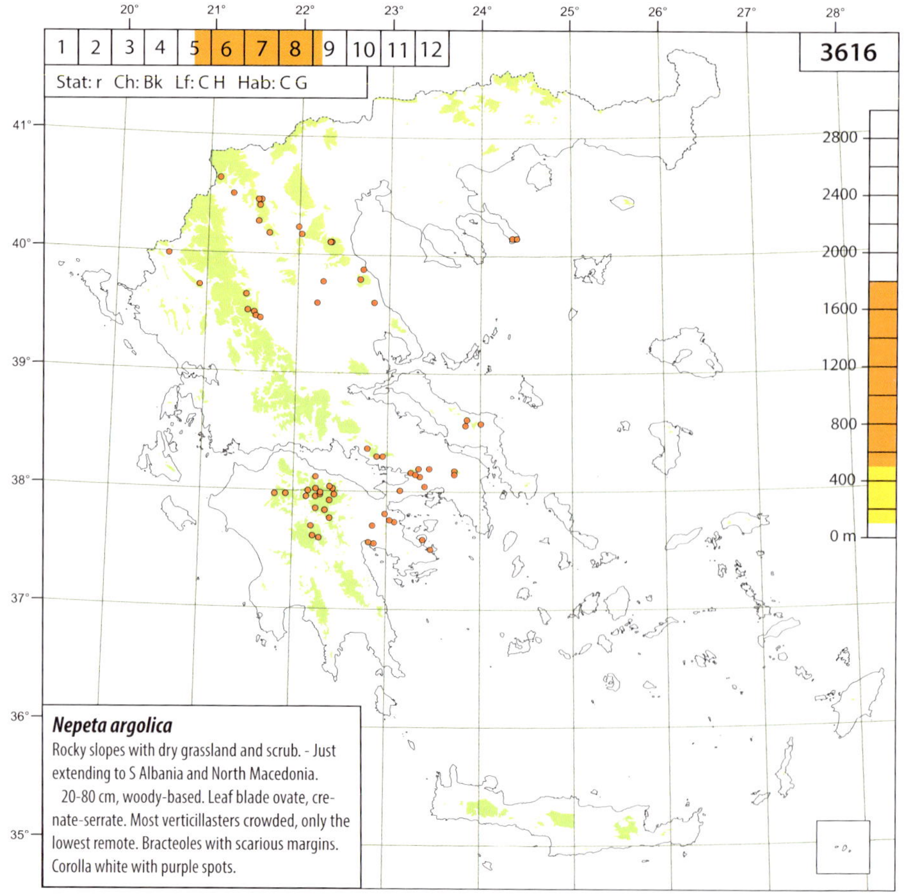

Nepeta argolica
Rocky slopes with dry grassland and scrub. - Just extending to S Albania and North Macedonia.
20-80 cm, woody-based. Leaf blade ovate, crenate-serrate. Most verticillasters crowded, only the lowest remote. Bracteoles with scarious margins. Corolla white with purple spots.

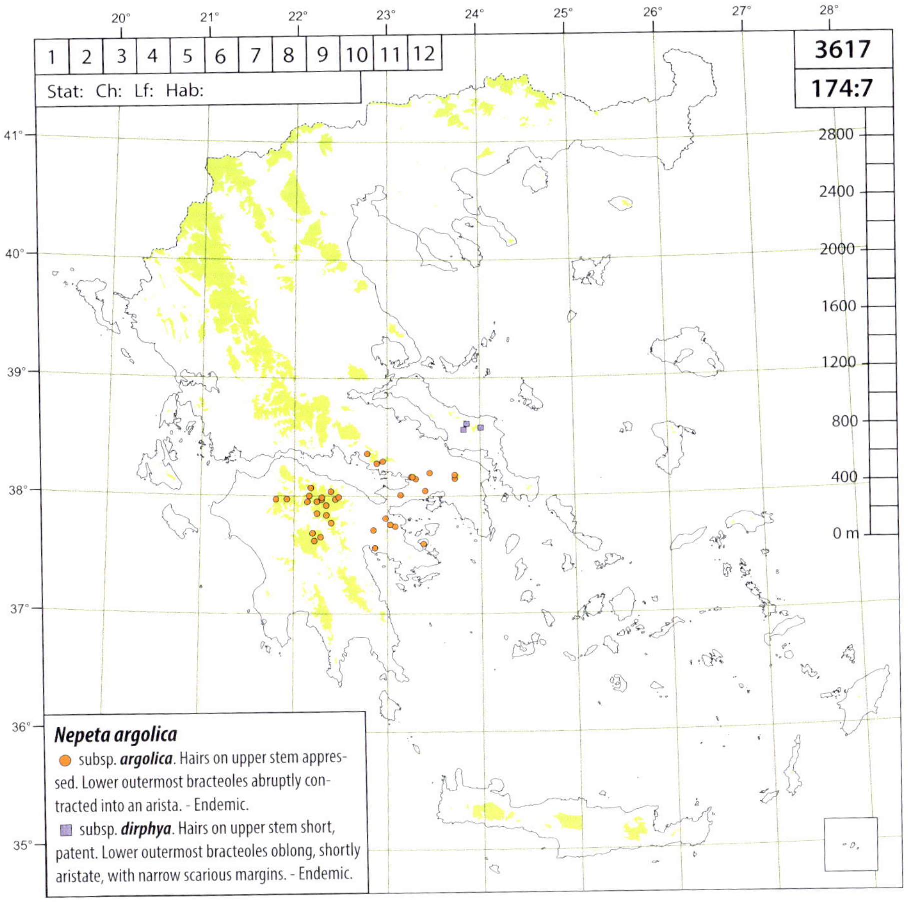

Nepeta argolica

● subsp. ***argolica***. Hairs on upper stem appressed. Lower outermost bracteoles abruptly contracted into an arista. - Endemic.

■ subsp. ***dirphya***. Hairs on upper stem short, patent. Lower outermost bracteoles oblong, shortly aristate, with narrow scarious margins. - Endemic.

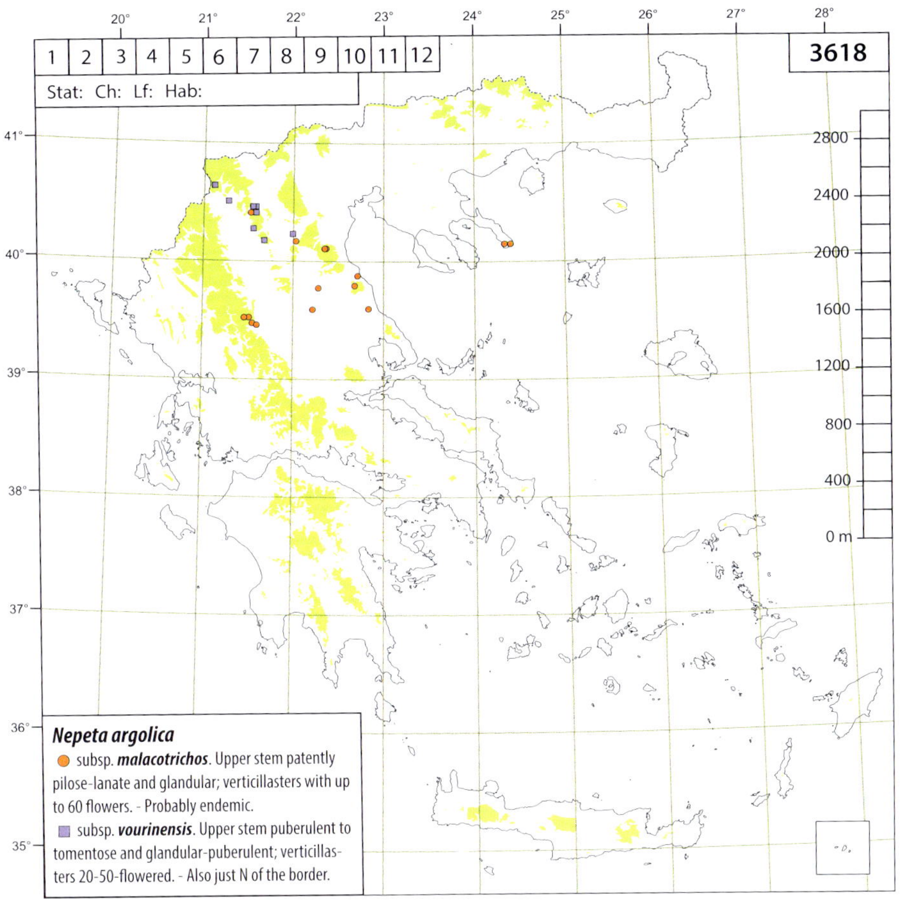

Nepeta argolica

● subsp. ***malacotrichos***. Upper stem patently pilose-lanate and glandular; verticillasters with up to 60 flowers. - Probably endemic.

■ subsp. ***vourinensis***. Upper stem puberulent to tomentose and glandular-puberulent; verticillasters 20-50-flowered. - Also just N of the border.

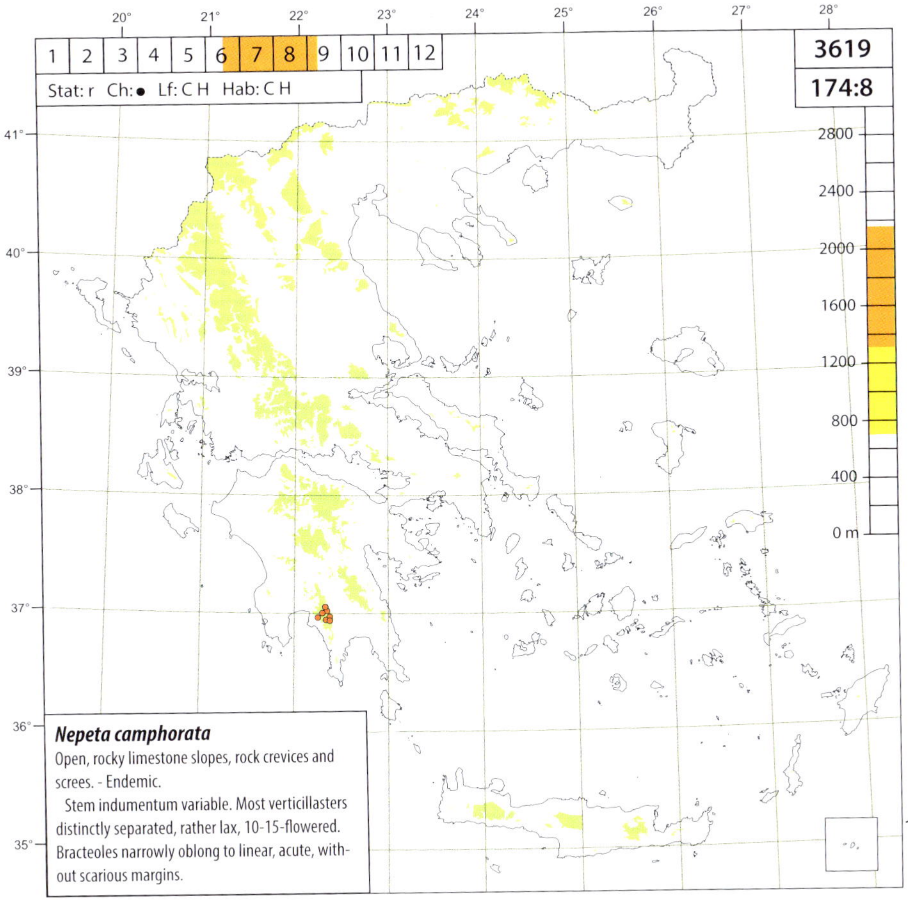

Nepeta camphorata

Open, rocky limestone slopes, rock crevices and screes. - Endemic.

Stem indumentum variable. Most verticillasters distinctly separated, rather lax, 10-15-flowered. Bracteoles narrowly oblong to linear, acute, without scarious margins.

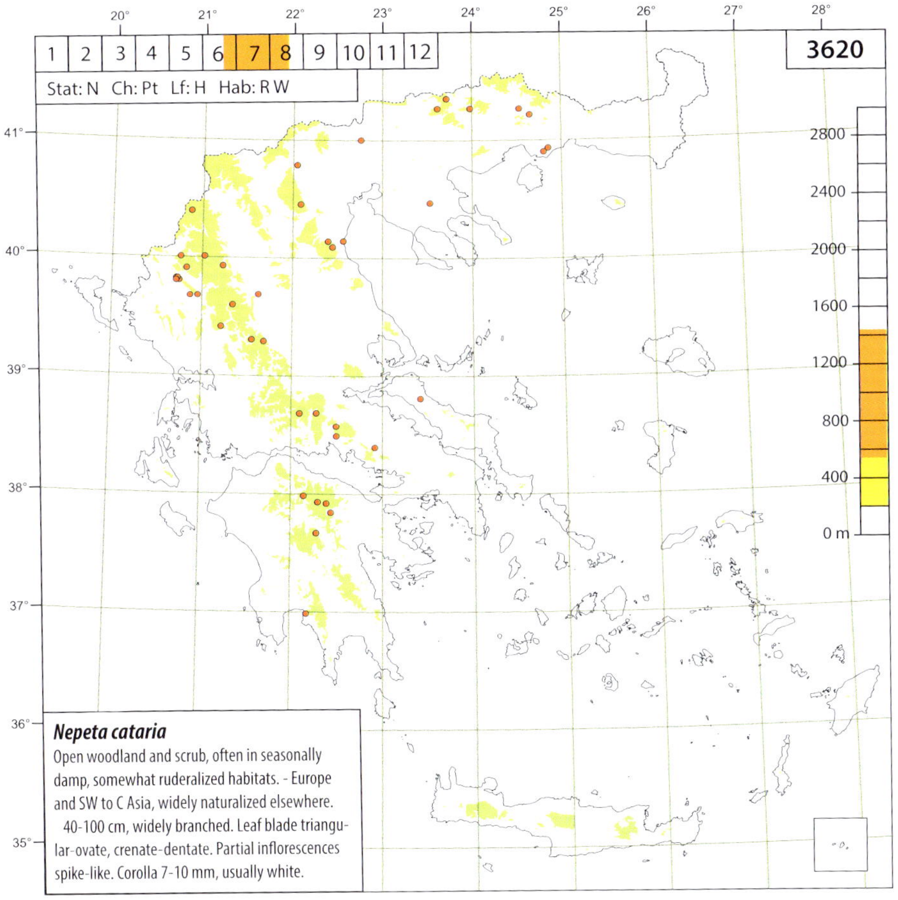

Nepeta cataria

Open woodland and scrub, often in seasonally damp, somewhat ruderalized habitats. - Europe and SW to C Asia, widely naturalized elsewhere.

40-100 cm, widely branched. Leaf blade triangular-ovate, crenate-dentate. Partial inflorescences spike-like. Corolla 7-10 mm, usually white.

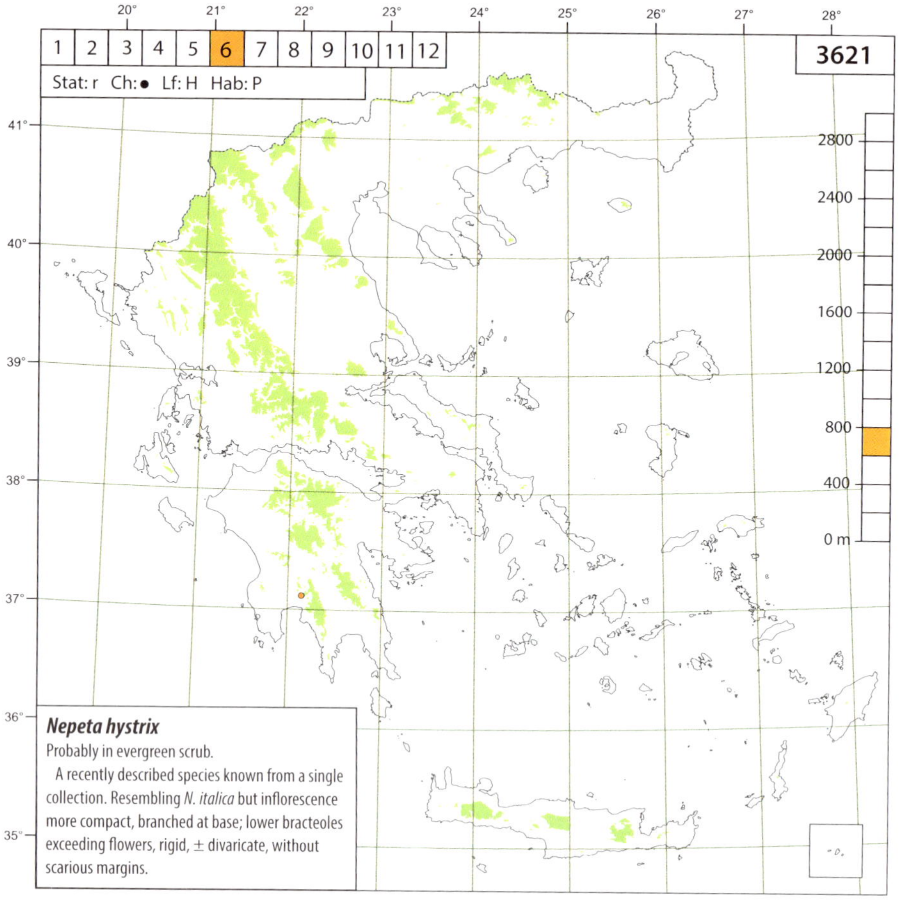

Nepeta hystrix

Probably in evergreen scrub.

A recently described species known from a single collection. Resembling *N. italica* but inflorescence more compact, branched at base; lower bracteoles exceeding flowers, rigid, ± divaricate, without scarious margins.

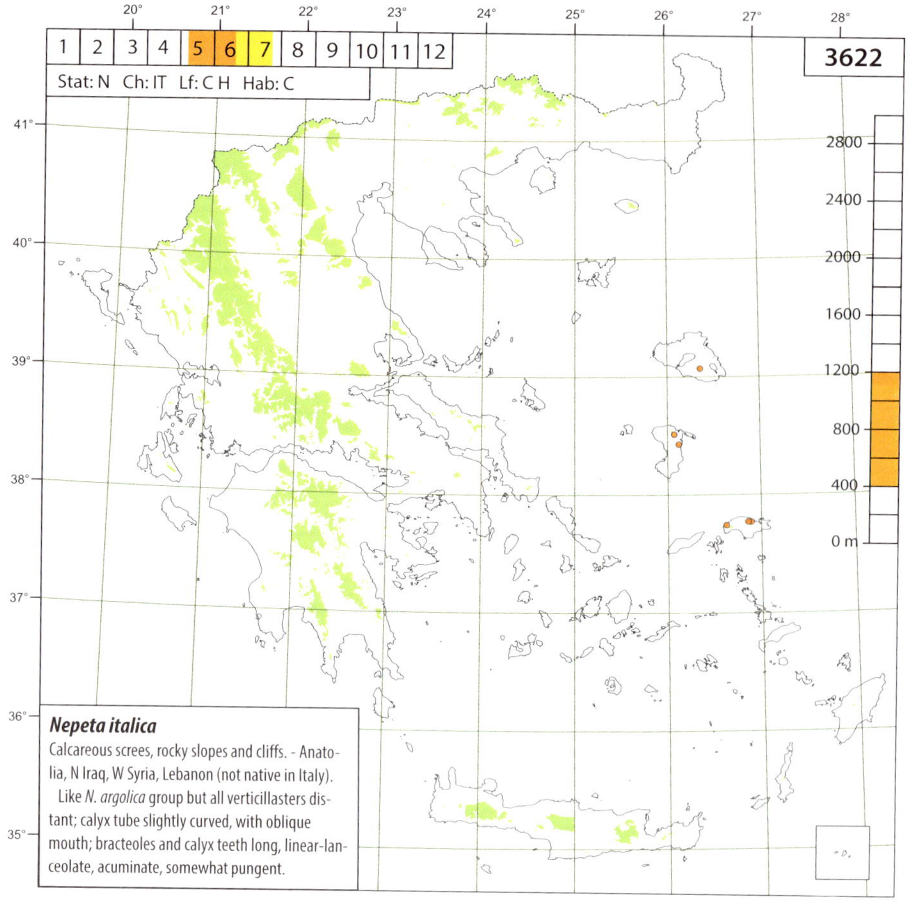

Nepeta italica

Calcareous screes, rocky slopes and cliffs. - Anatolia, N Iraq, W Syria, Lebanon (not native in Italy).

Like *N. argolica* group but all verticillasters distant; calyx tube slightly curved, with oblique mouth; bracteoles and calyx teeth long, linear-lanceolate, acuminate, somewhat pungent.

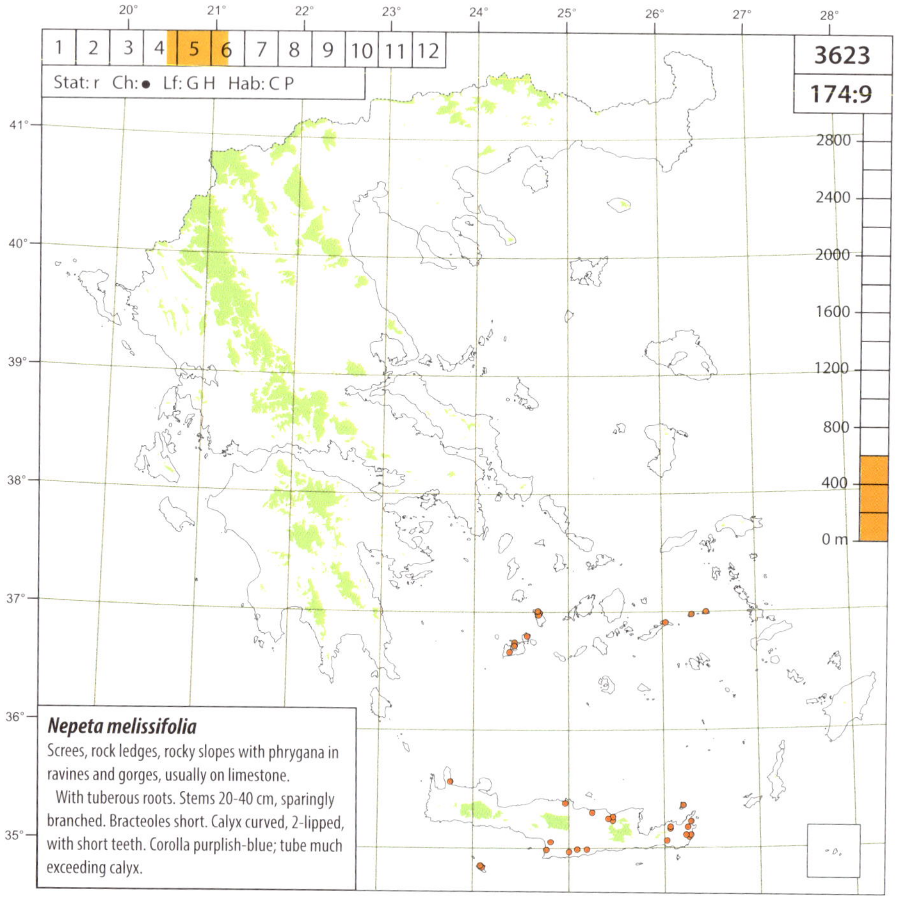

Nepeta melissifolia

Screes, rock ledges, rocky slopes with phrygana in ravines and gorges, usually on limestone.

With tuberous roots. Stems 20-40 cm, sparingly branched. Bracteoles short. Calyx curved, 2-lipped, with short teeth. Corolla purplish-blue; tube much exceeding calyx.

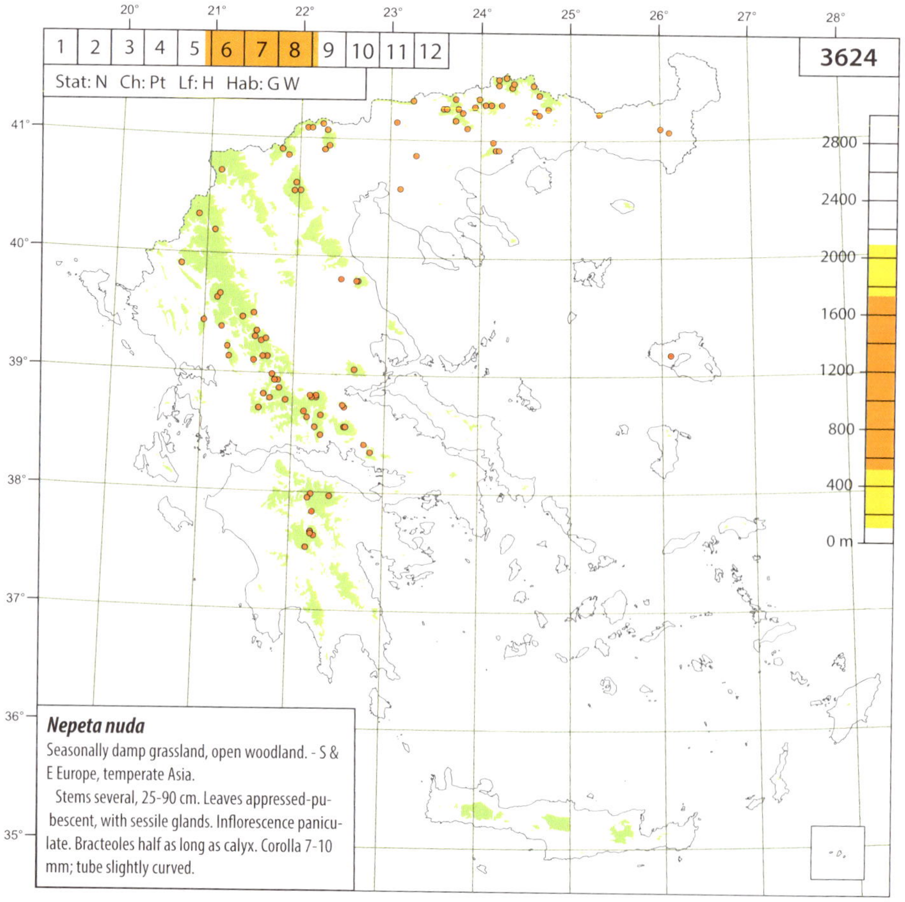

Nepeta nuda

Seasonally damp grassland, open woodland. - S & E Europe, temperate Asia.

Stems several, 25-90 cm. Leaves appressed-pubescent, with sessile glands. Inflorescence paniculate. Bracteoles half as long as calyx. Corolla 7-10 mm; tube slightly curved.

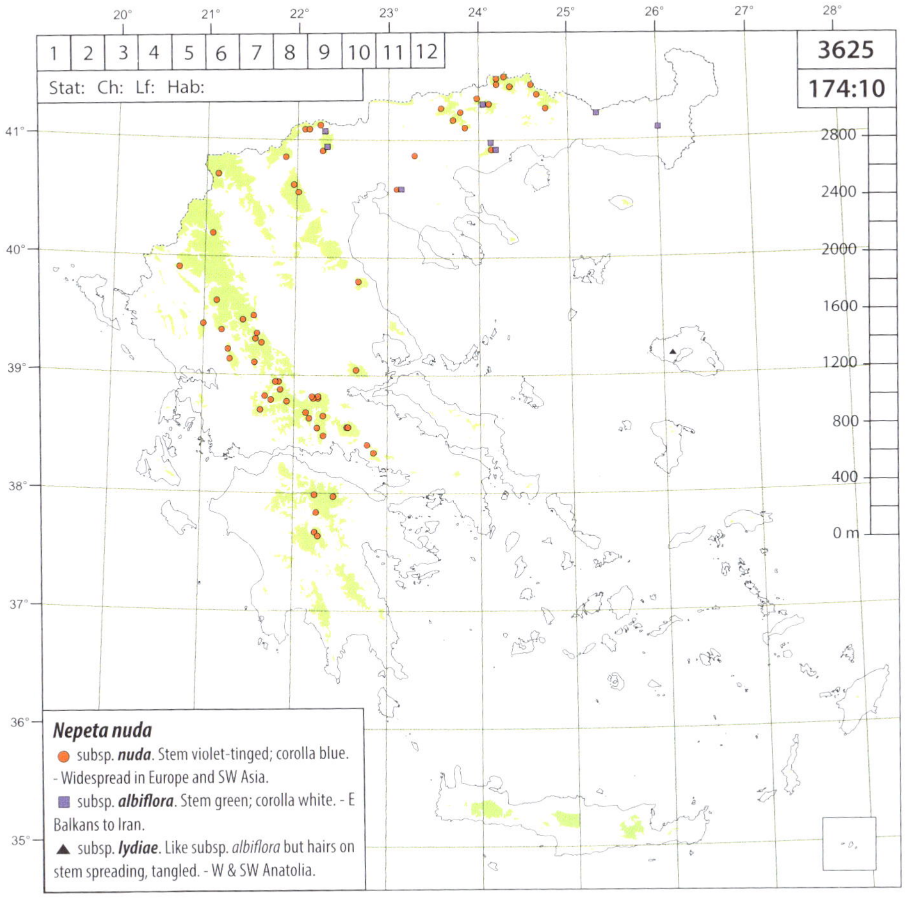

Nepeta nuda

● subsp. ***nuda***. Stem violet-tinged; corolla blue. - Widespread in Europe and SW Asia.

■ subsp. ***albiflora***. Stem green; corolla white. - E Balkans to Iran.

▲ subsp. ***lydiae***. Like subsp. *albiflora* but hairs on stem spreading, tangled. - W & SW Anatolia.

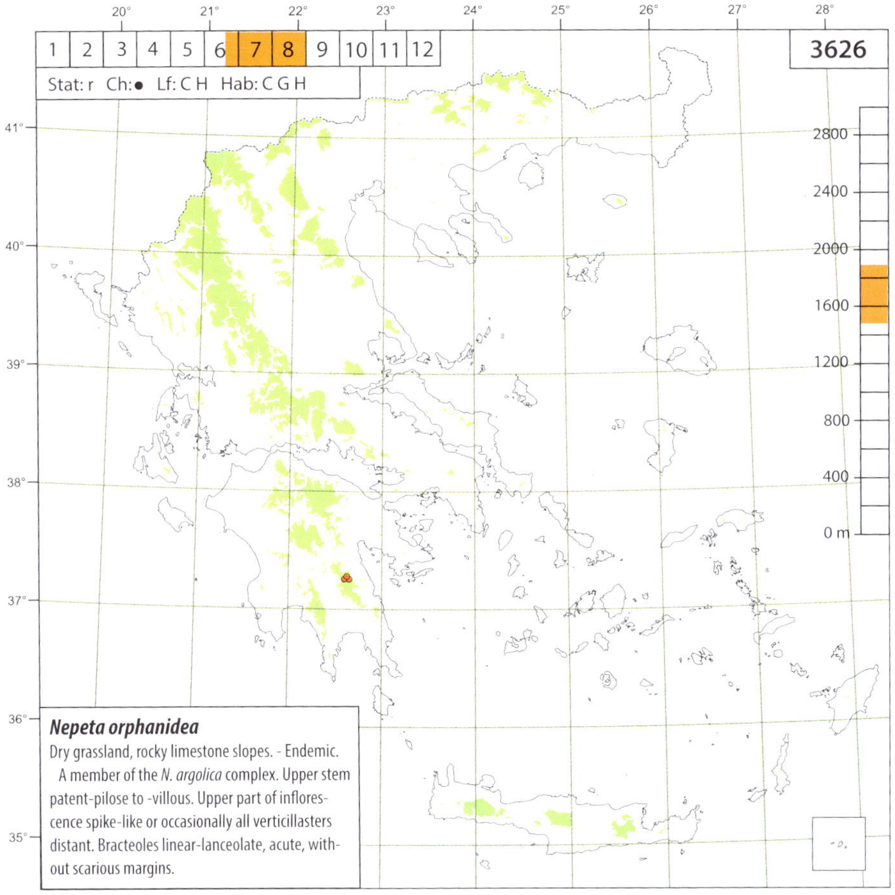

Nepeta orphanidea

Dry grassland, rocky limestone slopes. - Endemic.

A member of the *N. argolica* complex. Upper stem patent-pilose to -villous. Upper part of inflorescence spike-like or occasionally all verticillasters distant. Bracteoles linear-lanceolate, acute, without scarious margins.

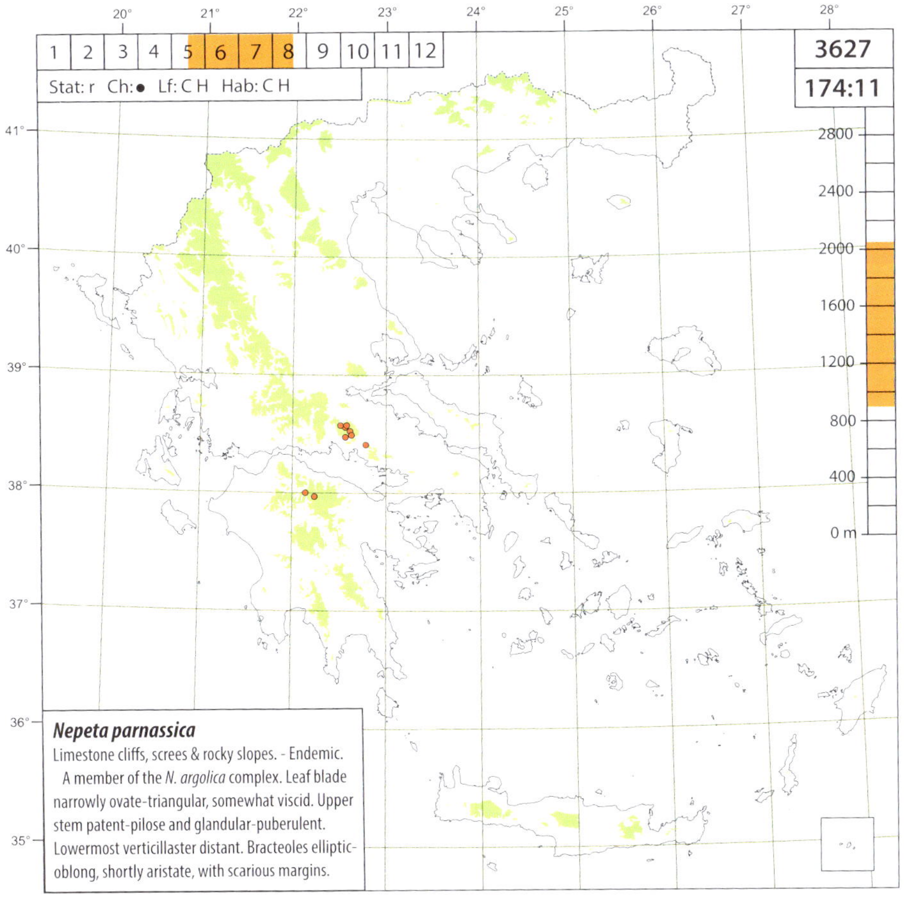

Nepeta parnassica

Limestone cliffs, screes & rocky slopes. - Endemic.

A member of the *N. argolica* complex. Leaf blade narrowly ovate-triangular, somewhat viscid. Upper stem patent-pilose and glandular-puberulent. Lowermost verticillaster distant. Bracteoles elliptic-oblong, shortly aristate, with scarious margins.

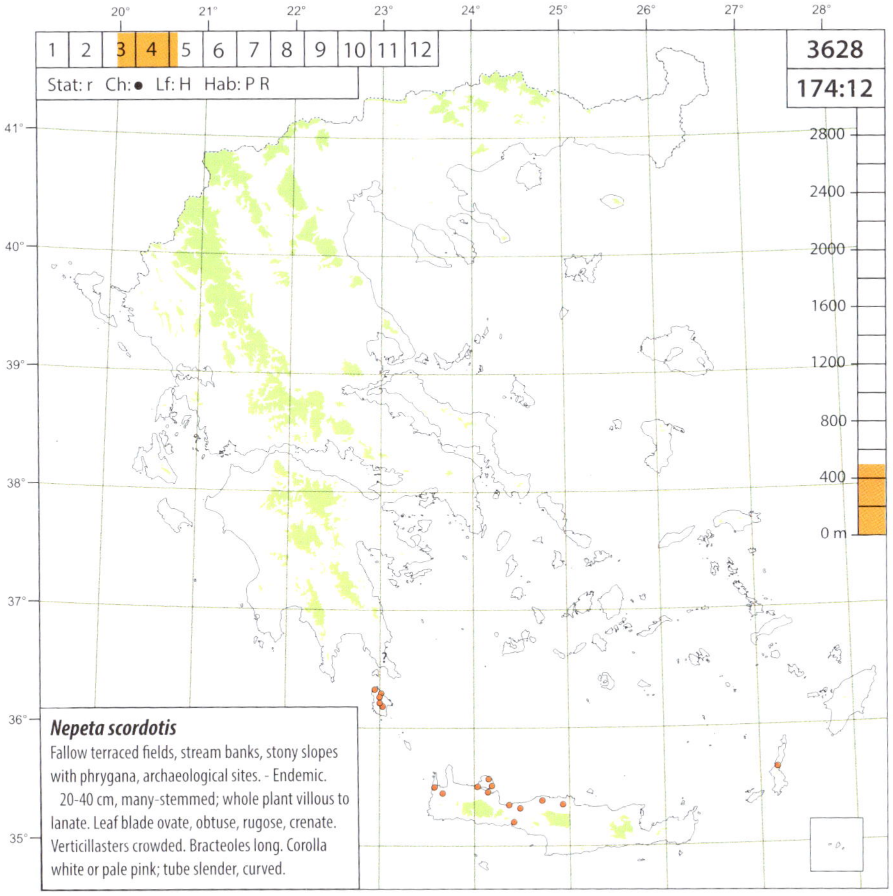

Nepeta scordotis

Fallow terraced fields, stream banks, stony slopes with phrygana, archaeological sites. - Endemic.

20-40 cm, many-stemmed; whole plant villous to lanate. Leaf blade ovate, obtuse, rugose, crenate. Verticillasters crowded. Bracteoles long. Corolla white or pale pink; tube slender, curved.

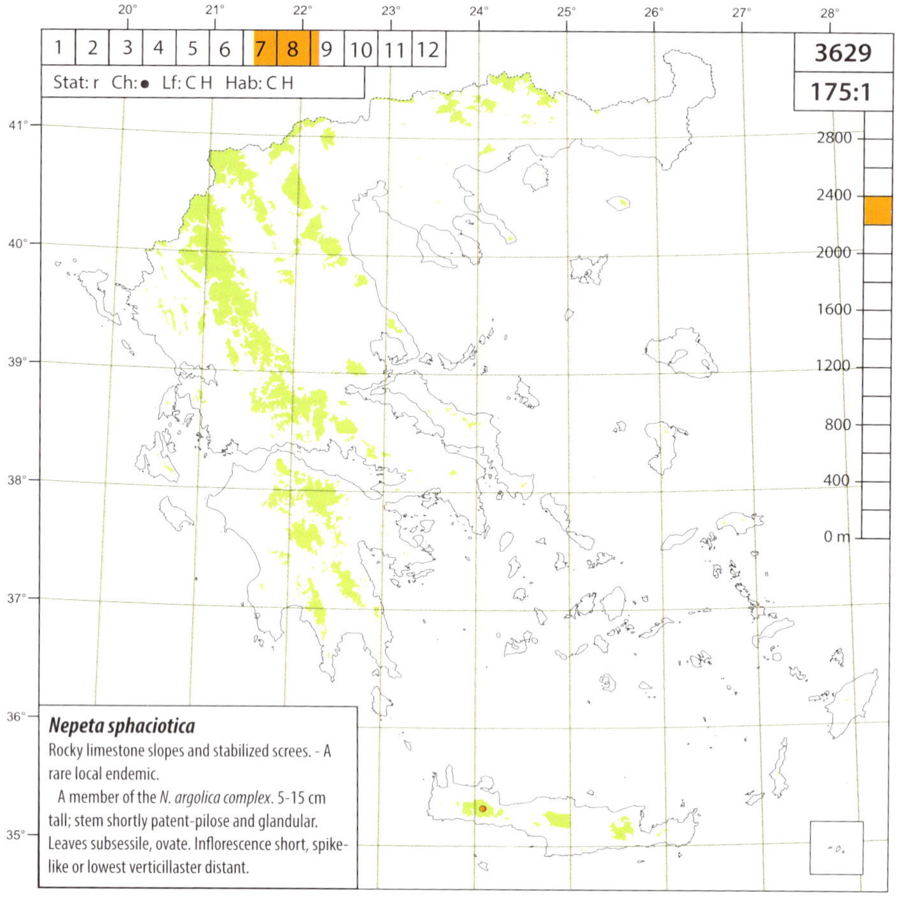

Nepeta sphaciotica
Rocky limestone slopes and stabilized screes. - A rare local endemic.
A member of the *N. argolica complex*. 5-15 cm tall; stem shortly patent-pilose and glandular. Leaves subsessile, ovate. Inflorescence short, spike-like or lowest verticillaster distant.

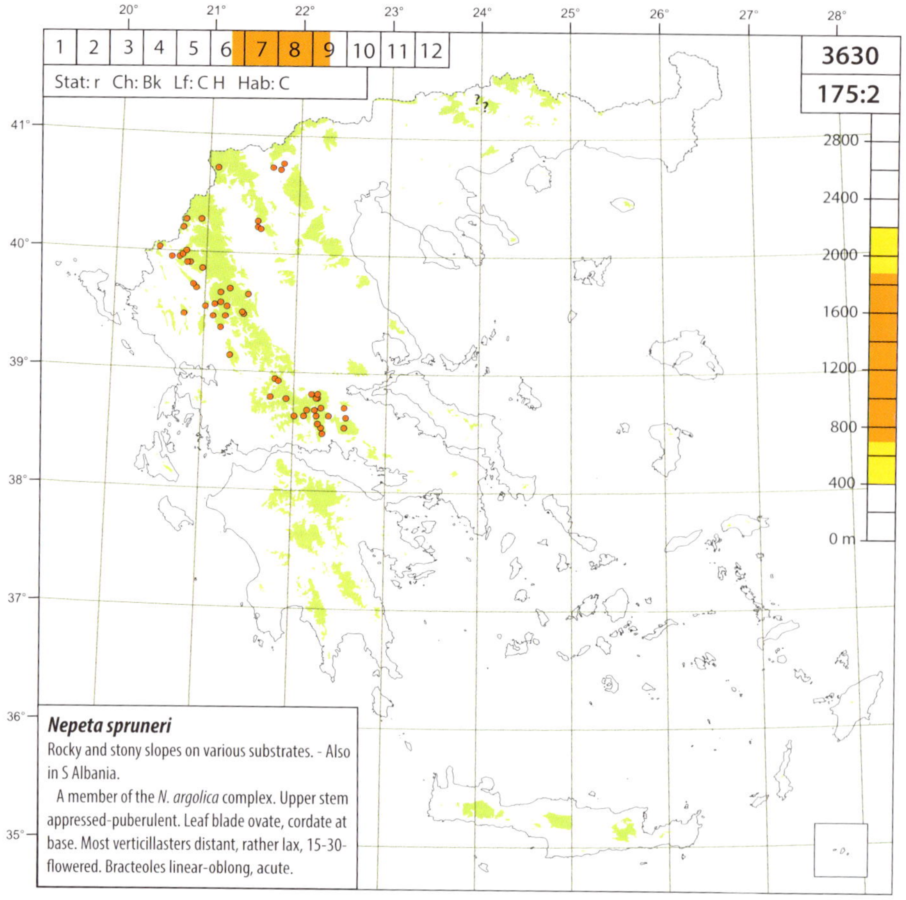

Nepeta spruneri
Rocky and stony slopes on various substrates. - Also in S Albania.
A member of the *N. argolica* complex. Upper stem appressed-puberulent. Leaf blade ovate, cordate at base. Most verticillasters distant, rather lax, 15-30-flowered. Bracteoles linear-oblong, acute.

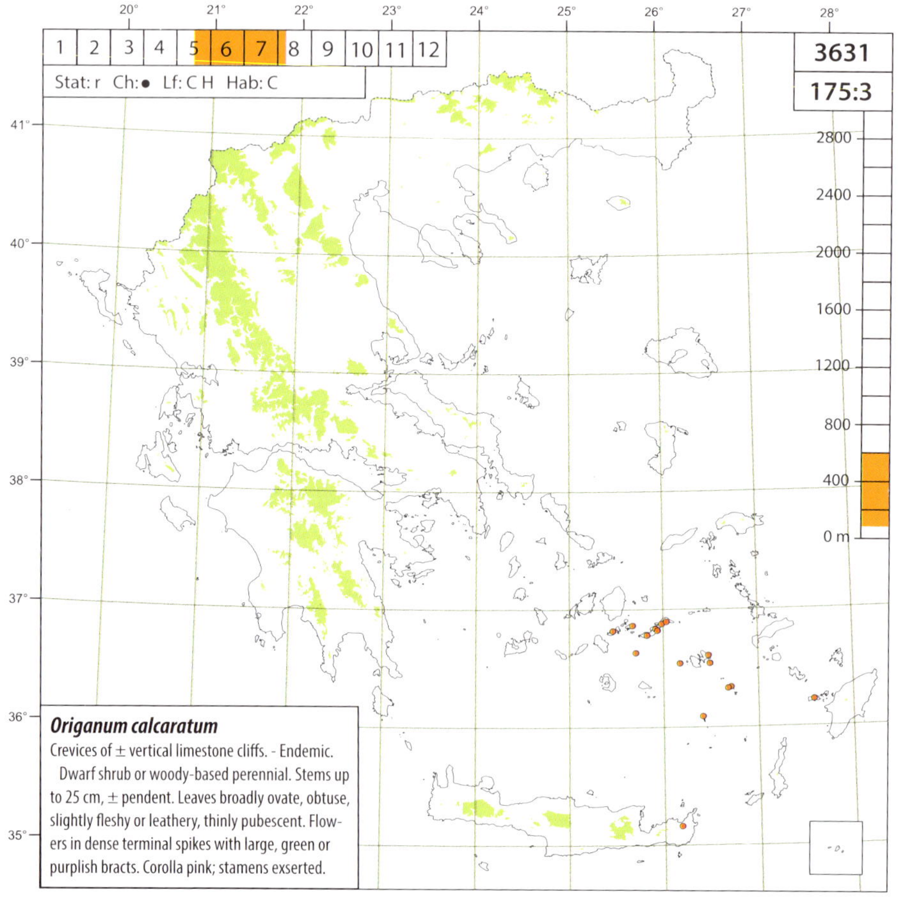

Origanum calcaratum
Crevices of ± vertical limestone cliffs. - Endemic.
Dwarf shrub or woody-based perennial. Stems up to 25 cm, ± pendent. Leaves broadly ovate, obtuse, slightly fleshy or leathery, thinly pubescent. Flowers in dense terminal spikes with large, green or purplish bracts. Corolla pink; stamens exserted.

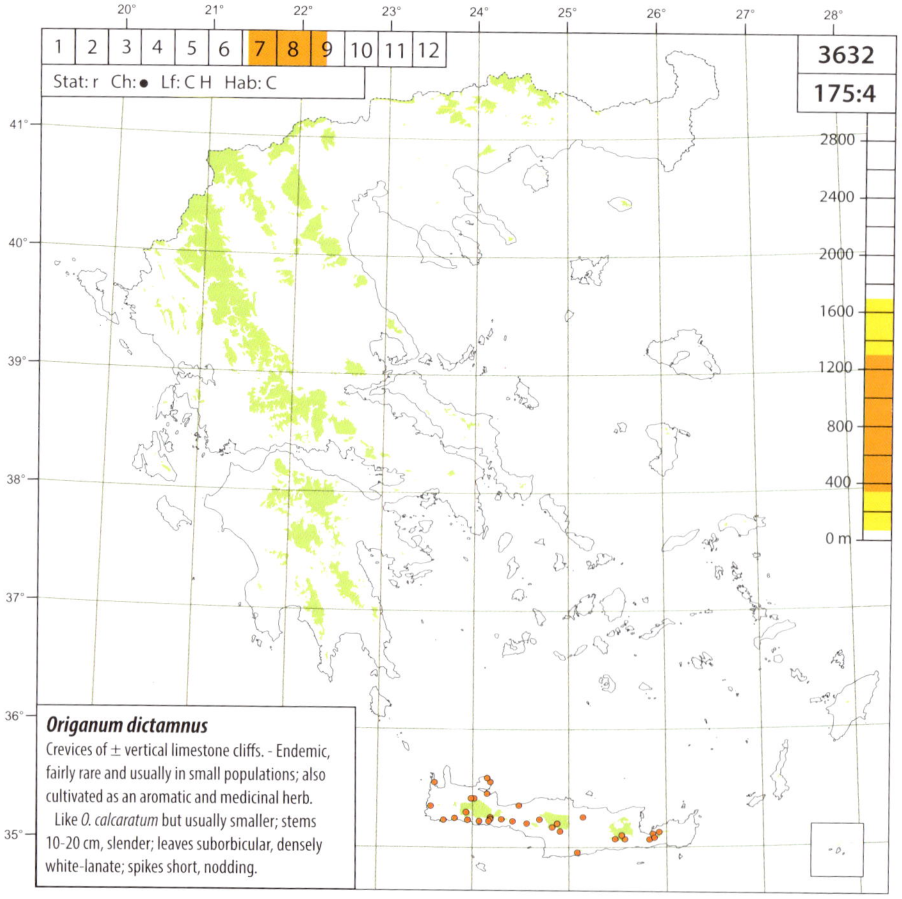

Origanum dictamnus
Crevices of ± vertical limestone cliffs. - Endemic, fairly rare and usually in small populations; also cultivated as an aromatic and medicinal herb.
Like *O. calcaratum* but usually smaller; stems 10-20 cm, slender; leaves suborbicular, densely white-lanate; spikes short, nodding.

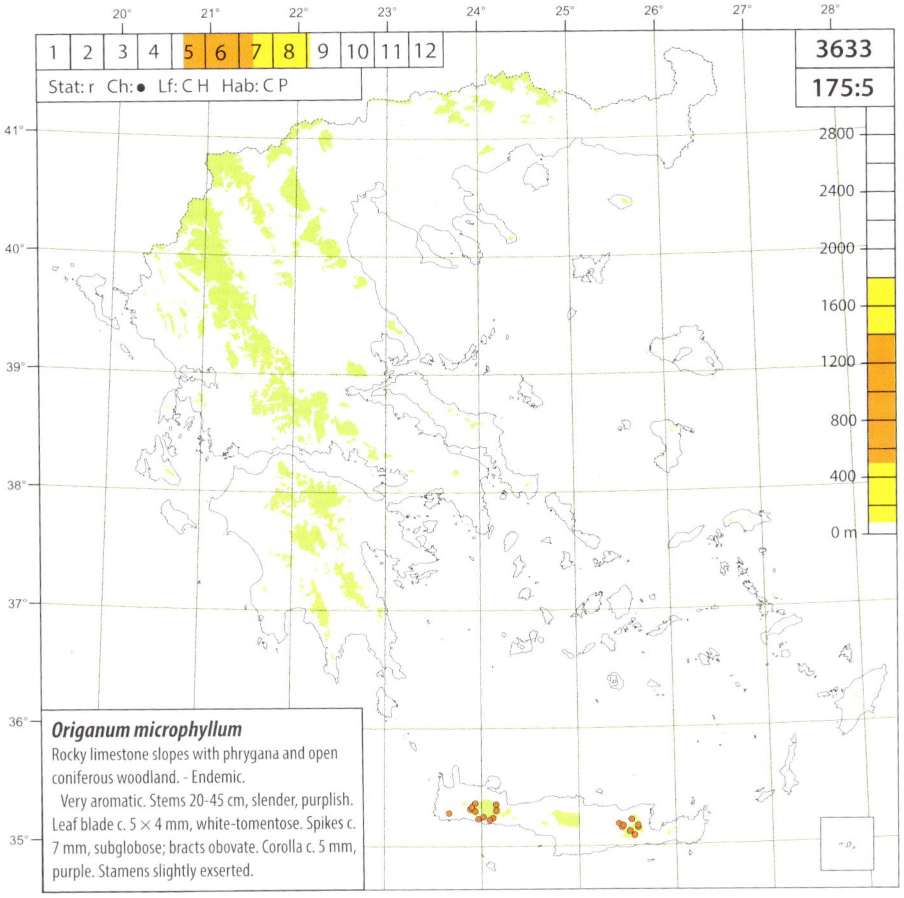

Origanum microphyllum

Rocky limestone slopes with phrygana and open coniferous woodland. - Endemic.

Very aromatic. Stems 20-45 cm, slender, purplish. Leaf blade c. 5 × 4 mm, white-tomentose. Spikes c. 7 mm, subglobose; bracts obovate. Corolla c. 5 mm, purple. Stamens slightly exserted.

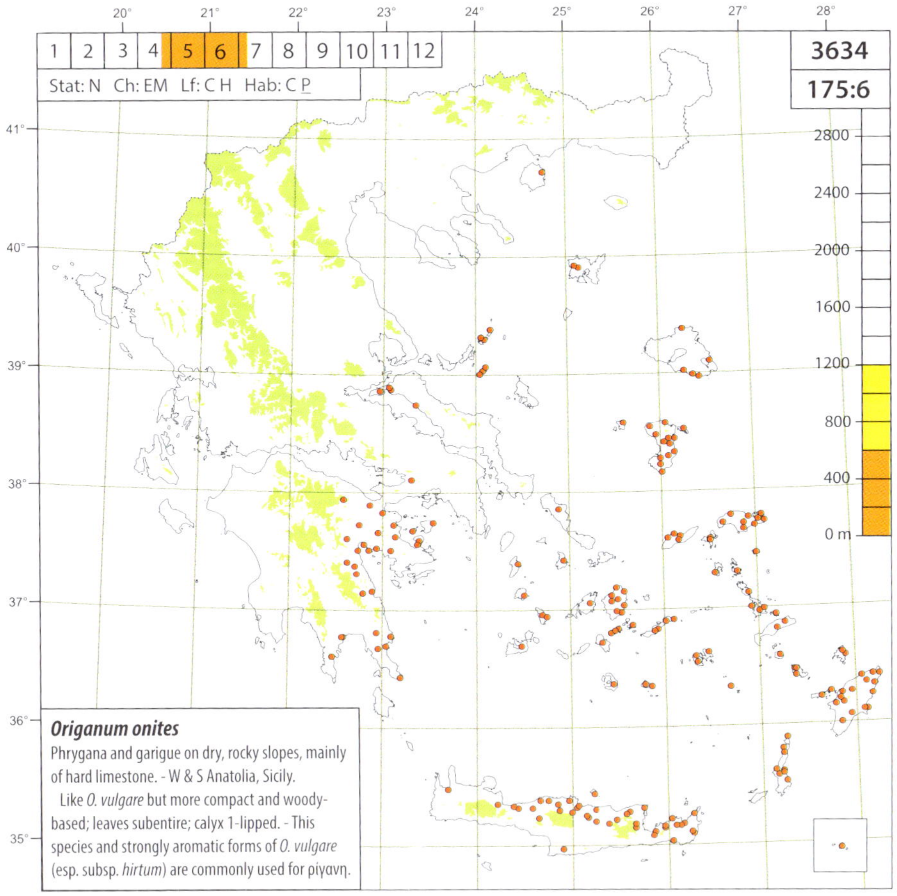

Origanum onites

Phrygana and garigue on dry, rocky slopes, mainly of hard limestone. - W & S Anatolia, Sicily.

Like *O. vulgare* but more compact and woody-based; leaves subentire; calyx 1-lipped. - This species and strongly aromatic forms of *O. vulgare* (esp. subsp. *hirtum*) are commonly used for ρίγανη.

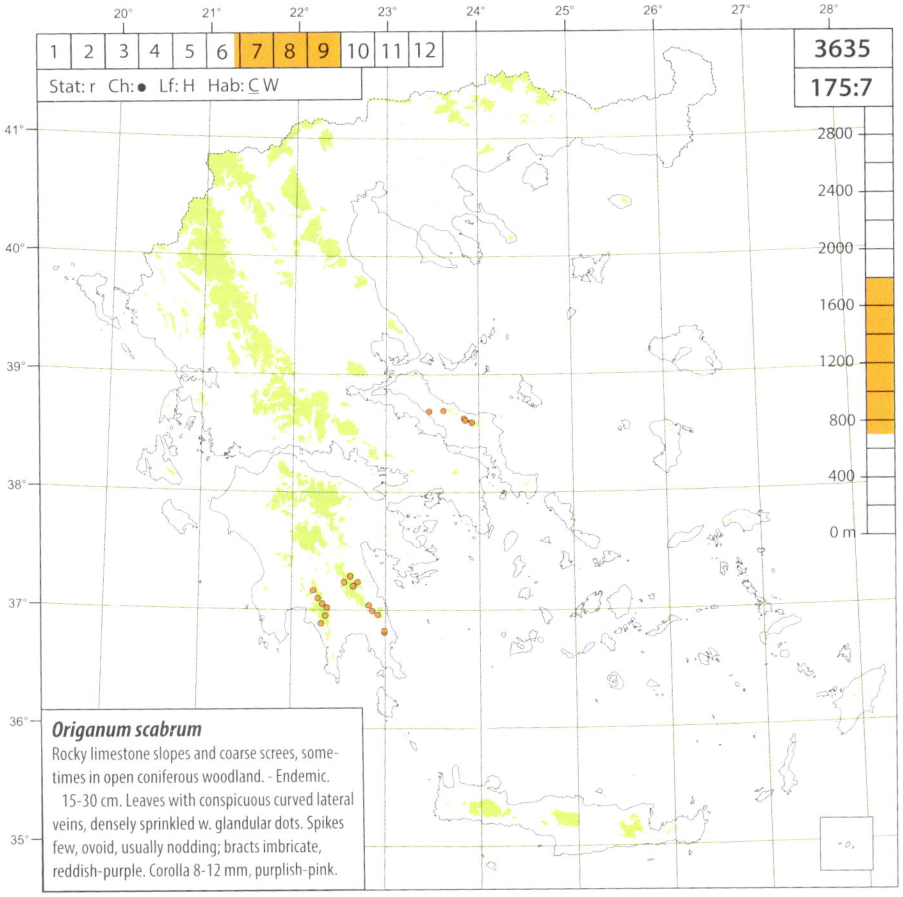

Origanum scabrum

Rocky limestone slopes and coarse screes, sometimes in open coniferous woodland. - Endemic.

15-30 cm. Leaves with conspicuous curved lateral veins, densely sprinkled w. glandular dots. Spikes few, ovoid, usually nodding; bracts imbricate, reddish-purple. Corolla 8-12 mm, purplish-pink.

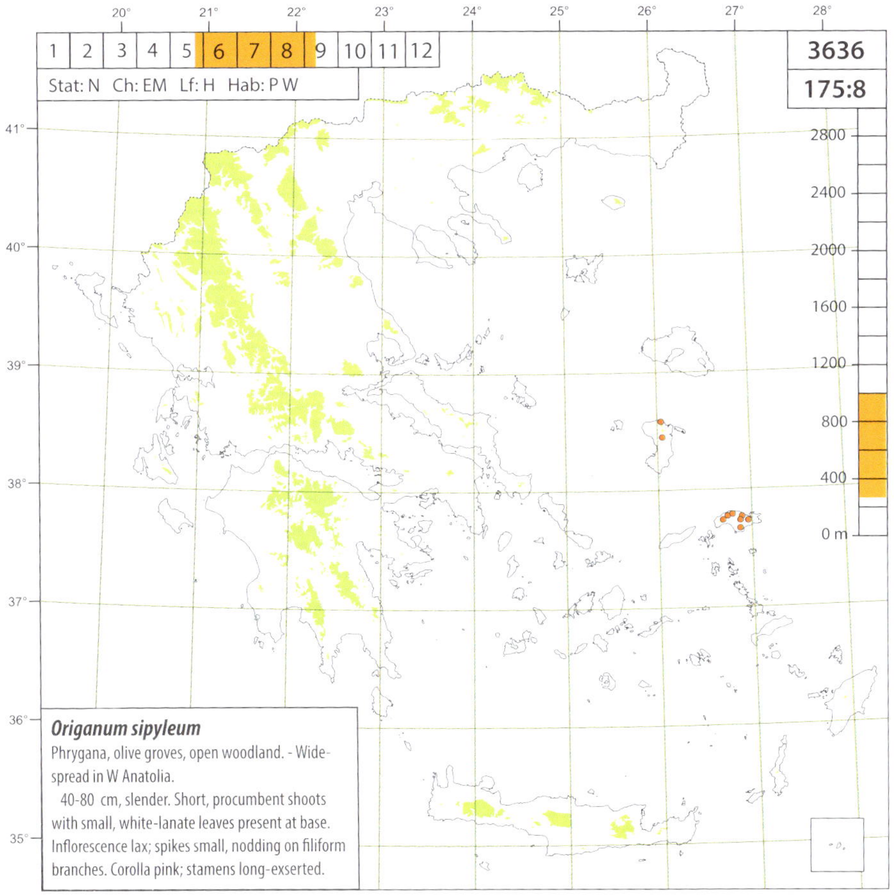

Origanum sipyleum

Phrygana, olive groves, open woodland. - Widespread in W Anatolia.

40-80 cm, slender. Short, procumbent shoots with small, white-lanate leaves present at base. Inflorescence lax; spikes small, nodding on filiform branches. Corolla pink; stamens long-exserted.

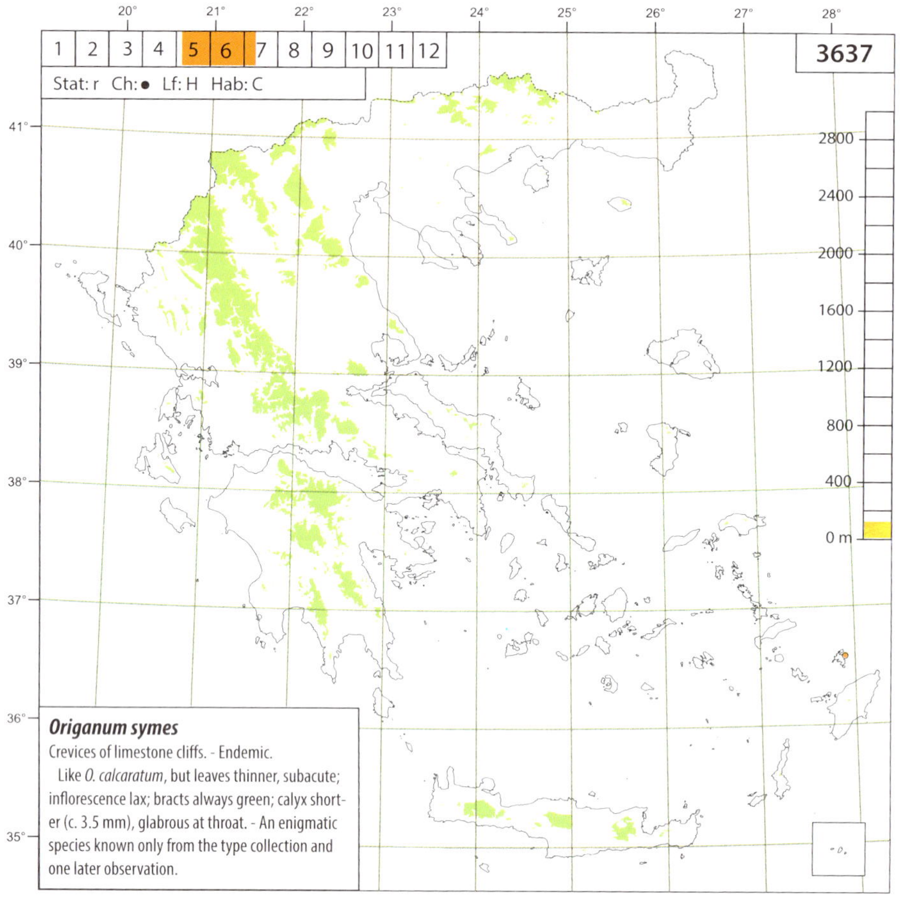

Origanum symes
Crevices of limestone cliffs. - Endemic.
Like *O. calcaratum*, but leaves thinner, subacute; inflorescence lax; bracts always green; calyx shorter (c. 3.5 mm), glabrous at throat. - An enigmatic species known only from the type collection and one later observation.

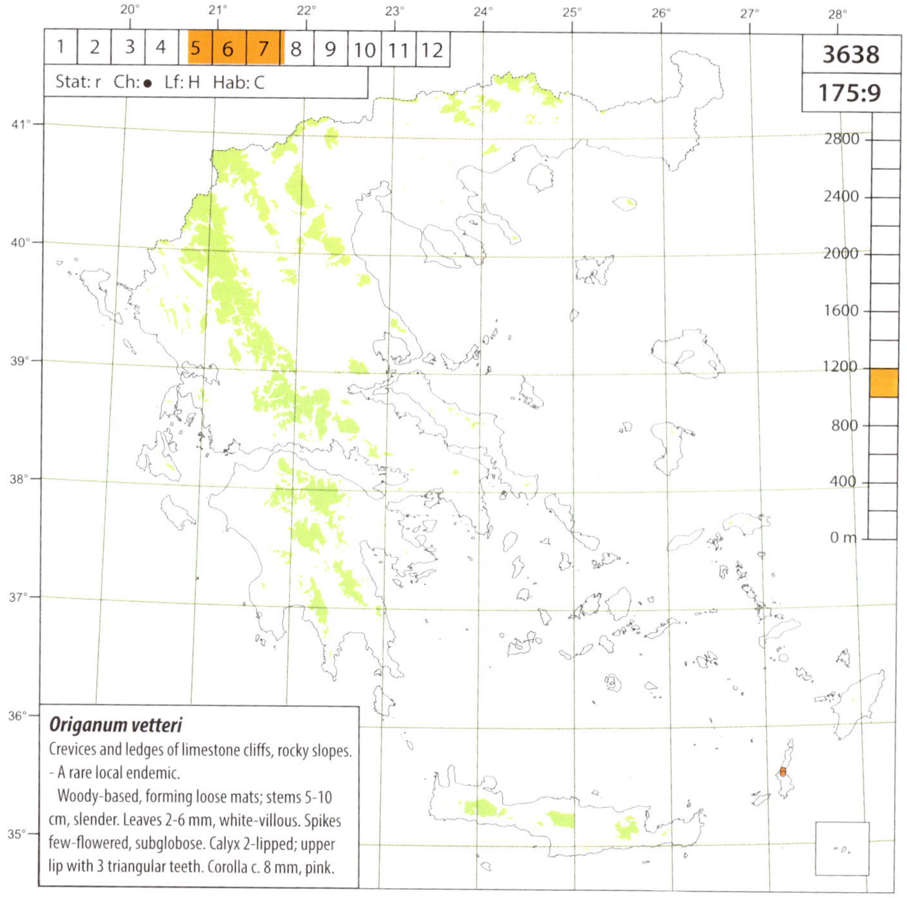

Origanum vetteri
Crevices and ledges of limestone cliffs, rocky slopes. - A rare local endemic.
Woody-based, forming loose mats; stems 5-10 cm, slender. Leaves 2-6 mm, white-villous. Spikes few-flowered, subglobose. Calyx 2-lipped; upper lip with 3 triangular teeth. Corolla c. 8 mm, pink.

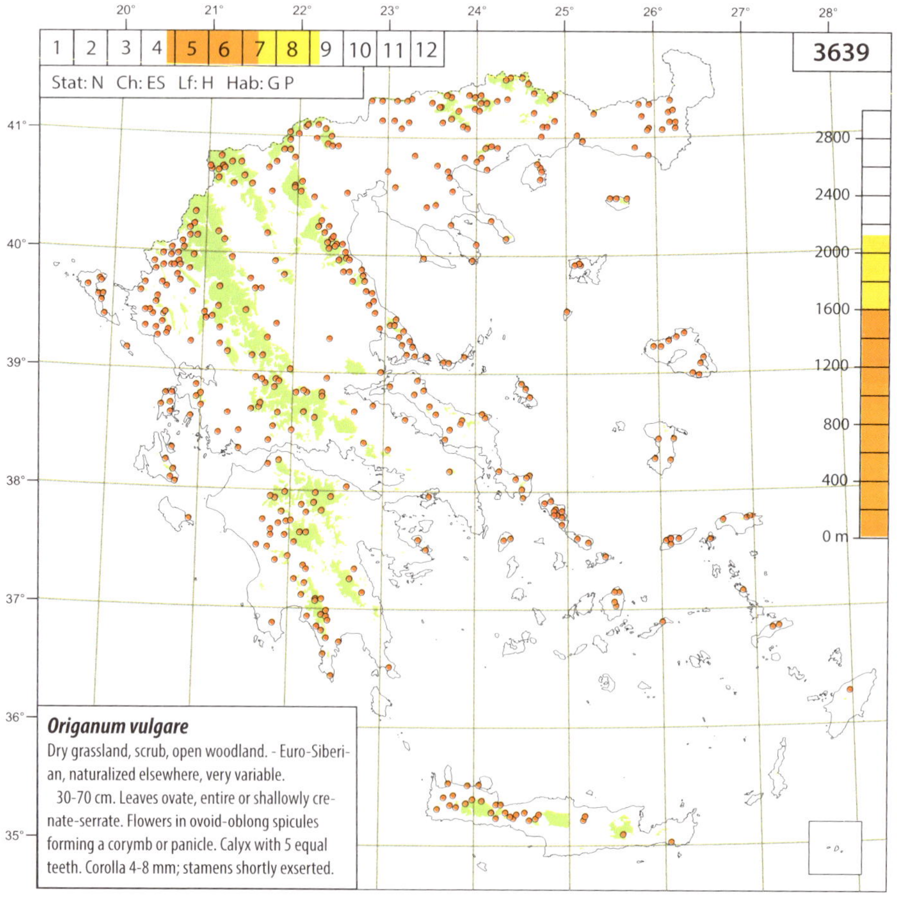

Origanum vulgare
Dry grassland, scrub, open woodland. - Euro-Siberian, naturalized elsewhere, very variable.
30-70 cm. Leaves ovate, entire or shallowly crenate-serrate. Flowers in ovoid-oblong spicules forming a corymb or panicle. Calyx with 5 equal teeth. Corolla 4-8 mm; stamens shortly exserted.

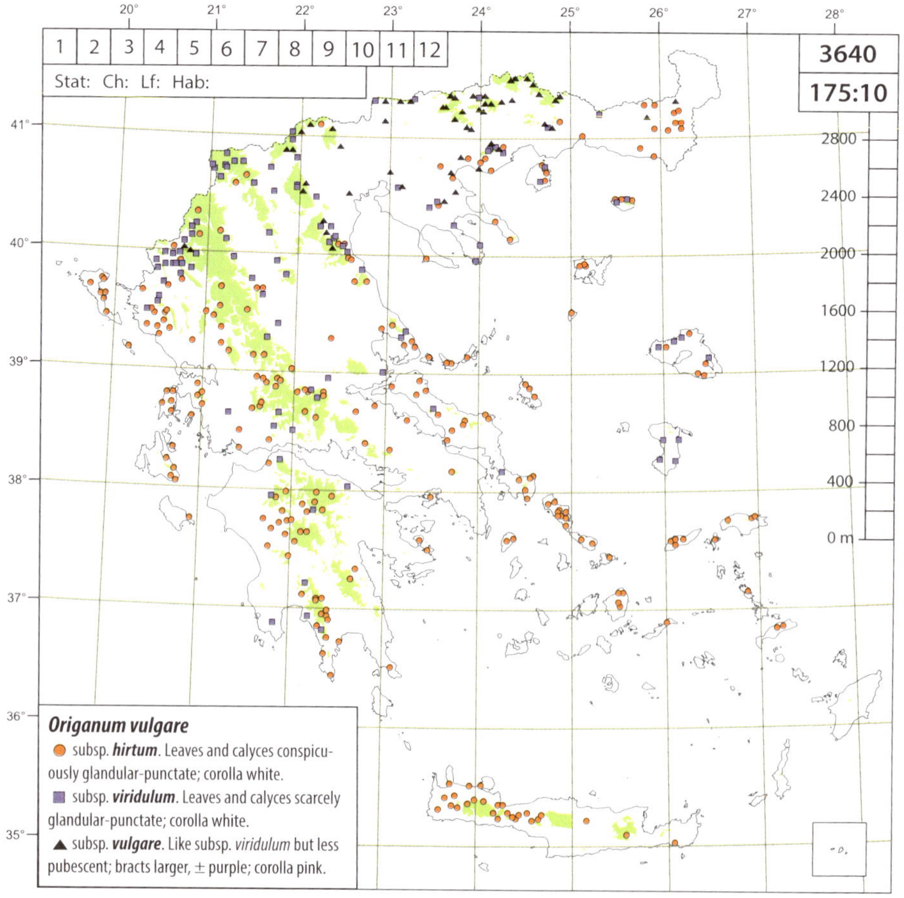

Origanum vulgare
● subsp. ***hirtum***. Leaves and calyces conspicuously glandular-punctate; corolla white.
■ subsp. ***viridulum***. Leaves and calyces scarcely glandular-punctate; corolla white.
▲ subsp. ***vulgare***. Like subsp. *viridulum* but less pubescent; bracts larger, ± purple; corolla pink.

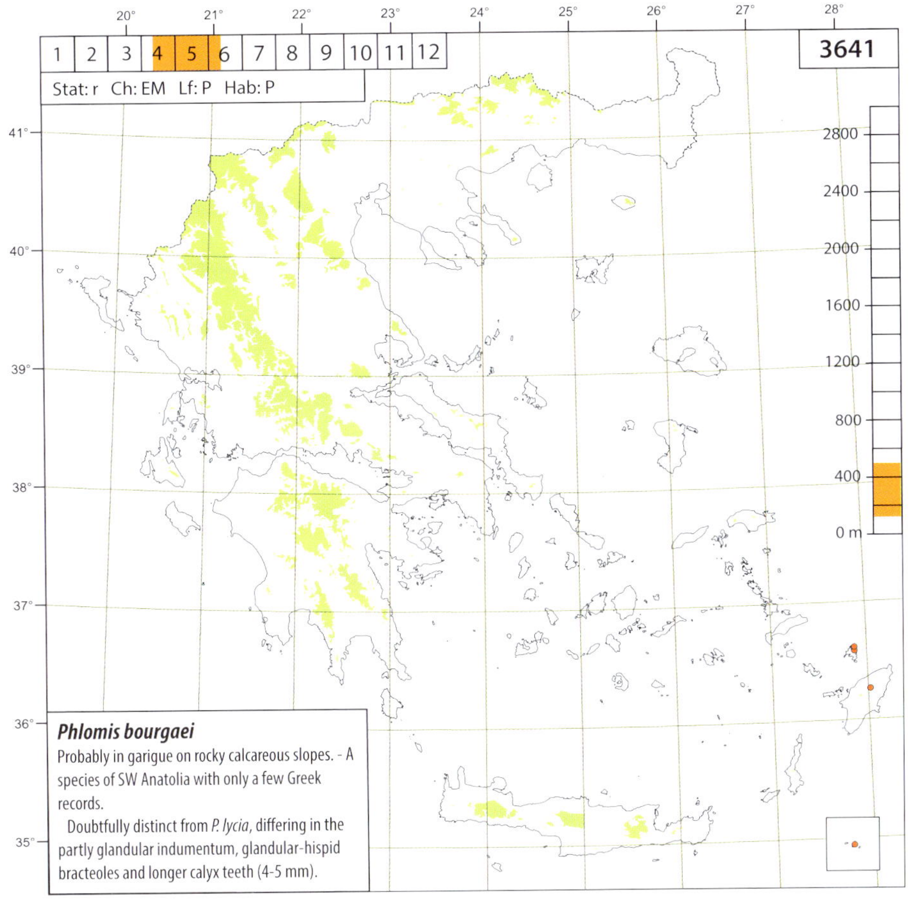

Phlomis bourgaei

Probably in garigue on rocky calcareous slopes. - A species of SW Anatolia with only a few Greek records.

Doubtfully distinct from *P. lycia*, differing in the partly glandular indumentum, glandular-hispid bracteoles and longer calyx teeth (4-5 mm).

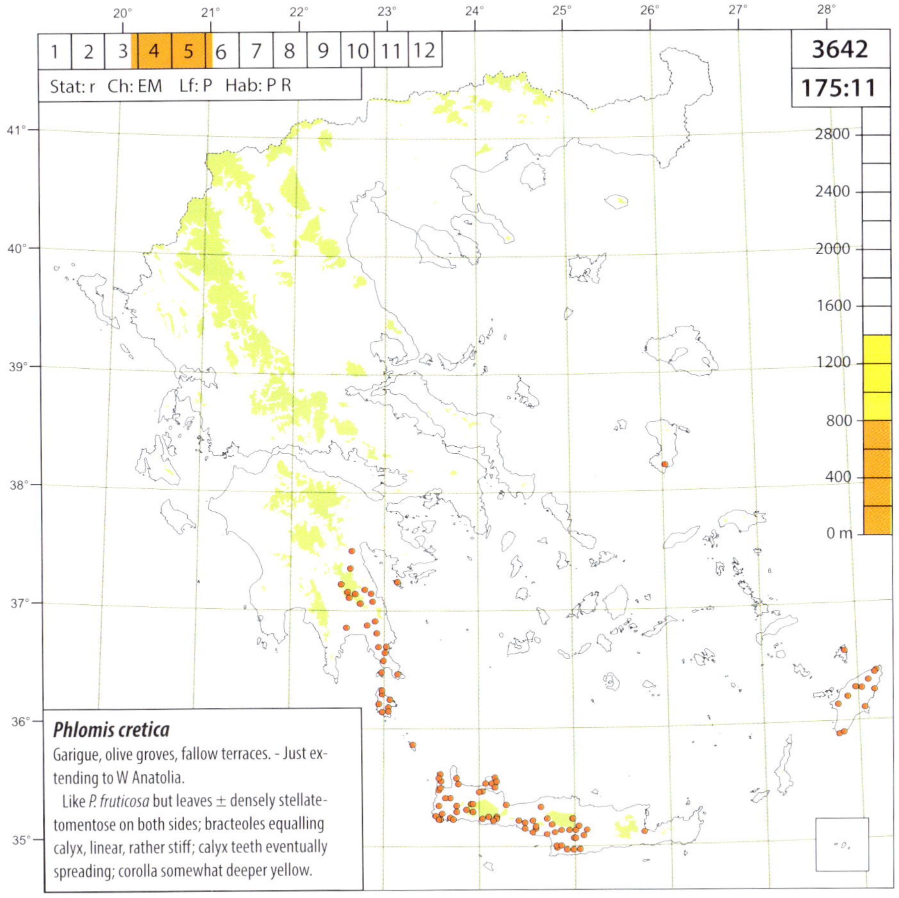

Phlomis cretica

Garigue, olive groves, fallow terraces. - Just extending to W Anatolia.

Like *P. fruticosa* but leaves ± densely stellate-tomentose on both sides; bracteoles equalling calyx, linear, rather stiff; calyx teeth eventually spreading; corolla somewhat deeper yellow.

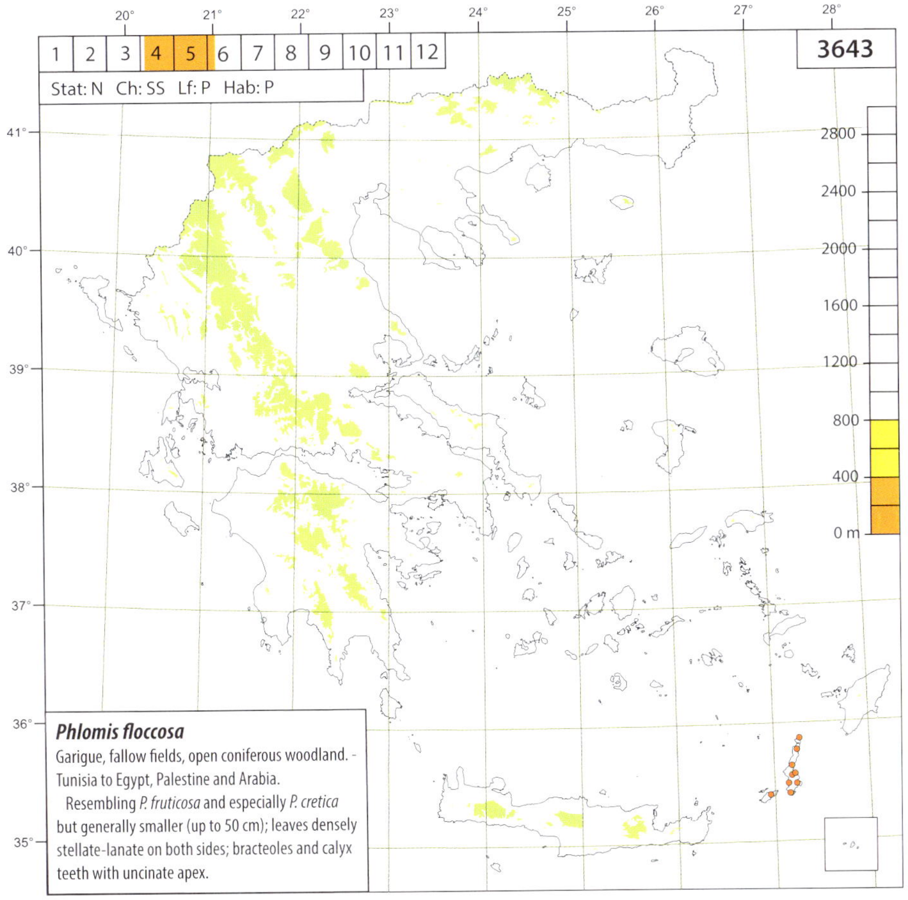

Phlomis floccosa

Garigue, fallow fields, open coniferous woodland. - Tunisia to Egypt, Palestine and Arabia.

Resembling *P. fruticosa* and especially *P. cretica* but generally smaller (up to 50 cm); leaves densely stellate-lanate on both sides; bracteoles and calyx teeth with uncinate apex.

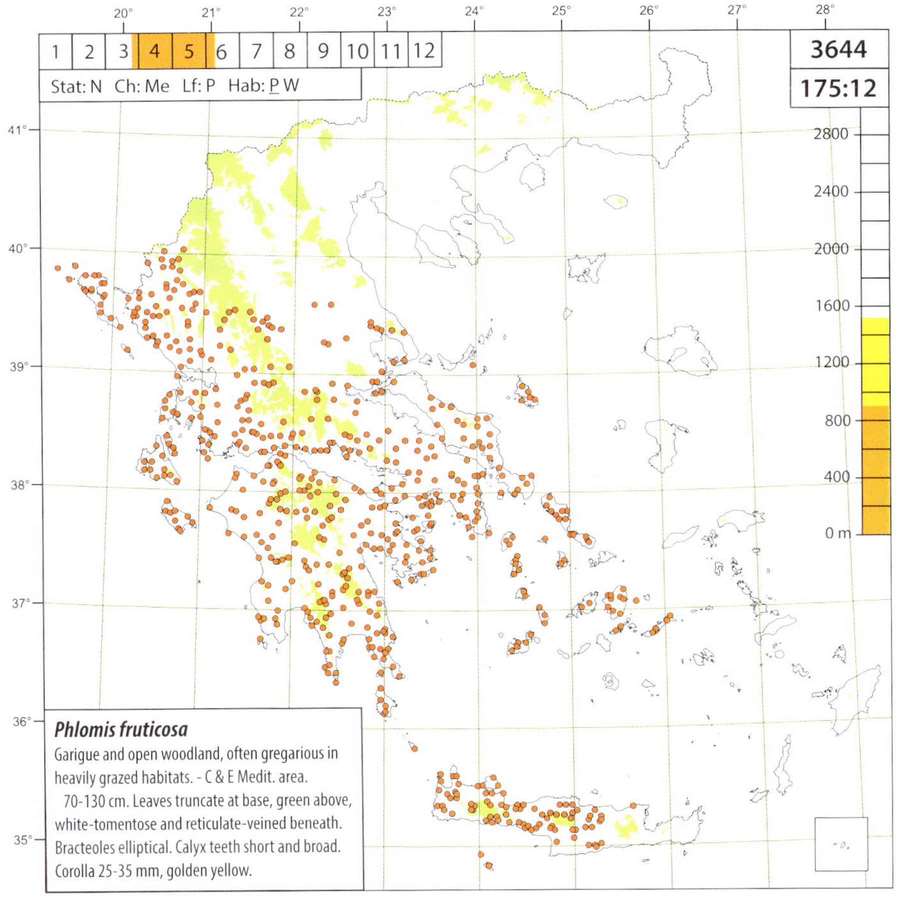

Phlomis fruticosa

Garigue and open woodland, often gregarious in heavily grazed habitats. - C & E Medit. area.

70-130 cm. Leaves truncate at base, green above, white-tomentose and reticulate-veined beneath. Bracteoles elliptical. Calyx teeth short and broad. Corolla 25-35 mm, golden yellow.

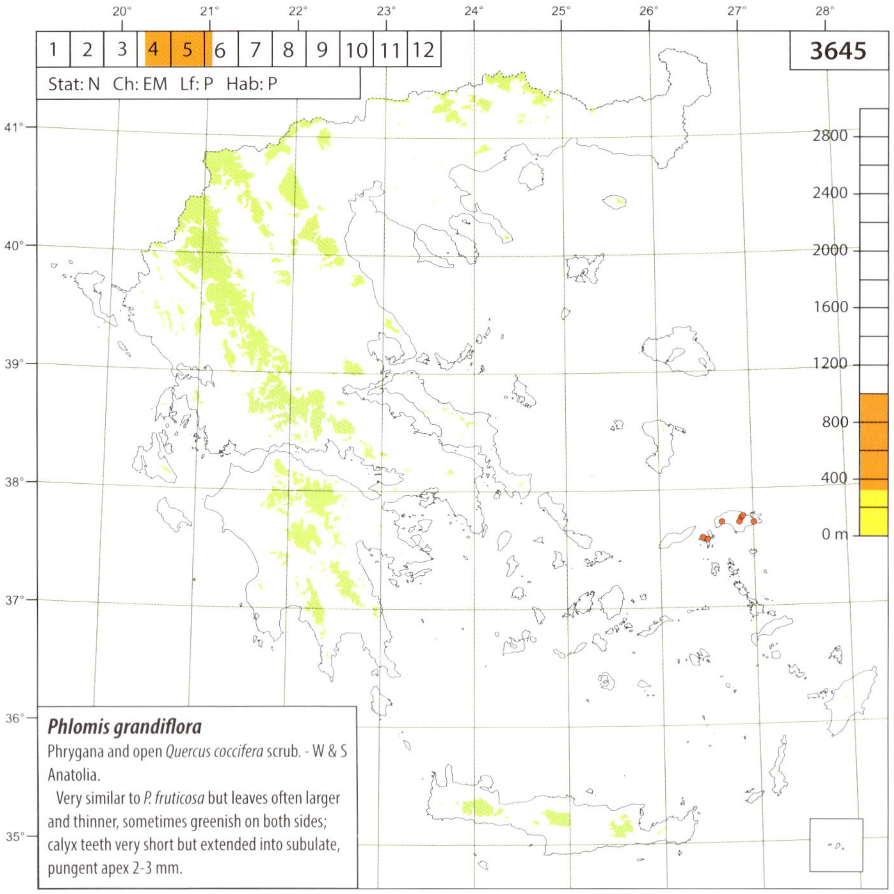

Phlomis grandiflora
Phrygana and open *Quercus coccifera* scrub. - W & S Anatolia.
Very similar to *P. fruticosa* but leaves often larger and thinner, sometimes greenish on both sides; calyx teeth very short but extended into subulate, pungent apex 2-3 mm.

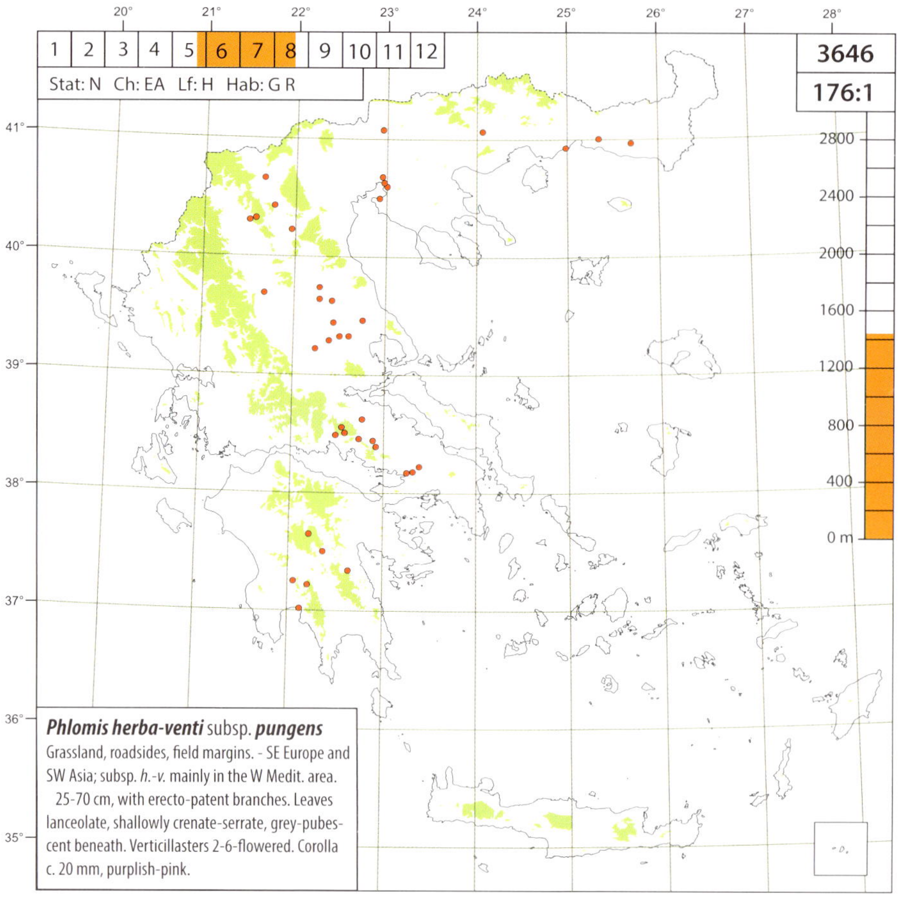

Phlomis herba-venti subsp. pungens
Grassland, roadsides, field margins. - SE Europe and SW Asia; subsp. *h.-v.* mainly in the W Medit. area.
25-70 cm, with erecto-patent branches. Leaves lanceolate, shallowly crenate-serrate, grey-pubescent beneath. Verticillasters 2-6-flowered. Corolla c. 20 mm, purplish-pink.

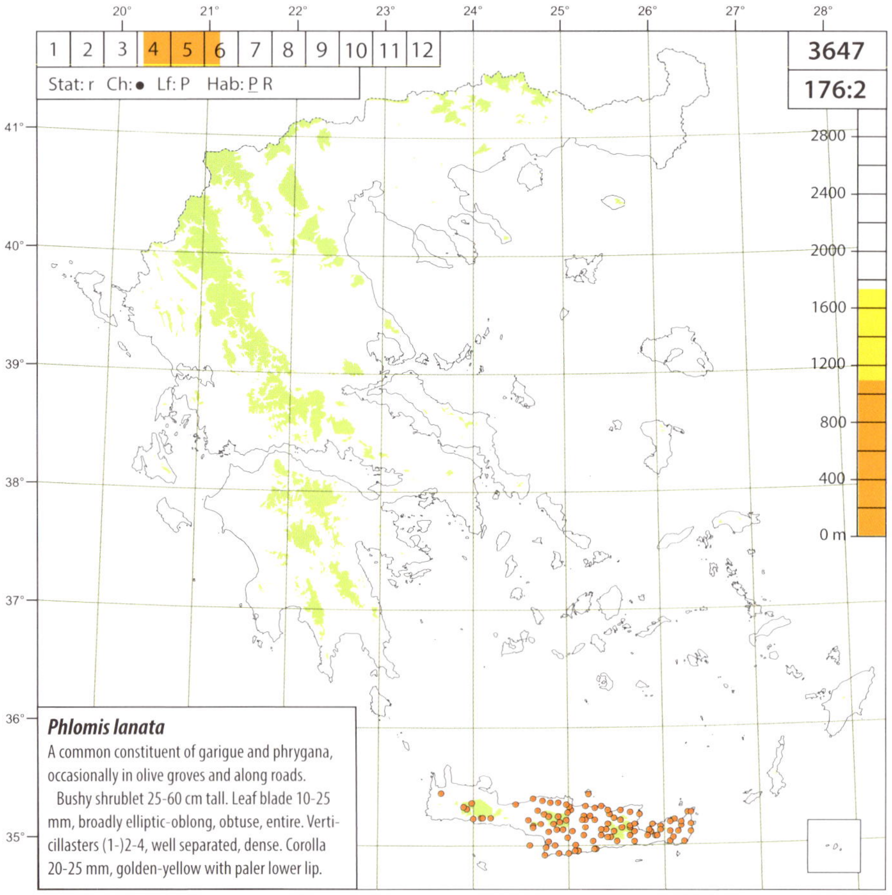

Phlomis lanata
A common constituent of garigue and phrygana, occasionally in olive groves and along roads.
Bushy shrublet 25-60 cm tall. Leaf blade 10-25 mm, broadly elliptic-oblong, obtuse, entire. Verticillasters (1-)2-4, well separated, dense. Corolla 20-25 mm, golden-yellow with paler lower lip.

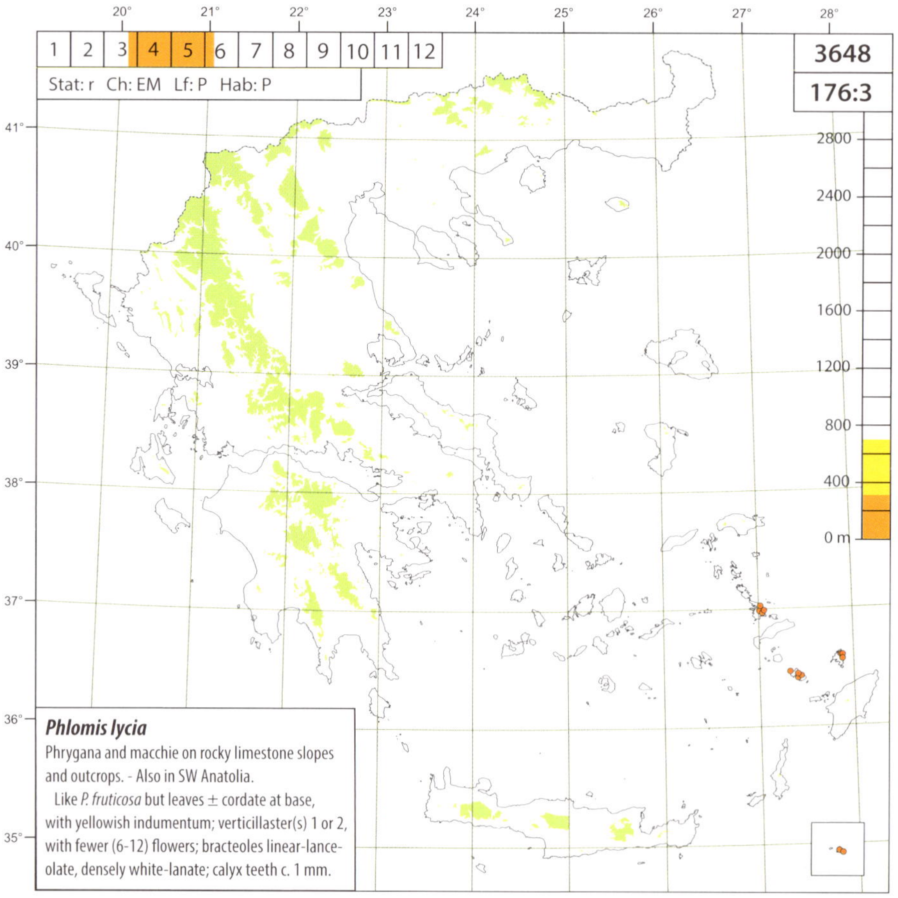

Phlomis lycia
Phrygana and macchie on rocky limestone slopes and outcrops. - Also in SW Anatolia.
Like *P. fruticosa* but leaves ± cordate at base, with yellowish indumentum; verticillaster(s) 1 or 2, with fewer (6-12) flowers; bracteoles linear-lanceolate, densely white-lanate; calyx teeth c. 1 mm.

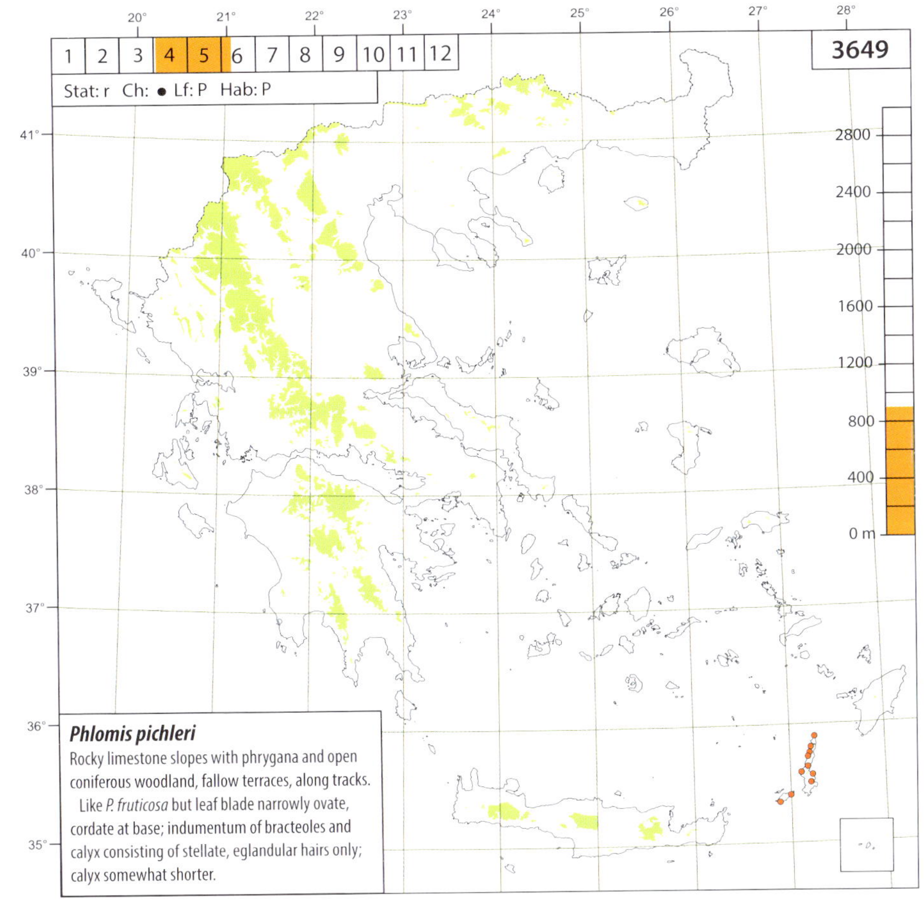

Phlomis pichleri
Rocky limestone slopes with phrygana and open coniferous woodland, fallow terraces, along tracks.
Like *P. fruticosa* but leaf blade narrowly ovate, cordate at base; indumentum of bracteoles and calyx consisting of stellate, eglandular hairs only; calyx somewhat shorter.

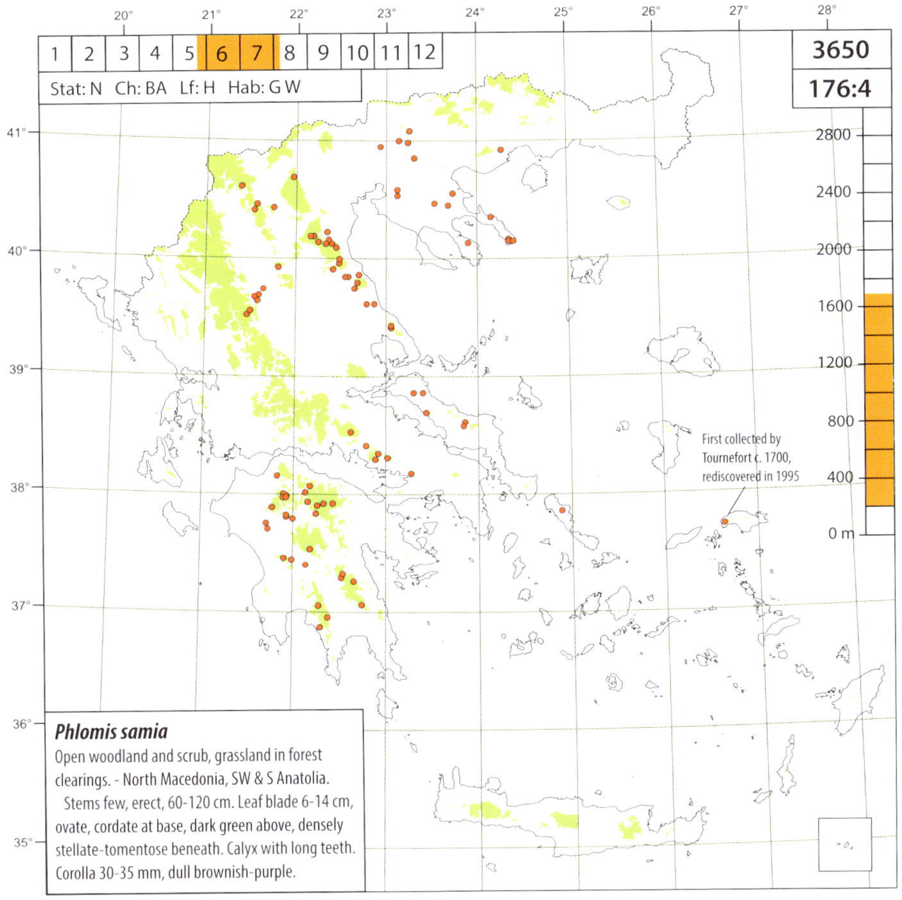

Phlomis samia
Open woodland and scrub, grassland in forest clearings. - North Macedonia, SW & S Anatolia.
Stems few, erect, 60-120 cm. Leaf blade 6-14 cm, ovate, cordate at base, dark green above, densely stellate-tomentose beneath. Calyx with long teeth. Corolla 30-35 mm, dull brownish-purple.

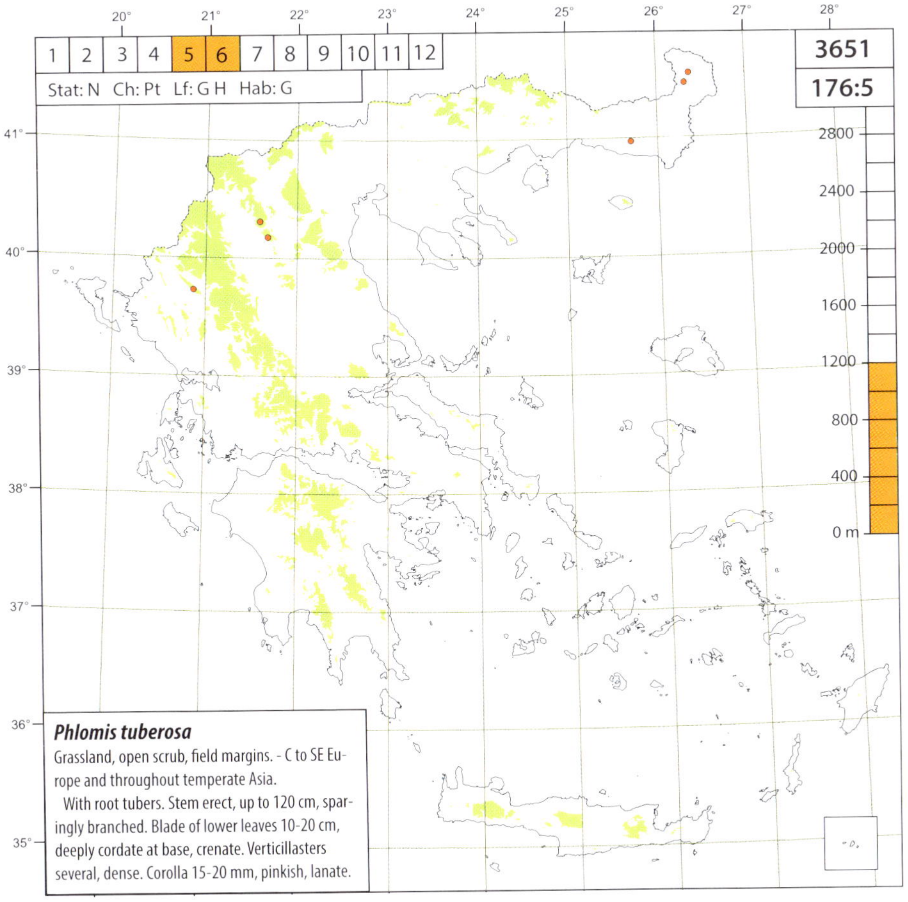

Phlomis tuberosa
Grassland, open scrub, field margins. - C to SE Europe and throughout temperate Asia.
With root tubers. Stem erect, up to 120 cm, sparingly branched. Blade of lower leaves 10-20 cm, deeply cordate at base, crenate. Verticillasters several, dense. Corolla 15-20 mm, pinkish, lanate.

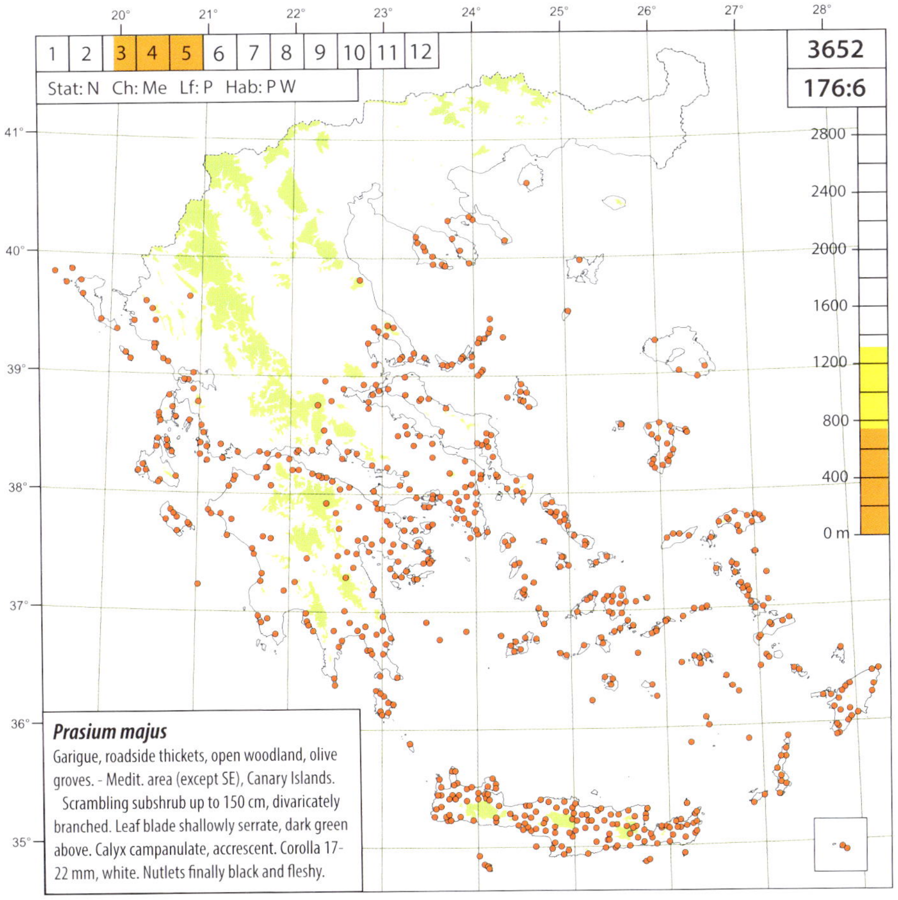

Prasium majus
Garigue, roadside thickets, open woodland, olive groves. - Medit. area (except SE), Canary Islands.
Scrambling subshrub up to 150 cm, divaricately branched. Leaf blade shallowly serrate, dark green above. Calyx campanulate, accrescent. Corolla 17-22 mm, white. Nutlets finally black and fleshy.

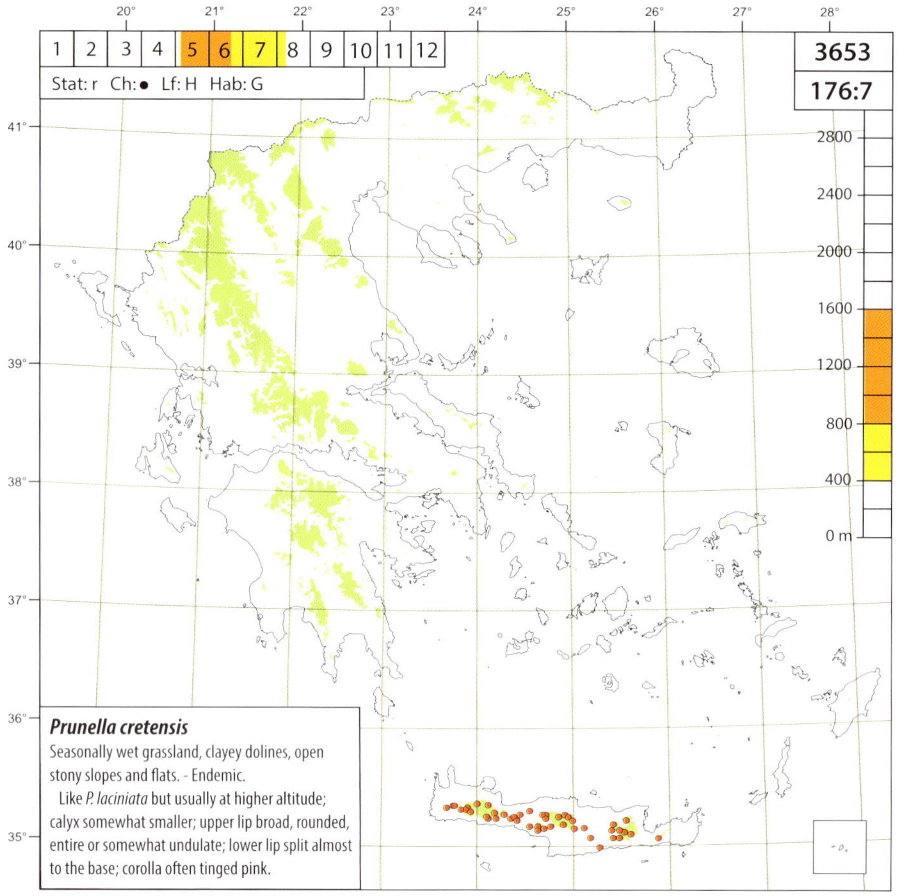

Prunella cretensis

Seasonally wet grassland, clayey dolines, open stony slopes and flats. - Endemic.

Like *P. laciniata* but usually at higher altitude; calyx somewhat smaller; upper lip broad, rounded, entire or somewhat undulate; lower lip split almost to the base; corolla often tinged pink.

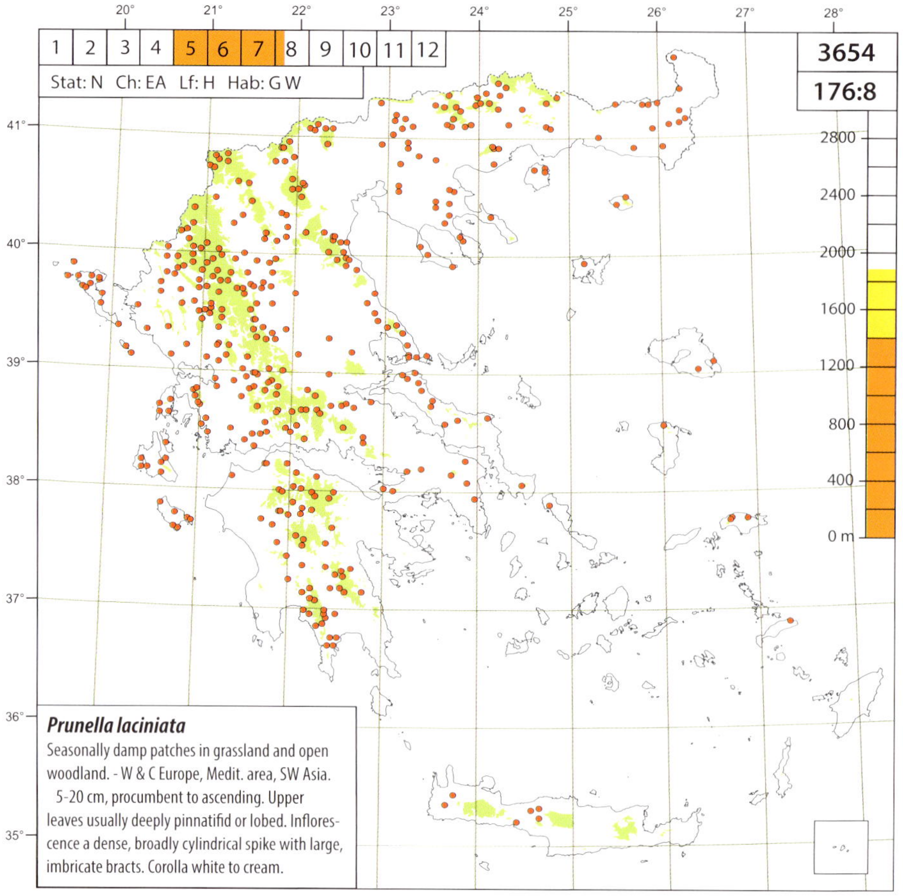

Prunella laciniata

Seasonally damp patches in grassland and open woodland. - W & C Europe, Medit. area, SW Asia.

5-20 cm, procumbent to ascending. Upper leaves usually deeply pinnatifid or lobed. Inflorescence a dense, broadly cylindrical spike with large, imbricate bracts. Corolla white to cream.

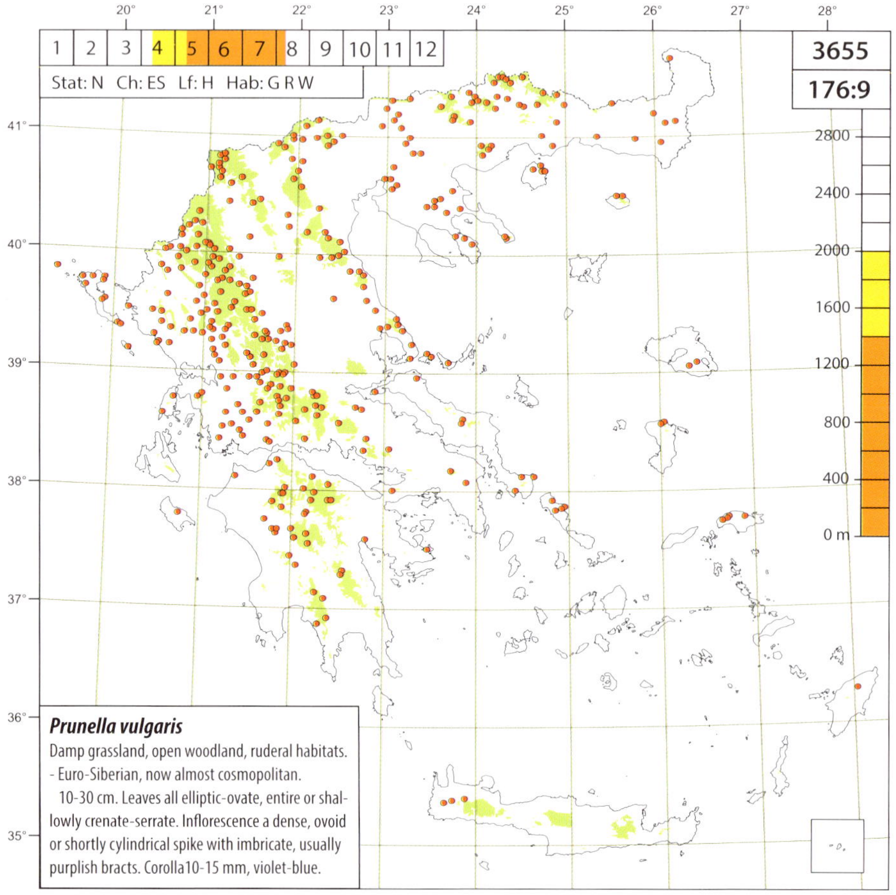

Prunella vulgaris

Damp grassland, open woodland, ruderal habitats. - Euro-Siberian, now almost cosmopolitan.

10-30 cm. Leaves all elliptic-ovate, entire or shallowly crenate-serrate. Inflorescence a dense, ovoid or shortly cylindrical spike with imbricate, usually purplish bracts. Corolla10-15 mm, violet-blue.

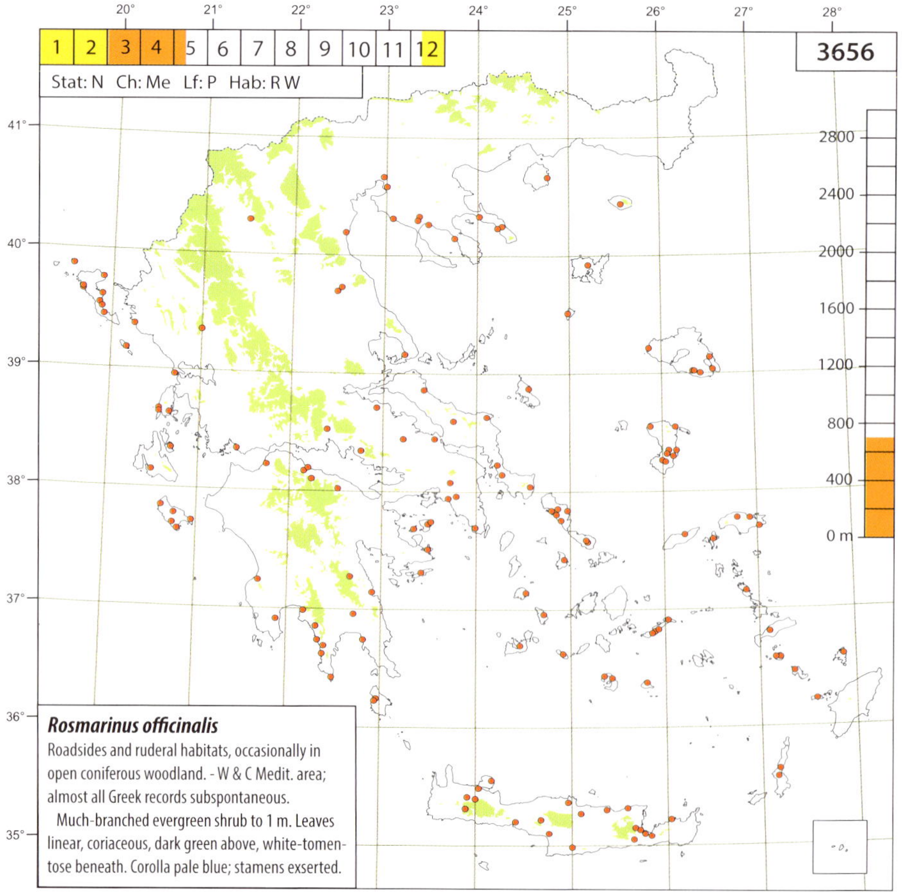

Rosmarinus officinalis

Roadsides and ruderal habitats, occasionally in open coniferous woodland. - W & C Medit. area; almost all Greek records subspontaneous.

Much-branched evergreen shrub to 1 m. Leaves linear, coriaceous, dark green above, white-tomentose beneath. Corolla pale blue; stamens exserted.

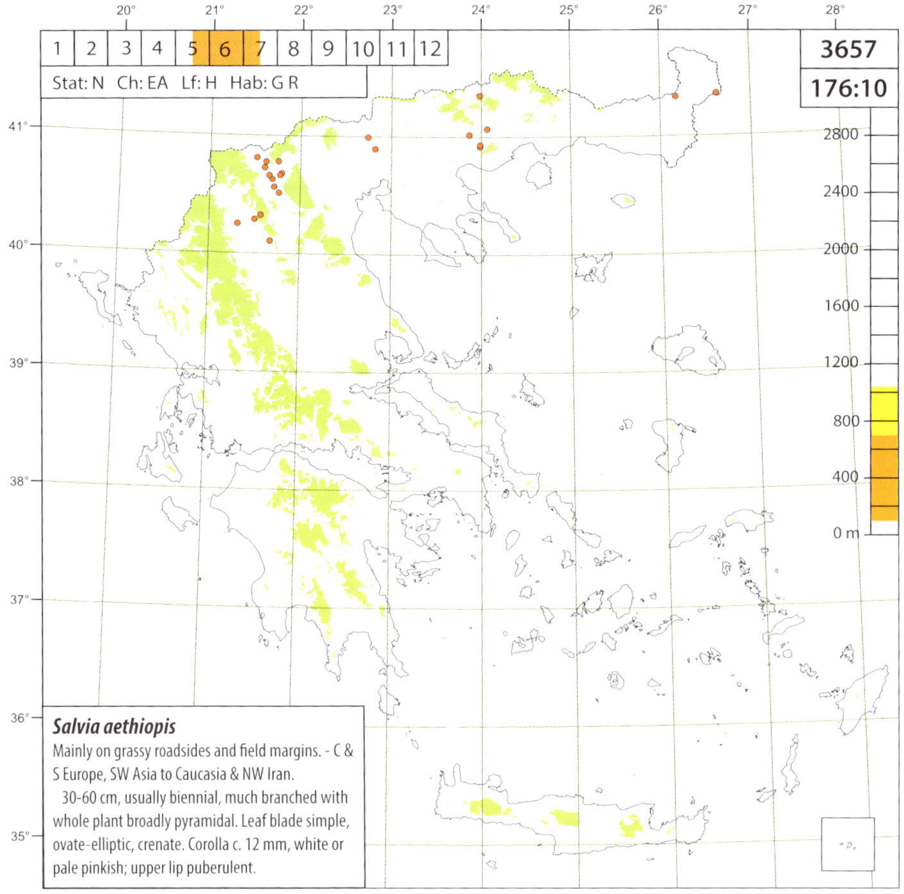

Salvia aethiopis
Mainly on grassy roadsides and field margins. - C & S Europe, SW Asia to Caucasia & NW Iran.
30-60 cm, usually biennial, much branched with whole plant broadly pyramidal. Leaf blade simple, ovate-elliptic, crenate. Corolla c. 12 mm, white or pale pinkish; upper lip puberulent.

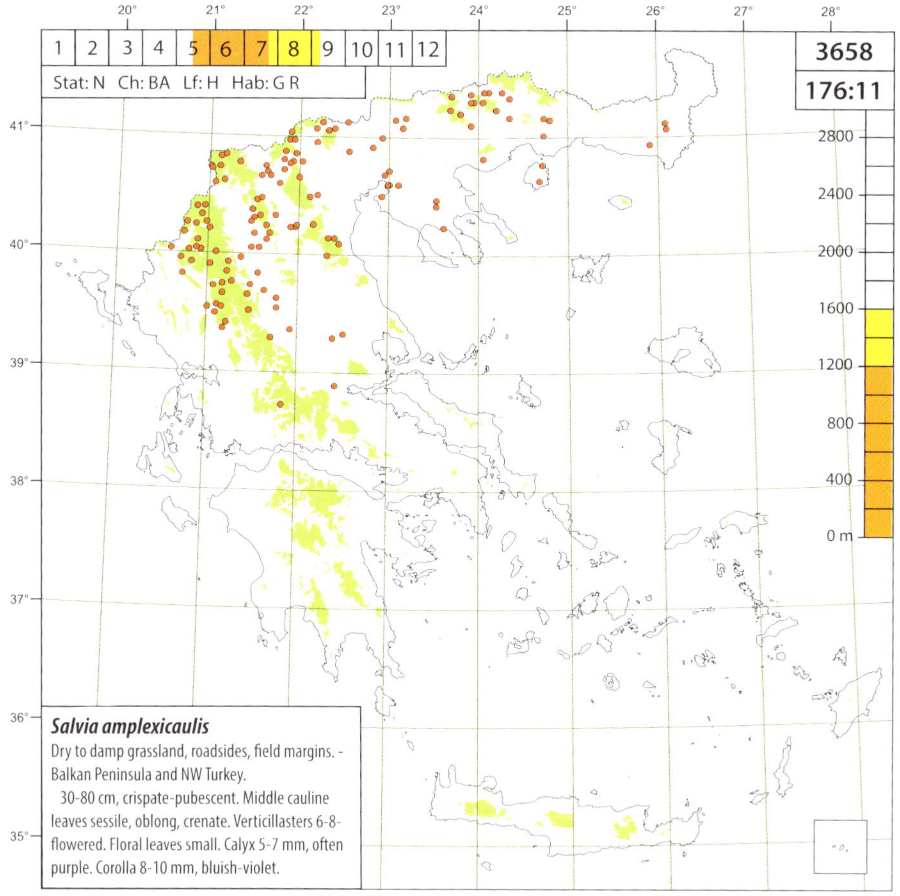

Salvia amplexicaulis
Dry to damp grassland, roadsides, field margins. - Balkan Peninsula and NW Turkey.
30-80 cm, crispate-pubescent. Middle cauline leaves sessile, oblong, crenate. Verticillasters 6-8-flowered. Floral leaves small. Calyx 5-7 mm, often purple. Corolla 8-10 mm, bluish-violet.

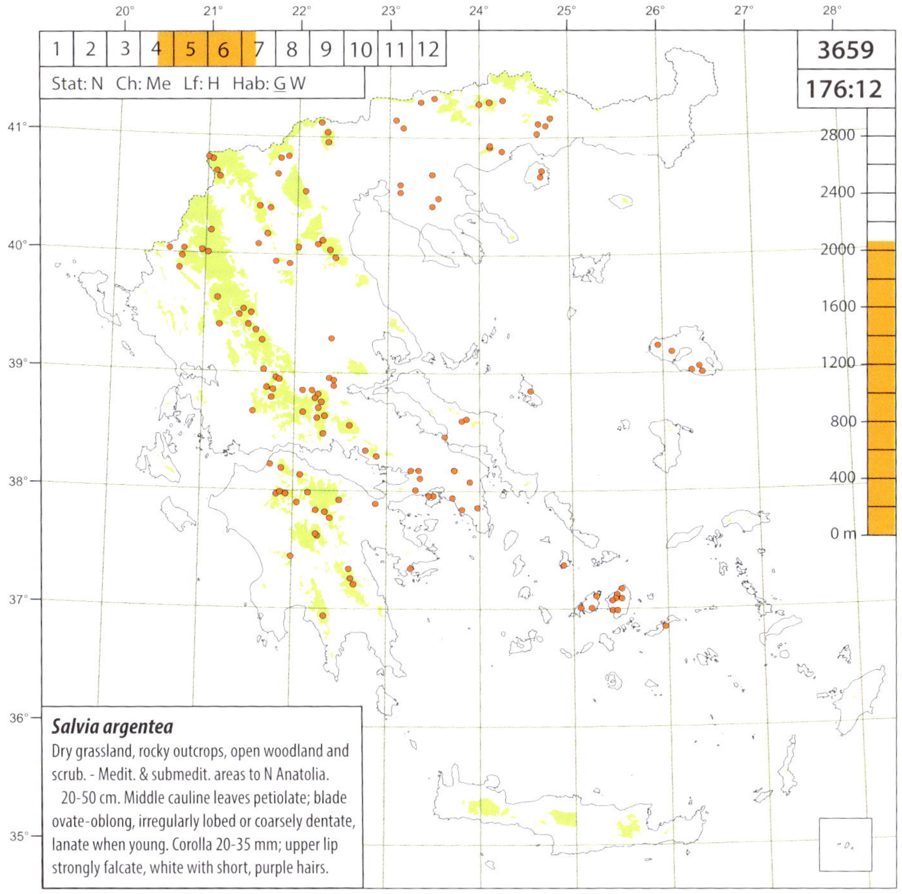

Salvia argentea
Dry grassland, rocky outcrops, open woodland and scrub. - Medit. & submedit. areas to N Anatolia.
20-50 cm. Middle cauline leaves petiolate; blade ovate-oblong, irregularly lobed or coarsely dentate, lanate when young. Corolla 20-35 mm; upper lip strongly falcate, white with short, purple hairs.

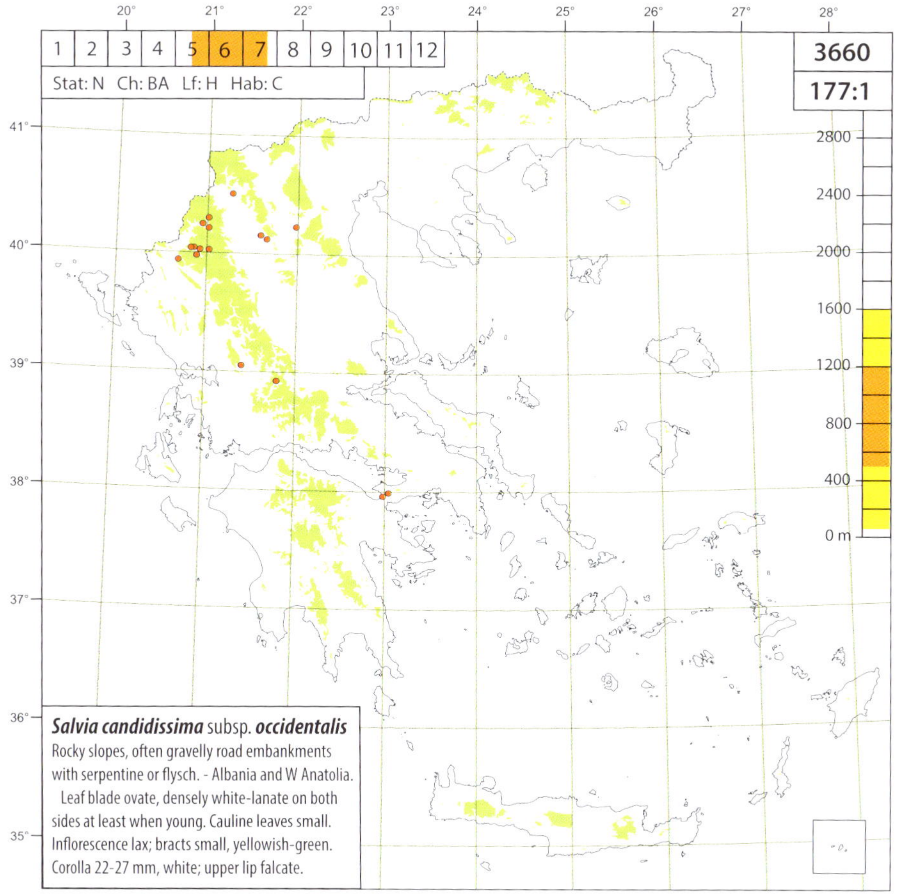

Salvia candidissima **subsp. *occidentalis***
Rocky slopes, often gravelly road embankments with serpentine or flysch. - Albania and W Anatolia.
Leaf blade ovate, densely white-lanate on both sides at least when young. Cauline leaves small. Inflorescence lax; bracts small, yellowish-green. Corolla 22-27 mm, white; upper lip falcate.

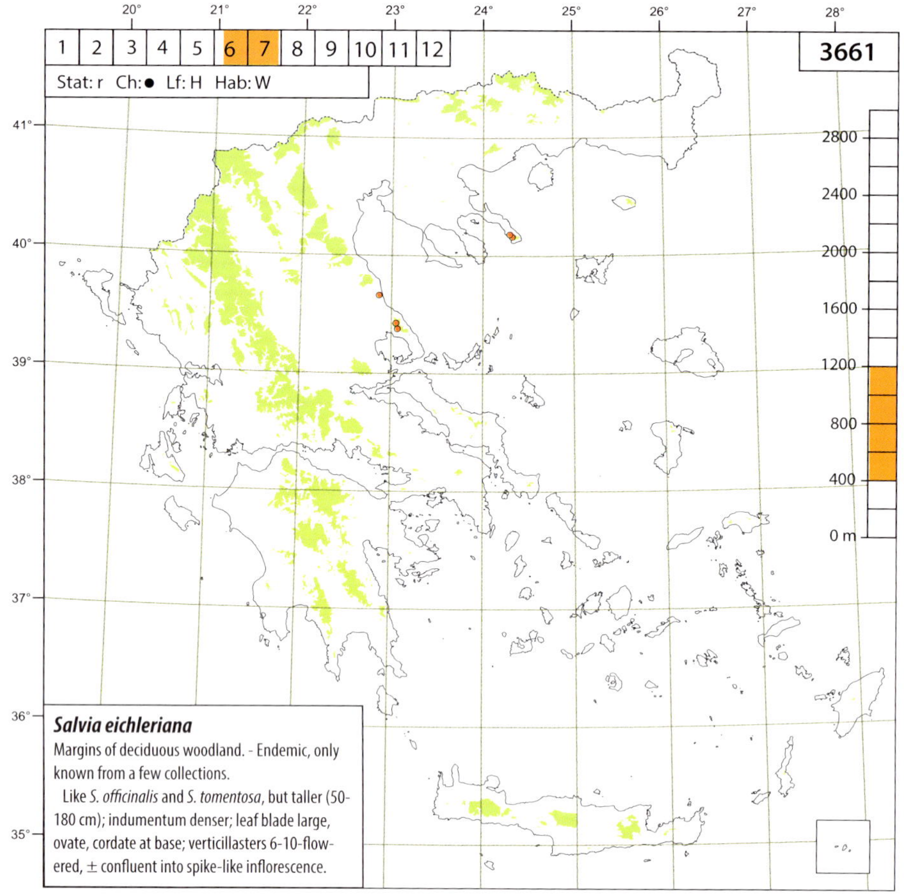

Salvia eichleriana

Margins of deciduous woodland. - Endemic, only known from a few collections.

Like *S. officinalis* and *S. tomentosa*, but taller (50-180 cm); indumentum denser; leaf blade large, ovate, cordate at base; verticillasters 6-10-flowered, ± confluent into spike-like inflorescence.

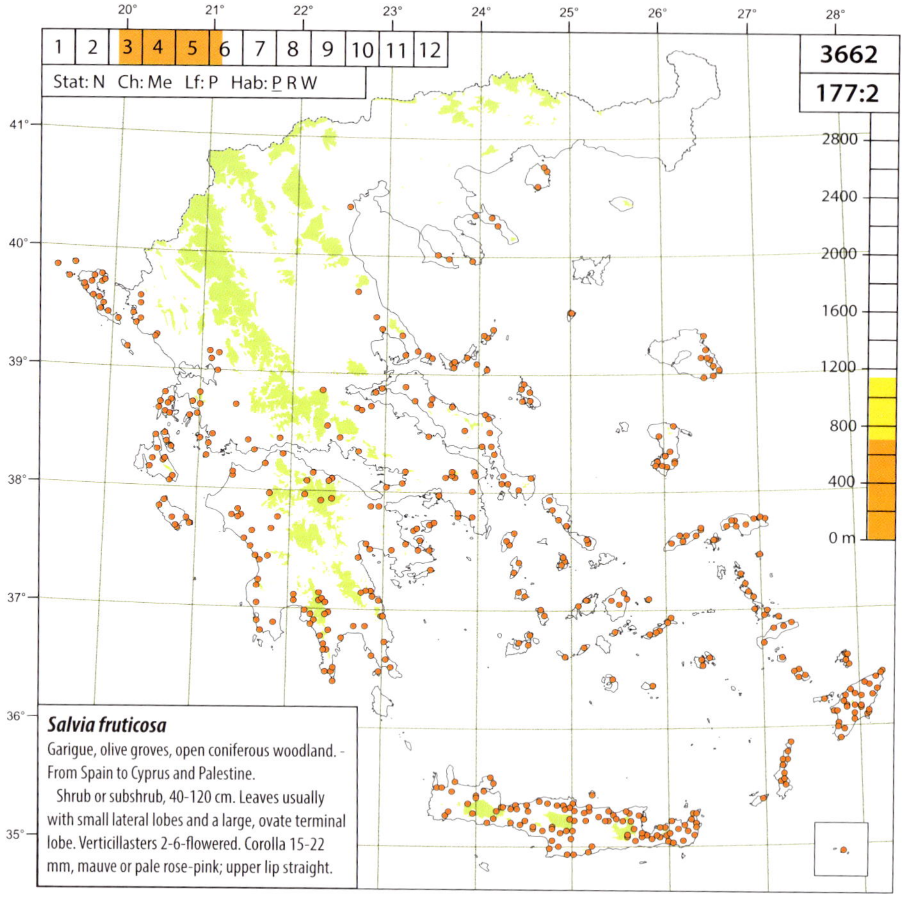

Salvia fruticosa

Garigue, olive groves, open coniferous woodland. - From Spain to Cyprus and Palestine.

Shrub or subshrub, 40-120 cm. Leaves usually with small lateral lobes and a large, ovate terminal lobe. Verticillasters 2-6-flowered. Corolla 15-22 mm, mauve or pale rose-pink; upper lip straight.

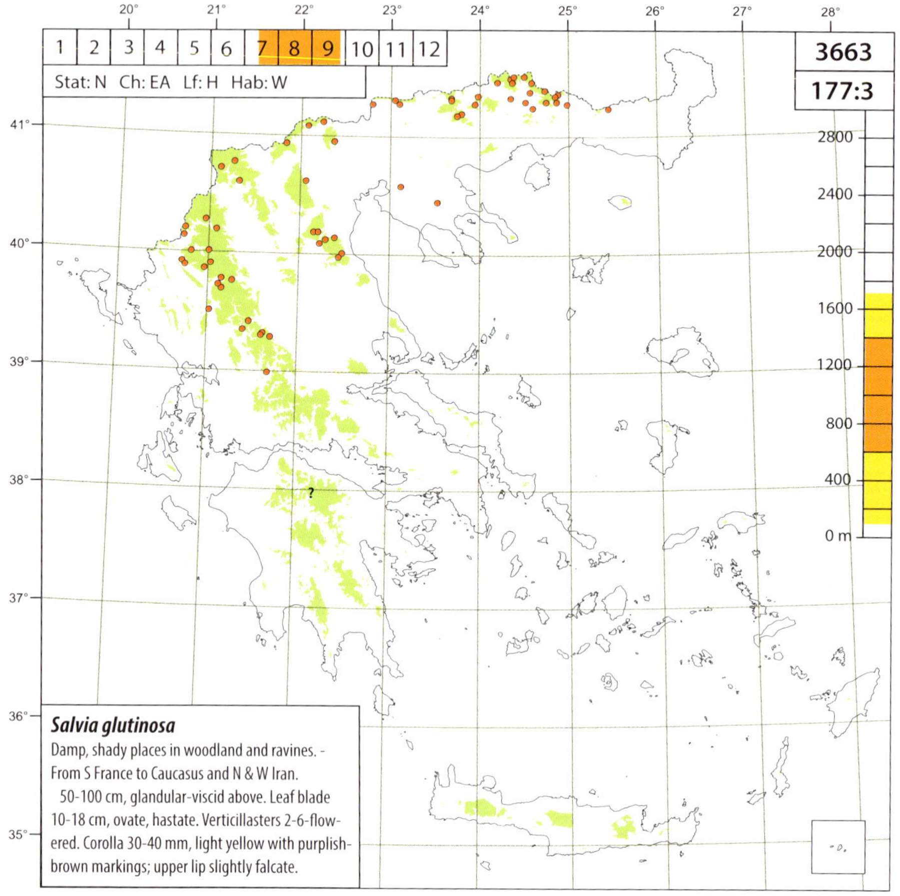

Salvia glutinosa

Damp, shady places in woodland and ravines. - From S France to Caucasus and N & W Iran.

50-100 cm, glandular-viscid above. Leaf blade 10-18 cm, ovate, hastate. Verticillasters 2-6-flowered. Corolla 30-40 mm, light yellow with purplish-brown markings; upper lip slightly falcate.

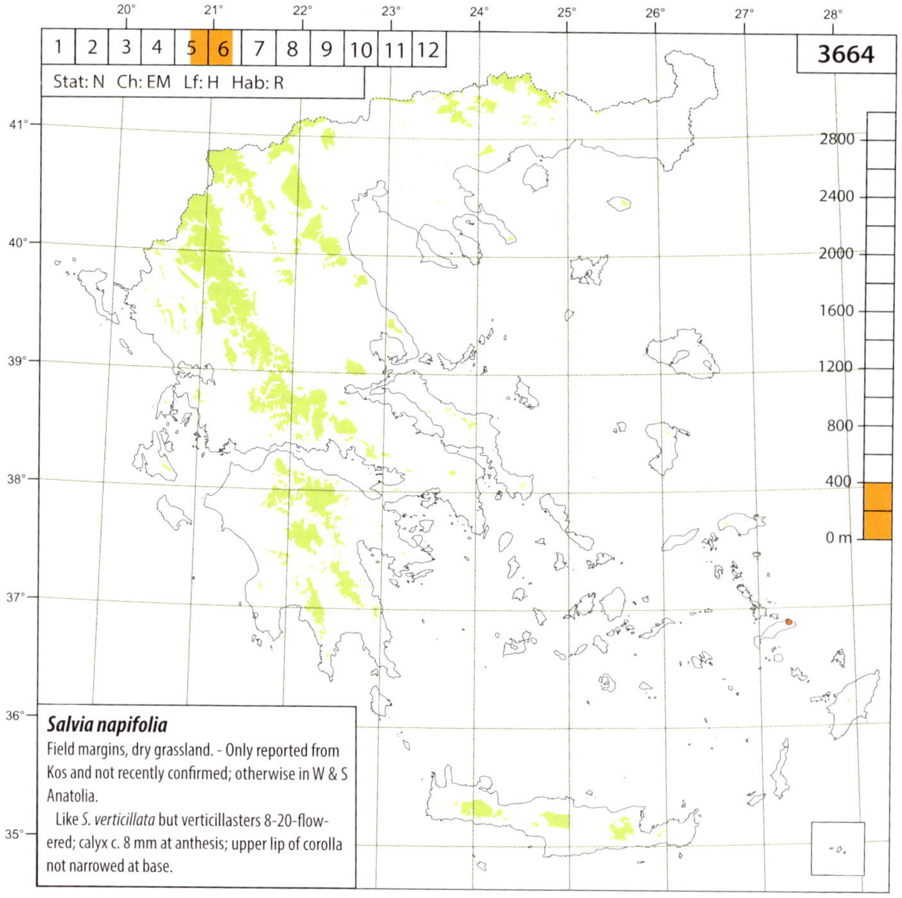

Salvia napifolia

Field margins, dry grassland. - Only reported from Kos and not recently confirmed; otherwise in W & S Anatolia.

Like *S. verticillata* but verticillasters 8-20-flowered; calyx c. 8 mm at anthesis; upper lip of corolla not narrowed at base.

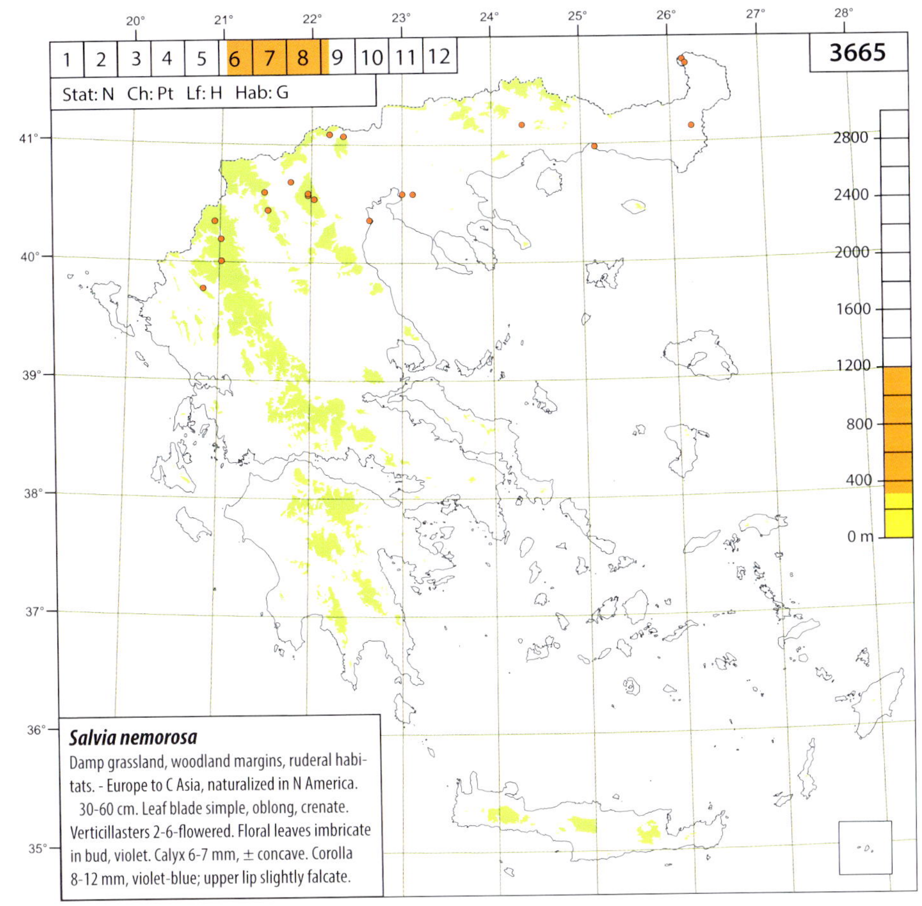

Salvia nemorosa
Damp grassland, woodland margins, ruderal habitats. - Europe to C Asia, naturalized in N America.
30-60 cm. Leaf blade simple, oblong, crenate. Verticillasters 2-6-flowered. Floral leaves imbricate in bud, violet. Calyx 6-7 mm, ± concave. Corolla 8-12 mm, violet-blue; upper lip slightly falcate.

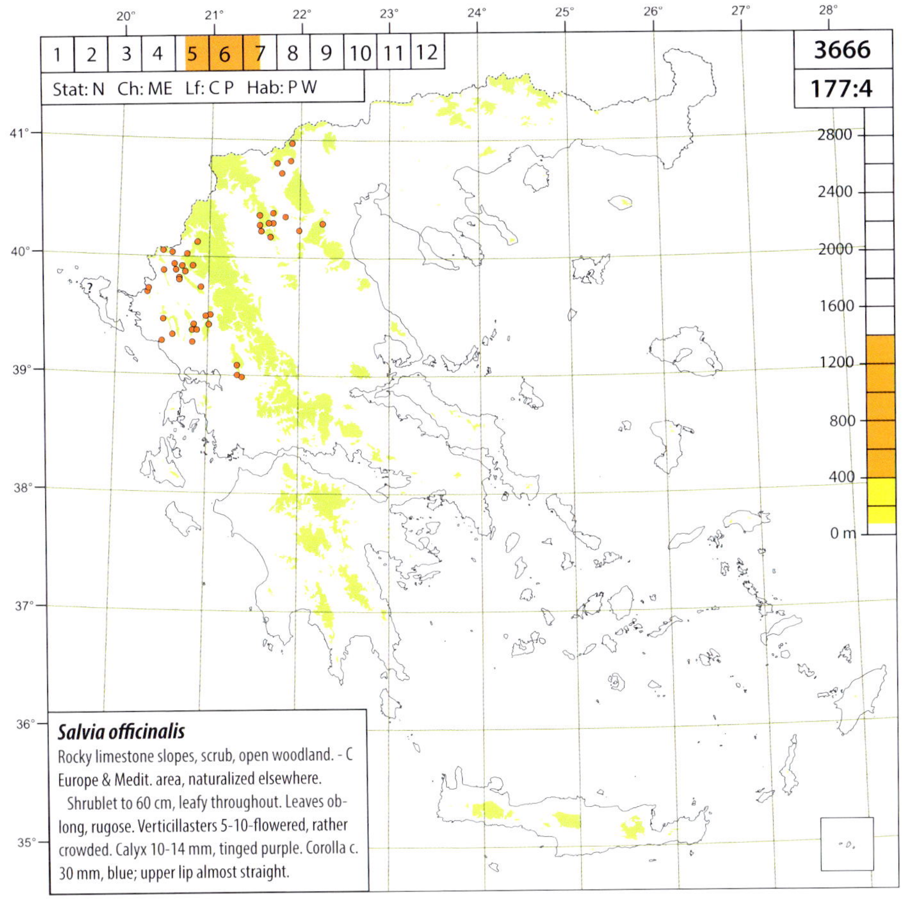

Salvia officinalis
Rocky limestone slopes, scrub, open woodland. - C Europe & Medit. area, naturalized elsewhere.
Shrublet to 60 cm, leafy throughout. Leaves oblong, rugose. Verticillasters 5-10-flowered, rather crowded. Calyx 10-14 mm, tinged purple. Corolla c. 30 mm, blue; upper lip almost straight.

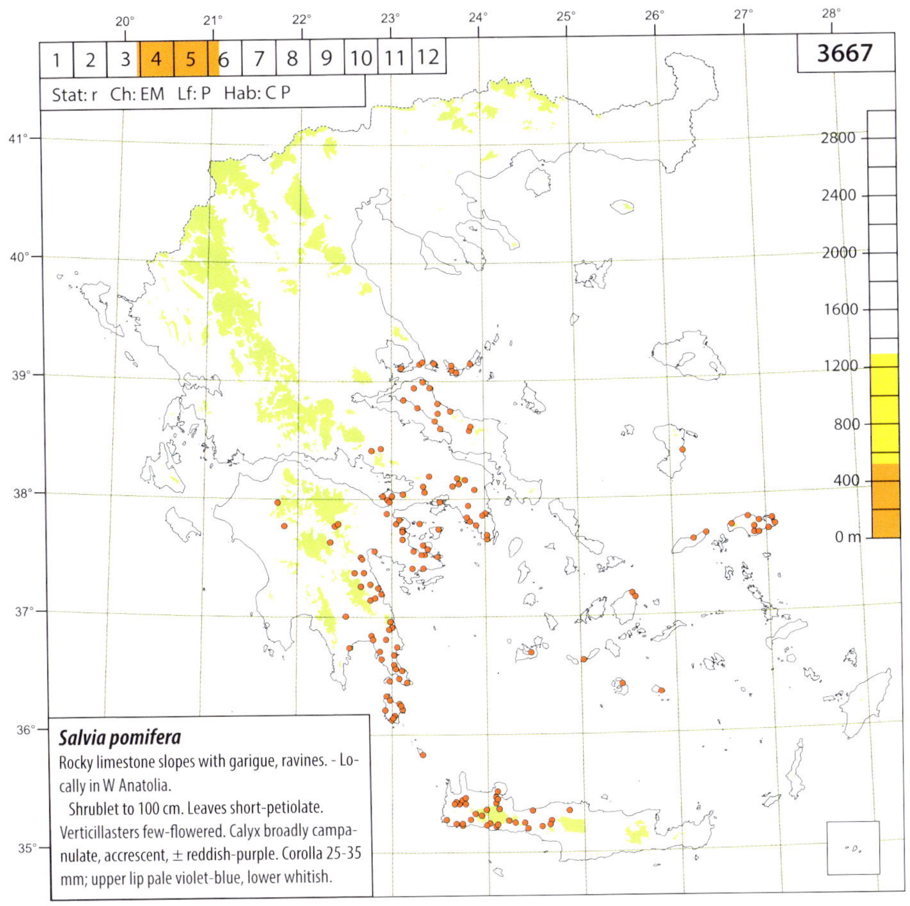

Salvia pomifera
Rocky limestone slopes with garigue, ravines. - Locally in W Anatolia.
Shrublet to 100 cm. Leaves short-petiolate. Verticillasters few-flowered. Calyx broadly campanulate, accrescent, ± reddish-purple. Corolla 25-35 mm; upper lip pale violet-blue, lower whitish.

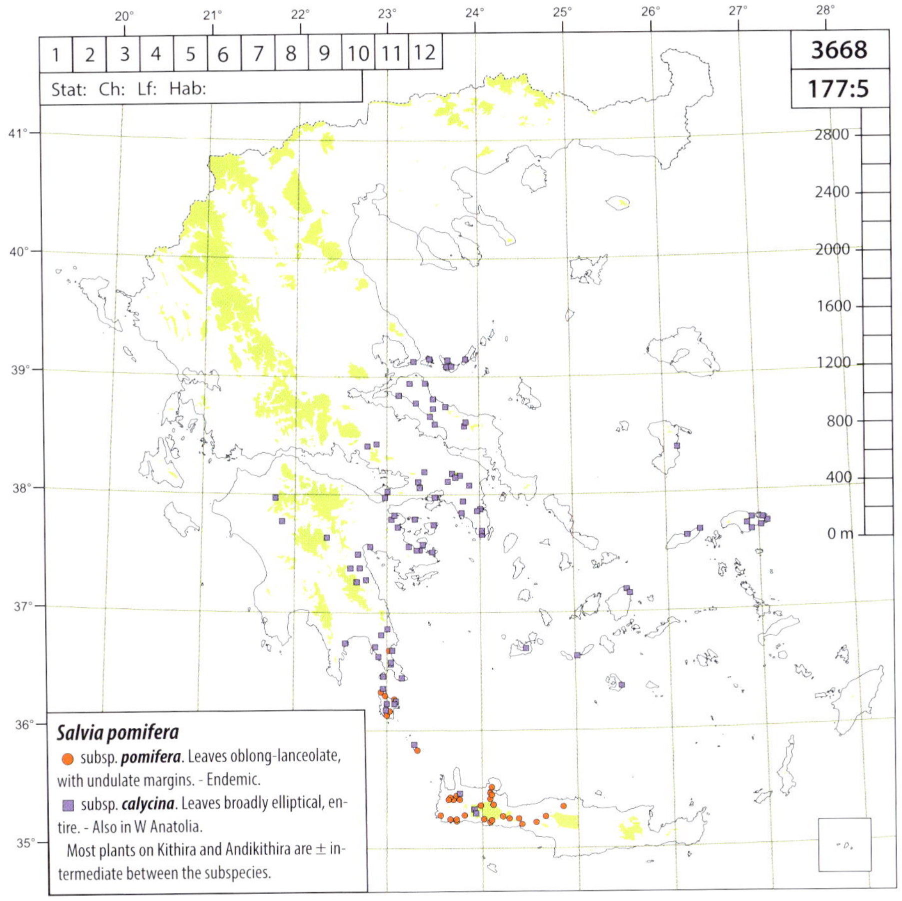

Salvia pomifera
● subsp. ***pomifera***. Leaves oblong-lanceolate, with undulate margins. - Endemic.
■ subsp. ***calycina***. Leaves broadly elliptical, entire. - Also in W Anatolia.
Most plants on Kithira and Andikithira are ± intermediate between the subspecies.

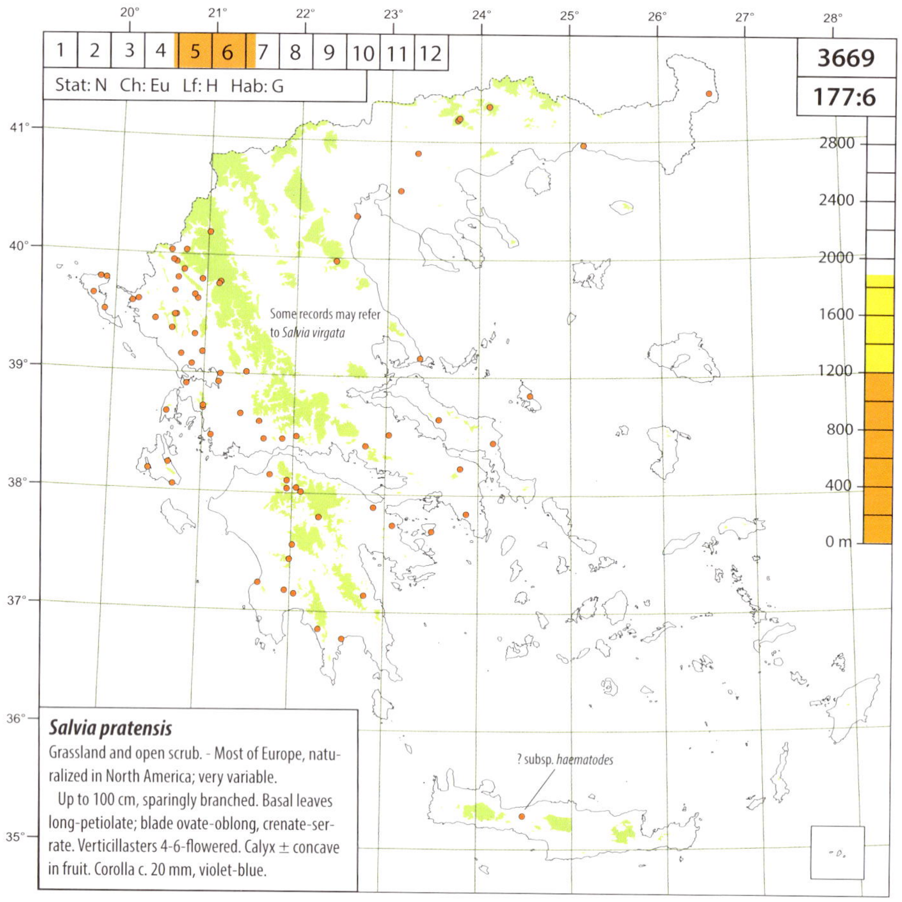

Salvia pratensis

Grassland and open scrub. - Most of Europe, naturalized in North America; very variable.

Up to 100 cm, sparingly branched. Basal leaves long-petiolate; blade ovate-oblong, crenate-serrate. Verticillasters 4-6-flowered. Calyx ± concave in fruit. Corolla c. 20 mm, violet-blue.

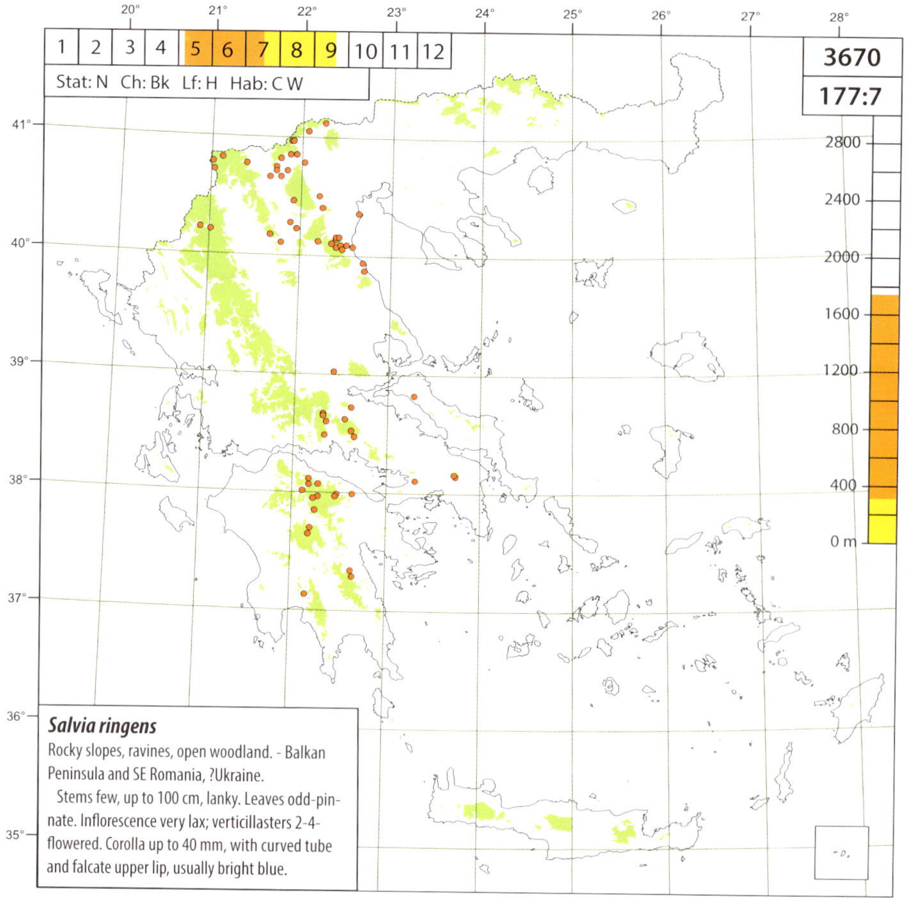

Salvia ringens

Rocky slopes, ravines, open woodland. - Balkan Peninsula and SE Romania, ?Ukraine.

Stems few, up to 100 cm, lanky. Leaves odd-pinnate. Inflorescence very lax; verticillasters 2-4-flowered. Corolla up to 40 mm, with curved tube and falcate upper lip, usually bright blue.

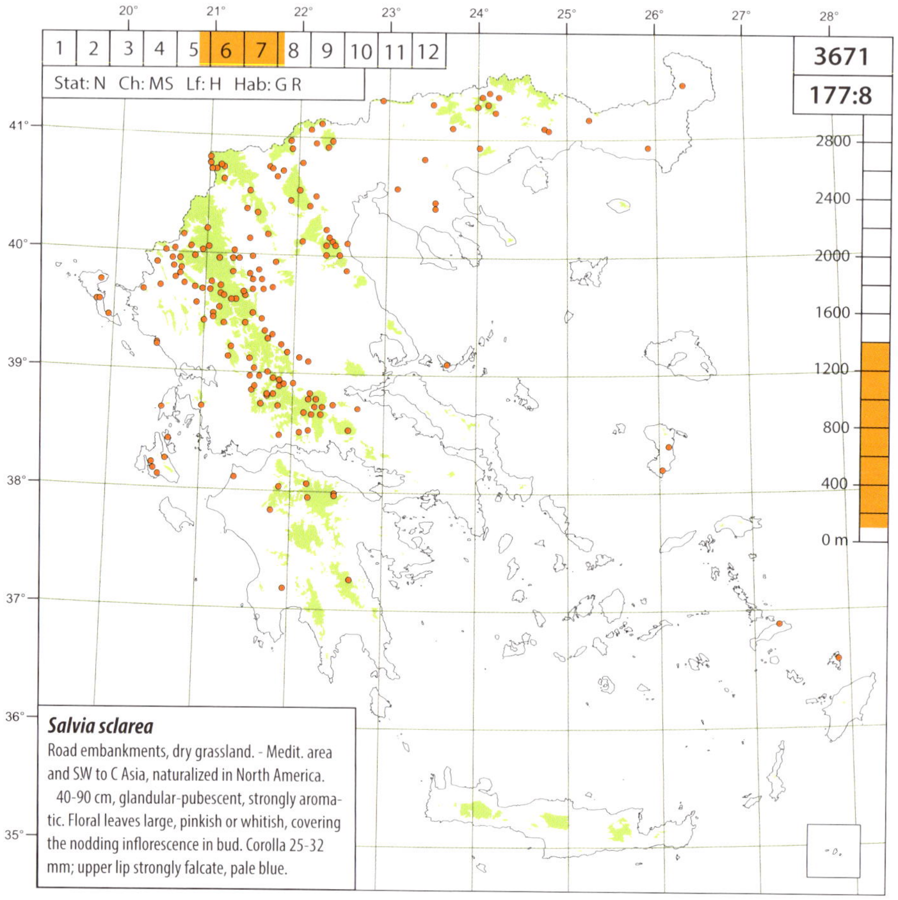

Salvia sclarea

Road embankments, dry grassland. - Medit. area and SW to C Asia, naturalized in North America.

40-90 cm, glandular-pubescent, strongly aromatic. Floral leaves large, pinkish or whitish, covering the nodding inflorescence in bud. Corolla 25-32 mm; upper lip strongly falcate, pale blue.

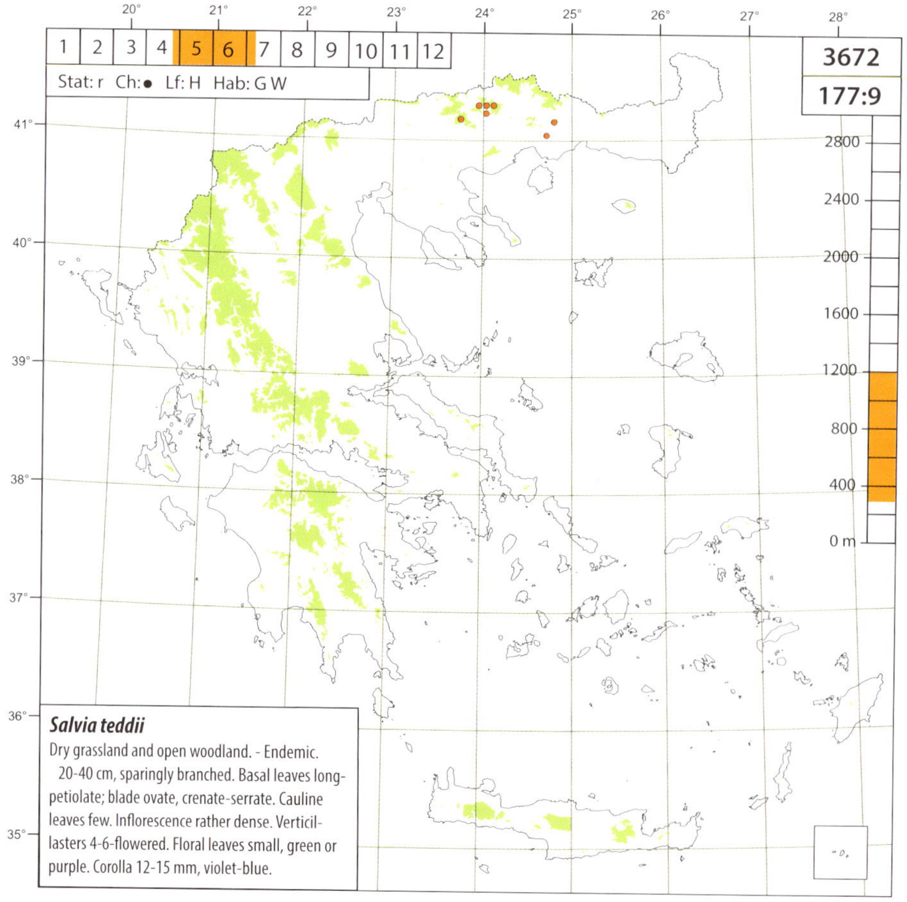

Salvia teddii

Dry grassland and open woodland. - Endemic.

20-40 cm, sparingly branched. Basal leaves long-petiolate; blade ovate, crenate-serrate. Cauline leaves few. Inflorescence rather dense. Verticillasters 4-6-flowered. Floral leaves small, green or purple. Corolla 12-15 mm, violet-blue.

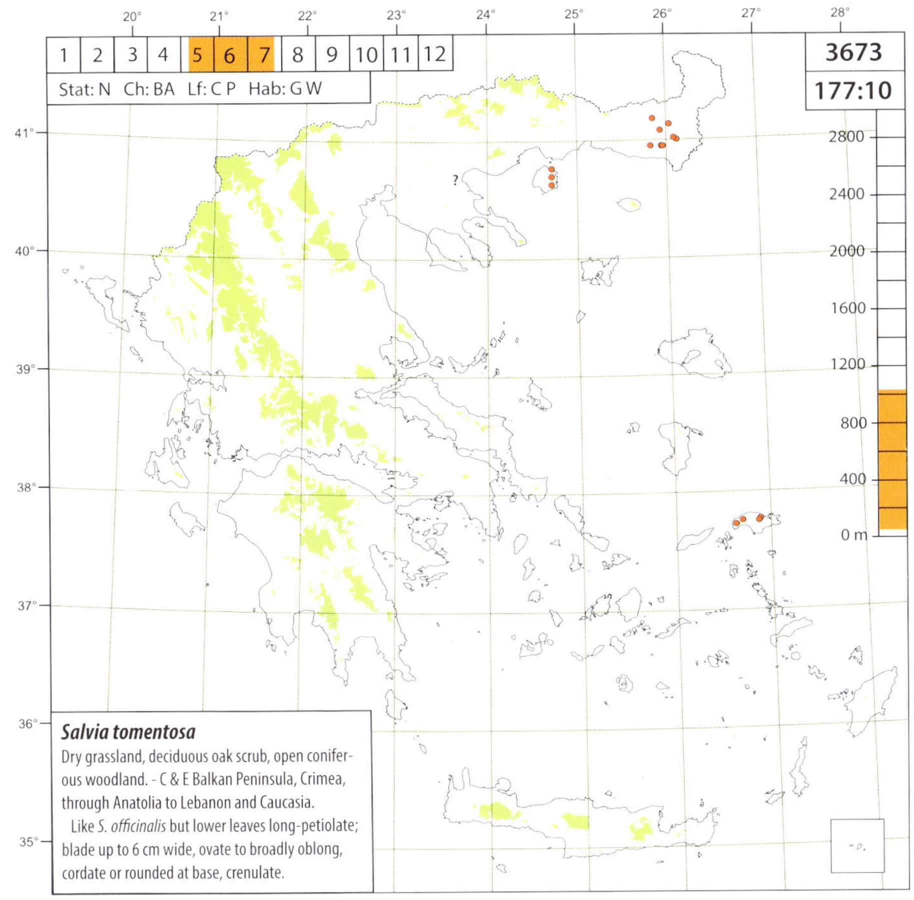

Salvia tomentosa
Dry grassland, deciduous oak scrub, open coniferous woodland. - C & E Balkan Peninsula, Crimea, through Anatolia to Lebanon and Caucasia.
Like *S. officinalis* but lower leaves long-petiolate; blade up to 6 cm wide, ovate to broadly oblong, cordate or rounded at base, crenulate.

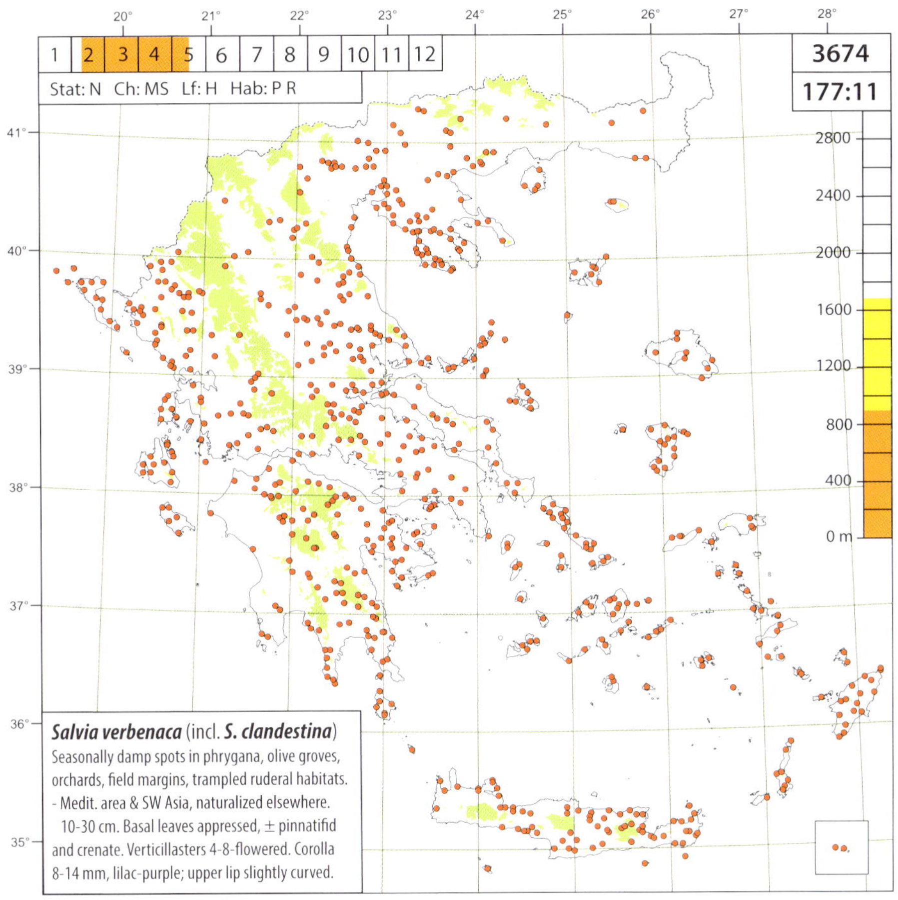

Salvia verbenaca (incl. ***S. clandestina***)
Seasonally damp spots in phrygana, olive groves, orchards, field margins, trampled ruderal habitats. - Medit. area & SW Asia, naturalized elsewhere.
10-30 cm. Basal leaves appressed, ± pinnatifid and crenate. Verticillasters 4-8-flowered. Corolla 8-14 mm, lilac-purple; upper lip slightly curved.

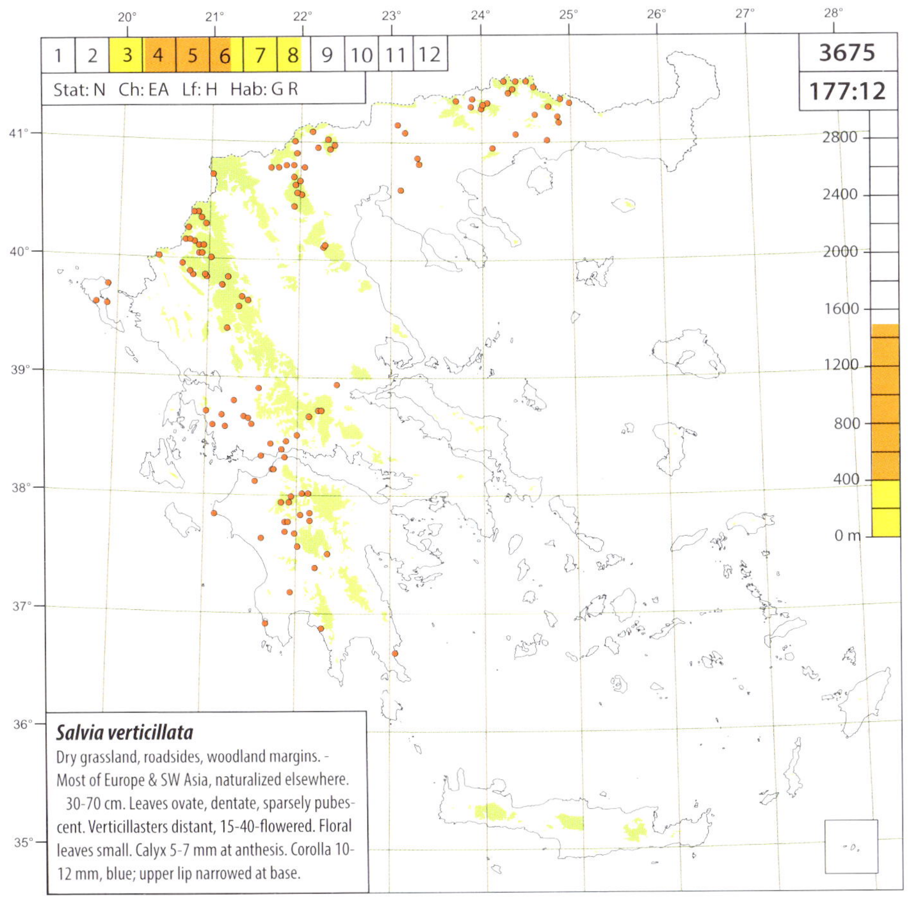

Salvia verticillata
Dry grassland, roadsides, woodland margins. - Most of Europe & SW Asia, naturalized elsewhere.
30-70 cm. Leaves ovate, dentate, sparsely pubescent. Verticillasters distant, 15-40-flowered. Floral leaves small. Calyx 5-7 mm at anthesis. Corolla 10-12 mm, blue; upper lip narrowed at base.

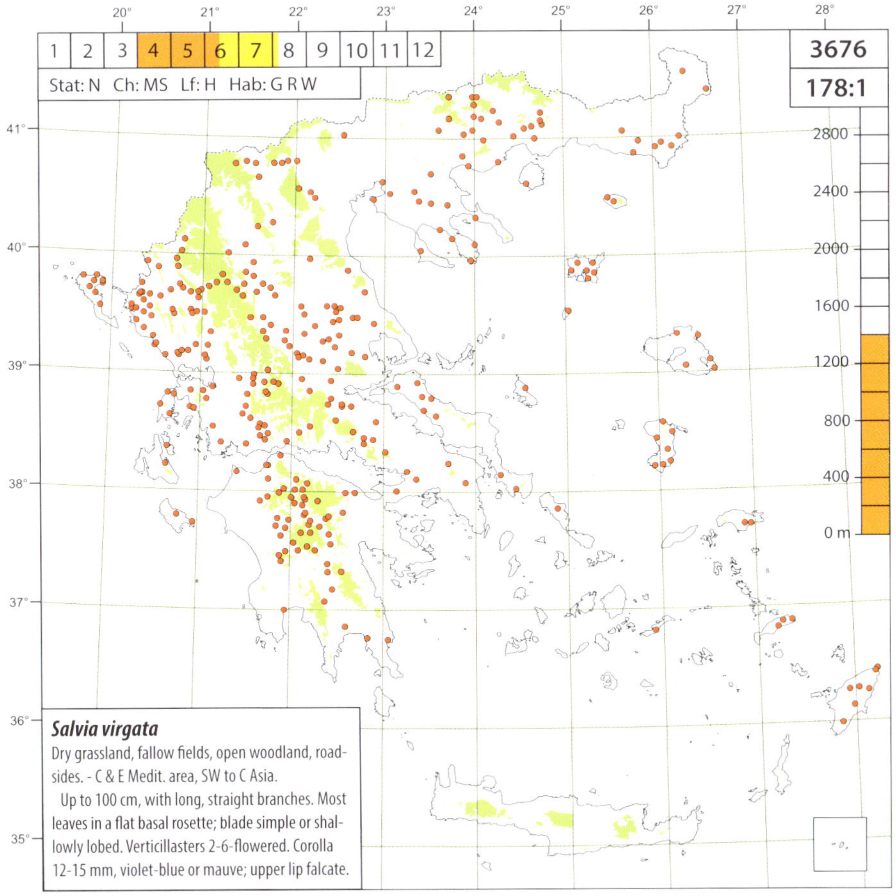

Salvia virgata
Dry grassland, fallow fields, open woodland, roadsides. - C & E Medit. area, SW to C Asia.
Up to 100 cm, with long, straight branches. Most leaves in a flat basal rosette; blade simple or shallowly lobed. Verticillasters 2-6-flowered. Corolla 12-15 mm, violet-blue or mauve; upper lip falcate.

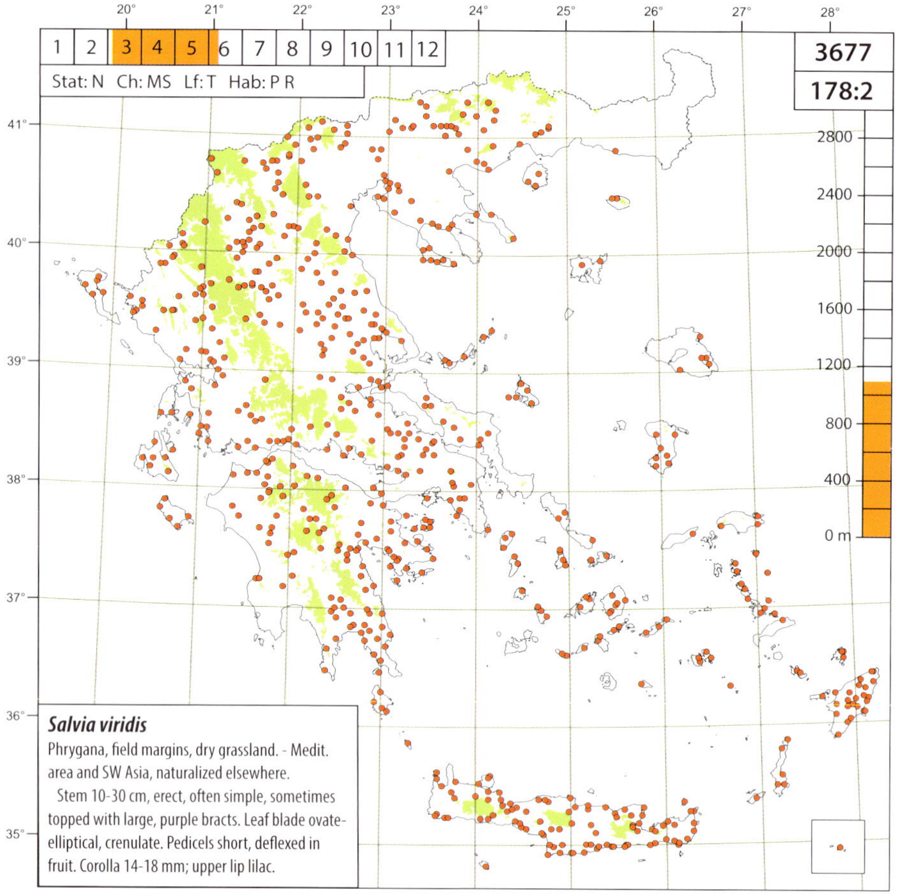

Salvia viridis

Phrygana, field margins, dry grassland. - Medit. area and SW Asia, naturalized elsewhere.

Stem 10-30 cm, erect, often simple, sometimes topped with large, purple bracts. Leaf blade ovate-elliptical, crenulate. Pedicels short, deflexed in fruit. Corolla 14-18 mm; upper lip lilac.

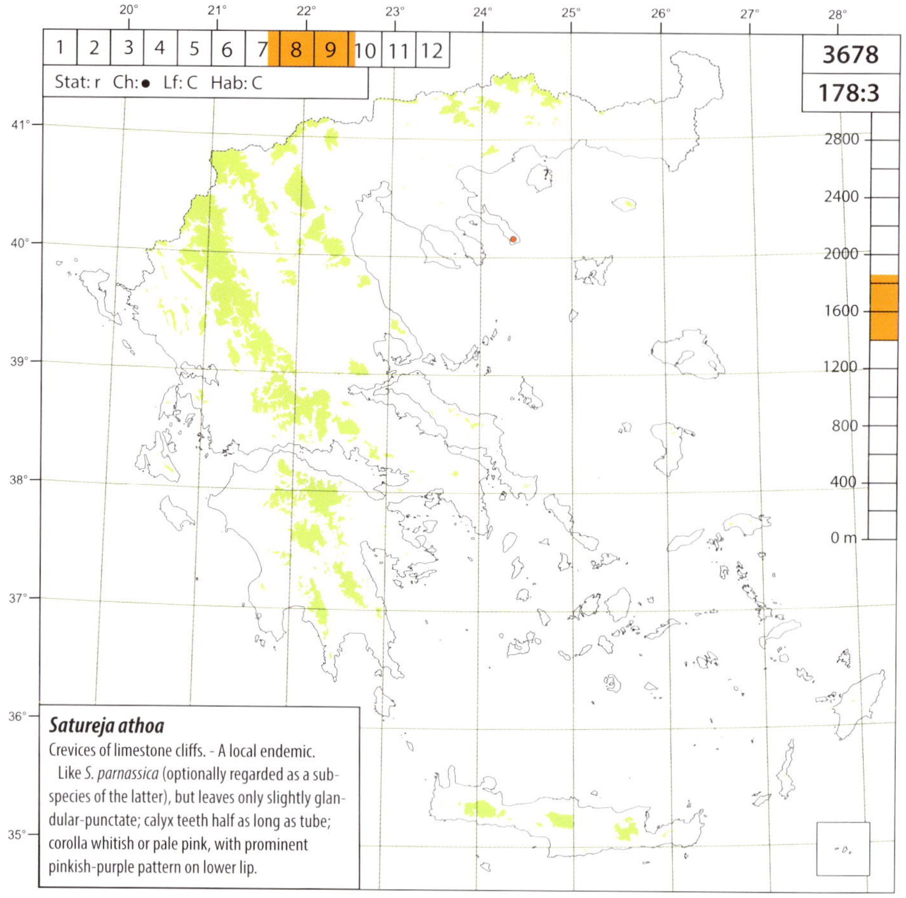

Satureja athoa

Crevices of limestone cliffs. - A local endemic.

Like *S. parnassica* (optionally regarded as a subspecies of the latter), but leaves only slightly glandular-punctate; calyx teeth half as long as tube; corolla whitish or pale pink, with prominent pinkish-purple pattern on lower lip.

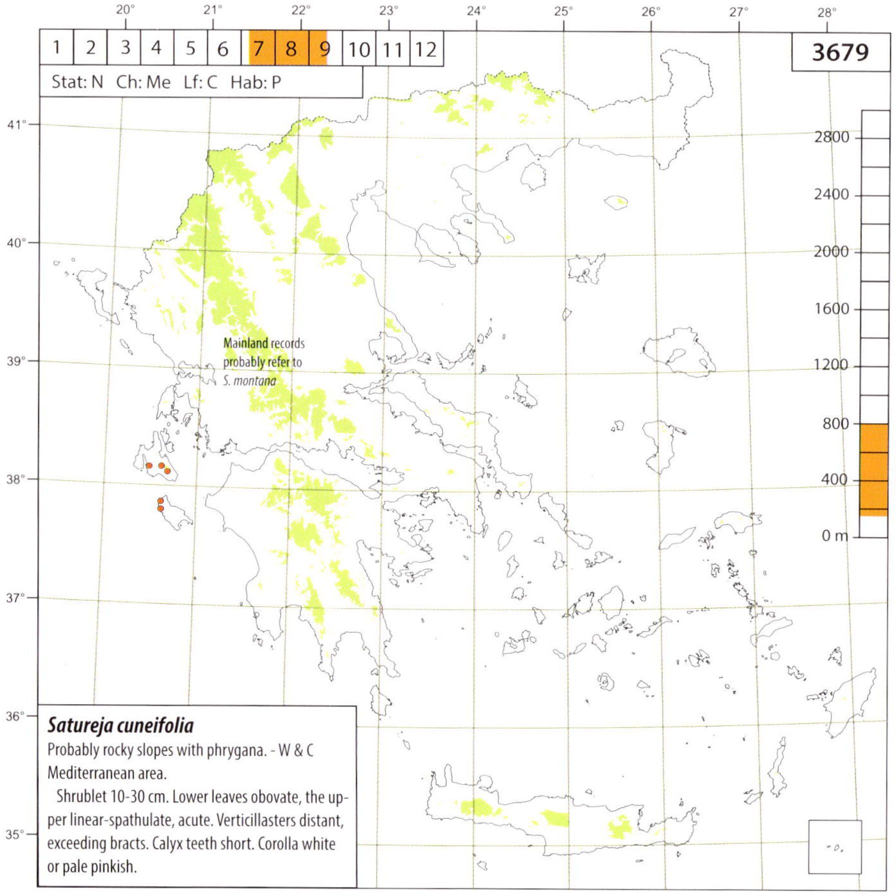

Satureja cuneifolia

Probably rocky slopes with phrygana. - W & C Mediterranean area.

Shrublet 10-30 cm. Lower leaves obovate, the upper linear-spathulate, acute. Verticillasters distant, exceeding bracts. Calyx teeth short. Corolla white or pale pinkish.

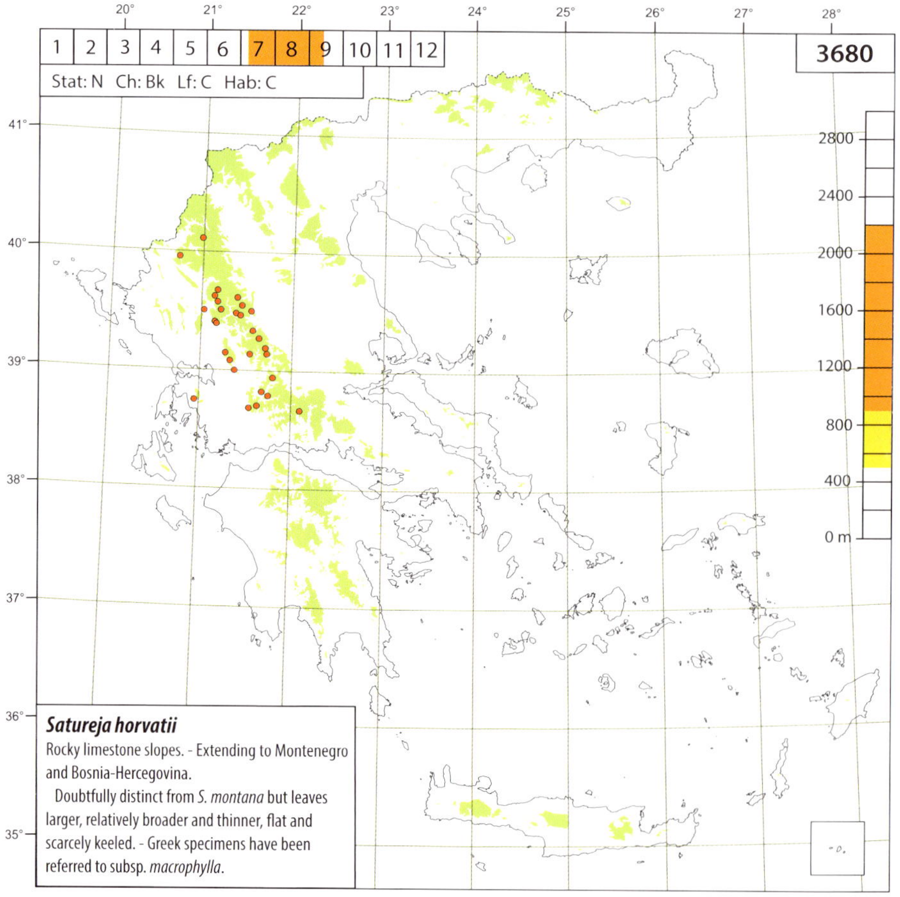

Satureja horvatii

Rocky limestone slopes. - Extending to Montenegro and Bosnia-Hercegovina.

Doubtfully distinct from *S. montana* but leaves larger, relatively broader and thinner, flat and scarcely keeled. - Greek specimens have been referred to subsp. *macrophylla*.

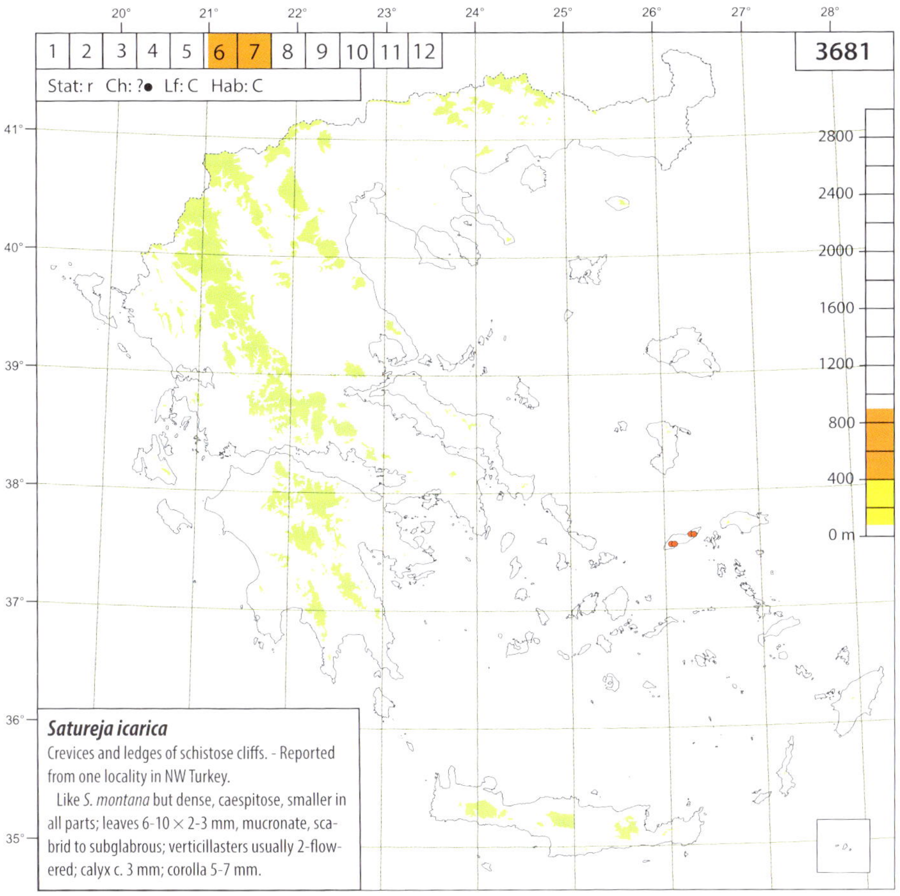

Satureja icarica
Crevices and ledges of schistose cliffs. - Reported from one locality in NW Turkey.
Like *S. montana* but dense, caespitose, smaller in all parts; leaves 6-10 × 2-3 mm, mucronate, scabrid to subglabrous; verticillasters usually 2-flowered; calyx c. 3 mm; corolla 5-7 mm.

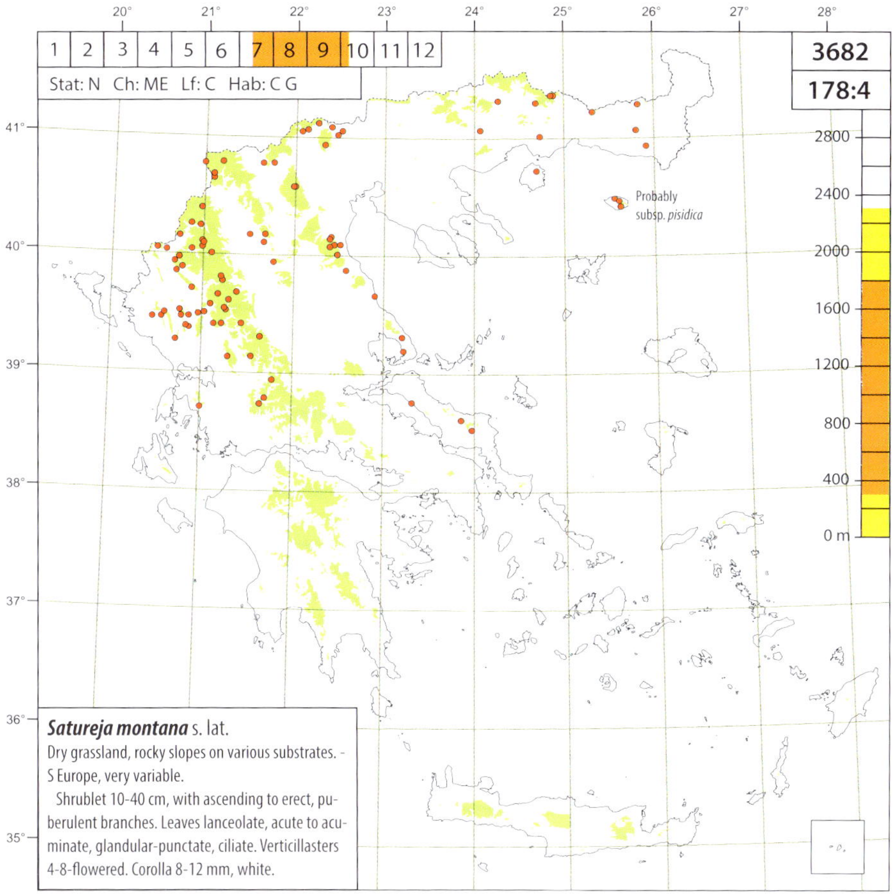

Satureja montana s. lat.
Dry grassland, rocky slopes on various substrates. - S Europe, very variable.
Shrublet 10-40 cm, with ascending to erect, puberulent branches. Leaves lanceolate, acute to acuminate, glandular-punctate, ciliate. Verticillasters 4-8-flowered. Corolla 8-12 mm, white.

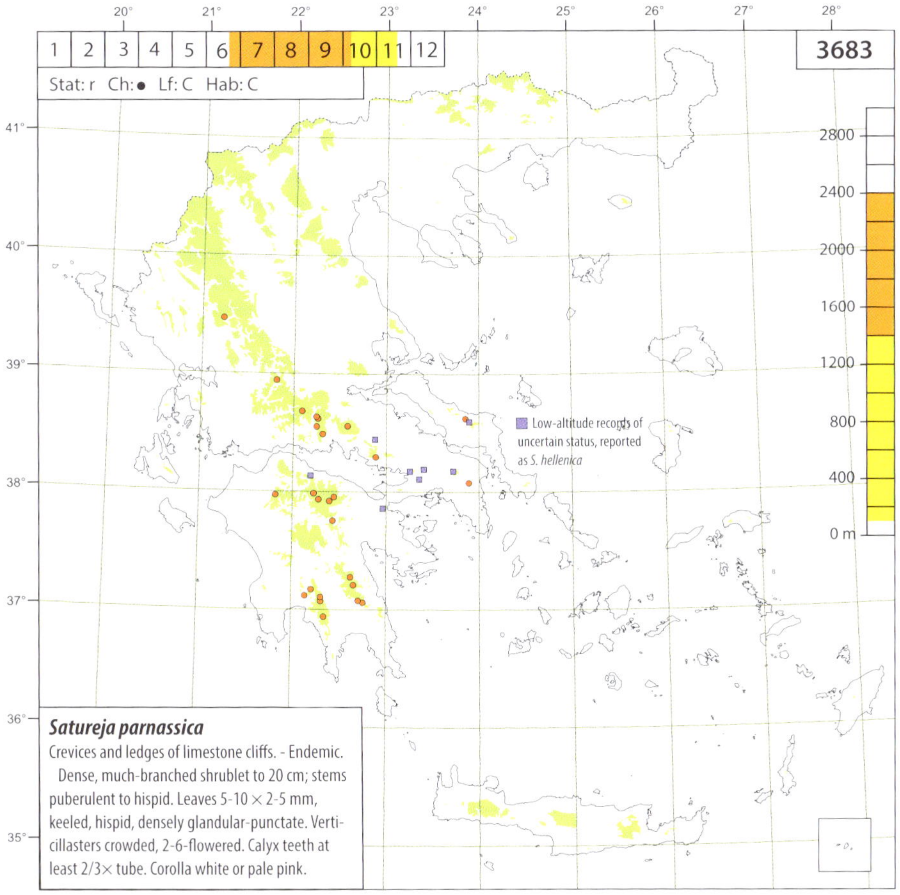

Satureja parnassica
Crevices and ledges of limestone cliffs. - Endemic.
Dense, much-branched shrublet to 20 cm; stems puberulent to hispid. Leaves 5-10 × 2-5 mm, keeled, hispid, densely glandular-punctate. Verticillasters crowded, 2-6-flowered. Calyx teeth at least 2/3× tube. Corolla white or pale pink.

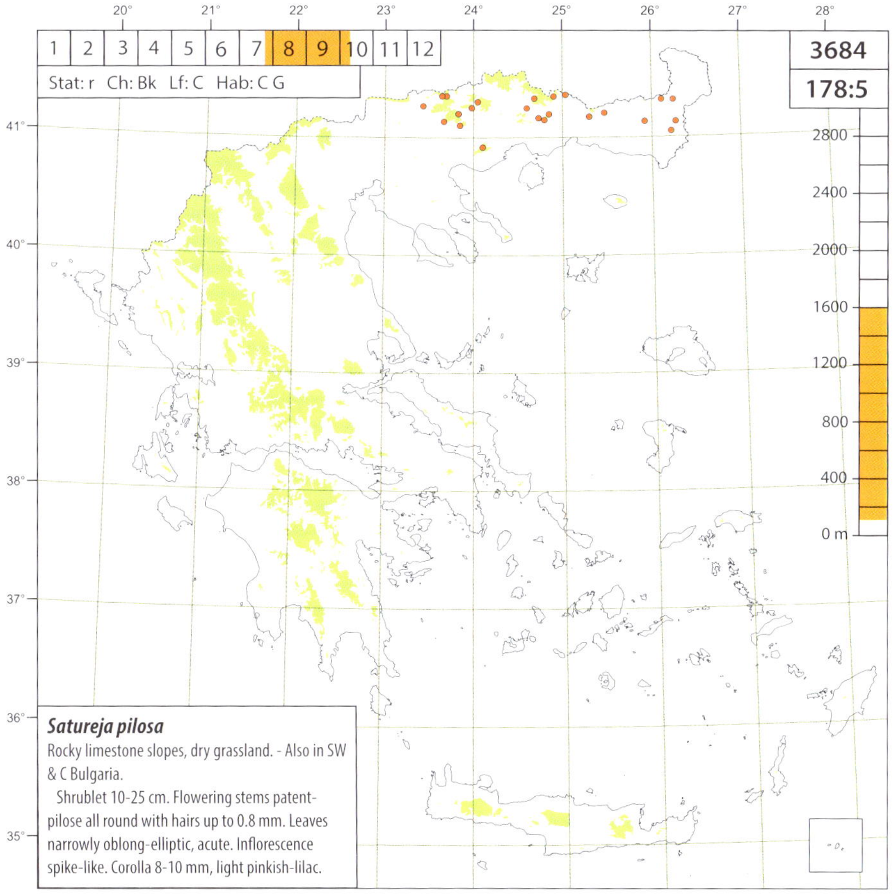

Satureja pilosa
Rocky limestone slopes, dry grassland. - Also in SW & C Bulgaria.
Shrublet 10-25 cm. Flowering stems patent-pilose all round with hairs up to 0.8 mm. Leaves narrowly oblong-elliptic, acute. Inflorescence spike-like. Corolla 8-10 mm, light pinkish-lilac.

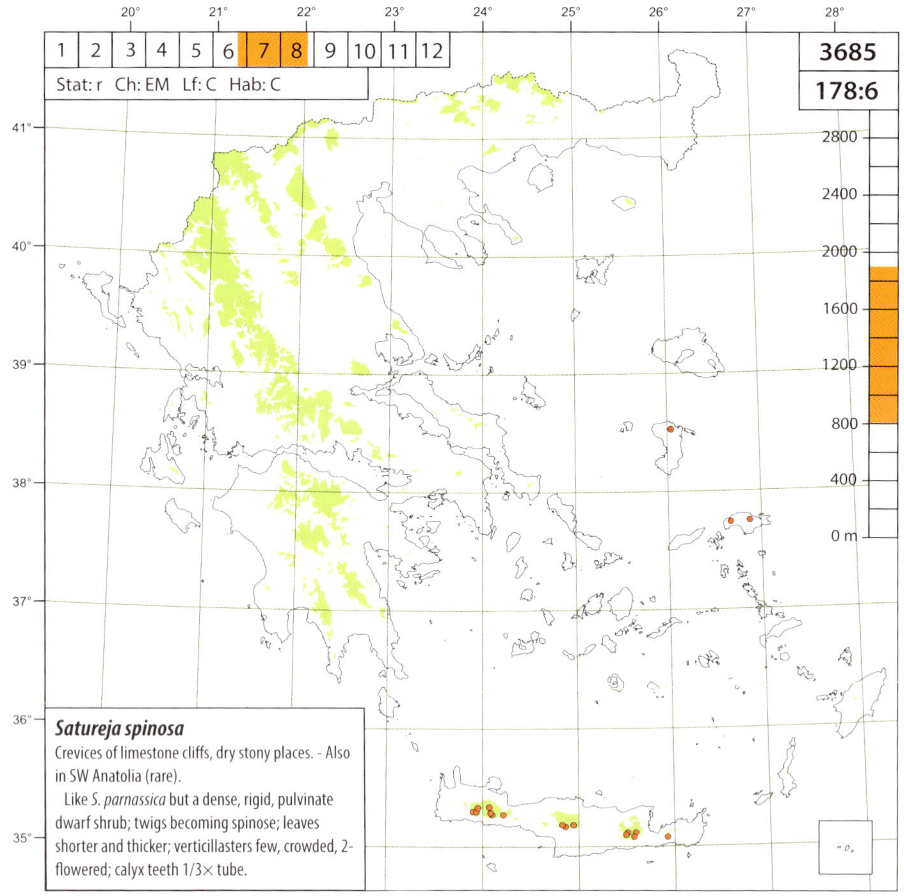

Satureja spinosa
Crevices of limestone cliffs, dry stony places. - Also in SW Anatolia (rare).
Like *S. parnassica* but a dense, rigid, pulvinate dwarf shrub; twigs becoming spinose; leaves shorter and thicker; verticillasters few, crowded, 2-flowered; calyx teeth 1/3× tube.

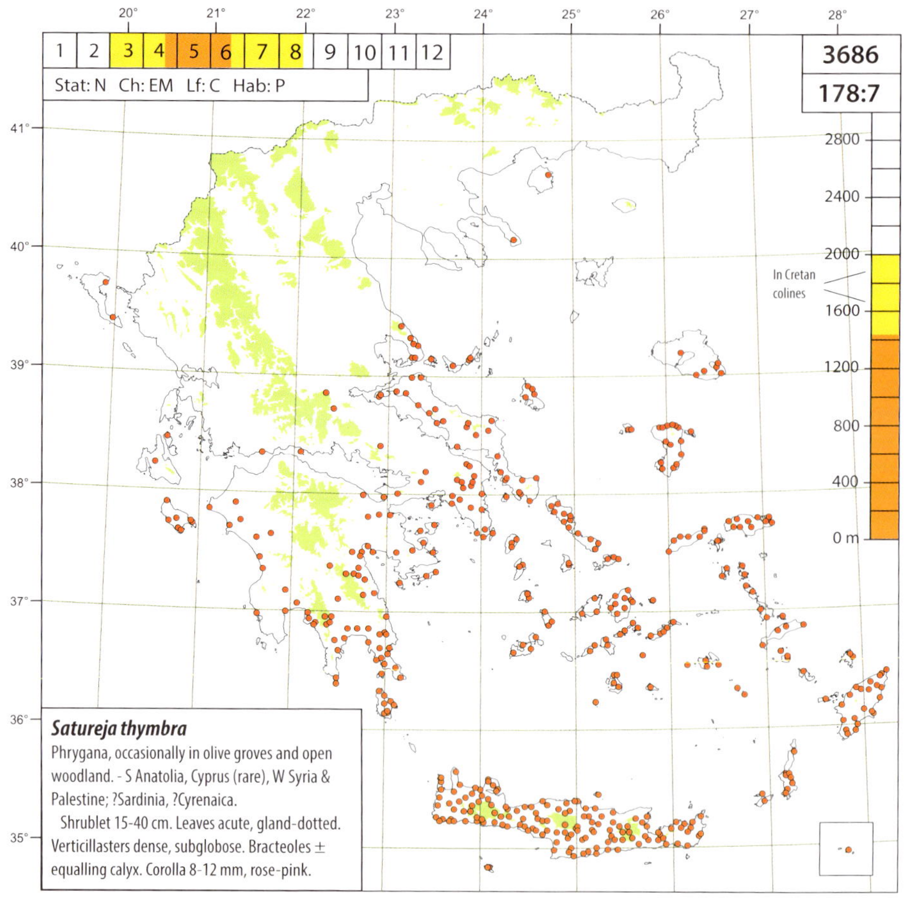

Satureja thymbra
Phrygana, occasionally in olive groves and open woodland. - S Anatolia, Cyprus (rare), W Syria & Palestine; ?Sardinia, ?Cyrenaica.
Shrublet 15-40 cm. Leaves acute, gland-dotted. Verticillasters dense, subglobose. Bracteoles ± equalling calyx. Corolla 8-12 mm, rose-pink.

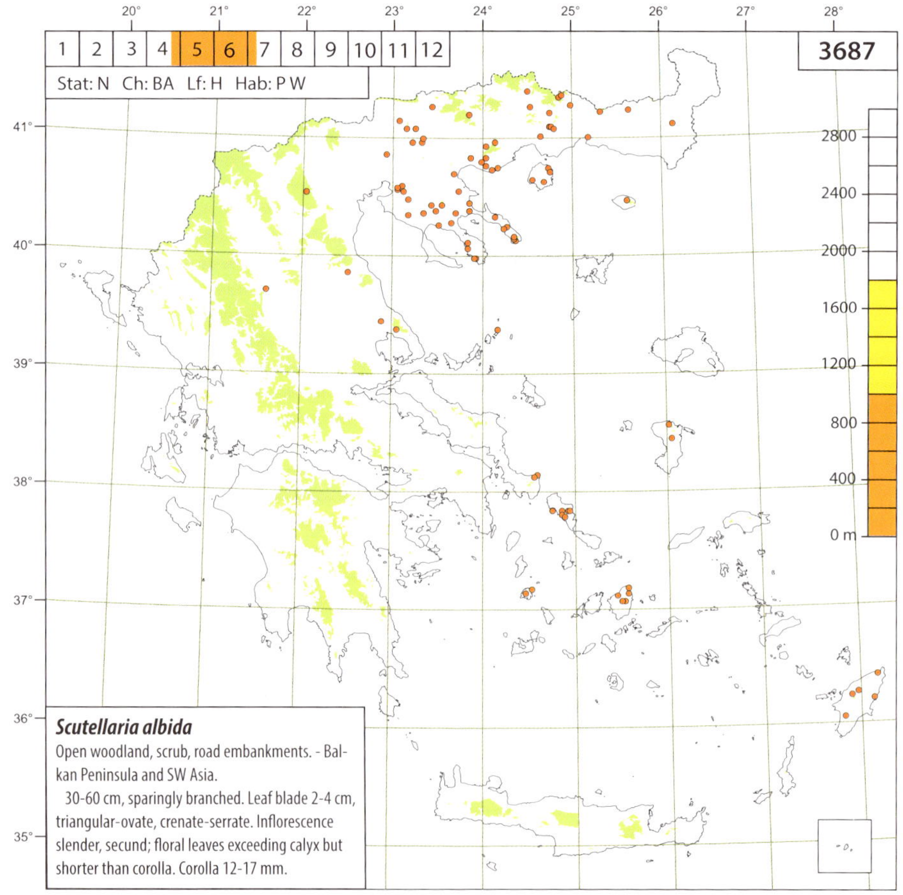

Scutellaria albida
Open woodland, scrub, road embankments. - Balkan Peninsula and SW Asia.
30-60 cm, sparingly branched. Leaf blade 2-4 cm, triangular-ovate, crenate-serrate. Inflorescence slender, secund; floral leaves exceeding calyx but shorter than corolla. Corolla 12-17 mm.

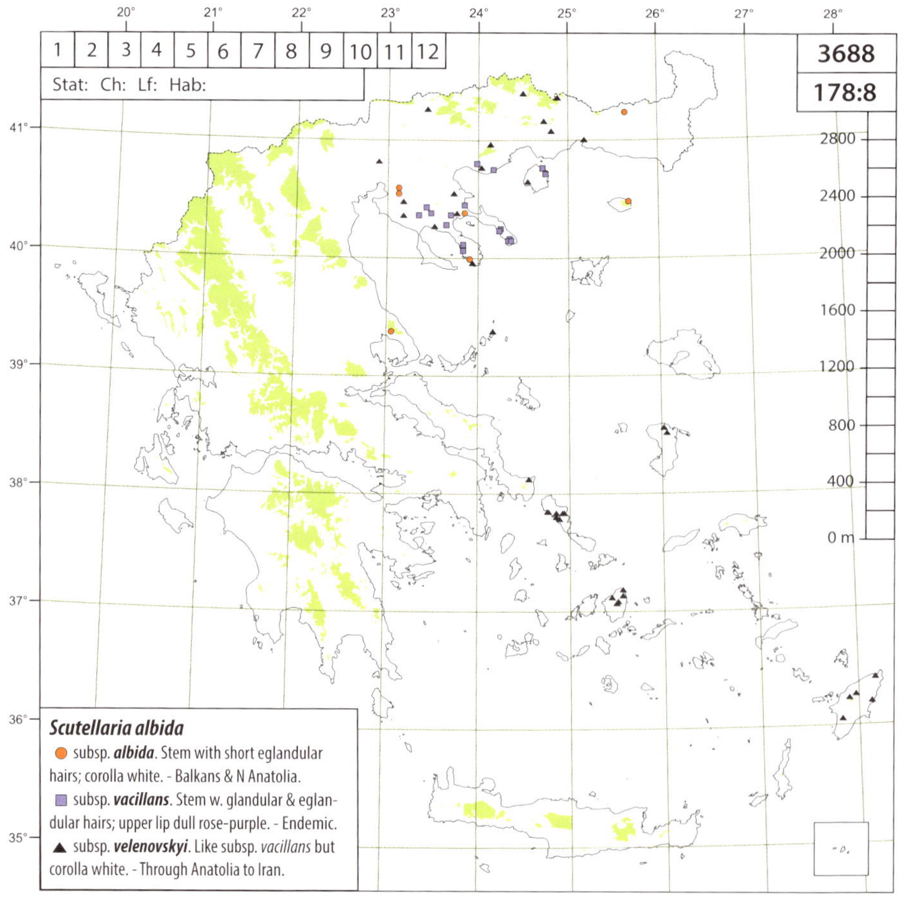

Scutellaria albida
● subsp. ***albida***. Stem with short eglandular hairs; corolla white. - Balkans & N Anatolia.
■ subsp. ***vacillans***. Stem w. glandular & eglandular hairs; upper lip dull rose-purple. - Endemic.
▲ subsp. ***velenovskyi***. Like subsp. *vacillans* but corolla white. - Through Anatolia to Iran.

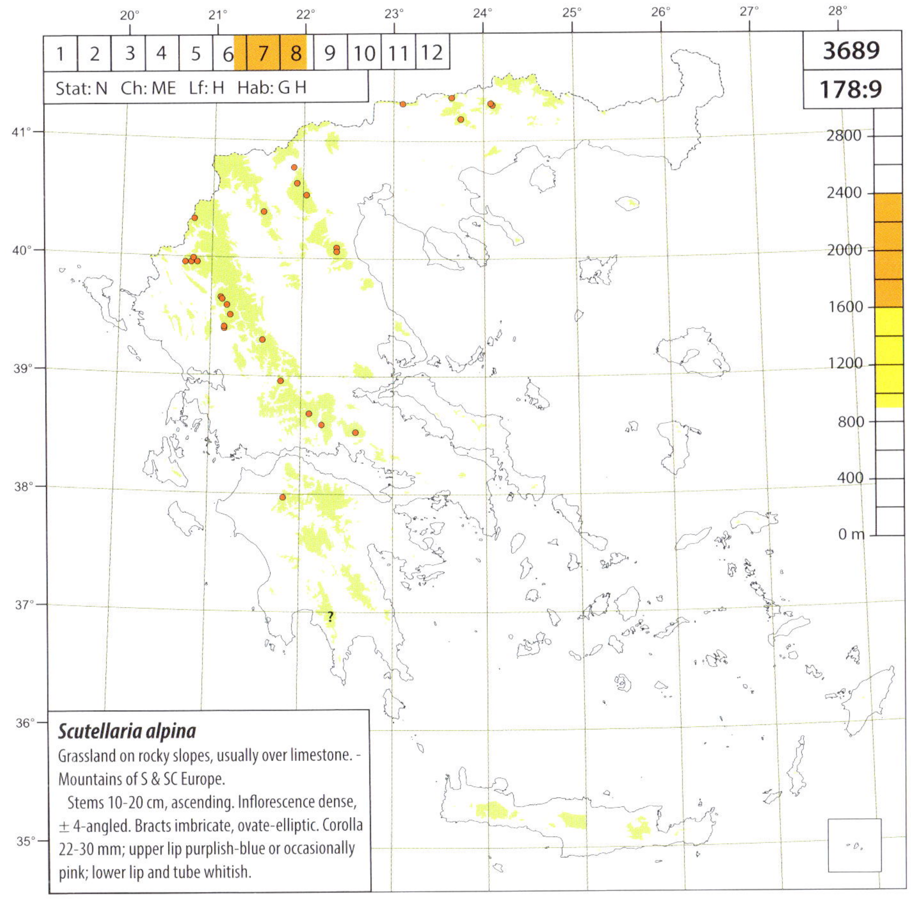

Scutellaria alpina

Grassland on rocky slopes, usually over limestone. - Mountains of S & SC Europe.

Stems 10-20 cm, ascending. Inflorescence dense, ± 4-angled. Bracts imbricate, ovate-elliptic. Corolla 22-30 mm; upper lip purplish-blue or occasionally pink; lower lip and tube whitish.

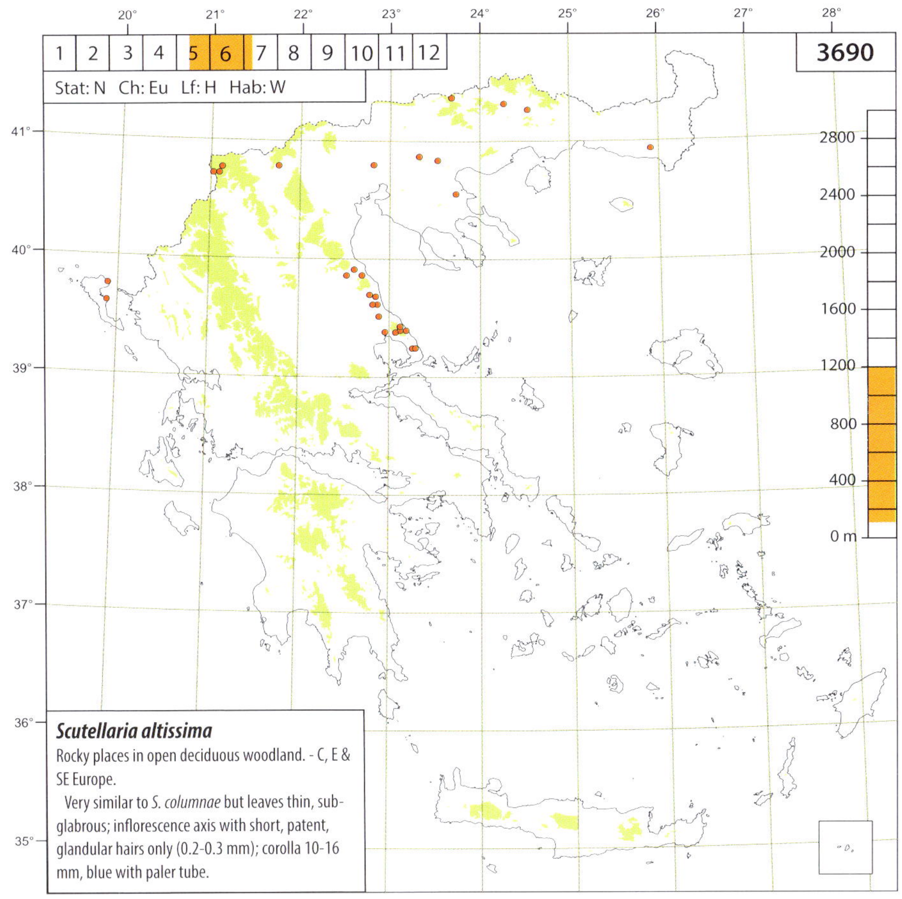

Scutellaria altissima

Rocky places in open deciduous woodland. - C, E & SE Europe.

Very similar to *S. columnae* but leaves thin, sub-glabrous; inflorescence axis with short, patent, glandular hairs only (0.2-0.3 mm); corolla 10-16 mm, blue with paler tube.

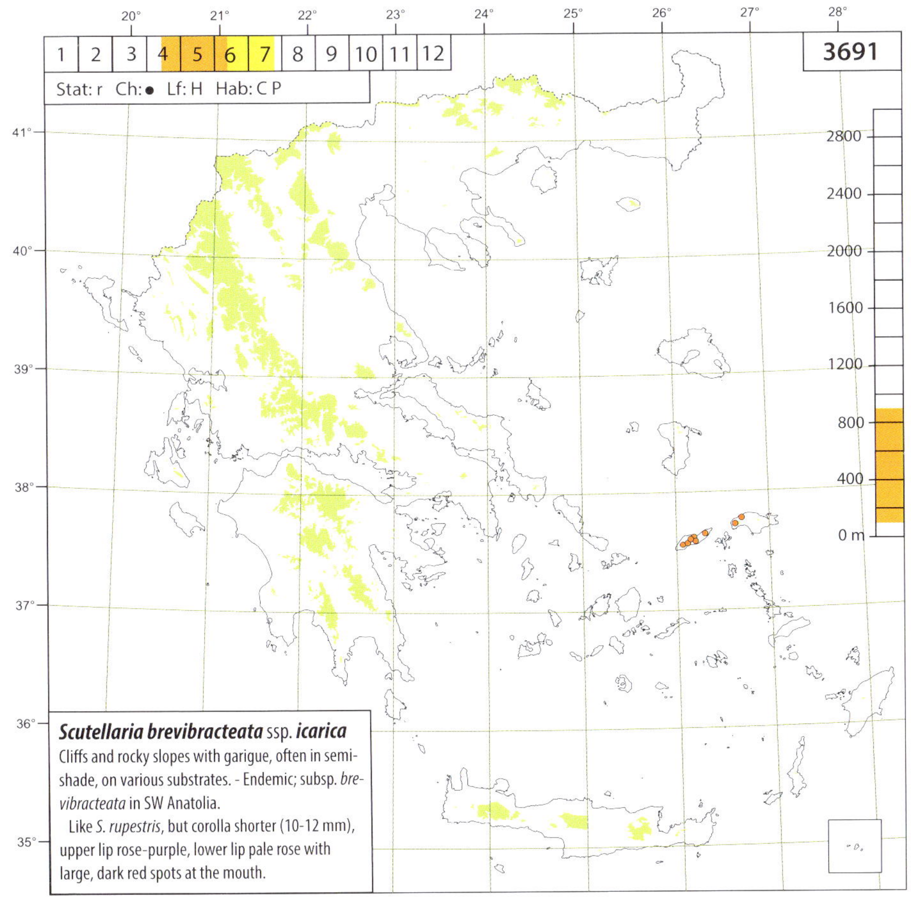

Scutellaria brevibracteata ssp. *icarica*

Cliffs and rocky slopes with garigue, often in semi-shade, on various substrates. - Endemic; subsp. *brevibracteata* in SW Anatolia.

Like *S. rupestris*, but corolla shorter (10-12 mm), upper lip rose-purple, lower lip pale rose with large, dark red spots at the mouth.

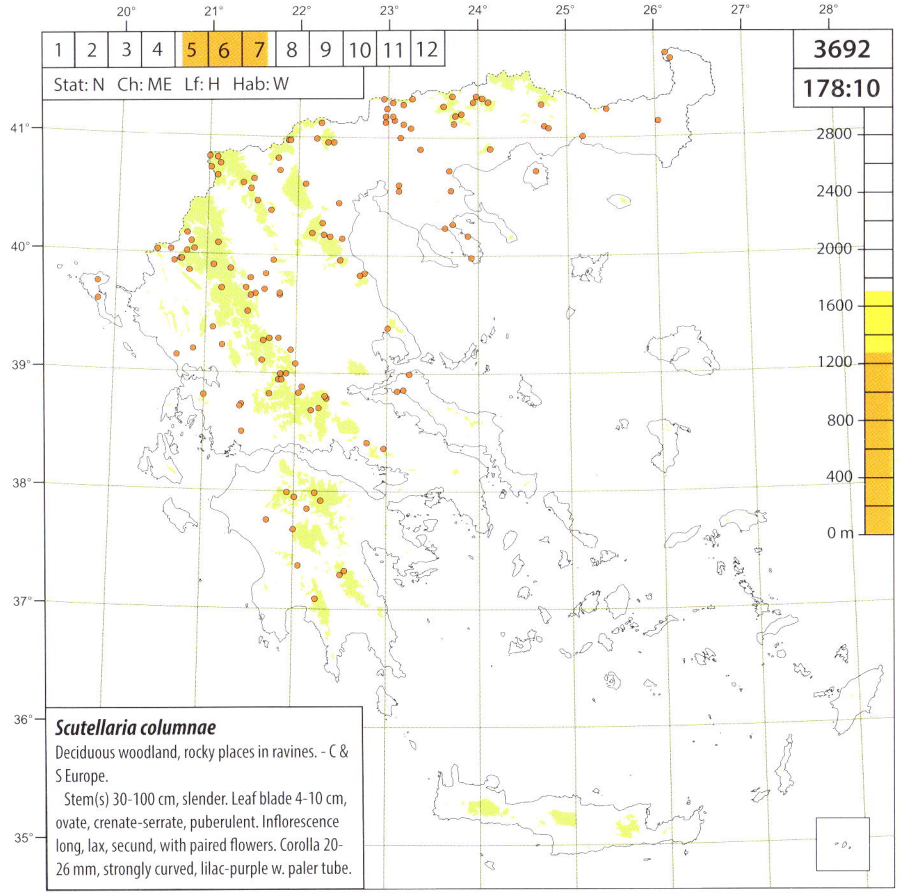

Scutellaria columnae

Deciduous woodland, rocky places in ravines. - C & S Europe.

Stem(s) 30-100 cm, slender. Leaf blade 4-10 cm, ovate, crenate-serrate, puberulent. Inflorescence long, lax, secund, with paired flowers. Corolla 20-26 mm, strongly curved, lilac-purple w. paler tube.

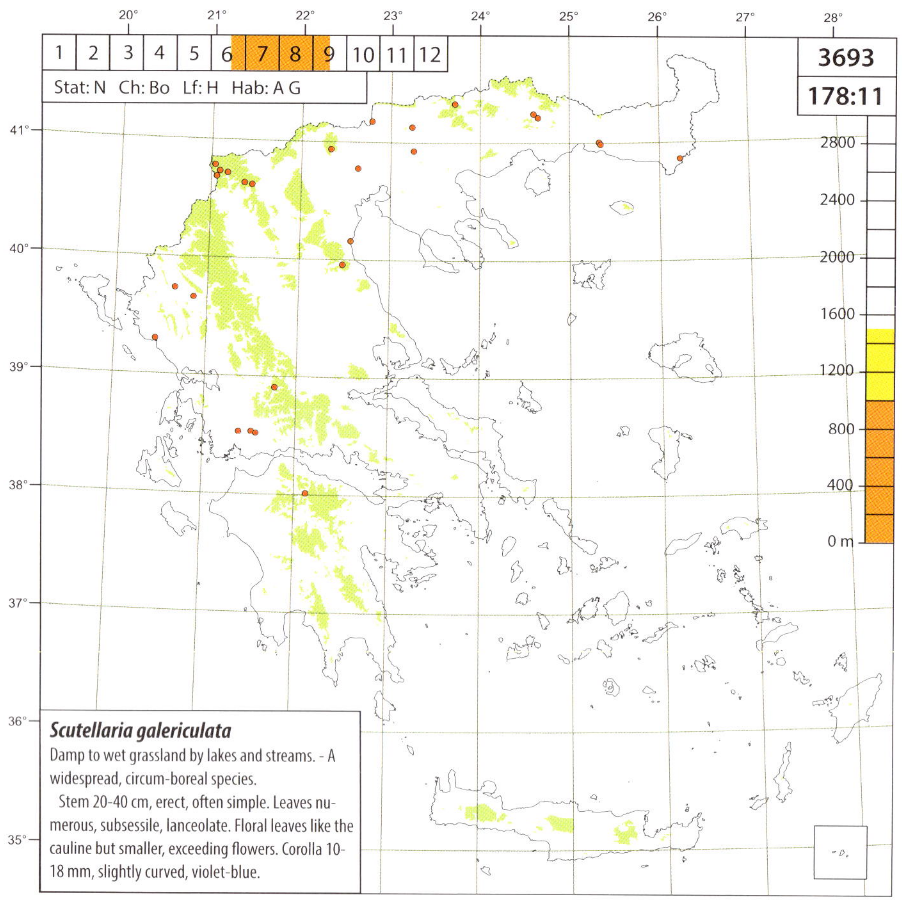

Scutellaria galericulata

Damp to wet grassland by lakes and streams. - A widespread, circum-boreal species.

Stem 20-40 cm, erect, often simple. Leaves numerous, subsessile, lanceolate. Floral leaves like the cauline but smaller, exceeding flowers. Corolla 10-18 mm, slightly curved, violet-blue.

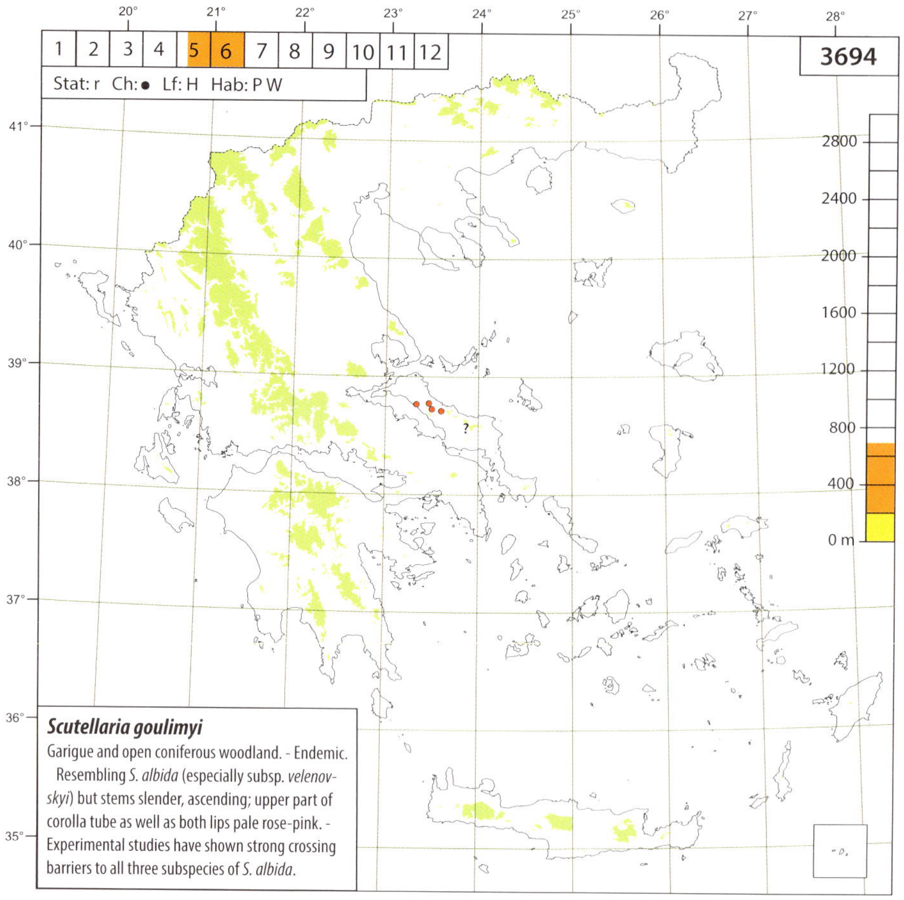

Scutellaria goulimyi

Garigue and open coniferous woodland. - Endemic.

Resembling *S. albida* (especially subsp. *velenovskyi*) but stems slender, ascending; upper part of corolla tube as well as both lips pale rose-pink. - Experimental studies have shown strong crossing barriers to all three subspecies of *S. albida*.

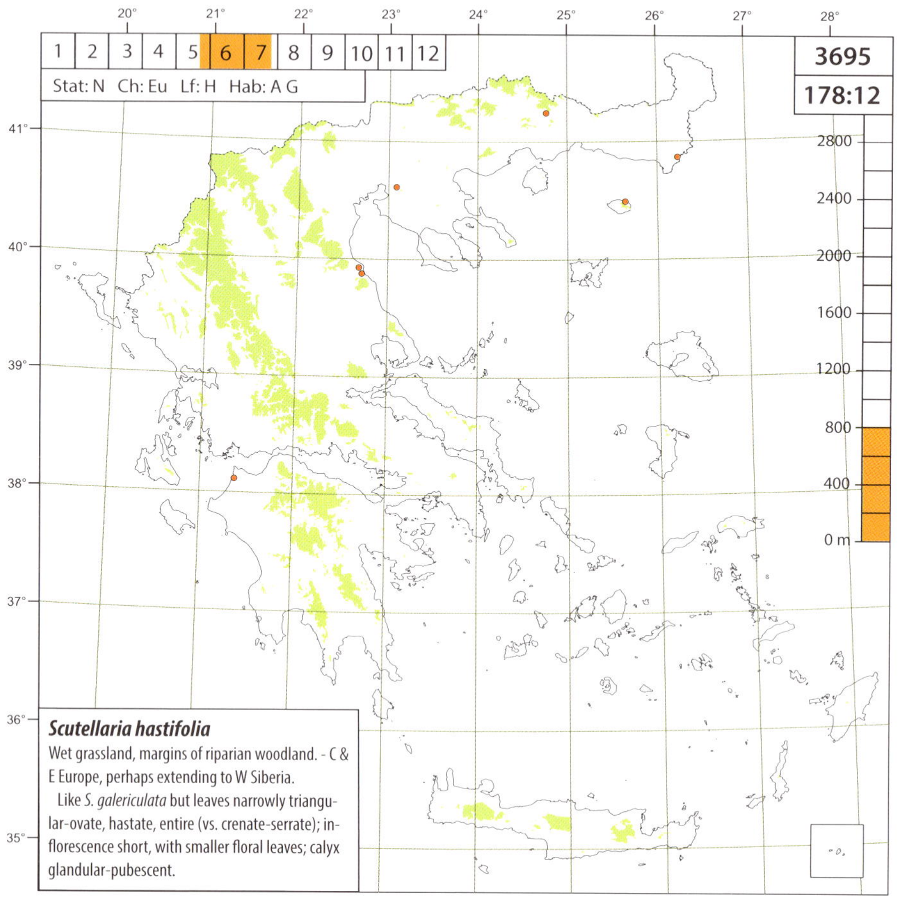

Scutellaria hastifolia

Wet grassland, margins of riparian woodland. - C & E Europe, perhaps extending to W Siberia.

Like *S. galericulata* but leaves narrowly triangular-ovate, hastate, entire (vs. crenate-serrate); inflorescence short, with smaller floral leaves; calyx glandular-pubescent.

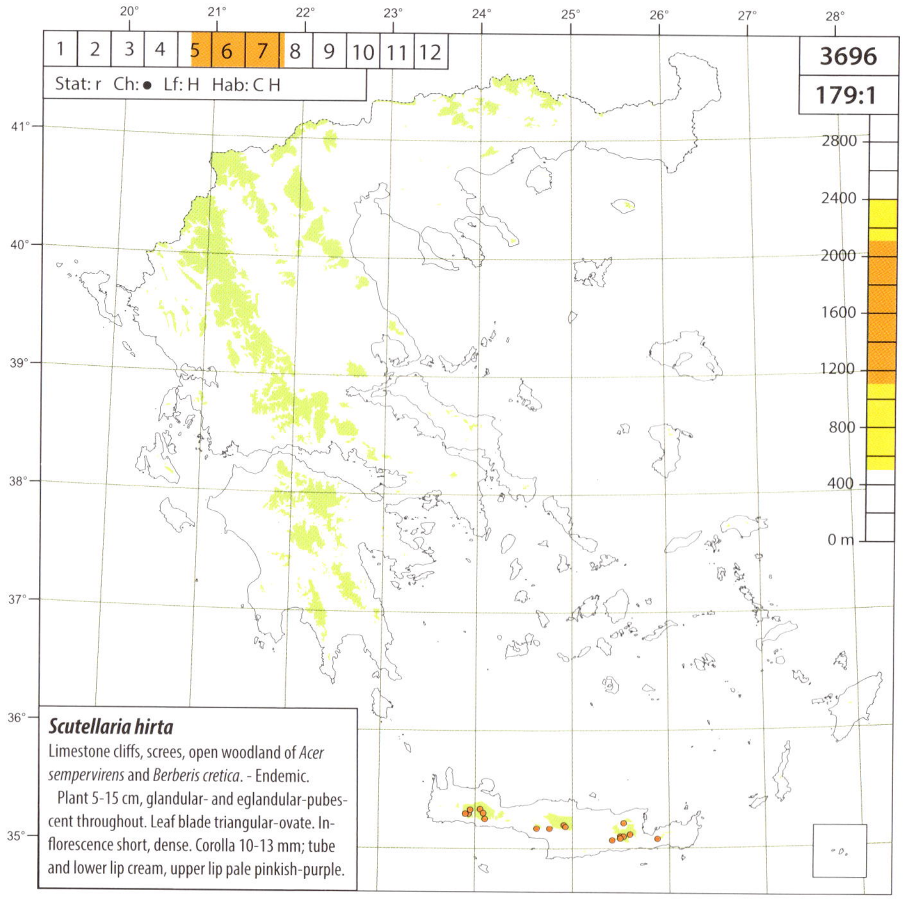

Scutellaria hirta

Limestone cliffs, screes, open woodland of *Acer sempervirens* and *Berberis cretica*. - Endemic.

Plant 5-15 cm, glandular- and eglandular-pubescent throughout. Leaf blade triangular-ovate. Inflorescence short, dense. Corolla 10-13 mm; tube and lower lip cream, upper lip pale pinkish-purple.

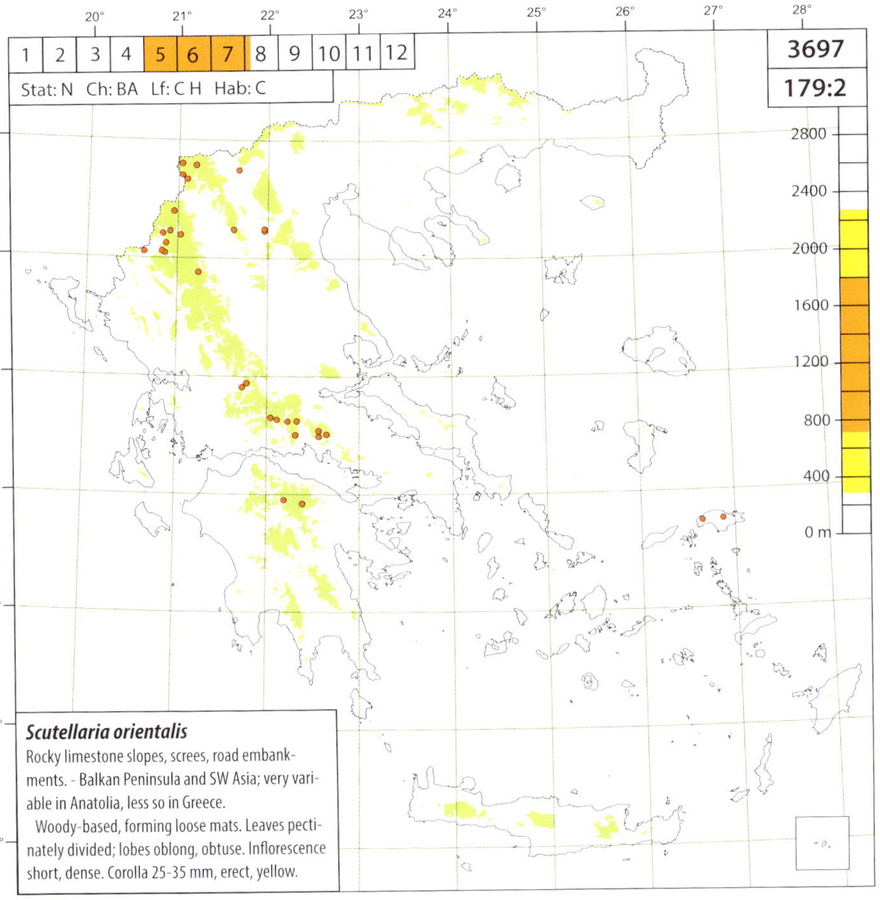

Scutellaria orientalis

Rocky limestone slopes, screes, road embankments. - Balkan Peninsula and SW Asia; very variable in Anatolia, less so in Greece.

Woody-based, forming loose mats. Leaves pectinately divided; lobes oblong, obtuse. Inflorescence short, dense. Corolla 25-35 mm, erect, yellow.

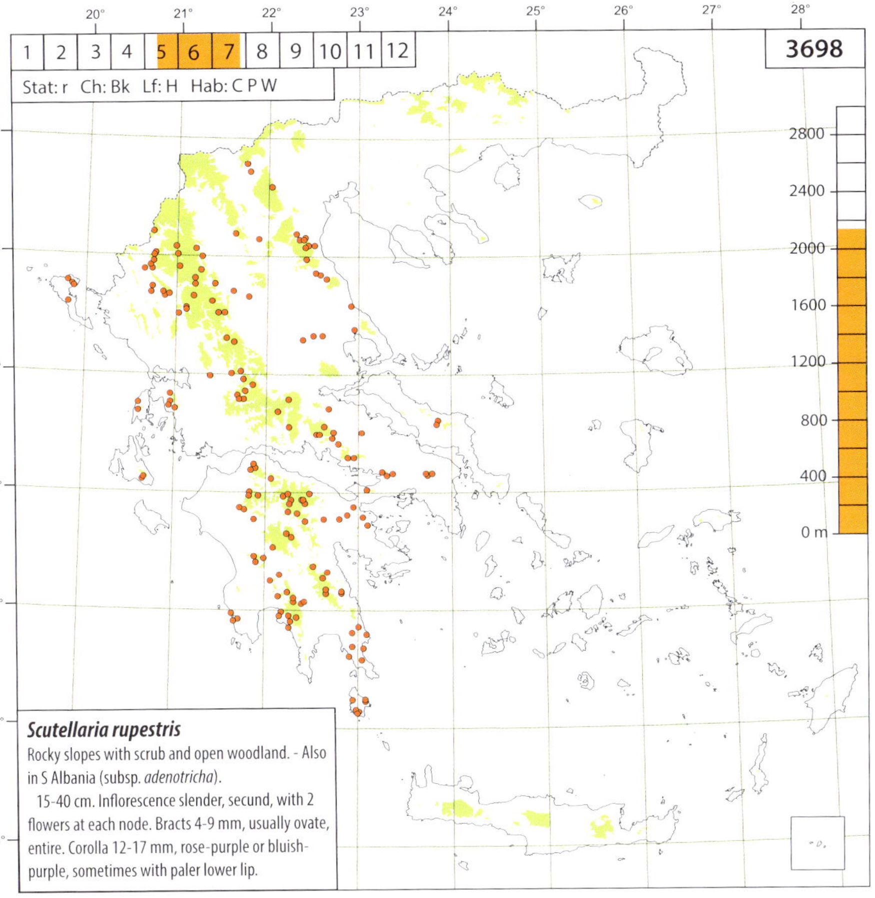

Scutellaria rupestris

Rocky slopes with scrub and open woodland. - Also in S Albania (subsp. *adenotricha*).

15-40 cm. Inflorescence slender, secund, with 2 flowers at each node. Bracts 4-9 mm, usually ovate, entire. Corolla 12-17 mm, rose-purple or bluish-purple, sometimes with paler lower lip.

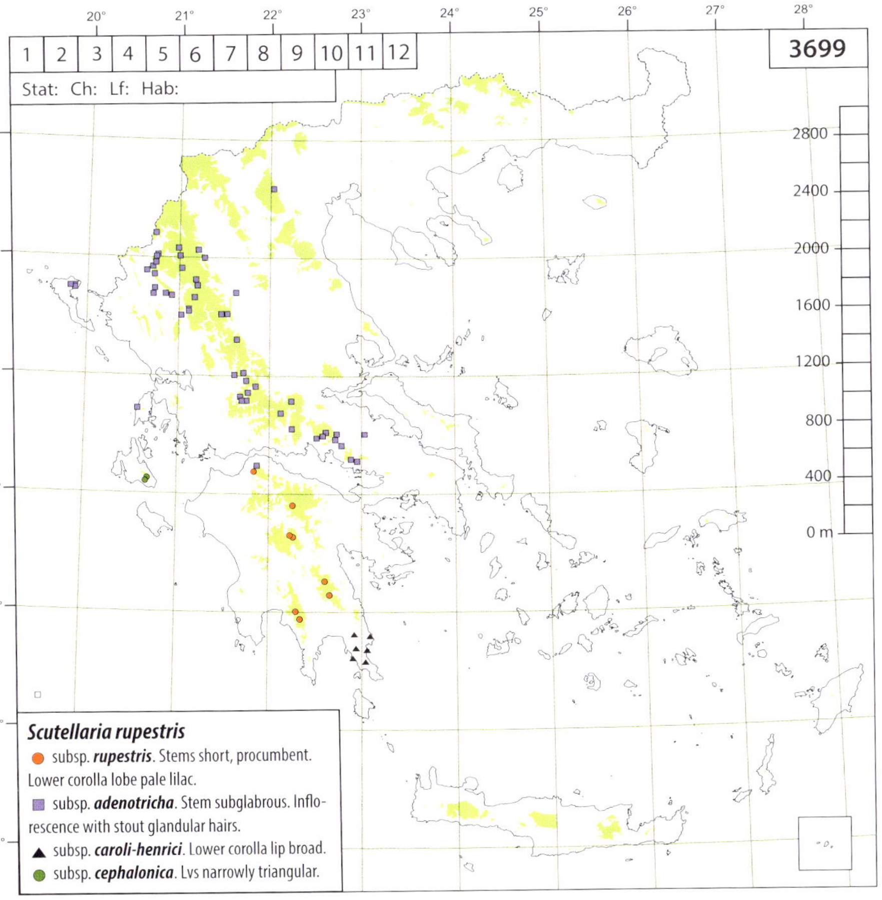

Scutellaria rupestris

● subsp. ***rupestris***. Stems short, procumbent. Lower corolla lobe pale lilac.

■ subsp. ***adenotricha***. Stem subglabrous. Inflorescence with stout glandular hairs.

▲ subsp. ***caroli-henrici***. Lower corolla lip broad.

● subsp. ***cephalonica***. Lvs narrowly triangular.

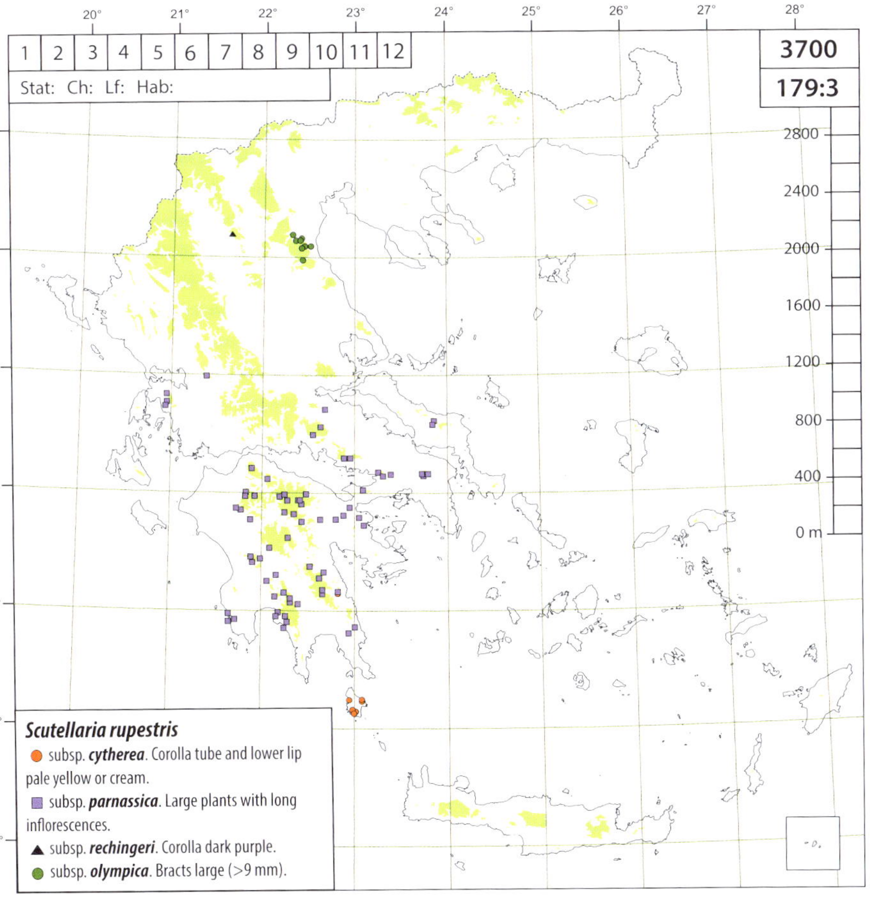

Scutellaria rupestris

● subsp. ***cytherea***. Corolla tube and lower lip pale yellow or cream.

■ subsp. ***parnassica***. Large plants with long inflorescences.

▲ subsp. ***rechingeri***. Corolla dark purple.

● subsp. ***olympica***. Bracts large (>9 mm).

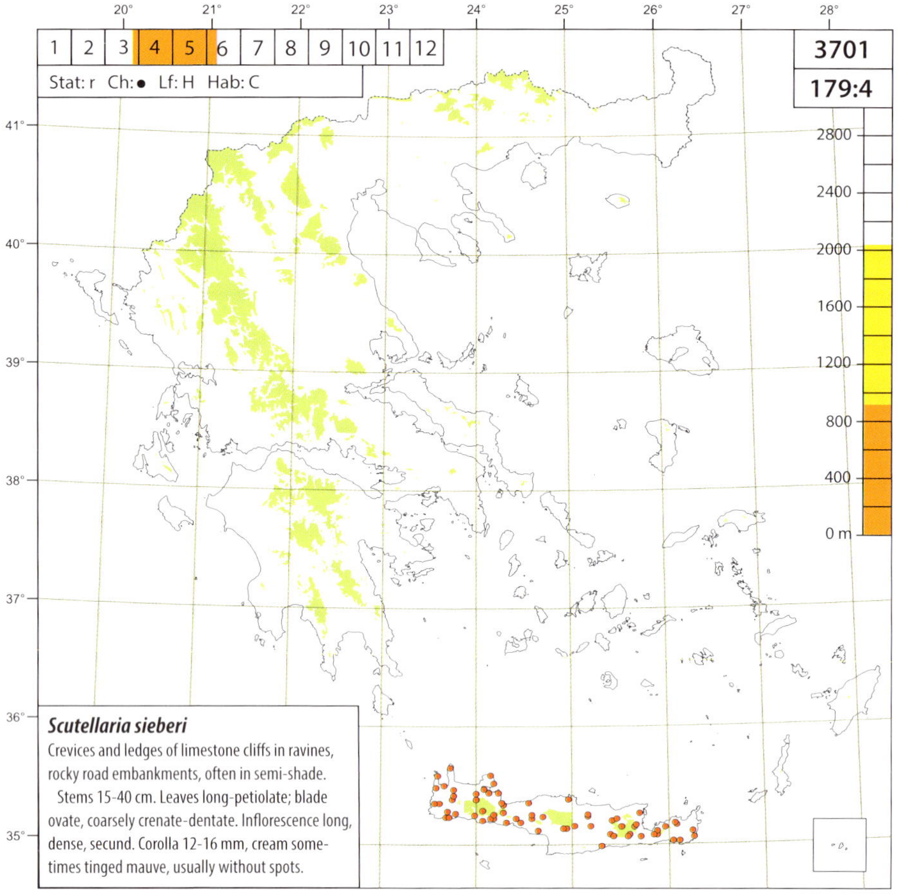

Scutellaria sieberi
Crevices and ledges of limestone cliffs in ravines, rocky road embankments, often in semi-shade.
Stems 15-40 cm. Leaves long-petiolate; blade ovate, coarsely crenate-dentate. Inflorescence long, dense, secund. Corolla 12-16 mm, cream sometimes tinged mauve, usually without spots.

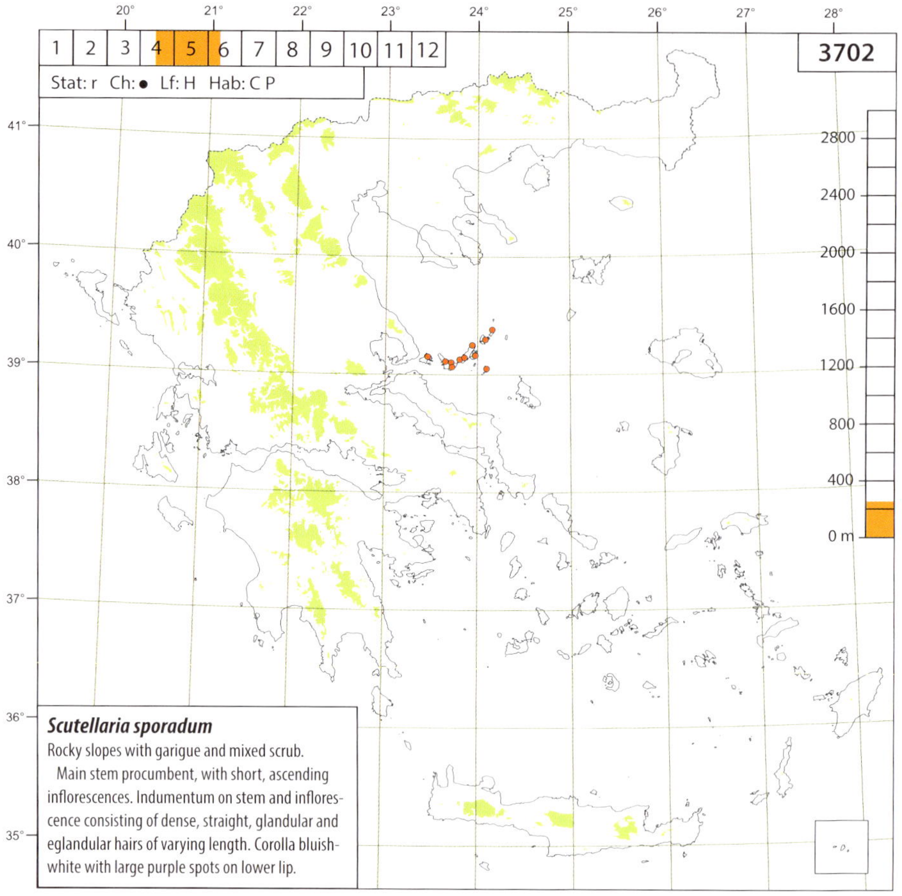

Scutellaria sporadum
Rocky slopes with garigue and mixed scrub.
Main stem procumbent, with short, ascending inflorescences. Indumentum on stem and inflorescence consisting of dense, straight, glandular and eglandular hairs of varying length. Corolla bluish-white with large purple spots on lower lip.

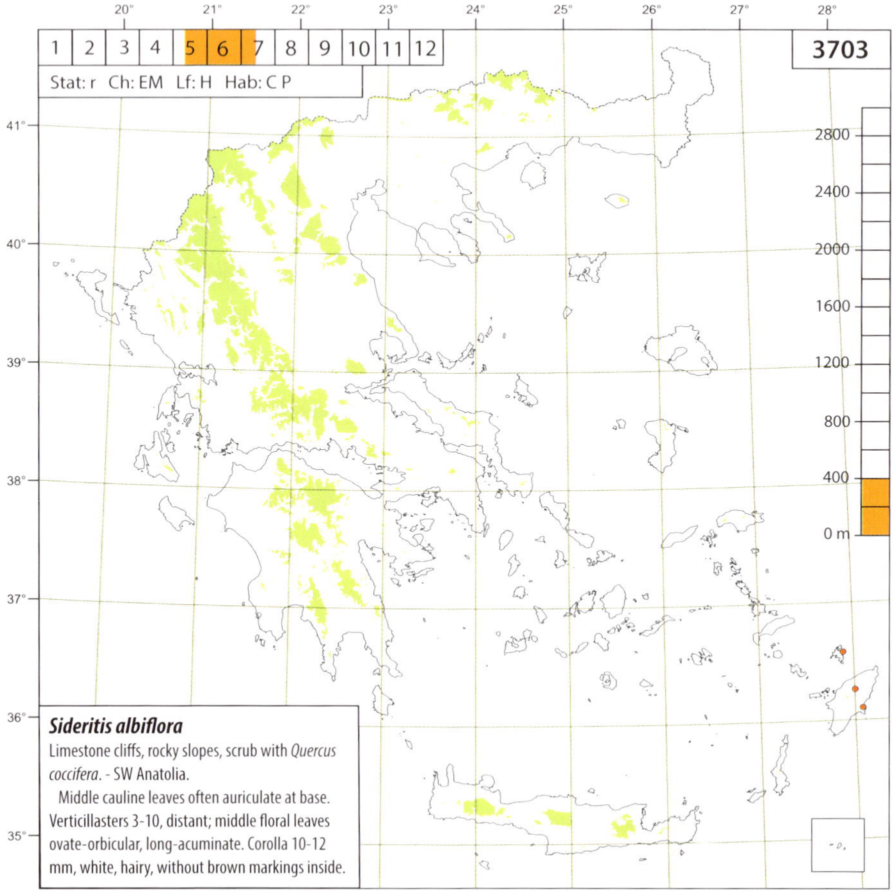

Sideritis albiflora
Limestone cliffs, rocky slopes, scrub with *Quercus coccifera*. - SW Anatolia.
Middle cauline leaves often auriculate at base. Verticillasters 3-10, distant; middle floral leaves ovate-orbicular, long-acuminate. Corolla 10-12 mm, white, hairy, without brown markings inside.

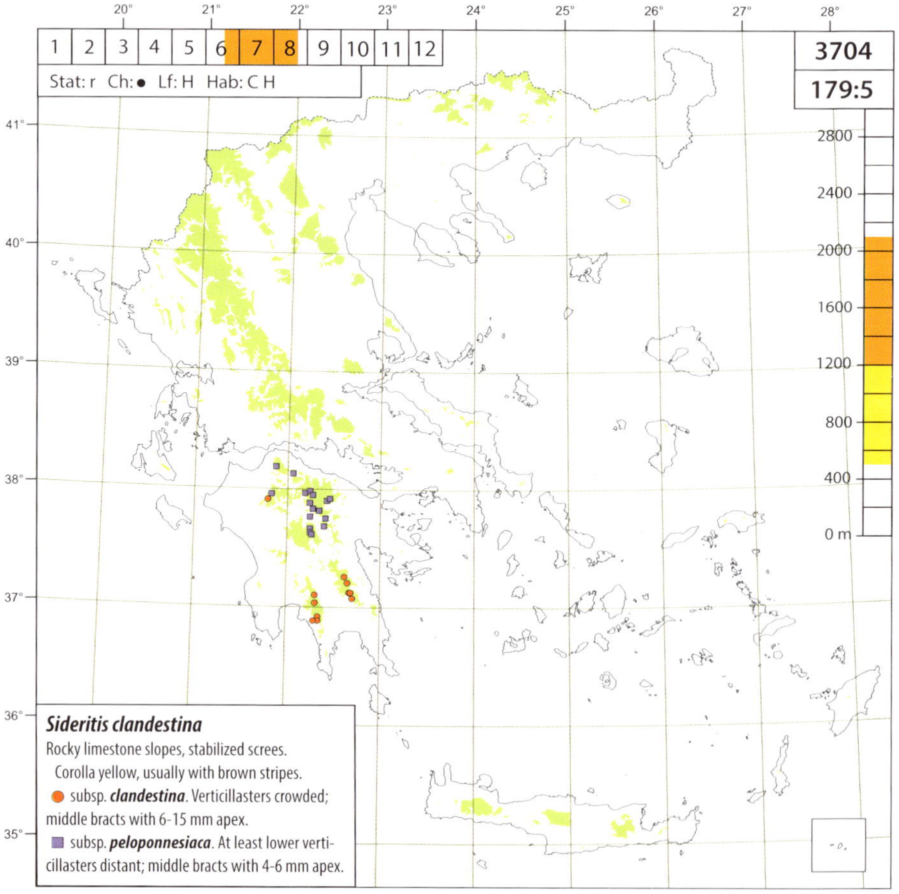

Sideritis clandestina
Rocky limestone slopes, stabilized screes.
Corolla yellow, usually with brown stripes.
● subsp. ***clandestina***. Verticillasters crowded; middle bracts with 6-15 mm apex.
■ subsp. ***peloponnesiaca***. At least lower verticillasters distant; middle bracts with 4-6 mm apex.

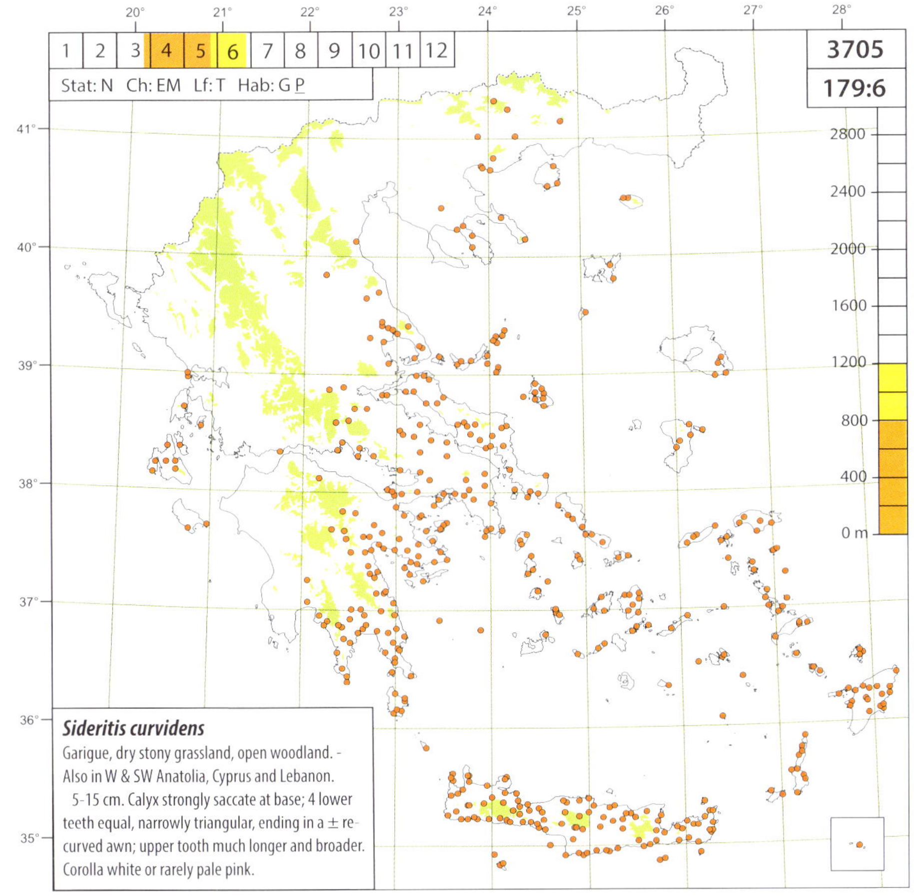

Sideritis curvidens
Garigue, dry stony grassland, open woodland. - Also in W & SW Anatolia, Cyprus and Lebanon.
5-15 cm. Calyx strongly saccate at base; 4 lower teeth equal, narrowly triangular, ending in a ± recurved awn; upper tooth much longer and broader. Corolla white or rarely pale pink.

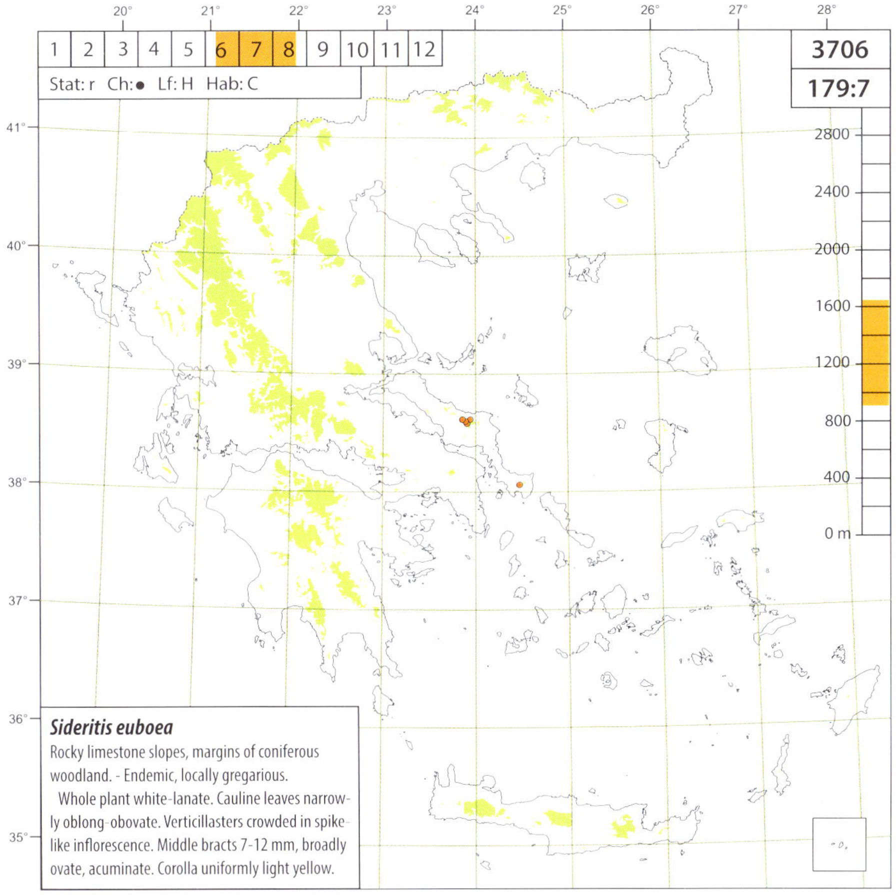

Sideritis euboea
Rocky limestone slopes, margins of coniferous woodland. - Endemic, locally gregarious.
Whole plant white-lanate. Cauline leaves narrowly oblong-obovate. Verticillasters crowded in spike-like inflorescence. Middle bracts 7-12 mm, broadly ovate, acuminate. Corolla uniformly light yellow.

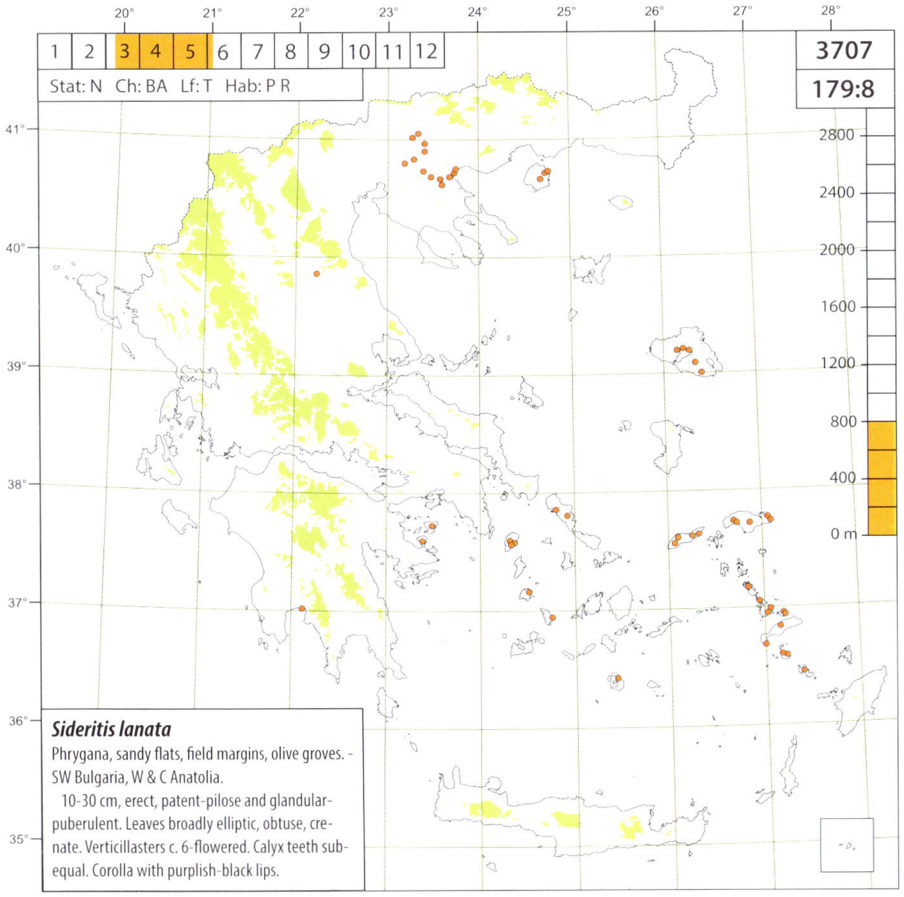

Sideritis lanata
Phrygana, sandy flats, field margins, olive groves. - SW Bulgaria, W & C Anatolia.
10-30 cm, erect, patent-pilose and glandular-puberulent. Leaves broadly elliptic, obtuse, crenate. Verticillasters c. 6-flowered. Calyx teeth subequal. Corolla with purplish-black lips.

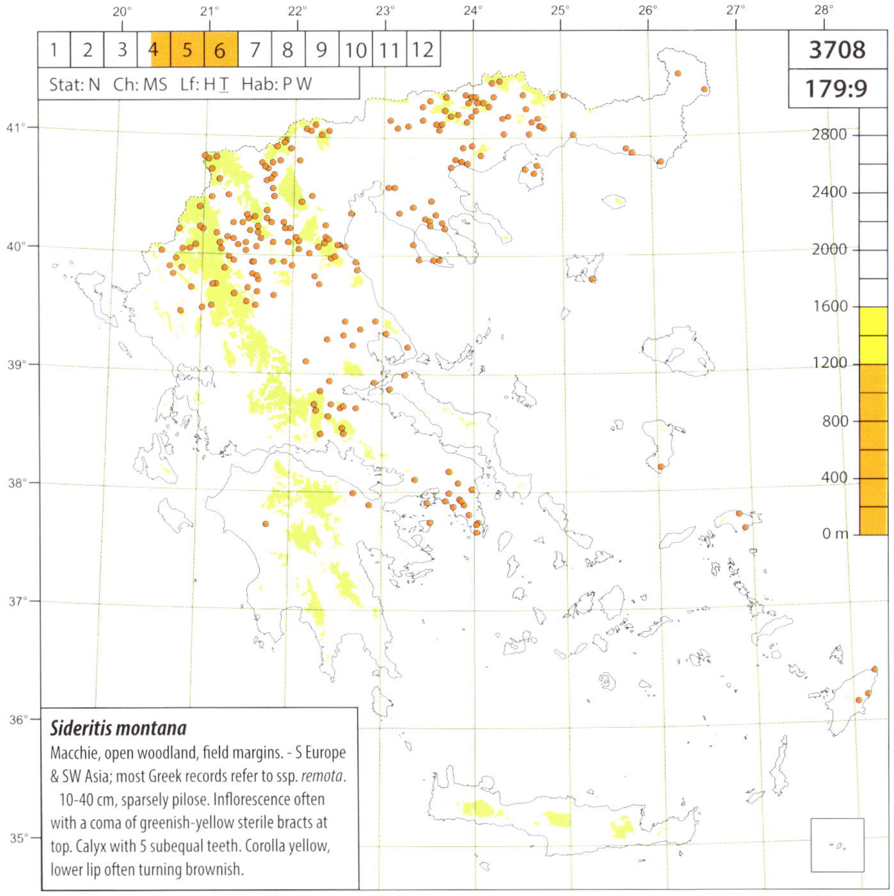

Sideritis montana
Macchie, open woodland, field margins. - S Europe & SW Asia; most Greek records refer to ssp. *remota*.
10-40 cm, sparsely pilose. Inflorescence often with a coma of greenish-yellow sterile bracts at top. Calyx with 5 subequal teeth. Corolla yellow, lower lip often turning brownish.

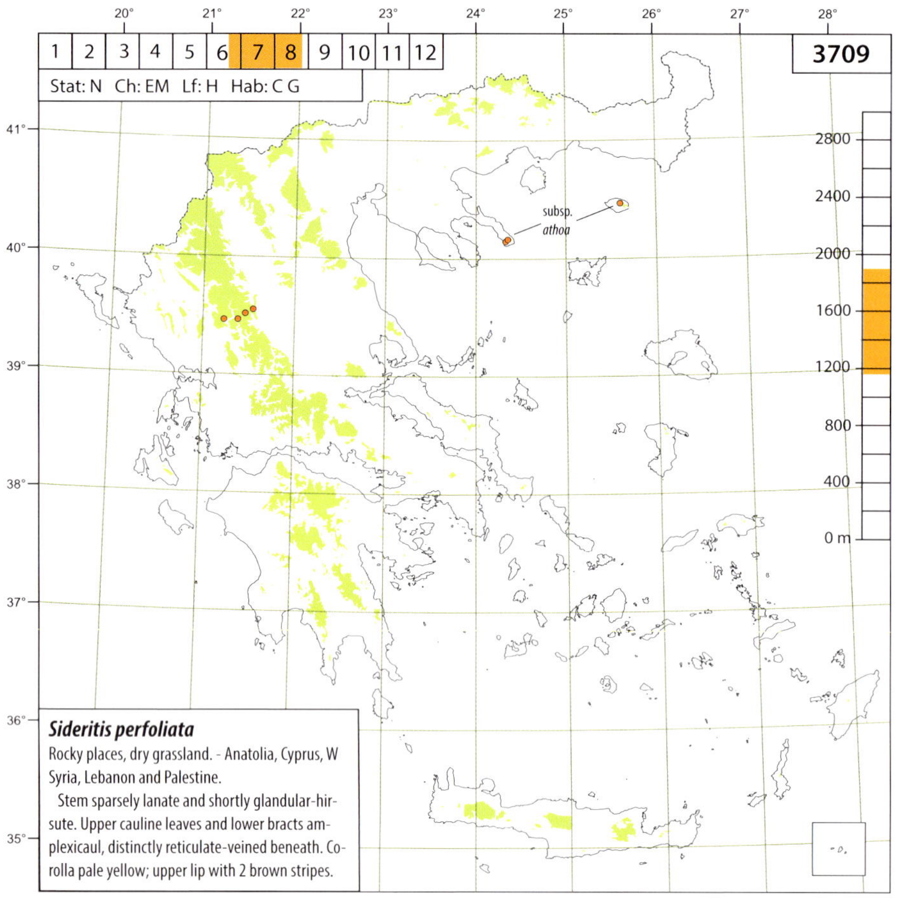

Sideritis perfoliata

Rocky places, dry grassland. - Anatolia, Cyprus, W Syria, Lebanon and Palestine.

Stem sparsely lanate and shortly glandular-hirsute. Upper cauline leaves and lower bracts amplexicaul, distinctly reticulate-veined beneath. Corolla pale yellow; upper lip with 2 brown stripes.

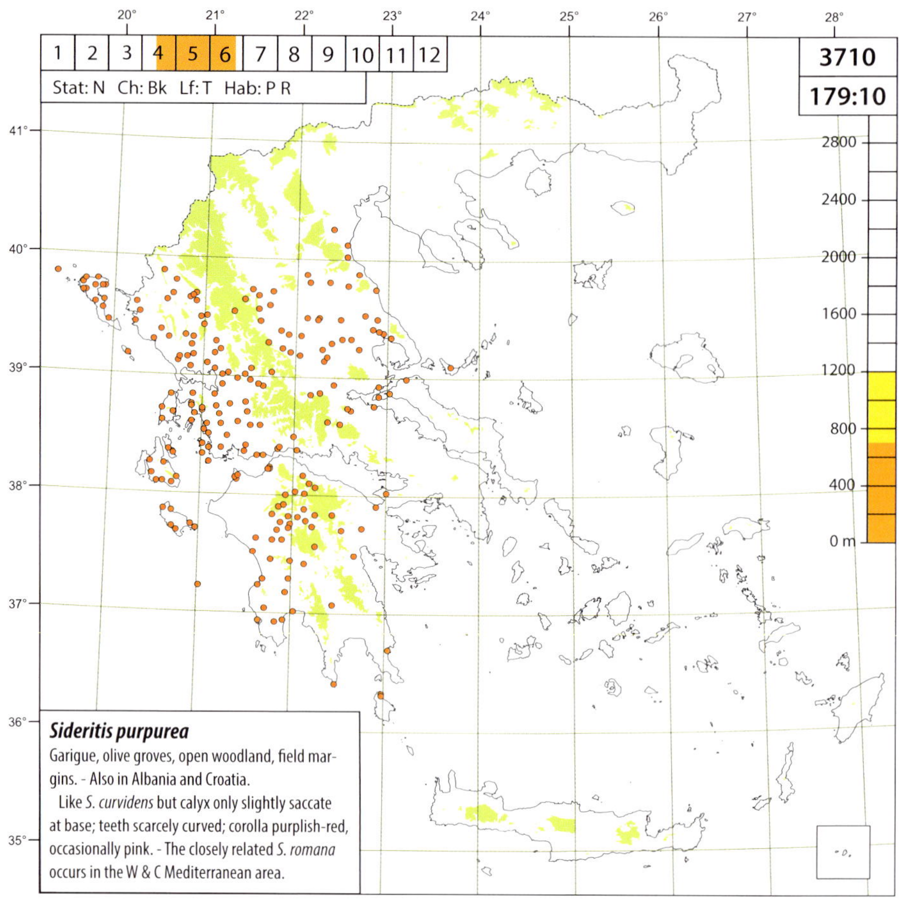

Sideritis purpurea

Garigue, olive groves, open woodland, field margins. - Also in Albania and Croatia.

Like *S. curvidens* but calyx only slightly saccate at base; teeth scarcely curved; corolla purplish-red, occasionally pink. - The closely related *S. romana* occurs in the W & C Mediterranean area.

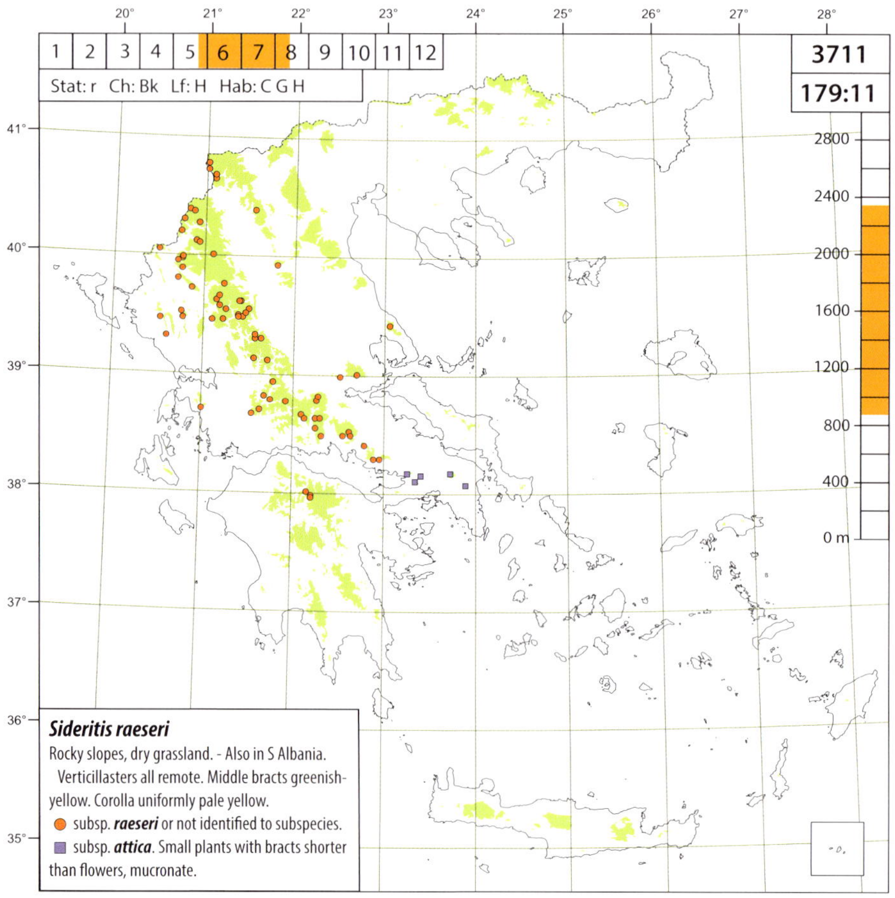

Sideritis raeseri

Rocky slopes, dry grassland. - Also in S Albania.

Verticillasters all remote. Middle bracts greenish-yellow. Corolla uniformly pale yellow.

● subsp. ***raeseri*** or not identified to subspecies.

■ subsp. ***attica***. Small plants with bracts shorter than flowers, mucronate.

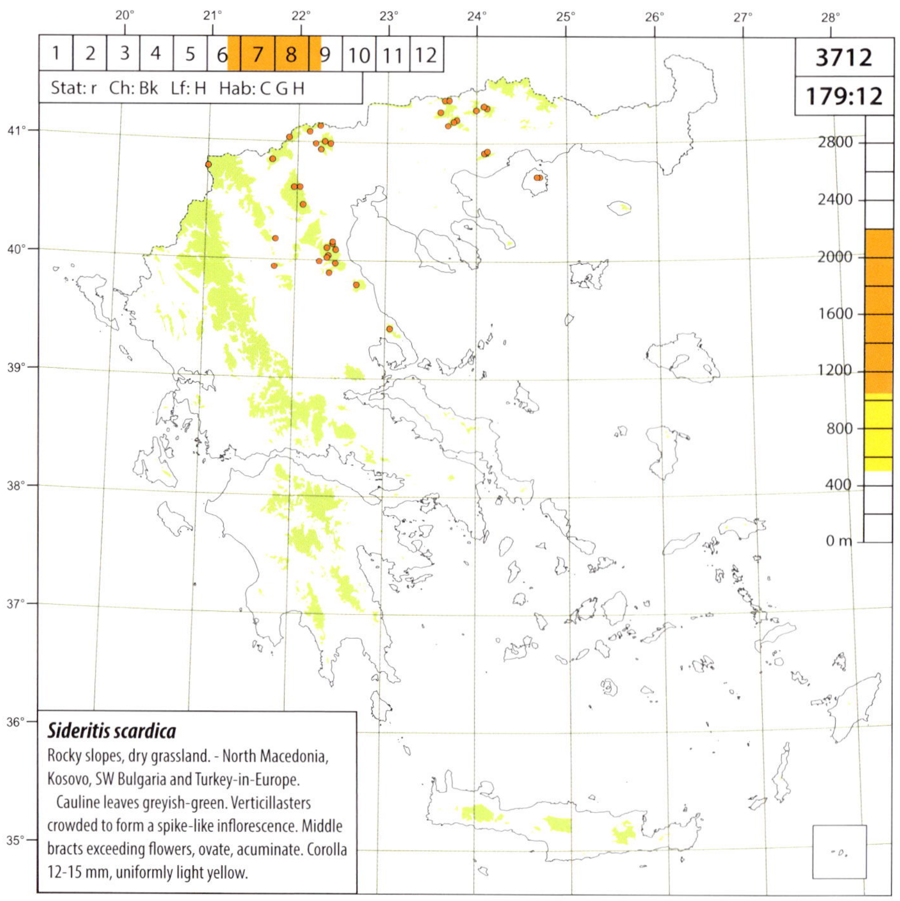

Sideritis scardica

Rocky slopes, dry grassland. - North Macedonia, Kosovo, SW Bulgaria and Turkey-in-Europe.

Cauline leaves greyish-green. Verticillasters crowded to form a spike-like inflorescence. Middle bracts exceeding flowers, ovate, acuminate. Corolla 12-15 mm, uniformly light yellow.

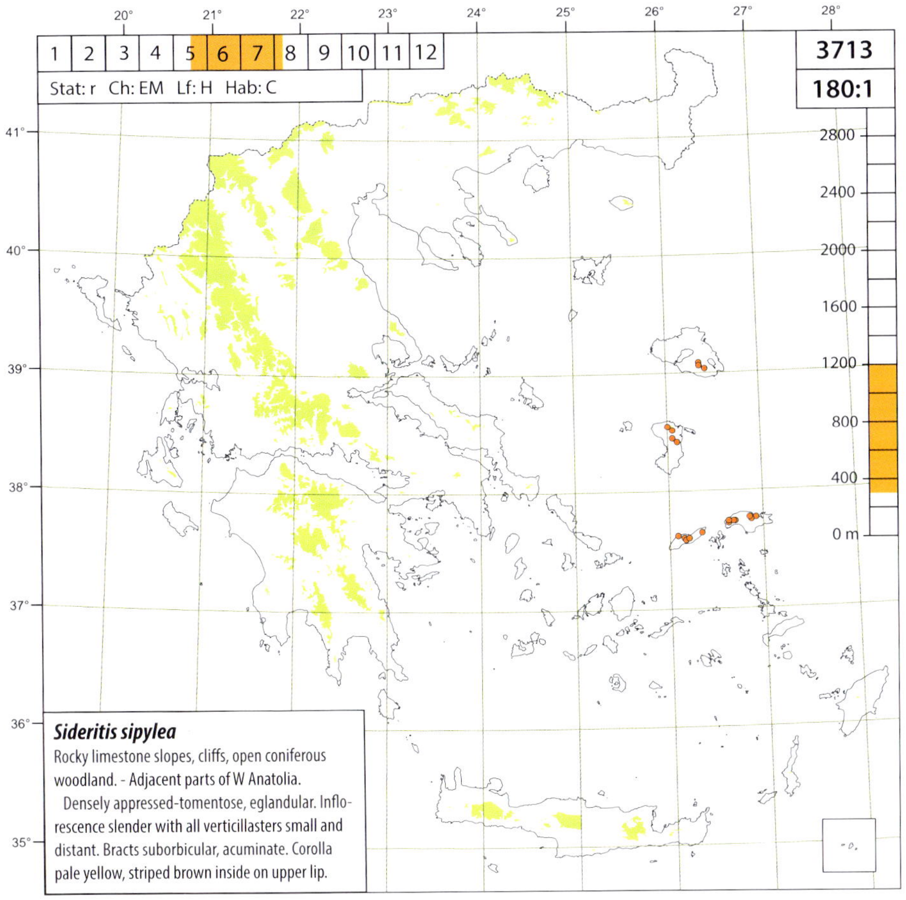

Sideritis sipylea
Rocky limestone slopes, cliffs, open coniferous woodland. - Adjacent parts of W Anatolia.
Densely appressed-tomentose, eglandular. Inflorescence slender with all verticillasters small and distant. Bracts suborbicular, acuminate. Corolla pale yellow, striped brown inside on upper lip.

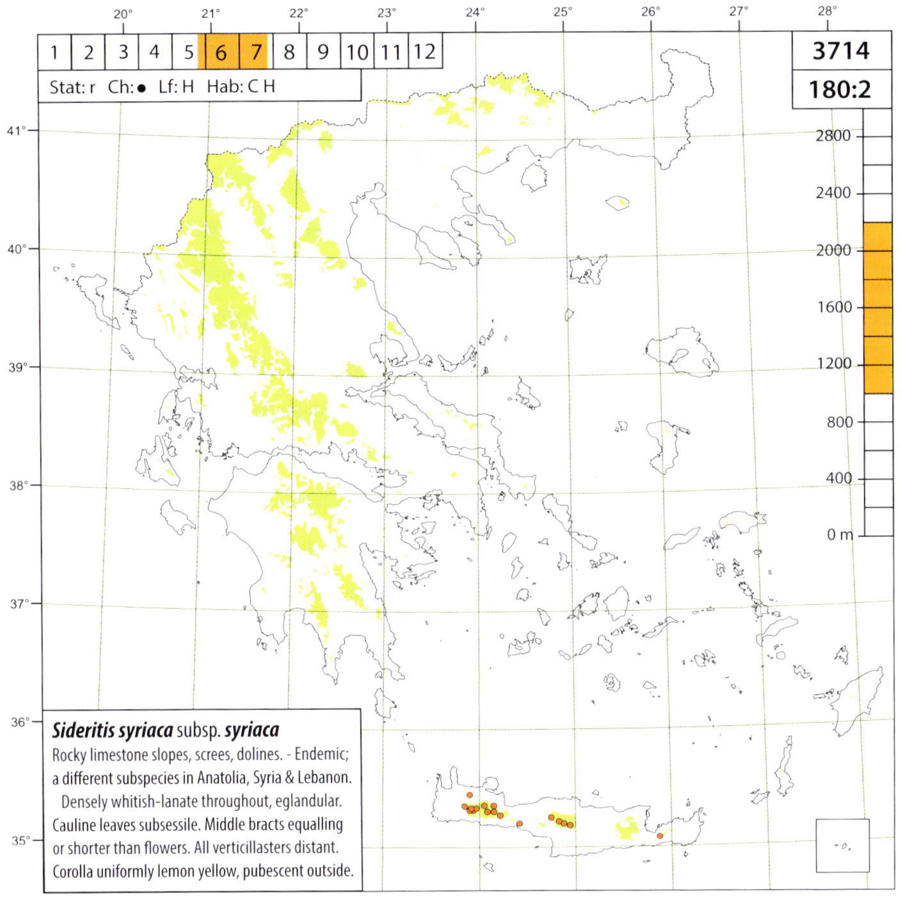

Sideritis syriaca* subsp. *syriaca
Rocky limestone slopes, screes, dolines. - Endemic; a different subspecies in Anatolia, Syria & Lebanon.
Densely whitish-lanate throughout, eglandular. Cauline leaves subsessile. Middle bracts equalling or shorter than flowers. All verticillasters distant. Corolla uniformly lemon yellow, pubescent outside.

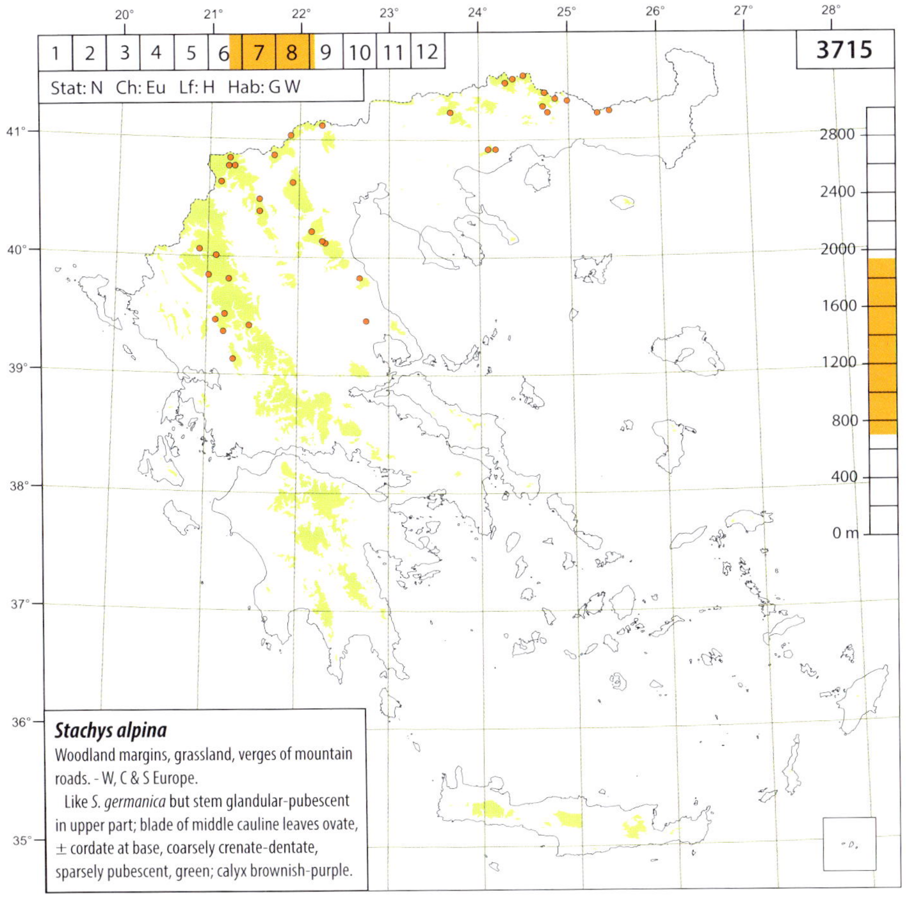

Stachys alpina
Woodland margins, grassland, verges of mountain roads. - W, C & S Europe.
Like *S. germanica* but stem glandular-pubescent in upper part; blade of middle cauline leaves ovate, ± cordate at base, coarsely crenate-dentate, sparsely pubescent, green; calyx brownish-purple.

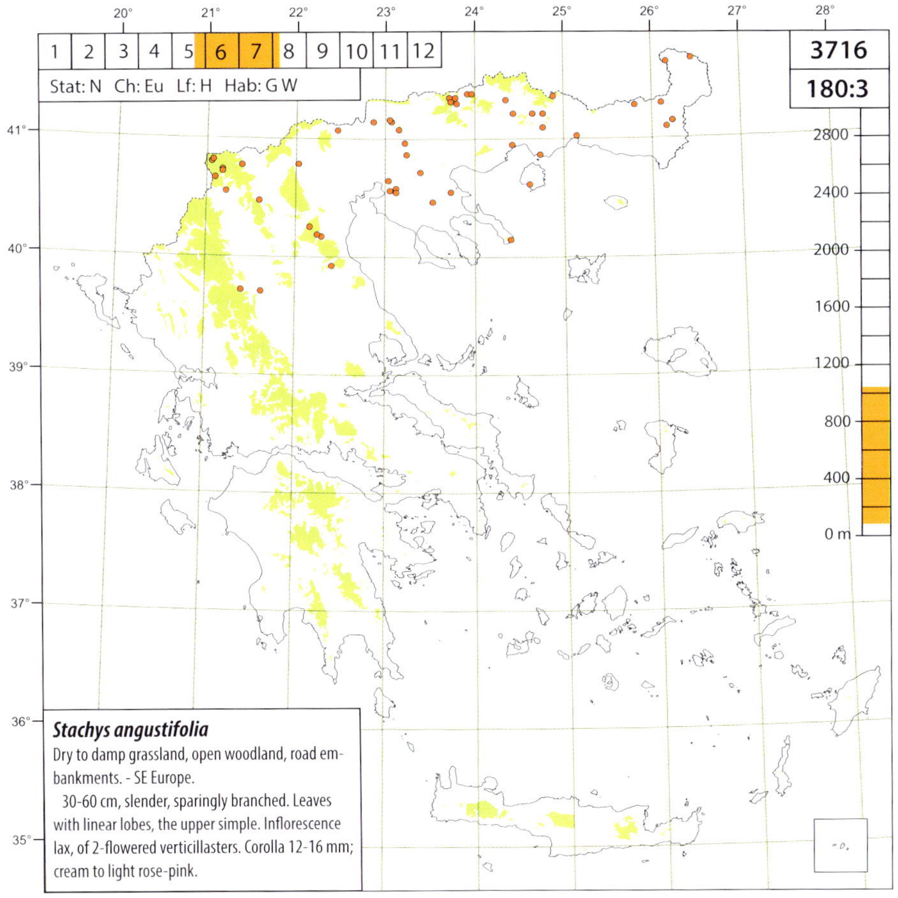

Stachys angustifolia
Dry to damp grassland, open woodland, road embankments. - SE Europe.
30-60 cm, slender, sparingly branched. Leaves with linear lobes, the upper simple. Inflorescence lax, of 2-flowered verticillasters. Corolla 12-16 mm; cream to light rose-pink.

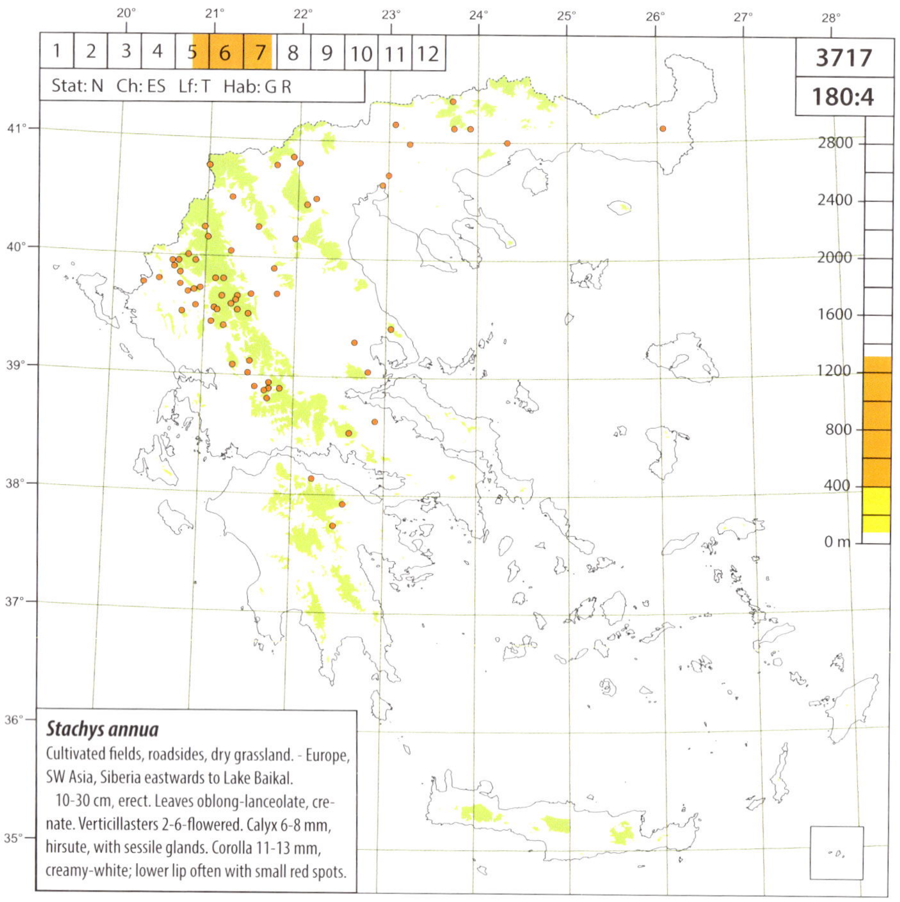

Stachys annua
Cultivated fields, roadsides, dry grassland. - Europe, SW Asia, Siberia eastwards to Lake Baikal.
10-30 cm, erect. Leaves oblong-lanceolate, crenate. Verticillasters 2-6-flowered. Calyx 6-8 mm, hirsute, with sessile glands. Corolla 11-13 mm, creamy-white; lower lip often with small red spots.

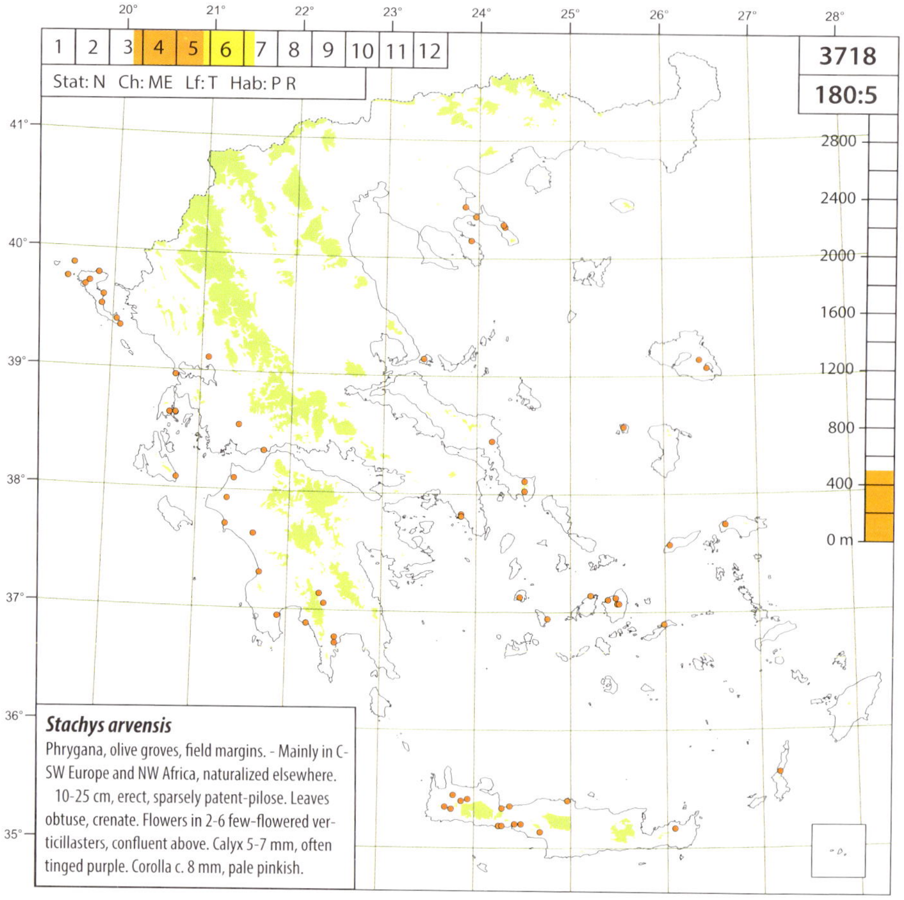

Stachys arvensis
Phrygana, olive groves, field margins. - Mainly in C-SW Europe and NW Africa, naturalized elsewhere.
10-25 cm, erect, sparsely patent-pilose. Leaves obtuse, crenate. Flowers in 2-6 few-flowered verticillasters, confluent above. Calyx 5-7 mm, often tinged purple. Corolla c. 8 mm, pale pinkish.

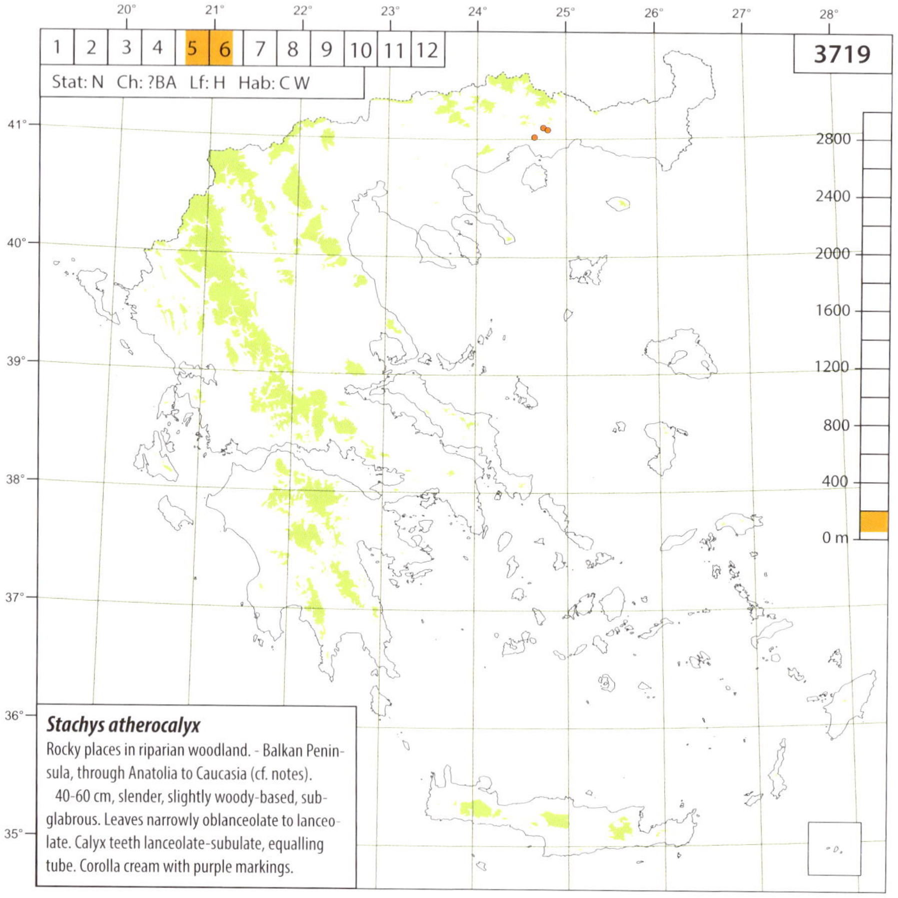

Stachys atherocalyx
Rocky places in riparian woodland. - Balkan Peninsula, through Anatolia to Caucasia (cf. notes).
40-60 cm, slender, slightly woody-based, subglabrous. Leaves narrowly oblanceolate to lanceolate. Calyx teeth lanceolate-subulate, equalling tube. Corolla cream with purple markings.

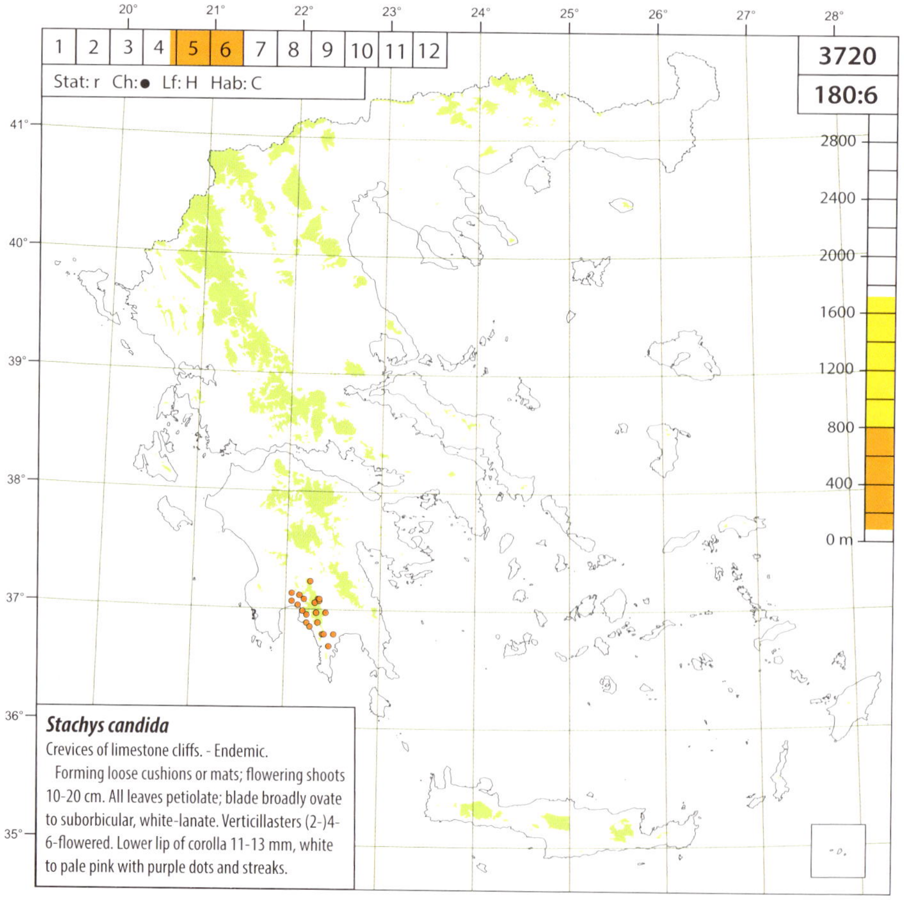

Stachys candida
Crevices of limestone cliffs. - Endemic.
Forming loose cushions or mats; flowering shoots 10-20 cm. All leaves petiolate; blade broadly ovate to suborbicular, white-lanate. Verticillasters (2-)4-6-flowered. Lower lip of corolla 11-13 mm, white to pale pink with purple dots and streaks.

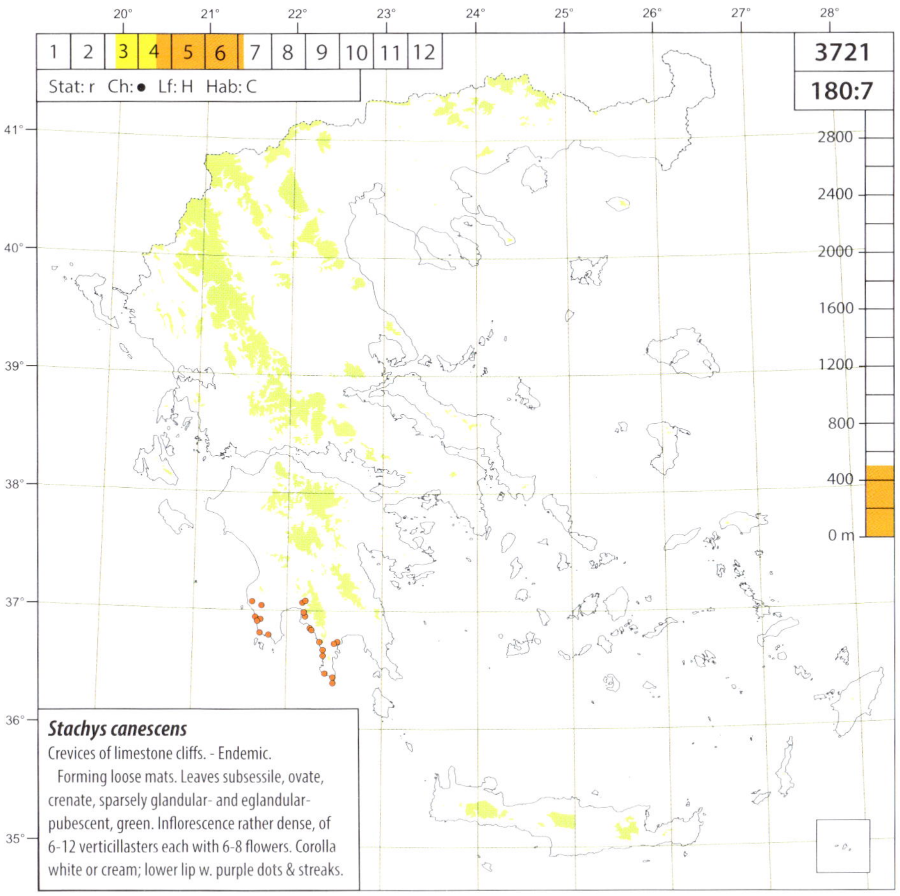

Stachys canescens
Crevices of limestone cliffs. - Endemic.
Forming loose mats. Leaves subsessile, ovate, crenate, sparsely glandular- and eglandular-pubescent, green. Inflorescence rather dense, of 6-12 verticillasters each with 6-8 flowers. Corolla white or cream; lower lip w. purple dots & streaks.

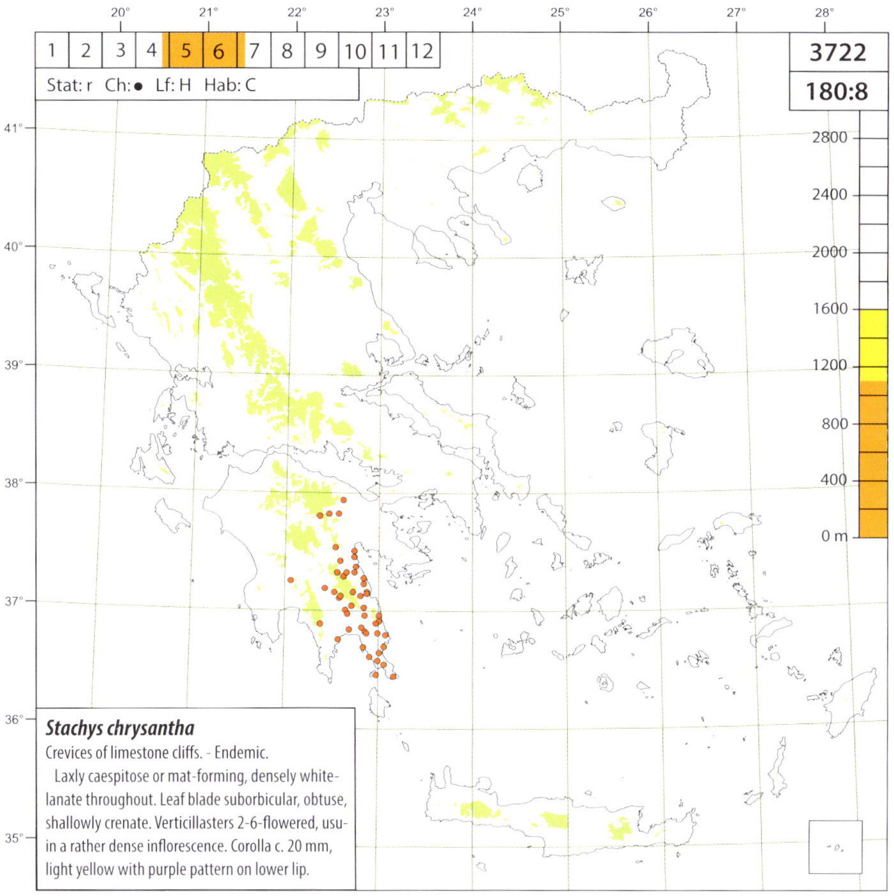

Stachys chrysantha
Crevices of limestone cliffs. - Endemic.
Laxly caespitose or mat-forming, densely white-lanate throughout. Leaf blade suborbicular, obtuse, shallowly crenate. Verticillasters 2-6-flowered, usu- in a rather dense inflorescence. Corolla c. 20 mm, light yellow with purple pattern on lower lip.

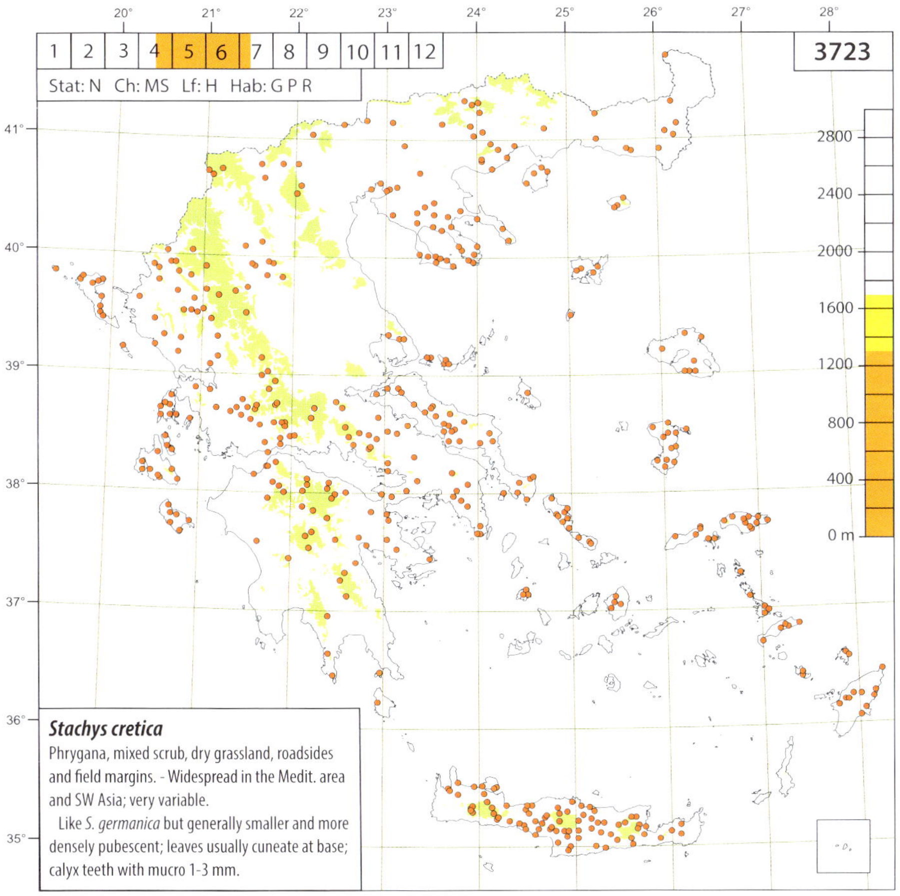

Stachys cretica
Phrygana, mixed scrub, dry grassland, roadsides and field margins. - Widespread in the Medit. area and SW Asia; very variable.
Like *S. germanica* but generally smaller and more densely pubescent; leaves usually cuneate at base; calyx teeth with mucro 1-3 mm.

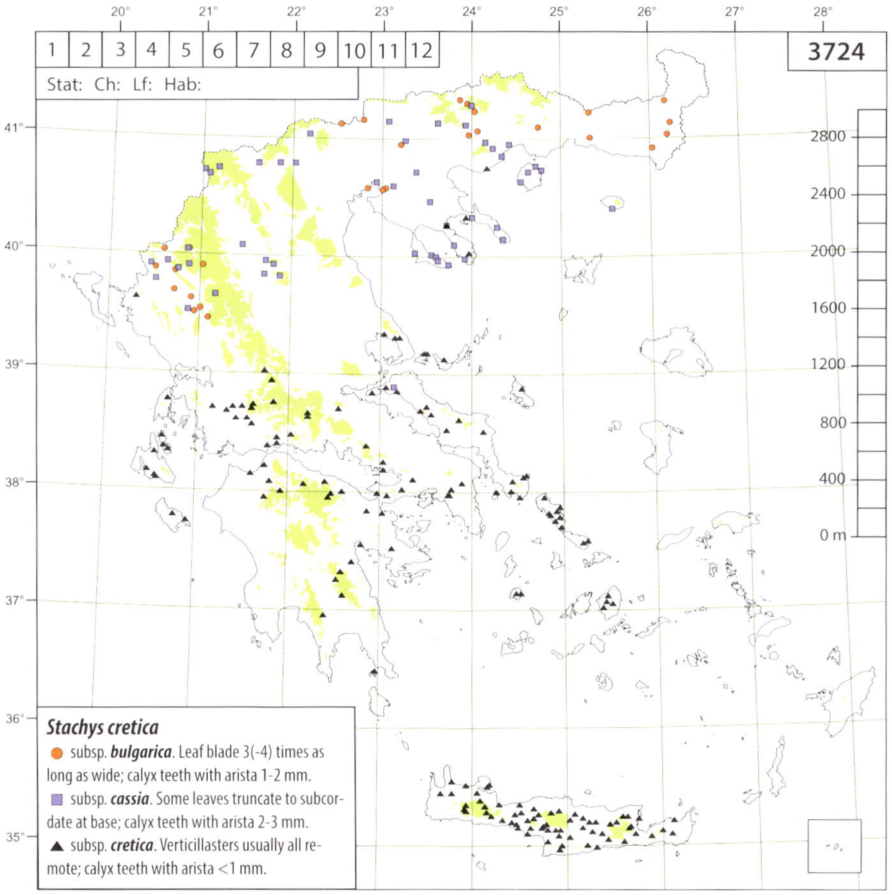

Stachys cretica
● subsp. ***bulgarica***. Leaf blade 3(-4) times as long as wide; calyx teeth with arista 1-2 mm.
■ subsp. ***cassia***. Some leaves truncate to subcordate at base; calyx teeth with arista 2-3 mm.
▲ subsp. ***cretica***. Verticillasters usually all remote; calyx teeth with arista <1 mm.

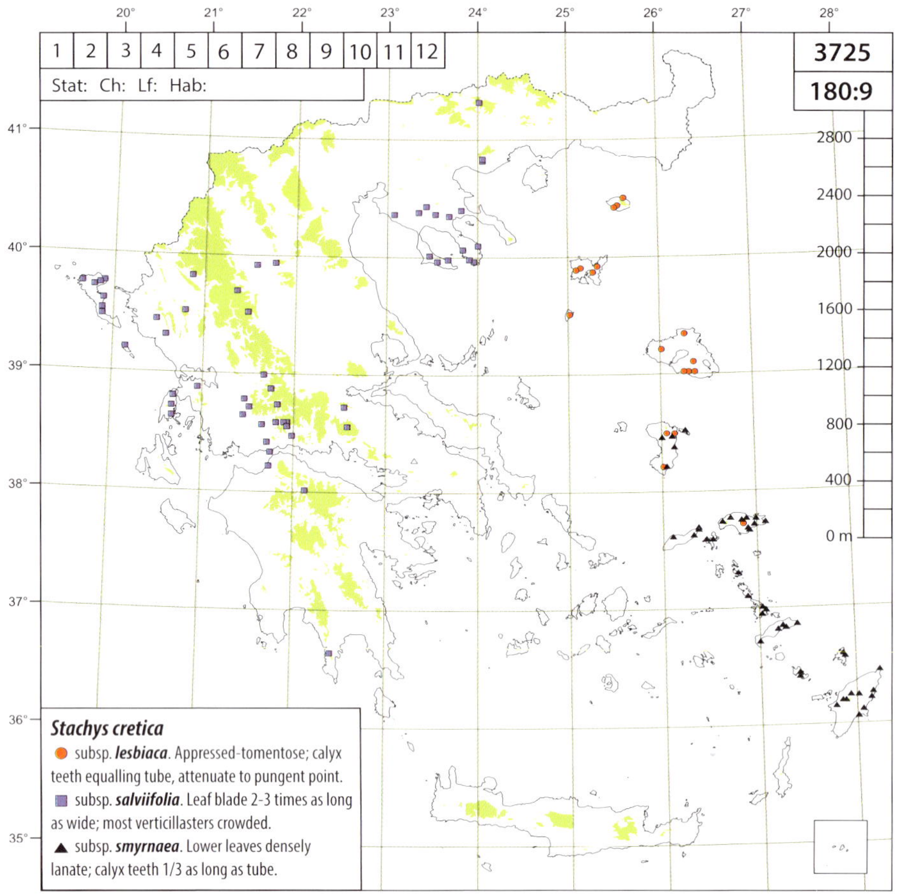

Stachys cretica

● subsp. ***lesbiaca***. Appressed-tomentose; calyx teeth equalling tube, attenuate to pungent point.

■ subsp. ***salviifolia***. Leaf blade 2-3 times as long as wide; most verticillasters crowded.

▲ subsp. ***smyrnaea***. Lower leaves densely lanate; calyx teeth 1/3 as long as tube.

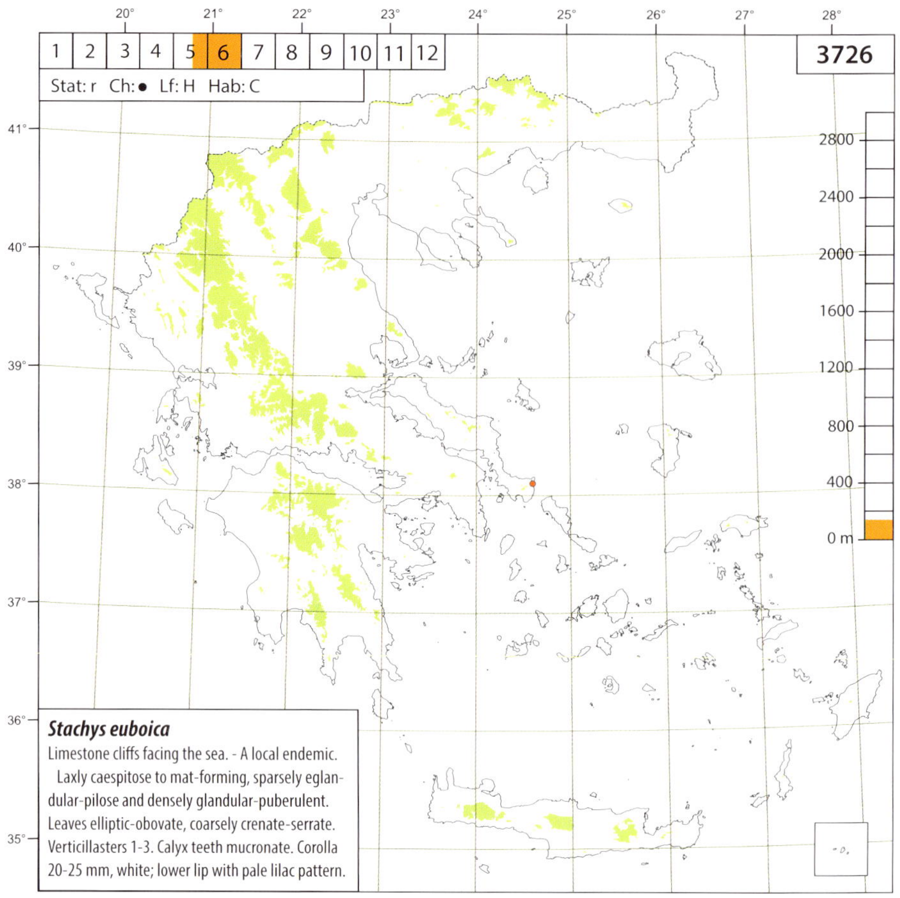

Stachys euboica

Limestone cliffs facing the sea. - A local endemic.

Laxly caespitose to mat-forming, sparsely eglandular-pilose and densely glandular-puberulent. Leaves elliptic-obovate, coarsely crenate-serrate. Verticillasters 1-3. Calyx teeth mucronate. Corolla 20-25 mm, white; lower lip with pale lilac pattern.

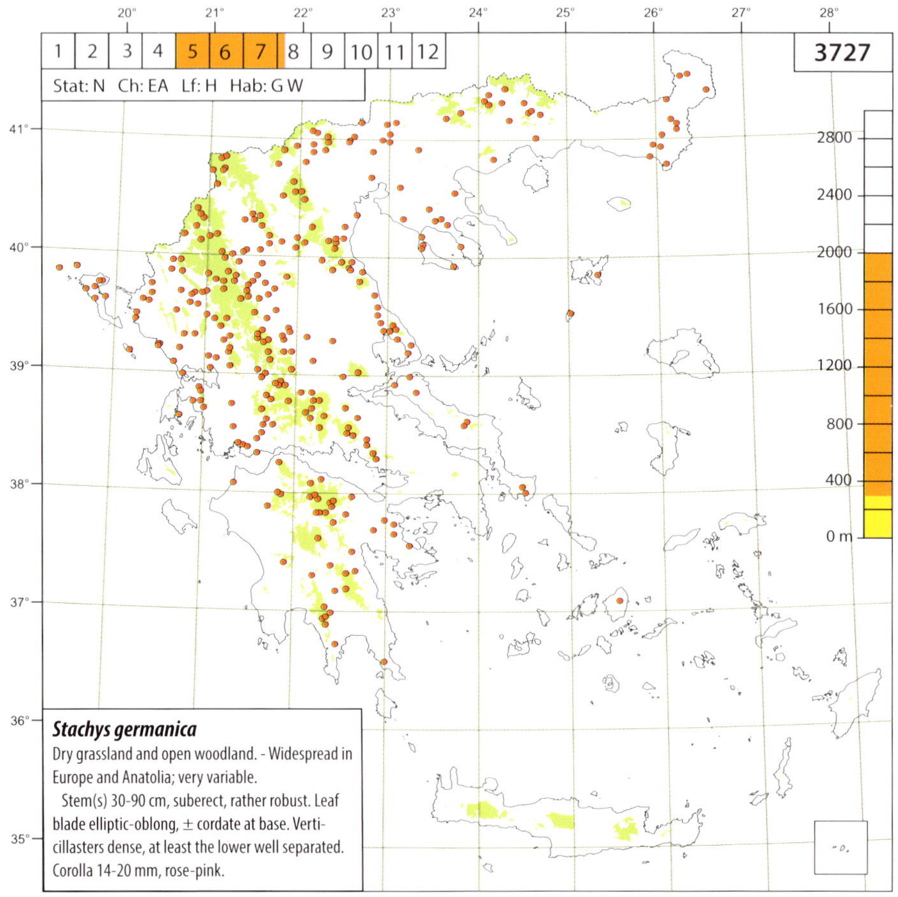

Stachys germanica

Dry grassland and open woodland. - Widespread in Europe and Anatolia; very variable.

Stem(s) 30-90 cm, suberect, rather robust. Leaf blade elliptic-oblong, ± cordate at base. Verticillasters dense, at least the lower well separated. Corolla 14-20 mm, rose-pink.

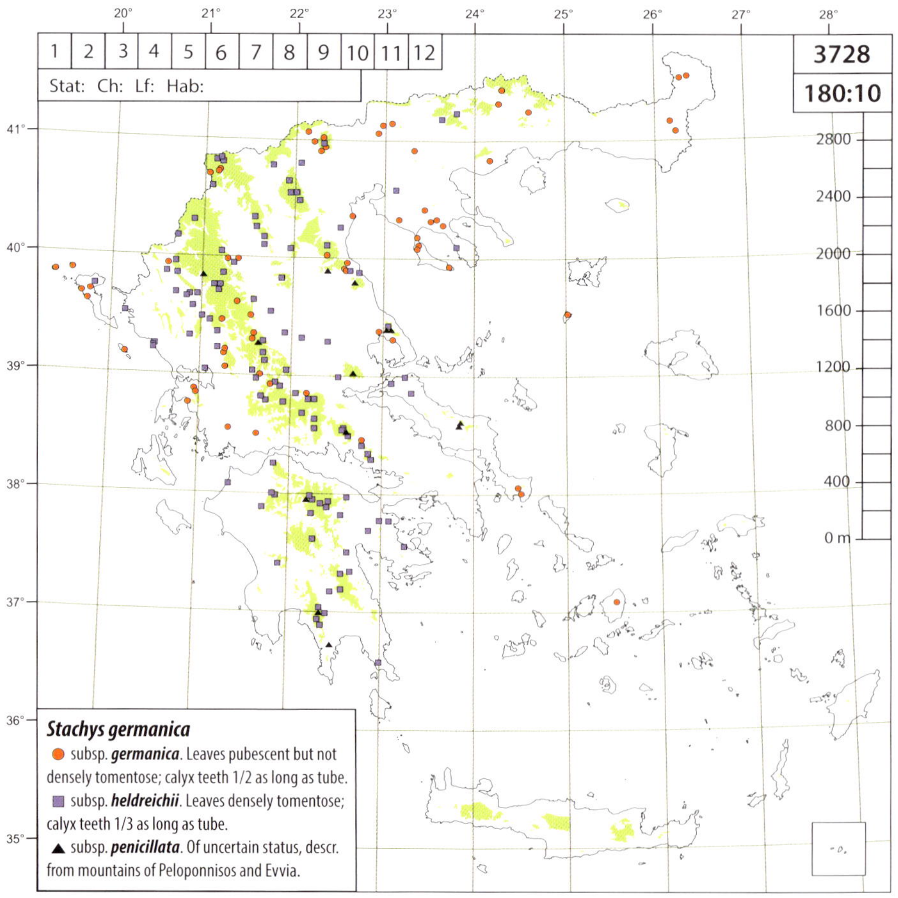

Stachys germanica

● subsp. ***germanica***. Leaves pubescent but not densely tomentose; calyx teeth 1/2 as long as tube.

■ subsp. ***heldreichii***. Leaves densely tomentose; calyx teeth 1/3 as long as tube.

▲ subsp. ***penicillata***. Of uncertain status, descr. from mountains of Peloponnisos and Evvia.

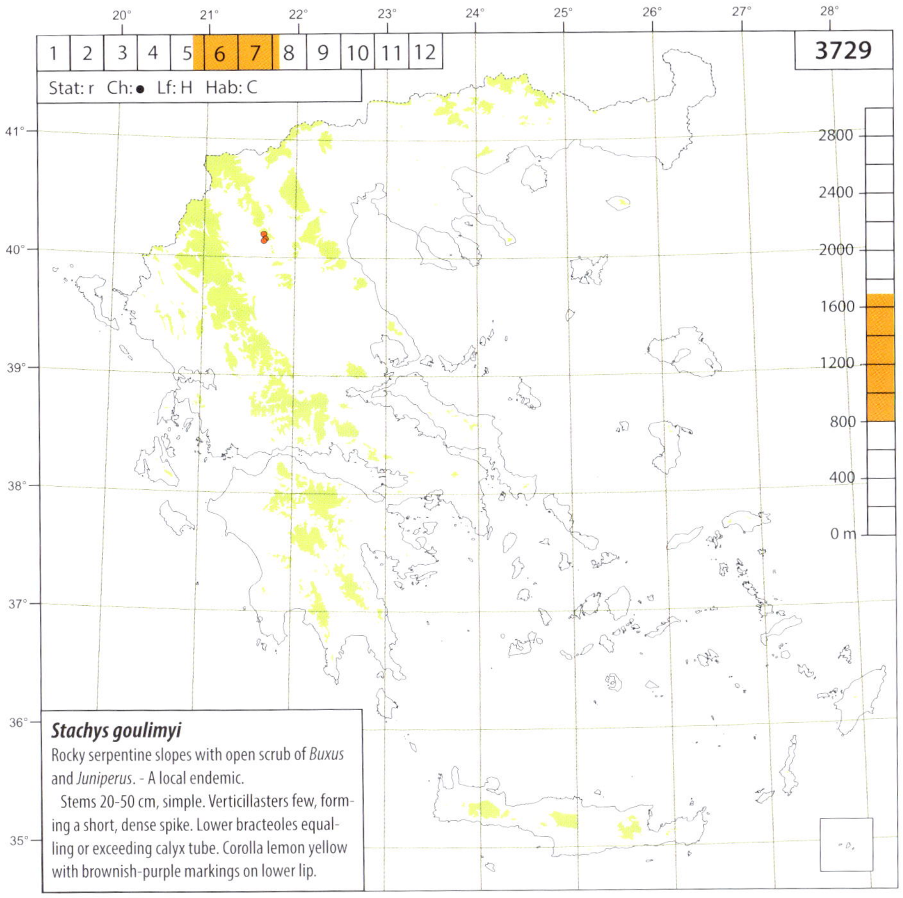

Stachys goulimyi

Rocky serpentine slopes with open scrub of *Buxus* and *Juniperus*. - A local endemic.

Stems 20-50 cm, simple. Verticillasters few, forming a short, dense spike. Lower bracteoles equalling or exceeding calyx tube. Corolla lemon yellow with brownish-purple markings on lower lip.

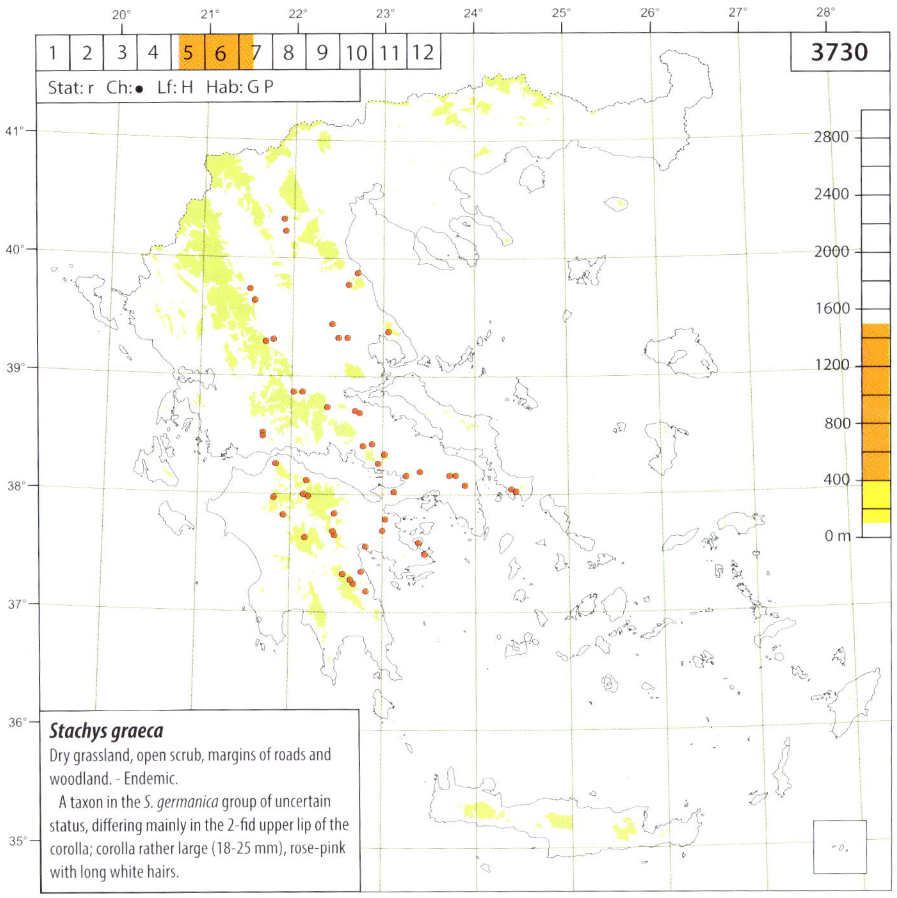

Stachys graeca

Dry grassland, open scrub, margins of roads and woodland. - Endemic.

A taxon in the *S. germanica* group of uncertain status, differing mainly in the 2-fid upper lip of the corolla; corolla rather large (18-25 mm), rose-pink with long white hairs.

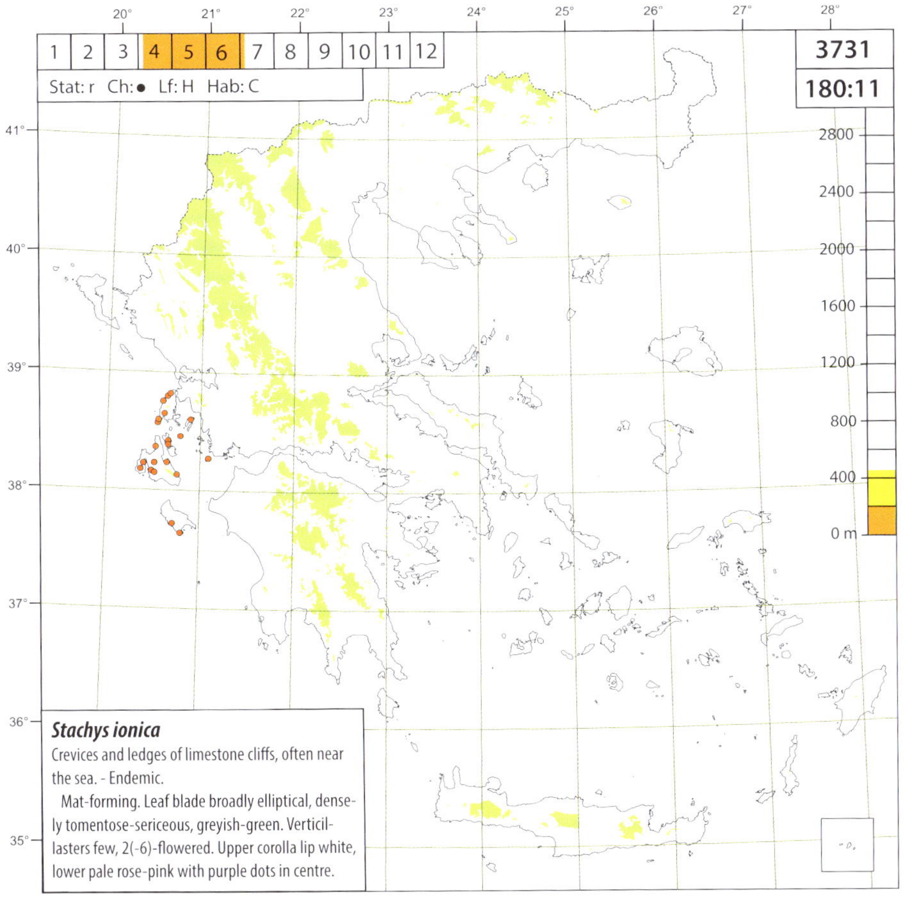

Stachys ionica

Crevices and ledges of limestone cliffs, often near the sea. - Endemic.

Mat-forming. Leaf blade broadly elliptical, densely tomentose-sericeous, greyish-green. Verticillasters few, 2(-6)-flowered. Upper corolla lip white, lower pale rose-pink with purple dots in centre.

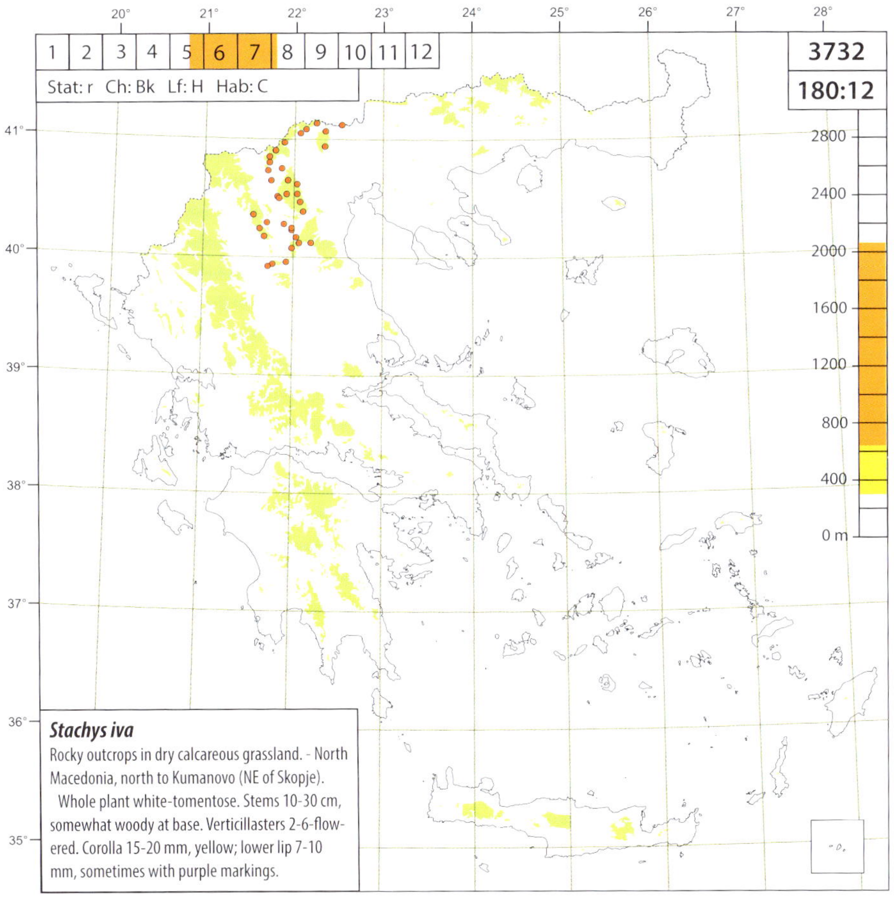

Stachys iva

Rocky outcrops in dry calcareous grassland. - North Macedonia, north to Kumanovo (NE of Skopje).

Whole plant white-tomentose. Stems 10-30 cm, somewhat woody at base. Verticillasters 2-6-flowered. Corolla 15-20 mm, yellow; lower lip 7-10 mm, sometimes with purple markings.

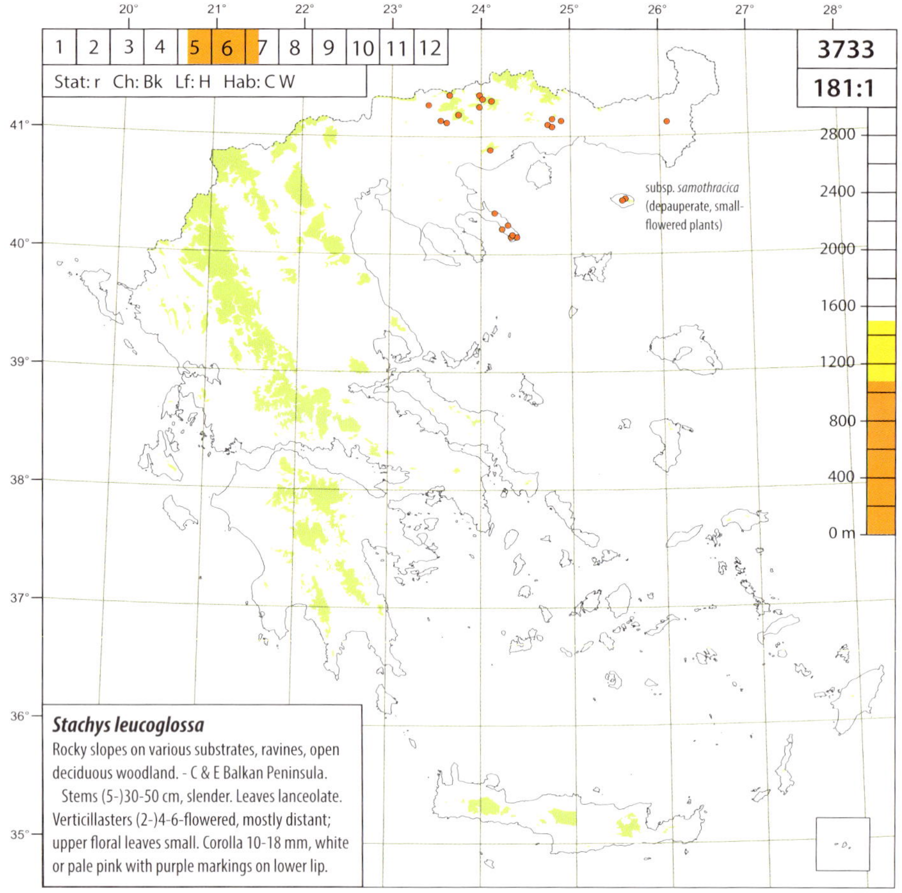

Stachys leucoglossa

Rocky slopes on various substrates, ravines, open deciduous woodland. - C & E Balkan Peninsula.

Stems (5-)30-50 cm, slender. Leaves lanceolate. Verticillasters (2-)4-6-flowered, mostly distant; upper floral leaves small. Corolla 10-18 mm, white or pale pink with purple markings on lower lip.

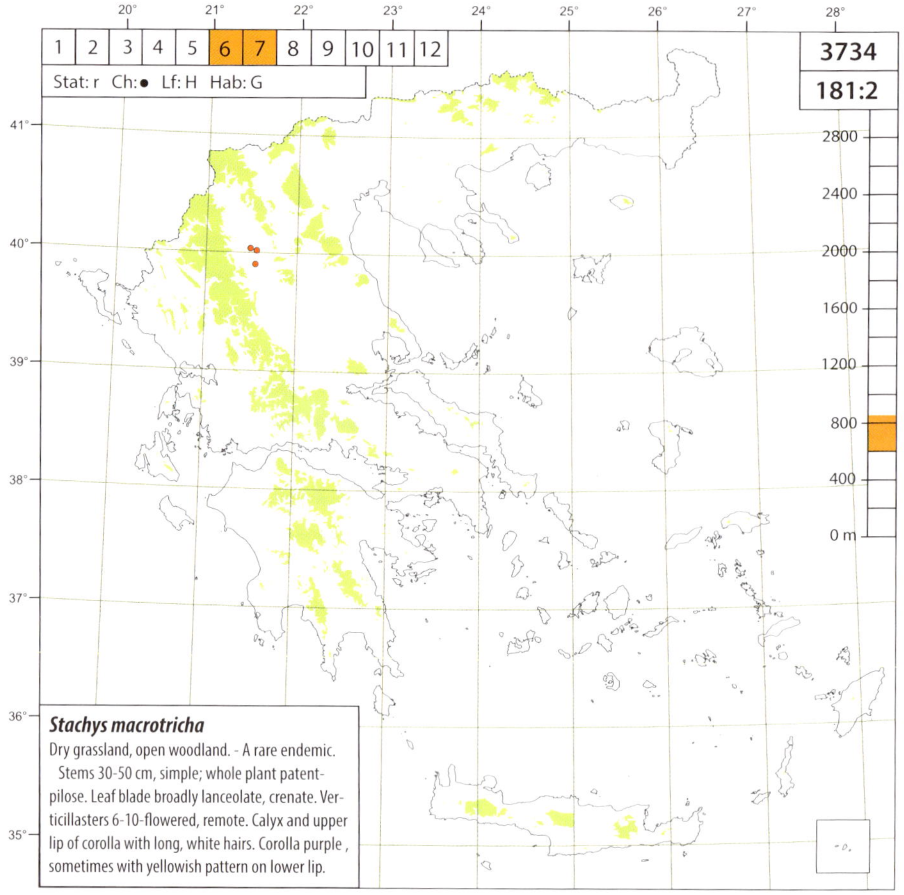

Stachys macrotricha

Dry grassland, open woodland. - A rare endemic.

Stems 30-50 cm, simple; whole plant patent-pilose. Leaf blade broadly lanceolate, crenate. Verticillasters 6-10-flowered, remote. Calyx and upper lip of corolla with long, white hairs. Corolla purple, sometimes with yellowish pattern on lower lip.

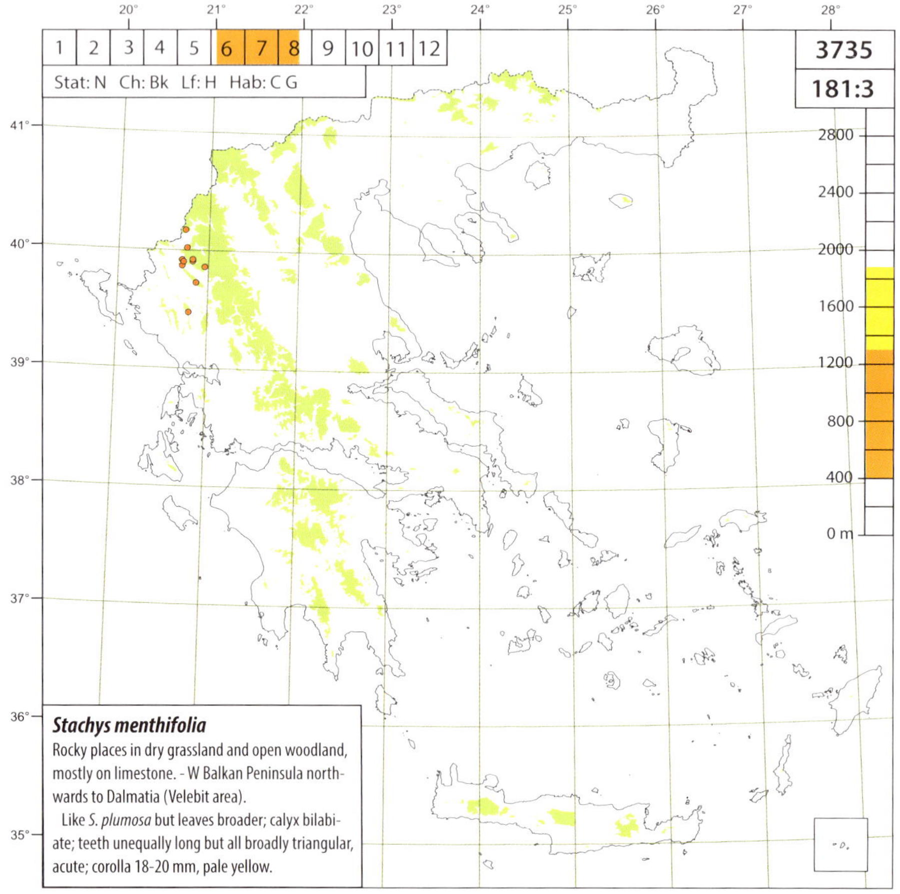

Stachys menthifolia

Rocky places in dry grassland and open woodland, mostly on limestone. - W Balkan Peninsula northwards to Dalmatia (Velebit area).

Like *S. plumosa* but leaves broader; calyx bilabiate; teeth unequally long but all broadly triangular, acute; corolla 18-20 mm, pale yellow.

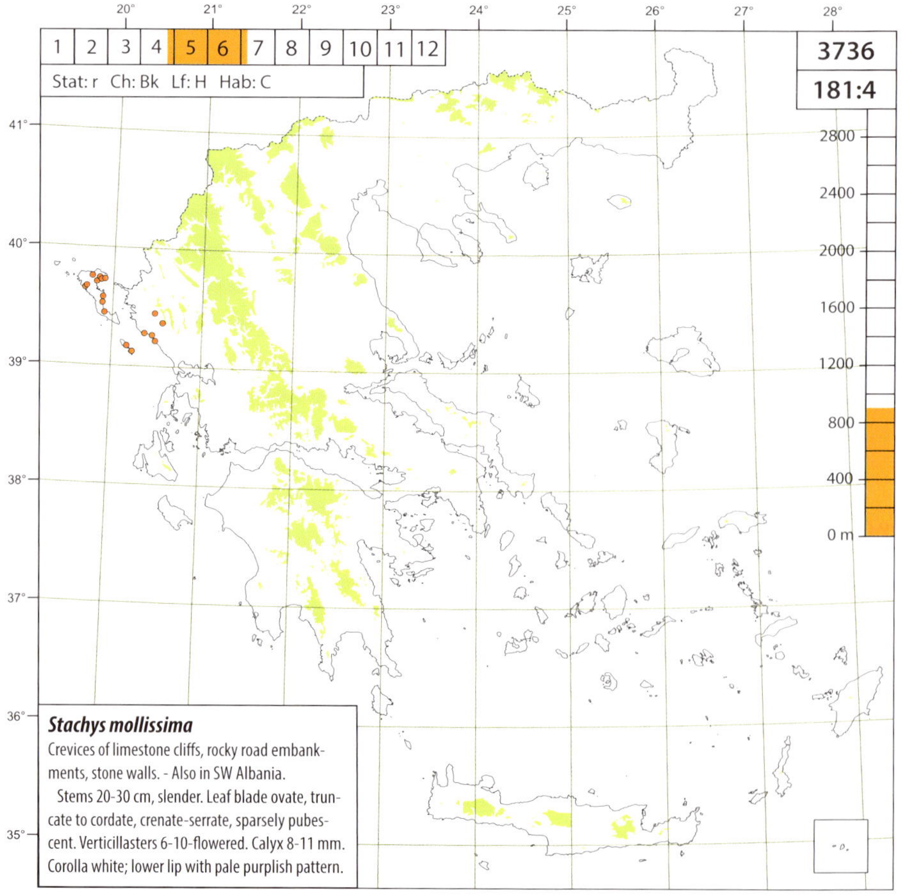

Stachys mollissima

Crevices of limestone cliffs, rocky road embankments, stone walls. - Also in SW Albania.

Stems 20-30 cm, slender. Leaf blade ovate, truncate to cordate, crenate-serrate, sparsely pubescent. Verticillasters 6-10-flowered. Calyx 8-11 mm. Corolla white; lower lip with pale purplish pattern.

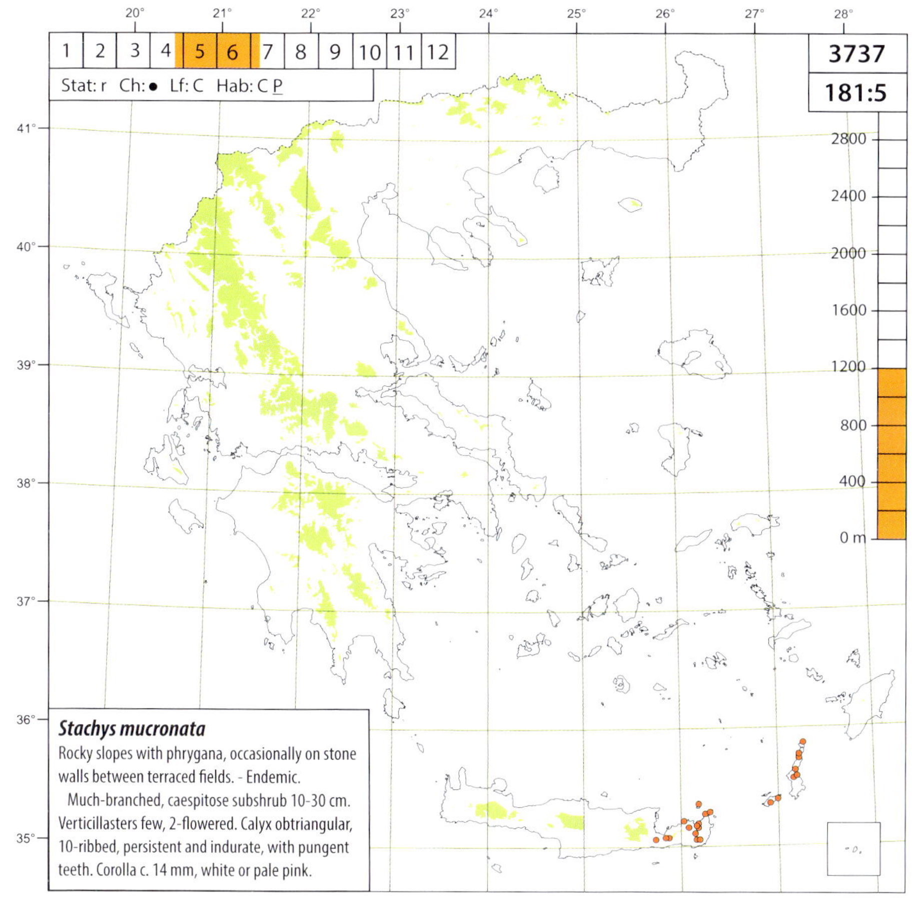

Stachys mucronata
Rocky slopes with phrygana, occasionally on stone walls between terraced fields. - Endemic.
Much-branched, caespitose subshrub 10-30 cm. Verticillasters few, 2-flowered. Calyx obtriangular, 10-ribbed, persistent and indurate, with pungent teeth. Corolla c. 14 mm, white or pale pink.

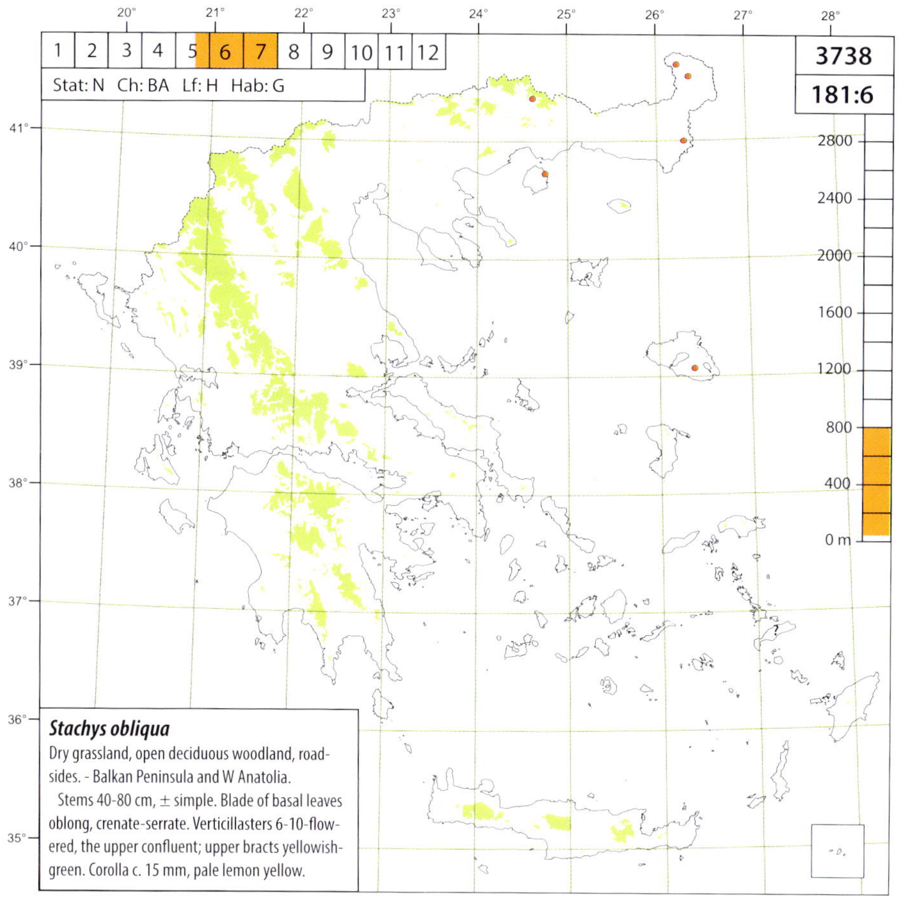

Stachys obliqua
Dry grassland, open deciduous woodland, roadsides. - Balkan Peninsula and W Anatolia.
Stems 40-80 cm, ± simple. Blade of basal leaves oblong, crenate-serrate. Verticillasters 6-10-flowered, the upper confluent; upper bracts yellowish-green. Corolla c. 15 mm, pale lemon yellow.

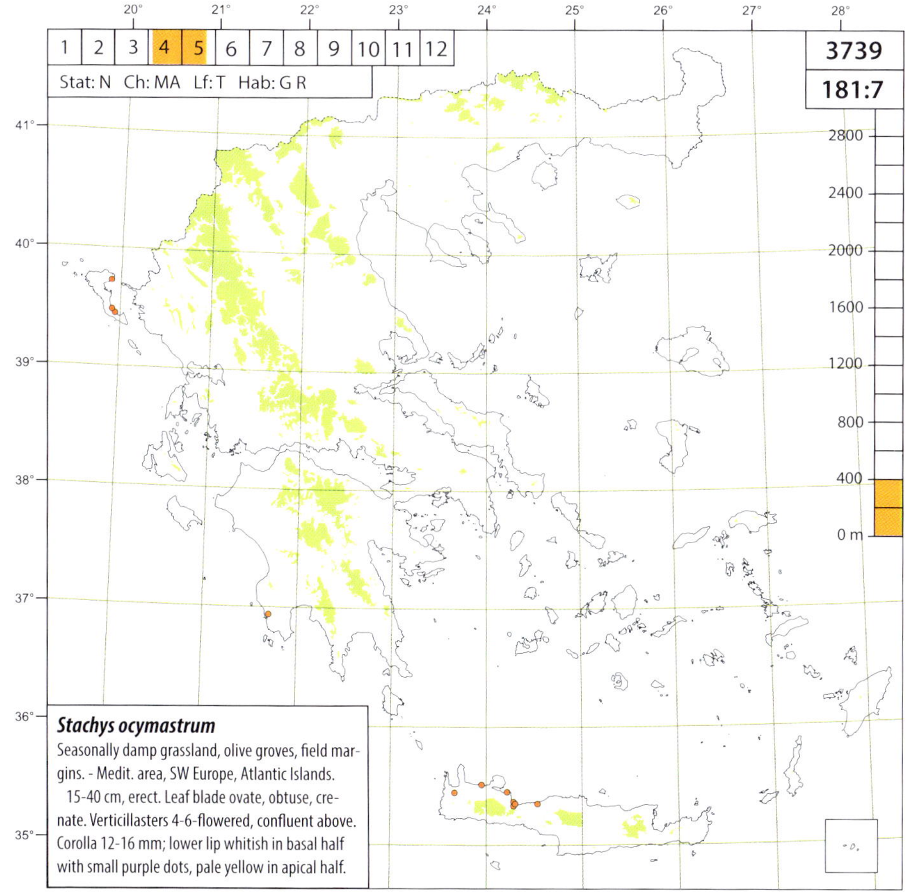

Stachys ocymastrum
Seasonally damp grassland, olive groves, field margins. - Medit. area, SW Europe, Atlantic Islands.
15-40 cm, erect. Leaf blade ovate, obtuse, crenate. Verticillasters 4-6-flowered, confluent above. Corolla 12-16 mm; lower lip whitish in basal half with small purple dots, pale yellow in apical half.

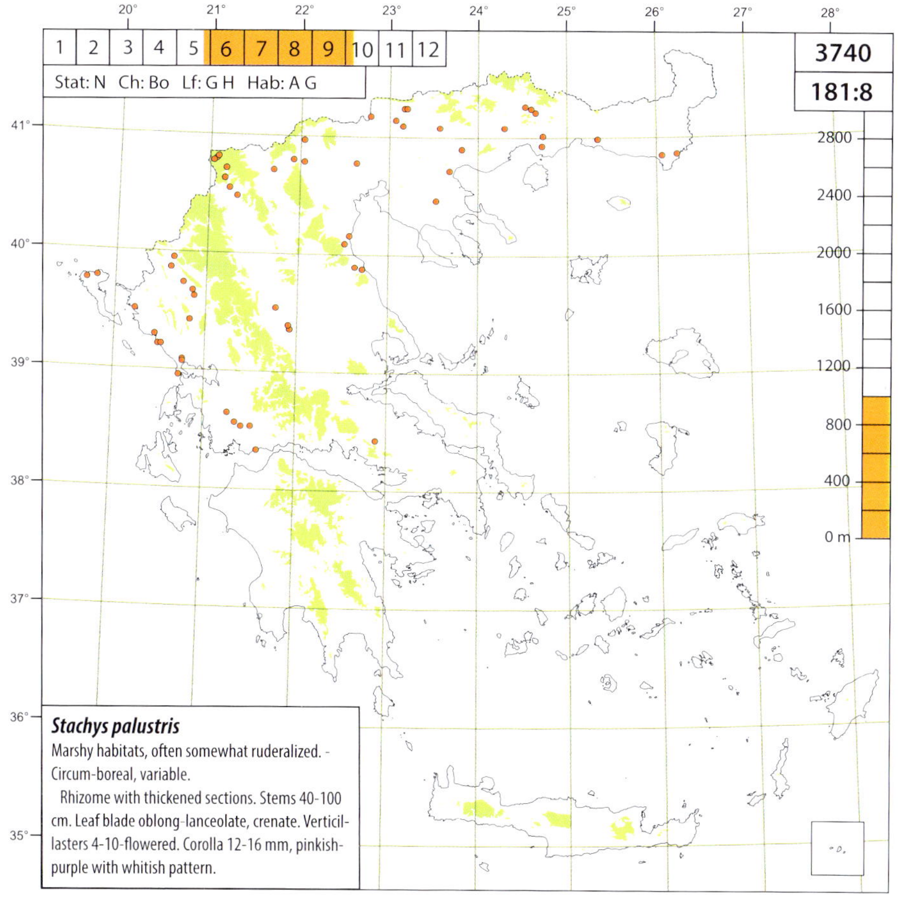

Stachys palustris
Marshy habitats, often somewhat ruderalized. - Circum-boreal, variable.
Rhizome with thickened sections. Stems 40-100 cm. Leaf blade oblong-lanceolate, crenate. Verticillasters 4-10-flowered. Corolla 12-16 mm, pinkish-purple with whitish pattern.

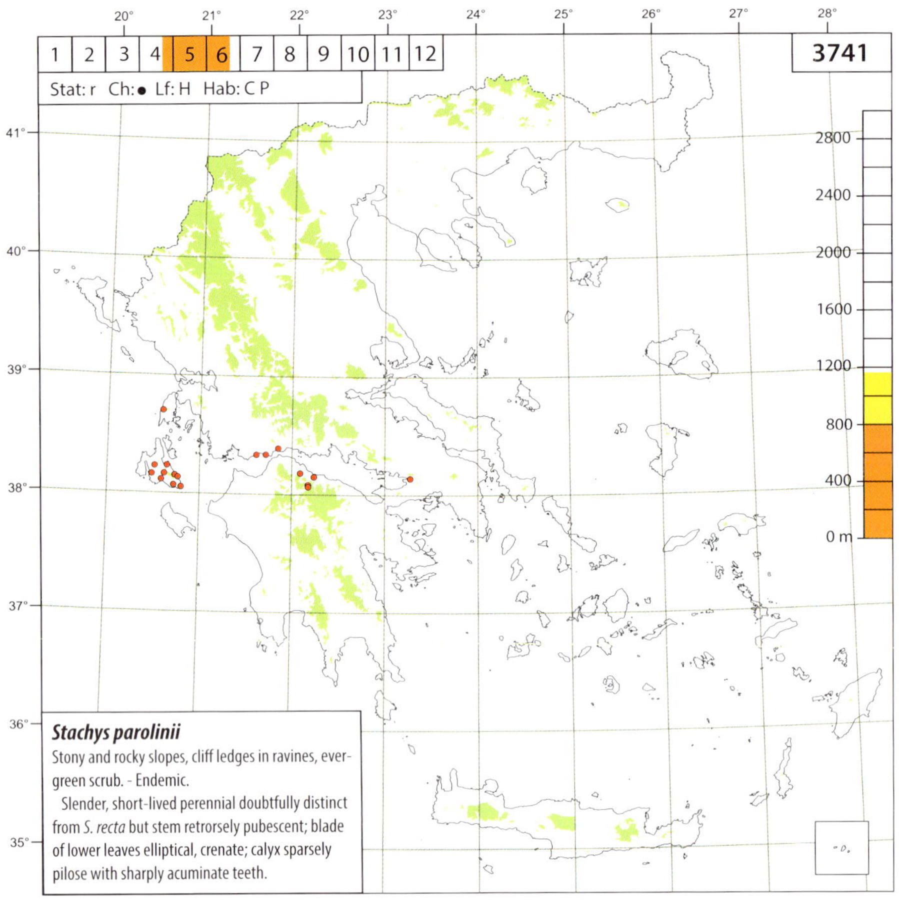

Stachys parolinii
Stony and rocky slopes, cliff ledges in ravines, evergreen scrub. - Endemic.
Slender, short-lived perennial doubtfully distinct from *S. recta* but stem retrorsely pubescent; blade of lower leaves elliptical, crenate; calyx sparsely pilose with sharply acuminate teeth.

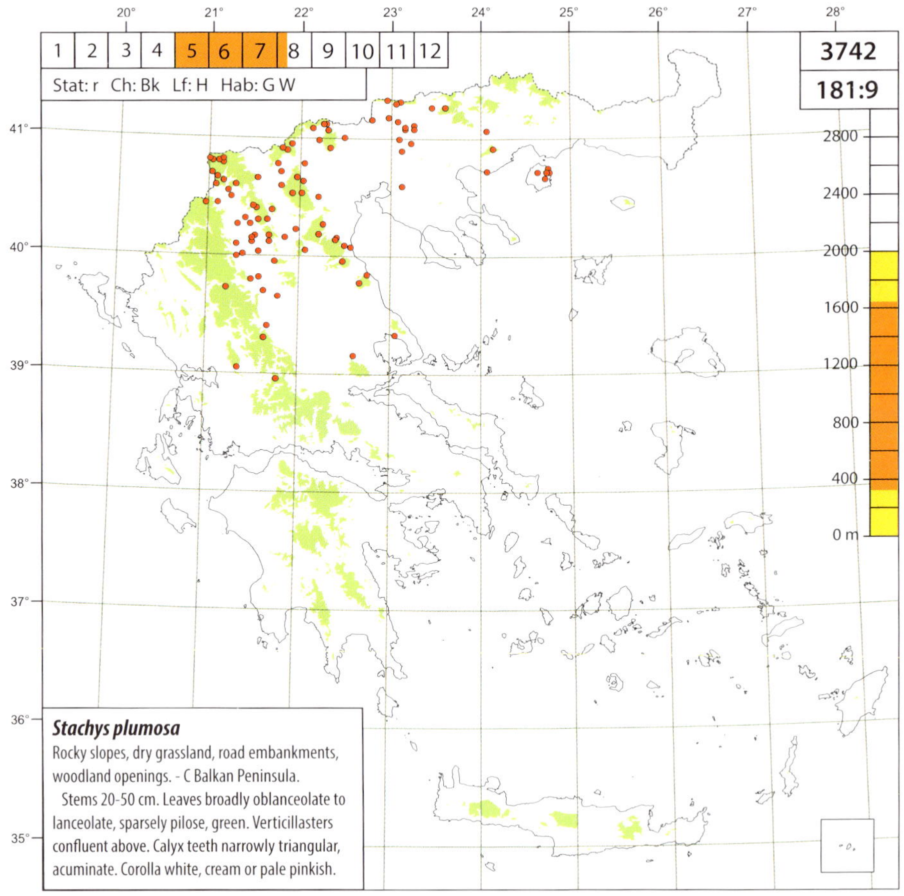

Stachys plumosa
Rocky slopes, dry grassland, road embankments, woodland openings. - C Balkan Peninsula.
Stems 20-50 cm. Leaves broadly oblanceolate to lanceolate, sparsely pilose, green. Verticillasters confluent above. Calyx teeth narrowly triangular, acuminate. Corolla white, cream or pale pinkish.

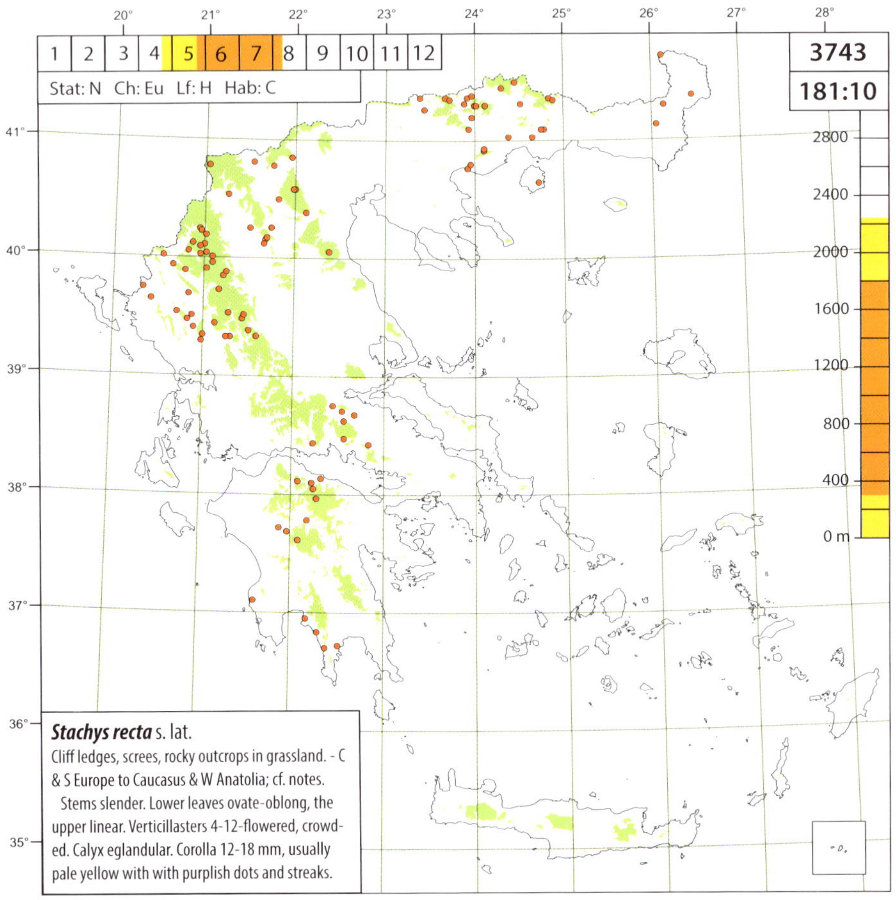

Stachys recta s. lat.
Cliff ledges, screes, rocky outcrops in grassland. - C & S Europe to Caucasus & W Anatolia; cf. notes.
Stems slender. Lower leaves ovate-oblong, the upper linear. Verticillasters 4-12-flowered, crowded. Calyx eglandular. Corolla 12-18 mm, usually pale yellow with with purplish dots and streaks.

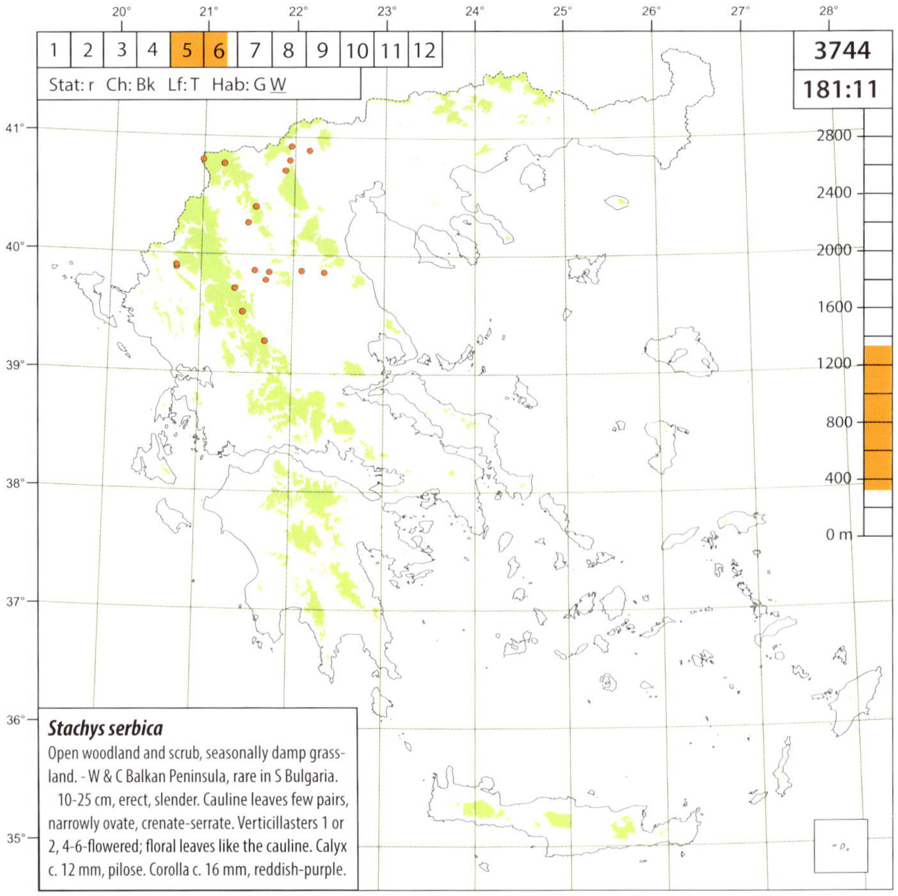

Stachys serbica
Open woodland and scrub, seasonally damp grassland. - W & C Balkan Peninsula, rare in S Bulgaria.
10-25 cm, erect, slender. Cauline leaves few pairs, narrowly ovate, crenate-serrate. Verticillasters 1 or 2, 4-6-flowered; floral leaves like the cauline. Calyx c. 12 mm, pilose. Corolla c. 16 mm, reddish-purple.

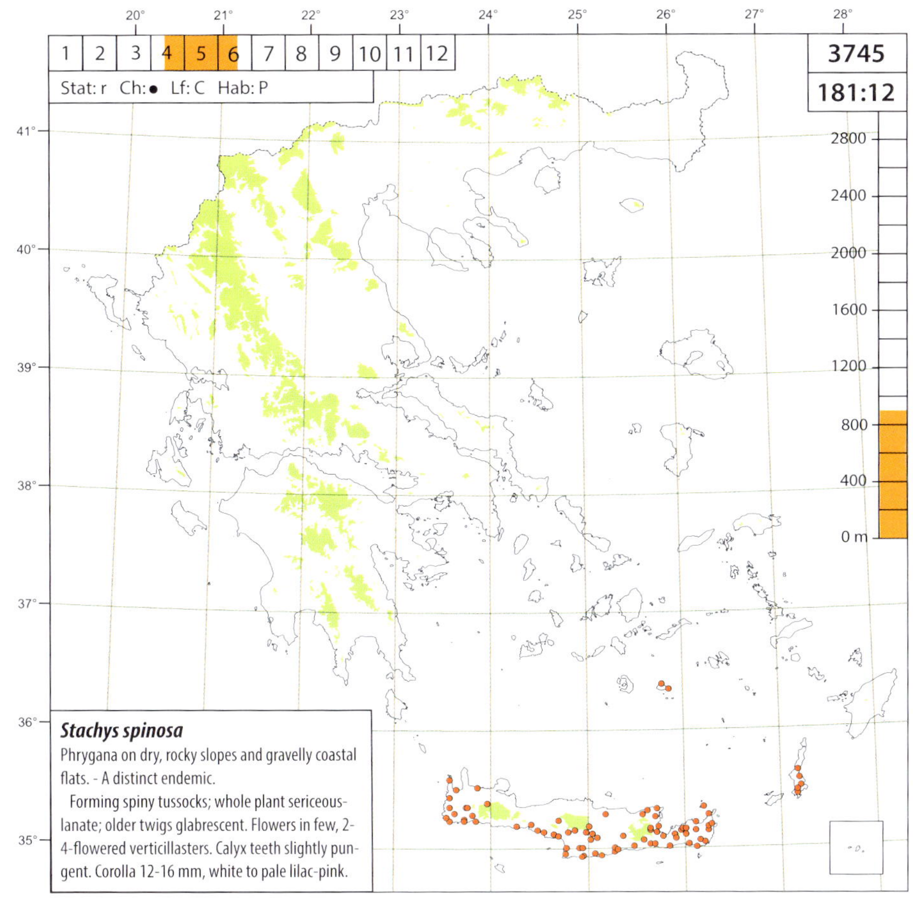

Stachys spinosa

Phrygana on dry, rocky slopes and gravelly coastal flats. - A distinct endemic.

Forming spiny tussocks; whole plant sericeous-lanate; older twigs glabrescent. Flowers in few, 2-4-flowered verticillasters. Calyx teeth slightly pungent. Corolla 12-16 mm, white to pale lilac-pink.

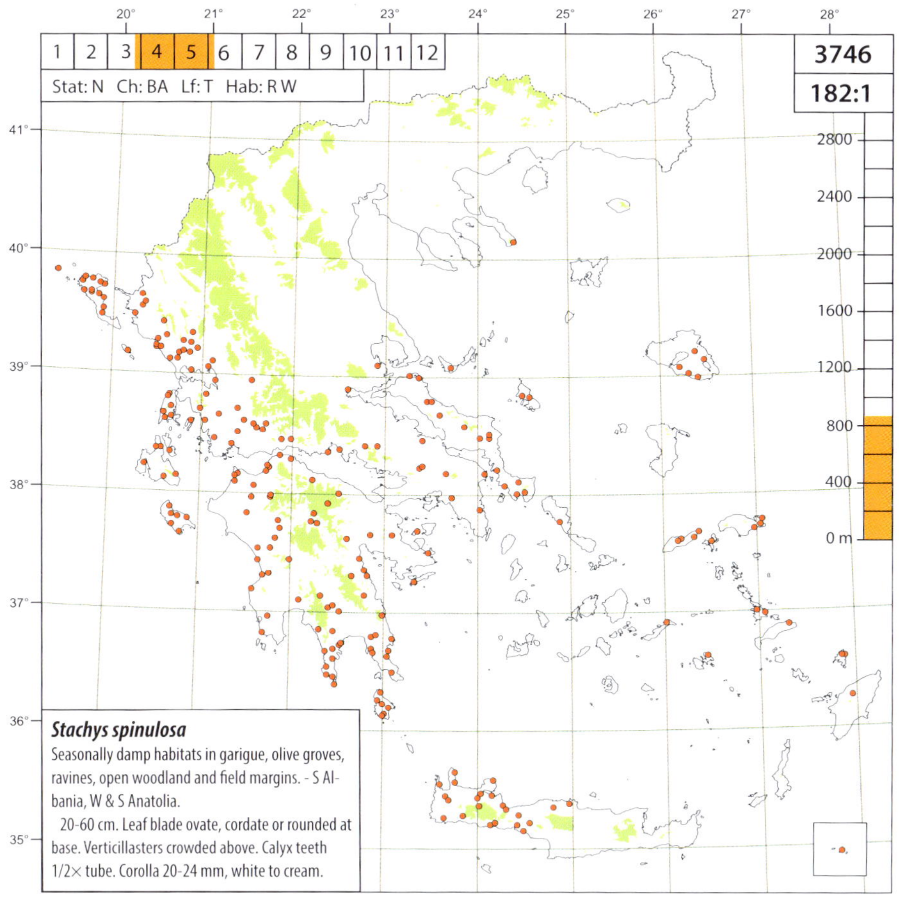

Stachys spinulosa

Seasonally damp habitats in garigue, olive groves, ravines, open woodland and field margins. - S Albania, W & S Anatolia.

20-60 cm. Leaf blade ovate, cordate or rounded at base. Verticillasters crowded above. Calyx teeth 1/2× tube. Corolla 20-24 mm, white to cream.

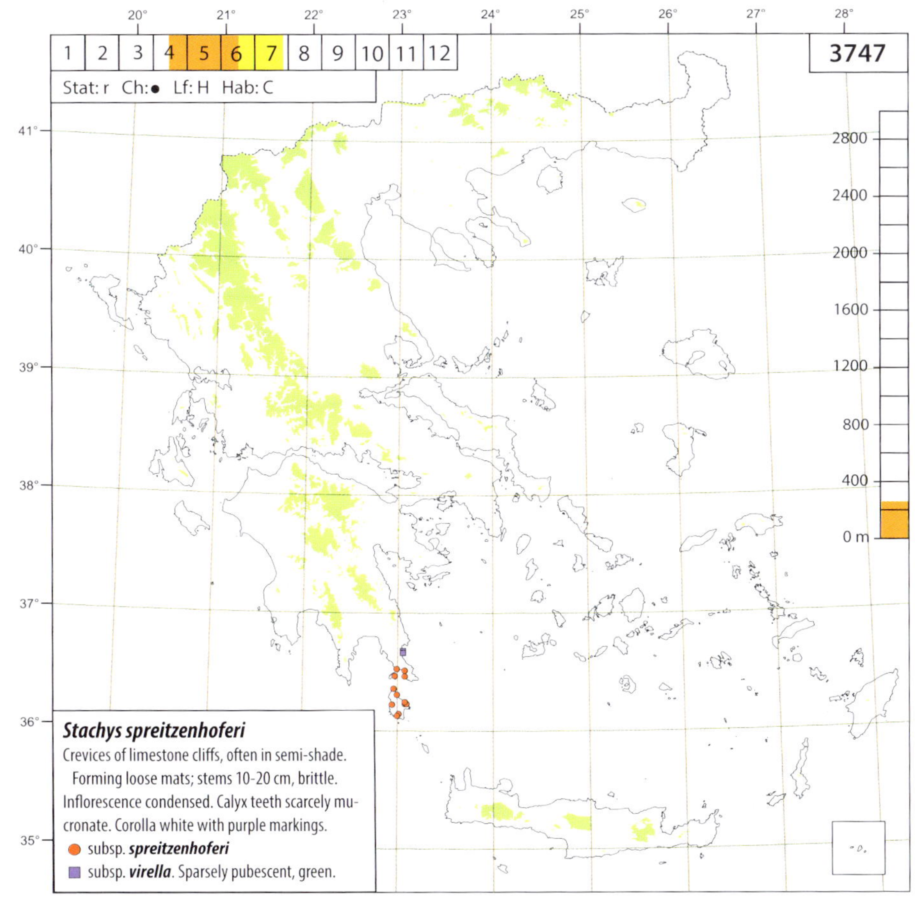

Stachys spreitzenhoferi

Crevices of limestone cliffs, often in semi-shade.

Forming loose mats; stems 10-20 cm, brittle. Inflorescence condensed. Calyx teeth scarcely mucronate. Corolla white with purple markings.

● subsp. ***spreitzenhoferi***

■ subsp. ***virella***. Sparsely pubescent, green.

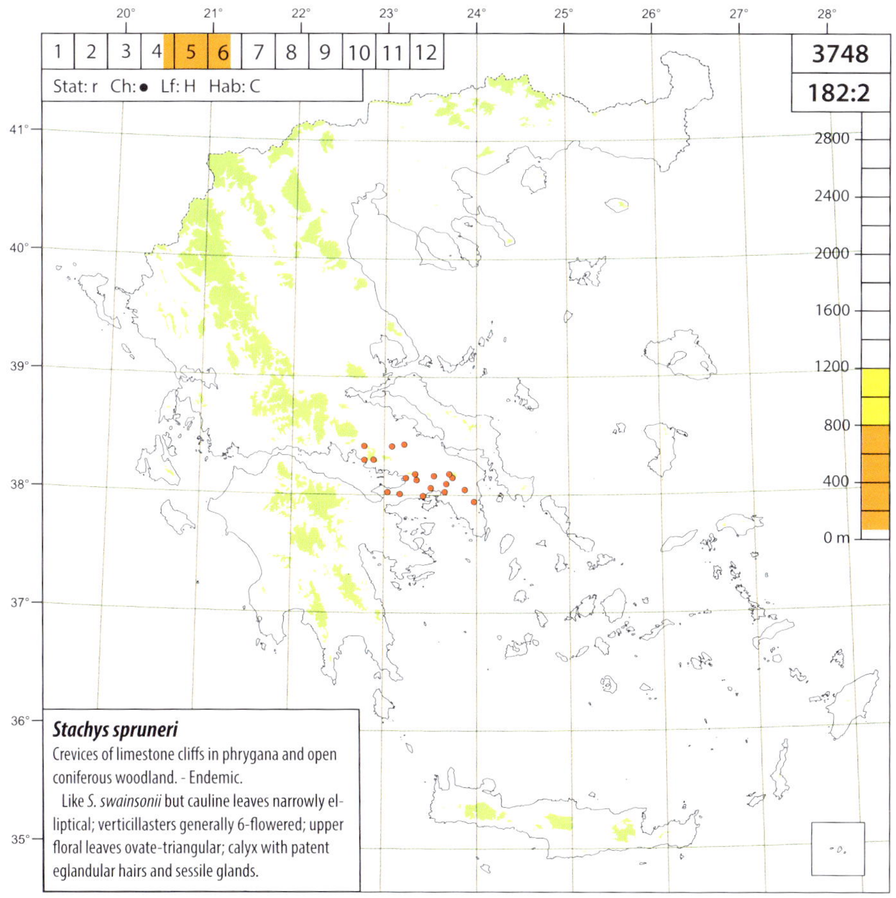

Stachys spruneri

Crevices of limestone cliffs in phrygana and open coniferous woodland. - Endemic.

Like *S. swainsonii* but cauline leaves narrowly elliptical; verticillasters generally 6-flowered; upper floral leaves ovate-triangular; calyx with patent eglandular hairs and sessile glands.

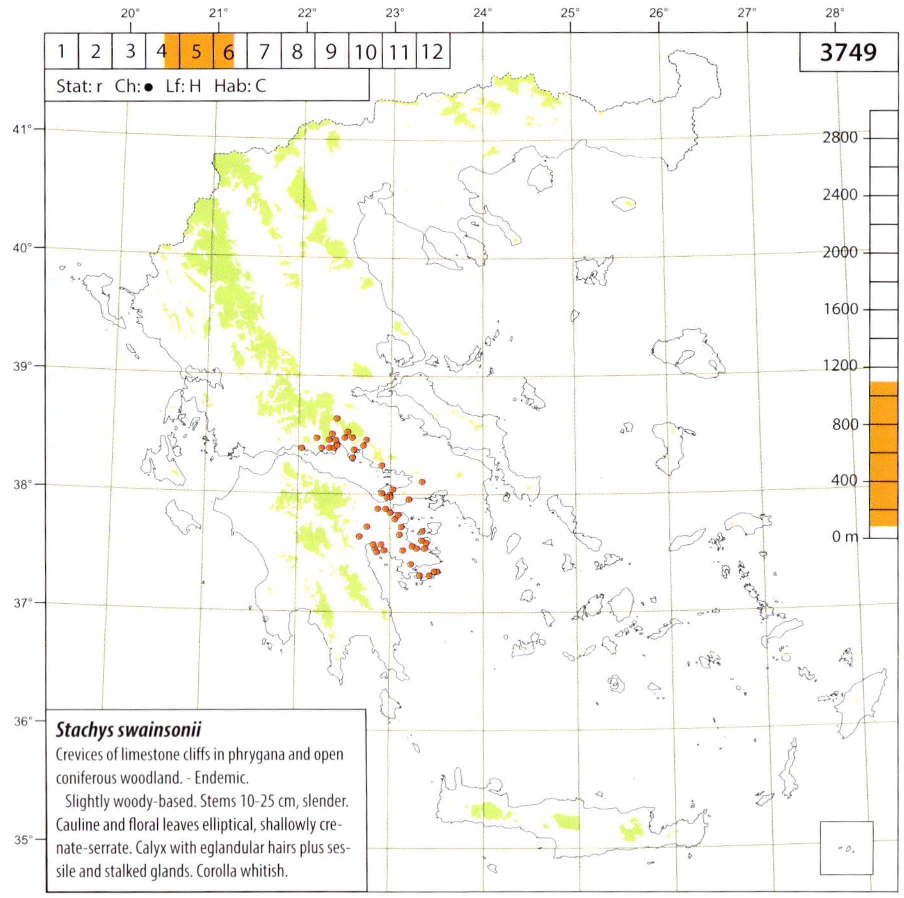

Stachys swainsonii
Crevices of limestone cliffs in phrygana and open coniferous woodland. - Endemic.
Slightly woody-based. Stems 10-25 cm, slender. Cauline and floral leaves elliptical, shallowly crenate-serrate. Calyx with eglandular hairs plus sessile and stalked glands. Corolla whitish.

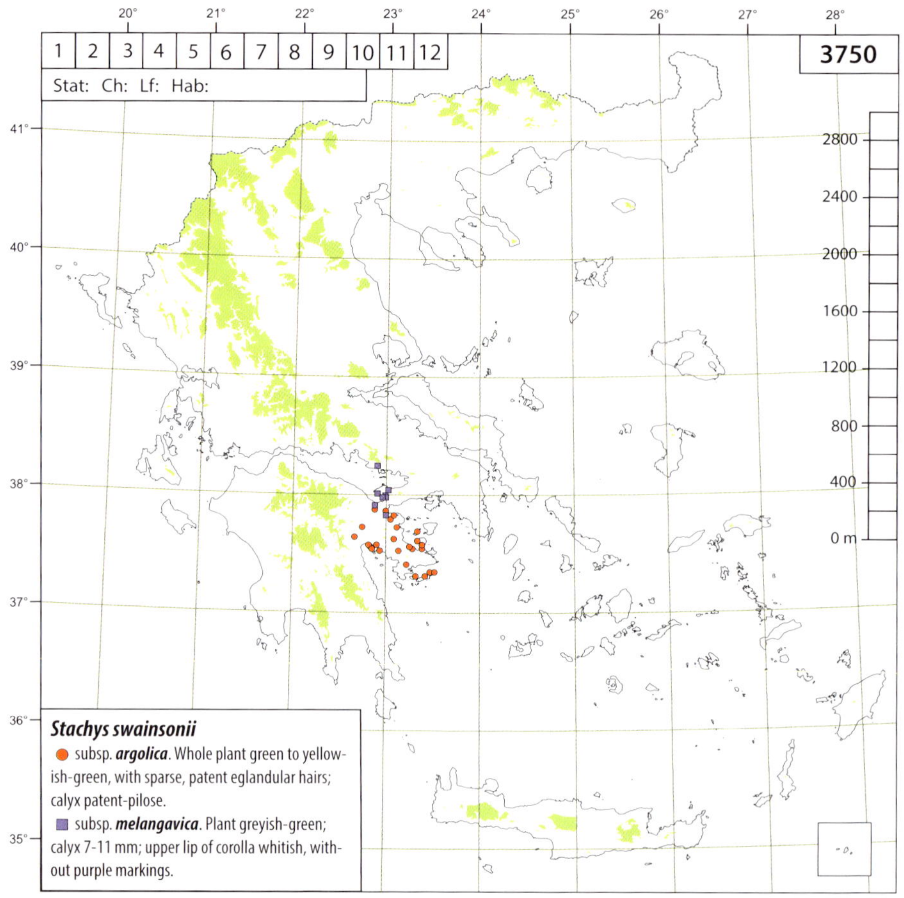

Stachys swainsonii
● subsp. ***argolica***. Whole plant green to yellowish-green, with sparse, patent eglandular hairs; calyx patent-pilose.
■ subsp. ***melangavica***. Plant greyish-green; calyx 7-11 mm; upper lip of corolla whitish, without purple markings.

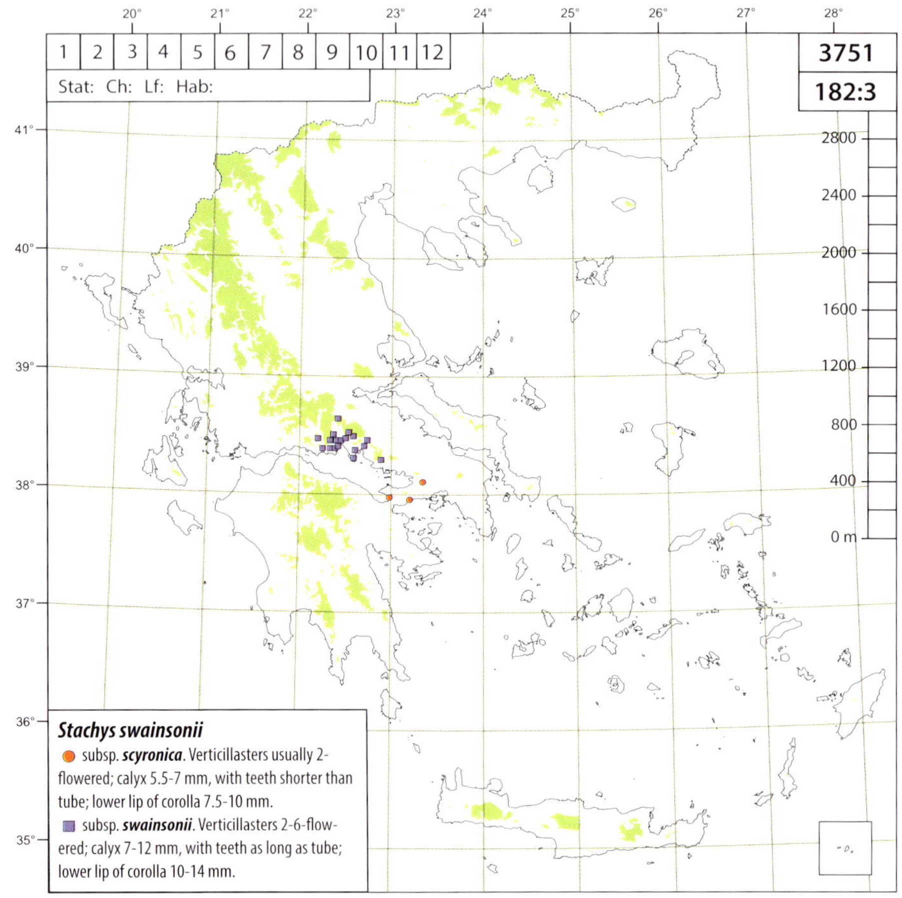

Stachys swainsonii
● subsp. ***scyronica***. Verticillasters usually 2-flowered; calyx 5.5-7 mm, with teeth shorter than tube; lower lip of corolla 7.5-10 mm.
■ subsp. ***swainsonii***. Verticillasters 2-6-flowered; calyx 7-12 mm, with teeth as long as tube; lower lip of corolla 10-14 mm.

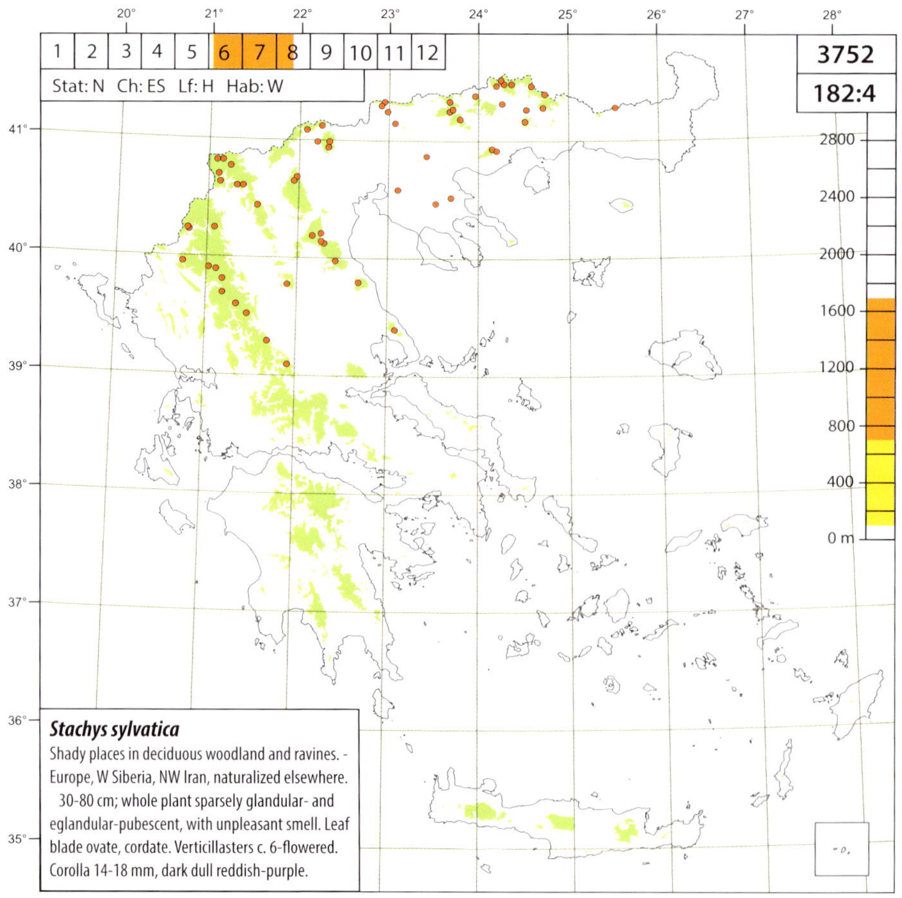

Stachys sylvatica
Shady places in deciduous woodland and ravines. - Europe, W Siberia, NW Iran, naturalized elsewhere.
30-80 cm; whole plant sparsely glandular- and eglandular-pubescent, with unpleasant smell. Leaf blade ovate, cordate. Verticillasters c. 6-flowered. Corolla 14-18 mm, dark dull reddish-purple.

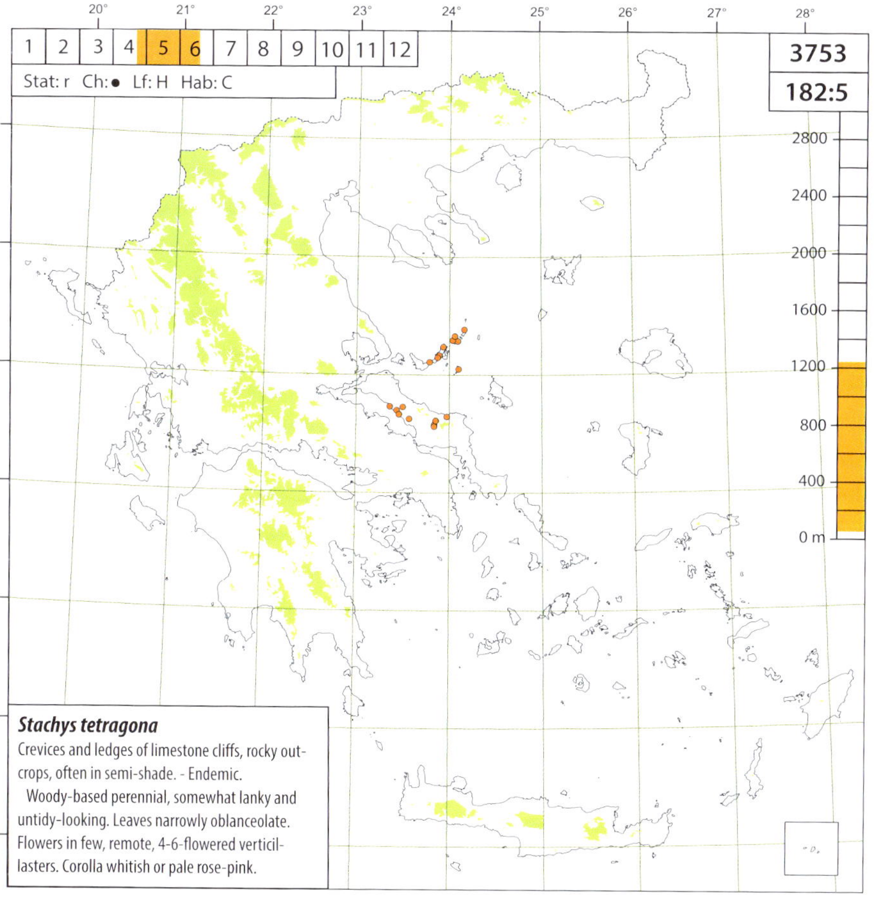

Stachys tetragona

Crevices and ledges of limestone cliffs, rocky outcrops, often in semi-shade. - Endemic.

Woody-based perennial, somewhat lanky and untidy-looking. Leaves narrowly oblanceolate. Flowers in few, remote, 4-6-flowered verticillasters. Corolla whitish or pale rose-pink.

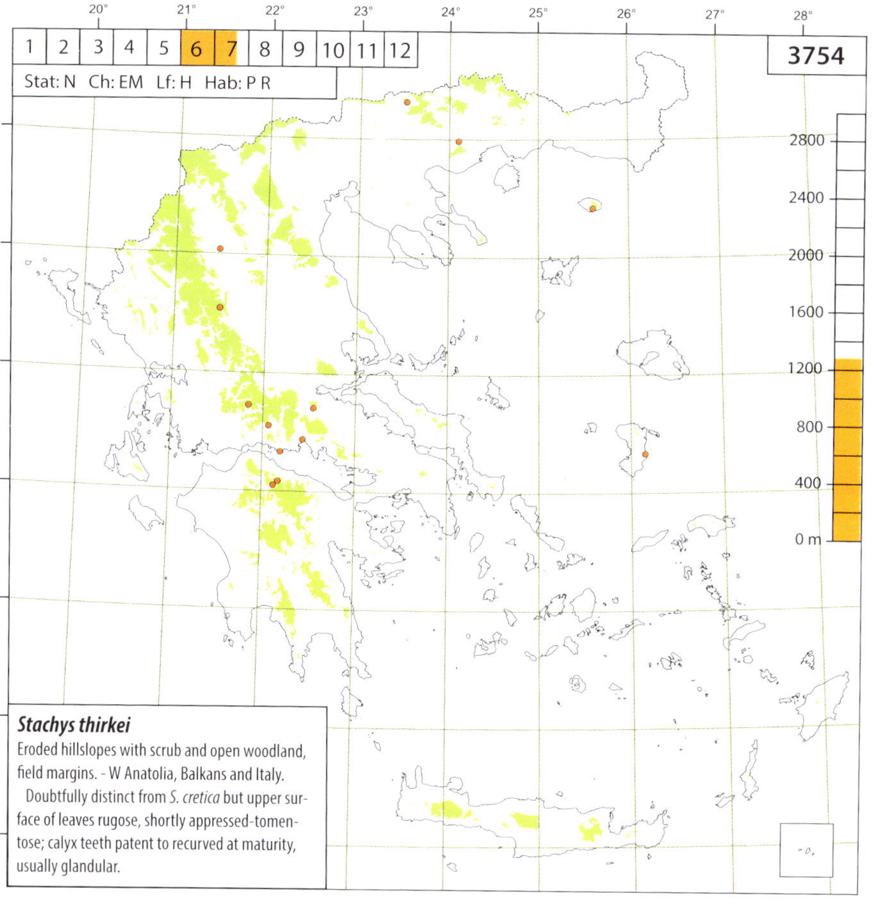

Stachys thirkei

Eroded hillslopes with scrub and open woodland, field margins. - W Anatolia, Balkans and Italy.

Doubtfully distinct from *S. cretica* but upper surface of leaves rugose, shortly appressed-tomentose; calyx teeth patent to recurved at maturity, usually glandular.

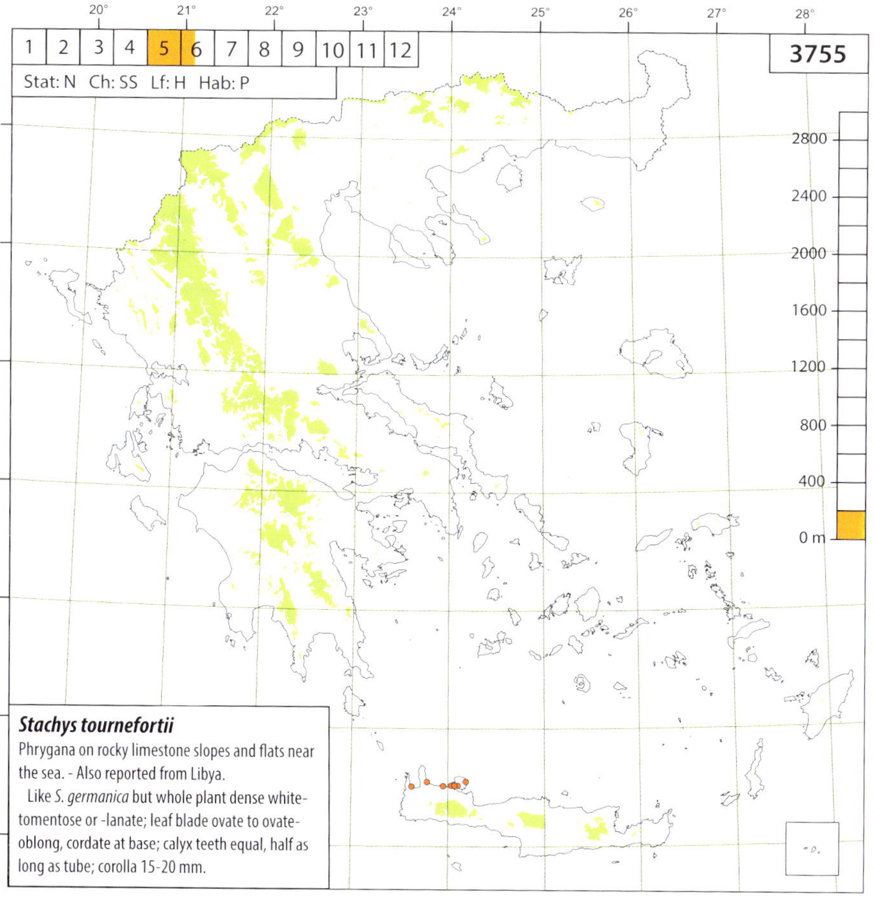

Stachys tournefortii

Phrygana on rocky limestone slopes and flats near the sea. - Also reported from Libya.

Like *S. germanica* but whole plant dense white-tomentose or -lanate; leaf blade ovate to ovate-oblong, cordate at base; calyx teeth equal, half as long as tube; corolla 15-20 mm.

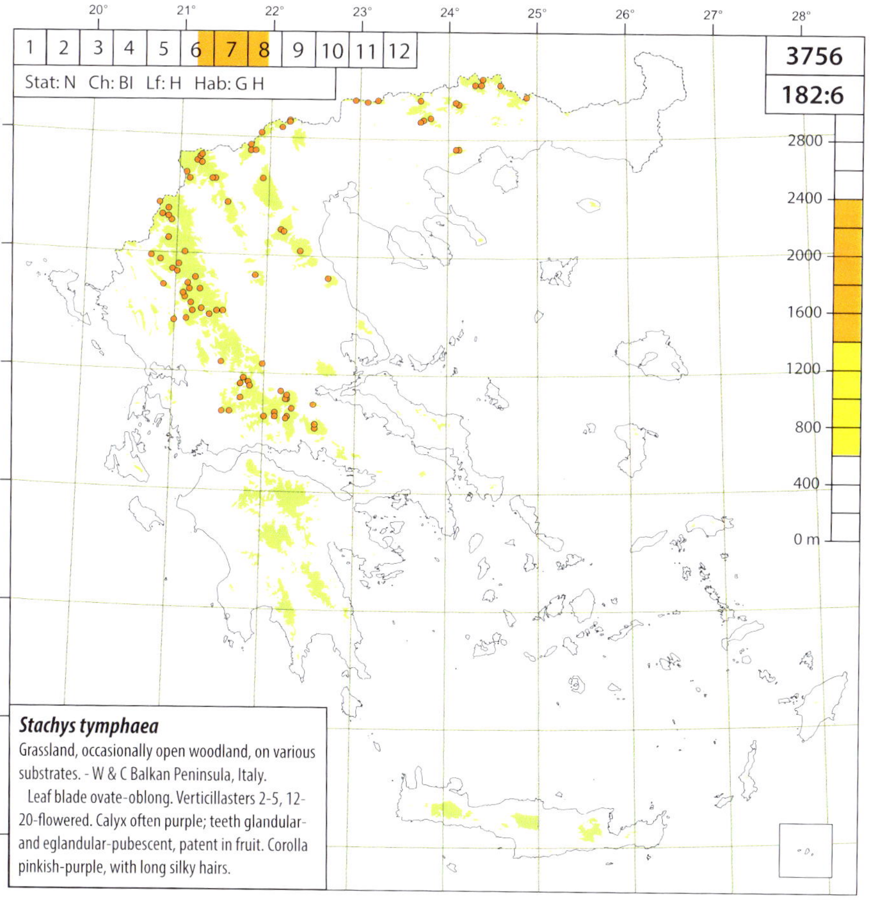

Stachys tymphaea

Grassland, occasionally open woodland, on various substrates. - W & C Balkan Peninsula, Italy.

Leaf blade ovate-oblong. Verticillasters 2-5, 12-20-flowered. Calyx often purple; teeth glandular- and eglandular-pubescent, patent in fruit. Corolla pinkish-purple, with long silky hairs.

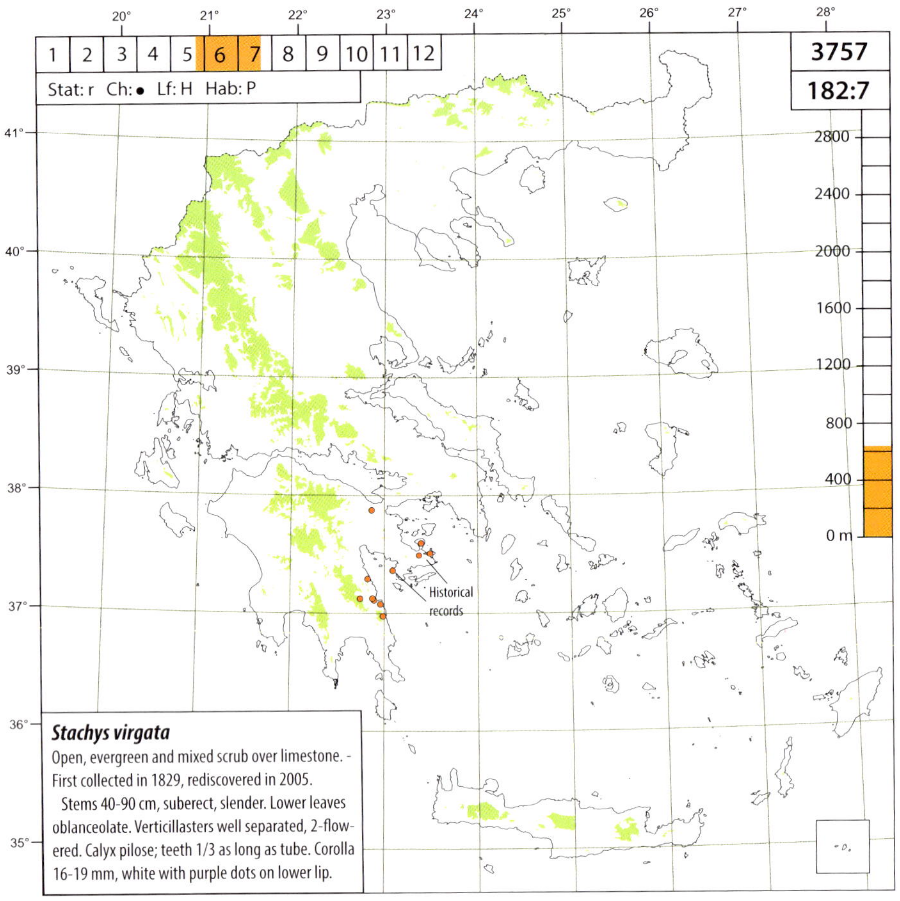

Stachys virgata

Open, evergreen and mixed scrub over limestone. - First collected in 1829, rediscovered in 2005.

Stems 40-90 cm, suberect, slender. Lower leaves oblanceolate. Verticillasters well separated, 2-flowered. Calyx pilose; teeth 1/3 as long as tube. Corolla 16-19 mm, white with purple dots on lower lip.

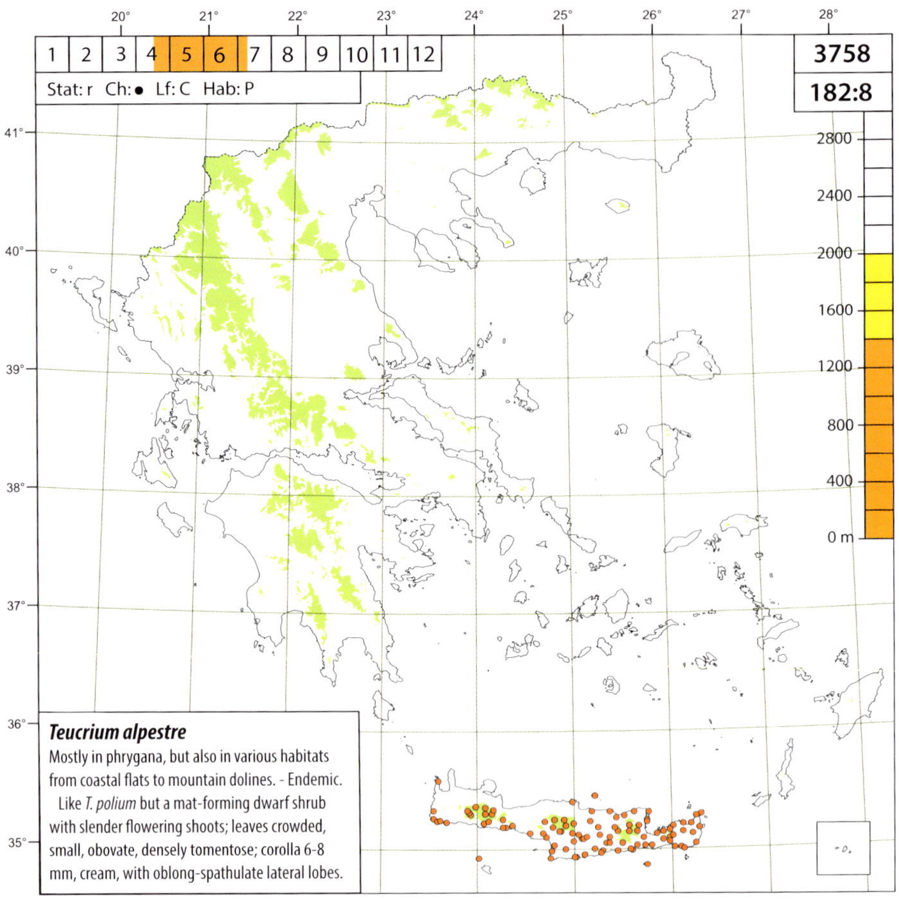

Teucrium alpestre

Mostly in phrygana, but also in various habitats from coastal flats to mountain dolines. - Endemic.

Like *T. polium* but a mat-forming dwarf shrub with slender flowering shoots; leaves crowded, small, obovate, densely tomentose; corolla 6-8 mm, cream, with oblong-spathulate lateral lobes.

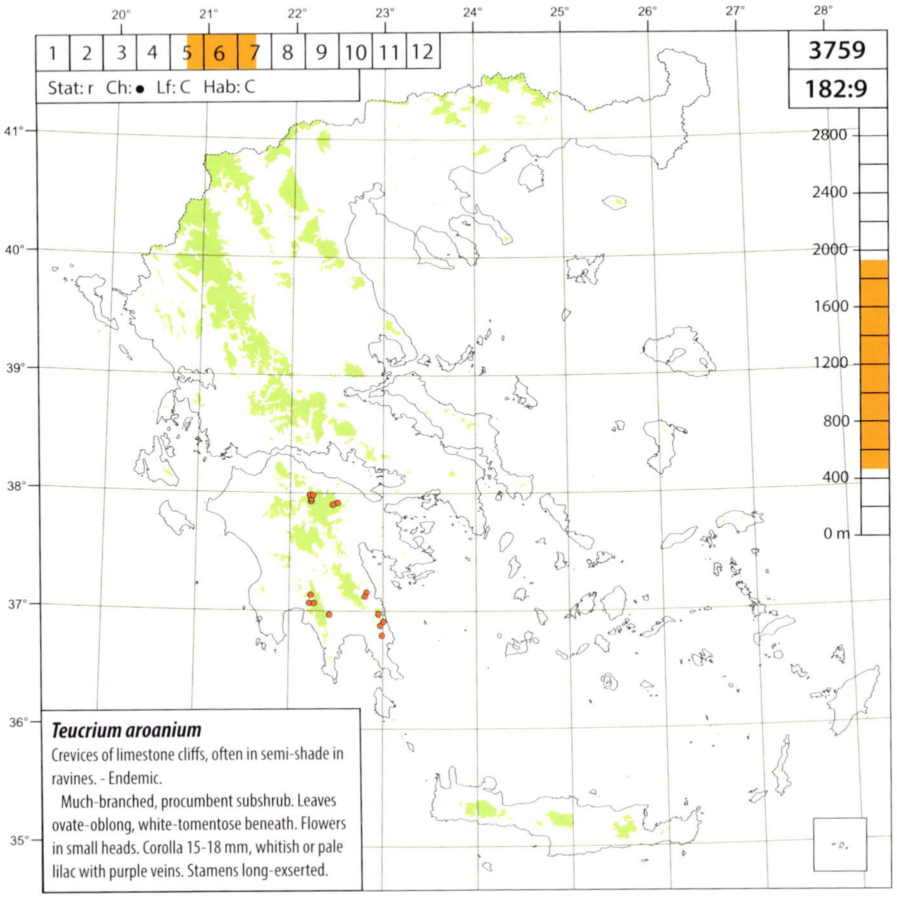

Teucrium aroanium

Crevices of limestone cliffs, often in semi-shade in ravines. - Endemic.

Much-branched, procumbent subshrub. Leaves ovate-oblong, white-tomentose beneath. Flowers in small heads. Corolla 15-18 mm, whitish or pale lilac with purple veins. Stamens long-exserted.

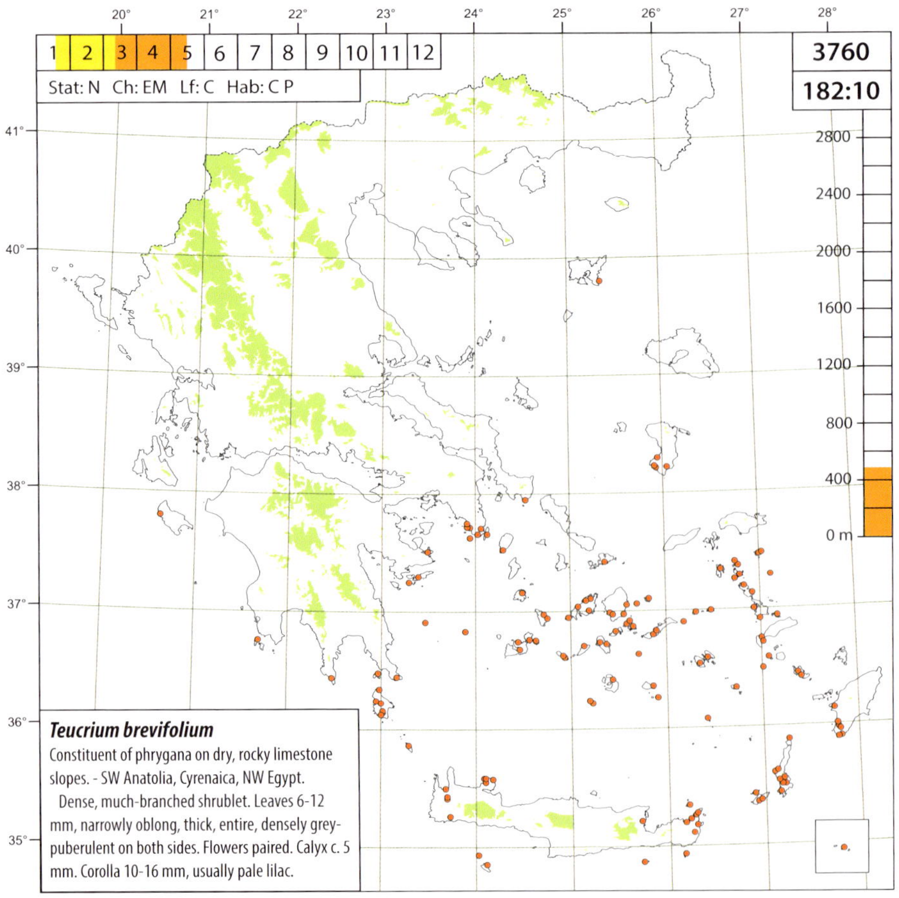

Teucrium brevifolium

Constituent of phrygana on dry, rocky limestone slopes. - SW Anatolia, Cyrenaica, NW Egypt.

Dense, much-branched shrublet. Leaves 6-12 mm, narrowly oblong, thick, entire, densely grey-puberulent on both sides. Flowers paired. Calyx c. 5 mm. Corolla 10-16 mm, usually pale lilac.

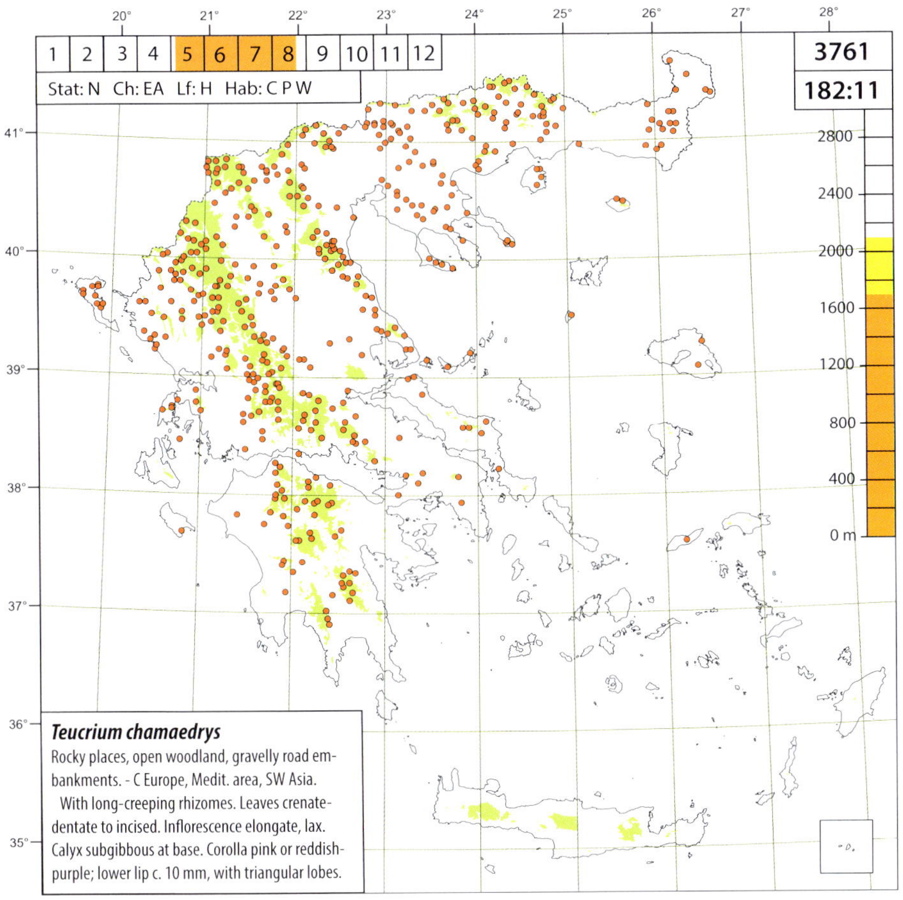

Teucrium chamaedrys

Rocky places, open woodland, gravelly road embankments. - C Europe, Medit. area, SW Asia.

With long-creeping rhizomes. Leaves crenate-dentate to incised. Inflorescence elongate, lax. Calyx subgibbous at base. Corolla pink or reddish-purple; lower lip c. 10 mm, with triangular lobes.

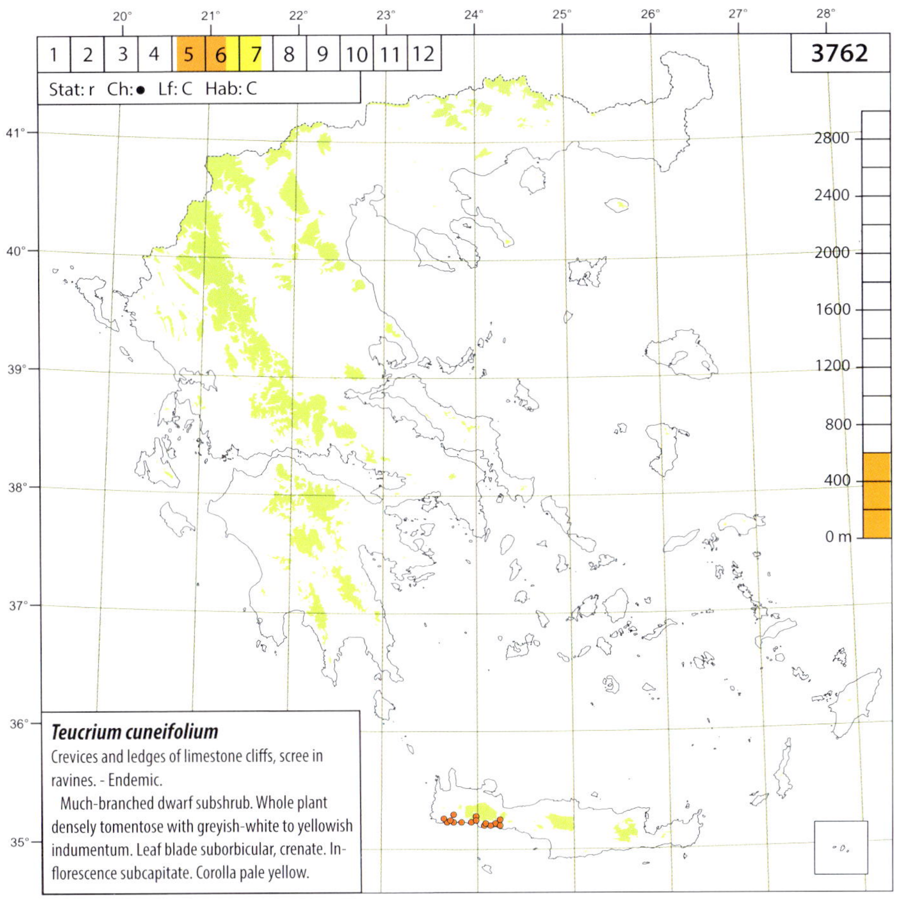

Teucrium cuneifolium

Crevices and ledges of limestone cliffs, scree in ravines. - Endemic.

Much-branched dwarf subshrub. Whole plant densely tomentose with greyish-white to yellowish indumentum. Leaf blade suborbicular, crenate. Inflorescence subcapitate. Corolla pale yellow.

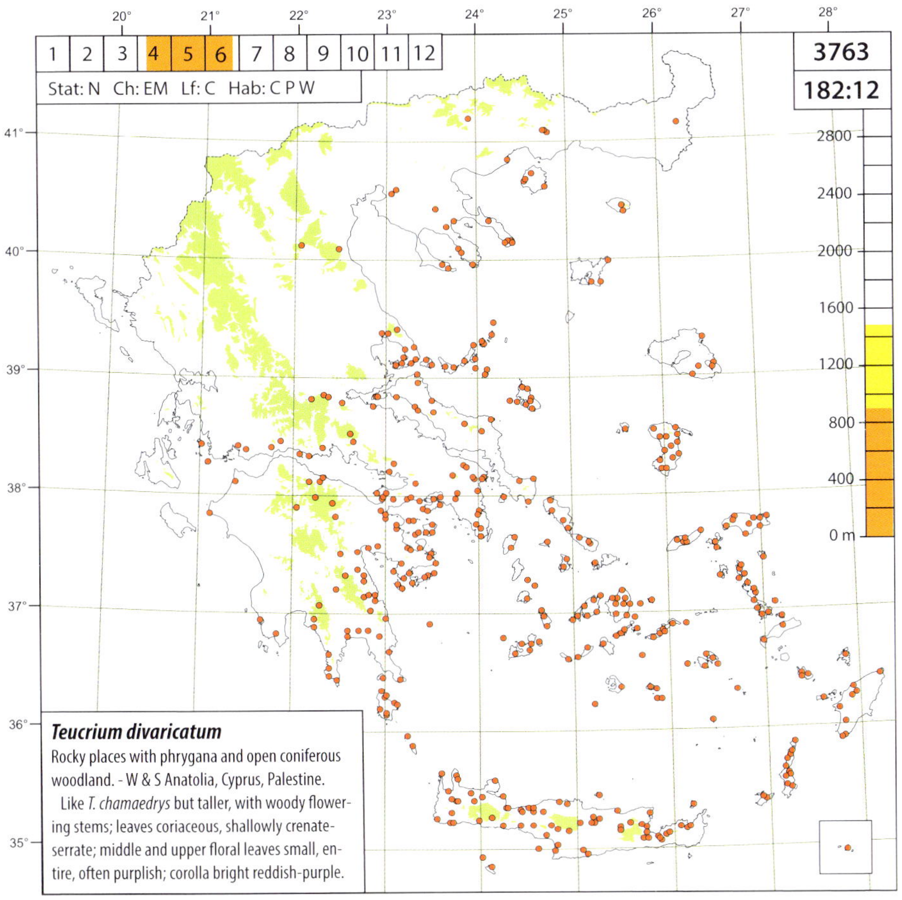

Teucrium divaricatum

Rocky places with phrygana and open coniferous woodland. - W & S Anatolia, Cyprus, Palestine.

Like *T. chamaedrys* but taller, with woody flowering stems; leaves coriaceous, shallowly crenate-serrate; middle and upper floral leaves small, entire, often purplish; corolla bright reddish-purple.

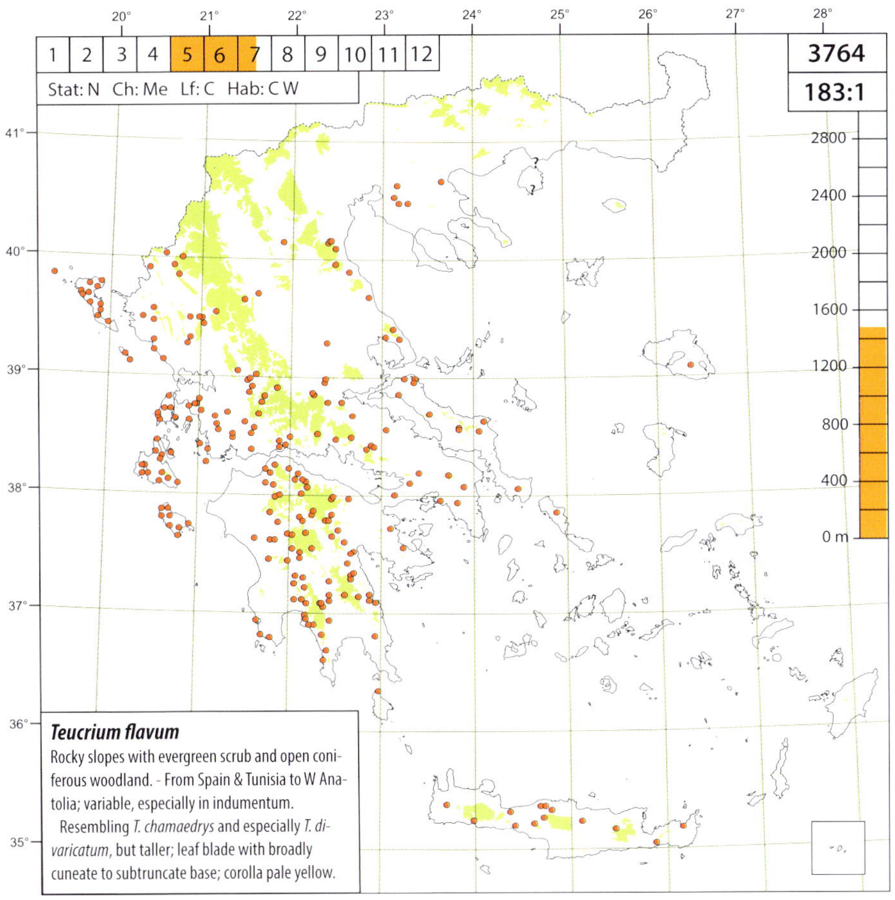

Teucrium flavum

Rocky slopes with evergreen scrub and open coniferous woodland. - From Spain & Tunisia to W Anatolia; variable, especially in indumentum.

Resembling *T. chamaedrys* and especially *T. divaricatum*, but taller; leaf blade with broadly cuneate to subtruncate base; corolla pale yellow.

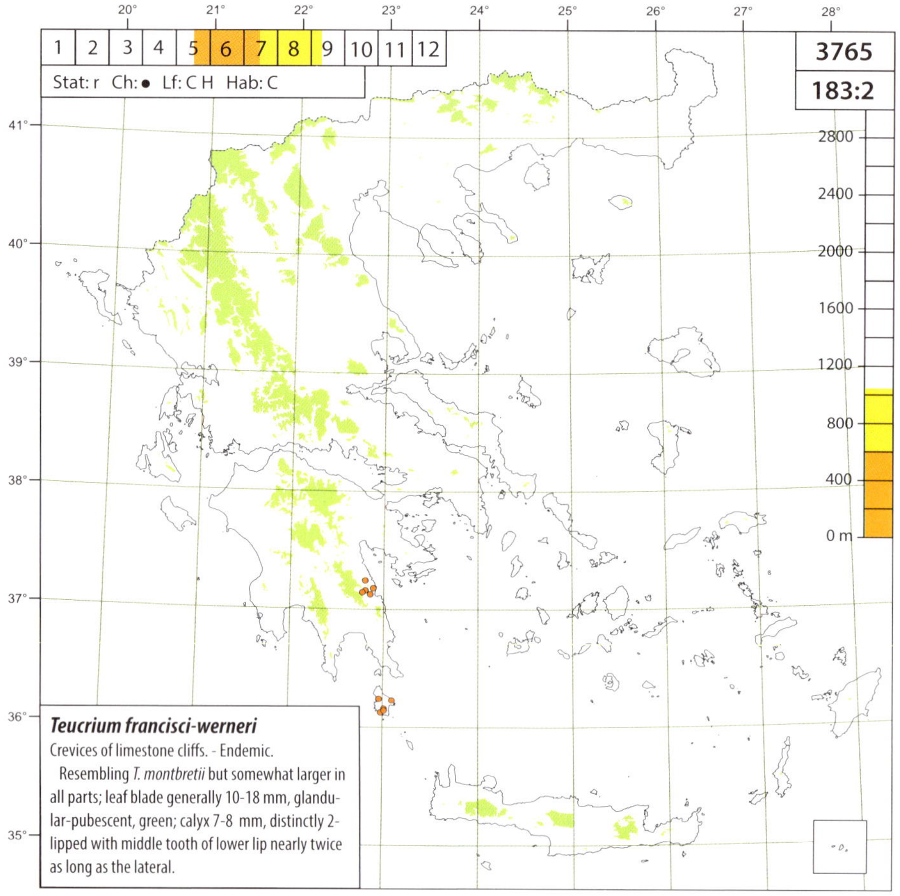

Teucrium francisci-werneri

Crevices of limestone cliffs. - Endemic.

Resembling *T. montbretii* but somewhat larger in all parts; leaf blade generally 10-18 mm, glandular-pubescent, green; calyx 7-8 mm, distinctly 2-lipped with middle tooth of lower lip nearly twice as long as the lateral.

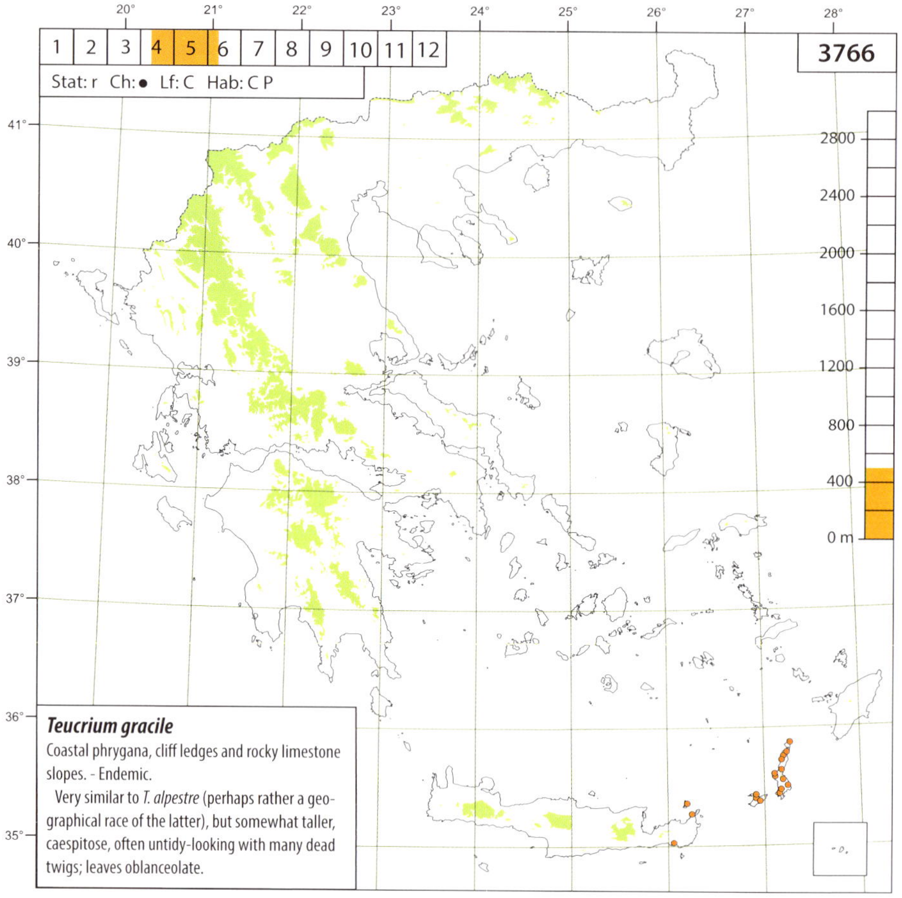

Teucrium gracile

Coastal phrygana, cliff ledges and rocky limestone slopes. - Endemic.

Very similar to *T. alpestre* (perhaps rather a geographical race of the latter), but somewhat taller, caespitose, often untidy-looking with many dead twigs; leaves oblanceolate.

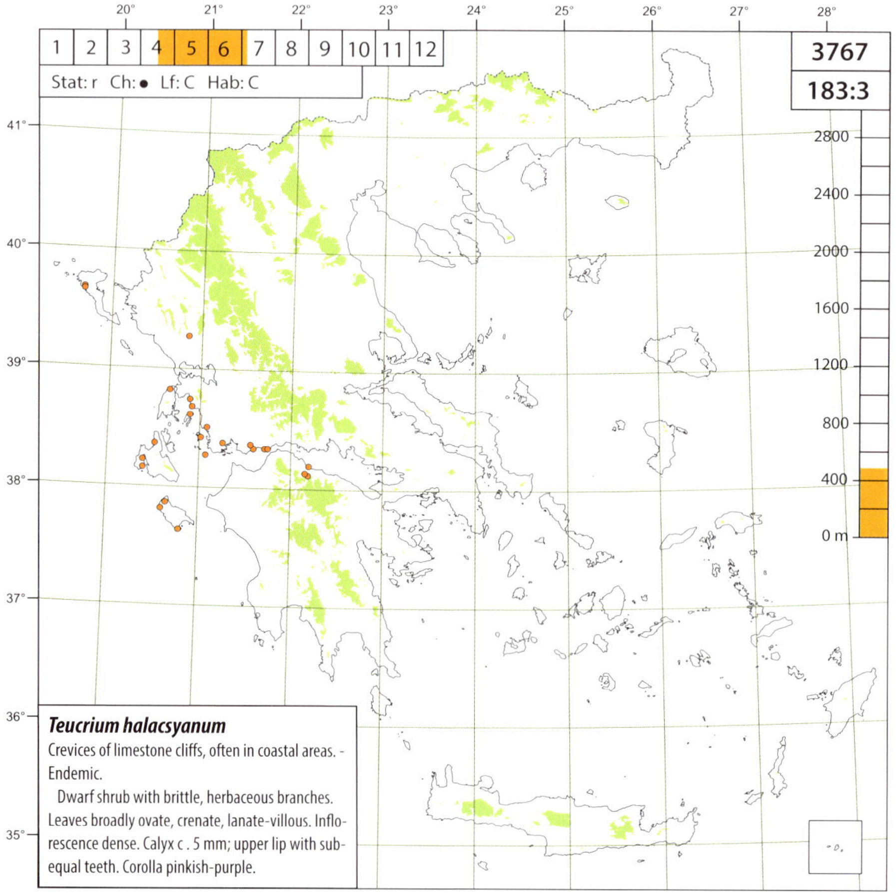

Teucrium halacsyanum

Crevices of limestone cliffs, often in coastal areas. - Endemic.

Dwarf shrub with brittle, herbaceous branches. Leaves broadly ovate, crenate, lanate-villous. Inflorescence dense. Calyx c. 5 mm; upper lip with subequal teeth. Corolla pinkish-purple.

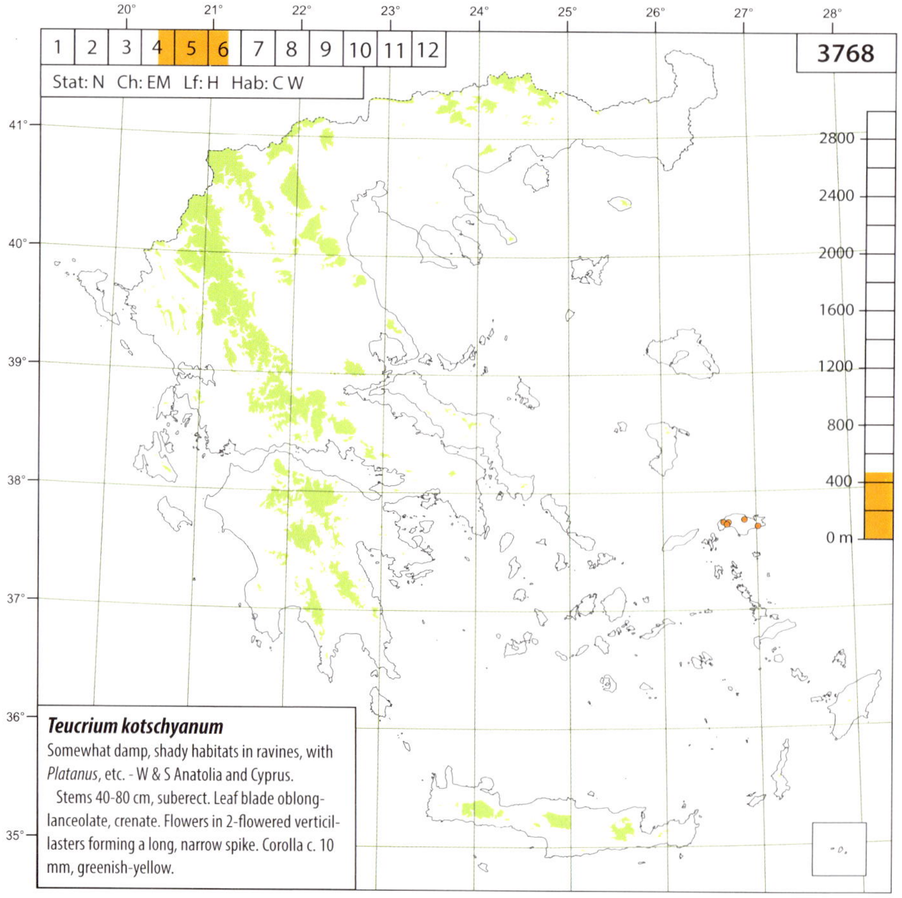

Teucrium kotschyanum

Somewhat damp, shady habitats in ravines, with *Platanus*, etc. - W & S Anatolia and Cyprus.

Stems 40-80 cm, suberect. Leaf blade oblong-lanceolate, crenate. Flowers in 2-flowered verticillasters forming a long, narrow spike. Corolla c. 10 mm, greenish-yellow.

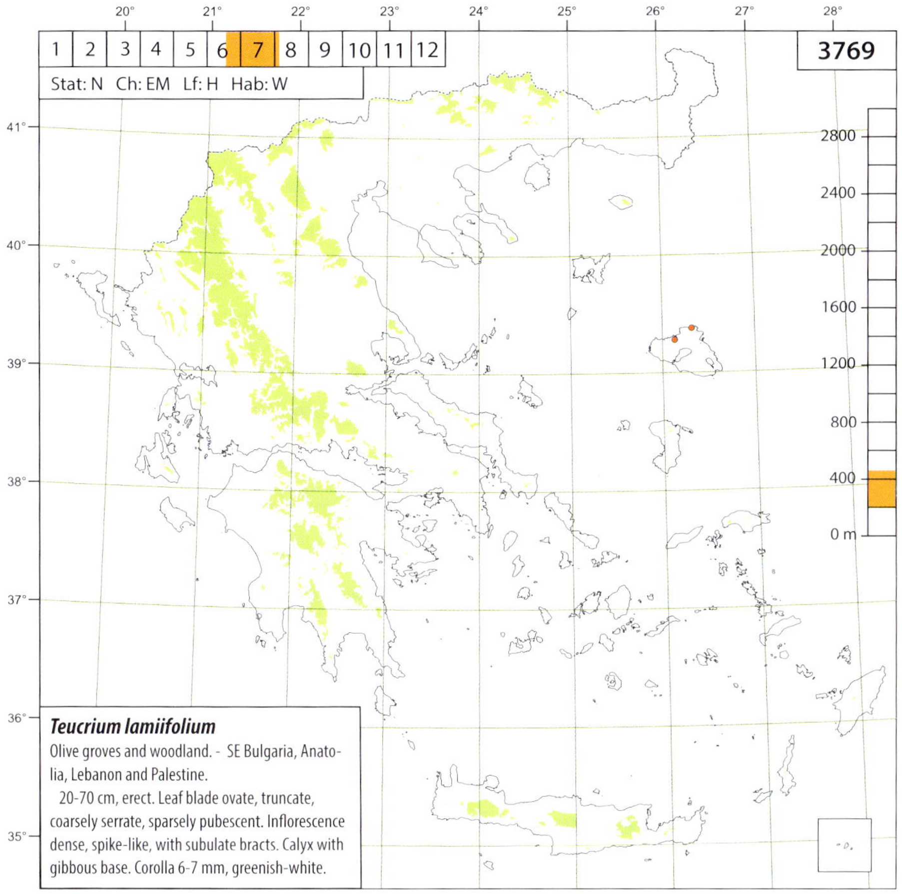

Teucrium lamiifolium
Olive groves and woodland. - SE Bulgaria, Anatolia, Lebanon and Palestine.
20-70 cm, erect. Leaf blade ovate, truncate, coarsely serrate, sparsely pubescent. Inflorescence dense, spike-like, with subulate bracts. Calyx with gibbous base. Corolla 6-7 mm, greenish-white.

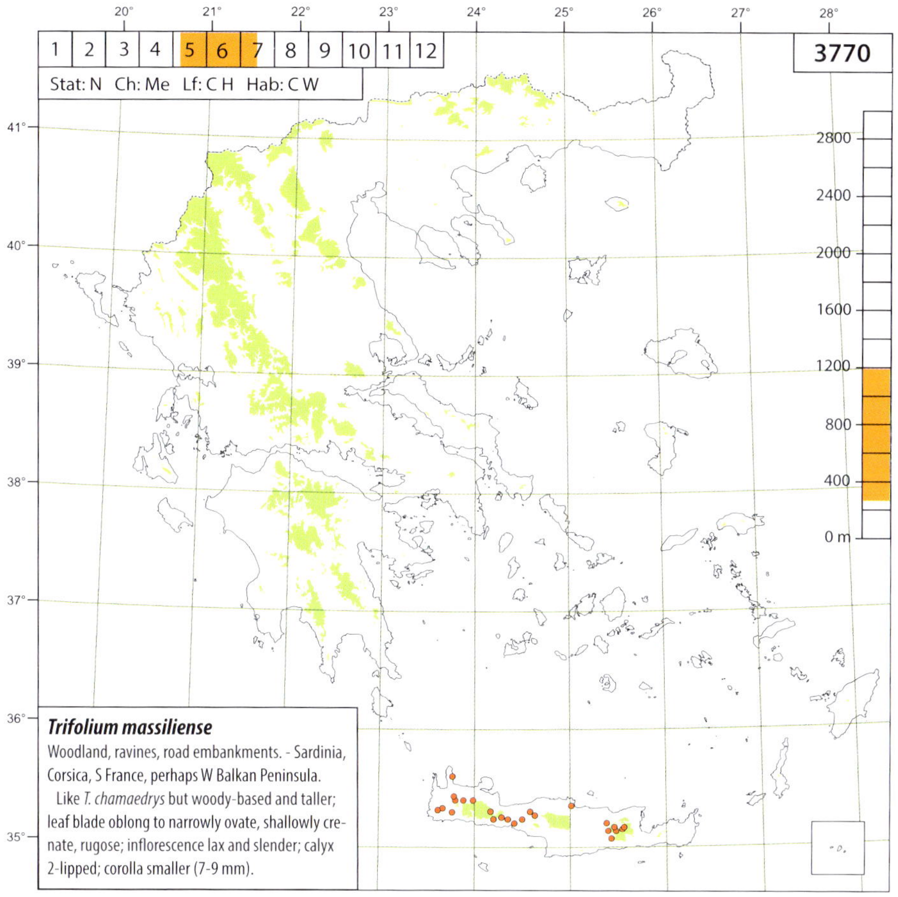

Trifolium massiliense
Woodland, ravines, road embankments. - Sardinia, Corsica, S France, perhaps W Balkan Peninsula.
Like *T. chamaedrys* but woody-based and taller; leaf blade oblong to narrowly ovate, shallowly crenate, rugose; inflorescence lax and slender; calyx 2-lipped; corolla smaller (7-9 mm).

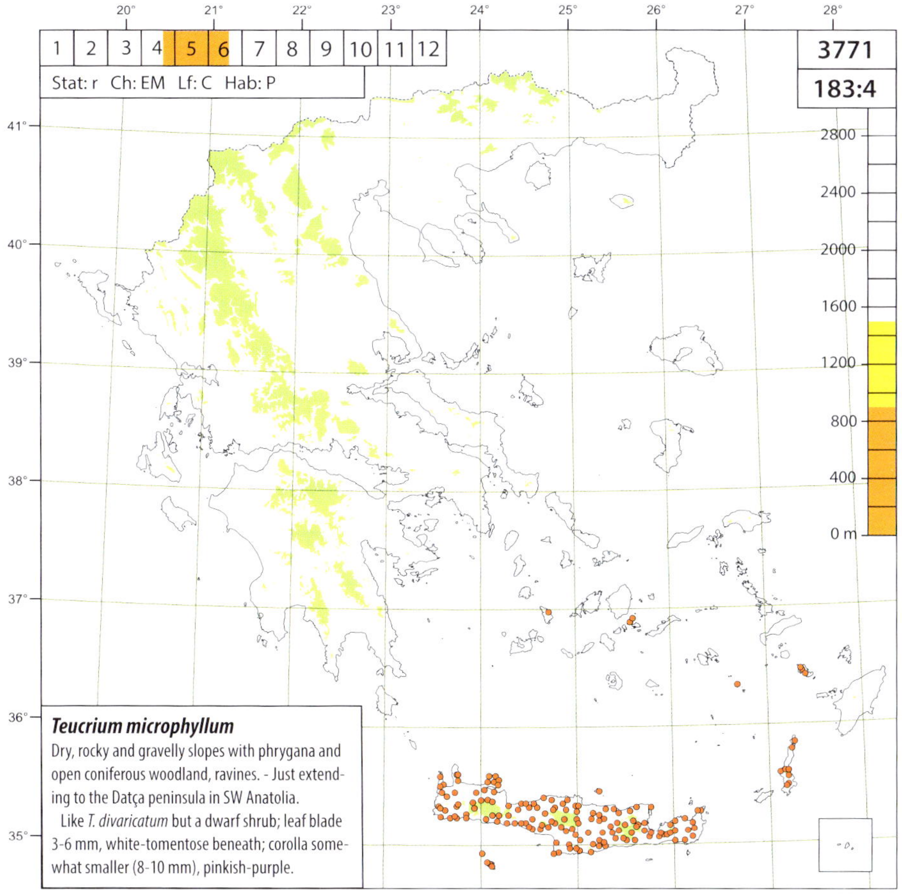

Teucrium microphyllum
Dry, rocky and gravelly slopes with phrygana and open coniferous woodland, ravines. - Just extending to the Datça peninsula in SW Anatolia.
Like *T. divaricatum* but a dwarf shrub; leaf blade 3-6 mm, white-tomentose beneath; corolla somewhat smaller (8-10 mm), pinkish-purple.

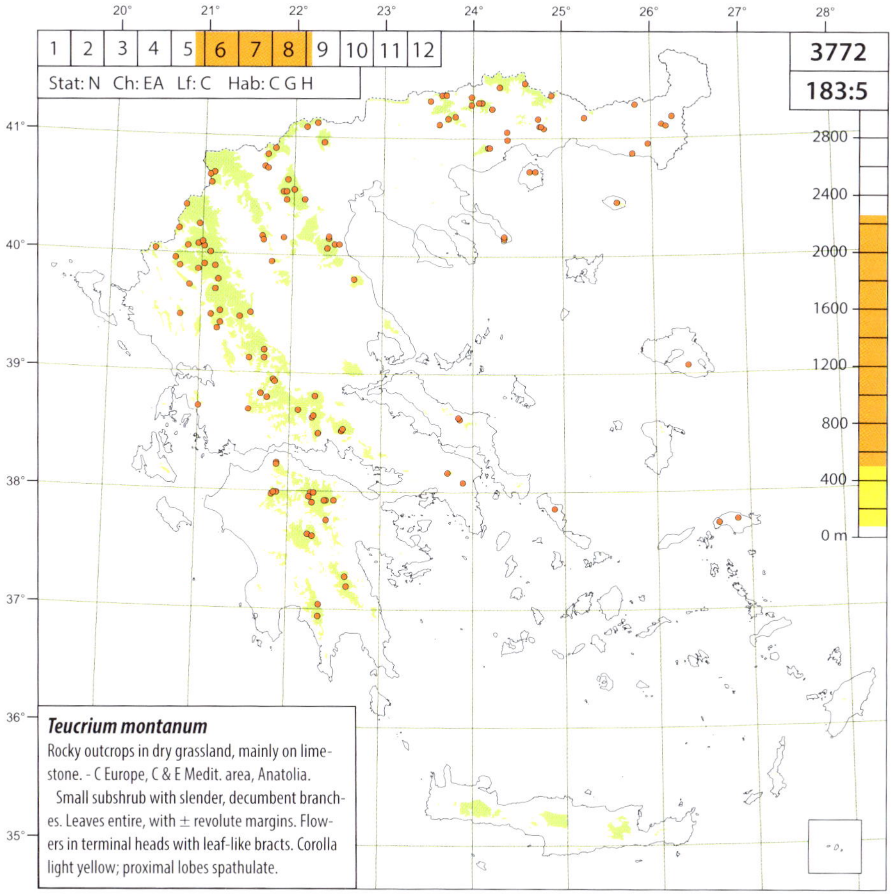

Teucrium montanum
Rocky outcrops in dry grassland, mainly on limestone. - C Europe, C & E Medit. area, Anatolia.
Small subshrub with slender, decumbent branches. Leaves entire, with ± revolute margins. Flowers in terminal heads with leaf-like bracts. Corolla light yellow; proximal lobes spathulate.

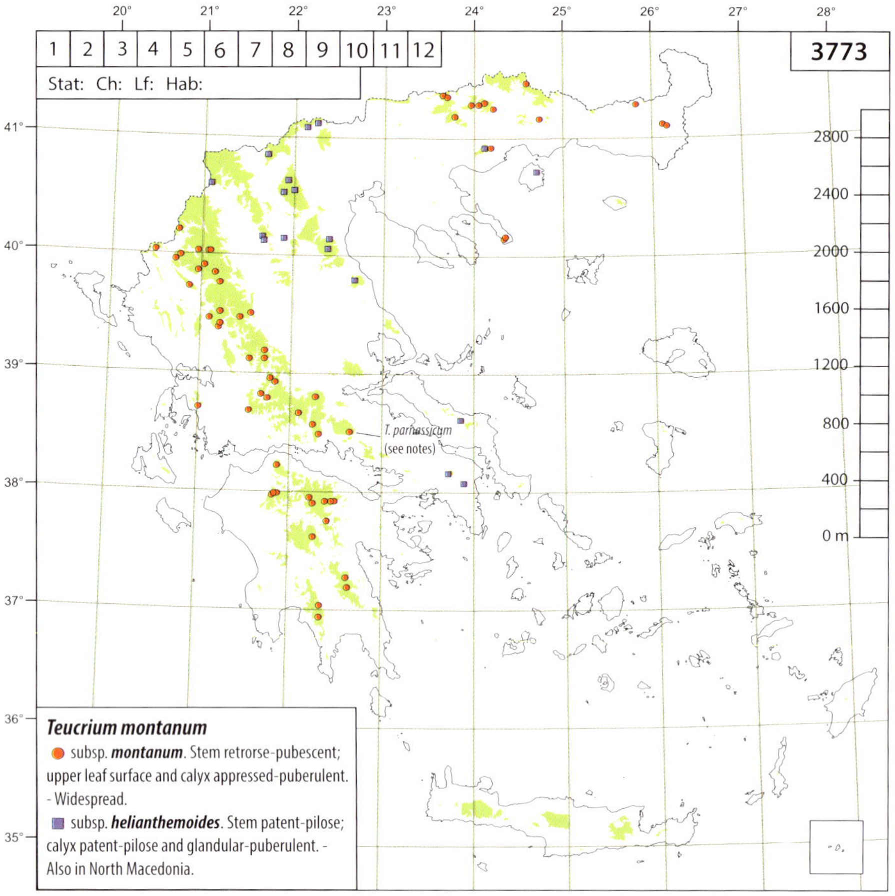

Teucrium montanum

● subsp. ***montanum***. Stem retrorse-pubescent; upper leaf surface and calyx appressed-puberulent. - Widespread.

■ subsp. ***helianthemoides***. Stem patent-pilose; calyx patent-pilose and glandular-puberulent. - Also in North Macedonia.

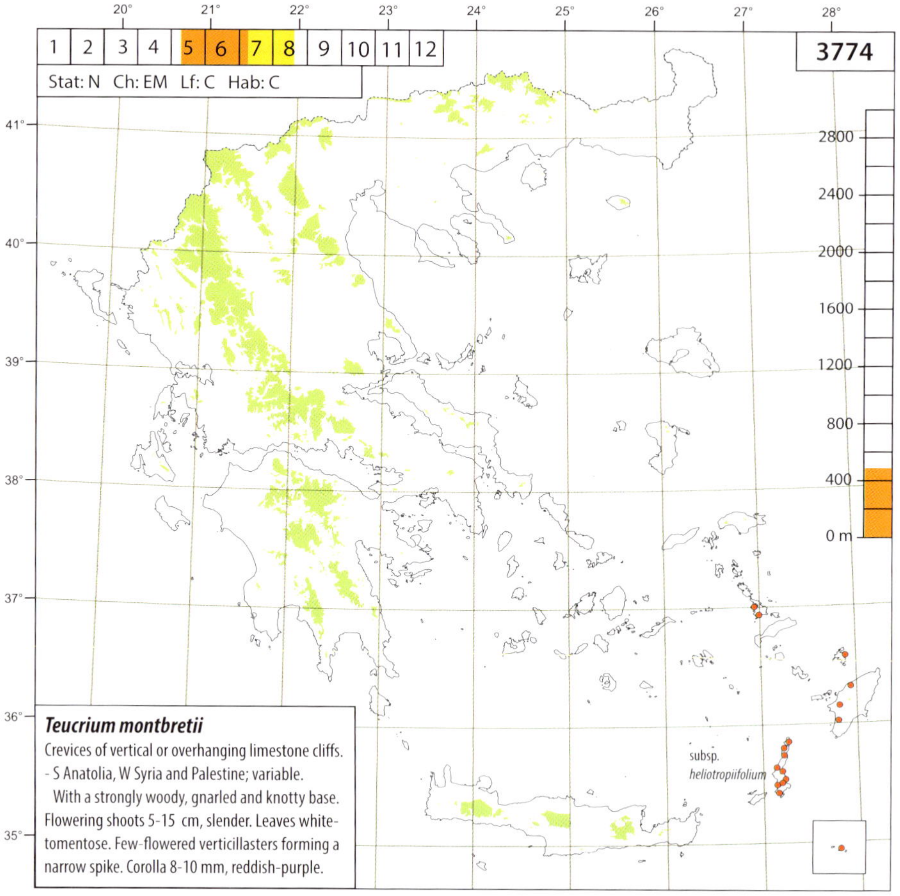

Teucrium montbretii

Crevices of vertical or overhanging limestone cliffs. - S Anatolia, W Syria and Palestine; variable.

With a strongly woody, gnarled and knotty base. Flowering shoots 5-15 cm, slender. Leaves white-tomentose. Few-flowered verticillasters forming a narrow spike. Corolla 8-10 mm, reddish-purple.

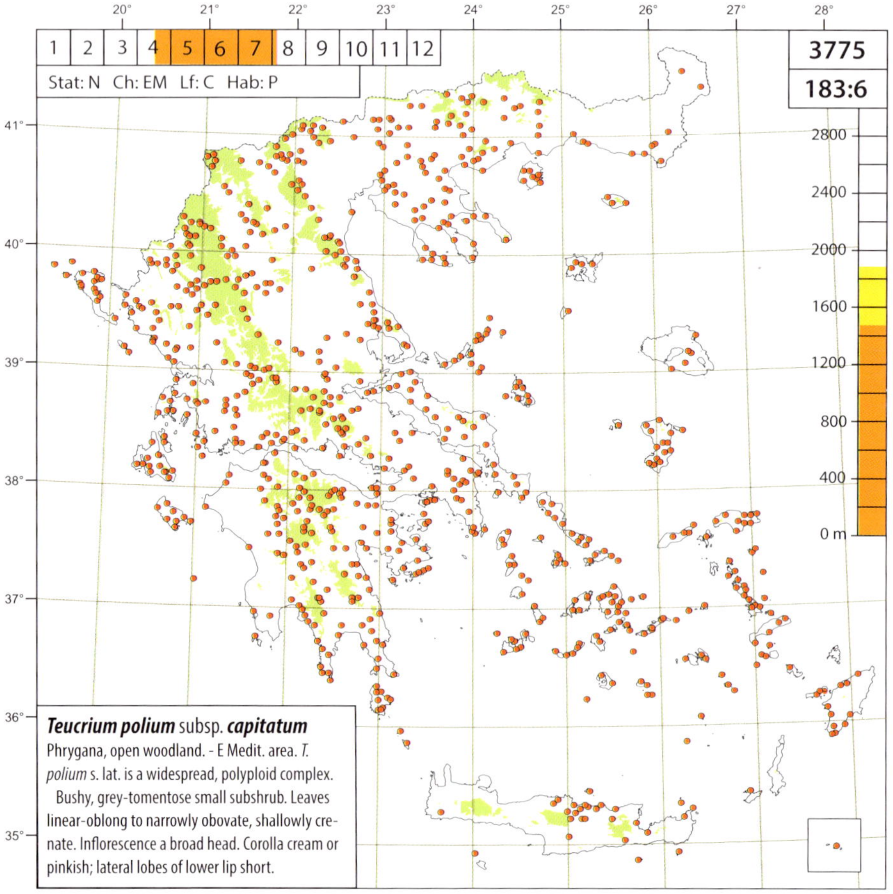

Teucrium polium subsp. ***capitatum***

Phrygana, open woodland. - E Medit. area. *T. polium* s. lat. is a widespread, polyploid complex.

Bushy, grey-tomentose small subshrub. Leaves linear-oblong to narrowly obovate, shallowly crenate. Inflorescence a broad head. Corolla cream or pinkish; lateral lobes of lower lip short.

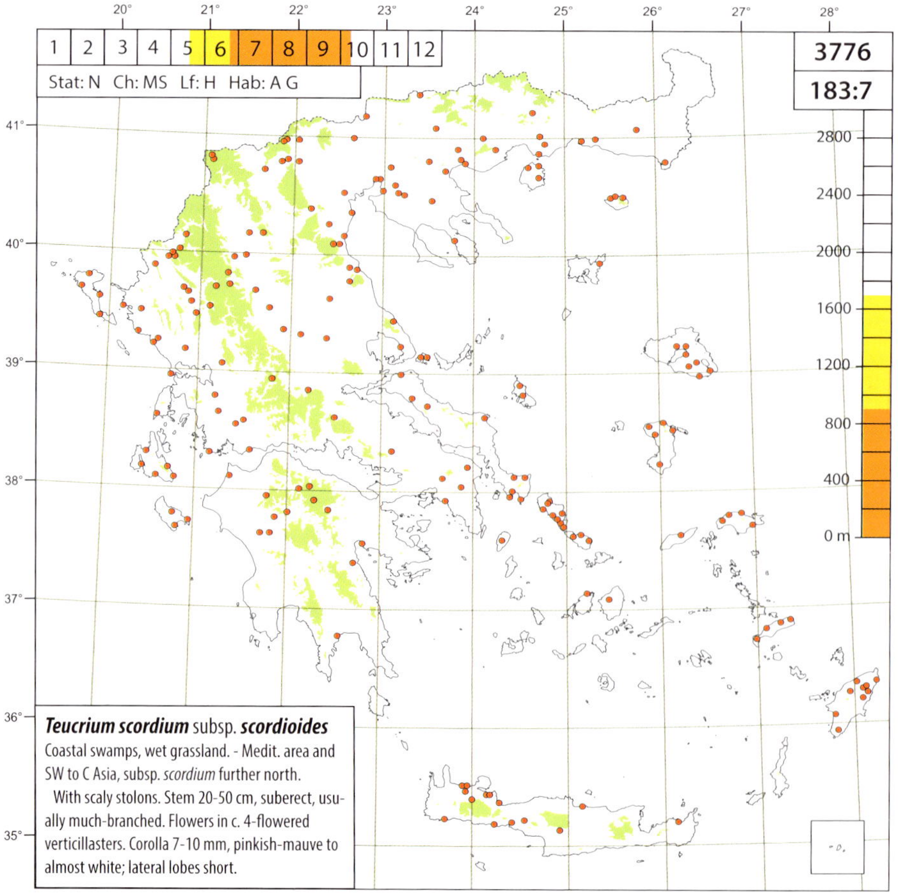

Teucrium scordium subsp. ***scordioides***

Coastal swamps, wet grassland. - Medit. area and SW to C Asia, subsp. *scordium* further north.

With scaly stolons. Stem 20-50 cm, suberect, usually much-branched. Flowers in c. 4-flowered verticillasters. Corolla 7-10 mm, pinkish-mauve to almost white; lateral lobes short.

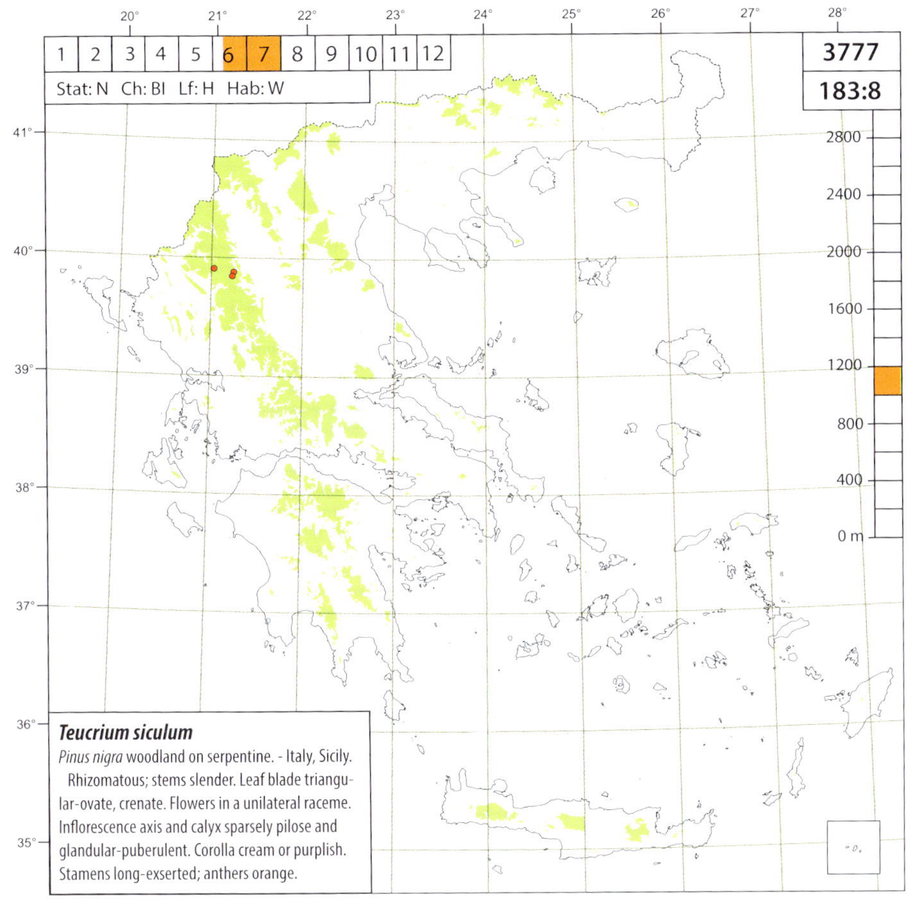

Teucrium siculum

Pinus nigra woodland on serpentine. - Italy, Sicily.

Rhizomatous; stems slender. Leaf blade triangular-ovate, crenate. Flowers in a unilateral raceme. Inflorescence axis and calyx sparsely pilose and glandular-puberulent. Corolla cream or purplish. Stamens long-exserted; anthers orange.

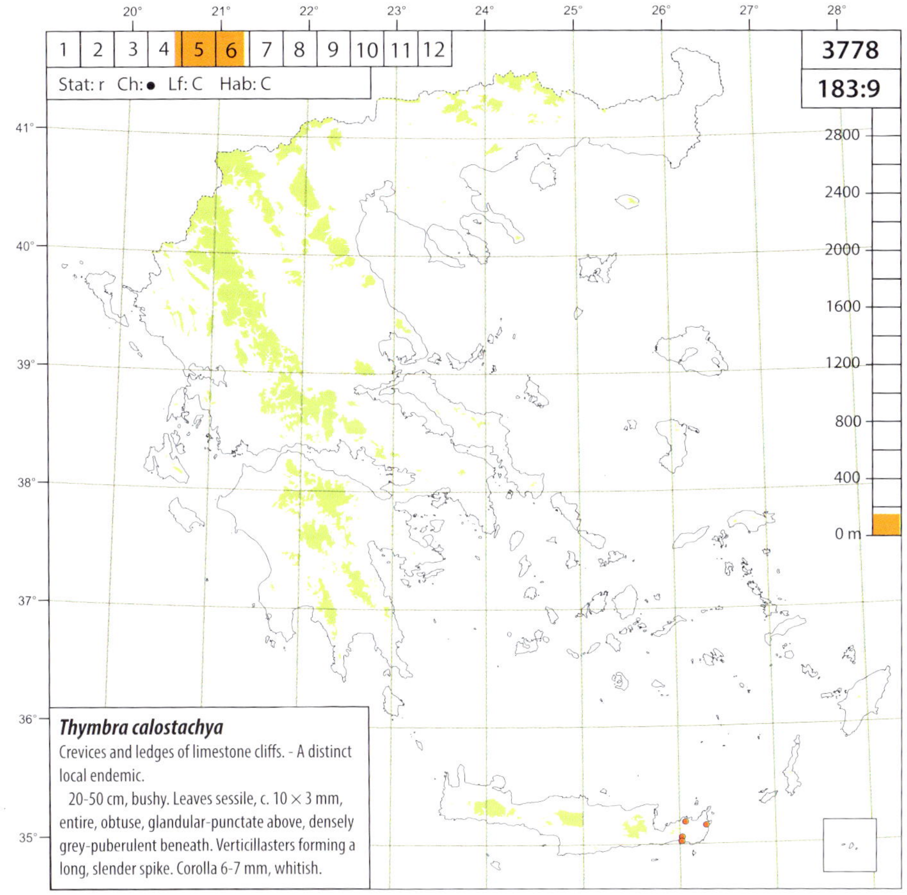

Thymbra calostachya

Crevices and ledges of limestone cliffs. - A distinct local endemic.

20-50 cm, bushy. Leaves sessile, c. 10 × 3 mm, entire, obtuse, glandular-punctate above, densely grey-puberulent beneath. Verticillasters forming a long, slender spike. Corolla 6-7 mm, whitish.

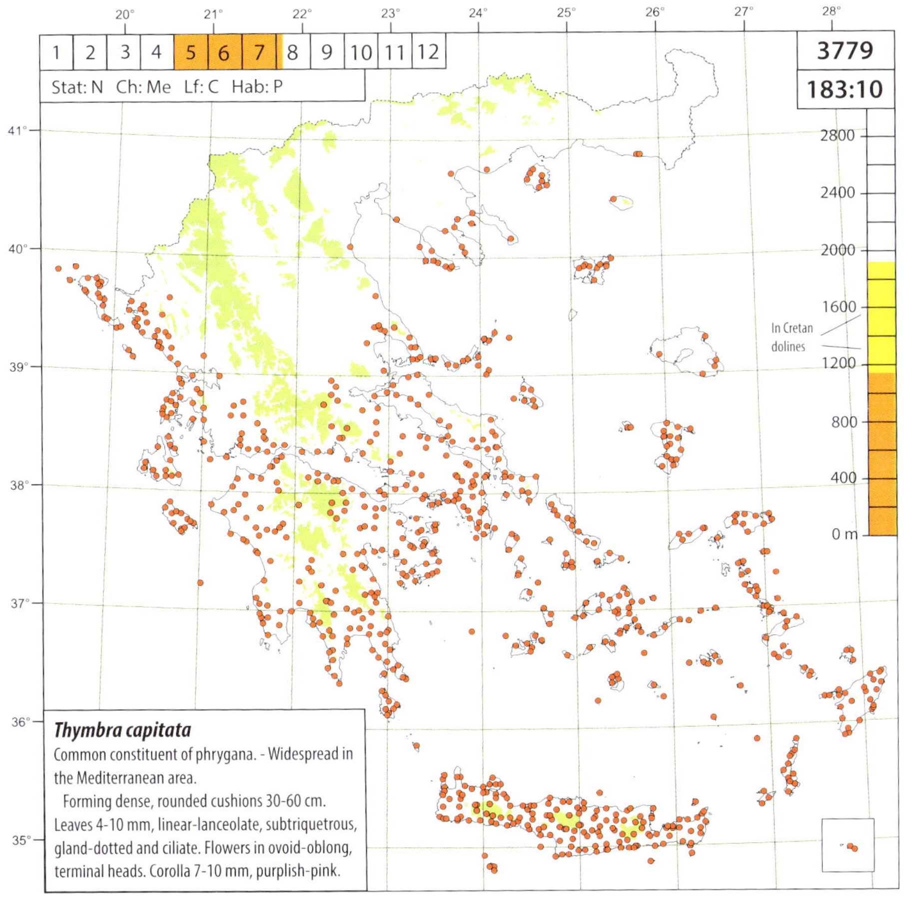

Thymbra capitata

Common constituent of phrygana. - Widespread in the Mediterranean area.

Forming dense, rounded cushions 30-60 cm. Leaves 4-10 mm, linear-lanceolate, subtriquetrous, gland-dotted and ciliate. Flowers in ovoid-oblong, terminal heads. Corolla 7-10 mm, purplish-pink.

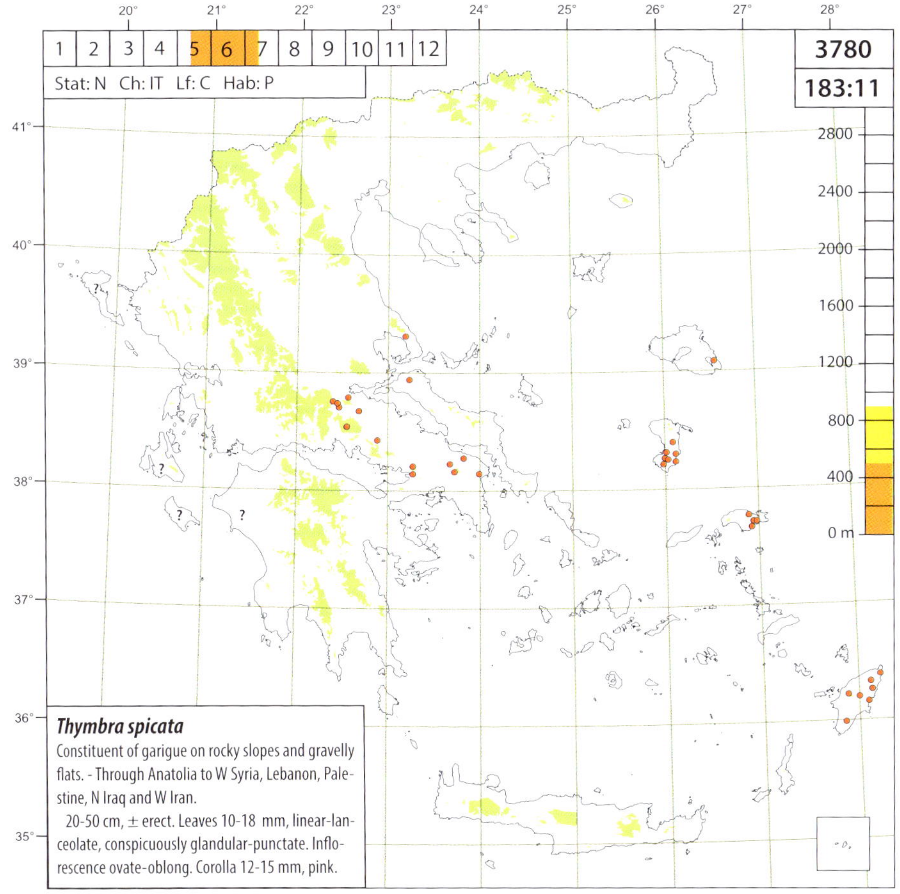

Thymbra spicata

Constituent of garigue on rocky slopes and gravelly flats. - Through Anatolia to W Syria, Lebanon, Palestine, N Iraq and W Iran.

20-50 cm, ± erect. Leaves 10-18 mm, linear-lanceolate, conspicuously glandular-punctate. Inflorescence ovate-oblong. Corolla 12-15 mm, pink.

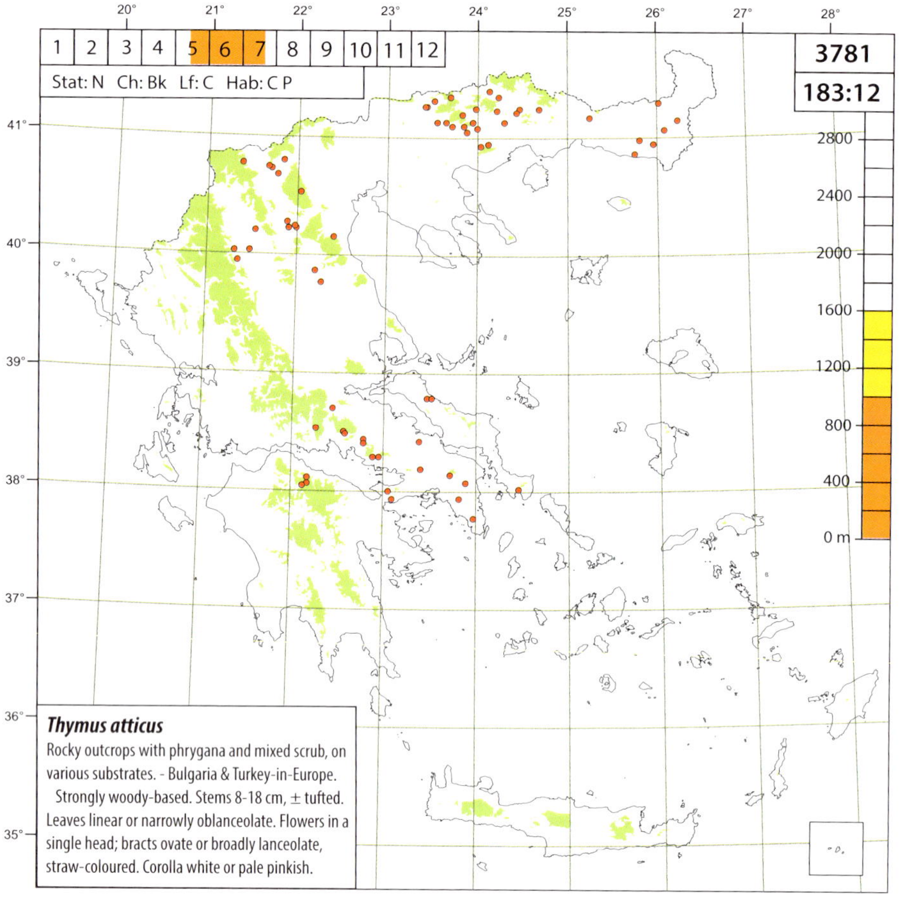

Thymus atticus

Rocky outcrops with phrygana and mixed scrub, on various substrates. - Bulgaria & Turkey-in-Europe.

Strongly woody-based. Stems 8-18 cm, ± tufted. Leaves linear or narrowly oblanceolate. Flowers in a single head; bracts ovate or broadly lanceolate, straw-coloured. Corolla white or pale pinkish.

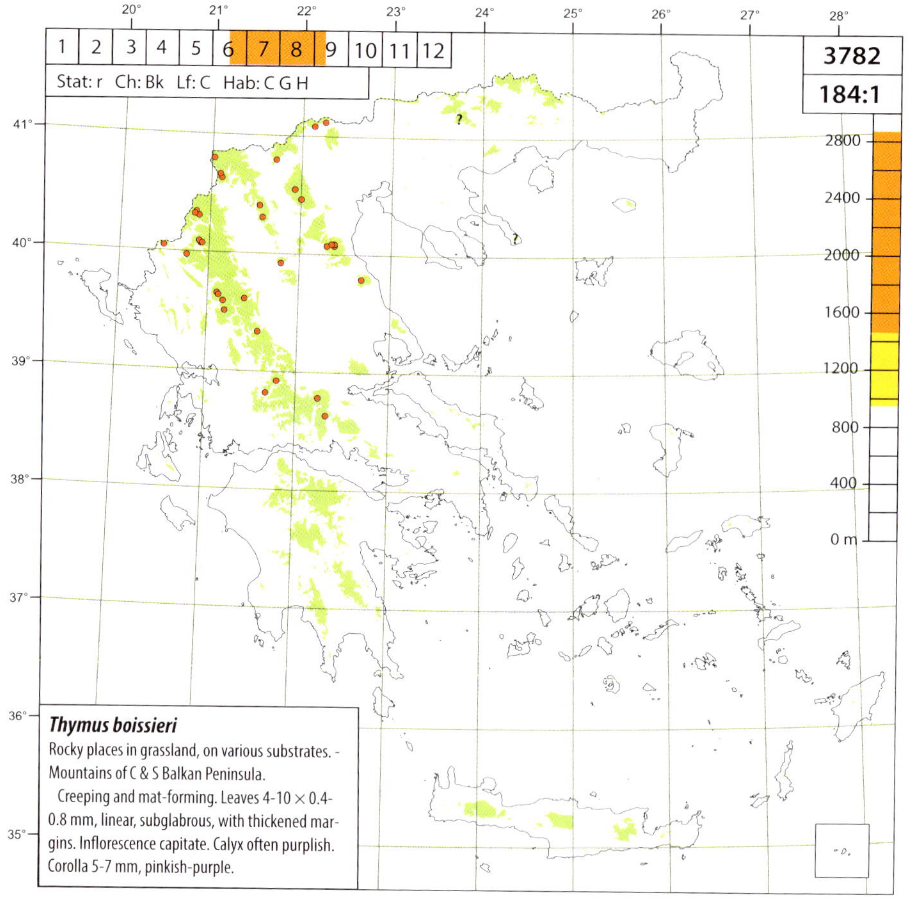

Thymus boissieri

Rocky places in grassland, on various substrates. - Mountains of C & S Balkan Peninsula.

Creeping and mat-forming. Leaves 4-10 × 0.4-0.8 mm, linear, subglabrous, with thickened margins. Inflorescence capitate. Calyx often purplish. Corolla 5-7 mm, pinkish-purple.

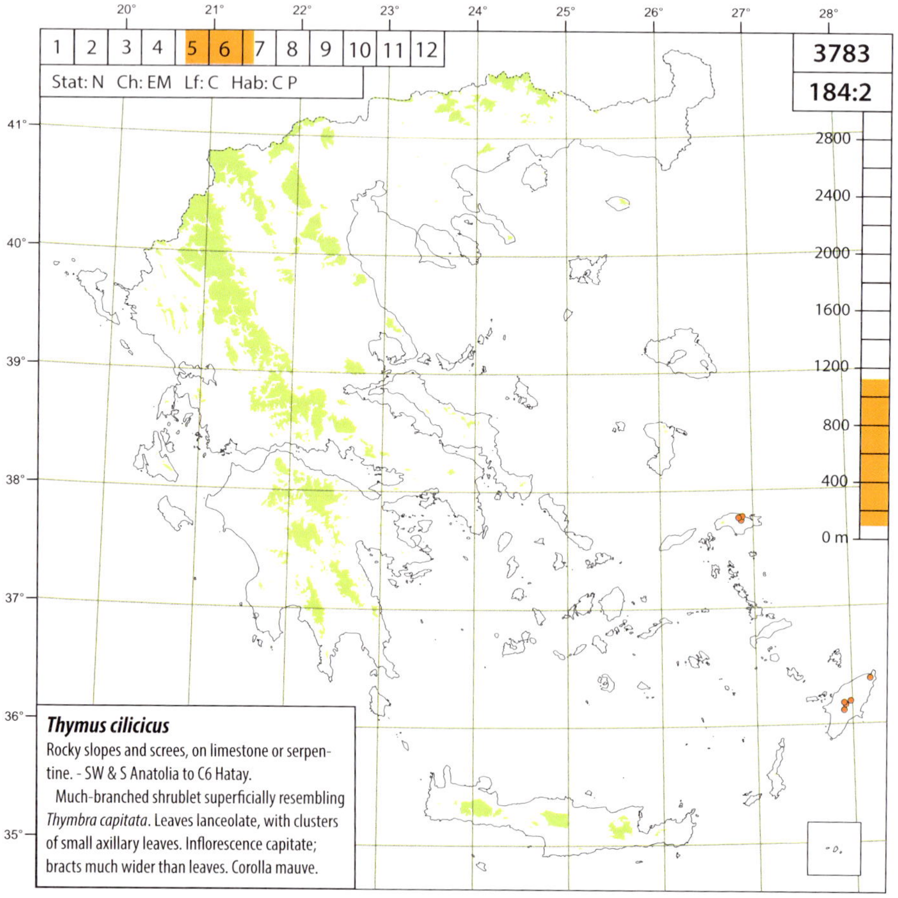

Thymus cilicicus

Rocky slopes and screes, on limestone or serpentine. - SW & S Anatolia to C6 Hatay.

Much-branched shrublet superficially resembling *Thymbra capitata*. Leaves lanceolate, with clusters of small axillary leaves. Inflorescence capitate; bracts much wider than leaves. Corolla mauve.

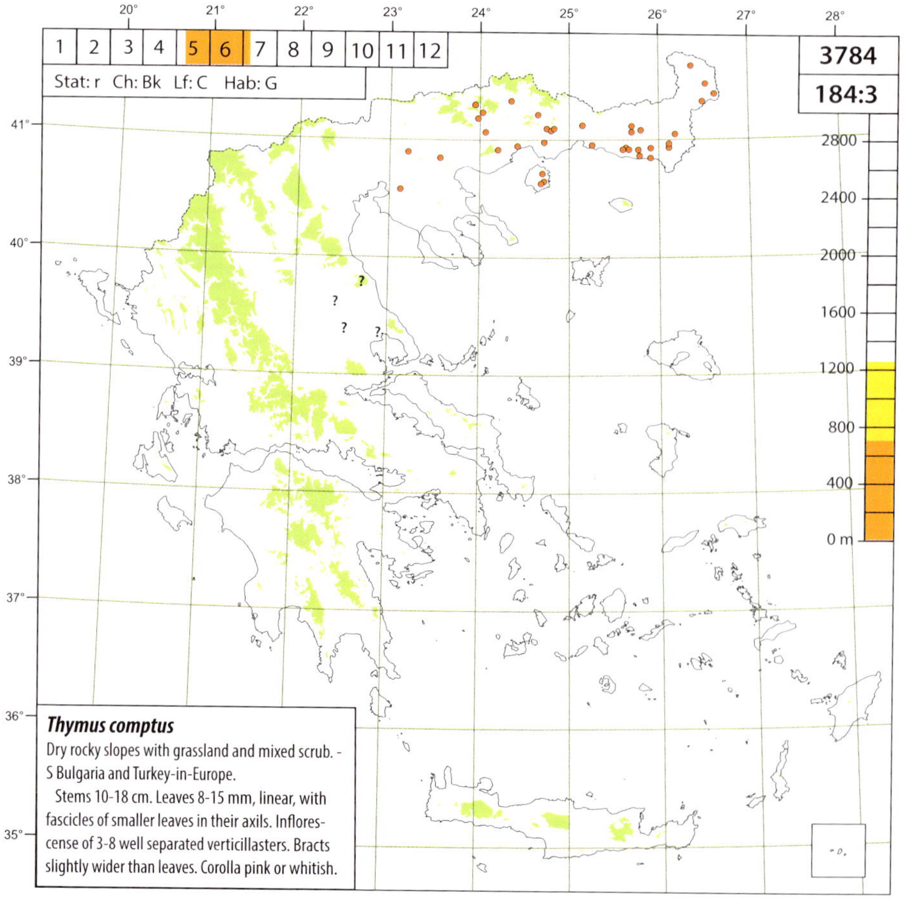

Thymus comptus

Dry rocky slopes with grassland and mixed scrub. - S Bulgaria and Turkey-in-Europe.

Stems 10-18 cm. Leaves 8-15 mm, linear, with fascicles of smaller leaves in their axils. Inflorescence of 3-8 well separated verticillasters. Bracts slightly wider than leaves. Corolla pink or whitish.

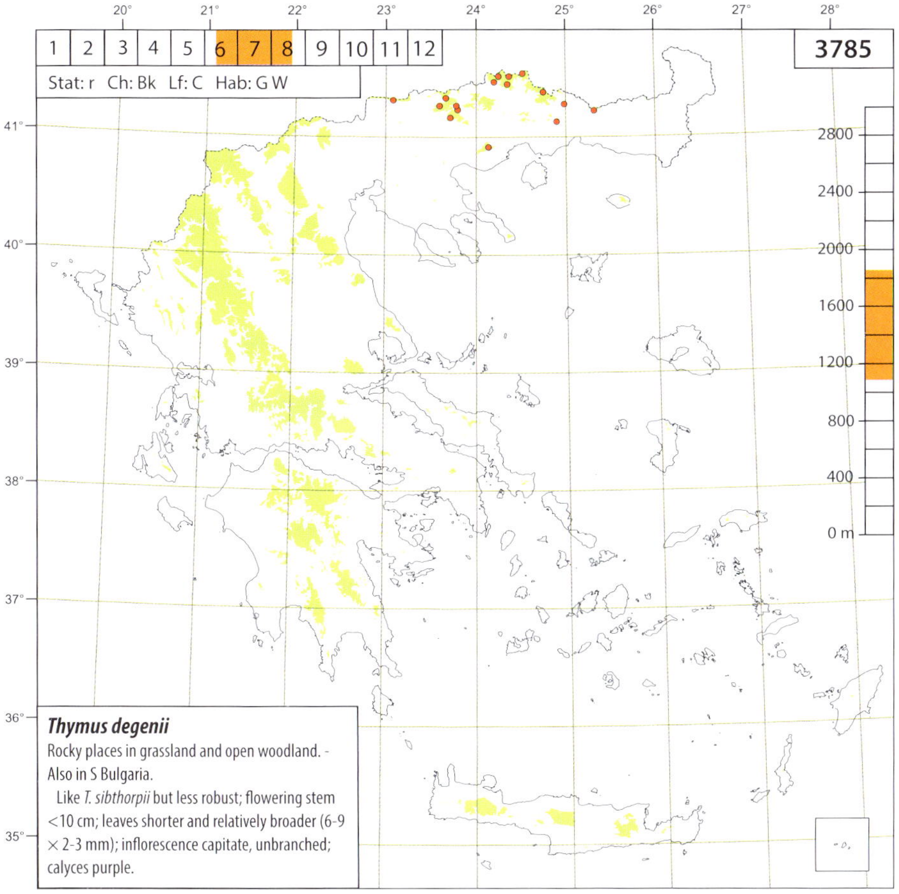

Thymus degenii
Rocky places in grassland and open woodland. - Also in S Bulgaria.
Like *T. sibthorpii* but less robust; flowering stem <10 cm; leaves shorter and relatively broader (6-9 × 2-3 mm); inflorescence capitate, unbranched; calyces purple.

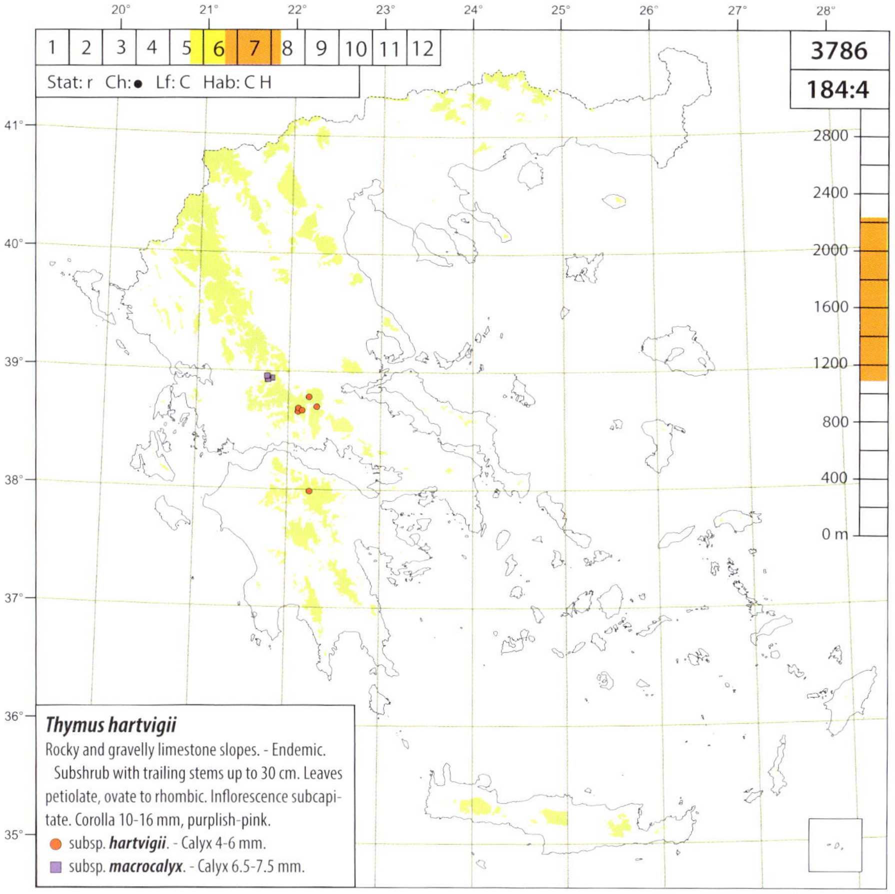

Thymus hartvigii
Rocky and gravelly limestone slopes. - Endemic.
Subshrub with trailing stems up to 30 cm. Leaves petiolate, ovate to rhombic. Inflorescence subcapitate. Corolla 10-16 mm, purplish-pink.
- ● subsp. ***hartvigii***. - Calyx 4-6 mm.
- ■ subsp. ***macrocalyx***. - Calyx 6.5-7.5 mm.

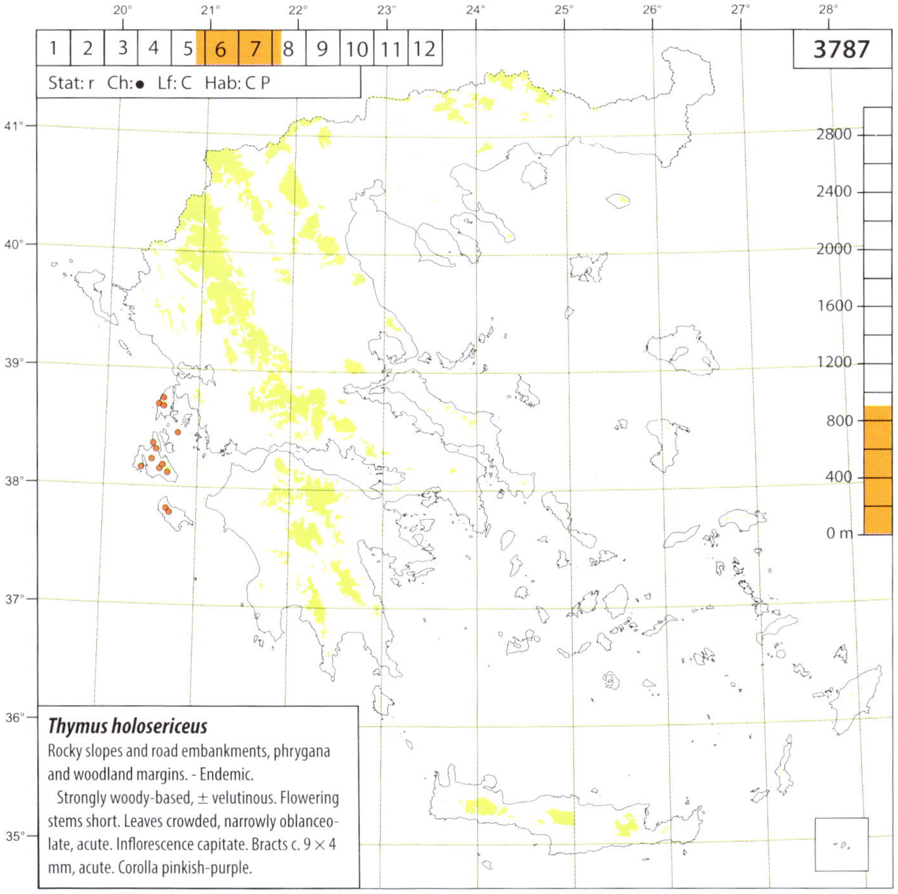

Thymus holosericeus
Rocky slopes and road embankments, phrygana and woodland margins. - Endemic.
Strongly woody-based, ± velutinous. Flowering stems short. Leaves crowded, narrowly oblanceolate, acute. Inflorescence capitate. Bracts c. 9 × 4 mm, acute. Corolla pinkish-purple.

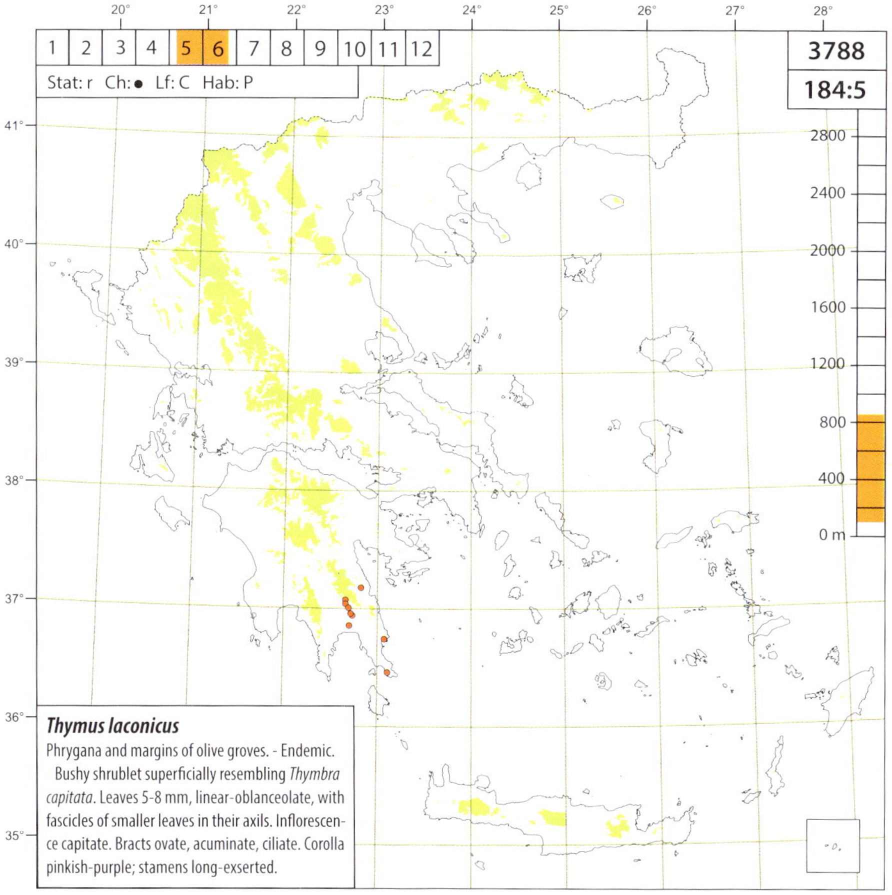

Thymus laconicus
Phrygana and margins of olive groves. - Endemic.
Bushy shrublet superficially resembling *Thymbra capitata*. Leaves 5-8 mm, linear-oblanceolate, with fascicles of smaller leaves in their axils. Inflorescence capitate. Bracts ovate, acuminate, ciliate. Corolla pinkish-purple; stamens long-exserted.

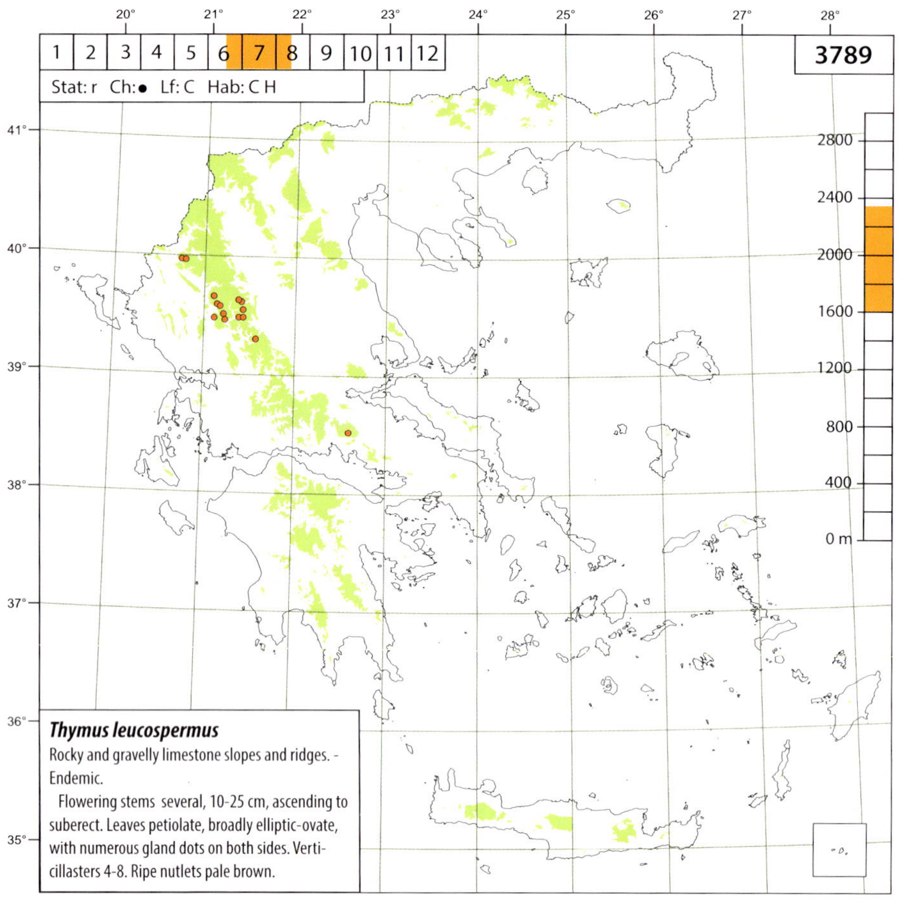

Thymus leucospermus
Rocky and gravelly limestone slopes and ridges. - Endemic.
Flowering stems several, 10-25 cm, ascending to suberect. Leaves petiolate, broadly elliptic-ovate, with numerous gland dots on both sides. Verticillasters 4-8. Ripe nutlets pale brown.

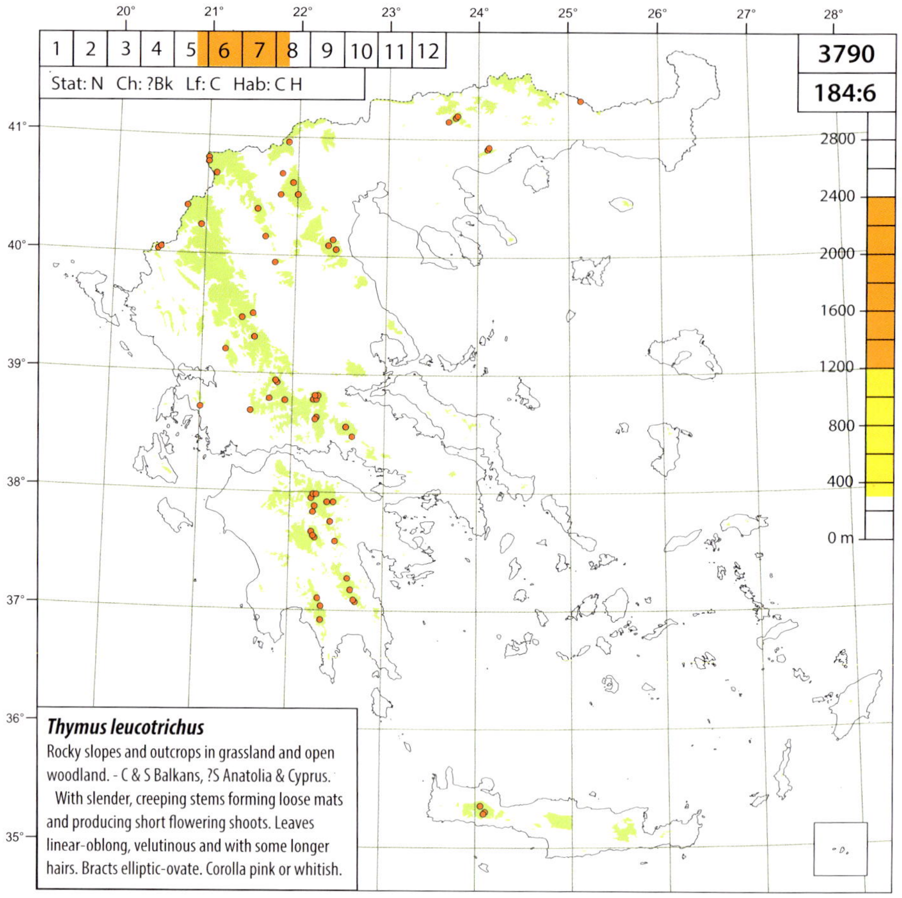

Thymus leucotrichus
Rocky slopes and outcrops in grassland and open woodland. - C & S Balkans, ?S Anatolia & Cyprus.
With slender, creeping stems forming loose mats and producing short flowering shoots. Leaves linear-oblong, velutinous and with some longer hairs. Bracts elliptic-ovate. Corolla pink or whitish.

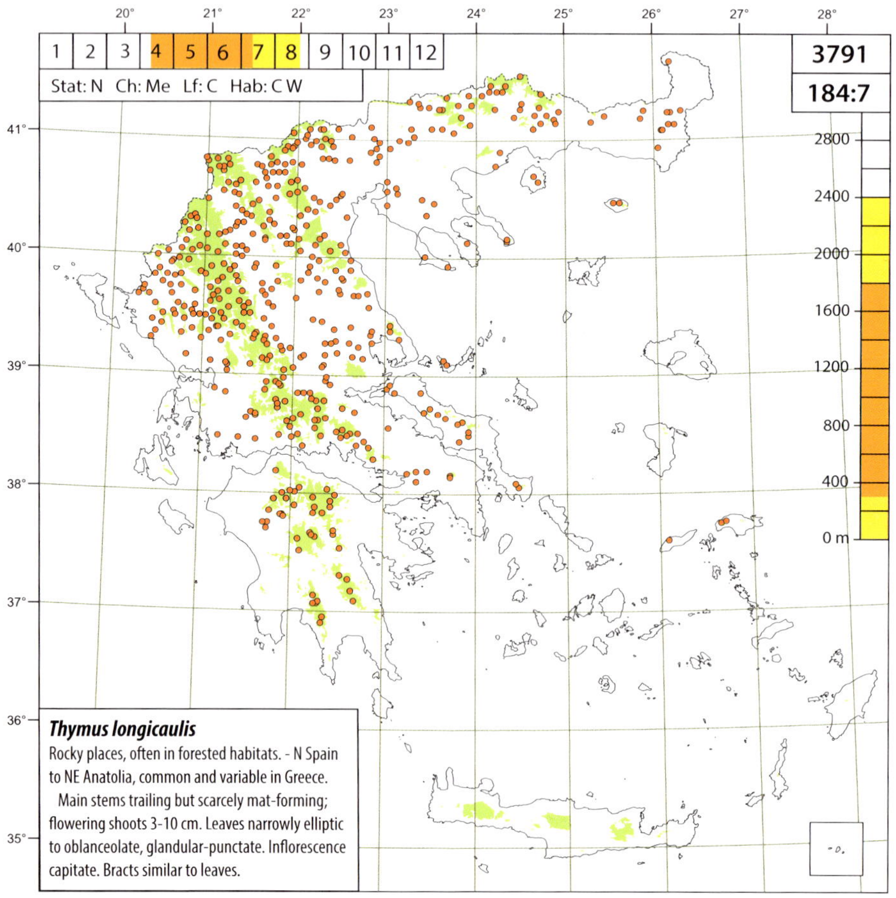

Thymus longicaulis
Rocky places, often in forested habitats. - N Spain to NE Anatolia, common and variable in Greece.
Main stems trailing but scarcely mat-forming; flowering shoots 3-10 cm. Leaves narrowly elliptic to oblanceolate, glandular-punctate. Inflorescence capitate. Bracts similar to leaves.

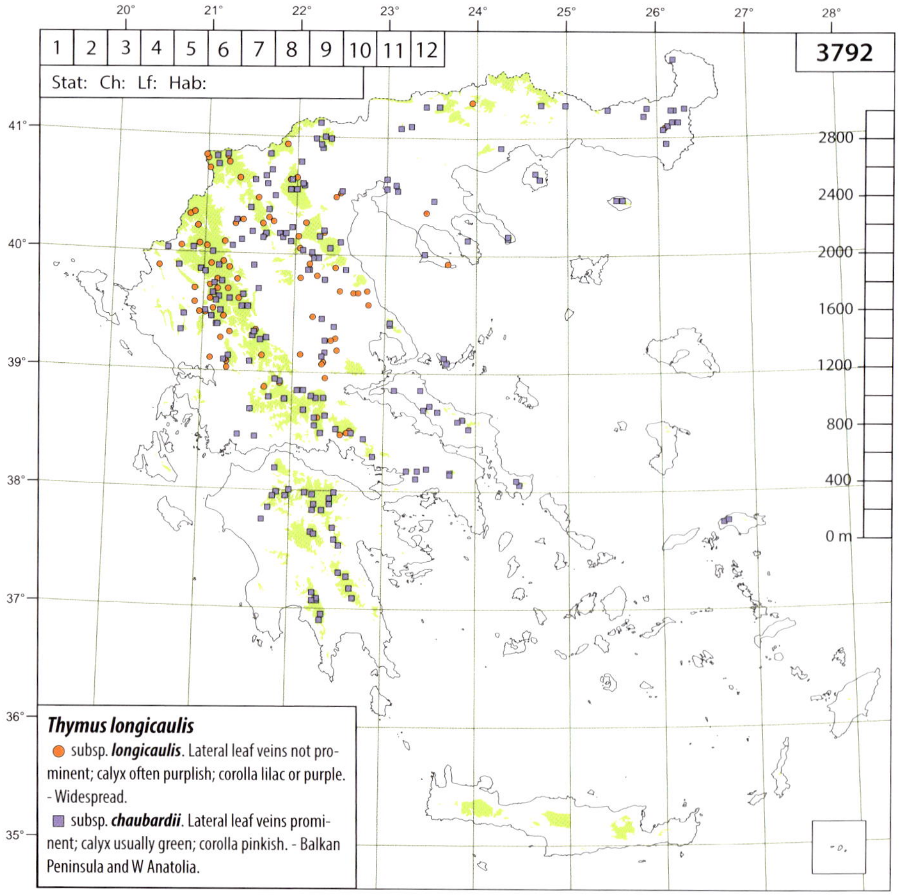

Thymus longicaulis
● subsp. ***longicaulis***. Lateral leaf veins not prominent; calyx often purplish; corolla lilac or purple. - Widespread.
■ subsp. ***chaubardii***. Lateral leaf veins prominent; calyx usually green; corolla pinkish. - Balkan Peninsula and W Anatolia.

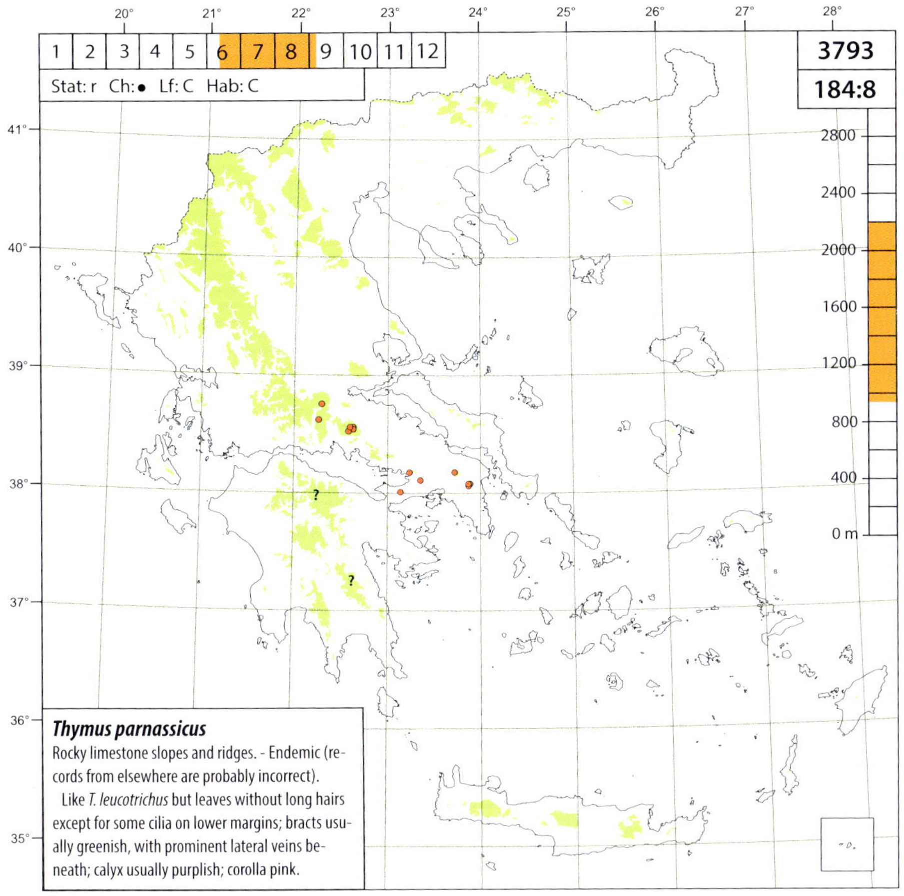

Thymus parnassicus
Rocky limestone slopes and ridges. - Endemic (records from elsewhere are probably incorrect).
Like *T. leucotrichus* but leaves without long hairs except for some cilia on lower margins; bracts usually greenish, with prominent lateral veins beneath; calyx usually purplish; corolla pink.

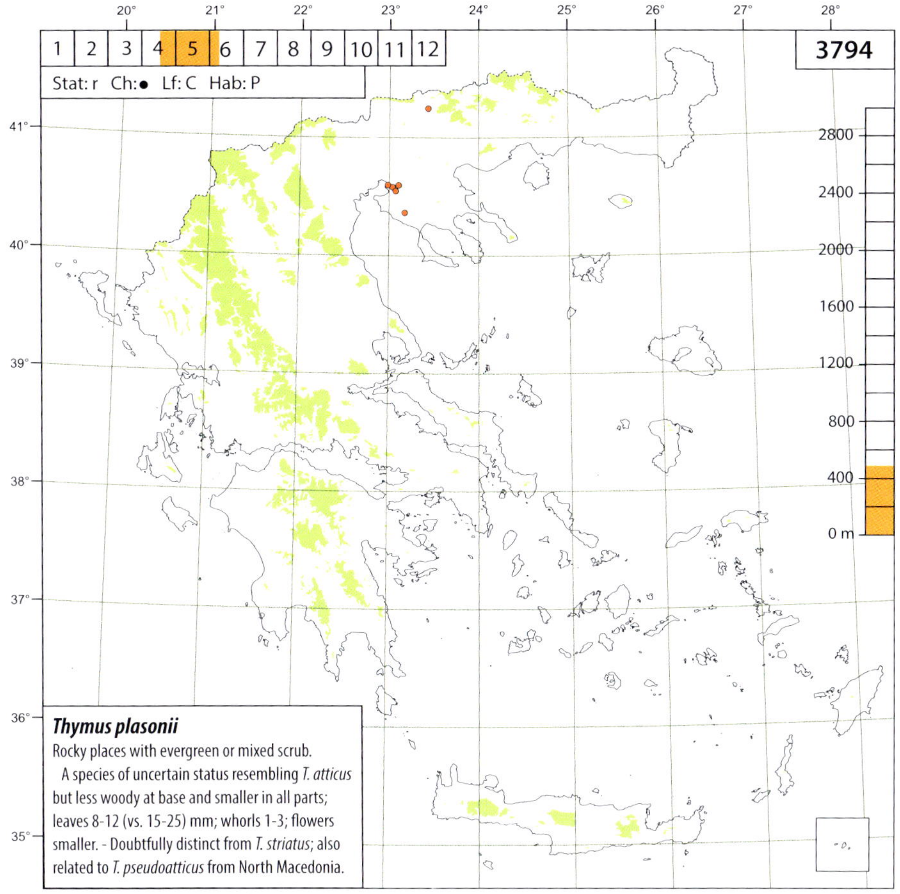

Thymus plasonii
Rocky places with evergreen or mixed scrub.
A species of uncertain status resembling *T. atticus* but less woody at base and smaller in all parts; leaves 8-12 (vs. 15-25) mm; whorls 1-3; flowers smaller. - Doubtfully distinct from *T. striatus*; also related to *T. pseudoatticus* from North Macedonia.

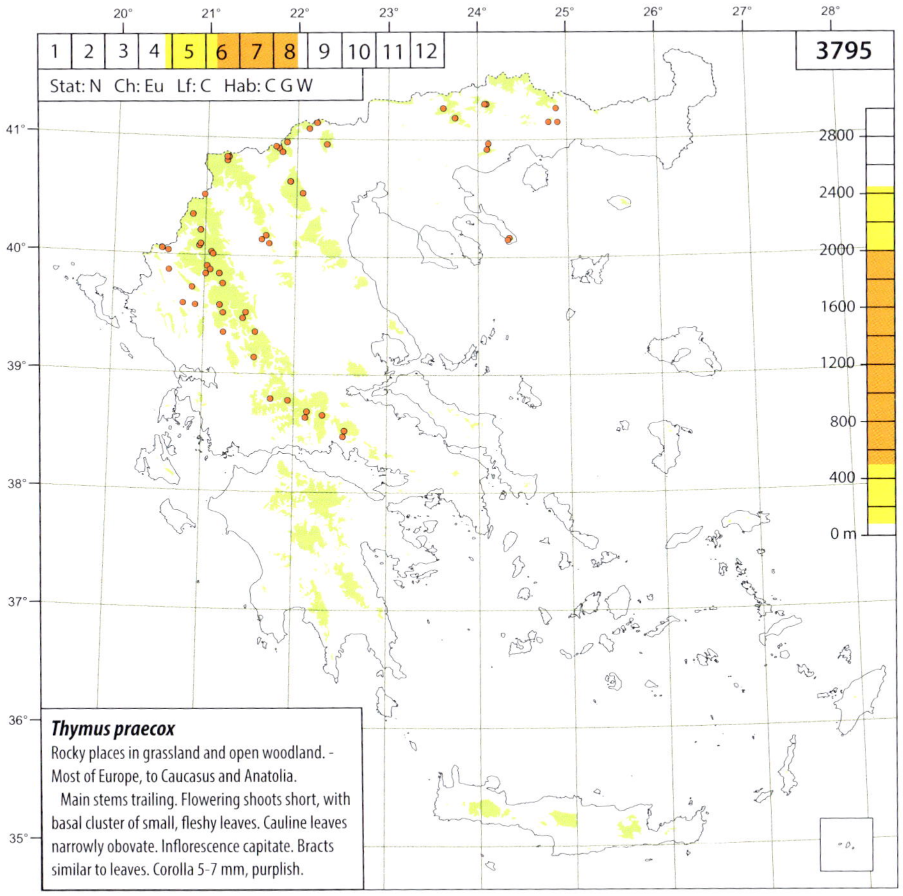

Thymus praecox
Rocky places in grassland and open woodland. - Most of Europe, to Caucasus and Anatolia.
Main stems trailing. Flowering shoots short, with basal cluster of small, fleshy leaves. Cauline leaves narrowly obovate. Inflorescence capitate. Bracts similar to leaves. Corolla 5-7 mm, purplish.

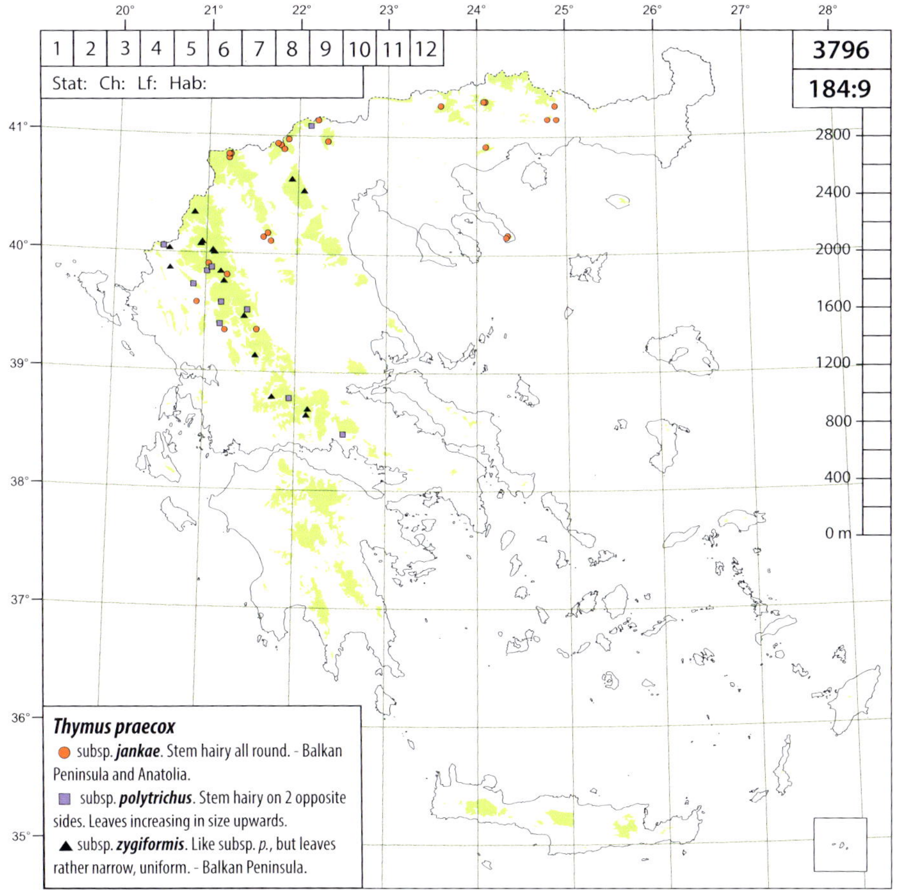

Thymus praecox
● subsp. ***jankae***. Stem hairy all round. - Balkan Peninsula and Anatolia.
■ subsp. ***polytrichus***. Stem hairy on 2 opposite sides. Leaves increasing in size upwards.
▲ subsp. ***zygiformis***. Like subsp. *p.*, but leaves rather narrow, uniform. - Balkan Peninsula.

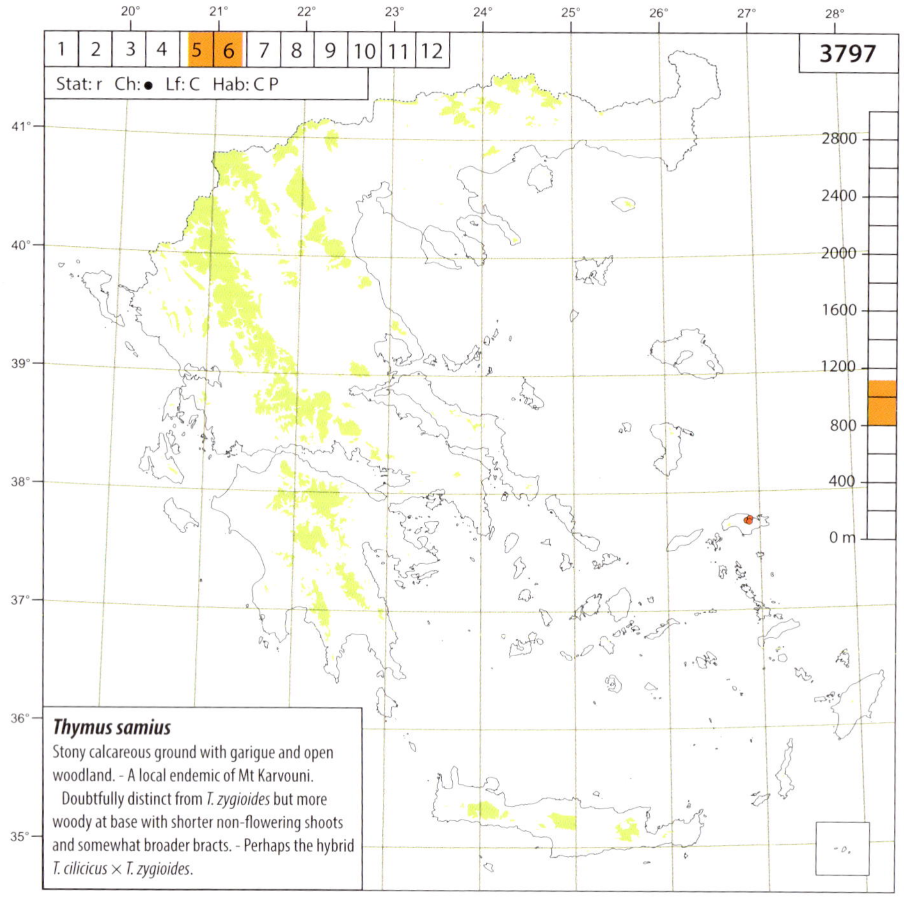

Thymus samius

Stony calcareous ground with garigue and open woodland. - A local endemic of Mt Karvouni.

Doubtfully distinct from *T. zygioides* but more woody at base with shorter non-flowering shoots and somewhat broader bracts. - Perhaps the hybrid *T. cilicicus* × *T. zygioides*.

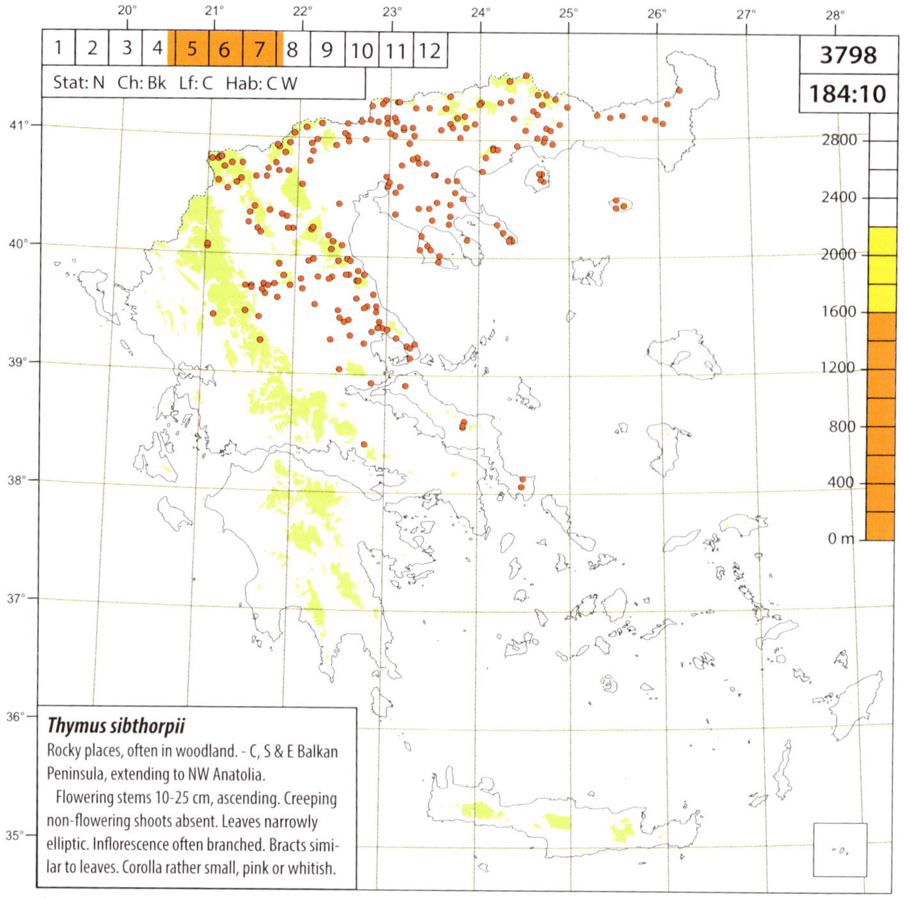

Thymus sibthorpii

Rocky places, often in woodland. - C, S & E Balkan Peninsula, extending to NW Anatolia.

Flowering stems 10-25 cm, ascending. Creeping non-flowering shoots absent. Leaves narrowly elliptic. Inflorescence often branched. Bracts similar to leaves. Corolla rather small, pink or whitish.

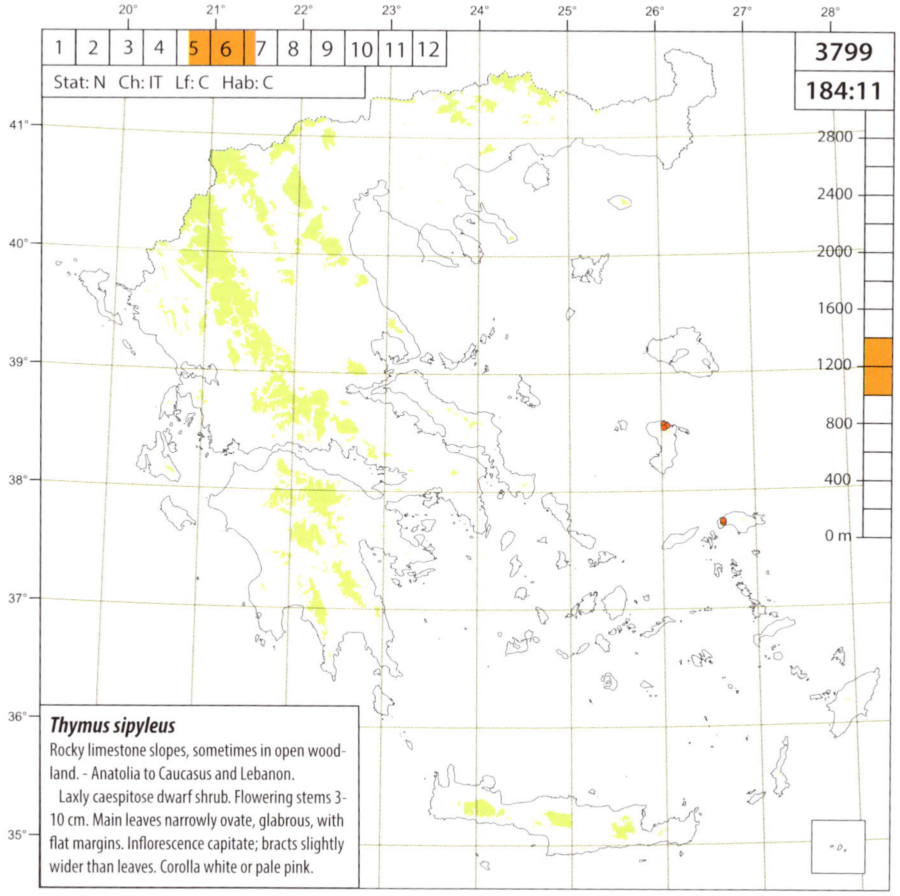

Thymus sipyleus

Rocky limestone slopes, sometimes in open woodland. - Anatolia to Caucasus and Lebanon.

Laxly caespitose dwarf shrub. Flowering stems 3-10 cm. Main leaves narrowly ovate, glabrous, with flat margins. Inflorescence capitate; bracts slightly wider than leaves. Corolla white or pale pink.

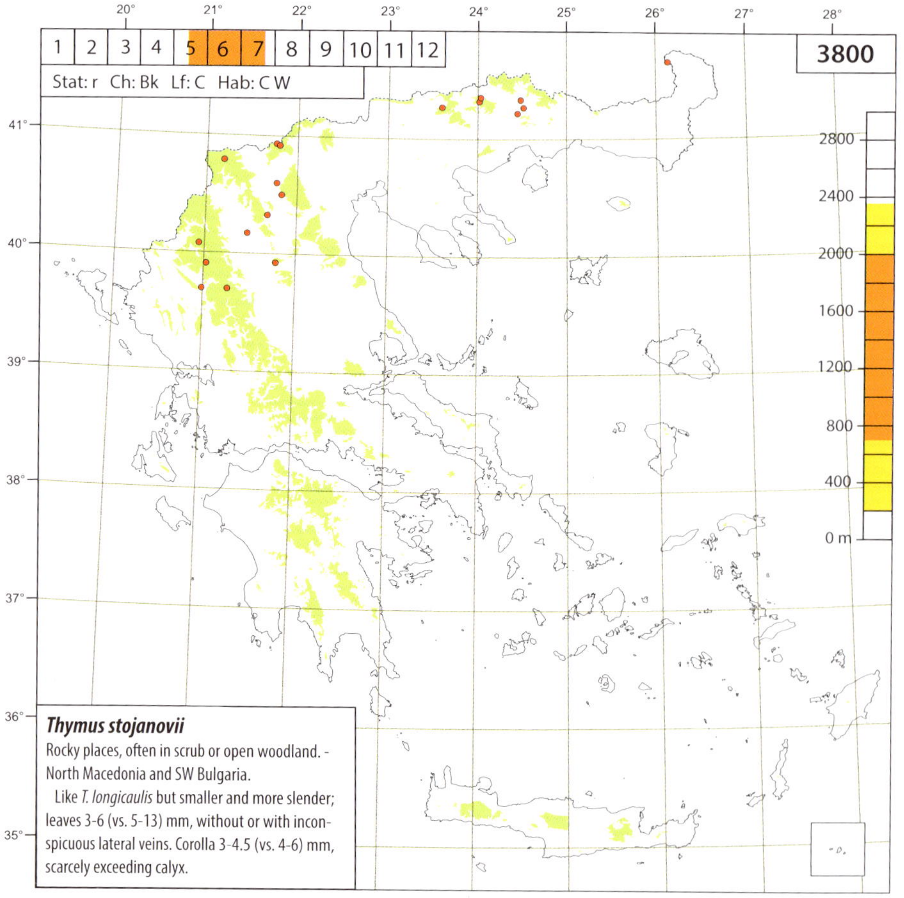

Thymus stojanovii

Rocky places, often in scrub or open woodland. - North Macedonia and SW Bulgaria.

Like *T. longicaulis* but smaller and more slender; leaves 3-6 (vs. 5-13) mm, without or with inconspicuous lateral veins. Corolla 3-4.5 (vs. 4-6) mm, scarcely exceeding calyx.

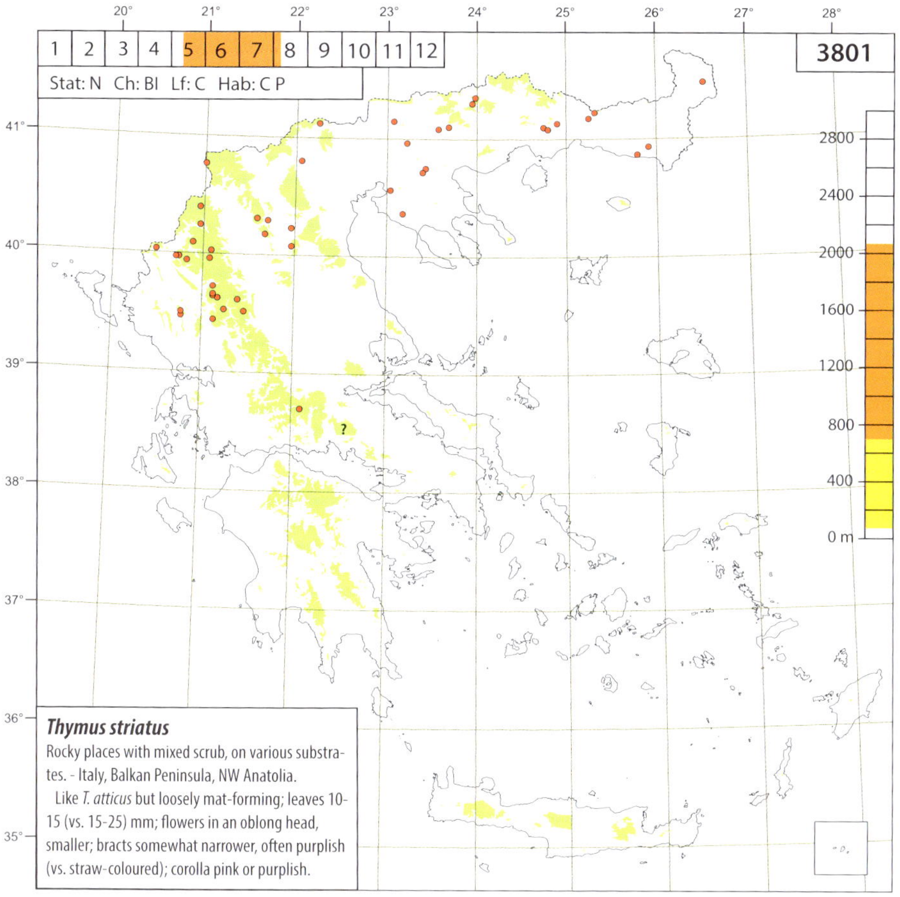

Thymus striatus
Rocky places with mixed scrub, on various substrates. - Italy, Balkan Peninsula, NW Anatolia.
Like *T. atticus* but loosely mat-forming; leaves 10-15 (vs. 15-25) mm; flowers in an oblong head, smaller; bracts somewhat narrower, often purplish (vs. straw-coloured); corolla pink or purplish.

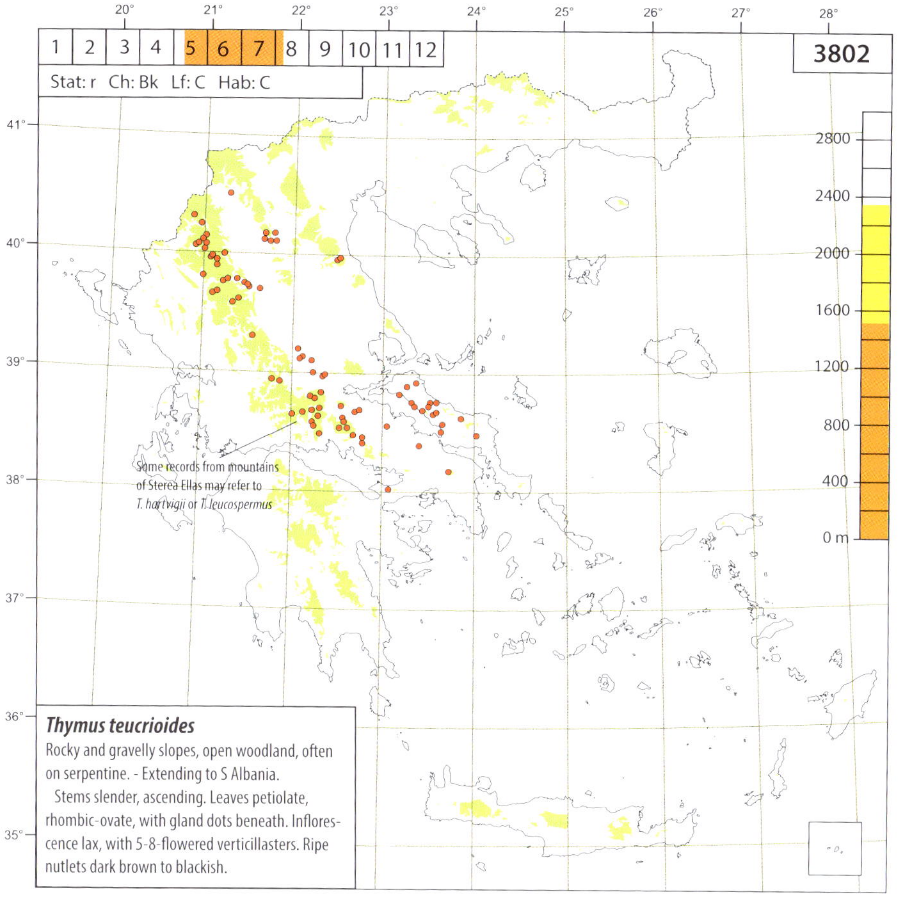

Thymus teucrioides
Rocky and gravelly slopes, open woodland, often on serpentine. - Extending to S Albania.
Stems slender, ascending. Leaves petiolate, rhombic-ovate, with gland dots beneath. Inflorescence lax, with 5-8-flowered verticillasters. Ripe nutlets dark brown to blackish.

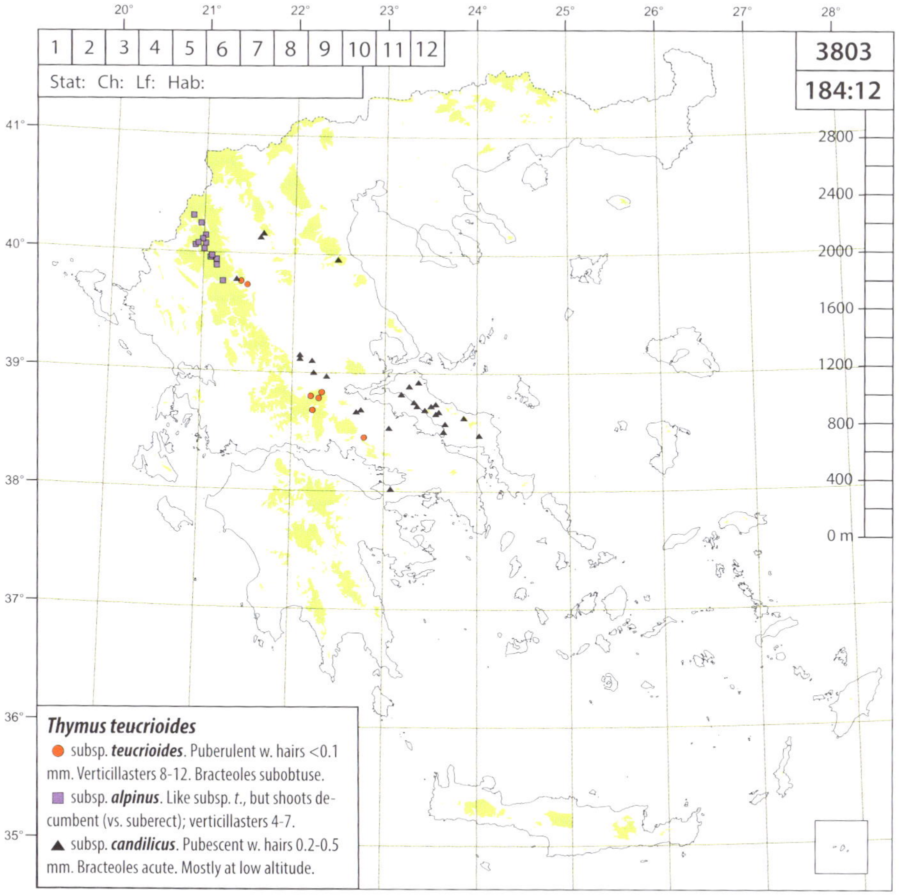

Thymus teucrioides
● subsp. ***teucrioides***. Puberulent w. hairs <0.1 mm. Verticillasters 8-12. Bracteoles subobtuse.
■ subsp. ***alpinus***. Like subsp. *t.*, but shoots decumbent (vs. suberect); verticillasters 4-7.
▲ subsp. ***candilicus***. Pubescent w. hairs 0.2-0.5 mm. Bracteoles acute. Mostly at low altitude.

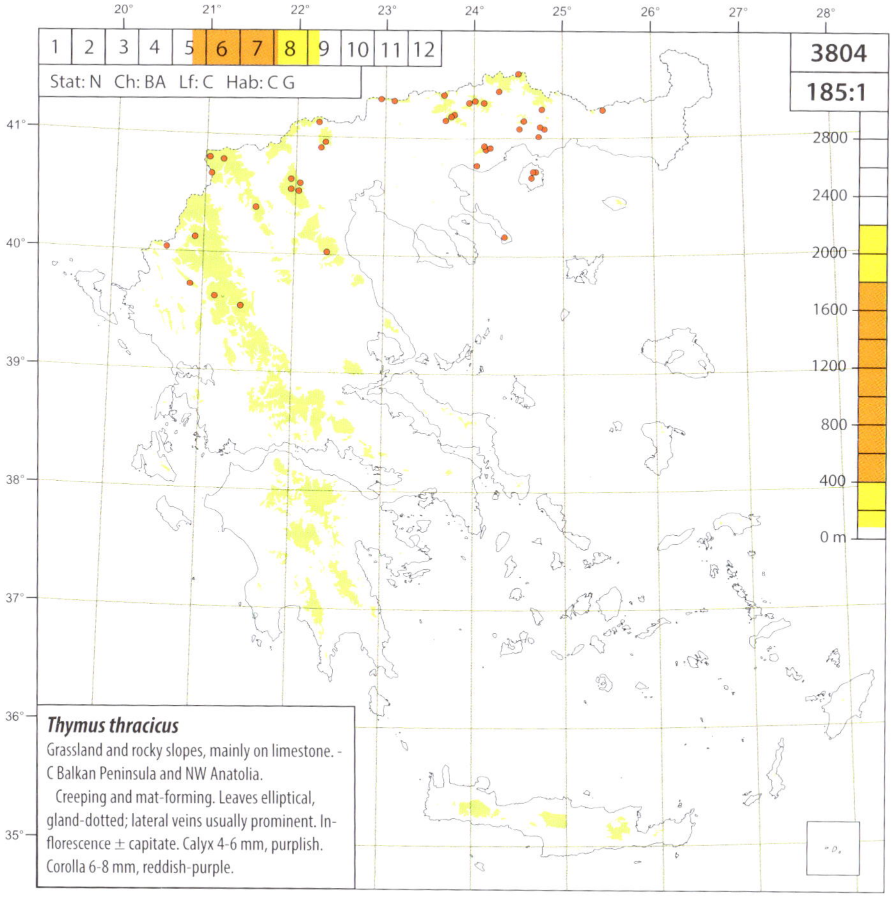

Thymus thracicus
Grassland and rocky slopes, mainly on limestone. - C Balkan Peninsula and NW Anatolia.
Creeping and mat-forming. Leaves elliptical, gland-dotted; lateral veins usually prominent. Inflorescence ± capitate. Calyx 4-6 mm, purplish. Corolla 6-8 mm, reddish-purple.

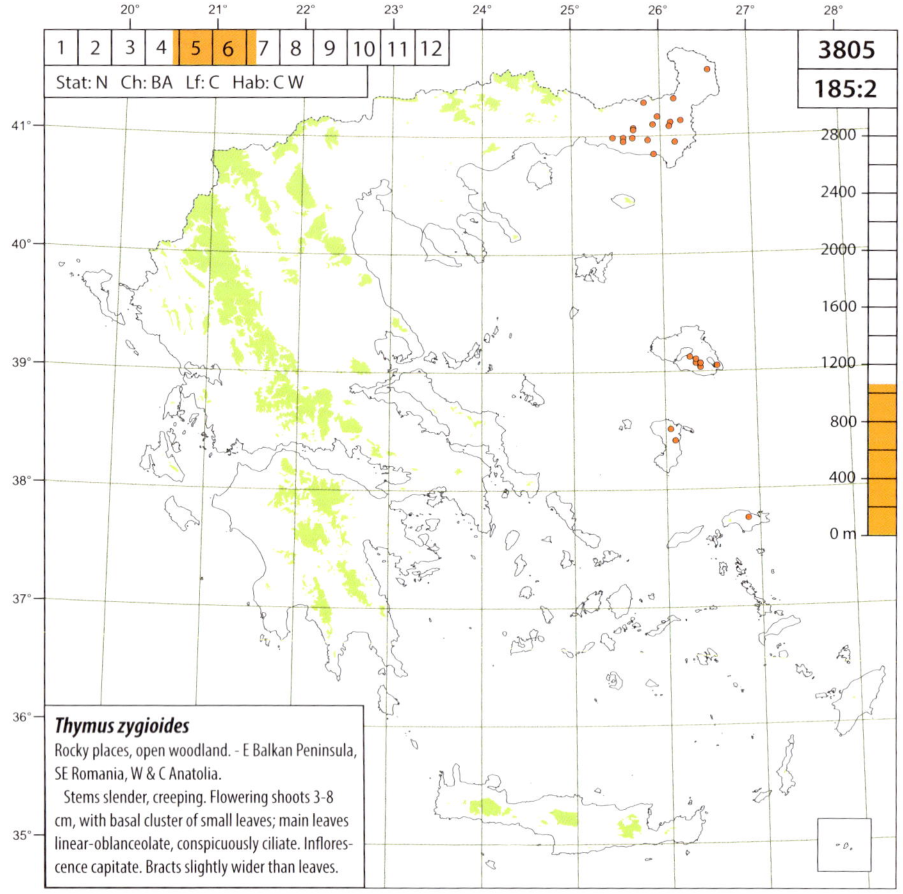

Thymus zygioides

Rocky places, open woodland. - E Balkan Peninsula, SE Romania, W & C Anatolia.

Stems slender, creeping. Flowering shoots 3-8 cm, with basal cluster of small leaves; main leaves linear-oblanceolate, conspicuously ciliate. Inflorescence capitate. Bracts slightly wider than leaves.

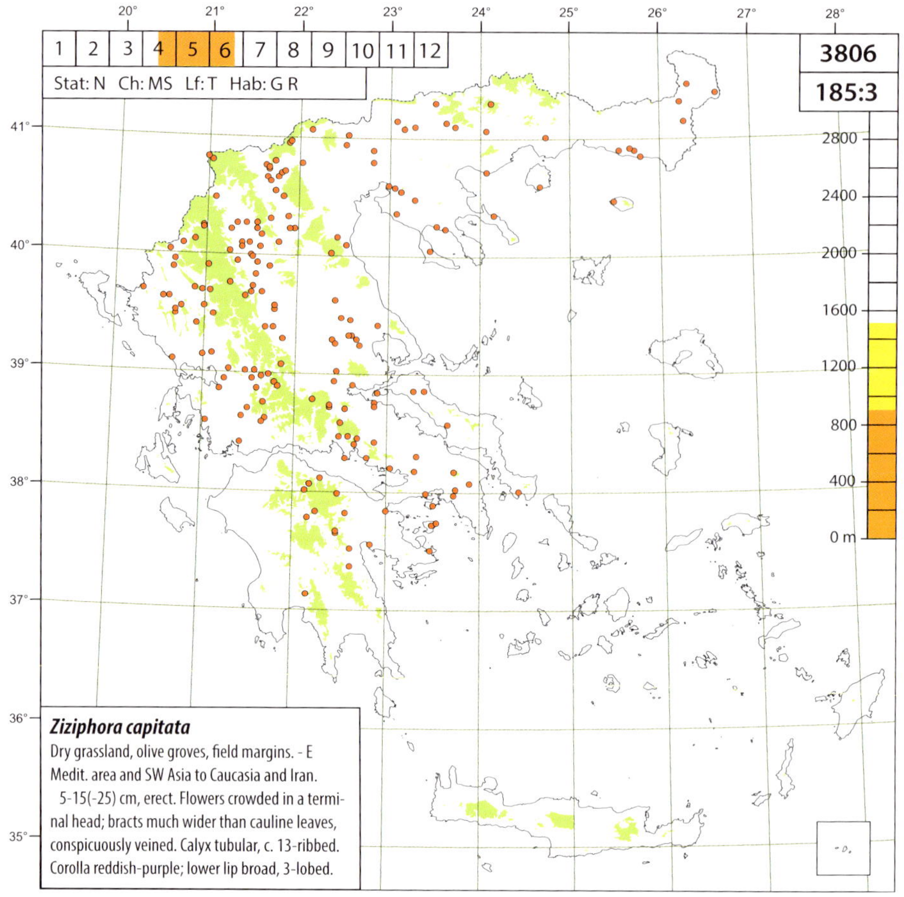

Ziziphora capitata

Dry grassland, olive groves, field margins. - E Medit. area and SW Asia to Caucasia and Iran.

5-15(-25) cm, erect. Flowers crowded in a terminal head; bracts much wider than cauline leaves, conspicuously veined. Calyx tubular, c. 13-ribbed. Corolla reddish-purple; lower lip broad, 3-lobed.

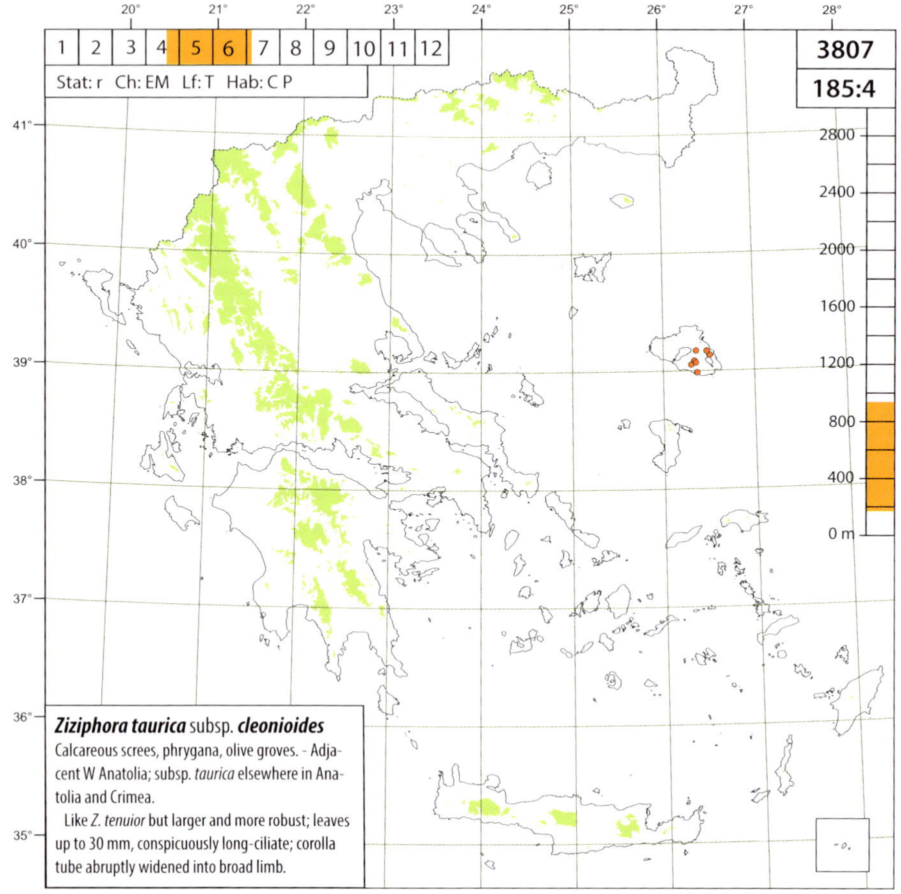

Ziziphora taurica subsp. *cleonioides*

Calcareous screes, phrygana, olive groves. - Adjacent W Anatolia; subsp. *taurica* elsewhere in Anatolia and Crimea.

Like *Z. tenuior* but larger and more robust; leaves up to 30 mm, conspicuously long-ciliate; corolla tube abruptly widened into broad limb.

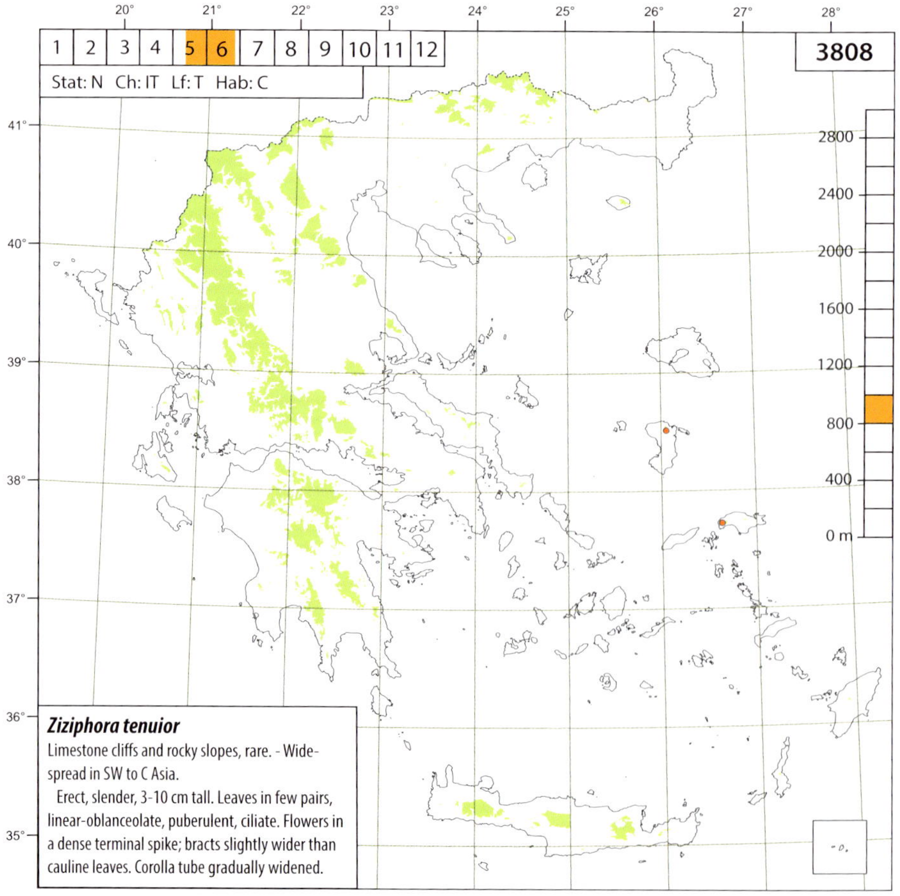

Ziziphora tenuior

Limestone cliffs and rocky slopes, rare. - Widespread in SW to C Asia.

Erect, slender, 3-10 cm tall. Leaves in few pairs, linear-oblanceolate, puberulent, ciliate. Flowers in a dense terminal spike; bracts slightly wider than cauline leaves. Corolla tube gradually widened.

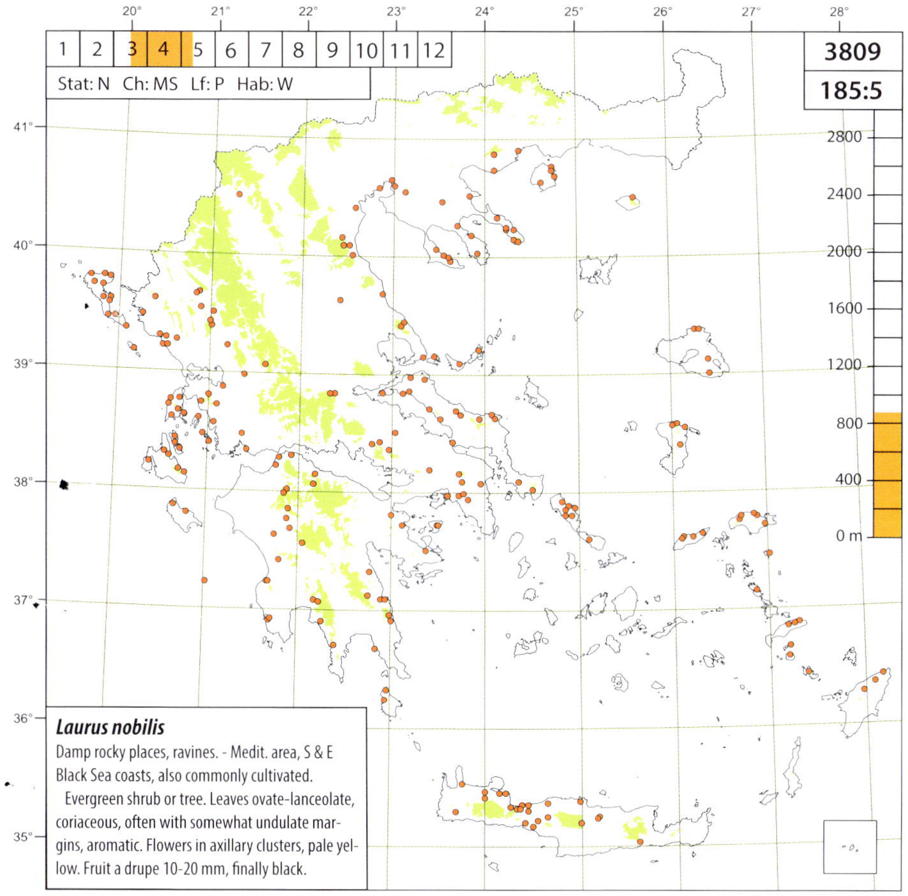

Laurus nobilis
Damp rocky places, ravines. - Medit. area, S & E Black Sea coasts, also commonly cultivated.
Evergreen shrub or tree. Leaves ovate-lanceolate, coriaceous, often with somewhat undulate margins, aromatic. Flowers in axillary clusters, pale yellow. Fruit a drupe 10-20 mm, finally black.

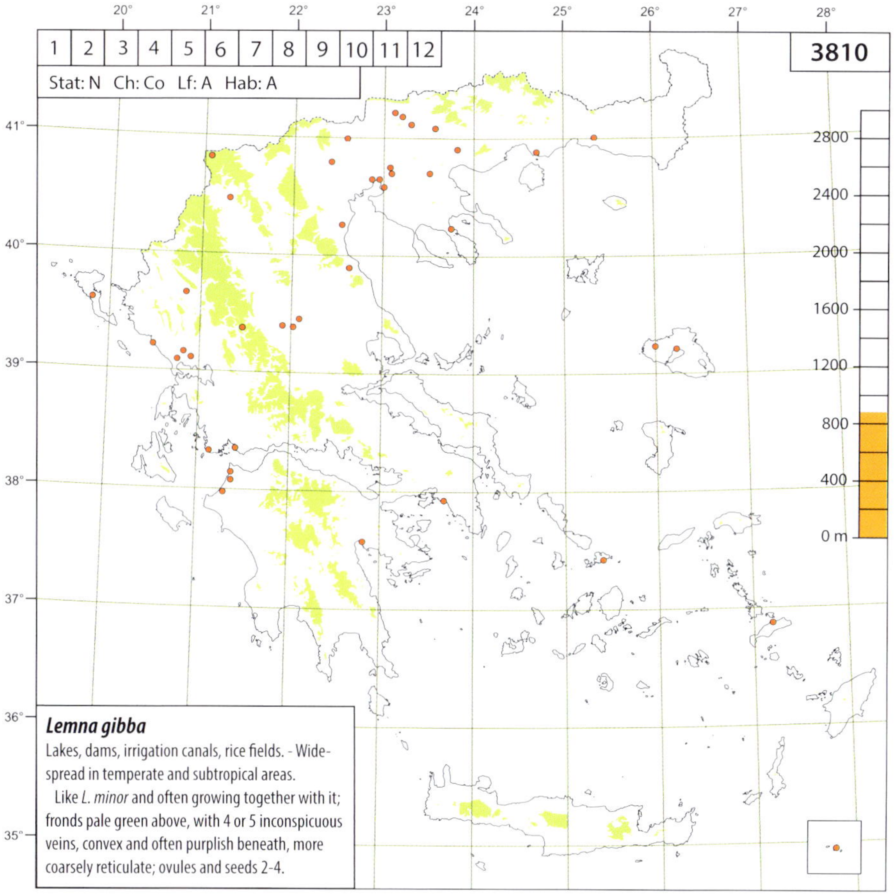

Lemna gibba
Lakes, dams, irrigation canals, rice fields. - Widespread in temperate and subtropical areas.
Like *L. minor* and often growing together with it; fronds pale green above, with 4 or 5 inconspicuous veins, convex and often purplish beneath, more coarsely reticulate; ovules and seeds 2-4.

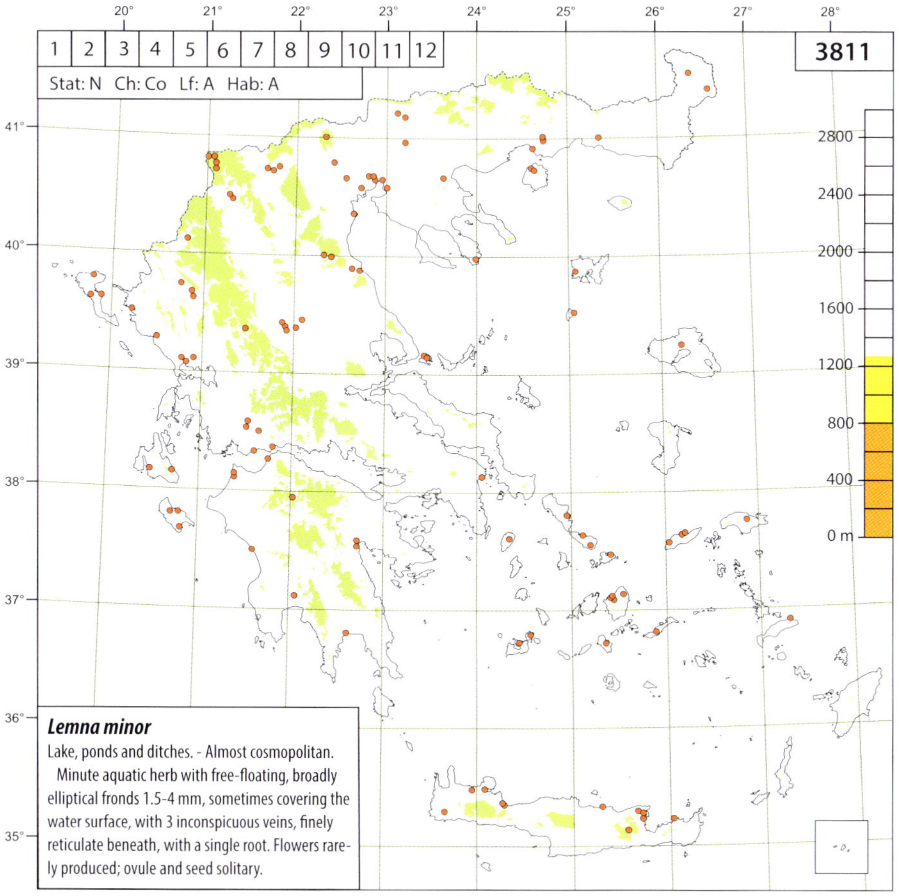

Lemna minor
Lake, ponds and ditches. - Almost cosmopolitan.
Minute aquatic herb with free-floating, broadly elliptical fronds 1.5-4 mm, sometimes covering the water surface, with 3 inconspicuous veins, finely reticulate beneath, with a single root. Flowers rarely produced; ovule and seed solitary.

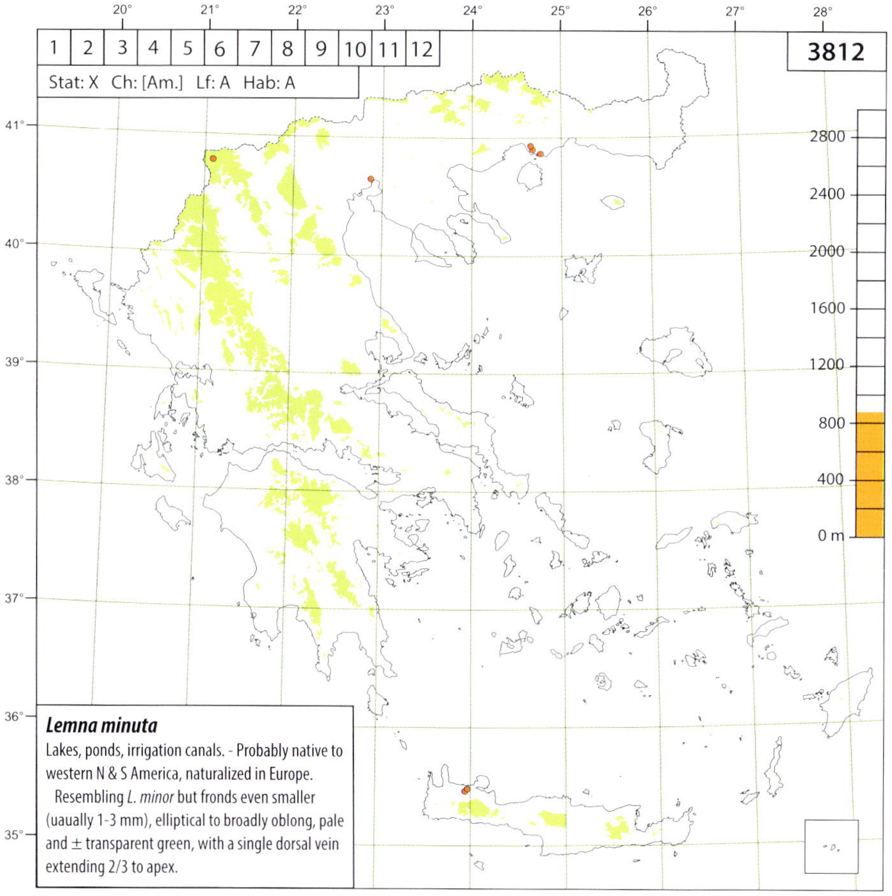

Lemna minuta
Lakes, ponds, irrigation canals. - Probably native to western N & S America, naturalized in Europe.
Resembling *L. minor* but fronds even smaller (uaually 1-3 mm), elliptical to broadly oblong, pale and ± transparent green, with a single dorsal vein extending 2/3 to apex.

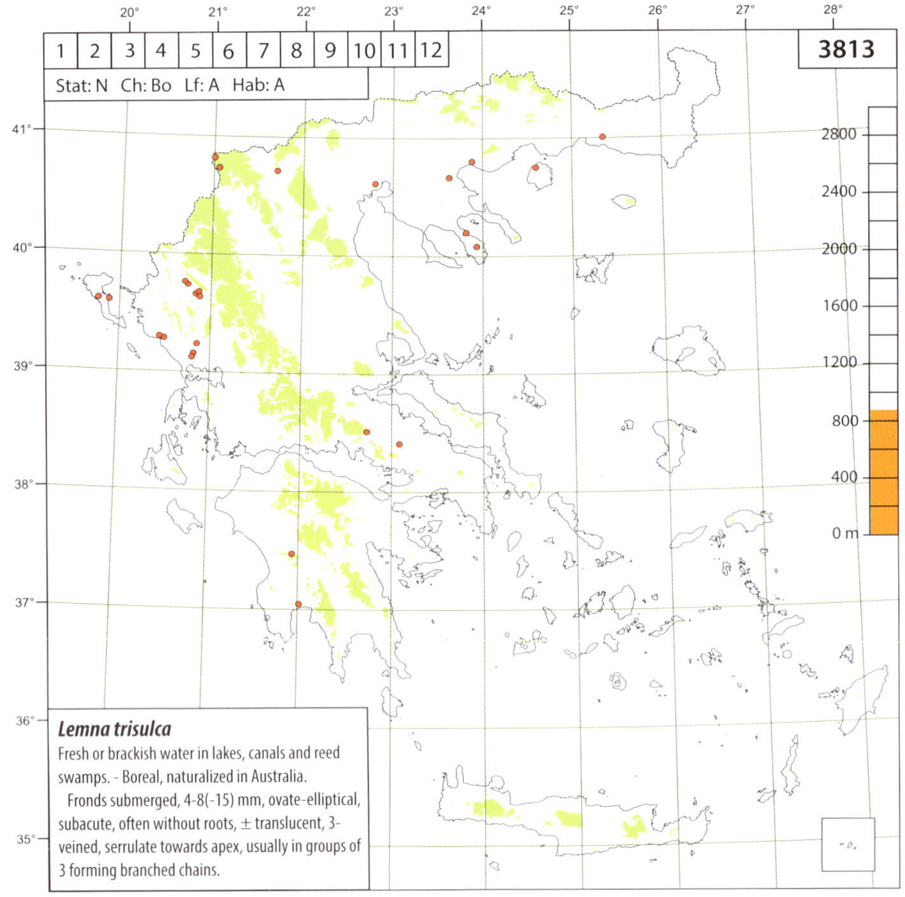

Lemna trisulca

Fresh or brackish water in lakes, canals and reed swamps. - Boreal, naturalized in Australia.

Fronds submerged, 4-8(-15) mm, ovate-elliptical, subacute, often without roots, ± translucent, 3-veined, serrulate towards apex, usually in groups of 3 forming branched chains.

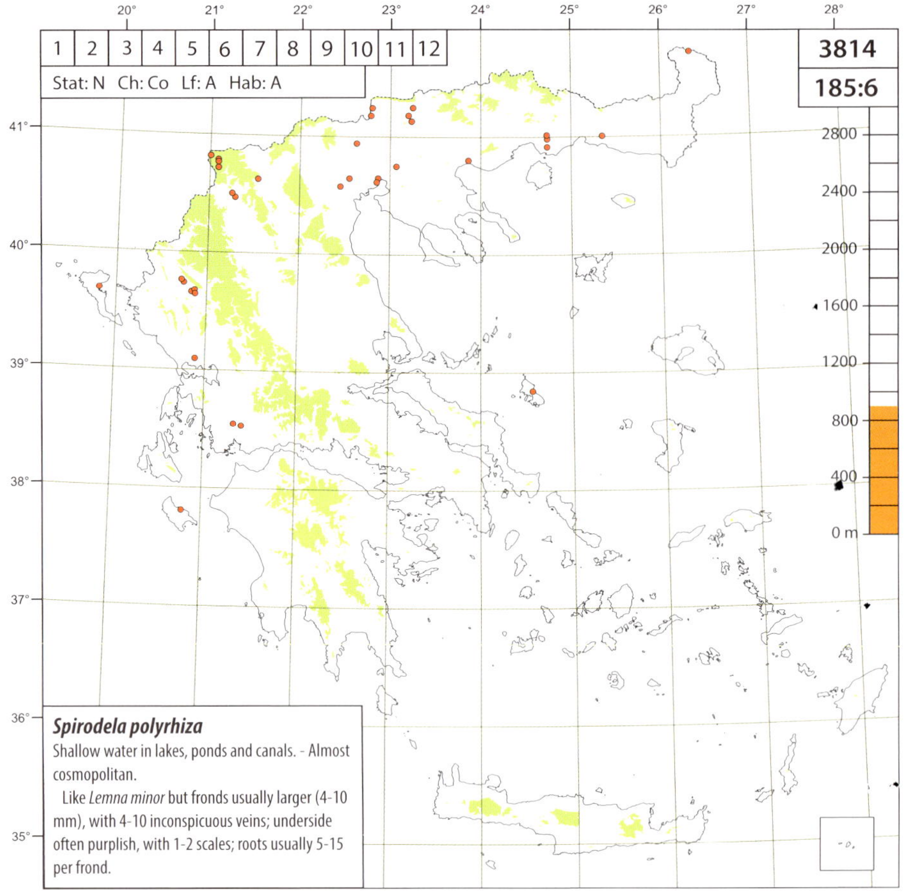

Spirodela polyrhiza

Shallow water in lakes, ponds and canals. - Almost cosmopolitan.

Like *Lemna minor* but fronds usually larger (4-10 mm), with 4-10 inconspicuous veins; underside often purplish, with 1-2 scales; roots usually 5-15 per frond.

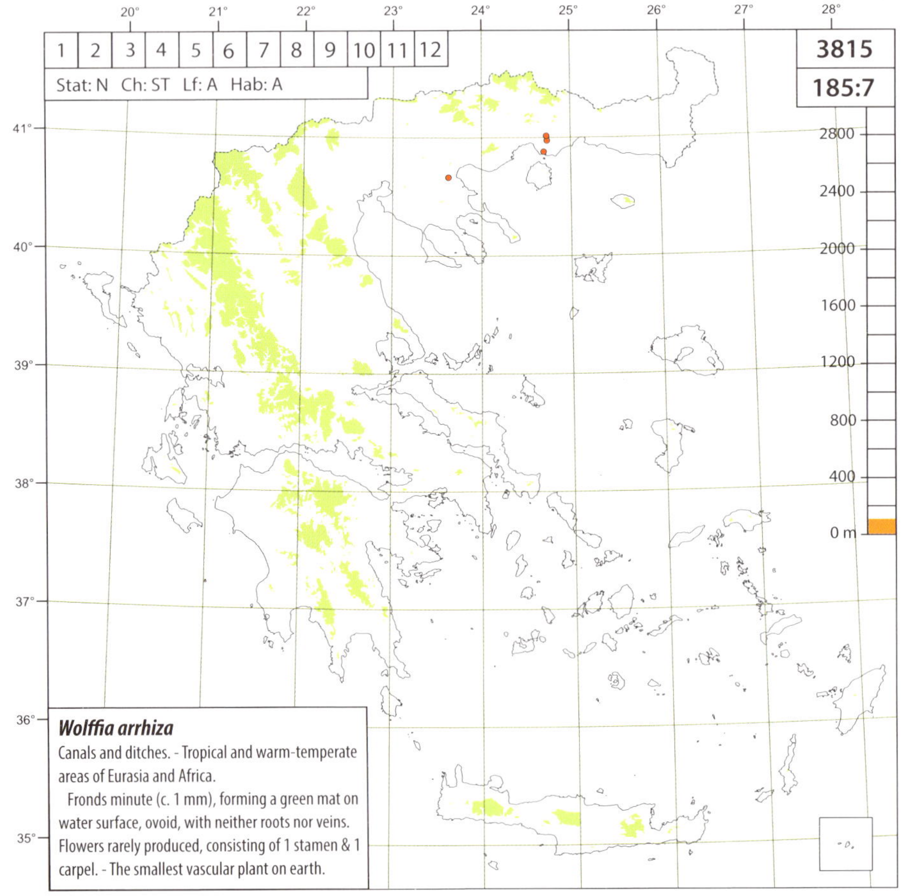

Wolffia arrhiza

Canals and ditches. - Tropical and warm-temperate areas of Eurasia and Africa.

Fronds minute (c. 1 mm), forming a green mat on water surface, ovoid, with neither roots nor veins. Flowers rarely produced, consisting of 1 stamen & 1 carpel. - The smallest vascular plant on earth.

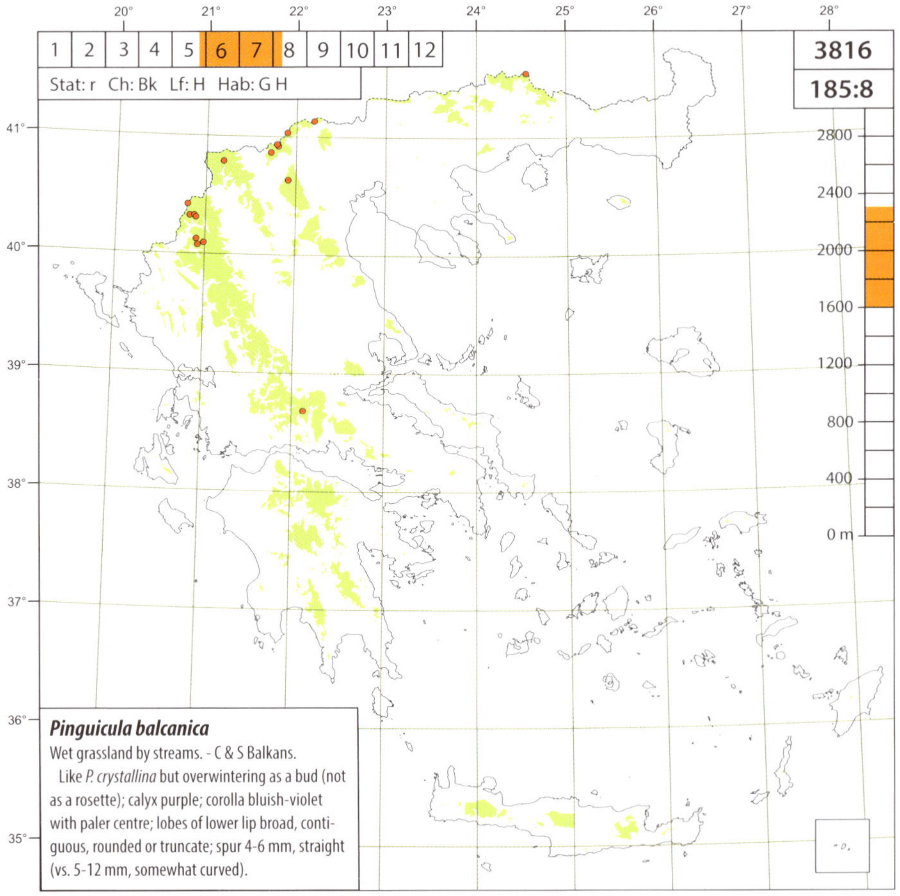

Pinguicula balcanica

Wet grassland by streams. - C & S Balkans.

Like *P. crystallina* but overwintering as a bud (not as a rosette); calyx purple; corolla bluish-violet with paler centre; lobes of lower lip broad, contiguous, rounded or truncate; spur 4-6 mm, straight (vs. 5-12 mm, somewhat curved).

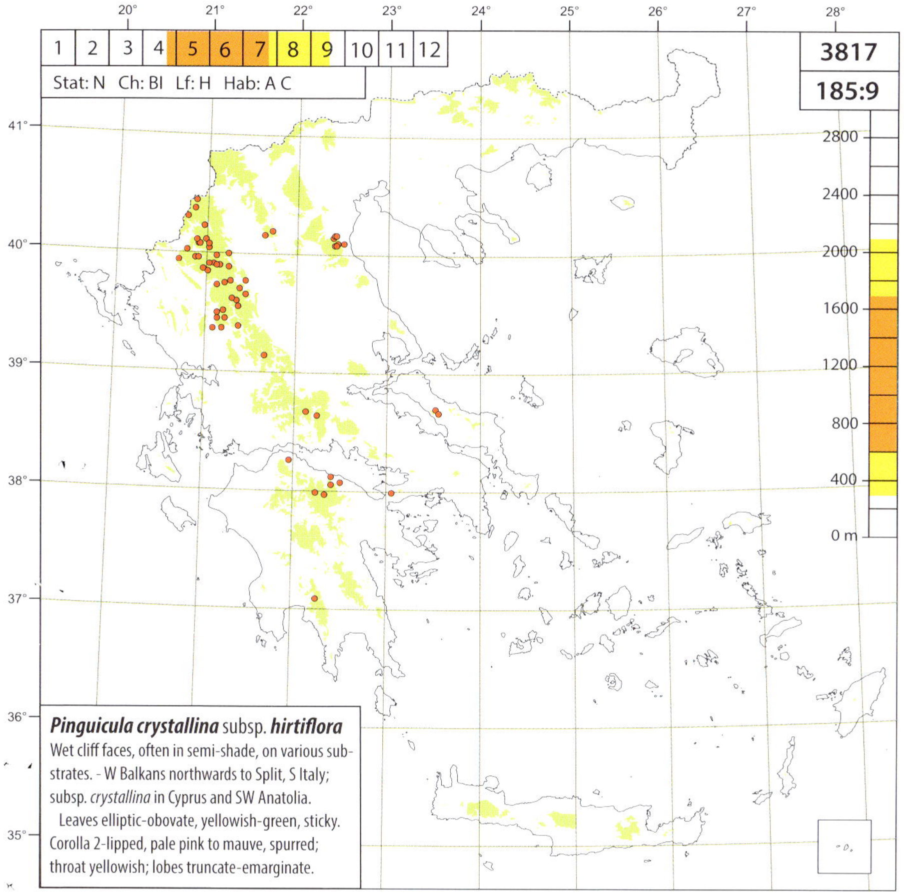

Pinguicula crystallina **subsp.** ***hirtiflora***
Wet cliff faces, often in semi-shade, on various substrates. - W Balkans northwards to Split, S Italy; subsp. *crystallina* in Cyprus and SW Anatolia.
Leaves elliptic-obovate, yellowish-green, sticky. Corolla 2-lipped, pale pink to mauve, spurred; throat yellowish; lobes truncate-emarginate.

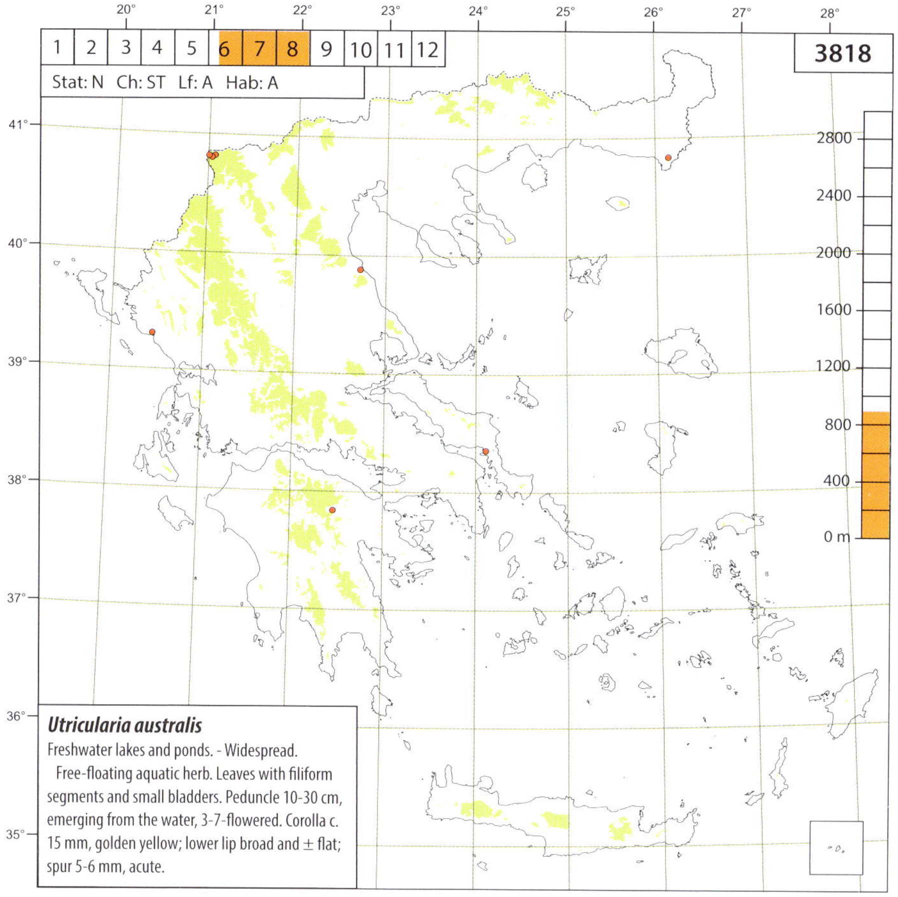

Utricularia australis
Freshwater lakes and ponds. - Widespread.
Free-floating aquatic herb. Leaves with filiform segments and small bladders. Peduncle 10-30 cm, emerging from the water, 3-7-flowered. Corolla c. 15 mm, golden yellow; lower lip broad and ± flat; spur 5-6 mm, acute.

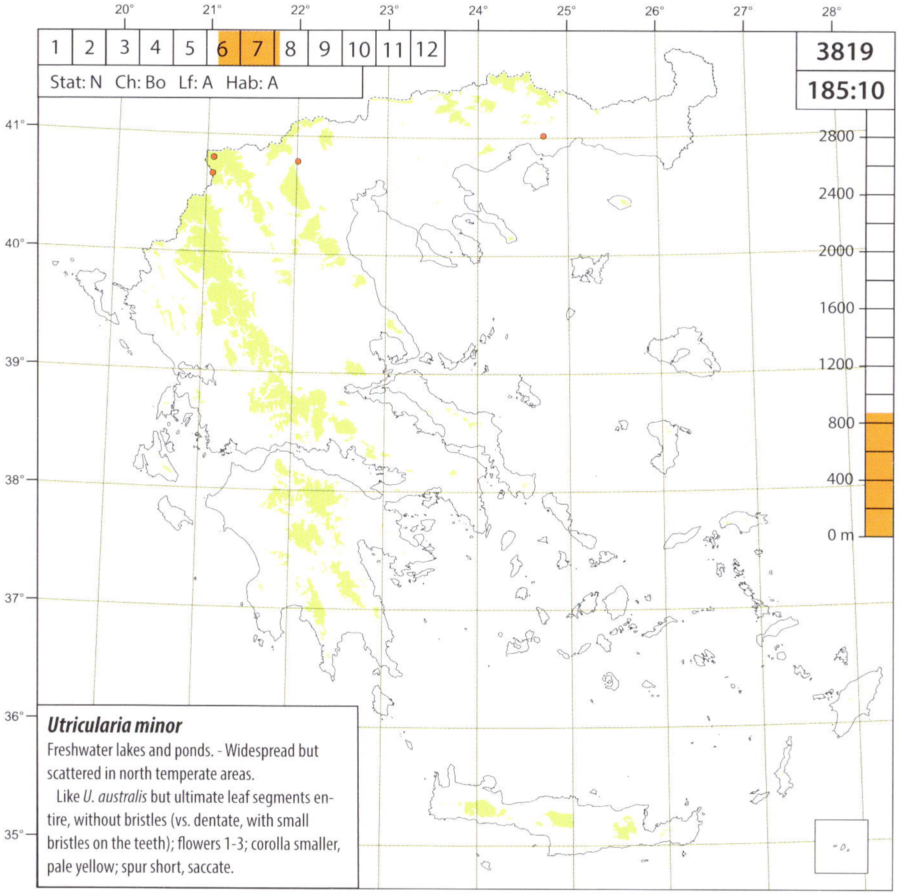

Utricularia minor
Freshwater lakes and ponds. - Widespread but scattered in north temperate areas.
Like *U. australis* but ultimate leaf segments entire, without bristles (vs. dentate, with small bristles on the teeth); flowers 1-3; corolla smaller, pale yellow; spur short, saccate.

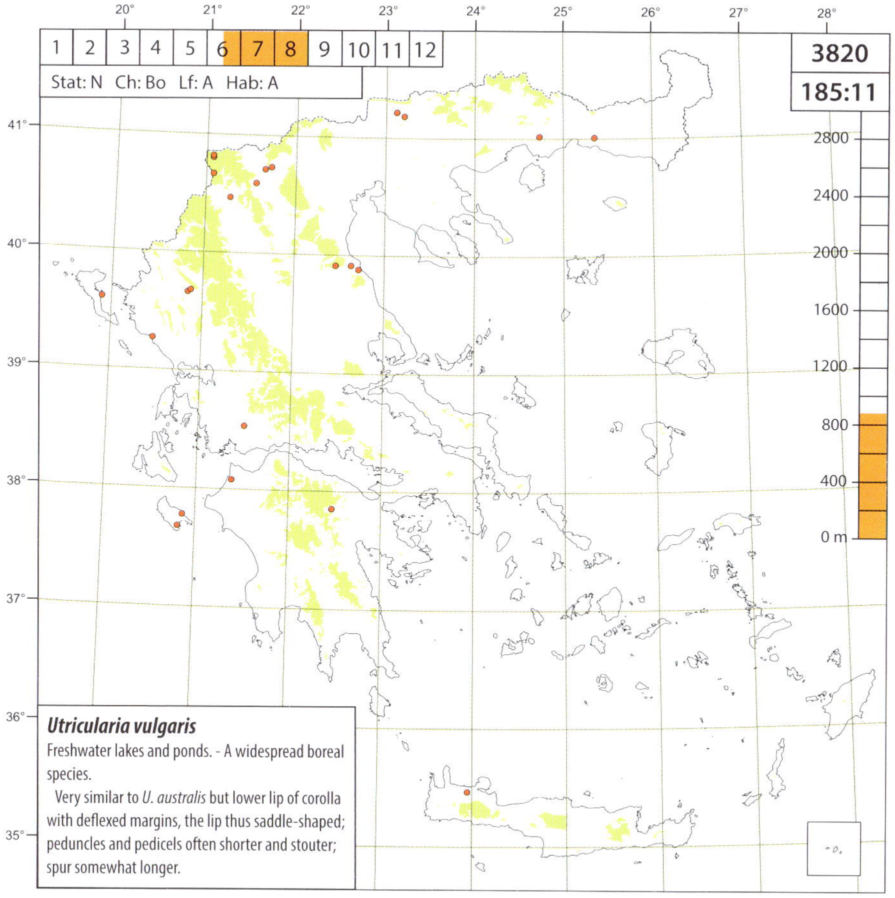

Utricularia vulgaris
Freshwater lakes and ponds. - A widespread boreal species.
Very similar to *U. australis* but lower lip of corolla with deflexed margins, the lip thus saddle-shaped; peduncles and pedicels often shorter and stouter; spur somewhat longer.

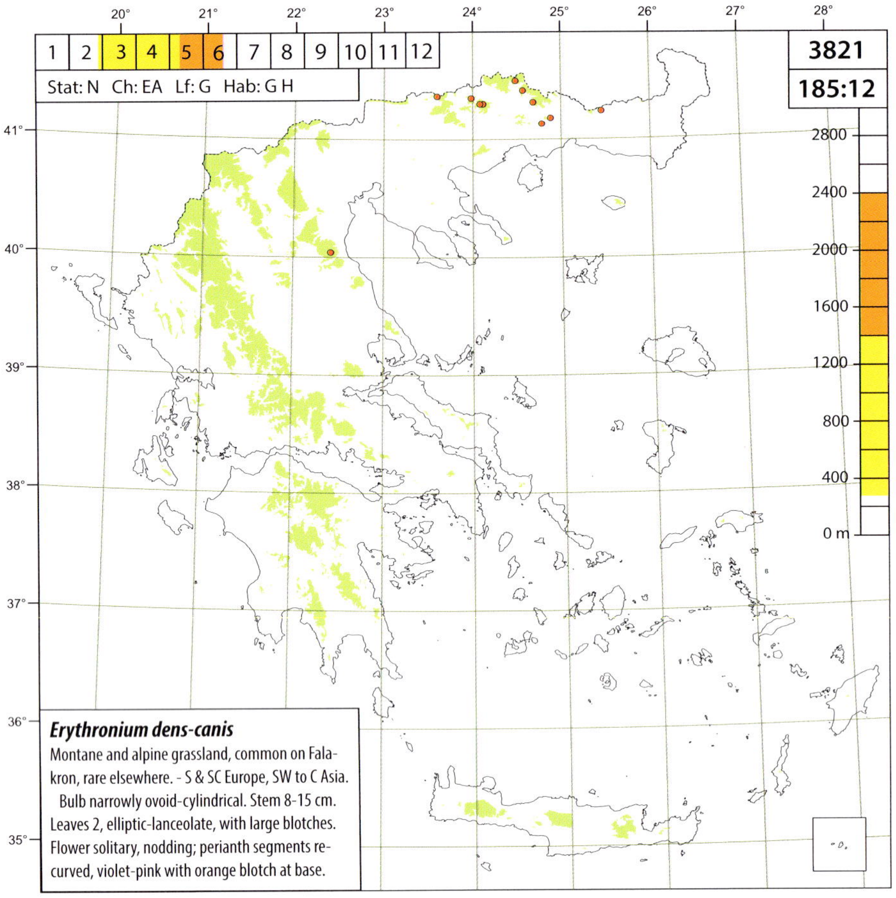

Erythronium dens-canis

Montane and alpine grassland, common on Falakron, rare elsewhere. - S & SC Europe, SW to C Asia.

Bulb narrowly ovoid-cylindrical. Stem 8-15 cm. Leaves 2, elliptic-lanceolate, with large blotches. Flower solitary, nodding; perianth segments recurved, violet-pink with orange blotch at base.

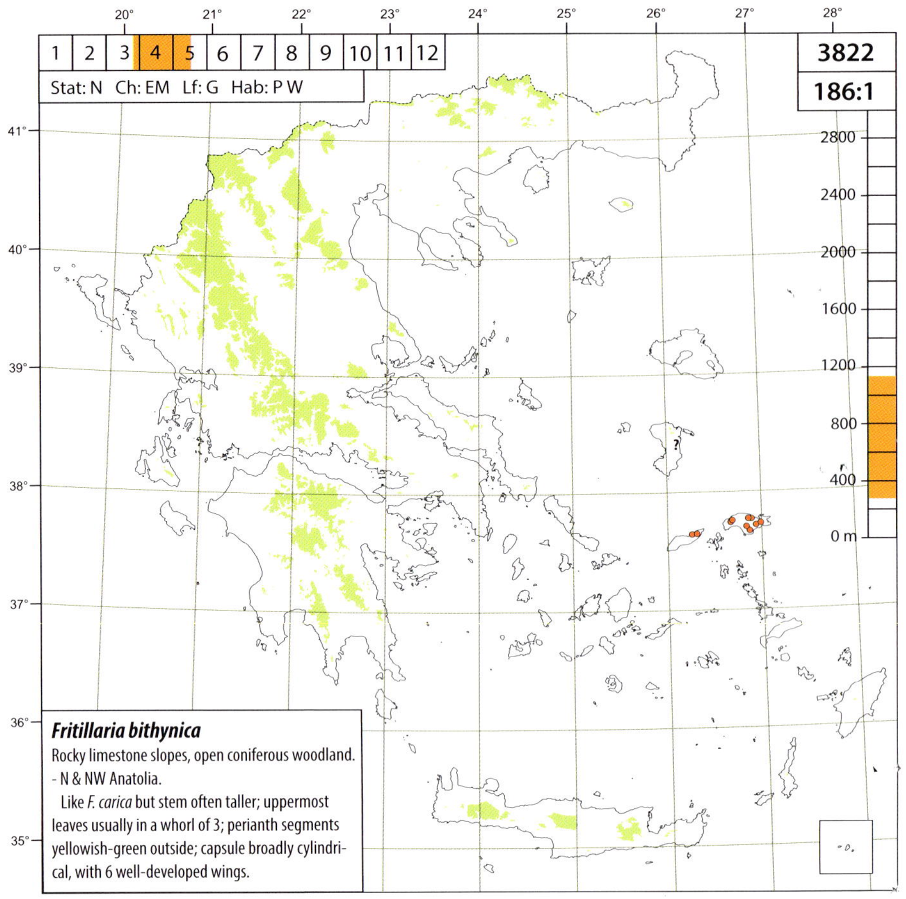

Fritillaria bithynica

Rocky limestone slopes, open coniferous woodland. - N & NW Anatolia.

Like *F. carica* but stem often taller; uppermost leaves usually in a whorl of 3; perianth segments yellowish-green outside; capsule broadly cylindrical, with 6 well-developed wings.

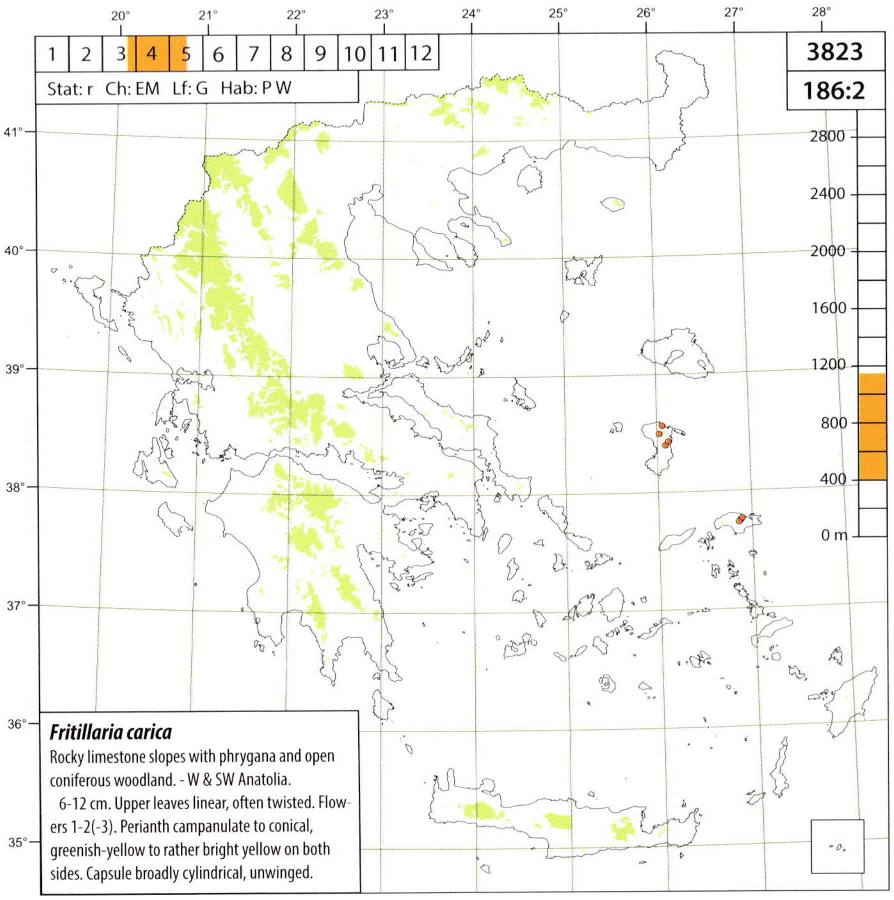

Fritillaria carica

Rocky limestone slopes with phrygana and open coniferous woodland. - W & SW Anatolia.

6-12 cm. Upper leaves linear, often twisted. Flowers 1-2(-3). Perianth campanulate to conical, greenish-yellow to rather bright yellow on both sides. Capsule broadly cylindrical, unwinged.

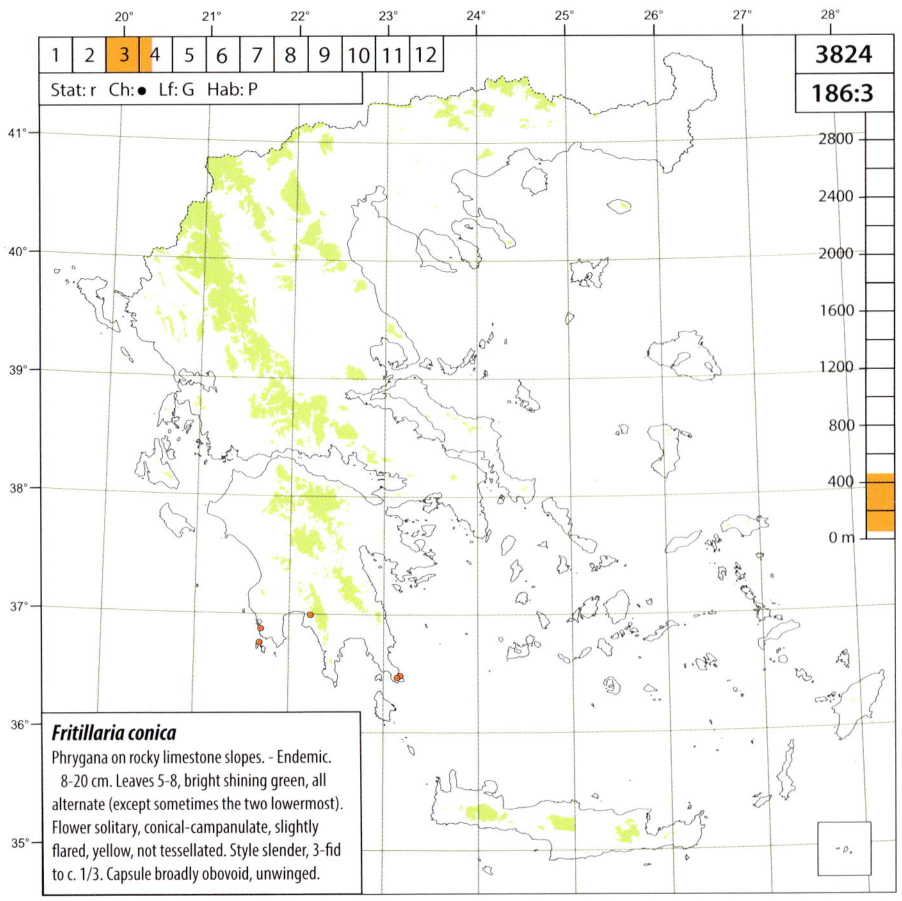

Fritillaria conica

Phrygana on rocky limestone slopes. - Endemic.

8-20 cm. Leaves 5-8, bright shining green, all alternate (except sometimes the two lowermost). Flower solitary, conical-campanulate, slightly flared, yellow, not tessellated. Style slender, 3-fid to c. 1/3. Capsule broadly obovoid, unwinged.

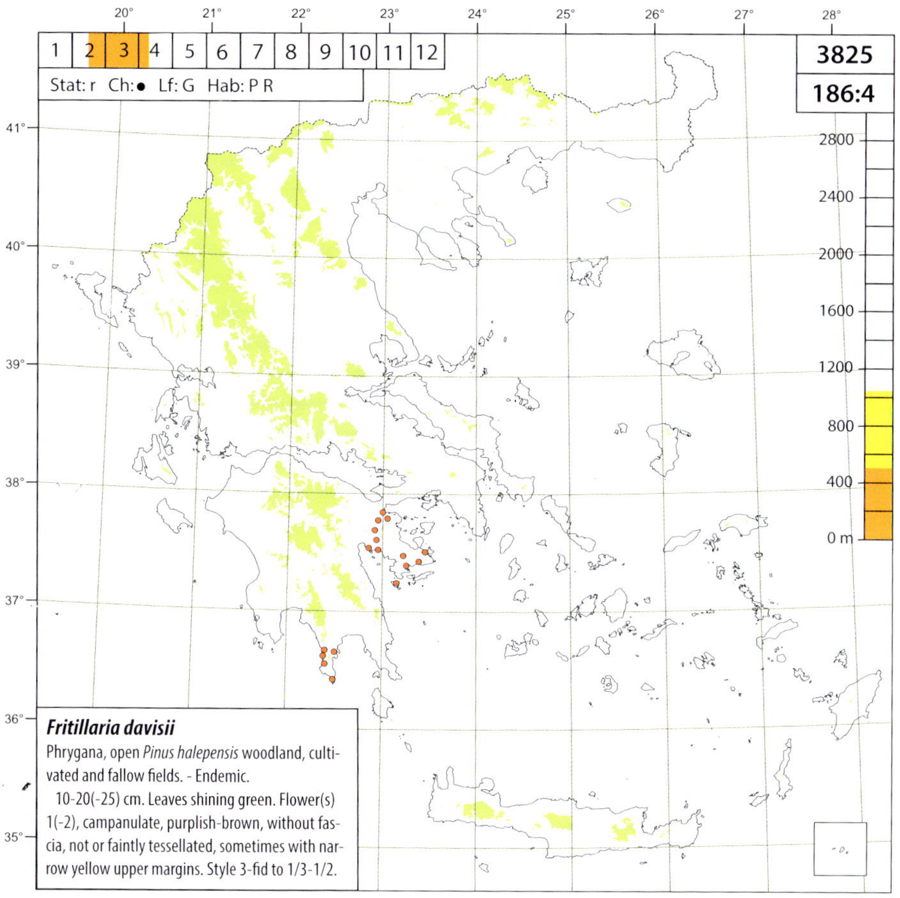

Fritillaria davisii

Phrygana, open *Pinus halepensis* woodland, cultivated and fallow fields. - Endemic.

10-20(-25) cm. Leaves shining green. Flower(s) 1(-2), campanulate, purplish-brown, without fascia, not or faintly tessellated, sometimes with narrow yellow upper margins. Style 3-fid to 1/3-1/2.

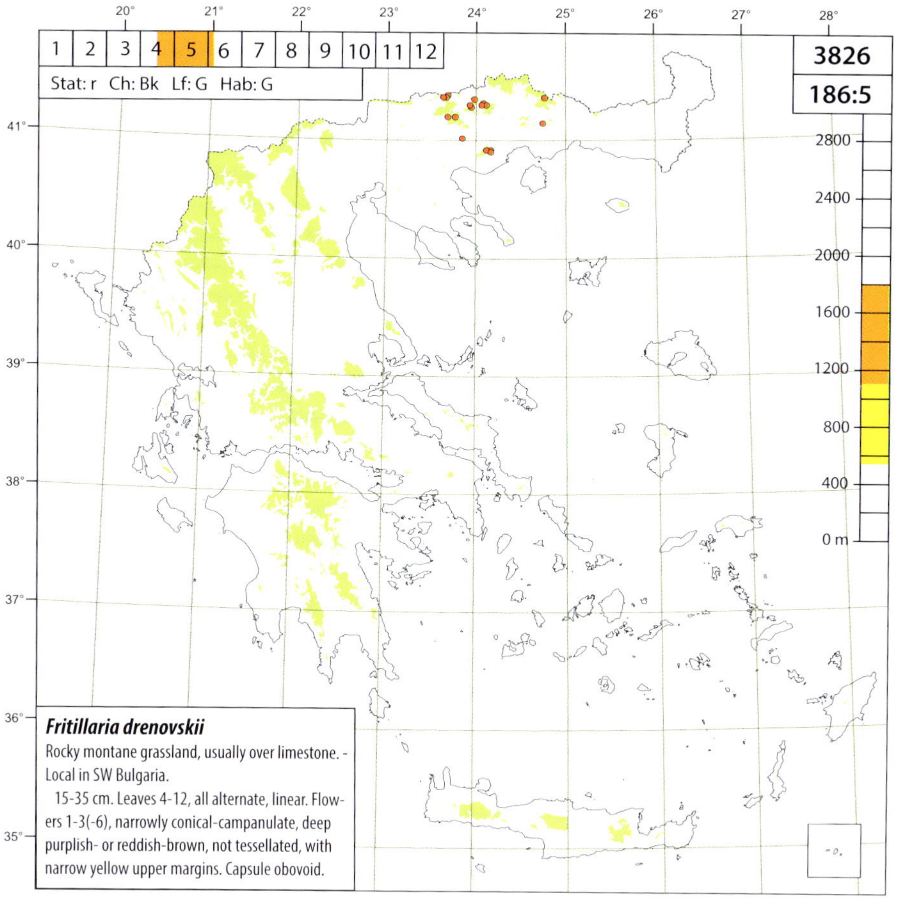

Fritillaria drenovskii

Rocky montane grassland, usually over limestone. - Local in SW Bulgaria.

15-35 cm. Leaves 4-12, all alternate, linear. Flowers 1-3(-6), narrowly conical-campanulate, deep purplish- or reddish-brown, not tessellated, with narrow yellow upper margins. Capsule obovoid.

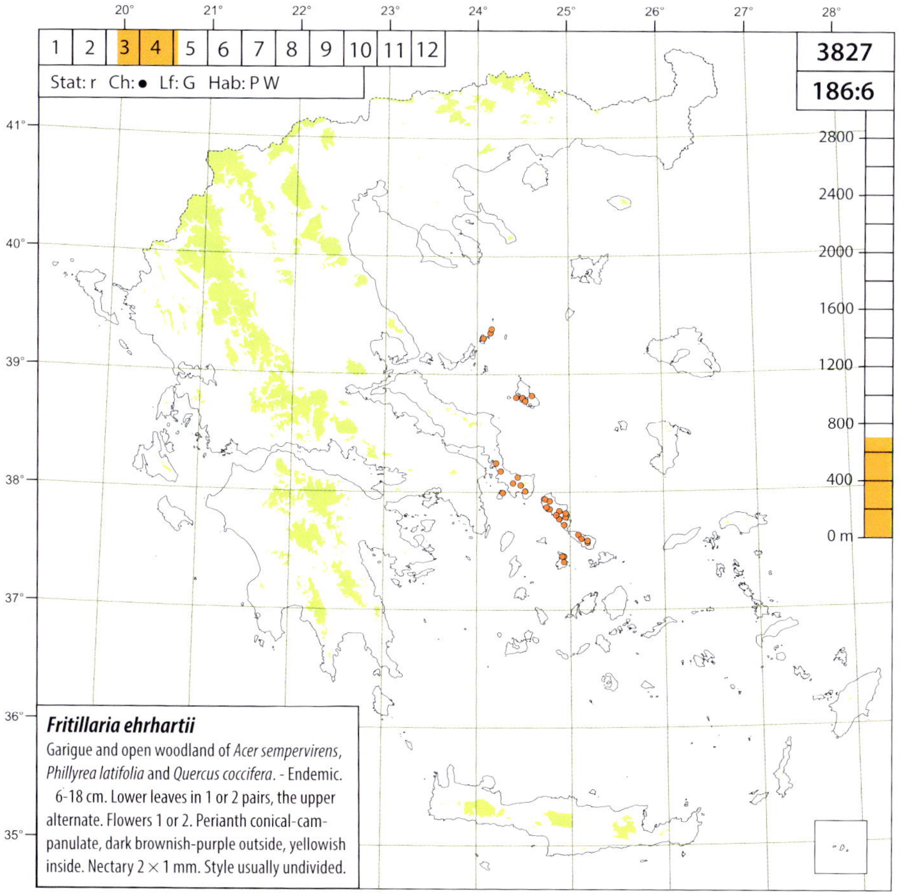

Fritillaria ehrhartii

Garigue and open woodland of *Acer sempervirens*, *Phillyrea latifolia* and *Quercus coccifera*. - Endemic.

6-18 cm. Lower leaves in 1 or 2 pairs, the upper alternate. Flowers 1 or 2. Perianth conical-campanulate, dark brownish-purple outside, yellowish inside. Nectary 2 × 1 mm. Style usually undivided.

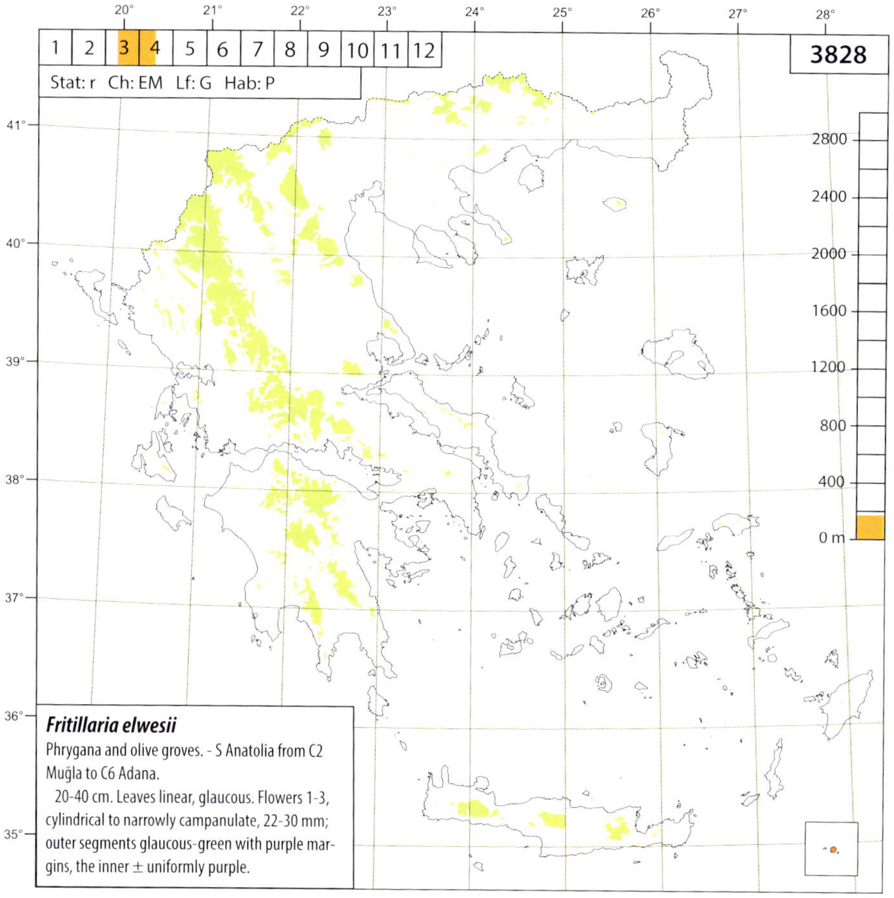

Fritillaria elwesii

Phrygana and olive groves. - S Anatolia from C2 Muğla to C6 Adana.

20-40 cm. Leaves linear, glaucous. Flowers 1-3, cylindrical to narrowly campanulate, 22-30 mm; outer segments glaucous-green with purple margins, the inner ± uniformly purple.

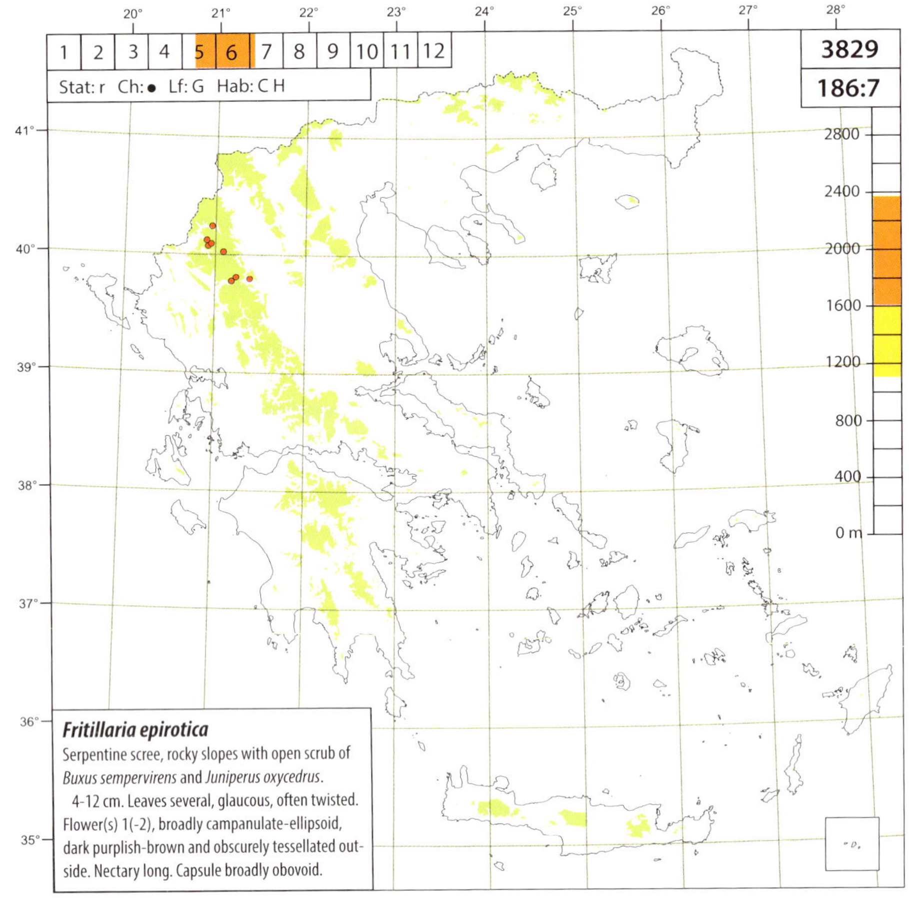

Fritillaria epirotica
Serpentine scree, rocky slopes with open scrub of *Buxus sempervirens* and *Juniperus oxycedrus*.
4-12 cm. Leaves several, glaucous, often twisted. Flower(s) 1(-2), broadly campanulate-ellipsoid, dark purplish-brown and obscurely tessellated outside. Nectary long. Capsule broadly obovoid.

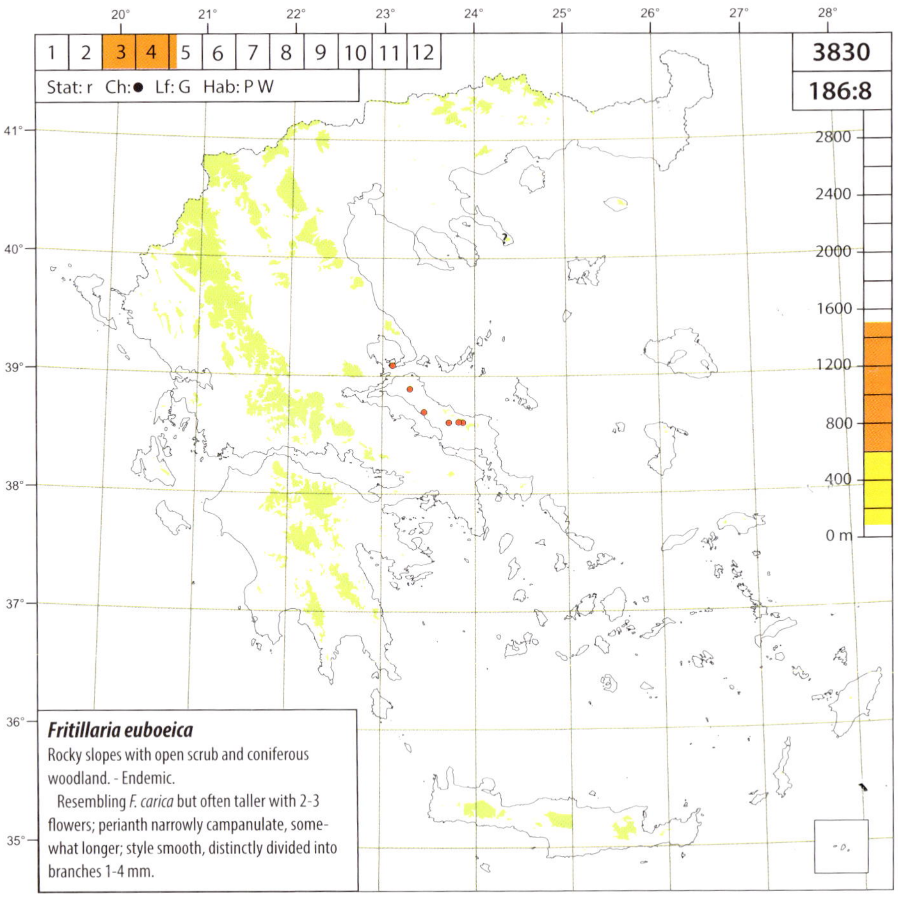

Fritillaria euboeica
Rocky slopes with open scrub and coniferous woodland. - Endemic.
Resembling *F. carica* but often taller with 2-3 flowers; perianth narrowly campanulate, somewhat longer; style smooth, distinctly divided into branches 1-4 mm.

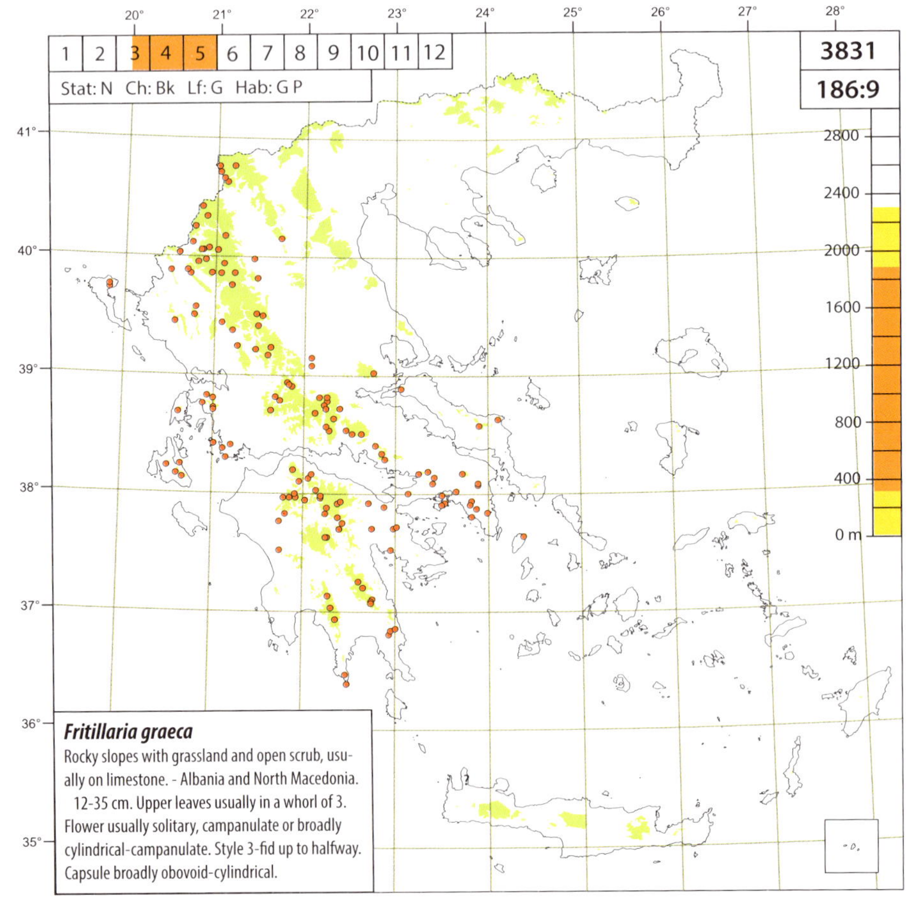

Fritillaria graeca
Rocky slopes with grassland and open scrub, usually on limestone. - Albania and North Macedonia.
12-35 cm. Upper leaves usually in a whorl of 3. Flower usually solitary, campanulate or broadly cylindrical-campanulate. Style 3-fid up to halfway. Capsule broadly obovoid-cylindrical.

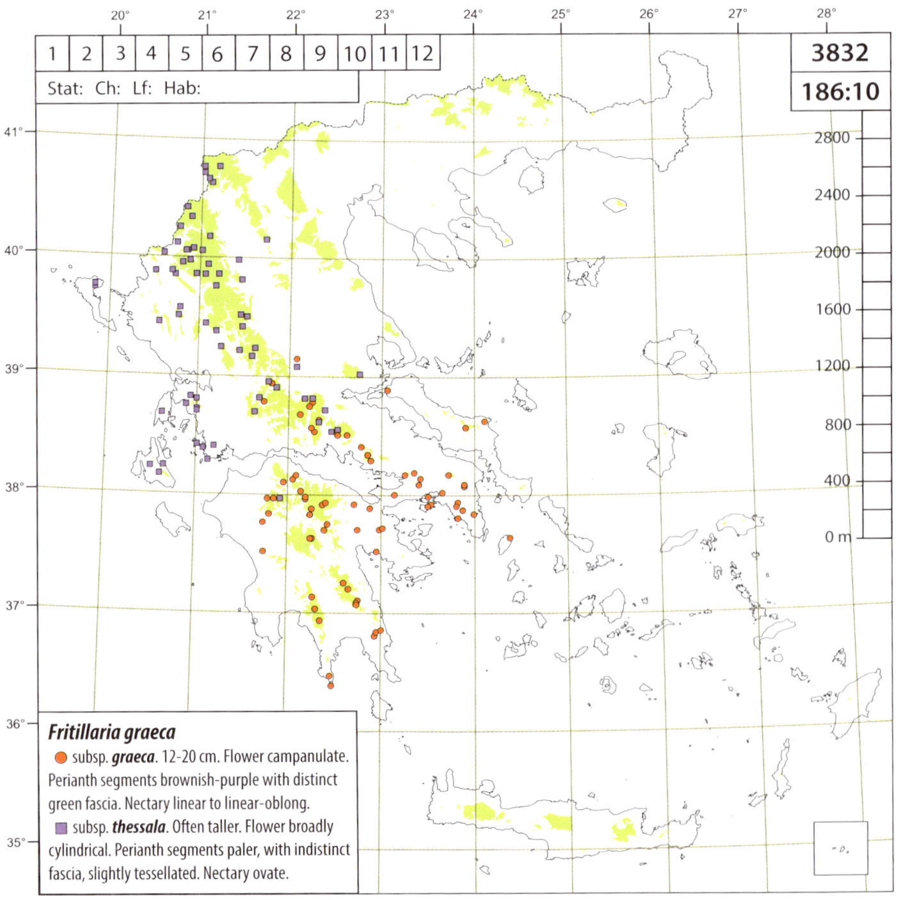

Fritillaria graeca
● subsp. ***graeca***. 12-20 cm. Flower campanulate. Perianth segments brownish-purple with distinct green fascia. Nectary linear to linear-oblong.
■ subsp. ***thessala***. Often taller. Flower broadly cylindrical. Perianth segments paler, with indistinct fascia, slightly tessellated. Nectary ovate.

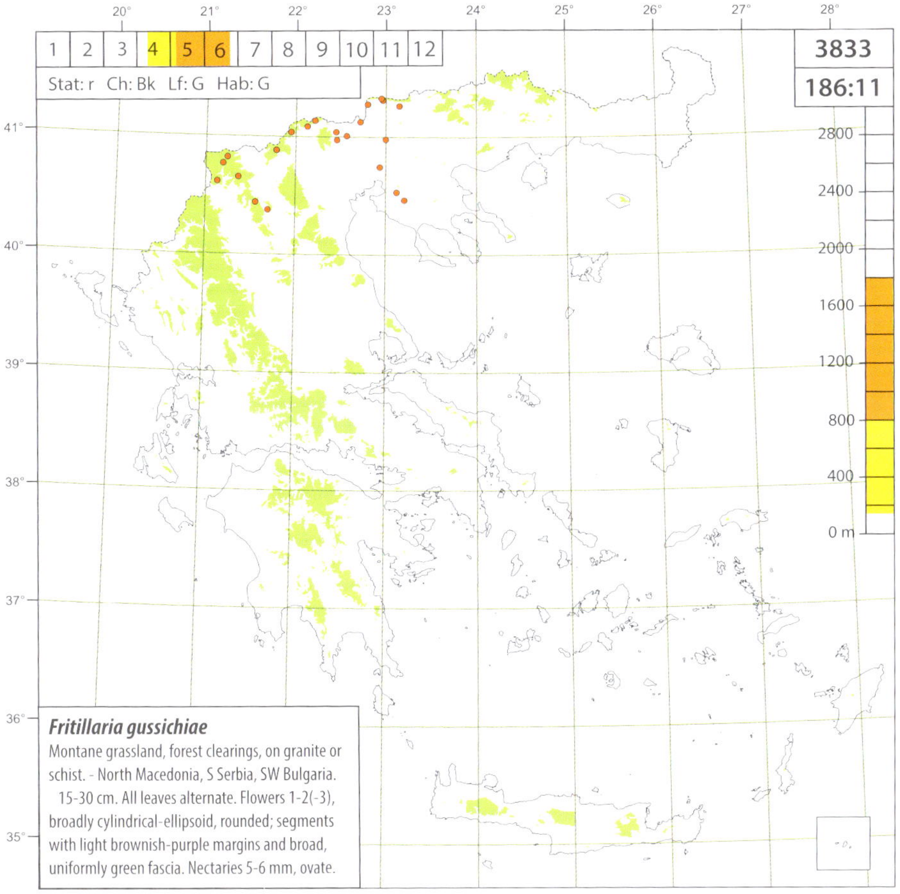

Fritillaria gussichiae
Montane grassland, forest clearings, on granite or schist. - North Macedonia, S Serbia, SW Bulgaria.
15-30 cm. All leaves alternate. Flowers 1-2(-3), broadly cylindrical-ellipsoid, rounded; segments with light brownish-purple margins and broad, uniformly green fascia. Nectaries 5-6 mm, ovate.

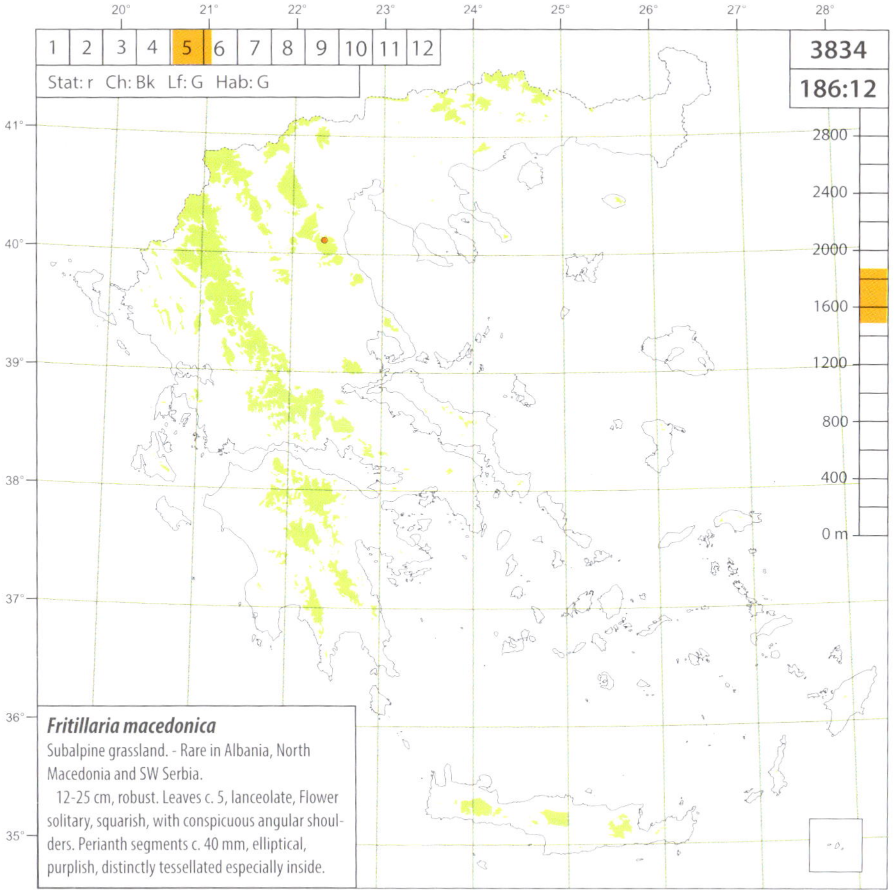

Fritillaria macedonica
Subalpine grassland. - Rare in Albania, North Macedonia and SW Serbia.
12-25 cm, robust. Leaves c. 5, lanceolate, Flower solitary, squarish, with conspicuous angular shoulders. Perianth segments c. 40 mm, elliptical, purplish, distinctly tessellated especially inside.

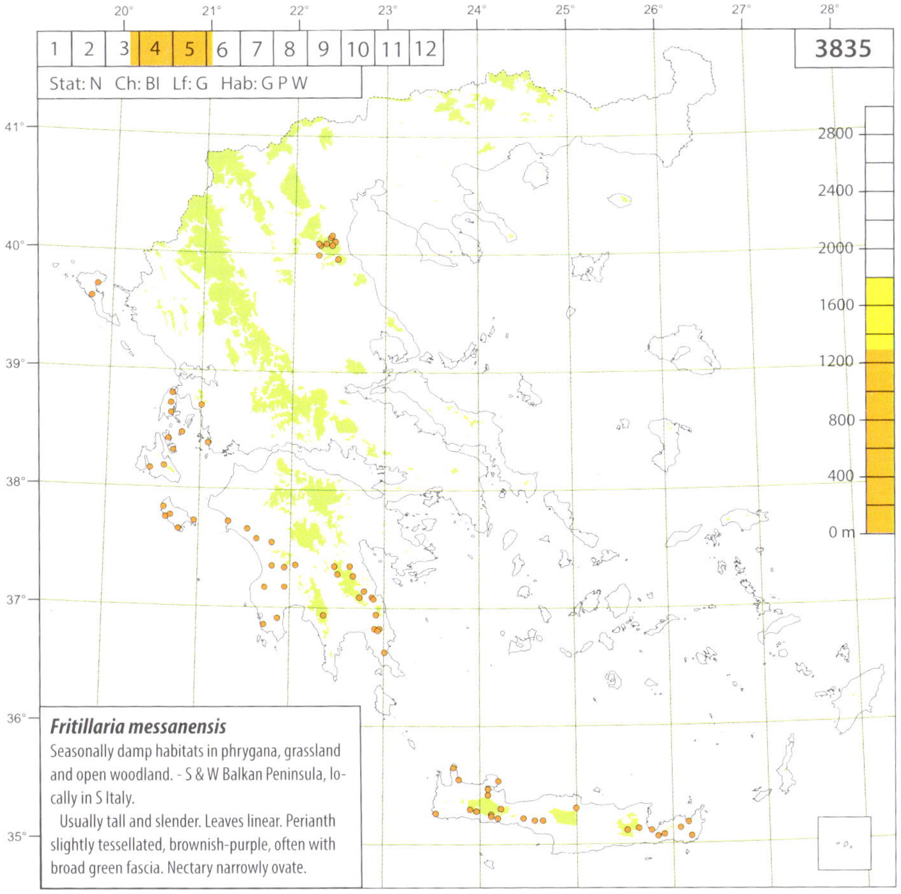

Fritillaria messanensis
Seasonally damp habitats in phrygana, grassland and open woodland. - S & W Balkan Peninsula, locally in S Italy.
Usually tall and slender. Leaves linear. Perianth slightly tessellated, brownish-purple, often with broad green fascia. Nectary narrowly ovate.

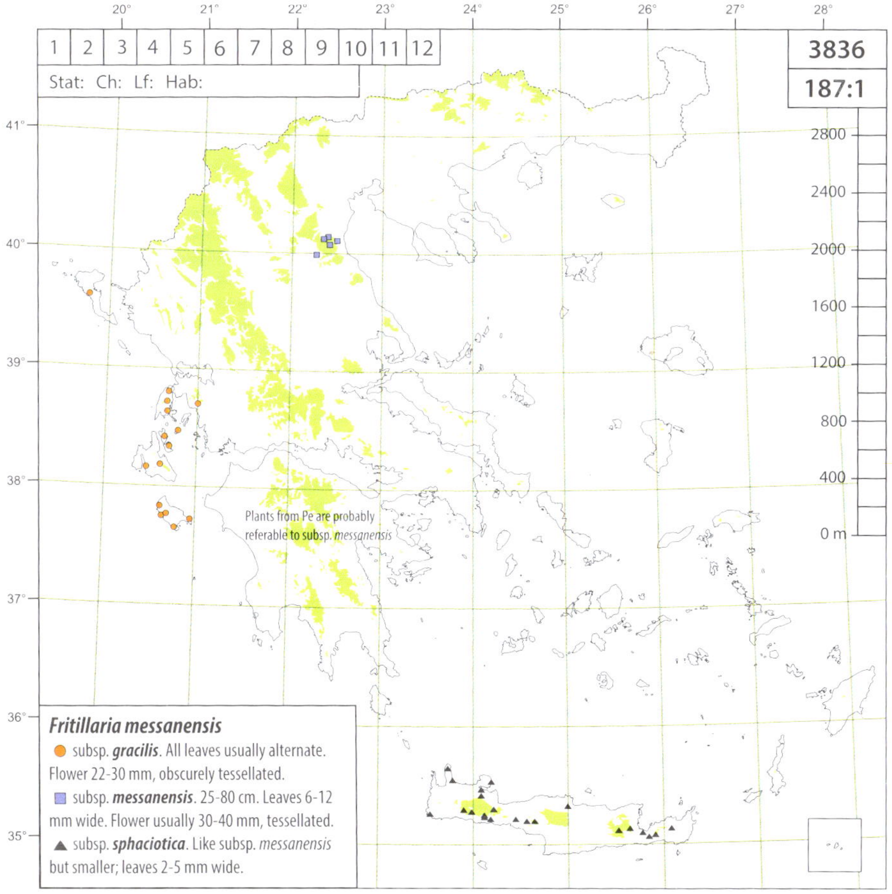

Fritillaria messanensis
● subsp. ***gracilis***. All leaves usually alternate. Flower 22-30 mm, obscurely tessellated.
■ subsp. ***messanensis***. 25-80 cm. Leaves 6-12 mm wide. Flower usually 30-40 mm, tessellated.
▲ subsp. ***sphaciotica***. Like subsp. *messanensis* but smaller; leaves 2-5 mm wide.

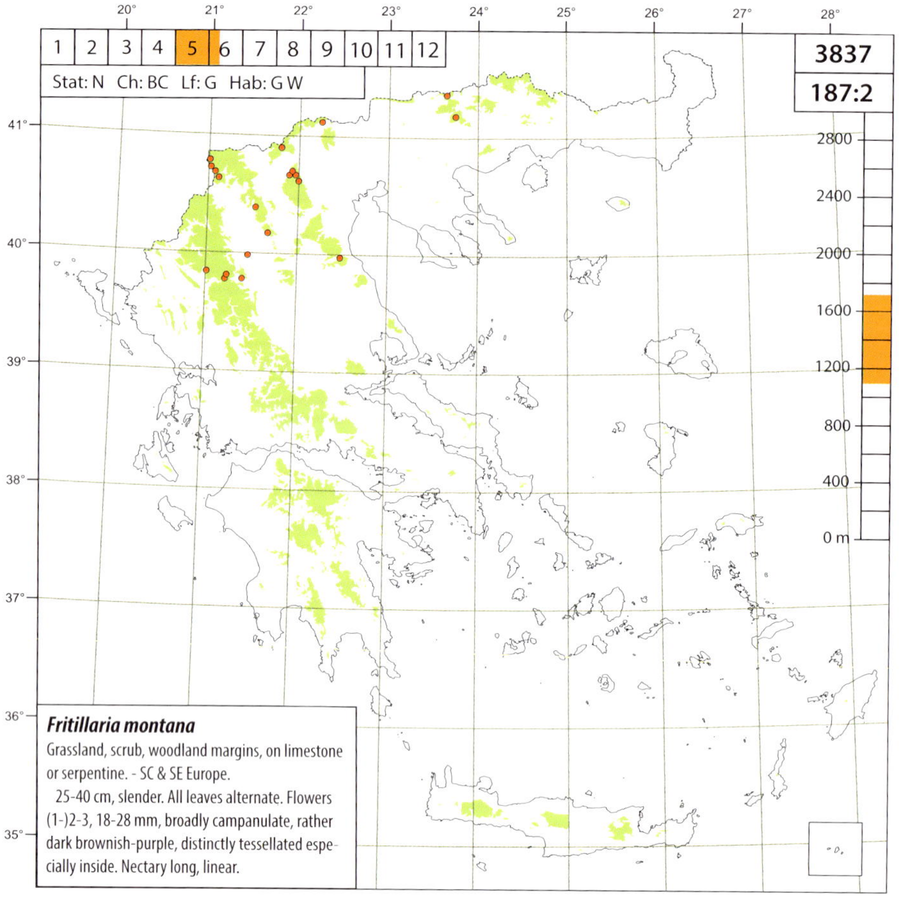

Fritillaria montana
Grassland, scrub, woodland margins, on limestone or serpentine. - SC & SE Europe.
25-40 cm, slender. All leaves alternate. Flowers (1-)2-3, 18-28 mm, broadly campanulate, rather dark brownish-purple, distinctly tessellated especially inside. Nectary long, linear.

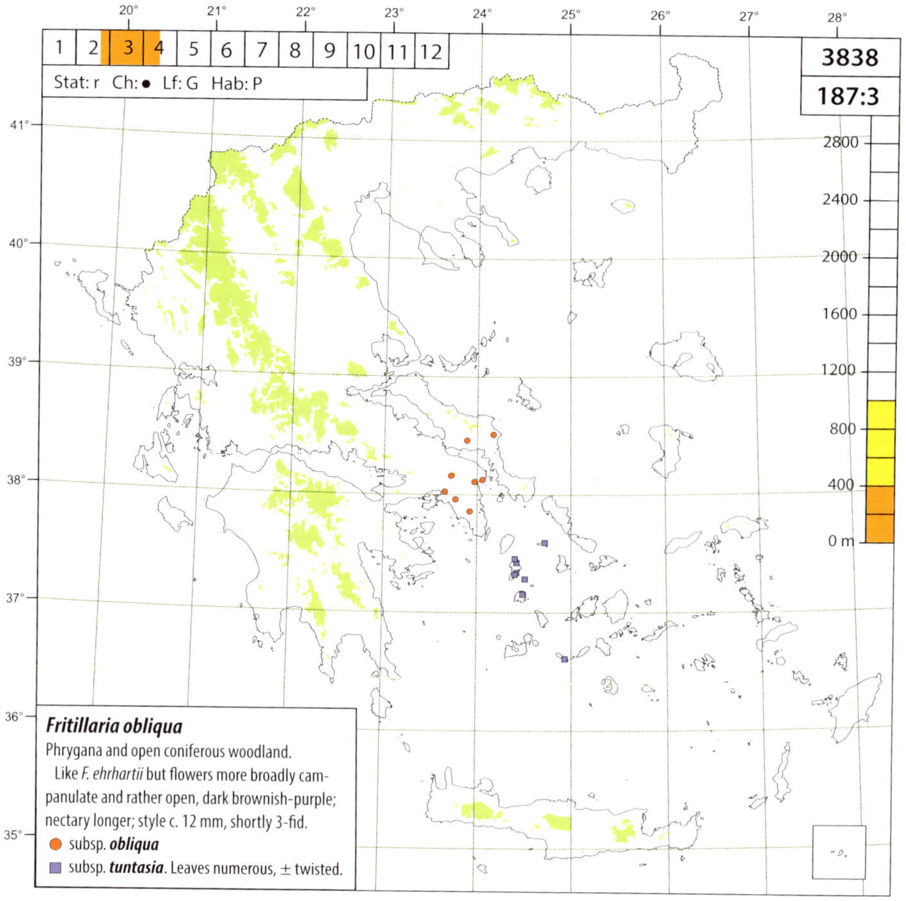

Fritillaria obliqua
Phrygana and open coniferous woodland.
Like *F. ehrhartii* but flowers more broadly campanulate and rather open, dark brownish-purple; nectary longer; style c. 12 mm, shortly 3-fid.
● subsp. ***obliqua***
■ subsp. ***tuntasia***. Leaves numerous, ± twisted.

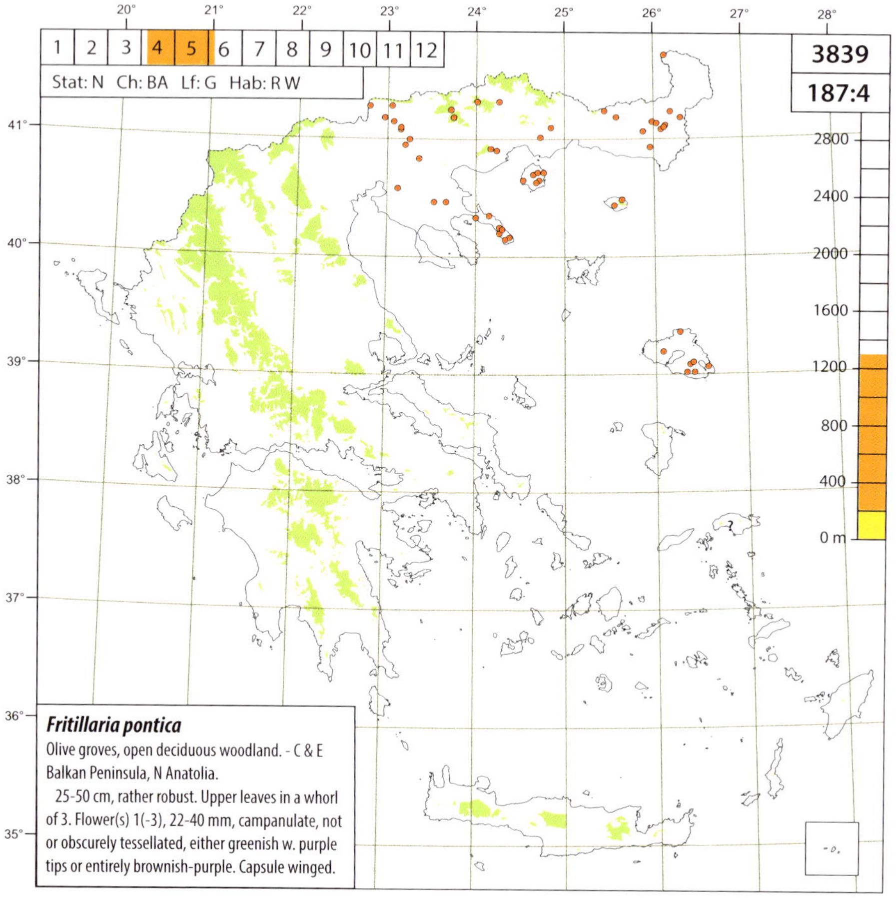

Fritillaria pontica
Olive groves, open deciduous woodland. - C & E Balkan Peninsula, N Anatolia.
25-50 cm, rather robust. Upper leaves in a whorl of 3. Flower(s) 1(-3), 22-40 mm, campanulate, not or obscurely tessellated, either greenish w. purple tips or entirely brownish-purple. Capsule winged.

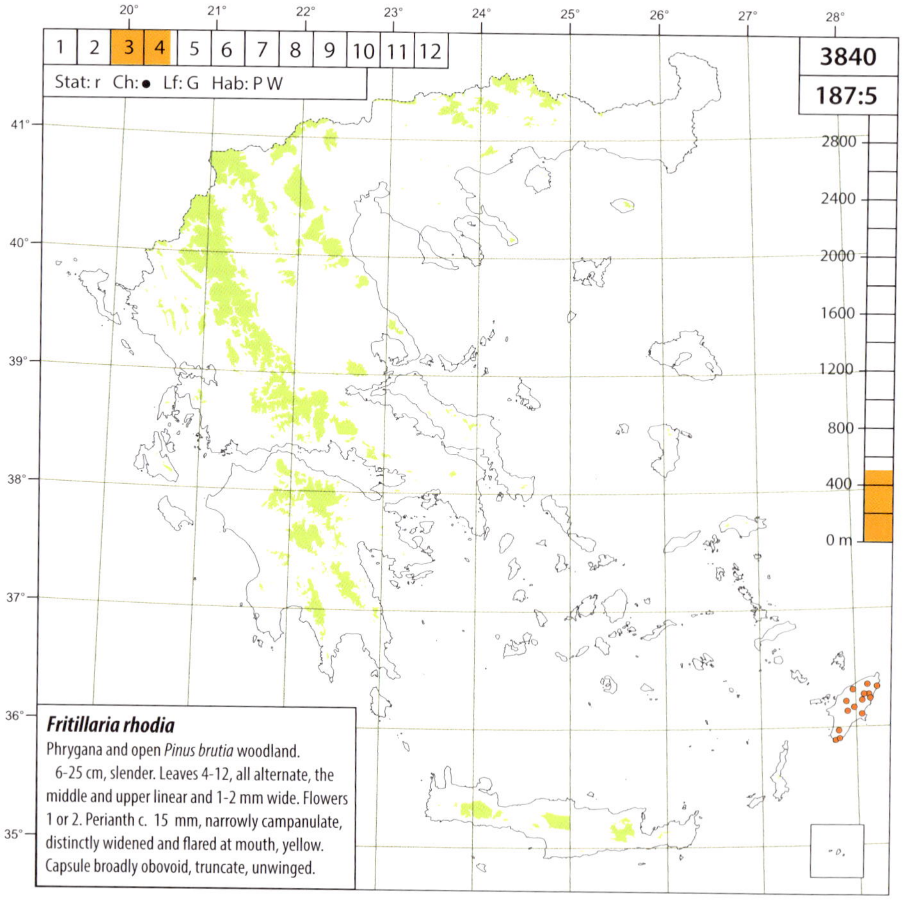

Fritillaria rhodia
Phrygana and open *Pinus brutia* woodland.
6-25 cm, slender. Leaves 4-12, all alternate, the middle and upper linear and 1-2 mm wide. Flowers 1 or 2. Perianth c. 15 mm, narrowly campanulate, distinctly widened and flared at mouth, yellow. Capsule broadly obovoid, truncate, unwinged.

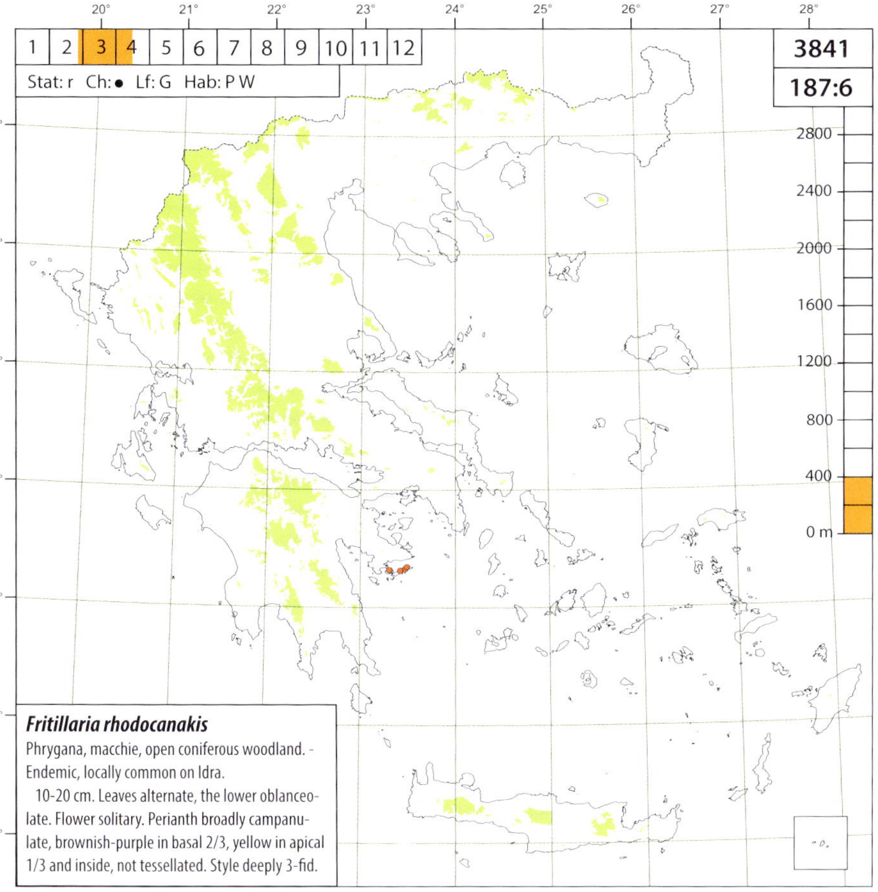

Fritillaria rhodocanakis
Phrygana, macchie, open coniferous woodland. - Endemic, locally common on Idra.
10-20 cm. Leaves alternate, the lower oblanceolate. Flower solitary. Perianth broadly campanulate, brownish-purple in basal 2/3, yellow in apical 1/3 and inside, not tessellated. Style deeply 3-fid.

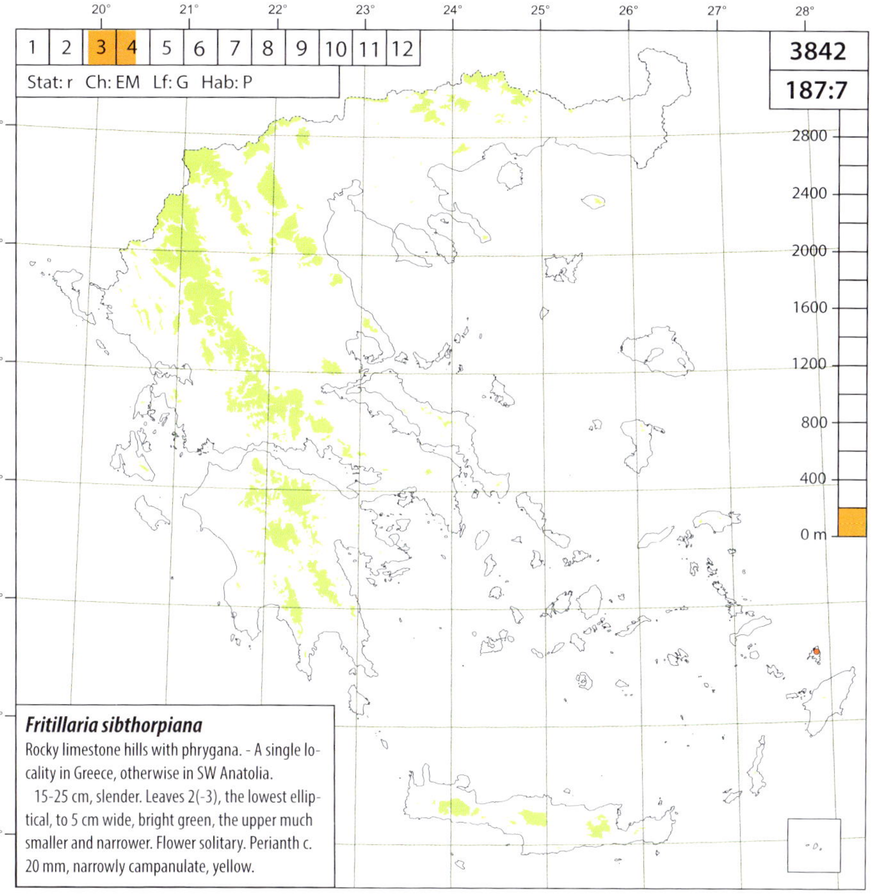

Fritillaria sibthorpiana
Rocky limestone hills with phrygana. - A single locality in Greece, otherwise in SW Anatolia.
15-25 cm, slender. Leaves 2(-3), the lowest elliptical, to 5 cm wide, bright green, the upper much smaller and narrower. Flower solitary. Perianth c. 20 mm, narrowly campanulate, yellow.

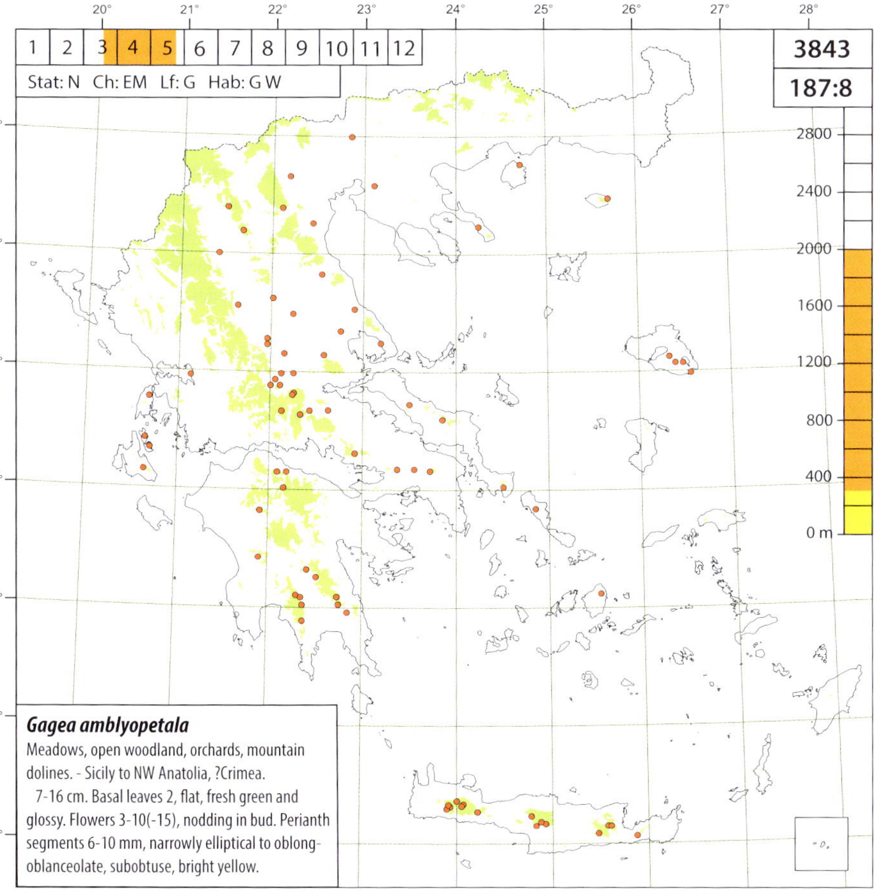

Gagea amblyopetala
Meadows, open woodland, orchards, mountain dolines. - Sicily to NW Anatolia, ?Crimea.
7-16 cm. Basal leaves 2, flat, fresh green and glossy. Flowers 3-10(-15), nodding in bud. Perianth segments 6-10 mm, narrowly elliptical to oblong-oblanceolate, subobtuse, bright yellow.

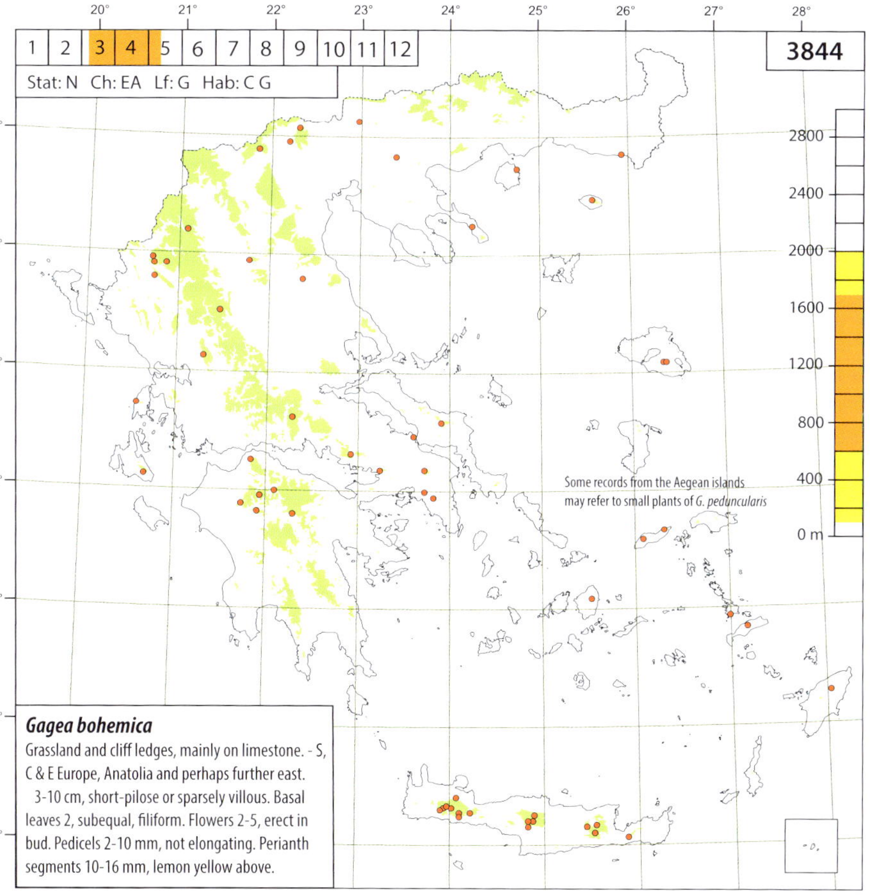

Gagea bohemica
Grassland and cliff ledges, mainly on limestone. - S, C & E Europe, Anatolia and perhaps further east.
3-10 cm, short-pilose or sparsely villous. Basal leaves 2, subequal, filiform. Flowers 2-5, erect in bud. Pedicels 2-10 mm, not elongating. Perianth segments 10-16 mm, lemon yellow above.

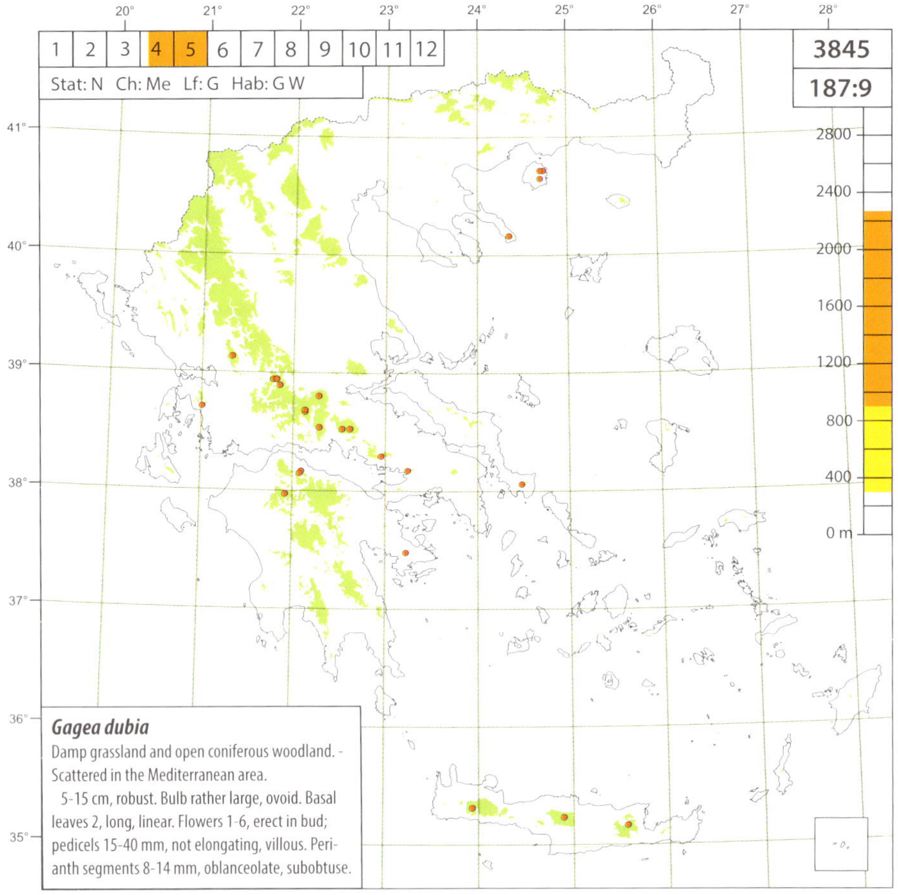

Gagea dubia

Damp grassland and open coniferous woodland. - Scattered in the Mediterranean area.

5-15 cm, robust. Bulb rather large, ovoid. Basal leaves 2, long, linear. Flowers 1-6, erect in bud; pedicels 15-40 mm, not elongating, villous. Perianth segments 8-14 mm, oblanceolate, subobtuse.

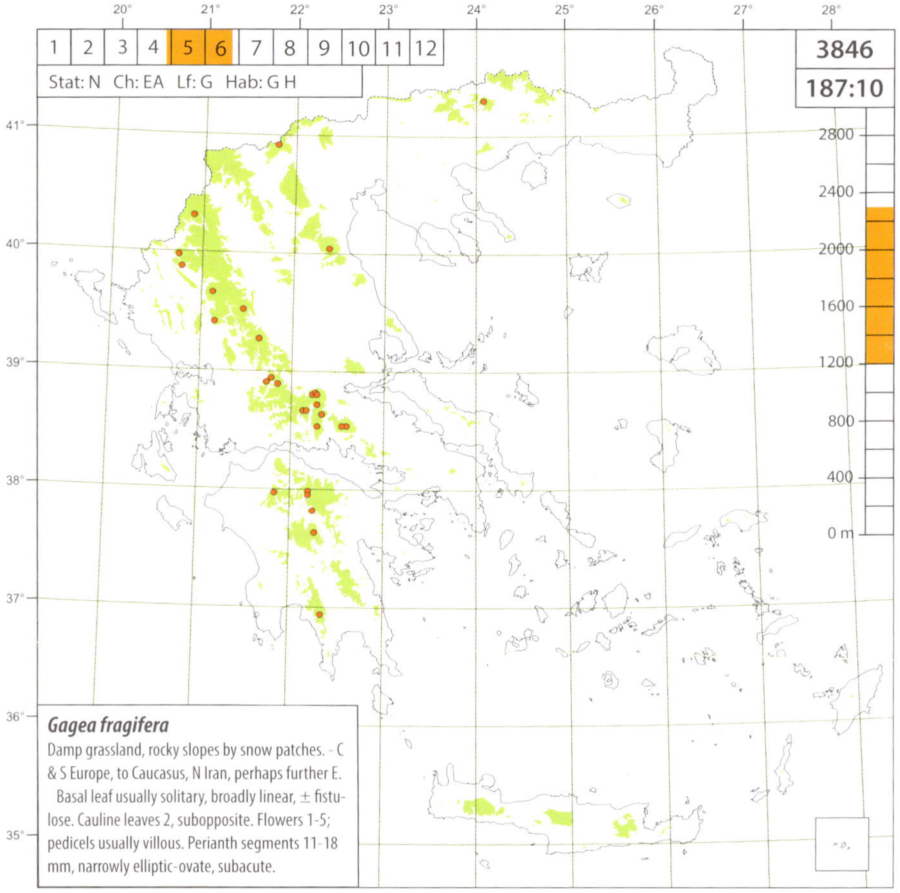

Gagea fragifera

Damp grassland, rocky slopes by snow patches. - C & S Europe, to Caucasus, N Iran, perhaps further E.

Basal leaf usually solitary, broadly linear, ± fistulose. Cauline leaves 2, subopposite. Flowers 1-5; pedicels usually villous. Perianth segments 11-18 mm, narrowly elliptic-ovate, subacute.

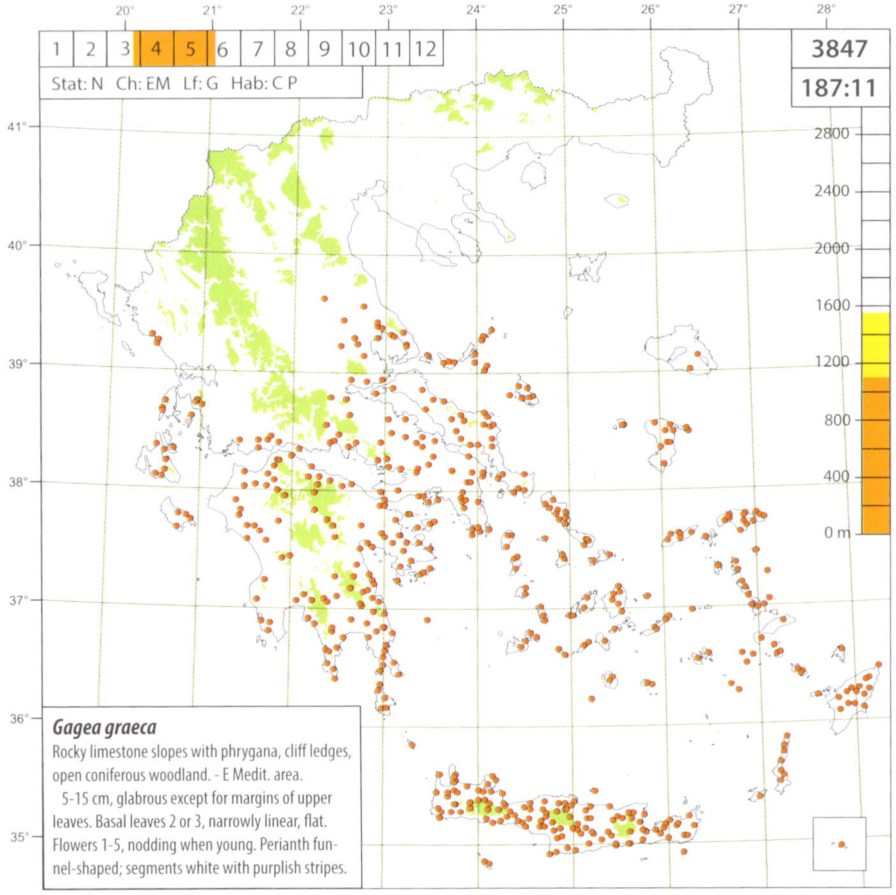

Gagea graeca

Rocky limestone slopes with phrygana, cliff ledges, open coniferous woodland. - E Medit. area.

5-15 cm, glabrous except for margins of upper leaves. Basal leaves 2 or 3, narrowly linear, flat. Flowers 1-5, nodding when young. Perianth funnel-shaped; segments white with purplish stripes.

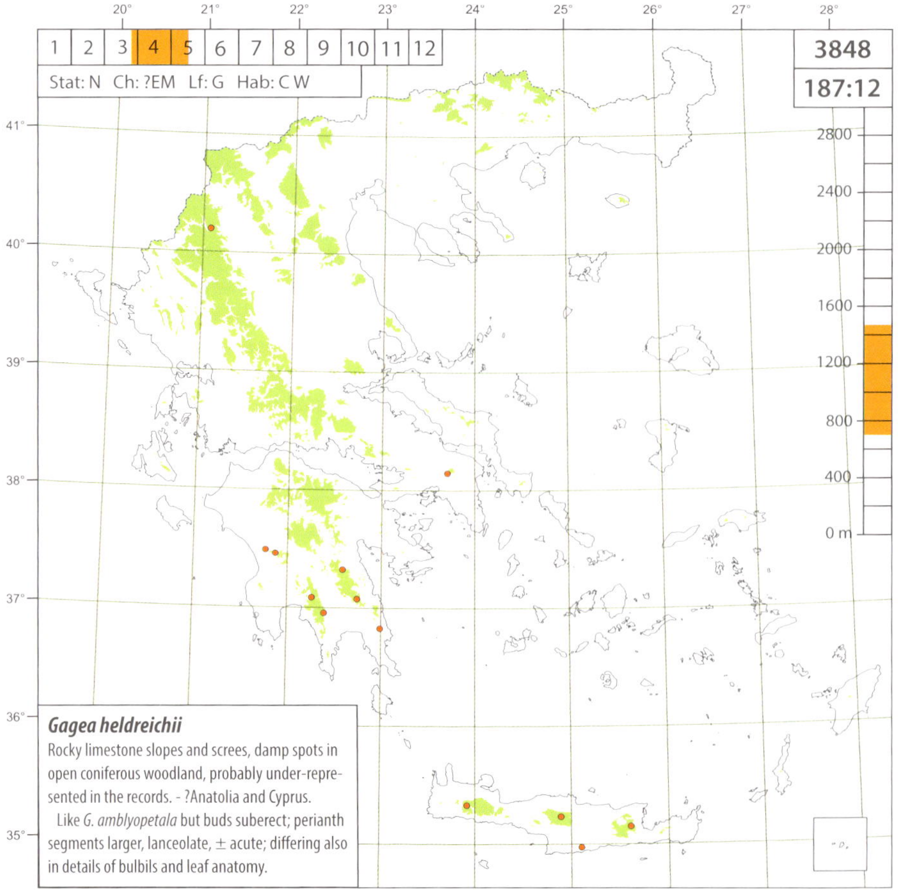

Gagea heldreichii

Rocky limestone slopes and screes, damp spots in open coniferous woodland, probably under-represented in the records. - ?Anatolia and Cyprus.

Like *G. amblyopetala* but buds suberect; perianth segments larger, lanceolate, ± acute; differing also in details of bulbils and leaf anatomy.

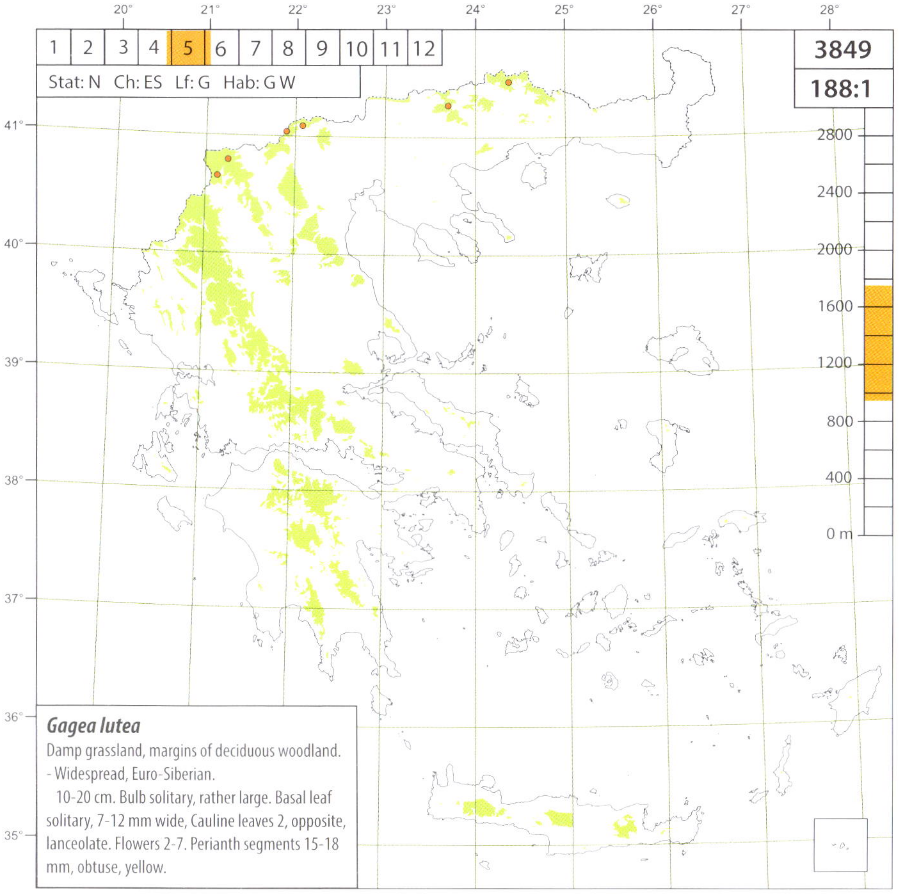

Gagea lutea
Damp grassland, margins of deciduous woodland. - Widespread, Euro-Siberian.
10-20 cm. Bulb solitary, rather large. Basal leaf solitary, 7-12 mm wide, Cauline leaves 2, opposite, lanceolate. Flowers 2-7. Perianth segments 15-18 mm, obtuse, yellow.

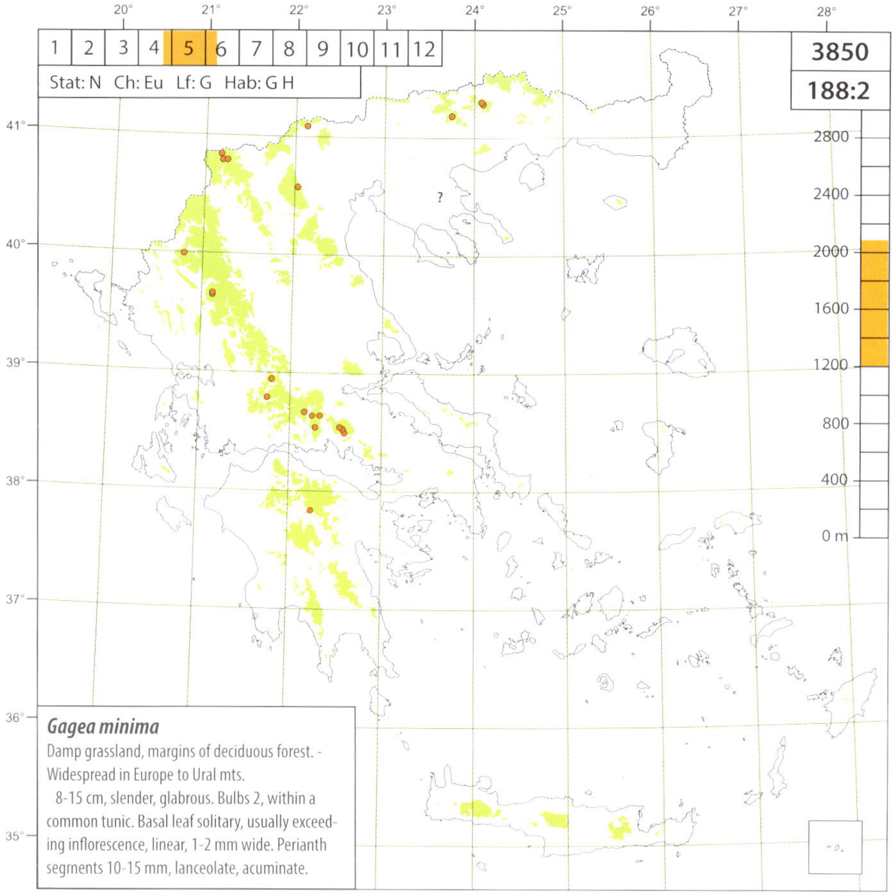

Gagea minima
Damp grassland, margins of deciduous forest. - Widespread in Europe to Ural mts.
8-15 cm, slender, glabrous. Bulbs 2, within a common tunic. Basal leaf solitary, usually exceeding inflorescence, linear, 1-2 mm wide. Perianth segments 10-15 mm, lanceolate, acuminate.

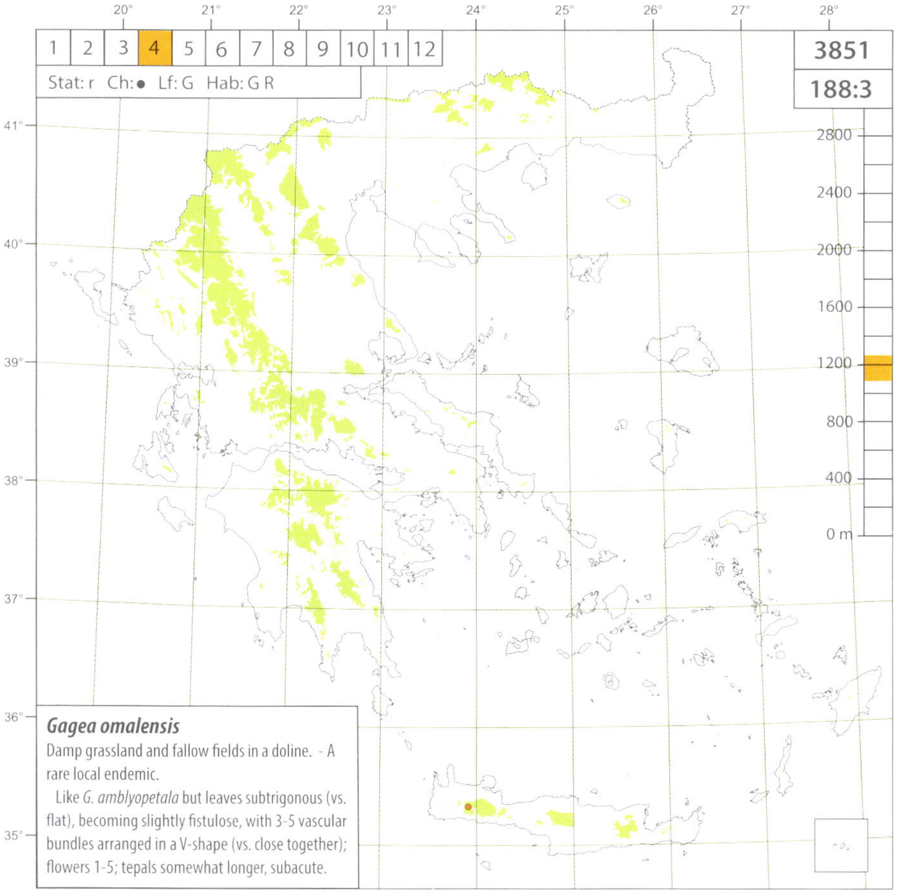

Gagea omalensis
Damp grassland and fallow fields in a doline. - A rare local endemic.
Like *G. amblyopetala* but leaves subtrigonous (vs. flat), becoming slightly fistulose, with 3-5 vascular bundles arranged in a V-shape (vs. close together); flowers 1-5; tepals somewhat longer, subacute.

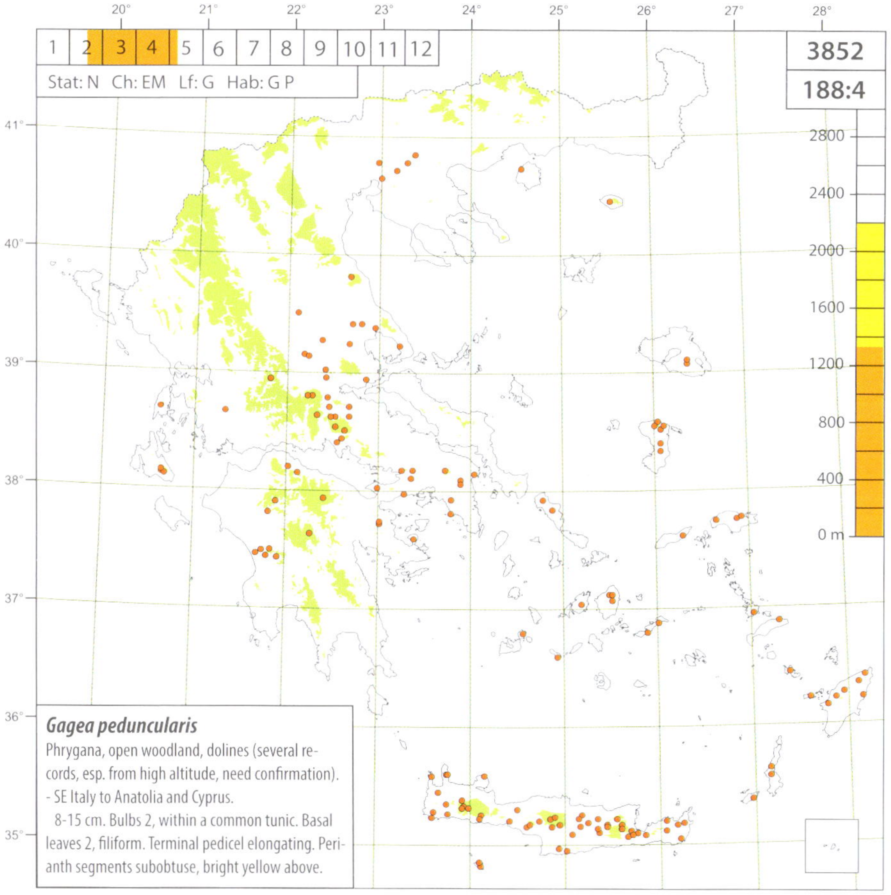

Gagea peduncularis
Phrygana, open woodland, dolines (several records, esp. from high altitude, need confirmation). - SE Italy to Anatolia and Cyprus.
8-15 cm. Bulbs 2, within a common tunic. Basal leaves 2, filiform. Terminal pedicel elongating. Perianth segments subobtuse, bright yellow above.

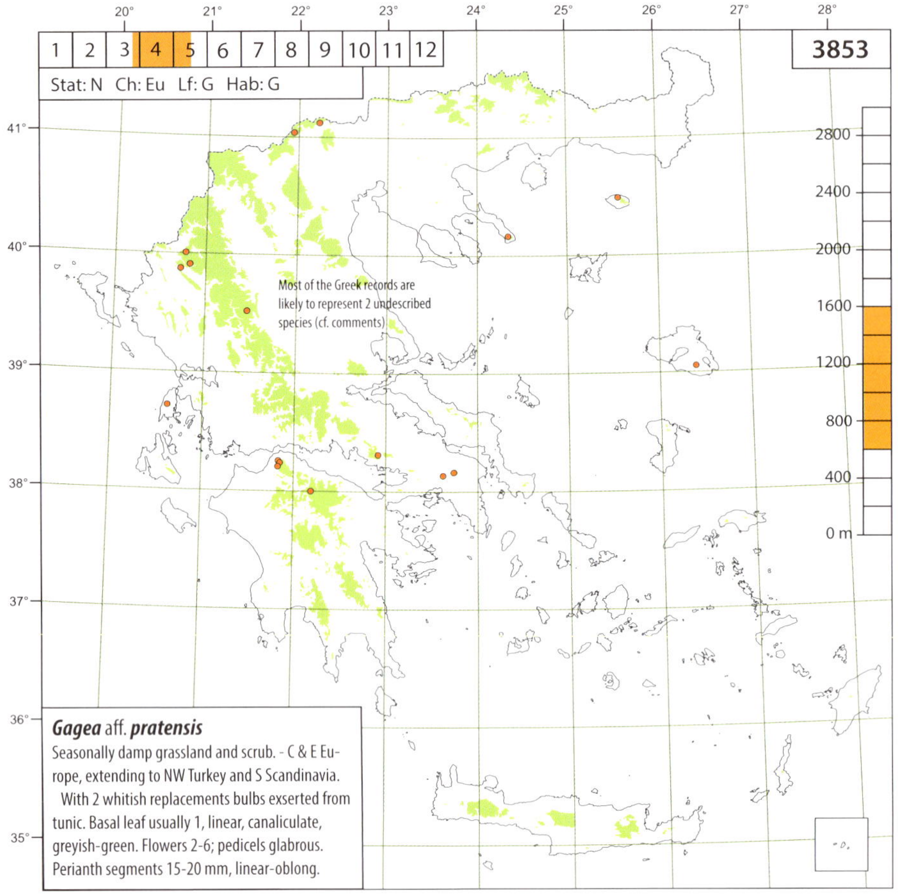

Gagea* aff. *pratensis

Seasonally damp grassland and scrub. - C & E Europe, extending to NW Turkey and S Scandinavia.

With 2 whitish replacements bulbs exserted from tunic. Basal leaf usually 1, linear, canaliculate, greyish-green. Flowers 2-6; pedicels glabrous. Perianth segments 15-20 mm, linear-oblong.

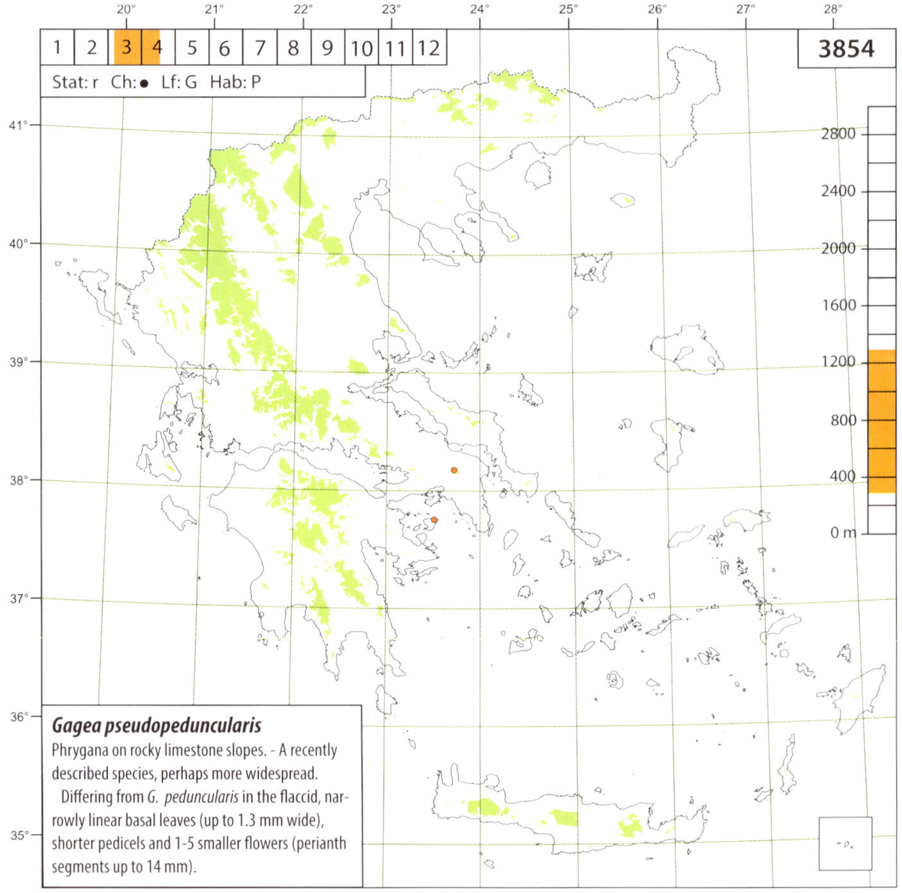

Gagea pseudopeduncularis

Phrygana on rocky limestone slopes. - A recently described species, perhaps more widespread.

Differing from *G. peduncularis* in the flaccid, narrowly linear basal leaves (up to 1.3 mm wide), shorter pedicels and 1-5 smaller flowers (perianth segments up to 14 mm).

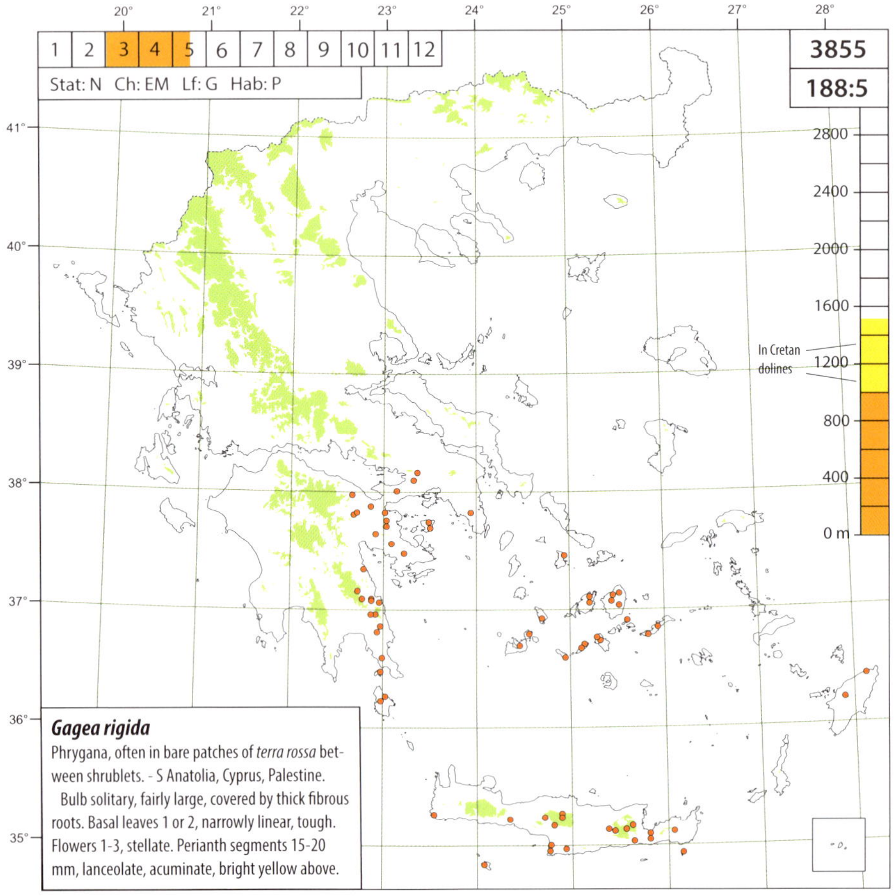

Gagea rigida

Phrygana, often in bare patches of *terra rossa* between shrublets. - S Anatolia, Cyprus, Palestine.

Bulb solitary, fairly large, covered by thick fibrous roots. Basal leaves 1 or 2, narrowly linear, tough. Flowers 1-3, stellate. Perianth segments 15-20 mm, lanceolate, acuminate, bright yellow above.

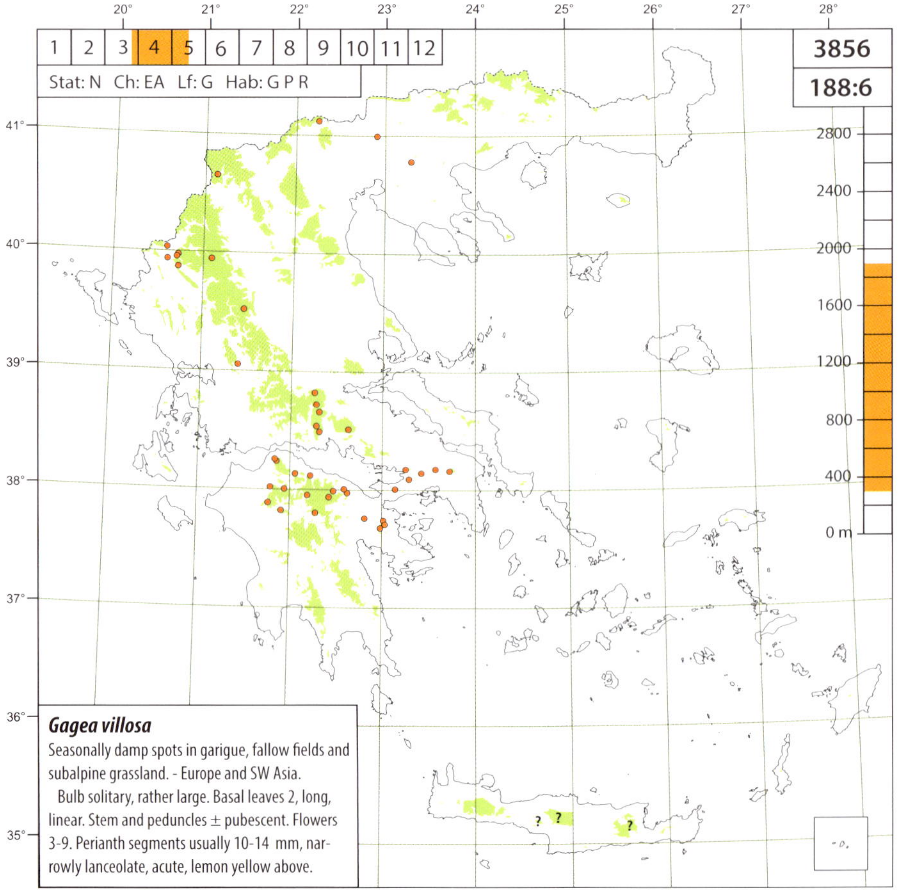

Gagea villosa

Seasonally damp spots in garigue, fallow fields and subalpine grassland. - Europe and SW Asia.

Bulb solitary, rather large. Basal leaves 2, long, linear. Stem and peduncles ± pubescent. Flowers 3-9. Perianth segments usually 10-14 mm, narrowly lanceolate, acute, lemon yellow above.

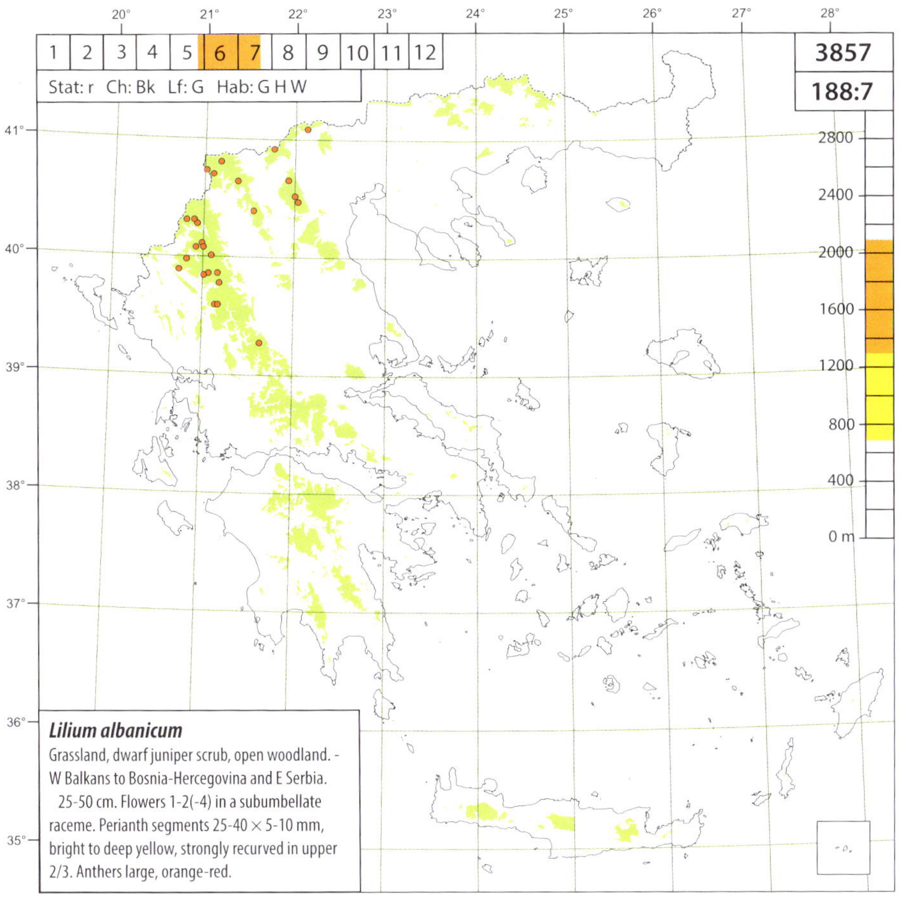

Lilium albanicum

Grassland, dwarf juniper scrub, open woodland. - W Balkans to Bosnia-Hercegovina and E Serbia.

25-50 cm. Flowers 1-2(-4) in a subumbellate raceme. Perianth segments 25-40 × 5-10 mm, bright to deep yellow, strongly recurved in upper 2/3. Anthers large, orange-red.

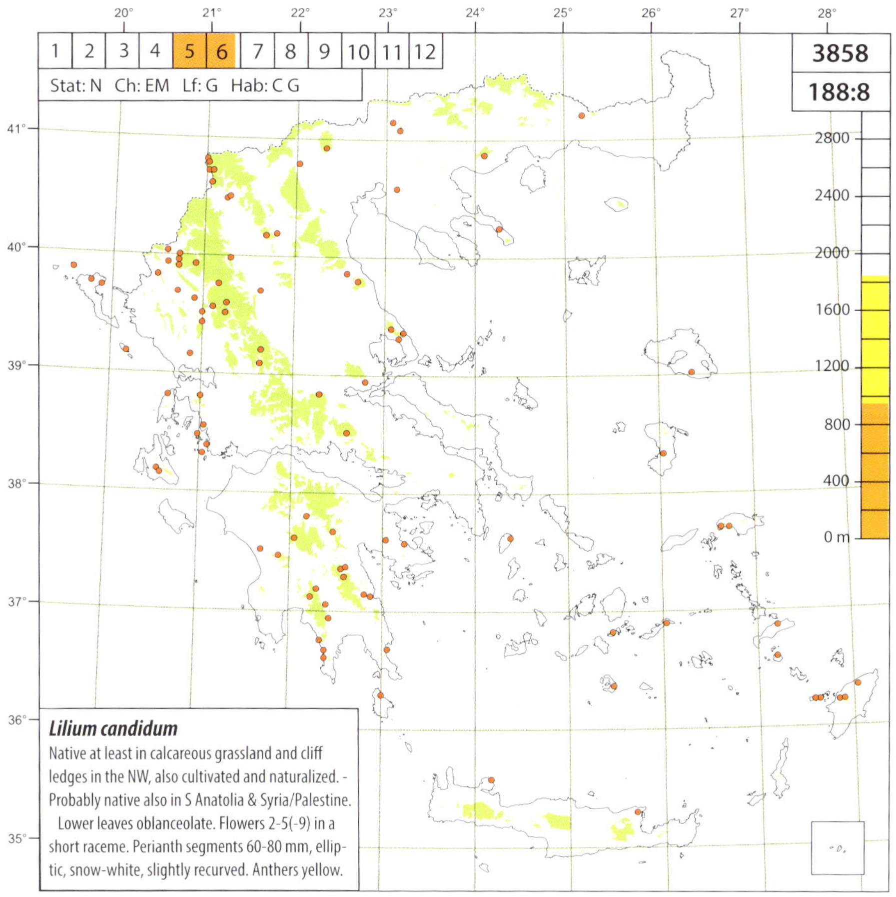

Lilium candidum

Native at least in calcareous grassland and cliff ledges in the NW, also cultivated and naturalized. - Probably native also in S Anatolia & Syria/Palestine.

Lower leaves oblanceolate. Flowers 2-5(-9) in a short raceme. Perianth segments 60-80 mm, elliptic, snow-white, slightly recurved. Anthers yellow.

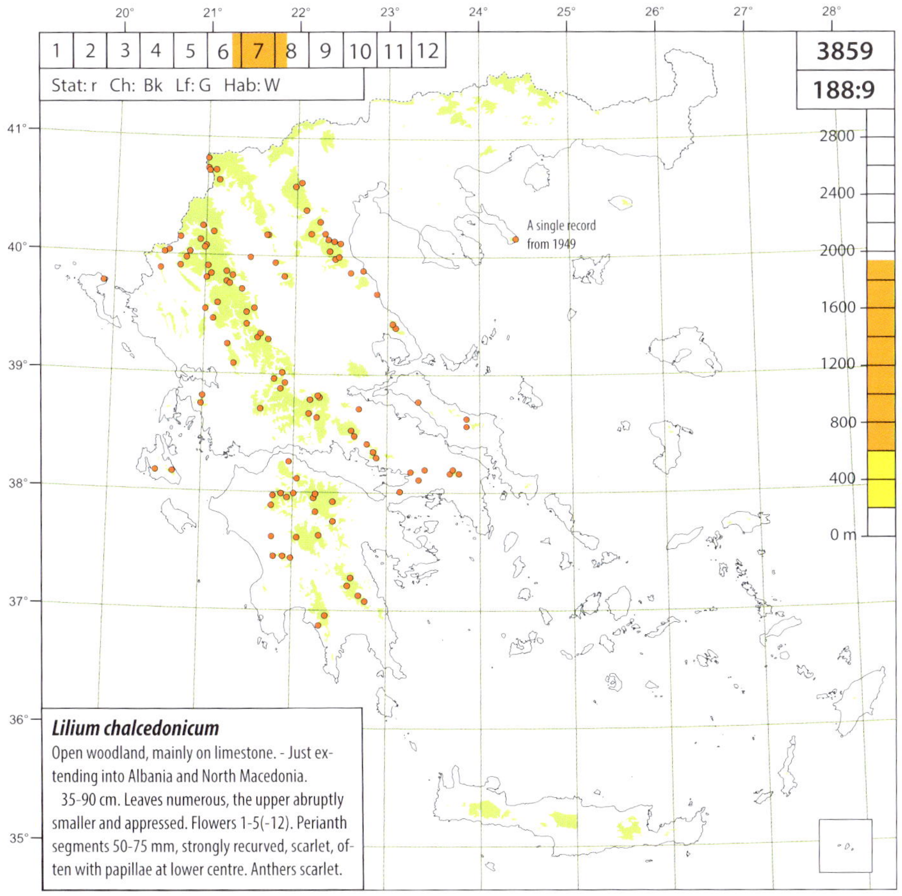

Lilium chalcedonicum

Open woodland, mainly on limestone. - Just extending into Albania and North Macedonia.

35-90 cm. Leaves numerous, the upper abruptly smaller and appressed. Flowers 1-5(-12). Perianth segments 50-75 mm, strongly recurved, scarlet, often with papillae at lower centre. Anthers scarlet.

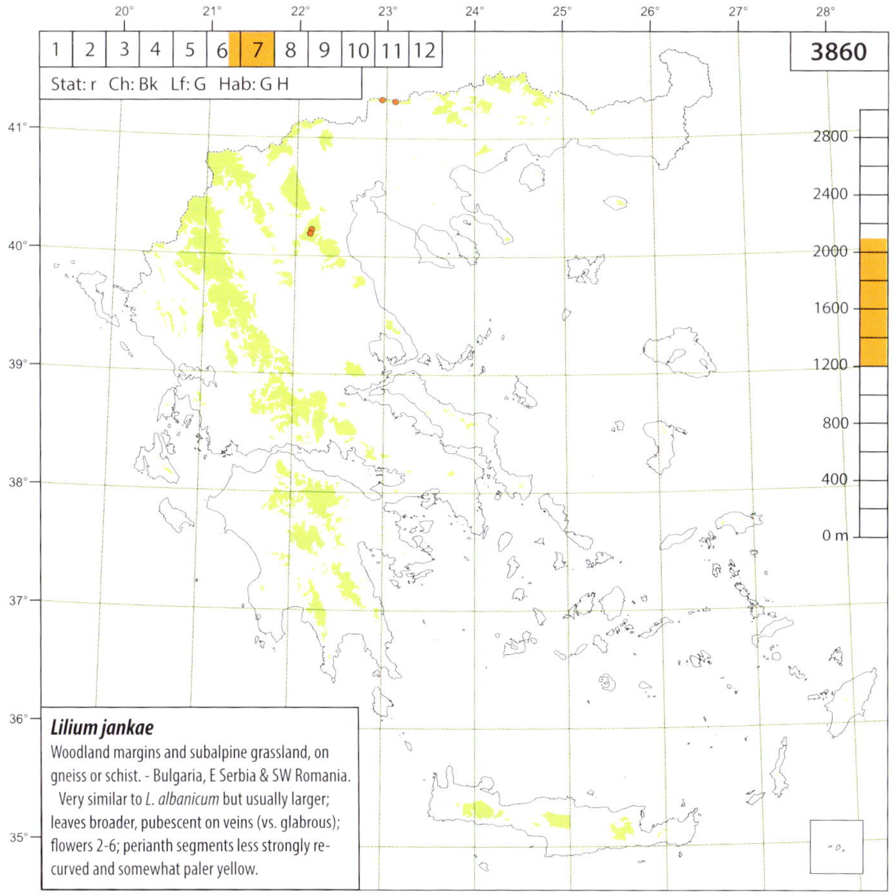

Lilium jankae

Woodland margins and subalpine grassland, on gneiss or schist. - Bulgaria, E Serbia & SW Romania.

Very similar to *L. albanicum* but usually larger; leaves broader, pubescent on veins (vs. glabrous); flowers 2-6; perianth segments less strongly recurved and somewhat paler yellow.

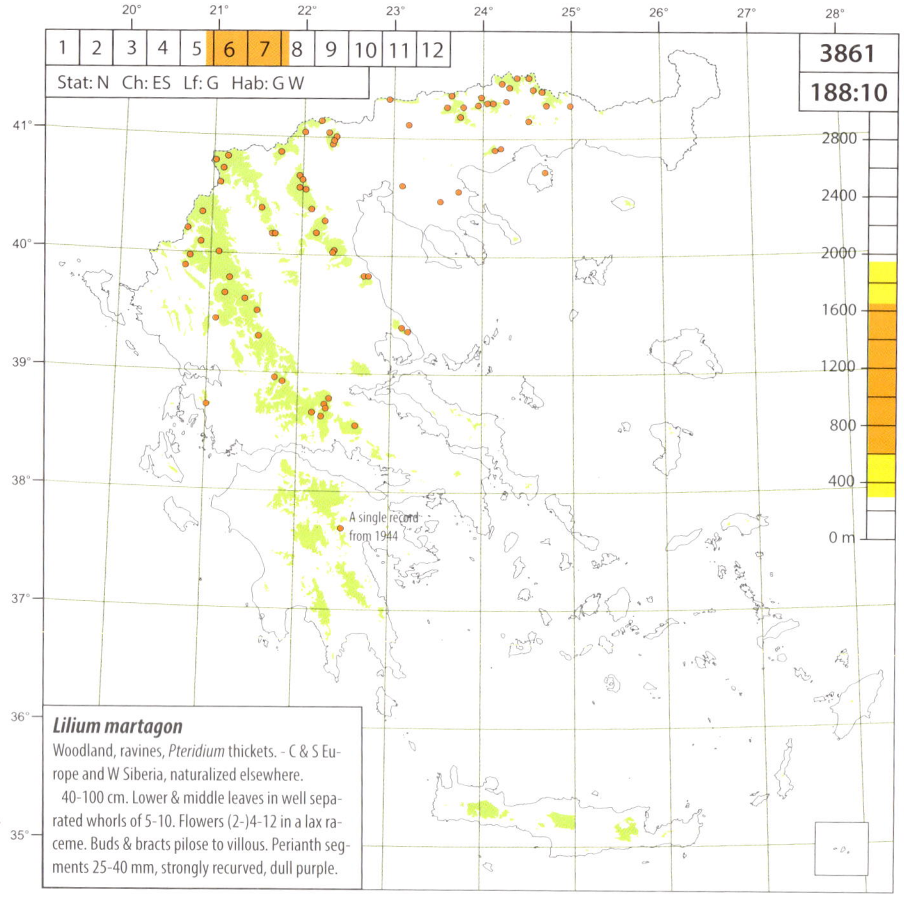

Lilium martagon

Woodland, ravines, *Pteridium* thickets. - C & S Europe and W Siberia, naturalized elsewhere.

40-100 cm. Lower & middle leaves in well separated whorls of 5-10. Flowers (2-)4-12 in a lax raceme. Buds & bracts pilose to villous. Perianth segments 25-40 mm, strongly recurved, dull purple.

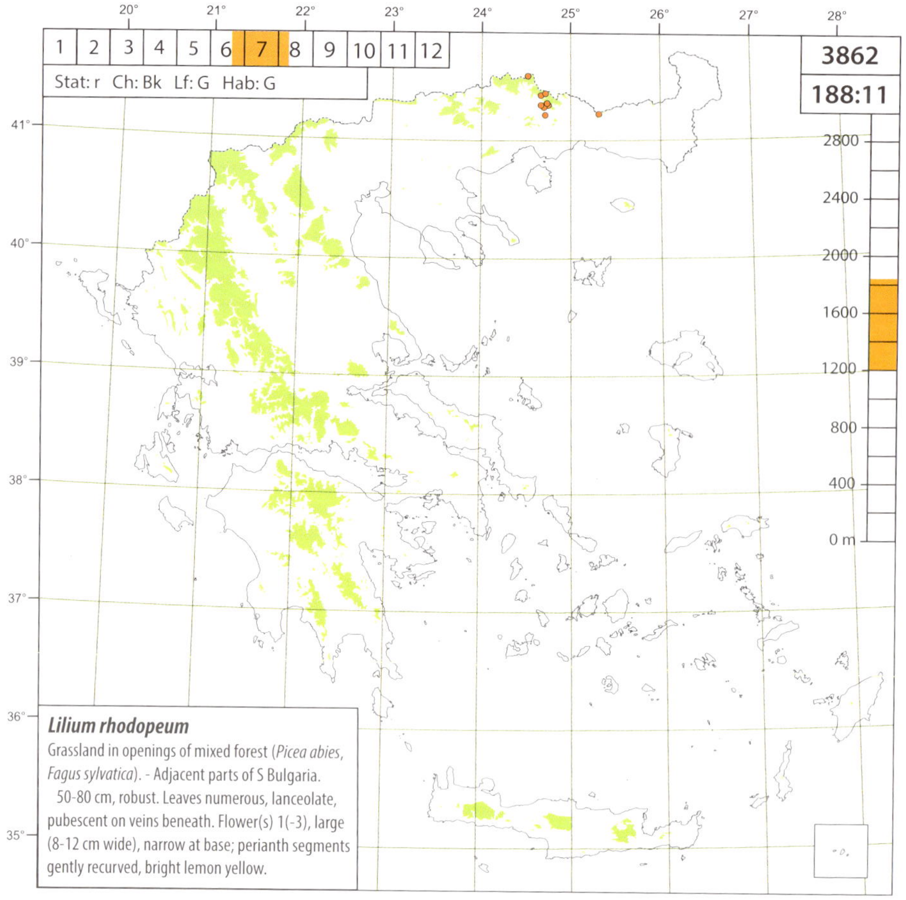

Lilium rhodopeum

Grassland in openings of mixed forest (*Picea abies*, *Fagus sylvatica*). - Adjacent parts of S Bulgaria.

50-80 cm, robust. Leaves numerous, lanceolate, pubescent on veins beneath. Flower(s) 1(-3), large (8-12 cm wide), narrow at base; perianth segments gently recurved, bright lemon yellow.

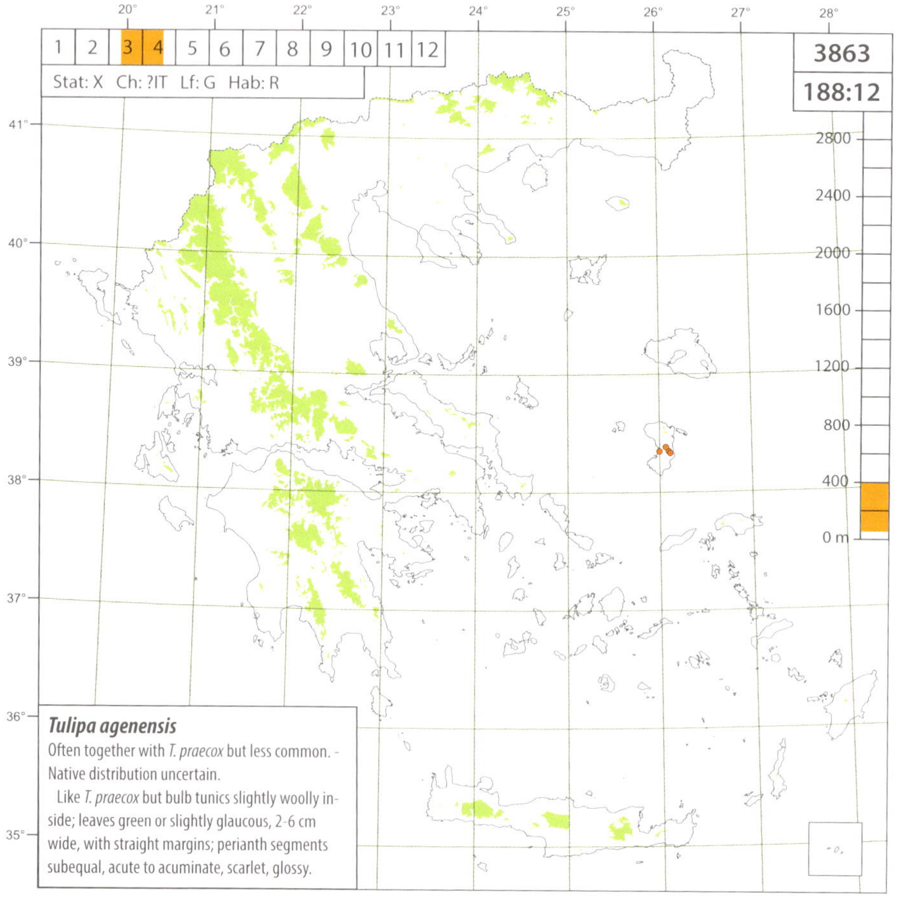

Tulipa agenensis

Often together with *T. praecox* but less common. - Native distribution uncertain.

Like *T. praecox* but bulb tunics slightly woolly inside; leaves green or slightly glaucous, 2-6 cm wide, with straight margins; perianth segments subequal, acute to acuminate, scarlet, glossy.

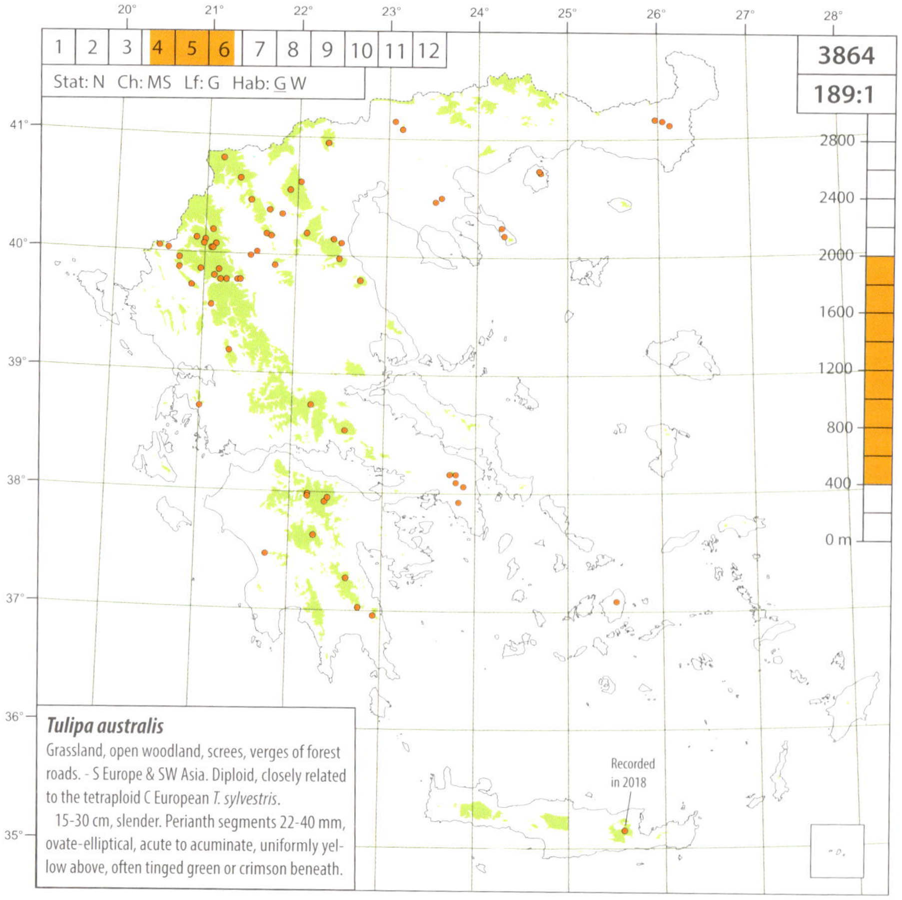

Tulipa australis

Grassland, open woodland, screes, verges of forest roads. - S Europe & SW Asia. Diploid, closely related to the tetraploid C European *T. sylvestris*.

15-30 cm, slender. Perianth segments 22-40 mm, ovate-elliptical, acute to acuminate, uniformly yellow above, often tinged green or crimson beneath.

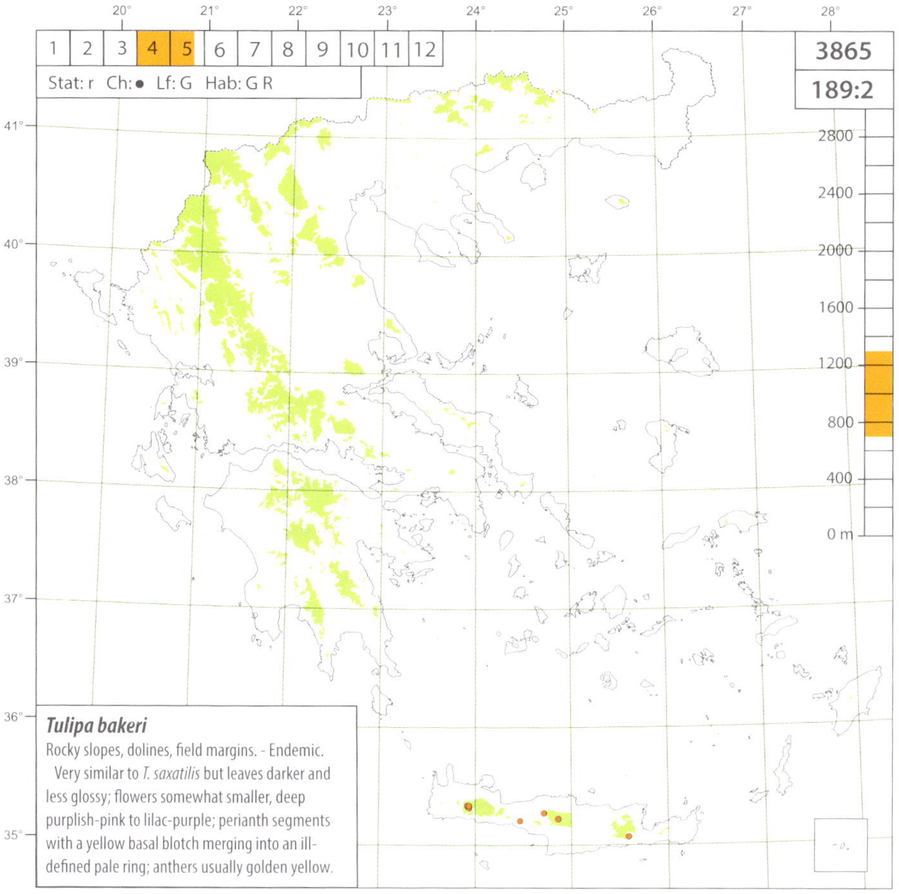

Tulipa bakeri
Rocky slopes, dolines, field margins. - Endemic.
Very similar to *T. saxatilis* but leaves darker and less glossy; flowers somewhat smaller, deep purplish-pink to lilac-purple; perianth segments with a yellow basal blotch merging into an ill-defined pale ring; anthers usually golden yellow.

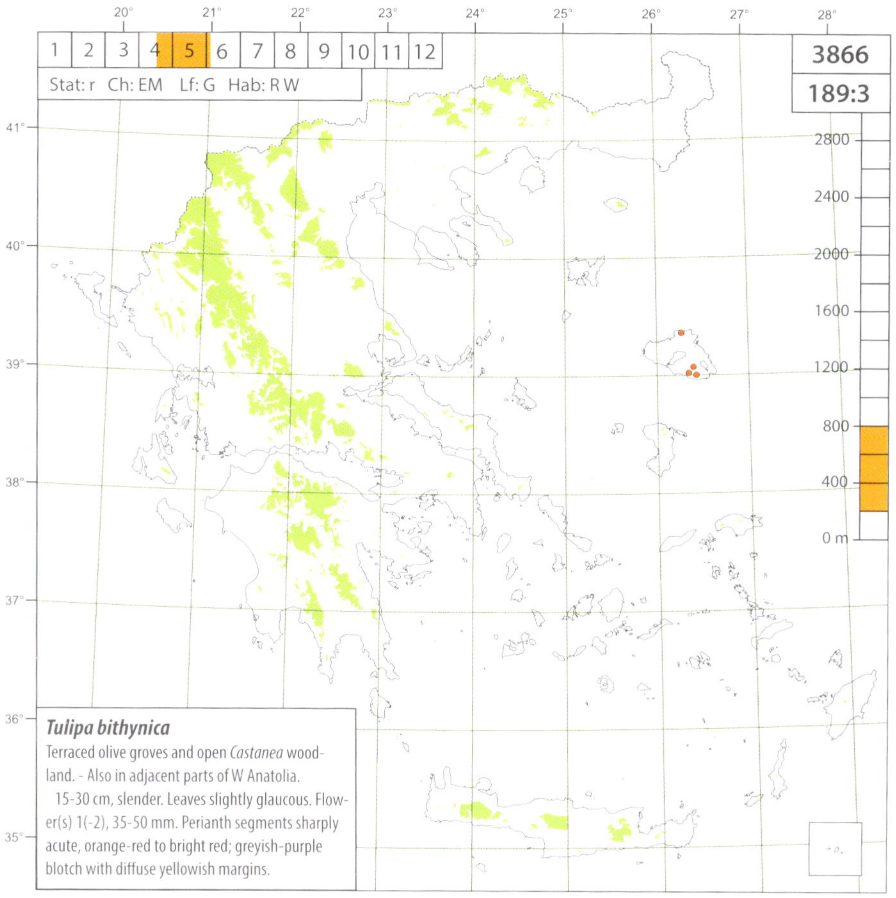

Tulipa bithynica
Terraced olive groves and open *Castanea* woodland. - Also in adjacent parts of W Anatolia.
15-30 cm, slender. Leaves slightly glaucous. Flower(s) 1(-2), 35-50 mm. Perianth segments sharply acute, orange-red to bright red; greyish-purple blotch with diffuse yellowish margins.

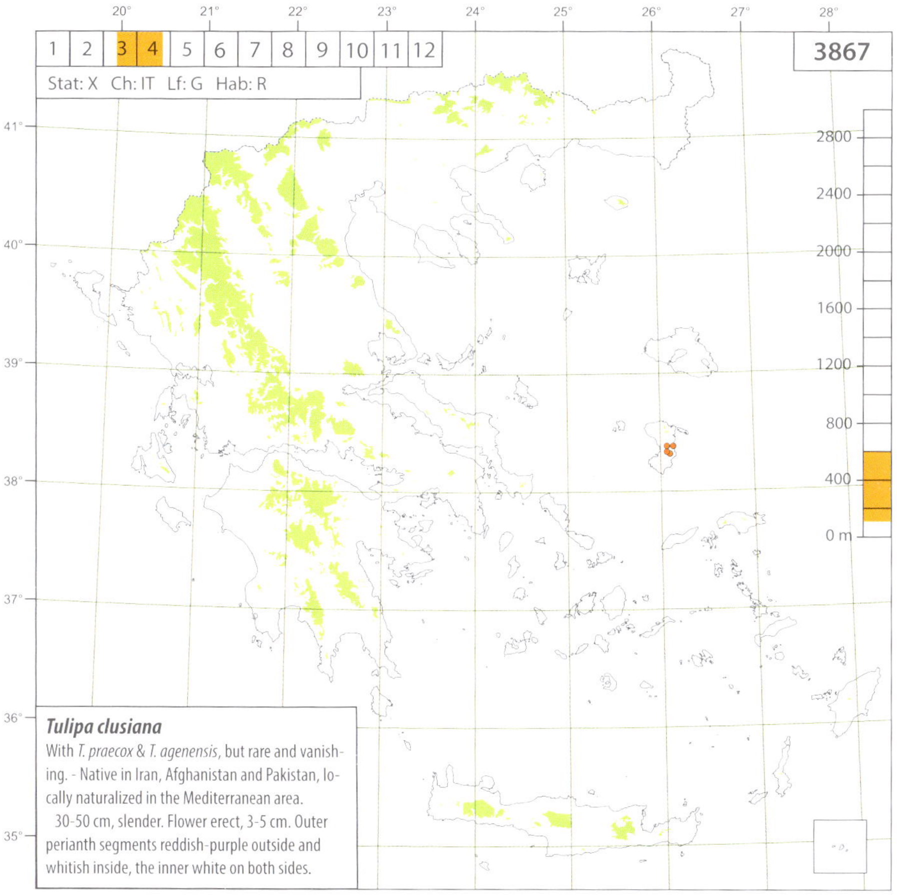

Tulipa clusiana
With *T. praecox* & *T. agenensis*, but rare and vanishing. - Native in Iran, Afghanistan and Pakistan, locally naturalized in the Mediterranean area.
30-50 cm, slender. Flower erect, 3-5 cm. Outer perianth segments reddish-purple outside and whitish inside, the inner white on both sides.

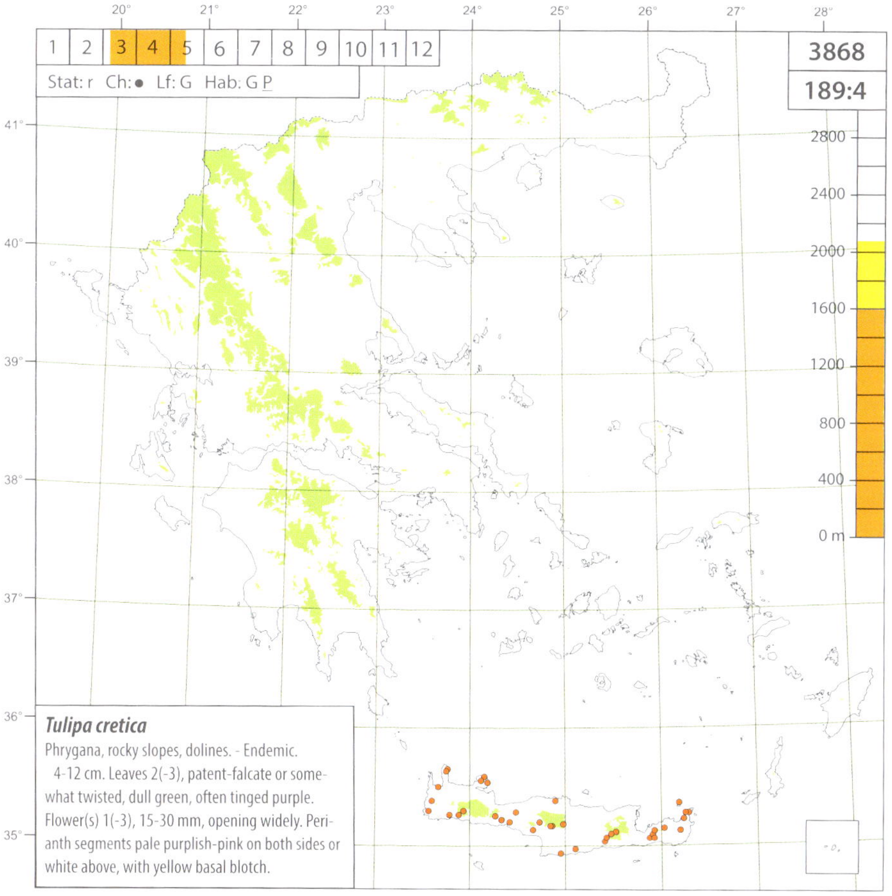

Tulipa cretica
Phrygana, rocky slopes, dolines. - Endemic.
4-12 cm. Leaves 2(-3), patent-falcate or somewhat twisted, dull green, often tinged purple. Flower(s) 1(-3), 15-30 mm, opening widely. Perianth segments pale purplish-pink on both sides or white above, with yellow basal blotch.

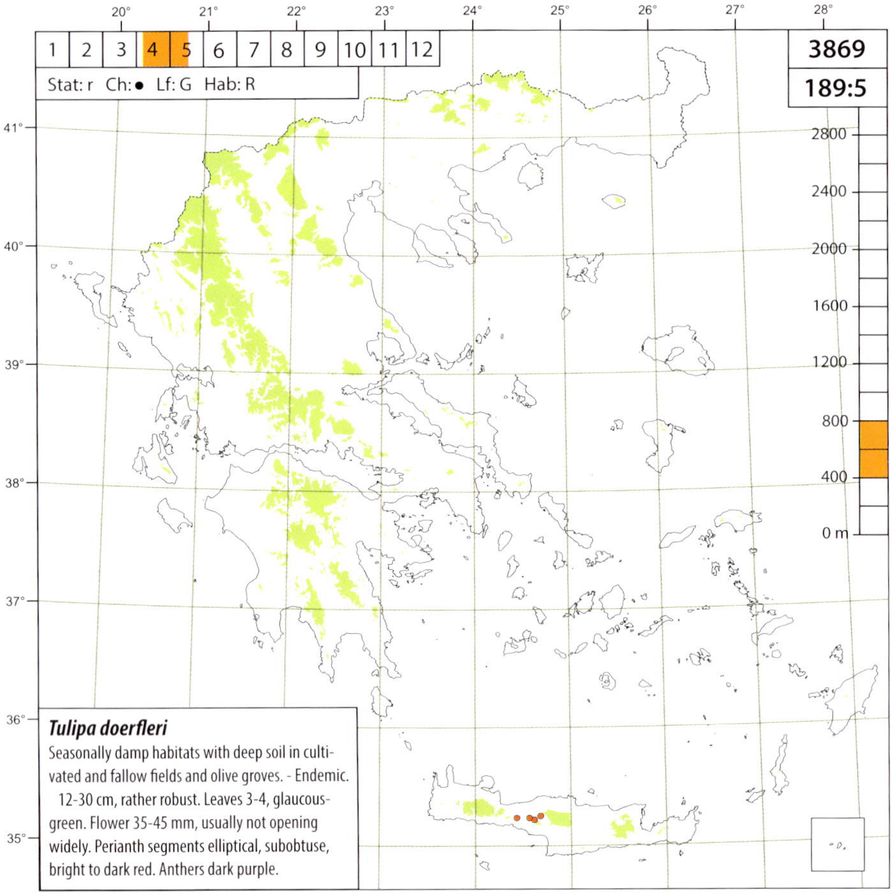

Tulipa doerfleri

Seasonally damp habitats with deep soil in cultivated and fallow fields and olive groves. - Endemic.

12-30 cm, rather robust. Leaves 3-4, glaucous-green. Flower 35-45 mm, usually not opening widely. Perianth segments elliptical, subobtuse, bright to dark red. Anthers dark purple.

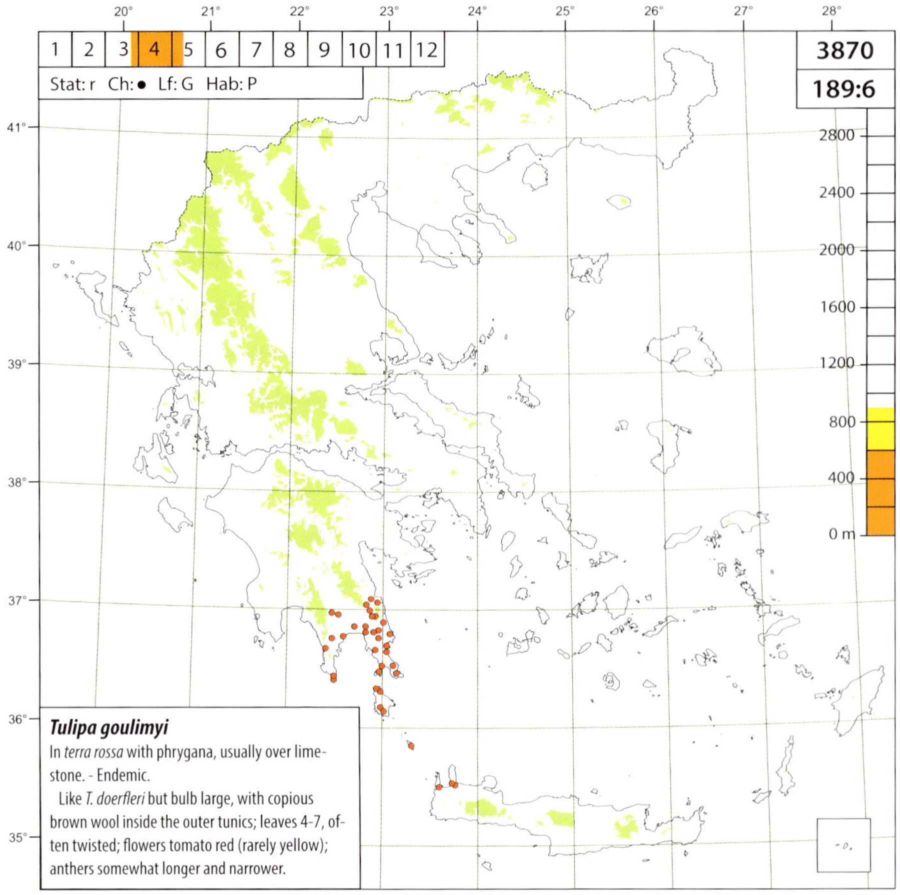

Tulipa goulimyi

In *terra rossa* with phrygana, usually over limestone. - Endemic.

Like *T. doerfleri* but bulb large, with copious brown wool inside the outer tunics; leaves 4-7, often twisted; flowers tomato red (rarely yellow); anthers somewhat longer and narrower.

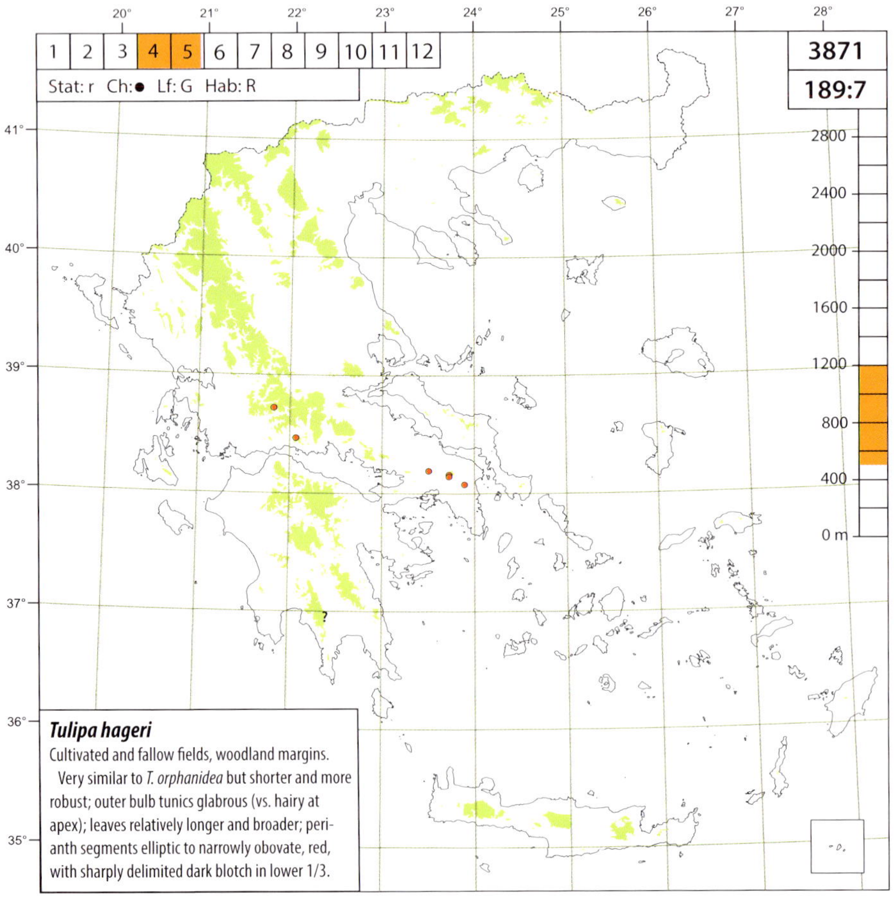

Tulipa hageri

Cultivated and fallow fields, woodland margins.

Very similar to *T. orphanidea* but shorter and more robust; outer bulb tunics glabrous (vs. hairy at apex); leaves relatively longer and broader; perianth segments elliptic to narrowly obovate, red, with sharply delimited dark blotch in lower 1/3.

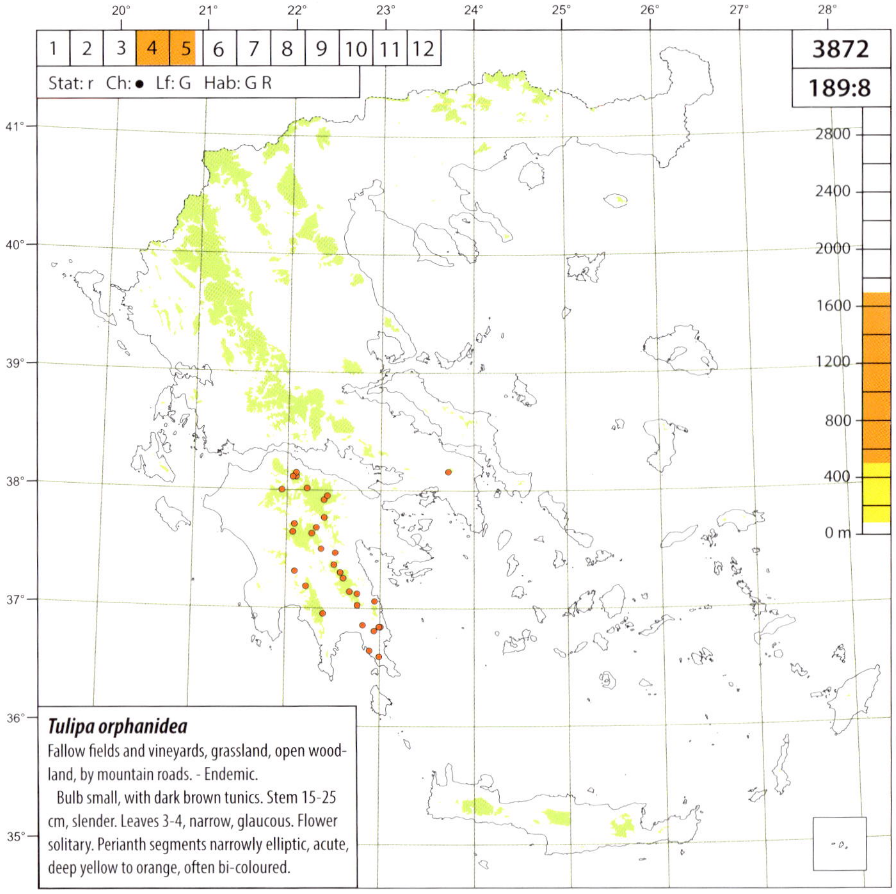

Tulipa orphanidea

Fallow fields and vineyards, grassland, open woodland, by mountain roads. - Endemic.

Bulb small, with dark brown tunics. Stem 15-25 cm, slender. Leaves 3-4, narrow, glaucous. Flower solitary. Perianth segments narrowly elliptic, acute, deep yellow to orange, often bi-coloured.

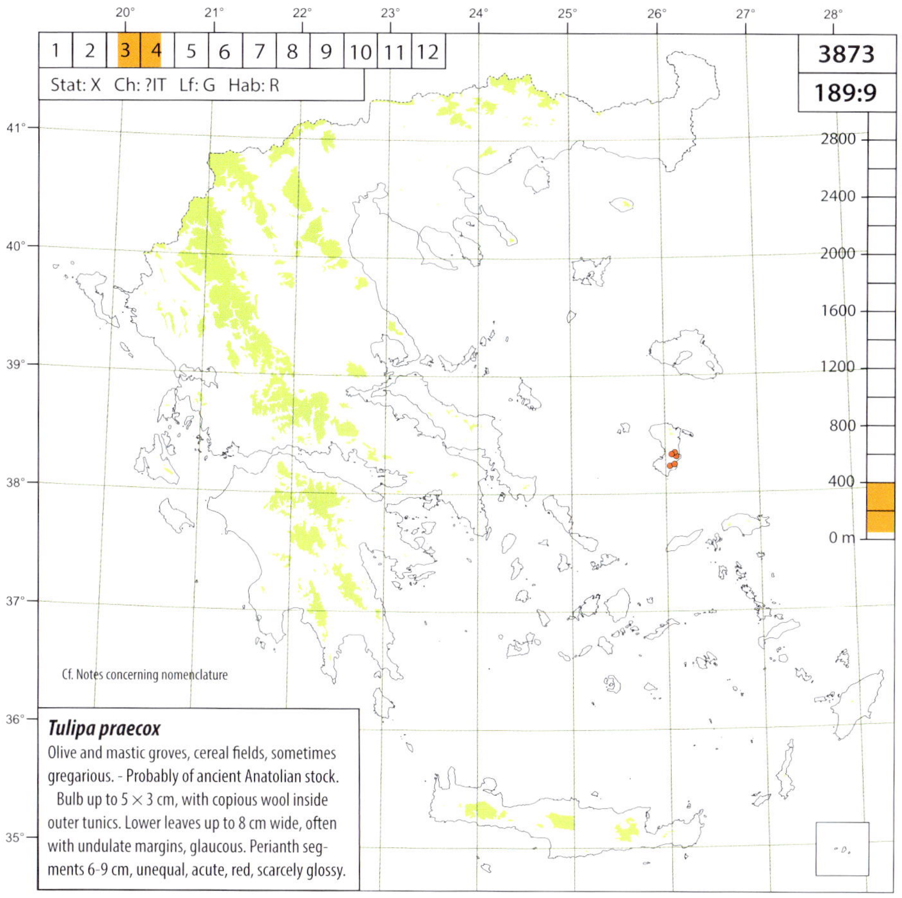

Tulipa praecox

Olive and mastic groves, cereal fields, sometimes gregarious. - Probably of ancient Anatolian stock.

Bulb up to 5 × 3 cm, with copious wool inside outer tunics. Lower leaves up to 8 cm wide, often with undulate margins, glaucous. Perianth segments 6-9 cm, unequal, acute, red, scarcely glossy.

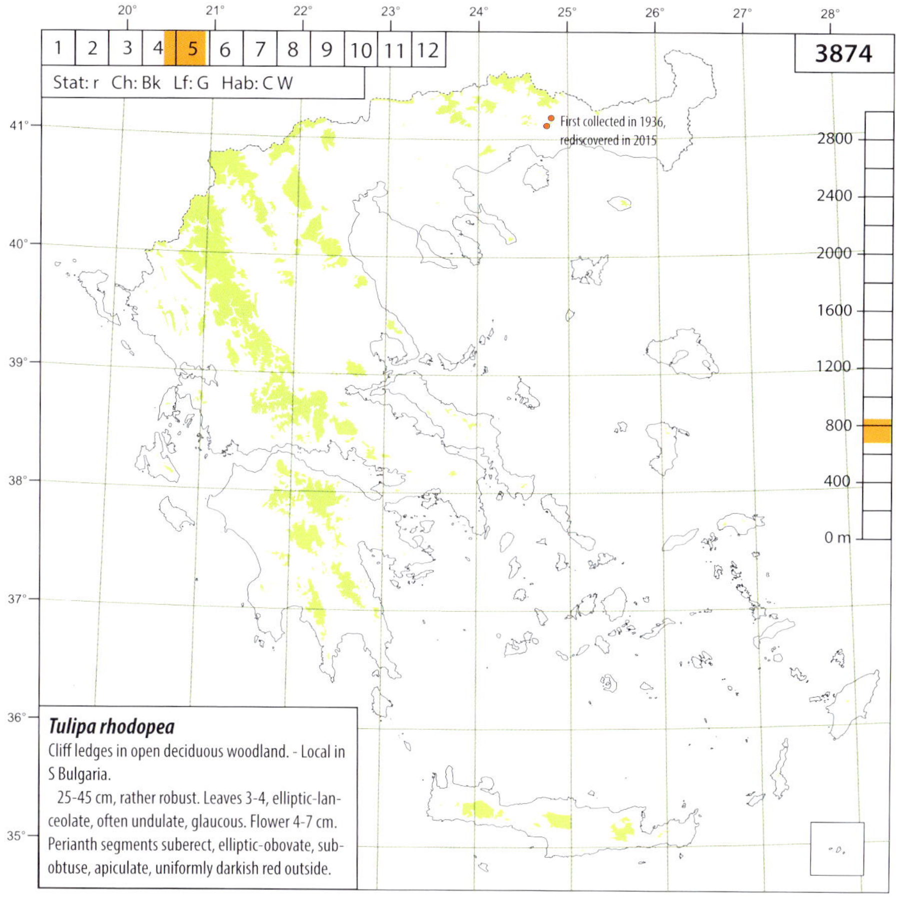

Tulipa rhodopea

Cliff ledges in open deciduous woodland. - Local in S Bulgaria.

25-45 cm, rather robust. Leaves 3-4, elliptic-lanceolate, often undulate, glaucous. Flower 4-7 cm. Perianth segments subereect, elliptic-obovate, subobtuse, apiculate, uniformly darkish red outside.

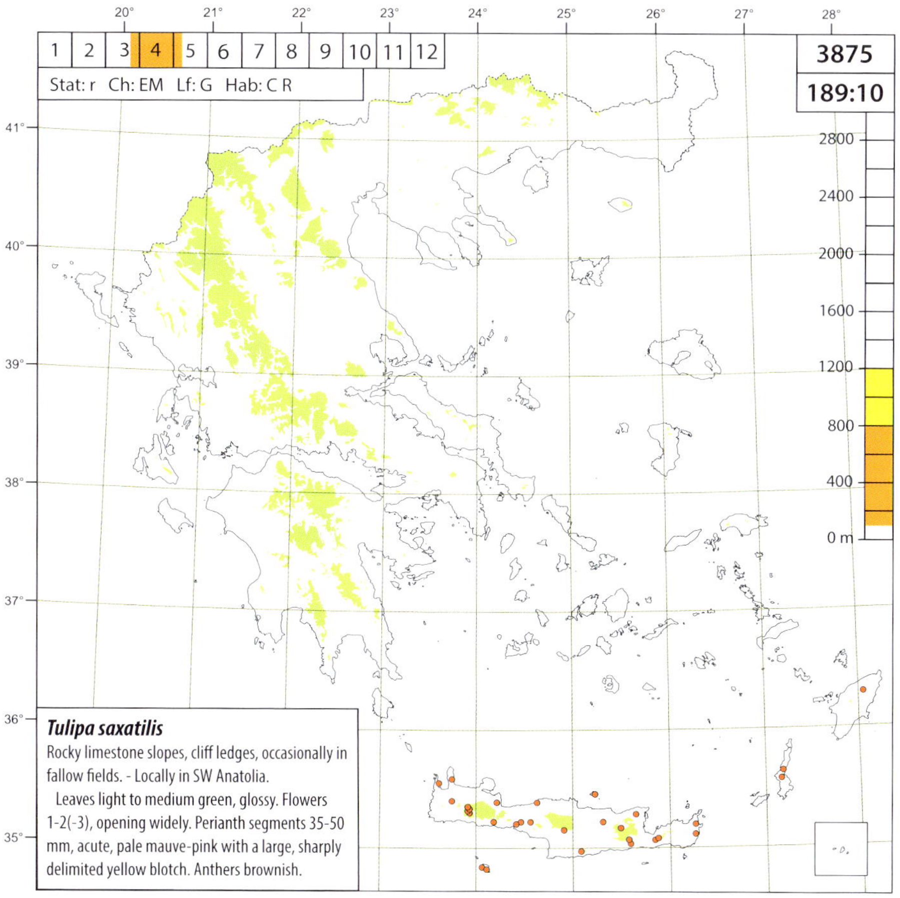

Tulipa saxatilis

Rocky limestone slopes, cliff ledges, occasionally in fallow fields. - Locally in SW Anatolia.

Leaves light to medium green, glossy. Flowers 1-2(-3), opening widely. Perianth segments 35-50 mm, acute, pale mauve-pink with a large, sharply delimited yellow blotch. Anthers brownish.

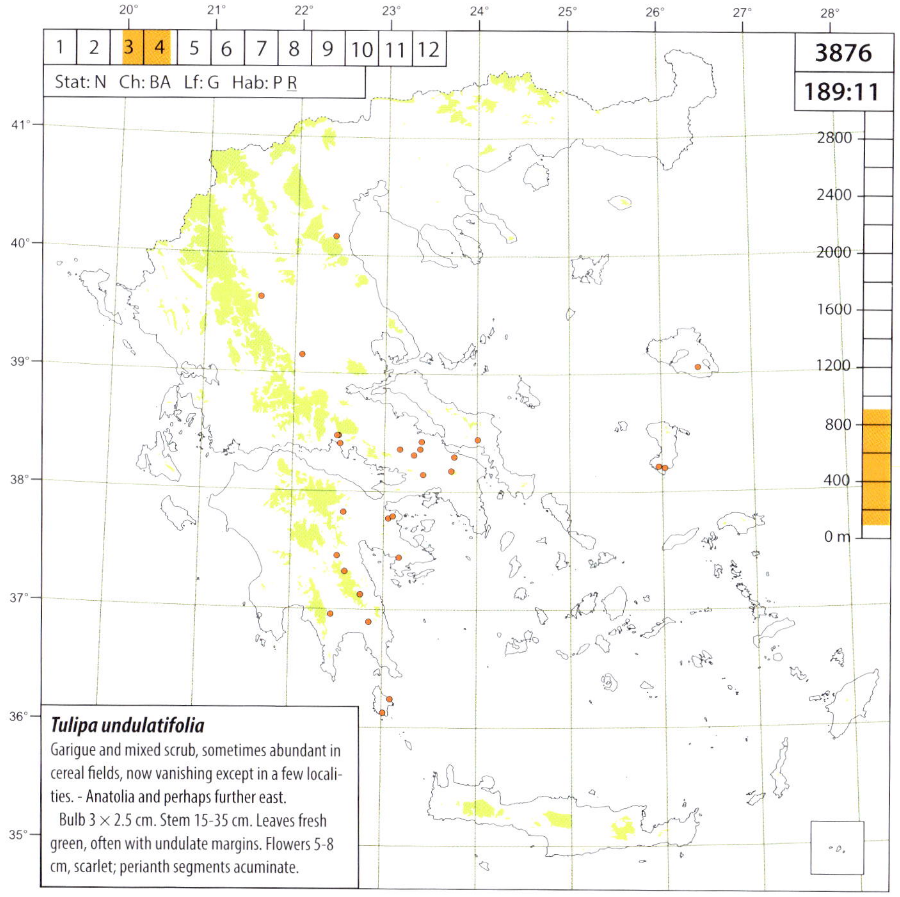

Tulipa undulatifolia

Garigue and mixed scrub, sometimes abundant in cereal fields, now vanishing except in a few localities. - Anatolia and perhaps further east.

Bulb 3 × 2.5 cm. Stem 15-35 cm. Leaves fresh green, often with undulate margins. Flowers 5-8 cm, scarlet; perianth segments acuminate.

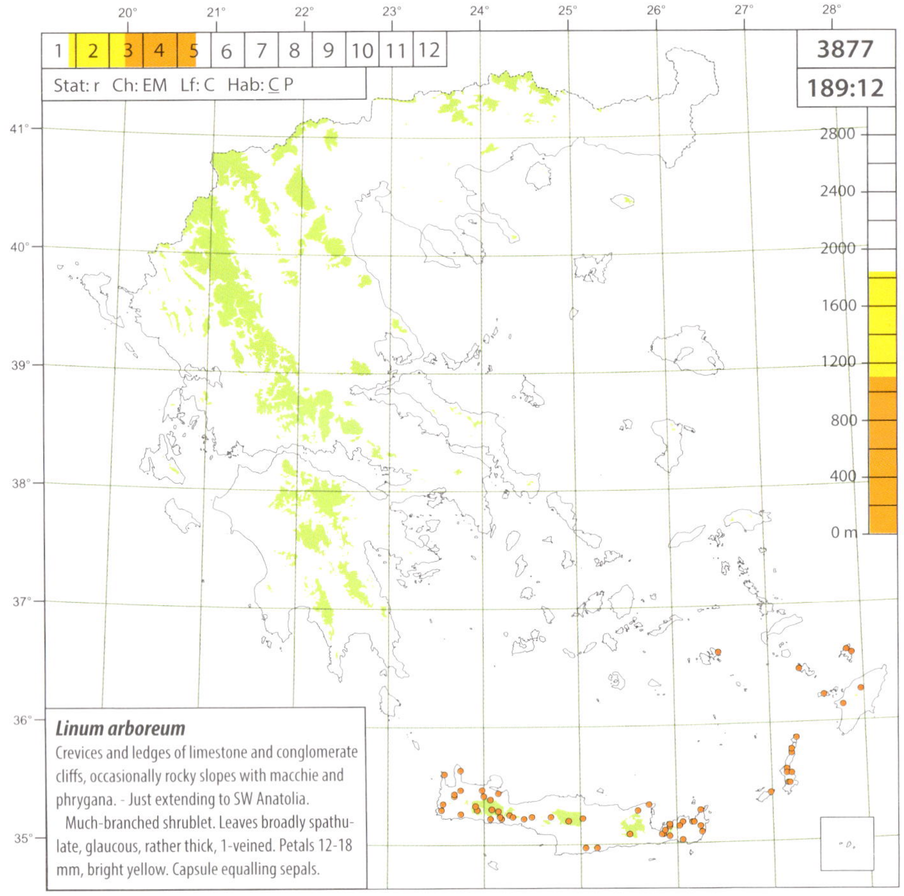

Linum arboreum

Crevices and ledges of limestone and conglomerate cliffs, occasionally rocky slopes with macchie and phrygana. - Just extending to SW Anatolia.

Much-branched shrublet. Leaves broadly spathulate, glaucous, rather thick, 1-veined. Petals 12-18 mm, bright yellow. Capsule equalling sepals.

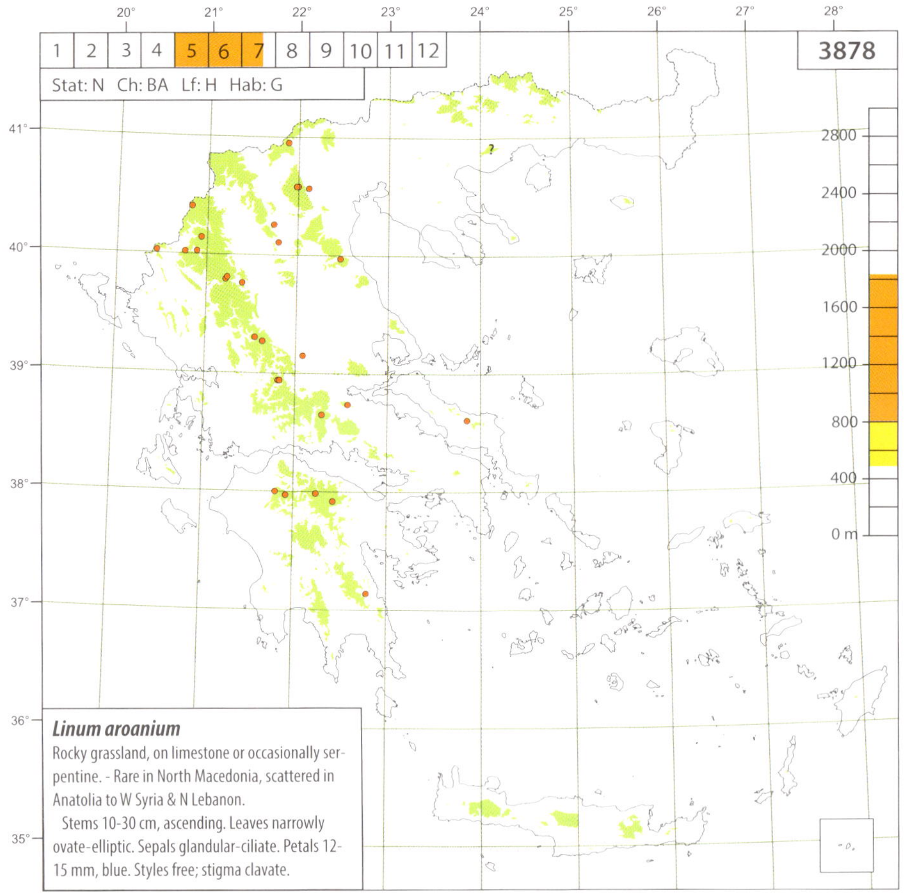

Linum aroanium

Rocky grassland, on limestone or occasionally serpentine. - Rare in North Macedonia, scattered in Anatolia to W Syria & N Lebanon.

Stems 10-30 cm, ascending. Leaves narrowly ovate-elliptic. Sepals glandular-ciliate. Petals 12-15 mm, blue. Styles free; stigma clavate.

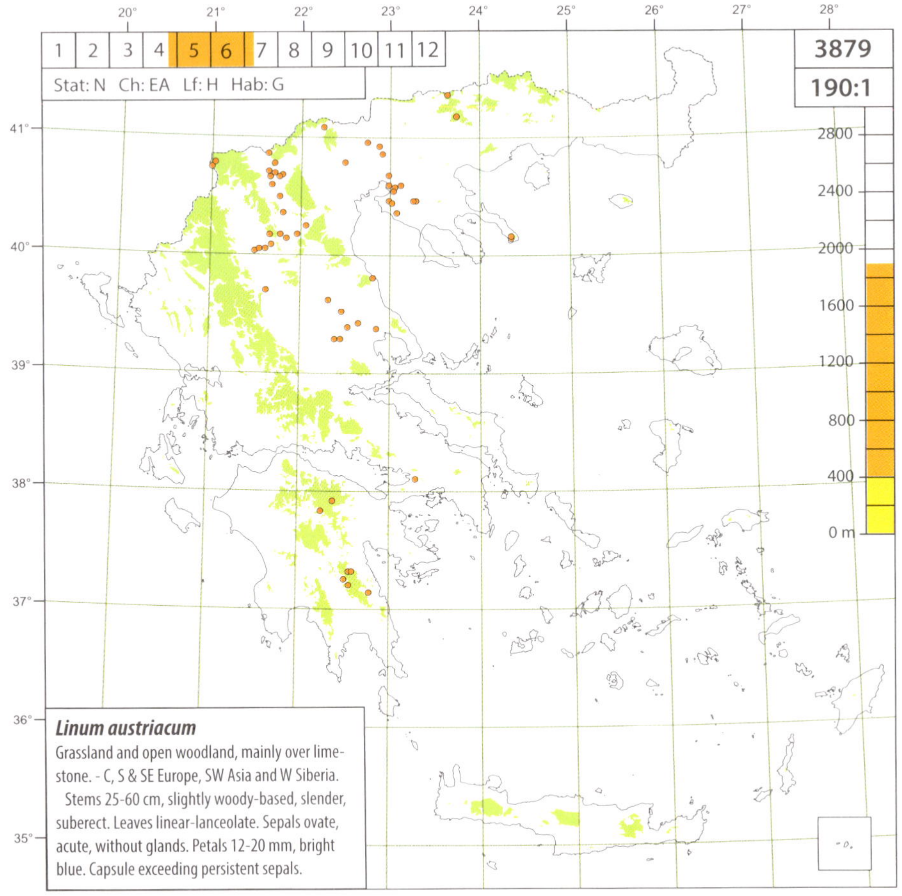

Linum austriacum

Grassland and open woodland, mainly over limestone. - C, S & SE Europe, SW Asia and W Siberia.

Stems 25-60 cm, slightly woody-based, slender, suberect. Leaves linear-lanceolate. Sepals ovate, acute, without glands. Petals 12-20 mm, bright blue. Capsule exceeding persistent sepals.

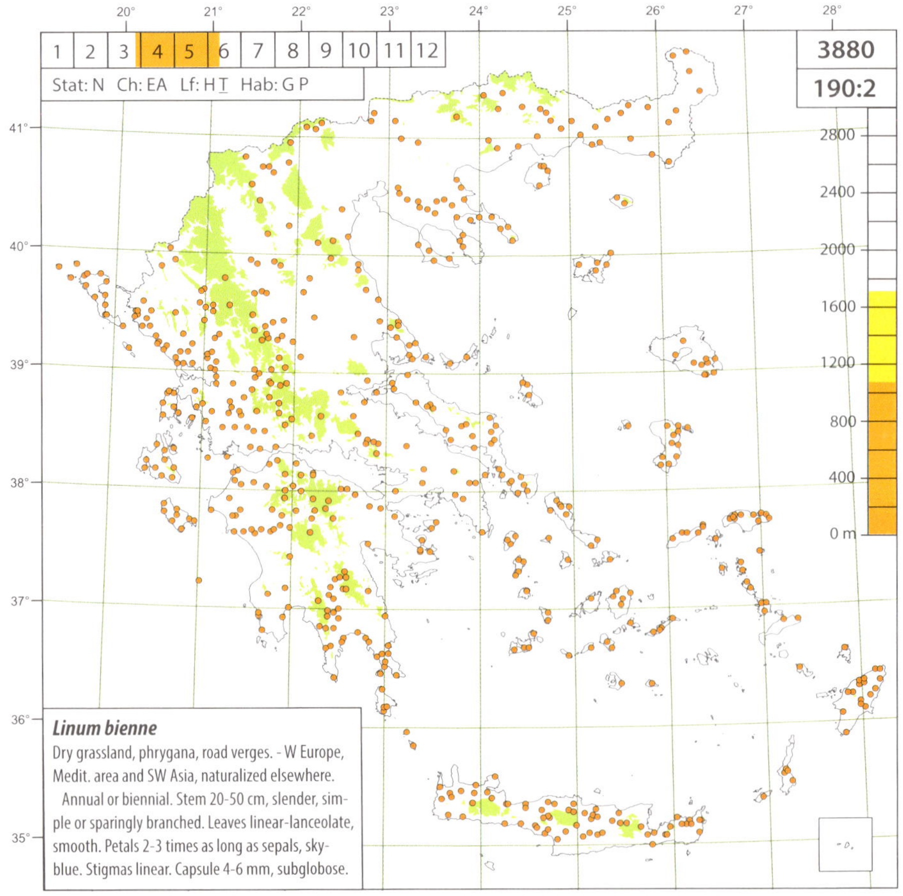

Linum bienne

Dry grassland, phrygana, road verges. - W Europe, Medit. area and SW Asia, naturalized elsewhere.

Annual or biennial. Stem 20-50 cm, slender, simple or sparingly branched. Leaves linear-lanceolate, smooth. Petals 2-3 times as long as sepals, sky-blue. Stigmas linear. Capsule 4-6 mm, subglobose.

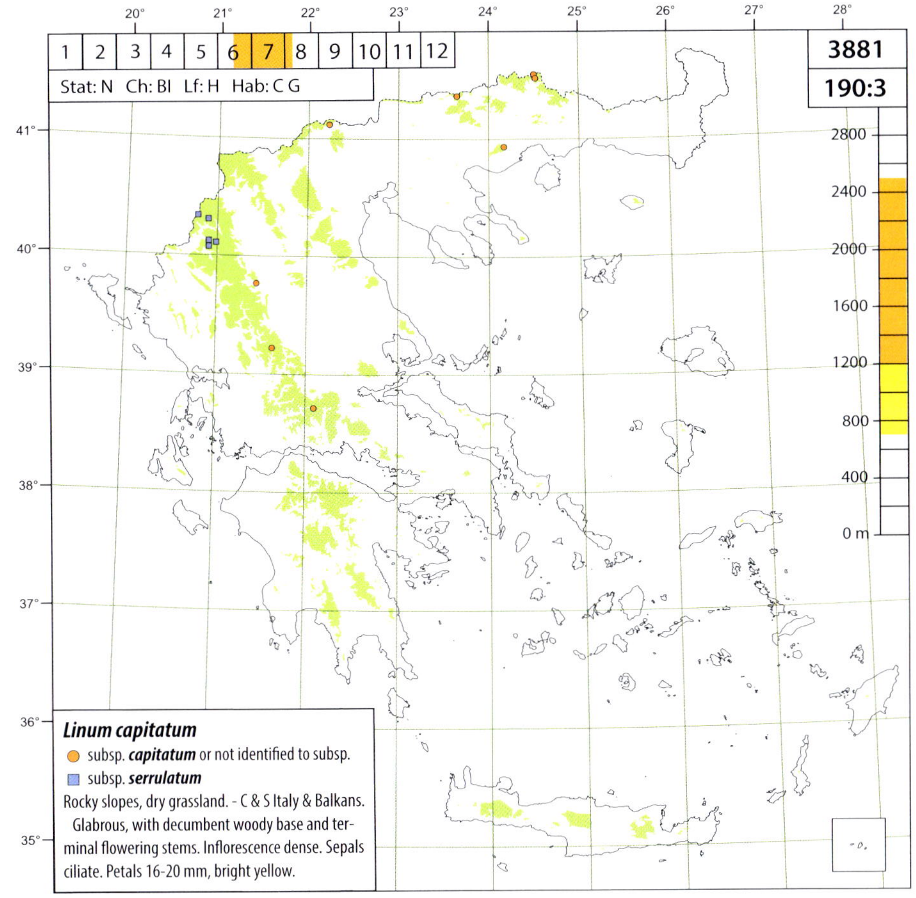

Linum capitatum
subsp. ***capitatum*** or not identified to subsp.
subsp. ***serrulatum***
Rocky slopes, dry grassland. - C & S Italy & Balkans.
Glabrous, with decumbent woody base and terminal flowering stems. Inflorescence dense. Sepals ciliate. Petals 16-20 mm, bright yellow.

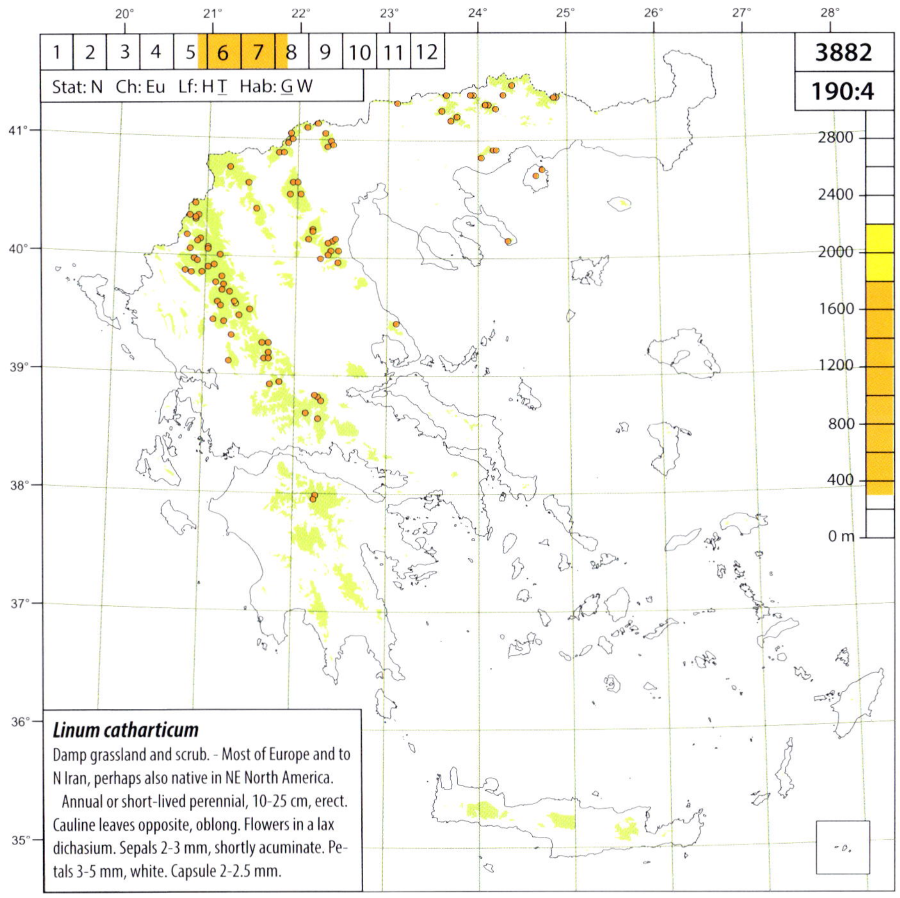

Linum catharticum
Damp grassland and scrub. - Most of Europe and to N Iran, perhaps also native in NE North America.
Annual or short-lived perennial, 10-25 cm, erect. Cauline leaves opposite, oblong. Flowers in a lax dichasium. Sepals 2-3 mm, shortly acuminate. Petals 3-5 mm, white. Capsule 2-2.5 mm.

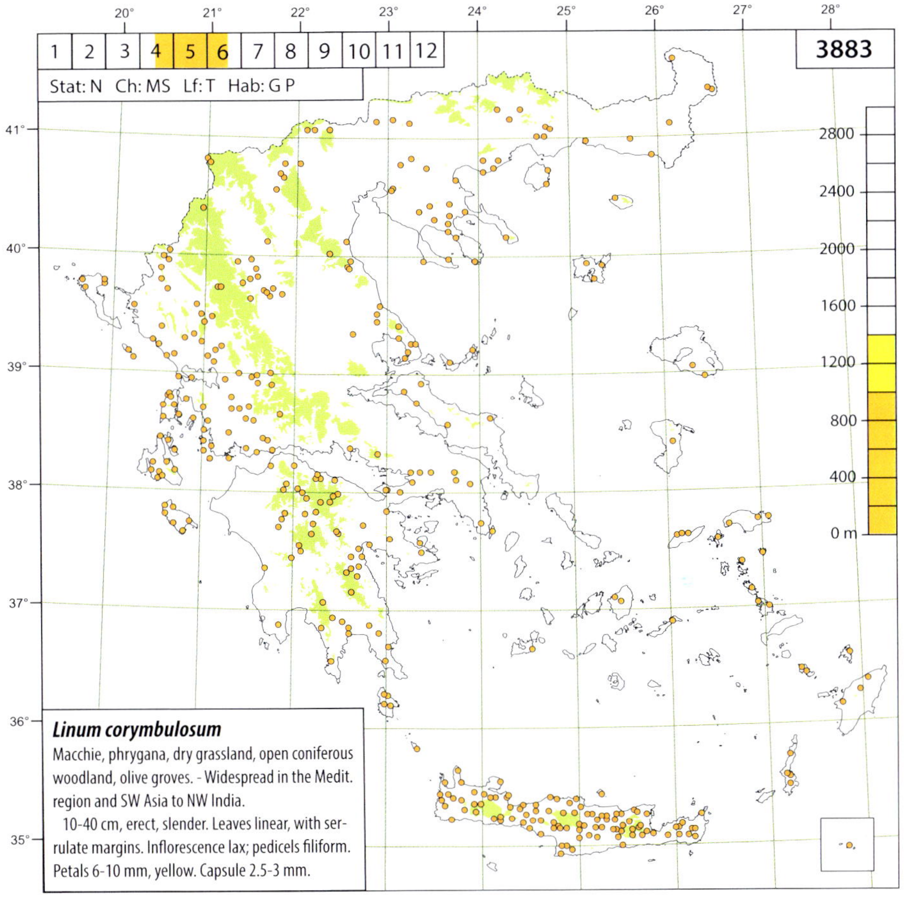

Linum corymbulosum
Macchie, phrygana, dry grassland, open coniferous woodland, olive groves. - Widespread in the Medit. region and SW Asia to NW India.
10-40 cm, erect, slender. Leaves linear, with serrulate margins. Inflorescence lax; pedicels filiform. Petals 6-10 mm, yellow. Capsule 2.5-3 mm.

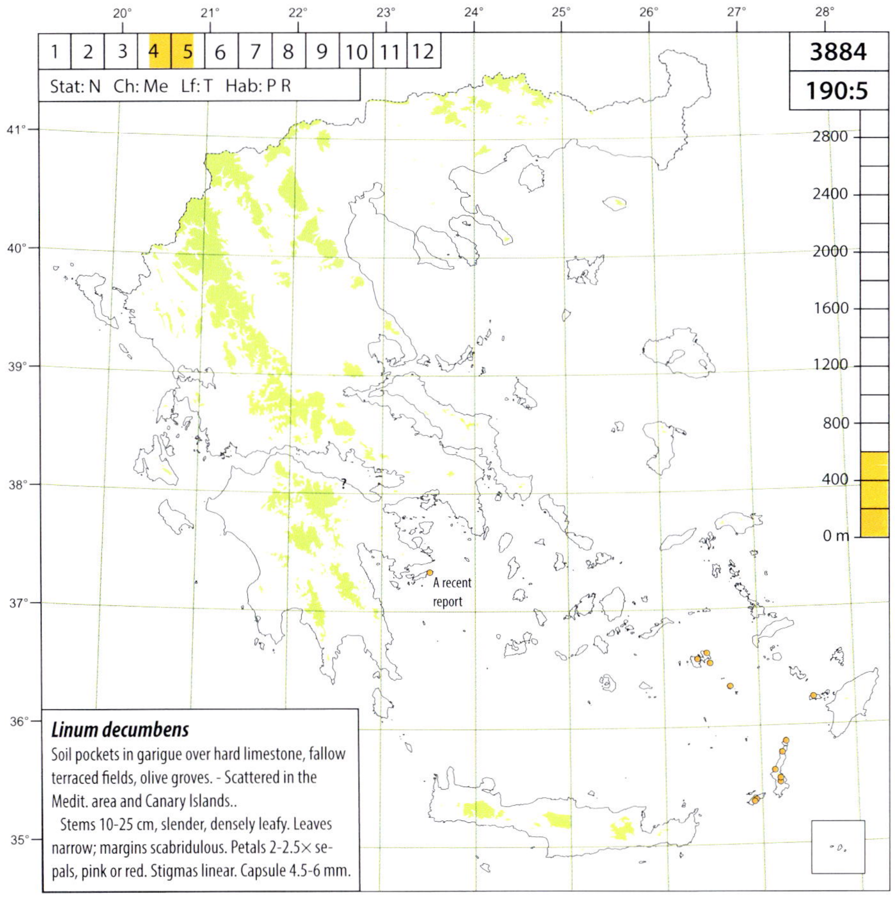

Linum decumbens
Soil pockets in garigue over hard limestone, fallow terraced fields, olive groves. - Scattered in the Medit. area and Canary Islands..
Stems 10-25 cm, slender, densely leafy. Leaves narrow; margins scabridulous. Petals 2-2.5× sepals, pink or red. Stigmas linear. Capsule 4.5-6 mm.

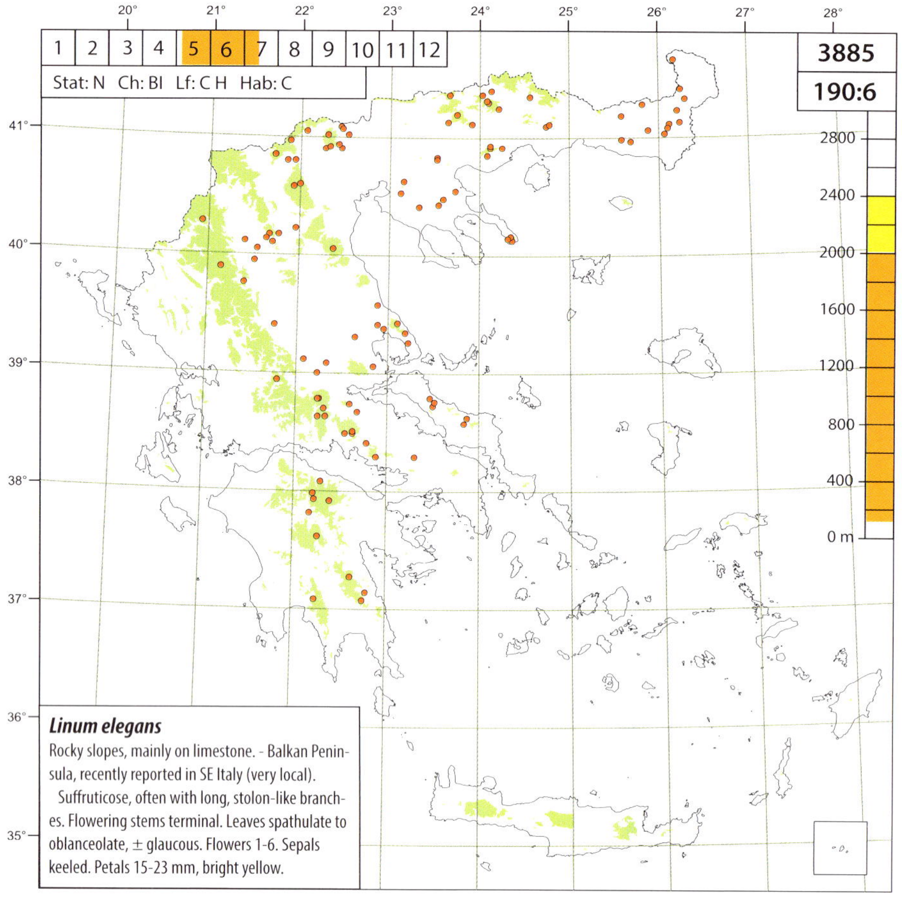

Linum elegans

Rocky slopes, mainly on limestone. - Balkan Peninsula, recently reported in SE Italy (very local).

Suffruticose, often with long, stolon-like branches. Flowering stems terminal. Leaves spathulate to oblanceolate, ± glaucous. Flowers 1-6. Sepals keeled. Petals 15-23 mm, bright yellow.

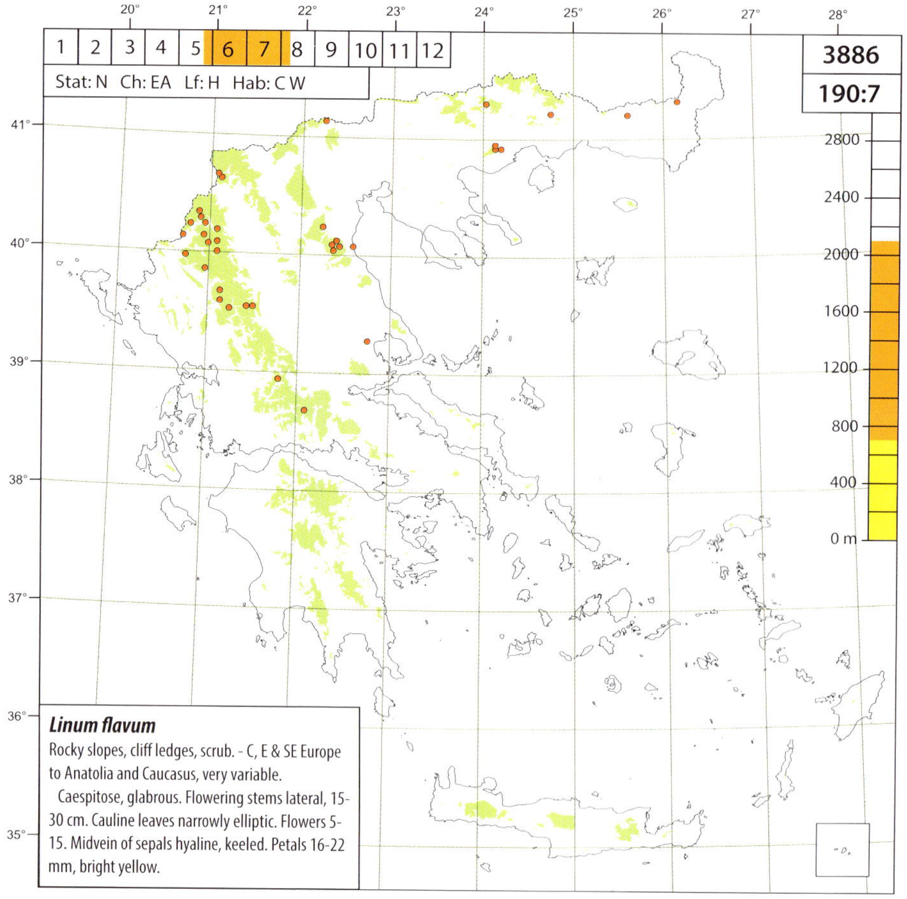

Linum flavum

Rocky slopes, cliff ledges, scrub. - C, E & SE Europe to Anatolia and Caucasus, very variable.

Caespitose, glabrous. Flowering stems lateral, 15-30 cm. Cauline leaves narrowly elliptic. Flowers 5-15. Midvein of sepals hyaline, keeled. Petals 16-22 mm, bright yellow.

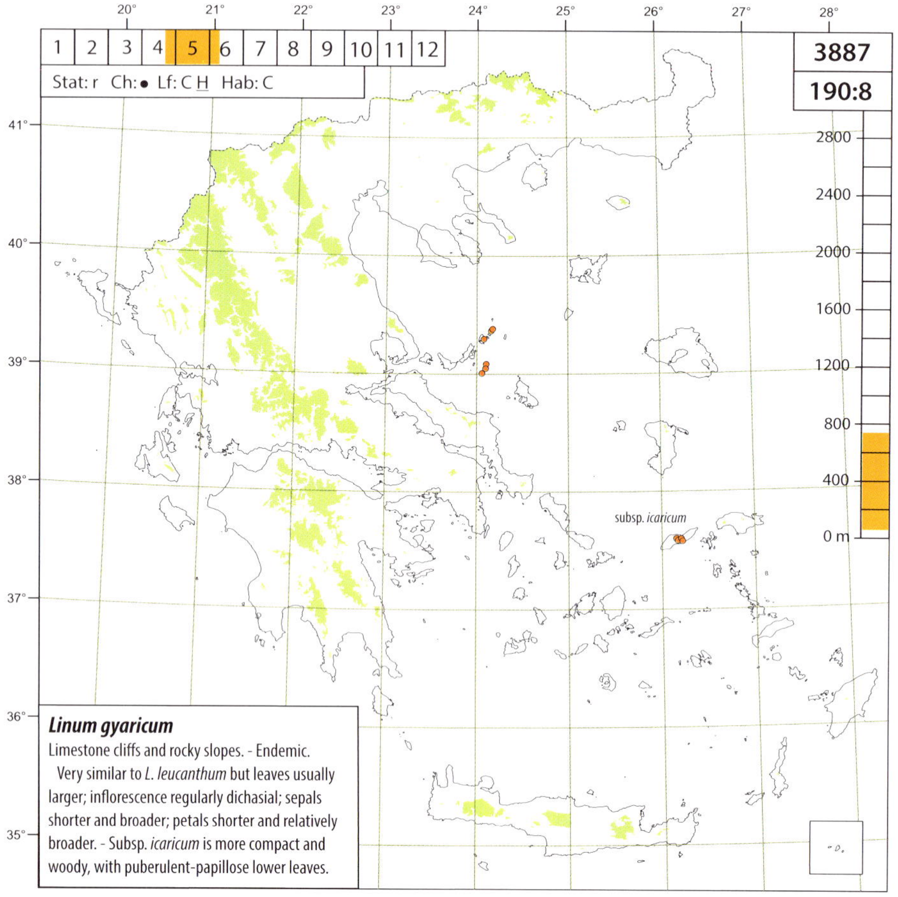

Linum gyaricum

Limestone cliffs and rocky slopes. - Endemic.

Very similar to *L. leucanthum* but leaves usually larger; inflorescence regularly dichasial; sepals shorter and broader; petals shorter and relatively broader. - Subsp. *icaricum* is more compact and woody, with puberulent-papillose lower leaves.

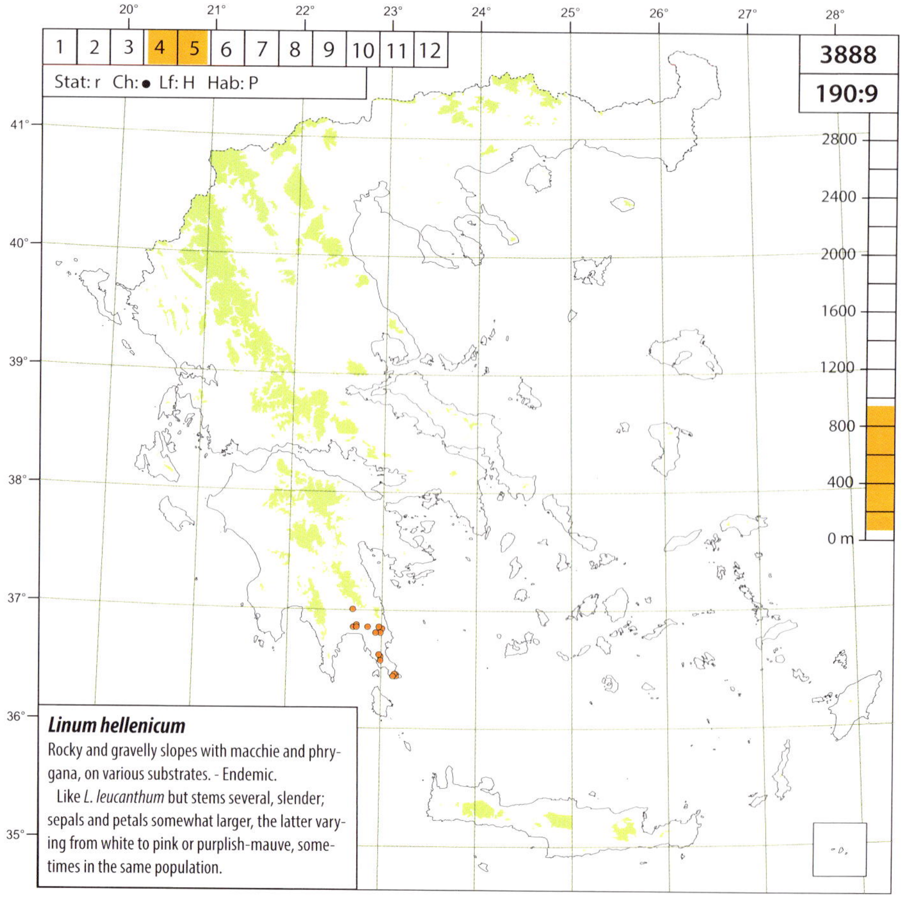

Linum hellenicum

Rocky and gravelly slopes with macchie and phrygana, on various substrates. - Endemic.

Like *L. leucanthum* but stems several, slender; sepals and petals somewhat larger, the latter varying from white to pink or purplish-mauve, sometimes in the same population.

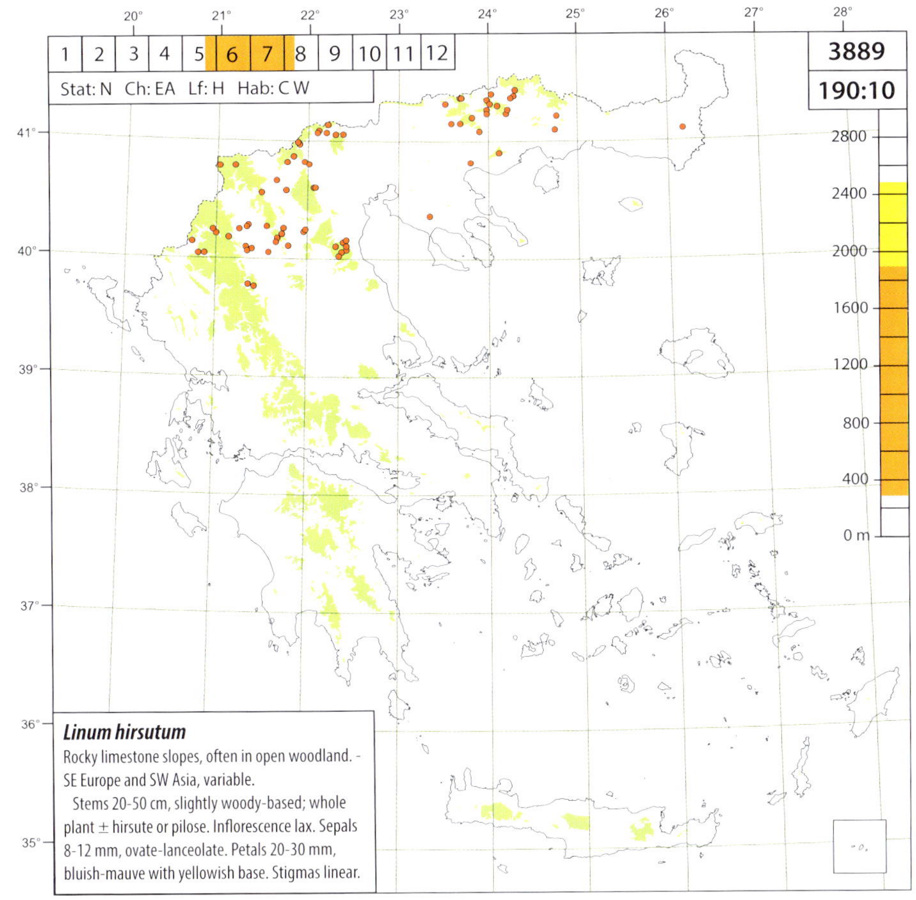

Linum hirsutum
Rocky limestone slopes, often in open woodland. - SE Europe and SW Asia, variable.
Stems 20-50 cm, slightly woody-based; whole plant ± hirsute or pilose. Inflorescence lax. Sepals 8-12 mm, ovate-lanceolate. Petals 20-30 mm, bluish-mauve with yellowish base. Stigmas linear.

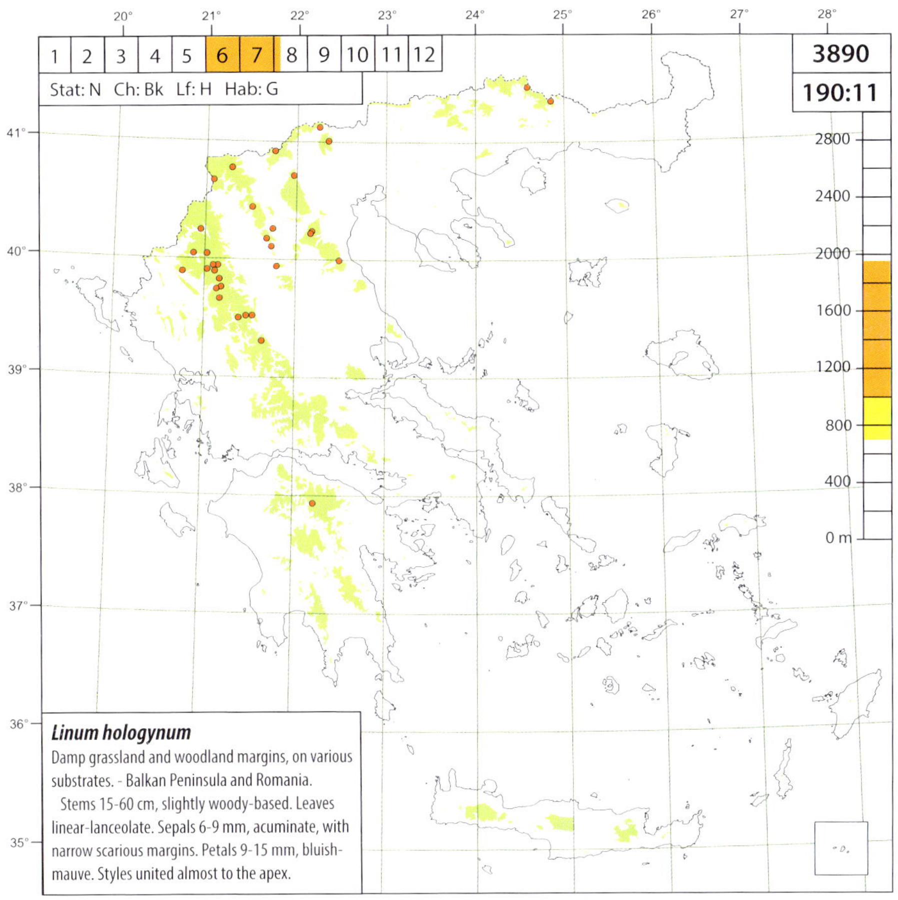

Linum hologynum
Damp grassland and woodland margins, on various substrates. - Balkan Peninsula and Romania.
Stems 15-60 cm, slightly woody-based. Leaves linear-lanceolate. Sepals 6-9 mm, acuminate, with narrow scarious margins. Petals 9-15 mm, bluish-mauve. Styles united almost to the apex.

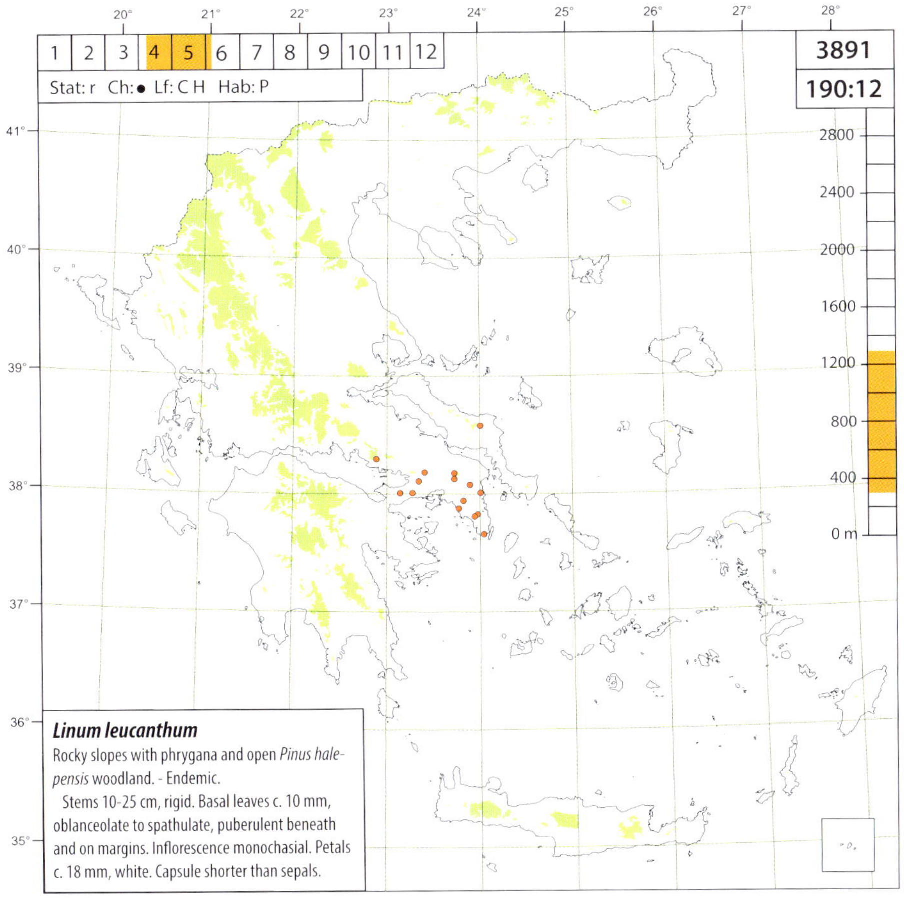

Linum leucanthum
Rocky slopes with phrygana and open *Pinus halepensis* woodland. - Endemic.
Stems 10-25 cm, rigid. Basal leaves c. 10 mm, oblanceolate to spathulate, puberulent beneath and on margins. Inflorescence monochasial. Petals c. 18 mm, white. Capsule shorter than sepals.

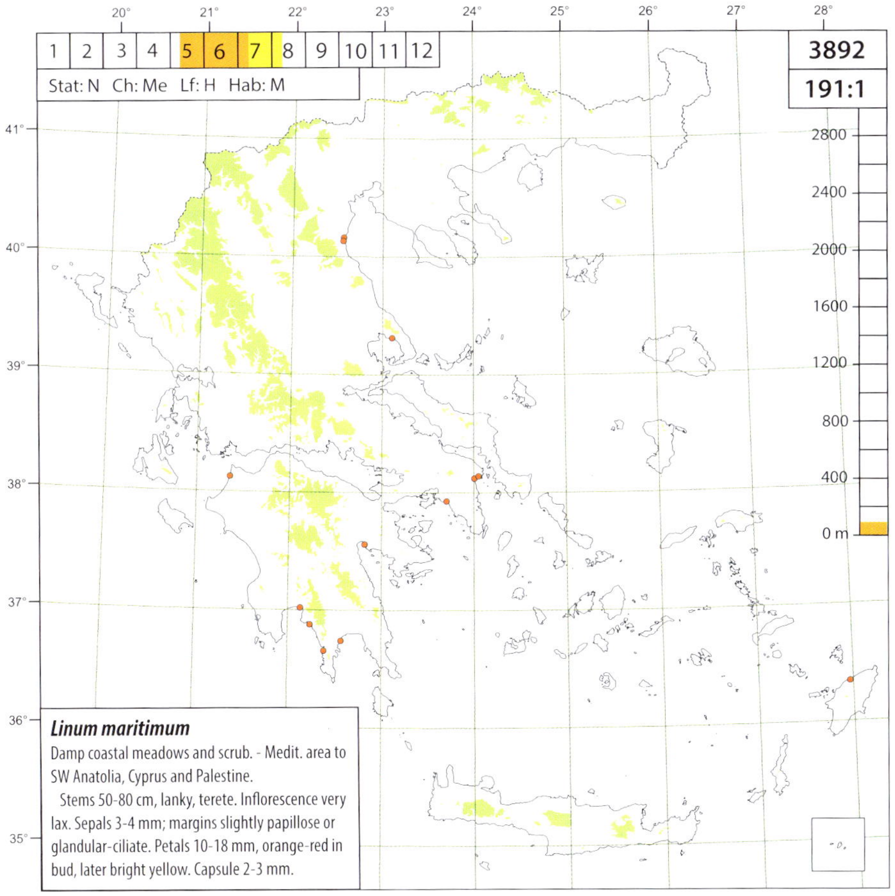

Linum maritimum
Damp coastal meadows and scrub. - Medit. area to SW Anatolia, Cyprus and Palestine.
Stems 50-80 cm, lanky, terete. Inflorescence very lax. Sepals 3-4 mm; margins slightly papillose or glandular-ciliate. Petals 10-18 mm, orange-red in bud, later bright yellow. Capsule 2-3 mm.

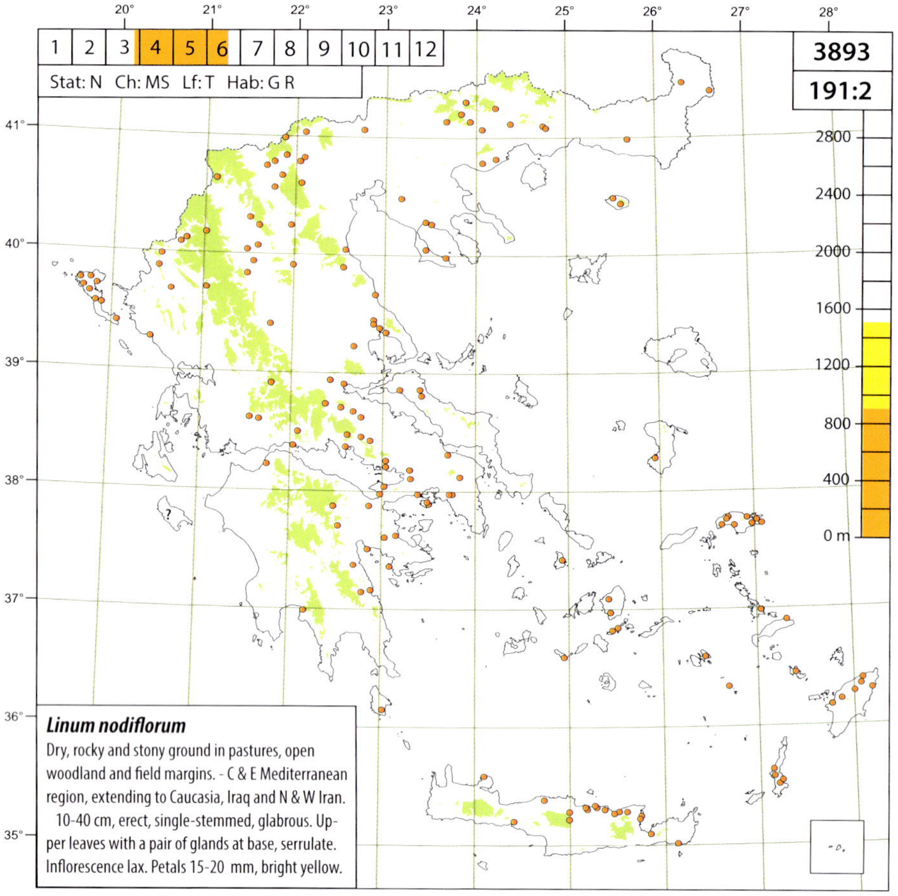

Linum nodiflorum
Dry, rocky and stony ground in pastures, open woodland and field margins. - C & E Mediterranean region, extending to Caucasia, Iraq and N & W Iran.
10-40 cm, erect, single-stemmed, glabrous. Upper leaves with a pair of glands at base, serrulate. Inflorescence lax. Petals 15-20 mm, bright yellow.

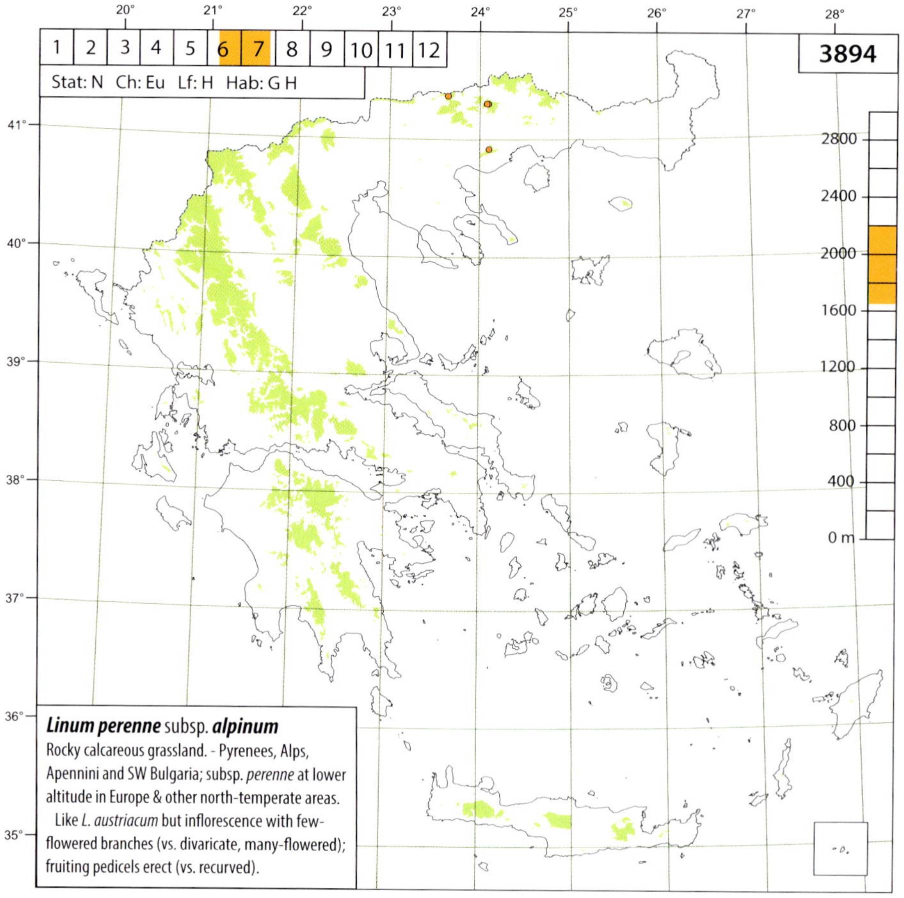

Linum perenne* subsp. *alpinum
Rocky calcareous grassland. - Pyrenees, Alps, Apennini and SW Bulgaria; subsp. *perenne* at lower altitude in Europe & other north-temperate areas.
Like *L. austriacum* but inflorescence with few-flowered branches (vs. divaricate, many-flowered); fruiting pedicels erect (vs. recurved).

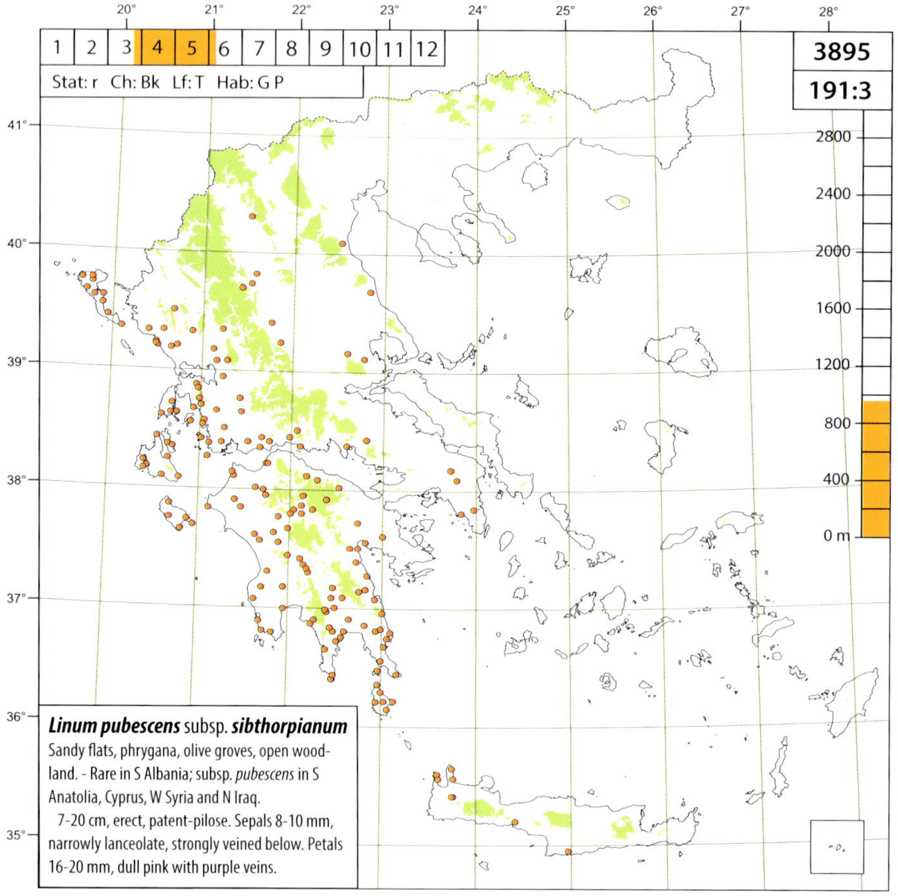

Linum pubescens* subsp. *sibthorpianum
Sandy flats, phrygana, olive groves, open woodland. - Rare in S Albania; subsp. *pubescens* in S Anatolia, Cyprus, W Syria and N Iraq.
7-20 cm, erect, patent-pilose. Sepals 8-10 mm, narrowly lanceolate, strongly veined below. Petals 16-20 mm, dull pink with purple veins.

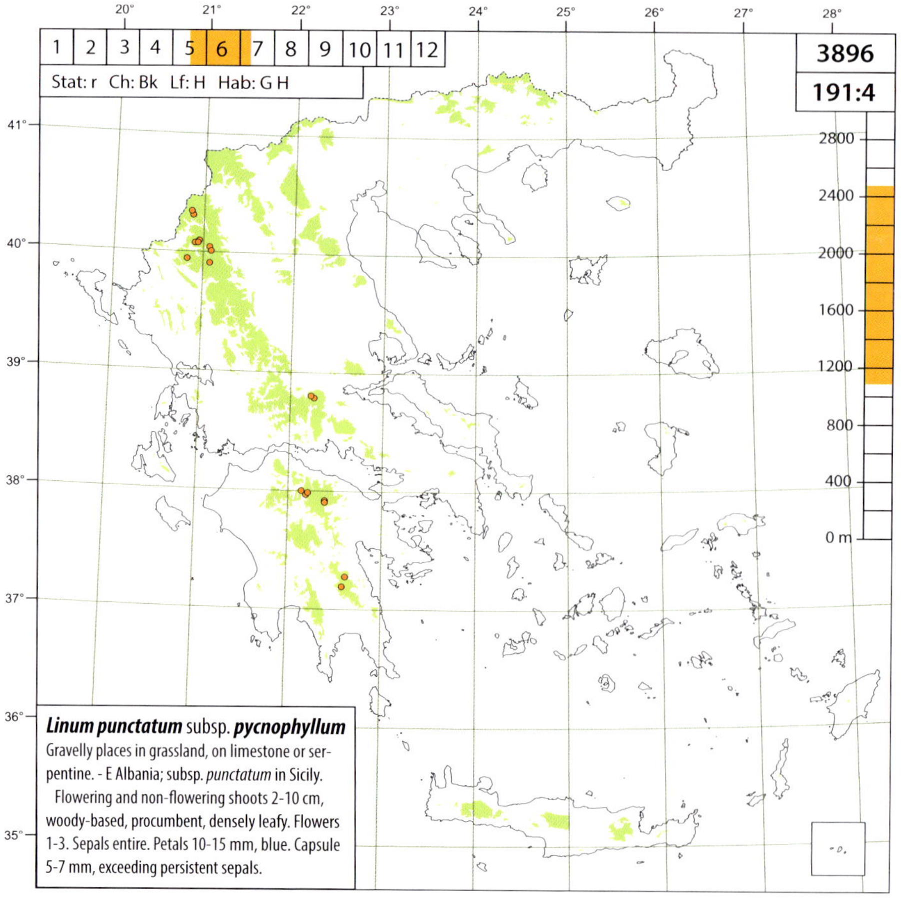

Linum punctatum* subsp. *pycnophyllum
Gravelly places in grassland, on limestone or serpentine. - E Albania; subsp. *punctatum* in Sicily.
Flowering and non-flowering shoots 2-10 cm, woody-based, procumbent, densely leafy. Flowers 1-3. Sepals entire. Petals 10-15 mm, blue. Capsule 5-7 mm, exceeding persistent sepals.

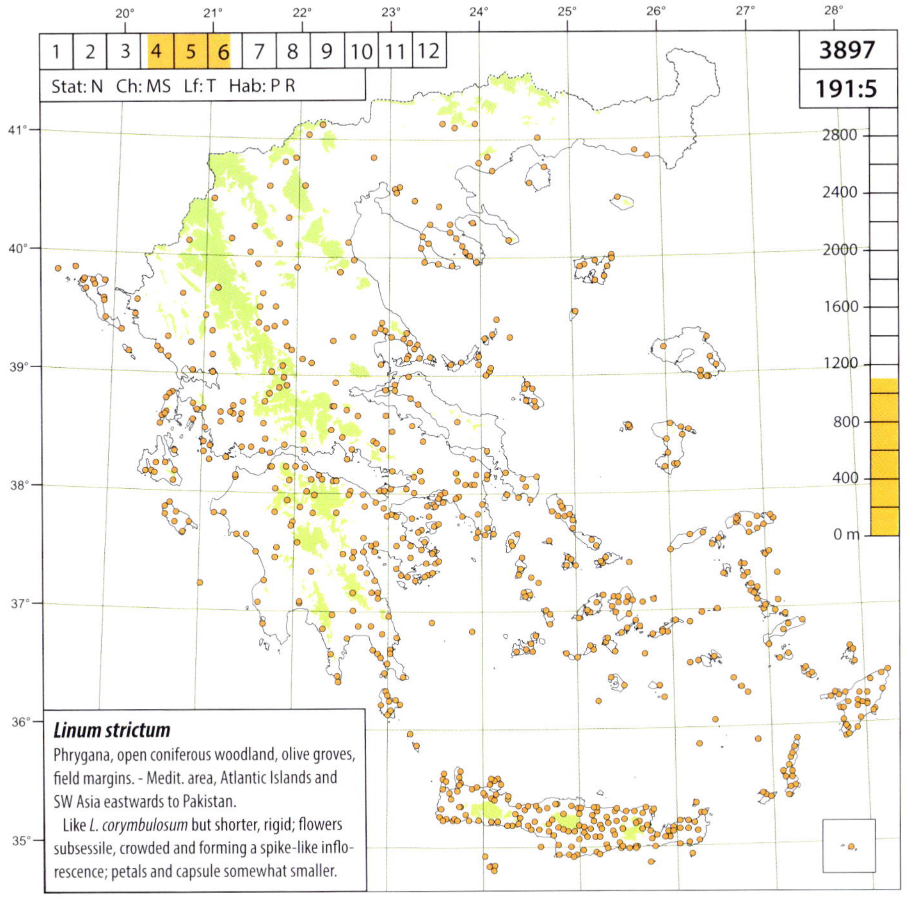

Linum strictum
Phrygana, open coniferous woodland, olive groves, field margins. - Medit. area, Atlantic Islands and SW Asia eastwards to Pakistan.
Like *L. corymbulosum* but shorter, rigid; flowers subsessile, crowded and forming a spike-like inflorescence; petals and capsule somewhat smaller.

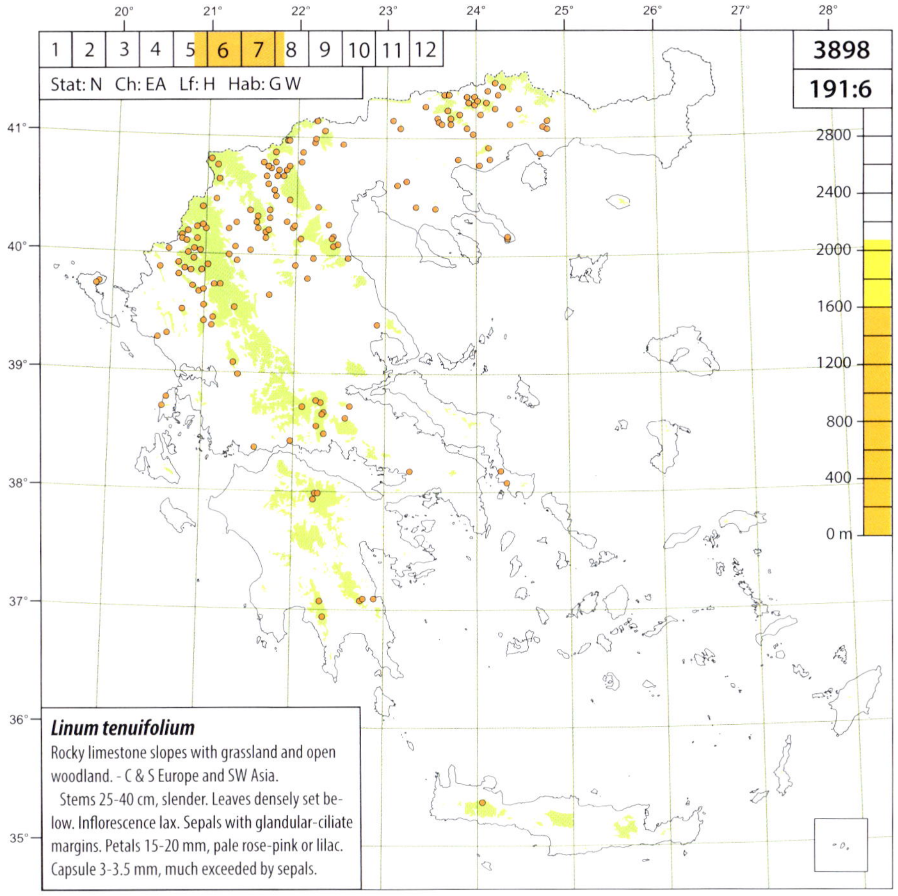

Linum tenuifolium
Rocky limestone slopes with grassland and open woodland. - C & S Europe and SW Asia.
Stems 25-40 cm, slender. Leaves densely set below. Inflorescence lax. Sepals with glandular-ciliate margins. Petals 15-20 mm, pale rose-pink or lilac. Capsule 3-3.5 mm, much exceeded by sepals.

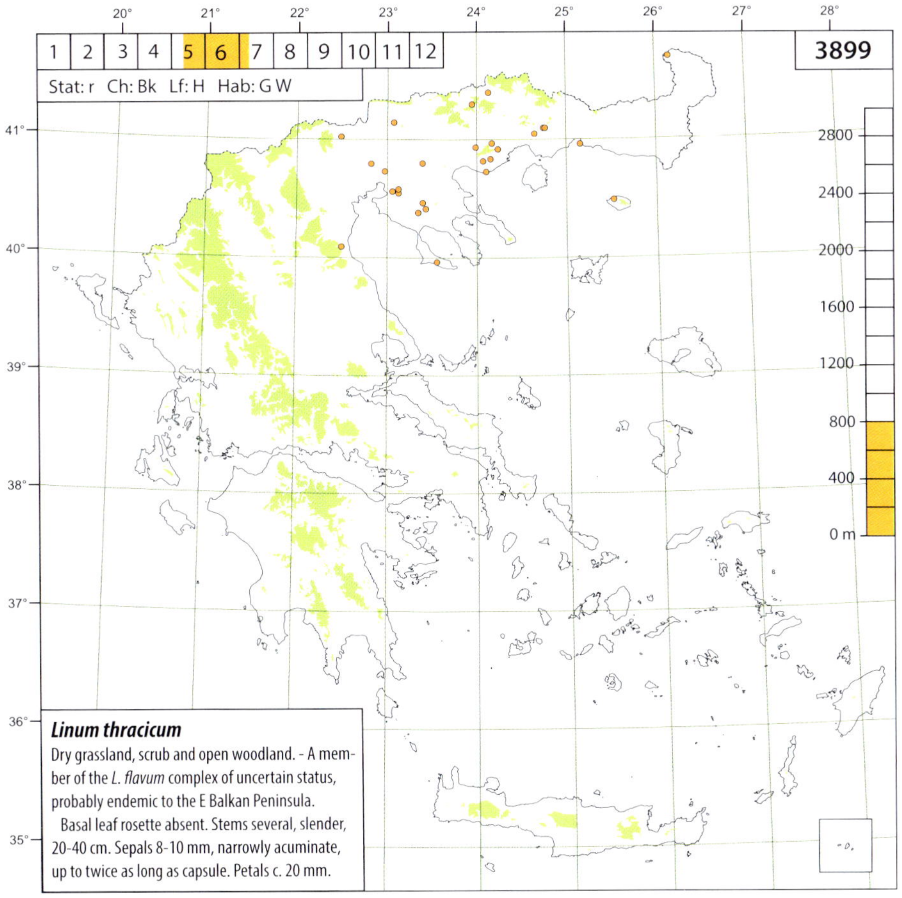

Linum thracicum
Dry grassland, scrub and open woodland. - A member of the *L. flavum* complex of uncertain status, probably endemic to the E Balkan Peninsula.
Basal leaf rosette absent. Stems several, slender, 20-40 cm. Sepals 8-10 mm, narrowly acuminate, up to twice as long as capsule. Petals c. 20 mm.

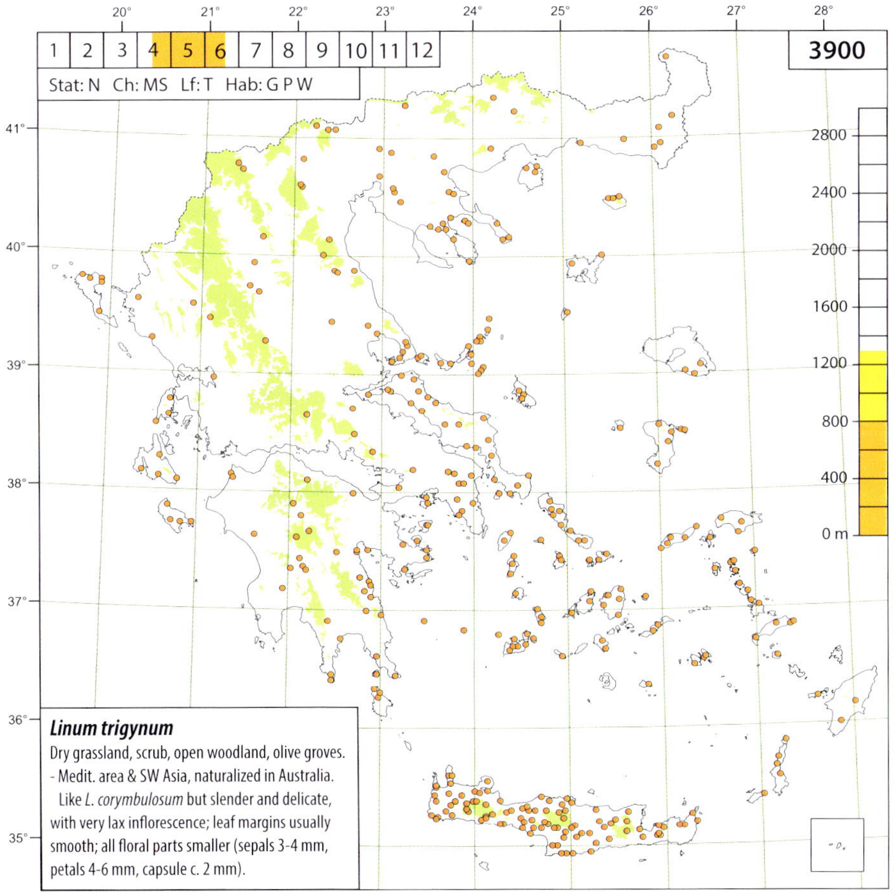

Linum trigynum
Dry grassland, scrub, open woodland, olive groves. - Medit. area & SW Asia, naturalized in Australia.
Like *L. corymbulosum* but slender and delicate, with very lax inflorescence; leaf margins usually smooth; all floral parts smaller (sepals 3-4 mm, petals 4-6 mm, capsule c. 2 mm).

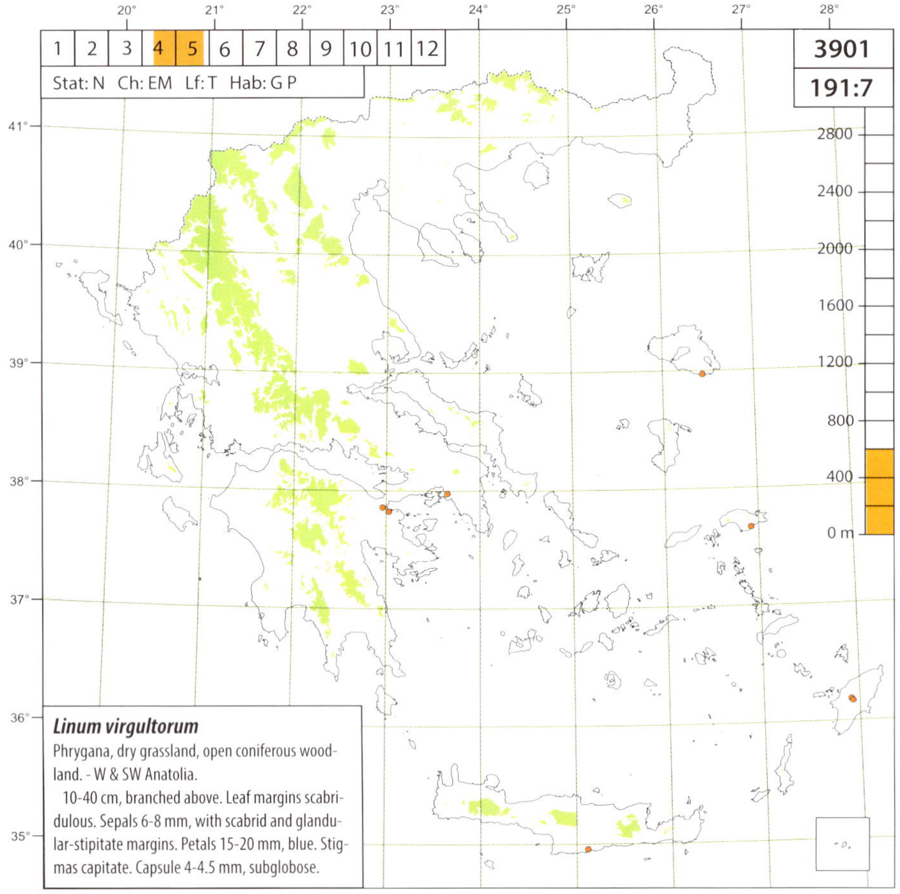

Linum virgultorum

Phrygana, dry grassland, open coniferous woodland. - W & SW Anatolia.

10-40 cm, branched above. Leaf margins scabridulous. Sepals 6-8 mm, with scabrid and glandular-stipitate margins. Petals 15-20 mm, blue. Stigmas capitate. Capsule 4-4.5 mm, subglobose.

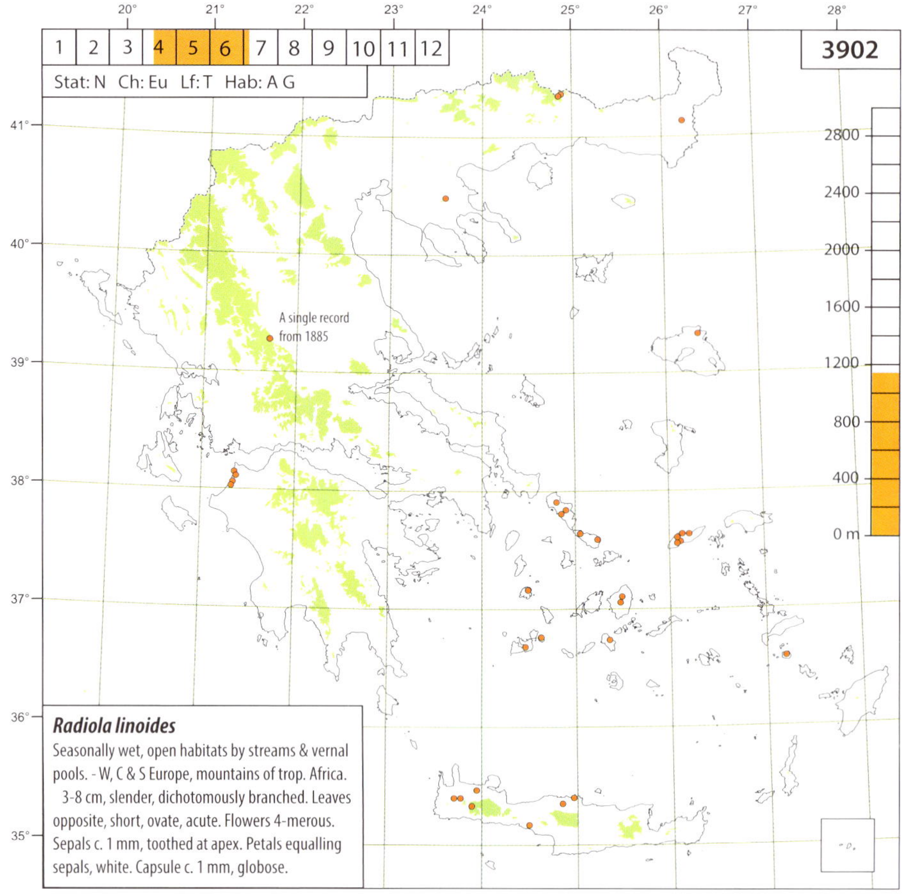

Radiola linoides

Seasonally wet, open habitats by streams & vernal pools. - W, C & S Europe, mountains of trop. Africa.

3-8 cm, slender, dichotomously branched. Leaves opposite, short, ovate, acute. Flowers 4-merous. Sepals c. 1 mm, toothed at apex. Petals equalling sepals, white. Capsule c. 1 mm, globose.

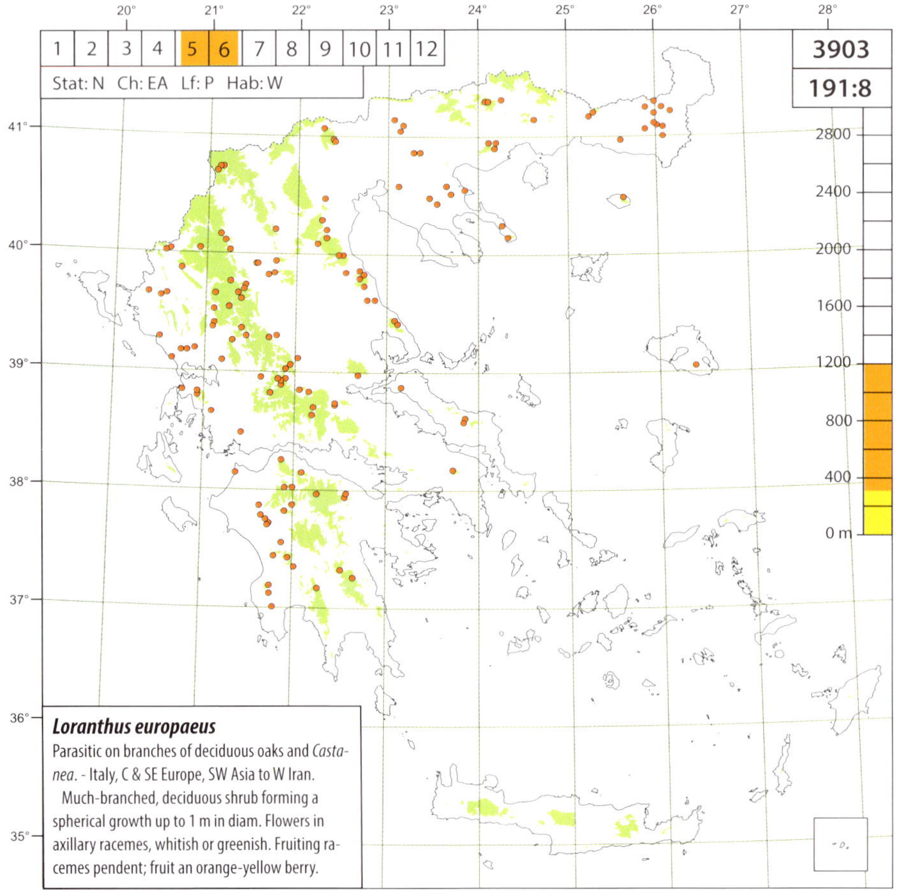

Loranthus europaeus

Parasitic on branches of deciduous oaks and *Castanea*. - Italy, C & SE Europe, SW Asia to W Iran.

Much-branched, deciduous shrub forming a spherical growth up to 1 m in diam. Flowers in axillary racemes, whitish or greenish. Fruiting racemes pendent; fruit an orange-yellow berry.

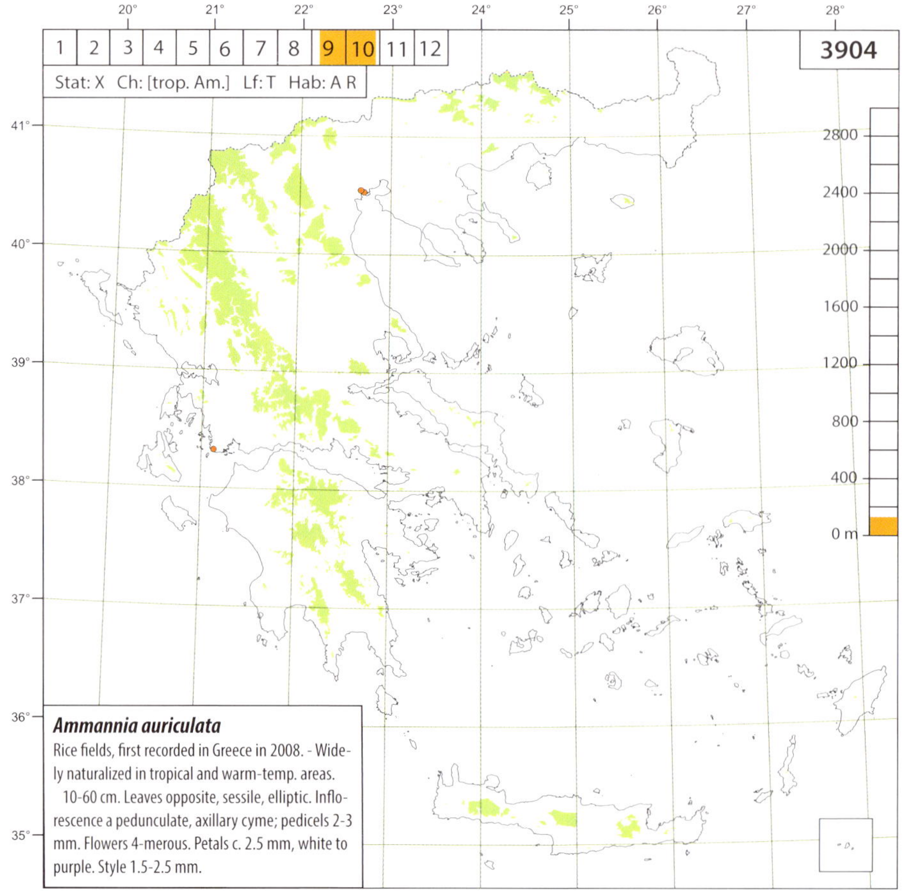

Ammannia auriculata

Rice fields, first recorded in Greece in 2008. - Widely naturalized in tropical and warm-temp. areas.

10-60 cm. Leaves opposite, sessile, elliptic. Inflorescence a pedunculate, axillary cyme; pedicels 2-3 mm. Flowers 4-merous. Petals c. 2.5 mm, white to purple. Style 1.5-2.5 mm.

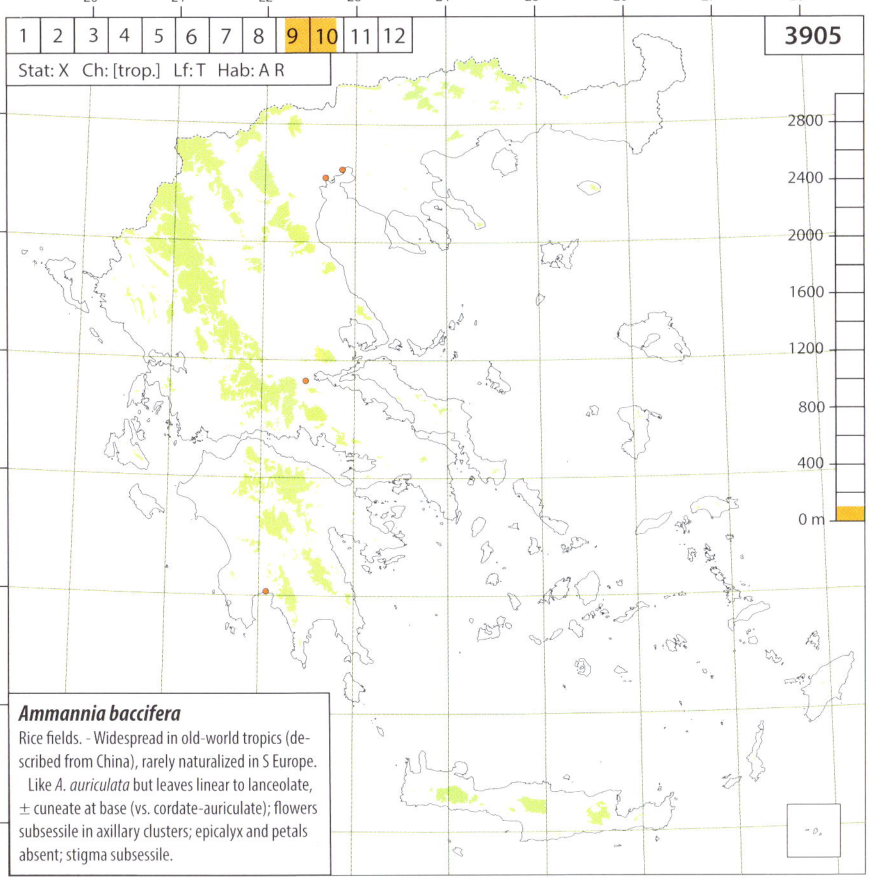

Ammannia baccifera

Rice fields. - Widespread in old-world tropics (described from China), rarely naturalized in S Europe.

Like *A. auriculata* but leaves linear to lanceolate, ± cuneate at base (vs. cordate-auriculate); flowers subsessile in axillary clusters; epicalyx and petals absent; stigma subsessile.

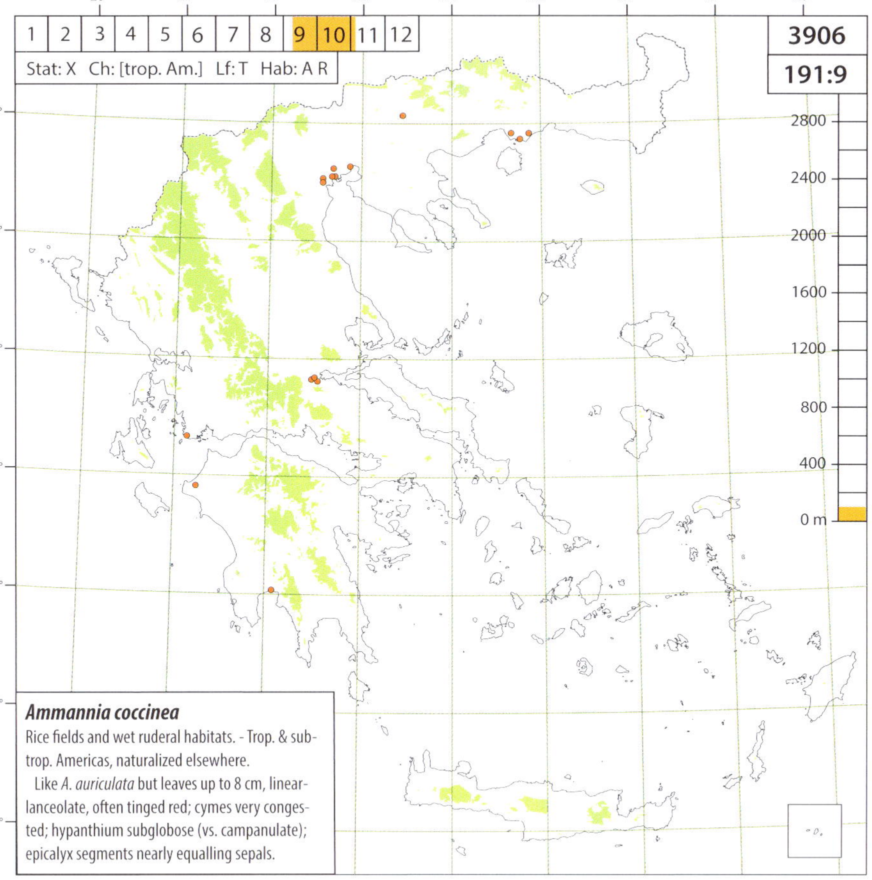

Ammannia coccinea

Rice fields and wet ruderal habitats. - Trop. & subtrop. Americas, naturalized elsewhere.

Like *A. auriculata* but leaves up to 8 cm, linear-lanceolate, often tinged red; cymes very congested; hypanthium subglobose (vs. campanulate); epicalyx segments nearly equalling sepals.

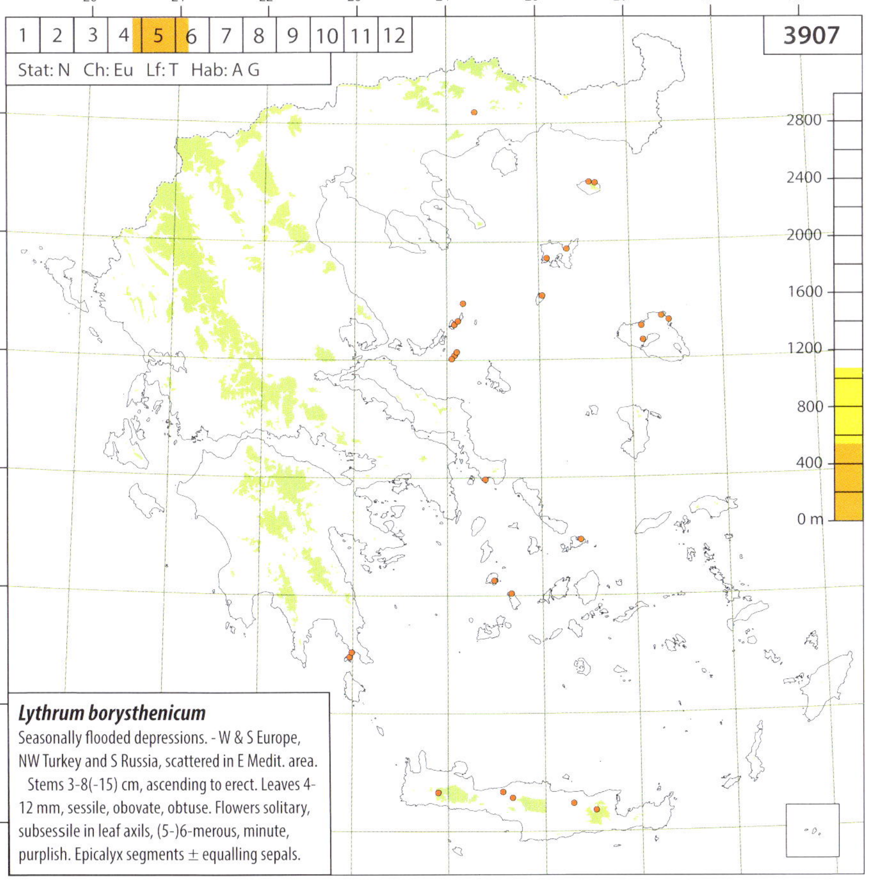

Lythrum borysthenicum

Seasonally flooded depressions. - W & S Europe, NW Turkey and S Russia, scattered in E Medit. area.

Stems 3-8(-15) cm, ascending to erect. Leaves 4-12 mm, sessile, obovate, obtuse. Flowers solitary, subsessile in leaf axils, (5-)6-merous, minute, purplish. Epicalyx segments ± equalling sepals.

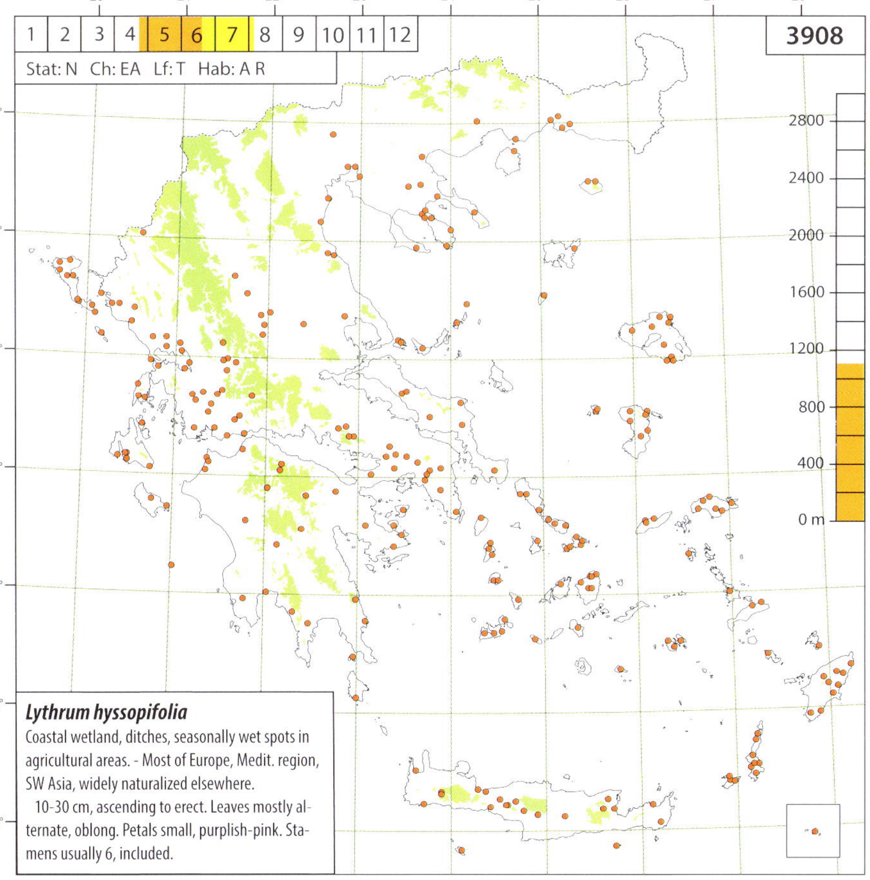

Lythrum hyssopifolia

Coastal wetland, ditches, seasonally wet spots in agricultural areas. - Most of Europe, Medit. region, SW Asia, widely naturalized elsewhere.

10-30 cm, ascending to erect. Leaves mostly alternate, oblong. Petals small, purplish-pink. Stamens usually 6, included.

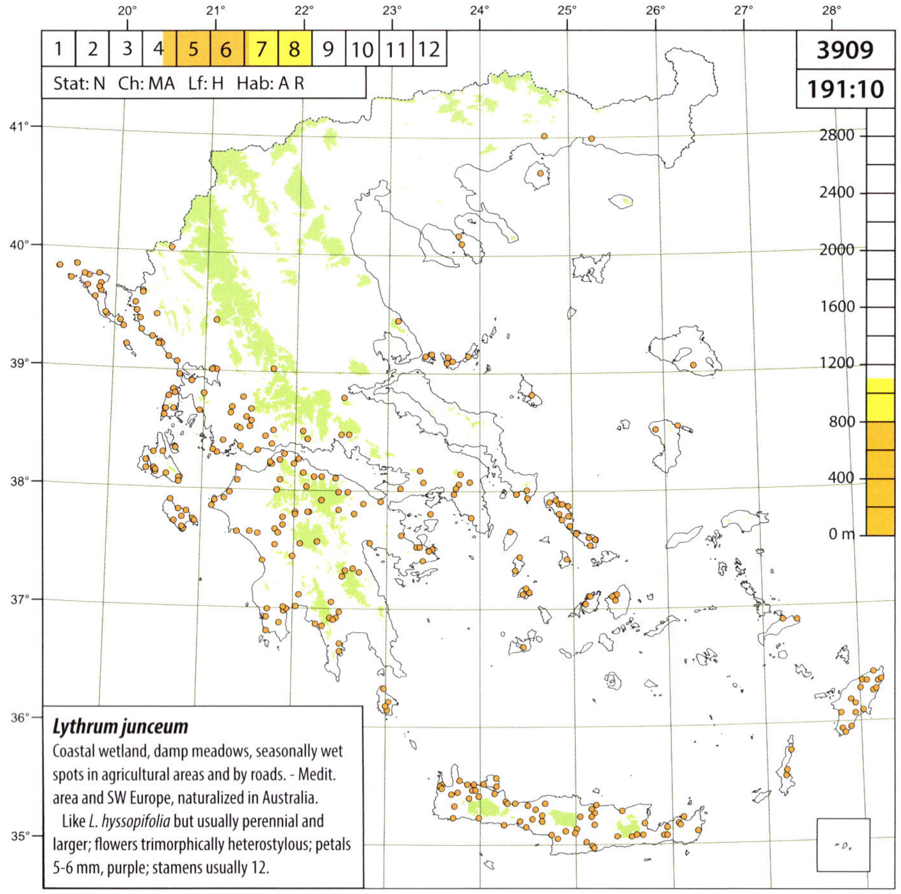

Lythrum junceum
Coastal wetland, damp meadows, seasonally wet spots in agricultural areas and by roads. - Medit. area and SW Europe, naturalized in Australia.
Like *L. hyssopifolia* but usually perennial and larger; flowers trimorphically heterostylous; petals 5-6 mm, purple; stamens usually 12.

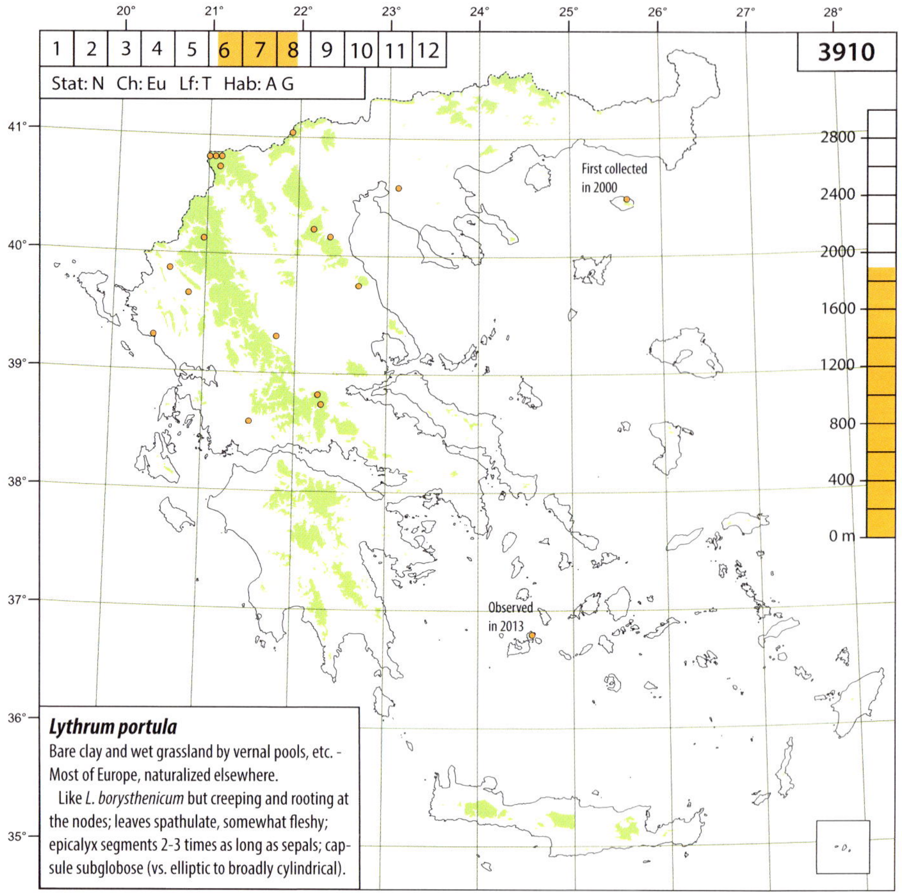

Lythrum portula
Bare clay and wet grassland by vernal pools, etc. - Most of Europe, naturalized elsewhere.
Like *L. borysthenicum* but creeping and rooting at the nodes; leaves spathulate, somewhat fleshy; epicalyx segments 2-3 times as long as sepals; capsule subglobose (vs. elliptic to broadly cylindrical).

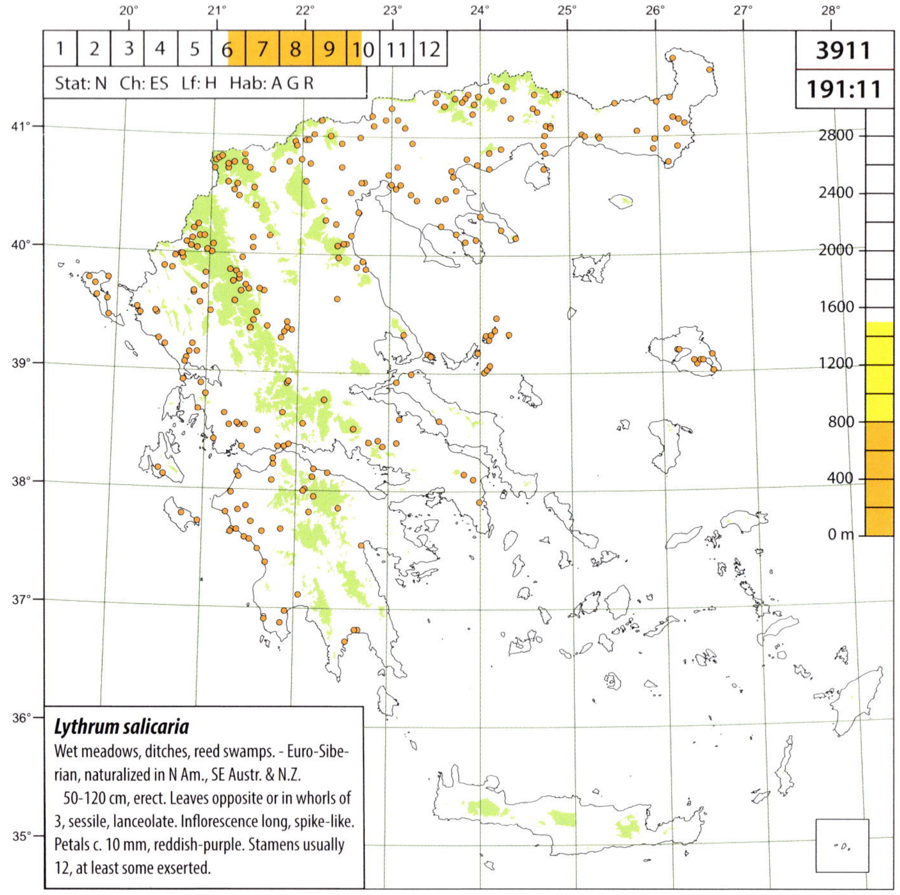

Lythrum salicaria
Wet meadows, ditches, reed swamps. - Euro-Siberian, naturalized in N Am., SE Austr. & N.Z.
50-120 cm, erect. Leaves opposite or in whorls of 3, sessile, lanceolate. Inflorescence long, spike-like. Petals c. 10 mm, reddish-purple. Stamens usually 12, at least some exserted.

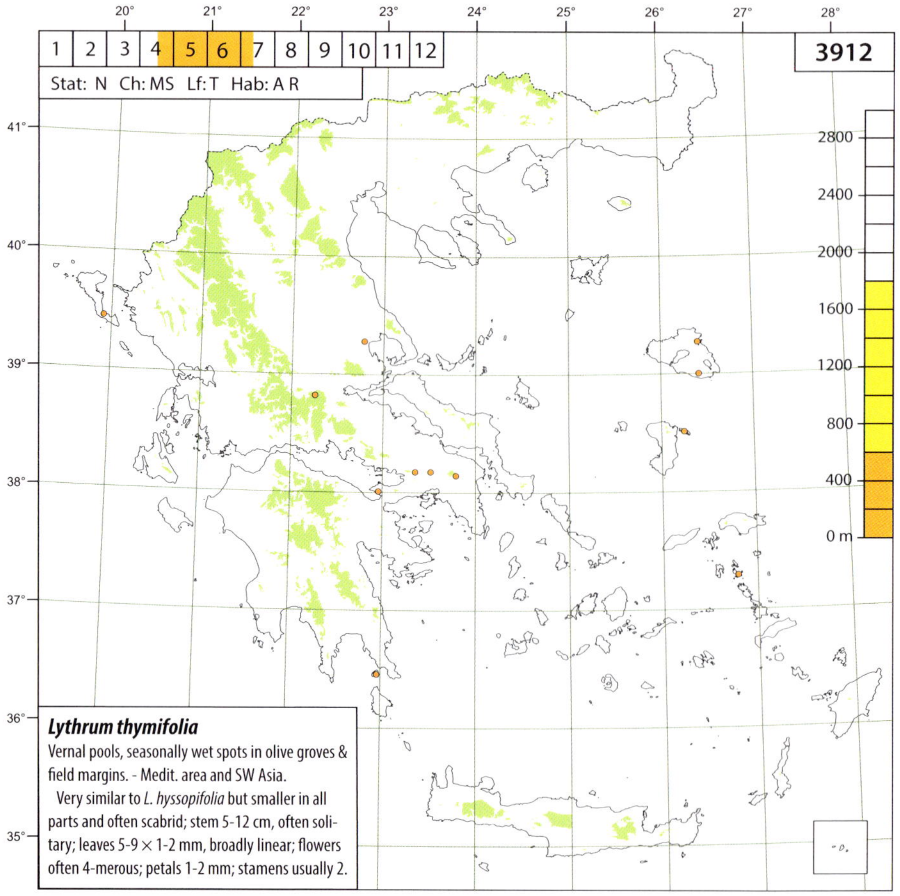

Lythrum thymifolia
Vernal pools, seasonally wet spots in olive groves & field margins. - Medit. area and SW Asia.
Very similar to *L. hyssopifolia* but smaller in all parts and often scabrid; stem 5-12 cm, often solitary; leaves 5-9 × 1-2 mm, broadly linear; flowers often 4-merous; petals 1-2 mm; stamens usually 2.

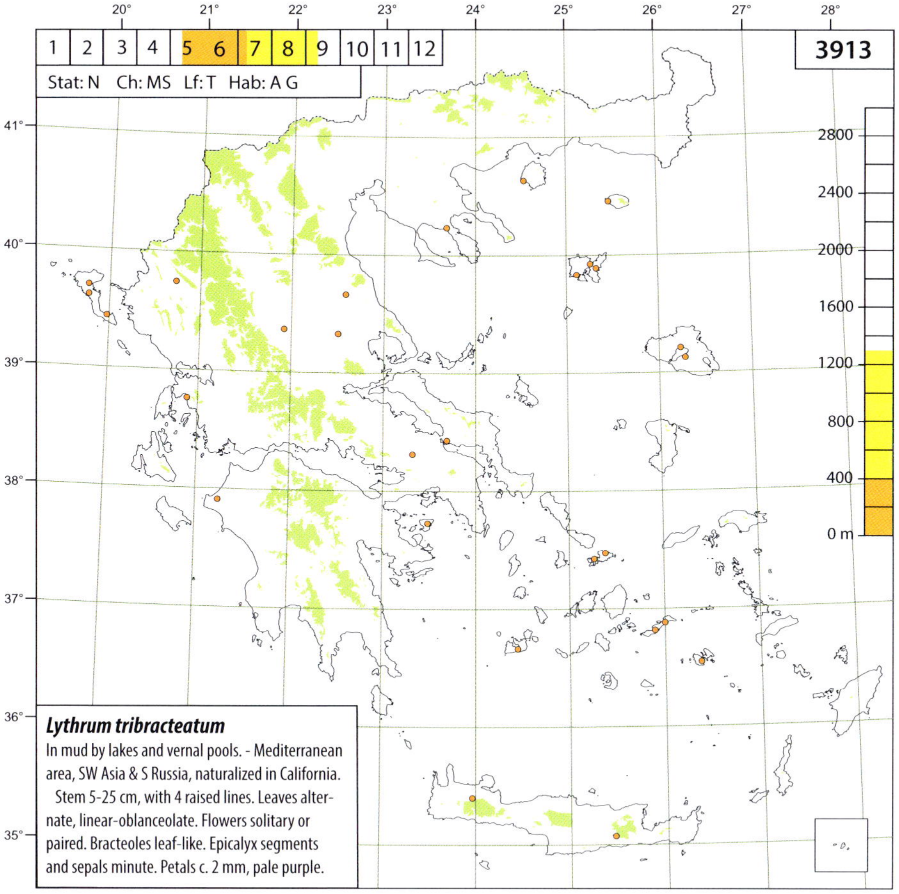

Lythrum tribracteatum

In mud by lakes and vernal pools. - Mediterranean area, SW Asia & S Russia, naturalized in California.

Stem 5-25 cm, with 4 raised lines. Leaves alternate, linear-oblanceolate. Flowers solitary or paired. Bracteoles leaf-like. Epicalyx segments and sepals minute. Petals c. 2 mm, pale purple.

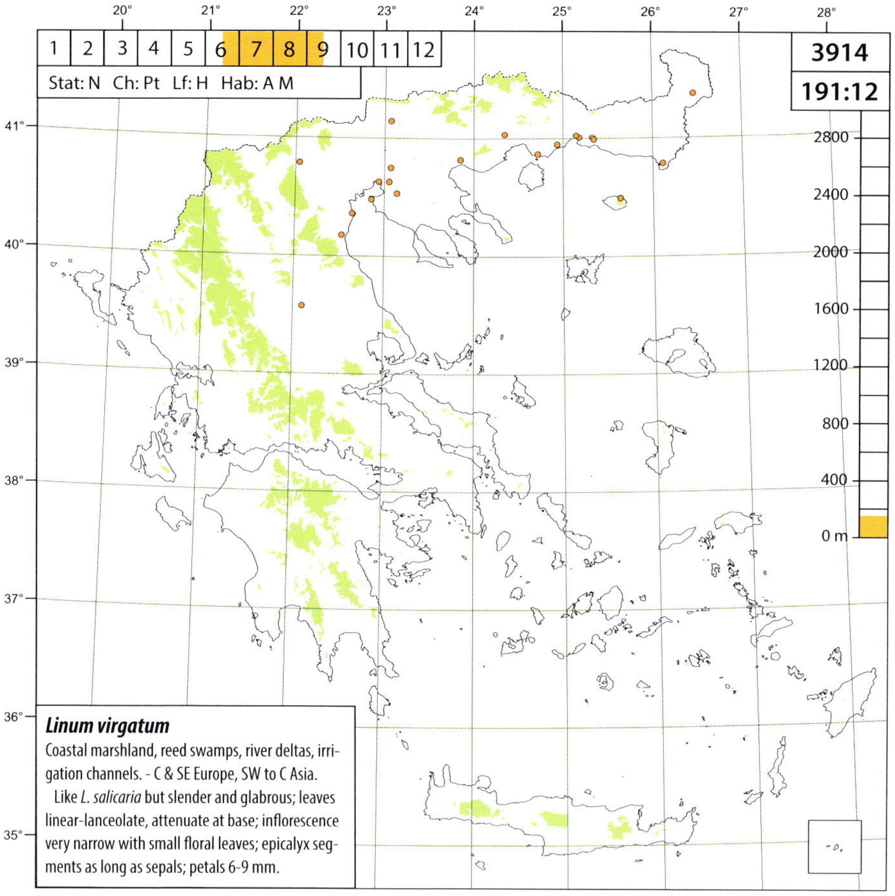

Linum virgatum

Coastal marshland, reed swamps, river deltas, irrigation channels. - C & SE Europe, SW to C Asia.

Like *L. salicaria* but slender and glabrous; leaves linear-lanceolate, attenuate at base; inflorescence very narrow with small floral leaves; epicalyx segments as long as sepals; petals 6-9 mm.

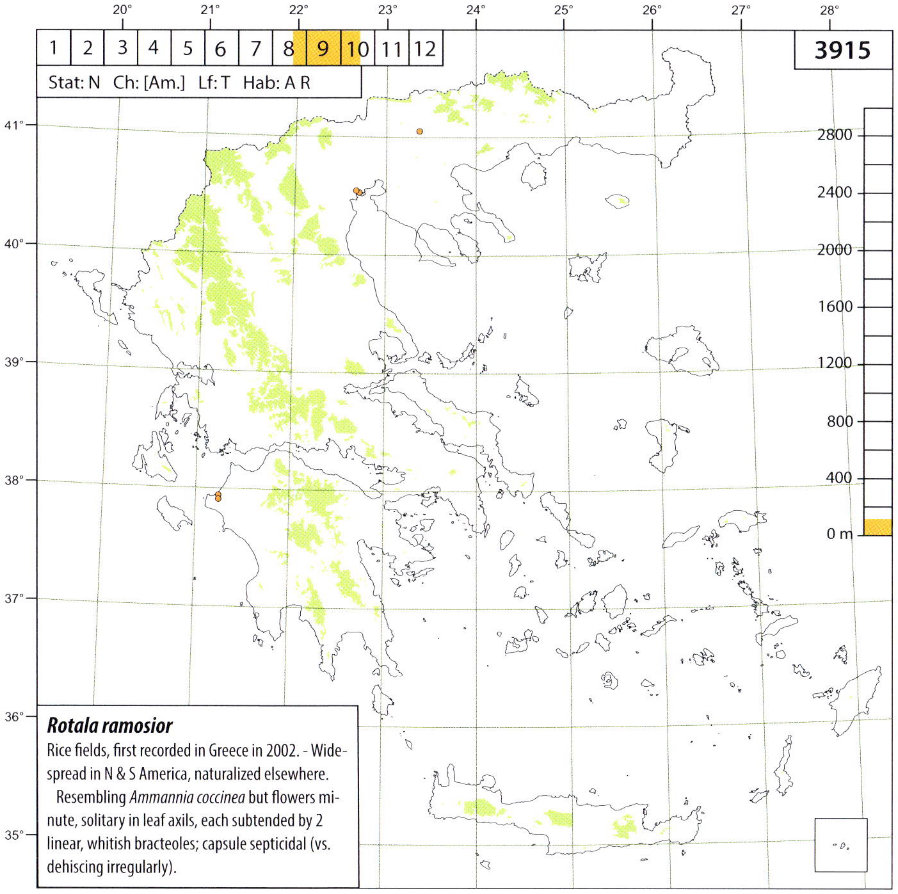

Rotala ramosior

Rice fields, first recorded in Greece in 2002. - Widespread in N & S America, naturalized elsewhere.

Resembling *Ammannia coccinea* but flowers minute, solitary in leaf axils, each subtended by 2 linear, whitish bracteoles; capsule septicidal (vs. dehiscing irregularly).

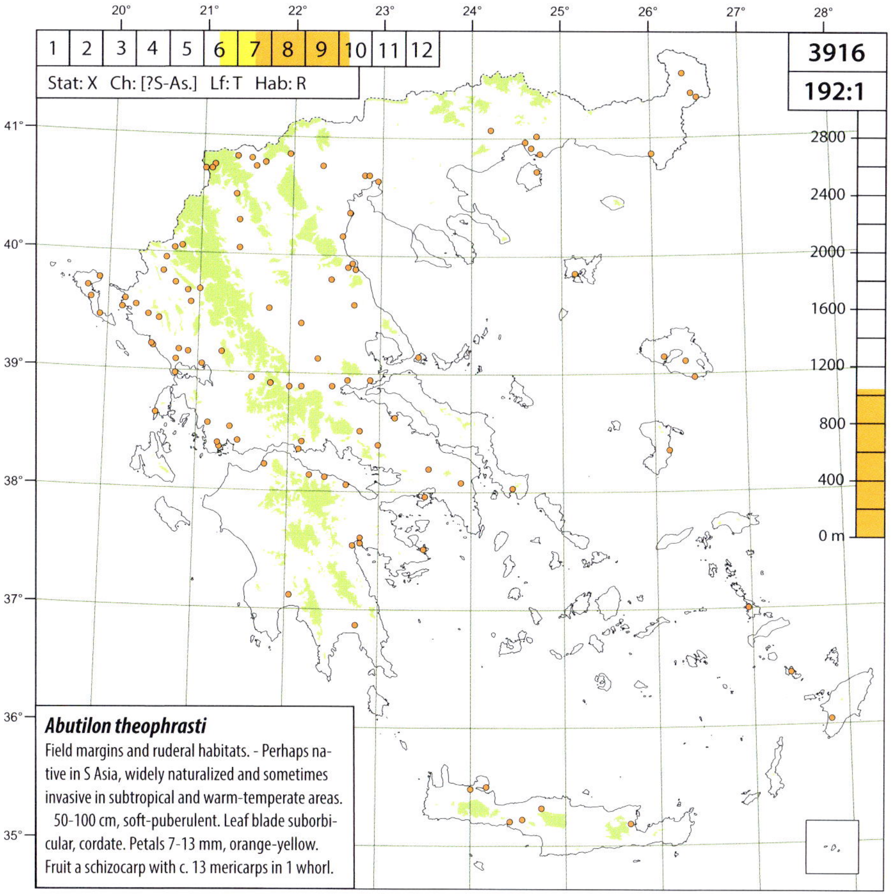

Abutilon theophrasti

Field margins and ruderal habitats. - Perhaps native in S Asia, widely naturalized and sometimes invasive in subtropical and warm-temperate areas.

50-100 cm, soft-puberulent. Leaf blade suborbicular, cordate. Petals 7-13 mm, orange-yellow. Fruit a schizocarp with c. 13 mericarps in 1 whorl.

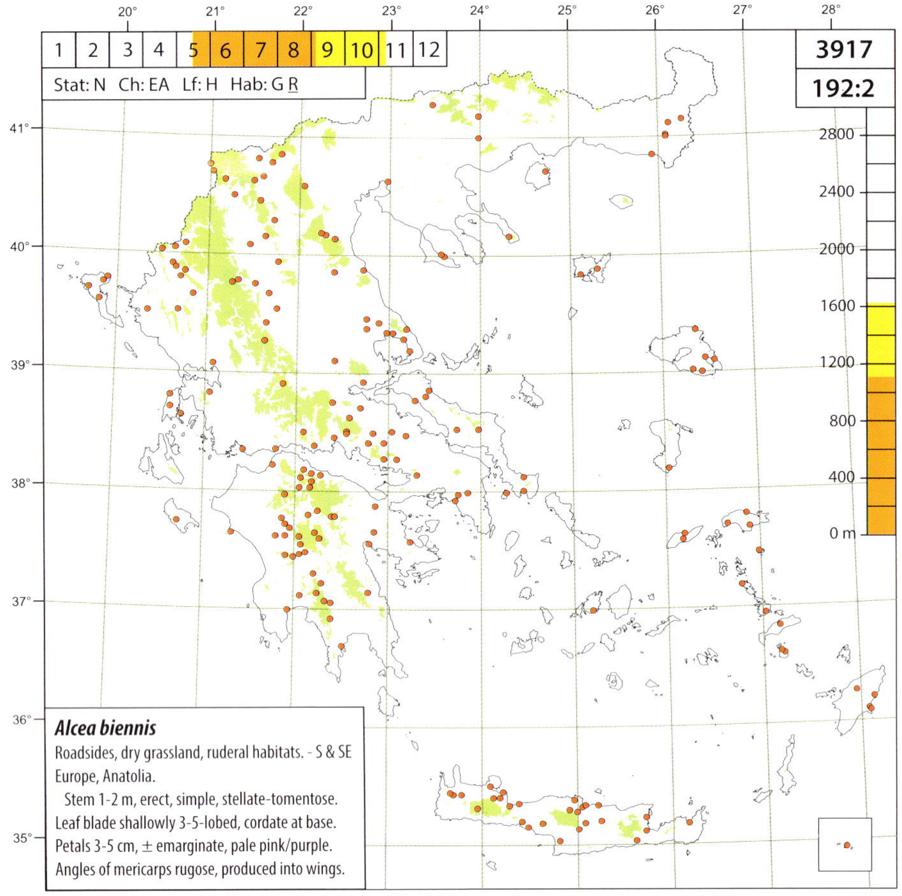

Alcea biennis
Roadsides, dry grassland, ruderal habitats. - S & SE Europe, Anatolia.
Stem 1-2 m, erect, simple, stellate-tomentose. Leaf blade shallowly 3-5-lobed, cordate at base. Petals 3-5 cm, ± emarginate, pale pink/purple. Angles of mericarps rugose, produced into wings.

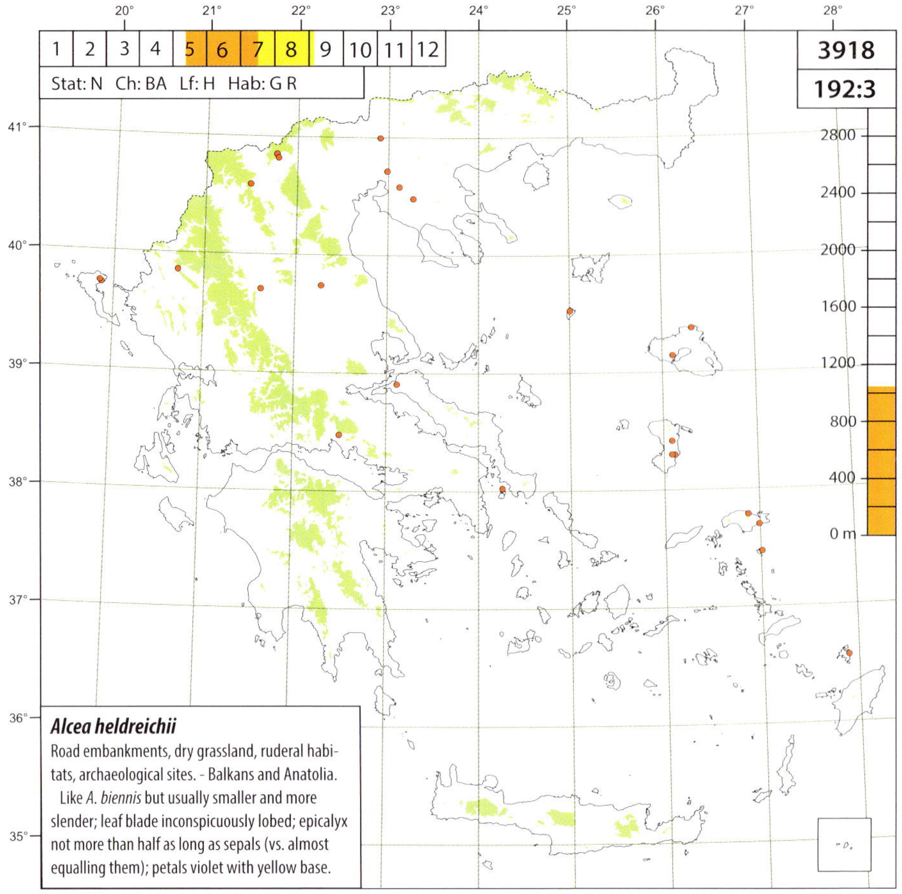

Alcea heldreichii
Road embankments, dry grassland, ruderal habitats, archaeological sites. - Balkans and Anatolia.
Like *A. biennis* but usually smaller and more slender; leaf blade inconspicuously lobed; epicalyx not more than half as long as sepals (vs. almost equalling them); petals violet with yellow base.

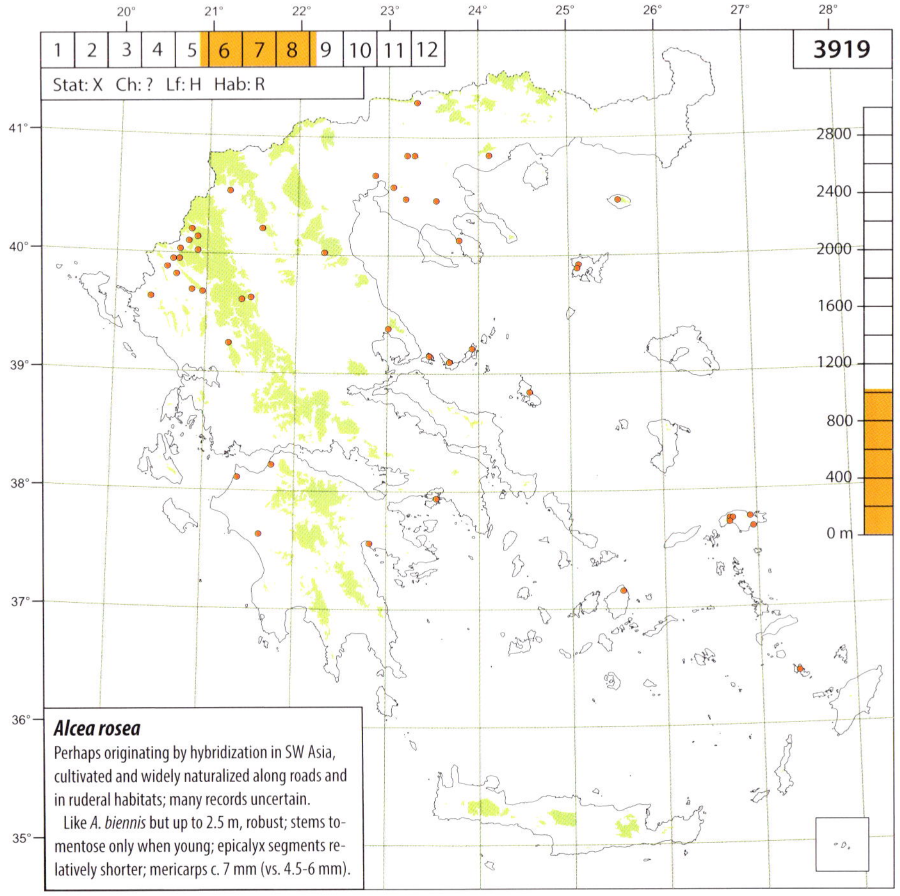

Alcea rosea
Perhaps originating by hybridization in SW Asia, cultivated and widely naturalized along roads and in ruderal habitats; many records uncertain.
Like *A. biennis* but up to 2.5 m, robust; stems tomentose only when young; epicalyx segments relatively shorter; mericarps c. 7 mm (vs. 4.5-6 mm).

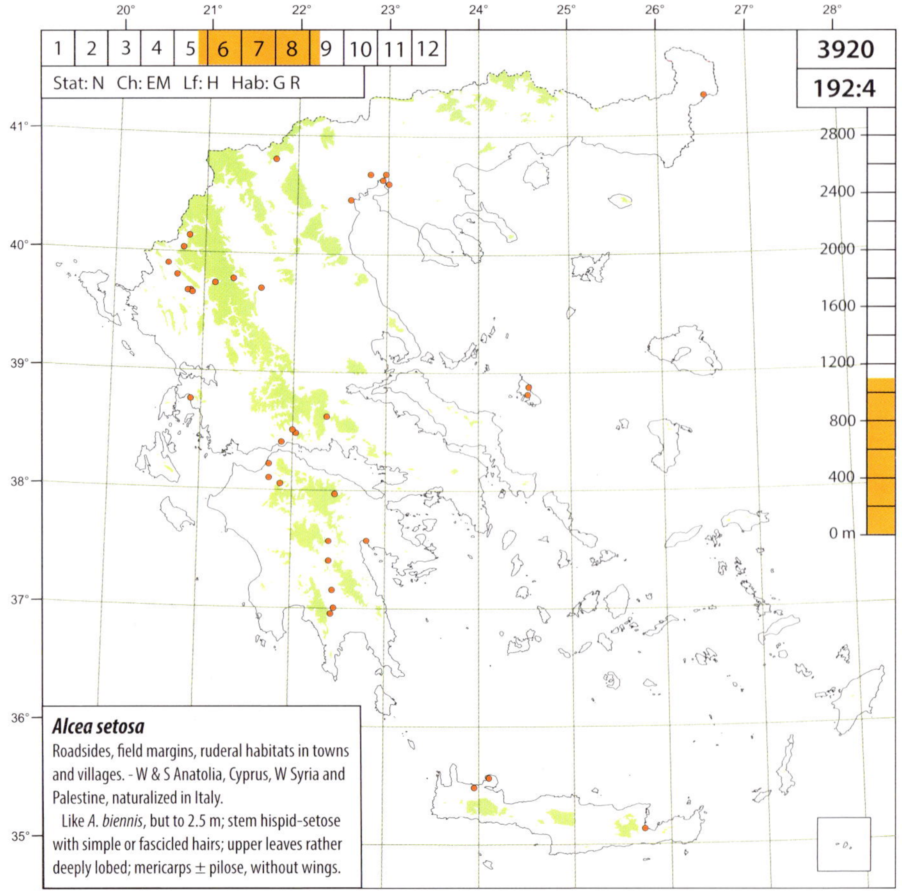

Alcea setosa
Roadsides, field margins, ruderal habitats in towns and villages. - W & S Anatolia, Cyprus, W Syria and Palestine, naturalized in Italy.
Like *A. biennis*, but to 2.5 m; stem hispid-setose with simple or fascicled hairs; upper leaves rather deeply lobed; mericarps ± pilose, without wings.

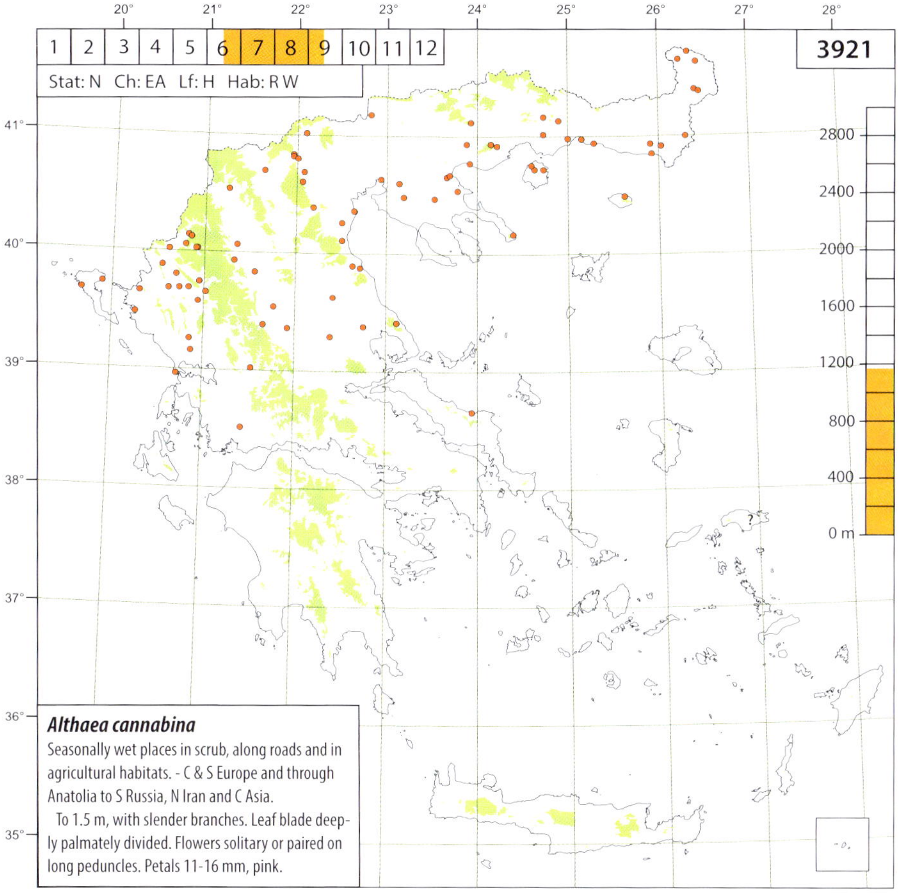

Althaea cannabina

Seasonally wet places in scrub, along roads and in agricultural habitats. - C & S Europe and through Anatolia to S Russia, N Iran and C Asia.

To 1.5 m, with slender branches. Leaf blade deeply palmately divided. Flowers solitary or paired on long peduncles. Petals 11-16 mm, pink.

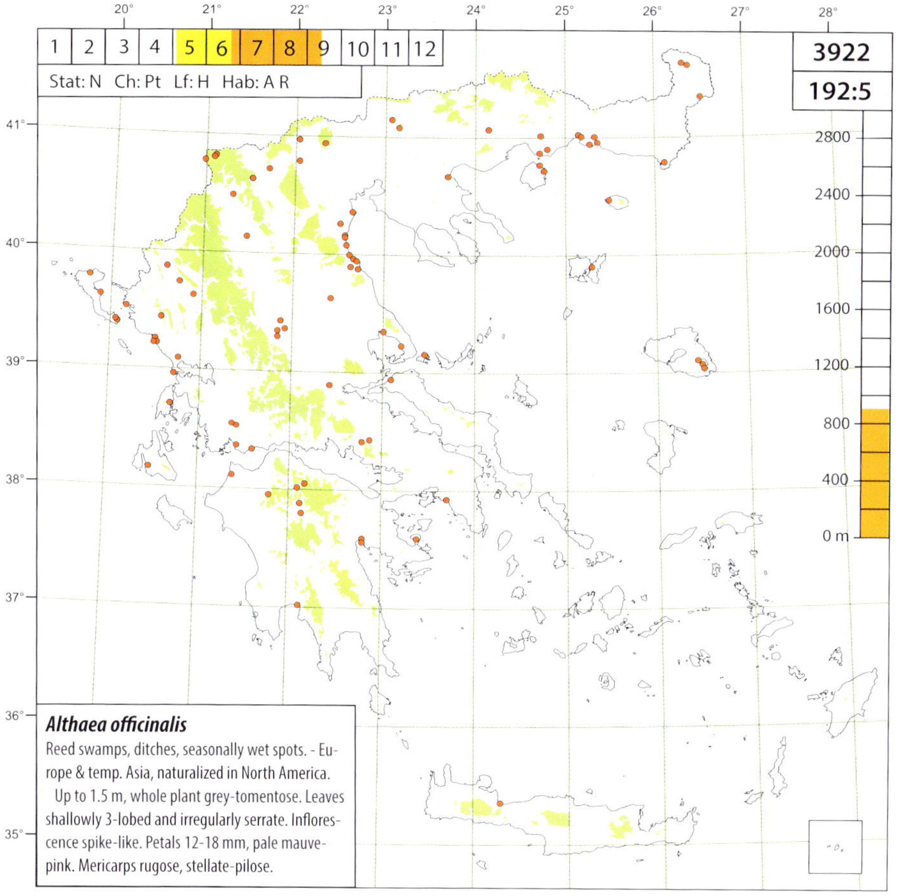

Althaea officinalis

Reed swamps, ditches, seasonally wet spots. - Europe & temp. Asia, naturalized in North America.

Up to 1.5 m, whole plant grey-tomentose. Leaves shallowly 3-lobed and irregularly serrate. Inflorescence spike-like. Petals 12-18 mm, pale mauve-pink. Mericarps rugose, stellate-pilose.

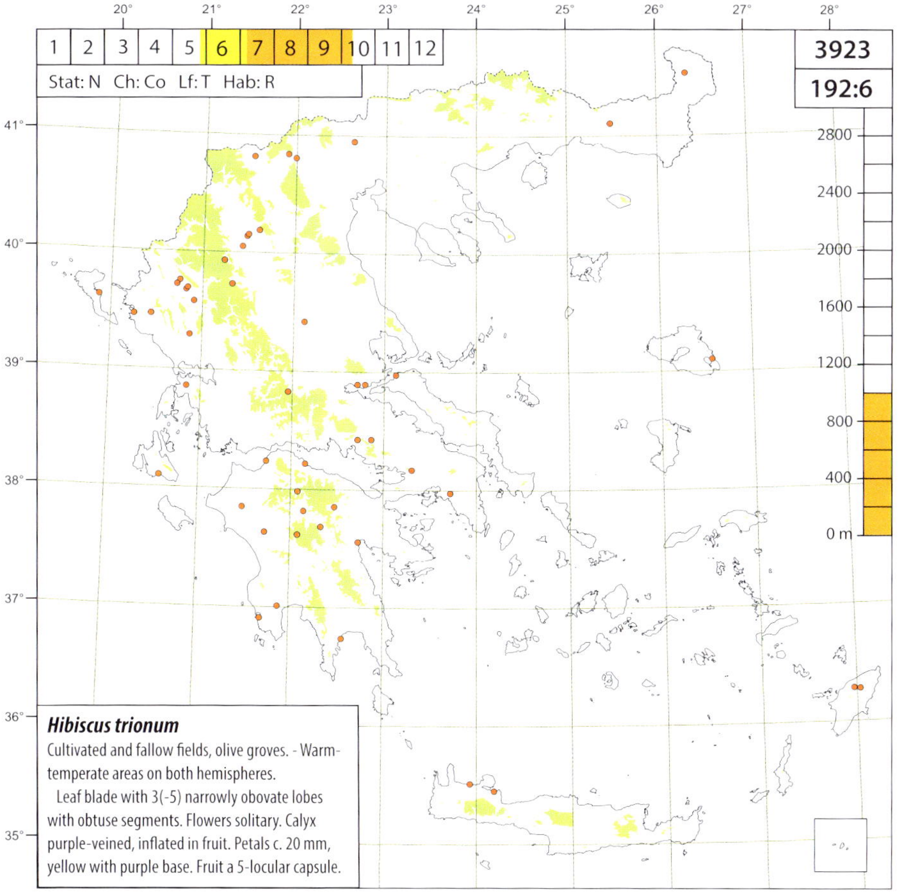

Hibiscus trionum

Cultivated and fallow fields, olive groves. - Warm-temperate areas on both hemispheres.

Leaf blade with 3(-5) narrowly obovate lobes with obtuse segments. Flowers solitary. Calyx purple-veined, inflated in fruit. Petals c. 20 mm, yellow with purple base. Fruit a 5-locular capsule.

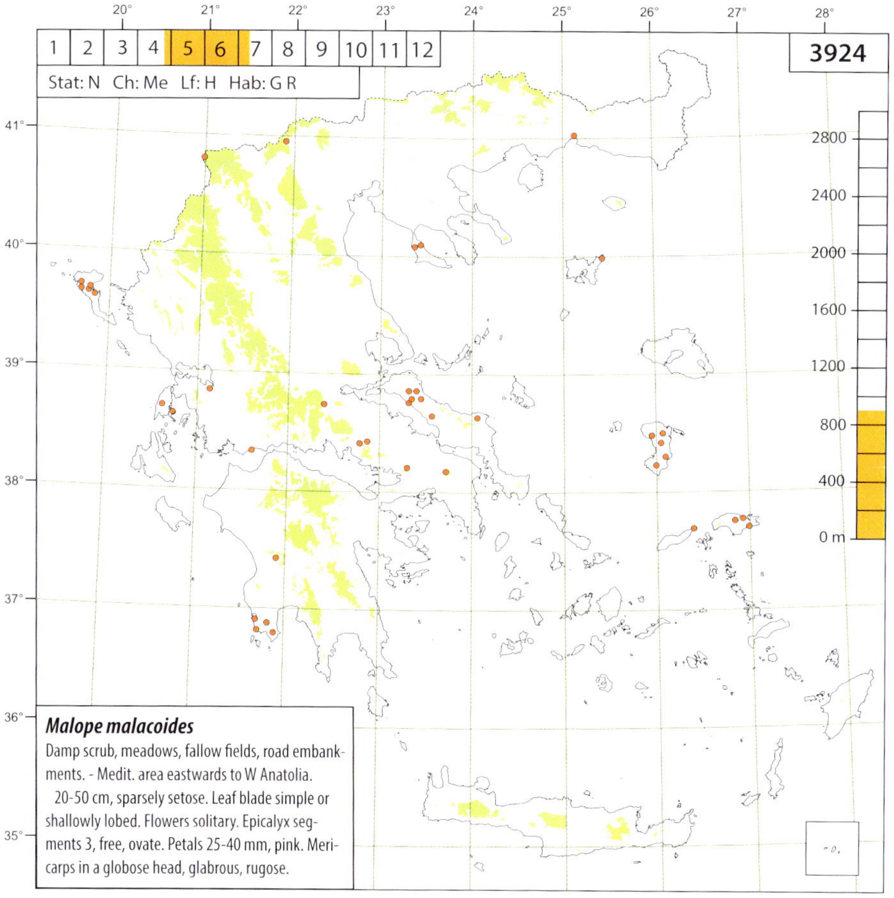

Malope malacoides

Damp scrub, meadows, fallow fields, road embankments. - Medit. area eastwards to W Anatolia.

20-50 cm, sparsely setose. Leaf blade simple or shallowly lobed. Flowers solitary. Epicalyx segments 3, free, ovate. Petals 25-40 mm, pink. Mericarps in a globose head, glabrous, rugose.

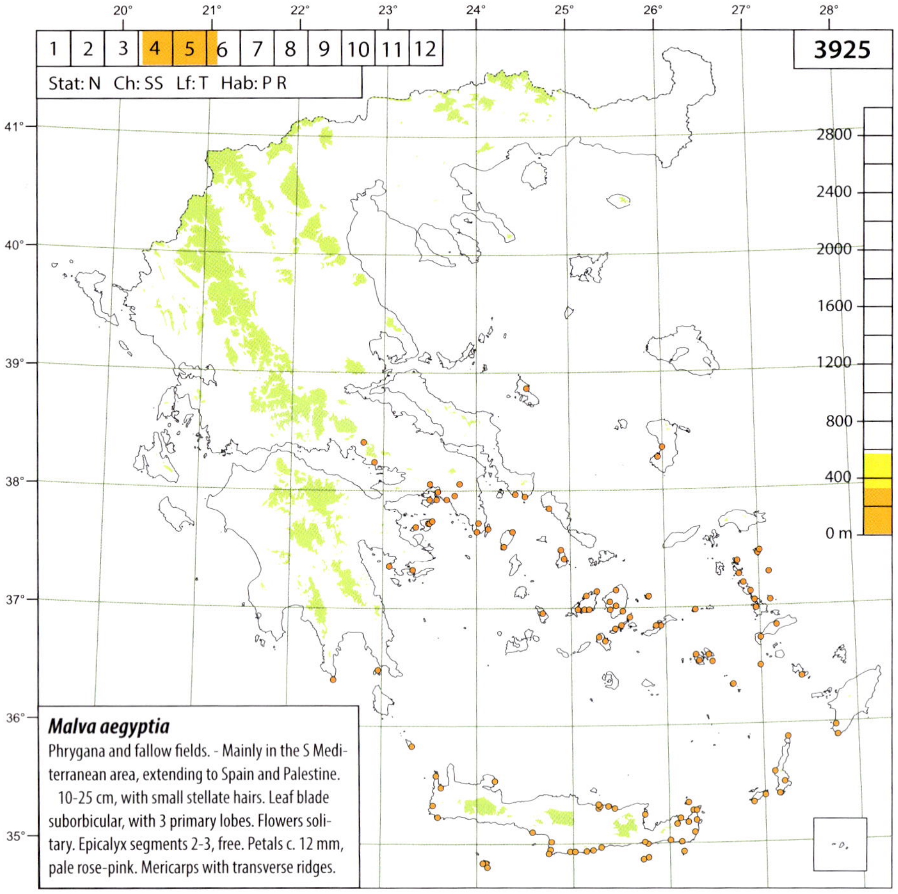

Malva aegyptia
Phrygana and fallow fields. - Mainly in the S Mediterranean area, extending to Spain and Palestine.
10-25 cm, with small stellate hairs. Leaf blade suborbicular, with 3 primary lobes. Flowers solitary. Epicalyx segments 2-3, free. Petals c. 12 mm, pale rose-pink. Mericarps with transverse ridges.

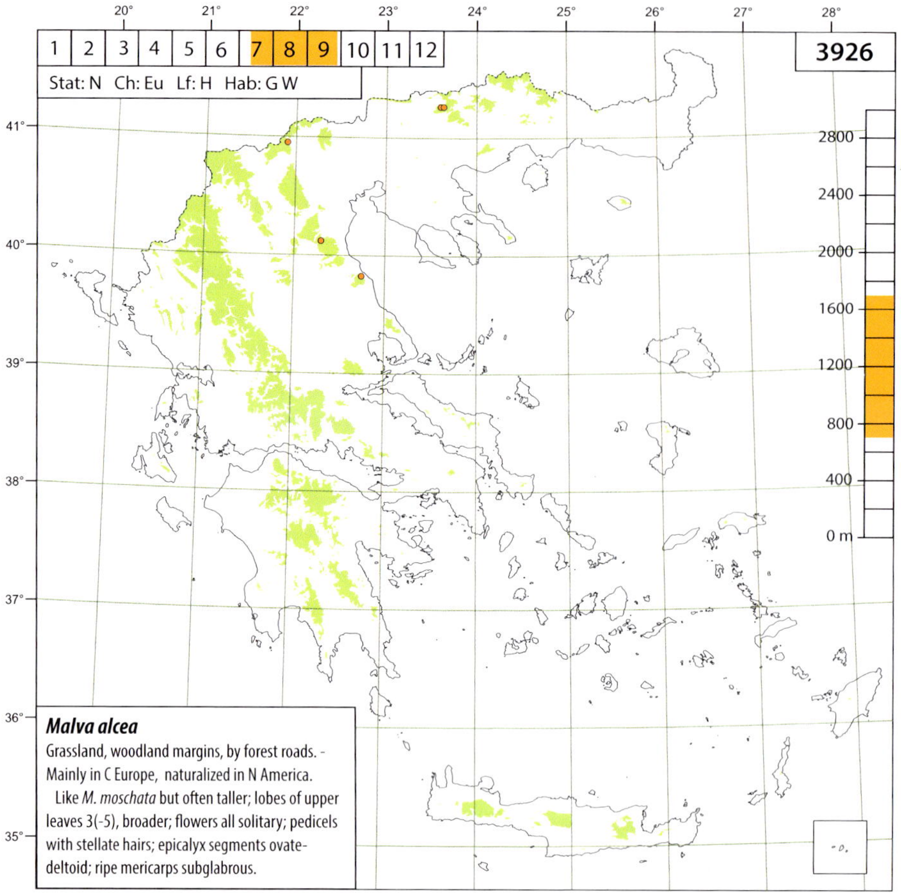

Malva alcea
Grassland, woodland margins, by forest roads. - Mainly in C Europe, naturalized in N America.
Like *M. moschata* but often taller; lobes of upper leaves 3(-5), broader; flowers all solitary; pedicels with stellate hairs; epicalyx segments ovate-deltoid; ripe mericarps subglabrous.

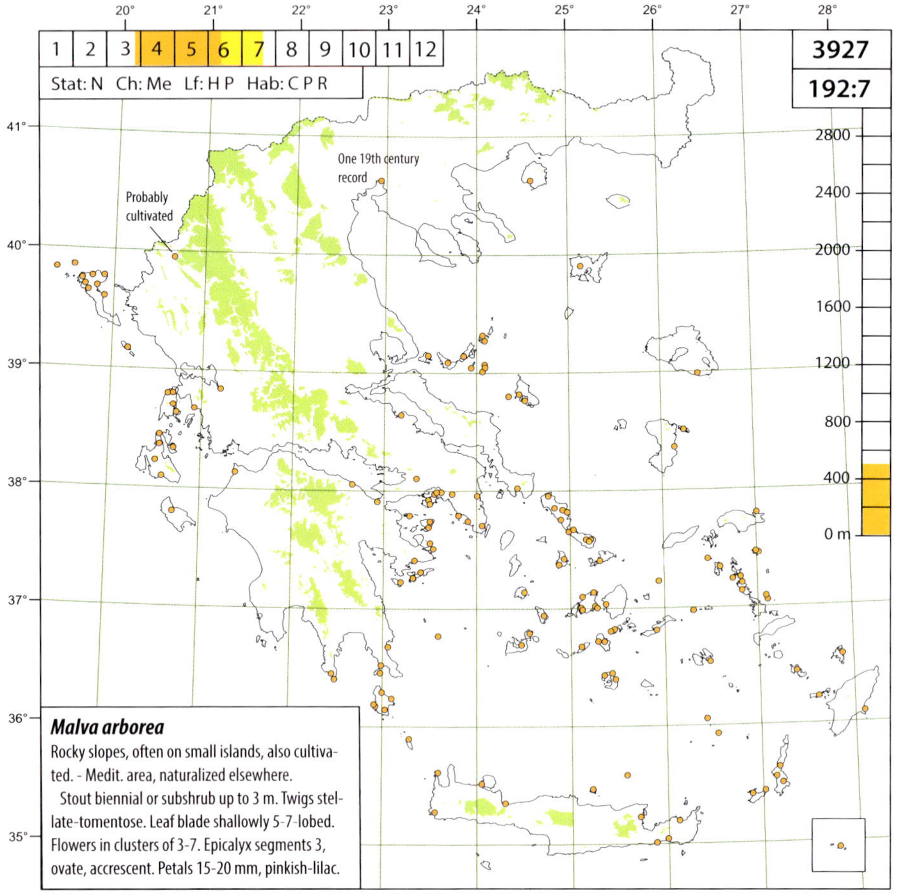

Malva arborea
Rocky slopes, often on small islands, also cultivated. - Medit. area, naturalized elsewhere.
Stout biennial or subshrub up to 3 m. Twigs stellate-tomentose. Leaf blade shallowly 5-7-lobed. Flowers in clusters of 3-7. Epicalyx segments 3, ovate, accrescent. Petals 15-20 mm, pinkish-lilac.

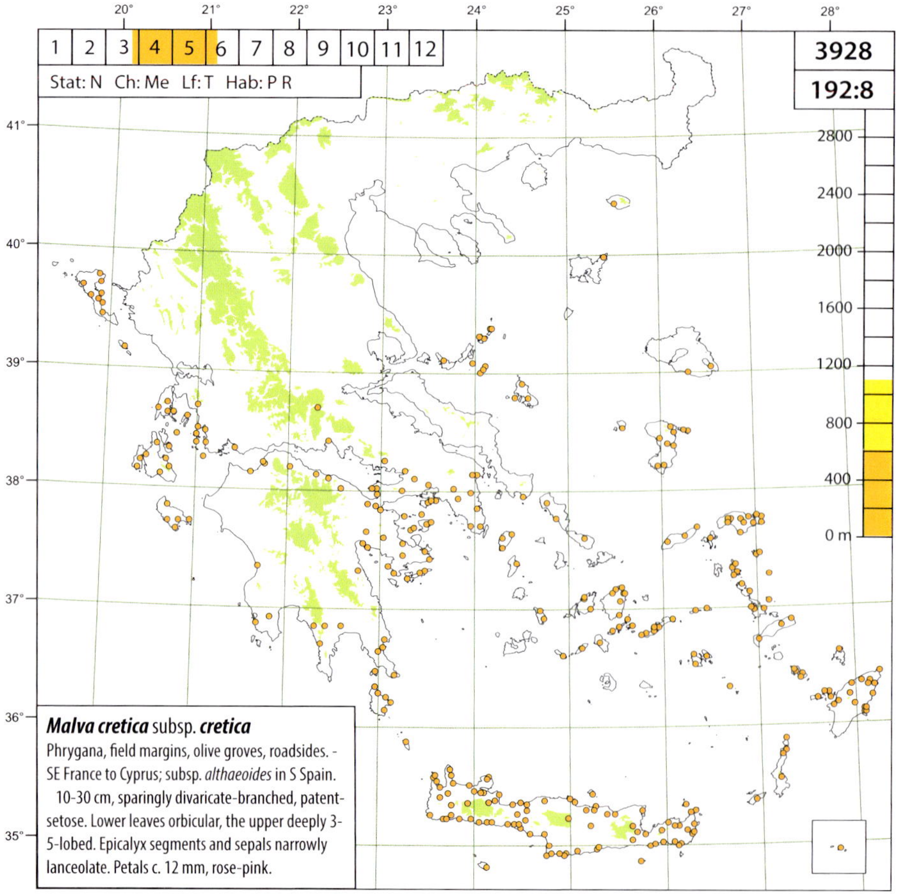

Malva cretica* subsp. *cretica
Phrygana, field margins, olive groves, roadsides. - SE France to Cyprus; subsp. *althaeoides* in S Spain.
10-30 cm, sparingly divaricate-branched, patent-setose. Lower leaves orbicular, the upper deeply 3-5-lobed. Epicalyx segments and sepals narrowly lanceolate. Petals c. 12 mm, rose-pink.

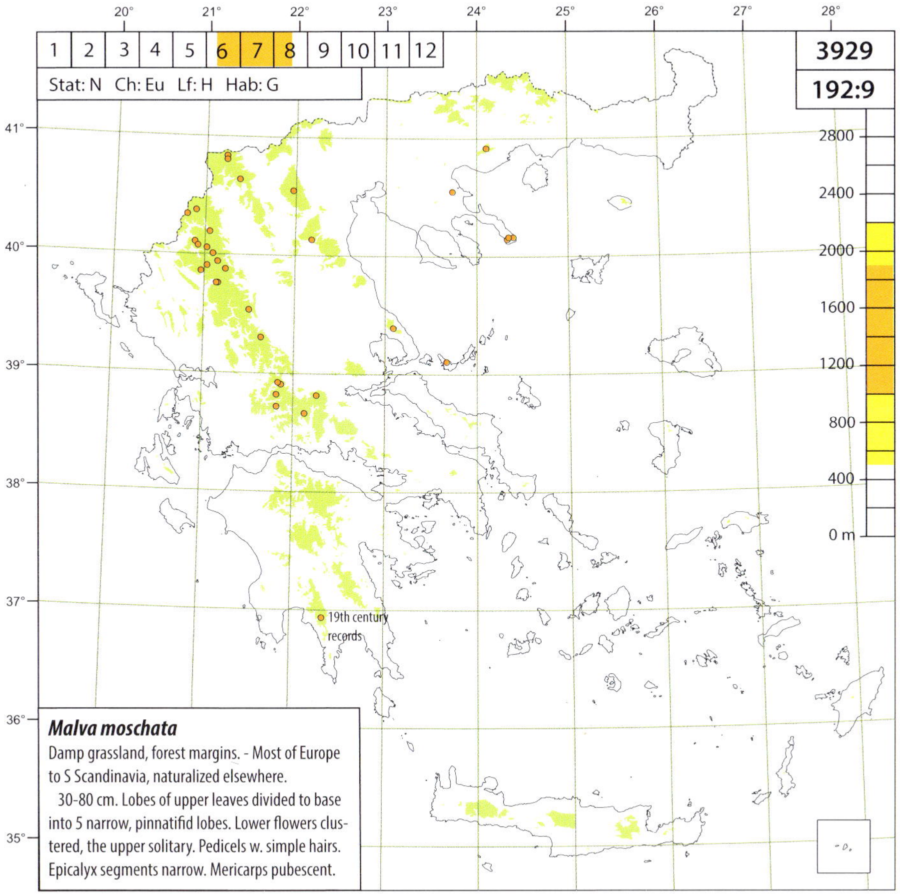

Malva moschata

Damp grassland, forest margins. - Most of Europe to S Scandinavia, naturalized elsewhere.

30-80 cm. Lobes of upper leaves divided to base into 5 narrow, pinnatifid lobes. Lower flowers clustered, the upper solitary. Pedicels w. simple hairs. Epicalyx segments narrow. Mericarps pubescent.

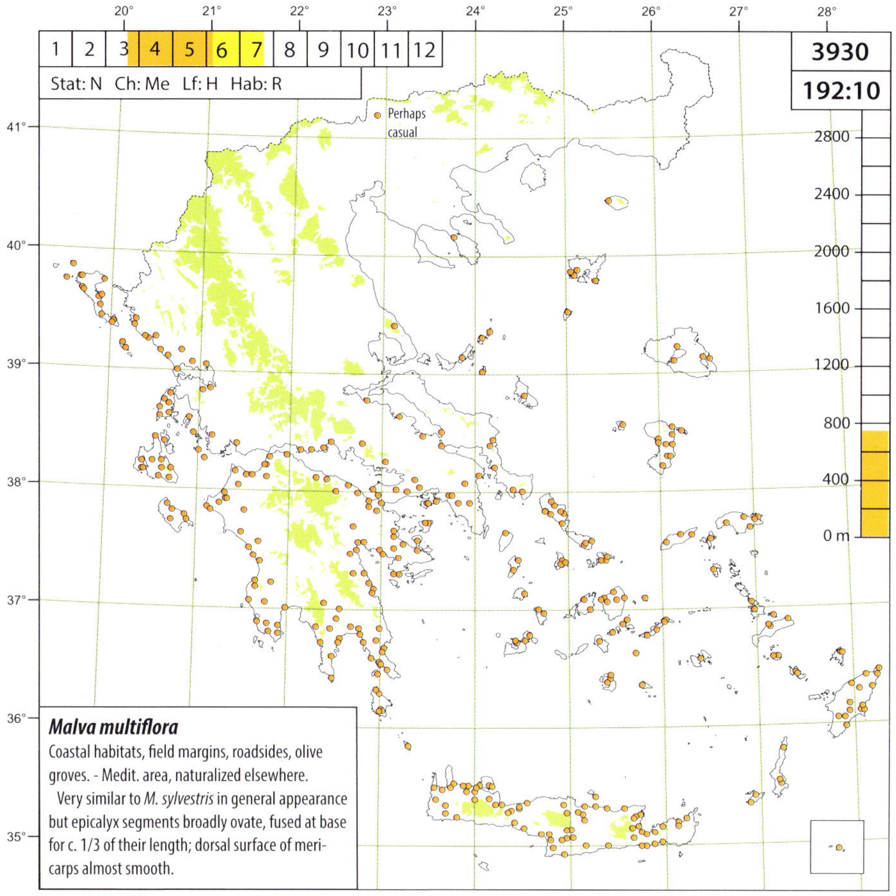

Malva multiflora

Coastal habitats, field margins, roadsides, olive groves. - Medit. area, naturalized elsewhere.

Very similar to *M. sylvestris* in general appearance but epicalyx segments broadly ovate, fused at base for c. 1/3 of their length; dorsal surface of mericarps almost smooth.

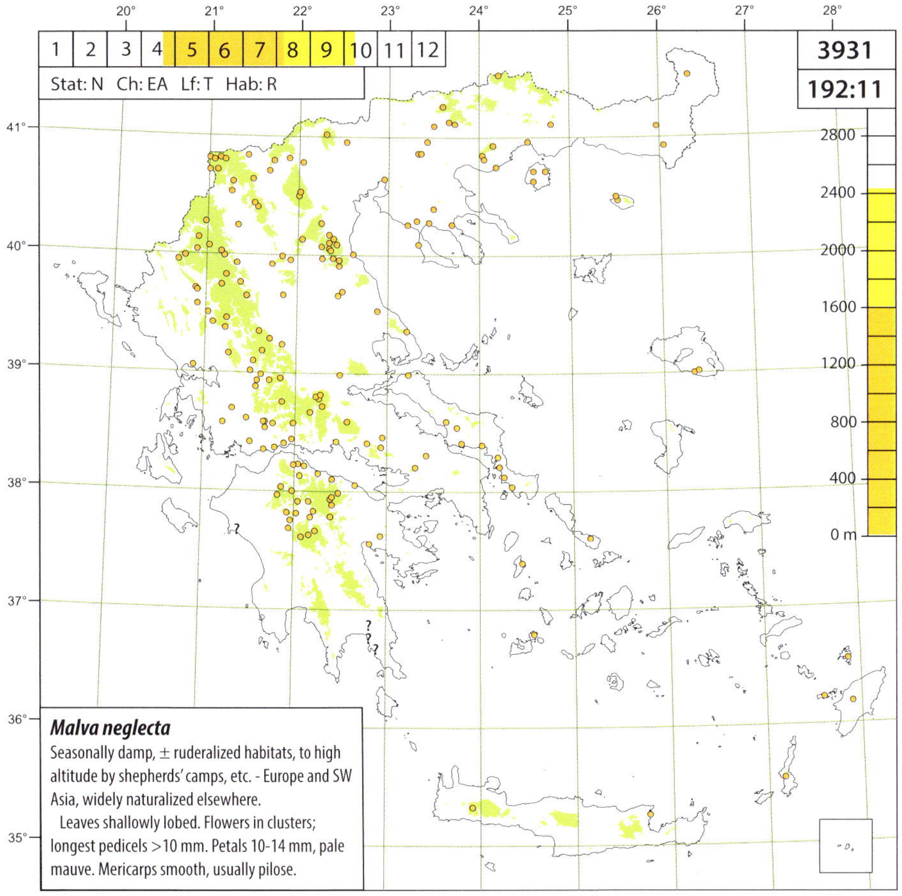

Malva neglecta

Seasonally damp, ± ruderalized habitats, to high altitude by shepherds' camps, etc. - Europe and SW Asia, widely naturalized elsewhere.

Leaves shallowly lobed. Flowers in clusters; longest pedicels >10 mm. Petals 10-14 mm, pale mauve. Mericarps smooth, usually pilose.

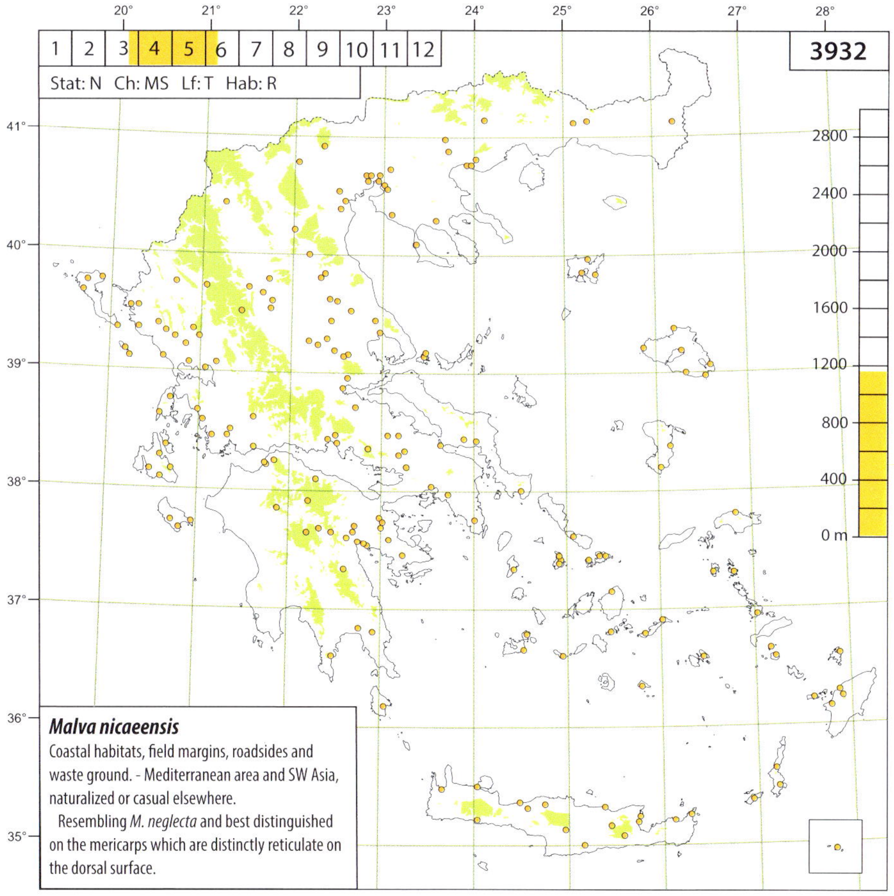

Malva nicaeensis

Coastal habitats, field margins, roadsides and waste ground. - Mediterranean area and SW Asia, naturalized or casual elsewhere.

Resembling *M. neglecta* and best distinguished on the mericarps which are distinctly reticulate on the dorsal surface.

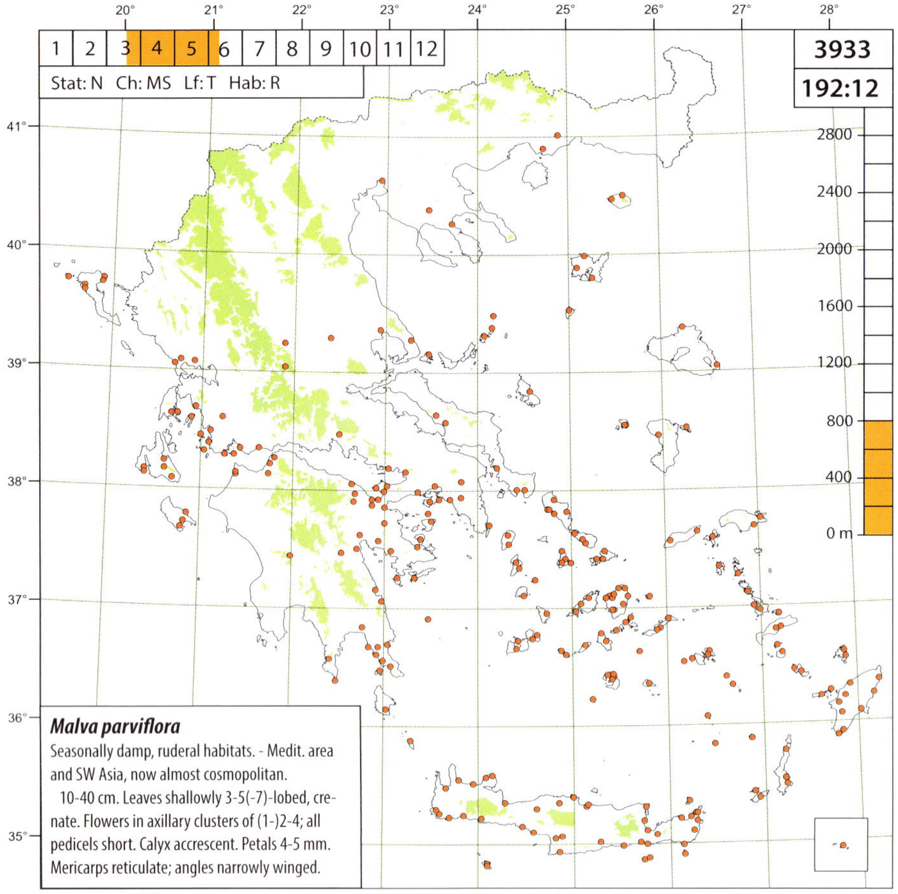

Malva parviflora

Seasonally damp, ruderal habitats. - Medit. area and SW Asia, now almost cosmopolitan.

10-40 cm. Leaves shallowly 3-5(-7)-lobed, crenate. Flowers in axillary clusters of (1-)2-4; all pedicels short. Calyx accrescent. Petals 4-5 mm. Mericarps reticulate; angles narrowly winged.

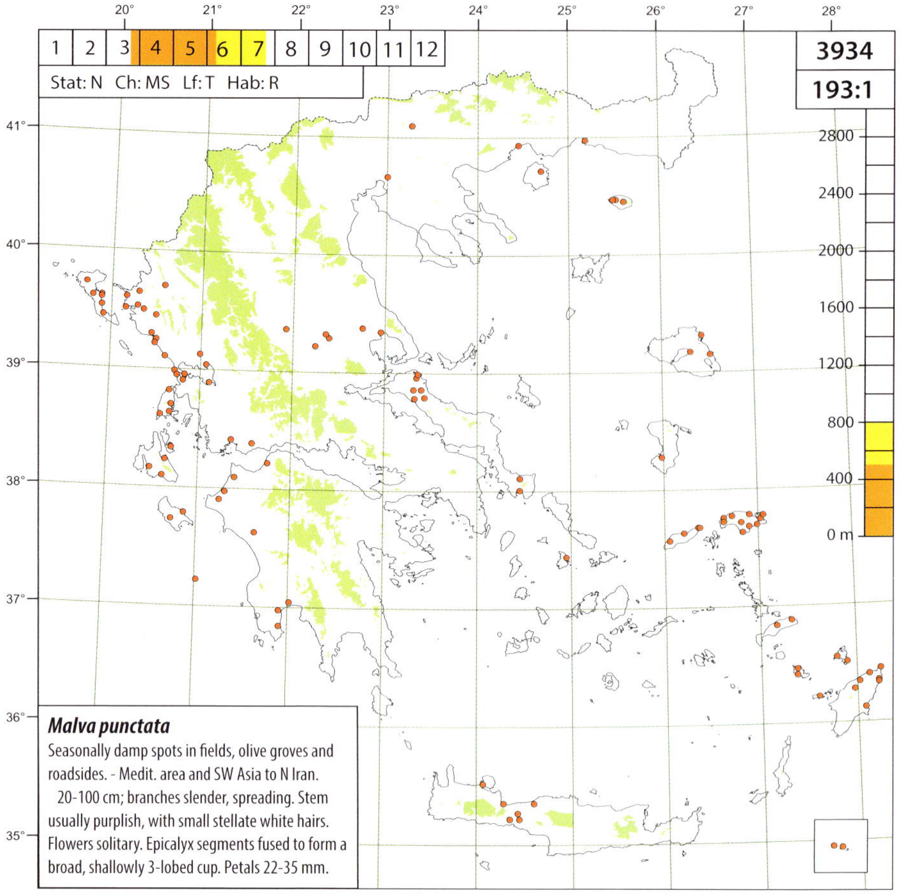

Malva punctata

Seasonally damp spots in fields, olive groves and roadsides. - Medit. area and SW Asia to N Iran.

20-100 cm; branches slender, spreading. Stem usually purplish, with small stellate white hairs. Flowers solitary. Epicalyx segments fused to form a broad, shallowly 3-lobed cup. Petals 22-35 mm.

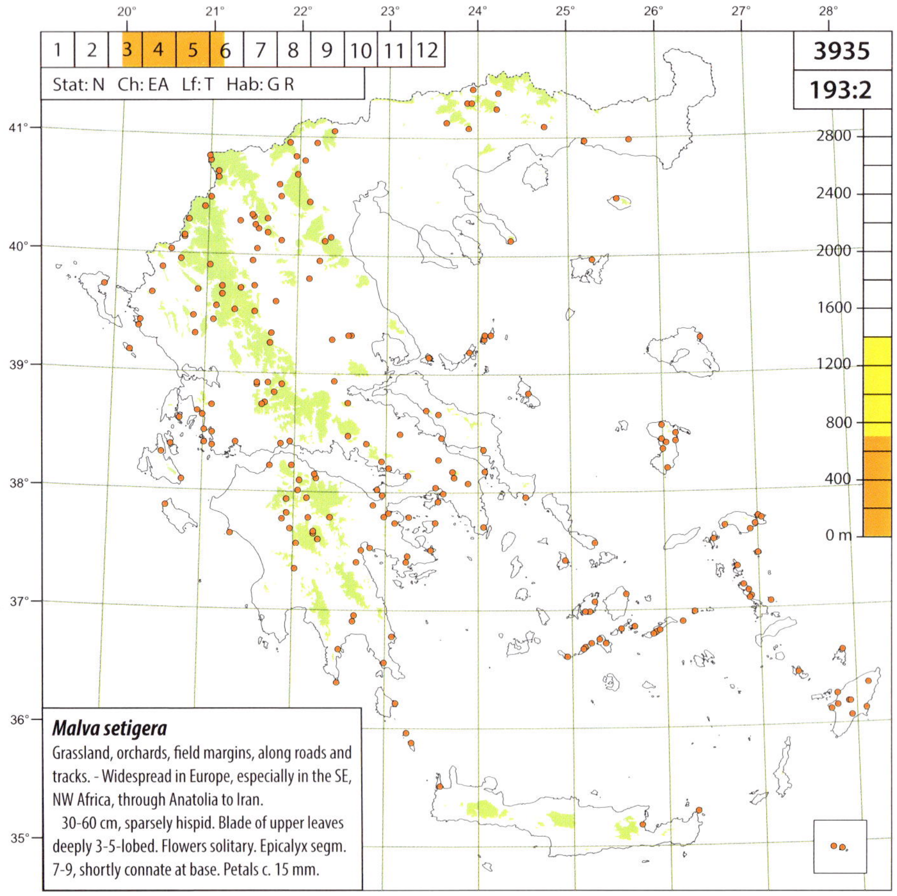

Malva setigera

Grassland, orchards, field margins, along roads and tracks. - Widespread in Europe, especially in the SE, NW Africa, through Anatolia to Iran.

30-60 cm, sparsely hispid. Blade of upper leaves deeply 3-5-lobed. Flowers solitary. Epicalyx segm. 7-9, shortly connate at base. Petals c. 15 mm.

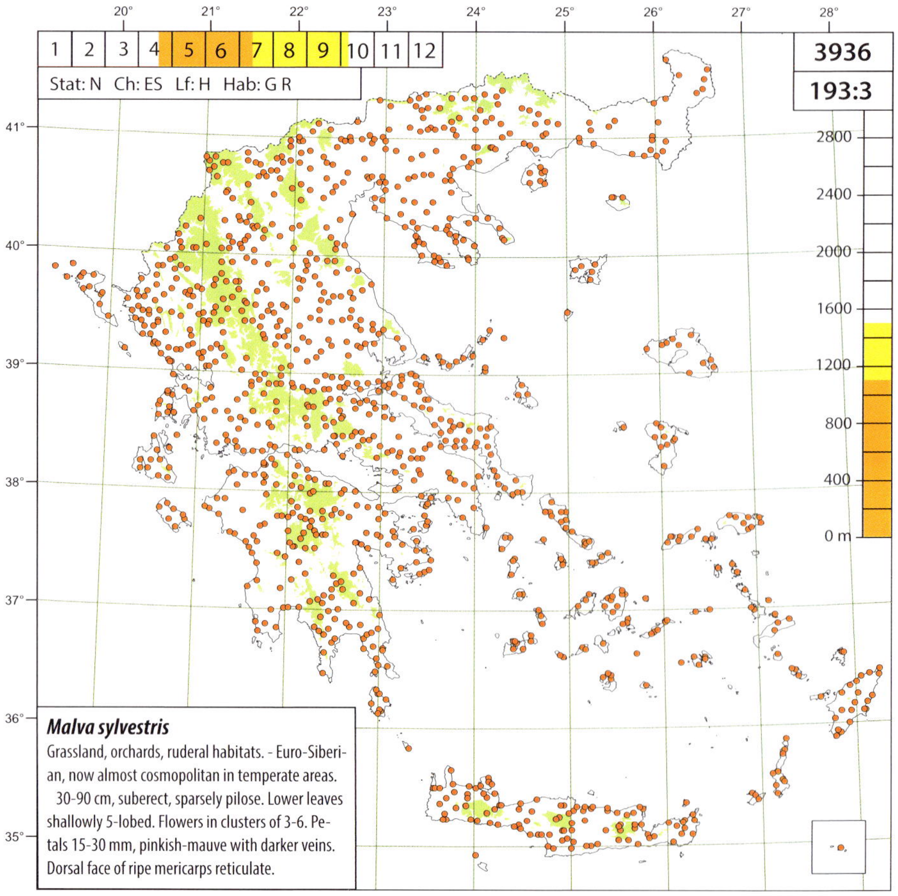

Malva sylvestris

Grassland, orchards, ruderal habitats. - Euro-Siberian, now almost cosmopolitan in temperate areas.

30-90 cm, suberect, sparsely pilose. Lower leaves shallowly 5-lobed. Flowers in clusters of 3-6. Petals 15-30 mm, pinkish-mauve with darker veins. Dorsal face of ripe mericarps reticulate.

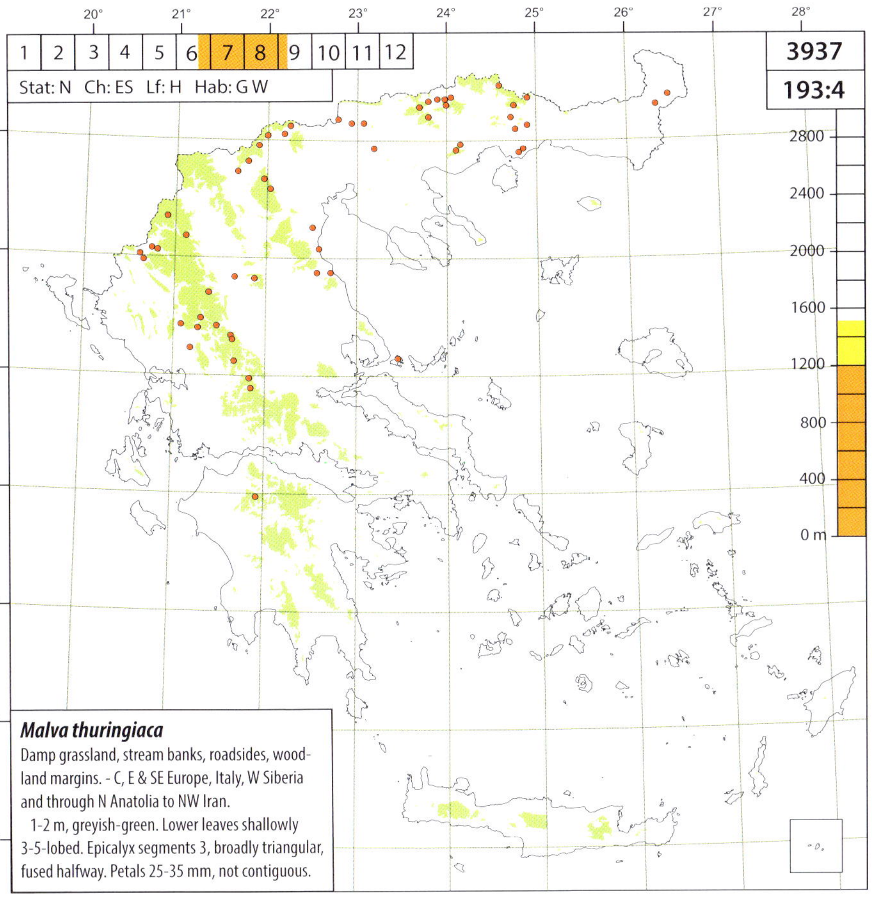

Malva thuringiaca

Damp grassland, stream banks, roadsides, woodland margins. - C, E & SE Europe, Italy, W Siberia and through N Anatolia to NW Iran.

1-2 m, greyish-green. Lower leaves shallowly 3-5-lobed. Epicalyx segments 3, broadly triangular, fused halfway. Petals 25-35 mm, not contiguous.

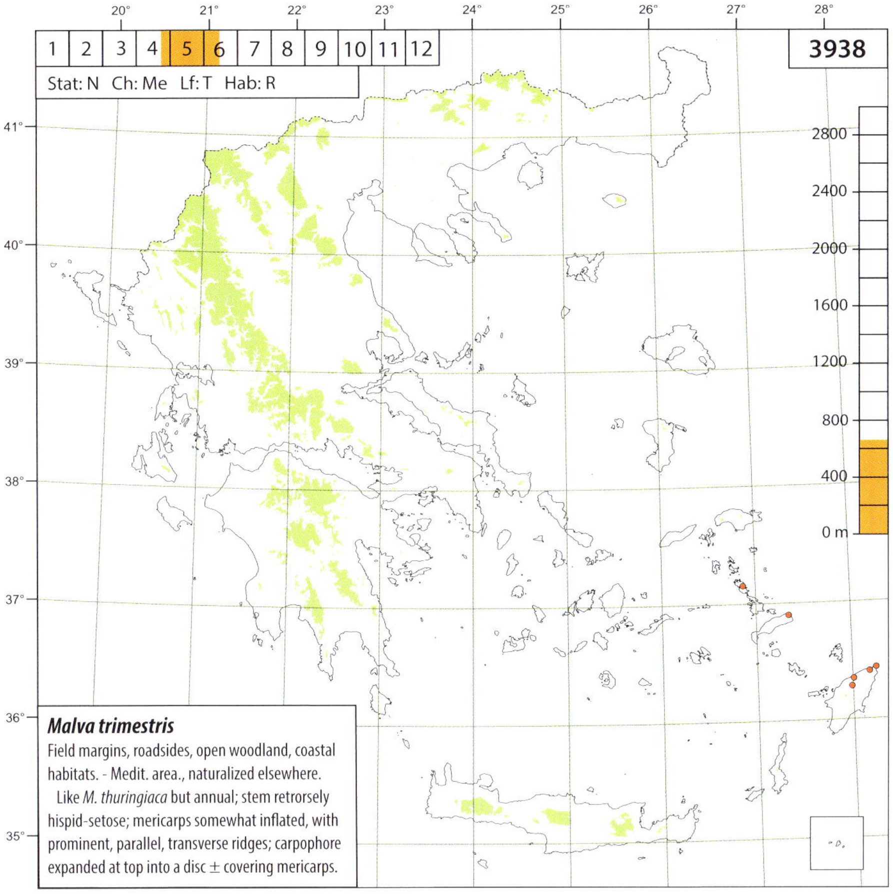

Malva trimestris

Field margins, roadsides, open woodland, coastal habitats. - Medit. area., naturalized elsewhere.

Like *M. thuringiaca* but annual; stem retrorsely hispid-setose; mericarps somewhat inflated, with prominent, parallel, transverse ridges; carpophore expanded at top into a disc ± covering mericarps.

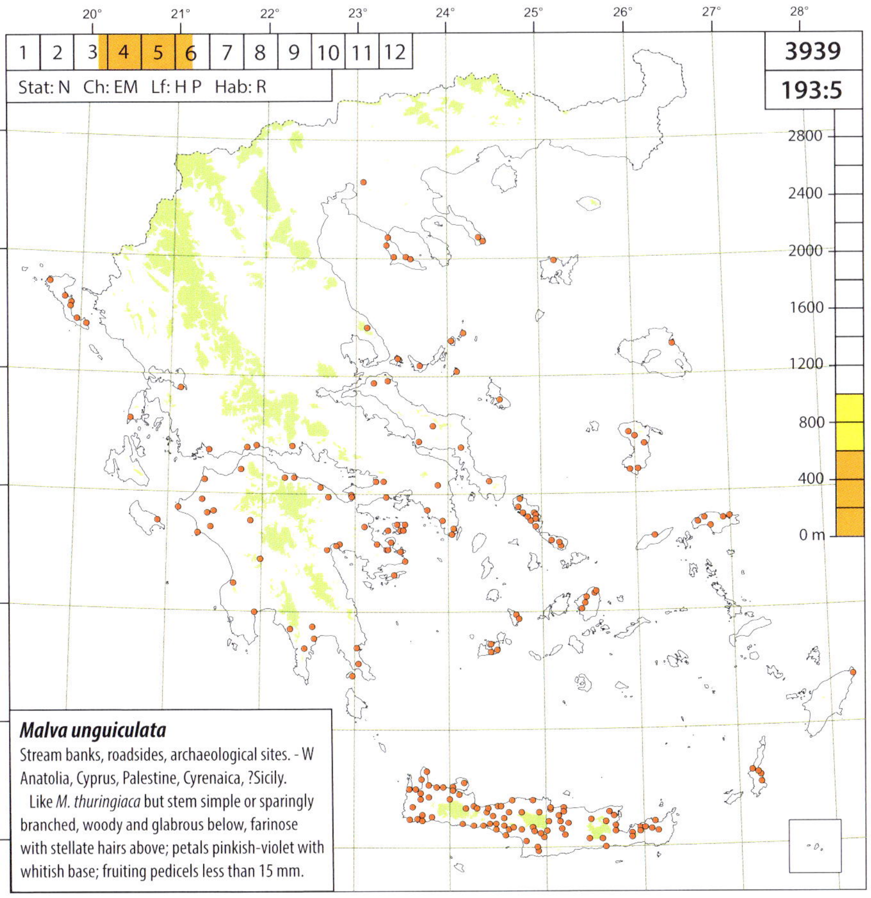

Malva unguiculata

Stream banks, roadsides, archaeological sites. - W Anatolia, Cyprus, Palestine, Cyrenaica, ?Sicily.

Like *M. thuringiaca* but stem simple or sparingly branched, woody and glabrous below, farinose with stellate hairs above; petals pinkish-violet with whitish base; fruiting pedicels less than 15 mm.

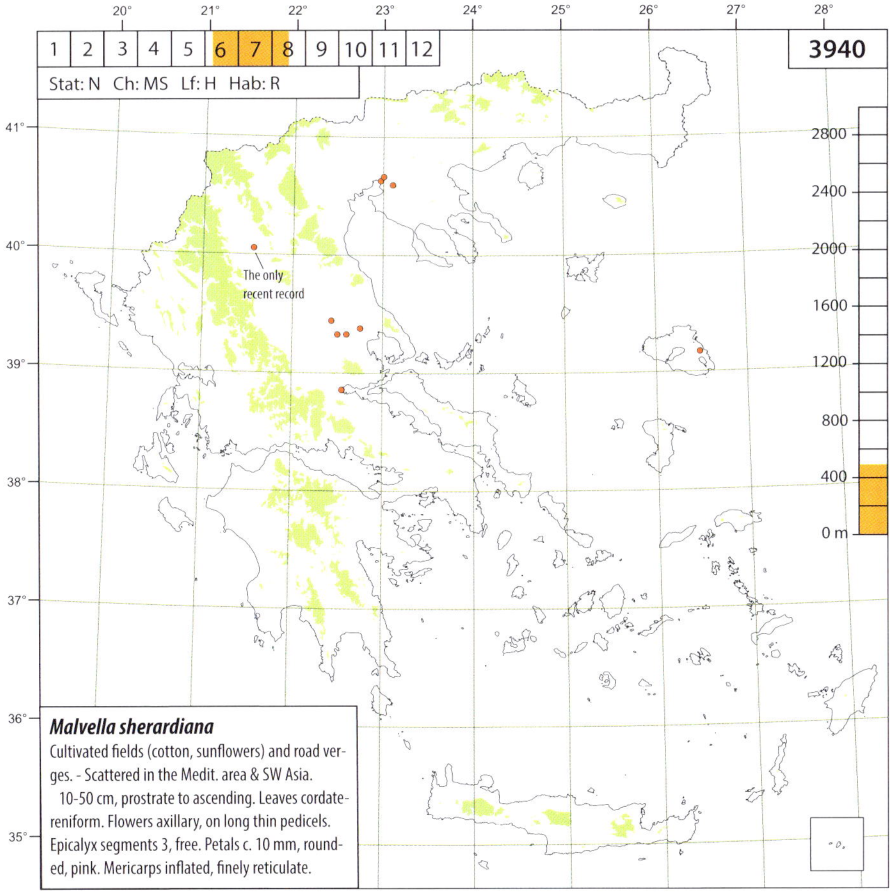

Malvella sherardiana

Cultivated fields (cotton, sunflowers) and road verges. - Scattered in the Medit. area & SW Asia.

10-50 cm, prostrate to ascending. Leaves cordate-reniform. Flowers axillary, on long thin pedicels. Epicalyx segments 3, free. Petals c. 10 mm, rounded, pink. Mericarps inflated, finely reticulate.

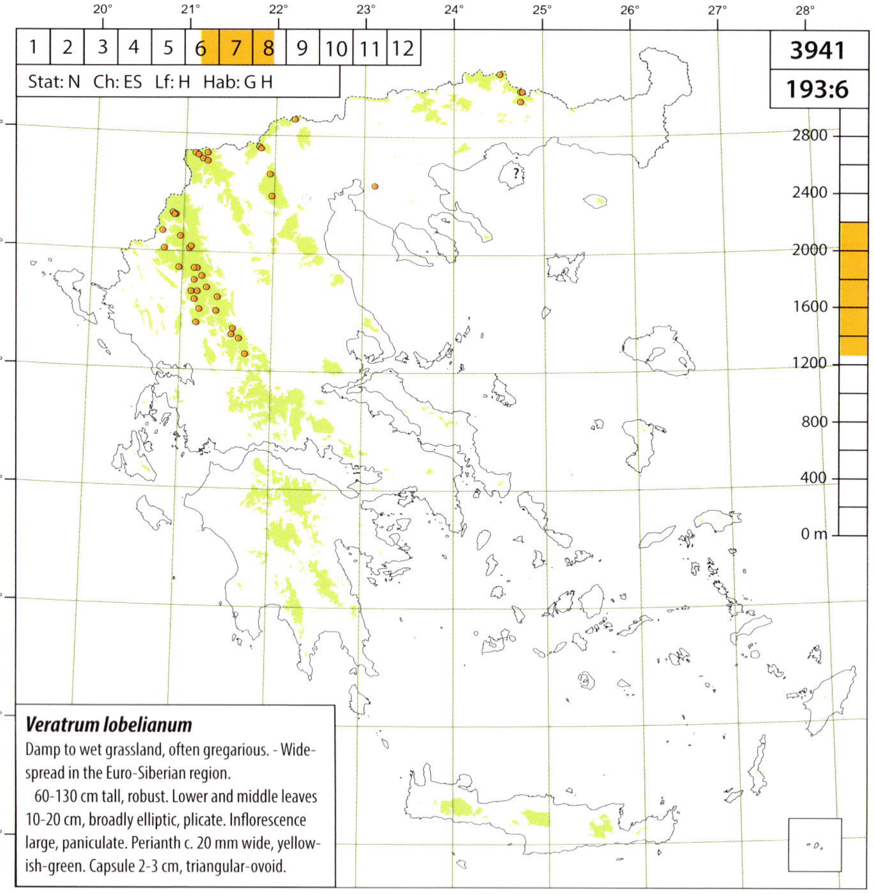

Veratrum lobelianum

Damp to wet grassland, often gregarious. - Widespread in the Euro-Siberian region.

60-130 cm tall, robust. Lower and middle leaves 10-20 cm, broadly elliptic, plicate. Inflorescence large, paniculate. Perianth c. 20 mm wide, yellowish-green. Capsule 2-3 cm, triangular-ovoid.

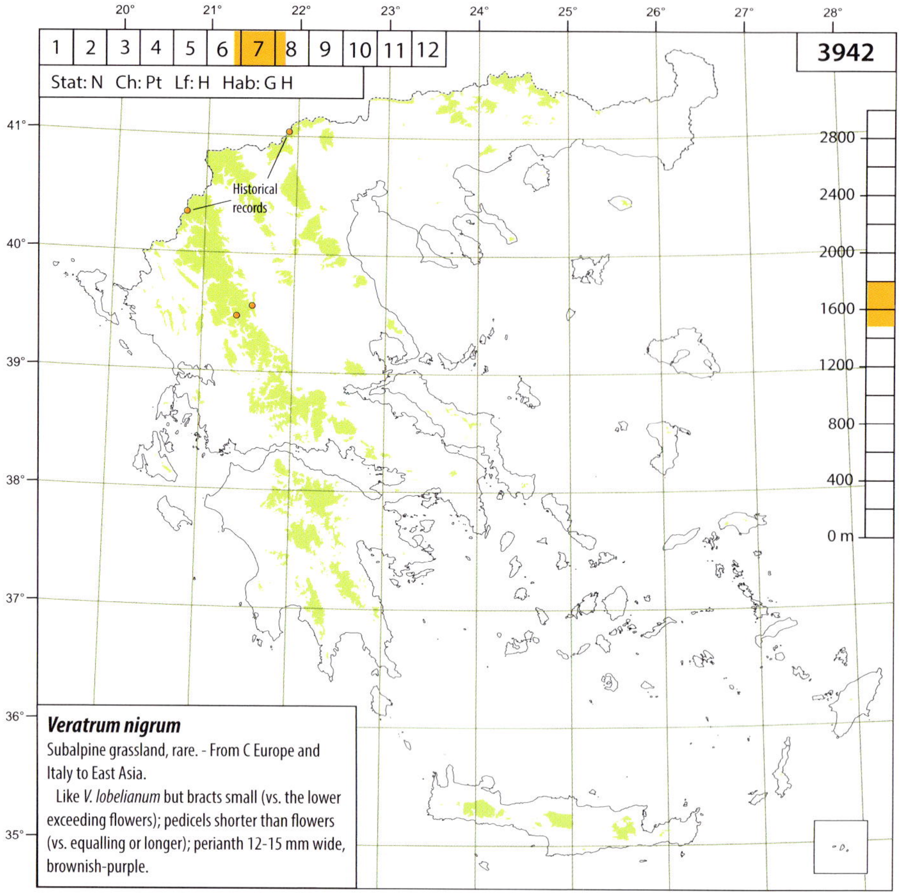

Veratrum nigrum

Subalpine grassland, rare. - From C Europe and Italy to East Asia.

Like *V. lobelianum* but bracts small (vs. the lower exceeding flowers); pedicels shorter than flowers (vs. equalling or longer); perianth 12-15 mm wide, brownish-purple.

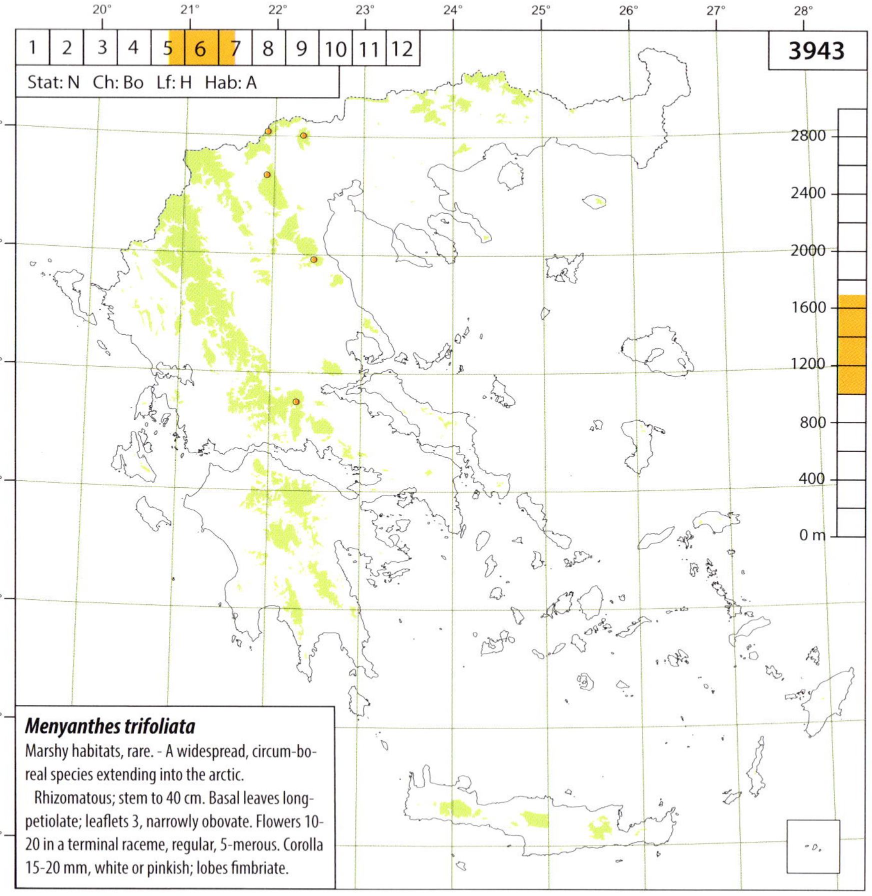

Menyanthes trifoliata

Marshy habitats, rare. - A widespread, circum-boreal species extending into the arctic.

Rhizomatous; stem to 40 cm. Basal leaves long-petiolate; leaflets 3, narrowly obovate. Flowers 10-20 in a terminal raceme, regular, 5-merous. Corolla 15-20 mm, white or pinkish; lobes fimbriate.

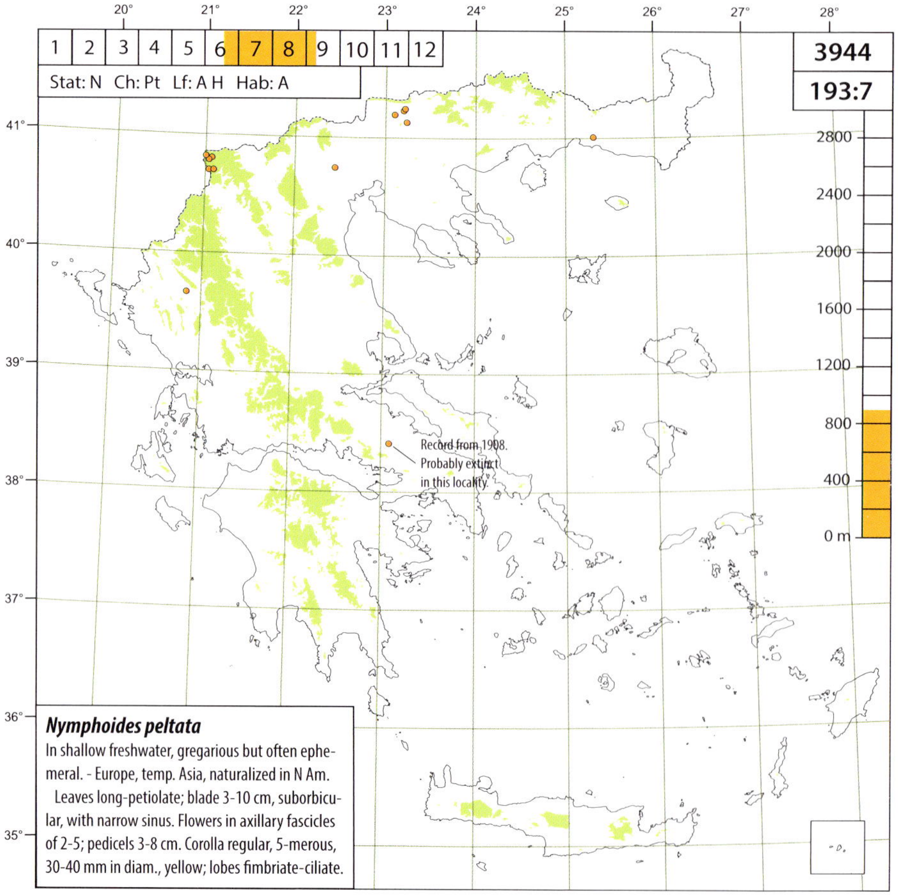

Nymphoides peltata

In shallow freshwater, gregarious but often ephemeral. - Europe, temp. Asia, naturalized in N Am.

Leaves long-petiolate; blade 3-10 cm, suborbicular, with narrow sinus. Flowers in axillary fascicles of 2-5; pedicels 3-8 cm. Corolla regular, 5-merous, 30-40 mm in diam., yellow; lobes fimbriate-ciliate.

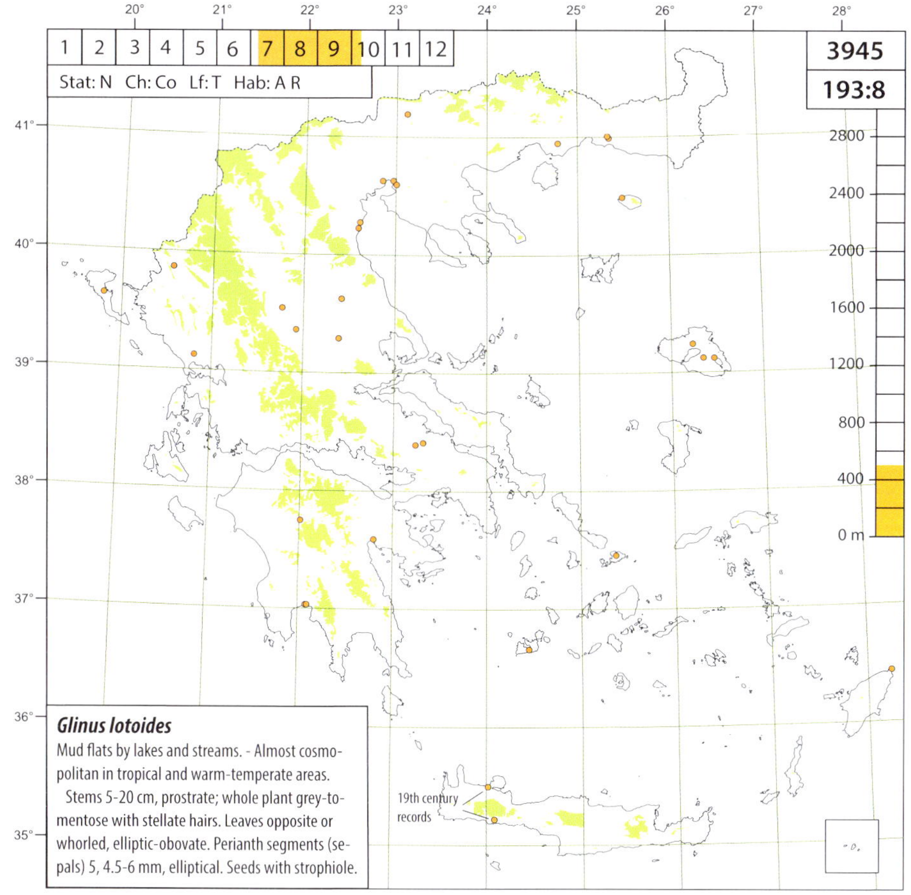

Glinus lotoides

Mud flats by lakes and streams. - Almost cosmopolitan in tropical and warm-temperate areas.

Stems 5-20 cm, prostrate; whole plant grey-tomentose with stellate hairs. Leaves opposite or whorled, elliptic-obovate. Perianth segments (sepals) 5, 4.5-6 mm, elliptical. Seeds with strophiole.

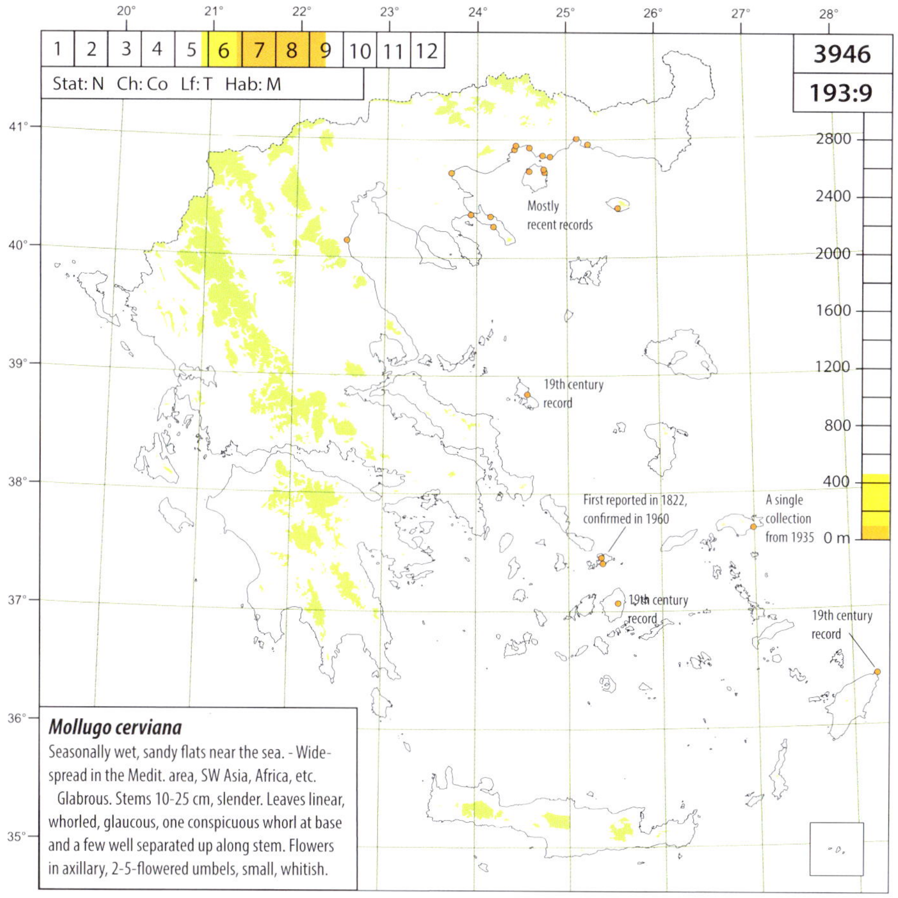

Mollugo cerviana

Seasonally wet, sandy flats near the sea. - Widespread in the Medit. area, SW Asia, Africa, etc.

Glabrous. Stems 10-25 cm, slender. Leaves linear, whorled, glaucous, one conspicuous whorl at base and a few well separated up along stem. Flowers in axillary, 2-5-flowered umbels, small, whitish.

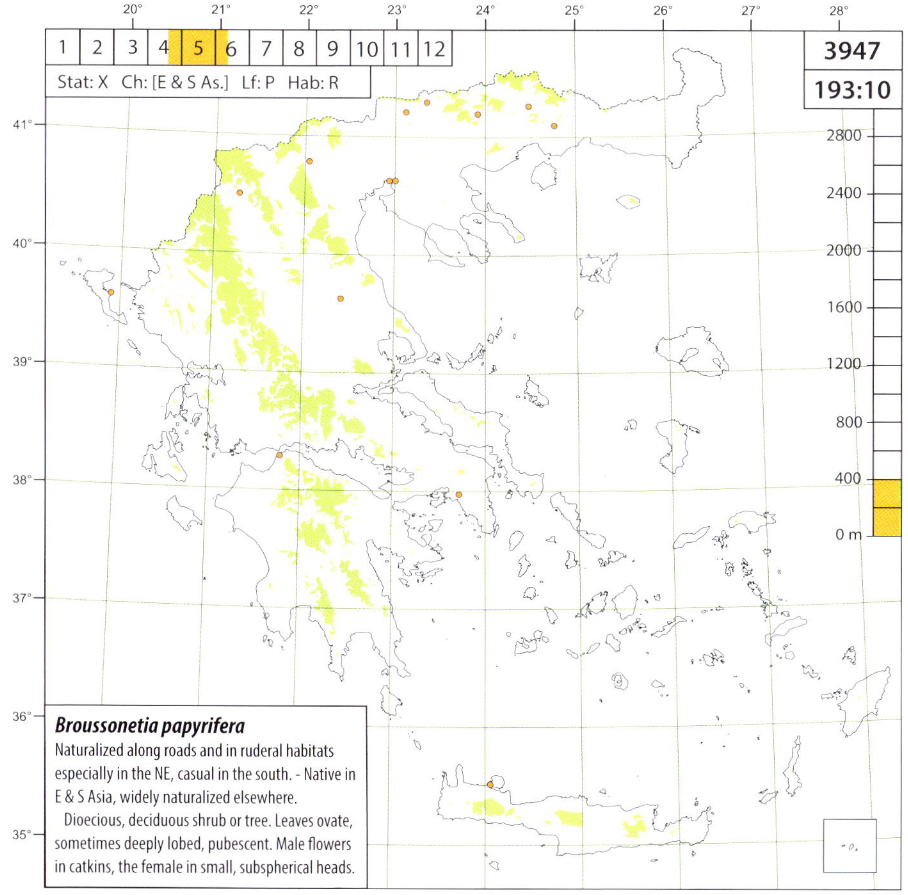

Broussonetia papyrifera

Naturalized along roads and in ruderal habitats especially in the NE, casual in the south. - Native in E & S Asia, widely naturalized elsewhere.

Dioecious, deciduous shrub or tree. Leaves ovate, sometimes deeply lobed, pubescent. Male flowers in catkins, the female in small, subspherical heads.

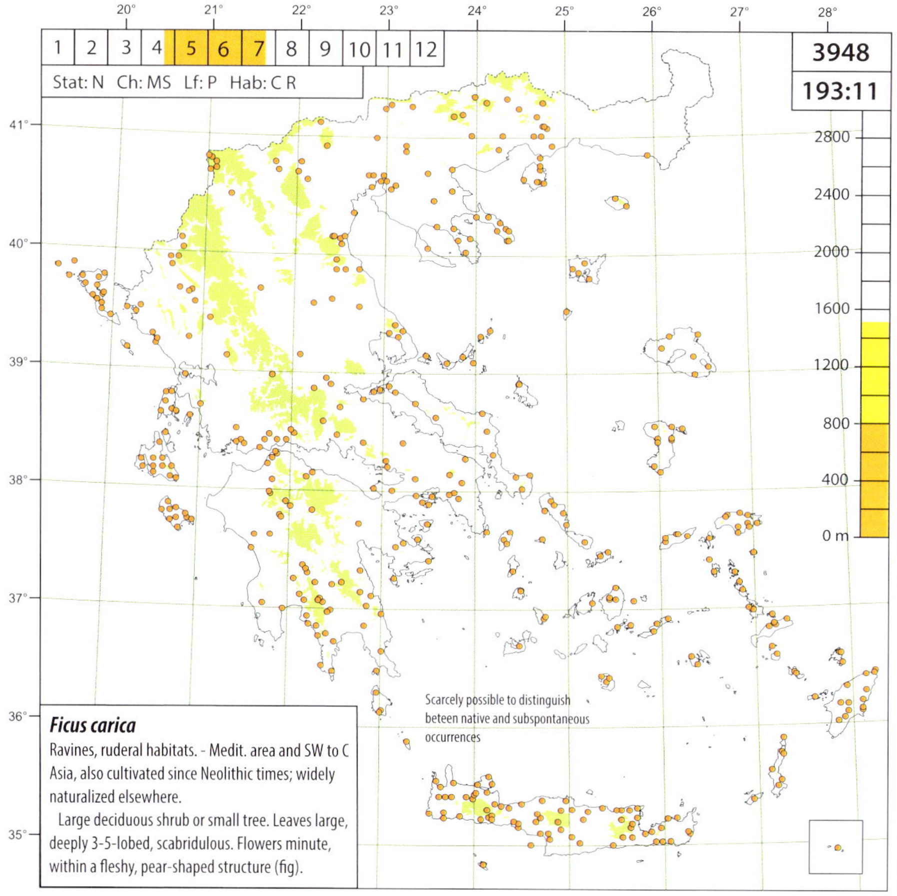

Ficus carica

Ravines, ruderal habitats. - Medit. area and SW to C Asia, also cultivated since Neolithic times; widely naturalized elsewhere.

Large deciduous shrub or small tree. Leaves large, deeply 3-5-lobed, scabridulous. Flowers minute, within a fleshy, pear-shaped structure (fig).

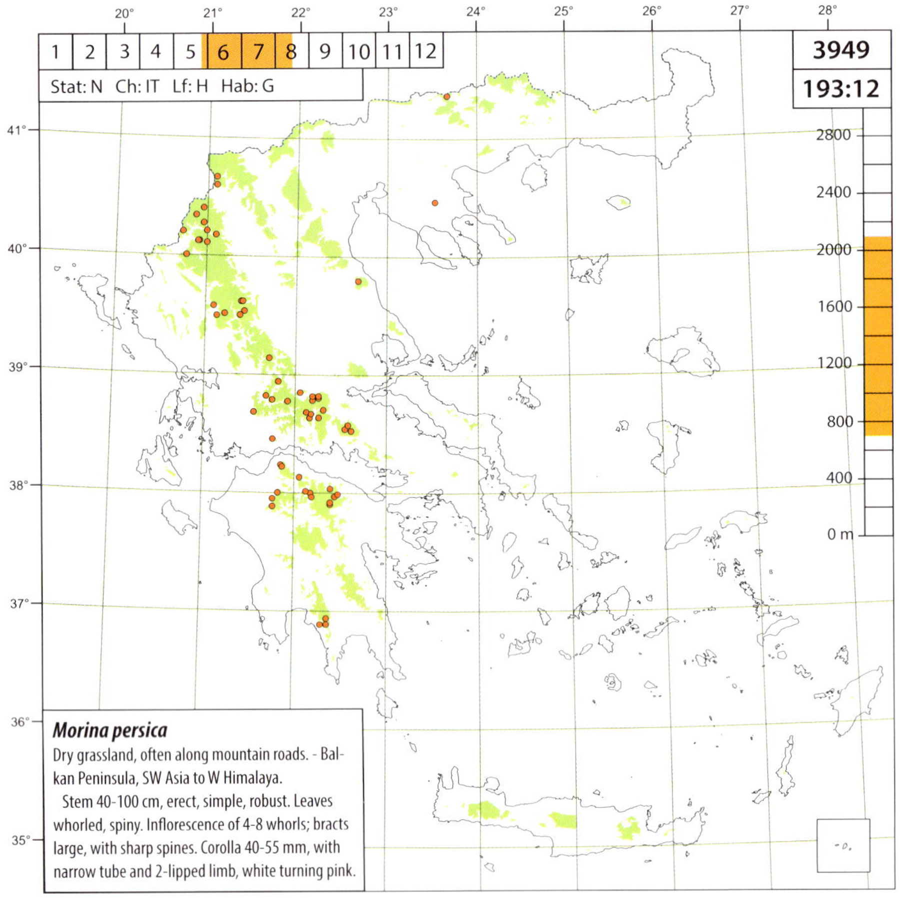

Morina persica

Dry grassland, often along mountain roads. - Balkan Peninsula, SW Asia to W Himalaya.

Stem 40-100 cm, erect, simple, robust. Leaves whorled, spiny. Inflorescence of 4-8 whorls; bracts large, with sharp spines. Corolla 40-55 mm, with narrow tube and 2-lipped limb, white turning pink.

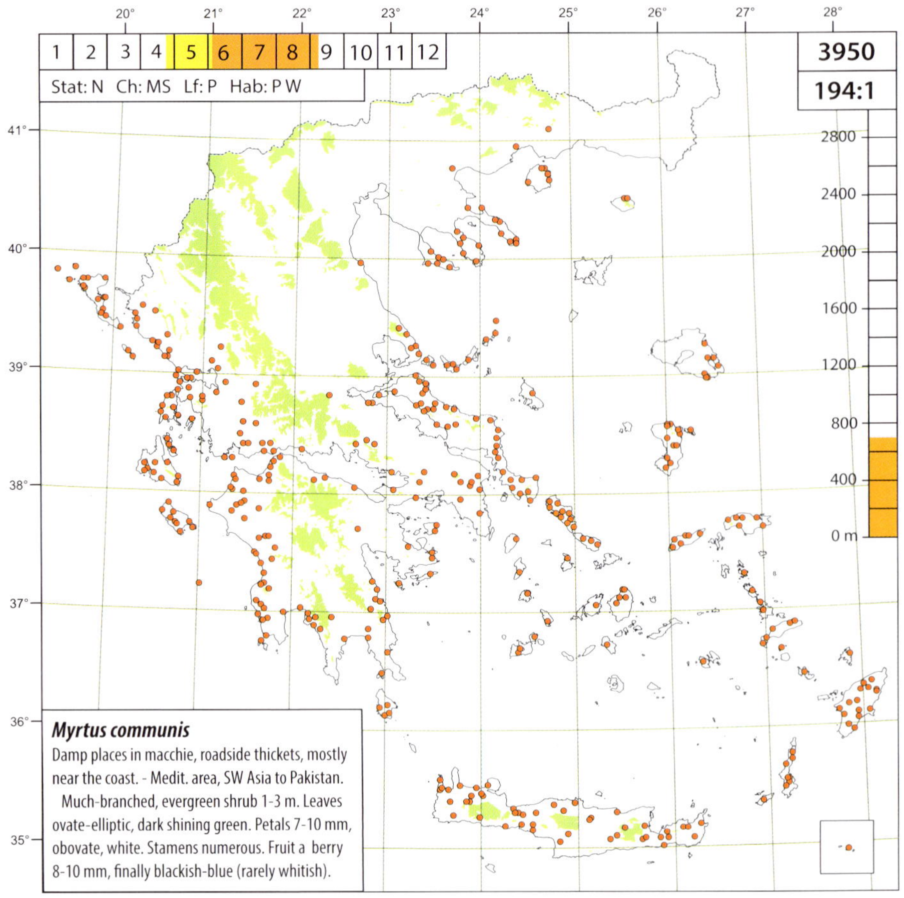

Myrtus communis

Damp places in macchie, roadside thickets, mostly near the coast. - Medit. area, SW Asia to Pakistan.

Much-branched, evergreen shrub 1-3 m. Leaves ovate-elliptic, dark shining green. Petals 7-10 mm, obovate, white. Stamens numerous. Fruit a berry 8-10 mm, finally blackish-blue (rarely whitish).

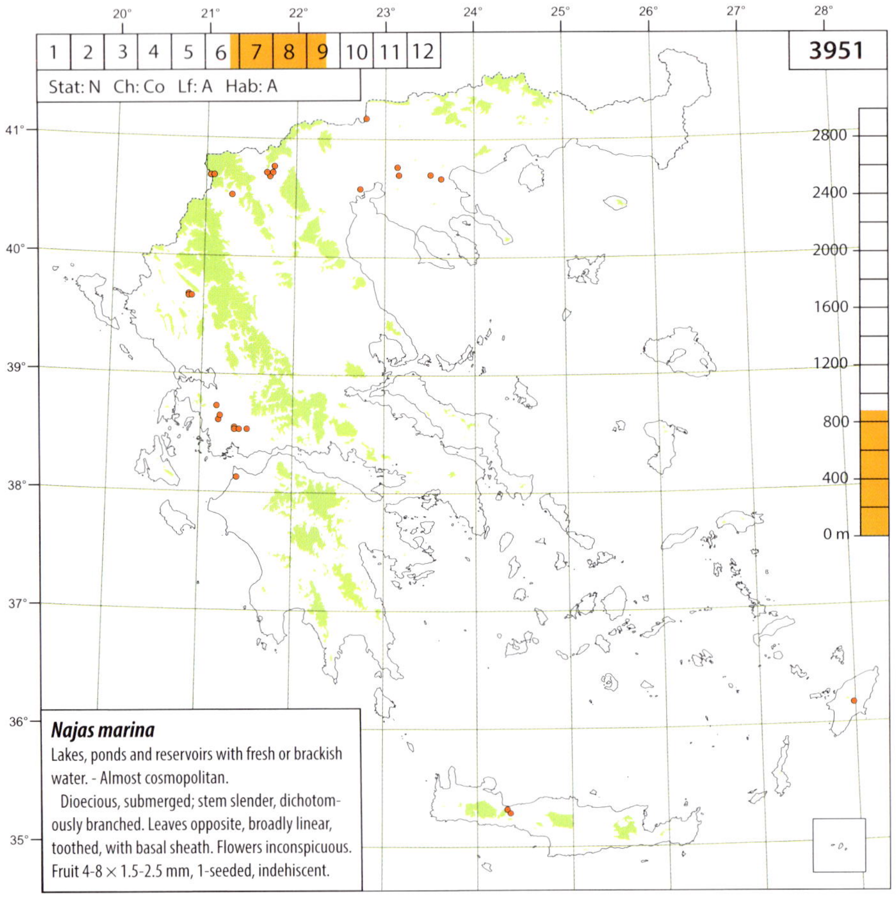

Najas marina

Lakes, ponds and reservoirs with fresh or brackish water. - Almost cosmopolitan.

Dioecious, submerged; stem slender, dichotomously branched. Leaves opposite, broadly linear, toothed, with basal sheath. Flowers inconspicuous. Fruit 4-8 × 1.5-2.5 mm, 1-seeded, indehiscent.

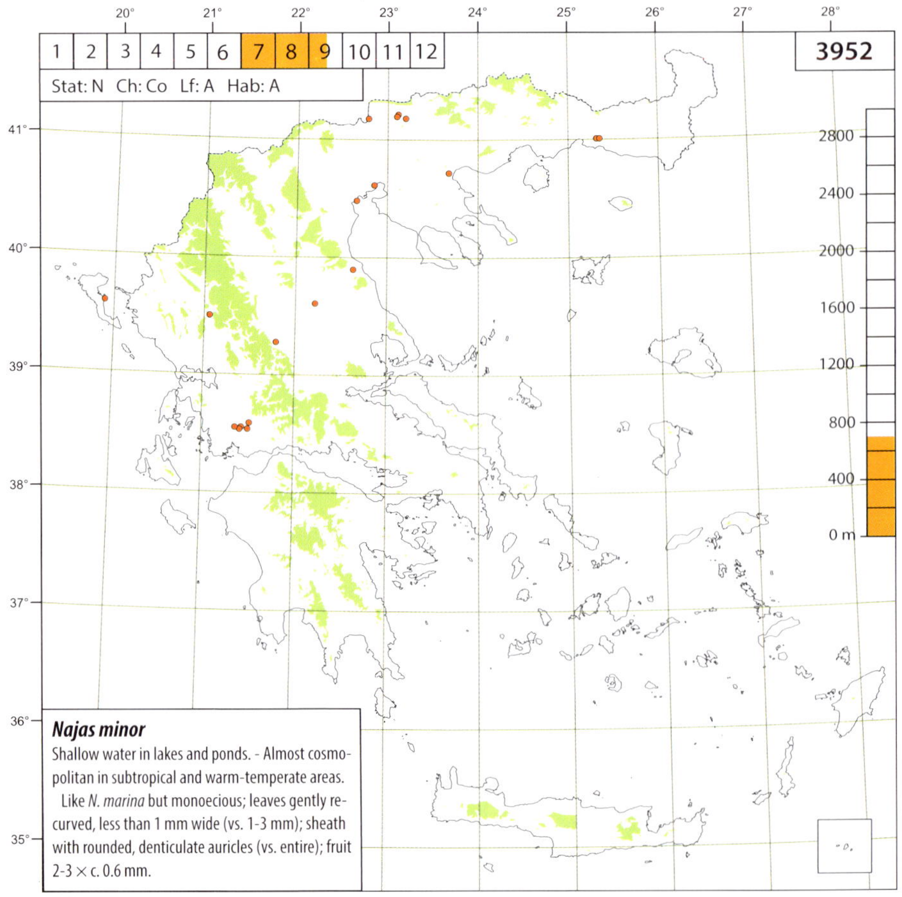

Najas minor

Shallow water in lakes and ponds. - Almost cosmopolitan in subtropical and warm-temperate areas.

Like *N. marina* but monoecious; leaves gently recurved, less than 1 mm wide (vs. 1-3 mm); sheath with rounded, denticulate auricles (vs. entire); fruit 2-3 × c. 0.6 mm.

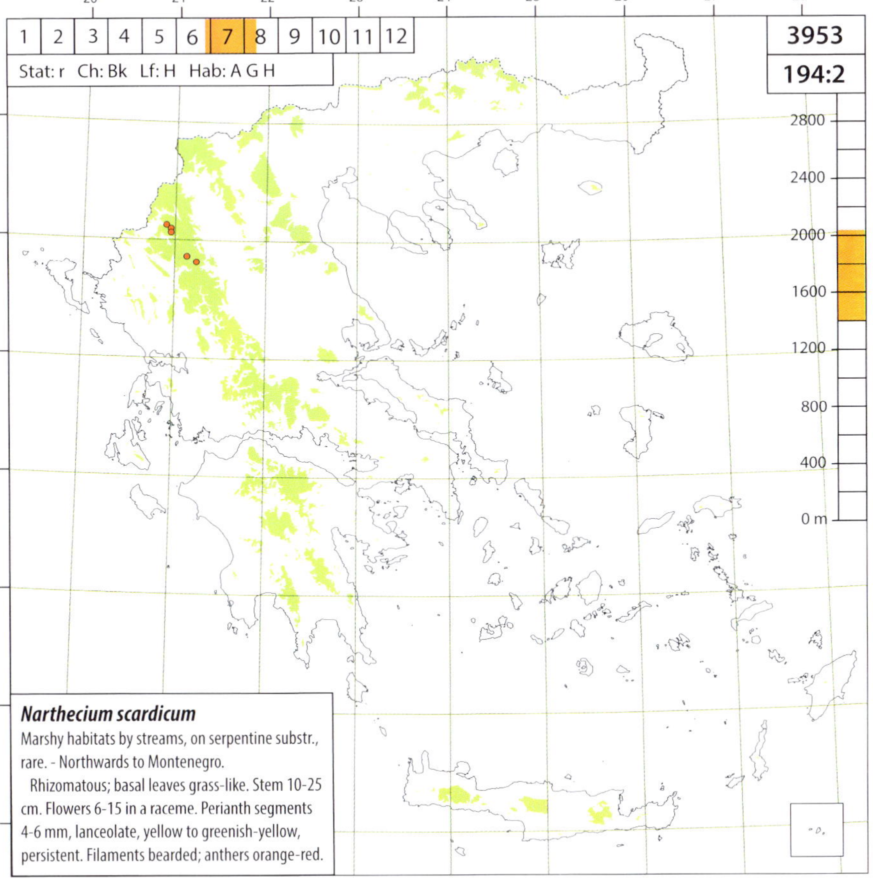

Narthecium scardicum

Marshy habitats by streams, on serpentine substr., rare. - Northwards to Montenegro.

Rhizomatous; basal leaves grass-like. Stem 10-25 cm. Flowers 6-15 in a raceme. Perianth segments 4-6 mm, lanceolate, yellow to greenish-yellow, persistent. Filaments bearded; anthers orange-red.

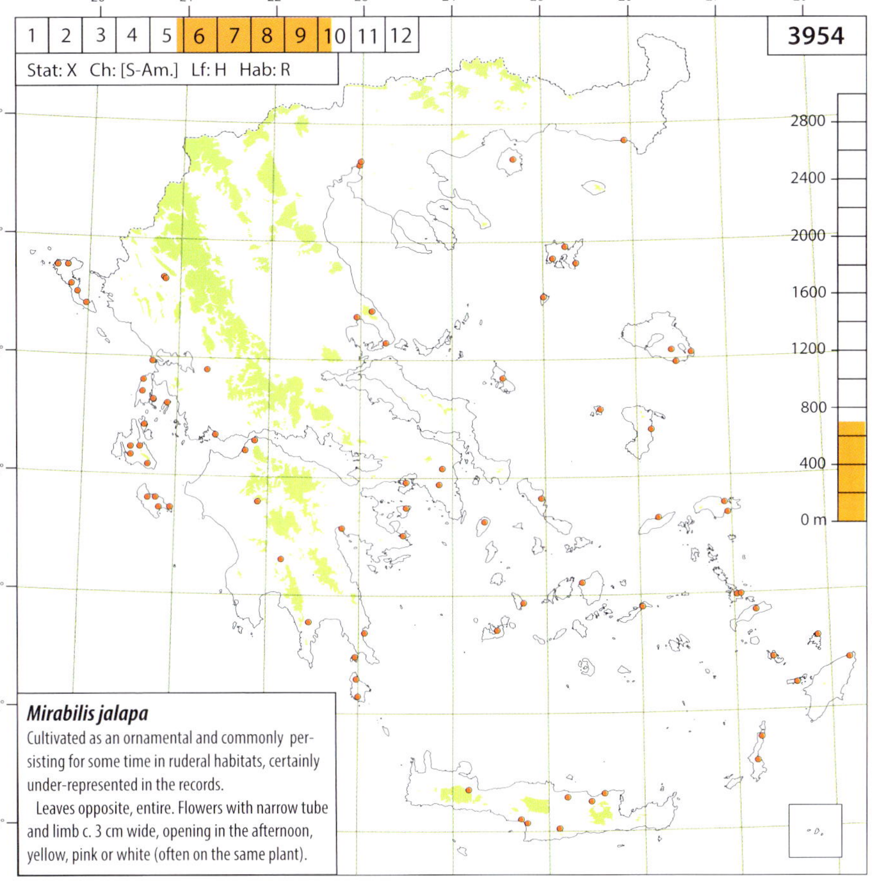

Mirabilis jalapa

Cultivated as an ornamental and commonly persisting for some time in ruderal habitats, certainly under-represented in the records.

Leaves opposite, entire. Flowers with narrow tube and limb c. 3 cm wide, opening in the afternoon, yellow, pink or white (often on the same plant).

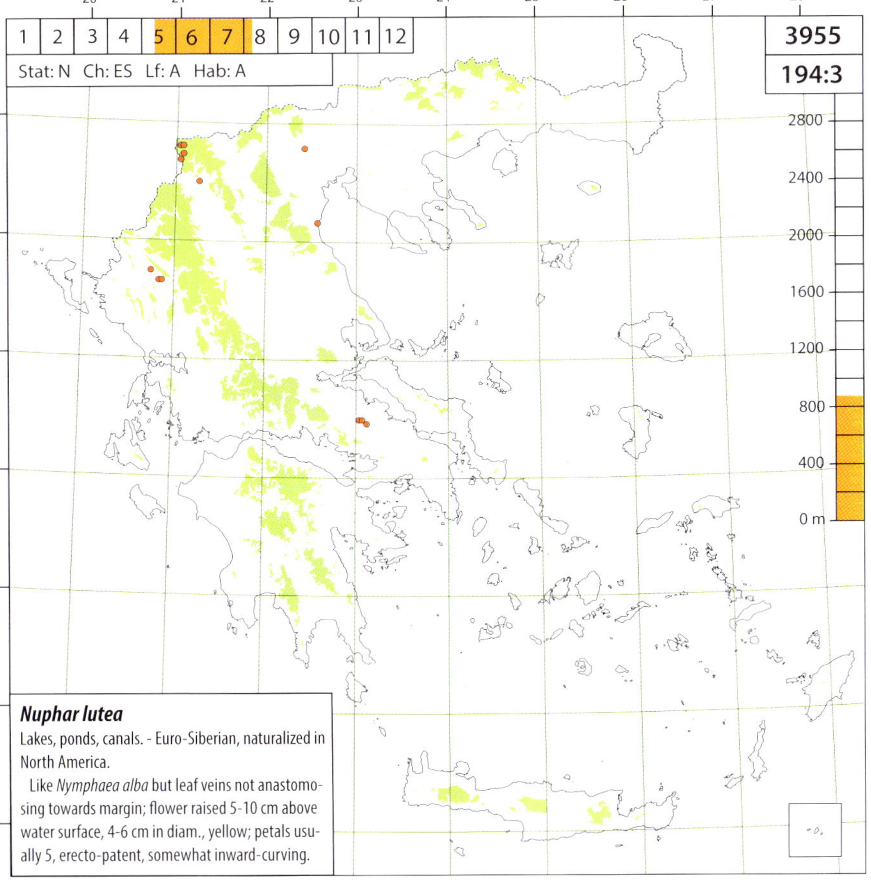

Nuphar lutea

Lakes, ponds, canals. - Euro-Siberian, naturalized in North America.

Like *Nymphaea alba* but leaf veins not anastomosing towards margin; flower raised 5-10 cm above water surface, 4-6 cm in diam., yellow; petals usually 5, erecto-patent, somewhat inward-curving.

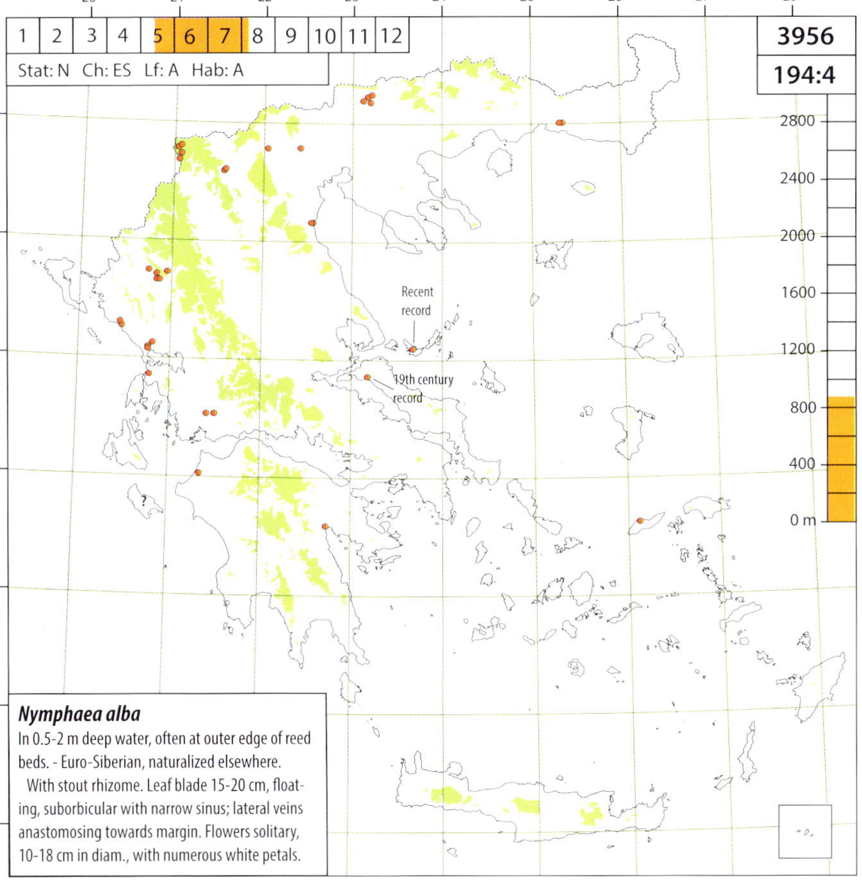

Nymphaea alba

In 0.5-2 m deep water, often at outer edge of reed beds. - Euro-Siberian, naturalized elsewhere.

With stout rhizome. Leaf blade 15-20 cm, floating, suborbicular with narrow sinus; lateral veins anastomosing towards margin. Flowers solitary, 10-18 cm in diam., with numerous white petals.

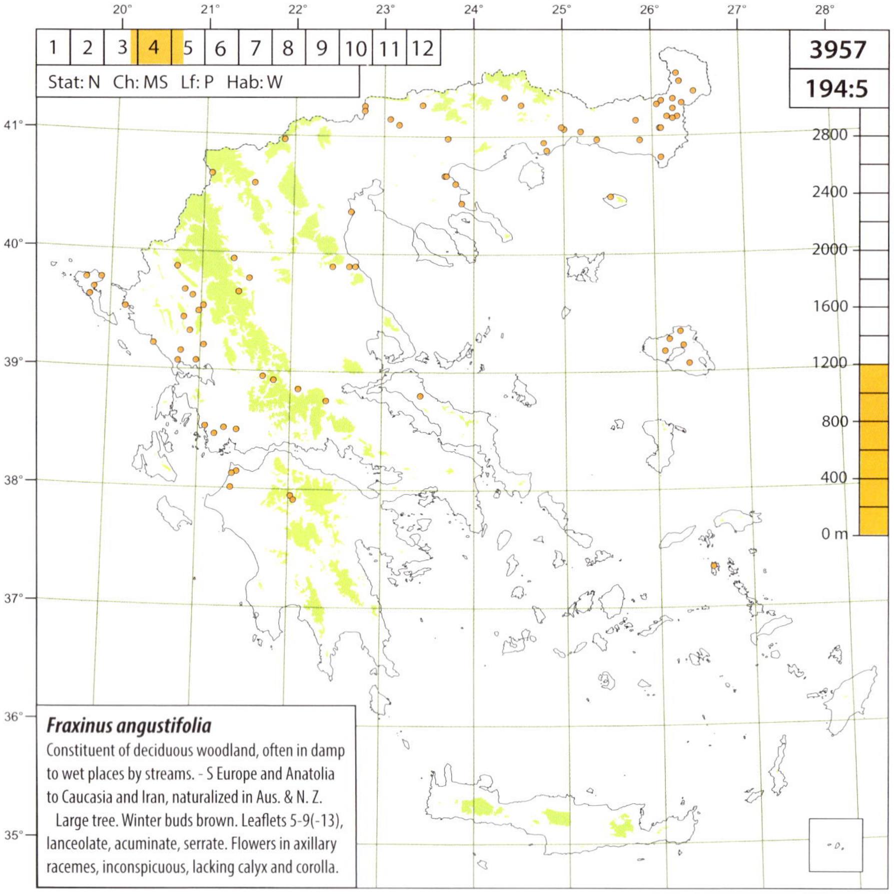

Fraxinus angustifolia

Constituent of deciduous woodland, often in damp to wet places by streams. - S Europe and Anatolia to Caucasia and Iran, naturalized in Aus. & N. Z.

Large tree. Winter buds brown. Leaflets 5-9(-13), lanceolate, acuminate, serrate. Flowers in axillary racemes, inconspicuous, lacking calyx and corolla.

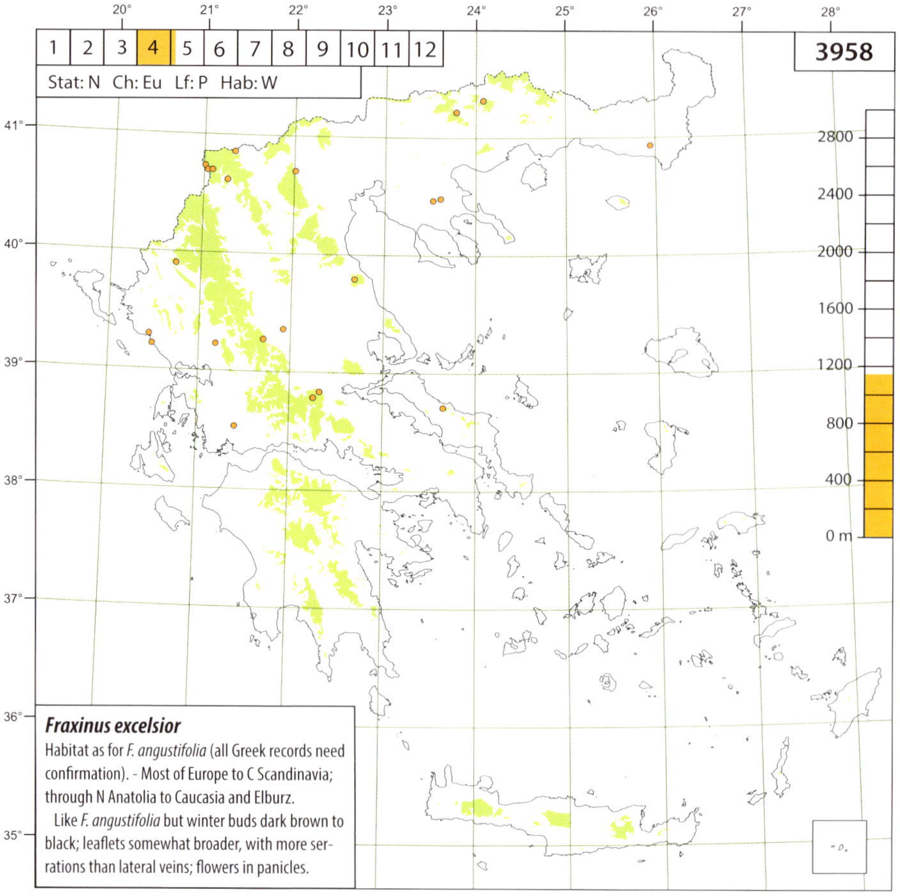

Fraxinus excelsior

Habitat as for *F. angustifolia* (all Greek records need confirmation). - Most of Europe to C Scandinavia; through N Anatolia to Caucasia and Elburz.

Like *F. angustifolia* but winter buds dark brown to black; leaflets somewhat broader, with more serrations than lateral veins; flowers in panicles.

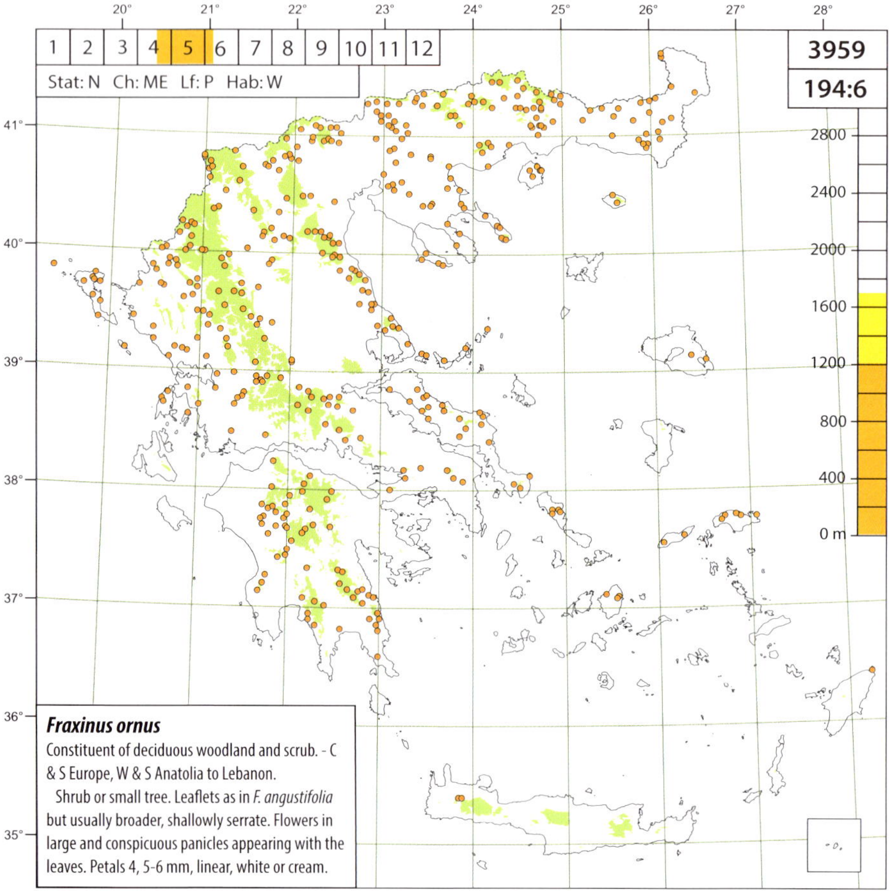

Fraxinus ornus

Constituent of deciduous woodland and scrub. - C & S Europe, W & S Anatolia to Lebanon.

Shrub or small tree. Leaflets as in *F. angustifolia* but usually broader, shallowly serrate. Flowers in large and conspicuous panicles appearing with the leaves. Petals 4, 5-6 mm, linear, white or cream.

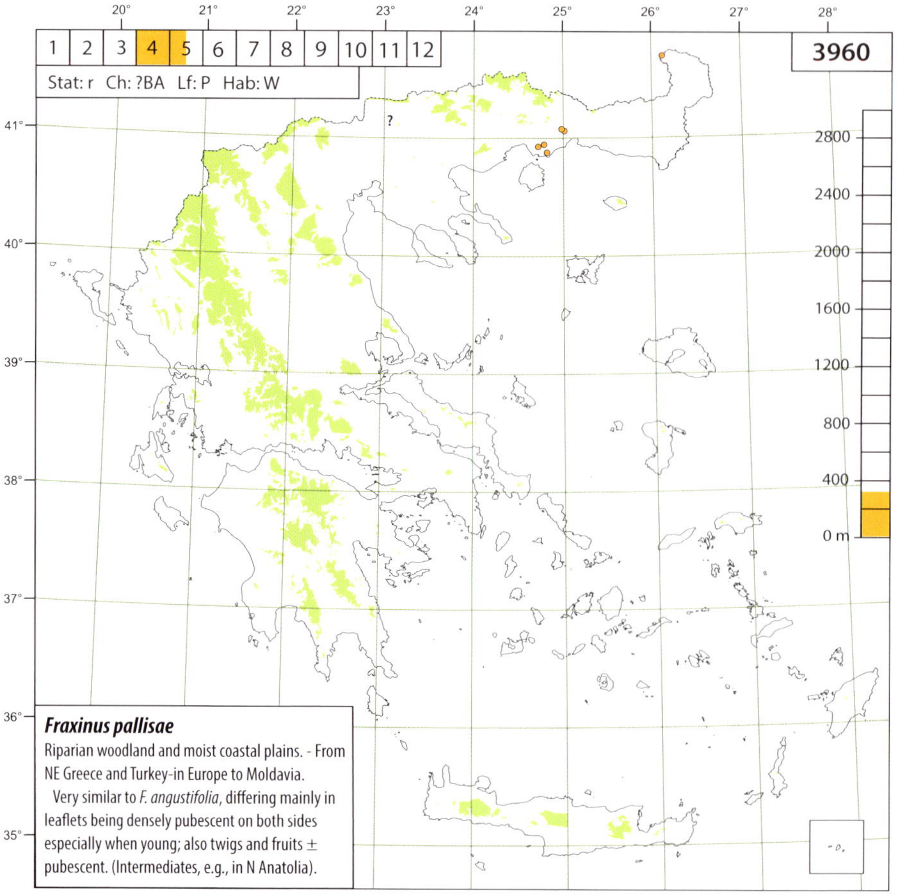

Fraxinus pallisae

Riparian woodland and moist coastal plains. - From NE Greece and Turkey-in Europe to Moldavia.

Very similar to *F. angustifolia*, differing mainly in leaflets being densely pubescent on both sides especially when young; also twigs and fruits ± pubescent. (Intermediates, e.g., in N Anatolia).

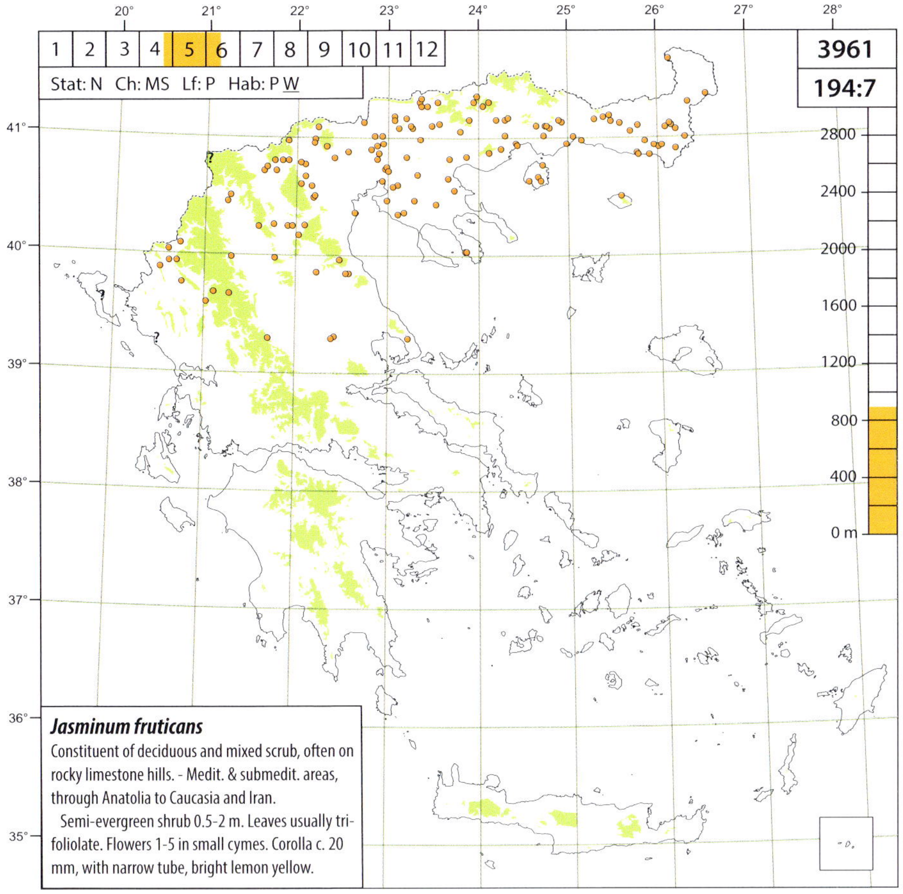

Jasminum fruticans
Constituent of deciduous and mixed scrub, often on rocky limestone hills. - Medit. & submedit. areas, through Anatolia to Caucasia and Iran.
Semi-evergreen shrub 0.5-2 m. Leaves usually trifoliolate. Flowers 1-5 in small cymes. Corolla c. 20 mm, with narrow tube, bright lemon yellow.

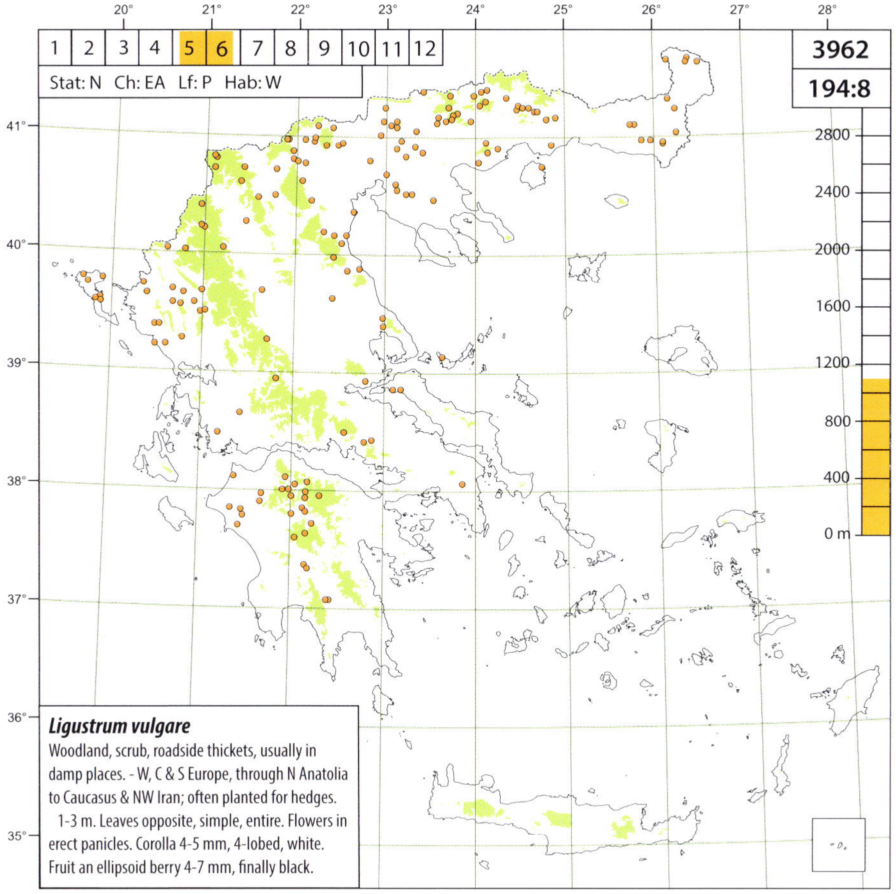

Ligustrum vulgare
Woodland, scrub, roadside thickets, usually in damp places. - W, C & S Europe, through N Anatolia to Caucasus & NW Iran; often planted for hedges.
1-3 m. Leaves opposite, simple, entire. Flowers in erect panicles. Corolla 4-5 mm, 4-lobed, white. Fruit an ellipsoid berry 4-7 mm, finally black.

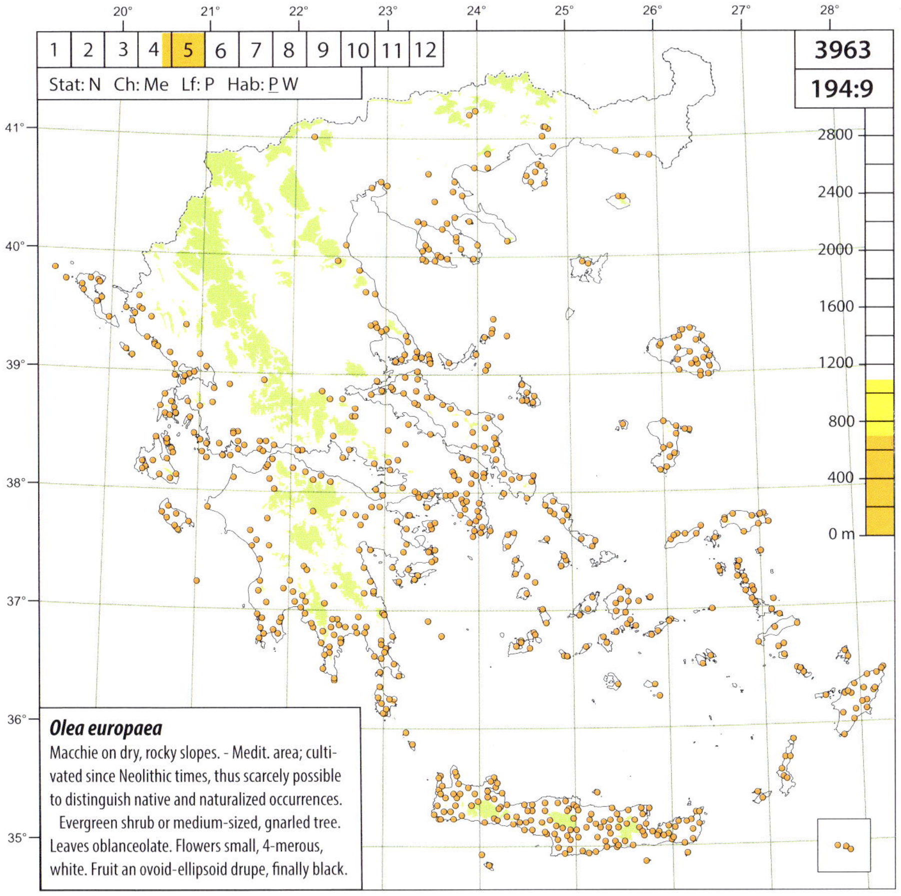

Olea europaea
Macchie on dry, rocky slopes. - Medit. area; cultivated since Neolithic times, thus scarcely possible to distinguish native and naturalized occurrences.
Evergreen shrub or medium-sized, gnarled tree. Leaves oblanceolate. Flowers small, 4-merous, white. Fruit an ovoid-ellipsoid drupe, finally black.

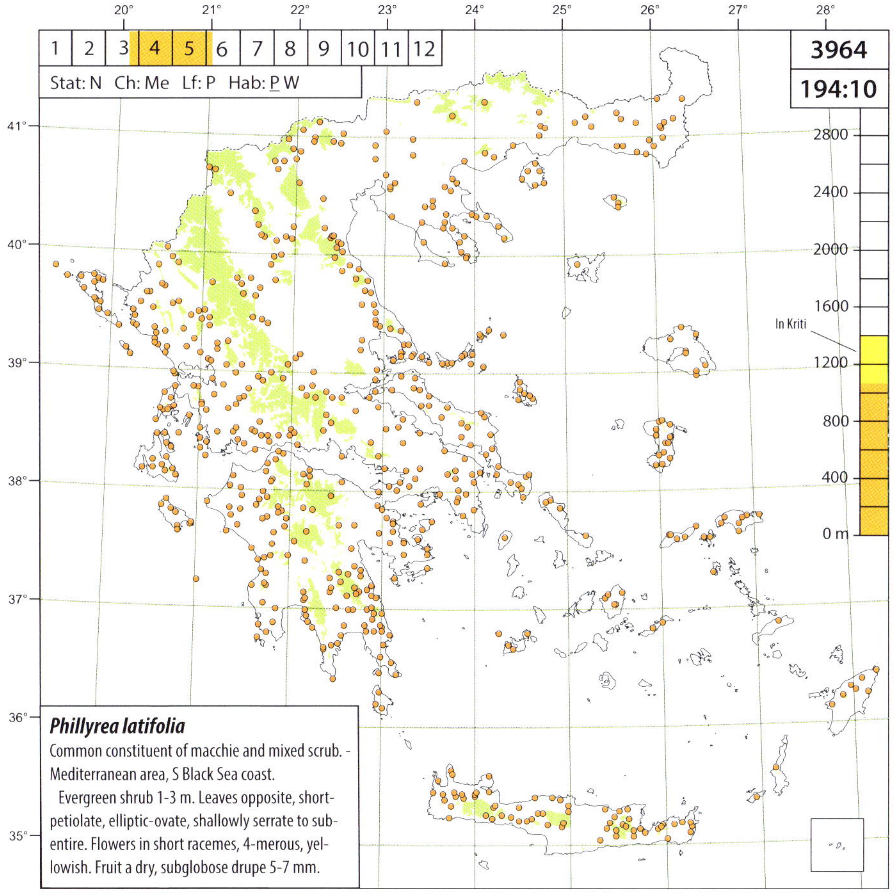

Phillyrea latifolia
Common constituent of macchie and mixed scrub. - Mediterranean area, S Black Sea coast.
Evergreen shrub 1-3 m. Leaves opposite, short-petiolate, elliptic-ovate, shallowly serrate to subentire. Flowers in short racemes, 4-merous, yellowish. Fruit a dry, subglobose drupe 5-7 mm.

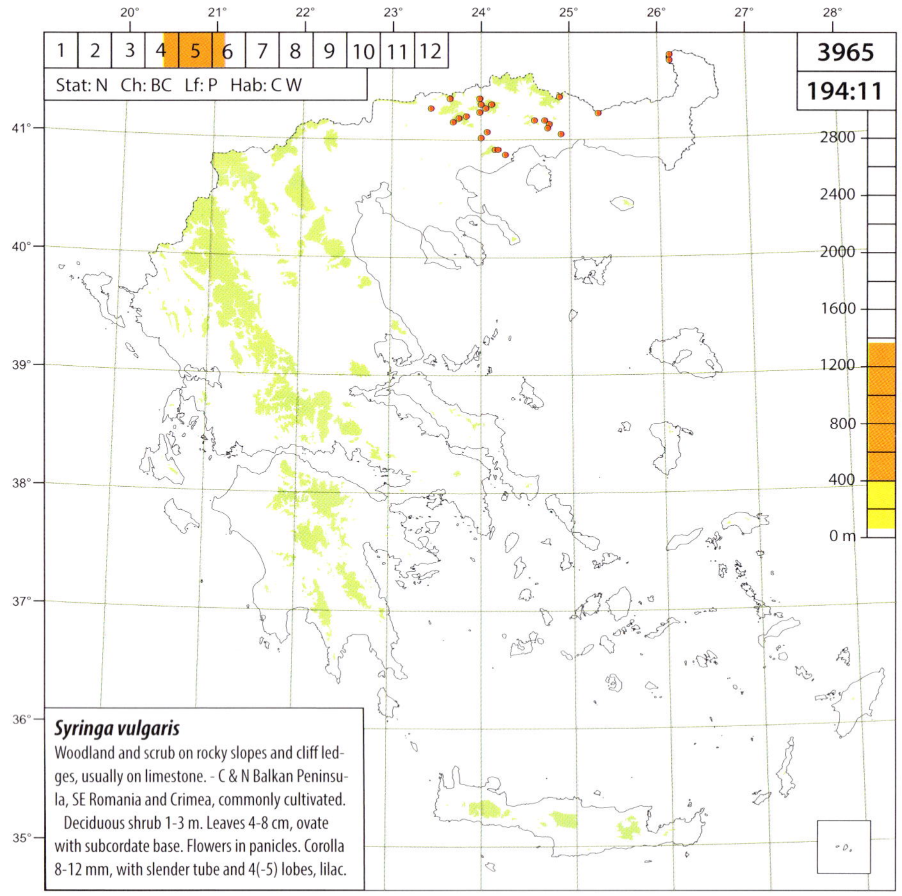

Syringa vulgaris

Woodland and scrub on rocky slopes and cliff ledges, usually on limestone. - C & N Balkan Peninsula, SE Romania and Crimea, commonly cultivated.

Deciduous shrub 1-3 m. Leaves 4-8 cm, ovate with subcordate base. Flowers in panicles. Corolla 8-12 mm, with slender tube and 4(-5) lobes, lilac.

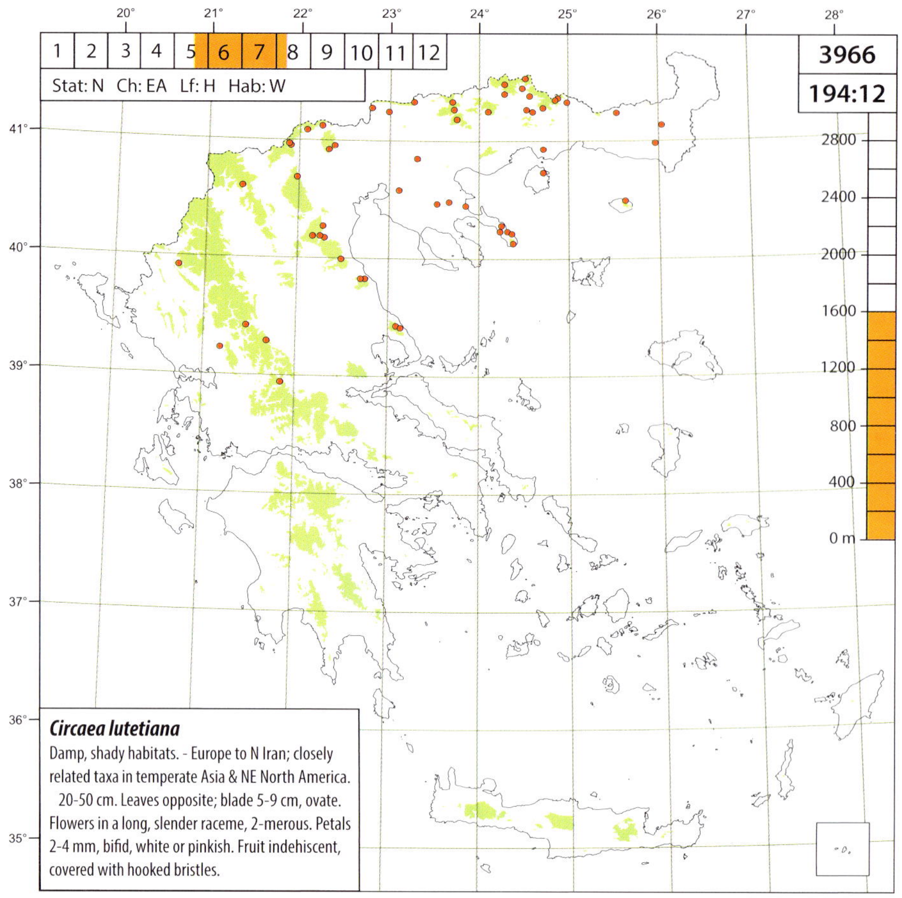

Circaea lutetiana

Damp, shady habitats. - Europe to N Iran; closely related taxa in temperate Asia & NE North America.

20-50 cm. Leaves opposite; blade 5-9 cm, ovate. Flowers in a long, slender raceme, 2-merous. Petals 2-4 mm, bifid, white or pinkish. Fruit indehiscent, covered with hooked bristles.

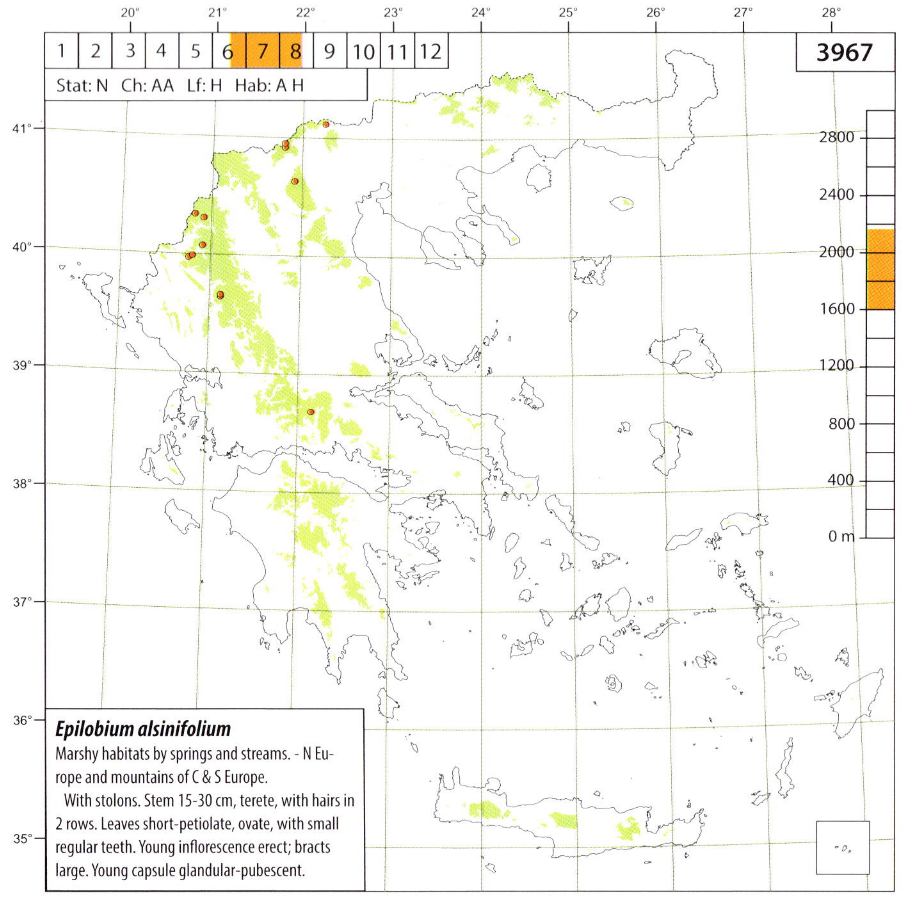

Epilobium alsinifolium

Marshy habitats by springs and streams. - N Europe and mountains of C & S Europe.

With stolons. Stem 15-30 cm, terete, with hairs in 2 rows. Leaves short-petiolate, ovate, with small regular teeth. Young inflorescence erect; bracts large. Young capsule glandular-pubescent.

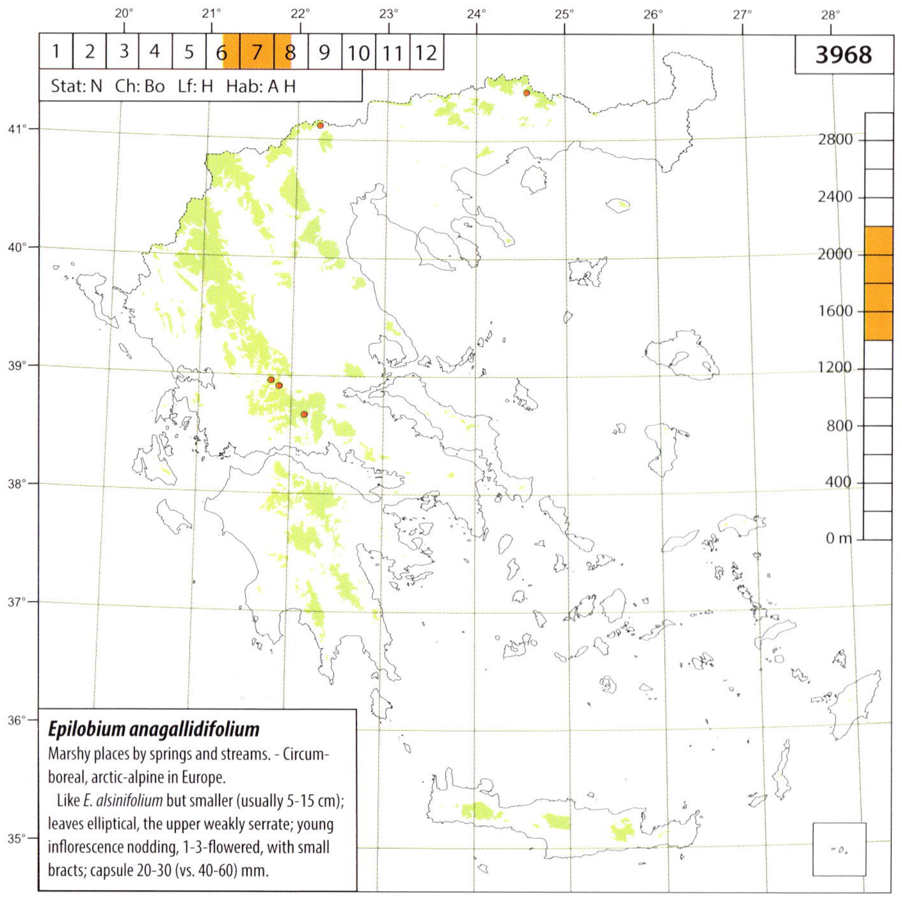

Epilobium anagallidifolium

Marshy places by springs and streams. - Circumboreal, arctic-alpine in Europe.

Like *E. alsinifolium* but smaller (usually 5-15 cm); leaves elliptical, the upper weakly serrate; young inflorescence nodding, 1-3-flowered, with small bracts; capsule 20-30 (vs. 40-60) mm.

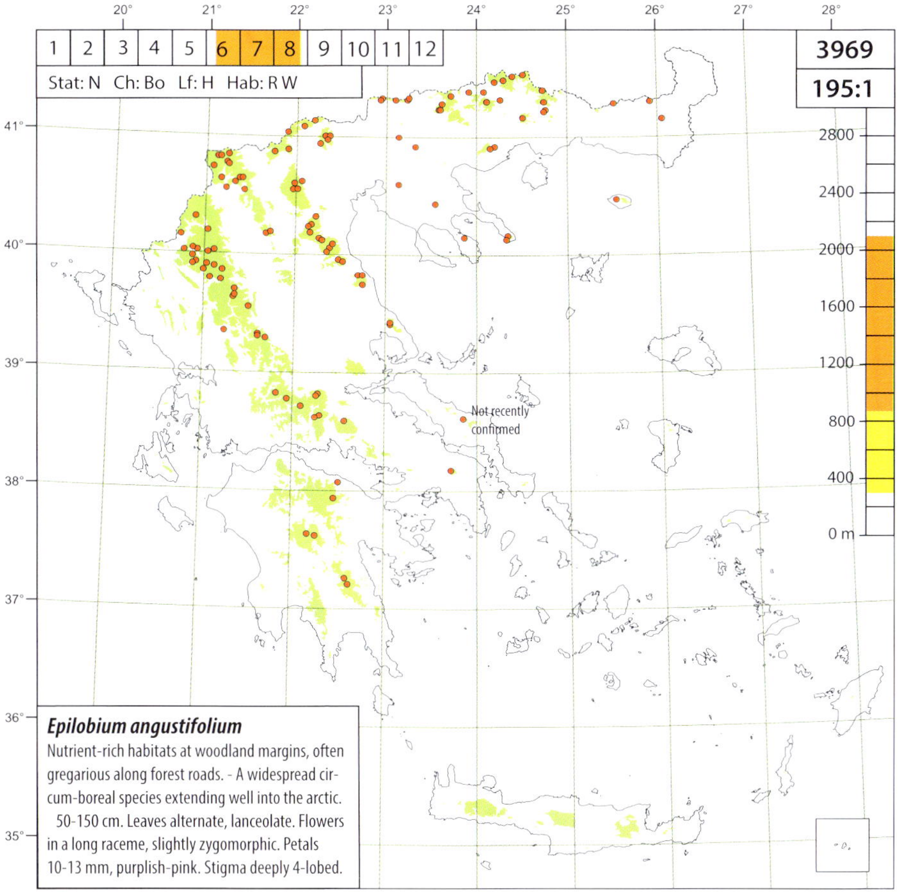

Epilobium angustifolium

Nutrient-rich habitats at woodland margins, often gregarious along forest roads. - A widespread circum-boreal species extending well into the arctic.
50-150 cm. Leaves alternate, lanceolate. Flowers in a long raceme, slightly zygomorphic. Petals 10-13 mm, purplish-pink. Stigma deeply 4-lobed.

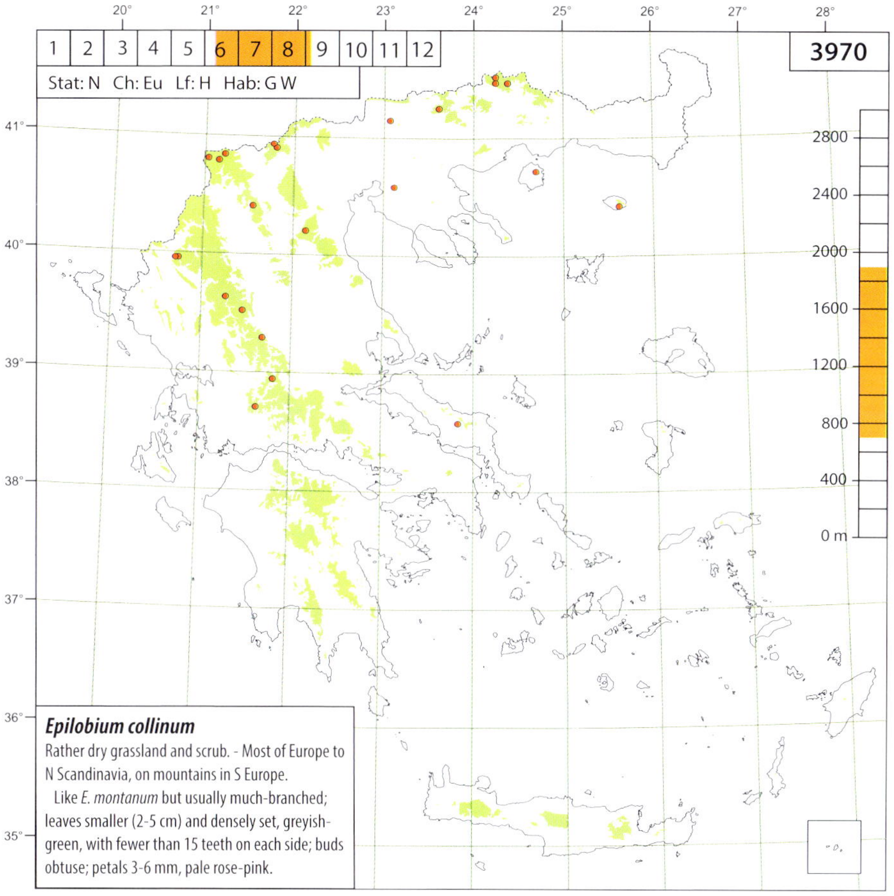

Epilobium collinum

Rather dry grassland and scrub. - Most of Europe to N Scandinavia, on mountains in S Europe.
Like *E. montanum* but usually much-branched; leaves smaller (2-5 cm) and densely set, greyish-green, with fewer than 15 teeth on each side; buds obtuse; petals 3-6 mm, pale rose-pink.

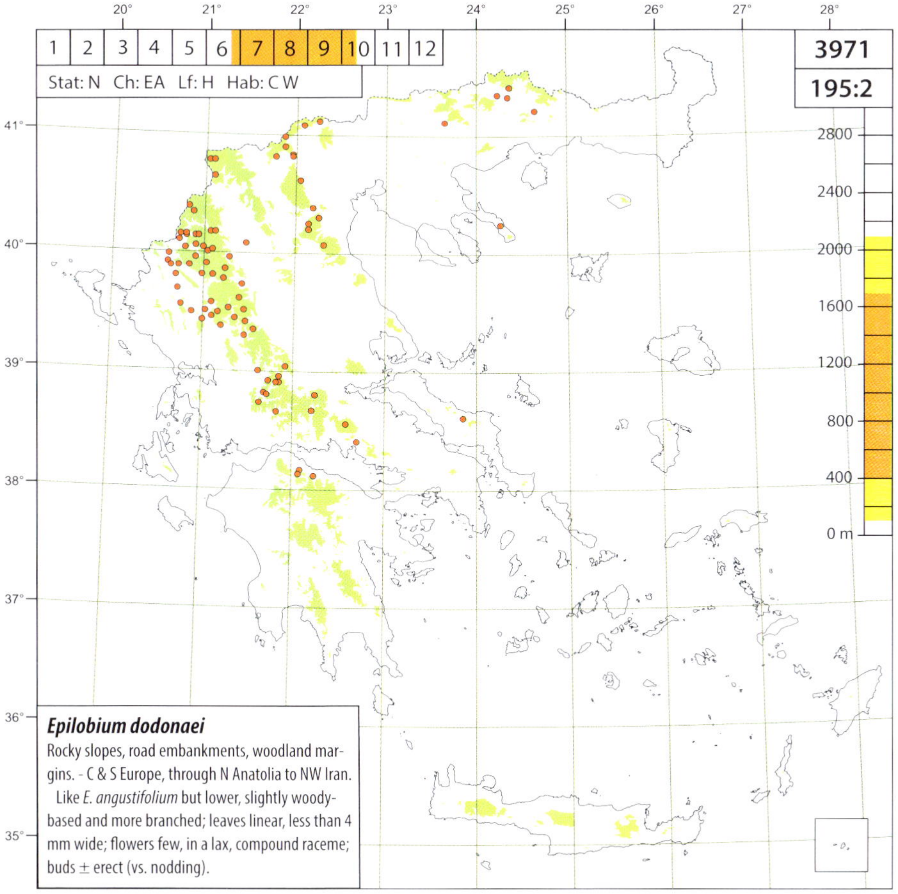

Epilobium dodonaei

Rocky slopes, road embankments, woodland margins. - C & S Europe, through N Anatolia to NW Iran.
Like *E. angustifolium* but lower, slightly woody-based and more branched; leaves linear, less than 4 mm wide; flowers few, in a lax, compound raceme; buds ± erect (vs. nodding).

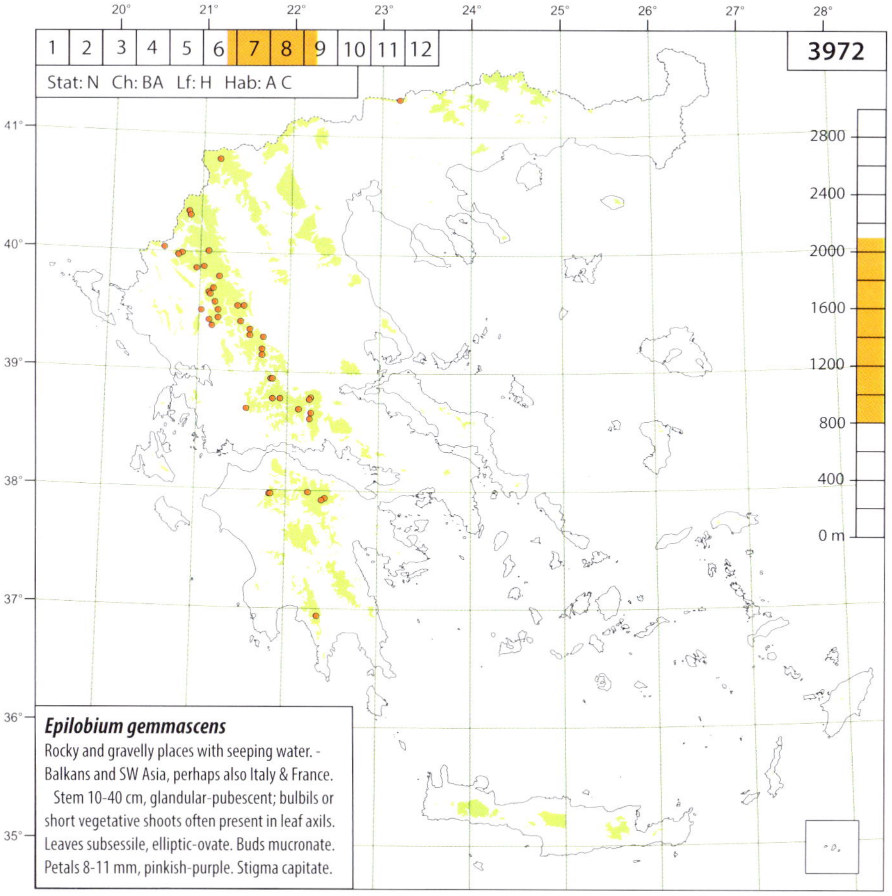

Epilobium gemmascens

Rocky and gravelly places with seeping water. - Balkans and SW Asia, perhaps also Italy & France.
Stem 10-40 cm, glandular-pubescent; bulbils or short vegetative shoots often present in leaf axils. Leaves subsessile, elliptic-ovate. Buds mucronate. Petals 8-11 mm, pinkish-purple. Stigma capitate.

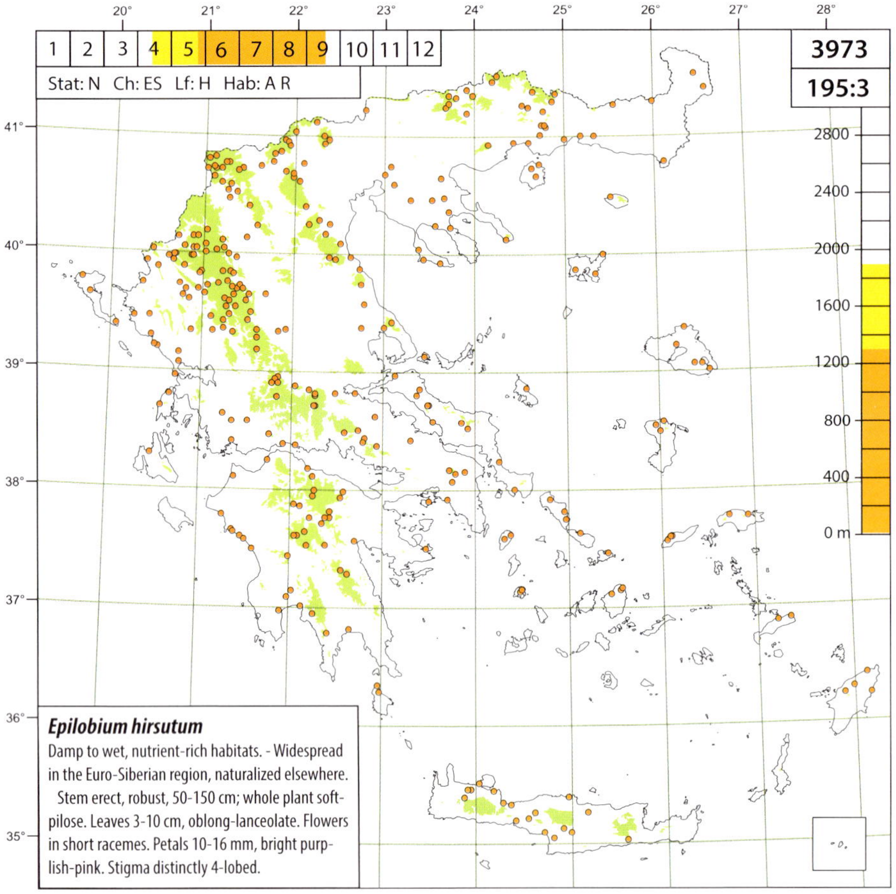

Epilobium hirsutum

Damp to wet, nutrient-rich habitats. - Widespread in the Euro-Siberian region, naturalized elsewhere.

Stem erect, robust, 50-150 cm; whole plant soft-pilose. Leaves 3-10 cm, oblong-lanceolate. Flowers in short racemes. Petals 10-16 mm, bright purplish-pink. Stigma distinctly 4-lobed.

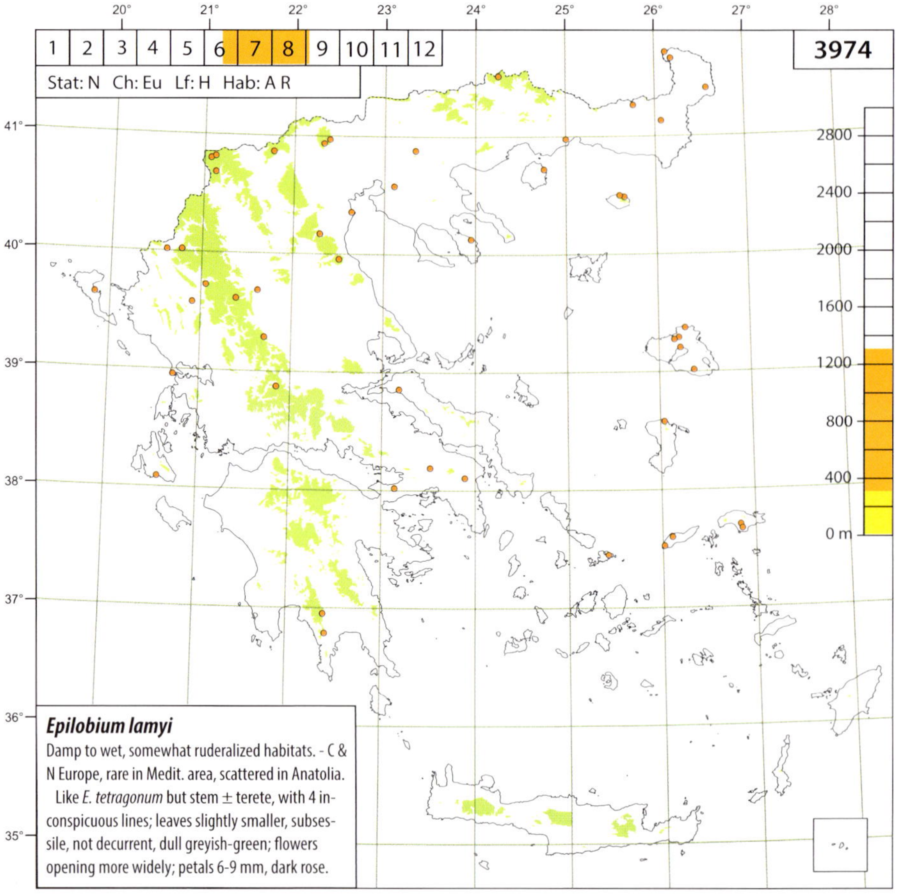

Epilobium lamyi

Damp to wet, somewhat ruderalized habitats. - C & N Europe, rare in Medit. area, scattered in Anatolia.

Like *E. tetragonum* but stem ± terete, with 4 inconspicuous lines; leaves slightly smaller, subsessile, not decurrent, dull greyish-green; flowers opening more widely; petals 6-9 mm, dark rose.

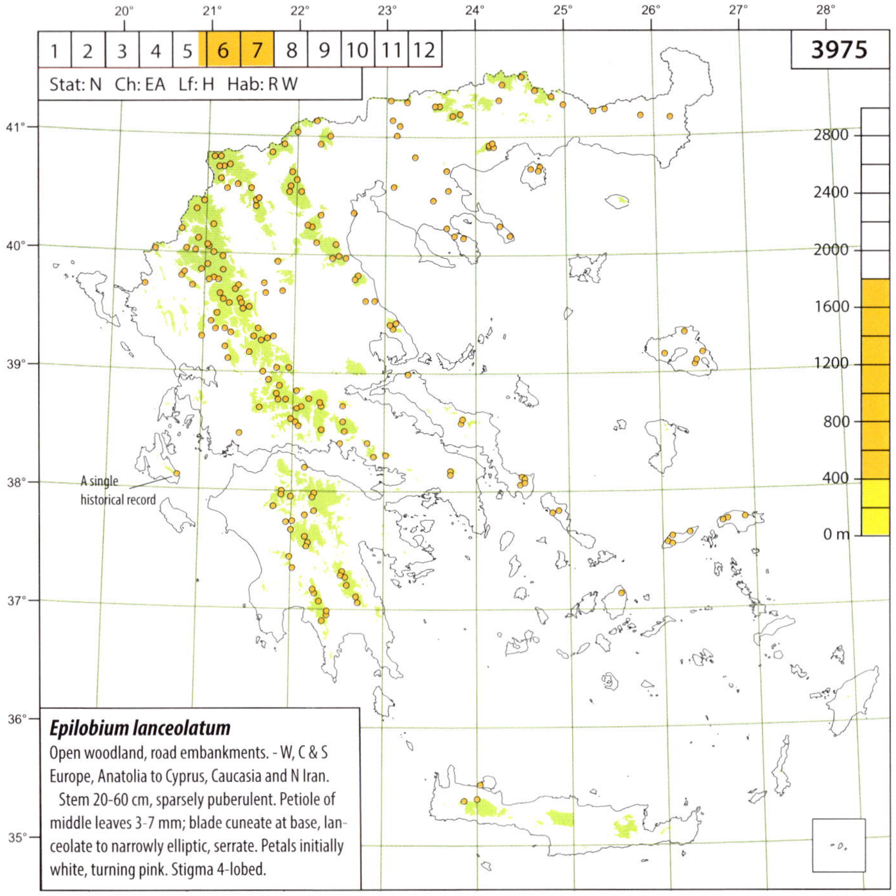

Epilobium lanceolatum

Open woodland, road embankments. - W, C & S Europe, Anatolia to Cyprus, Caucasia and N Iran.

Stem 20-60 cm, sparsely puberulent. Petiole of middle leaves 3-7 mm; blade cuneate at base, lanceolate to narrowly elliptic, serrate. Petals initially white, turning pink. Stigma 4-lobed.

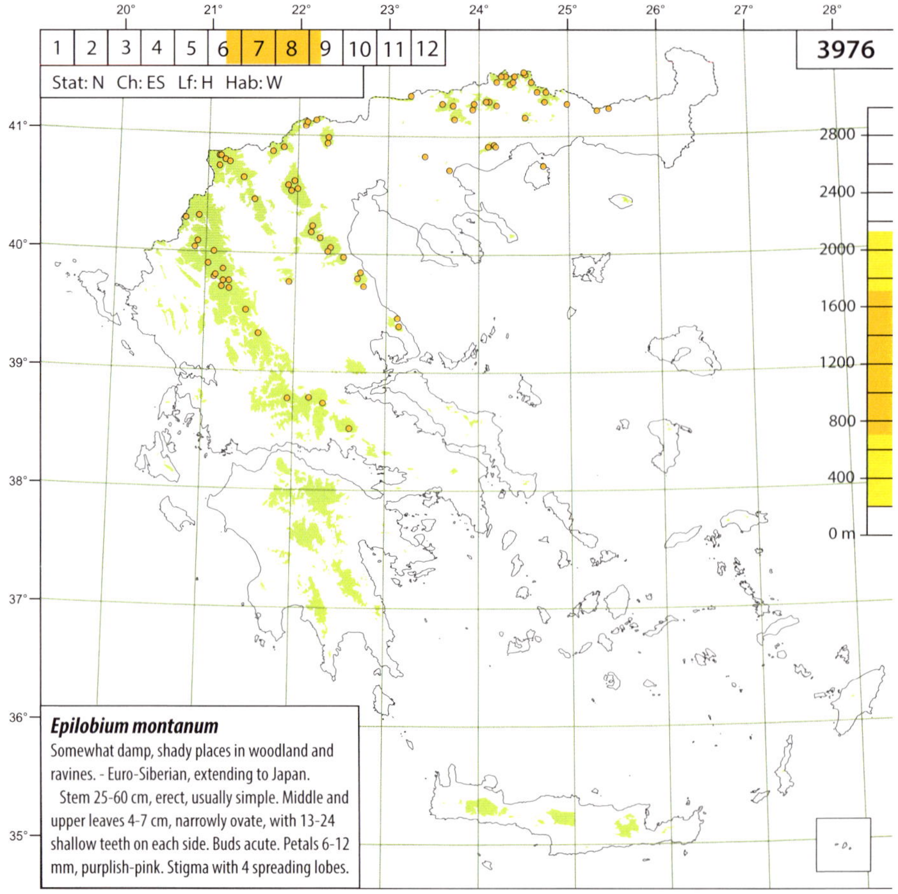

Epilobium montanum

Somewhat damp, shady places in woodland and ravines. - Euro-Siberian, extending to Japan.

Stem 25-60 cm, erect, usually simple. Middle and upper leaves 4-7 cm, narrowly ovate, with 13-24 shallow teeth on each side. Buds acute. Petals 6-12 mm, purplish-pink. Stigma with 4 spreading lobes.

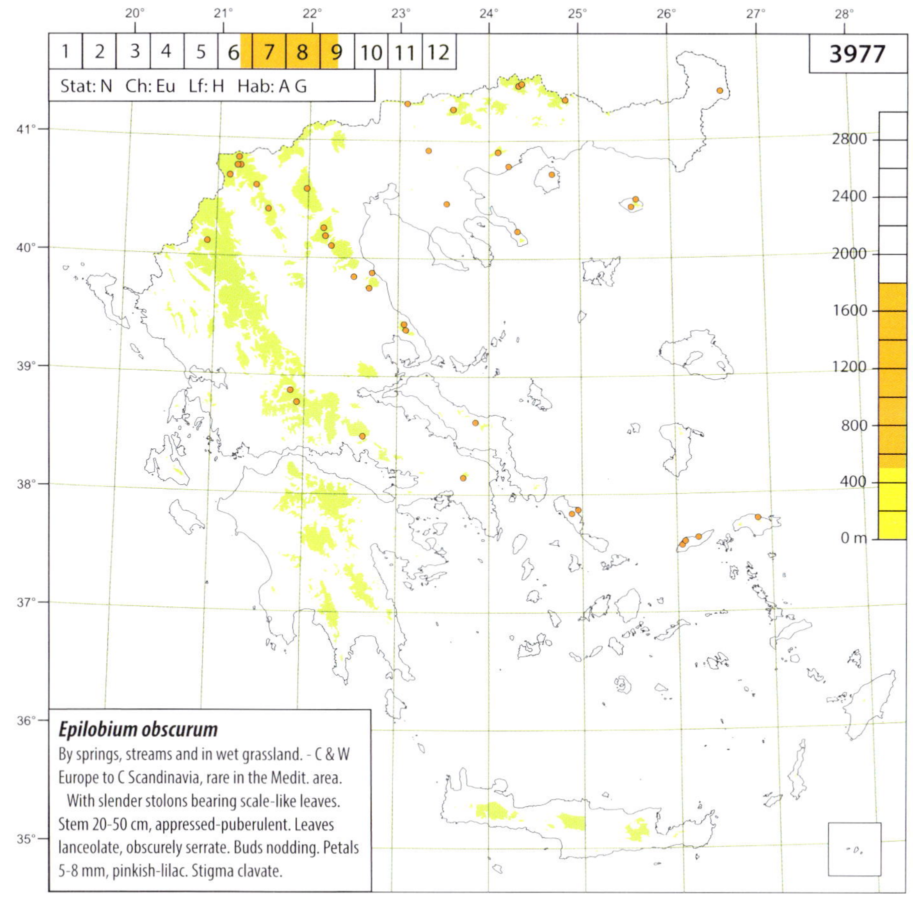

Epilobium obscurum

By springs, streams and in wet grassland. - C & W Europe to C Scandinavia, rare in the Medit. area.

With slender stolons bearing scale-like leaves. Stem 20-50 cm, appressed-puberulent. Leaves lanceolate, obscurely serrate. Buds nodding. Petals 5-8 mm, pinkish-lilac. Stigma clavate.

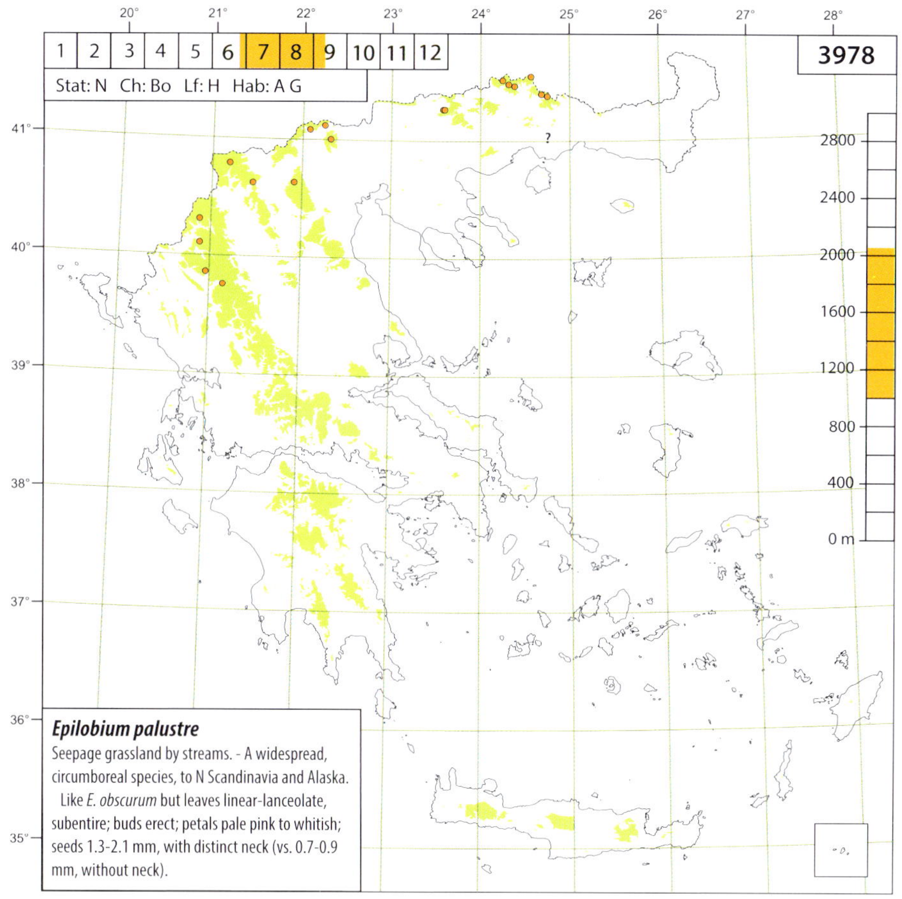

Epilobium palustre

Seepage grassland by streams. - A widespread, circumboreal species, to N Scandinavia and Alaska.

Like *E. obscurum* but leaves linear-lanceolate, subentire; buds erect; petals pale pink to whitish; seeds 1.3-2.1 mm, with distinct neck (vs. 0.7-0.9 mm, without neck).

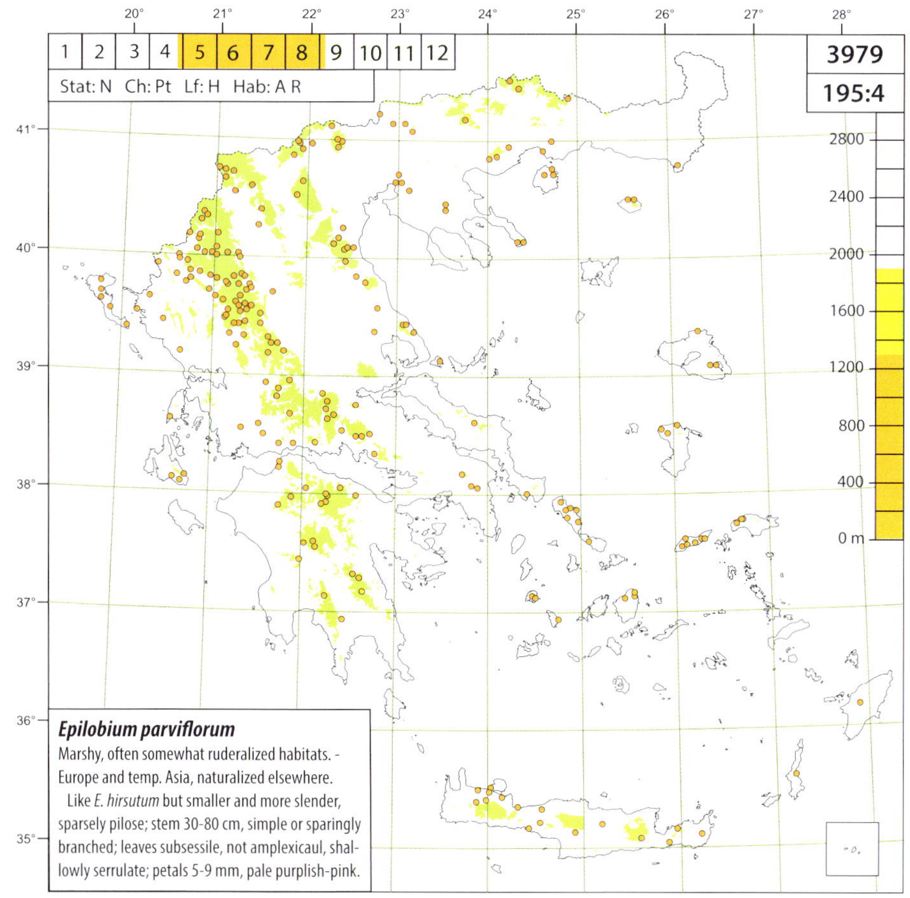

Epilobium parviflorum

Marshy, often somewhat ruderalized habitats. - Europe and temp. Asia, naturalized elsewhere.

Like *E. hirsutum* but smaller and more slender, sparsely pilose; stem 30-80 cm, simple or sparingly branched; leaves subsessile, not amplexicaul, shallowly serrulate; petals 5-9 mm, pale purplish-pink.

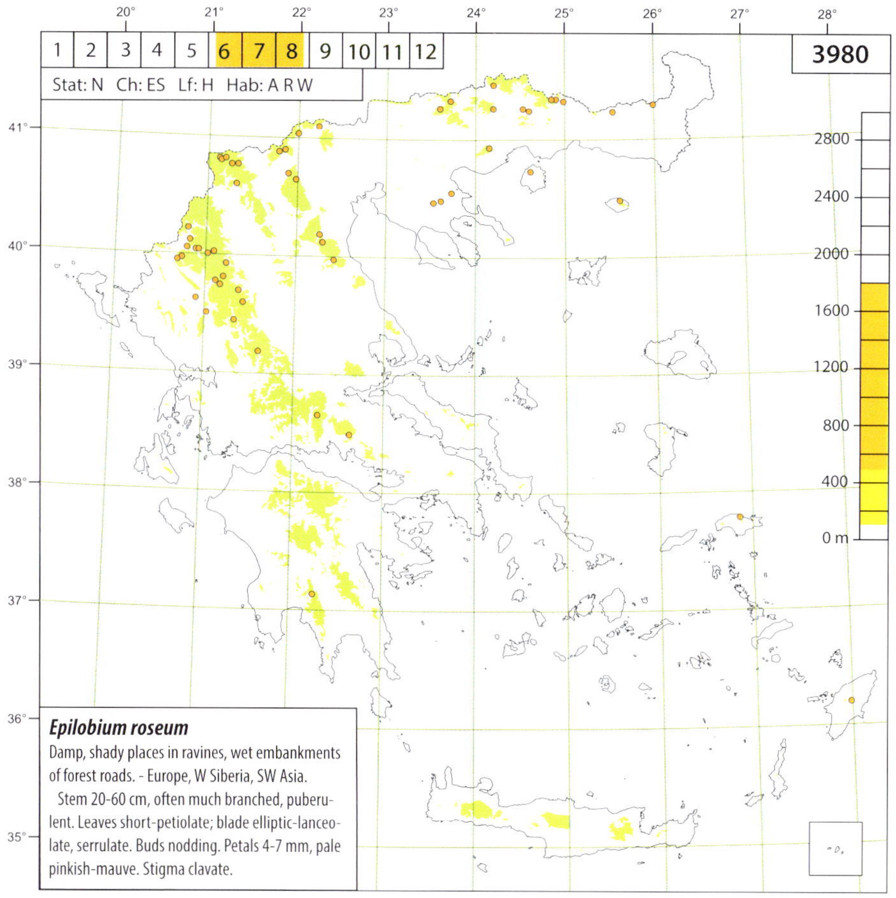

Epilobium roseum

Damp, shady places in ravines, wet embankments of forest roads. - Europe, W Siberia, SW Asia.

Stem 20-60 cm, often much branched, puberulent. Leaves short-petiolate; blade elliptic-lanceolate, serrulate. Buds nodding. Petals 4-7 mm, pale pinkish-mauve. Stigma clavate.

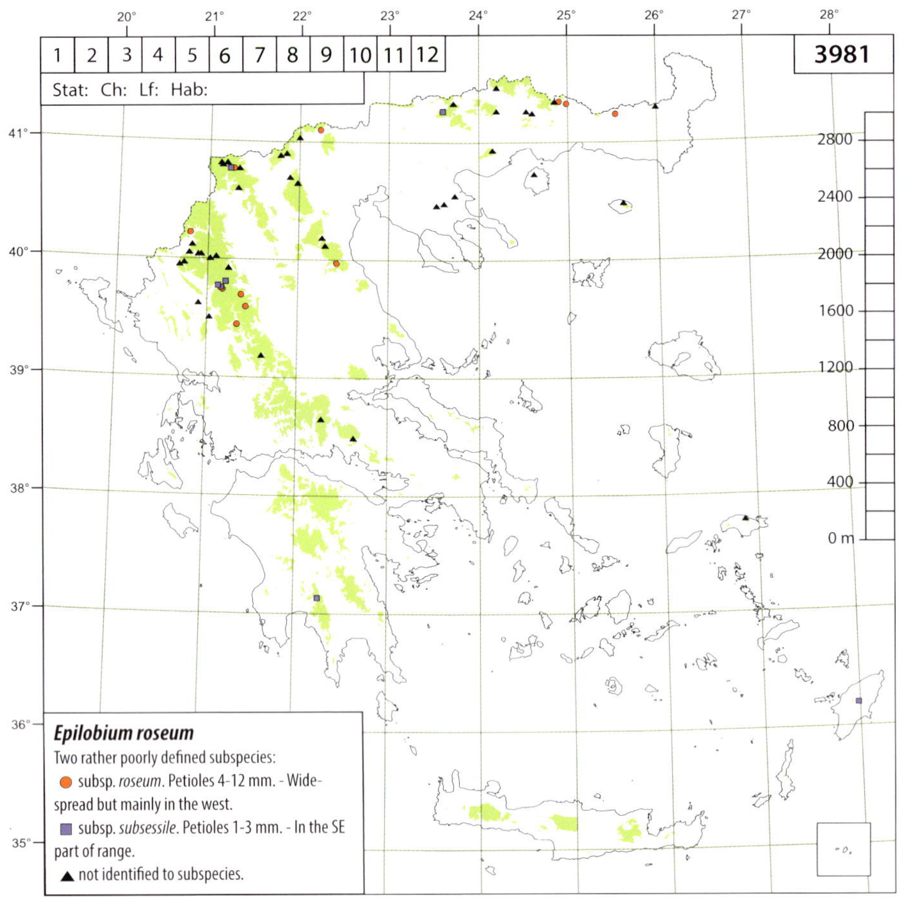

Epilobium roseum

Two rather poorly defined subspecies:

● subsp. *roseum*. Petioles 4-12 mm. - Widespread but mainly in the west.

■ subsp. *subsessile*. Petioles 1-3 mm. - In the SE part of range.

▲ not identified to subspecies.

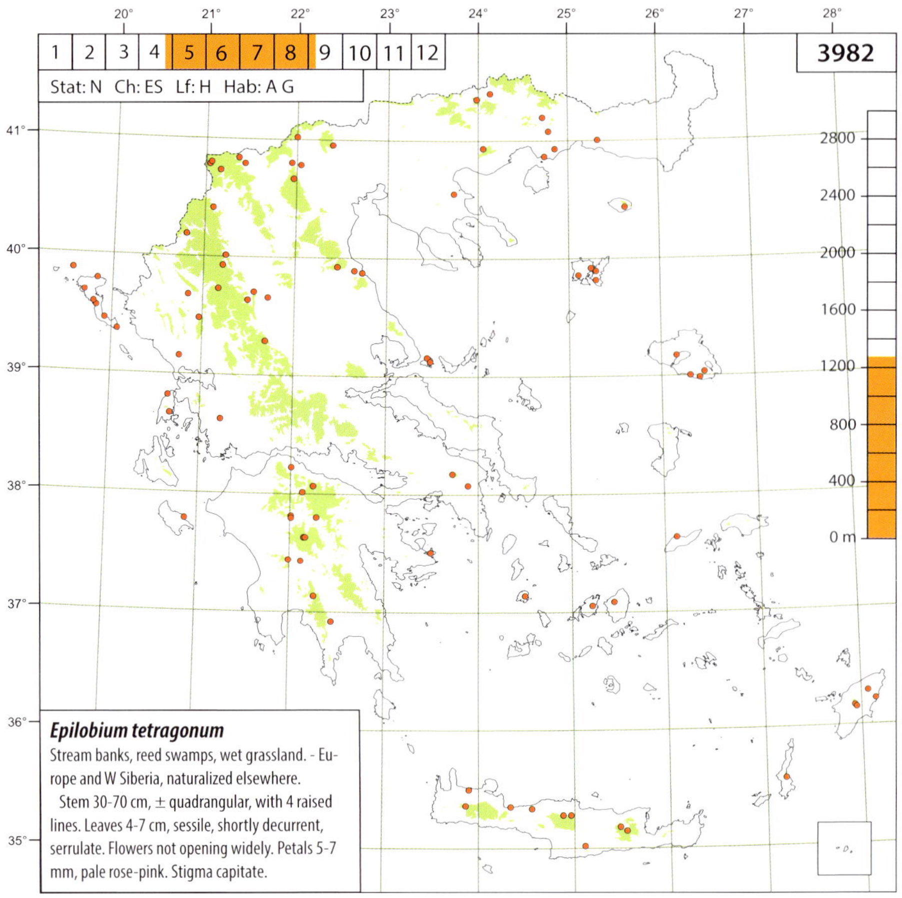

Epilobium tetragonum

Stream banks, reed swamps, wet grassland. - Europe and W Siberia, naturalized elsewhere.

Stem 30-70 cm, ± quadrangular, with 4 raised lines. Leaves 4-7 cm, sessile, shortly decurrent, serrulate. Flowers not opening widely. Petals 5-7 mm, pale rose-pink. Stigma capitate.

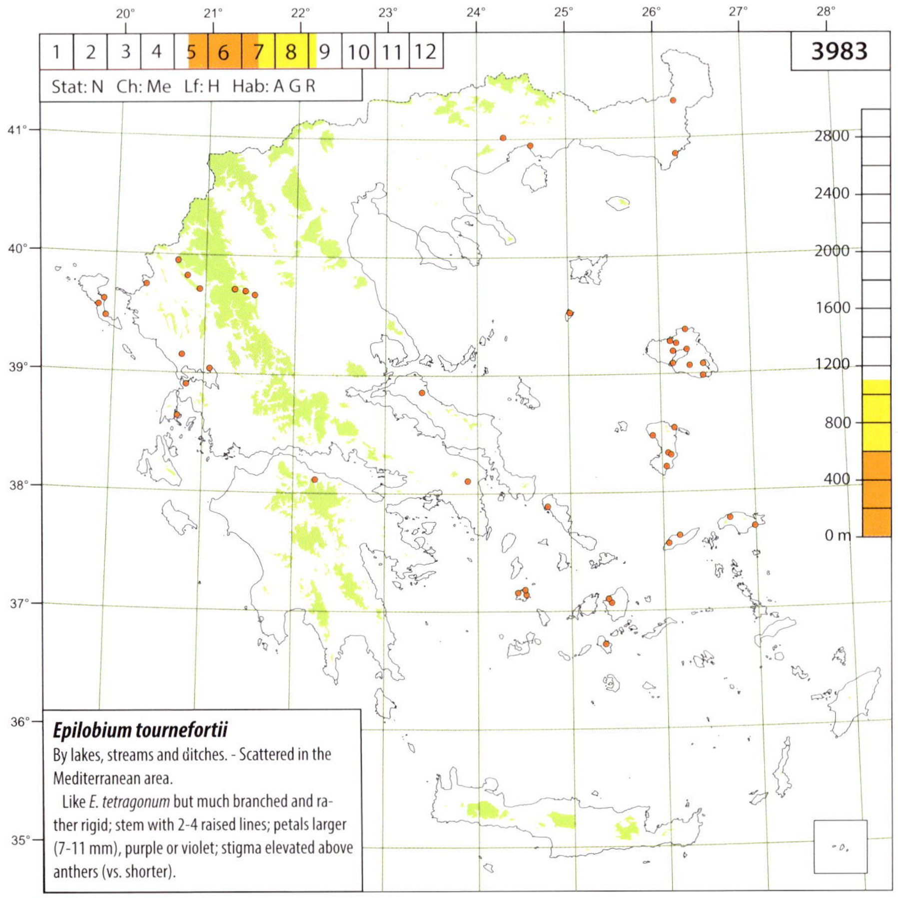

Epilobium tournefortii

By lakes, streams and ditches. - Scattered in the Mediterranean area.

Like *E. tetragonum* but much branched and rather rigid; stem with 2-4 raised lines; petals larger (7-11 mm), purple or violet; stigma elevated above anthers (vs. shorter).

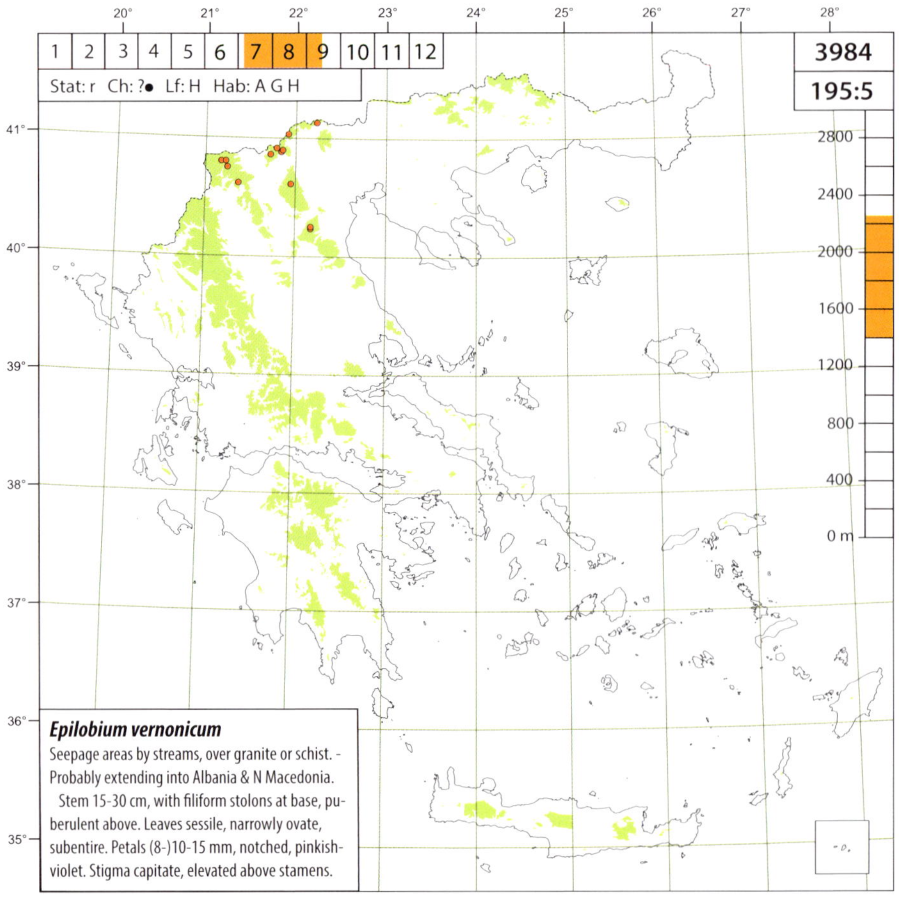

Epilobium vernonicum

Seepage areas by streams, over granite or schist. - Probably extending into Albania & N Macedonia.

Stem 15-30 cm, with filiform stolons at base, puberulent above. Leaves sessile, narrowly ovate, subentire. Petals (8-)10-15 mm, notched, pinkish-violet. Stigma capitate, elevated above stamens.

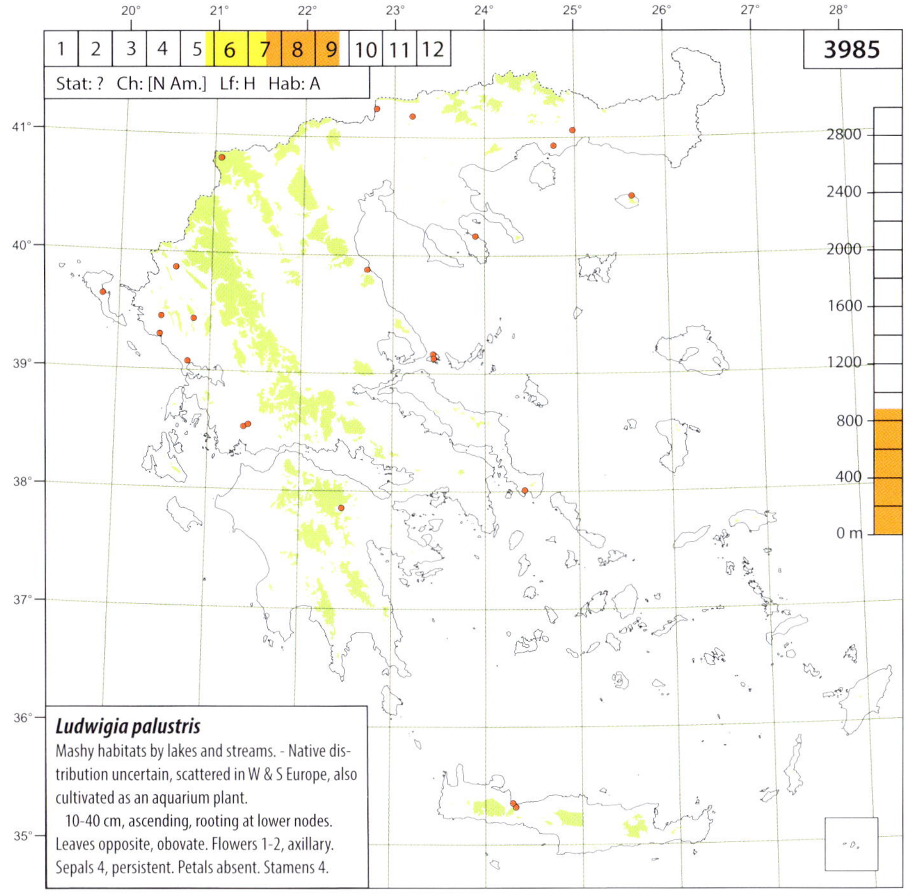

Ludwigia palustris
Mashy habitats by lakes and streams. - Native distribution uncertain, scattered in W & S Europe, also cultivated as an aquarium plant.
10-40 cm, ascending, rooting at lower nodes. Leaves opposite, obovate. Flowers 1-2, axillary. Sepals 4, persistent. Petals absent. Stamens 4.

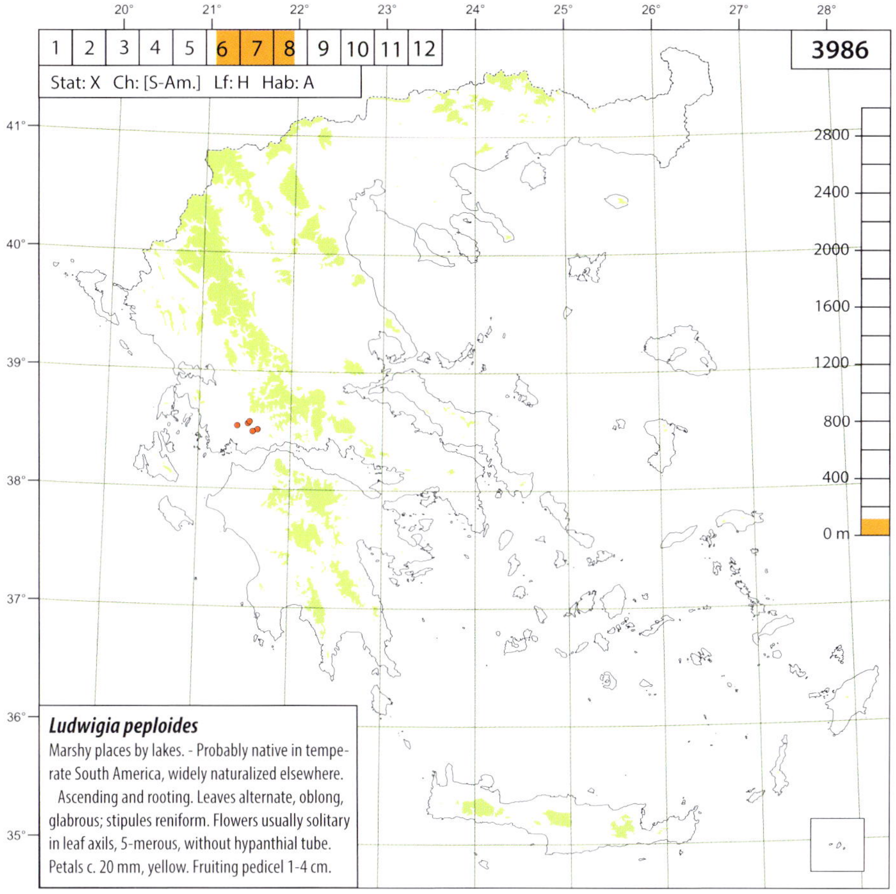

Ludwigia peploides
Marshy places by lakes. - Probably native in temperate South America, widely naturalized elsewhere.
Ascending and rooting. Leaves alternate, oblong, glabrous; stipules reniform. Flowers usually solitary in leaf axils, 5-merous, without hypanthial tube. Petals c. 20 mm, yellow. Fruiting pedicel 1-4 cm.

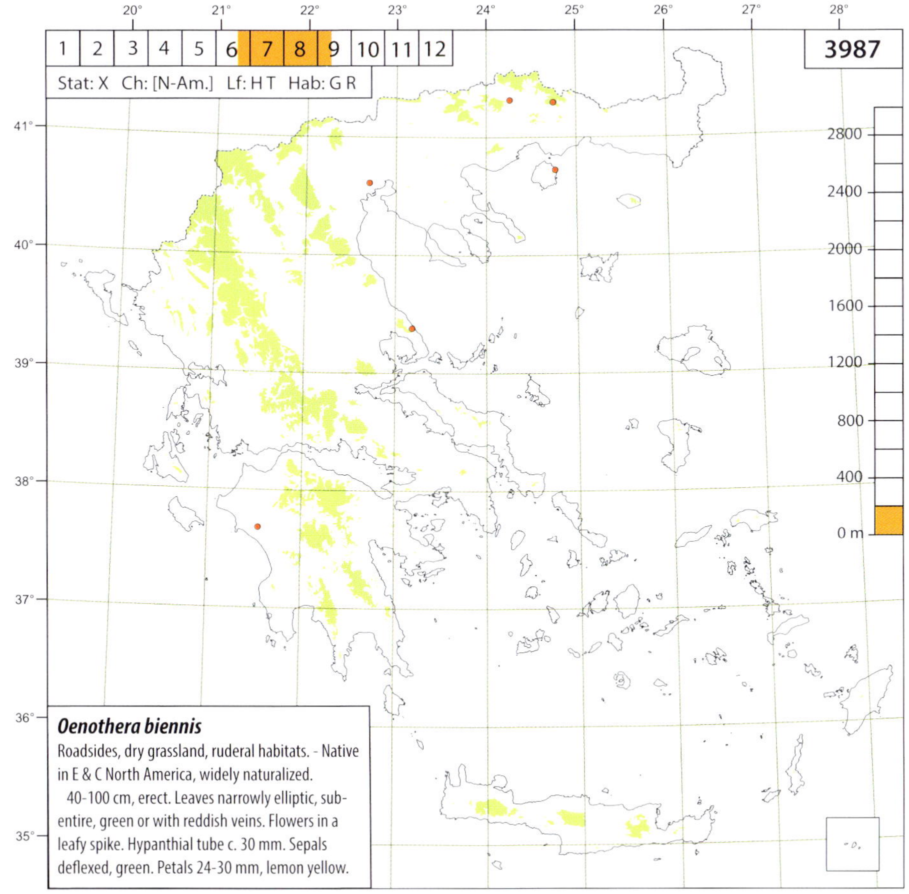

Oenothera biennis
Roadsides, dry grassland, ruderal habitats. - Native in E & C North America, widely naturalized.
40-100 cm, erect. Leaves narrowly elliptic, subentire, green or with reddish veins. Flowers in a leafy spike. Hypanthial tube c. 30 mm. Sepals deflexed, green. Petals 24-30 mm, lemon yellow.

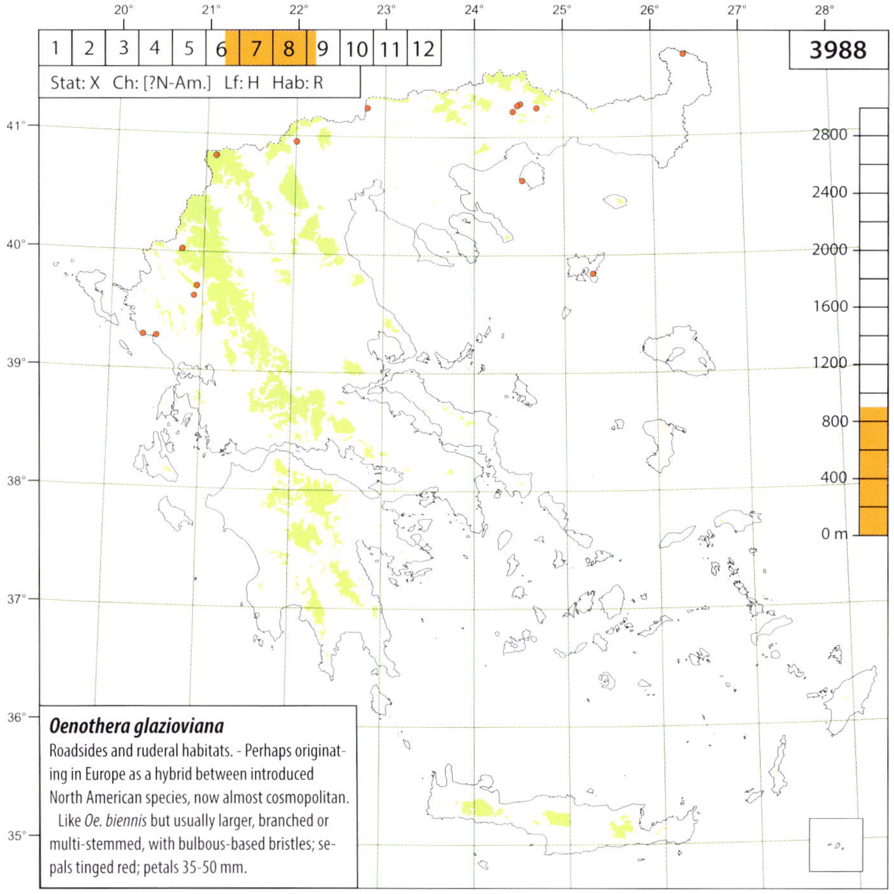

Oenothera glazioviana
Roadsides and ruderal habitats. - Perhaps originating in Europe as a hybrid between introduced North American species, now almost cosmopolitan.
Like *Oe. biennis* but usually larger, branched or multi-stemmed, with bulbous-based bristles; sepals tinged red; petals 35-50 mm.

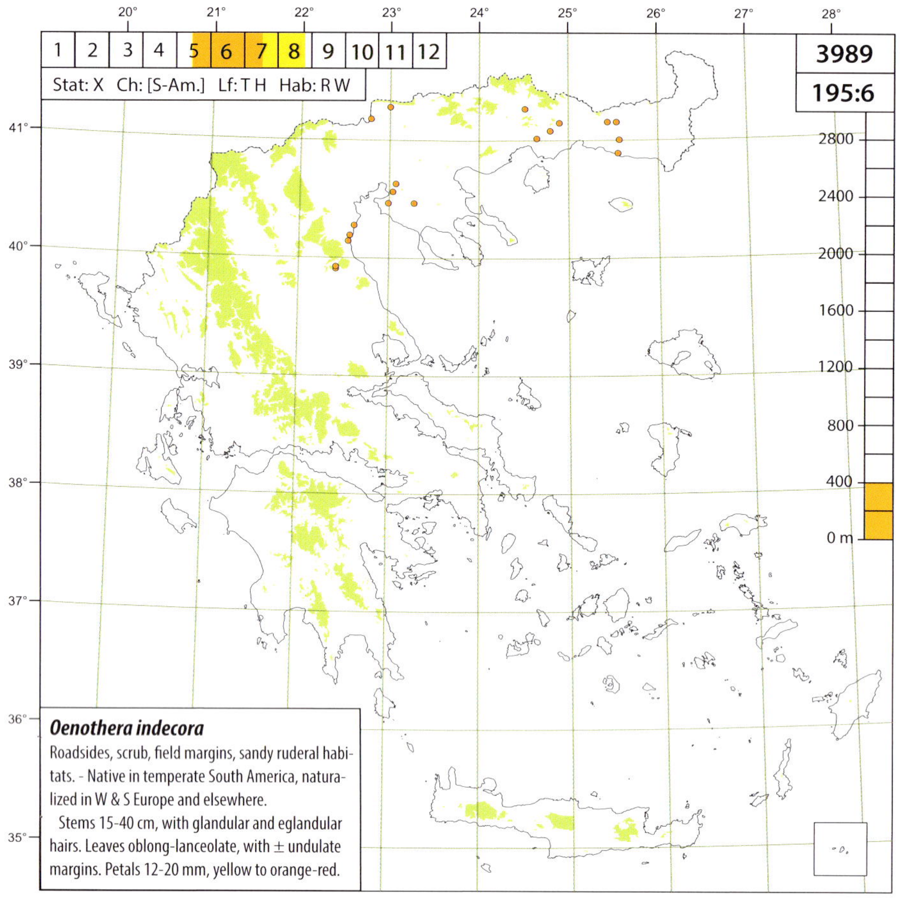

Oenothera indecora

Roadsides, scrub, field margins, sandy ruderal habitats. - Native in temperate South America, naturalized in W & S Europe and elsewhere.

Stems 15-40 cm, with glandular and eglandular hairs. Leaves oblong-lanceolate, with ± undulate margins. Petals 12-20 mm, yellow to orange-red.

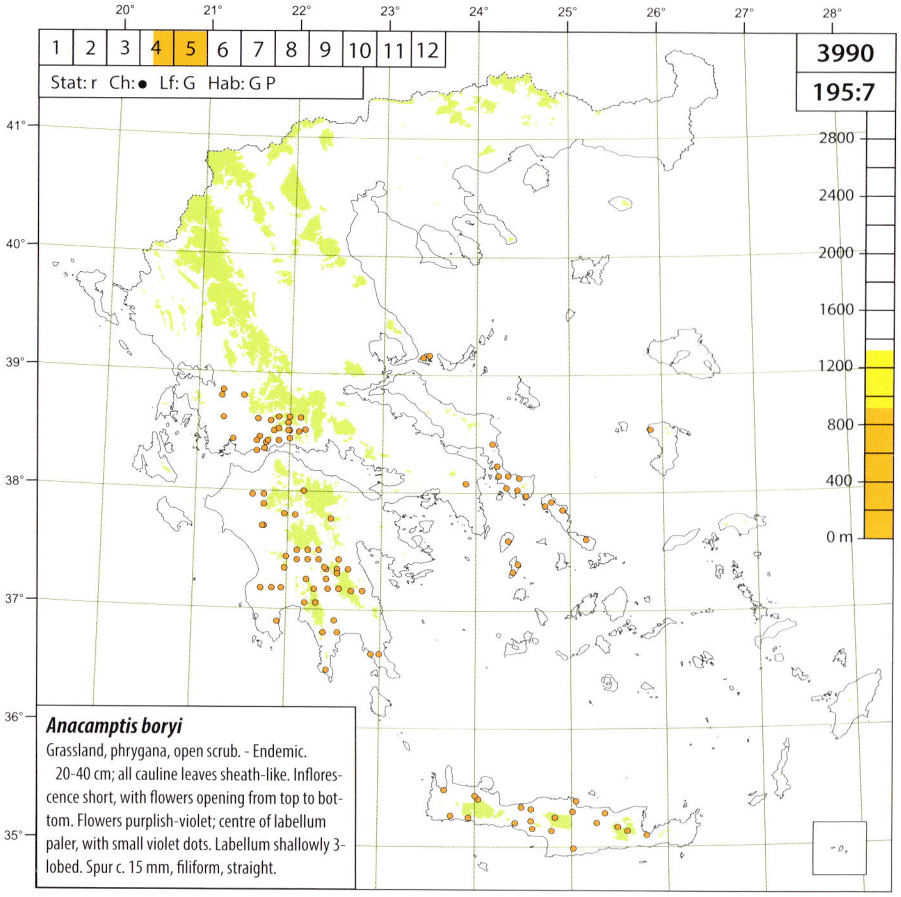

Anacamptis boryi

Grassland, phrygana, open scrub. - Endemic.

20-40 cm; all cauline leaves sheath-like. Inflorescence short, with flowers opening from top to bottom. Flowers purplish-violet; centre of labellum paler, with small violet dots. Labellum shallowly 3-lobed. Spur c. 15 mm, filiform, straight.

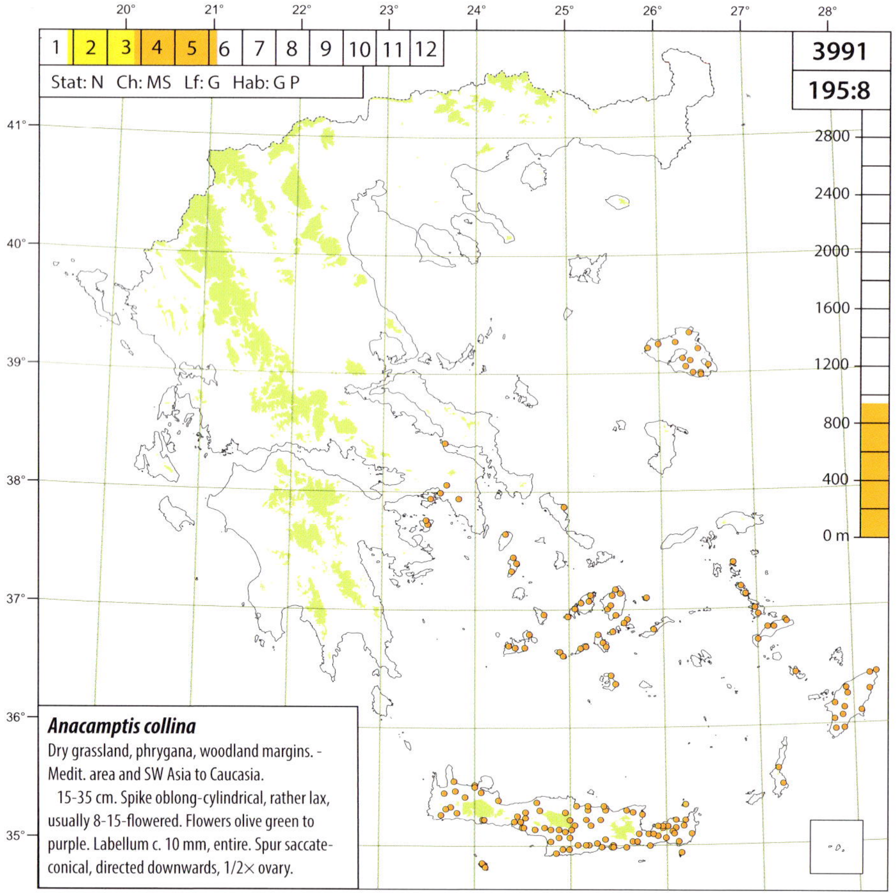

Anacamptis collina

Dry grassland, phrygana, woodland margins. - Medit. area and SW Asia to Caucasia.

15-35 cm. Spike oblong-cylindrical, rather lax, usually 8-15-flowered. Flowers olive green to purple. Labellum c. 10 mm, entire. Spur saccate-conical, directed downwards, 1/2× ovary.

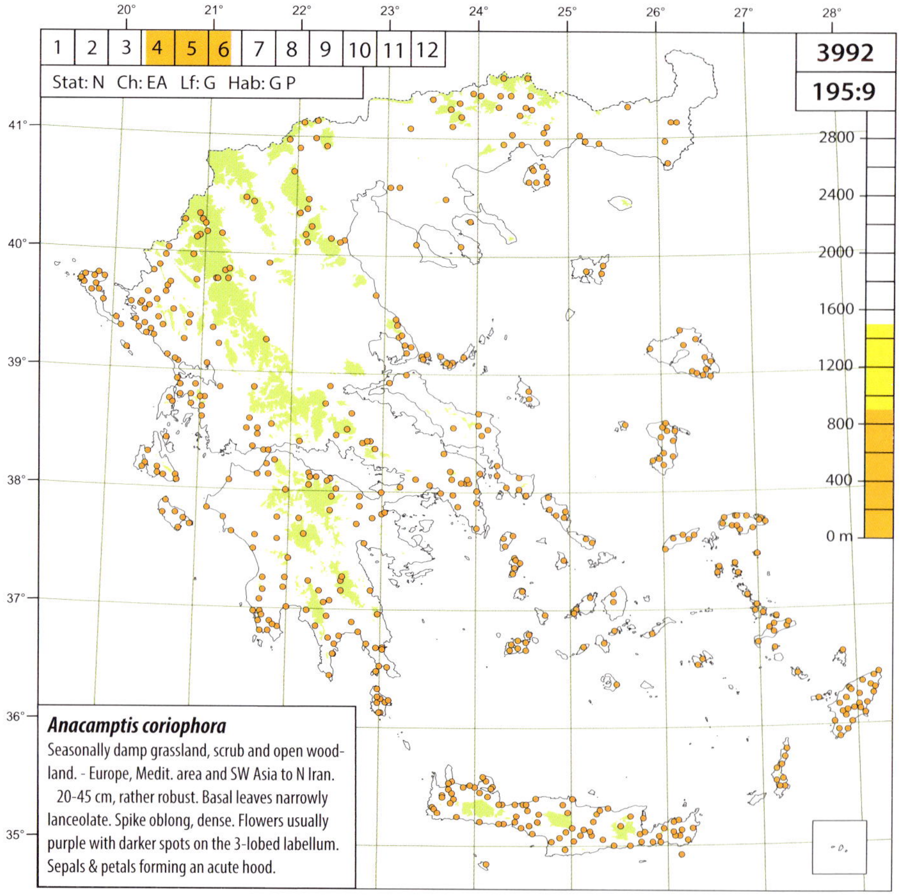

Anacamptis coriophora

Seasonally damp grassland, scrub and open woodland. - Europe, Medit. area and SW Asia to N Iran.

20-45 cm, rather robust. Basal leaves narrowly lanceolate. Spike oblong, dense. Flowers usually purple with darker spots on the 3-lobed labellum. Sepals & petals forming an acute hood.

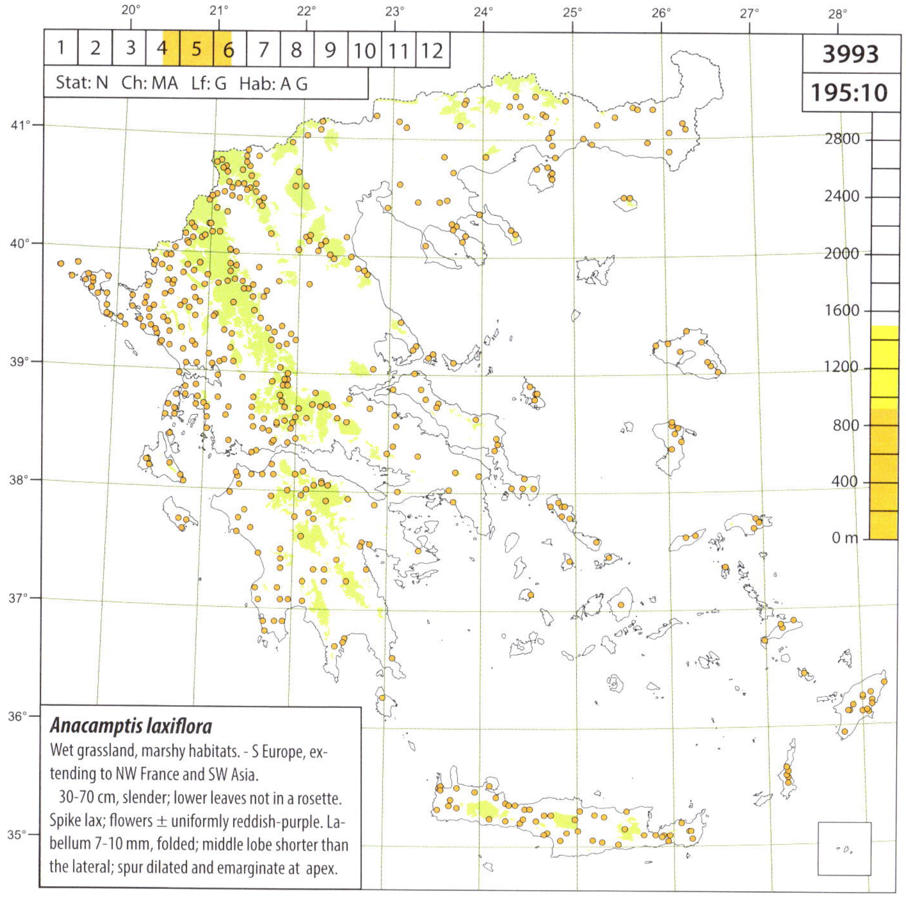

Anacamptis laxiflora
Wet grassland, marshy habitats. - S Europe, extending to NW France and SW Asia.
30-70 cm, slender; lower leaves not in a rosette. Spike lax; flowers ± uniformly reddish-purple. Labellum 7-10 mm, folded; middle lobe shorter than the lateral; spur dilated and emarginate at apex.

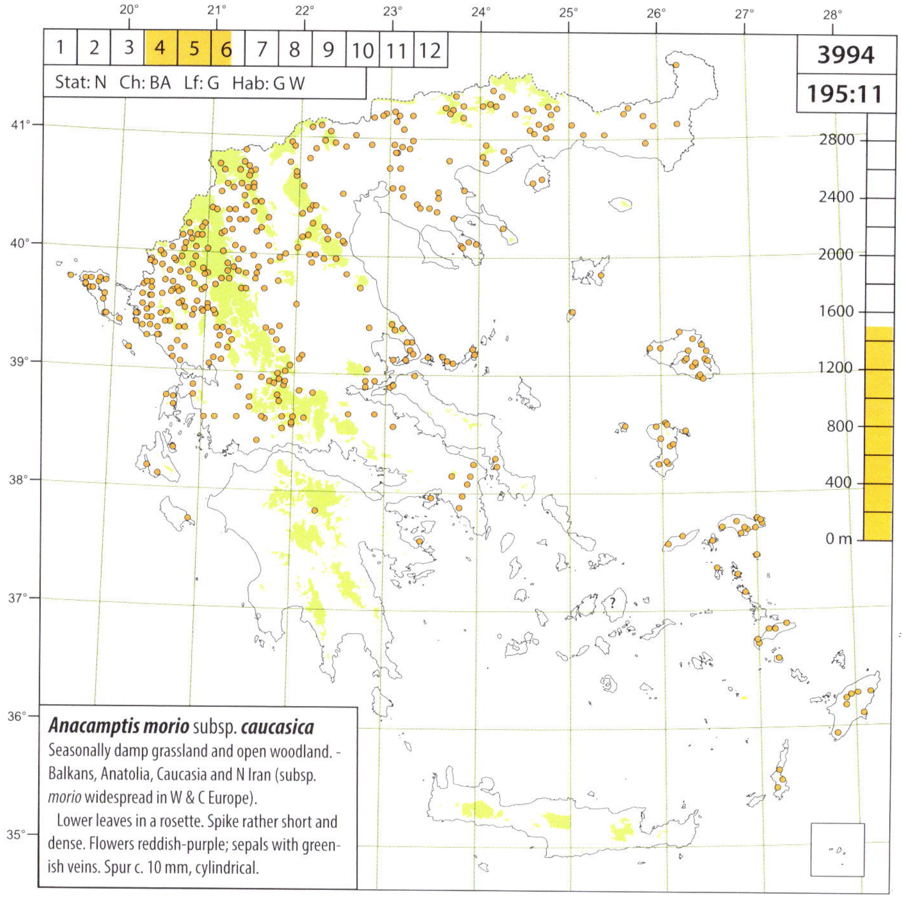

Anacamptis morio subsp. ***caucasica***
Seasonally damp grassland and open woodland. - Balkans, Anatolia, Caucasia and N Iran (subsp. *morio* widespread in W & C Europe).
Lower leaves in a rosette. Spike rather short and dense. Flowers reddish-purple; sepals with greenish veins. Spur c. 10 mm, cylindrical.

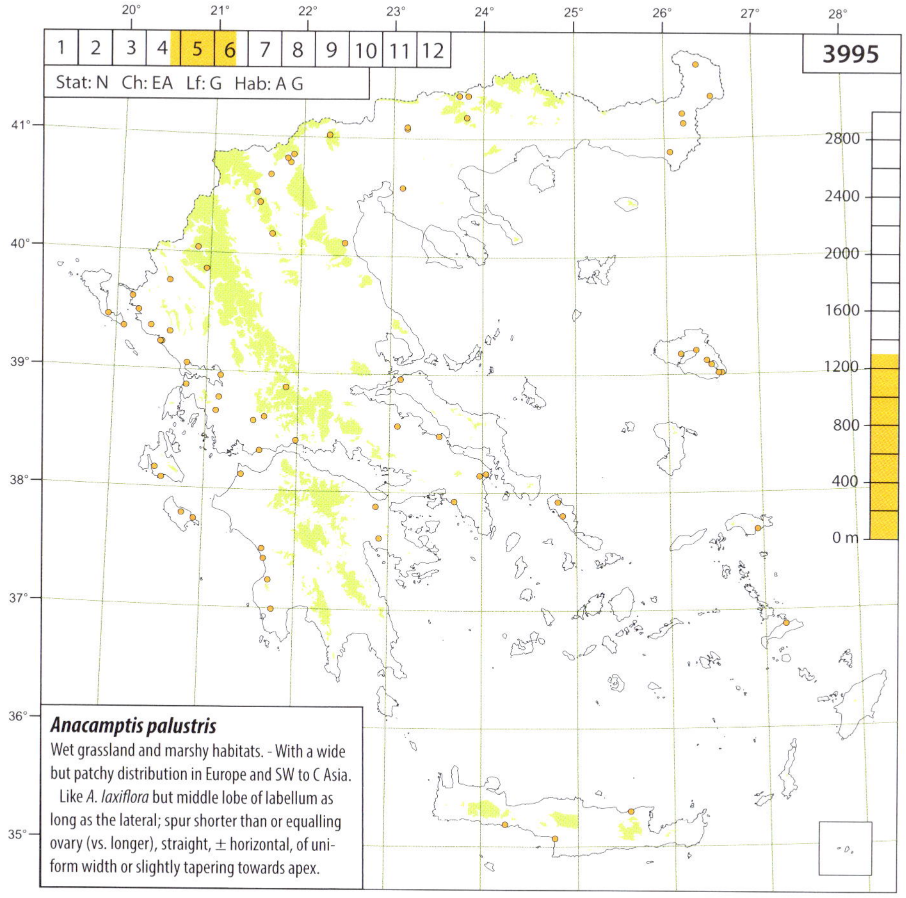

Anacamptis palustris
Wet grassland and marshy habitats. - With a wide but patchy distribution in Europe and SW to C Asia.
Like *A. laxiflora* but middle lobe of labellum as long as the lateral; spur shorter than or equalling ovary (vs. longer), straight, ± horizontal, of uniform width or slightly tapering towards apex.

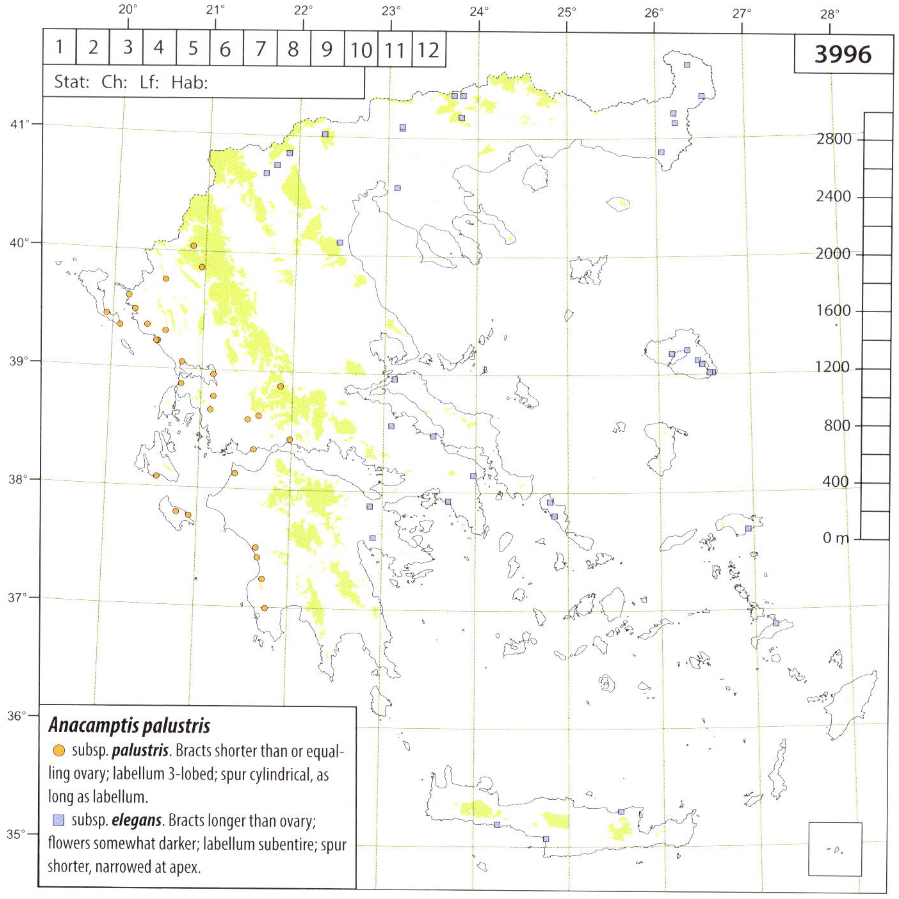

Anacamptis palustris
● subsp. ***palustris***. Bracts shorter than or equalling ovary; labellum 3-lobed; spur cylindrical, as long as labellum.
■ subsp. ***elegans***. Bracts longer than ovary; flowers somewhat darker; labellum subentire; spur shorter, narrowed at apex.

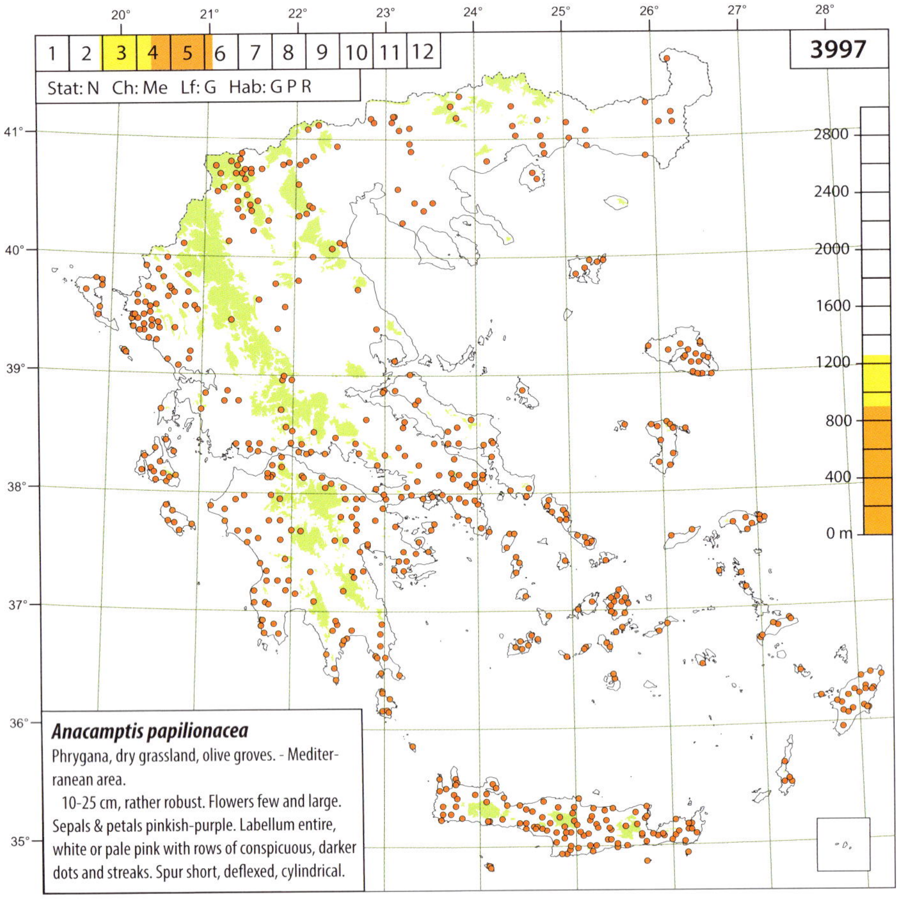

Anacamptis papilionacea
Phrygana, dry grassland, olive groves. - Mediterranean area.
10-25 cm, rather robust. Flowers few and large. Sepals & petals pinkish-purple. Labellum entire, white or pale pink with rows of conspicuous, darker dots and streaks. Spur short, deflexed, cylindrical.

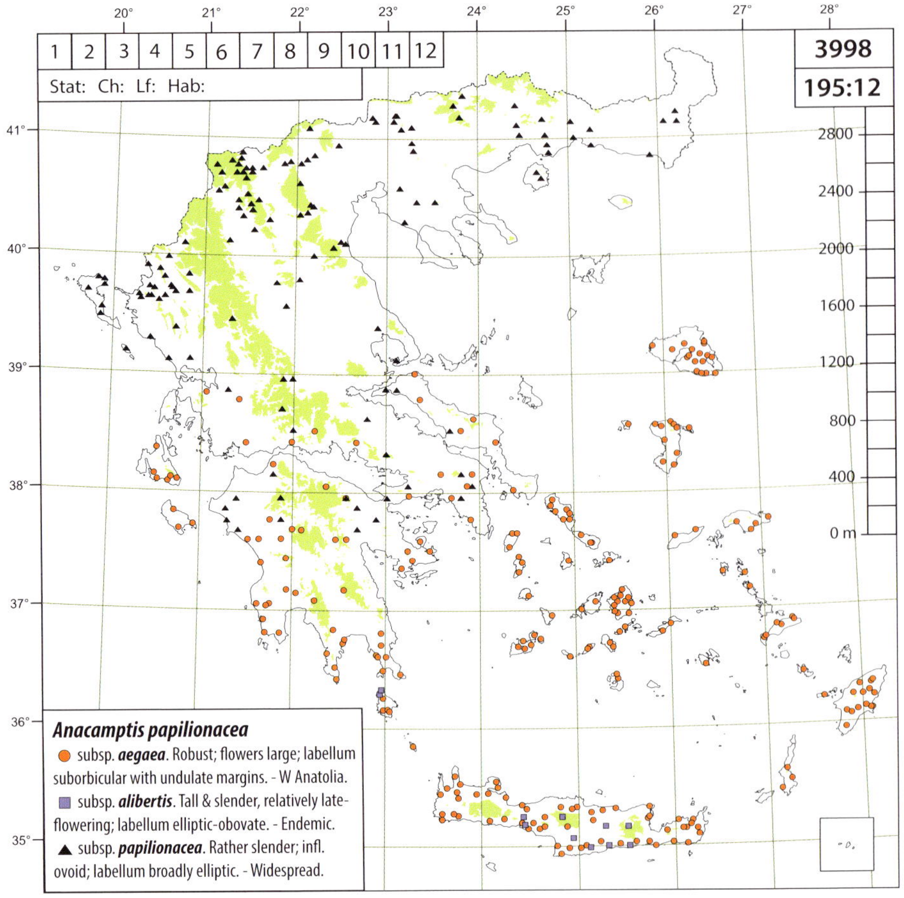

Anacamptis papilionacea
● subsp. ***aegaea***. Robust; flowers large; labellum suborbicular with undulate margins. - W Anatolia.
■ subsp. ***alibertis***. Tall & slender, relatively late-flowering; labellum elliptic-obovate. - Endemic.
▲ subsp. ***papilionacea***. Rather slender; infl. ovoid; labellum broadly elliptic. - Widespread.

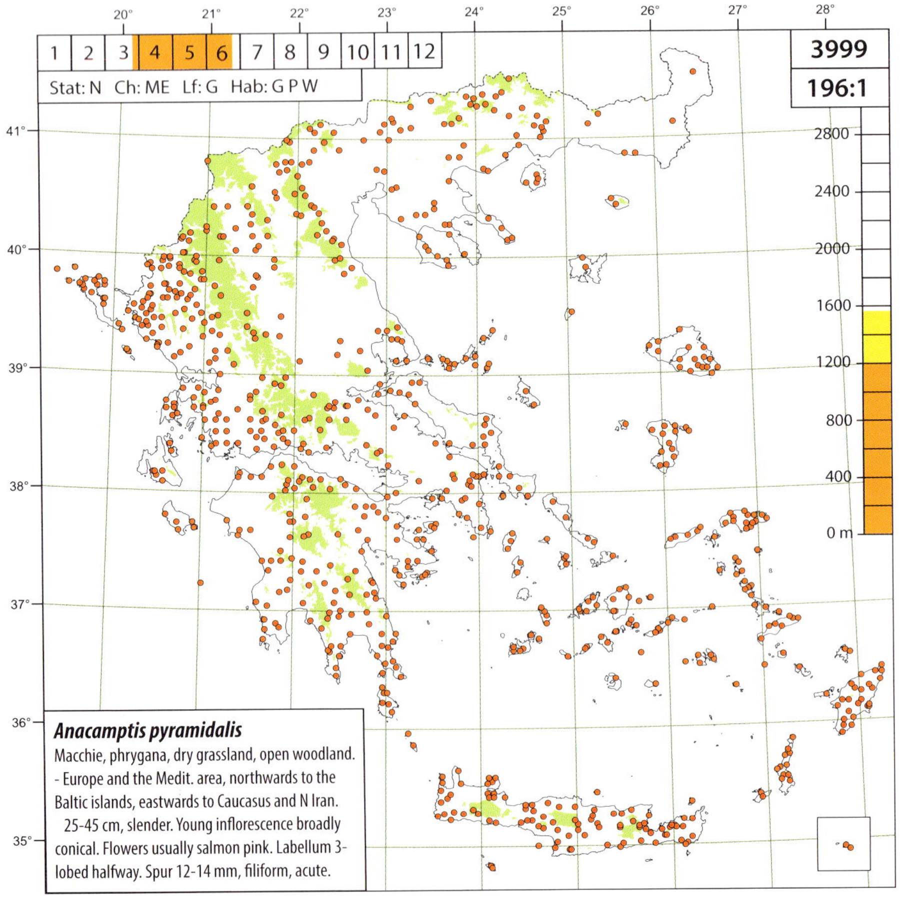

Anacamptis pyramidalis
Macchie, phrygana, dry grassland, open woodland. - Europe and the Medit. area, northwards to the Baltic islands, eastwards to Caucasus and N Iran.
25-45 cm, slender. Young inflorescence broadly conical. Flowers usually salmon pink. Labellum 3-lobed halfway. Spur 12-14 mm, filiform, acute.

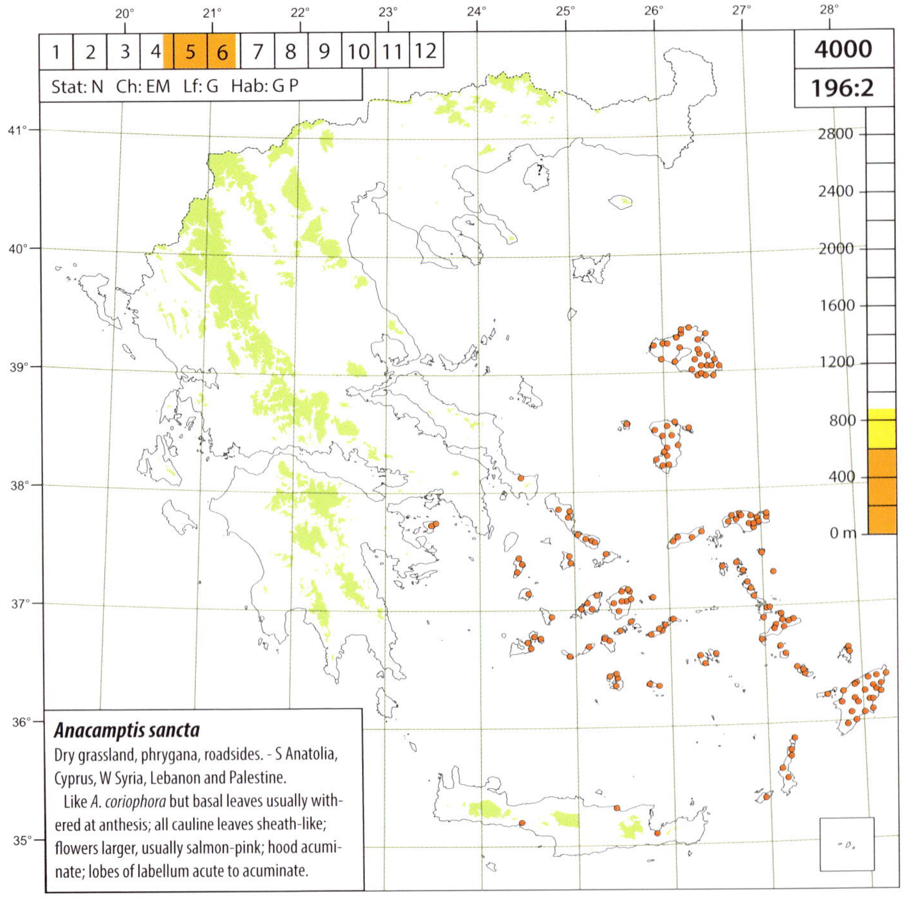

Anacamptis sancta
Dry grassland, phrygana, roadsides. - S Anatolia, Cyprus, W Syria, Lebanon and Palestine.
Like *A. coriophora* but basal leaves usually withered at anthesis; all cauline leaves sheath-like; flowers larger, usually salmon-pink; hood acuminate; lobes of labellum acute to acuminate.

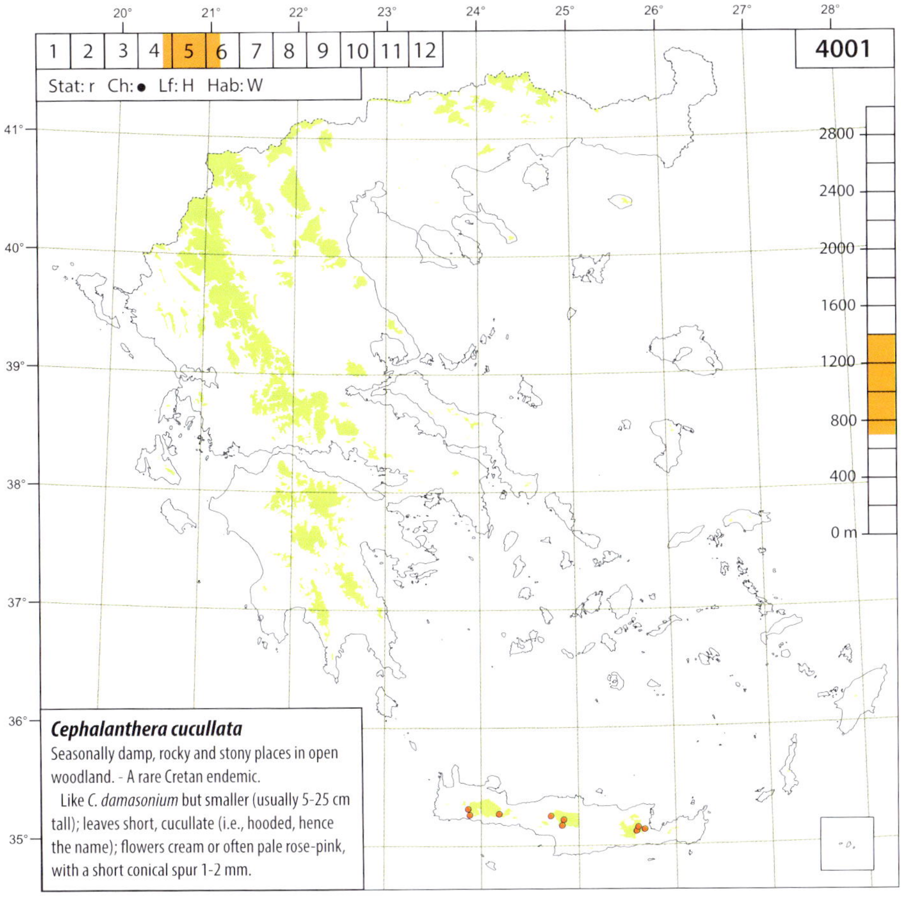

Cephalanthera cucullata

Seasonally damp, rocky and stony places in open woodland. - A rare Cretan endemic.

Like *C. damasonium* but smaller (usually 5-25 cm tall); leaves short, cucullate (i.e., hooded, hence the name); flowers cream or often pale rose-pink, with a short conical spur 1-2 mm.

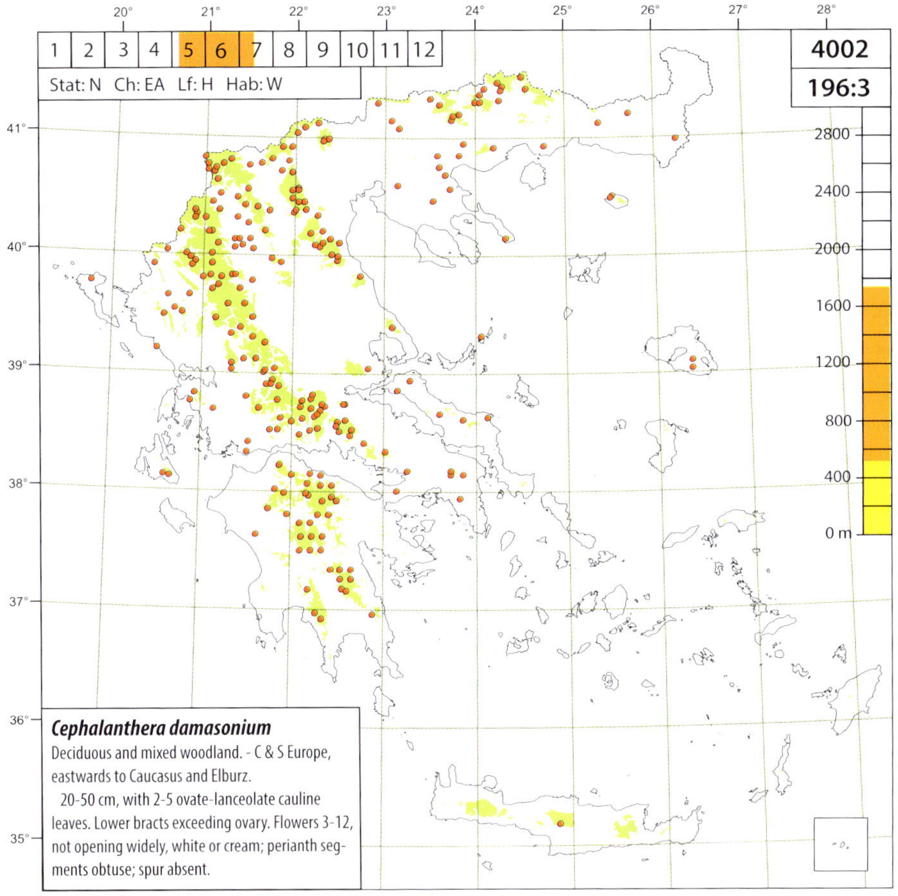

Cephalanthera damasonium

Deciduous and mixed woodland. - C & S Europe, eastwards to Caucasus and Elburz.

20-50 cm, with 2-5 ovate-lanceolate cauline leaves. Lower bracts exceeding ovary. Flowers 3-12, not opening widely, white or cream; perianth segments obtuse; spur absent.

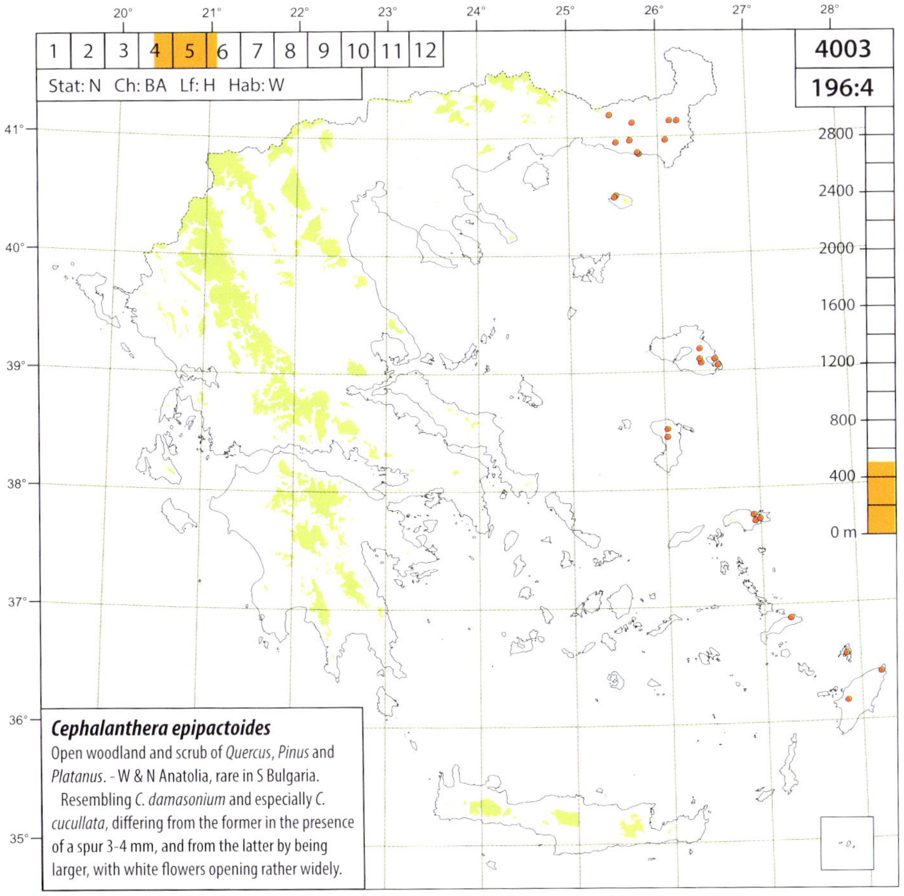

Cephalanthera epipactoides

Open woodland and scrub of *Quercus*, *Pinus* and *Platanus*. - W & N Anatolia, rare in S Bulgaria.

Resembling *C. damasonium* and especially *C. cucullata*, differing from the former in the presence of a spur 3-4 mm, and from the latter by being larger, with white flowers opening rather widely.

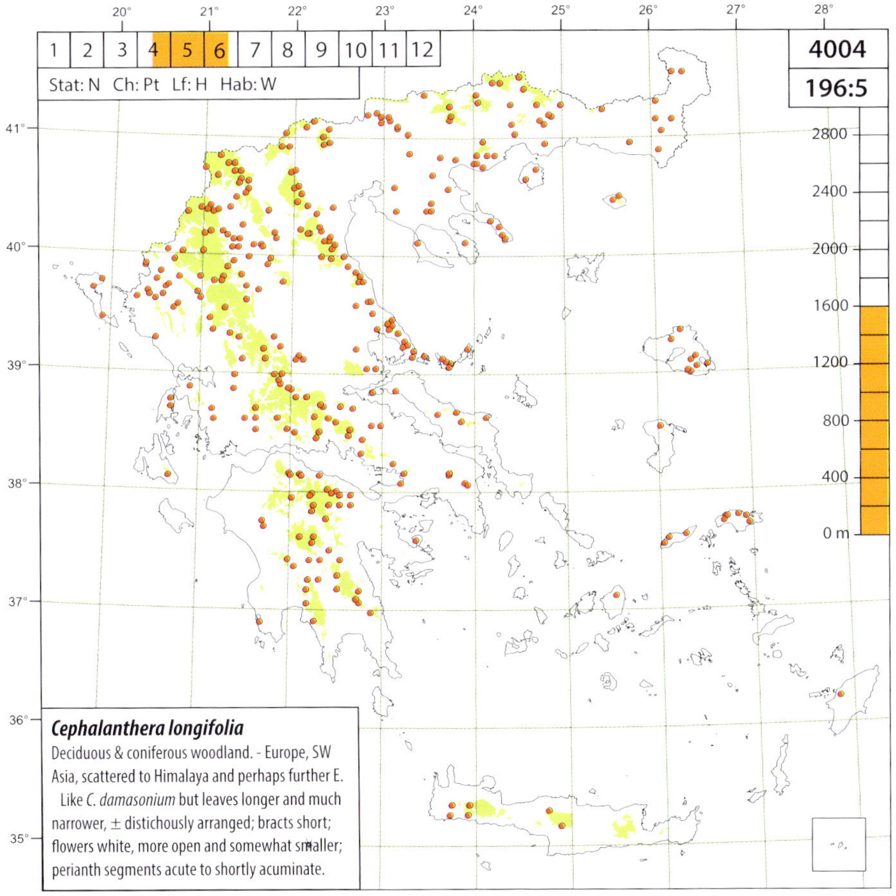

Cephalanthera longifolia

Deciduous & coniferous woodland. - Europe, SW Asia, scattered to Himalaya and perhaps further E.

Like *C. damasonium* but leaves longer and much narrower, ± distichously arranged; bracts short; flowers white, more open and somewhat smaller; perianth segments acute to shortly acuminate.

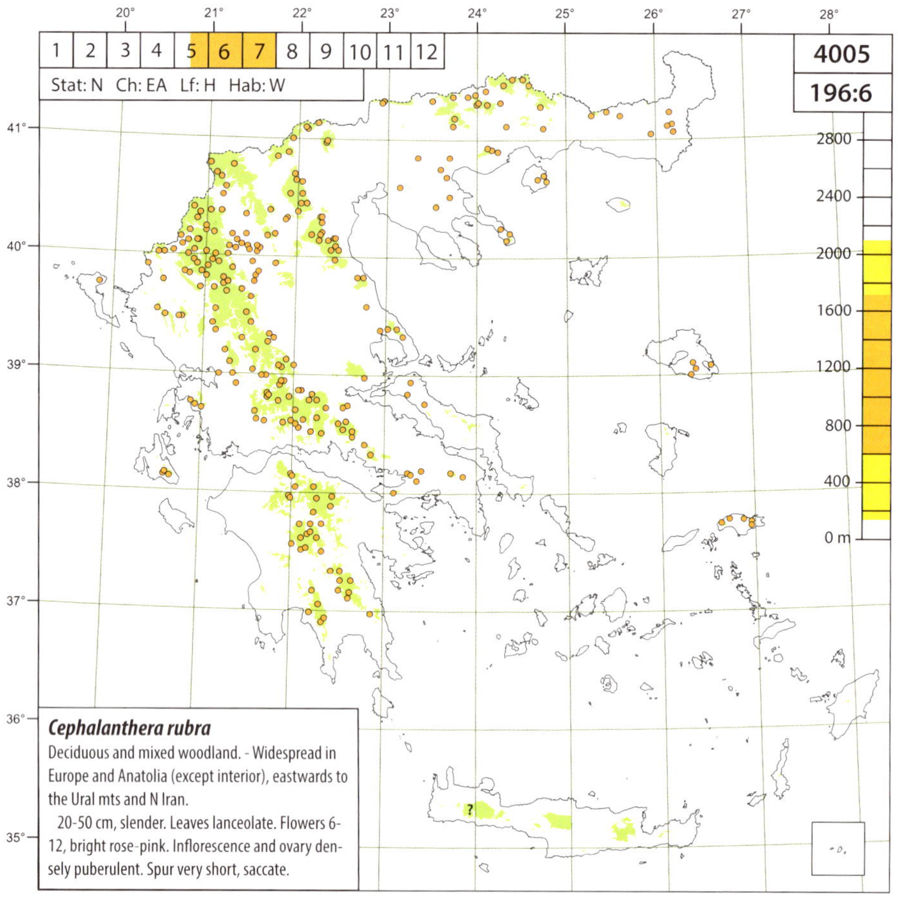

Cephalanthera rubra
Deciduous and mixed woodland. - Widespread in Europe and Anatolia (except interior), eastwards to the Ural mts and N Iran.
20-50 cm, slender. Leaves lanceolate. Flowers 6-12, bright rose-pink. Inflorescence and ovary densely puberulent. Spur very short, saccate.

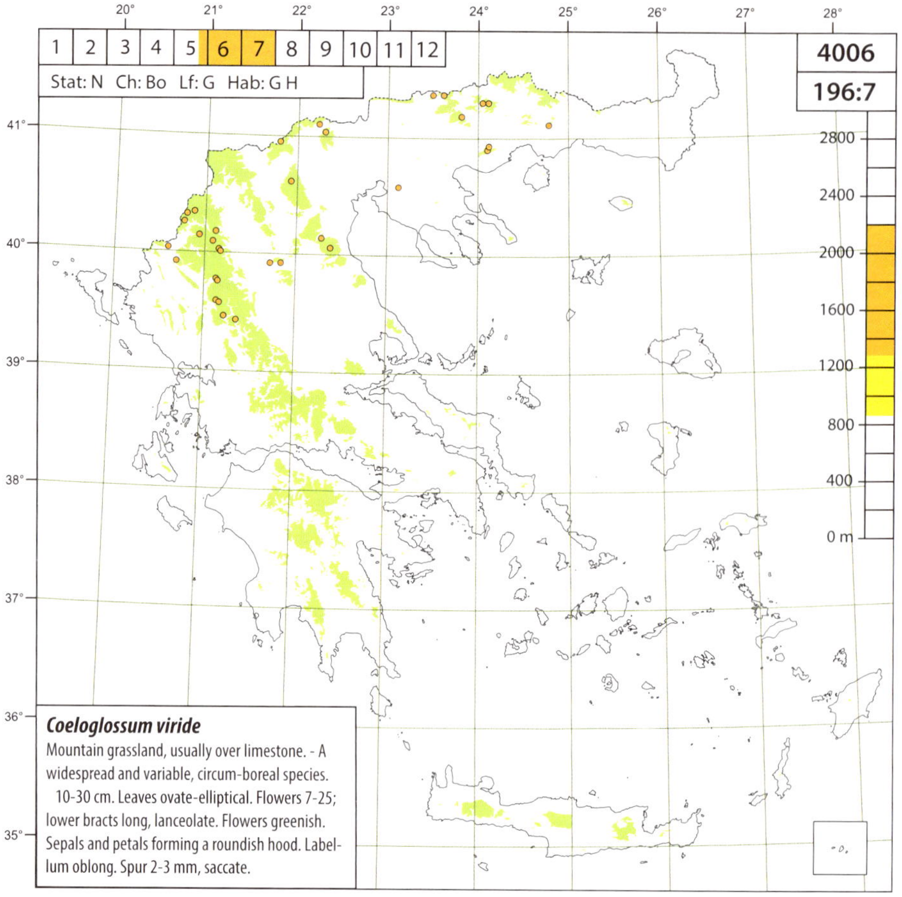

Coeloglossum viride
Mountain grassland, usually over limestone. - A widespread and variable, circum-boreal species.
10-30 cm. Leaves ovate-elliptical. Flowers 7-25; lower bracts long, lanceolate. Flowers greenish. Sepals and petals forming a roundish hood. Labellum oblong. Spur 2-3 mm, saccate.

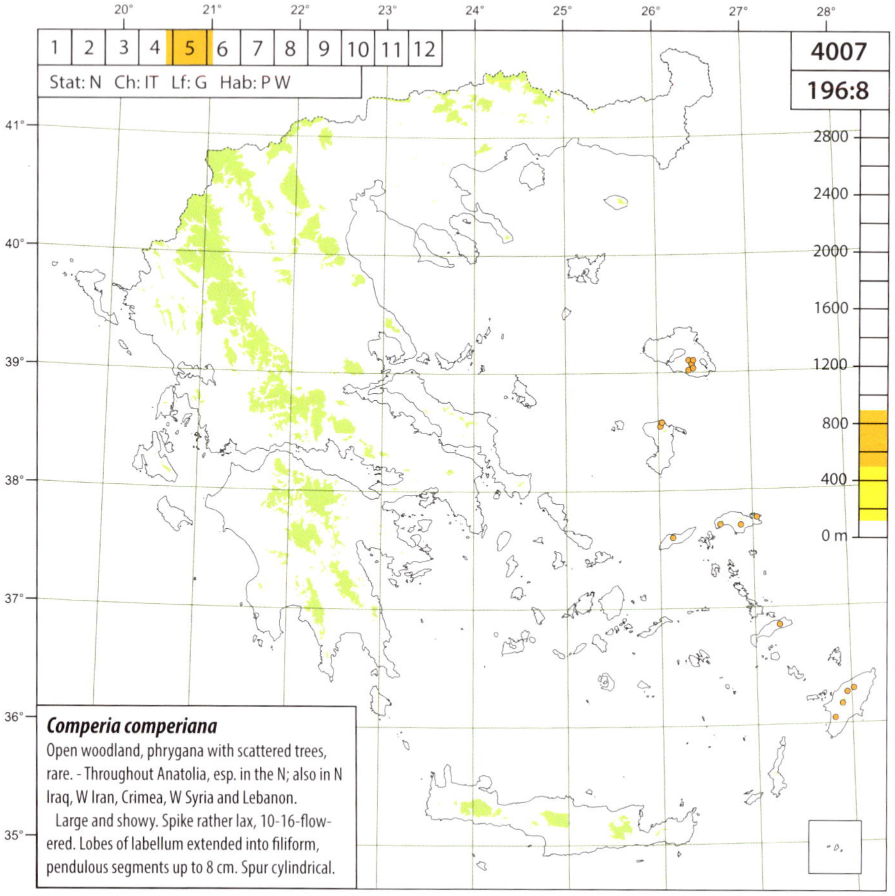

Comperia comperiana
Open woodland, phrygana with scattered trees, rare. - Throughout Anatolia, esp. in the N; also in N Iraq, W Iran, Crimea, W Syria and Lebanon.
Large and showy. Spike rather lax, 10-16-flowered. Lobes of labellum extended into filiform, pendulous segments up to 8 cm. Spur cylindrical.

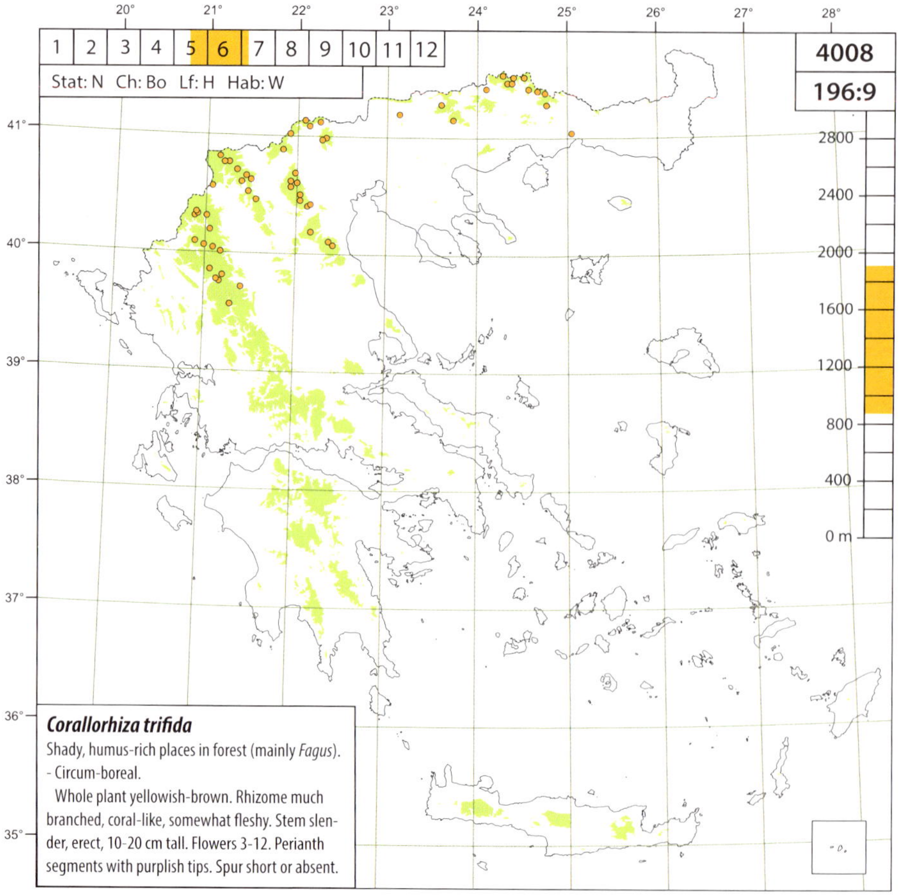

Corallorhiza trifida
Shady, humus-rich places in forest (mainly *Fagus*). - Circum-boreal.
Whole plant yellowish-brown. Rhizome much branched, coral-like, somewhat fleshy. Stem slender, erect, 10-20 cm tall. Flowers 3-12. Perianth segments with purplish tips. Spur short or absent.

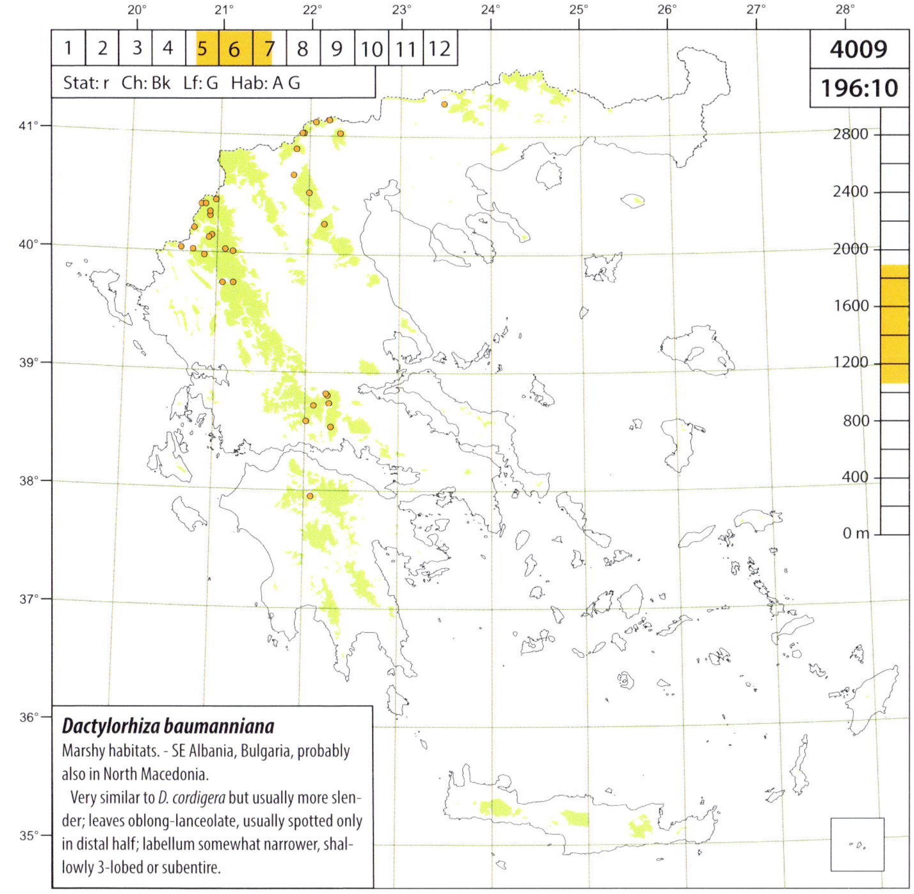

Dactylorhiza baumanniana

Marshy habitats. - SE Albania, Bulgaria, probably also in North Macedonia.

Very similar to *D. cordigera* but usually more slender; leaves oblong-lanceolate, usually spotted only in distal half; labellum somewhat narrower, shallowly 3-lobed or subentire.

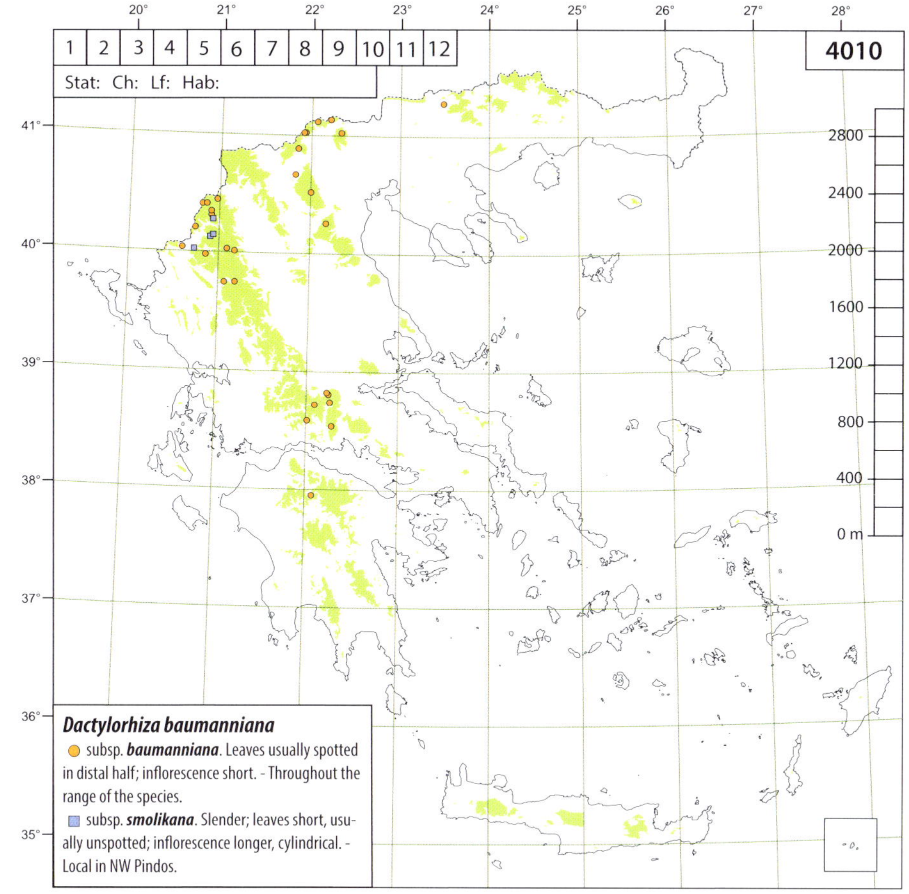

Dactylorhiza baumanniana

● subsp. ***baumanniana***. Leaves usually spotted in distal half; inflorescence short. - Throughout the range of the species.

■ subsp. ***smolikana***. Slender; leaves short, usually unspotted; inflorescence longer, cylindrical. - Local in NW Pindos.

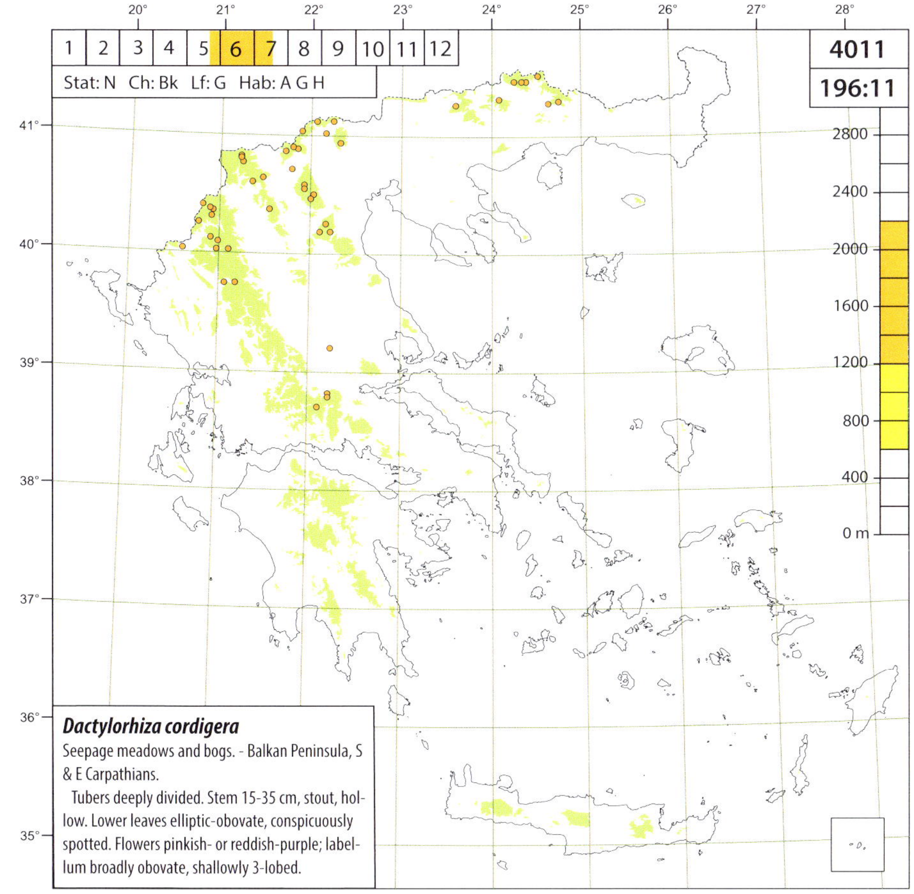

Dactylorhiza cordigera

Seepage meadows and bogs. - Balkan Peninsula, S & E Carpathians.

Tubers deeply divided. Stem 15-35 cm, stout, hollow. Lower leaves elliptic-obovate, conspicuously spotted. Flowers pinkish- or reddish-purple; labellum broadly obovate, shallowly 3-lobed.

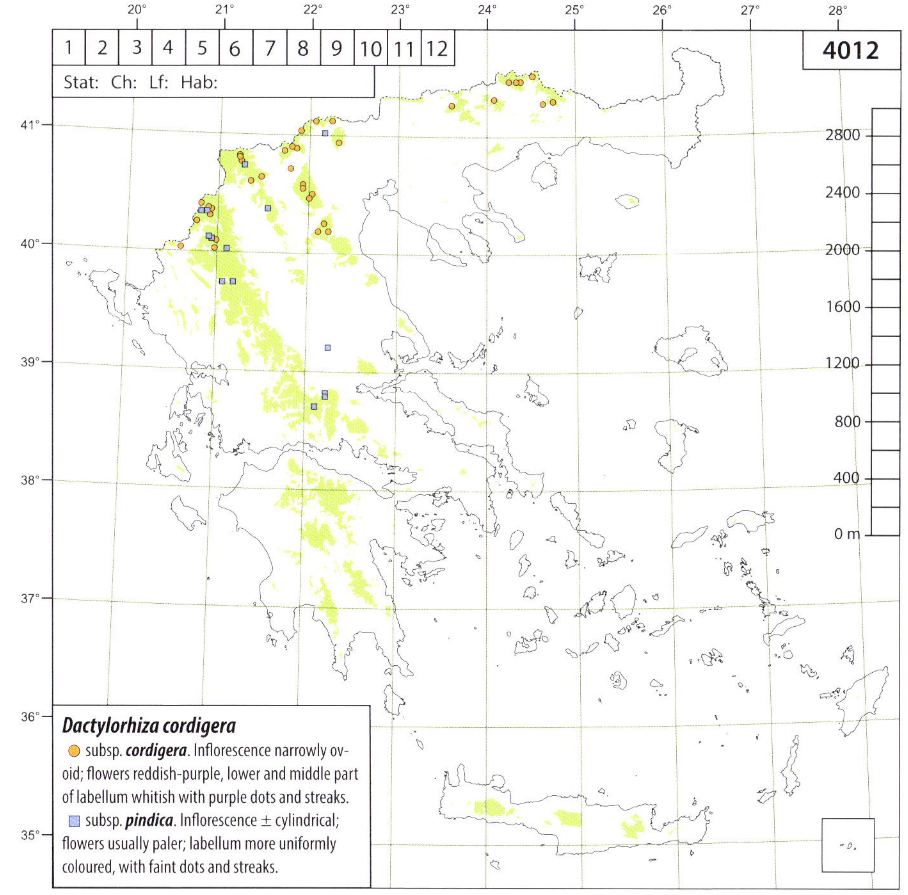

Dactylorhiza cordigera

● subsp. ***cordigera***. Inflorescence narrowly ovoid; flowers reddish-purple, lower and middle part of labellum whitish with purple dots and streaks.

■ subsp. ***pindica***. Inflorescence ± cylindrical; flowers usually paler; labellum more uniformly coloured, with faint dots and streaks.

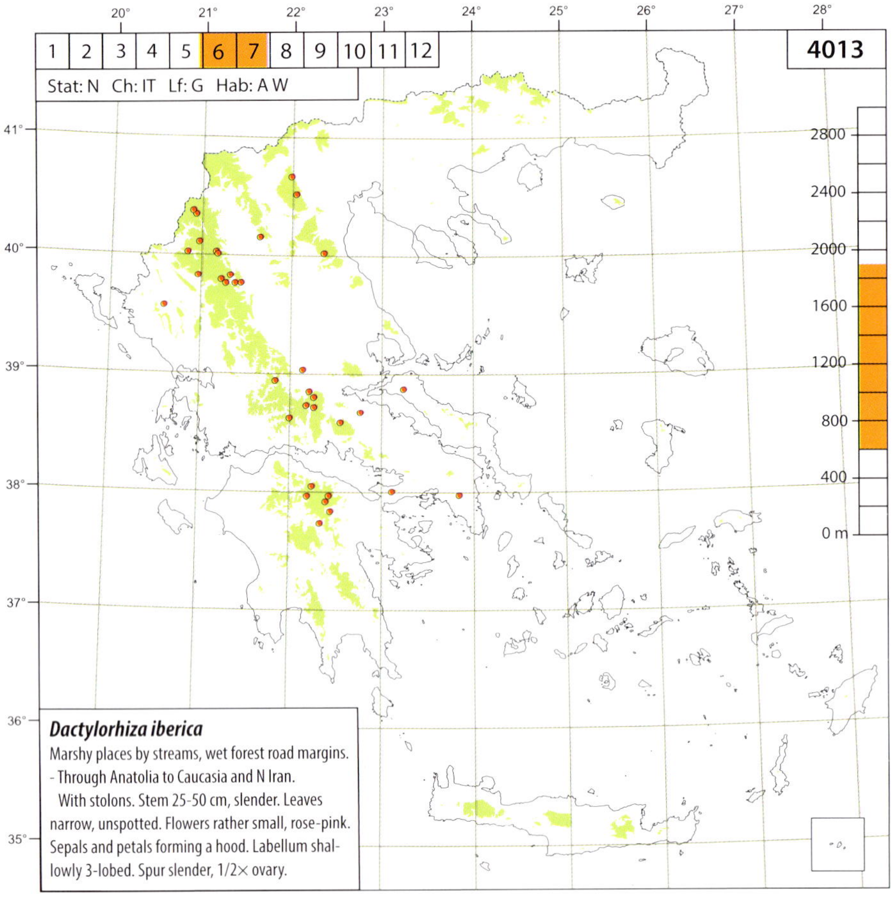

Dactylorhiza iberica
Marshy places by streams, wet forest road margins. - Through Anatolia to Caucasia and N Iran.
With stolons. Stem 25-50 cm, slender. Leaves narrow, unspotted. Flowers rather small, rose-pink. Sepals and petals forming a hood. Labellum shallowly 3-lobed. Spur slender, 1/2× ovary.

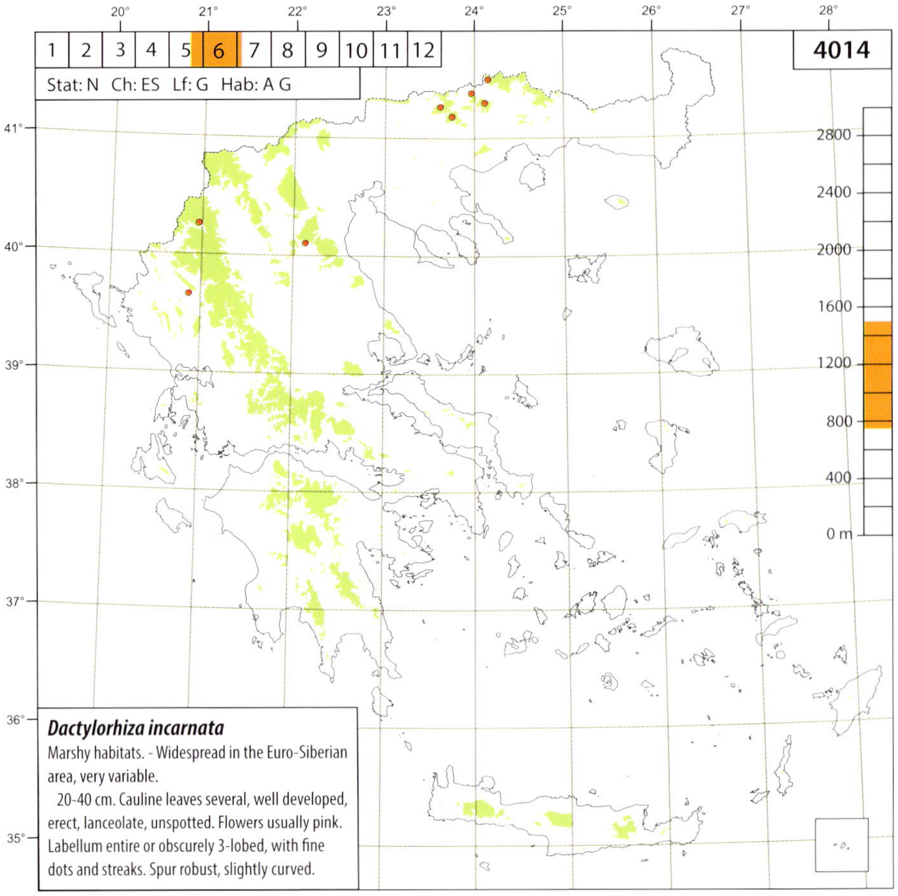

Dactylorhiza incarnata
Marshy habitats. - Widespread in the Euro-Siberian area, very variable.
20-40 cm. Cauline leaves several, well developed, erect, lanceolate, unspotted. Flowers usually pink. Labellum entire or obscurely 3-lobed, with fine dots and streaks. Spur robust, slightly curved.

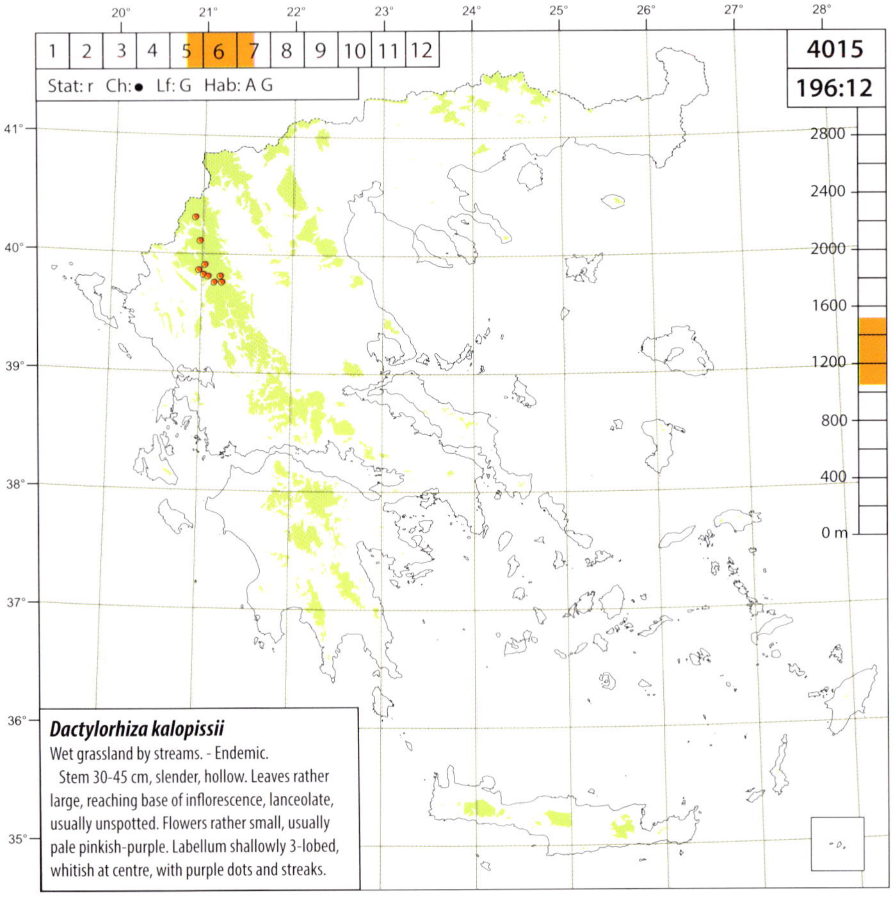

Dactylorhiza kalopissii
Wet grassland by streams. - Endemic.
Stem 30-45 cm, slender, hollow. Leaves rather large, reaching base of inflorescence, lanceolate, usually unspotted. Flowers rather small, usually pale pinkish-purple. Labellum shallowly 3-lobed, whitish at centre, with purple dots and streaks.

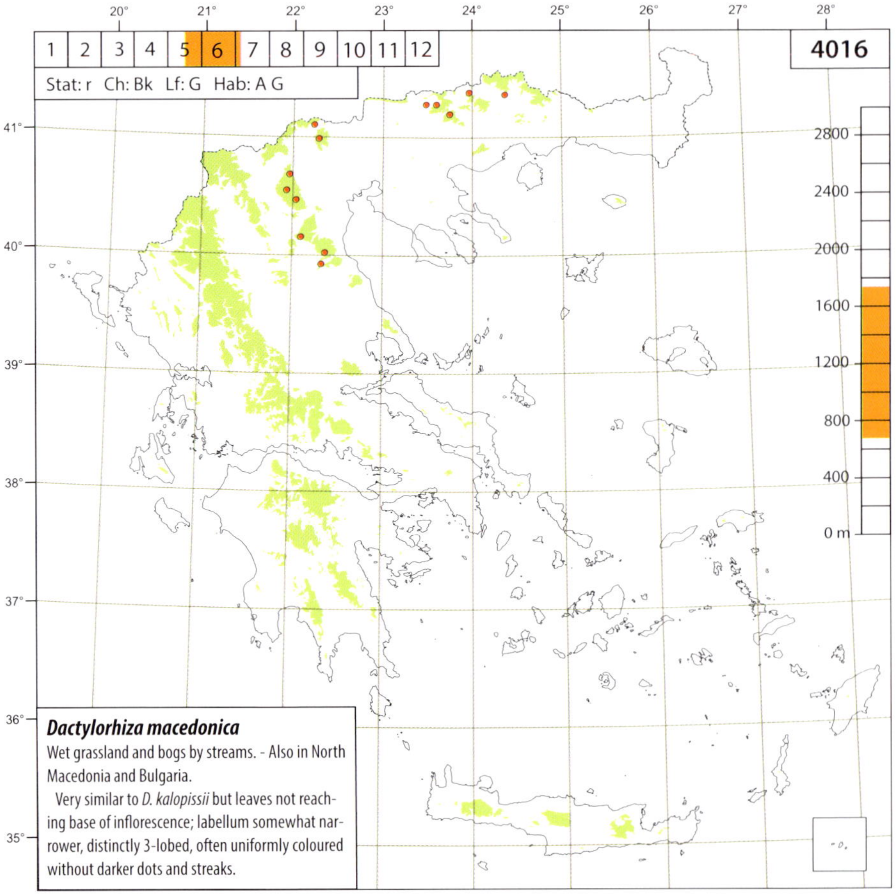

Dactylorhiza macedonica
Wet grassland and bogs by streams. - Also in North Macedonia and Bulgaria.
Very similar to *D. kalopissii* but leaves not reaching base of inflorescence; labellum somewhat narrower, distinctly 3-lobed, often uniformly coloured without darker dots and streaks.

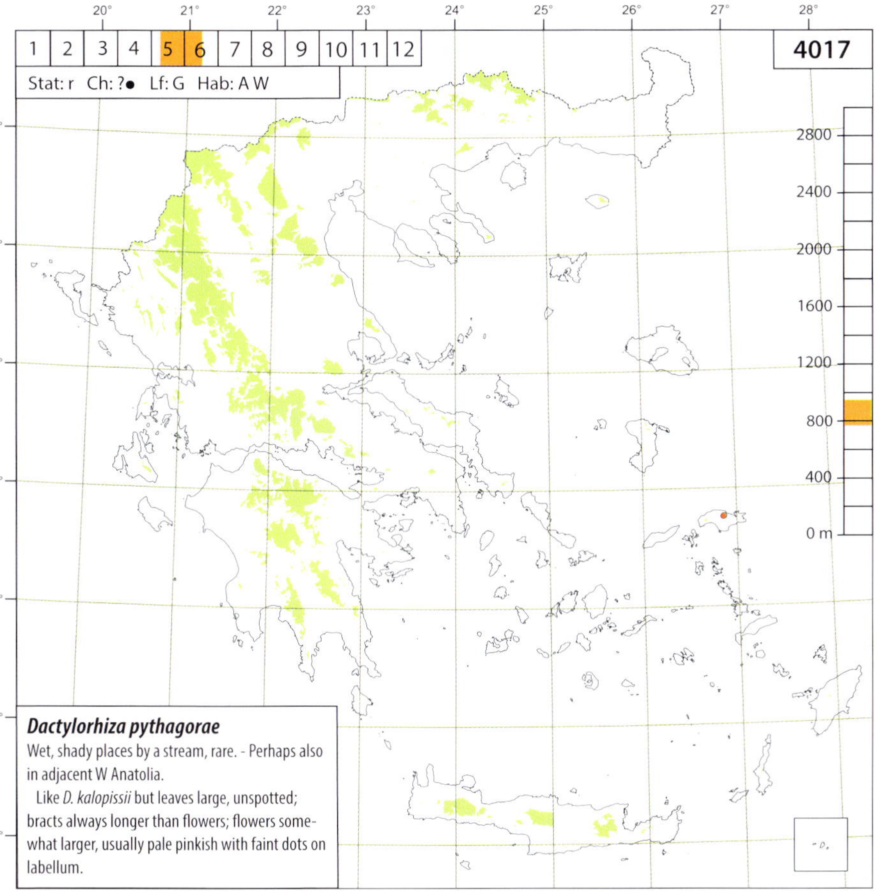

Dactylorhiza pythagorae

Wet, shady places by a stream, rare. - Perhaps also in adjacent W Anatolia.

Like *D. kalopissii* but leaves large, unspotted; bracts always longer than flowers; flowers somewhat larger, usually pale pinkish with faint dots on labellum.

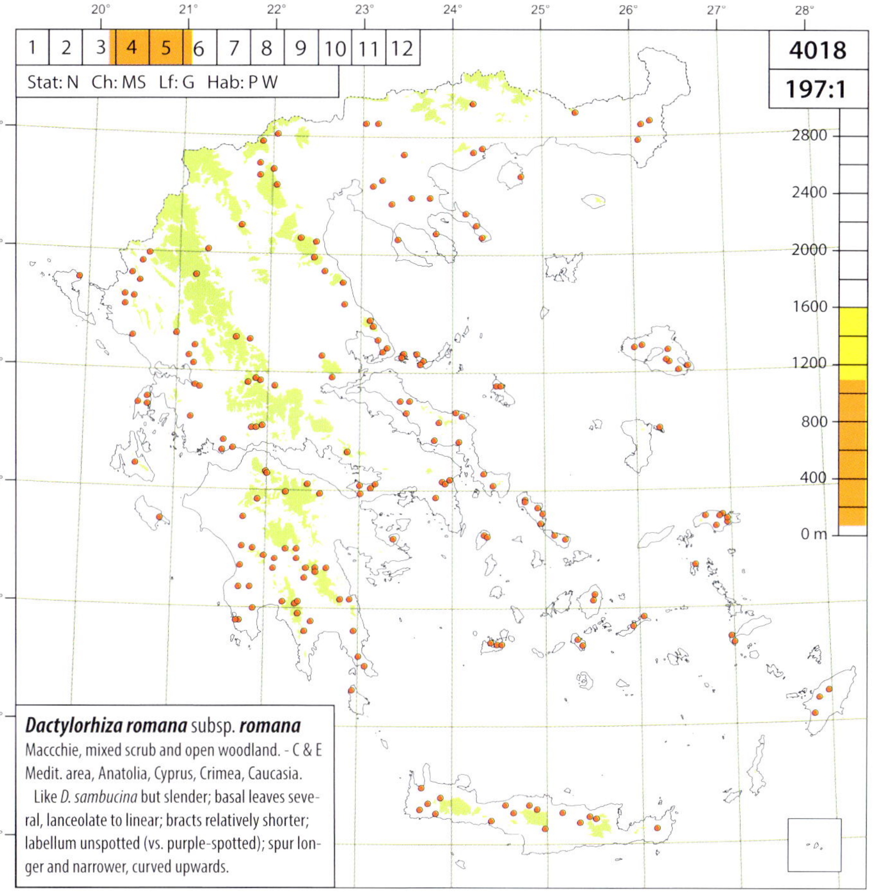

Dactylorhiza romana* subsp. *romana

Macchie, mixed scrub and open woodland. - C & E Medit. area, Anatolia, Cyprus, Crimea, Caucasia.

Like *D. sambucina* but slender; basal leaves several, lanceolate to linear; bracts relatively shorter; labellum unspotted (vs. purple-spotted); spur longer and narrower, curved upwards.

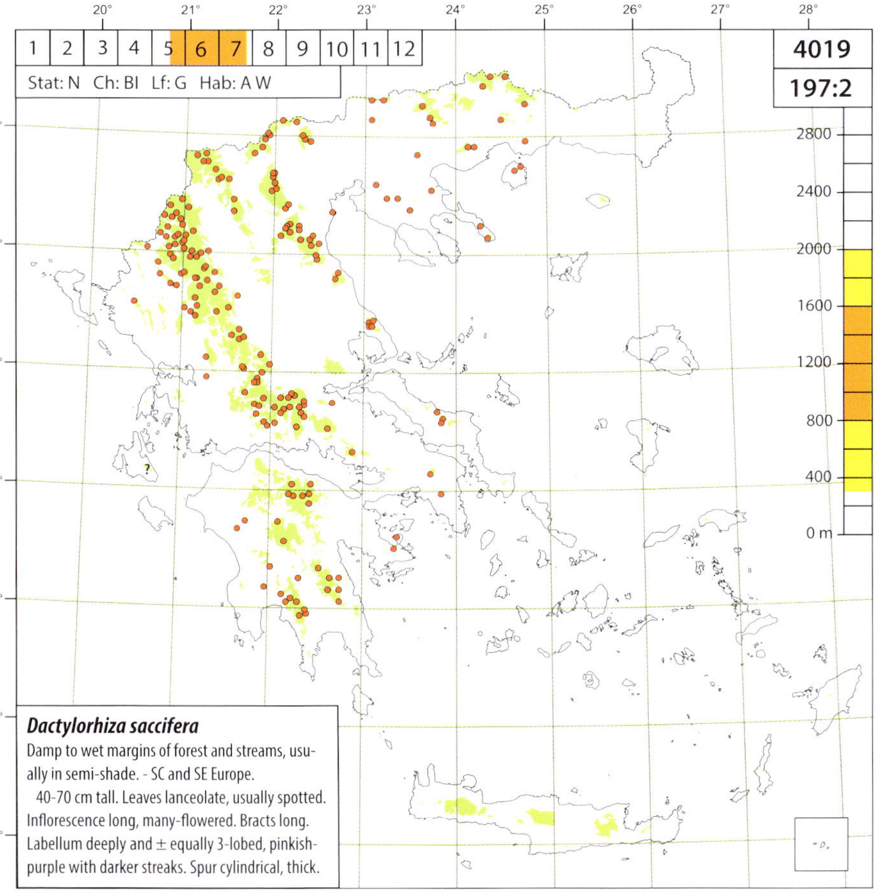

Dactylorhiza saccifera

Damp to wet margins of forest and streams, usually in semi-shade. - SC and SE Europe.

40-70 cm tall. Leaves lanceolate, usually spotted. Inflorescence long, many-flowered. Bracts long. Labellum deeply and ± equally 3-lobed, pinkish-purple with darker streaks. Spur cylindrical, thick.

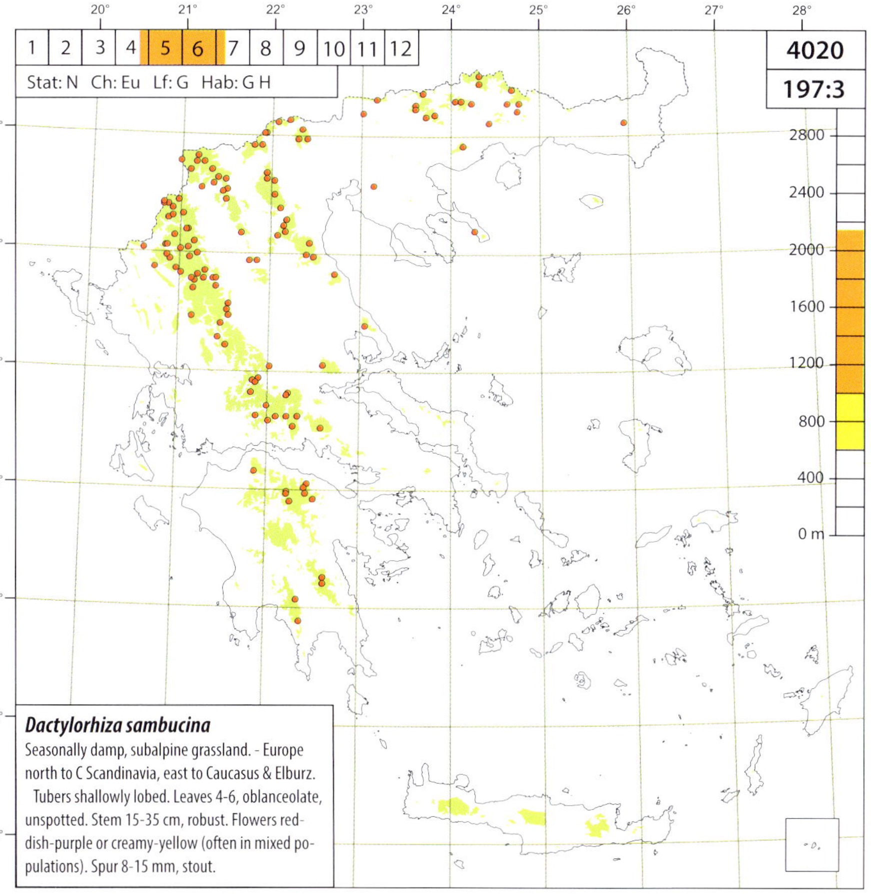

Dactylorhiza sambucina

Seasonally damp, subalpine grassland. - Europe north to C Scandinavia, east to Caucasus & Elburz.

Tubers shallowly lobed. Leaves 4-6, oblanceolate, unspotted. Stem 15-35 cm, robust. Flowers reddish-purple or creamy-yellow (often in mixed populations). Spur 8-15 mm, stout.

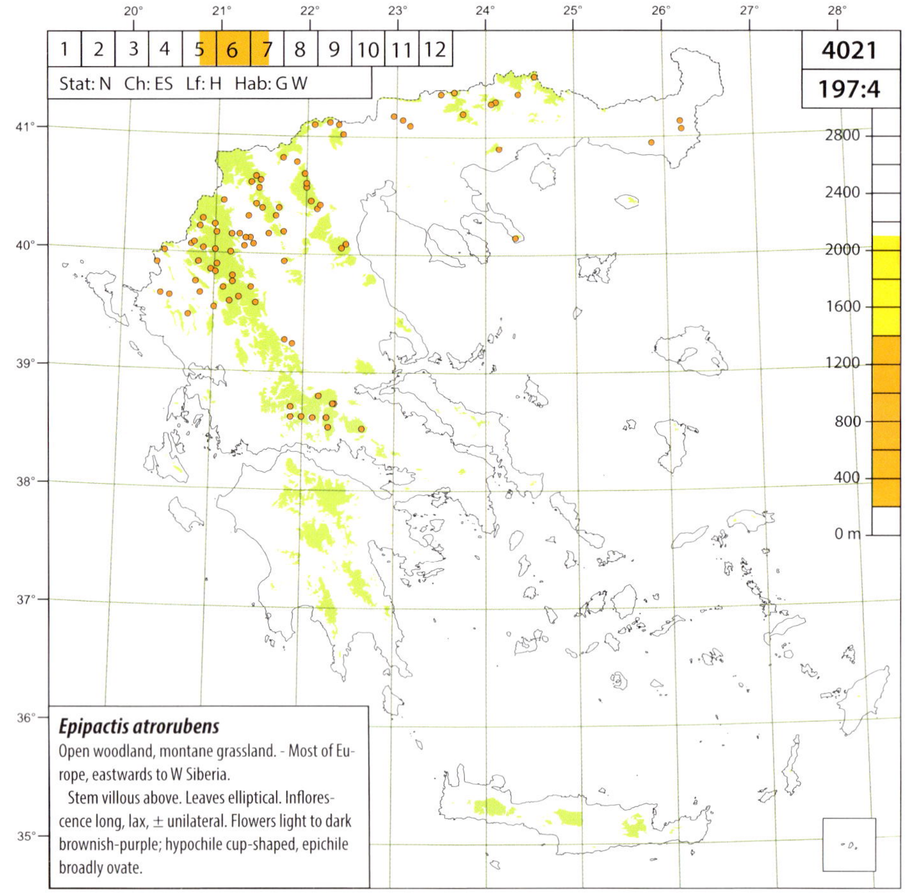

Epipactis atrorubens
Open woodland, montane grassland. - Most of Europe, eastwards to W Siberia.
Stem villous above. Leaves elliptical. Inflorescence long, lax, ± unilateral. Flowers light to dark brownish-purple; hypochile cup-shaped, epichile broadly ovate.

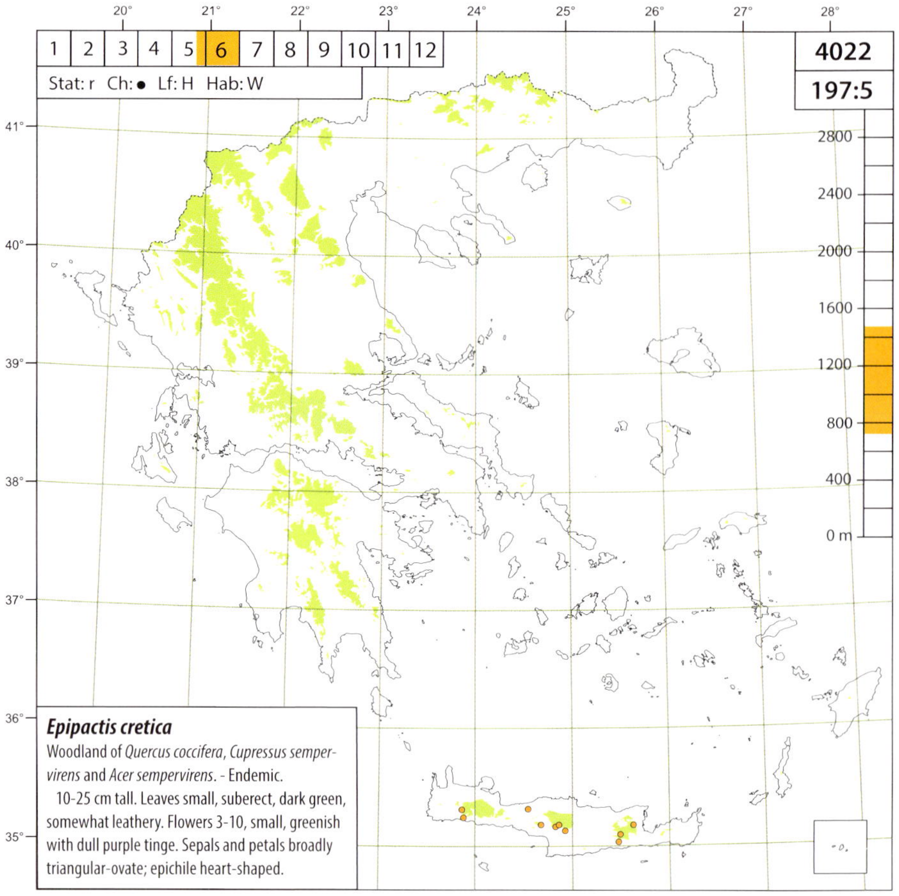

Epipactis cretica
Woodland of *Quercus coccifera*, *Cupressus sempervirens* and *Acer sempervirens*. - Endemic.
10-25 cm tall. Leaves small, suberect, dark green, somewhat leathery. Flowers 3-10, small, greenish with dull purple tinge. Sepals and petals broadly triangular-ovate; epichile heart-shaped.

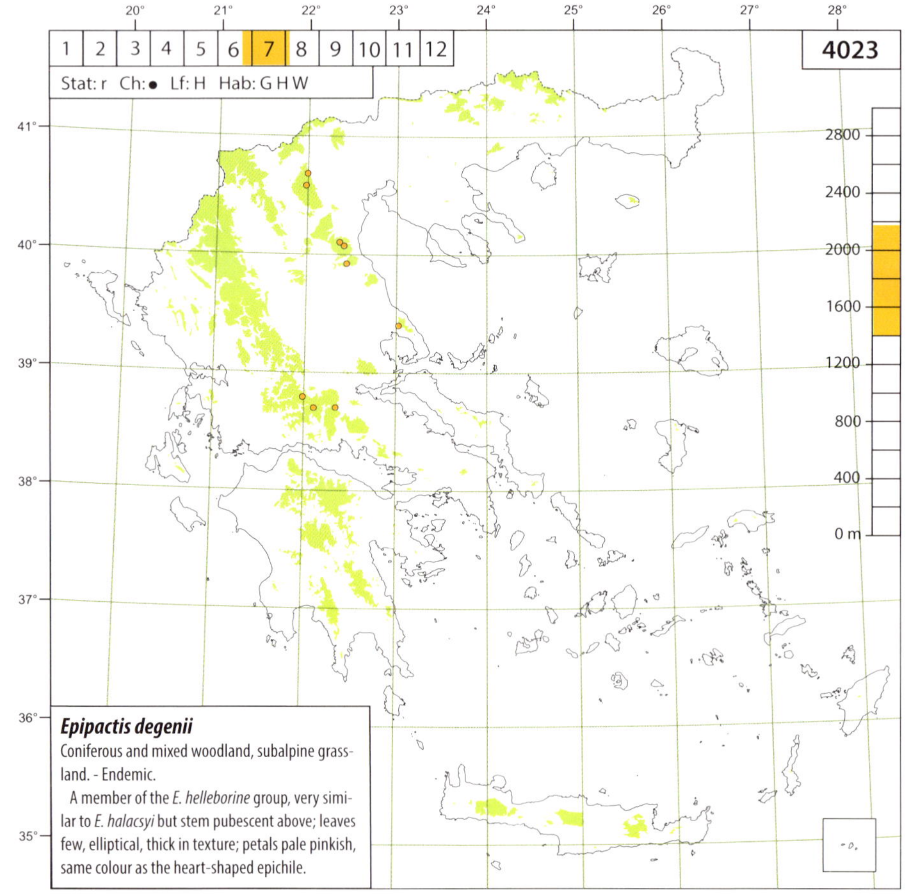

Epipactis degenii
Coniferous and mixed woodland, subalpine grassland. - Endemic.
A member of the *E. helleborine* group, very similar to *E. halacsyi* but stem pubescent above; leaves few, elliptical, thick in texture; petals pale pinkish, same colour as the heart-shaped epichile.

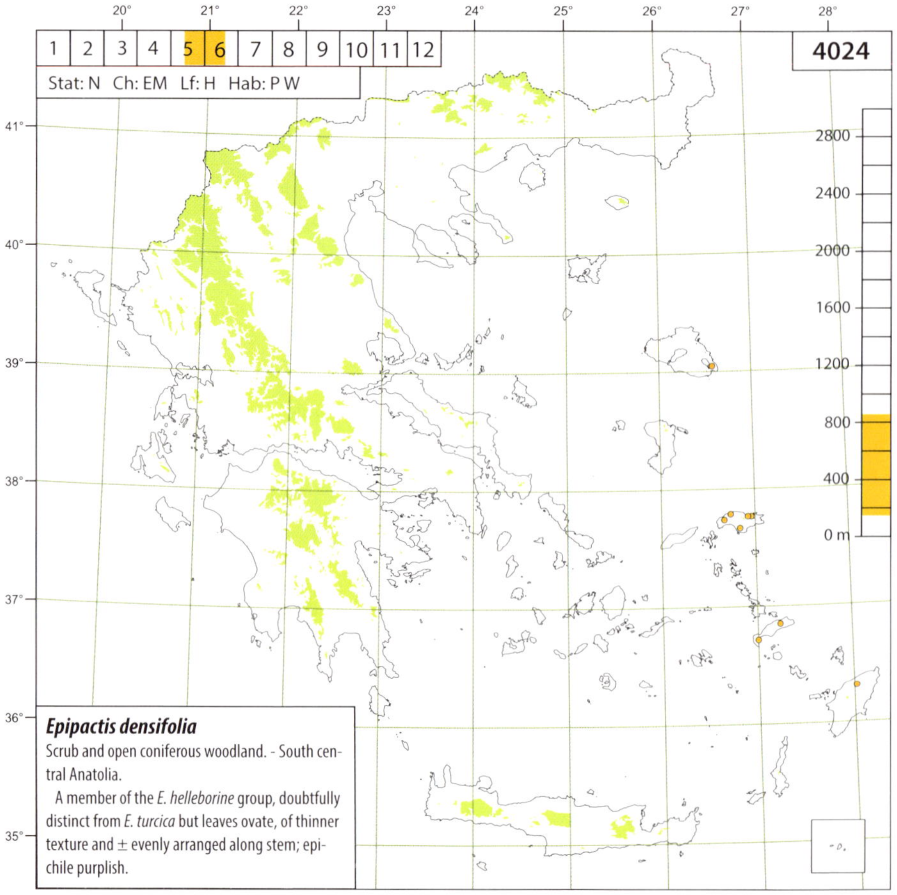

Epipactis densifolia
Scrub and open coniferous woodland. - South central Anatolia.
A member of the *E. helleborine* group, doubtfully distinct from *E. turcica* but leaves ovate, of thinner texture and ± evenly arranged along stem; epichile purplish.

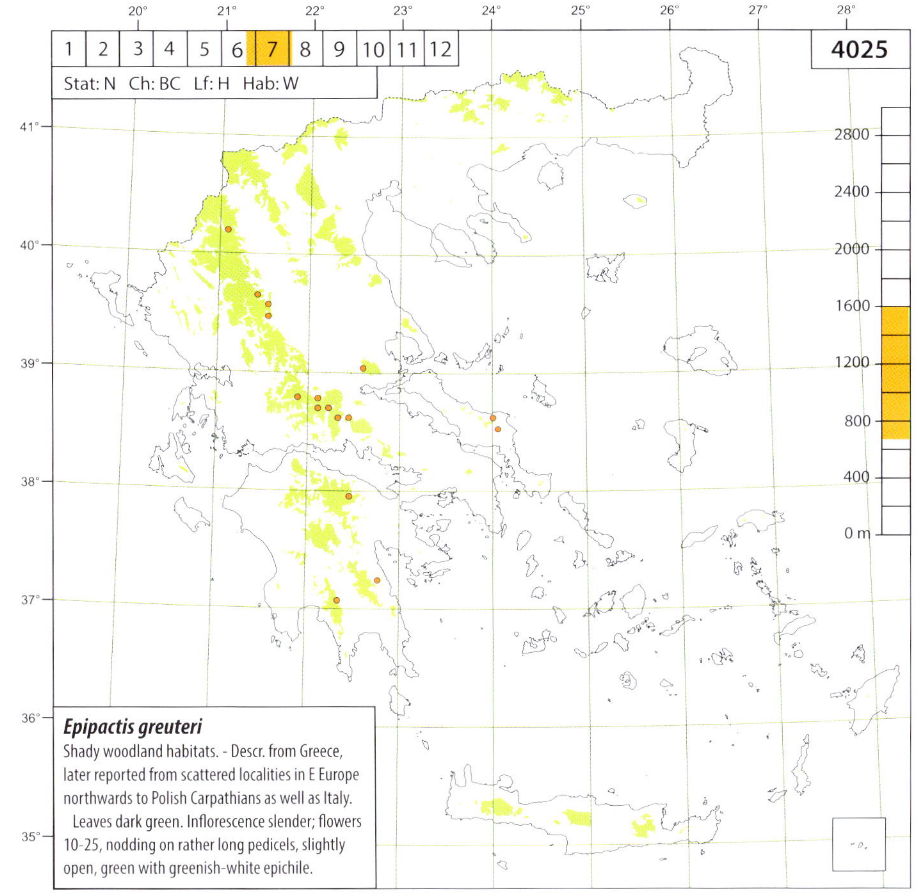

Epipactis greuteri
Shady woodland habitats. - Descr. from Greece, later reported from scattered localities in E Europe northwards to Polish Carpathians as well as Italy.
Leaves dark green. Inflorescence slender; flowers 10-25, nodding on rather long pedicels, slightly open, green with greenish-white epichile.

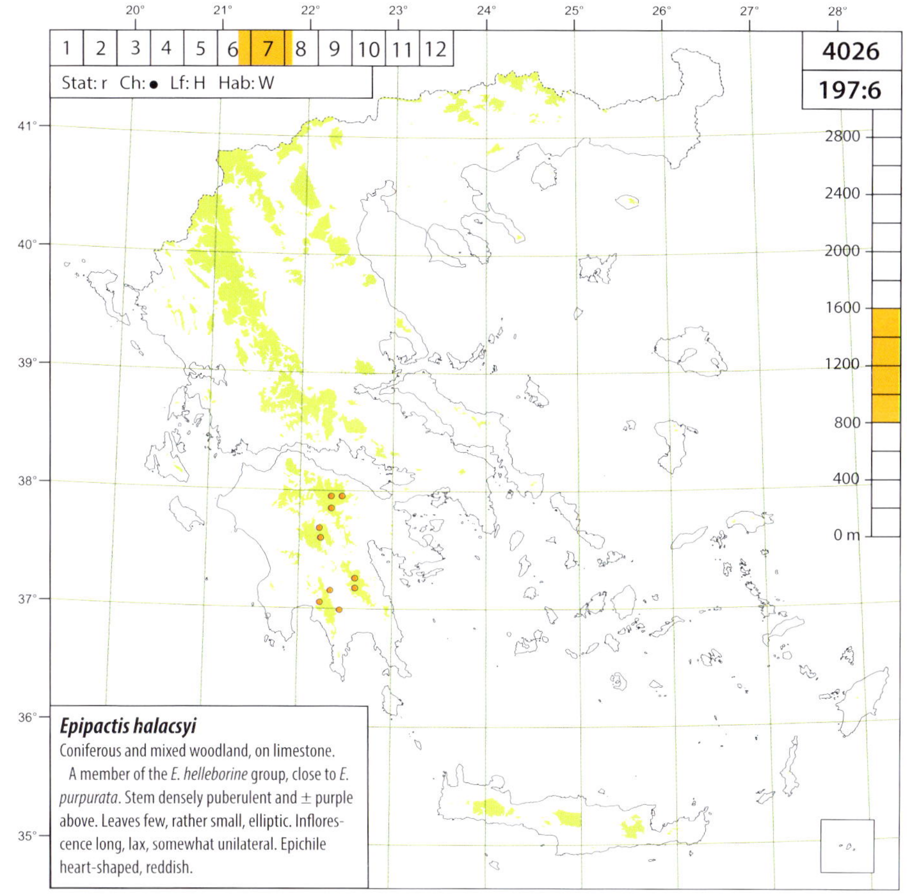

Epipactis halacsyi
Coniferous and mixed woodland, on limestone.
A member of the *E. helleborine* group, close to *E. purpurata*. Stem densely puberulent and ± purple above. Leaves few, rather small, elliptic. Inflorescence long, lax, somewhat unilateral. Epichile heart-shaped, reddish.

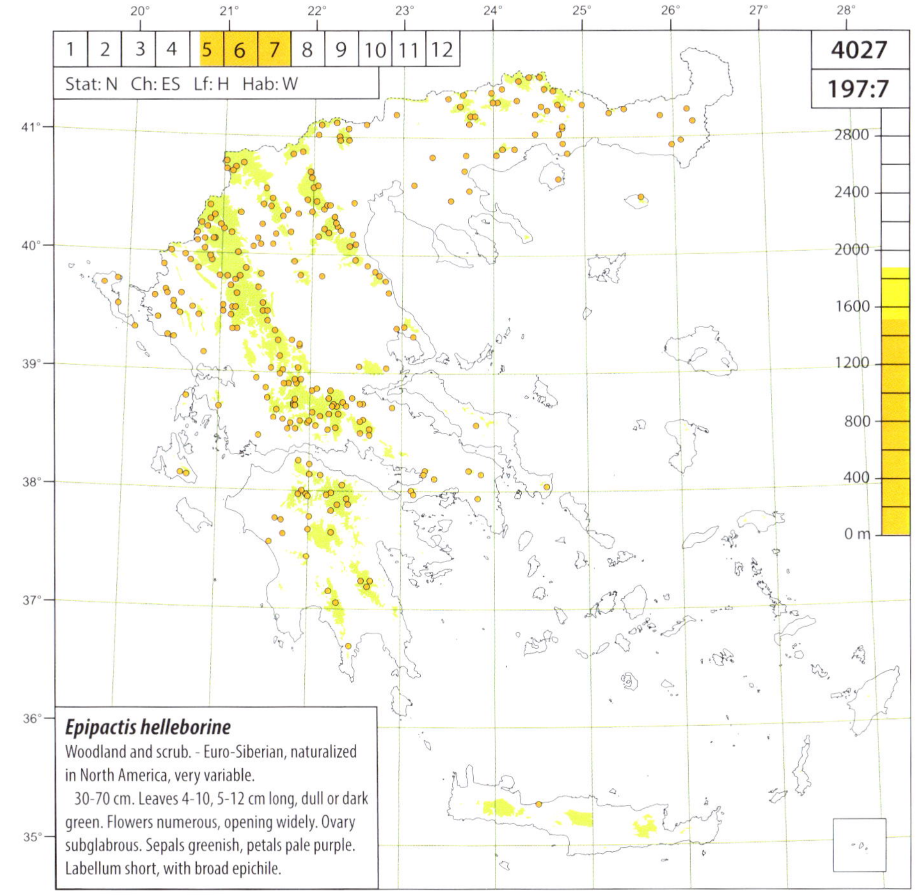

Epipactis helleborine
Woodland and scrub. - Euro-Siberian, naturalized in North America, very variable.
30-70 cm. Leaves 4-10, 5-12 cm long, dull or dark green. Flowers numerous, opening widely. Ovary subglabrous. Sepals greenish, petals pale purple. Labellum short, with broad epichile.

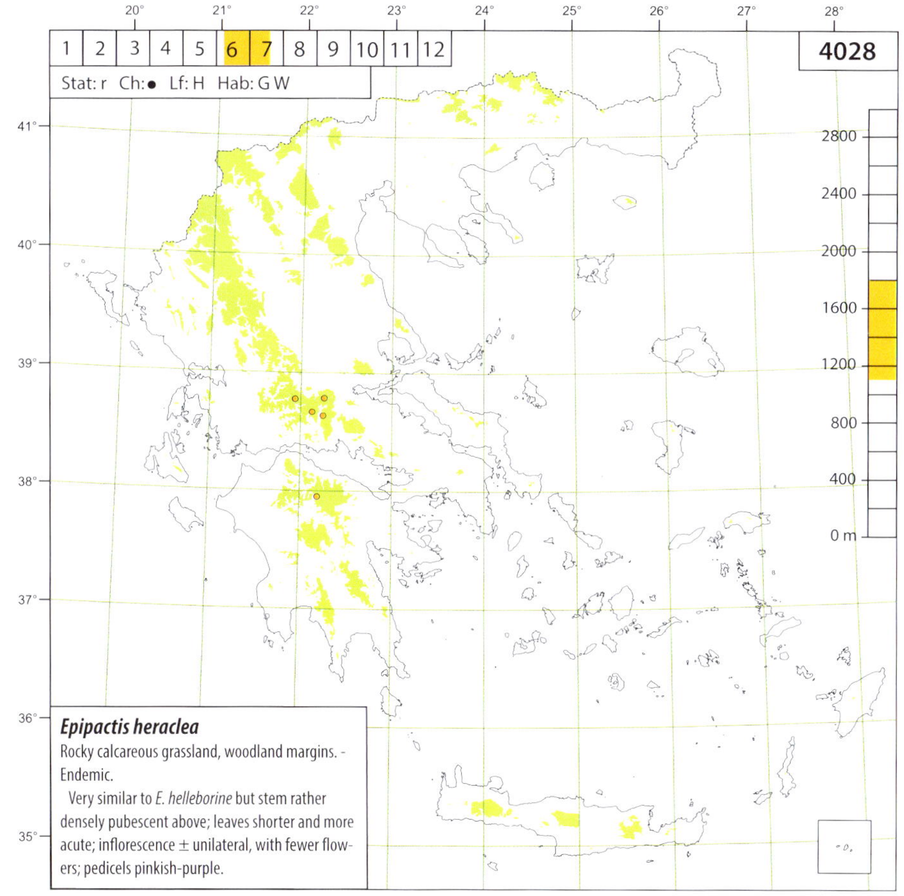

Epipactis heraclea
Rocky calcareous grassland, woodland margins. - Endemic.
Very similar to *E. helleborine* but stem rather densely pubescent above; leaves shorter and more acute; inflorescence ± unilateral, with fewer flowers; pedicels pinkish-purple.

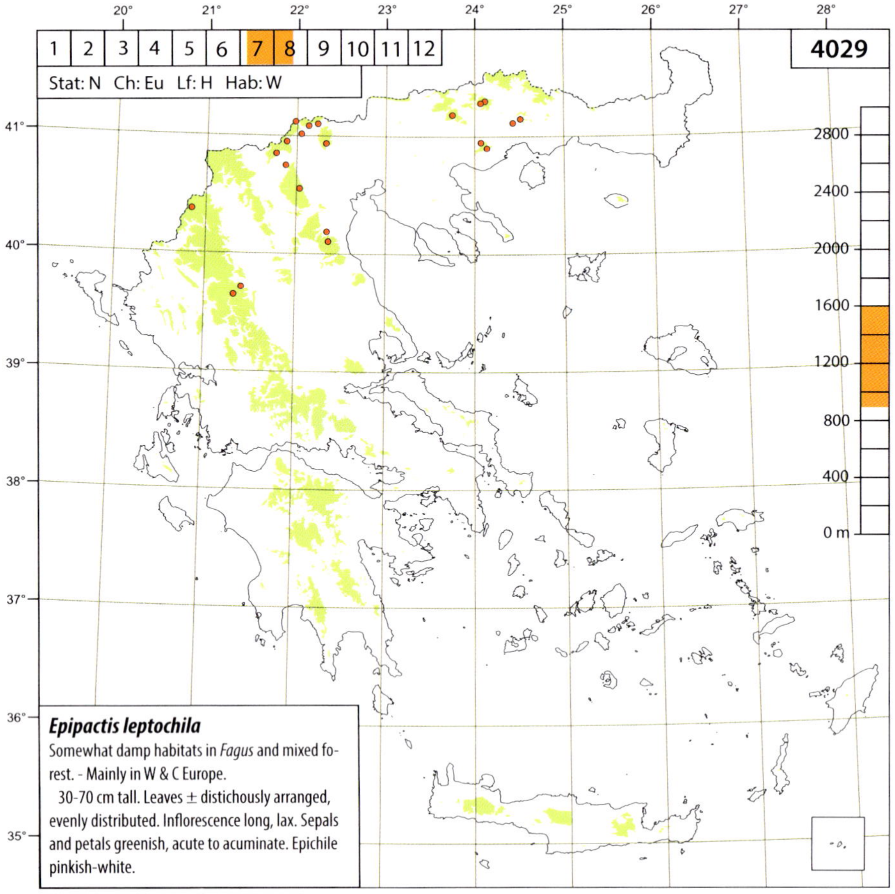

Epipactis leptochila

Somewhat damp habitats in *Fagus* and mixed forest. - Mainly in W & C Europe.

30-70 cm tall. Leaves ± distichously arranged, evenly distributed. Inflorescence long, lax. Sepals and petals greenish, acute to acuminate. Epichile pinkish-white.

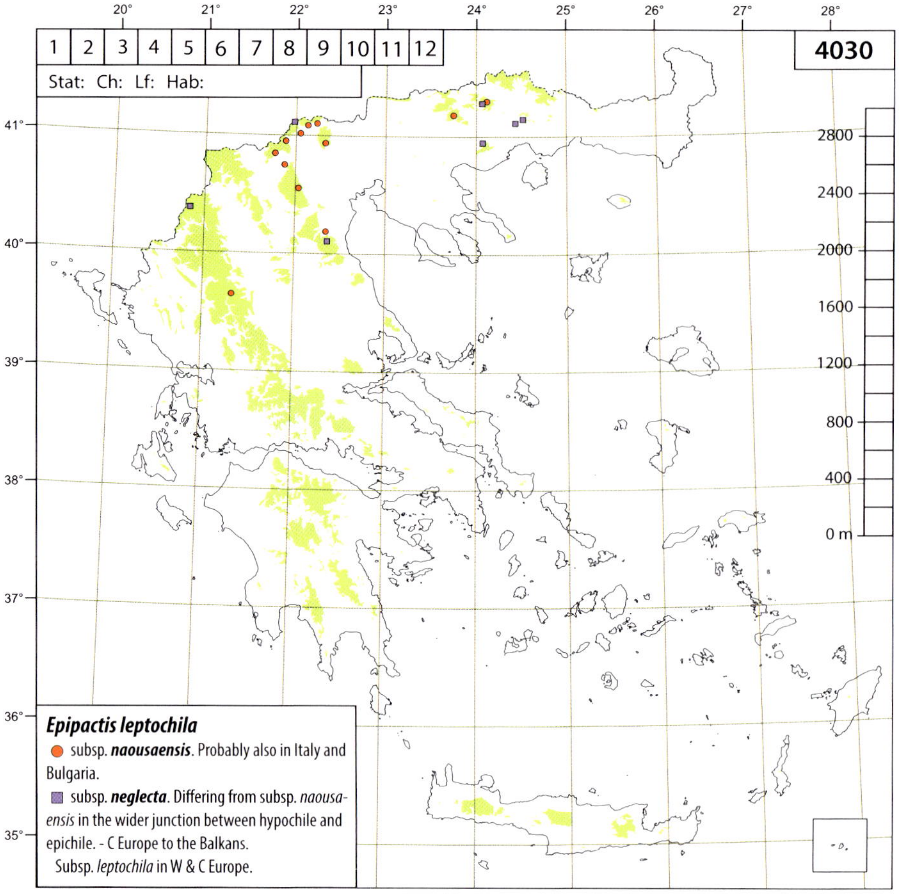

Epipactis leptochila

● subsp. ***naousaensis***. Probably also in Italy and Bulgaria.

■ subsp. ***neglecta***. Differing from subsp. *naousaensis* in the wider junction between hypochile and epichile. - C Europe to the Balkans.

Subsp. *leptochila* in W & C Europe.

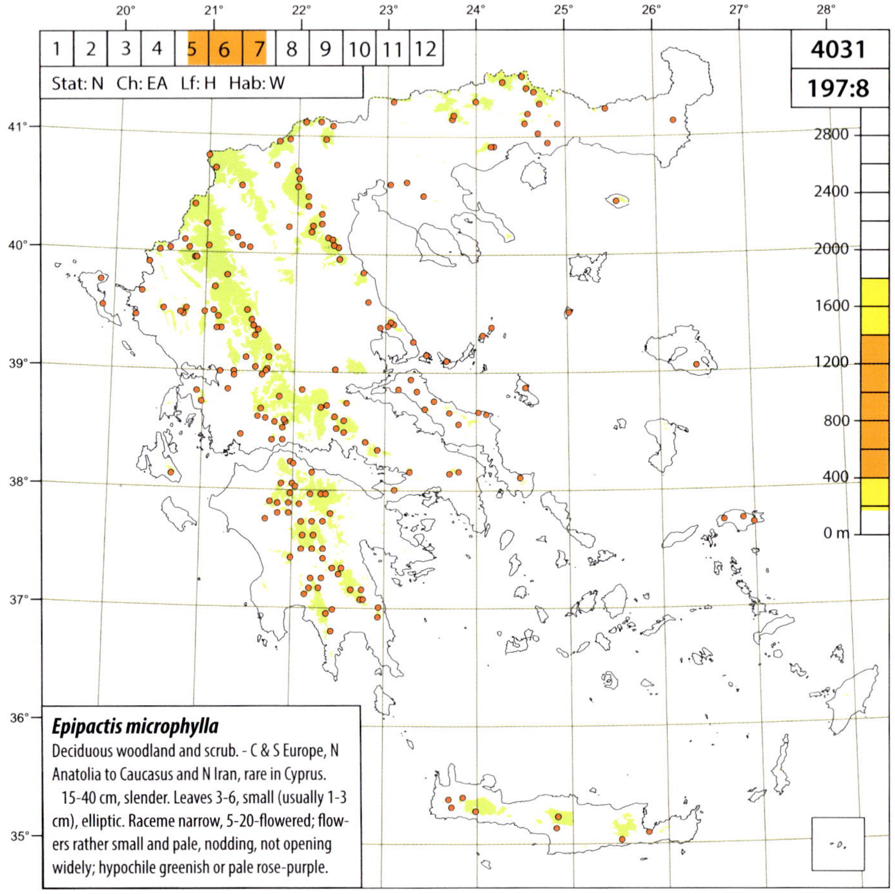

Epipactis microphylla

Deciduous woodland and scrub. - C & S Europe, N Anatolia to Caucasus and N Iran, rare in Cyprus.

15-40 cm, slender. Leaves 3-6, small (usually 1-3 cm), elliptic. Raceme narrow, 5-20-flowered; flowers rather small and pale, nodding, not opening widely; hypochile greenish or pale rose-purple.

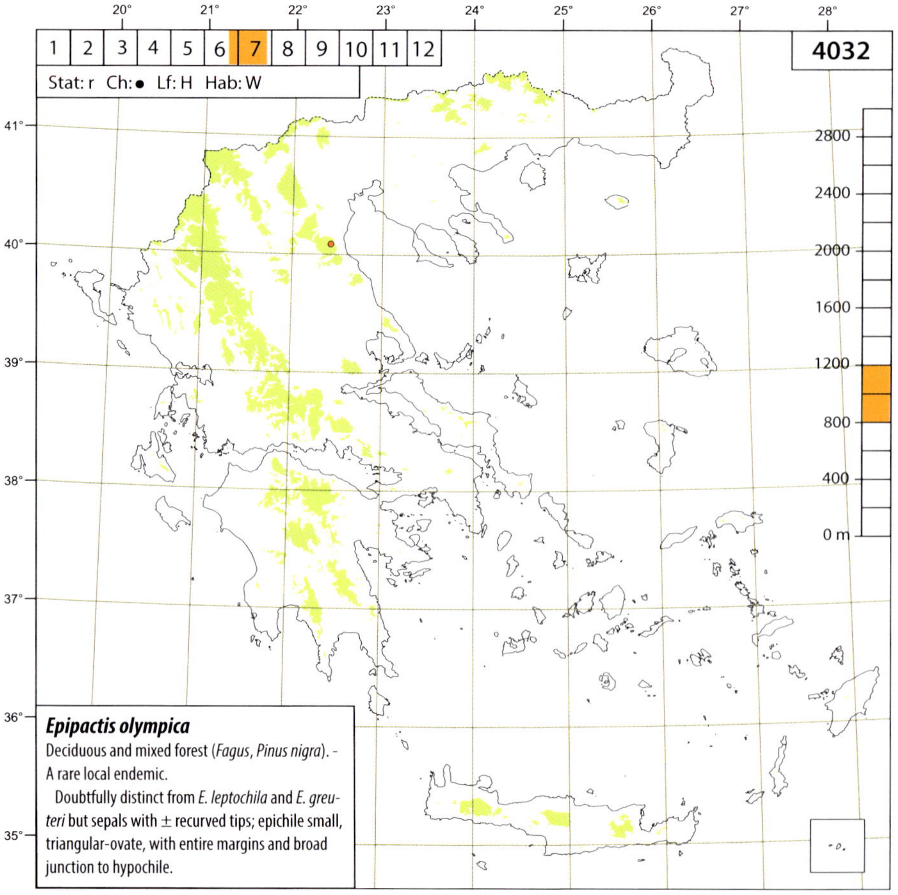

Epipactis olympica

Deciduous and mixed forest (*Fagus*, *Pinus nigra*). - A rare local endemic.

Doubtfully distinct from *E. leptochila* and *E. greuteri* but sepals with ± recurved tips; epichile small, triangular-ovate, with entire margins and broad junction to hypochile.

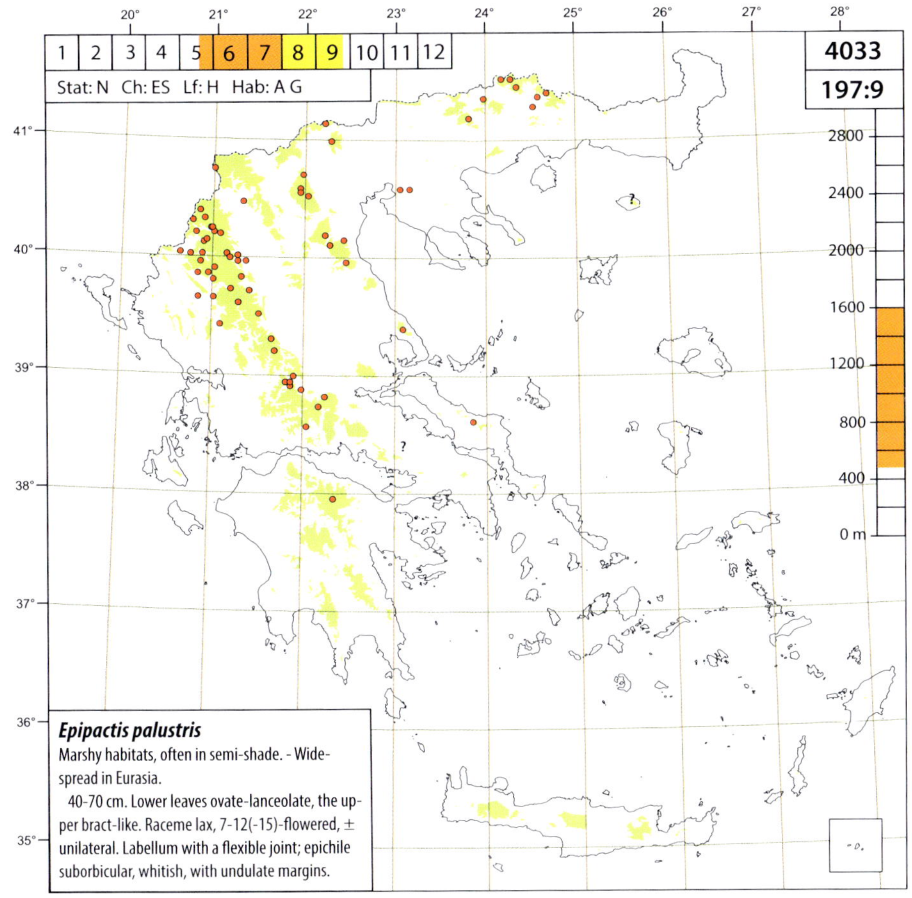

Epipactis palustris
Marshy habitats, often in semi-shade. - Widespread in Eurasia.
40-70 cm. Lower leaves ovate-lanceolate, the upper bract-like. Raceme lax, 7-12(-15)-flowered, ± unilateral. Labellum with a flexible joint; epichile suborbicular, whitish, with undulate margins.

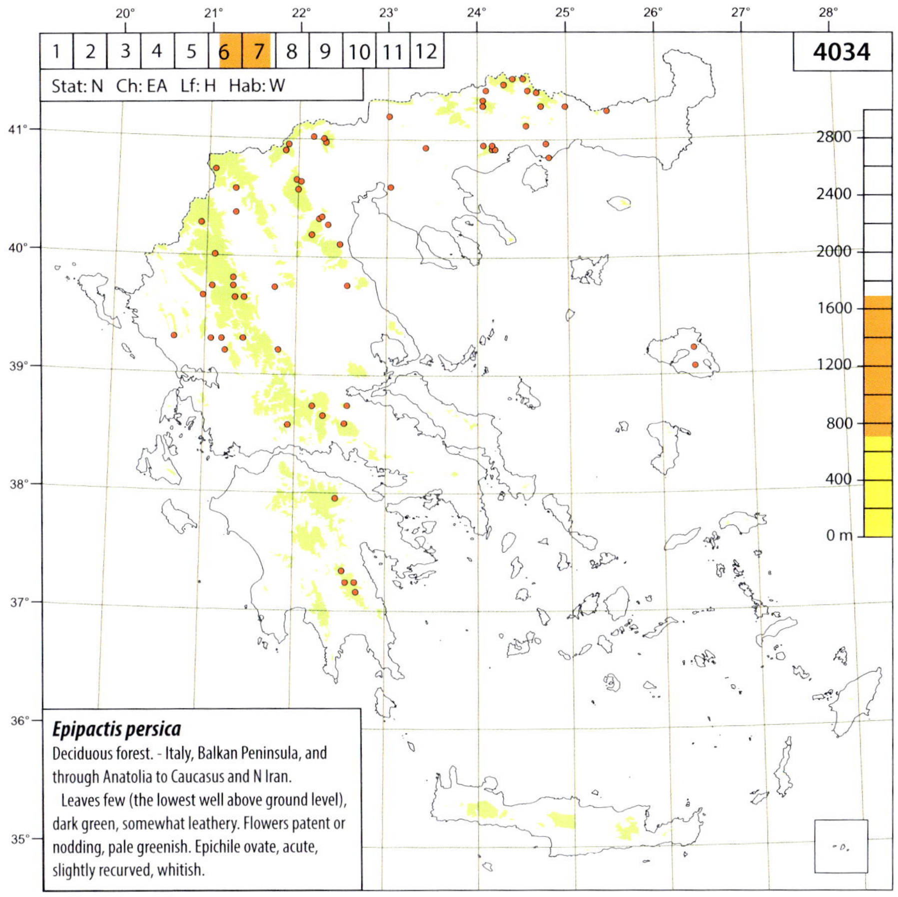

Epipactis persica
Deciduous forest. - Italy, Balkan Peninsula, and through Anatolia to Caucasus and N Iran.
Leaves few (the lowest well above ground level), dark green, somewhat leathery. Flowers patent or nodding, pale greenish. Epichile ovate, acute, slightly recurved, whitish.

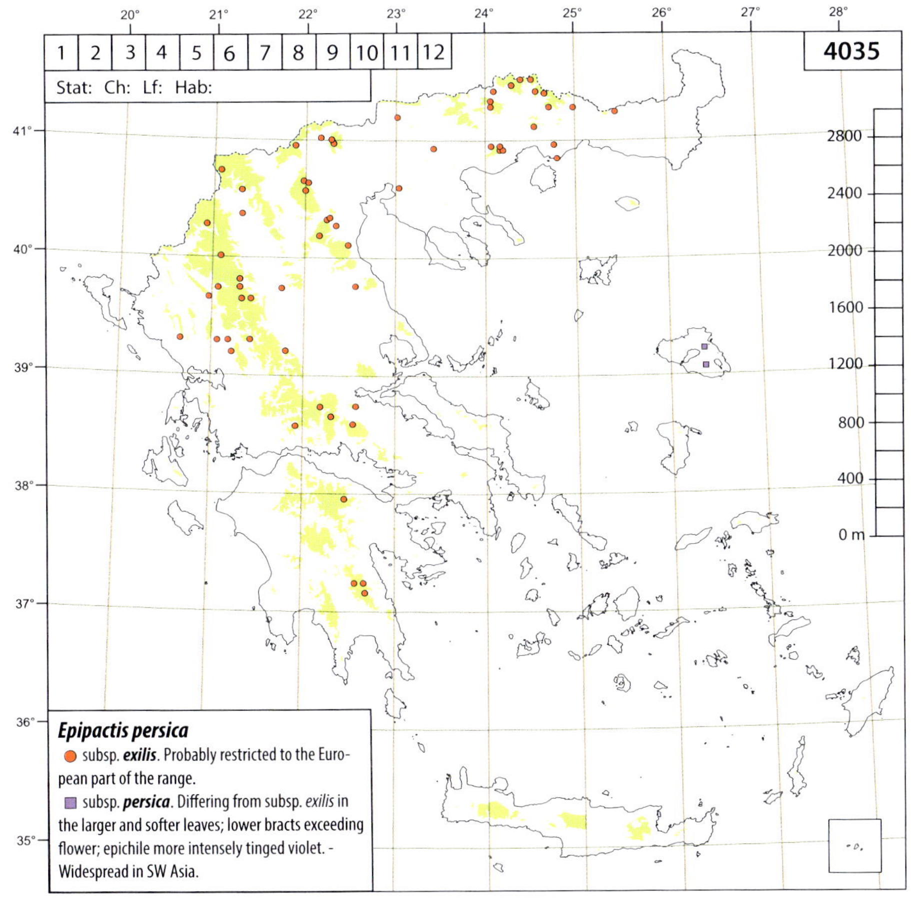

Epipactis persica
● subsp. ***exilis***. Probably restricted to the European part of the range.
■ subsp. ***persica***. Differing from subsp. *exilis* in the larger and softer leaves; lower bracts exceeding flower; epichile more intensely tinged violet. - Widespread in SW Asia.

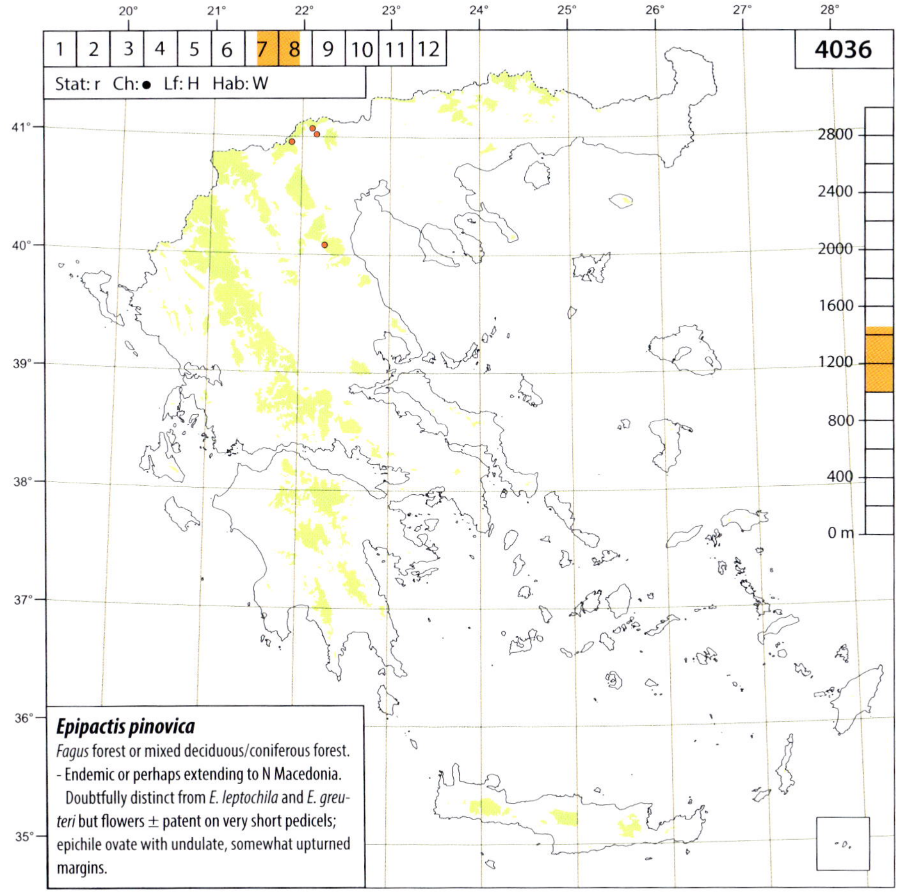

Epipactis pinovica
Fagus forest or mixed deciduous/coniferous forest. - Endemic or perhaps extending to N Macedonia.
Doubtfully distinct from *E. leptochila* and *E. greuteri* but flowers ± patent on very short pedicels; epichile ovate with undulate, somewhat upturned margins.

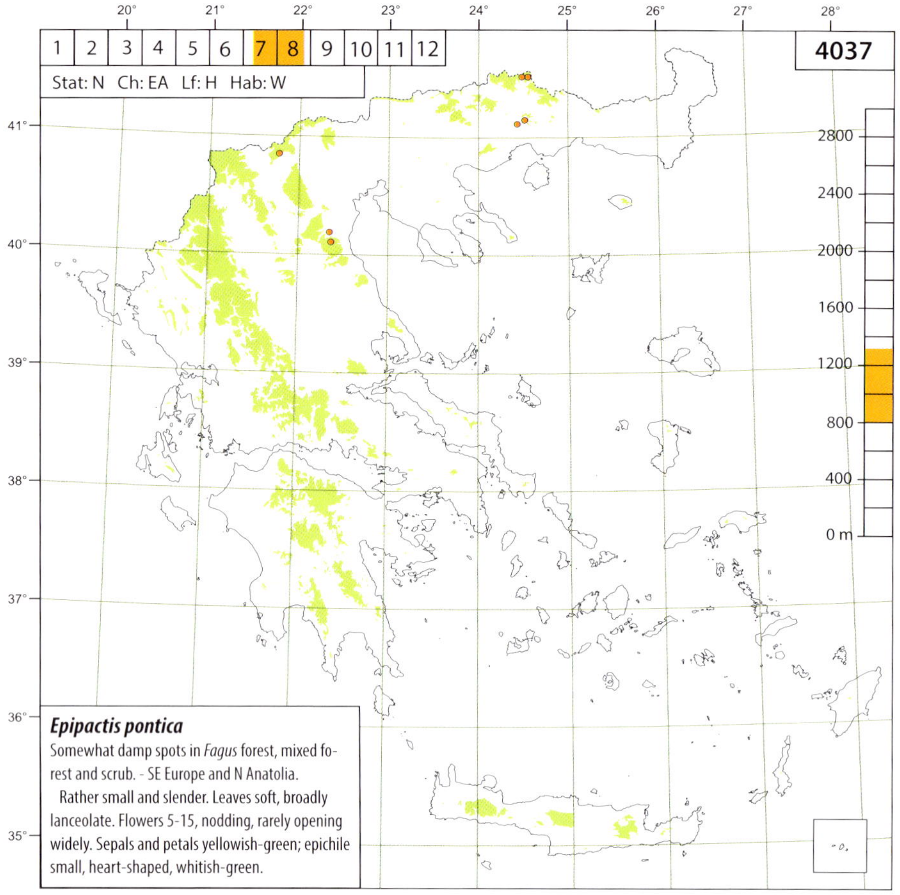

Epipactis pontica

Somewhat damp spots in *Fagus* forest, mixed forest and scrub. - SE Europe and N Anatolia.

Rather small and slender. Leaves soft, broadly lanceolate. Flowers 5-15, nodding, rarely opening widely. Sepals and petals yellowish-green; epichile small, heart-shaped, whitish-green.

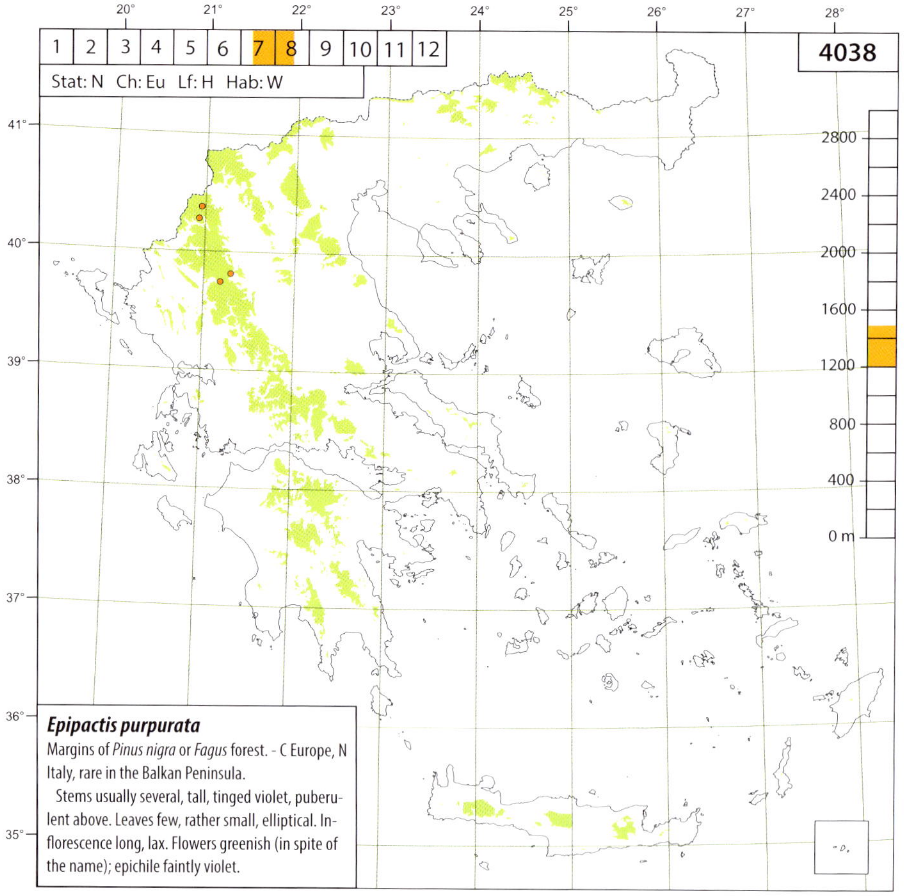

Epipactis purpurata

Margins of *Pinus nigra* or *Fagus* forest. - C Europe, N Italy, rare in the Balkan Peninsula.

Stems usually several, tall, tinged violet, puberulent above. Leaves few, rather small, elliptical. Inflorescence long, lax. Flowers greenish (in spite of the name); epichile faintly violet.

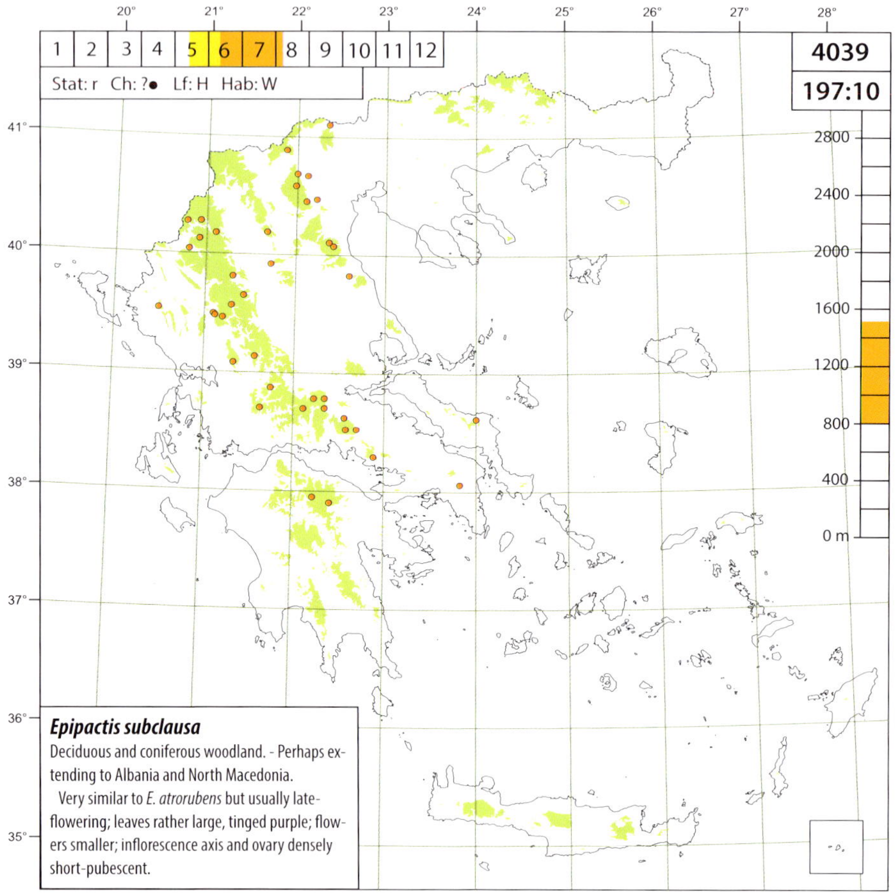

Epipactis subclausa

Deciduous and coniferous woodland. - Perhaps extending to Albania and North Macedonia.

Very similar to *E. atrorubens* but usually late-flowering; leaves rather large, tinged purple; flowers smaller; inflorescence axis and ovary densely short-pubescent.

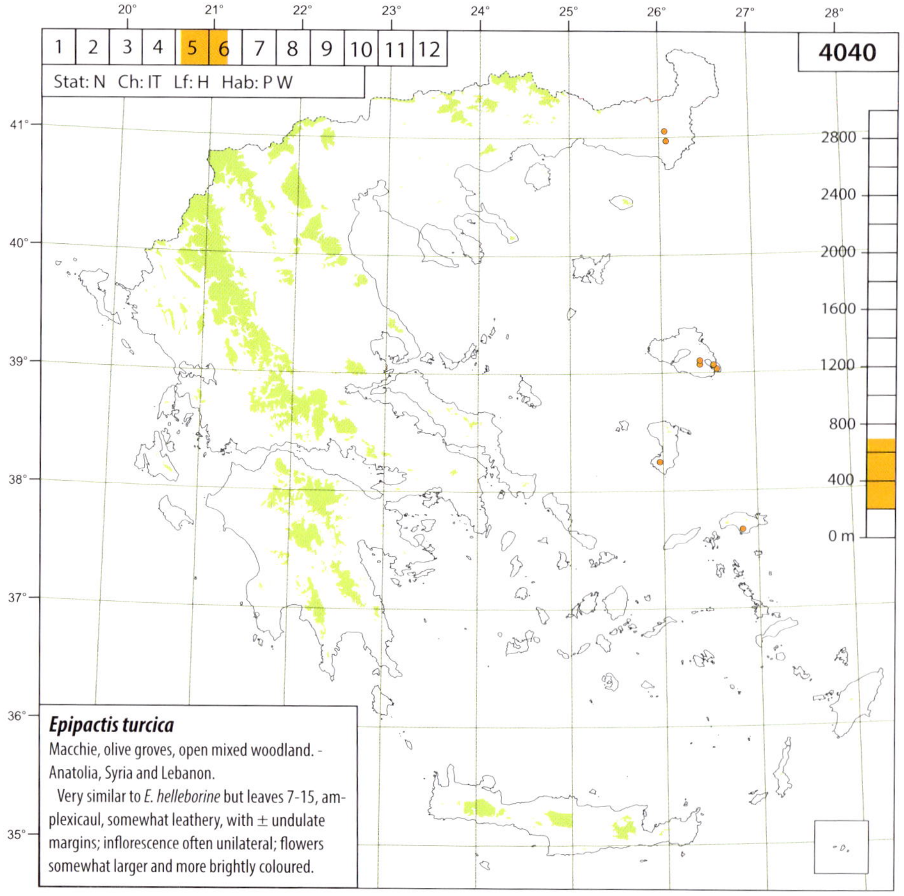

Epipactis turcica

Macchie, olive groves, open mixed woodland. - Anatolia, Syria and Lebanon.

Very similar to *E. helleborine* but leaves 7-15, amplexicaul, somewhat leathery, with ± undulate margins; inflorescence often unilateral; flowers somewhat larger and more brightly coloured.

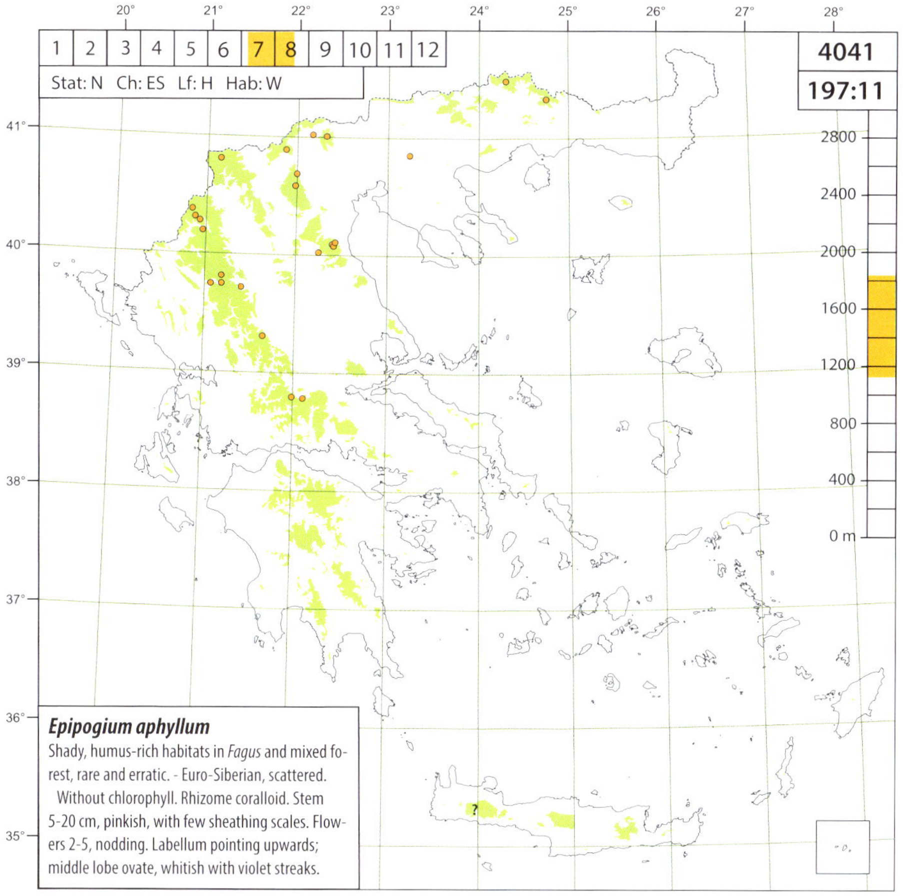

Epipogium aphyllum

Shady, humus-rich habitats in *Fagus* and mixed forest, rare and erratic. - Euro-Siberian, scattered.

Without chlorophyll. Rhizome coralloid. Stem 5-20 cm, pinkish, with few sheathing scales. Flowers 2-5, nodding. Labellum pointing upwards; middle lobe ovate, whitish with violet streaks.

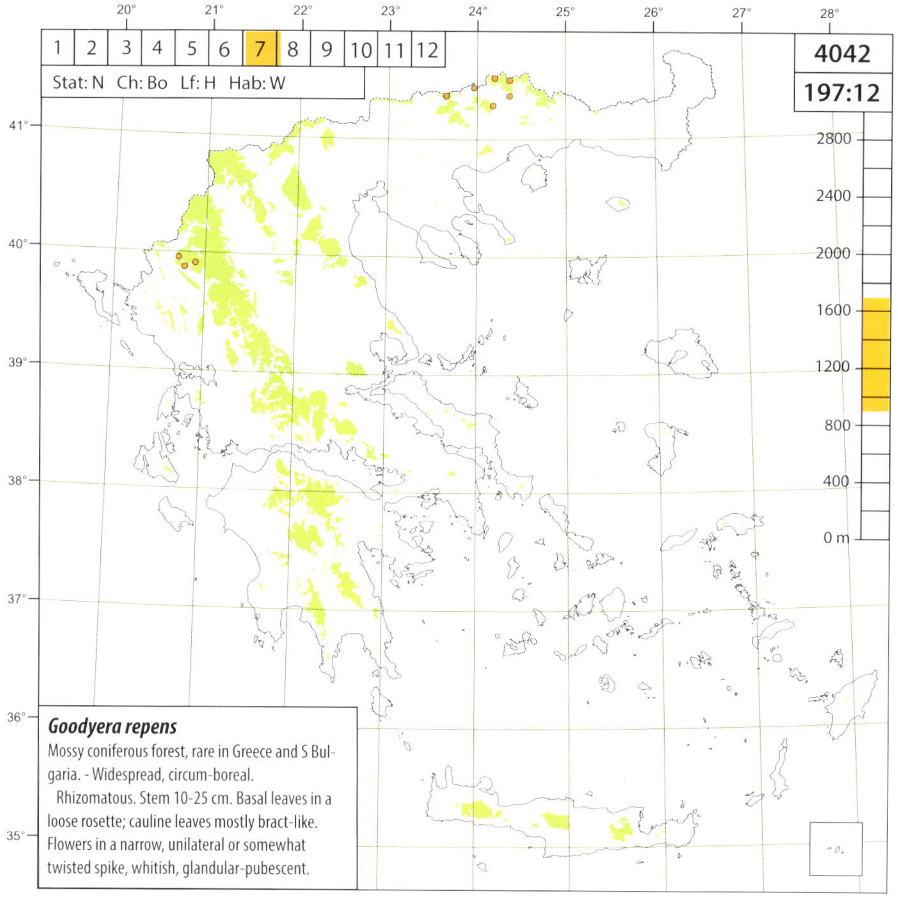

Goodyera repens

Mossy coniferous forest, rare in Greece and S Bulgaria. - Widespread, circum-boreal.

Rhizomatous. Stem 10-25 cm. Basal leaves in a loose rosette; cauline leaves mostly bract-like. Flowers in a narrow, unilateral or somewhat twisted spike, whitish, glandular-pubescent.

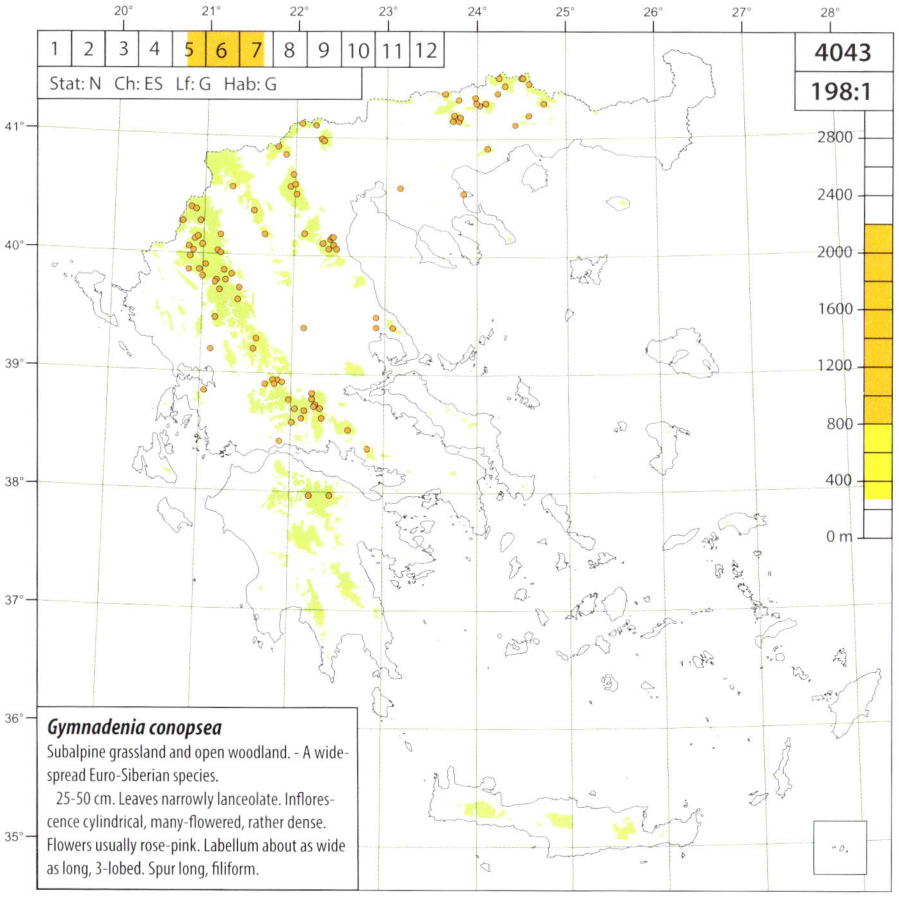

Gymnadenia conopsea

Subalpine grassland and open woodland. - A widespread Euro-Siberian species.

25-50 cm. Leaves narrowly lanceolate. Inflorescence cylindrical, many-flowered, rather dense. Flowers usually rose-pink. Labellum about as wide as long, 3-lobed. Spur long, filiform.

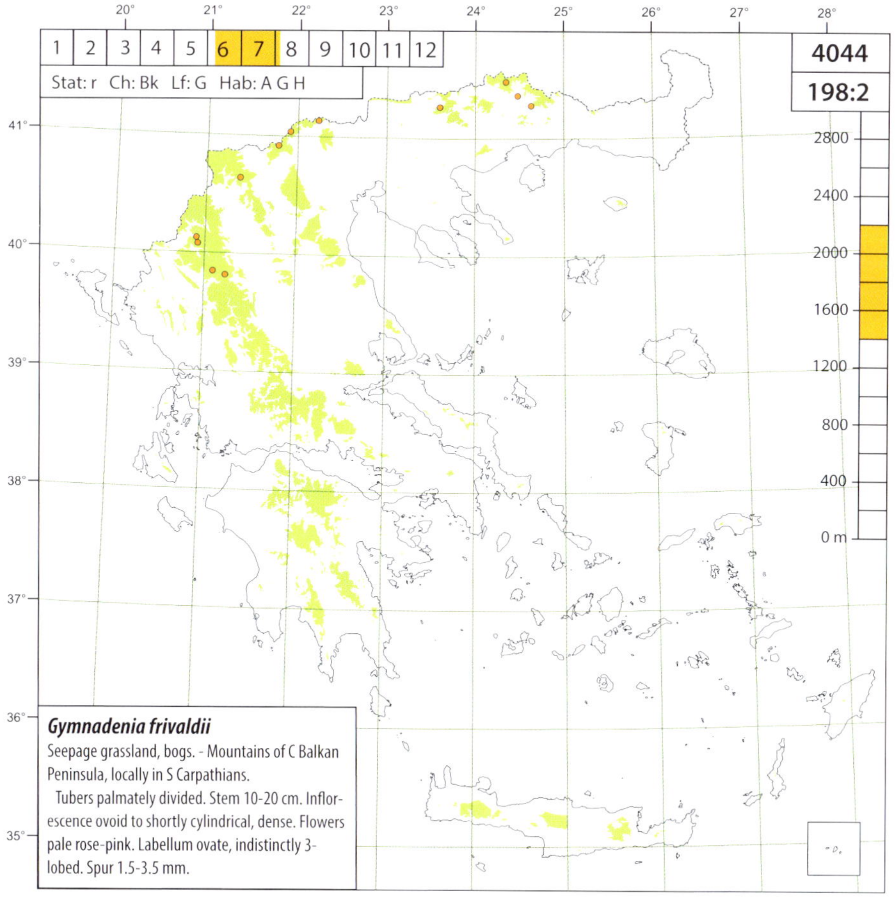

Gymnadenia frivaldii

Seepage grassland, bogs. - Mountains of C Balkan Peninsula, locally in S Carpathians.

Tubers palmately divided. Stem 10-20 cm. Inflorescence ovoid to shortly cylindrical, dense. Flowers pale rose-pink. Labellum ovate, indistinctly 3-lobed. Spur 1.5-3.5 mm.

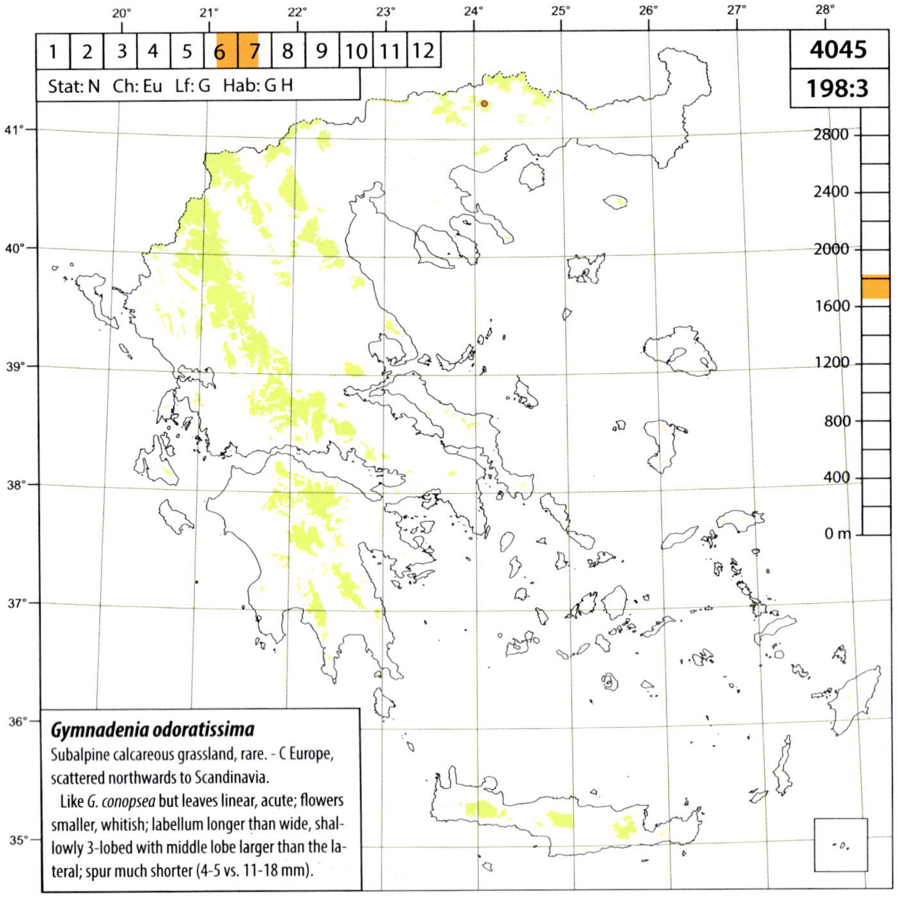

Gymnadenia odoratissima

Subalpine calcareous grassland, rare. - C Europe, scattered northwards to Scandinavia.

Like *G. conopsea* but leaves linear, acute; flowers smaller, whitish; labellum longer than wide, shallowly 3-lobed with middle lobe larger than the lateral; spur much shorter (4-5 vs. 11-18 mm).

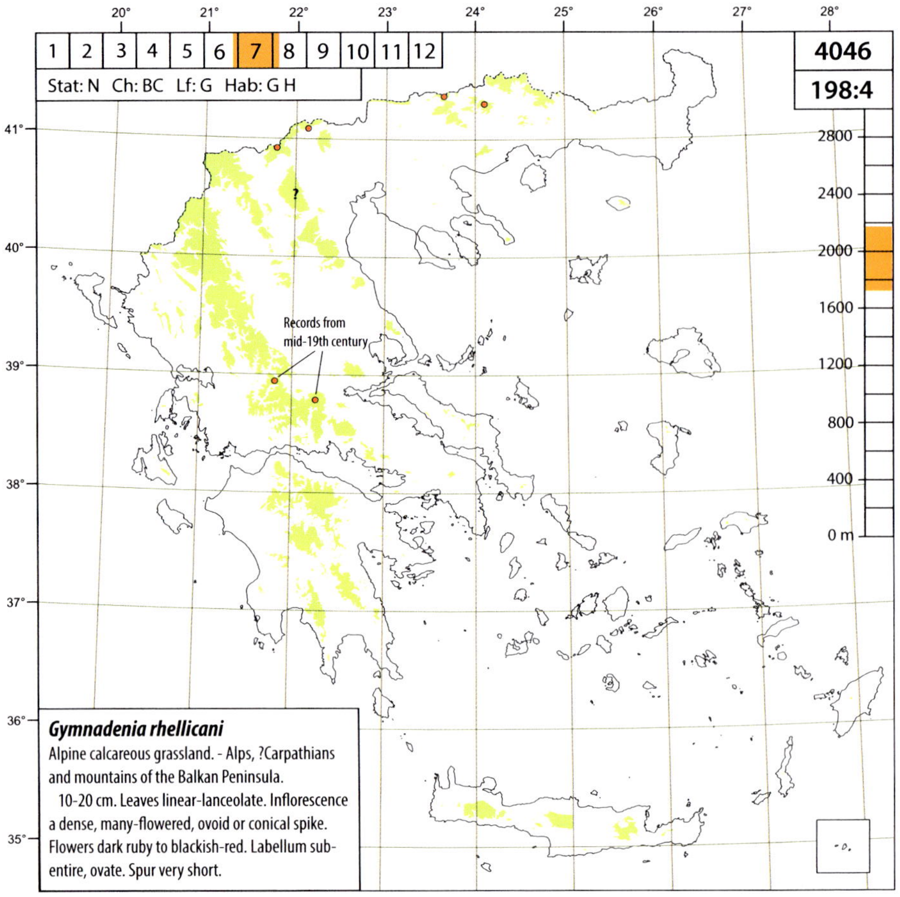

Gymnadenia rhellicani

Alpine calcareous grassland. - Alps, ?Carpathians and mountains of the Balkan Peninsula.

10-20 cm. Leaves linear-lanceolate. Inflorescence a dense, many-flowered, ovoid or conical spike. Flowers dark ruby to blackish-red. Labellum subentire, ovate. Spur very short.

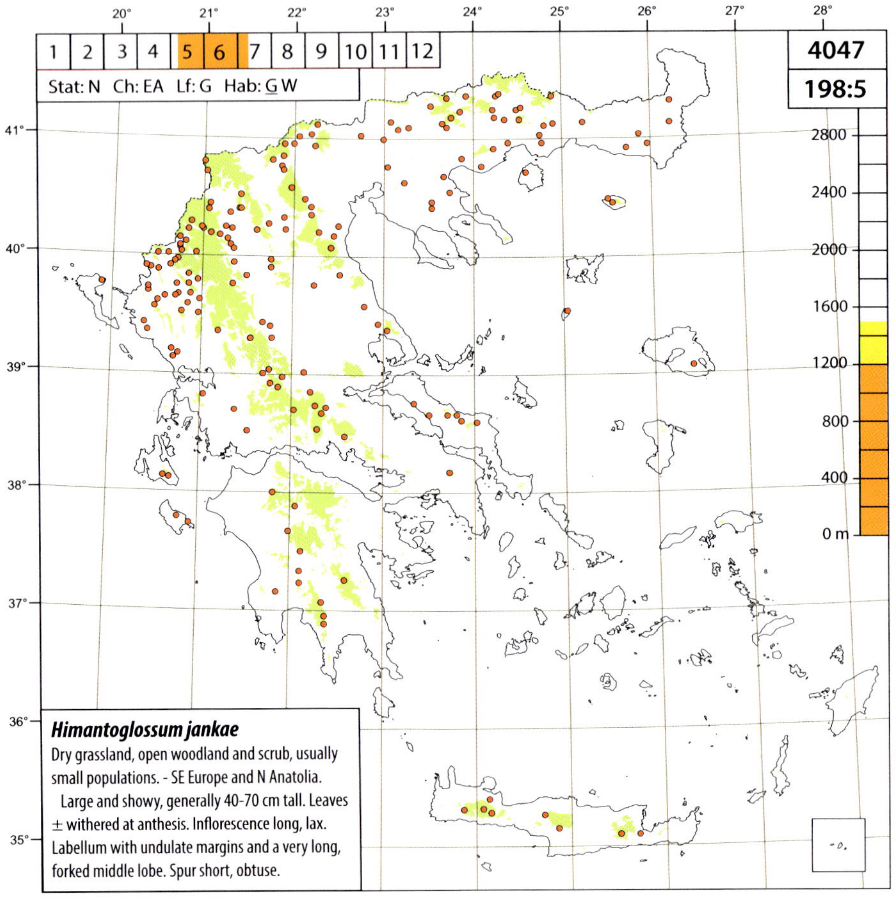

Himantoglossum jankae

Dry grassland, open woodland and scrub, usually small populations. - SE Europe and N Anatolia.

Large and showy, generally 40-70 cm tall. Leaves ± withered at anthesis. Inflorescence long, lax. Labellum with undulate margins and a very long, forked middle lobe. Spur short, obtuse.

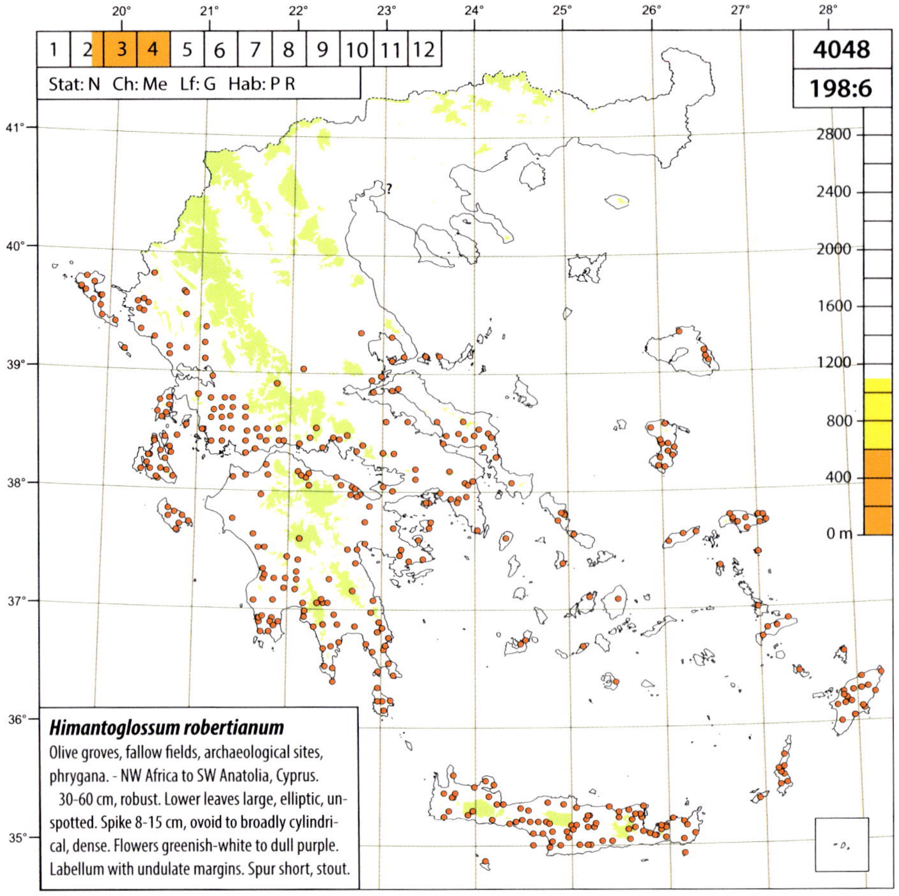

Himantoglossum robertianum

Olive groves, fallow fields, archaeological sites, phrygana. - NW Africa to SW Anatolia, Cyprus.

30-60 cm, robust. Lower leaves large, elliptic, unspotted. Spike 8-15 cm, ovoid to broadly cylindrical, dense. Flowers greenish-white to dull purple. Labellum with undulate margins. Spur short, stout.

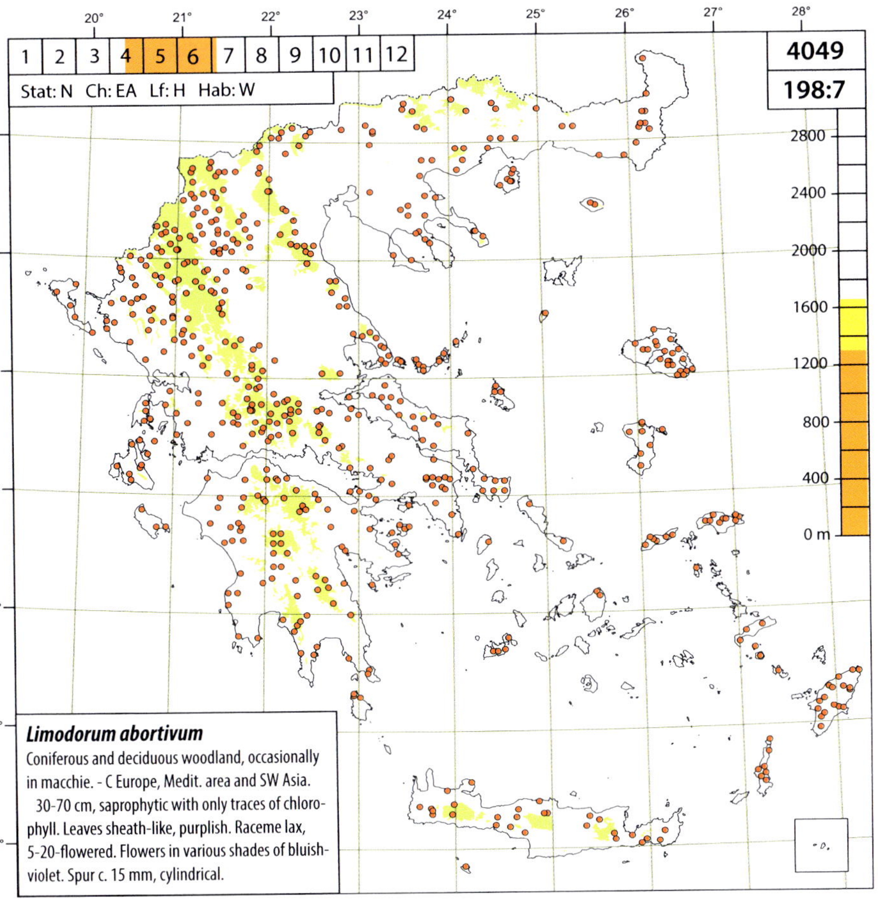

Limodorum abortivum

Coniferous and deciduous woodland, occasionally in macchie. - C Europe, Medit. area and SW Asia.

30-70 cm, saprophytic with only traces of chlorophyll. Leaves sheath-like, purplish. Raceme lax, 5-20-flowered. Flowers in various shades of bluish-violet. Spur c. 15 mm, cylindrical.

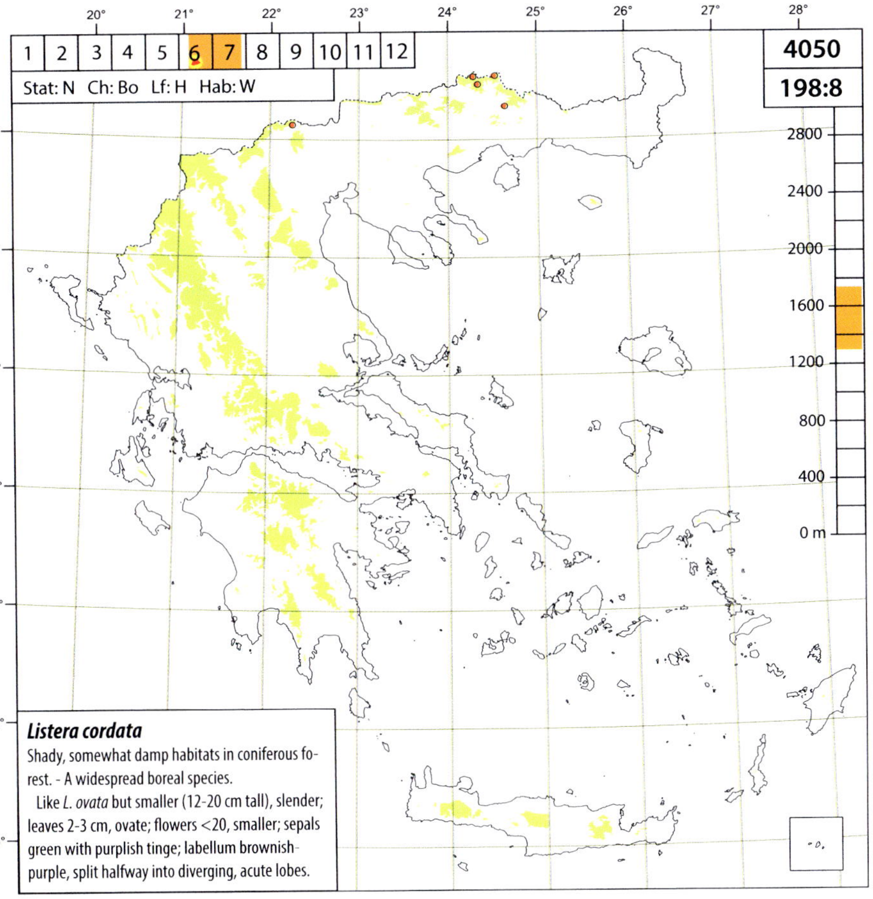

Listera cordata

Shady, somewhat damp habitats in coniferous forest. - A widespread boreal species.

Like *L. ovata* but smaller (12-20 cm tall), slender; leaves 2-3 cm, ovate; flowers <20, smaller; sepals green with purplish tinge; labellum brownish-purple, split halfway into diverging, acute lobes.

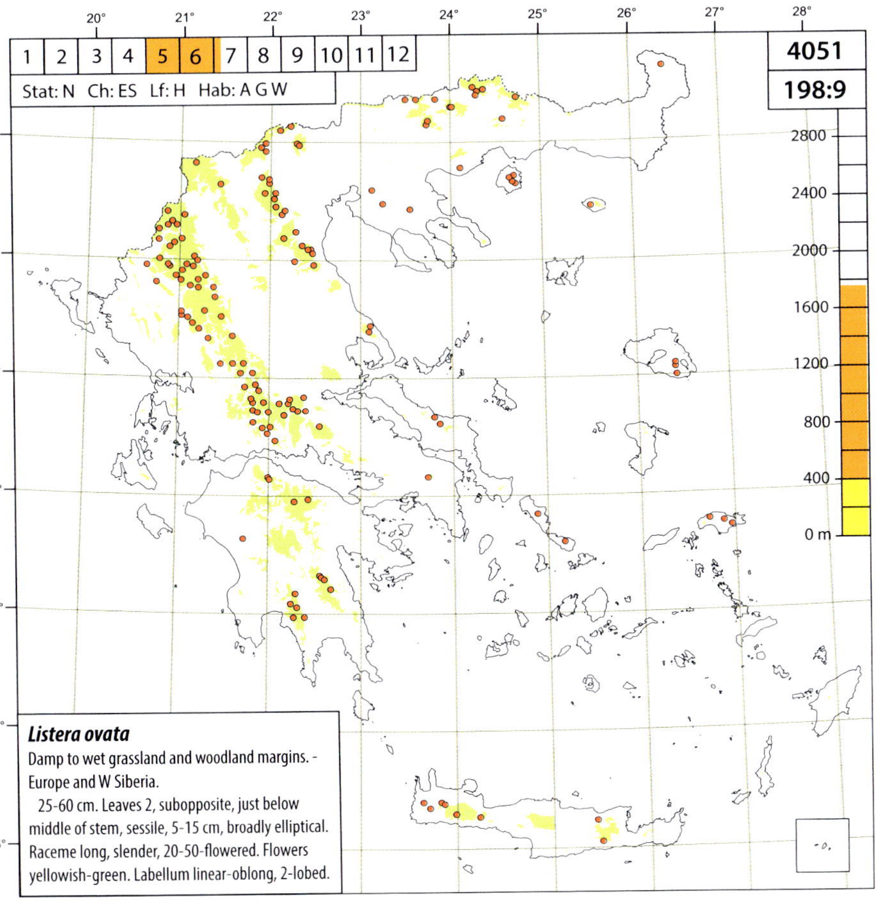

Listera ovata

Damp to wet grassland and woodland margins. - Europe and W Siberia.

25-60 cm. Leaves 2, subopposite, just below middle of stem, sessile, 5-15 cm, broadly elliptical. Raceme long, slender, 20-50-flowered. Flowers yellowish-green. Labellum linear-oblong, 2-lobed.

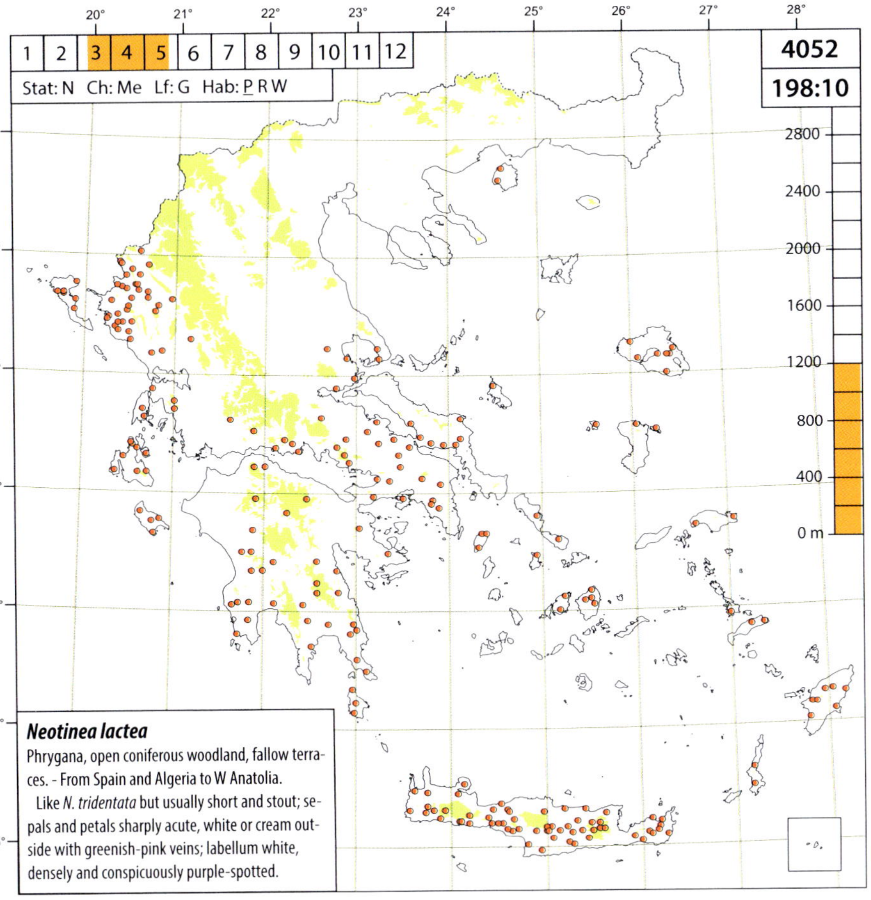

Neotinea lactea

Phrygana, open coniferous woodland, fallow terraces. - From Spain and Algeria to W Anatolia.

Like *N. tridentata* but usually short and stout; sepals and petals sharply acute, white or cream outside with greenish-pink veins; labellum white, densely and conspicuously purple-spotted.

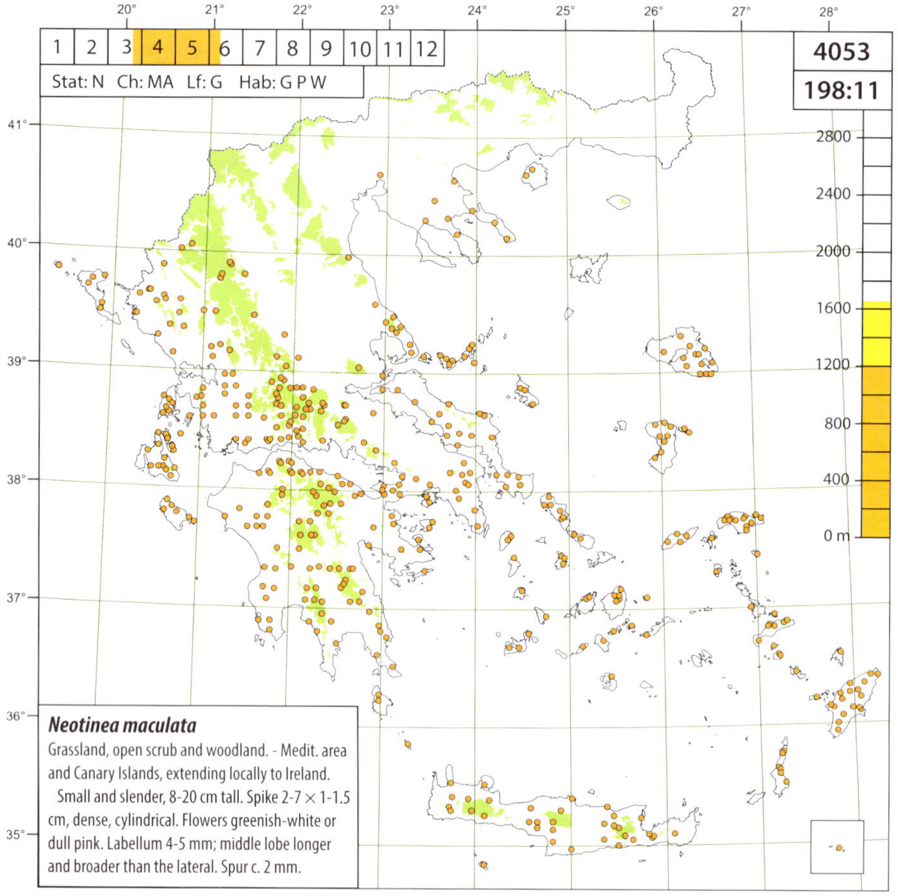

Neotinea maculata
Grassland, open scrub and woodland. - Medit. area and Canary Islands, extending locally to Ireland.
Small and slender, 8-20 cm tall. Spike 2-7 × 1-1.5 cm, dense, cylindrical. Flowers greenish-white or dull pink. Labellum 4-5 mm; middle lobe longer and broader than the lateral. Spur c. 2 mm.

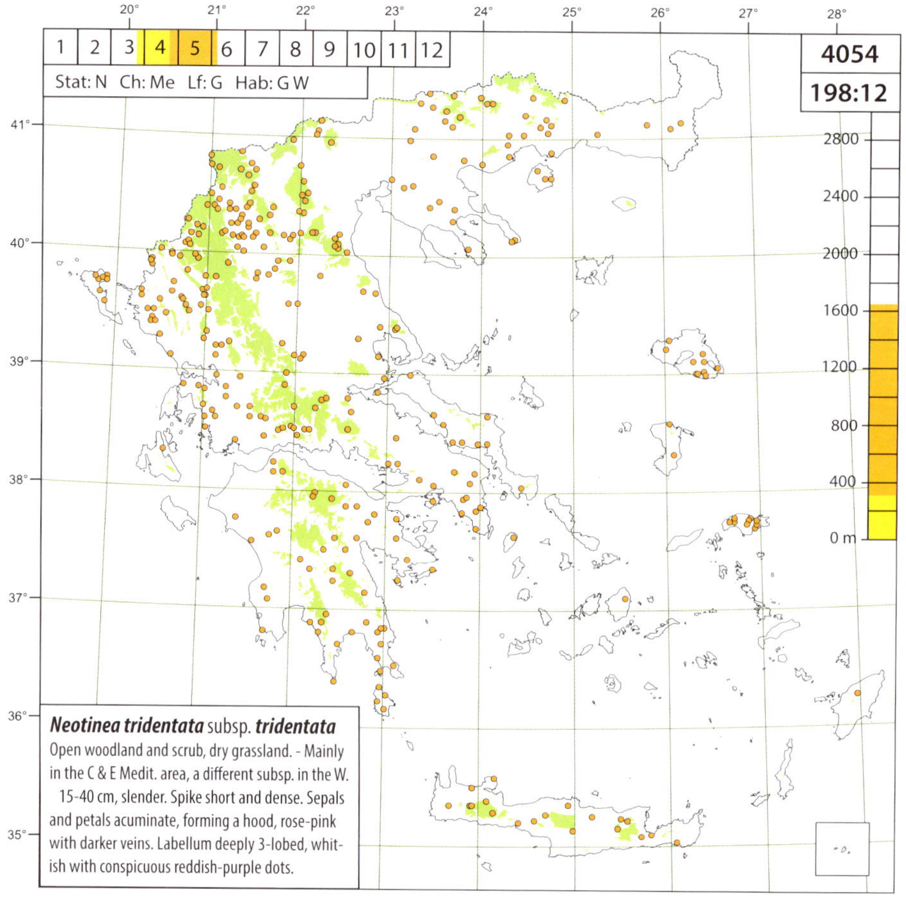

Neotinea tridentata* subsp. *tridentata
Open woodland and scrub, dry grassland. - Mainly in the C & E Medit. area, a different subsp. in the W.
15-40 cm, slender. Spike short and dense. Sepals and petals acuminate, forming a hood, rose-pink with darker veins. Labellum deeply 3-lobed, whitish with conspicuous reddish-purple dots.

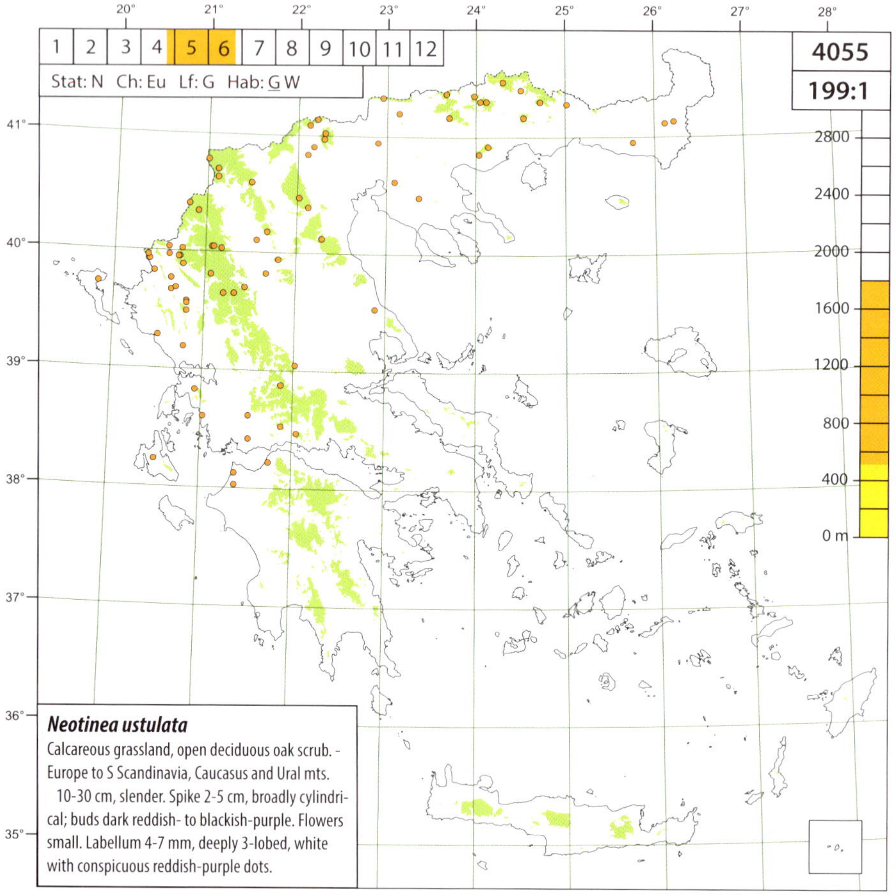

Neotinea ustulata
Calcareous grassland, open deciduous oak scrub. - Europe to S Scandinavia, Caucasus and Ural mts.
10-30 cm, slender. Spike 2-5 cm, broadly cylindrical; buds dark reddish- to blackish-purple. Flowers small. Labellum 4-7 mm, deeply 3-lobed, white with conspicuous reddish-purple dots.

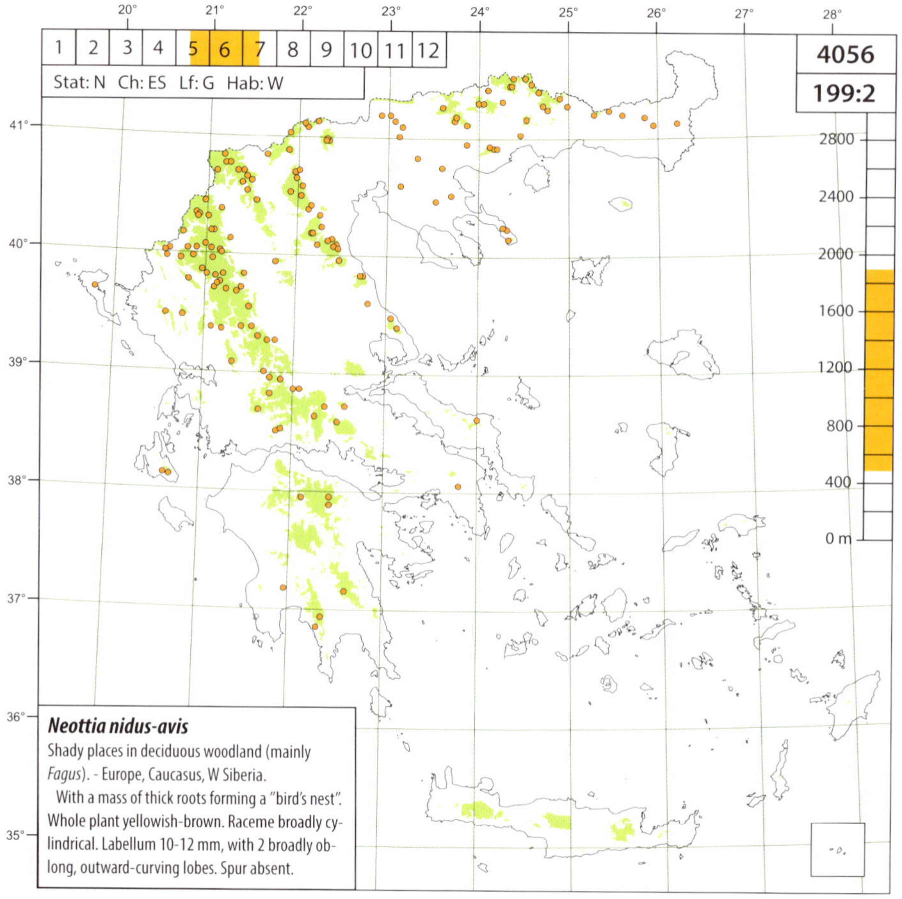

Neottia nidus-avis
Shady places in deciduous woodland (mainly *Fagus*). - Europe, Caucasus, W Siberia.
With a mass of thick roots forming a "bird's nest". Whole plant yellowish-brown. Raceme broadly cylindrical. Labellum 10-12 mm, with 2 broadly oblong, outward-curving lobes. Spur absent.

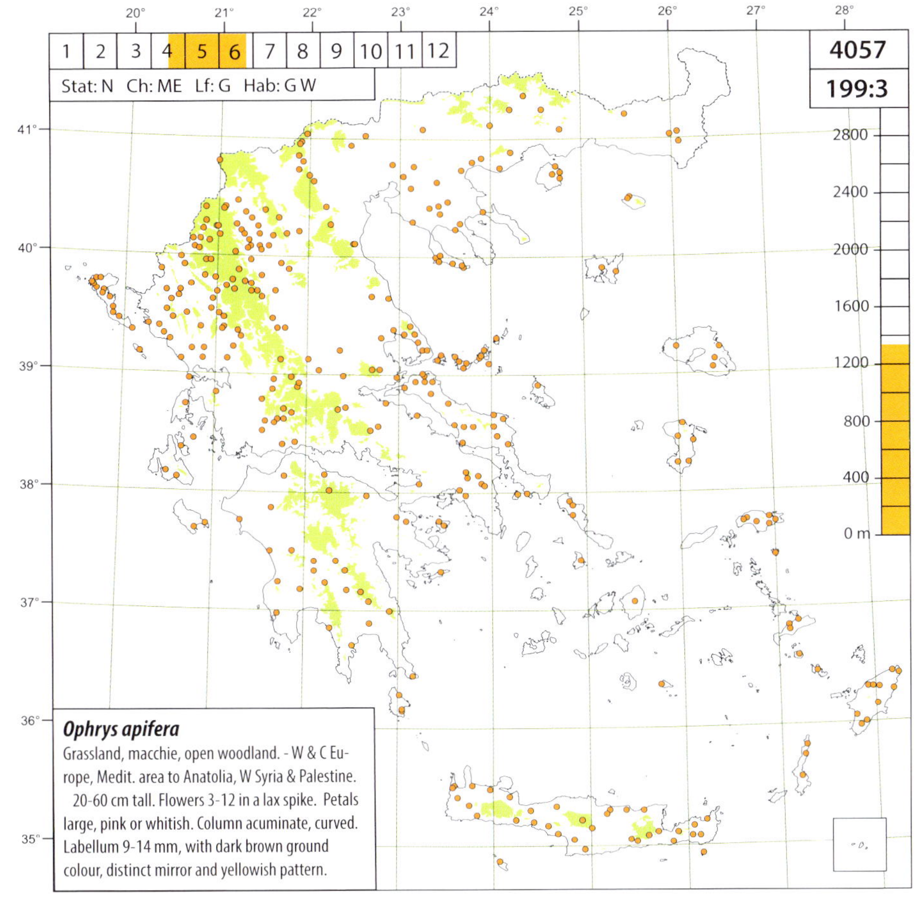

Ophrys apifera

Grassland, macchie, open woodland. - W & C Europe, Medit. area to Anatolia, W Syria & Palestine.
20-60 cm tall. Flowers 3-12 in a lax spike. Petals large, pink or whitish. Column acuminate, curved. Labellum 9-14 mm, with dark brown ground colour, distinct mirror and yellowish pattern.

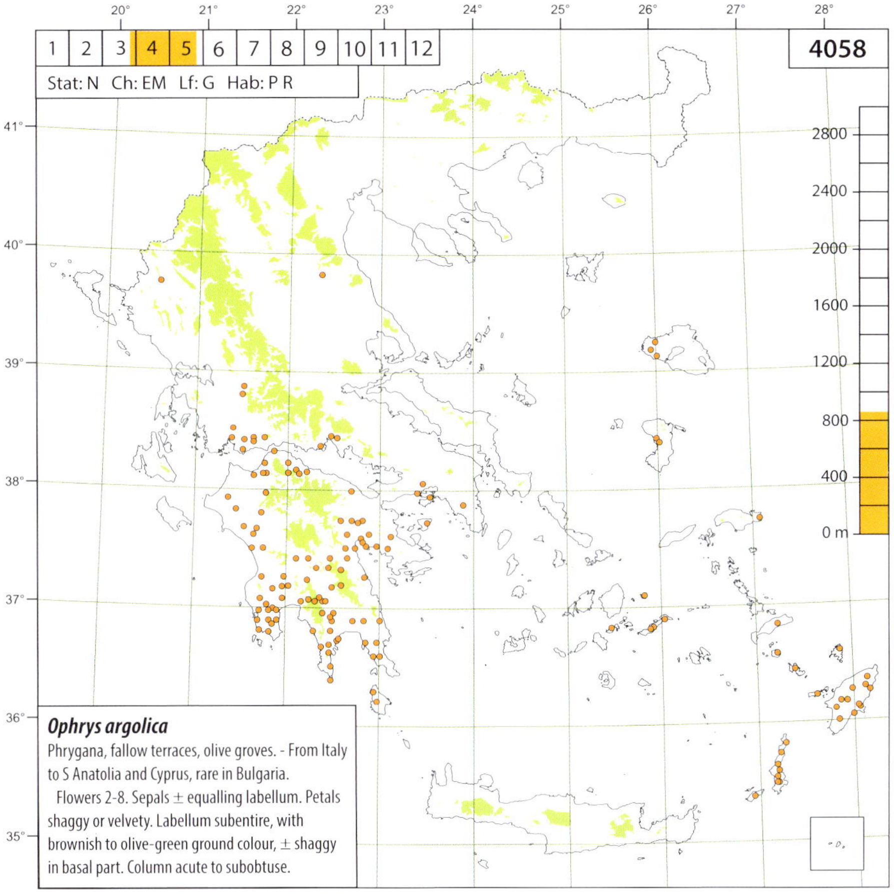

Ophrys argolica

Phrygana, fallow terraces, olive groves. - From Italy to S Anatolia and Cyprus, rare in Bulgaria.
Flowers 2-8. Sepals ± equalling labellum. Petals shaggy or velvety. Labellum subentire, with brownish to olive-green ground colour, ± shaggy in basal part. Column acute to subobtuse.

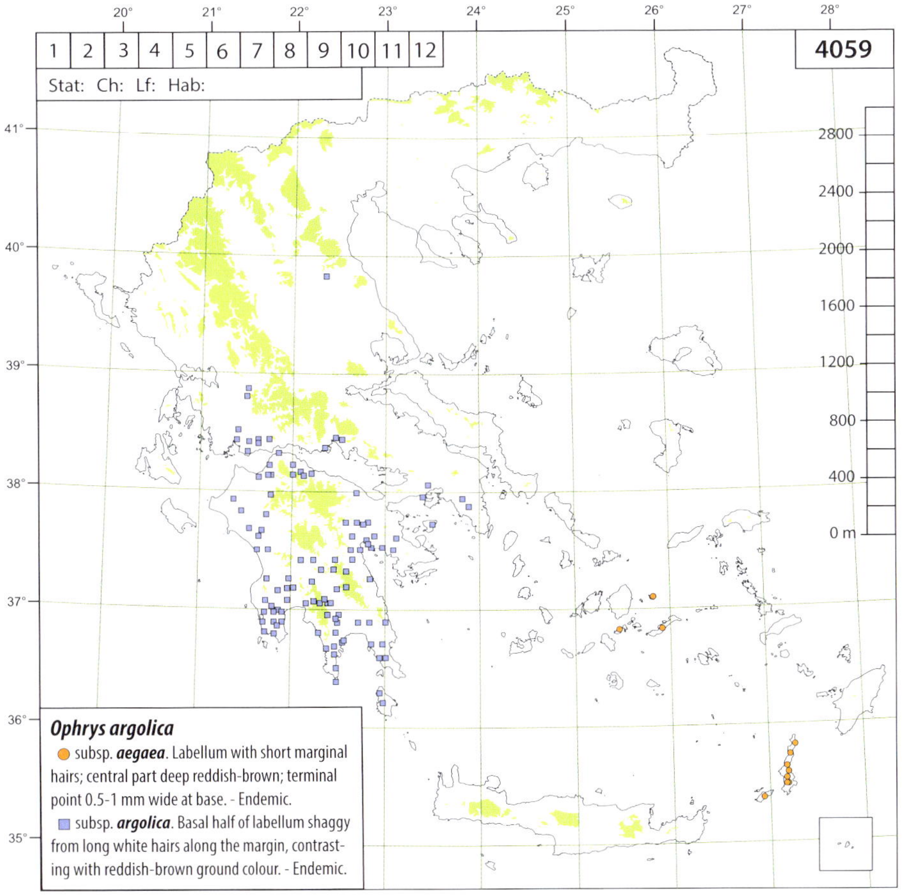

Ophrys argolica

● subsp. ***aegaea***. Labellum with short marginal hairs; central part deep reddish-brown; terminal point 0.5-1 mm wide at base. - Endemic.
■ subsp. ***argolica***. Basal half of labellum shaggy from long white hairs along the margin, contrasting with reddish-brown ground colour. - Endemic.

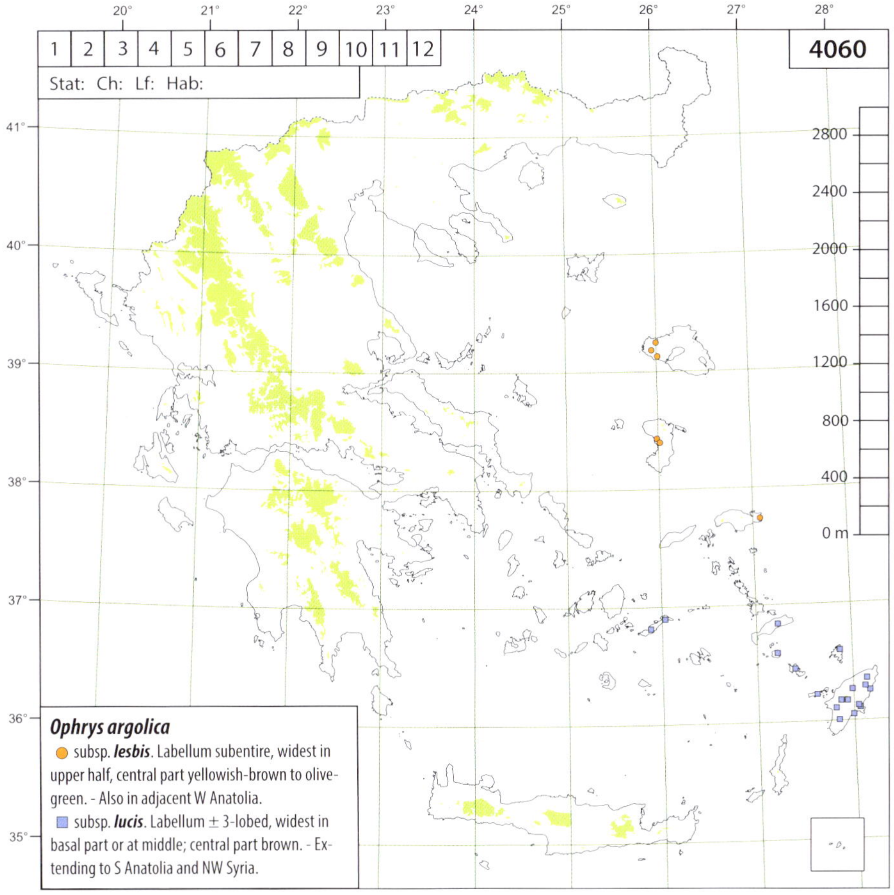

Ophrys argolica

● subsp. ***lesbis***. Labellum subentire, widest in upper half, central part yellowish-brown to olive-green. - Also in adjacent W Anatolia.
■ subsp. ***lucis***. Labellum ± 3-lobed, widest in basal part or at middle; central part brown. - Extending to S Anatolia and NW Syria.

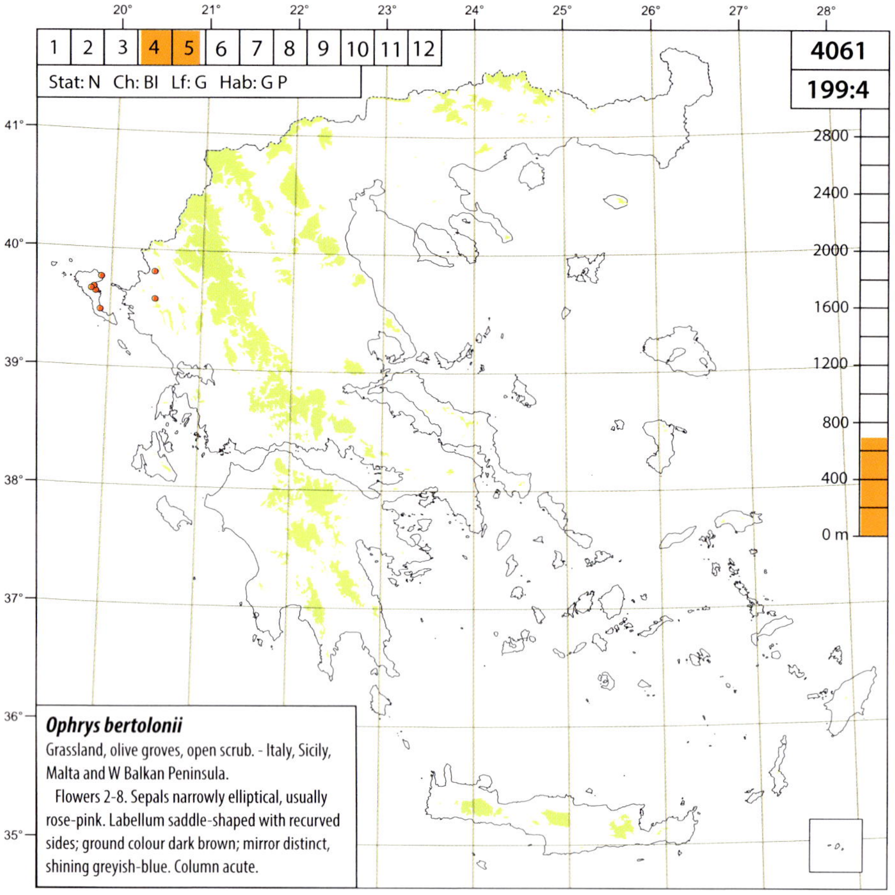

Ophrys bertolonii

Grassland, olive groves, open scrub. - Italy, Sicily, Malta and W Balkan Peninsula.

Flowers 2-8. Sepals narrowly elliptical, usually rose-pink. Labellum saddle-shaped with recurved sides; ground colour dark brown; mirror distinct, shining greyish-blue. Column acute.

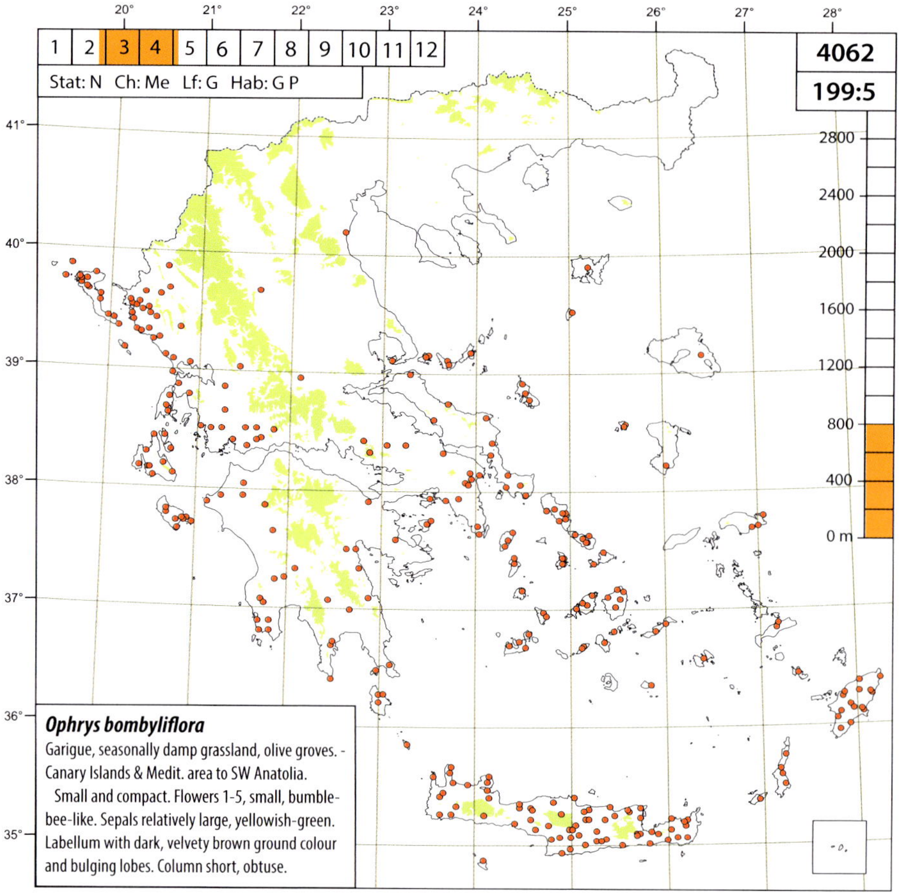

Ophrys bombyliflora

Garigue, seasonally damp grassland, olive groves. - Canary Islands & Medit. area to SW Anatolia.

Small and compact. Flowers 1-5, small, bumble-bee-like. Sepals relatively large, yellowish-green. Labellum with dark, velvety brown ground colour and bulging lobes. Column short, obtuse.

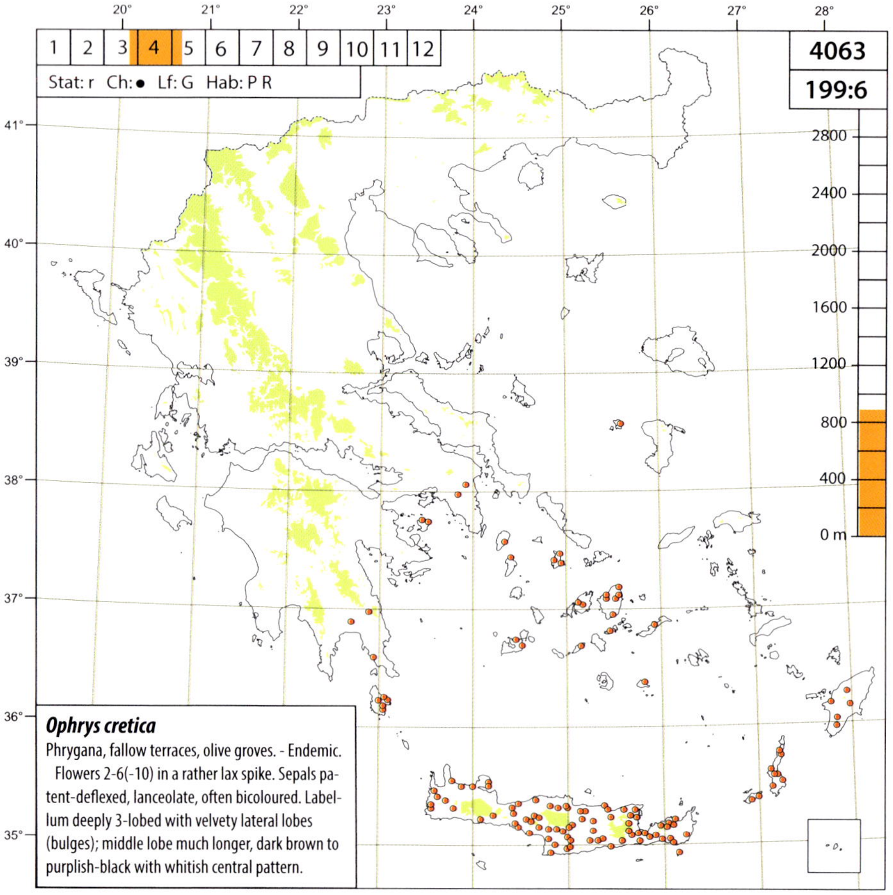

Ophrys cretica

Phrygana, fallow terraces, olive groves. - Endemic.

Flowers 2-6(-10) in a rather lax spike. Sepals patent-deflexed, lanceolate, often bicoloured. Labellum deeply 3-lobed with velvety lateral lobes (bulges); middle lobe much longer, dark brown to purplish-black with whitish central pattern.

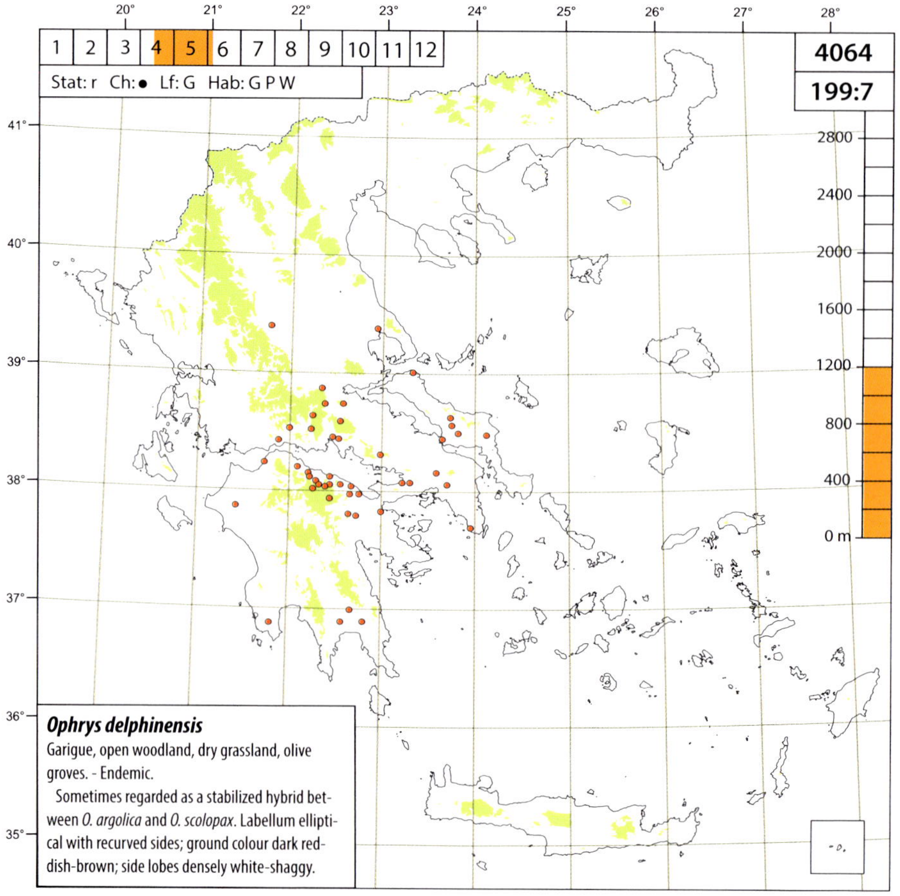

Ophrys delphinensis

Garigue, open woodland, dry grassland, olive groves. - Endemic.

Sometimes regarded as a stabilized hybrid between *O. argolica* and *O. scolopax*. Labellum elliptical with recurved sides; ground colour dark reddish-brown; side lobes densely white-shaggy.

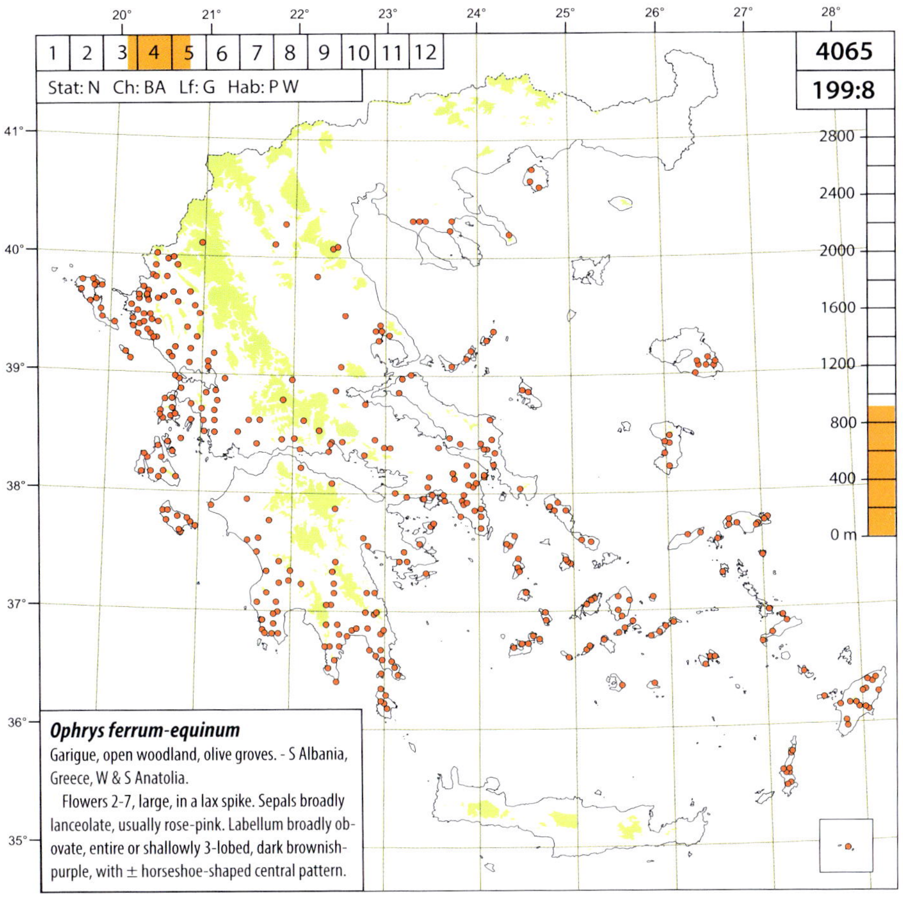

Ophrys ferrum-equinum
Garigue, open woodland, olive groves. - S Albania, Greece, W & S Anatolia.
Flowers 2-7, large, in a lax spike. Sepals broadly lanceolate, usually rose-pink. Labellum broadly obovate, entire or shallowly 3-lobed, dark brownish-purple, with ± horseshoe-shaped central pattern.

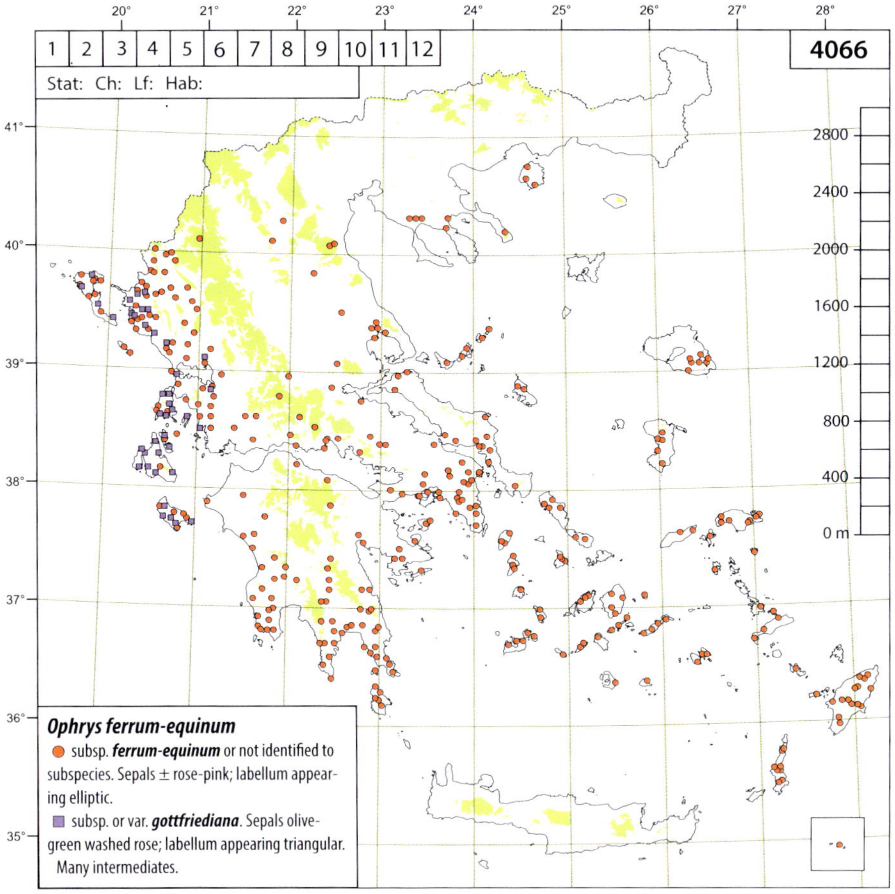

Ophrys ferrum-equinum
● subsp. ***ferrum-equinum*** or not identified to subspecies. Sepals ± rose-pink; labellum appearing elliptic.
■ subsp. or var. ***gottfriediana***. Sepals olive-green washed rose; labellum appearing triangular. Many intermediates.

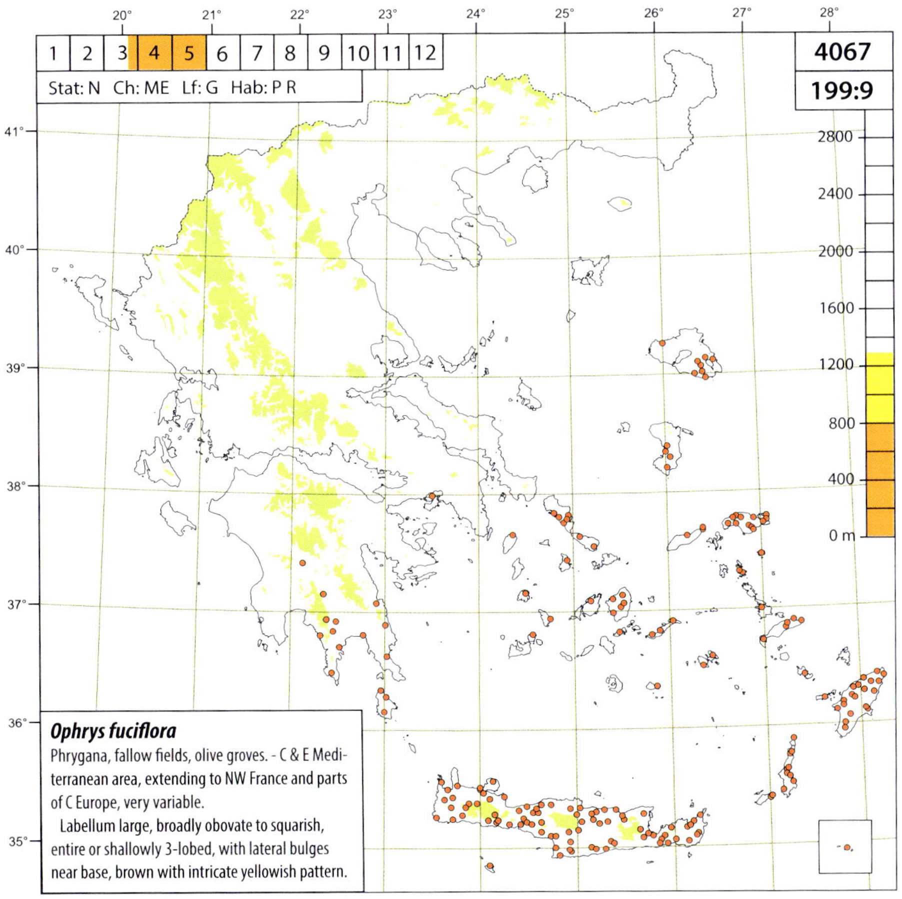

Ophrys fuciflora
Phrygana, fallow fields, olive groves. - C & E Mediterranean area, extending to NW France and parts of C Europe, very variable.
Labellum large, broadly obovate to squarish, entire or shallowly 3-lobed, with lateral bulges near base, brown with intricate yellowish pattern.

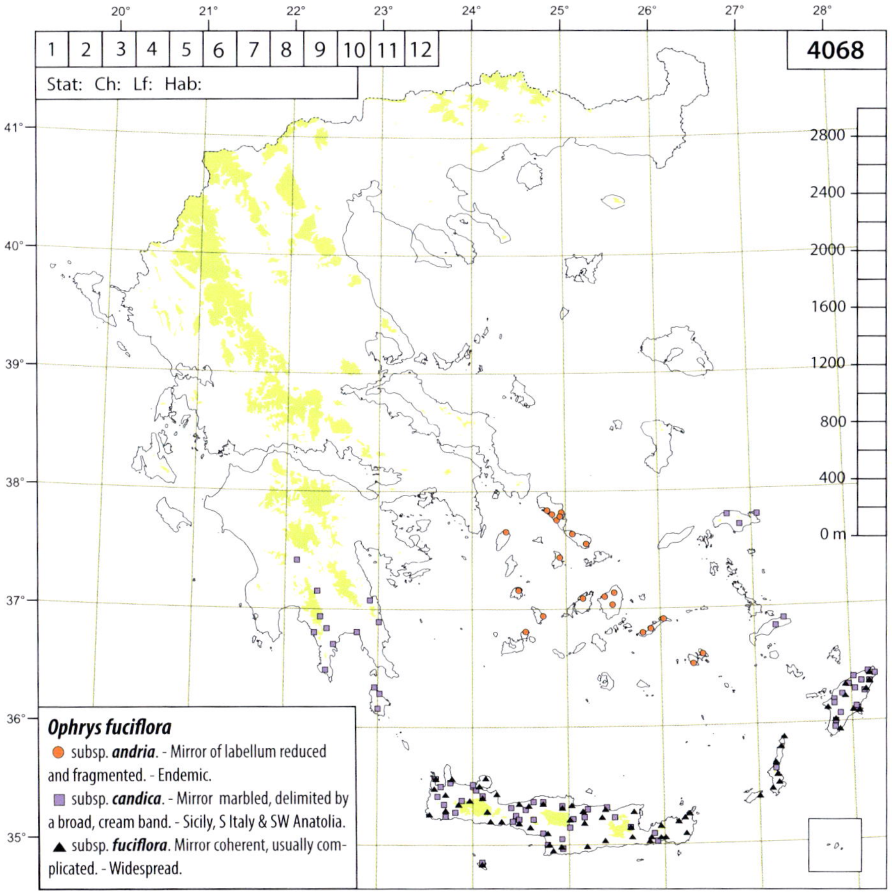

Ophrys fuciflora
● subsp. ***andria***. - Mirror of labellum reduced and fragmented. - Endemic.
■ subsp. ***candica***. - Mirror marbled, delimited by a broad, cream band. - Sicily, S Italy & SW Anatolia.
▲ subsp. ***fuciflora***. Mirror coherent, usually complicated. - Widespread.

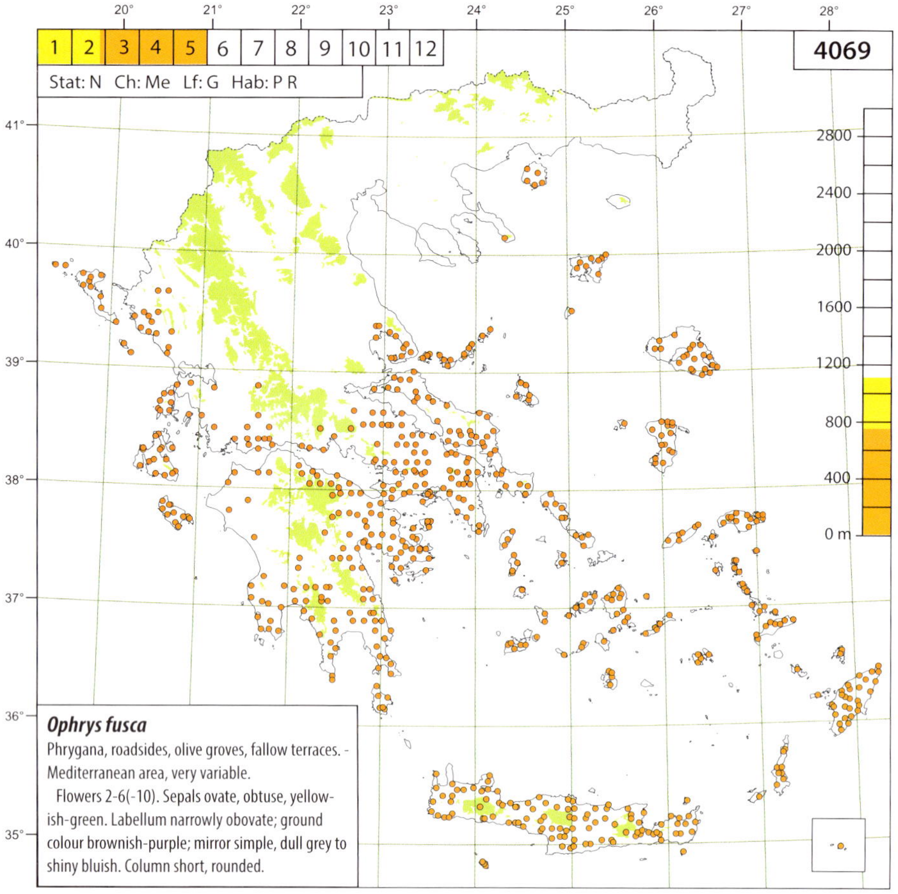

Ophrys fusca
Phrygana, roadsides, olive groves, fallow terraces. - Mediterranean area, very variable.
Flowers 2-6(-10). Sepals ovate, obtuse, yellowish-green. Labellum narrowly obovate; ground colour brownish-purple; mirror simple, dull grey to shiny bluish. Column short, rounded.

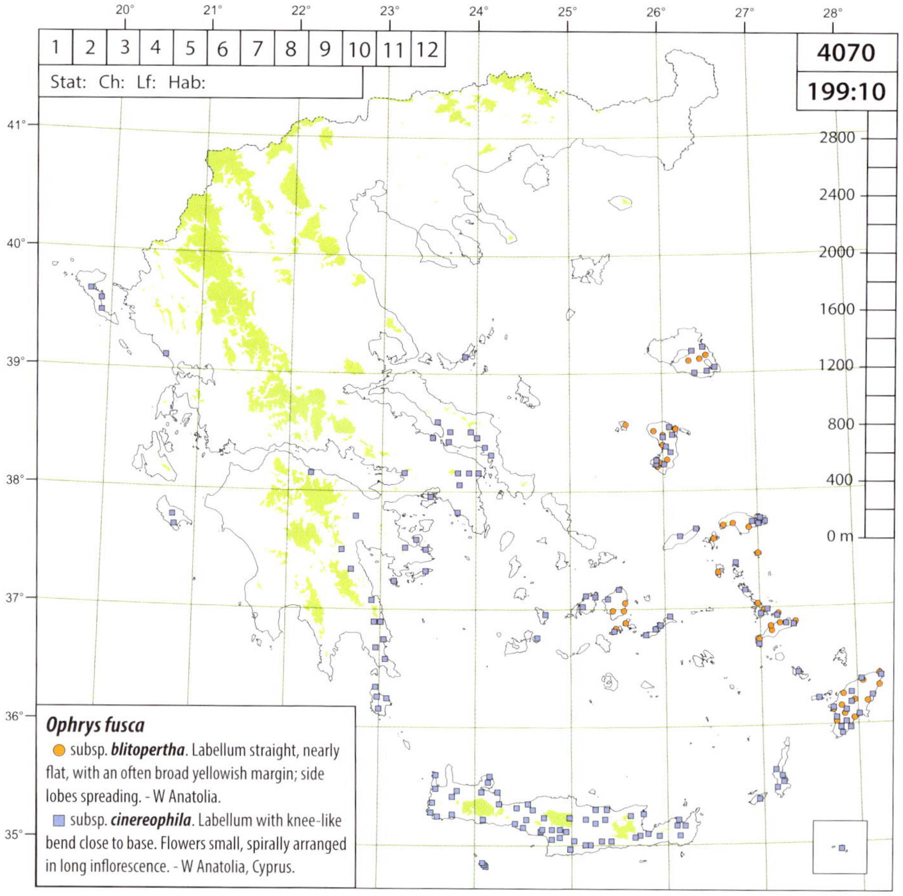

Ophrys fusca
● subsp. ***blitopertha***. Labellum straight, nearly flat, with an often broad yellowish margin; side lobes spreading. - W Anatolia.
■ subsp. ***cinereophila***. Labellum with knee-like bend close to base. Flowers small, spirally arranged in long inflorescence. - W Anatolia, Cyprus.

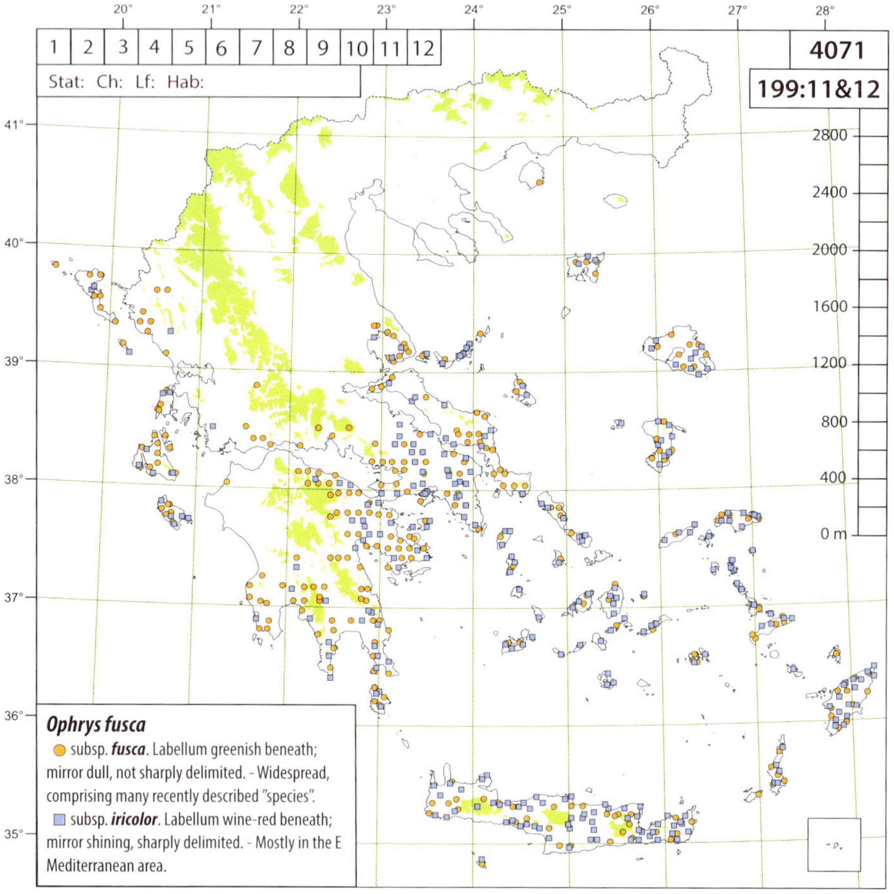

Ophrys fusca
● subsp. ***fusca***. Labellum greenish beneath; mirror dull, not sharply delimited. - Widespread, comprising many recently described "species".
■ subsp. ***iricolor***. Labellum wine-red beneath; mirror shining, sharply delimited. - Mostly in the E Mediterranean area.

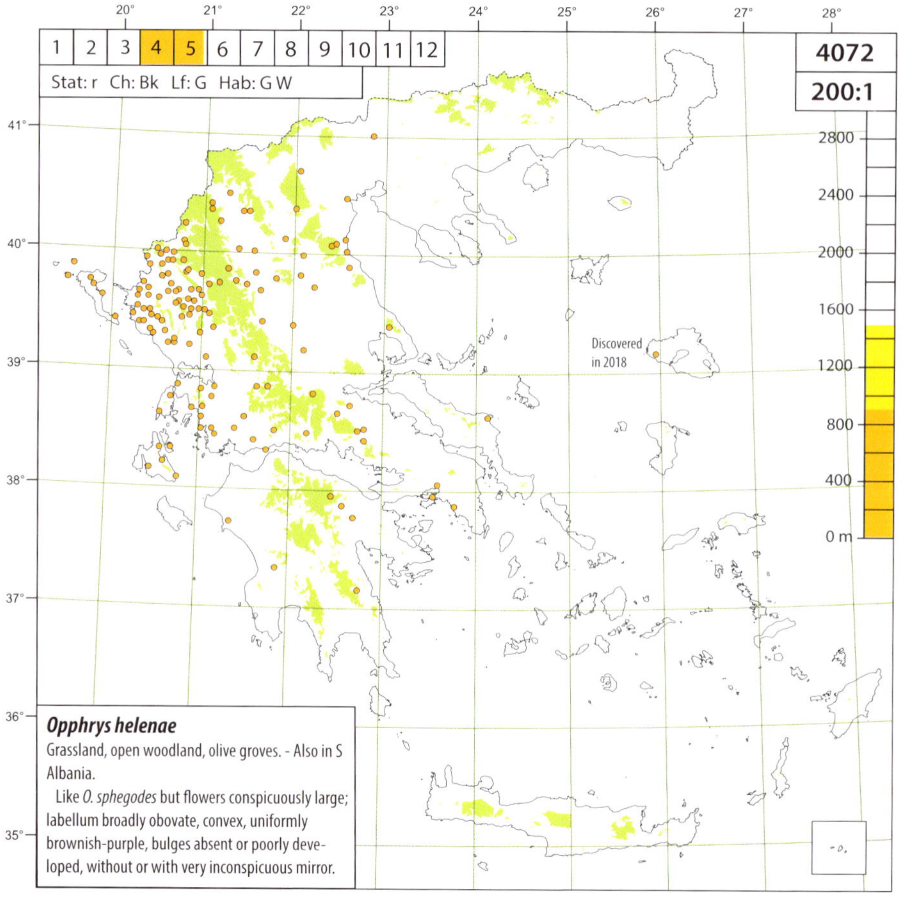

Opphrys helenae
Grassland, open woodland, olive groves. - Also in S Albania.
Like *O. sphegodes* but flowers conspicuously large; labellum broadly obovate, convex, uniformly brownish-purple, bulges absent or poorly developed, without or with very inconspicuous mirror.

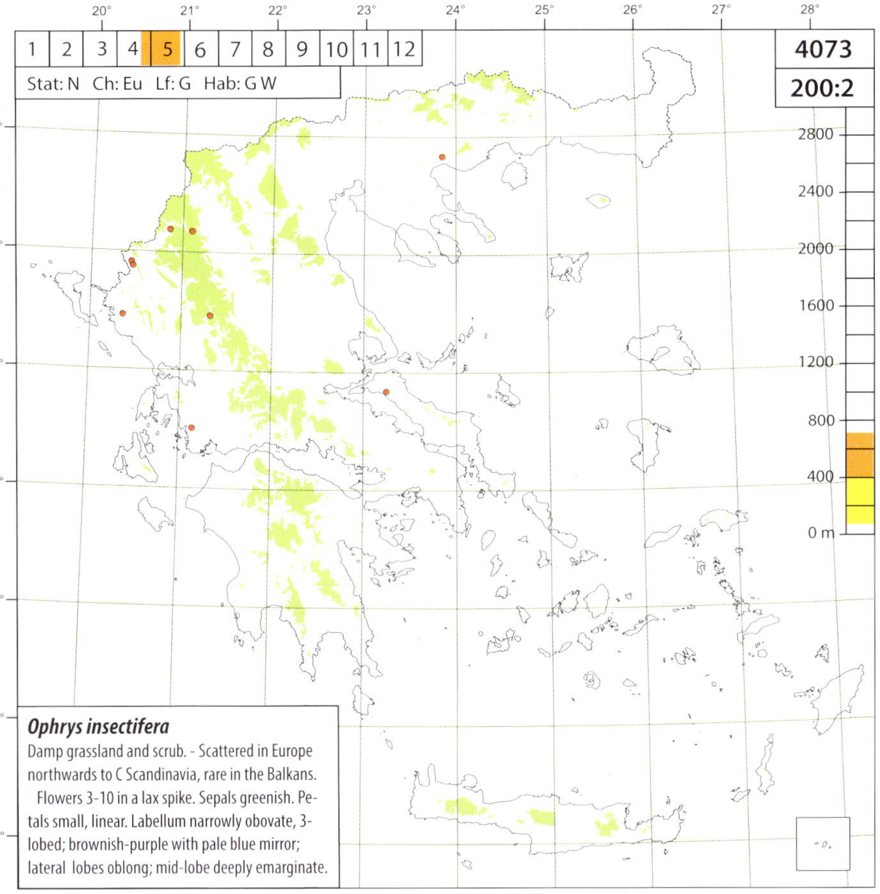

Ophrys insectifera

Damp grassland and scrub. - Scattered in Europe northwards to C Scandinavia, rare in the Balkans.

Flowers 3-10 in a lax spike. Sepals greenish. Petals small, linear. Labellum narrowly obovate, 3-lobed; brownish-purple with pale blue mirror; lateral lobes oblong; mid-lobe deeply emarginate.

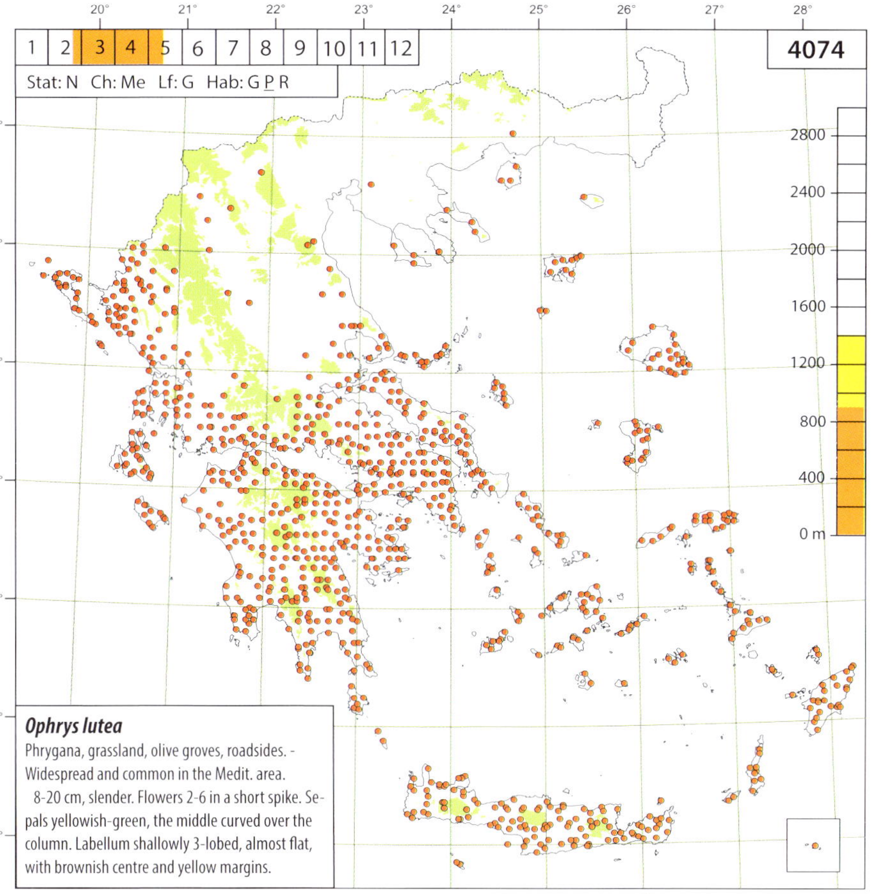

Ophrys lutea

Phrygana, grassland, olive groves, roadsides. - Widespread and common in the Medit. area.

8-20 cm, slender. Flowers 2-6 in a short spike. Sepals yellowish-green, the middle curved over the column. Labellum shallowly 3-lobed, almost flat, with brownish centre and yellow margins.

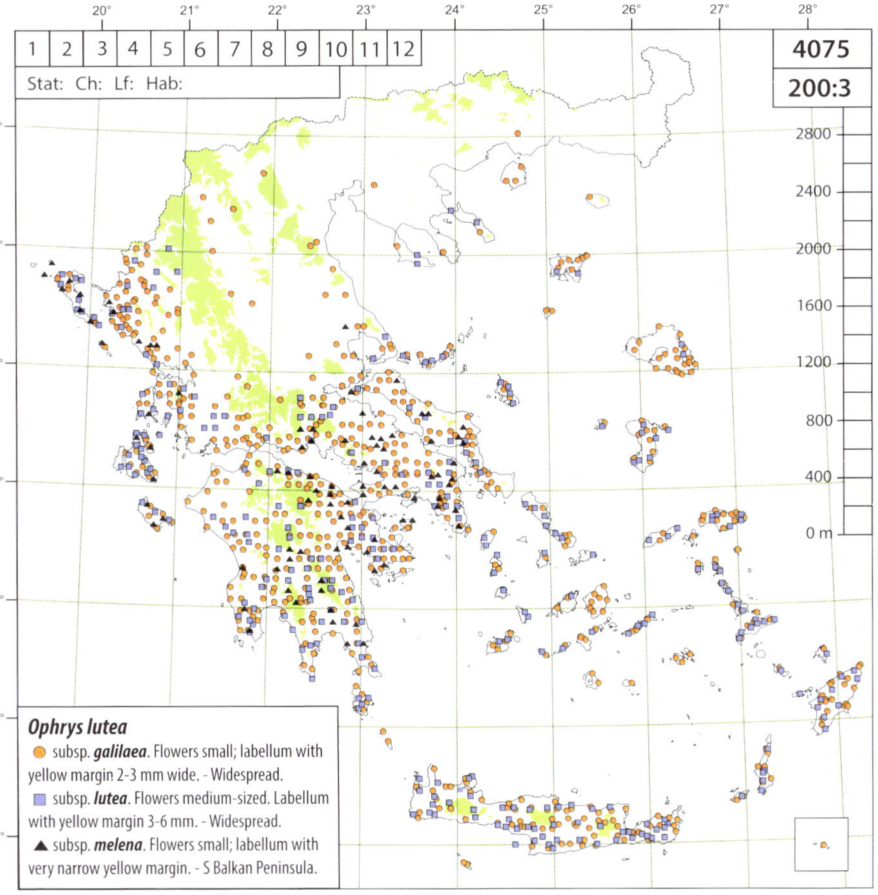

Ophrys lutea

● subsp. ***galilaea***. Flowers small; labellum with yellow margin 2-3 mm wide. - Widespread.

■ subsp. ***lutea***. Flowers medium-sized. Labellum with yellow margin 3-6 mm. - Widespread.

▲ subsp. ***melena***. Flowers small; labellum with very narrow yellow margin. - S Balkan Peninsula.

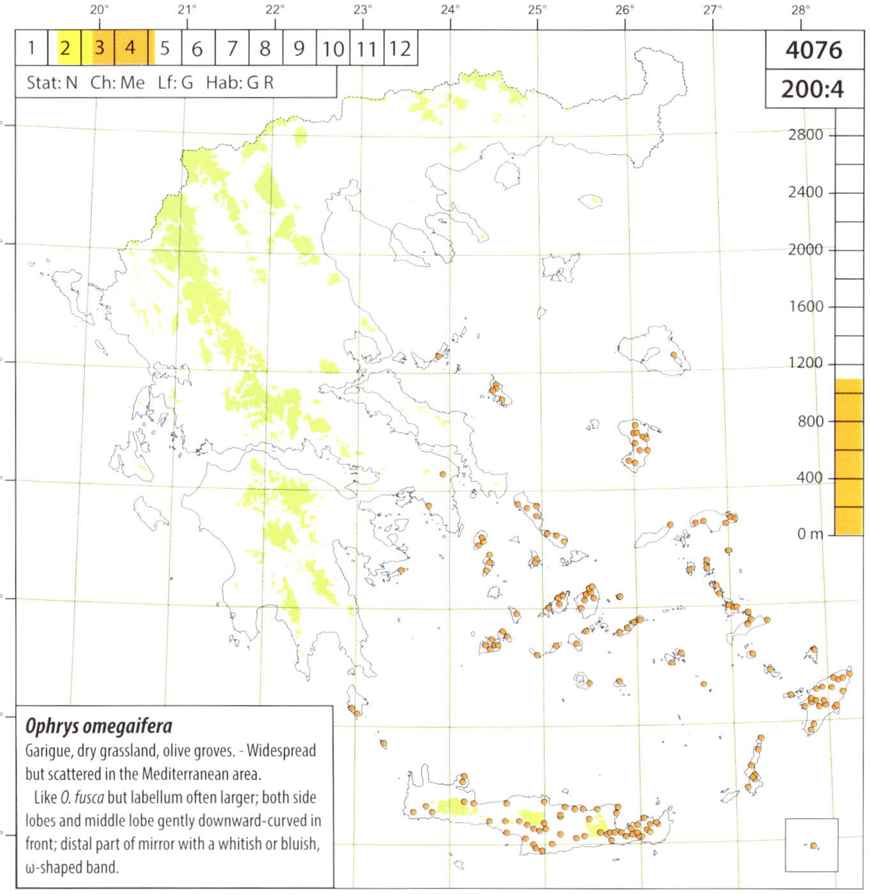

Ophrys omegaifera

Garigue, dry grassland, olive groves. - Widespread but scattered in the Mediterranean area.

Like *O. fusca* but labellum often larger; both side lobes and middle lobe gently downward-curved in front; distal part of mirror with a whitish or bluish, ω-shaped band.

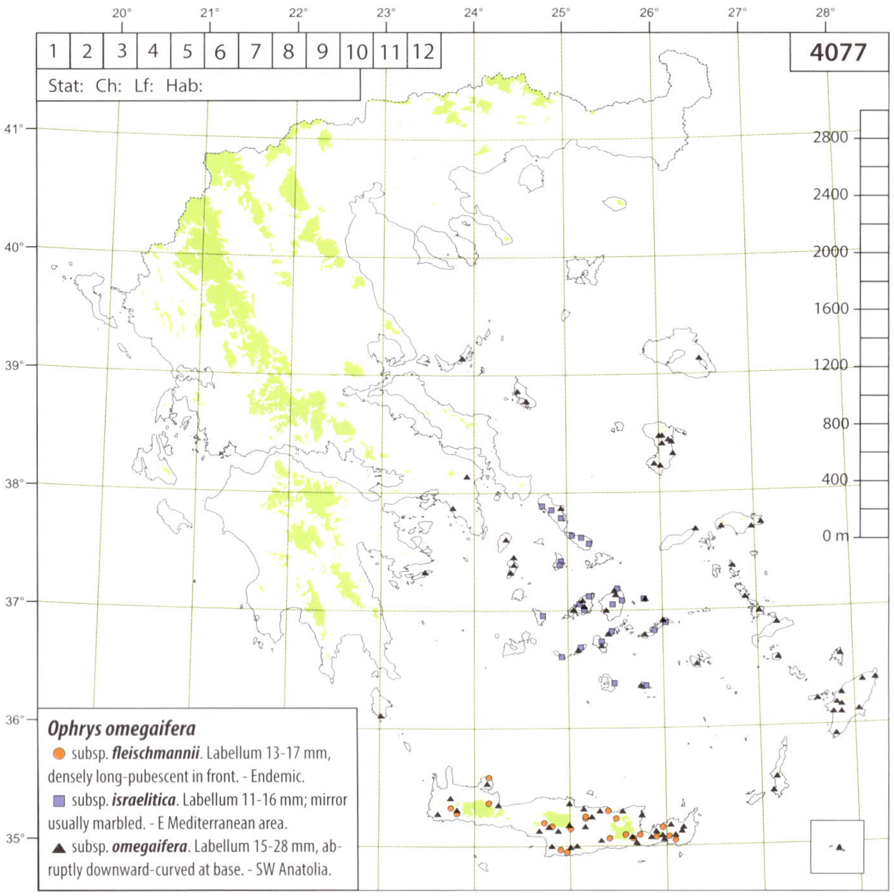

Ophrys omegaifera
● subsp. ***fleischmannii***. Labellum 13-17 mm, densely long-pubescent in front. - Endemic.
■ subsp. ***israelitica***. Labellum 11-16 mm; mirror usually marbled. - E Mediterranean area.
▲ subsp. ***omegaifera***. Labellum 15-28 mm, abruptly downward-curved at base. - SW Anatolia.

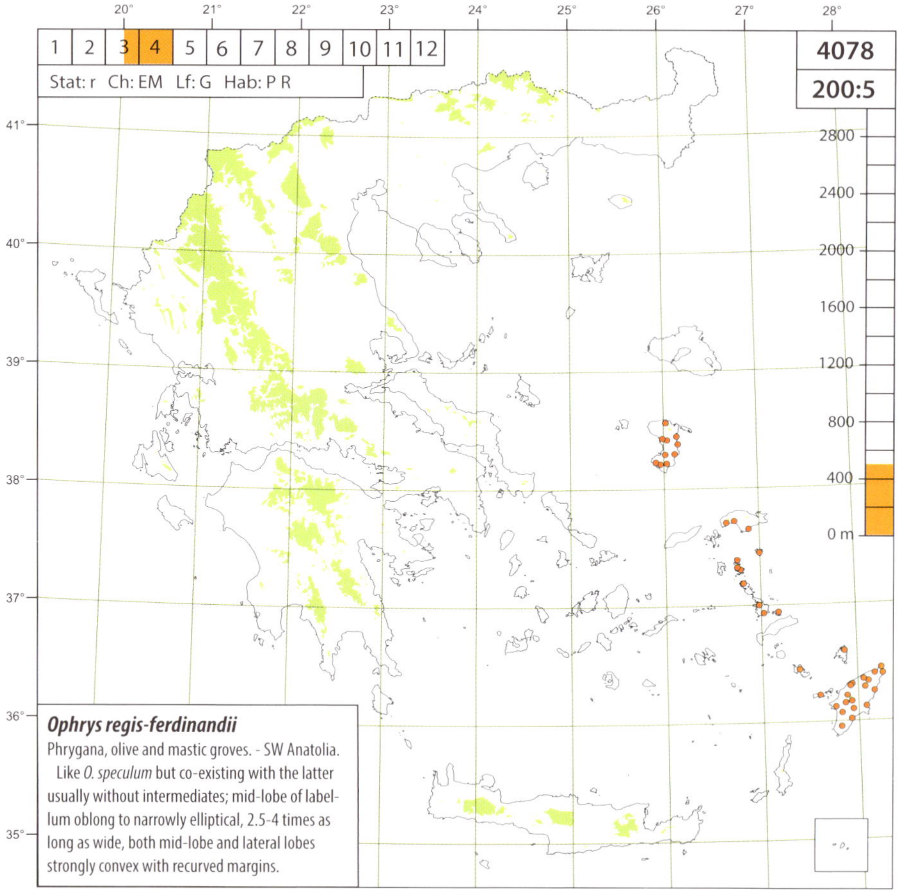

Ophrys regis-ferdinandii
Phrygana, olive and mastic groves. - SW Anatolia.
Like *O. speculum* but co-existing with the latter usually without intermediates; mid-lobe of labellum oblong to narrowly elliptical, 2.5-4 times as long as wide, both mid-lobe and lateral lobes strongly convex with recurved margins.

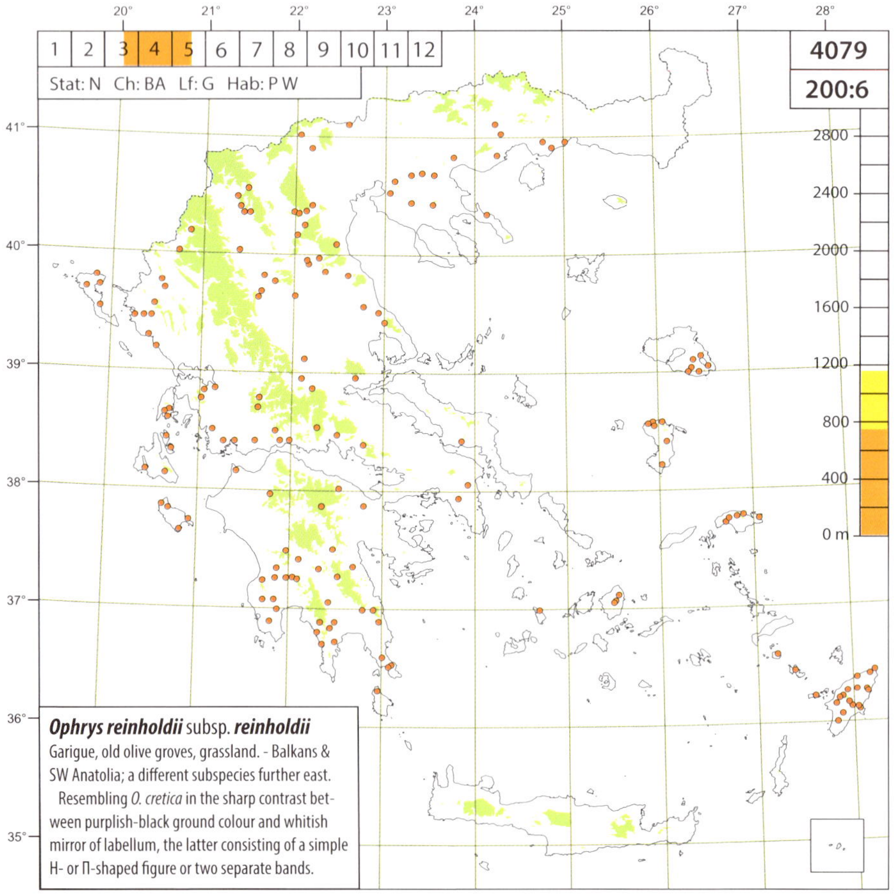

Ophrys reinholdii subsp. ***reinholdii***
Garigue, old olive groves, grassland. - Balkans & SW Anatolia; a different subspecies further east.
Resembling *O. cretica* in the sharp contrast between purplish-black ground colour and whitish mirror of labellum, the latter consisting of a simple H- or Π-shaped figure or two separate bands.

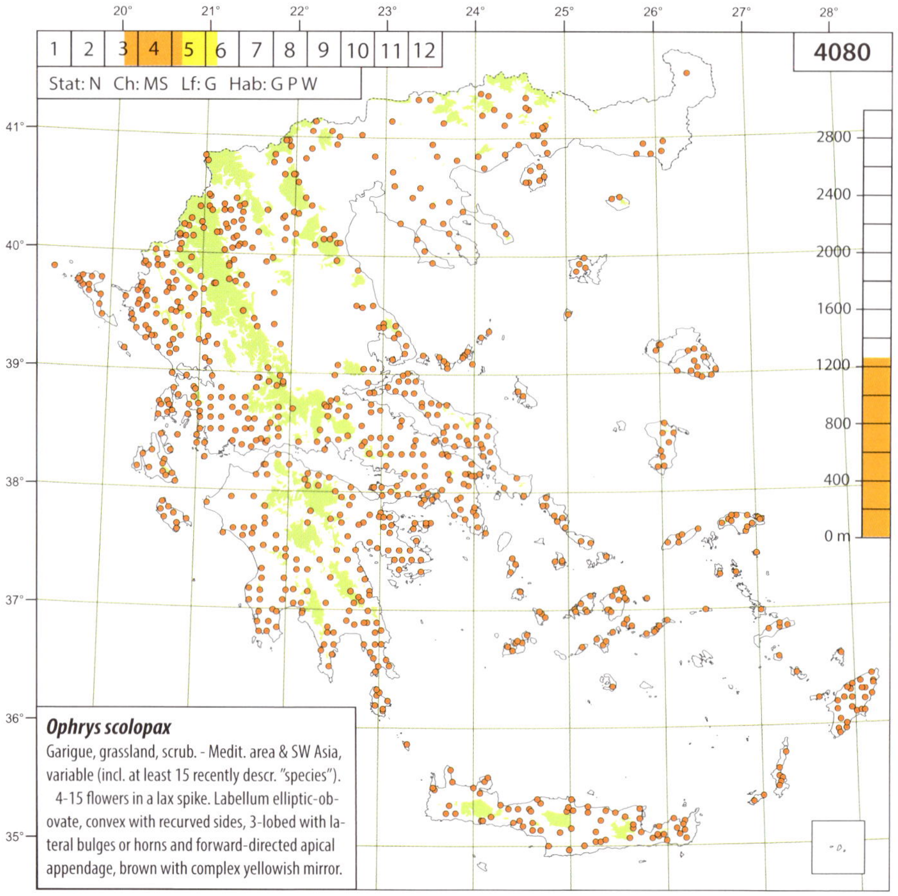

Ophrys scolopax
Garigue, grassland, scrub. - Medit. area & SW Asia, variable (incl. at least 15 recently descr. "species").
4-15 flowers in a lax spike. Labellum elliptic-obovate, convex with recurved sides, 3-lobed with lateral bulges or horns and forward-directed apical appendage, brown with complex yellowish mirror.

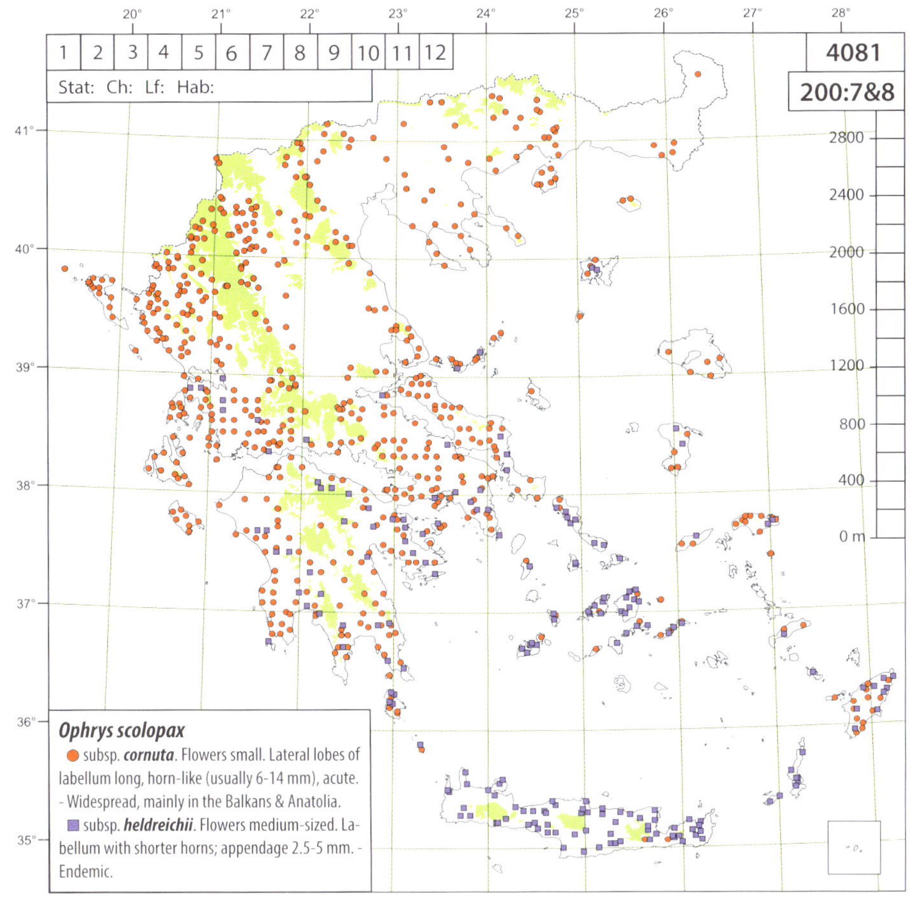

Ophrys scolopax
● subsp. ***cornuta***. Flowers small. Lateral lobes of labellum long, horn-like (usually 6-14 mm), acute. - Widespread, mainly in the Balkans & Anatolia.
■ subsp. ***heldreichii***. Flowers medium-sized. Labellum with shorter horns; appendage 2.5-5 mm. - Endemic.

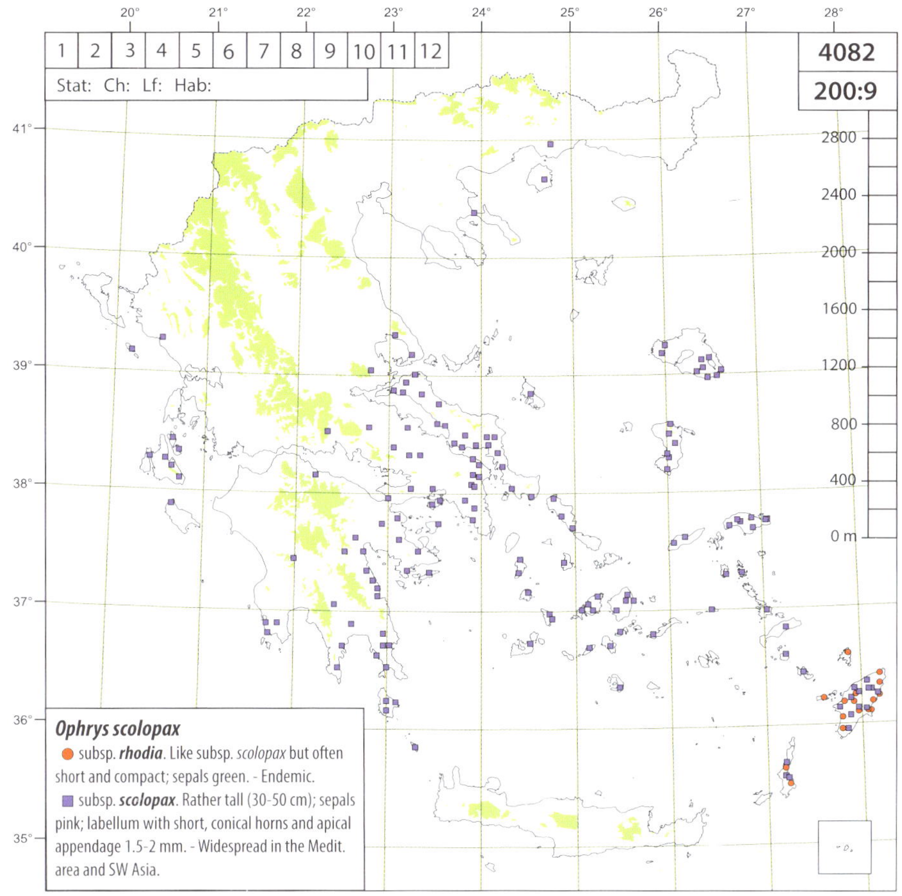

Ophrys scolopax
● subsp. ***rhodia***. Like subsp. *scolopax* but often short and compact; sepals green. - Endemic.
■ subsp. ***scolopax***. Rather tall (30-50 cm); sepals pink; labellum with short, conical horns and apical appendage 1.5-2 mm. - Widespread in the Medit. area and SW Asia.

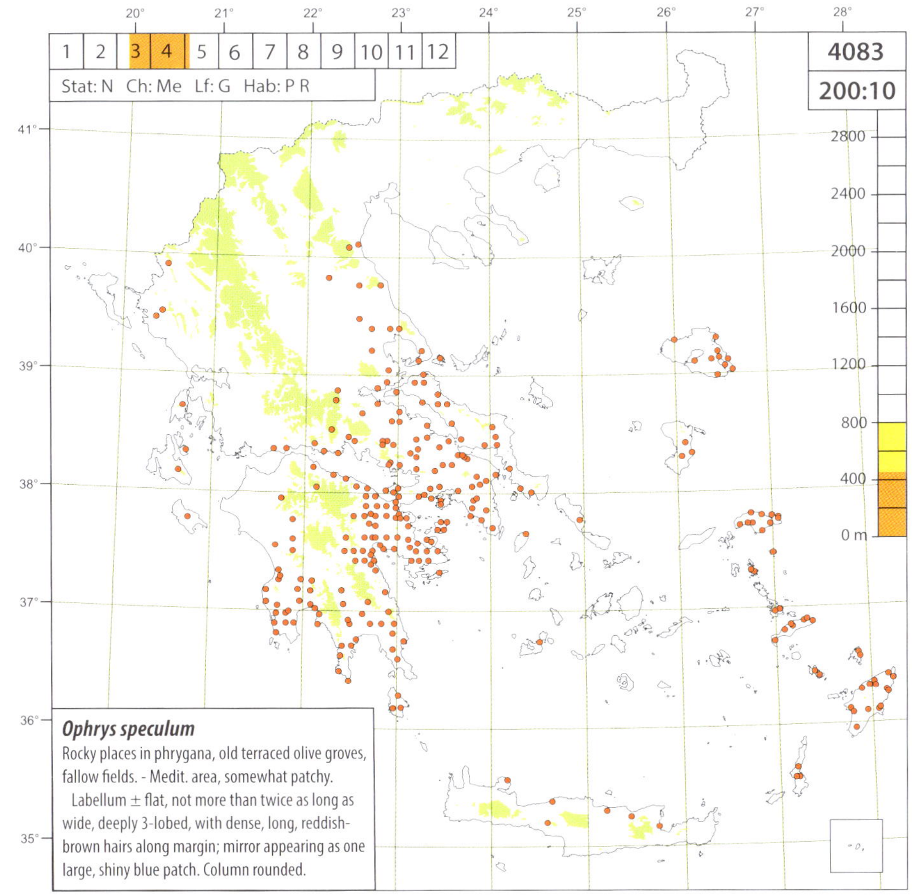

Ophrys speculum
Rocky places in phrygana, old terraced olive groves, fallow fields. - Medit. area, somewhat patchy.
Labellum ± flat, not more than twice as long as wide, deeply 3-lobed, with dense, long, reddish-brown hairs along margin; mirror appearing as one large, shiny blue patch. Column rounded.

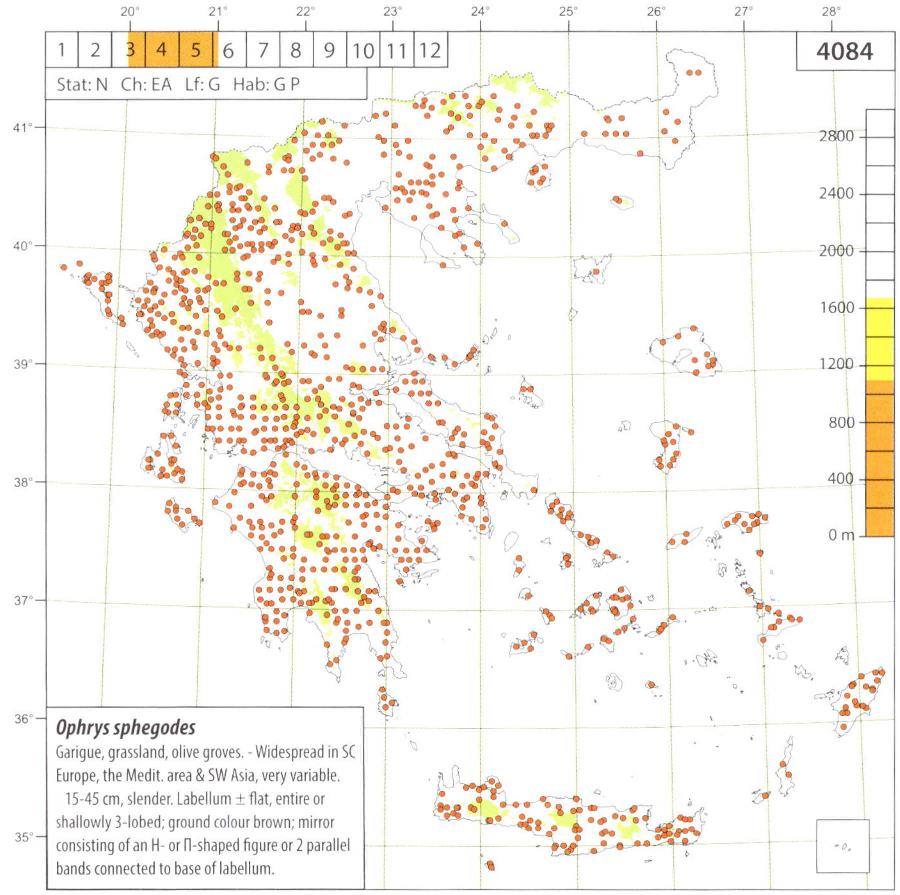

Ophrys sphegodes
Garigue, grassland, olive groves. - Widespread in SC Europe, the Medit. area & SW Asia, very variable.
15-45 cm, slender. Labellum ± flat, entire or shallowly 3-lobed; ground colour brown; mirror consisting of an H- or Π-shaped figure or 2 parallel bands connected to base of labellum.

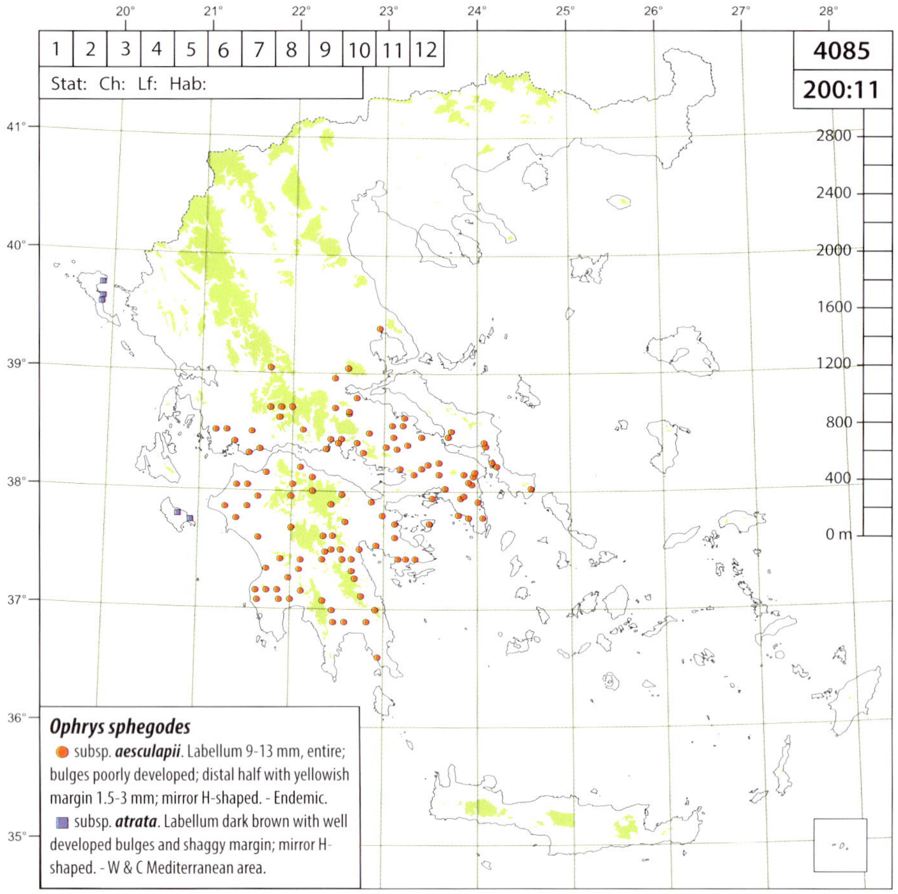

Ophrys sphegodes
- subsp. ***aesculapii***. Labellum 9-13 mm, entire; bulges poorly developed; distal half with yellowish margin 1.5-3 mm; mirror H-shaped. - Endemic.
- subsp. ***atrata***. Labellum dark brown with well developed bulges and shaggy margin; mirror H-shaped. - W & C Mediterranean area.

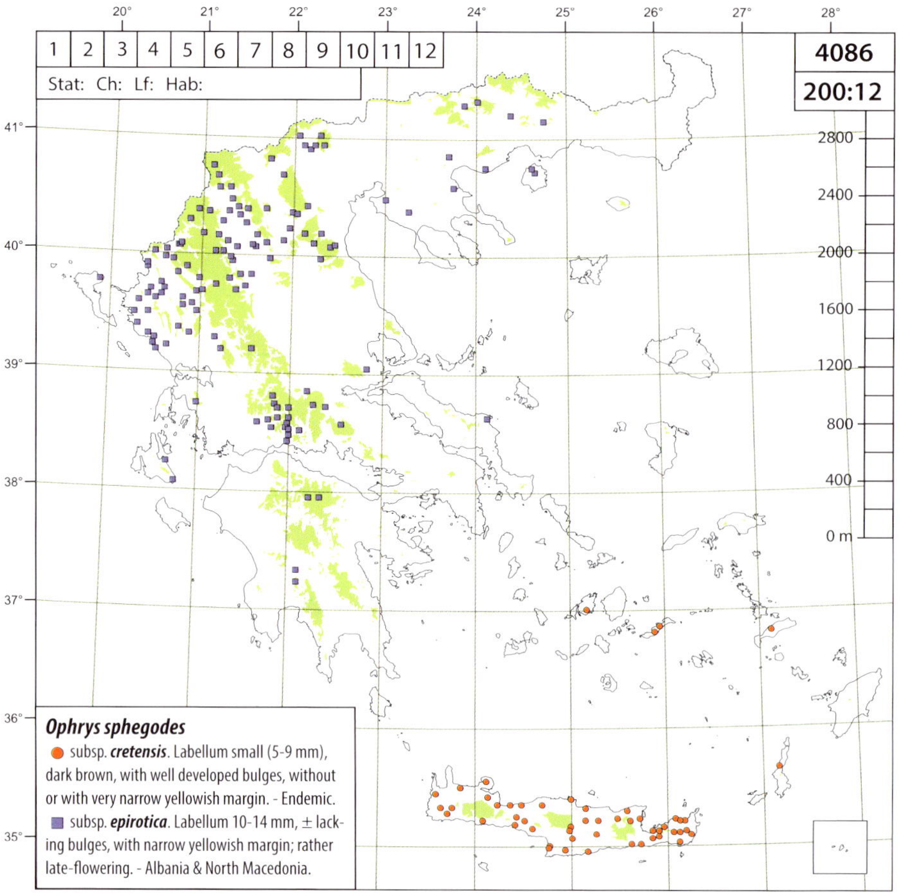

Ophrys sphegodes
- subsp. ***cretensis***. Labellum small (5-9 mm), dark brown, with well developed bulges, without or with very narrow yellowish margin. - Endemic.
- subsp. ***epirotica***. Labellum 10-14 mm, ± lacking bulges, with narrow yellowish margin; rather late-flowering. - Albania & North Macedonia.

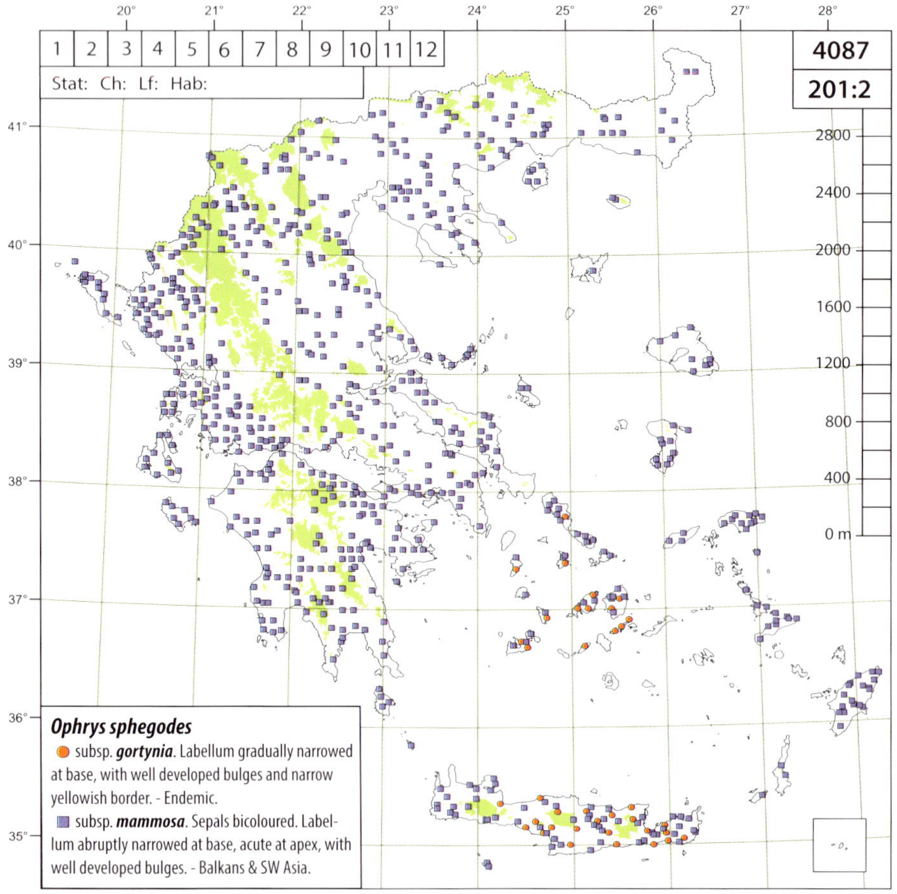

Ophrys sphegodes
- subsp. ***gortynia***. Labellum gradually narrowed at base, with well developed bulges and narrow yellowish border. - Endemic.
- subsp. ***mammosa***. Sepals bicoloured. Labellum abruptly narrowed at base, acute at apex, with well developed bulges. - Balkans & SW Asia.

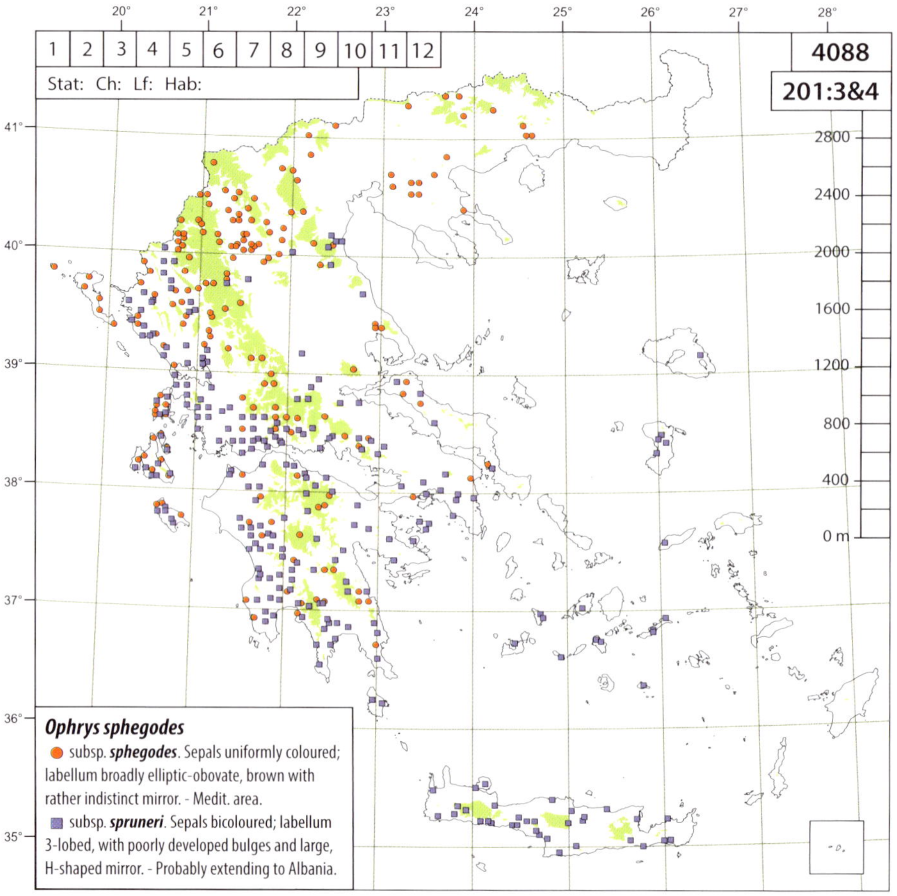

Ophrys sphegodes
- subsp. ***sphegodes***. Sepals uniformly coloured; labellum broadly elliptic-obovate, brown with rather indistinct mirror. - Medit. area.
- subsp. ***spruneri***. Sepals bicoloured; labellum 3-lobed, with poorly developed bulges and large, H-shaped mirror. - Probably extending to Albania.

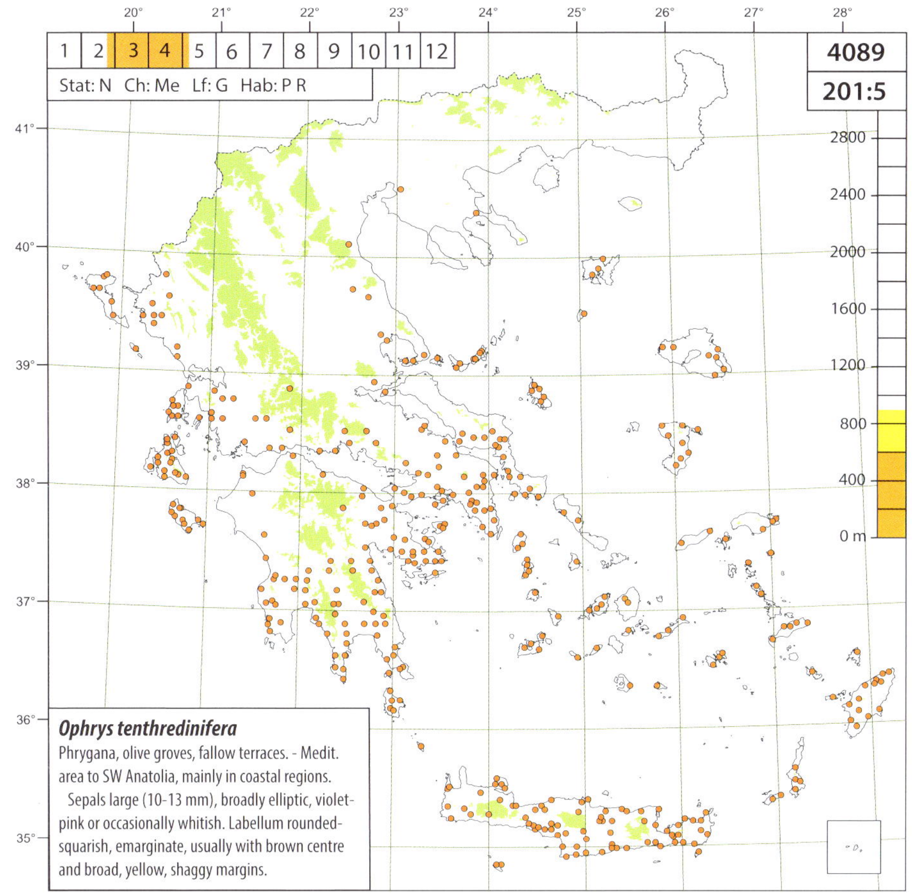

Ophrys tenthredinifera

Phrygana, olive groves, fallow terraces. - Medit. area to SW Anatolia, mainly in coastal regions.

Sepals large (10-13 mm), broadly elliptic, violet-pink or occasionally whitish. Labellum rounded-squarish, emarginate, usually with brown centre and broad, yellow, shaggy margins.

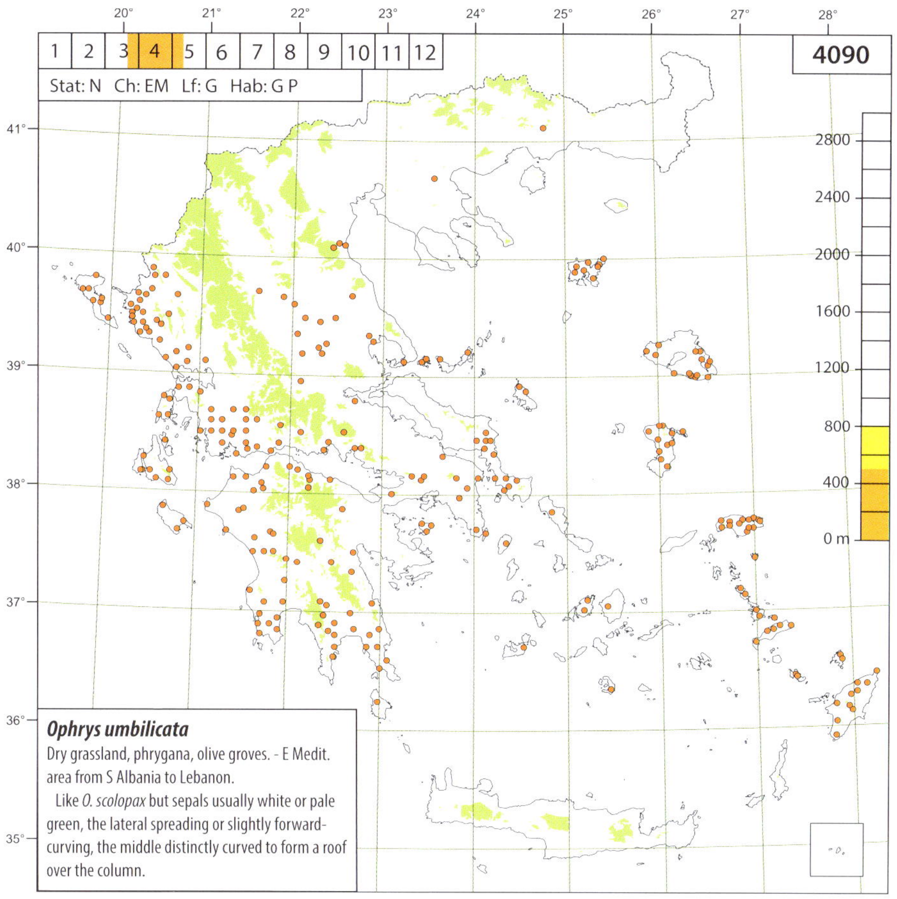

Ophrys umbilicata

Dry grassland, phrygana, olive groves. - E Medit. area from S Albania to Lebanon.

Like *O. scolopax* but sepals usually white or pale green, the lateral spreading or slightly forward-curving, the middle distinctly curved to form a roof over the column.

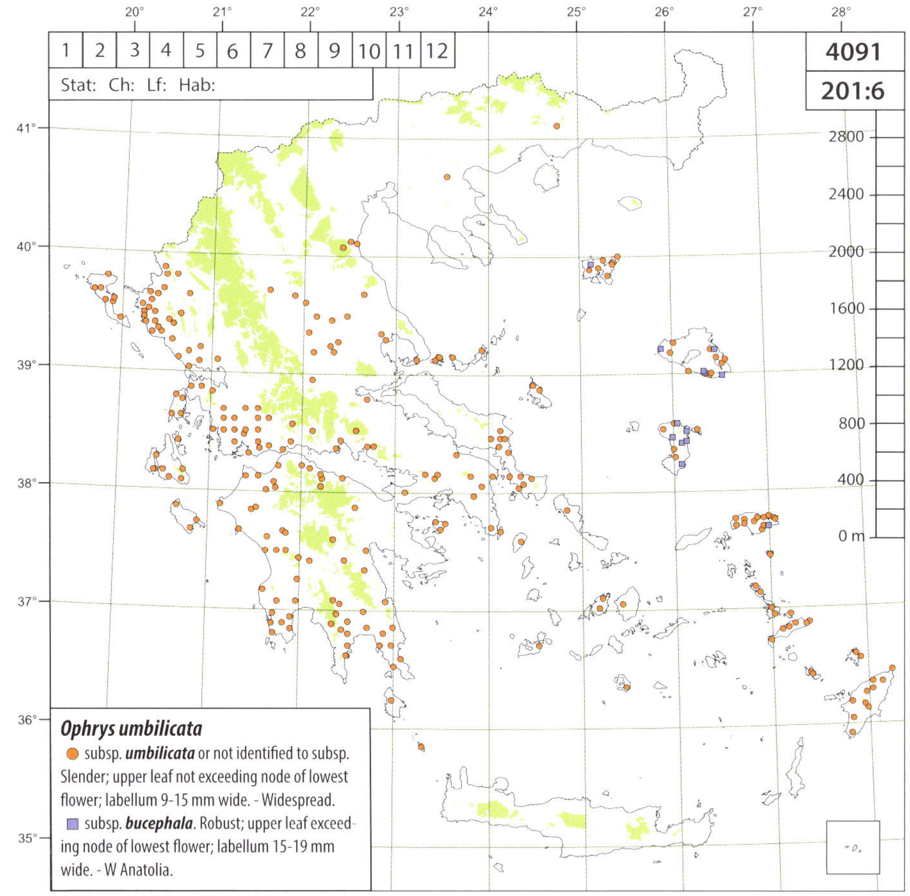

Ophrys umbilicata

● subsp. ***umbilicata*** or not identified to subsp. Slender; upper leaf not exceeding node of lowest flower; labellum 9-15 mm wide. - Widespread.

■ subsp. ***bucephala***. Robust; upper leaf exceeding node of lowest flower; labellum 15-19 mm wide. - W Anatolia.

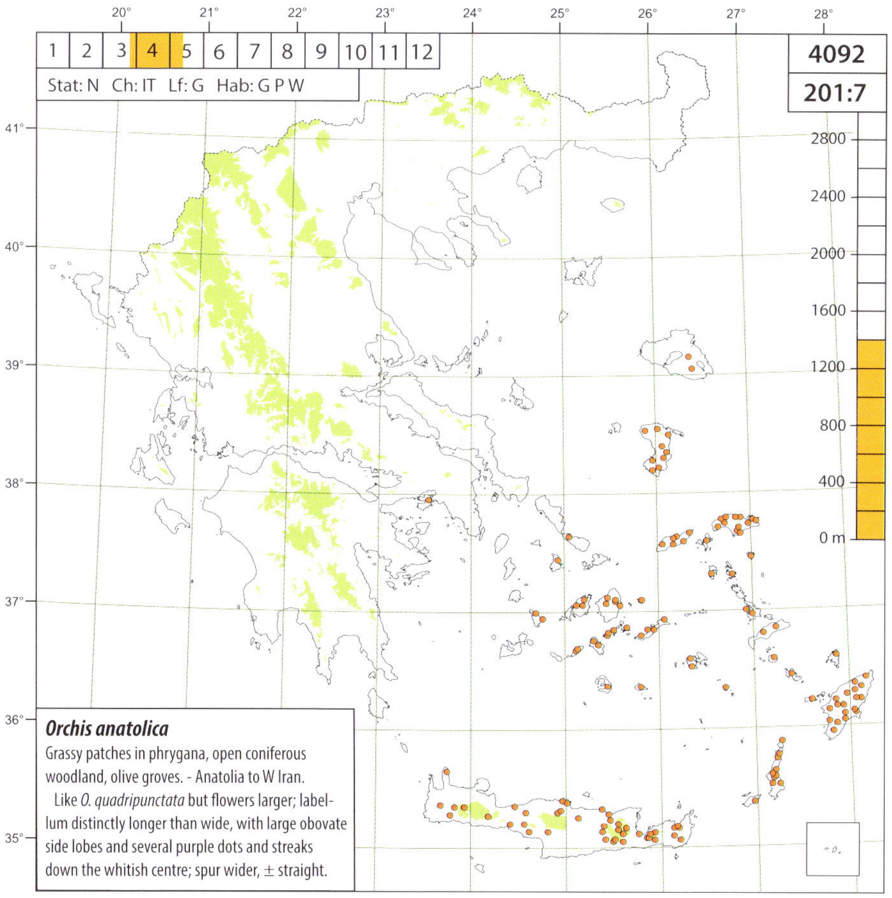

Orchis anatolica

Grassy patches in phrygana, open coniferous woodland, olive groves. - Anatolia to W Iran.

Like *O. quadripunctata* but flowers larger; labellum distinctly longer than wide, with large obovate side lobes and several purple dots and streaks down the whitish centre; spur wider, ± straight.

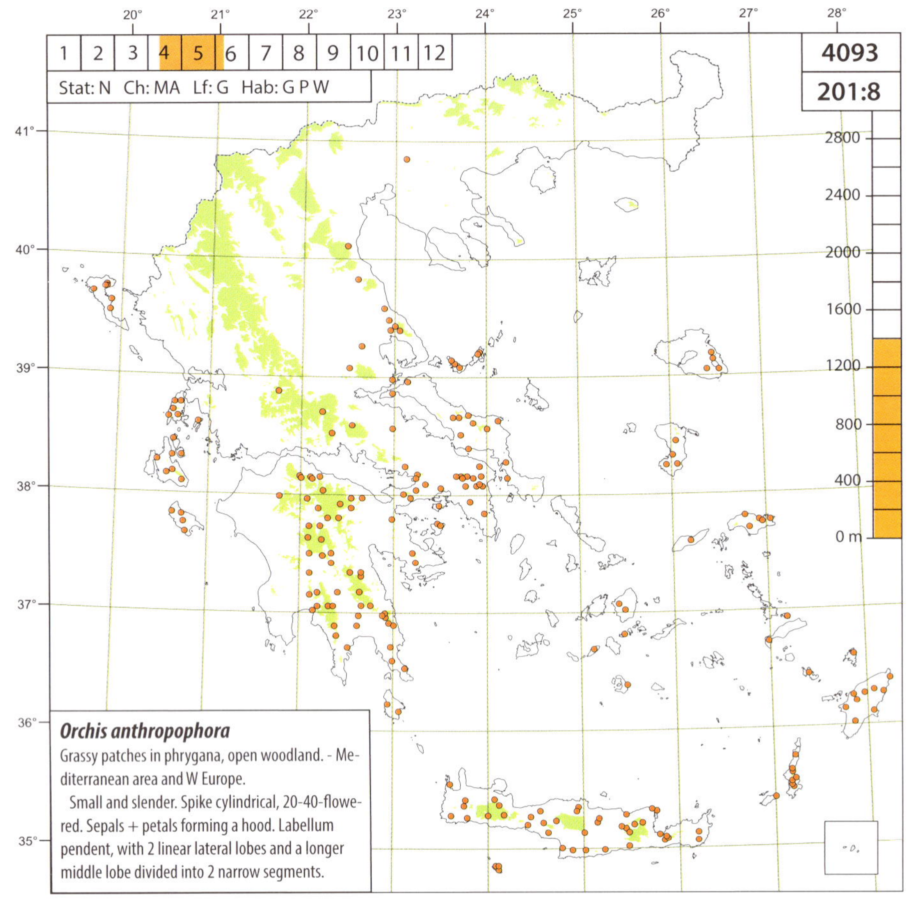

Orchis anthropophora

Grassy patches in phrygana, open woodland. - Mediterranean area and W Europe.

Small and slender. Spike cylindrical, 20-40-flowered. Sepals + petals forming a hood. Labellum pendent, with 2 linear lateral lobes and a longer middle lobe divided into 2 narrow segments.

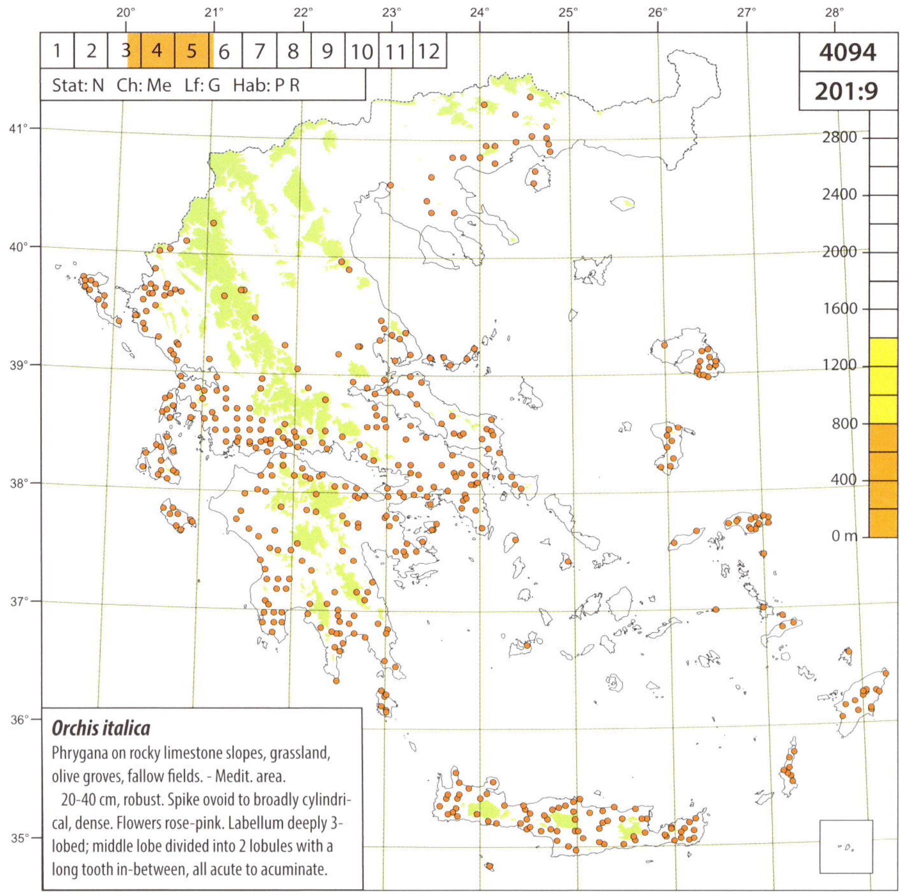

Orchis italica

Phrygana on rocky limestone slopes, grassland, olive groves, fallow fields. - Medit. area.

20-40 cm, robust. Spike ovoid to broadly cylindrical, dense. Flowers rose-pink. Labellum deeply 3-lobed; middle lobe divided into 2 lobules with a long tooth in-between, all acute to acuminate.

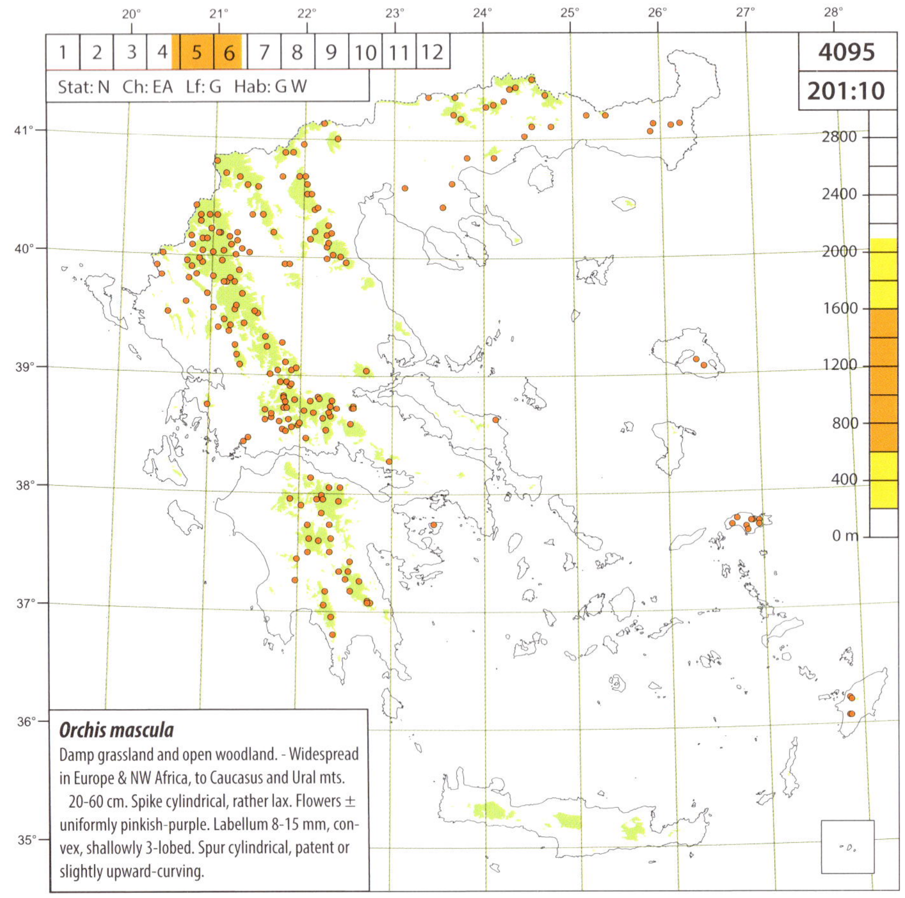

Orchis mascula

Damp grassland and open woodland. - Widespread in Europe & NW Africa, to Caucasus and Ural mts.

20-60 cm. Spike cylindrical, rather lax. Flowers ± uniformly pinkish-purple. Labellum 8-15 mm, convex, shallowly 3-lobed. Spur cylindrical, patent or slightly upward-curving.

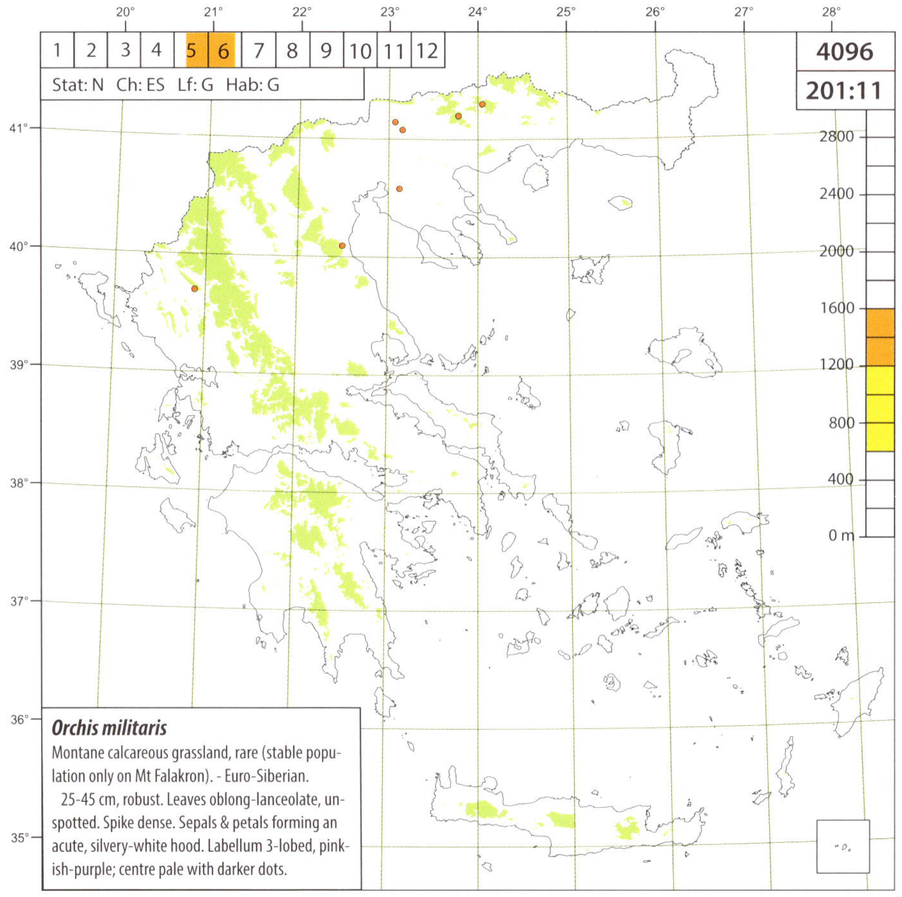

Orchis militaris

Montane calcareous grassland, rare (stable population only on Mt Falakron). - Euro-Siberian.

25-45 cm, robust. Leaves oblong-lanceolate, unspotted. Spike dense. Sepals & petals forming an acute, silvery-white hood. Labellum 3-lobed, pinkish-purple; centre pale with darker dots.

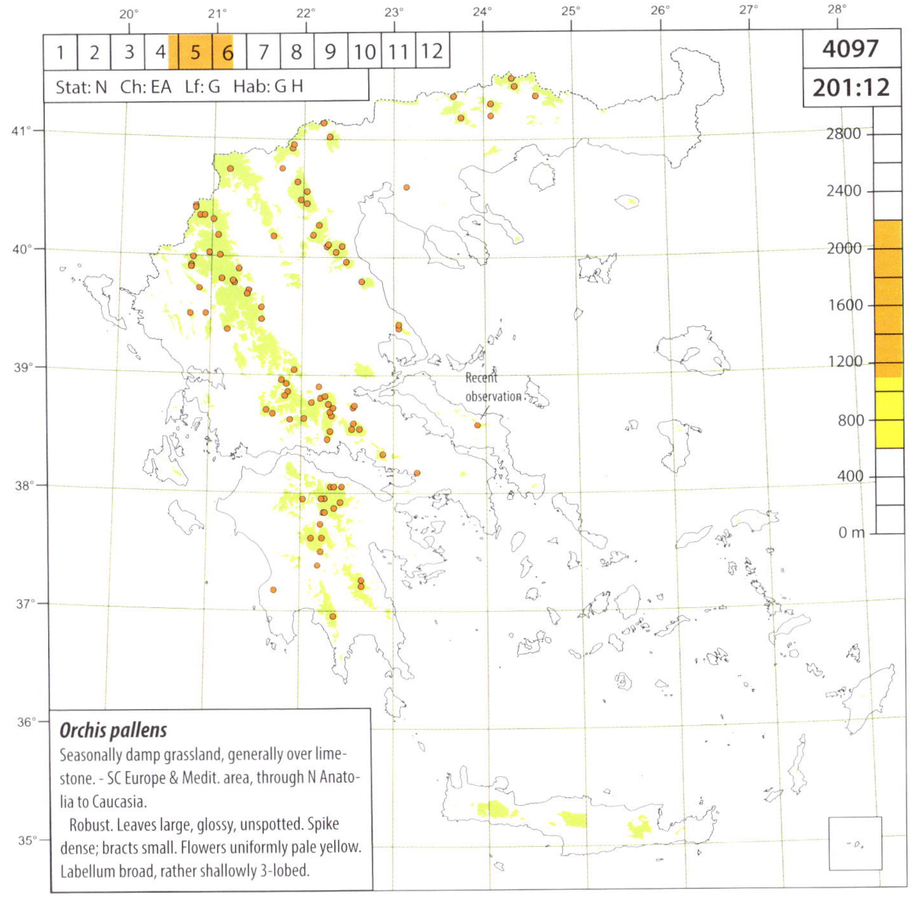

Orchis pallens
Seasonally damp grassland, generally over limestone. - SC Europe & Medit. area, through N Anatolia to Caucasia.
Robust. Leaves large, glossy, unspotted. Spike dense; bracts small. Flowers uniformly pale yellow. Labellum broad, rather shallowly 3-lobed.

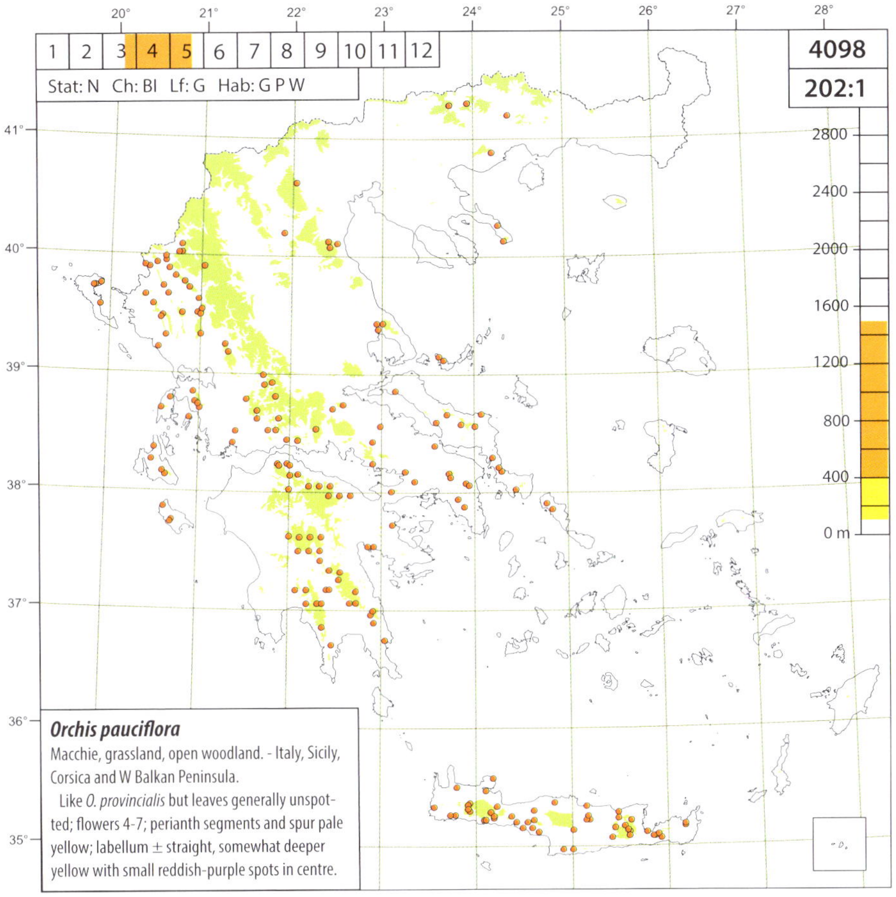

Orchis pauciflora
Macchie, grassland, open woodland. - Italy, Sicily, Corsica and W Balkan Peninsula.
Like *O. provincialis* but leaves generally unspotted; flowers 4-7; perianth segments and spur pale yellow; labellum ± straight, somewhat deeper yellow with small reddish-purple spots in centre.

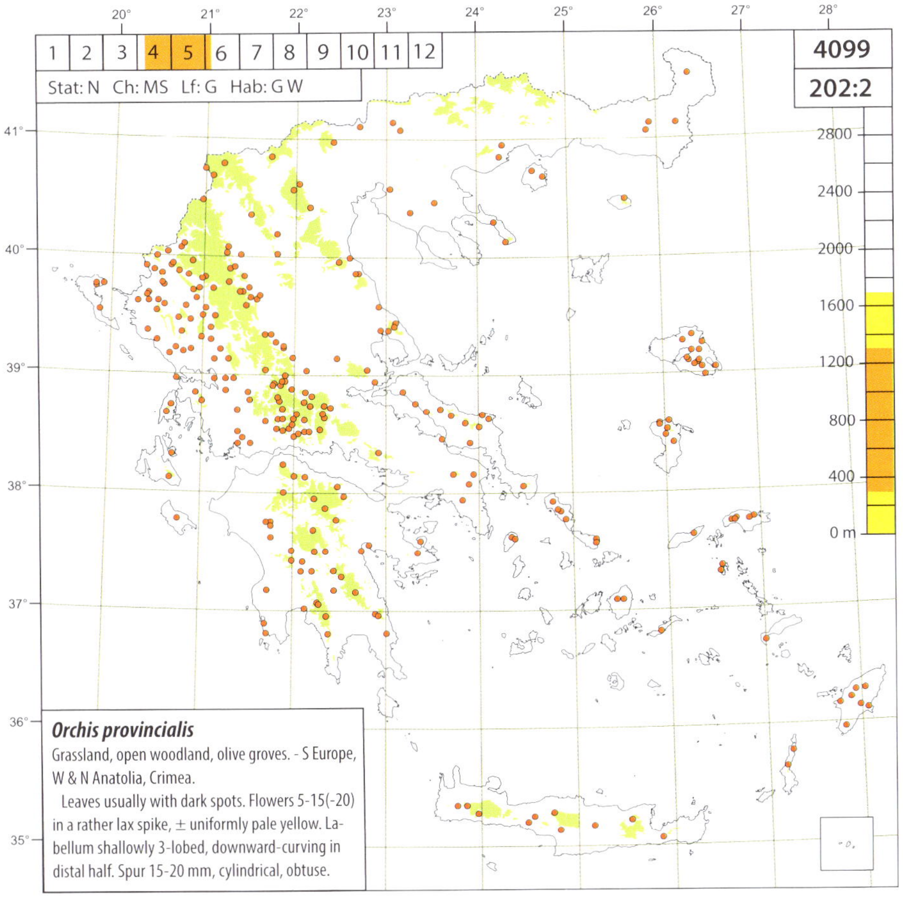

Orchis provincialis
Grassland, open woodland, olive groves. - S Europe, W & N Anatolia, Crimea.
Leaves usually with dark spots. Flowers 5-15(-20) in a rather lax spike, ± uniformly pale yellow. Labellum shallowly 3-lobed, downward-curving in distal half. Spur 15-20 mm, cylindrical, obtuse.

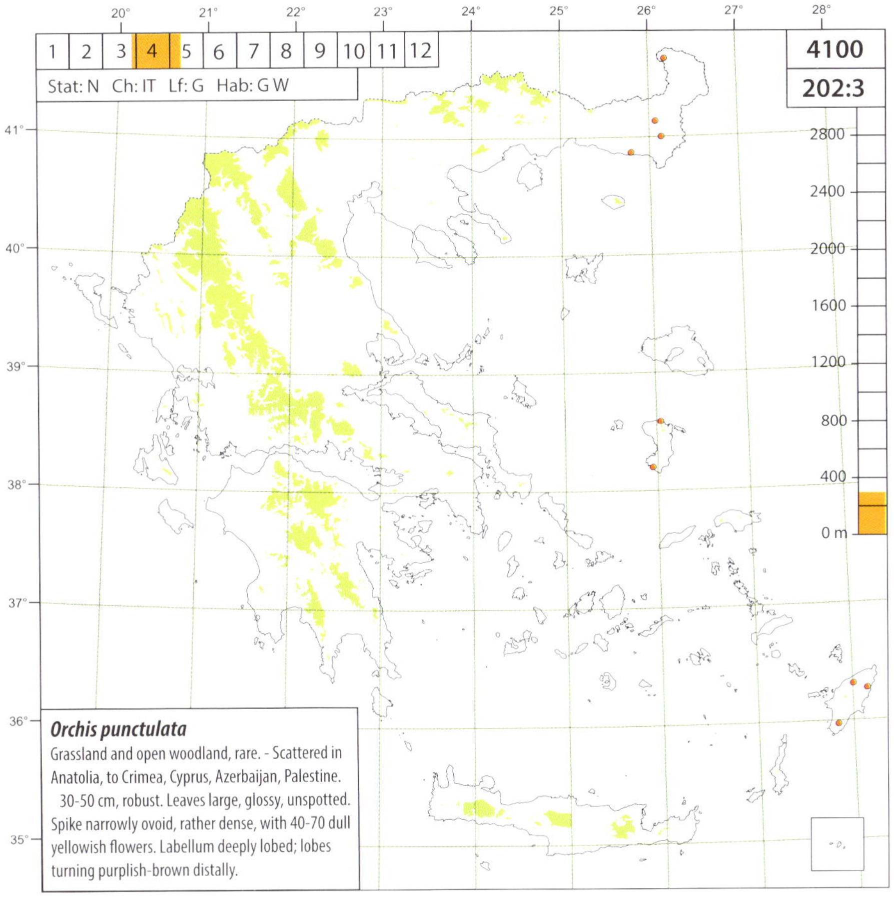

Orchis punctulata
Grassland and open woodland, rare. - Scattered in Anatolia, to Crimea, Cyprus, Azerbaijan, Palestine.
30-50 cm, robust. Leaves large, glossy, unspotted. Spike narrowly ovoid, rather dense, with 40-70 dull yellowish flowers. Labellum deeply lobed; lobes turning purplish-brown distally.

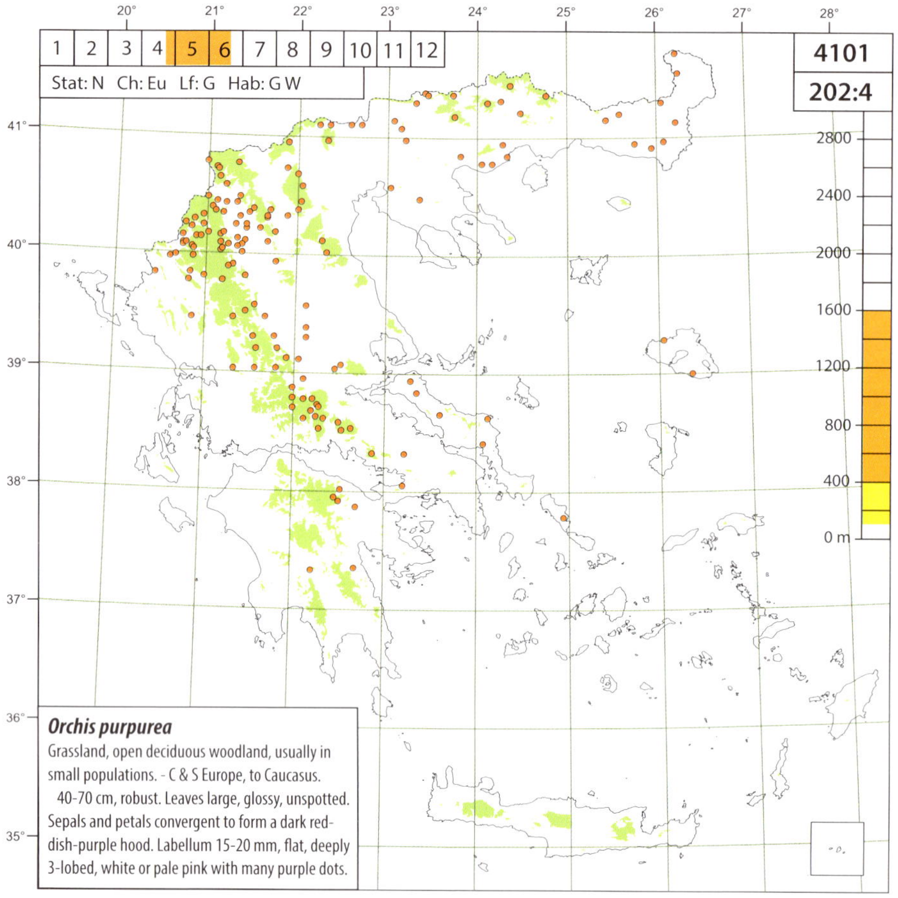

Orchis purpurea

Grassland, open deciduous woodland, usually in small populations. - C & S Europe, to Caucasus.

40-70 cm, robust. Leaves large, glossy, unspotted. Sepals and petals convergent to form a dark red-dish-purple hood. Labellum 15-20 mm, flat, deeply 3-lobed, white or pale pink with many purple dots.

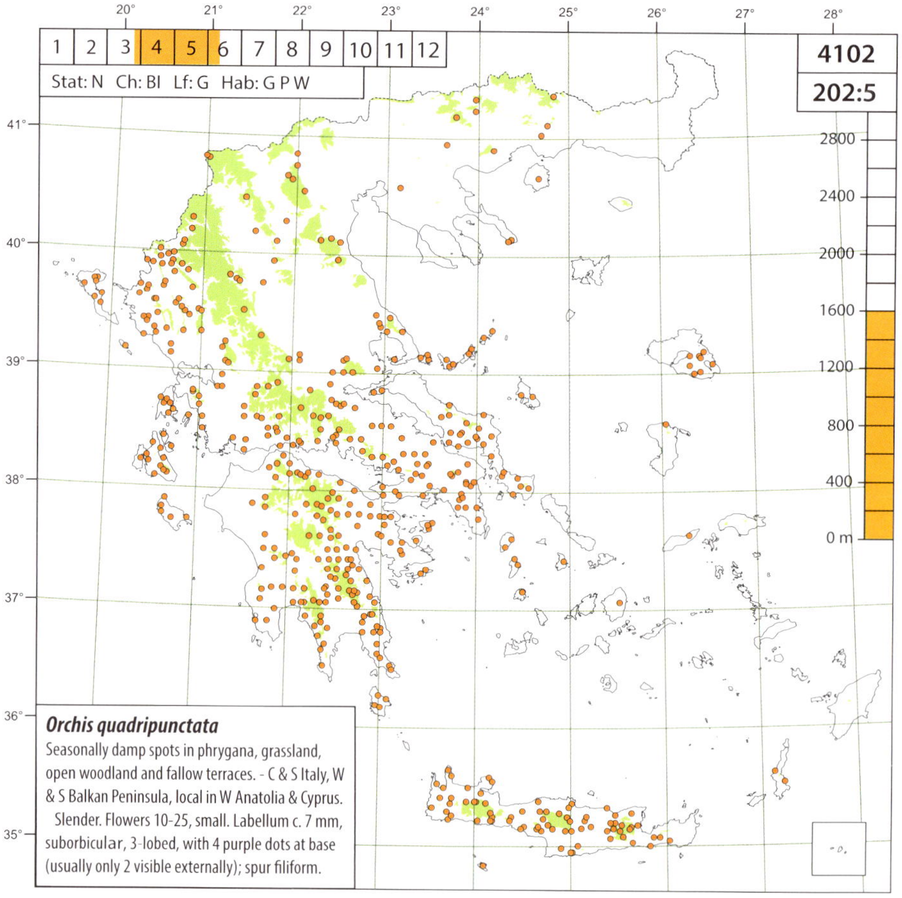

Orchis quadripunctata

Seasonally damp spots in phrygana, grassland, open woodland and fallow terraces. - C & S Italy, W & S Balkan Peninsula, local in W Anatolia & Cyprus.

Slender. Flowers 10-25, small. Labellum c. 7 mm, suborbicular, 3-lobed, with 4 purple dots at base (usually only 2 visible externally); spur filiform.

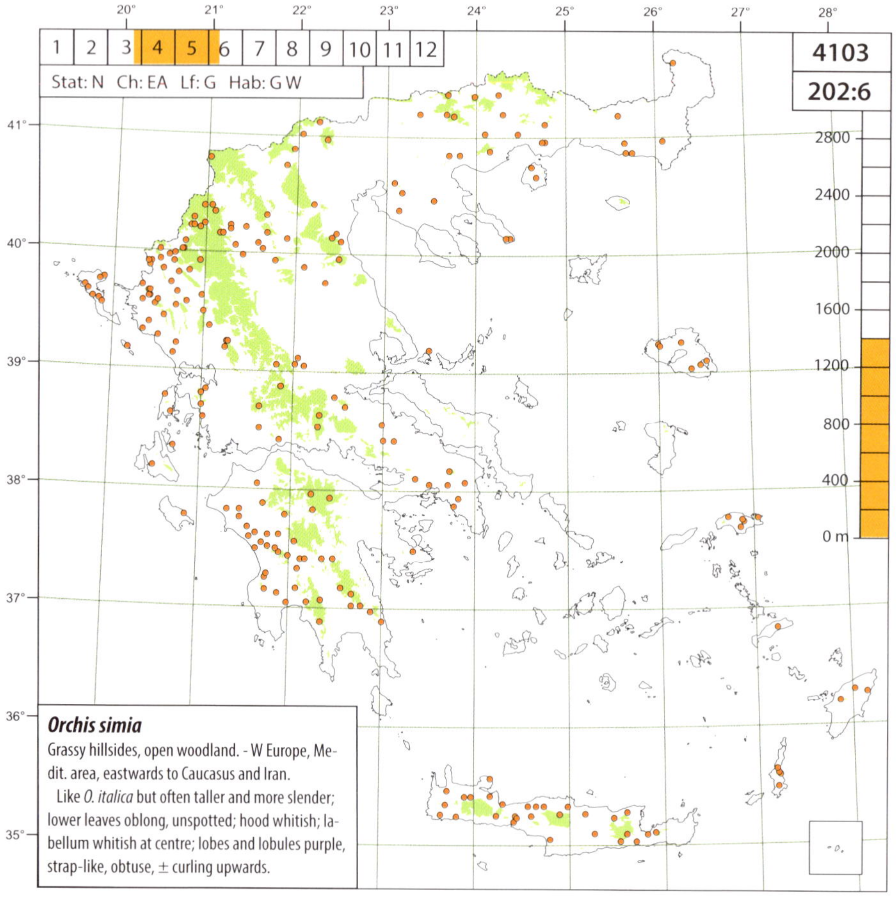

Orchis simia

Grassy hillsides, open woodland. - W Europe, Me-dit. area, eastwards to Caucasus and Iran.

Like *O. italica* but often taller and more slender; lower leaves oblong, unspotted; hood whitish; la-bellum whitish at centre; lobes and lobules purple, strap-like, obtuse, ± curling upwards.

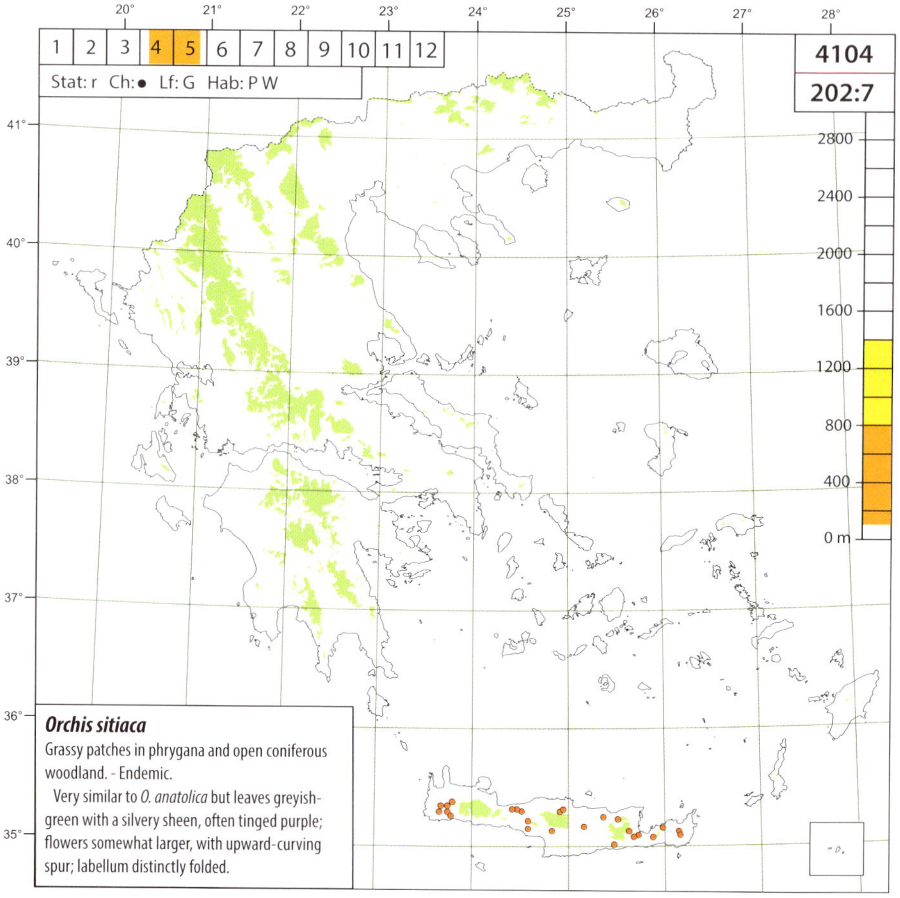

Orchis sitiaca

Grassy patches in phrygana and open coniferous woodland. - Endemic.

Very similar to *O. anatolica* but leaves greyish-green with a silvery sheen, often tinged purple; flowers somewhat larger, with upward-curving spur; labellum distinctly folded.

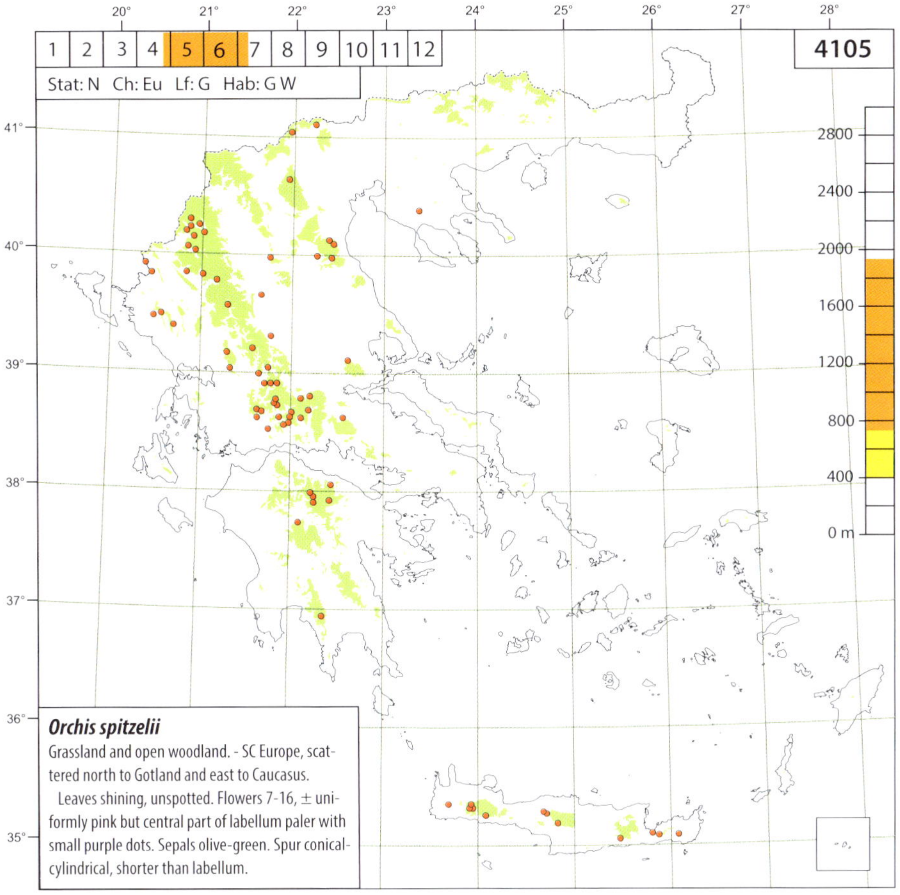

Orchis spitzelii

Grassland and open woodland. - SC Europe, scattered north to Gotland and east to Caucasus.

Leaves shining, unspotted. Flowers 7-16, ± uniformly pink but central part of labellum paler with small purple dots. Sepals olive-green. Spur conical-cylindrical, shorter than labellum.

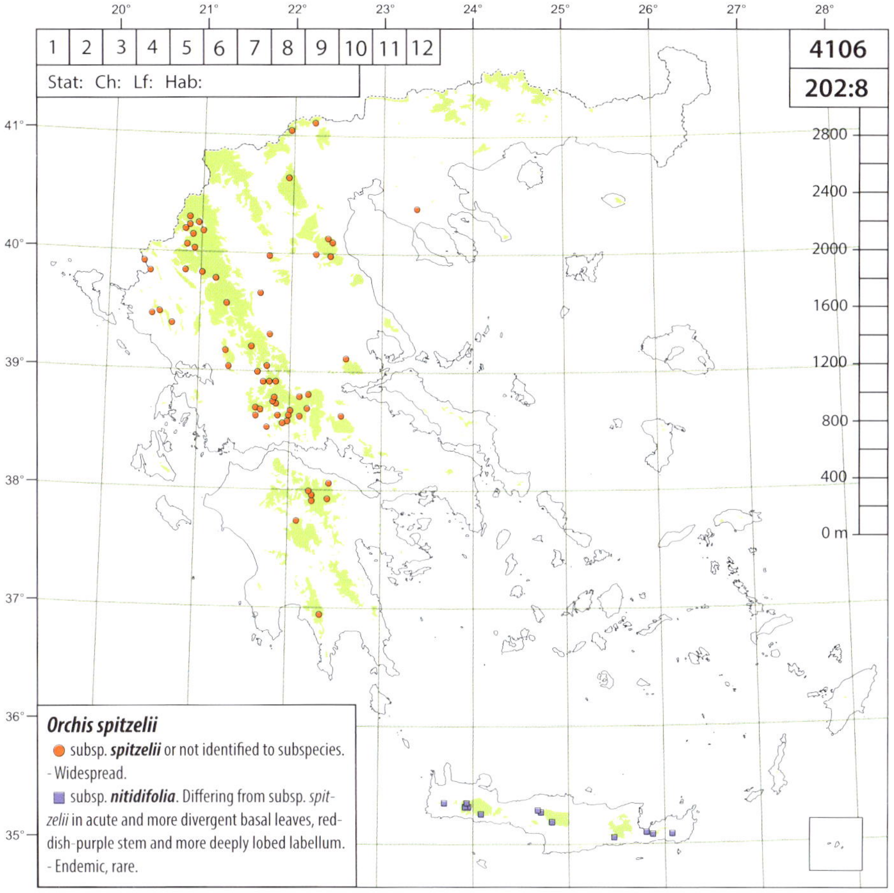

Orchis spitzelii

● subsp. ***spitzelii*** or not identified to subspecies. - Widespread.

■ subsp. ***nitidifolia***. Differing from subsp. *spitzelii* in acute and more divergent basal leaves, reddish-purple stem and more deeply lobed labellum. - Endemic, rare.

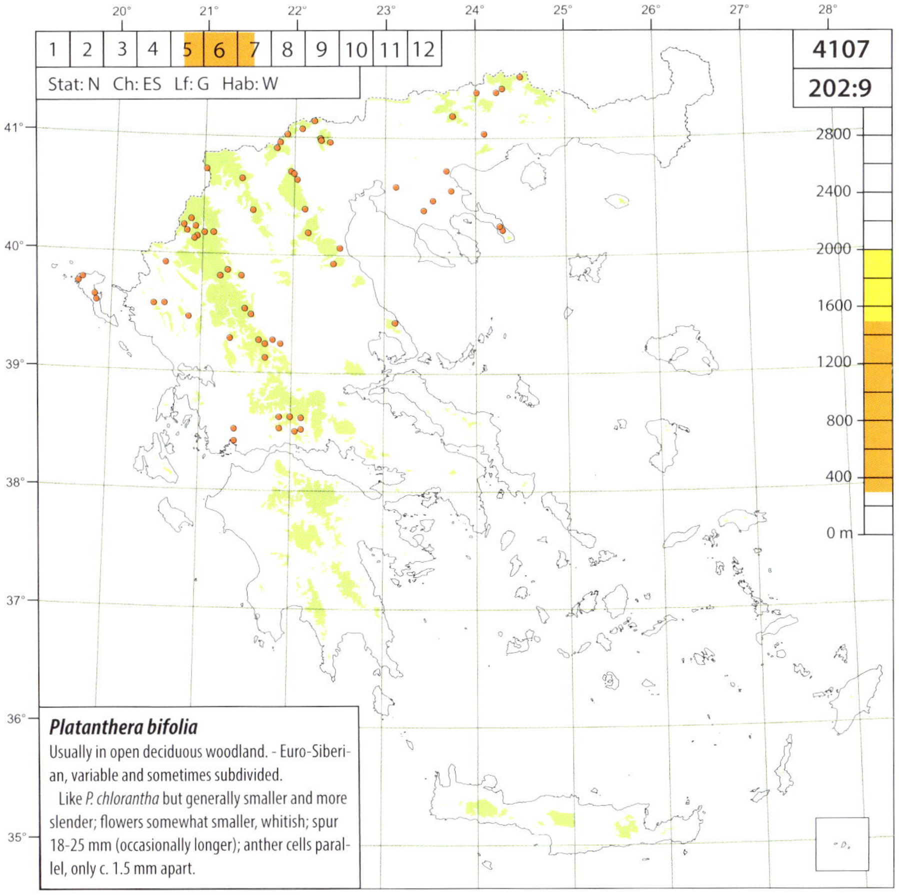

Platanthera bifolia

Usually in open deciduous woodland. - Euro-Siberian, variable and sometimes subdivided.

Like *P. chlorantha* but generally smaller and more slender; flowers somewhat smaller, whitish; spur 18-25 mm (occasionally longer); anther cells parallel, only c. 1.5 mm apart.

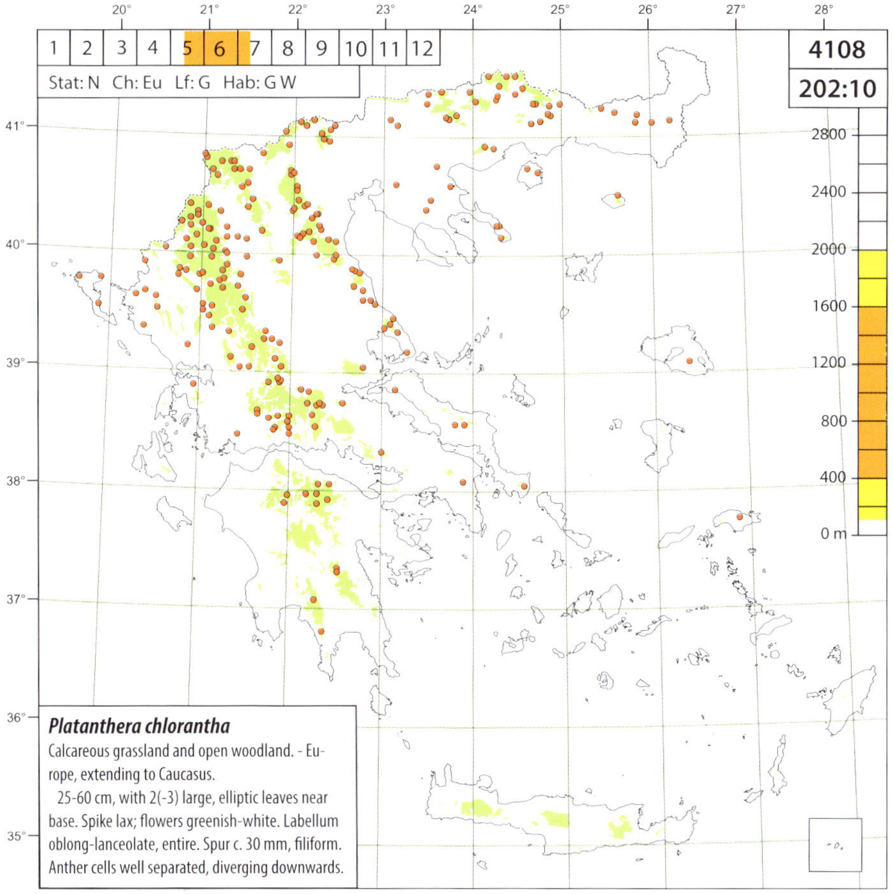

Platanthera chlorantha

Calcareous grassland and open woodland. - Europe, extending to Caucasus.

25-60 cm, with 2(-3) large, elliptic leaves near base. Spike lax; flowers greenish-white. Labellum oblong-lanceolate, entire. Spur c. 30 mm, filiform. Anther cells well separated, diverging downwards.

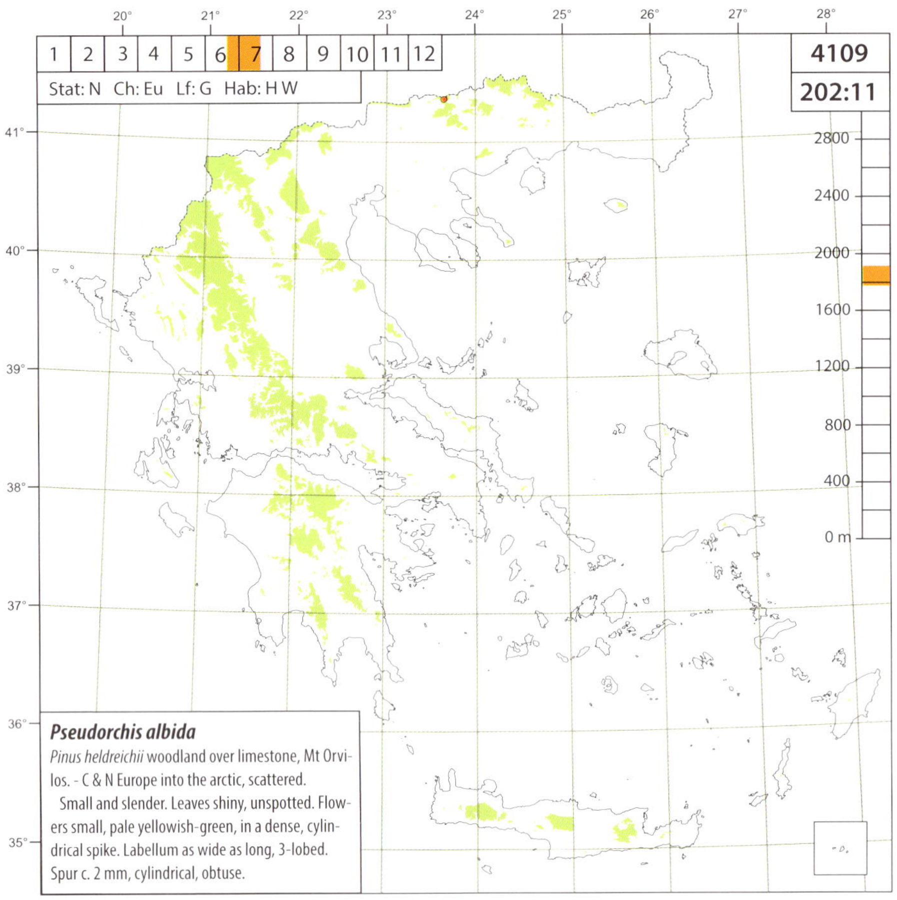

Pseudorchis albida

Pinus heldreichii woodland over limestone, Mt Orvilos. - C & N Europe into the arctic, scattered.

Small and slender. Leaves shiny, unspotted. Flowers small, pale yellowish-green, in a dense, cylindrical spike. Labellum as wide as long, 3-lobed. Spur c. 2 mm, cylindrical, obtuse.

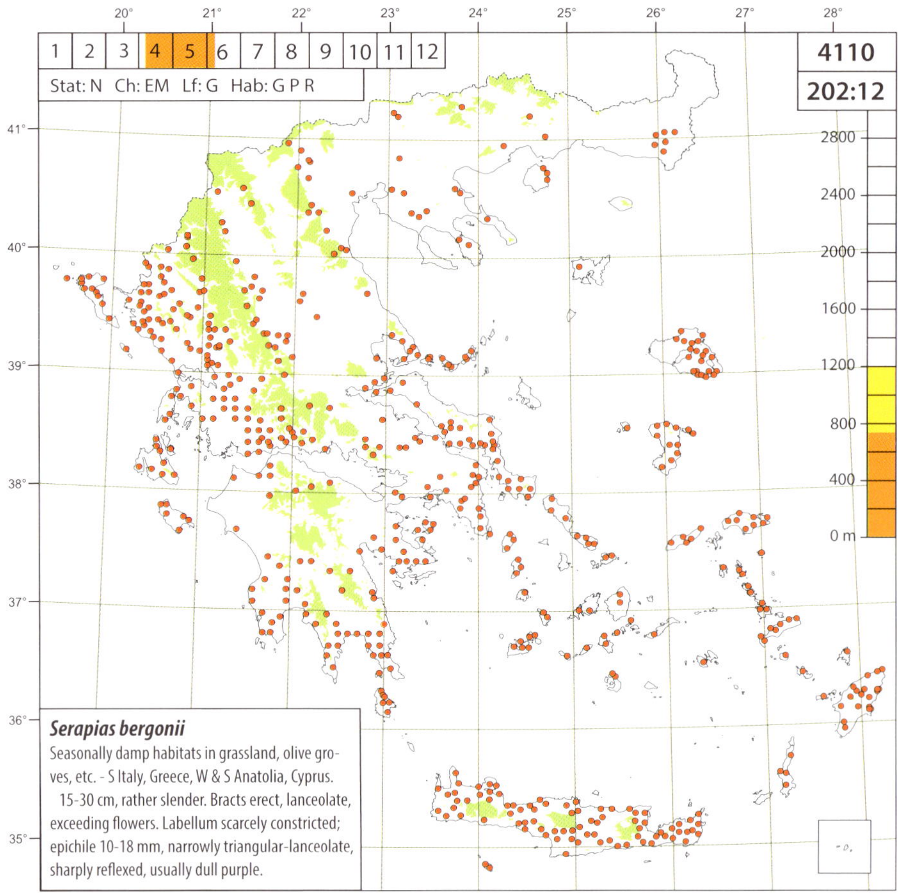

Serapias bergonii

Seasonally damp habitats in grassland, olive groves, etc. - S Italy, Greece, W & S Anatolia, Cyprus.

15-30 cm, rather slender. Bracts erect, lanceolate, exceeding flowers. Labellum scarcely constricted; epichile 10-18 mm, narrowly triangular-lanceolate, sharply reflexed, usually dull purple.

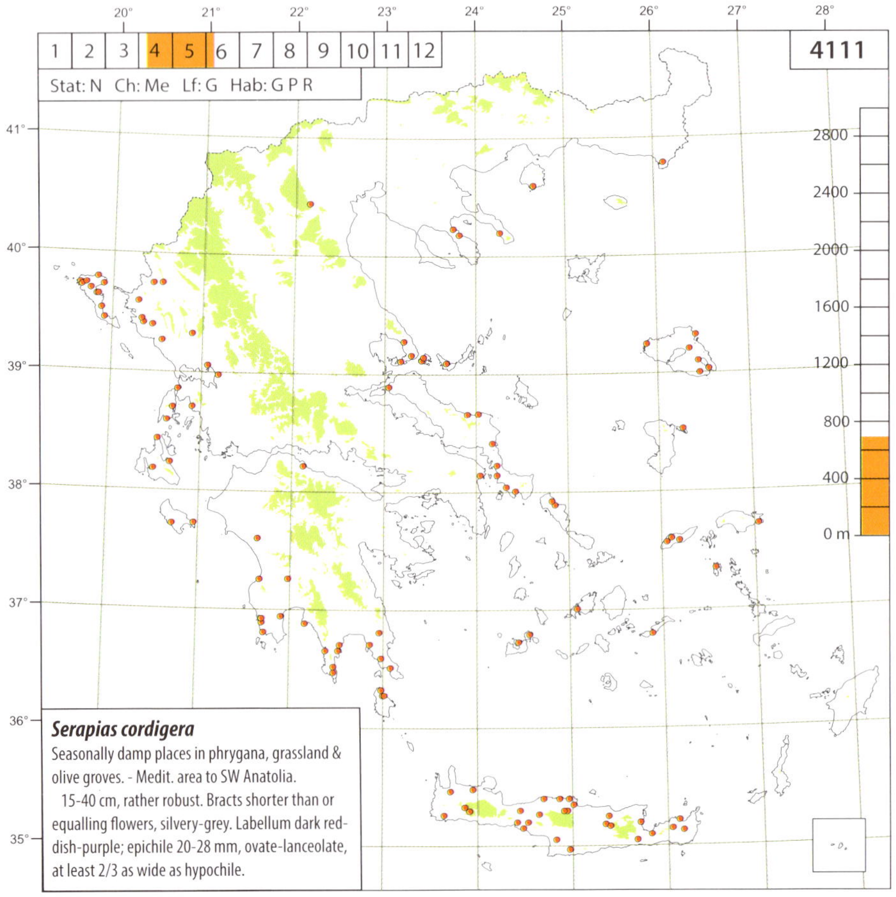

Serapias cordigera

Seasonally damp places in phrygana, grassland & olive groves. - Medit. area to SW Anatolia.

15-40 cm, rather robust. Bracts shorter than or equalling flowers, silvery-grey. Labellum dark reddish-purple; epichile 20-28 mm, ovate-lanceolate, at least 2/3 as wide as hypochile.

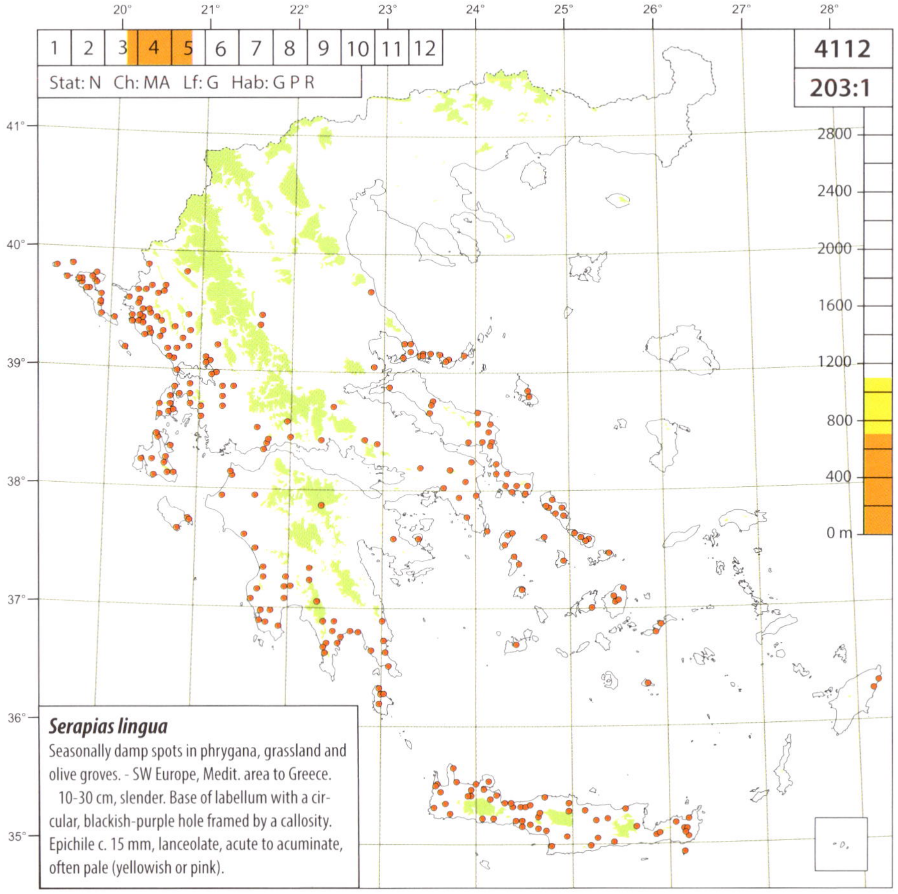

Serapias lingua

Seasonally damp spots in phrygana, grassland and olive groves. - SW Europe, Medit. area to Greece.

10-30 cm, slender. Base of labellum with a circular, blackish-purple hole framed by a callosity. Epichile c. 15 mm, lanceolate, acute to acuminate, often pale (yellowish or pink).

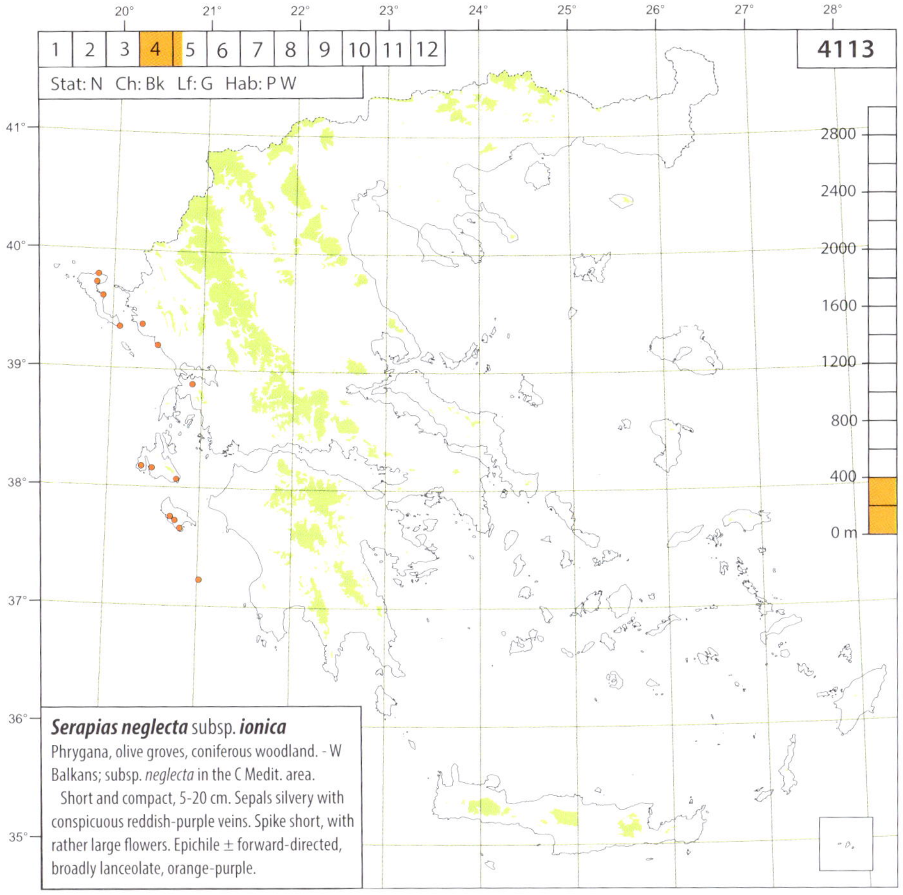

Serapias neglecta* subsp. *ionica
Phrygana, olive groves, coniferous woodland. - W Balkans; subsp. *neglecta* in the C Medit. area.
Short and compact, 5-20 cm. Sepals silvery with conspicuous reddish-purple veins. Spike short, with rather large flowers. Epichile ± forward-directed, broadly lanceolate, orange-purple.

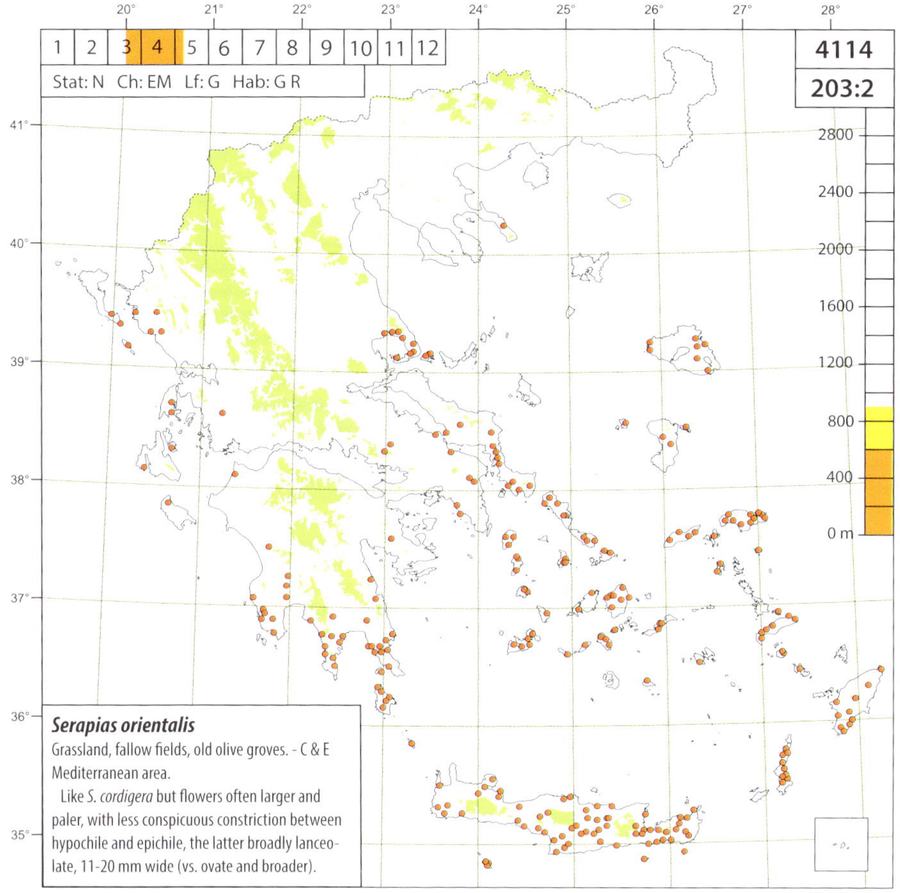

Serapias orientalis
Grassland, fallow fields, old olive groves. - C & E Mediterranean area.
Like *S. cordigera* but flowers often larger and paler, with less conspicuous constriction between hypochile and epichile, the latter broadly lanceolate, 11-20 mm wide (vs. ovate and broader).

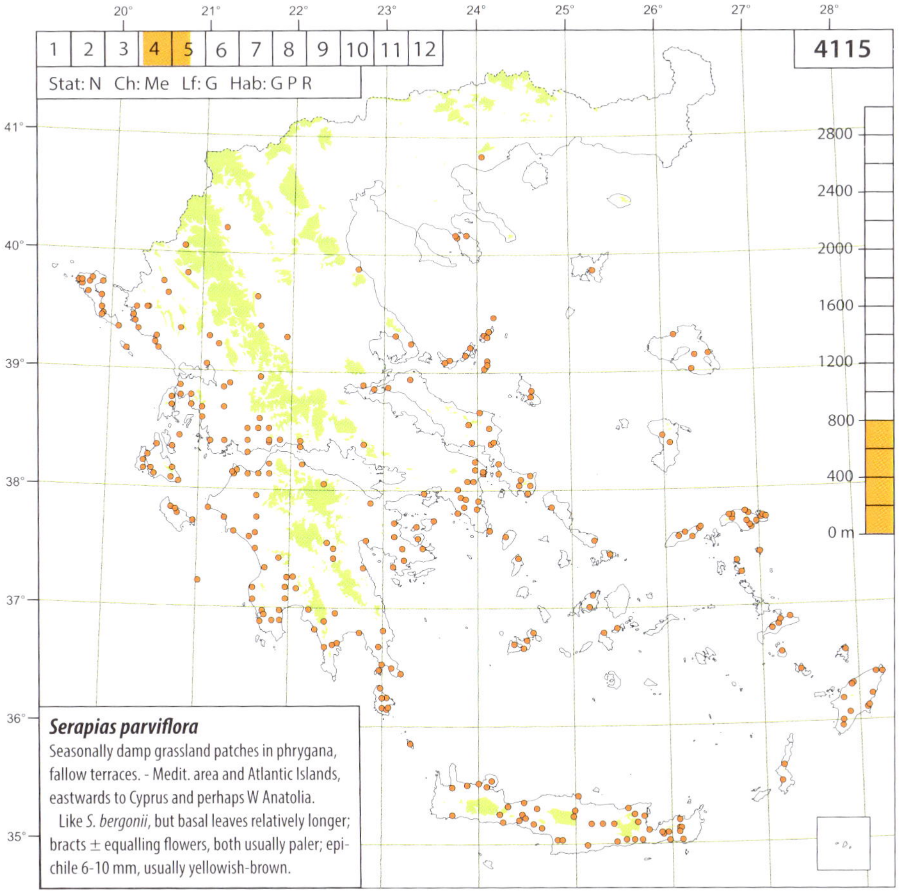

Serapias parviflora
Seasonally damp grassland patches in phrygana, fallow terraces. - Medit. area and Atlantic Islands, eastwards to Cyprus and perhaps W Anatolia.
Like *S. bergonii*, but basal leaves relatively longer; bracts ± equalling flowers, both usually paler; epichile 6-10 mm, usually yellowish-brown.

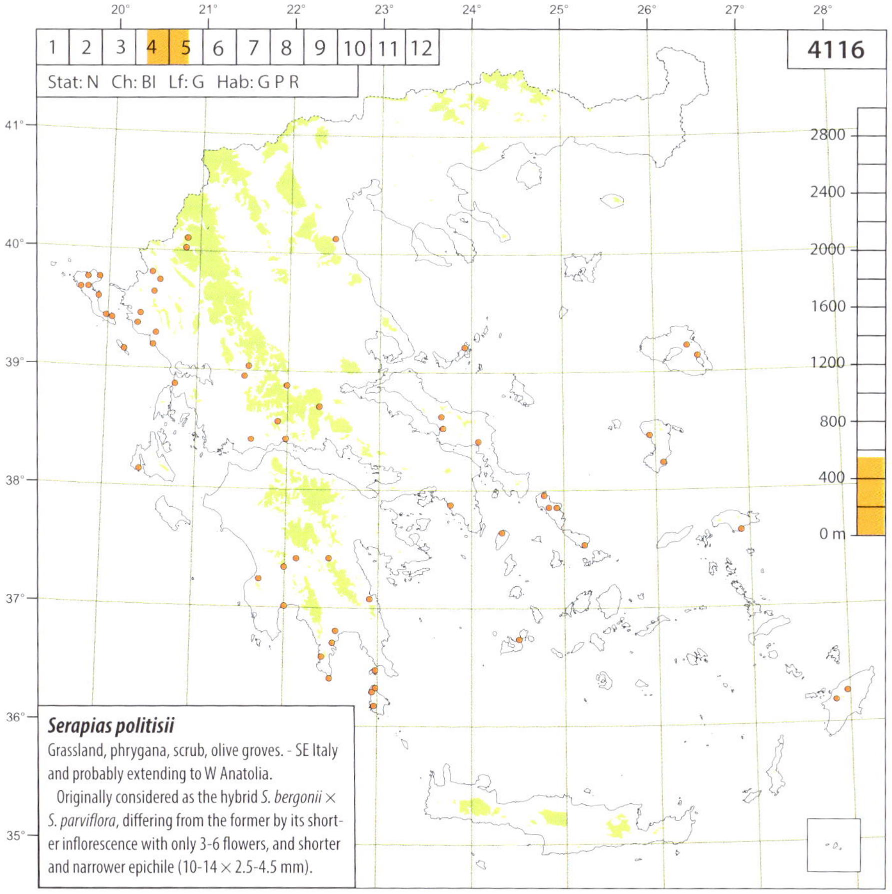

Serapias politisii
Grassland, phrygana, scrub, olive groves. - SE Italy and probably extending to W Anatolia.
Originally considered as the hybrid *S. bergonii* × *S. parviflora*, differing from the former by its shorter inflorescence with only 3-6 flowers, and shorter and narrower epichile (10-14 × 2.5-4.5 mm).

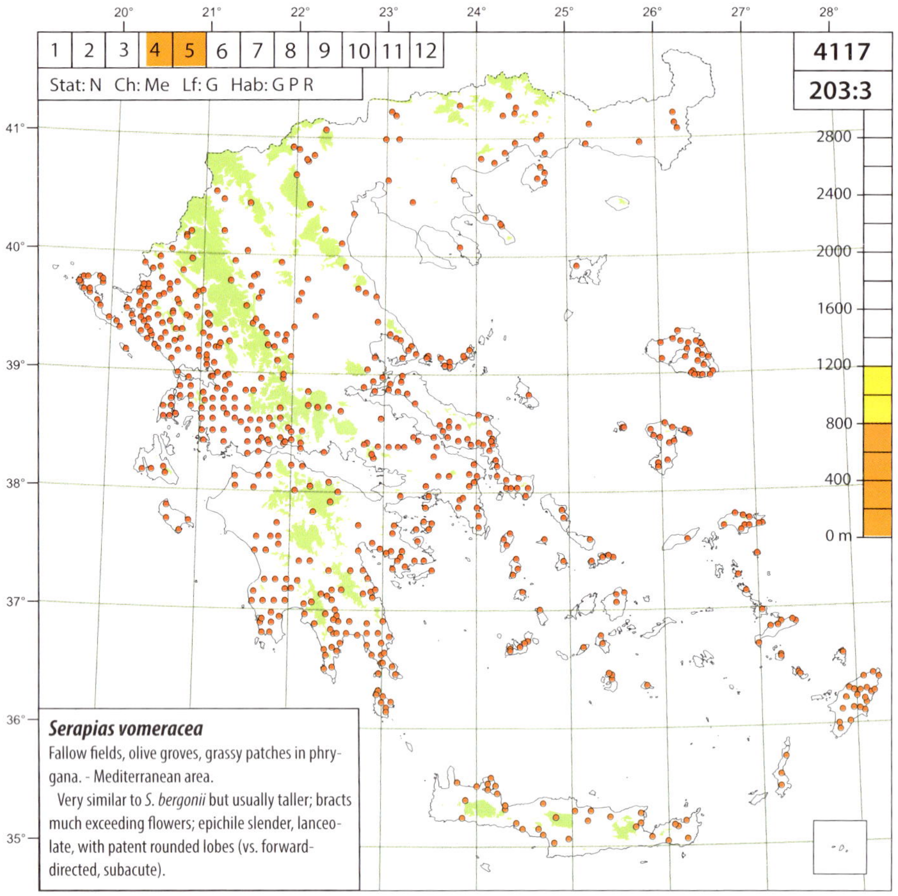

Serapias vomeracea
Fallow fields, olive groves, grassy patches in phrygana. - Mediterranean area.
Very similar to *S. bergonii* but usually taller; bracts much exceeding flowers; epichile slender, lanceolate, with patent rounded lobes (vs. forward-directed, subacute).

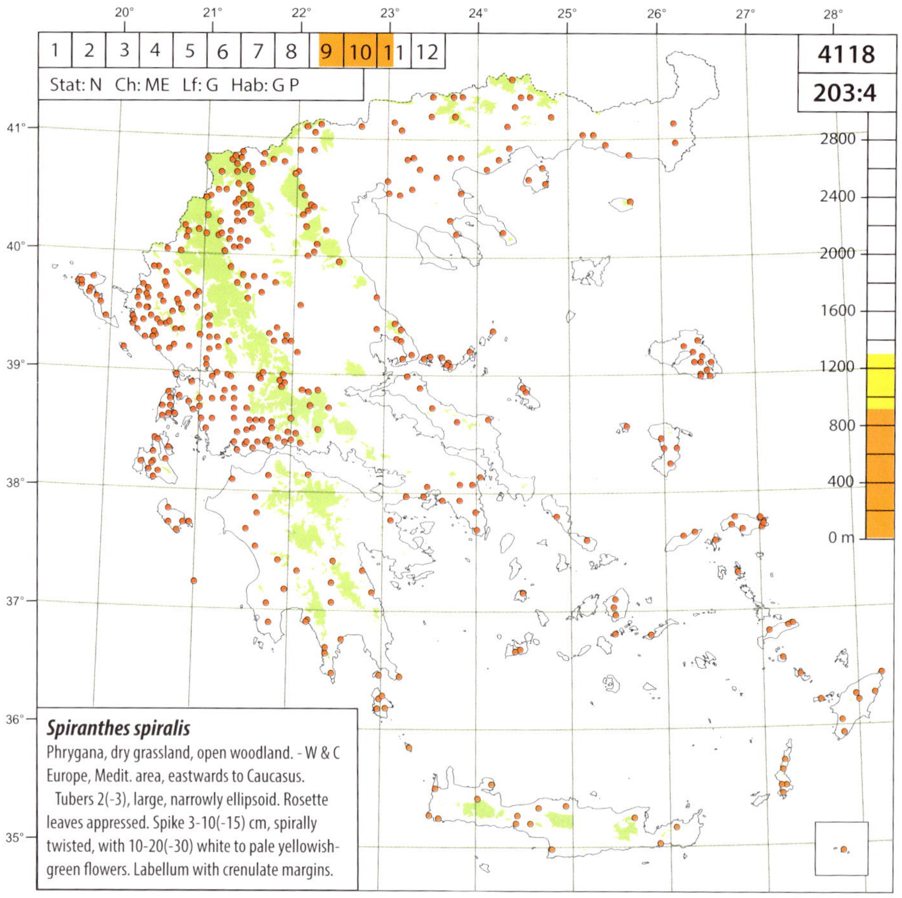

Spiranthes spiralis
Phrygana, dry grassland, open woodland. - W & C Europe, Medit. area, eastwards to Caucasus.
Tubers 2(-3), large, narrowly ellipsoid. Rosette leaves appressed. Spike 3-10(-15) cm, spirally twisted, with 10-20(-30) white to pale yellowish-green flowers. Labellum with crenulate margins.

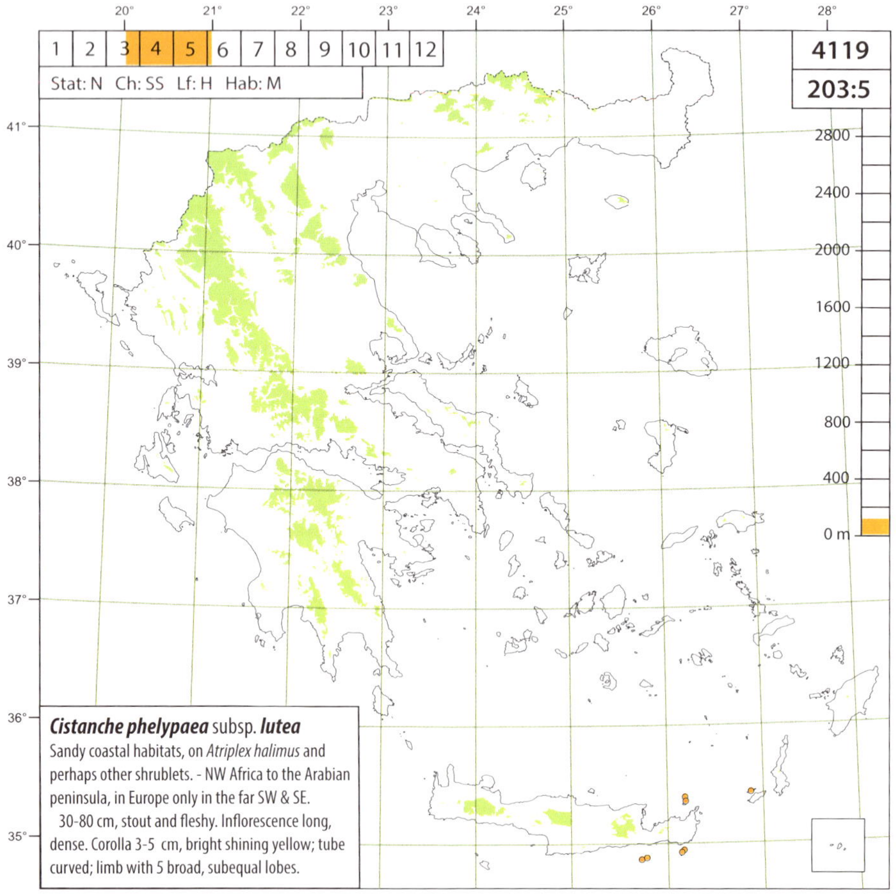

Cistanche phelypaea subsp. ***lutea***
Sandy coastal habitats, on *Atriplex halimus* and perhaps other shrublets. - NW Africa to the Arabian peninsula, in Europe only in the far SW & SE.
30-80 cm, stout and fleshy. Inflorescence long, dense. Corolla 3-5 cm, bright shining yellow; tube curved; limb with 5 broad, subequal lobes.

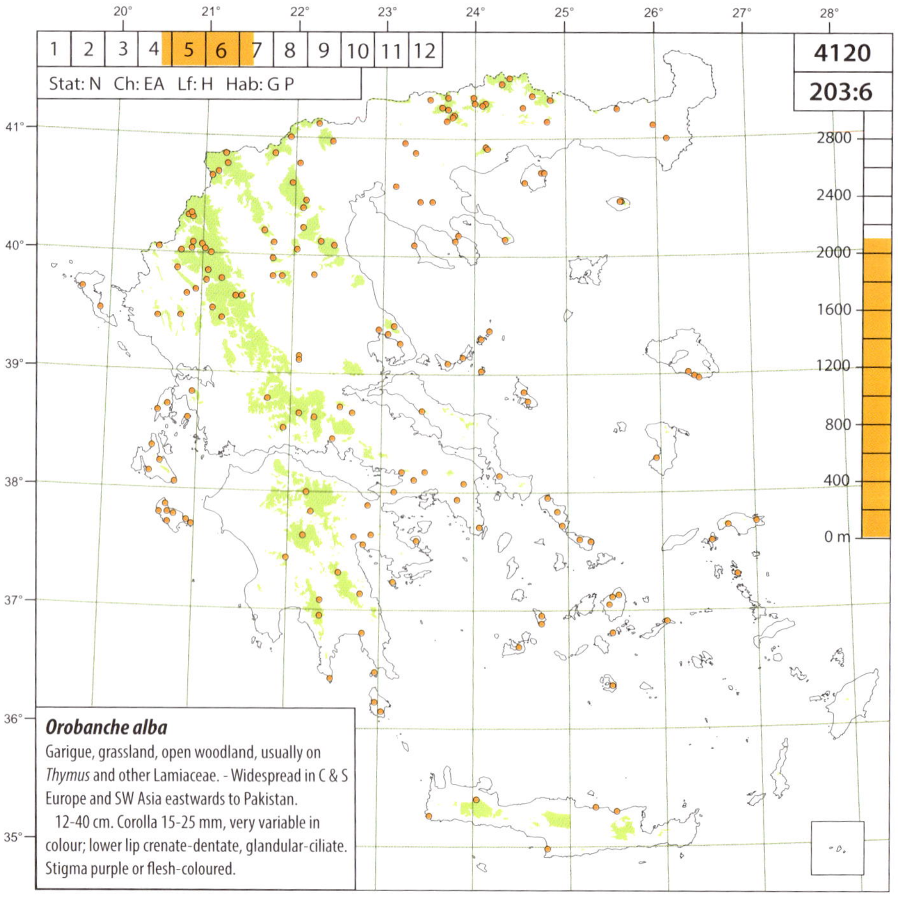

Orobanche alba
Garigue, grassland, open woodland, usually on *Thymus* and other Lamiaceae. - Widespread in C & S Europe and SW Asia eastwards to Pakistan.
12-40 cm. Corolla 15-25 mm, very variable in colour; lower lip crenate-dentate, glandular-ciliate. Stigma purple or flesh-coloured.

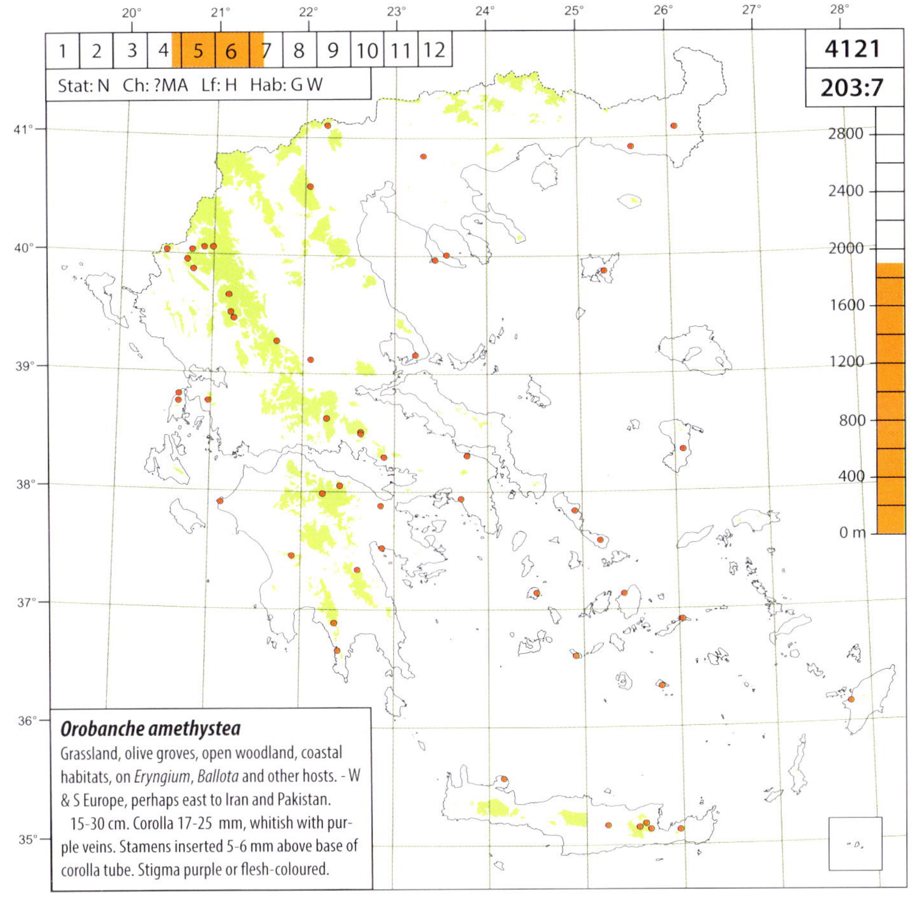

Orobanche amethystea

Grassland, olive groves, open woodland, coastal habitats, on *Eryngium*, *Ballota* and other hosts. - W & S Europe, perhaps east to Iran and Pakistan.

15-30 cm. Corolla 17-25 mm, whitish with purple veins. Stamens inserted 5-6 mm above base of corolla tube. Stigma purple or flesh-coloured.

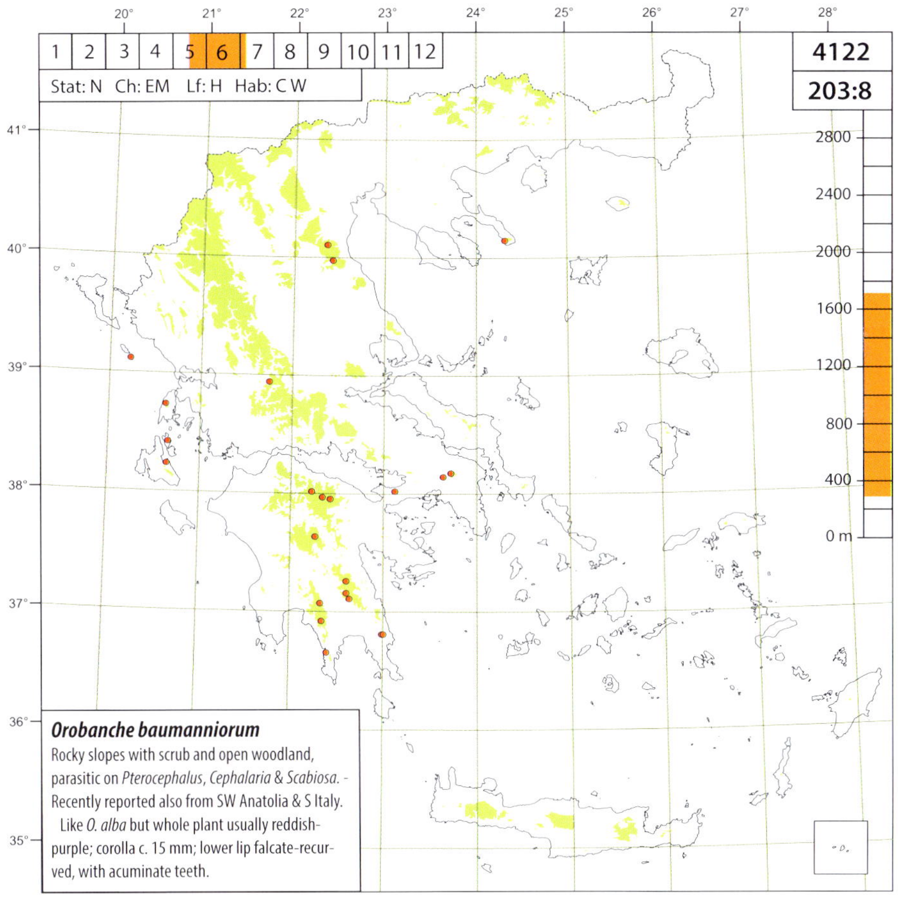

Orobanche baumanniorum

Rocky slopes with scrub and open woodland, parasitic on *Pterocephalus*, *Cephalaria* & *Scabiosa*. - Recently reported also from SW Anatolia & S Italy.

Like *O. alba* but whole plant usually reddish-purple; corolla c. 15 mm; lower lip falcate-recurved, with acuminate teeth.

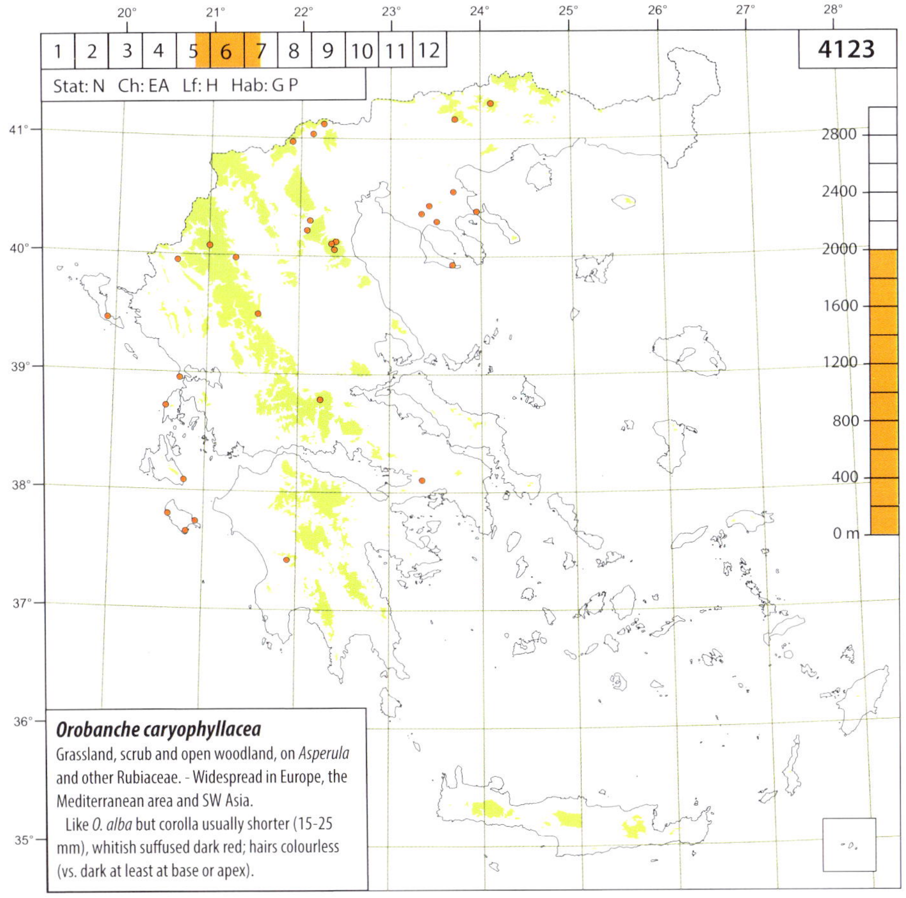

Orobanche caryophyllacea

Grassland, scrub and open woodland, on *Asperula* and other Rubiaceae. - Widespread in Europe, the Mediterranean area and SW Asia.

Like *O. alba* but corolla usually shorter (15-25 mm), whitish suffused dark red; hairs colourless (vs. dark at least at base or apex).

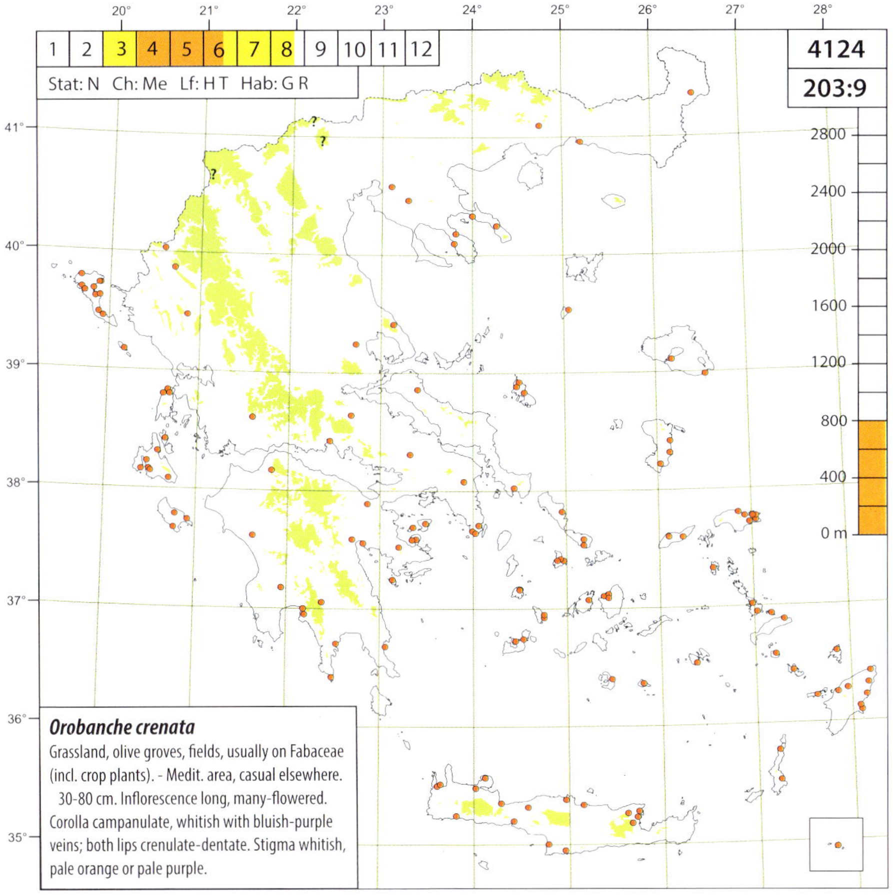

Orobanche crenata

Grassland, olive groves, fields, usually on Fabaceae (incl. crop plants). - Medit. area, casual elsewhere.

30-80 cm. Inflorescence long, many-flowered. Corolla campanulate, whitish with bluish-purple veins; both lips crenulate-dentate. Stigma whitish, pale orange or pale purple.

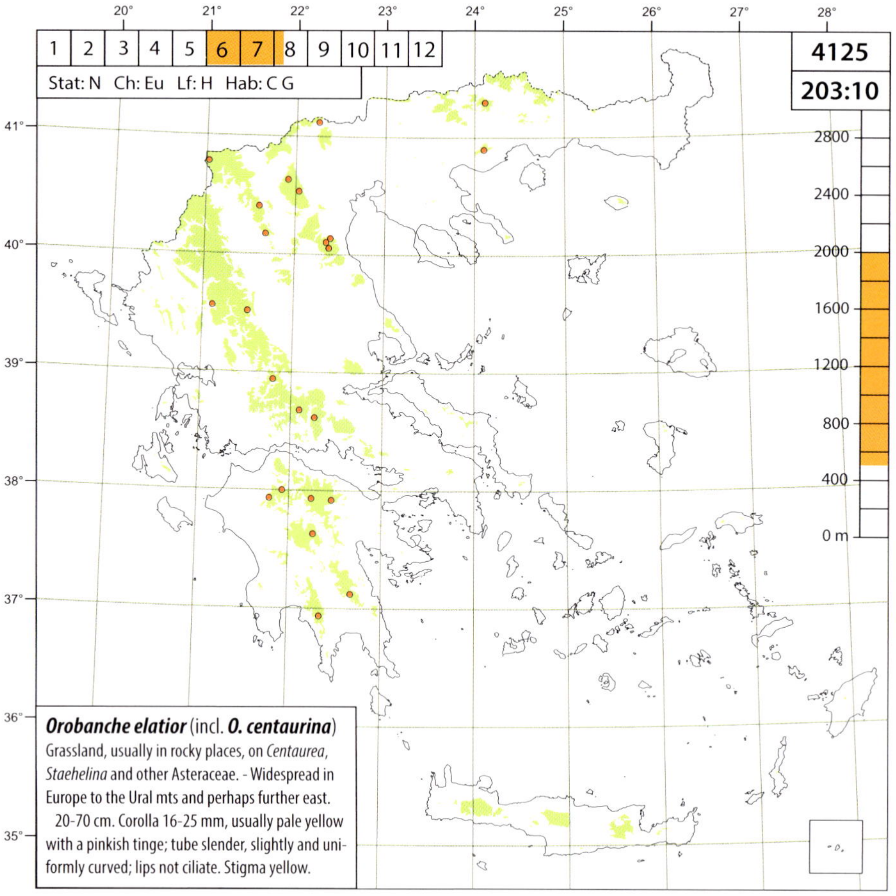

Orobanche elatior (incl. ***O. centaurina***)
Grassland, usually in rocky places, on *Centaurea*, *Staehelina* and other Asteraceae. - Widespread in Europe to the Ural mts and perhaps further east.
20-70 cm. Corolla 16-25 mm, usually pale yellow with a pinkish tinge; tube slender, slightly and uniformly curved; lips not ciliate. Stigma yellow.

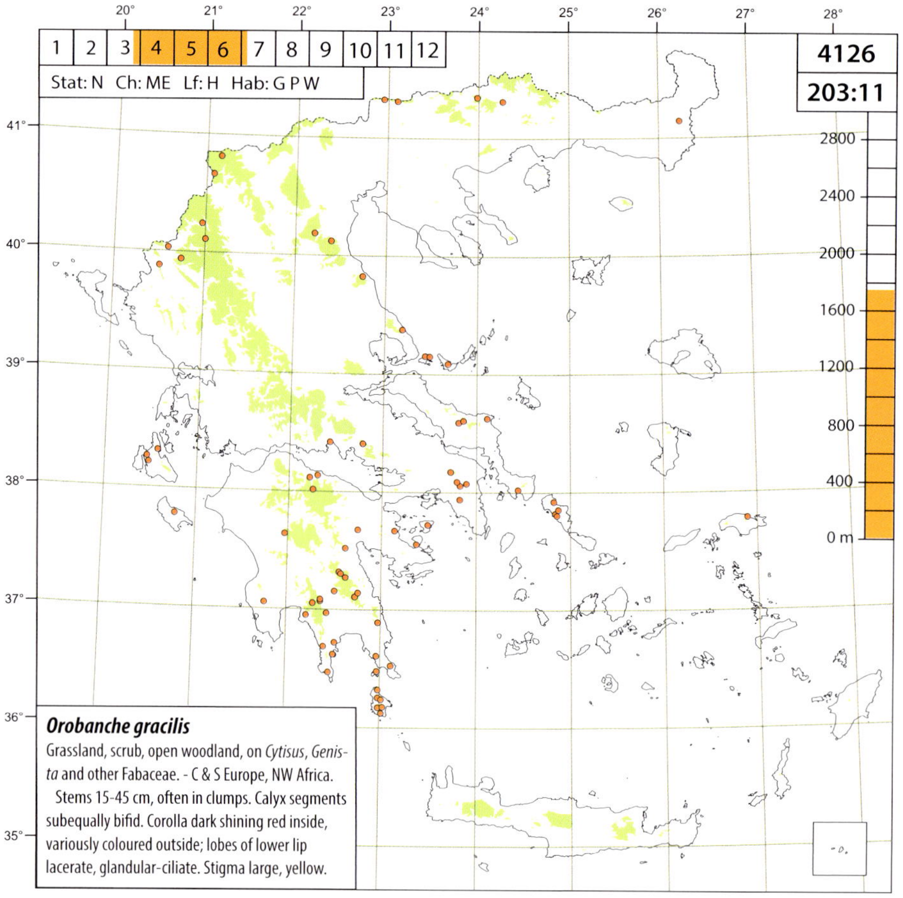

Orobanche gracilis
Grassland, scrub, open woodland, on *Cytisus*, *Genista* and other Fabaceae. - C & S Europe, NW Africa.
Stems 15-45 cm, often in clumps. Calyx segments subequally bifid. Corolla dark shining red inside, variously coloured outside; lobes of lower lip lacerate, glandular-ciliate. Stigma large, yellow.

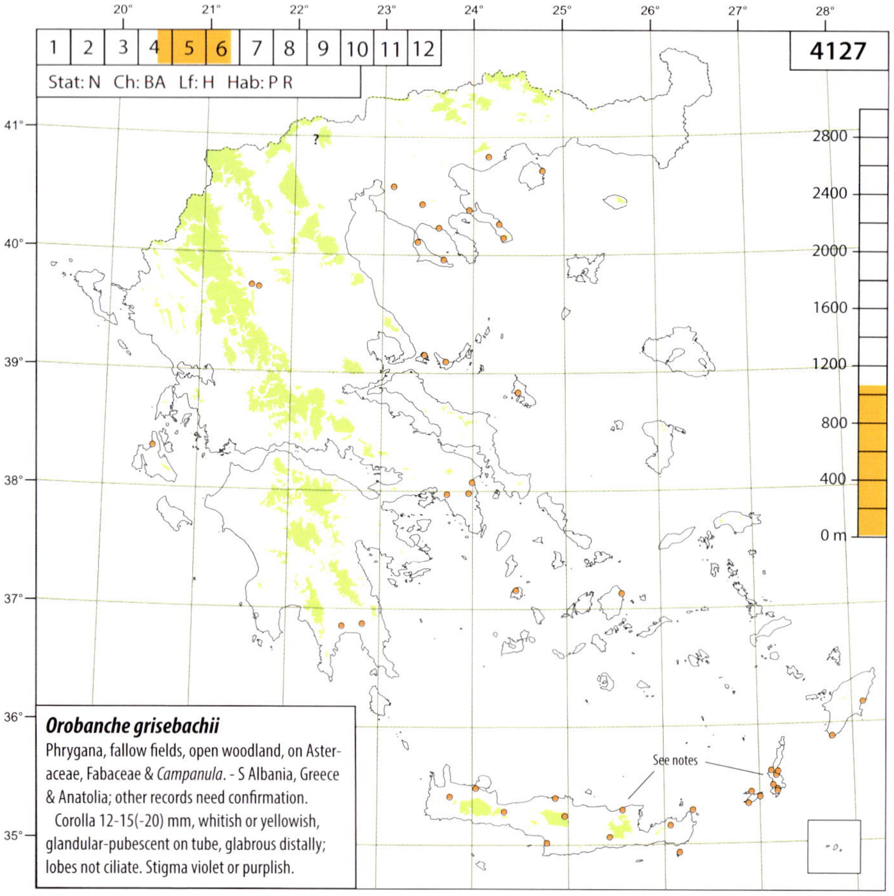

Orobanche grisebachii
Phrygana, fallow fields, open woodland, on Asteraceae, Fabaceae & *Campanula*. - S Albania, Greece & Anatolia; other records need confirmation.
Corolla 12-15(-20) mm, whitish or yellowish, glandular-pubescent on tube, glabrous distally; lobes not ciliate. Stigma violet or purplish.

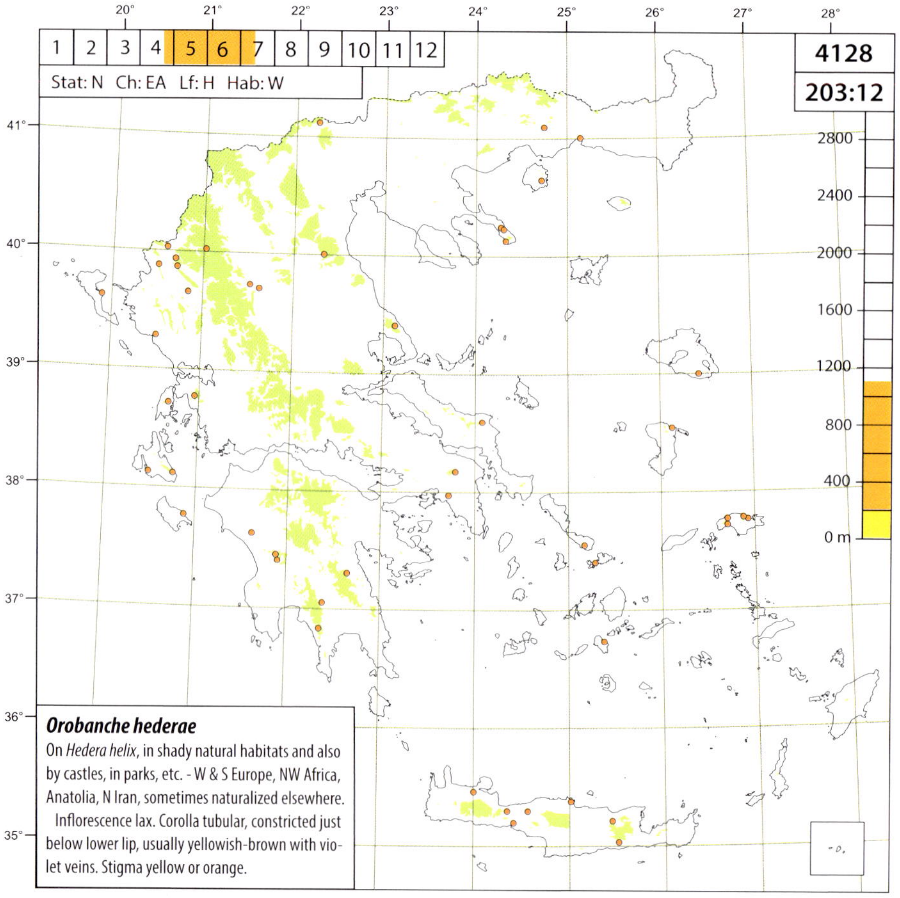

Orobanche hederae
On *Hedera helix*, in shady natural habitats and also by castles, in parks, etc. - W & S Europe, NW Africa, Anatolia, N Iran, sometimes naturalized elsewhere.
Inflorescence lax. Corolla tubular, constricted just below lower lip, usually yellowish-brown with violet veins. Stigma yellow or orange.

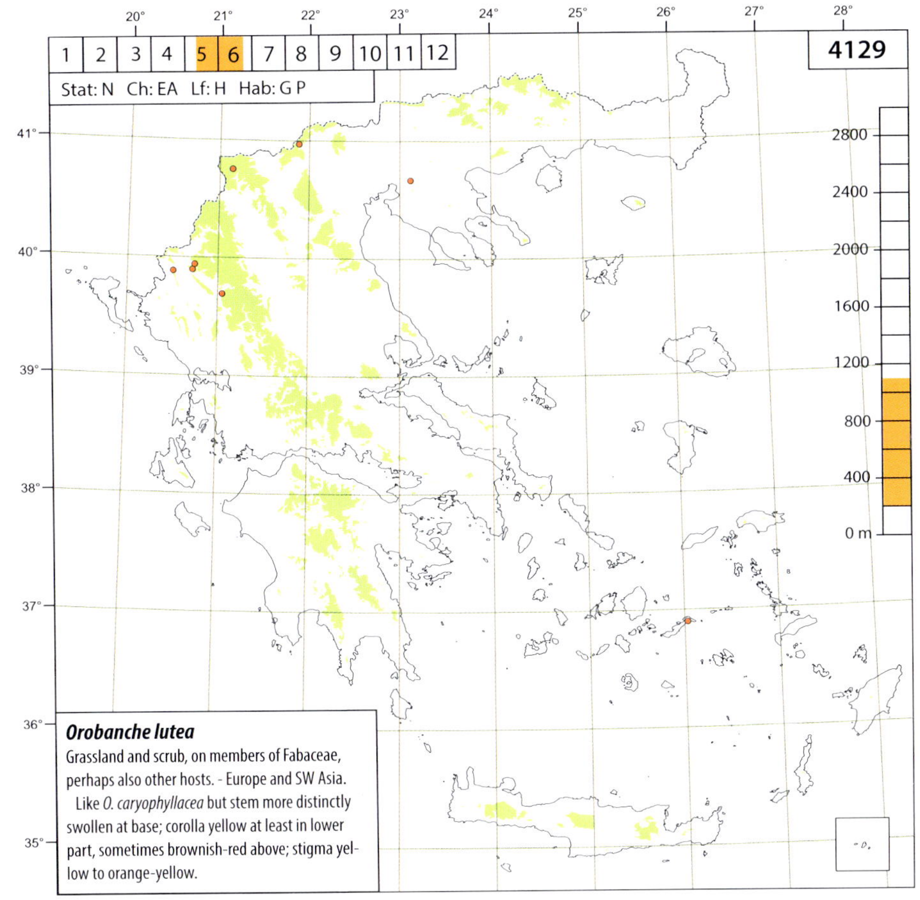

Orobanche lutea

Grassland and scrub, on members of Fabaceae, perhaps also other hosts. - Europe and SW Asia.

Like *O. caryophyllacea* but stem more distinctly swollen at base; corolla yellow at least in lower part, sometimes brownish-red above; stigma yellow to orange-yellow.

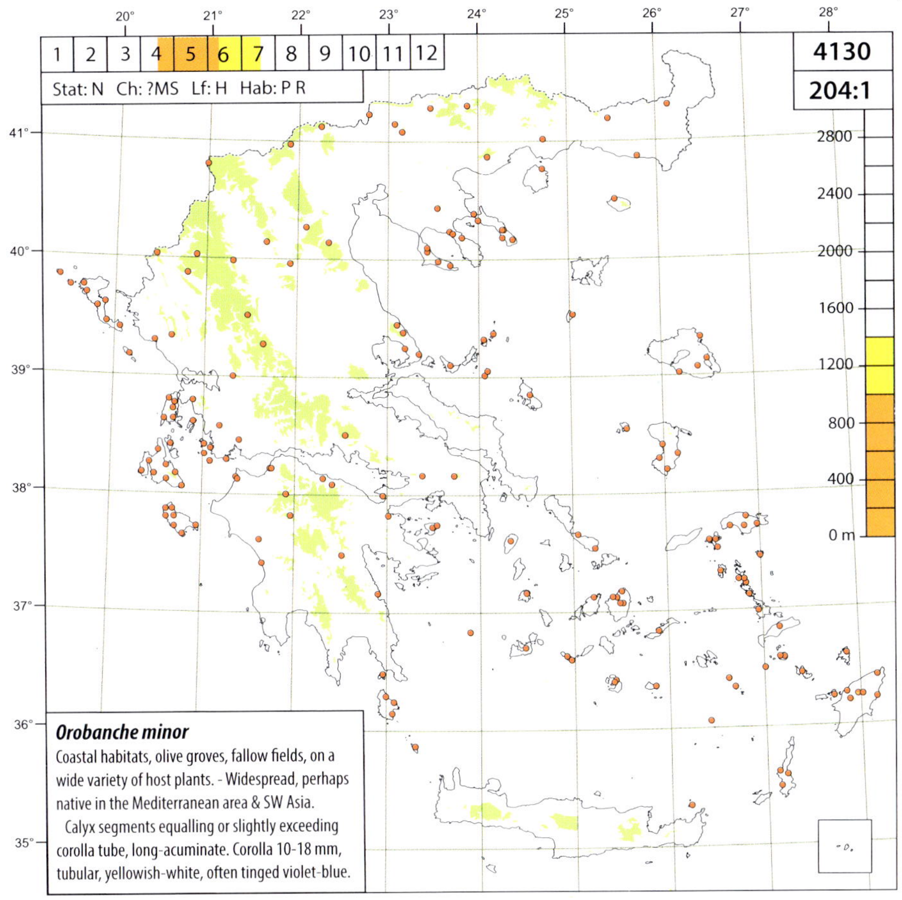

Orobanche minor

Coastal habitats, olive groves, fallow fields, on a wide variety of host plants. - Widespread, perhaps native in the Mediterranean area & SW Asia.

Calyx segments equalling or slightly exceeding corolla tube, long-acuminate. Corolla 10-18 mm, tubular, yellowish-white, often tinged violet-blue.

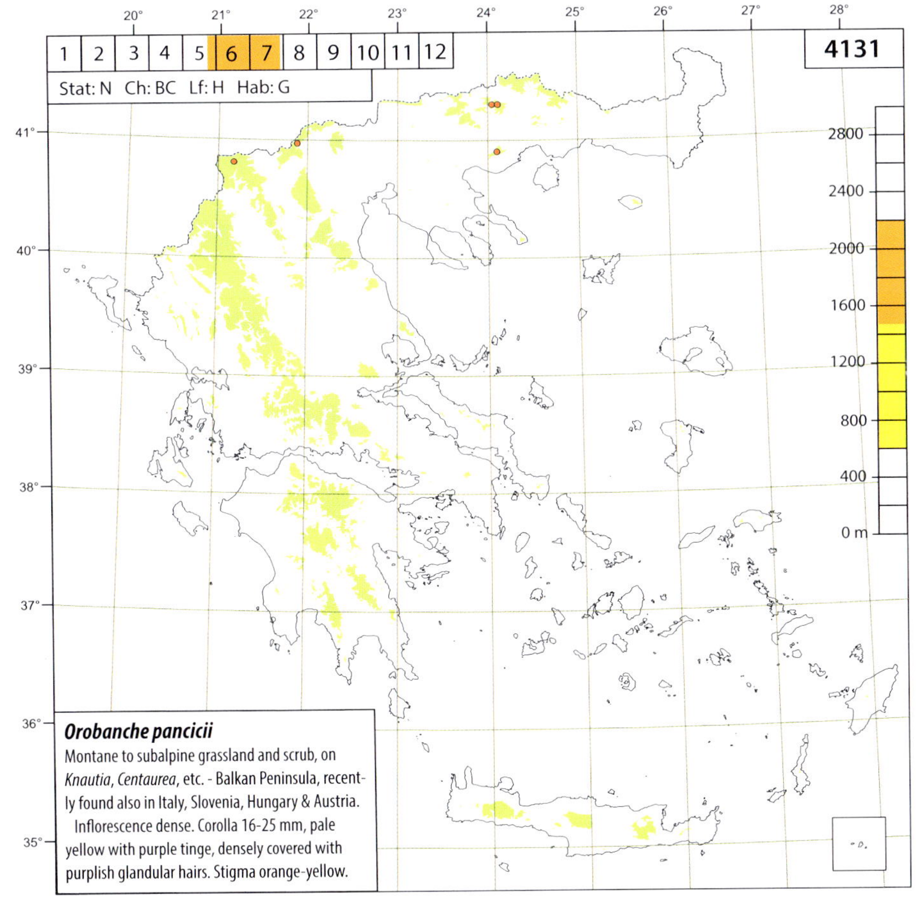

Orobanche pancicii

Montane to subalpine grassland and scrub, on *Knautia*, *Centaurea*, etc. - Balkan Peninsula, recently found also in Italy, Slovenia, Hungary & Austria.

Inflorescence dense. Corolla 16-25 mm, pale yellow with purple tinge, densely covered with purplish glandular hairs. Stigma orange-yellow.

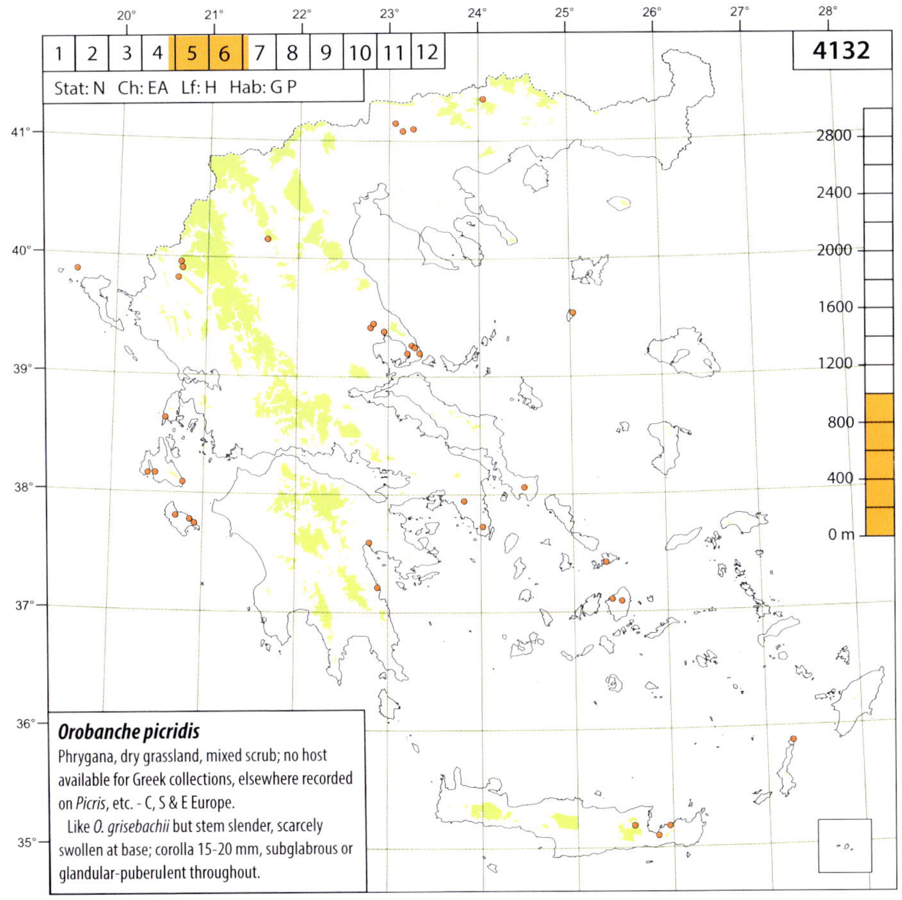

Orobanche picridis

Phrygana, dry grassland, mixed scrub; no host available for Greek collections, elsewhere recorded on *Picris*, etc. - C, S & E Europe.

Like *O. grisebachii* but stem slender, scarcely swollen at base; corolla 15-20 mm, subglabrous or glandular-puberulent throughout.

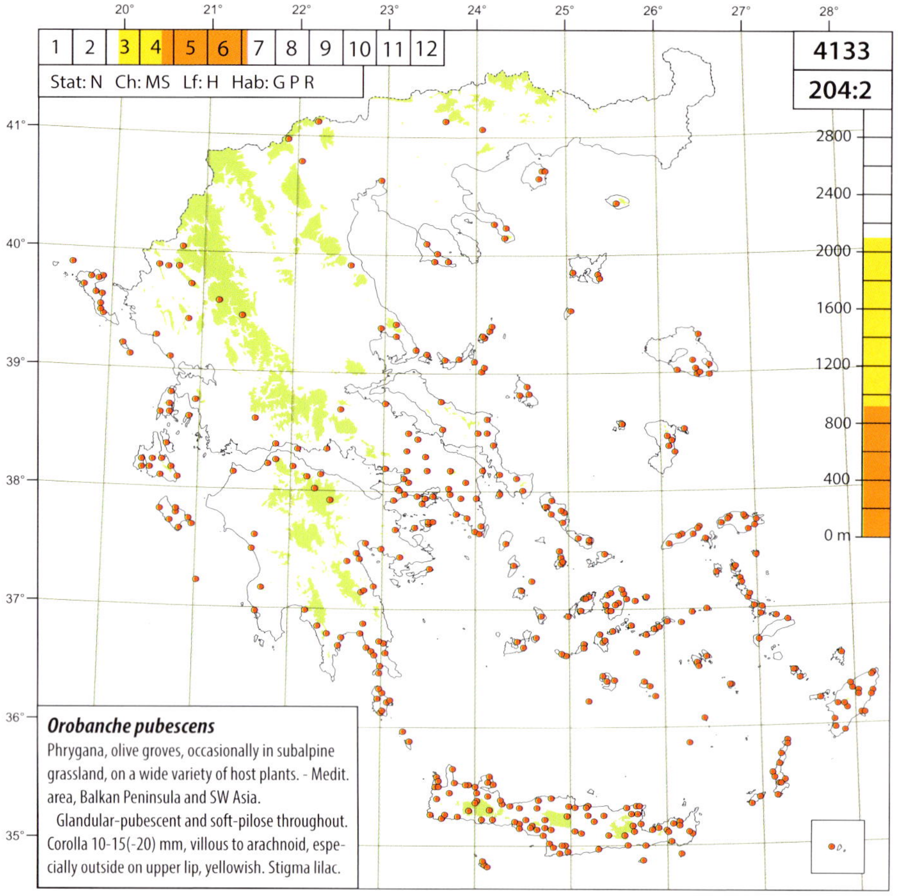

Orobanche pubescens
Phrygana, olive groves, occasionally in subalpine grassland, on a wide variety of host plants. - Medit. area, Balkan Peninsula and SW Asia.
Glandular-pubescent and soft-pilose throughout. Corolla 10-15(-20) mm, villous to arachnoid, especially outside on upper lip, yellowish. Stigma lilac.

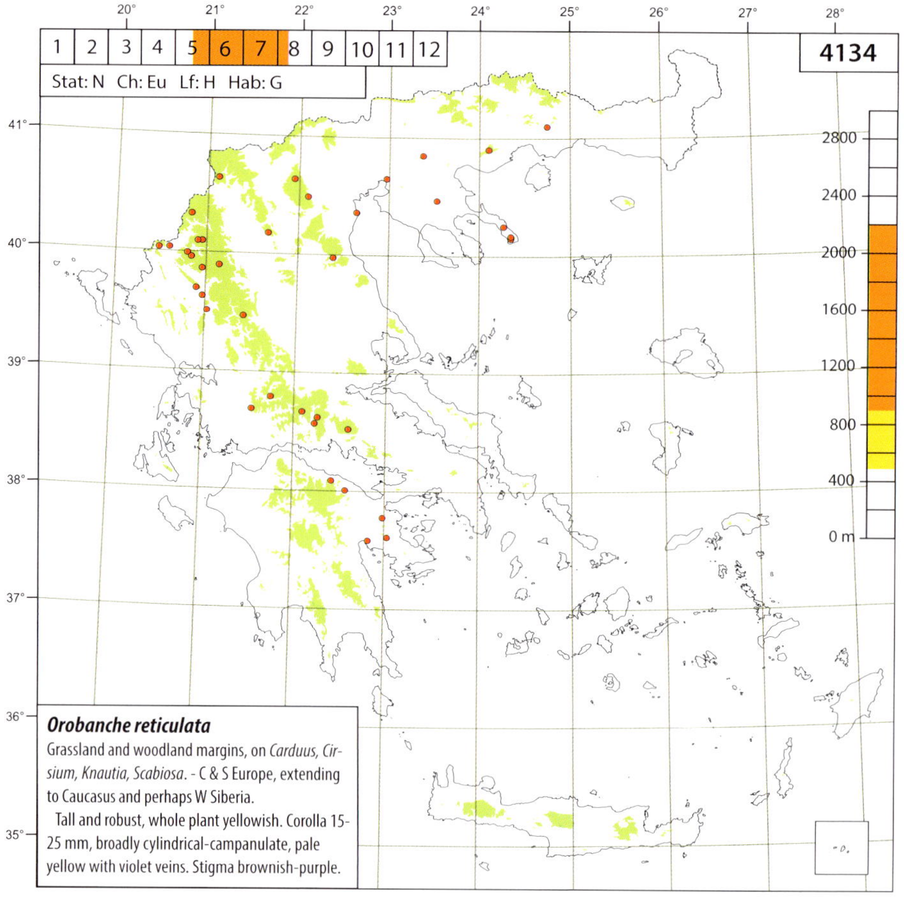

Orobanche reticulata
Grassland and woodland margins, on *Carduus*, *Cirsium*, *Knautia*, *Scabiosa*. - C & S Europe, extending to Caucasus and perhaps W Siberia.
Tall and robust, whole plant yellowish. Corolla 15-25 mm, broadly cylindrical-campanulate, pale yellow with violet veins. Stigma brownish-purple.

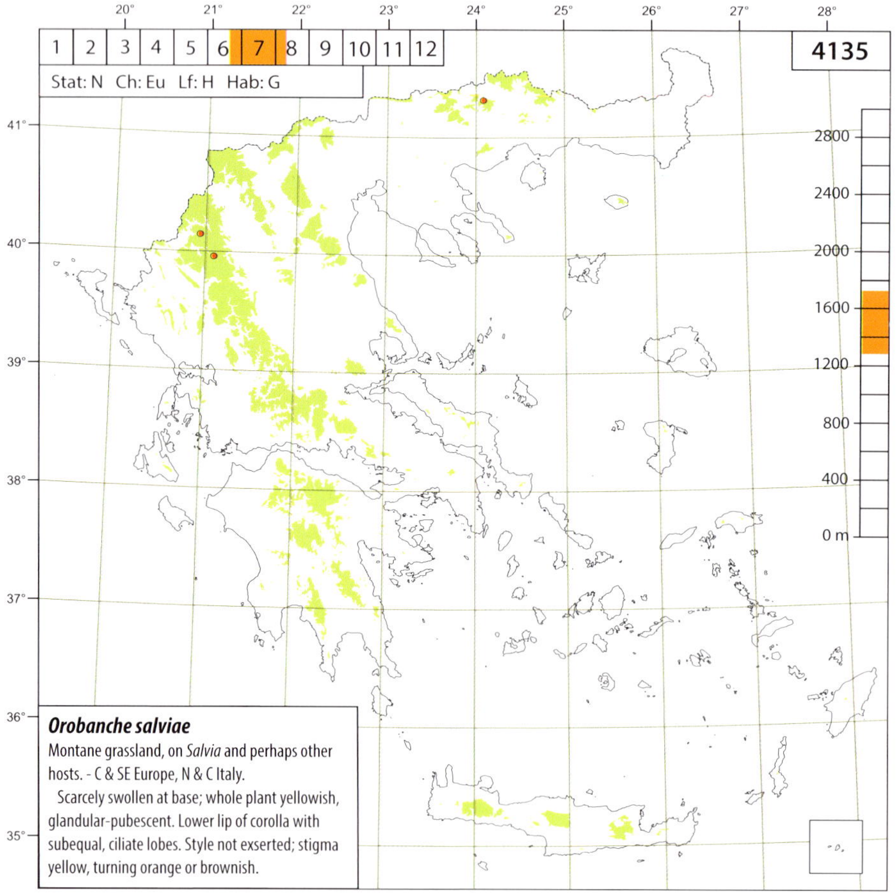

Orobanche salviae
Montane grassland, on *Salvia* and perhaps other hosts. - C & SE Europe, N & C Italy.
Scarcely swollen at base; whole plant yellowish, glandular-pubescent. Lower lip of corolla with subequal, ciliate lobes. Style not exserted; stigma yellow, turning orange or brownish.

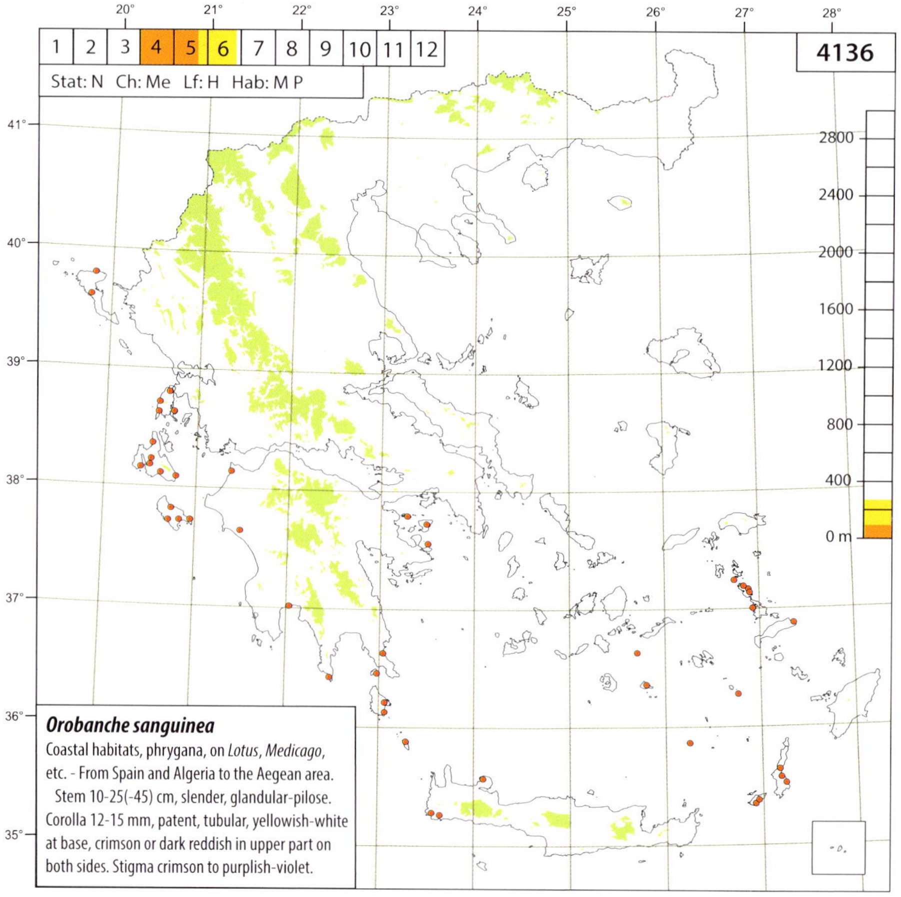

Orobanche sanguinea
Coastal habitats, phrygana, on *Lotus*, *Medicago*, etc. - From Spain and Algeria to the Aegean area.
Stem 10-25(-45) cm, slender, glandular-pilose. Corolla 12-15 mm, patent, tubular, yellowish-white at base, crimson or dark reddish in upper part on both sides. Stigma crimson to purplish-violet.

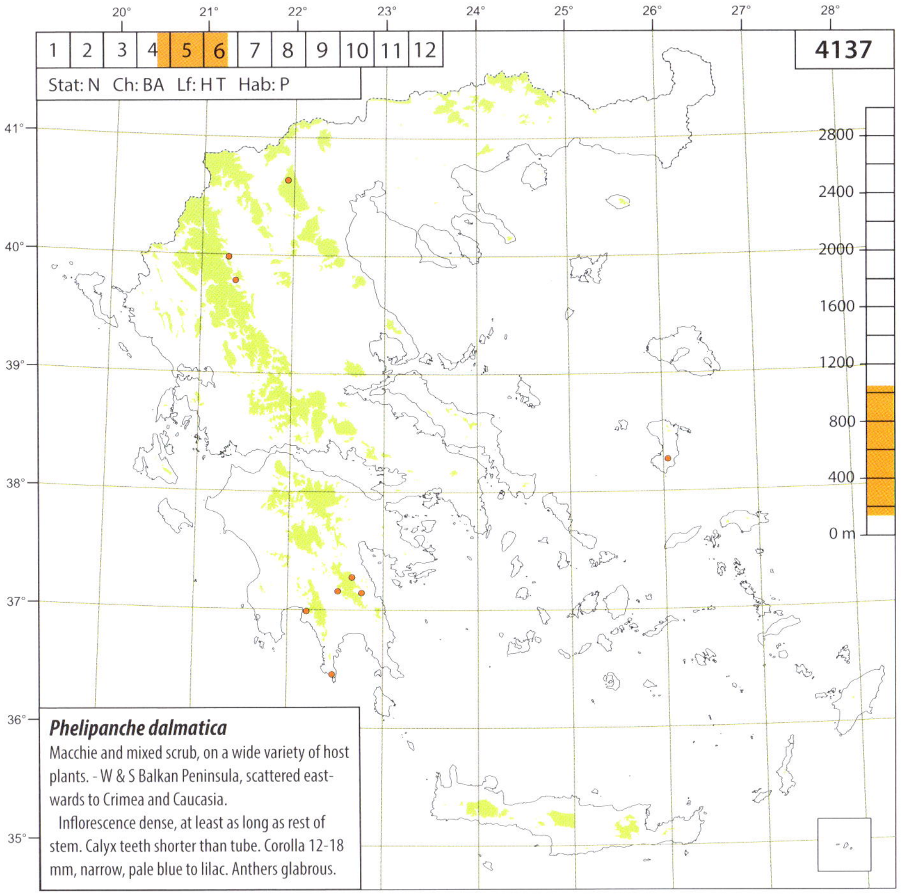

Phelipanche dalmatica

Macchie and mixed scrub, on a wide variety of host plants. - W & S Balkan Peninsula, scattered eastwards to Crimea and Caucasia.

Inflorescence dense, at least as long as rest of stem. Calyx teeth shorter than tube. Corolla 12-18 mm, narrow, pale blue to lilac. Anthers glabrous.

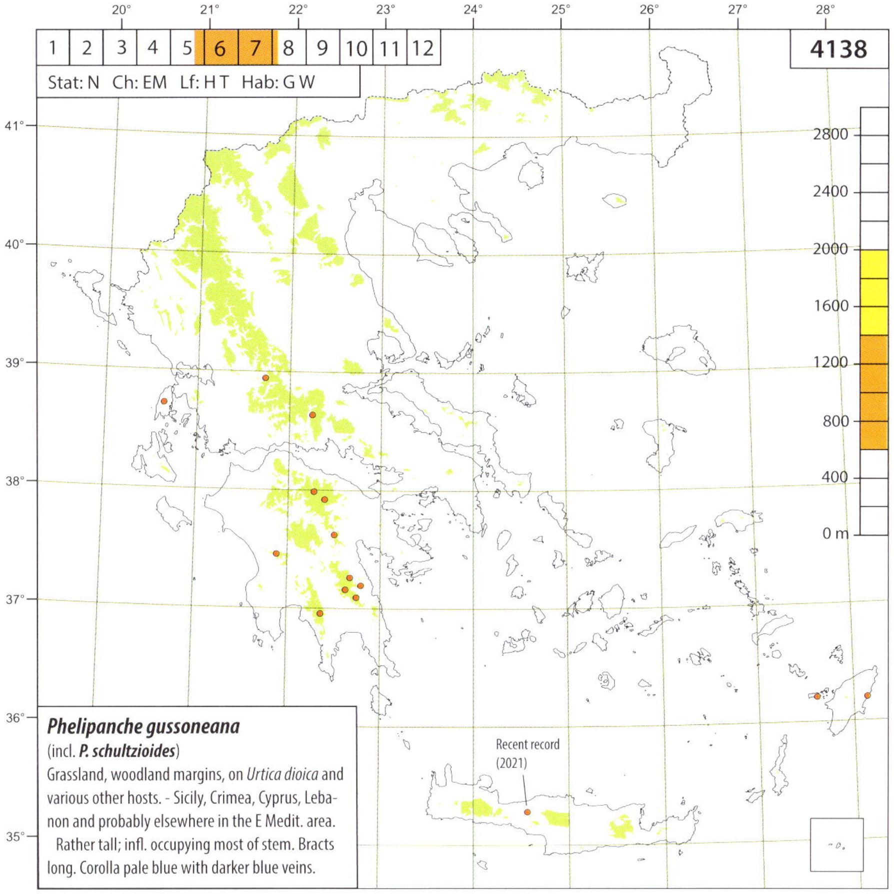

Phelipanche gussoneana

(incl. ***P. schultzioides***)

Grassland, woodland margins, on *Urtica dioica* and various other hosts. - Sicily, Crimea, Cyprus, Lebanon and probably elsewhere in the E Medit. area.

Rather tall; infl. occupying most of stem. Bracts long. Corolla pale blue with darker blue veins.

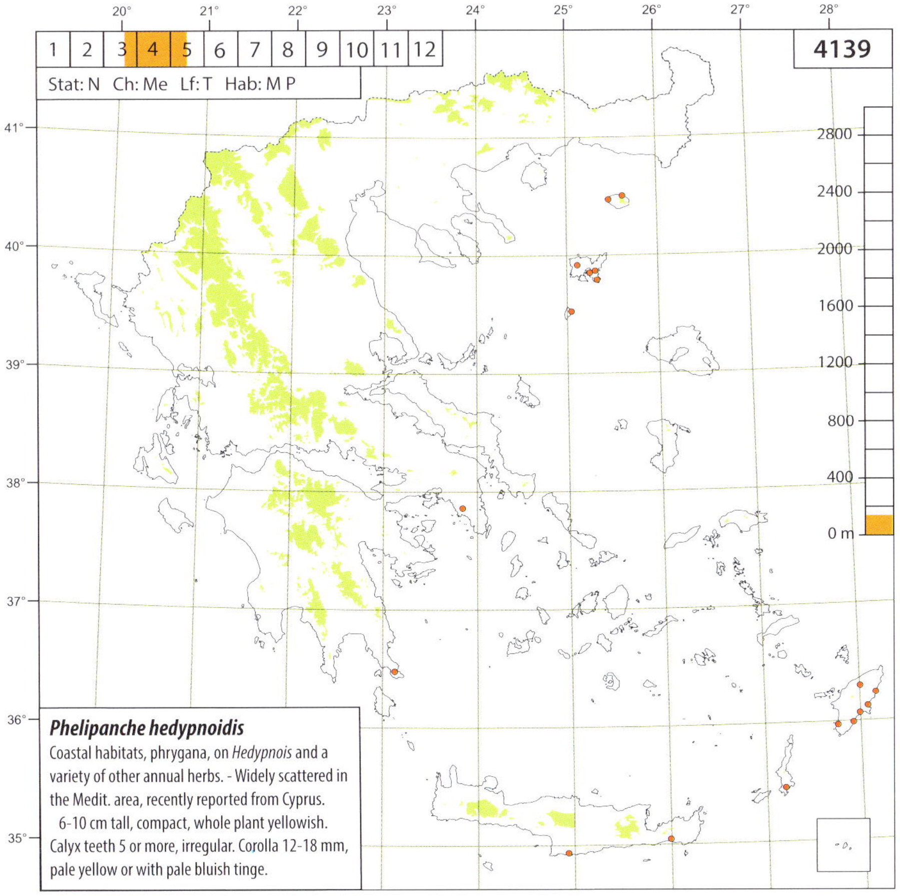

Phelipanche hedypnoidis

Coastal habitats, phrygana, on *Hedypnois* and a variety of other annual herbs. - Widely scattered in the Medit. area, recently reported from Cyprus.

6-10 cm tall, compact, whole plant yellowish. Calyx teeth 5 or more, irregular. Corolla 12-18 mm, pale yellow or with pale bluish tinge.

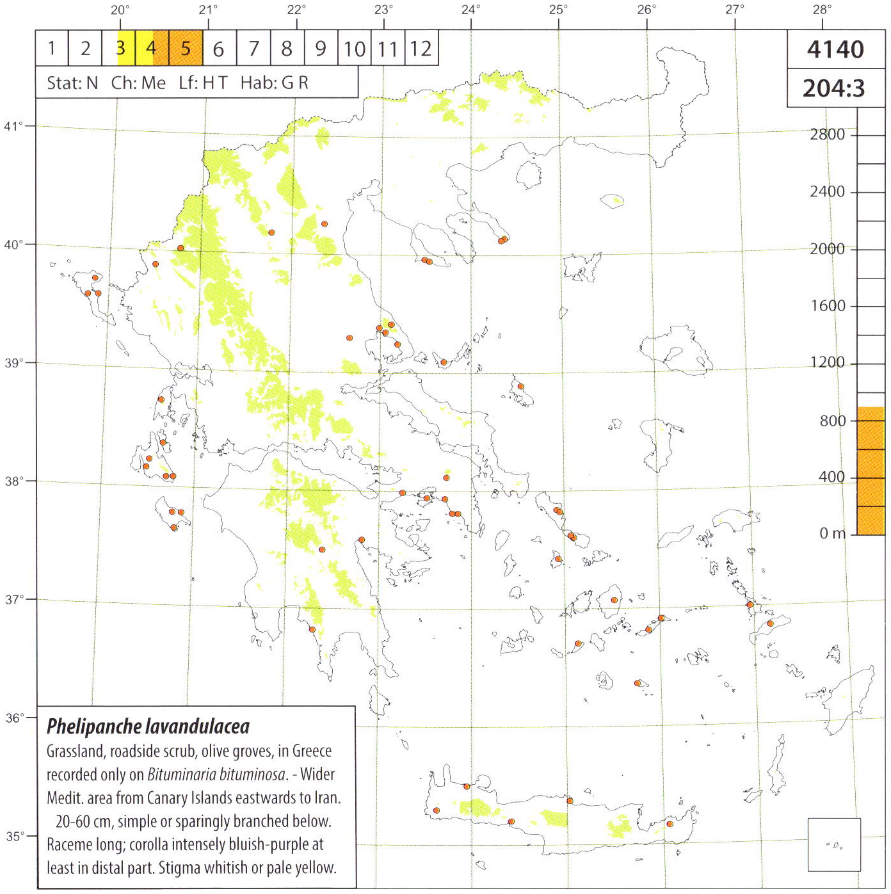

Phelipanche lavandulacea

Grassland, roadside scrub, olive groves, in Greece recorded only on *Bituminaria bituminosa*. - Wider Medit. area from Canary Islands eastwards to Iran.

20-60 cm, simple or sparingly branched below. Raceme long; corolla intensely bluish-purple at least in distal part. Stigma whitish or pale yellow.

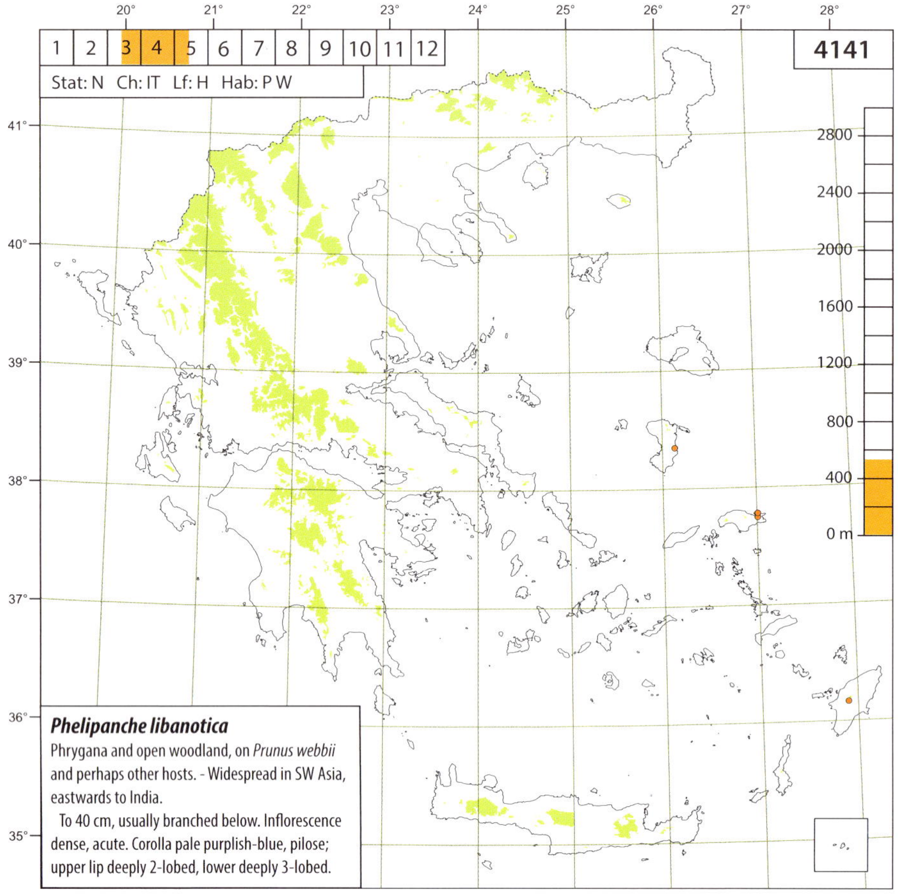

Phelipanche libanotica
Phrygana and open woodland, on *Prunus webbii* and perhaps other hosts. - Widespread in SW Asia, eastwards to India.
To 40 cm, usually branched below. Inflorescence dense, acute. Corolla pale purplish-blue, pilose; upper lip deeply 2-lobed, lower deeply 3-lobed.

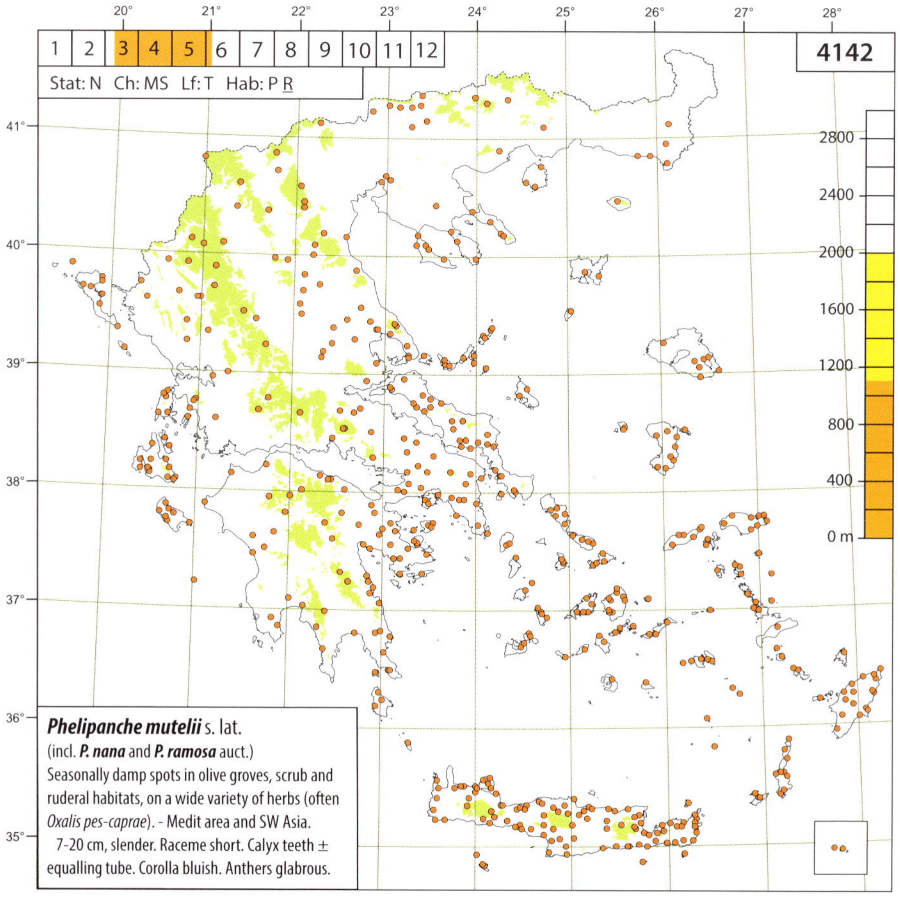

Phelipanche mutelii s. lat.
(incl. ***P. nana*** and ***P. ramosa*** auct.)
Seasonally damp spots in olive groves, scrub and ruderal habitats, on a wide variety of herbs (often *Oxalis pes-caprae*). - Medit area and SW Asia.
7-20 cm, slender. Raceme short. Calyx teeth ± equalling tube. Corolla bluish. Anthers glabrous.

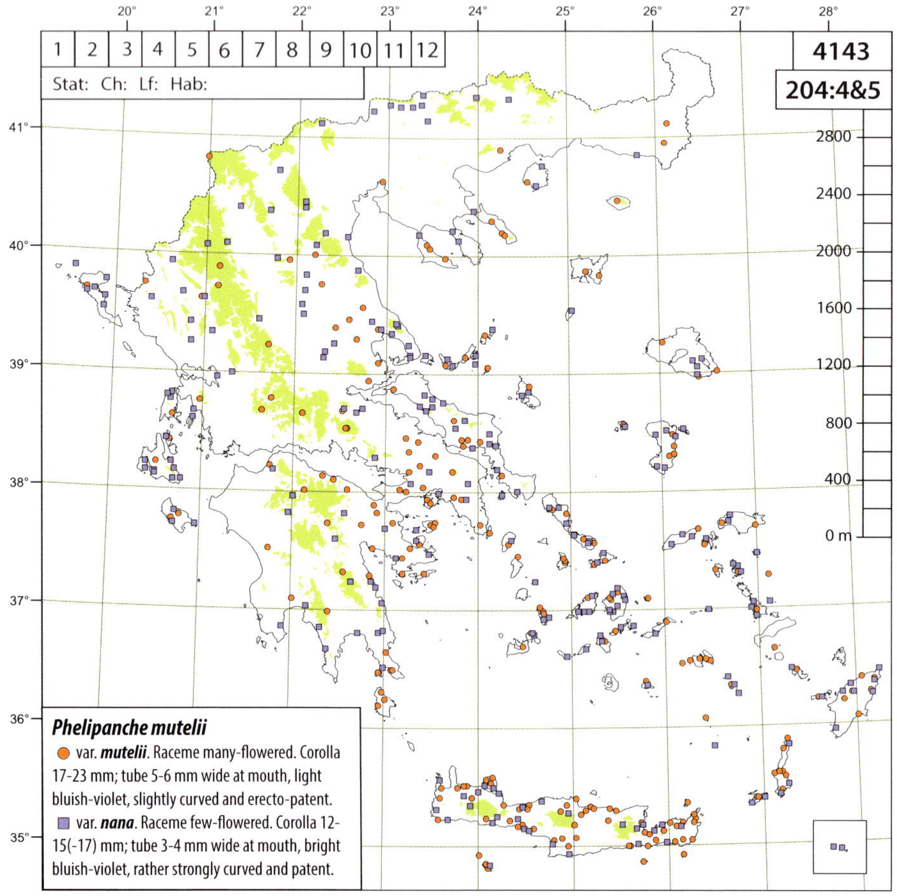

Phelipanche mutelii
● var. ***mutelii***. Raceme many-flowered. Corolla 17-23 mm; tube 5-6 mm wide at mouth, light bluish-violet, slightly curved and erecto-patent.
■ var. ***nana***. Raceme few-flowered. Corolla 12-15(-17) mm; tube 3-4 mm wide at mouth, bright bluish-violet, rather strongly curved and patent.

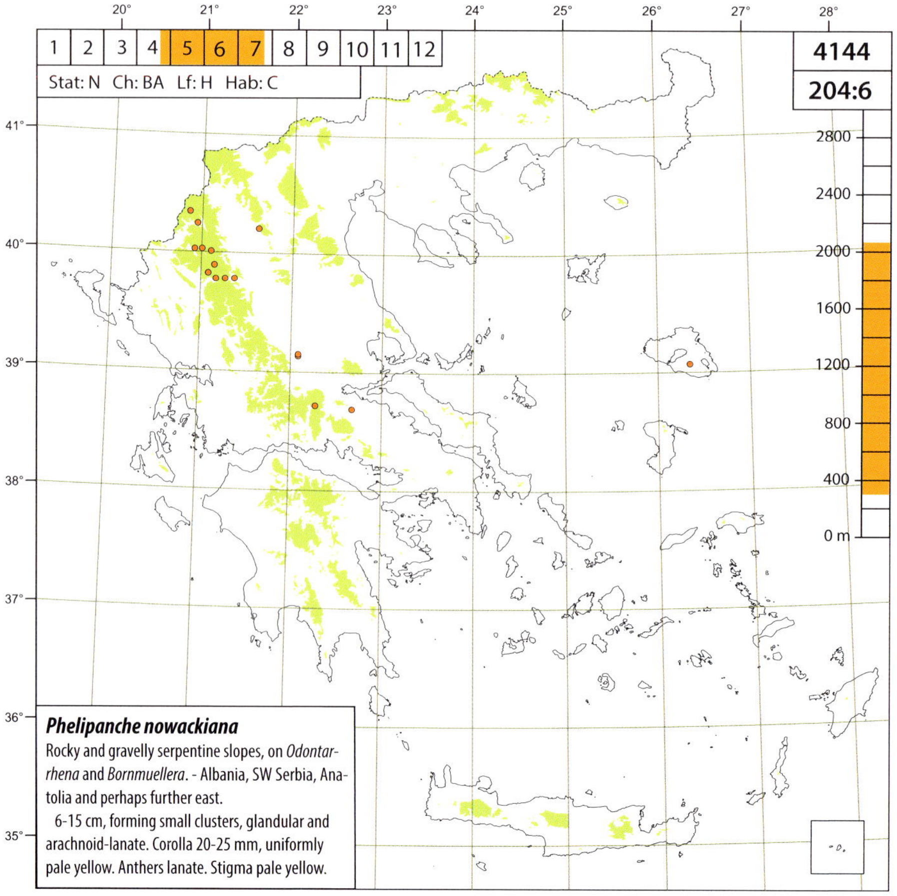

Phelipanche nowackiana
Rocky and gravelly serpentine slopes, on *Odontarrhena* and *Bornmuellera*. - Albania, SW Serbia, Anatolia and perhaps further east.
6-15 cm, forming small clusters, glandular and arachnoid-lanate. Corolla 20-25 mm, uniformly pale yellow. Anthers lanate. Stigma pale yellow.

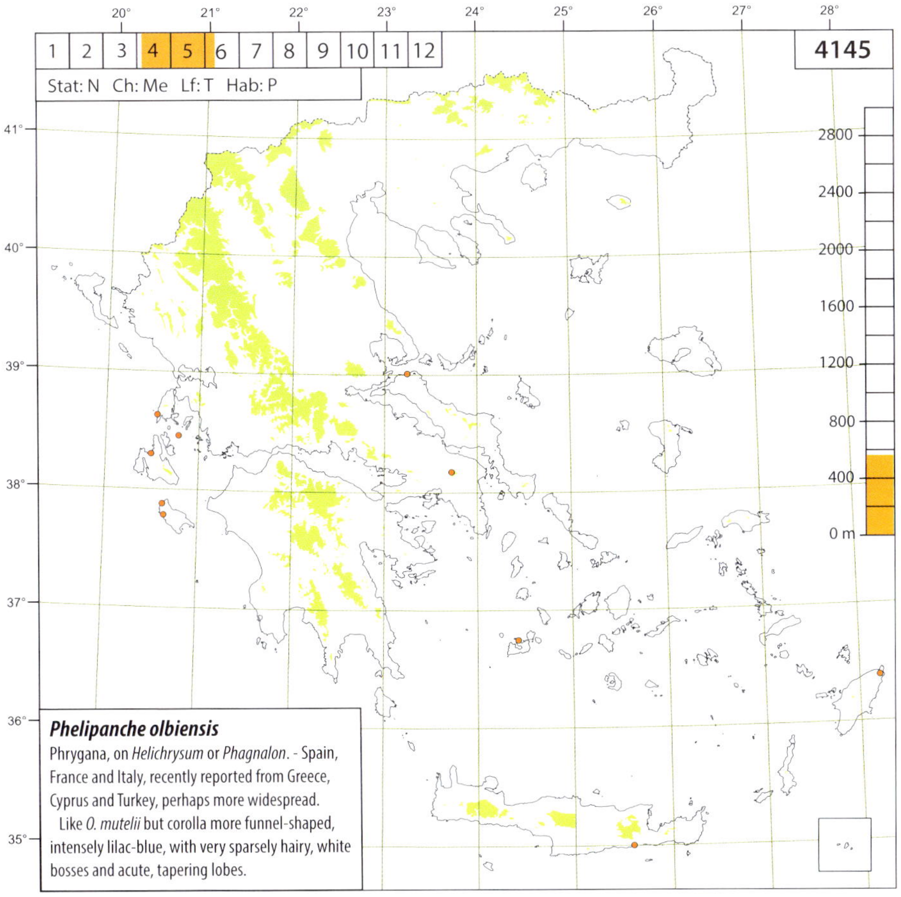

Phelipanche olbiensis

Phrygana, on *Helichrysum* or *Phagnalon*. - Spain, France and Italy, recently reported from Greece, Cyprus and Turkey, perhaps more widespread.

Like *O. mutelii* but corolla more funnel-shaped, intensely lilac-blue, with very sparsely hairy, white bosses and acute, tapering lobes.

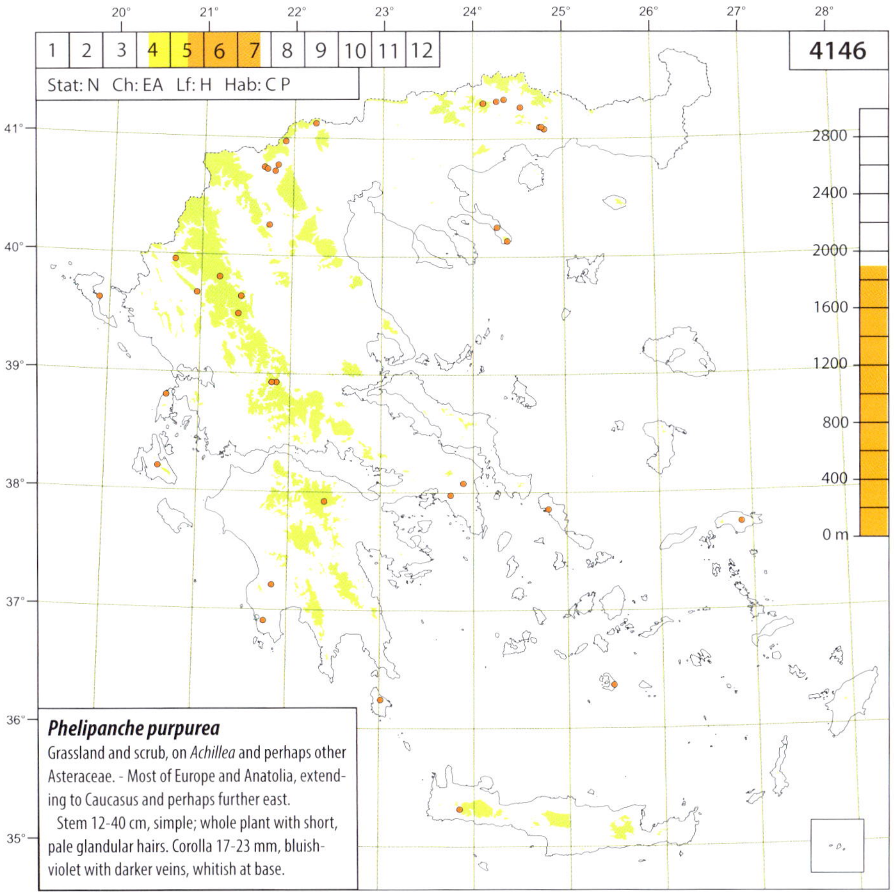

Phelipanche purpurea

Grassland and scrub, on *Achillea* and perhaps other Asteraceae. - Most of Europe and Anatolia, extending to Caucasus and perhaps further east.

Stem 12-40 cm, simple; whole plant with short, pale glandular hairs. Corolla 17-23 mm, bluish-violet with darker veins, whitish at base.

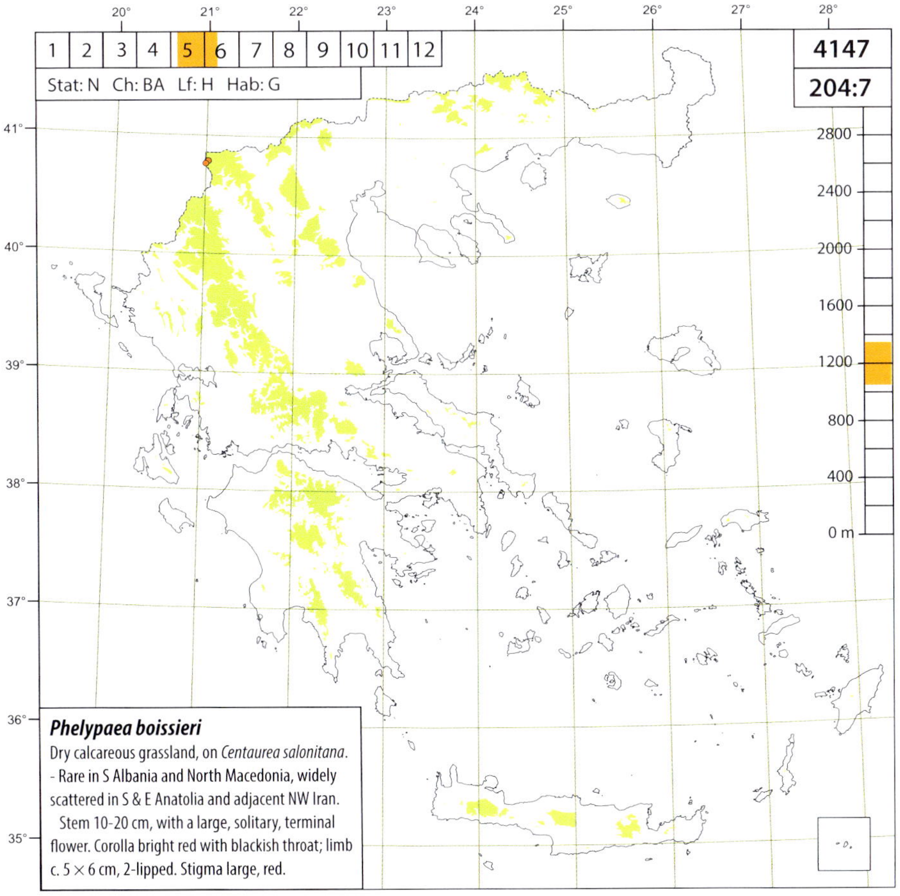

Phelypaea boissieri

Dry calcareous grassland, on *Centaurea salonitana*. - Rare in S Albania and North Macedonia, widely scattered in S & E Anatolia and adjacent NW Iran.

Stem 10-20 cm, with a large, solitary, terminal flower. Corolla bright red with blackish throat; limb c. 5 × 6 cm, 2-lipped. Stigma large, red.

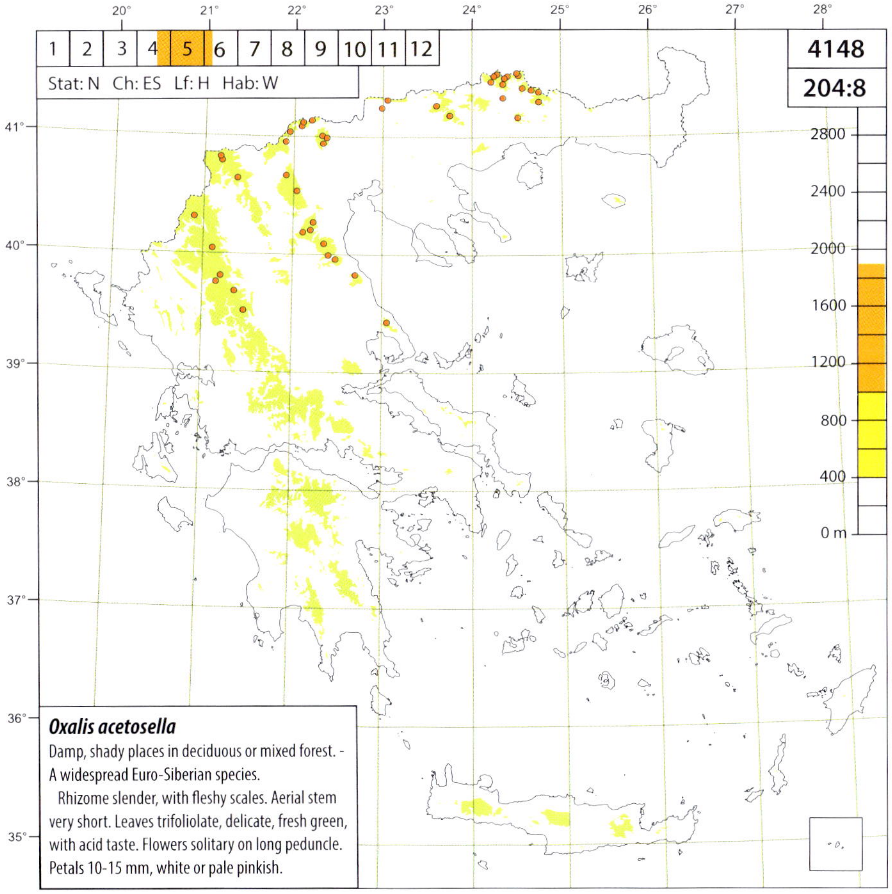

Oxalis acetosella

Damp, shady places in deciduous or mixed forest. - A widespread Euro-Siberian species.

Rhizome slender, with fleshy scales. Aerial stem very short. Leaves trifoliolate, delicate, fresh green, with acid taste. Flowers solitary on long peduncle. Petals 10-15 mm, white or pale pinkish.

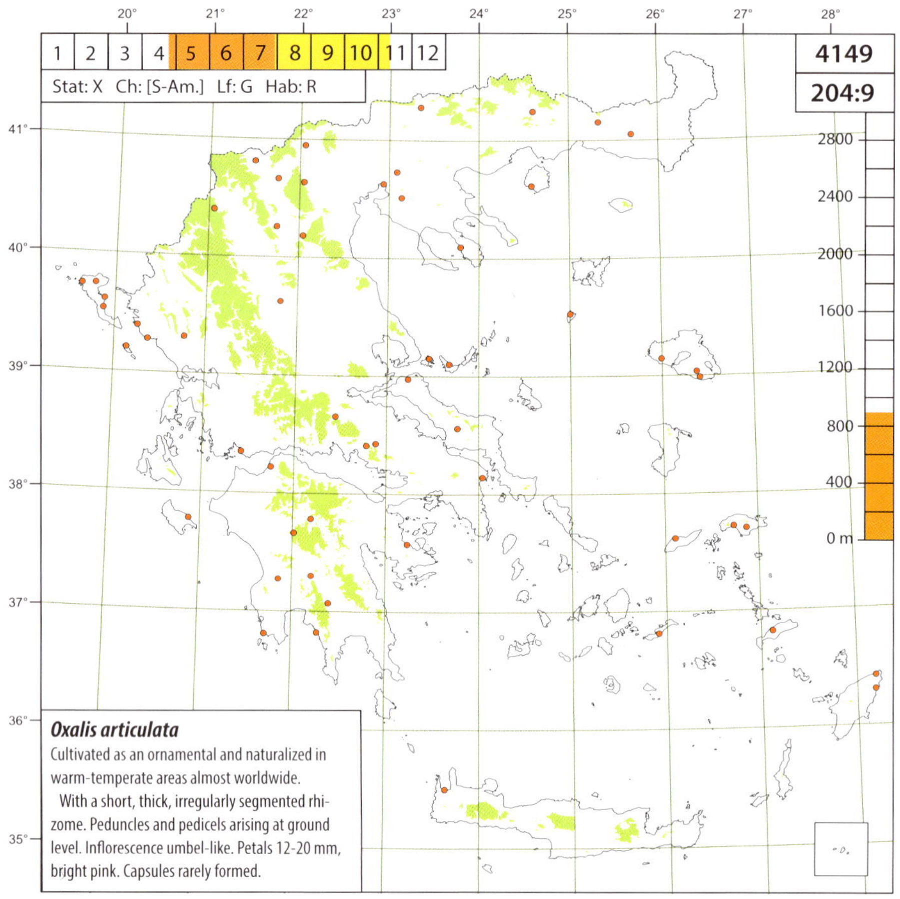

Oxalis articulata

Cultivated as an ornamental and naturalized in warm-temperate areas almost worldwide.

With a short, thick, irregularly segmented rhizome. Peduncles and pedicels arising at ground level. Inflorescence umbel-like. Petals 12-20 mm, bright pink. Capsules rarely formed.

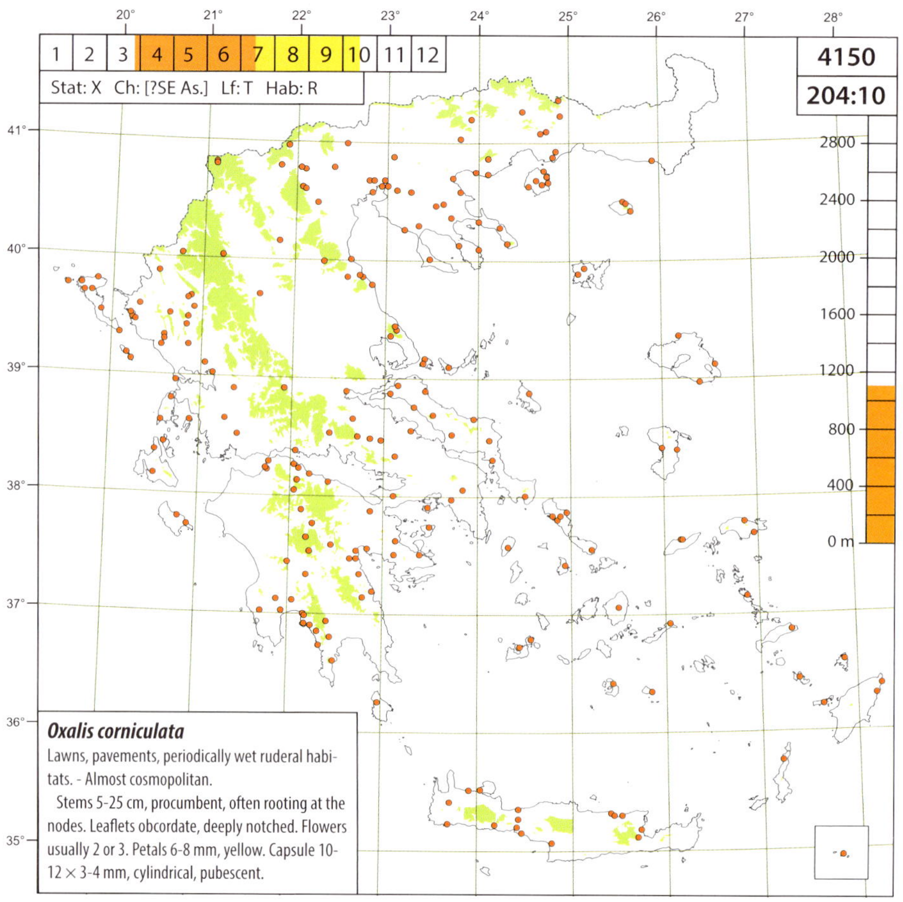

Oxalis corniculata

Lawns, pavements, periodically wet ruderal habitats. - Almost cosmopolitan.

Stems 5-25 cm, procumbent, often rooting at the nodes. Leaflets obcordate, deeply notched. Flowers usually 2 or 3. Petals 6-8 mm, yellow. Capsule 10-12 × 3-4 mm, cylindrical, pubescent.

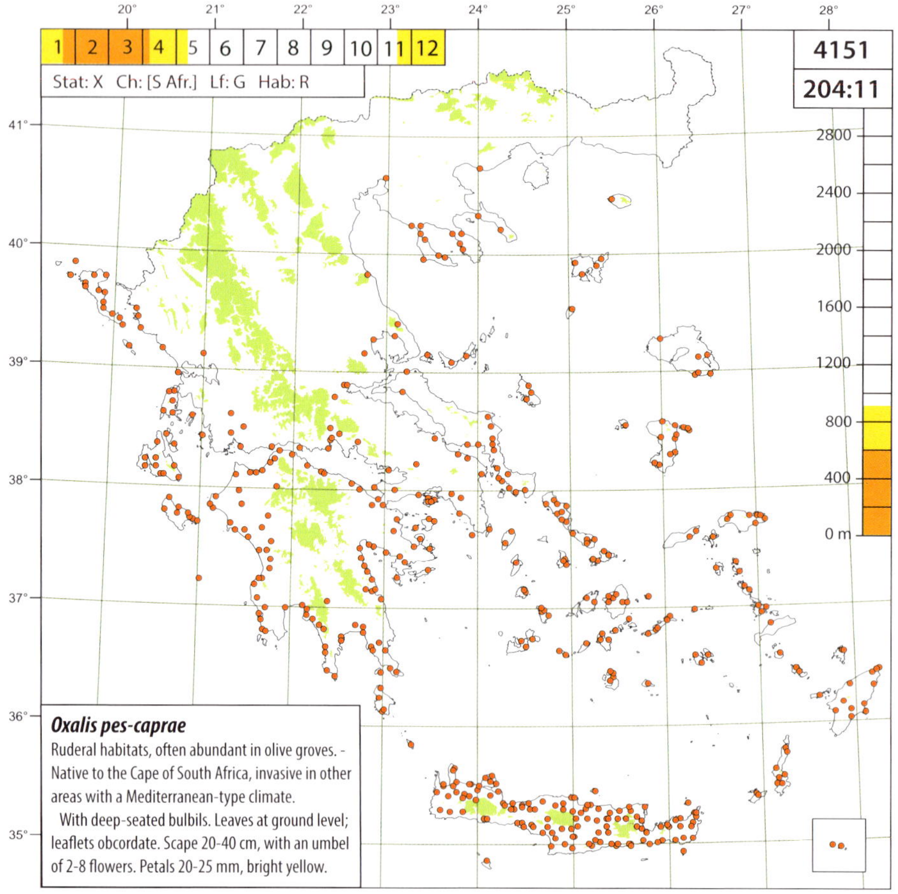

Oxalis pes-caprae

Ruderal habitats, often abundant in olive groves. - Native to the Cape of South Africa, invasive in other areas with a Mediterranean-type climate.

With deep-seated bulbils. Leaves at ground level; leaflets obcordate. Scape 20-40 cm, with an umbel of 2-8 flowers. Petals 20-25 mm, bright yellow.

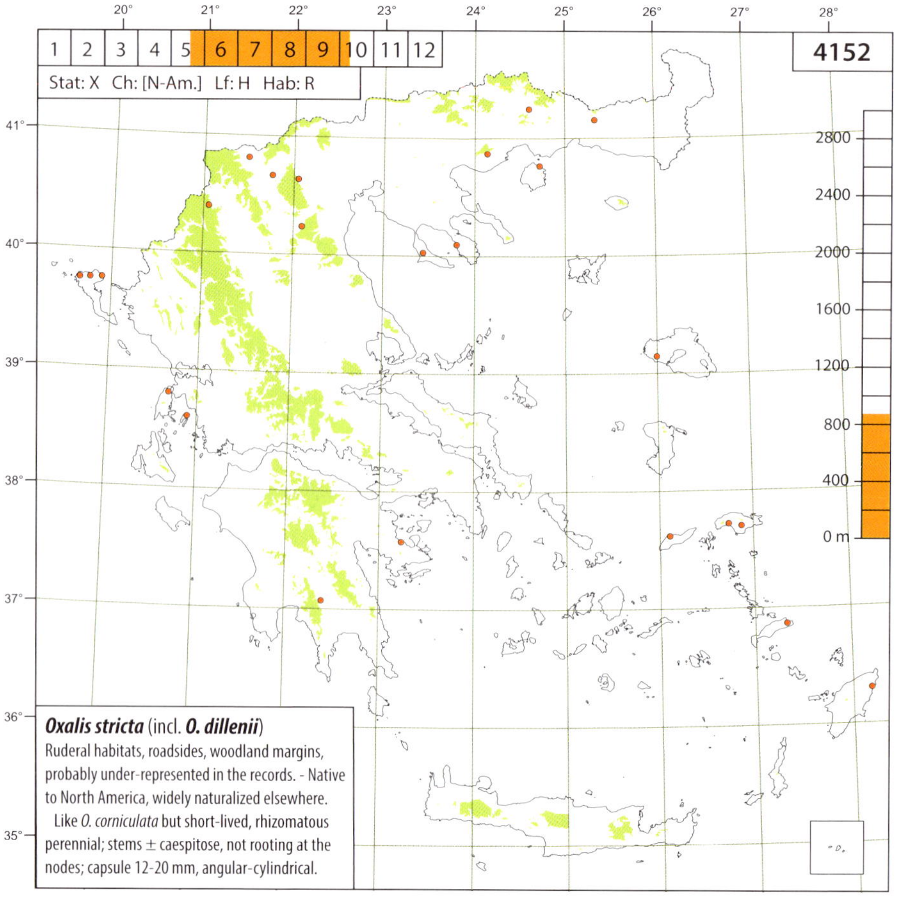

Oxalis stricta (incl. *O. dillenii*)

Ruderal habitats, roadsides, woodland margins, probably under-represented in the records. - Native to North America, widely naturalized elsewhere.

Like *O. corniculata* but short-lived, rhizomatous perennial; stems ± caespitose, not rooting at the nodes; capsule 12-20 mm, angular-cylindrical.

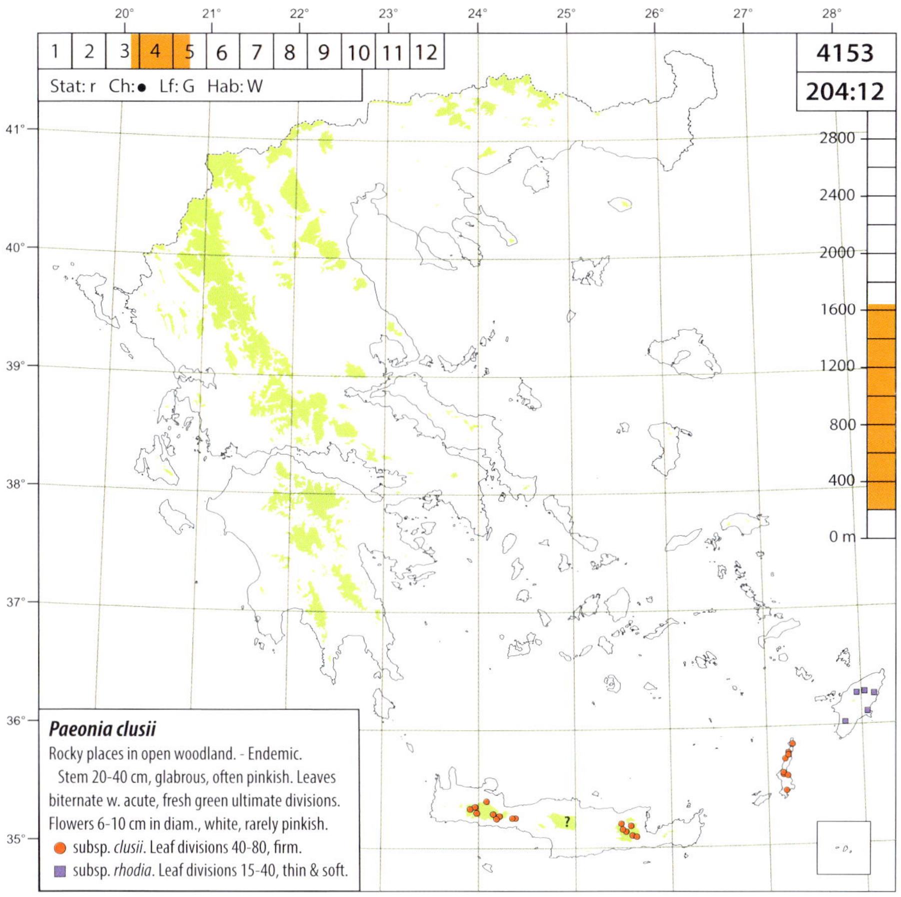

Paeonia clusii
Rocky places in open woodland. - Endemic.
Stem 20-40 cm, glabrous, often pinkish. Leaves biternate w. acute, fresh green ultimate divisions. Flowers 6-10 cm in diam., white, rarely pinkish.
● subsp. *clusii*. Leaf divisions 40-80, firm.
■ subsp. *rhodia*. Leaf divisions 15-40, thin & soft.

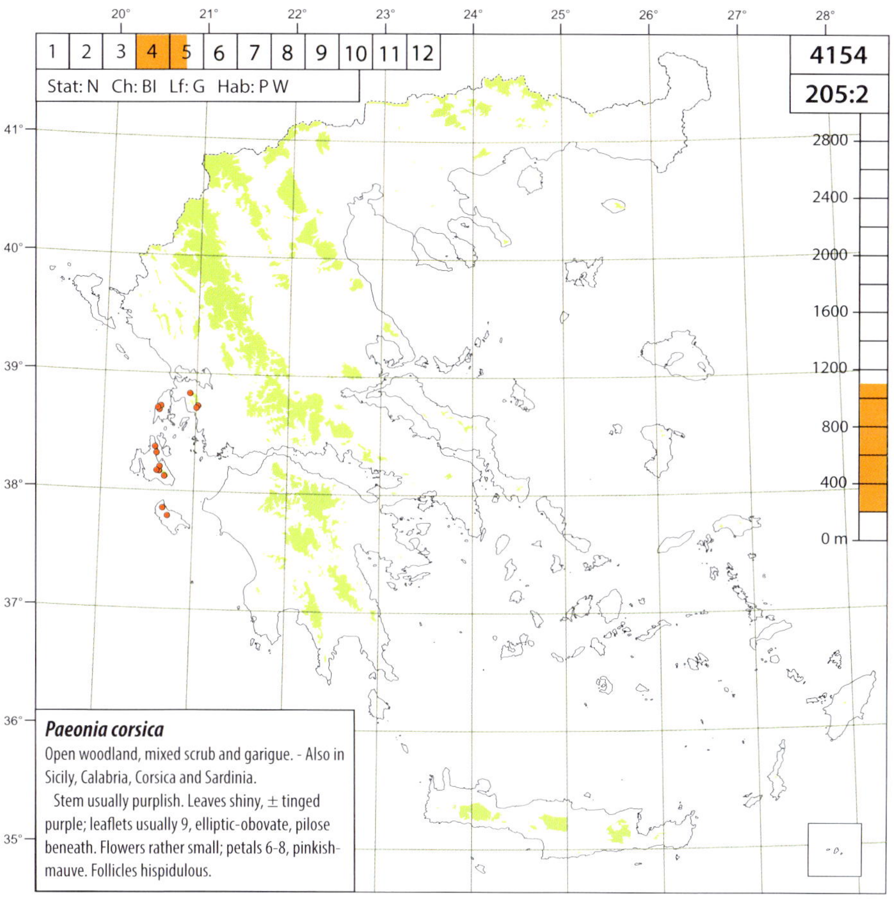

Paeonia corsica
Open woodland, mixed scrub and garigue. - Also in Sicily, Calabria, Corsica and Sardinia.
Stem usually purplish. Leaves shiny, ± tinged purple; leaflets usually 9, elliptic-obovate, pilose beneath. Flowers rather small; petals 6-8, pinkish-mauve. Follicles hispidulous.

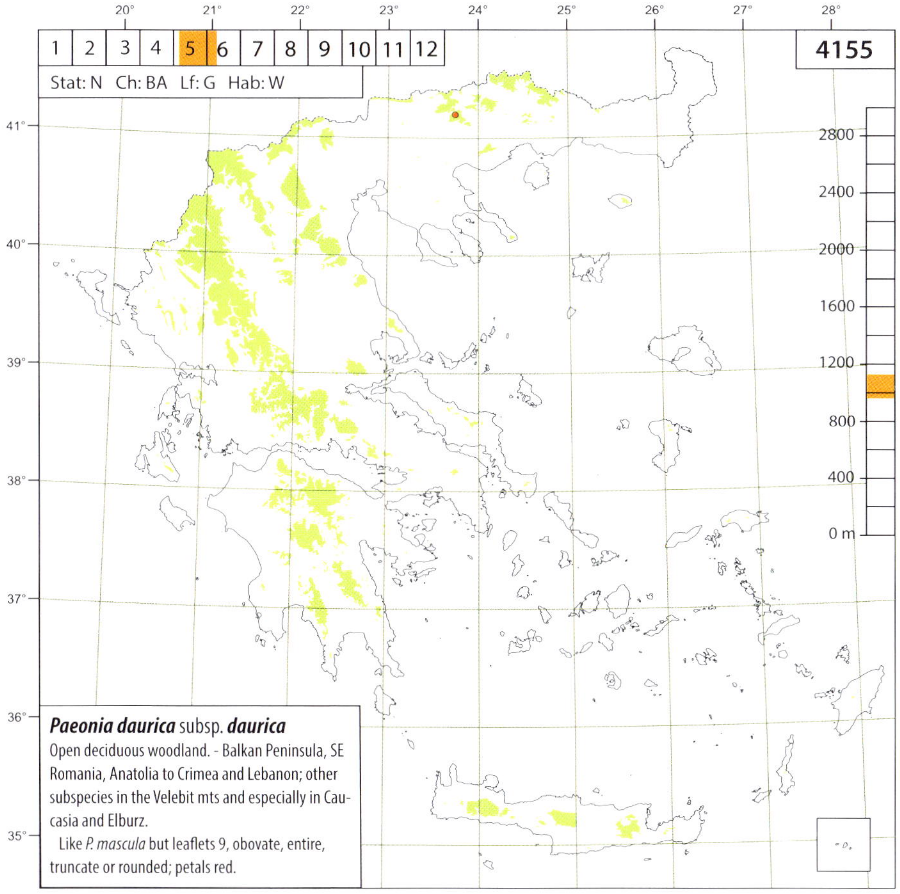

Paeonia daurica* subsp. *daurica
Open deciduous woodland. - Balkan Peninsula, SE Romania, Anatolia to Crimea and Lebanon; other subspecies in the Velebit mts and especially in Caucasia and Elburz.
Like *P. mascula* but leaflets 9, obovate, entire, truncate or rounded; petals red.

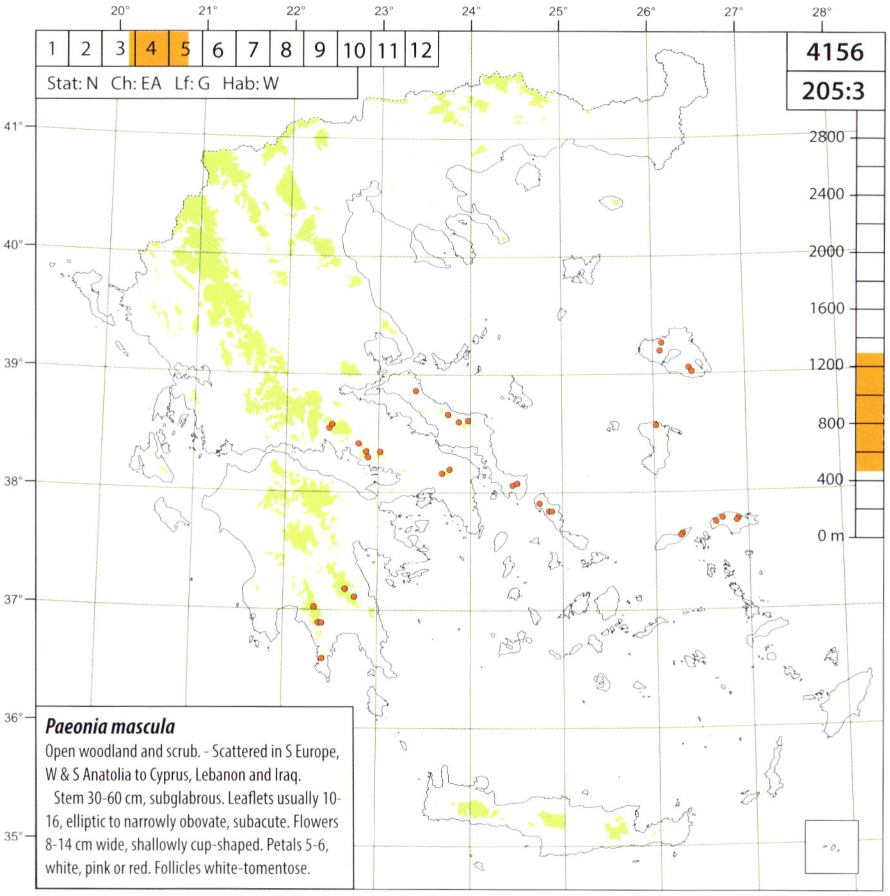

Paeonia mascula
Open woodland and scrub. - Scattered in S Europe, W & S Anatolia to Cyprus, Lebanon and Iraq.
Stem 30-60 cm, subglabrous. Leaflets usually 10-16, elliptic to narrowly obovate, subacute. Flowers 8-14 cm wide, shallowly cup-shaped. Petals 5-6, white, pink or red. Follicles white-tomentose.

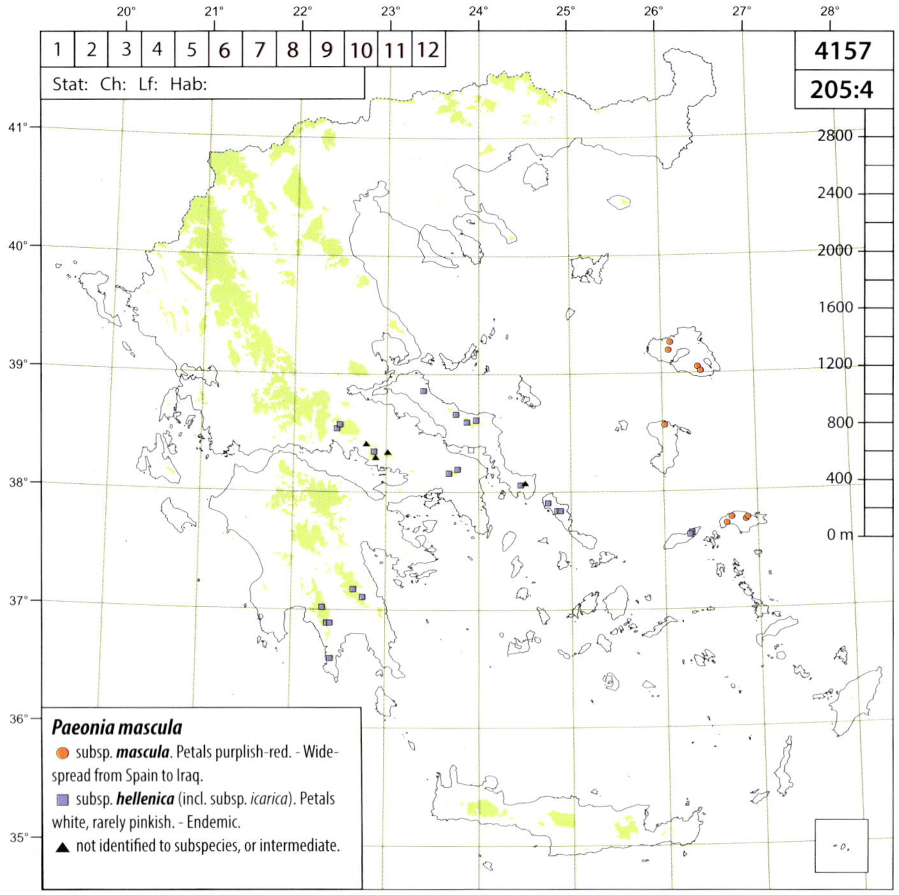

Paeonia mascula

● subsp. ***mascula***. Petals purplish-red. - Widespread from Spain to Iraq.

■ subsp. ***hellenica*** (incl. subsp. *icarica*). Petals white, rarely pinkish. - Endemic.

▲ not identified to subspecies, or intermediate.

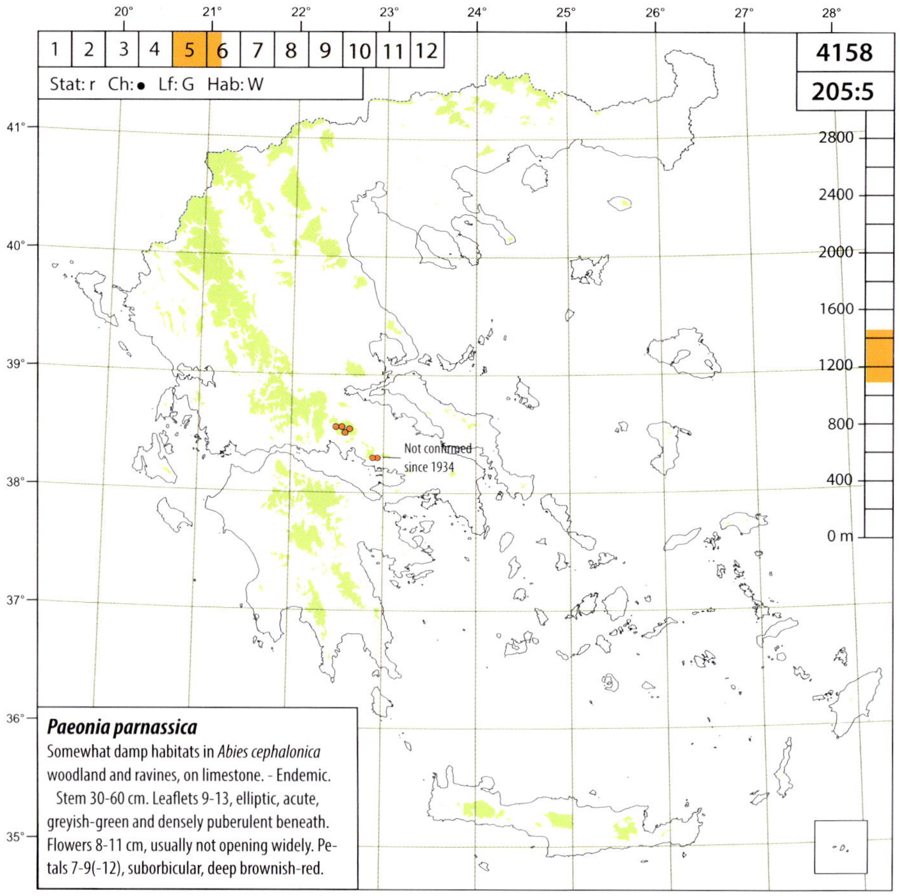

Paeonia parnassica

Somewhat damp habitats in *Abies cephalonica* woodland and ravines, on limestone. - Endemic.

Stem 30-60 cm. Leaflets 9-13, elliptic, acute, greyish-green and densely puberulent beneath. Flowers 8-11 cm, usually not opening widely. Petals 7-9(-12), suborbicular, deep brownish-red.

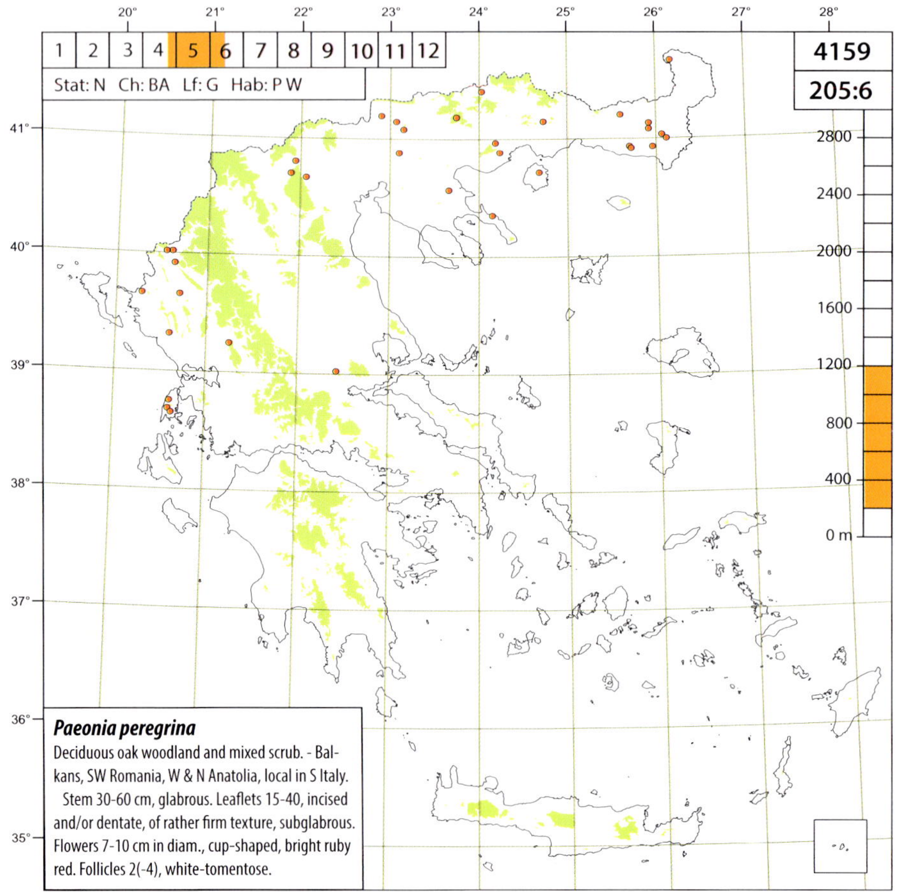

Paeonia peregrina

Deciduous oak woodland and mixed scrub. - Balkans, SW Romania, W & N Anatolia, local in S Italy.

Stem 30-60 cm, glabrous. Leaflets 15-40, incised and/or dentate, of rather firm texture, subglabrous. Flowers 7-10 cm in diam., cup-shaped, bright ruby red. Follicles 2(-4), white-tomentose.

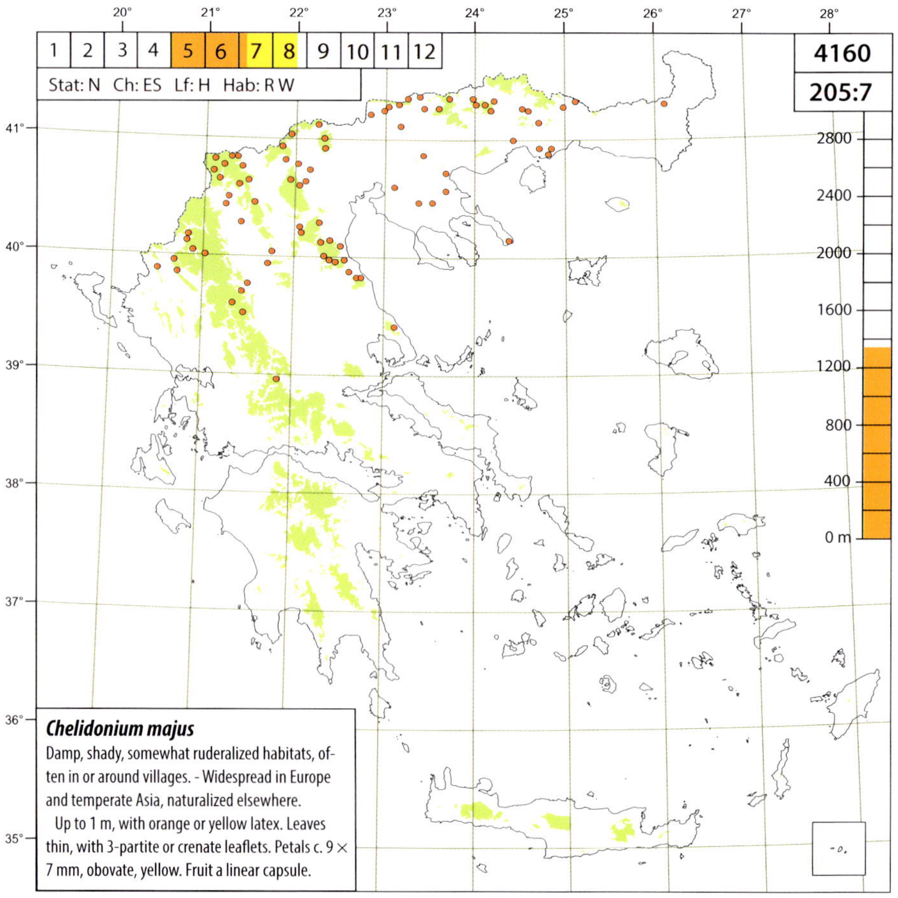

Chelidonium majus

Damp, shady, somewhat ruderalized habitats, often in or around villages. - Widespread in Europe and temperate Asia, naturalized elsewhere.

Up to 1 m, with orange or yellow latex. Leaves thin, with 3-partite or crenate leaflets. Petals c. 9 × 7 mm, obovate, yellow. Fruit a linear capsule.

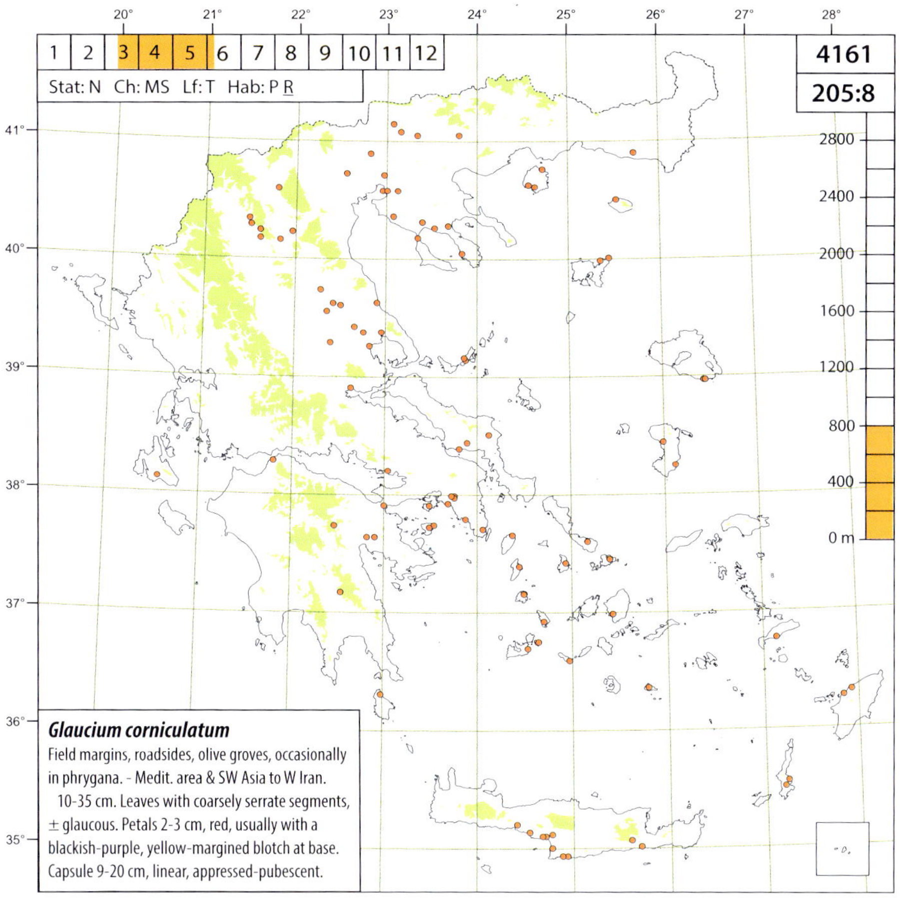

Glaucium corniculatum

Field margins, roadsides, olive groves, occasionally in phrygana. - Medit. area & SW Asia to W Iran.

10-35 cm. Leaves with coarsely serrate segments, ± glaucous. Petals 2-3 cm, red, usually with a blackish-purple, yellow-margined blotch at base. Capsule 9-20 cm, linear, appressed-pubescent.

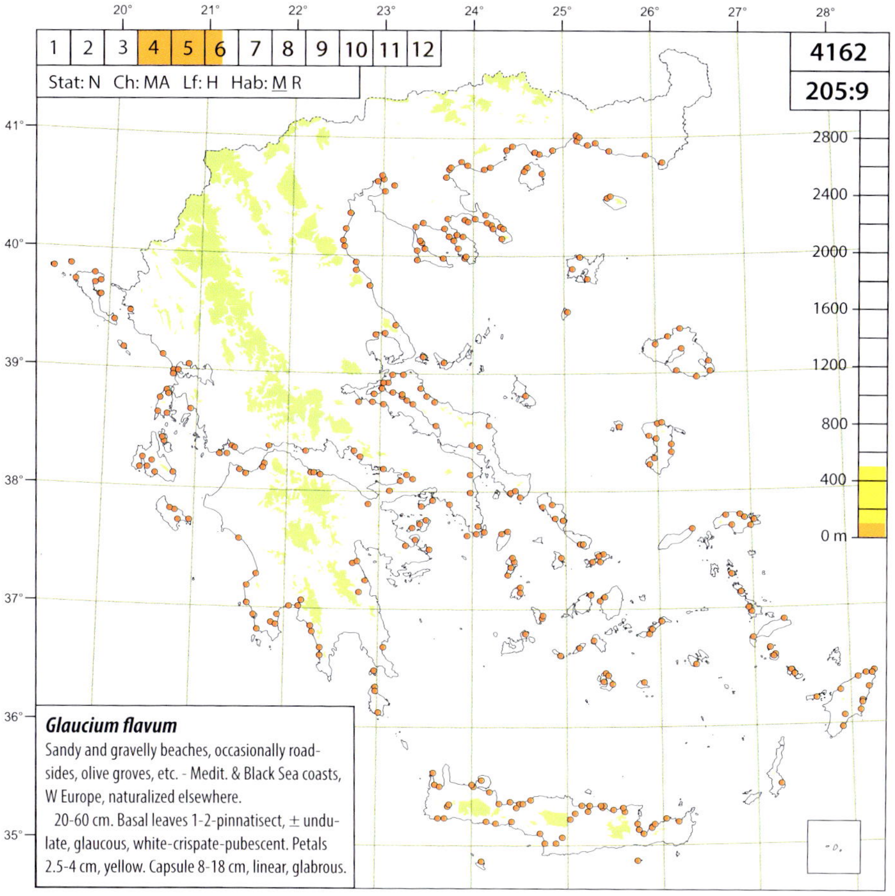

Glaucium flavum

Sandy and gravelly beaches, occasionally roadsides, olive groves, etc. - Medit. & Black Sea coasts, W Europe, naturalized elsewhere.

20-60 cm. Basal leaves 1-2-pinnatisect, ± undulate, glaucous, white-crispate-pubescent. Petals 2.5-4 cm, yellow. Capsule 8-18 cm, linear, glabrous.

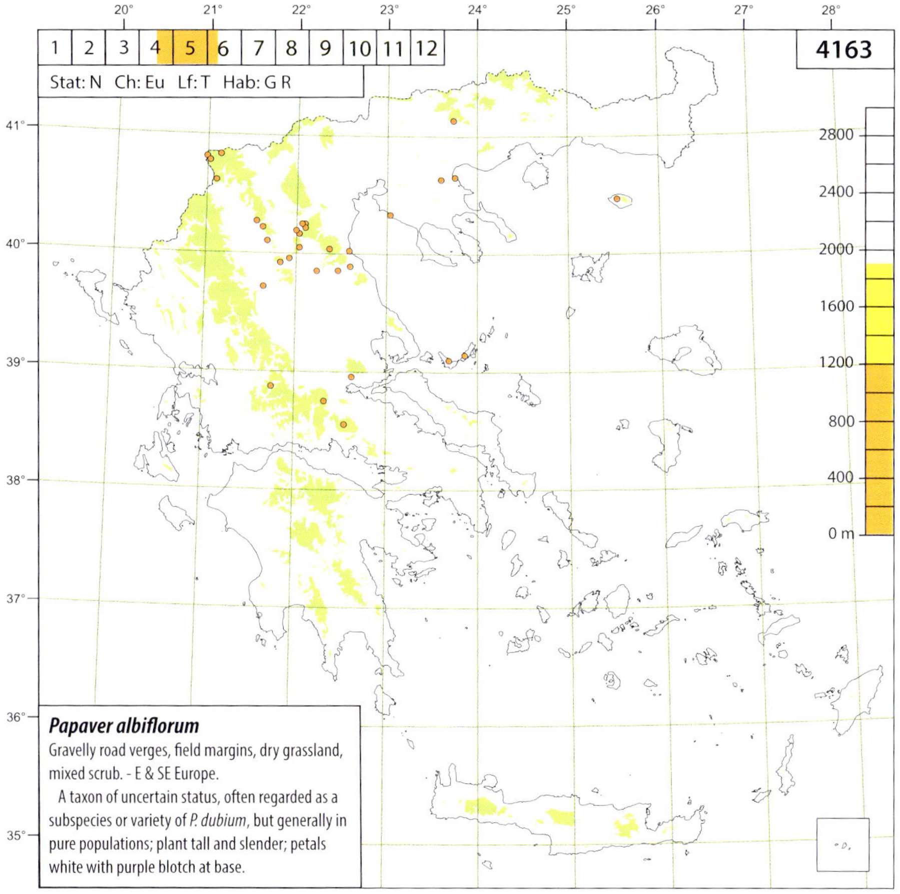

Papaver albiflorum

Gravelly road verges, field margins, dry grassland, mixed scrub. - E & SE Europe.

A taxon of uncertain status, often regarded as a subspecies or variety of *P. dubium*, but generally in pure populations; plant tall and slender; petals white with purple blotch at base.

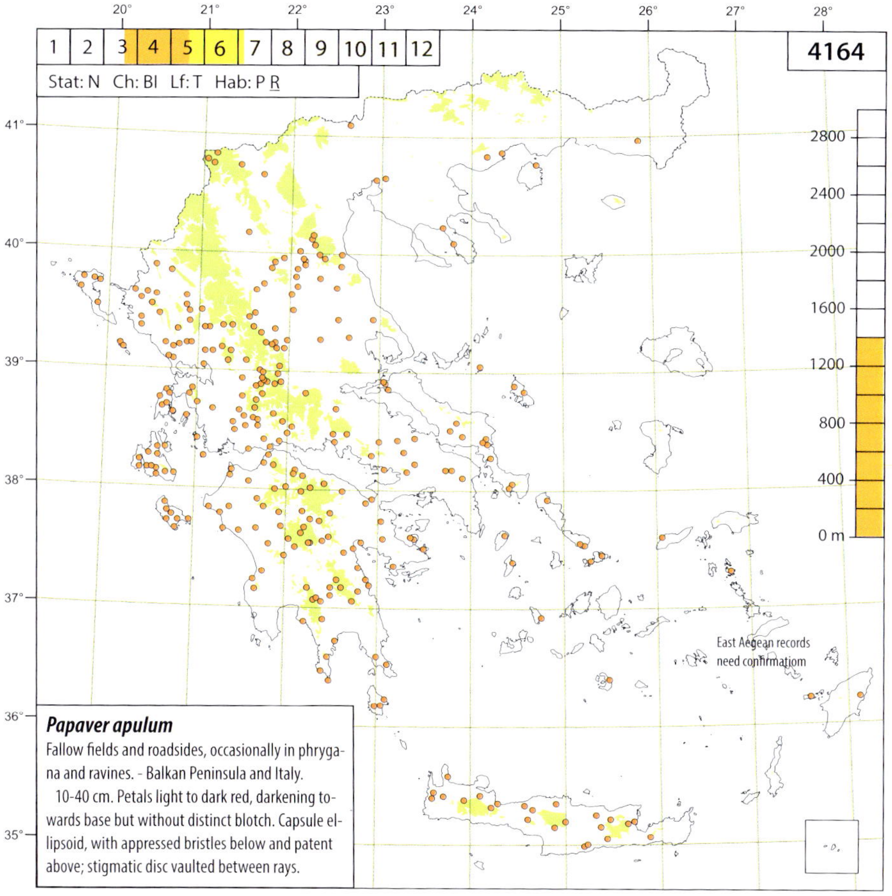

Papaver apulum

Fallow fields and roadsides, occasionally in phrygana and ravines. - Balkan Peninsula and Italy.

10-40 cm. Petals light to dark red, darkening towards base but without distinct blotch. Capsule ellipsoid, with appressed bristles below and patent above; stigmatic disc vaulted between rays.

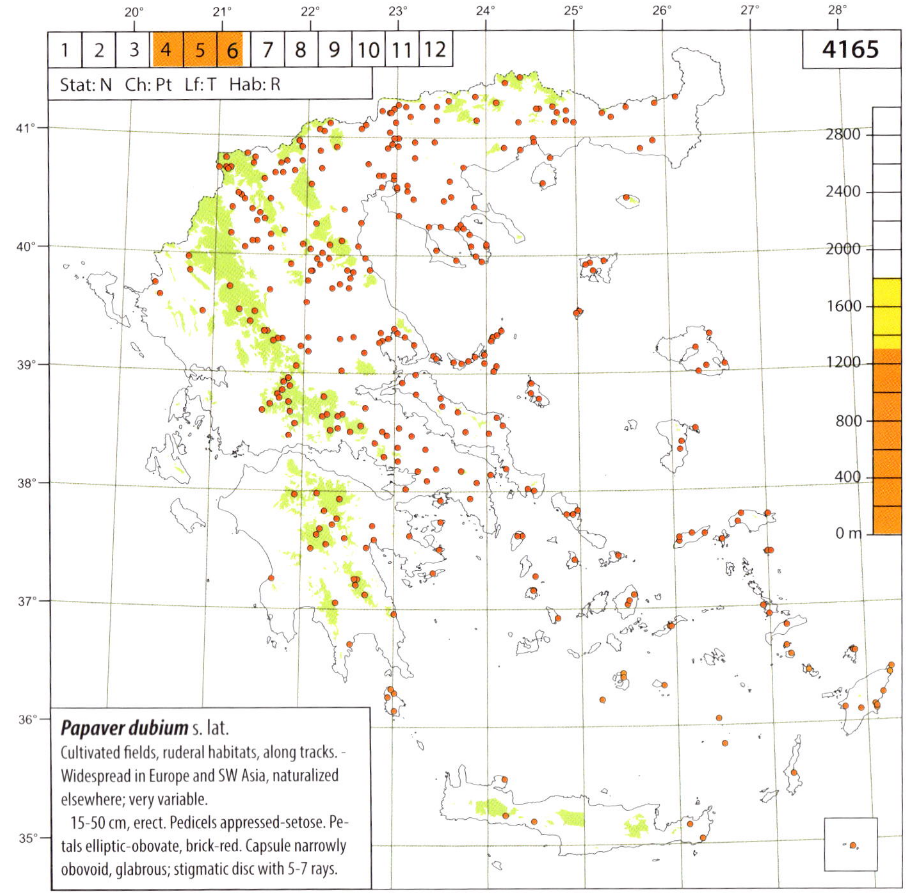

Papaver dubium s. lat.
Cultivated fields, ruderal habitats, along tracks. - Widespread in Europe and SW Asia, naturalized elsewhere; very variable.
15-50 cm, erect. Pedicels appressed-setose. Petals elliptic-obovate, brick-red. Capsule narrowly obovoid, glabrous; stigmatic disc with 5-7 rays.

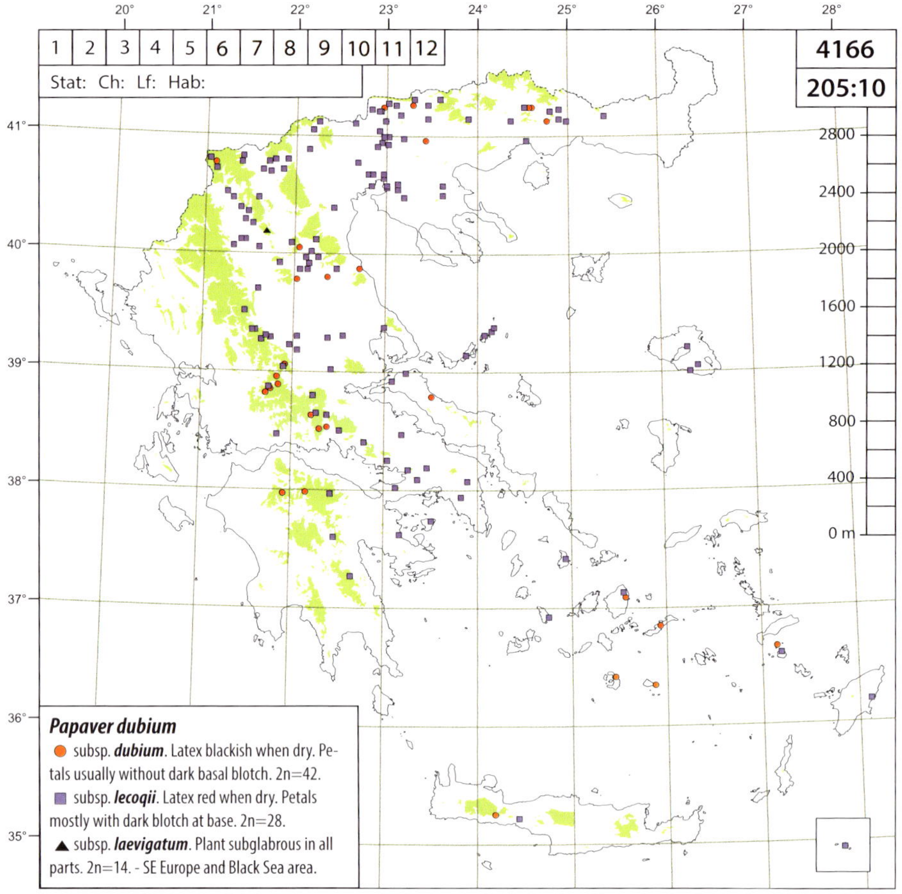

Papaver dubium
● subsp. ***dubium***. Latex blackish when dry. Petals usually without dark basal blotch. 2n=42.
■ subsp. ***lecoqii***. Latex red when dry. Petals mostly with dark blotch at base. 2n=28.
▲ subsp. ***laevigatum***. Plant subglabrous in all parts. 2n=14. - SE Europe and Black Sea area.

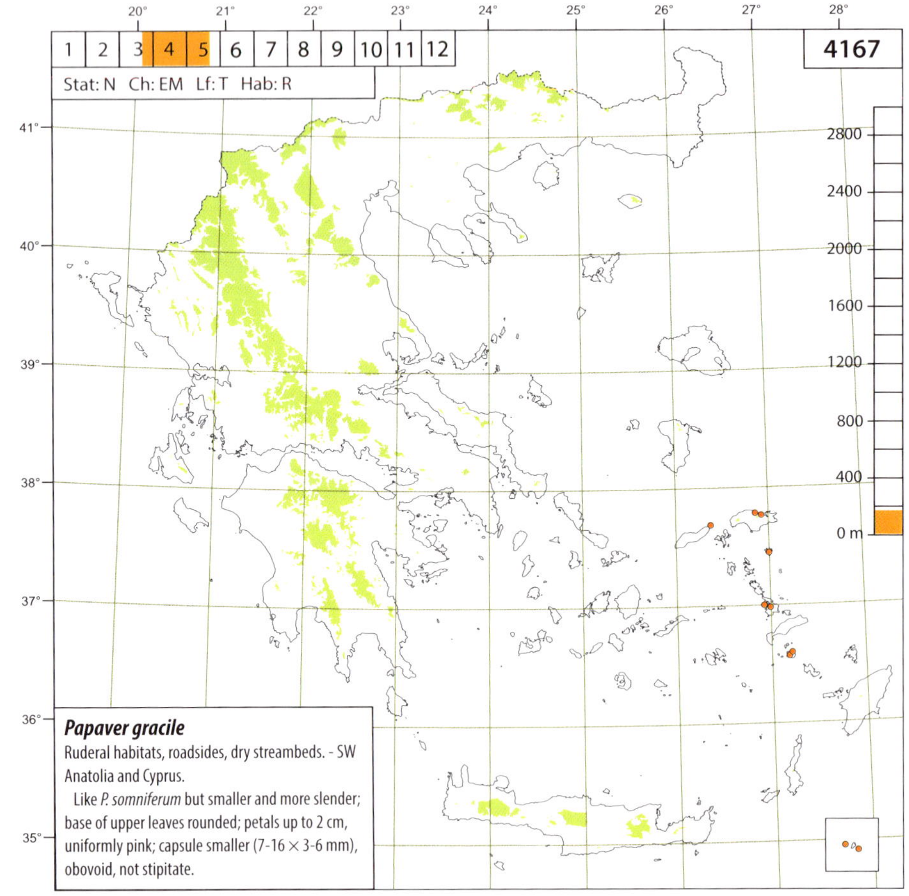

Papaver gracile
Ruderal habitats, roadsides, dry streambeds. - SW Anatolia and Cyprus.
Like *P. somniferum* but smaller and more slender; base of upper leaves rounded; petals up to 2 cm, uniformly pink; capsule smaller (7-16 × 3-6 mm), obovoid, not stipitate.

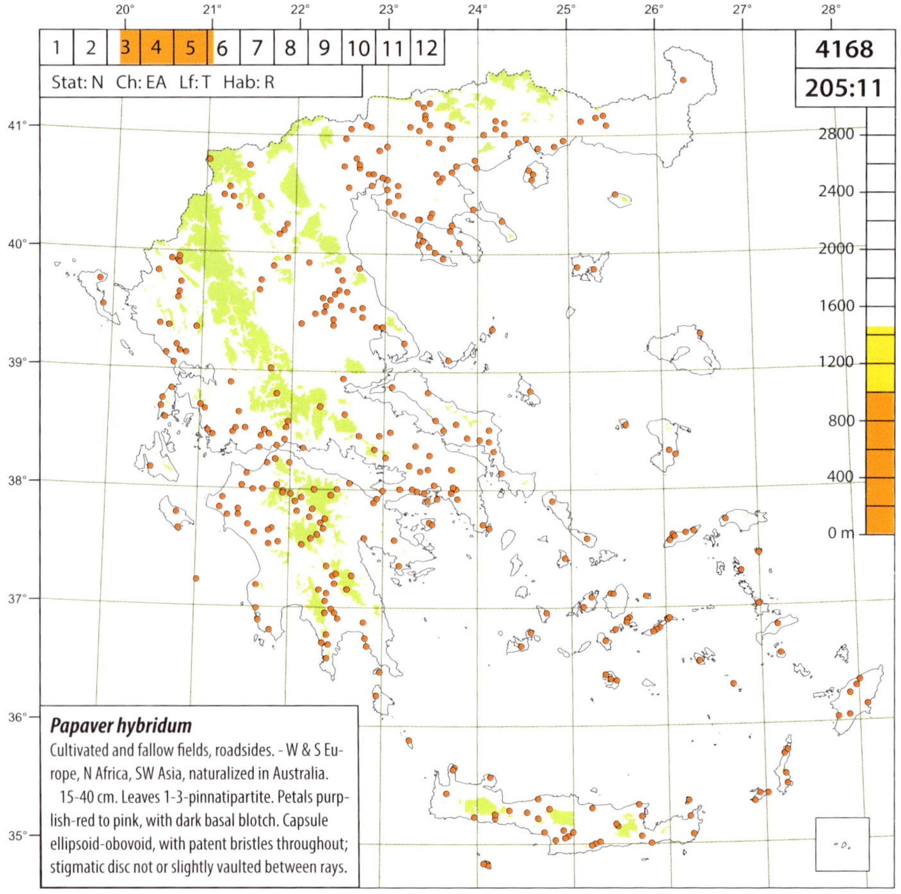

Papaver hybridum
Cultivated and fallow fields, roadsides. - W & S Europe, N Africa, SW Asia, naturalized in Australia.
15-40 cm. Leaves 1-3-pinnatipartite. Petals purplish-red to pink, with dark basal blotch. Capsule ellipsoid-obovoid, with patent bristles throughout; stigmatic disc not or slightly vaulted between rays.

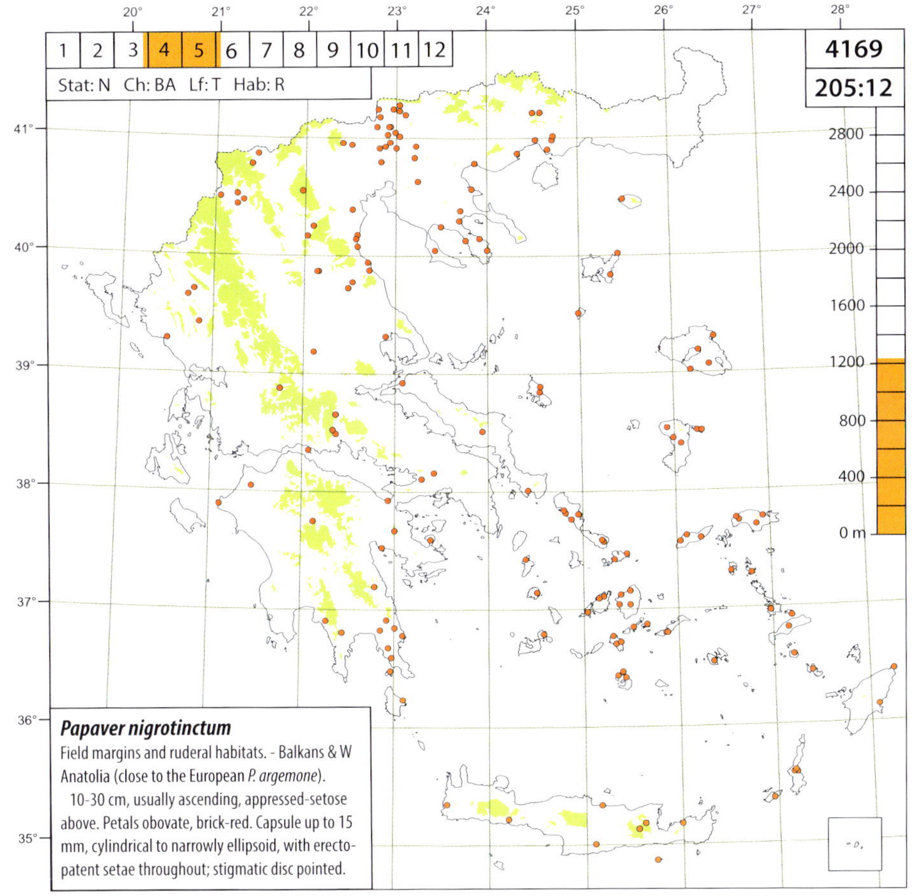

Papaver nigrotinctum
Field margins and ruderal habitats. - Balkans & W Anatolia (close to the European *P. argemone*).
10-30 cm, usually ascending, appressed-setose above. Petals obovate, brick-red. Capsule up to 15 mm, cylindrical to narrowly ellipsoid, with erecto-patent setae throughout; stigmatic disc pointed.

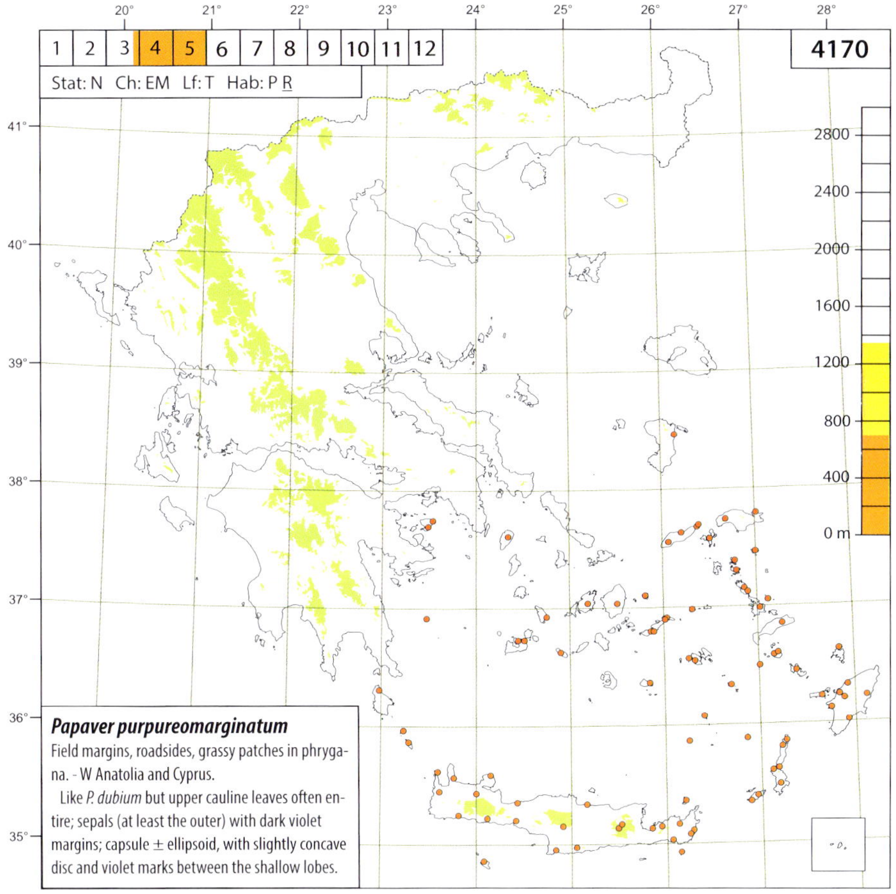

Papaver purpureomarginatum
Field margins, roadsides, grassy patches in phrygana. - W Anatolia and Cyprus.
Like *P. dubium* but upper cauline leaves often entire; sepals (at least the outer) with dark violet margins; capsule ± ellipsoid, with slightly concave disc and violet marks between the shallow lobes.

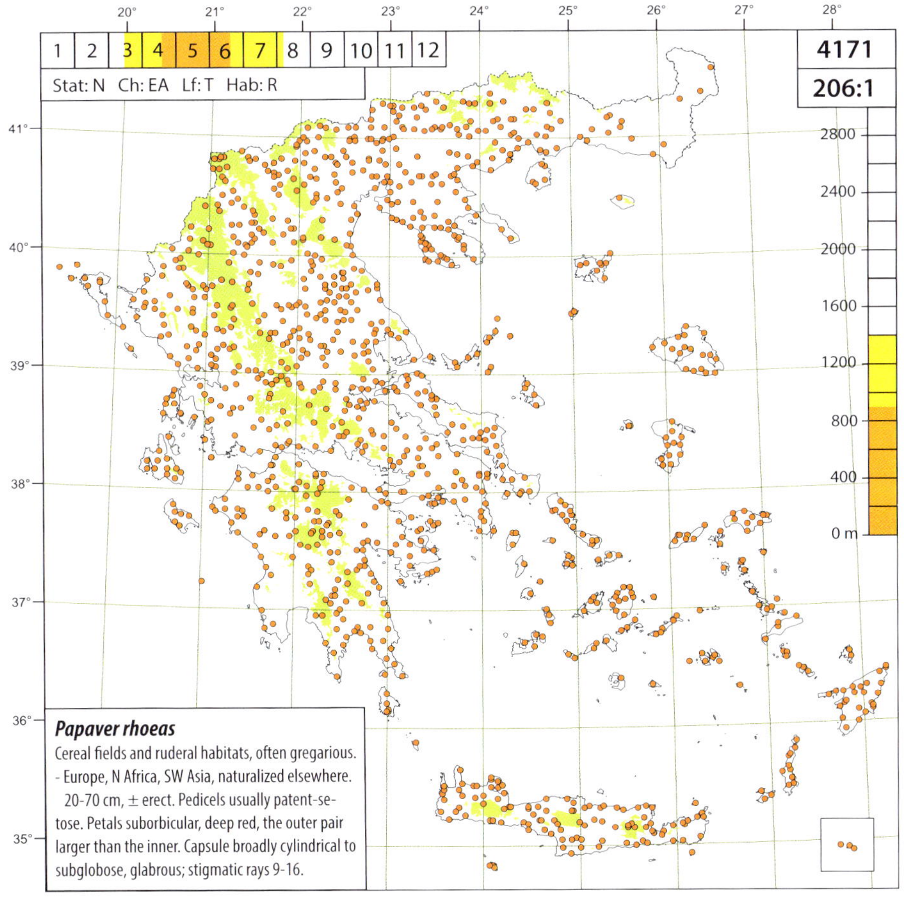

Papaver rhoeas
Cereal fields and ruderal habitats, often gregarious. - Europe, N Africa, SW Asia, naturalized elsewhere.
20-70 cm, ± erect. Pedicels usually patent-setose. Petals suborbicular, deep red, the outer pair larger than the inner. Capsule broadly cylindrical to subglobose, glabrous; stigmatic rays 9-16.

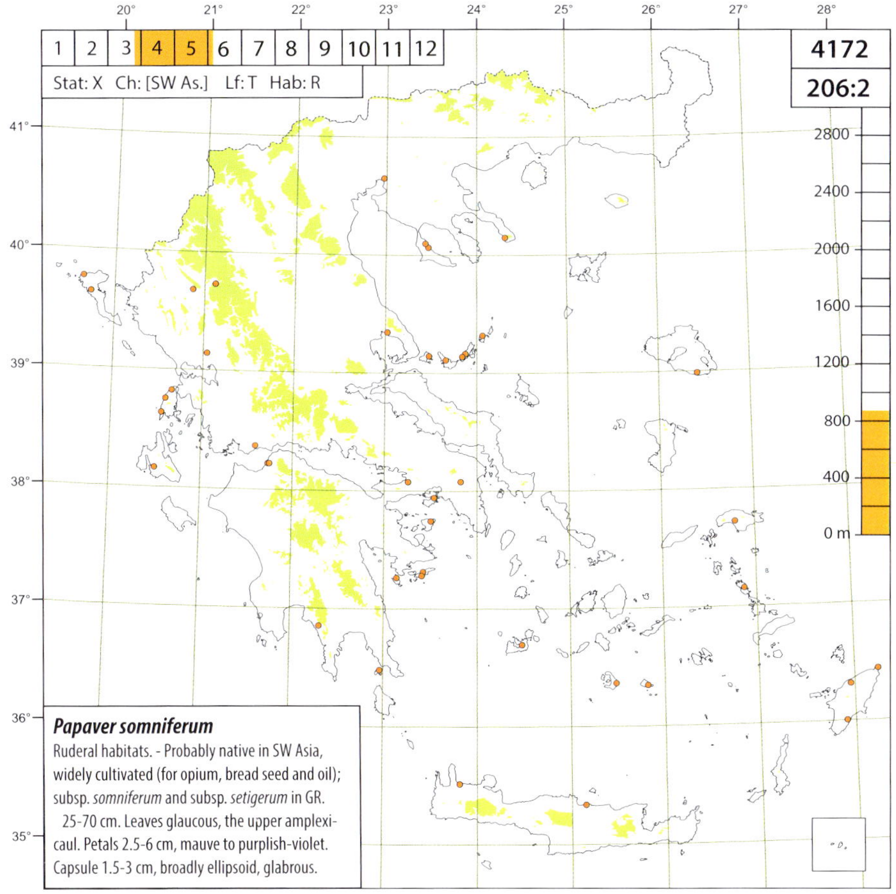

Papaver somniferum
Ruderal habitats. - Probably native in SW Asia, widely cultivated (for opium, bread seed and oil); subsp. *somniferum* and subsp. *setigerum* in GR.
25-70 cm. Leaves glaucous, the upper amplexicaul. Petals 2.5-6 cm, mauve to purplish-violet. Capsule 1.5-3 cm, broadly ellipsoid, glabrous.

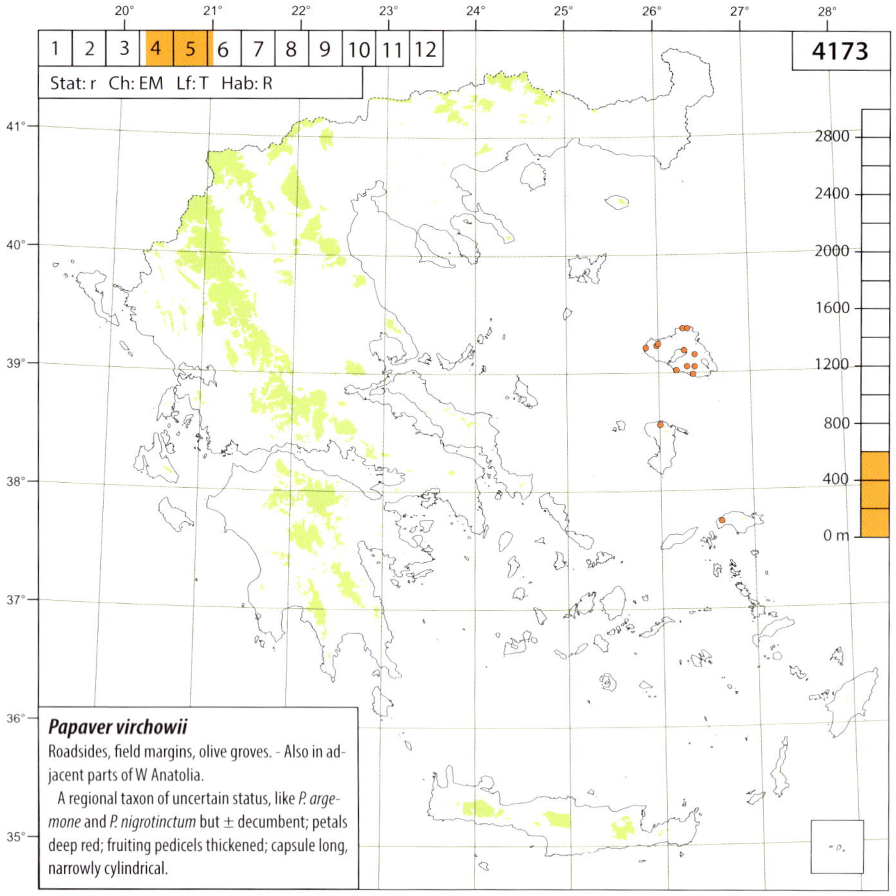

Papaver virchowii
Roadsides, field margins, olive groves. - Also in adjacent parts of W Anatolia.
A regional taxon of uncertain status, like *P. argemone* and *P. nigrotinctum* but ± decumbent; petals deep red; fruiting pedicels thickened; capsule long, narrowly cylindrical.

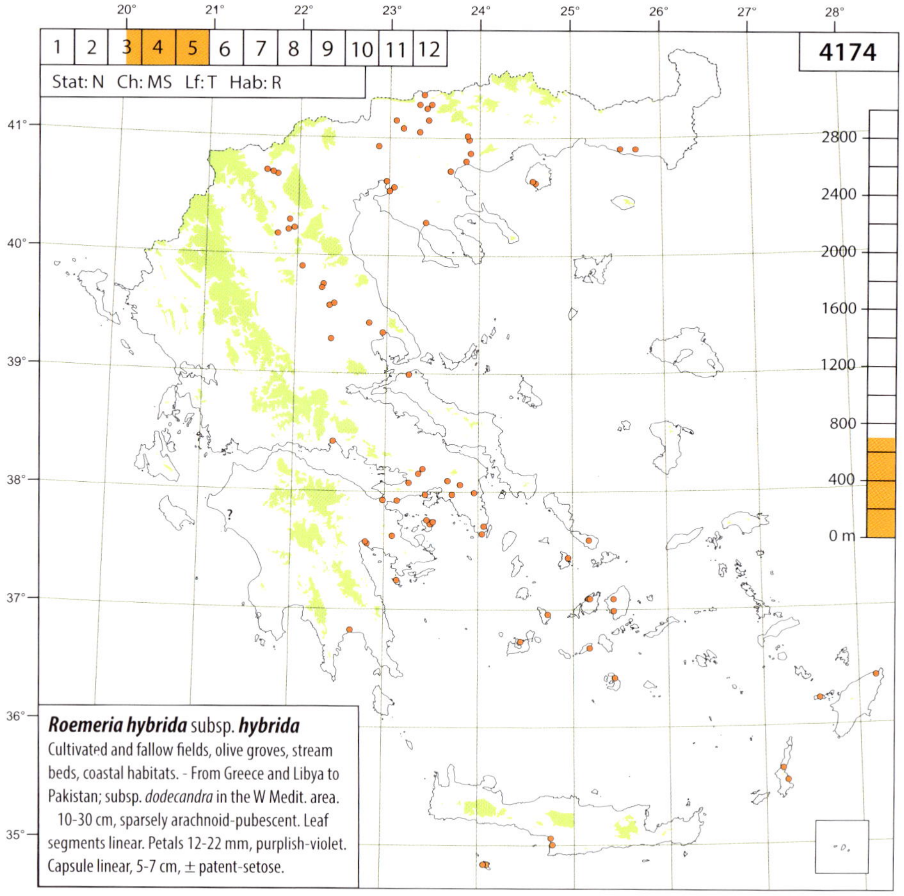

Roemeria hybrida subsp. ***hybrida***
Cultivated and fallow fields, olive groves, stream beds, coastal habitats. - From Greece and Libya to Pakistan; subsp. *dodecandra* in the W Medit. area.
10-30 cm, sparsely arachnoid-pubescent. Leaf segments linear. Petals 12-22 mm, purplish-violet. Capsule linear, 5-7 cm, ± patent-setose.

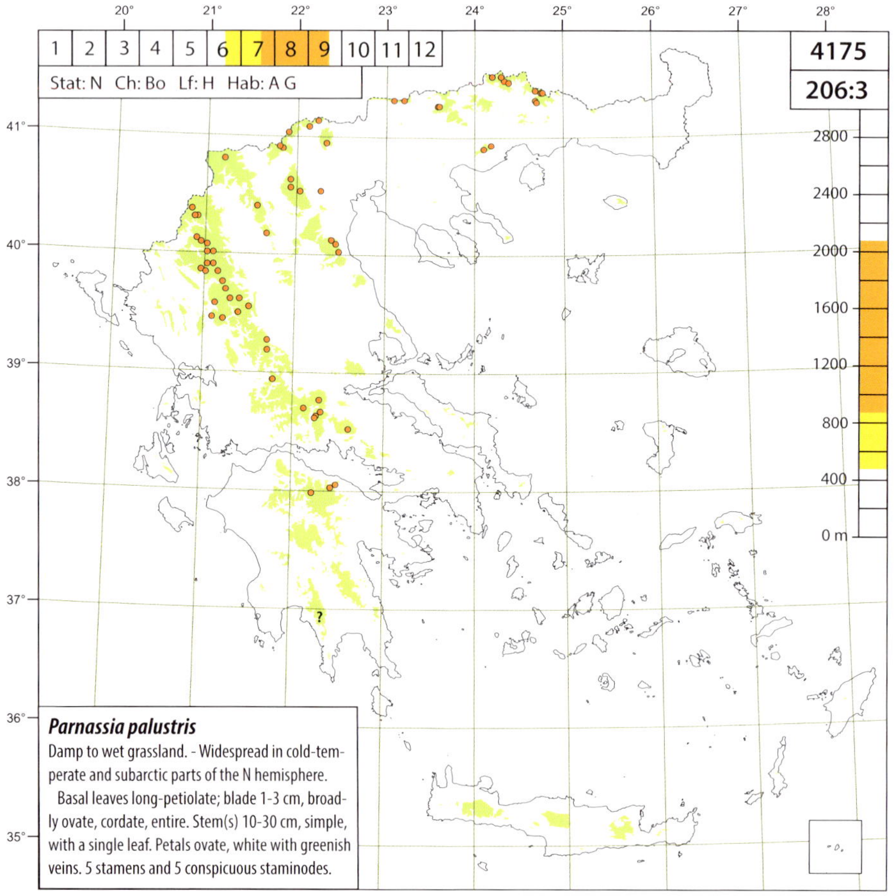

Parnassia palustris
Damp to wet grassland. - Widespread in cold-temperate and subarctic parts of the N hemisphere.
Basal leaves long-petiolate; blade 1-3 cm, broadly ovate, cordate, entire. Stem(s) 10-30 cm, simple, with a single leaf. Petals ovate, white with greenish veins. 5 stamens and 5 conspicuous staminodes.

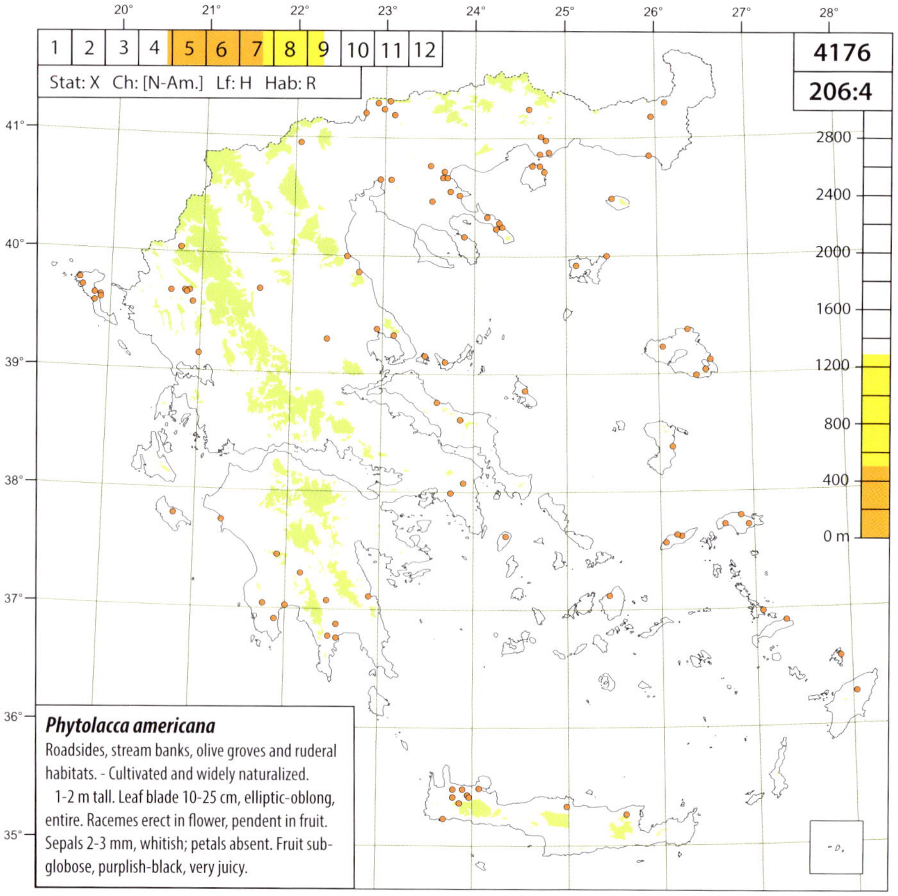

Phytolacca americana
Roadsides, stream banks, olive groves and ruderal habitats. - Cultivated and widely naturalized.
1-2 m tall. Leaf blade 10-25 cm, elliptic-oblong, entire. Racemes erect in flower, pendent in fruit. Sepals 2-3 mm, whitish; petals absent. Fruit subglobose, purplish-black, very juicy.

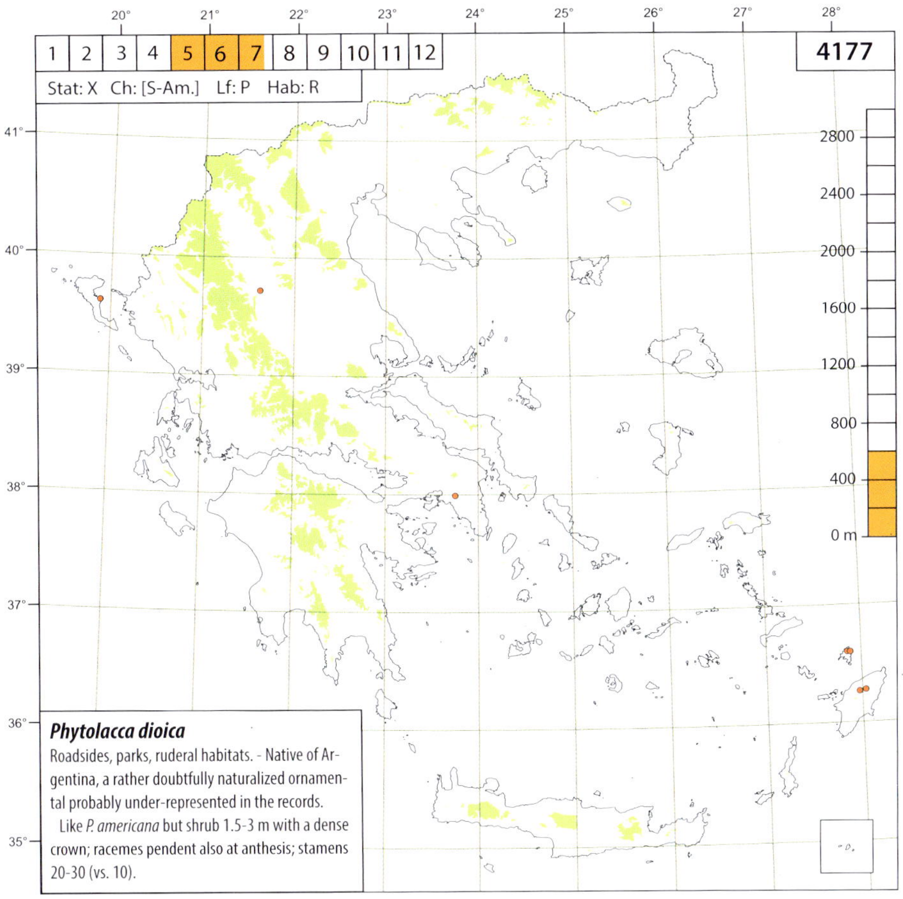

Phytolacca dioica

Roadsides, parks, ruderal habitats. - Native of Argentina, a rather doubtfully naturalized ornamental probably under-represented in the records.

Like *P. americana* but shrub 1.5-3 m with a dense crown; racemes pendent also at anthesis; stamens 20-30 (vs. 10).

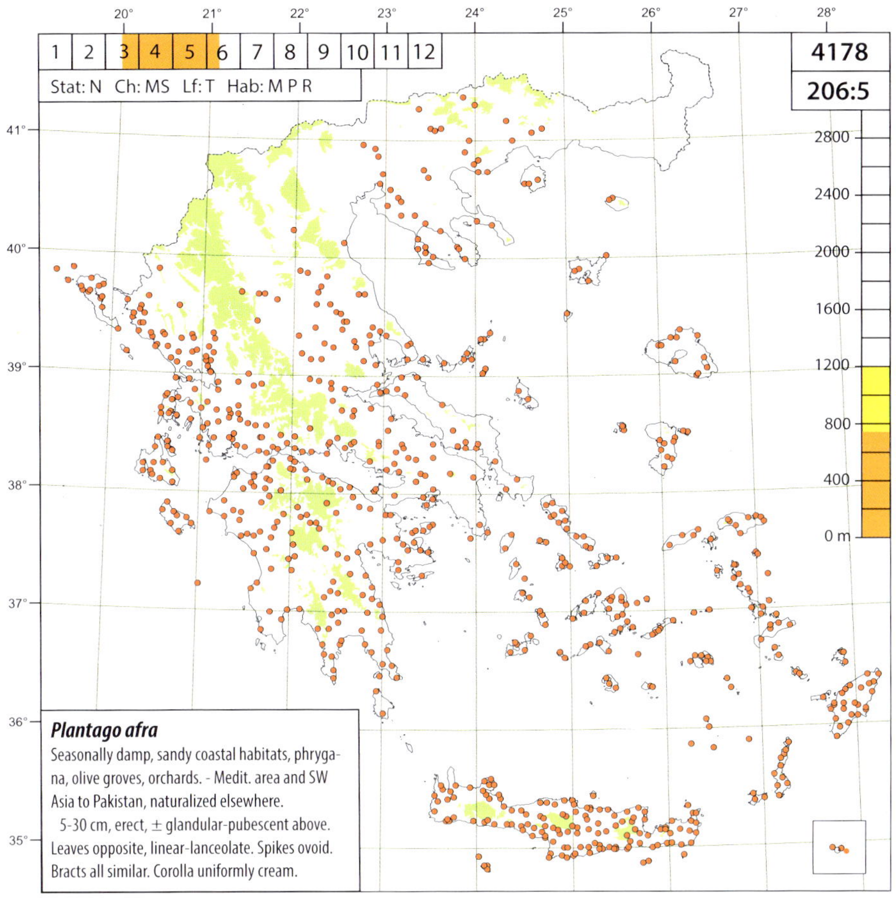

Plantago afra

Seasonally damp, sandy coastal habitats, phrygana, olive groves, orchards. - Medit. area and SW Asia to Pakistan, naturalized elsewhere.

5-30 cm, erect, ± glandular-pubescent above. Leaves opposite, linear-lanceolate. Spikes ovoid. Bracts all similar. Corolla uniformly cream.

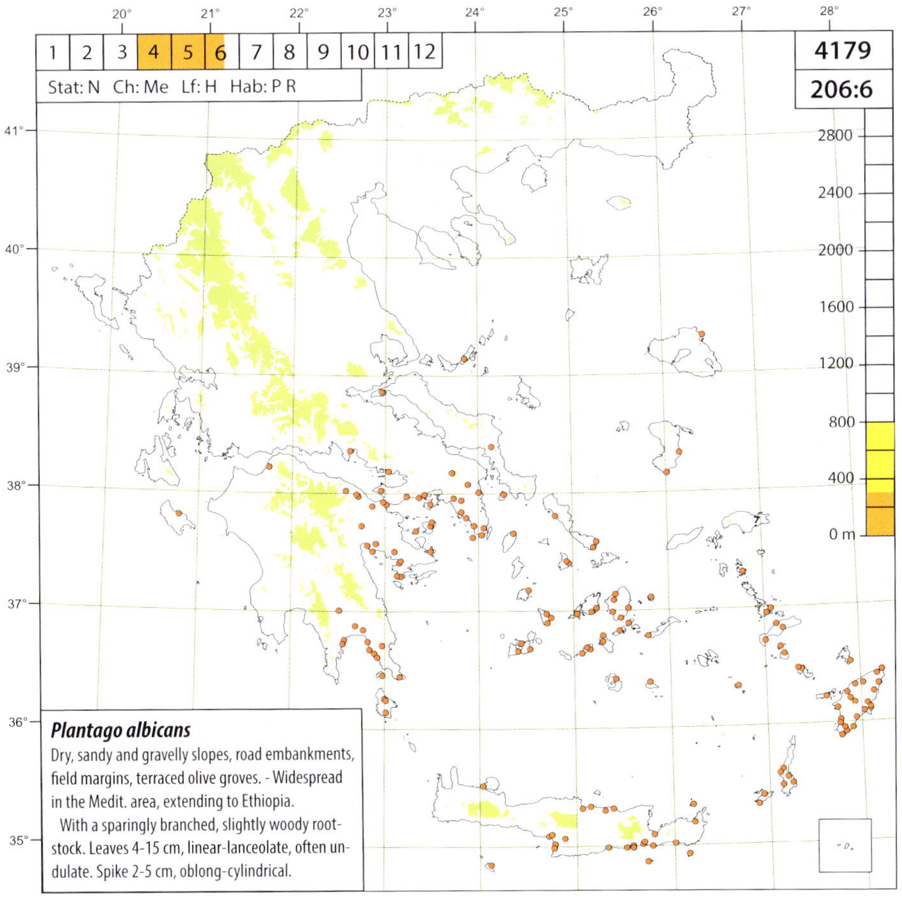

Plantago albicans

Dry, sandy and gravelly slopes, road embankments, field margins, terraced olive groves. - Widespread in the Medit. area, extending to Ethiopia.

With a sparingly branched, slightly woody rootstock. Leaves 4-15 cm, linear-lanceolate, often undulate. Spike 2-5 cm, oblong-cylindrical.

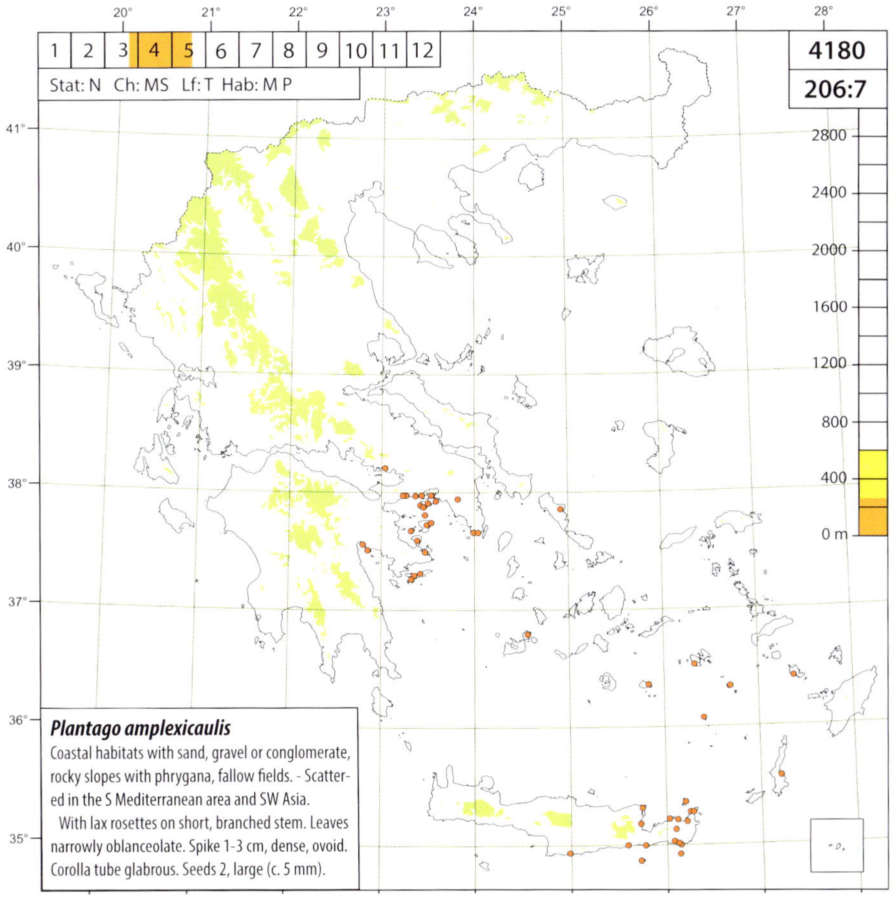

Plantago amplexicaulis

Coastal habitats with sand, gravel or conglomerate, rocky slopes with phrygana, fallow fields. - Scattered in the S Mediterranean area and SW Asia.

With lax rosettes on short, branched stem. Leaves narrowly oblanceolate. Spike 1-3 cm, dense, ovoid. Corolla tube glabrous. Seeds 2, large (c. 5 mm).

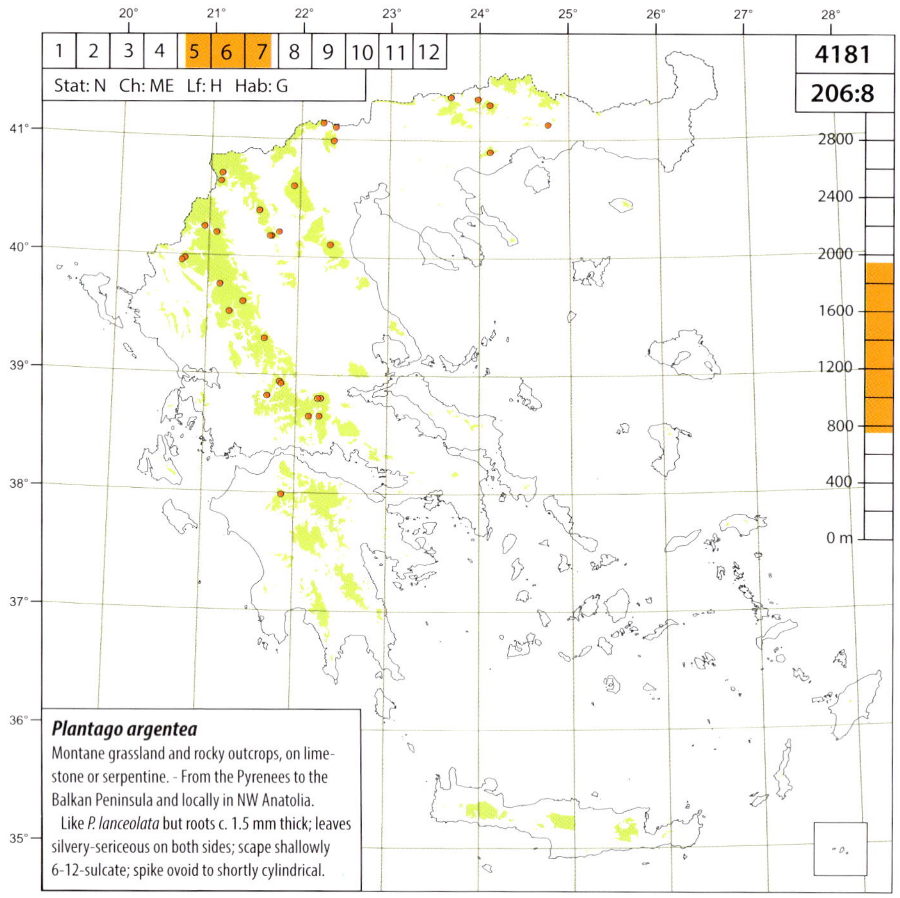

Plantago argentea

Montane grassland and rocky outcrops, on limestone or serpentine. - From the Pyrenees to the Balkan Peninsula and locally in NW Anatolia.

Like *P. lanceolata* but roots c. 1.5 mm thick; leaves silvery-sericeous on both sides; scape shallowly 6-12-sulcate; spike ovoid to shortly cylindrical.

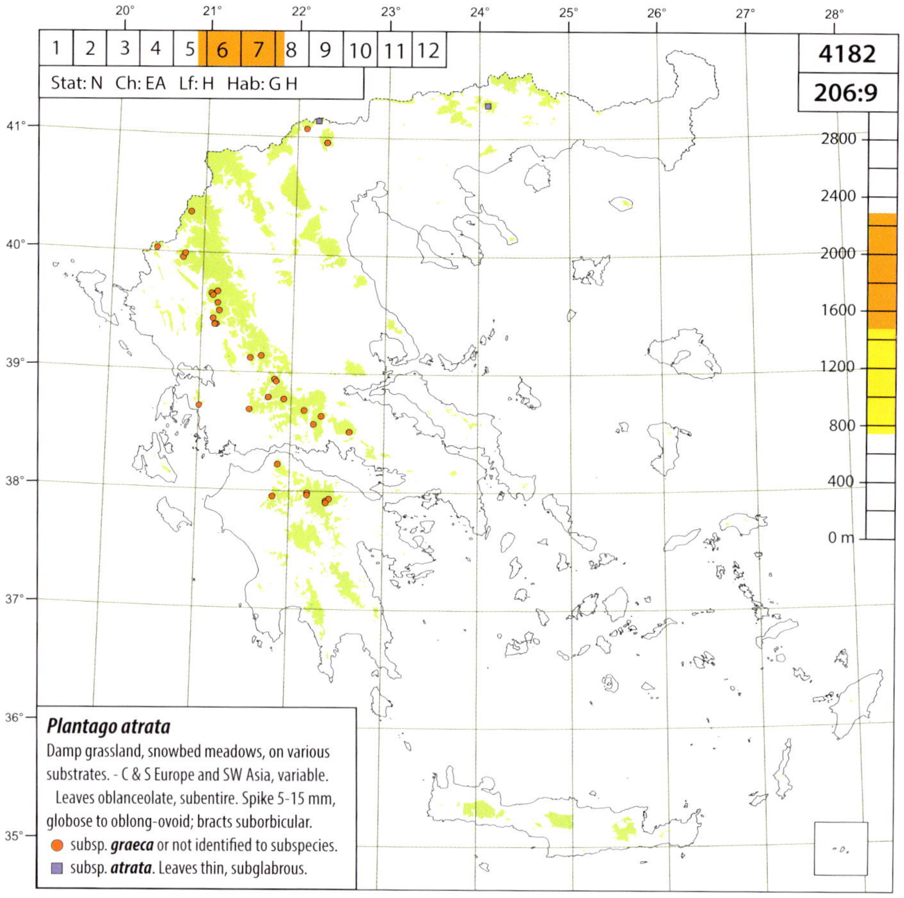

Plantago atrata

Damp grassland, snowbed meadows, on various substrates. - C & S Europe and SW Asia, variable.

Leaves oblanceolate, subentire. Spike 5-15 mm, globose to oblong-ovoid; bracts suborbicular.

● subsp. ***graeca*** or not identified to subspecies.

■ subsp. ***atrata***. Leaves thin, subglabrous.

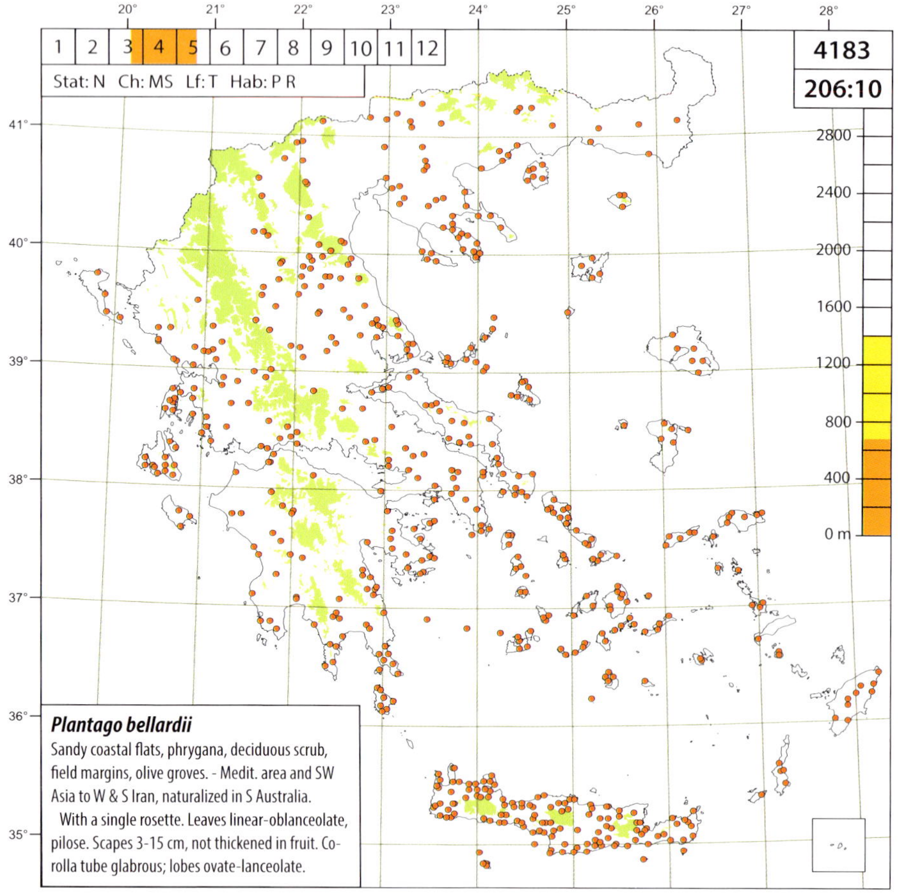

Plantago bellardii

Sandy coastal flats, phrygana, deciduous scrub, field margins, olive groves. - Medit. area and SW Asia to W & S Iran, naturalized in S Australia.

With a single rosette. Leaves linear-oblanceolate, pilose. Scapes 3-15 cm, not thickened in fruit. Corolla tube glabrous; lobes ovate-lanceolate.

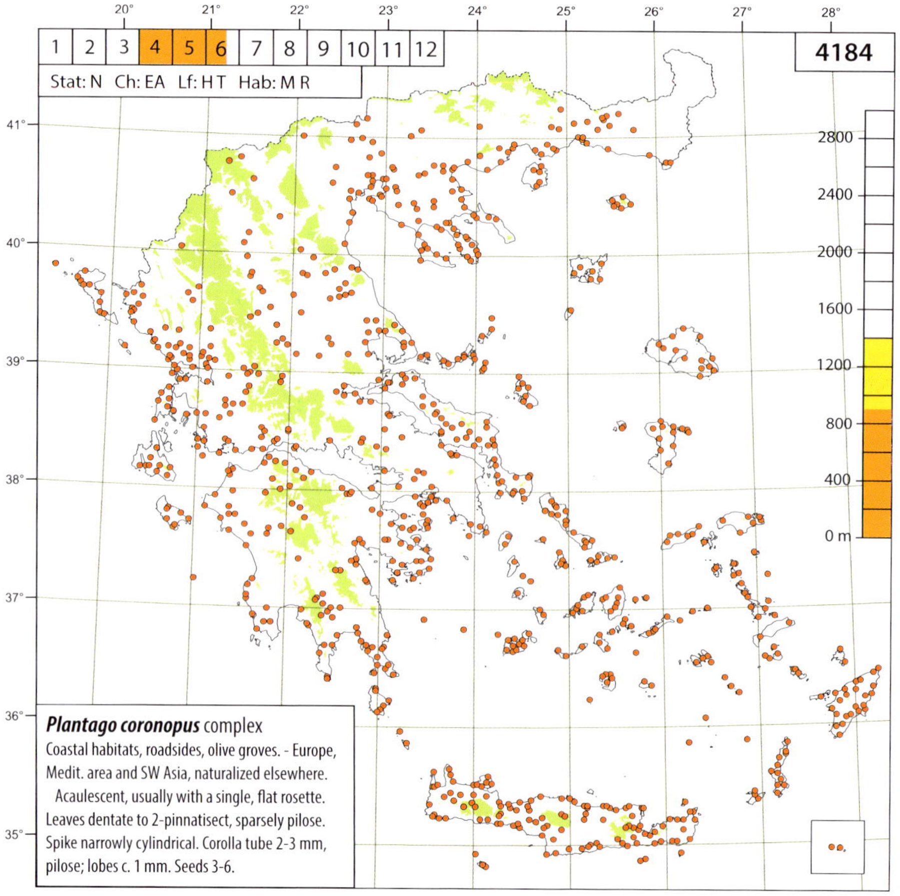

Plantago coronopus complex

Coastal habitats, roadsides, olive groves. - Europe, Medit. area and SW Asia, naturalized elsewhere.

Acaulescent, usually with a single, flat rosette. Leaves dentate to 2-pinnatisect, sparsely pilose. Spike narrowly cylindrical. Corolla tube 2-3 mm, pilose; lobes c. 1 mm. Seeds 3-6.

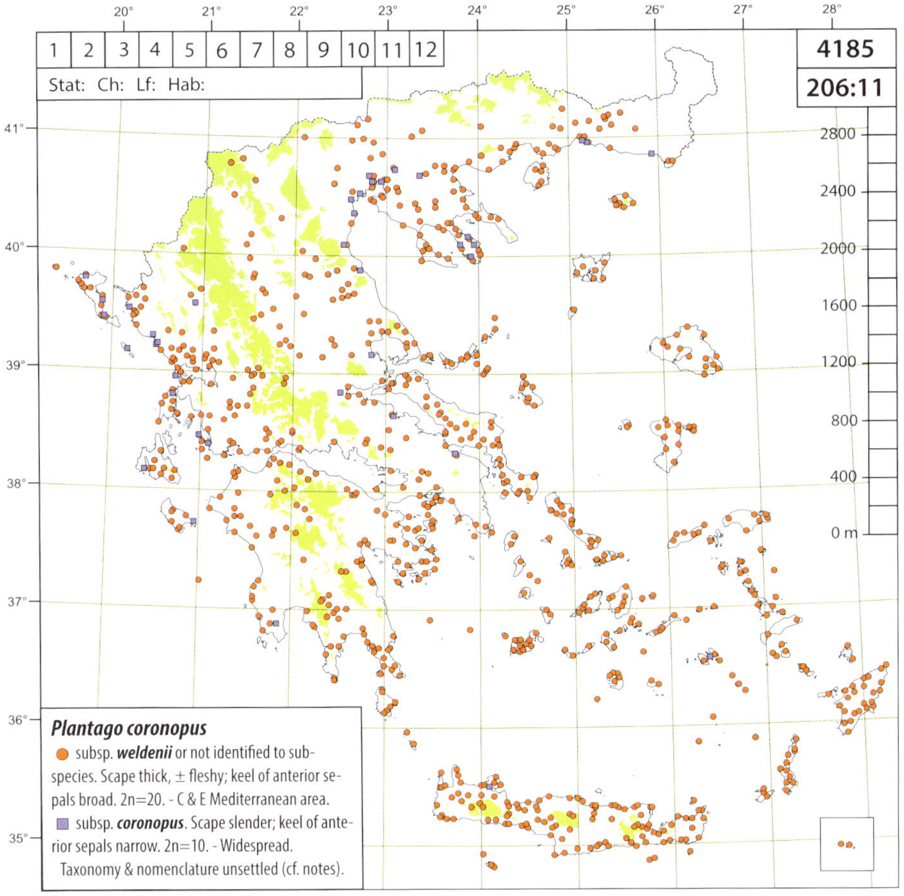

Plantago coronopus

● subsp. ***weldenii*** or not identified to subspecies. Scape thick, ± fleshy; keel of anterior sepals broad. 2n=20. - C & E Mediterranean area.

■ subsp. ***coronopus***. Scape slender; keel of anterior sepals narrow. 2n=10. - Widespread.

Taxonomy & nomenclature unsettled (cf. notes).

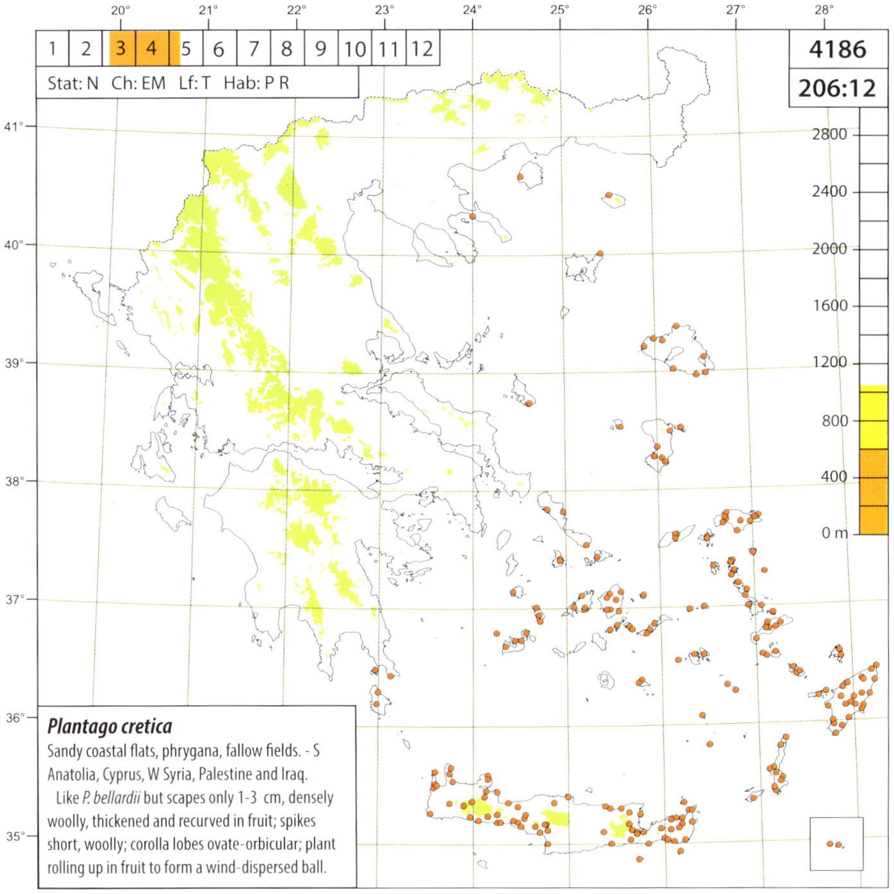

Plantago cretica

Sandy coastal flats, phrygana, fallow fields. - S Anatolia, Cyprus, W Syria, Palestine and Iraq.

Like *P. bellardii* but scapes only 1-3 cm, densely woolly, thickened and recurved in fruit; spikes short, woolly; corolla lobes ovate-orbicular; plant rolling up in fruit to form a wind-dispersed ball.

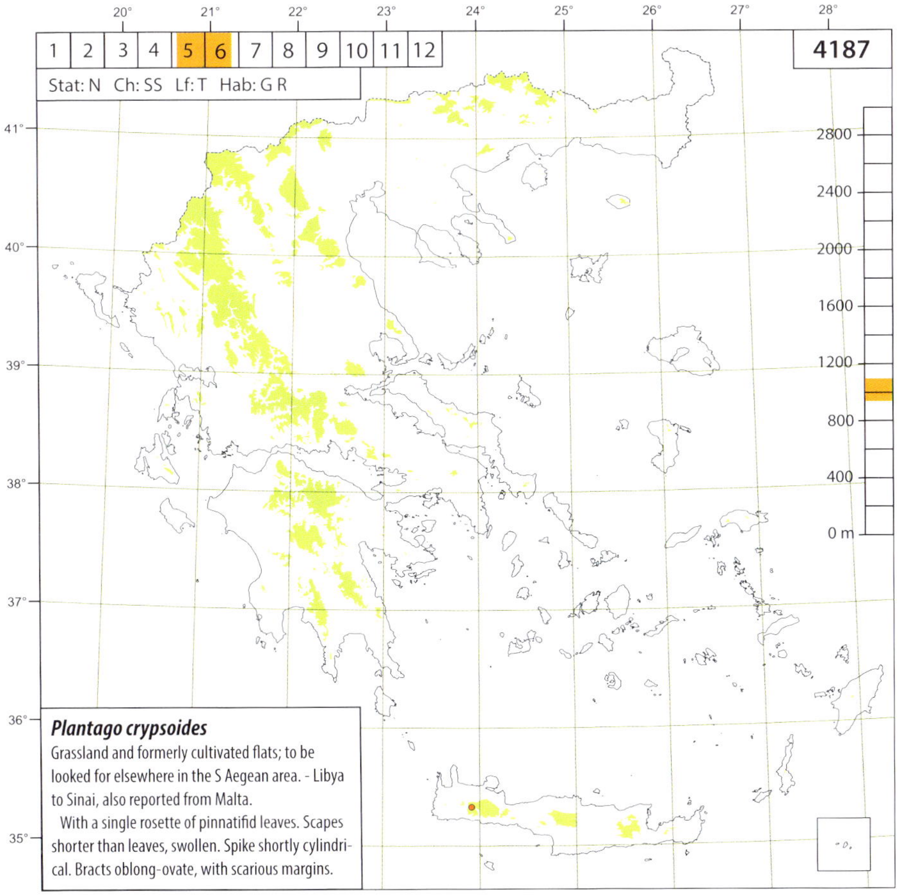

Plantago crypsoides

Grassland and formerly cultivated flats; to be looked for elsewhere in the S Aegean area. - Libya to Sinai, also reported from Malta.

With a single rosette of pinnatifid leaves. Scapes shorter than leaves, swollen. Spike shortly cylindrical. Bracts oblong-ovate, with scarious margins.

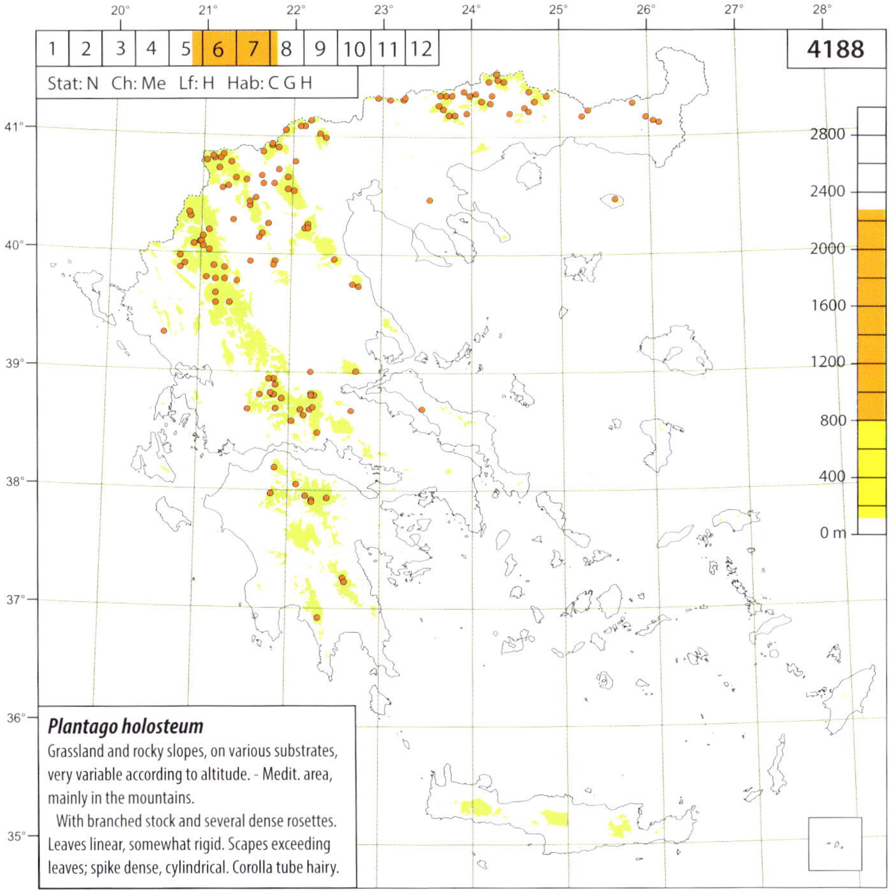

Plantago holosteum

Grassland and rocky slopes, on various substrates, very variable according to altitude. - Medit. area, mainly in the mountains.

With branched stock and several dense rosettes. Leaves linear, somewhat rigid. Scapes exceeding leaves; spike dense, cylindrical. Corolla tube hairy.

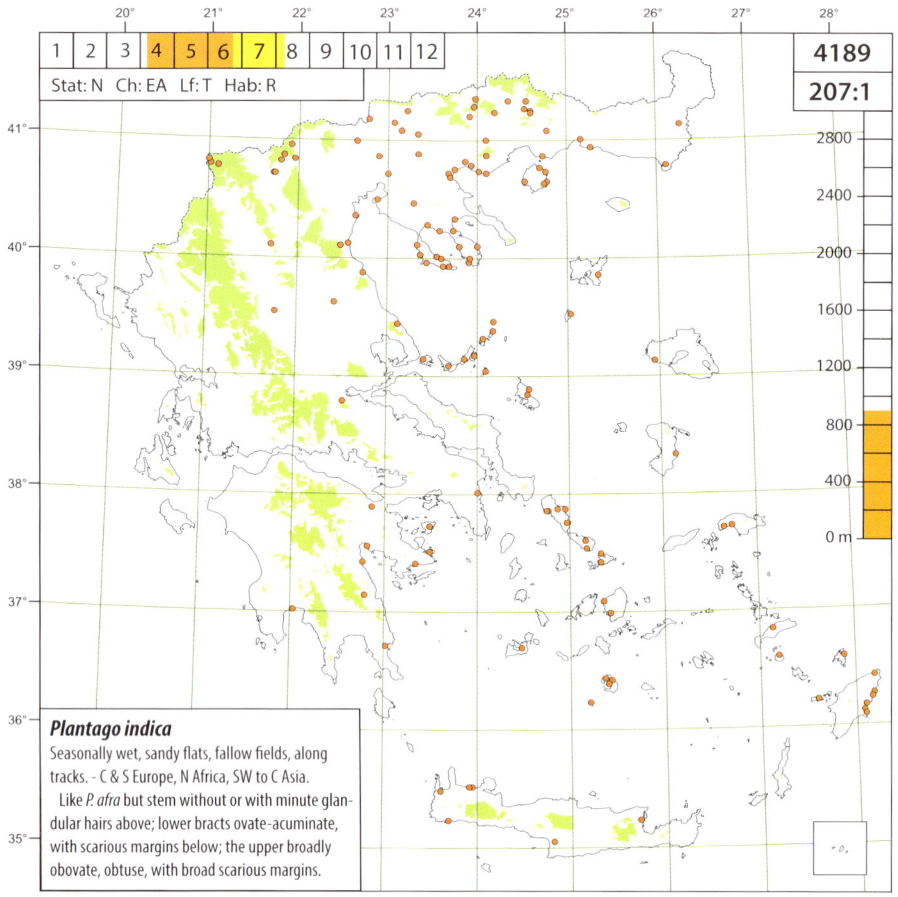

Plantago indica

Seasonally wet, sandy flats, fallow fields, along tracks. - C & S Europe, N Africa, SW to C Asia.

Like *P. afra* but stem without or with minute glandular hairs above; lower bracts ovate-acuminate, with scarious margins below; the upper broadly obovate, obtuse, with broad scarious margins.

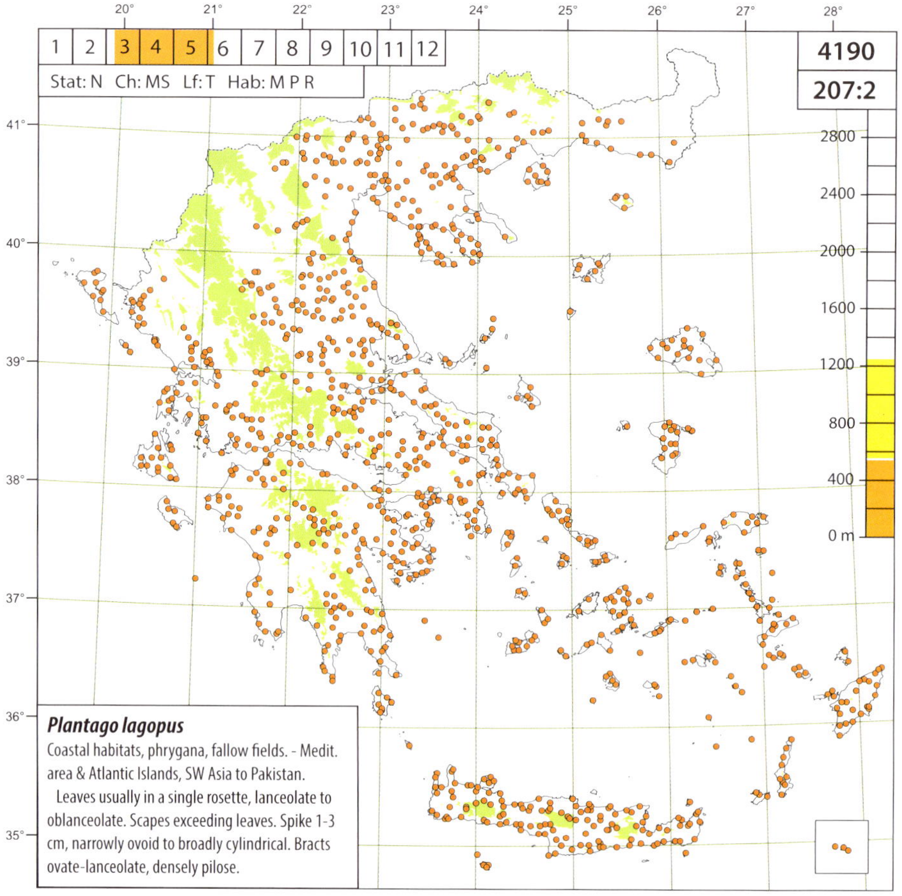

Plantago lagopus

Coastal habitats, phrygana, fallow fields. - Medit. area & Atlantic Islands, SW Asia to Pakistan.

Leaves usually in a single rosette, lanceolate to oblanceolate. Scapes exceeding leaves. Spike 1-3 cm, narrowly ovoid to broadly cylindrical. Bracts ovate-lanceolate, densely pilose.

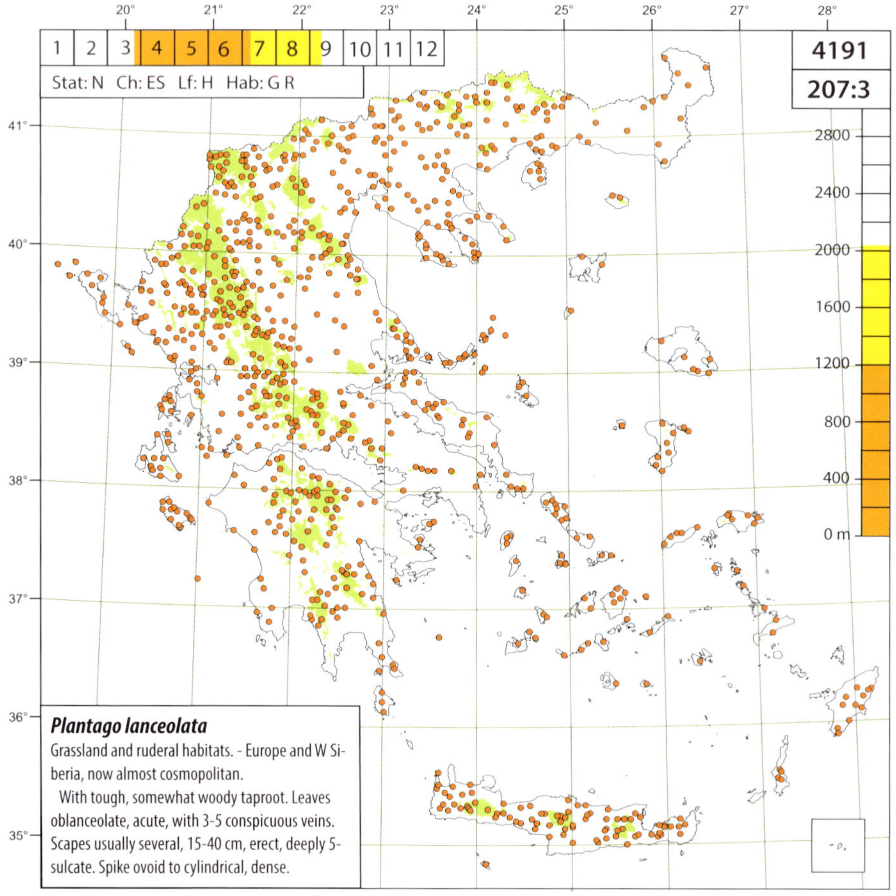

Plantago lanceolata

Grassland and ruderal habitats. - Europe and W Siberia, now almost cosmopolitan.

With tough, somewhat woody taproot. Leaves oblanceolate, acute, with 3-5 conspicuous veins. Scapes usually several, 15-40 cm, erect, deeply 5-sulcate. Spike ovoid to cylindrical, dense.

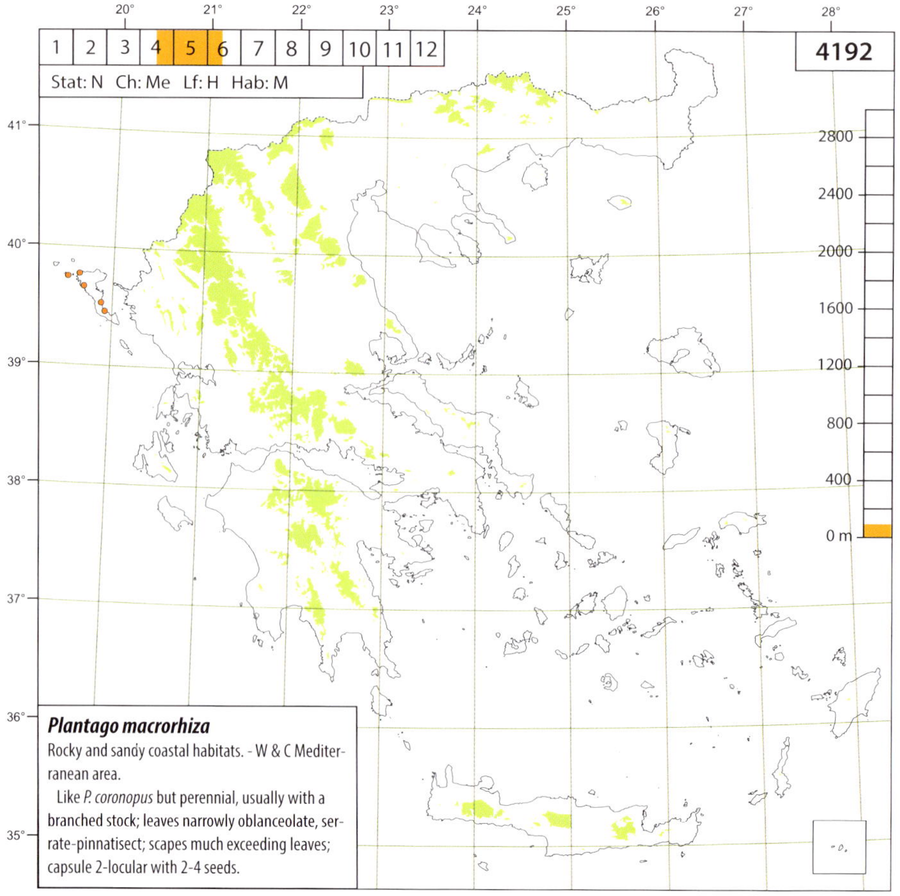

Plantago macrorhiza

Rocky and sandy coastal habitats. - W & C Mediterranean area.

Like *P. coronopus* but perennial, usually with a branched stock; leaves narrowly oblanceolate, serrate-pinnatisect; scapes much exceeding leaves; capsule 2-locular with 2-4 seeds.

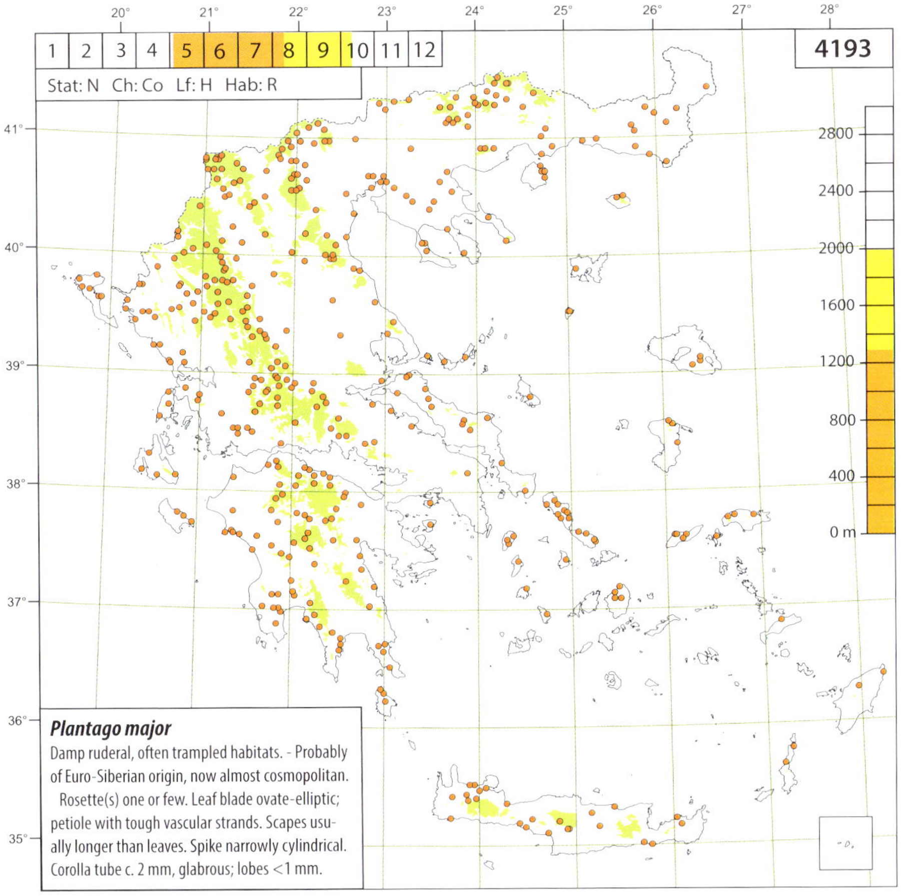

Plantago major
Damp ruderal, often trampled habitats. - Probably of Euro-Siberian origin, now almost cosmopolitan.
Rosette(s) one or few. Leaf blade ovate-elliptic; petiole with tough vascular strands. Scapes usually longer than leaves. Spike narrowly cylindrical. Corolla tube c. 2 mm, glabrous; lobes <1 mm.

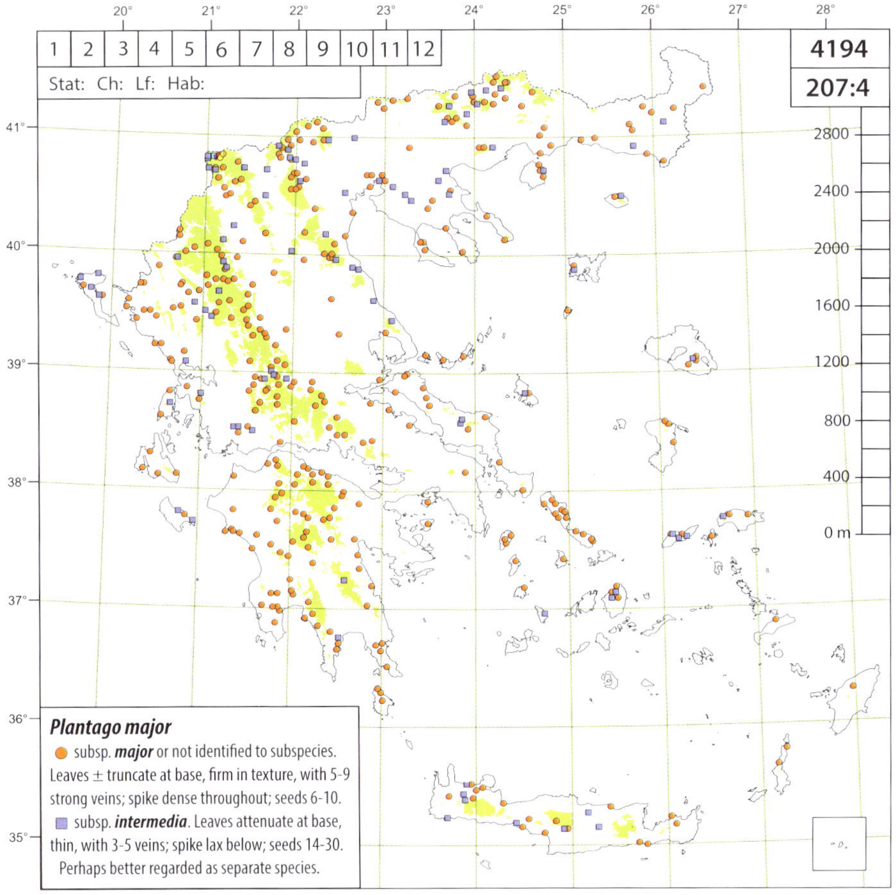

Plantago major
● subsp. ***major*** or not identified to subspecies. Leaves ± truncate at base, firm in texture, with 5-9 strong veins; spike dense throughout; seeds 6-10.
■ subsp. ***intermedia***. Leaves attenuate at base, thin, with 3-5 veins; spike lax below; seeds 14-30. Perhaps better regarded as separate species.

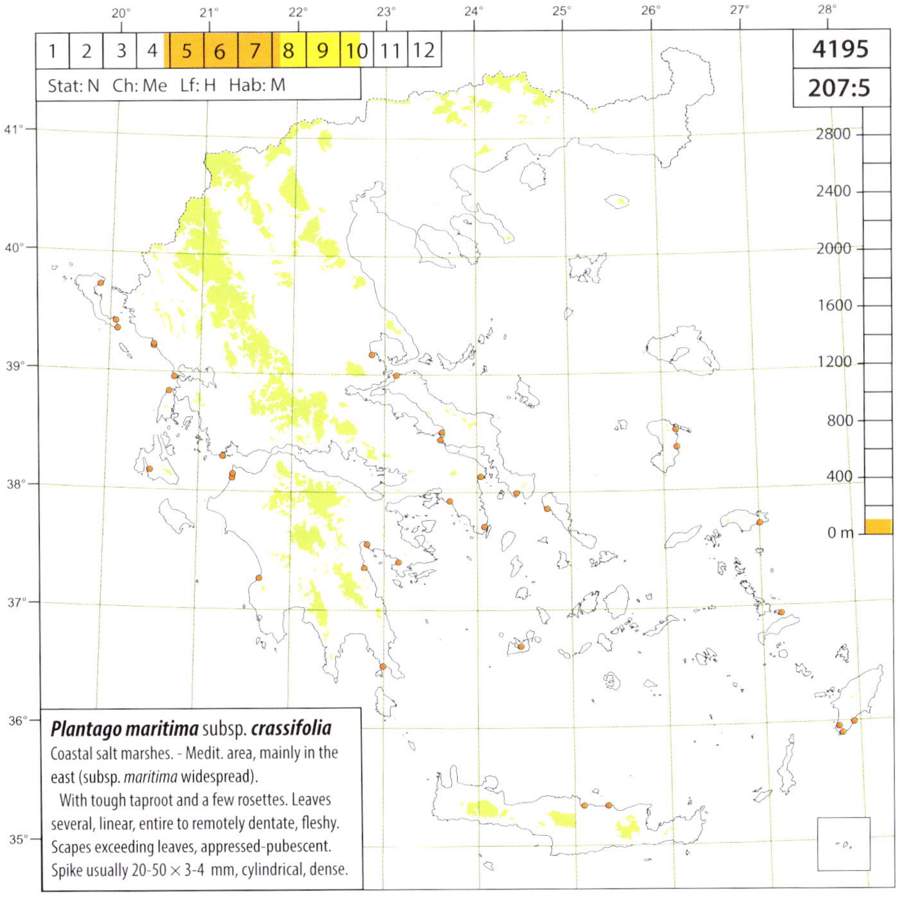

Plantago maritima subsp. ***crassifolia***
Coastal salt marshes. - Medit. area, mainly in the east (subsp. *maritima* widespread).
With tough taproot and a few rosettes. Leaves several, linear, entire to remotely dentate, fleshy. Scapes exceeding leaves, appressed-pubescent. Spike usually 20-50 × 3-4 mm, cylindrical, dense.

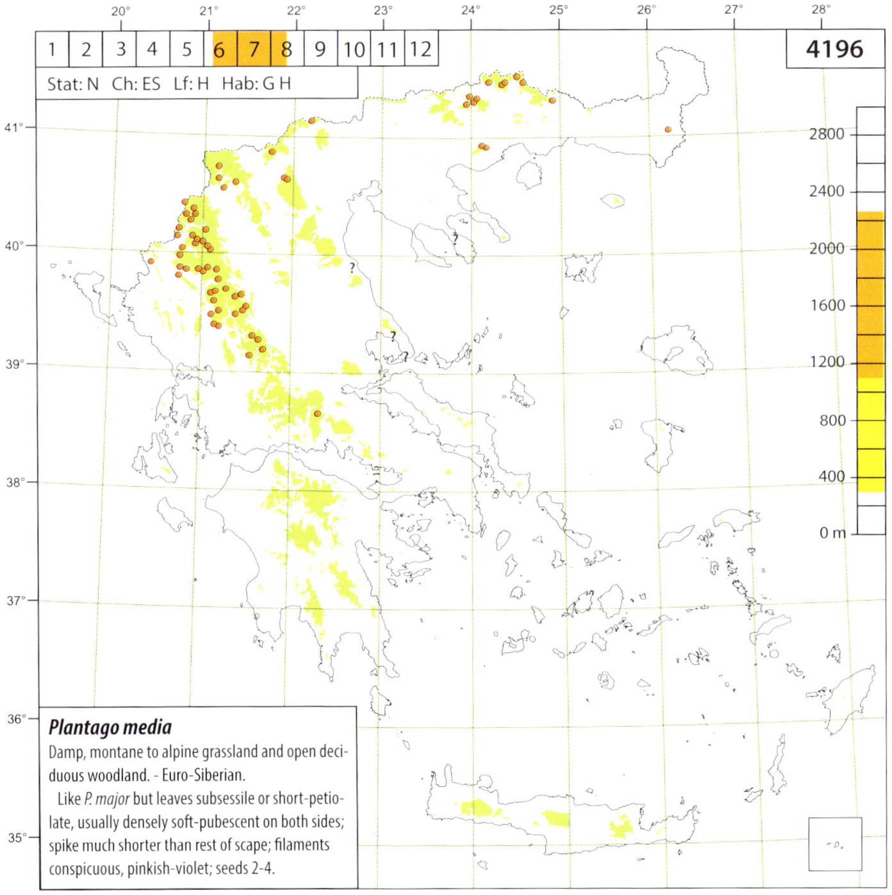

Plantago media
Damp, montane to alpine grassland and open deciduous woodland. - Euro-Siberian.
Like *P. major* but leaves subsessile or short-petiolate, usually densely soft-pubescent on both sides; spike much shorter than rest of scape; filaments conspicuous, pinkish-violet; seeds 2-4.

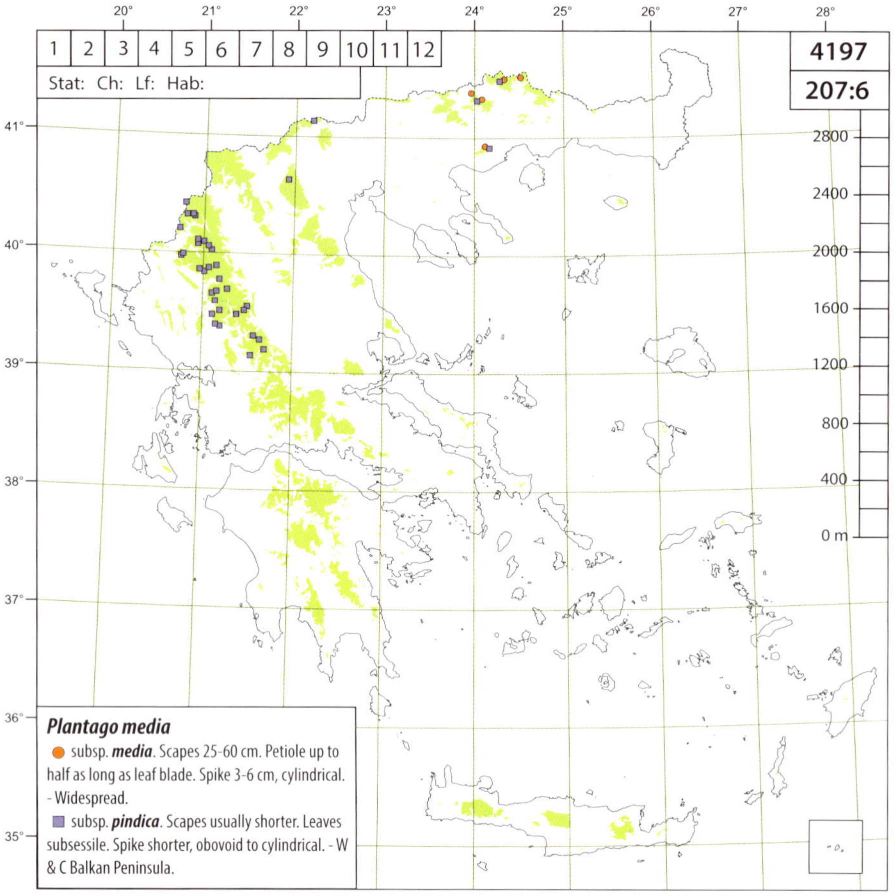

Plantago media

● subsp. ***media***. Scapes 25-60 cm. Petiole up to half as long as leaf blade. Spike 3-6 cm, cylindrical. - Widespread.

■ subsp. ***pindica***. Scapes usually shorter. Leaves subsessile. Spike shorter, obovoid to cylindrical. - W & C Balkan Peninsula.

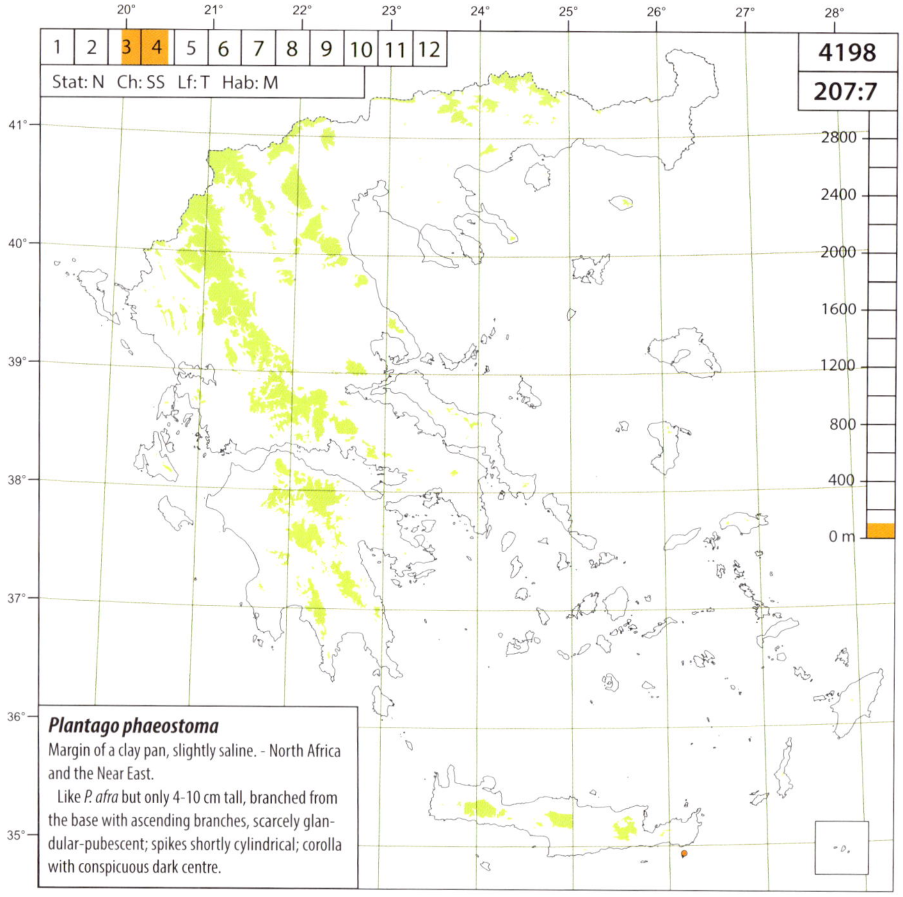

Plantago phaeostoma

Margin of a clay pan, slightly saline. - North Africa and the Near East.

Like *P. afra* but only 4-10 cm tall, branched from the base with ascending branches, scarcely glandular-pubescent; spikes shortly cylindrical; corolla with conspicuous dark centre.

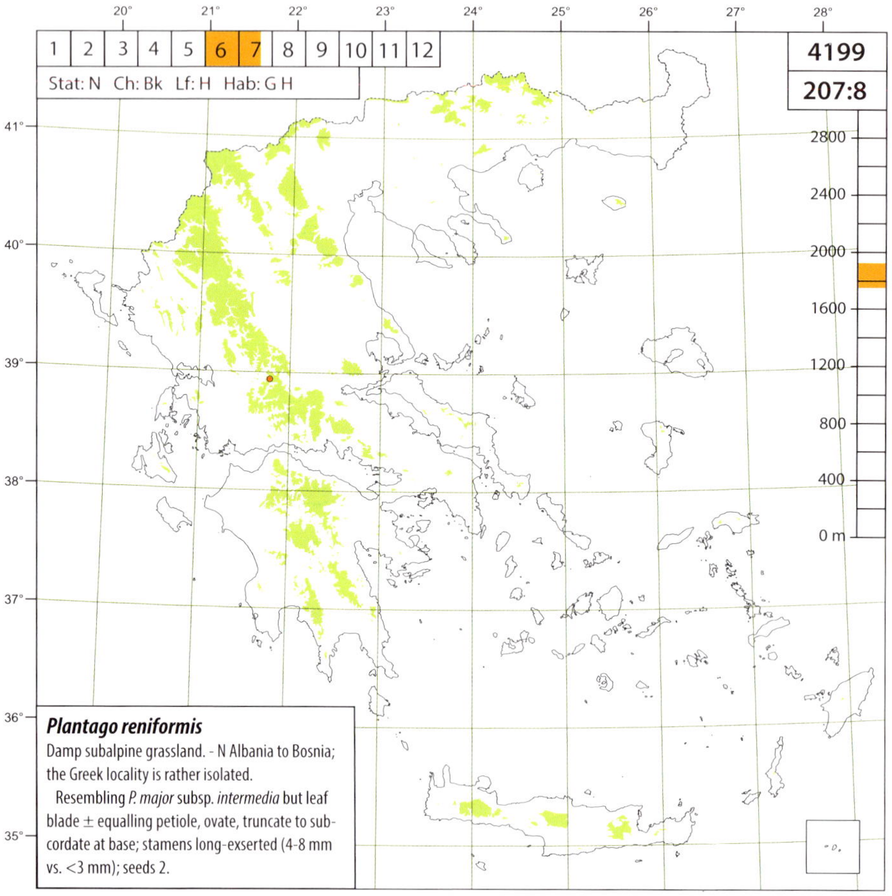

Plantago reniformis

Damp subalpine grassland. - N Albania to Bosnia; the Greek locality is rather isolated.

Resembling *P. major* subsp. *intermedia* but leaf blade ± equalling petiole, ovate, truncate to subcordate at base; stamens long-exserted (4-8 mm vs. <3 mm); seeds 2.

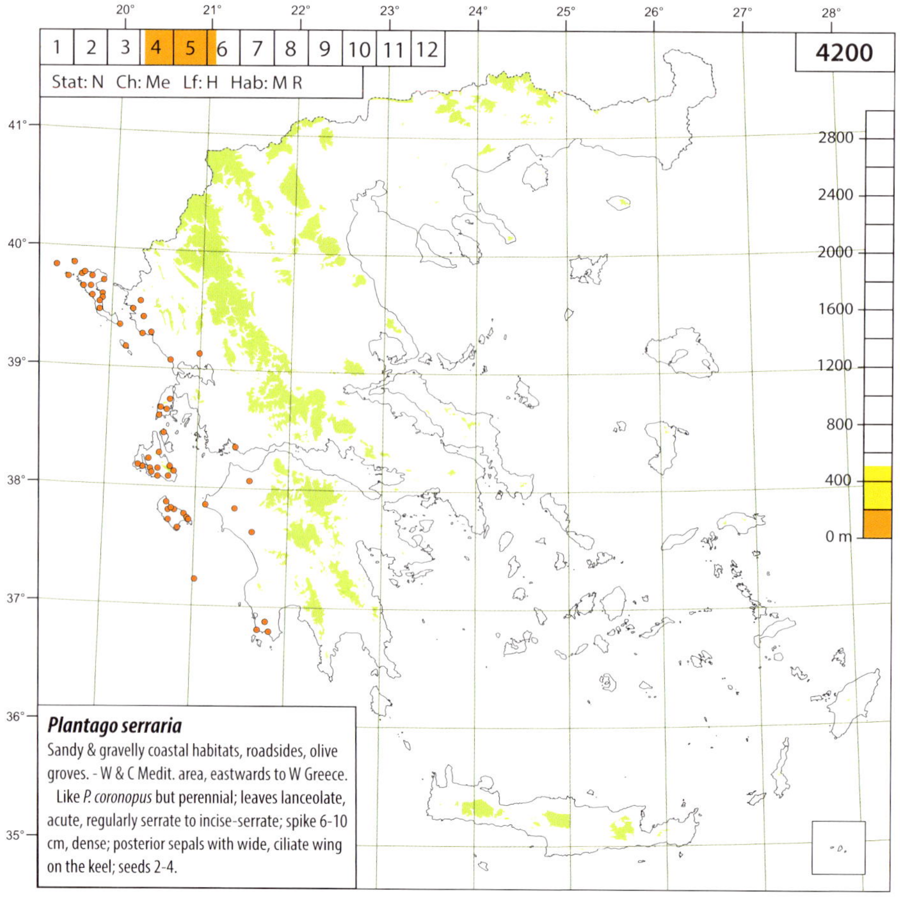

Plantago serraria

Sandy & gravelly coastal habitats, roadsides, olive groves. - W & C Medit. area, eastwards to W Greece.

Like *P. coronopus* but perennial; leaves lanceolate, acute, regularly serrate to incise-serrate; spike 6-10 cm, dense; posterior sepals with wide, ciliate wing on the keel; seeds 2-4.

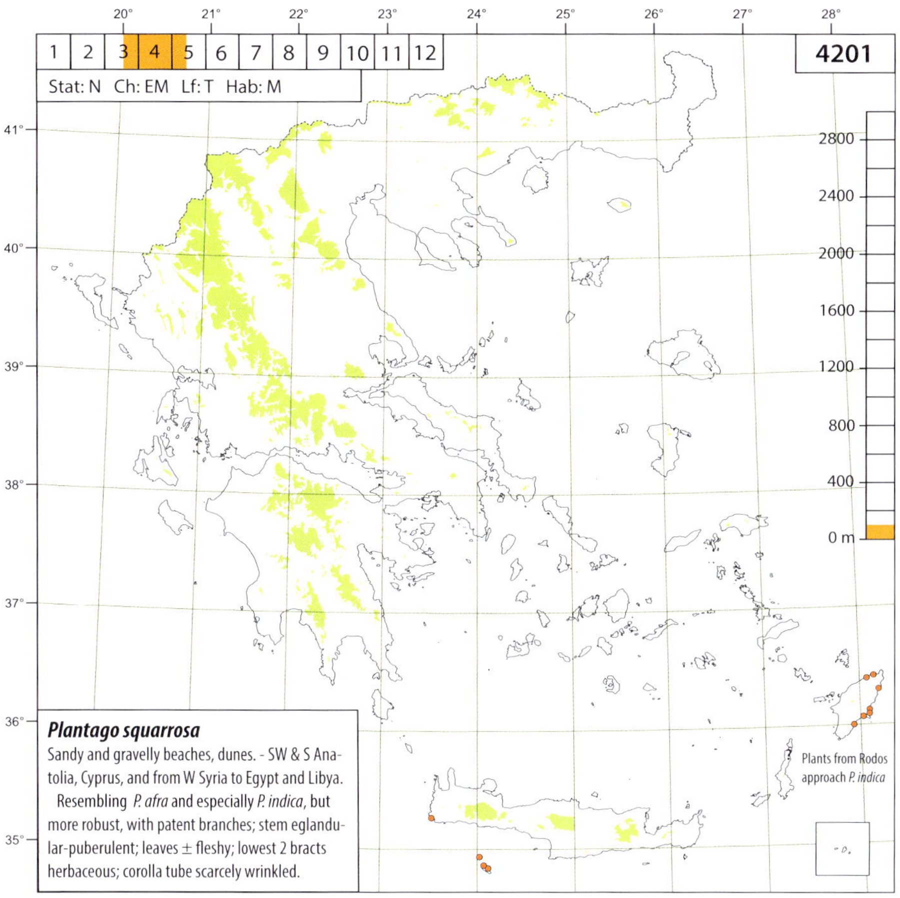

Plantago squarrosa

Sandy and gravelly beaches, dunes. - SW & S Anatolia, Cyprus, and from W Syria to Egypt and Libya.

Resembling *P. afra* and especially *P. indica*, but more robust, with patent branches; stem eglandular-puberulent; leaves ± fleshy; lowest 2 bracts herbaceous; corolla tube scarcely wrinkled.

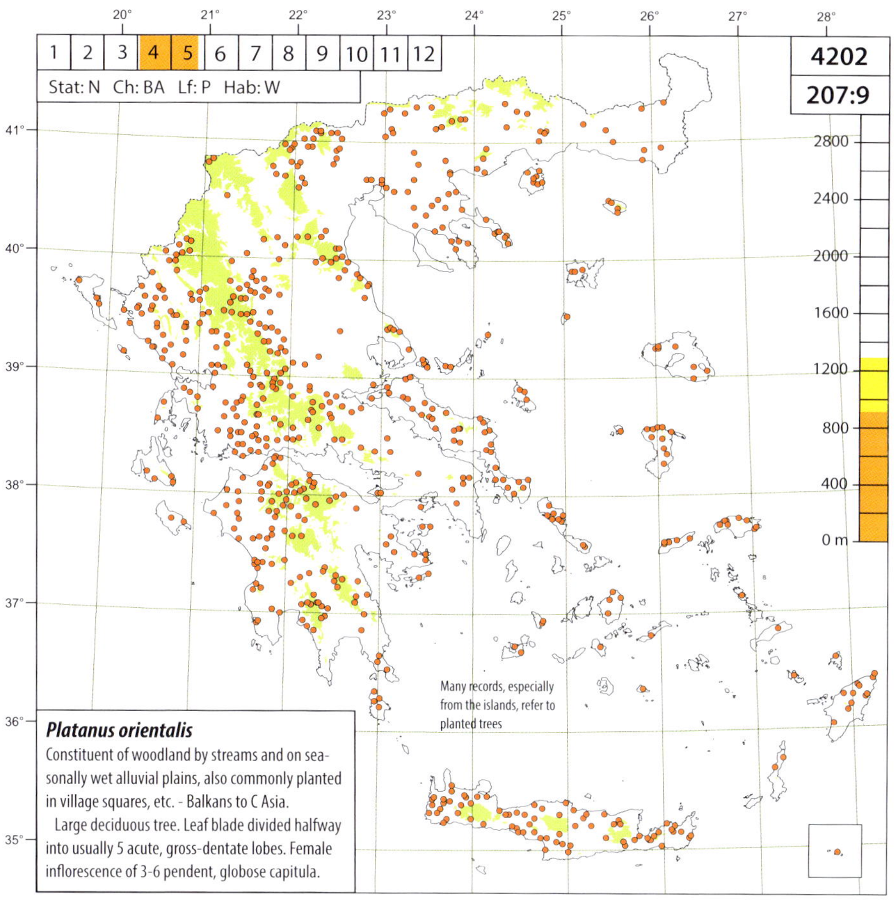

Platanus orientalis

Constituent of woodland by streams and on seasonally wet alluvial plains, also commonly planted in village squares, etc. - Balkans to C Asia.

Large deciduous tree. Leaf blade divided halfway into usually 5 acute, gross-dentate lobes. Female inflorescence of 3-6 pendent, globose capitula.

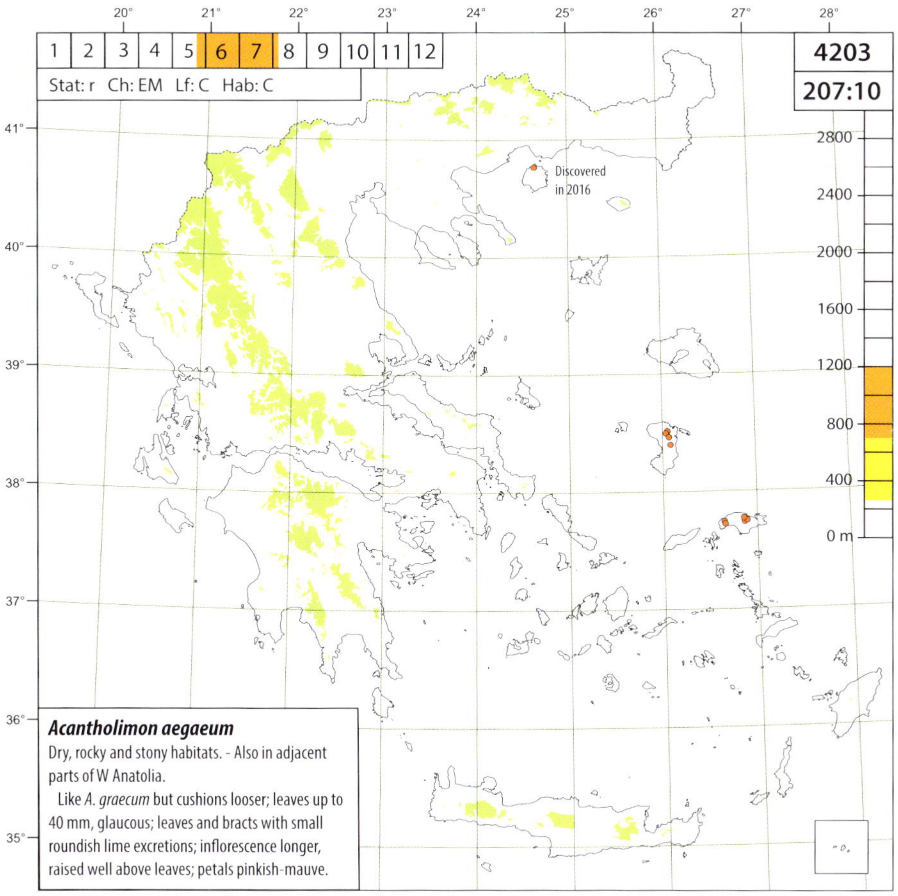

Acantholimon aegaeum

Dry, rocky and stony habitats. - Also in adjacent parts of W Anatolia.

Like *A. graecum* but cushions looser; leaves up to 40 mm, glaucous; leaves and bracts with small roundish lime excretions; inflorescence longer, raised well above leaves; petals pinkish-mauve.

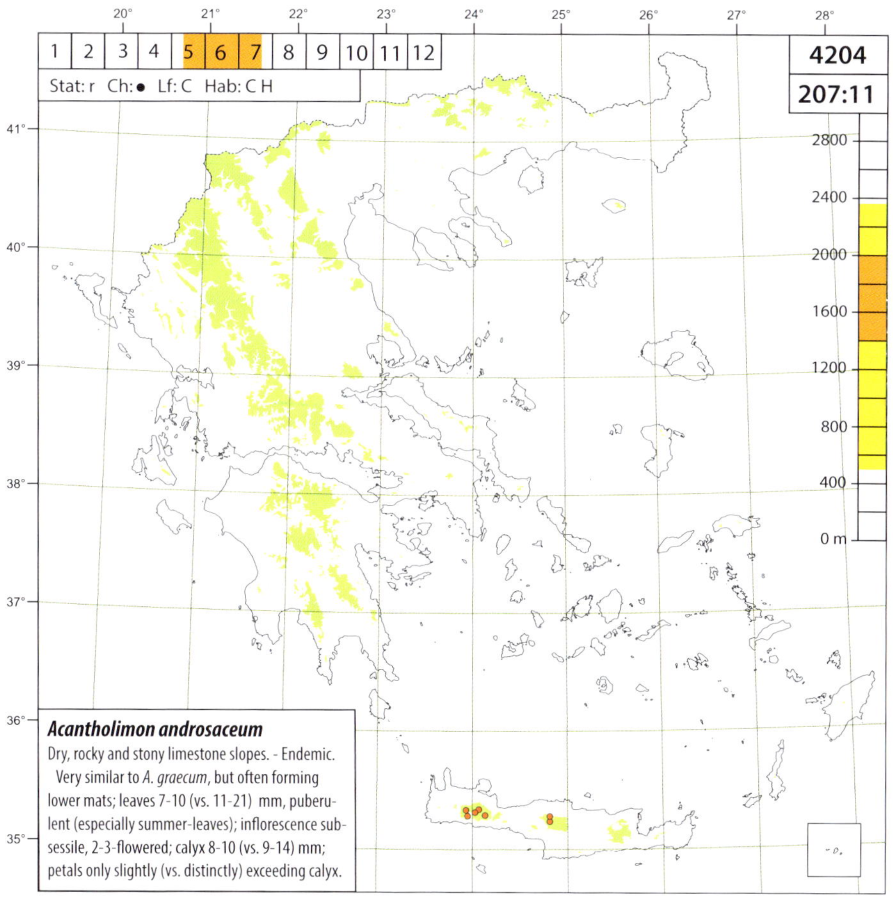

Acantholimon androsaceum

Dry, rocky and stony limestone slopes. - Endemic.

Very similar to *A. graecum*, but often forming lower mats; leaves 7-10 (vs. 11-21) mm, puberulent (especially summer-leaves); inflorescence subsessile, 2-3-flowered; calyx 8-10 (vs. 9-14) mm; petals only slightly (vs. distinctly) exceeding calyx.

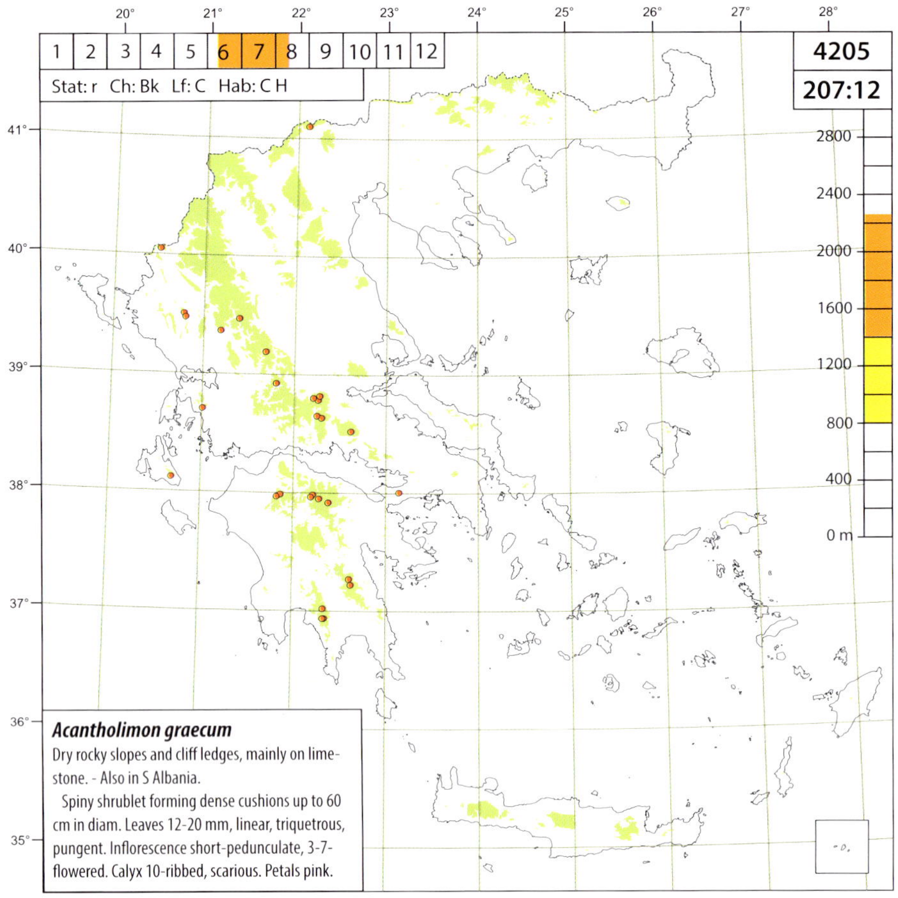

Acantholimon graecum
Dry rocky slopes and cliff ledges, mainly on limestone. - Also in S Albania.
Spiny shrublet forming dense cushions up to 60 cm in diam. Leaves 12-20 mm, linear, triquetrous, pungent. Inflorescence short-pedunculate, 3-7-flowered. Calyx 10-ribbed, scarious. Petals pink.

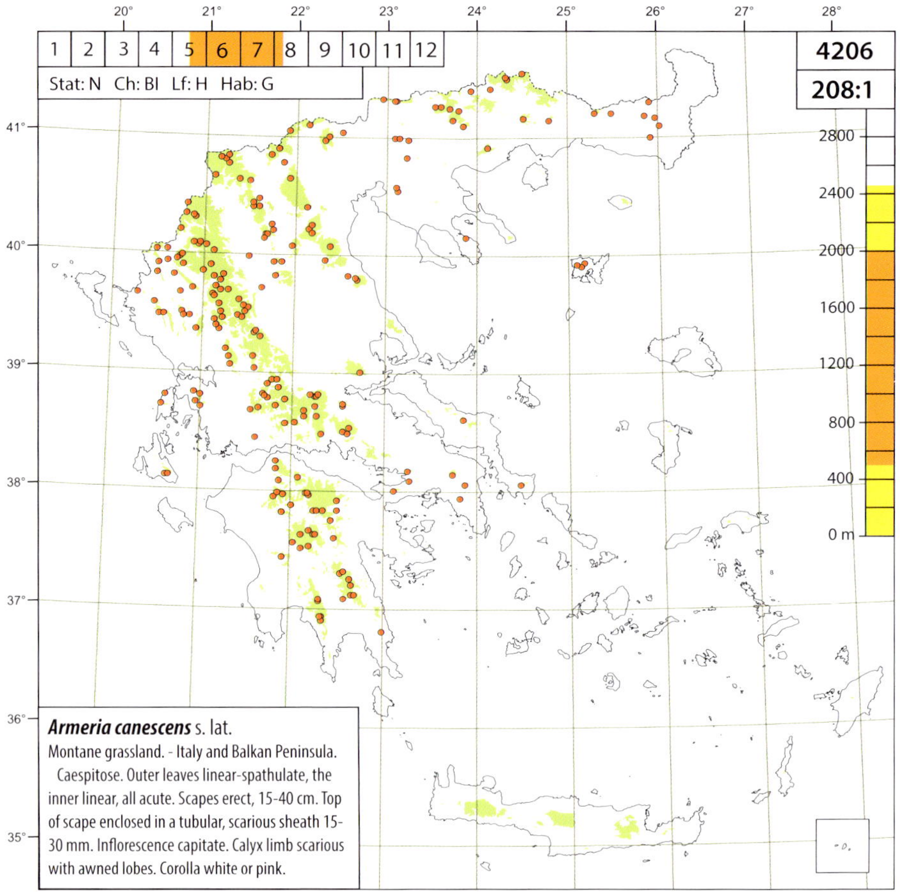

Armeria canescens s. lat.
Montane grassland. - Italy and Balkan Peninsula.
Caespitose. Outer leaves linear-spathulate, the inner linear, all acute. Scapes erect, 15-40 cm. Top of scape enclosed in a tubular, scarious sheath 15-30 mm. Inflorescence capitate. Calyx limb scarious with awned lobes. Corolla white or pink.

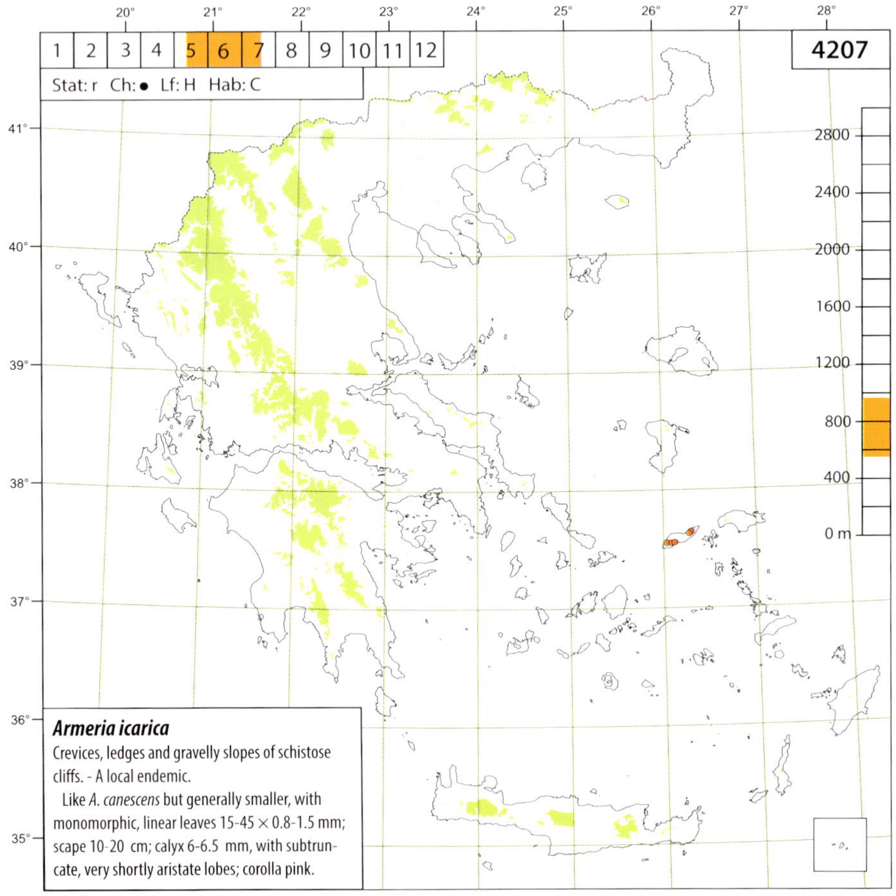

Armeria icarica
Crevices, ledges and gravelly slopes of schistose cliffs. - A local endemic.
Like *A. canescens* but generally smaller, with monomorphic, linear leaves 15-45 × 0.8-1.5 mm; scape 10-20 cm; calyx 6-6.5 mm, with subtruncate, very shortly aristate lobes; corolla pink.

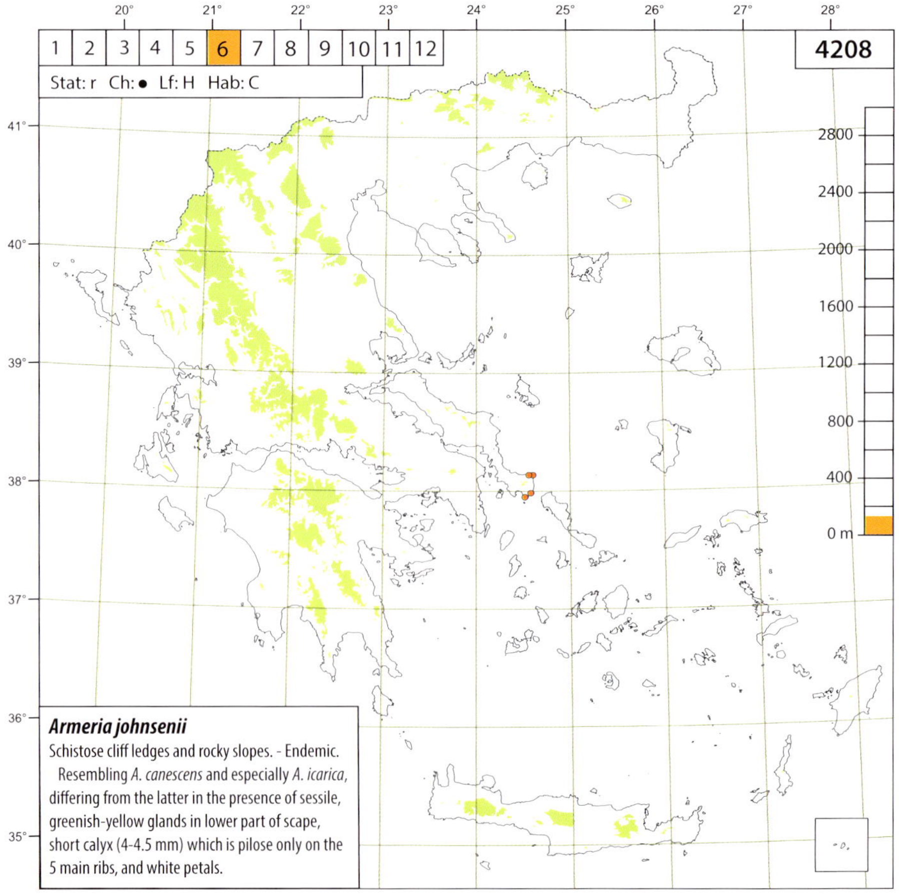

Armeria johnsenii
Schistose cliff ledges and rocky slopes. - Endemic.
Resembling *A. canescens* and especially *A. icarica*, differing from the latter in the presence of sessile, greenish-yellow glands in lower part of scape, short calyx (4-4.5 mm) which is pilose only on the 5 main ribs, and white petals.

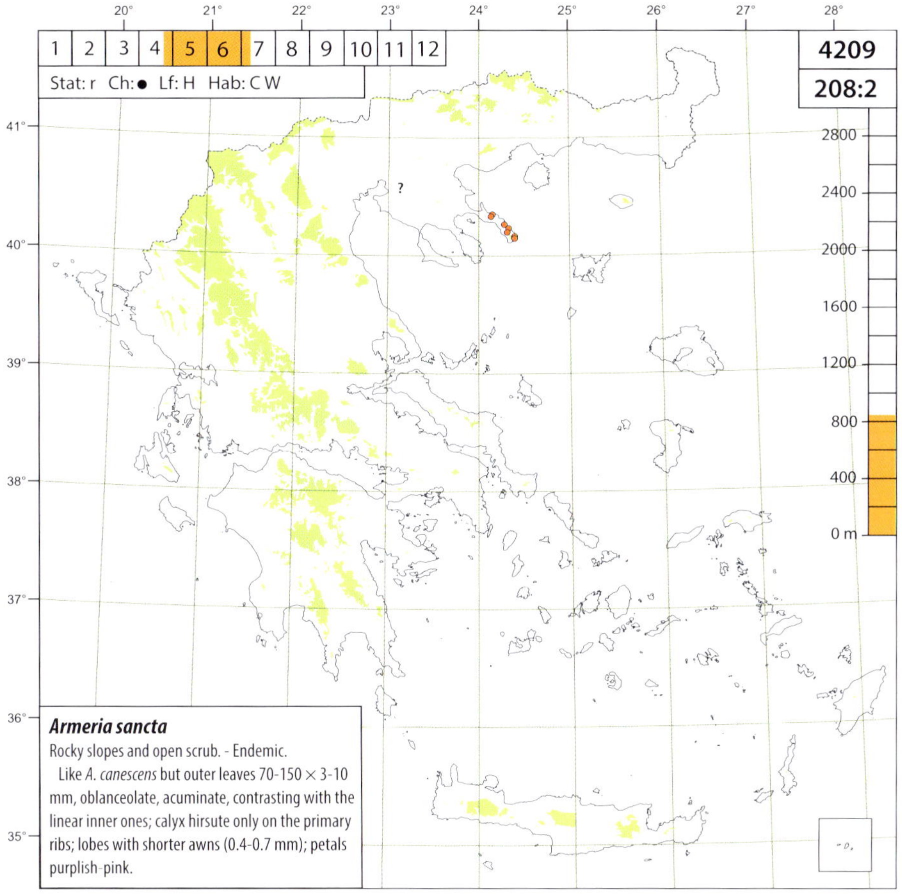

Armeria sancta

Rocky slopes and open scrub. - Endemic.

Like *A. canescens* but outer leaves 70-150 × 3-10 mm, oblanceolate, acuminate, contrasting with the linear inner ones; calyx hirsute only on the primary ribs; lobes with shorter awns (0.4-0.7 mm); petals purplish-pink.

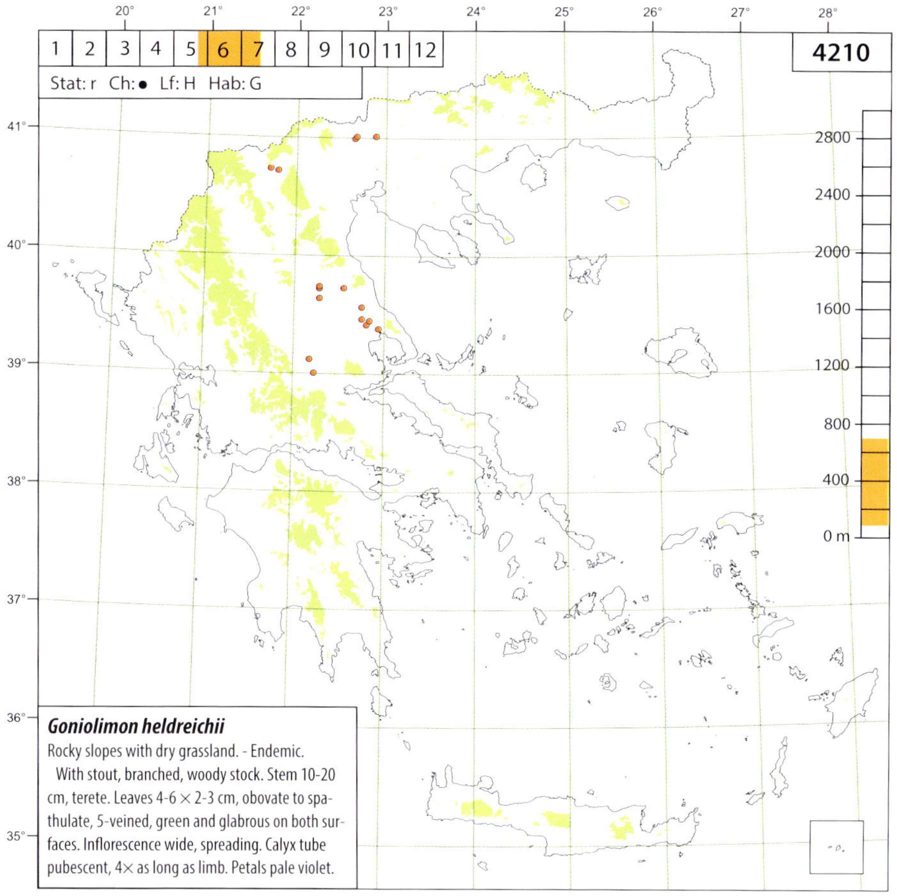

Goniolimon heldreichii

Rocky slopes with dry grassland. - Endemic.

With stout, branched, woody stock. Stem 10-20 cm, terete. Leaves 4-6 × 2-3 cm, obovate to spathulate, 5-veined, green and glabrous on both surfaces. Inflorescence wide, spreading. Calyx tube pubescent, 4× as long as limb. Petals pale violet.

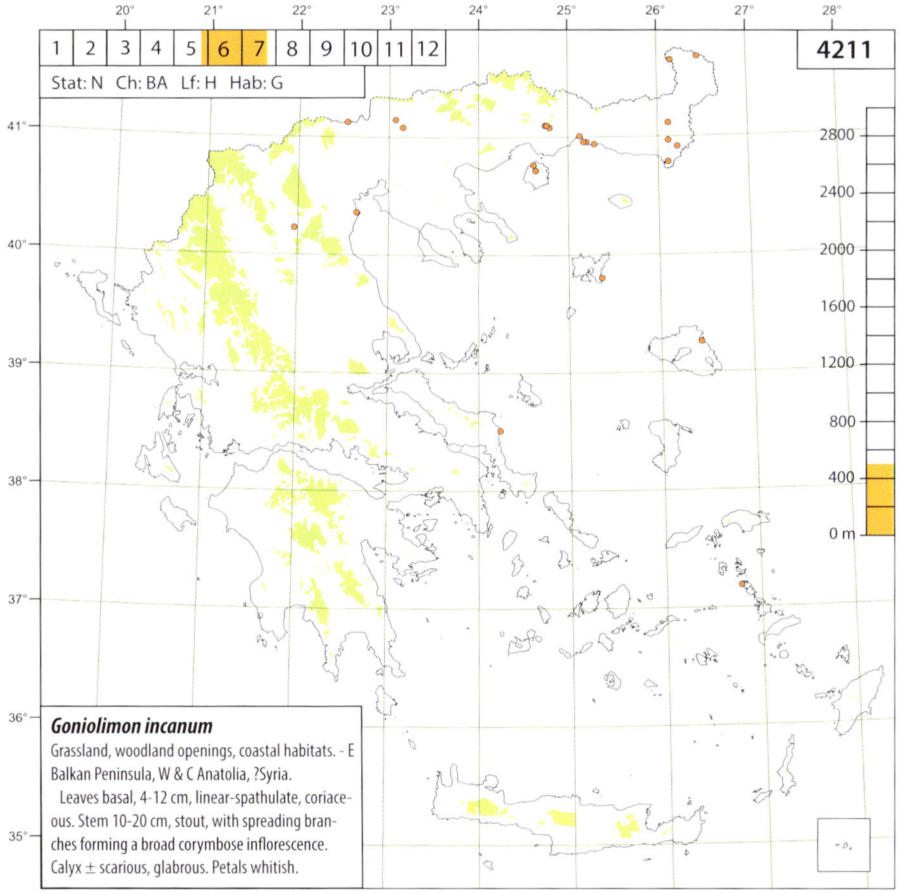

Goniolimon incanum

Grassland, woodland openings, coastal habitats. - E Balkan Peninsula, W & C Anatolia, ?Syria.

Leaves basal, 4-12 cm, linear-spathulate, coriaceous. Stem 10-20 cm, stout, with spreading branches forming a broad corymbose inflorescence. Calyx ± scarious, glabrous. Petals whitish.

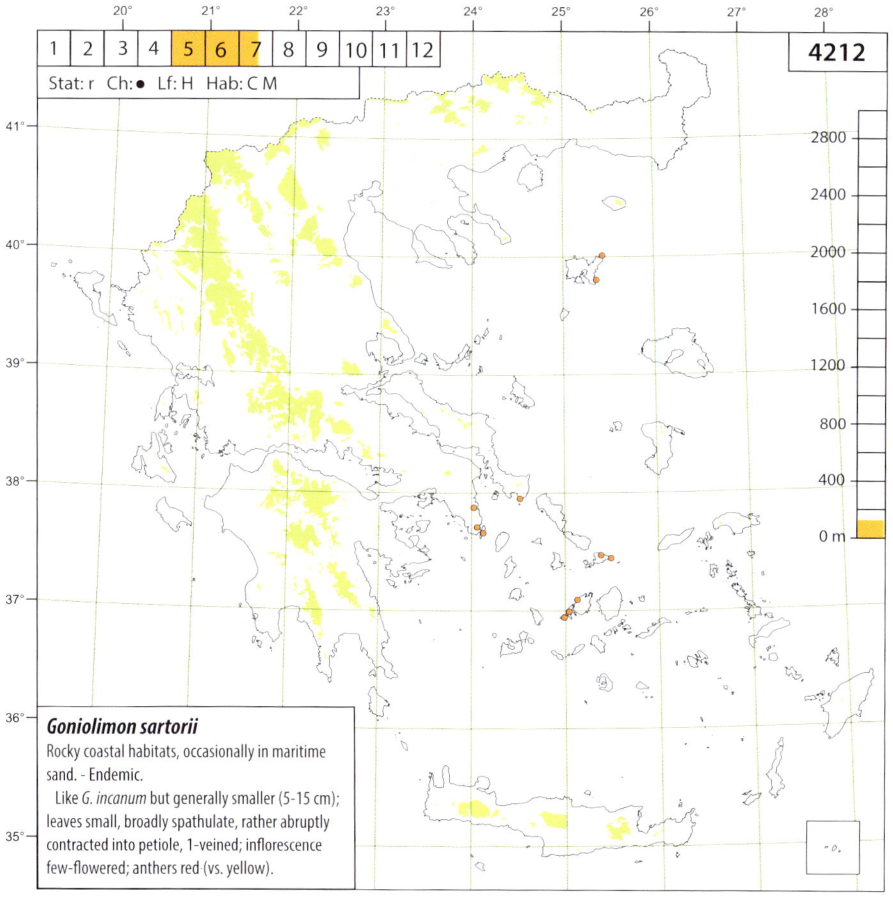

Goniolimon sartorii

Rocky coastal habitats, occasionally in maritime sand. - Endemic.

Like *G. incanum* but generally smaller (5-15 cm); leaves small, broadly spathulate, rather abruptly contracted into petiole, 1-veined; inflorescence few-flowered; anthers red (vs. yellow).

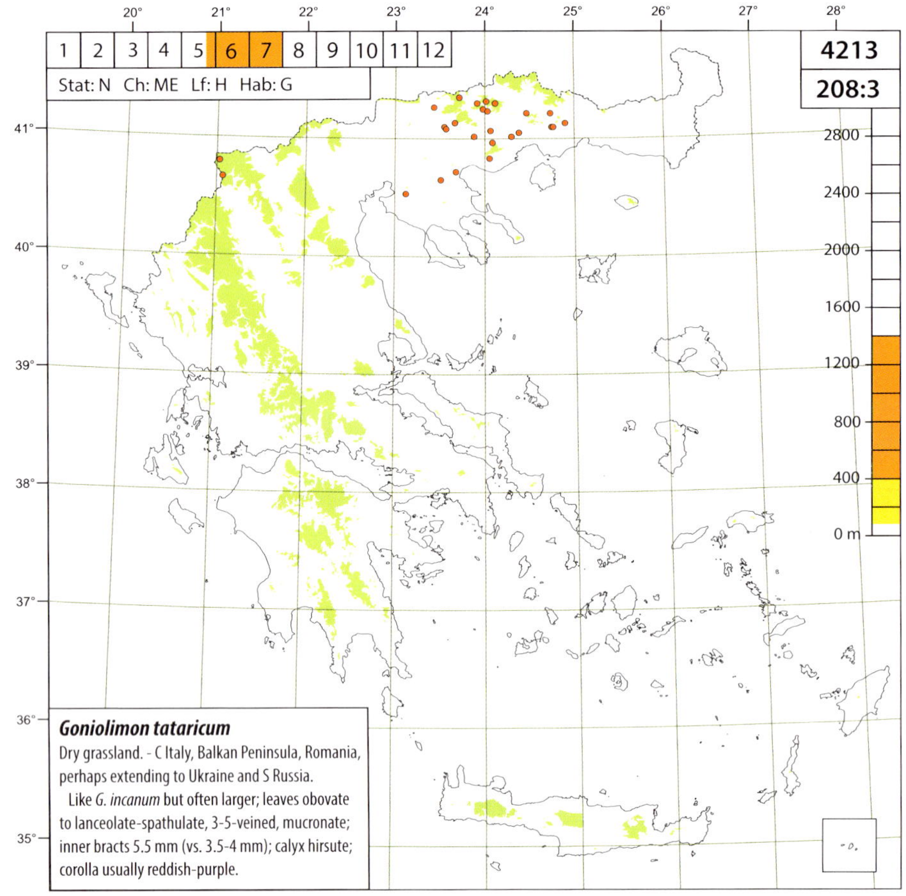

Goniolimon tataricum

Dry grassland. - C Italy, Balkan Peninsula, Romania, perhaps extending to Ukraine and S Russia.

Like *G. incanum* but often larger; leaves obovate to lanceolate-spathulate, 3-5-veined, mucronate; inner bracts 5.5 mm (vs. 3.5-4 mm); calyx hirsute; corolla usually reddish-purple.

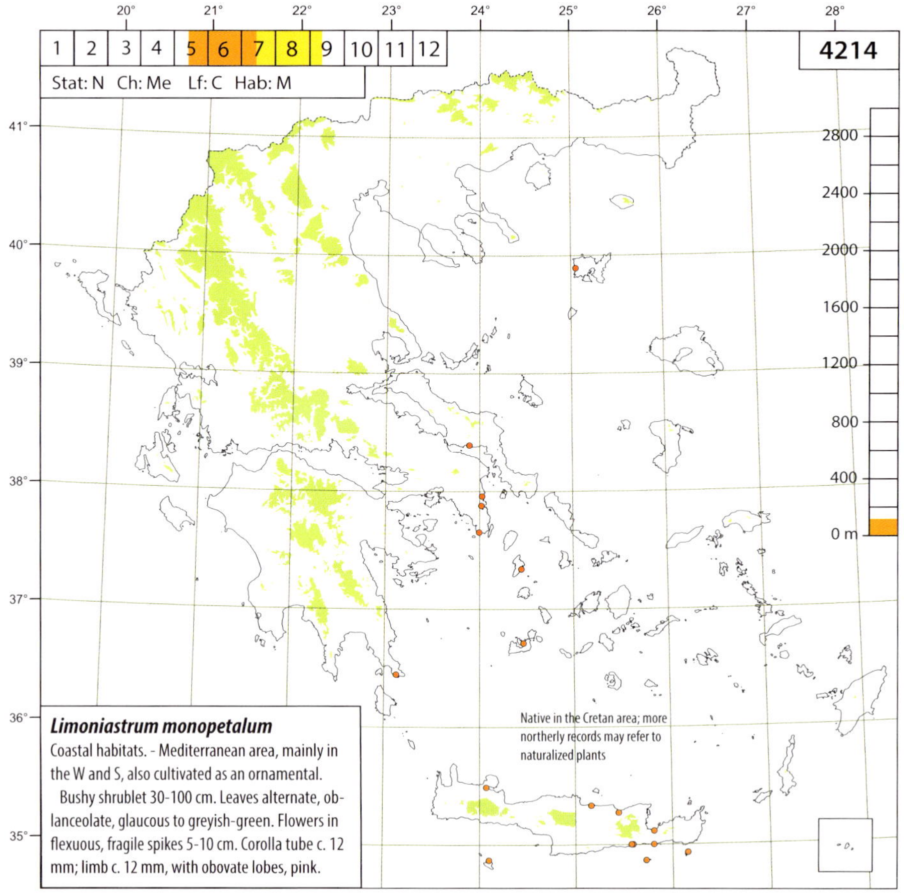

Limoniastrum monopetalum

Coastal habitats. - Mediterranean area, mainly in the W and S, also cultivated as an ornamental.

Bushy shrublet 30-100 cm. Leaves alternate, ob-lanceolate, glaucous to greyish-green. Flowers in flexuous, fragile spikes 5-10 cm. Corolla tube c. 12 mm; limb c. 12 mm, with obovate lobes, pink.

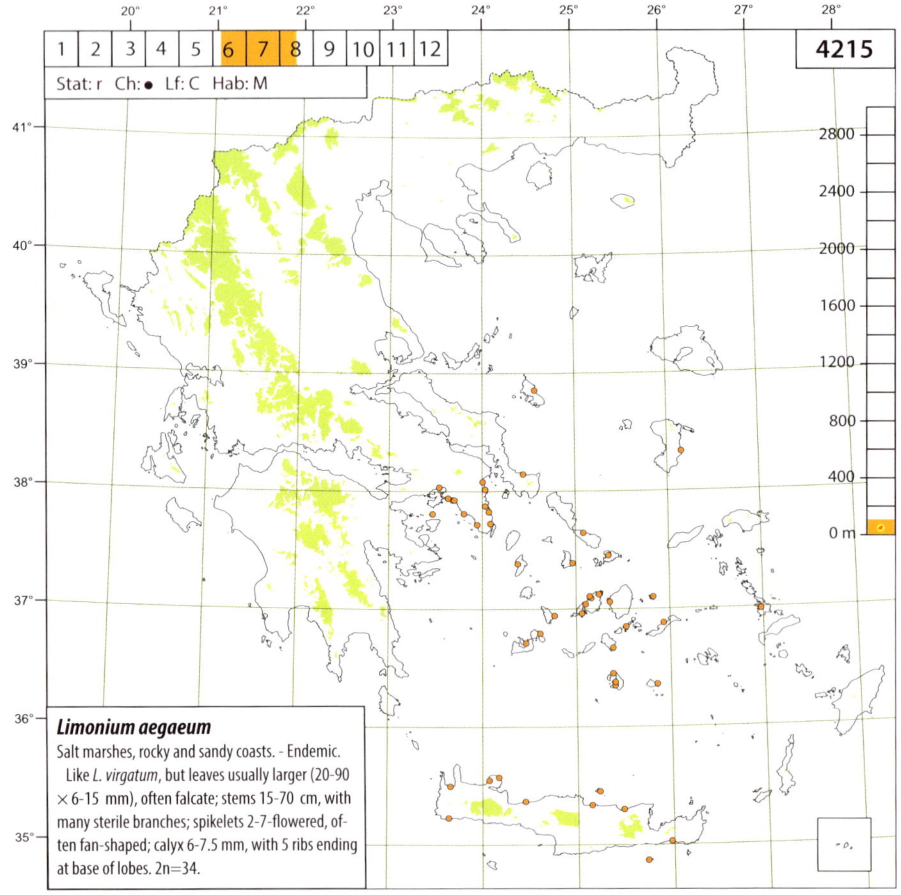

Limonium aegaeum

Salt marshes, rocky and sandy coasts. - Endemic.

Like *L. virgatum*, but leaves usually larger (20-90 × 6-15 mm), often falcate; stems 15-70 cm, with many sterile branches; spikelets 2-7-flowered, often fan-shaped; calyx 6-7.5 mm, with 5 ribs ending at base of lobes. 2n=34.

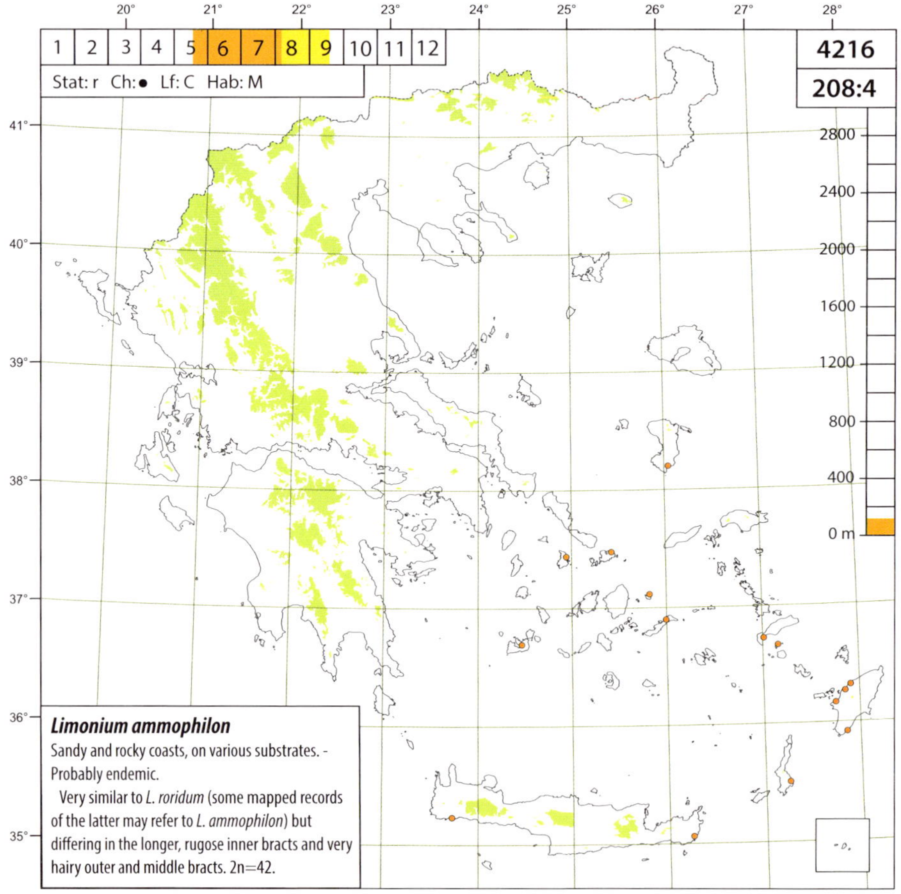

Limonium ammophilon

Sandy and rocky coasts, on various substrates. - Probably endemic.

Very similar to *L. roridum* (some mapped records of the latter may refer to *L. ammophilon*) but differing in the longer, rugose inner bracts and very hairy outer and middle bracts. 2n=42.

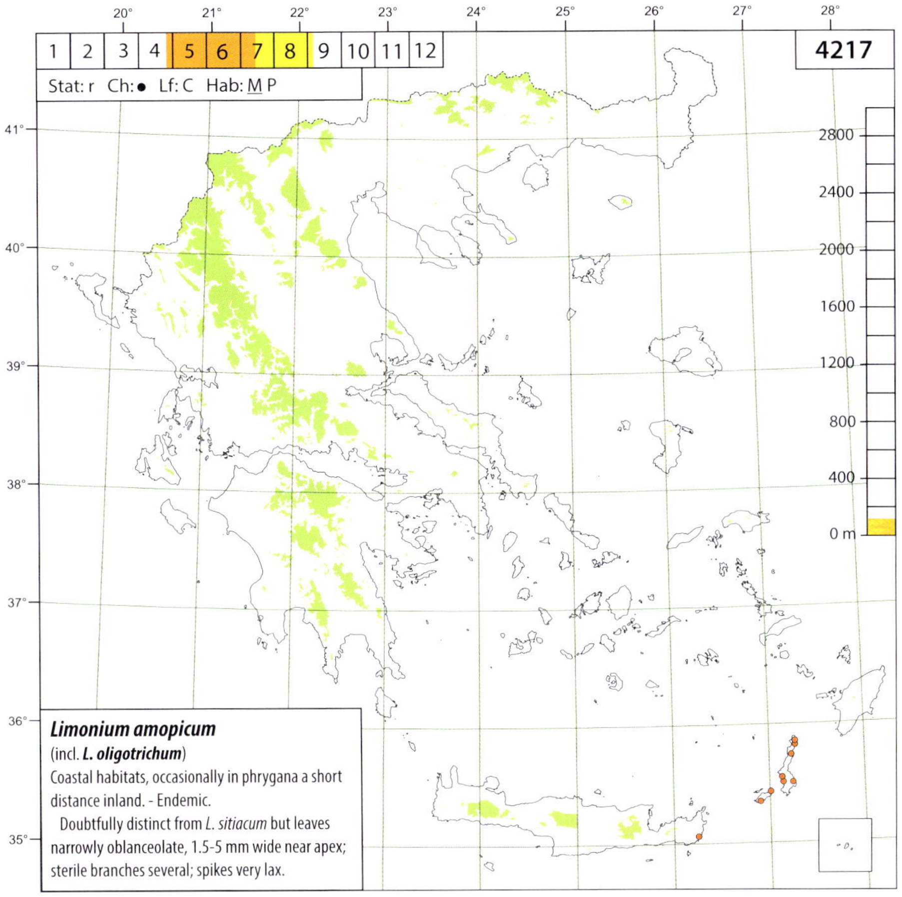

Limonium amopicum
(incl. ***L. oligotrichum***)
Coastal habitats, occasionally in phrygana a short distance inland. - Endemic.
Doubtfully distinct from *L. sitiacum* but leaves narrowly oblanceolate, 1.5-5 mm wide near apex; sterile branches several; spikes very lax.

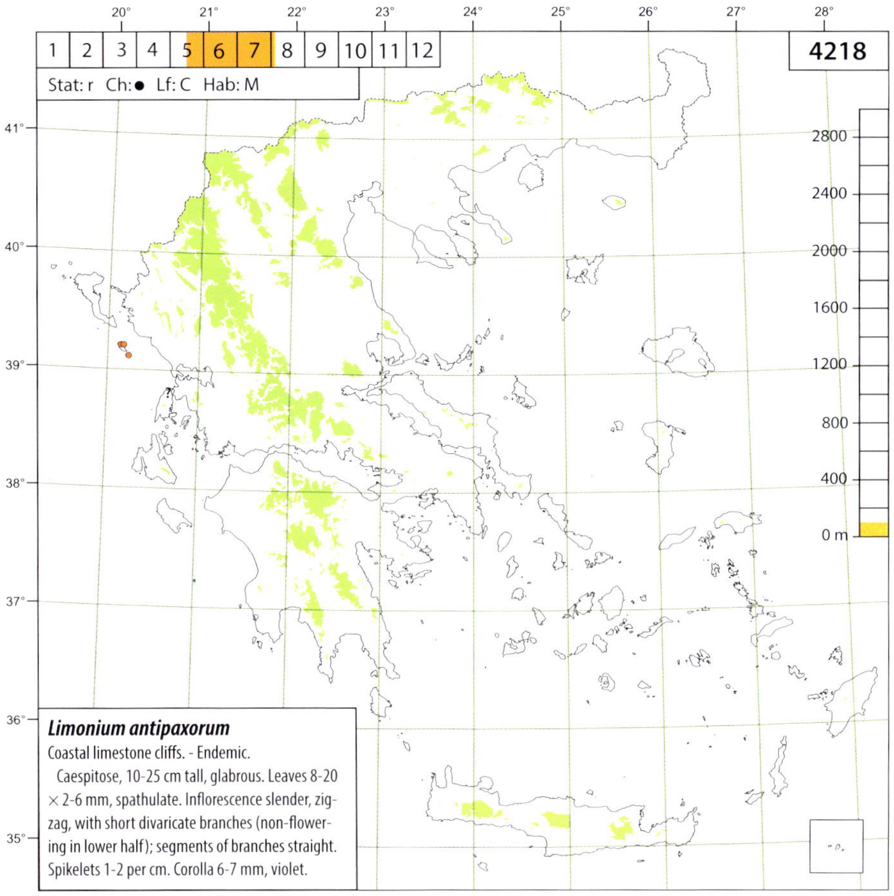

Limonium antipaxorum
Coastal limestone cliffs. - Endemic.
Caespitose, 10-25 cm tall, glabrous. Leaves 8-20 × 2-6 mm, spathulate. Inflorescence slender, zig-zag, with short divaricate branches (non-flowering in lower half); segments of branches straight. Spikelets 1-2 per cm. Corolla 6-7 mm, violet.

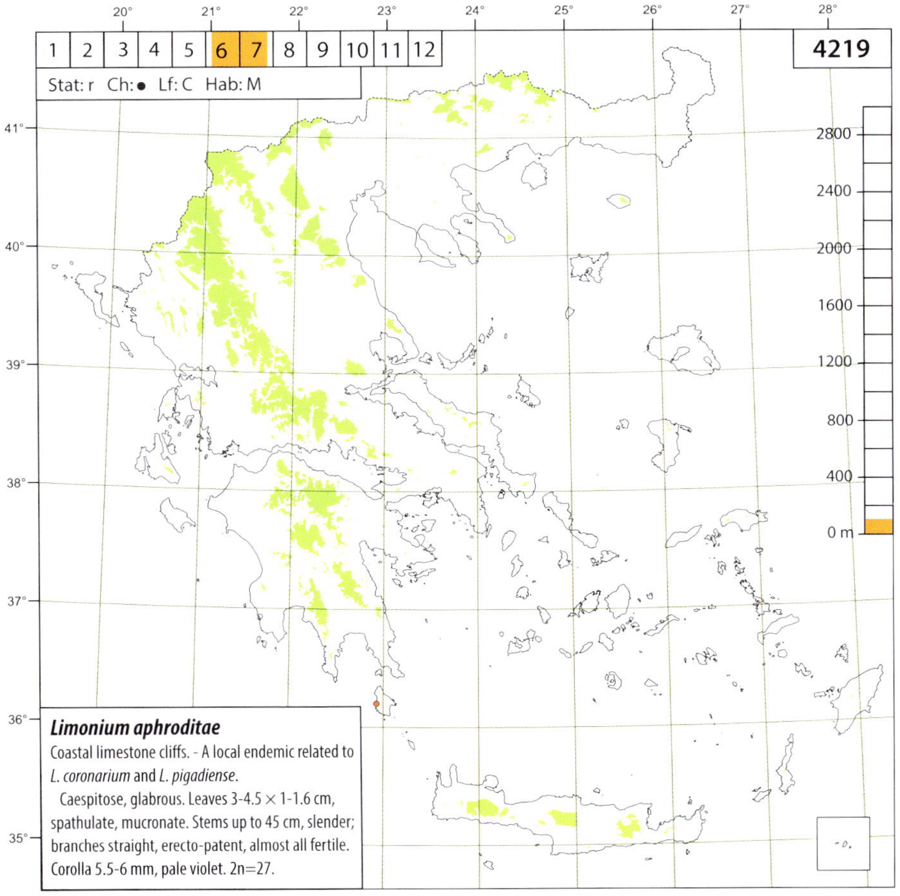

Limonium aphroditae
Coastal limestone cliffs. - A local endemic related to *L. coronarium* and *L. pigadiense*.
Caespitose, glabrous. Leaves 3-4.5 × 1-1.6 cm, spathulate, mucronate. Stems up to 45 cm, slender; branches straight, erecto-patent, almost all fertile. Corolla 5.5-6 mm, pale violet. 2n=27.

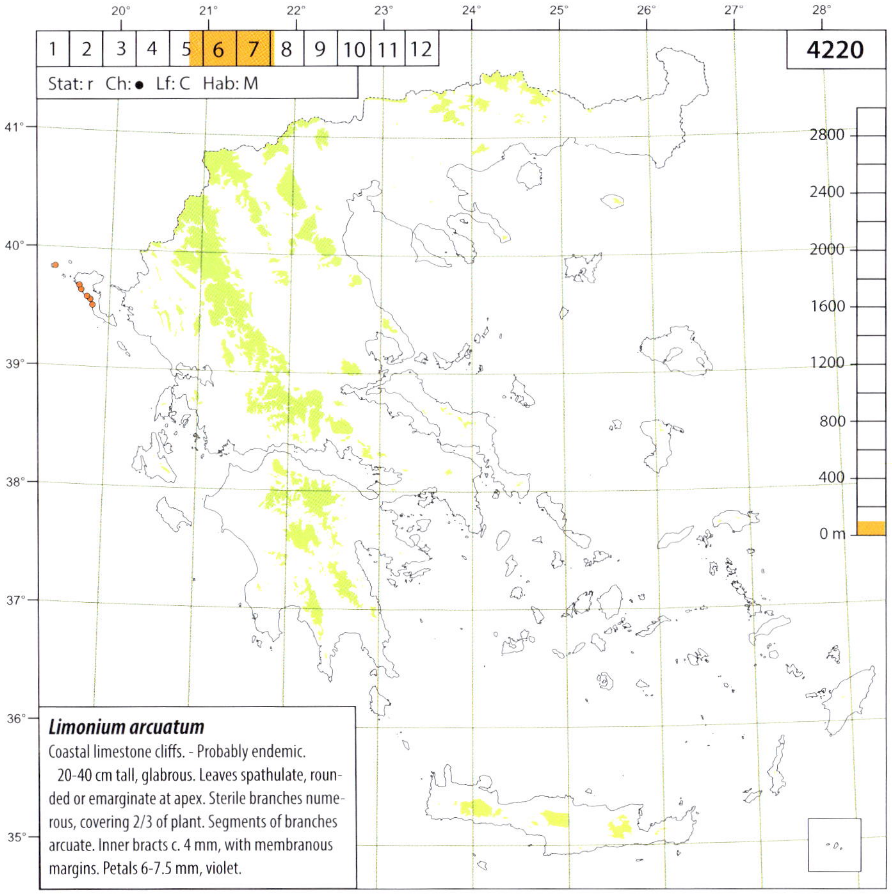

Limonium arcuatum
Coastal limestone cliffs. - Probably endemic.
20-40 cm tall, glabrous. Leaves spathulate, rounded or emarginate at apex. Sterile branches numerous, covering 2/3 of plant. Segments of branches arcuate. Inner bracts c. 4 mm, with membranous margins. Petals 6-7.5 mm, violet.

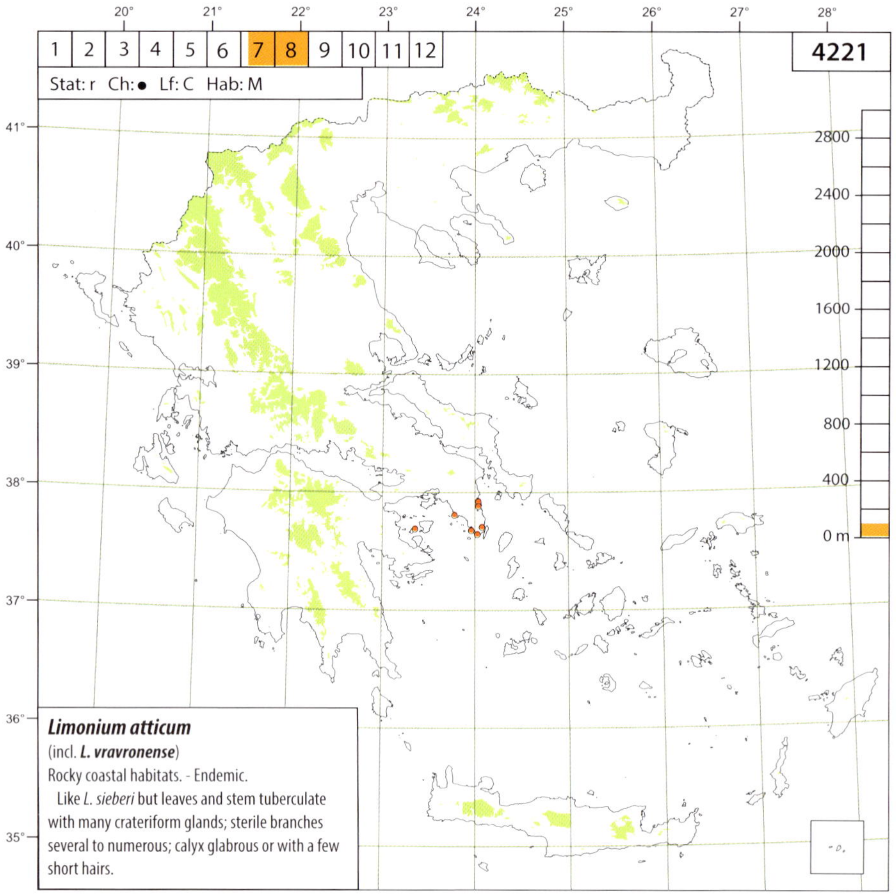

Limonium atticum
(incl. ***L. vravronense***)
Rocky coastal habitats. - Endemic.
Like *L. sieberi* but leaves and stem tuberculate with many crateriform glands; sterile branches several to numerous; calyx glabrous or with a few short hairs.

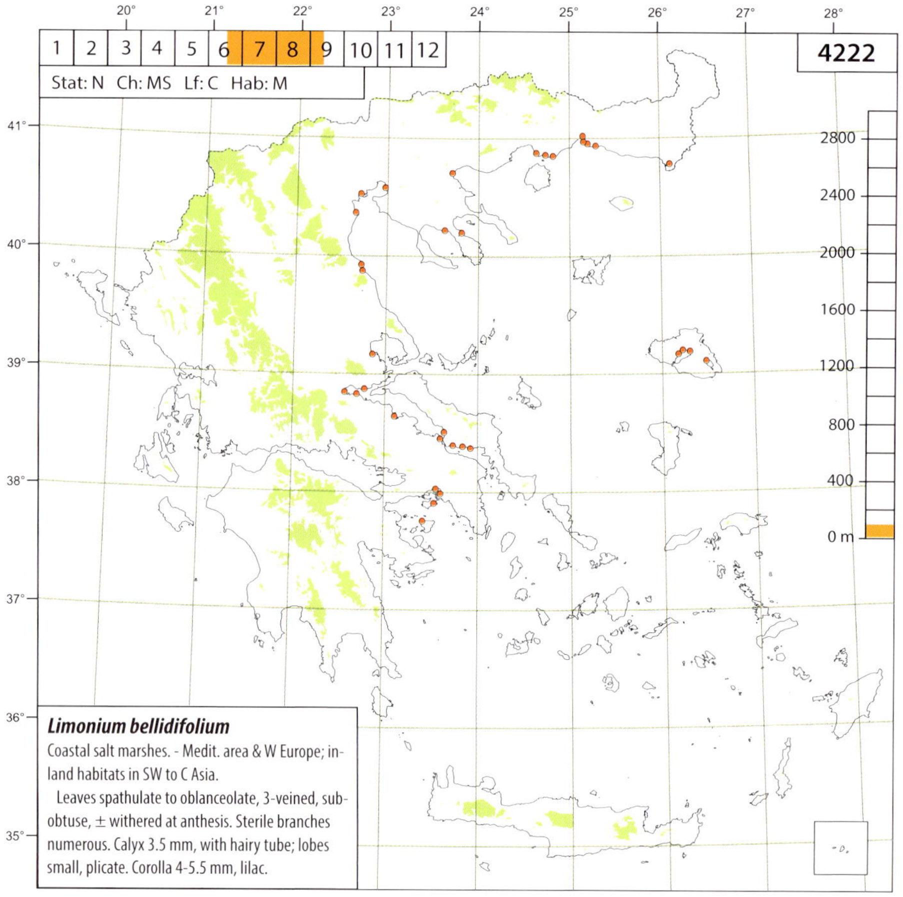

Limonium bellidifolium
Coastal salt marshes. - Medit. area & W Europe; inland habitats in SW to C Asia.
Leaves spathulate to oblanceolate, 3-veined, subobtuse, ± withered at anthesis. Sterile branches numerous. Calyx 3.5 mm, with hairy tube; lobes small, plicate. Corolla 4-5.5 mm, lilac.

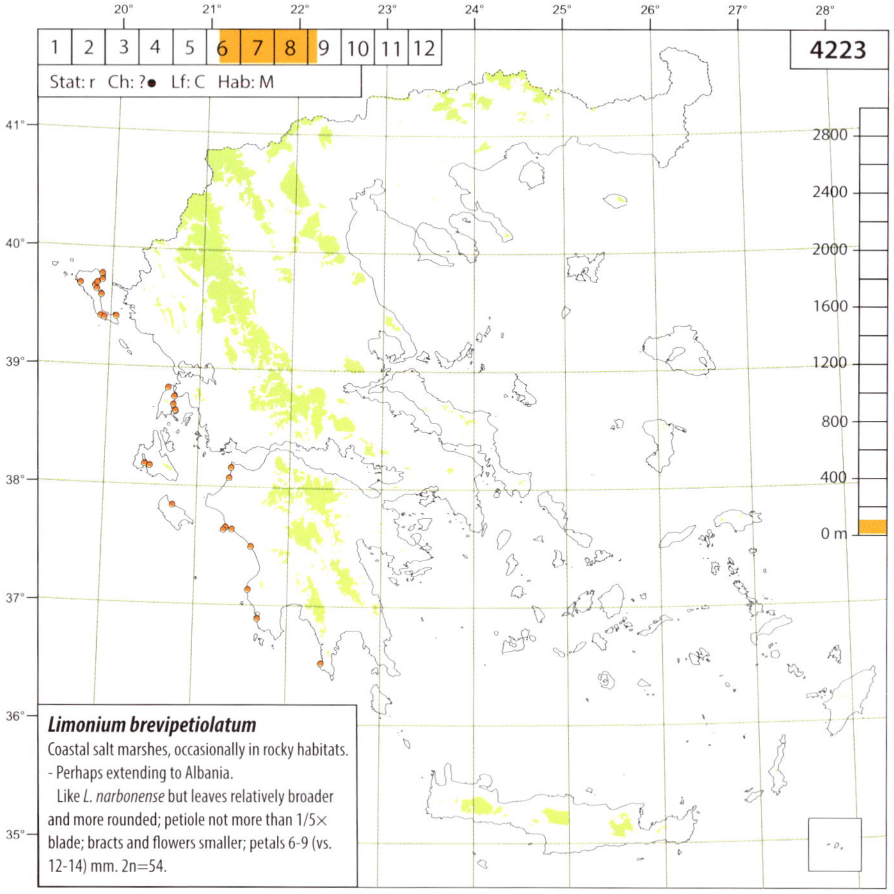

Limonium brevipetiolatum
Coastal salt marshes, occasionally in rocky habitats. - Perhaps extending to Albania.
Like *L. narbonense* but leaves relatively broader and more rounded; petiole not more than 1/5× blade; bracts and flowers smaller; petals 6-9 (vs. 12-14) mm. 2n=54.

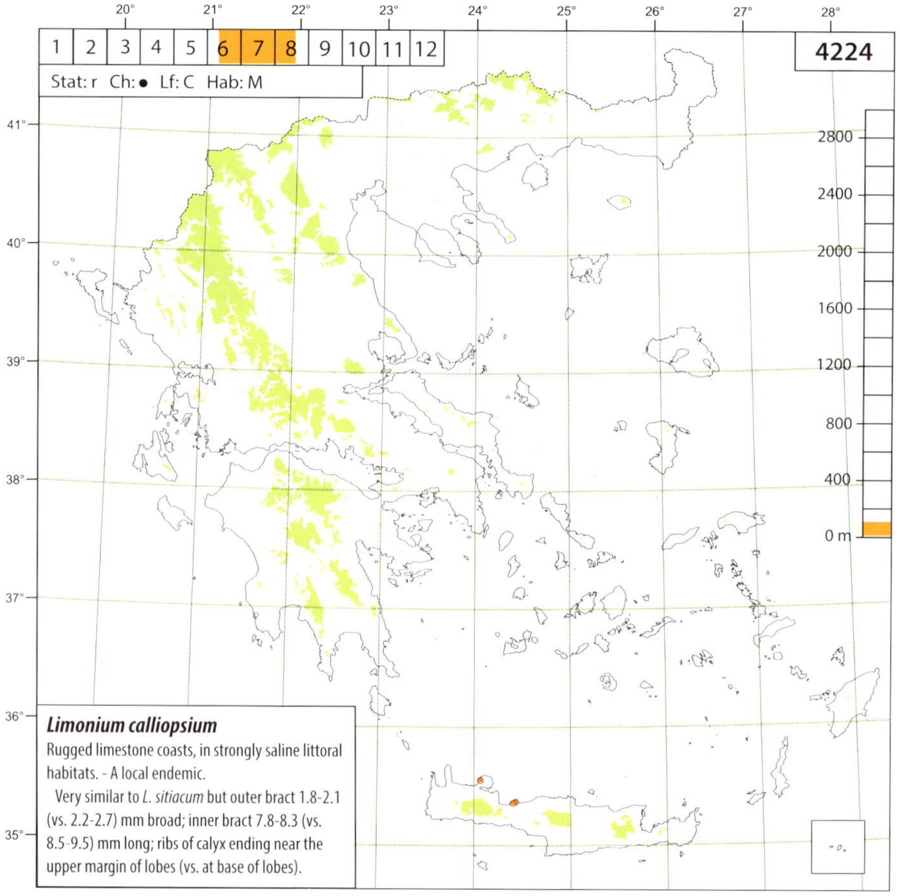

Limonium calliopsium
Rugged limestone coasts, in strongly saline littoral habitats. - A local endemic.
Very similar to *L. sitiacum* but outer bract 1.8-2.1 (vs. 2.2-2.7) mm broad; inner bract 7.8-8.3 (vs. 8.5-9.5) mm long; ribs of calyx ending near the upper margin of lobes (vs. at base of lobes).

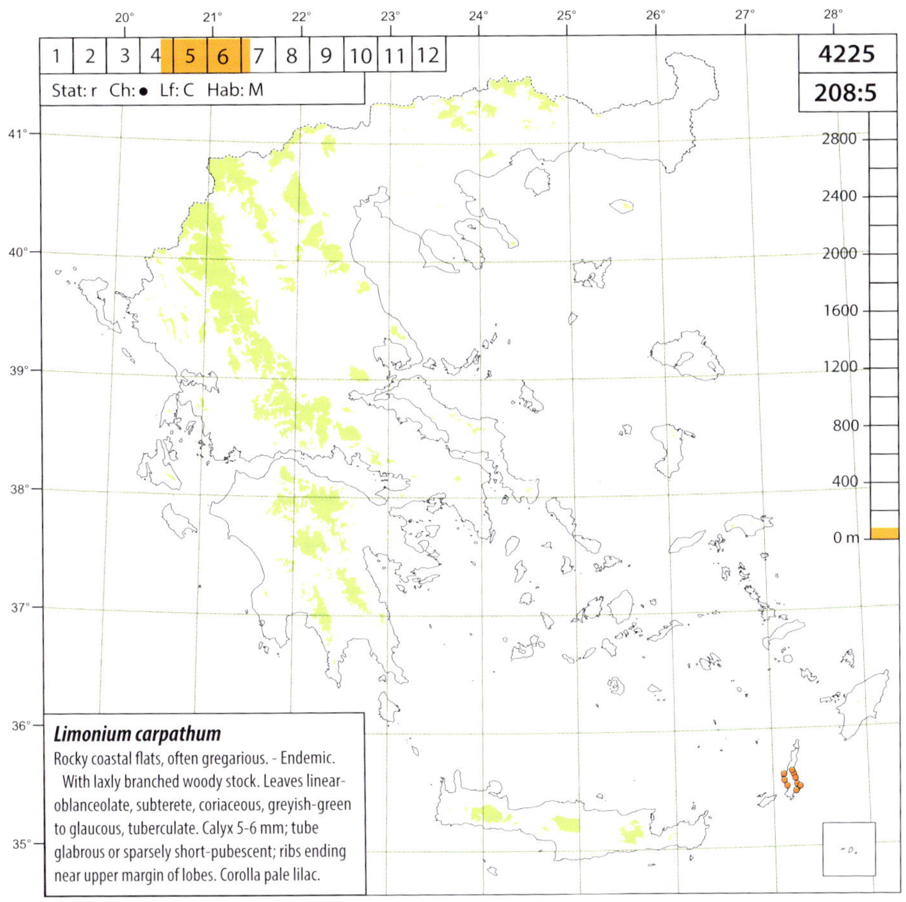

Limonium carpathum
Rocky coastal flats, often gregarious. - Endemic.
With laxly branched woody stock. Leaves linear-oblanceolate, subterete, coriaceous, greyish-green to glaucous, tuberculate. Calyx 5-6 mm; tube glabrous or sparsely short-pubescent; ribs ending near upper margin of lobes. Corolla pale lilac.

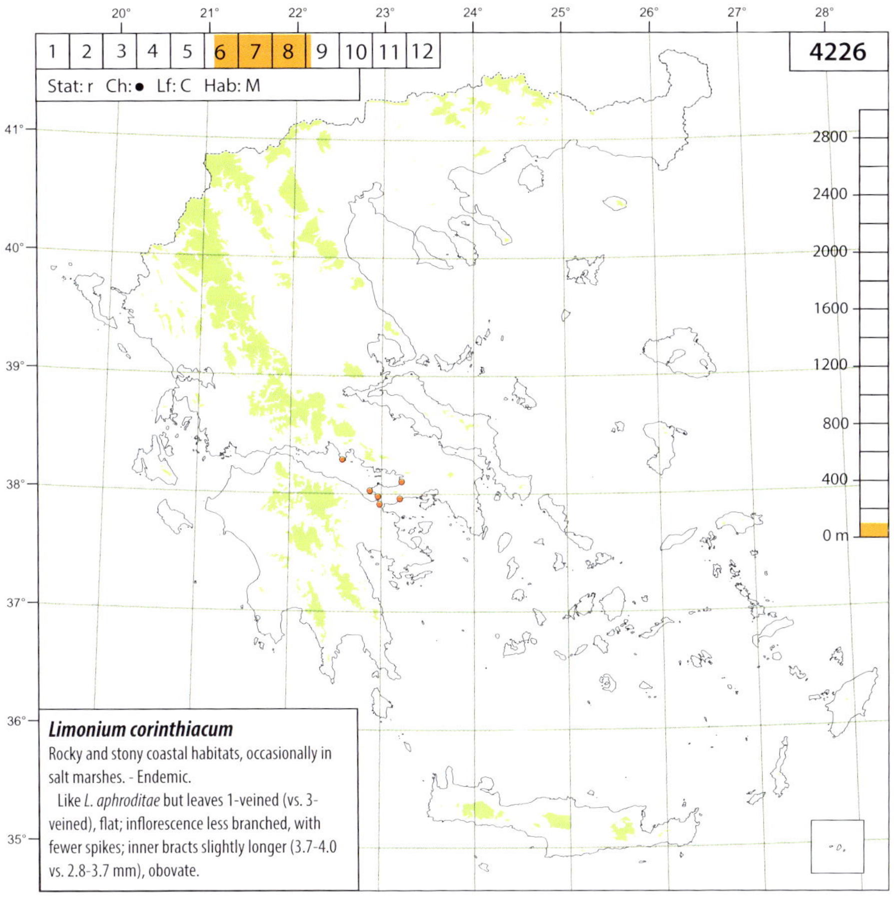

Limonium corinthiacum
Rocky and stony coastal habitats, occasionally in salt marshes. - Endemic.
Like *L. aphroditae* but leaves 1-veined (vs. 3-veined), flat; inflorescence less branched, with fewer spikes; inner bracts slightly longer (3.7-4.0 vs. 2.8-3.7 mm), obovate.

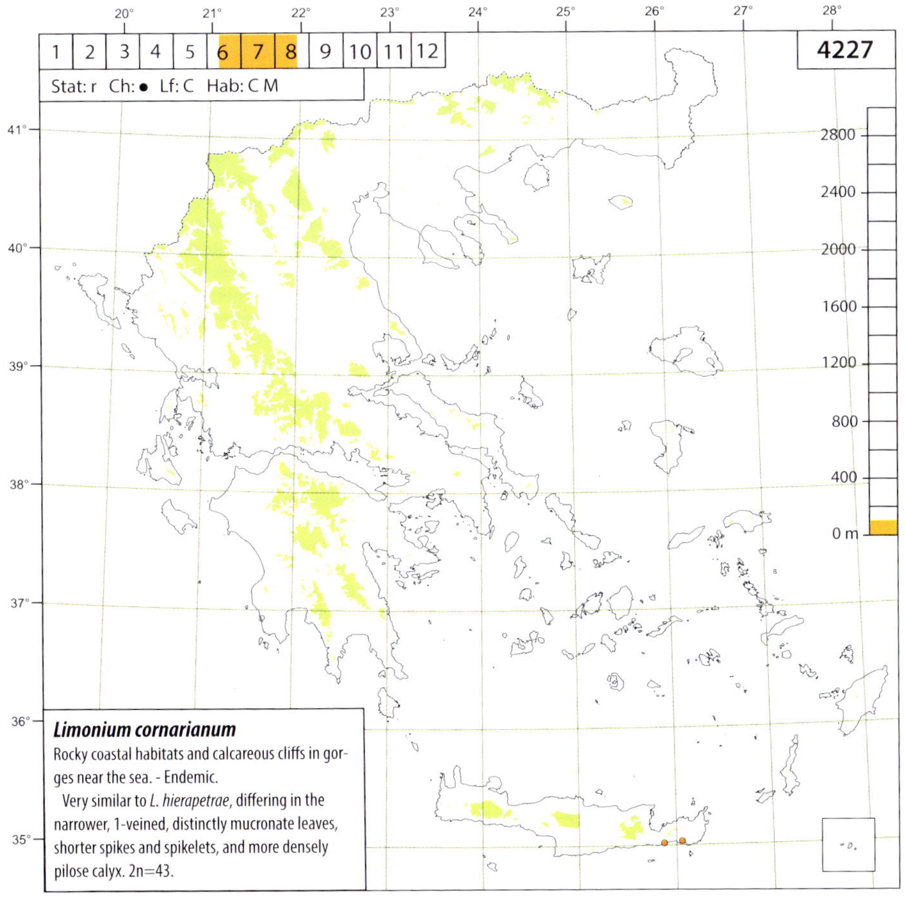

Limonium cornarianum
Rocky coastal habitats and calcareous cliffs in gorges near the sea. - Endemic.
Very similar to *L. hierapetrae*, differing in the narrower, 1-veined, distinctly mucronate leaves, shorter spikes and spikelets, and more densely pilose calyx. 2n=43.

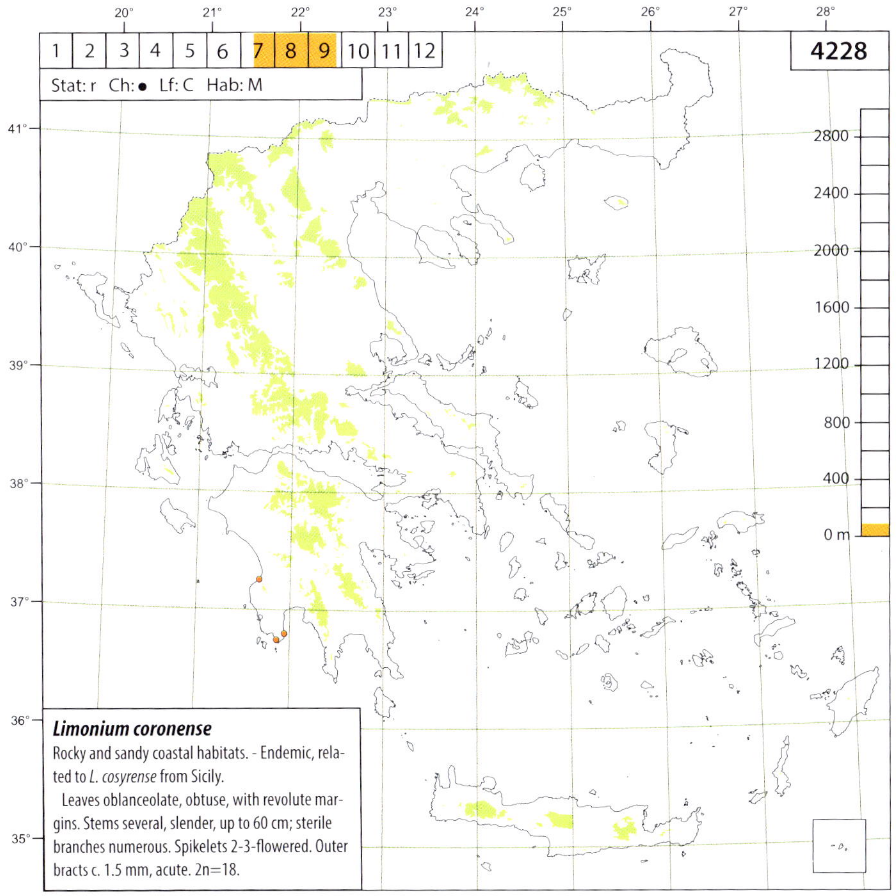

Limonium coronense
Rocky and sandy coastal habitats. - Endemic, related to *L. cosyrense* from Sicily.
Leaves oblanceolate, obtuse, with revolute margins. Stems several, slender, up to 60 cm; sterile branches numerous. Spikelets 2-3-flowered. Outer bracts c. 1.5 mm, acute. 2n=18.

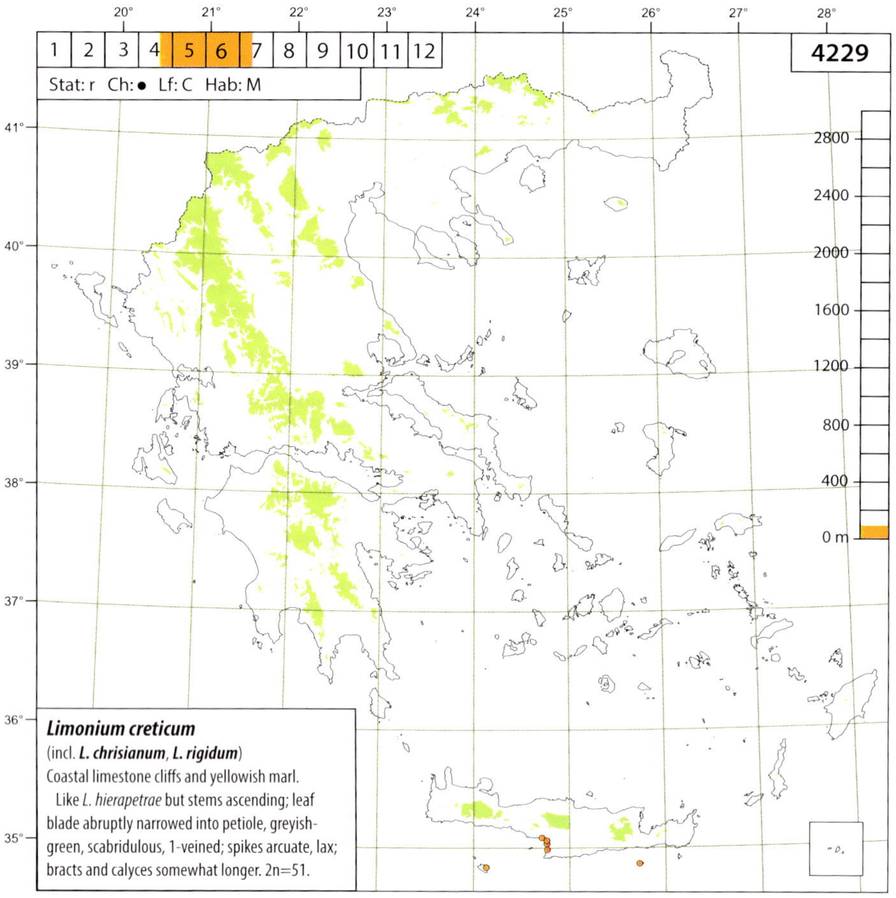

Limonium creticum
(incl. ***L. chrisianum***, ***L. rigidum***)
Coastal limestone cliffs and yellowish marl.
Like *L. hierapetrae* but stems ascending; leaf blade abruptly narrowed into petiole, greyish-green, scabridulous, 1-veined; spikes arcuate, lax; bracts and calyces somewhat longer. 2n=51.

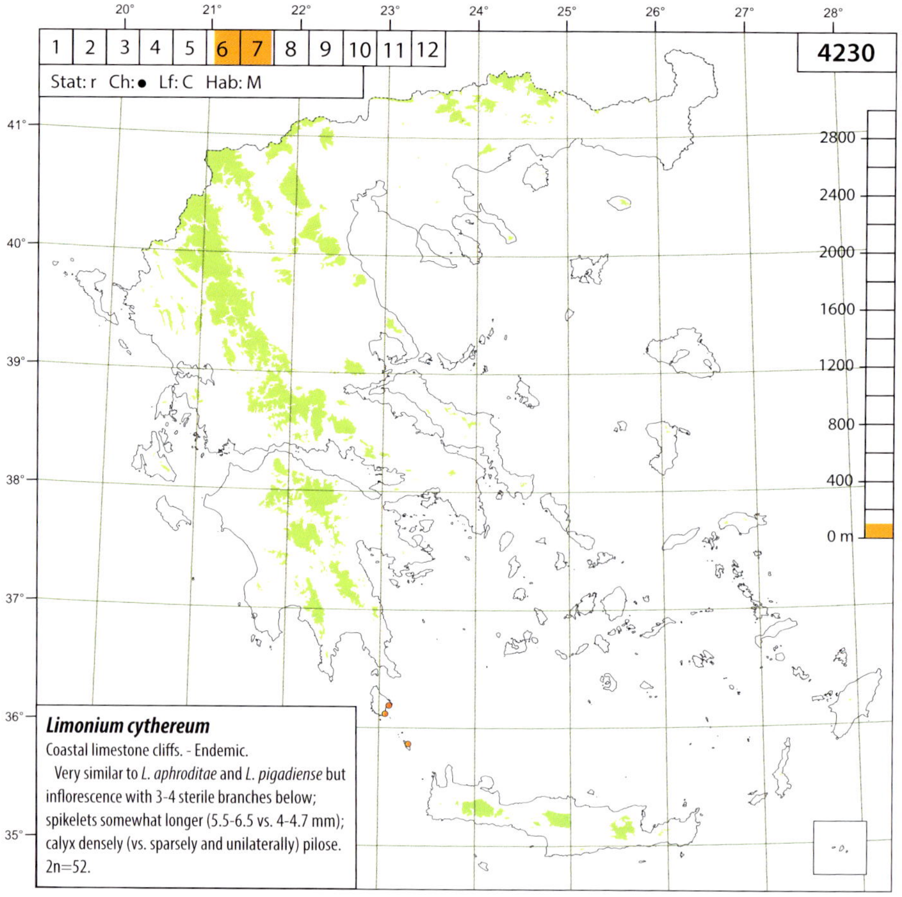

Limonium cythereum
Coastal limestone cliffs. - Endemic.
Very similar to *L. aphroditae* and *L. pigadiense* but inflorescence with 3-4 sterile branches below; spikelets somewhat longer (5.5-6.5 vs. 4-4.7 mm); calyx densely (vs. sparsely and unilaterally) pilose. 2n=52.

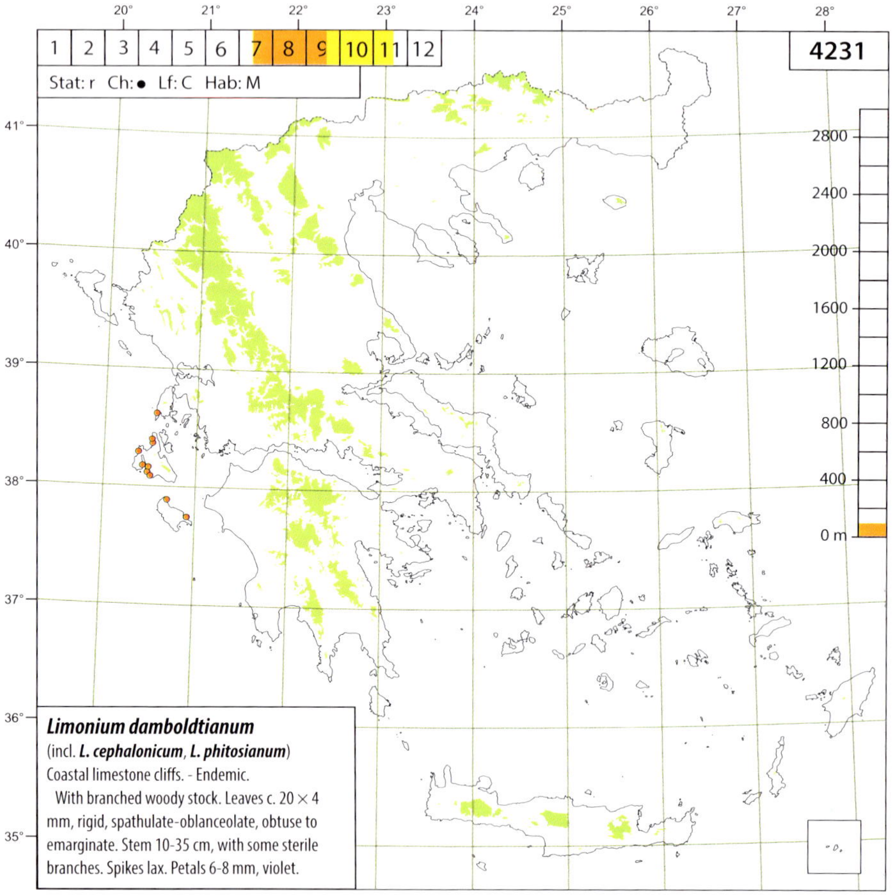

Limonium damboldtianum
(incl. ***L. cephalonicum***, ***L. phitosianum***)
Coastal limestone cliffs. - Endemic.
With branched woody stock. Leaves c. 20 × 4 mm, rigid, spathulate-oblanceolate, obtuse to emarginate. Stem 10-35 cm, with some sterile branches. Spikes lax. Petals 6-8 mm, violet.

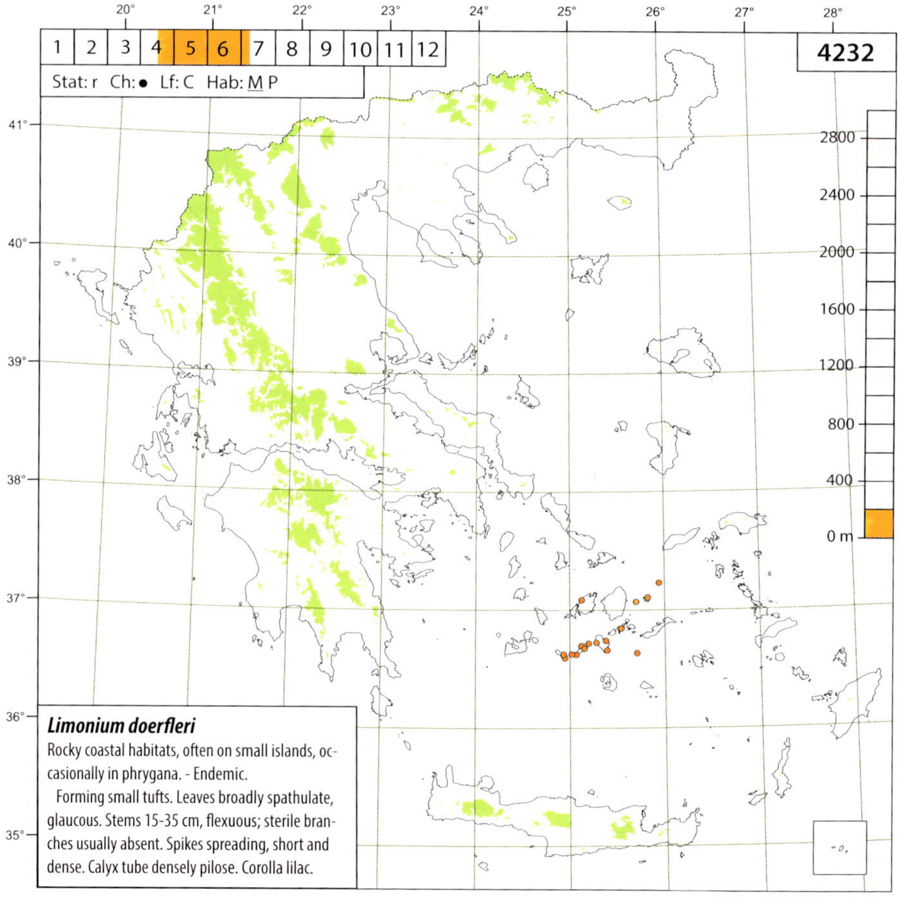

Limonium doerfleri
Rocky coastal habitats, often on small islands, occasionally in phrygana. - Endemic.
Forming small tufts. Leaves broadly spathulate, glaucous. Stems 15-35 cm, flexuous; sterile branches usually absent. Spikes spreading, short and dense. Calyx tube densely pilose. Corolla lilac.

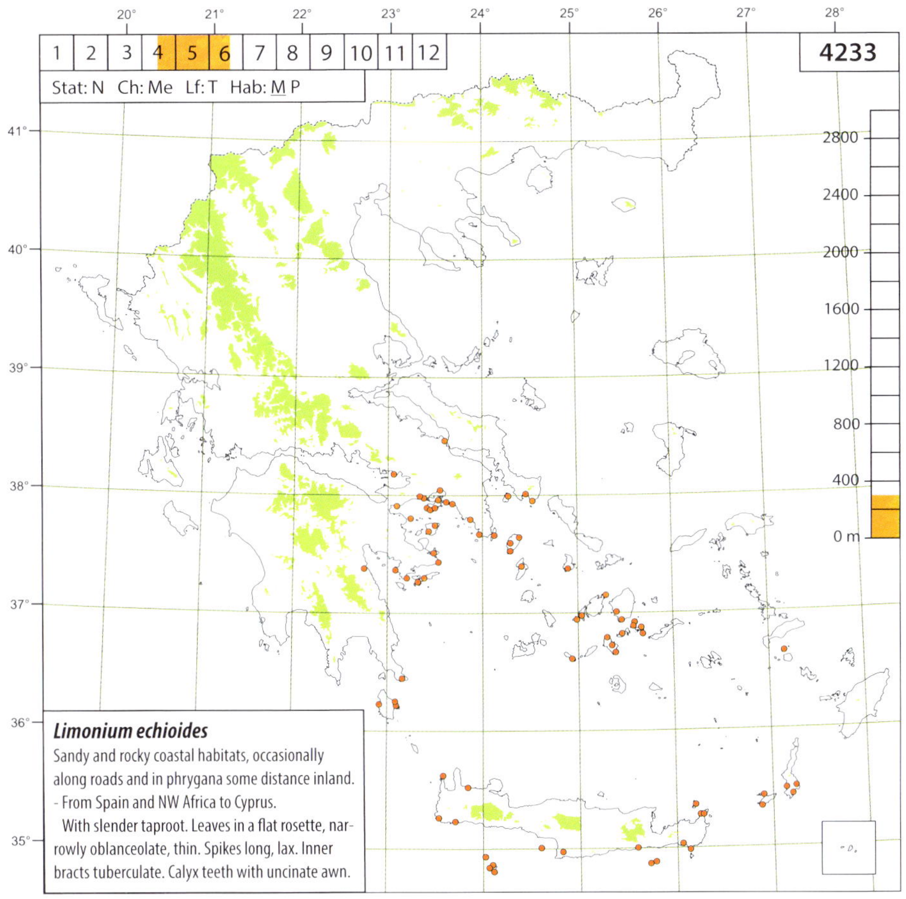

Limonium echioides

Sandy and rocky coastal habitats, occasionally along roads and in phrygana some distance inland. - From Spain and NW Africa to Cyprus.

With slender taproot. Leaves in a flat rosette, narrowly oblanceolate, thin. Spikes long, lax. Inner bracts tuberculate. Calyx teeth with uncinate awn.

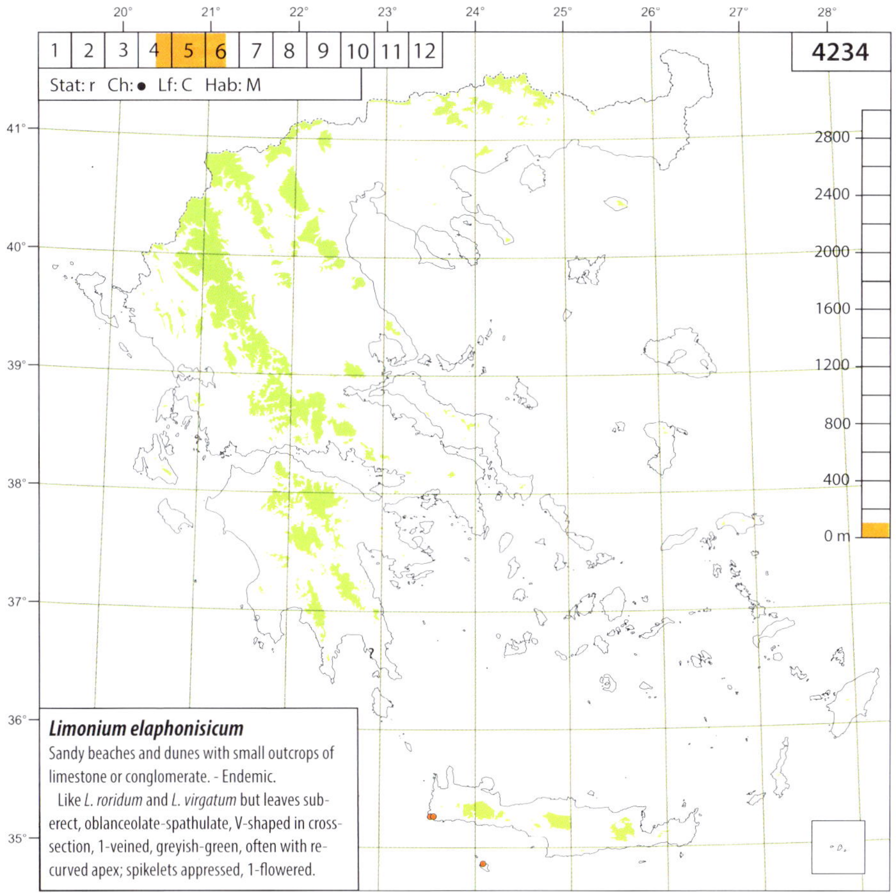

Limonium elaphonisicum

Sandy beaches and dunes with small outcrops of limestone or conglomerate. - Endemic.

Like *L. roridum* and *L. virgatum* but leaves suberect, oblanceolate-spathulate, V-shaped in cross-section, 1-veined, greyish-green, often with recurved apex; spikelets appressed, 1-flowered.

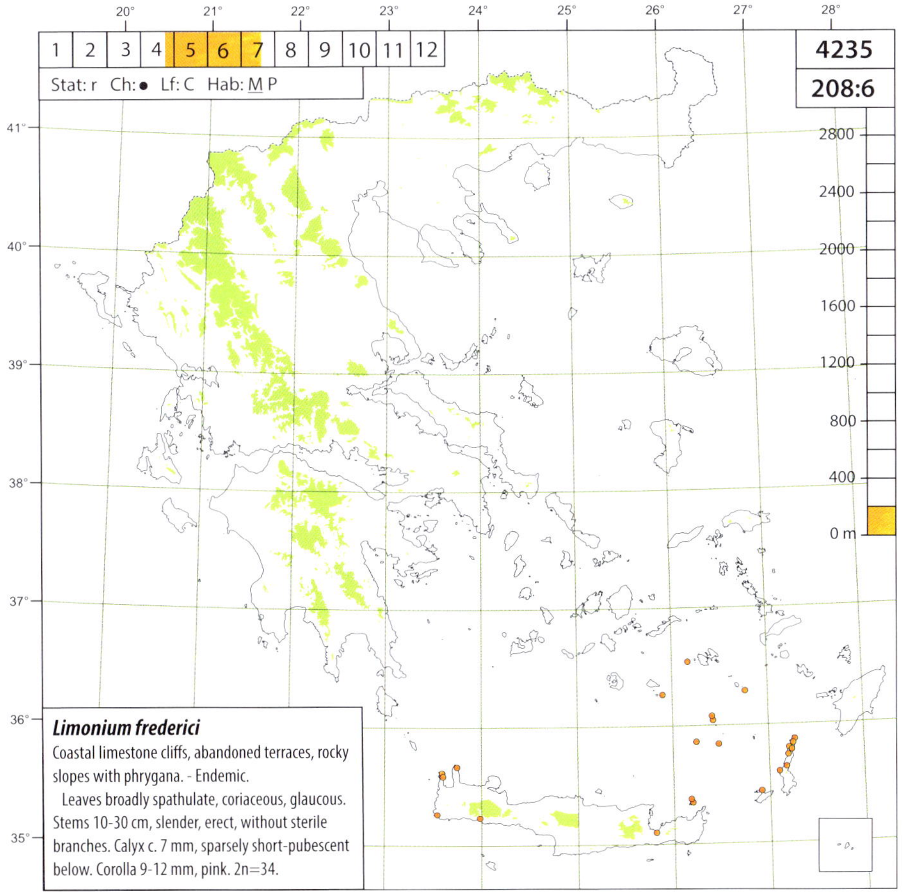

Limonium frederici

Coastal limestone cliffs, abandoned terraces, rocky slopes with phrygana. - Endemic.

Leaves broadly spathulate, coriaceous, glaucous. Stems 10-30 cm, slender, erect, without sterile branches. Calyx c. 7 mm, sparsely short-pubescent below. Corolla 9-12 mm, pink. 2n=34.

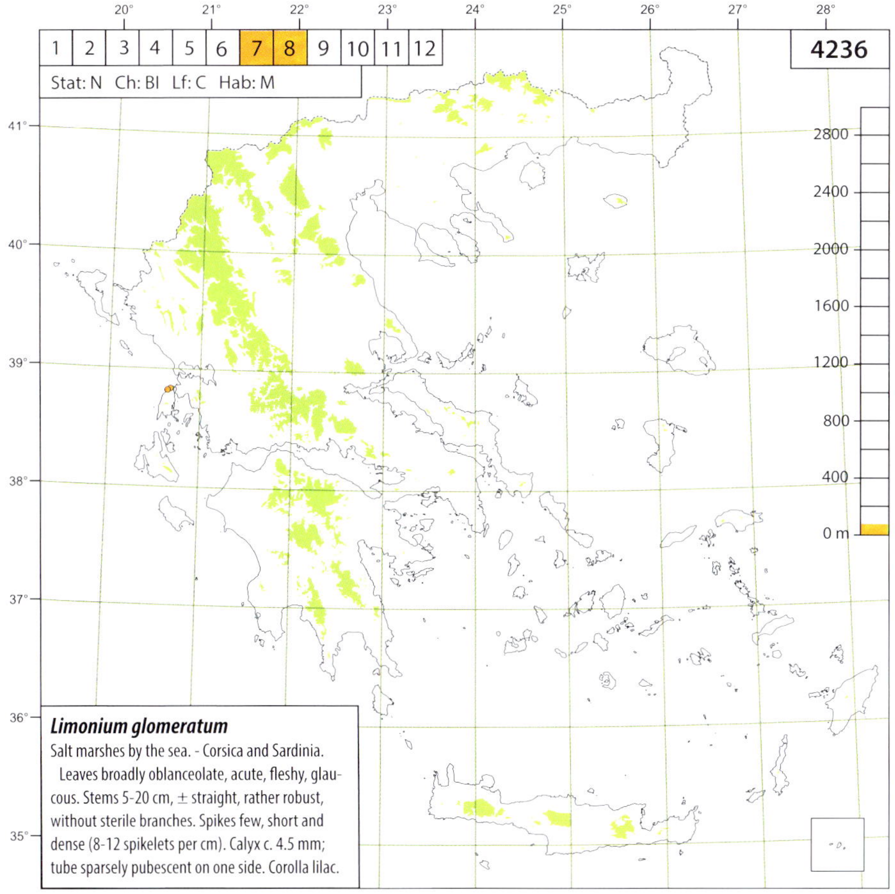

Limonium glomeratum

Salt marshes by the sea. - Corsica and Sardinia.

Leaves broadly oblanceolate, acute, fleshy, glaucous. Stems 5-20 cm, ± straight, rather robust, without sterile branches. Spikes few, short and dense (8-12 spikelets per cm). Calyx c. 4.5 mm; tube sparsely pubescent on one side. Corolla lilac.

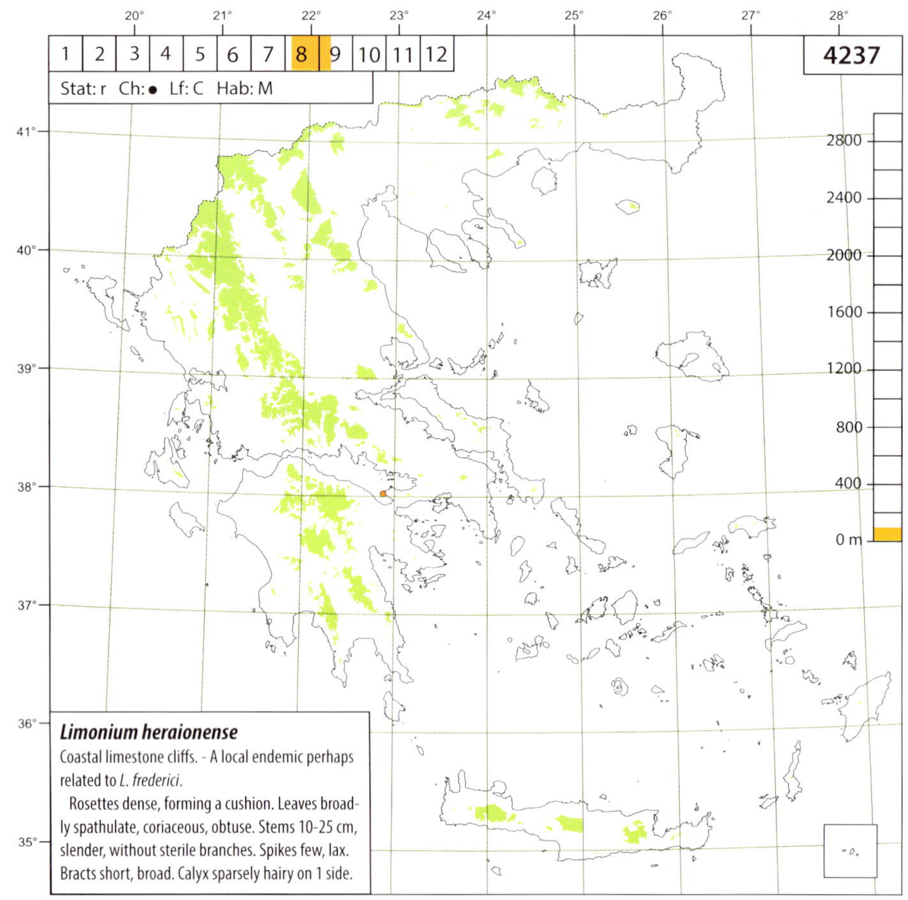

Limonium heraionense
Coastal limestone cliffs. - A local endemic perhaps related to *L. frederici*.
Rosettes dense, forming a cushion. Leaves broadly spathulate, coriaceous, obtuse. Stems 10-25 cm, slender, without sterile branches. Spikes few, lax. Bracts short, broad. Calyx sparsely hairy on 1 side.

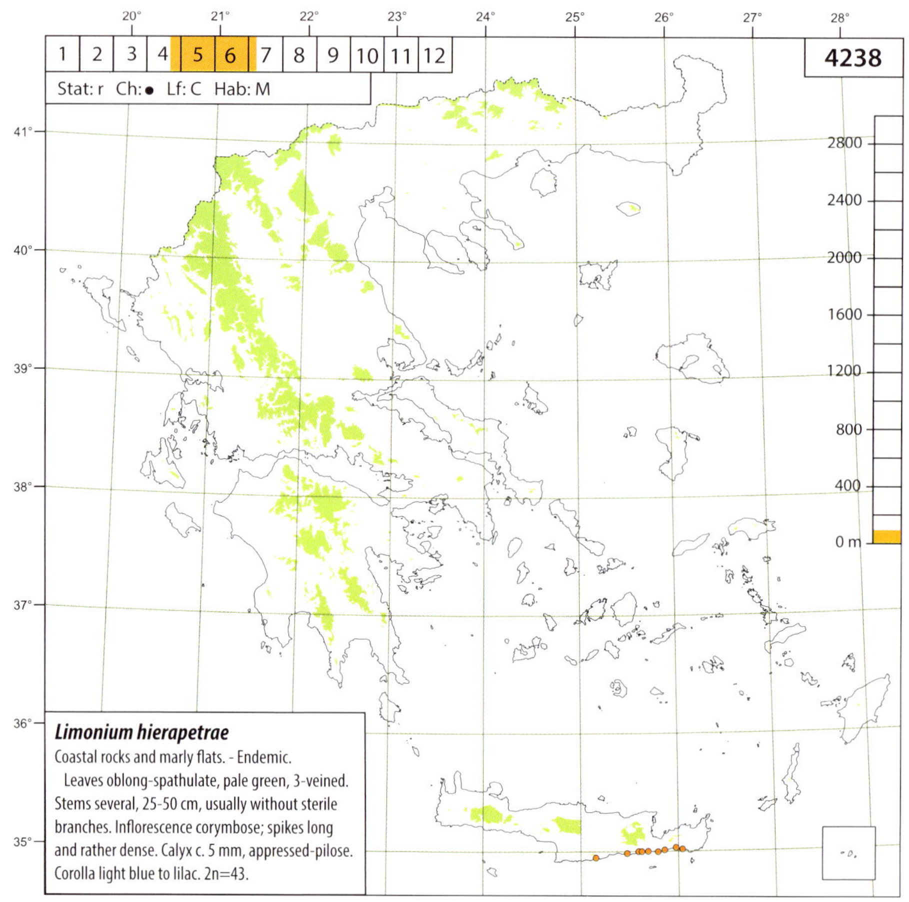

Limonium hierapetrae
Coastal rocks and marly flats. - Endemic.
Leaves oblong-spathulate, pale green, 3-veined. Stems several, 25-50 cm, usually without sterile branches. Inflorescence corymbose; spikes long and rather dense. Calyx c. 5 mm, appressed-pilose. Corolla light blue to lilac. 2n=43.

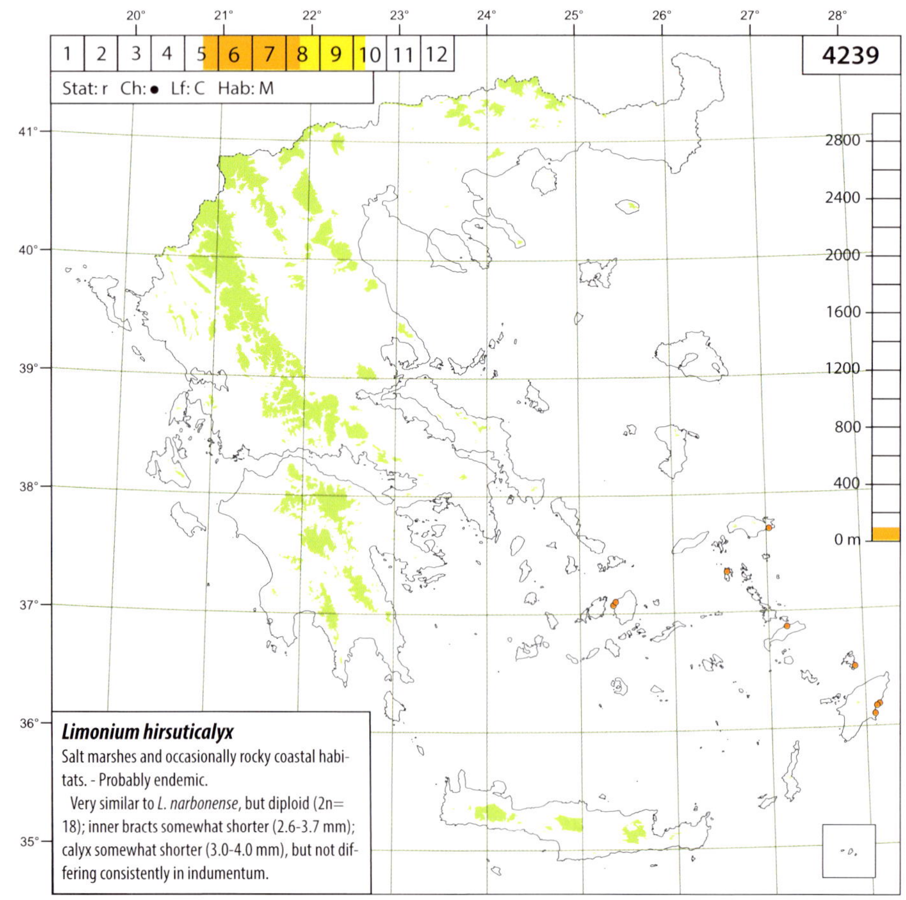

Limonium hirsuticalyx
Salt marshes and occasionally rocky coastal habitats. - Probably endemic.
Very similar to *L. narbonense*, but diploid (2n=18); inner bracts somewhat shorter (2.6-3.7 mm); calyx somewhat shorter (3.0-4.0 mm), but not differing consistently in indumentum.

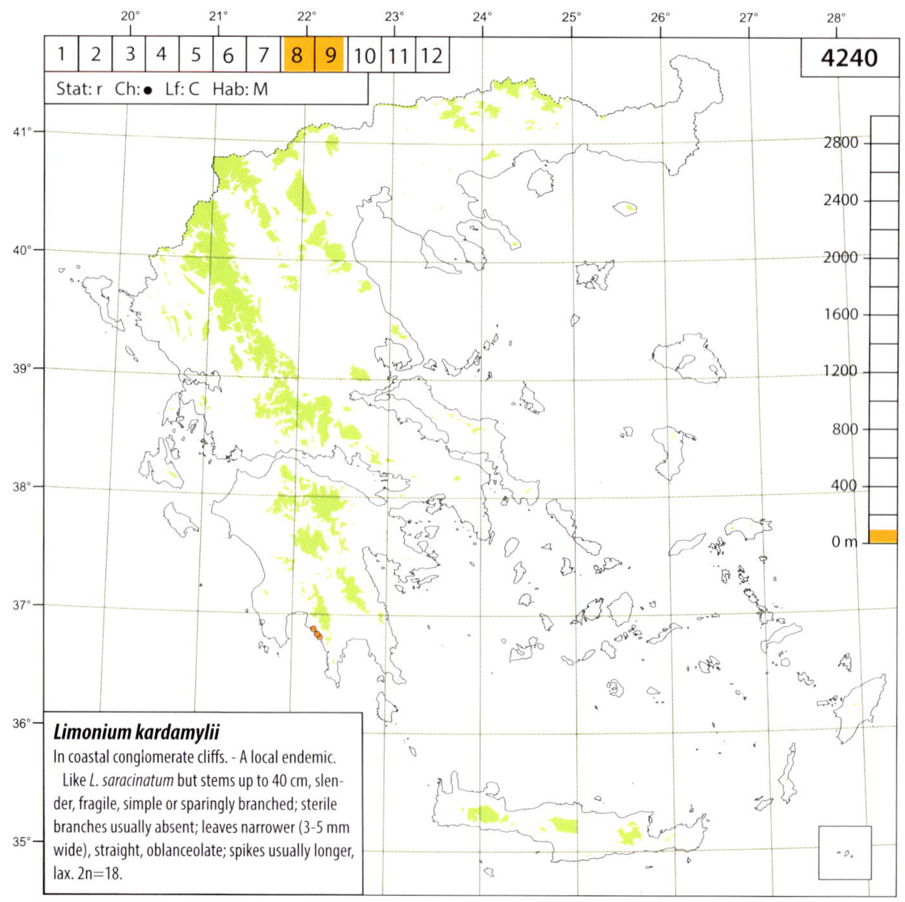

Limonium kardamylii
In coastal conglomerate cliffs. - A local endemic.
Like *L. saracinatum* but stems up to 40 cm, slender, fragile, simple or sparingly branched; sterile branches usually absent; leaves narrower (3-5 mm wide), straight, oblanceolate; spikes usually longer, lax. 2n=18.

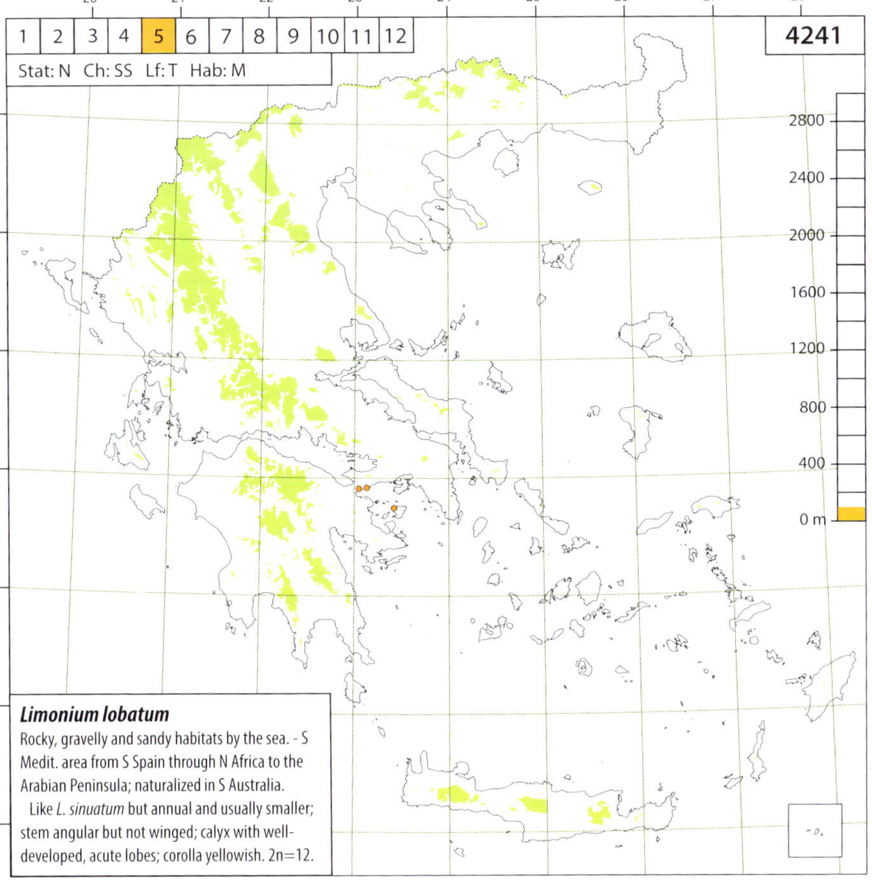

Limonium lobatum

Rocky, gravelly and sandy habitats by the sea. - S Medit. area from S Spain through N Africa to the Arabian Peninsula; naturalized in S Australia.

Like *L. sinuatum* but annual and usually smaller; stem angular but not winged; calyx with well-developed, acute lobes; corolla yellowish. 2n=12.

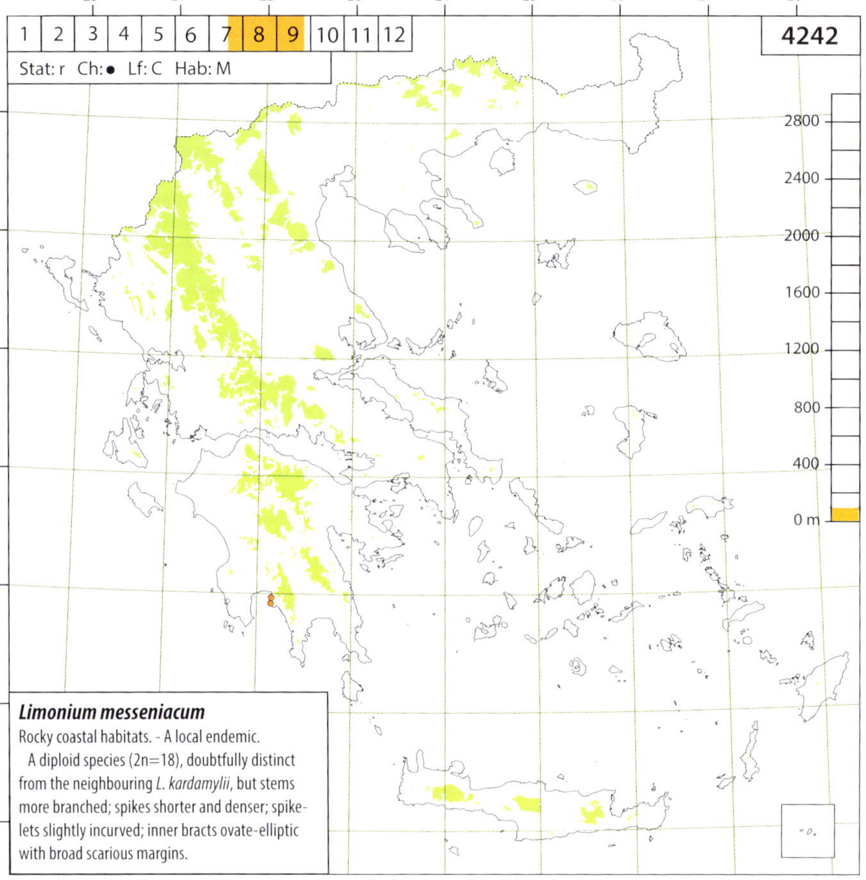

Limonium messeniacum

Rocky coastal habitats. - A local endemic.

A diploid species (2n=18), doubtfully distinct from the neighbouring *L. kardamylii*, but stems more branched; spikes shorter and denser; spikelets slightly incurved; inner bracts ovate-elliptic with broad scarious margins.

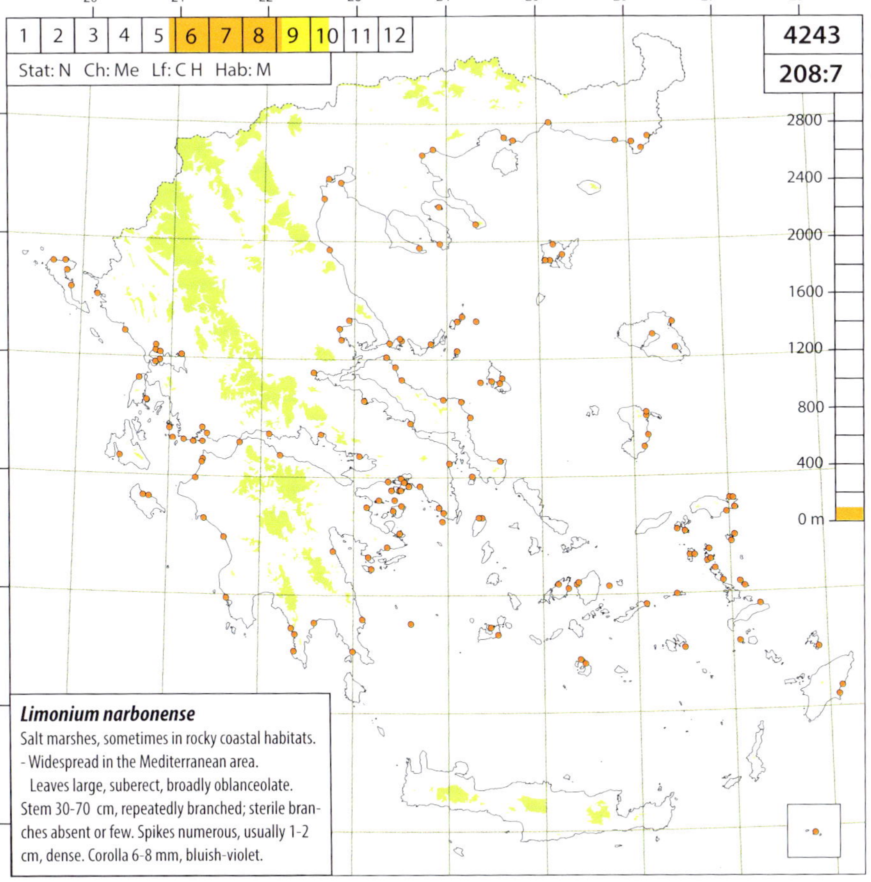

Limonium narbonense

Salt marshes, sometimes in rocky coastal habitats. - Widespread in the Mediterranean area.

Leaves large, suberect, broadly oblanceolate. Stem 30-70 cm, repeatedly branched; sterile branches absent or few. Spikes numerous, usually 1-2 cm, dense. Corolla 6-8 mm, bluish-violet.

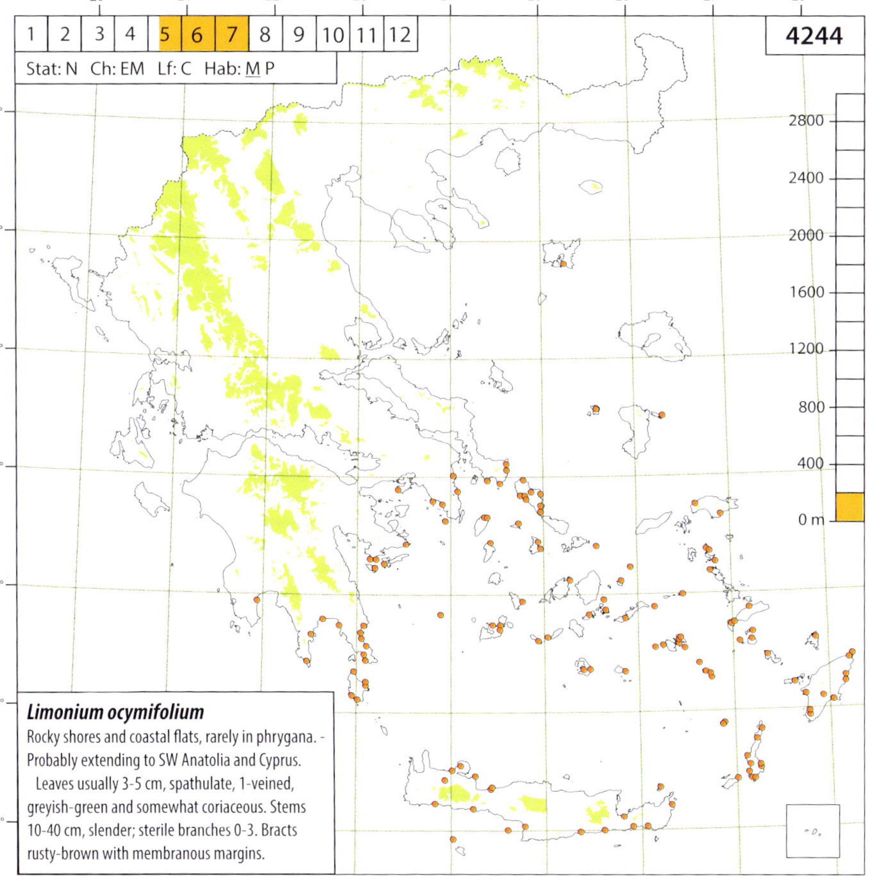

Limonium ocymifolium

Rocky shores and coastal flats, rarely in phrygana. - Probably extending to SW Anatolia and Cyprus.

Leaves usually 3-5 cm, spathulate, 1-veined, greyish-green and somewhat coriaceous. Stems 10-40 cm, slender; sterile branches 0-3. Bracts rusty-brown with membranous margins.

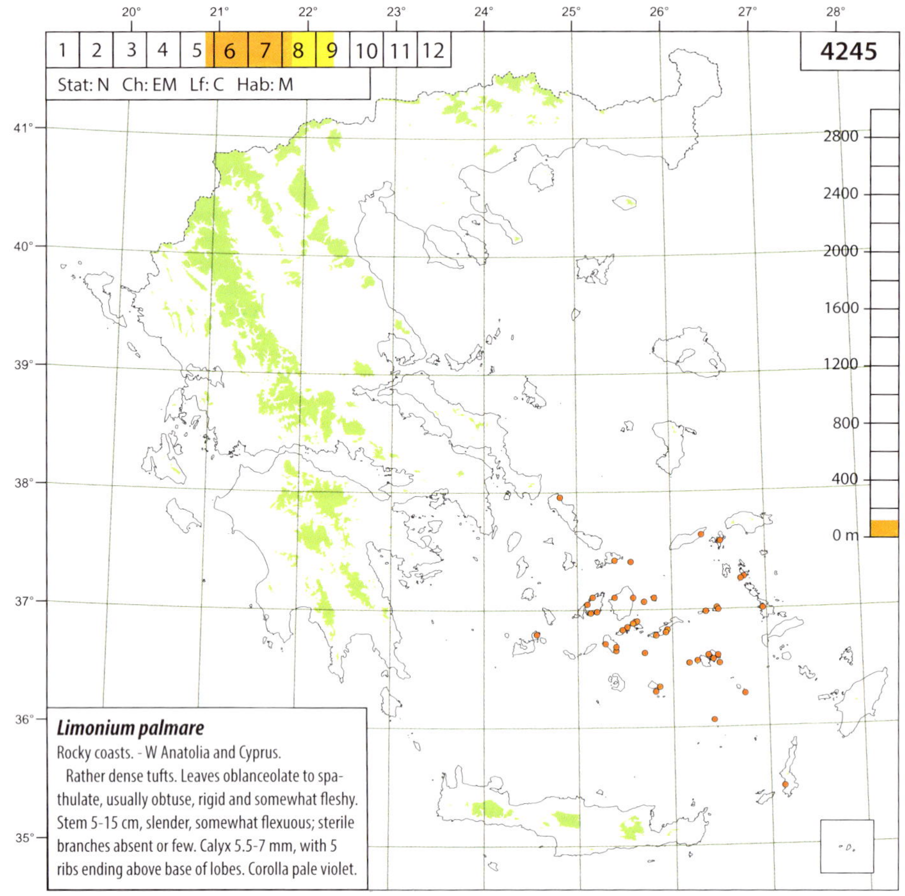

Limonium palmare

Rocky coasts. - W Anatolia and Cyprus.

Rather dense tufts. Leaves oblanceolate to spathulate, usually obtuse, rigid and somewhat fleshy. Stem 5-15 cm, slender, somewhat flexuous; sterile branches absent or few. Calyx 5.5-7 mm, with 5 ribs ending above base of lobes. Corolla pale violet.

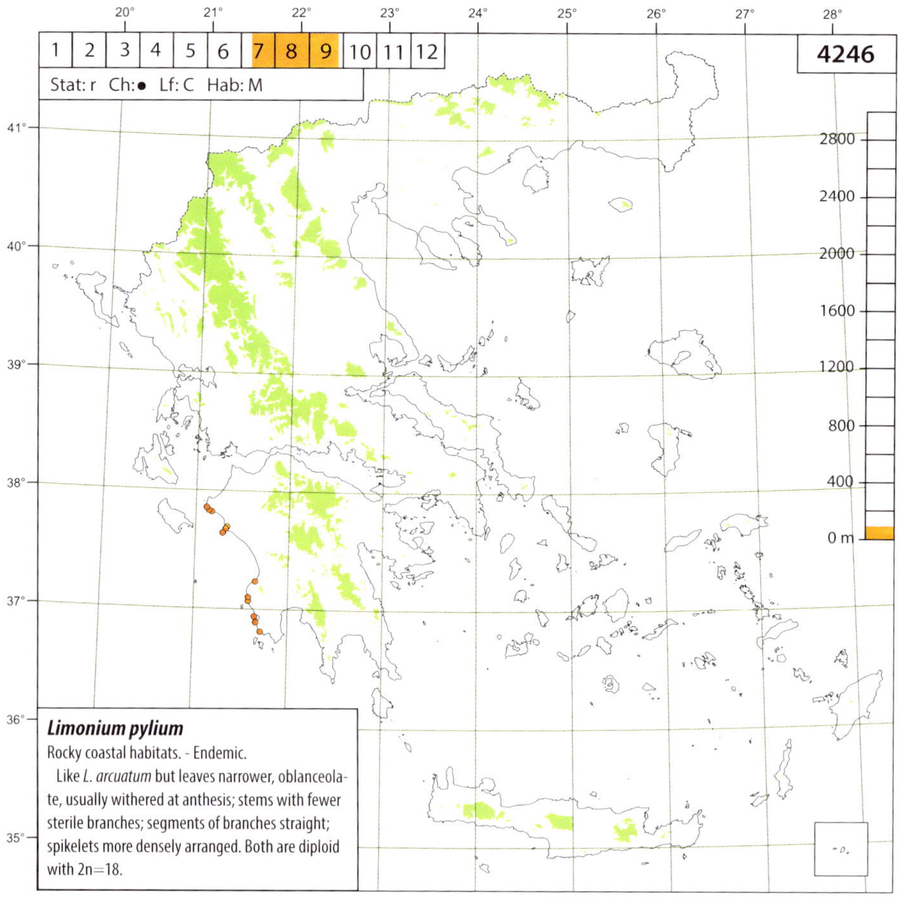

Limonium pylium

Rocky coastal habitats. - Endemic.

Like *L. arcuatum* but leaves narrower, oblanceolate, usually withered at anthesis; stems with fewer sterile branches; segments of branches straight; spikelets more densely arranged. Both are diploid with 2n=18.

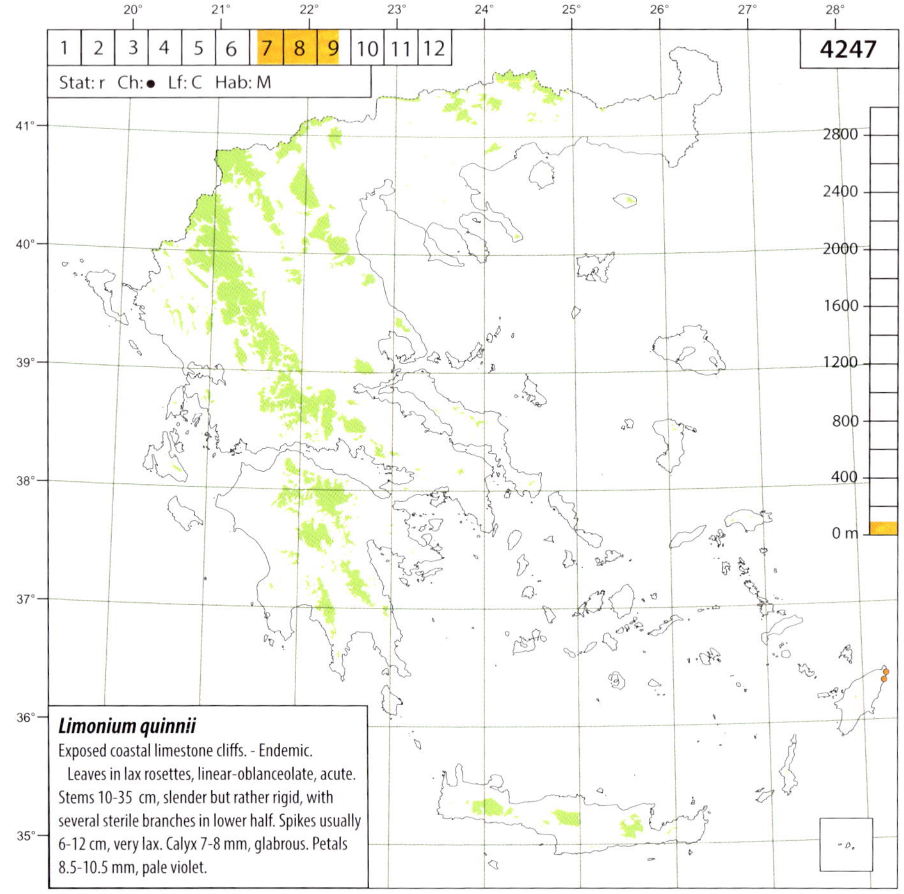

Limonium quinnii

Exposed coastal limestone cliffs. - Endemic.

Leaves in lax rosettes, linear-oblanceolate, acute. Stems 10-35 cm, slender but rather rigid, with several sterile branches in lower half. Spikes usually 6-12 cm, very lax. Calyx 7-8 mm, glabrous. Petals 8.5-10.5 mm, pale violet.

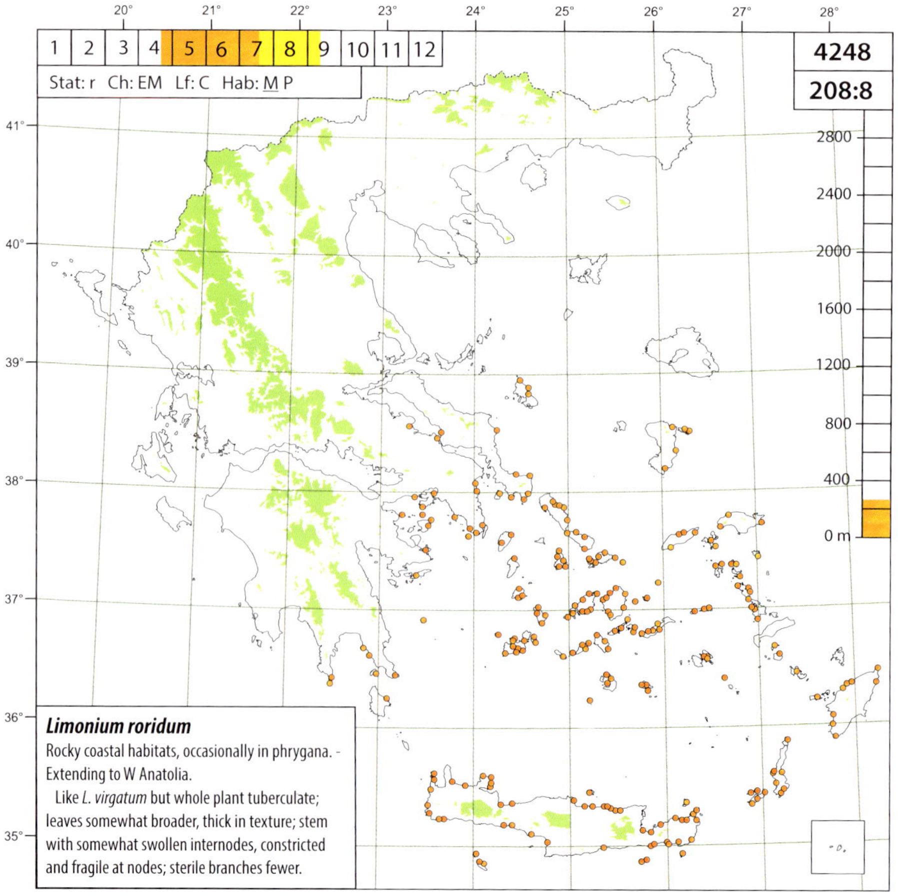

Limonium roridum

Rocky coastal habitats, occasionally in phrygana. - Extending to W Anatolia.

Like *L. virgatum* but whole plant tuberculate; leaves somewhat broader, thick in texture; stem with somewhat swollen internodes, constricted and fragile at nodes; sterile branches fewer.

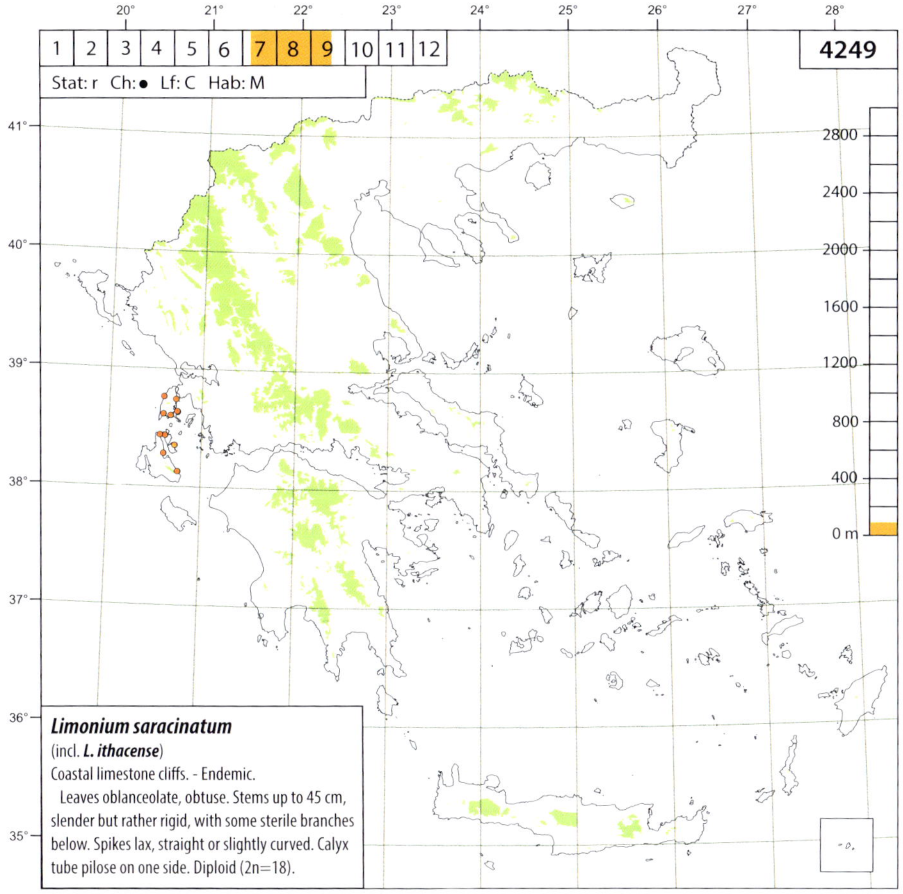

Limonium saracinatum

(incl. ***L. ithacense***)
Coastal limestone cliffs. - Endemic.
Leaves oblanceolate, obtuse. Stems up to 45 cm, slender but rather rigid, with some sterile branches below. Spikes lax, straight or slightly curved. Calyx tube pilose on one side. Diploid (2n=18).

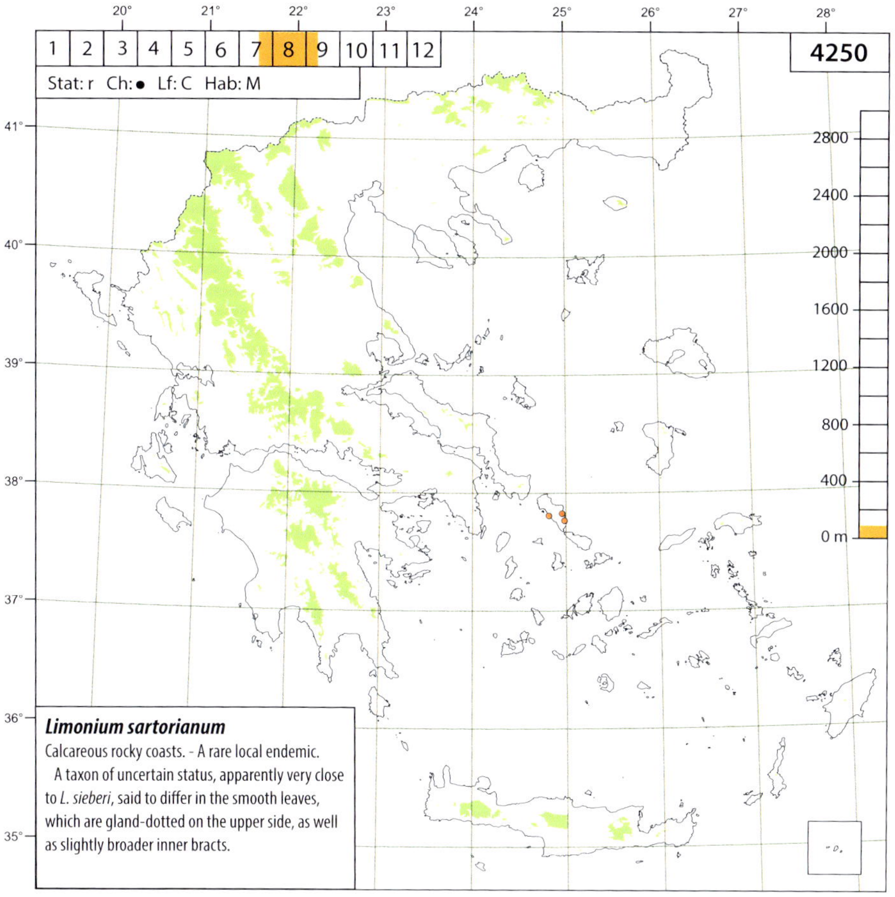

Limonium sartorianum

Calcareous rocky coasts. - A rare local endemic.
A taxon of uncertain status, apparently very close to *L. sieberi*, said to differ in the smooth leaves, which are gland-dotted on the upper side, as well as slightly broader inner bracts.

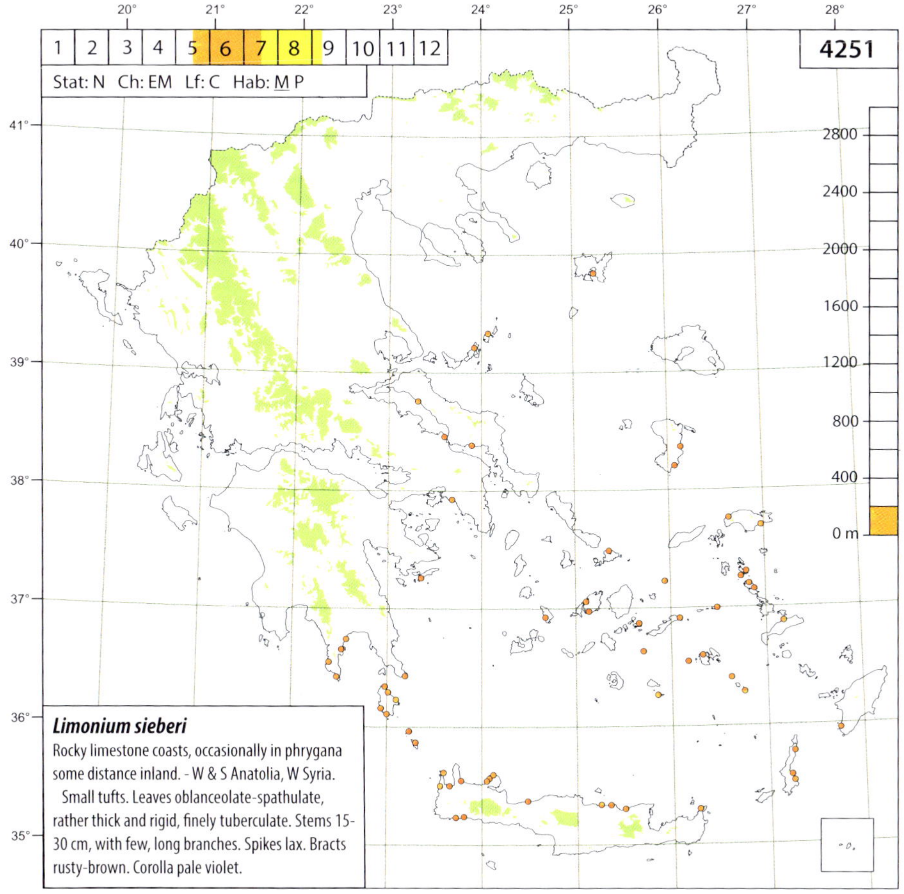

Limonium sieberi

Rocky limestone coasts, occasionally in phrygana some distance inland. - W & S Anatolia, W Syria.
Small tufts. Leaves oblanceolate-spathulate, rather thick and rigid, finely tuberculate. Stems 15-30 cm, with few, long branches. Spikes lax. Bracts rusty-brown. Corolla pale violet.

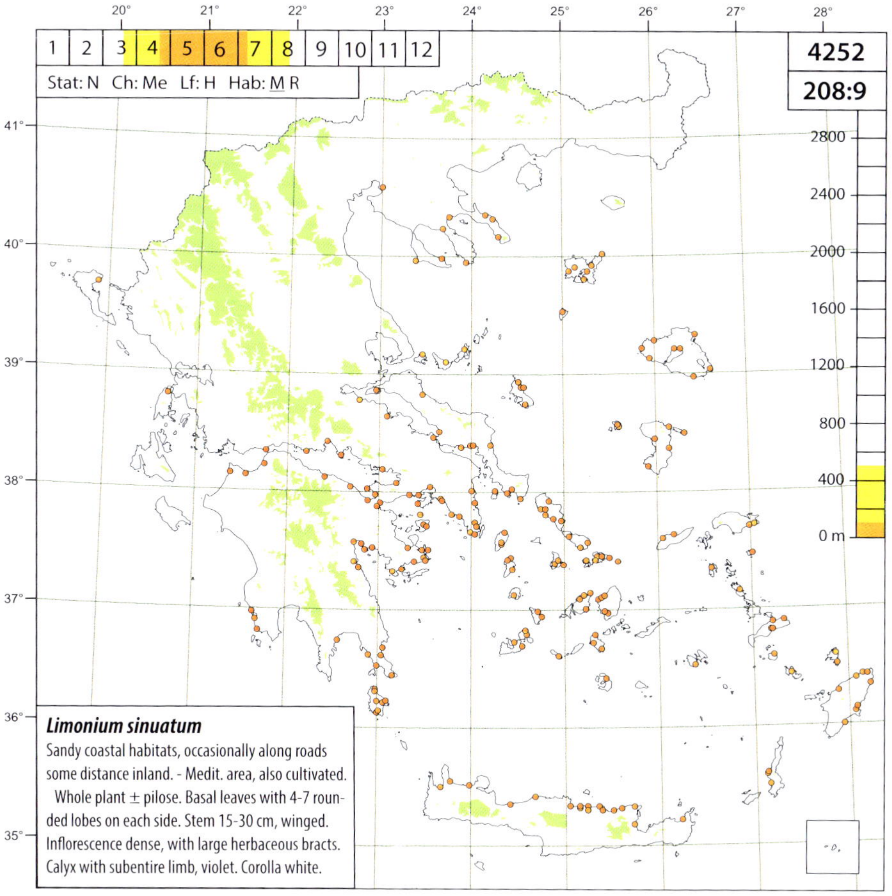

Limonium sinuatum

Sandy coastal habitats, occasionally along roads some distance inland. - Medit. area, also cultivated.
Whole plant ± pilose. Basal leaves with 4-7 rounded lobes on each side. Stem 15-30 cm, winged. Inflorescence dense, with large herbaceous bracts. Calyx with subentire limb, violet. Corolla white.

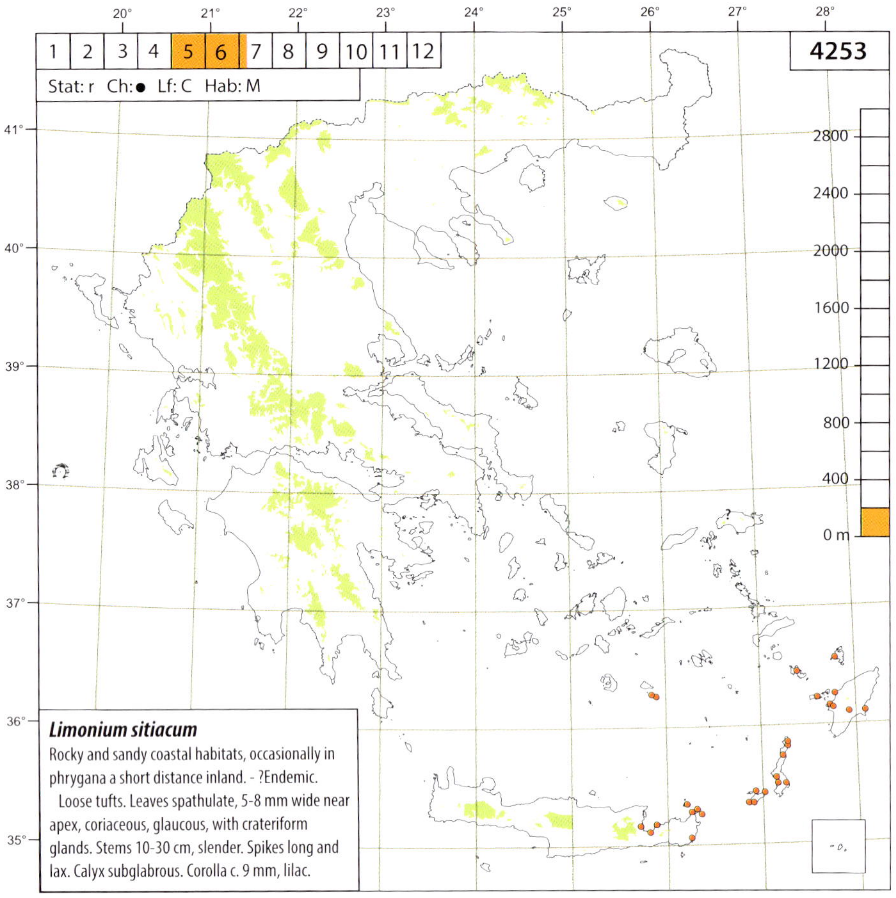

Limonium sitiacum

Rocky and sandy coastal habitats, occasionally in phrygana a short distance inland. - ?Endemic.

Loose tufts. Leaves spathulate, 5-8 mm wide near apex, coriaceous, glaucous, with crateriform glands. Stems 10-30 cm, slender. Spikes long and lax. Calyx subglabrous. Corolla c. 9 mm, lilac.

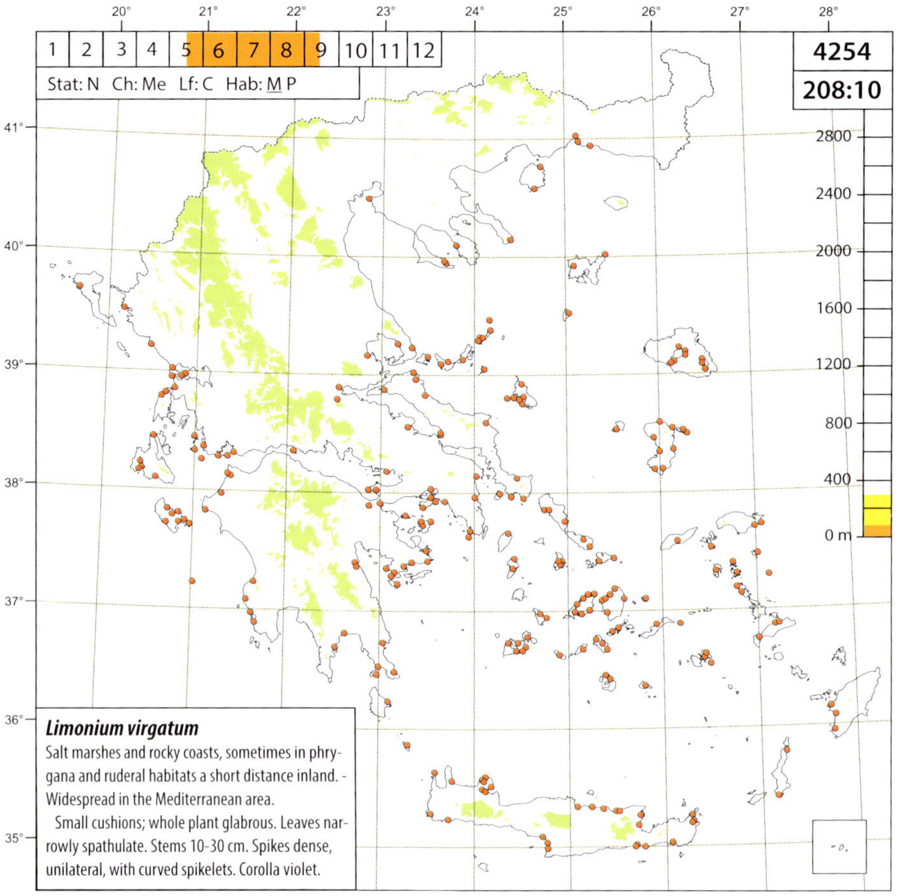

Limonium virgatum

Salt marshes and rocky coasts, sometimes in phrygana and ruderal habitats a short distance inland. - Widespread in the Mediterranean area.

Small cushions; whole plant glabrous. Leaves narrowly spathulate. Stems 10-30 cm. Spikes dense, unilateral, with curved spikelets. Corolla violet.

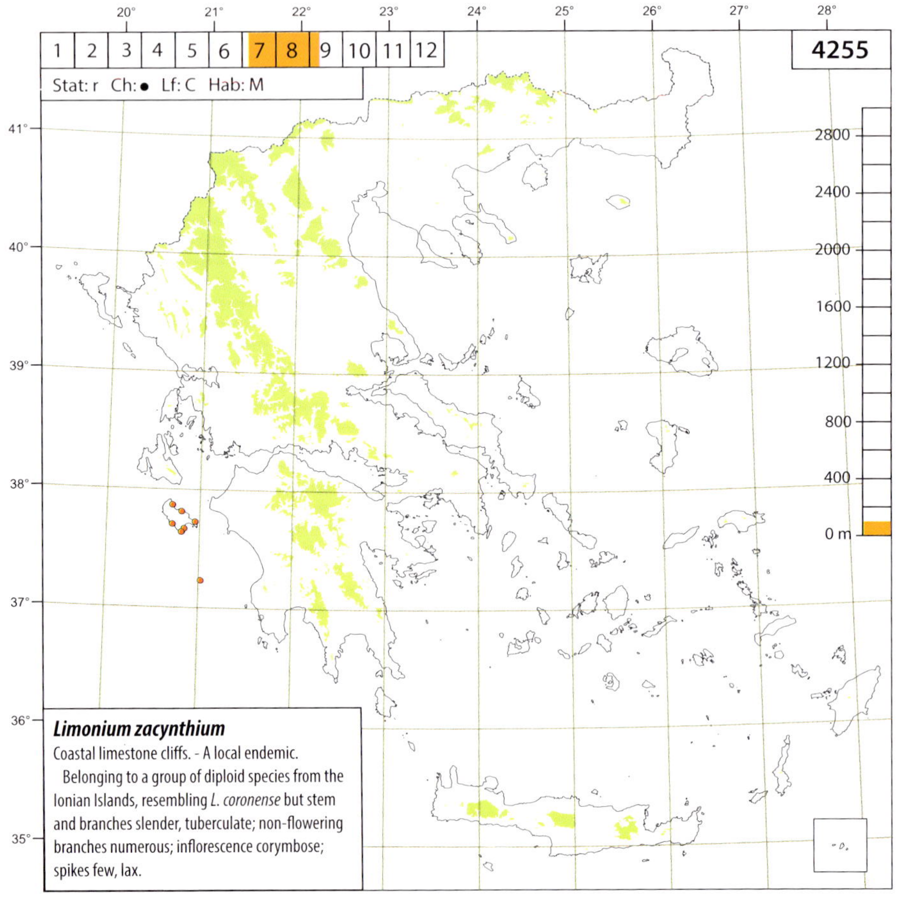

Limonium zacynthium

Coastal limestone cliffs. - A local endemic.

Belonging to a group of diploid species from the Ionian Islands, resembling *L. coronense* but stem and branches slender, tuberculate; non-flowering branches numerous; inflorescence corymbose; spikes few, lax.

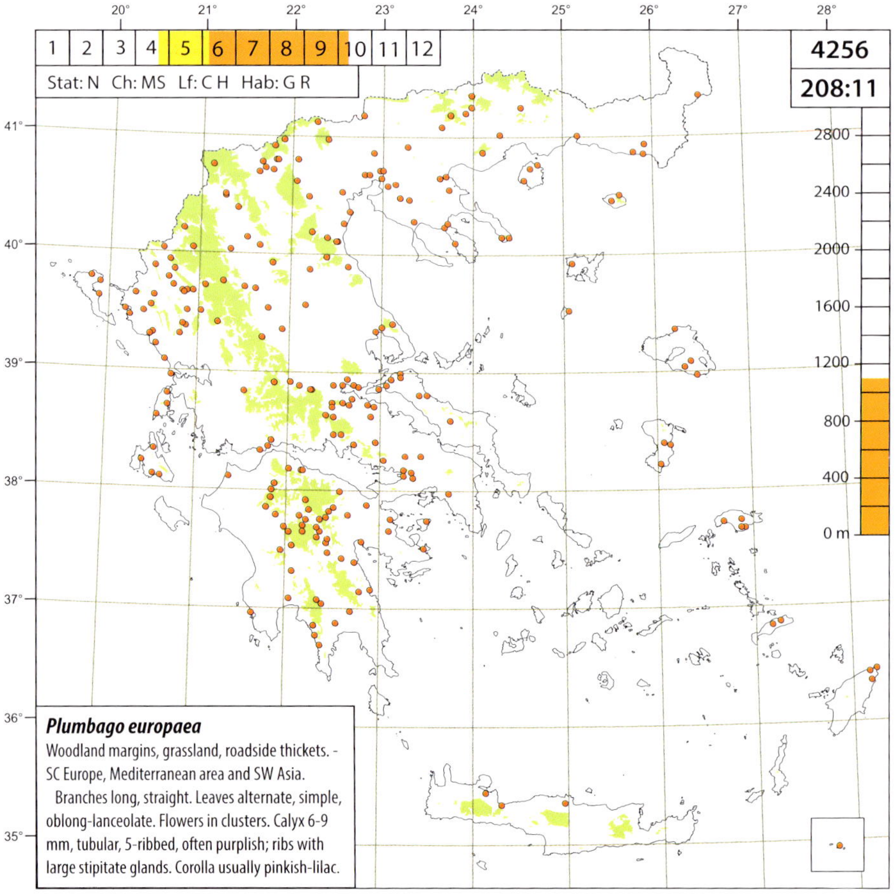

Plumbago europaea

Woodland margins, grassland, roadside thickets. - SC Europe, Mediterranean area and SW Asia.

Branches long, straight. Leaves alternate, simple, oblong-lanceolate. Flowers in clusters. Calyx 6-9 mm, tubular, 5-ribbed, often purplish; ribs with large stipitate glands. Corolla usually pinkish-lilac.

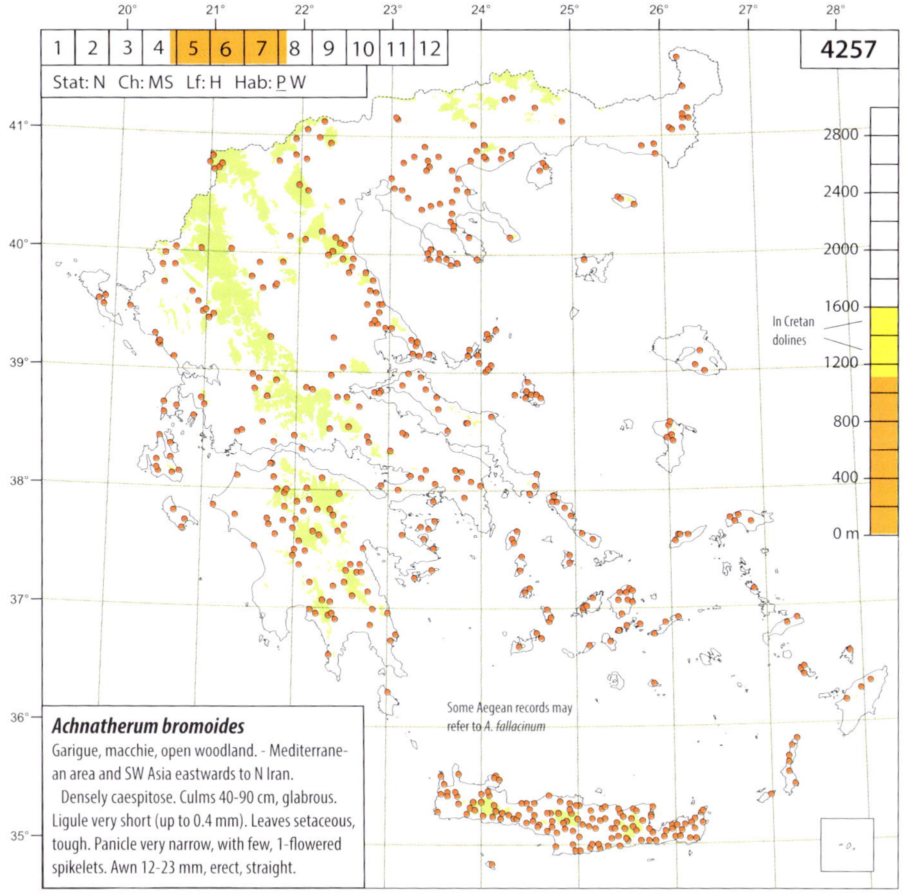

Achnatherum bromoides

Garigue, macchie, open woodland. - Mediterranean area and SW Asia eastwards to N Iran.

Densely caespitose. Culms 40-90 cm, glabrous. Ligule very short (up to 0.4 mm). Leaves setaceous, tough. Panicle very narrow, with few, 1-flowered spikelets. Awn 12-23 mm, erect, straight.

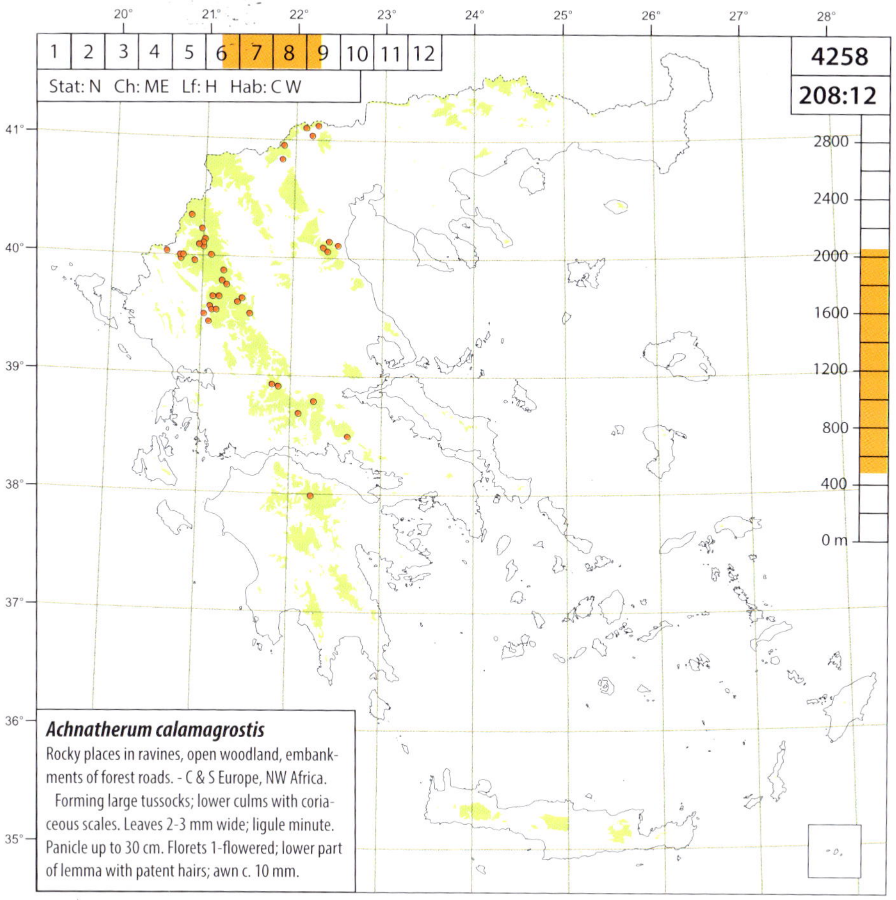

Achnatherum calamagrostis

Rocky places in ravines, open woodland, embankments of forest roads. - C & S Europe, NW Africa.

Forming large tussocks; lower culms with coriaceous scales. Leaves 2-3 mm wide; ligule minute. Panicle up to 30 cm. Florets 1-flowered; lower part of lemma with patent hairs; awn c. 10 mm.

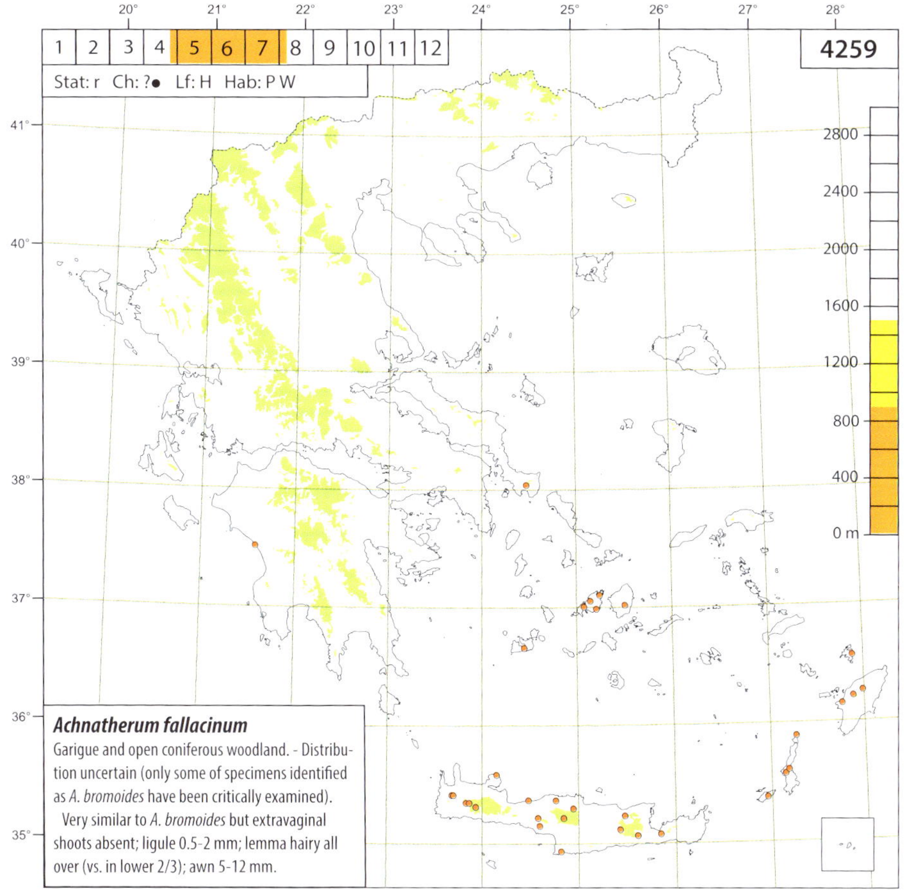

Achnatherum fallacinum

Garigue and open coniferous woodland. - Distribution uncertain (only some of specimens identified as *A. bromoides* have been critically examined).

Very similar to *A. bromoides* but extravaginal shoots absent; ligule 0.5-2 mm; lemma hairy all over (vs. in lower 2/3); awn 5-12 mm.

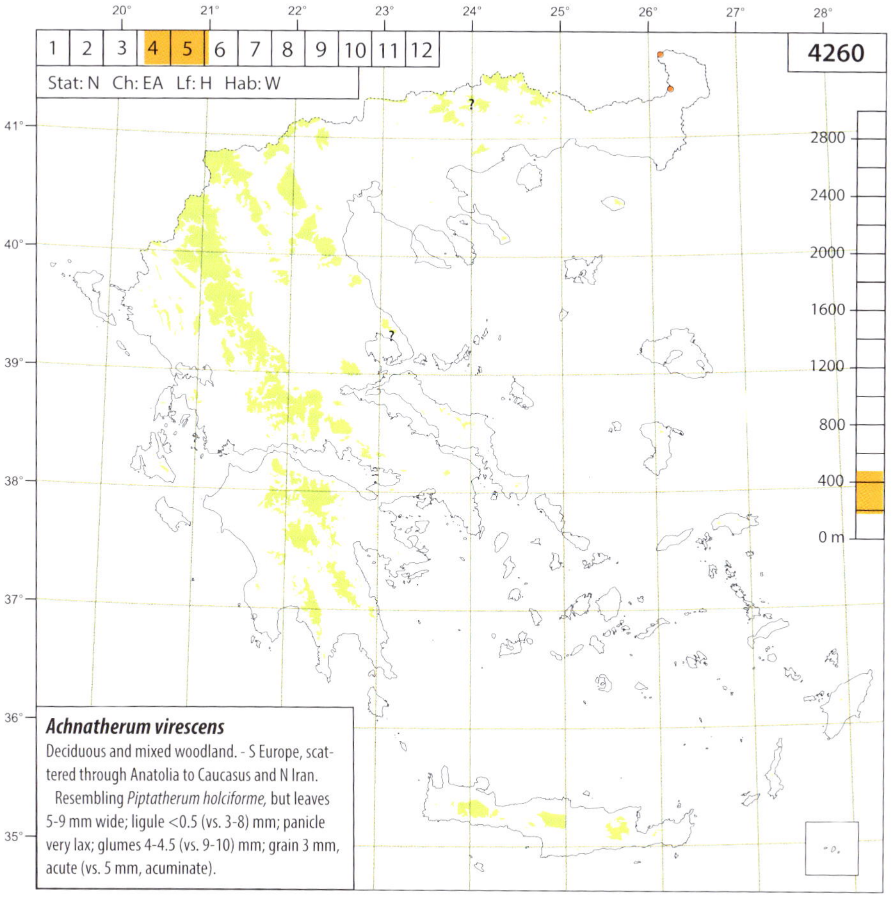

Achnatherum virescens

Deciduous and mixed woodland. - S Europe, scattered through Anatolia to Caucasus and N Iran.

Resembling *Piptatherum holciforme*, but leaves 5-9 mm wide; ligule <0.5 (vs. 3-8) mm; panicle very lax; glumes 4-4.5 (vs. 9-10) mm; grain 3 mm, acute (vs. 5 mm, acuminate).

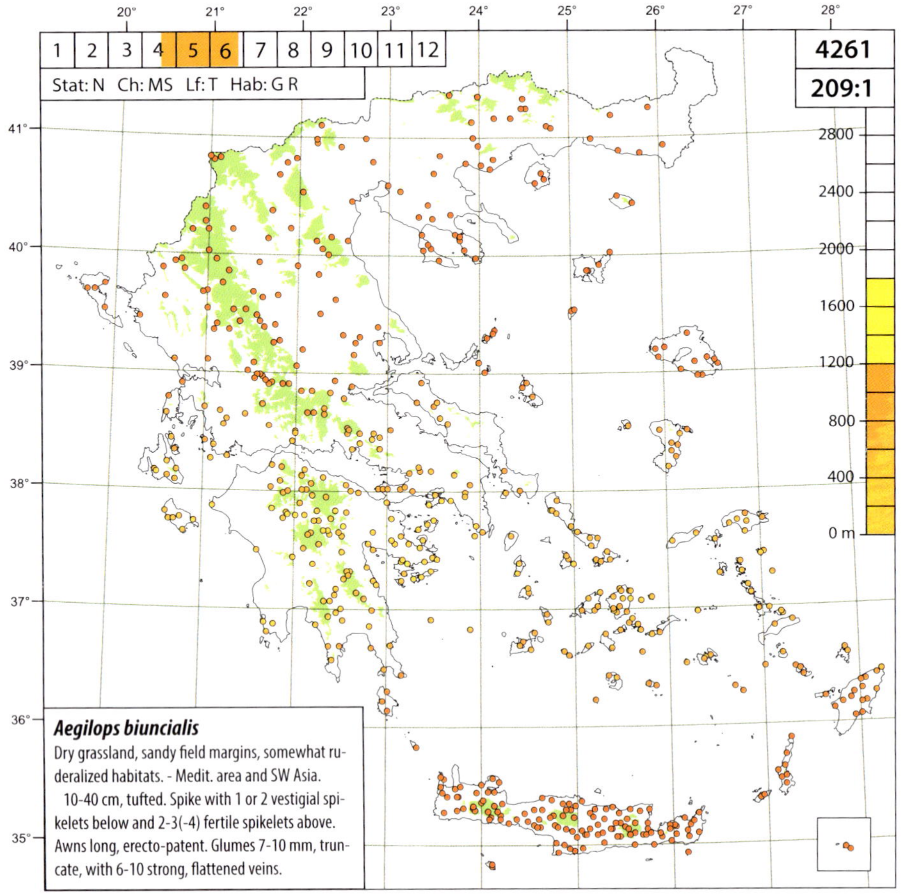

Aegilops biuncialis

Dry grassland, sandy field margins, somewhat ruderalized habitats. - Medit. area and SW Asia.

10-40 cm, tufted. Spike with 1 or 2 vestigial spikelets below and 2-3(-4) fertile spikelets above. Awns long, erecto-patent. Glumes 7-10 mm, truncate, with 6-10 strong, flattened veins.

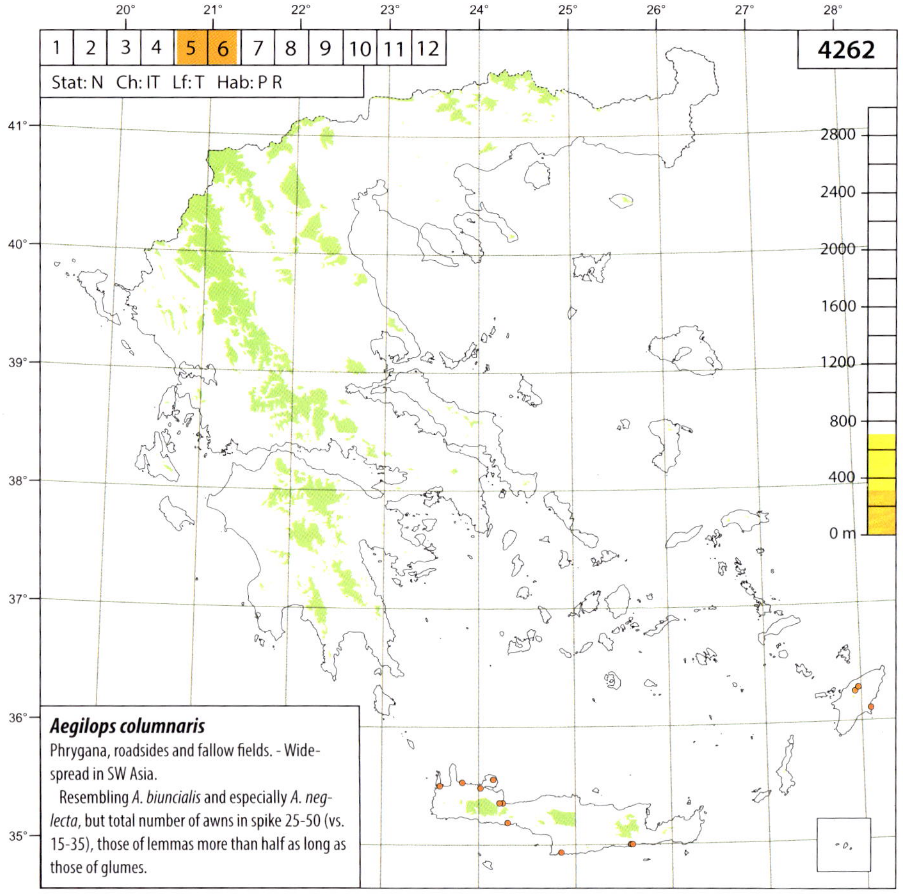

Aegilops columnaris

Phrygana, roadsides and fallow fields. - Widespread in SW Asia.

Resembling *A. biuncialis* and especially *A. neglecta*, but total number of awns in spike 25-50 (vs. 15-35), those of lemmas more than half as long as those of glumes.

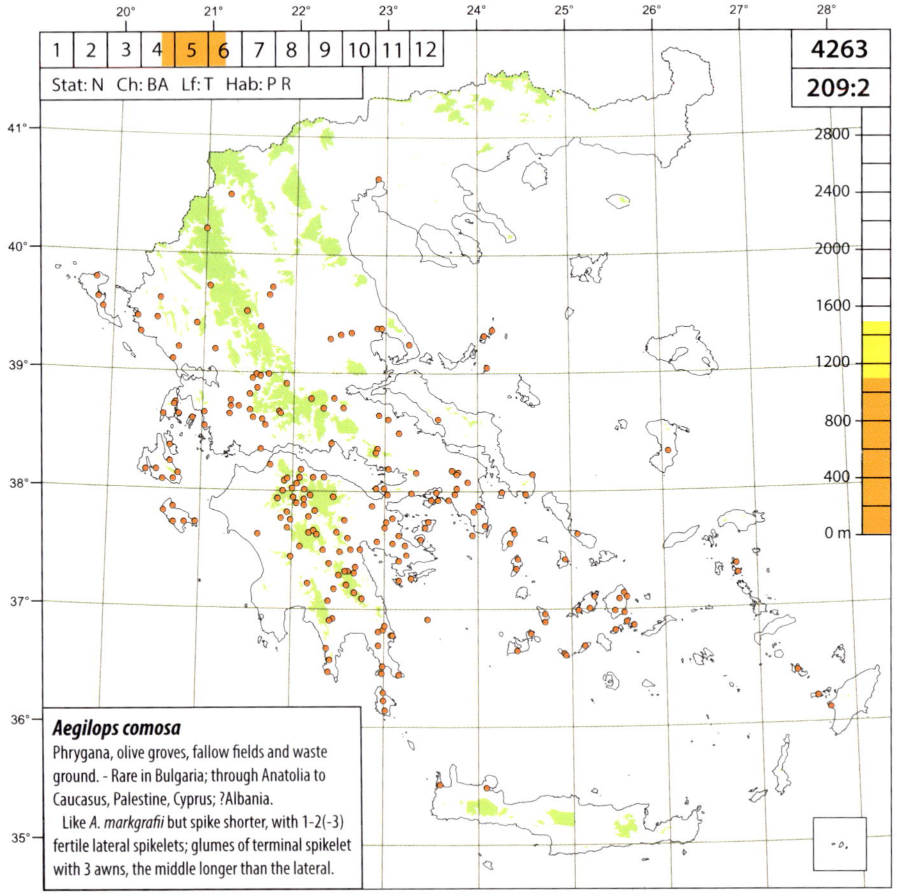

Aegilops comosa

Phrygana, olive groves, fallow fields and waste ground. - Rare in Bulgaria; through Anatolia to Caucasus, Palestine, Cyprus; ?Albania.

Like *A. markgrafii* but spike shorter, with 1-2(-3) fertile lateral spikelets; glumes of terminal spikelet with 3 awns, the middle longer than the lateral.

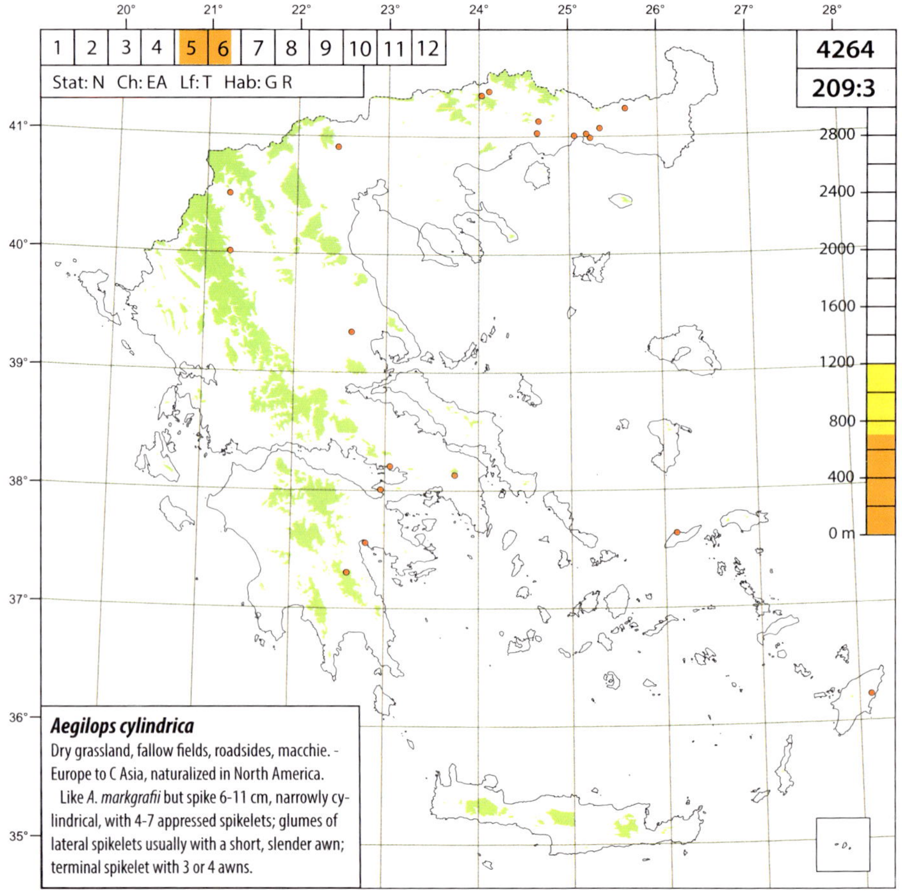

Aegilops cylindrica

Dry grassland, fallow fields, roadsides, macchie. - Europe to C Asia, naturalized in North America.

Like *A. markgrafii* but spike 6-11 cm, narrowly cylindrical, with 4-7 appressed spikelets; glumes of lateral spikelets usually with a short, slender awn; terminal spikelet with 3 or 4 awns.

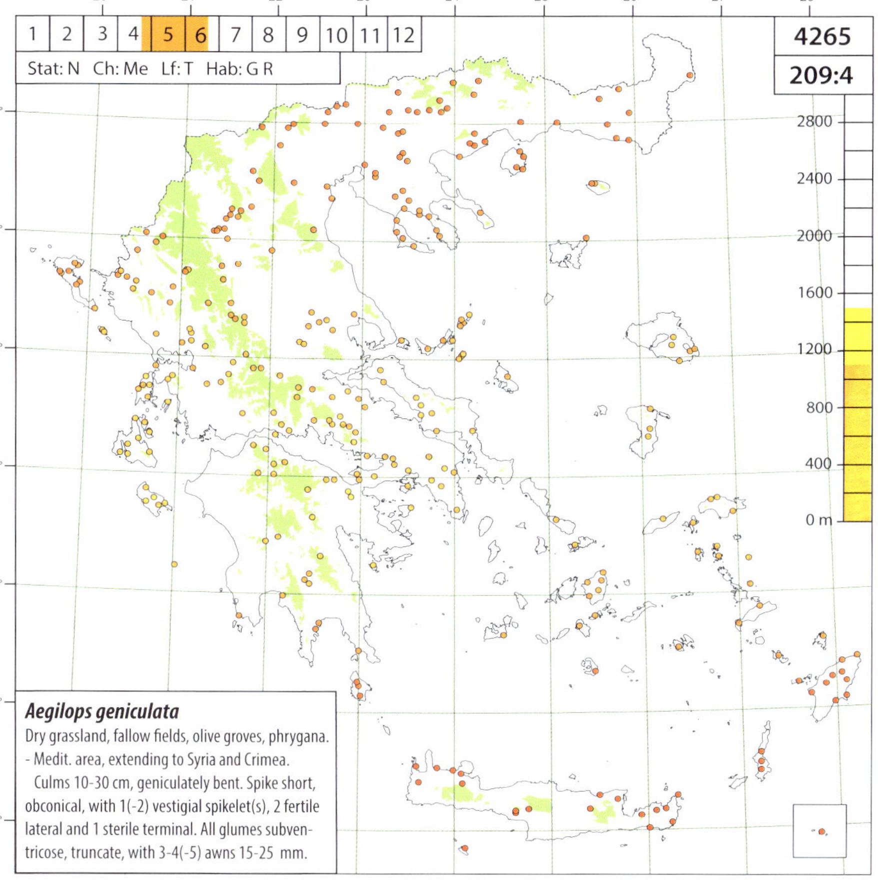

Aegilops geniculata
Dry grassland, fallow fields, olive groves, phrygana. - Medit. area, extending to Syria and Crimea.
Culms 10-30 cm, geniculately bent. Spike short, obconical, with 1(-2) vestigial spikelet(s), 2 fertile lateral and 1 sterile terminal. All glumes subventricose, truncate, with 3-4(-5) awns 15-25 mm.

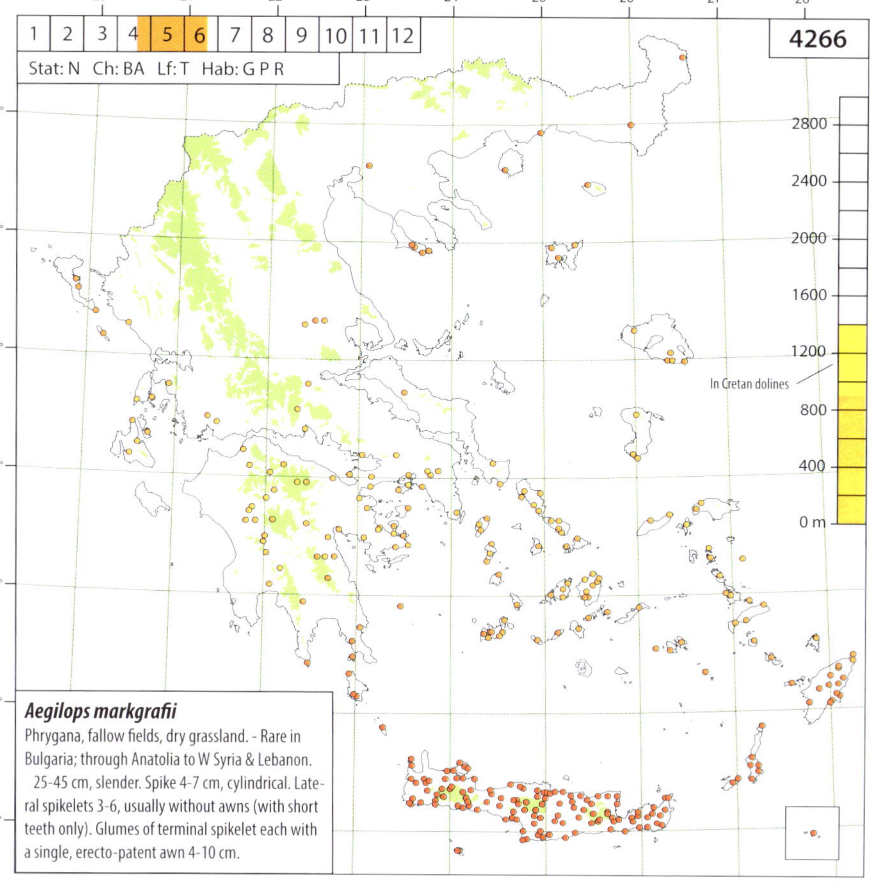

Aegilops markgrafii
Phrygana, fallow fields, dry grassland. - Rare in Bulgaria; through Anatolia to W Syria & Lebanon.
25-45 cm, slender. Spike 4-7 cm, cylindrical. Lateral spikelets 3-6, usually without awns (with short teeth only). Glumes of terminal spikelet each with a single, erecto-patent awn 4-10 cm.

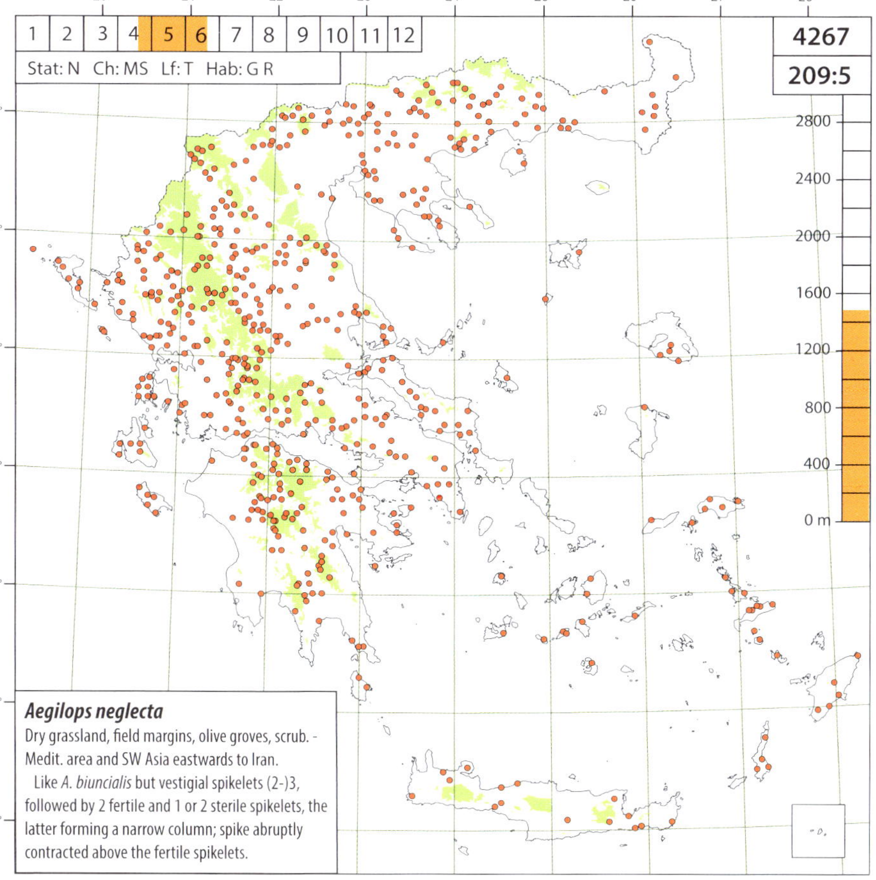

Aegilops neglecta
Dry grassland, field margins, olive groves, scrub. - Medit. area and SW Asia eastwards to Iran.
Like _A. biuncialis_ but vestigial spikelets (2-)3, followed by 2 fertile and 1 or 2 sterile spikelets, the latter forming a narrow column; spike abruptly contracted above the fertile spikelets.

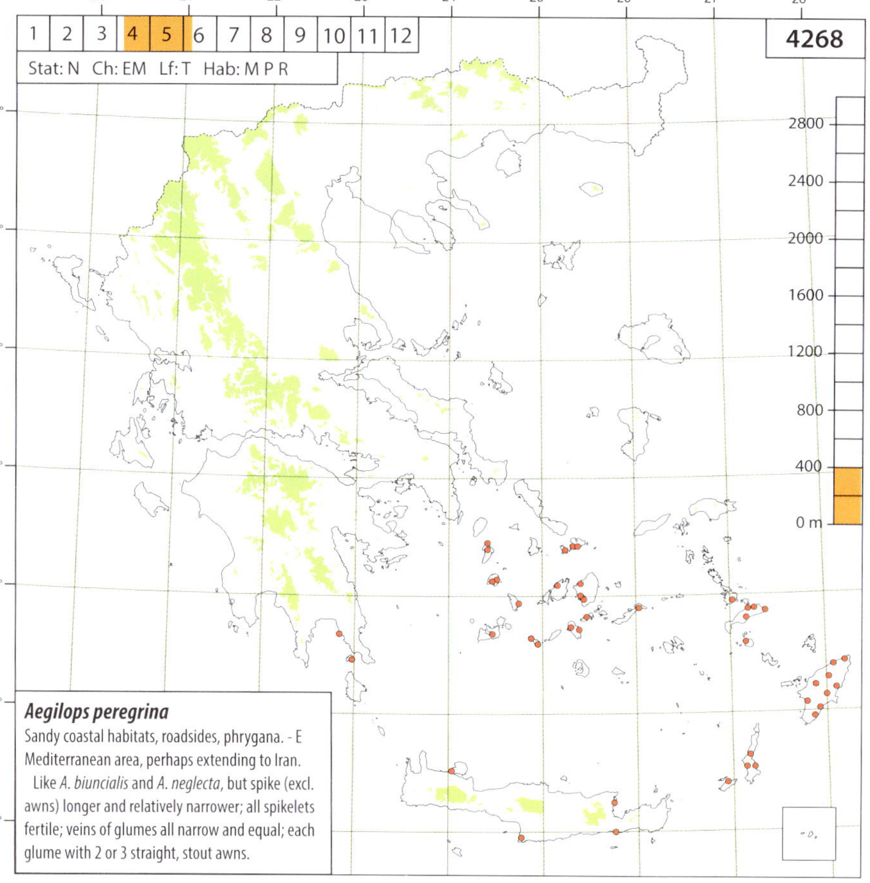

Aegilops peregrina
Sandy coastal habitats, roadsides, phrygana. - E Mediterranean area, perhaps extending to Iran.
Like _A. biuncialis_ and _A. neglecta_, but spike (excl. awns) longer and relatively narrower; all spikelets fertile; veins of glumes all narrow and equal; each glume with 2 or 3 straight, stout awns.

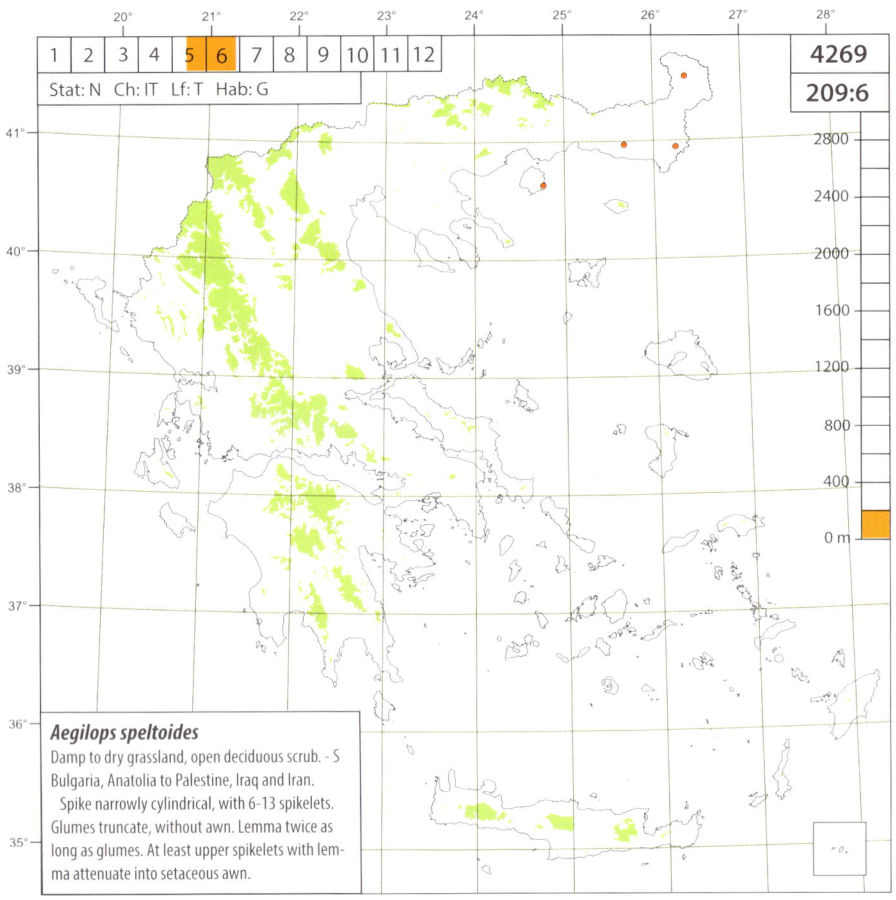

Aegilops speltoides

Damp to dry grassland, open deciduous scrub. - S Bulgaria, Anatolia to Palestine, Iraq and Iran.

Spike narrowly cylindrical, with 6-13 spikelets. Glumes truncate, without awn. Lemma twice as long as glumes. At least upper spikelets with lemma attenuate into setaceous awn.

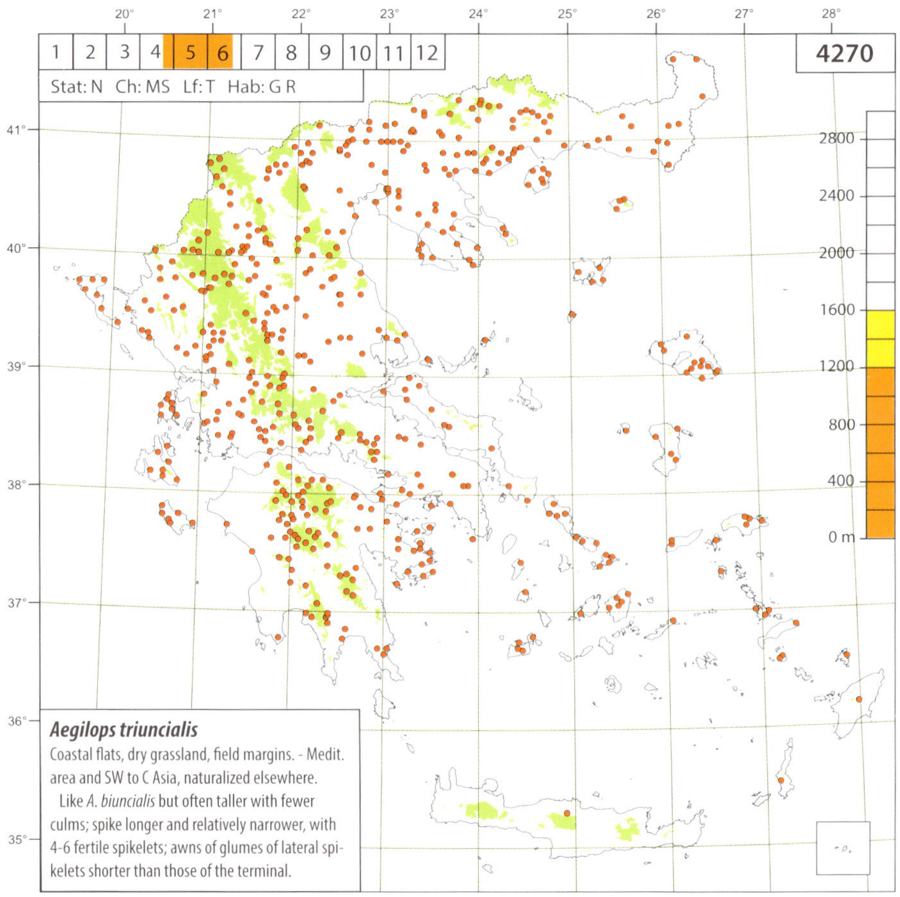

Aegilops triuncialis

Coastal flats, dry grassland, field margins. - Medit. area and SW to C Asia, naturalized elsewhere.

Like *A. biuncialis* but often taller with fewer culms; spike longer and relatively narrower, with 4-6 fertile spikelets; awns of glumes of lateral spikelets shorter than those of the terminal.

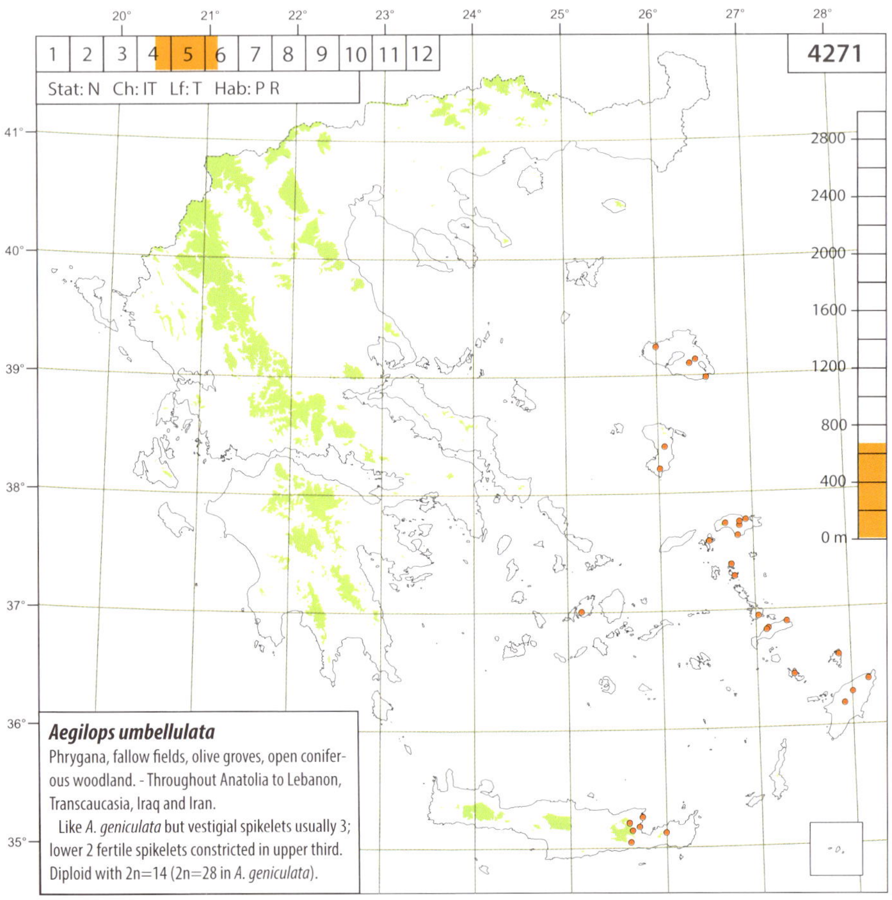

Aegilops umbellulata

Phrygana, fallow fields, olive groves, open coniferous woodland. - Throughout Anatolia to Lebanon, Transcaucasia, Iraq and Iran.

Like *A. geniculata* but vestigial spikelets usually 3; lower 2 fertile spikelets constricted in upper third. Diploid with 2n=14 (2n=28 in *A. geniculata*).

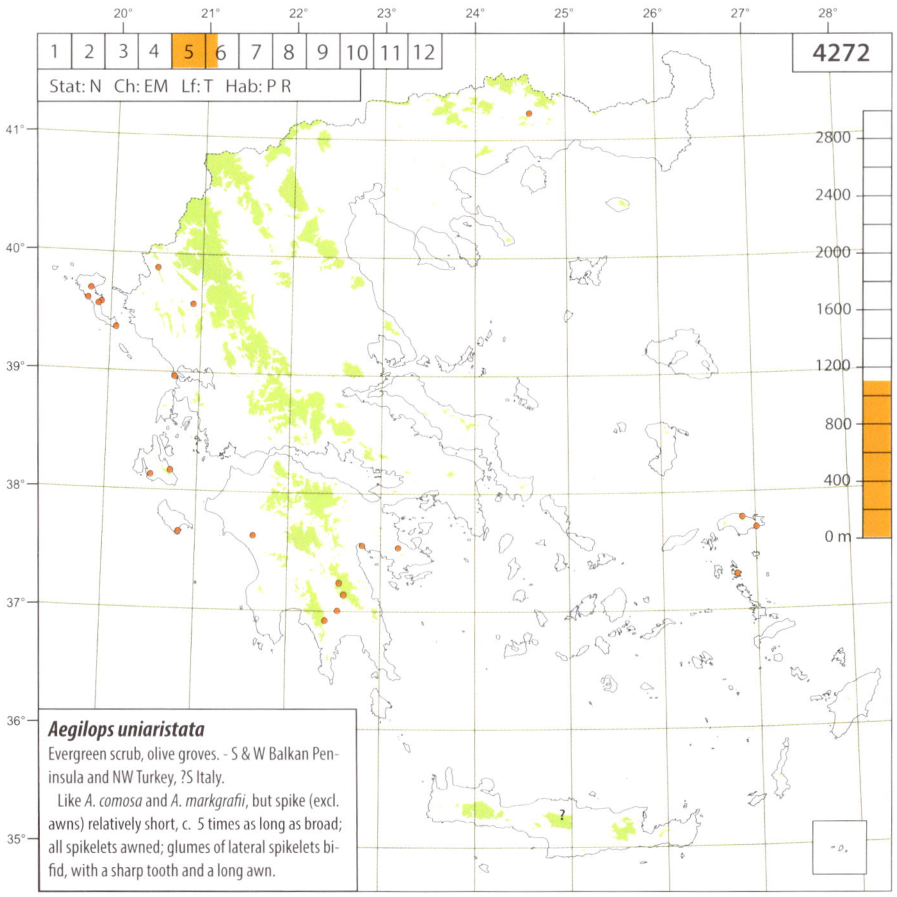

Aegilops uniaristata

Evergreen scrub, olive groves. - S & W Balkan Peninsula and NW Turkey, ?S Italy.

Like *A. comosa* and *A. markgrafii*, but spike (excl. awns) relatively short, c. 5 times as long as broad; all spikelets awned; glumes of lateral spikelets bifid, with a sharp tooth and a long awn.

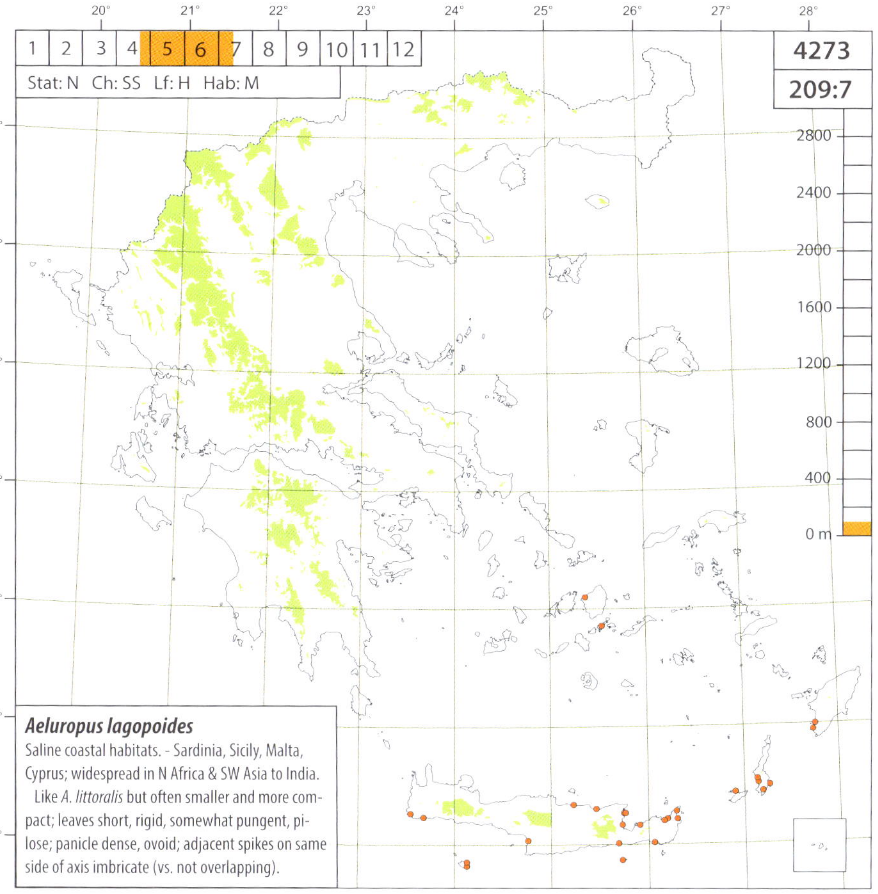

Aeluropus lagopoides
Saline coastal habitats. - Sardinia, Sicily, Malta, Cyprus; widespread in N Africa & SW Asia to India.
Like *A. littoralis* but often smaller and more compact; leaves short, rigid, somewhat pungent, pilose; panicle dense, ovoid; adjacent spikes on same side of axis imbricate (vs. not overlapping).

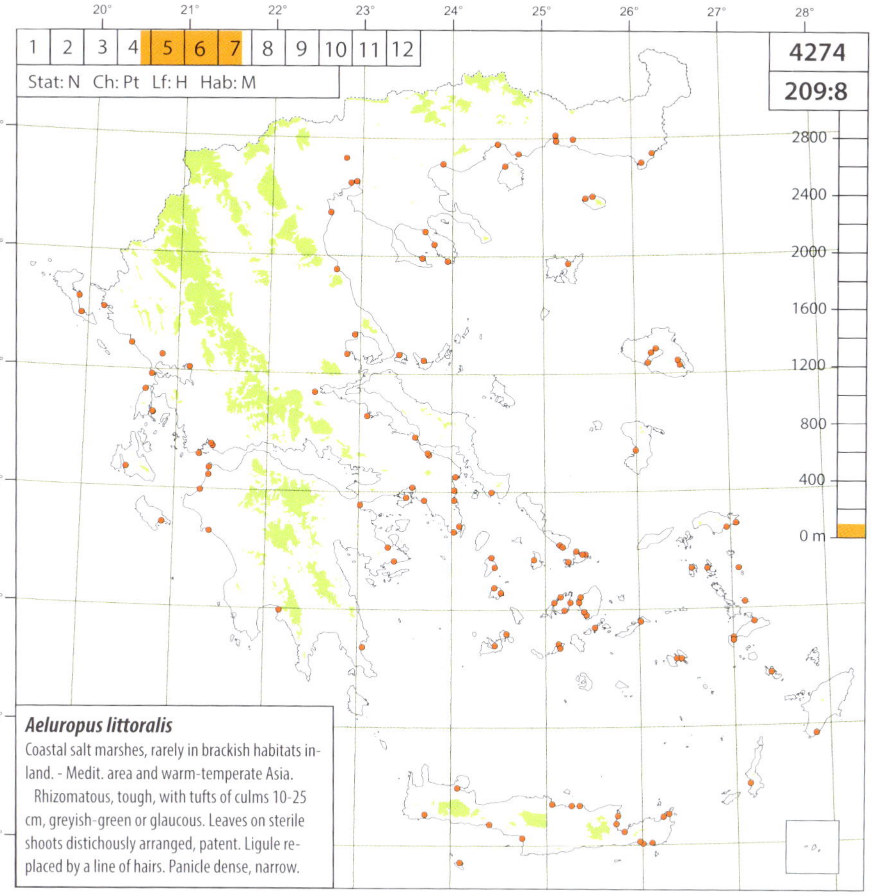

Aeluropus littoralis
Coastal salt marshes, rarely in brackish habitats inland. - Medit. area and warm-temperate Asia.
Rhizomatous, tough, with tufts of culms 10-25 cm, greyish-green or glaucous. Leaves on sterile shoots distichously arranged, patent. Ligule replaced by a line of hairs. Panicle dense, narrow.

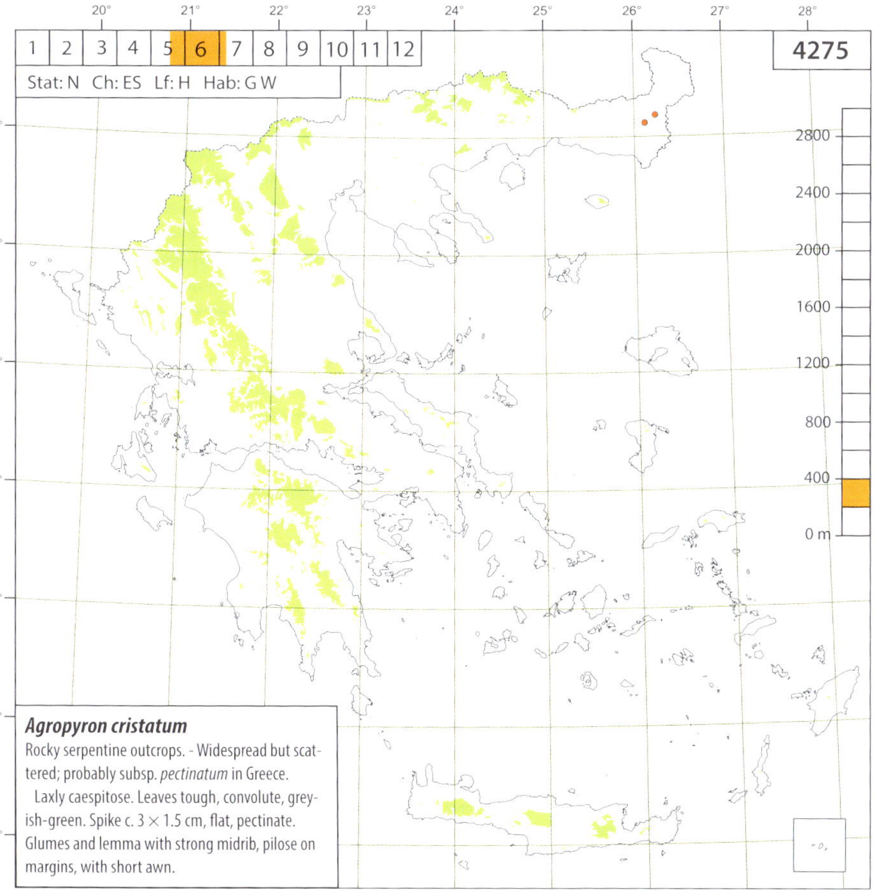

Agropyron cristatum
Rocky serpentine outcrops. - Widespread but scattered; probably subsp. *pectinatum* in Greece.
Laxly caespitose. Leaves tough, convolute, greyish-green. Spike c. 3 × 1.5 cm, flat, pectinate. Glumes and lemma with strong midrib, pilose on margins, with short awn.

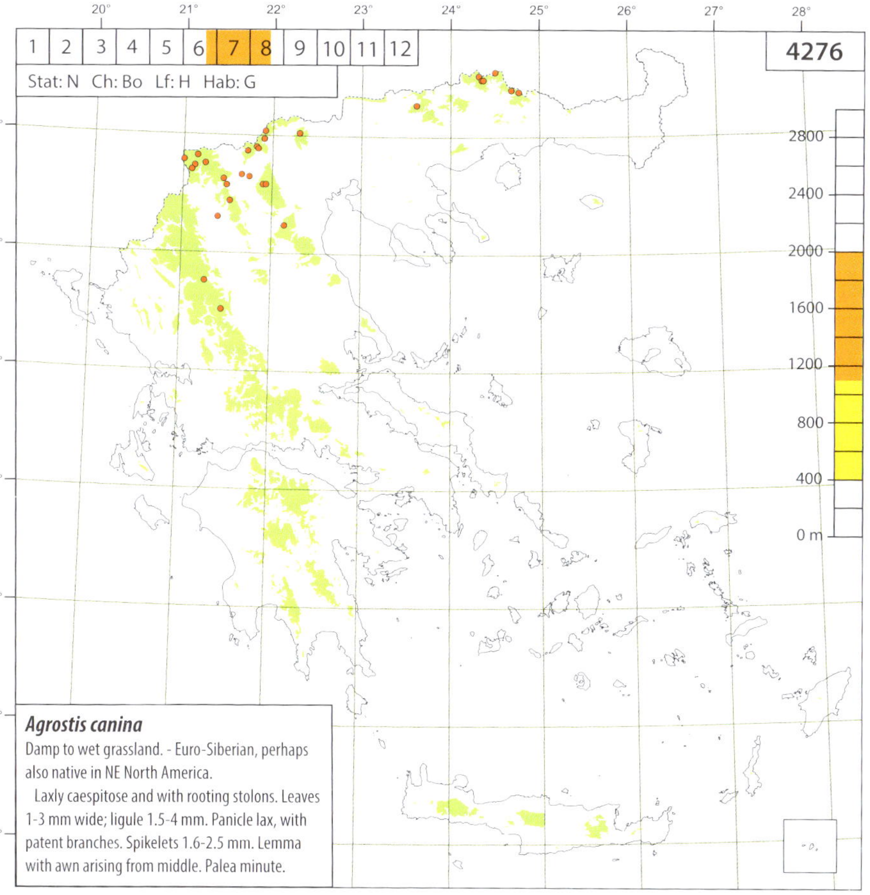

Agrostis canina
Damp to wet grassland. - Euro-Siberian, perhaps also native in NE North America.
Laxly caespitose and with rooting stolons. Leaves 1-3 mm wide; ligule 1.5-4 mm. Panicle lax, with patent branches. Spikelets 1.6-2.5 mm. Lemma with awn arising from middle. Palea minute.

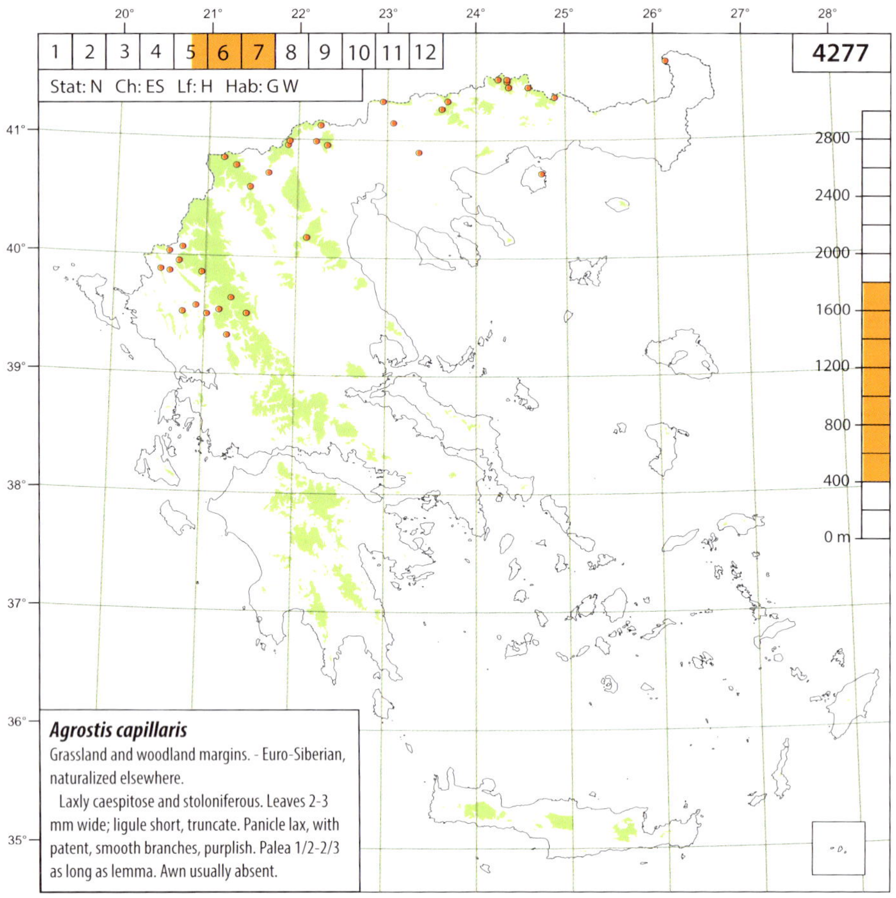

Agrostis capillaris
Grassland and woodland margins. - Euro-Siberian, naturalized elsewhere.
Laxly caespitose and stoloniferous. Leaves 2-3 mm wide; ligule short, truncate. Panicle lax, with patent, smooth branches, purplish. Palea 1/2-2/3 as long as lemma. Awn usually absent.

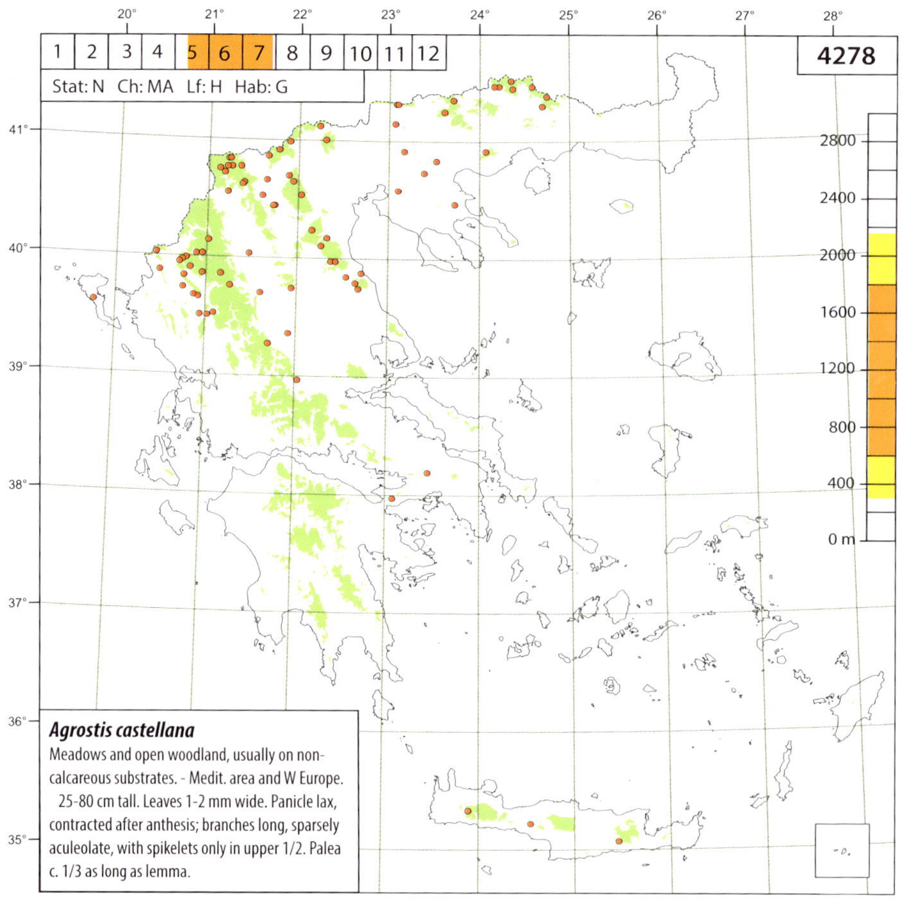

Agrostis castellana
Meadows and open woodland, usually on non-calcareous substrates. - Medit. area and W Europe.
25-80 cm tall. Leaves 1-2 mm wide. Panicle lax, contracted after anthesis; branches long, sparsely aculeolate, with spikelets only in upper 1/2. Palea c. 1/3 as long as lemma.

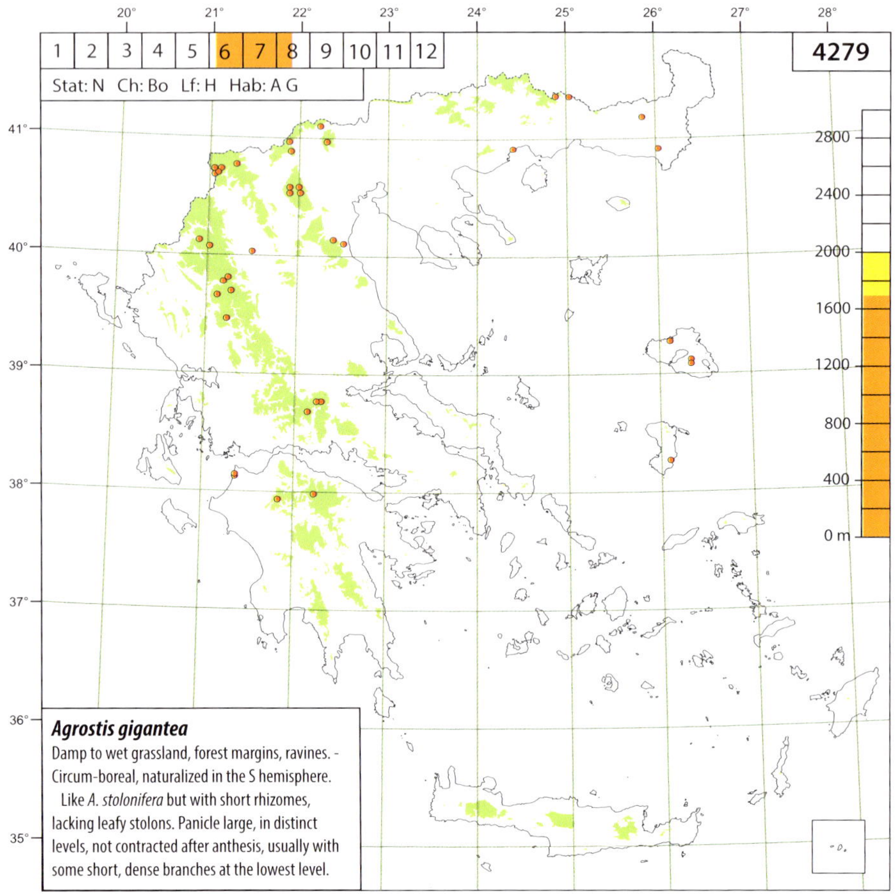

Agrostis gigantea
Damp to wet grassland, forest margins, ravines. - Circum-boreal, naturalized in the S hemisphere.
Like *A. stolonifera* but with short rhizomes, lacking leafy stolons. Panicle large, in distinct levels, not contracted after anthesis, usually with some short, dense branches at the lowest level.

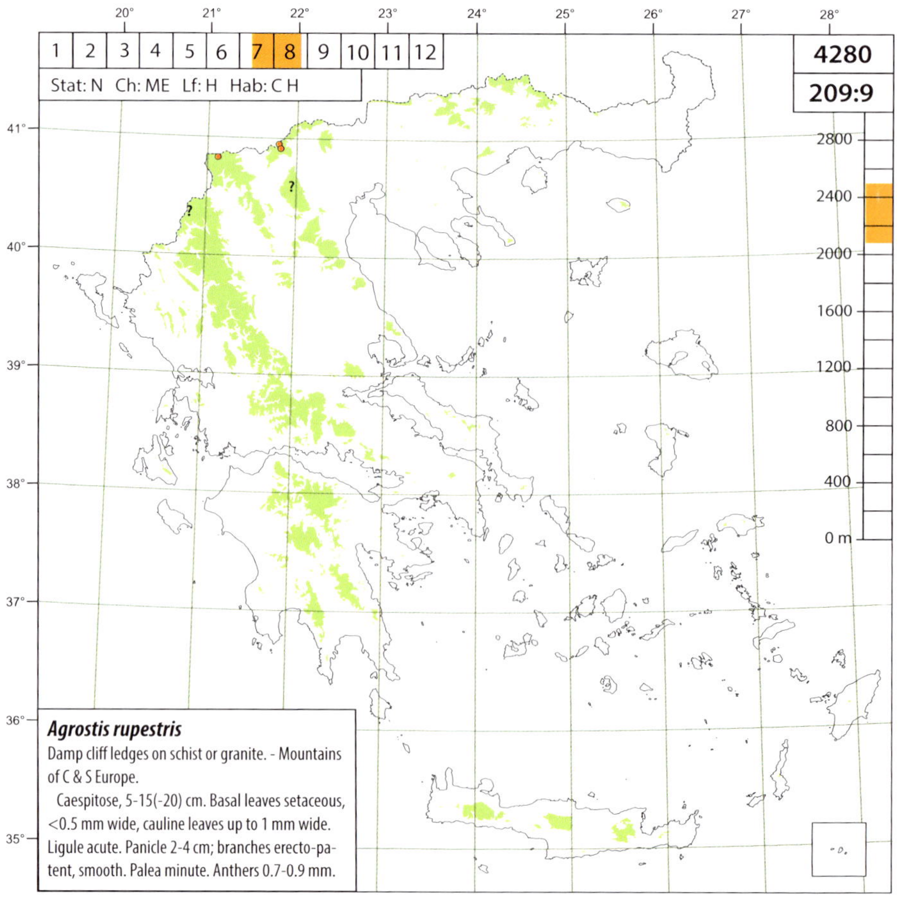

Agrostis rupestris
Damp cliff ledges on schist or granite. - Mountains of C & S Europe.
Caespitose, 5-15(-20) cm. Basal leaves setaceous, <0.5 mm wide, cauline leaves up to 1 mm wide. Ligule acute. Panicle 2-4 cm; branches erecto-patent, smooth. Palea minute. Anthers 0.7-0.9 mm.

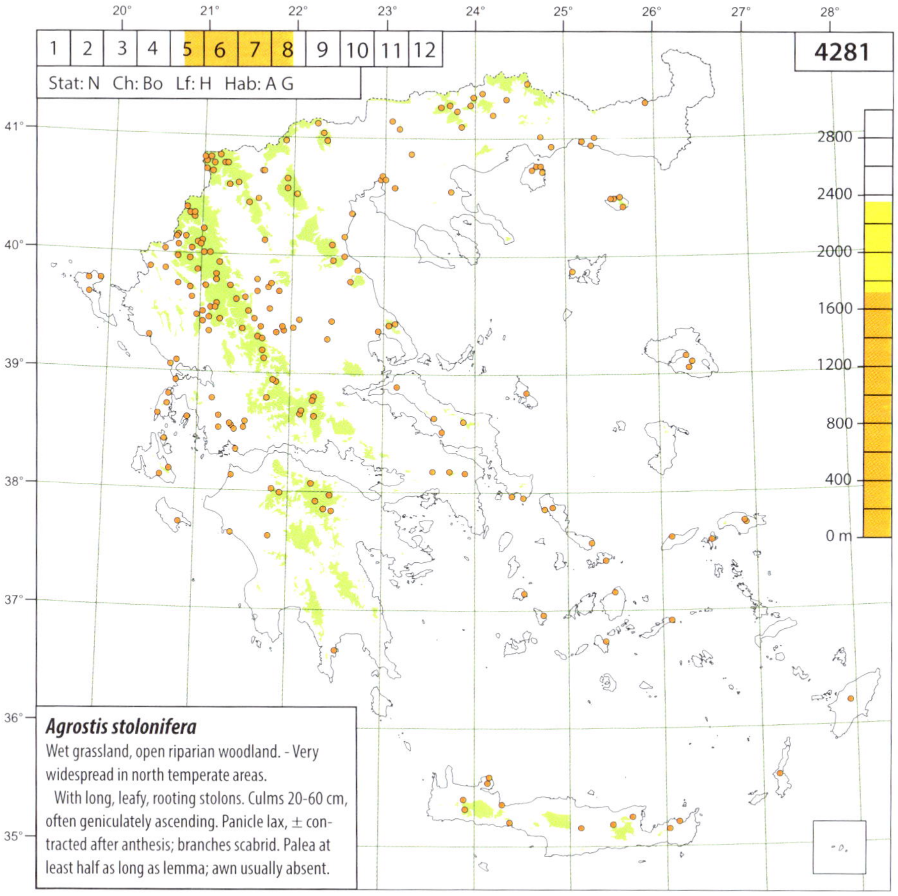

Agrostis stolonifera

Wet grassland, open riparian woodland. - Very widespread in north temperate areas.

With long, leafy, rooting stolons. Culms 20-60 cm, often geniculately ascending. Panicle lax, ± contracted after anthesis; branches scabrid. Palea at least half as long as lemma; awn usually absent.

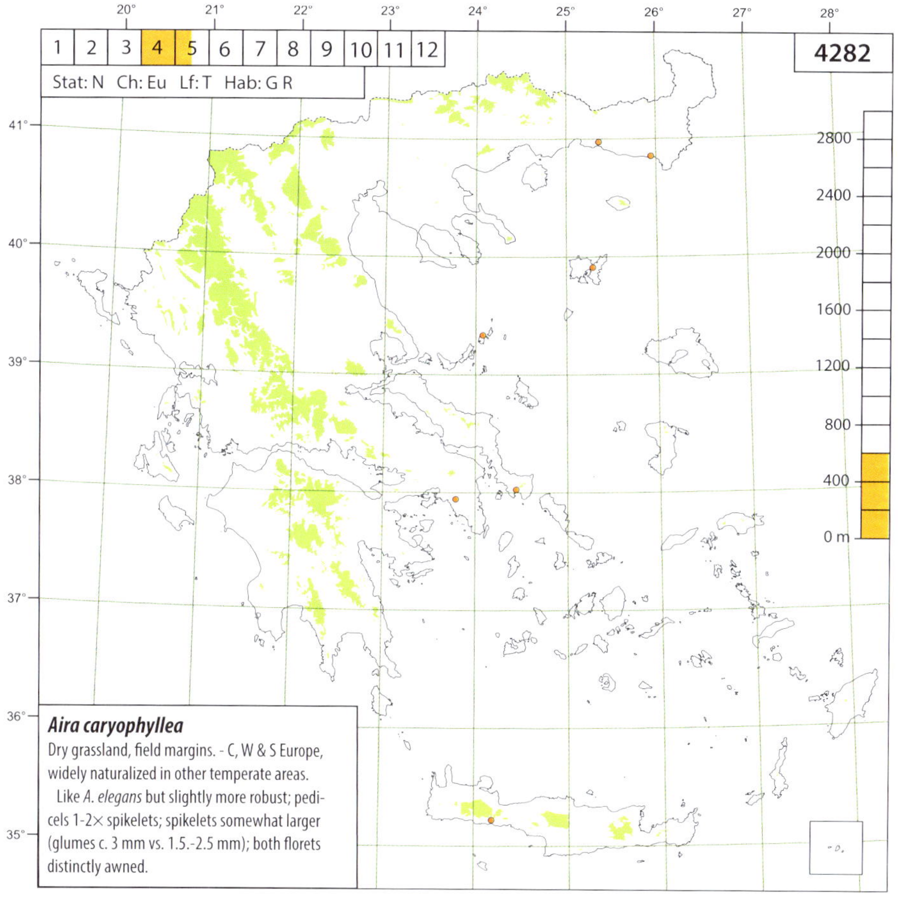

Aira caryophyllea

Dry grassland, field margins. - C, W & S Europe, widely naturalized in other temperate areas.

Like *A. elegans* but slightly more robust; pedicels 1-2× spikelets; spikelets somewhat larger (glumes c. 3 mm vs. 1.5.-2.5 mm); both florets distinctly awned.

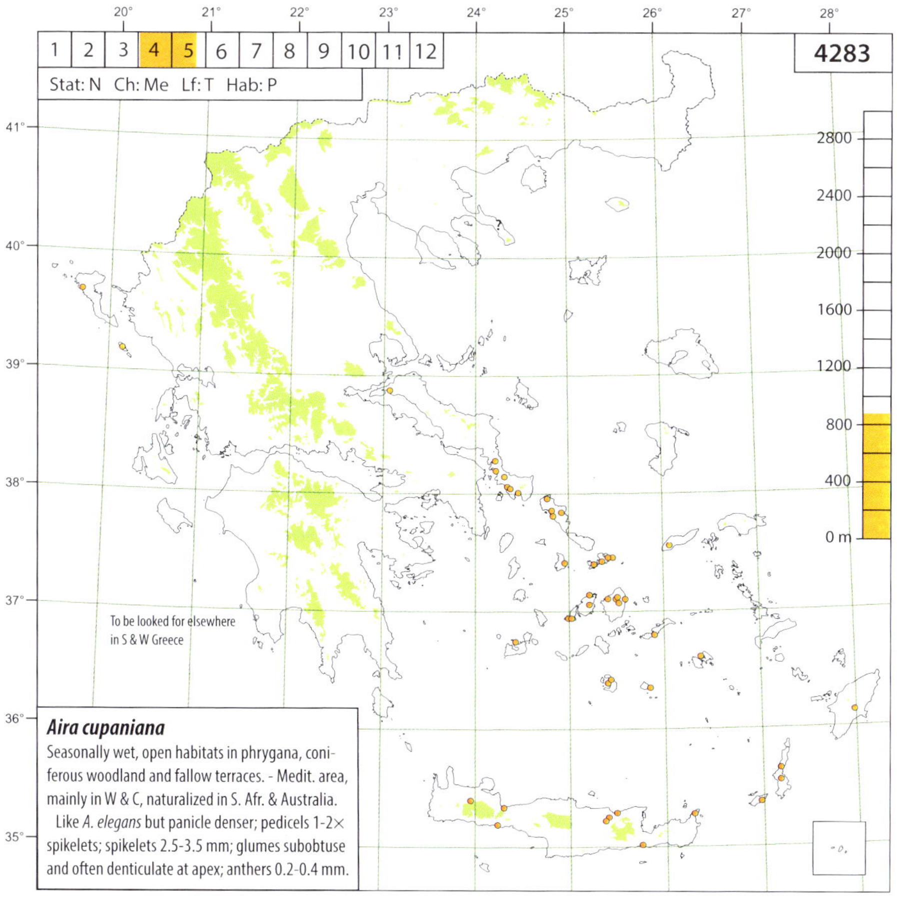

Aira cupaniana

Seasonally wet, open habitats in phrygana, coniferous woodland and fallow terraces. - Medit. area, mainly in W & C, naturalized in S. Afr. & Australia.

Like *A. elegans* but panicle denser; pedicels 1-2× spikelets; spikelets 2.5-3.5 mm; glumes subobtuse and often denticulate at apex; anthers 0.2-0.4 mm.

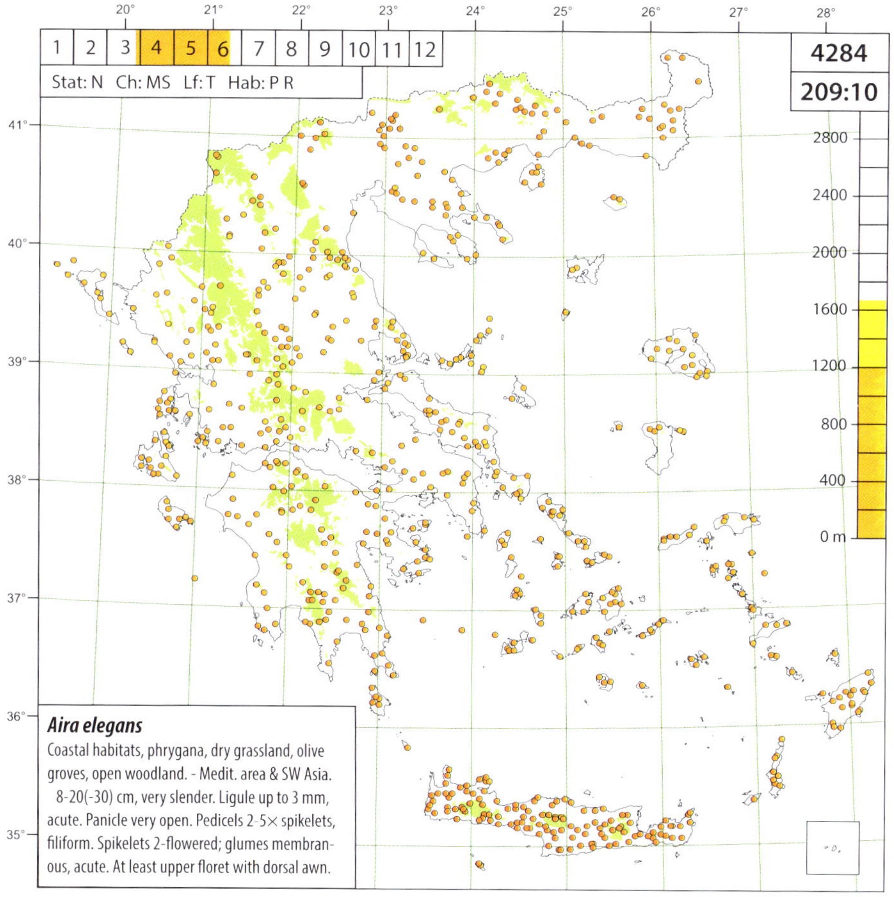

Aira elegans

Coastal habitats, phrygana, dry grassland, olive groves, open woodland. - Medit. area & SW Asia.

8-20(-30) cm, very slender. Ligule up to 3 mm, acute. Panicle very open. Pedicels 2-5× spikelets, filiform. Spikelets 2-flowered; glumes membranous, acute. At least upper floret with dorsal awn.

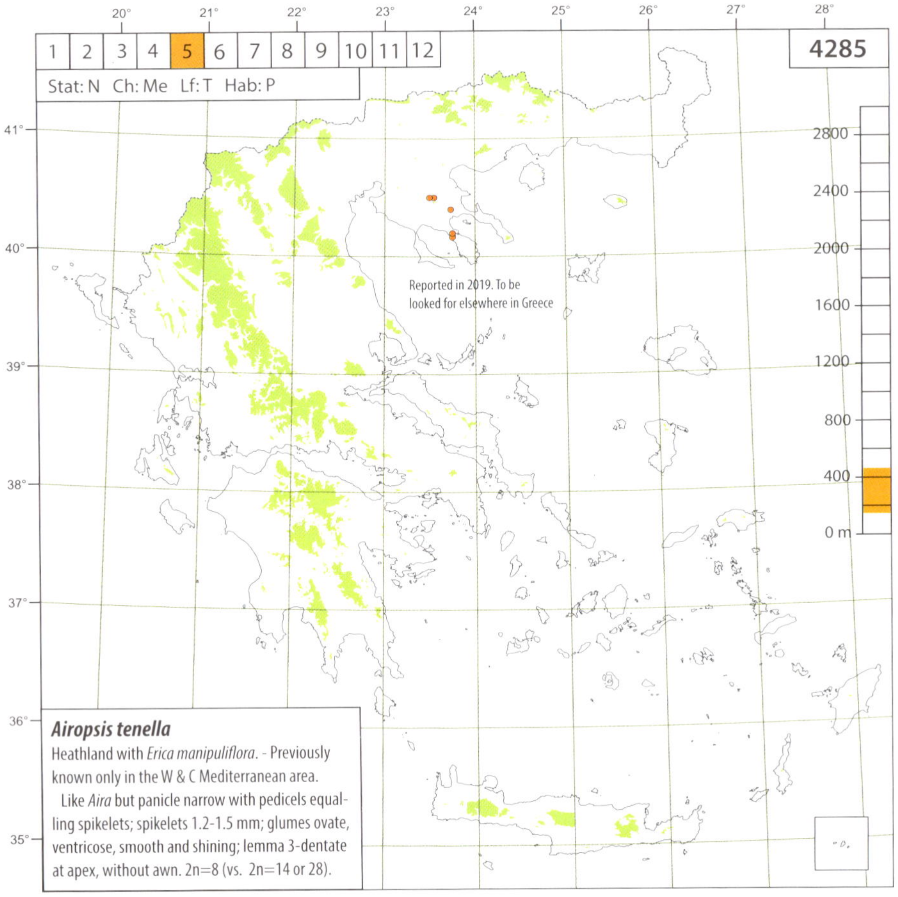

Airopsis tenella

Heathland with *Erica manipuliflora*. - Previously known only in the W & C Mediterranean area.

Like *Aira* but panicle narrow with pedicels equalling spikelets; spikelets 1.2-1.5 mm; glumes ovate, ventricose, smooth and shining; lemma 3-dentate at apex, without awn. 2n=8 (vs. 2n=14 or 28).

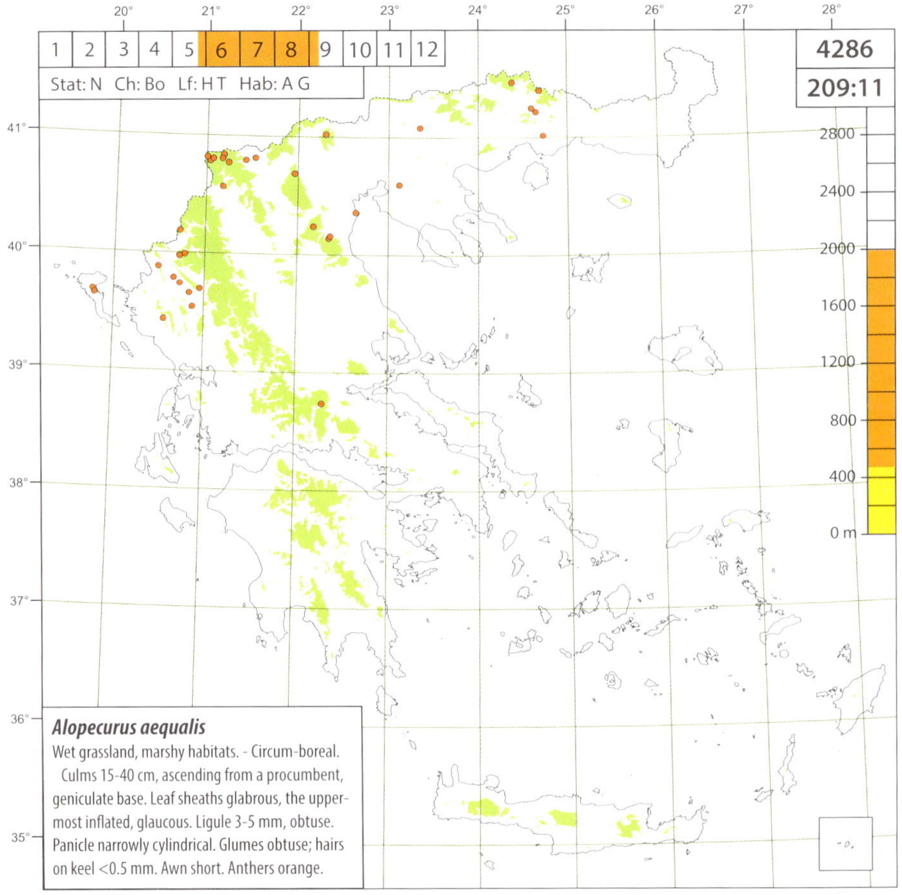

Alopecurus aequalis

Wet grassland, marshy habitats. - Circum-boreal.

Culms 15-40 cm, ascending from a procumbent, geniculate base. Leaf sheaths glabrous, the uppermost inflated, glaucous. Ligule 3-5 mm, obtuse. Panicle narrowly cylindrical. Glumes obtuse; hairs on keel <0.5 mm. Awn short. Anthers orange.

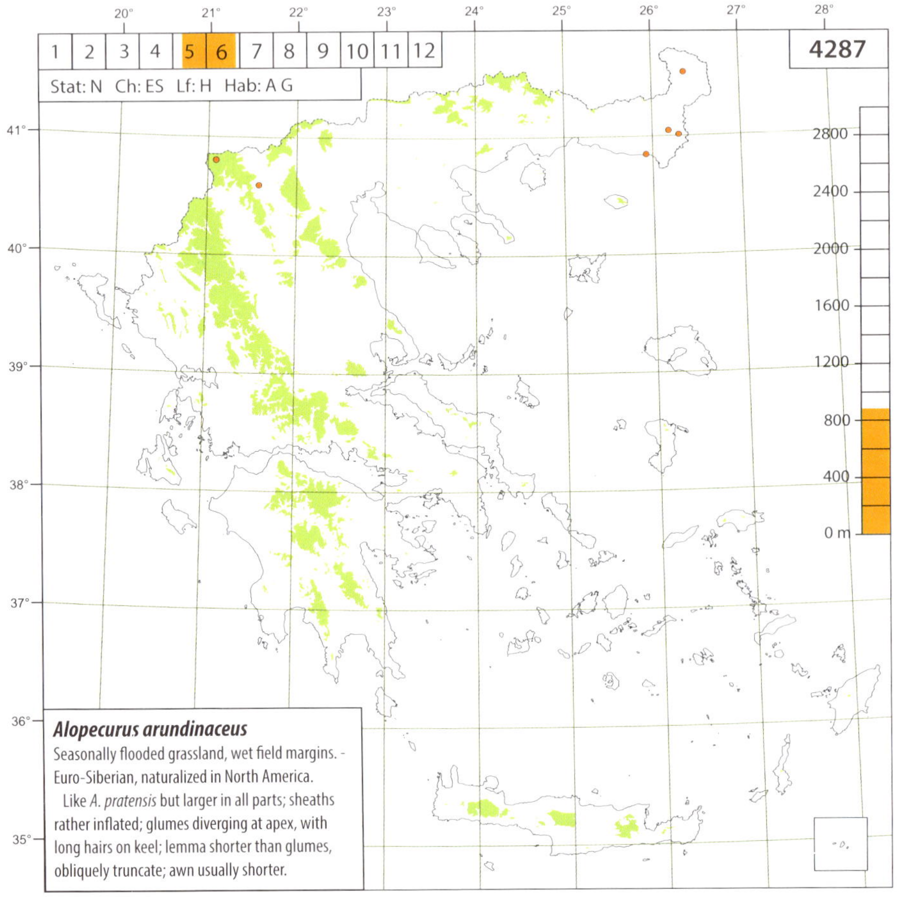

Alopecurus arundinaceus

Seasonally flooded grassland, wet field margins. - Euro-Siberian, naturalized in North America.

Like *A. pratensis* but larger in all parts; sheaths rather inflated; glumes diverging at apex, with long hairs on keel; lemma shorter than glumes, obliquely truncate; awn usually shorter.

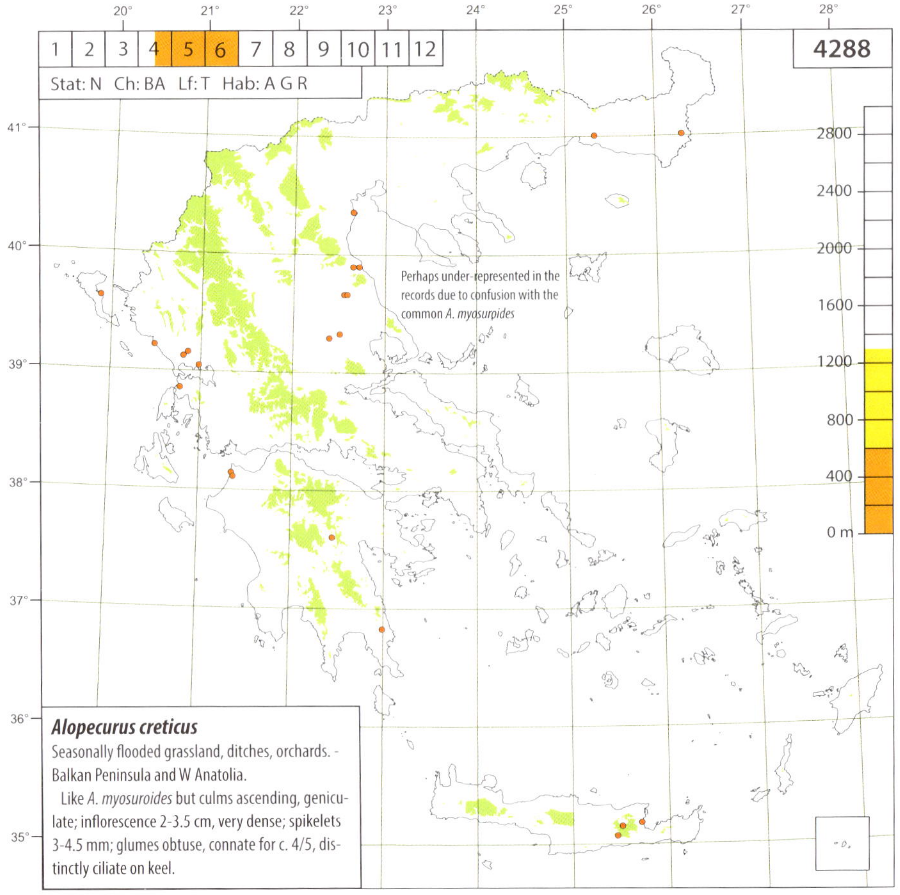

Alopecurus creticus

Seasonally flooded grassland, ditches, orchards. - Balkan Peninsula and W Anatolia.

Like *A. myosuroides* but culms ascending, geniculate; inflorescence 2-3.5 cm, very dense; spikelets 3-4.5 mm; glumes obtuse, connate for c. 4/5, distinctly ciliate on keel.

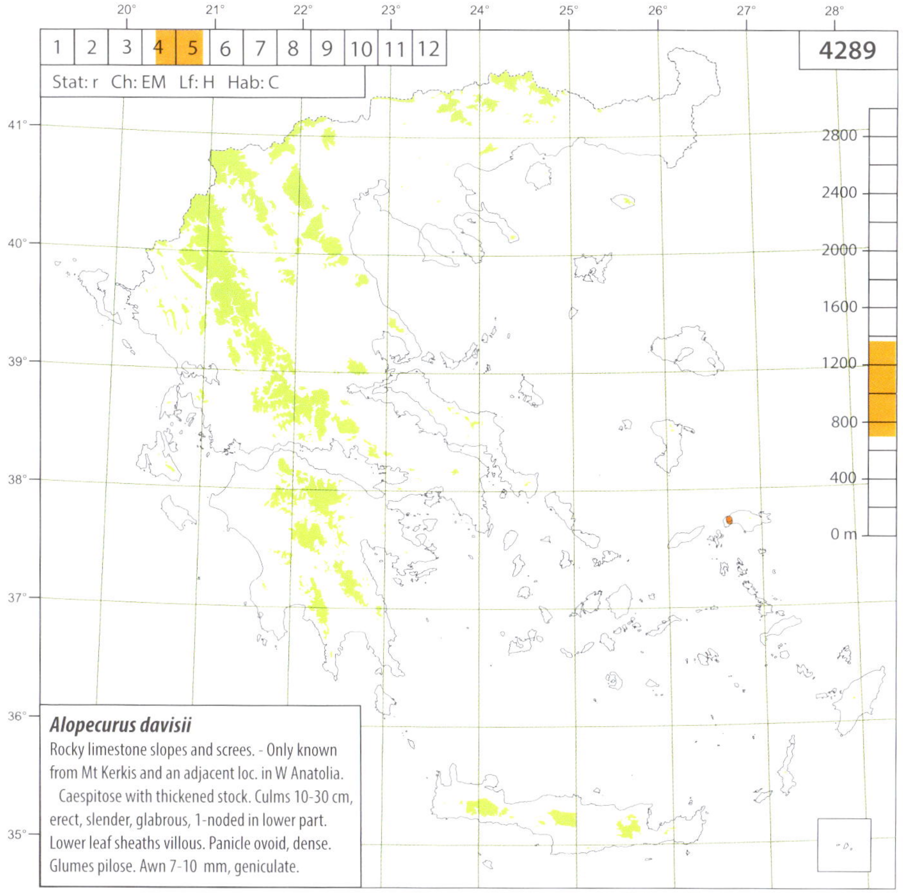

Alopecurus davisii
Rocky limestone slopes and screes. - Only known from Mt Kerkis and an adjacent loc. in W Anatolia.
Caespitose with thickened stock. Culms 10-30 cm, erect, slender, glabrous, 1-noded in lower part. Lower leaf sheaths villous. Panicle ovoid, dense. Glumes pilose. Awn 7-10 mm, geniculate.

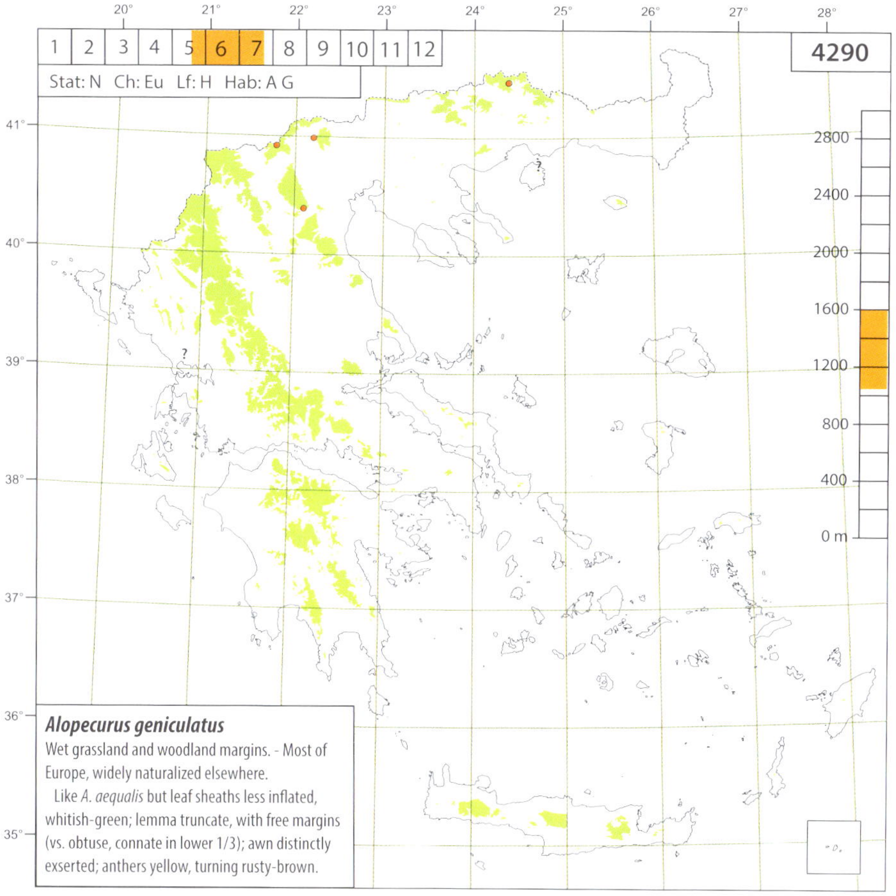

Alopecurus geniculatus
Wet grassland and woodland margins. - Most of Europe, widely naturalized elsewhere.
Like *A. aequalis* but leaf sheaths less inflated, whitish-green; lemma truncate, with free margins (vs. obtuse, connate in lower 1/3); awn distinctly exserted; anthers yellow, turning rusty-brown.

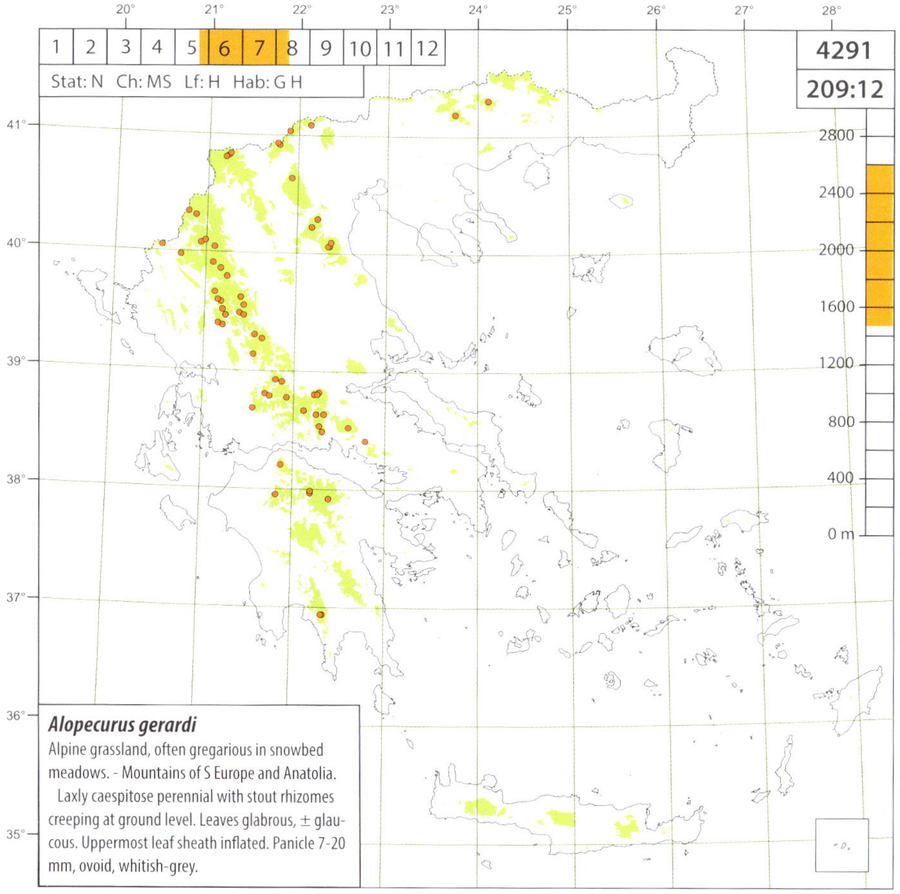

Alopecurus gerardi
Alpine grassland, often gregarious in snowbed meadows. - Mountains of S Europe and Anatolia.
Laxly caespitose perennial with stout rhizomes creeping at ground level. Leaves glabrous, ± glaucous. Uppermost leaf sheath inflated. Panicle 7-20 mm, ovoid, whitish-grey.

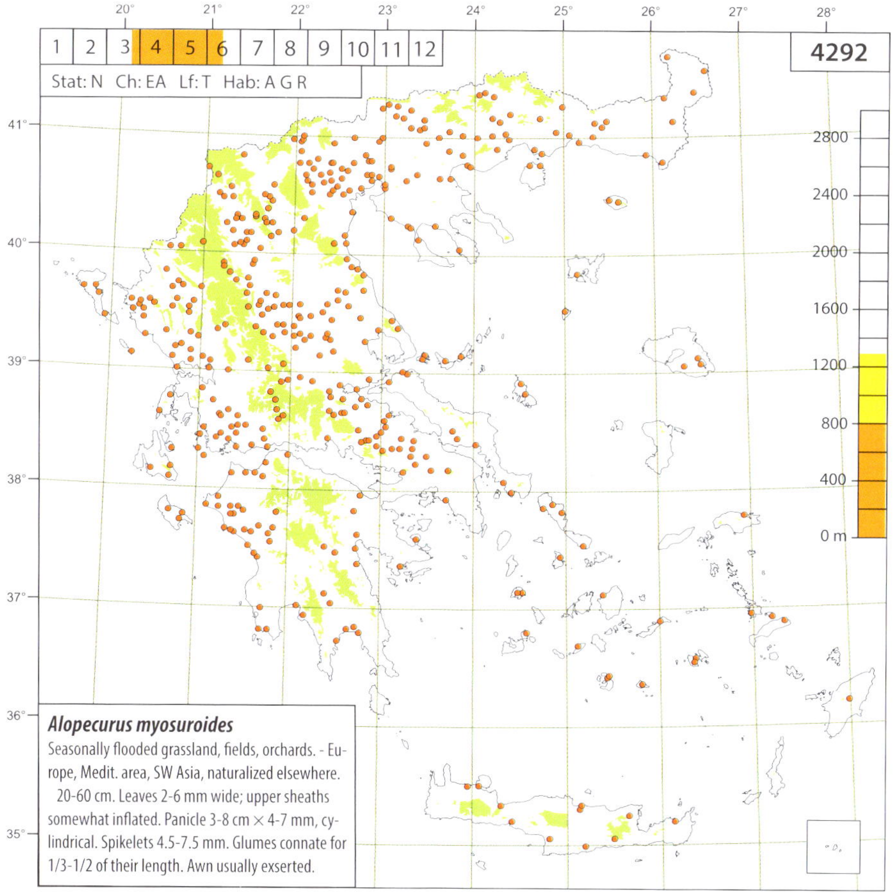

Alopecurus myosuroides
Seasonally flooded grassland, fields, orchards. - Europe, Medit. area, SW Asia, naturalized elsewhere.
20-60 cm. Leaves 2-6 mm wide; upper sheaths somewhat inflated. Panicle 3-8 cm × 4-7 mm, cylindrical. Spikelets 4.5-7.5 mm. Glumes connate for 1/3-1/2 of their length. Awn usually exserted.

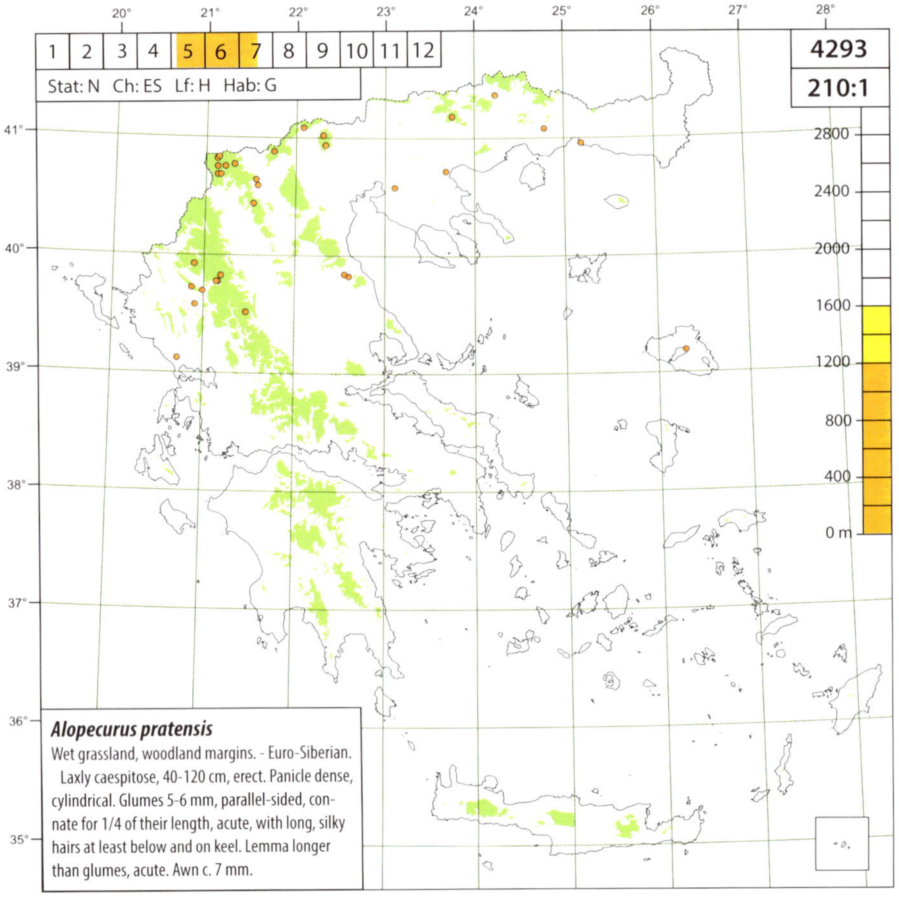

Alopecurus pratensis

Wet grassland, woodland margins. - Euro-Siberian.

Laxly caespitose, 40-120 cm, erect. Panicle dense, cylindrical. Glumes 5-6 mm, parallel-sided, connate for 1/4 of their length, acute, with long, silky hairs at least below and on keel. Lemma longer than glumes, acute. Awn c. 7 mm.

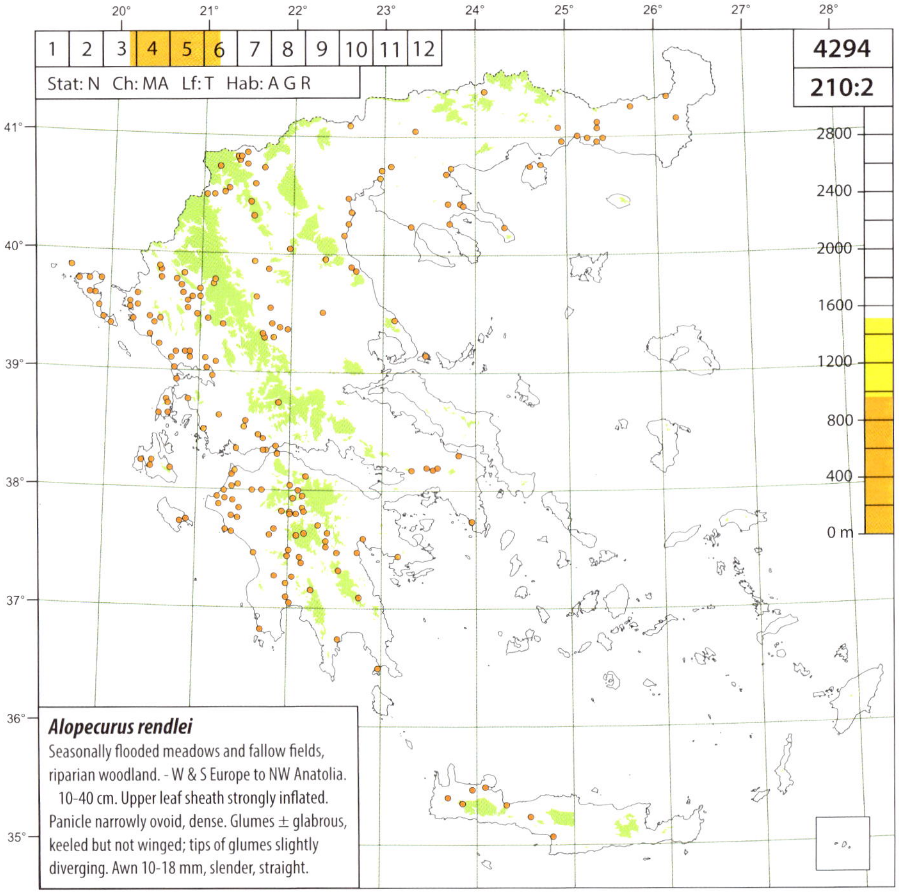

Alopecurus rendlei

Seasonally flooded meadows and fallow fields, riparian woodland. - W & S Europe to NW Anatolia.

10-40 cm. Upper leaf sheath strongly inflated. Panicle narrowly ovoid, dense. Glumes ± glabrous, keeled but not winged; tips of glumes slightly diverging. Awn 10-18 mm, slender, straight.

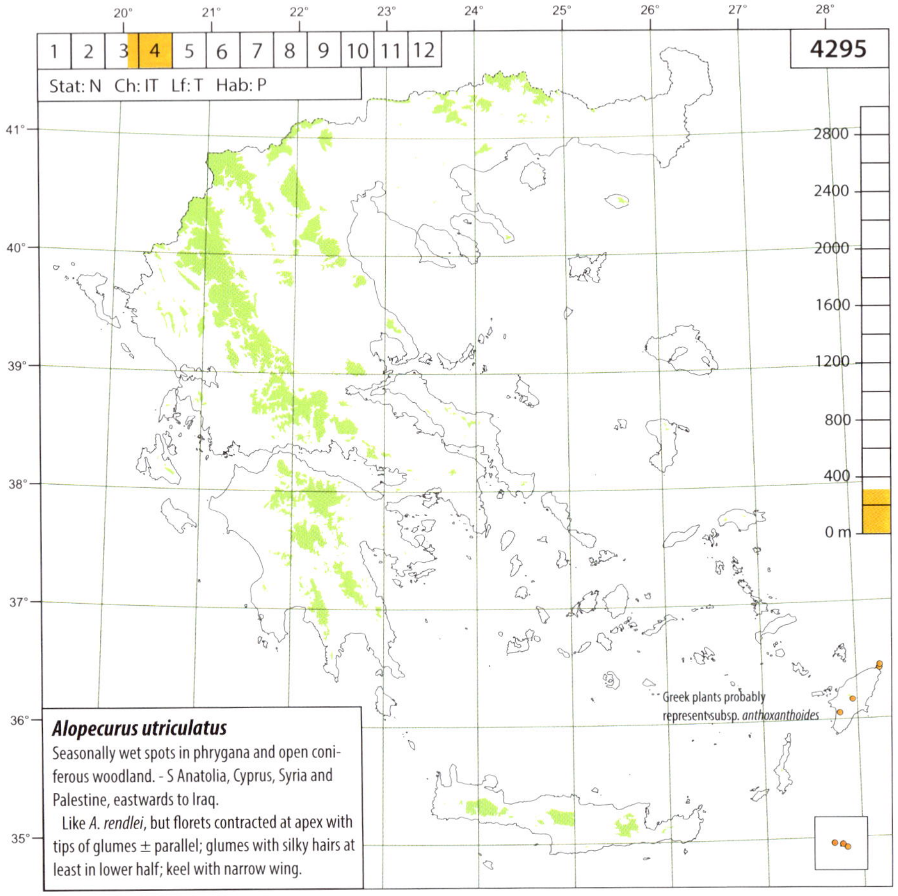

Alopecurus utriculatus

Seasonally wet spots in phrygana and open coniferous woodland. - S Anatolia, Cyprus, Syria and Palestine, eastwards to Iraq.

Like *A. rendlei*, but florets contracted at apex with tips of glumes ± parallel; glumes with silky hairs at least in lower half; keel with narrow wing.

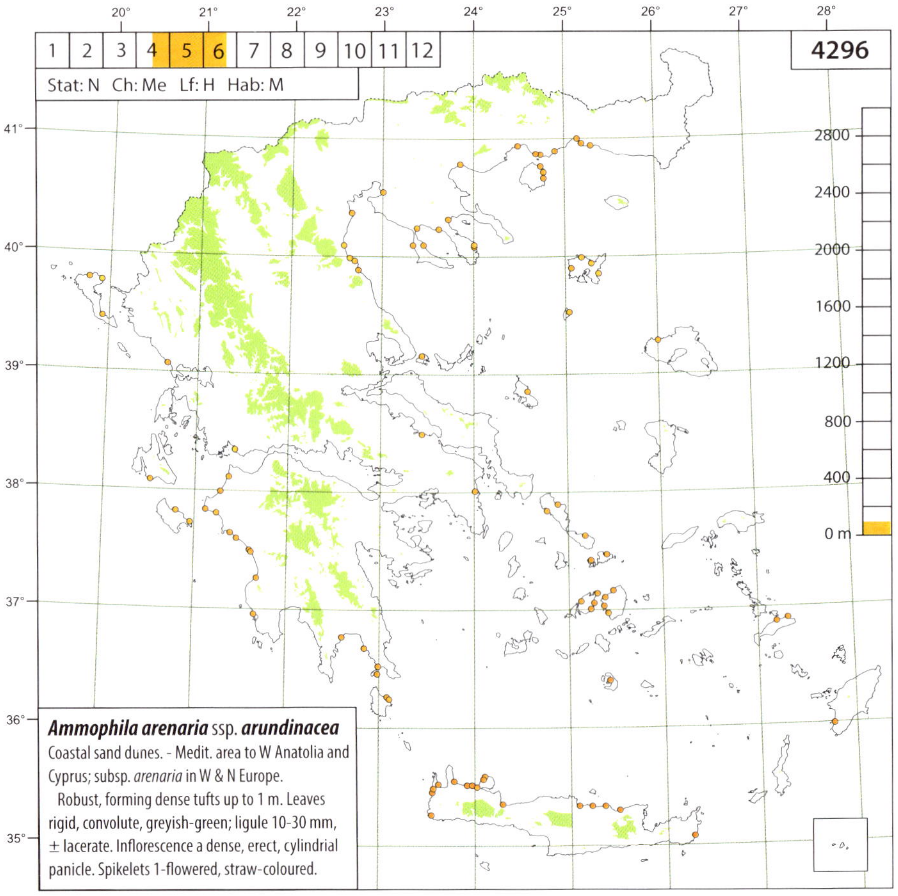

Ammophila arenaria ssp. ***arundinacea***

Coastal sand dunes. - Medit. area to W Anatolia and Cyprus; subsp. *arenaria* in W & N Europe.

Robust, forming dense tufts up to 1 m. Leaves rigid, convolute, greyish-green; ligule 10-30 mm, ± lacerate. Inflorescence a dense, erect, cylindrial panicle. Spikelets 1-flowered, straw-coloured.

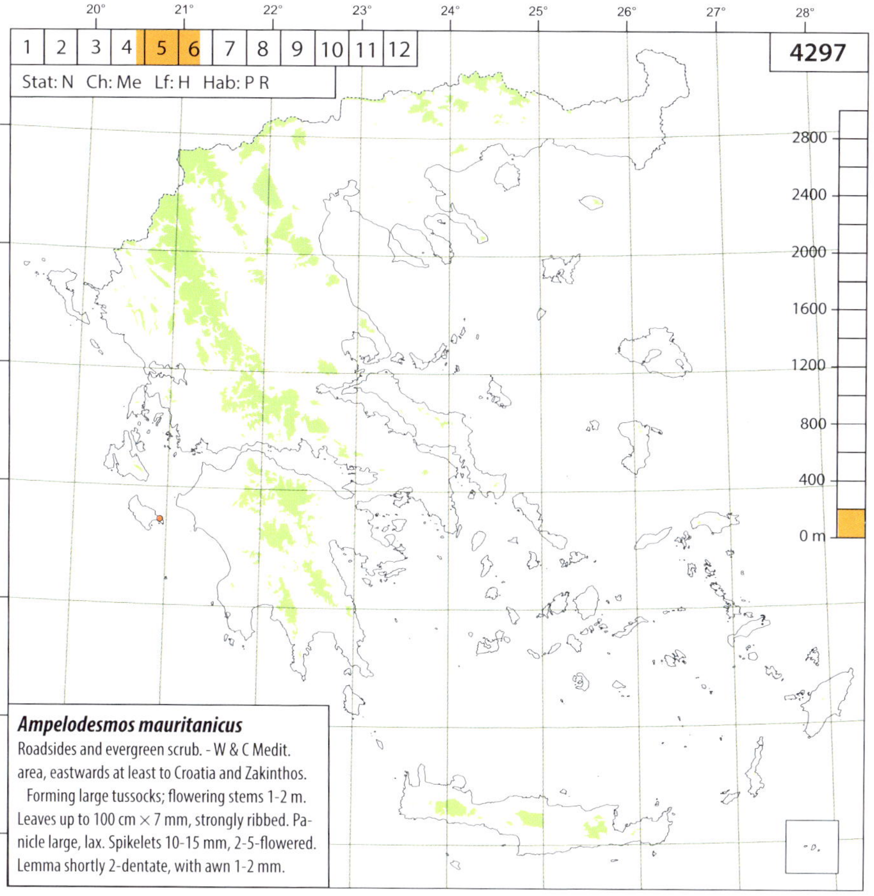

Ampelodesmos mauritanicus

Roadsides and evergreen scrub. - W & C Medit. area, eastwards at least to Croatia and Zakinthos.

Forming large tussocks; flowering stems 1-2 m. Leaves up to 100 cm × 7 mm, strongly ribbed. Panicle large, lax. Spikelets 10-15 mm, 2-5-flowered. Lemma shortly 2-dentate, with awn 1-2 mm.

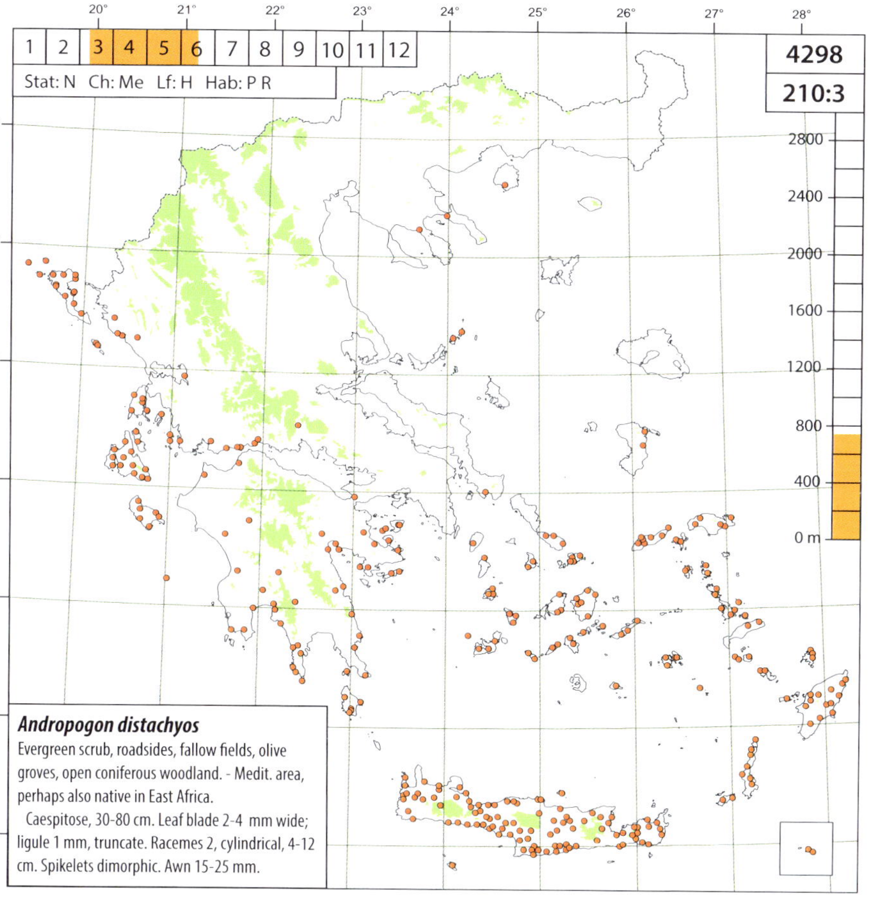

Andropogon distachyos

Evergreen scrub, roadsides, fallow fields, olive groves, open coniferous woodland. - Medit. area, perhaps also native in East Africa.

Caespitose, 30-80 cm. Leaf blade 2-4 mm wide; ligule 1 mm, truncate. Racemes 2, cylindrical, 4-12 cm. Spikelets dimorphic. Awn 15-25 mm.

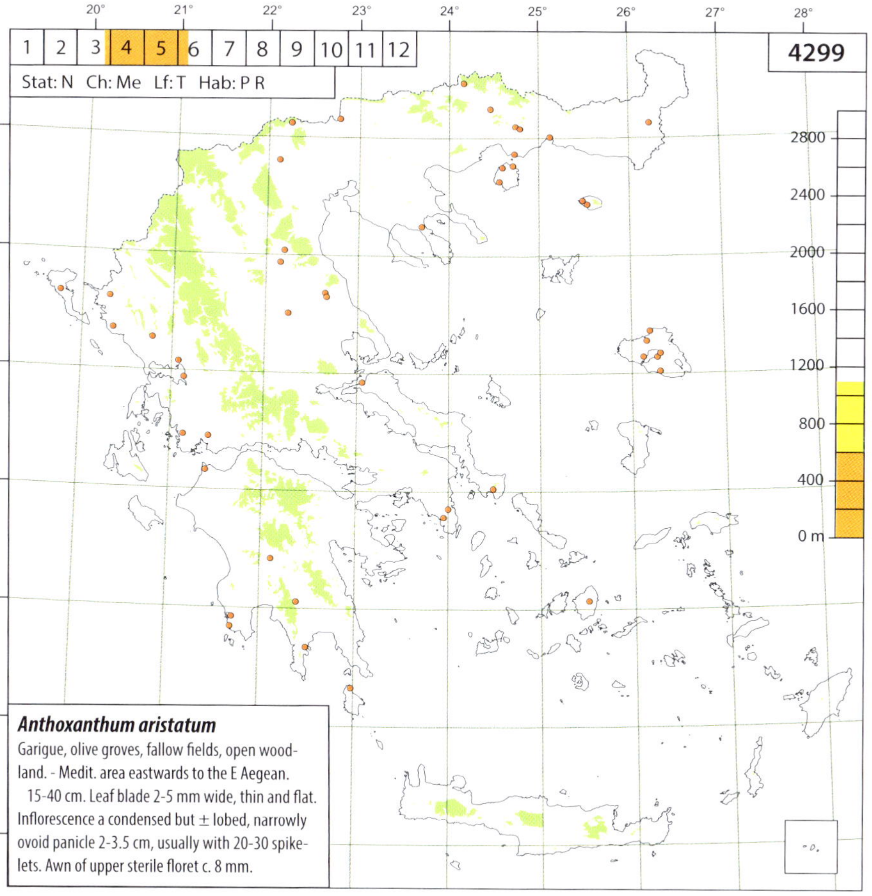

Anthoxanthum aristatum

Garigue, olive groves, fallow fields, open woodland. - Medit. area eastwards to the E Aegean.

15-40 cm. Leaf blade 2-5 mm wide, thin and flat. Inflorescence a condensed but ± lobed, narrowly ovoid panicle 2-3.5 cm, usually with 20-30 spikelets. Awn of upper sterile floret c. 8 mm.

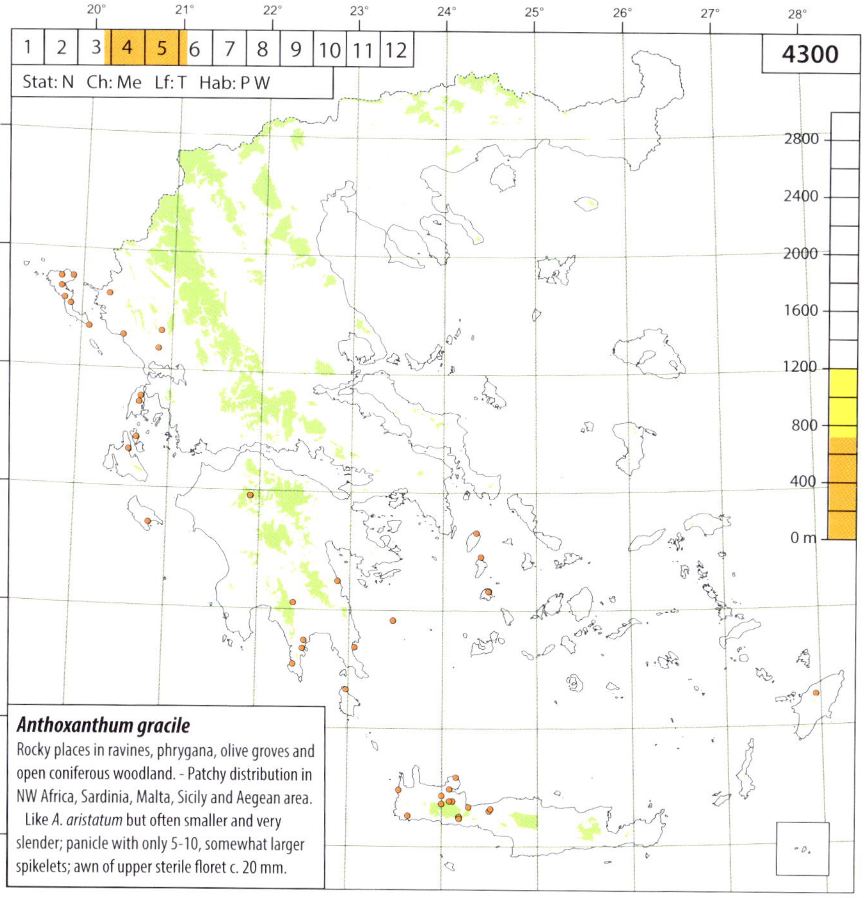

Anthoxanthum gracile

Rocky places in ravines, phrygana, olive groves and open coniferous woodland. - Patchy distribution in NW Africa, Sardinia, Malta, Sicily and Aegean area.

Like *A. aristatum* but often smaller and very slender; panicle with only 5-10, somewhat larger spikelets; awn of upper sterile floret c. 20 mm.

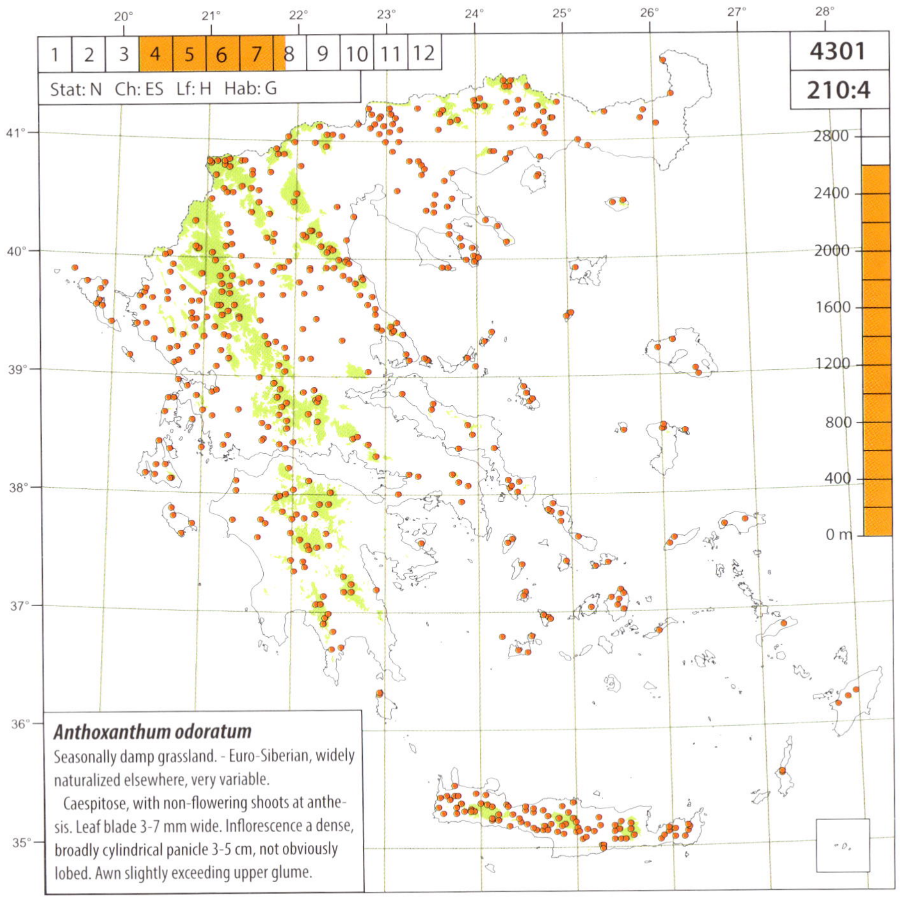

Anthoxanthum odoratum

Seasonally damp grassland. - Euro-Siberian, widely naturalized elsewhere, very variable.

Caespitose, with non-flowering shoots at anthesis. Leaf blade 3-7 mm wide. Inflorescence a dense, broadly cylindrical panicle 3-5 cm, not obviously lobed. Awn slightly exceeding upper glume.

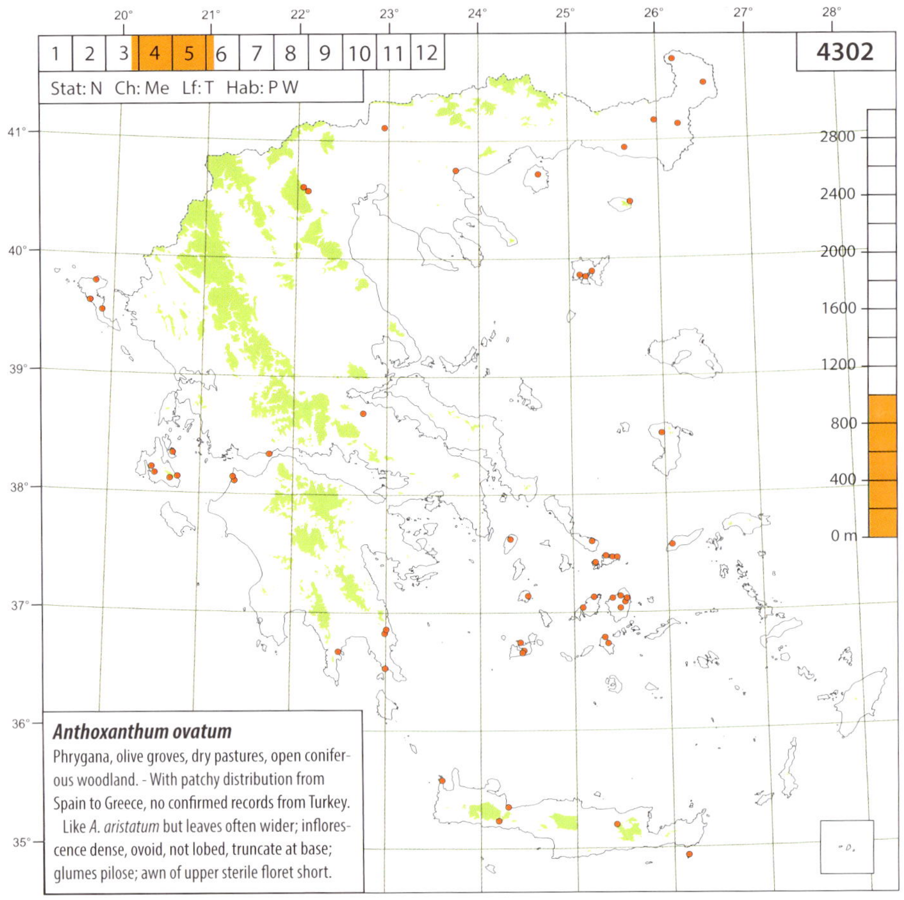

Anthoxanthum ovatum

Phrygana, olive groves, dry pastures, open coniferous woodland. - With patchy distribution from Spain to Greece, no confirmed records from Turkey.

Like *A. aristatum* but leaves often wider; inflorescence dense, ovoid, not lobed, truncate at base; glumes pilose; awn of upper sterile floret short.

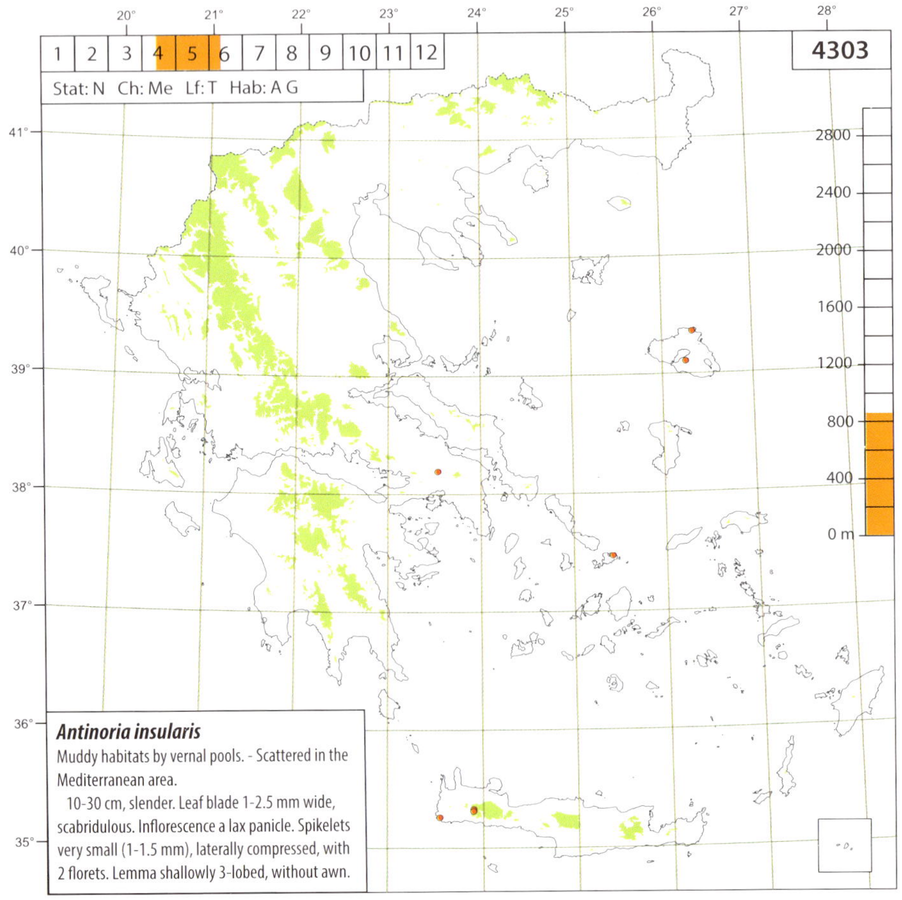

Antinoria insularis

Muddy habitats by vernal pools. - Scattered in the Mediterranean area.

10-30 cm, slender. Leaf blade 1-2.5 mm wide, scabridulous. Inflorescence a lax panicle. Spikelets very small (1-1.5 mm), laterally compressed, with 2 florets. Lemma shallowly 3-lobed, without awn.

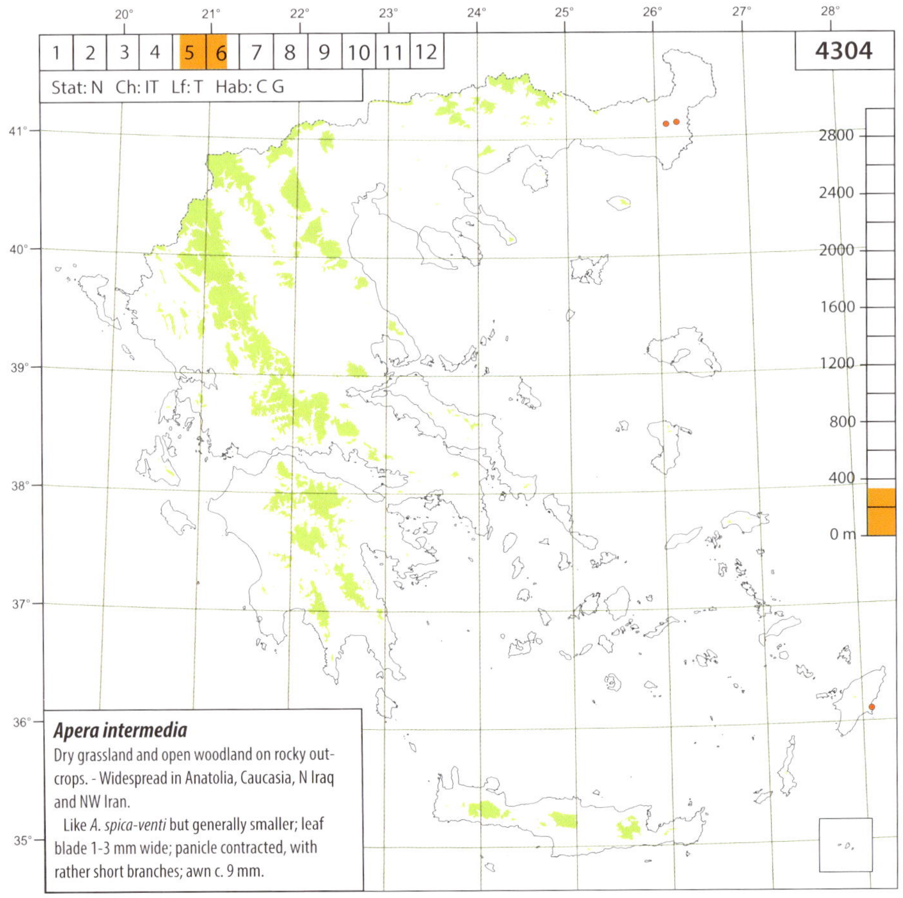

Apera intermedia

Dry grassland and open woodland on rocky outcrops. - Widespread in Anatolia, Caucasia, N Iraq and NW Iran.

Like *A. spica-venti* but generally smaller; leaf blade 1-3 mm wide; panicle contracted, with rather short branches; awn c. 9 mm.

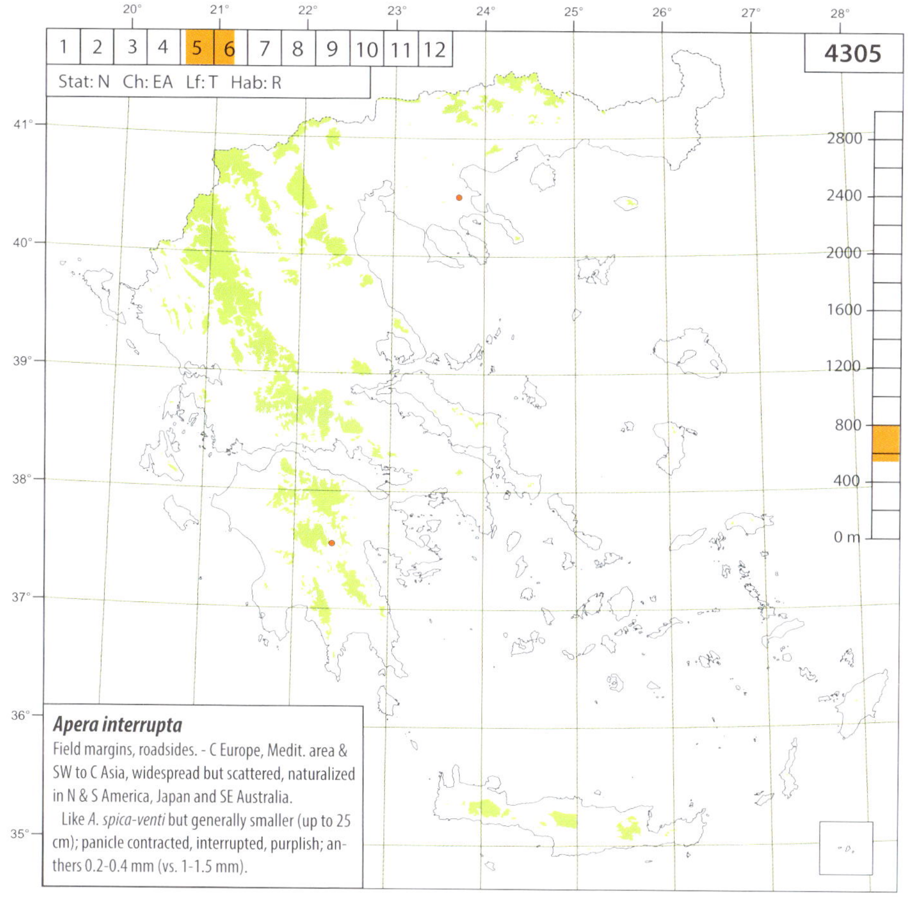

Apera interrupta

Field margins, roadsides. - C Europe, Medit. area & SW to C Asia, widespread but scattered, naturalized in N & S America, Japan and SE Australia.

Like *A. spica-venti* but generally smaller (up to 25 cm); panicle contracted, interrupted, purplish; anthers 0.2-0.4 mm (vs. 1-1.5 mm).

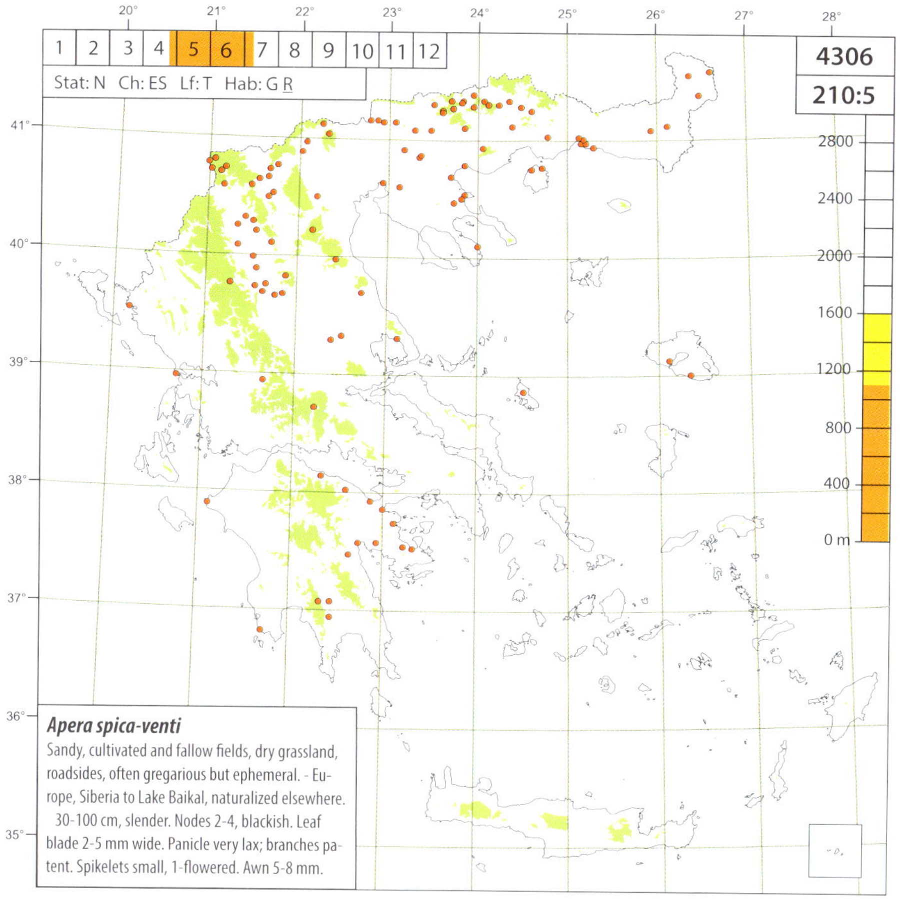

Apera spica-venti

Sandy, cultivated and fallow fields, dry grassland, roadsides, often gregarious but ephemeral. - Europe, Siberia to Lake Baikal, naturalized elsewhere.

30-100 cm, slender. Nodes 2-4, blackish. Leaf blade 2-5 mm wide. Panicle very lax; branches patent. Spikelets small, 1-flowered. Awn 5-8 mm.

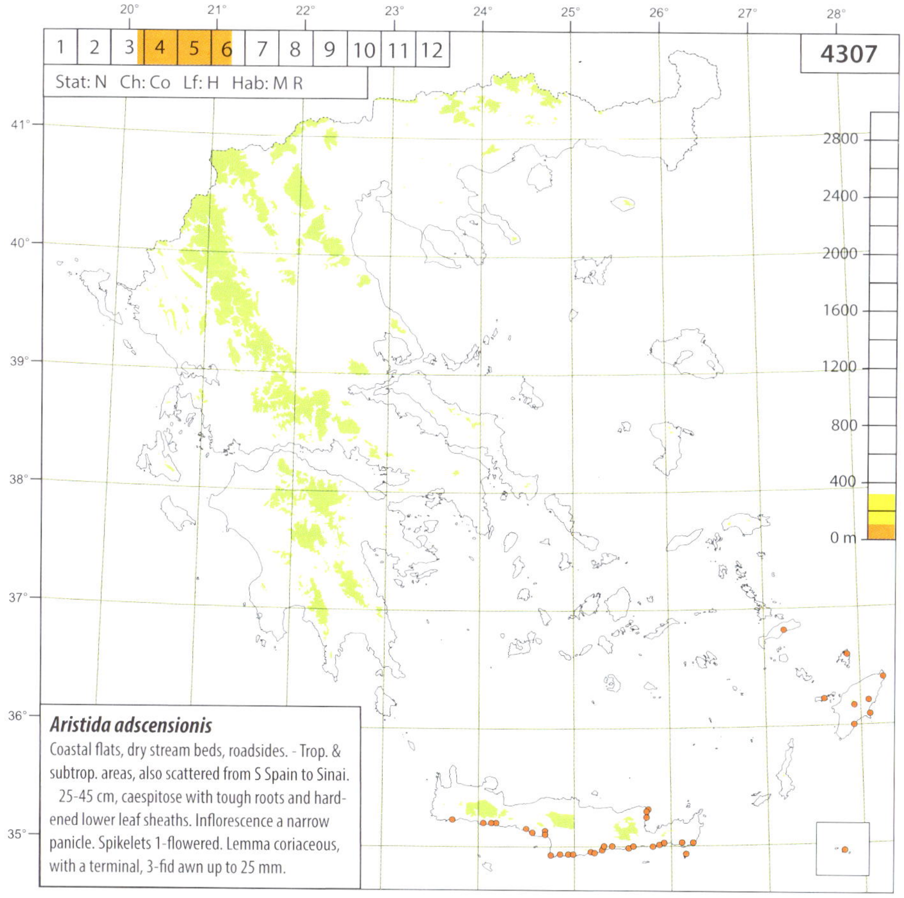

Aristida adscensionis

Coastal flats, dry stream beds, roadsides. - Trop. & subtrop. areas, also scattered from S Spain to Sinai.

25-45 cm, caespitose with tough roots and hardened lower leaf sheaths. Inflorescence a narrow panicle. Spikelets 1-flowered. Lemma coriaceous, with a terminal, 3-fid awn up to 25 mm.

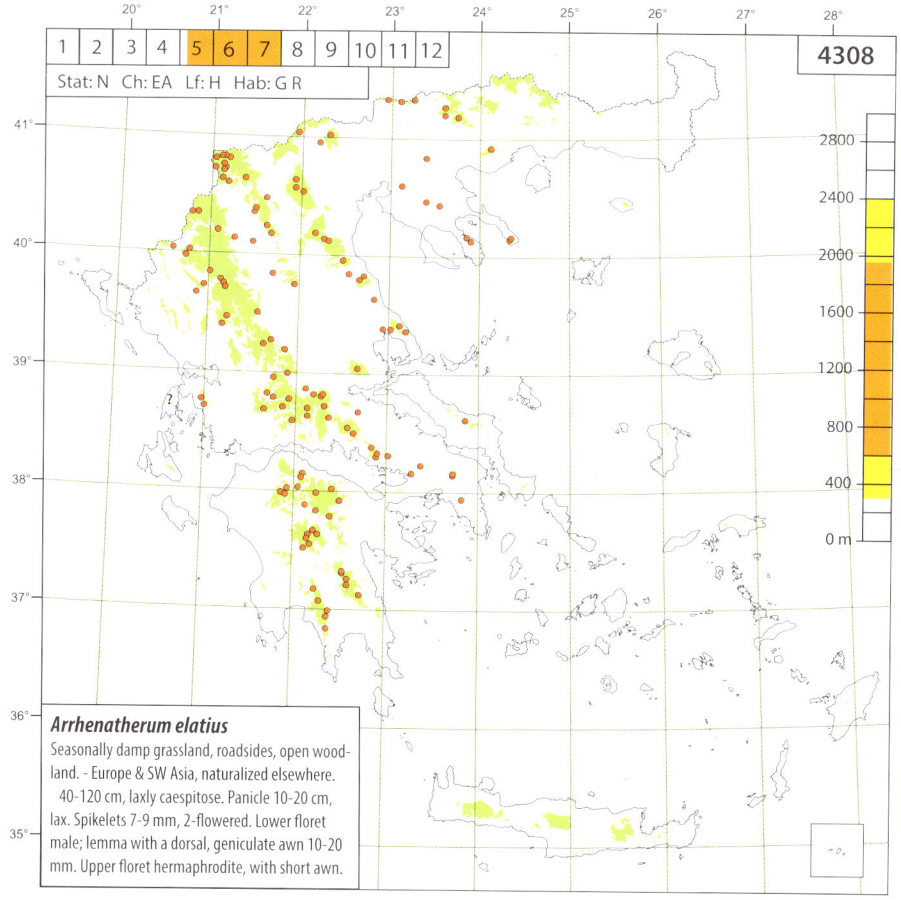

Arrhenatherum elatius

Seasonally damp grassland, roadsides, open woodland. - Europe & SW Asia, naturalized elsewhere.

40-120 cm, laxly caespitose. Panicle 10-20 cm, lax. Spikelets 7-9 mm, 2-flowered. Lower floret male; lemma with a dorsal, geniculate awn 10-20 mm. Upper floret hermaphrodite, with short awn.

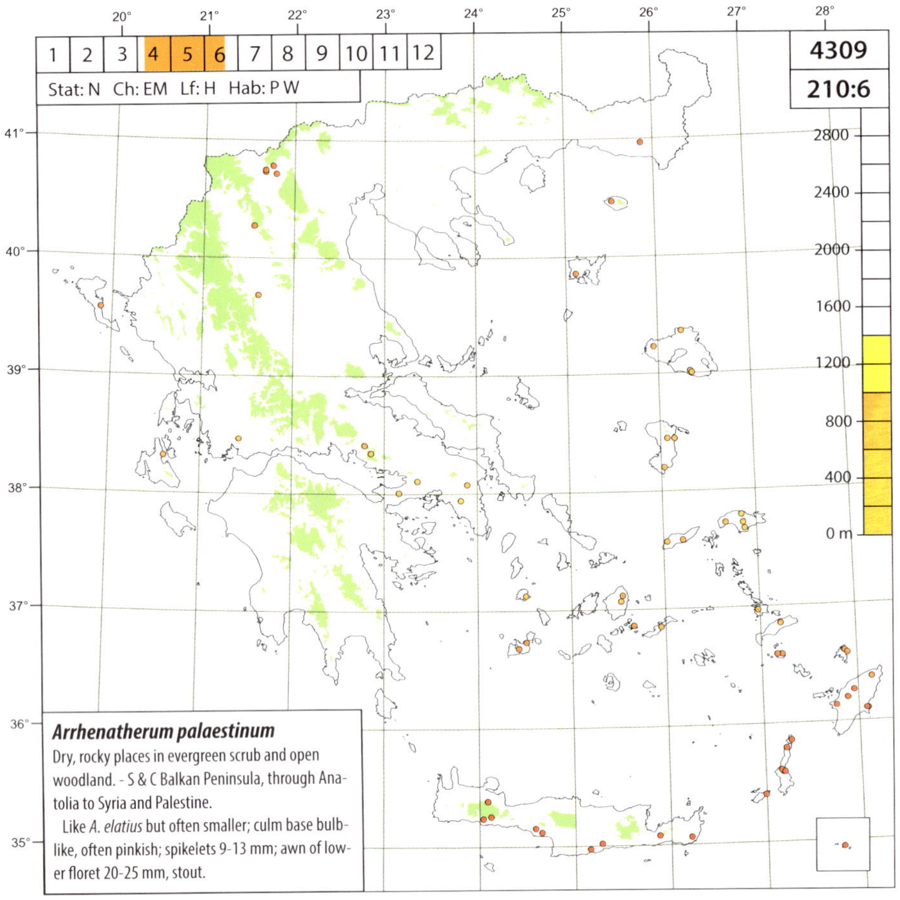

Arrhenatherum palaestinum

Dry, rocky places in evergreen scrub and open woodland. - S & C Balkan Peninsula, through Anatolia to Syria and Palestine.

Like *A. elatius* but often smaller; culm base bulb-like, often pinkish; spikelets 9-13 mm; awn of lower floret 20-25 mm, stout.

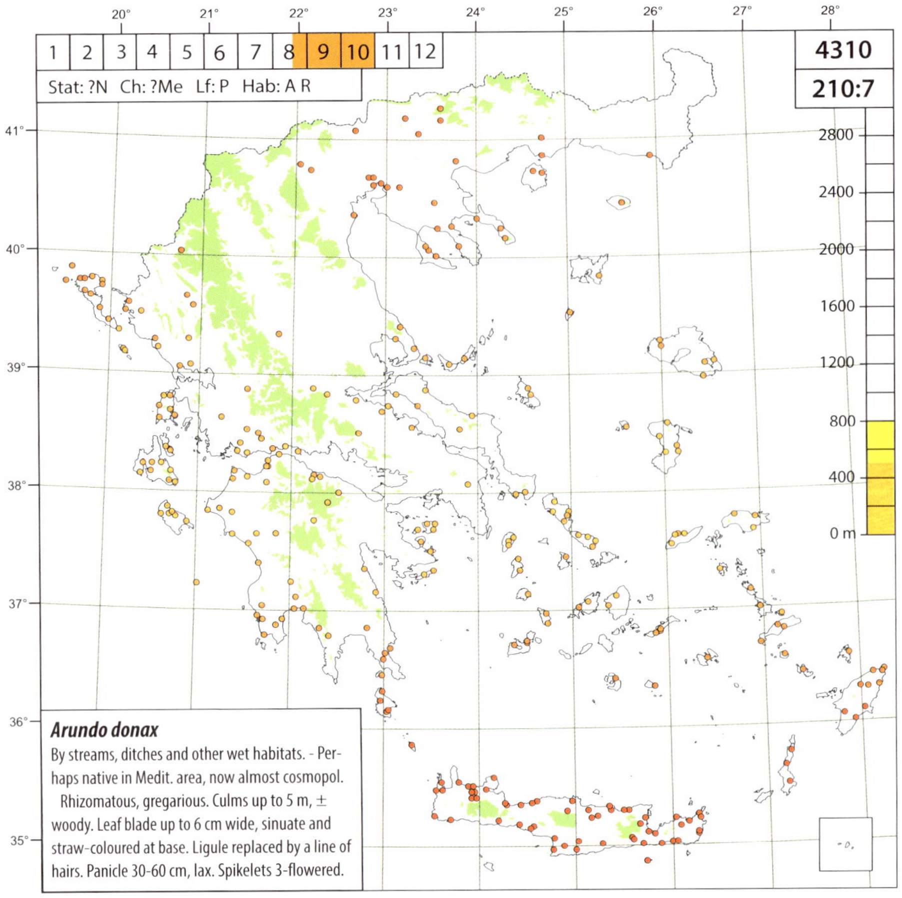

Arundo donax

By streams, ditches and other wet habitats. - Perhaps native in Medit. area, now almost cosmopol.

Rhizomatous, gregarious. Culms up to 5 m, ± woody. Leaf blade up to 6 cm wide, sinuate and straw-coloured at base. Ligule replaced by a line of hairs. Panicle 30-60 cm, lax. Spikelets 3-flowered.

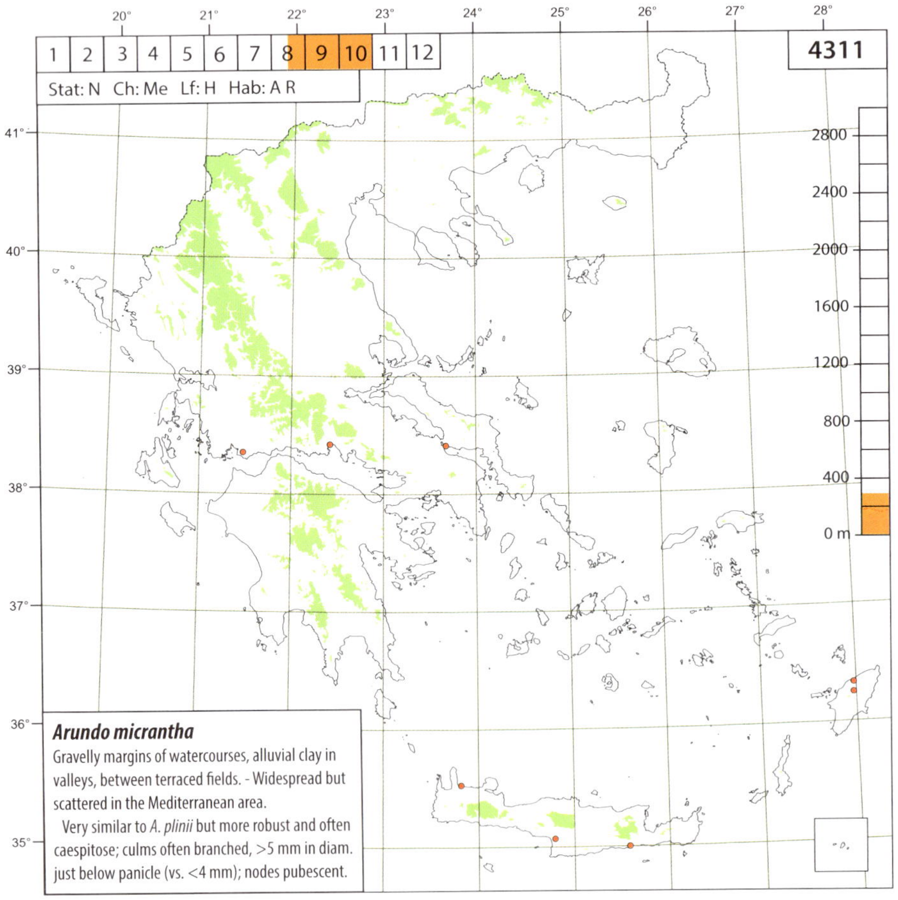

Arundo micrantha

Gravelly margins of watercourses, alluvial clay in valleys, between terraced fields. - Widespread but scattered in the Mediterranean area.

Very similar to *A. plinii* but more robust and often caespitose; culms often branched, >5 mm in diam. just below panicle (vs. <4 mm); nodes pubescent.

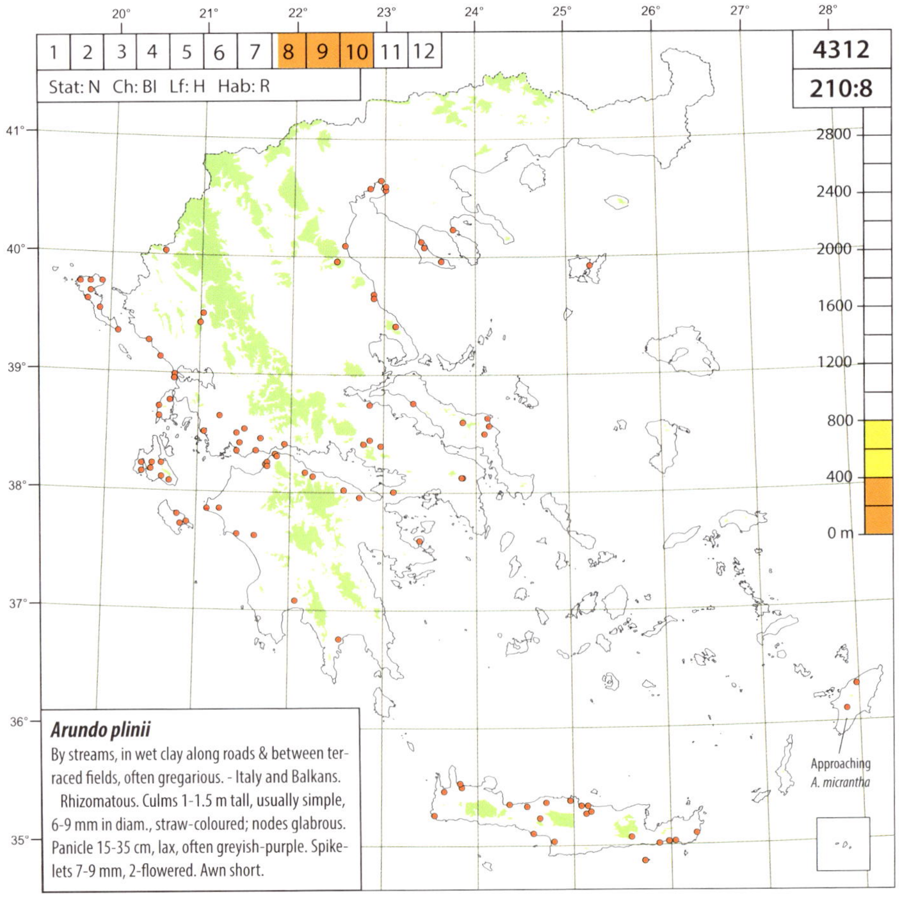

Arundo plinii

By streams, in wet clay along roads & between terraced fields, often gregarious. - Italy and Balkans.

Rhizomatous. Culms 1-1.5 m tall, usually simple, 6-9 mm in diam., straw-coloured; nodes glabrous. Panicle 15-35 cm, lax, often greyish-purple. Spikelets 7-9 mm, 2-flowered. Awn short.

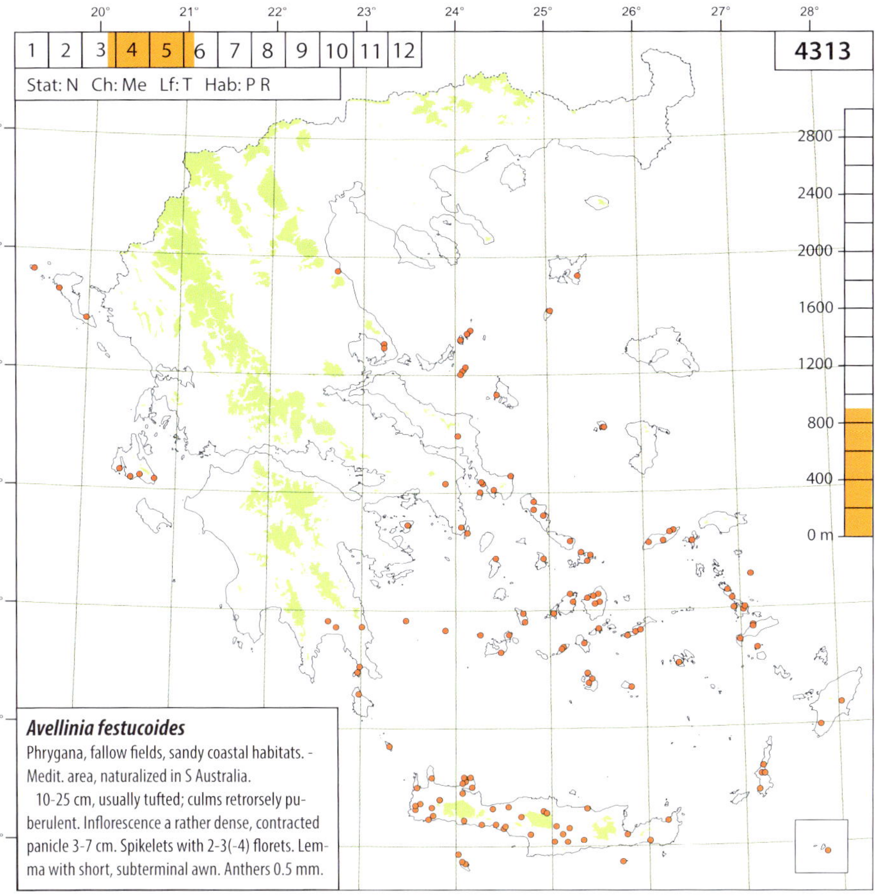

Avellinia festucoides

Phrygana, fallow fields, sandy coastal habitats. - Medit. area, naturalized in S Australia.

10-25 cm, usually tufted; culms retrorsely puberulent. Inflorescence a rather dense, contracted panicle 3-7 cm. Spikelets with 2-3(-4) florets. Lemma with short, subterminal awn. Anthers 0.5 mm.

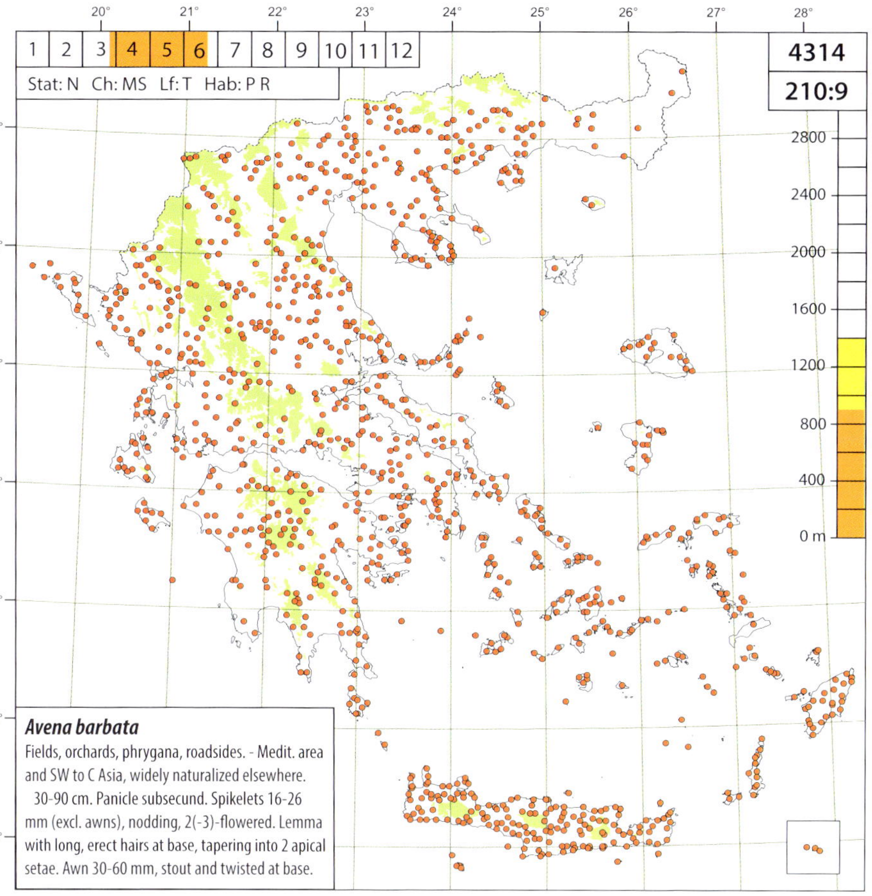

Avena barbata

Fields, orchards, phrygana, roadsides. - Medit. area and SW to C Asia, widely naturalized elsewhere.

30-90 cm. Panicle subsecund. Spikelets 16-26 mm (excl. awns), nodding, 2(-3)-flowered. Lemma with long, erect hairs at base, tapering into 2 apical setae. Awn 30-60 mm, stout and twisted at base.

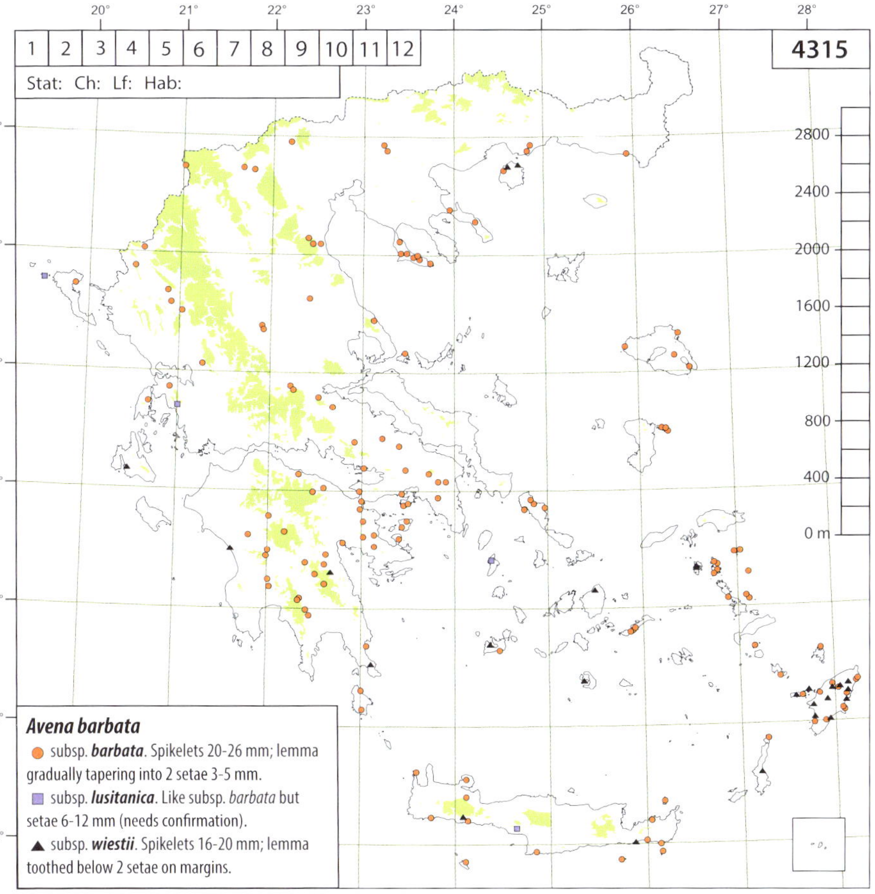

Avena barbata

● subsp. ***barbata***. Spikelets 20-26 mm; lemma gradually tapering into 2 setae 3-5 mm.

■ subsp. ***lusitanica***. Like subsp. *barbata* but setae 6-12 mm (needs confirmation).

▲ subsp. ***wiestii***. Spikelets 16-20 mm; lemma toothed below 2 setae on margins.

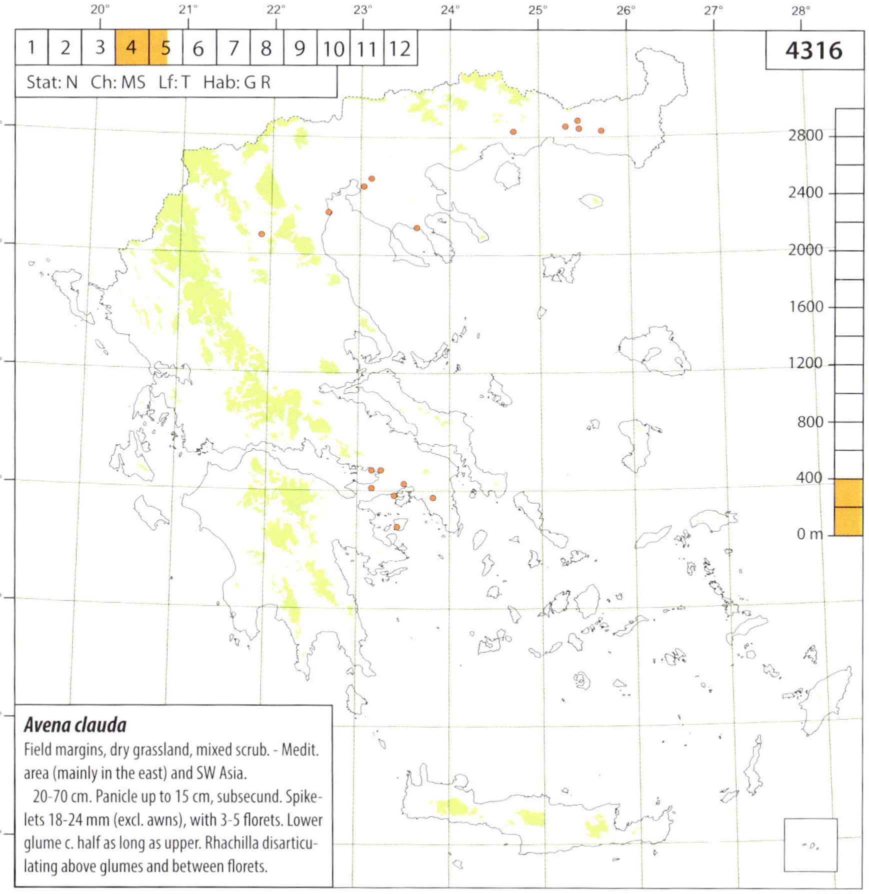

Avena clauda

Field margins, dry grassland, mixed scrub. - Medit. area (mainly in the east) and SW Asia.

20-70 cm. Panicle up to 15 cm, subsecund. Spikelets 18-24 mm (excl. awns), with 3-5 florets. Lower glume c. half as long as upper. Rhachilla disarticulating above glumes and between florets.

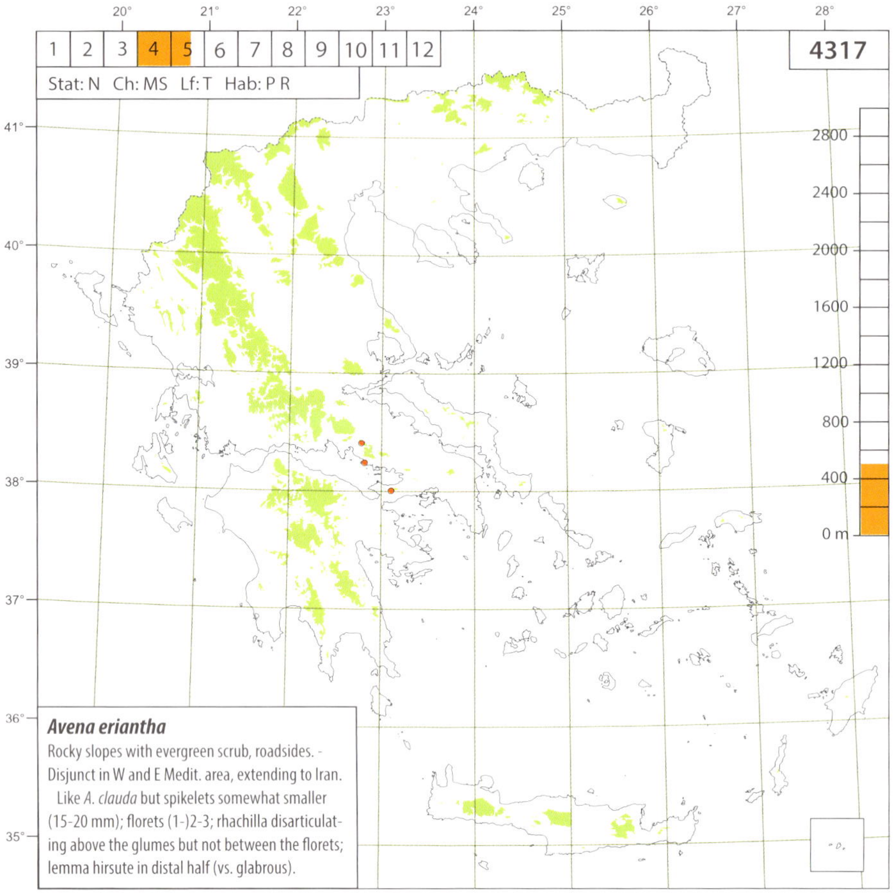

Avena eriantha
Rocky slopes with evergreen scrub, roadsides. - Disjunct in W and E Medit. area, extending to Iran.
Like *A. clauda* but spikelets somewhat smaller (15-20 mm); florets (1-)2-3; rhachilla disarticulating above the glumes but not between the florets; lemma hirsute in distal half (vs. glabrous).

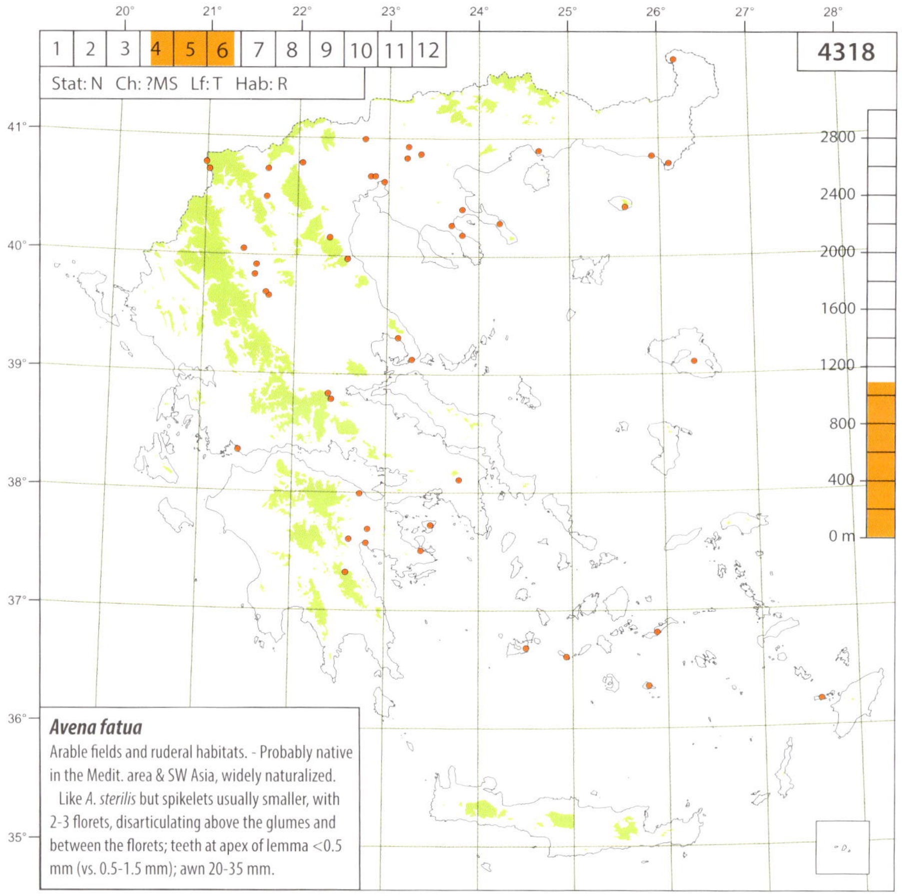

Avena fatua
Arable fields and ruderal habitats. - Probably native in the Medit. area & SW Asia, widely naturalized.
Like *A. sterilis* but spikelets usually smaller, with 2-3 florets, disarticulating above the glumes and between the florets; teeth at apex of lemma <0.5 mm (vs. 0.5-1.5 mm); awn 20-35 mm.

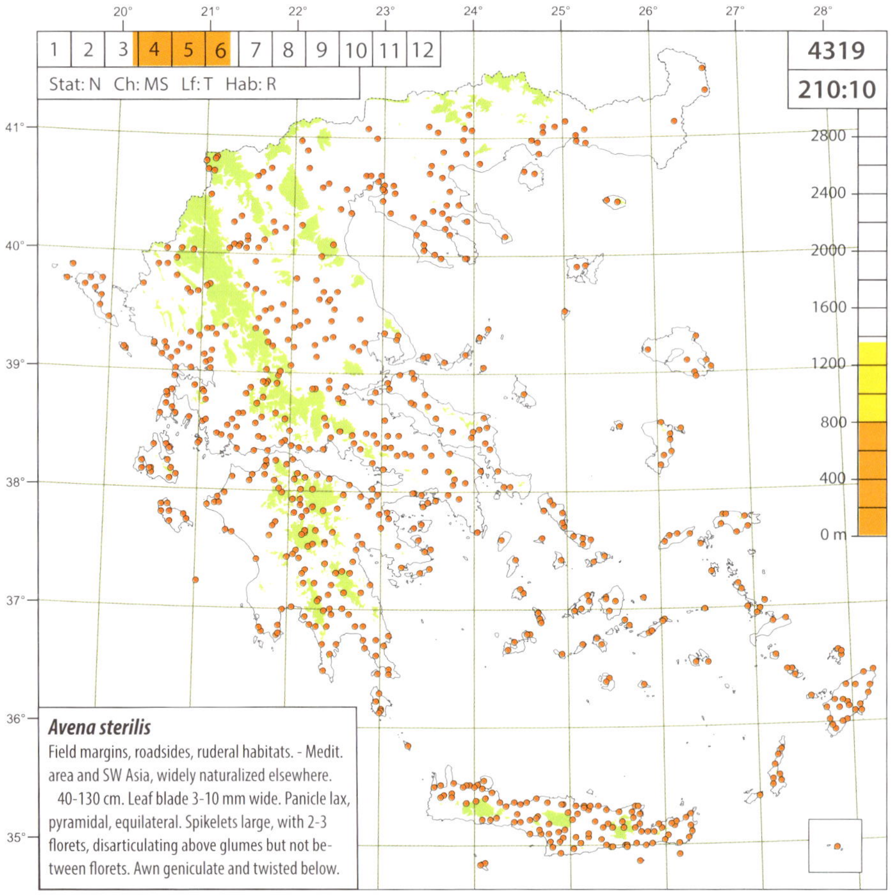

Avena sterilis
Field margins, roadsides, ruderal habitats. - Medit. area and SW Asia, widely naturalized elsewhere.
40-130 cm. Leaf blade 3-10 mm wide. Panicle lax, pyramidal, equilateral. Spikelets large, with 2-3 florets, disarticulating above glumes but not between florets. Awn geniculate and twisted below.

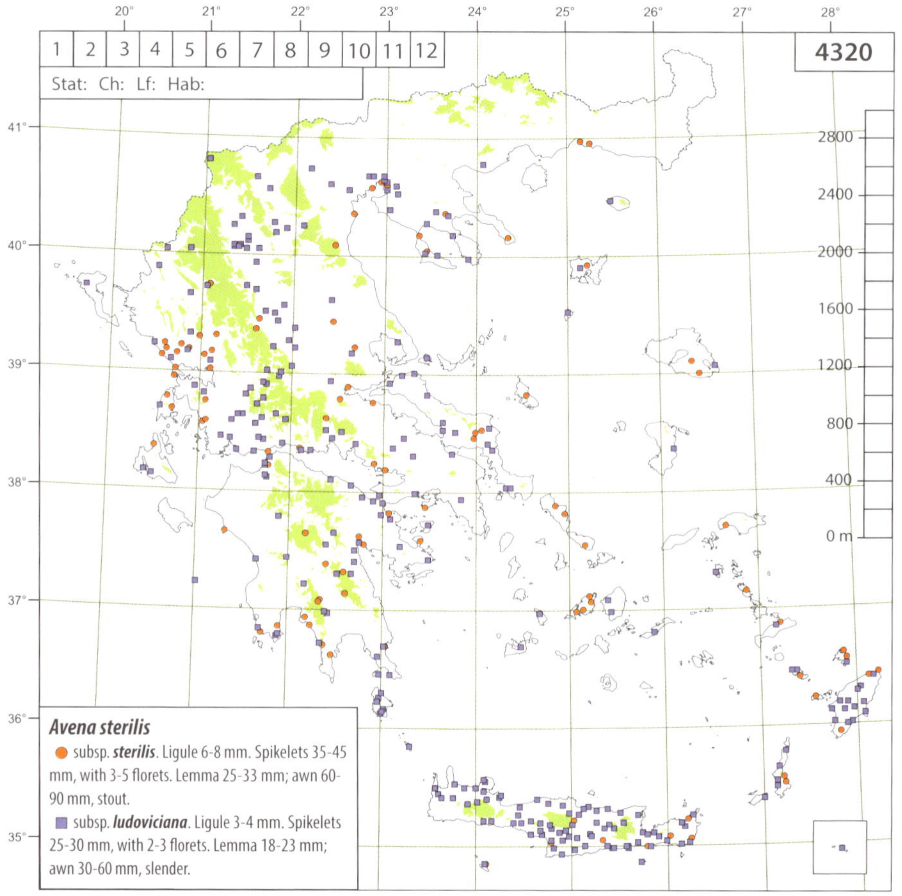

Avena sterilis
● subsp. ***sterilis***. Ligule 6-8 mm. Spikelets 35-45 mm, with 3-5 florets. Lemma 25-33 mm; awn 60-90 mm, stout.
■ subsp. ***ludoviciana***. Ligule 3-4 mm. Spikelets 25-30 mm, with 2-3 florets. Lemma 18-23 mm; awn 30-60 mm, slender.

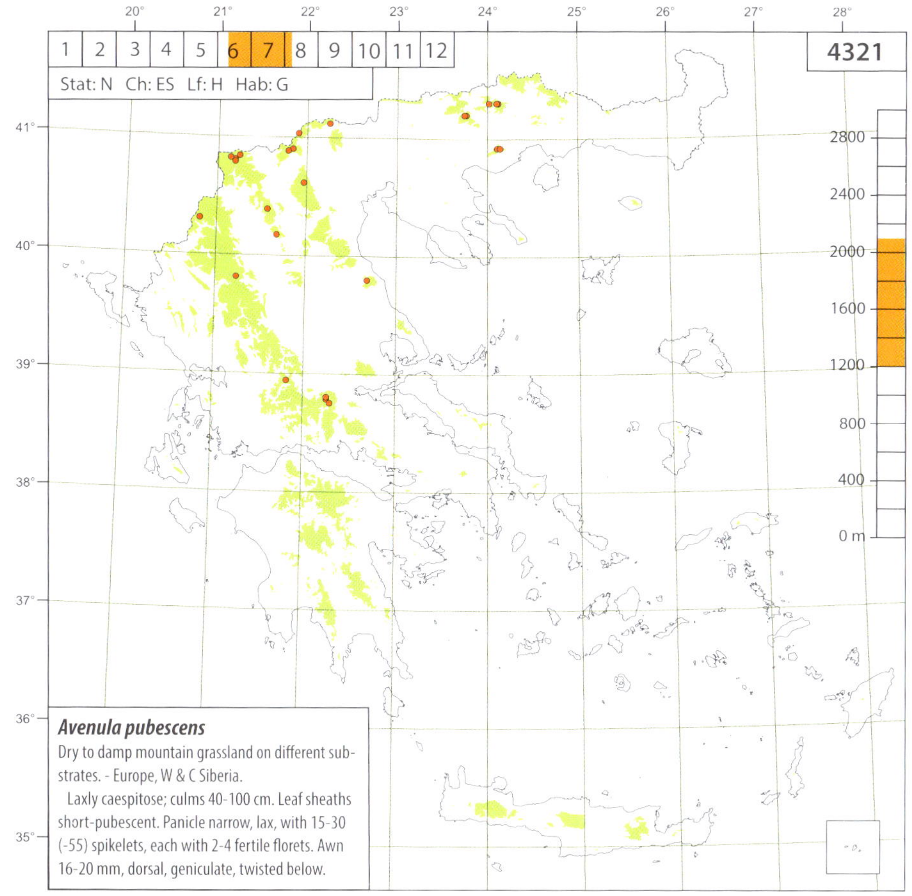

Avenula pubescens

Dry to damp mountain grassland on different substrates. - Europe, W & C Siberia.

Laxly caespitose; culms 40-100 cm. Leaf sheaths short-pubescent. Panicle narrow, lax, with 15-30 (-55) spikelets, each with 2-4 fertile florets. Awn 16-20 mm, dorsal, geniculate, twisted below.

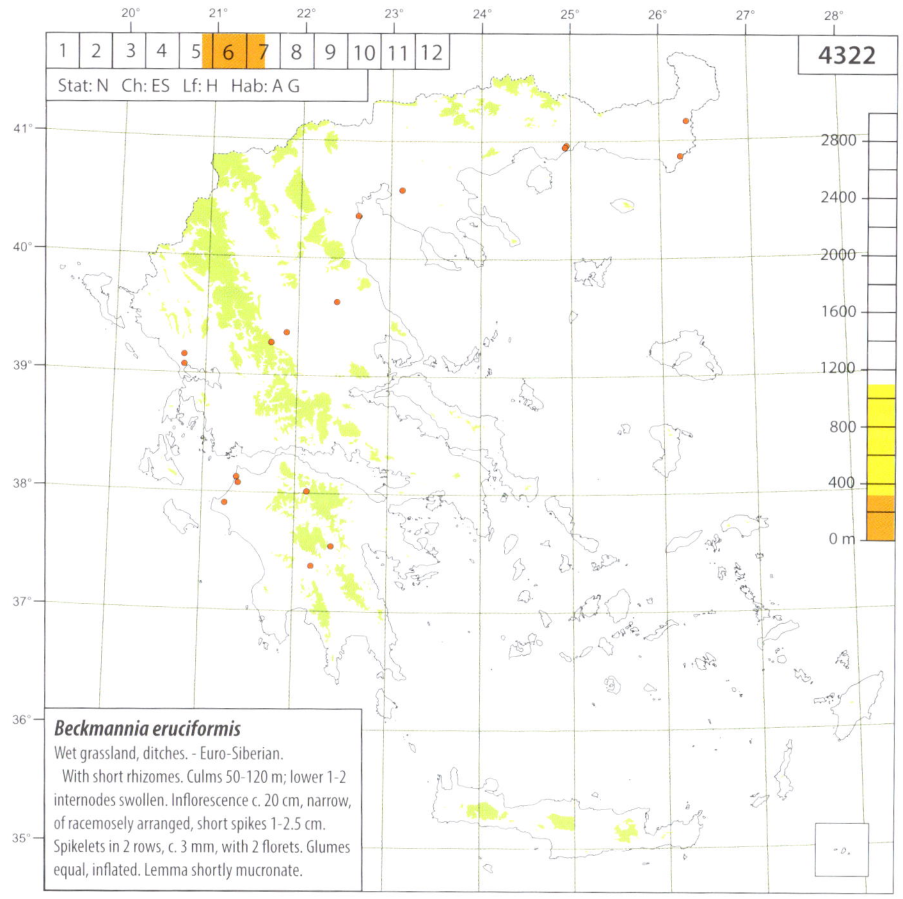

Beckmannia eruciformis

Wet grassland, ditches. - Euro-Siberian.

With short rhizomes. Culms 50-120 m; lower 1-2 internodes swollen. Inflorescence c. 20 cm, narrow, of racemosely arranged, short spikes 1-2.5 cm. Spikelets in 2 rows, c. 3 mm, with 2 florets. Glumes equal, inflated. Lemma shortly mucronate.

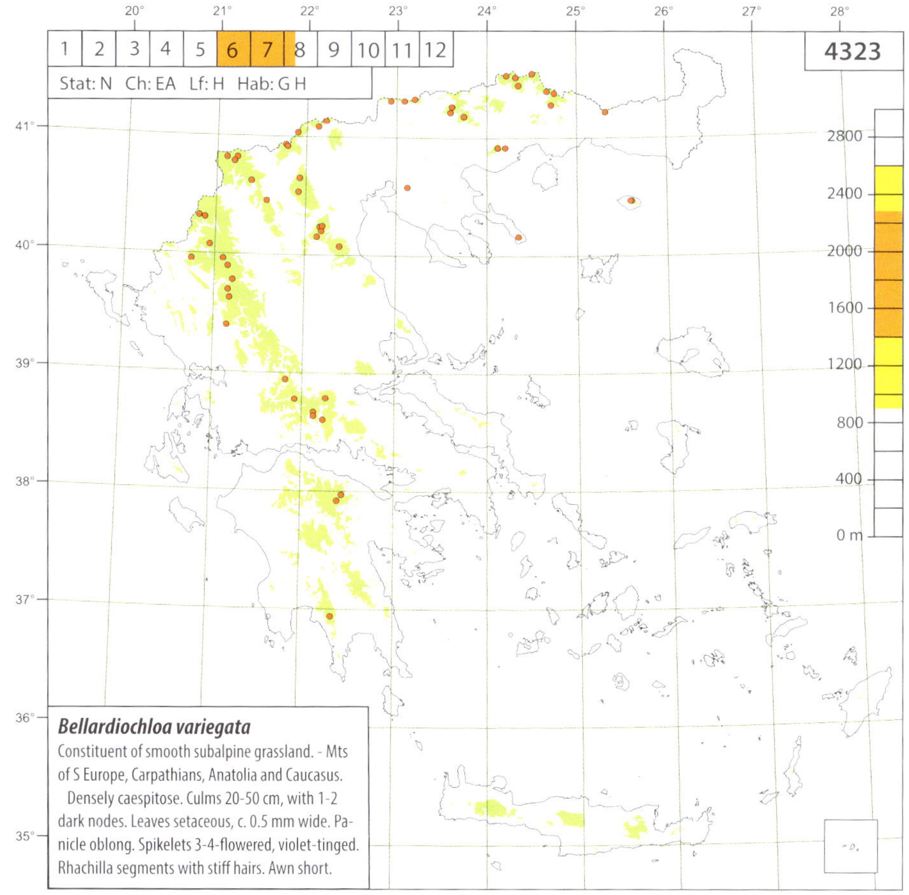

Bellardiochloa variegata

Constituent of smooth subalpine grassland. - Mts of S Europe, Carpathians, Anatolia and Caucasus.

Densely caespitose. Culms 20-50 cm, with 1-2 dark nodes. Leaves setaceous, c. 0.5 mm wide. Panicle oblong. Spikelets 3-4-flowered, violet-tinged. Rhachilla segments with stiff hairs. Awn short.

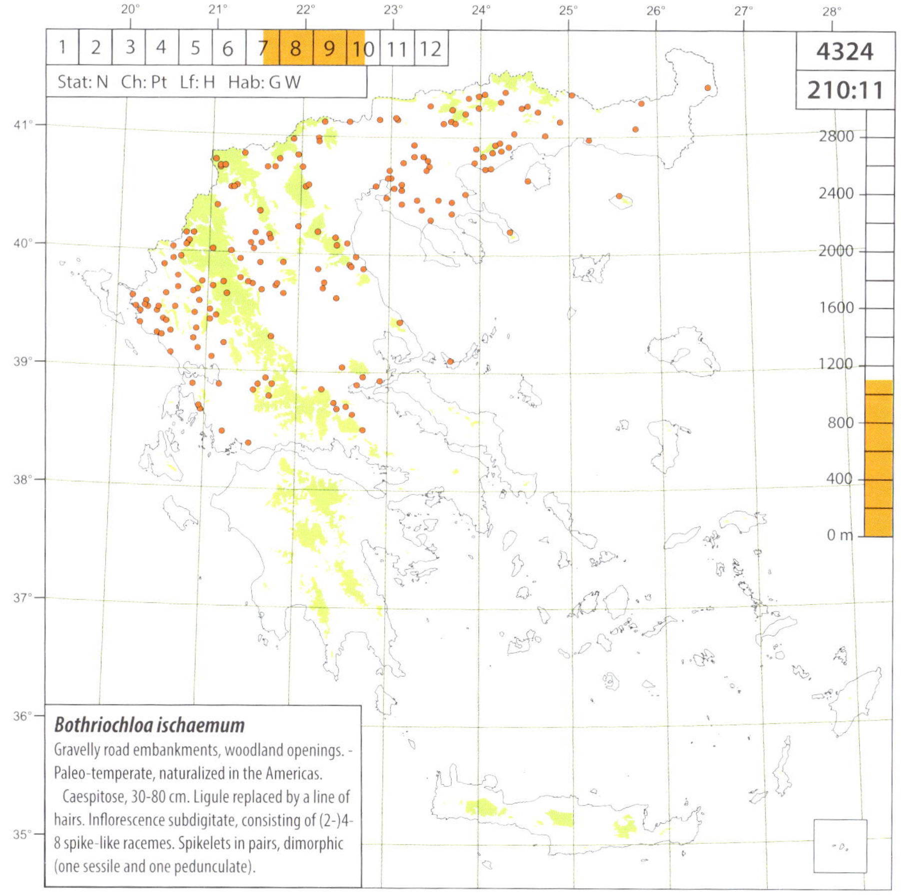

Bothriochloa ischaemum

Gravelly road embankments, woodland openings. - Paleo-temperate, naturalized in the Americas.

Caespitose, 30-80 cm. Ligule replaced by a line of hairs. Inflorescence subdigitate, consisting of (2-)4-8 spike-like racemes. Spikelets in pairs, dimorphic (one sessile and one pedunculate).

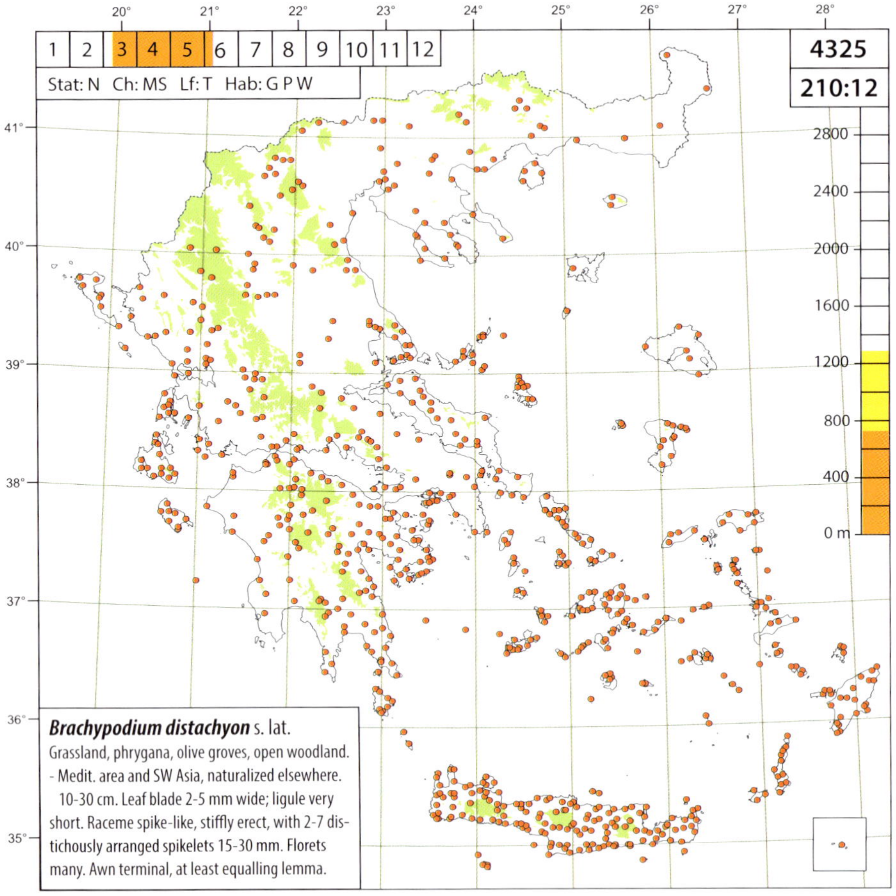

Brachypodium distachyon **s. lat.**
Grassland, phrygana, olive groves, open woodland. - Medit. area and SW Asia, naturalized elsewhere.
10-30 cm. Leaf blade 2-5 mm wide; ligule very short. Raceme spike-like, stiffly erect, with 2-7 distichously arranged spikelets 15-30 mm. Florets many. Awn terminal, at least equalling lemma.

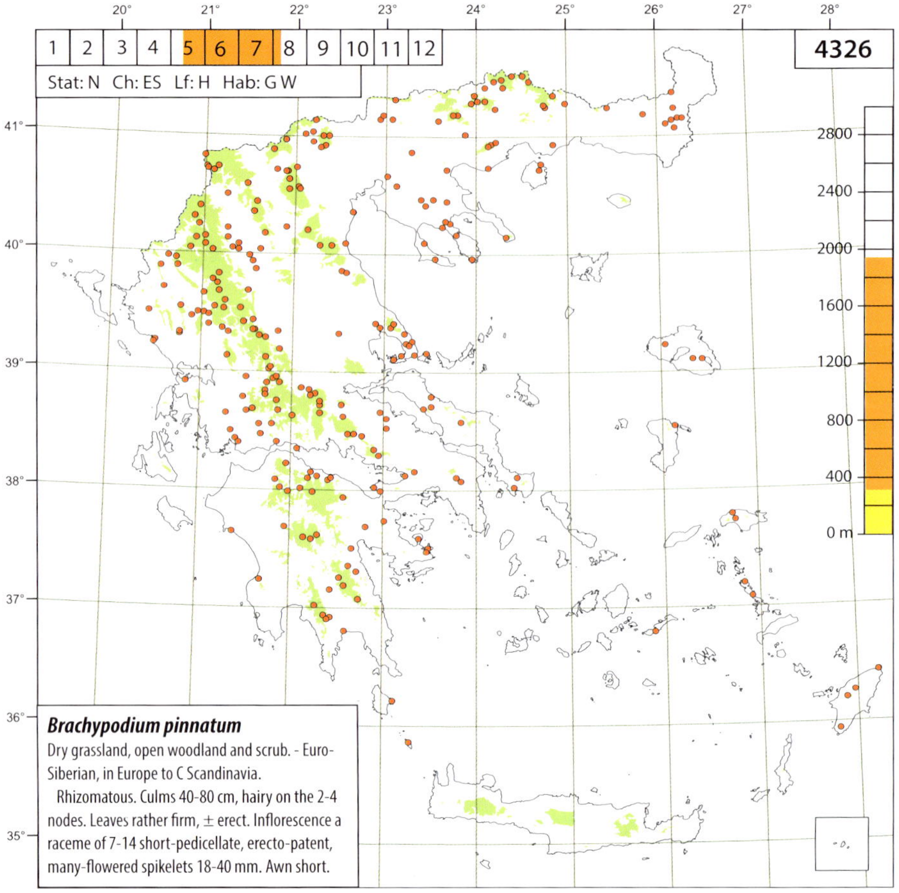

Brachypodium pinnatum
Dry grassland, open woodland and scrub. - Euro-Siberian, in Europe to C Scandinavia.
Rhizomatous. Culms 40-80 cm, hairy on the 2-4 nodes. Leaves rather firm, ± erect. Inflorescence a raceme of 7-14 short-pedicellate, erecto-patent, many-flowered spikelets 18-40 mm. Awn short.

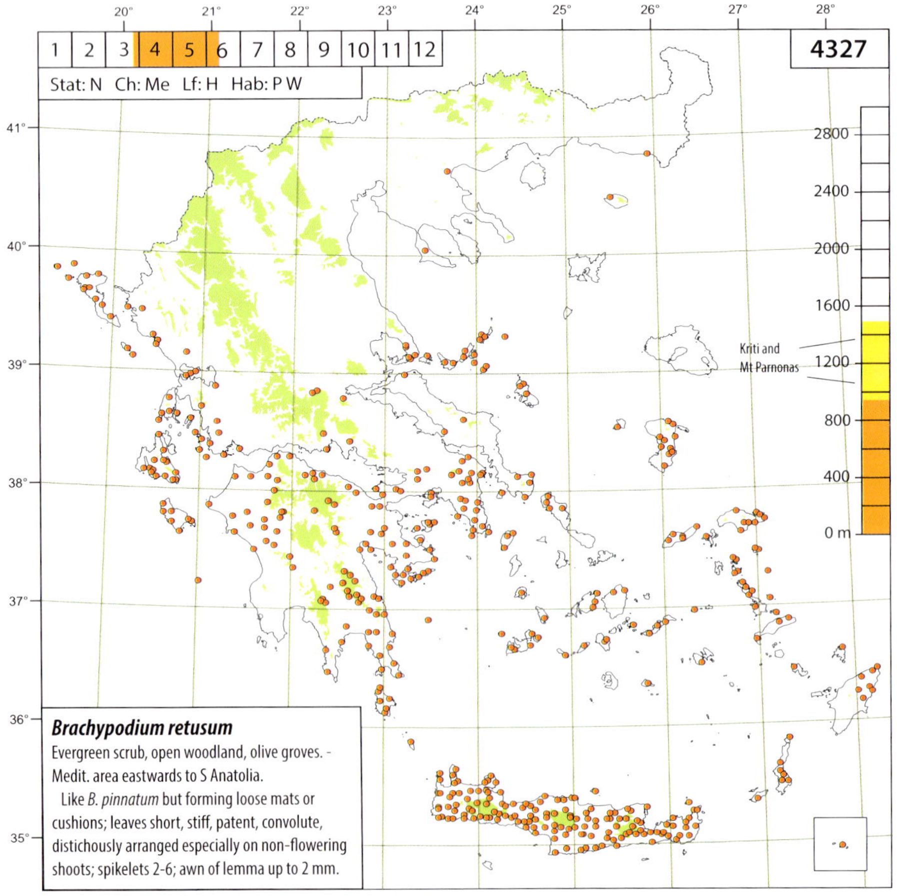

Brachypodium retusum
Evergreen scrub, open woodland, olive groves. - Medit. area eastwards to S Anatolia.
Like *B. pinnatum* but forming loose mats or cushions; leaves short, stiff, patent, convolute, distichously arranged especially on non-flowering shoots; spikelets 2-6; awn of lemma up to 2 mm.

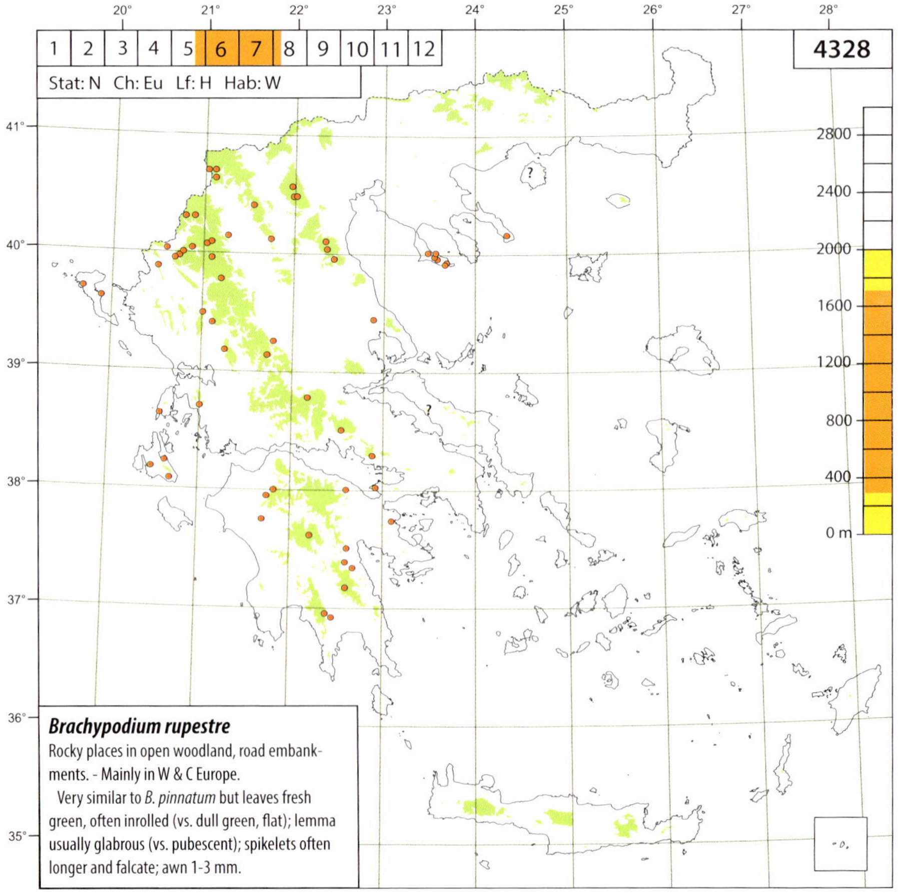

Brachypodium rupestre
Rocky places in open woodland, road embankments. - Mainly in W & C Europe.
Very similar to *B. pinnatum* but leaves fresh green, often inrolled (vs. dull green, flat); lemma usually glabrous (vs. pubescent); spikelets often longer and falcate; awn 1-3 mm.

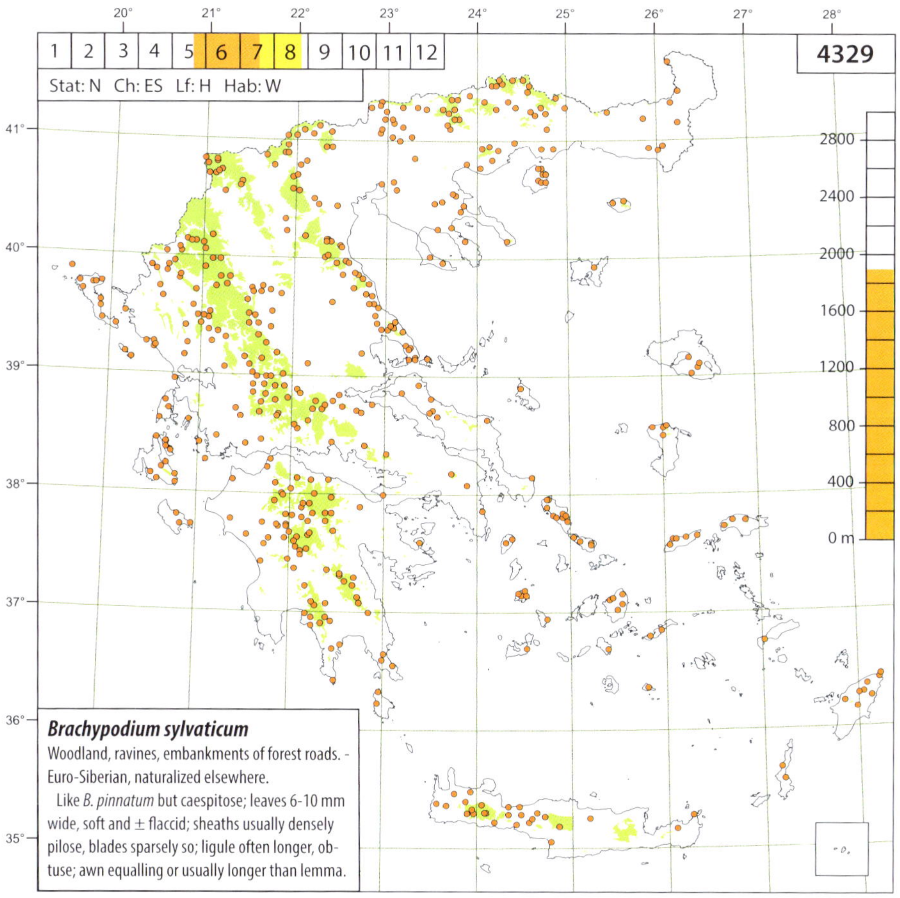

Brachypodium sylvaticum
Woodland, ravines, embankments of forest roads. - Euro-Siberian, naturalized elsewhere.
Like *B. pinnatum* but caespitose; leaves 6-10 mm wide, soft and ± flaccid; sheaths usually densely pilose, blades sparsely so; ligule often longer, obtuse; awn equalling or usually longer than lemma.

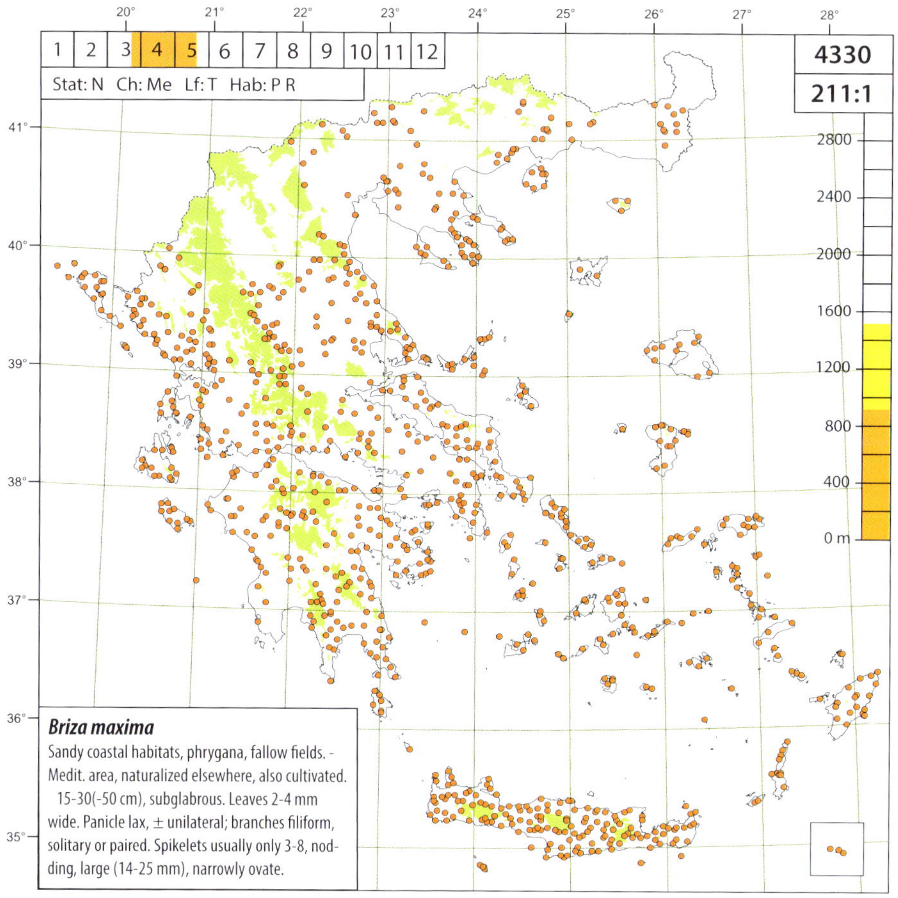

Briza maxima
Sandy coastal habitats, phrygana, fallow fields. - Medit. area, naturalized elsewhere, also cultivated.
15-30(-50 cm), subglabrous. Leaves 2-4 mm wide. Panicle lax, ± unilateral; branches filiform, solitary or paired. Spikelets usually only 3-8, nodding, large (14-25 mm), narrowly ovate.

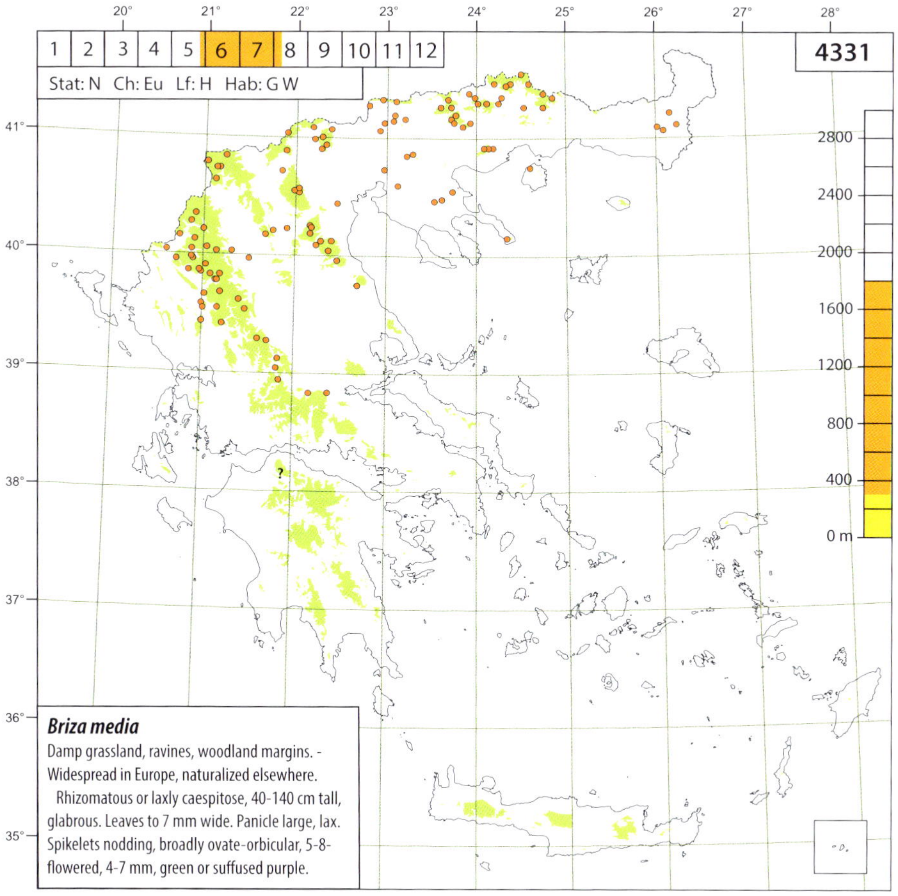

Briza media
Damp grassland, ravines, woodland margins. - Widespread in Europe, naturalized elsewhere.
Rhizomatous or laxly caespitose, 40-140 cm tall, glabrous. Leaves to 7 mm wide. Panicle large, lax. Spikelets nodding, broadly ovate-orbicular, 5-8-flowered, 4-7 mm, green or suffused purple.

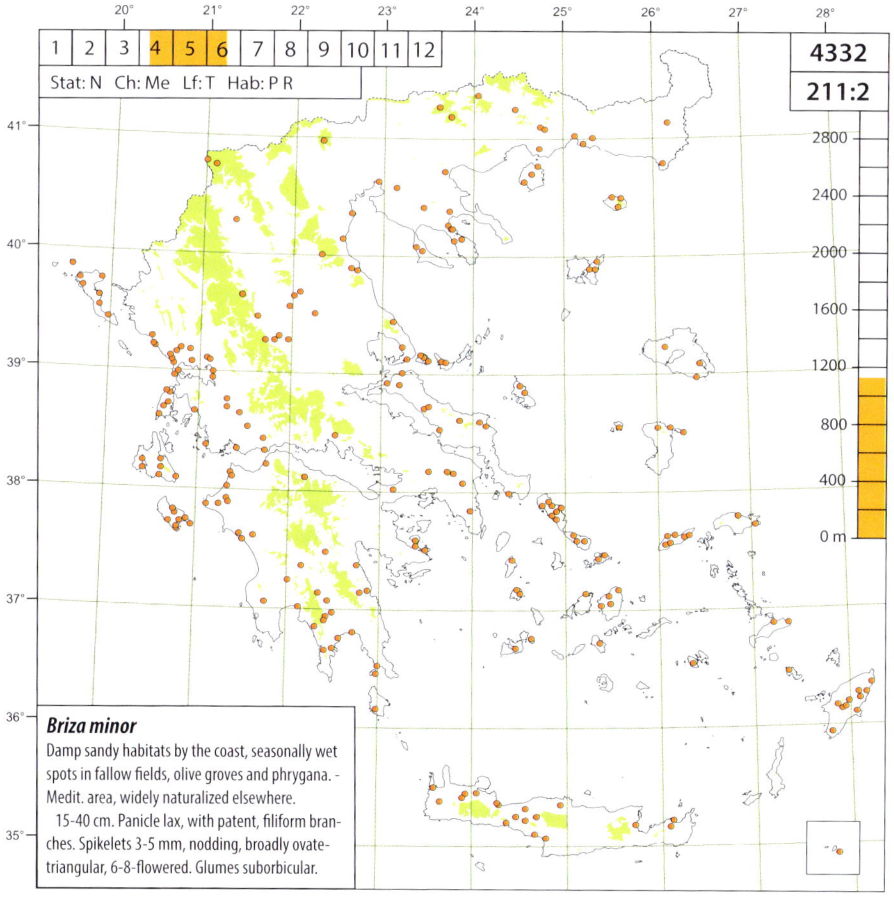

Briza minor
Damp sandy habitats by the coast, seasonally wet spots in fallow fields, olive groves and phrygana. - Medit. area, widely naturalized elsewhere.
15-40 cm. Panicle lax, with patent, filiform branches. Spikelets 3-5 mm, nodding, broadly ovate-triangular, 6-8-flowered. Glumes suborbicular.

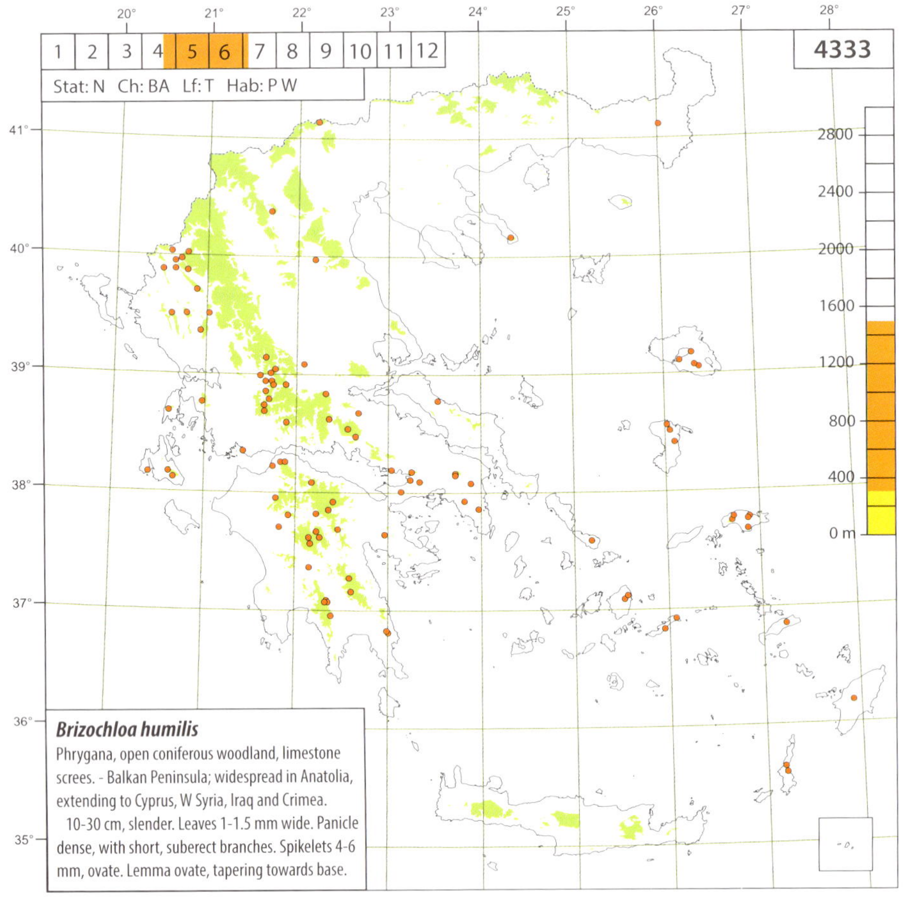

Brizochloa humilis
Phrygana, open coniferous woodland, limestone screes. - Balkan Peninsula; widespread in Anatolia, extending to Cyprus, W Syria, Iraq and Crimea.
10-30 cm, slender. Leaves 1-1.5 mm wide. Panicle dense, with short, suberect branches. Spikelets 4-6 mm, ovate. Lemma ovate, tapering towards base.

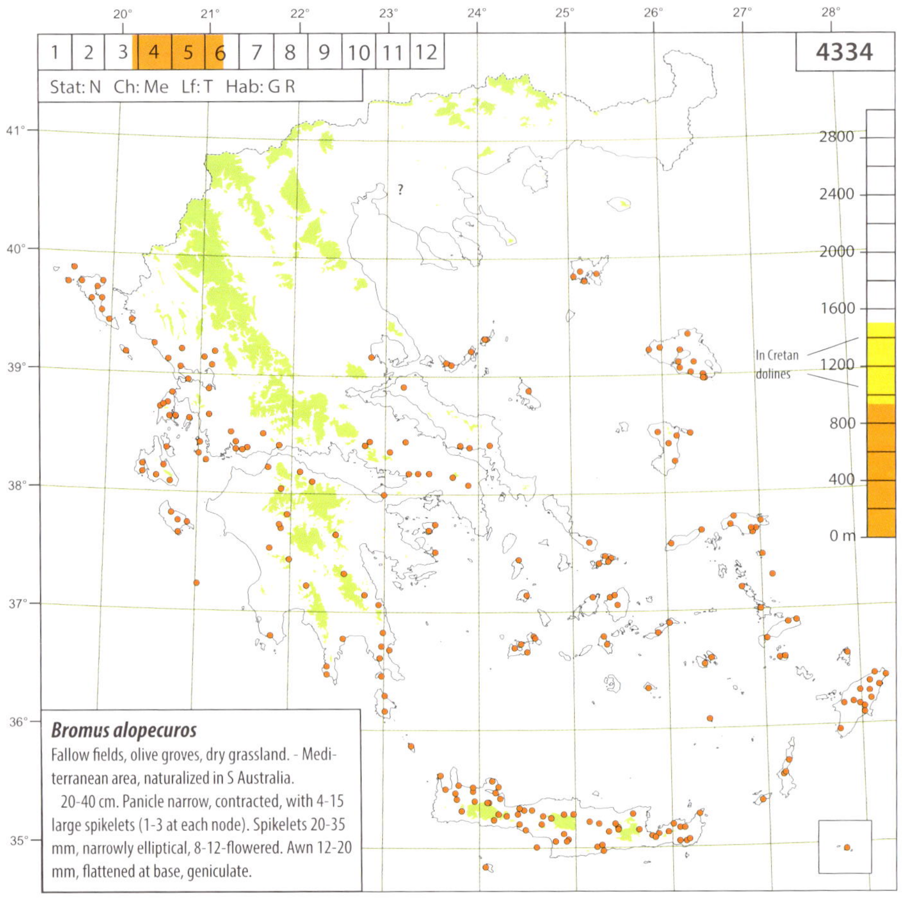

Bromus alopecuros
Fallow fields, olive groves, dry grassland. - Mediterranean area, naturalized in S Australia.
20-40 cm. Panicle narrow, contracted, with 4-15 large spikelets (1-3 at each node). Spikelets 20-35 mm, narrowly elliptical, 8-12-flowered. Awn 12-20 mm, flattened at base, geniculate.

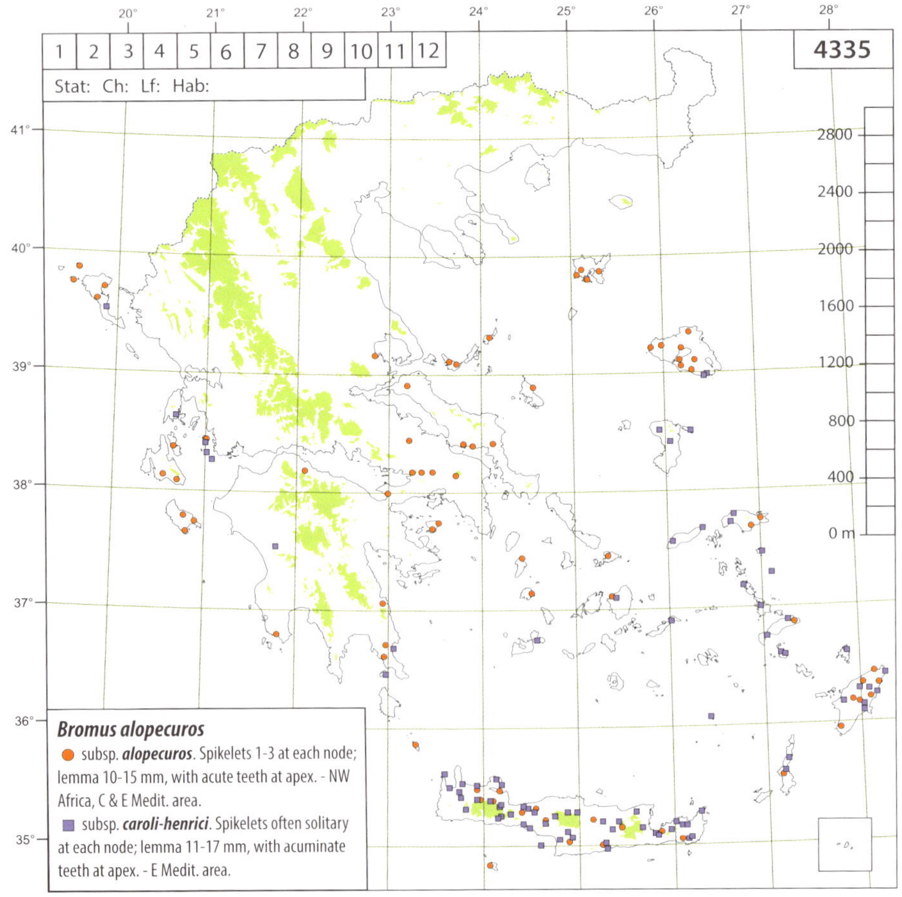

Bromus alopecuros
● subsp. ***alopecuros***. Spikelets 1-3 at each node; lemma 10-15 mm, with acute teeth at apex. - NW Africa, C & E Medit. area.
■ subsp. ***caroli-henrici***. Spikelets often solitary at each node; lemma 11-17 mm, with acuminate teeth at apex. - E Medit. area.

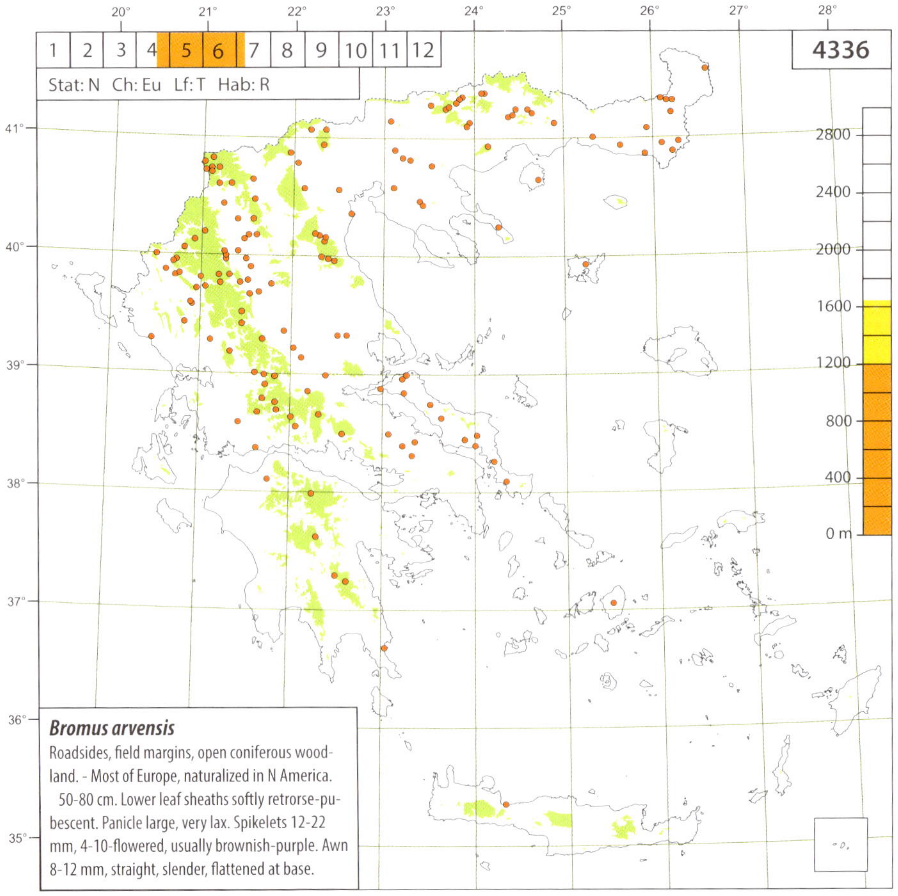

Bromus arvensis
Roadsides, field margins, open coniferous woodland. - Most of Europe, naturalized in N America.
50-80 cm. Lower leaf sheaths softly retrorse-pubescent. Panicle large, very lax. Spikelets 12-22 mm, 4-10-flowered, usually brownish-purple. Awn 8-12 mm, straight, slender, flattened at base.

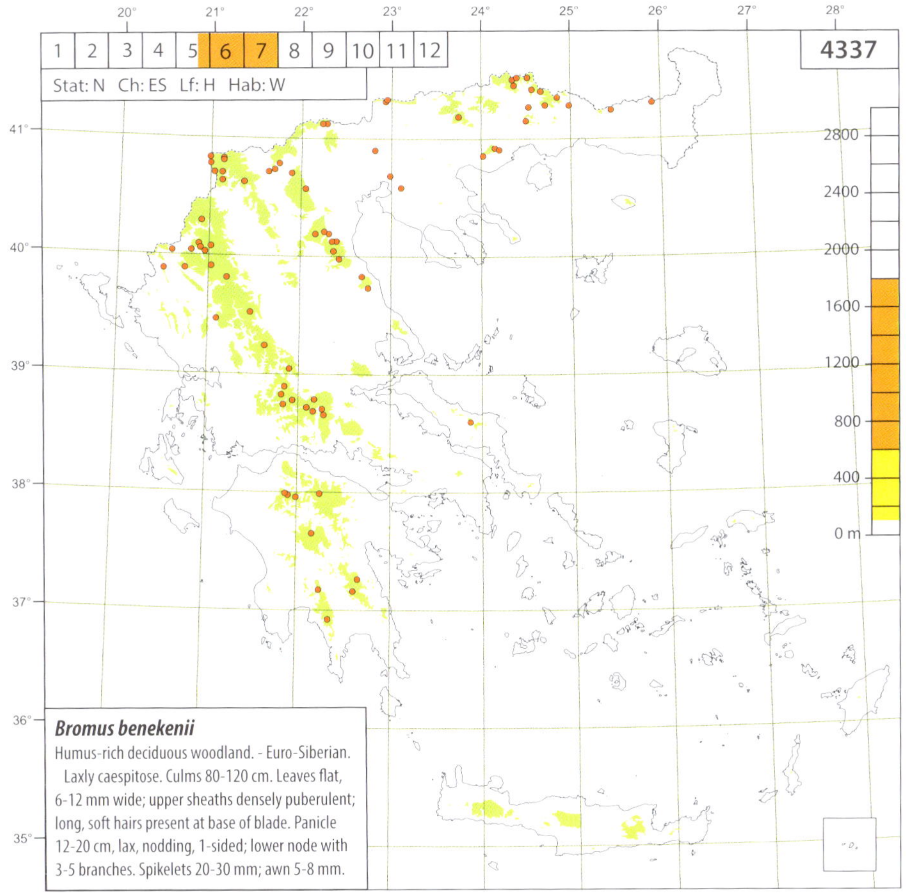

Bromus benekenii
Humus-rich deciduous woodland. - Euro-Siberian.
Laxly caespitose. Culms 80-120 cm. Leaves flat, 6-12 mm wide; upper sheaths densely puberulent; long, soft hairs present at base of blade. Panicle 12-20 cm, lax, nodding, 1-sided; lower node with 3-5 branches. Spikelets 20-30 mm; awn 5-8 mm.

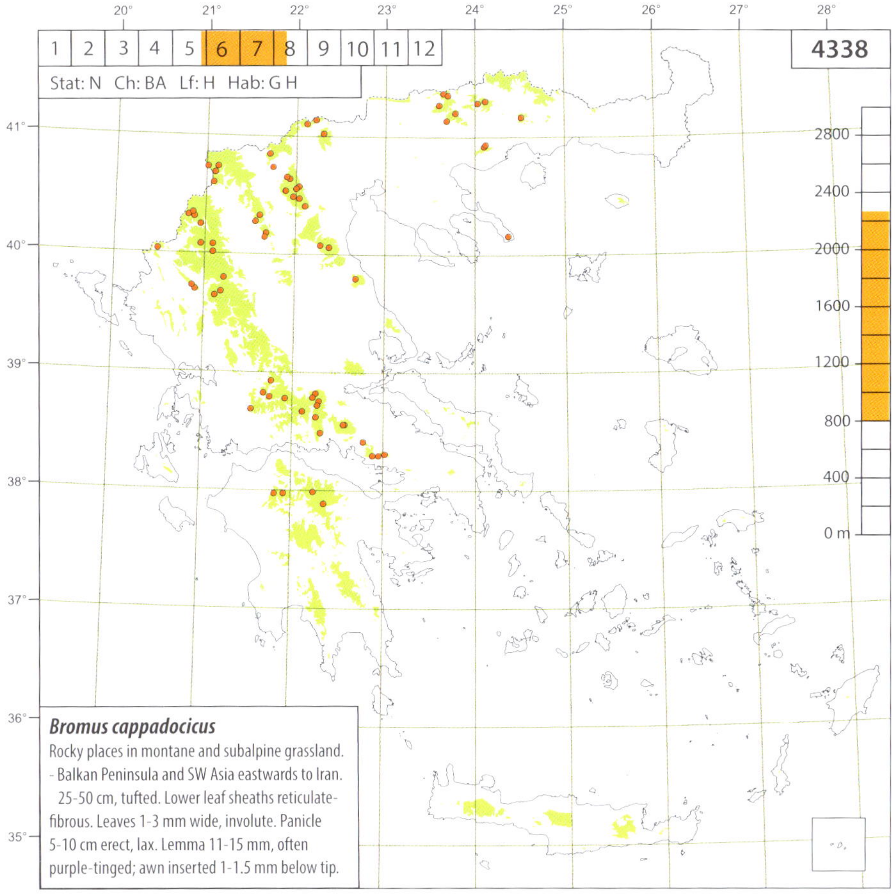

Bromus cappadocicus
Rocky places in montane and subalpine grassland. - Balkan Peninsula and SW Asia eastwards to Iran.
25-50 cm, tufted. Lower leaf sheaths reticulate-fibrous. Leaves 1-3 mm wide, involute. Panicle 5-10 cm erect, lax. Lemma 11-15 mm, often purple-tinged; awn inserted 1-1.5 mm below tip.

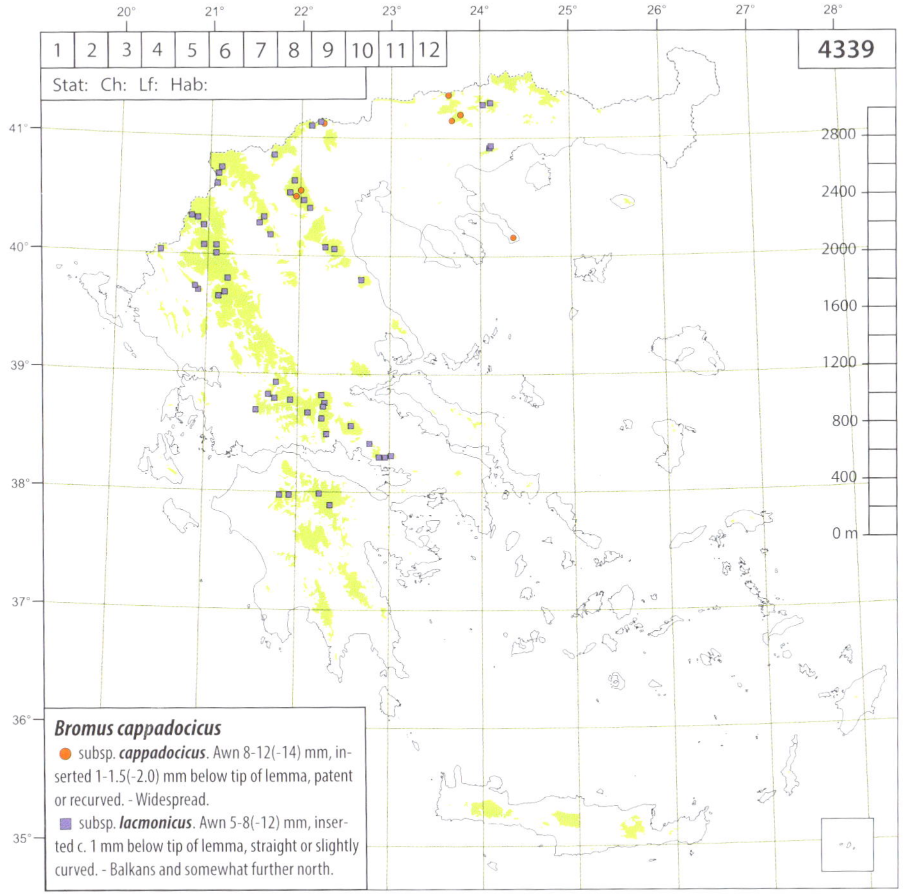

Bromus cappadocicus
● subsp. ***cappadocicus***. Awn 8-12(-14) mm, inserted 1-1.5(-2.0) mm below tip of lemma, patent or recurved. - Widespread.
■ subsp. ***lacmonicus***. Awn 5-8(-12) mm, inserted c. 1 mm below tip of lemma, straight or slightly curved. - Balkans and somewhat further north.

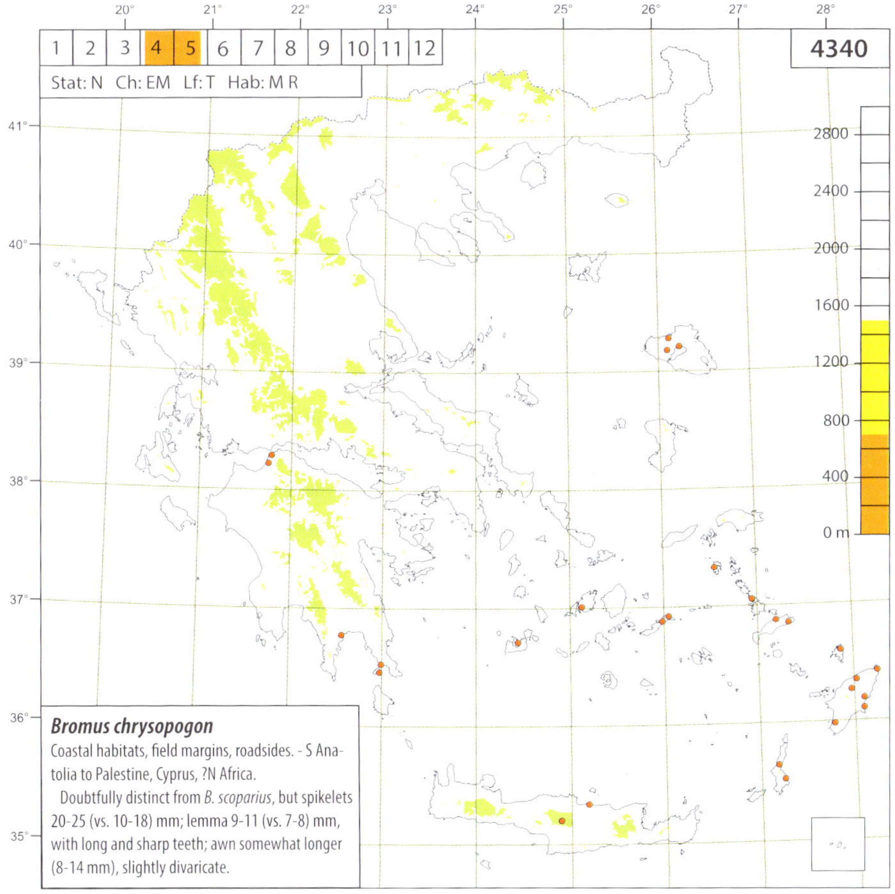

Bromus chrysopogon
Coastal habitats, field margins, roadsides. - S Anatolia to Palestine, Cyprus, ?N Africa.
Doubtfully distinct from *B. scoparius*, but spikelets 20-25 (vs. 10-18) mm; lemma 9-11 (vs. 7-8) mm, with long and sharp teeth; awn somewhat longer (8-14 mm), slightly divaricate.

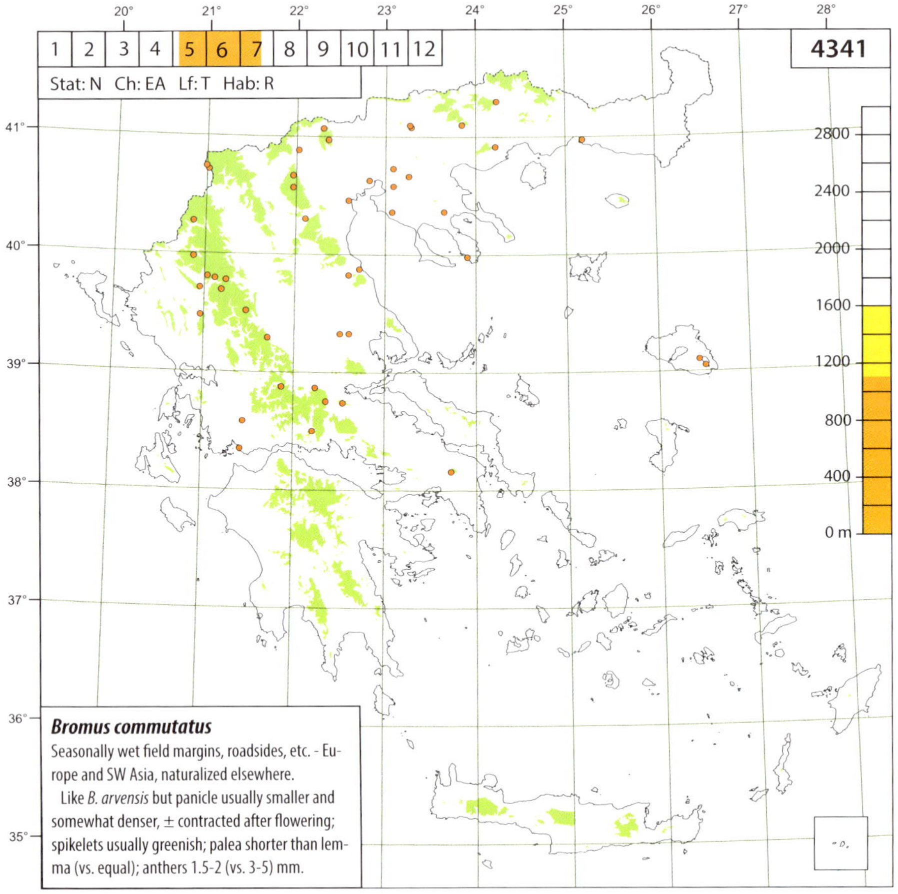

Bromus commutatus

Seasonally wet field margins, roadsides, etc. - Europe and SW Asia, naturalized elsewhere.

Like *B. arvensis* but panicle usually smaller and somewhat denser, ± contracted after flowering; spikelets usually greenish; palea shorter than lemma (vs. equal); anthers 1.5-2 (vs. 3-5) mm.

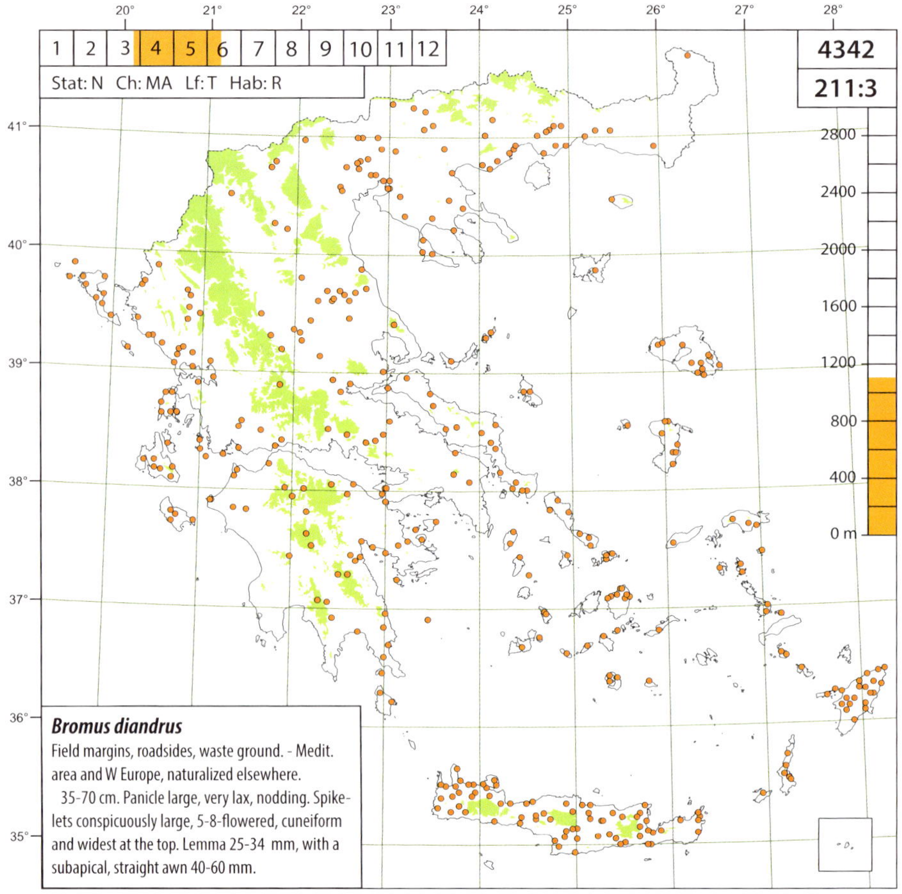

Bromus diandrus

Field margins, roadsides, waste ground. - Medit. area and W Europe, naturalized elsewhere.

35-70 cm. Panicle large, very lax, nodding. Spikelets conspicuously large, 5-8-flowered, cuneiform and widest at the top. Lemma 25-34 mm, with a subapical, straight awn 40-60 mm.

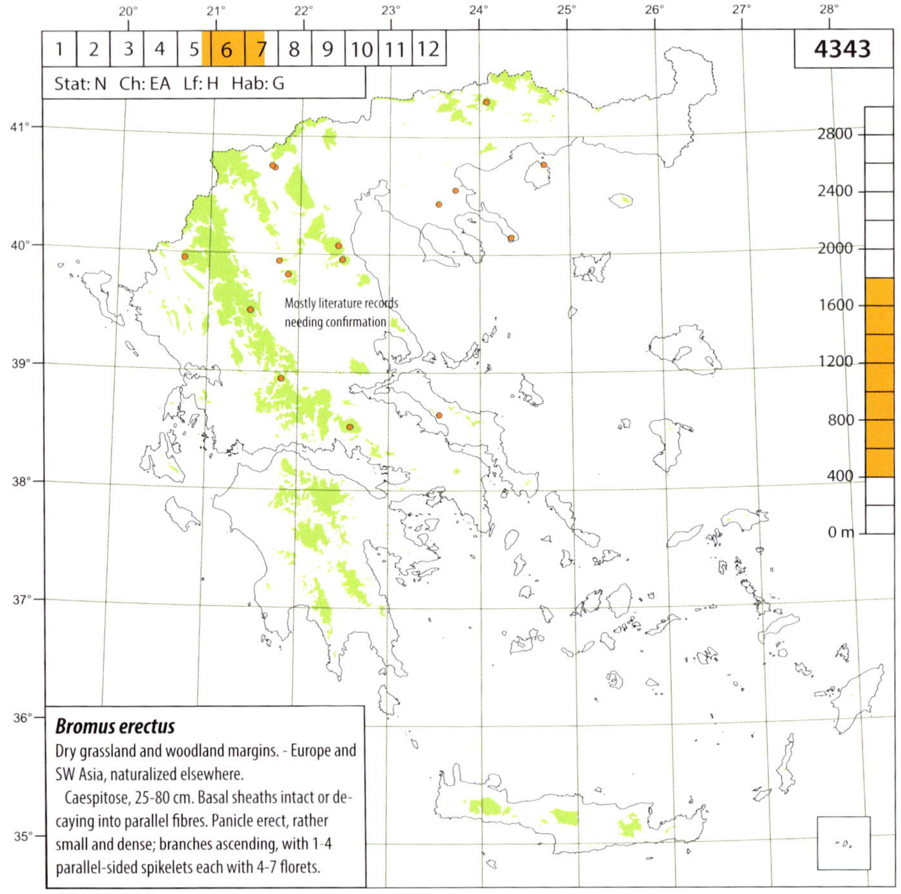

Bromus erectus

Dry grassland and woodland margins. - Europe and SW Asia, naturalized elsewhere.

Caespitose, 25-80 cm. Basal sheaths intact or decaying into parallel fibres. Panicle erect, rather small and dense; branches ascending, with 1-4 parallel-sided spikelets each with 4-7 florets.

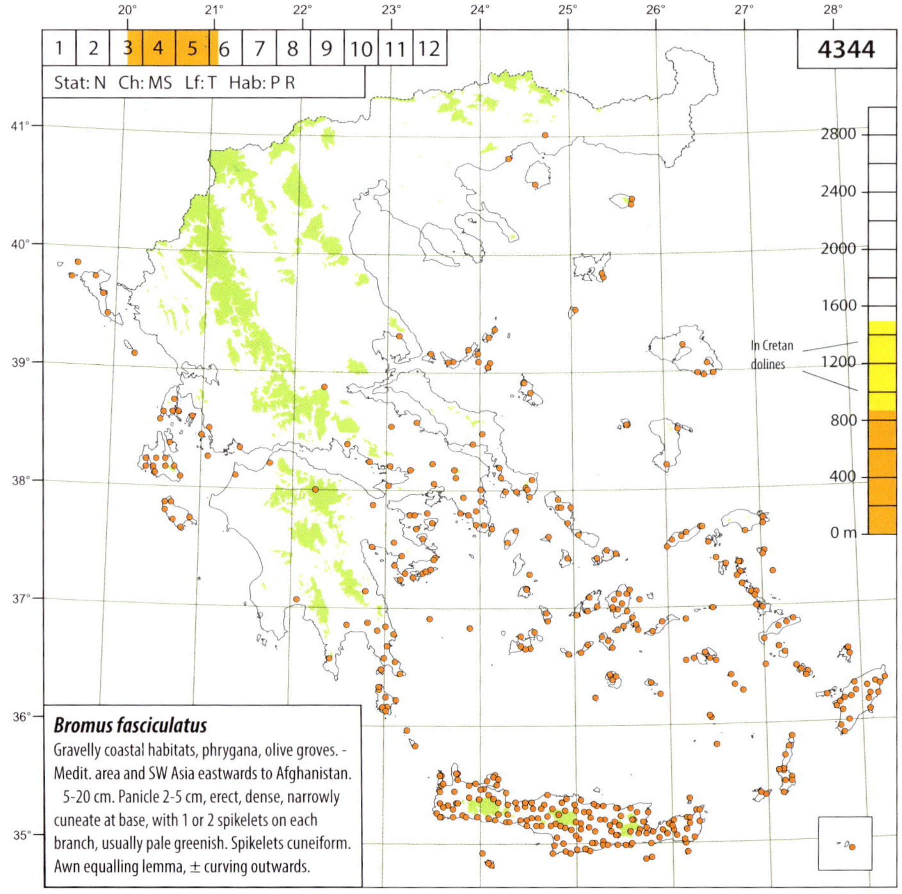

Bromus fasciculatus

Gravelly coastal habitats, phrygana, olive groves. - Medit. area and SW Asia eastwards to Afghanistan.

5-20 cm. Panicle 2-5 cm, erect, dense, narrowly cuneate at base, with 1 or 2 spikelets on each branch, usually pale greenish. Spikelets cuneiform. Awn equalling lemma, ± curving outwards.

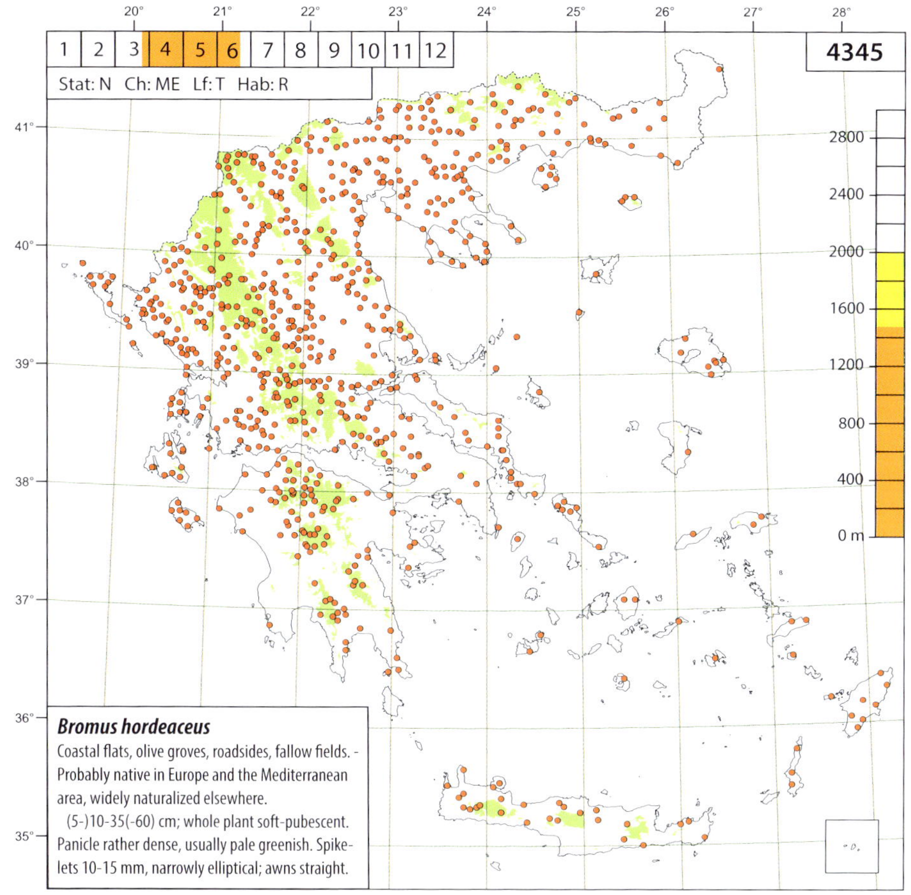

Bromus hordeaceus
Coastal flats, olive groves, roadsides, fallow fields. - Probably native in Europe and the Mediterranean area, widely naturalized elsewhere.
(5-)10-35(-60) cm; whole plant soft-pubescent. Panicle rather dense, usually pale greenish. Spikelets 10-15 mm, narrowly elliptical; awns straight.

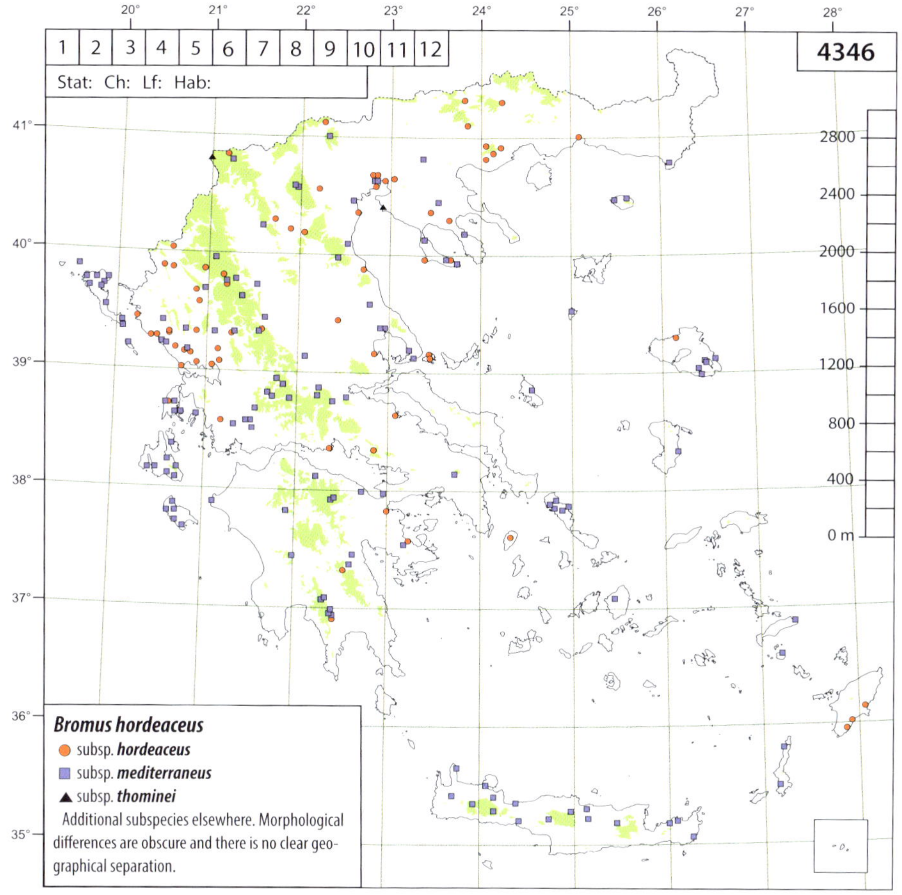

Bromus hordeaceus
subsp. ***hordeaceus***
subsp. ***mediterraneus***
subsp. ***thominei***
Additional subspecies elsewhere. Morphological differences are obscure and there is no clear geographical separation.

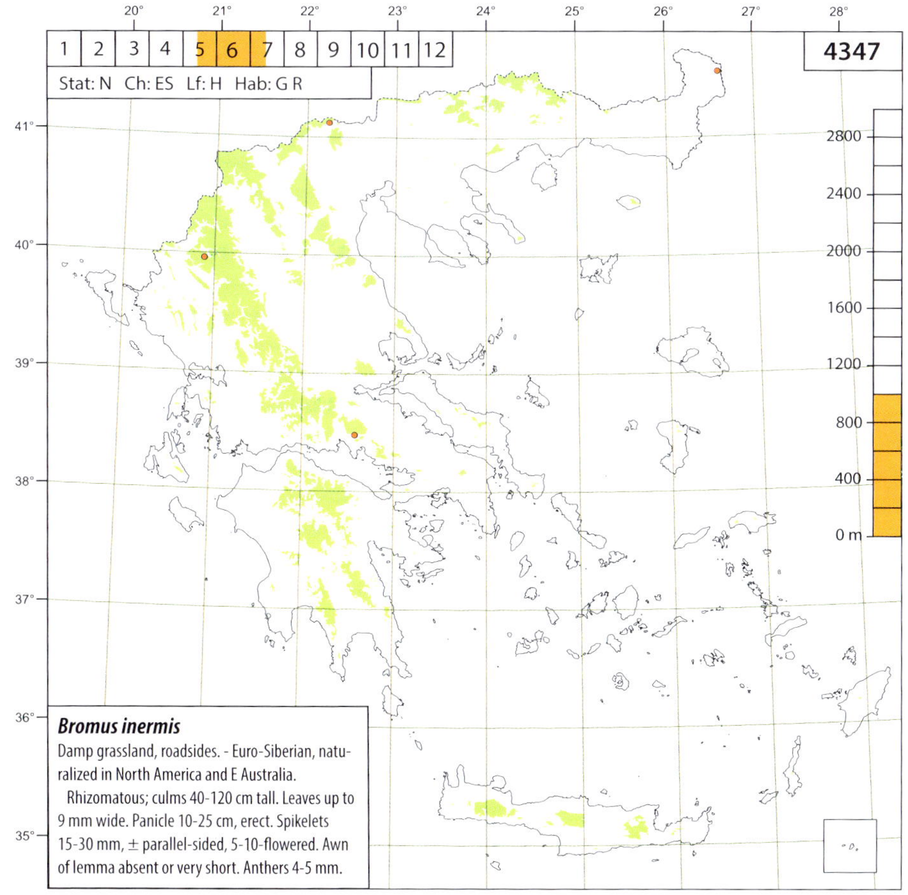

Bromus inermis
Damp grassland, roadsides. - Euro-Siberian, naturalized in North America and E Australia.
Rhizomatous; culms 40-120 cm tall. Leaves up to 9 mm wide. Panicle 10-25 cm, erect. Spikelets 15-30 mm, ± parallel-sided, 5-10-flowered. Awn of lemma absent or very short. Anthers 4-5 mm.

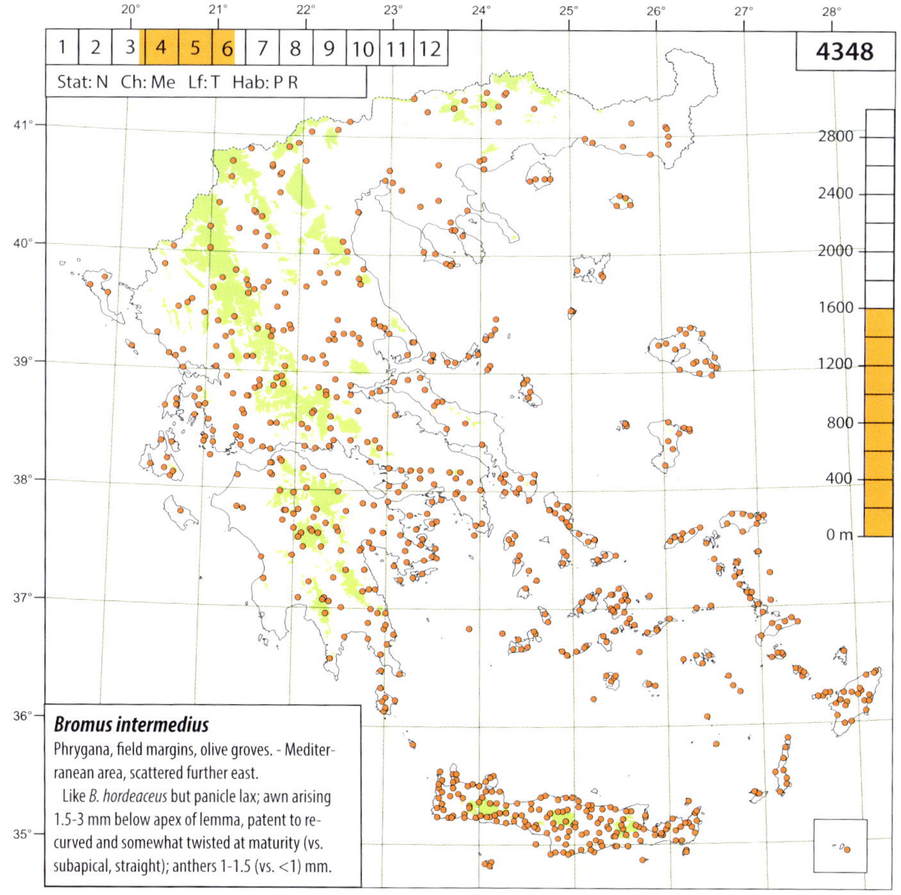

Bromus intermedius
Phrygana, field margins, olive groves. - Mediterranean area, scattered further east.
Like *B. hordeaceus* but panicle lax; awn arising 1.5-3 mm below apex of lemma, patent to recurved and somewhat twisted at maturity (vs. subapical, straight); anthers 1-1.5 (vs. <1) mm.

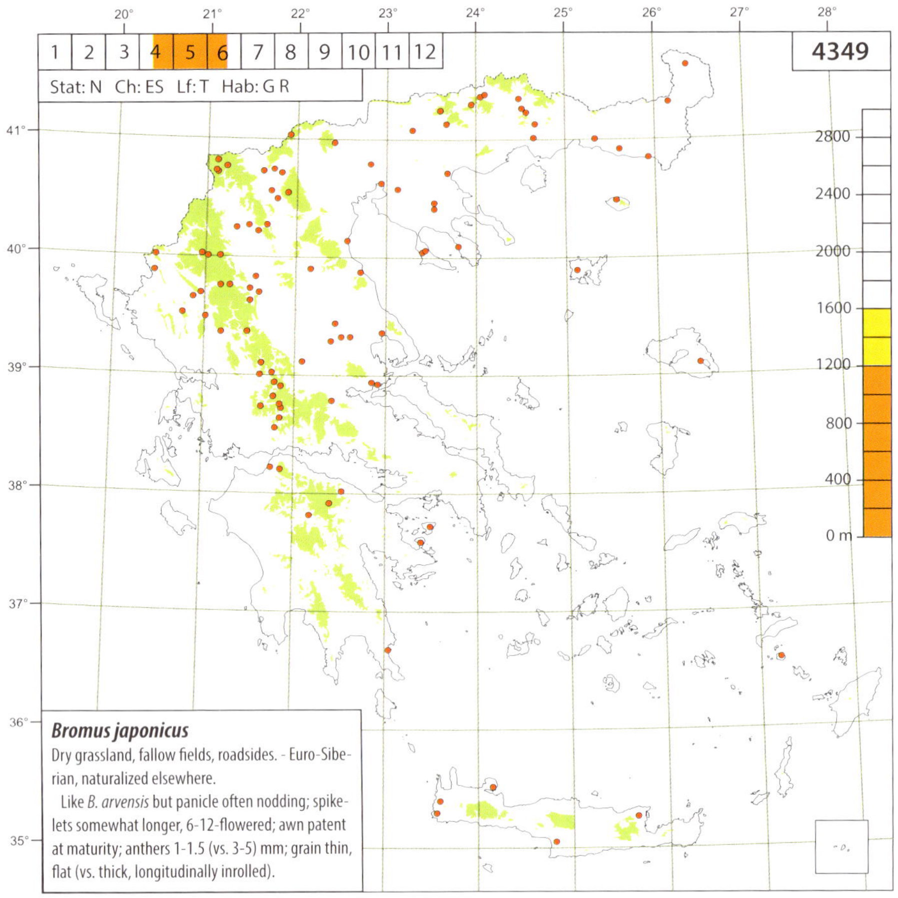

Bromus japonicus

Dry grassland, fallow fields, roadsides. - Euro-Siberian, naturalized elsewhere.

Like *B. arvensis* but panicle often nodding; spikelets somewhat longer, 6-12-flowered; awn patent at maturity; anthers 1-1.5 (vs. 3-5) mm; grain thin, flat (vs. thick, longitudinally inrolled).

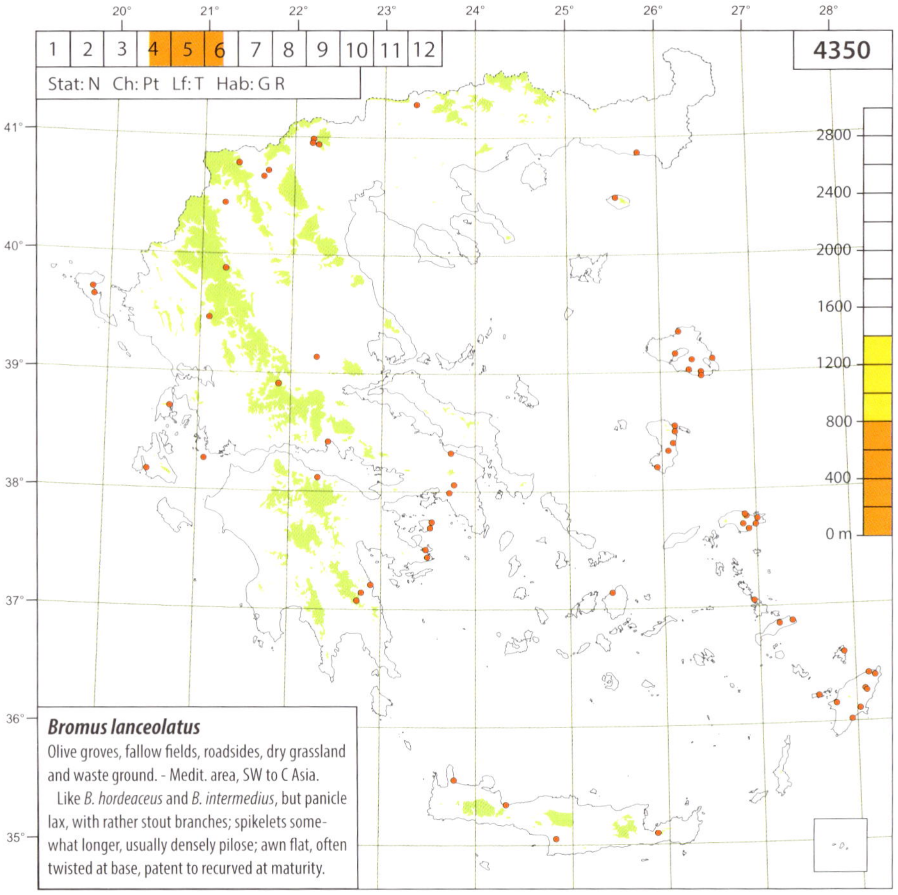

Bromus lanceolatus

Olive groves, fallow fields, roadsides, dry grassland and waste ground. - Medit. area, SW to C Asia.

Like *B. hordeaceus* and *B. intermedius*, but panicle lax, with rather stout branches; spikelets somewhat longer, usually densely pilose; awn flat, often twisted at base, patent to recurved at maturity.

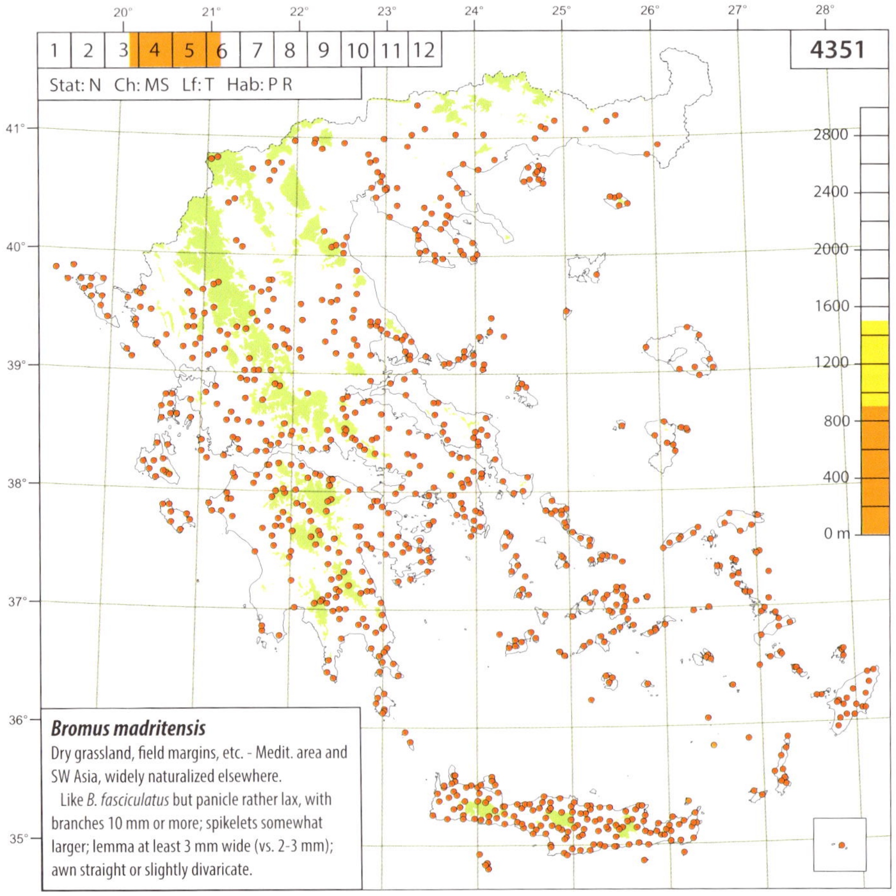

Bromus madritensis

Dry grassland, field margins, etc. - Medit. area and SW Asia, widely naturalized elsewhere.

Like *B. fasciculatus* but panicle rather lax, with branches 10 mm or more; spikelets somewhat larger; lemma at least 3 mm wide (vs. 2-3 mm); awn straight or slightly divaricate.

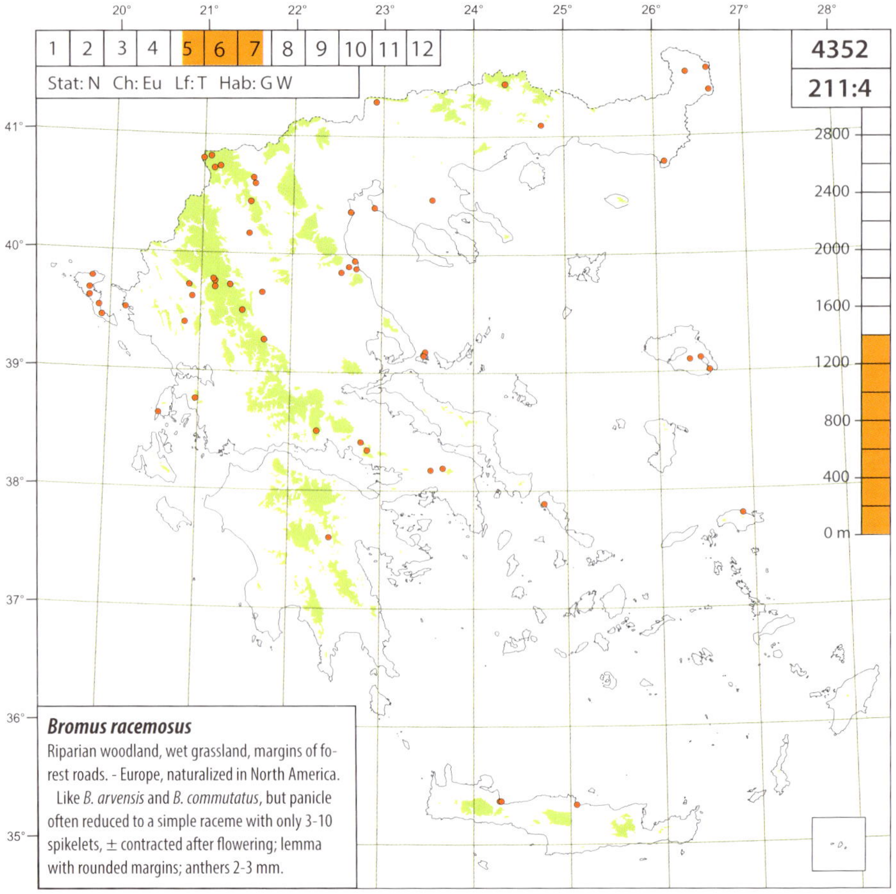

Bromus racemosus

Riparian woodland, wet grassland, margins of forest roads. - Europe, naturalized in North America.

Like *B. arvensis* and *B. commutatus*, but panicle often reduced to a simple raceme with only 3-10 spikelets, ± contracted after flowering; lemma with rounded margins; anthers 2-3 mm.

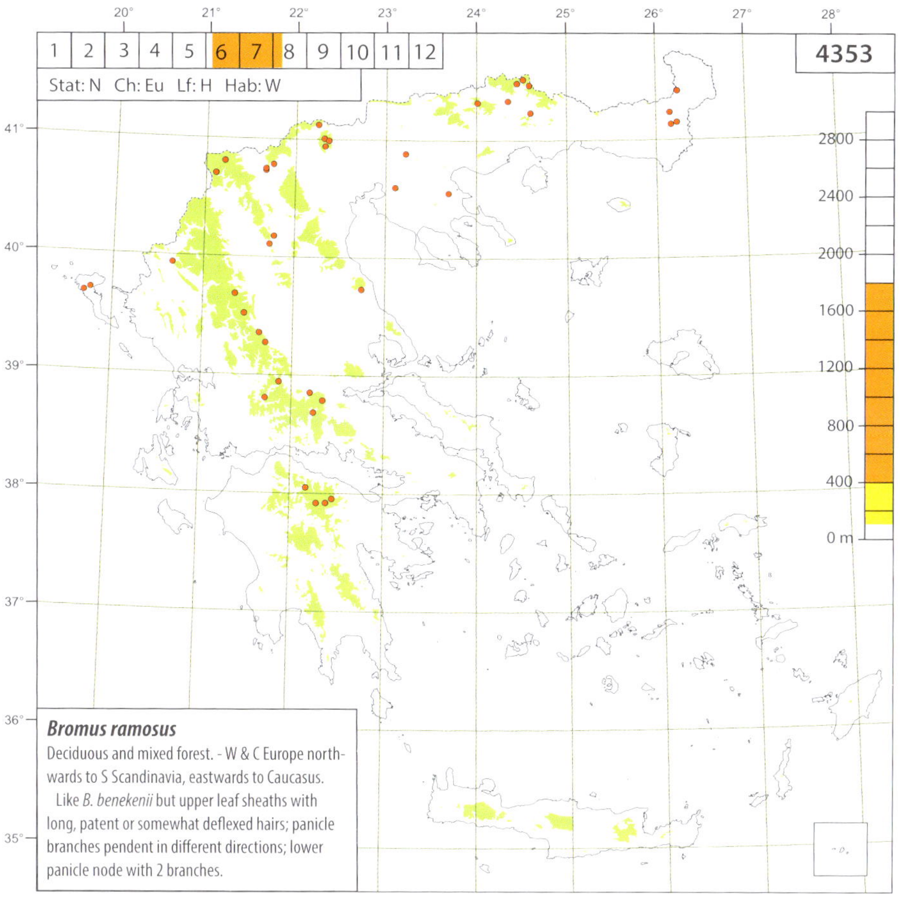

Bromus ramosus
Deciduous and mixed forest. - W & C Europe northwards to S Scandinavia, eastwards to Caucasus.
Like *B. benekenii* but upper leaf sheaths with long, patent or somewhat deflexed hairs; panicle branches pendent in different directions; lower panicle node with 2 branches.

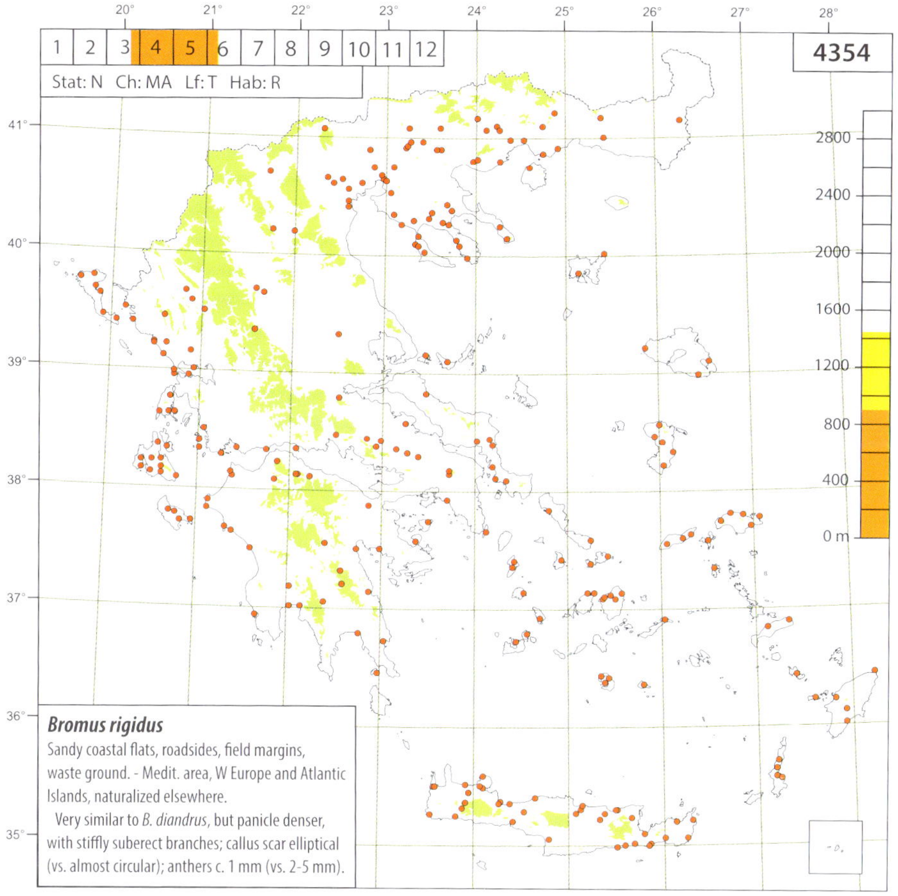

Bromus rigidus
Sandy coastal flats, roadsides, field margins, waste ground. - Medit. area, W Europe and Atlantic Islands, naturalized elsewhere.
Very similar to *B. diandrus*, but panicle denser, with stiffly suberect branches; callus scar elliptical (vs. almost circular); anthers c. 1 mm (vs. 2-5 mm).

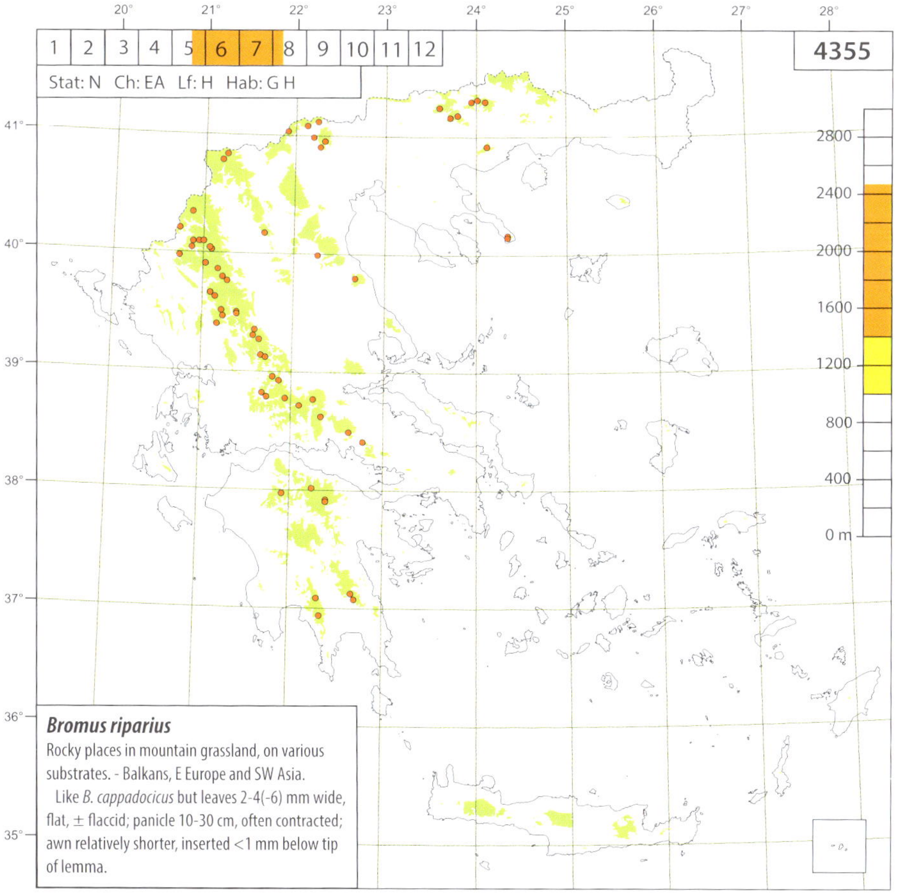

Bromus riparius
Rocky places in mountain grassland, on various substrates. - Balkans, E Europe and SW Asia.
Like *B. cappadocicus* but leaves 2-4(-6) mm wide, flat, ± flaccid; panicle 10-30 cm, often contracted; awn relatively shorter, inserted <1 mm below tip of lemma.

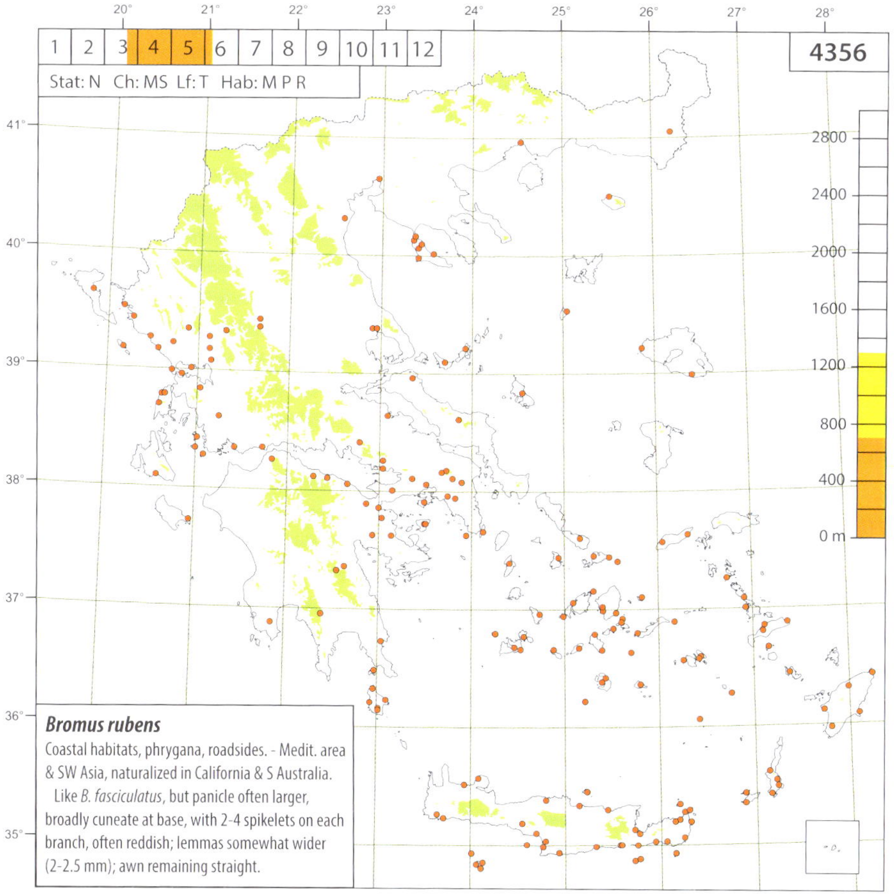

Bromus rubens
Coastal habitats, phrygana, roadsides. - Medit. area & SW Asia, naturalized in California & S Australia.
Like *B. fasciculatus*, but panicle often larger, broadly cuneate at base, with 2-4 spikelets on each branch, often reddish; lemmas somewhat wider (2-2.5 mm); awn remaining straight.

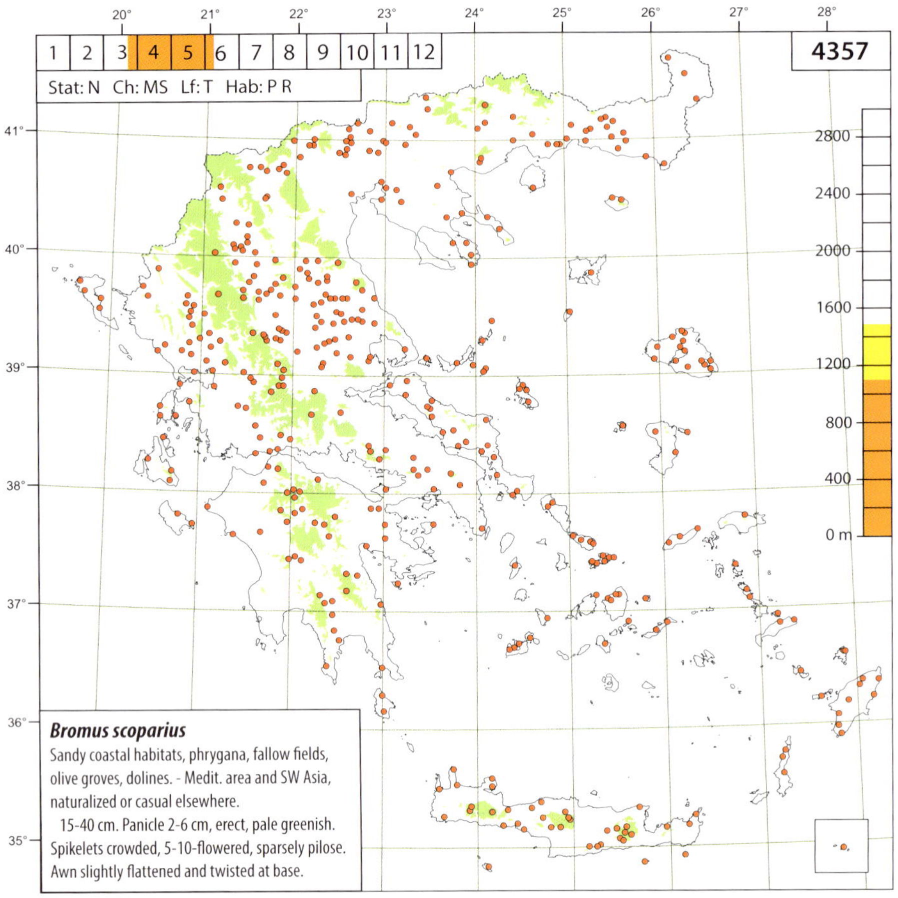

Bromus scoparius

Sandy coastal habitats, phrygana, fallow fields, olive groves, dolines. - Medit. area and SW Asia, naturalized or casual elsewhere.

15-40 cm. Panicle 2-6 cm, erect, pale greenish. Spikelets crowded, 5-10-flowered, sparsely pilose. Awn slightly flattened and twisted at base.

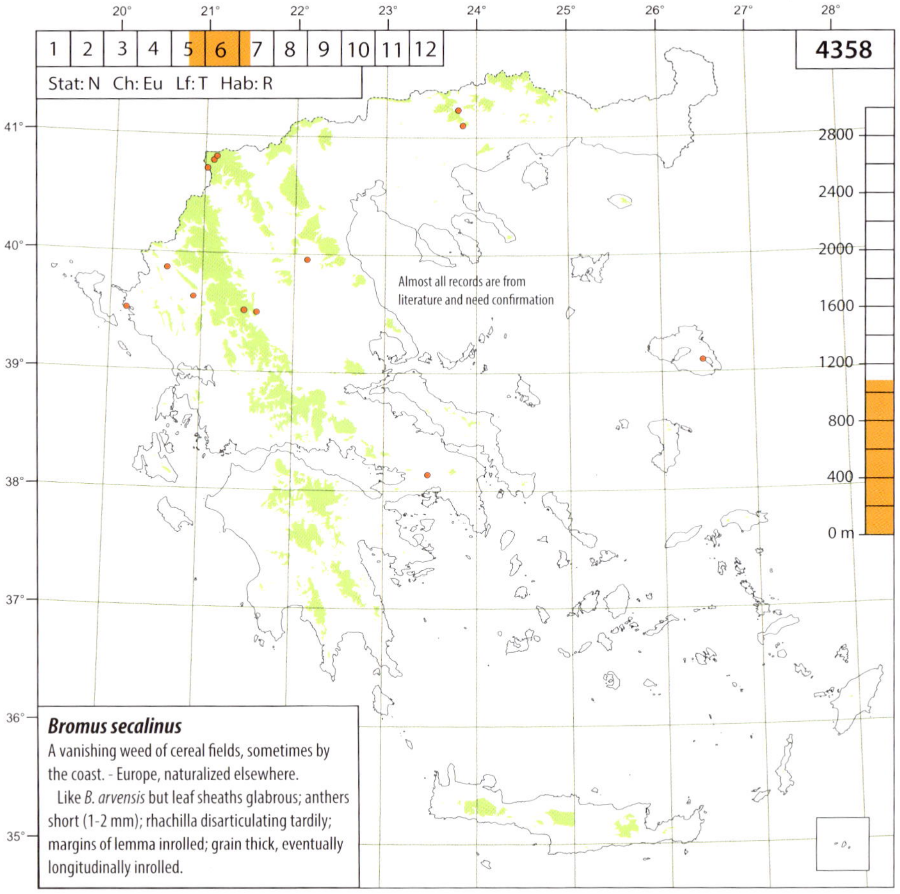

Bromus secalinus

A vanishing weed of cereal fields, sometimes by the coast. - Europe, naturalized elsewhere.

Like *B. arvensis* but leaf sheaths glabrous; anthers short (1-2 mm); rhachilla disarticulating tardily; margins of lemma inrolled; grain thick, eventually longitudinally inrolled.

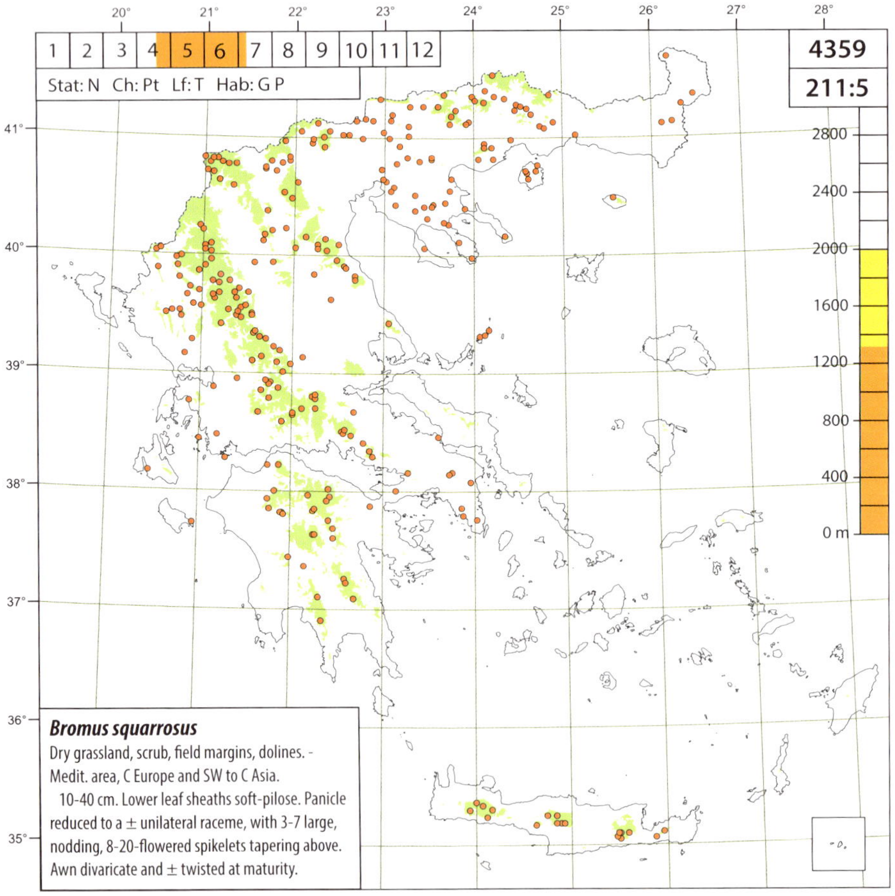

Bromus squarrosus

Dry grassland, scrub, field margins, dolines. - Medit. area, C Europe and SW to C Asia.

10-40 cm. Lower leaf sheaths soft-pilose. Panicle reduced to a ± unilateral raceme, with 3-7 large, nodding, 8-20-flowered spikelets tapering above. Awn divaricate and ± twisted at maturity.

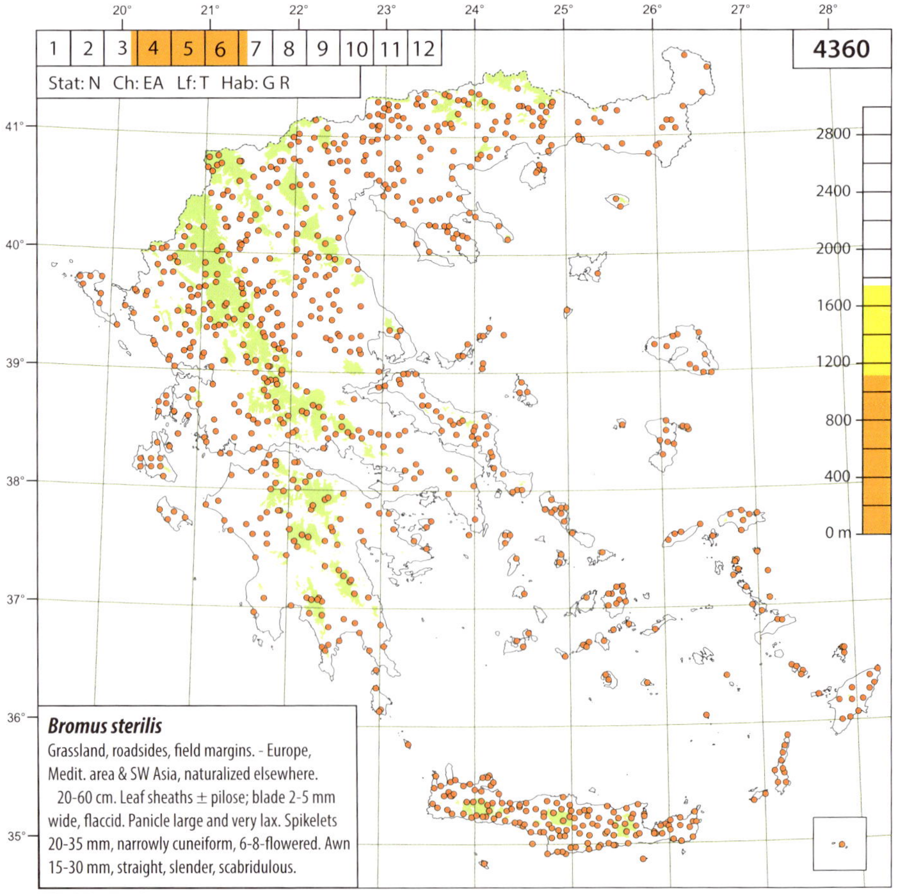

Bromus sterilis

Grassland, roadsides, field margins. - Europe, Medit. area & SW Asia, naturalized elsewhere.

20-60 cm. Leaf sheaths ± pilose; blade 2-5 mm wide, flaccid. Panicle large and very lax. Spikelets 20-35 mm, narrowly cuneiform, 6-8-flowered. Awn 15-30 mm, straight, slender, scabridulous.

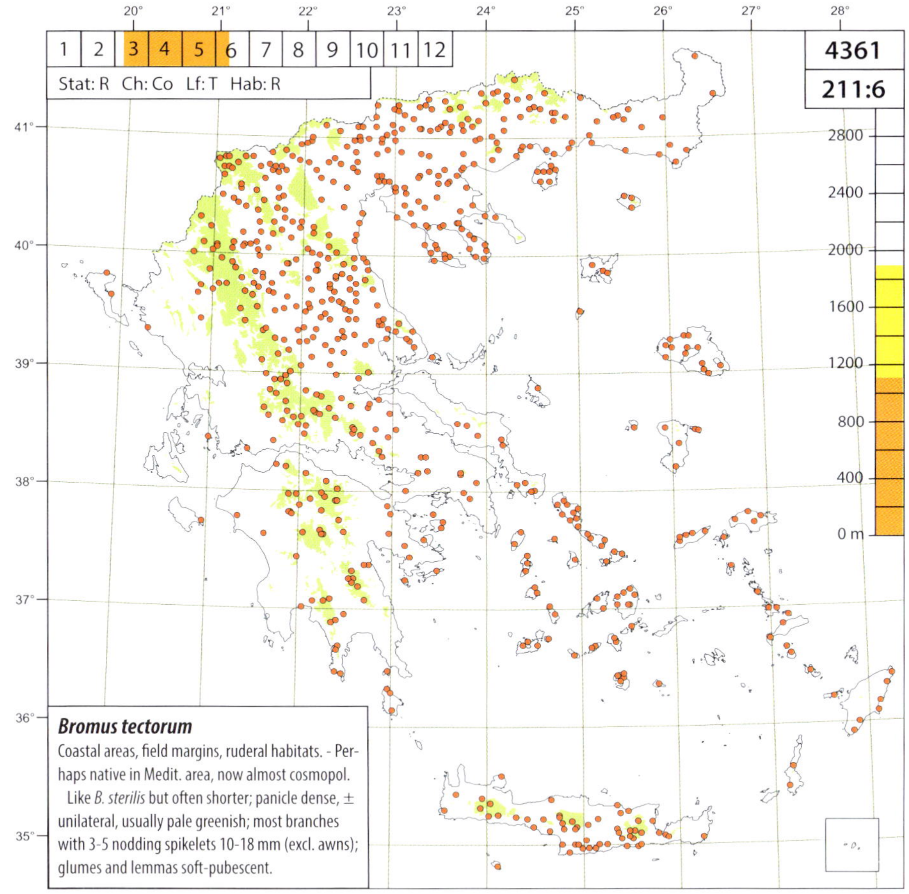

Bromus tectorum
Coastal areas, field margins, ruderal habitats. - Perhaps native in Medit. area, now almost cosmopol.
Like *B. sterilis* but often shorter; panicle dense, ± unilateral, usually pale greenish; most branches with 3-5 nodding spikelets 10-18 mm (excl. awns); glumes and lemmas soft-pubescent.

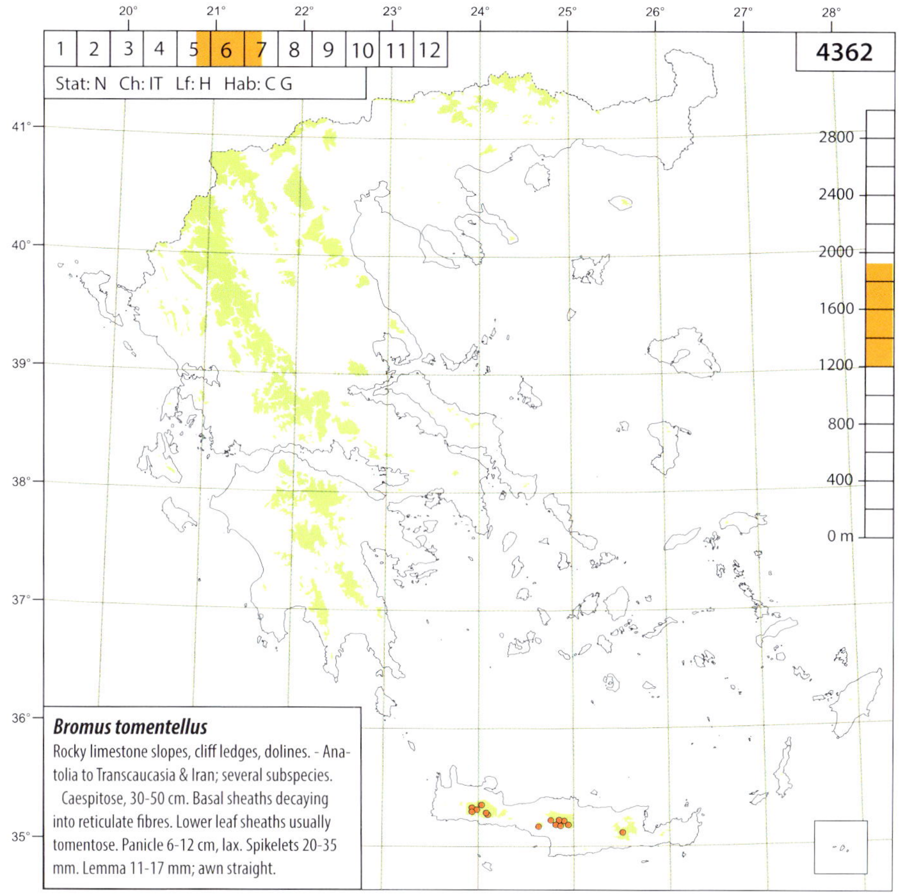

Bromus tomentellus
Rocky limestone slopes, cliff ledges, dolines. - Anatolia to Transcaucasia & Iran; several subspecies.
Caespitose, 30-50 cm. Basal sheaths decaying into reticulate fibres. Lower leaf sheaths usually tomentose. Panicle 6-12 cm, lax. Spikelets 20-35 mm. Lemma 11-17 mm; awn straight.

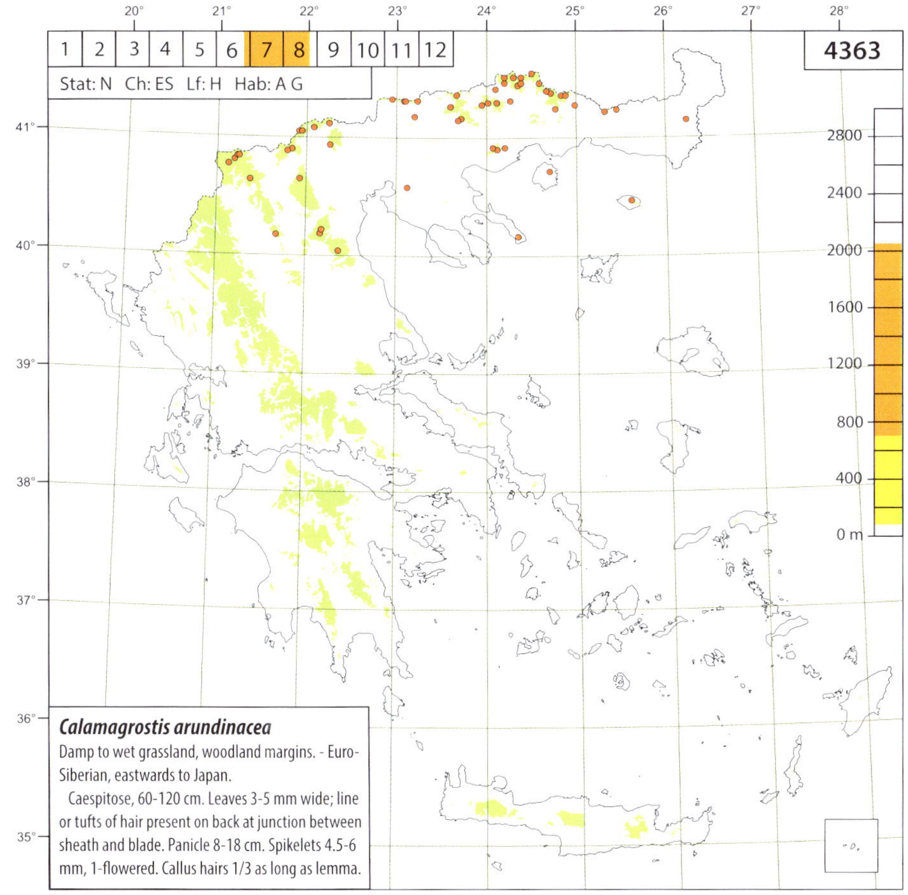

Calamagrostis arundinacea
Damp to wet grassland, woodland margins. - Euro-Siberian, eastwards to Japan.
Caespitose, 60-120 cm. Leaves 3-5 mm wide; line or tufts of hair present on back at junction between sheath and blade. Panicle 8-18 cm. Spikelets 4.5-6 mm, 1-flowered. Callus hairs 1/3 as long as lemma.

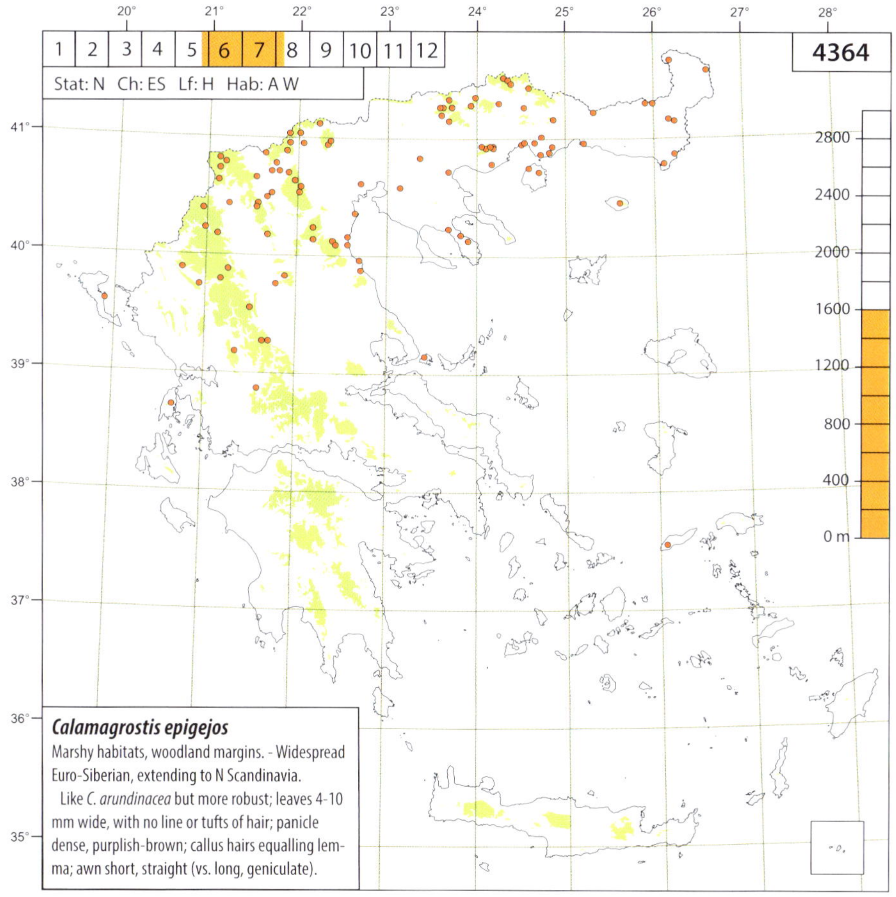

Calamagrostis epigejos
Marshy habitats, woodland margins. - Widespread Euro-Siberian, extending to N Scandinavia.
Like *C. arundinacea* but more robust; leaves 4-10 mm wide, with no line or tufts of hair; panicle dense, purplish-brown; callus hairs equalling lemma; awn short, straight (vs. long, geniculate).

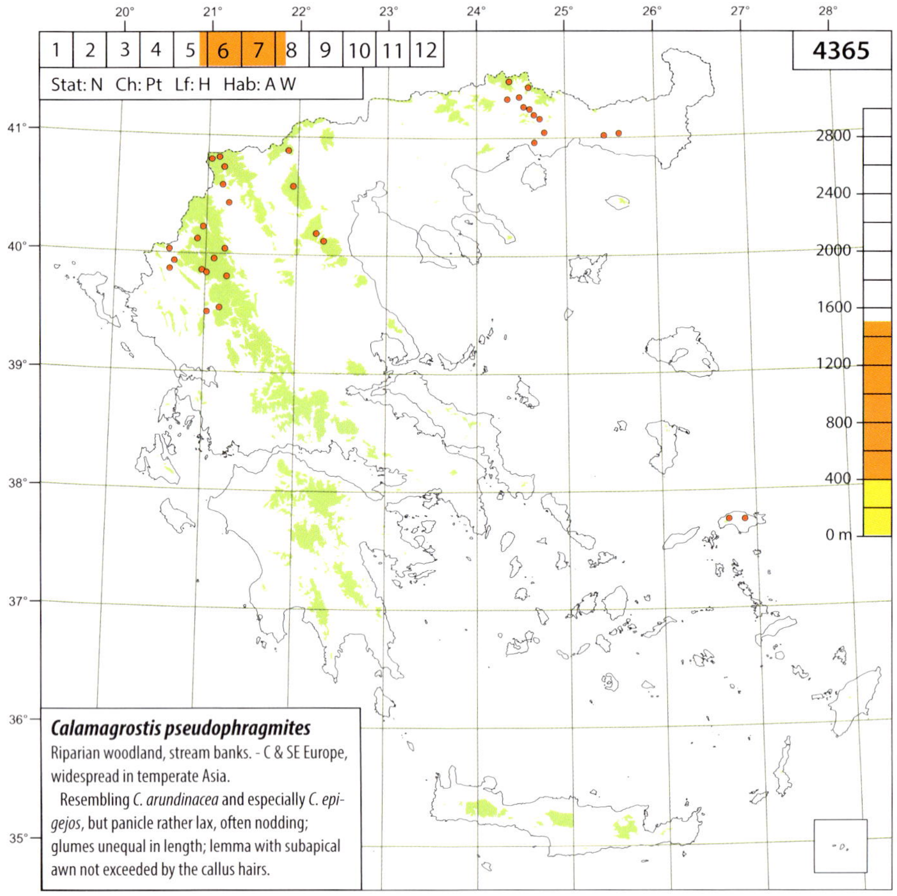

Calamagrostis pseudophragmites
Riparian woodland, stream banks. - C & SE Europe, widespread in temperate Asia.
Resembling *C. arundinacea* and especially *C. epigejos*, but panicle rather lax, often nodding; glumes unequal in length; lemma with subapical awn not exceeded by the callus hairs.

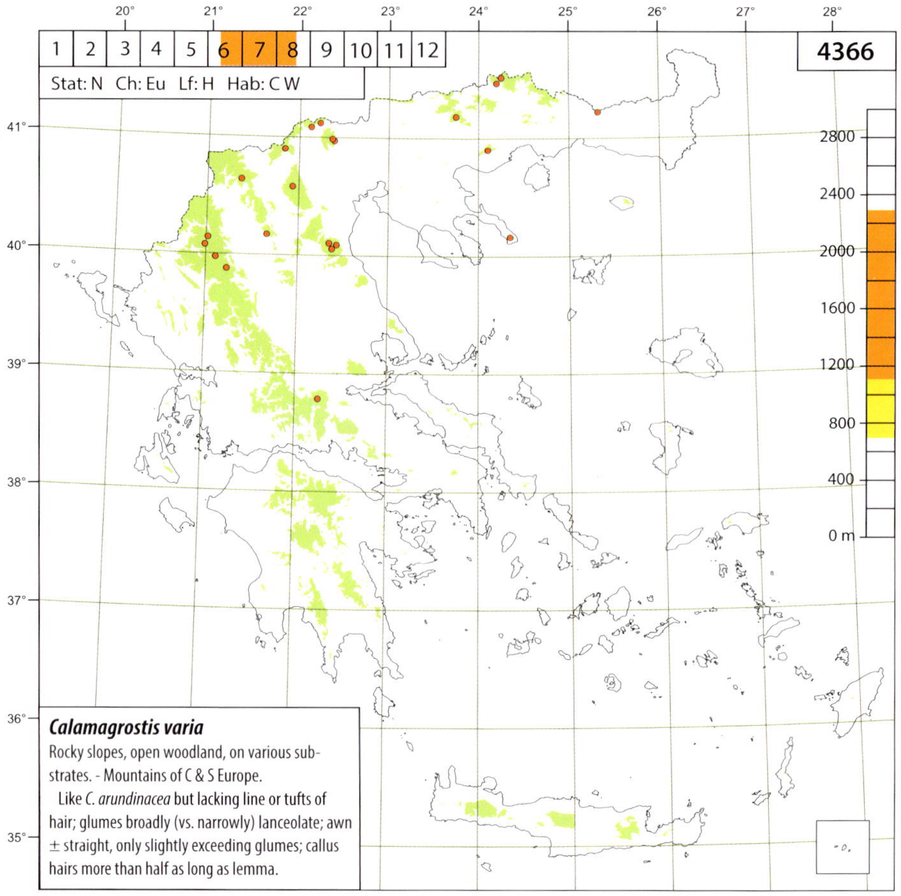

Calamagrostis varia
Rocky slopes, open woodland, on various substrates. - Mountains of C & S Europe.
Like *C. arundinacea* but lacking line or tufts of hair; glumes broadly (vs. narrowly) lanceolate; awn ± straight, only slightly exceeding glumes; callus hairs more than half as long as lemma.

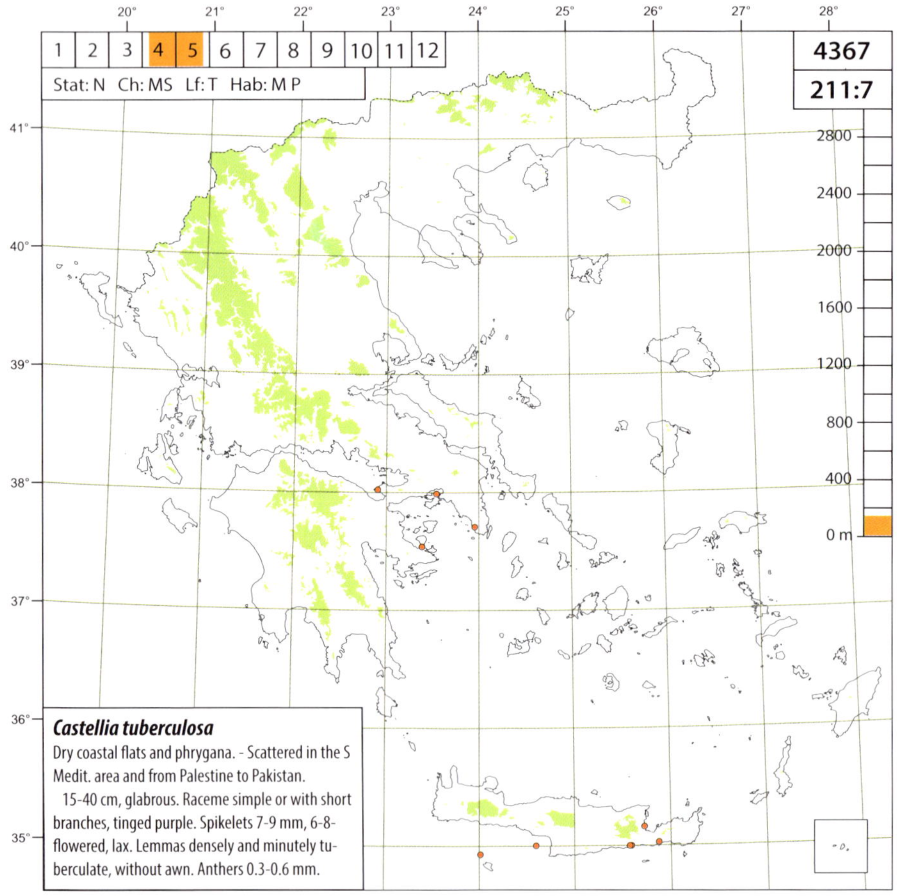

Castellia tuberculosa
Dry coastal flats and phrygana. - Scattered in the S Medit. area and from Palestine to Pakistan.
15-40 cm, glabrous. Raceme simple or with short branches, tinged purple. Spikelets 7-9 mm, 6-8-flowered, lax. Lemmas densely and minutely tuberculate, without awn. Anthers 0.3-0.6 mm.

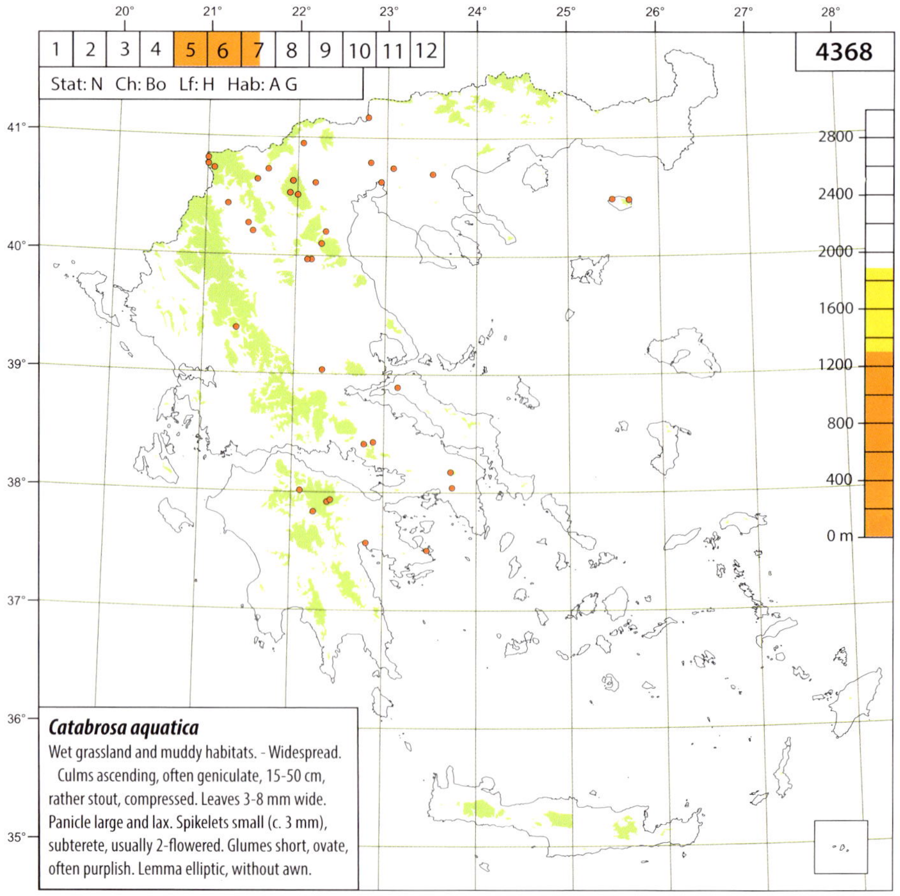

Catabrosa aquatica
Wet grassland and muddy habitats. - Widespread.
Culms ascending, often geniculate, 15-50 cm, rather stout, compressed. Leaves 3-8 mm wide. Panicle large and lax. Spikelets small (c. 3 mm), subterete, usually 2-flowered. Glumes short, ovate, often purplish. Lemma elliptic, without awn.

4369

Stat: N Ch: Me Lf: T Hab: M R

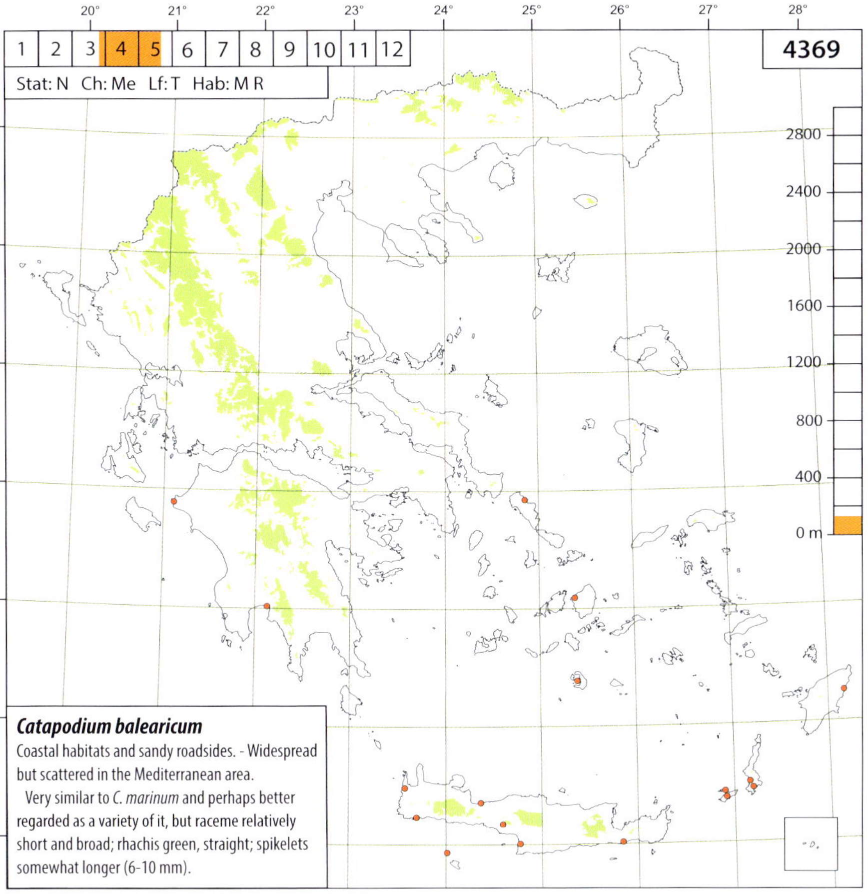

Catapodium balearicum

Coastal habitats and sandy roadsides. - Widespread but scattered in the Mediterranean area.

Very similar to *C. marinum* and perhaps better regarded as a variety of it, but raceme relatively short and broad; rhachis green, straight; spikelets somewhat longer (6-10 mm).

4370

211:8

Stat: N Ch: MA Lf: T Hab: M

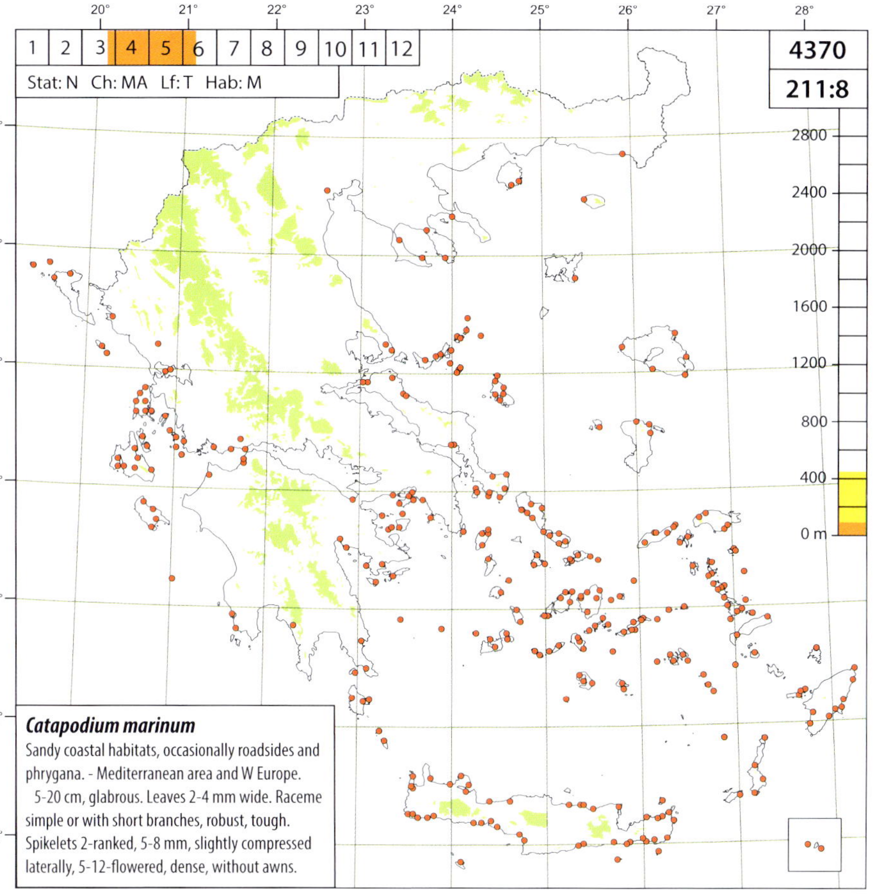

Catapodium marinum

Sandy coastal habitats, occasionally roadsides and phrygana. - Mediterranean area and W Europe.

5-20 cm, glabrous. Leaves 2-4 mm wide. Raceme simple or with short branches, robust, tough. Spikelets 2-ranked, 5-8 mm, slightly compressed laterally, 5-12-flowered, dense, without awns.

4371

Stat: N Ch: EA Lf: T Hab: M P R

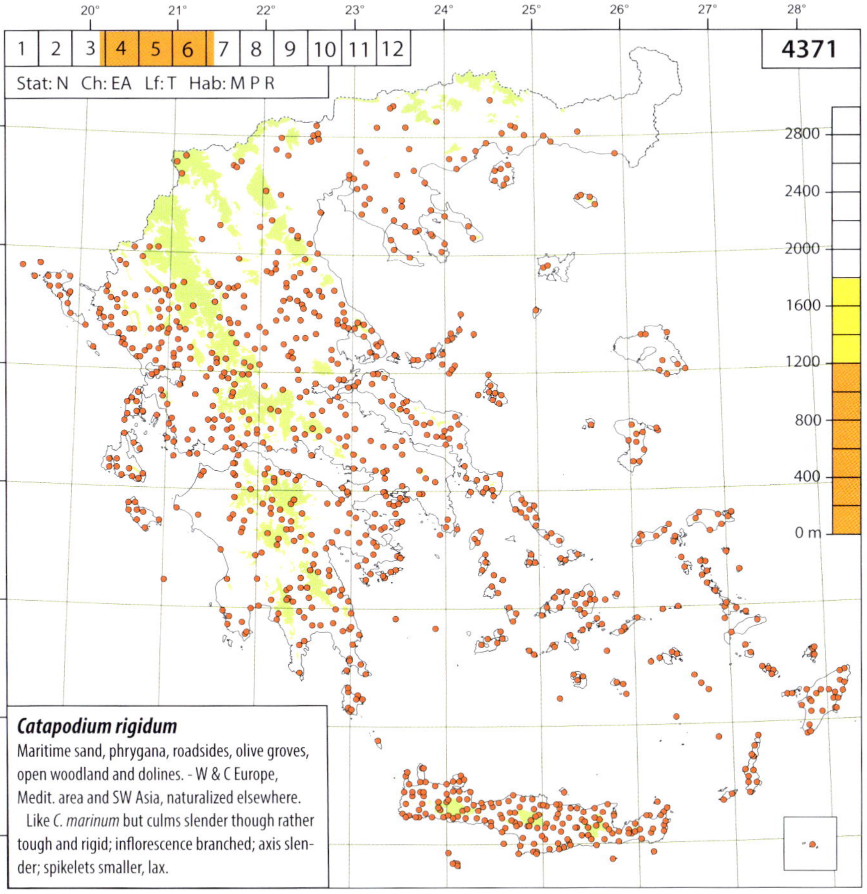

Catapodium rigidum

Maritime sand, phrygana, roadsides, olive groves, open woodland and dolines. - W & C Europe, Medit. area and SW Asia, naturalized elsewhere.

Like *C. marinum* but culms slender though rather tough and rigid; inflorescence branched; axis slender; spikelets smaller, lax.

4372

Stat: X Ch: [trop.] Lf: H Hab: C R

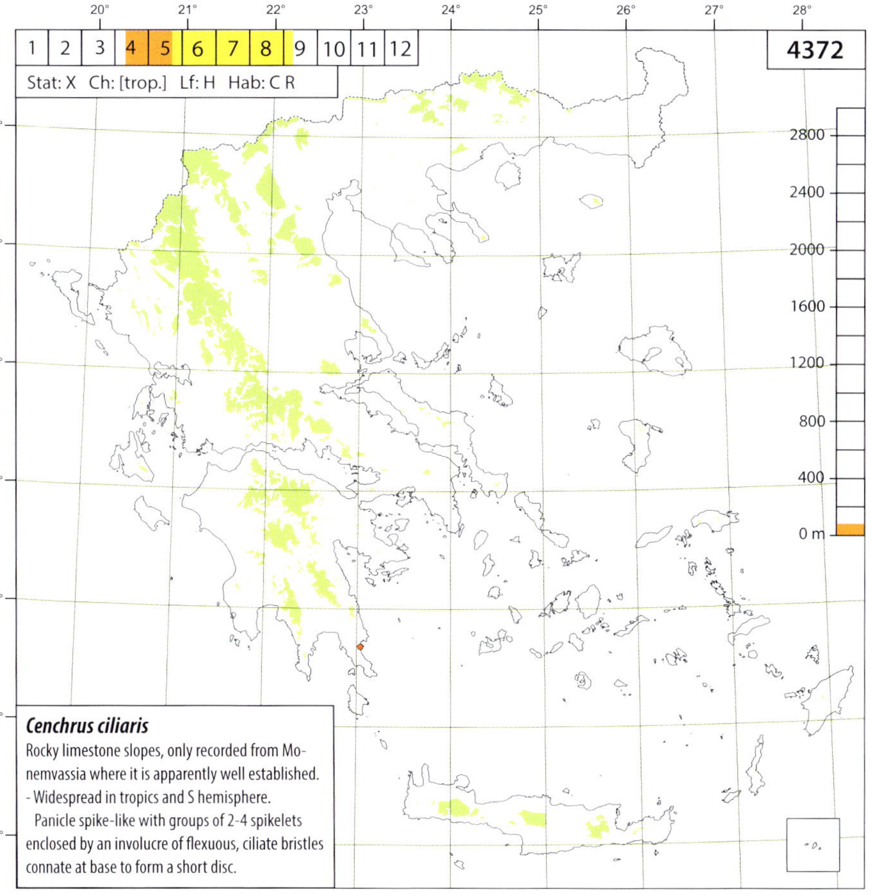

Cenchrus ciliaris

Rocky limestone slopes, only recorded from Monemvassia where it is apparently well established. - Widespread in tropics and S hemisphere.

Panicle spike-like with groups of 2-4 spikelets enclosed by an involucre of flexuous, ciliate bristles connate at base to form a short disc.

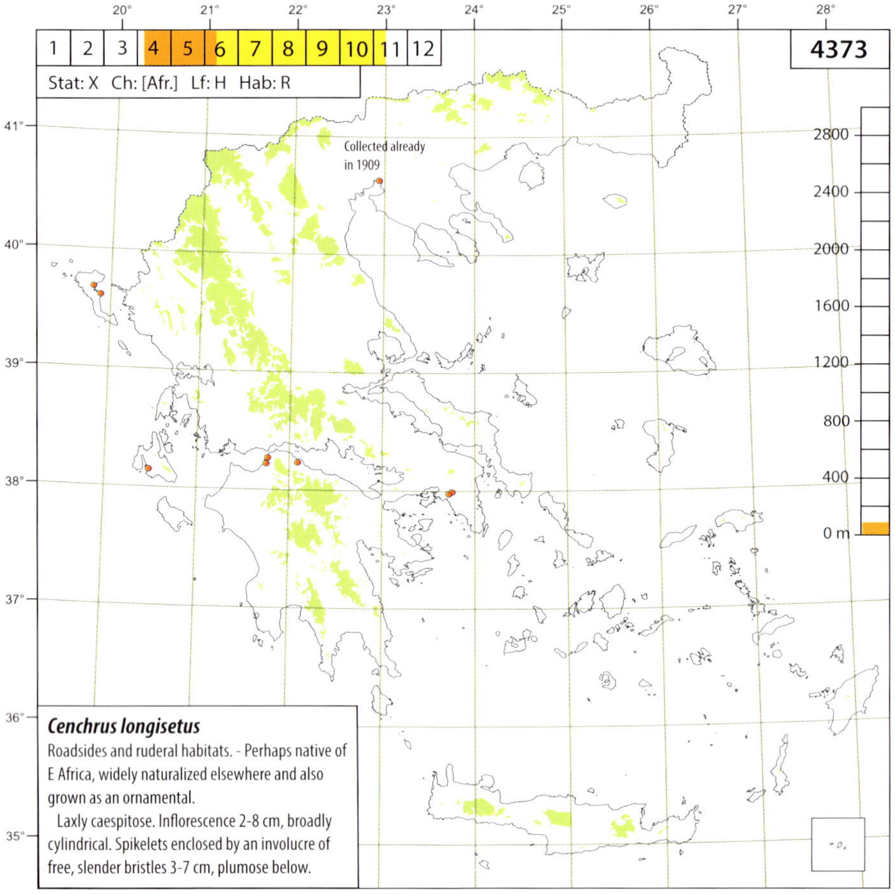

Cenchrus longisetus
Roadsides and ruderal habitats. - Perhaps native of E Africa, widely naturalized elsewhere and also grown as an ornamental.
Laxly caespitose. Inflorescence 2-8 cm, broadly cylindrical. Spikelets enclosed by an involucre of free, slender bristles 3-7 cm, plumose below.

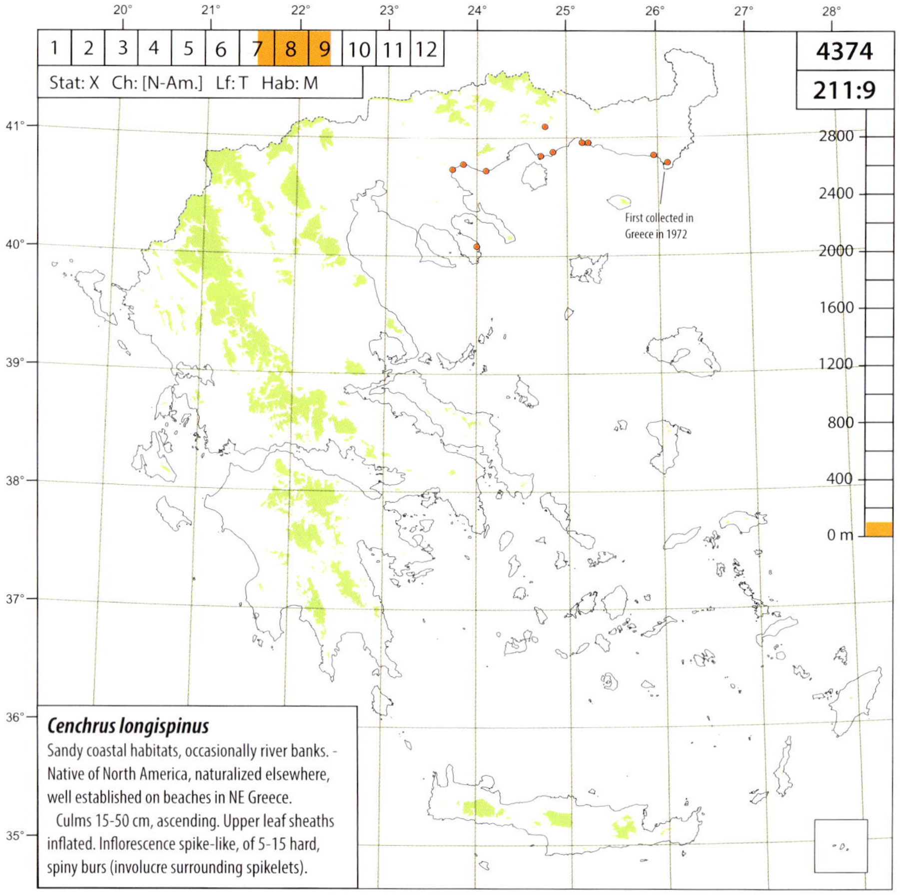

Cenchrus longispinus
Sandy coastal habitats, occasionally river banks. - Native of North America, naturalized elsewhere, well established on beaches in NE Greece.
Culms 15-50 cm, ascending. Upper leaf sheaths inflated. Inflorescence spike-like, of 5-15 hard, spiny burs (involucre surrounding spikelets).

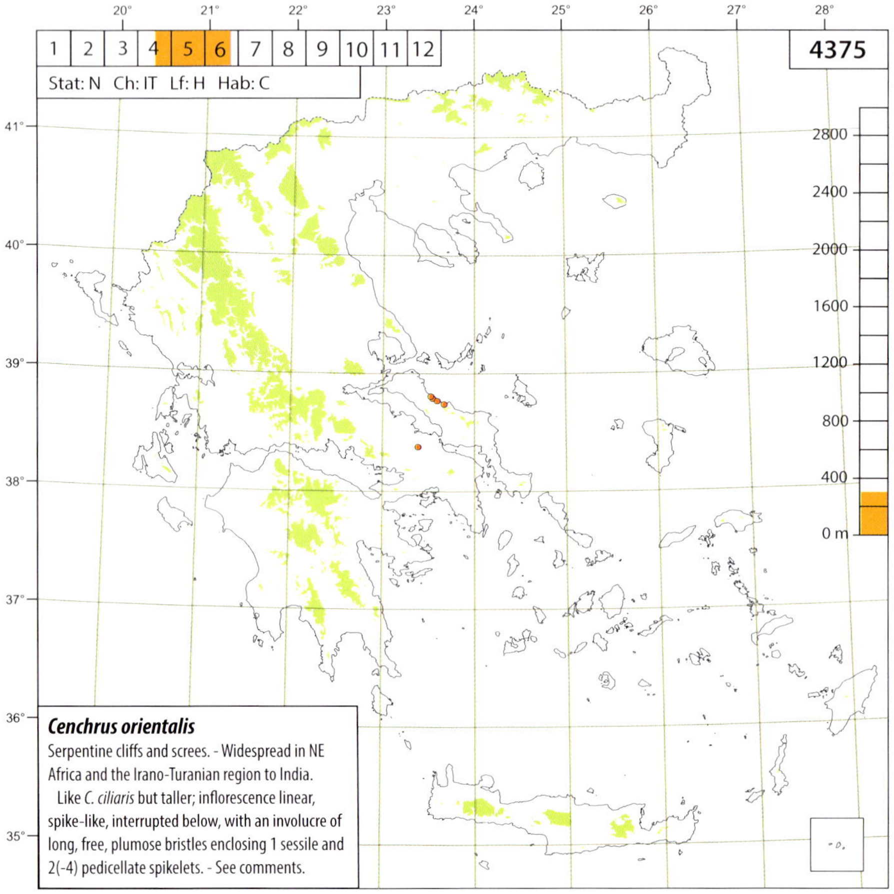

Cenchrus orientalis
Serpentine cliffs and screes. - Widespread in NE Africa and the Irano-Turanian region to India.
Like *C. ciliaris* but taller; inflorescence linear, spike-like, interrupted below, with an involucre of long, free, plumose bristles enclosing 1 sessile and 2(-4) pedicellate spikelets. - See comments.

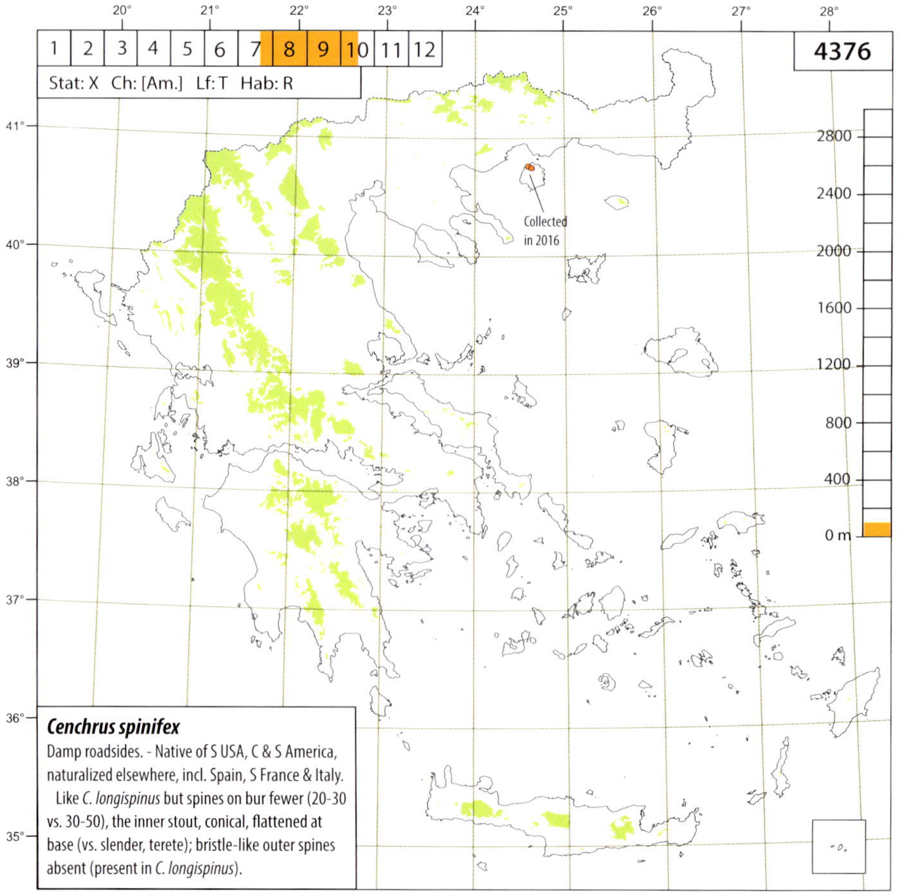

Cenchrus spinifex
Damp roadsides. - Native of S USA, C & S America, naturalized elsewhere, incl. Spain, S France & Italy.
Like *C. longispinus* but spines on bur fewer (20-30 vs. 30-50), the inner stout, conical, flattened at base (vs. slender, terete); bristle-like outer spines absent (present in *C. longispinus*).

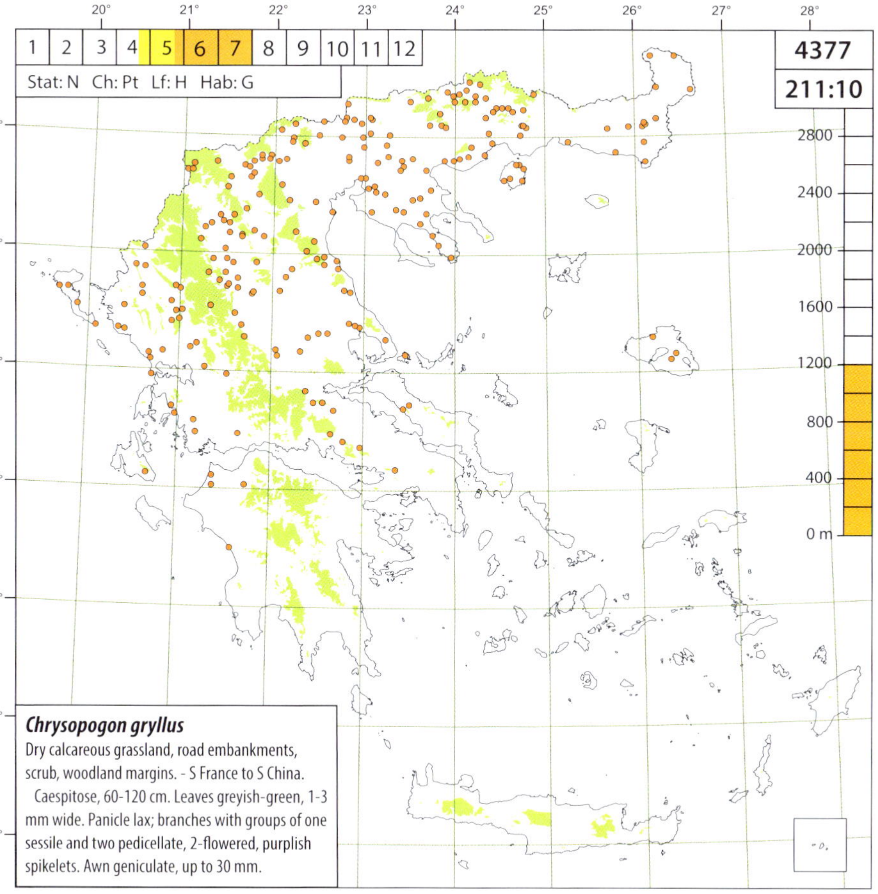

Chrysopogon gryllus

Dry calcareous grassland, road embankments, scrub, woodland margins. - S France to S China.
Caespitose, 60-120 cm. Leaves greyish-green, 1-3 mm wide. Panicle lax; branches with groups of one sessile and two pedicellate, 2-flowered, purplish spikelets. Awn geniculate, up to 30 mm.

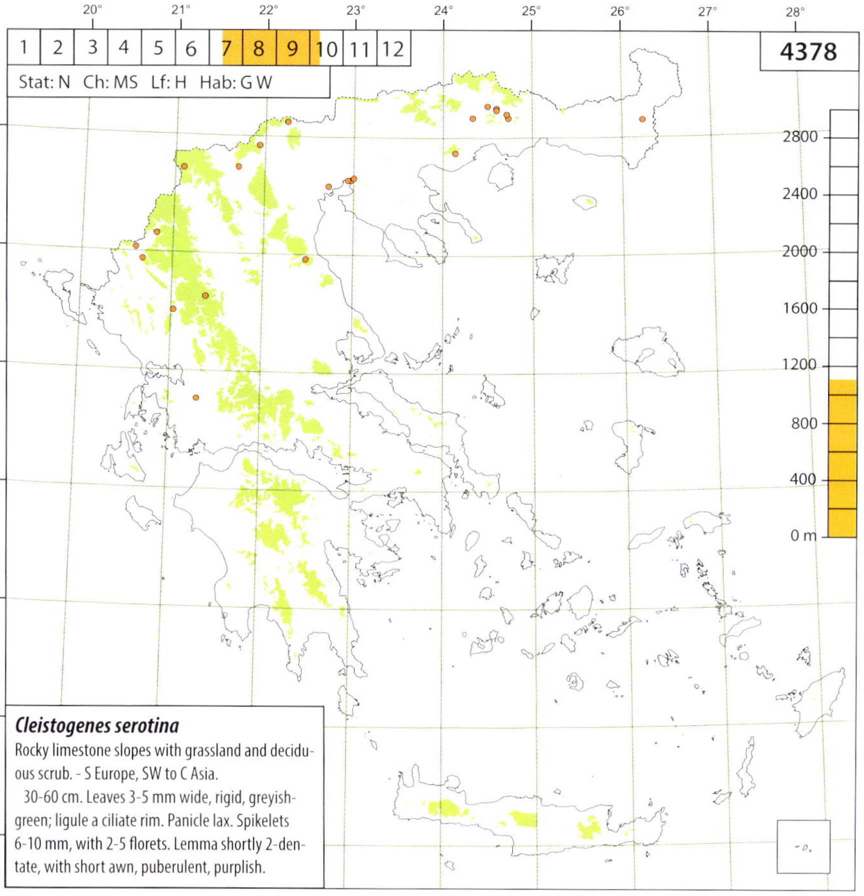

Cleistogenes serotina

Rocky limestone slopes with grassland and deciduous scrub. - S Europe, SW to C Asia.
30-60 cm. Leaves 3-5 mm wide, rigid, greyish-green; ligule a ciliate rim. Panicle lax. Spikelets 6-10 mm, with 2-5 florets. Lemma shortly 2-dentate, with short awn, puberulent, purplish.

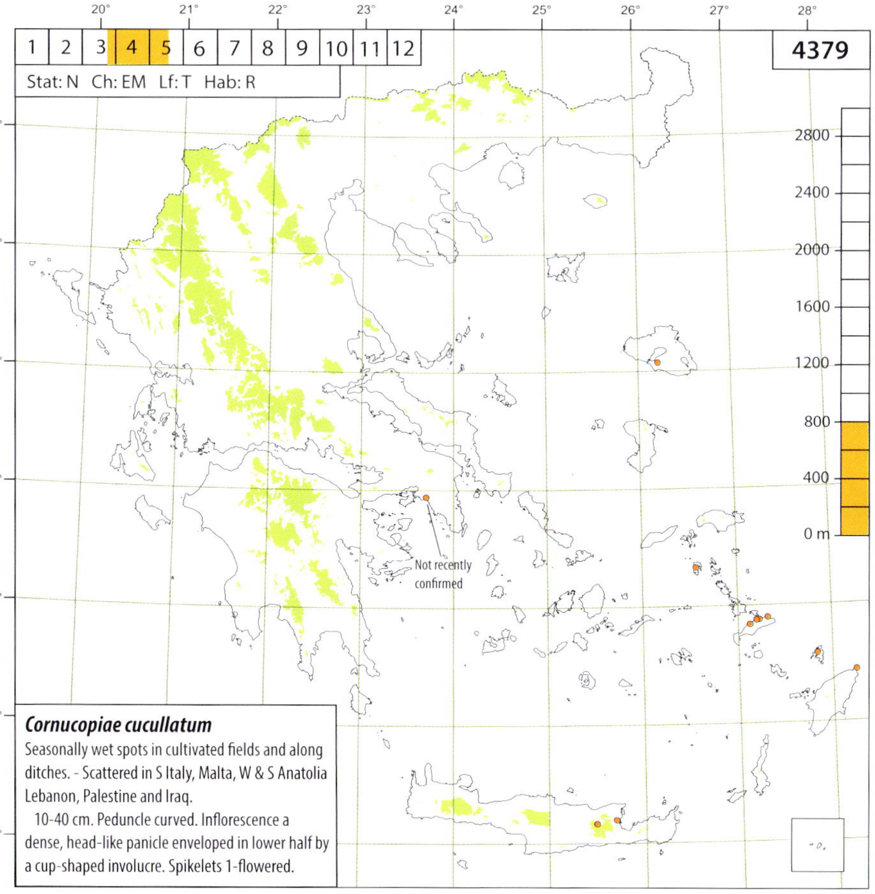

Cornucopiae cucullatum

Seasonally wet spots in cultivated fields and along ditches. - Scattered in S Italy, Malta, W & S Anatolia Lebanon, Palestine and Iraq.
10-40 cm. Peduncle curved. Inflorescence a dense, head-like panicle enveloped in lower half by a cup-shaped involucre. Spikelets 1-flowered.

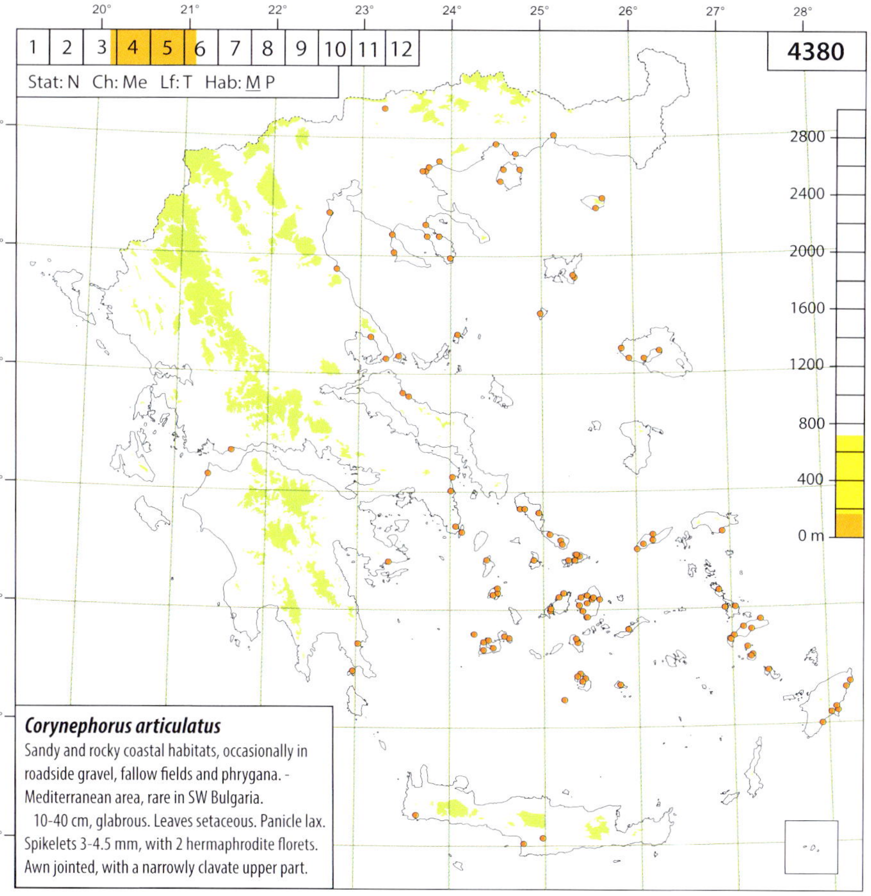

Corynephorus articulatus

Sandy and rocky coastal habitats, occasionally in roadside gravel, fallow fields and phrygana. - Mediterranean area, rare in SW Bulgaria.
10-40 cm, glabrous. Leaves setaceous. Panicle lax. Spikelets 3-4.5 mm, with 2 hermaphrodite florets. Awn jointed, with a narrowly clavate upper part.

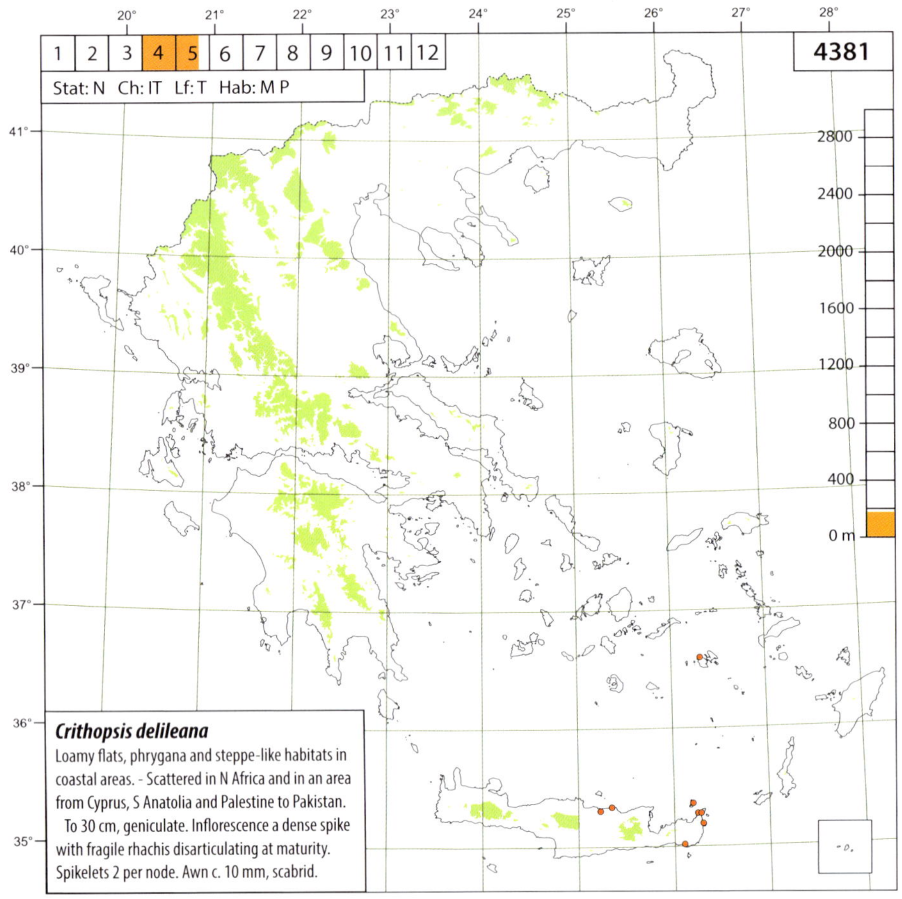

Crithopsis delileana
Loamy flats, phrygana and steppe-like habitats in coastal areas. - Scattered in N Africa and in an area from Cyprus, S Anatolia and Palestine to Pakistan.
To 30 cm, geniculate. Inflorescence a dense spike with fragile rhachis disarticulating at maturity. Spikelets 2 per node. Awn c. 10 mm, scabrid.

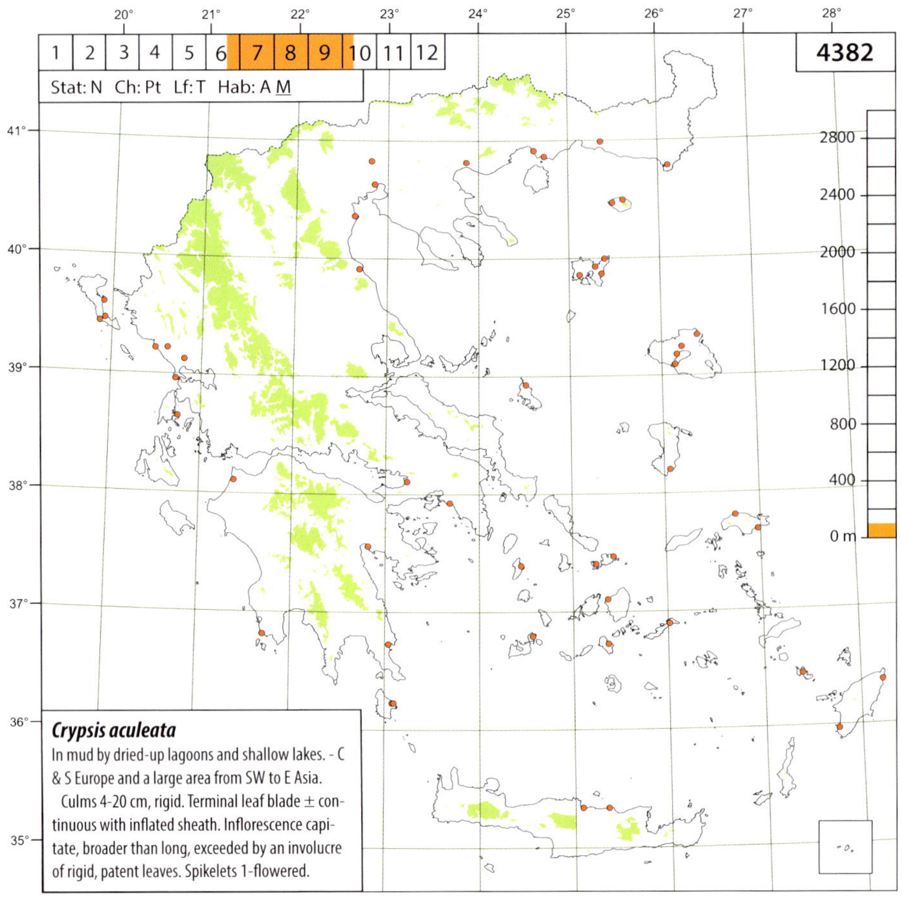

Crypsis aculeata
In mud by dried-up lagoons and shallow lakes. - C & S Europe and a large area from SW to E Asia.
Culms 4-20 cm, rigid. Terminal leaf blade ± continuous with inflated sheath. Inflorescence capitate, broader than long, exceeded by an involucre of rigid, patent leaves. Spikelets 1-flowered.

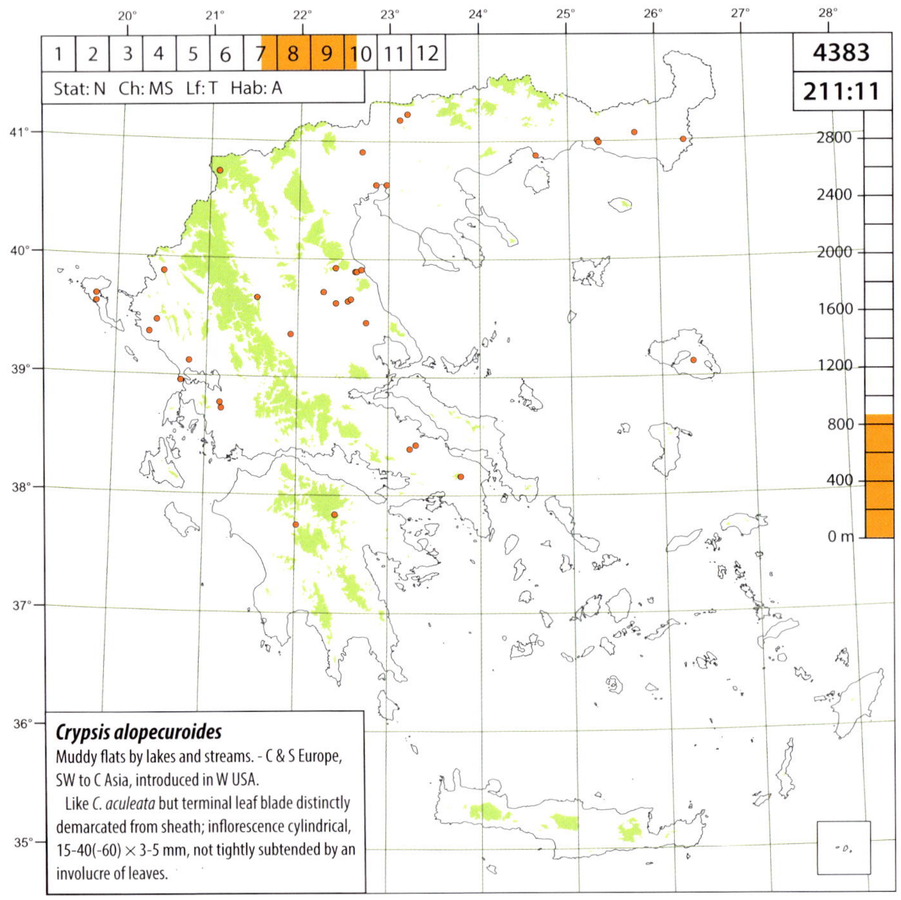

Crypsis alopecuroides
Muddy flats by lakes and streams. - C & S Europe, SW to C Asia, introduced in W USA.
Like *C. aculeata* but terminal leaf blade distinctly demarcated from sheath; inflorescence cylindrical, 15-40(-60) × 3-5 mm, not tightly subtended by an involucre of leaves.

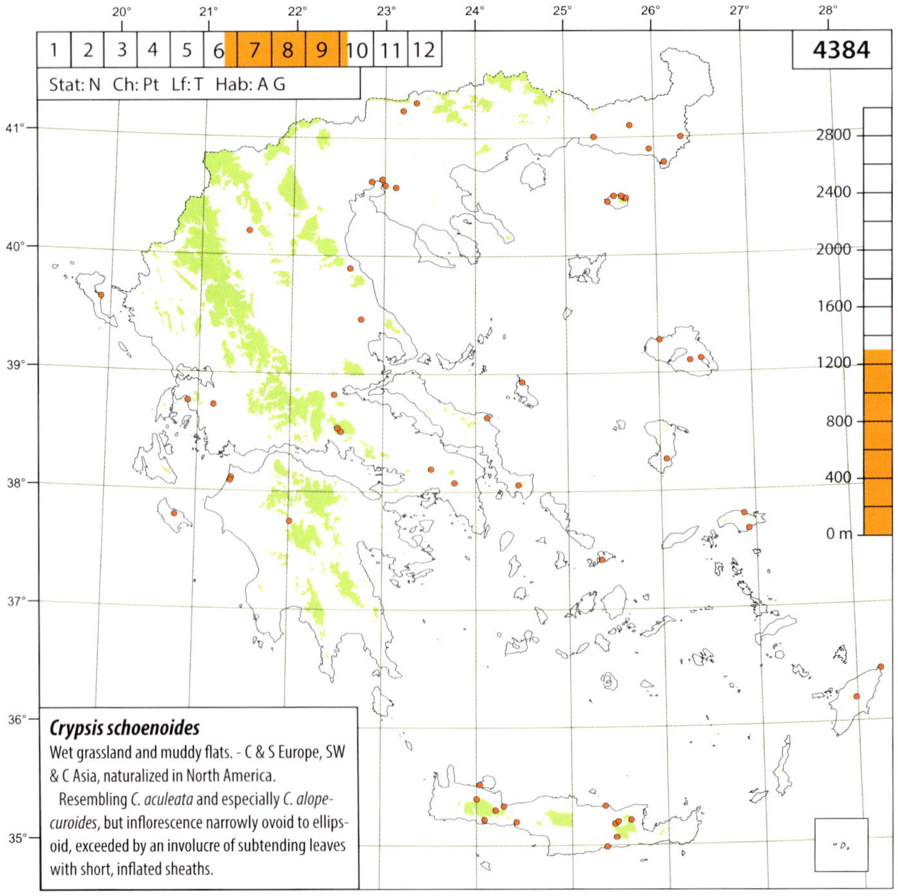

Crypsis schoenoides
Wet grassland and muddy flats. - C & S Europe, SW & C Asia, naturalized in North America.
Resembling *C. aculeata* and especially *C. alopecuroides*, but inflorescence narrowly ovoid to ellipsoid, exceeded by an involucre of subtending leaves with short, inflated sheaths.

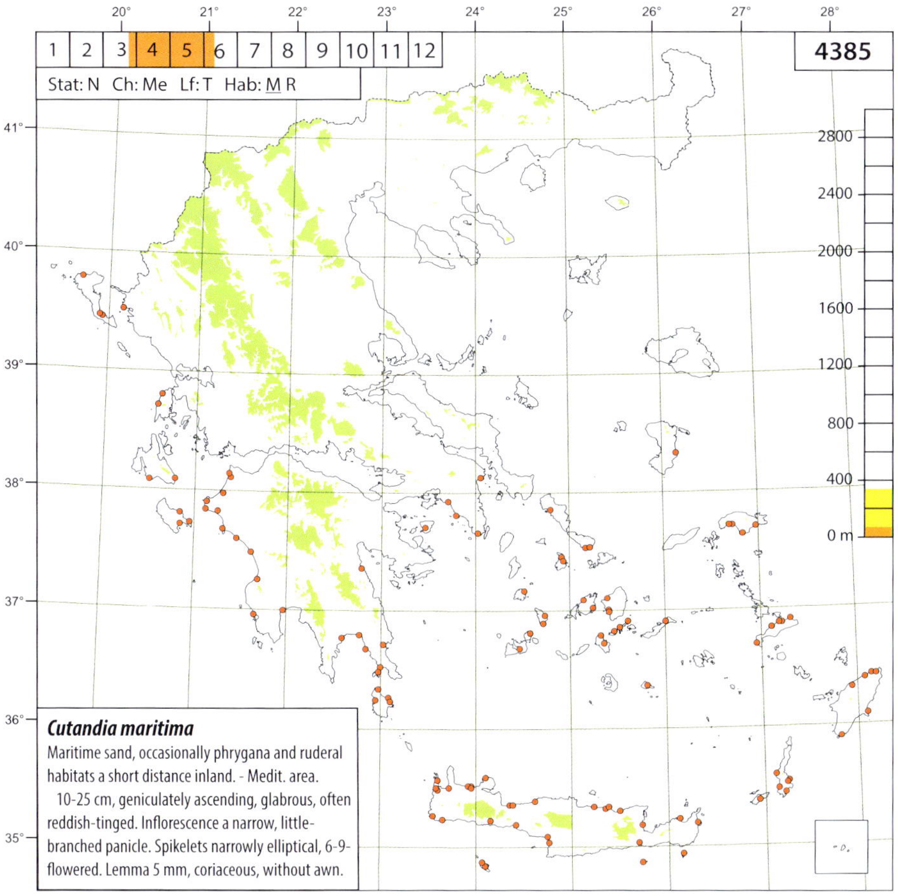

Cutandia maritima

Maritime sand, occasionally phrygana and ruderal habitats a short distance inland. - Medit. area.

10-25 cm, geniculately ascending, glabrous, often reddish-tinged. Inflorescence a narrow, little-branched panicle. Spikelets narrowly elliptical, 6-9-flowered. Lemma 5 mm, coriaceous, without awn.

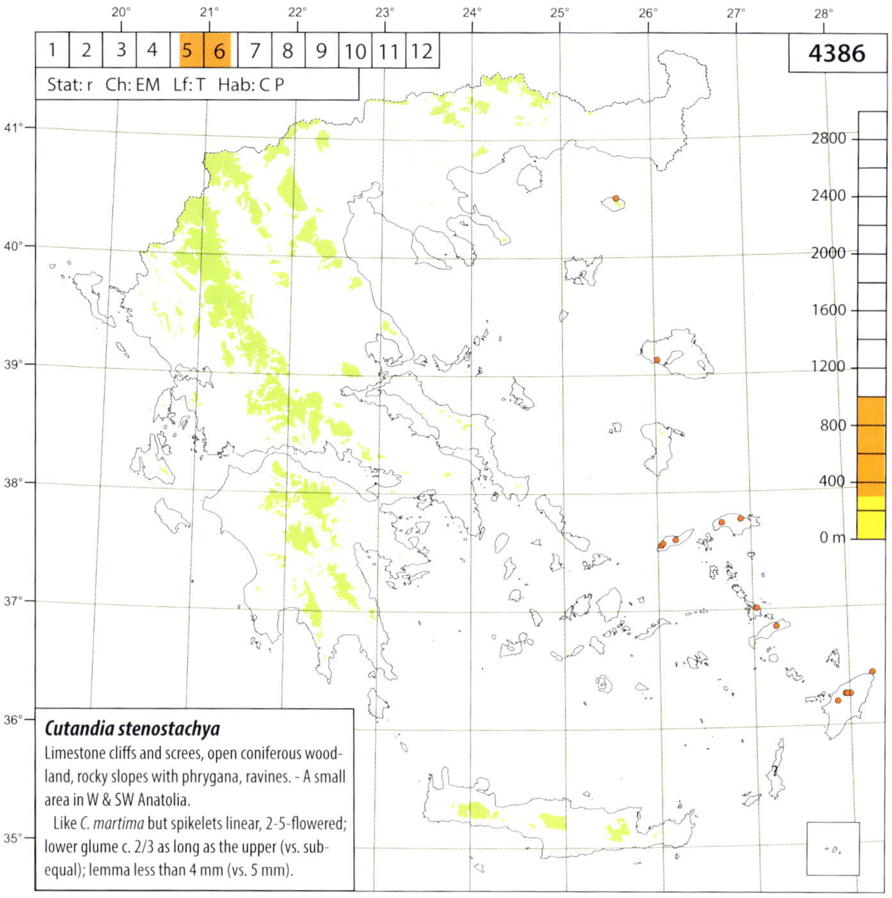

Cutandia stenostachya

Limestone cliffs and screes, open coniferous woodland, rocky slopes with phrygana, ravines. - A small area in W & SW Anatolia.

Like *C. martima* but spikelets linear, 2-5-flowered; lower glume c. 2/3 as long as the upper (vs. sub-equal); lemma less than 4 mm (vs. 5 mm).

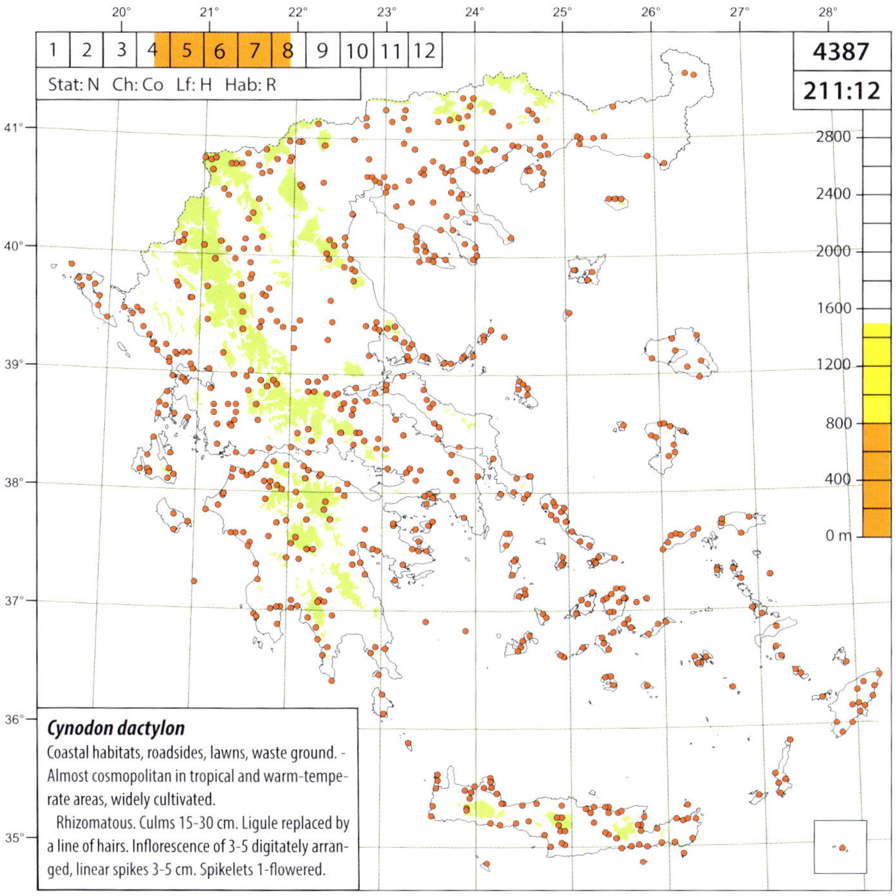

Cynodon dactylon

Coastal habitats, roadsides, lawns, waste ground. - Almost cosmopolitan in tropical and warm-temperate areas, widely cultivated.

Rhizomatous. Culms 15-30 cm. Ligule replaced by a line of hairs. Inflorescence of 3-5 digitately arranged, linear spikes 3-5 cm. Spikelets 1-flowered.

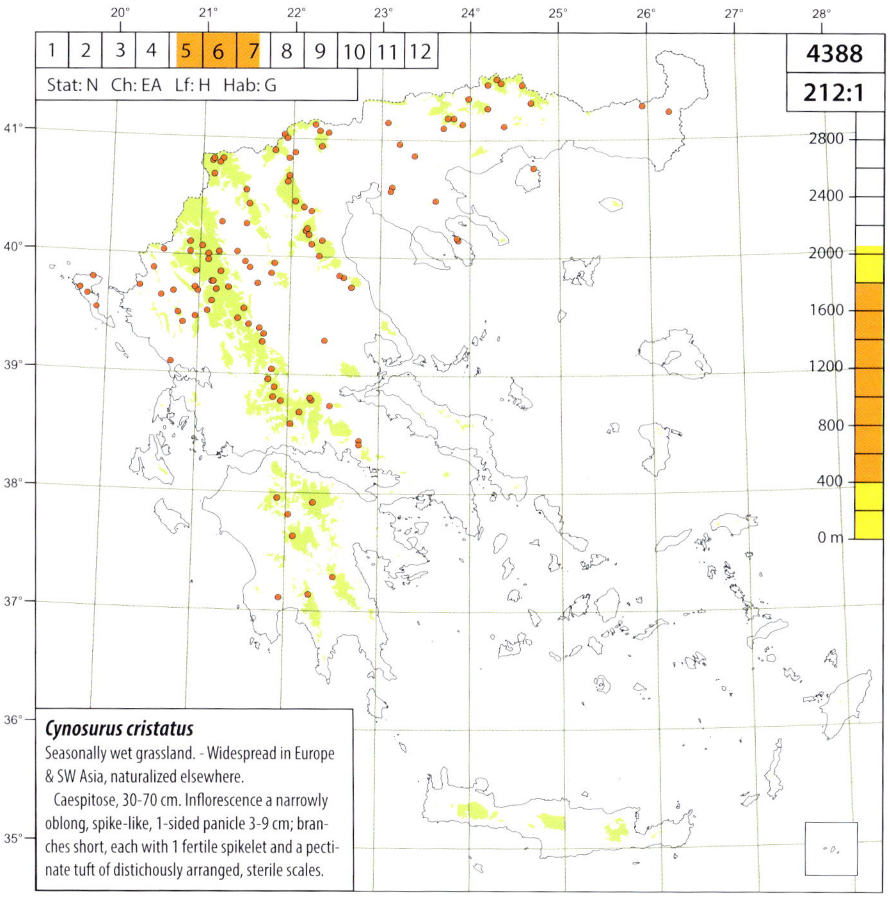

Cynosurus cristatus

Seasonally wet grassland. - Widespread in Europe & SW Asia, naturalized elsewhere.

Caespitose, 30-70 cm. Inflorescence a narrowly oblong, spike-like, 1-sided panicle 3-9 cm; branches short, each with 1 fertile spikelet and a pectinate tuft of distichously arranged, sterile scales.

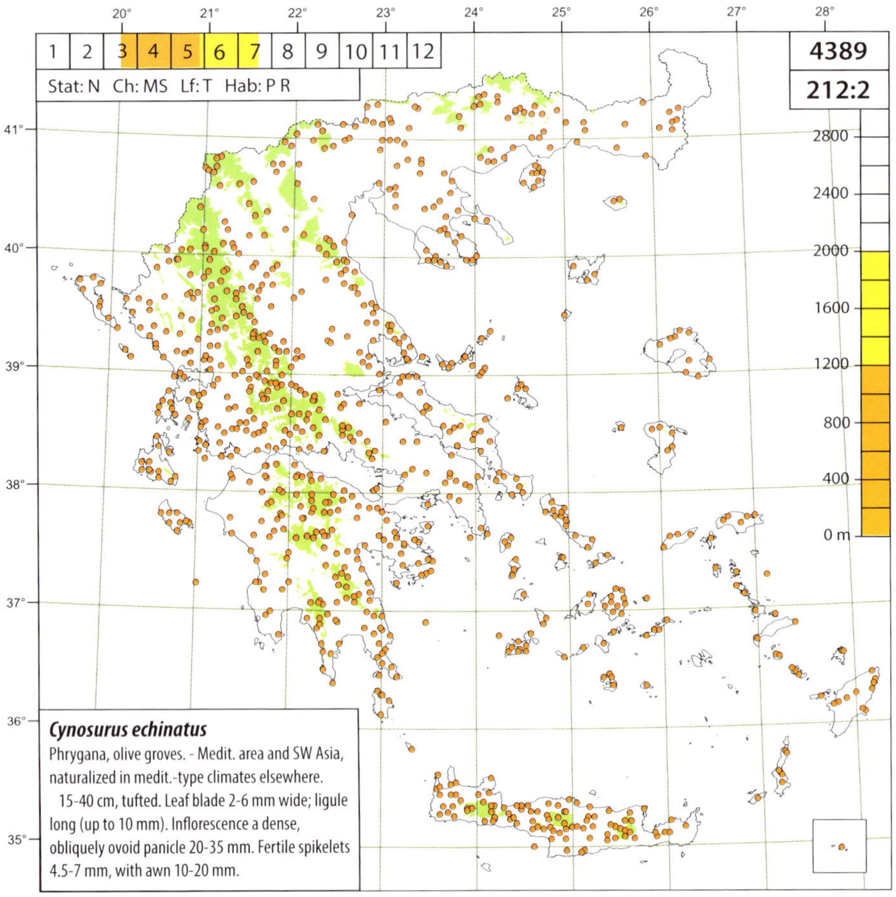

Cynosurus echinatus
Phrygana, olive groves. - Medit. area and SW Asia, naturalized in medit.-type climates elsewhere.
15-40 cm, tufted. Leaf blade 2-6 mm wide; ligule long (up to 10 mm). Inflorescence a dense, obliquely ovoid panicle 20-35 mm. Fertile spikelets 4.5-7 mm, with awn 10-20 mm.

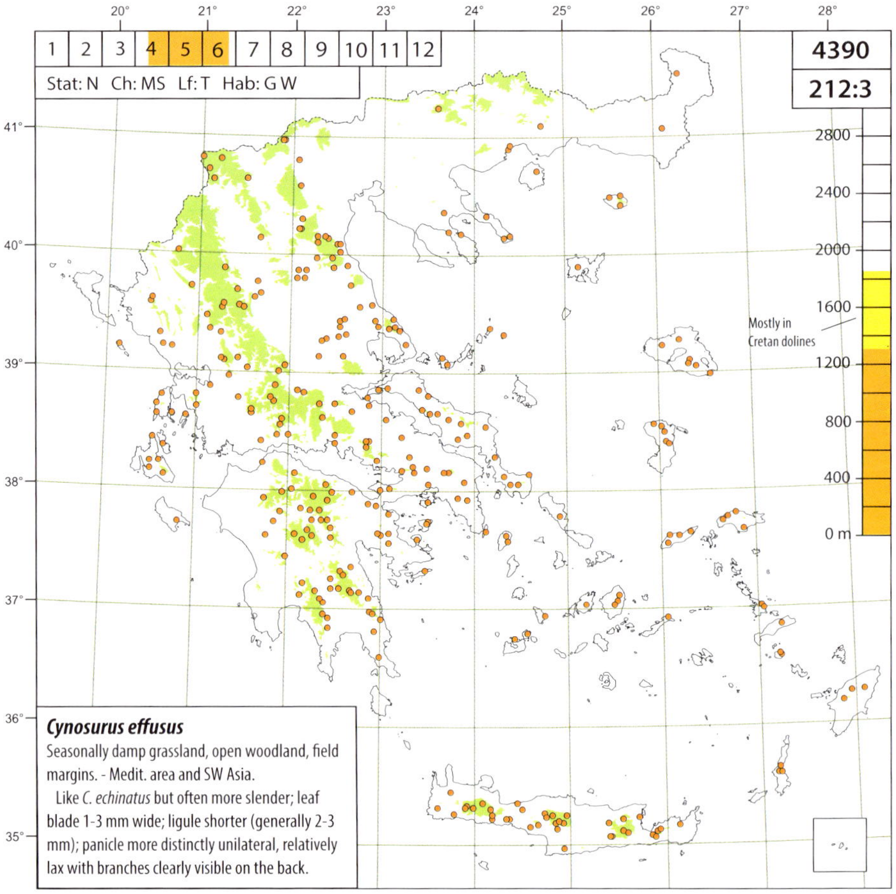

Cynosurus effusus
Seasonally damp grassland, open woodland, field margins. - Medit. area and SW Asia.
Like *C. echinatus* but often more slender; leaf blade 1-3 mm wide; ligule shorter (generally 2-3 mm); panicle more distinctly unilateral, relatively lax with branches clearly visible on the back.

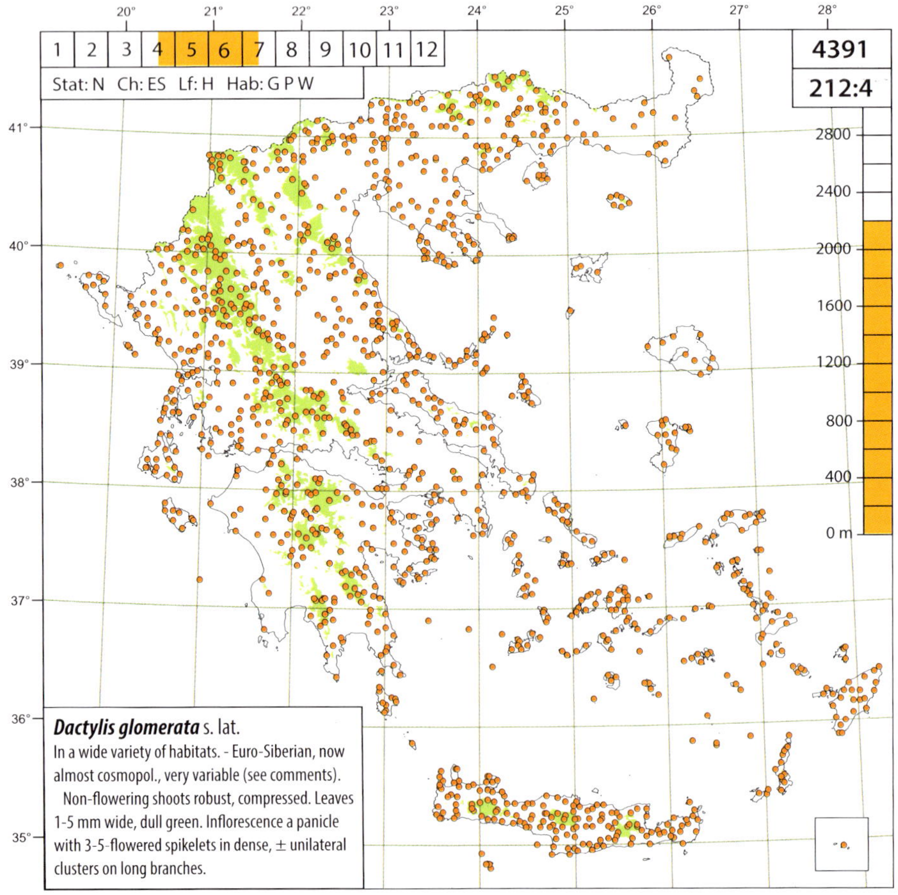

Dactylis glomerata s. lat.
In a wide variety of habitats. - Euro-Siberian, now almost cosmopol., very variable (see comments).
Non-flowering shoots robust, compressed. Leaves 1-5 mm wide, dull green. Inflorescence a panicle with 3-5-flowered spikelets in dense, ± unilateral clusters on long branches.

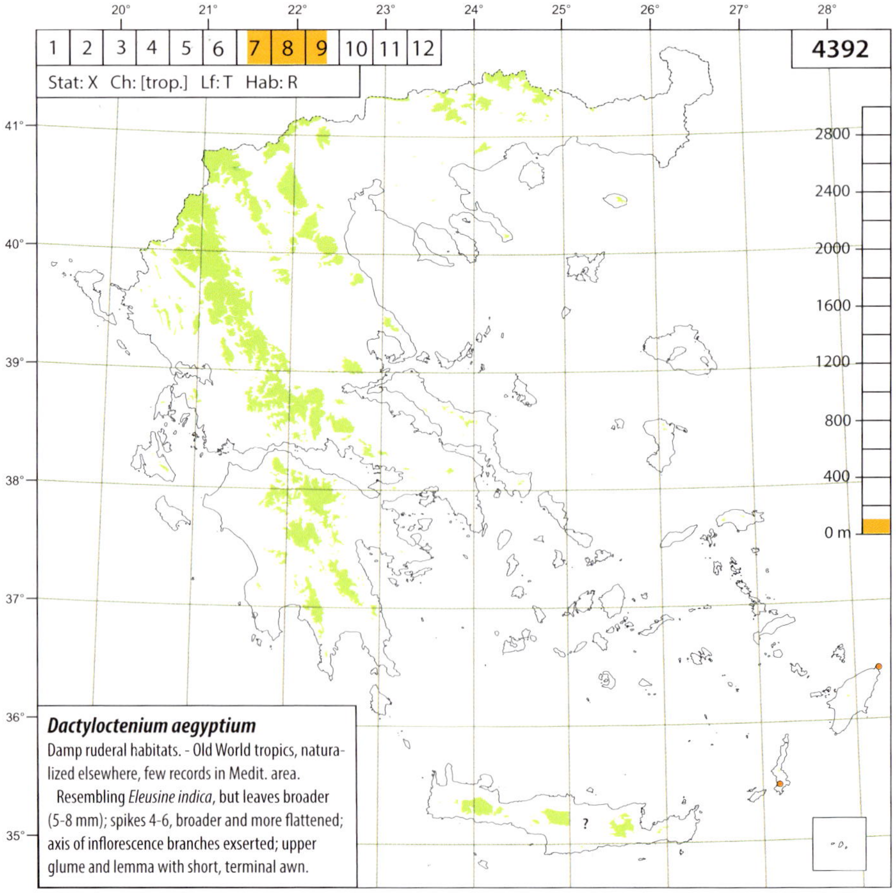

Dactyloctenium aegyptium
Damp ruderal habitats. - Old World tropics, naturalized elsewhere, few records in Medit. area.
Resembling *Eleusine indica*, but leaves broader (5-8 mm); spikes 4-6, broader and more flattened; axis of inflorescence branches exserted; upper glume and lemma with short, terminal awn.

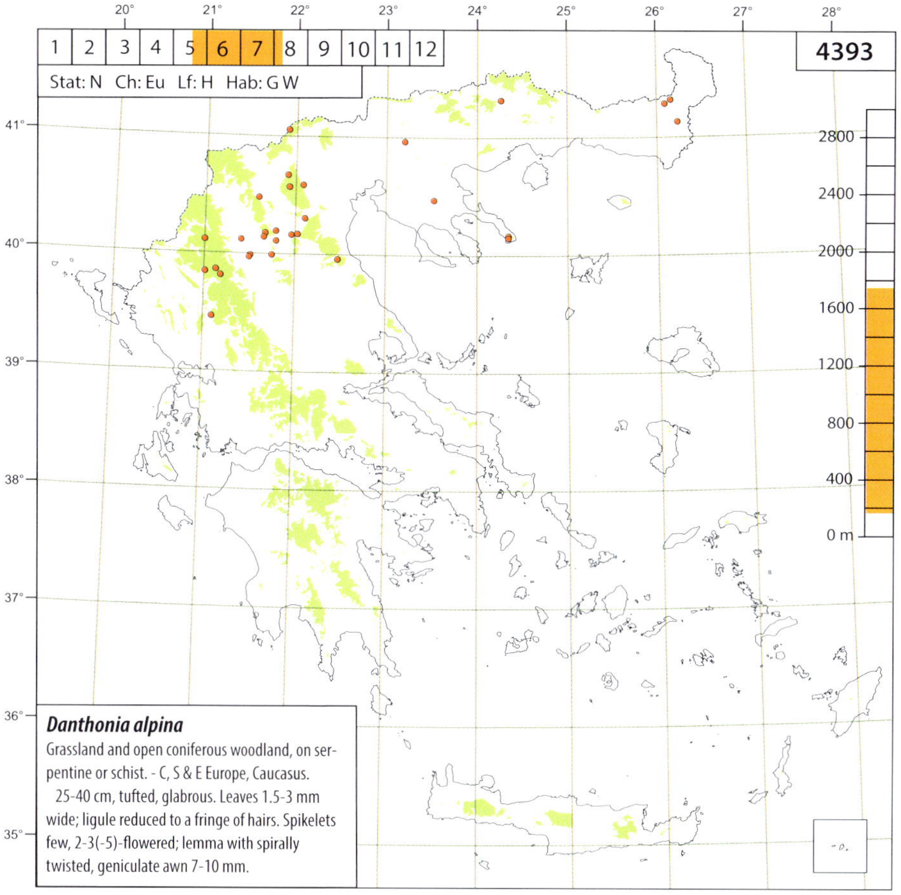

Danthonia alpina
Grassland and open coniferous woodland, on serpentine or schist. - C, S & E Europe, Caucasus.
25-40 cm, tufted, glabrous. Leaves 1.5-3 mm wide; ligule reduced to a fringe of hairs. Spikelets few, 2-3(-5)-flowered; lemma with spirally twisted, geniculate awn 7-10 mm.

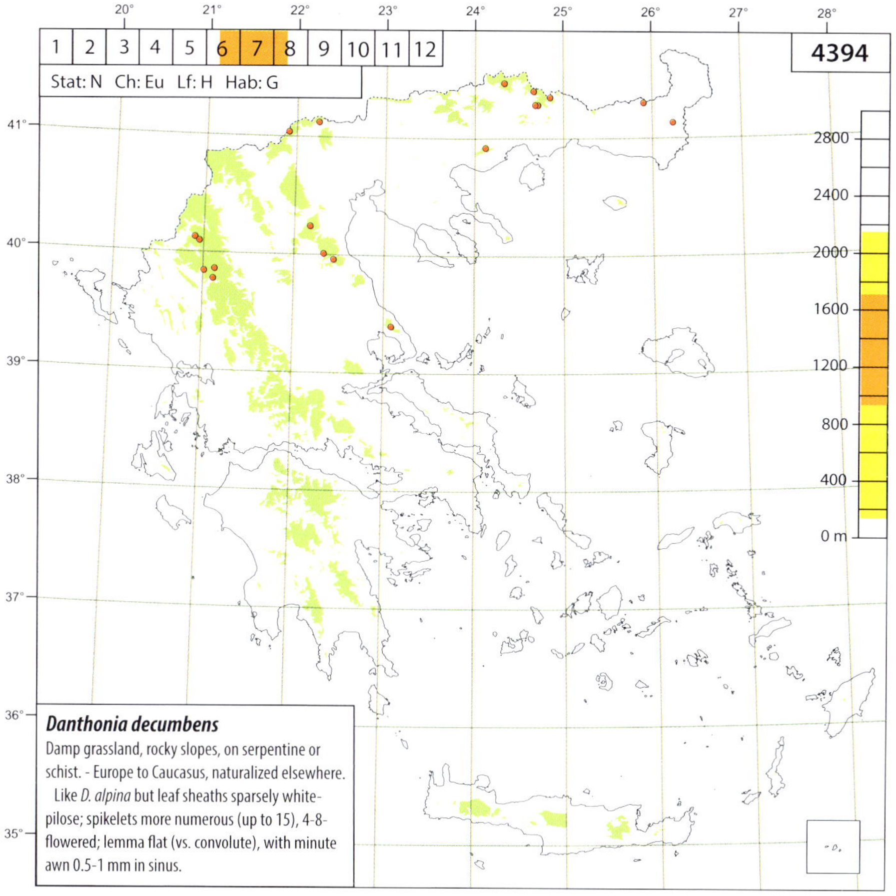

Danthonia decumbens
Damp grassland, rocky slopes, on serpentine or schist. - Europe to Caucasus, naturalized elsewhere.
Like *D. alpina* but leaf sheaths sparsely white-pilose; spikelets more numerous (up to 15), 4-8-flowered; lemma flat (vs. convolute), with minute awn 0.5-1 mm in sinus.

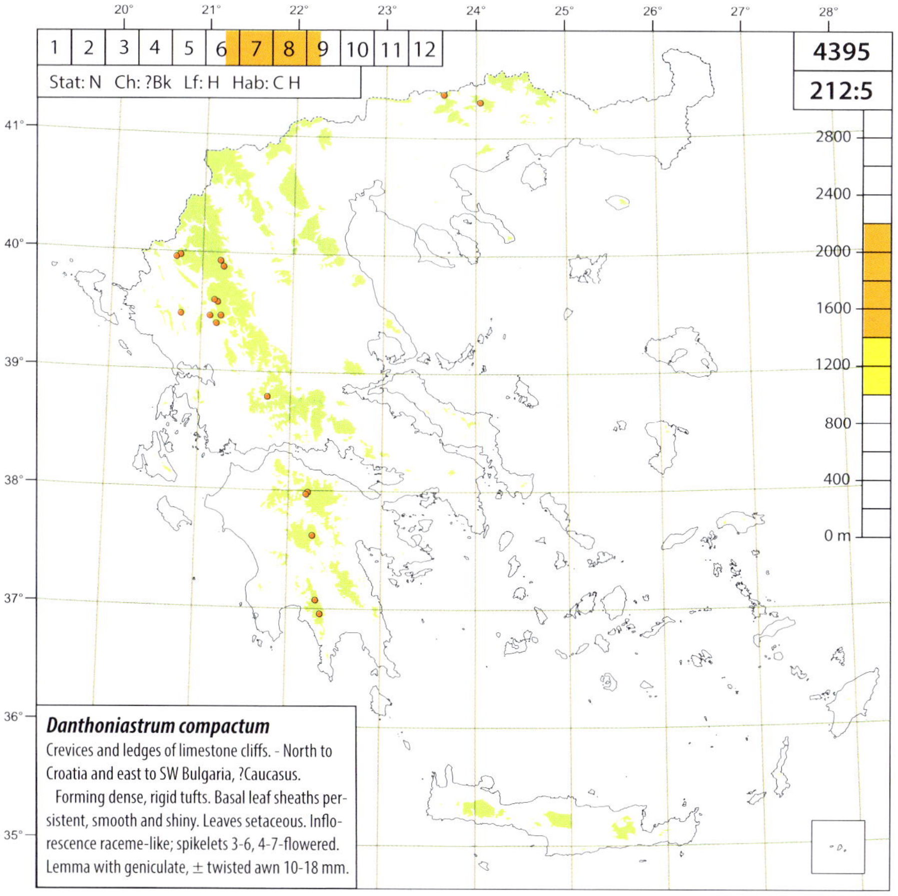

Danthoniastrum compactum
Crevices and ledges of limestone cliffs. - North to Croatia and east to SW Bulgaria, ?Caucasus.
Forming dense, rigid tufts. Basal leaf sheaths persistent, smooth and shiny. Leaves setaceous. Inflorescence raceme-like; spikelets 3-6, 4-7-flowered. Lemma with geniculate, ± twisted awn 10-18 mm.

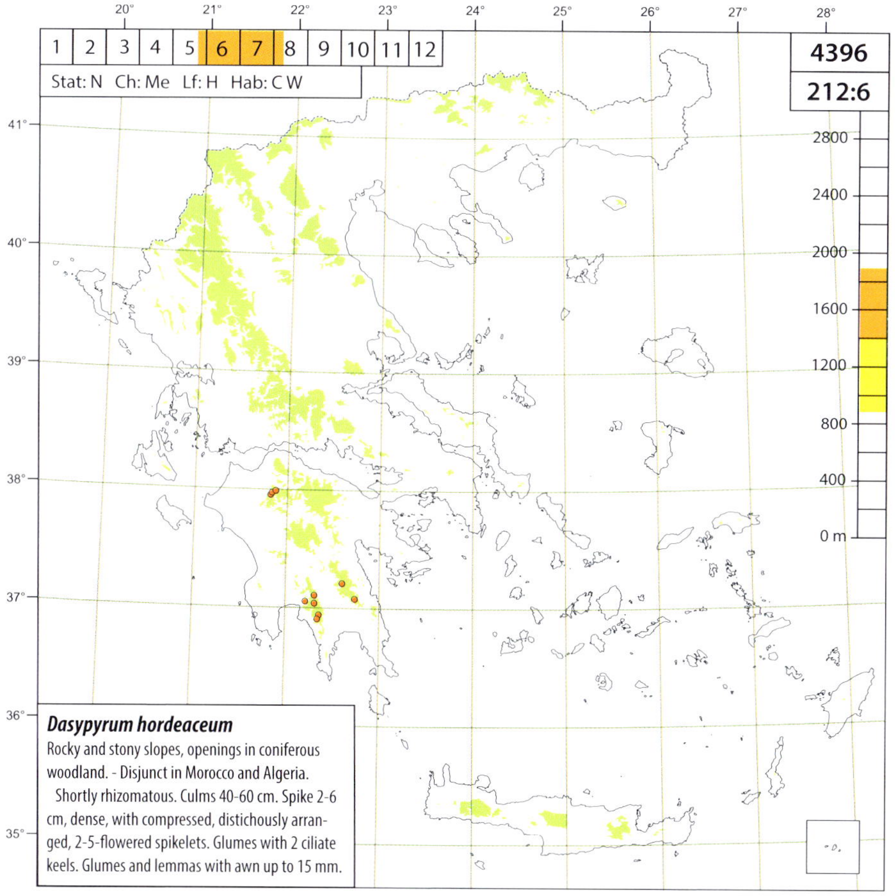

Dasypyrum hordeaceum
Rocky and stony slopes, openings in coniferous woodland. - Disjunct in Morocco and Algeria.
Shortly rhizomatous. Culms 40-60 cm. Spike 2-6 cm, dense, with compressed, distichously arranged, 2-5-flowered spikelets. Glumes with 2 ciliate keels. Glumes and lemmas with awn up to 15 mm.

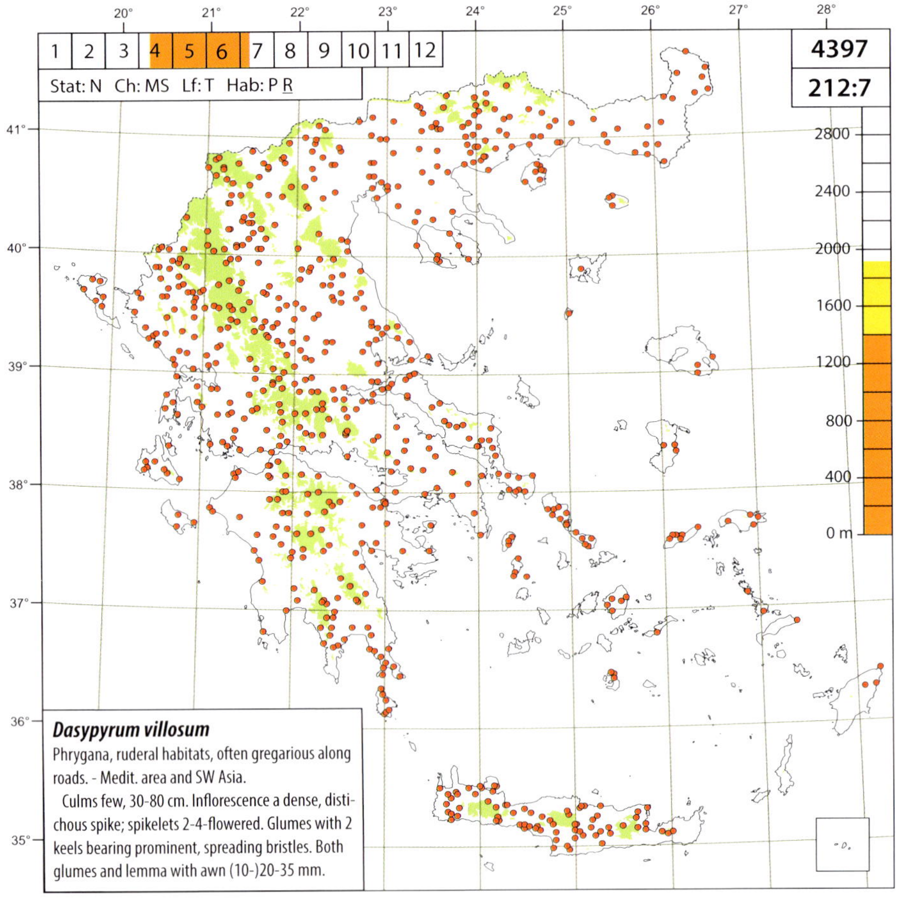

Dasypyrum villosum

Phrygana, ruderal habitats, often gregarious along roads. - Medit. area and SW Asia.

Culms few, 30-80 cm. Inflorescence a dense, distichous spike; spikelets 2-4-flowered. Glumes with 2 keels bearing prominent, spreading bristles. Both glumes and lemma with awn (10-)20-35 mm.

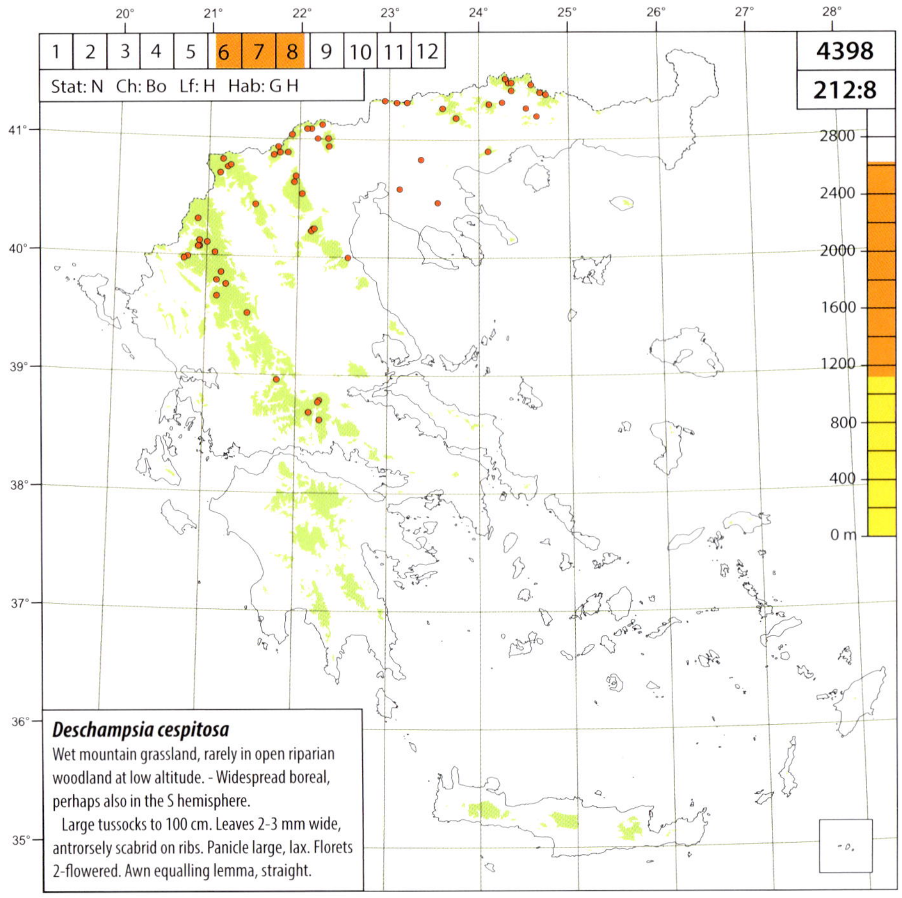

Deschampsia cespitosa

Wet mountain grassland, rarely in open riparian woodland at low altitude. - Widespread boreal, perhaps also in the S hemisphere.

Large tussocks to 100 cm. Leaves 2-3 mm wide, antrorsely scabrid on ribs. Panicle large, lax. Florets 2-flowered. Awn equalling lemma, straight.

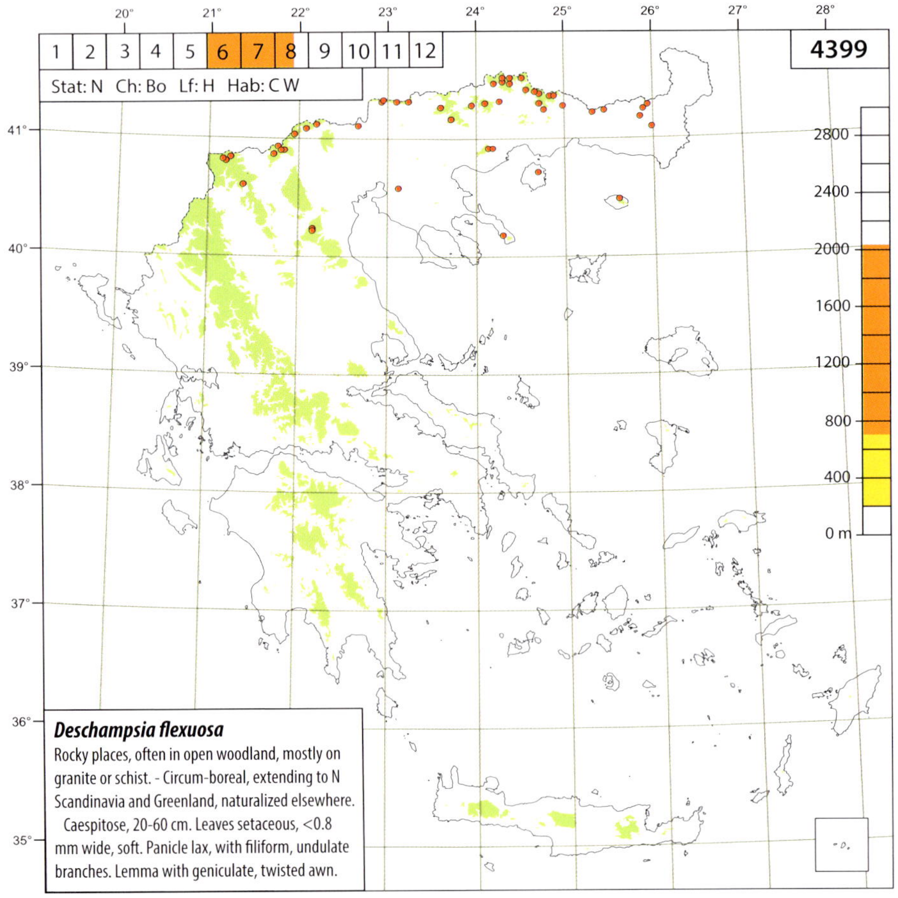

Deschampsia flexuosa

Rocky places, often in open woodland, mostly on granite or schist. - Circum-boreal, extending to N Scandinavia and Greenland, naturalized elsewhere.

Caespitose, 20-60 cm. Leaves setaceous, <0.8 mm wide, soft. Panicle lax, with filiform, undulate branches. Lemma with geniculate, twisted awn.

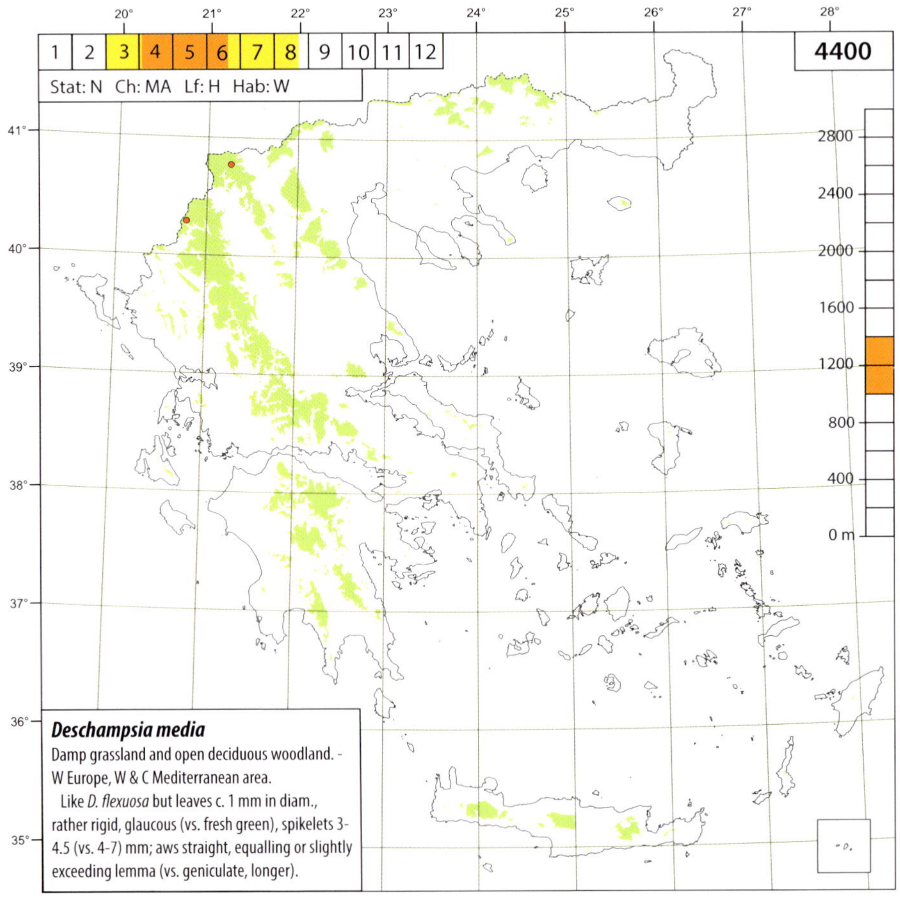

Deschampsia media

Damp grassland and open deciduous woodland. - W Europe, W & C Mediterranean area.

Like *D. flexuosa* but leaves c. 1 mm in diam., rather rigid, glaucous (vs. fresh green), spikelets 3-4.5 (vs. 4-7) mm; aws straight, equalling or slightly exceeding lemma (vs. geniculate, longer).

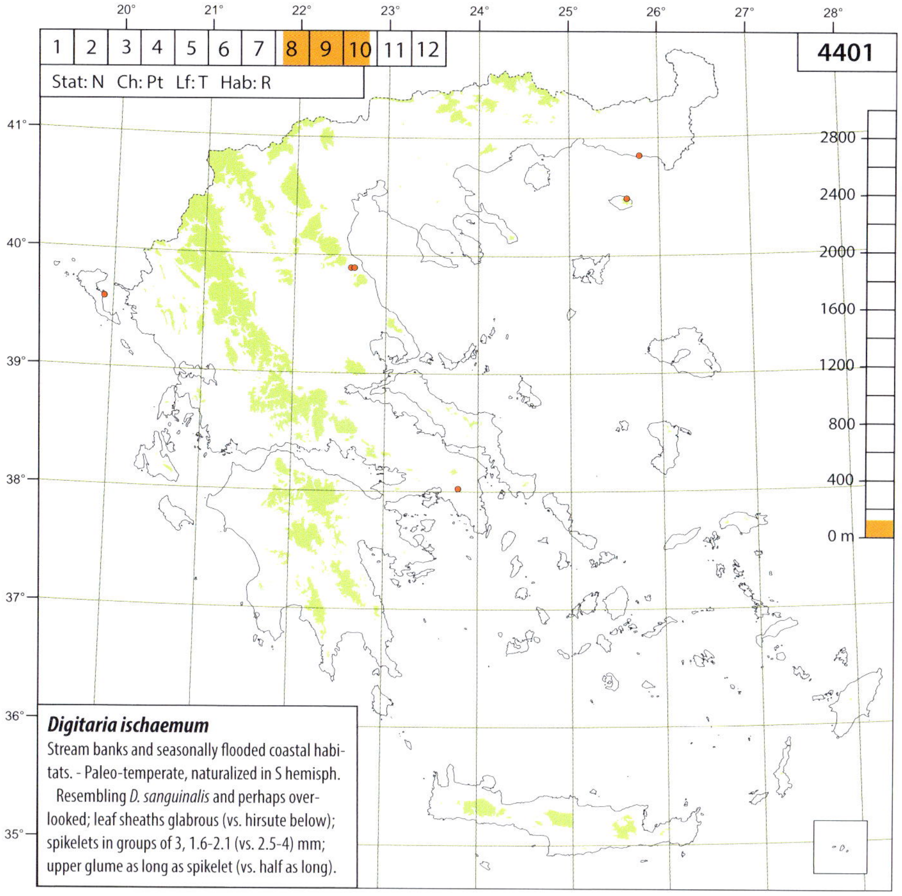

Digitaria ischaemum
Stream banks and seasonally flooded coastal habitats. - Paleo-temperate, naturalized in S hemisph.
Resembling *D. sanguinalis* and perhaps overlooked; leaf sheaths glabrous (vs. hirsute below); spikelets in groups of 3, 1.6-2.1 (vs. 2.5-4) mm; upper glume as long as spikelet (vs. half as long).

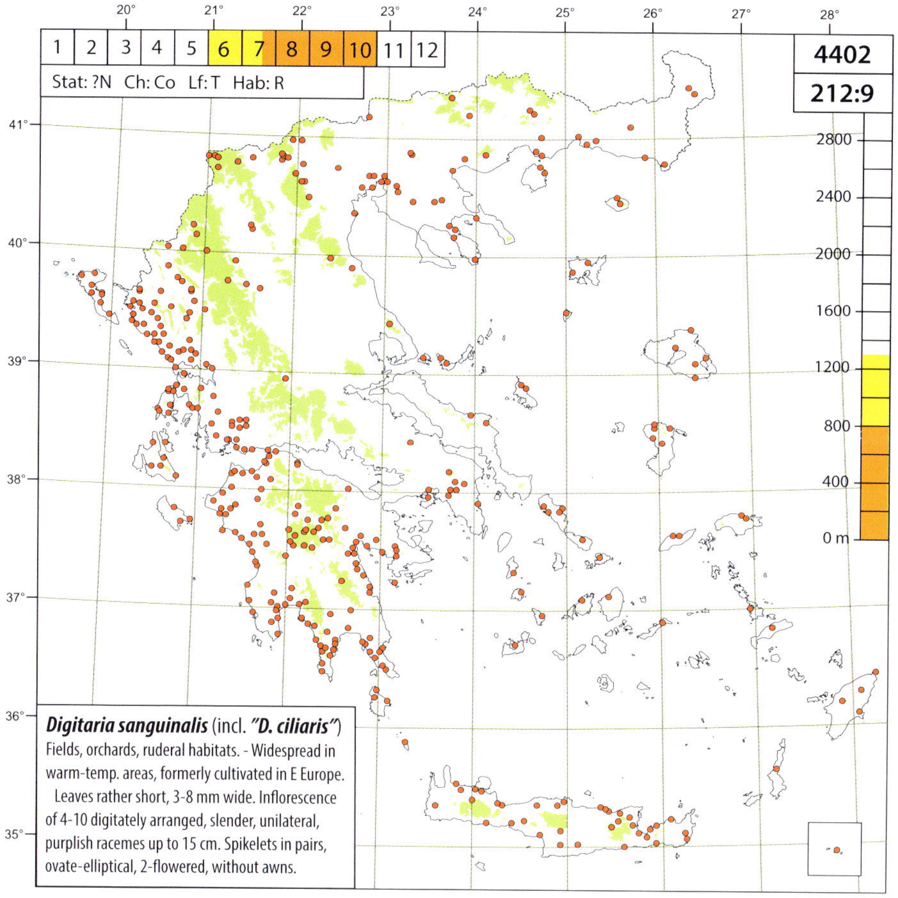

Digitaria sanguinalis (incl. ***"D. ciliaris"***)
Fields, orchards, ruderal habitats. - Widespread in warm-temp. areas, formerly cultivated in E Europe.
Leaves rather short, 3-8 mm wide. Inflorescence of 4-10 digitately arranged, slender, unilateral, purplish racemes up to 15 cm. Spikelets in pairs, ovate-elliptical, 2-flowered, without awns.

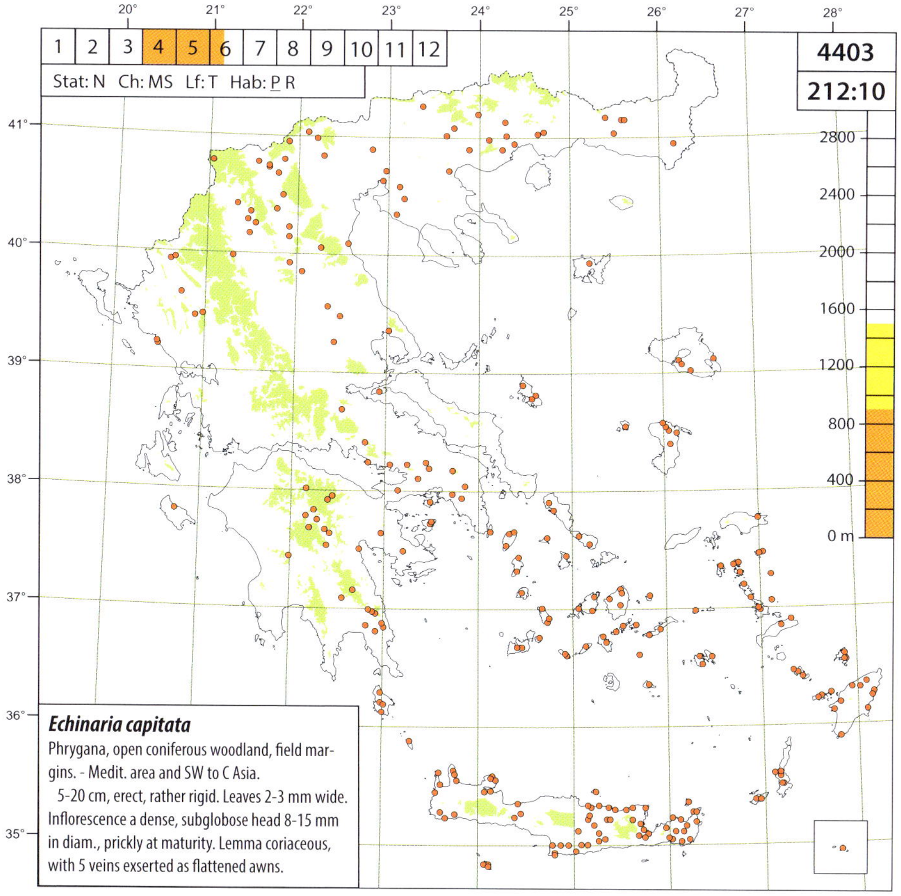

Echinaria capitata
Phrygana, open coniferous woodland, field margins. - Medit. area and SW to C Asia.
5-20 cm, erect, rather rigid. Leaves 2-3 mm wide. Inflorescence a dense, subglobose head 8-15 mm in diam., prickly at maturity. Lemma coriaceous, with 5 veins exserted as flattened awns.

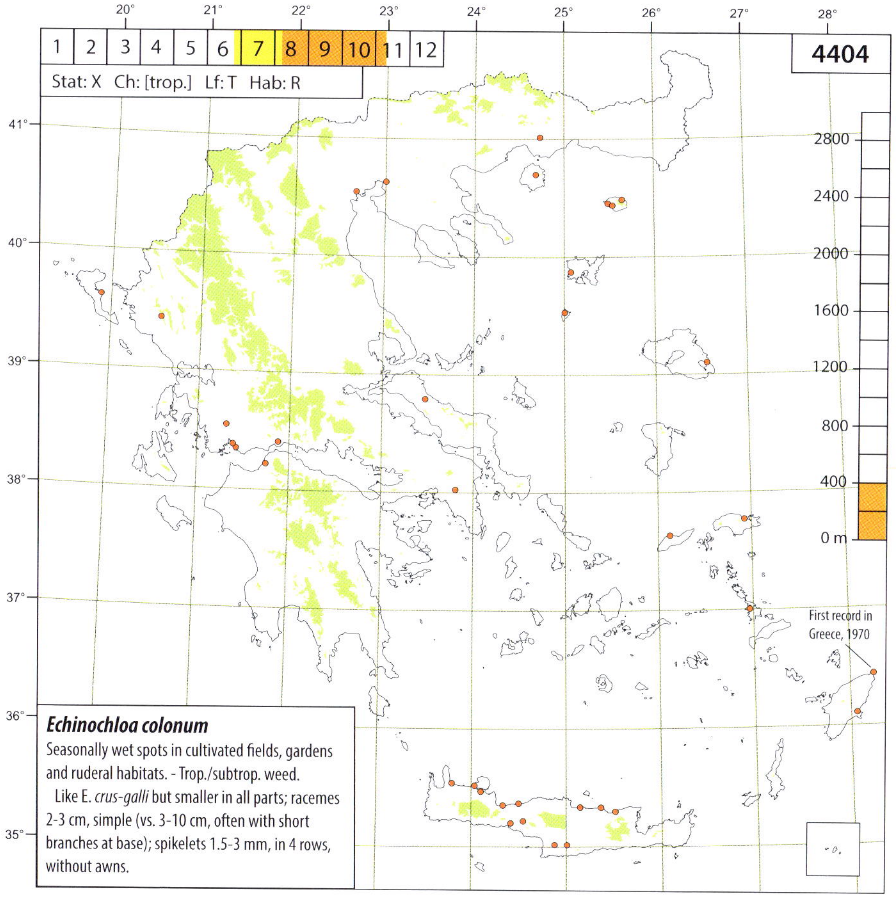

Echinochloa colonum
Seasonally wet spots in cultivated fields, gardens and ruderal habitats. - Trop./subtrop. weed.
Like *E. crus-galli* but smaller in all parts; racemes 2-3 cm, simple (vs. 3-10 cm, often with short branches at base); spikelets 1.5-3 mm, in 4 rows, without awns.

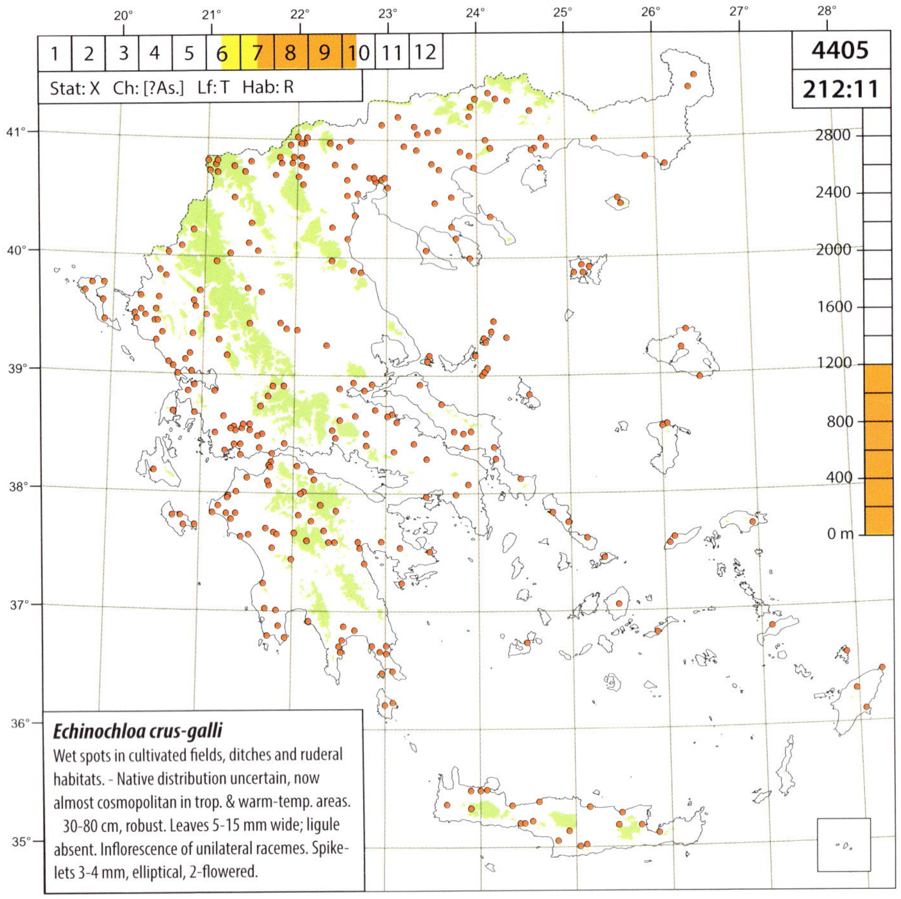

Echinochloa crus-galli
Wet spots in cultivated fields, ditches and ruderal habitats. - Native distribution uncertain, now almost cosmopolitan in trop. & warm-temp. areas.
30-80 cm, robust. Leaves 5-15 mm wide; ligule absent. Inflorescence of unilateral racemes. Spikelets 3-4 mm, elliptical, 2-flowered.

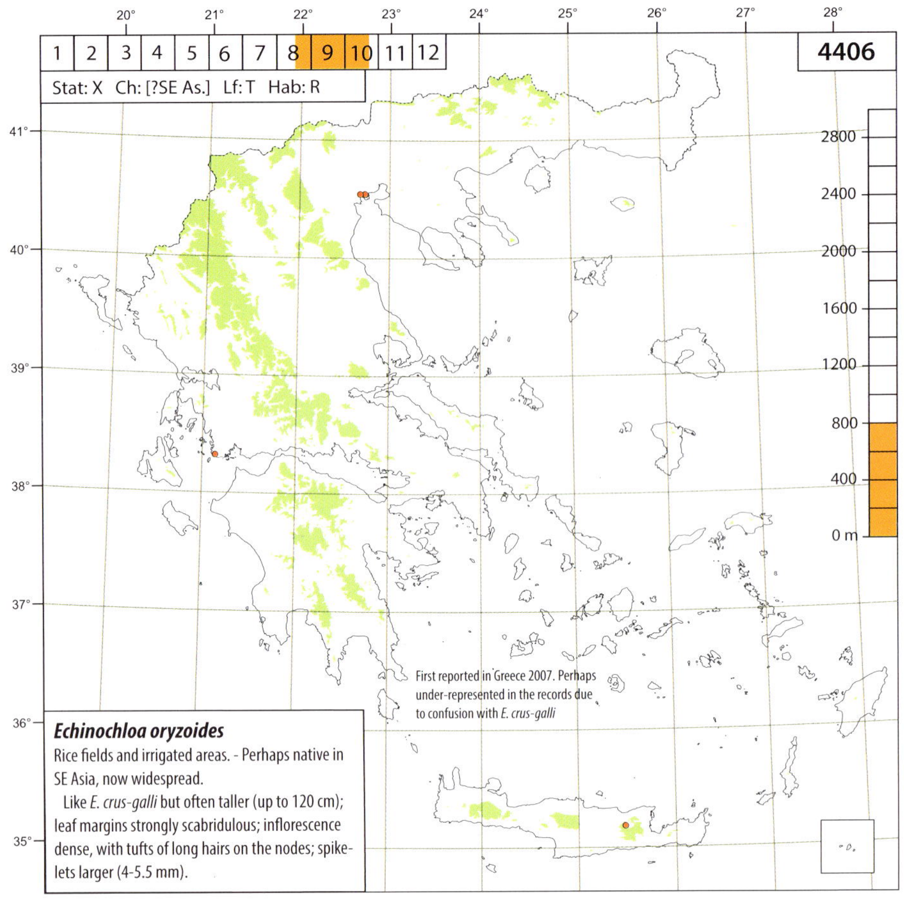

Echinochloa oryzoides
Rice fields and irrigated areas. - Perhaps native in SE Asia, now widespread.
Like *E. crus-galli* but often taller (up to 120 cm); leaf margins strongly scabridulous; inflorescence dense, with tufts of long hairs on the nodes; spikelets larger (4-5.5 mm).

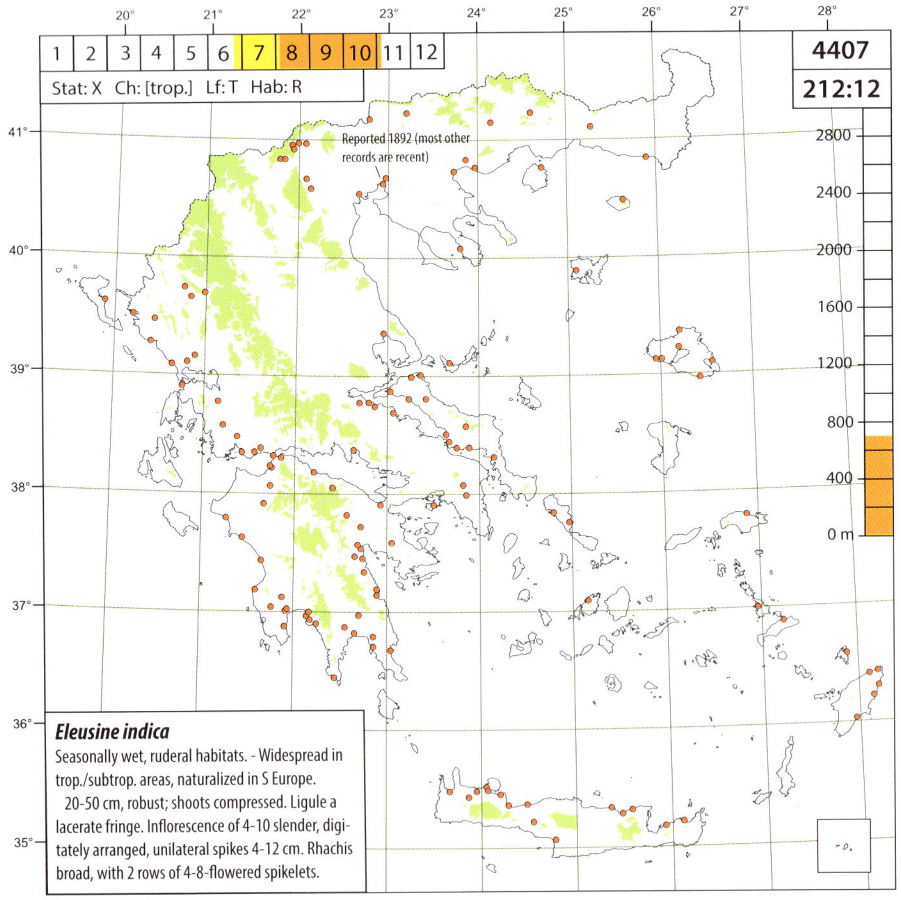

Eleusine indica
Seasonally wet, ruderal habitats. - Widespread in trop./subtrop. areas, naturalized in S Europe.
20-50 cm, robust; shoots compressed. Ligule a lacerate fringe. Inflorescence of 4-10 slender, digitately arranged, unilateral spikes 4-12 cm. Rhachis broad, with 2 rows of 4-8-flowered spikelets.

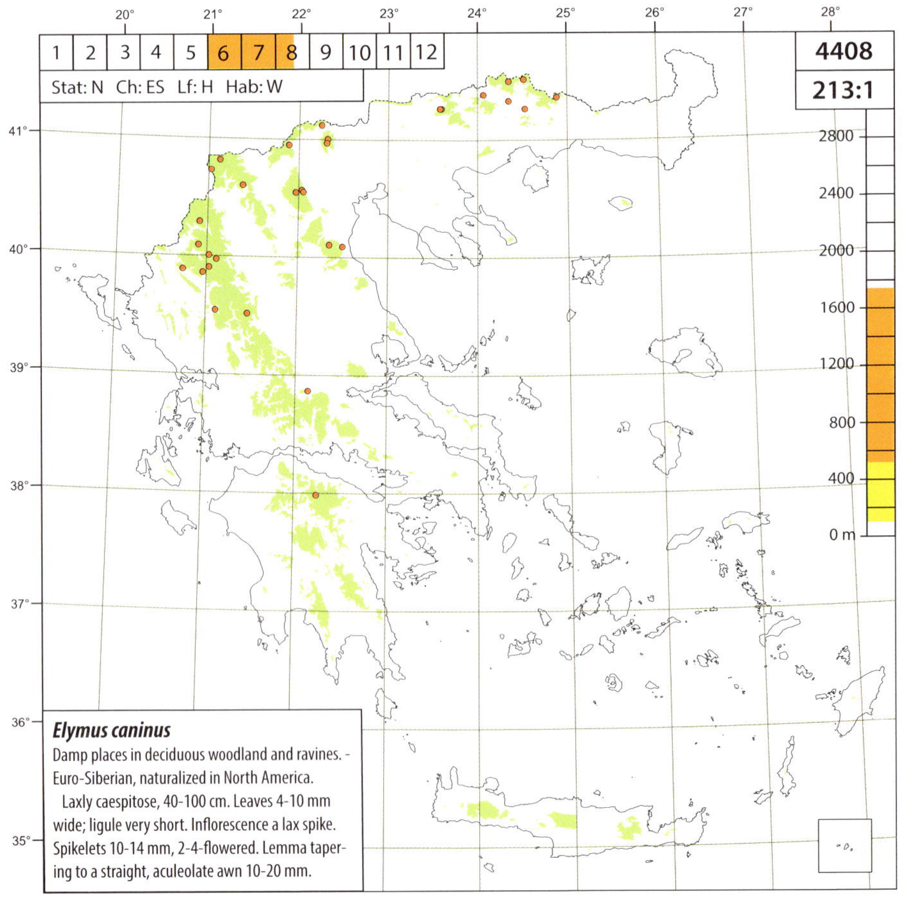

Elymus caninus
Damp places in deciduous woodland and ravines. - Euro-Siberian, naturalized in North America.
Laxly caespitose, 40-100 cm. Leaves 4-10 mm wide; ligule very short. Inflorescence a lax spike. Spikelets 10-14 mm, 2-4-flowered. Lemma tapering to a straight, aculeolate awn 10-20 mm.

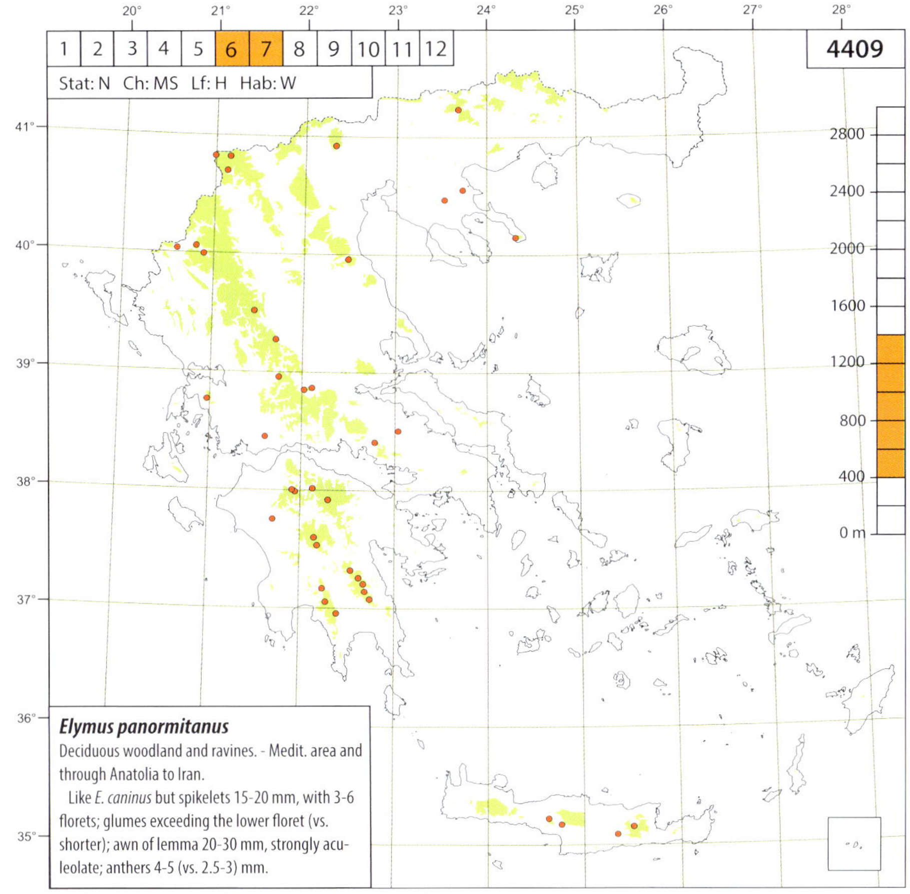

Elymus panormitanus
Deciduous woodland and ravines. - Medit. area and through Anatolia to Iran.
Like *E. caninus* but spikelets 15-20 mm, with 3-6 florets; glumes exceeding the lower floret (vs. shorter); awn of lemma 20-30 mm, strongly aculeolate; anthers 4-5 (vs. 2.5-3) mm.

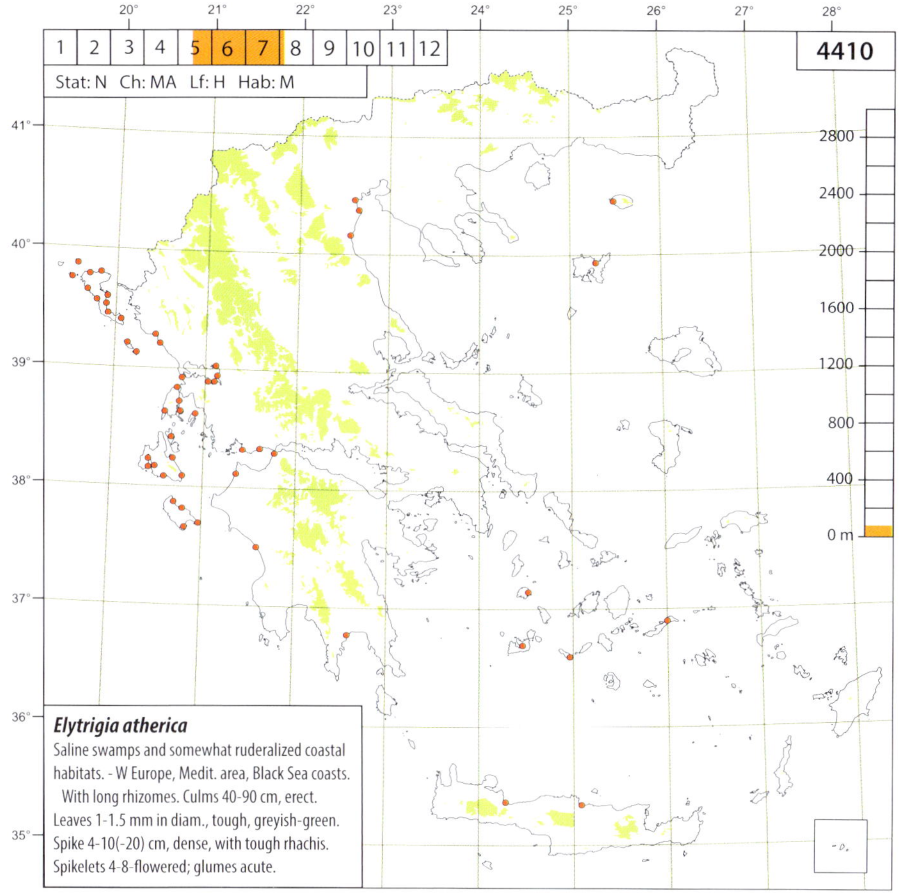

Elytrigia atherica
Saline swamps and somewhat ruderalized coastal habitats. - W Europe, Medit. area, Black Sea coasts.
With long rhizomes. Culms 40-90 cm, erect. Leaves 1-1.5 mm in diam., tough, greyish-green. Spike 4-10(-20) cm, dense, with tough rhachis. Spikelets 4-8-flowered; glumes acute.

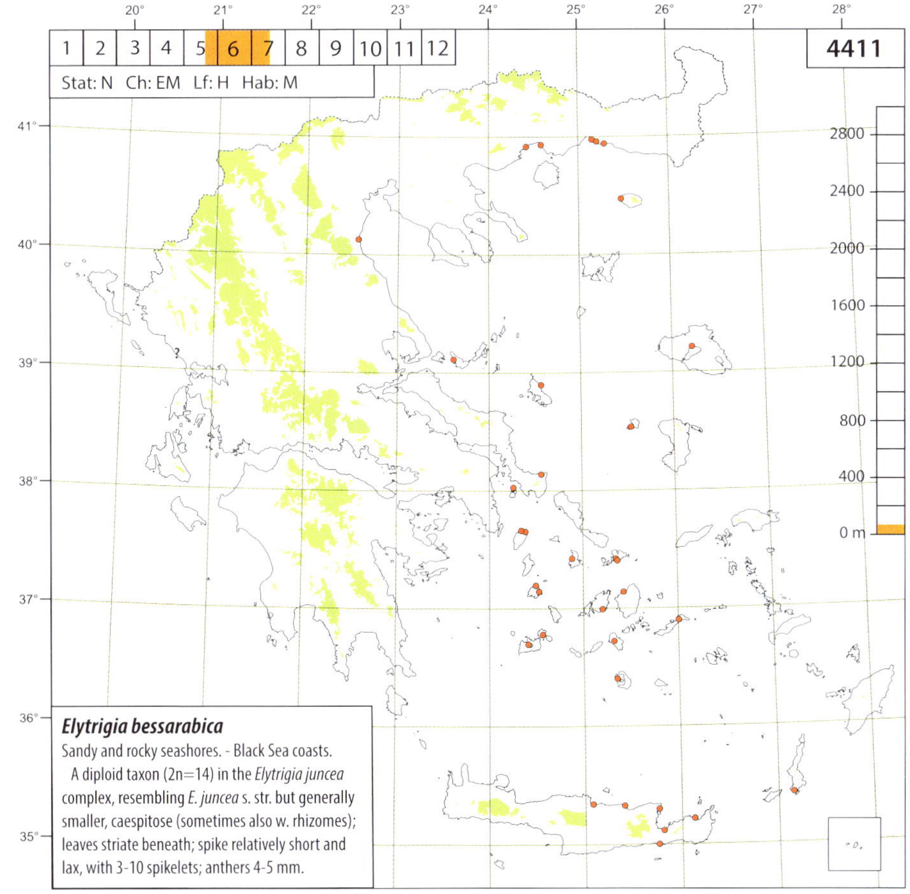

Elytrigia bessarabica
Sandy and rocky seashores. - Black Sea coasts.
A diploid taxon (2n=14) in the *Elytrigia juncea* complex, resembling *E. juncea* s. str. but generally smaller, caespitose (sometimes also w. rhizomes); leaves striate beneath; spike relatively short and lax, with 3-10 spikelets; anthers 4-5 mm.

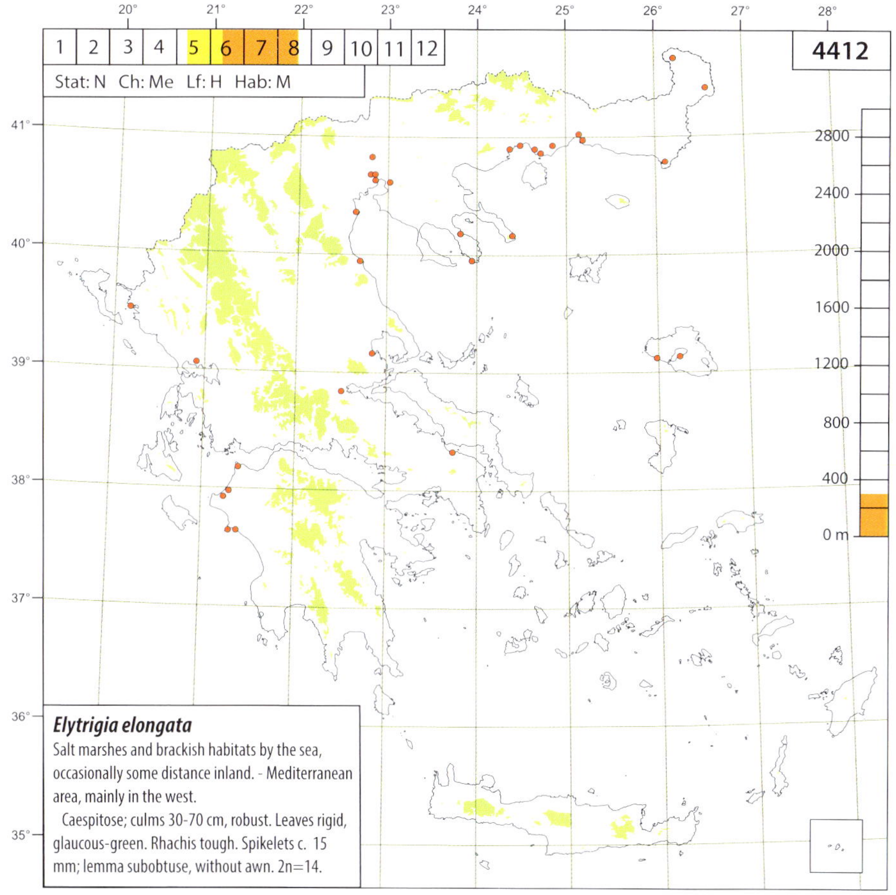

Elytrigia elongata
Salt marshes and brackish habitats by the sea, occasionally some distance inland. - Mediterranean area, mainly in the west.
Caespitose; culms 30-70 cm, robust. Leaves rigid, glaucous-green. Rhachis tough. Spikelets c. 15 mm; lemma subobtuse, without awn. 2n=14.

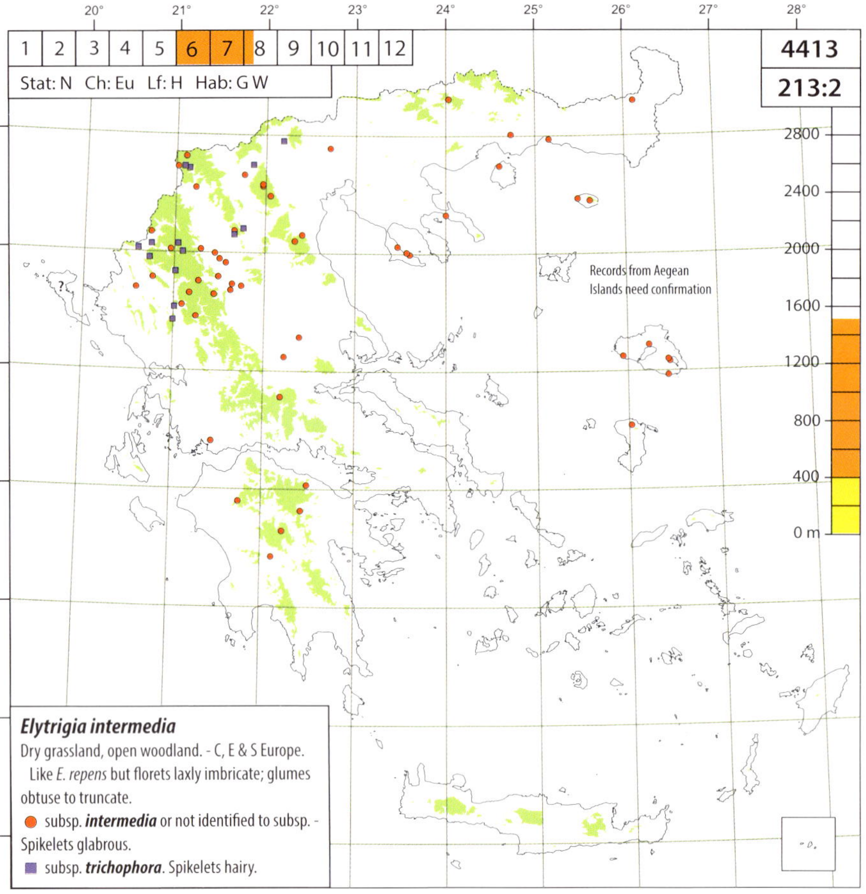

Elytrigia intermedia
Dry grassland, open woodland. - C, E & S Europe.
Like *E. repens* but florets laxly imbricate; glumes obtuse to truncate.
● subsp. ***intermedia*** or not identified to subsp. - Spikelets glabrous.
■ subsp. ***trichophora***. Spikelets hairy.

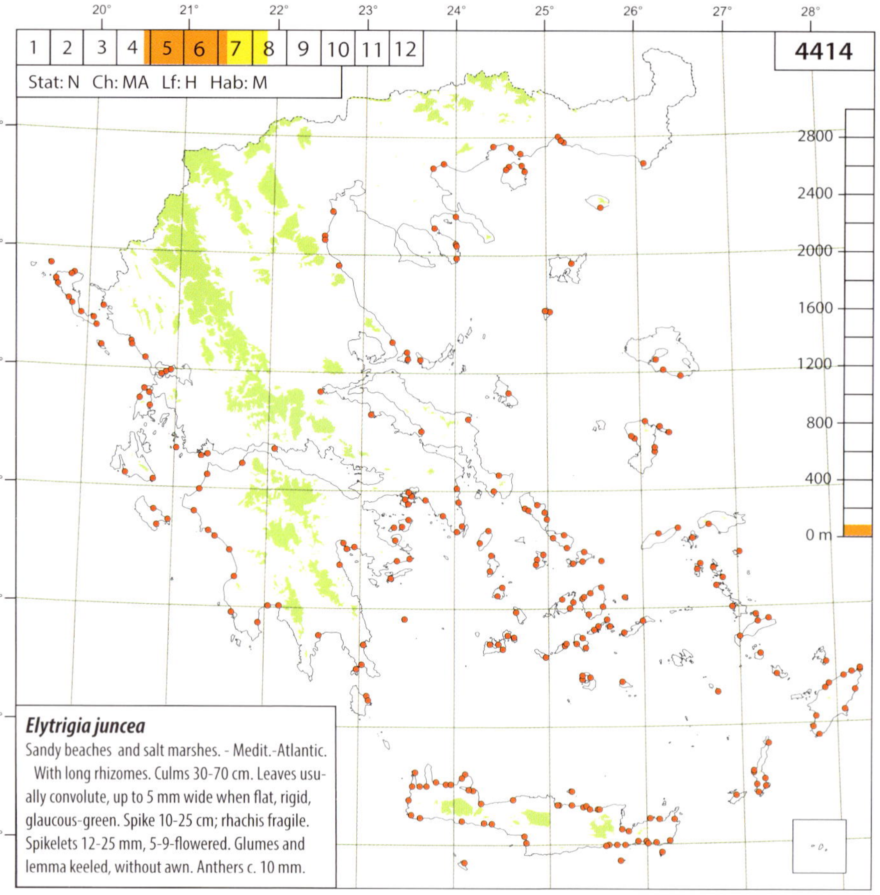

Elytrigia juncea
Sandy beaches and salt marshes. - Medit.-Atlantic.
With long rhizomes. Culms 30-70 cm. Leaves usually convolute, up to 5 mm wide when flat, rigid, glaucous-green. Spike 10-25 cm; rhachis fragile. Spikelets 12-25 mm, 5-9-flowered. Glumes and lemma keeled, without awn. Anthers c. 10 mm.

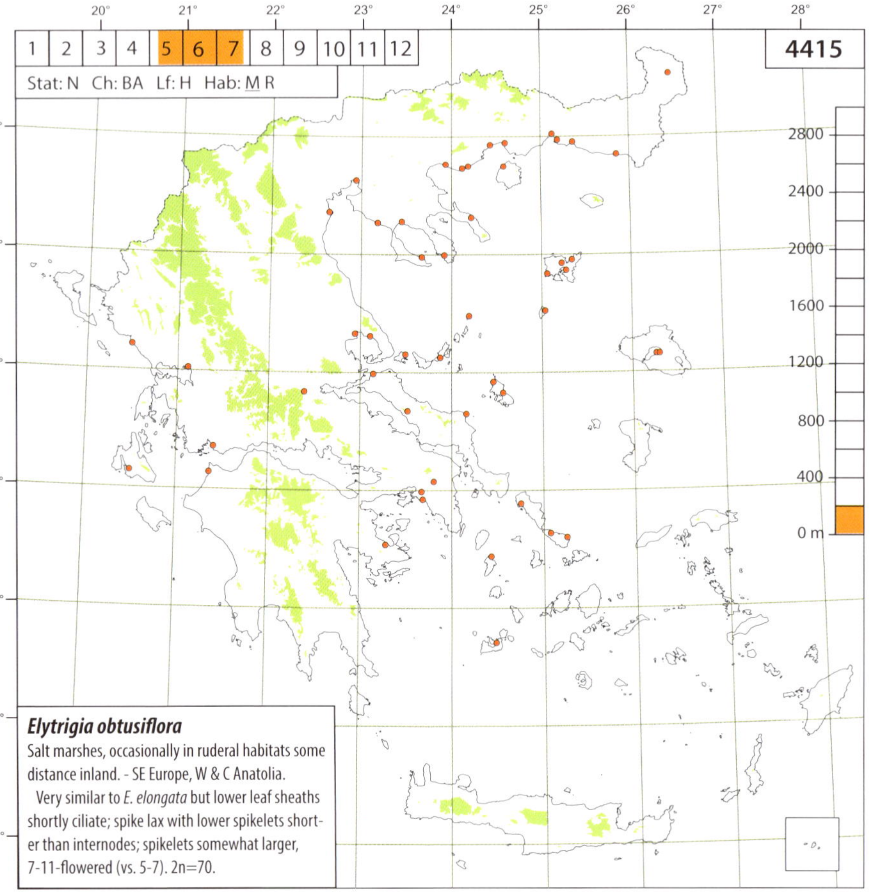

Elytrigia obtusiflora
Salt marshes, occasionally in ruderal habitats some distance inland. - SE Europe, W & C Anatolia.
Very similar to *E. elongata* but lower leaf sheaths shortly ciliate; spike lax with lower spikelets shorter than internodes; spikelets somewhat larger, 7-11-flowered (vs. 5-7). 2n=70.

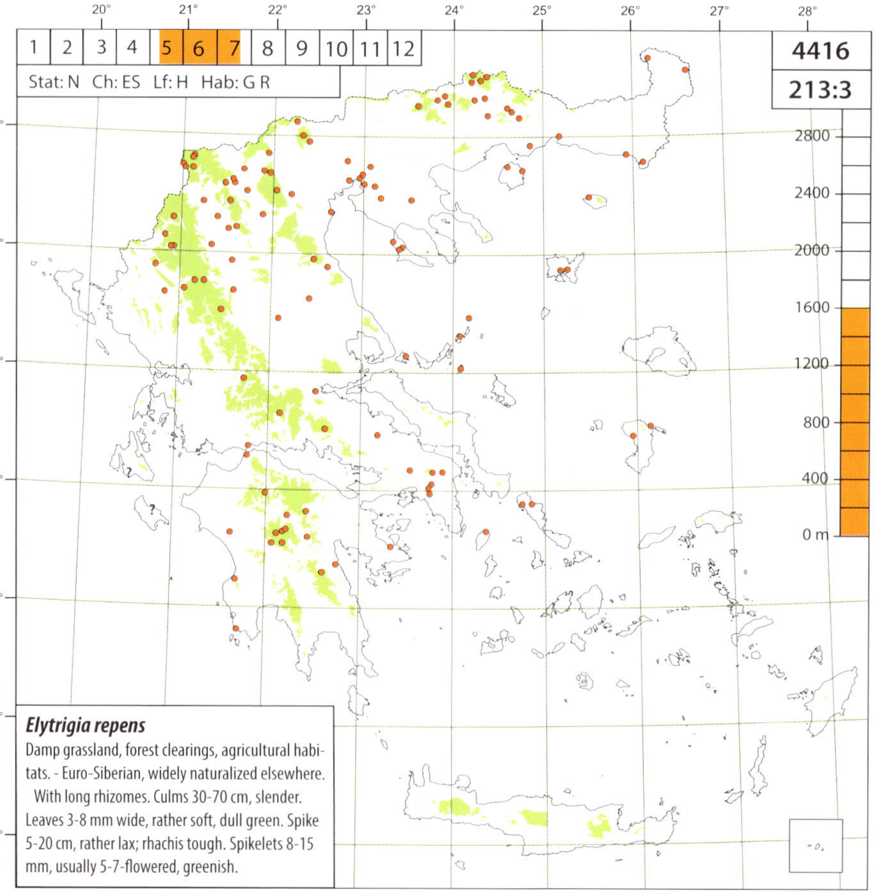

Elytrigia repens
Damp grassland, forest clearings, agricultural habitats. - Euro-Siberian, widely naturalized elsewhere.
With long rhizomes. Culms 30-70 cm, slender. Leaves 3-8 mm wide, rather soft, dull green. Spike 5-20 cm, rather lax; rhachis tough. Spikelets 8-15 mm, usually 5-7-flowered, greenish.

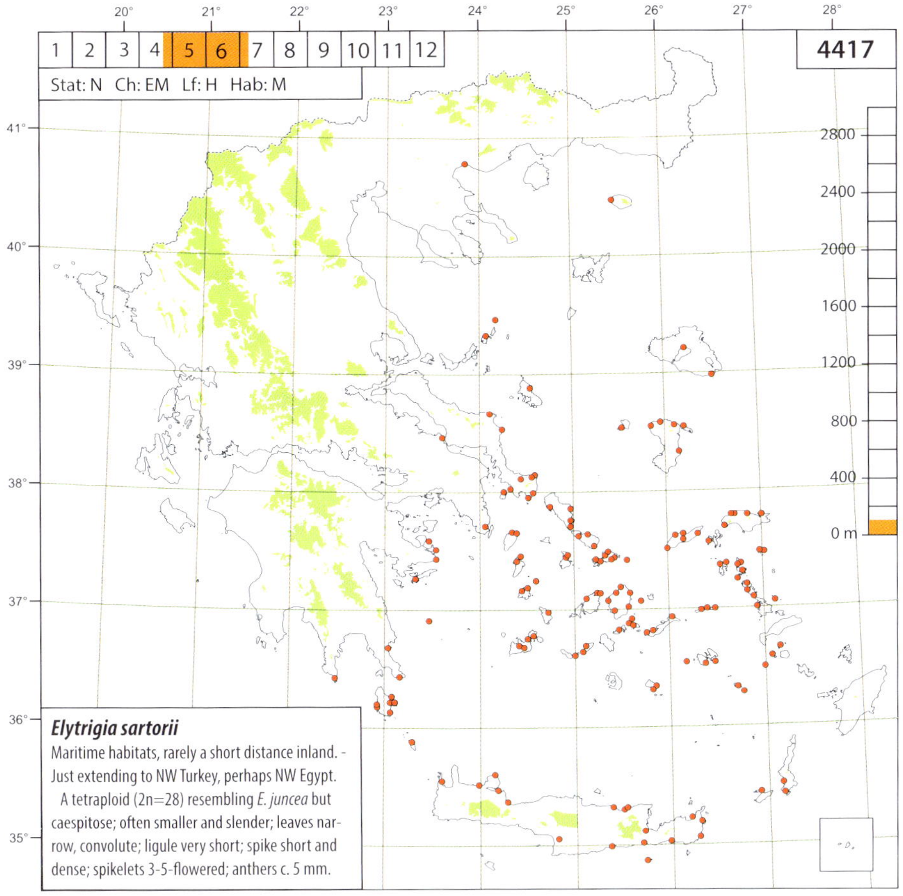

Elytrigia sartorii

Maritime habitats, rarely a short distance inland. - Just extending to NW Turkey, perhaps NW Egypt.

A tetraploid (2n=28) resembling *E. juncea* but caespitose; often smaller and slender; leaves narrow, convolute; ligule very short; spike short and dense; spikelets 3-5-flowered; anthers c. 5 mm.

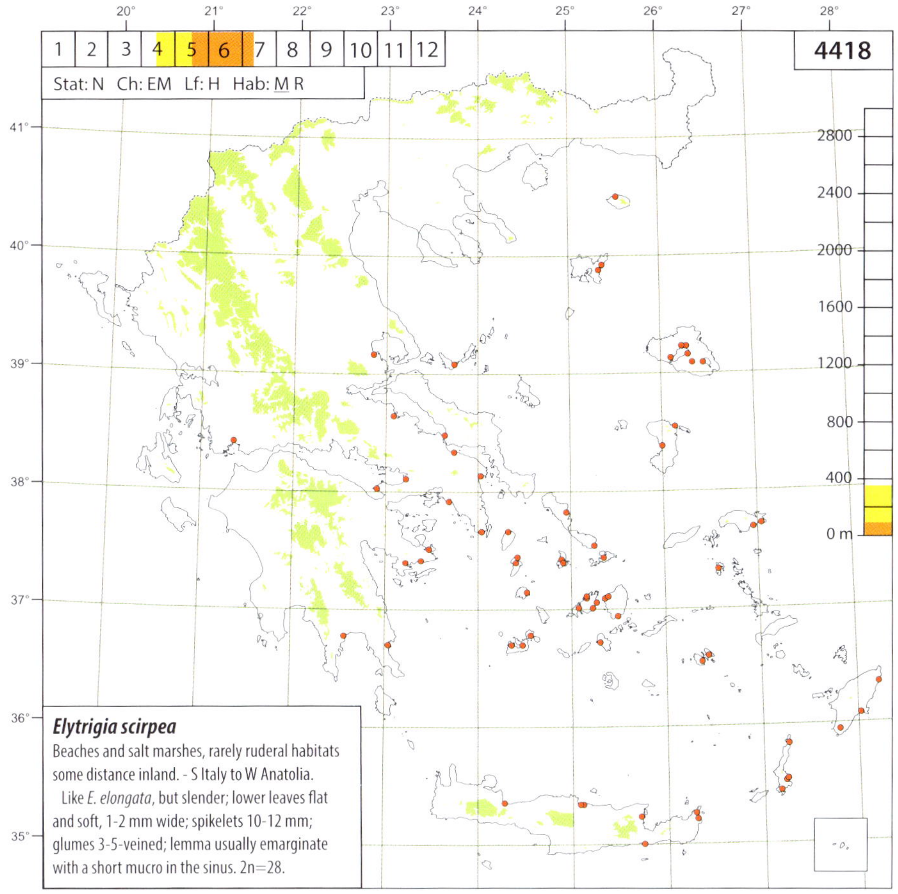

Elytrigia scirpea

Beaches and salt marshes, rarely ruderal habitats some distance inland. - S Italy to W Anatolia.

Like *E. elongata*, but slender; lower leaves flat and soft, 1-2 mm wide; spikelets 10-12 mm; glumes 3-5-veined; lemma usually emarginate with a short mucro in the sinus. 2n=28.

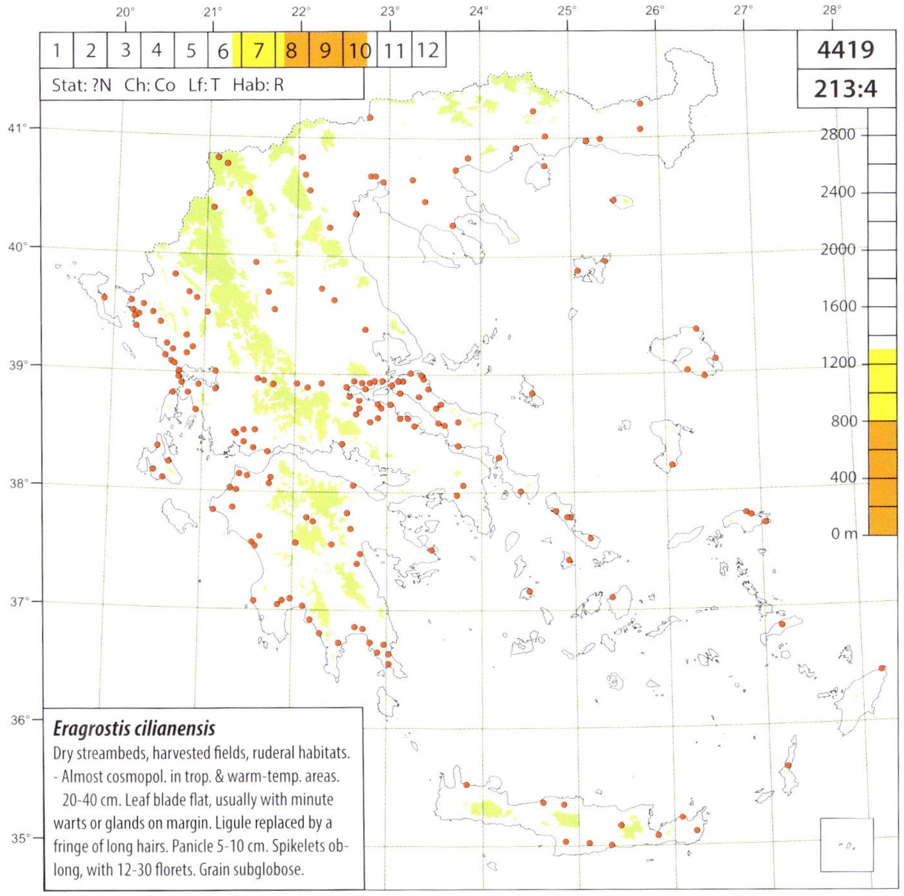

Eragrostis cilianensis

Dry streambeds, harvested fields, ruderal habitats. - Almost cosmopol. in trop. & warm-temp. areas.

20-40 cm. Leaf blade flat, usually with minute warts or glands on margin. Ligule replaced by a fringe of long hairs. Panicle 5-10 cm. Spikelets oblong, with 12-30 florets. Grain subglobose.

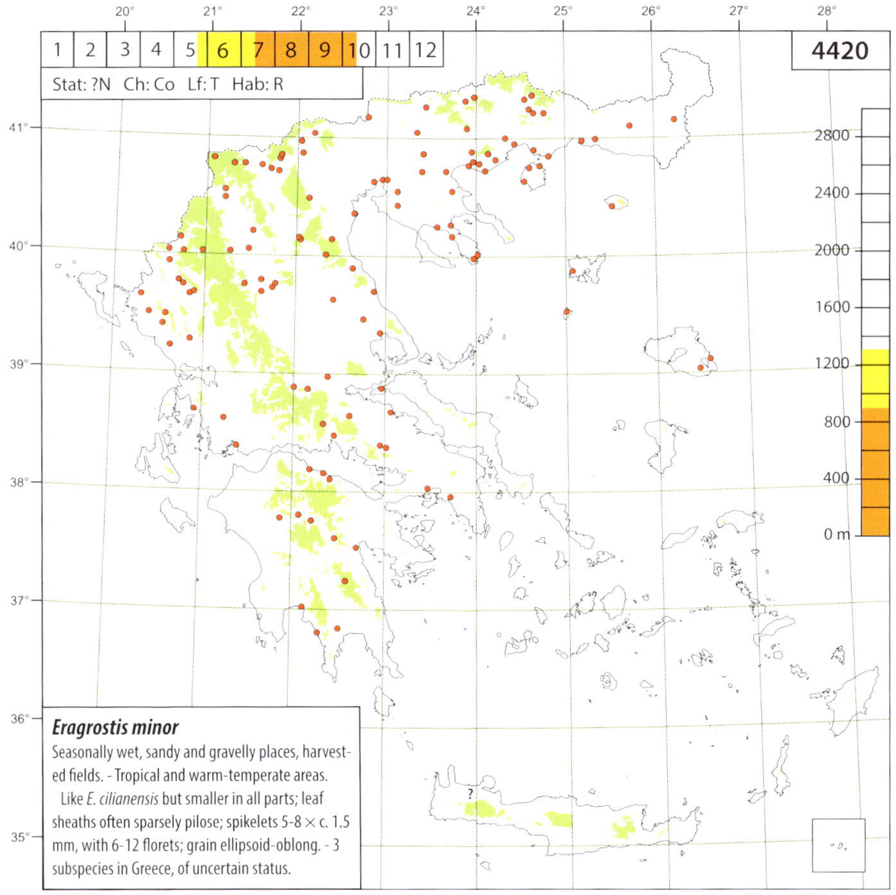

Eragrostis minor

Seasonally wet, sandy and gravelly places, harvested fields. - Tropical and warm-temperate areas.

Like *E. cilianensis* but smaller in all parts; leaf sheaths often sparsely pilose; spikelets 5-8 × c. 1.5 mm, with 6-12 florets; grain ellipsoid-oblong. - 3 subspecies in Greece, of uncertain status.

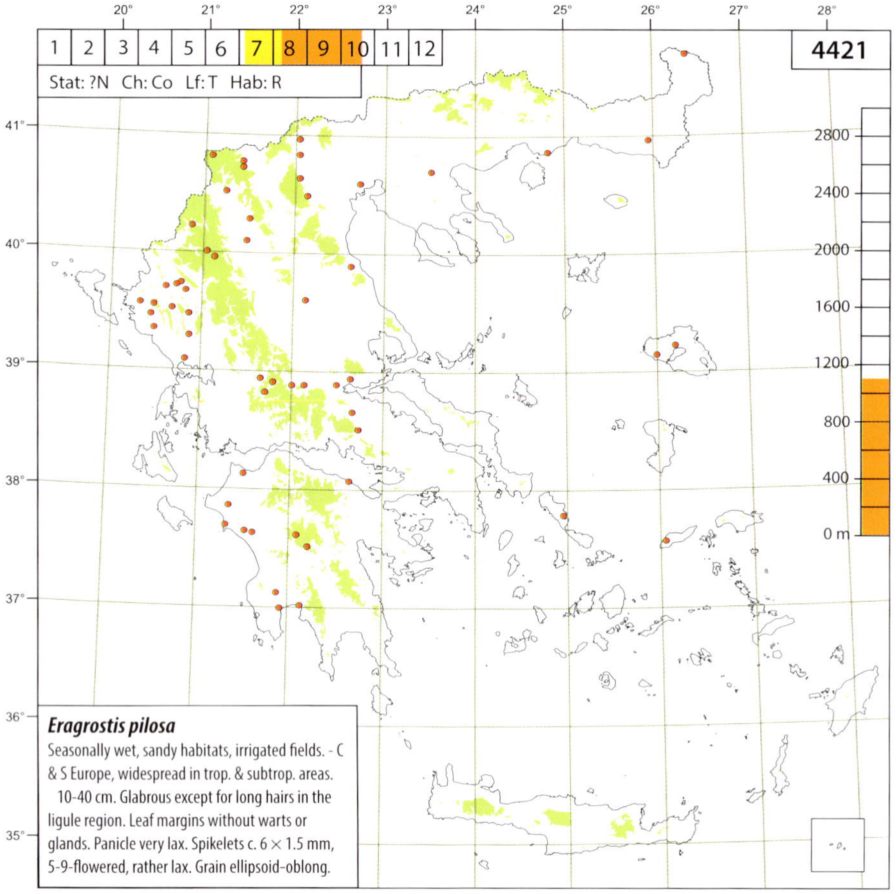

Eragrostis pilosa

Seasonally wet, sandy habitats, irrigated fields. - C & S Europe, widespread in trop. & subtrop. areas.

10-40 cm. Glabrous except for long hairs in the ligule region. Leaf margins without warts or glands. Panicle very lax. Spikelets c. 6 × 1.5 mm, 5-9-flowered, rather lax. Grain ellipsoid-oblong.

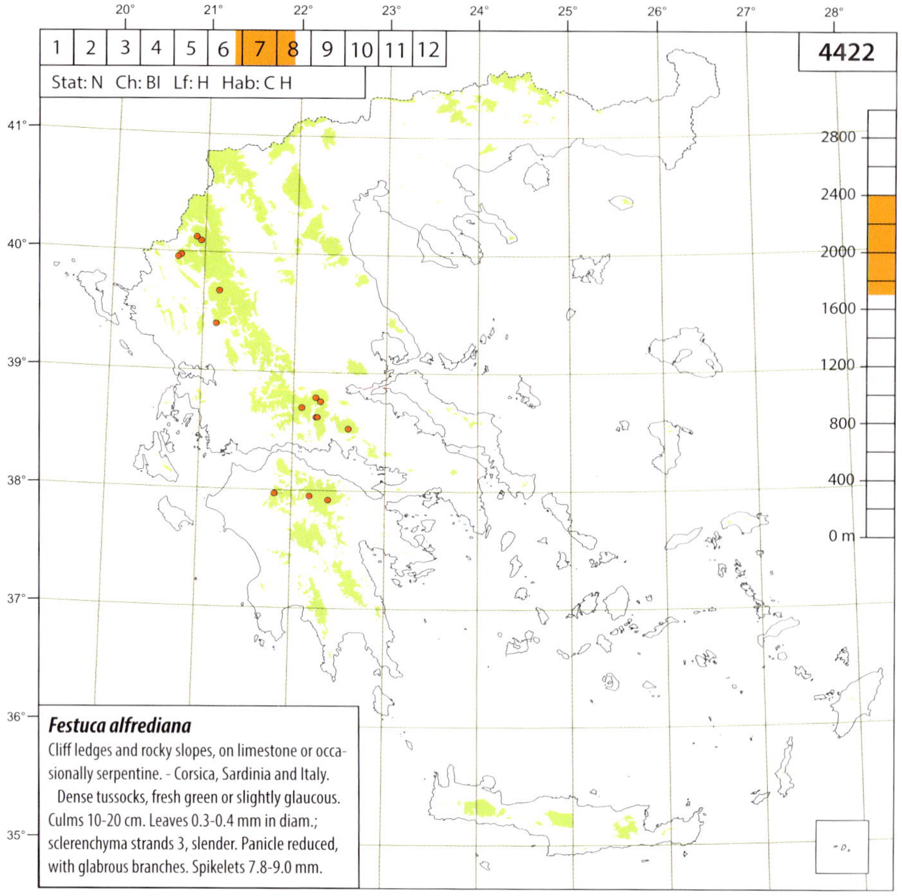

Festuca alfrediana

Cliff ledges and rocky slopes, on limestone or occasionally serpentine. - Corsica, Sardinia and Italy.

Dense tussocks, fresh green or slightly glaucous. Culms 10-20 cm. Leaves 0.3-0.4 mm in diam.; sclerenchyma strands 3, slender. Panicle reduced, with glabrous branches. Spikelets 7.8-9.0 mm.

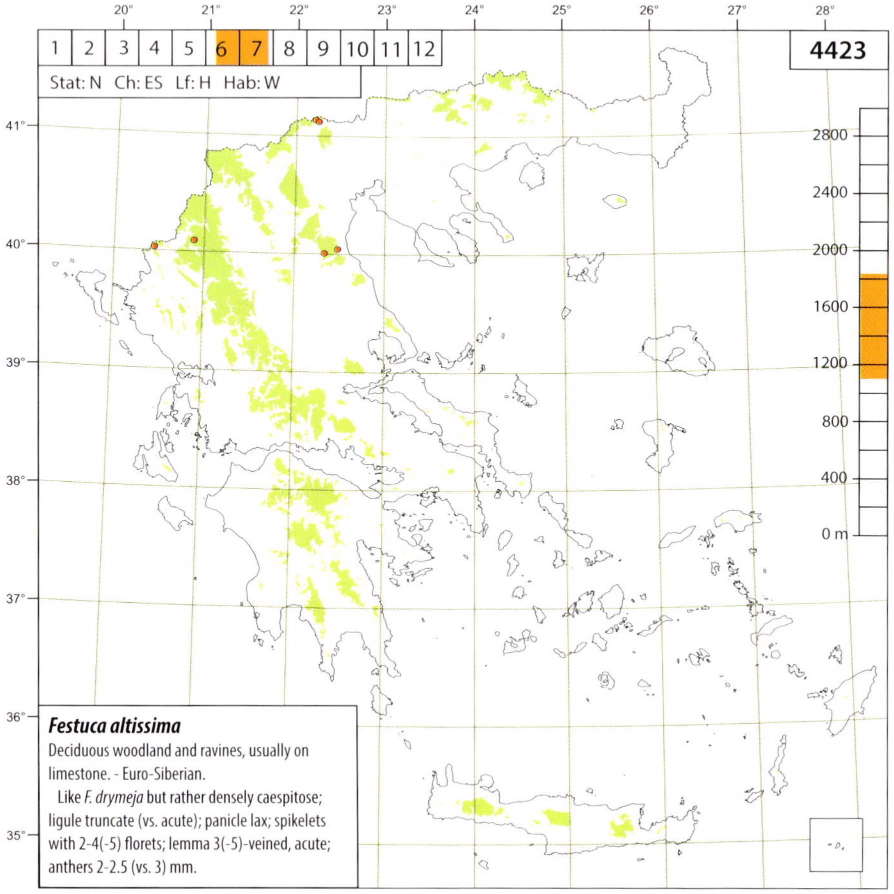

Festuca altissima

Deciduous woodland and ravines, usually on limestone. - Euro-Siberian.

Like *F. drymeja* but rather densely caespitose; ligule truncate (vs. acute); panicle lax; spikelets with 2-4(-5) florets; lemma 3(-5)-veined, acute; anthers 2-2.5 (vs. 3) mm.

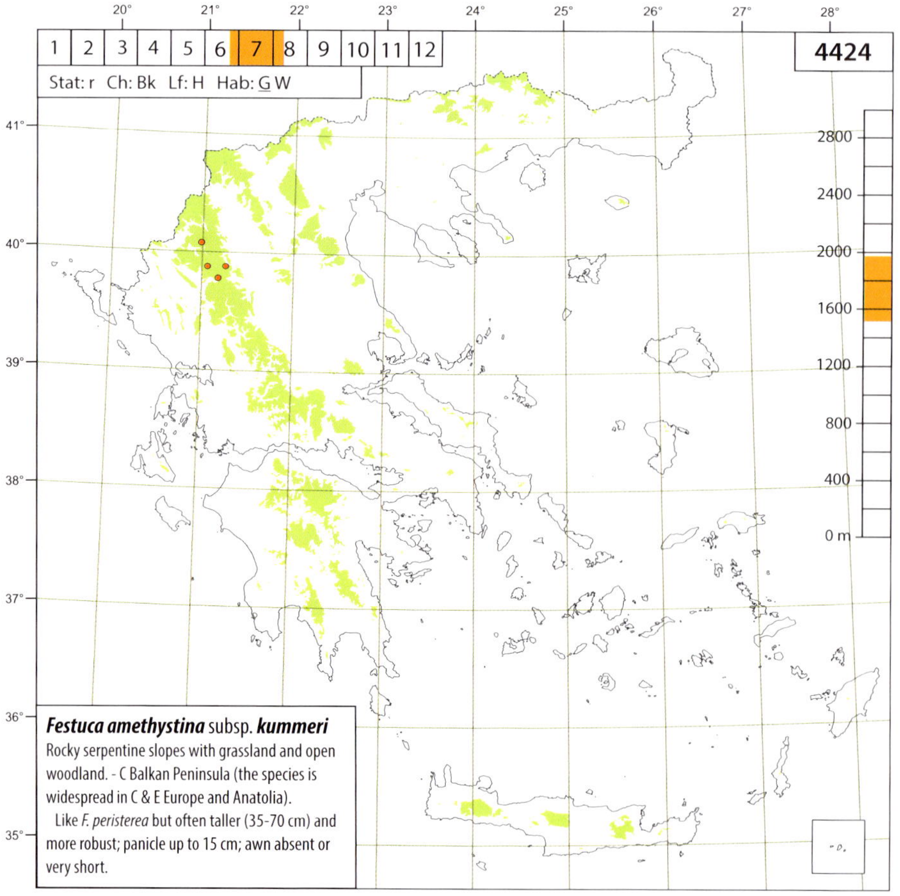

Festuca amethystina subsp. ***kummeri***

Rocky serpentine slopes with grassland and open woodland. - C Balkan Peninsula (the species is widespread in C & E Europe and Anatolia).

Like *F. peristerea* but often taller (35-70 cm) and more robust; panicle up to 15 cm; awn absent or very short.

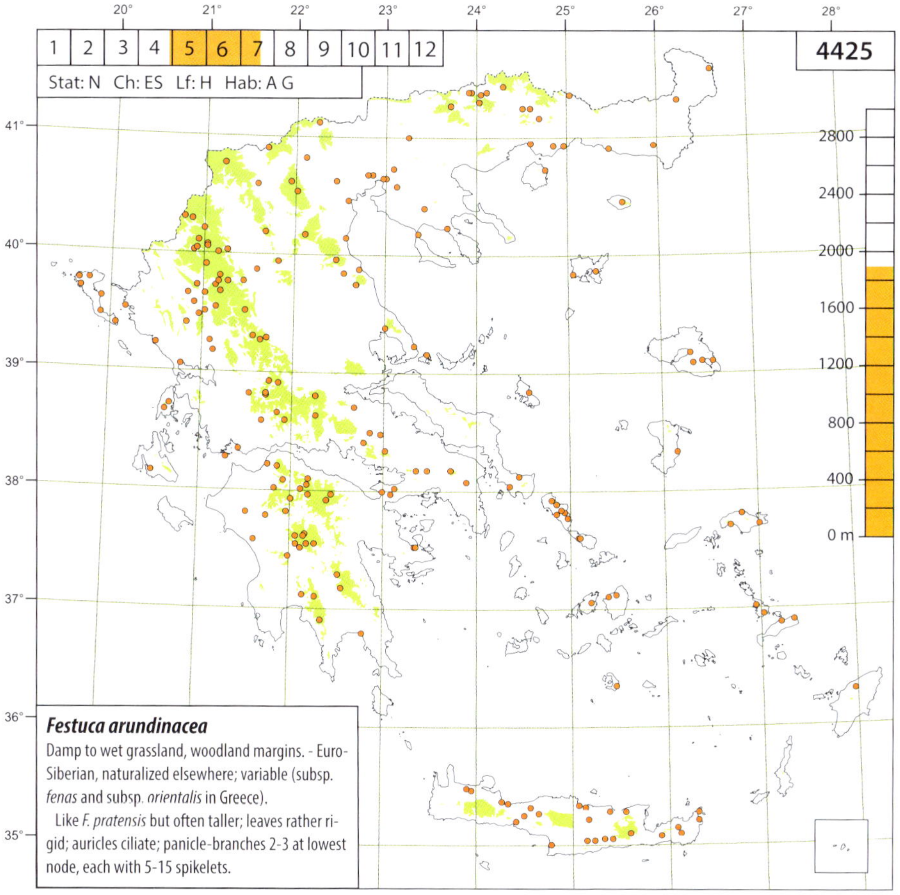

Festuca arundinacea

Damp to wet grassland, woodland margins. - Euro-Siberian, naturalized elsewhere; variable (subsp. *fenas* and subsp. *orientalis* in Greece).

Like *F. pratensis* but often taller; leaves rather rigid; auricles ciliate; panicle-branches 2-3 at lowest node, each with 5-15 spikelets.

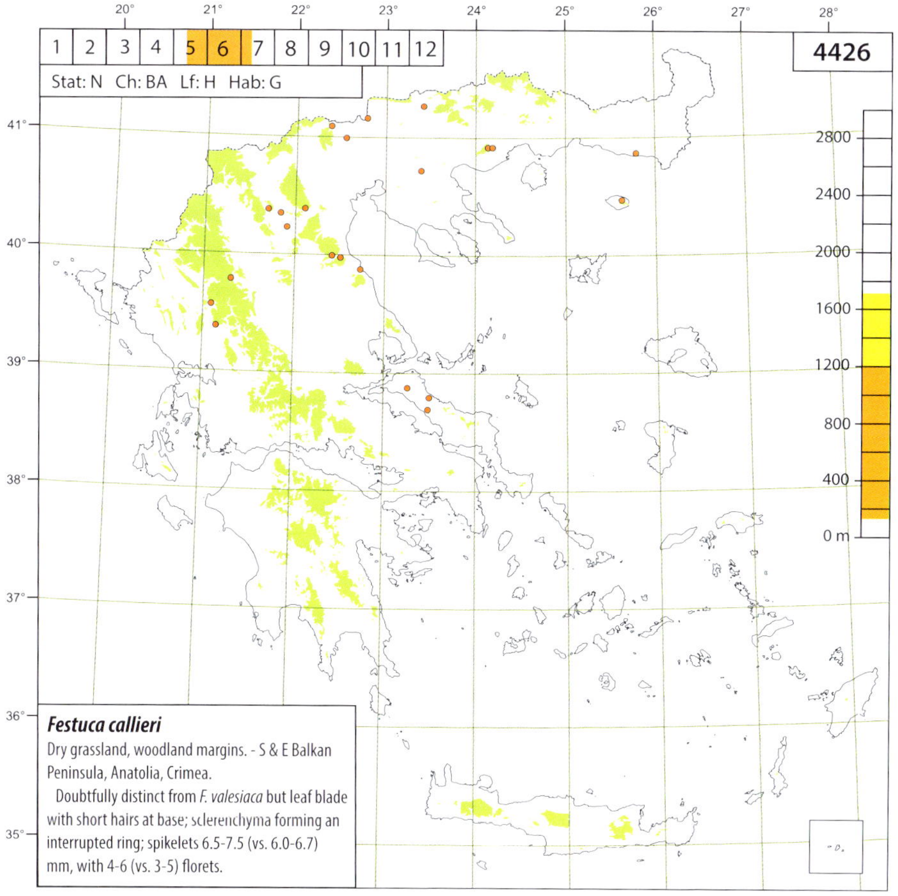

Festuca callieri

Dry grassland, woodland margins. - S & E Balkan Peninsula, Anatolia, Crimea.

Doubtfully distinct from *F. valesiaca* but leaf blade with short hairs at base; sclerenchyma forming an interrupted ring; spikelets 6.5-7.5 (vs. 6.0-6.7) mm, with 4-6 (vs. 3-5) florets.

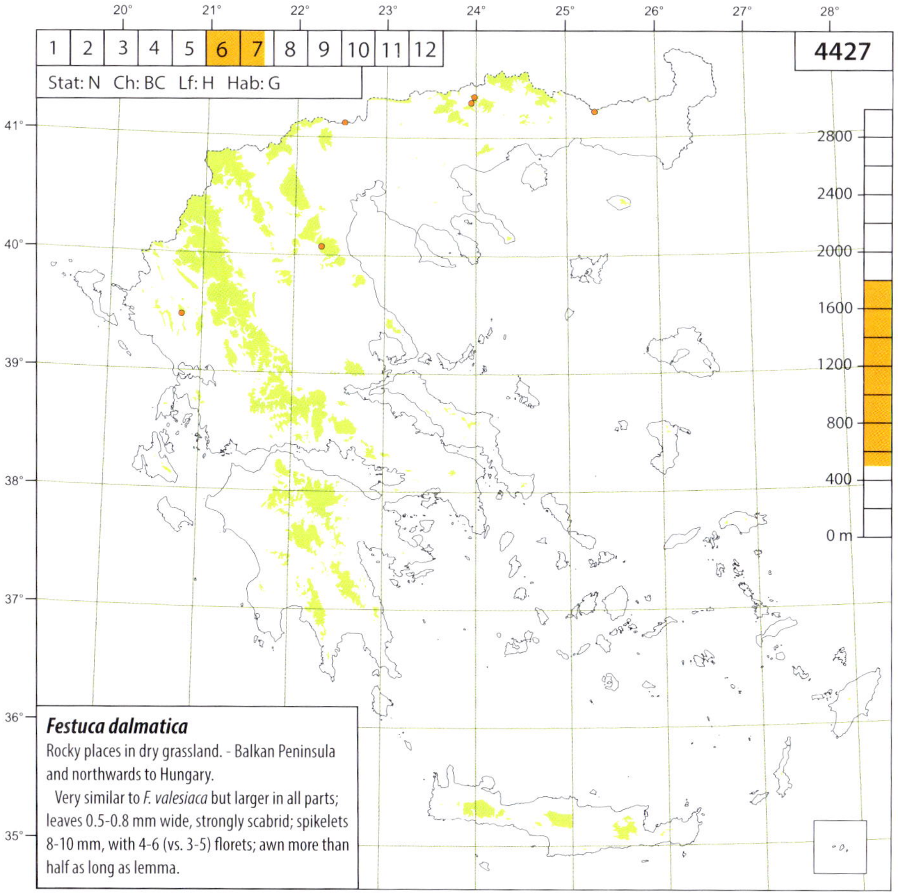

Festuca dalmatica

Rocky places in dry grassland. - Balkan Peninsula and northwards to Hungary.

Very similar to *F. valesiaca* but larger in all parts; leaves 0.5-0.8 mm wide, strongly scabrid; spikelets 8-10 mm, with 4-6 (vs. 3-5) florets; awn more than half as long as lemma.

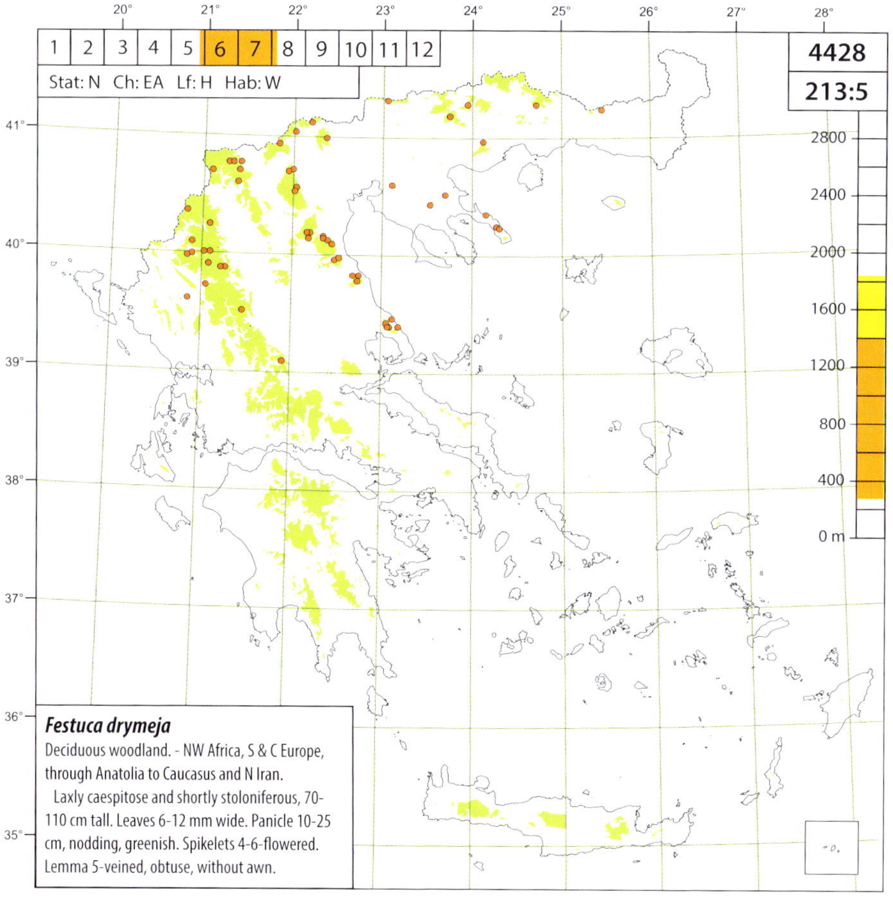

Festuca drymeja

Deciduous woodland. - NW Africa, S & C Europe, through Anatolia to Caucasus and N Iran.

Laxly caespitose and shortly stoloniferous, 70-110 cm tall. Leaves 6-12 mm wide. Panicle 10-25 cm, nodding, greenish. Spikelets 4-6-flowered. Lemma 5-veined, obtuse, without awn.

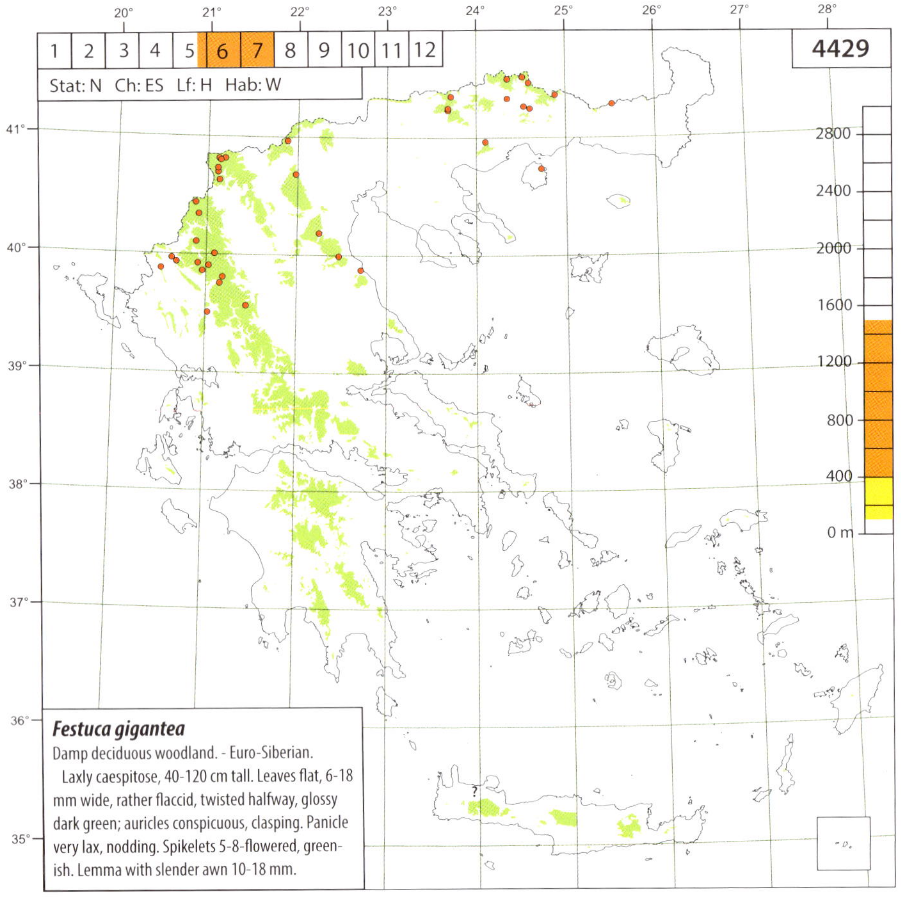

Festuca gigantea

Damp deciduous woodland. - Euro-Siberian.

Laxly caespitose, 40-120 cm tall. Leaves flat, 6-18 mm wide, rather flaccid, twisted halfway, glossy dark green; auricles conspicuous, clasping. Panicle very lax, nodding. Spikelets 5-8-flowered, greenish. Lemma with slender awn 10-18 mm.

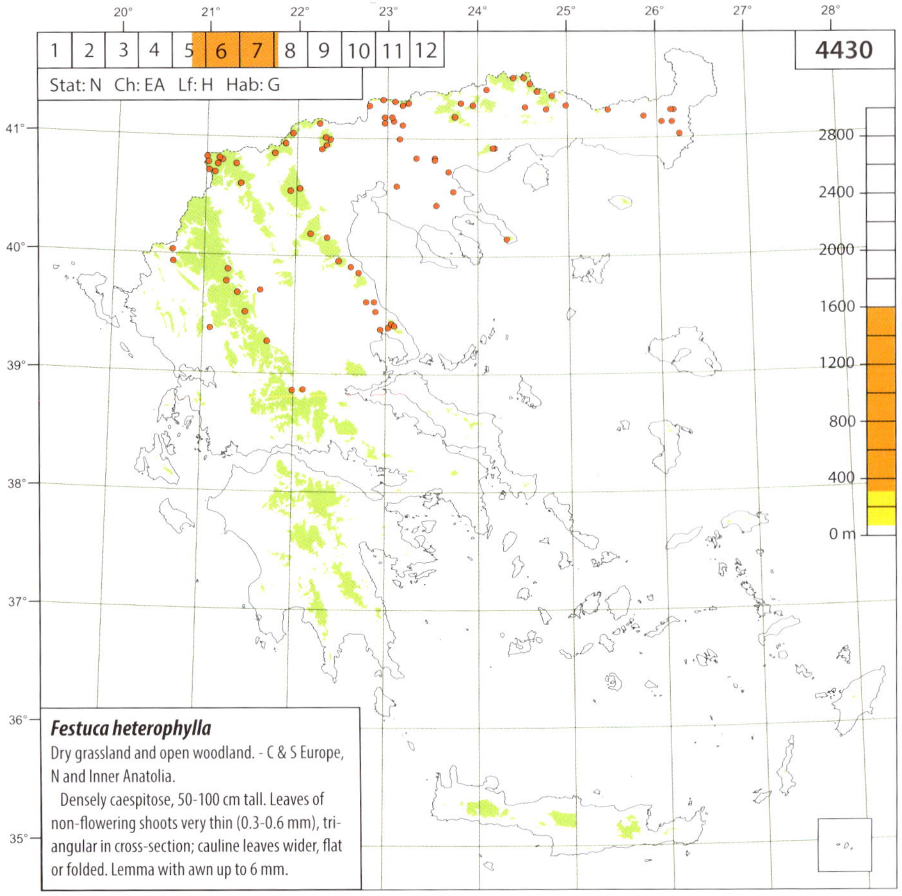

Festuca heterophylla

Dry grassland and open woodland. - C & S Europe, N and Inner Anatolia.

Densely caespitose, 50-100 cm tall. Leaves of non-flowering shoots very thin (0.3-0.6 mm), triangular in cross-section; cauline leaves wider, flat or folded. Lemma with awn up to 6 mm.

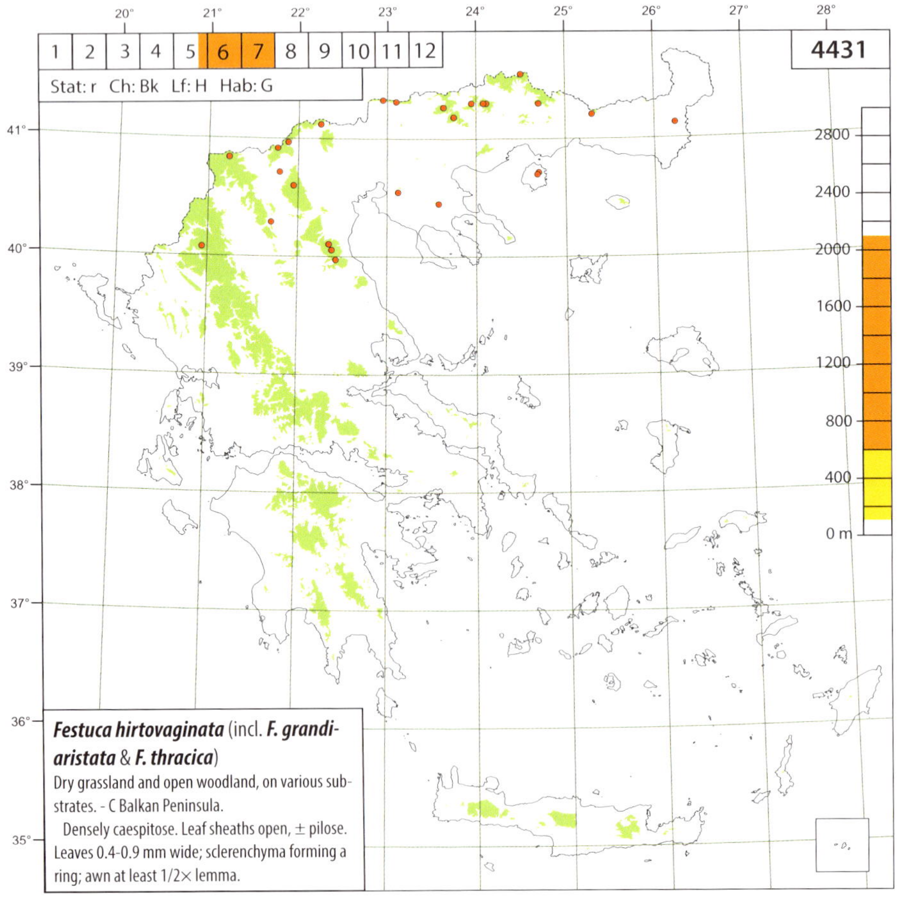

Festuca hirtovaginata (incl. ***F. grandi-aristata*** & ***F. thracica***)

Dry grassland and open woodland, on various substrates. - C Balkan Peninsula.

Densely caespitose. Leaf sheaths open, ± pilose. Leaves 0.4-0.9 mm wide; sclerenchyma forming a ring; awn at least 1/2× lemma.

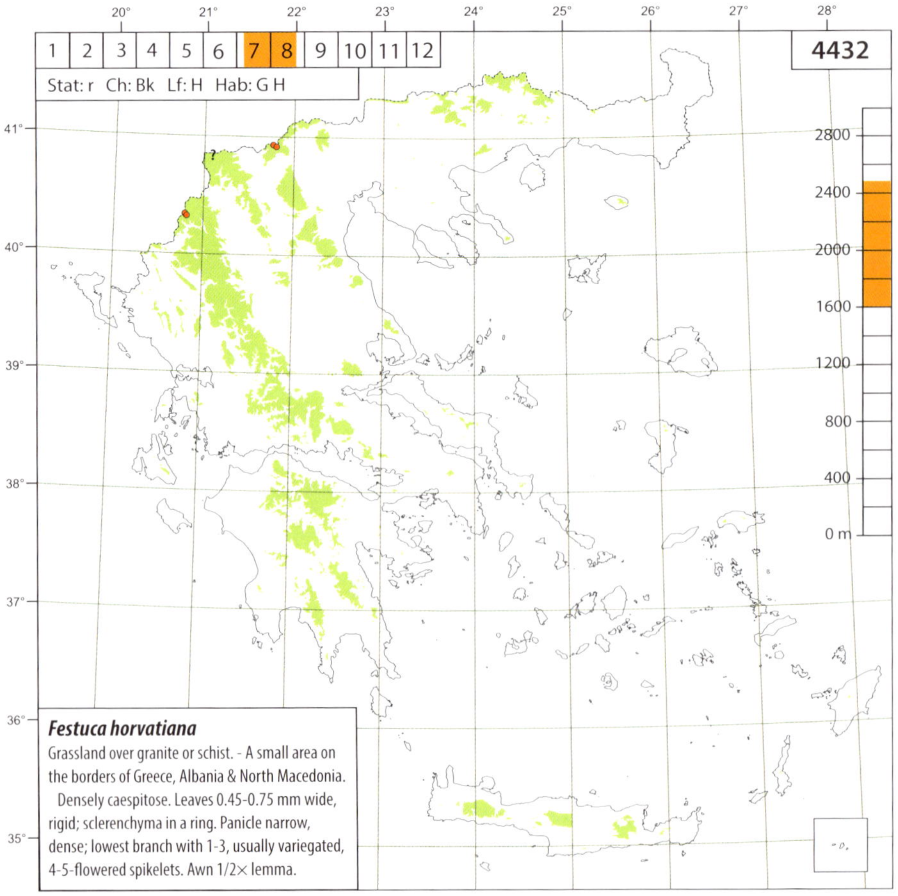

Festuca horvatiana

Grassland over granite or schist. - A small area on the borders of Greece, Albania & North Macedonia.

Densely caespitose. Leaves 0.45-0.75 mm wide, rigid; sclerenchyma in a ring. Panicle narrow, dense; lowest branch with 1-3, usually variegated, 4-5-flowered spikelets. Awn 1/2× lemma.

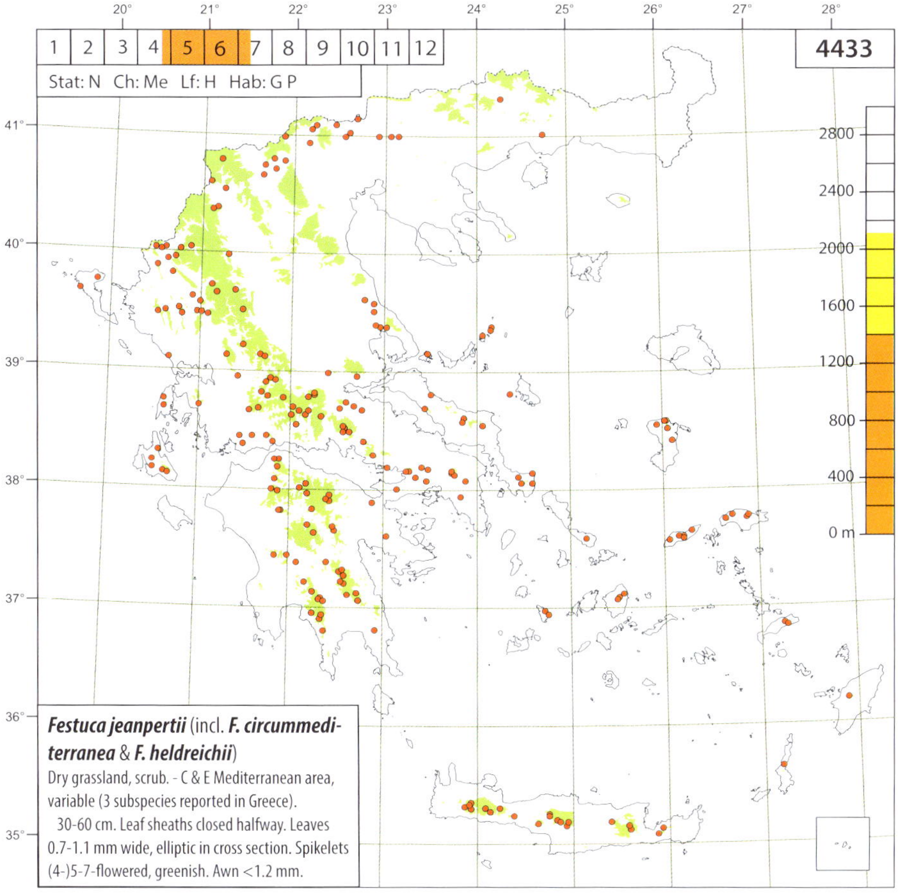

Festuca jeanpertii (incl. ***F. circummediterranea*** & ***F. heldreichii***)
Dry grassland, scrub. - C & E Mediterranean area, variable (3 subspecies reported in Greece).
30-60 cm. Leaf sheaths closed halfway. Leaves 0.7-1.1 mm wide, elliptic in cross section. Spikelets (4-)5-7-flowered, greenish. Awn <1.2 mm.

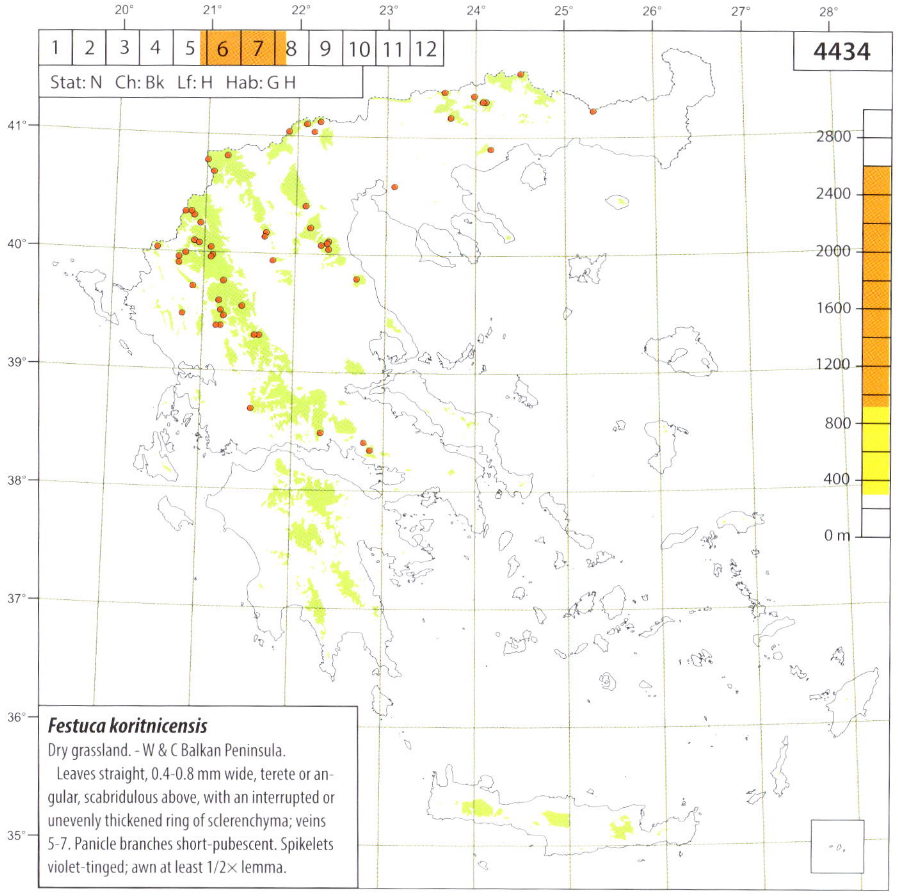

Festuca koritnicensis
Dry grassland. - W & C Balkan Peninsula.
Leaves straight, 0.4-0.8 mm wide, terete or angular, scabridulous above, with an interrupted or unevenly thickened ring of sclerenchyma; veins 5-7. Panicle branches short-pubescent. Spikelets violet-tinged; awn at least 1/2× lemma.

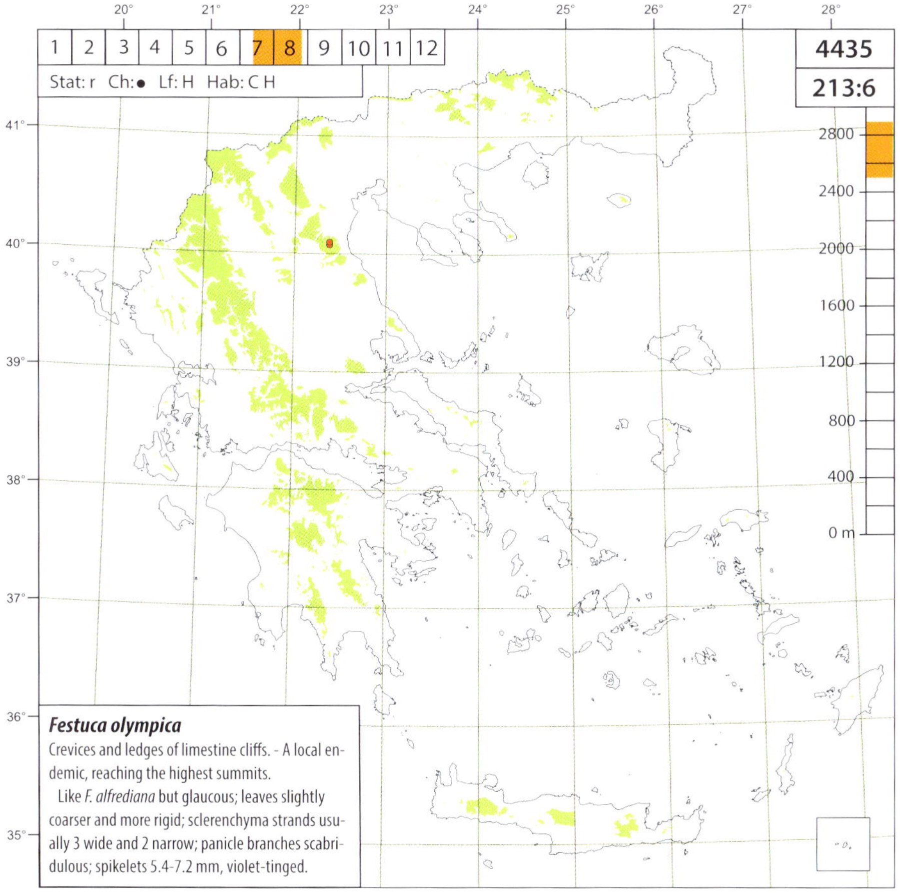

Festuca olympica
Crevices and ledges of limestone cliffs. - A local endemic, reaching the highest summits.
Like *F. alfrediana* but glaucous; leaves slightly coarser and more rigid; sclerenchyma strands usually 3 wide and 2 narrow; panicle branches scabridulous; spikelets 5.4-7.2 mm, violet-tinged.

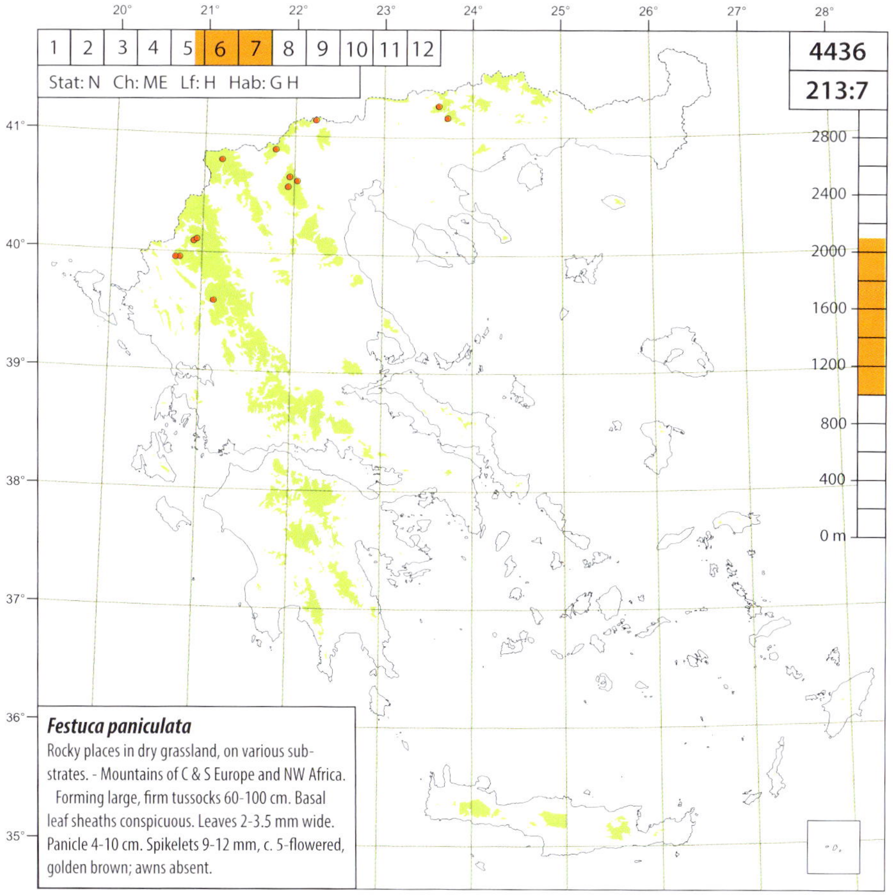

Festuca paniculata
Rocky places in dry grassland, on various substrates. - Mountains of C & S Europe and NW Africa.
Forming large, firm tussocks 60-100 cm. Basal leaf sheaths conspicuous. Leaves 2-3.5 mm wide. Panicle 4-10 cm. Spikelets 9-12 mm, c. 5-flowered, golden brown; awns absent.

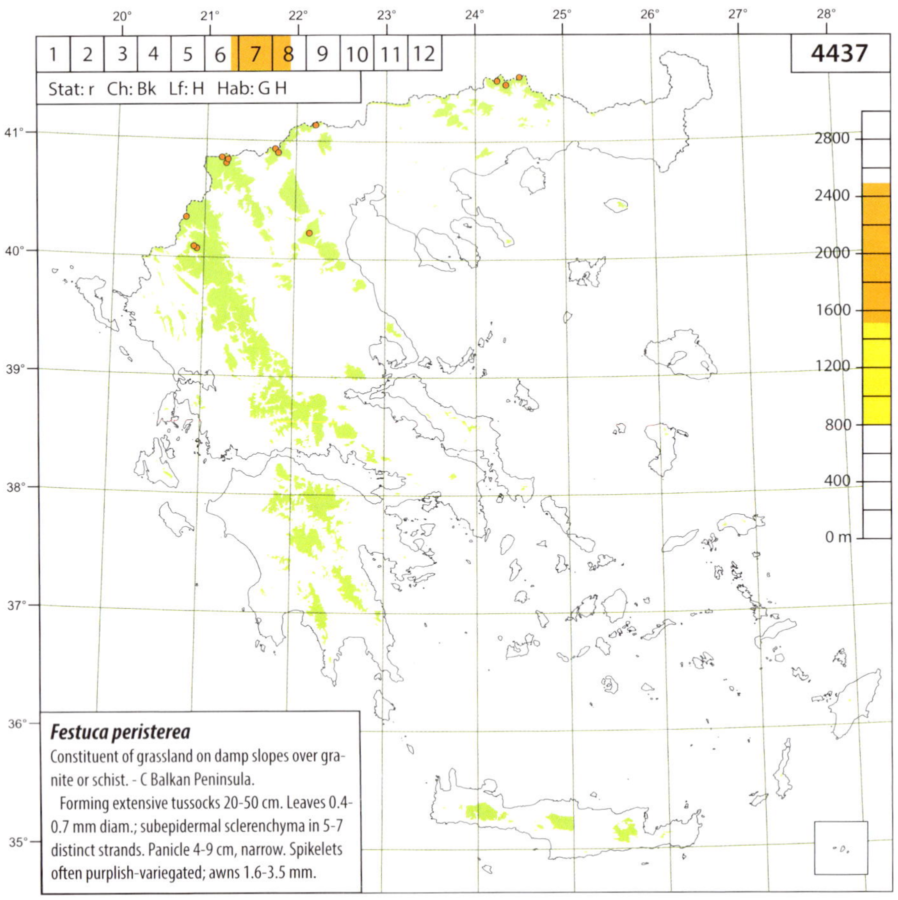

Festuca peristerea

Constituent of grassland on damp slopes over granite or schist. - C Balkan Peninsula.

Forming extensive tussocks 20-50 cm. Leaves 0.4-0.7 mm diam.; subepidermal sclerenchyma in 5-7 distinct strands. Panicle 4-9 cm, narrow. Spikelets often purplish-variegated; awns 1.6-3.5 mm.

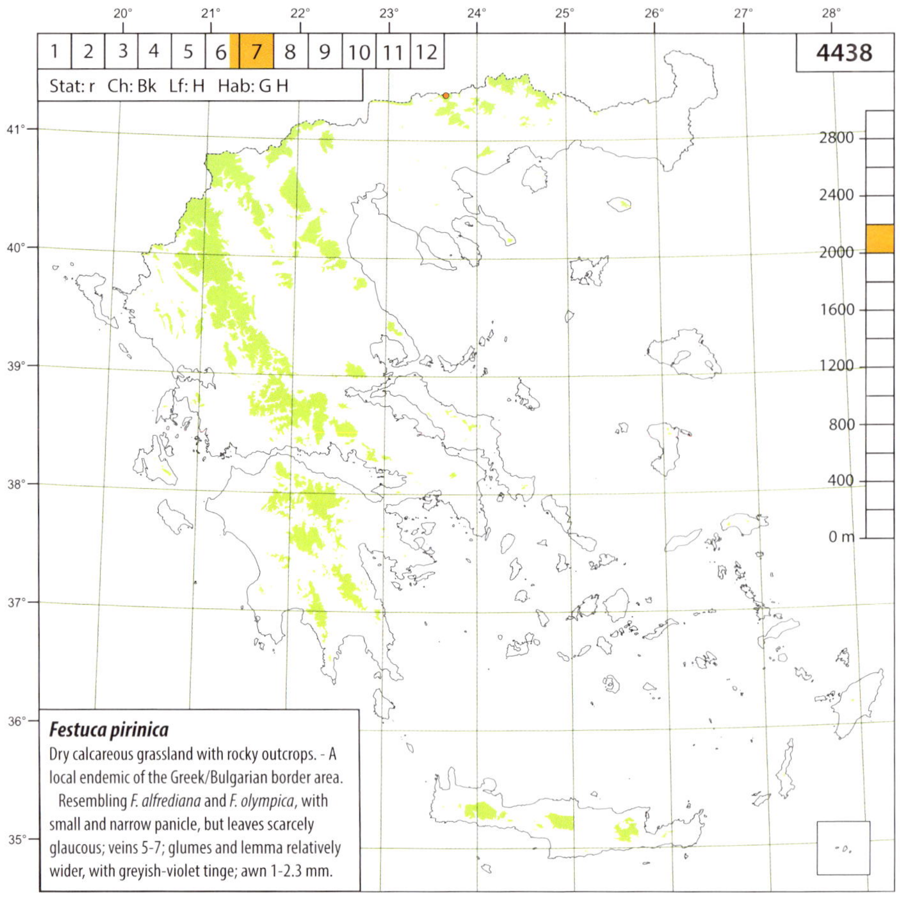

Festuca pirinica

Dry calcareous grassland with rocky outcrops. - A local endemic of the Greek/Bulgarian border area.

Resembling *F. alfrediana* and *F. olympica*, with small and narrow panicle, but leaves scarcely glaucous; veins 5-7; glumes and lemma relatively wider, with greyish-violet tinge; awn 1-2.3 mm.

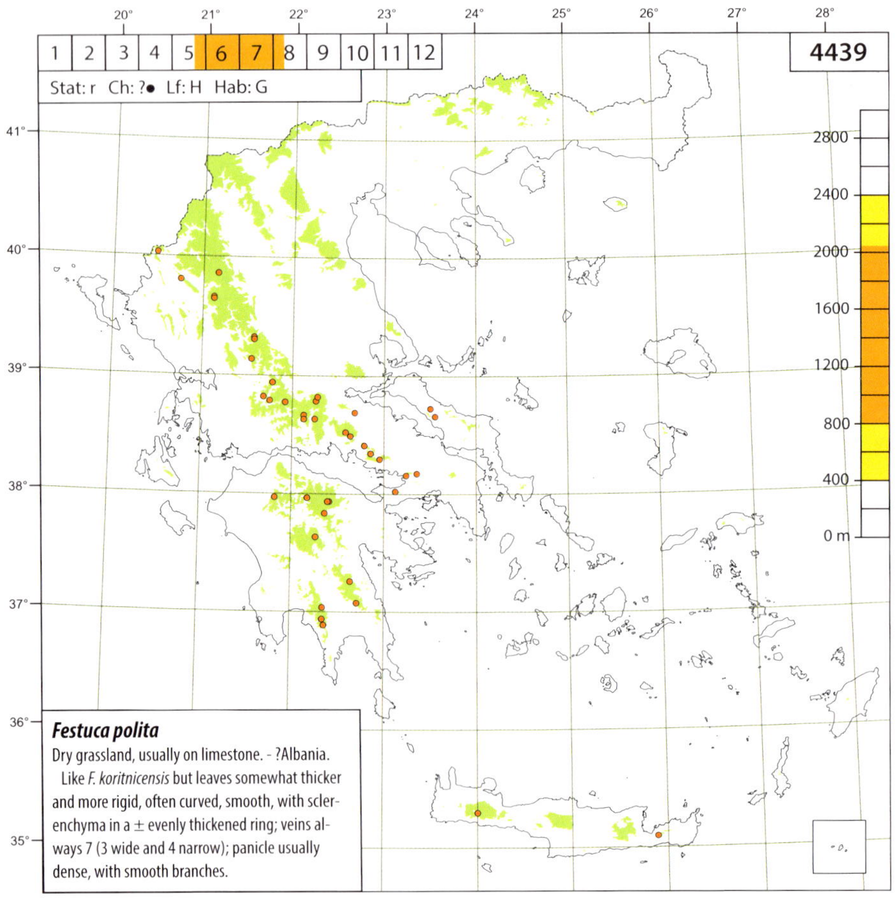

Festuca polita

Dry grassland, usually on limestone. - ?Albania.

Like *F. koritnicensis* but leaves somewhat thicker and more rigid, often curved, smooth, with sclerenchyma in a ± evenly thickened ring; veins always 7 (3 wide and 4 narrow); panicle usually dense, with smooth branches.

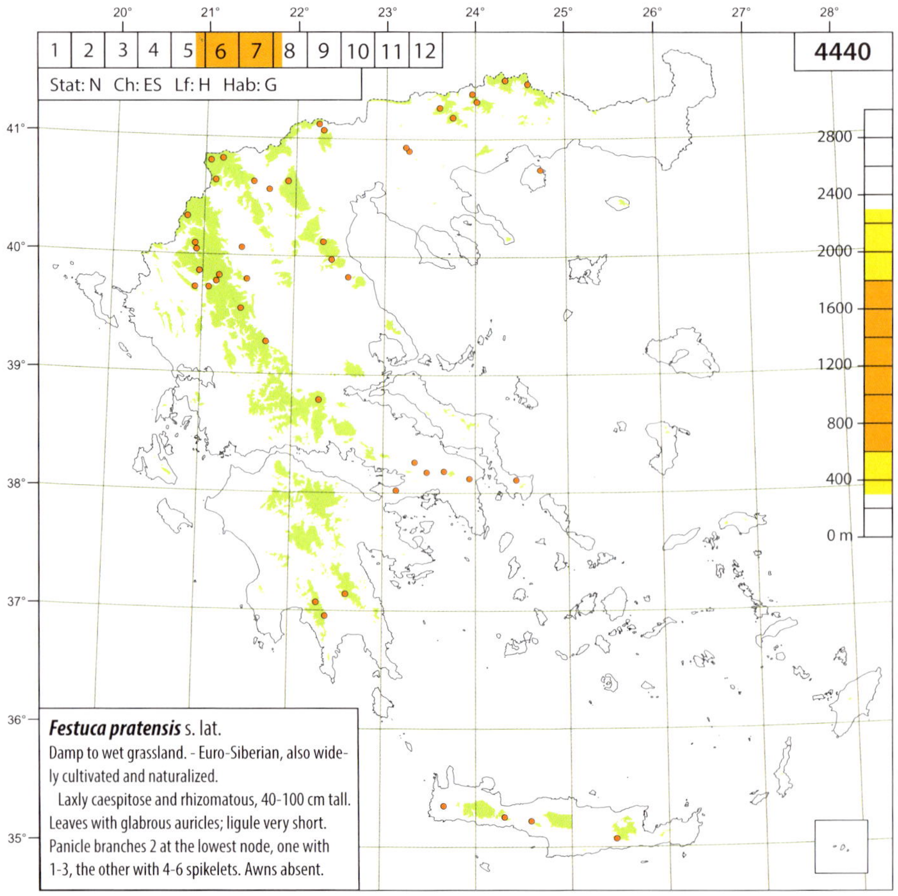

Festuca pratensis s. lat.

Damp to wet grassland. - Euro-Siberian, also widely cultivated and naturalized.

Laxly caespitose and rhizomatous, 40-100 cm tall. Leaves with glabrous auricles; ligule very short. Panicle branches 2 at the lowest node, one with 1-3, the other with 4-6 spikelets. Awns absent.

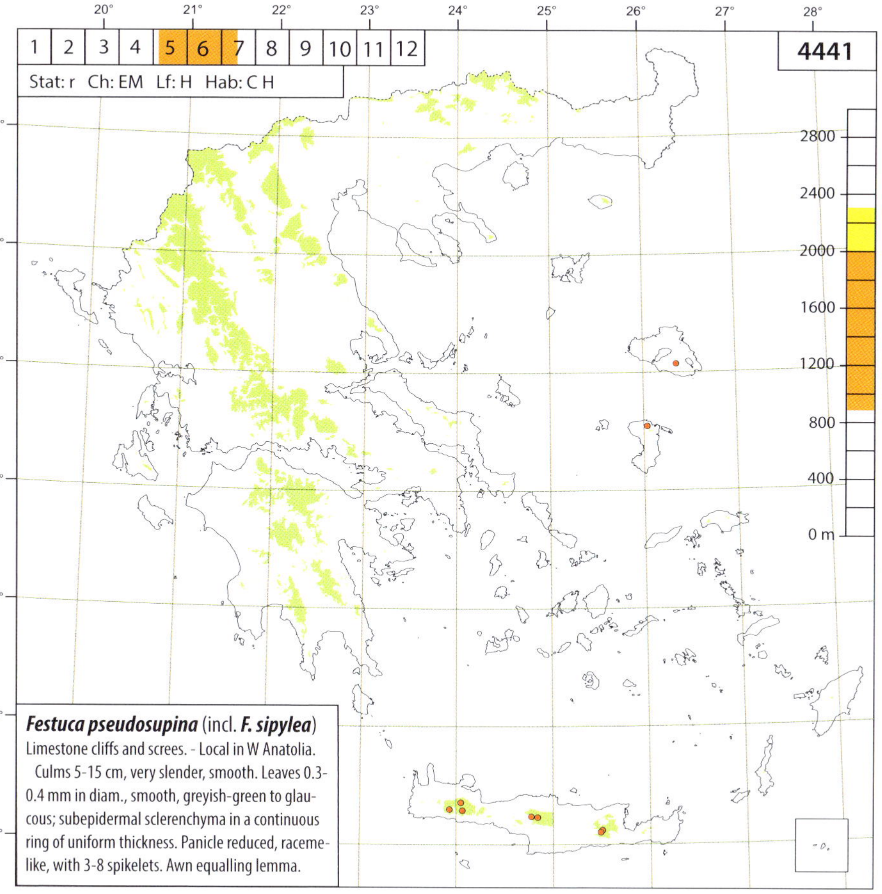

Festuca pseudosupina (incl. **F. sipylea**)
Limestone cliffs and screes. - Local in W Anatolia.
Culms 5-15 cm, very slender, smooth. Leaves 0.3-0.4 mm in diam., smooth, greyish-green to glaucous; subepidermal sclerenchyma in a continuous ring of uniform thickness. Panicle reduced, raceme-like, with 3-8 spikelets. Awn equalling lemma.

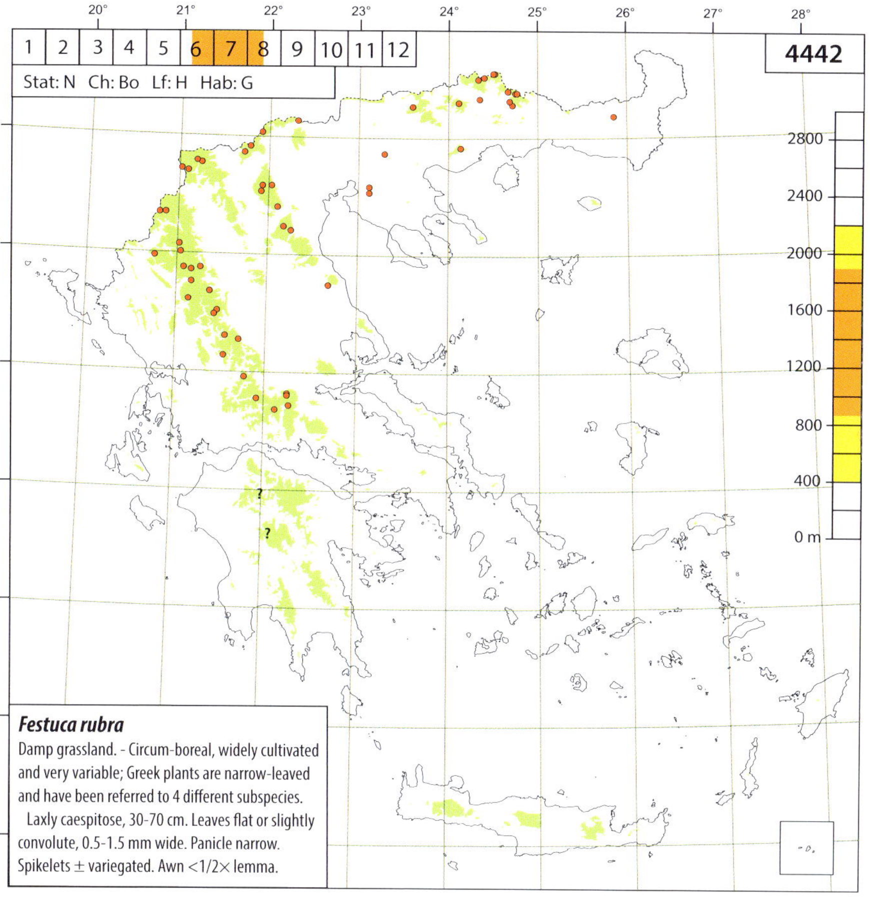

Festuca rubra
Damp grassland. - Circum-boreal, widely cultivated and very variable; Greek plants are narrow-leaved and have been referred to 4 different subspecies.
Laxly caespitose, 30-70 cm. Leaves flat or slightly convolute, 0.5-1.5 mm wide. Panicle narrow. Spikelets ± variegated. Awn <1/2× lemma.

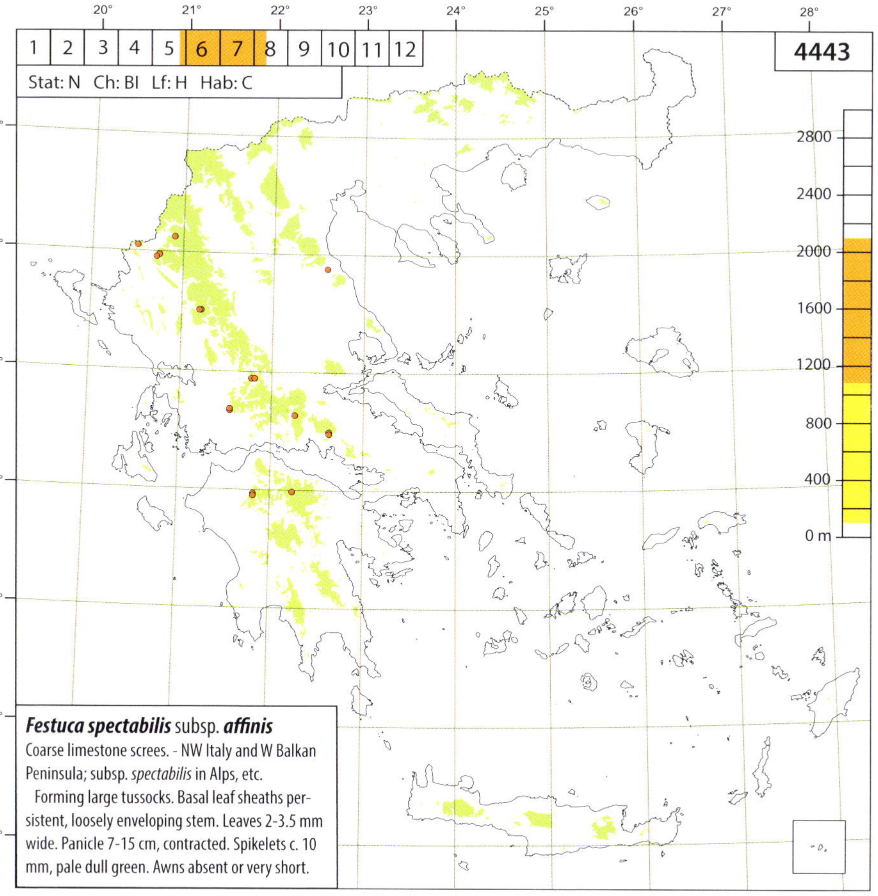

Festuca spectabilis subsp. **affinis**
Coarse limestone screes. - NW Italy and W Balkan Peninsula; subsp. *spectabilis* in Alps, etc.
Forming large tussocks. Basal leaf sheaths persistent, loosely enveloping stem. Leaves 2-3.5 mm wide. Panicle 7-15 cm, contracted. Spikelets c. 10 mm, pale dull green. Awns absent or very short.

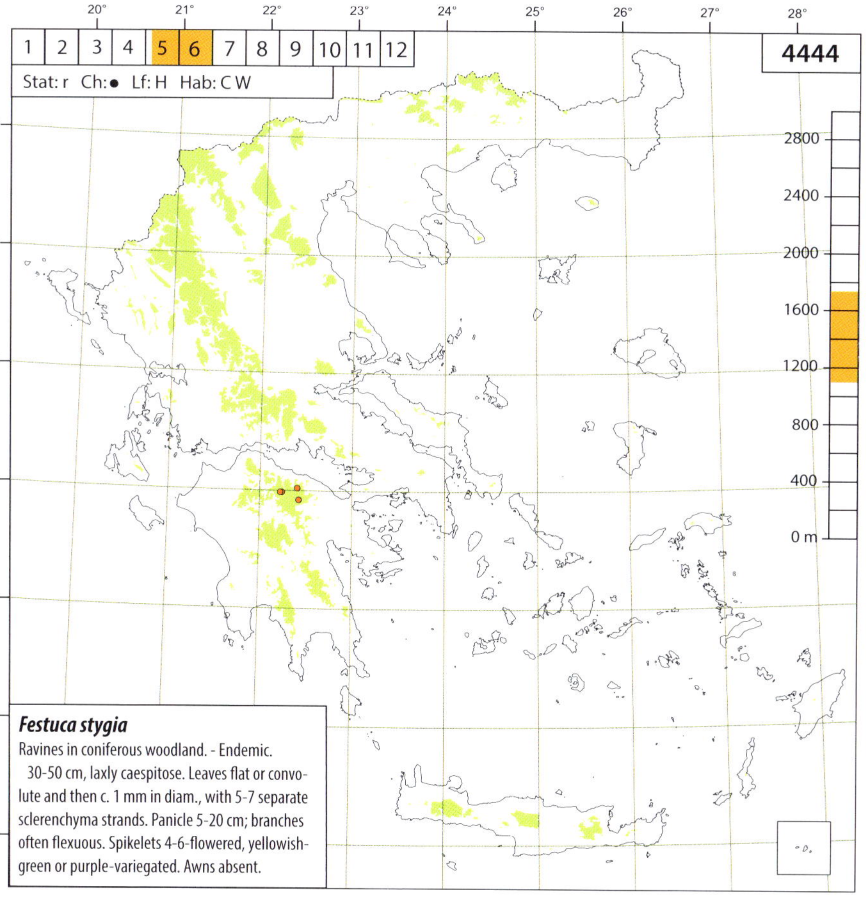

Festuca stygia
Ravines in coniferous woodland. - Endemic.
30-50 cm, laxly caespitose. Leaves flat or convolute and then c. 1 mm in diam., with 5-7 separate sclerenchyma strands. Panicle 5-20 cm; branches often flexuous. Spikelets 4-6-flowered, yellowish-green or purple-variegated. Awns absent.

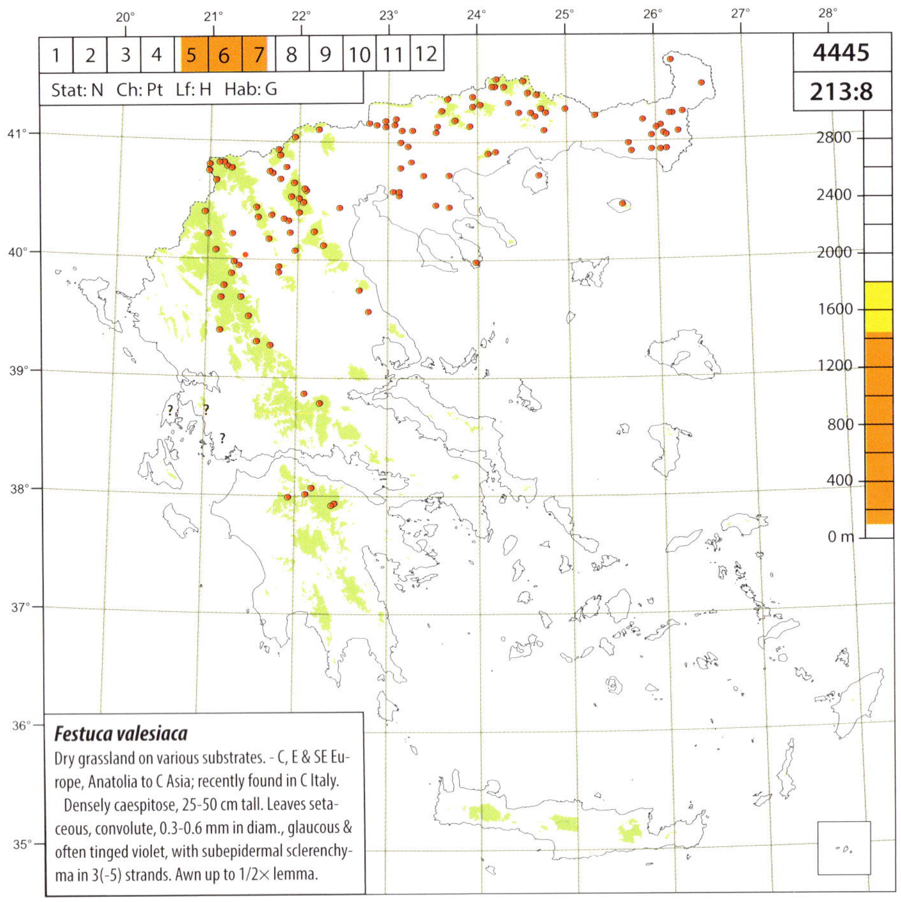

Festuca valesiaca
Dry grassland on various substrates. - C, E & SE Europe, Anatolia to C Asia; recently found in C Italy.
Densely caespitose, 25-50 cm tall. Leaves setaceous, convolute, 0.3-0.6 mm in diam., glaucous & often tinged violet, with subepidermal sclerenchyma in 3(-5) strands. Awn up to 1/2× lemma.

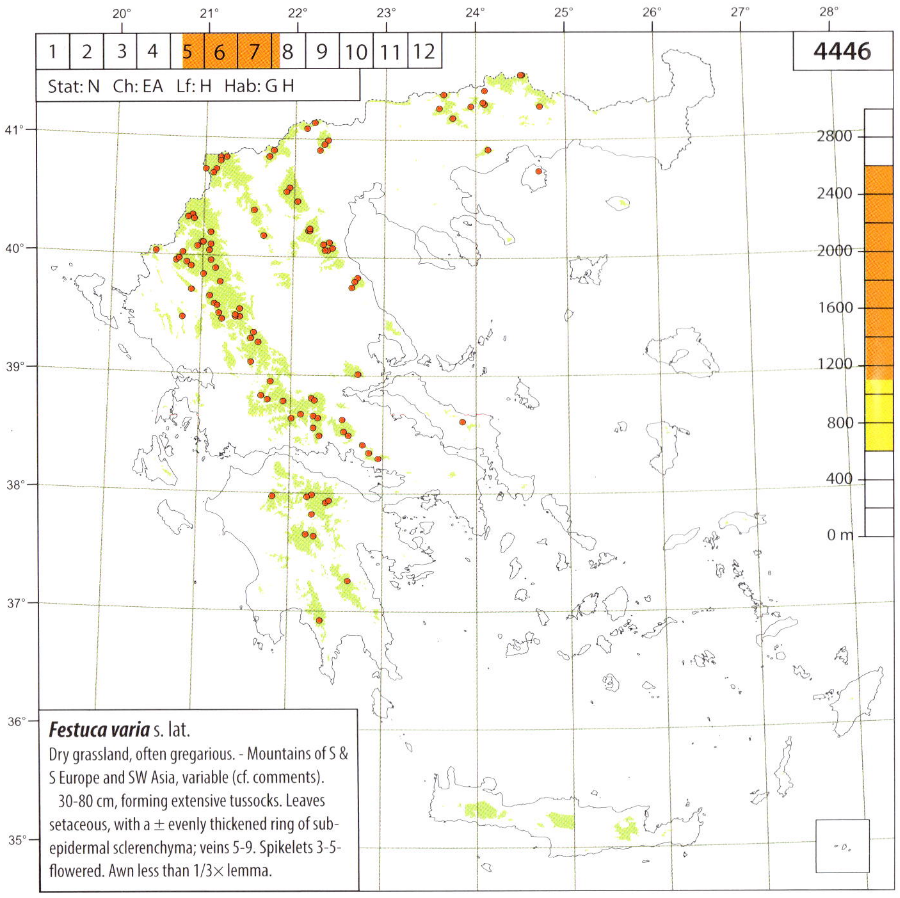

Festuca varia s. lat.
Dry grassland, often gregarious. - Mountains of S & S Europe and SW Asia, variable (cf. comments).
30-80 cm, forming extensive tussocks. Leaves setaceous, with a ± evenly thickened ring of subepidermal sclerenchyma; veins 5-9. Spikelets 3-5-flowered. Awn less than 1/3× lemma.

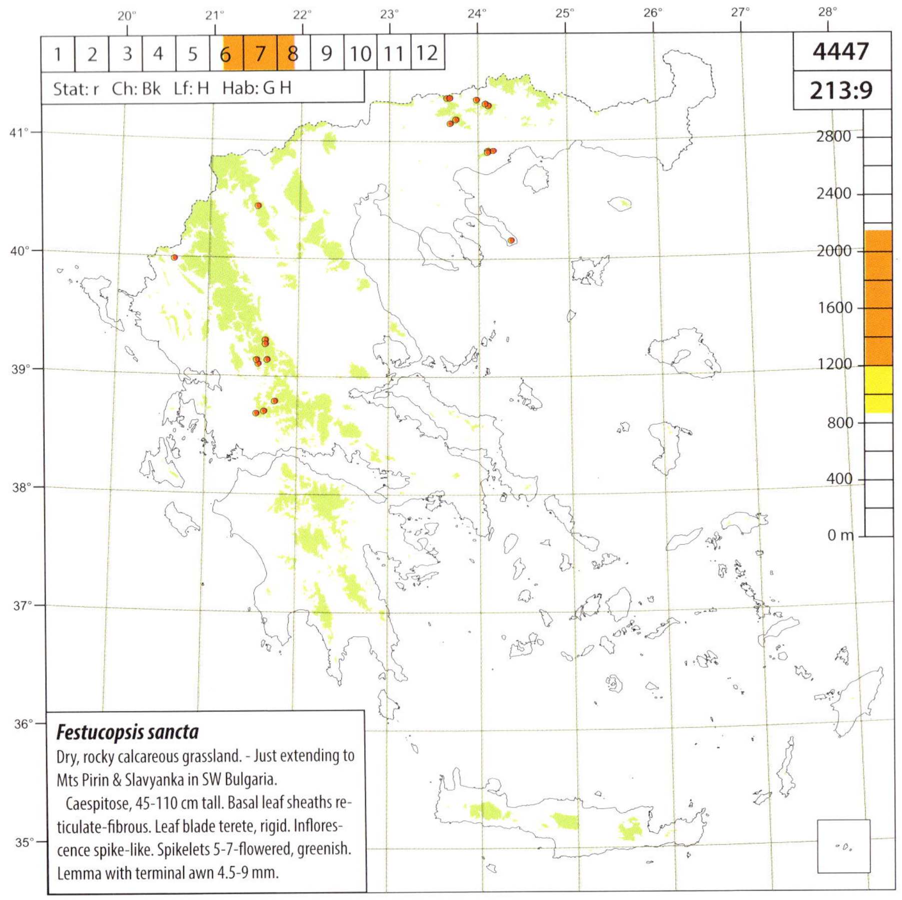

Festucopsis sancta
Dry, rocky calcareous grassland. - Just extending to Mts Pirin & Slavyanka in SW Bulgaria.
Caespitose, 45-110 cm tall. Basal leaf sheaths reticulate-fibrous. Leaf blade terete, rigid. Inflorescence spike-like. Spikelets 5-7-flowered, greenish. Lemma with terminal awn 4.5-9 mm.

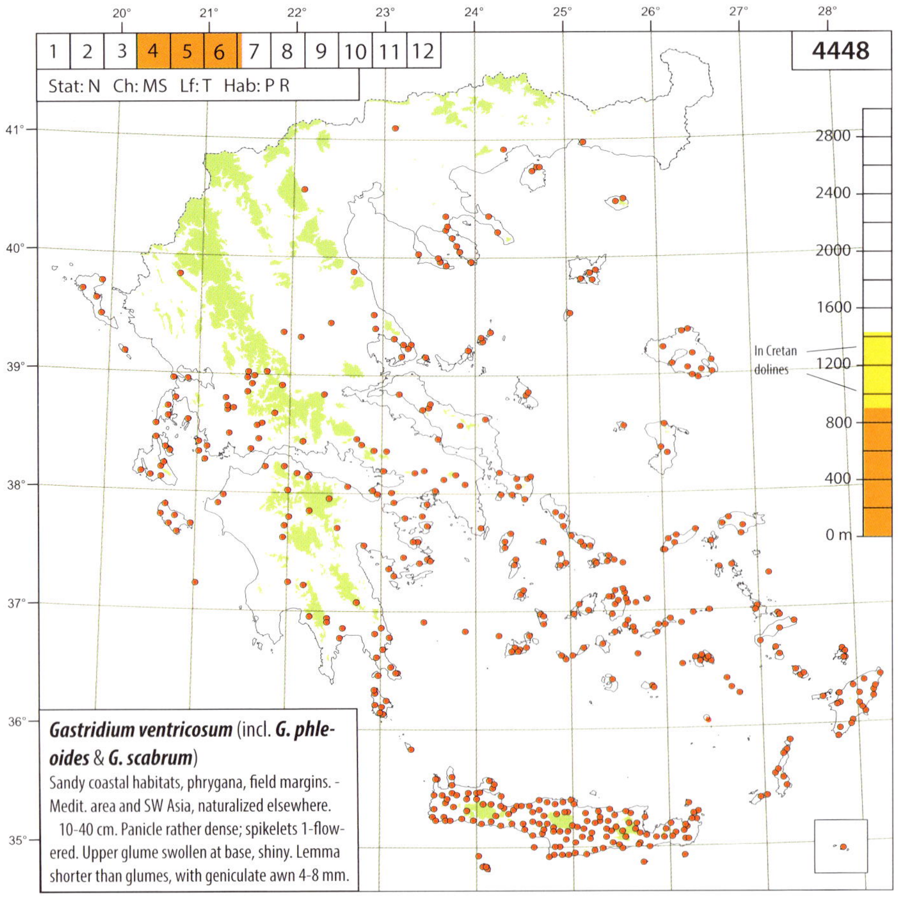

Gastridium ventricosum (incl. ***G. phleoides*** & ***G. scabrum***)
Sandy coastal habitats, phrygana, field margins. - Medit. area and SW Asia, naturalized elsewhere.
10-40 cm. Panicle rather dense; spikelets 1-flowered. Upper glume swollen at base, shiny. Lemma shorter than glumes, with geniculate awn 4-8 mm.

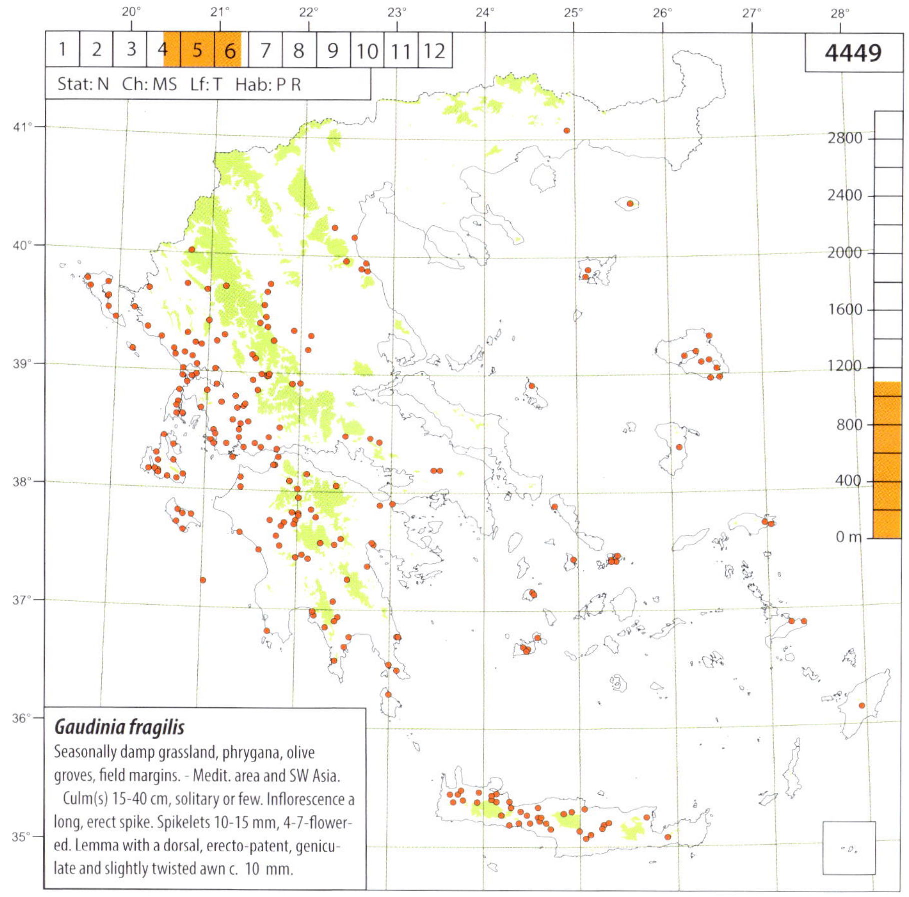

Gaudinia fragilis

Seasonally damp grassland, phrygana, olive groves, field margins. - Medit. area and SW Asia.

Culm(s) 15-40 cm, solitary or few. Inflorescence a long, erect spike. Spikelets 10-15 mm, 4-7-flowered. Lemma with a dorsal, erecto-patent, geniculate and slightly twisted awn c. 10 mm.

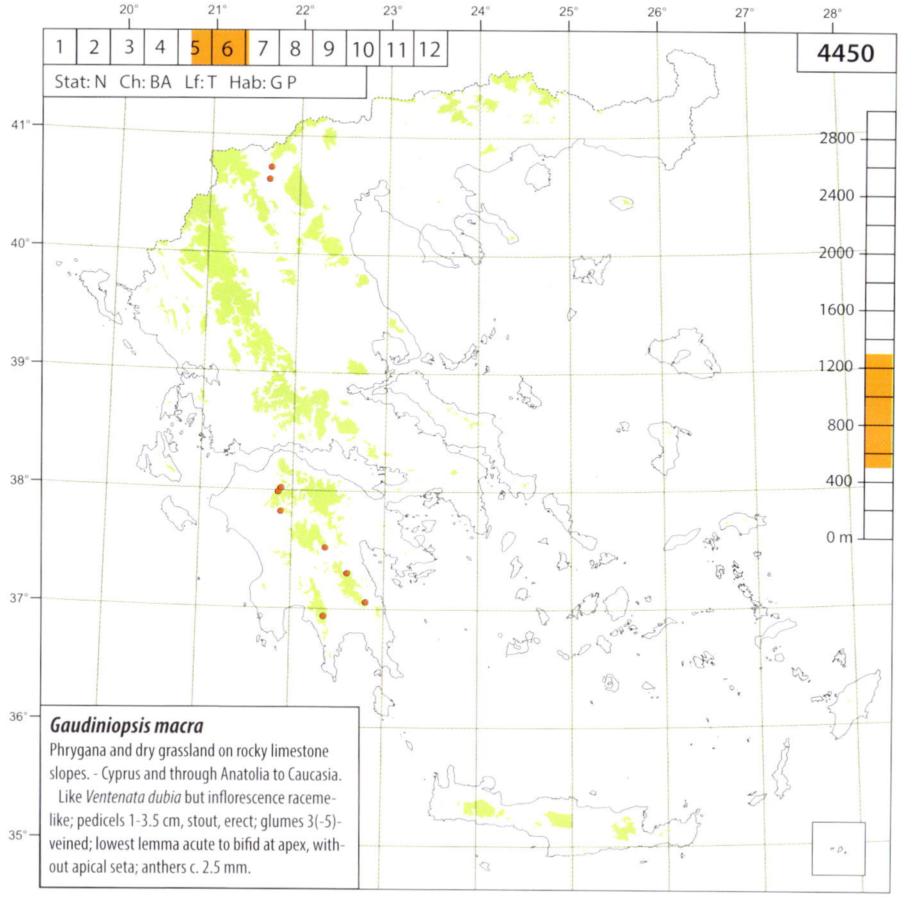

Gaudiniopsis macra

Phrygana and dry grassland on rocky limestone slopes. - Cyprus and through Anatolia to Caucasia.

Like *Ventenata dubia* but inflorescence raceme-like; pedicels 1-3.5 cm, stout, erect; glumes 3(-5)-veined; lowest lemma acute to bifid at apex, without apical seta; anthers c. 2.5 mm.

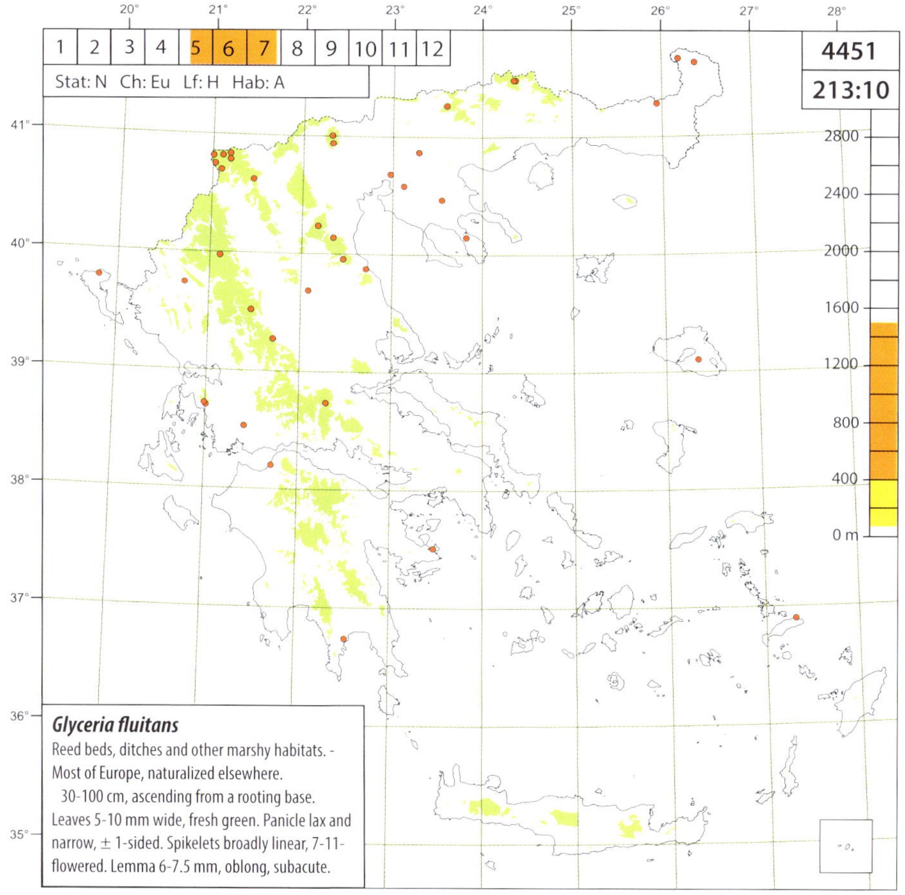

Glyceria fluitans

Reed beds, ditches and other marshy habitats. - Most of Europe, naturalized elsewhere.

30-100 cm, ascending from a rooting base. Leaves 5-10 mm wide, fresh green. Panicle lax and narrow, ± 1-sided. Spikelets broadly linear, 7-11-flowered. Lemma 6-7.5 mm, oblong, subacute.

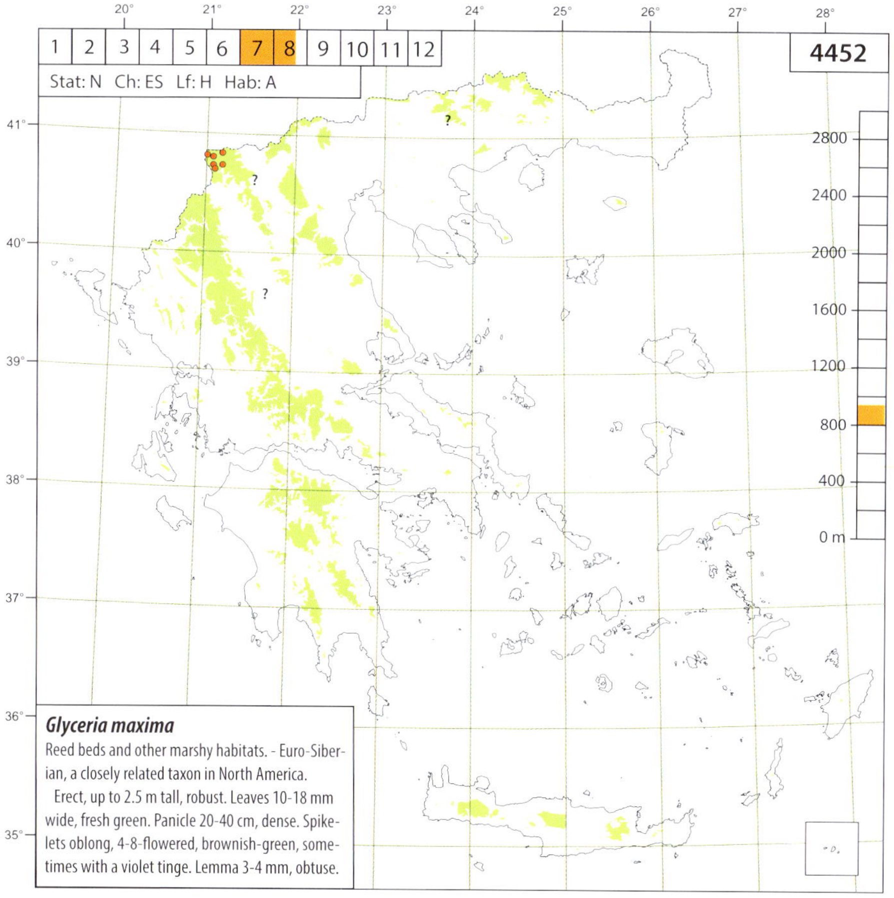

Glyceria maxima

Reed beds and other marshy habitats. - Euro-Siberian, a closely related taxon in North America.

Erect, up to 2.5 m tall, robust. Leaves 10-18 mm wide, fresh green. Panicle 20-40 cm, dense. Spikelets oblong, 4-8-flowered, brownish-green, sometimes with a violet tinge. Lemma 3-4 mm, obtuse.

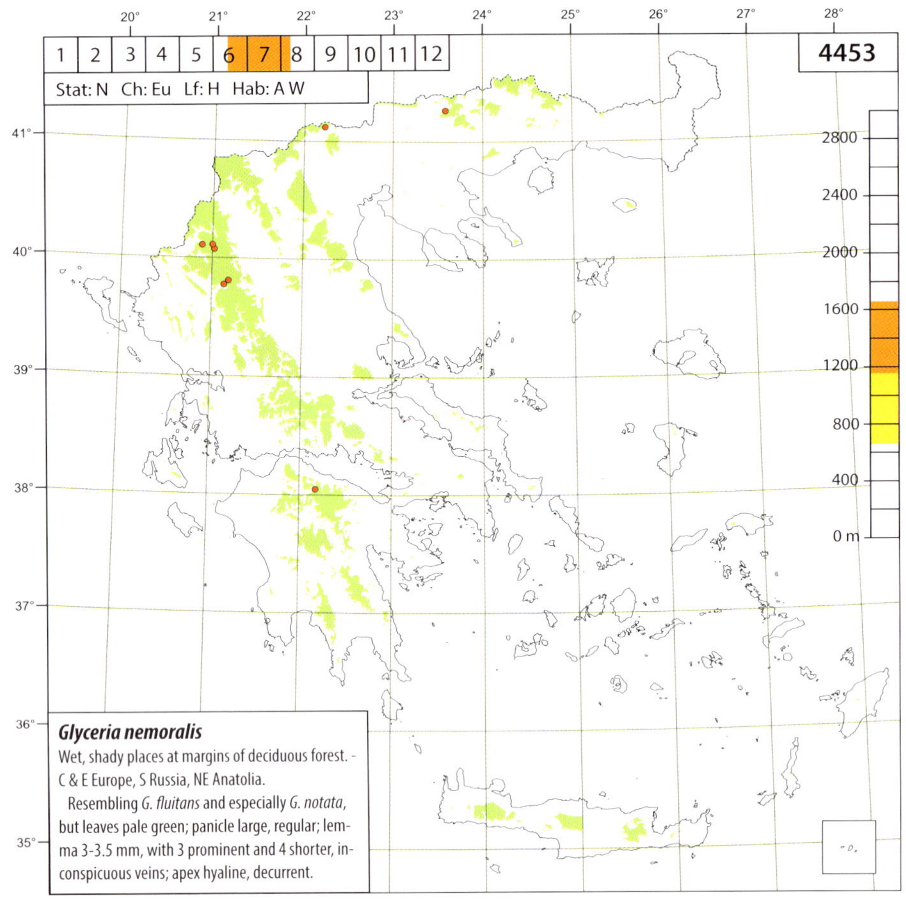

Glyceria nemoralis

Wet, shady places at margins of deciduous forest. – C & E Europe, S Russia, NE Anatolia.

Resembling *G. fluitans* and especially *G. notata*, but leaves pale green; panicle large, regular; lemma 3-3.5 mm, with 3 prominent and 4 shorter, inconspicuous veins; apex hyaline, decurrent.

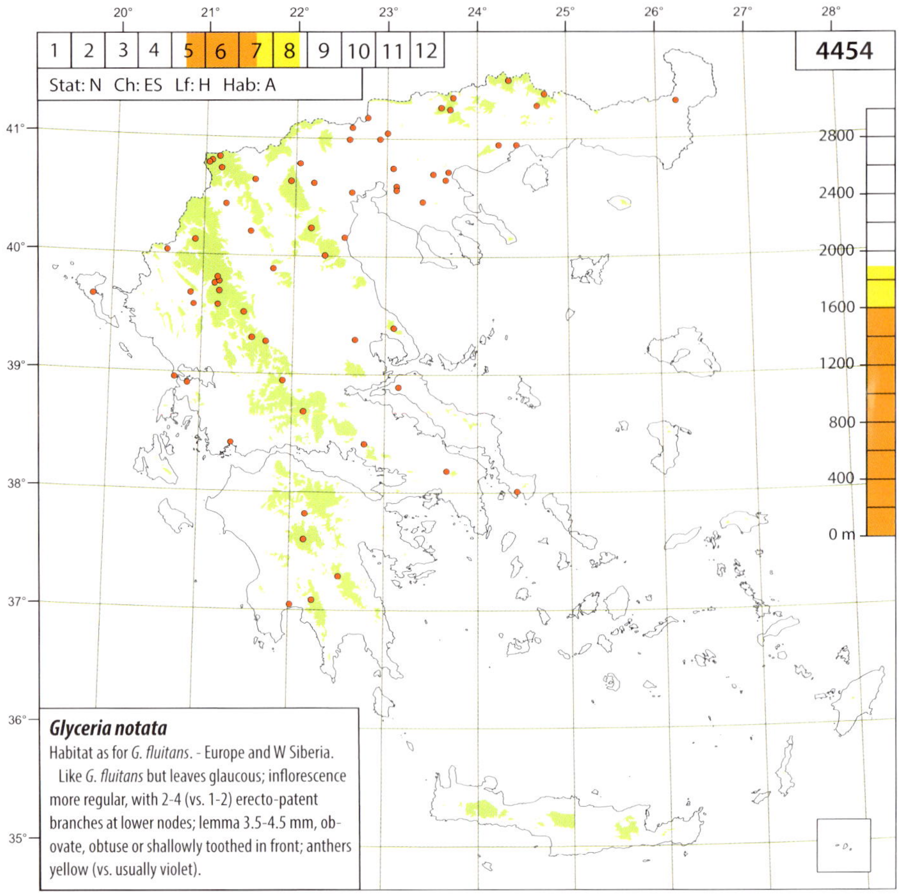

Glyceria notata

Habitat as for *G. fluitans*. – Europe and W Siberia.

Like *G. fluitans* but leaves glaucous; inflorescence more regular, with 2-4 (vs. 1-2) erecto-patent branches at lower nodes; lemma 3.5-4.5 mm, obovate, obtuse or shallowly toothed in front; anthers yellow (vs. usually violet).

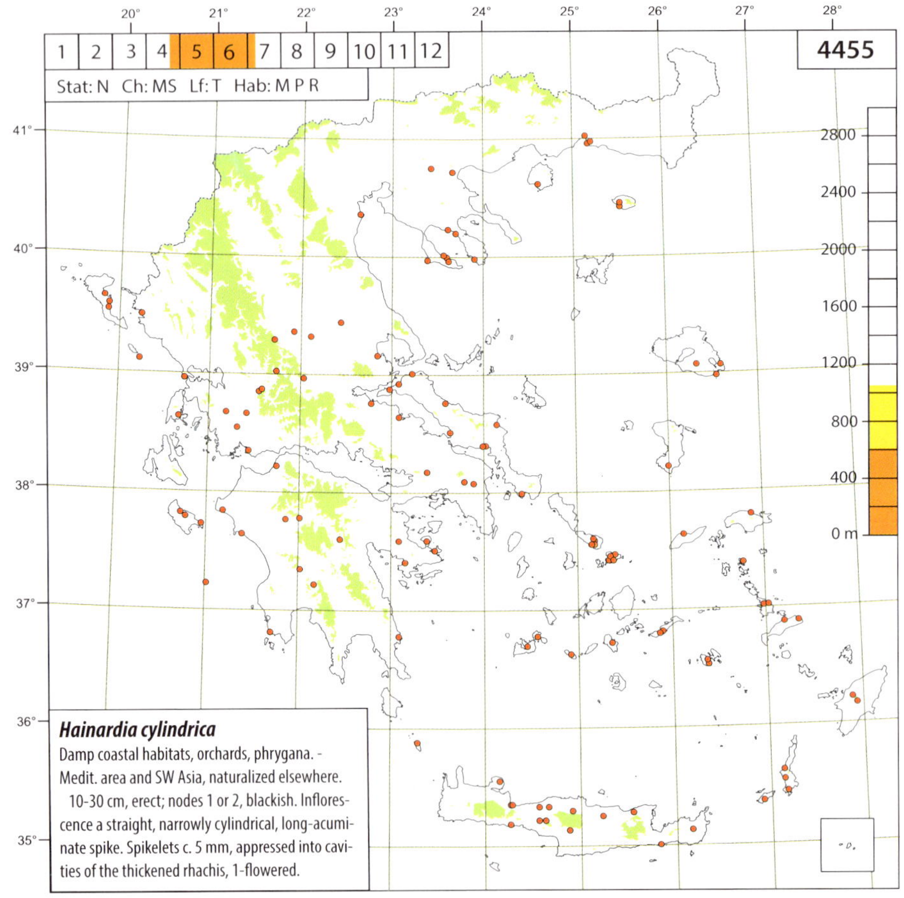

Hainardia cylindrica

Damp coastal habitats, orchards, phrygana. – Medit. area and SW Asia, naturalized elsewhere.

10-30 cm, erect; nodes 1 or 2, blackish. Inflorescence a straight, narrowly cylindrical, long-acuminate spike. Spikelets c. 5 mm, appressed into cavities of the thickened rhachis, 1-flowered.

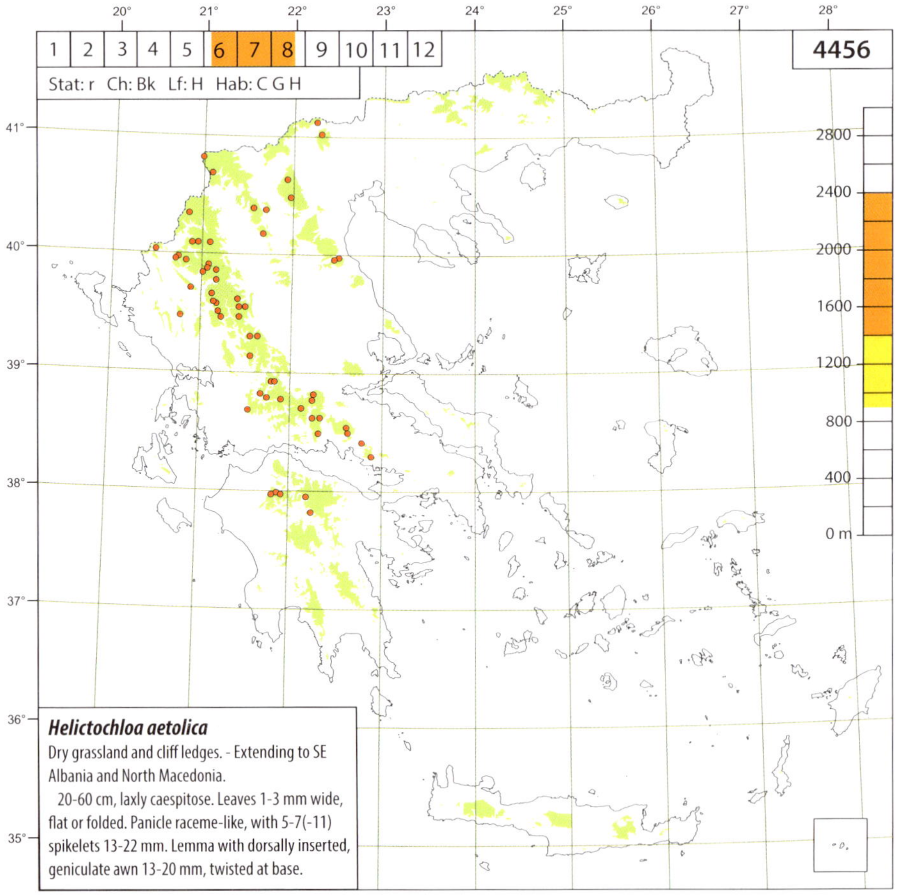

Helictochloa aetolica

Dry grassland and cliff ledges. – Extending to SE Albania and North Macedonia.

20-60 cm, laxly caespitose. Leaves 1-3 mm wide, flat or folded. Panicle raceme-like, with 5-7(-11) spikelets 13-22 mm. Lemma with dorsally inserted, geniculate awn 13-20 mm, twisted at base.

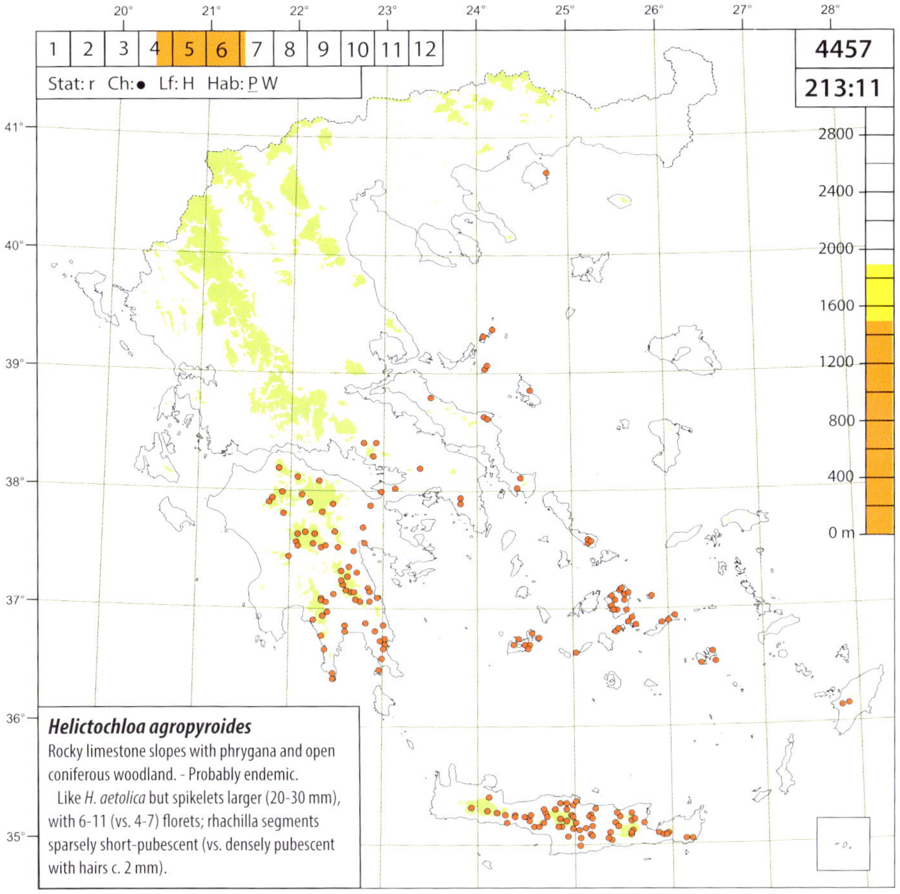

Helictochloa agropyroides

Rocky limestone slopes with phrygana and open coniferous woodland. - Probably endemic.

Like *H. aetolica* but spikelets larger (20-30 mm), with 6-11 (vs. 4-7) florets; rhachilla segments sparsely short-pubescent (vs. densely pubescent with hairs c. 2 mm).

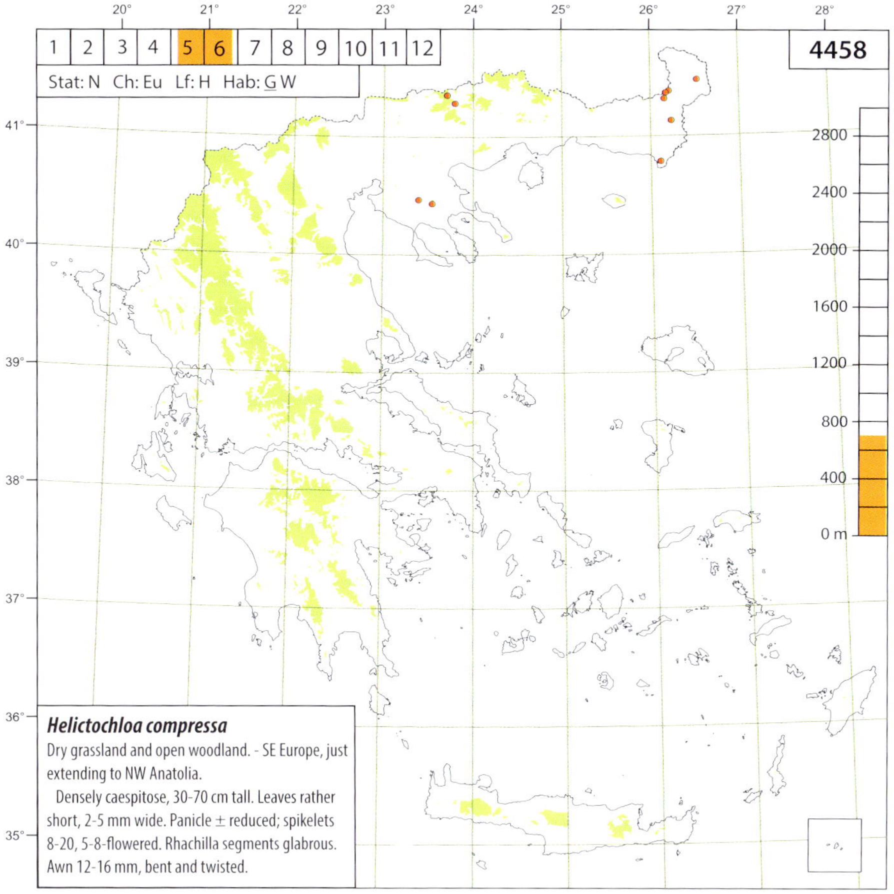

Helictochloa compressa

Dry grassland and open woodland. - SE Europe, just extending to NW Anatolia.

Densely caespitose, 30-70 cm tall. Leaves rather short, 2-5 mm wide. Panicle ± reduced; spikelets 8-20, 5-8-flowered. Rhachilla segments glabrous. Awn 12-16 mm, bent and twisted.

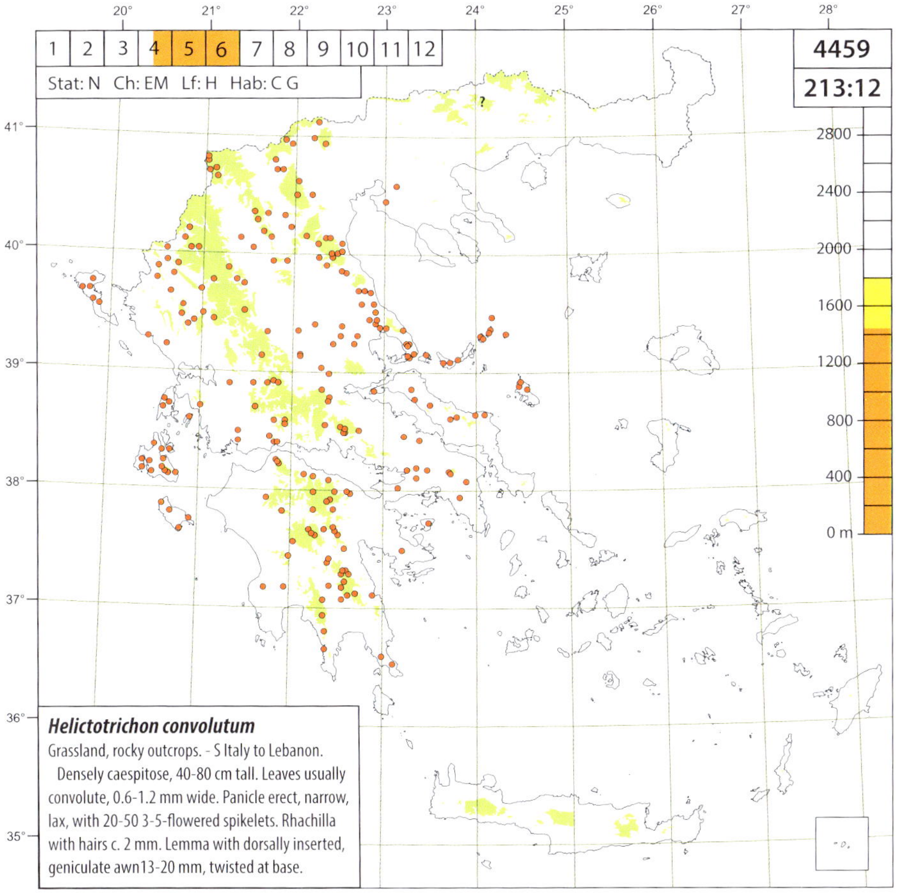

Helictotrichon convolutum

Grassland, rocky outcrops. - S Italy to Lebanon.

Densely caespitose, 40-80 cm tall. Leaves usually convolute, 0.6-1.2 mm wide. Panicle erect, narrow, lax, with 20-50 3-5-flowered spikelets. Rhachilla with hairs c. 2 mm. Lemma with dorsally inserted, geniculate awn13-20 mm, twisted at base.

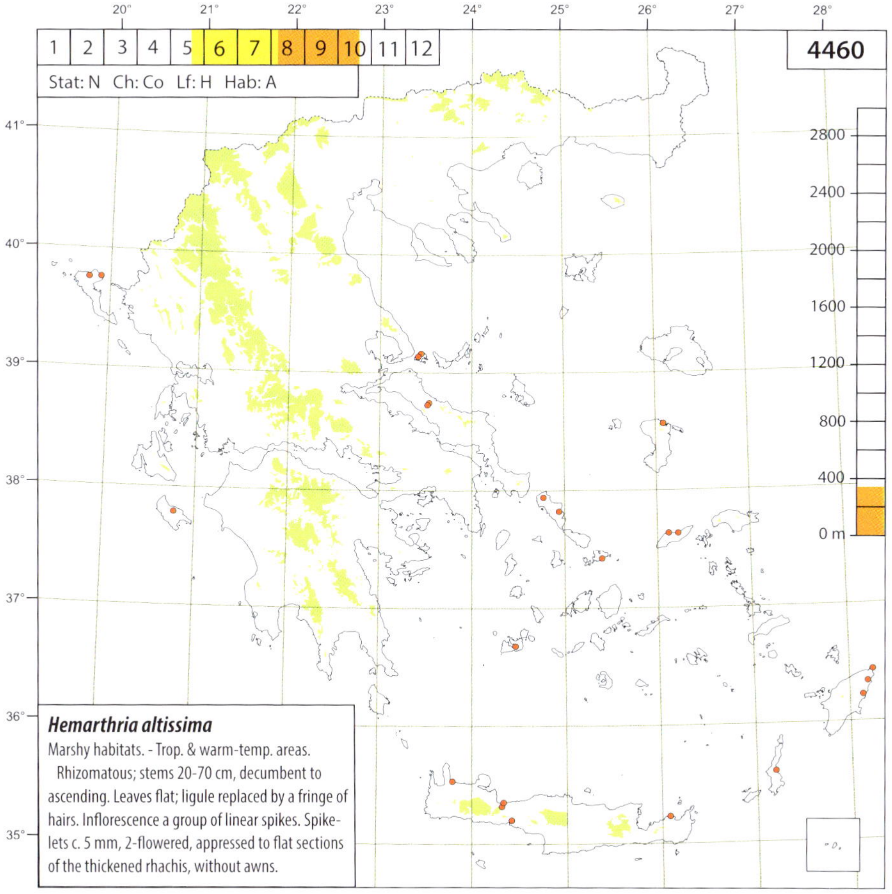

Hemarthria altissima

Marshy habitats. - Trop. & warm-temp. areas.

Rhizomatous; stems 20-70 cm, decumbent to ascending. Leaves flat; ligule replaced by a fringe of hairs. Inflorescence a group of linear spikes. Spikelets c. 5 mm, 2-flowered, appressed to flat sections of the thickened rhachis, without awns.

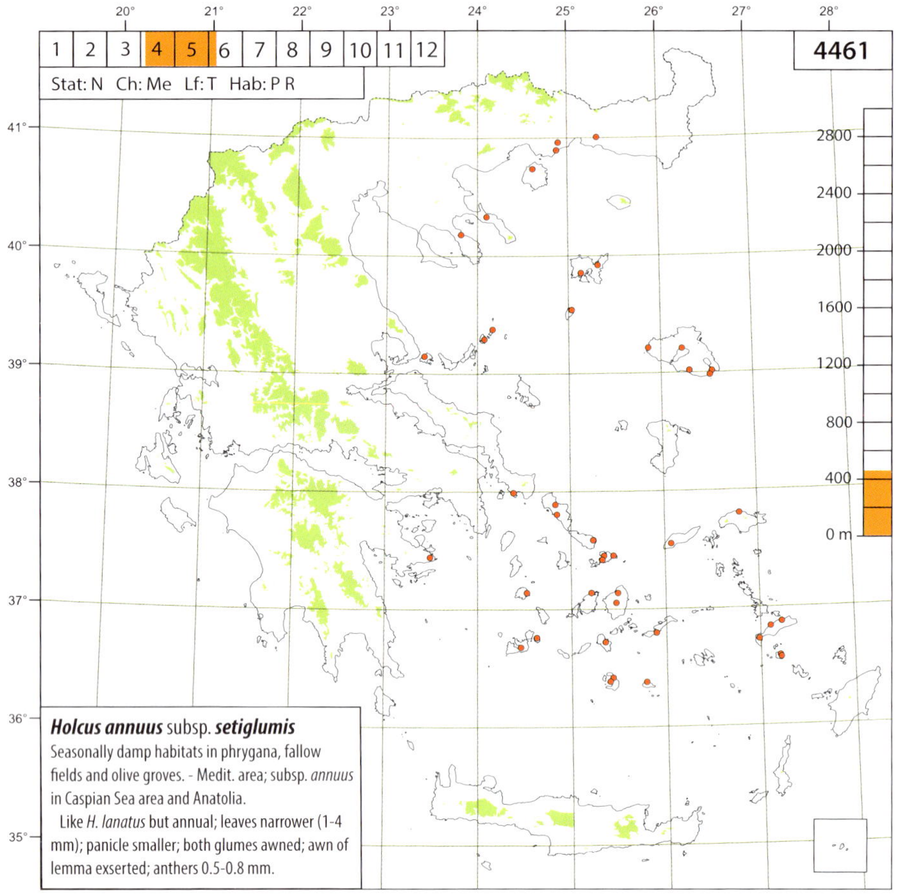

Holcus annuus subsp. ***setiglumis***
Seasonally damp habitats in phrygana, fallow fields and olive groves. - Medit. area; subsp. *annuus* in Caspian Sea area and Anatolia.
Like *H. lanatus* but annual; leaves narrower (1-4 mm); panicle smaller; both glumes awned; awn of lemma exserted; anthers 0.5-0.8 mm.

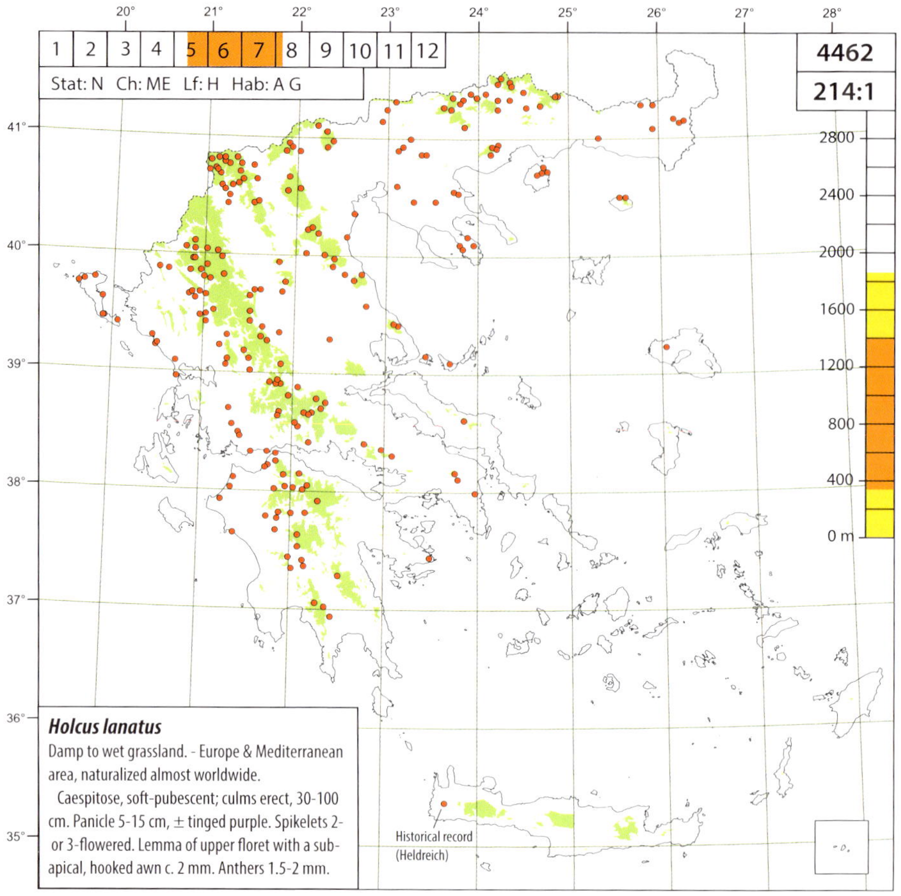

Holcus lanatus
Damp to wet grassland. - Europe & Mediterranean area, naturalized almost worldwide.
Caespitose, soft-pubescent; culms erect, 30-100 cm. Panicle 5-15 cm, ± tinged purple. Spikelets 2- or 3-flowered. Lemma of upper floret with a sub-apical, hooked awn c. 2 mm. Anthers 1.5-2 mm.

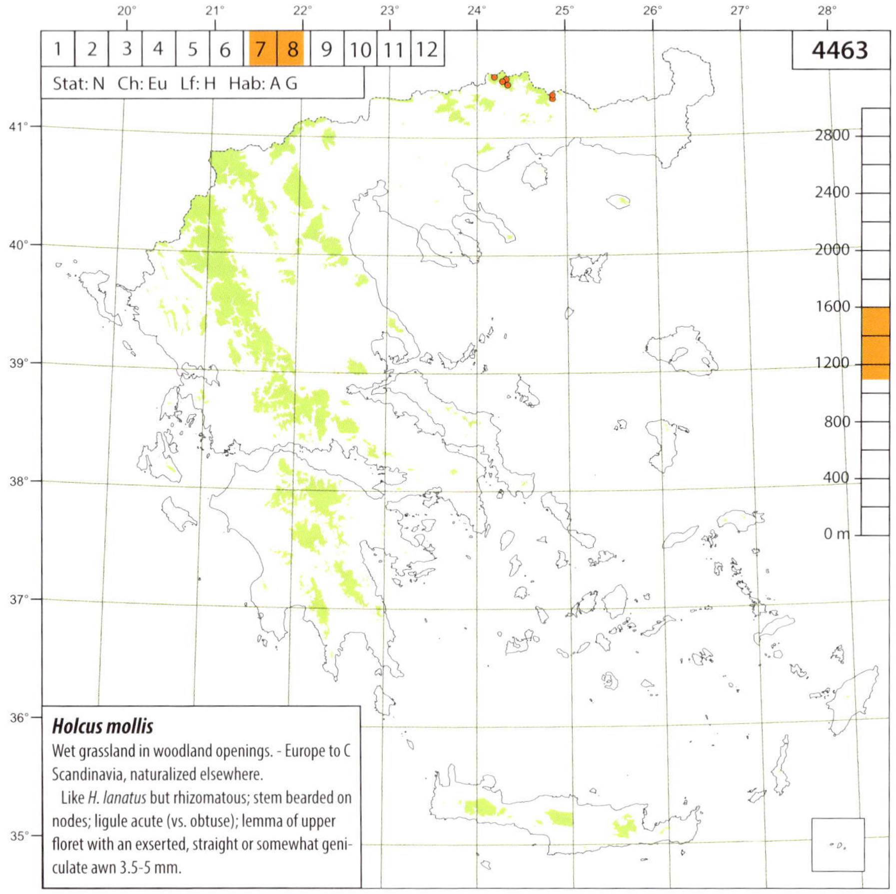

Holcus mollis
Wet grassland in woodland openings. - Europe to C Scandinavia, naturalized elsewhere.
Like *H. lanatus* but rhizomatous; stem bearded on nodes; ligule acute (vs. obtuse); lemma of upper floret with an exserted, straight or somewhat geniculate awn 3.5-5 mm.

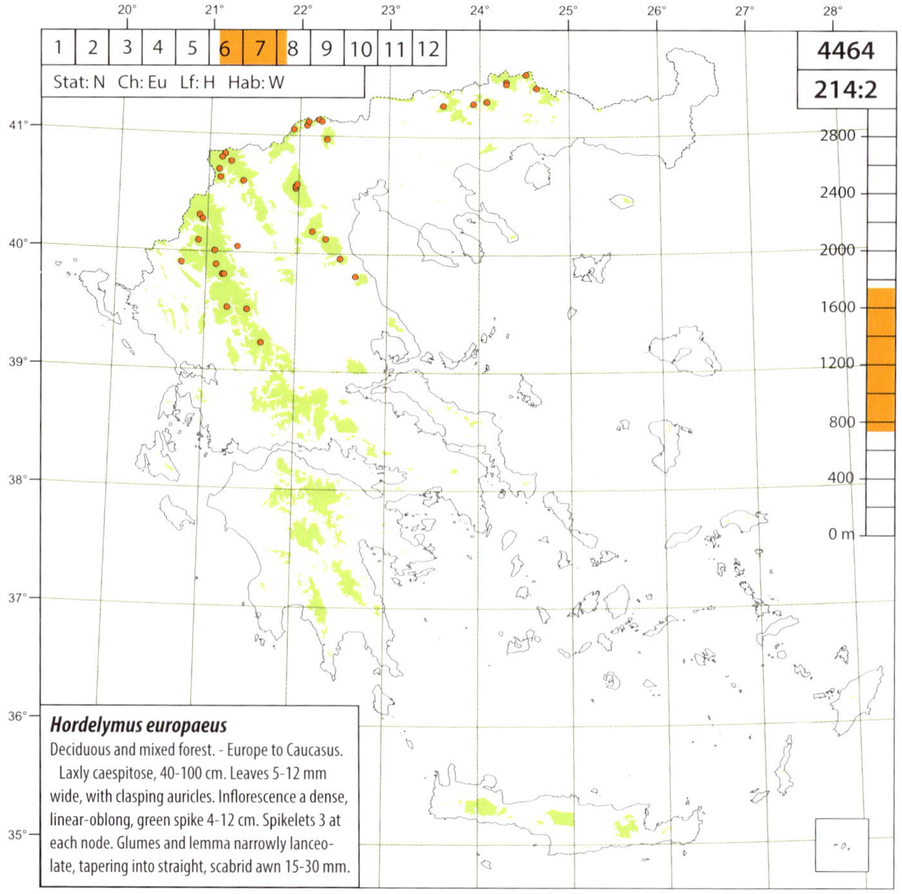

Hordelymus europaeus
Deciduous and mixed forest. - Europe to Caucasus.
Laxly caespitose, 40-100 cm. Leaves 5-12 mm wide, with clasping auricles. Inflorescence a dense, linear-oblong, green spike 4-12 cm. Spikelets 3 at each node. Glumes and lemma narrowly lanceolate, tapering into straight, scabrid awn 15-30 mm.

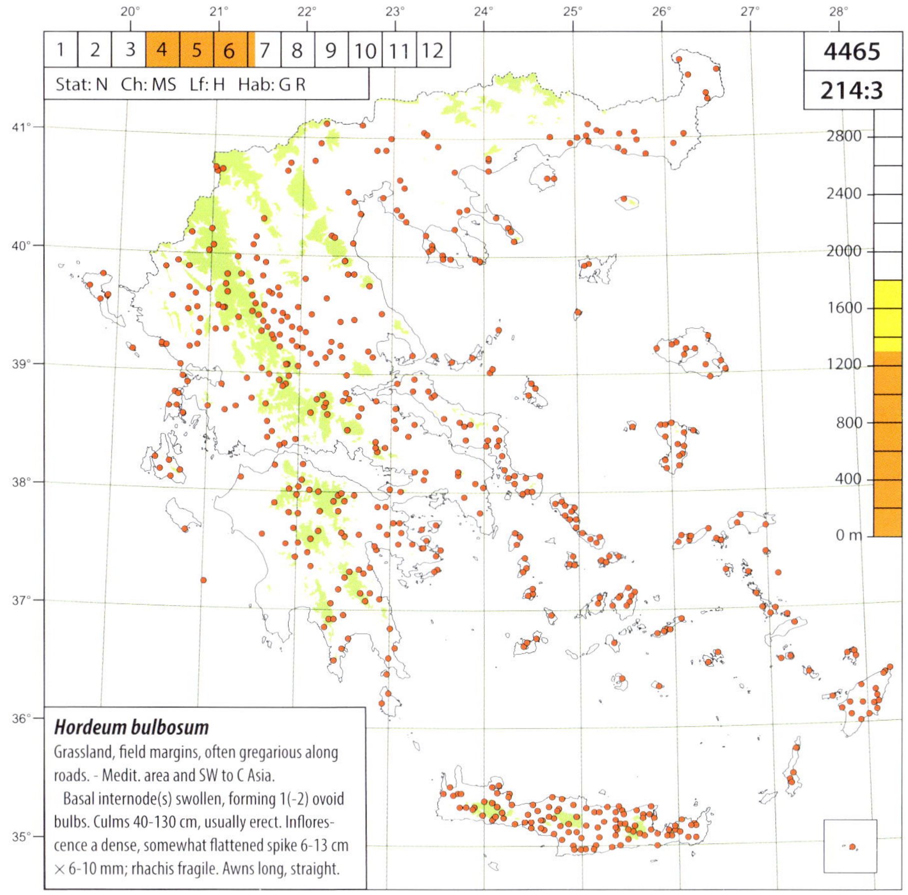

Hordeum bulbosum

Grassland, field margins, often gregarious along roads. - Medit. area and SW to C Asia.

Basal internode(s) swollen, forming 1(-2) ovoid bulbs. Culms 40-130 cm, usually erect. Inflorescence a dense, somewhat flattened spike 6-13 cm × 6-10 mm; rhachis fragile. Awns long, straight.

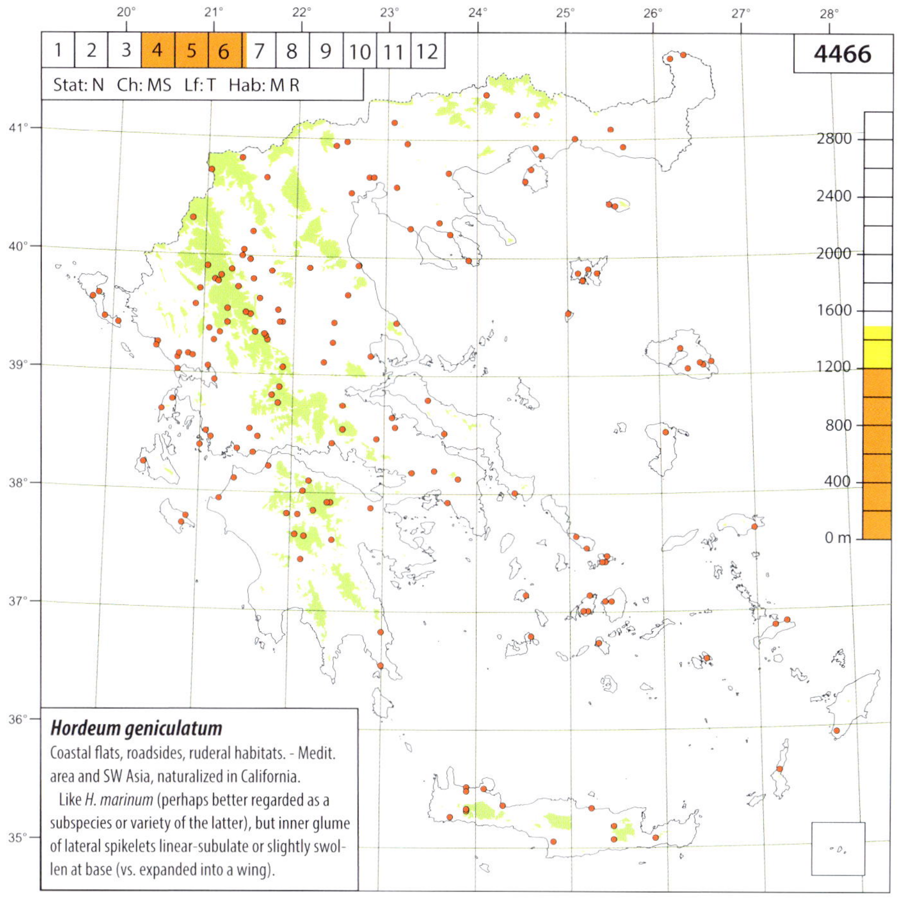

Hordeum geniculatum

Coastal flats, roadsides, ruderal habitats. - Medit. area and SW Asia, naturalized in California.

Like *H. marinum* (perhaps better regarded as a subspecies or variety of the latter), but inner glume of lateral spikelets linear-subulate or slightly swollen at base (vs. expanded into a wing).

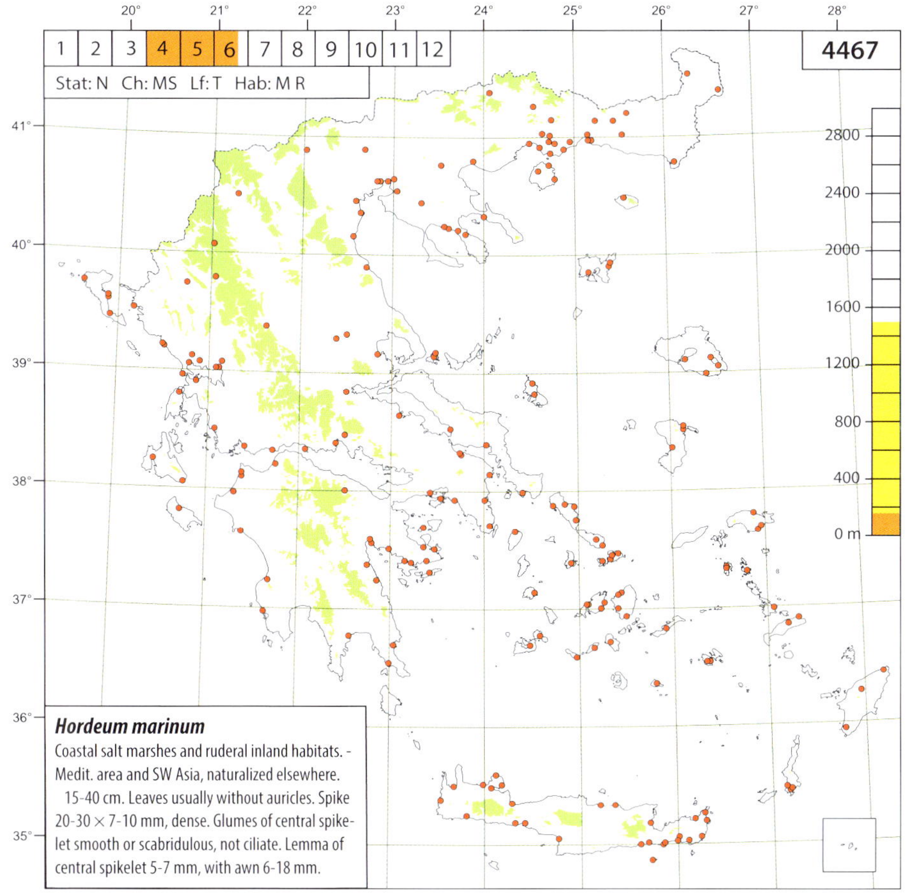

Hordeum marinum

Coastal salt marshes and ruderal inland habitats. - Medit. area and SW Asia, naturalized elsewhere.

15-40 cm. Leaves usually without auricles. Spike 20-30 × 7-10 mm, dense. Glumes of central spikelet smooth or scabridulous, not ciliate. Lemma of central spikelet 5-7 mm, with awn 6-18 mm.

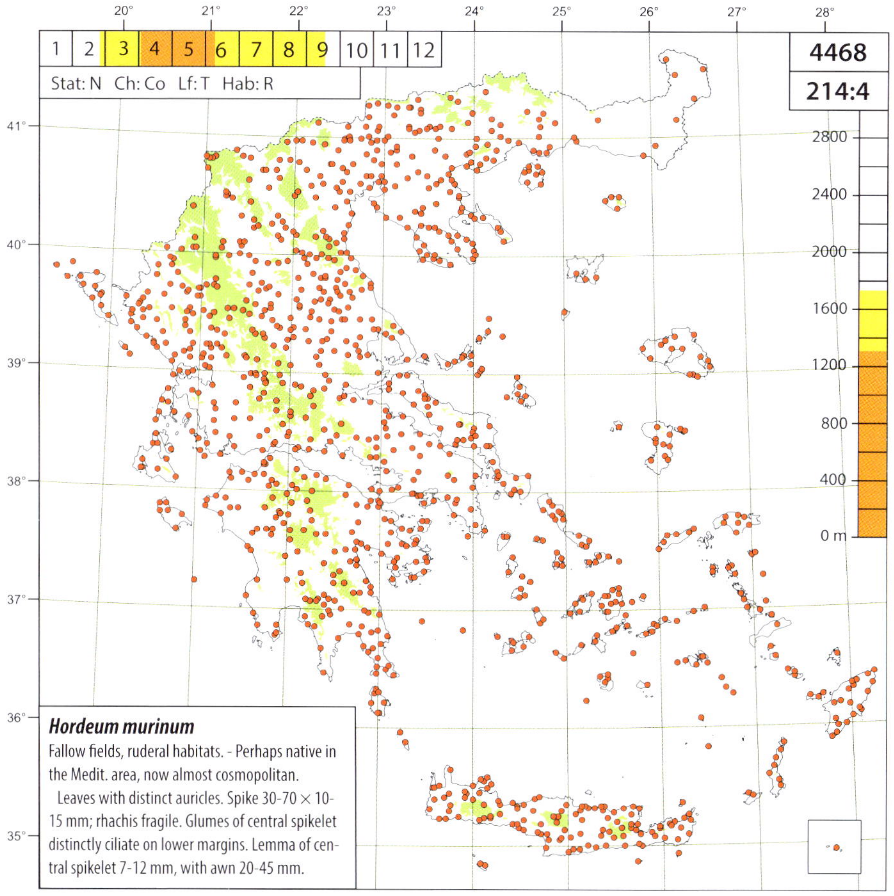

Hordeum murinum

Fallow fields, ruderal habitats. - Perhaps native in the Medit. area, now almost cosmopolitan.

Leaves with distinct auricles. Spike 30-70 × 10-15 mm; rhachis fragile. Glumes of central spikelet distinctly ciliate on lower margins. Lemma of central spikelet 7-12 mm, with awn 20-45 mm.

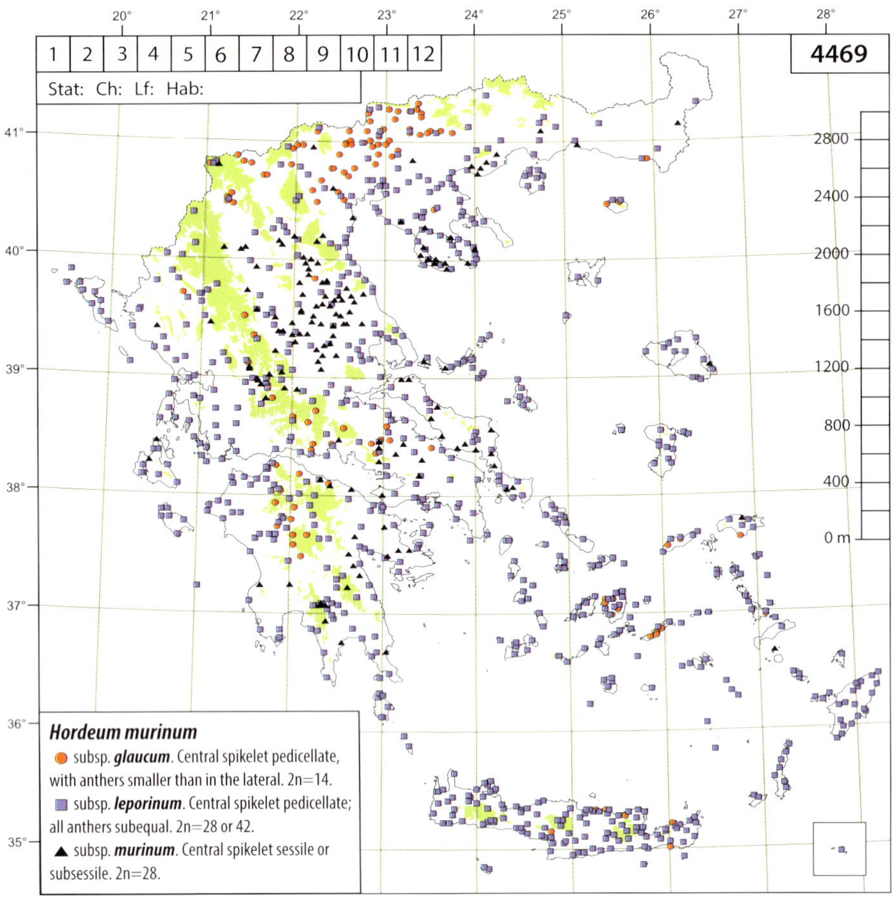

Hordeum murinum

● subsp. ***glaucum***. Central spikelet pedicellate, with anthers smaller than in the lateral. 2n=14.
■ subsp. ***leporinum***. Central spikelet pedicellate; all anthers subequal. 2n=28 or 42.
▲ subsp. ***murinum***. Central spikelet sessile or subsessile. 2n=28.

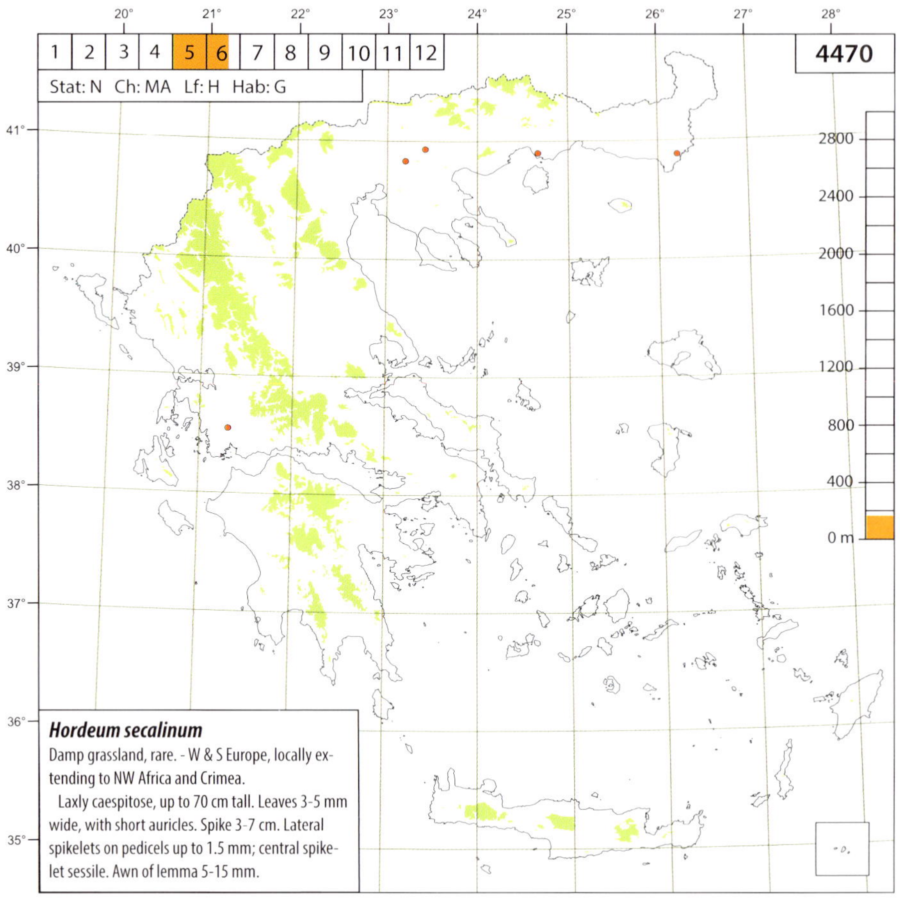

Hordeum secalinum

Damp grassland, rare. - W & S Europe, locally extending to NW Africa and Crimea.

Laxly caespitose, up to 70 cm tall. Leaves 3-5 mm wide, with short auricles. Spike 3-7 cm. Lateral spikelets on pedicels up to 1.5 mm; central spikelet sessile. Awn of lemma 5-15 mm.

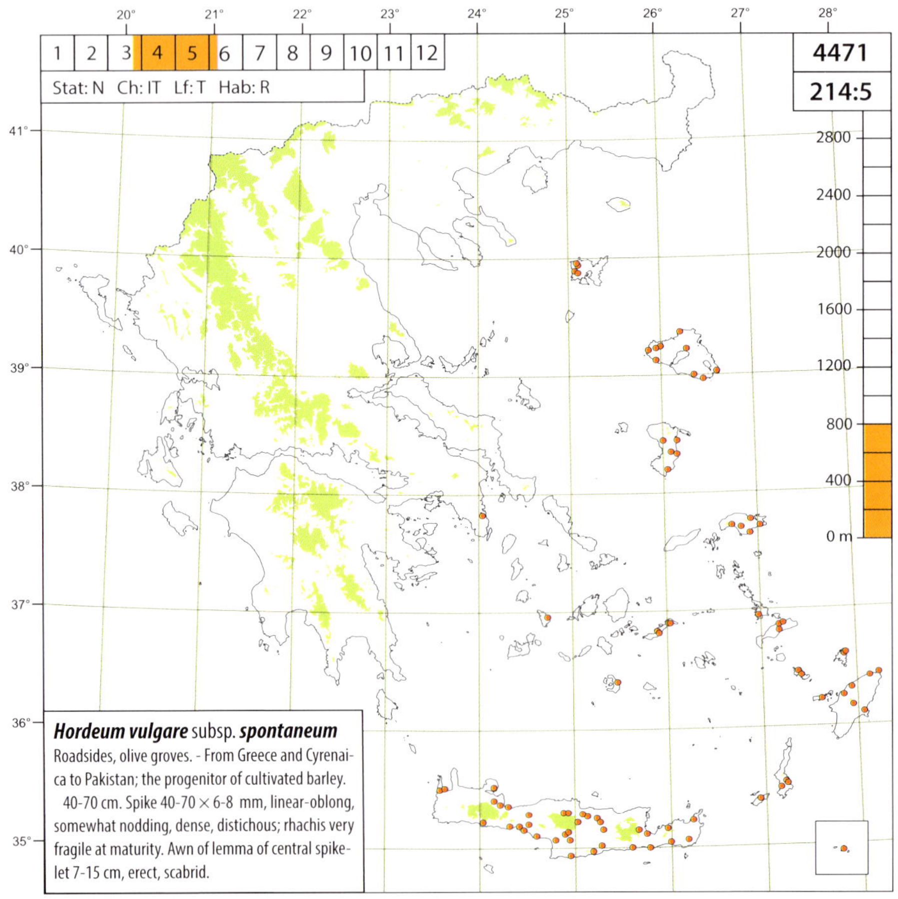

Hordeum vulgare **subsp.** ***spontaneum***

Roadsides, olive groves. - From Greece and Cyrenaica to Pakistan; the progenitor of cultivated barley.

40-70 cm. Spike 40-70 × 6-8 mm, linear-oblong, somewhat nodding, dense, distichous; rhachis very fragile at maturity. Awn of lemma of central spikelet 7-15 cm, erect, scabrid.

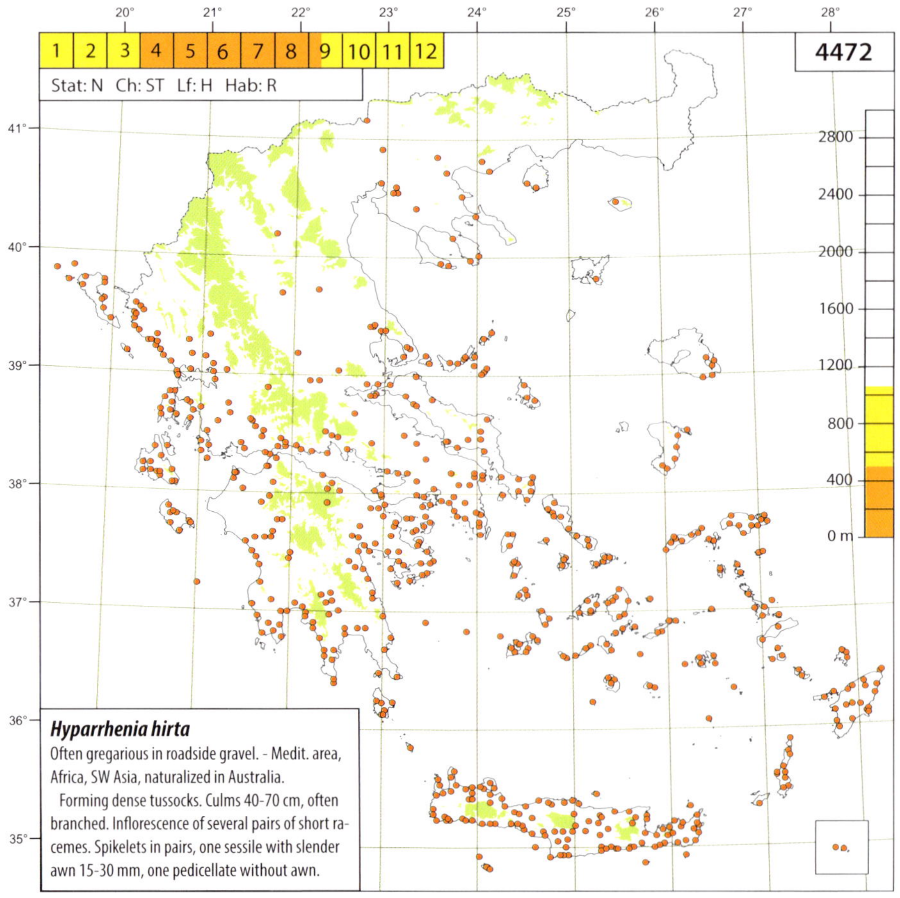

Hyparrhenia hirta

Often gregarious in roadside gravel. - Medit. area, Africa, SW Asia, naturalized in Australia.

Forming dense tussocks. Culms 40-70 cm, often branched. Inflorescence of several pairs of short racemes. Spikelets in pairs, one sessile with slender awn 15-30 mm, one pedicellate without awn.

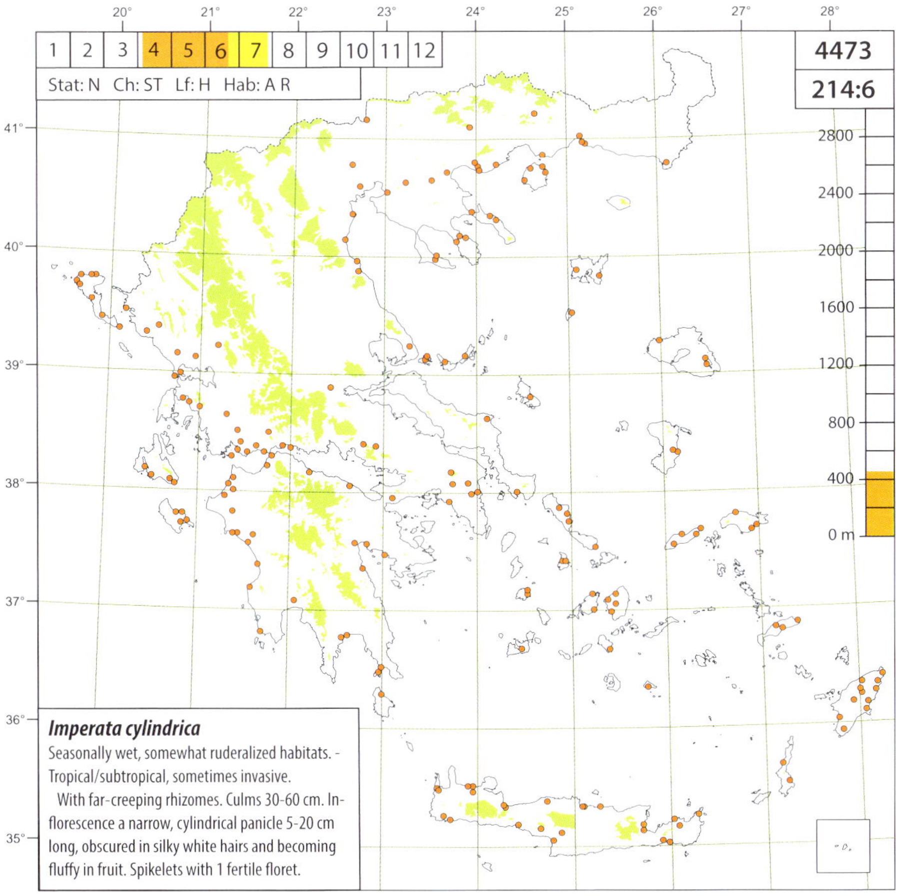

Imperata cylindrica

Seasonally wet, somewhat ruderalized habitats. - Tropical/subtropical, sometimes invasive.

With far-creeping rhizomes. Culms 30-60 cm. Inflorescence a narrow, cylindrical panicle 5-20 cm long, obscured in silky white hairs and becoming fluffy in fruit. Spikelets with 1 fertile floret.

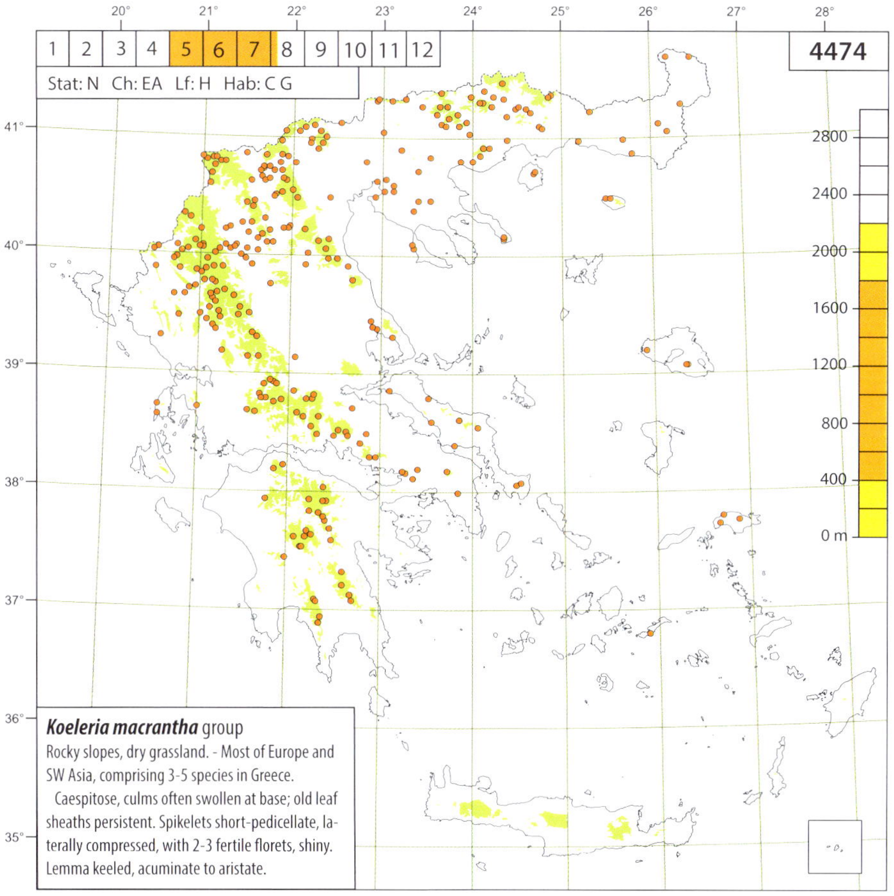

Koeleria macrantha **group**

Rocky slopes, dry grassland. - Most of Europe and SW Asia, comprising 3-5 species in Greece.

Caespitose, culms often swollen at base; old leaf sheaths persistent. Spikelets short-pedicellate, laterally compressed, with 2-3 fertile florets, shiny. Lemma keeled, acuminate to aristate.

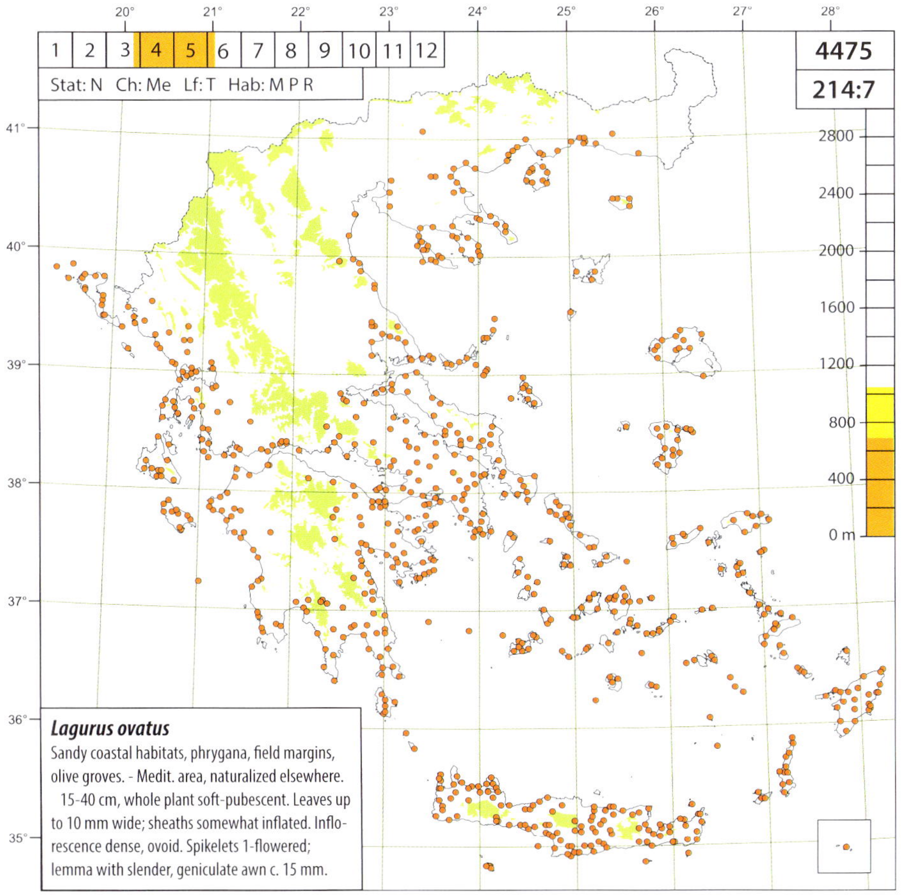

Lagurus ovatus

Sandy coastal habitats, phrygana, field margins, olive groves. - Medit. area, naturalized elsewhere.

15-40 cm, whole plant soft-pubescent. Leaves up to 10 mm wide; sheaths somewhat inflated. Inflorescence dense, ovoid. Spikelets 1-flowered; lemma with slender, geniculate awn c. 15 mm.

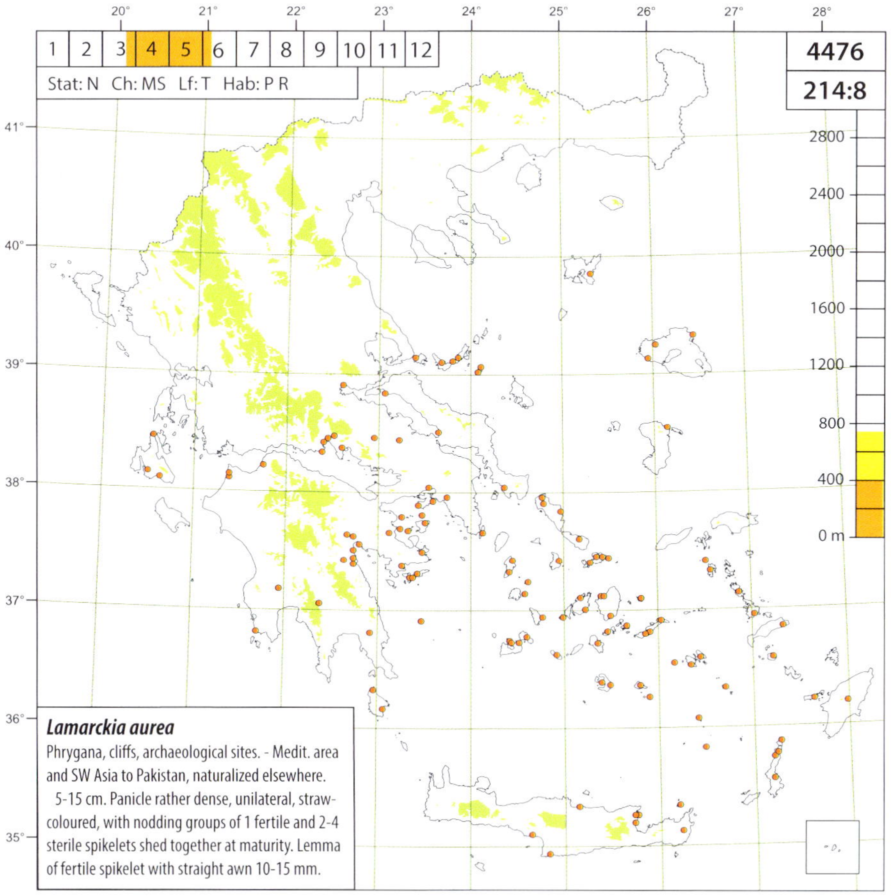

Lamarckia aurea

Phrygana, cliffs, archaeological sites. - Medit. area and SW Asia to Pakistan, naturalized elsewhere.

5-15 cm. Panicle rather dense, unilateral, straw-coloured, with nodding groups of 1 fertile and 2-4 sterile spikelets shed together at maturity. Lemma of fertile spikelet with straight awn 10-15 mm.

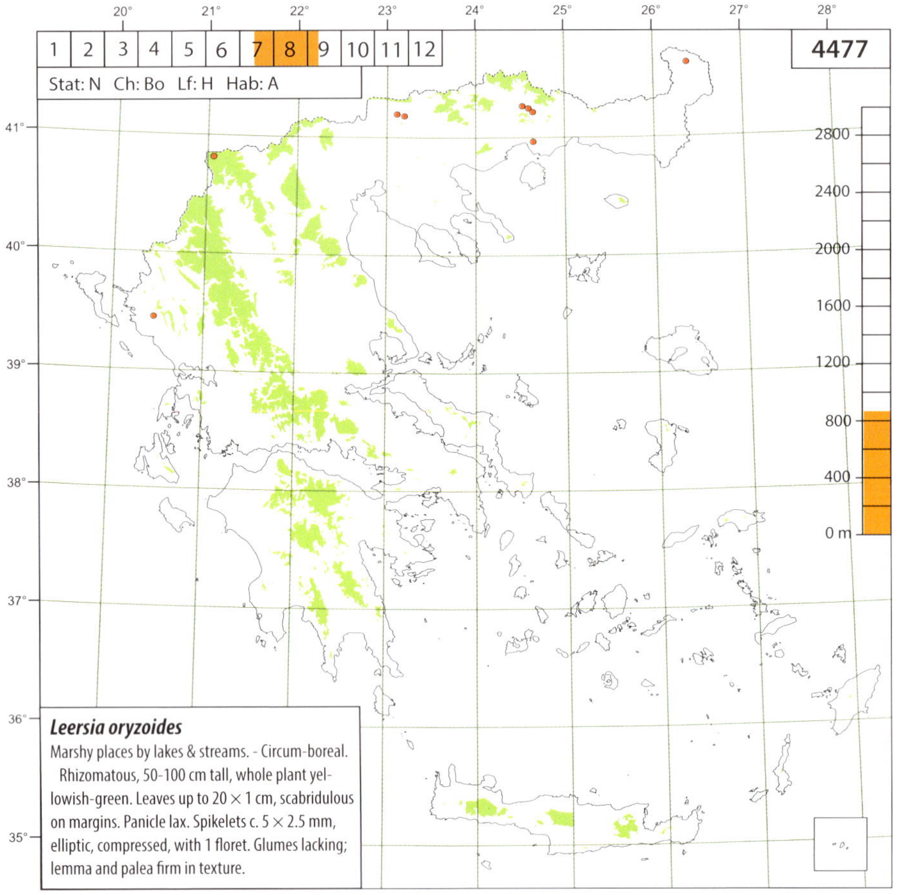

Leersia oryzoides

Marshy places by lakes & streams. - Circum-boreal.

Rhizomatous, 50-100 cm tall, whole plant yellowish-green. Leaves up to 20×1 cm, scabridulous on margins. Panicle lax. Spikelets c. 5×2.5 mm, elliptic, compressed, with 1 floret. Glumes lacking; lemma and palea firm in texture.

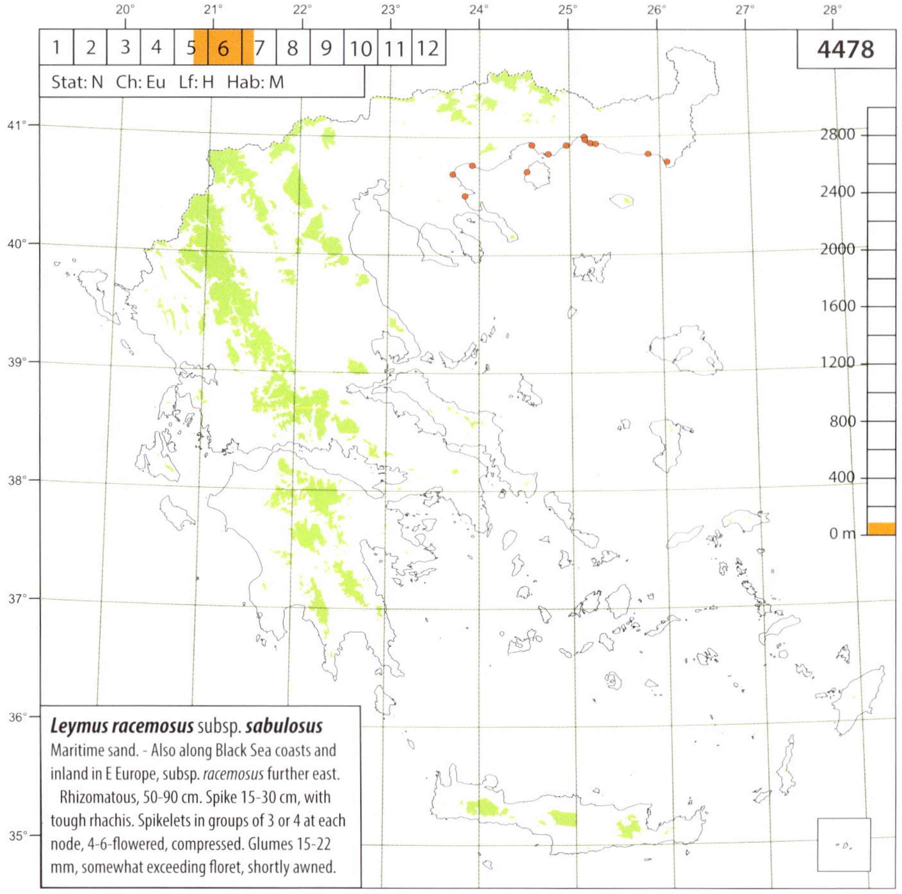

Leymus racemosus subsp. ***sabulosus***

Maritime sand. - Also along Black Sea coasts and inland in E Europe, subsp. *racemosus* further east.

Rhizomatous, 50-90 cm. Spike 15-30 cm, with tough rhachis. Spikelets in groups of 3 or 4 at each node, 4-6-flowered, compressed. Glumes 15-22 mm, somewhat exceeding floret, shortly awned.

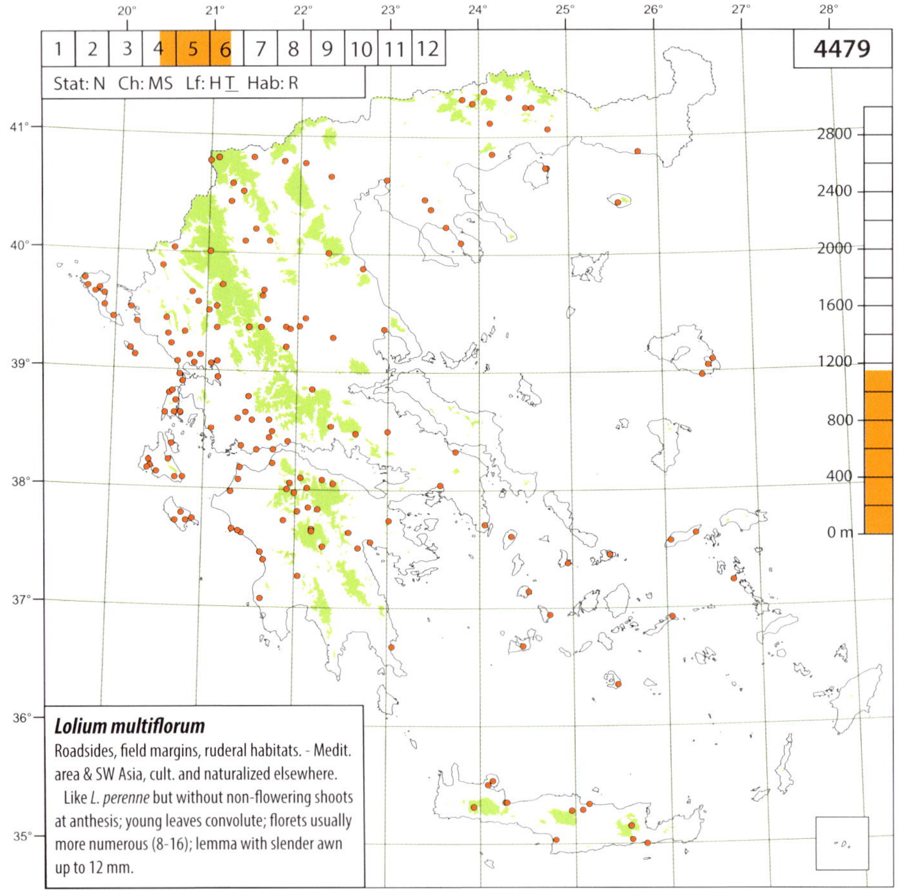

Lolium multiflorum

Roadsides, field margins, ruderal habitats. - Medit. area & SW Asia, cult. and naturalized elsewhere.

Like *L. perenne* but without non-flowering shoots at anthesis; young leaves convolute; florets usually more numerous (8-16); lemma with slender awn up to 12 mm.

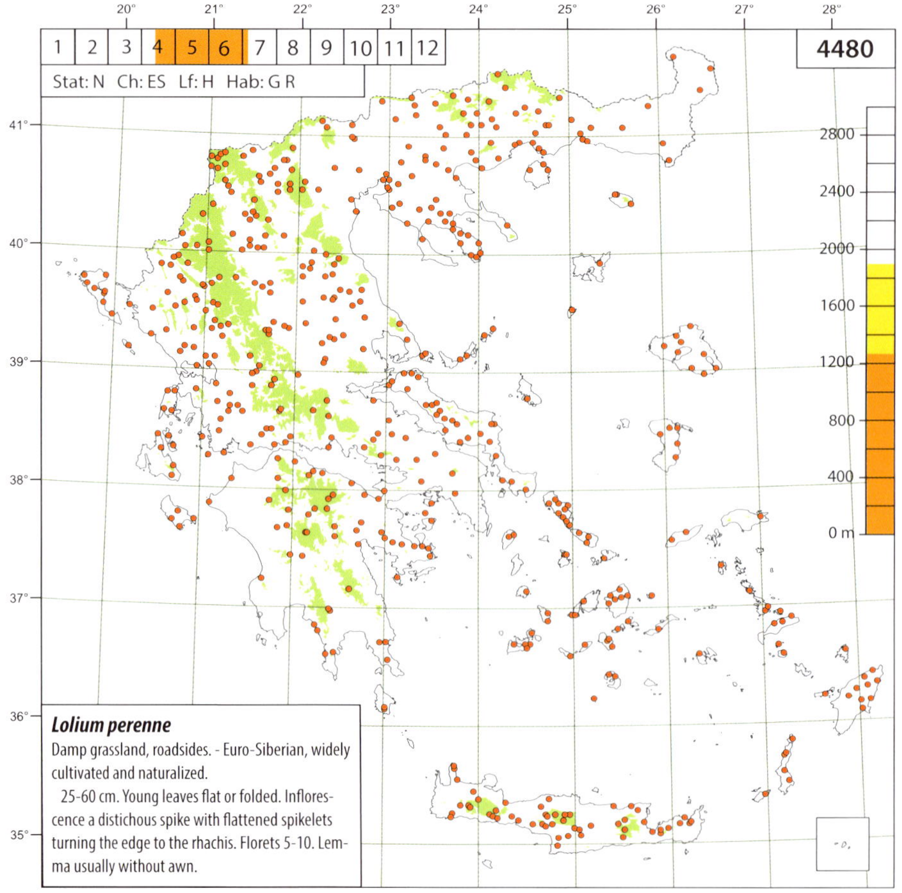

Lolium perenne

Damp grassland, roadsides. - Euro-Siberian, widely cultivated and naturalized.

25-60 cm. Young leaves flat or folded. Inflorescence a distichous spike with flattened spikelets turning the edge to the rhachis. Florets 5-10. Lemma usually without awn.

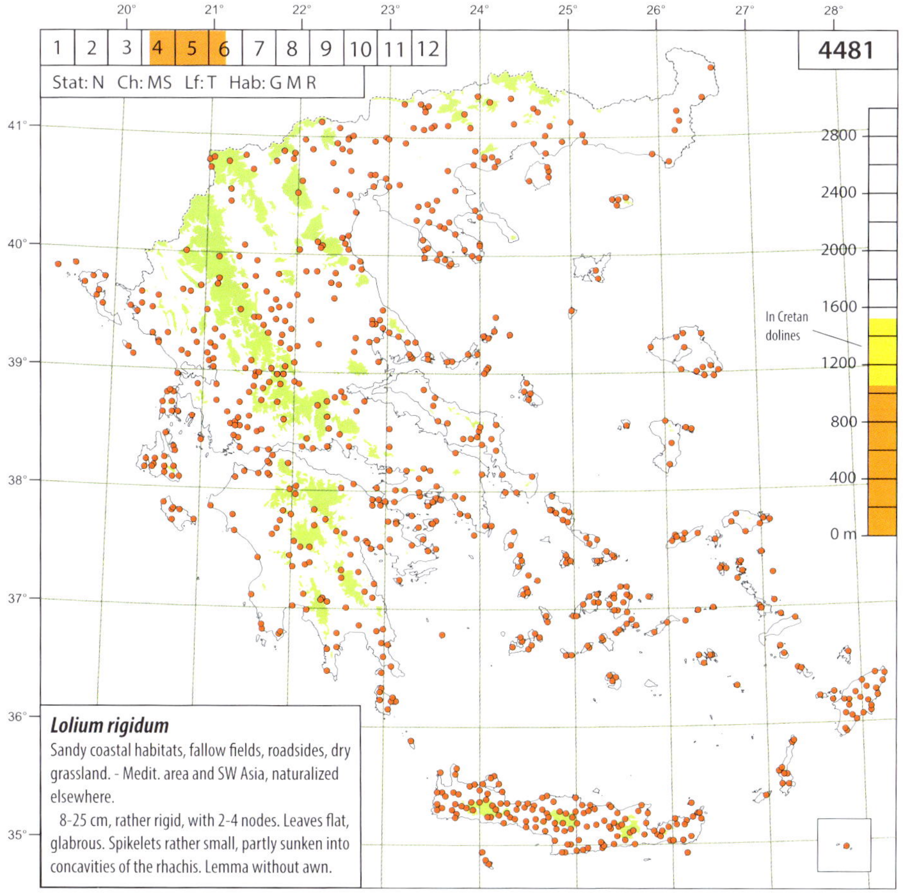

Lolium rigidum

Sandy coastal habitats, fallow fields, roadsides, dry grassland. - Medit. area and SW Asia, naturalized elsewhere.

8-25 cm, rather rigid, with 2-4 nodes. Leaves flat, glabrous. Spikelets rather small, partly sunken into concavities of the rhachis. Lemma without awn.

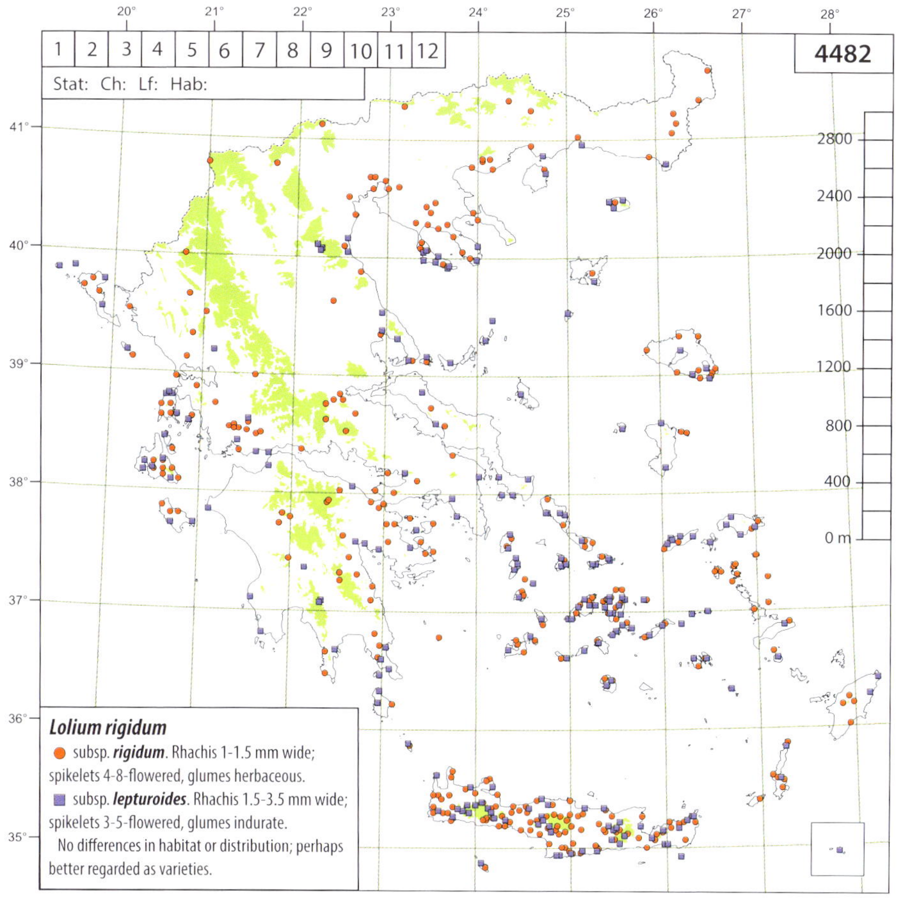

Lolium rigidum

● subsp. ***rigidum***. Rhachis 1-1.5 mm wide; spikelets 4-8-flowered, glumes herbaceous.

■ subsp. ***lepturoides***. Rhachis 1.5-3.5 mm wide; spikelets 3-5-flowered, glumes indurate.

No differences in habitat or distribution; perhaps better regarded as varieties.

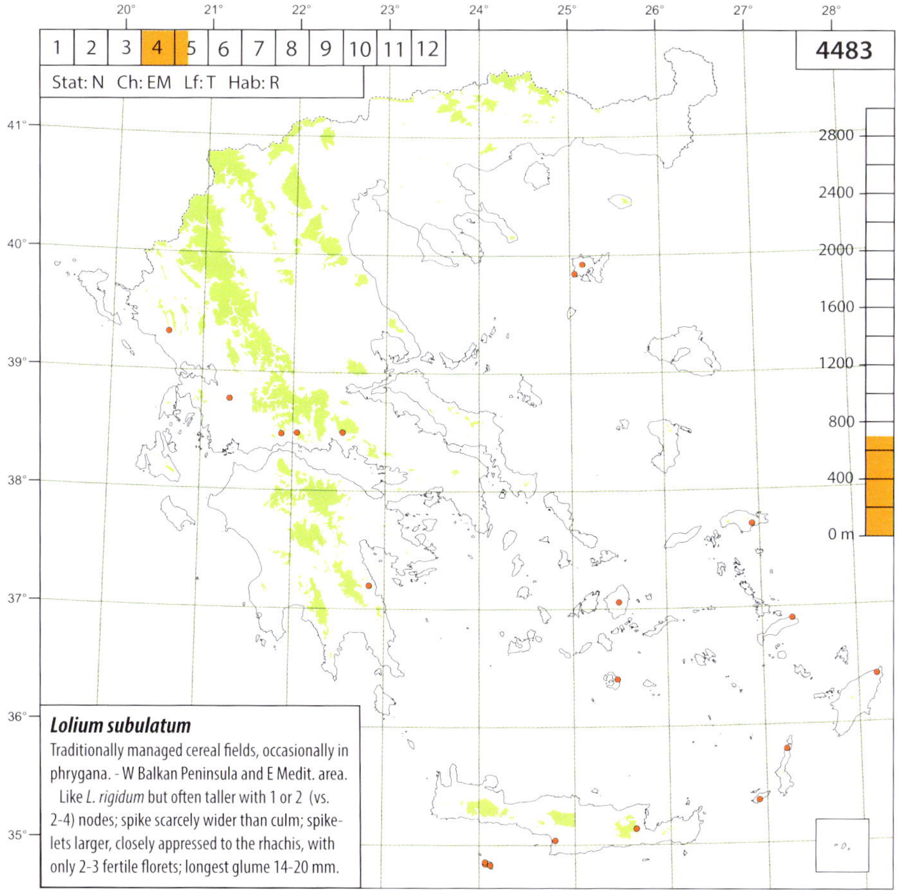

Lolium subulatum

Traditionally managed cereal fields, occasionally in phrygana. - W Balkan Peninsula and E Medit. area.

Like *L. rigidum* but often taller with 1 or 2 (vs. 2-4) nodes; spike scarcely wider than culm; spikelets larger, closely appressed to the rhachis, with only 2-3 fertile florets; longest glume 14-20 mm.

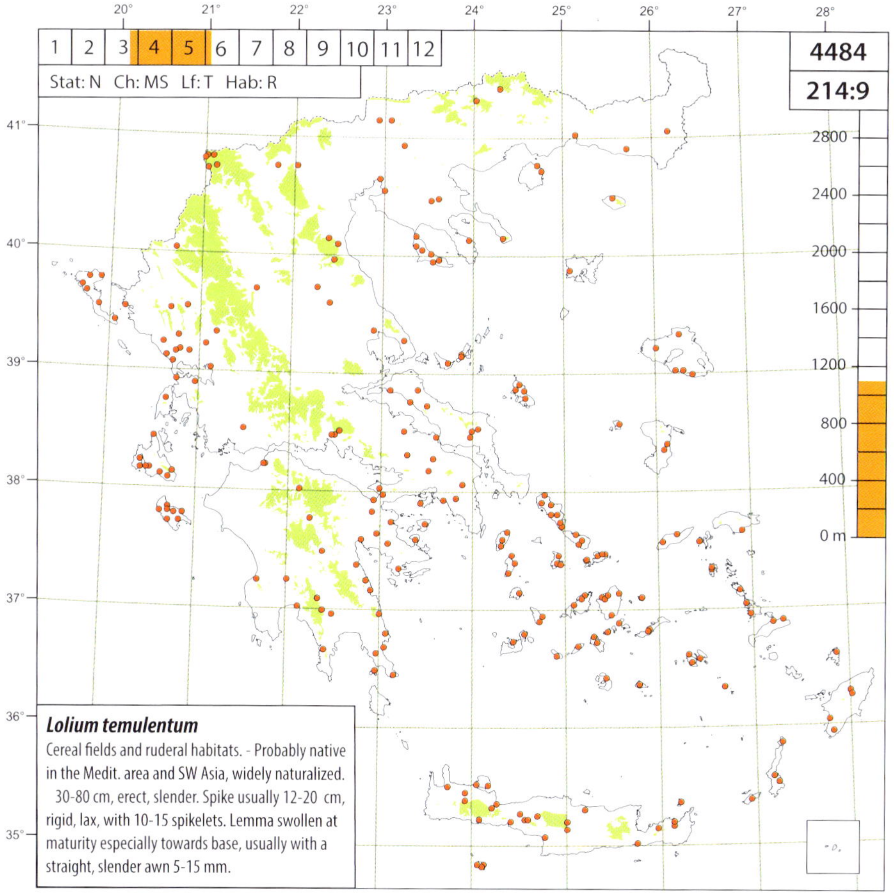

Lolium temulentum

Cereal fields and ruderal habitats. - Probably native in the Medit. area and SW Asia, widely naturalized.

30-80 cm, erect, slender. Spike usually 12-20 cm, rigid, lax, with 10-15 spikelets. Lemma swollen at maturity especially towards base, usually with a straight, slender awn 5-15 mm.

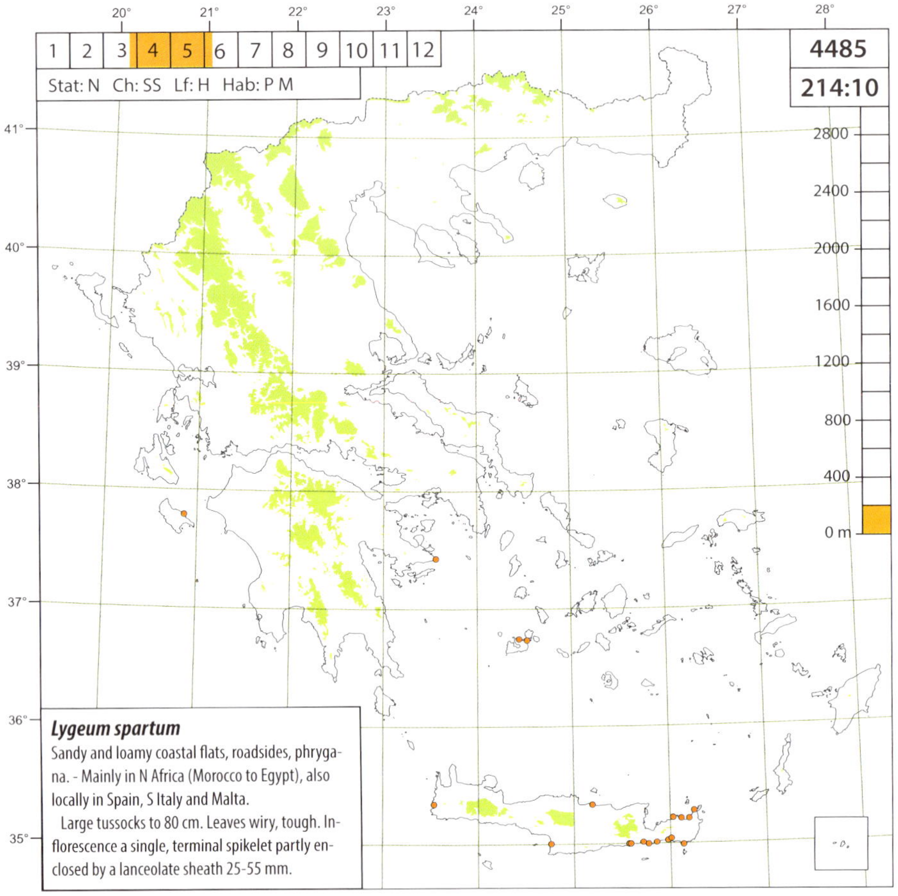

Lygeum spartum
Sandy and loamy coastal flats, roadsides, phrygana. - Mainly in N Africa (Morocco to Egypt), also locally in Spain, S Italy and Malta.
Large tussocks to 80 cm. Leaves wiry, tough. Inflorescence a single, terminal spikelet partly enclosed by a lanceolate sheath 25-55 mm.

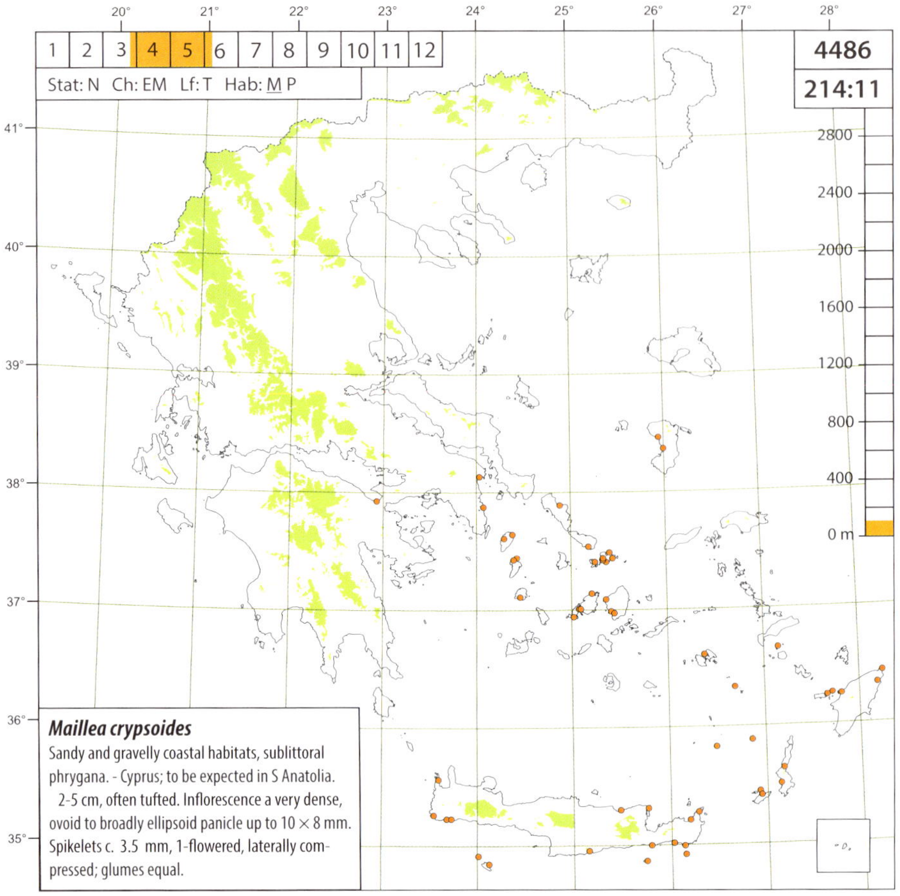

Maillea crypsoides
Sandy and gravelly coastal habitats, sublittoral phrygana. - Cyprus; to be expected in S Anatolia.
2-5 cm, often tufted. Inflorescence a very dense, ovoid to broadly ellipsoid panicle up to 10 × 8 mm. Spikelets c. 3.5 mm, 1-flowered, laterally compressed; glumes equal.

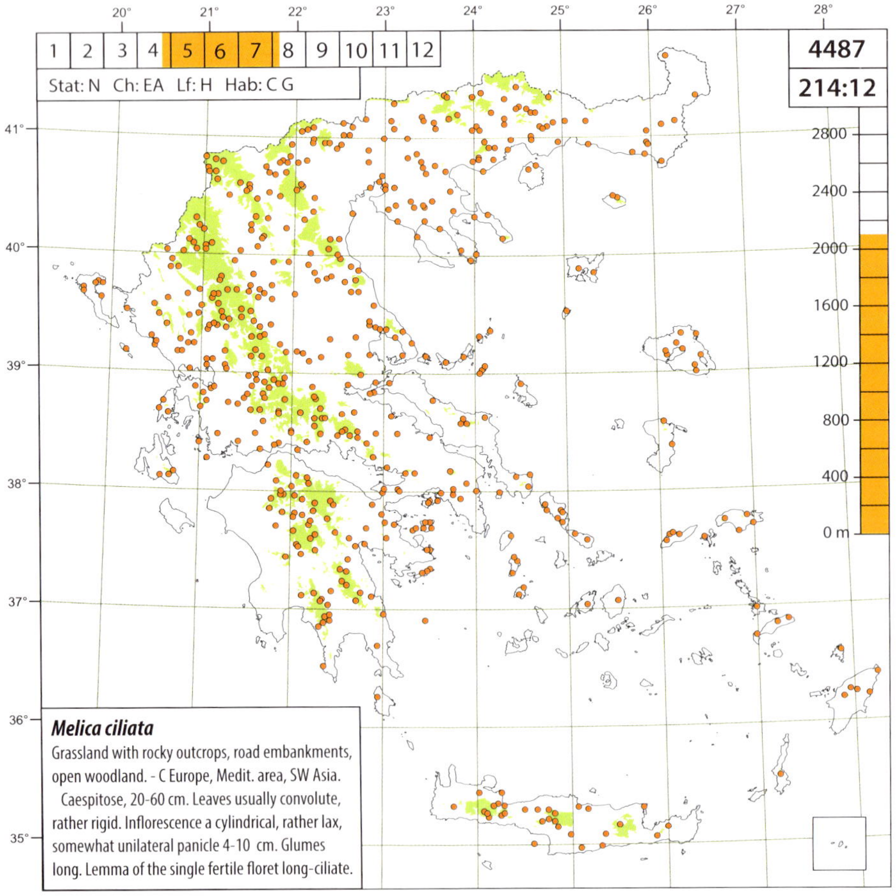

Melica ciliata
Grassland with rocky outcrops, road embankments, open woodland. - C Europe, Medit. area, SW Asia.
Caespitose, 20-60 cm. Leaves usually convolute, rather rigid. Inflorescence a cylindrical, rather lax, somewhat unilateral panicle 4-10 cm. Glumes long. Lemma of the single fertile floret long-ciliate.

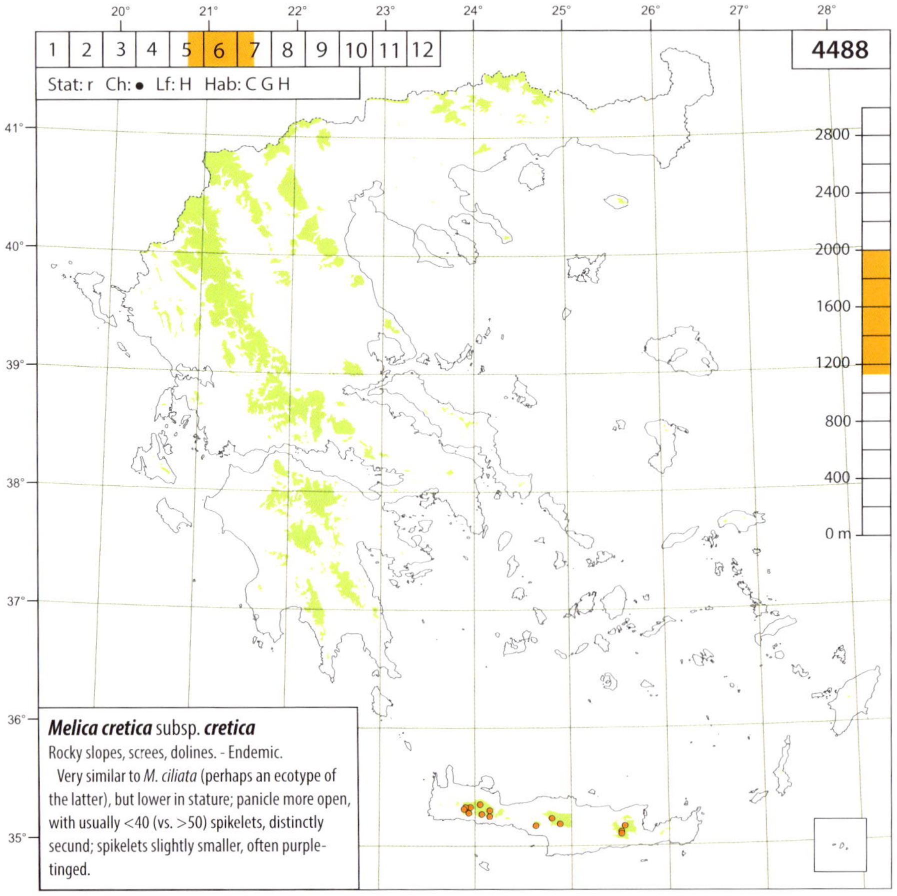

Melica cretica subsp. ***cretica***
Rocky slopes, screes, dolines. - Endemic.
Very similar to *M. ciliata* (perhaps an ecotype of the latter), but lower in stature; panicle more open, with usually <40 (vs. >50) spikelets, distinctly secund; spikelets slightly smaller, often purple-tinged.

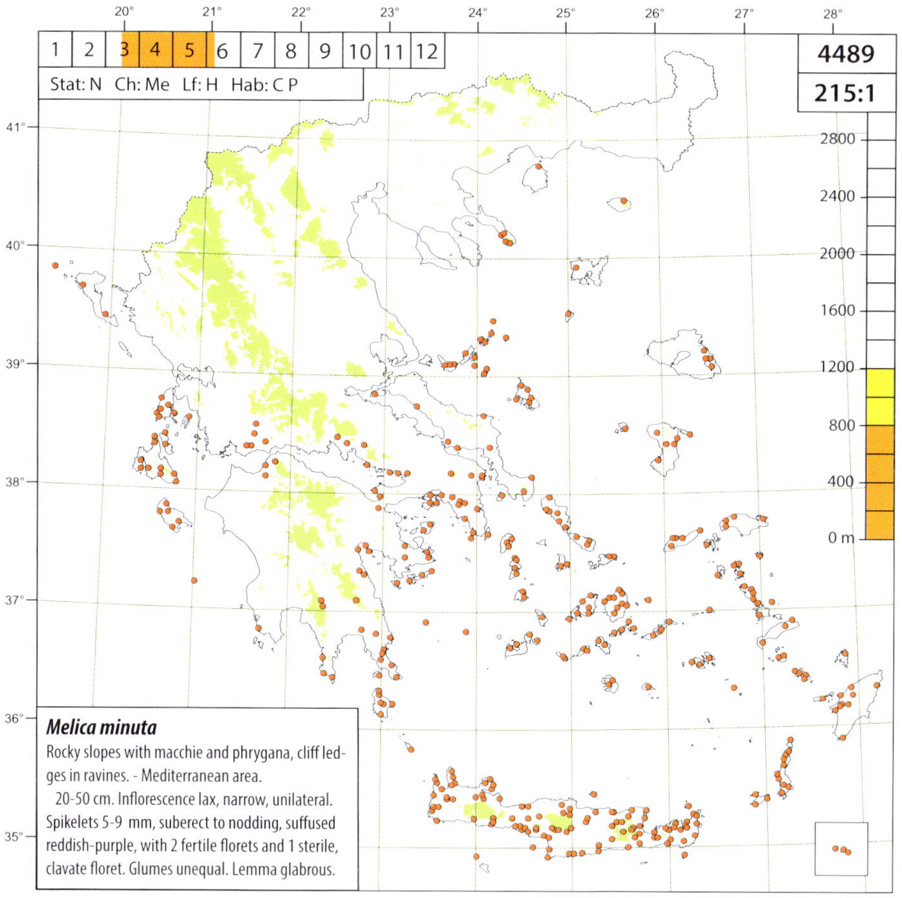

Melica minuta
Rocky slopes with macchie and phrygana, cliff ledges in ravines. - Mediterranean area.
20-50 cm. Inflorescence lax, narrow, unilateral. Spikelets 5-9 mm, suberect to nodding, suffused reddish-purple, with 2 fertile florets and 1 sterile, clavate floret. Glumes unequal. Lemma glabrous.

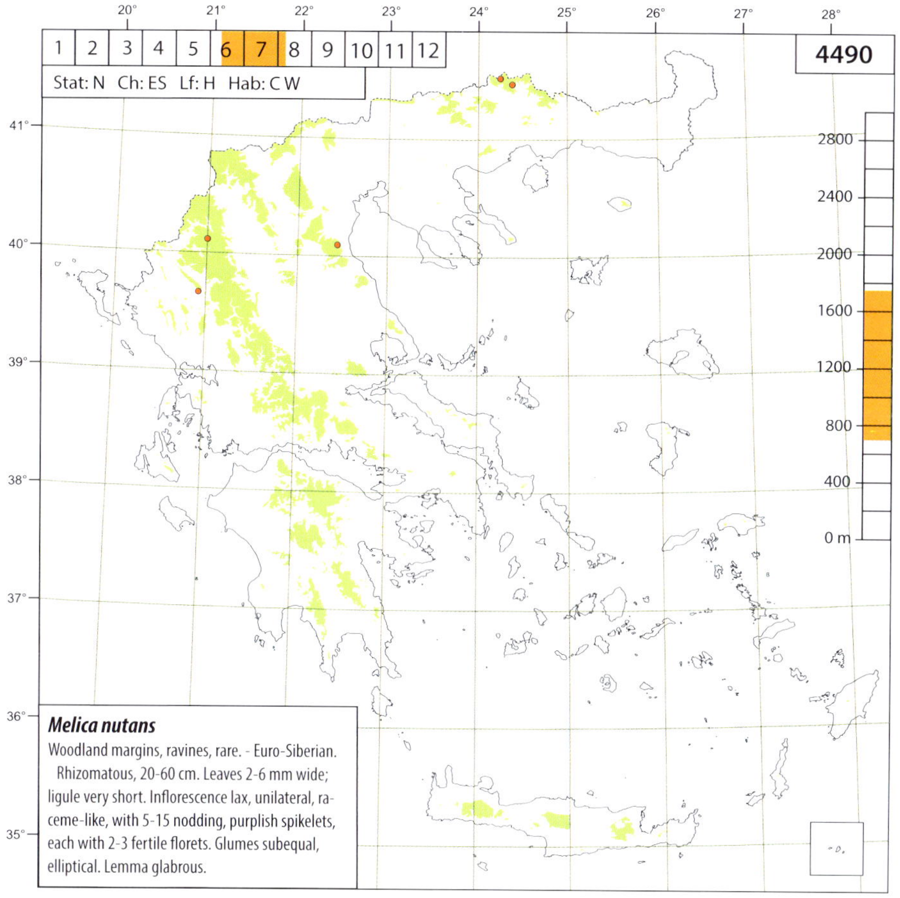

Melica nutans
Woodland margins, ravines, rare. - Euro-Siberian.
Rhizomatous, 20-60 cm. Leaves 2-6 mm wide; ligule very short. Inflorescence lax, unilateral, raceme-like, with 5-15 nodding, purplish spikelets, each with 2-3 fertile florets. Glumes subequal, elliptical. Lemma glabrous.

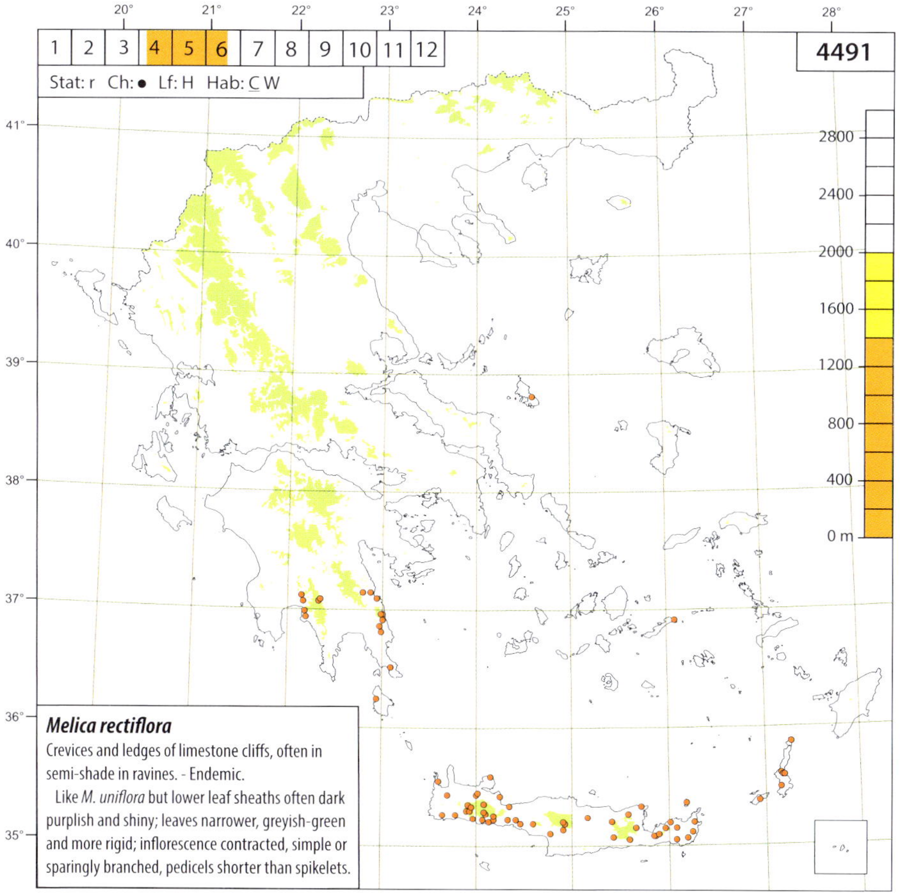

Melica rectiflora
Crevices and ledges of limestone cliffs, often in semi-shade in ravines. - Endemic.
Like *M. uniflora* but lower leaf sheaths often dark purplish and shiny; leaves narrower, greyish-green and more rigid; inflorescence contracted, simple or sparingly branched, pedicels shorter than spikelets.

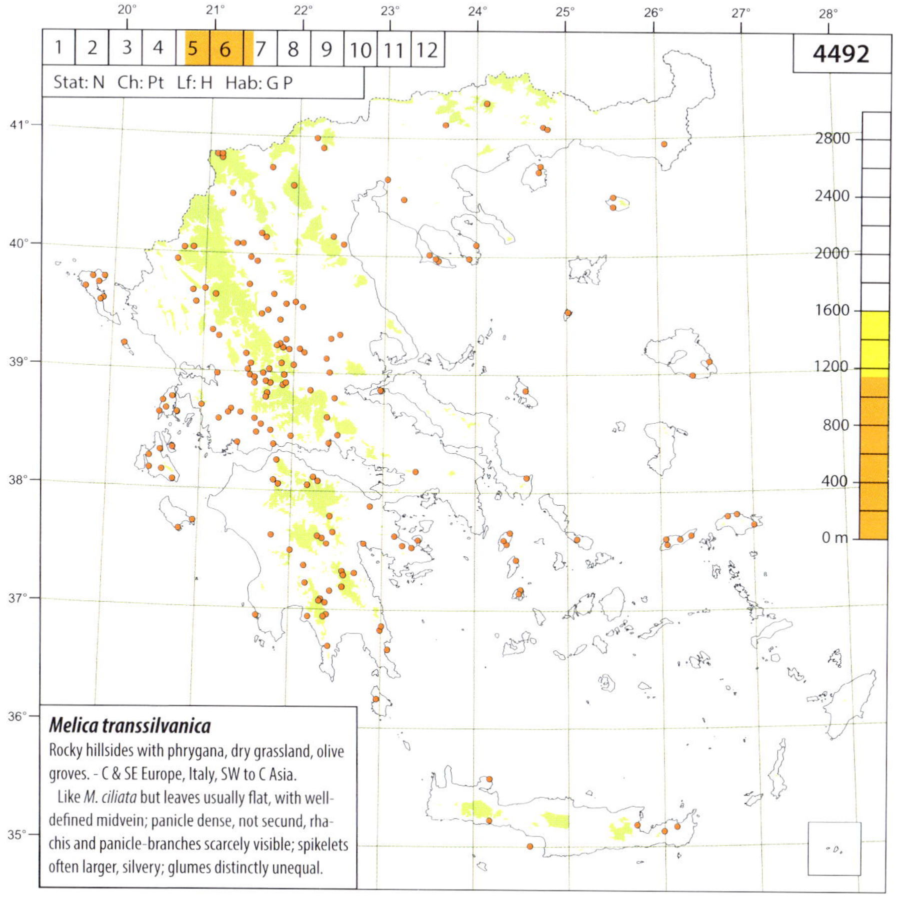

Melica transsilvanica
Rocky hillsides with phrygana, dry grassland, olive groves. - C & SE Europe, Italy, SW to C Asia.
Like *M. ciliata* but leaves usually flat, with well-defined midvein; panicle dense, not secund, rhachis and panicle-branches scarcely visible; spikelets often larger, silvery; glumes distinctly unequal.

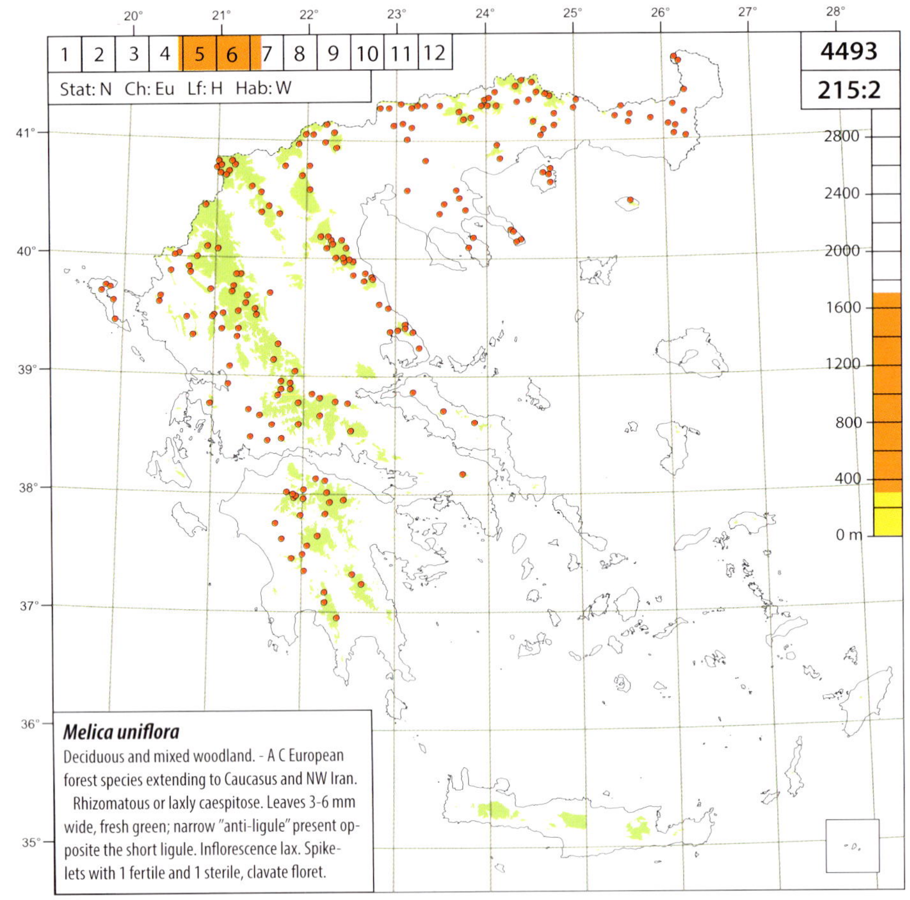

Melica uniflora

Deciduous and mixed woodland. - A C European forest species extending to Caucasus and NW Iran.
Rhizomatous or laxly caespitose. Leaves 3-6 mm wide, fresh green; narrow "anti-ligule" present opposite the short ligule. Inflorescence lax. Spikelets with 1 fertile and 1 sterile, clavate floret.

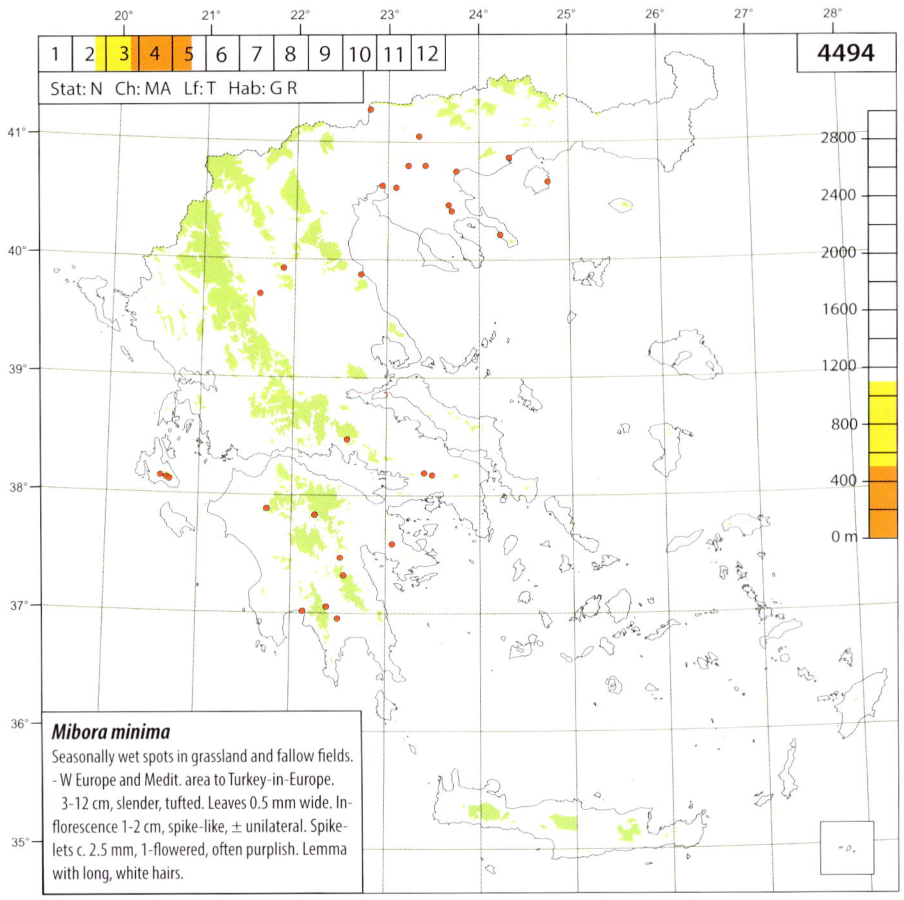

Mibora minima

Seasonally wet spots in grassland and fallow fields. - W Europe and Medit. area to Turkey-in-Europe.
3-12 cm, slender, tufted. Leaves 0.5 mm wide. Inflorescence 1-2 cm, spike-like, ± unilateral. Spikelets c. 2.5 mm, 1-flowered, often purplish. Lemma with long, white hairs.

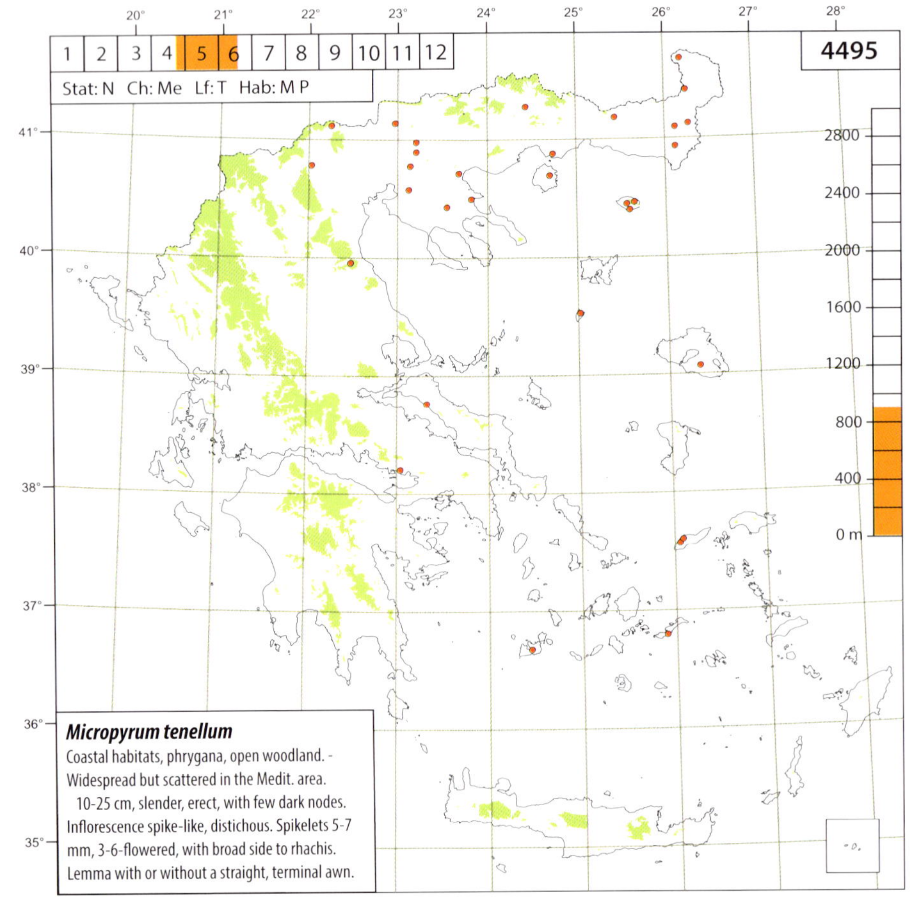

Micropyrum tenellum

Coastal habitats, phrygana, open woodland. - Widespread but scattered in the Medit. area.
10-25 cm, slender, erect, with few dark nodes. Inflorescence spike-like, distichous. Spikelets 5-7 mm, 3-6-flowered, with broad side to rhachis. Lemma with or without a straight, terminal awn.

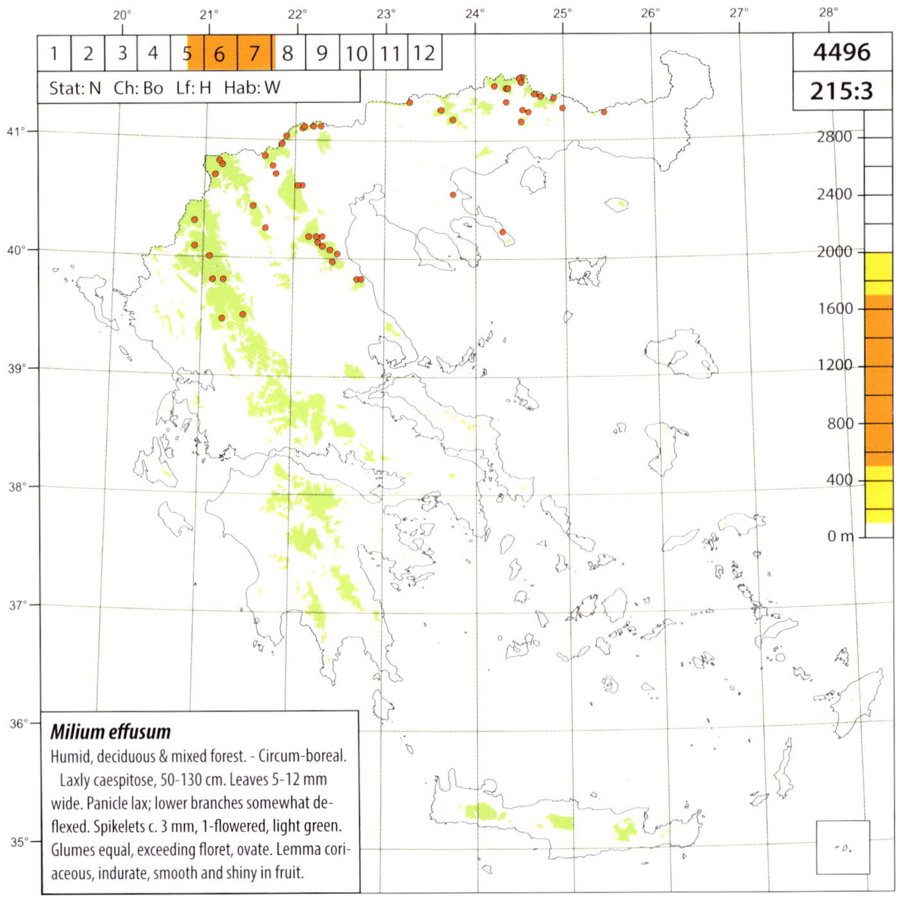

Milium effusum

Humid, deciduous & mixed forest. - Circum-boreal.
Laxly caespitose, 50-130 cm. Leaves 5-12 mm wide. Panicle lax; lower branches somewhat deflexed. Spikelets c. 3 mm, 1-flowered, light green. Glumes equal, exceeding floret, ovate. Lemma coriaceous, indurate, smooth and shiny in fruit.

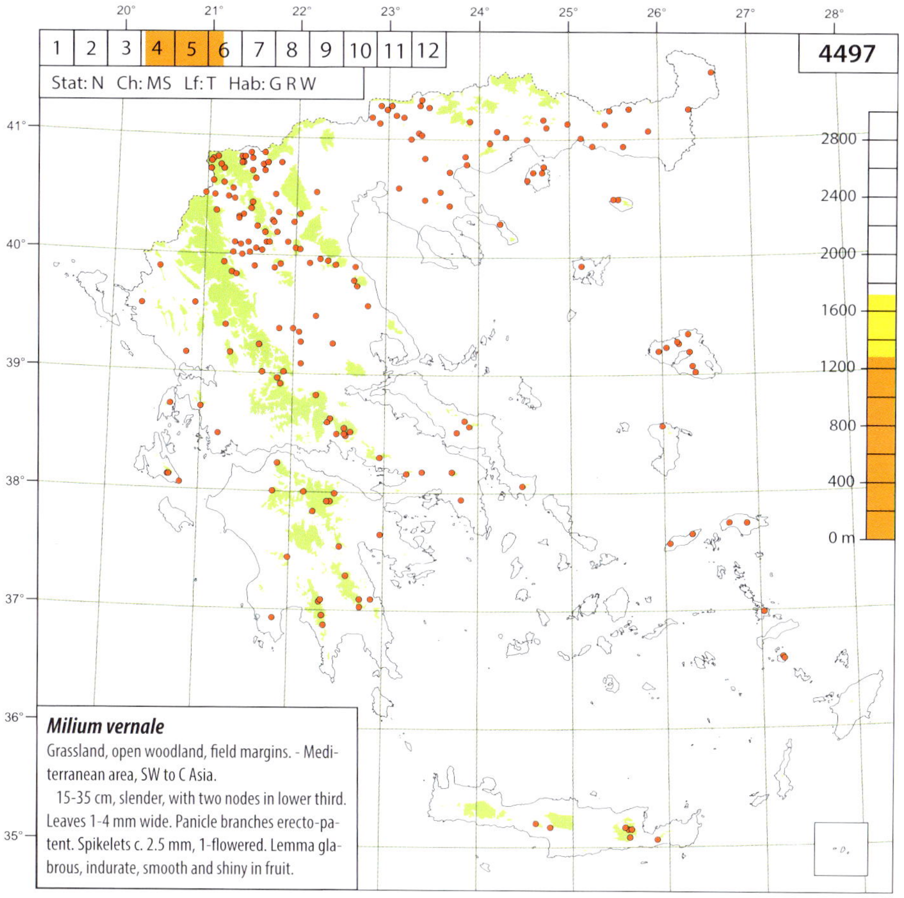

Milium vernale
Grassland, open woodland, field margins. - Mediterranean area, SW to C Asia.
15-35 cm, slender, with two nodes in lower third. Leaves 1-4 mm wide. Panicle branches erecto-patent. Spikelets c. 2.5 mm, 1-flowered. Lemma glabrous, indurate, smooth and shiny in fruit.

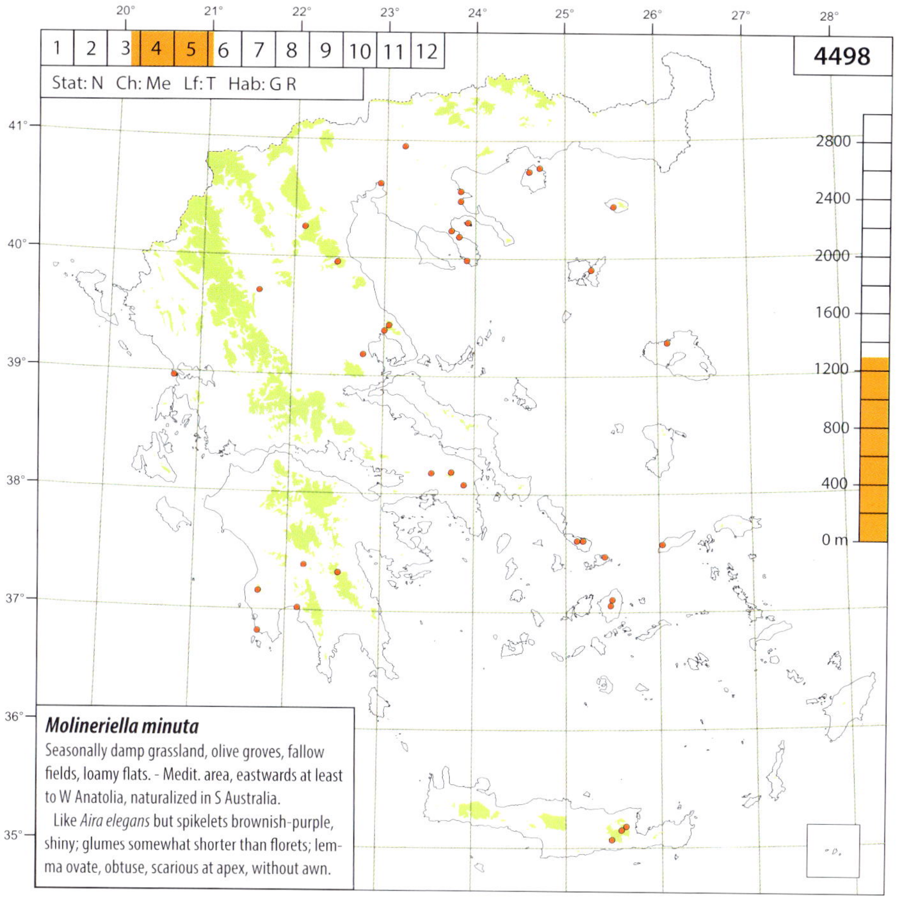

Molineriella minuta
Seasonally damp grassland, olive groves, fallow fields, loamy flats. - Medit. area, eastwards at least to W Anatolia, naturalized in S Australia.
Like *Aira elegans* but spikelets brownish-purple, shiny; glumes somewhat shorter than florets; lemma ovate, obtuse, scarious at apex, without awn.

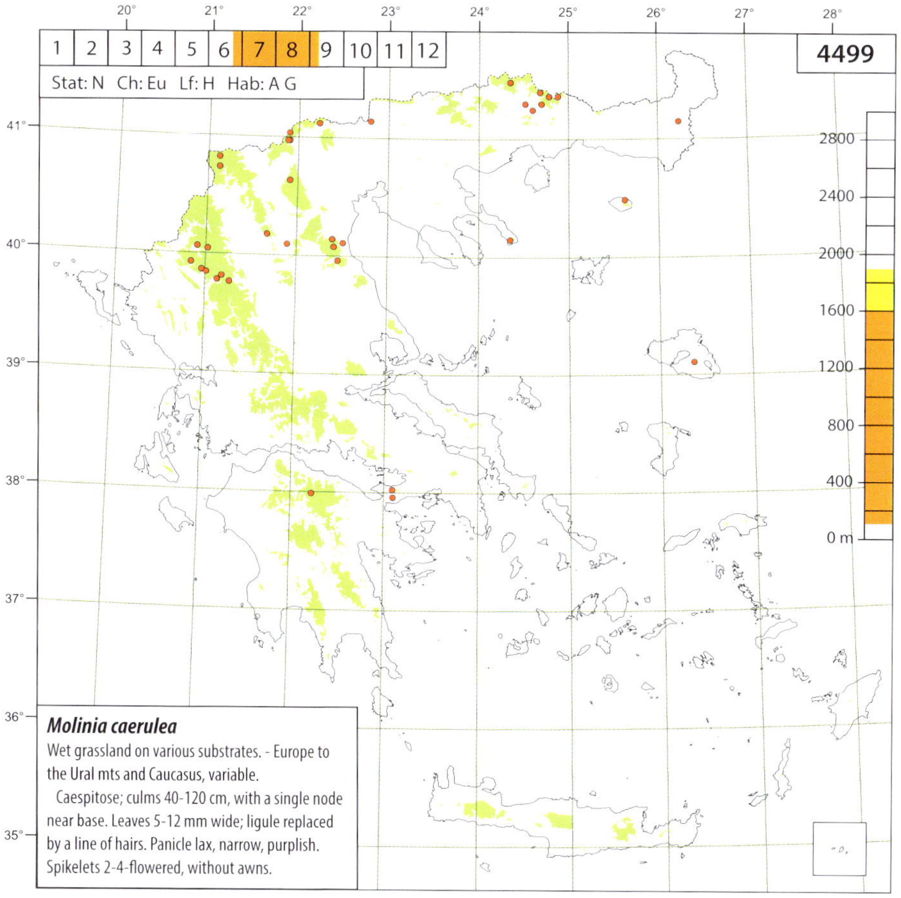

Molinia caerulea
Wet grassland on various substrates. - Europe to the Ural mts and Caucasus, variable.
Caespitose; culms 40-120 cm, with a single node near base. Leaves 5-12 mm wide; ligule replaced by a line of hairs. Panicle lax, narrow, purplish. Spikelets 2-4-flowered, without awns.

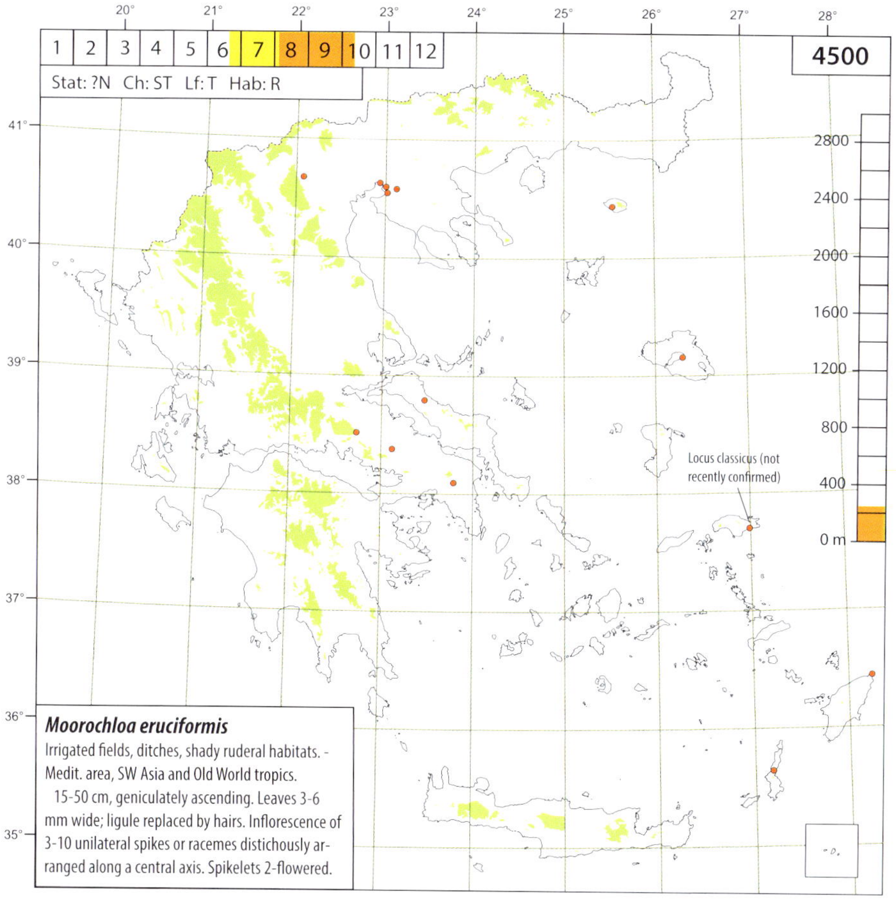

Moorochloa eruciformis
Irrigated fields, ditches, shady ruderal habitats. - Medit. area, SW Asia and Old World tropics.
15-50 cm, geniculately ascending. Leaves 3-6 mm wide; ligule replaced by hairs. Inflorescence of 3-10 unilateral spikes or racemes distichously arranged along a central axis. Spikelets 2-flowered.

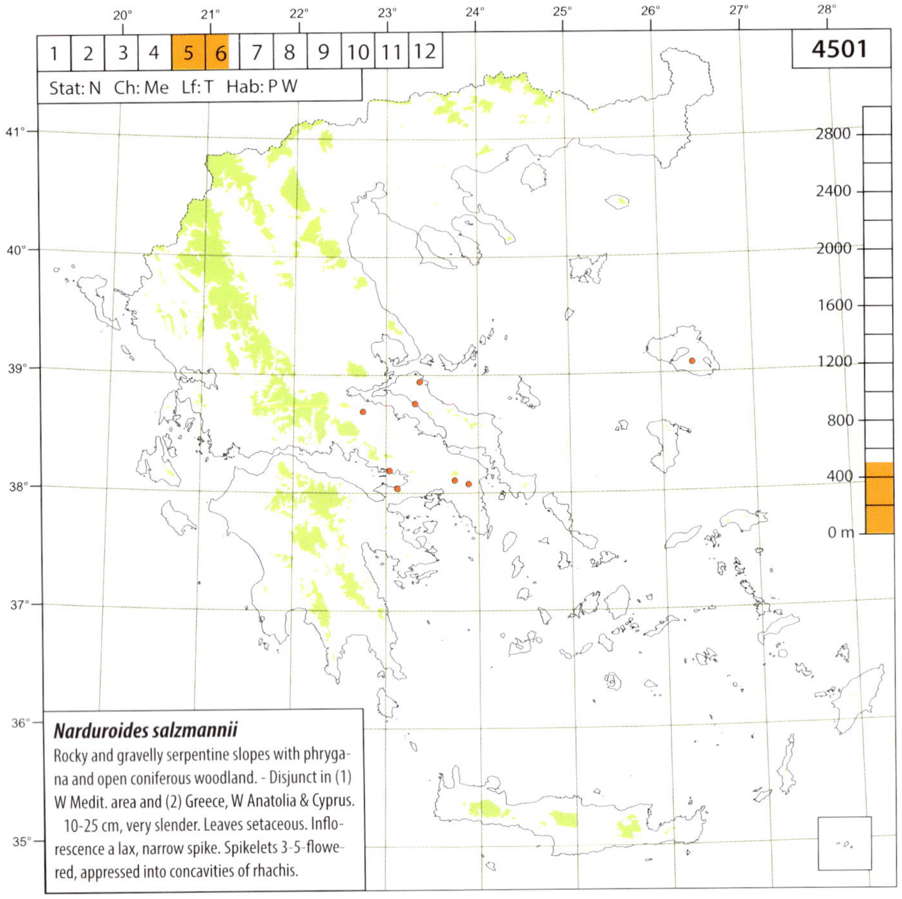

Narduroides salzmannii
Rocky and gravelly serpentine slopes with phrygana and open coniferous woodland. - Disjunct in (1) W Medit. area and (2) Greece, W Anatolia & Cyprus.
10-25 cm, very slender. Leaves setaceous. Inflorescence a lax, narrow spike. Spikelets 3-5-flowered, appressed into concavities of rhachis.

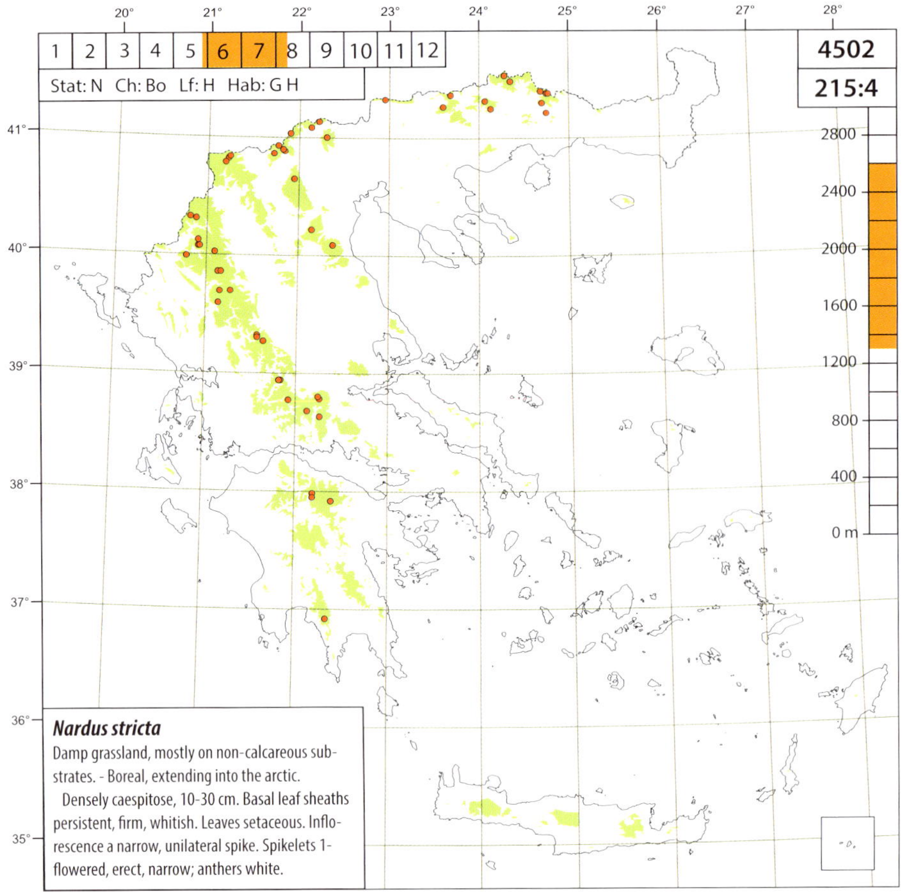

Nardus stricta
Damp grassland, mostly on non-calcareous substrates. - Boreal, extending into the arctic.
Densely caespitose, 10-30 cm. Basal leaf sheaths persistent, firm, whitish. Leaves setaceous. Inflorescence a narrow, unilateral spike. Spikelets 1-flowered, erect, narrow; anthers white.

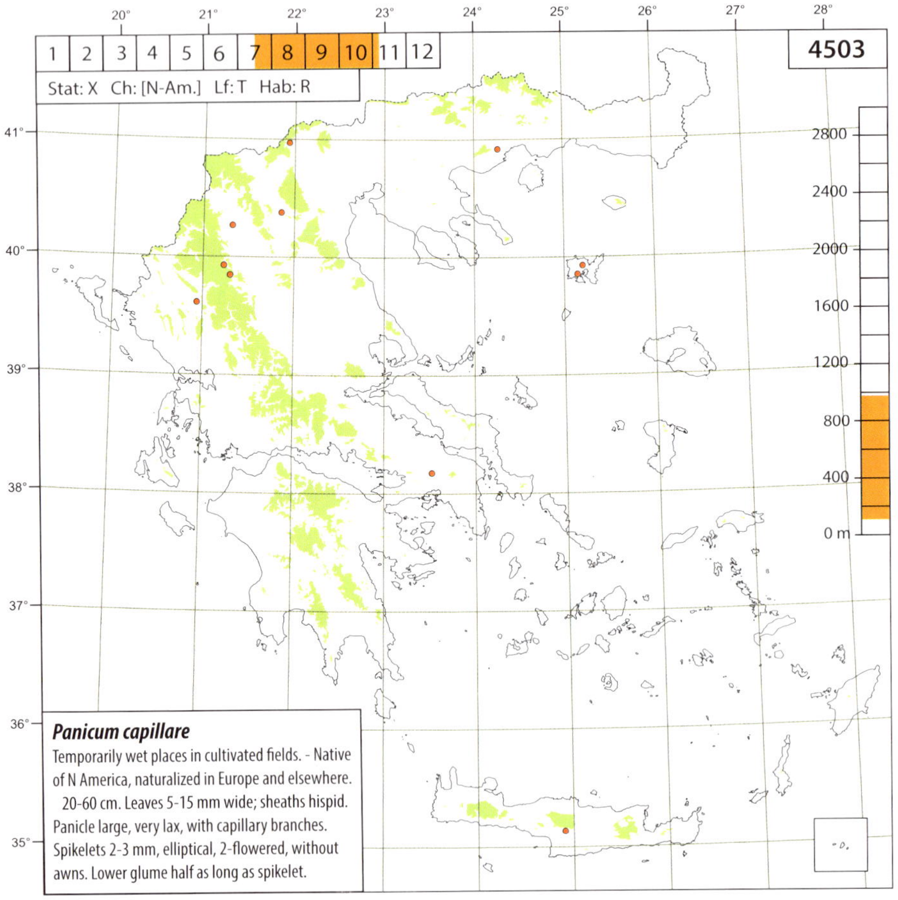

Panicum capillare
Temporarily wet places in cultivated fields. - Native of N America, naturalized in Europe and elsewhere.
20-60 cm. Leaves 5-15 mm wide; sheaths hispid. Panicle large, very lax, with capillary branches. Spikelets 2-3 mm, elliptical, 2-flowered, without awns. Lower glume half as long as spikelet.

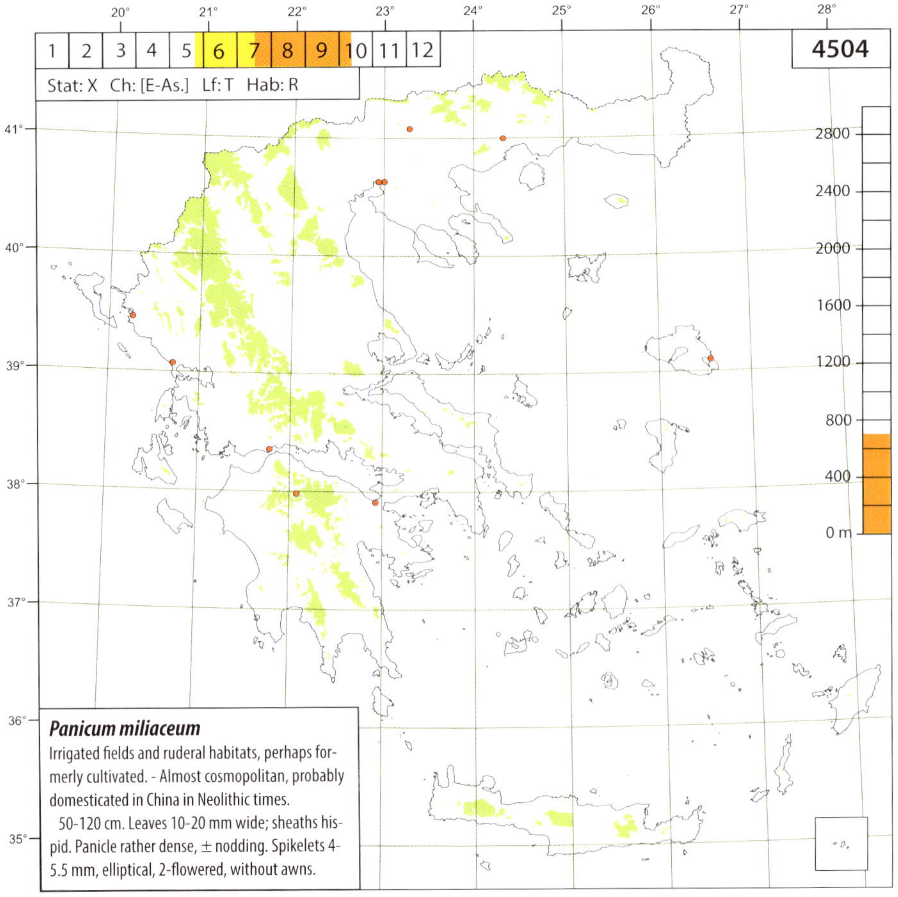

Panicum miliaceum
Irrigated fields and ruderal habitats, perhaps formerly cultivated. - Almost cosmopolitan, probably domesticated in China in Neolithic times.
50-120 cm. Leaves 10-20 mm wide; sheaths hispid. Panicle rather dense, ± nodding. Spikelets 4-5.5 mm, elliptical, 2-flowered, without awns.

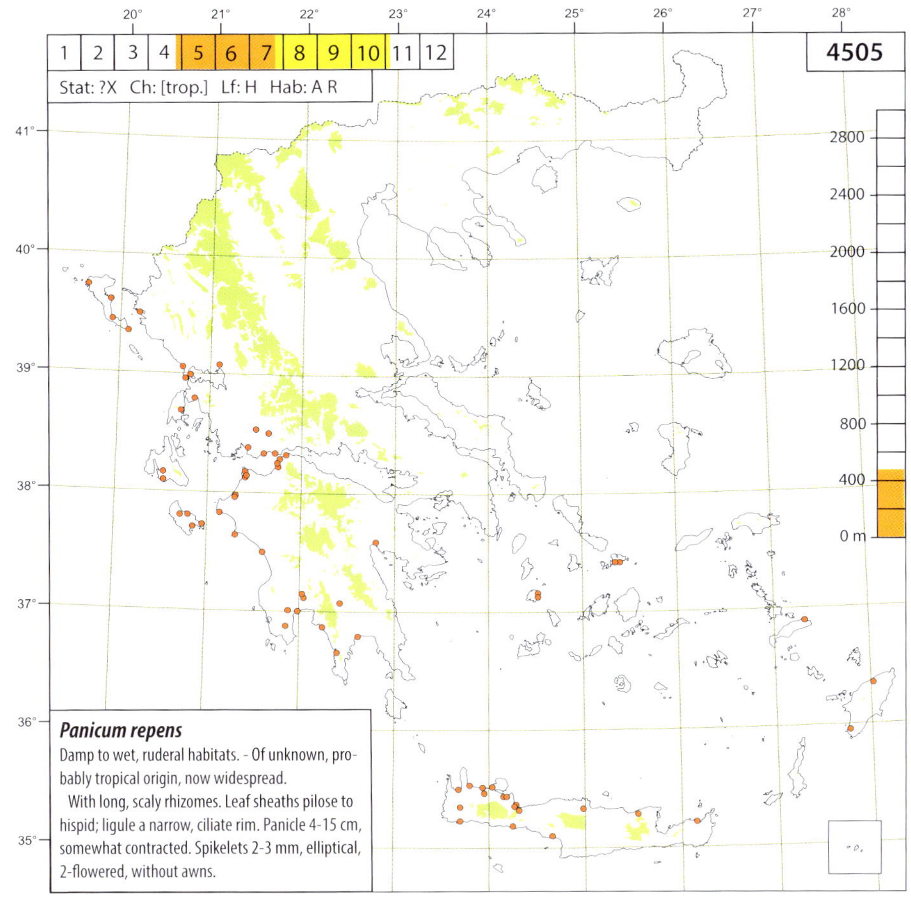

Panicum repens
Damp to wet, ruderal habitats. - Of unknown, probably tropical origin, now widespread.
With long, scaly rhizomes. Leaf sheaths pilose to hispid; ligule a narrow, ciliate rim. Panicle 4-15 cm, somewhat contracted. Spikelets 2-3 mm, elliptical, 2-flowered, without awns.

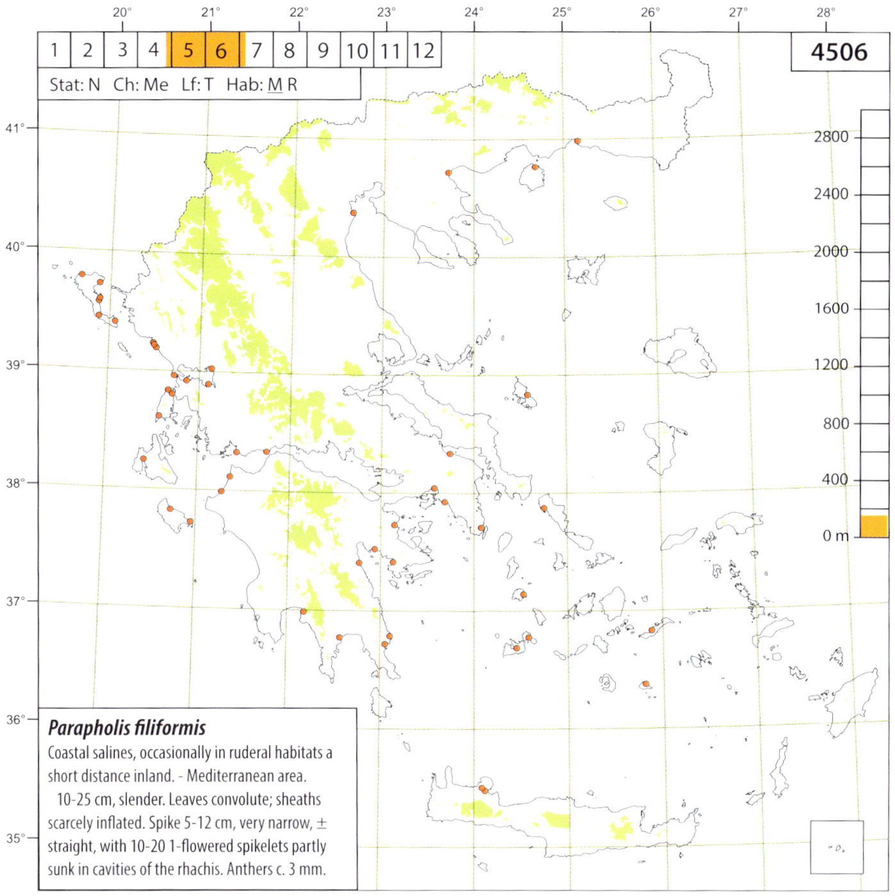

Parapholis filiformis
Coastal salines, occasionally in ruderal habitats a short distance inland. - Mediterranean area.
10-25 cm, slender. Leaves convolute; sheaths scarcely inflated. Spike 5-12 cm, very narrow, ± straight, with 10-20 1-flowered spikelets partly sunk in cavities of the rhachis. Anthers c. 3 mm.

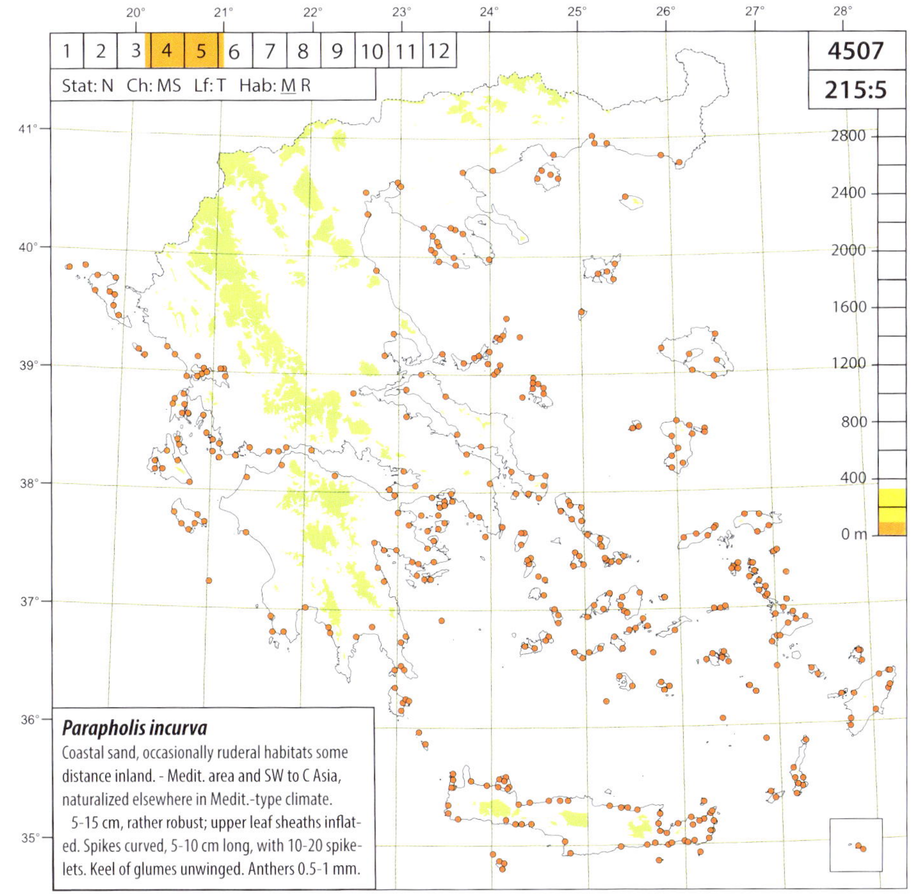

Parapholis incurva
Coastal sand, occasionally ruderal habitats some distance inland. - Medit. area and SW to C Asia, naturalized elsewhere in Medit.-type climate.
5-15 cm, rather robust; upper leaf sheaths inflated. Spikes curved, 5-10 cm long, with 10-20 spikelets. Keel of glumes unwinged. Anthers 0.5-1 mm.

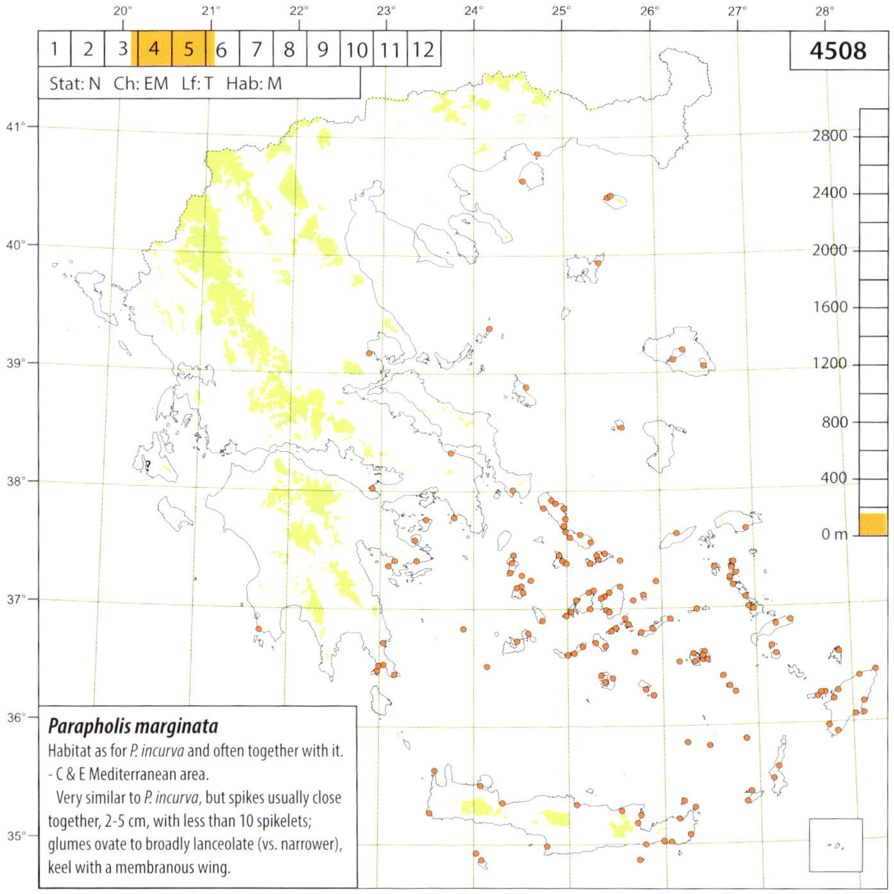

Parapholis marginata
Habitat as for *P. incurva* and often together with it. - C & E Mediterranean area.
Very similar to *P. incurva*, but spikes usually close together, 2-5 cm, with less than 10 spikelets; glumes ovate to broadly lanceolate (vs. narrower), keel with a membranous wing.

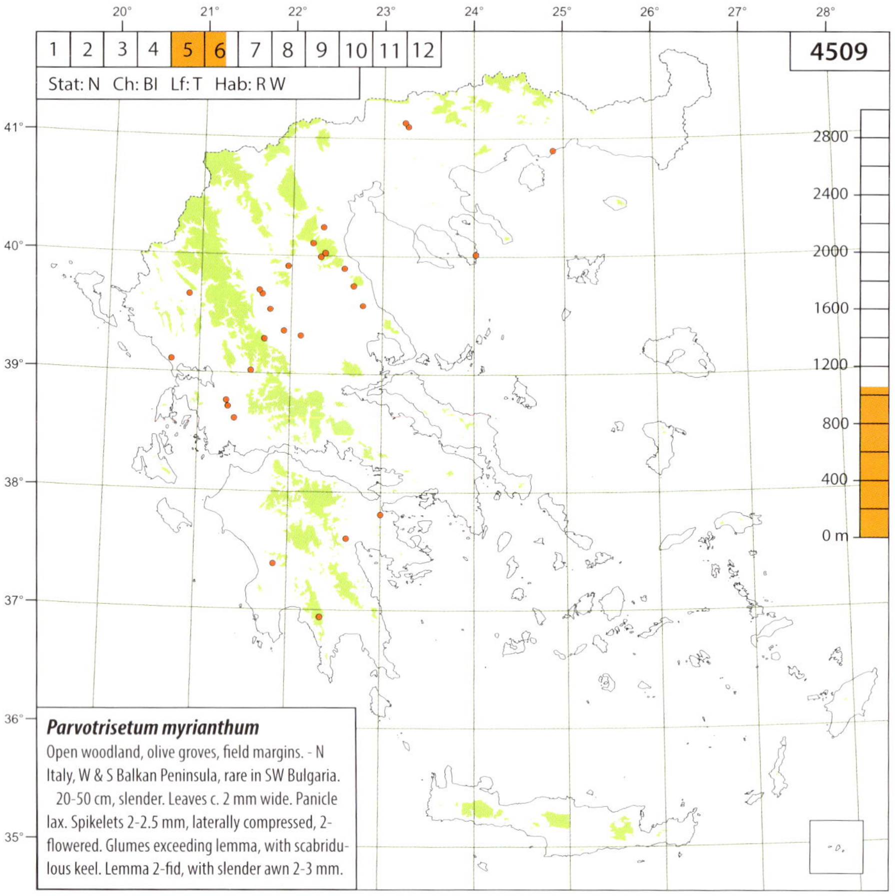

Parvotrisetum myrianthum

Open woodland, olive groves, field margins. - N Italy, W & S Balkan Peninsula, rare in SW Bulgaria.

20-50 cm, slender. Leaves c. 2 mm wide. Panicle lax. Spikelets 2-2.5 mm, laterally compressed, 2-flowered. Glumes exceeding lemma, with scabridulous keel. Lemma 2-fid, with slender awn 2-3 mm.

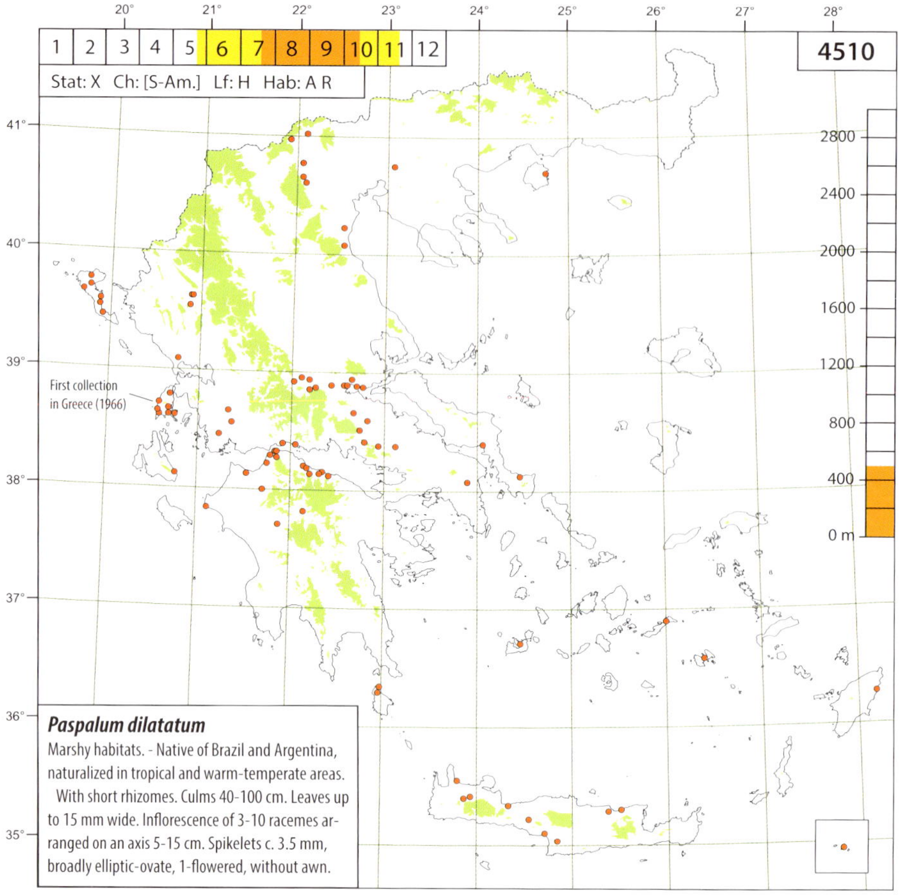

Paspalum dilatatum

Marshy habitats. - Native of Brazil and Argentina, naturalized in tropical and warm-temperate areas.

With short rhizomes. Culms 40-100 cm. Leaves up to 15 mm wide. Inflorescence of 3-10 racemes arranged on an axis 5-15 cm. Spikelets c. 3.5 mm, broadly elliptic-ovate, 1-flowered, without awn.

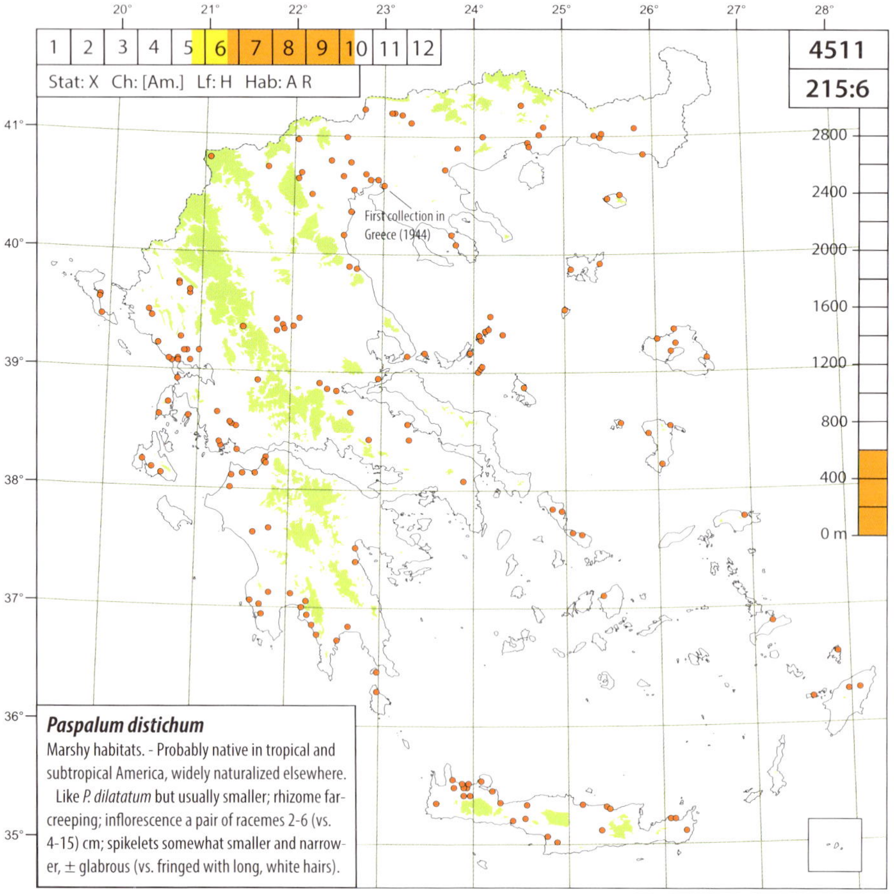

Paspalum distichum

Marshy habitats. - Probably native in tropical and subtropical America, widely naturalized elsewhere.

Like *P. dilatatum* but usually smaller; rhizome far-creeping; inflorescence a pair of racemes 2-6 (vs. 4-15) cm; spikelets somewhat smaller and narrower, ± glabrous (vs. fringed with long, white hairs).

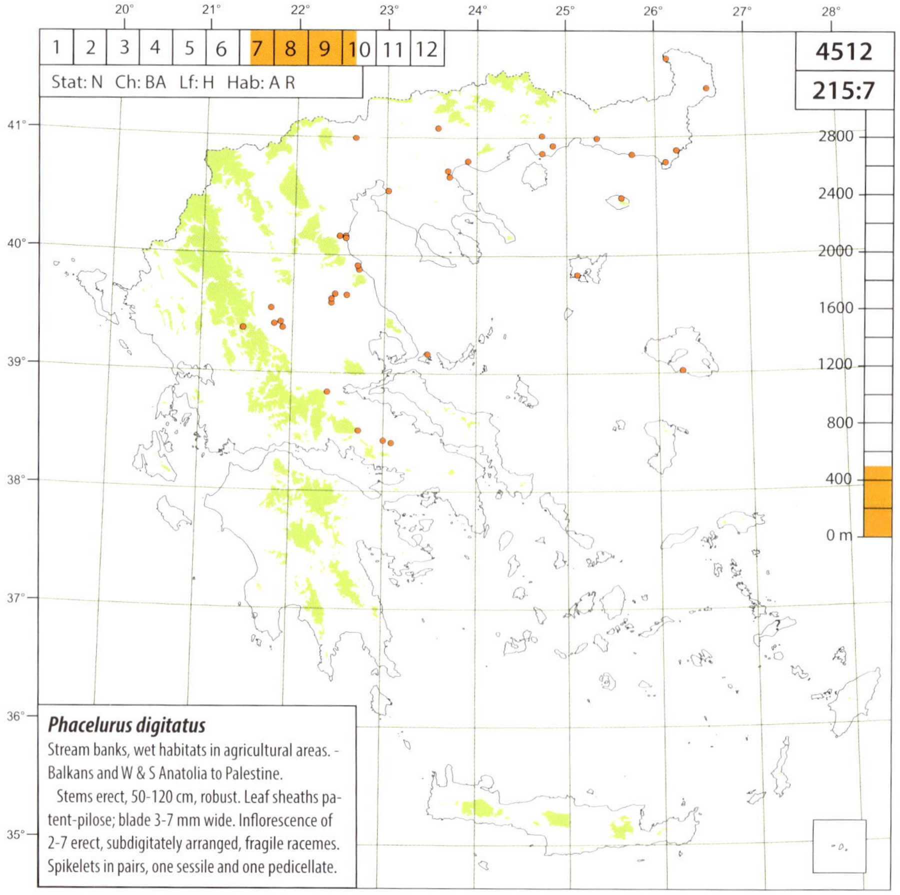

Phacelurus digitatus

Stream banks, wet habitats in agricultural areas. - Balkans and W & S Anatolia to Palestine.

Stems erect, 50-120 cm, robust. Leaf sheaths patent-pilose; blade 3-7 mm wide. Inflorescence of 2-7 erect, subdigitately arranged, fragile racemes. Spikelets in pairs, one sessile and one pedicellate.

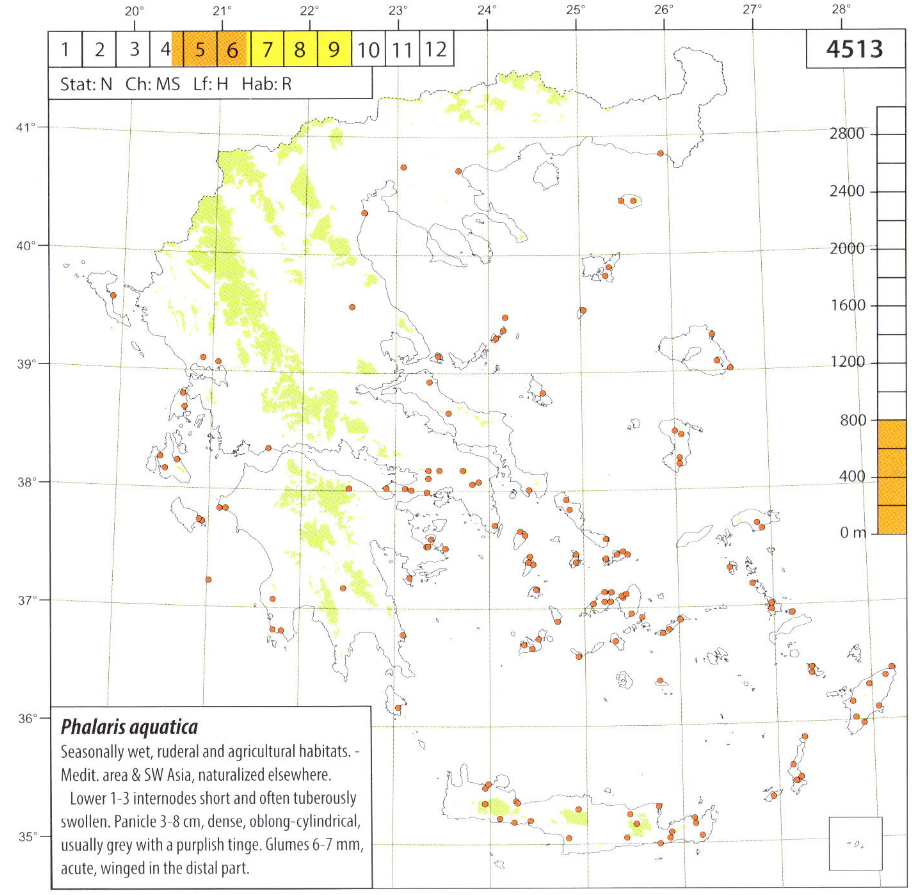

Phalaris aquatica

Seasonally wet, ruderal and agricultural habitats. - Medit. area & SW Asia, naturalized elsewhere.

Lower 1-3 internodes short and often tuberously swollen. Panicle 3-8 cm, dense, oblong-cylindrical, usually grey with a purplish tinge. Glumes 6-7 mm, acute, winged in the distal part.

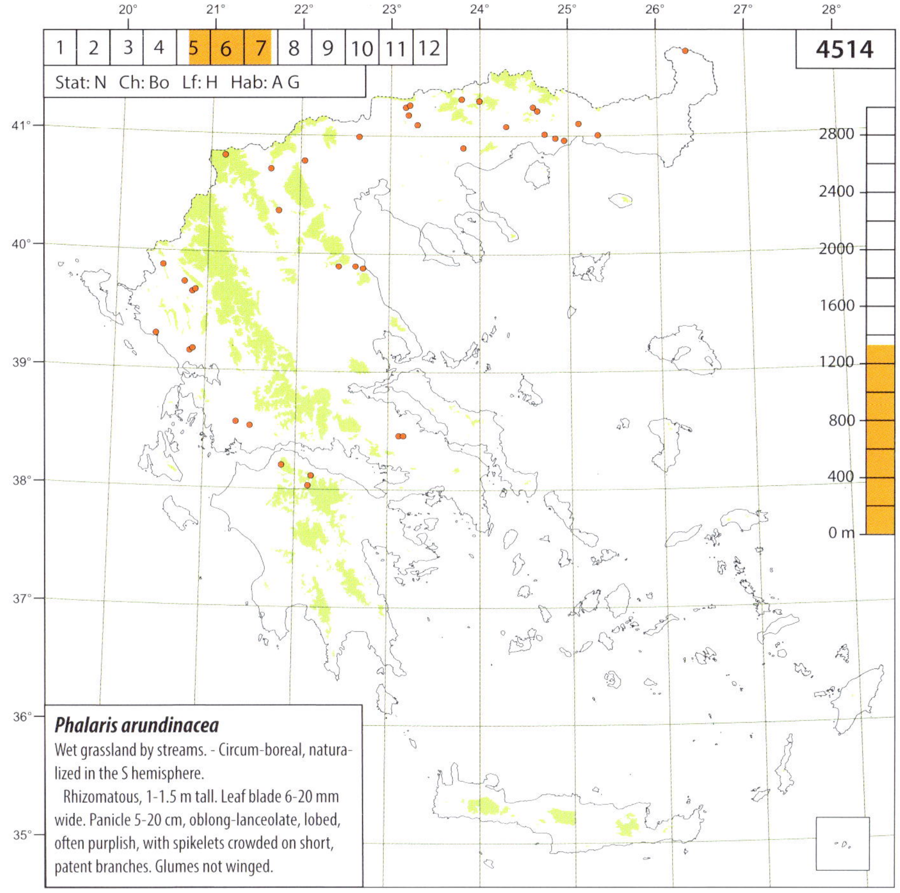

Phalaris arundinacea

Wet grassland by streams. - Circum-boreal, naturalized in the S hemisphere.

Rhizomatous, 1-1.5 m tall. Leaf blade 6-20 mm wide. Panicle 5-20 cm, oblong-lanceolate, lobed, often purplish, with spikelets crowded on short, patent branches. Glumes not winged.

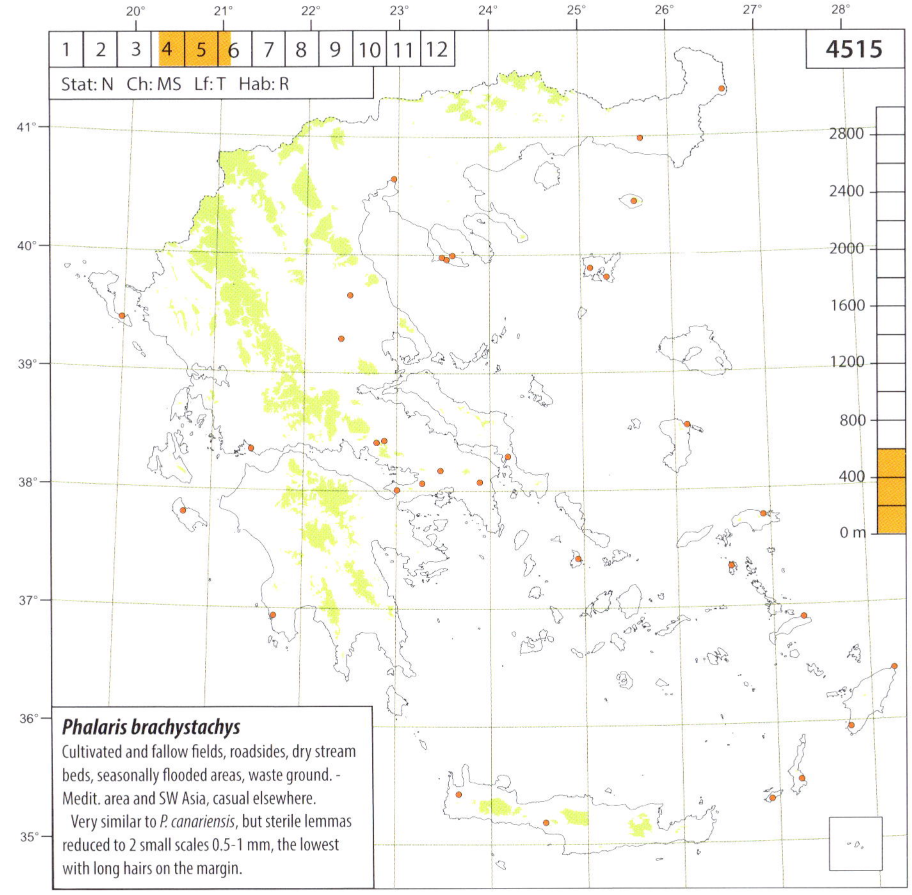

Phalaris brachystachys

Cultivated and fallow fields, roadsides, dry stream beds, seasonally flooded areas, waste ground. - Medit. area and SW Asia, casual elsewhere.

Very similar to *P. canariensis*, but sterile lemmas reduced to 2 small scales 0.5-1 mm, the lowest with long hairs on the margin.

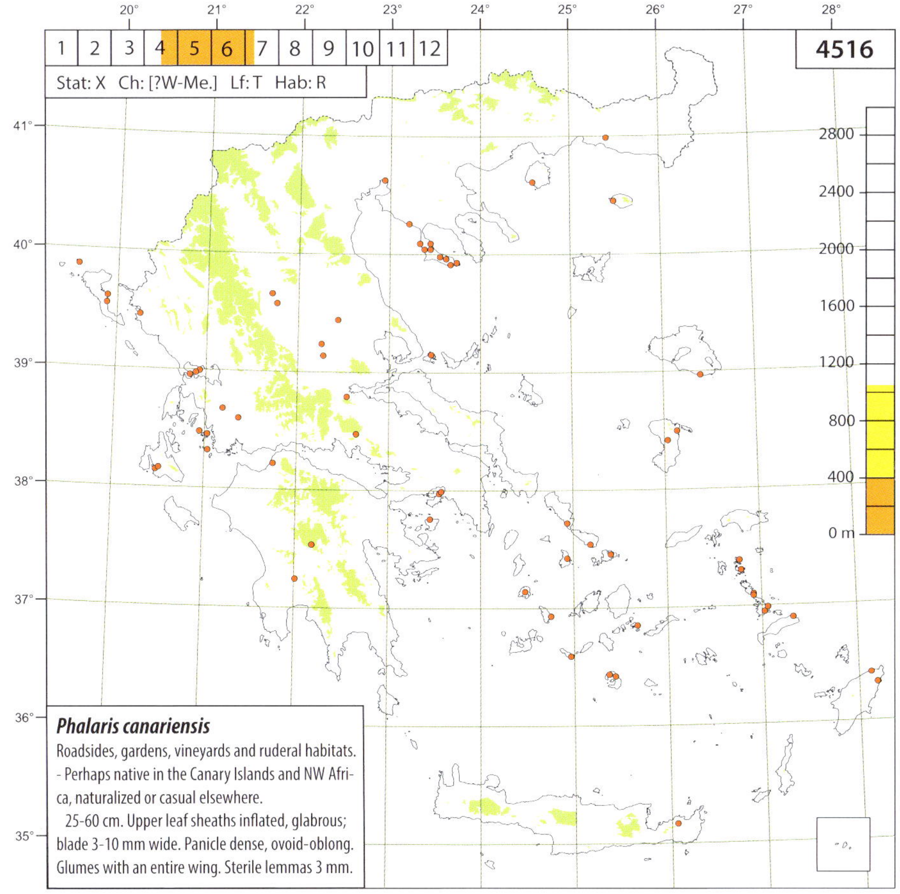

Phalaris canariensis

Roadsides, gardens, vineyards and ruderal habitats. - Perhaps native in the Canary Islands and NW Africa, naturalized or casual elsewhere.

25-60 cm. Upper leaf sheaths inflated, glabrous; blade 3-10 mm wide. Panicle dense, ovoid-oblong. Glumes with an entire wing. Sterile lemmas 3 mm.

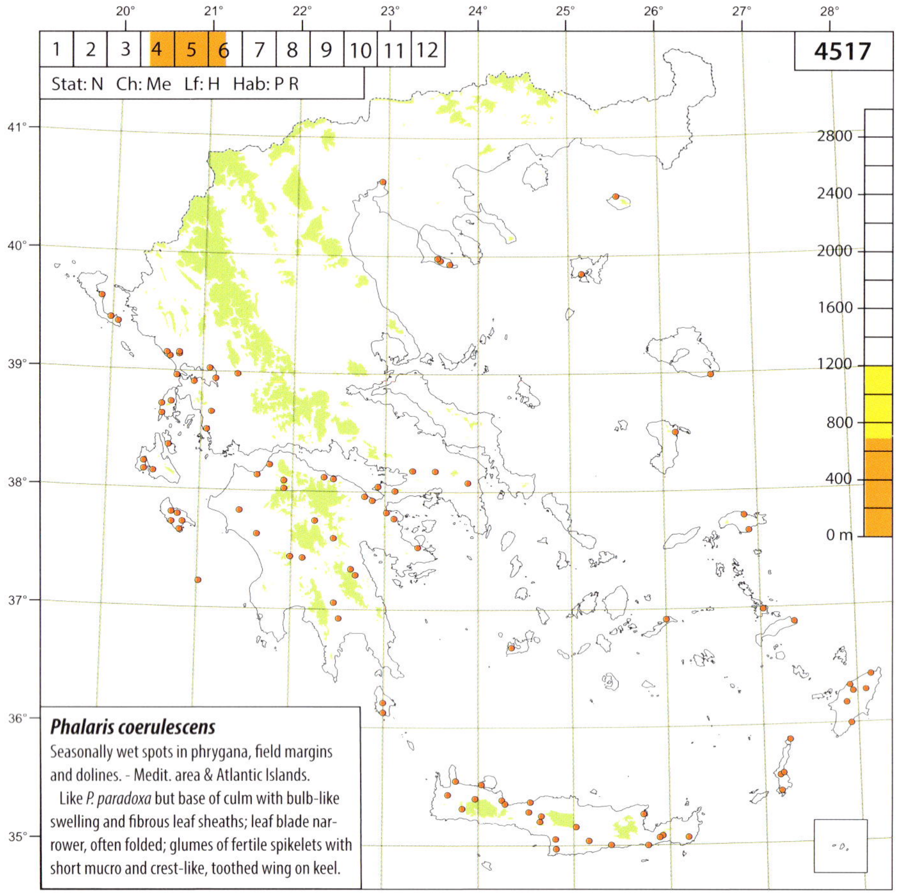

Phalaris coerulescens
Seasonally wet spots in phrygana, field margins and dolines. - Medit. area & Atlantic Islands.
Like *P. paradoxa* but base of culm with bulb-like swelling and fibrous leaf sheaths; leaf blade narrower, often folded; glumes of fertile spikelets with short mucro and crest-like, toothed wing on keel.

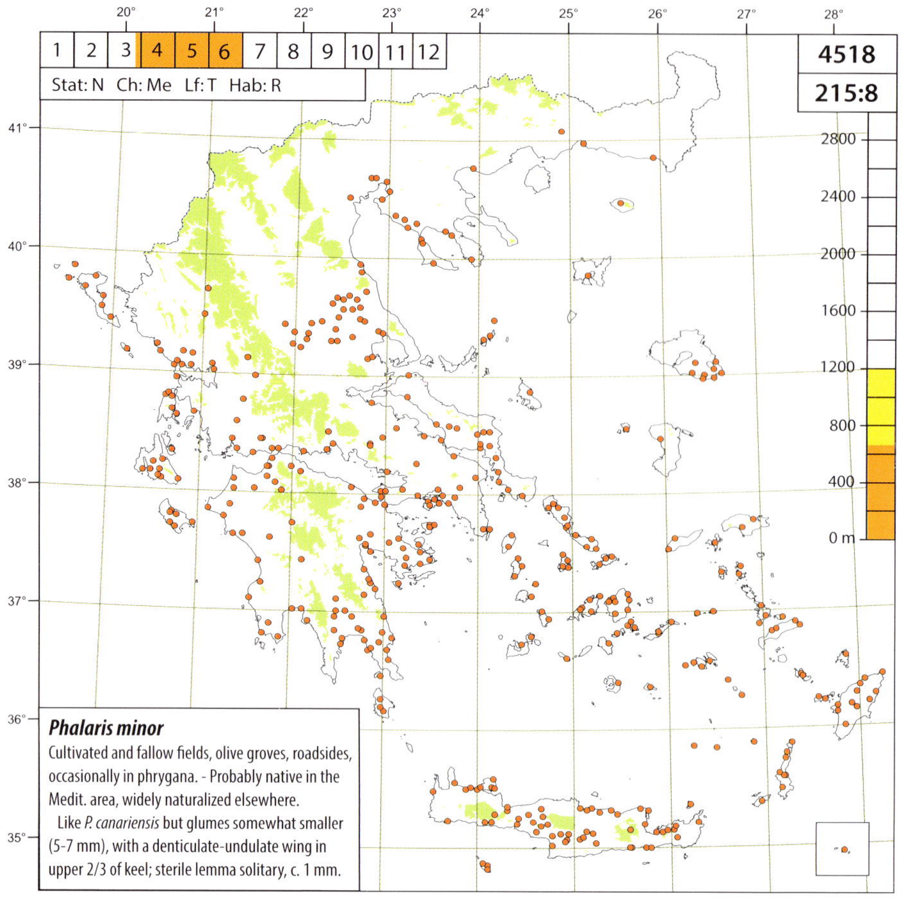

Phalaris minor
Cultivated and fallow fields, olive groves, roadsides, occasionally in phrygana. - Probably native in the Medit. area, widely naturalized elsewhere.
Like *P. canariensis* but glumes somewhat smaller (5-7 mm), with a denticulate-undulate wing in upper 2/3 of keel; sterile lemma solitary, c. 1 mm.

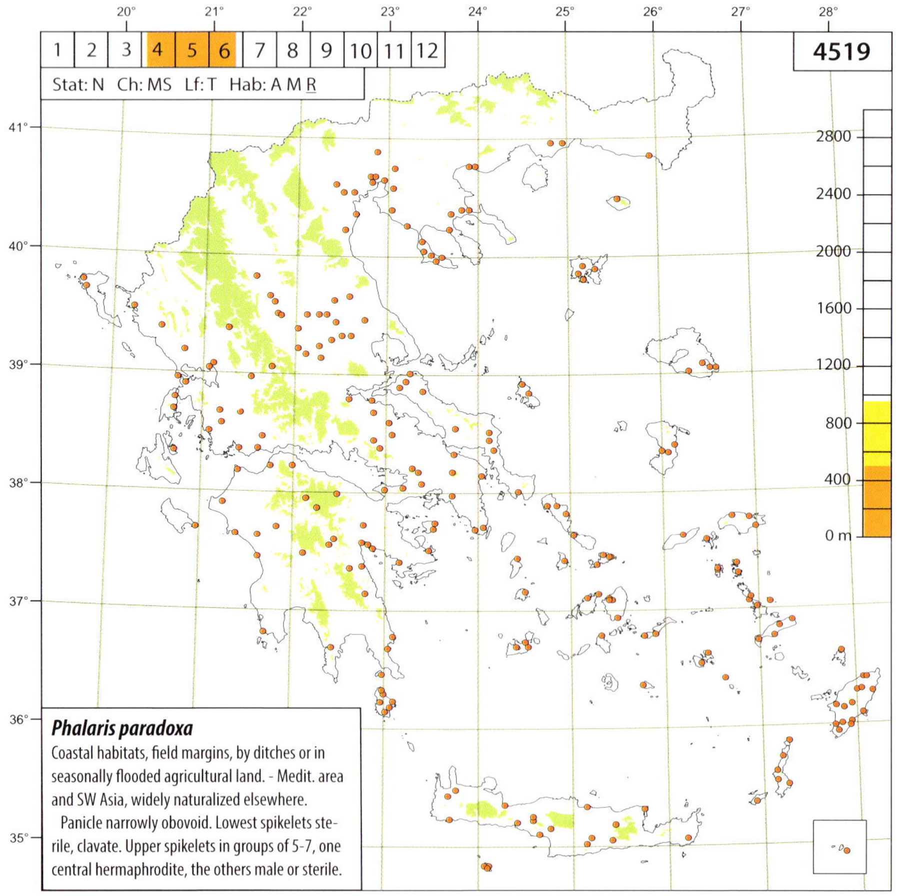

Phalaris paradoxa
Coastal habitats, field margins, by ditches or in seasonally flooded agricultural land. - Medit. area and SW Asia, widely naturalized elsewhere.
Panicle narrowly obovoid. Lowest spikelets sterile, clavate. Upper spikelets in groups of 5-7, one central hermaphrodite, the others male or sterile.

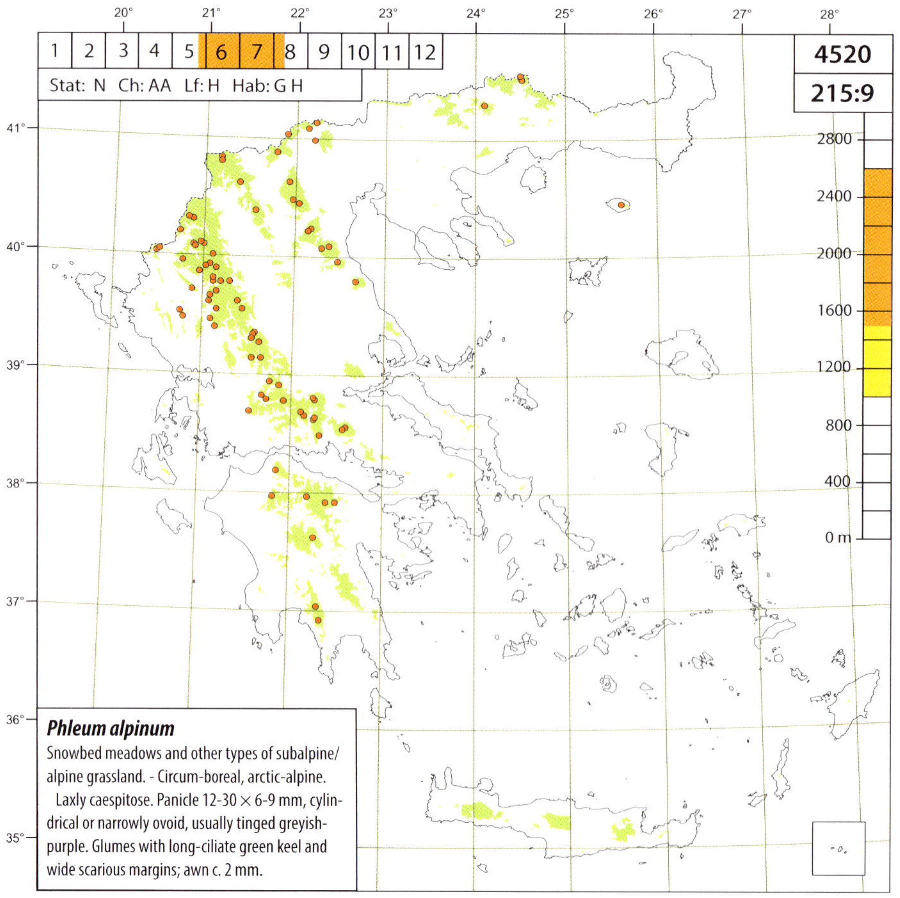

Phleum alpinum
Snowbed meadows and other types of subalpine/alpine grassland. - Circum-boreal, arctic-alpine.
Laxly caespitose. Panicle 12-30 × 6-9 mm, cylindrical or narrowly ovoid, usually tinged greyish-purple. Glumes with long-ciliate green keel and wide scarious margins; awn c. 2 mm.

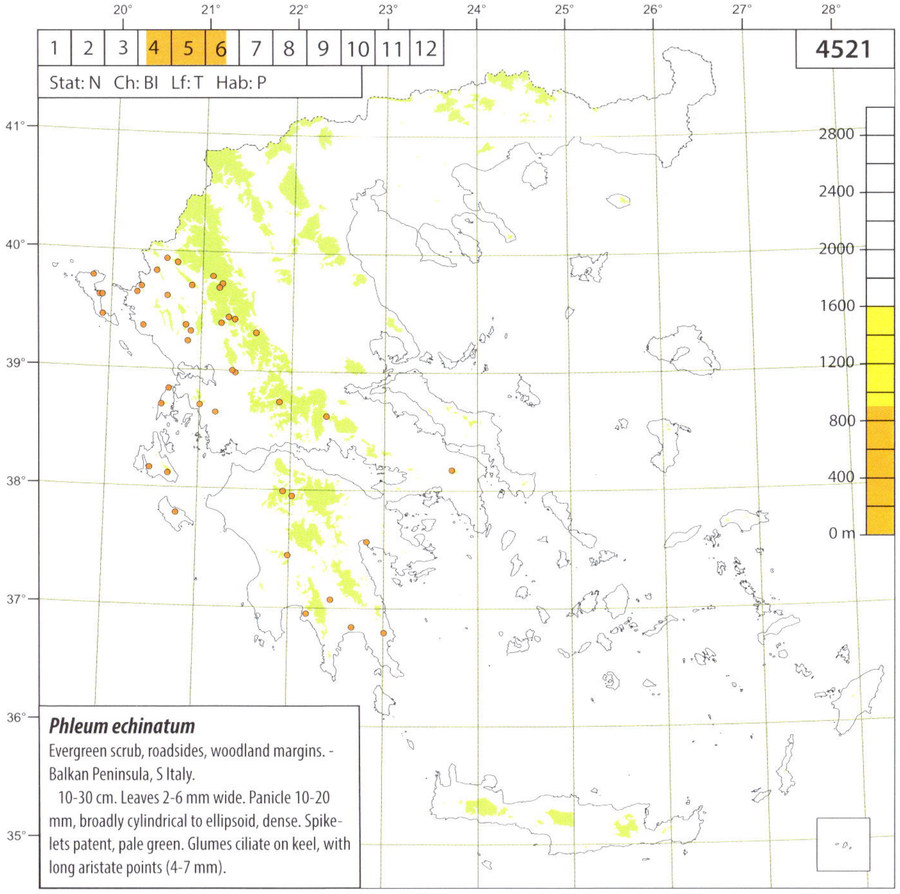

Phleum echinatum
Evergreen scrub, roadsides, woodland margins. - Balkan Peninsula, S Italy.
10-30 cm. Leaves 2-6 mm wide. Panicle 10-20 mm, broadly cylindrical to ellipsoid, dense. Spikelets patent, pale green. Glumes ciliate on keel, with long aristate points (4-7 mm).

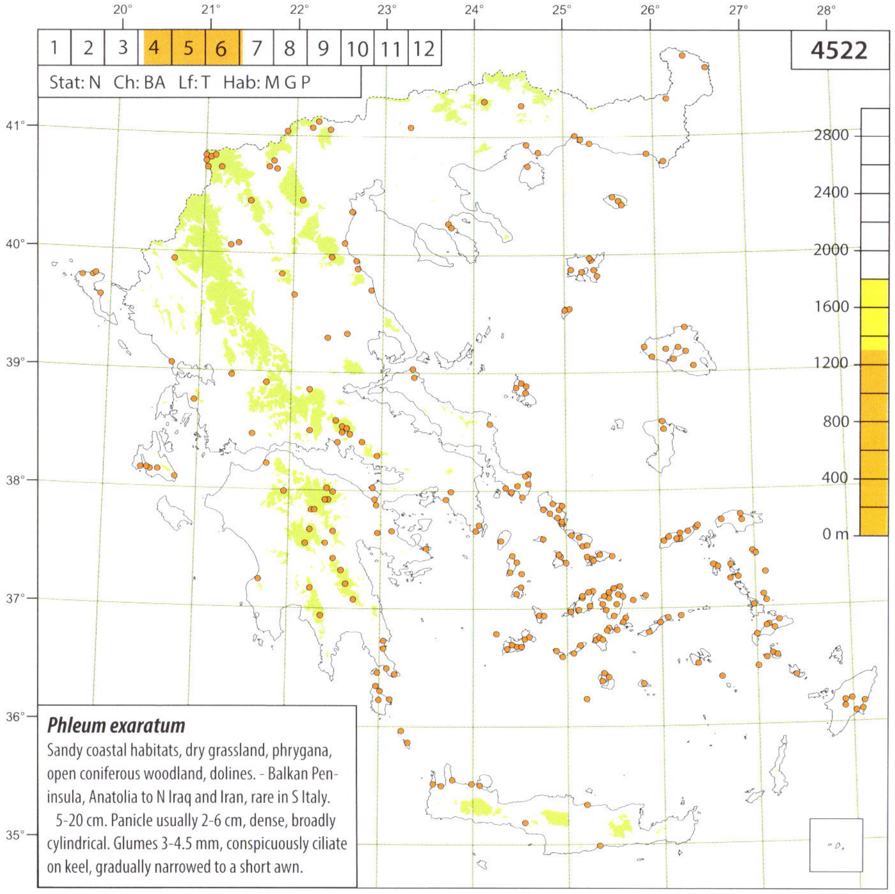

Phleum exaratum
Sandy coastal habitats, dry grassland, phrygana, open coniferous woodland, dolines. - Balkan Peninsula, Anatolia to N Iraq and Iran, rare in S Italy.
5-20 cm. Panicle usually 2-6 cm, dense, broadly cylindrical. Glumes 3-4.5 mm, conspicuously ciliate on keel, gradually narrowed to a short awn.

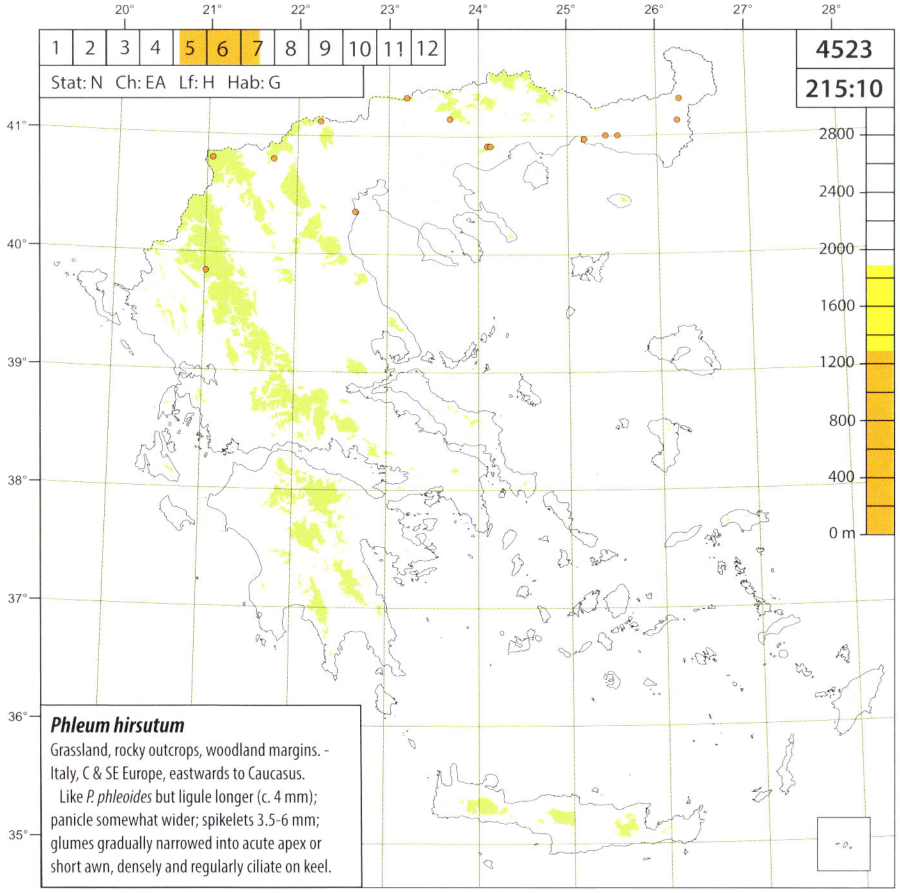

Phleum hirsutum
Grassland, rocky outcrops, woodland margins. - Italy, C & SE Europe, eastwards to Caucasus.
Like *P. phleoides* but ligule longer (c. 4 mm); panicle somewhat wider; spikelets 3.5-6 mm; glumes gradually narrowed into acute apex or short awn, densely and regularly ciliate on keel.

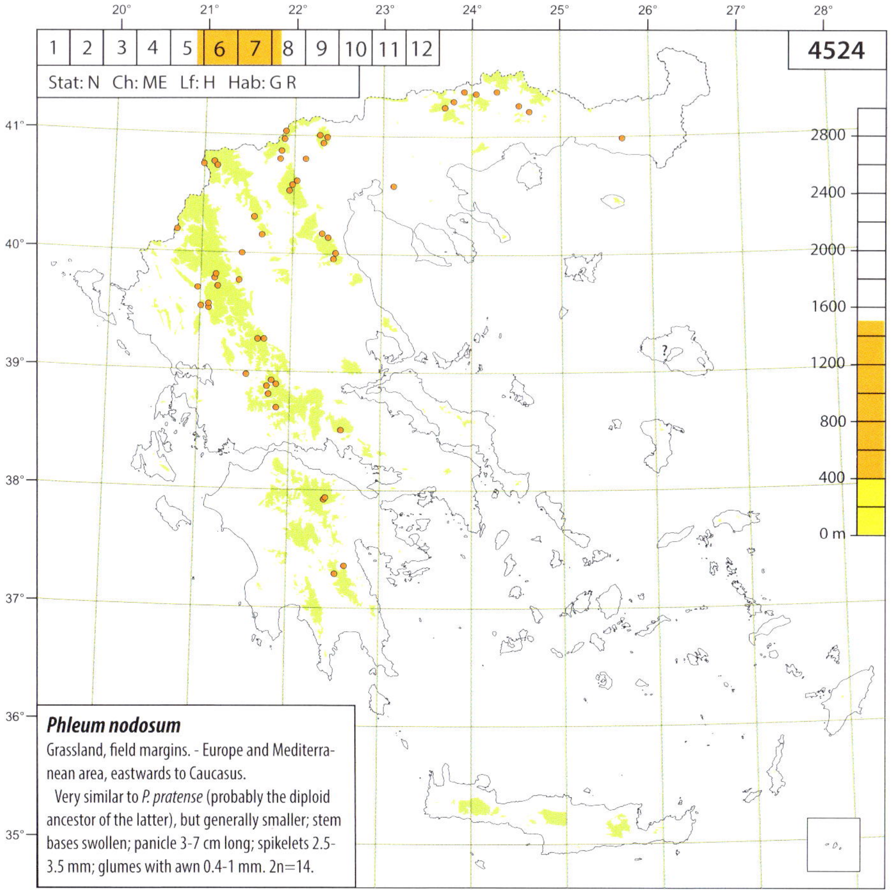

Phleum nodosum
Grassland, field margins. - Europe and Mediterranean area, eastwards to Caucasus.
Very similar to *P. pratense* (probably the diploid ancestor of the latter), but generally smaller; stem bases swollen; panicle 3-7 cm long; spikelets 2.5-3.5 mm; glumes with awn 0.4-1 mm. 2n=14.

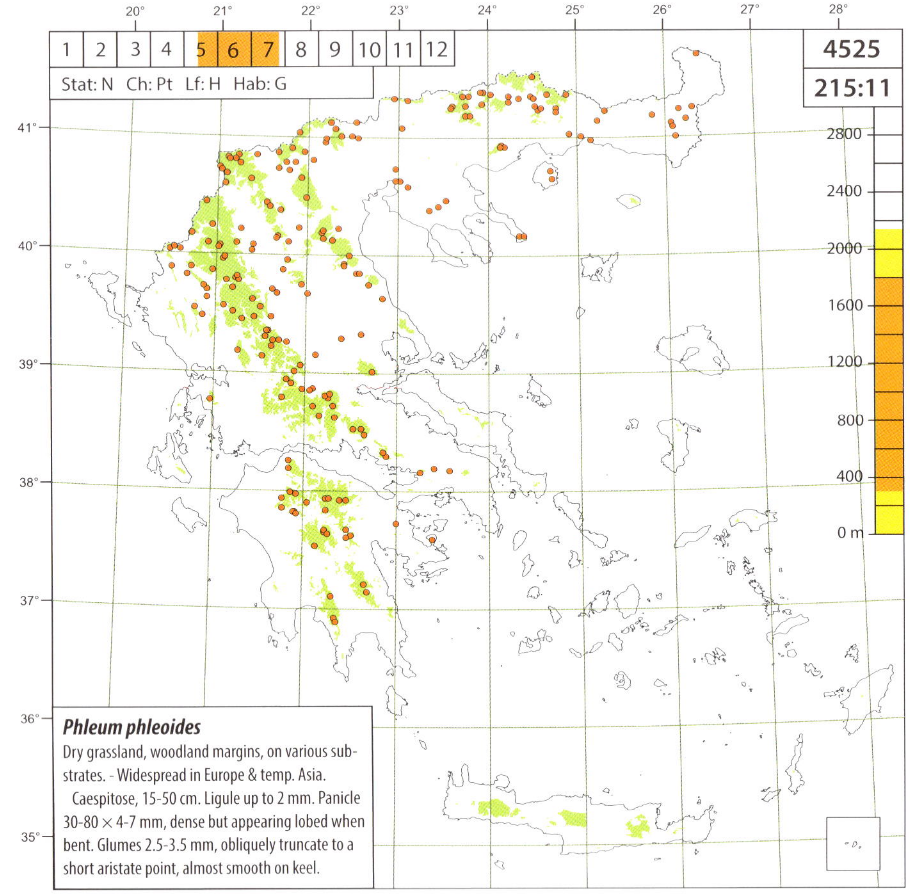

Phleum phleoides

Dry grassland, woodland margins, on various substrates. - Widespread in Europe & temp. Asia.

Caespitose, 15-50 cm. Ligule up to 2 mm. Panicle 30-80 × 4-7 mm, dense but appearing lobed when bent. Glumes 2.5-3.5 mm, obliquely truncate to a short aristate point, almost smooth on keel.

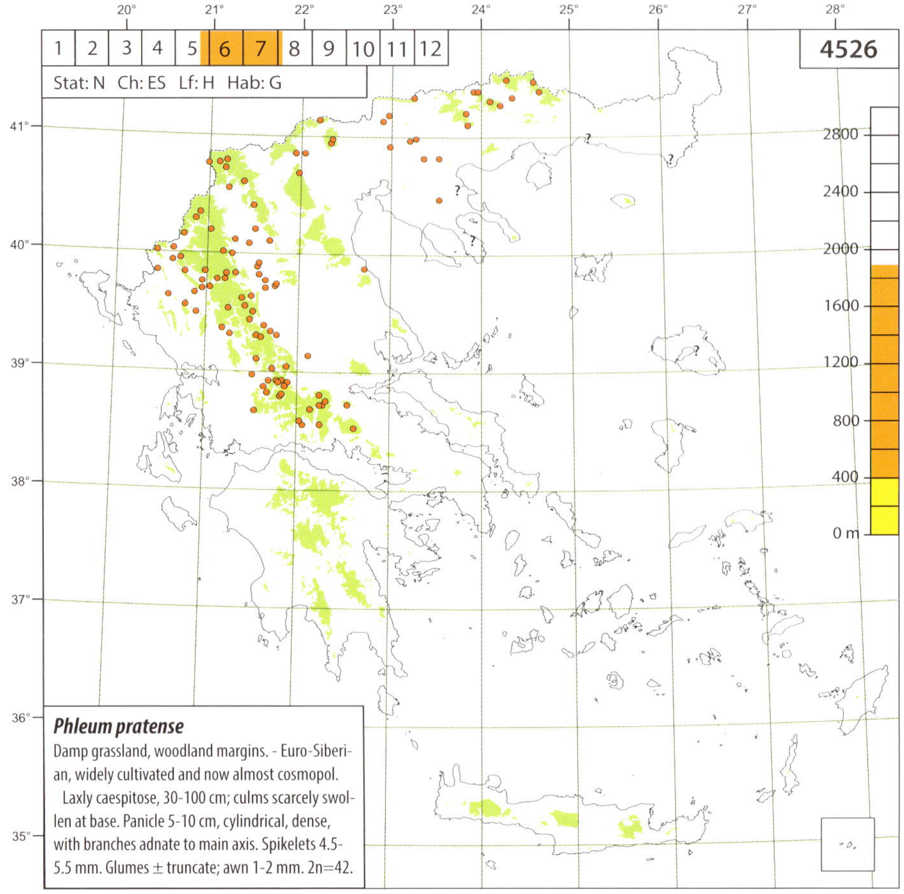

Phleum pratense

Damp grassland, woodland margins. - Euro-Siberian, widely cultivated and now almost cosmopol.

Laxly caespitose, 30-100 cm; culms scarcely swollen at base. Panicle 5-10 cm, cylindrical, dense, with branches adnate to main axis. Spikelets 4.5-5.5 mm. Glumes ± truncate; awn 1-2 mm. 2n=42.

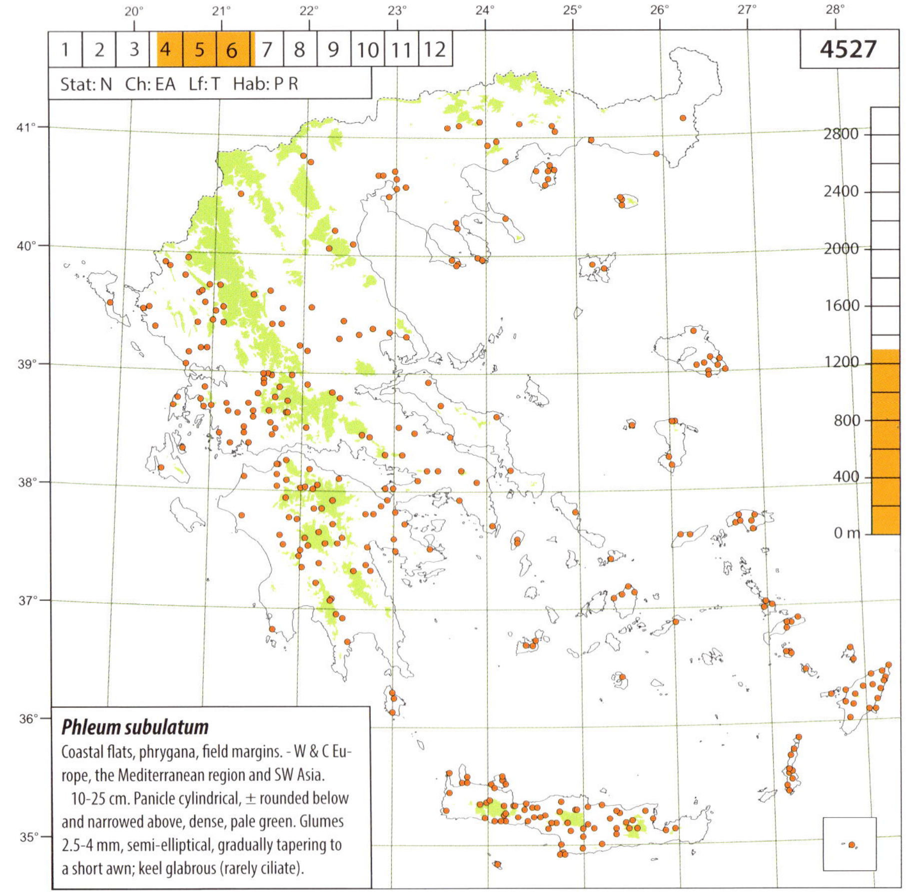

Phleum subulatum

Coastal flats, phrygana, field margins. - W & C Europe, the Mediterranean region and SW Asia.

10-25 cm. Panicle cylindrical, ± rounded below and narrowed above, dense, pale green. Glumes 2.5-4 mm, semi-elliptical, gradually tapering to a short awn; keel glabrous (rarely ciliate).

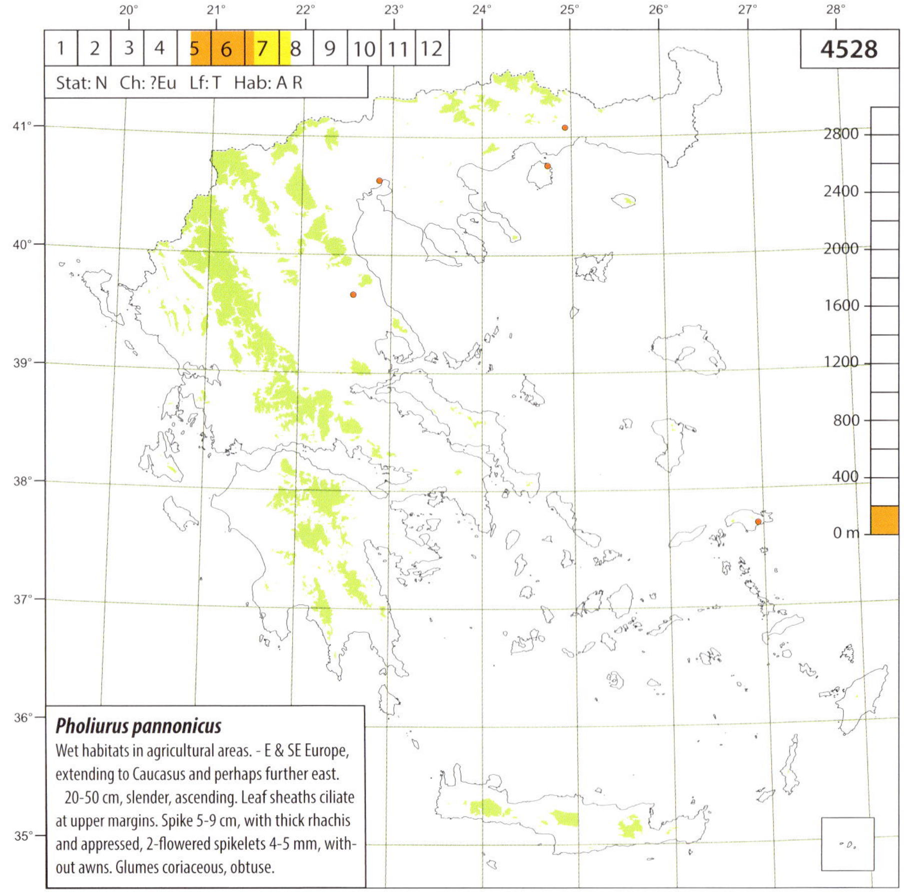

Pholiurus pannonicus

Wet habitats in agricultural areas. - E & SE Europe, extending to Caucasus and perhaps further east.

20-50 cm, slender, ascending. Leaf sheaths ciliate at upper margins. Spike 5-9 cm, with thick rhachis and appressed, 2-flowered spikelets 4-5 mm, without awns. Glumes coriaceous, obtuse.

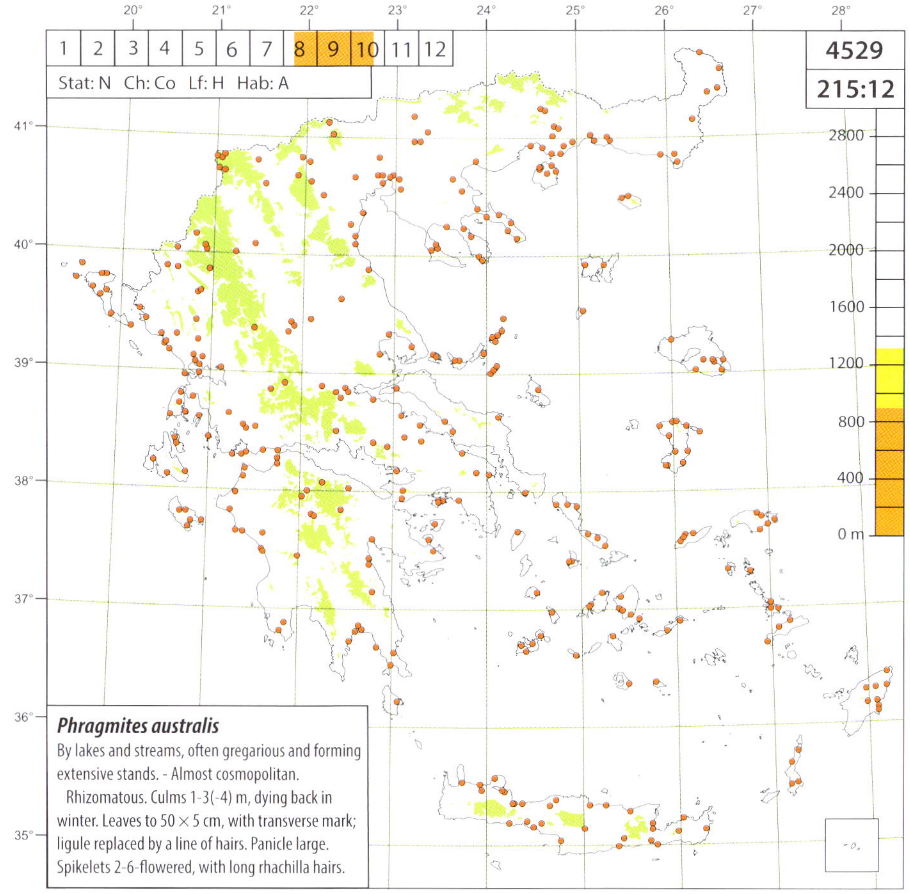

Phragmites australis

By lakes and streams, often gregarious and forming extensive stands. - Almost cosmopolitan.

Rhizomatous. Culms 1-3(-4) m, dying back in winter. Leaves to 50 × 5 cm, with transverse mark; ligule replaced by a line of hairs. Panicle large. Spikelets 2-6-flowered, with long rhachilla hairs.

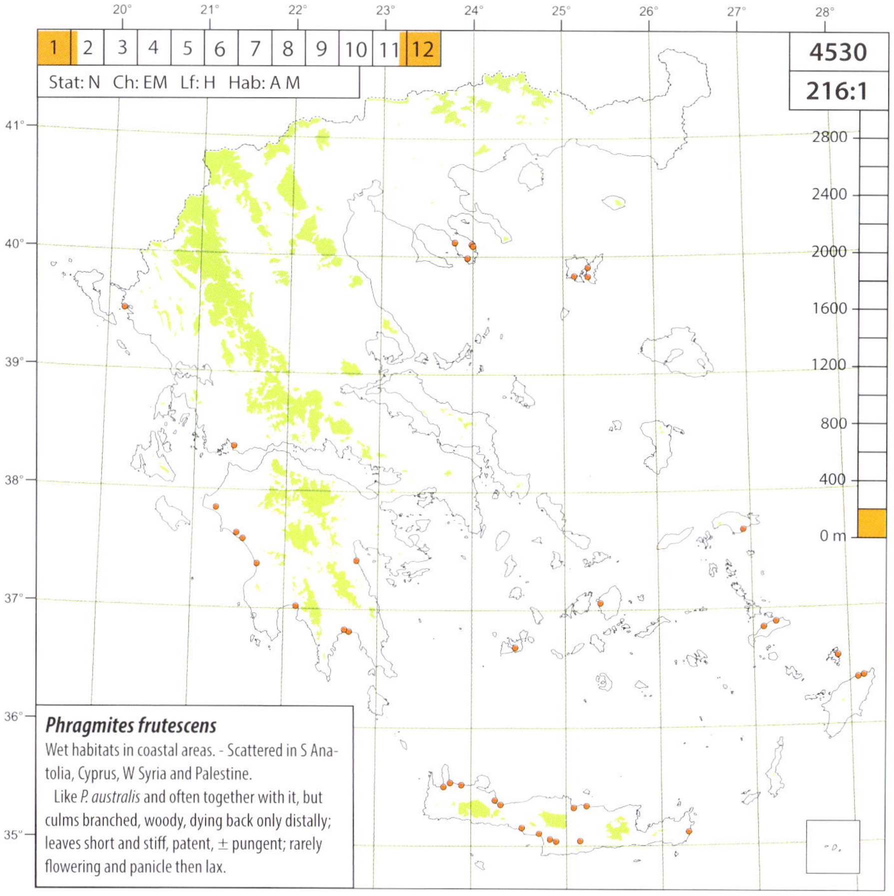

Phragmites frutescens

Wet habitats in coastal areas. - Scattered in S Anatolia, Cyprus, W Syria and Palestine.

Like *P. australis* and often together with it, but culms branched, woody, dying back only distally; leaves short and stiff, patent, ± pungent; rarely flowering and panicle then lax.

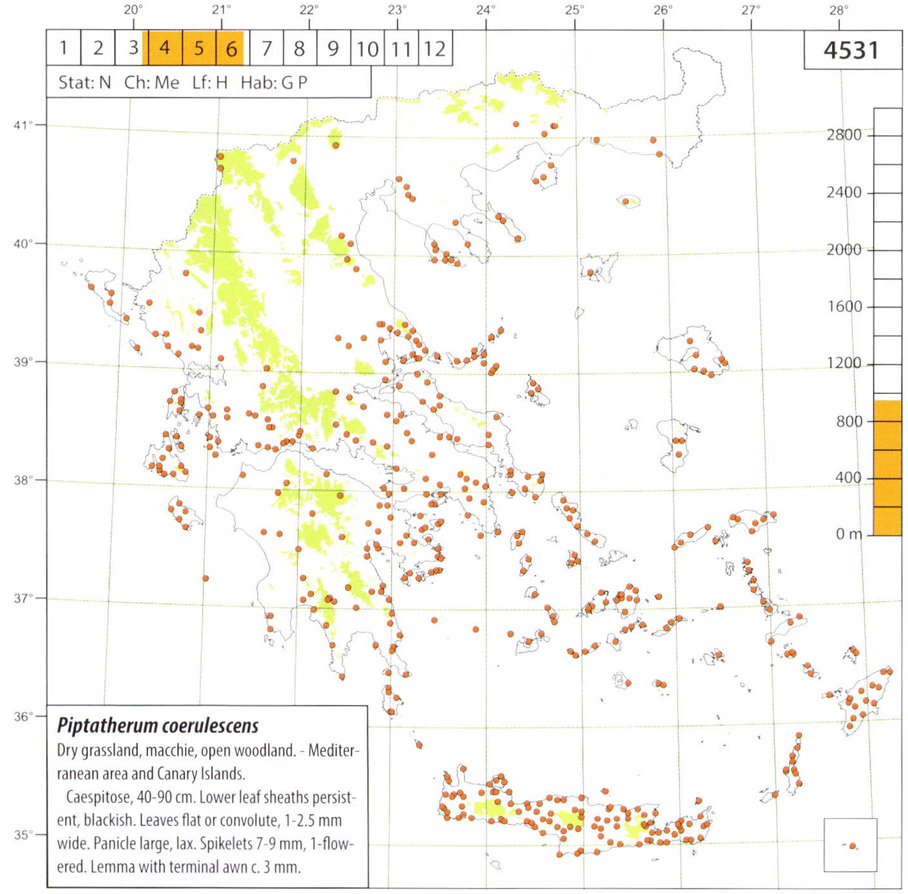

Piptatherum coerulescens

Dry grassland, macchie, open woodland. - Mediterranean area and Canary Islands.

Caespitose, 40-90 cm. Lower leaf sheaths persistent, blackish. Leaves flat or convolute, 1-2.5 mm wide. Panicle large, lax. Spikelets 7-9 mm, 1-flowered. Lemma with terminal awn c. 3 mm.

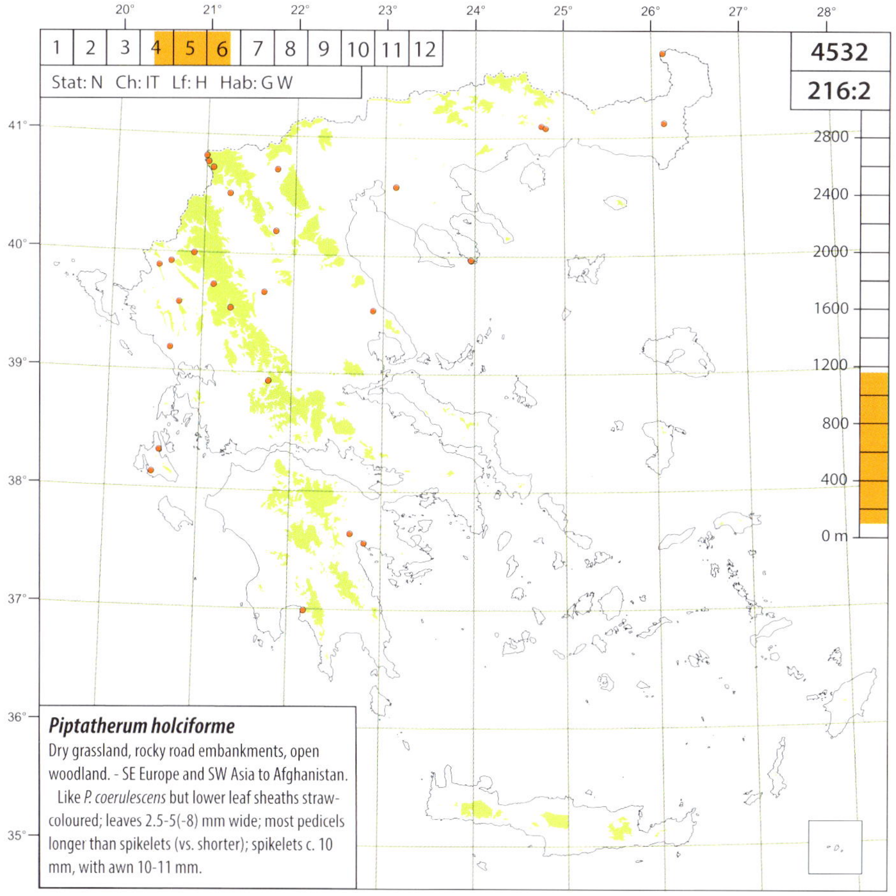

Piptatherum holciforme

Dry grassland, rocky road embankments, open woodland. - SE Europe and SW Asia to Afghanistan.

Like *P. coerulescens* but lower leaf sheaths straw-coloured; leaves 2.5-5(-8) mm wide; most pedicels longer than spikelets (vs. shorter); spikelets c. 10 mm, with awn 10-11 mm.

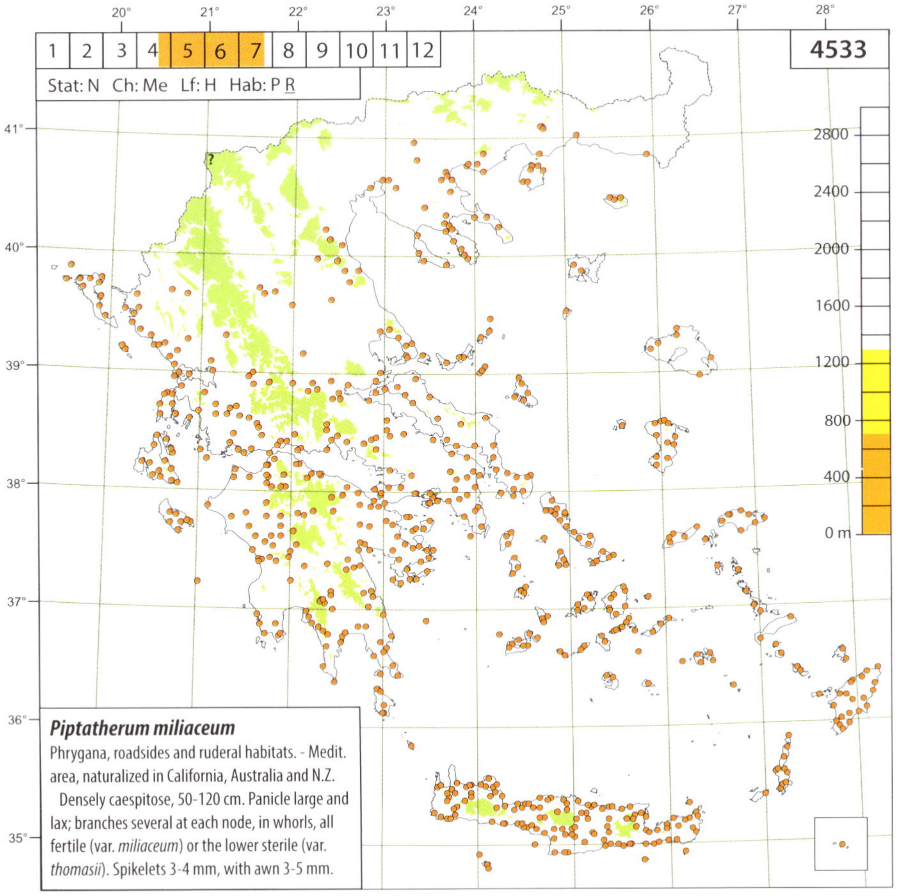

Piptatherum miliaceum

Phrygana, roadsides and ruderal habitats. - Medit. area, naturalized in California, Australia and N.Z.

Densely caespitose, 50-120 cm. Panicle large and lax; branches several at each node, in whorls, all fertile (var. *miliaceum*) or the lower sterile (var. *thomasii*). Spikelets 3-4 mm, with awn 3-5 mm.

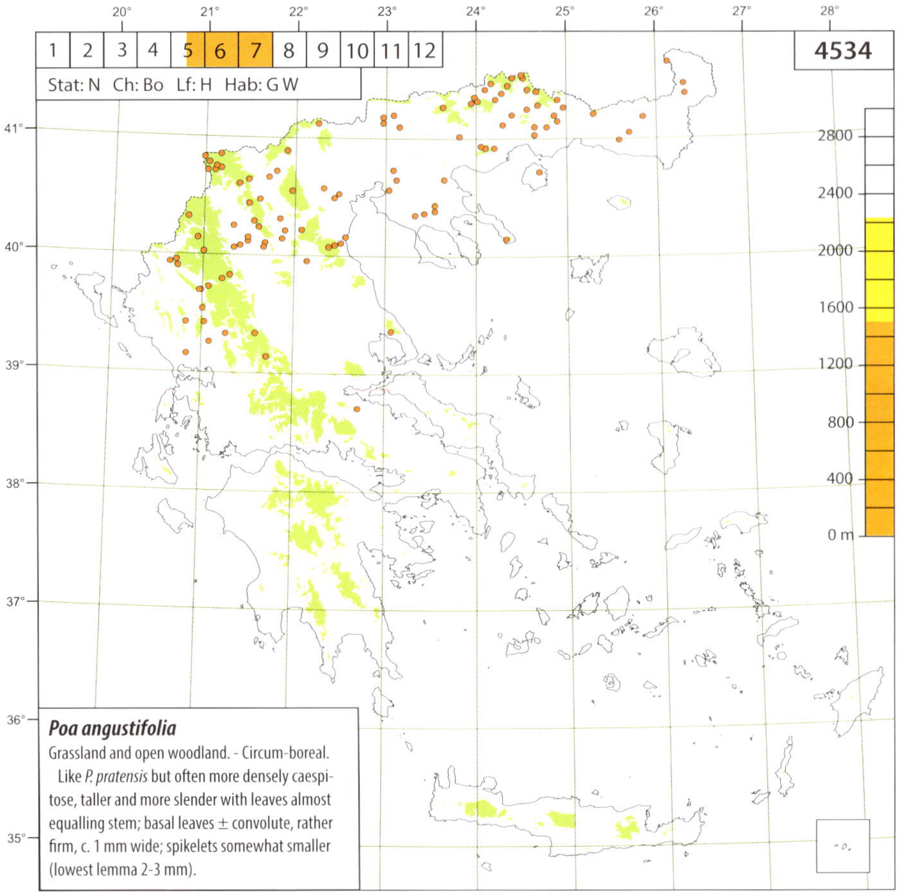

Poa angustifolia

Grassland and open woodland. - Circum-boreal.

Like *P. pratensis* but often more densely caespitose, taller and more slender with leaves almost equalling stem; basal leaves ± convolute, rather firm, c. 1 mm wide; spikelets somewhat smaller (lowest lemma 2-3 mm).

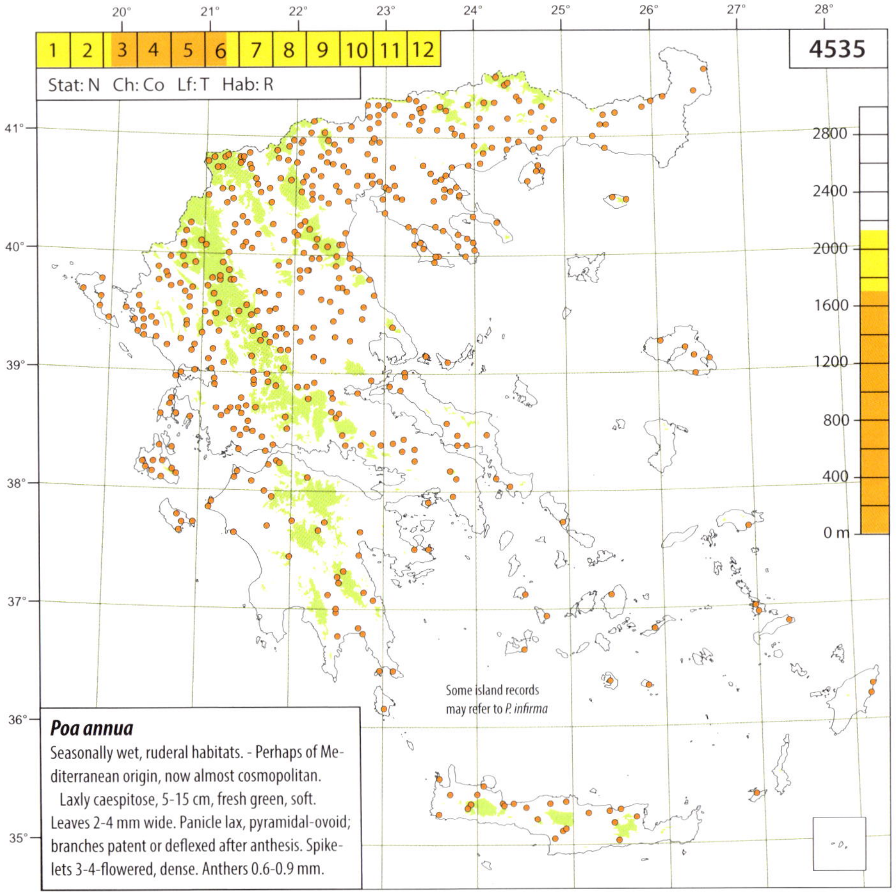

Poa annua

Seasonally wet, ruderal habitats. - Perhaps of Mediterranean origin, now almost cosmopolitan.

Laxly caespitose, 5-15 cm, fresh green, soft. Leaves 2-4 mm wide. Panicle lax, pyramidal-ovoid; branches patent or deflexed after anthesis. Spikelets 3-4-flowered, dense. Anthers 0.6-0.9 mm.

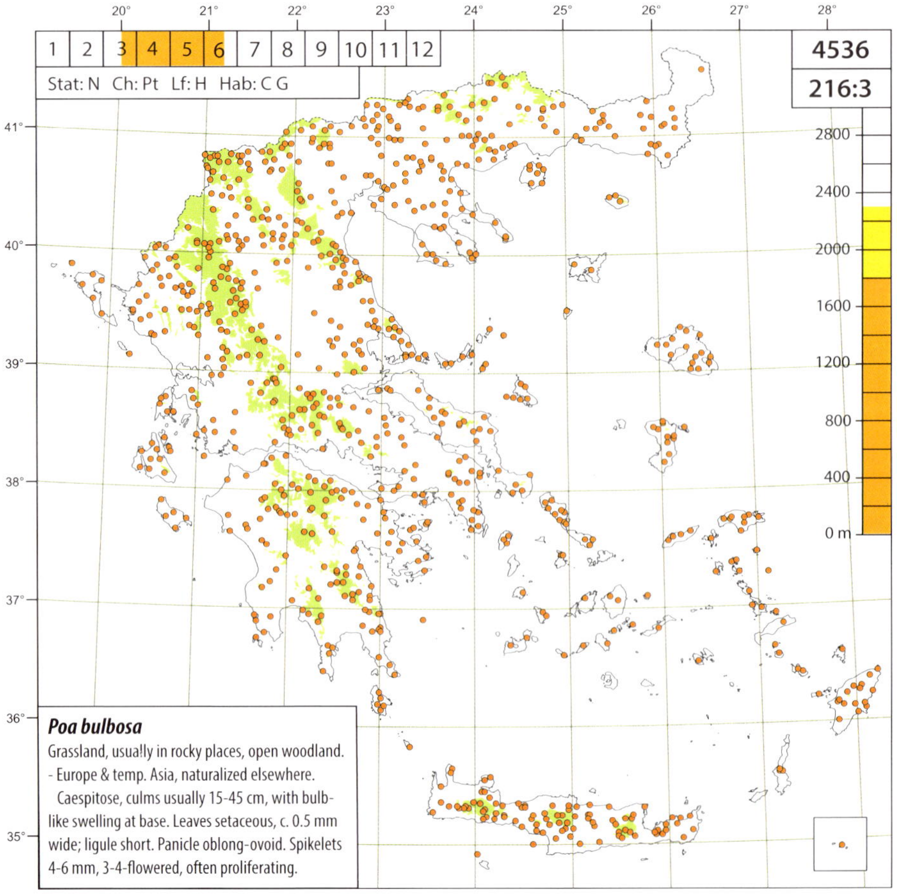

Poa bulbosa

Grassland, usually in rocky places, open woodland. - Europe & temp. Asia, naturalized elsewhere.

Caespitose, culms usually 15-45 cm, with bulb-like swelling at base. Leaves setaceous, c. 0.5 mm wide; ligule short. Panicle oblong-ovoid. Spikelets 4-6 mm, 3-4-flowered, often proliferating.

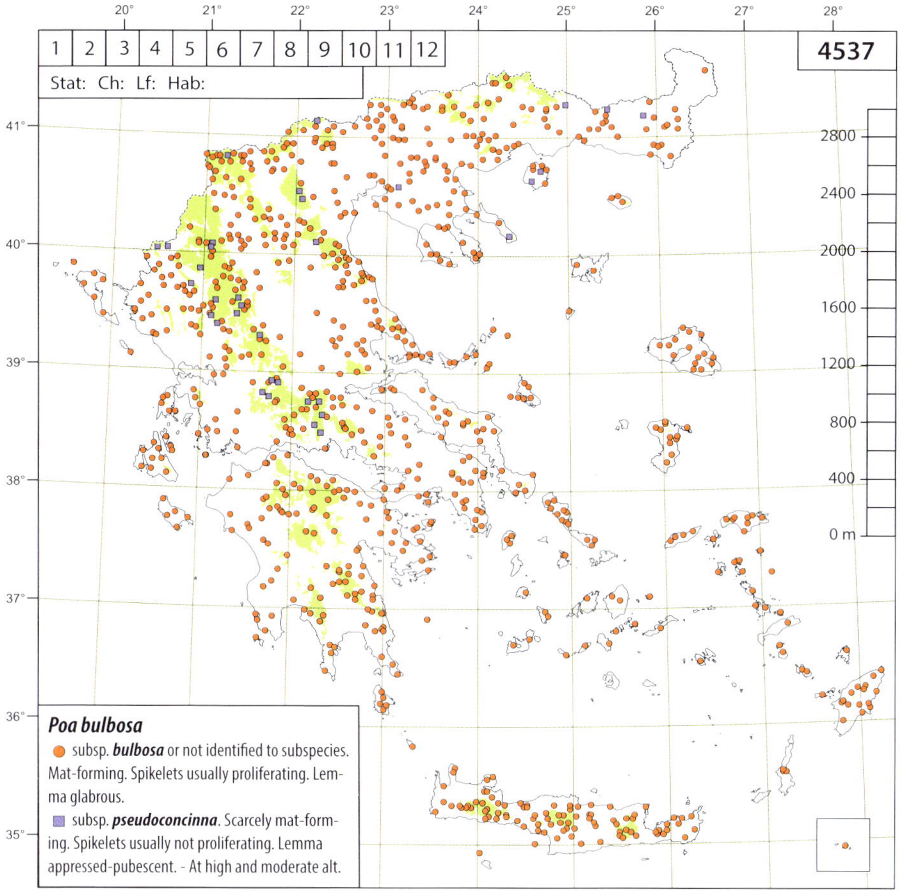

Poa bulbosa

● subsp. ***bulbosa*** or not identified to subspecies. Mat-forming. Spikelets usually proliferating. Lemma glabrous.

■ subsp. ***pseudoconcinna***. Scarcely mat-forming. Spikelets usually not proliferating. Lemma appressed-pubescent. - At high and moderate alt.

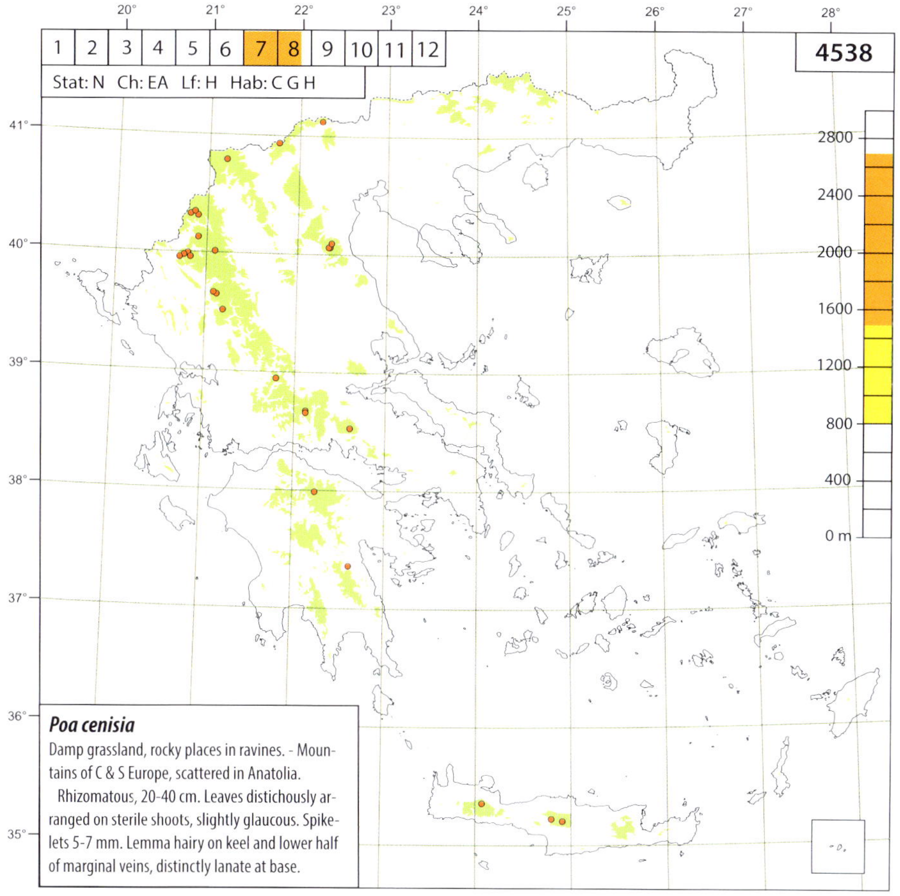

Poa cenisia

Damp grassland, rocky places in ravines. - Mountains of C & S Europe, scattered in Anatolia.

Rhizomatous, 20-40 cm. Leaves distichously arranged on sterile shoots, slightly glaucous. Spikelets 5-7 mm. Lemma hairy on keel and lower half of marginal veins, distinctly lanate at base.

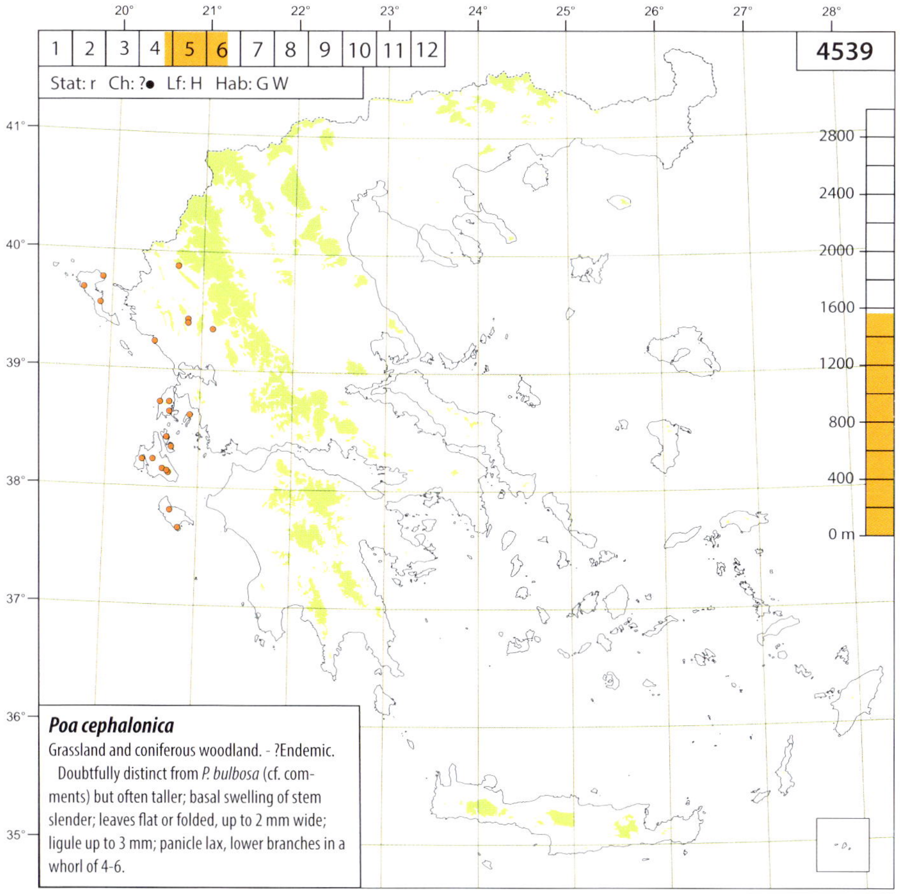

Poa cephalonica

Grassland and coniferous woodland. - ?Endemic.

Doubtfully distinct from *P. bulbosa* (cf. comments) but often taller; basal swelling of stem slender; leaves flat or folded, up to 2 mm wide; ligule up to 3 mm; panicle lax, lower branches in a whorl of 4-6.

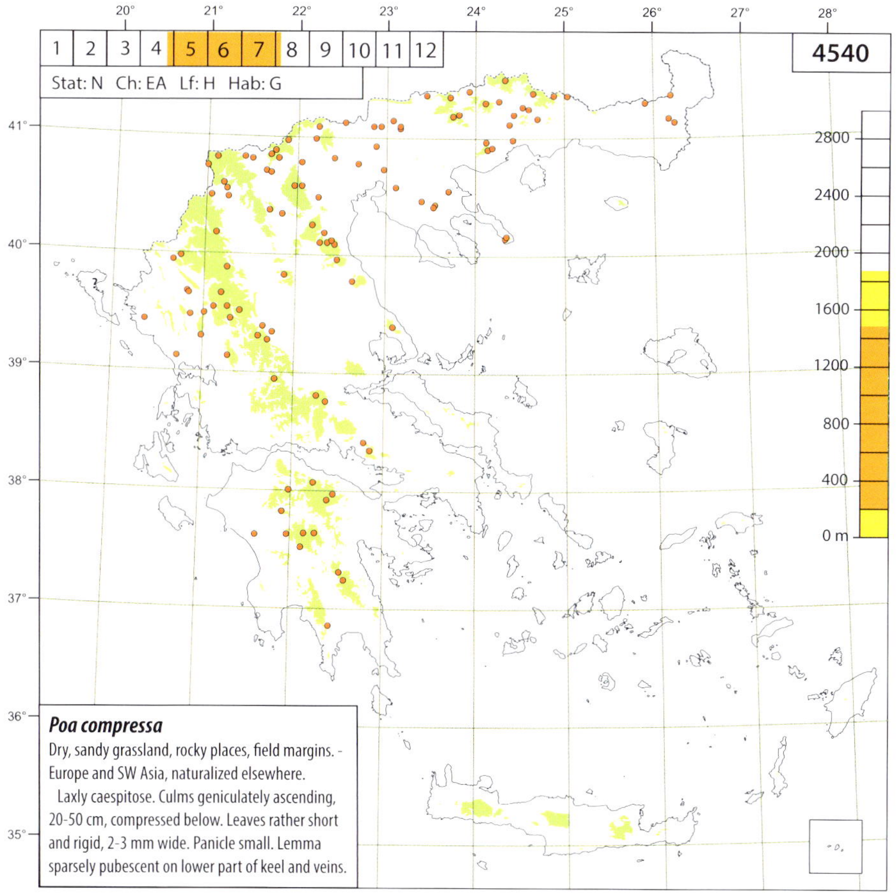

Poa compressa

Dry, sandy grassland, rocky places, field margins. - Europe and SW Asia, naturalized elsewhere.

Laxly caespitose. Culms geniculately ascending, 20-50 cm, compressed below. Leaves rather short and rigid, 2-3 mm wide. Panicle small. Lemma sparsely pubescent on lower part of keel and veins.

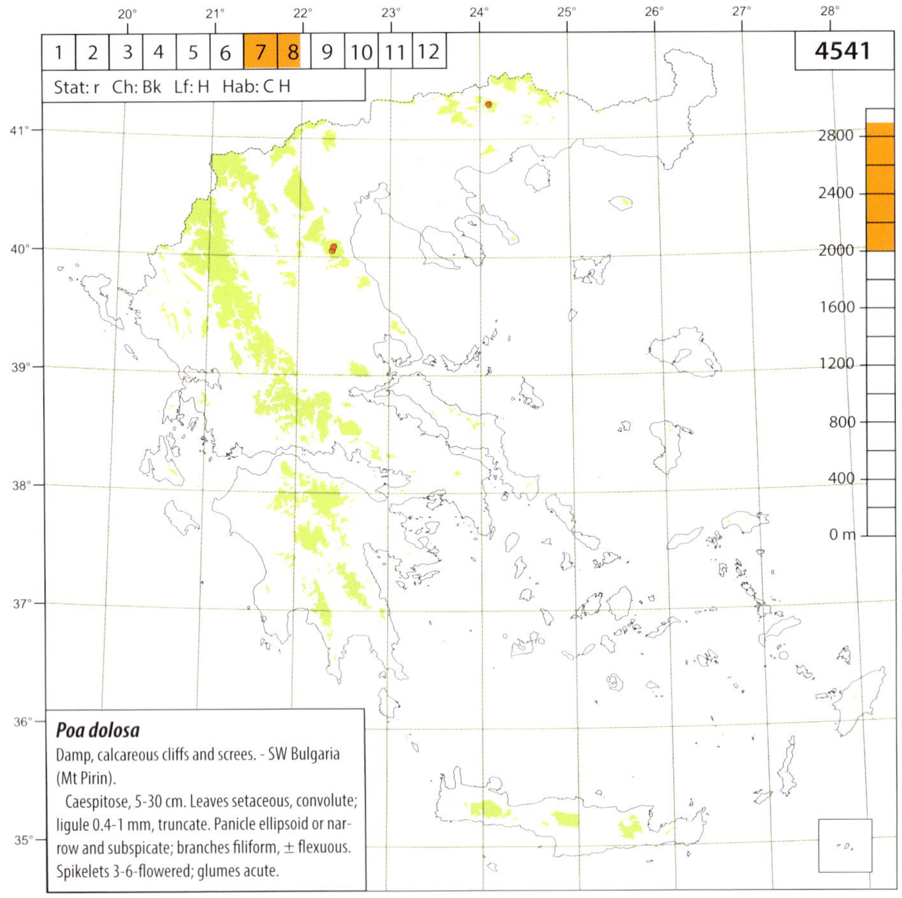

Poa dolosa
Damp, calcareous cliffs and screes. - SW Bulgaria (Mt Pirin).
Caespitose, 5-30 cm. Leaves setaceous, convolute; ligule 0.4-1 mm, truncate. Panicle ellipsoid or narrow and subspicate; branches filiform, ± flexuous. Spikelets 3-6-flowered; glumes acute.

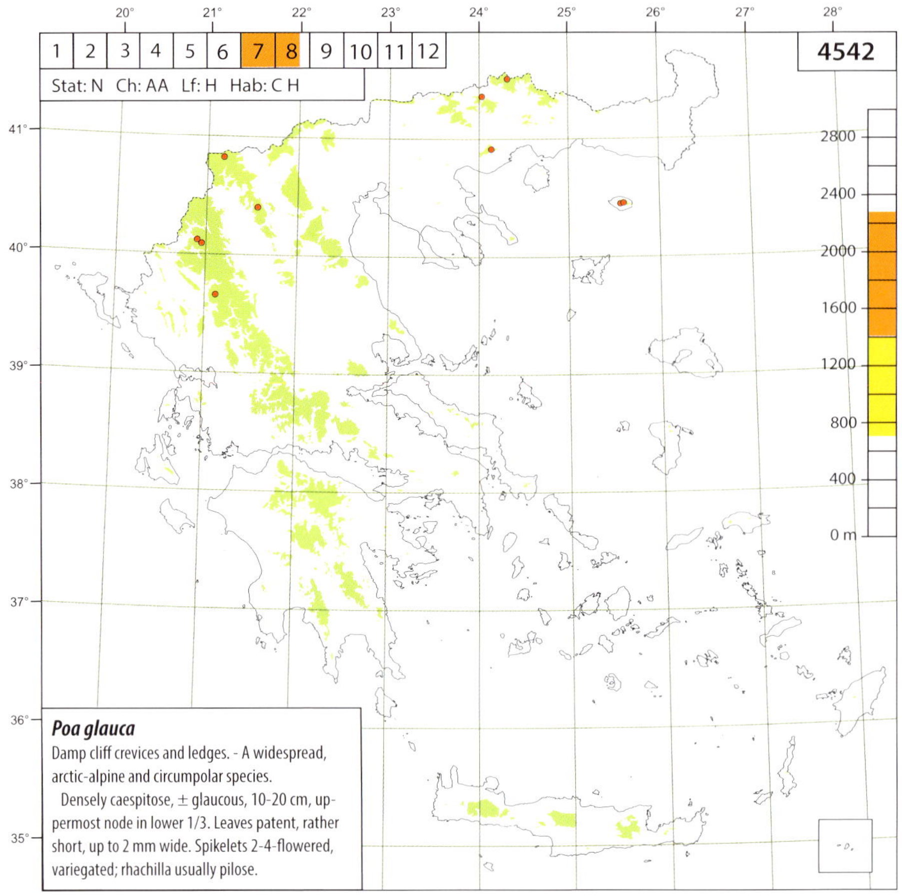

Poa glauca
Damp cliff crevices and ledges. - A widespread, arctic-alpine and circumpolar species.
Densely caespitose, ± glaucous, 10-20 cm, uppermost node in lower 1/3. Leaves patent, rather short, up to 2 mm wide. Spikelets 2-4-flowered, variegated; rhachilla usually pilose.

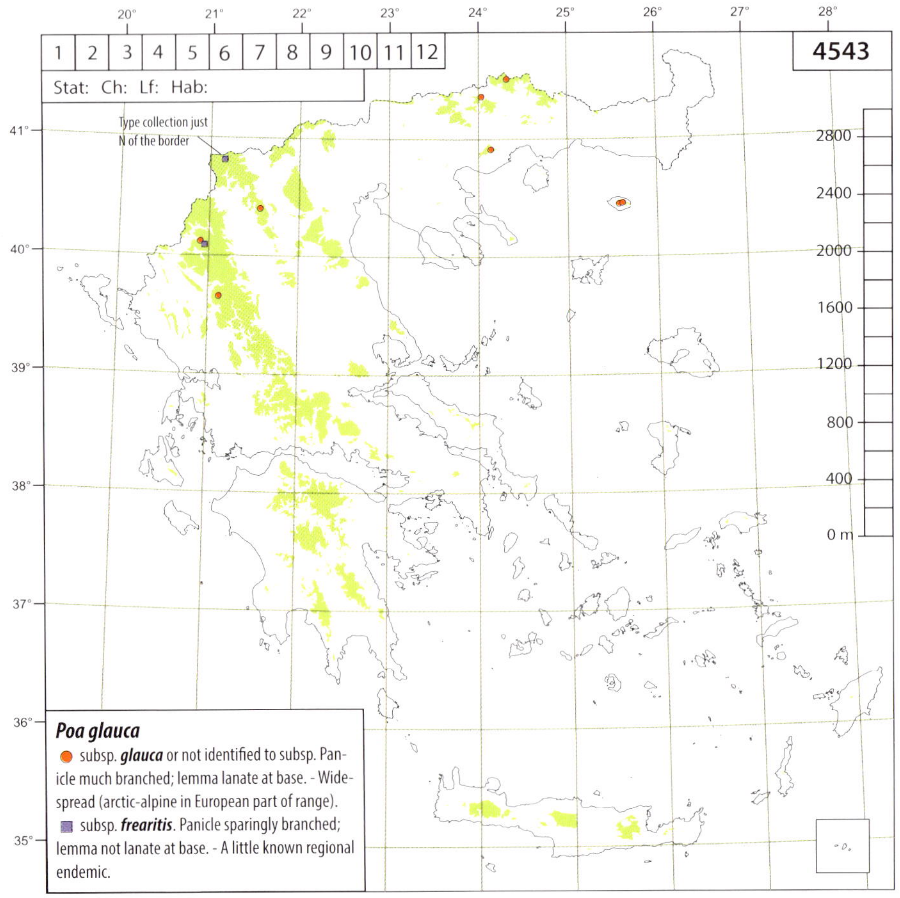

Poa glauca
● subsp. ***glauca*** or not identified to subsp. Panicle much branched; lemma lanate at base. - Widespread (arctic-alpine in European part of range).
■ subsp. ***frearitis***. Panicle sparingly branched; lemma not lanate at base. - A little known regional endemic.

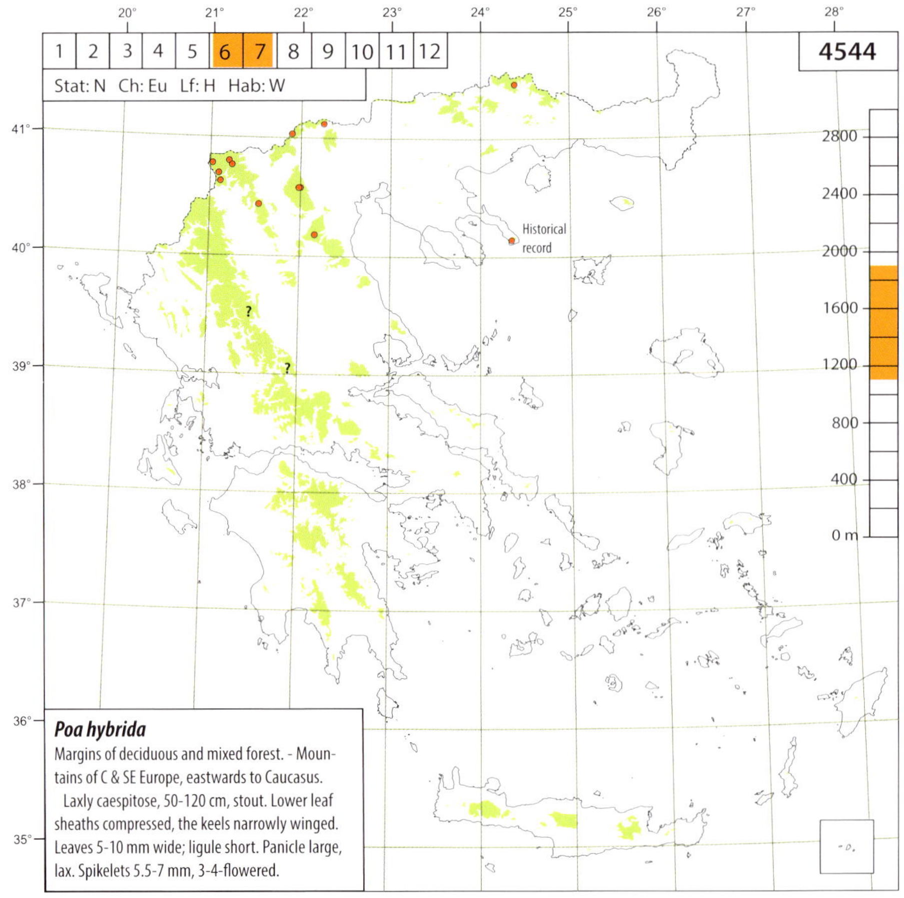

Poa hybrida
Margins of deciduous and mixed forest. - Mountains of C & SE Europe, eastwards to Caucasus.
Laxly caespitose, 50-120 cm, stout. Lower leaf sheaths compressed, the keels narrowly winged. Leaves 5-10 mm wide; ligule short. Panicle large, lax. Spikelets 5.5-7 mm, 3-4-flowered.

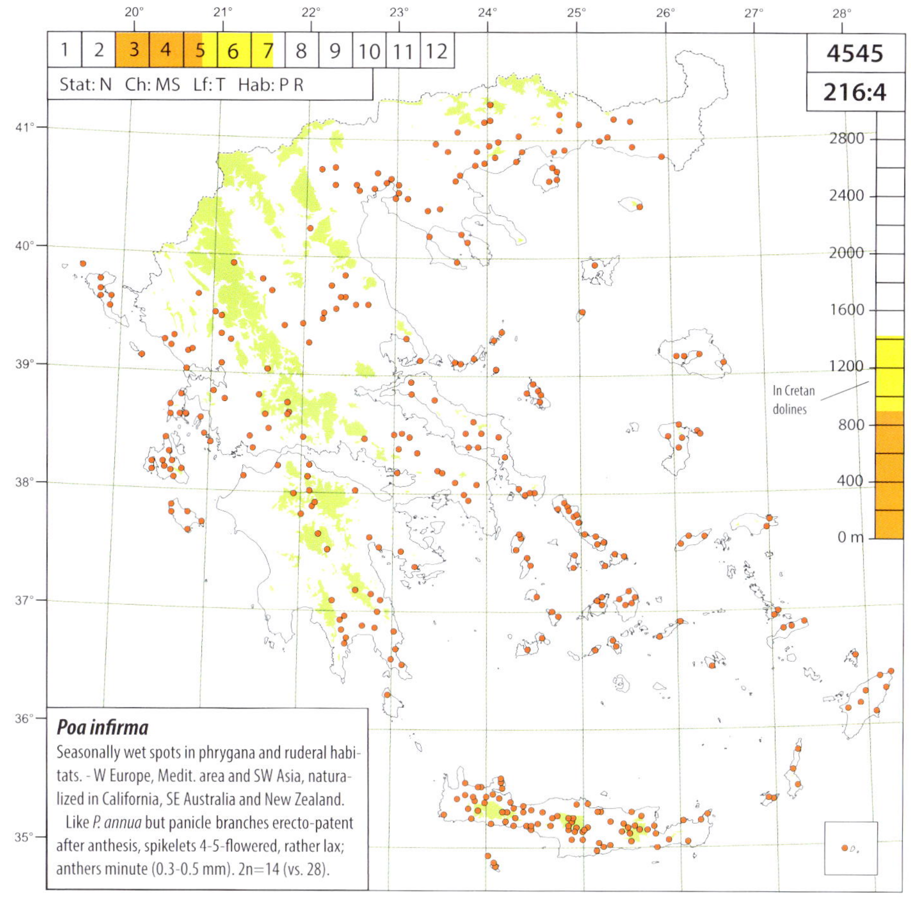

Poa infirma

Seasonally wet spots in phrygana and ruderal habitats. - W Europe, Medit. area and SW Asia, naturalized in California, SE Australia and New Zealand.

Like *P. annua* but panicle branches erecto-patent after anthesis, spikelets 4-5-flowered, rather lax; anthers minute (0.3-0.5 mm). 2n=14 (vs. 28).

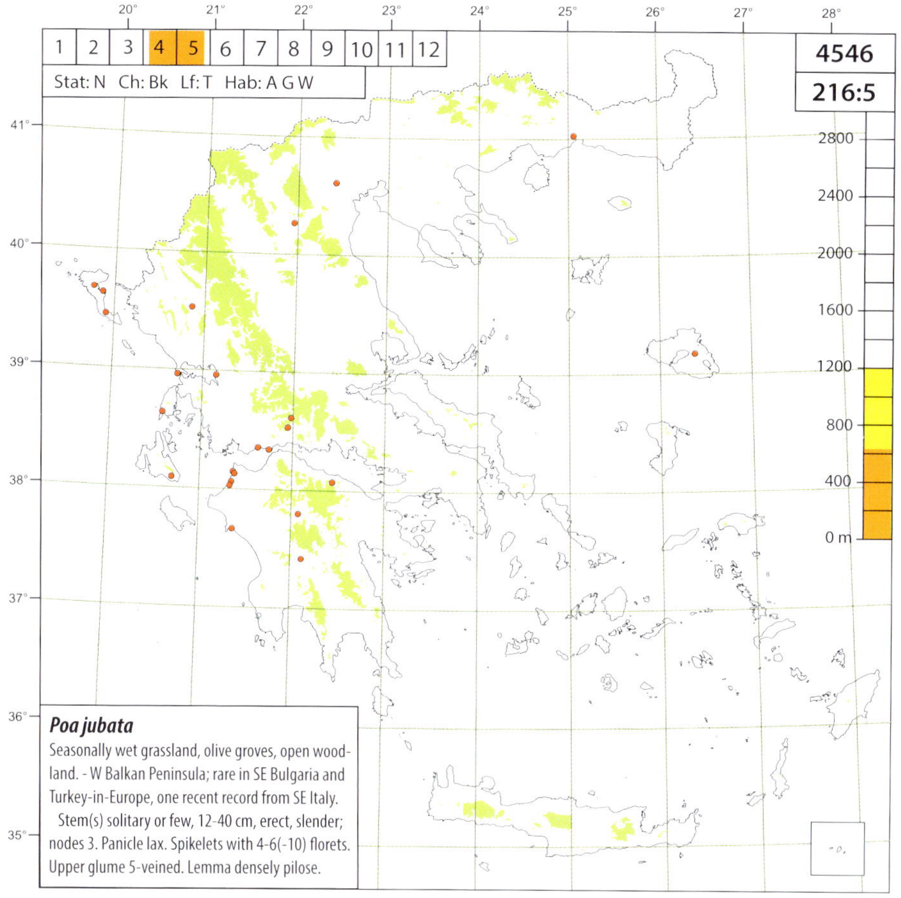

Poa jubata

Seasonally wet grassland, olive groves, open woodland. - W Balkan Peninsula; rare in SE Bulgaria and Turkey-in-Europe, one recent record from SE Italy.

Stem(s) solitary or few, 12-40 cm, erect, slender; nodes 3. Panicle lax. Spikelets with 4-6(-10) florets. Upper glume 5-veined. Lemma densely pilose.

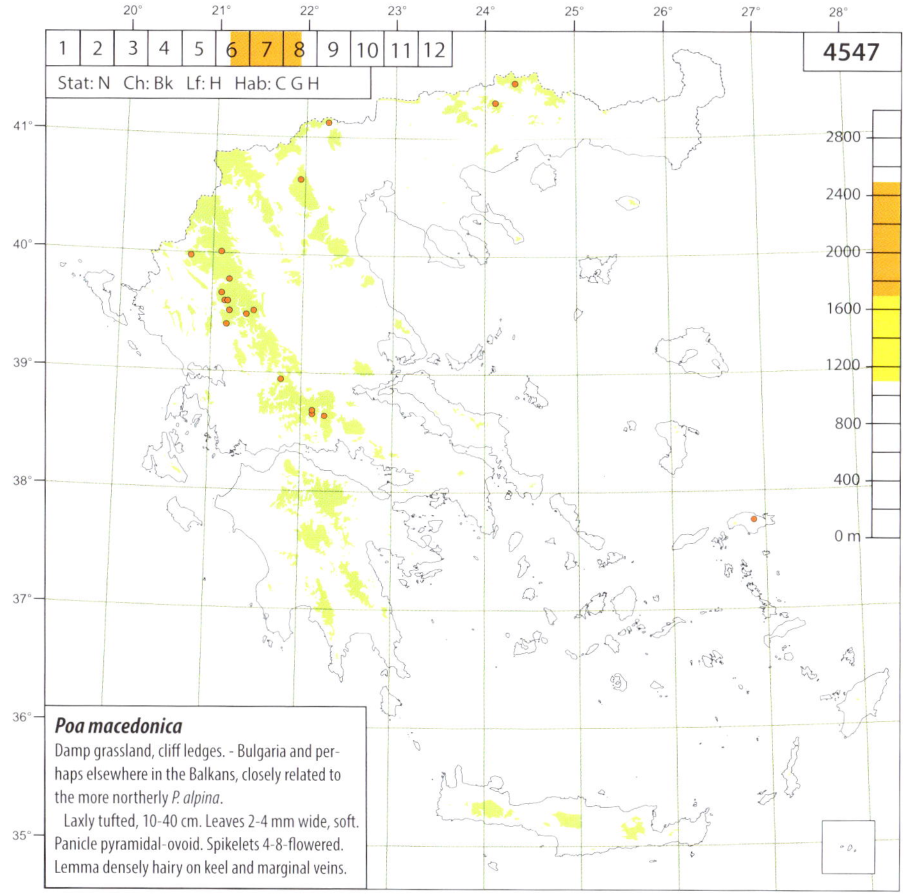

Poa macedonica

Damp grassland, cliff ledges. - Bulgaria and perhaps elsewhere in the Balkans, closely related to the more northerly *P. alpina*.

Laxly tufted, 10-40 cm. Leaves 2-4 mm wide, soft. Panicle pyramidal-ovoid. Spikelets 4-8-flowered. Lemma densely hairy on keel and marginal veins.

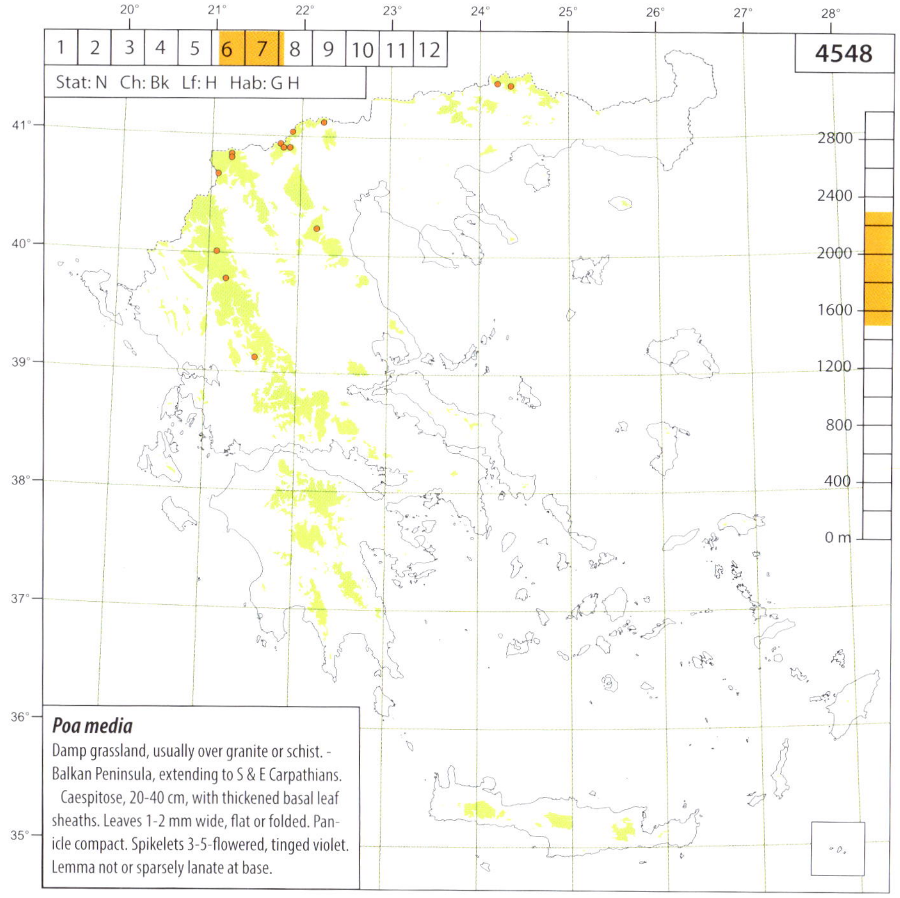

Poa media

Damp grassland, usually over granite or schist. - Balkan Peninsula, extending to S & E Carpathians.

Caespitose, 20-40 cm, with thickened basal leaf sheaths. Leaves 1-2 mm wide, flat or folded. Panicle compact. Spikelets 3-5-flowered, tinged violet. Lemma not or sparsely lanate at base.

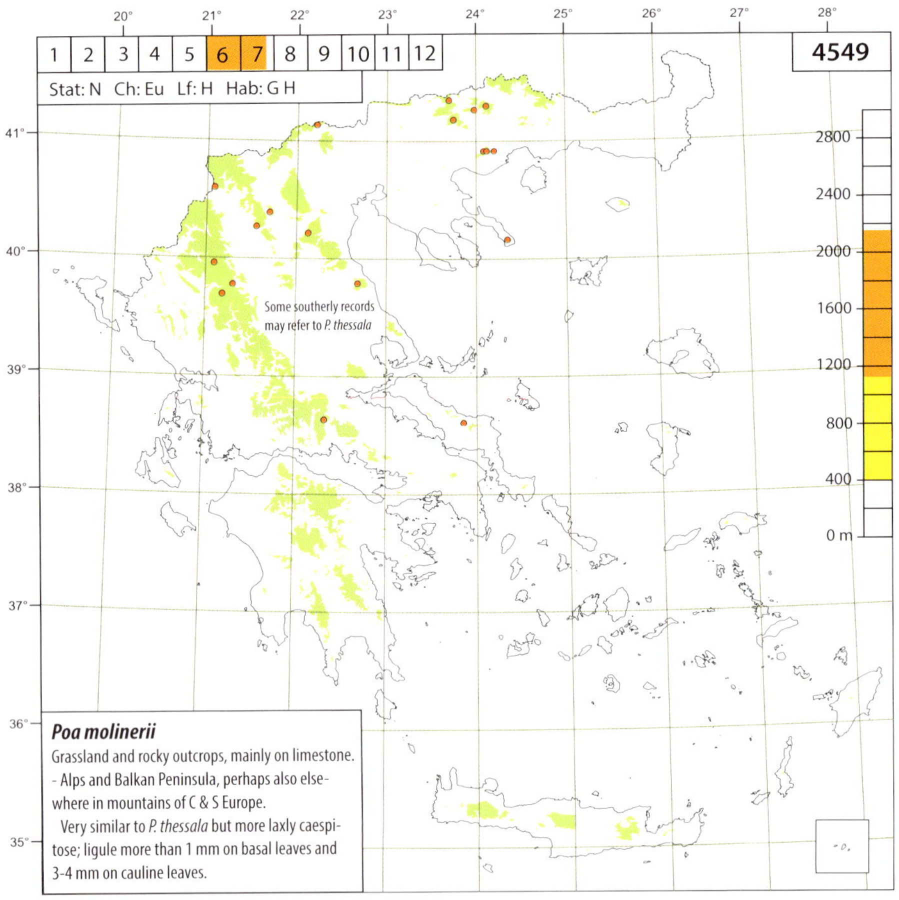

Poa molinerii

Grassland and rocky outcrops, mainly on limestone. - Alps and Balkan Peninsula, perhaps also elsewhere in mountains of C & S Europe.

Very similar to *P. thessala* but more laxly caespitose; ligule more than 1 mm on basal leaves and 3-4 mm on cauline leaves.

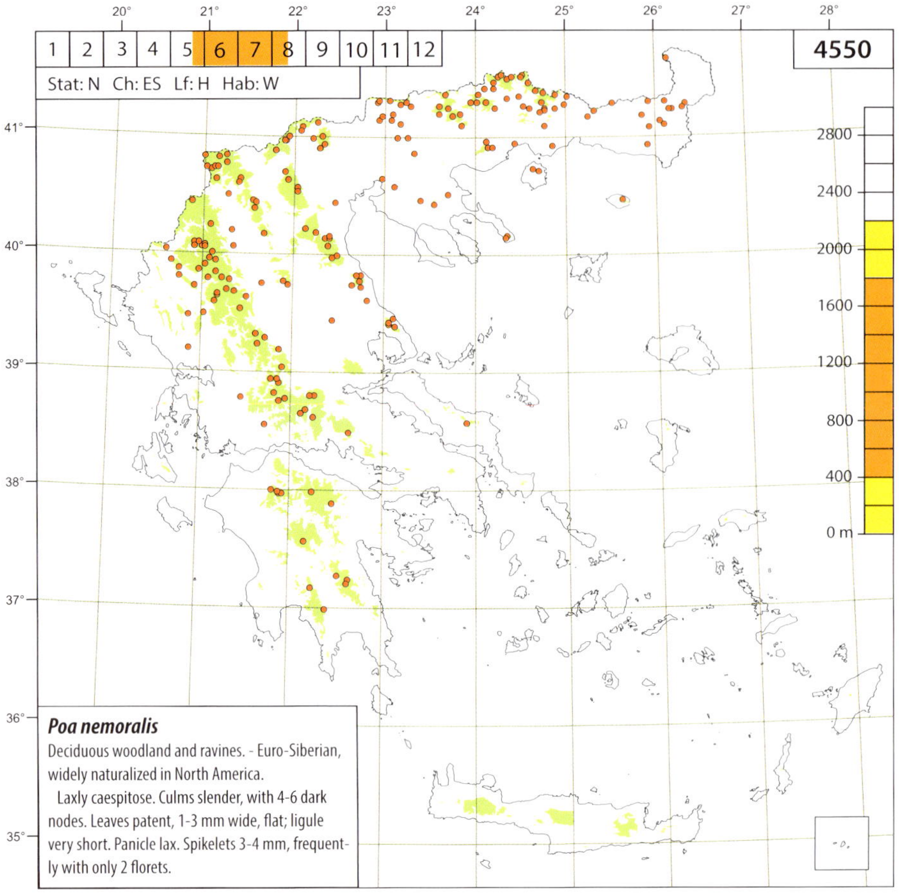

Poa nemoralis

Deciduous woodland and ravines. - Euro-Siberian, widely naturalized in North America.

Laxly caespitose. Culms slender, with 4-6 dark nodes. Leaves patent, 1-3 mm wide, flat; ligule very short. Panicle lax. Spikelets 3-4 mm, frequently with only 2 florets.

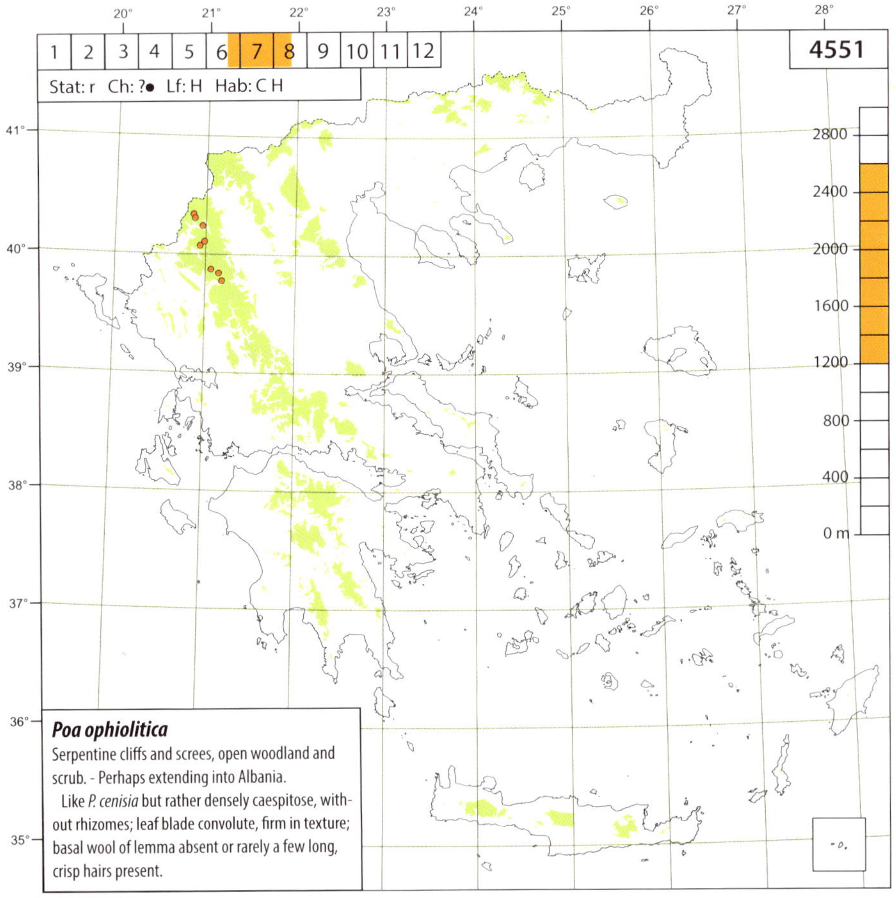

Poa ophiolitica

Serpentine cliffs and screes, open woodland and scrub. - Perhaps extending into Albania.

Like *P. cenisia* but rather densely caespitose, without rhizomes; leaf blade convolute, firm in texture; basal wool of lemma absent or rarely a few long, crisp hairs present.

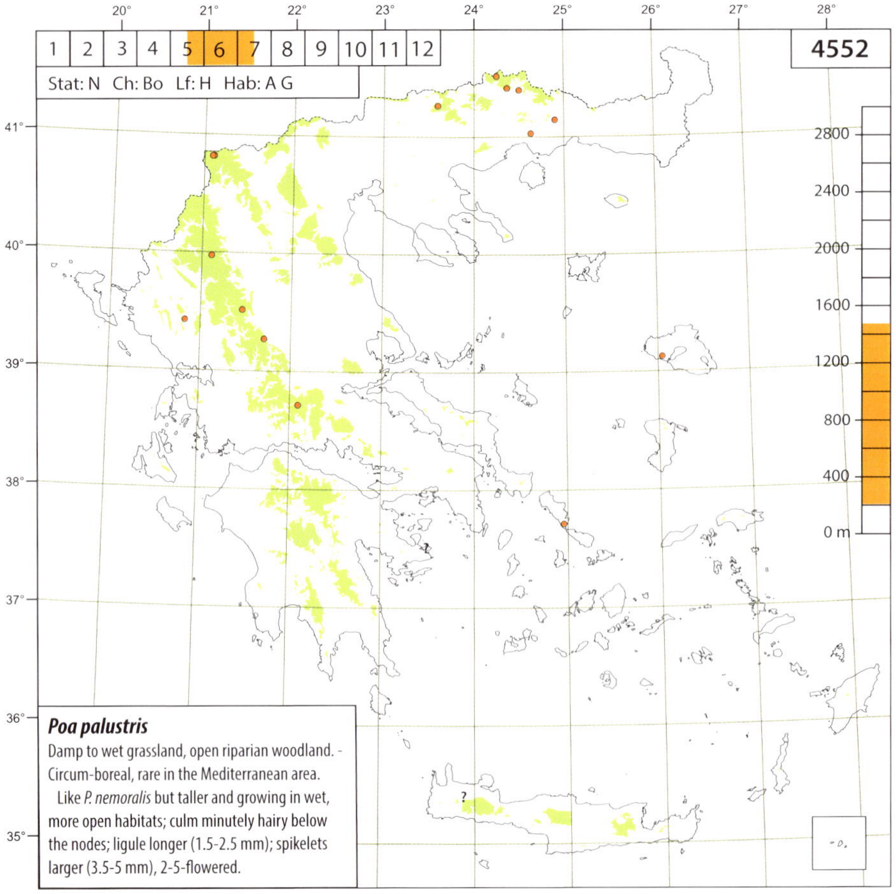

Poa palustris

Damp to wet grassland, open riparian woodland. - Circum-boreal, rare in the Mediterranean area.

Like *P. nemoralis* but taller and growing in wet, more open habitats; culm minutely hairy below the nodes; ligule longer (1.5-2.5 mm); spikelets larger (3.5-5 mm), 2-5-flowered.

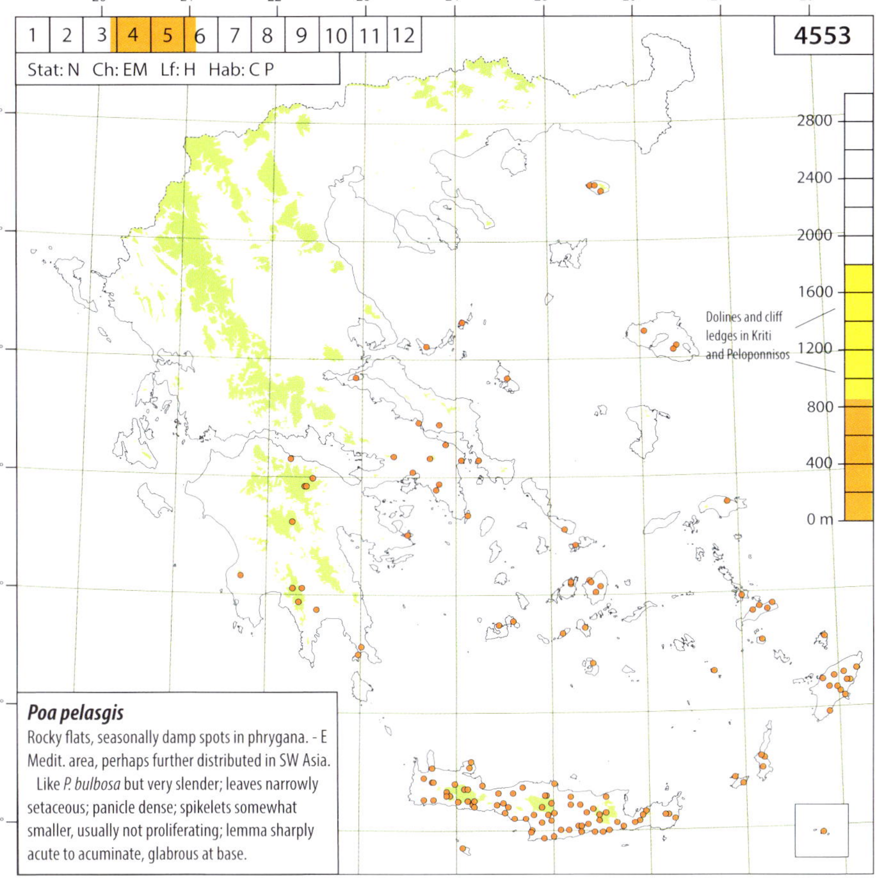

Poa pelasgis
Rocky flats, seasonally damp spots in phrygana. - E Medit. area, perhaps further distributed in SW Asia.
Like *P. bulbosa* but very slender; leaves narrowly setaceous; panicle dense; spikelets somewhat smaller, usually not proliferating; lemma sharply acute to acuminate, glabrous at base.

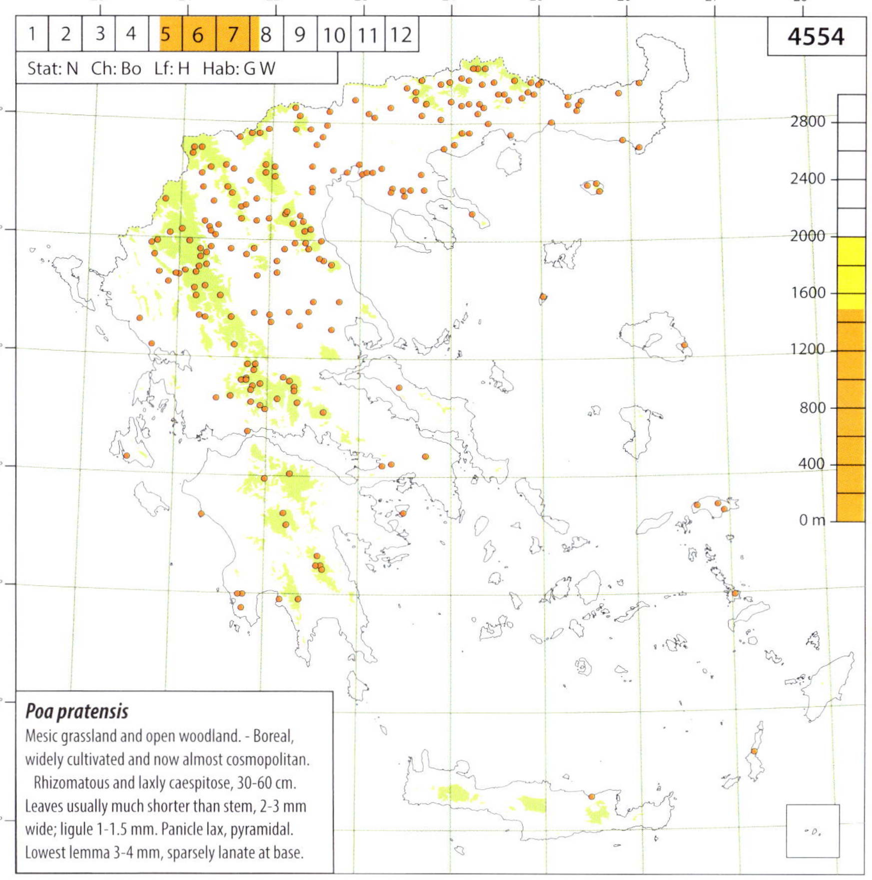

Poa pratensis
Mesic grassland and open woodland. - Boreal, widely cultivated and now almost cosmopolitan.
Rhizomatous and laxly caespitose, 30-60 cm. Leaves usually much shorter than stem, 2-3 mm wide; ligule 1-1.5 mm. Panicle lax, pyramidal. Lowest lemma 3-4 mm, sparsely lanate at base.

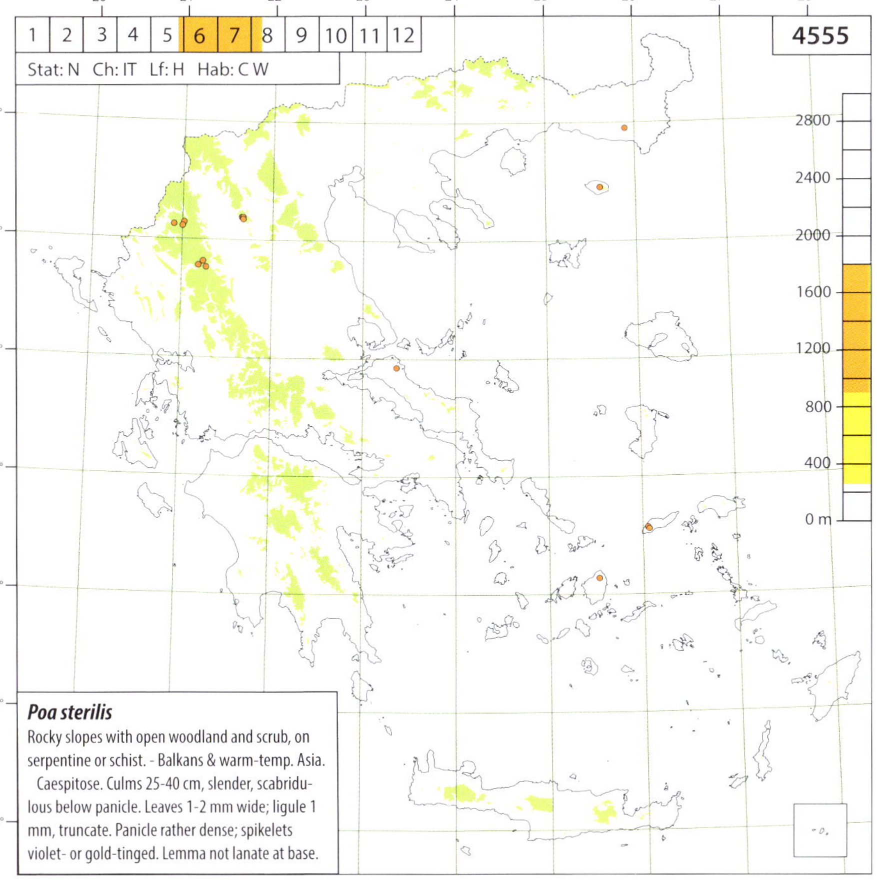

Poa sterilis
Rocky slopes with open woodland and scrub, on serpentine or schist. - Balkans & warm-temp. Asia.
Caespitose. Culms 25-40 cm, slender, scabridulous below panicle. Leaves 1-2 mm wide; ligule 1 mm, truncate. Panicle rather dense; spikelets violet- or gold-tinged. Lemma not lanate at base.

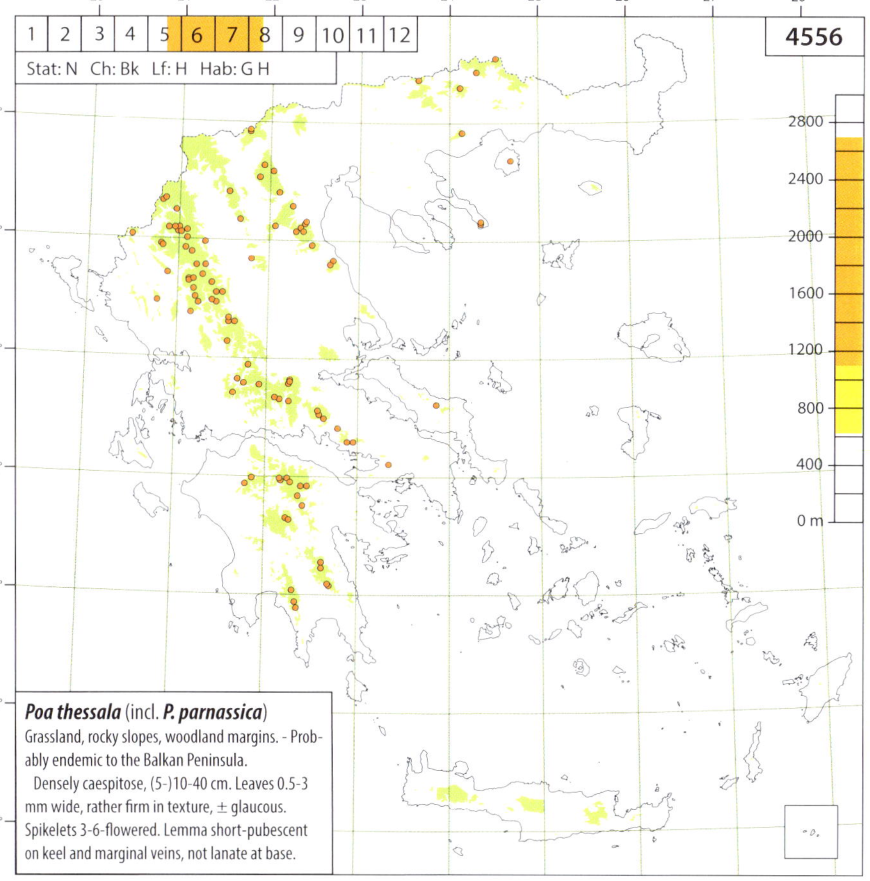

Poa thessala (incl. ***P. parnassica***)
Grassland, rocky slopes, woodland margins. - Probably endemic to the Balkan Peninsula.
Densely caespitose, (5-)10-40 cm. Leaves 0.5-3 mm wide, rather firm in texture, ± glaucous. Spikelets 3-6-flowered. Lemma short-pubescent on keel and marginal veins, not lanate at base.

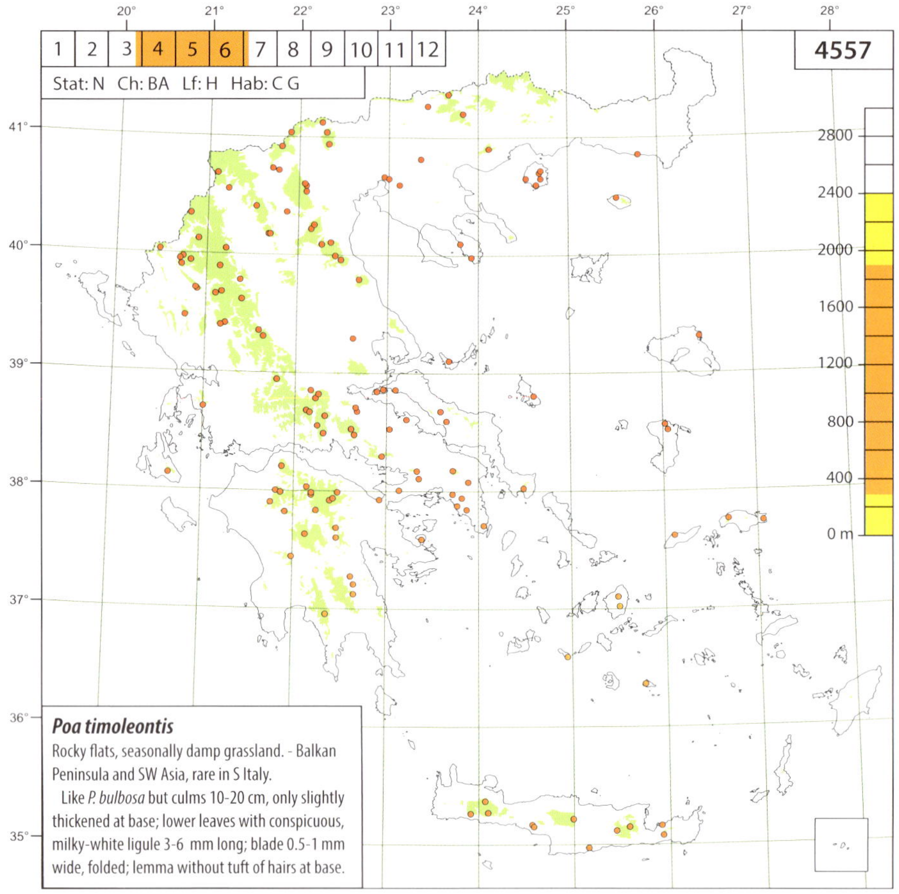

Poa timoleontis

Rocky flats, seasonally damp grassland. - Balkan Peninsula and SW Asia, rare in S Italy.

Like *P. bulbosa* but culms 10-20 cm, only slightly thickened at base; lower leaves with conspicuous, milky-white ligule 3-6 mm long; blade 0.5-1 mm wide, folded; lemma without tuft of hairs at base.

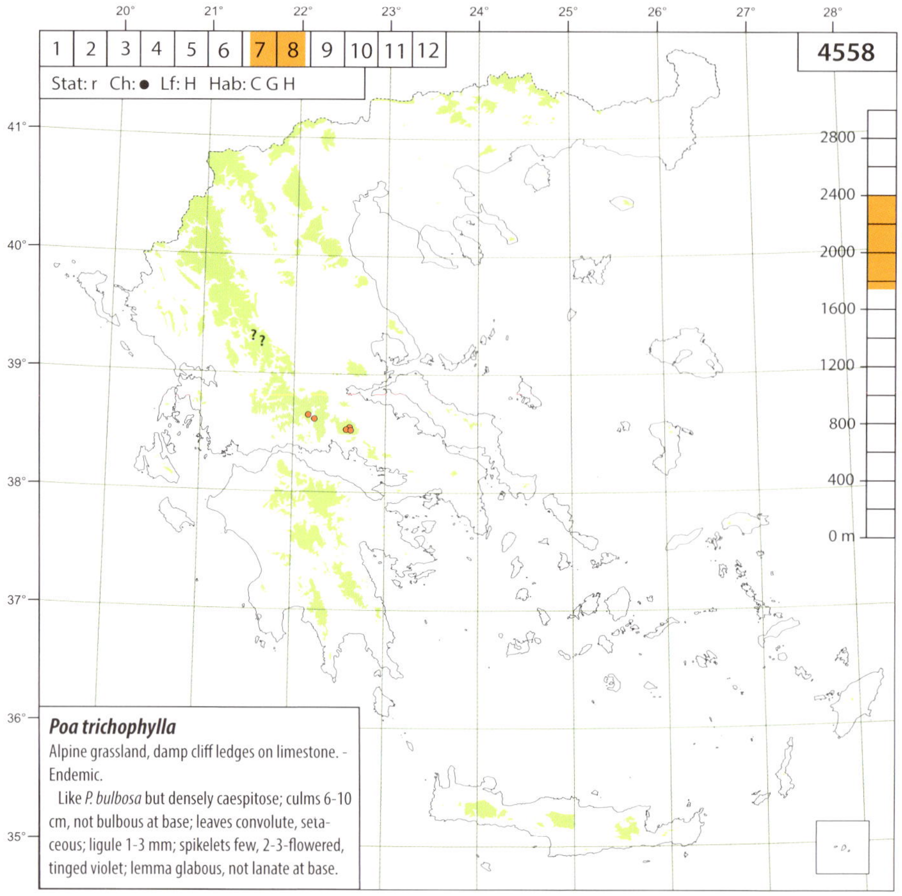

Poa trichophylla

Alpine grassland, damp cliff ledges on limestone. - Endemic.

Like *P. bulbosa* but densely caespitose; culms 6-10 cm, not bulbous at base; leaves convolute, setaceous; ligule 1-3 mm; spikelets few, 2-3-flowered, tinged violet; lemma glabous, not lanate at base.

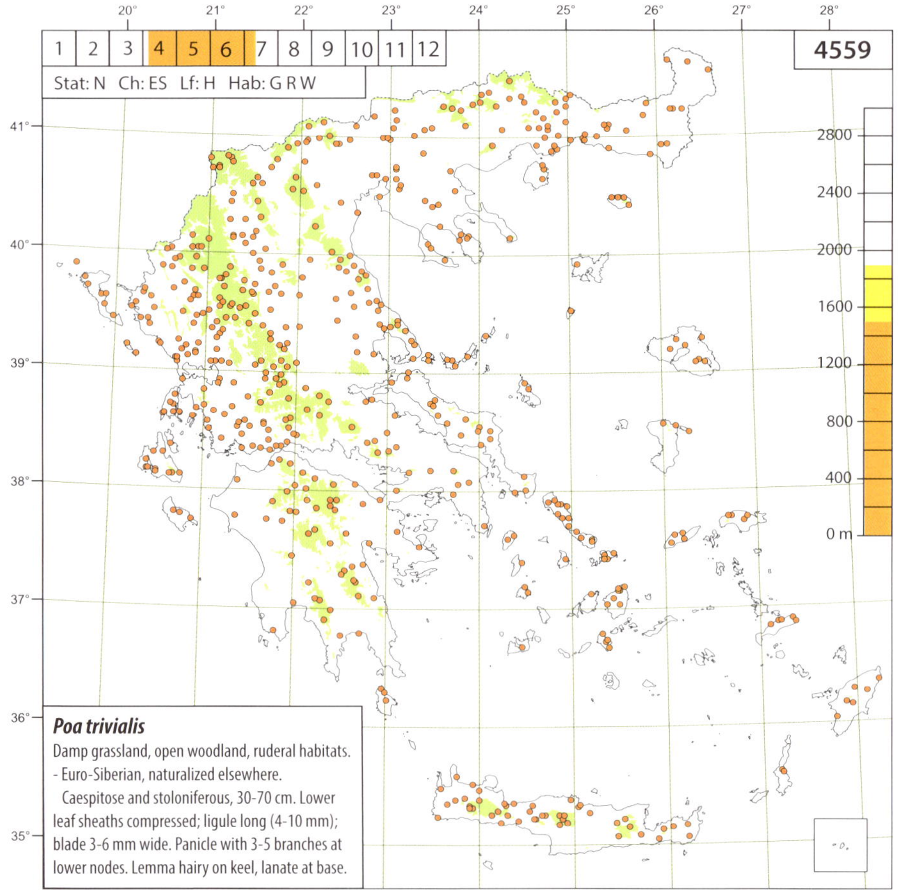

Poa trivialis

Damp grassland, open woodland, ruderal habitats. - Euro-Siberian, naturalized elsewhere.

Caespitose and stoloniferous, 30-70 cm. Lower leaf sheaths compressed; ligule long (4-10 mm); blade 3-6 mm wide. Panicle with 3-5 branches at lower nodes. Lemma hairy on keel, lanate at base.

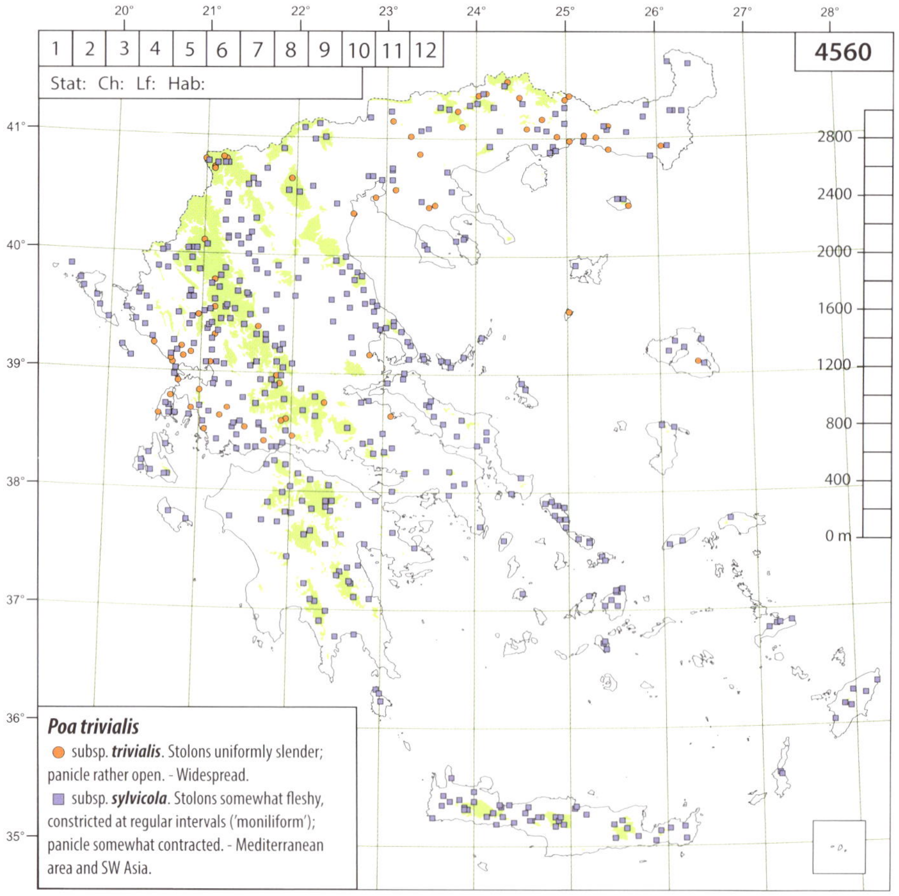

Poa trivialis

● subsp. ***trivialis***. Stolons uniformly slender; panicle rather open. - Widespread.

■ subsp. ***sylvicola***. Stolons somewhat fleshy, constricted at regular intervals ('moniliform'); panicle somewhat contracted. - Mediterranean area and SW Asia.

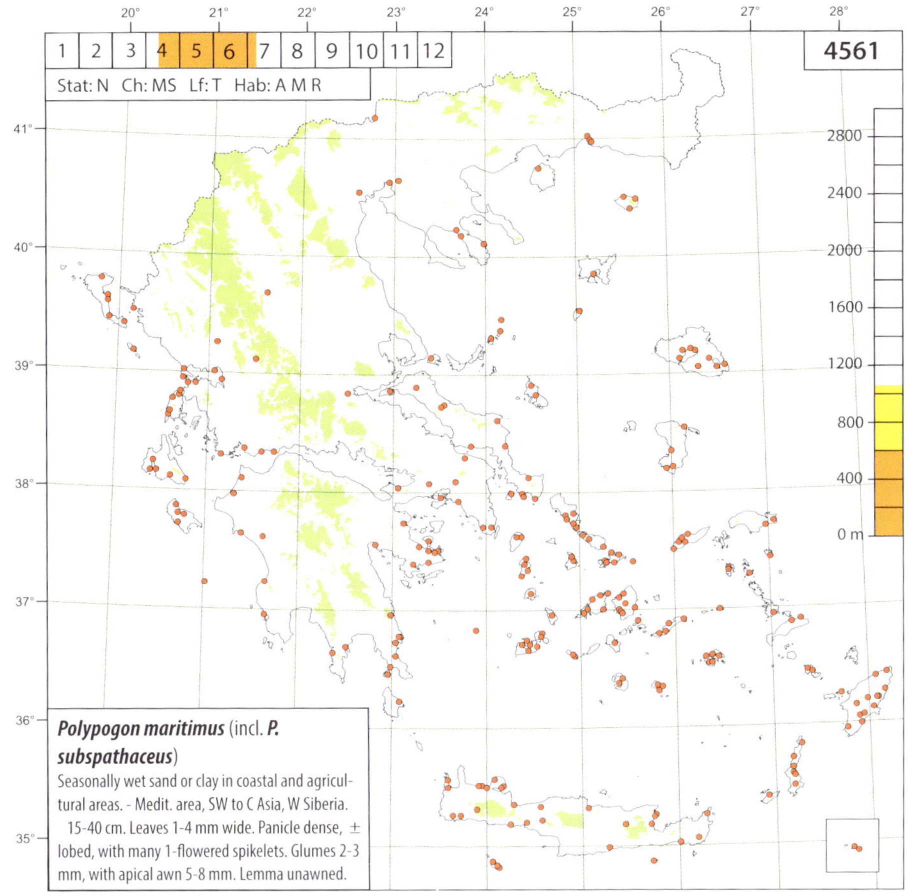

Polypogon maritimus (incl. ***P. subspathaceus***)

Seasonally wet sand or clay in coastal and agricultural areas. - Medit. area, SW to C Asia, W Siberia.

15-40 cm. Leaves 1-4 mm wide. Panicle dense, ± lobed, with many 1-flowered spikelets. Glumes 2-3 mm, with apical awn 5-8 mm. Lemma unawned.

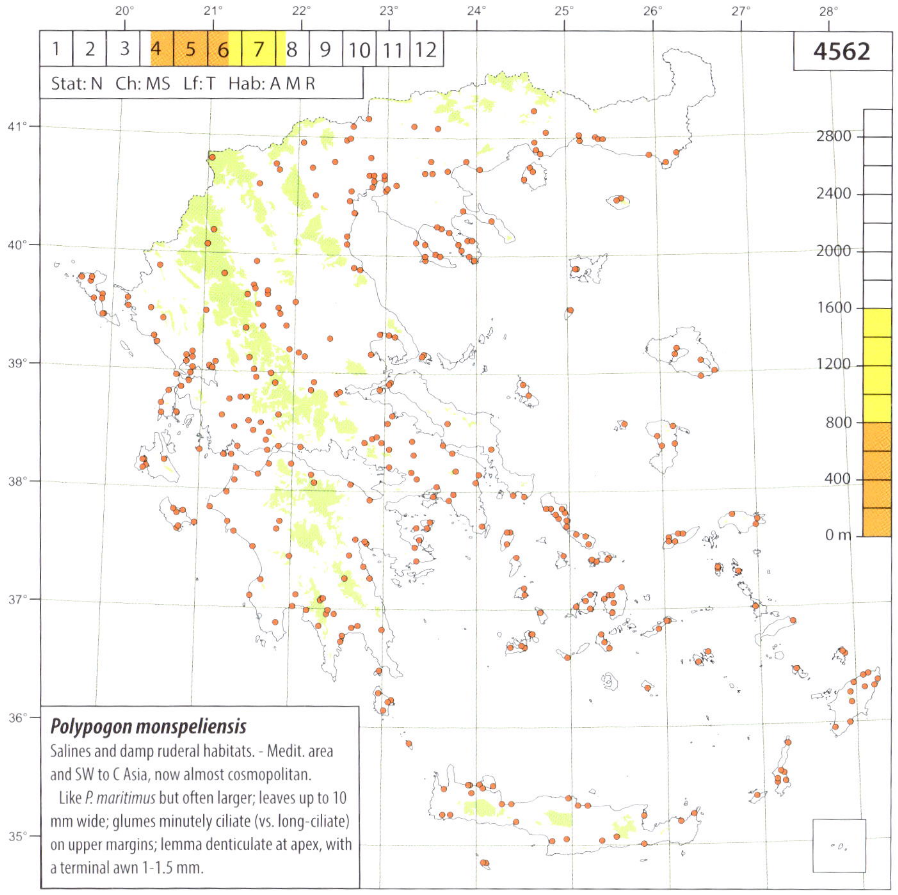

Polypogon monspeliensis

Salines and damp ruderal habitats. - Medit. area and SW to C Asia, now almost cosmopolitan.

Like *P. maritimus* but often larger; leaves up to 10 mm wide; glumes minutely ciliate (vs. long-ciliate) on upper margins; lemma denticulate at apex, with a terminal awn 1-1.5 mm.

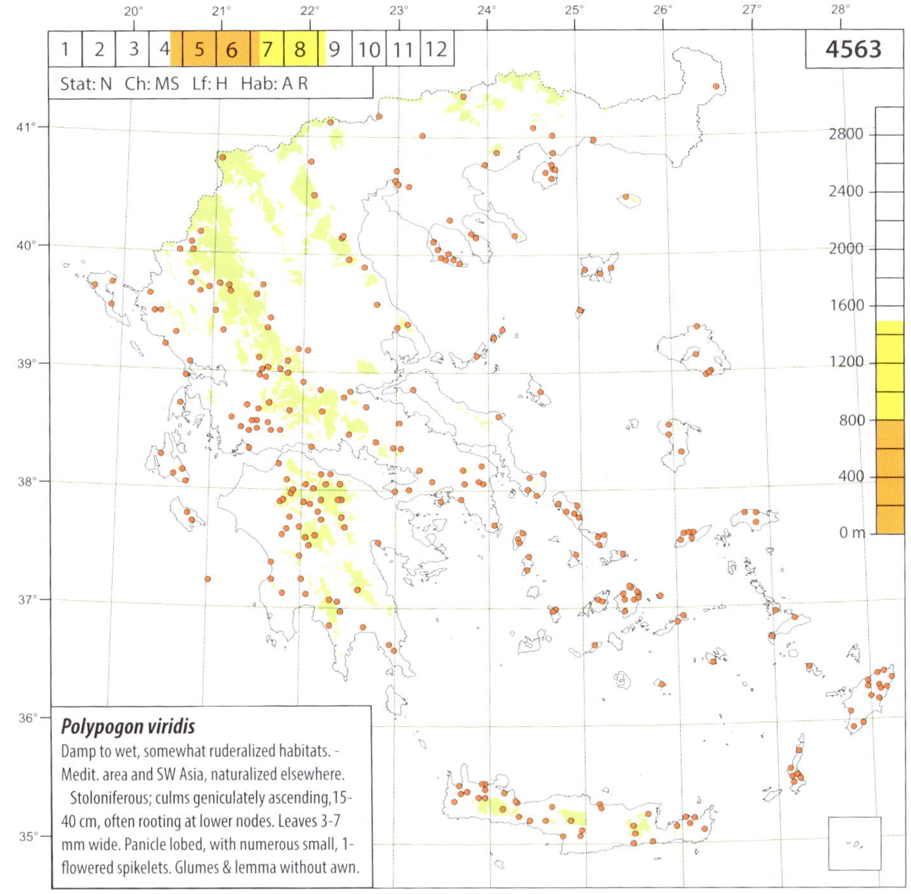

Polypogon viridis

Damp to wet, somewhat ruderalized habitats. - Medit. area and SW Asia, naturalized elsewhere.

Stoloniferous; culms geniculately ascending, 15-40 cm, often rooting at lower nodes. Leaves 3-7 mm wide. Panicle lobed, with numerous small, 1-flowered spikelets. Glumes & lemma without awn.

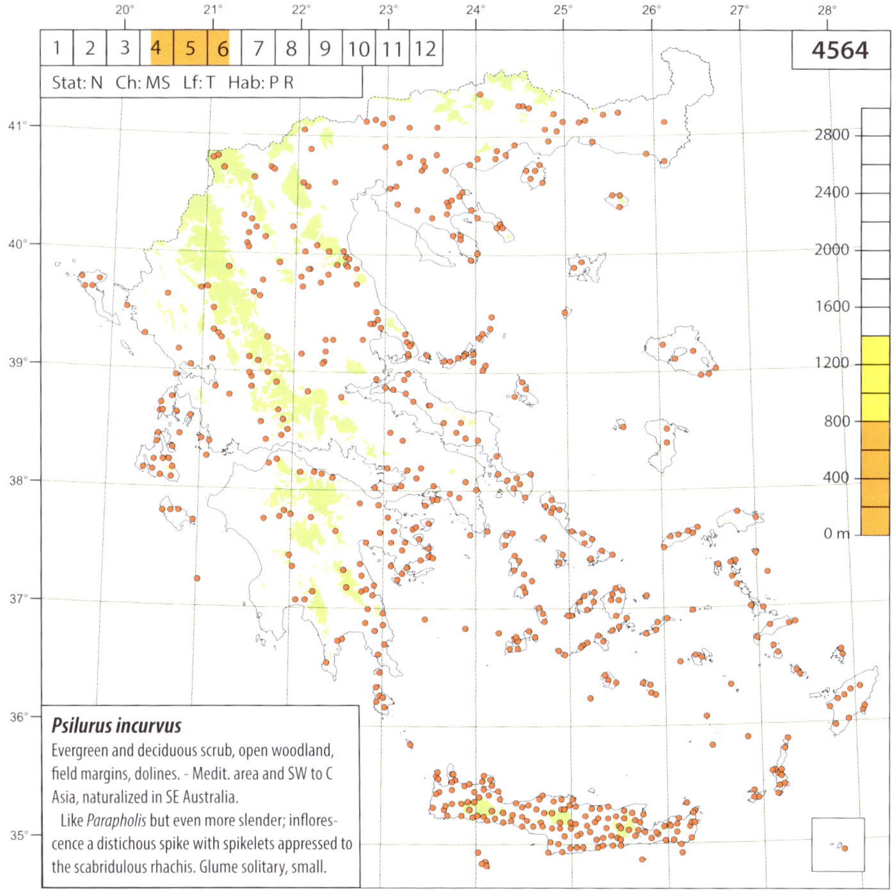

Psilurus incurvus

Evergreen and deciduous scrub, open woodland, field margins, dolines. - Medit. area and SW to C Asia, naturalized in SE Australia.

Like *Parapholis* but even more slender; inflorescence a distichous spike with spikelets appressed to the scabridulous rhachis. Glume solitary, small.

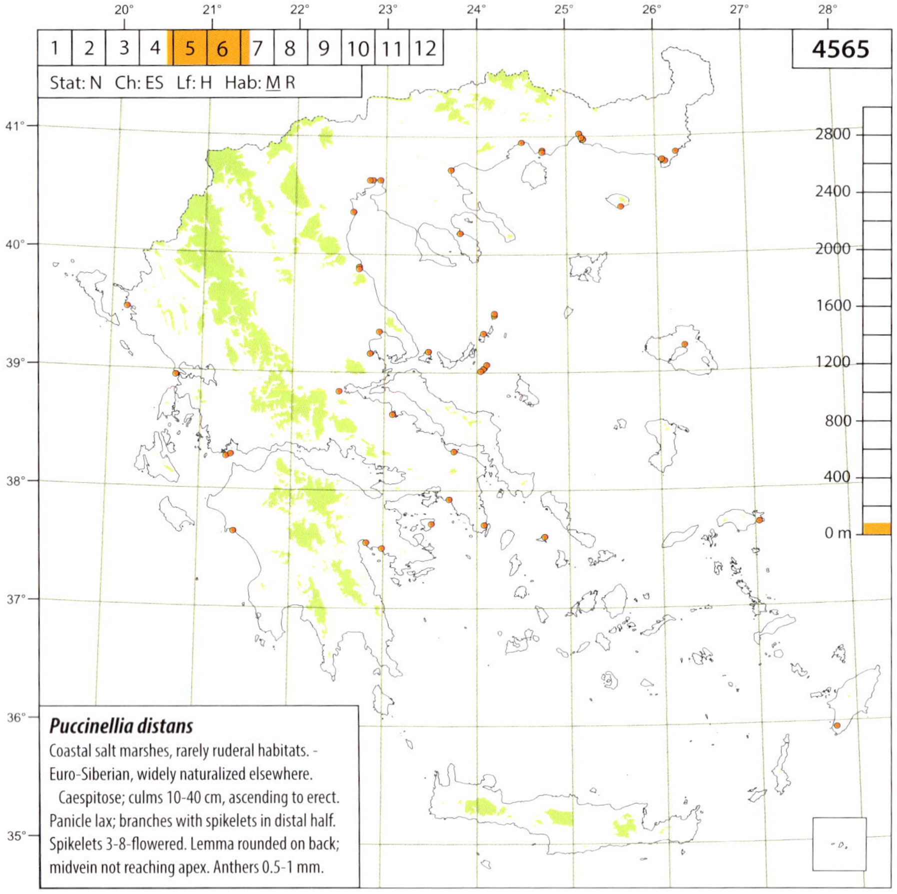

Puccinellia distans
Coastal salt marshes, rarely ruderal habitats. - Euro-Siberian, widely naturalized elsewhere.
Caespitose; culms 10-40 cm, ascending to erect. Panicle lax; branches with spikelets in distal half. Spikelets 3-8-flowered. Lemma rounded on back; midvein not reaching apex. Anthers 0.5-1 mm.

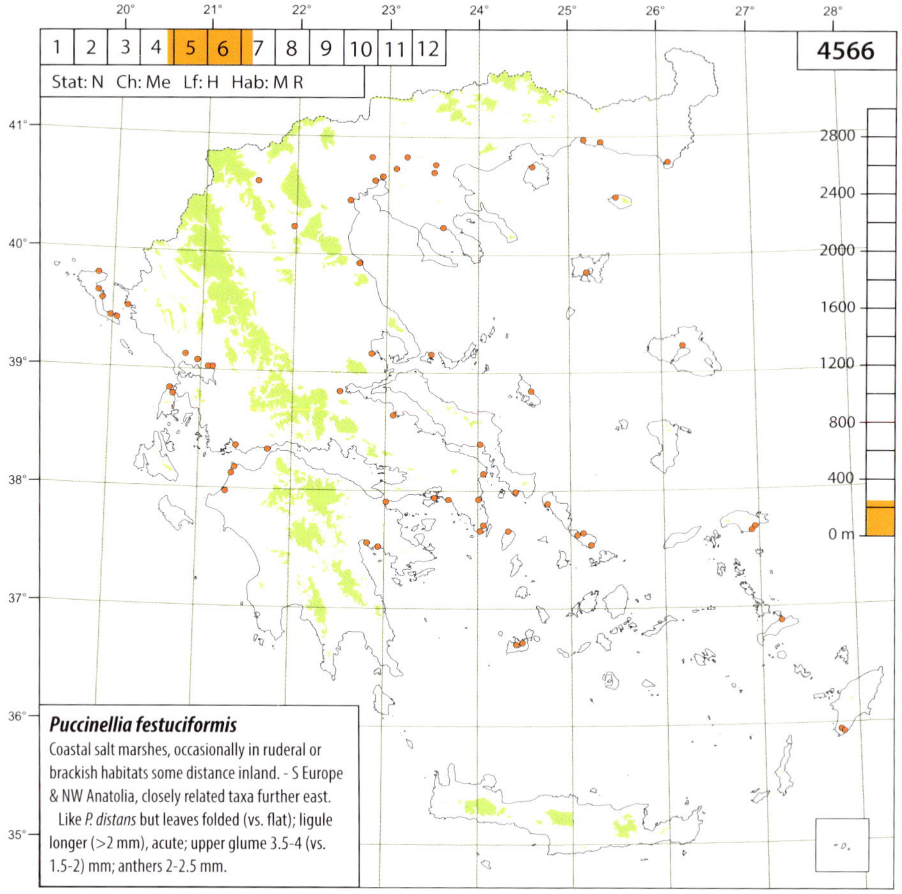

Puccinellia festuciformis
Coastal salt marshes, occasionally in ruderal or brackish habitats some distance inland. - S Europe & NW Anatolia, closely related taxa further east.
Like *P. distans* but leaves folded (vs. flat); ligule longer (>2 mm), acute; upper glume 3.5-4 (vs. 1.5-2) mm; anthers 2-2.5 mm.

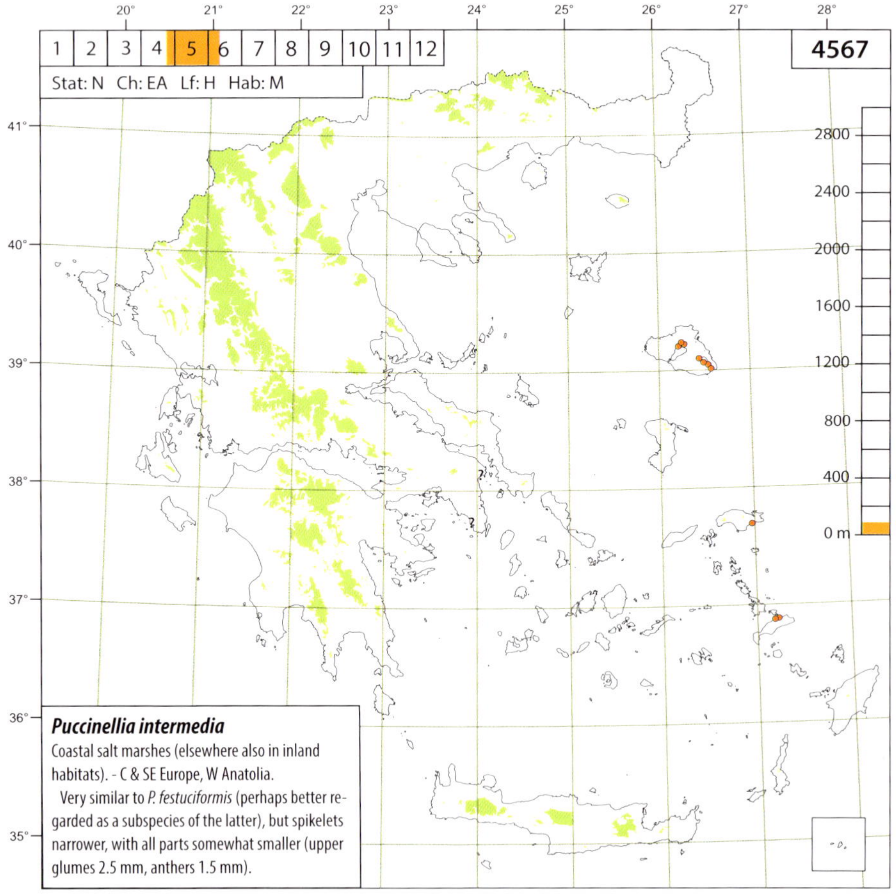

Puccinellia intermedia
Coastal salt marshes (elsewhere also in inland habitats). - C & SE Europe, W Anatolia.
Very similar to *P. festuciformis* (perhaps better regarded as a subspecies of the latter), but spikelets narrower, with all parts somewhat smaller (upper glumes 2.5 mm, anthers 1.5 mm).

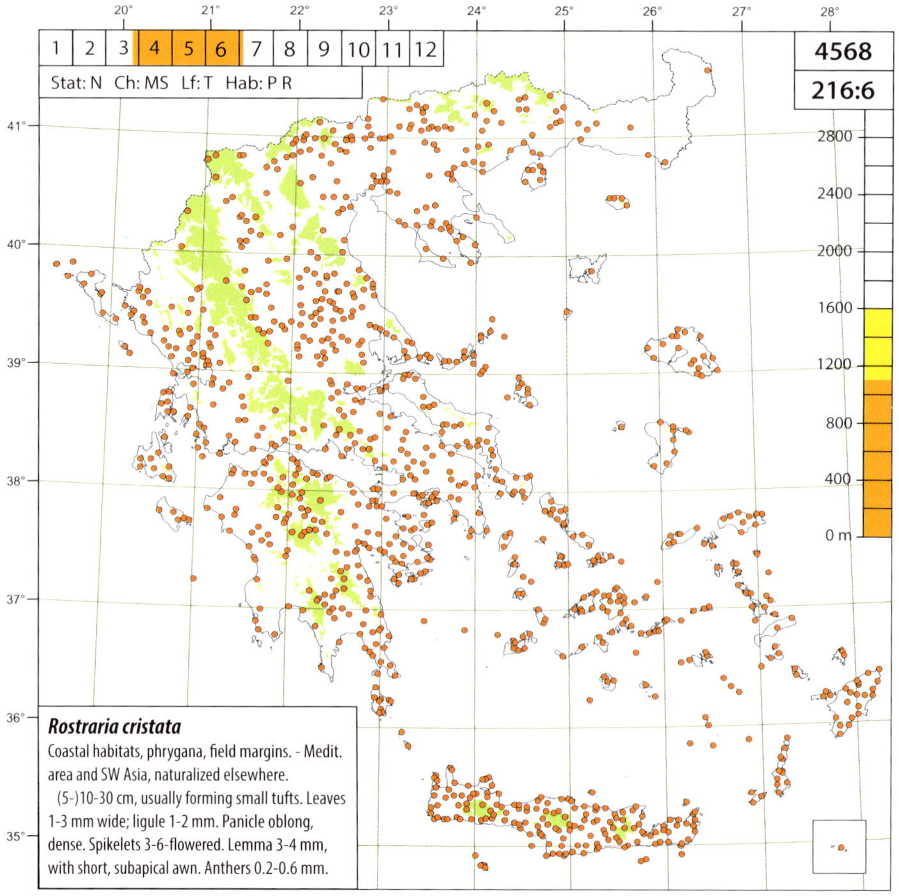

Rostraria cristata
Coastal habitats, phrygana, field margins. - Medit. area and SW Asia, naturalized elsewhere.
(5-)10-30 cm, usually forming small tufts. Leaves 1-3 mm wide; ligule 1-2 mm. Panicle oblong, dense. Spikelets 3-6-flowered. Lemma 3-4 mm, with short, subapical awn. Anthers 0.2-0.6 mm.

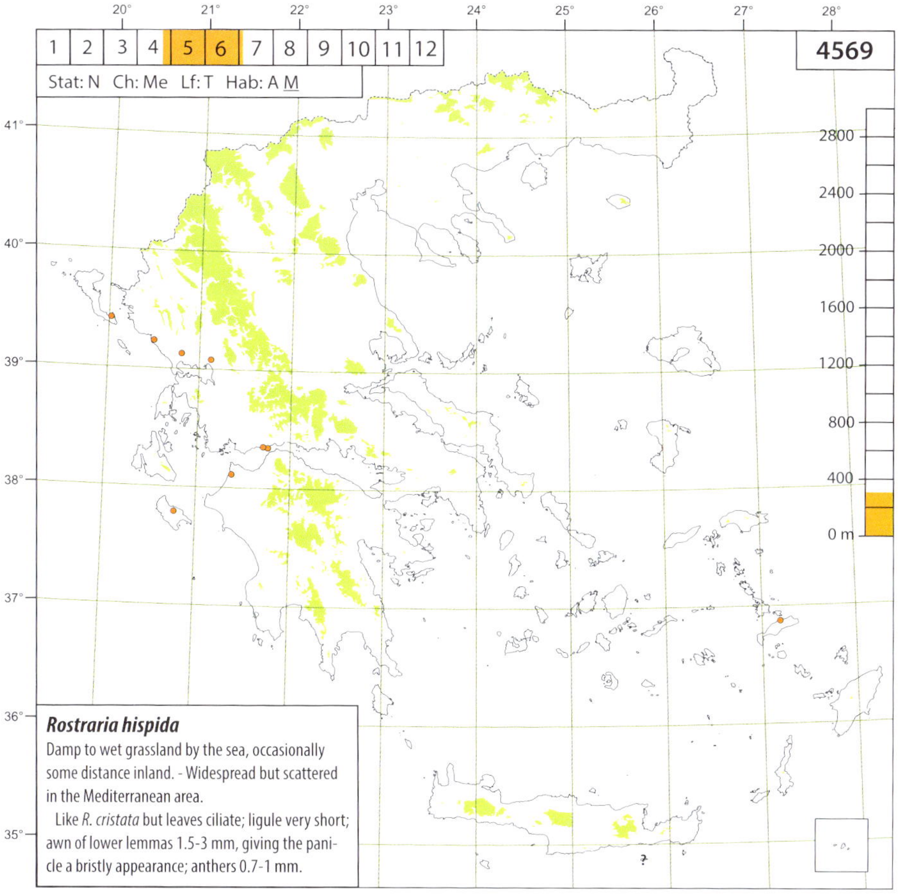

Rostraria hispida
Damp to wet grassland by the sea, occasionally some distance inland. - Widespread but scattered in the Mediterranean area.
Like *R. cristata* but leaves ciliate; ligule very short; awn of lower lemmas 1.5-3 mm, giving the panicle a bristly appearance; anthers 0.7-1 mm.

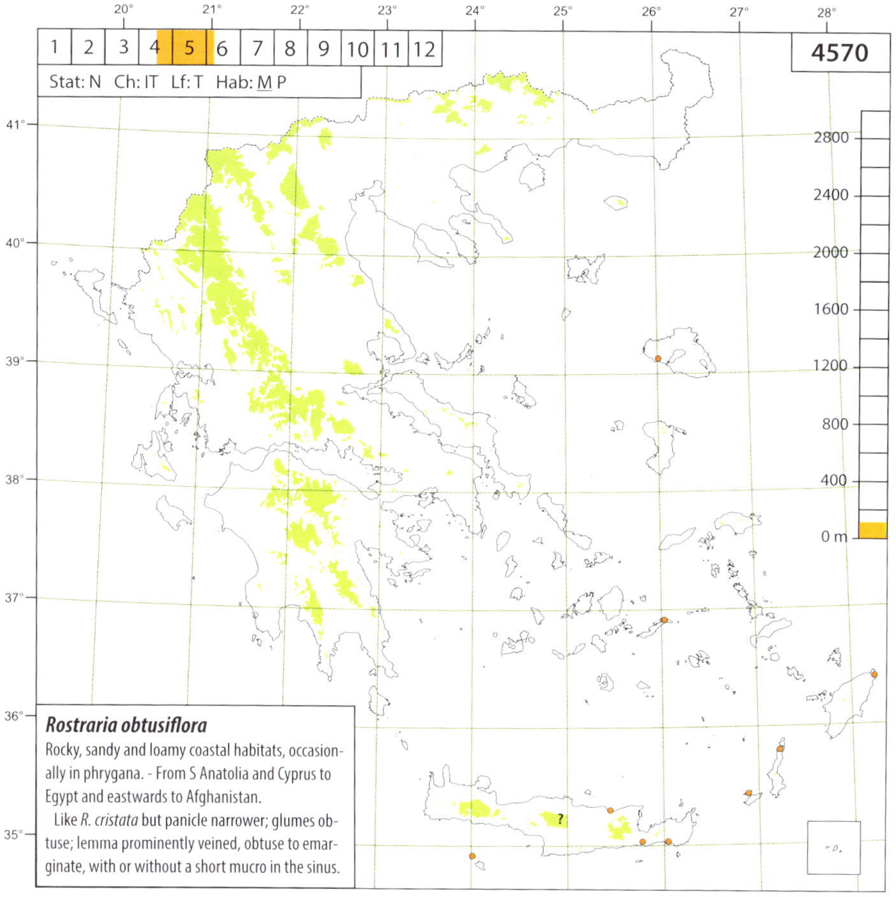

Rostraria obtusiflora
Rocky, sandy and loamy coastal habitats, occasionally in phrygana. - From S Anatolia and Cyprus to Egypt and eastwards to Afghanistan.
Like *R. cristata* but panicle narrower; glumes obtuse; lemma prominently veined, obtuse to emarginate, with or without a short mucro in the sinus.

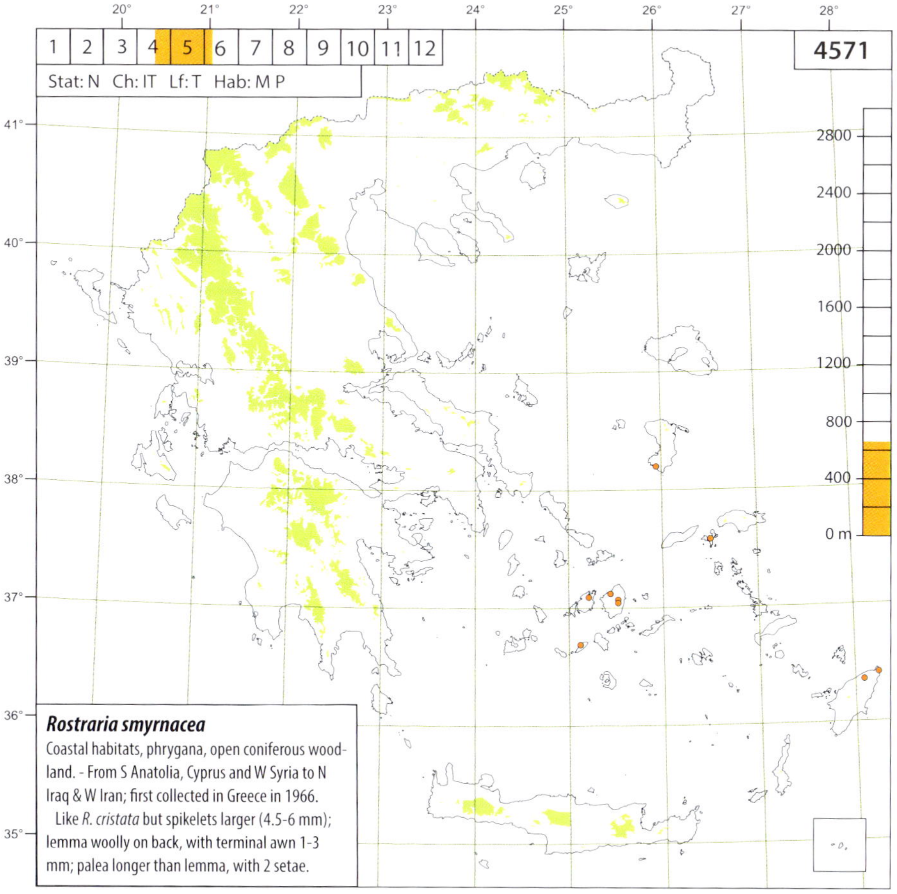

Rostraria smyrnacea
Coastal habitats, phrygana, open coniferous woodland. - From S Anatolia, Cyprus and W Syria to N Iraq & W Iran; first collected in Greece in 1966.
Like *R. cristata* but spikelets larger (4.5-6 mm); lemma woolly on back, with terminal awn 1-3 mm; palea longer than lemma, with 2 setae.

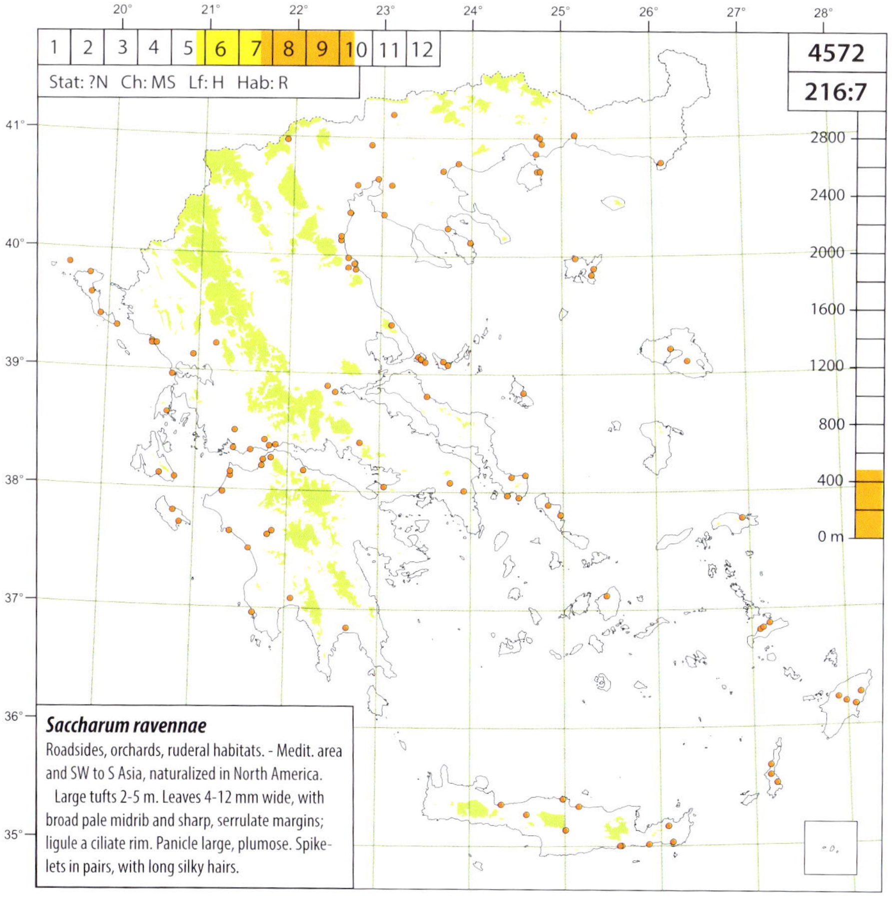

Saccharum ravennae
Roadsides, orchards, ruderal habitats. - Medit. area and SW to S Asia, naturalized in North America.
Large tufts 2-5 m. Leaves 4-12 mm wide, with broad pale midrib and sharp, serrulate margins; ligule a ciliate rim. Panicle large, plumose. Spikelets in pairs, with long silky hairs.

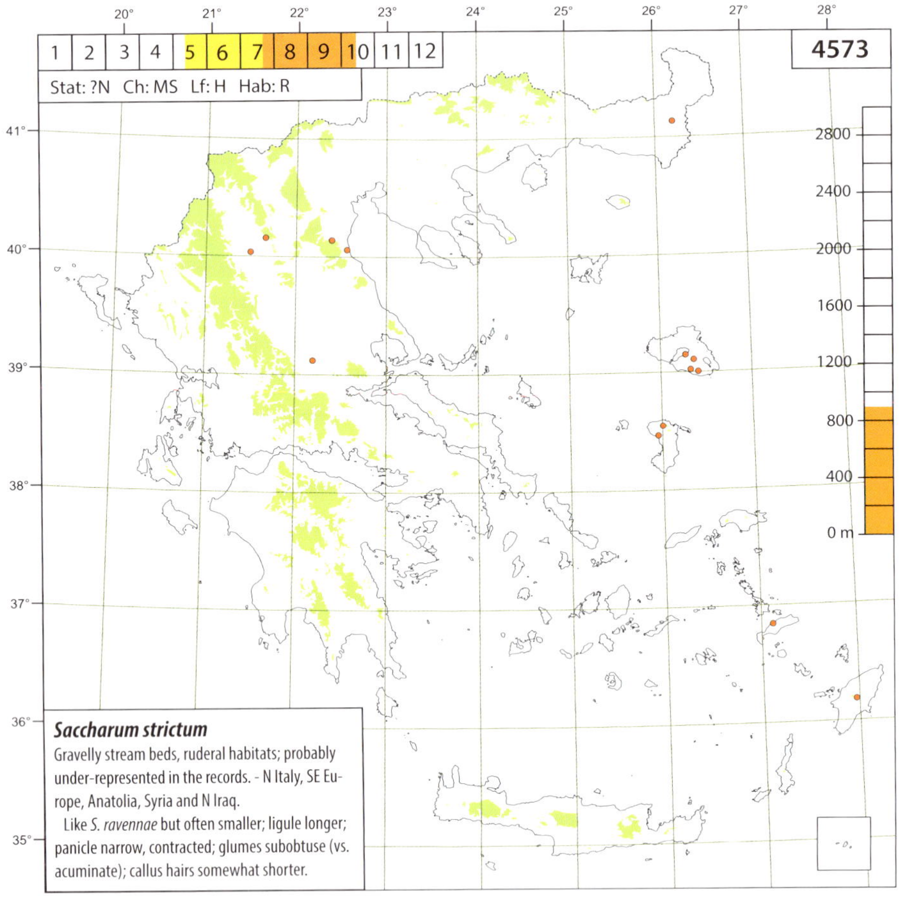

Saccharum strictum

Gravelly stream beds, ruderal habitats; probably under-represented in the records. - N Italy, SE Europe, Anatolia, Syria and N Iraq.

Like *S. ravennae* but often smaller; ligule longer; panicle narrow, contracted; glumes subobtuse (vs. acuminate); callus hairs somewhat shorter.

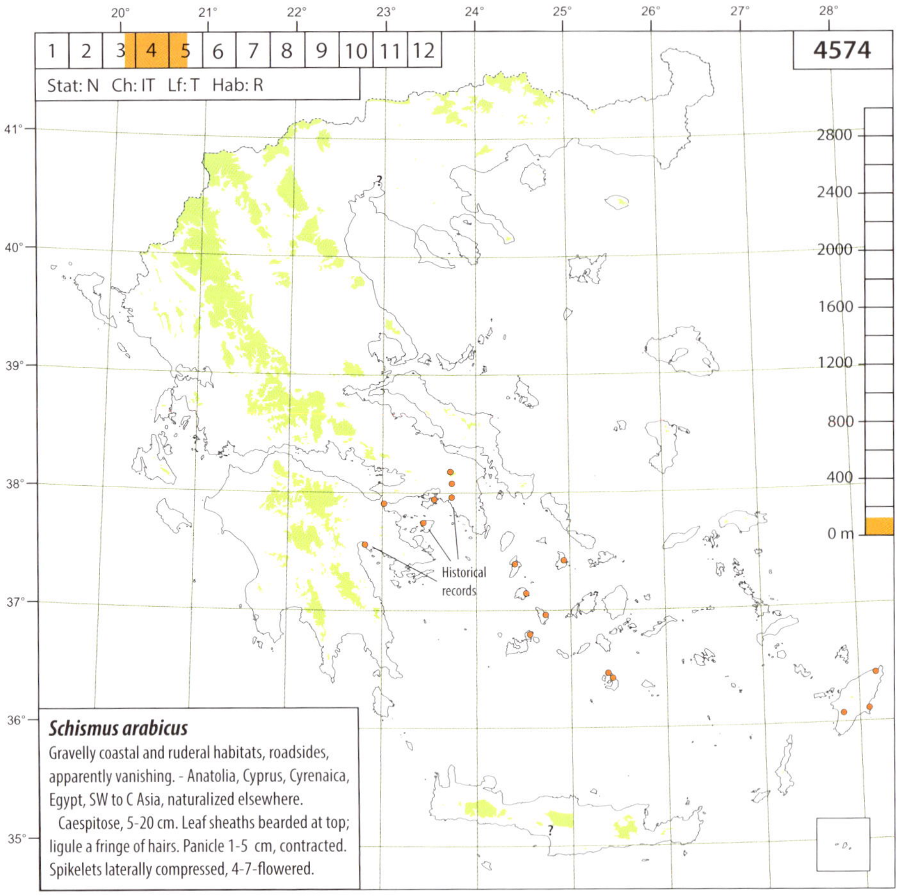

Schismus arabicus

Gravelly coastal and ruderal habitats, roadsides, apparently vanishing. - Anatolia, Cyprus, Cyrenaica, Egypt, SW to C Asia, naturalized elsewhere.

Caespitose, 5-20 cm. Leaf sheaths bearded at top; ligule a fringe of hairs. Panicle 1-5 cm, contracted. Spikelets laterally compressed, 4-7-flowered.

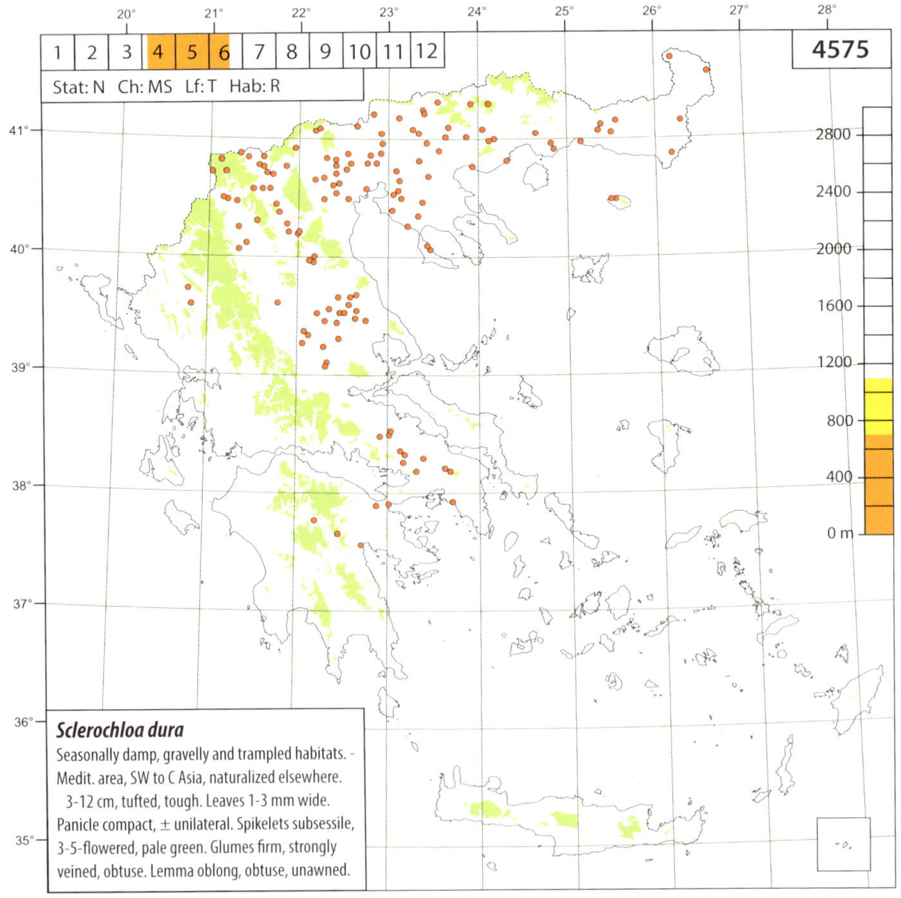

Sclerochloa dura

Seasonally damp, gravelly and trampled habitats. - Medit. area, SW to C Asia, naturalized elsewhere.

3-12 cm, tufted, tough. Leaves 1-3 mm wide. Panicle compact, ± unilateral. Spikelets subsessile, 3-5-flowered, pale green. Glumes firm, strongly veined, obtuse. Lemma oblong, obtuse, unawned.

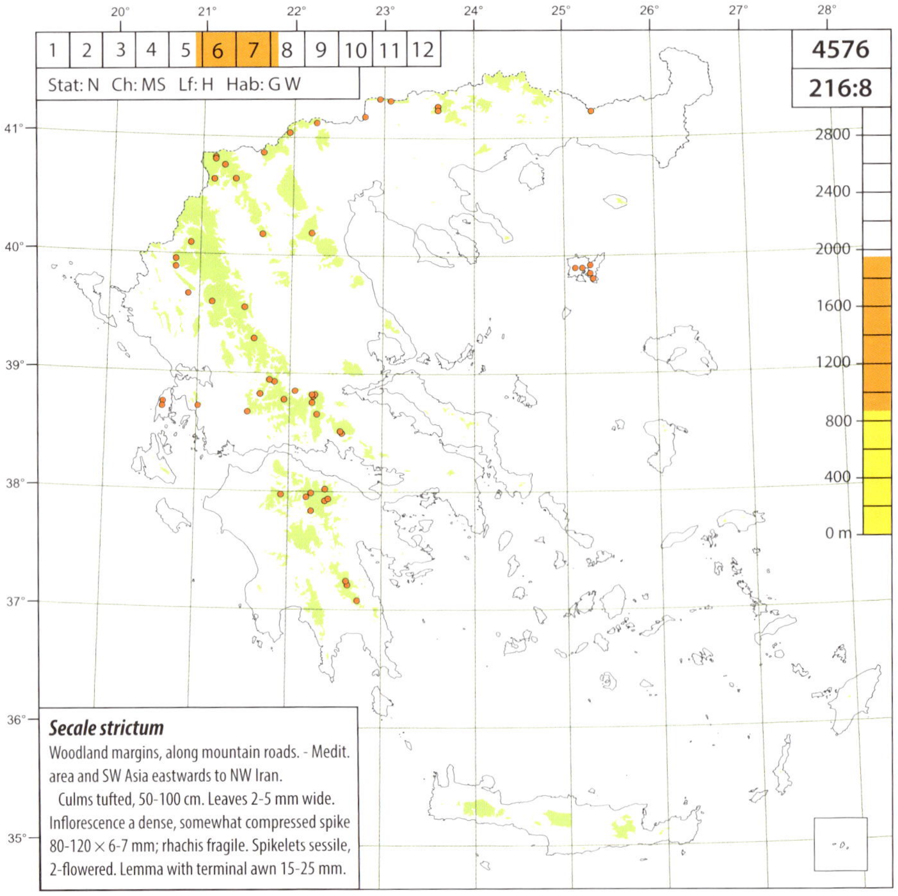

Secale strictum

Woodland margins, along mountain roads. - Medit. area and SW Asia eastwards to NW Iran.

Culms tufted, 50-100 cm. Leaves 2-5 mm wide. Inflorescence a dense, somewhat compressed spike 80-120 × 6-7 mm; rhachis fragile. Spikelets sessile, 2-flowered. Lemma with terminal awn 15-25 mm.

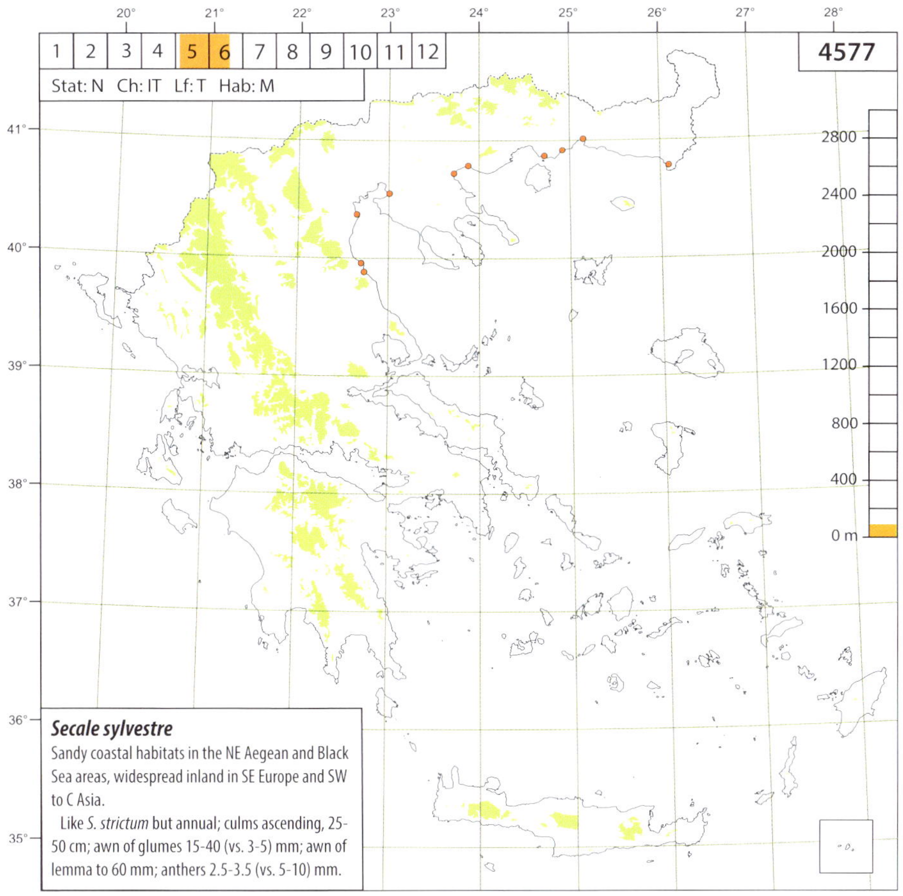

Secale sylvestre
Sandy coastal habitats in the NE Aegean and Black Sea areas, widespread inland in SE Europe and SW to C Asia.
Like *S. strictum* but annual; culms ascending, 25-50 cm; awn of glumes 15-40 (vs. 3-5) mm; awn of lemma to 60 mm; anthers 2.5-3.5 (vs. 5-10) mm.

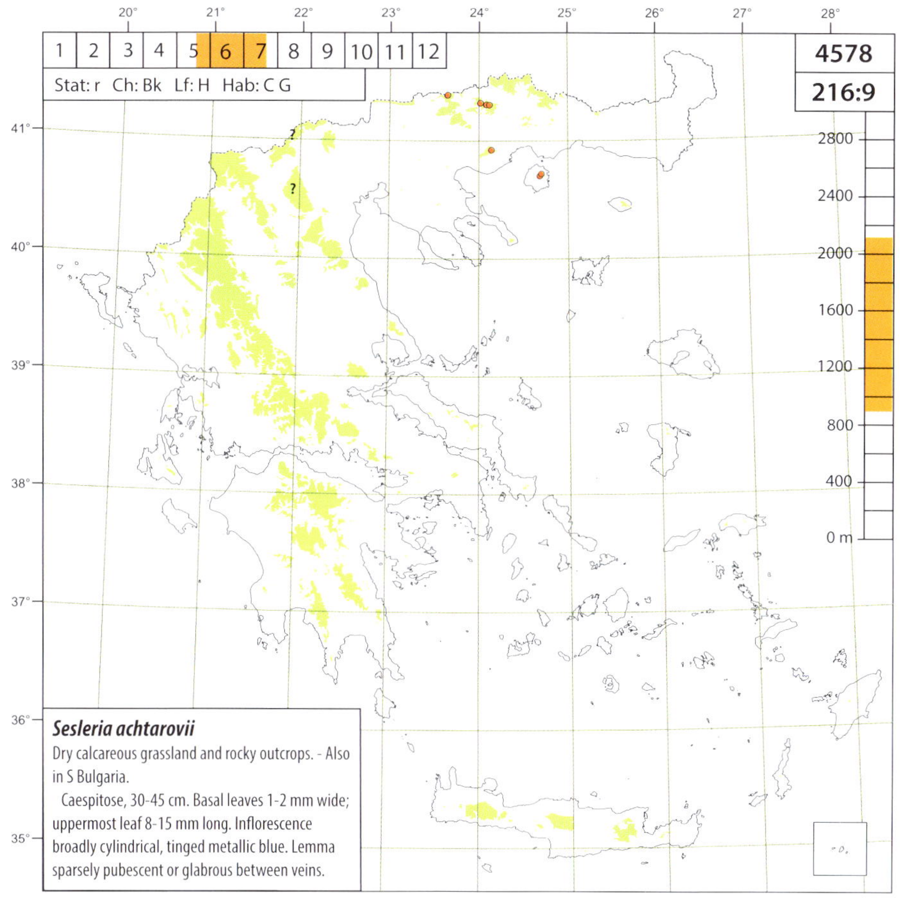

Sesleria achtarovii
Dry calcareous grassland and rocky outcrops. - Also in S Bulgaria.
Caespitose, 30-45 cm. Basal leaves 1-2 mm wide; uppermost leaf 8-15 mm long. Inflorescence broadly cylindrical, tinged metallic blue. Lemma sparsely pubescent or glabrous between veins.

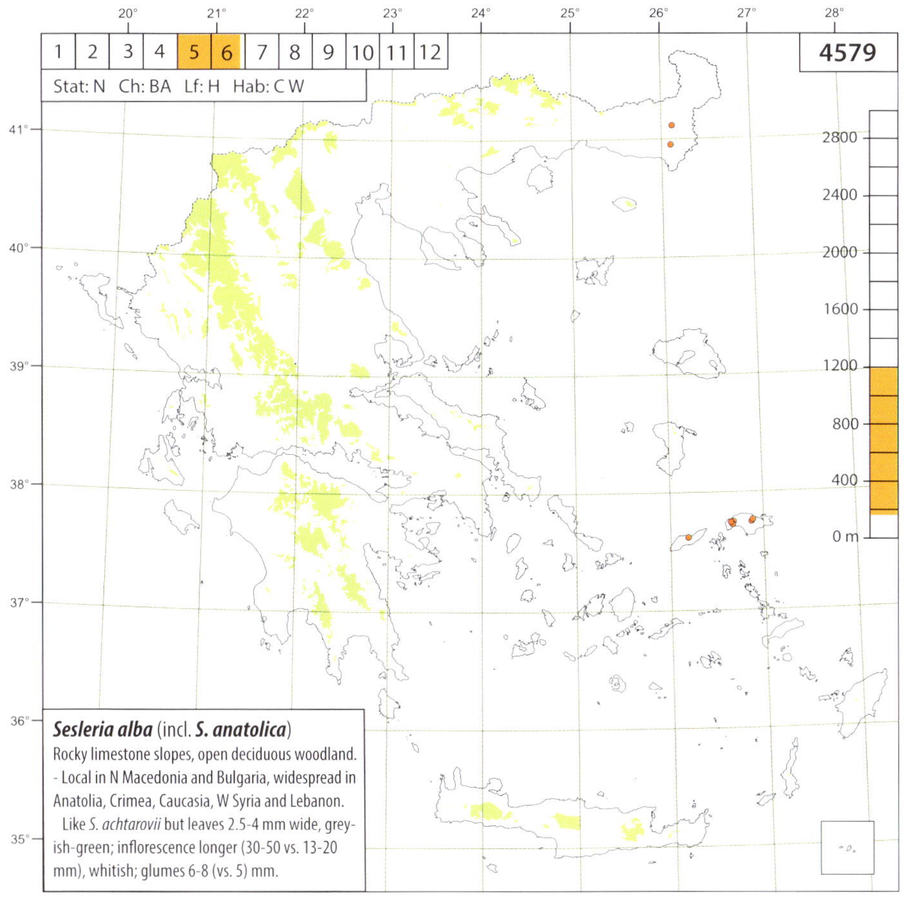

Sesleria alba (incl. ***S. anatolica***)
Rocky limestone slopes, open deciduous woodland. - Local in N Macedonia and Bulgaria, widespread in Anatolia, Crimea, Caucasia, W Syria and Lebanon.
Like *S. achtarovii* but leaves 2.5-4 mm wide, grey-ish-green; inflorescence longer (30-50 vs. 13-20 mm), whitish; glumes 6-8 (vs. 5) mm.

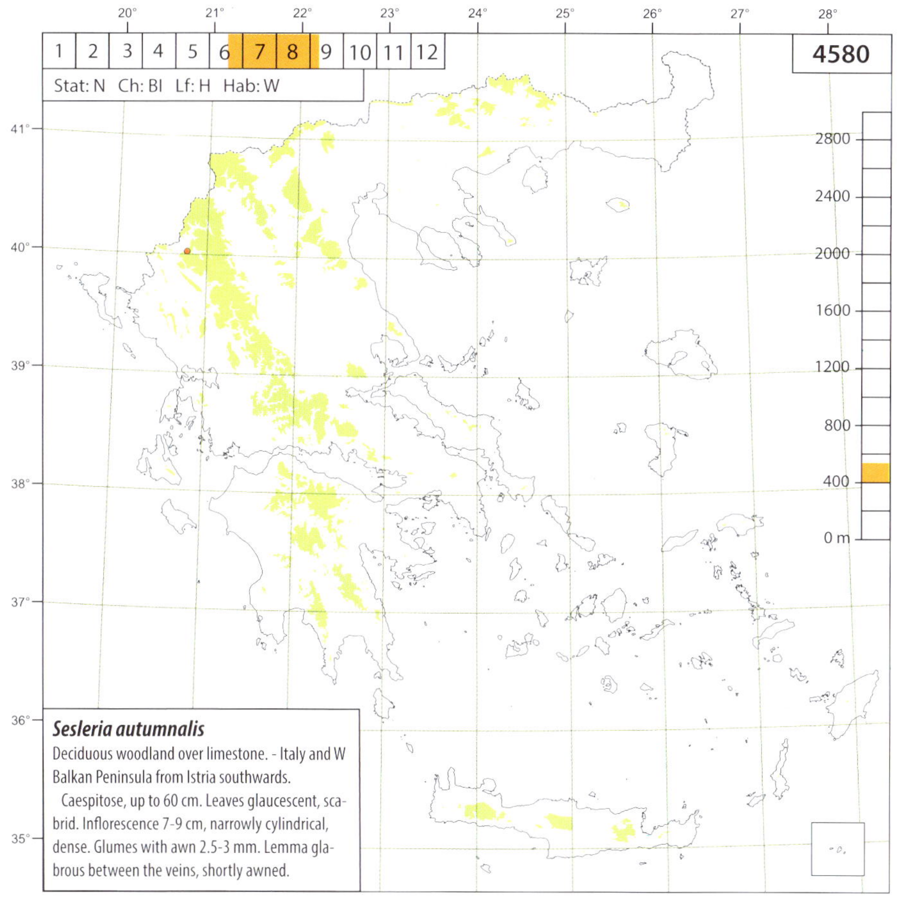

Sesleria autumnalis
Deciduous woodland over limestone. - Italy and W Balkan Peninsula from Istria southwards.
Caespitose, up to 60 cm. Leaves glaucescent, sca-brid. Inflorescence 7-9 cm, narrowly cylindrical, dense. Glumes with awn 2.5-3 mm. Lemma gla-brous between the veins, shortly awned.

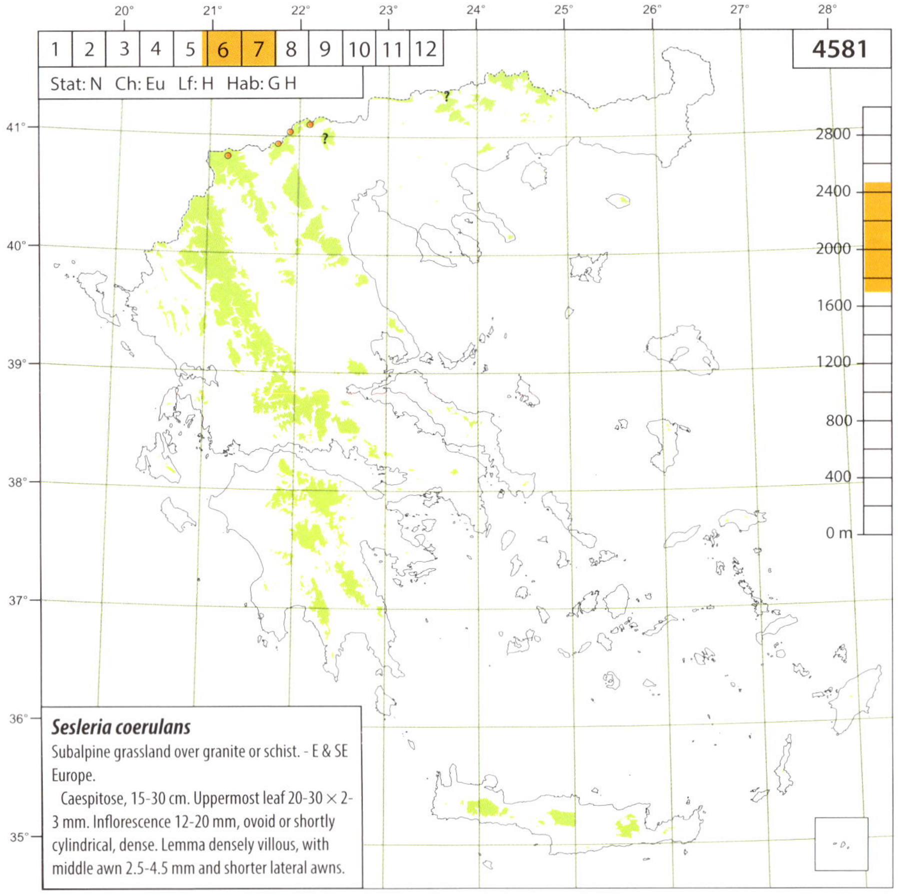

Sesleria coerulans
Subalpine grassland over granite or schist. - E & SE Europe.
Caespitose, 15-30 cm. Uppermost leaf 20-30 × 2-3 mm. Inflorescence 12-20 mm, ovoid or shortly cylindrical, dense. Lemma densely villous, with middle awn 2.5-4.5 mm and shorter lateral awns.

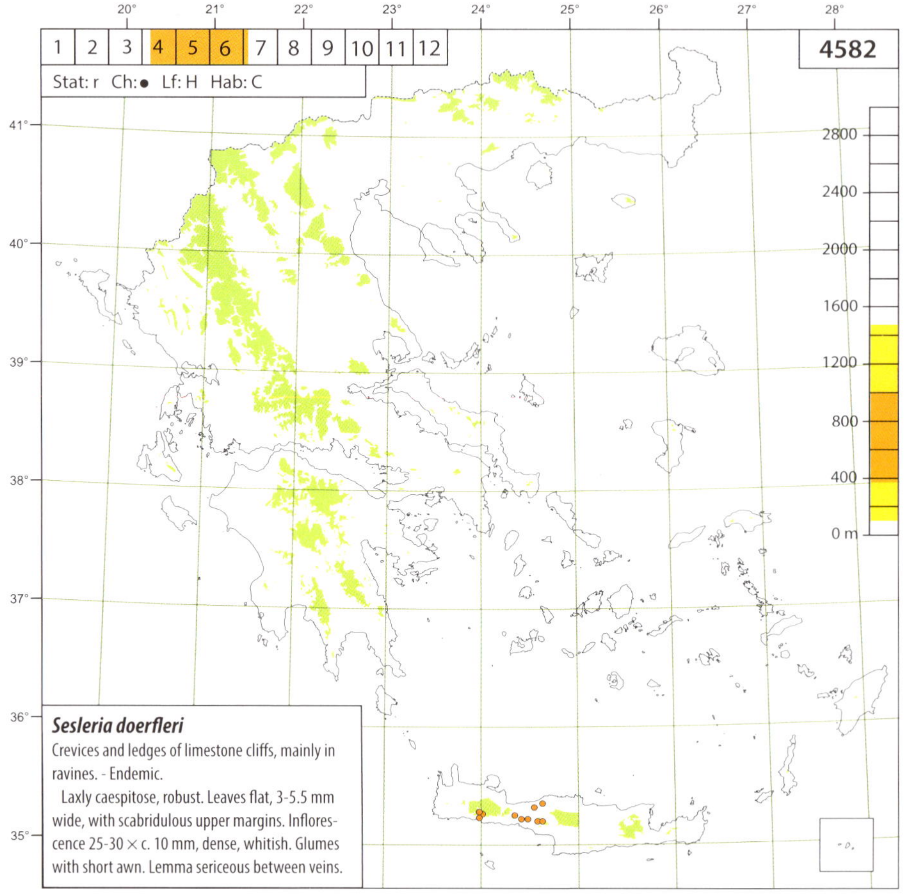

Sesleria doerfleri
Crevices and ledges of limestone cliffs, mainly in ravines. - Endemic.
Laxly caespitose, robust. Leaves flat, 3-5.5 mm wide, with scabridulous upper margins. Inflorescence 25-30 × c. 10 mm, dense, whitish. Glumes with short awn. Lemma sericeous between veins.

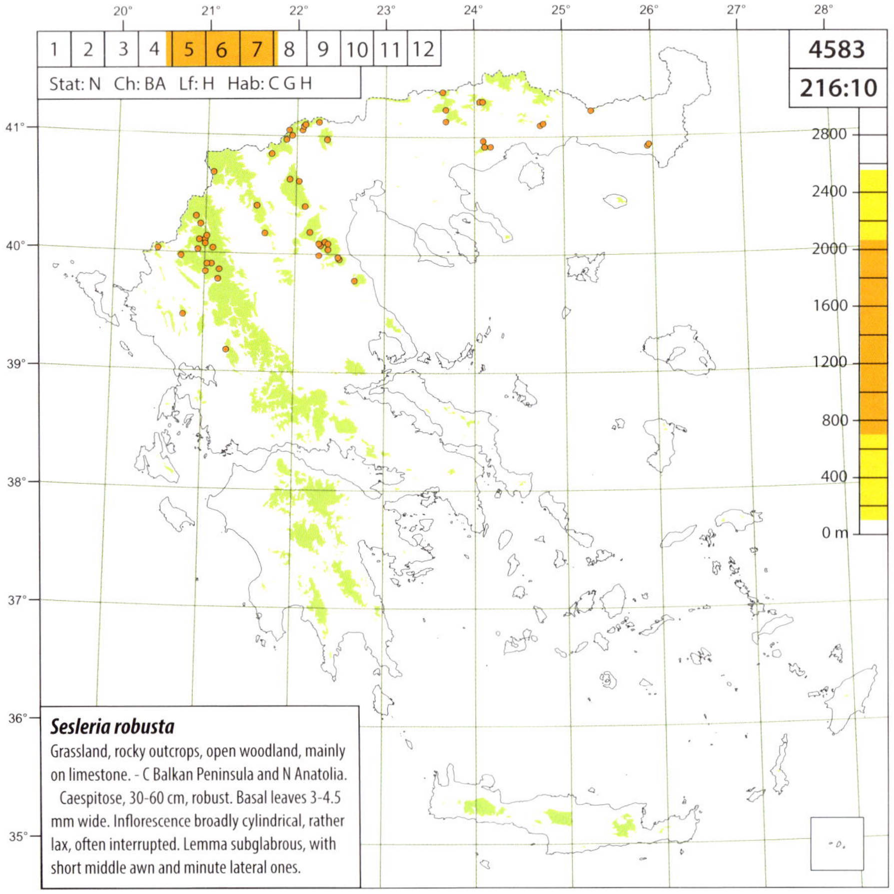

Sesleria robusta
Grassland, rocky outcrops, open woodland, mainly on limestone. - C Balkan Peninsula and N Anatolia.
Caespitose, 30-60 cm, robust. Basal leaves 3-4.5 mm wide. Inflorescence broadly cylindrical, rather lax, often interrupted. Lemma subglabrous, with short middle awn and minute lateral ones.

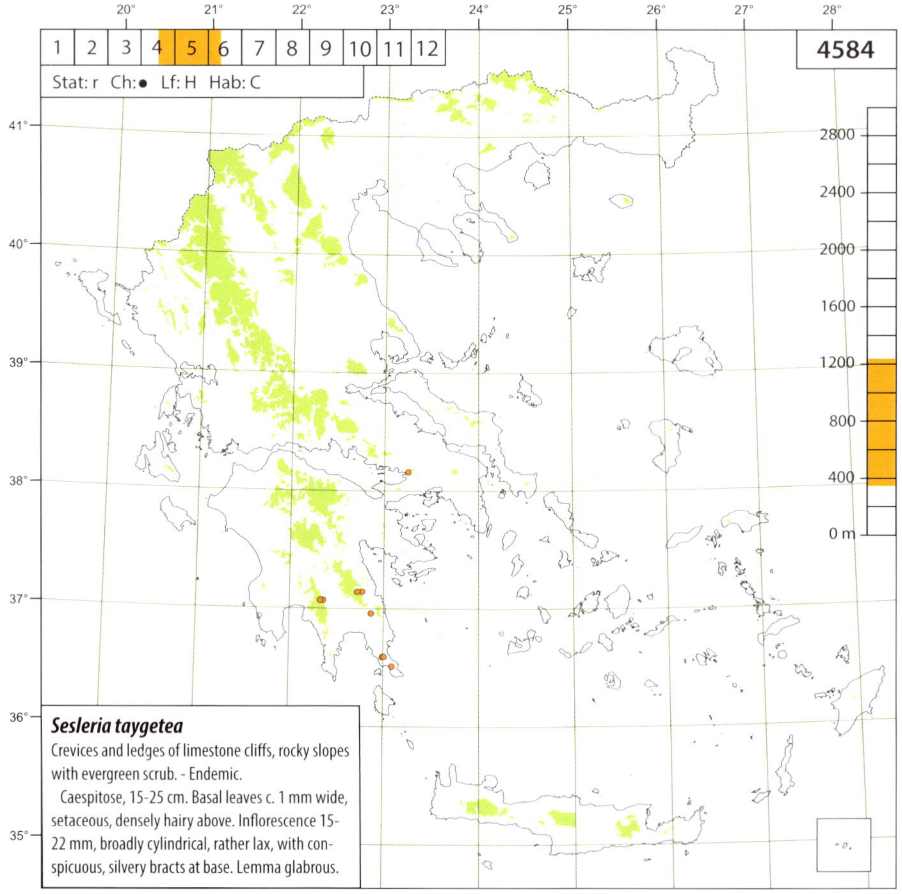

Sesleria taygetea
Crevices and ledges of limestone cliffs, rocky slopes with evergreen scrub. - Endemic.
Caespitose, 15-25 cm. Basal leaves c. 1 mm wide, setaceous, densely hairy above. Inflorescence 15-22 mm, broadly cylindrical, rather lax, with conspicuous, silvery bracts at base. Lemma glabrous.

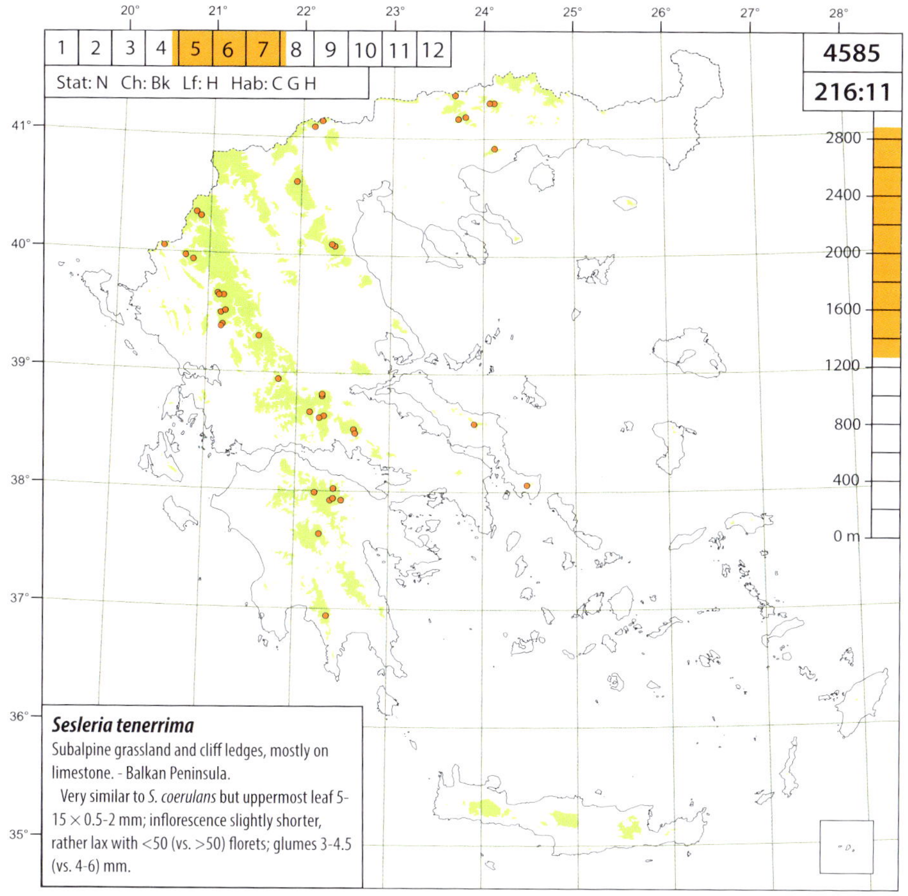

Sesleria tenerrima
Subalpine grassland and cliff ledges, mostly on limestone. - Balkan Peninsula.
Very similar to *S. coerulans* but uppermost leaf 5-15 × 0.5-2 mm; inflorescence slightly shorter, rather lax with <50 (vs. >50) florets; glumes 3-4.5 (vs. 4-6) mm.

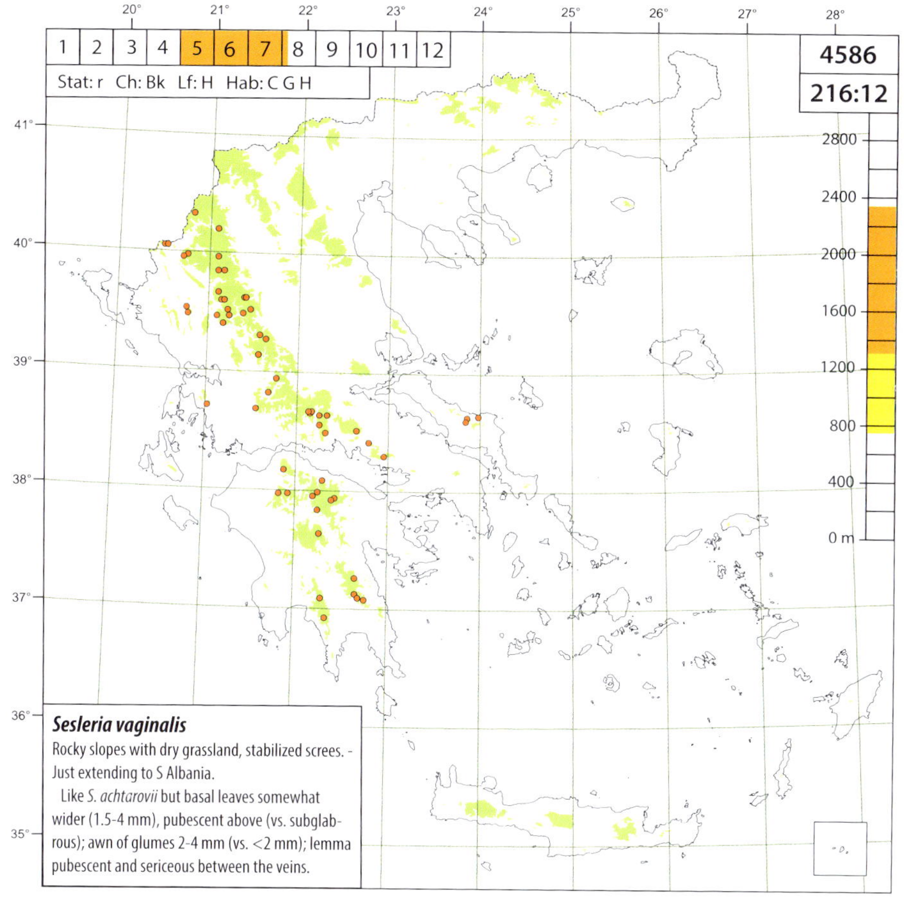

Sesleria vaginalis
Rocky slopes with dry grassland, stabilized screes. - Just extending to S Albania.
Like *S. achtarovii* but basal leaves somewhat wider (1.5-4 mm), pubescent above (vs. subglabrous); awn of glumes 2-4 mm (vs. <2 mm); lemma pubescent and sericeous between the veins.

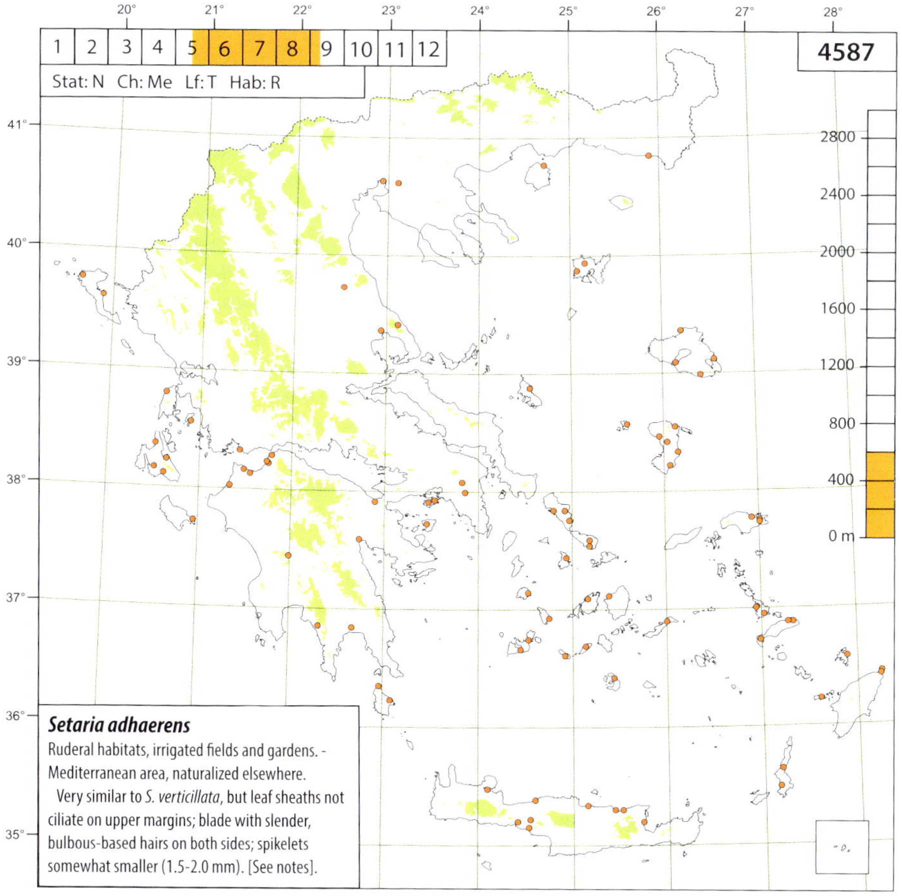

Setaria adhaerens
Ruderal habitats, irrigated fields and gardens. - Mediterranean area, naturalized elsewhere.
Very similar to *S. verticillata*, but leaf sheaths not ciliate on upper margins; blade with slender, bulbous-based hairs on both sides; spikelets somewhat smaller (1.5-2.0 mm). [See notes].

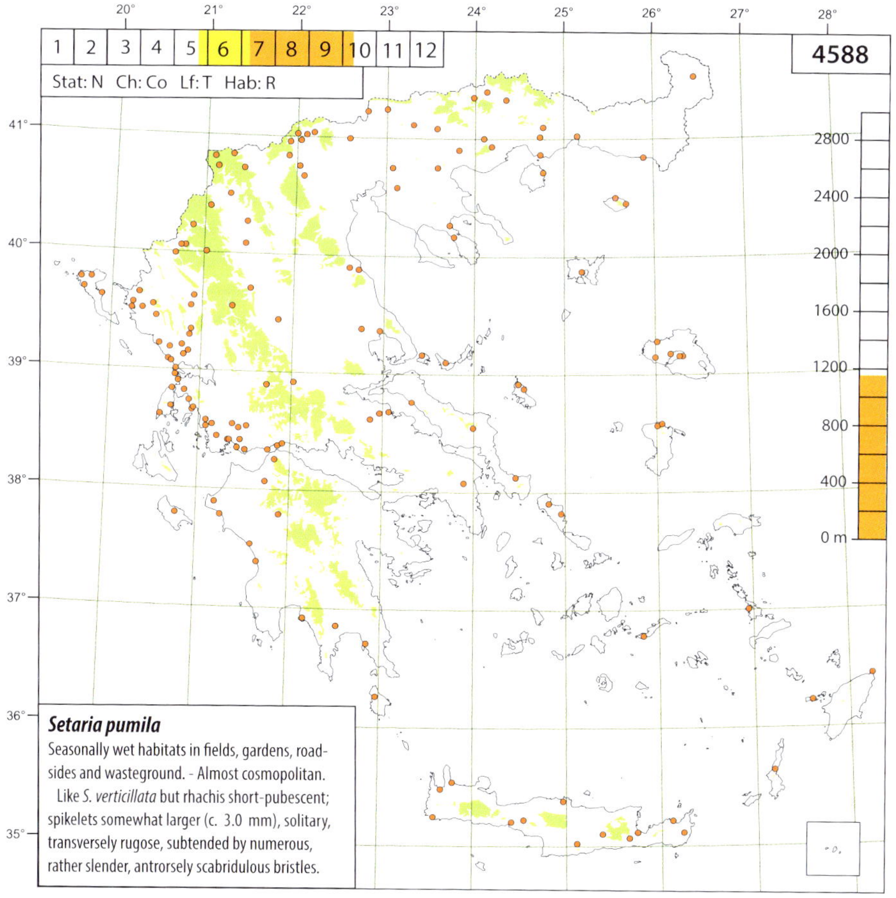

Setaria pumila
Seasonally wet habitats in fields, gardens, roadsides and wasteground. - Almost cosmopolitan.
Like *S. verticillata* but rhachis short-pubescent; spikelets somewhat larger (c. 3.0 mm), solitary, transversely rugose, subtended by numerous, rather slender, antrorsely scabridulous bristles.

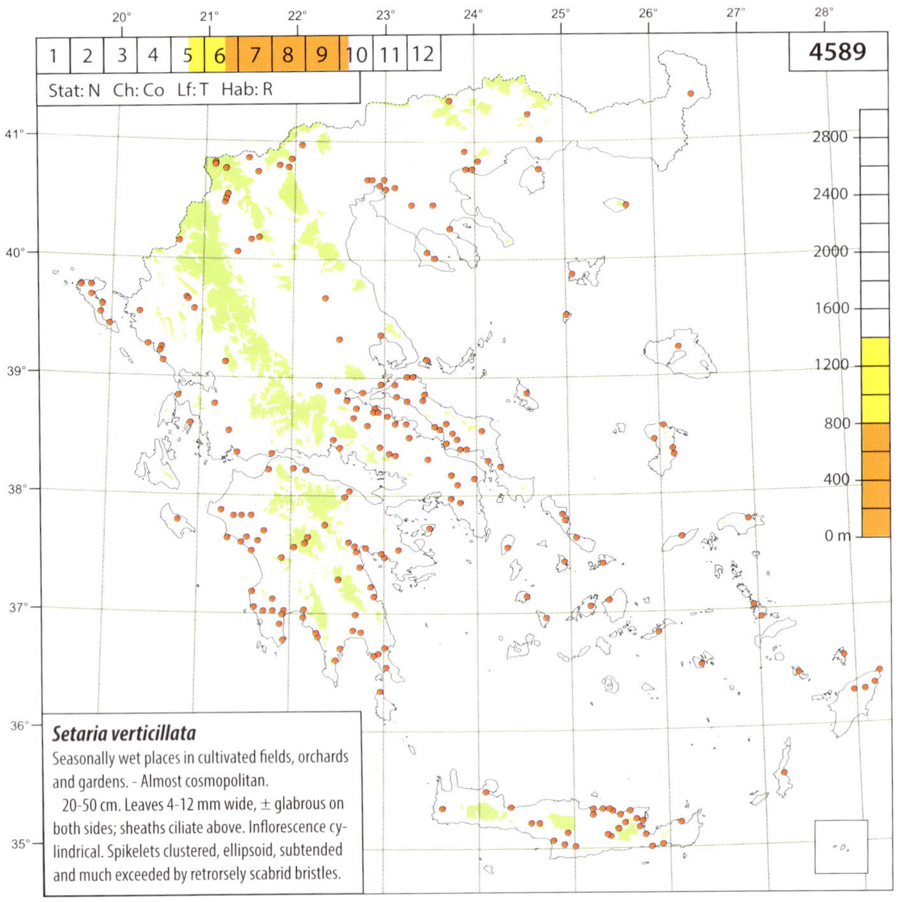

Setaria verticillata

Seasonally wet places in cultivated fields, orchards and gardens. - Almost cosmopolitan.

20-50 cm. Leaves 4-12 mm wide, ± glabrous on both sides; sheaths ciliate above. Inflorescence cylindrical. Spikelets clustered, ellipsoid, subtended and much exceeded by retrorsely scabrid bristles.

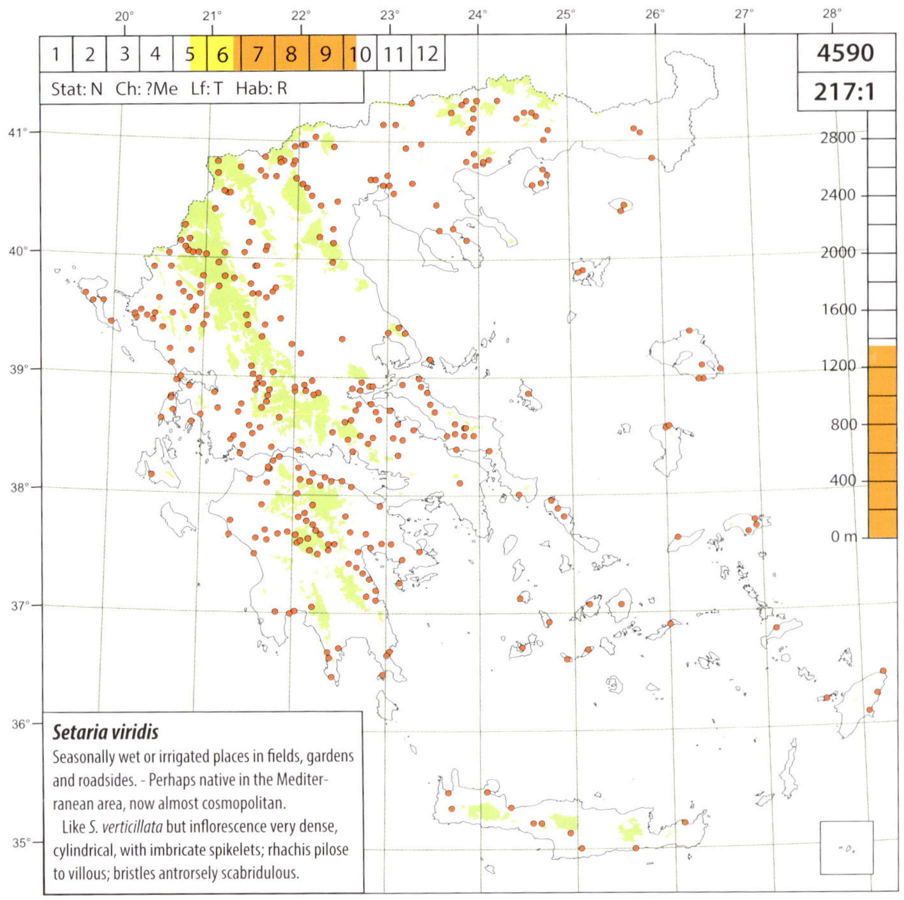

Setaria viridis

Seasonally wet or irrigated places in fields, gardens and roadsides. - Perhaps native in the Mediterranean area, now almost cosmopolitan.

Like *S. verticillata* but inflorescence very dense, cylindrical, with imbricate spikelets; rhachis pilose to villous; bristles antrorsely scabridulous.

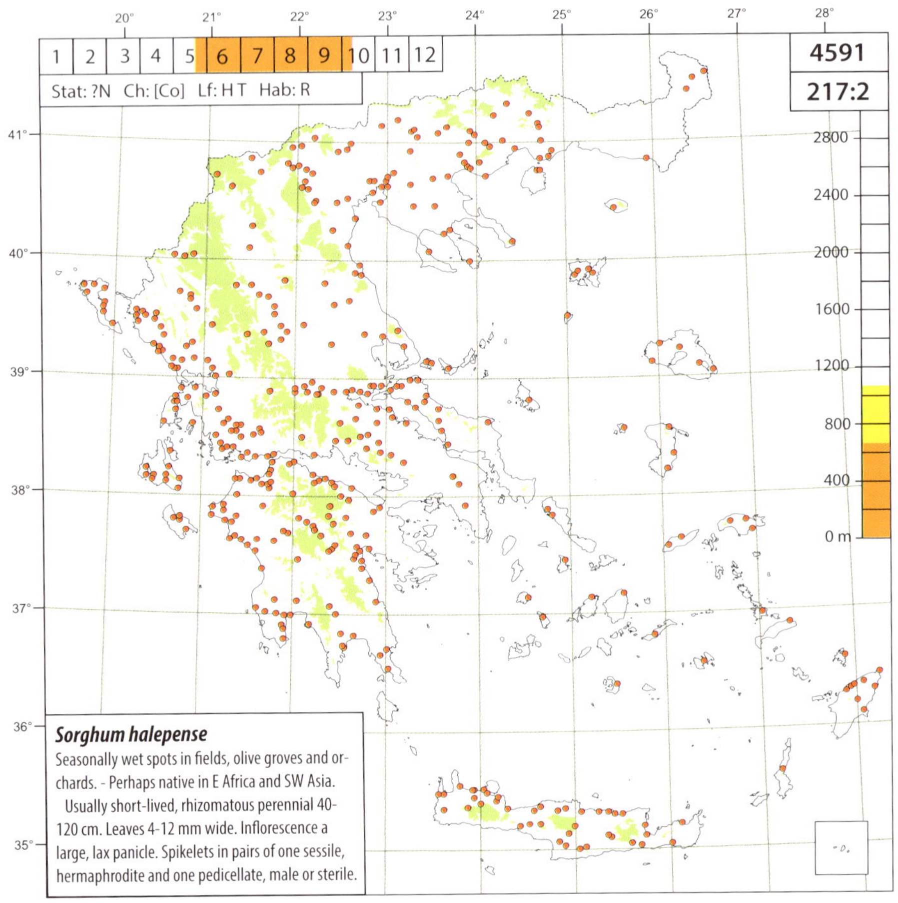

Sorghum halepense

Seasonally wet spots in fields, olive groves and orchards. - Perhaps native in E Africa and SW Asia.

Usually short-lived, rhizomatous perennial 40-120 cm. Leaves 4-12 mm wide. Inflorescence a large, lax panicle. Spikelets in pairs of one sessile, hermaphrodite and one pedicellate, male or sterile.

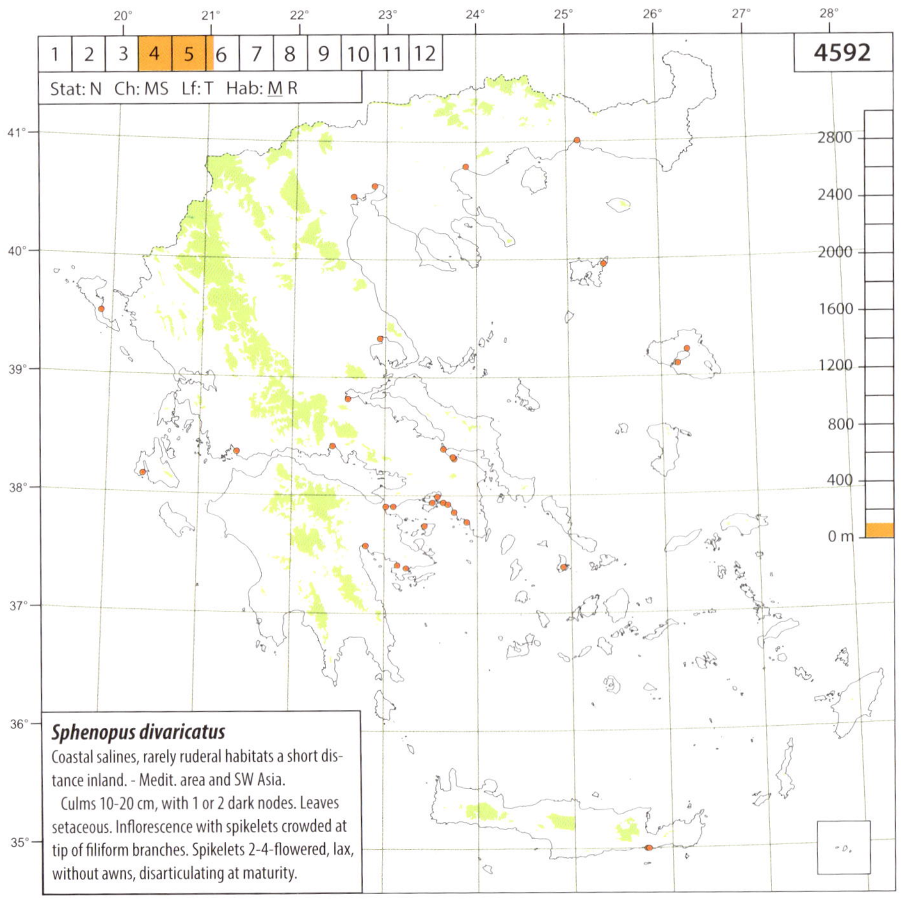

Sphenopus divaricatus

Coastal salines, rarely ruderal habitats a short distance inland. - Medit. area and SW Asia.

Culms 10-20 cm, with 1 or 2 dark nodes. Leaves setaceous. Inflorescence with spikelets crowded at tip of filiform branches. Spikelets 2-4-flowered, lax, without awns, disarticulating at maturity.

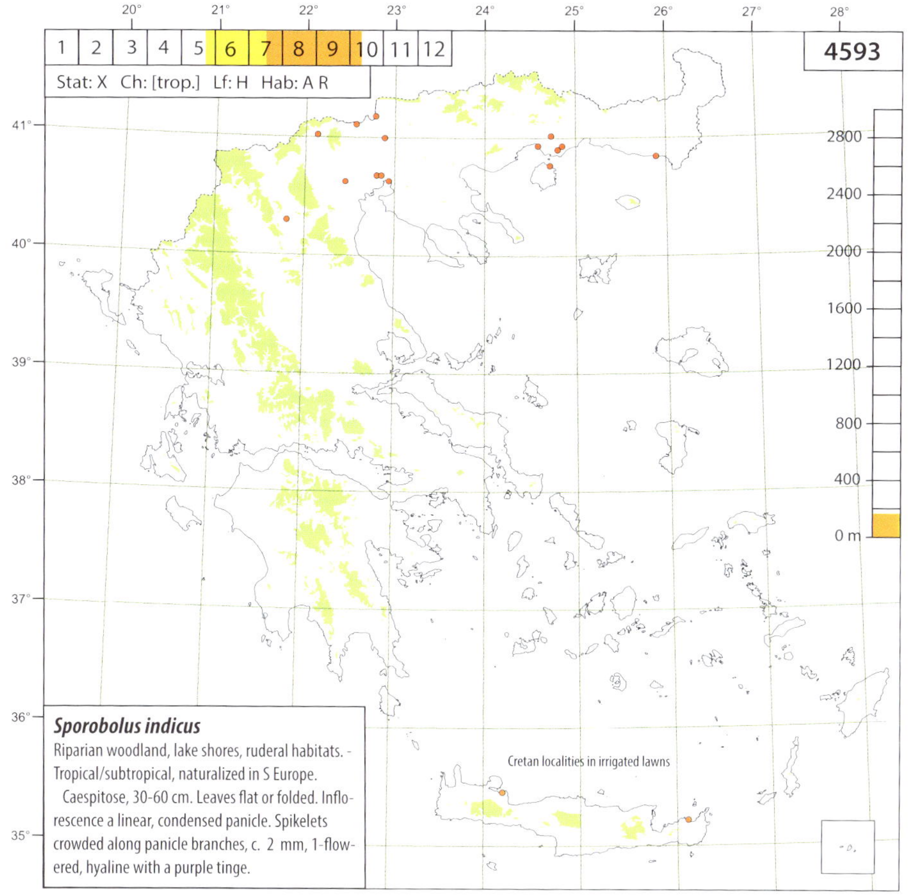

Sporobolus indicus

Riparian woodland, lake shores, ruderal habitats. - Tropical/subtropical, naturalized in S Europe.

Caespitose, 30-60 cm. Leaves flat or folded. Inflorescence a linear, condensed panicle. Spikelets crowded along panicle branches, c. 2 mm, 1-flowered, hyaline with a purple tinge.

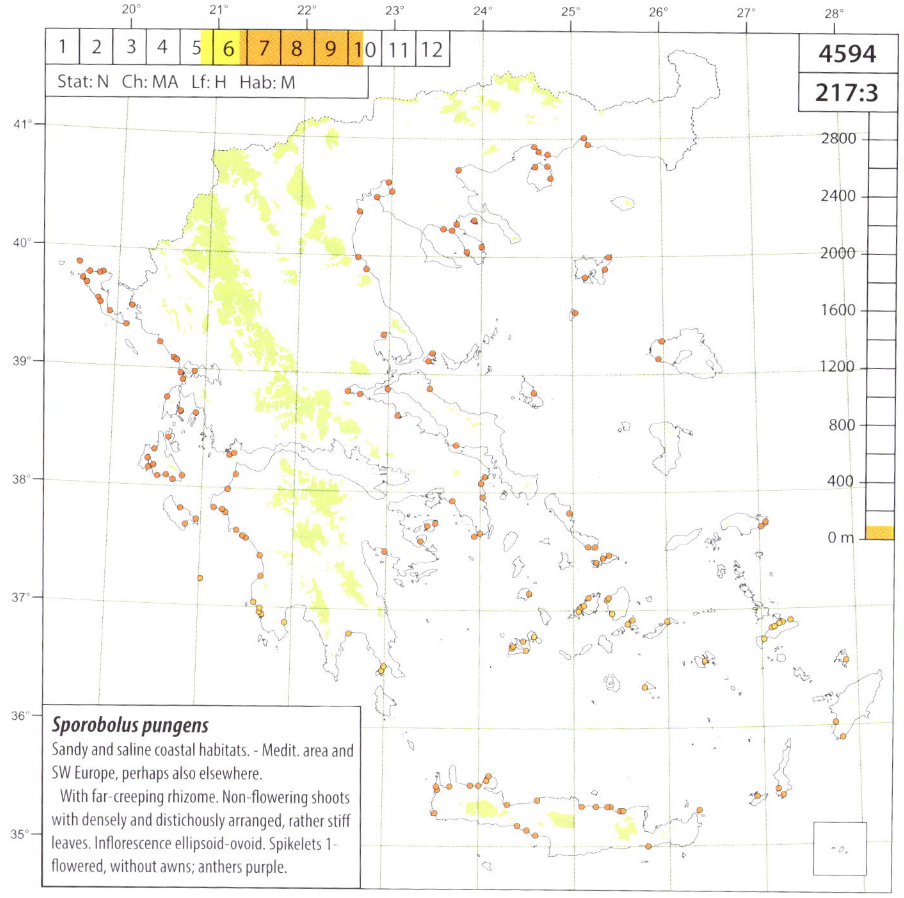

Sporobolus pungens

Sandy and saline coastal habitats. - Medit. area and SW Europe, perhaps also elsewhere.

With far-creeping rhizome. Non-flowering shoots with densely and distichously arranged, rather stiff leaves. Inflorescence ellipsoid-ovoid. Spikelets 1-flowered, without awns; anthers purple.

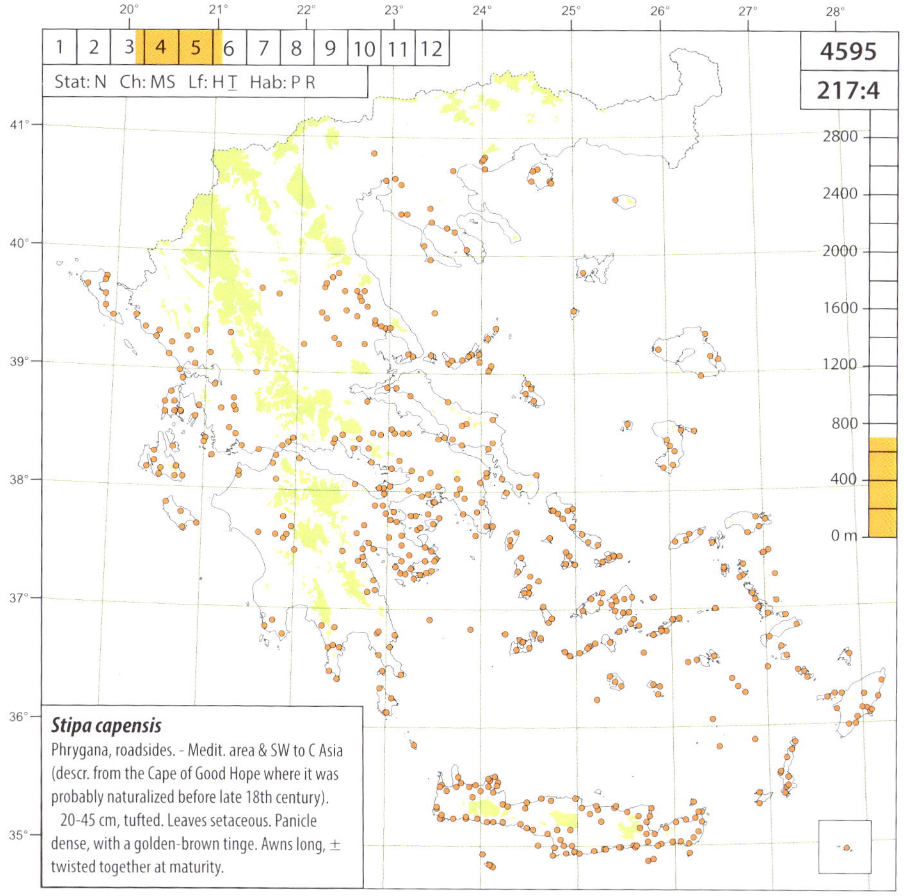

Stipa capensis

Phrygana, roadsides. - Medit. area & SW to C Asia (descr. from the Cape of Good Hope where it was probably naturalized before late 18th century).

20-45 cm, tufted. Leaves setaceous. Panicle dense, with a golden-brown tinge. Awns long, ± twisted together at maturity.

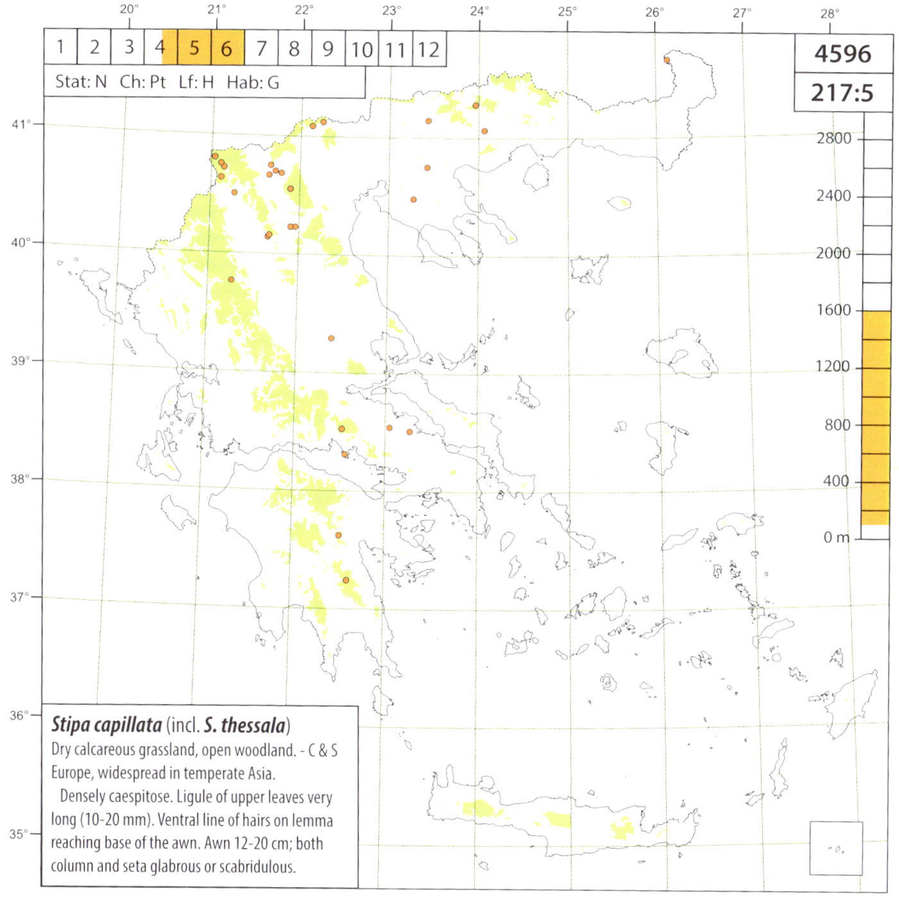

Stipa capillata (incl. ***S. thessala***)

Dry calcareous grassland, open woodland. - C & S Europe, widespread in temperate Asia.

Densely caespitose. Ligule of upper leaves very long (10-20 mm). Ventral line of hairs on lemma reaching base of the awn. Awn 12-20 cm; both column and seta glabrous or scabridulous.

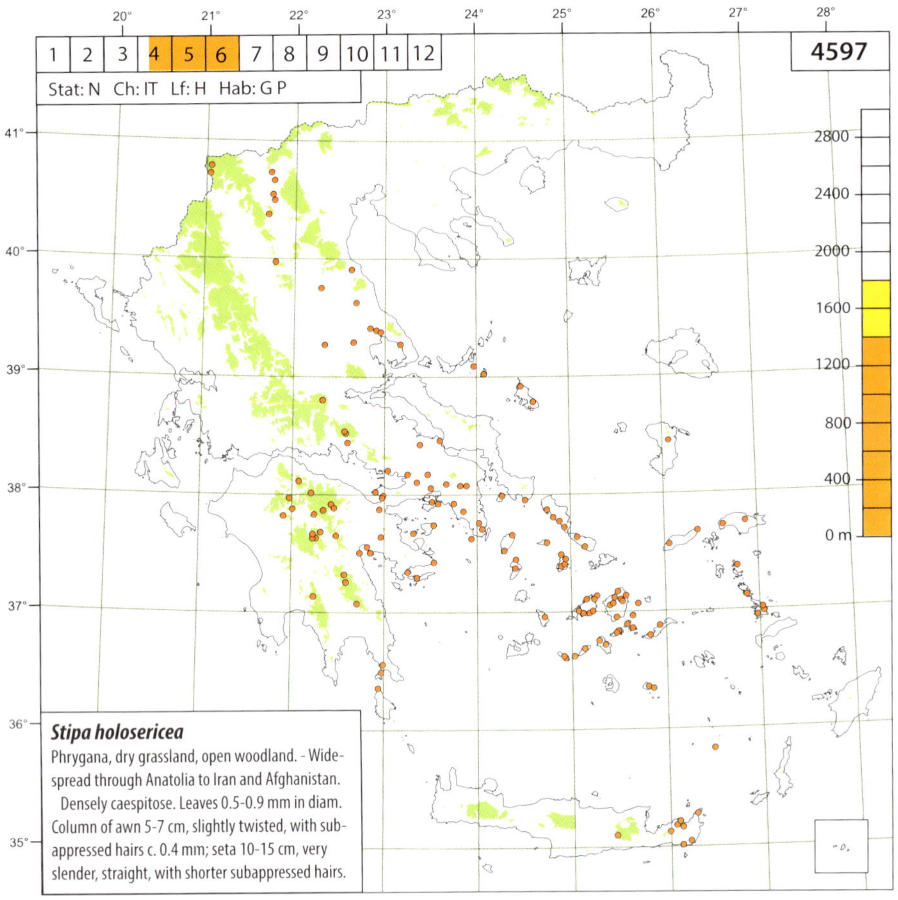

Stipa holosericea
Phrygana, dry grassland, open woodland. - Widespread through Anatolia to Iran and Afghanistan.
Densely caespitose. Leaves 0.5-0.9 mm in diam. Column of awn 5-7 cm, slightly twisted, with subappressed hairs c. 0.4 mm; seta 10-15 cm, very slender, straight, with shorter subappressed hairs.

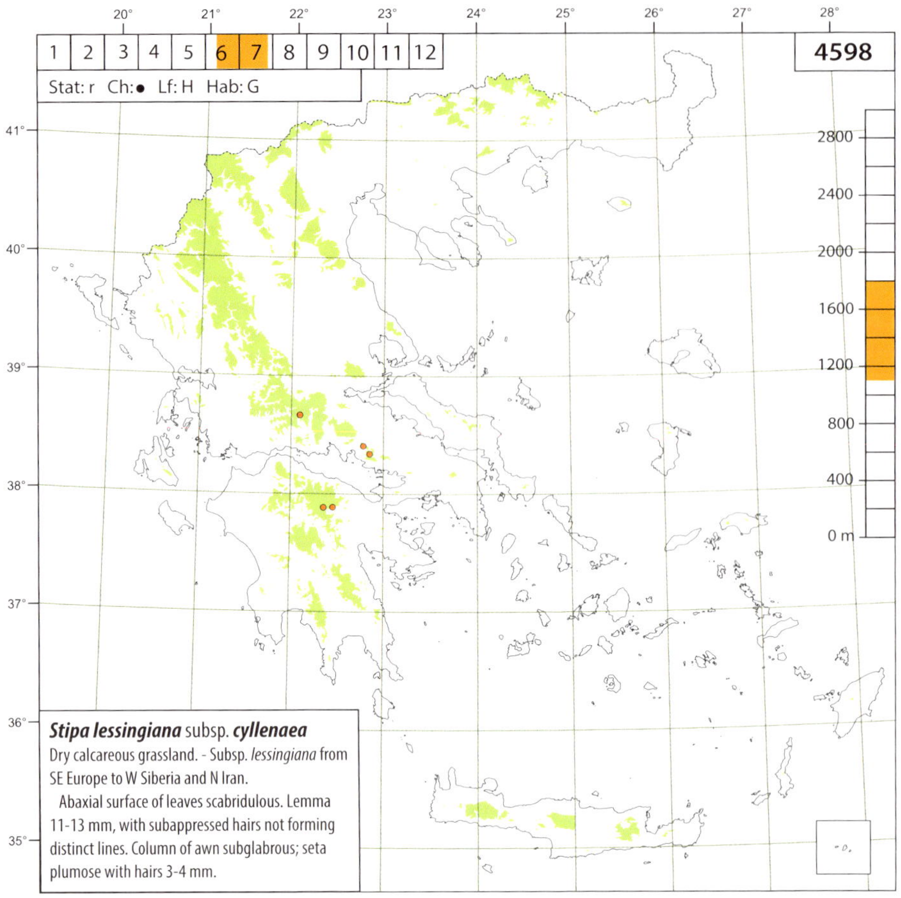

Stipa lessingiana* subsp. *cyllenaea
Dry calcareous grassland. - Subsp. *lessingiana* from SE Europe to W Siberia and N Iran.
Abaxial surface of leaves scabridulous. Lemma 11-13 mm, with subappressed hairs not forming distinct lines. Column of awn subglabrous; seta plumose with hairs 3-4 mm.

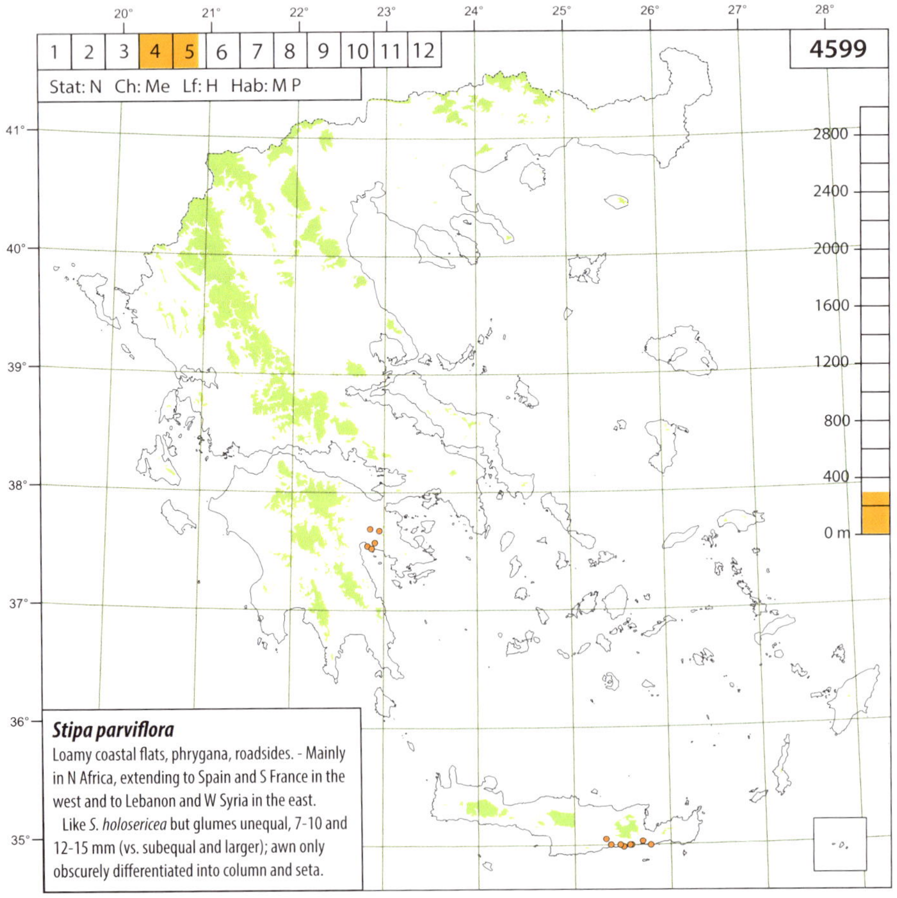

Stipa parviflora
Loamy coastal flats, phrygana, roadsides. - Mainly in N Africa, extending to Spain and S France in the west and to Lebanon and W Syria in the east.
Like *S. holosericea* but glumes unequal, 7-10 and 12-15 mm (vs. subequal and larger); awn only obscurely differentiated into column and seta.

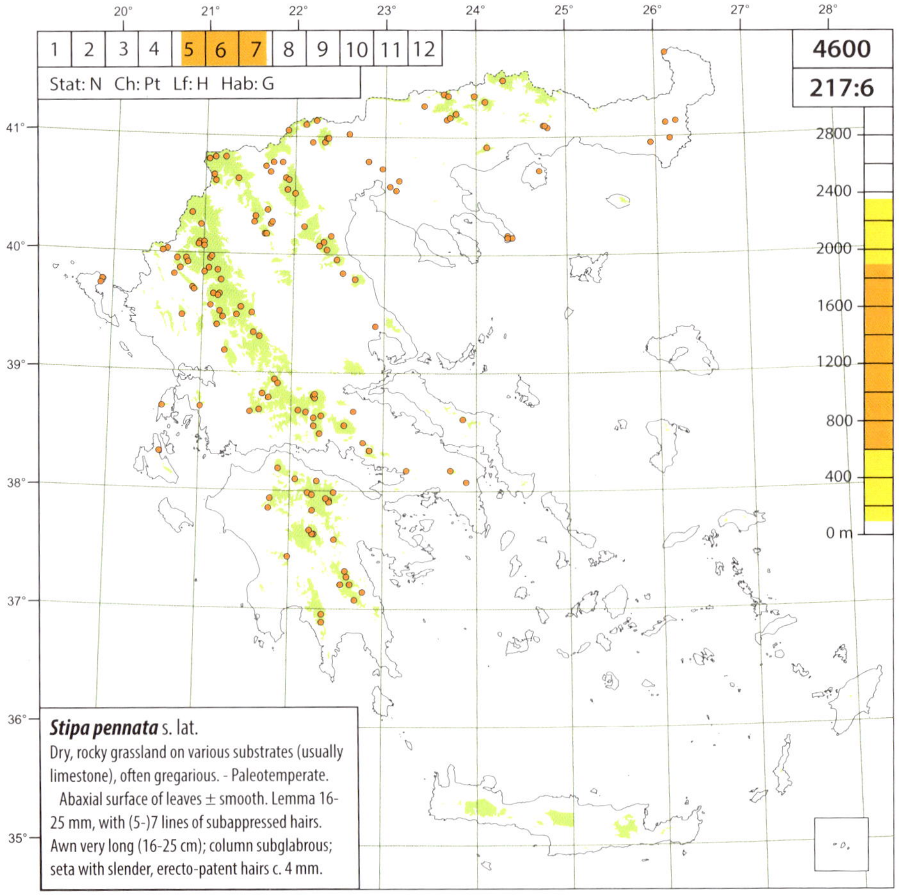

***Stipa pennata* s. lat.**
Dry, rocky grassland on various substrates (usually limestone), often gregarious. - Paleotemperate.
Abaxial surface of leaves ± smooth. Lemma 16-25 mm, with (5-)7 lines of subappressed hairs. Awn very long (16-25 cm); column subglabrous; seta with slender, erecto-patent hairs c. 4 mm.

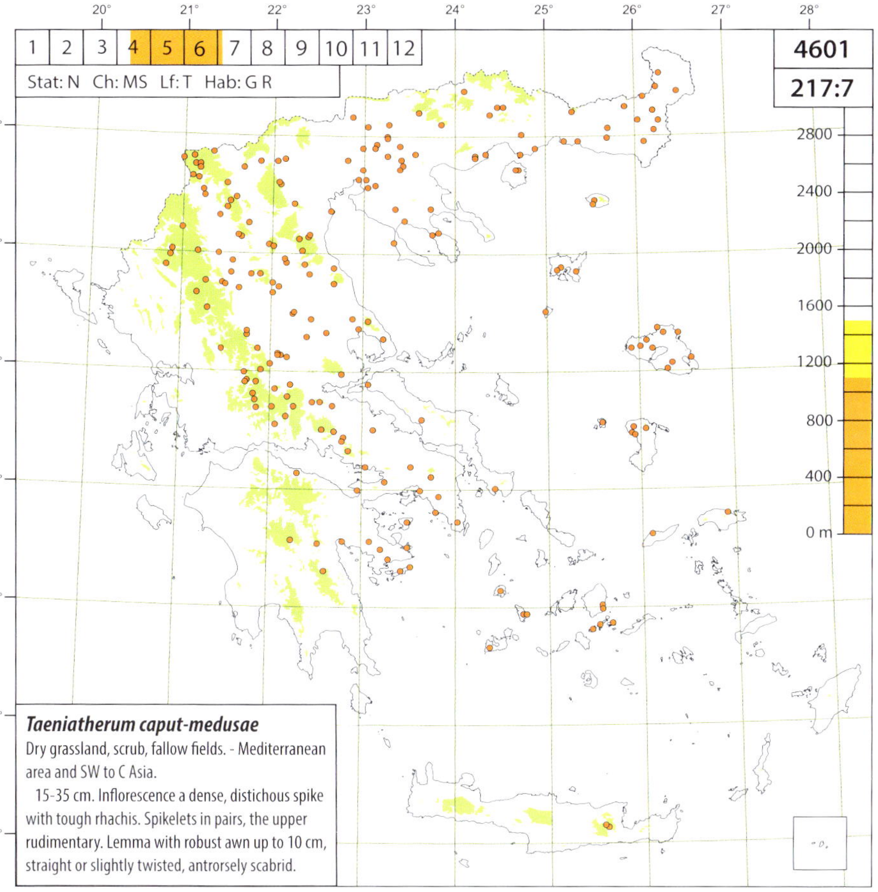

Taeniatherum caput-medusae
Dry grassland, scrub, fallow fields. - Mediterranean area and SW to C Asia.
15-35 cm. Inflorescence a dense, distichous spike with tough rhachis. Spikelets in pairs, the upper rudimentary. Lemma with robust awn up to 10 cm, straight or slightly twisted, antrorsely scabrid.

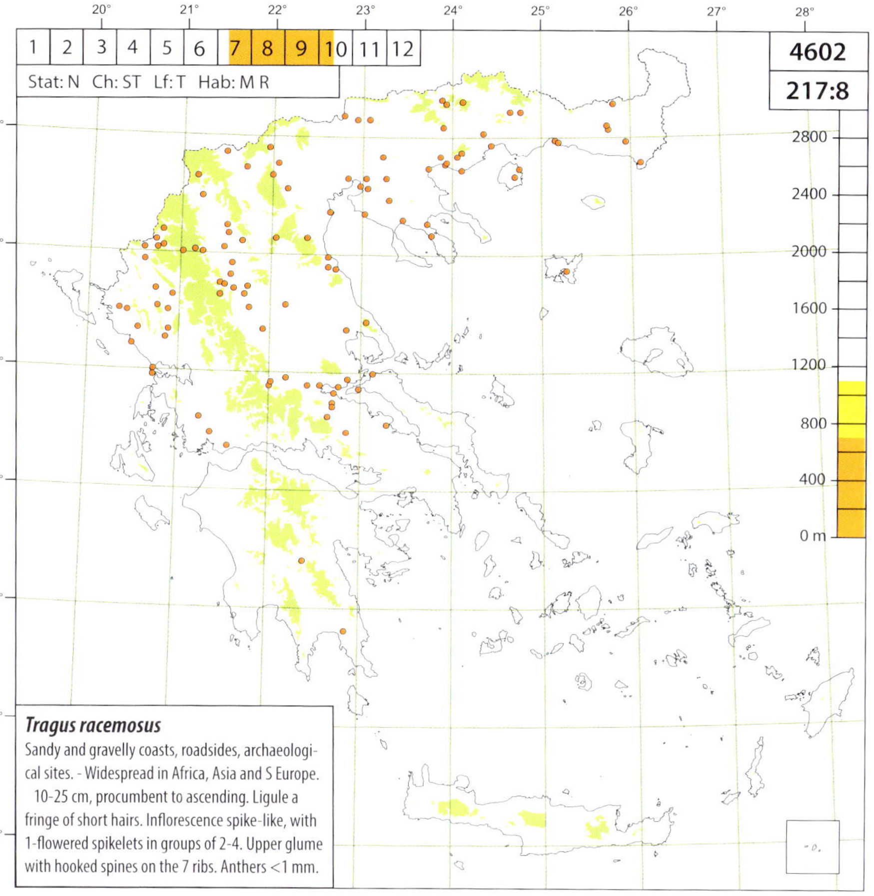

Tragus racemosus
Sandy and gravelly coasts, roadsides, archaeological sites. - Widespread in Africa, Asia and S Europe.
10-25 cm, procumbent to ascending. Ligule a fringe of short hairs. Inflorescence spike-like, with 1-flowered spikelets in groups of 2-4. Upper glume with hooked spines on the 7 ribs. Anthers <1 mm.

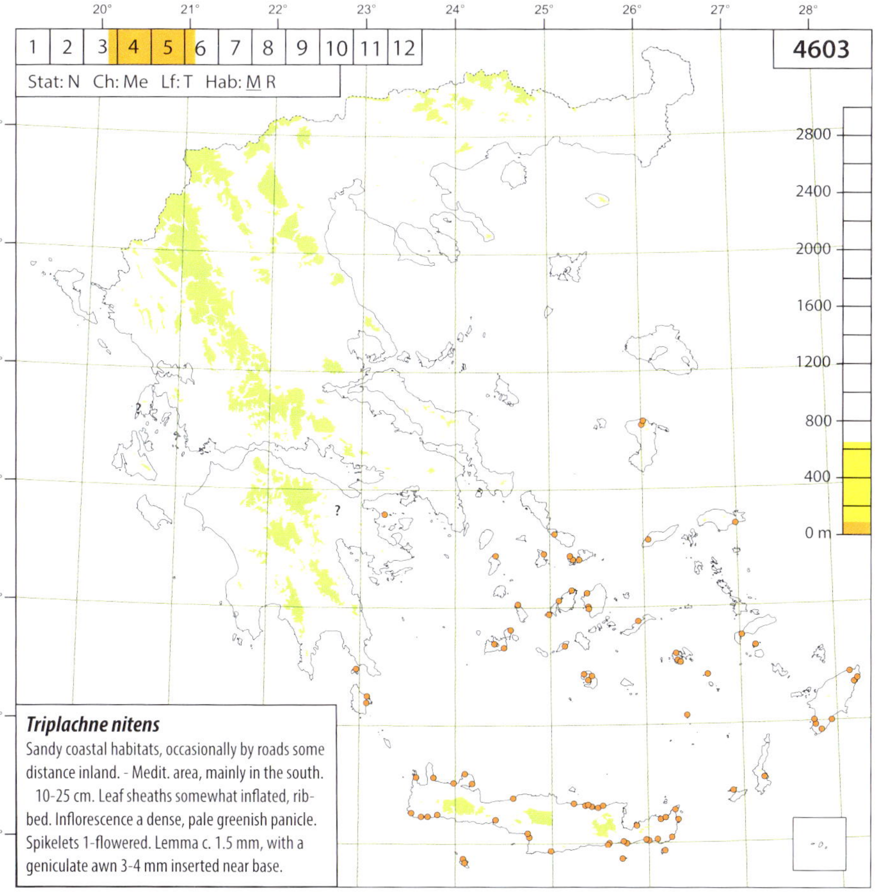

Triplachne nitens
Sandy coastal habitats, occasionally by roads some distance inland. - Medit. area, mainly in the south.
10-25 cm. Leaf sheaths somewhat inflated, ribbed. Inflorescence a dense, pale greenish panicle. Spikelets 1-flowered. Lemma c. 1.5 mm, with a geniculate awn 3-4 mm inserted near base.

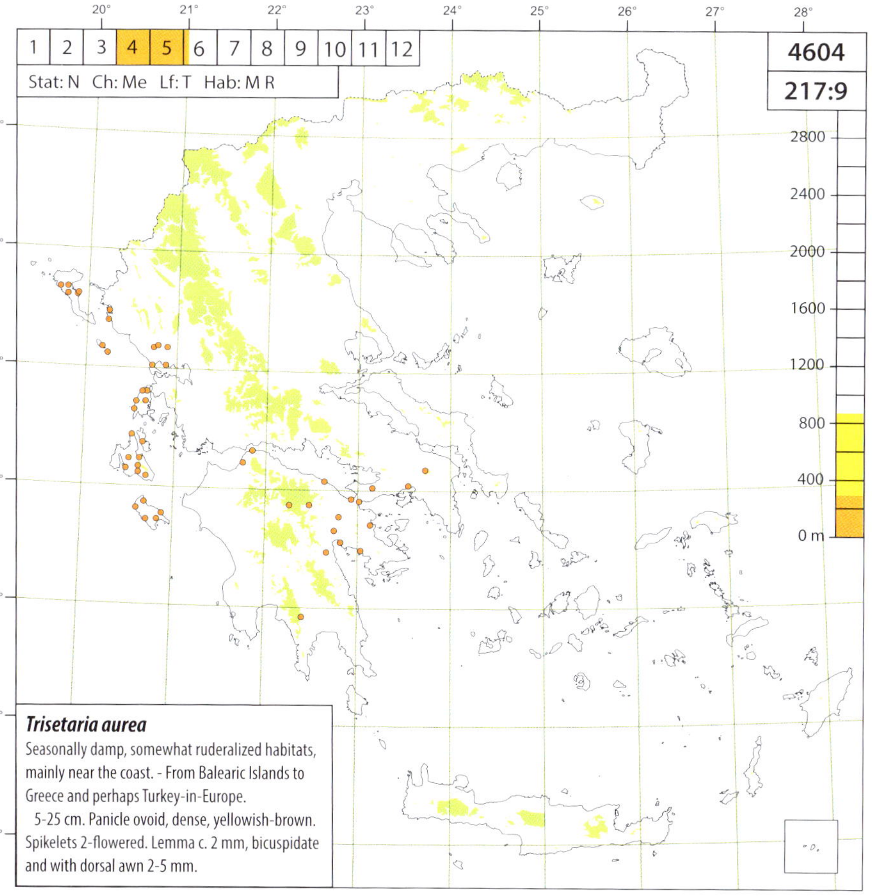

Trisetaria aurea
Seasonally damp, somewhat ruderalized habitats, mainly near the coast. - From Balearic Islands to Greece and perhaps Turkey-in-Europe.
5-25 cm. Panicle ovoid, dense, yellowish-brown. Spikelets 2-flowered. Lemma c. 2 mm, bicuspidate and with dorsal awn 2-5 mm.

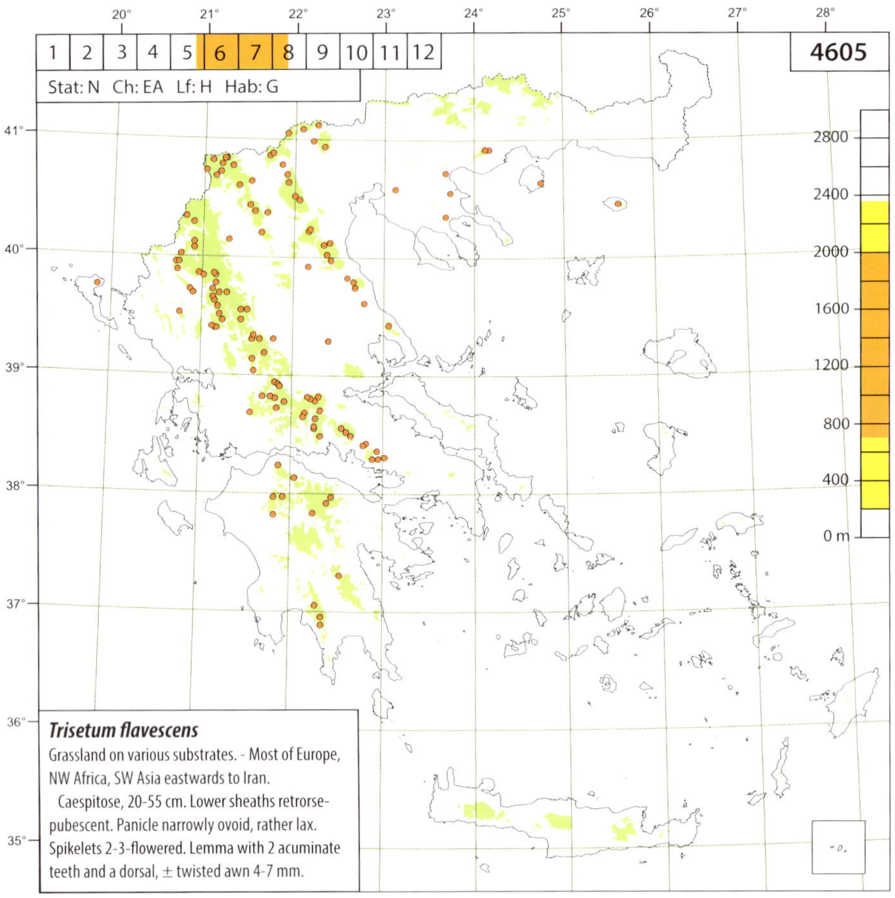

Trisetum flavescens

Grassland on various substrates. - Most of Europe, NW Africa, SW Asia eastwards to Iran.

Caespitose, 20-55 cm. Lower sheaths retrorse-pubescent. Panicle narrowly ovoid, rather lax. Spikelets 2-3-flowered. Lemma with 2 acuminate teeth and a dorsal, ± twisted awn 4-7 mm.

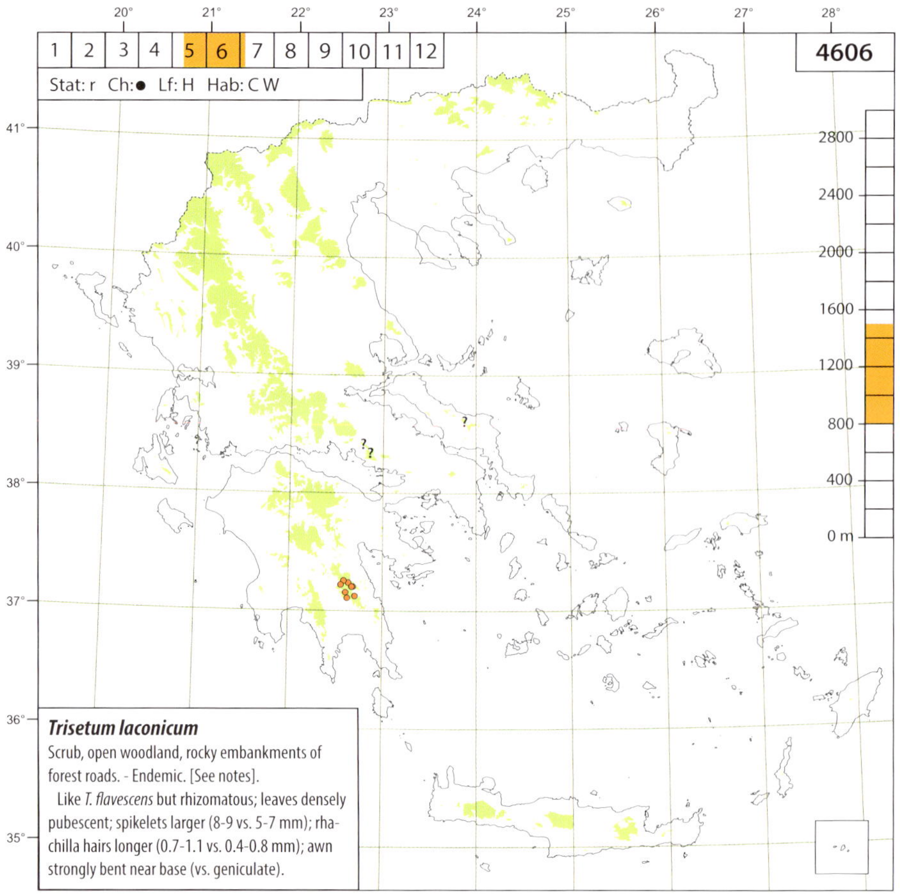

Trisetum laconicum

Scrub, open woodland, rocky embankments of forest roads. - Endemic. [See notes].

Like *T. flavescens* but rhizomatous; leaves densely pubescent; spikelets larger (8-9 vs. 5-7 mm); rhachilla hairs longer (0.7-1.1 vs. 0.4-0.8 mm); awn strongly bent near base (vs. geniculate).

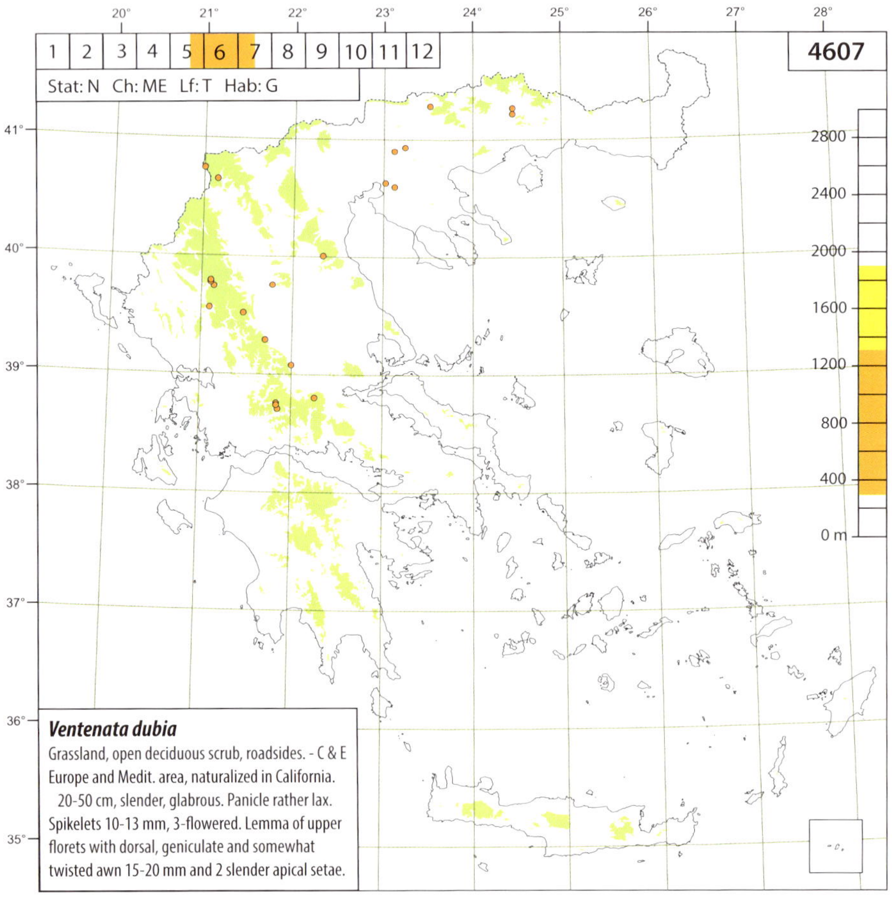

Ventenata dubia

Grassland, open deciduous scrub, roadsides. - C & E Europe and Medit. area, naturalized in California.

20-50 cm, slender, glabrous. Panicle rather lax. Spikelets 10-13 mm, 3-flowered. Lemma of upper florets with dorsal, geniculate and somewhat twisted awn 15-20 mm and 2 slender apical setae.

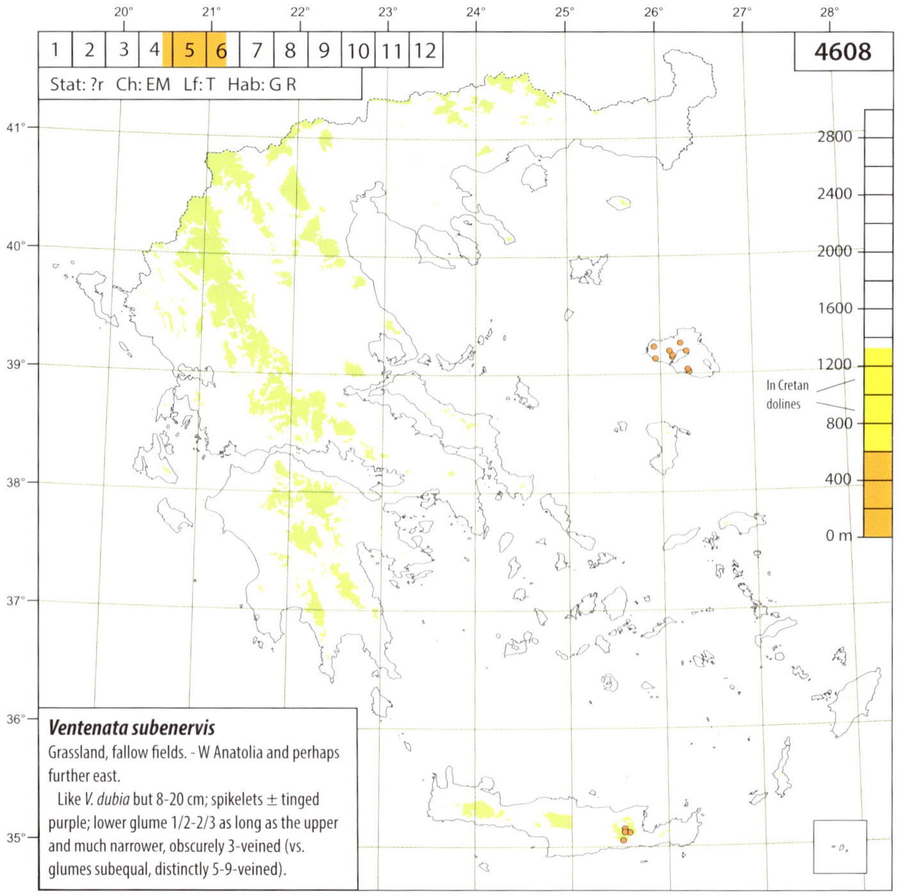

Ventenata subenervis

Grassland, fallow fields. - W Anatolia and perhaps further east.

Like *V. dubia* but 8-20 cm; spikelets ± tinged purple; lower glume 1/2-2/3 as long as the upper and much narrower, obscurely 3-veined (vs. glumes subequal, distinctly 5-9-veined).

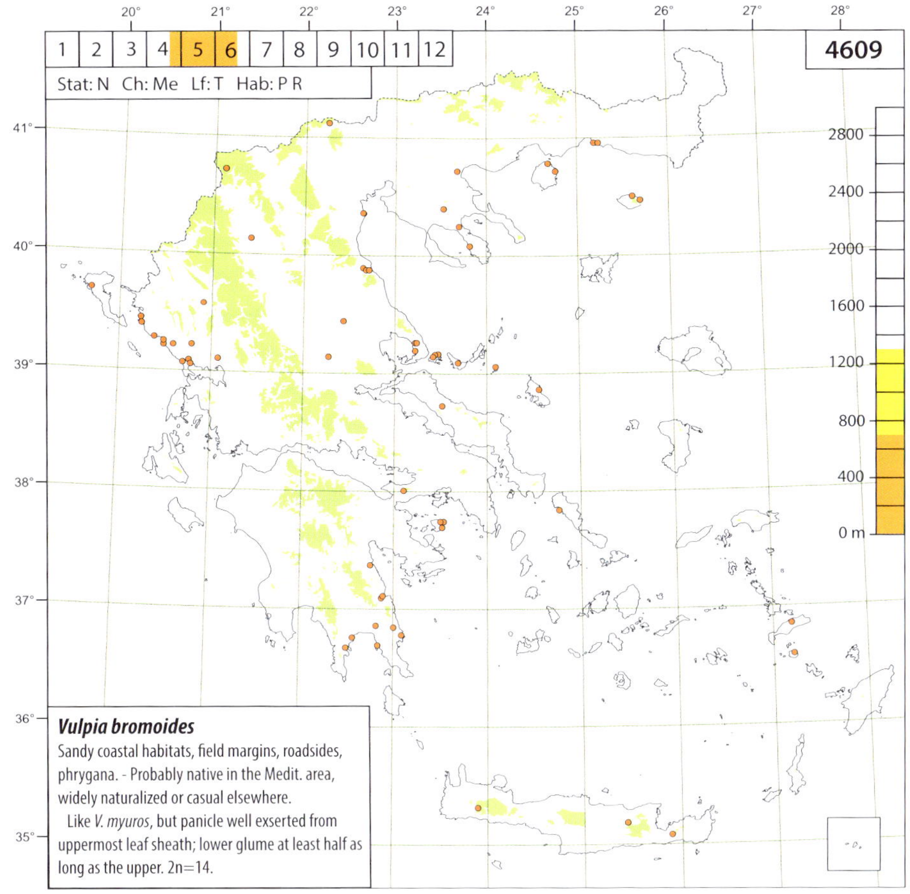

Vulpia bromoides
Sandy coastal habitats, field margins, roadsides, phrygana. - Probably native in the Medit. area, widely naturalized or casual elsewhere.
Like *V. myuros*, but panicle well exserted from uppermost leaf sheath; lower glume at least half as long as the upper. 2n=14.

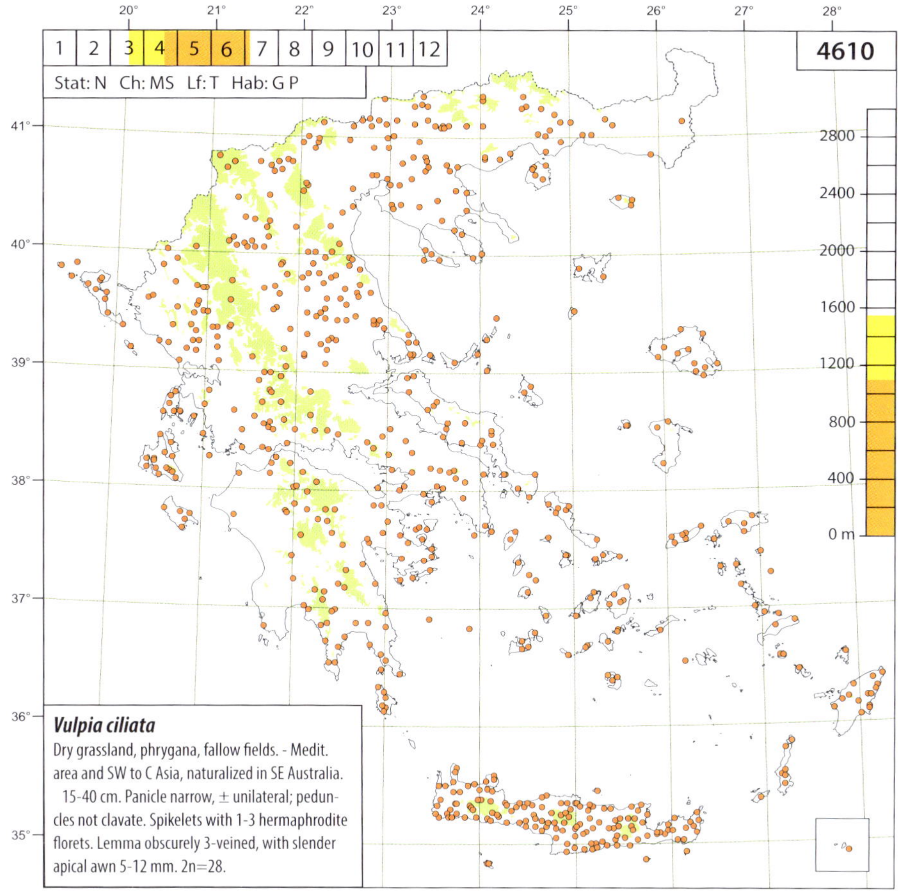

Vulpia ciliata
Dry grassland, phrygana, fallow fields. - Medit. area and SW to C Asia, naturalized in SE Australia.
15-40 cm. Panicle narrow, ± unilateral; peduncles not clavate. Spikelets with 1-3 hermaphrodite florets. Lemma obscurely 3-veined, with slender apical awn 5-12 mm. 2n=28.

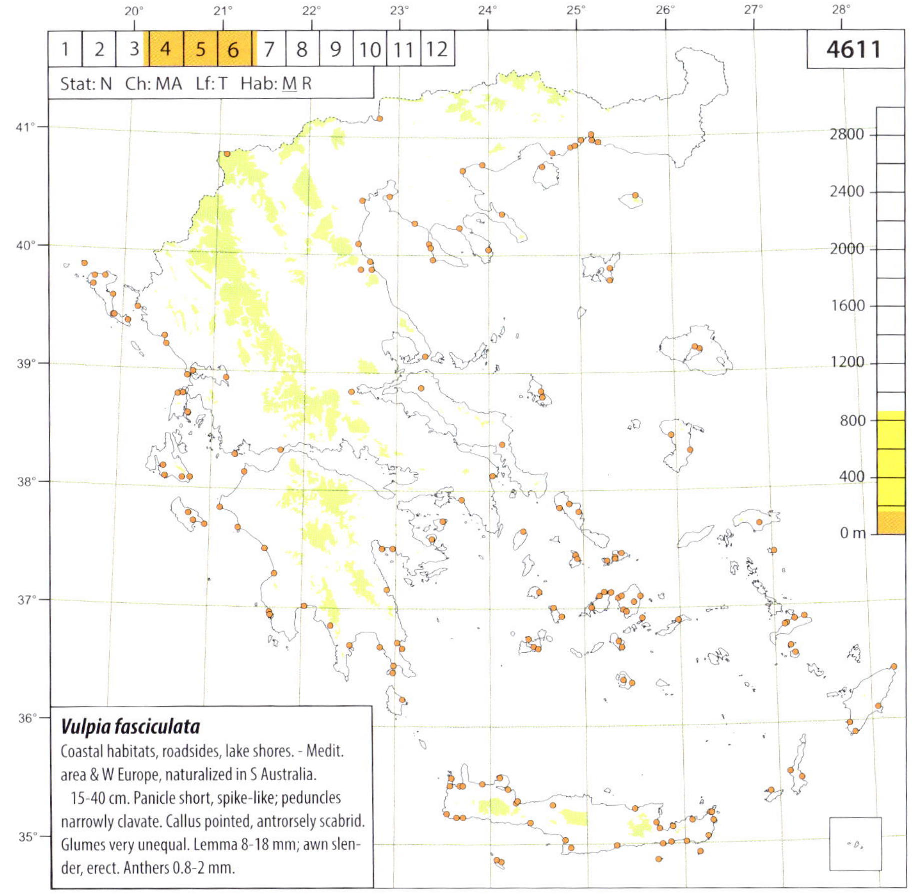

Vulpia fasciculata
Coastal habitats, roadsides, lake shores. - Medit. area & W Europe, naturalized in S Australia.
15-40 cm. Panicle short, spike-like; peduncles narrowly clavate. Callus pointed, antrorsely scabrid. Glumes very unequal. Lemma 8-18 mm; awn slender, erect. Anthers 0.8-2 mm.

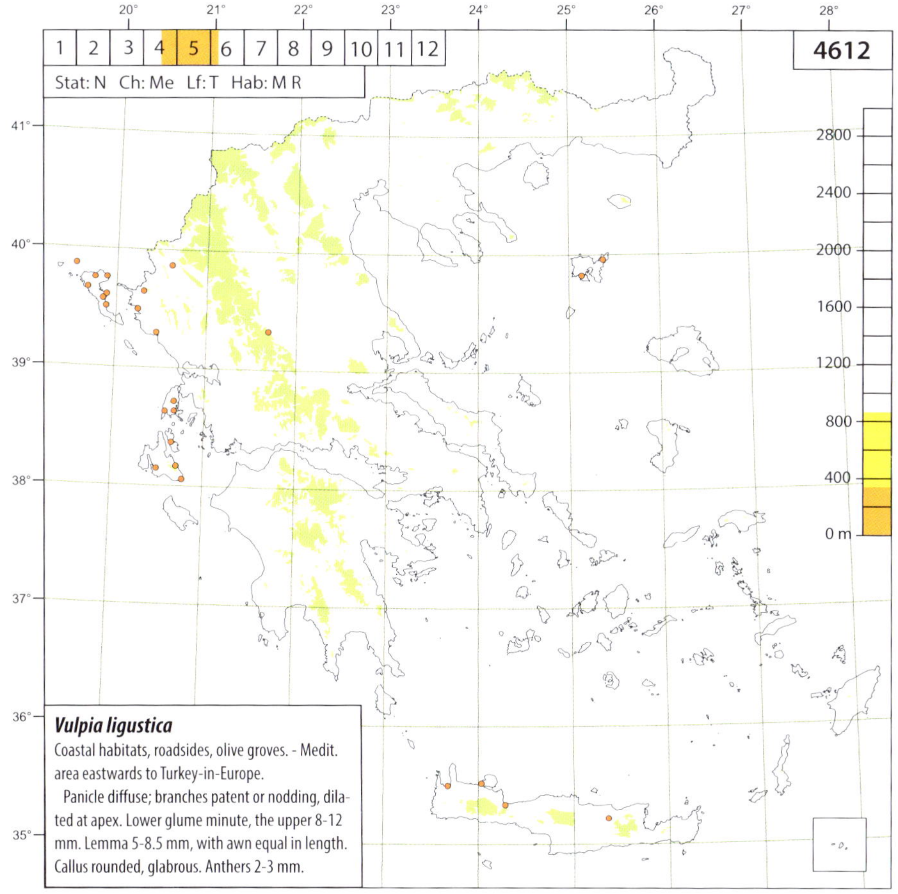

Vulpia ligustica
Coastal habitats, roadsides, olive groves. - Medit. area eastwards to Turkey-in-Europe.
Panicle diffuse; branches patent or nodding, dilated at apex. Lower glume minute, the upper 8-12 mm. Lemma 5-8.5 mm, with awn equal in length. Callus rounded, glabrous. Anthers 2-3 mm.

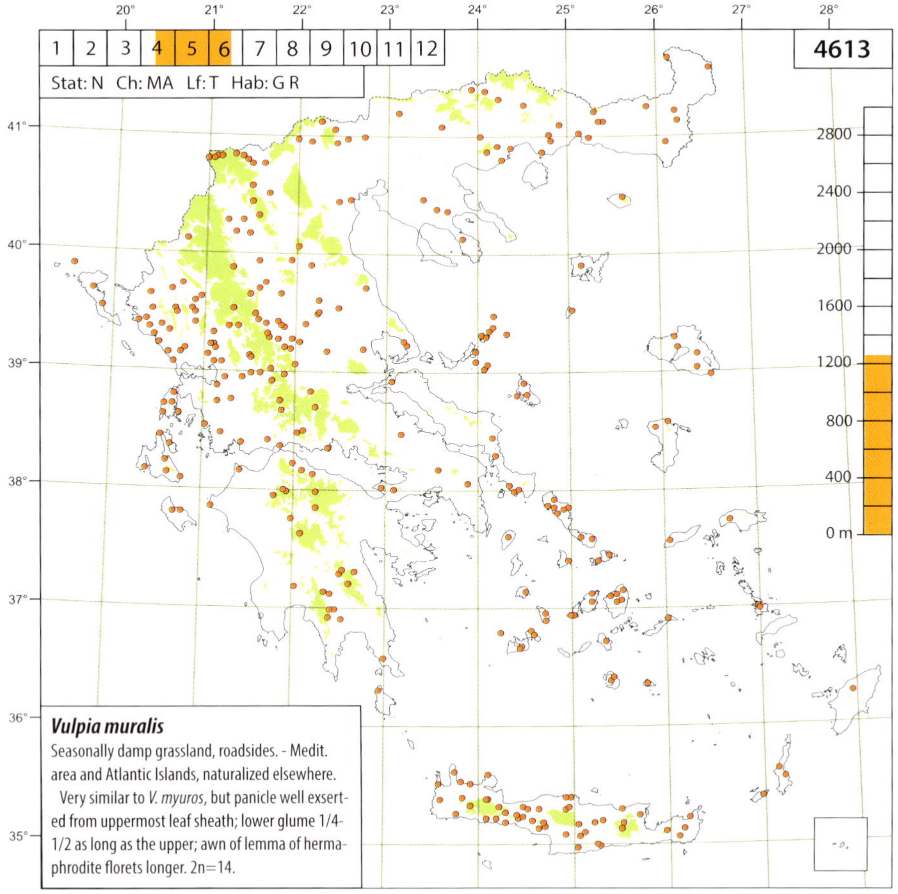

Vulpia muralis
Seasonally damp grassland, roadsides. - Medit. area and Atlantic Islands, naturalized elsewhere.
Very similar to *V. myuros*, but panicle well exserted from uppermost leaf sheath; lower glume 1/4-1/2 as long as the upper; awn of lemma of hermaphrodite florets longer. 2n=14.

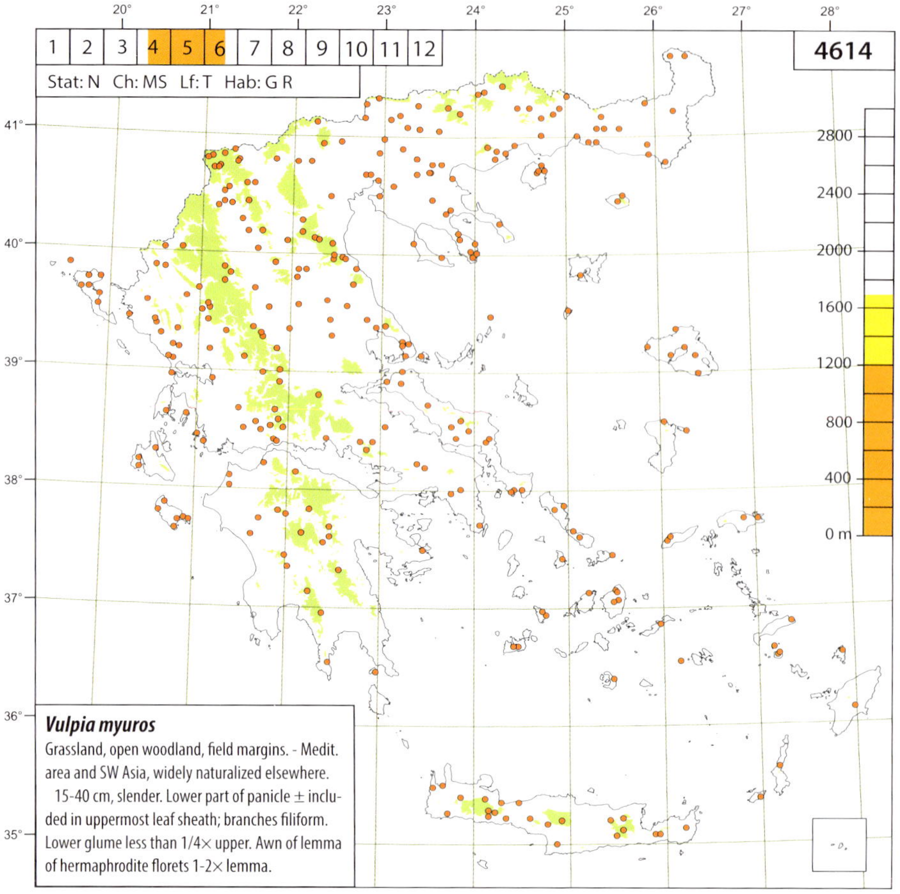

Vulpia myuros
Grassland, open woodland, field margins. - Medit. area and SW Asia, widely naturalized elsewhere.
15-40 cm, slender. Lower part of panicle ± included in uppermost leaf sheath; branches filiform. Lower glume less than 1/4× upper. Awn of lemma of hermaphrodite florets 1-2× lemma.

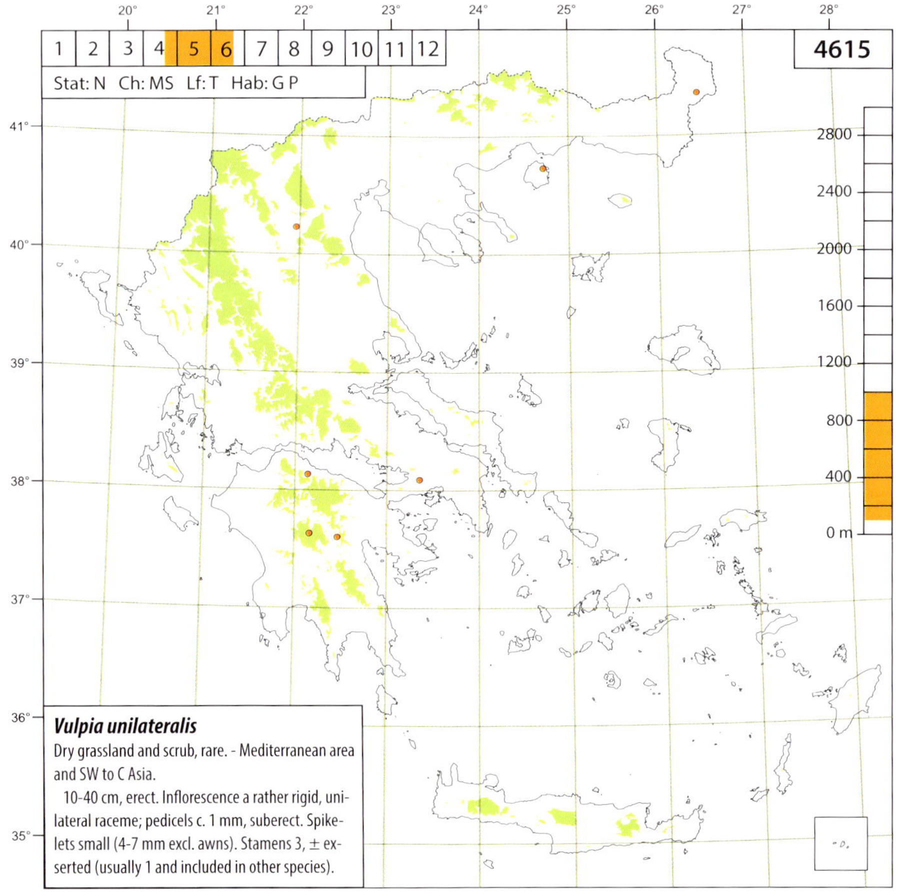

Vulpia unilateralis
Dry grassland and scrub, rare. - Mediterranean area and SW to C Asia.
10-40 cm, erect. Inflorescence a rather rigid, unilateral raceme; pedicels c. 1 mm, suberect. Spikelets small (4-7 mm excl. awns). Stamens 3, ± exserted (usually 1 and included in other species).

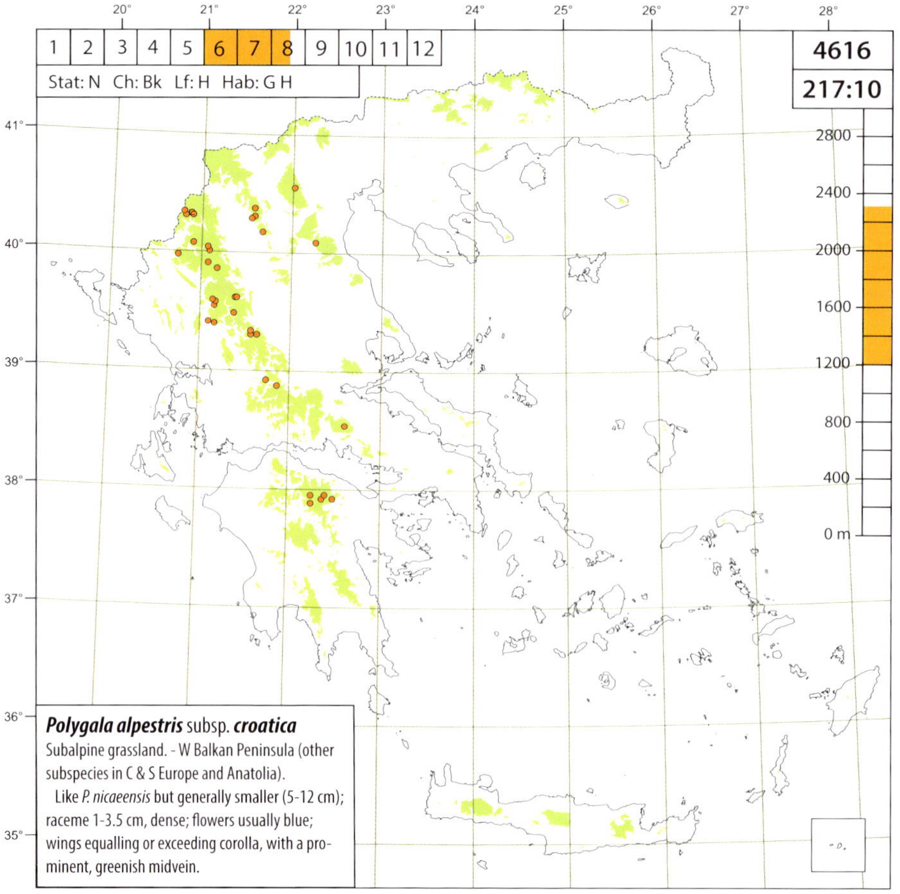

Polygala alpestris* subsp. *croatica
Subalpine grassland. - W Balkan Peninsula (other subspecies in C & S Europe and Anatolia).
Like *P. nicaeensis* but generally smaller (5-12 cm); raceme 1-3.5 cm, dense; flowers usually blue; wings equalling or exceeding corolla, with a prominent, greenish midvein.

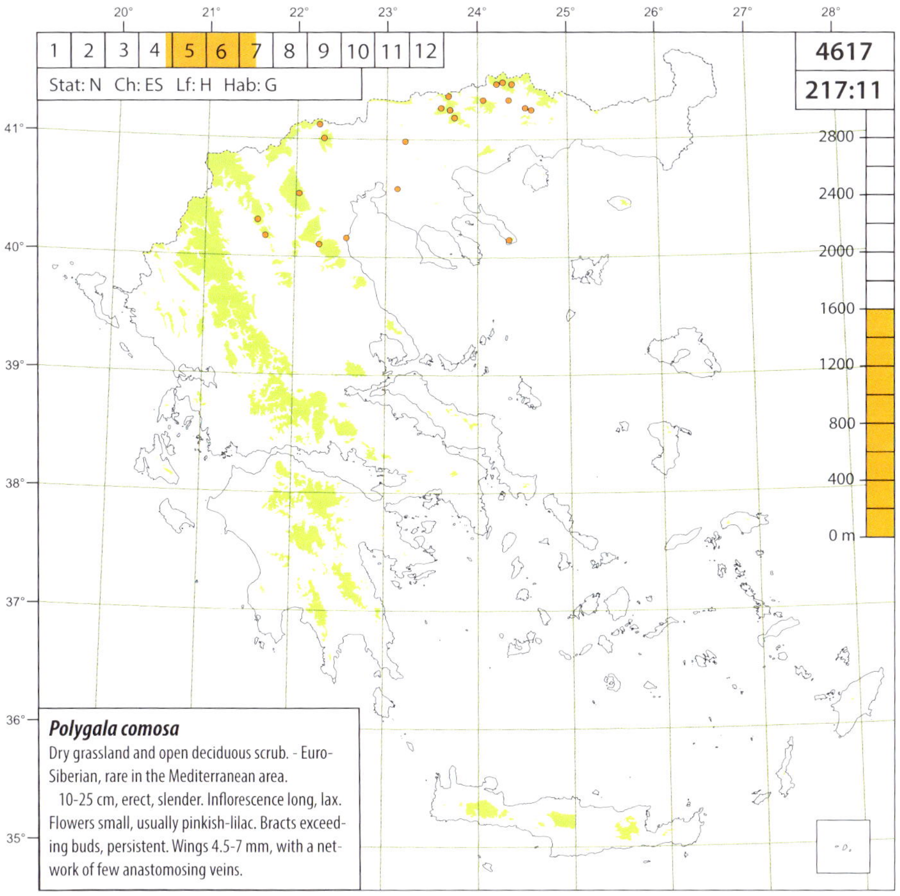

Polygala comosa

Dry grassland and open deciduous scrub. - Euro-Siberian, rare in the Mediterranean area.

10-25 cm, erect, slender. Inflorescence long, lax. Flowers small, usually pinkish-lilac. Bracts exceeding buds, persistent. Wings 4.5-7 mm, with a network of few anastomosing veins.

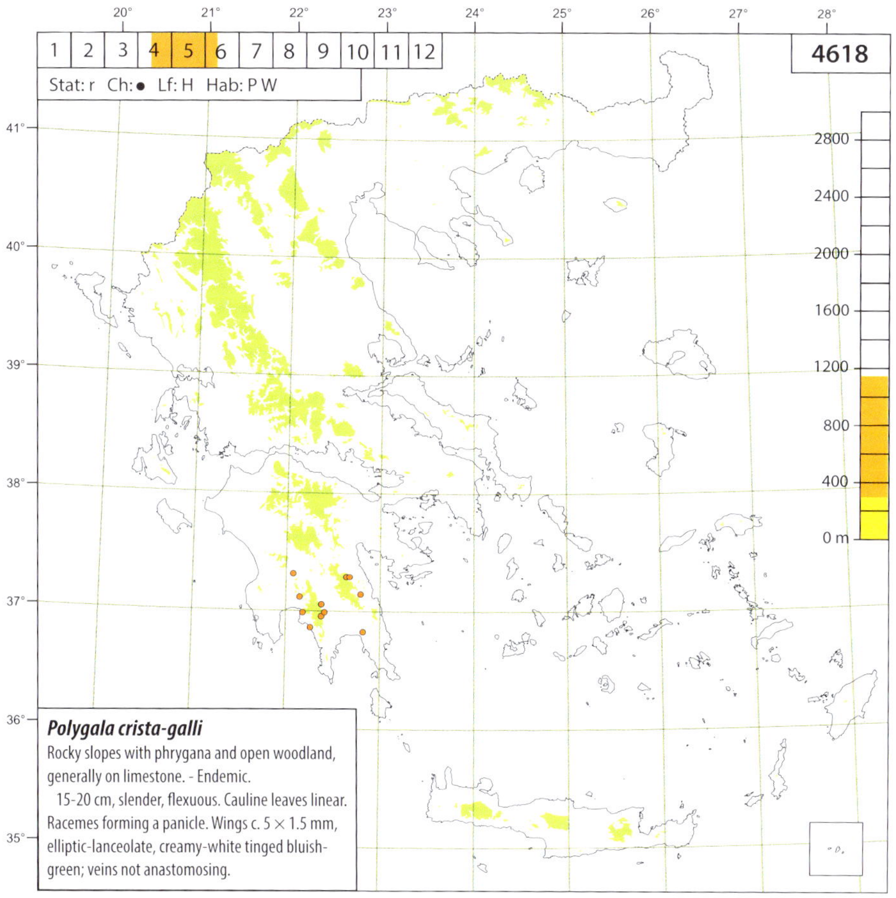

Polygala crista-galli

Rocky slopes with phrygana and open woodland, generally on limestone. - Endemic.

15-20 cm, slender, flexuous. Cauline leaves linear. Racemes forming a panicle. Wings c. 5 × 1.5 mm, elliptic-lanceolate, creamy-white tinged bluish-green; veins not anastomosing.

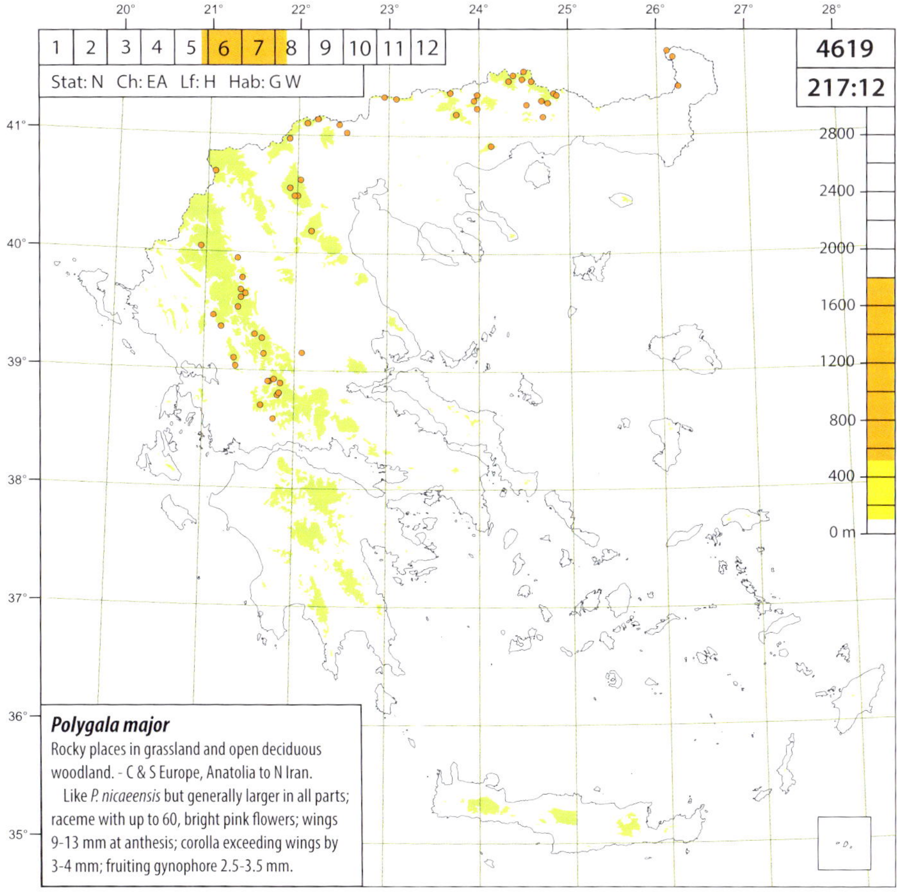

Polygala major

Rocky places in grassland and open deciduous woodland. - C & S Europe, Anatolia to N Iran.

Like *P. nicaeensis* but generally larger in all parts; raceme with up to 60, bright pink flowers; wings 9-13 mm at anthesis; corolla exceeding wings by 3-4 mm; fruiting gynophore 2.5-3.5 mm.

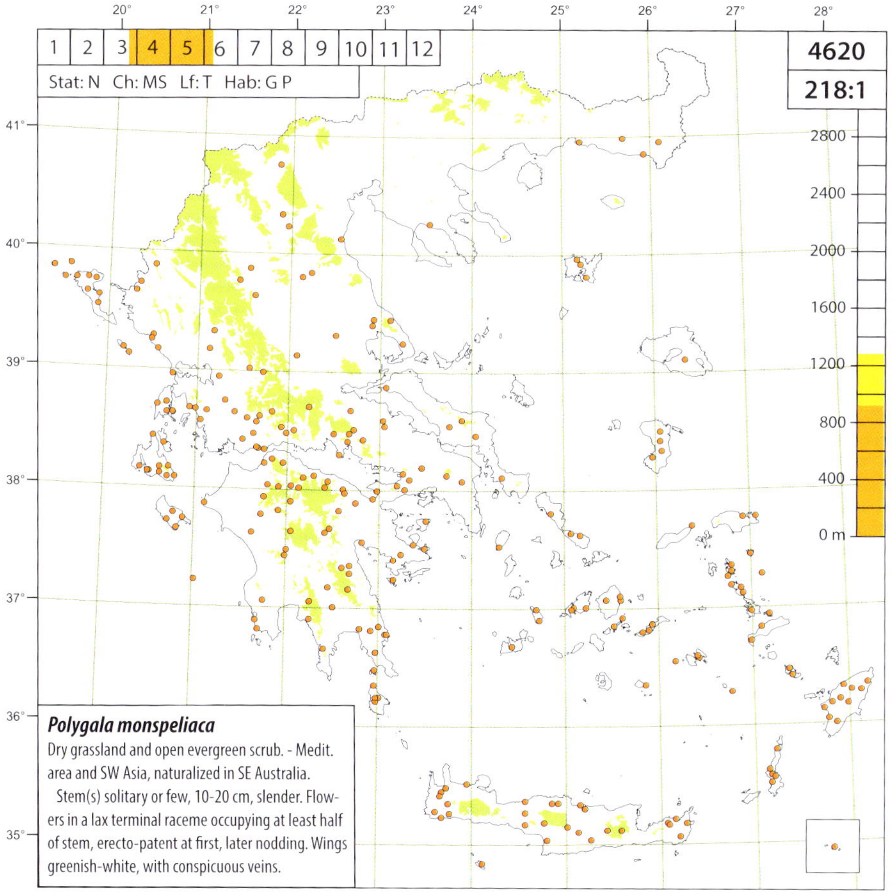

Polygala monspeliaca

Dry grassland and open evergreen scrub. - Medit. area and SW Asia, naturalized in SE Australia.

Stem(s) solitary or few, 10-20 cm, slender. Flowers in a lax terminal raceme occupying at least half of stem, erecto-patent at first, later nodding. Wings greenish-white, with conspicuous veins.

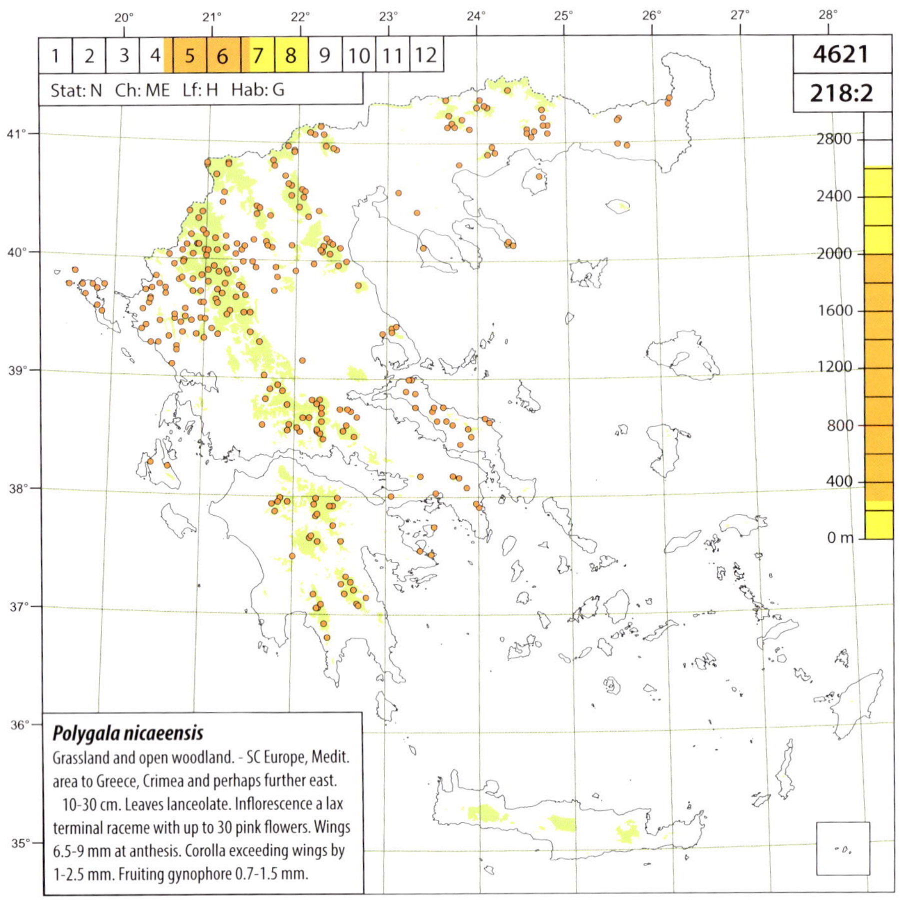

Polygala nicaeensis

Grassland and open woodland. - SC Europe, Medit. area to Greece, Crimea and perhaps further east.

10-30 cm. Leaves lanceolate. Inflorescence a lax terminal raceme with up to 30 pink flowers. Wings 6.5-9 mm at anthesis. Corolla exceeding wings by 1-2.5 mm. Fruiting gynophore 0.7-1.5 mm.

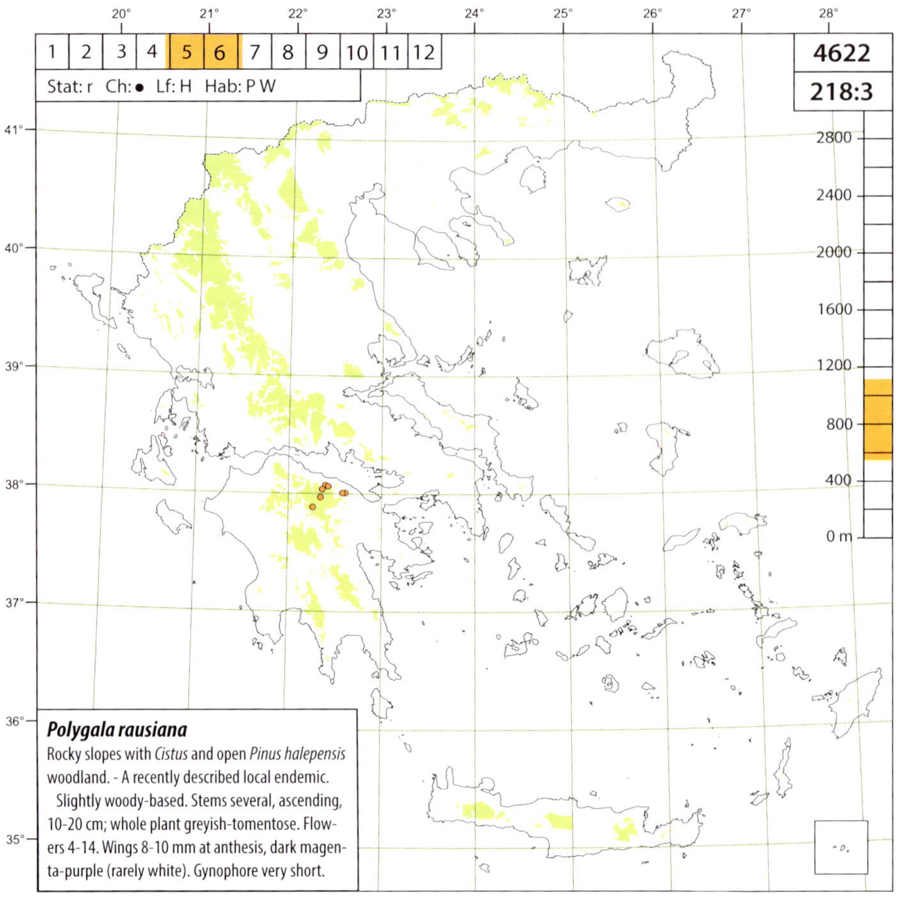

Polygala rausiana

Rocky slopes with *Cistus* and open *Pinus halepensis* woodland. - A recently described local endemic.

Slightly woody-based. Stems several, ascending, 10-20 cm; whole plant greyish-tomentose. Flowers 4-14. Wings 8-10 mm at anthesis, dark magenta-purple (rarely white). Gynophore very short.

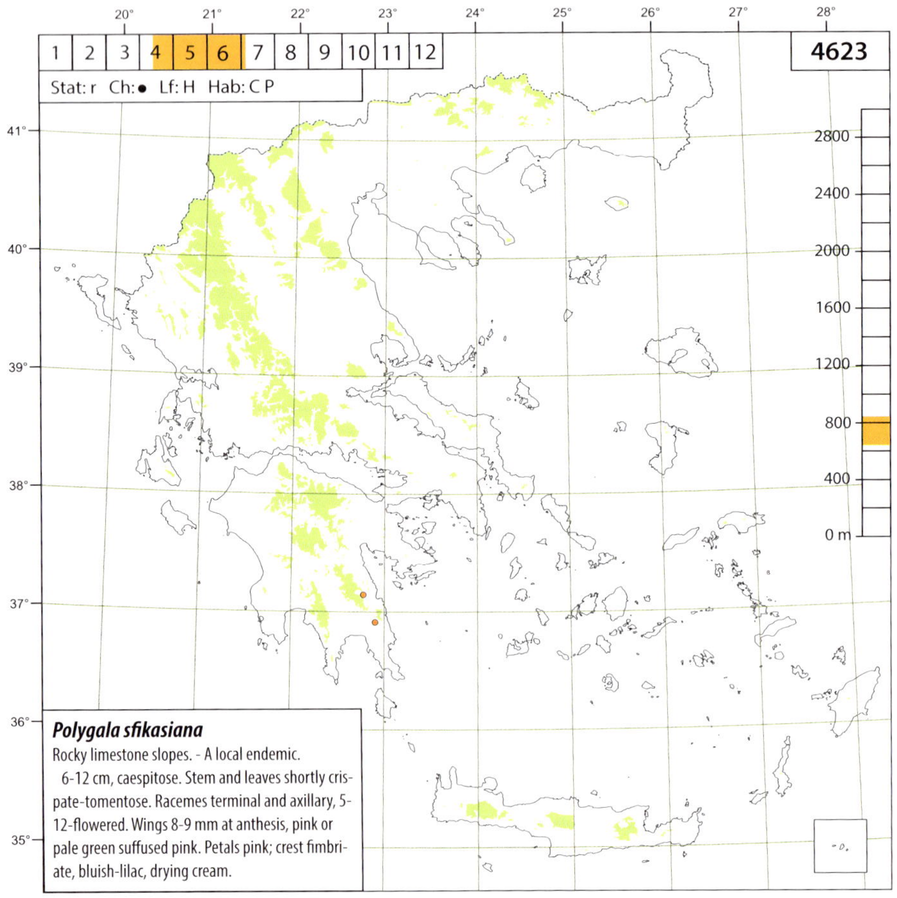

Polygala sfikasiana

Rocky limestone slopes. - A local endemic.

6-12 cm, caespitose. Stem and leaves shortly crispate-tomentose. Racemes terminal and axillary, 5-12-flowered. Wings 8-9 mm at anthesis, pink or pale green suffused pink. Petals pink; crest fimbriate, bluish-lilac, drying cream.

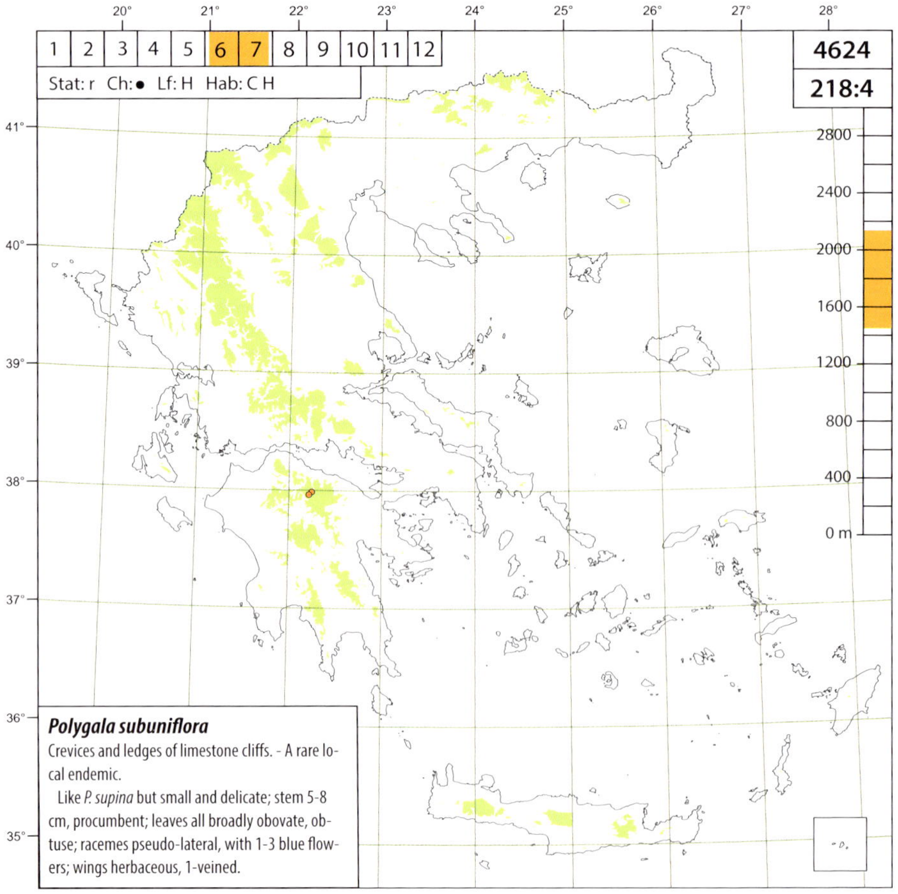

Polygala subuniflora

Crevices and ledges of limestone cliffs. - A rare local endemic.

Like *P. supina* but small and delicate; stem 5-8 cm, procumbent; leaves all broadly obovate, obtuse; racemes pseudo-lateral, with 1-3 blue flowers; wings herbaceous, 1-veined.

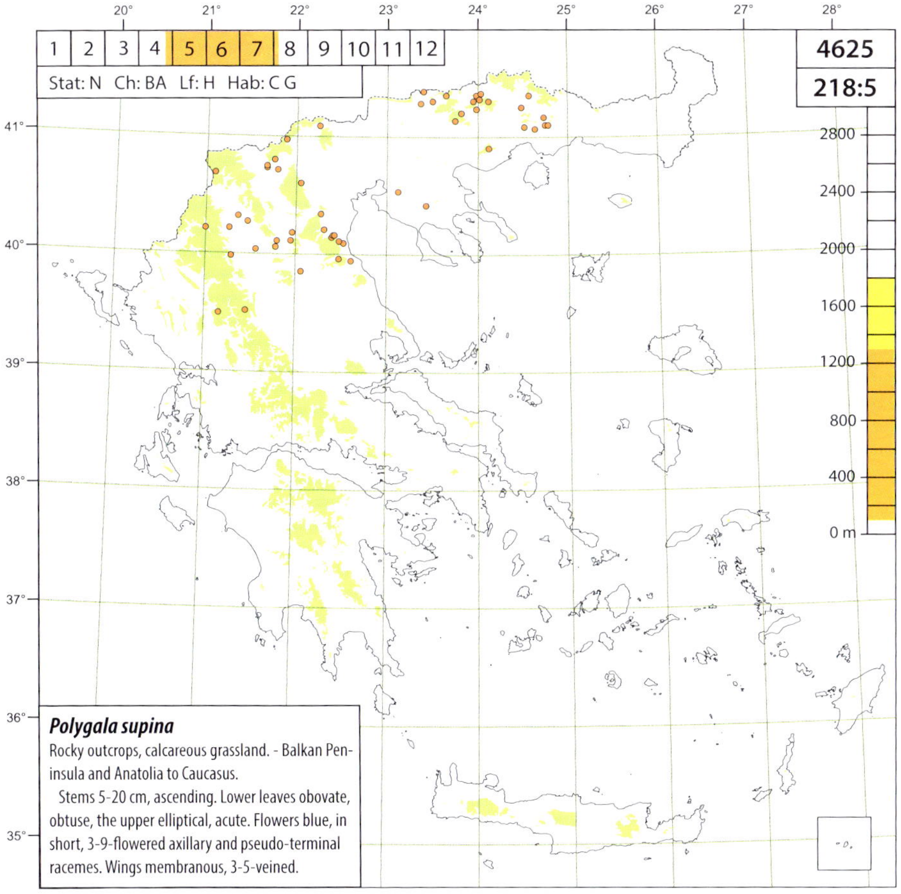

Polygala supina
Rocky outcrops, calcareous grassland. - Balkan Peninsula and Anatolia to Caucasus.
Stems 5-20 cm, ascending. Lower leaves obovate, obtuse, the upper elliptical, acute. Flowers blue, in short, 3-9-flowered axillary and pseudo-terminal racemes. Wings membranous, 3-5-veined.

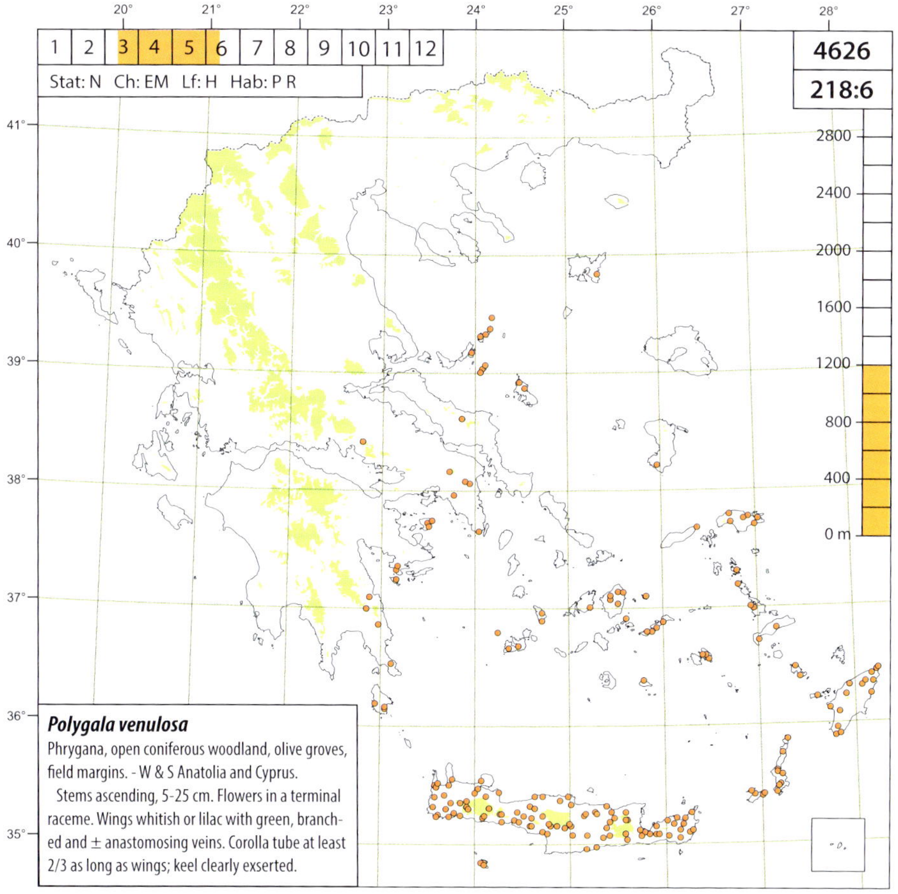

Polygala venulosa
Phrygana, open coniferous woodland, olive groves, field margins. - W & S Anatolia and Cyprus.
Stems ascending, 5-25 cm. Flowers in a terminal raceme. Wings whitish or lilac with green, branched and ± anastomosing veins. Corolla tube at least 2/3 as long as wings; keel clearly exserted.

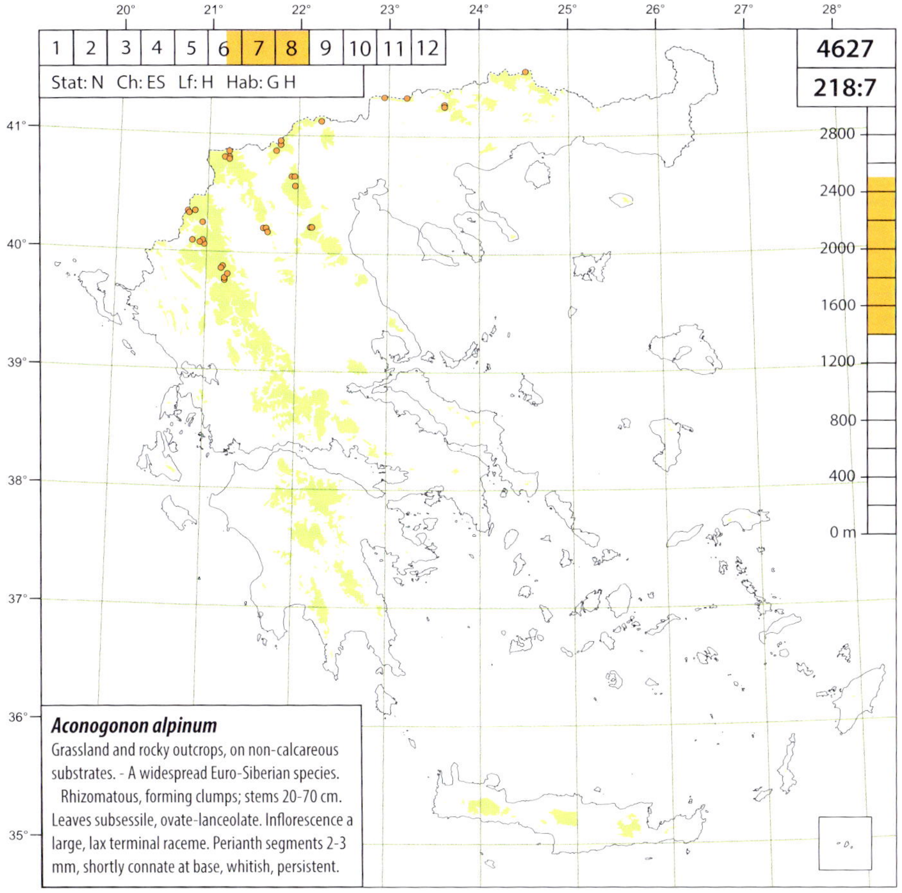

Aconogonon alpinum
Grassland and rocky outcrops, on non-calcareous substrates. - A widespread Euro-Siberian species.
Rhizomatous, forming clumps; stems 20-70 cm. Leaves subsessile, ovate-lanceolate. Inflorescence a large, lax terminal raceme. Perianth segments 2-3 mm, shortly connate at base, whitish, persistent.

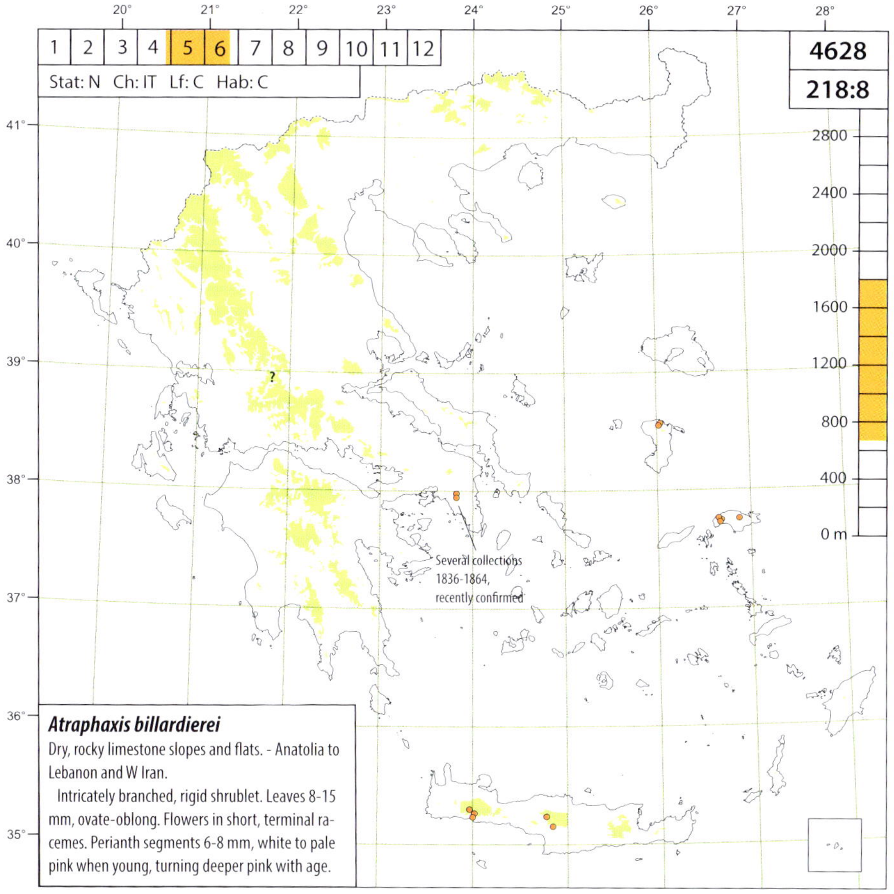

Atraphaxis billardierei
Dry, rocky limestone slopes and flats. - Anatolia to Lebanon and W Iran.
Intricately branched, rigid shrublet. Leaves 8-15 mm, ovate-oblong. Flowers in short, terminal racemes. Perianth segments 6-8 mm, white to pale pink when young, turning deeper pink with age.

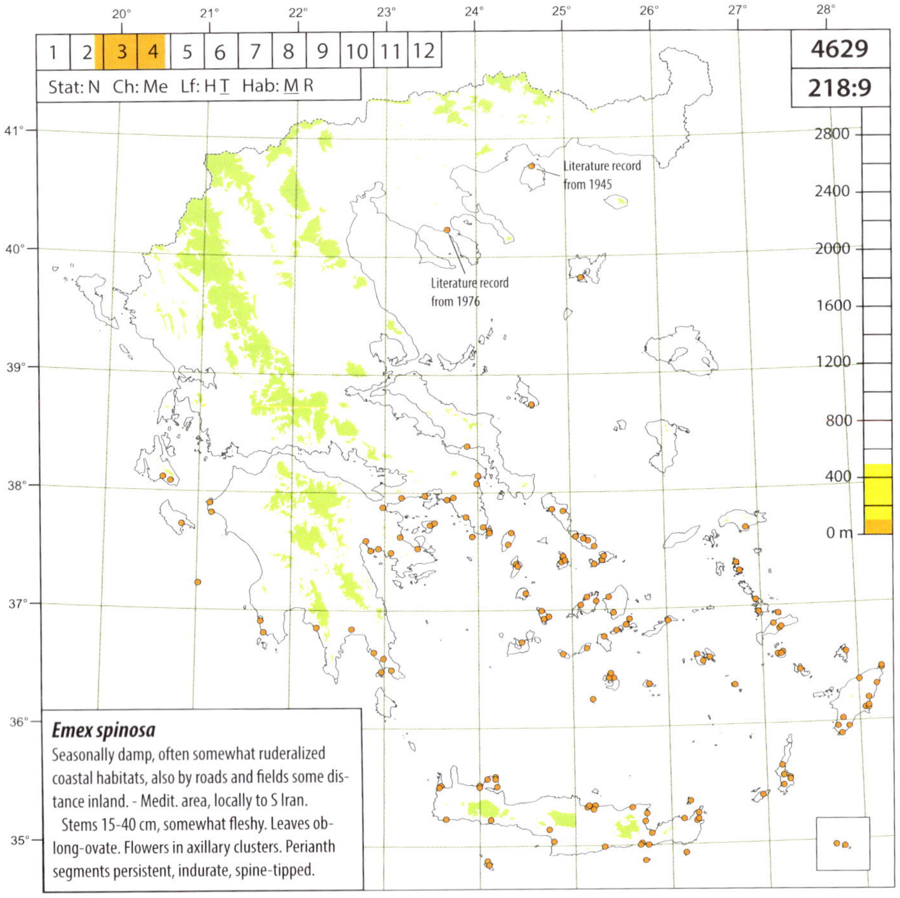

Emex spinosa
Seasonally damp, often somewhat ruderalized coastal habitats, also by roads and fields some distance inland. - Medit. area, locally to S Iran.
Stems 15-40 cm, somewhat fleshy. Leaves oblong-ovate. Flowers in axillary clusters. Perianth segments persistent, indurate, spine-tipped.

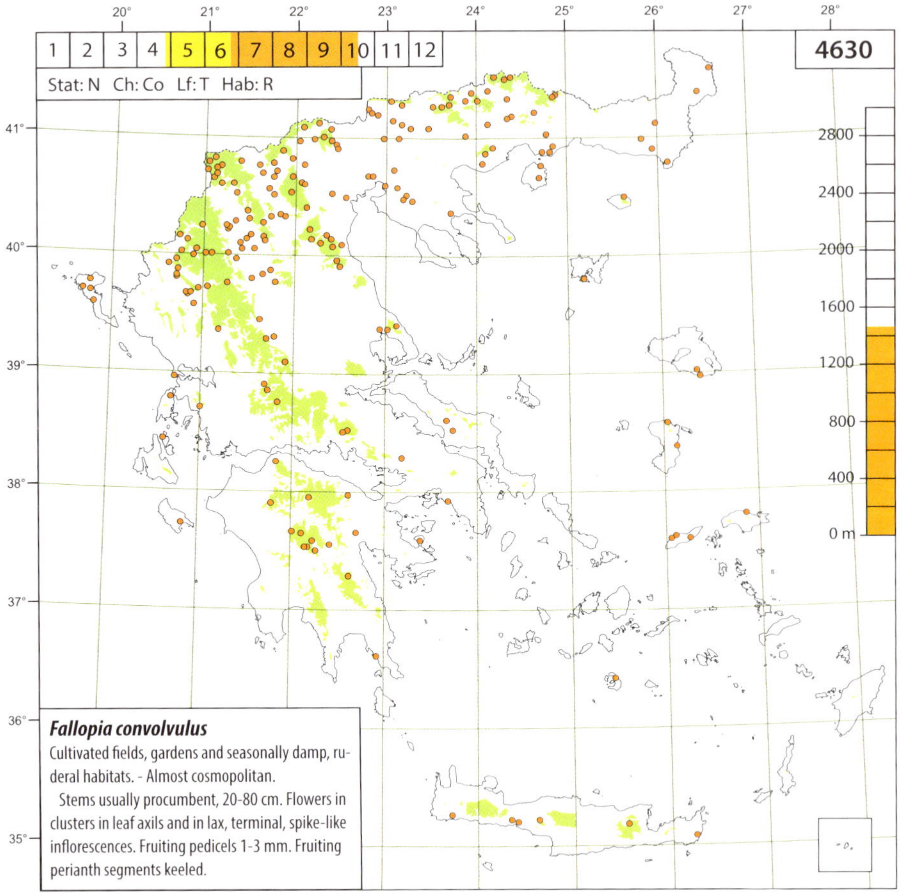

Fallopia convolvulus
Cultivated fields, gardens and seasonally damp, ruderal habitats. - Almost cosmopolitan.
Stems usually procumbent, 20-80 cm. Flowers in clusters in leaf axils and in lax, terminal, spike-like inflorescences. Fruiting pedicels 1-3 mm. Fruiting perianth segments keeled.

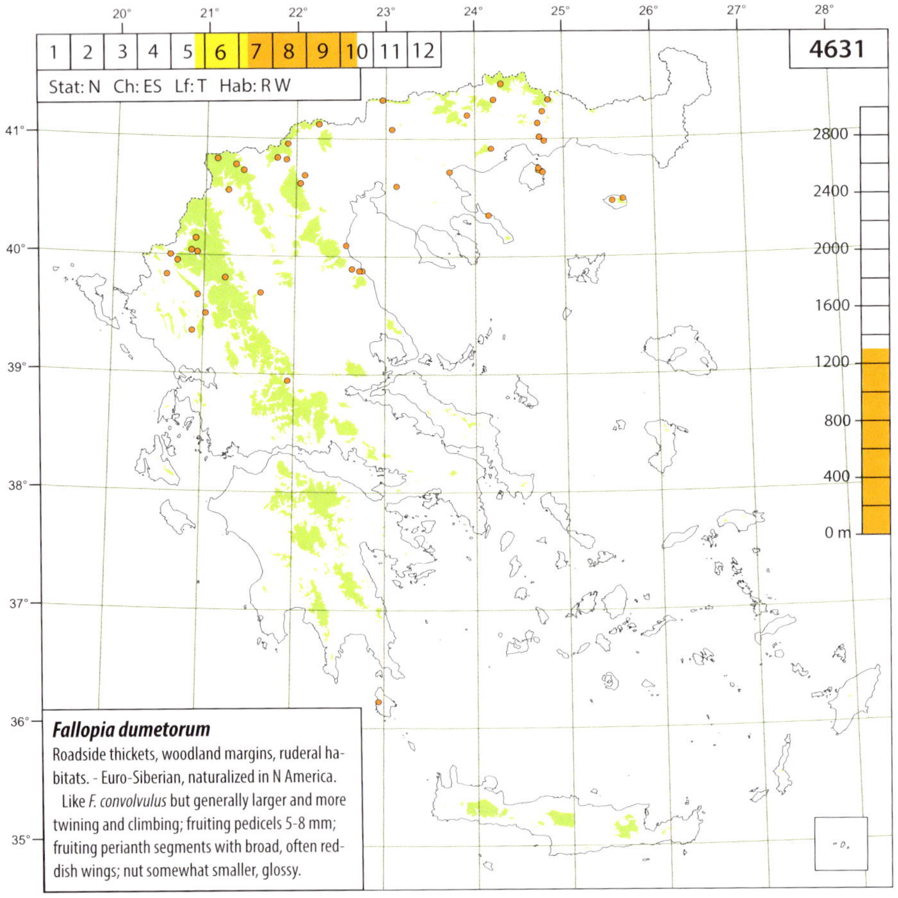

Fallopia dumetorum
Roadside thickets, woodland margins, ruderal habitats. - Euro-Siberian, naturalized in N America.
Like *F. convolvulus* but generally larger and more twining and climbing; fruiting pedicels 5-8 mm; fruiting perianth segments with broad, often reddish wings; nut somewhat smaller, glossy.

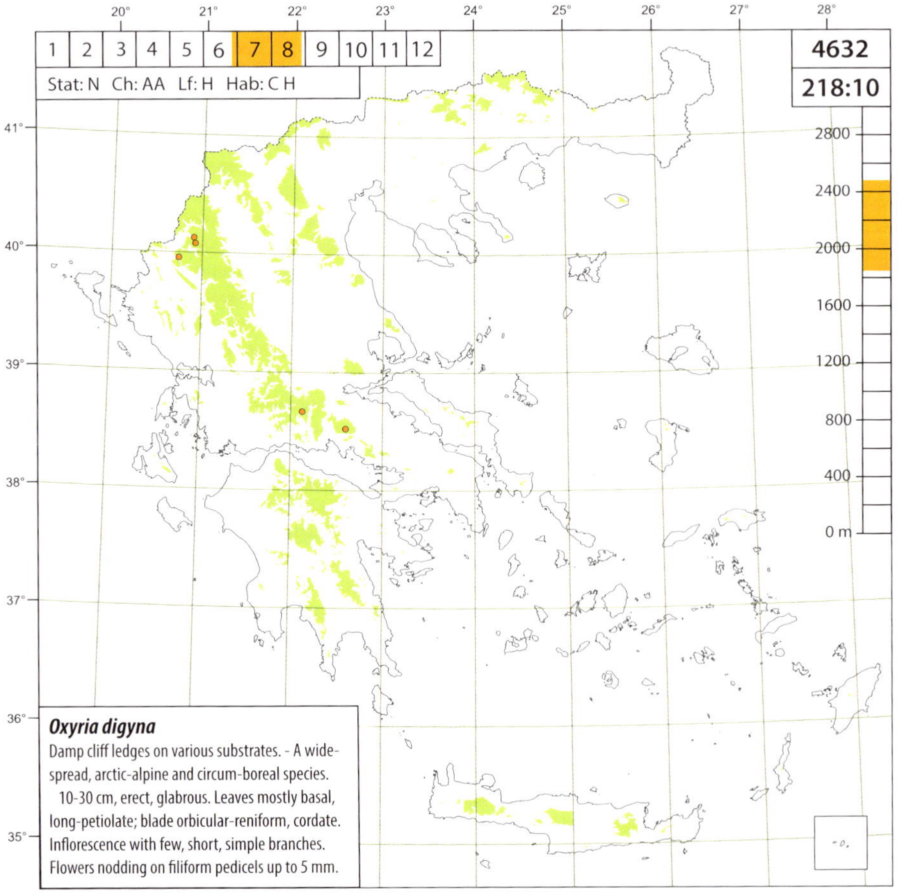

Oxyria digyna
Damp cliff ledges on various substrates. - A widespread, arctic-alpine and circum-boreal species.
10-30 cm, erect, glabrous. Leaves mostly basal, long-petiolate; blade orbicular-reniform, cordate. Inflorescence with few, short, simple branches. Flowers nodding on filiform pedicels up to 5 mm.

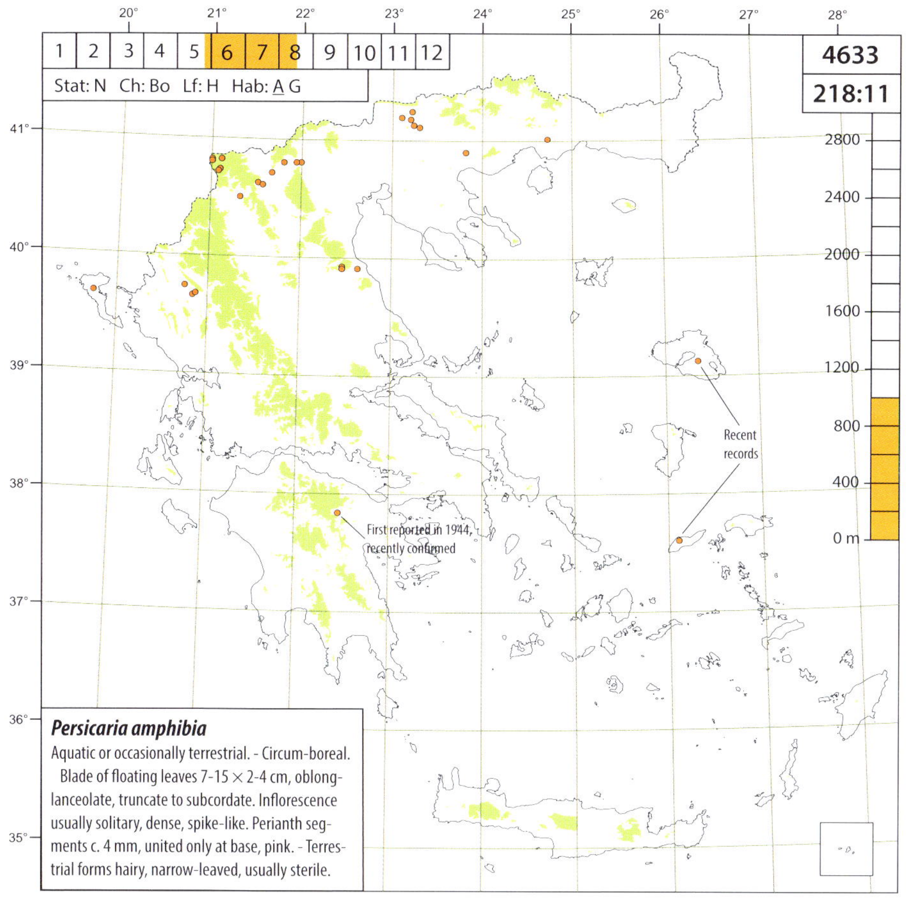

Persicaria amphibia

Aquatic or occasionally terrestrial. - Circum-boreal.

Blade of floating leaves 7-15 × 2-4 cm, oblong-lanceolate, truncate to subcordate. Inflorescence usually solitary, dense, spike-like. Perianth segments c. 4 mm, united only at base, pink. - Terrestrial forms hairy, narrow-leaved, usually sterile.

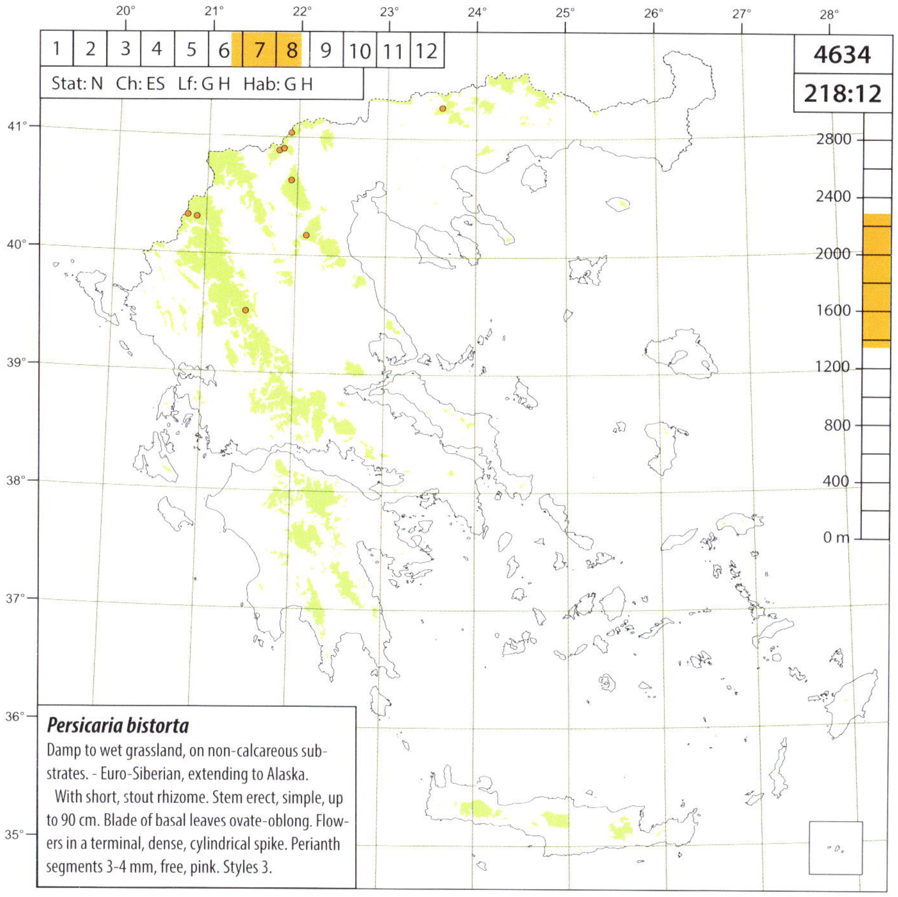

Persicaria bistorta

Damp to wet grassland, on non-calcareous substrates. - Euro-Siberian, extending to Alaska.

With short, stout rhizome. Stem erect, simple, up to 90 cm. Blade of basal leaves ovate-oblong. Flowers in a terminal, dense, cylindrical spike. Perianth segments 3-4 mm, free, pink. Styles 3.

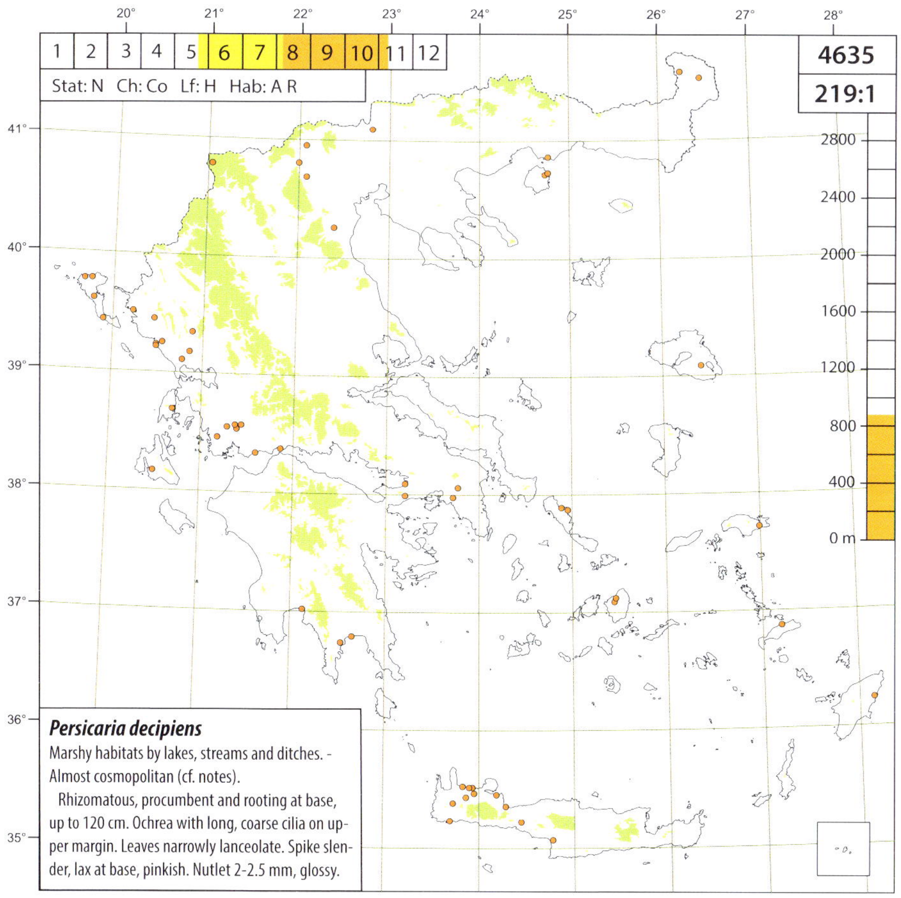

Persicaria decipiens

Marshy habitats by lakes, streams and ditches. - Almost cosmopolitan (cf. notes).

Rhizomatous, procumbent and rooting at base, up to 120 cm. Ochrea with long, coarse cilia on upper margin. Leaves narrowly lanceolate. Spike slender, lax at base, pinkish. Nutlet 2-2.5 mm, glossy.

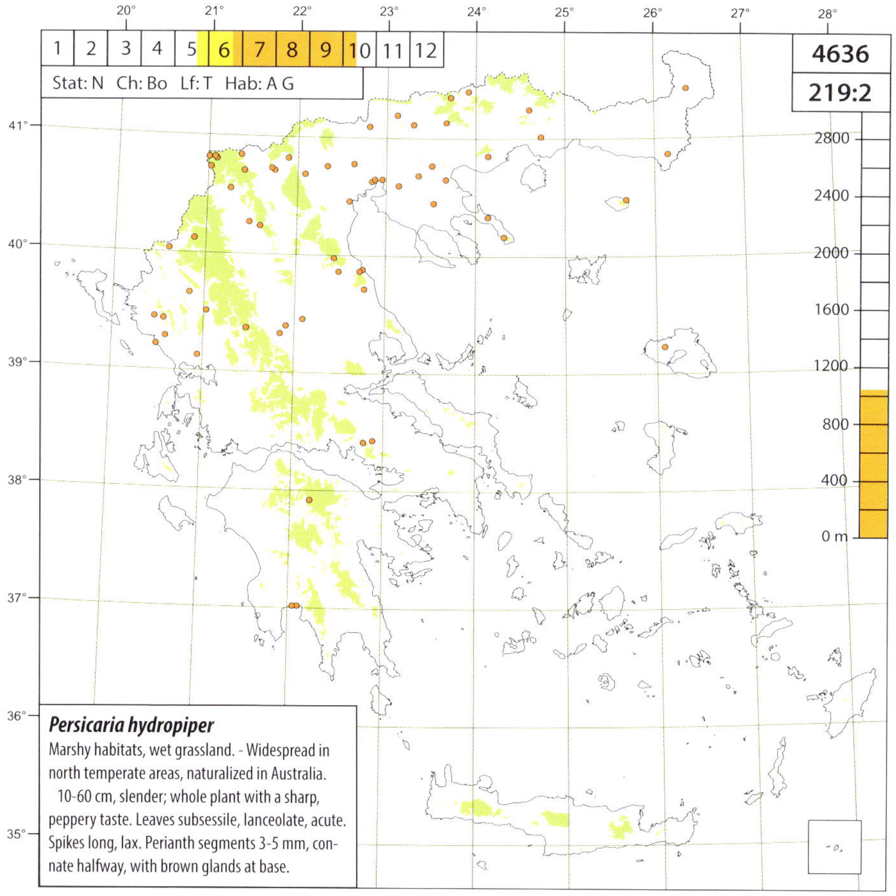

Persicaria hydropiper

Marshy habitats, wet grassland. - Widespread in north temperate areas, naturalized in Australia.

10-60 cm, slender; whole plant with a sharp, peppery taste. Leaves subsessile, lanceolate, acute. Spikes long, lax. Perianth segments 3-5 mm, connate halfway, with brown glands at base.

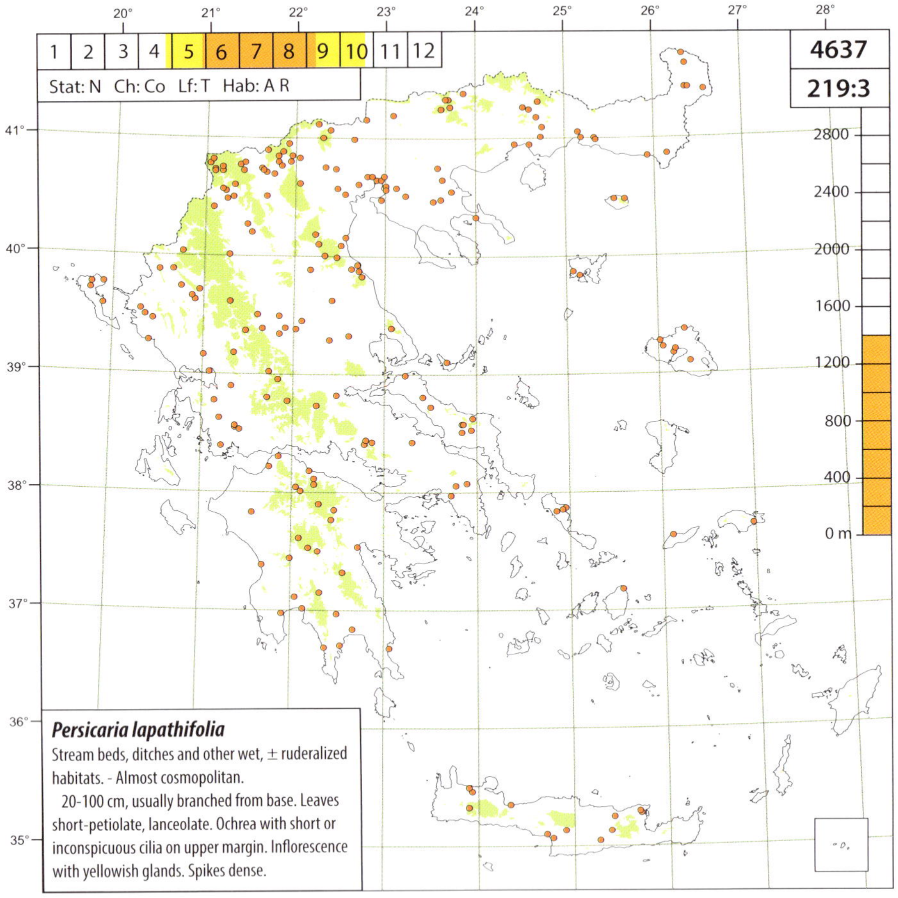

Persicaria lapathifolia

Stream beds, ditches and other wet, ± ruderalized habitats. - Almost cosmopolitan.

20-100 cm, usually branched from base. Leaves short-petiolate, lanceolate. Ochrea with short or inconspicuous cilia on upper margin. Inflorescence with yellowish glands. Spikes dense.

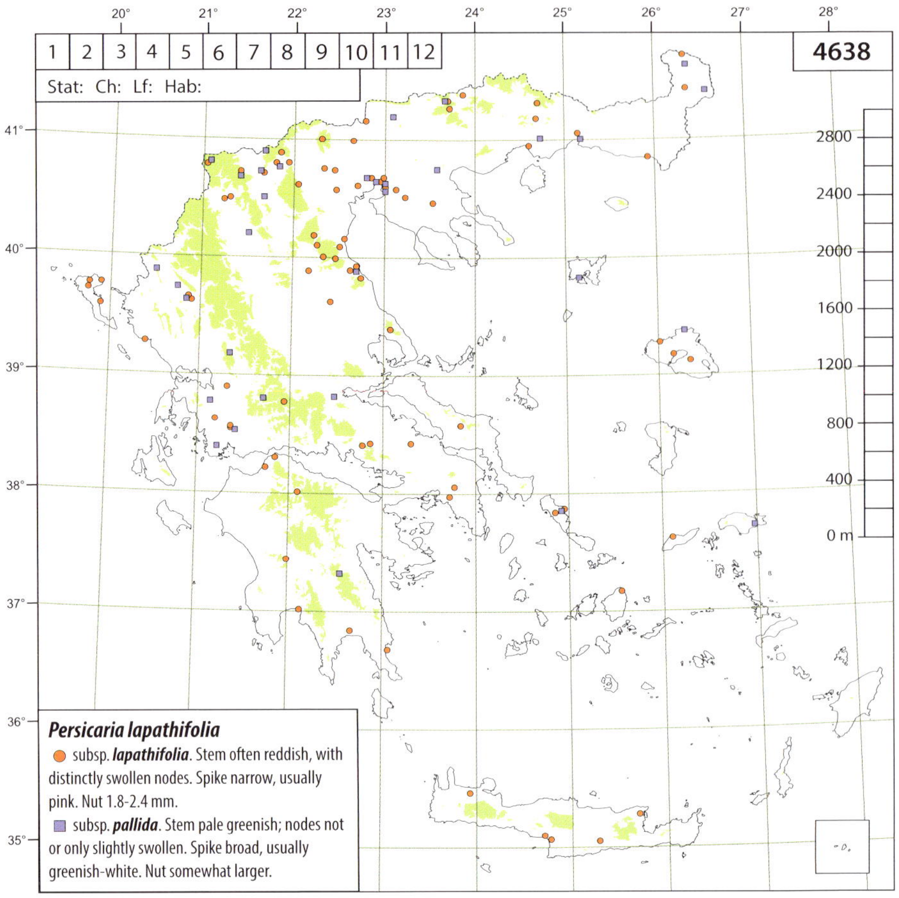

Persicaria lapathifolia

● subsp. ***lapathifolia***. Stem often reddish, with distinctly swollen nodes. Spike narrow, usually pink. Nut 1.8-2.4 mm.

■ subsp. ***pallida***. Stem pale greenish; nodes not or only slightly swollen. Spike broad, usually greenish-white. Nut somewhat larger.

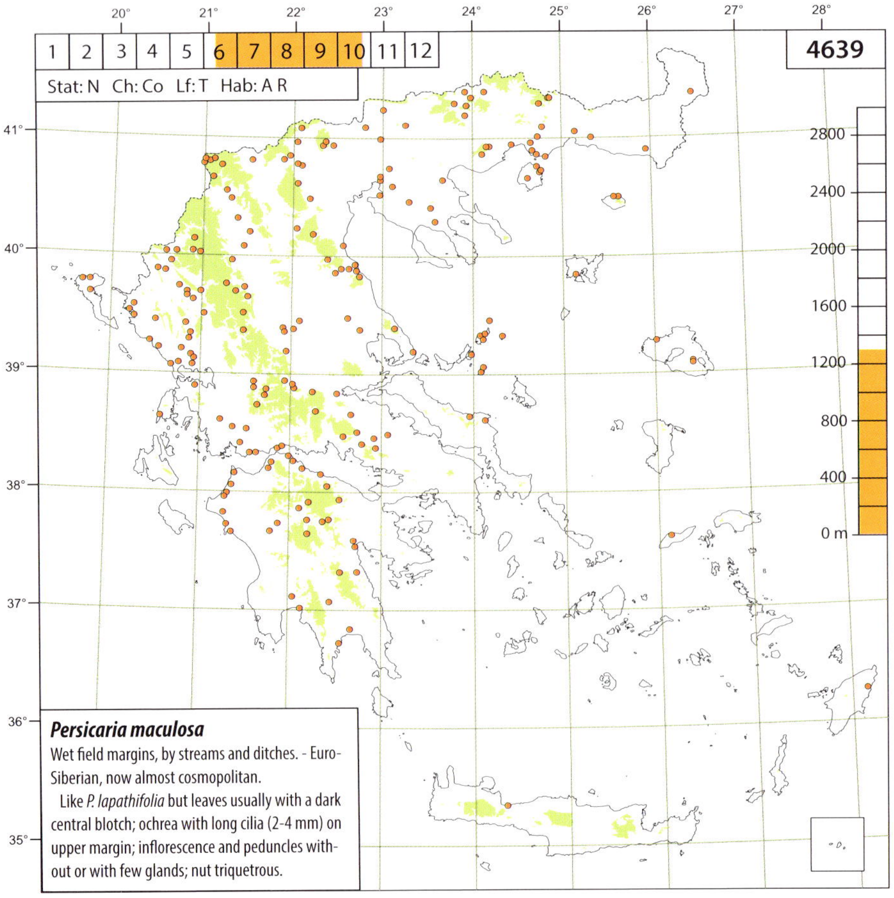

Persicaria maculosa

Wet field margins, by streams and ditches. - Euro-Siberian, now almost cosmopolitan.

Like *P. lapathifolia* but leaves usually with a dark central blotch; ochrea with long cilia (2-4 mm) on upper margin; inflorescence and peduncles without or with few glands; nut triquetrous.

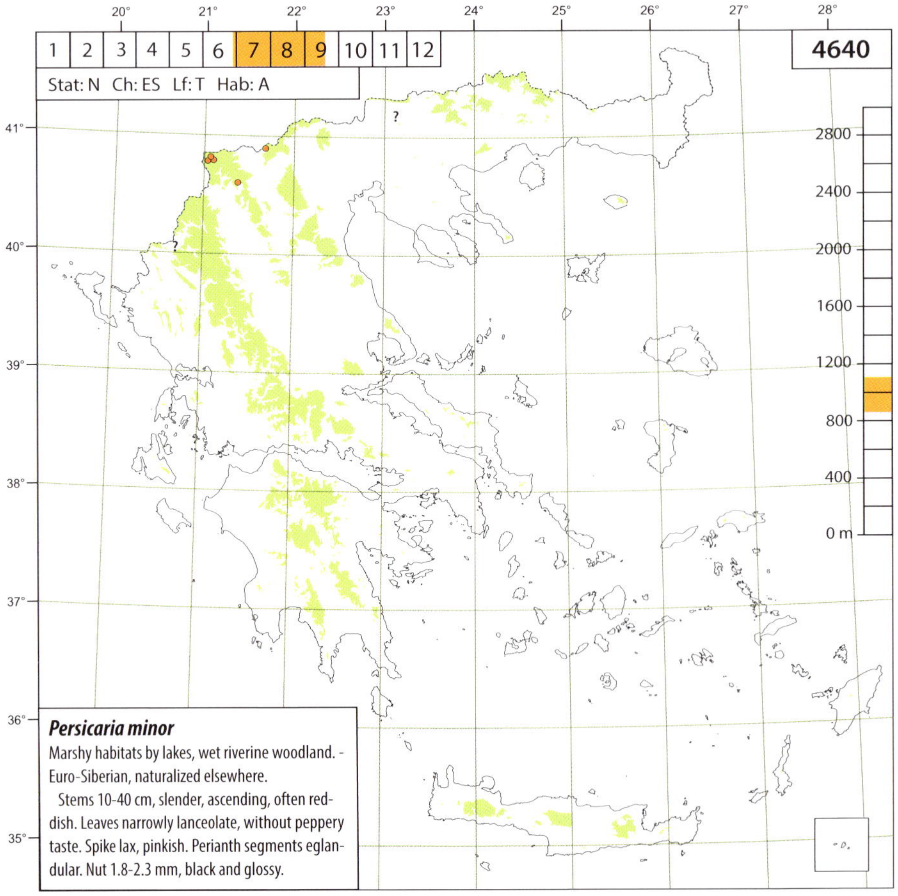

Persicaria minor

Marshy habitats by lakes, wet riverine woodland. - Euro-Siberian, naturalized elsewhere.

Stems 10-40 cm, slender, ascending, often reddish. Leaves narrowly lanceolate, without peppery taste. Spike lax, pinkish. Perianth segments eglandular. Nut 1.8-2.3 mm, black and glossy.

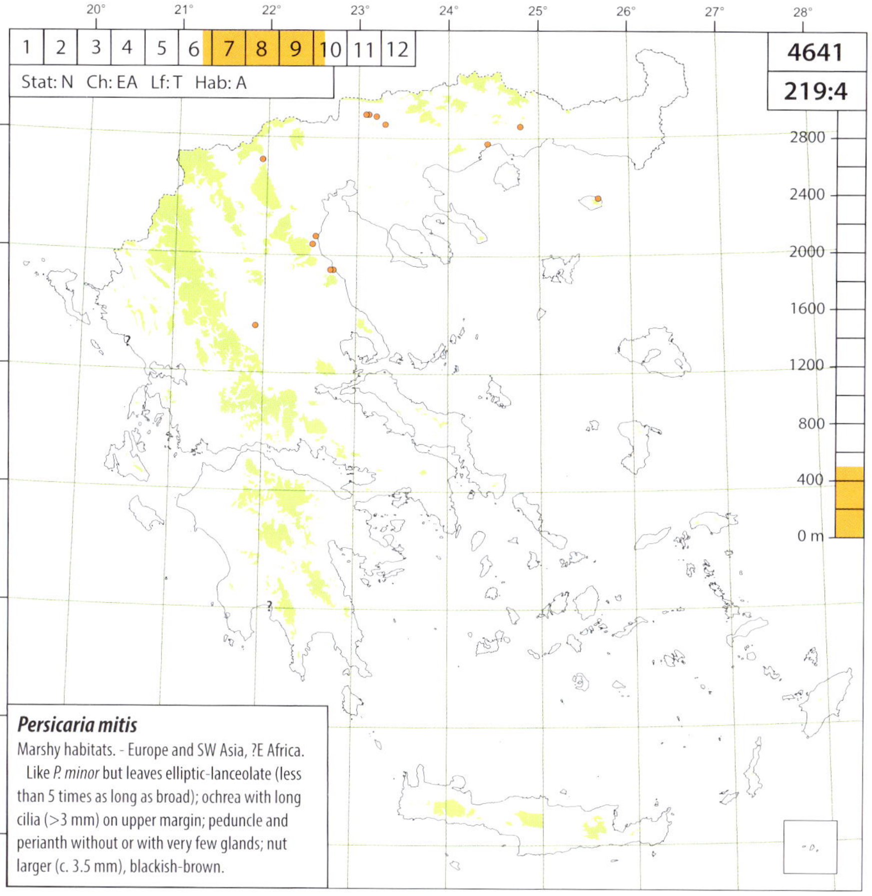

Persicaria mitis
Marshy habitats. - Europe and SW Asia, ?E Africa.
Like *P. minor* but leaves elliptic-lanceolate (less than 5 times as long as broad); ochrea with long cilia (>3 mm) on upper margin; peduncle and perianth without or with very few glands; nut larger (c. 3.5 mm), blackish-brown.

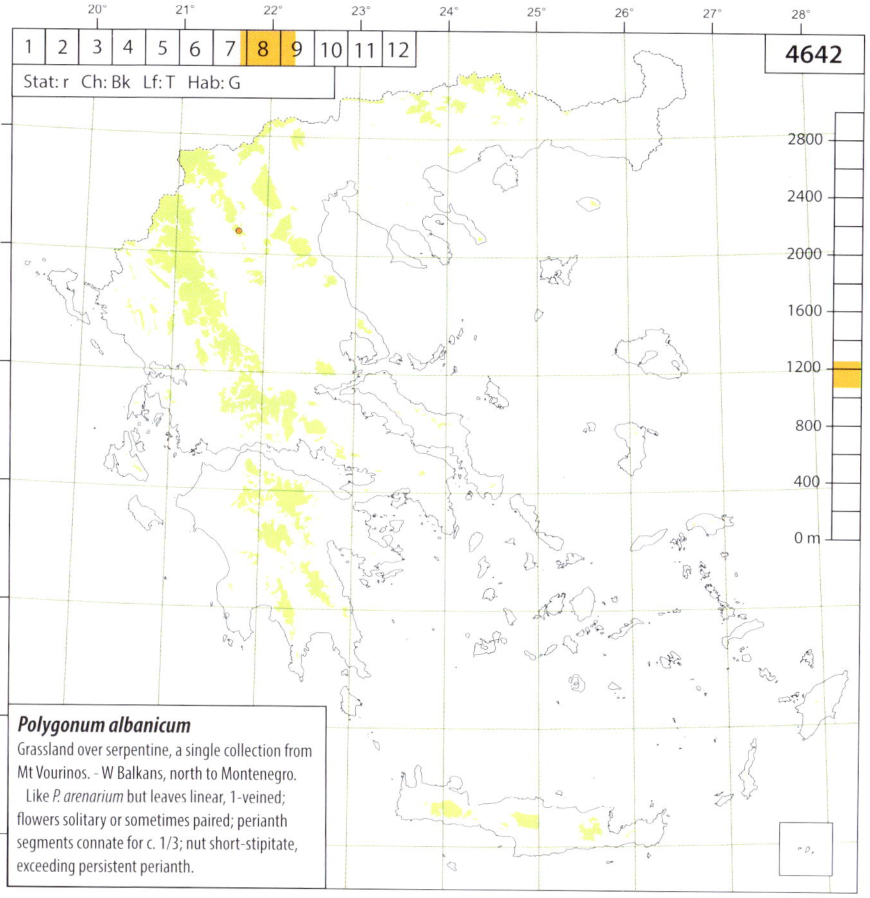

Polygonum albanicum
Grassland over serpentine, a single collection from Mt Vourinos. - W Balkans, north to Montenegro.
Like *P. arenarium* but leaves linear, 1-veined; flowers solitary or sometimes paired; perianth segments connate for c. 1/3; nut short-stipitate, exceeding persistent perianth.

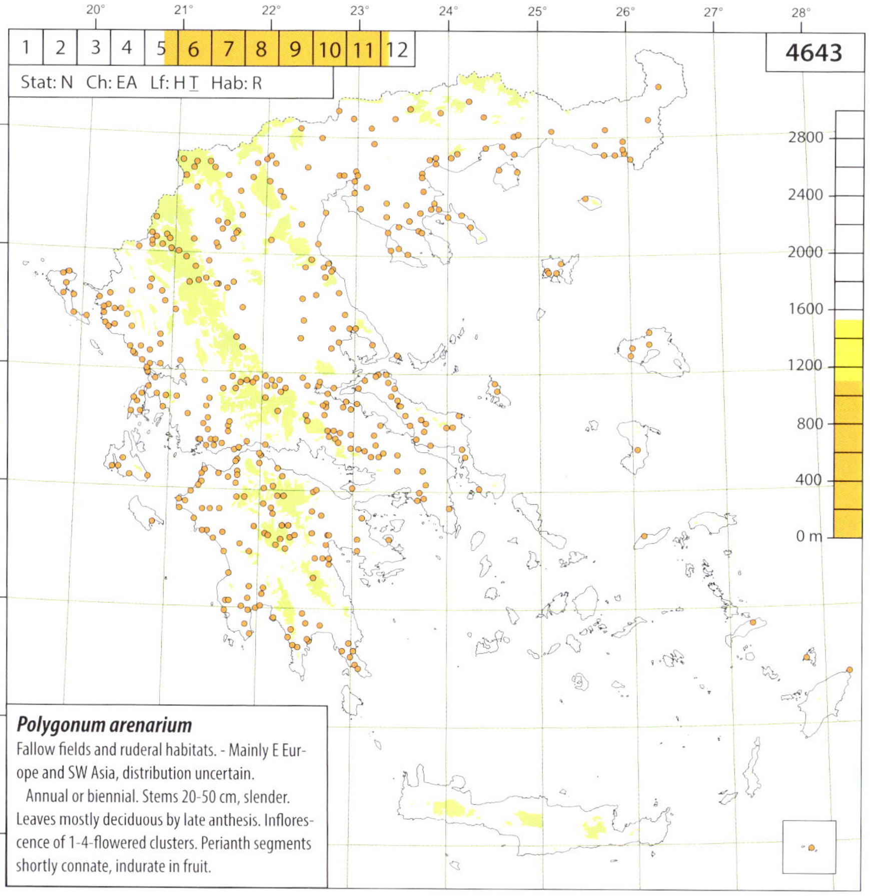

Polygonum arenarium
Fallow fields and ruderal habitats. - Mainly E Europe and SW Asia, distribution uncertain.
Annual or biennial. Stems 20-50 cm, slender. Leaves mostly deciduous by late anthesis. Inflorescence of 1-4-flowered clusters. Perianth segments shortly connate, indurate in fruit.

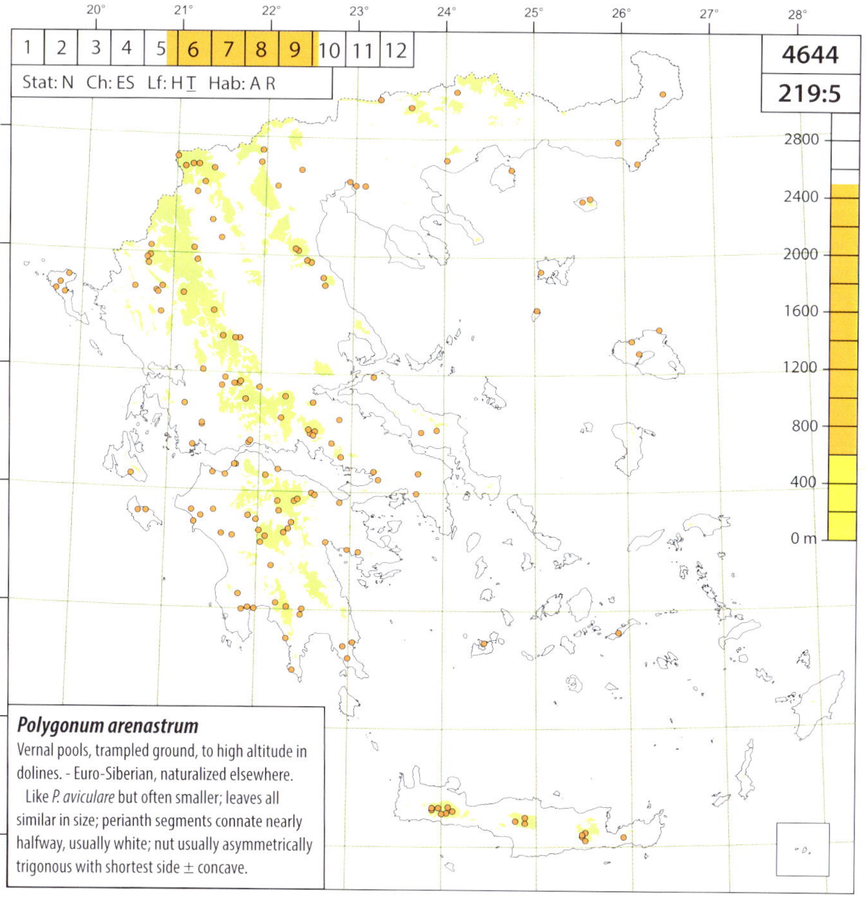

Polygonum arenastrum
Vernal pools, trampled ground, to high altitude in dolines. - Euro-Siberian, naturalized elsewhere.
Like *P. aviculare* but often smaller; leaves all similar in size; perianth segments connate nearly halfway, usually white; nut usually asymmetrically trigonous with shortest side ± concave.

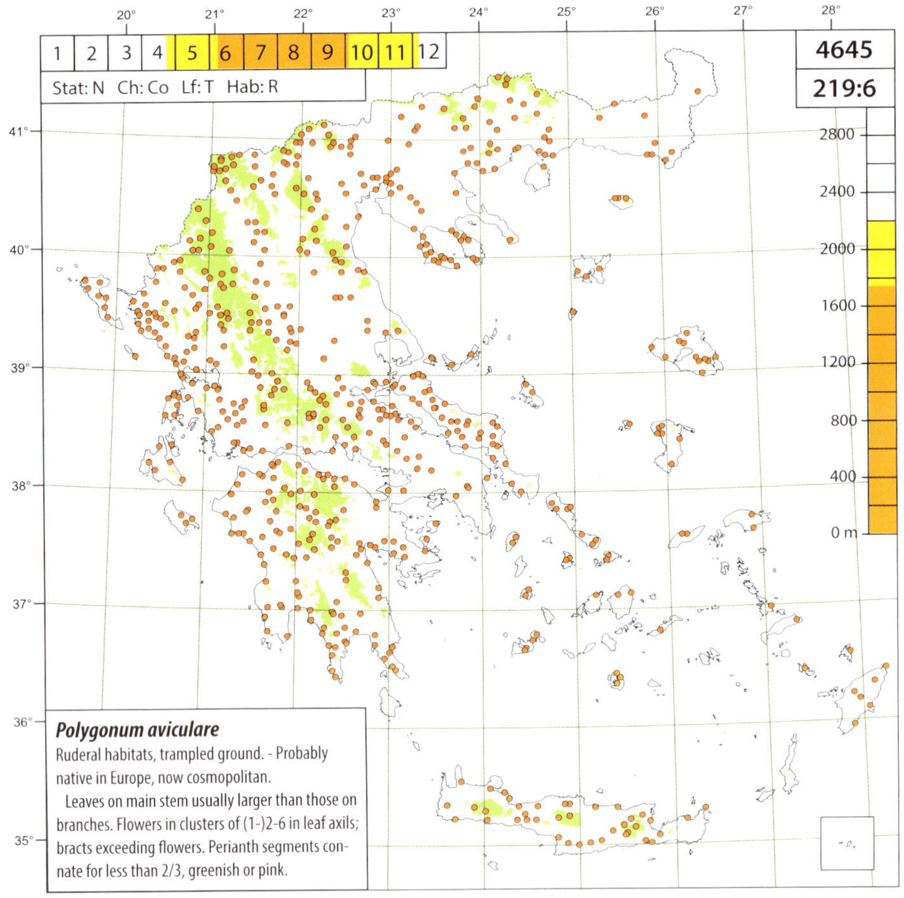

Polygonum aviculare
Ruderal habitats, trampled ground. - Probably native in Europe, now cosmopolitan.
Leaves on main stem usually larger than those on branches. Flowers in clusters of (1-)2-6 in leaf axils; bracts exceeding flowers. Perianth segments connate for less than 2/3, greenish or pink.

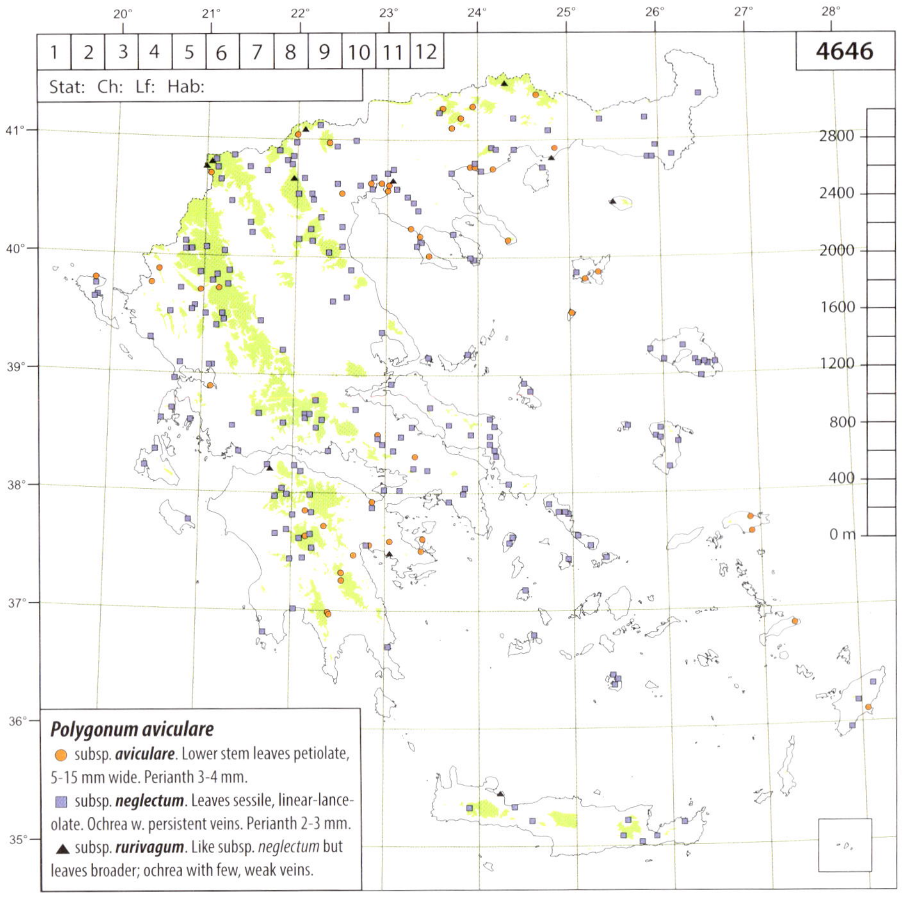

Polygonum aviculare
● subsp. ***aviculare***. Lower stem leaves petiolate, 5-15 mm wide. Perianth 3-4 mm.
■ subsp. ***neglectum***. Leaves sessile, linear-lanceolate. Ochrea w. persistent veins. Perianth 2-3 mm.
▲ subsp. ***rurivagum***. Like subsp. *neglectum* but leaves broader; ochrea with few, weak veins.

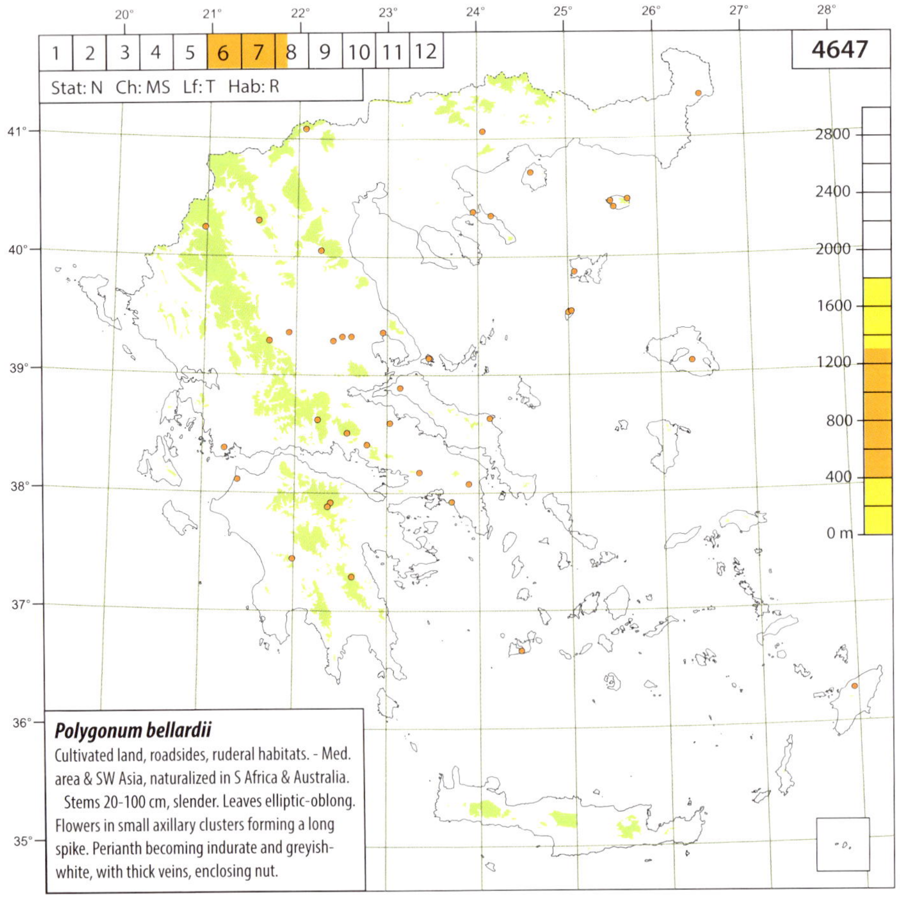

Polygonum bellardii
Cultivated land, roadsides, ruderal habitats. - Med. area & SW Asia, naturalized in S Africa & Australia.
Stems 20-100 cm, slender. Leaves elliptic-oblong. Flowers in small axillary clusters forming a long spike. Perianth becoming indurate and greyish-white, with thick veins, enclosing nut.

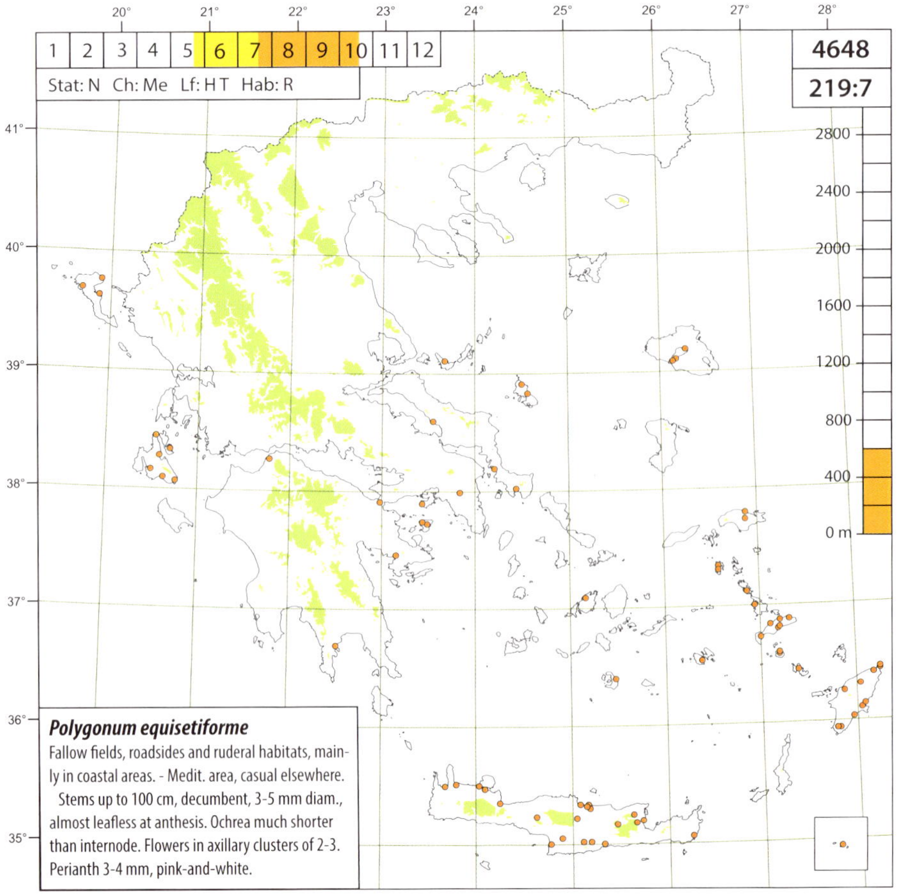

Polygonum equisetiforme
Fallow fields, roadsides and ruderal habitats, mainly in coastal areas. - Medit. area, casual elsewhere.
Stems up to 100 cm, decumbent, 3-5 mm diam., almost leafless at anthesis. Ochrea much shorter than internode. Flowers in axillary clusters of 2-3. Perianth 3-4 mm, pink-and-white.

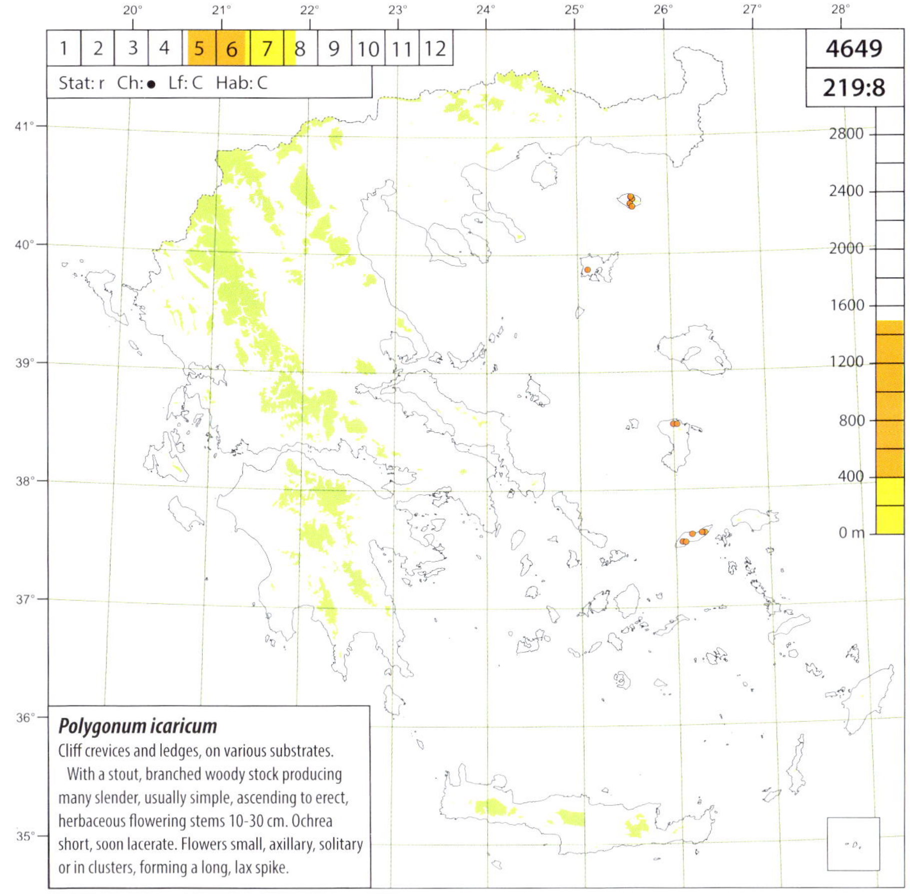

Polygonum icaricum
Cliff crevices and ledges, on various substrates.
With a stout, branched woody stock producing many slender, usually simple, ascending to erect, herbaceous flowering stems 10-30 cm. Ochrea short, soon lacerate. Flowers small, axillary, solitary or in clusters, forming a long, lax spike.

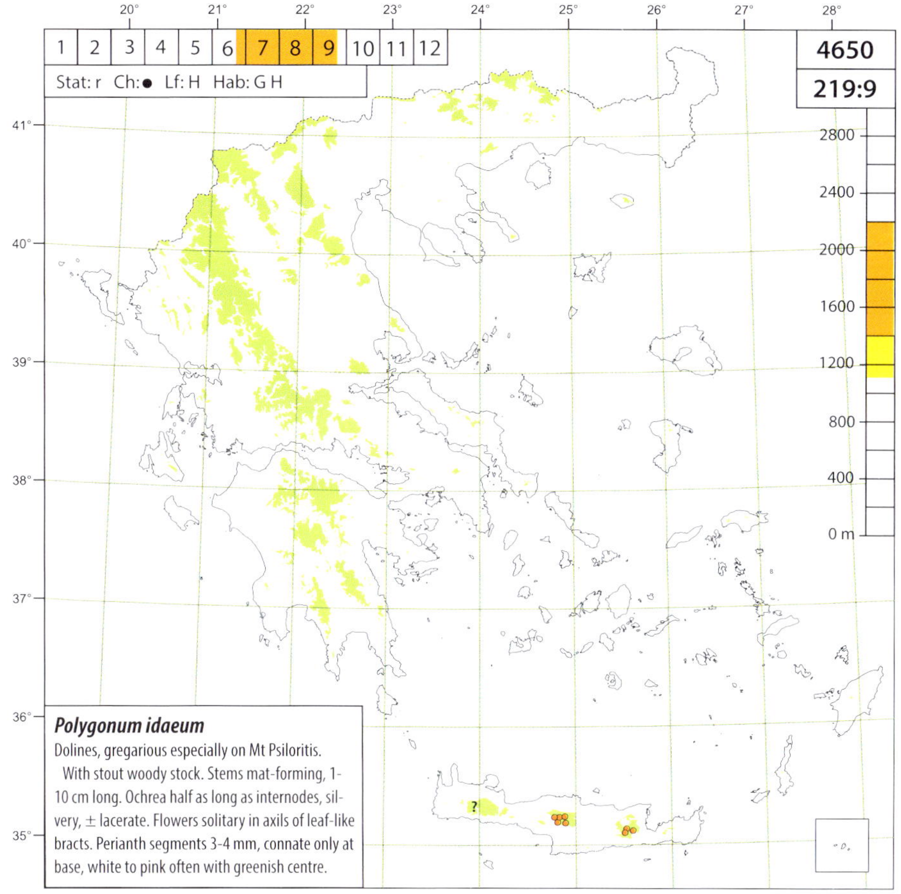

Polygonum idaeum
Dolines, gregarious especially on Mt Psiloritis.
With stout woody stock. Stems mat-forming, 1-10 cm long. Ochrea half as long as internodes, silvery, ± lacerate. Flowers solitary in axils of leaf-like bracts. Perianth segments 3-4 mm, connate only at base, white to pink often with greenish centre.

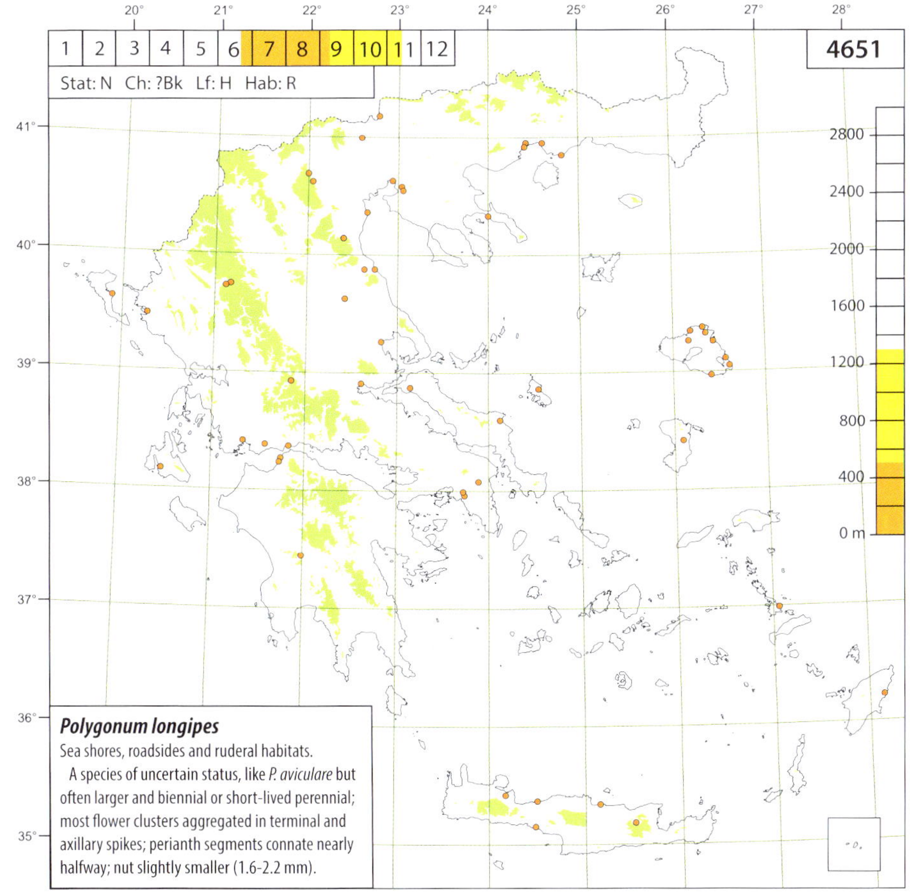

Polygonum longipes
Sea shores, roadsides and ruderal habitats.
A species of uncertain status, like *P. aviculare* but often larger and biennial or short-lived perennial; most flower clusters aggregated in terminal and axillary spikes; perianth segments connate nearly halfway; nut slightly smaller (1.6-2.2 mm).

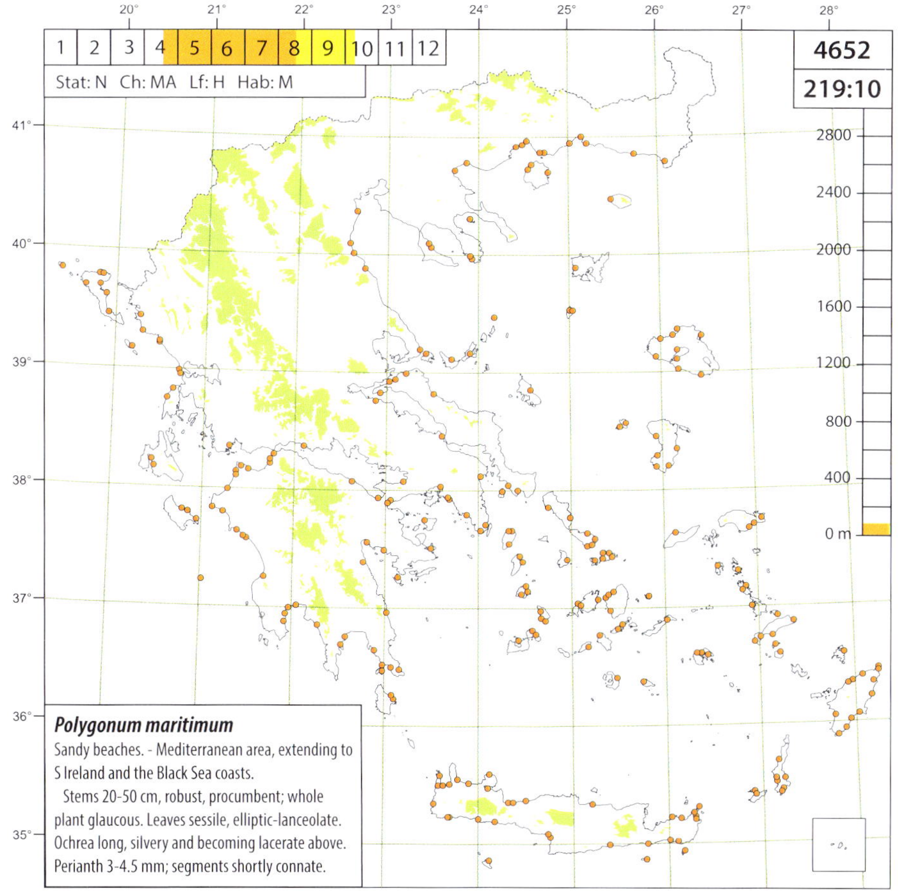

Polygonum maritimum
Sandy beaches. - Mediterranean area, extending to S Ireland and the Black Sea coasts.
Stems 20-50 cm, robust, procumbent; whole plant glaucous. Leaves sessile, elliptic-lanceolate. Ochrea long, silvery and becoming lacerate above. Perianth 3-4.5 mm; segments shortly connate.

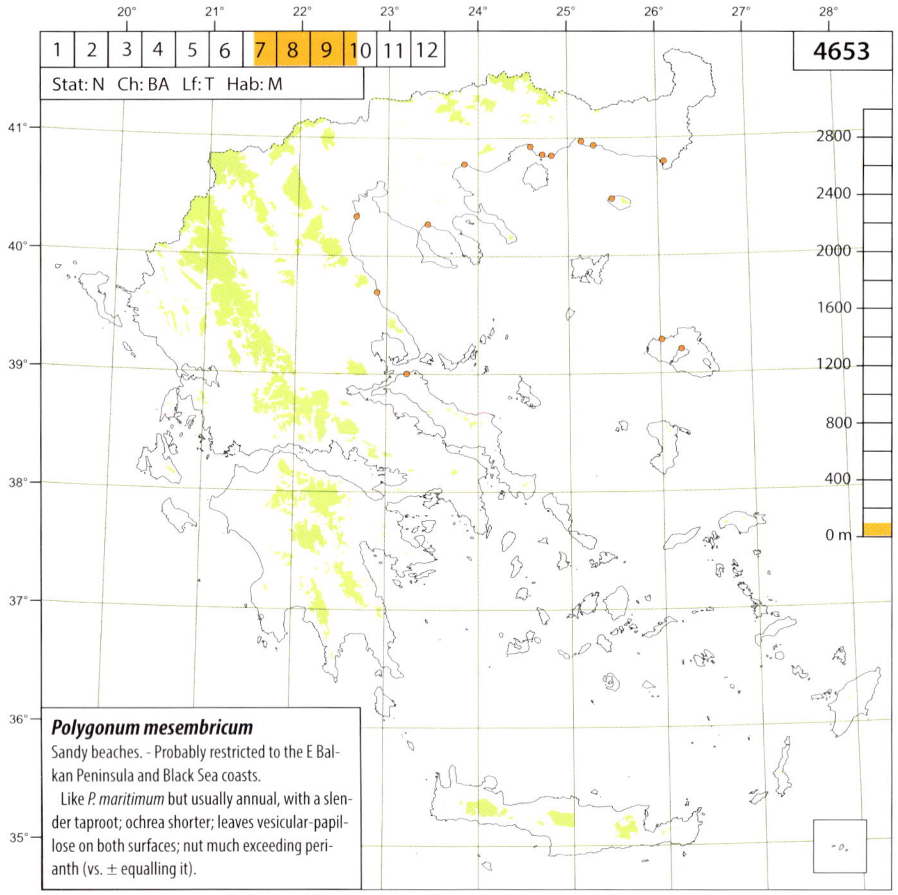

Polygonum mesembricum

Sandy beaches. - Probably restricted to the E Balkan Peninsula and Black Sea coasts.

Like *P. maritimum* but usually annual, with a slender taproot; ochrea shorter; leaves vesicular-papillose on both surfaces; nut much exceeding perianth (vs. ± equalling it).

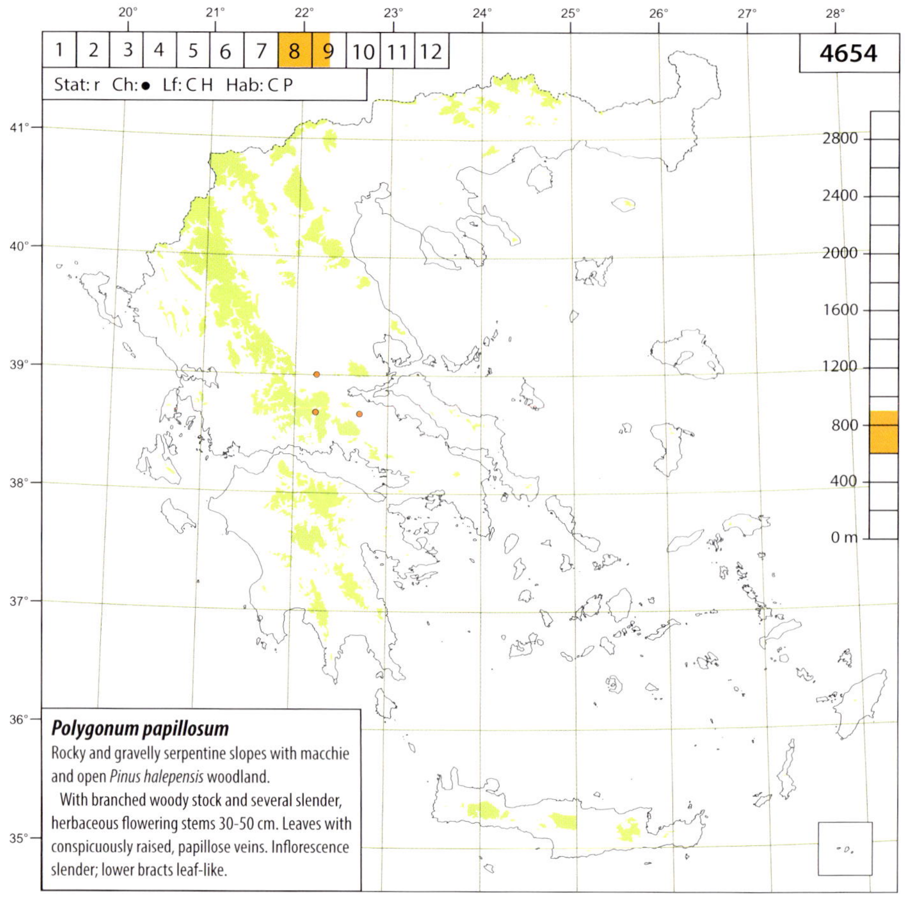

Polygonum papillosum

Rocky and gravelly serpentine slopes with macchie and open *Pinus halepensis* woodland.

With branched woody stock and several slender, herbaceous flowering stems 30-50 cm. Leaves with conspicuously raised, papillose veins. Inflorescence slender; lower bracts leaf-like.

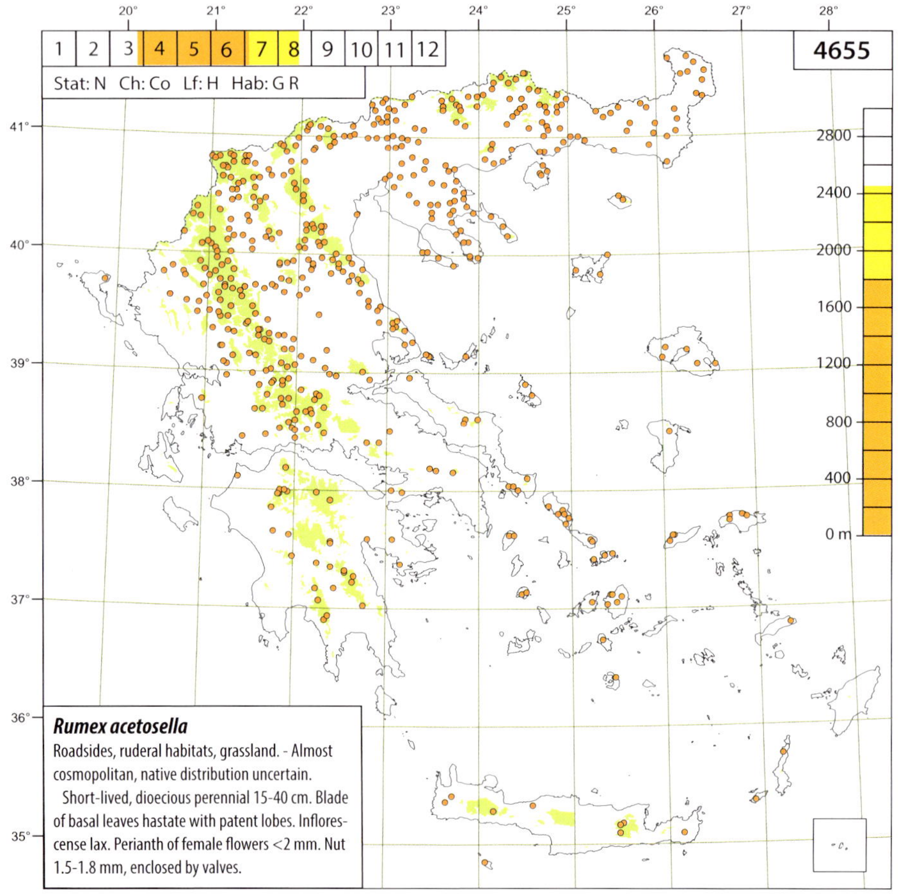

Rumex acetosella

Roadsides, ruderal habitats, grassland. - Almost cosmopolitan, native distribution uncertain.

Short-lived, dioecious perennial 15-40 cm. Blade of basal leaves hastate with patent lobes. Inflorescence lax. Perianth of female flowers <2 mm. Nut 1.5-1.8 mm, enclosed by valves.

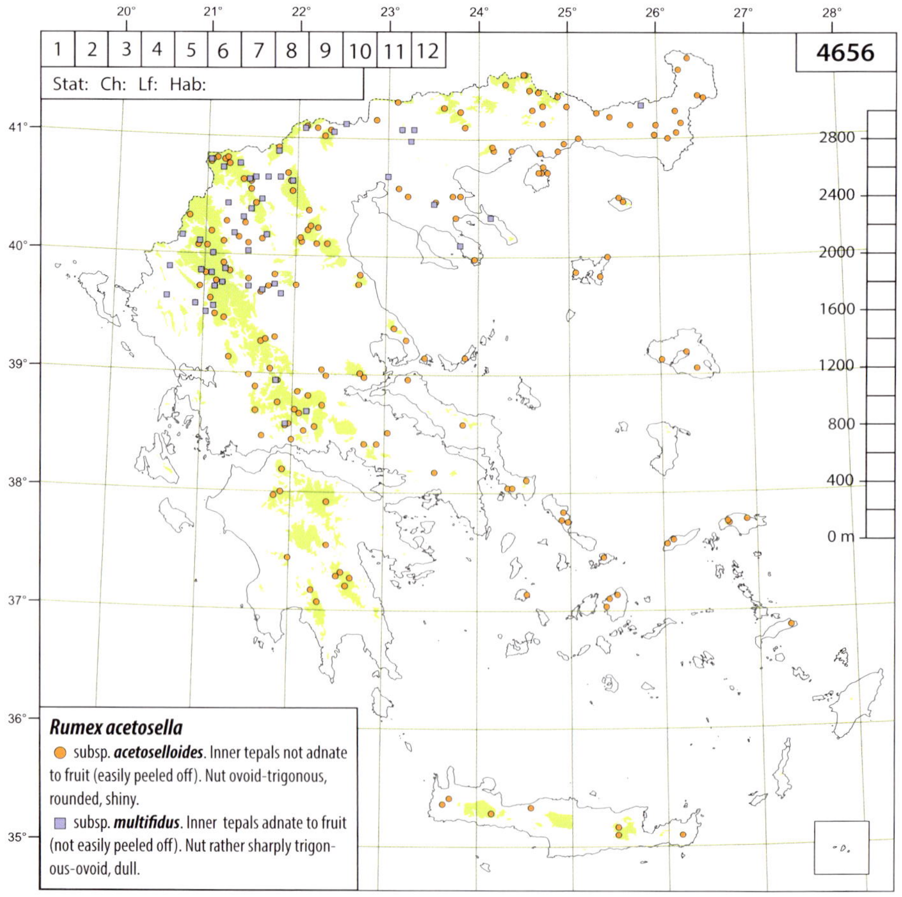

Rumex acetosella

● subsp. ***acetoselloides***. Inner tepals not adnate to fruit (easily peeled off). Nut ovoid-trigonous, rounded, shiny.

■ subsp. ***multifidus***. Inner tepals adnate to fruit (not easily peeled off). Nut rather sharply trigon-ous-ovoid, dull.

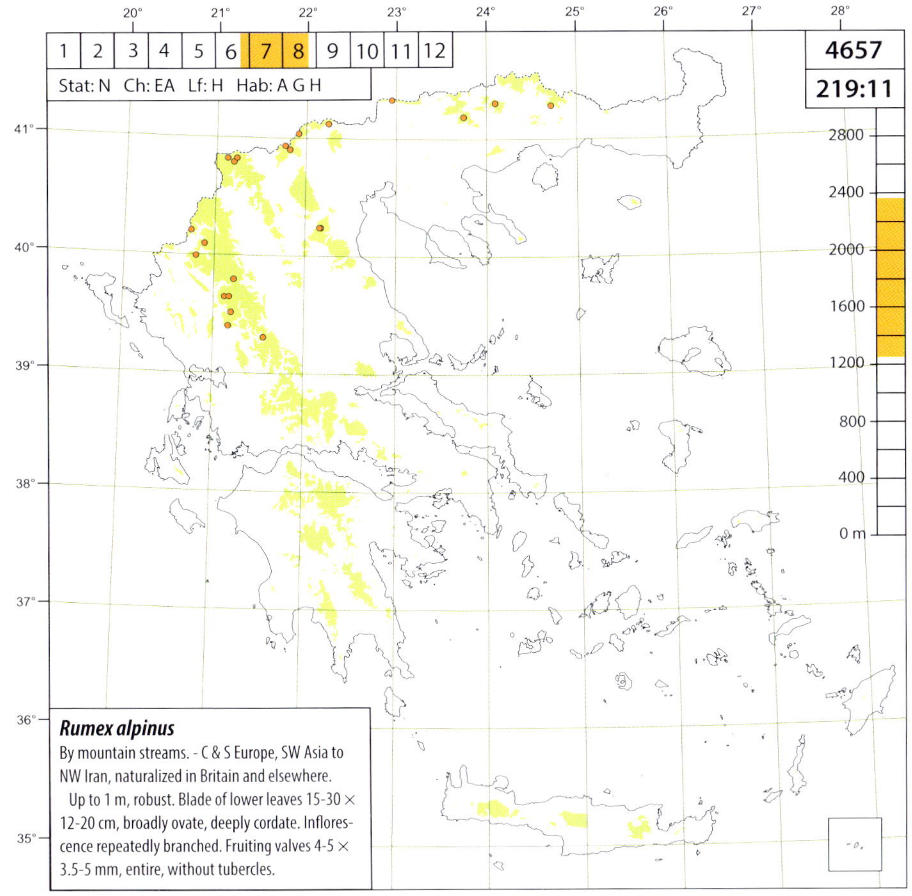

Rumex alpinus

By mountain streams. - C & S Europe, SW Asia to NW Iran, naturalized in Britain and elsewhere.

Up to 1 m, robust. Blade of lower leaves 15-30 × 12-20 cm, broadly ovate, deeply cordate. Inflorescence repeatedly branched. Fruiting valves 4-5 × 3.5-5 mm, entire, without tubercles.

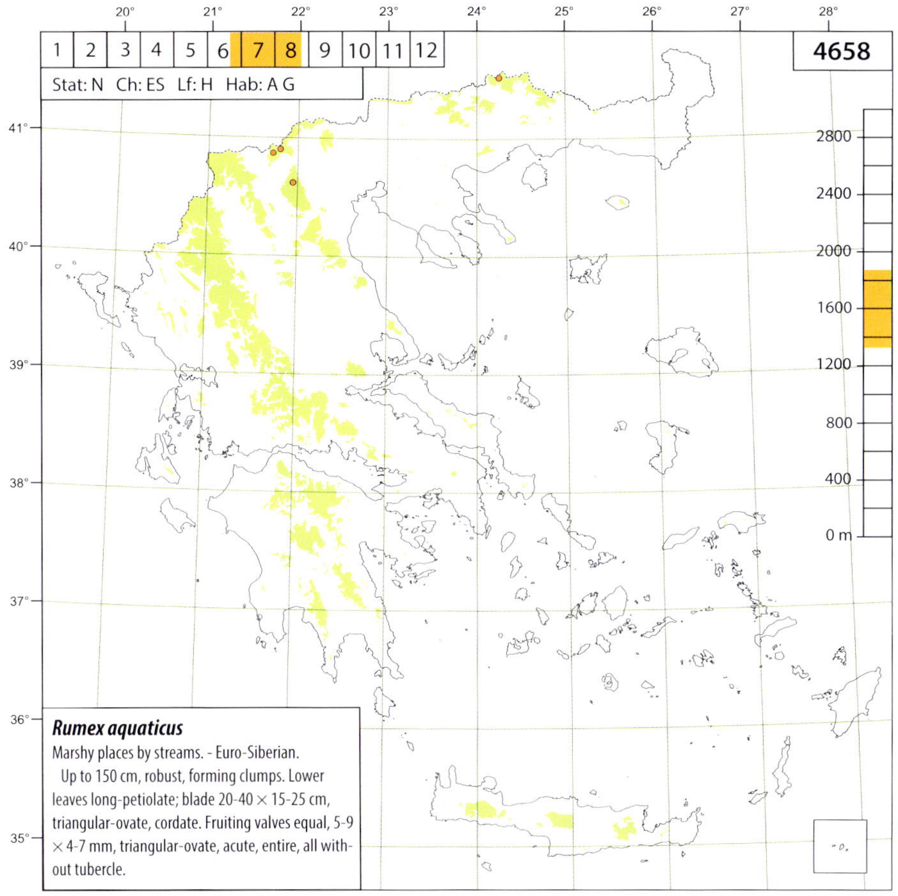

Rumex aquaticus

Marshy places by streams. - Euro-Siberian.

Up to 150 cm, robust, forming clumps. Lower leaves long-petiolate; blade 20-40 × 15-25 cm, triangular-ovate, cordate. Fruiting valves equal, 5-9 × 4-7 mm, triangular-ovate, acute, entire, all without tubercle.

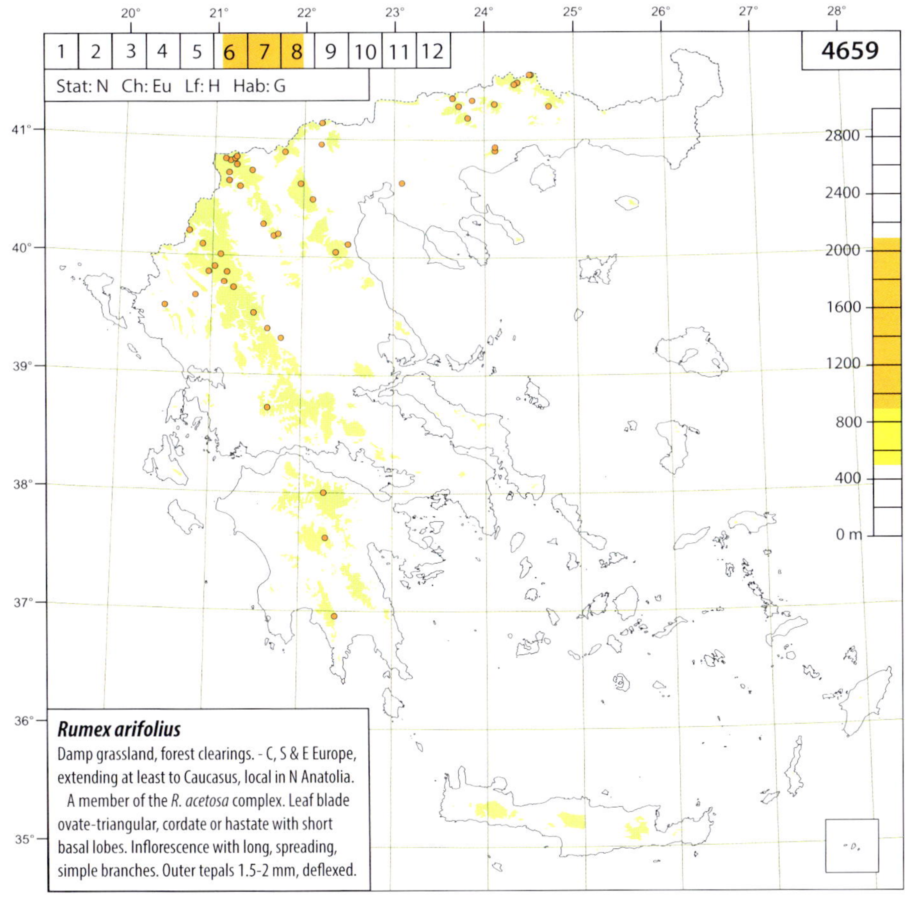

Rumex arifolius

Damp grassland, forest clearings. - C, S & E Europe, extending at least to Caucasus, local in N Anatolia.

A member of the *R. acetosa* complex. Leaf blade ovate-triangular, cordate or hastate with short basal lobes. Inflorescence with long, spreading, simple branches. Outer tepals 1.5-2 mm, deflexed.

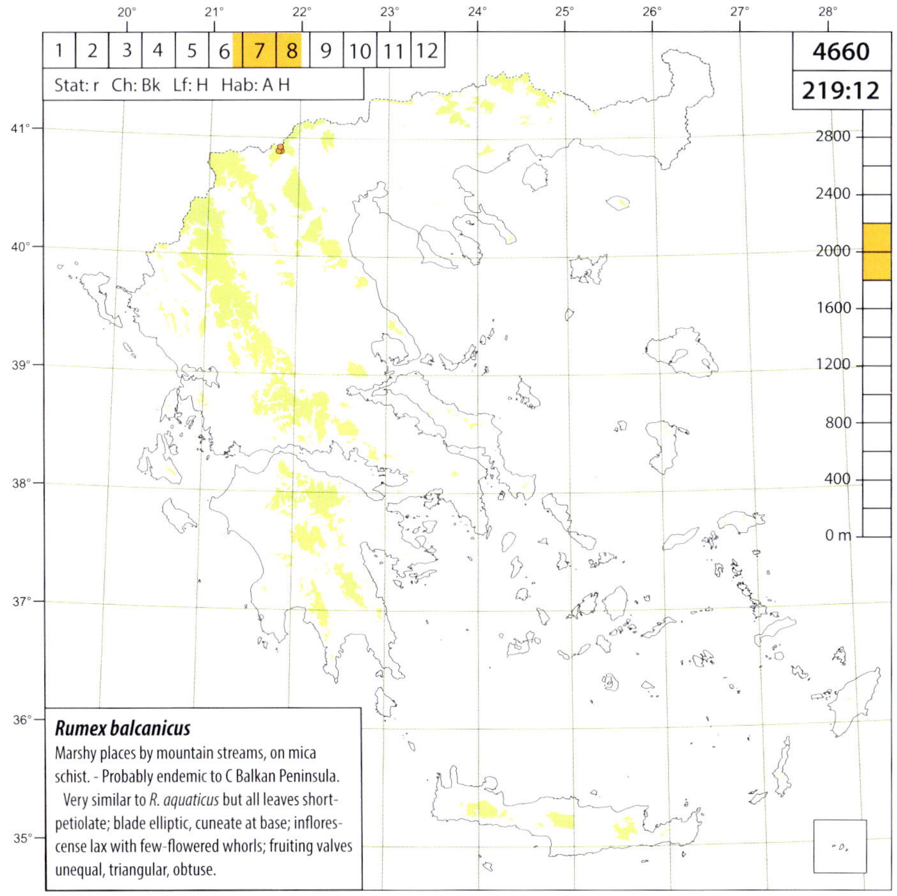

Rumex balcanicus

Marshy places by mountain streams, on mica schist. - Probably endemic to C Balkan Peninsula.

Very similar to *R. aquaticus* but all leaves short-petiolate; blade elliptic, cuneate at base; inflorescence lax with few-flowered whorls; fruiting valves unequal, triangular, obtuse.

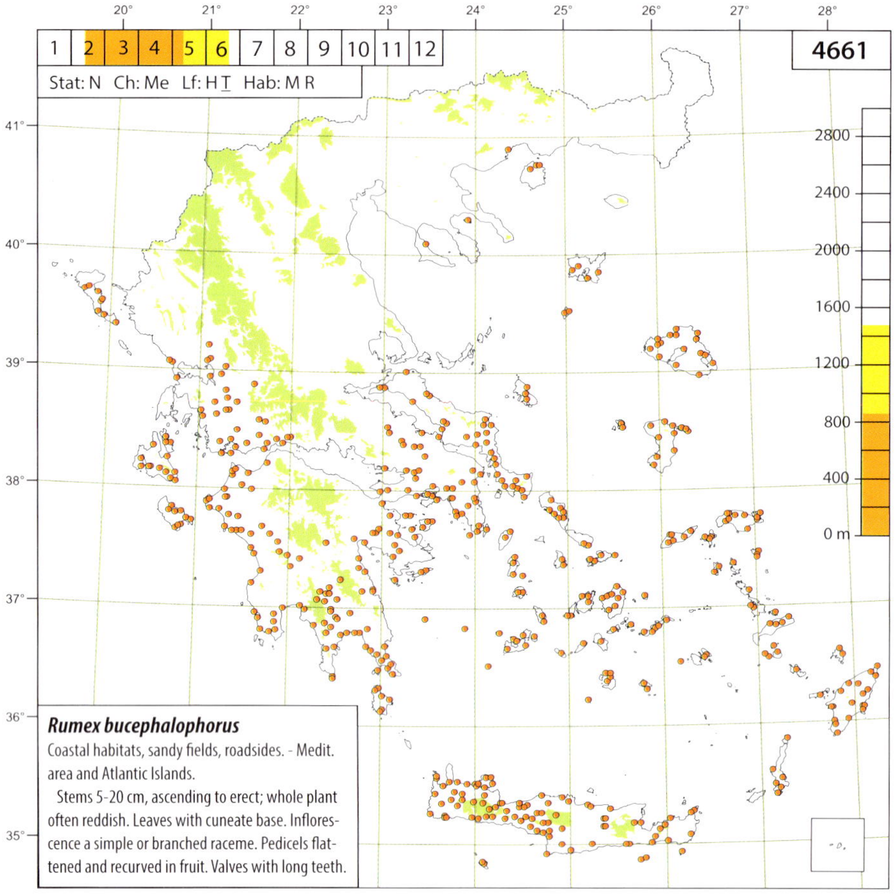
1 2 3 4 5 6 7 8 9 10 11 12
4661
Stat: N Ch: Me Lf: H T Hab: M R
2800
2400
2000
1600
1200
800
400
0 m
Rumex bucephalophorus
Coastal habitats, sandy fields, roadsides. - Medit. area and Atlantic Islands.
Stems 5-20 cm, ascending to erect; whole plant often reddish. Leaves with cuneate base. Inflorescence a simple or branched raceme. Pedicels flattened and recurved in fruit. Valves with long teeth.

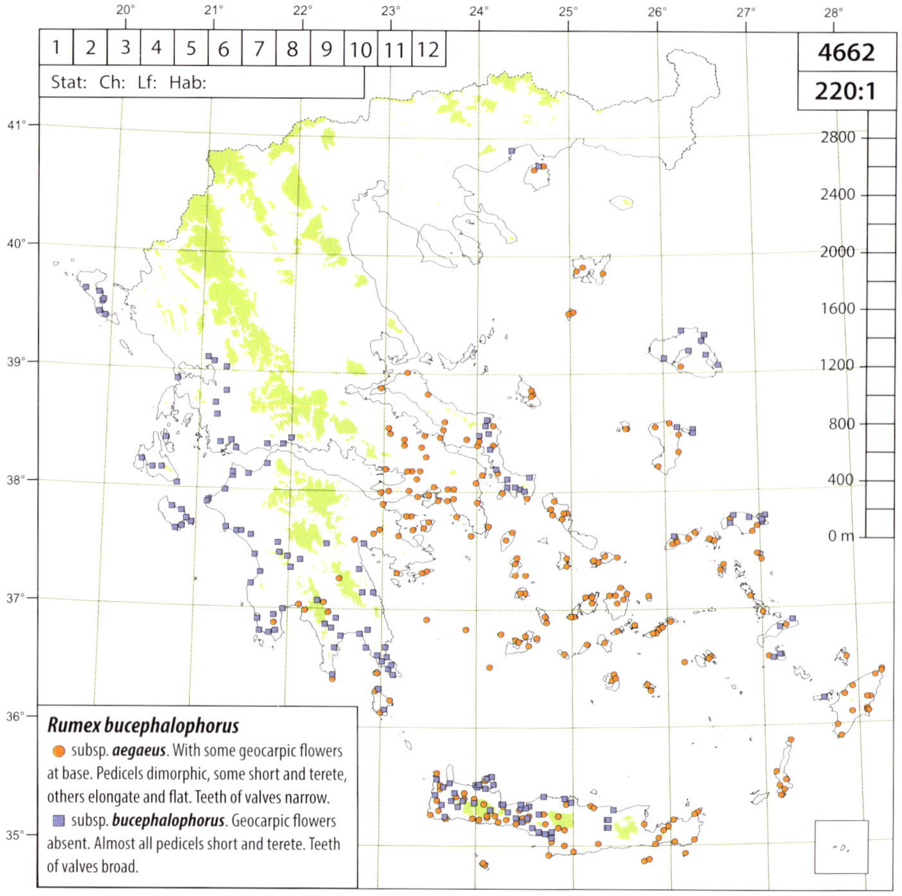
1 2 3 4 5 6 7 8 9 10 11 12
4662
220:1
Stat: Ch: Lf: Hab:
2800
2400
2000
1600
1200
800
400
0 m
Rumex bucephalophorus
subsp. **_aegaeus_**. With some geocarpic flowers at base. Pedicels dimorphic, some short and terete, others elongate and flat. Teeth of valves narrow.
subsp. **_bucephalophorus_**. Geocarpic flowers absent. Almost all pedicels short and terete. Teeth of valves broad.

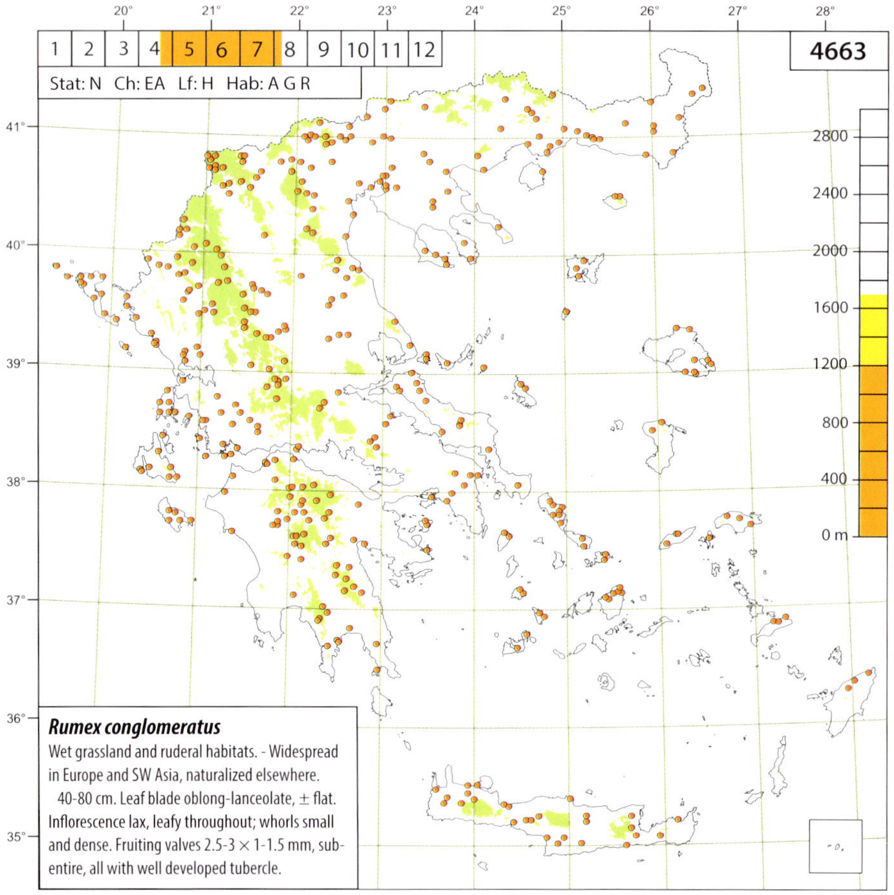
1 2 3 4 5 6 7 8 9 10 11 12
4663
Stat: N Ch: EA Lf: H Hab: A G R
2800
2400
2000
1600
1200
800
400
0 m
Rumex conglomeratus
Wet grassland and ruderal habitats. - Widespread in Europe and SW Asia, naturalized elsewhere.
40-80 cm. Leaf blade oblong-lanceolate, ± flat. Inflorescence lax, leafy throughout; whorls small and dense. Fruiting valves 2.5-3 × 1-1.5 mm, subentire, all with well developed tubercle.

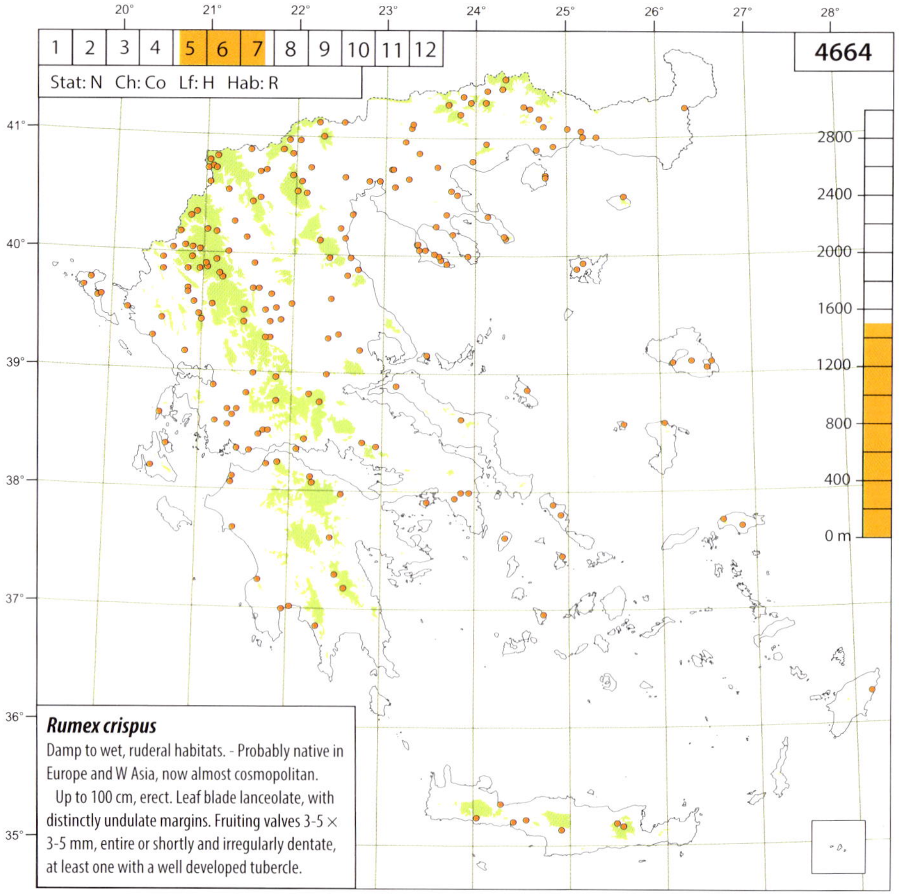
1 2 3 4 5 6 7 8 9 10 11 12
4664
Stat: N Ch: Co Lf: H Hab: R
2800
2400
2000
1600
1200
800
400
0 m
Rumex crispus
Damp to wet, ruderal habitats. - Probably native in Europe and W Asia, now almost cosmopolitan.
Up to 100 cm, erect. Leaf blade lanceolate, with distinctly undulate margins. Fruiting valves 3-5 × 3-5 mm, entire or shortly and irregularly dentate, at least one with a well developed tubercle.

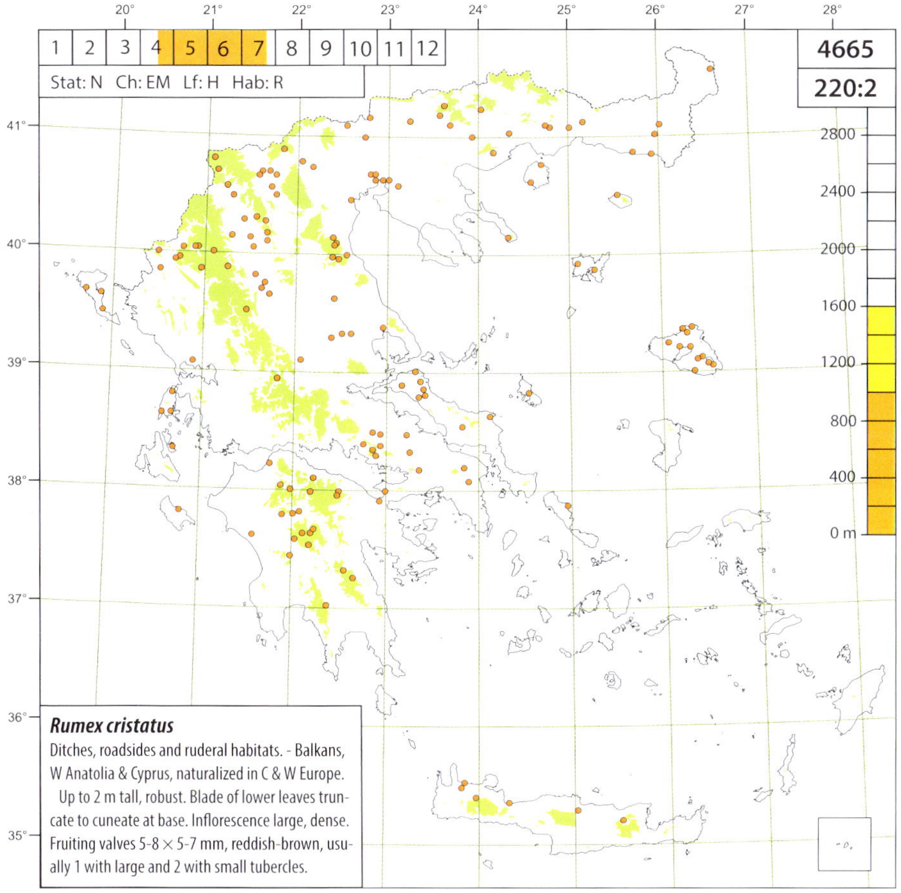

Rumex cristatus

Ditches, roadsides and ruderal habitats. - Balkans, W Anatolia & Cyprus, naturalized in C & W Europe.

Up to 2 m tall, robust. Blade of lower leaves truncate to cuneate at base. Inflorescence large, dense. Fruiting valves 5-8 × 5-7 mm, reddish-brown, usually 1 with large and 2 with small tubercles.

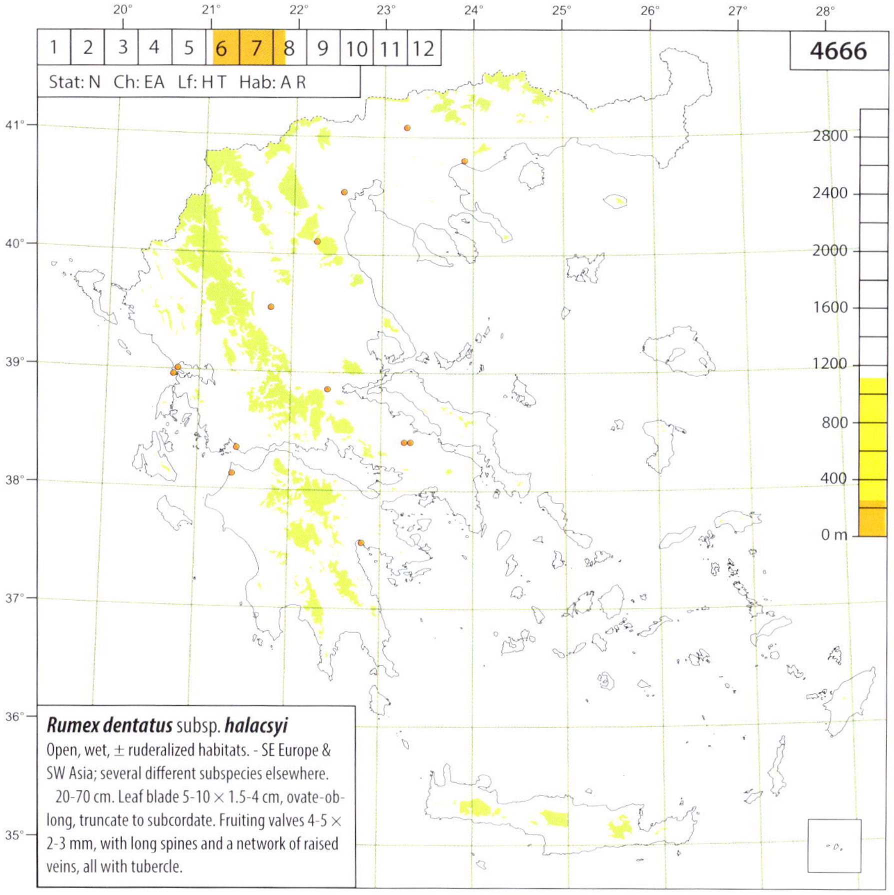

Rumex dentatus subsp. ***halacsyi***

Open, wet, ± ruderalized habitats. - SE Europe & SW Asia; several different subspecies elsewhere.

20-70 cm. Leaf blade 5-10 × 1.5-4 cm, ovate-oblong, truncate to subcordate. Fruiting valves 4-5 × 2-3 mm, with long spines and a network of raised veins, all with tubercle.

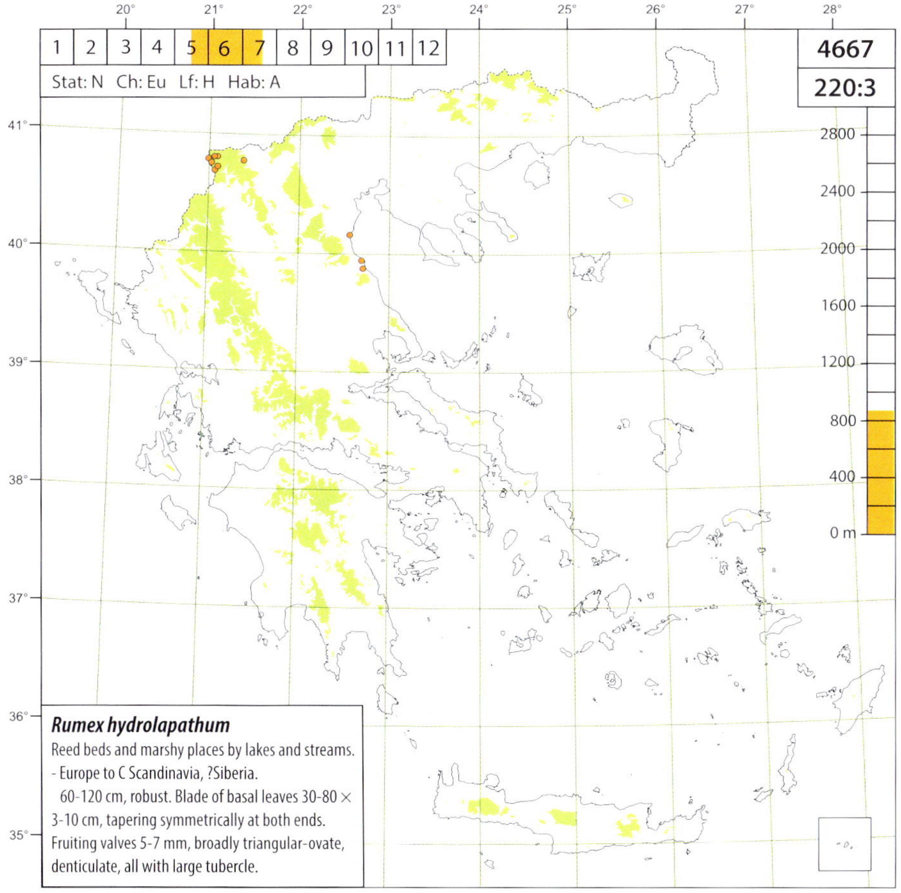

Rumex hydrolapathum

Reed beds and marshy places by lakes and streams. - Europe to C Scandinavia, ?Siberia.

60-120 cm, robust. Blade of basal leaves 30-80 × 3-10 cm, tapering symmetrically at both ends. Fruiting valves 5-7 mm, broadly triangular-ovate, denticulate, all with large tubercle.

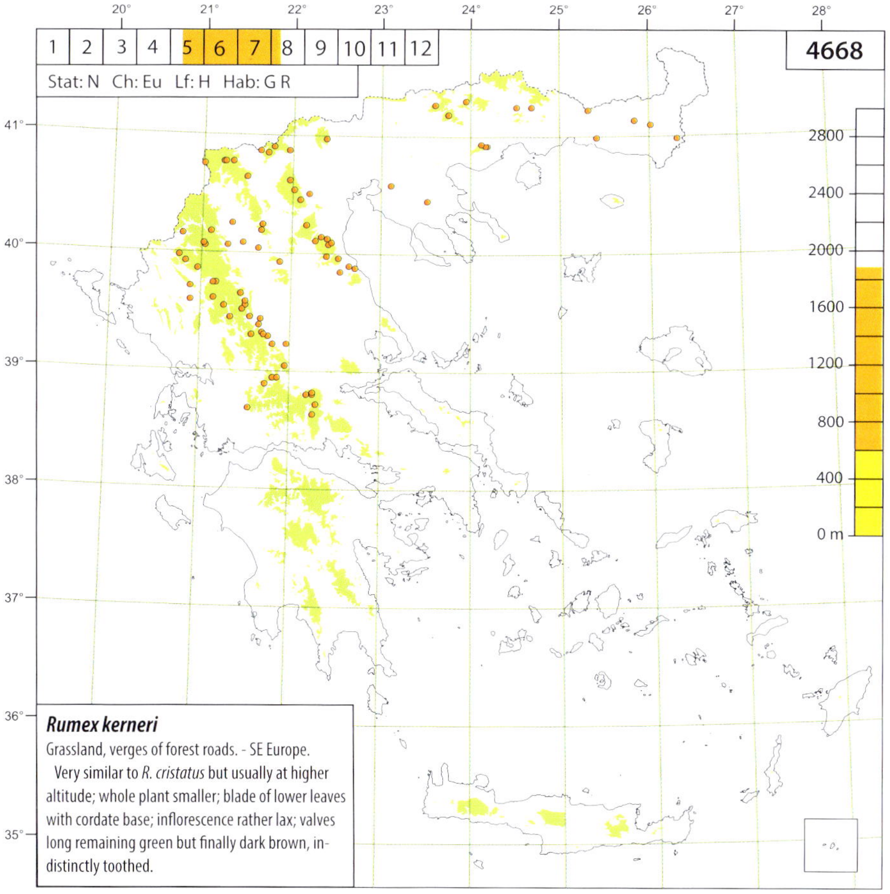

Rumex kerneri

Grassland, verges of forest roads. - SE Europe.

Very similar to *R. cristatus* but usually at higher altitude; whole plant smaller; blade of lower leaves with cordate base; inflorescence rather lax; valves long remaining green but finally dark brown, indistinctly toothed.

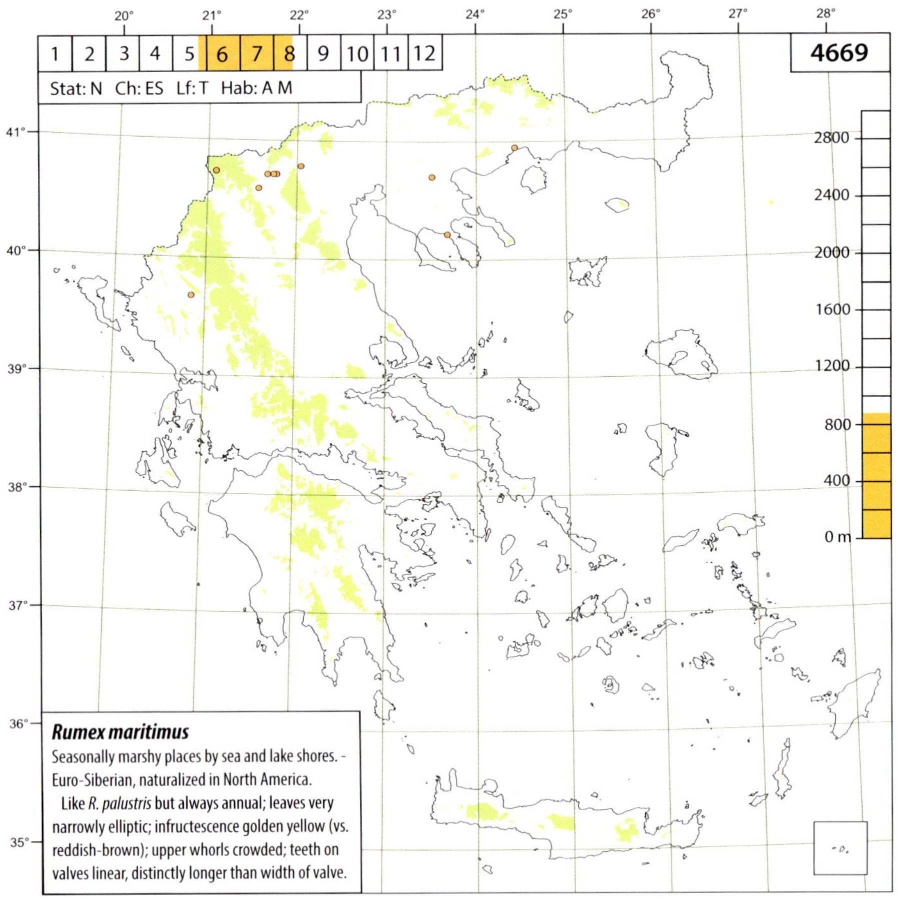

Rumex maritimus

Seasonally marshy places by sea and lake shores. - Euro-Siberian, naturalized in North America.

Like *R. palustris* but always annual; leaves very narrowly elliptic; infructescence golden yellow (vs. reddish-brown); upper whorls crowded; teeth on valves linear, distinctly longer than width of valve.

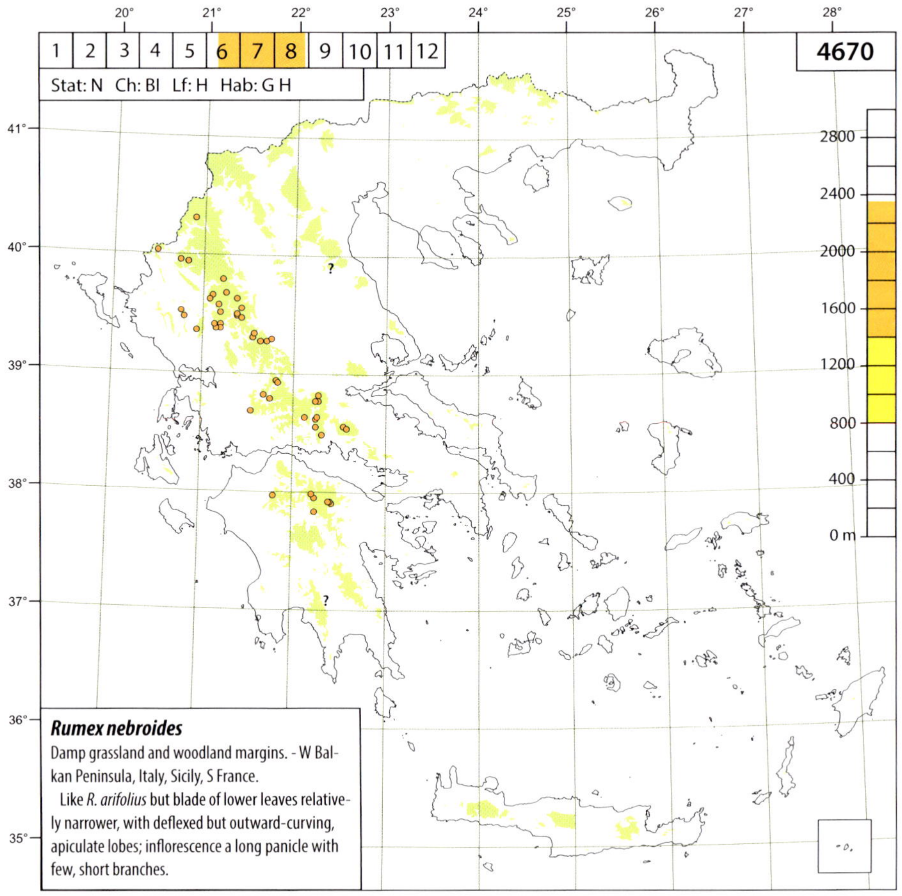

Rumex nebroides

Damp grassland and woodland margins. - W Balkan Peninsula, Italy, Sicily, S France.

Like *R. arifolius* but blade of lower leaves relatively narrower, with deflexed but outward-curving, apiculate lobes; inflorescence a long panicle with few, short branches.

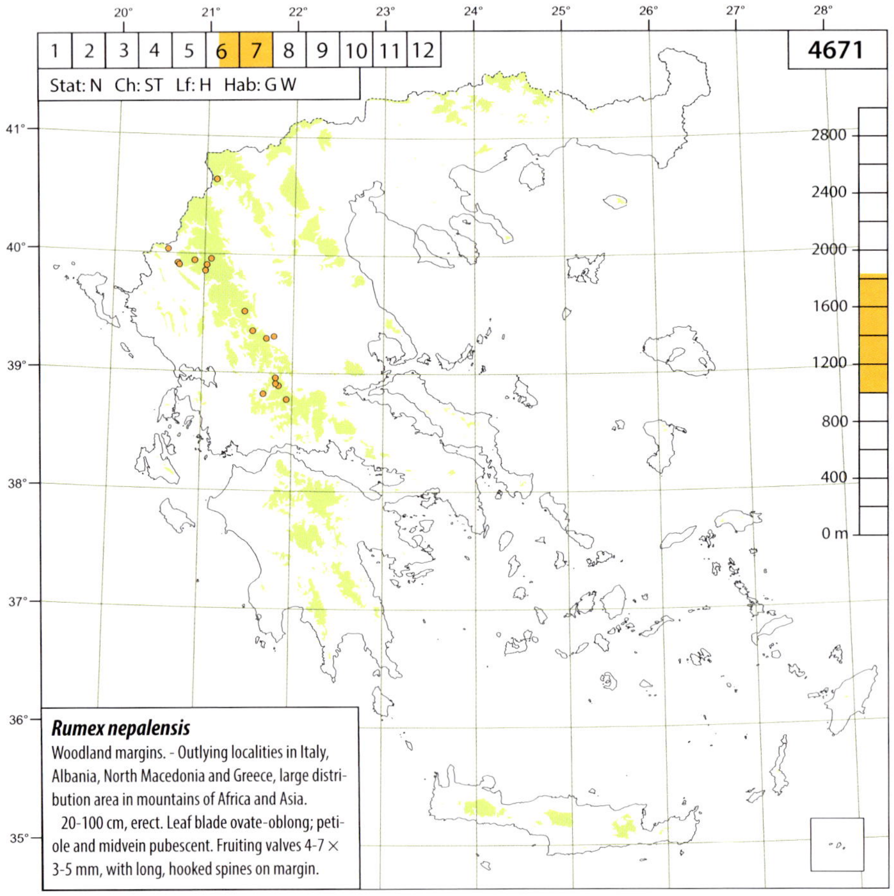

Rumex nepalensis

Woodland margins. - Outlying localities in Italy, Albania, North Macedonia and Greece, large distribution area in mountains of Africa and Asia.

20-100 cm, erect. Leaf blade ovate-oblong; petiole and midvein pubescent. Fruiting valves 4-7 × 3-5 mm, with long, hooked spines on margin.

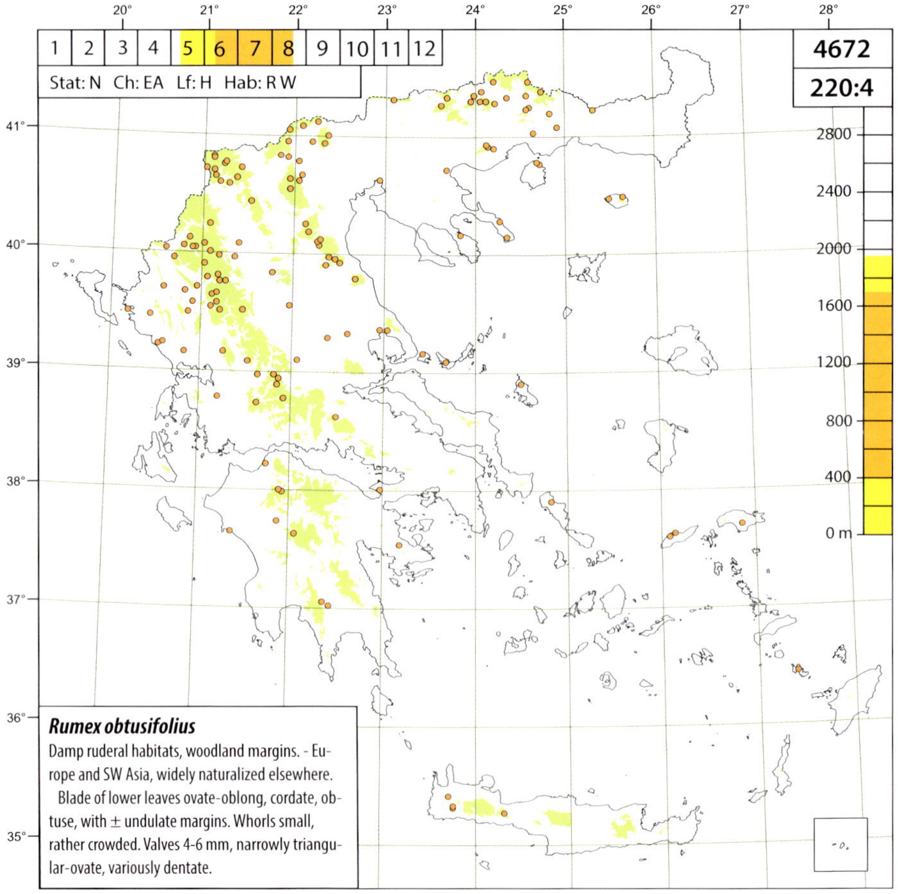

Rumex obtusifolius

Damp ruderal habitats, woodland margins. - Europe and SW Asia, widely naturalized elsewhere.

Blade of lower leaves ovate-oblong, cordate, obtuse, with ± undulate margins. Whorls small, rather crowded. Valves 4-6 mm, narrowly triangular-ovate, variously dentate.

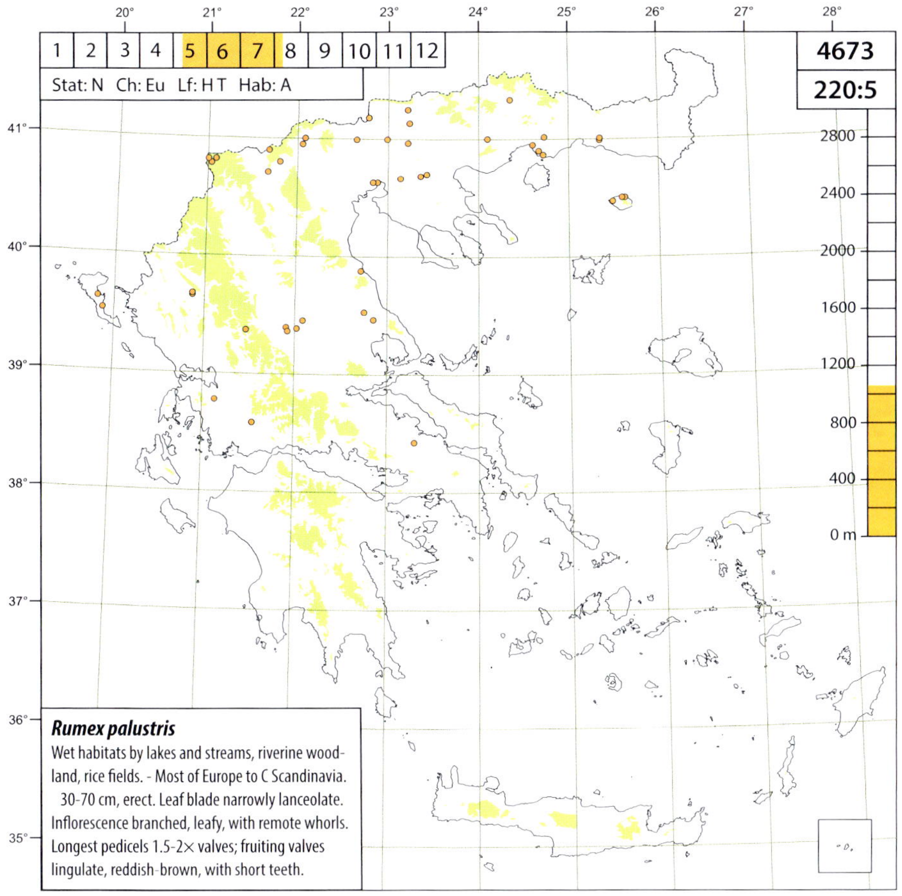

Rumex palustris
Wet habitats by lakes and streams, riverine woodland, rice fields. - Most of Europe to C Scandinavia.
30-70 cm, erect. Leaf blade narrowly lanceolate. Inflorescence branched, leafy, with remote whorls. Longest pedicels 1.5-2× valves; fruiting valves lingulate, reddish-brown, with short teeth.

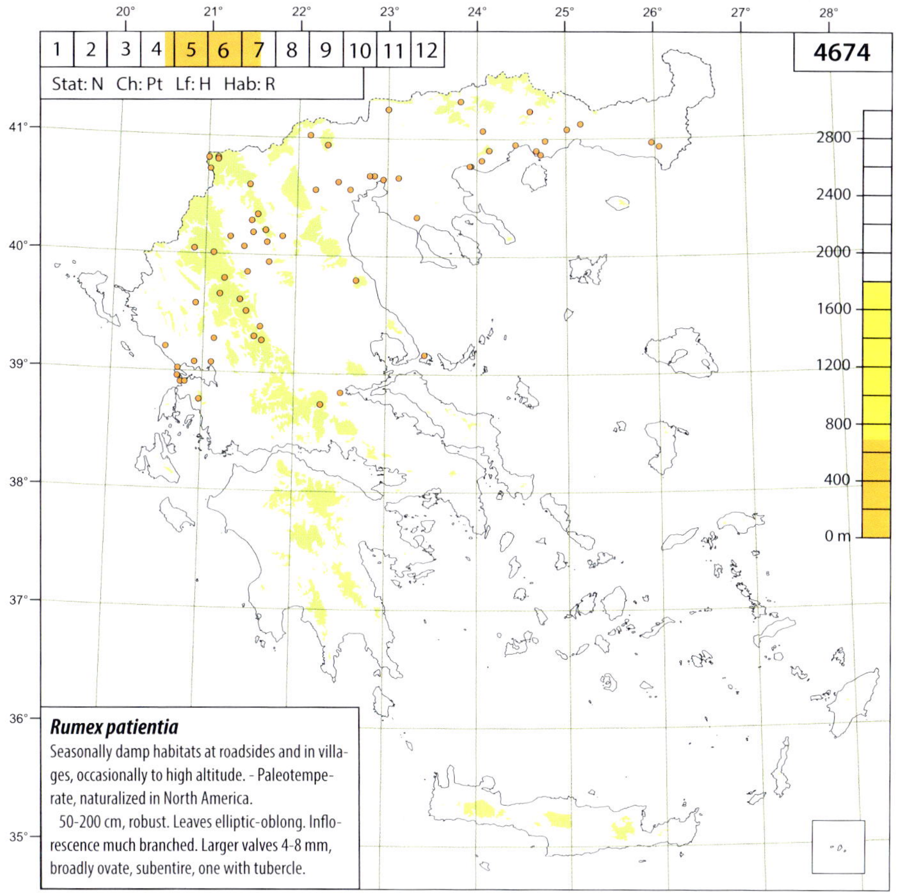

Rumex patientia
Seasonally damp habitats at roadsides and in villages, occasionally to high altitude. - Paleotemperate, naturalized in North America.
50-200 cm, robust. Leaves elliptic-oblong. Inflorescence much branched. Larger valves 4-8 mm, broadly ovate, subentire, one with tubercle.

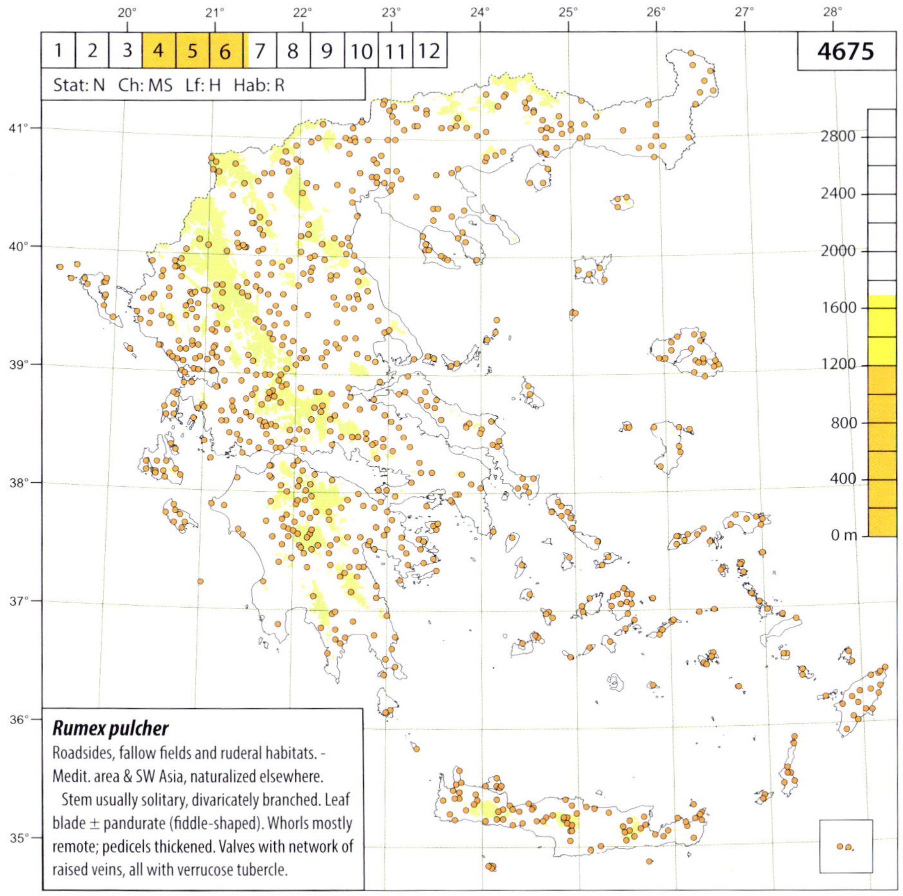

Rumex pulcher
Roadsides, fallow fields and ruderal habitats. - Medit. area & SW Asia, naturalized elsewhere.
Stem usually solitary, divaricately branched. Leaf blade ± pandurate (fiddle-shaped). Whorls mostly remote; pedicels thickened. Valves with network of raised veins, all with verrucose tubercle.

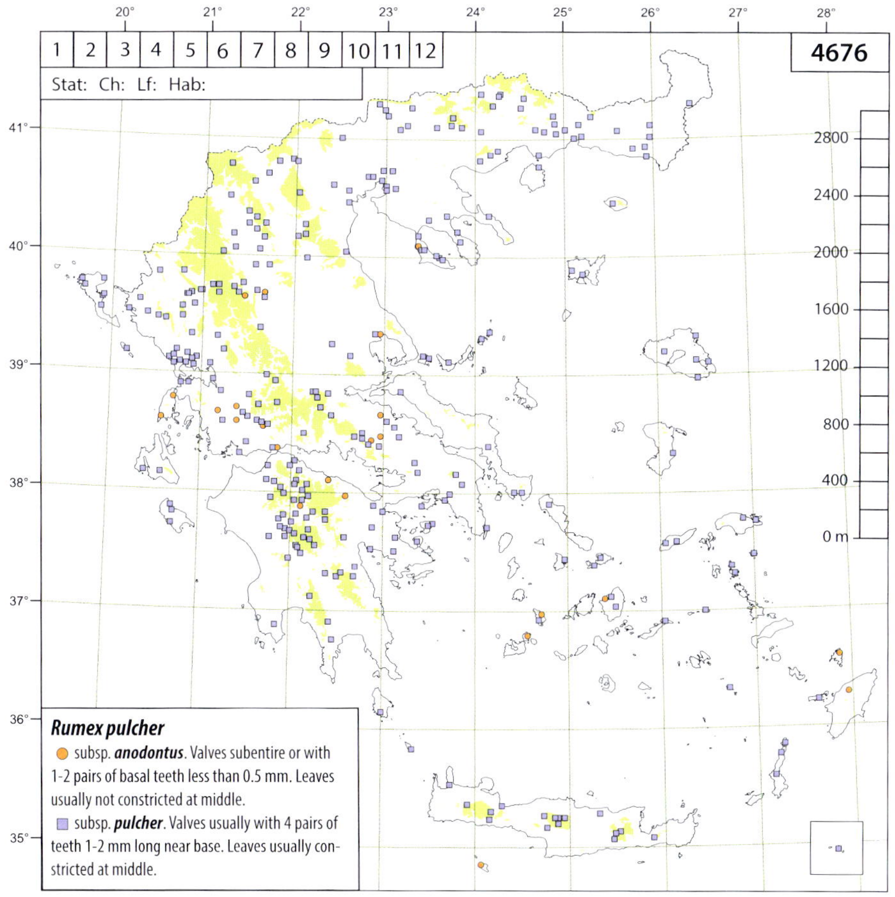

Rumex pulcher
● subsp. ***anodontus***. Valves subentire or with 1-2 pairs of basal teeth less than 0.5 mm. Leaves usually not constricted at middle.
■ subsp. ***pulcher***. Valves usually with 4 pairs of teeth 1-2 mm long near base. Leaves usually constricted at middle.

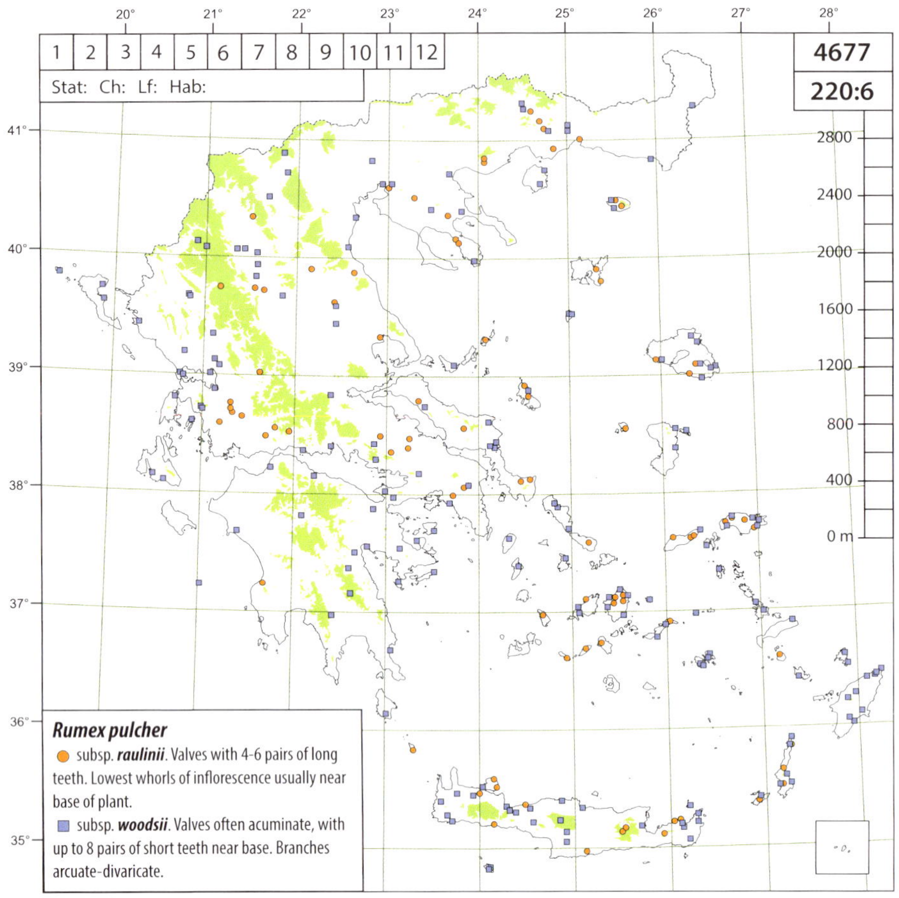

Rumex pulcher

● subsp. ***raulinii***. Valves with 4-6 pairs of long teeth. Lowest whorls of inflorescence usually near base of plant.

■ subsp. ***woodsii***. Valves often acuminate, with up to 8 pairs of short teeth near base. Branches arcuate-divaricate.

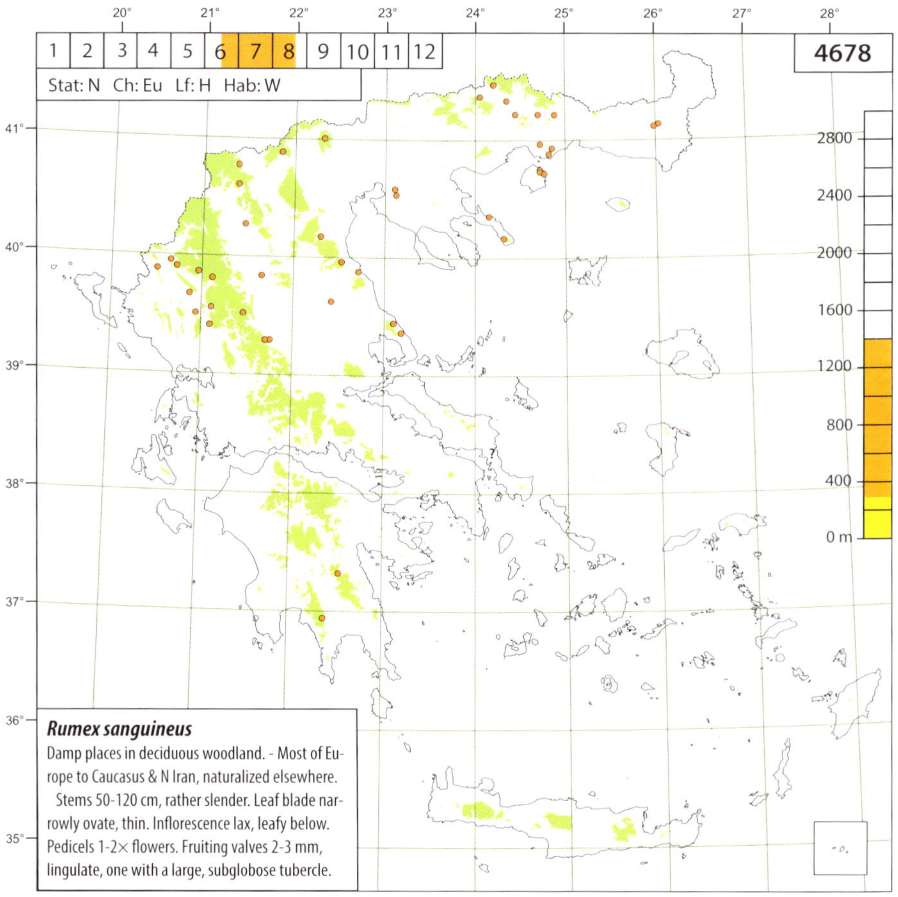

Rumex sanguineus

Damp places in deciduous woodland. - Most of Europe to Caucasus & N Iran, naturalized elsewhere.

Stems 50-120 cm, rather slender. Leaf blade narrowly ovate, thin. Inflorescence lax, leafy below. Pedicels 1-2× flowers. Fruiting valves 2-3 mm, lingulate, one with a large, subglobose tubercle.

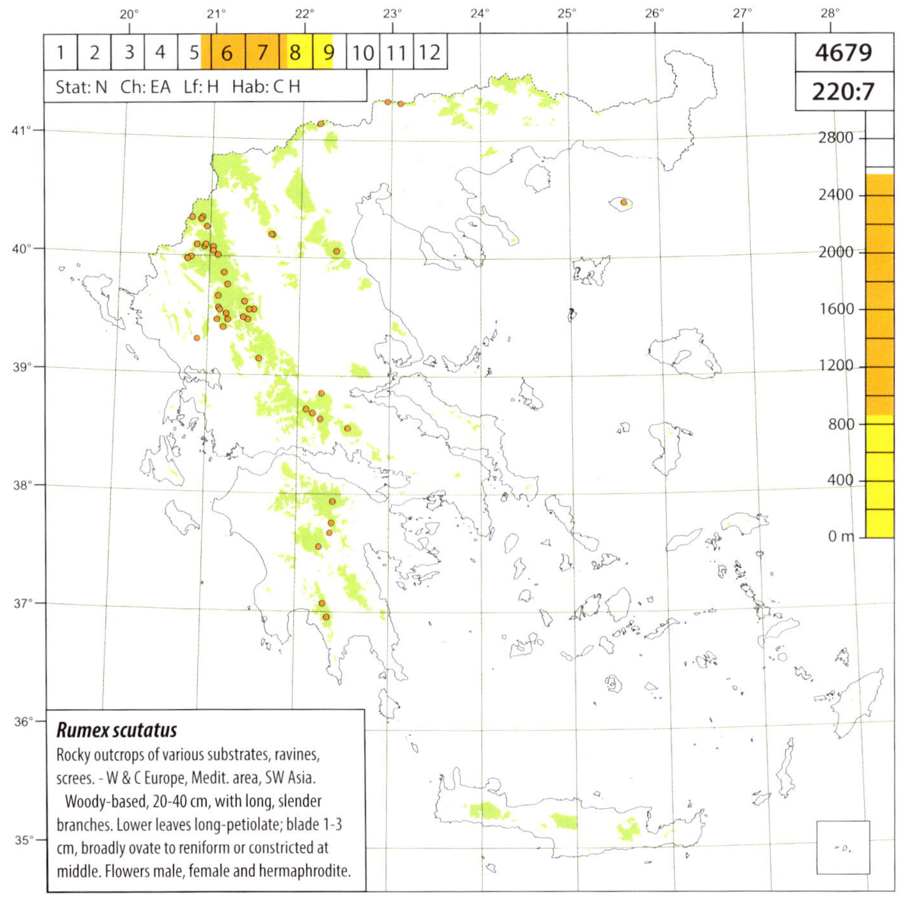

Rumex scutatus

Rocky outcrops of various substrates, ravines, screes. - W & C Europe, Medit. area, SW Asia.

Woody-based, 20-40 cm, with long, slender branches. Lower leaves long-petiolate; blade 1-3 cm, broadly ovate to reniform or constricted at middle. Flowers male, female and hermaphrodite.

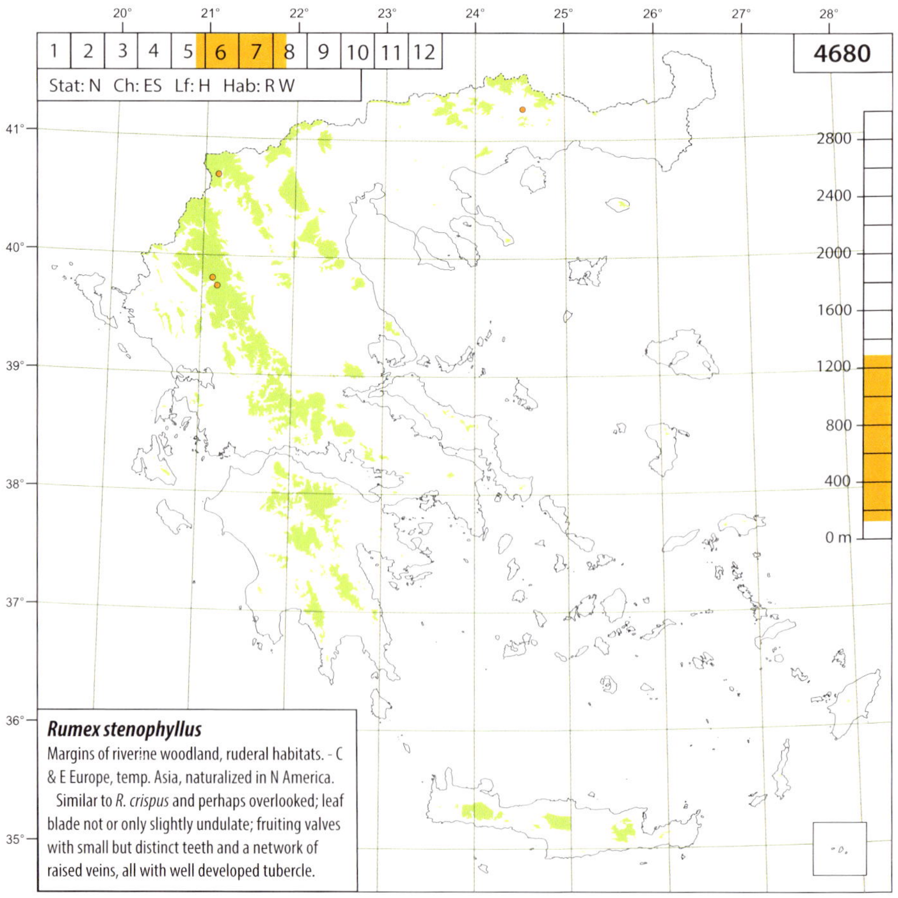

Rumex stenophyllus

Margins of riverine woodland, ruderal habitats. - C & E Europe, temp. Asia, naturalized in N America.

Similar to *R. crispus* and perhaps overlooked; leaf blade not or only slightly undulate; fruiting valves with small but distinct teeth and a network of raised veins, all with well developed tubercle.

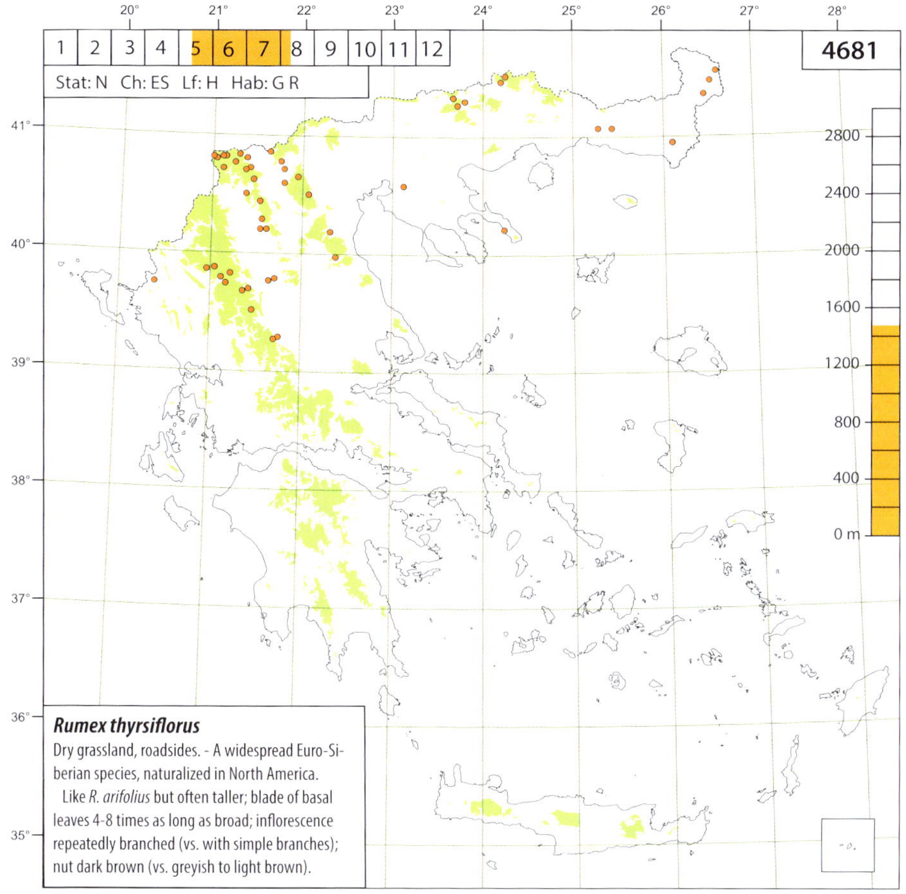

Rumex thyrsiflorus
Dry grassland, roadsides. - A widespread Euro-Siberian species, naturalized in North America.
Like *R. arifolius* but often taller; blade of basal leaves 4-8 times as long as broad; inflorescence repeatedly branched (vs. with simple branches); nut dark brown (vs. greyish to light brown).

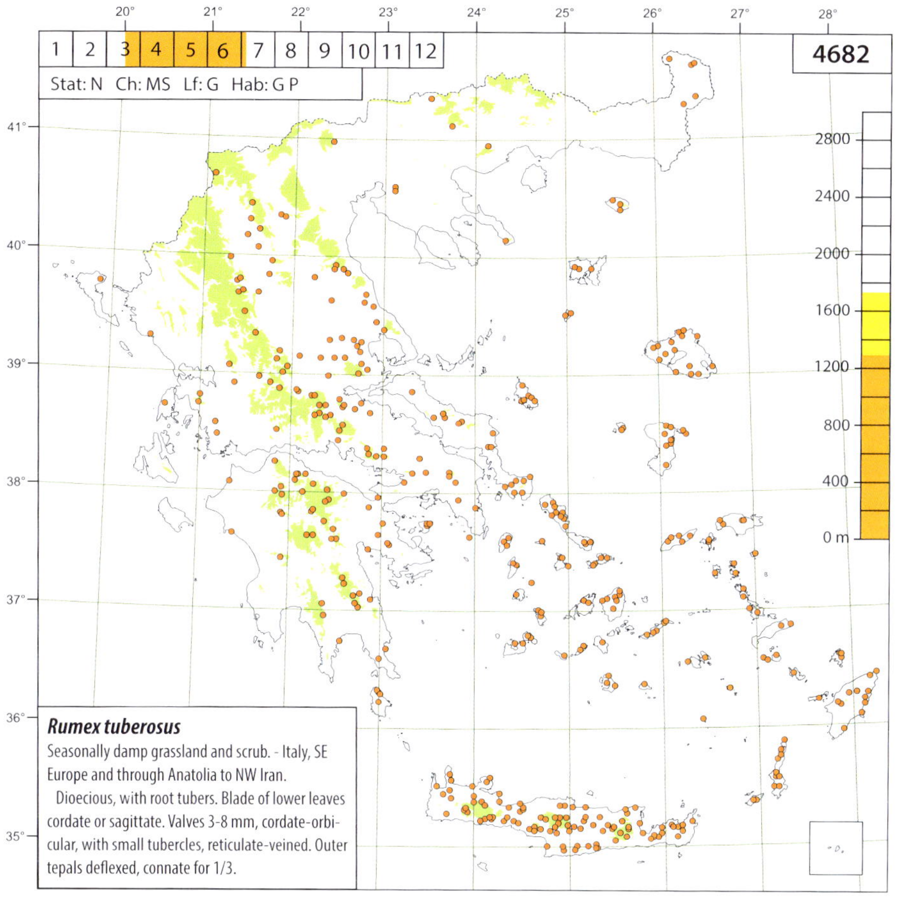

Rumex tuberosus
Seasonally damp grassland and scrub. - Italy, SE Europe and through Anatolia to NW Iran.
Dioecious, with root tubers. Blade of lower leaves cordate or sagittate. Valves 3-8 mm, cordate-orbicular, with small tubercles, reticulate-veined. Outer tepals deflexed, connate for 1/3.

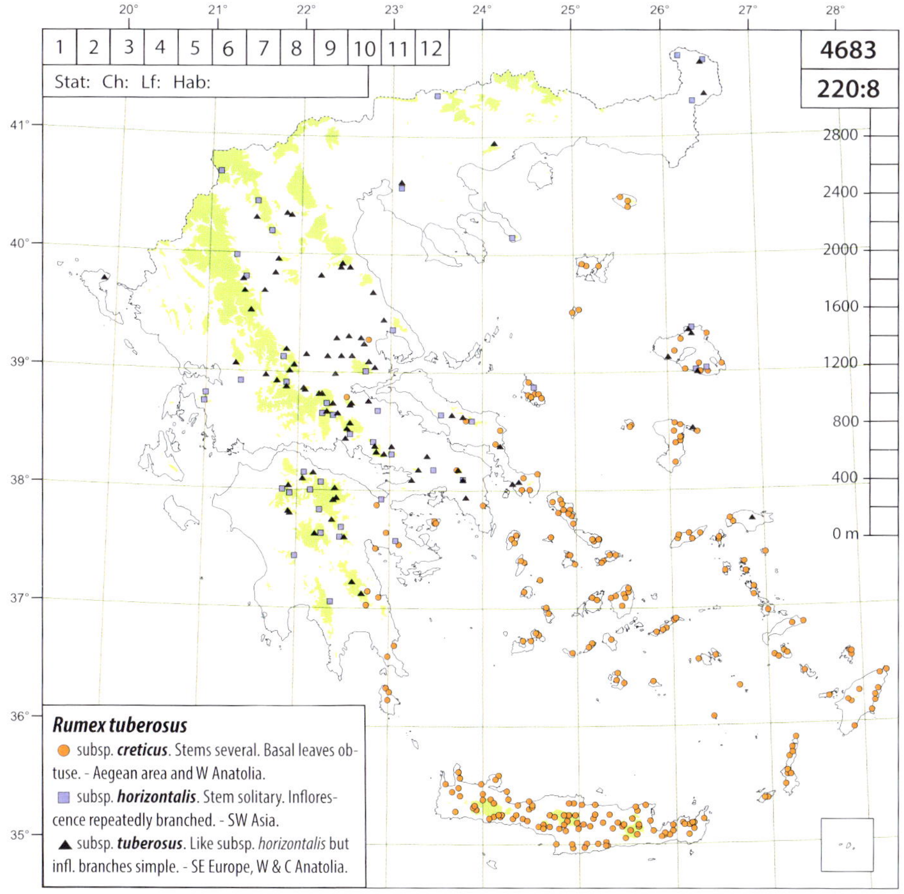

Rumex tuberosus
● subsp. ***creticus***. Stems several. Basal leaves obtuse. - Aegean area and W Anatolia.
■ subsp. ***horizontalis***. Stem solitary. Inflorescence repeatedly branched. - SW Asia.
▲ subsp. ***tuberosus***. Like subsp. *horizontalis* but infl. branches simple. - SE Europe, W & C Anatolia.

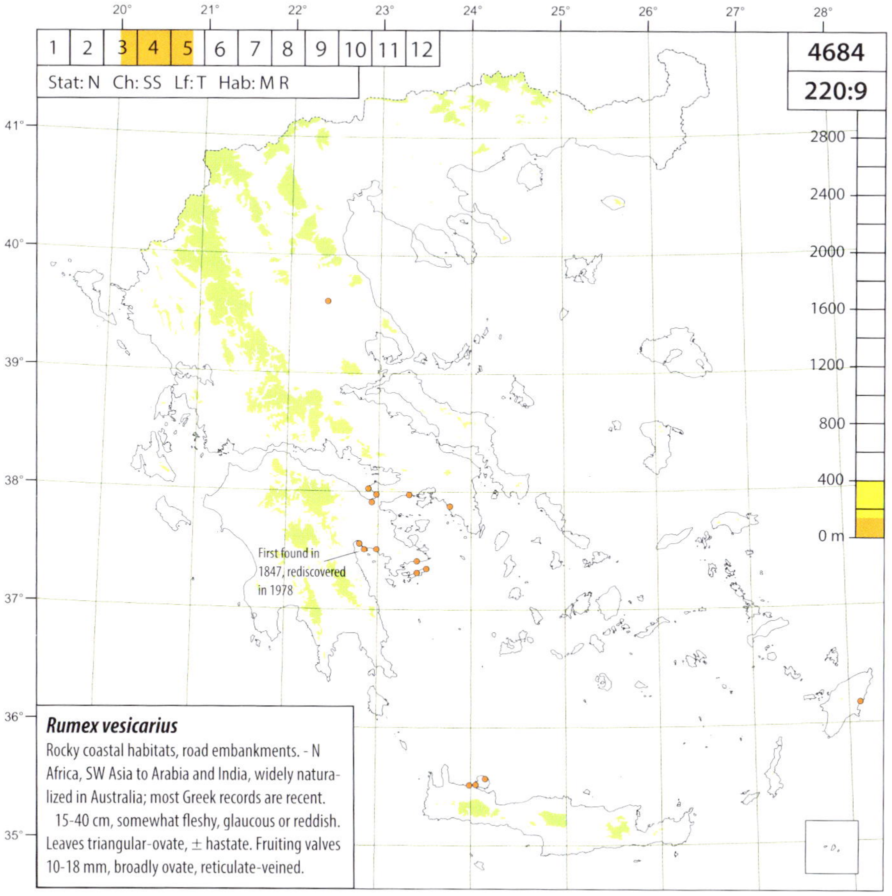

Rumex vesicarius
Rocky coastal habitats, road embankments. - N Africa, SW Asia to Arabia and India, widely naturalized in Australia; most Greek records are recent.
15-40 cm, somewhat fleshy, glaucous or reddish. Leaves triangular-ovate, ± hastate. Fruiting valves 10-18 mm, broadly ovate, reticulate-veined.

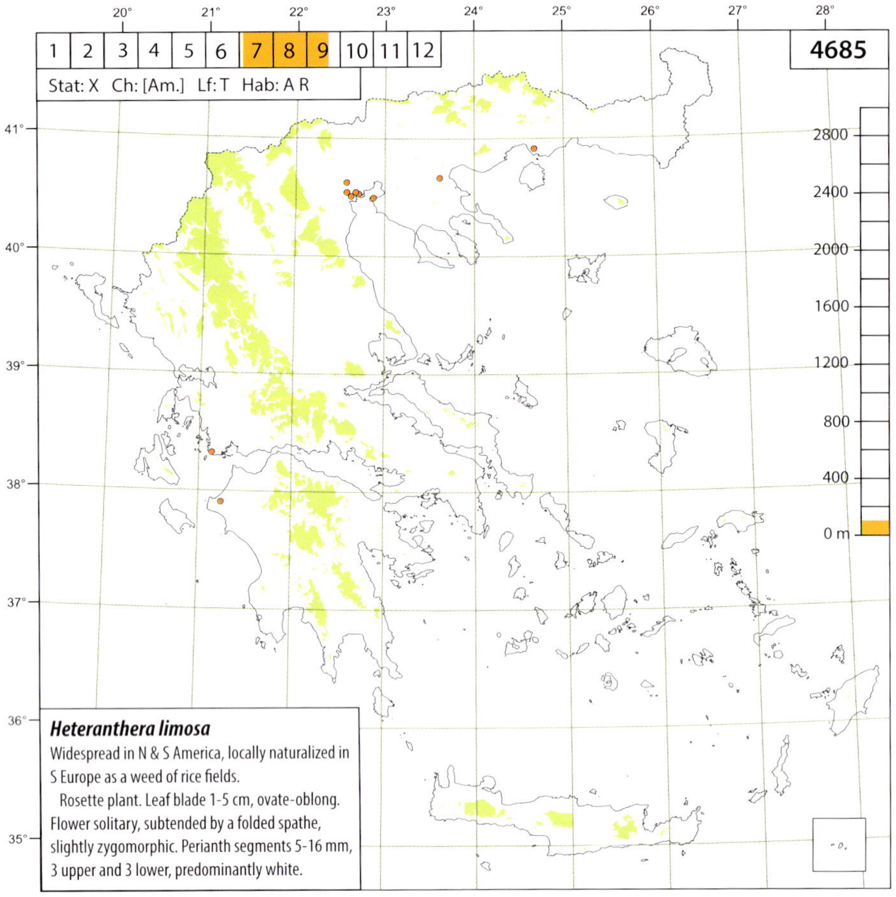

Heteranthera limosa

Widespread in N & S America, locally naturalized in S Europe as a weed of rice fields.

Rosette plant. Leaf blade 1-5 cm, ovate-oblong. Flower solitary, subtended by a folded spathe, slightly zygomorphic. Perianth segments 5-16 mm, 3 upper and 3 lower, predominantly white.

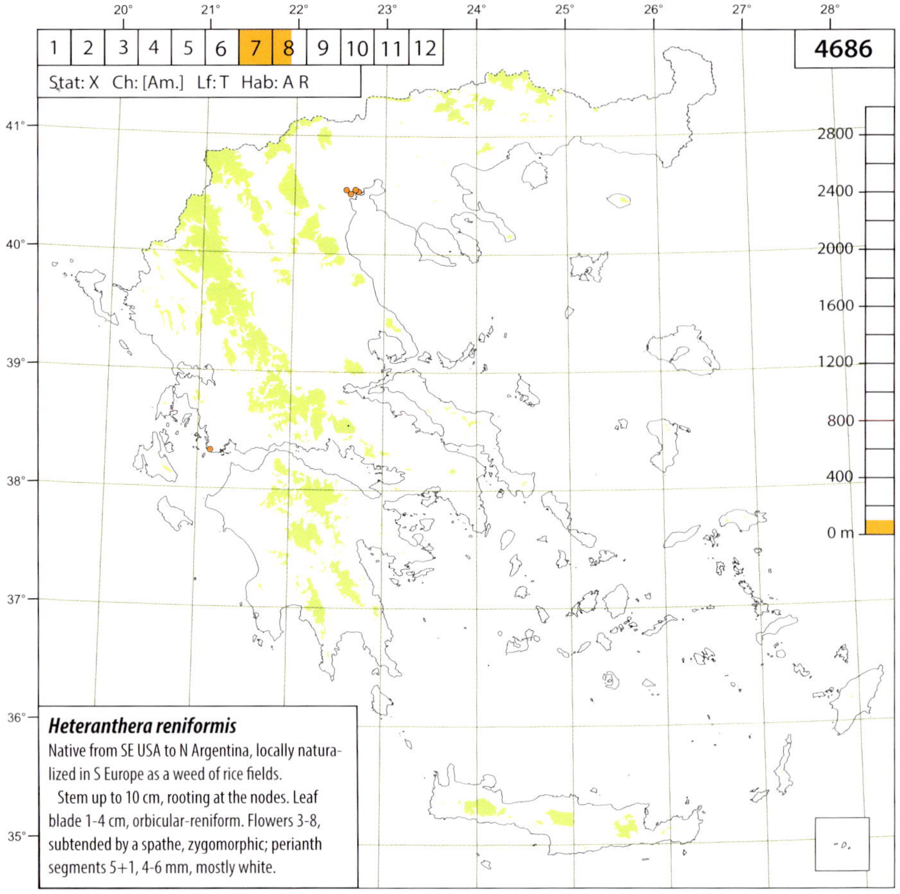

Heteranthera reniformis

Native from SE USA to N Argentina, locally naturalized in S Europe as a weed of rice fields.

Stem up to 10 cm, rooting at the nodes. Leaf blade 1-4 cm, orbicular-reniform. Flowers 3-8, subtended by a spathe, zygomorphic; perianth segments 5+1, 4-6 mm, mostly white.

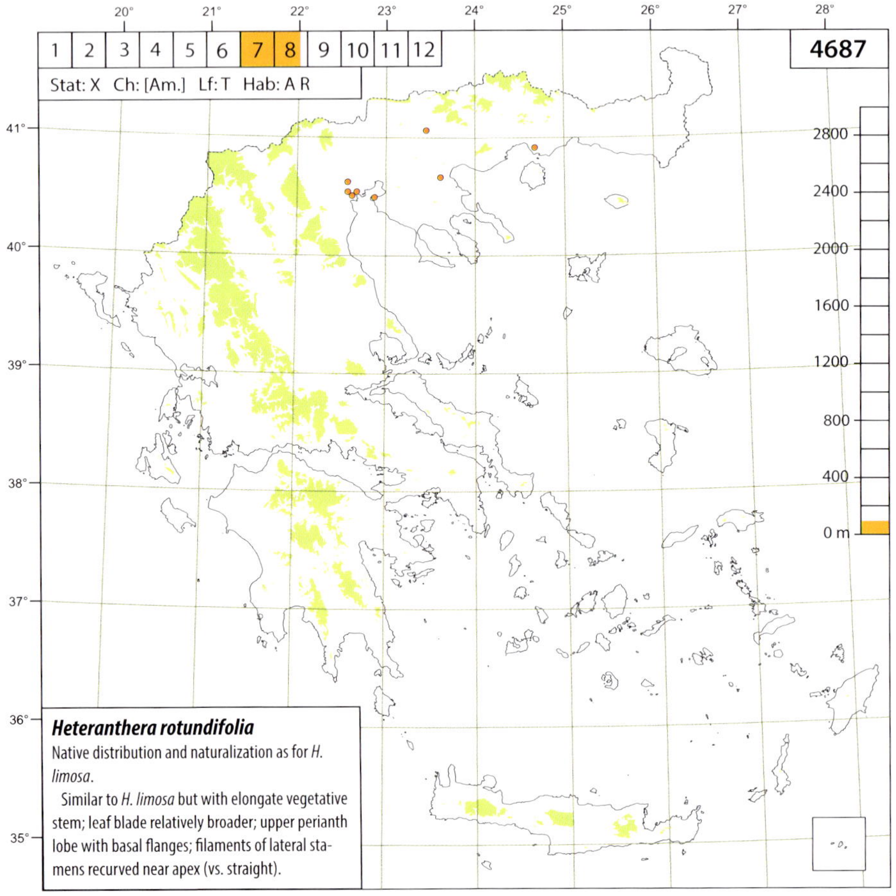

Heteranthera rotundifolia

Native distribution and naturalization as for *H. limosa*.

Similar to *H. limosa* but with elongate vegetative stem; leaf blade relatively broader; upper perianth lobe with basal flanges; filaments of lateral stamens recurved near apex (vs. straight).

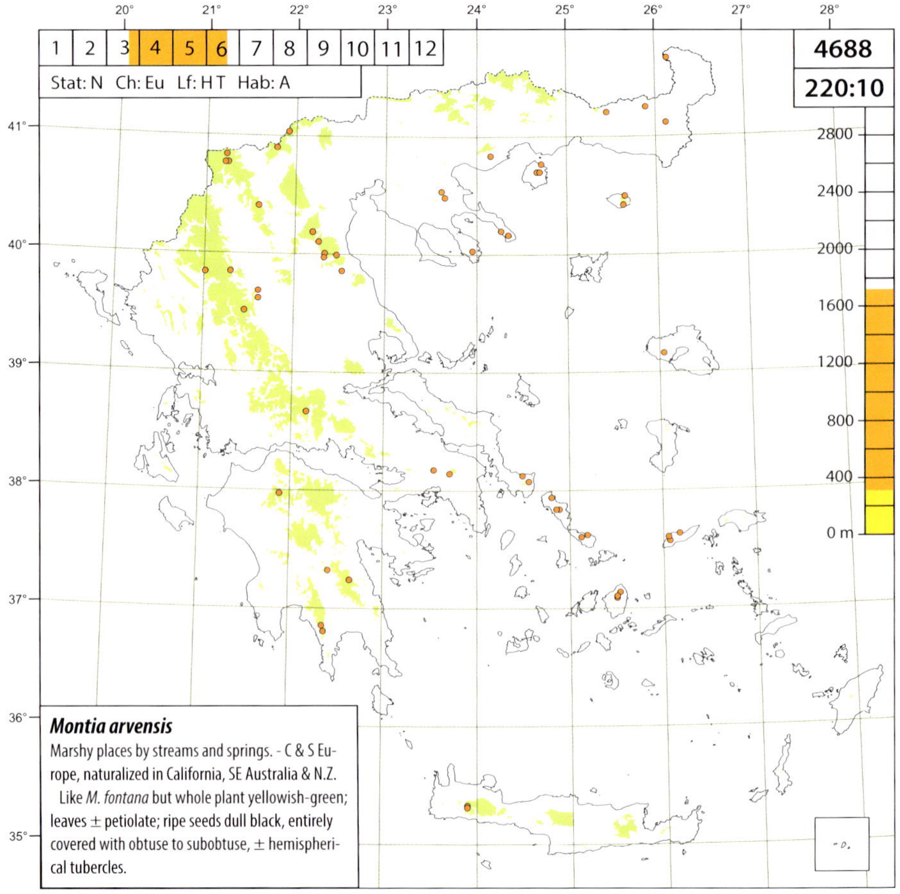

Montia arvensis

Marshy places by streams and springs. - C & S Europe, naturalized in California, SE Australia & N.Z.

Like *M. fontana* but whole plant yellowish-green; leaves ± petiolate; ripe seeds dull black, entirely covered with obtuse to subobtuse, ± hemispherical tubercles.

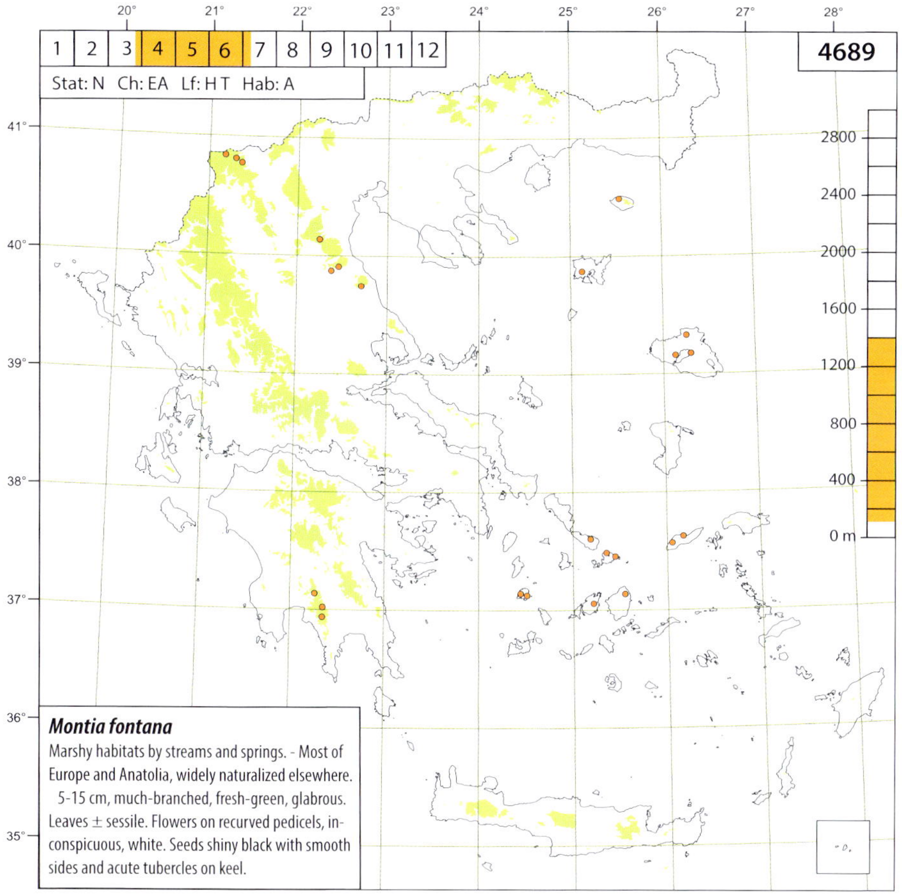

Montia fontana
Marshy habitats by streams and springs. - Most of Europe and Anatolia, widely naturalized elsewhere.
5-15 cm, much-branched, fresh-green, glabrous. Leaves ± sessile. Flowers on recurved pedicels, inconspicuous, white. Seeds shiny black with smooth sides and acute tubercles on keel.

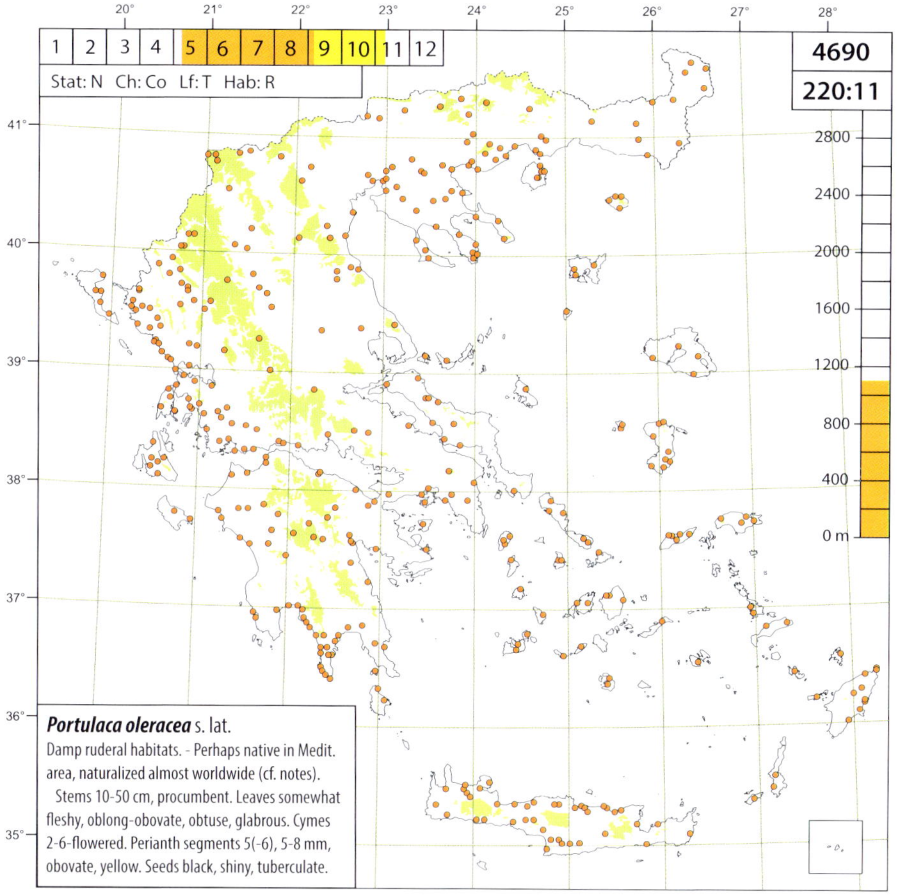

Portulaca oleracea s. lat.
Damp ruderal habitats. - Perhaps native in Medit. area, naturalized almost worldwide (cf. notes).
Stems 10-50 cm, procumbent. Leaves somewhat fleshy, oblong-obovate, obtuse, glabrous. Cymes 2-6-flowered. Perianth segments 5(-6), 5-8 mm, obovate, yellow. Seeds black, shiny, tuberculate.

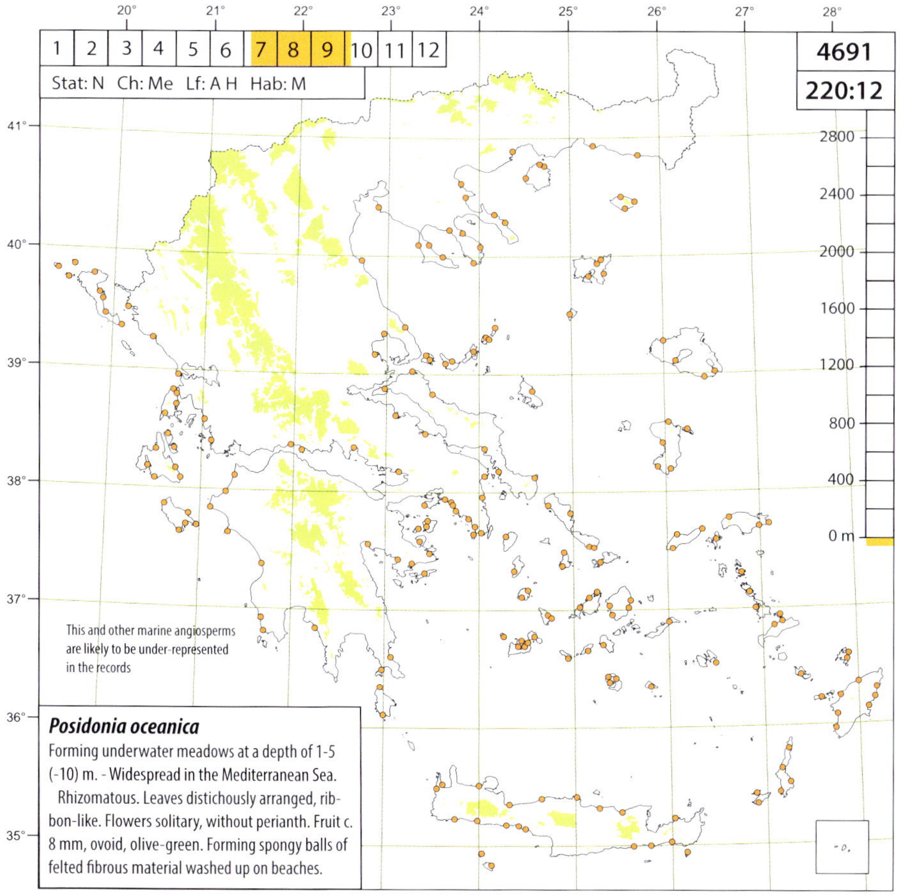

This and other marine angiosperms are likely to be under-represented in the records

Posidonia oceanica
Forming underwater meadows at a depth of 1-5 (-10) m. - Widespread in the Mediterranean Sea.
Rhizomatous. Leaves distichously arranged, ribbon-like. Flowers solitary, without perianth. Fruit c. 8 mm, ovoid, olive-green. Forming spongy balls of felted fibrous material washed up on beaches.

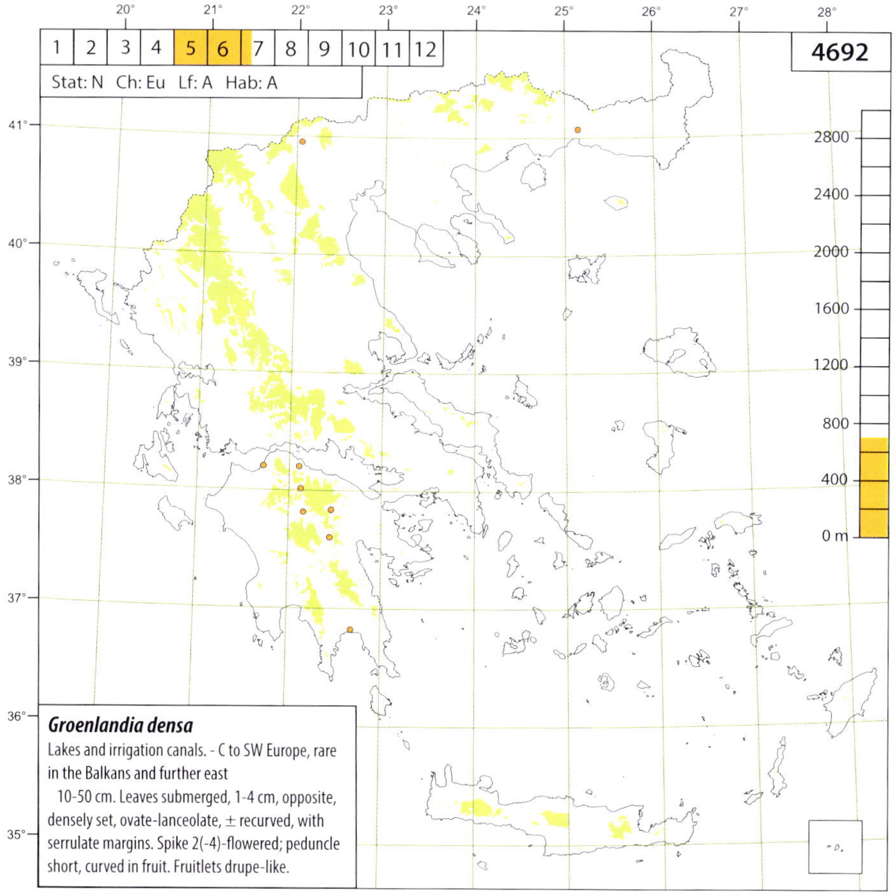

Groenlandia densa
Lakes and irrigation canals. - C to SW Europe, rare in the Balkans and further east
10-50 cm. Leaves submerged, 1-4 cm, opposite, densely set, ovate-lanceolate, ± recurved, with serrulate margins. Spike 2(-4)-flowered; peduncle short, curved in fruit. Fruitlets drupe-like.

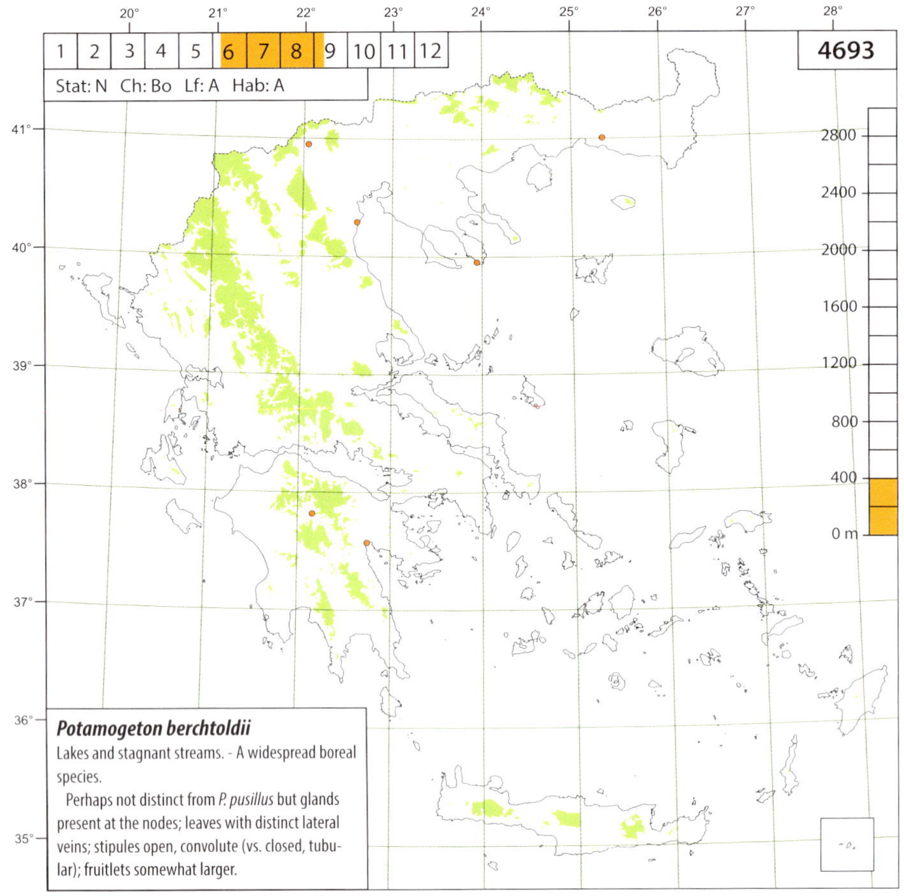

Potamogeton berchtoldii

Lakes and stagnant streams. - A widespread boreal species.

Perhaps not distinct from *P. pusillus* but glands present at the nodes; leaves with distinct lateral veins; stipules open, convolute (vs. closed, tubular); fruitlets somewhat larger.

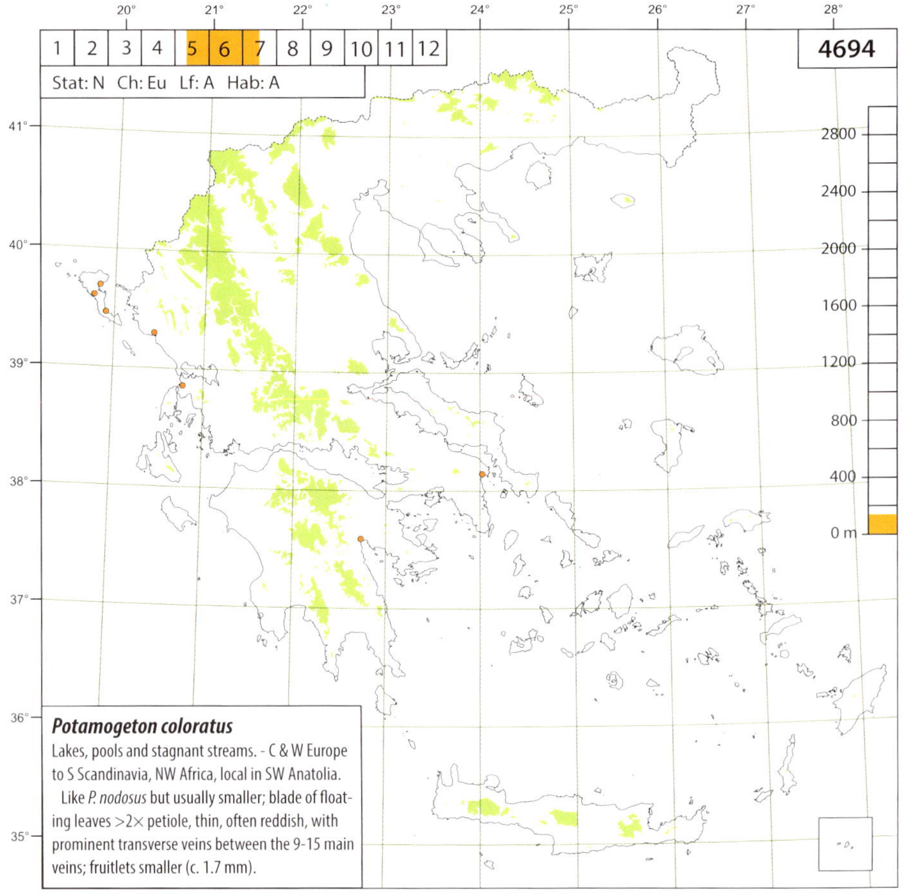

Potamogeton coloratus

Lakes, pools and stagnant streams. - C & W Europe to S Scandinavia, NW Africa, local in SW Anatolia.

Like *P. nodosus* but usually smaller; blade of floating leaves >2× petiole, thin, often reddish, with prominent transverse veins between the 9-15 main veins; fruitlets smaller (c. 1.7 mm).

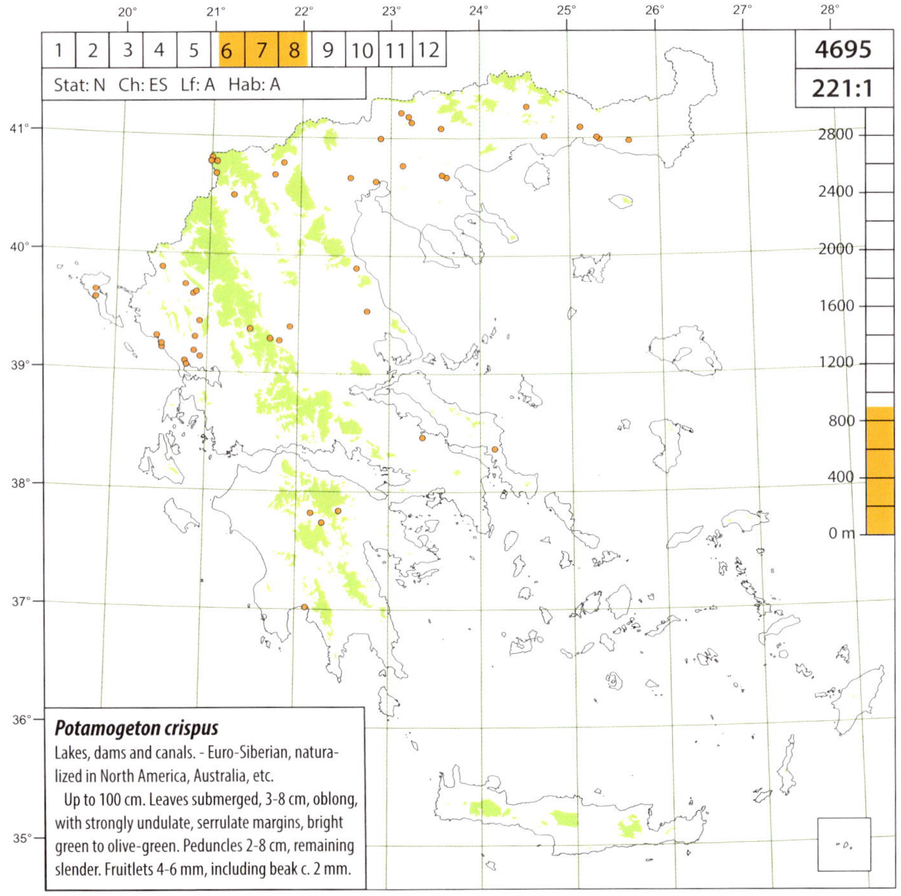

Potamogeton crispus

Lakes, dams and canals. - Euro-Siberian, naturalized in North America, Australia, etc.

Up to 100 cm. Leaves submerged, 3-8 cm, oblong, with strongly undulate, serrulate margins, bright green to olive-green. Peduncles 2-8 cm, remaining slender. Fruitlets 4-6 mm, including beak c. 2 mm.

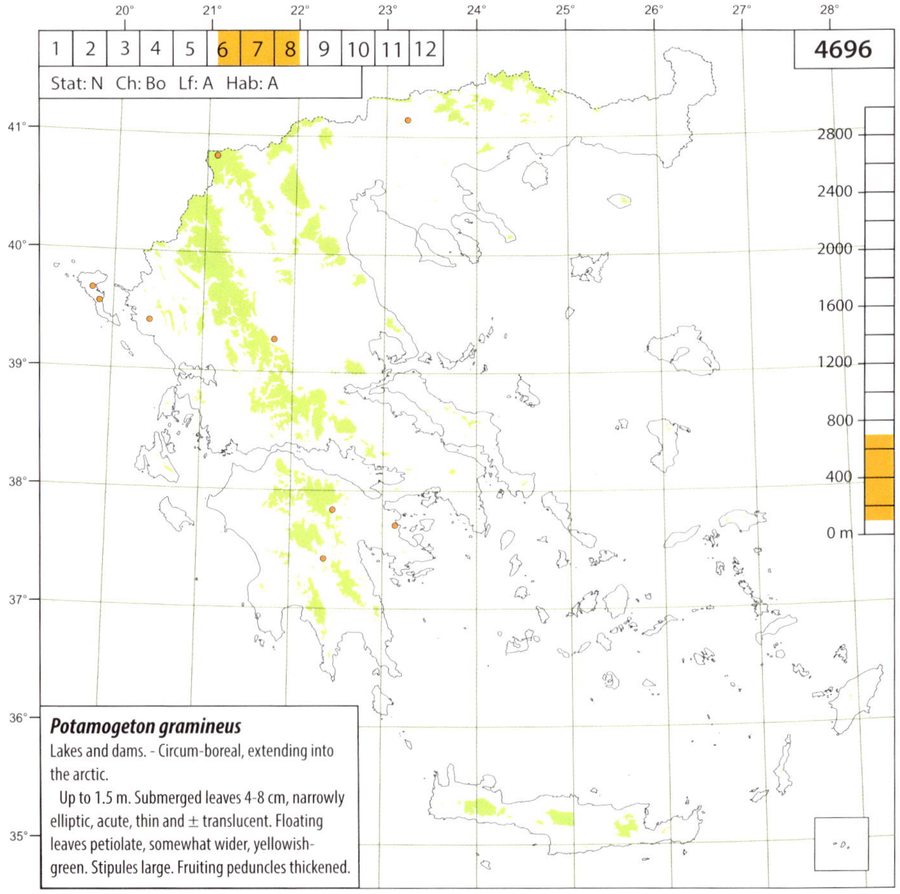

Potamogeton gramineus

Lakes and dams. - Circum-boreal, extending into the arctic.

Up to 1.5 m. Submerged leaves 4-8 cm, narrowly elliptic, acute, thin and ± translucent. Floating leaves petiolate, somewhat wider, yellowish-green. Stipules large. Fruiting peduncles thickened.

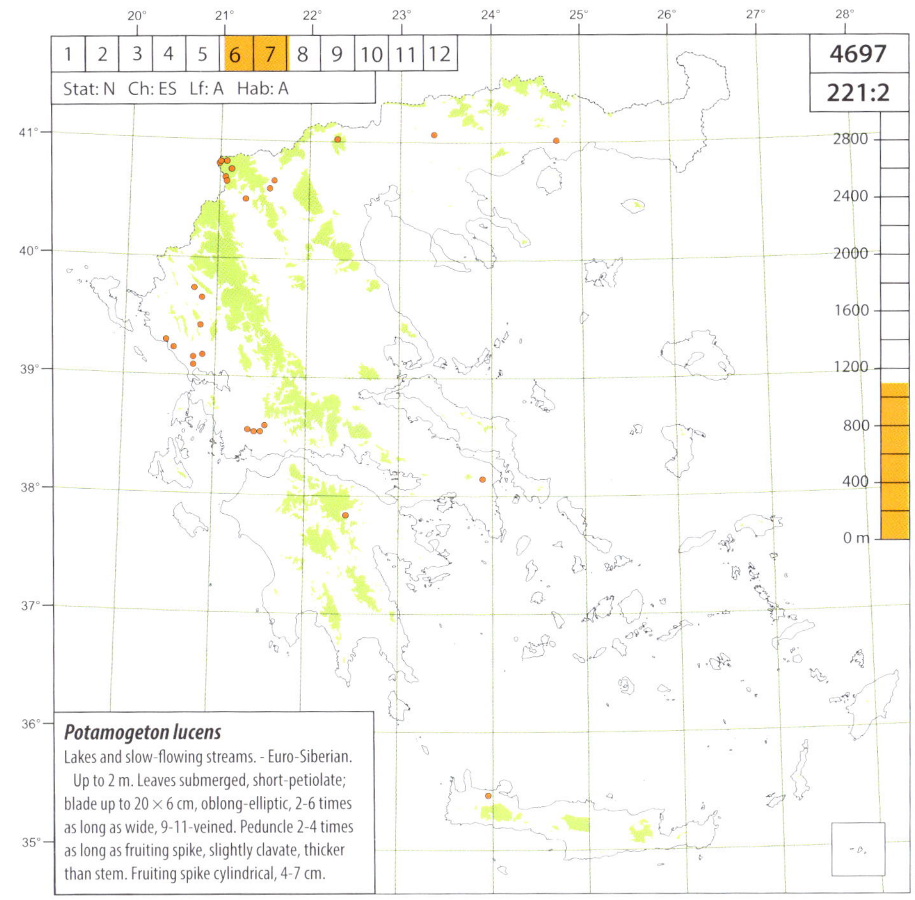

Potamogeton lucens

Lakes and slow-flowing streams. - Euro-Siberian.

Up to 2 m. Leaves submerged, short-petiolate; blade up to 20 × 6 cm, oblong-elliptic, 2-6 times as long as wide, 9-11-veined. Peduncle 2-4 times as long as fruiting spike, slightly clavate, thicker than stem. Fruiting spike cylindrical, 4-7 cm.

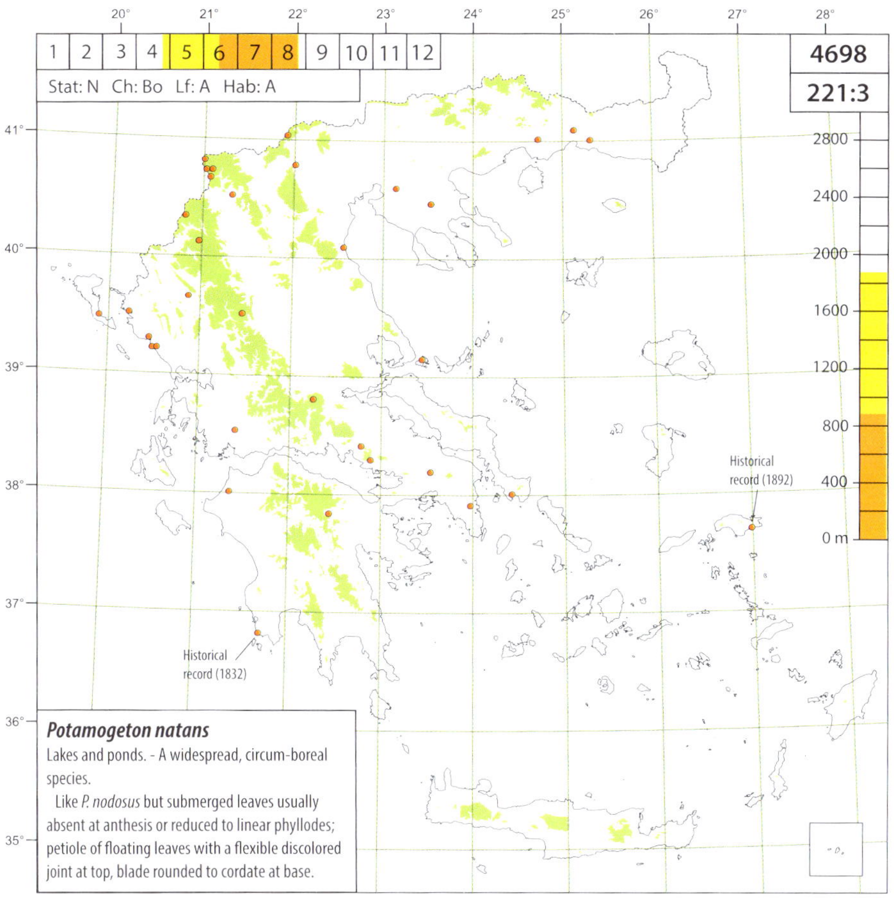

Potamogeton natans

Lakes and ponds. - A widespread, circum-boreal species.

Like *P. nodosus* but submerged leaves usually absent at anthesis or reduced to linear phyllodes; petiole of floating leaves with a flexible discolored joint at top, blade rounded to cordate at base.

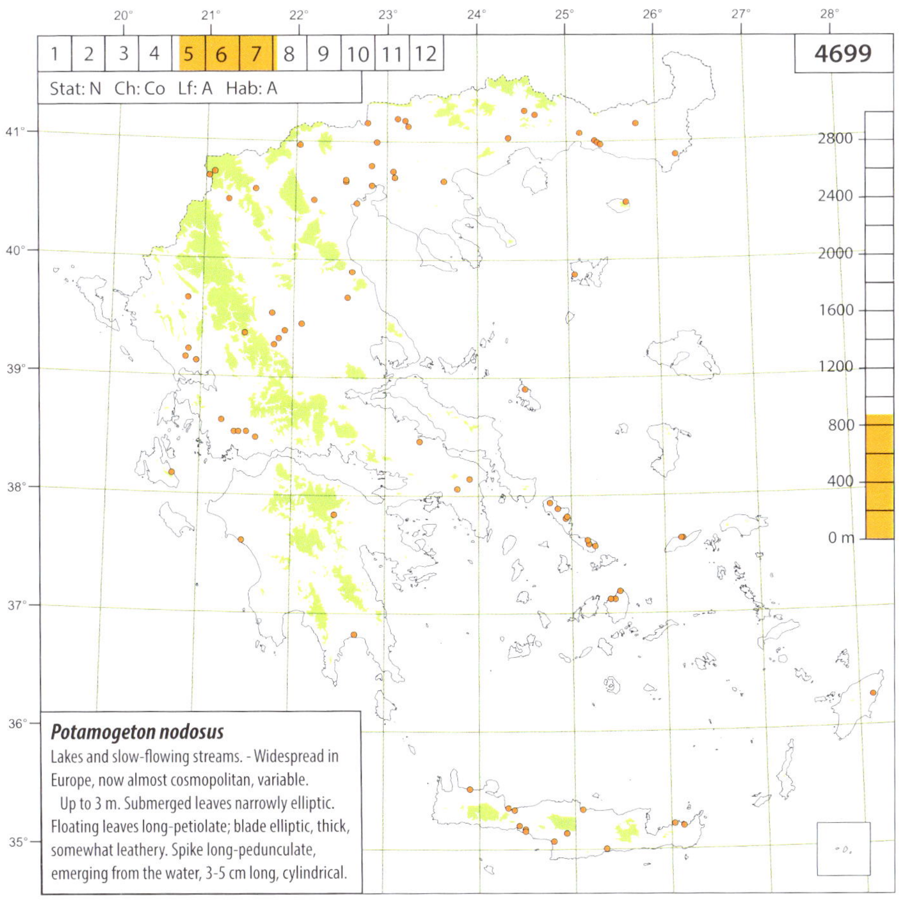

Potamogeton nodosus

Lakes and slow-flowing streams. - Widespread in Europe, now almost cosmopolitan, variable.

Up to 3 m. Submerged leaves narrowly elliptic. Floating leaves long-petiolate; blade elliptic, thick, somewhat leathery. Spike long-pedunculate, emerging from the water, 3-5 cm long, cylindrical.

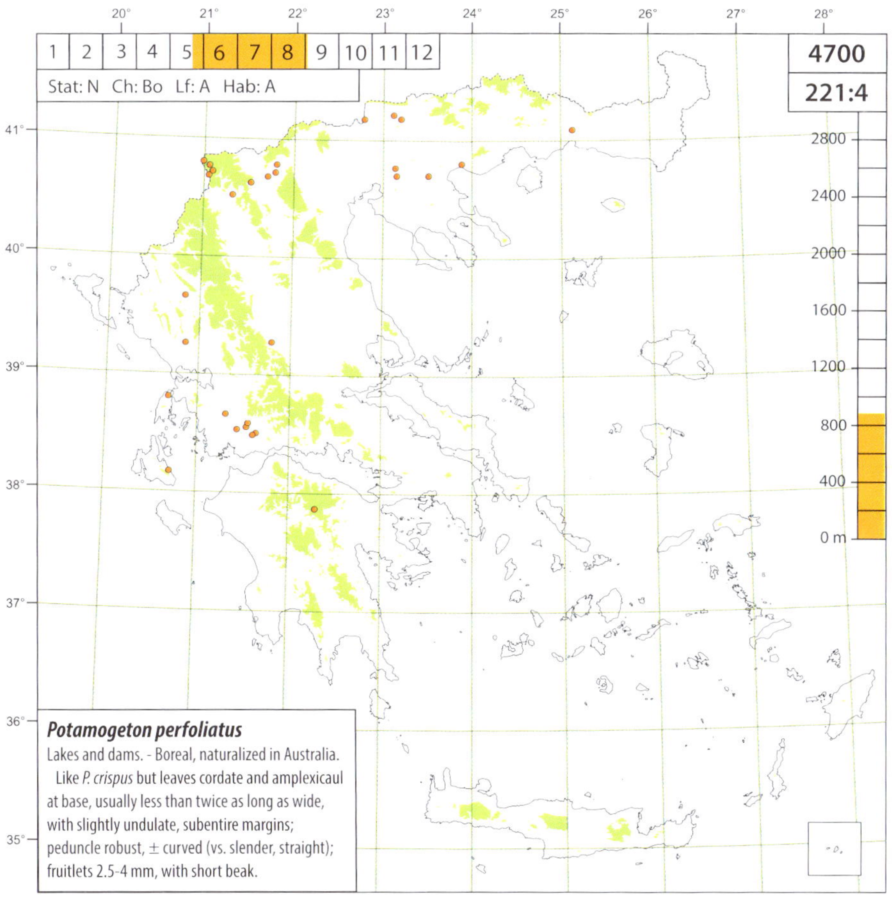

Potamogeton perfoliatus

Lakes and dams. - Boreal, naturalized in Australia.

Like *P. crispus* but leaves cordate and amplexicaul at base, usually less than twice as long as wide, with slightly undulate, subentire margins; peduncle robust, ± curved (vs. slender, straight); fruitlets 2.5-4 mm, with short beak.

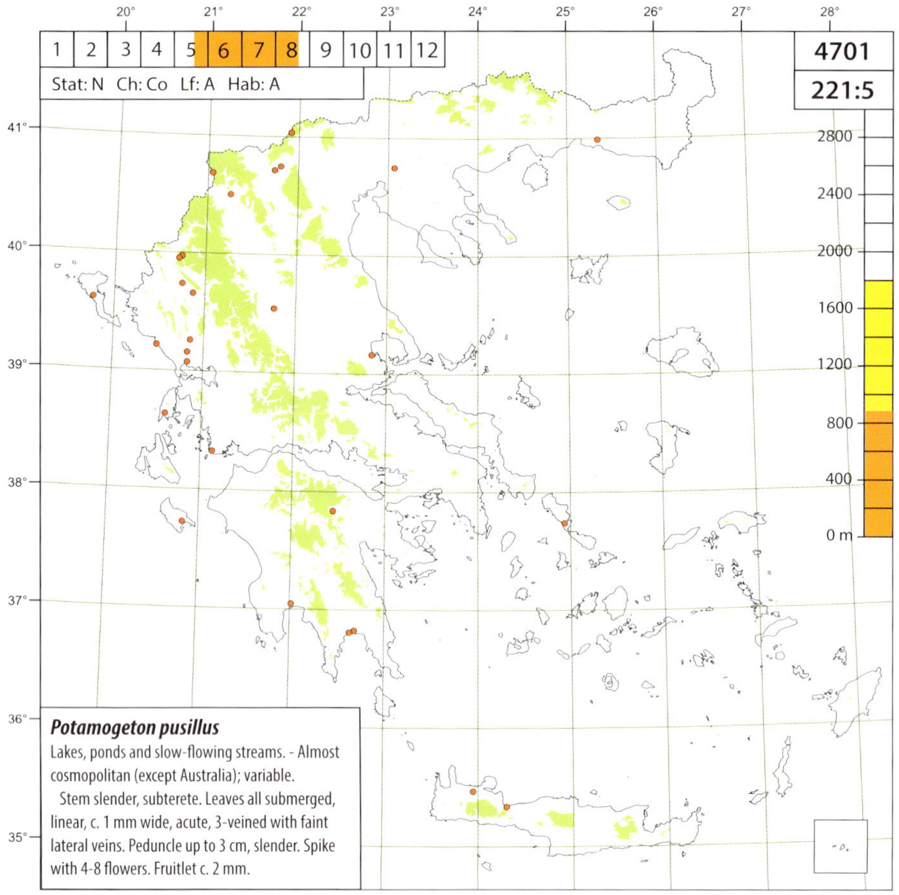

Potamogeton pusillus
Lakes, ponds and slow-flowing streams. - Almost cosmopolitan (except Australia); variable.
Stem slender, subterete. Leaves all submerged, linear, c. 1 mm wide, acute, 3-veined with faint lateral veins. Peduncle up to 3 cm, slender. Spike with 4-8 flowers. Fruitlet c. 2 mm.

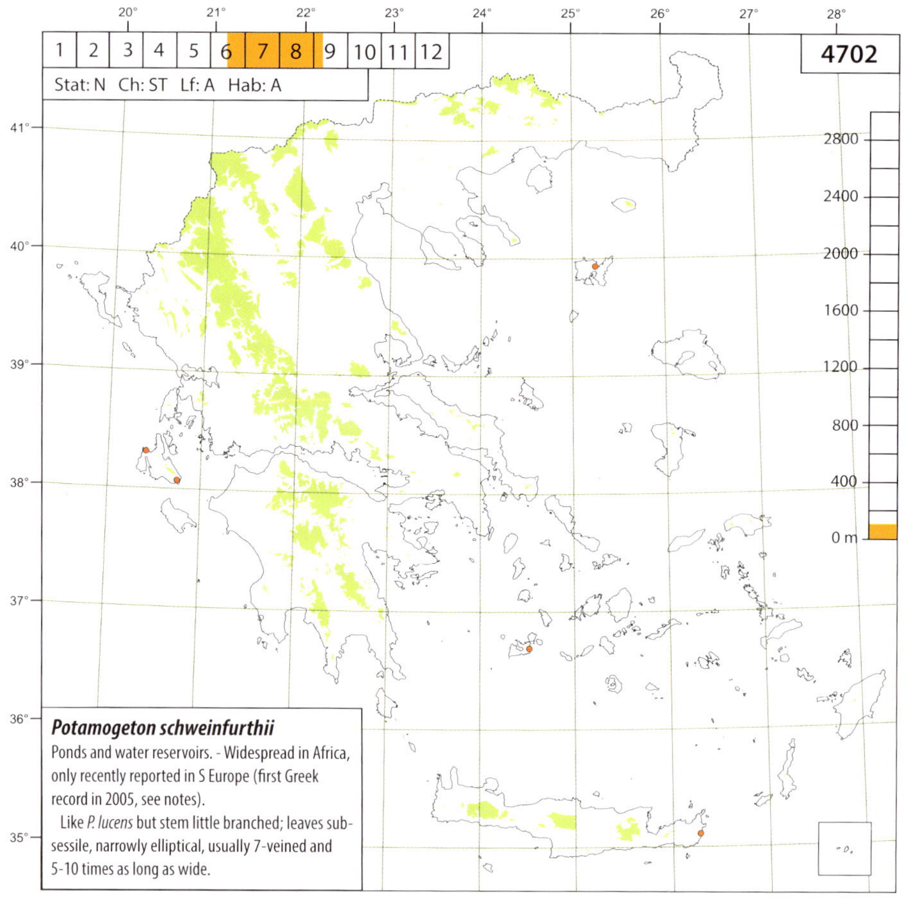

Potamogeton schweinfurthii
Ponds and water reservoirs. - Widespread in Africa, only recently reported in S Europe (first Greek record in 2005, see notes).
Like *P. lucens* but stem little branched; leaves sub-sessile, narrowly elliptical, usually 7-veined and 5-10 times as long as wide.

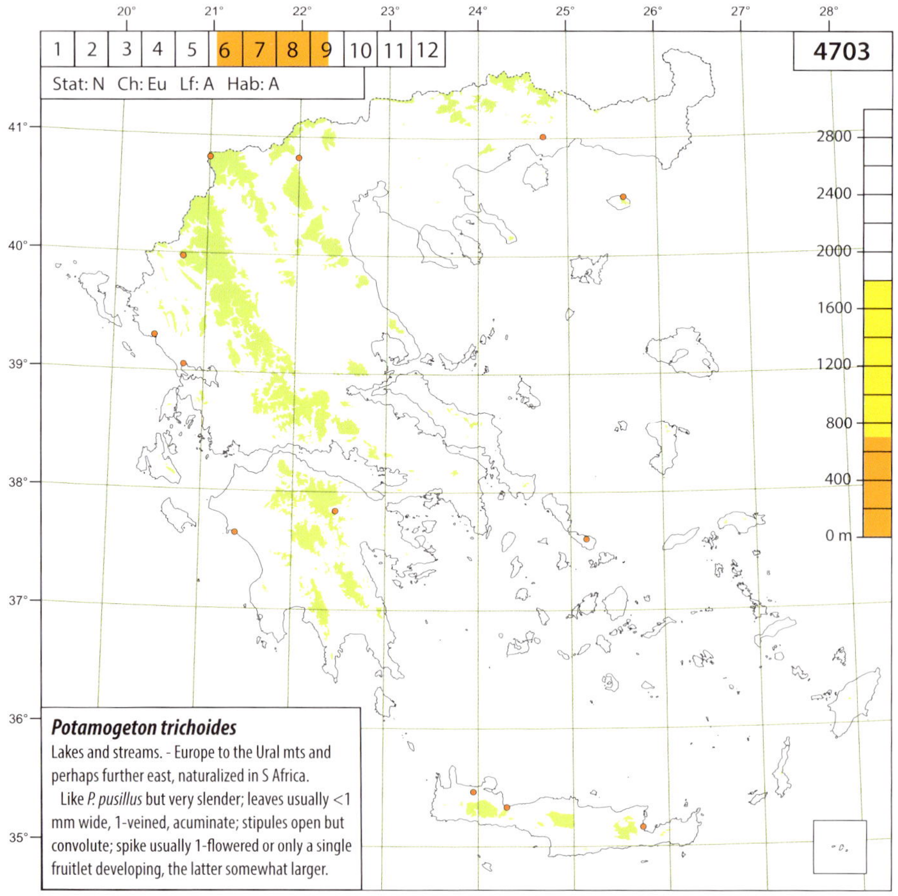

Potamogeton trichoides
Lakes and streams. - Europe to the Ural mts and perhaps further east, naturalized in S Africa.
Like *P. pusillus* but very slender; leaves usually <1 mm wide, 1-veined, acuminate; stipules open but convolute; spike usually 1-flowered or only a single fruitlet developing, the latter somewhat larger.

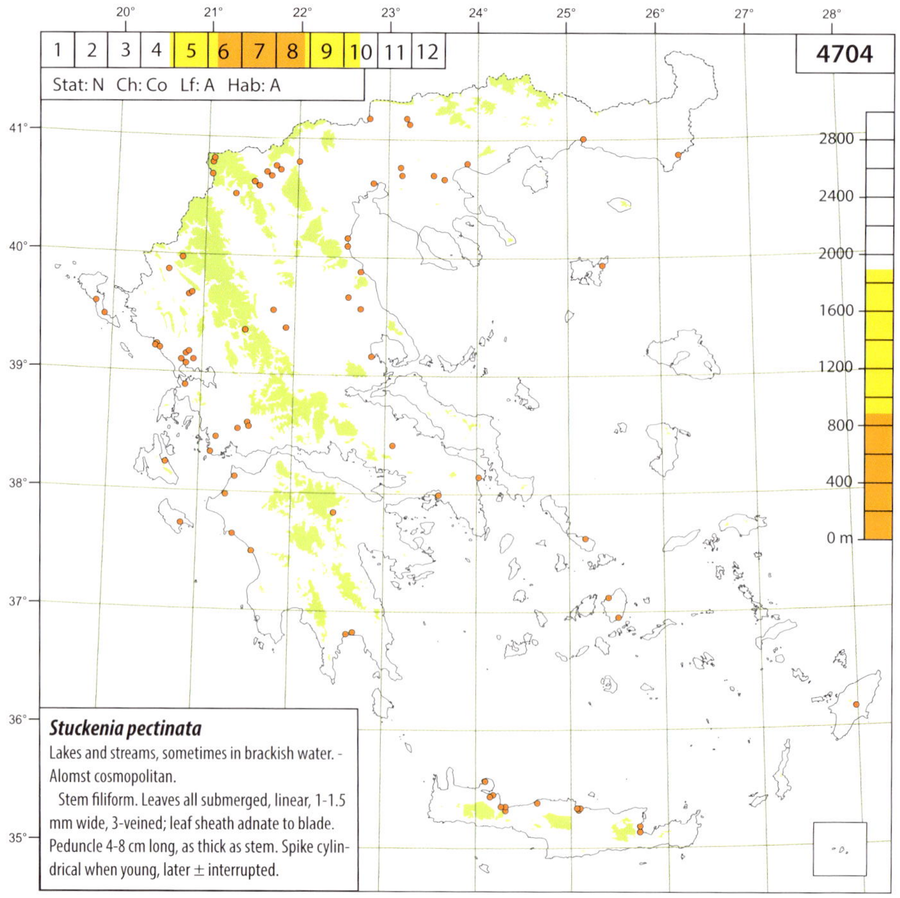

Stuckenia pectinata
Lakes and streams, sometimes in brackish water. - Alomst cosmopolitan.
Stem filiform. Leaves all submerged, linear, 1-1.5 mm wide, 3-veined; leaf sheath adnate to blade. Peduncle 4-8 cm long, as thick as stem. Spike cylindrical when young, later ± interrupted.

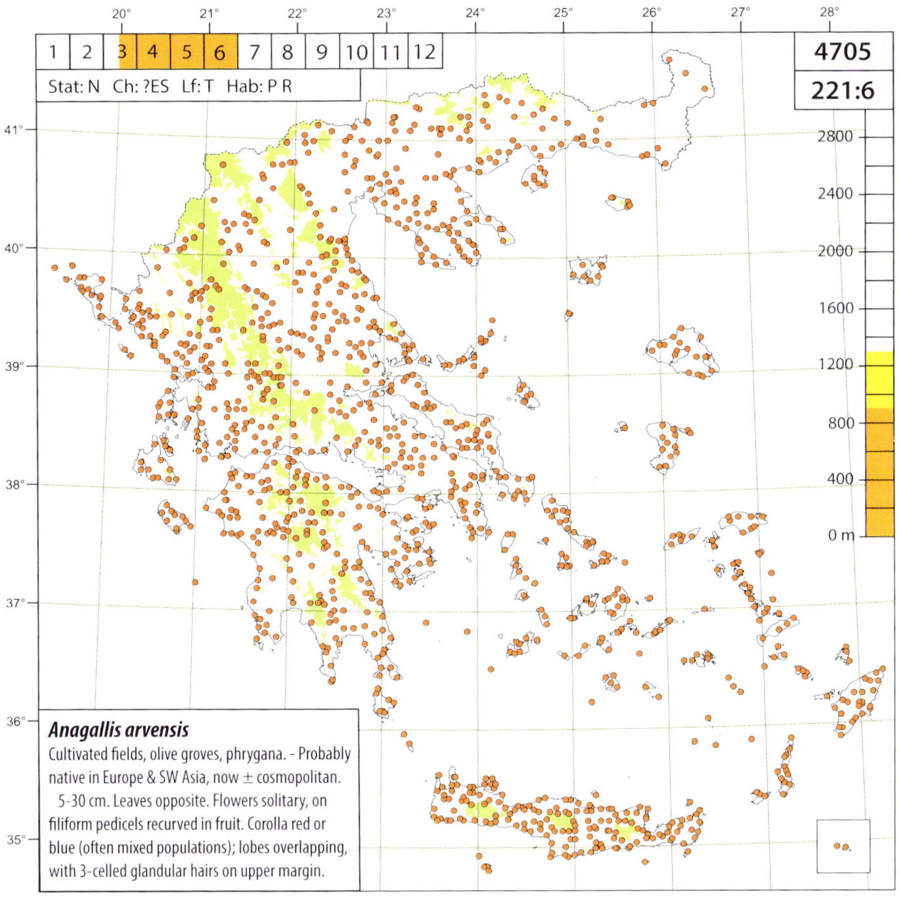

Anagallis arvensis

Cultivated fields, olive groves, phrygana. - Probably native in Europe & SW Asia, now ± cosmopolitan.

5-30 cm. Leaves opposite. Flowers solitary, on filiform pedicels recurved in fruit. Corolla red or blue (often mixed populations); lobes overlapping, with 3-celled glandular hairs on upper margin.

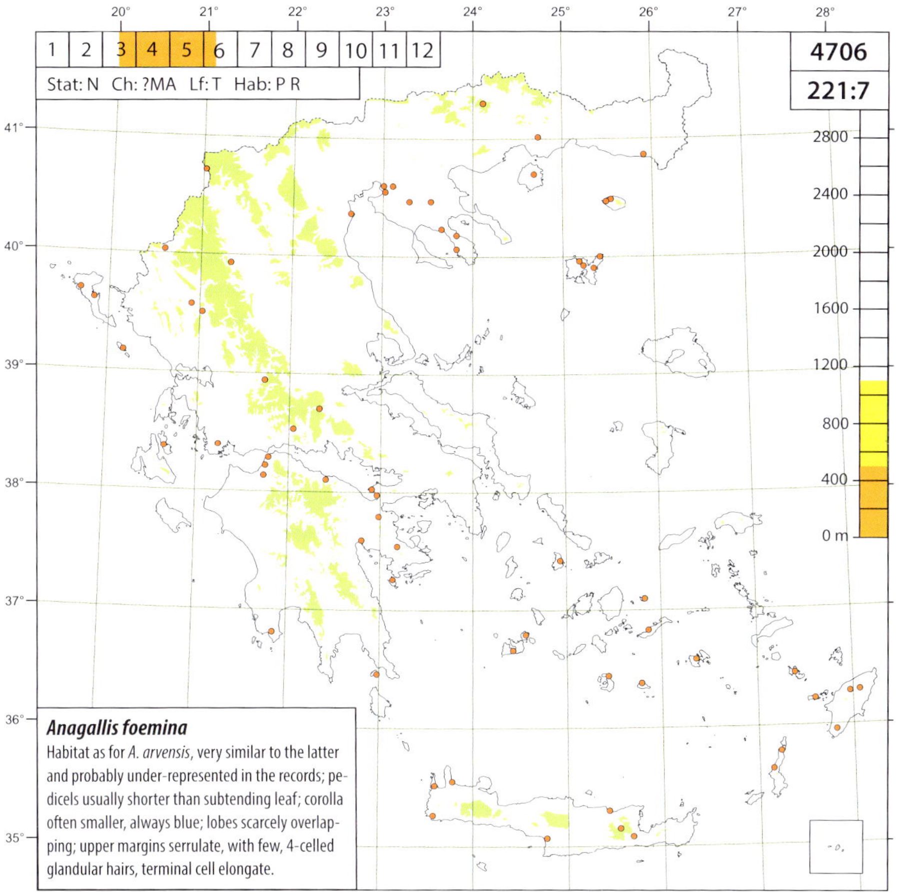

Anagallis foemina

Habitat as for *A. arvensis*, very similar to the latter and probably under-represented in the records; pedicels usually shorter than subtending leaf; corolla often smaller, always blue; lobes scarcely overlapping; upper margins serrulate, with few, 4-celled glandular hairs, terminal cell elongate.

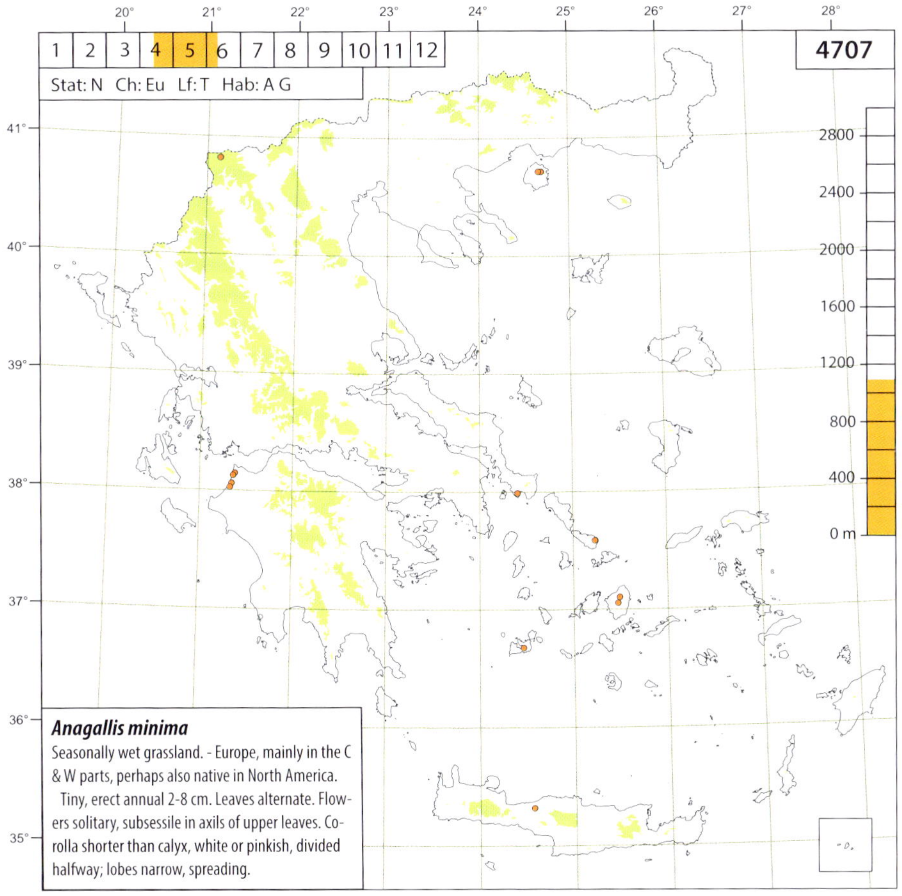

Anagallis minima

Seasonally wet grassland. - Europe, mainly in the C & W parts, perhaps also native in North America.

Tiny, erect annual 2-8 cm. Leaves alternate. Flowers solitary, subsessile in axils of upper leaves. Corolla shorter than calyx, white or pinkish, divided halfway; lobes narrow, spreading.

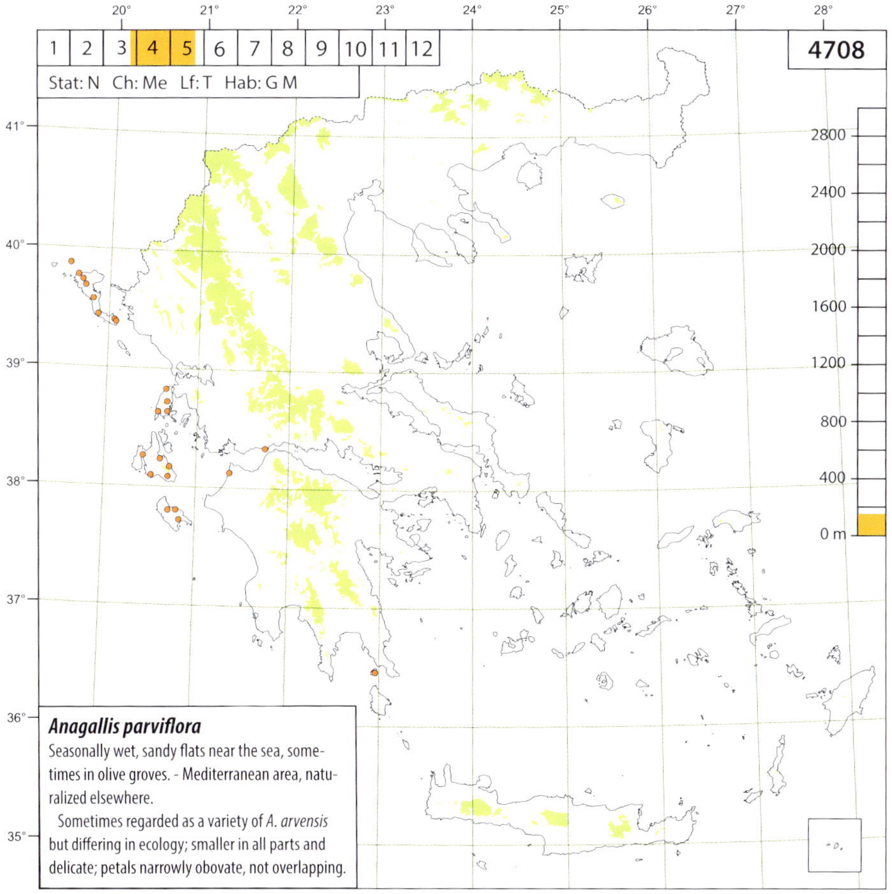

Anagallis parviflora

Seasonally wet, sandy flats near the sea, sometimes in olive groves. - Mediterranean area, naturalized elsewhere.

Sometimes regarded as a variety of *A. arvensis* but differing in ecology; smaller in all parts and delicate; petals narrowly obovate, not overlapping.

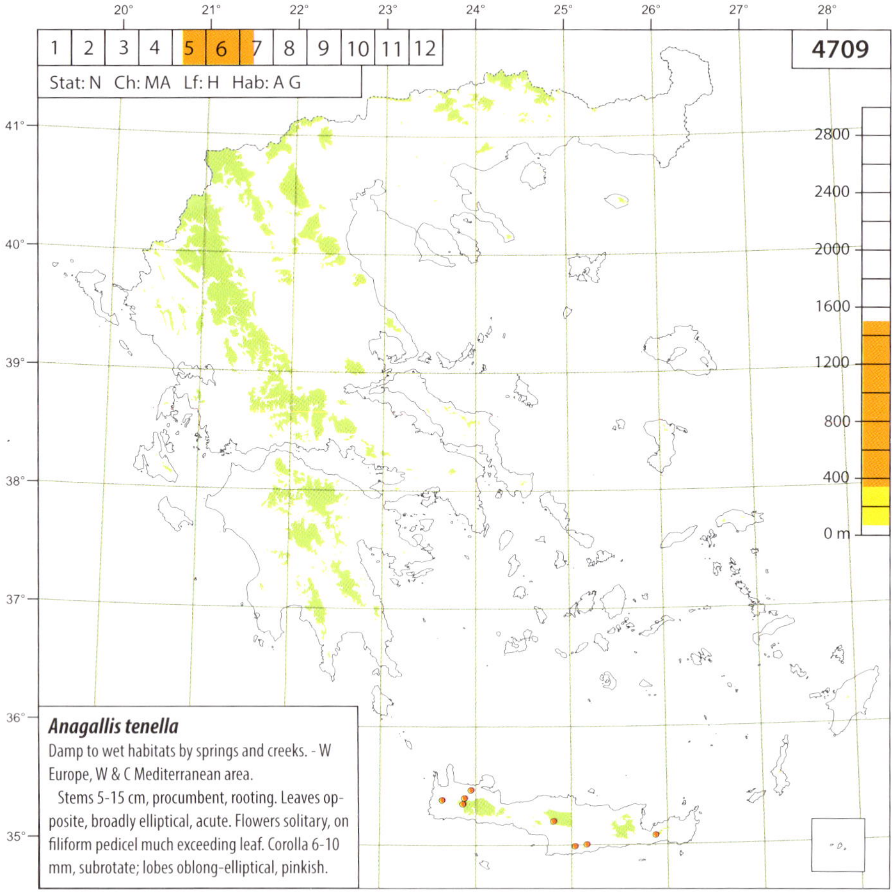

Anagallis tenella

Damp to wet habitats by springs and creeks. - W Europe, W & C Mediterranean area.

Stems 5-15 cm, procumbent, rooting. Leaves opposite, broadly elliptical, acute. Flowers solitary, on filiform pedicel much exceeding leaf. Corolla 6-10 mm, subrotate; lobes oblong-elliptical, pinkish.

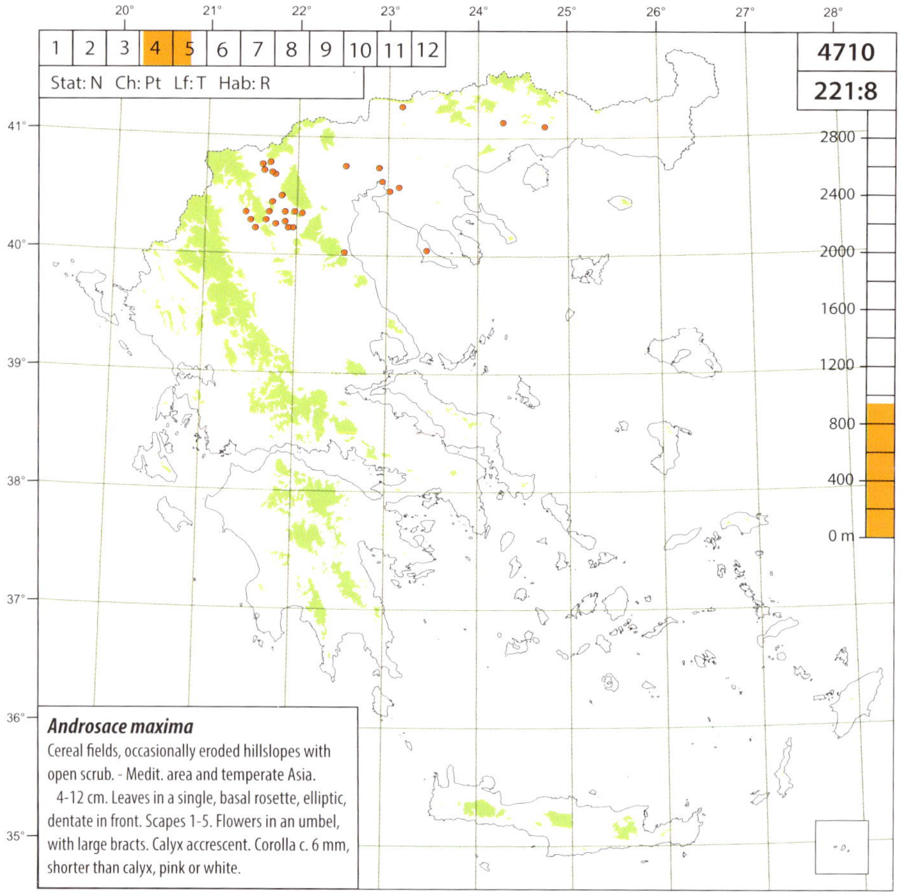

Androsace maxima

Cereal fields, occasionally eroded hillslopes with open scrub. - Medit. area and temperate Asia.

4-12 cm. Leaves in a single, basal rosette, elliptic, dentate in front. Scapes 1-5. Flowers in an umbel, with large bracts. Calyx accrescent. Corolla c. 6 mm, shorter than calyx, pink or white.

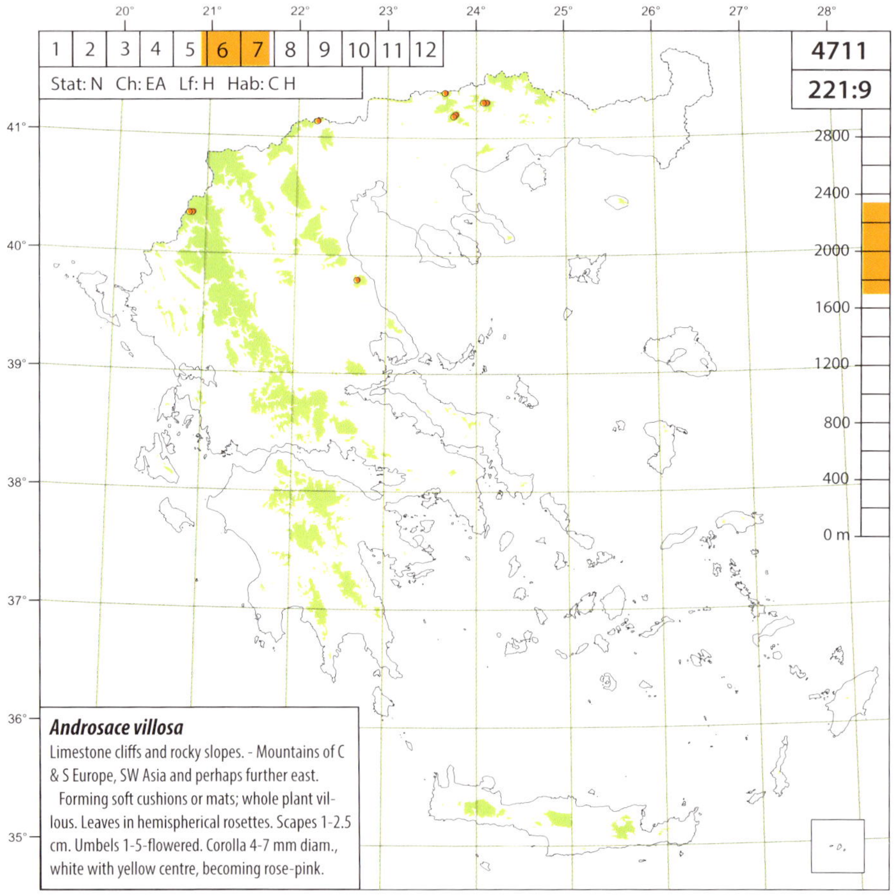

Androsace villosa

Limestone cliffs and rocky slopes. - Mountains of C & S Europe, SW Asia and perhaps further east.

Forming soft cushions or mats; whole plant villous. Leaves in hemispherical rosettes. Scapes 1-2.5 cm. Umbels 1-5-flowered. Corolla 4-7 mm diam., white with yellow centre, becoming rose-pink.

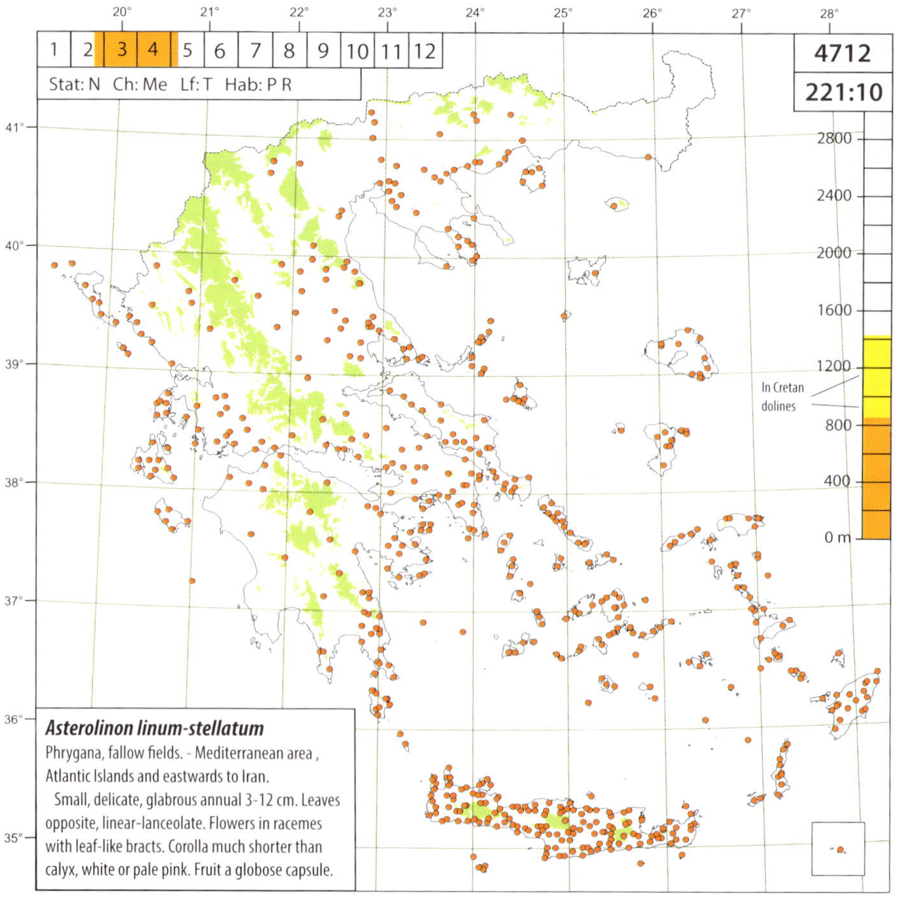

Asterolinon linum-stellatum

Phrygana, fallow fields. - Mediterranean area, Atlantic Islands and eastwards to Iran.

Small, delicate, glabrous annual 3-12 cm. Leaves opposite, linear-lanceolate. Flowers in racemes with leaf-like bracts. Corolla much shorter than calyx, white or pale pink. Fruit a globose capsule.

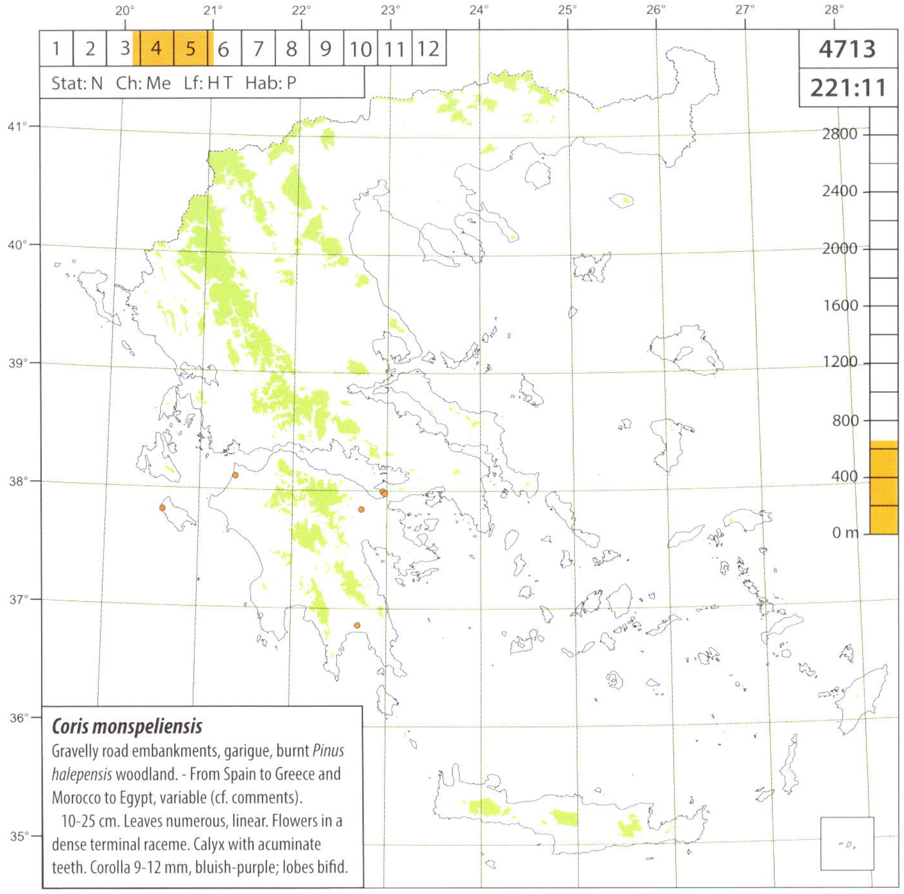

Coris monspeliensis

Gravelly road embankments, garigue, burnt *Pinus halepensis* woodland. - From Spain to Greece and Morocco to Egypt, variable (cf. comments).

10-25 cm. Leaves numerous, linear. Flowers in a dense terminal raceme. Calyx with acuminate teeth. Corolla 9-12 mm, bluish-purple; lobes bifid.

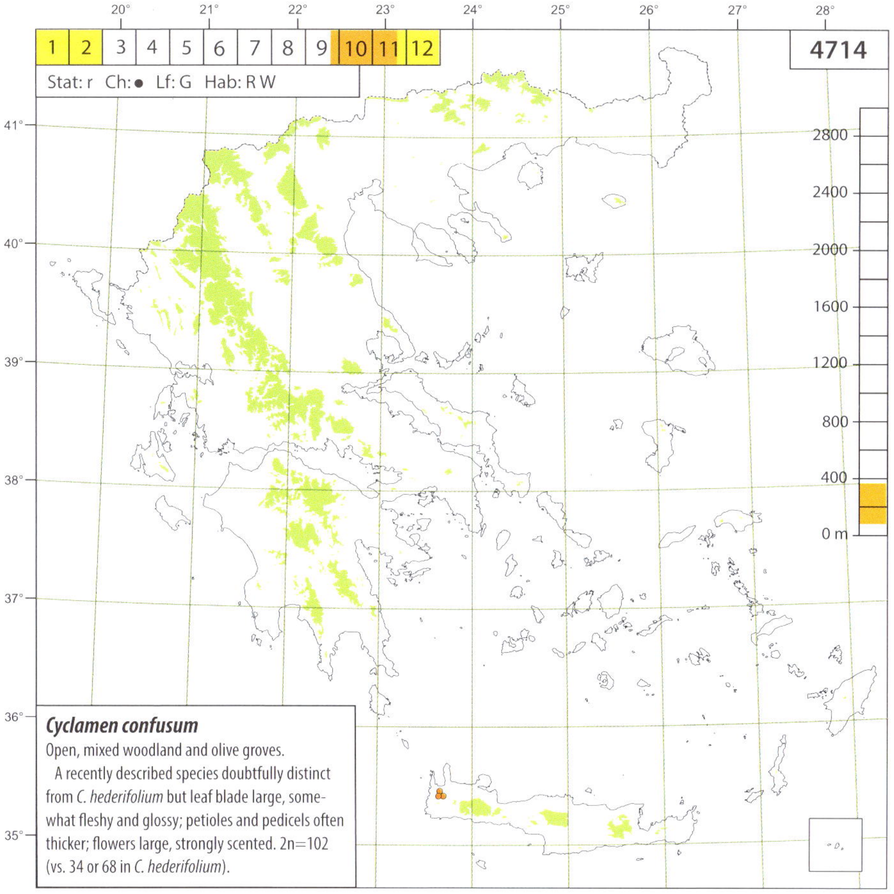

Cyclamen confusum

Open, mixed woodland and olive groves.

A recently described species doubtfully distinct from *C. hederifolium* but leaf blade large, somewhat fleshy and glossy; petioles and pedicels often thicker; flowers large, strongly scented. 2n=102 (vs. 34 or 68 in *C. hederifolium*).

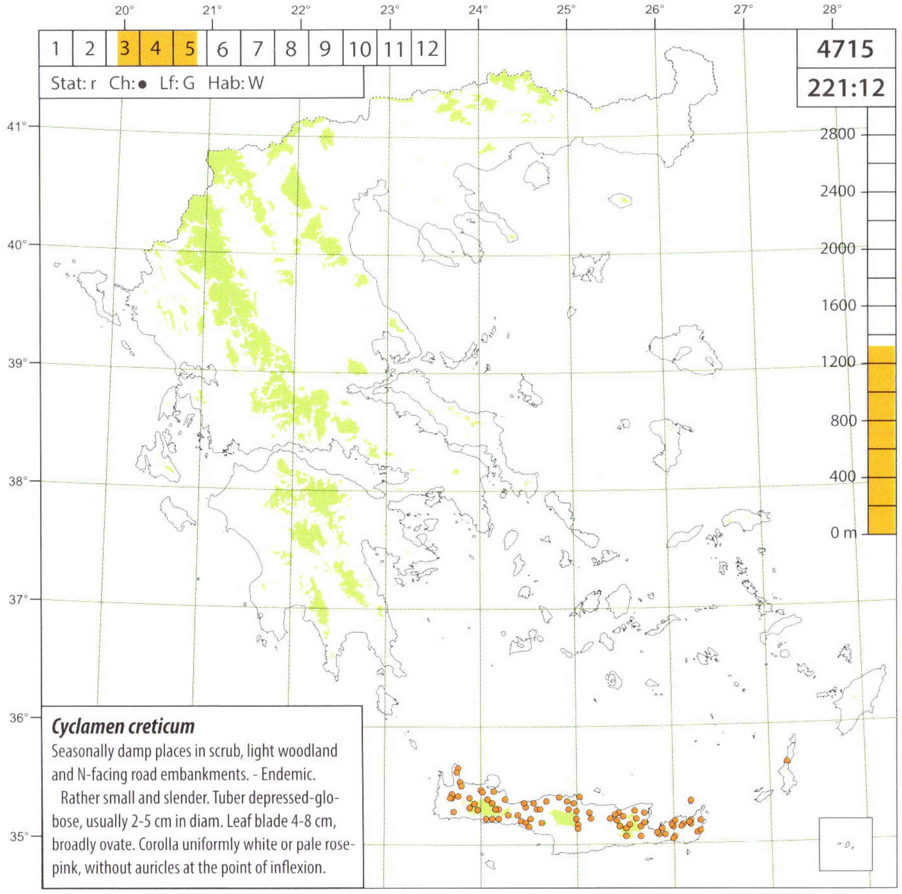

Cyclamen creticum

Seasonally damp places in scrub, light woodland and N-facing road embankments. - Endemic.

Rather small and slender. Tuber depressed-globose, usually 2-5 cm in diam. Leaf blade 4-8 cm, broadly ovate. Corolla uniformly white or pale rose-pink, without auricles at the point of inflexion.

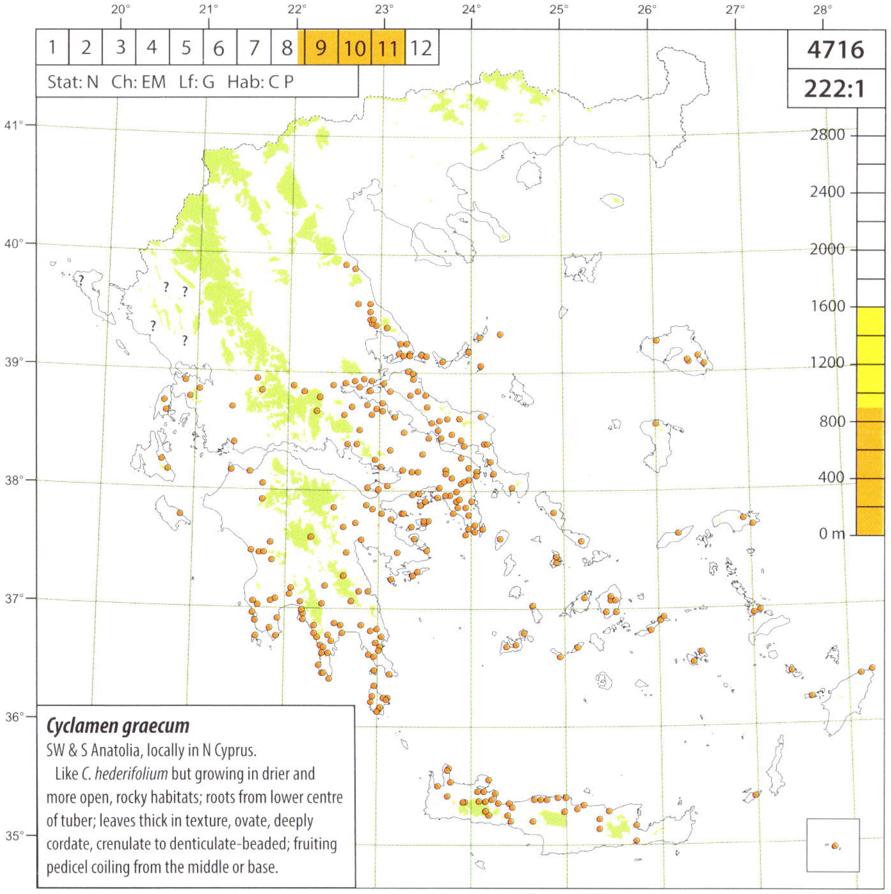

Cyclamen graecum

SW & S Anatolia, locally in N Cyprus.

Like *C. hederifolium* but growing in drier and more open, rocky habitats; roots from lower centre of tuber; leaves thick in texture, ovate, deeply cordate, crenulate to denticulate-beaded; fruiting pedicel coiling from the middle or base.

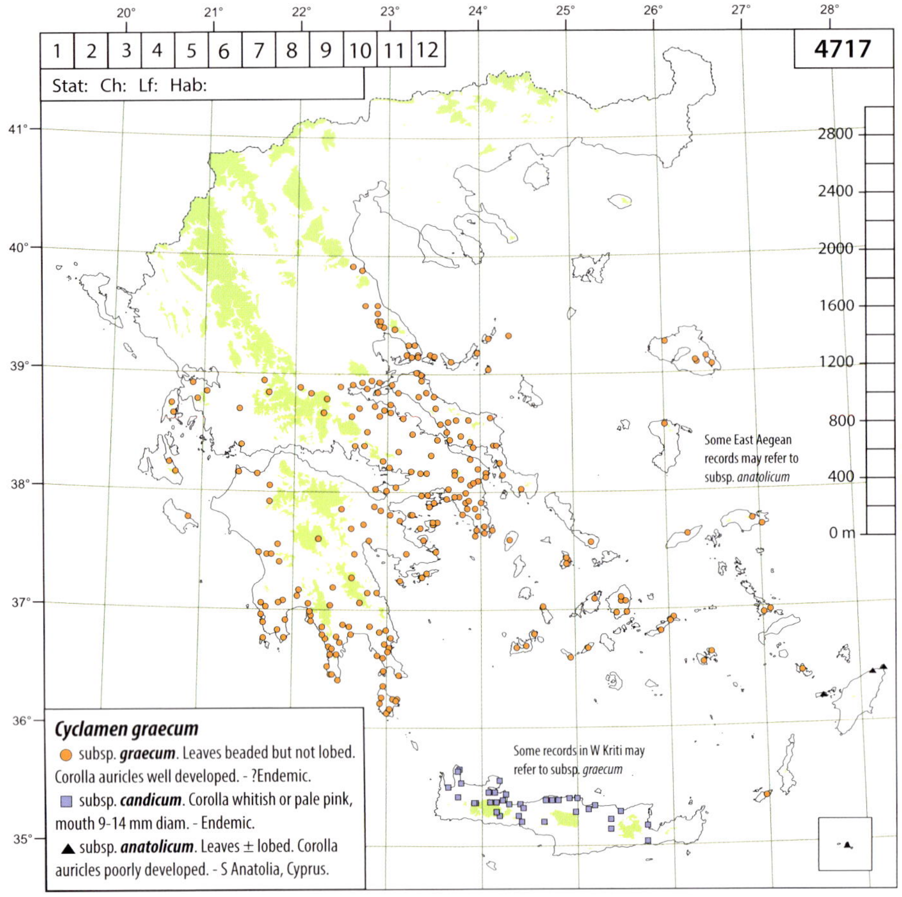

Cyclamen graecum
● subsp. ***graecum***. Leaves beaded but not lobed. Corolla auricles well developed. - ?Endemic.
■ subsp. ***candicum***. Corolla whitish or pale pink, mouth 9-14 mm diam. - Endemic.
▲ subsp. ***anatolicum***. Leaves ± lobed. Corolla auricles poorly developed. - S Anatolia, Cyprus.

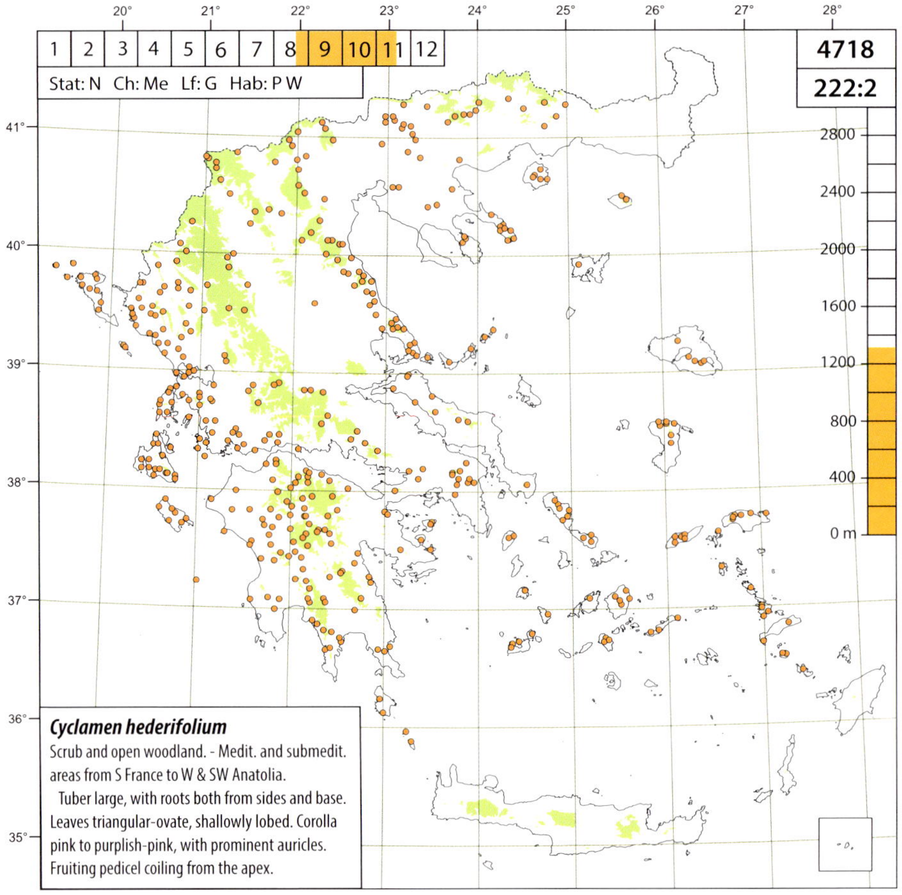

Cyclamen hederifolium
Scrub and open woodland. - Medit. and submedit. areas from S France to W & SW Anatolia.
Tuber large, with roots from both sides and base. Leaves triangular-ovate, shallowly lobed. Corolla pink to purplish-pink, with prominent auricles. Fruiting pedicel coiling from the apex.

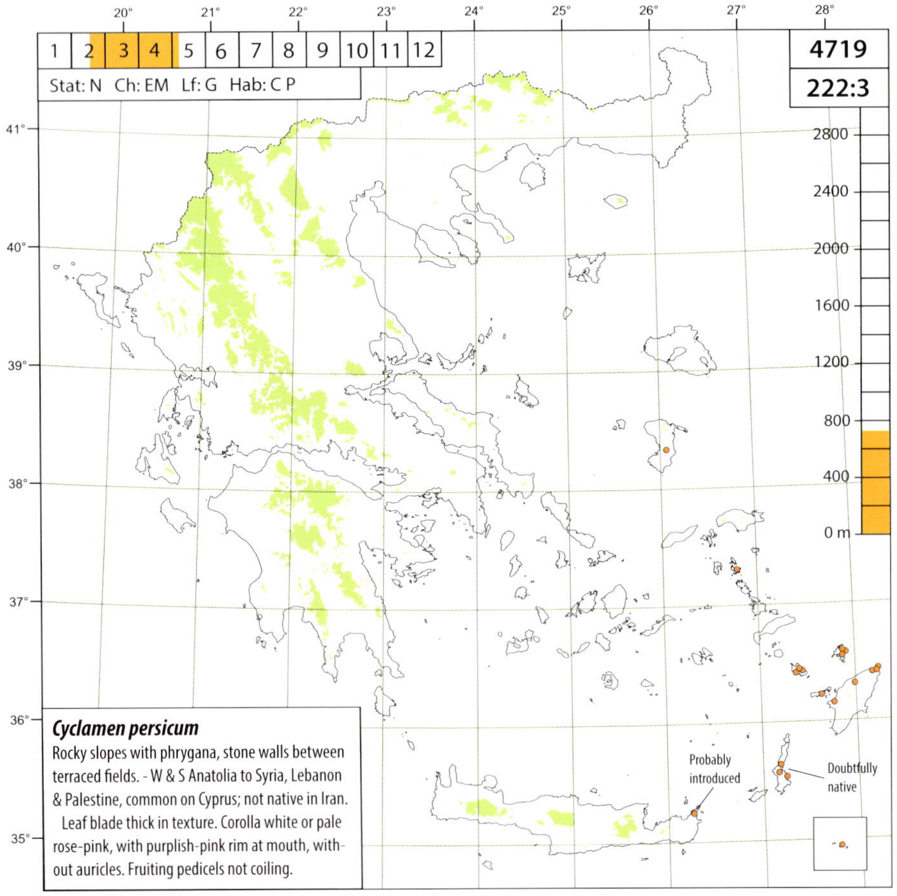

Cyclamen persicum
Rocky slopes with phrygana, stone walls between terraced fields. - W & S Anatolia to Syria, Lebanon & Palestine, common on Cyprus; not native in Iran.
Leaf blade thick in texture. Corolla white or pale rose-pink, with purplish-pink rim at mouth, without auricles. Fruiting pedicels not coiling.

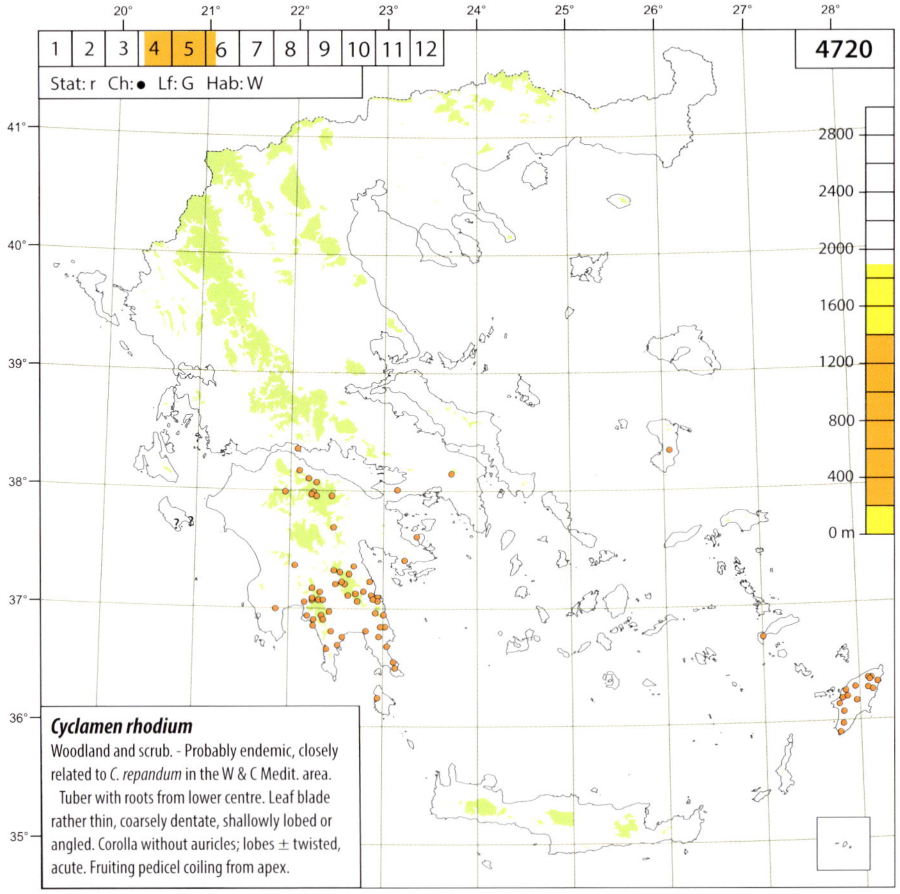

Cyclamen rhodium
Woodland and scrub. - Probably endemic, closely related to *C. repandum* in the W & C Medit. area.
Tuber with roots from lower centre. Leaf blade rather thin, coarsely dentate, shallowly lobed or angled. Corolla without auricles; lobes ± twisted, acute. Fruiting pedicel coiling from apex.

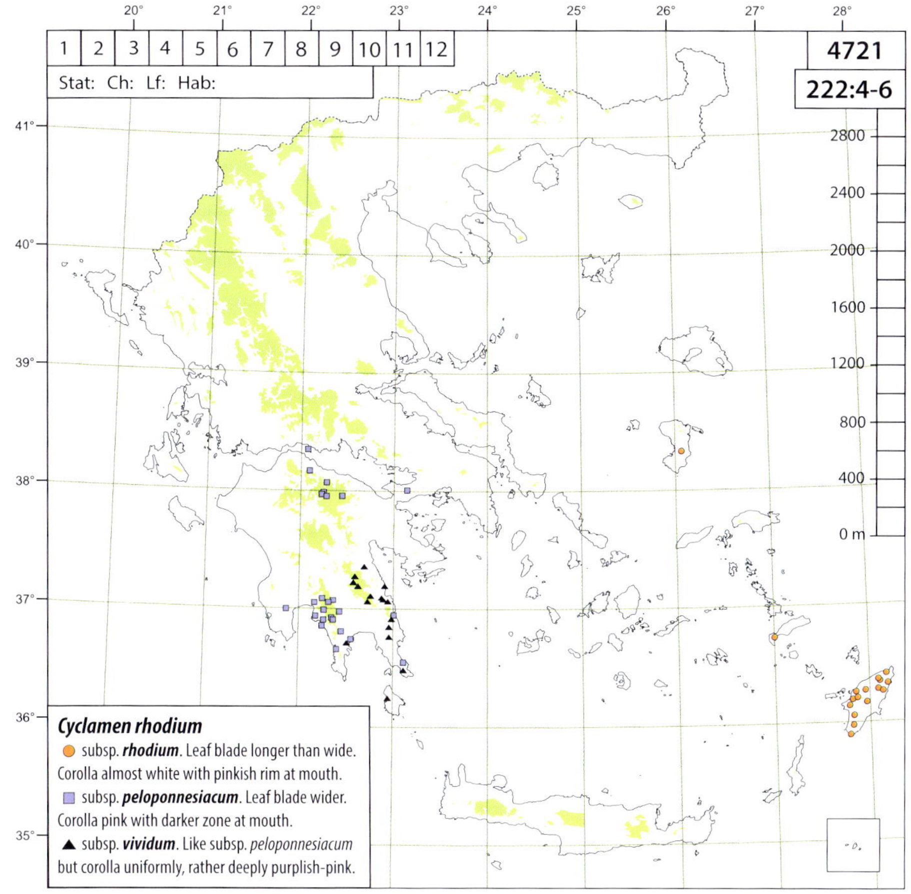

Cyclamen rhodium
● subsp. ***rhodium***. Leaf blade longer than wide. Corolla almost white with pinkish rim at mouth.
■ subsp. ***peloponnesiacum***. Leaf blade wider. Corolla pink with darker zone at mouth.
▲ subsp. ***vividum***. Like subsp. *peloponnesiacum* but corolla uniformly, rather deeply purplish-pink.

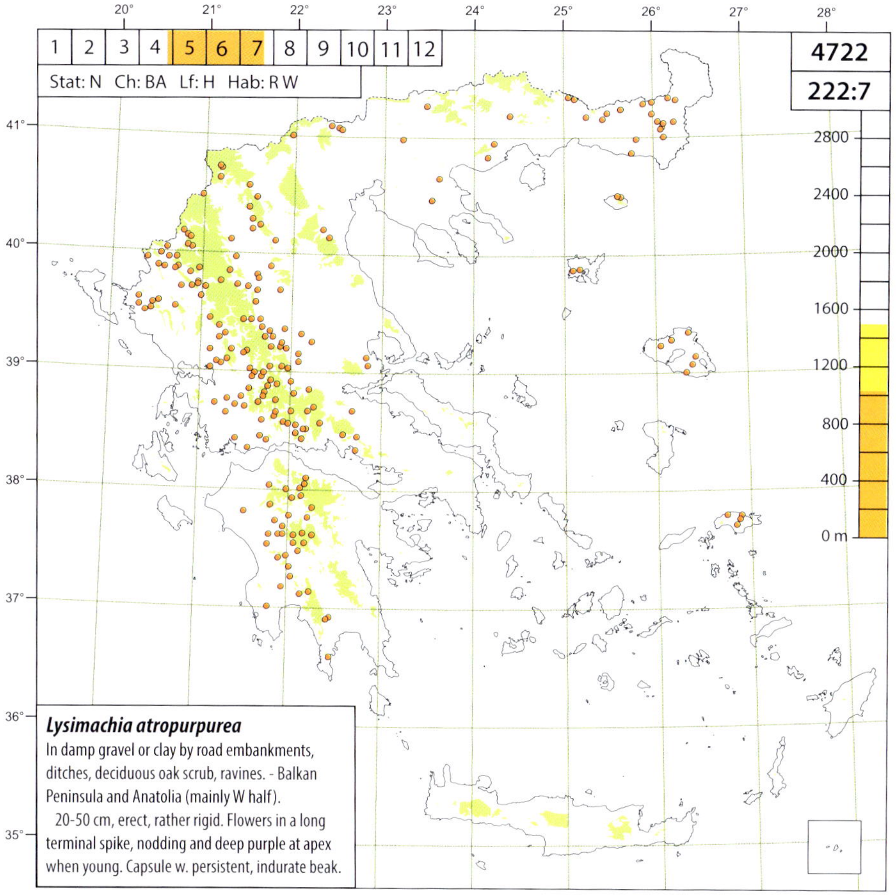

Lysimachia atropurpurea
In damp gravel or clay by road embankments, ditches, deciduous oak scrub, ravines. - Balkan Peninsula and Anatolia (mainly W half).
20-50 cm, erect, rather rigid. Flowers in a long terminal spike, nodding and deep purple at apex when young. Capsule w. persistent, indurate beak.

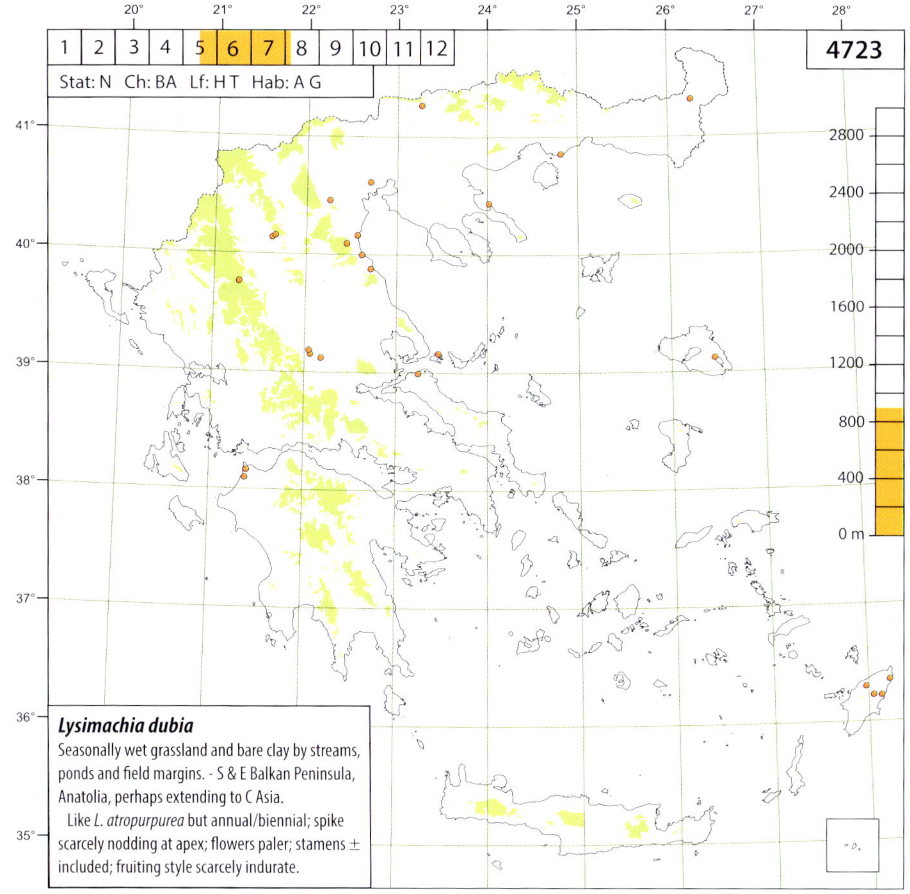

Lysimachia dubia
Seasonally wet grassland and bare clay by streams, ponds and field margins. - S & E Balkan Peninsula, Anatolia, perhaps extending to C Asia.
Like *L. atropurpurea* but annual/biennial; spike scarcely nodding at apex; flowers paler; stamens ± included; fruiting style scarcely indurate.

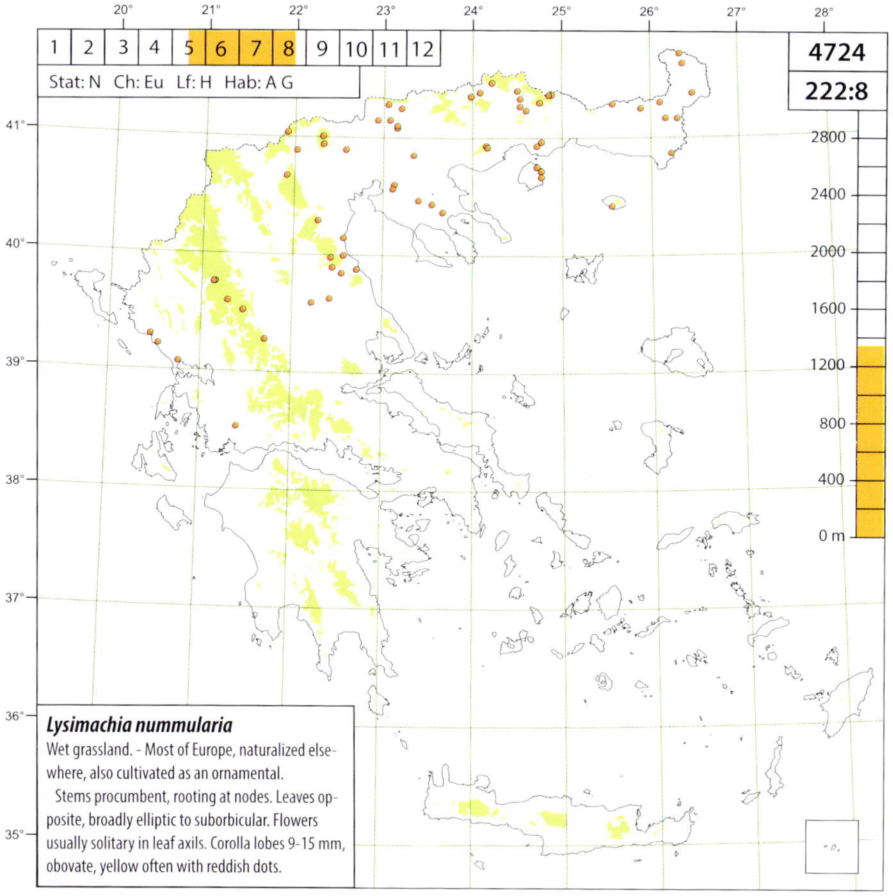

Lysimachia nummularia
Wet grassland. - Most of Europe, naturalized elsewhere, also cultivated as an ornamental.
Stems procumbent, rooting at nodes. Leaves opposite, broadly elliptic to suborbicular. Flowers usually solitary in leaf axils. Corolla lobes 9-15 mm, obovate, yellow often with reddish dots.

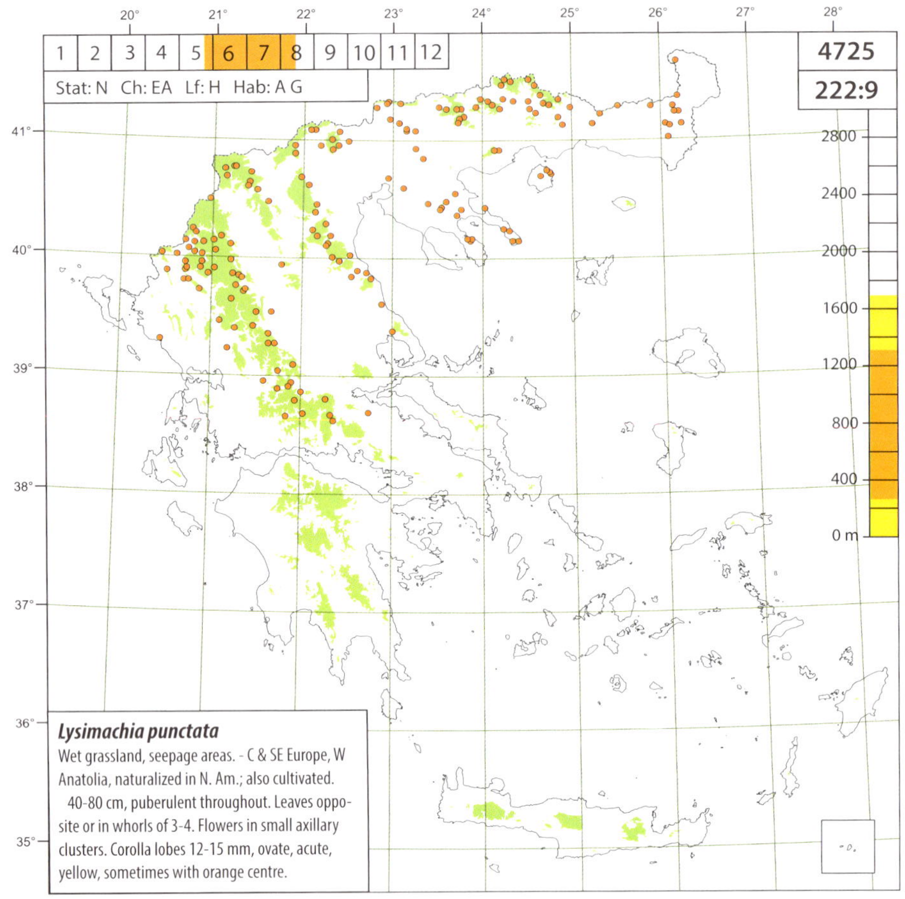

Lysimachia punctata
Wet grassland, seepage areas. - C & SE Europe, W Anatolia, naturalized in N. Am.; also cultivated.
40-80 cm, puberulent throughout. Leaves opposite or in whorls of 3-4. Flowers in small axillary clusters. Corolla lobes 12-15 mm, ovate, acute, yellow, sometimes with orange centre.

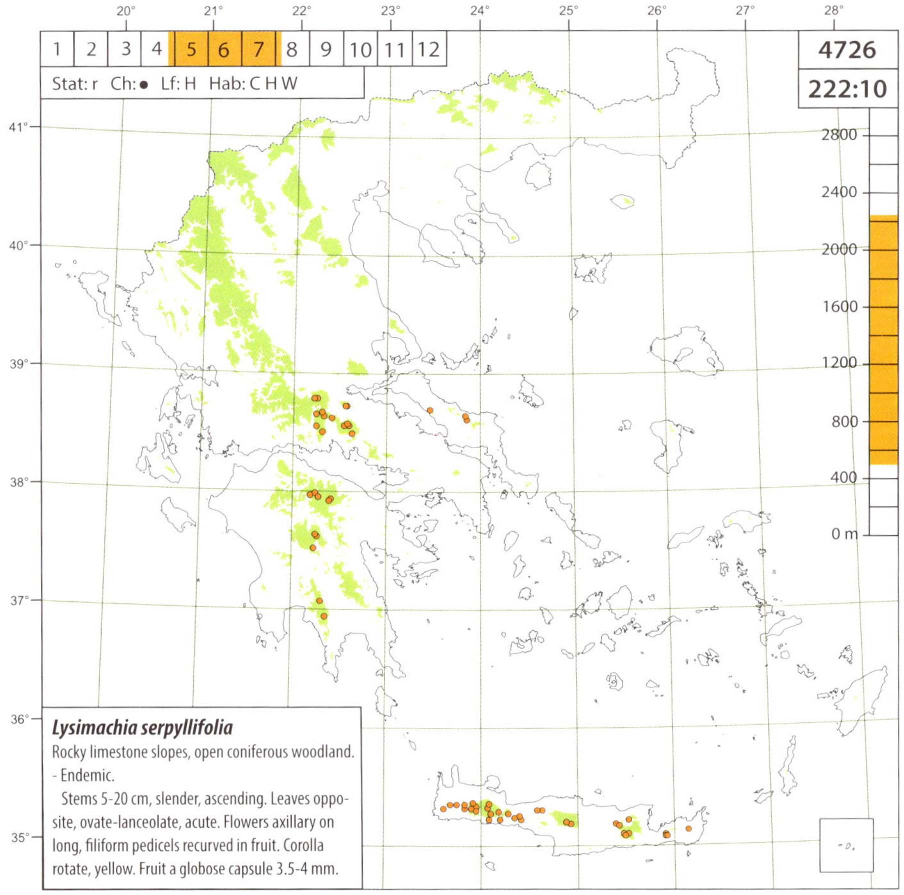

Lysimachia serpyllifolia
Rocky limestone slopes, open coniferous woodland. - Endemic.
Stems 5-20 cm, slender, ascending. Leaves opposite, ovate-lanceolate, acute. Flowers axillary on long, filiform pedicels recurved in fruit. Corolla rotate, yellow. Fruit a globose capsule 3.5-4 mm.

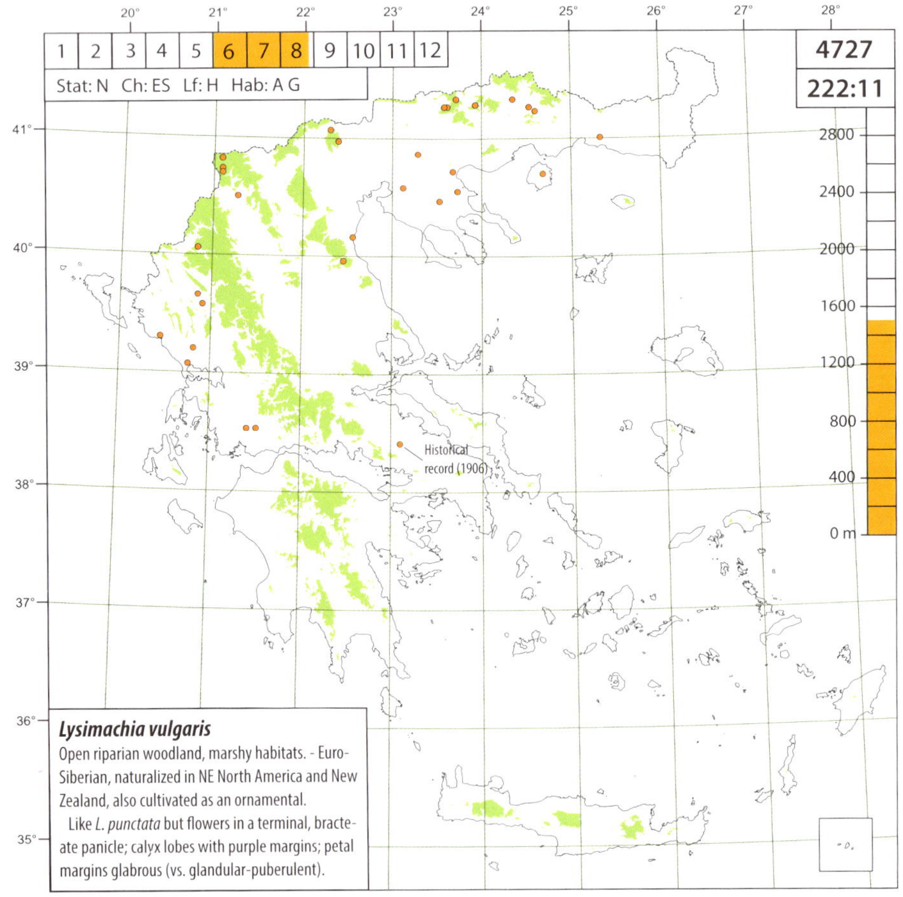

Lysimachia vulgaris
Open riparian woodland, marshy habitats. - Euro-Siberian, naturalized in NE North America and New Zealand, also cultivated as an ornamental.
Like *L. punctata* but flowers in a terminal, bracteate panicle; calyx lobes with purple margins; petal margins glabrous (vs. glandular-puberulent).

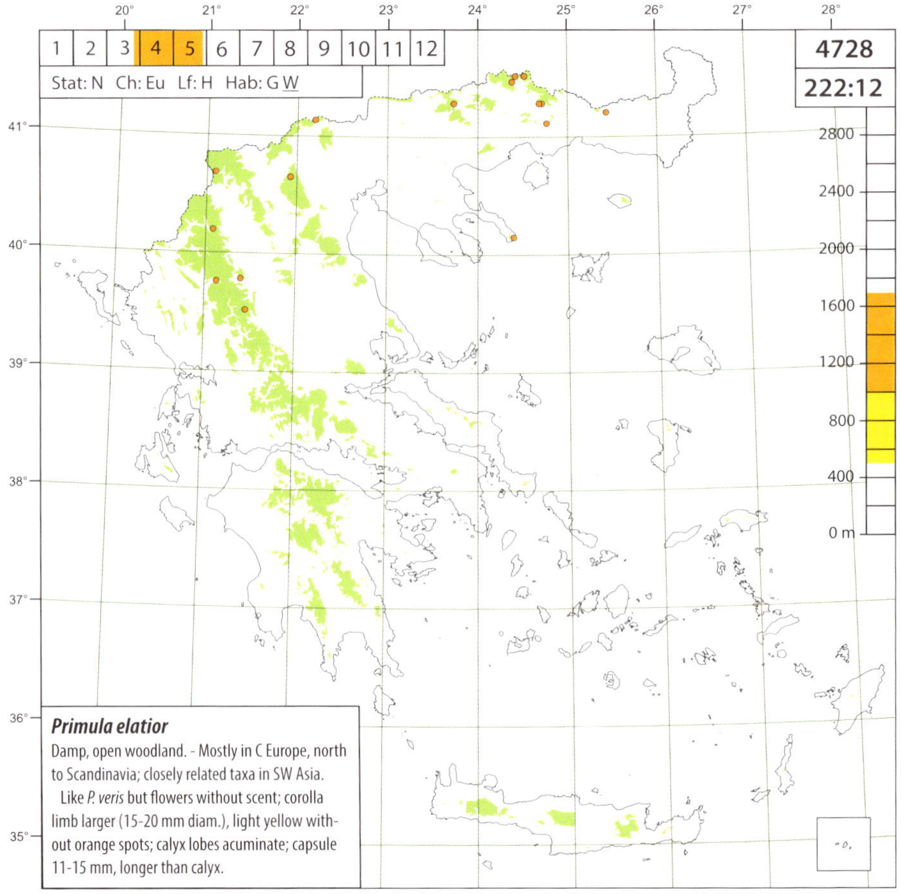

Primula elatior
Damp, open woodland. - Mostly in C Europe, north to Scandinavia; closely related taxa in SW Asia.
Like *P. veris* but flowers without scent; corolla limb larger (15-20 mm diam.), light yellow without orange spots; calyx lobes acuminate; capsule 11-15 mm, longer than calyx.

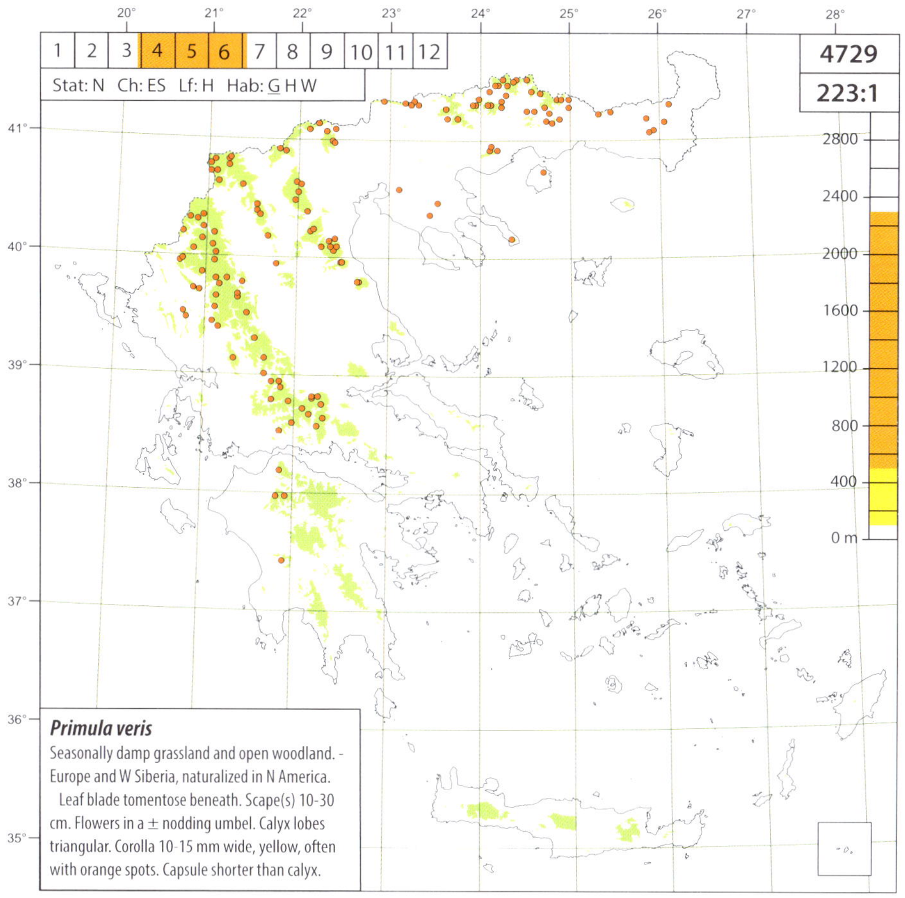

Primula veris
Seasonally damp grassland and open woodland. - Europe and W Siberia, naturalized in N America.
Leaf blade tomentose beneath. Scape(s) 10-30 cm. Flowers in a ± nodding umbel. Calyx lobes triangular. Corolla 10-15 mm wide, yellow, often with orange spots. Capsule shorter than calyx.

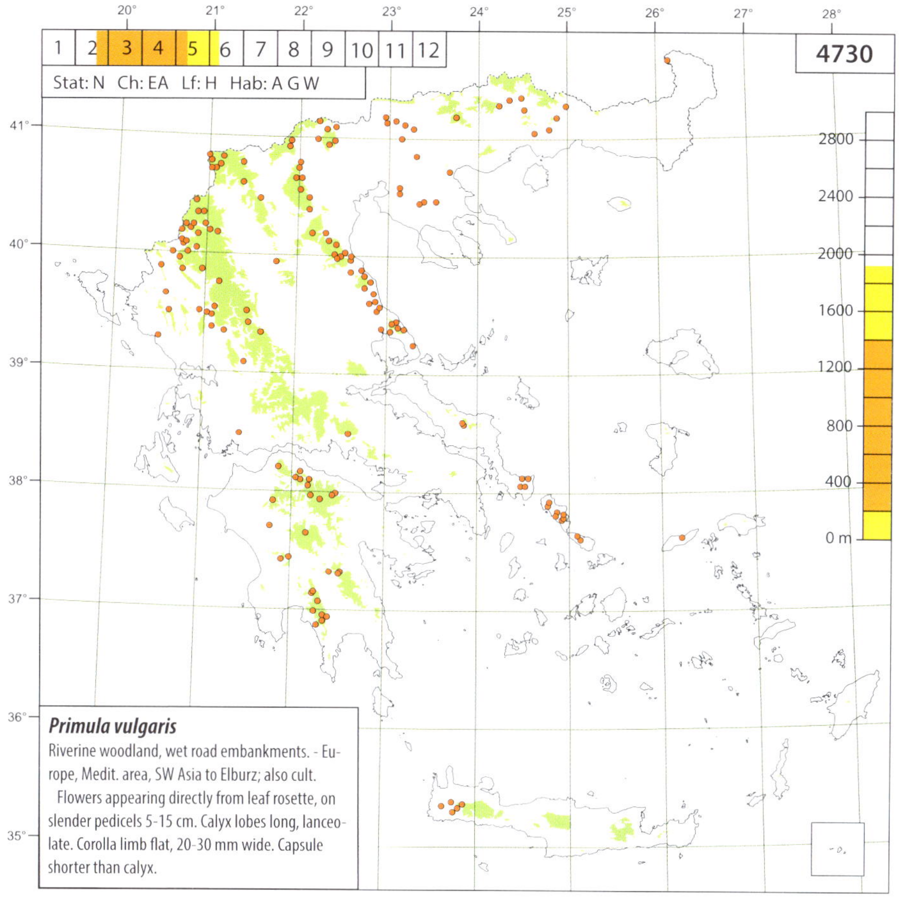

Primula vulgaris
Riverine woodland, wet road embankments. - Europe, Medit. area, SW Asia to Elburz; also cult.
Flowers appearing directly from leaf rosette, on slender pedicels 5-15 cm. Calyx lobes long, lanceolate. Corolla limb flat, 20-30 mm wide. Capsule shorter than calyx.

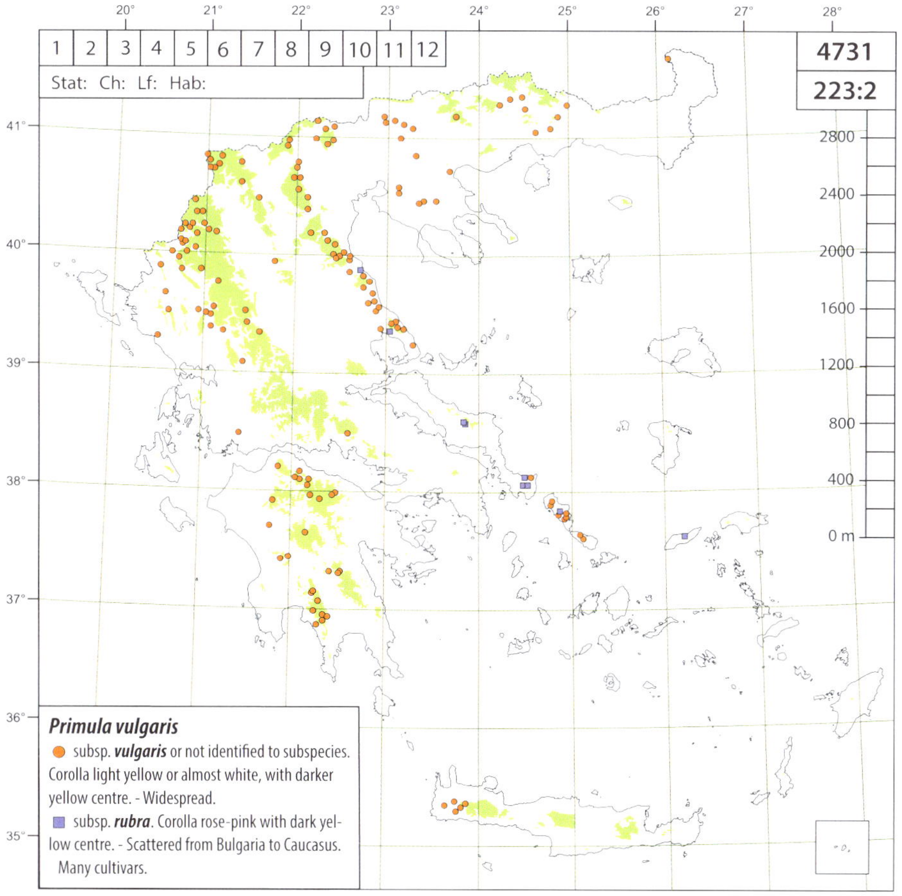

Primula vulgaris
● subsp. ***vulgaris*** or not identified to subspecies. Corolla light yellow or almost white, with darker yellow centre. - Widespread.
■ subsp. ***rubra***. Corolla rose-pink with dark yellow centre. - Scattered from Bulgaria to Caucasus. Many cultivars.

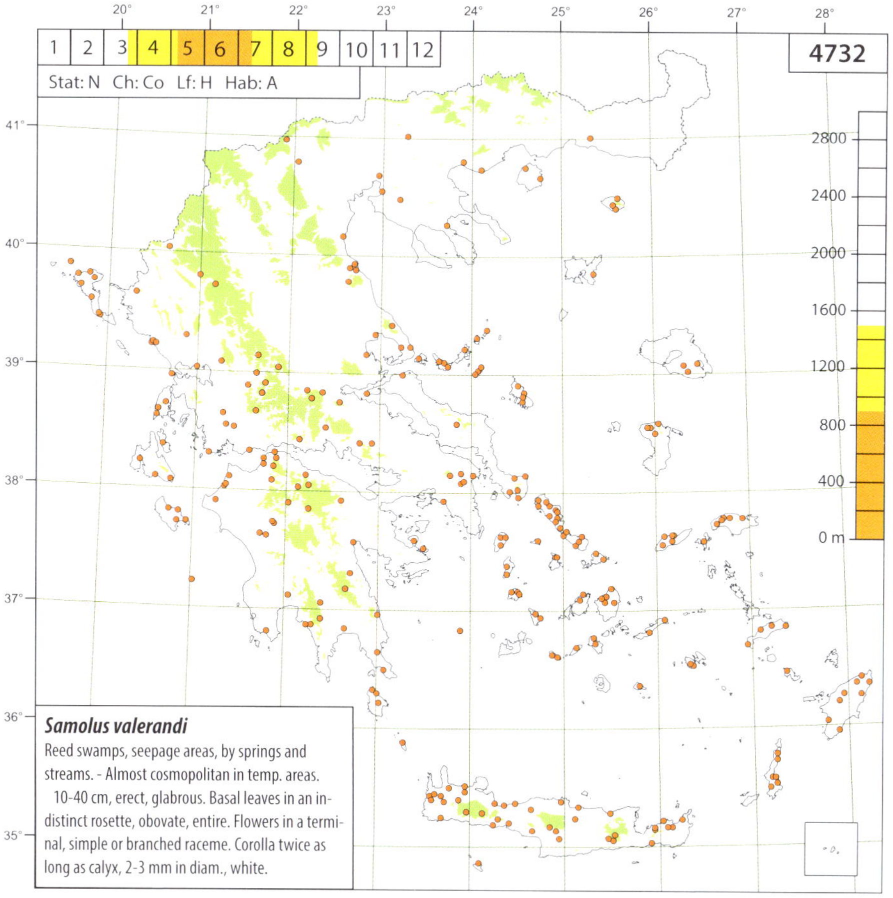

Samolus valerandi
Reed swamps, seepage areas, by springs and streams. - Almost cosmopolitan in temp. areas.
10-40 cm, erect, glabrous. Basal leaves in an indistinct rosette, obovate, entire. Flowers in a terminal, simple or branched raceme. Corolla twice as long as calyx, 2-3 mm in diam., white.

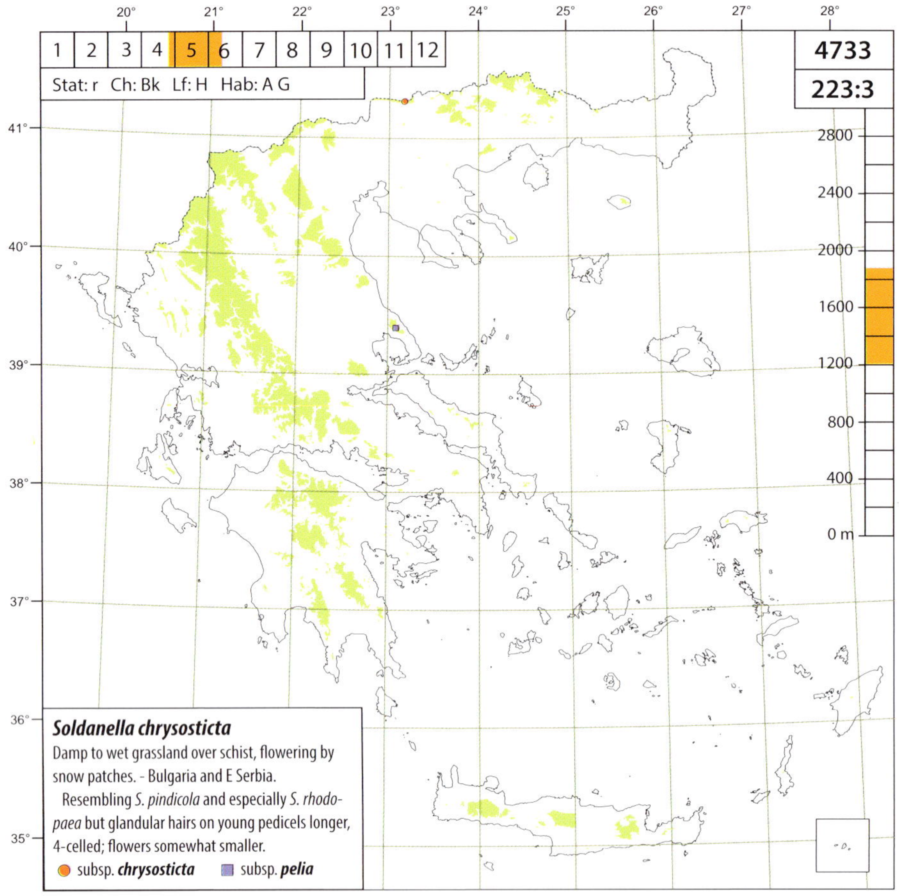

Soldanella chrysosticta

Damp to wet grassland over schist, flowering by snow patches. - Bulgaria and E Serbia.

Resembling *S. pindicola* and especially *S. rhodopaea* but glandular hairs on young pedicels longer, 4-celled; flowers somewhat smaller.

● subsp. ***chrysosticta*** ■ subsp. ***pelia***

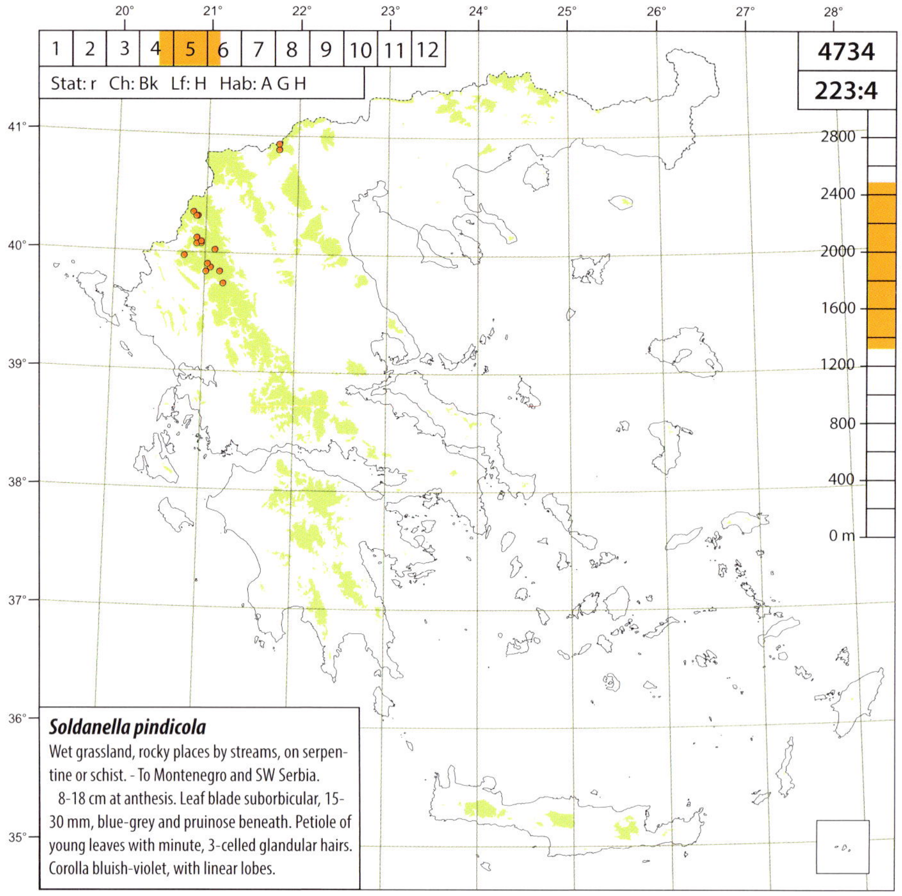

Soldanella pindicola

Wet grassland, rocky places by streams, on serpentine or schist. - To Montenegro and SW Serbia.

8-18 cm at anthesis. Leaf blade suborbicular, 15-30 mm, blue-grey and pruinose beneath. Petiole of young leaves with minute, 3-celled glandular hairs. Corolla bluish-violet, with linear lobes.

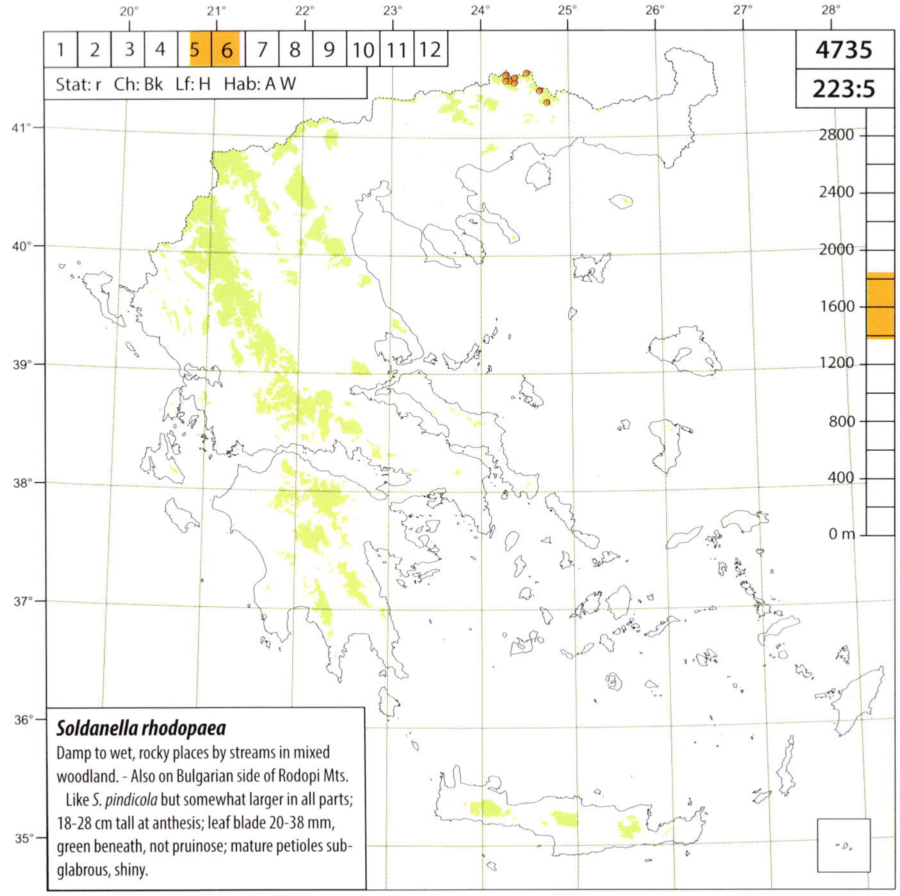

Soldanella rhodopaea

Damp to wet, rocky places by streams in mixed woodland. - Also on Bulgarian side of Rodopi Mts.

Like *S. pindicola* but somewhat larger in all parts; 18-28 cm tall at anthesis; leaf blade 20-38 mm, green beneath, not pruinose; mature petioles subglabrous, shiny.

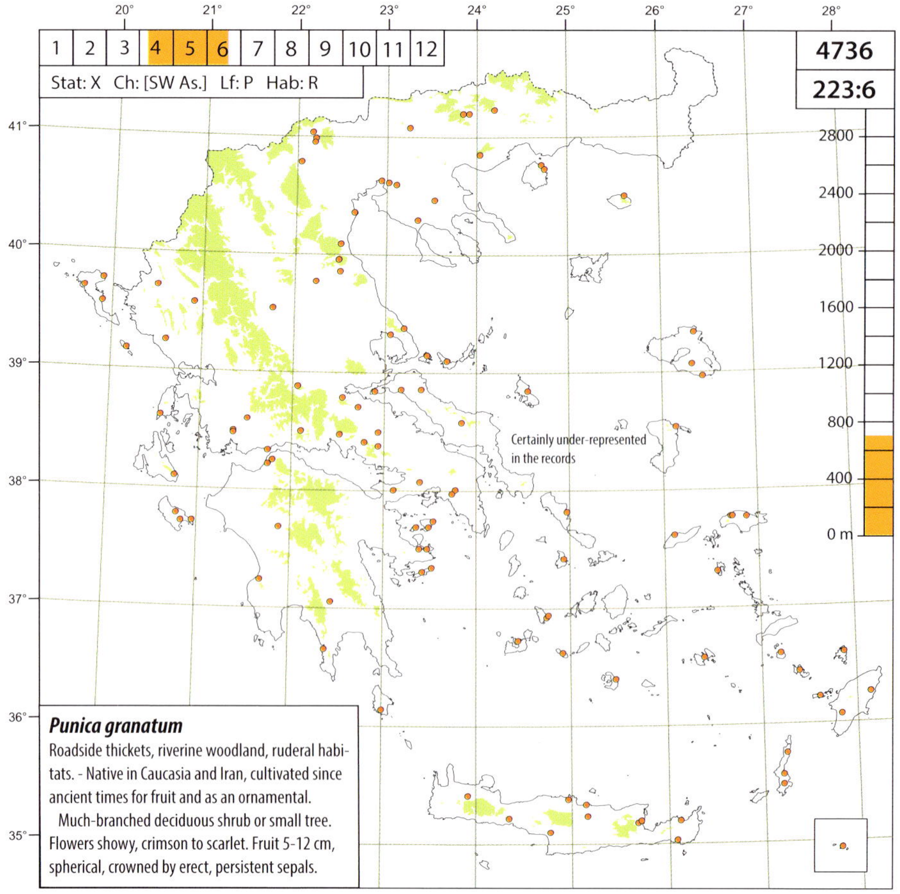

Punica granatum

Roadside thickets, riverine woodland, ruderal habitats. - Native in Caucasia and Iran, cultivated since ancient times for fruit and as an ornamental.

Much-branched deciduous shrub or small tree. Flowers showy, crimson to scarlet. Fruit 5-12 cm, spherical, crowned by erect, persistent sepals.

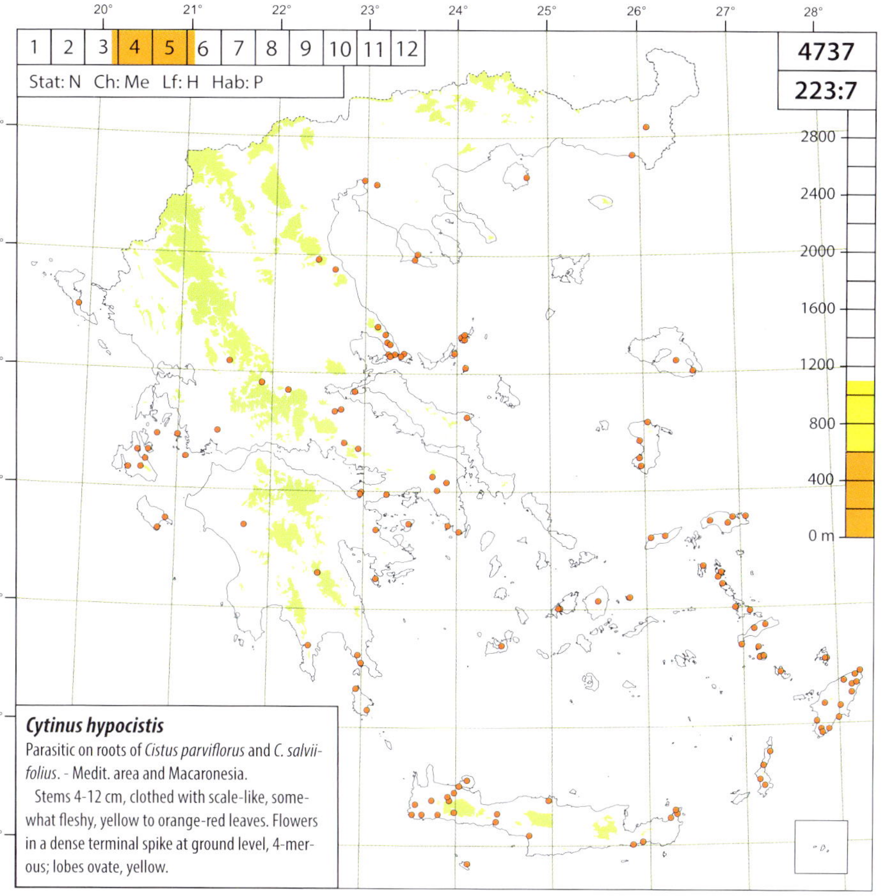

Cytinus hypocistis

Parasitic on roots of *Cistus parviflorus* and *C. salvii-folius*. - Medit. area and Macaronesia.

Stems 4-12 cm, clothed with scale-like, somewhat fleshy, yellow to orange-red leaves. Flowers in a dense terminal spike at ground level, 4-merous; lobes ovate, yellow.

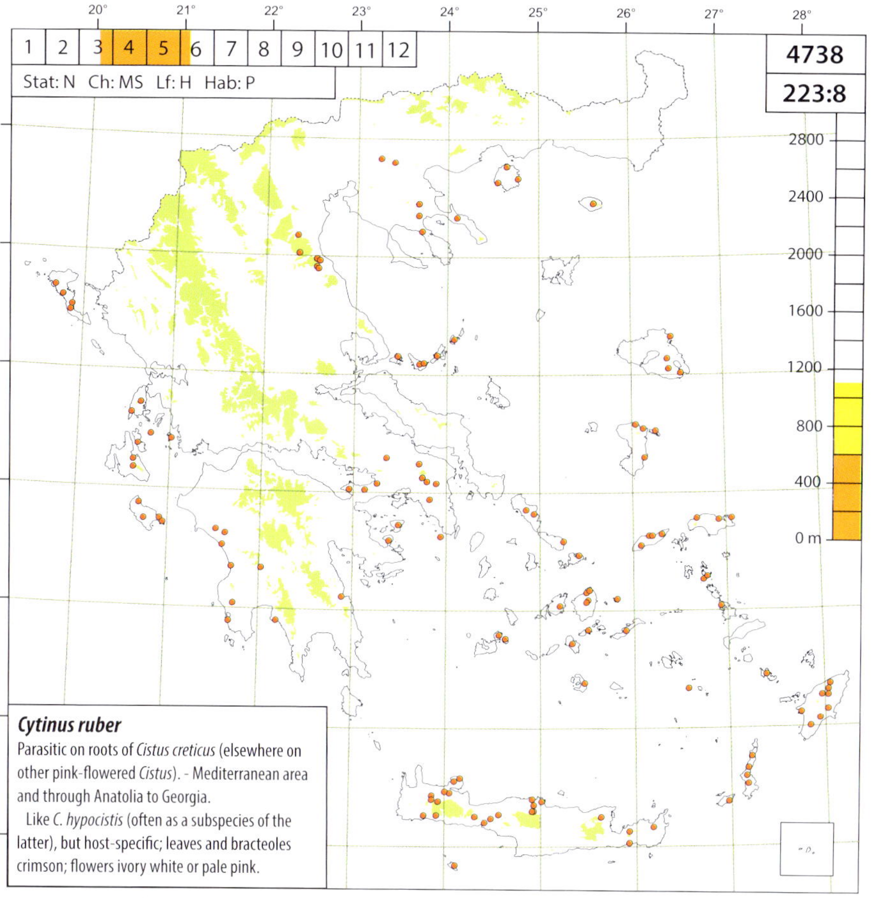

Cytinus ruber

Parasitic on roots of *Cistus creticus* (elsewhere on other pink-flowered *Cistus*). - Mediterranean area and through Anatolia to Georgia.

Like *C. hypocistis* (often as a subspecies of the latter), but host-specific; leaves and bracteoles crimson; flowers ivory white or pale pink.

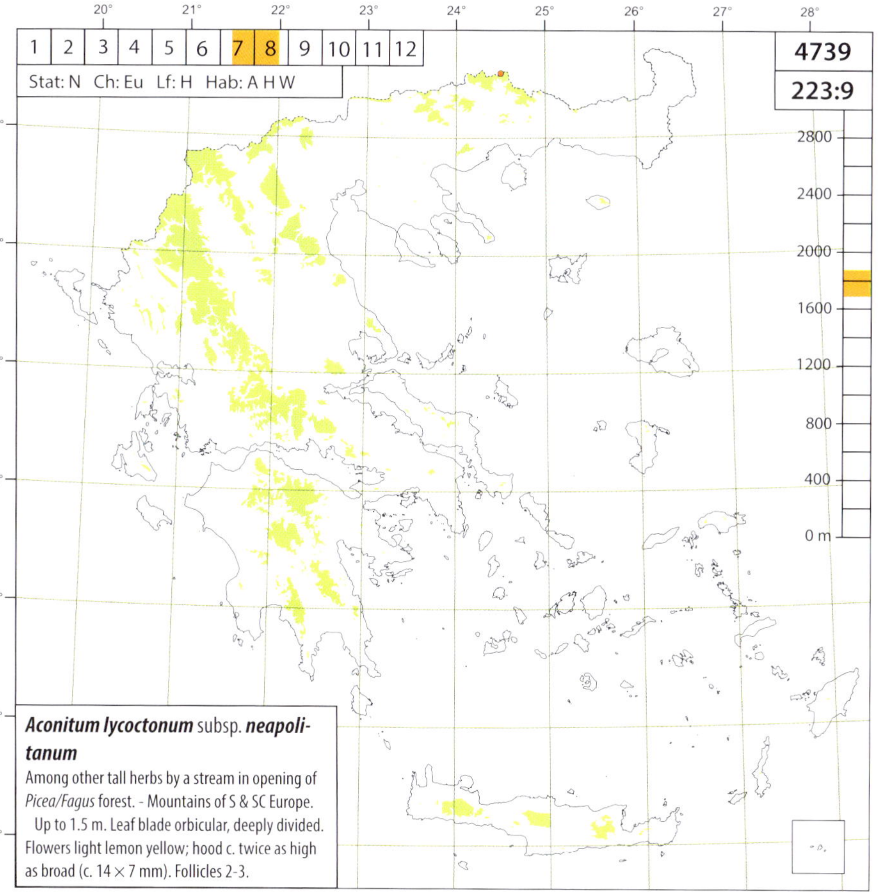

Aconitum lycoctonum subsp. *neapolitanum*

Among other tall herbs by a stream in opening of *Picea/Fagus* forest. - Mountains of S & SC Europe.

Up to 1.5 m. Leaf blade orbicular, deeply divided. Flowers light lemon yellow; hood c. twice as high as broad (c. 14 × 7 mm). Follicles 2-3.

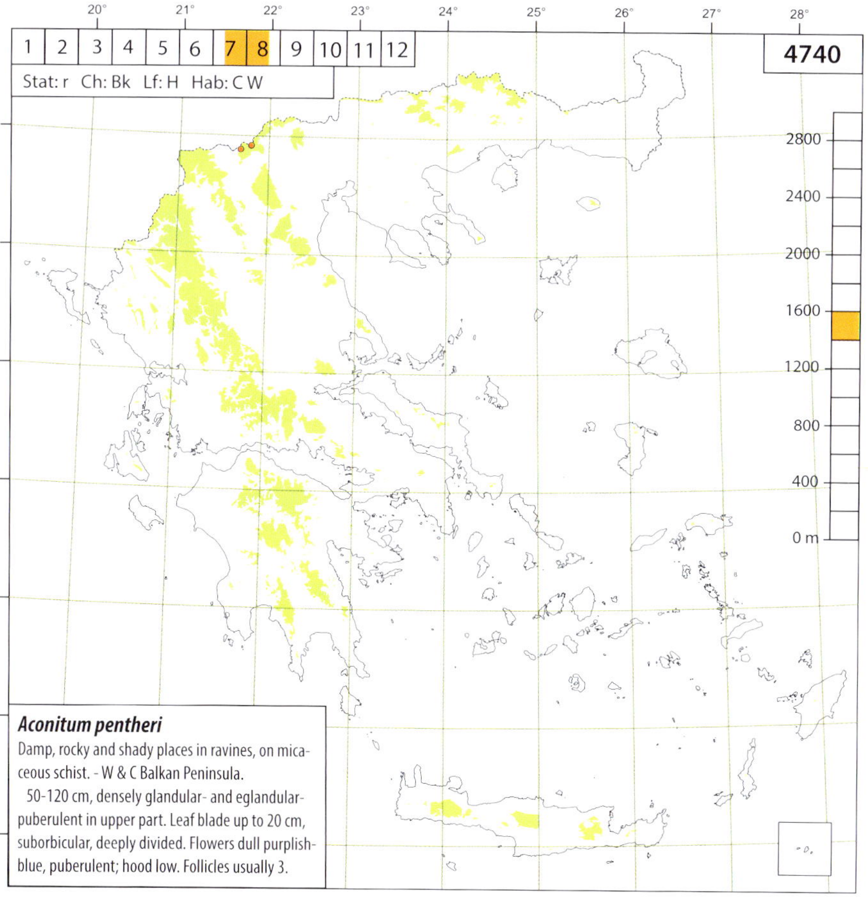

Aconitum pentheri

Damp, rocky and shady places in ravines, on micaceous schist. - W & C Balkan Peninsula.

50-120 cm, densely glandular- and eglandular-puberulent in upper part. Leaf blade up to 20 cm, suborbicular, deeply divided. Flowers dull purplish-blue, puberulent; hood low. Follicles usually 3.

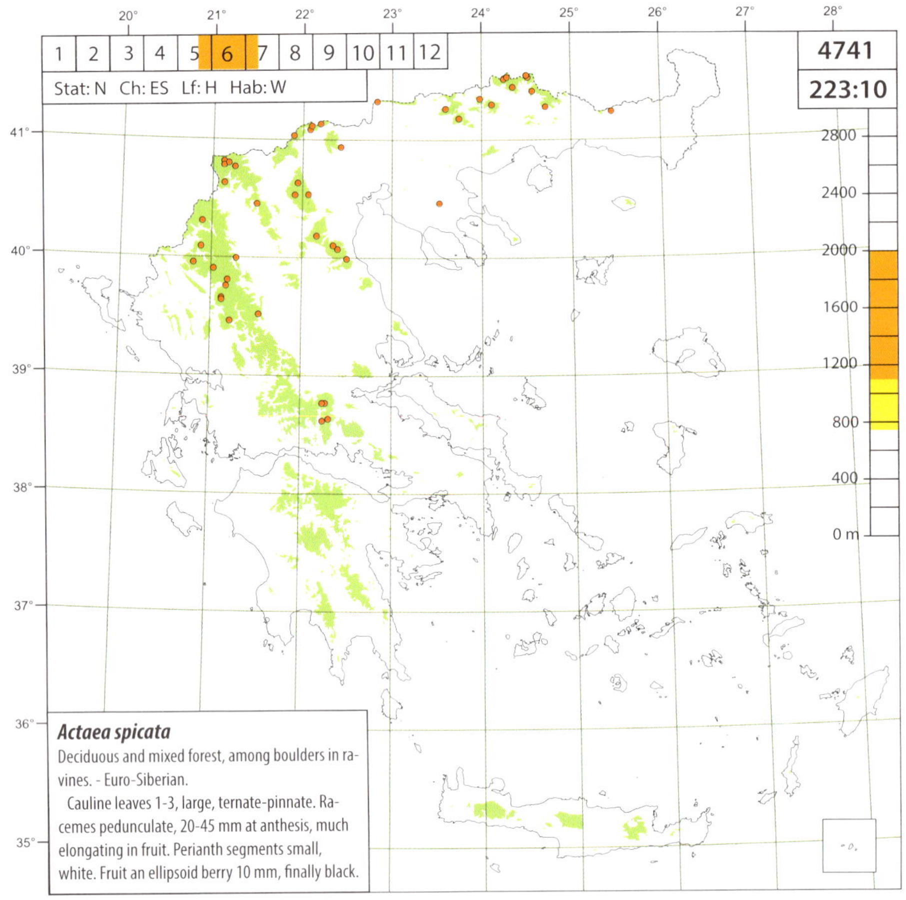

Actaea spicata
Deciduous and mixed forest, among boulders in ravines. - Euro-Siberian.

Cauline leaves 1-3, large, ternate-pinnate. Racemes pedunculate, 20-45 mm at anthesis, much elongating in fruit. Perianth segments small, white. Fruit an ellipsoid berry 10 mm, finally black.

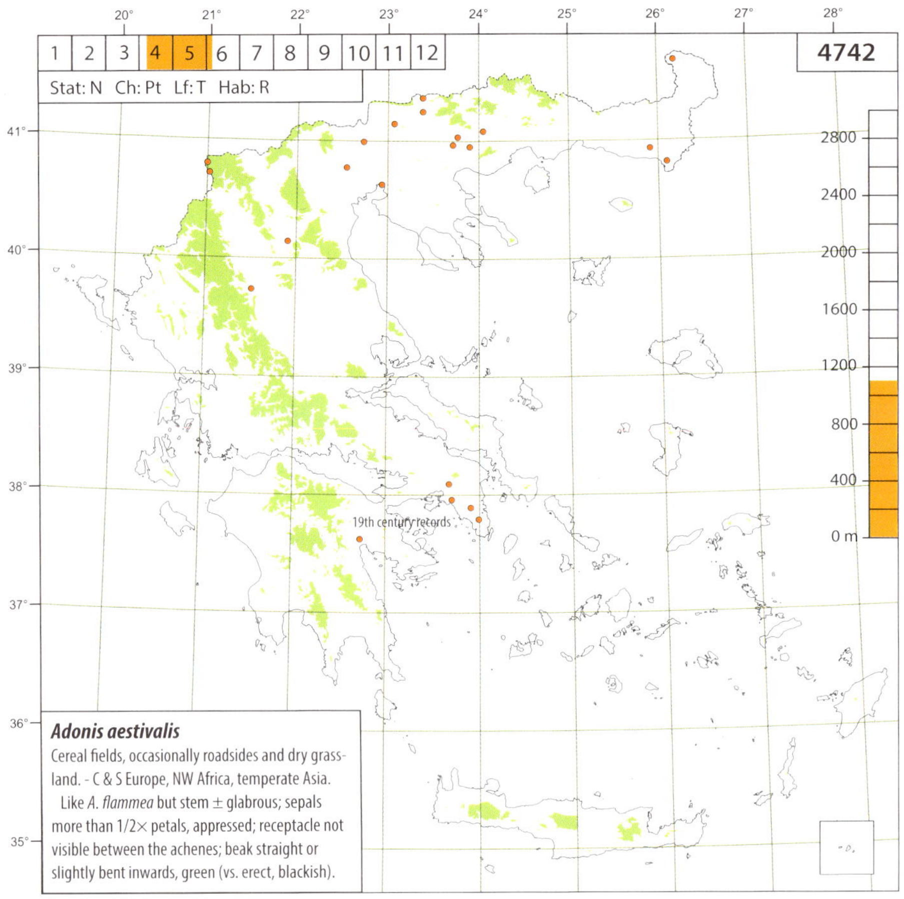

Adonis aestivalis
Cereal fields, occasionally roadsides and dry grassland. - C & S Europe, NW Africa, temperate Asia.

Like *A. flammea* but stem ± glabrous; sepals more than 1/2× petals, appressed; receptacle not visible between the achenes; beak straight or slightly bent inwards, green (vs. erect, blackish).

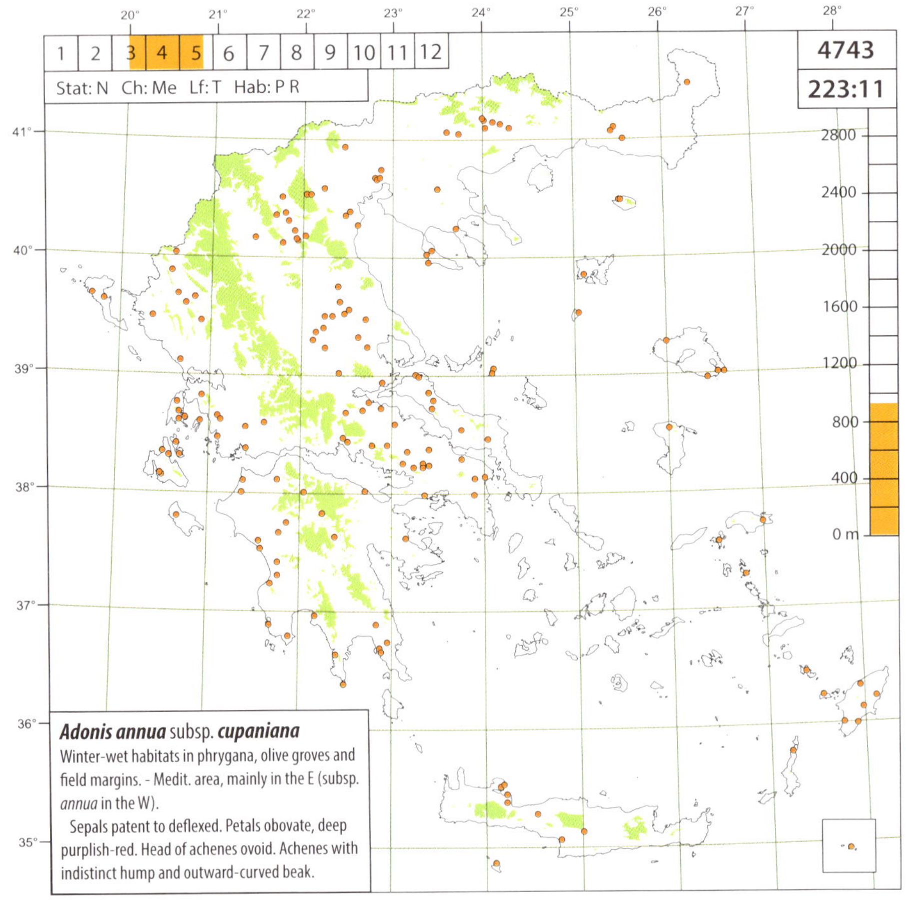

Adonis annua* subsp. *cupaniana
Winter-wet habitats in phrygana, olive groves and field margins. - Medit. area, mainly in the E (subsp. *annua* in the W).

Sepals patent to deflexed. Petals obovate, deep purplish-red. Head of achenes ovoid. Achenes with indistinct hump and outward-curved beak.

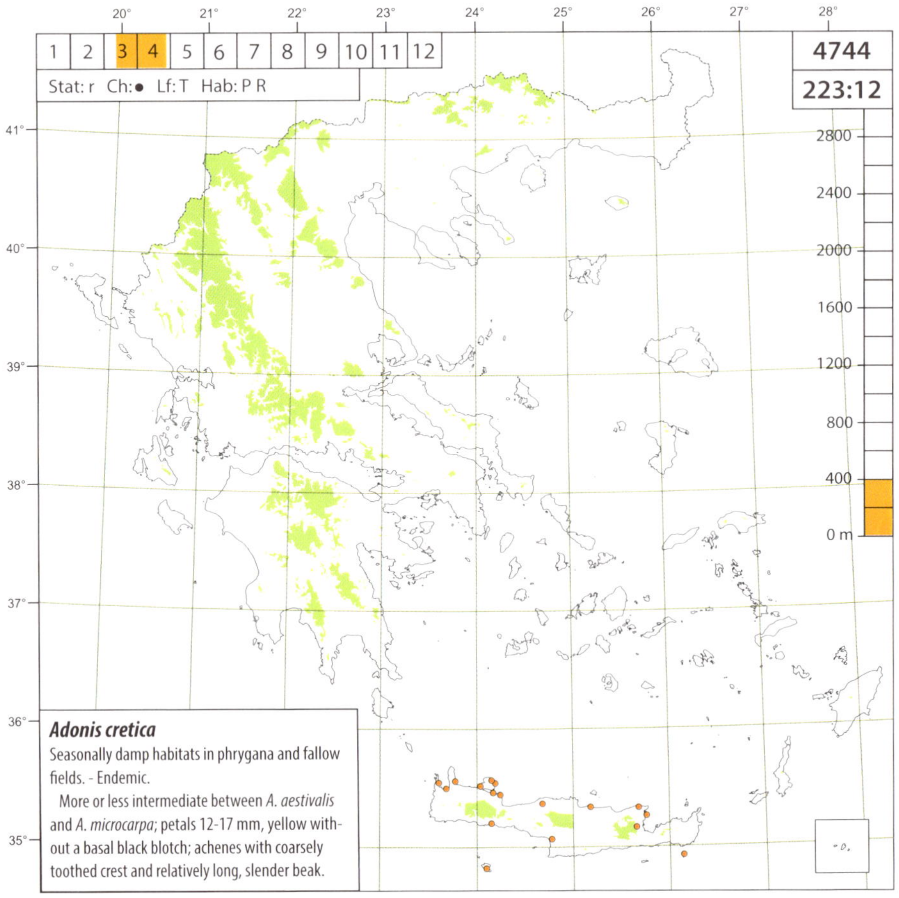

Adonis cretica
Seasonally damp habitats in phrygana and fallow fields. - Endemic.

More or less intermediate between *A. aestivalis* and *A. microcarpa*; petals 12-17 mm, yellow without a basal black blotch; achenes with coarsely toothed crest and relatively long, slender beak.

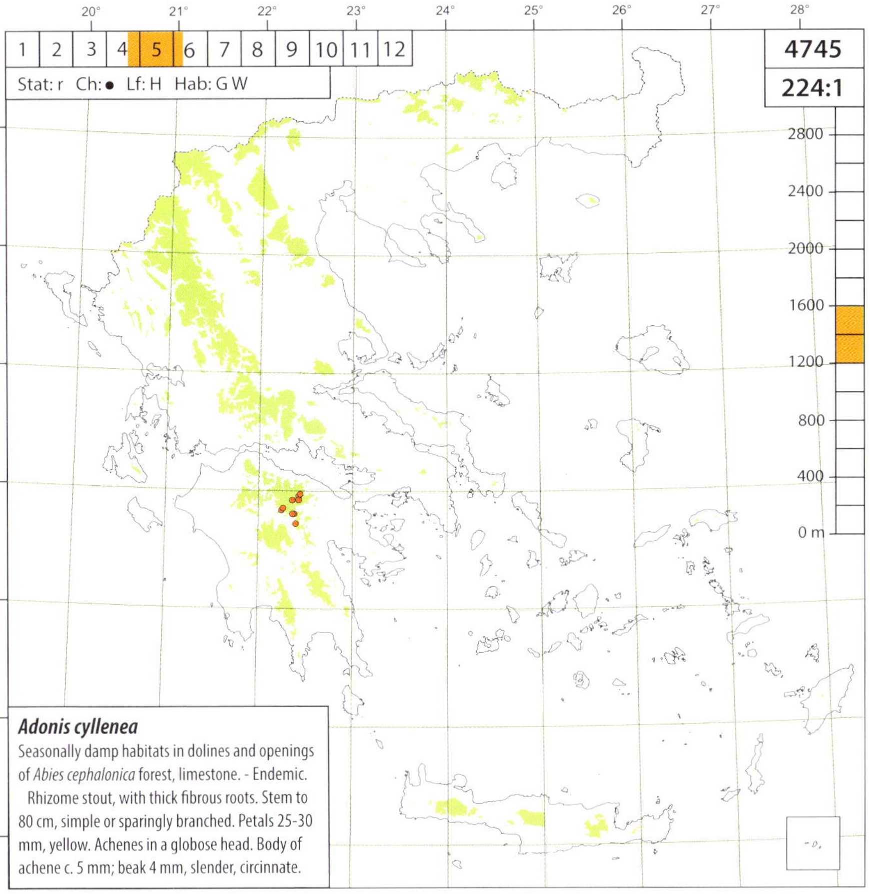

Adonis cyllenea

Seasonally damp habitats in dolines and openings of *Abies cephalonica* forest, limestone. - Endemic.

Rhizome stout, with thick fibrous roots. Stem to 80 cm, simple or sparingly branched. Petals 25-30 mm, yellow. Achenes in a globose head. Body of achene c. 5 mm; beak 4 mm, slender, circinnate.

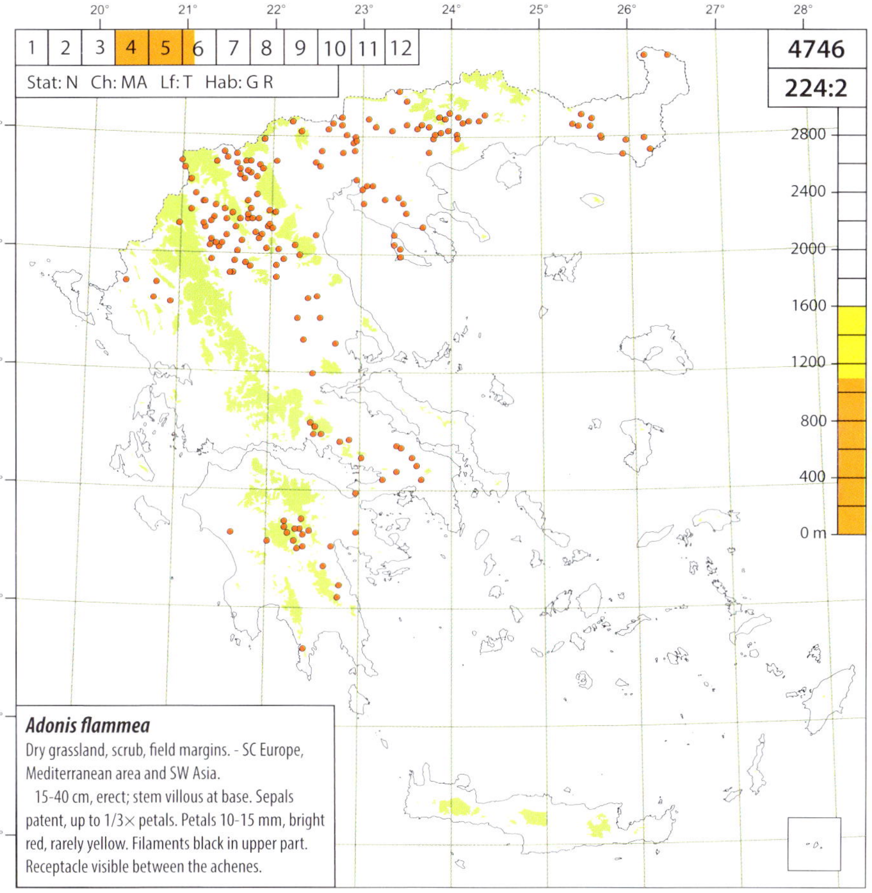

Adonis flammea

Dry grassland, scrub, field margins. - SC Europe, Mediterranean area and SW Asia.

15-40 cm, erect; stem villous at base. Sepals patent, up to 1/3× petals. Petals 10-15 mm, bright red, rarely yellow. Filaments black in upper part. Receptacle visible between the achenes.

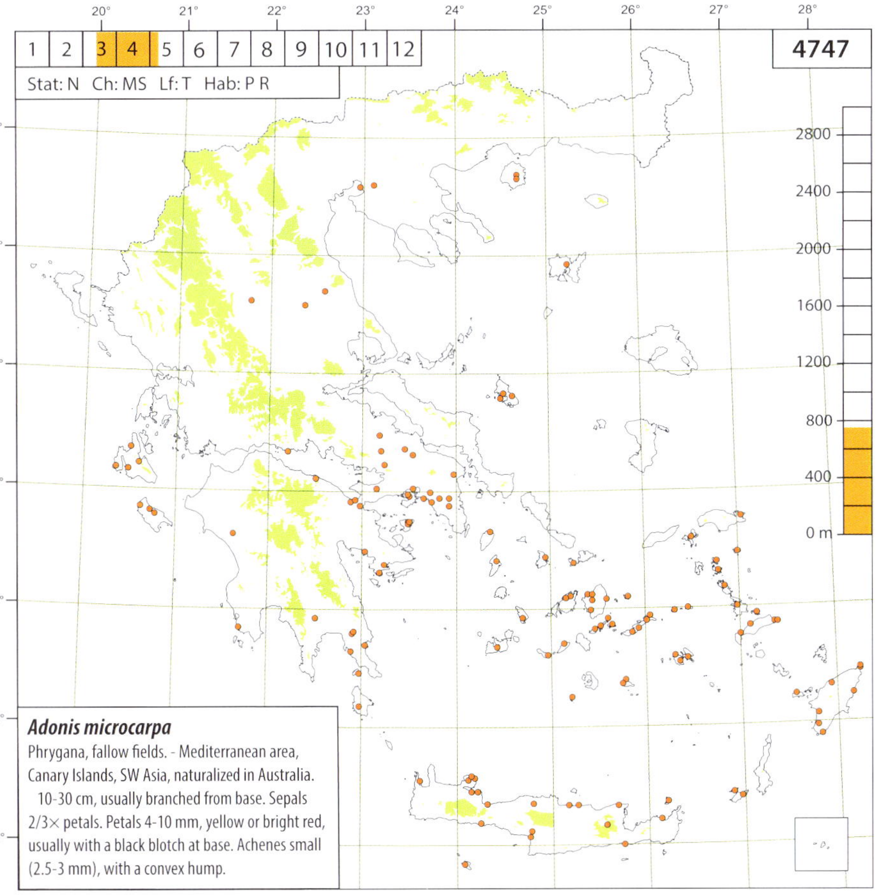

Adonis microcarpa

Phrygana, fallow fields. - Mediterranean area, Canary Islands, SW Asia, naturalized in Australia.

10-30 cm, usually branched from base. Sepals 2/3× petals. Petals 4-10 mm, yellow or bright red, usually with a black blotch at base. Achenes small (2.5-3 mm), with a convex hump.

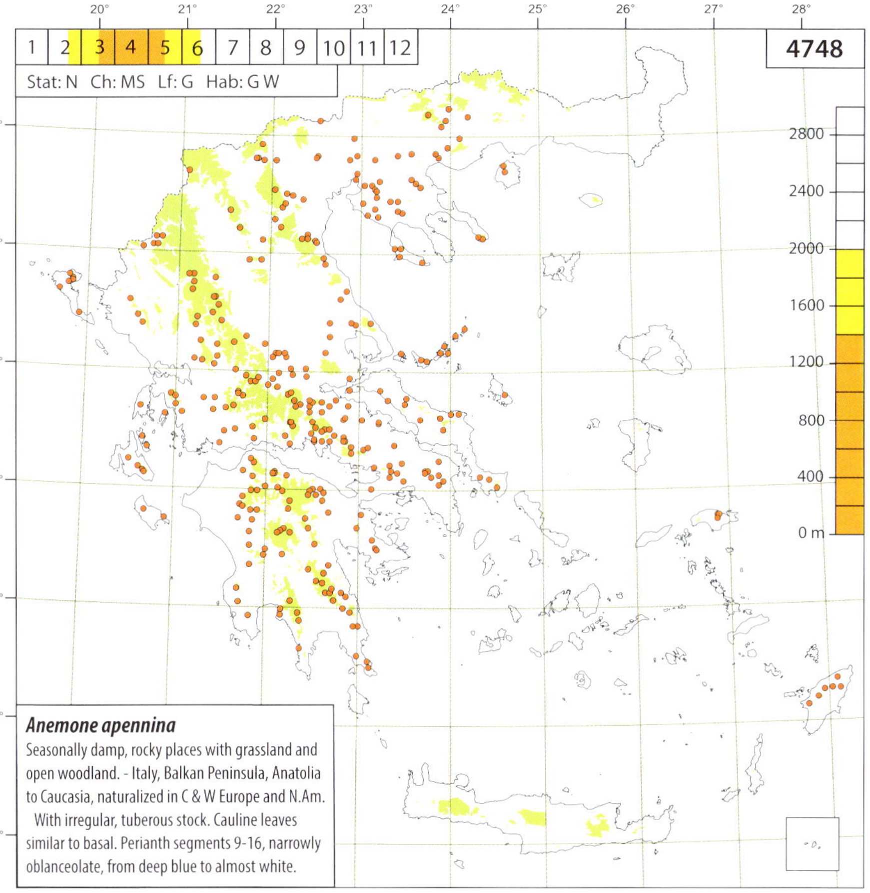

Anemone apennina

Seasonally damp, rocky places with grassland and open woodland. - Italy, Balkan Peninsula, Anatolia to Caucasia, naturalized in C & W Europe and N.Am.

With irregular, tuberous stock. Cauline leaves similar to basal. Perianth segments 9-16, narrowly oblanceolate, from deep blue to almost white.

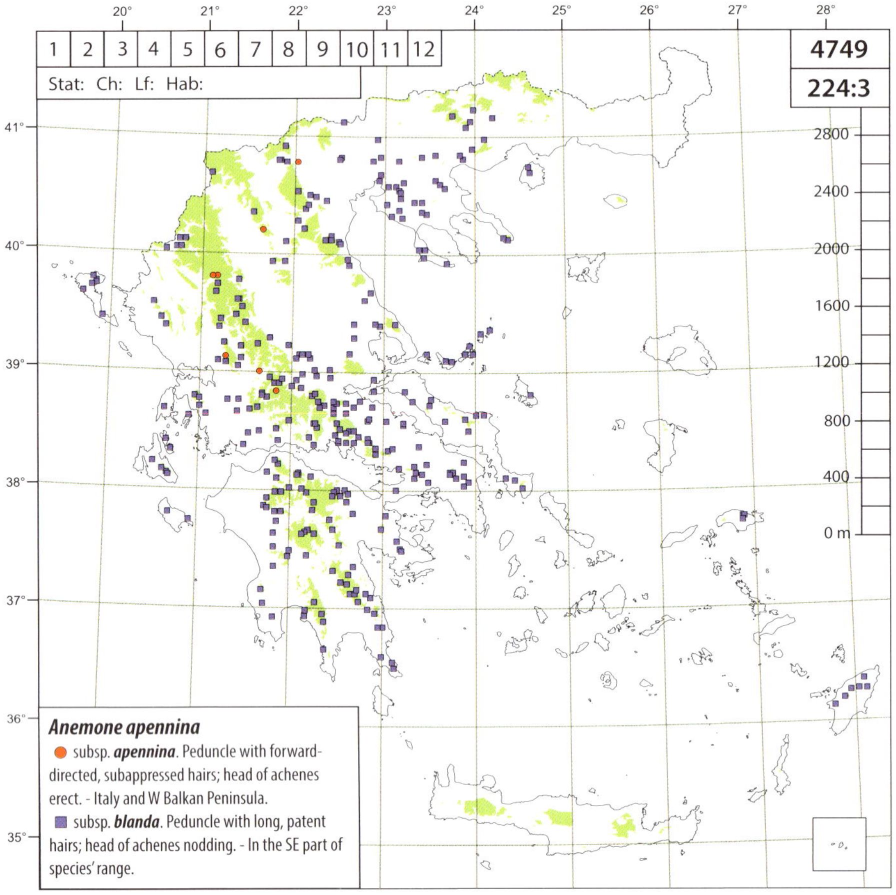

Anemone apennina
● subsp. ***apennina***. Peduncle with forward-directed, subappressed hairs; head of achenes erect. - Italy and W Balkan Peninsula.
■ subsp. ***blanda***. Peduncle with long, patent hairs; head of achenes nodding. - In the SE part of species' range.

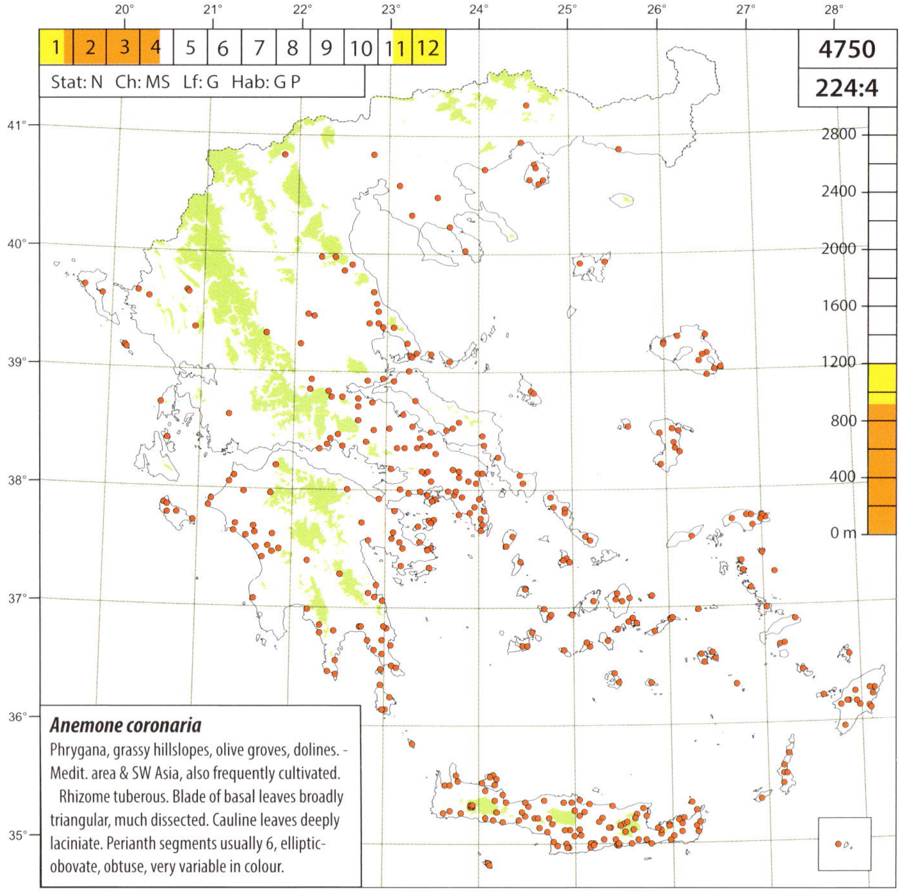

Anemone coronaria
Phrygana, grassy hillslopes, olive groves, dolines. - Medit. area & SW Asia, also frequently cultivated.
Rhizome tuberous. Blade of basal leaves broadly triangular, much dissected. Cauline leaves deeply laciniate. Perianth segments usually 6, elliptic-obovate, obtuse, very variable in colour.

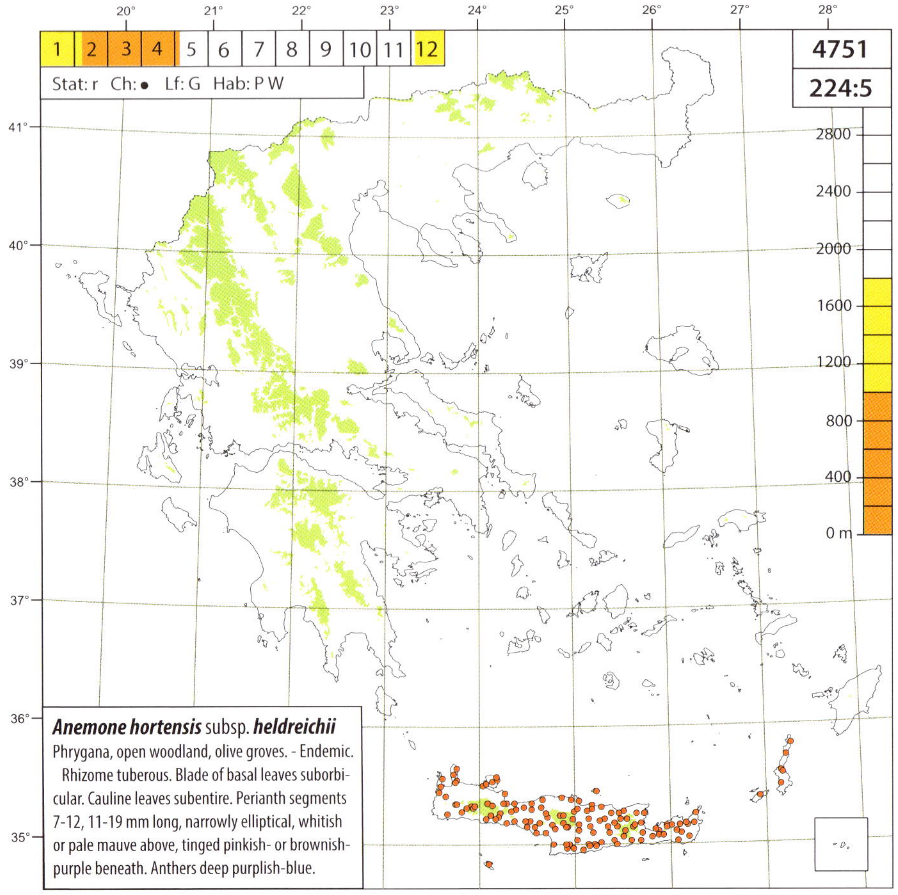

Anemone hortensis subsp. ***heldreichii***
Phrygana, open woodland, olive groves. - Endemic.
Rhizome tuberous. Blade of basal leaves suborbicular. Cauline leaves subentire. Perianth segments 7-12, 11-19 mm long, narrowly elliptical, whitish or pale mauve above, tinged pinkish- or brownish-purple beneath. Anthers deep purplish-blue.

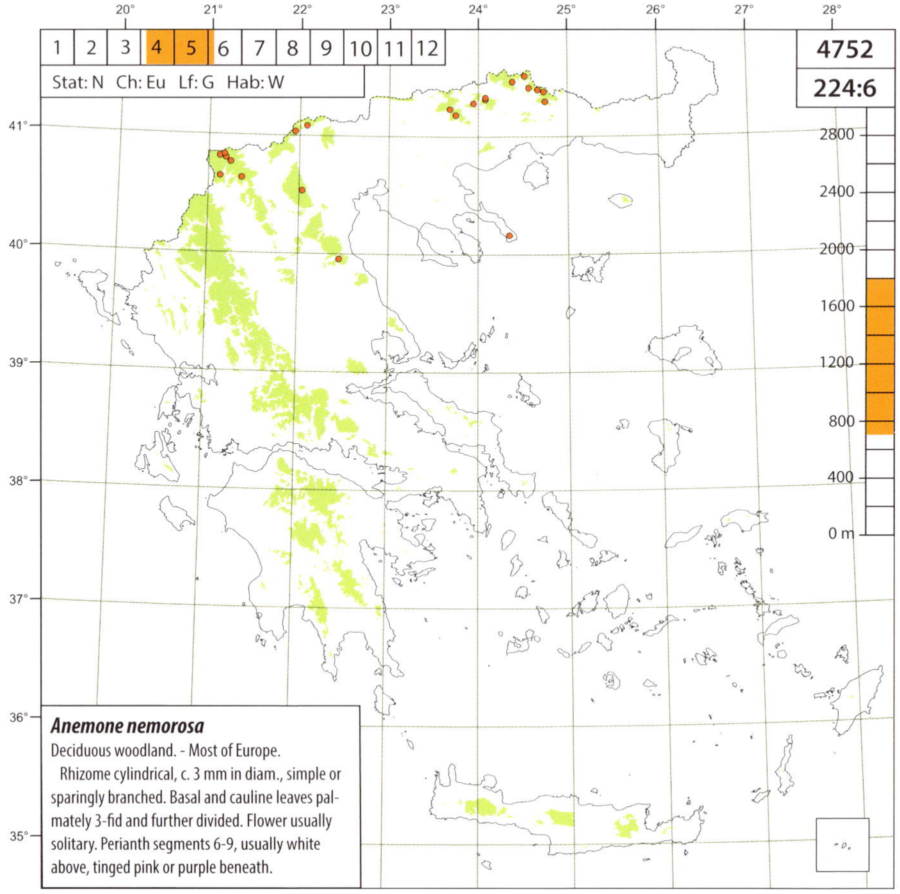

Anemone nemorosa
Deciduous woodland. - Most of Europe.
Rhizome cylindrical, c. 3 mm in diam., simple or sparingly branched. Basal and cauline leaves palmately 3-fid and further divided. Flower usually solitary. Perianth segments 6-9, usually white above, tinged pink or purple beneath.

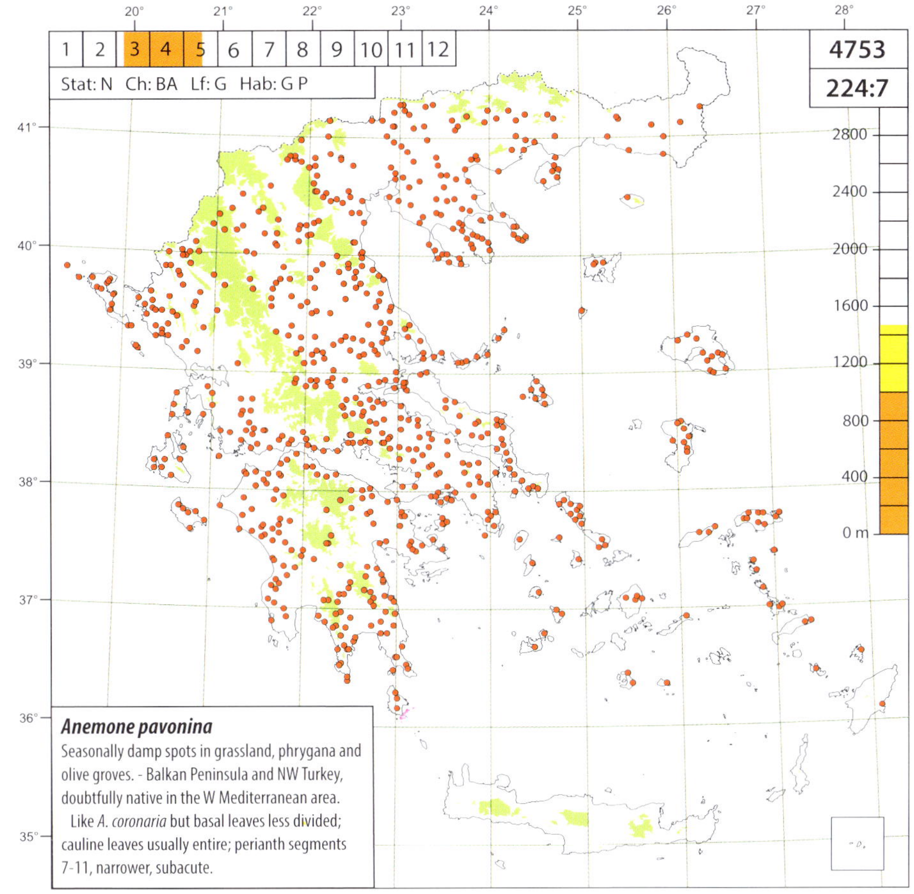

Anemone pavonina

Seasonally damp spots in grassland, phrygana and olive groves. - Balkan Peninsula and NW Turkey, doubtfully native in the W Mediterranean area.

Like *A. coronaria* but basal leaves less divided; cauline leaves usually entire; perianth segments 7-11, narrower, subacute.

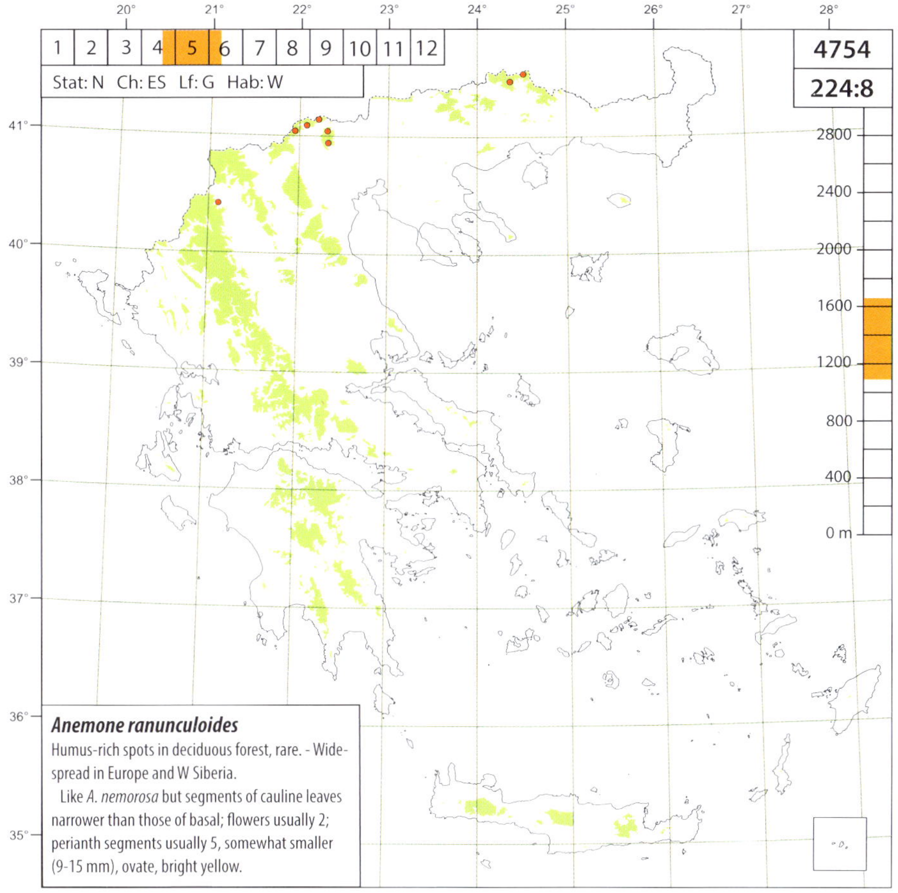

Anemone ranunculoides

Humus-rich spots in deciduous forest, rare. - Widespread in Europe and W Siberia.

Like *A. nemorosa* but segments of cauline leaves narrower than those of basal; flowers usually 2; perianth segments usually 5, somewhat smaller (9-15 mm), ovate, bright yellow.

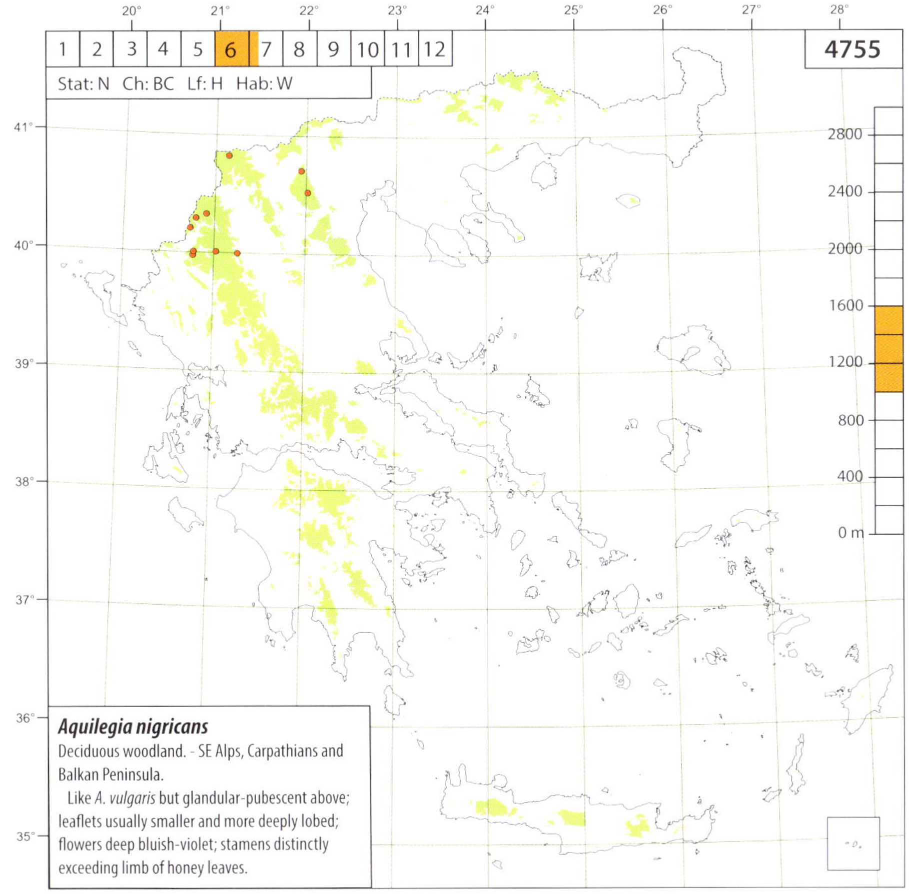

Aquilegia nigricans

Deciduous woodland. - SE Alps, Carpathians and Balkan Peninsula.

Like *A. vulgaris* but glandular-pubescent above; leaflets usually smaller and more deeply lobed; flowers deep bluish-violet; stamens distinctly exceeding limb of honey leaves.

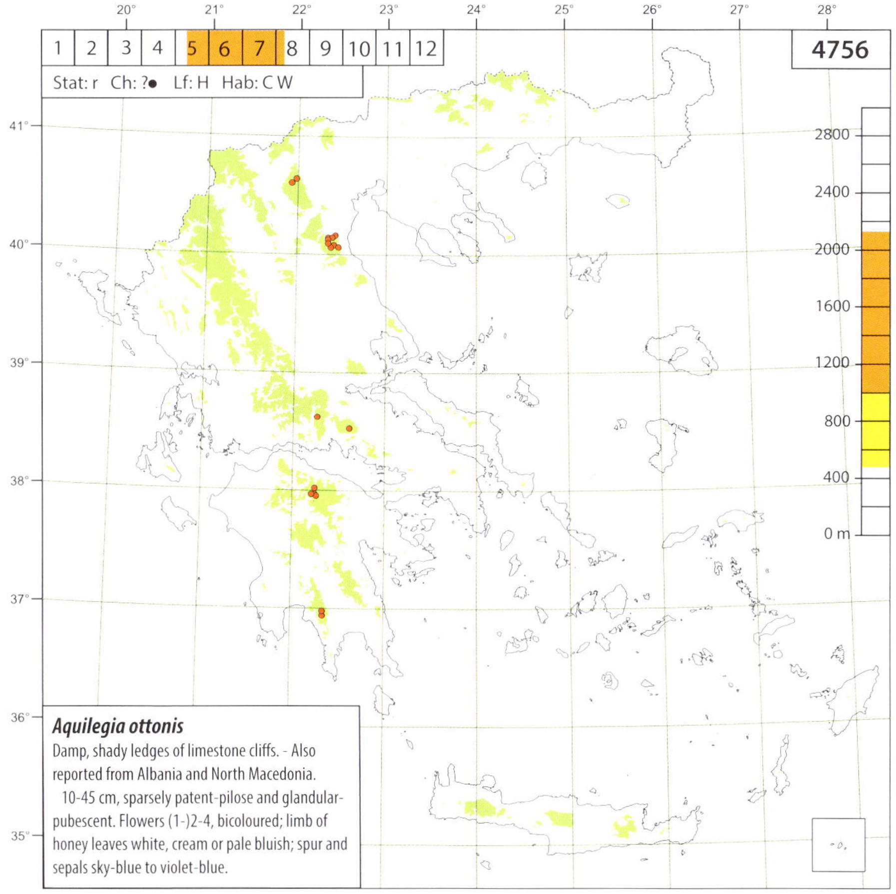

Aquilegia ottonis

Damp, shady ledges of limestone cliffs. - Also reported from Albania and North Macedonia.

10-45 cm, sparsely patent-pilose and glandular-pubescent. Flowers (1-)2-4, bicoloured; limb of honey leaves white, cream or pale bluish; spur and sepals sky-blue to violet-blue.

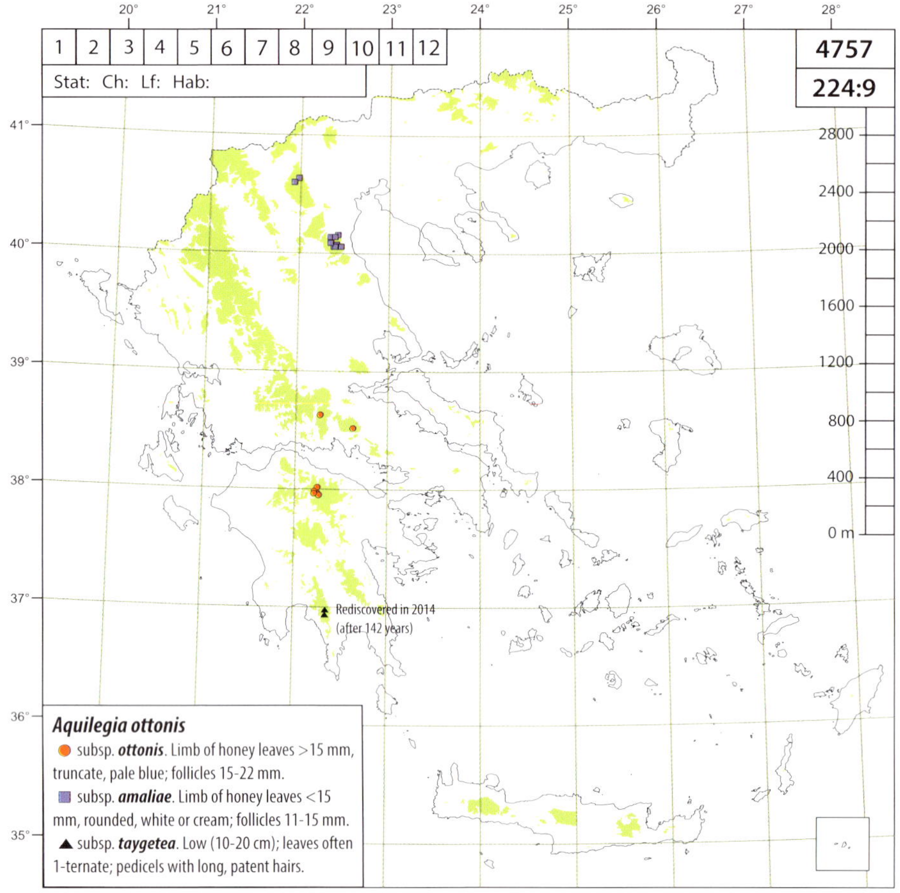

Aquilegia ottonis

● subsp. ***ottonis***. Limb of honey leaves >15 mm, truncate, pale blue; follicles 15-22 mm.
■ subsp. ***amaliae***. Limb of honey leaves <15 mm, rounded, white or cream; follicles 11-15 mm.
▲ subsp. ***taygetea***. Low (10-20 cm); leaves often 1-ternate; pedicels with long, patent hairs.

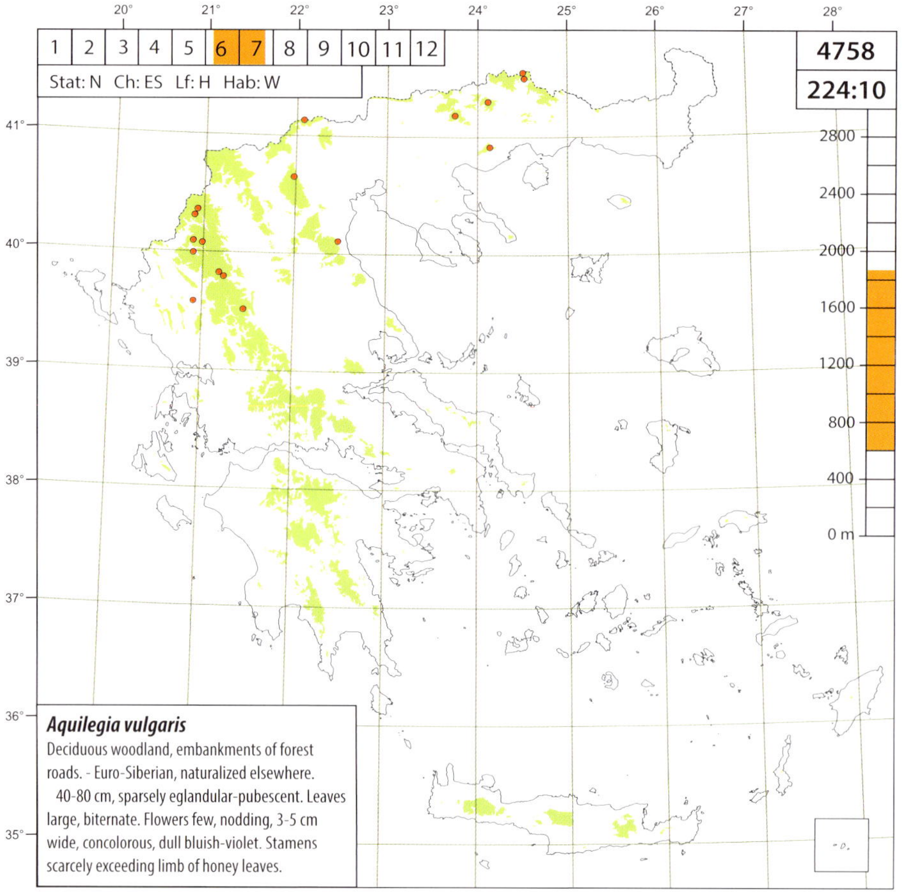

Aquilegia vulgaris

Deciduous woodland, embankments of forest roads. - Euro-Siberian, naturalized elsewhere.

40-80 cm, sparsely eglandular-pubescent. Leaves large, biternate. Flowers few, nodding, 3-5 cm wide, concolorous, dull bluish-violet. Stamens scarcely exceeding limb of honey leaves.

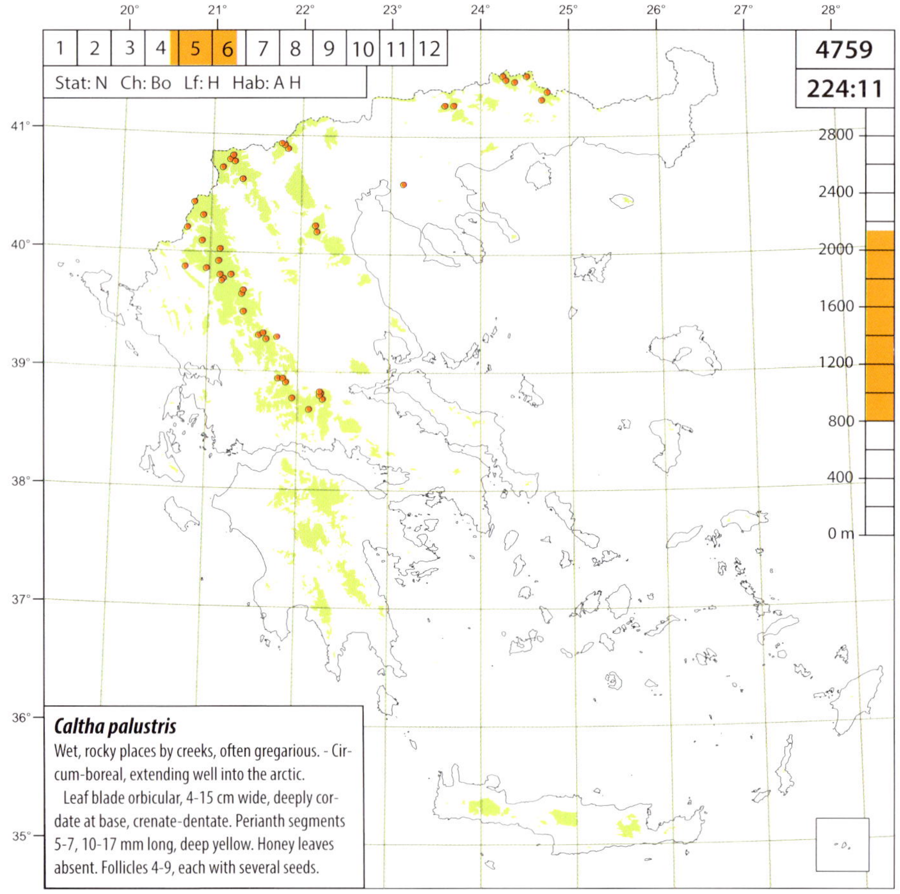

Caltha palustris

Wet, rocky places by creeks, often gregarious. - Circum-boreal, extending well into the arctic.

Leaf blade orbicular, 4-15 cm wide, deeply cordate at base, crenate-dentate. Perianth segments 5-7, 10-17 mm long, deep yellow. Honey leaves absent. Follicles 4-9, each with several seeds.

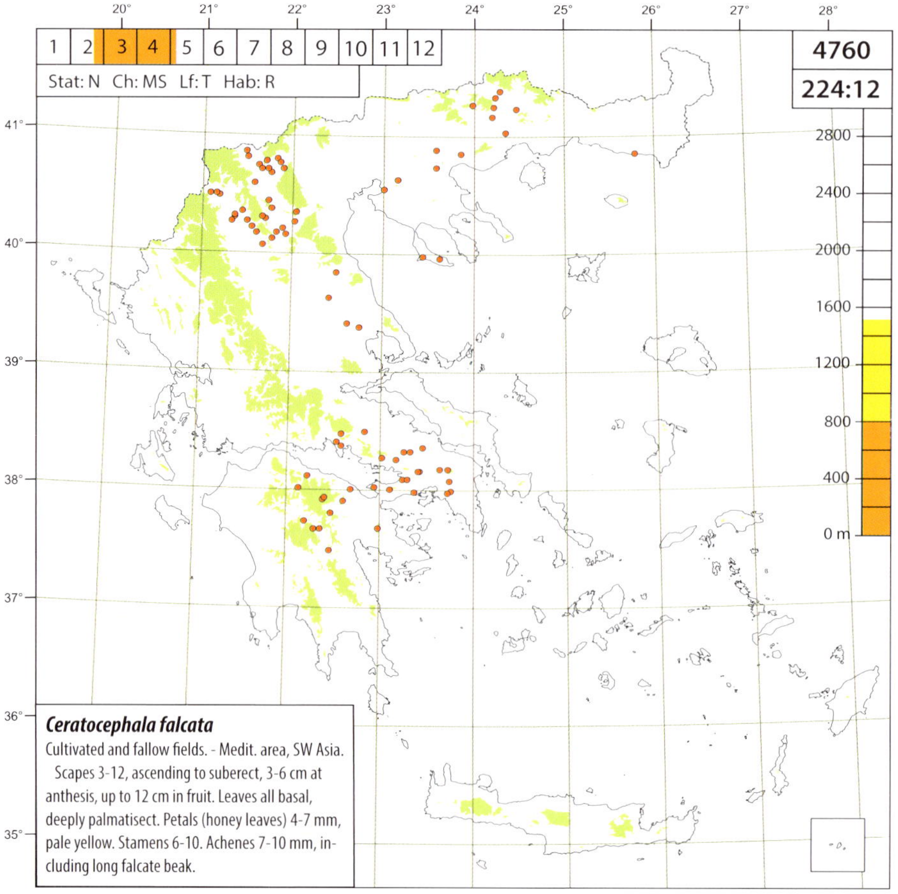

Ceratocephala falcata

Cultivated and fallow fields. - Medit. area, SW Asia.

Scapes 3-12, ascending to suberect, 3-6 cm at anthesis, up to 12 cm in fruit. Leaves all basal, deeply palmatisect. Petals (honey leaves) 4-7 mm, pale yellow. Stamens 6-10. Achenes 7-10 mm, including long falcate beak.

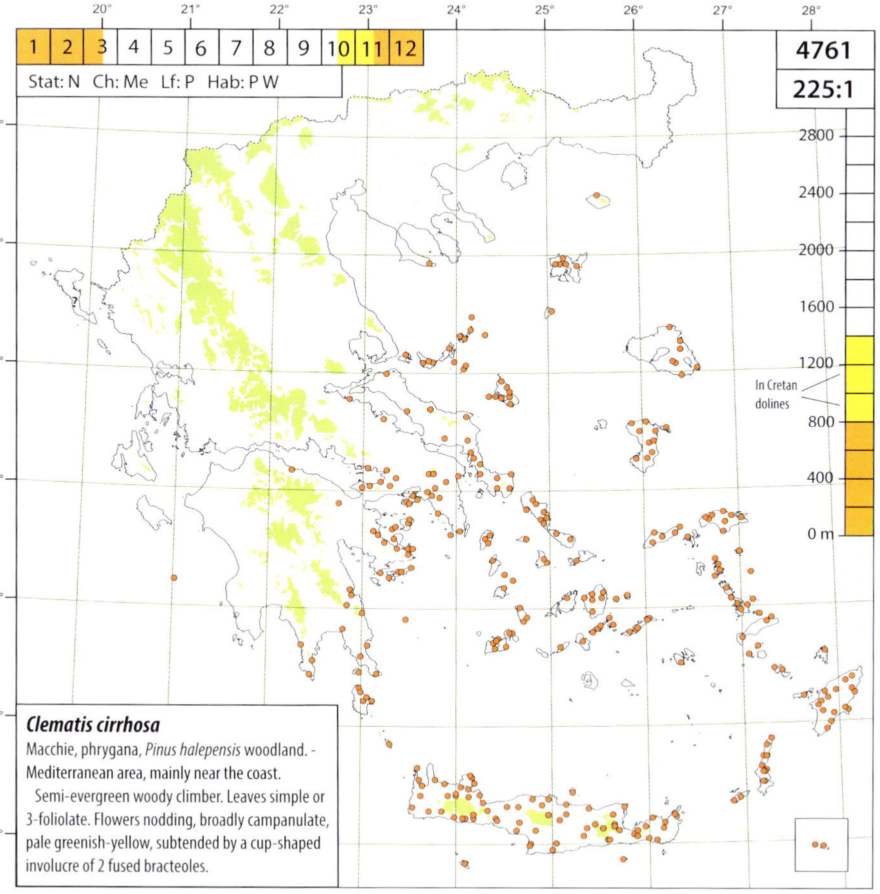

Clematis cirrhosa

Macchie, phrygana, *Pinus halepensis* woodland. - Mediterranean area, mainly near the coast.

Semi-evergreen woody climber. Leaves simple or 3-foliolate. Flowers nodding, broadly campanulate, pale greenish-yellow, subtended by a cup-shaped involucre of 2 fused bracteoles.

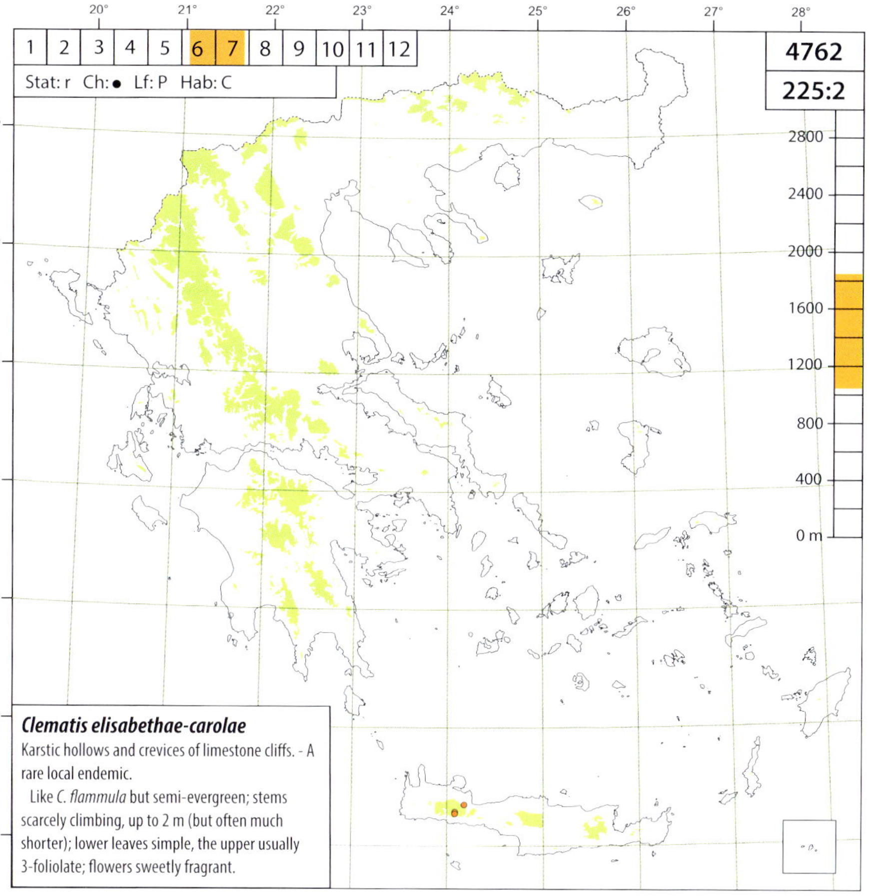

Clematis elisabethae-carolae

Karstic hollows and crevices of limestone cliffs. - A rare local endemic.

Like *C. flammula* but semi-evergreen; stems scarcely climbing, up to 2 m (but often much shorter); lower leaves simple, the upper usually 3-foliolate; flowers sweetly fragrant.

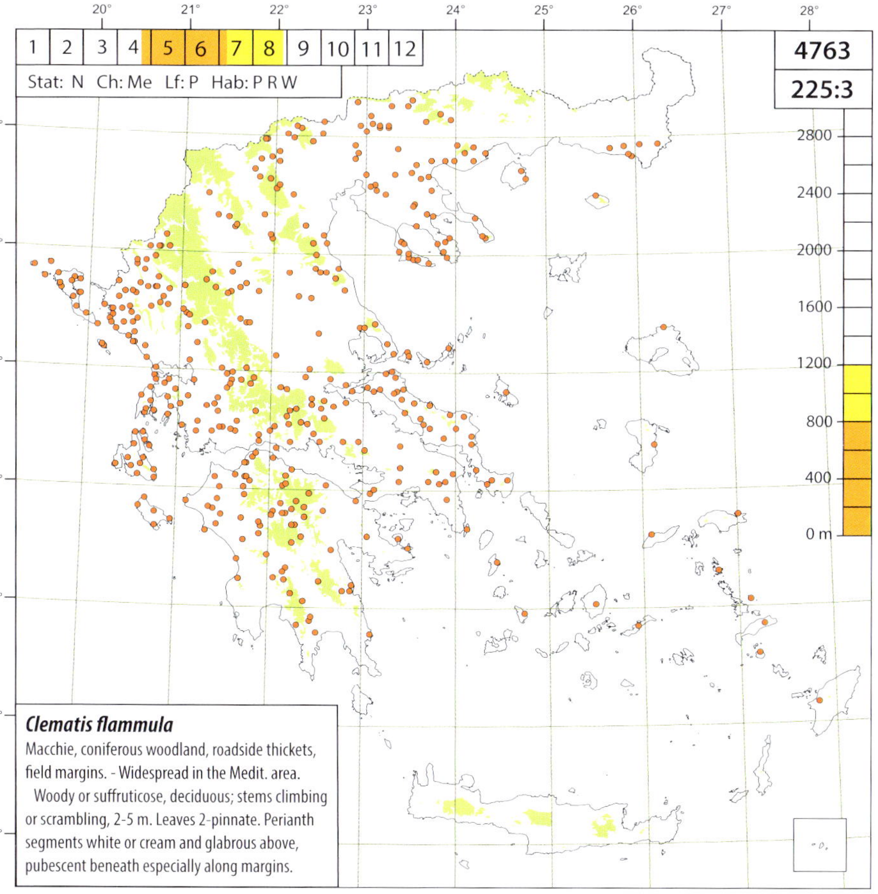

Clematis flammula

Macchie, coniferous woodland, roadside thickets, field margins. - Widespread in the Medit. area.

Woody or suffruticose, deciduous; stems climbing or scrambling, 2-5 m. Leaves 2-pinnate. Perianth segments white or cream and glabrous above, pubescent beneath especially along margins.

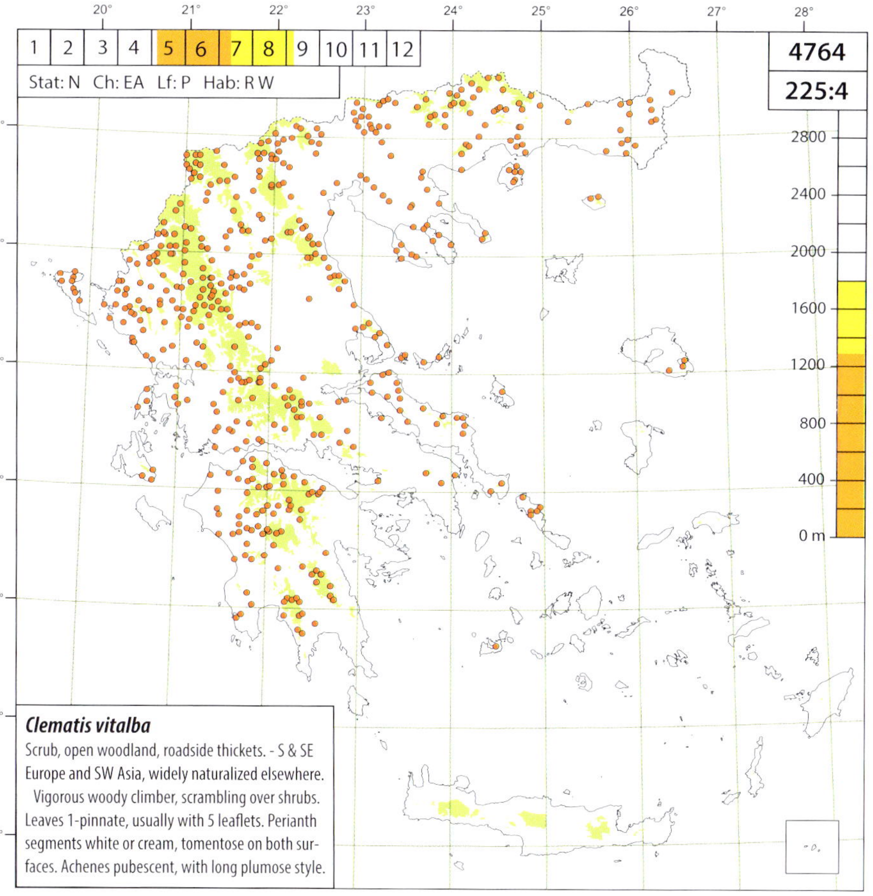

Clematis vitalba

Scrub, open woodland, roadside thickets. - S & SE Europe and SW Asia, widely naturalized elsewhere.

Vigorous woody climber, scrambling over shrubs. Leaves 1-pinnate, usually with 5 leaflets. Perianth segments white or cream, tomentose on both surfaces. Achenes pubescent, with long plumose style.

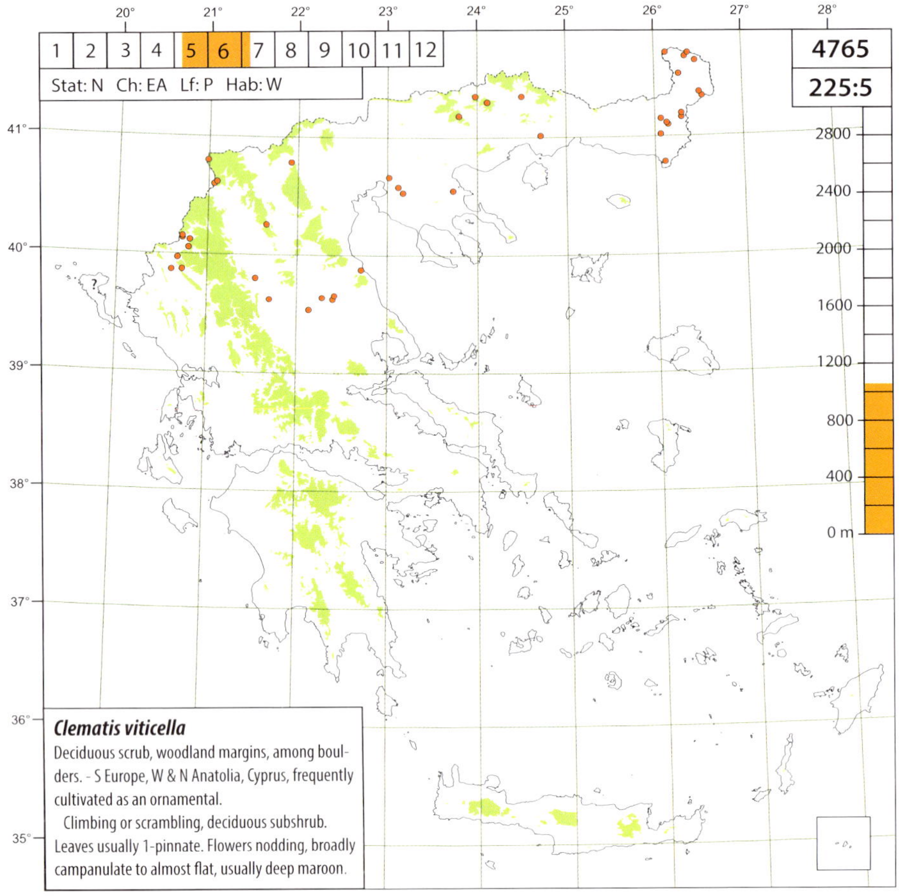

Clematis viticella

Deciduous scrub, woodland margins, among boulders. - S Europe, W & N Anatolia, Cyprus, frequently cultivated as an ornamental.

Climbing or scrambling, deciduous subshrub. Leaves usually 1-pinnate. Flowers nodding, broadly campanulate to almost flat, usually deep maroon.

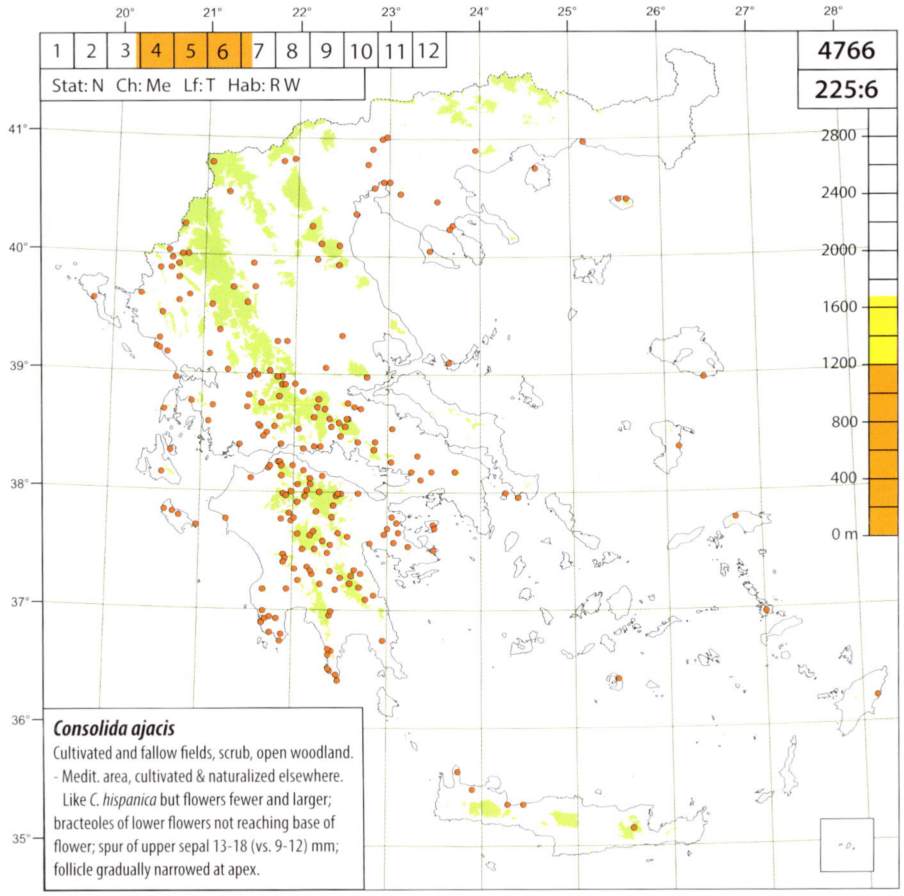

Consolida ajacis

Cultivated and fallow fields, scrub, open woodland. - Medit. area, cultivated & naturalized elsewhere.

Like *C. hispanica* but flowers fewer and larger; bracteoles of lower flowers not reaching base of flower; spur of upper sepal 13-18 (vs. 9-12) mm; follicle gradually narrowed at apex.

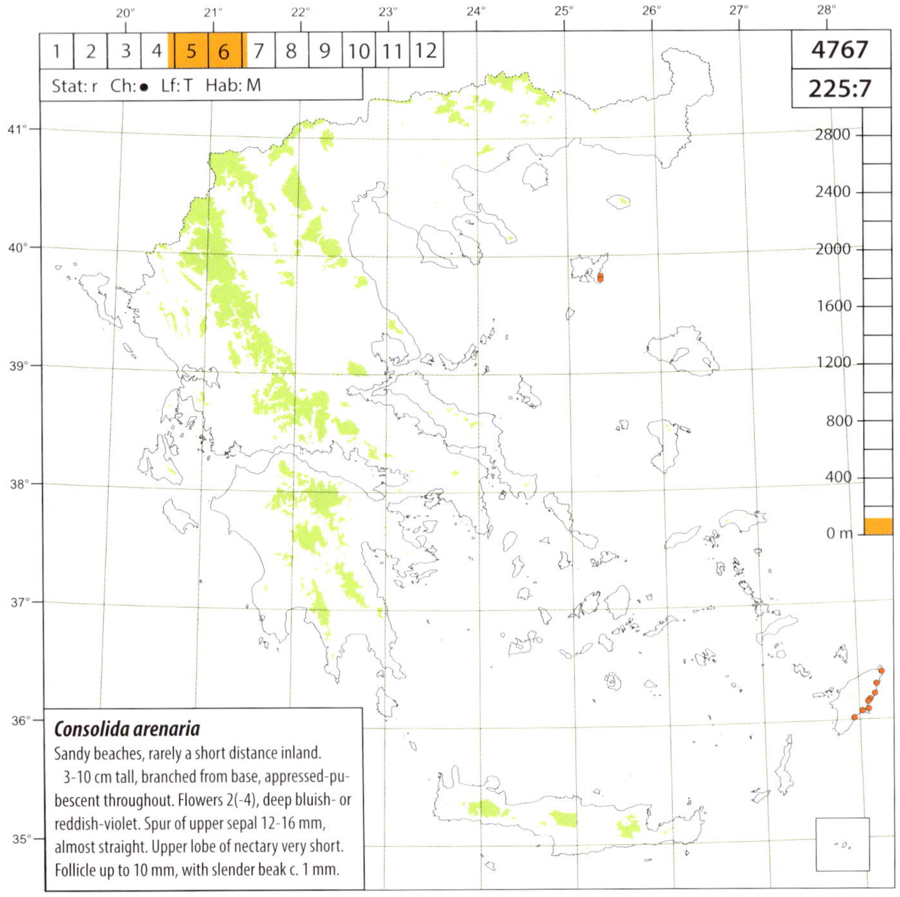

Consolida arenaria

Sandy beaches, rarely a short distance inland.

3-10 cm tall, branched from base, appressed-pubescent throughout. Flowers 2(-4), deep bluish- or reddish-violet. Spur of upper sepal 12-16 mm, almost straight. Upper lobe of nectary very short. Follicle up to 10 mm, with slender beak c. 1 mm.

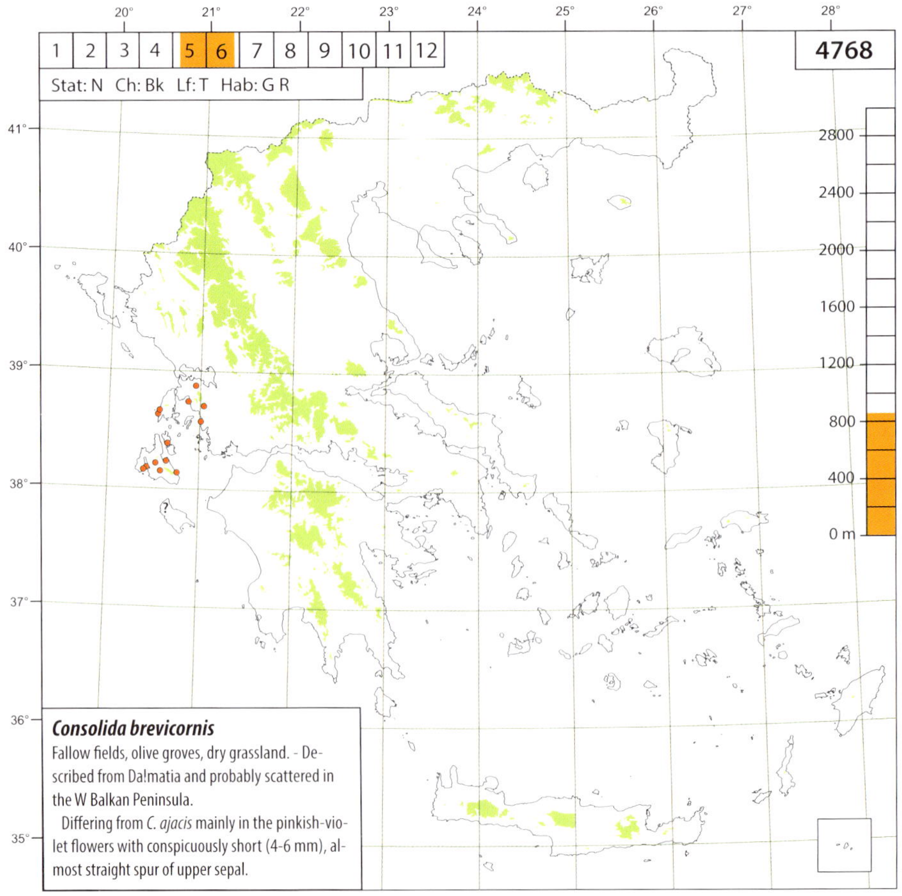

Consolida brevicornis

Fallow fields, olive groves, dry grassland. - Described from Dalmatia and probably scattered in the W Balkan Peninsula.

Differing from *C. ajacis* mainly in the pinkish-violet flowers with conspicuously short (4-6 mm), almost straight spur of upper sepal.

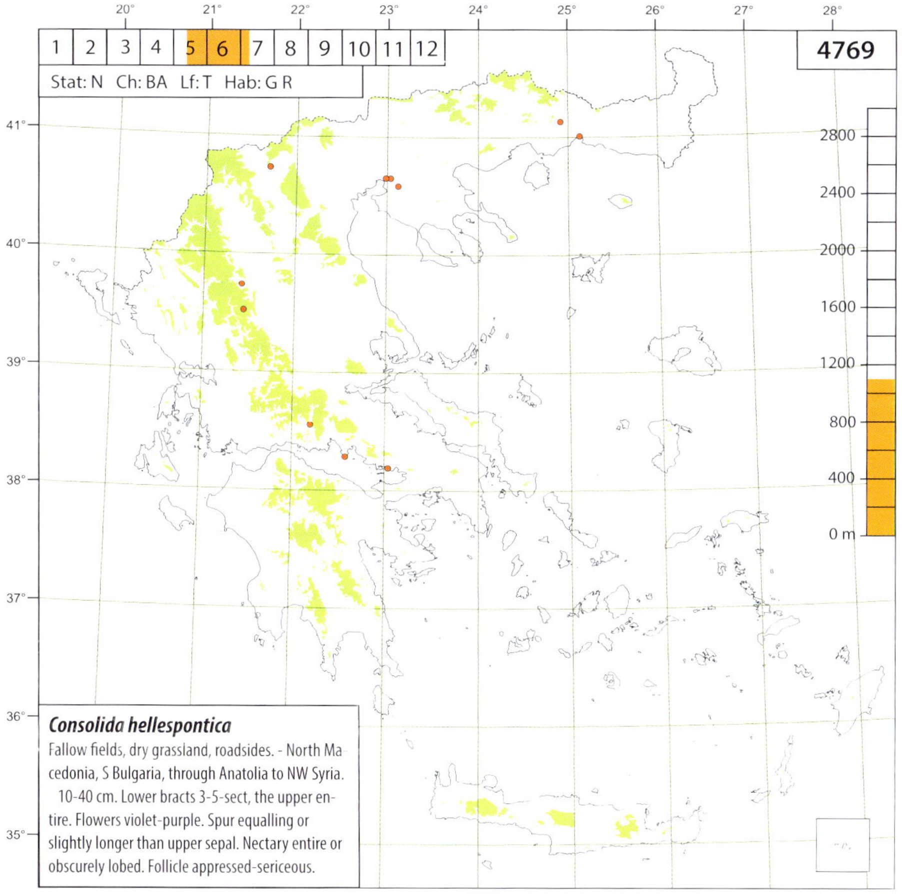

Consolida hellespontica

Fallow fields, dry grassland, roadsides. - North Macedonia, S Bulgaria, through Anatolia to NW Syria.

10-40 cm. Lower bracts 3-5-sect, the upper entire. Flowers violet-purple. Spur equalling or slightly longer than upper sepal. Nectary entire or obscurely lobed. Follicle appressed-sericeous.

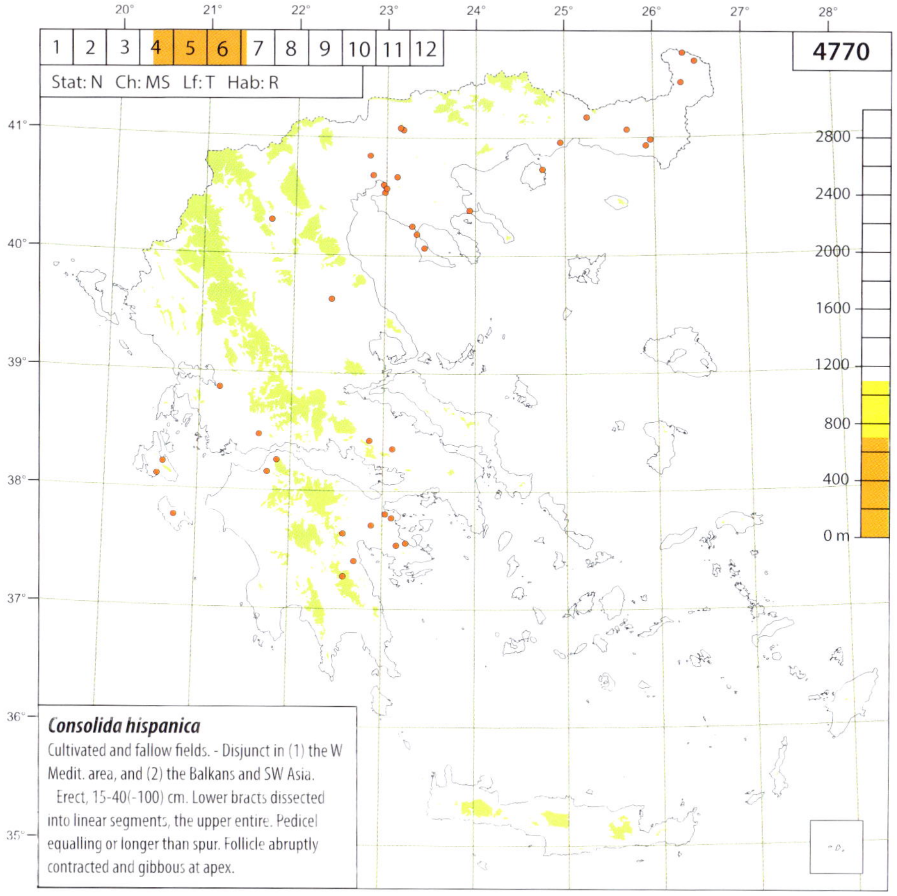

Consolida hispanica

Cultivated and fallow fields. - Disjunct in (1) the W Medit. area, and (2) the Balkans and SW Asia.

Erect, 15-40(-100) cm. Lower bracts dissected into linear segments, the upper entire. Pedicel equalling or longer than spur. Follicle abruptly contracted and gibbous at apex.

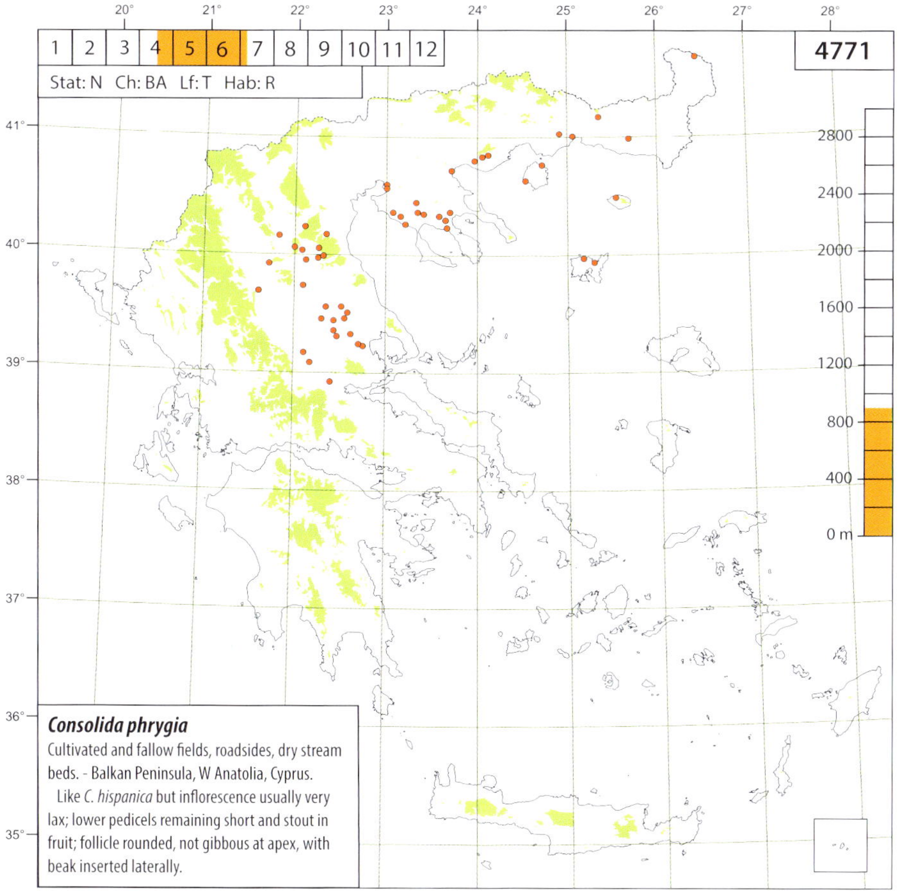

Consolida phrygia

Cultivated and fallow fields, roadsides, dry stream beds. - Balkan Peninsula, W Anatolia, Cyprus.

Like *C. hispanica* but inflorescence usually very lax; lower pedicels remaining short and stout in fruit; follicle rounded, not gibbous at apex, with beak inserted laterally.

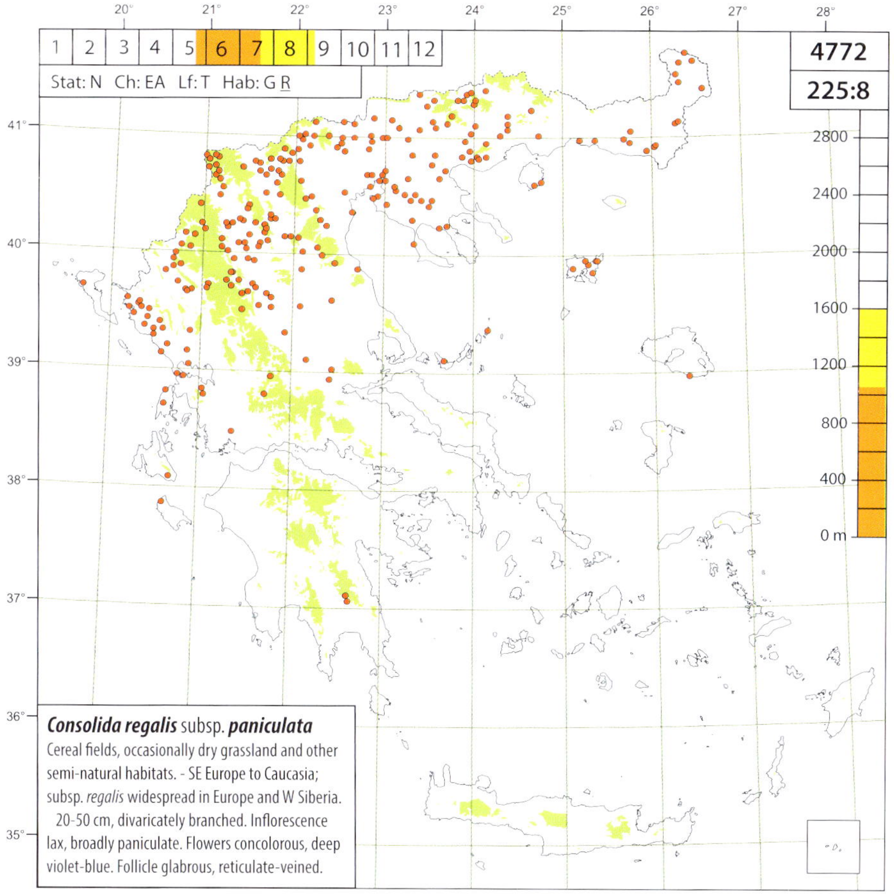

Consolida regalis subsp. ***paniculata***

Cereal fields, occasionally dry grassland and other semi-natural habitats. - SE Europe to Caucasia; subsp. *regalis* widespread in Europe and W Siberia.

20-50 cm, divaricately branched. Inflorescence lax, broadly paniculate. Flowers concolorous, deep violet-blue. Follicle glabrous, reticulate-veined.

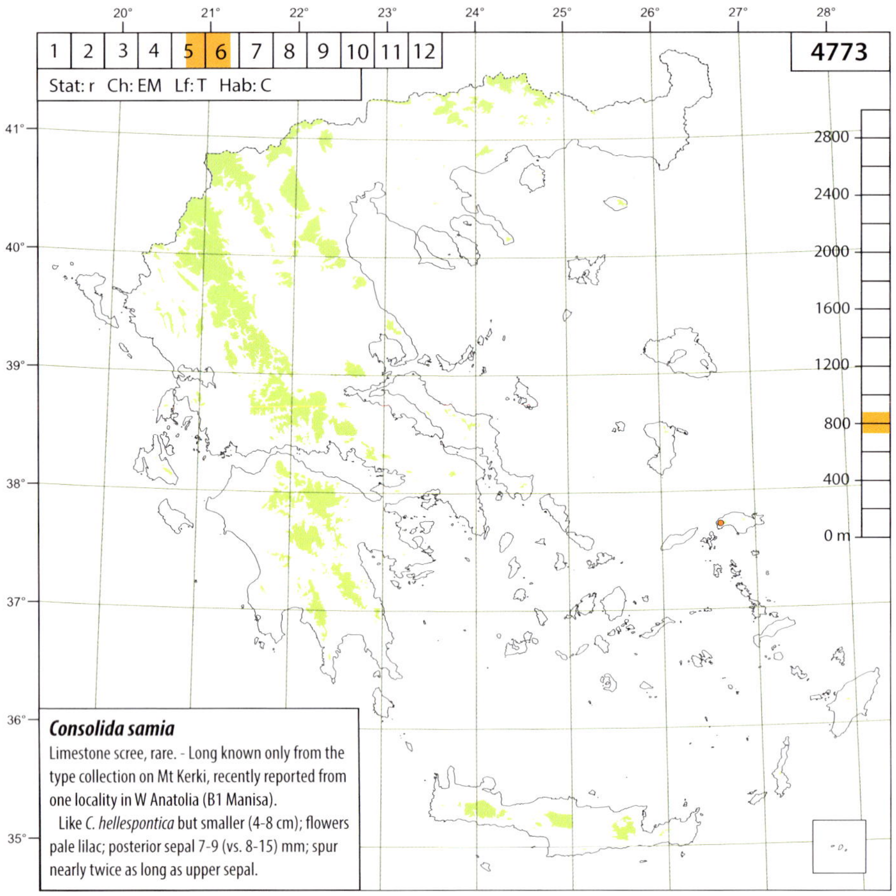

Consolida samia

Limestone scree, rare. - Long known only from the type collection on Mt Kerki, recently reported from one locality in W Anatolia (B1 Manisa).

Like *C. hellespontica* but smaller (4-8 cm); flowers pale lilac; posterior sepal 7-9 (vs. 8-15) mm; spur nearly twice as long as upper sepal.

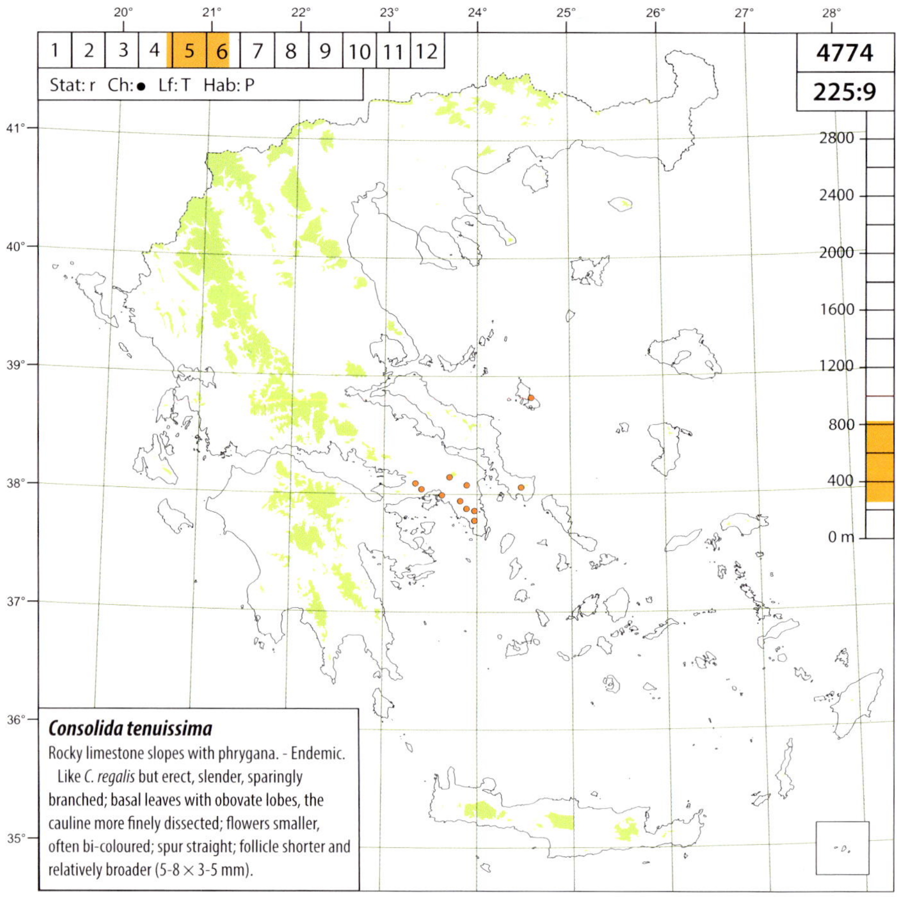

Consolida tenuissima

Rocky limestone slopes with phrygana. - Endemic.

Like *C. regalis* but erect, slender, sparingly branched; basal leaves with obovate lobes, the cauline more finely dissected; flowers smaller, often bi-coloured; spur straight; follicle shorter and relatively broader (5-8 × 3-5 mm).

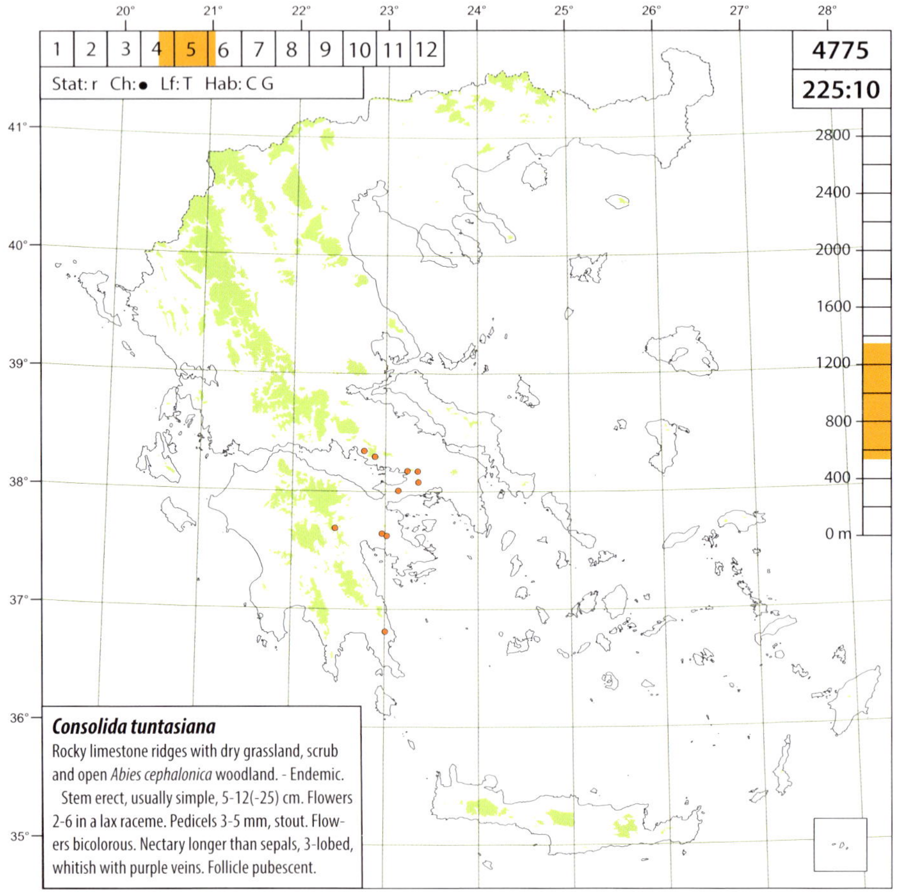

Consolida tuntasiana

Rocky limestone ridges with dry grassland, scrub and open *Abies cephalonica* woodland. - Endemic.

Stem erect, usually simple, 5-12(-25) cm. Flowers 2-6 in a lax raceme. Pedicels 3-5 mm, stout. Flowers bicolorous. Nectary longer than sepals, 3-lobed, whitish with purple veins. Follicle pubescent.

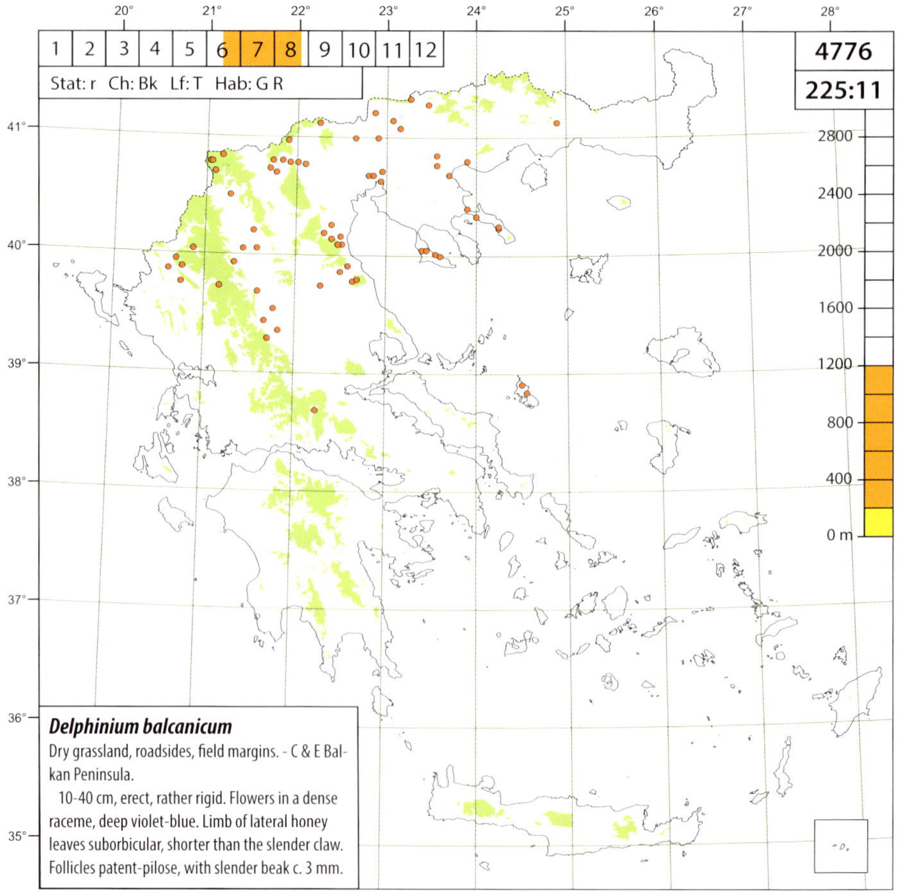

Delphinium balcanicum

Dry grassland, roadsides, field margins. - C & E Balkan Peninsula.

10-40 cm, erect, rather rigid. Flowers in a dense raceme, deep violet-blue. Limb of lateral honey leaves suborbicular, shorter than the slender claw. Follicles patent-pilose, with slender beak c. 3 mm.

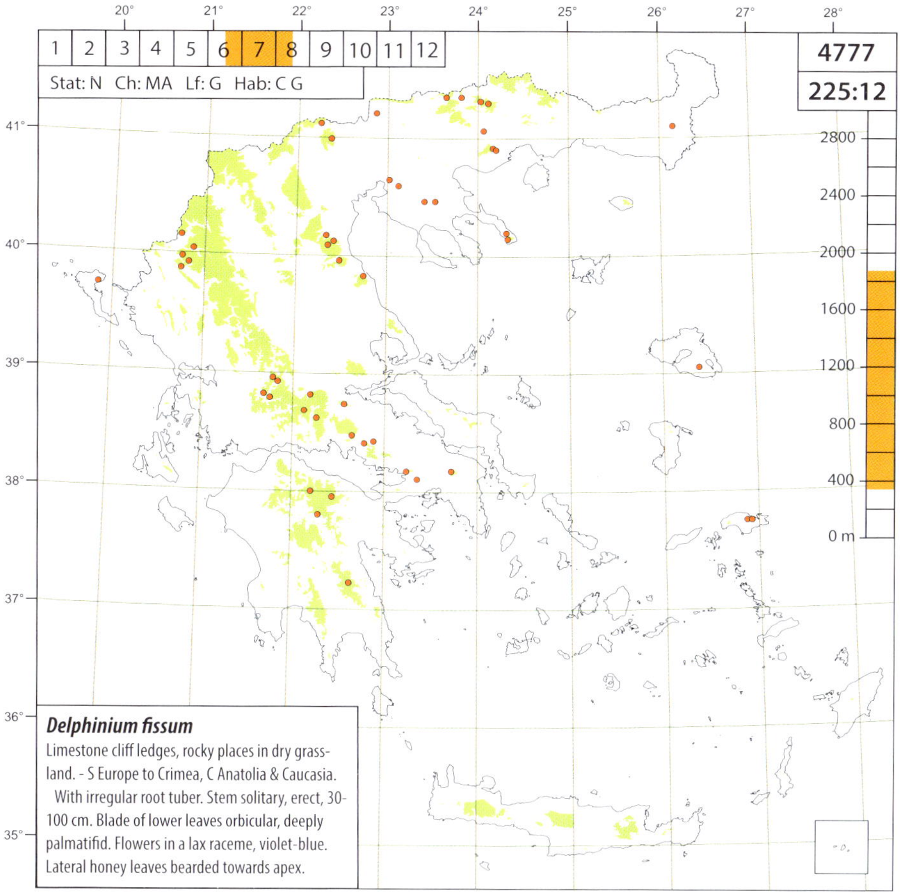

Delphinium fissum
Limestone cliff ledges, rocky places in dry grassland. - S Europe to Crimea, C Anatolia & Caucasia.
With irregular root tuber. Stem solitary, erect, 30-100 cm. Blade of lower leaves orbicular, deeply palmatifid. Flowers in a lax raceme, violet-blue. Lateral honey leaves bearded towards apex.

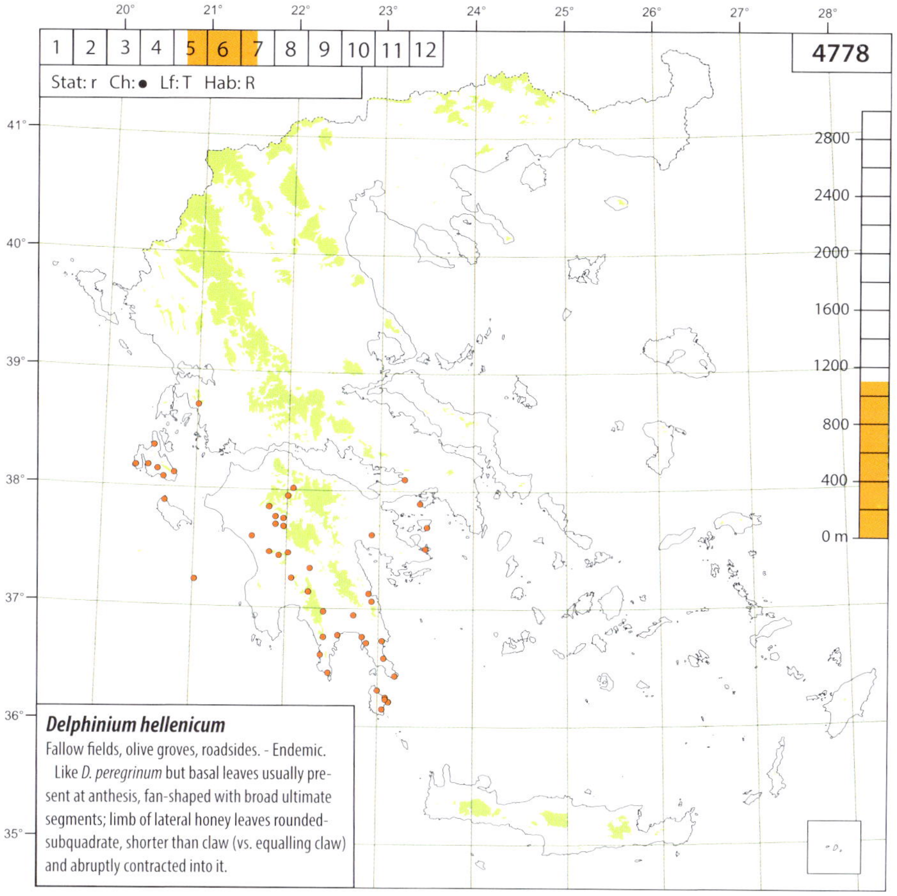

Delphinium hellenicum
Fallow fields, olive groves, roadsides. - Endemic.
Like *D. peregrinum* but basal leaves usually present at anthesis, fan-shaped with broad ultimate segments; limb of lateral honey leaves rounded-subquadrate, shorter than claw (vs. equalling claw) and abruptly contracted into it.

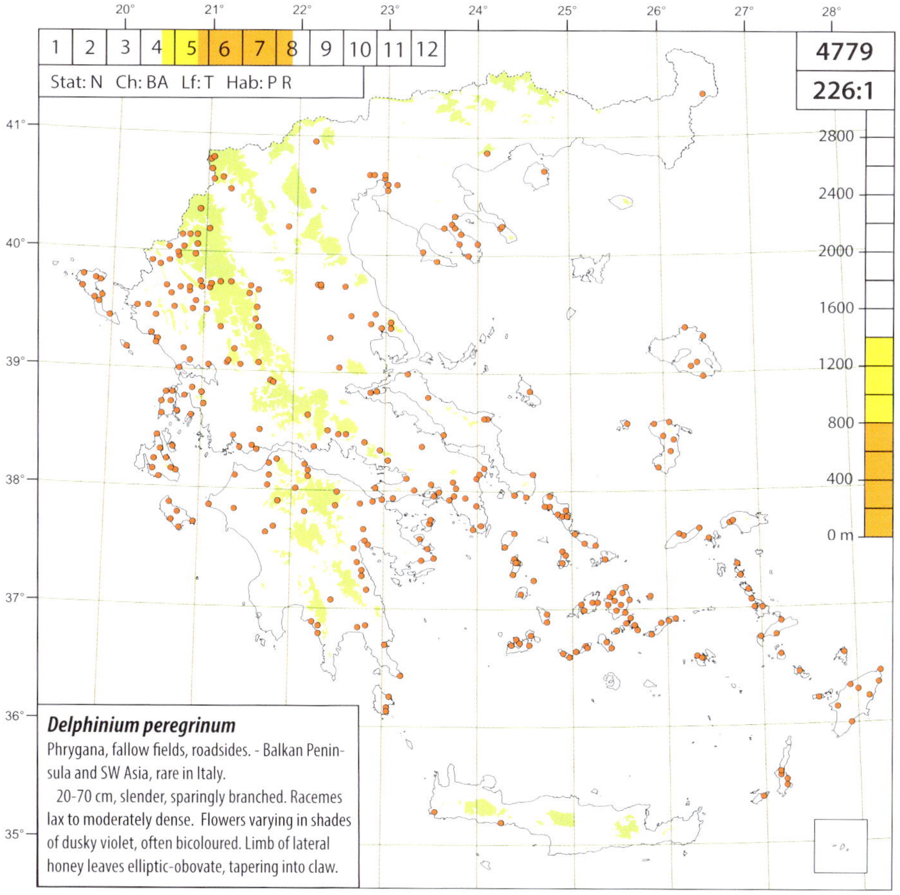

Delphinium peregrinum
Phrygana, fallow fields, roadsides. - Balkan Peninsula and SW Asia, rare in Italy.
20-70 cm, slender, sparingly branched. Racemes lax to moderately dense. Flowers varying in shades of dusky violet, often bicoloured. Limb of lateral honey leaves elliptic-obovate, tapering into claw.

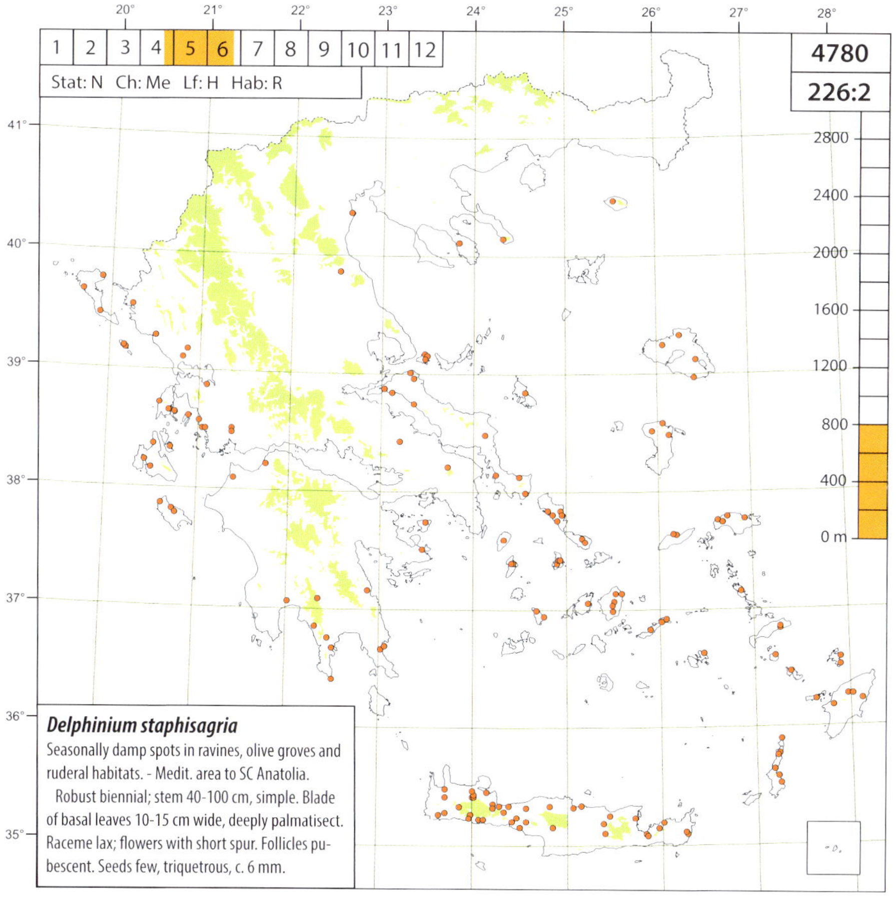

Delphinium staphisagria
Seasonally damp spots in ravines, olive groves and ruderal habitats. - Medit. area to SC Anatolia.
Robust biennial; stem 40-100 cm, simple. Blade of basal leaves 10-15 cm wide, deeply palmatisect. Raceme lax; flowers with short spur. Follicles pubescent. Seeds few, triquetrous, c. 6 mm.

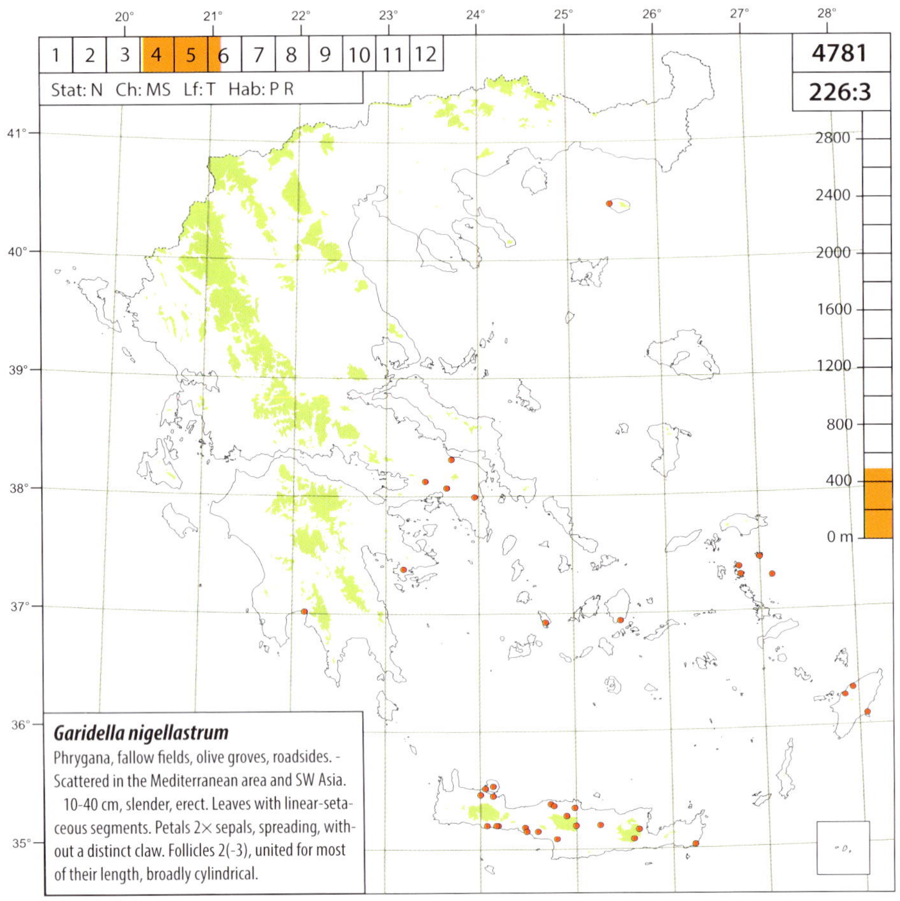

Garidella nigellastrum

Phrygana, fallow fields, olive groves, roadsides. - Scattered in the Mediterranean area and SW Asia.

10-40 cm, slender, erect. Leaves with linear-setaceous segments. Petals 2× sepals, spreading, without a distinct claw. Follicles 2(-3), united for most of their length, broadly cylindrical.

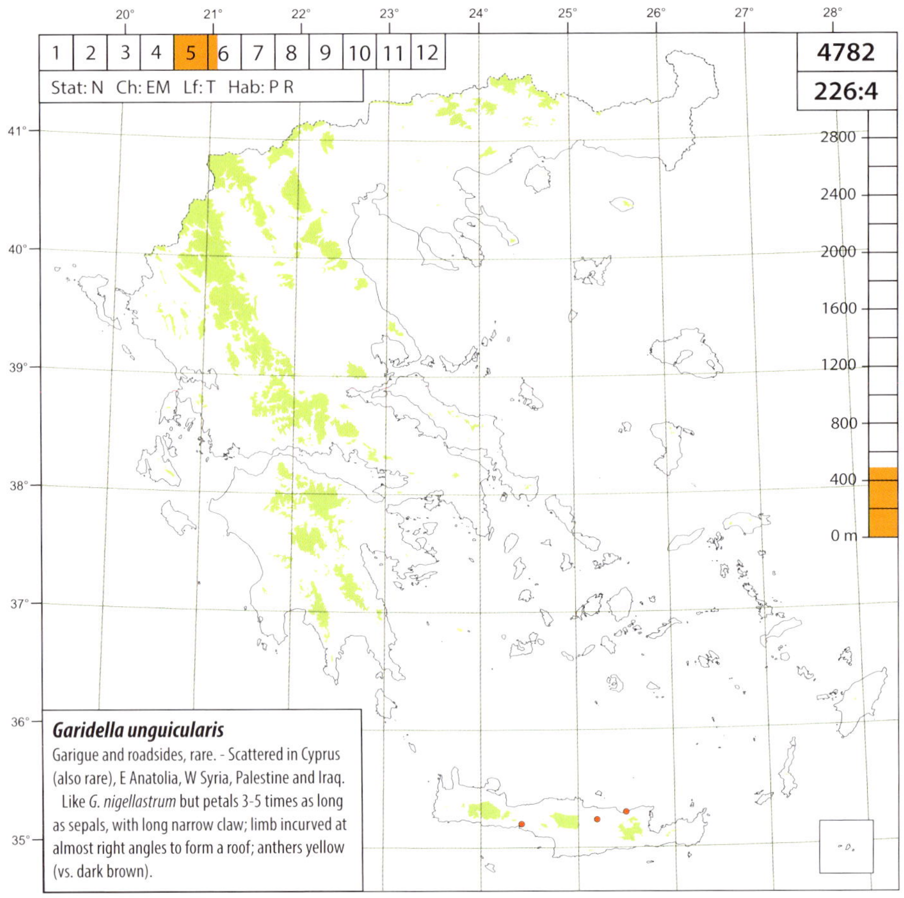

Garidella unguicularis

Garigue and roadsides, rare. - Scattered in Cyprus (also rare), E Anatolia, W Syria, Palestine and Iraq.

Like *G. nigellastrum* but petals 3-5 times as long as sepals, with long narrow claw; limb incurved at almost right angles to form a roof; anthers yellow (vs. dark brown).

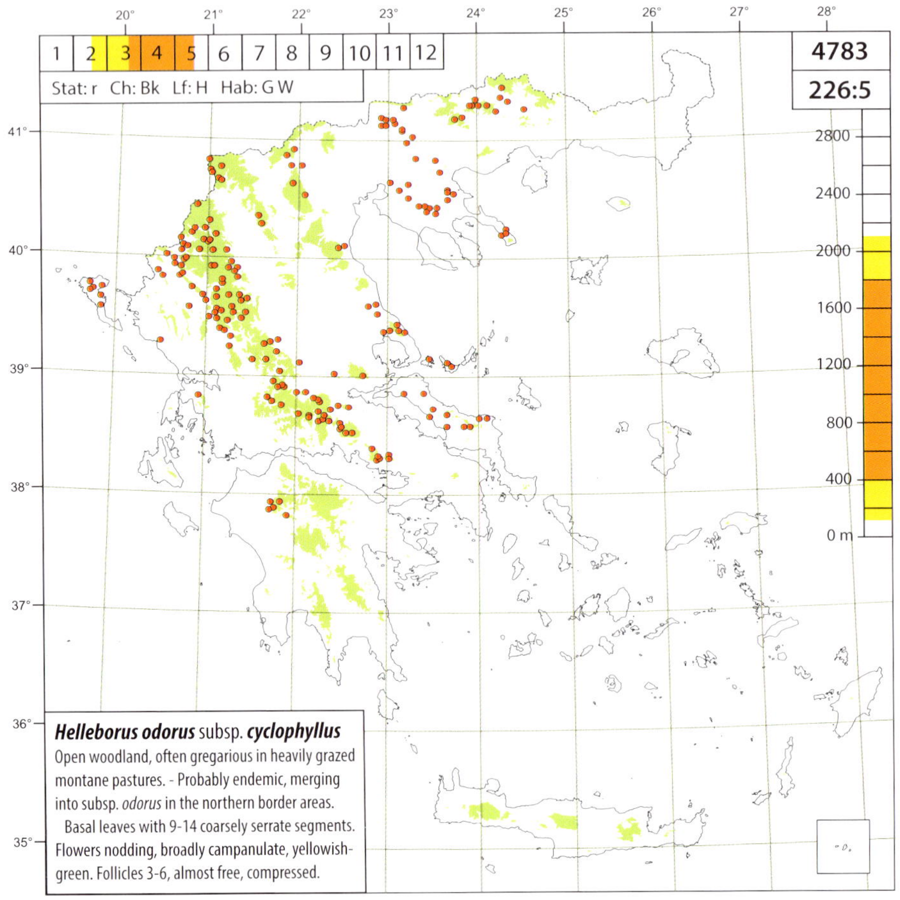

Helleborus odorus subsp. **cyclophyllus**

Open woodland, often gregarious in heavily grazed montane pastures. - Probably endemic, merging into subsp. *odorus* in the northern border areas.

Basal leaves with 9-14 coarsely serrate segments. Flowers nodding, broadly campanulate, yellowish-green. Follicles 3-6, almost free, compressed.

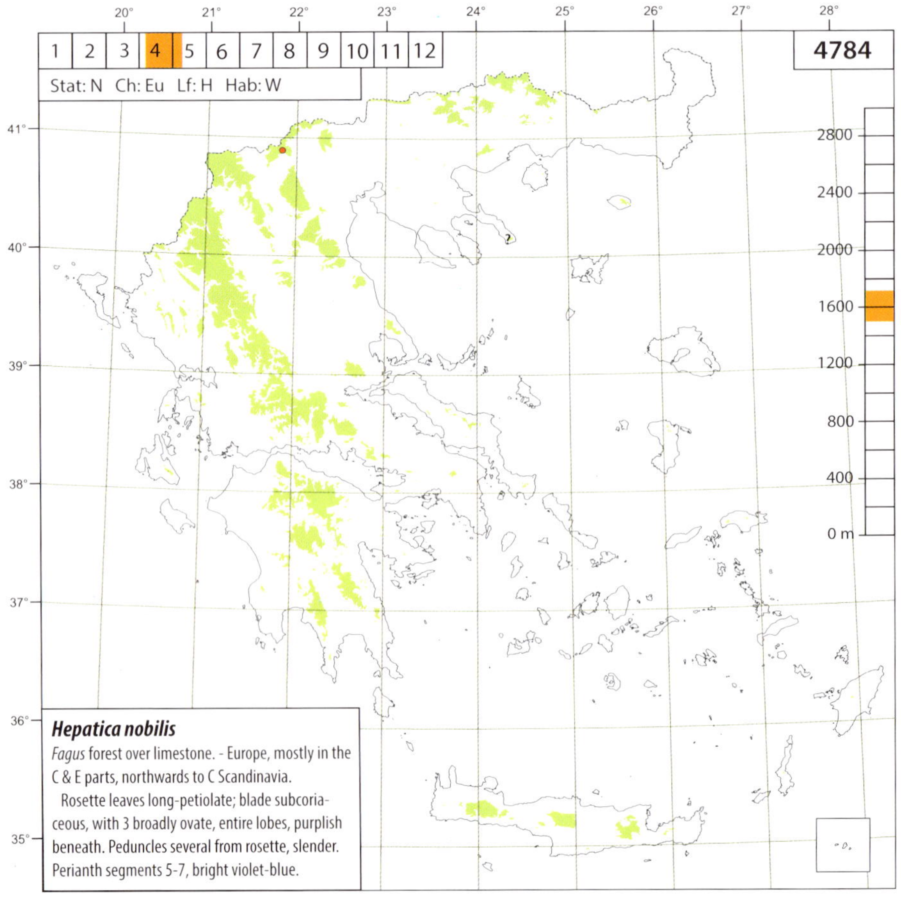

Hepatica nobilis

Fagus forest over limestone. - Europe, mostly in the C & E parts, northwards to C Scandinavia.

Rosette leaves long-petiolate; blade subcoriaceous, with 3 broadly ovate, entire lobes, purplish beneath. Peduncles several from rosette, slender. Perianth segments 5-7, bright violet-blue.

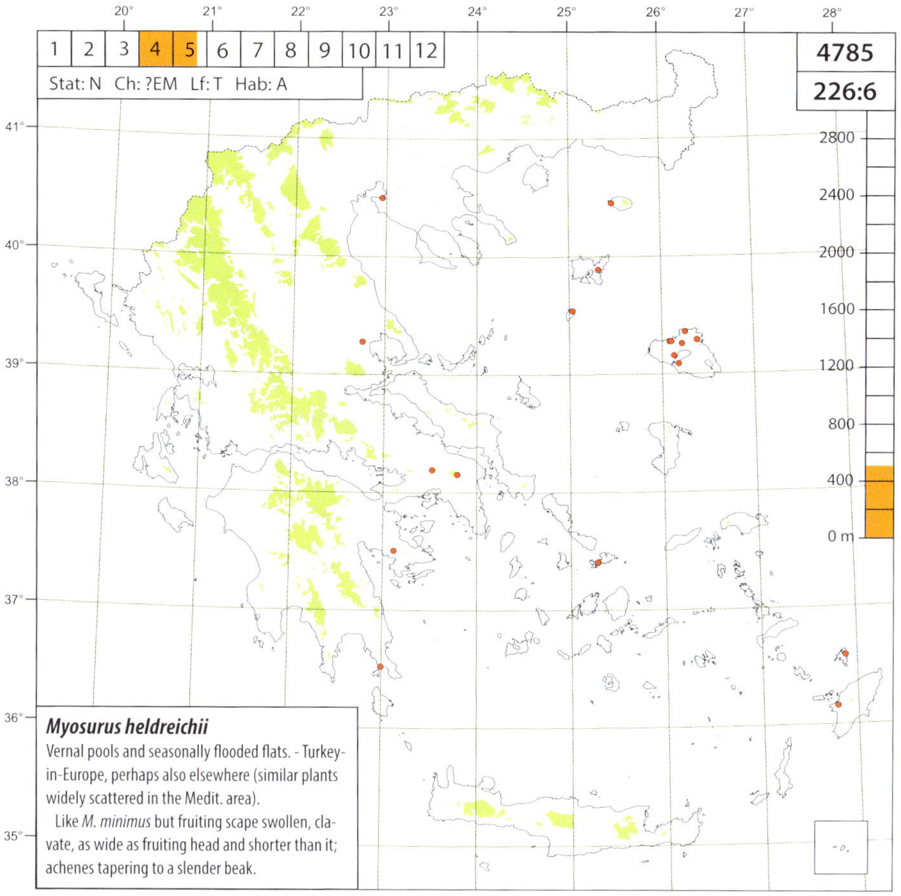

Myosurus heldreichii

Vernal pools and seasonally flooded flats. - Turkey-in-Europe, perhaps also elsewhere (similar plants widely scattered in the Medit. area).

Like *M. minimus* but fruiting scape swollen, clavate, as wide as fruiting head and shorter than it; achenes tapering to a slender beak.

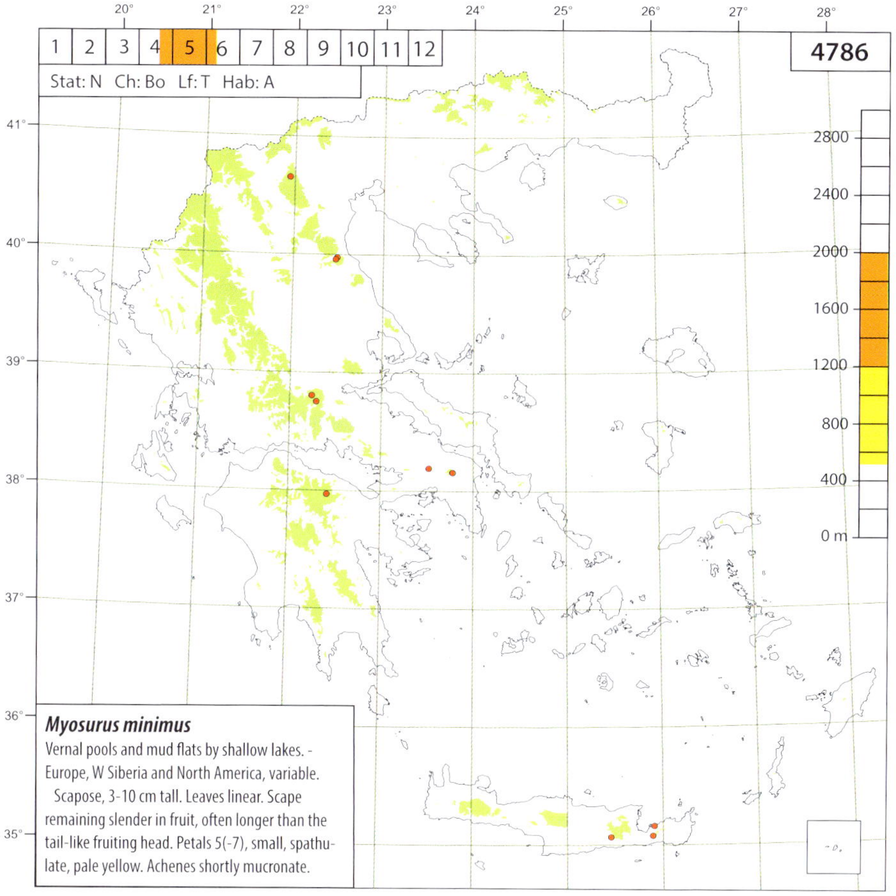

Myosurus minimus

Vernal pools and mud flats by shallow lakes. - Europe, W Siberia and North America, variable.

Scapose, 3-10 cm tall. Leaves linear. Scape remaining slender in fruit, often longer than the tail-like fruiting head. Petals 5(-7), small, spathulate, pale yellow. Achenes shortly mucronate.

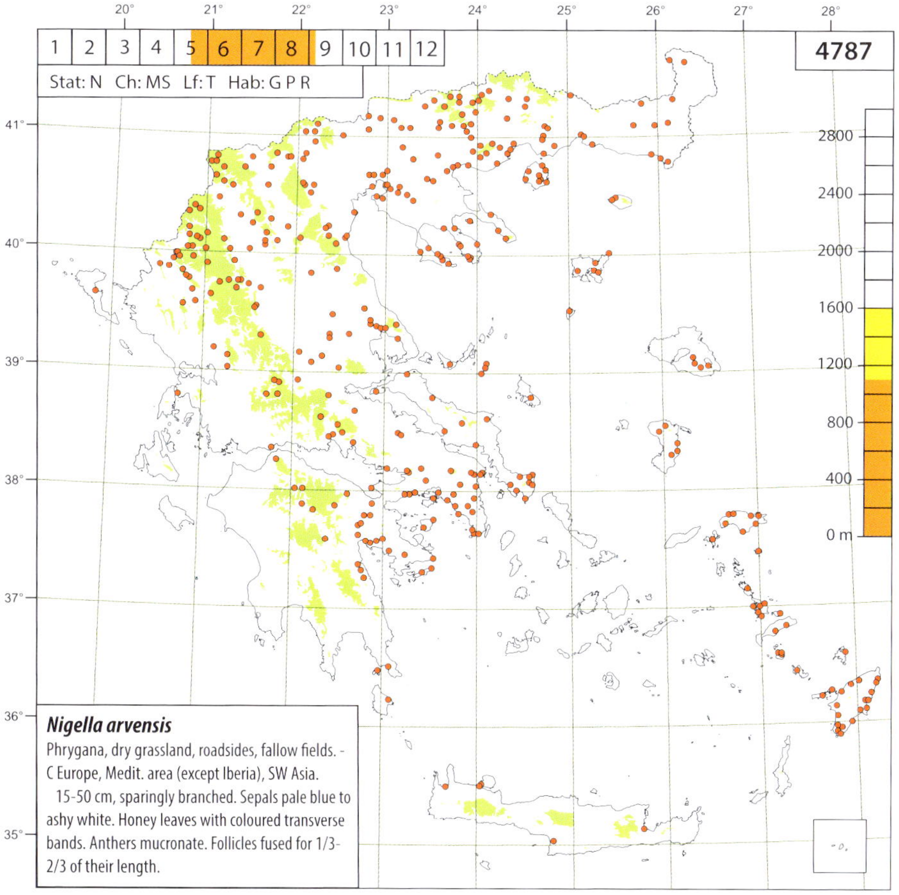

Nigella arvensis

Phrygana, dry grassland, roadsides, fallow fields. - C Europe, Medit. area (except Iberia), SW Asia.

15-50 cm, sparingly branched. Sepals pale blue to ashy white. Honey leaves with coloured transverse bands. Anthers mucronate. Follicles fused for 1/3-2/3 of their length.

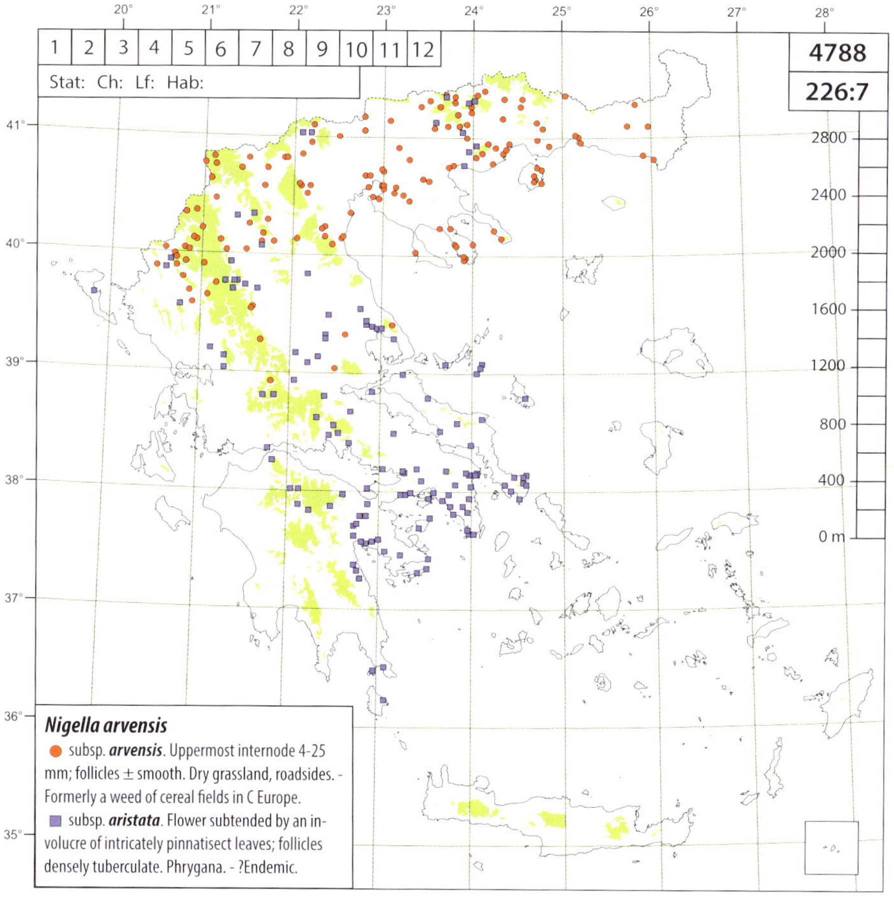

Nigella arvensis

● subsp. ***arvensis***. Uppermost internode 4-25 mm; follicles ± smooth. Dry grassland, roadsides. - Formerly a weed of cereal fields in C Europe.

■ subsp. ***aristata***. Flower subtended by an involucre of intricately pinnatisect leaves; follicles densely tuberculate. Phrygana. - ?Endemic.

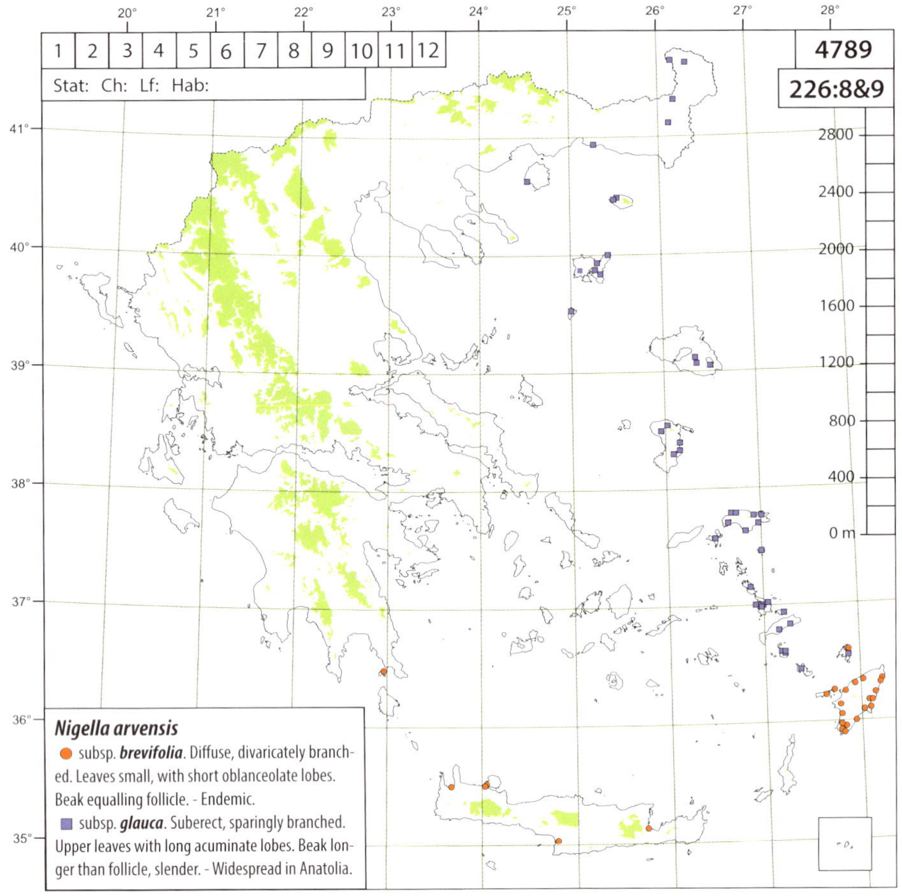

Nigella arvensis

● subsp. ***brevifolia***. Diffuse, divaricately branched. Leaves small, with short oblanceolate lobes. Beak equalling follicle. - Endemic.

■ subsp. ***glauca***. Suberect, sparingly branched. Upper leaves with long acuminate lobes. Beak longer than follicle, slender. - Widespread in Anatolia.

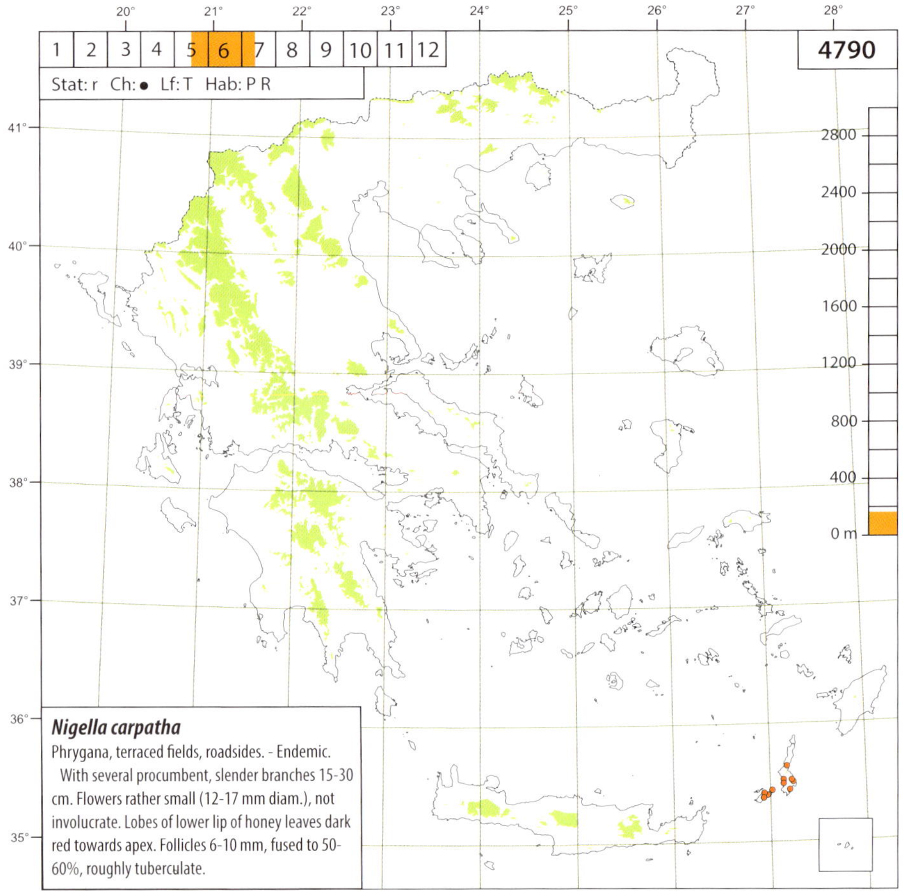

Nigella carpatha

Phrygana, terraced fields, roadsides. - Endemic.

With several procumbent, slender branches 15-30 cm. Flowers rather small (12-17 mm diam.), not involucrate. Lobes of lower lip of honey leaves dark red towards apex. Follicles 6-10 mm, fused to 50-60%, roughly tuberculate.

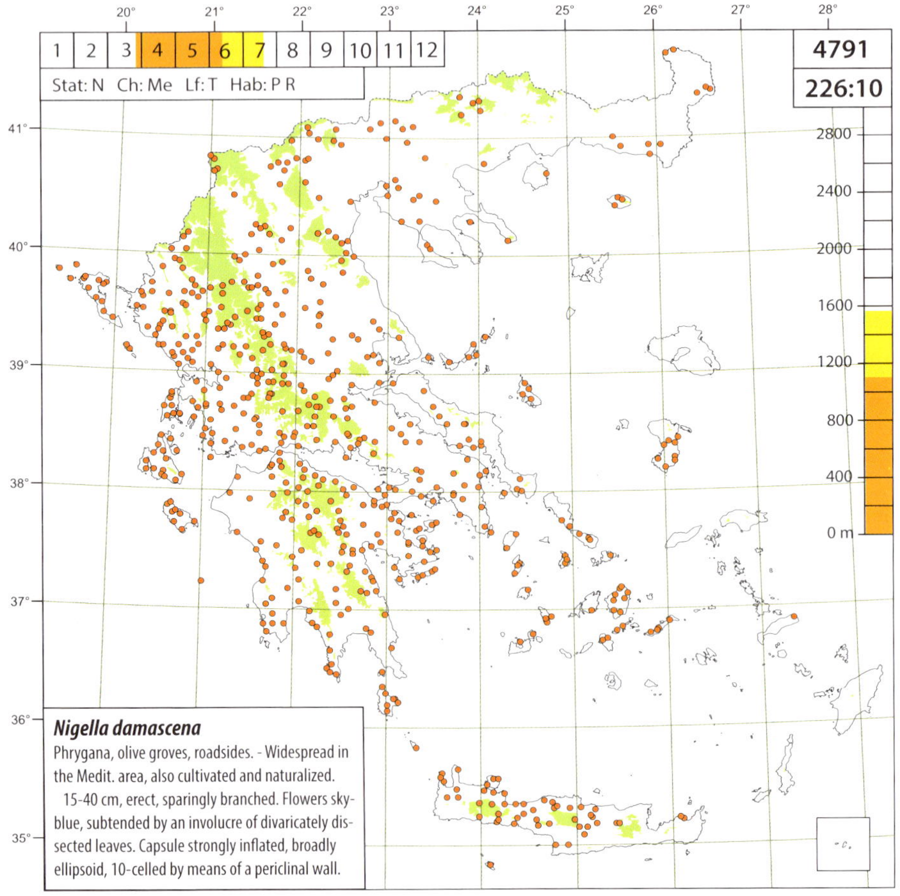

Nigella damascena

Phrygana, olive groves, roadsides. - Widespread in the Medit. area, also cultivated and naturalized.

15-40 cm, erect, sparingly branched. Flowers sky-blue, subtended by an involucre of divaricately dissected leaves. Capsule strongly inflated, broadly ellipsoid, 10-celled by means of a periclinal wall.

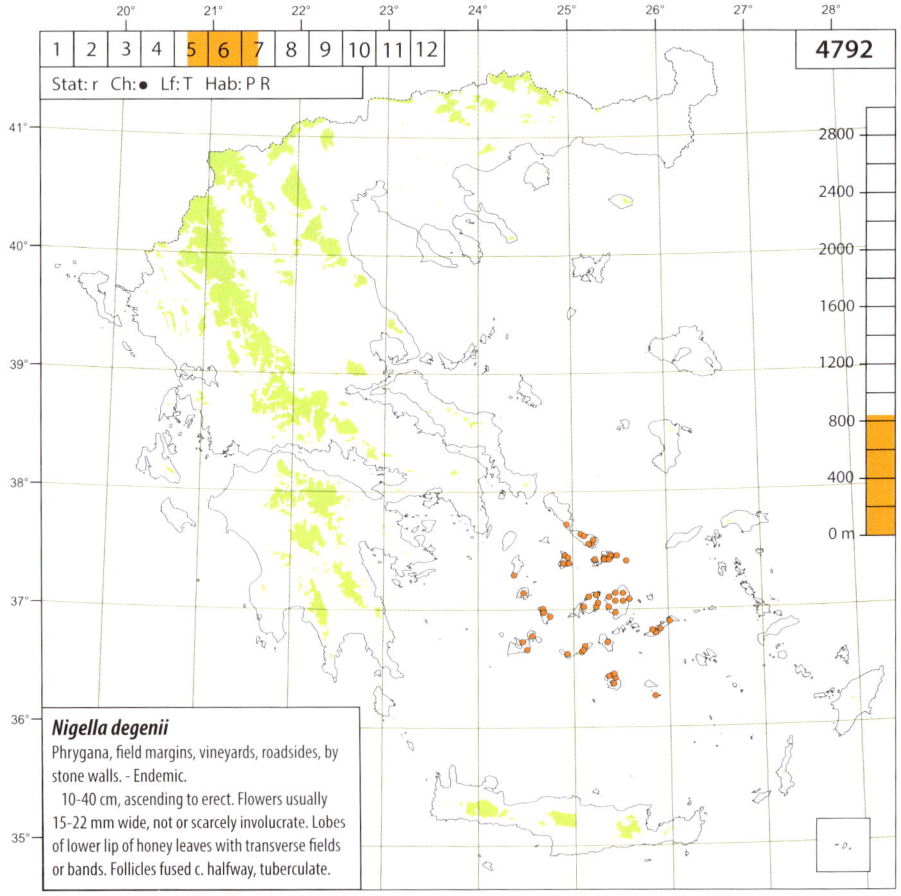

Nigella degenii

Phrygana, field margins, vineyards, roadsides, by stone walls. - Endemic.

10-40 cm, ascending to erect. Flowers usually 15-22 mm wide, not or scarcely involucrate. Lobes of lower lip of honey leaves with transverse fields or bands. Follicles fused c. halfway, tuberculate.

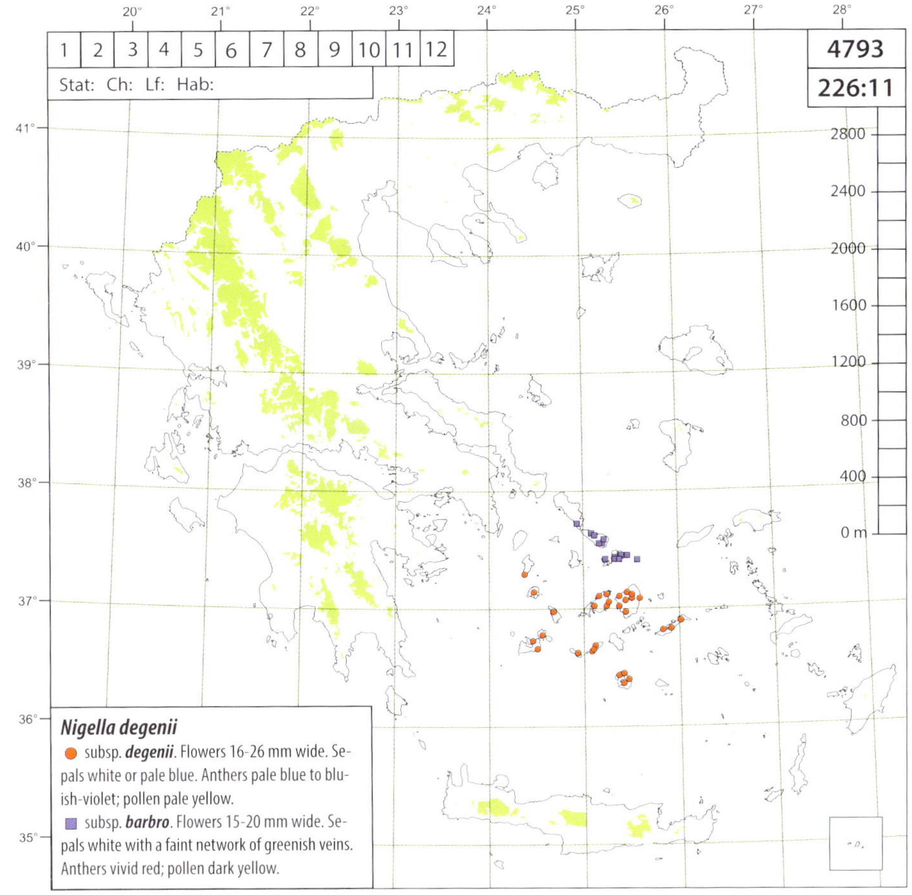

Nigella degenii

● subsp. ***degenii***. Flowers 16-26 mm wide. Sepals white or pale blue. Anthers pale blue to bluish-violet; pollen pale yellow.

■ subsp. ***barbro***. Flowers 15-20 mm wide. Sepals white with a faint network of greenish veins. Anthers vivid red; pollen dark yellow.

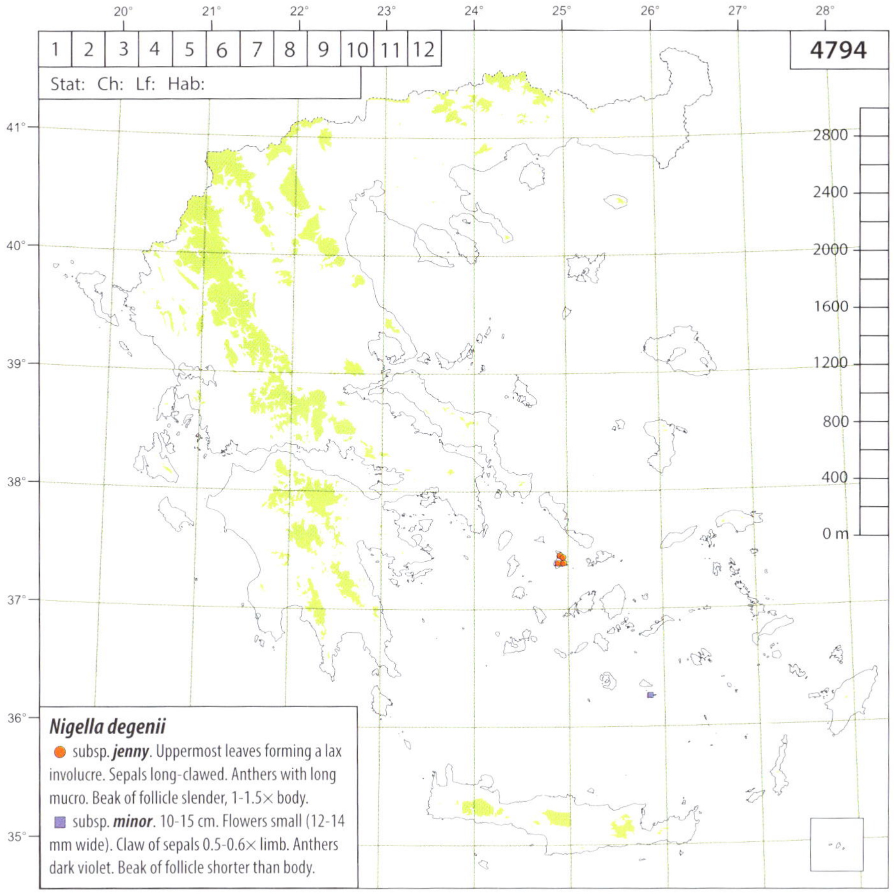

Nigella degenii

● subsp. ***jenny***. Uppermost leaves forming a lax involucre. Sepals long-clawed. Anthers with long mucro. Beak of follicle slender, 1-1.5× body.

■ subsp. ***minor***. 10-15 cm. Flowers small (12-14 mm wide). Claw of sepals 0.5-0.6× limb. Anthers dark violet. Beak of follicle shorter than body.

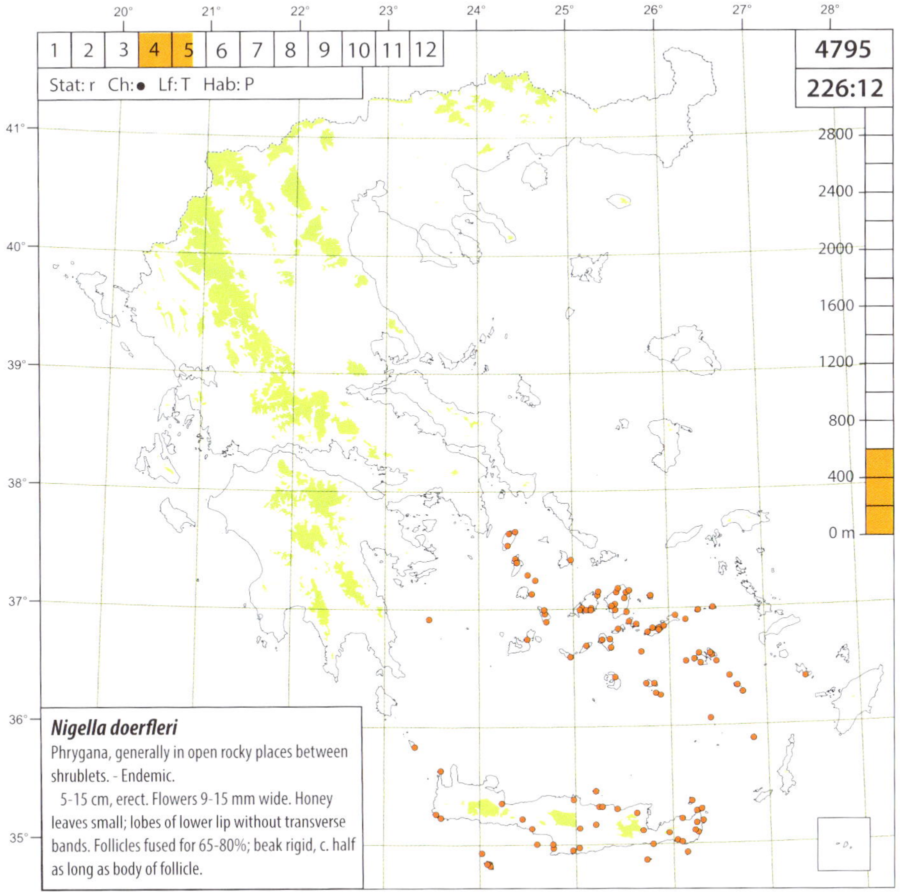

Nigella doerfleri

Phrygana, generally in open rocky places between shrublets. - Endemic.

5-15 cm, erect. Flowers 9-15 mm wide. Honey leaves small; lobes of lower lip without transverse bands. Follicles fused for 65-80%; beak rigid, c. half as long as body of follicle.

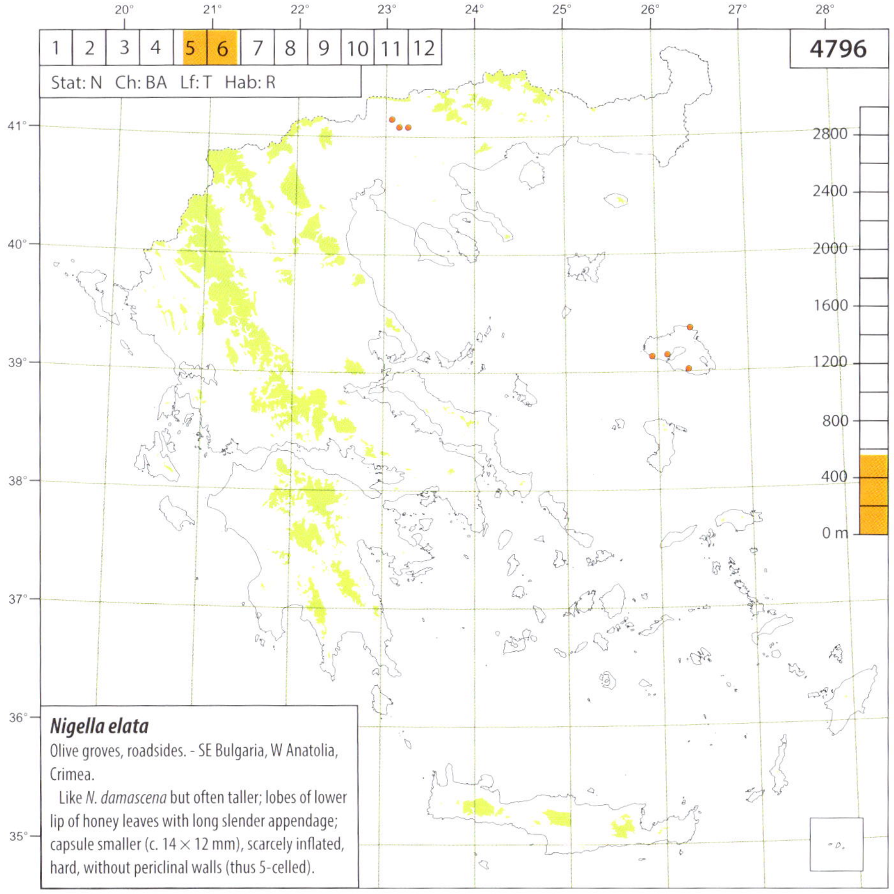

Nigella elata

Olive groves, roadsides. - SE Bulgaria, W Anatolia, Crimea.

Like *N. damascena* but often taller; lobes of lower lip of honey leaves with long slender appendage; capsule smaller (c. 14 × 12 mm), scarcely inflated, hard, without periclinal walls (thus 5-celled).

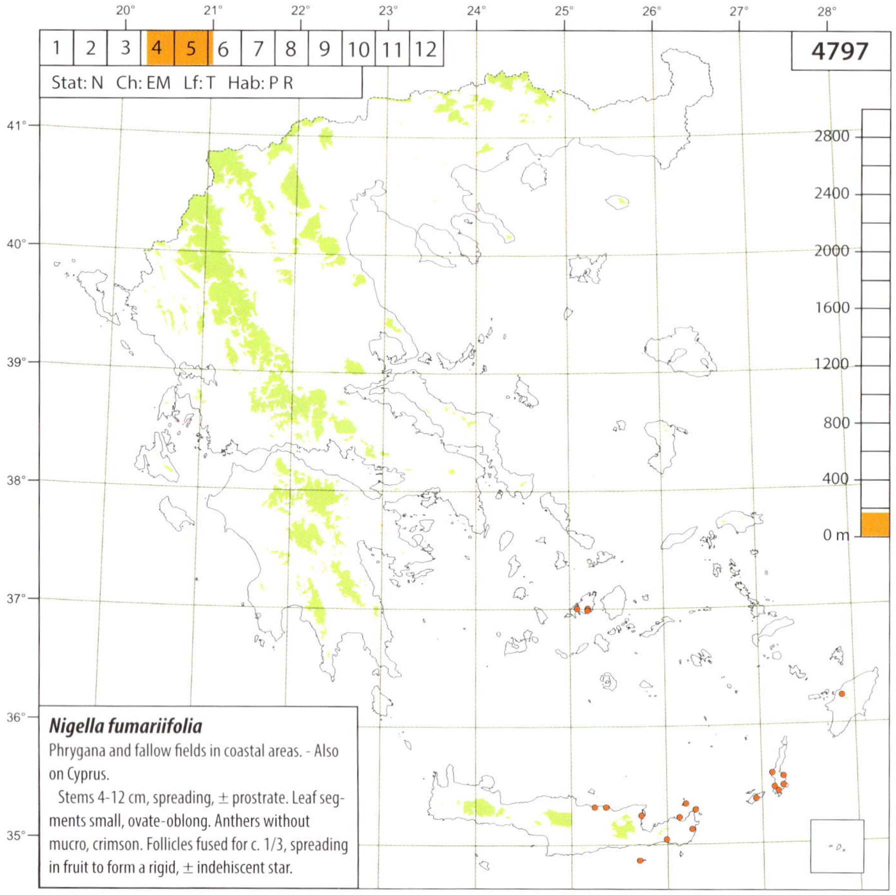

Nigella fumariifolia
Phrygana and fallow fields in coastal areas. - Also on Cyprus.
Stems 4-12 cm, spreading, ± prostrate. Leaf segments small, ovate-oblong. Anthers without mucro, crimson. Follicles fused for c. 1/3, spreading in fruit to form a rigid, ± indehiscent star.

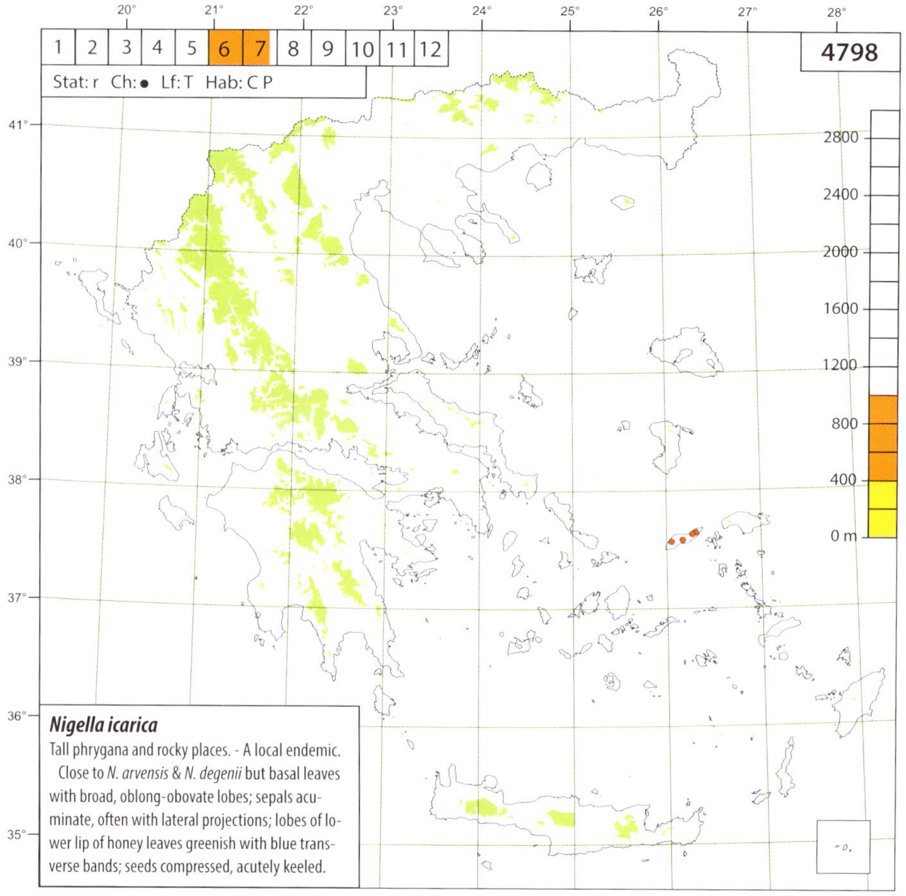

Nigella icarica
Tall phrygana and rocky places. - A local endemic.
Close to *N. arvensis* & *N. degenii* but basal leaves with broad, oblong-obovate lobes; sepals acuminate, often with lateral projections; lobes of lower lip of honey leaves greenish with blue transverse bands; seeds compressed, acutely keeled.

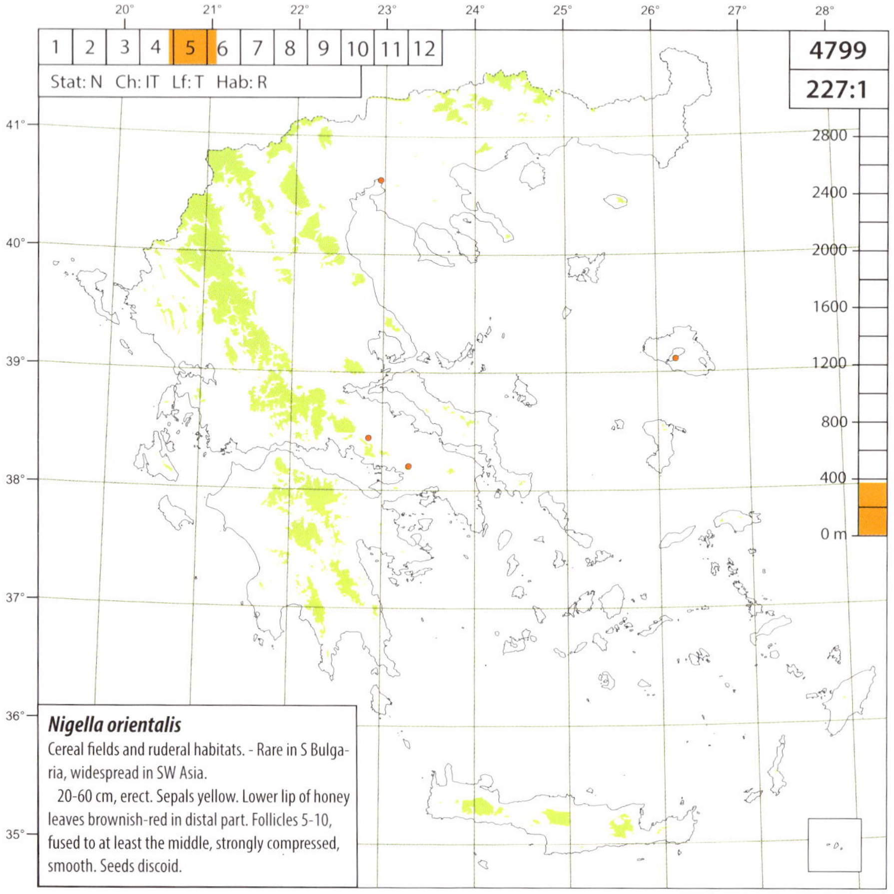

Nigella orientalis
Cereal fields and ruderal habitats. - Rare in S Bulgaria, widespread in SW Asia.
20-60 cm, erect. Sepals yellow. Lower lip of honey leaves brownish-red in distal part. Follicles 5-10, fused to at least the middle, strongly compressed, smooth. Seeds discoid.

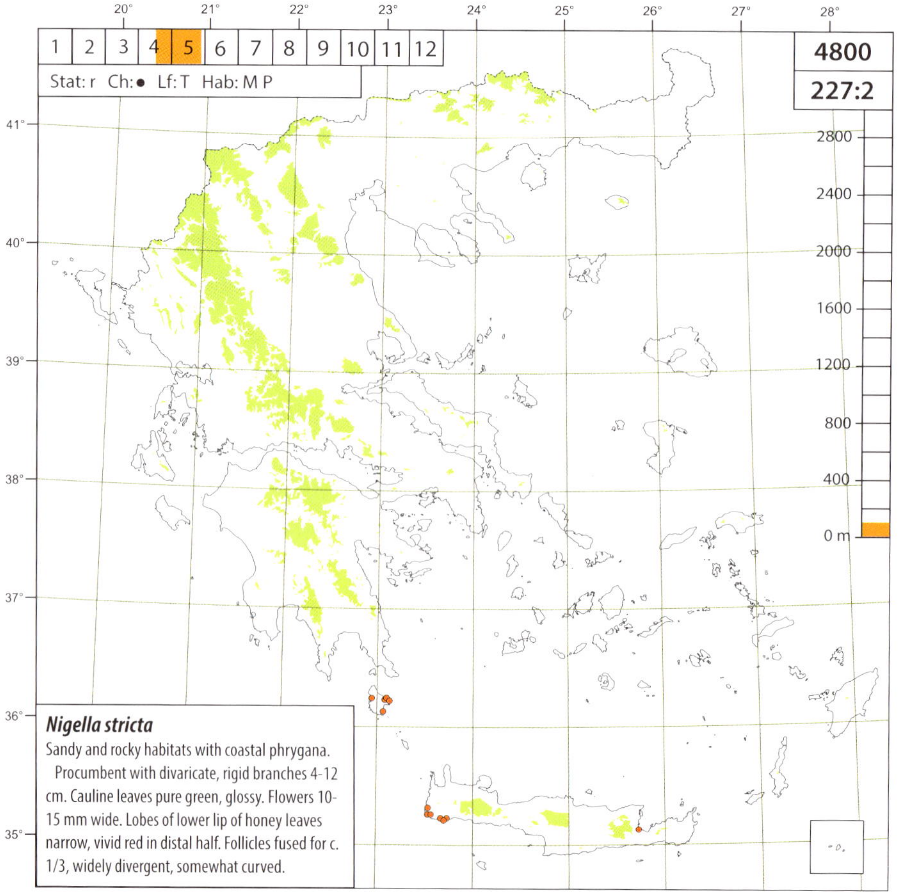

Nigella stricta
Sandy and rocky habitats with coastal phrygana.
Procumbent with divaricate, rigid branches 4-12 cm. Cauline leaves pure green, glossy. Flowers 10-15 mm wide. Lobes of lower lip of honey leaves narrow, vivid red in distal half. Follicles fused for c. 1/3, widely divergent, somewhat curved.

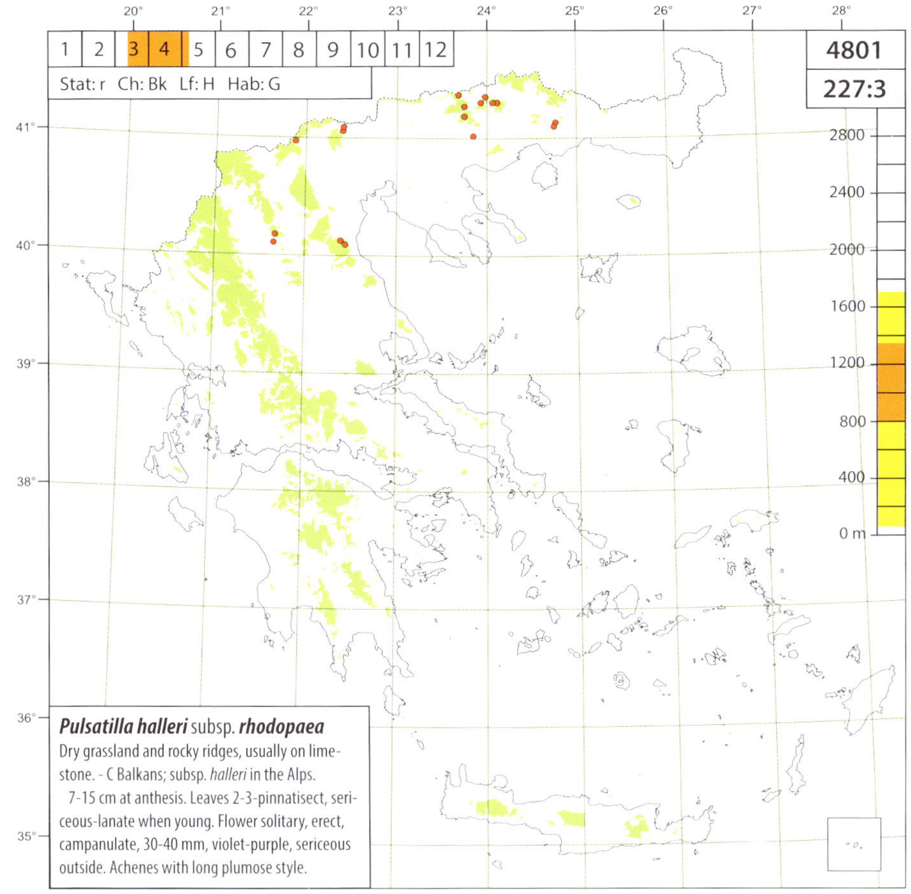

Pulsatilla halleri **subsp.** ***rhodopaea***
Dry grassland and rocky ridges, usually on limestone. - C Balkans; subsp. *halleri* in the Alps.
7-15 cm at anthesis. Leaves 2-3-pinnatisect, sericeous-lanate when young. Flower solitary, erect, campanulate, 30-40 mm, violet-purple, sericeous outside. Achenes with long plumose style.

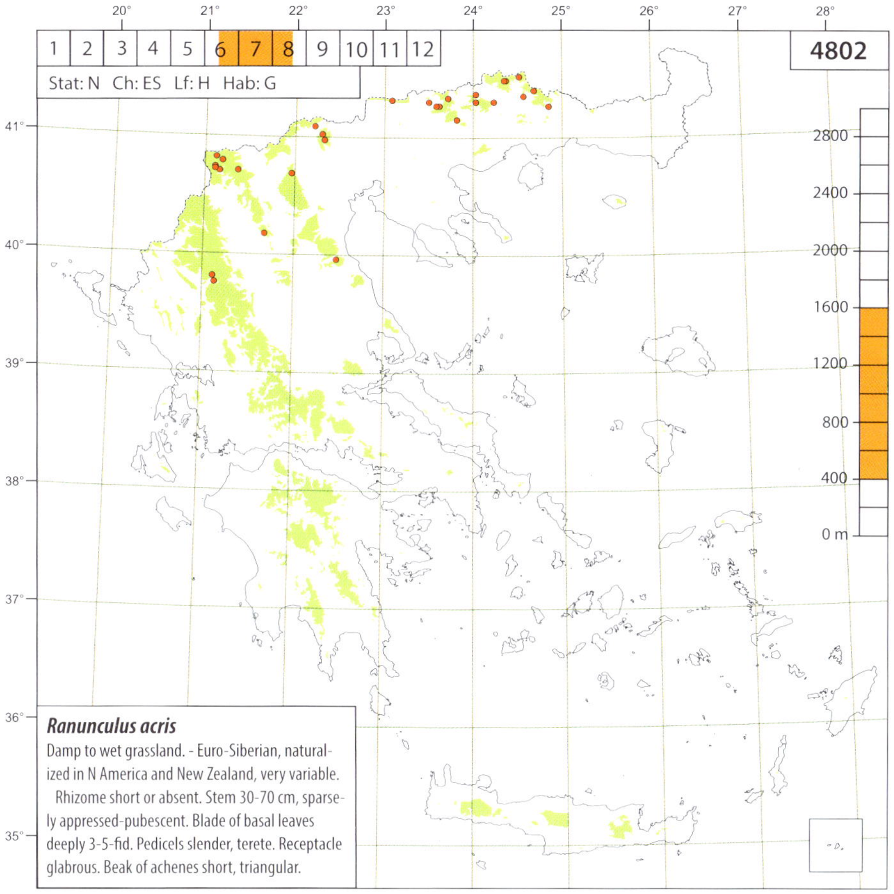

Ranunculus acris
Damp to wet grassland. - Euro-Siberian, naturalized in N America and New Zealand, very variable.
Rhizome short or absent. Stem 30-70 cm, sparsely appressed-pubescent. Blade of basal leaves deeply 3-5-fid. Pedicels slender, terete. Receptacle glabrous. Beak of achenes short, triangular.

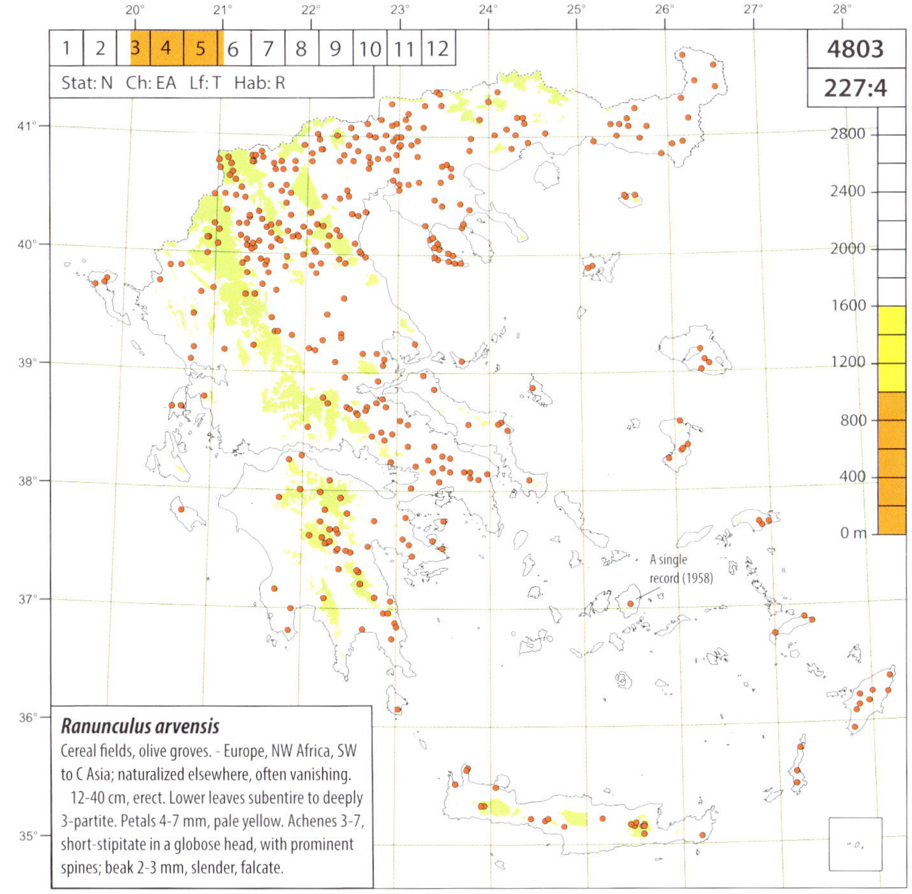

Ranunculus arvensis
Cereal fields, olive groves. - Europe, NW Africa, SW to C Asia; naturalized elsewhere, often vanishing.
12-40 cm, erect. Lower leaves subentire to deeply 3-partite. Petals 4-7 mm, pale yellow. Achenes 3-7, short-stipitate in a globose head, with prominent spines; beak 2-3 mm, slender, falcate.

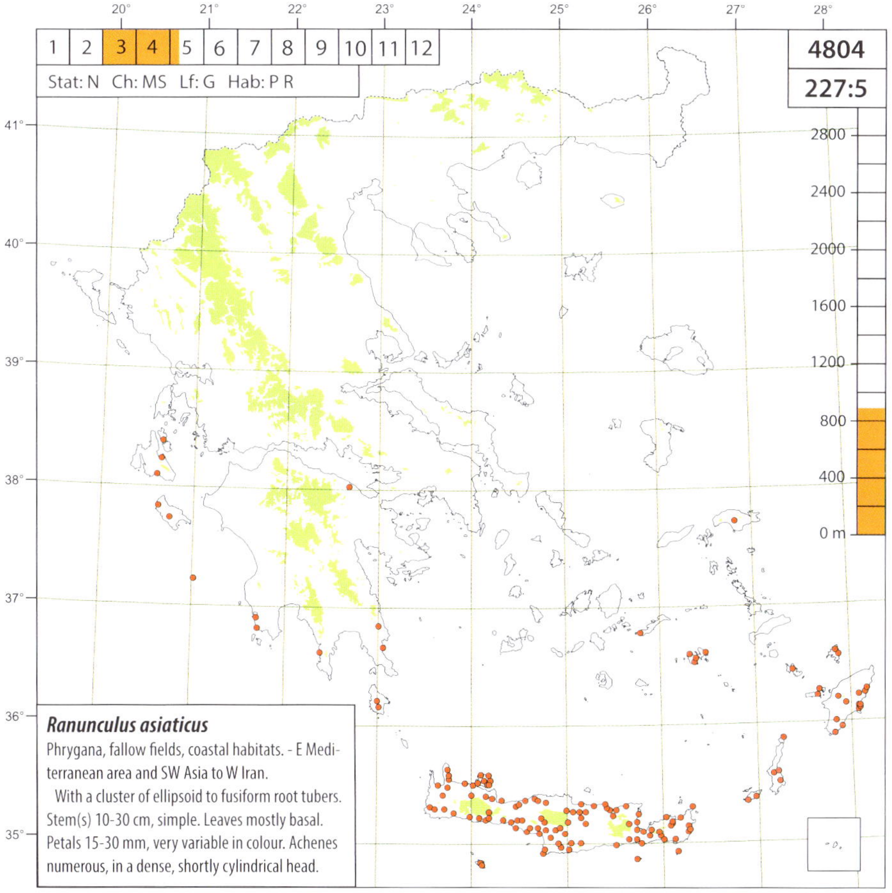

Ranunculus asiaticus
Phrygana, fallow fields, coastal habitats. - E Mediterranean area and SW Asia to W Iran.
With a cluster of ellipsoid to fusiform root tubers. Stem(s) 10-30 cm, simple. Leaves mostly basal. Petals 15-30 mm, very variable in colour. Achenes numerous, in a dense, shortly cylindrical head.

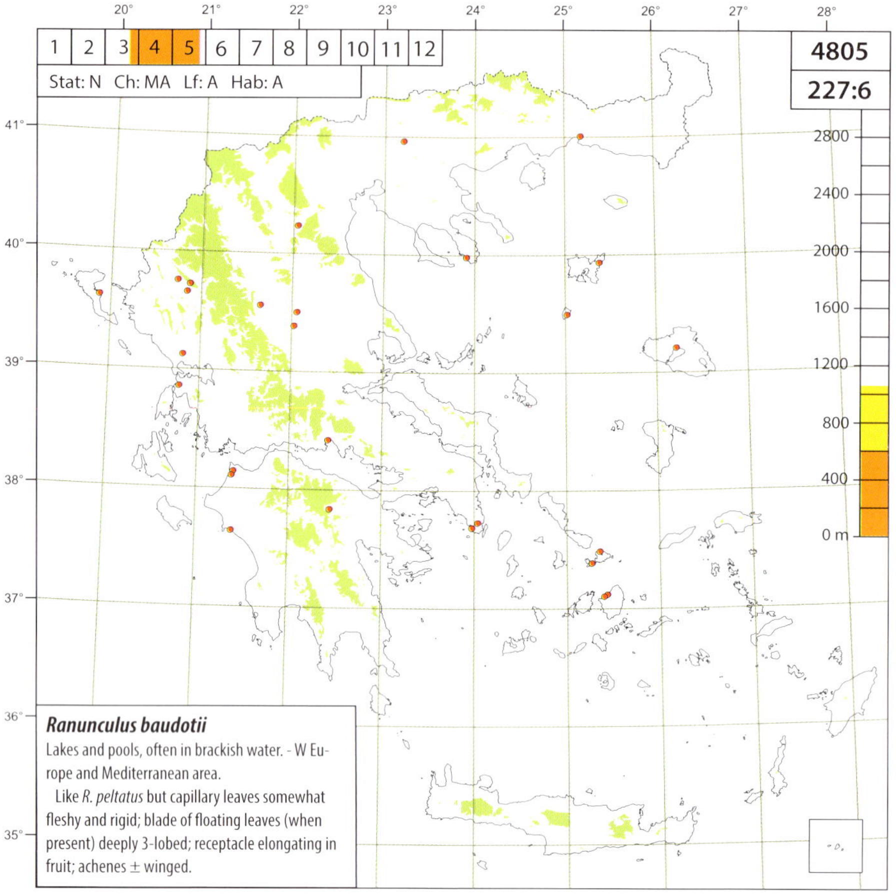

Ranunculus baudotii
Lakes and pools, often in brackish water. - W Europe and Mediterranean area.
Like *R. peltatus* but capillary leaves somewhat fleshy and rigid; blade of floating leaves (when present) deeply 3-lobed; receptacle elongating in fruit; achenes ± winged.

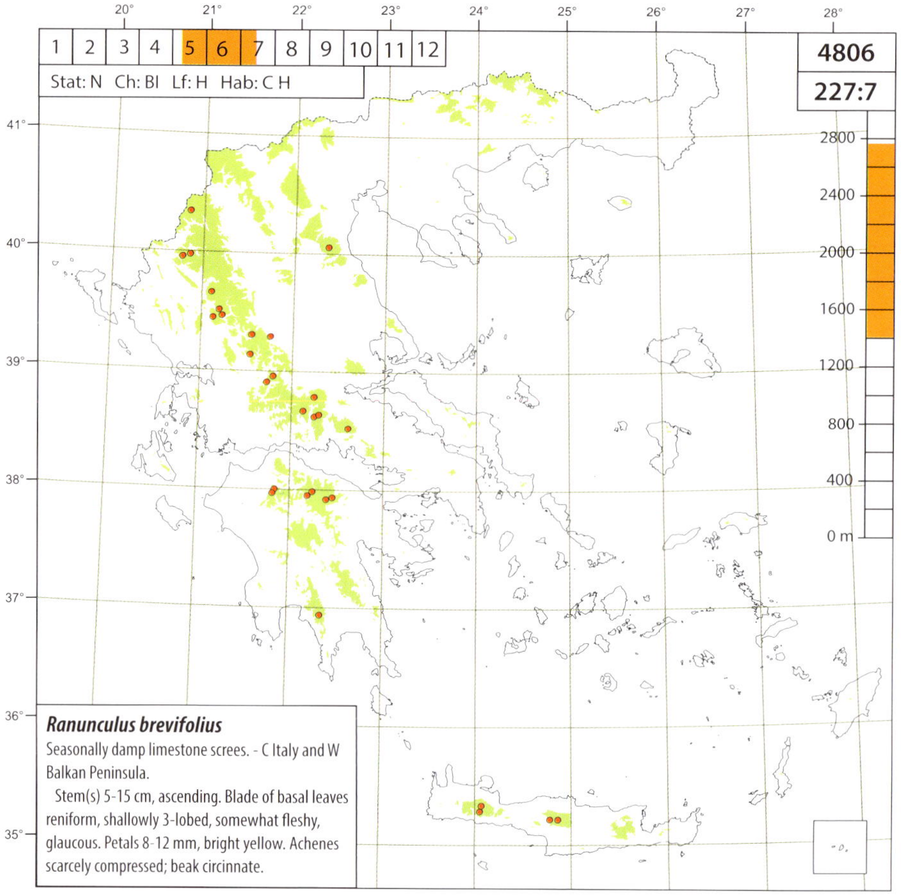

Ranunculus brevifolius
Seasonally damp limestone screes. - C Italy and W Balkan Peninsula.
Stem(s) 5-15 cm, ascending. Blade of basal leaves reniform, shallowly 3-lobed, somewhat fleshy, glaucous. Petals 8-12 mm, bright yellow. Achenes scarcely compressed; beak circinnate.

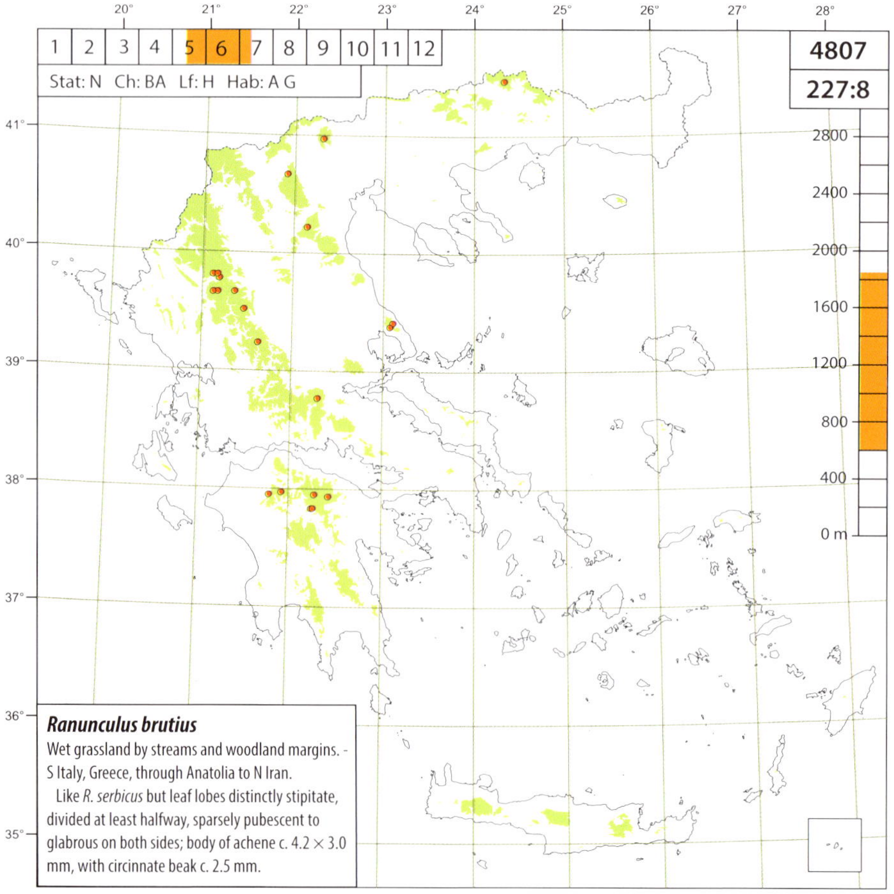

Ranunculus brutius
Wet grassland by streams and woodland margins. - S Italy, Greece, through Anatolia to N Iran.
Like *R. serbicus* but leaf lobes distinctly stipitate, divided at least halfway, sparsely pubescent to glabrous on both sides; body of achene c. 4.2 × 3.0 mm, with circinnate beak c. 2.5 mm.

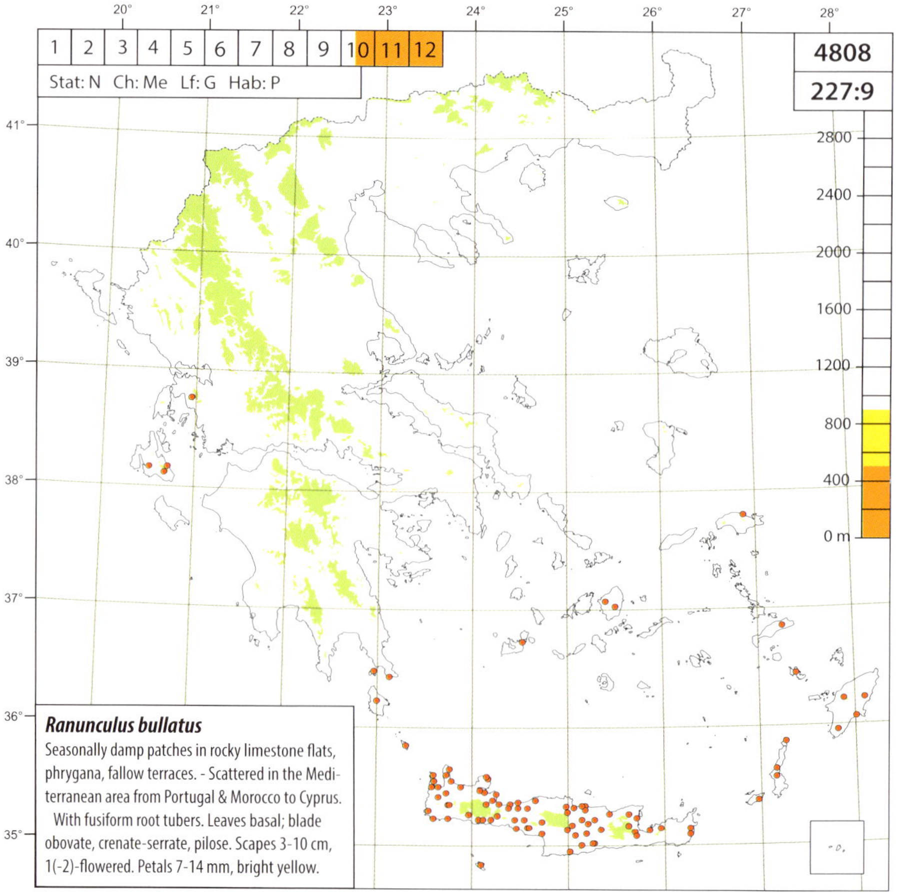

Ranunculus bullatus
Seasonally damp patches in rocky limestone flats, phrygana, fallow terraces. - Scattered in the Mediterranean area from Portugal & Morocco to Cyprus.
With fusiform root tubers. Leaves basal; blade obovate, crenate-serrate, pilose. Scapes 3-10 cm, 1(-2)-flowered. Petals 7-14 mm, bright yellow.

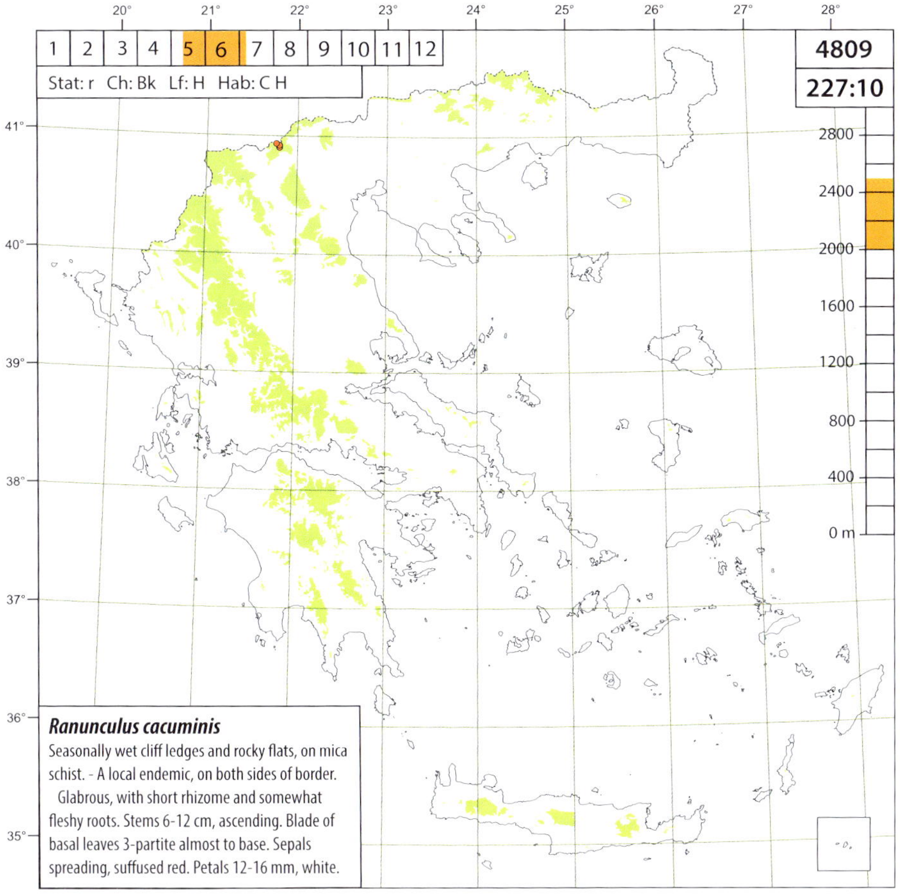

Ranunculus cacuminis

Seasonally wet cliff ledges and rocky flats, on mica schist. - A local endemic, on both sides of border.

Glabrous, with short rhizome and somewhat fleshy roots. Stems 6-12 cm, ascending. Blade of basal leaves 3-partite almost to base. Sepals spreading, suffused red. Petals 12-16 mm, white.

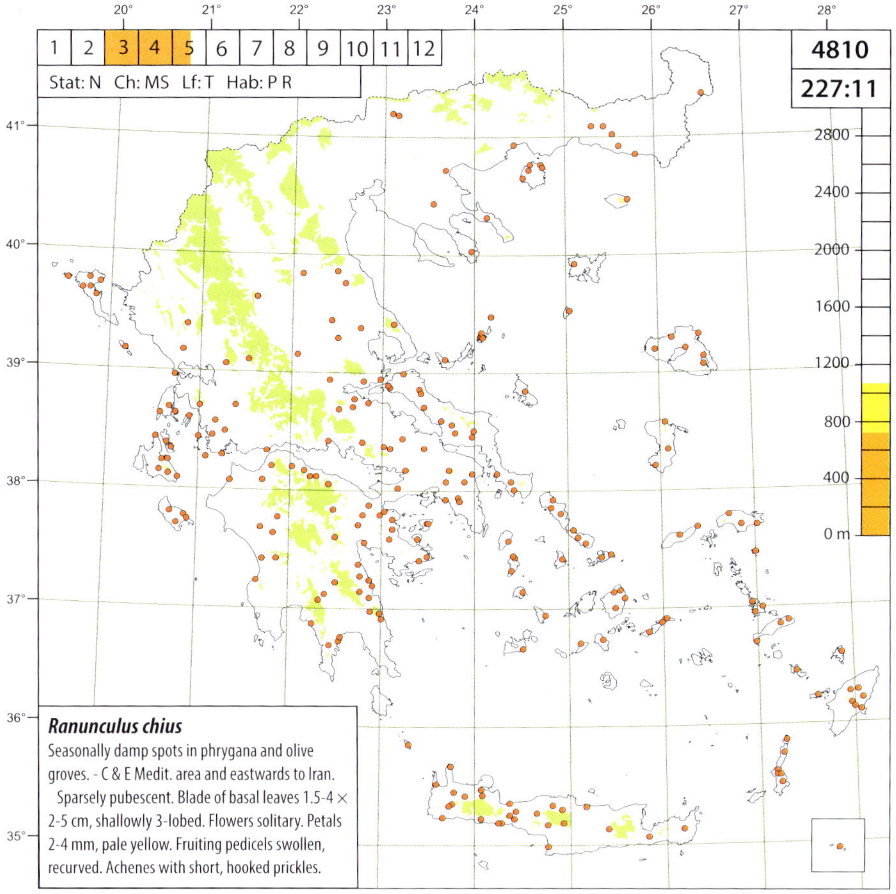

Ranunculus chius

Seasonally damp spots in phrygana and olive groves. - C & E Medit. area and eastwards to Iran.

Sparsely pubescent. Blade of basal leaves 1.5-4 × 2-5 cm, shallowly 3-lobed. Flowers solitary. Petals 2-4 mm, pale yellow. Fruiting pedicels swollen, recurved. Achenes with short, hooked prickles.

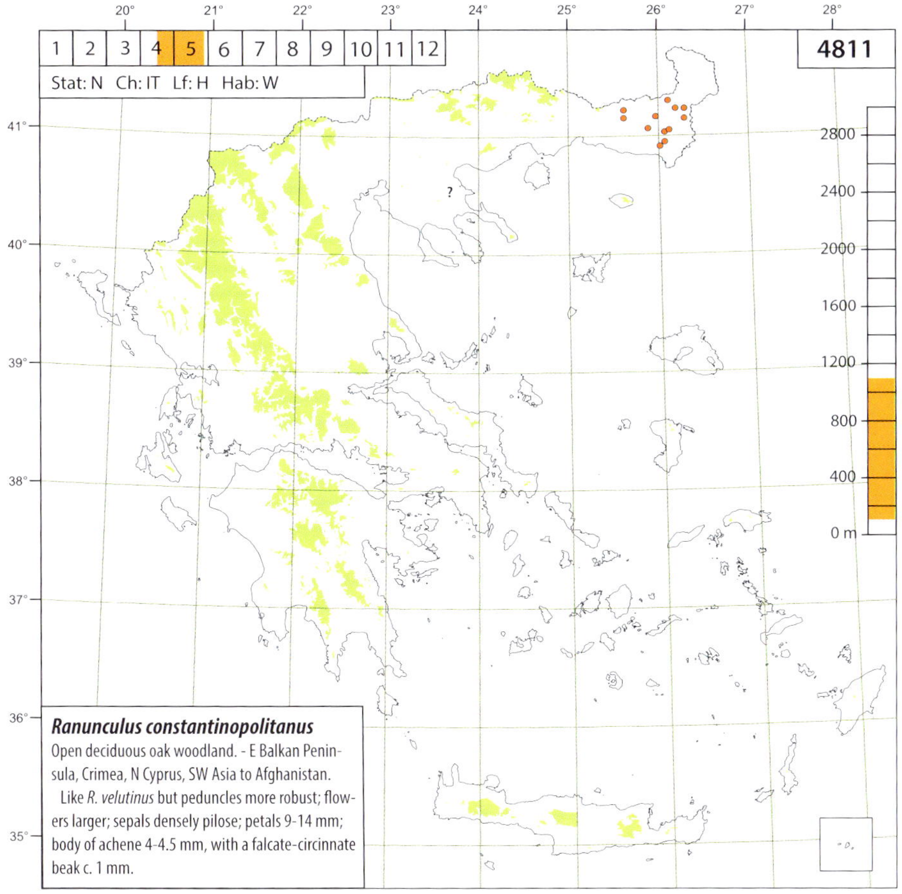

Ranunculus constantinopolitanus

Open deciduous oak woodland. - E Balkan Peninsula, Crimea, N Cyprus, SW Asia to Afghanistan.

Like *R. velutinus* but peduncles more robust; flowers larger; sepals densely pilose; petals 9-14 mm; body of achene 4-4.5 mm, with a falcate-circinnate beak c. 1 mm.

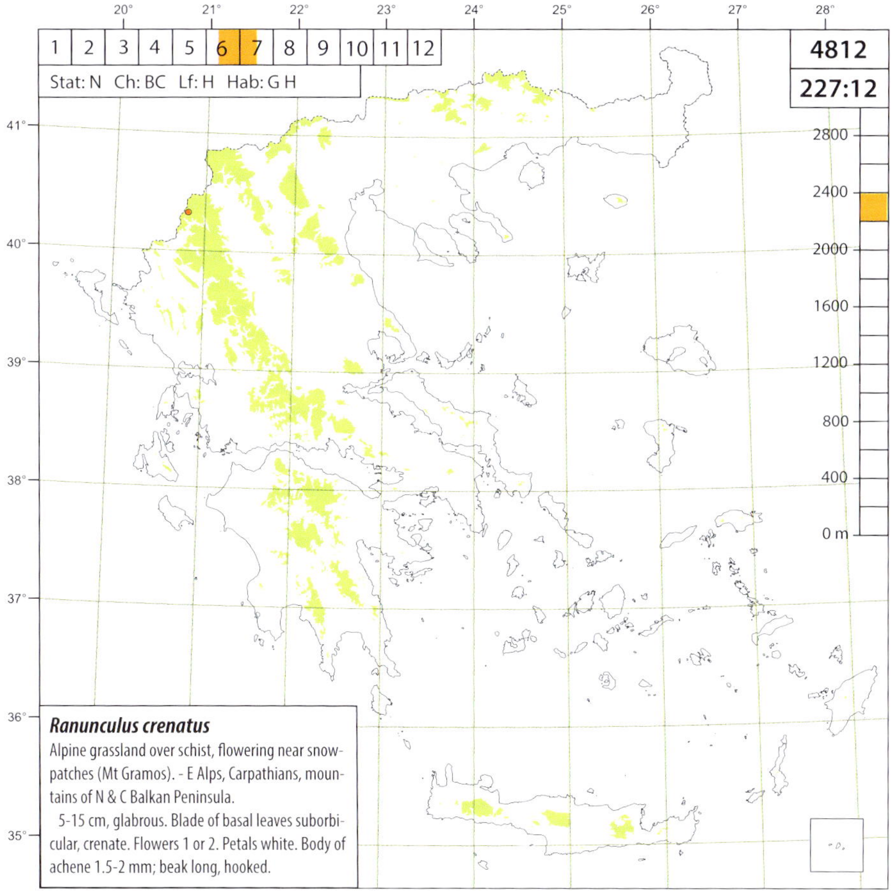

Ranunculus crenatus

Alpine grassland over schist, flowering near snow-patches (Mt Gramos). - E Alps, Carpathians, mountains of N & C Balkan Peninsula.

5-15 cm, glabrous. Blade of basal leaves suborbicular, crenate. Flowers 1 or 2. Petals white. Body of achene 1.5-2 mm; beak long, hooked.

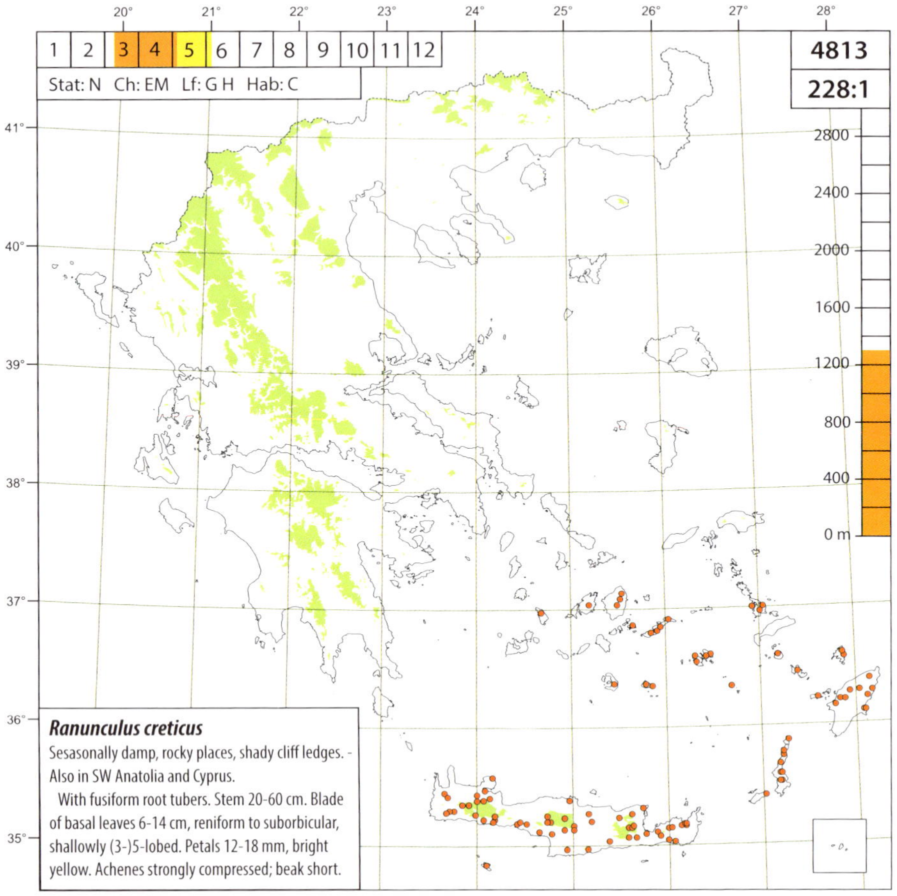

Ranunculus creticus
Seasonally damp, rocky places, shady cliff ledges. - Also in SW Anatolia and Cyprus.
With fusiform root tubers. Stem 20-60 cm. Blade of basal leaves 6-14 cm, reniform to suborbicular, shallowly (3-)5-lobed. Petals 12-18 mm, bright yellow. Achenes strongly compressed; beak short.

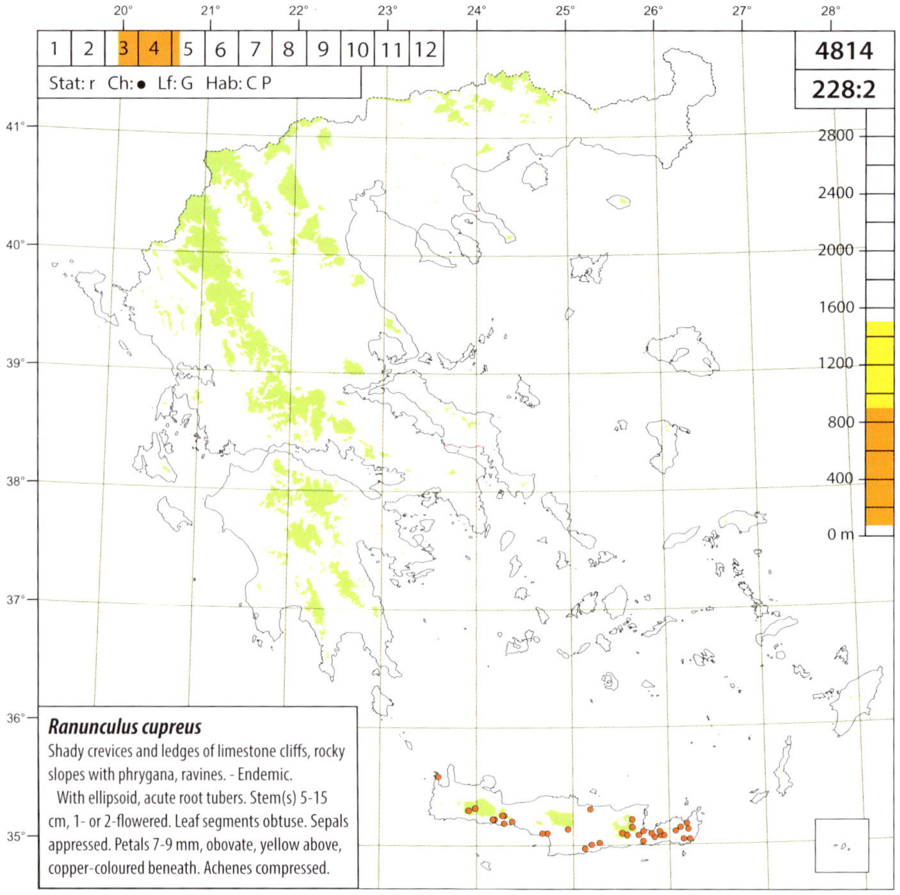

Ranunculus cupreus
Shady crevices and ledges of limestone cliffs, rocky slopes with phrygana, ravines. - Endemic.
With ellipsoid, acute root tubers. Stem(s) 5-15 cm, 1- or 2-flowered. Leaf segments obtuse. Sepals appressed. Petals 7-9 mm, obovate, yellow above, copper-coloured beneath. Achenes compressed.

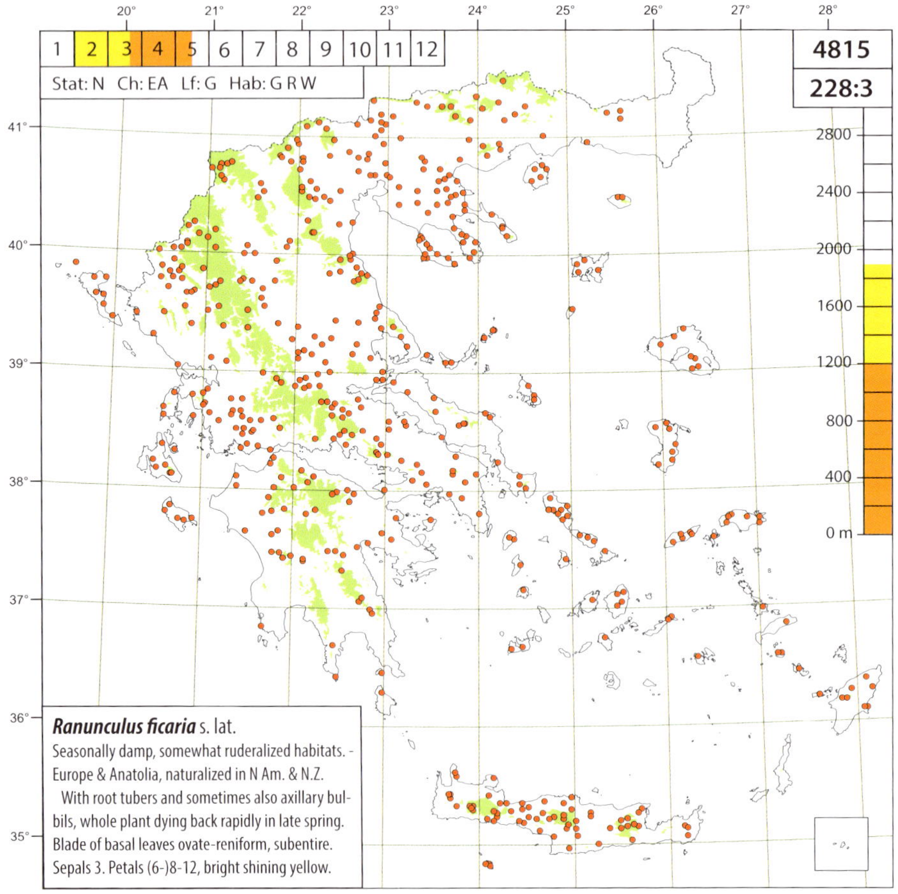

Ranunculus ficaria s. lat.
Seasonally damp, somewhat ruderalized habitats. - Europe & Anatolia, naturalized in N Am. & N.Z.
With root tubers and sometimes also axillary bulbils, whole plant dying back rapidly in late spring. Blade of basal leaves ovate-reniform, subentire. Sepals 3. Petals (6-)8-12, bright shining yellow.

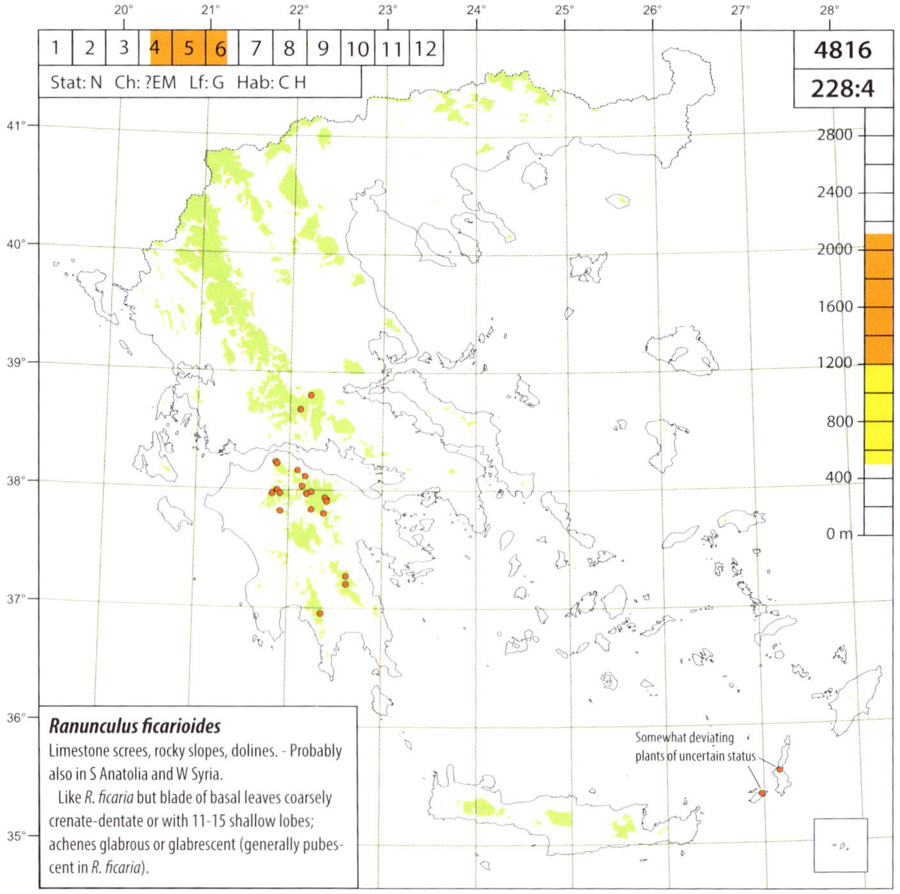

Ranunculus ficarioides
Limestone screes, rocky slopes, dolines. - Probably also in S Anatolia and W Syria.
Like *R. ficaria* but blade of basal leaves coarsely crenate-dentate or with 11-15 shallow lobes; achenes glabrous or glabrescent (generally pubescent in *R. ficaria*).

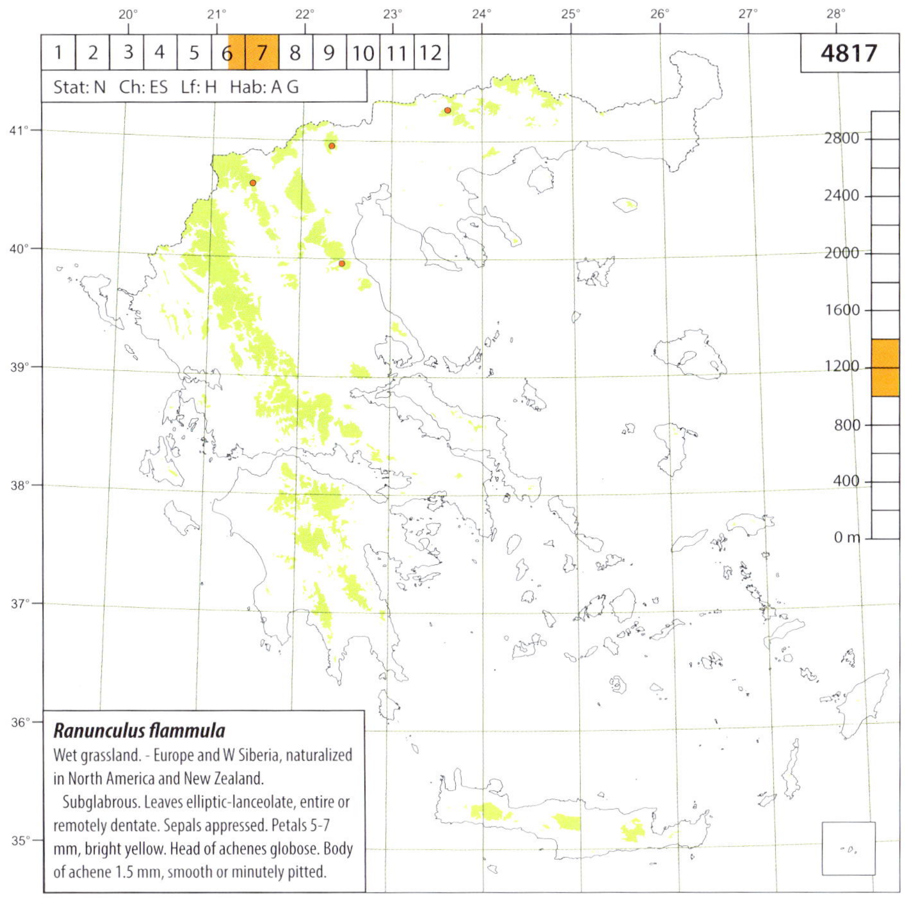

Ranunculus flammula

Wet grassland. - Europe and W Siberia, naturalized in North America and New Zealand.

Subglabrous. Leaves elliptic-lanceolate, entire or remotely dentate. Sepals appressed. Petals 5-7 mm, bright yellow. Head of achenes globose. Body of achene 1.5 mm, smooth or minutely pitted.

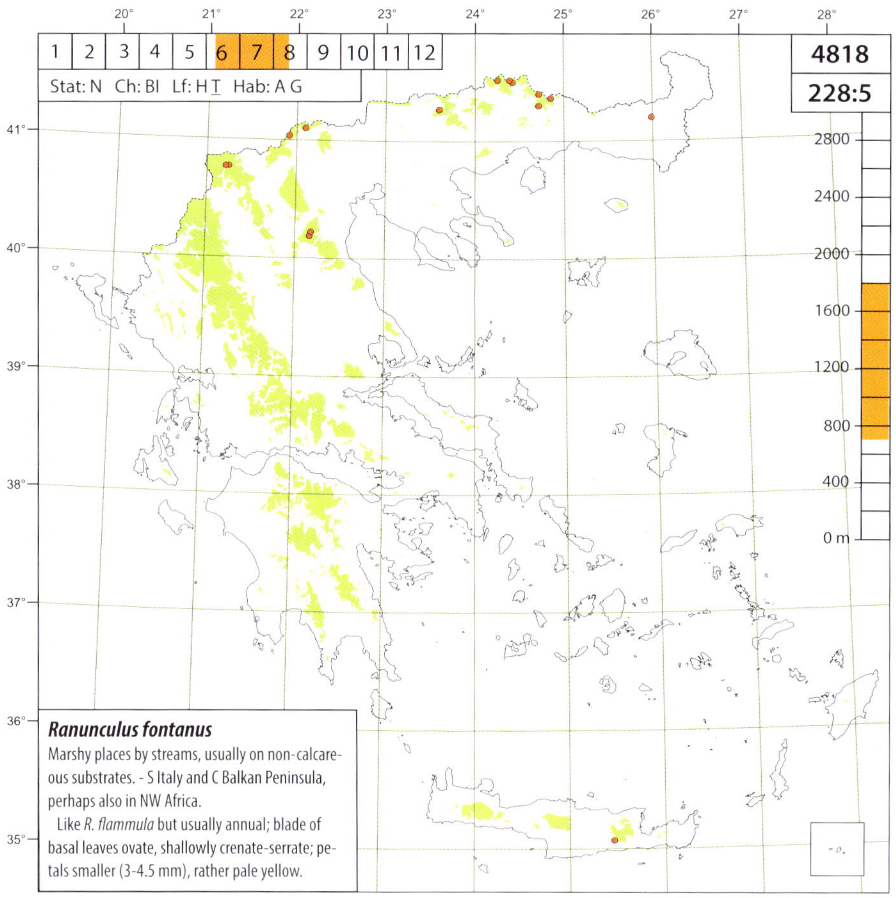

Ranunculus fontanus

Marshy places by streams, usually on non-calcareous substrates. - S Italy and C Balkan Peninsula, perhaps also in NW Africa.

Like *R. flammula* but usually annual; blade of basal leaves ovate, shallowly crenate-serrate; petals smaller (3-4.5 mm), rather pale yellow.

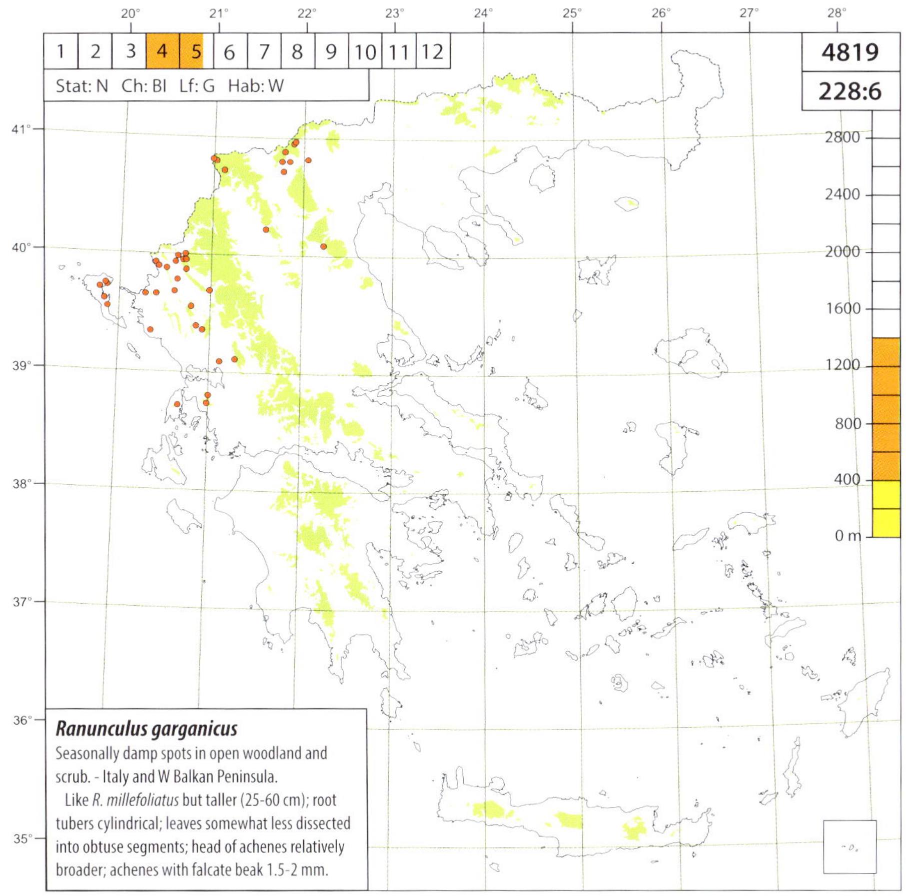

Ranunculus garganicus

Seasonally damp spots in open woodland and scrub. - Italy and W Balkan Peninsula.

Like *R. millefoliatus* but taller (25-60 cm); root tubers cylindrical; leaves somewhat less dissected into obtuse segments; head of achenes relatively broader; achenes with falcate beak 1.5-2 mm.

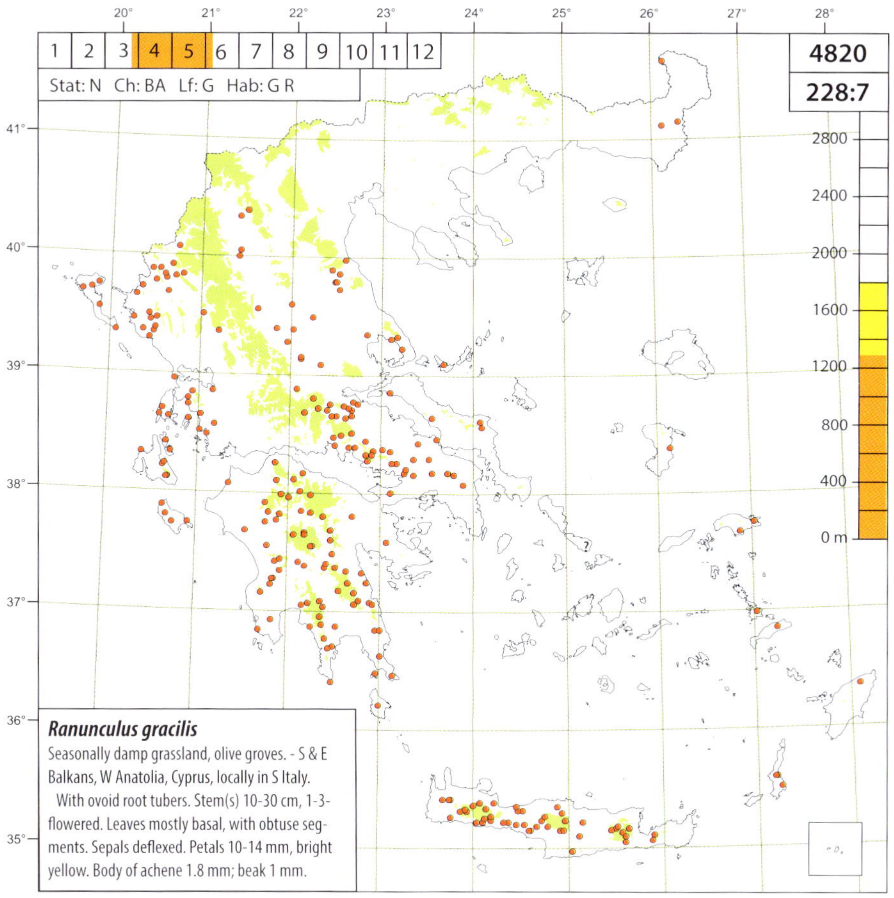

Ranunculus gracilis

Seasonally damp grassland, olive groves. - S & E Balkans, W Anatolia, Cyprus, locally in S Italy.

With ovoid root tubers. Stem(s) 10-30 cm, 1-3-flowered. Leaves mostly basal, with obtuse segments. Sepals deflexed. Petals 10-14 mm, bright yellow. Body of achene 1.8 mm; beak 1 mm.

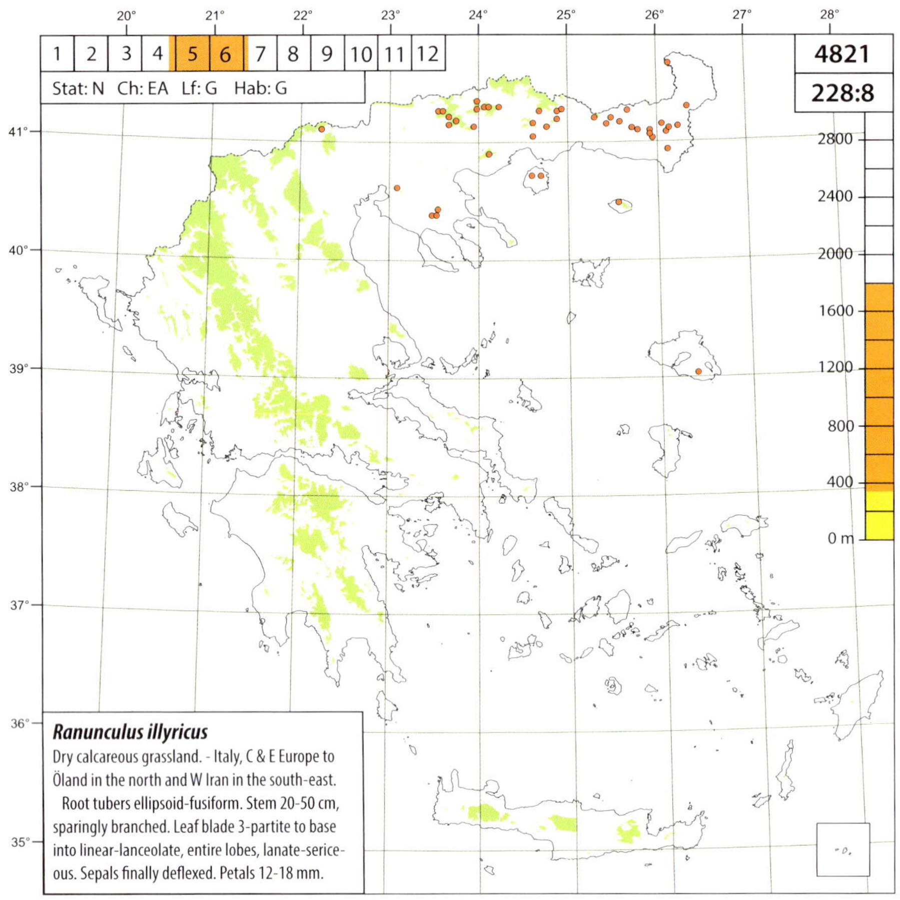

Ranunculus illyricus

Dry calcareous grassland. - Italy, C & E Europe to Öland in the north and W Iran in the south-east.

Root tubers ellipsoid-fusiform. Stem 20-50 cm, sparingly branched. Leaf blade 3-partite to base into linear-lanceolate, entire lobes, lanate-sericeous. Sepals finally deflexed. Petals 12-18 mm.

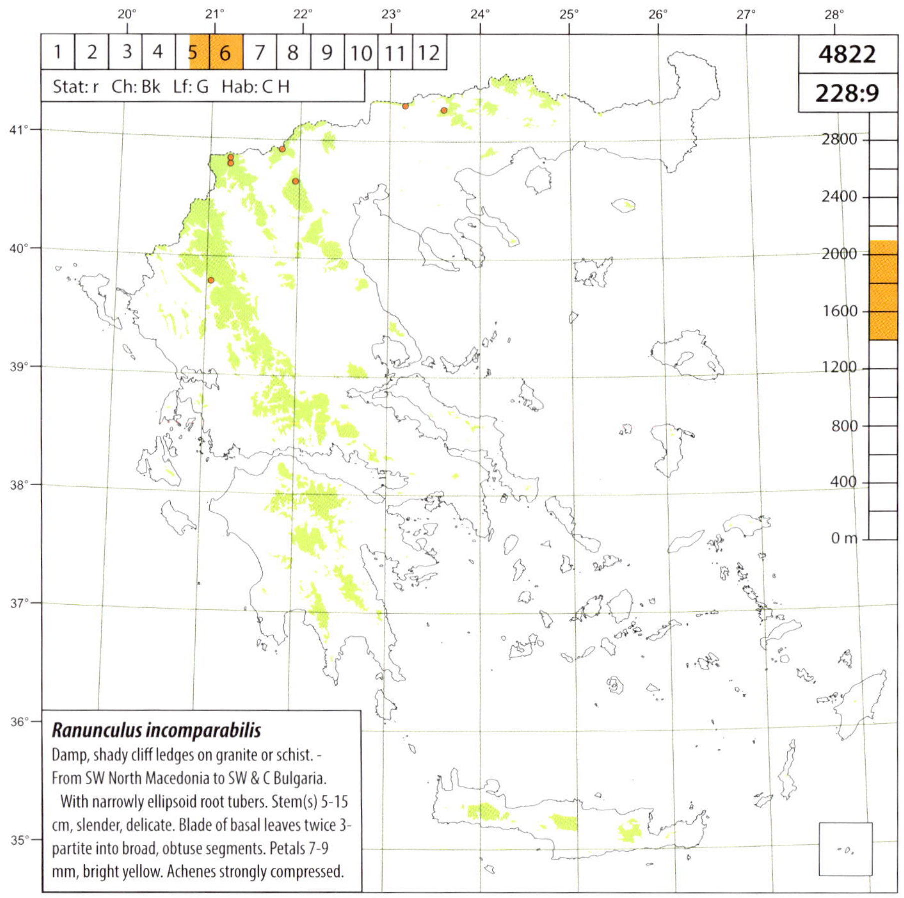

Ranunculus incomparabilis

Damp, shady cliff ledges on granite or schist. - From SW North Macedonia to SW & C Bulgaria.

With narrowly ellipsoid root tubers. Stem(s) 5-15 cm, slender, delicate. Blade of basal leaves twice 3-partite into broad, obtuse segments. Petals 7-9 mm, bright yellow. Achenes strongly compressed.

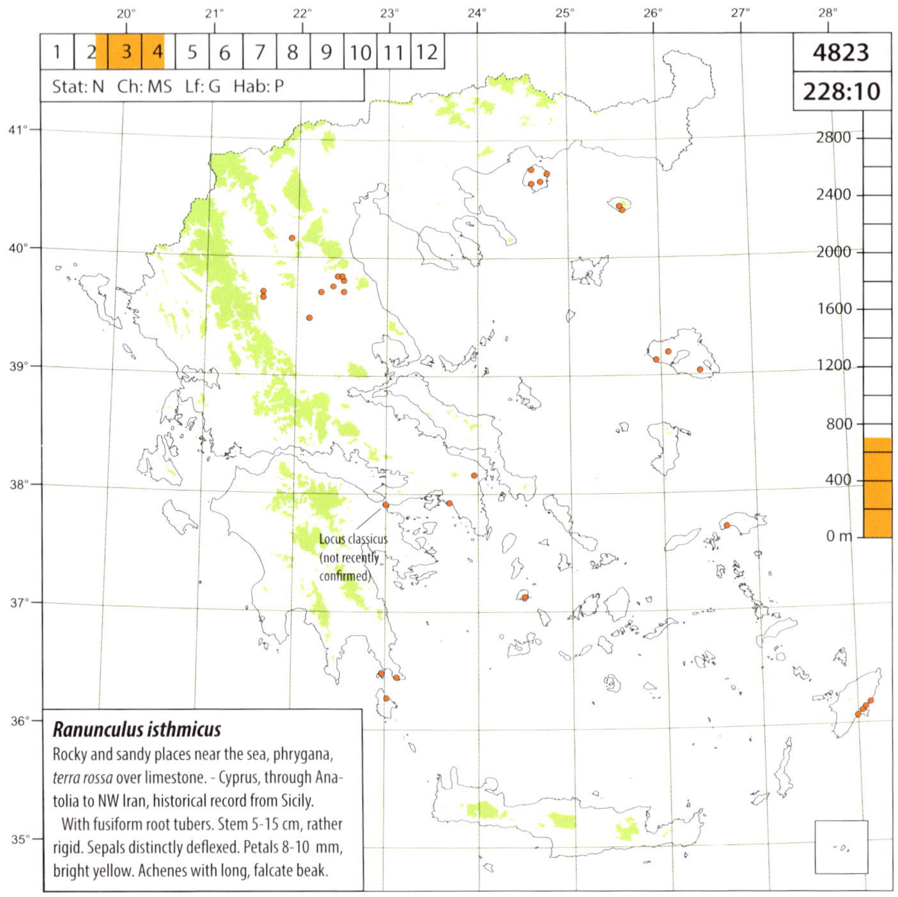

Ranunculus isthmicus

Rocky and sandy places near the sea, phrygana, *terra rossa* over limestone. - Cyprus, through Anatolia to NW Iran, historical record from Sicily.

With fusiform root tubers. Stem 5-15 cm, rather rigid. Sepals distinctly deflexed. Petals 8-10 mm, bright yellow. Achenes with long, falcate beak.

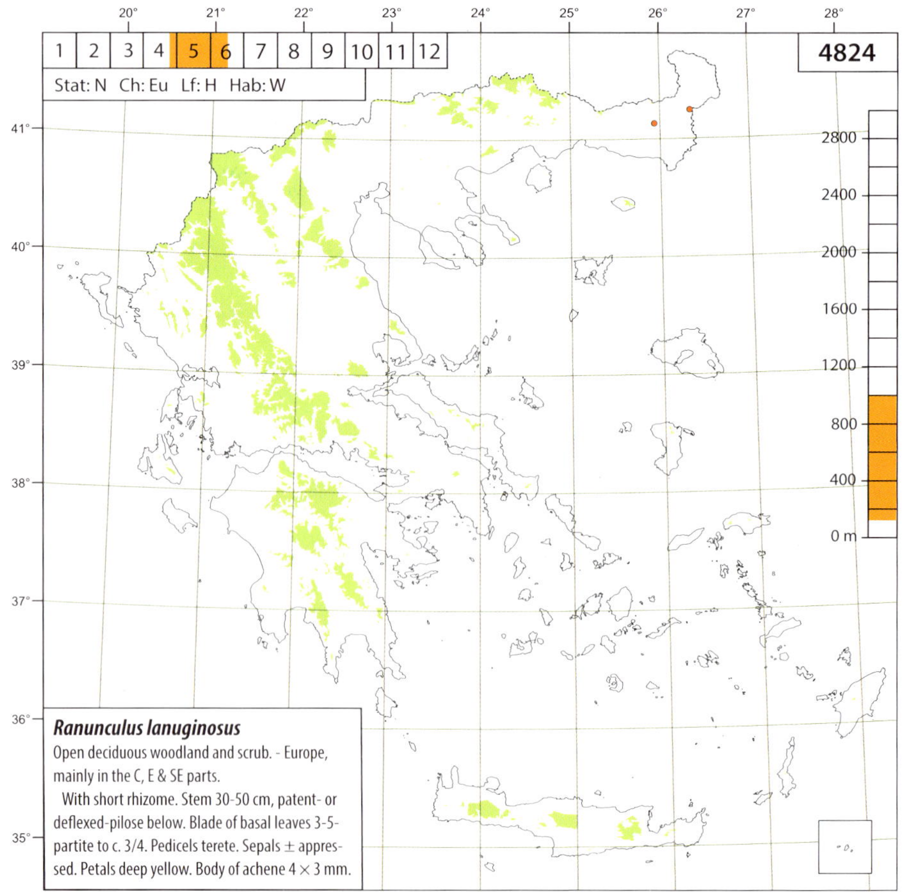

Ranunculus lanuginosus

Open deciduous woodland and scrub. - Europe, mainly in the C, E & SE parts.

With short rhizome. Stem 30-50 cm, patent- or deflexed-pilose below. Blade of basal leaves 3-5-partite to c. 3/4. Pedicels terete. Sepals ± appressed. Petals deep yellow. Body of achene 4 × 3 mm.

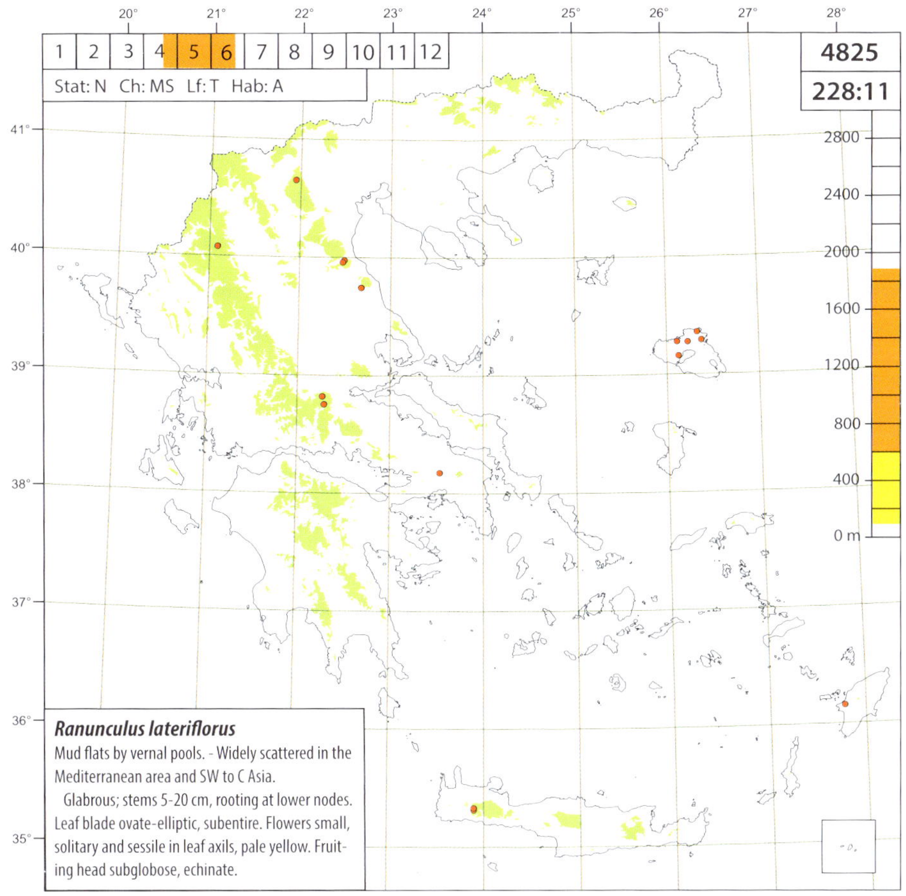

Ranunculus lateriflorus

Mud flats by vernal pools. - Widely scattered in the Mediterranean area and SW to C Asia.

Glabrous; stems 5-20 cm, rooting at lower nodes. Leaf blade ovate-elliptic, subentire. Flowers small, solitary and sessile in leaf axils, pale yellow. Fruiting head subglobose, echinate.

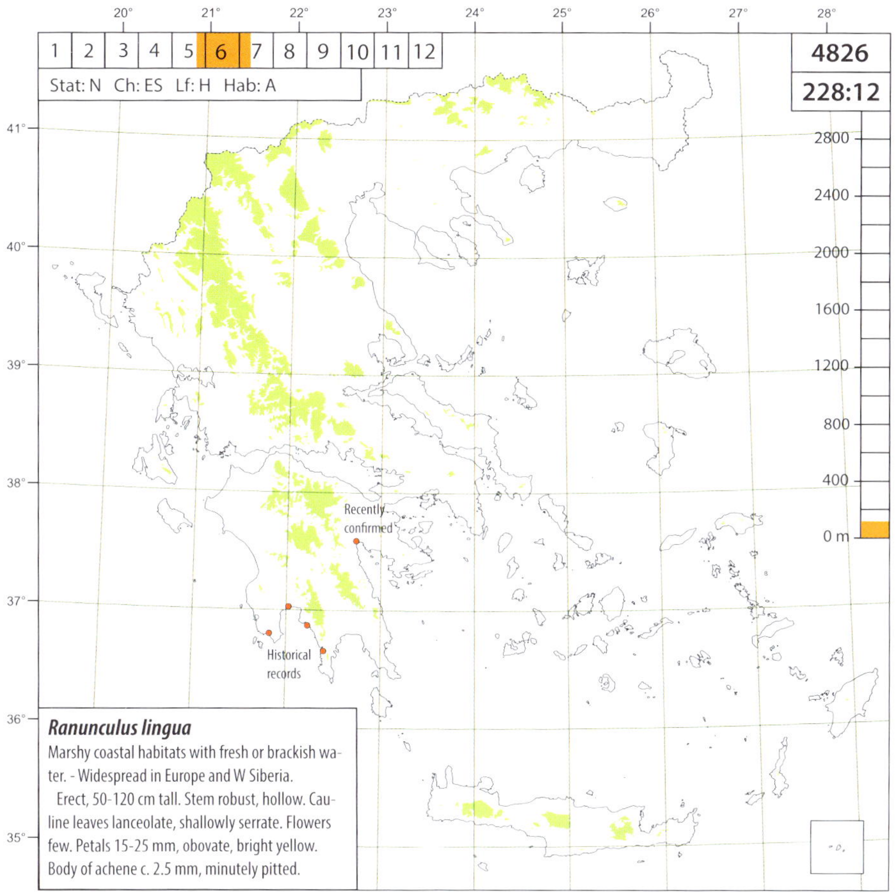

Ranunculus lingua

Marshy coastal habitats with fresh or brackish water. - Widespread in Europe and W Siberia.

Erect, 50-120 cm tall. Stem robust, hollow. Cauline leaves lanceolate, shallowly serrate. Flowers few. Petals 15-25 mm, obovate, bright yellow. Body of achene c. 2.5 mm, minutely pitted.

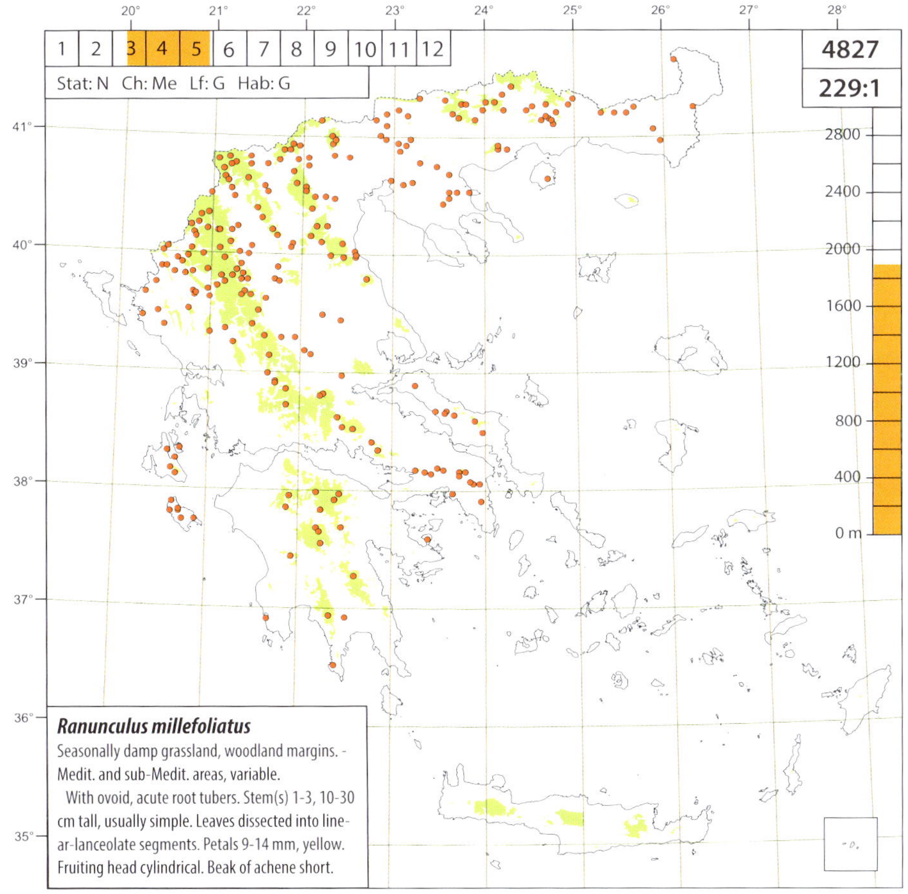

Ranunculus millefoliatus

Seasonally damp grassland, woodland margins. - Medit. and sub-Medit. areas, variable.

With ovoid, acute root tubers. Stem(s) 1-3, 10-30 cm tall, usually simple. Leaves dissected into linear-lanceolate segments. Petals 9-14 mm, yellow. Fruiting head cylindrical. Beak of achene short.

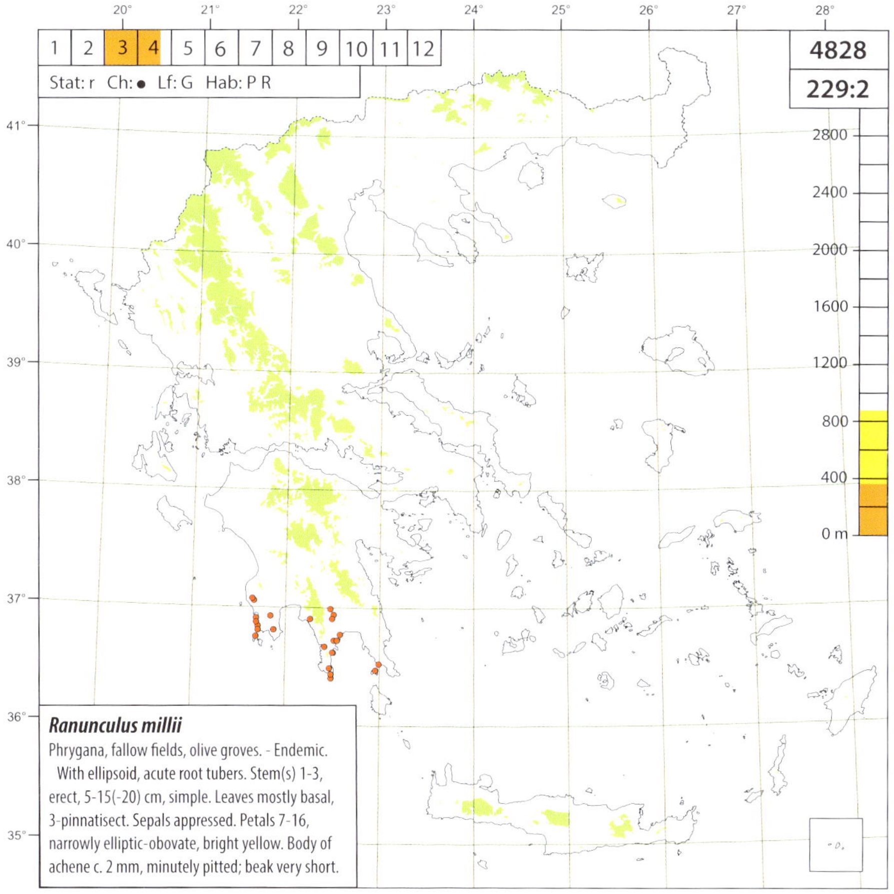

Ranunculus millii

Phrygana, fallow fields, olive groves. - Endemic.

With ellipsoid, acute root tubers. Stem(s) 1-3, erect, 5-15(-20) cm, simple. Leaves mostly basal, 3-pinnatisect. Sepals appressed. Petals 7-16, narrowly elliptic-obovate, bright yellow. Body of achene c. 2 mm, minutely pitted; beak very short.

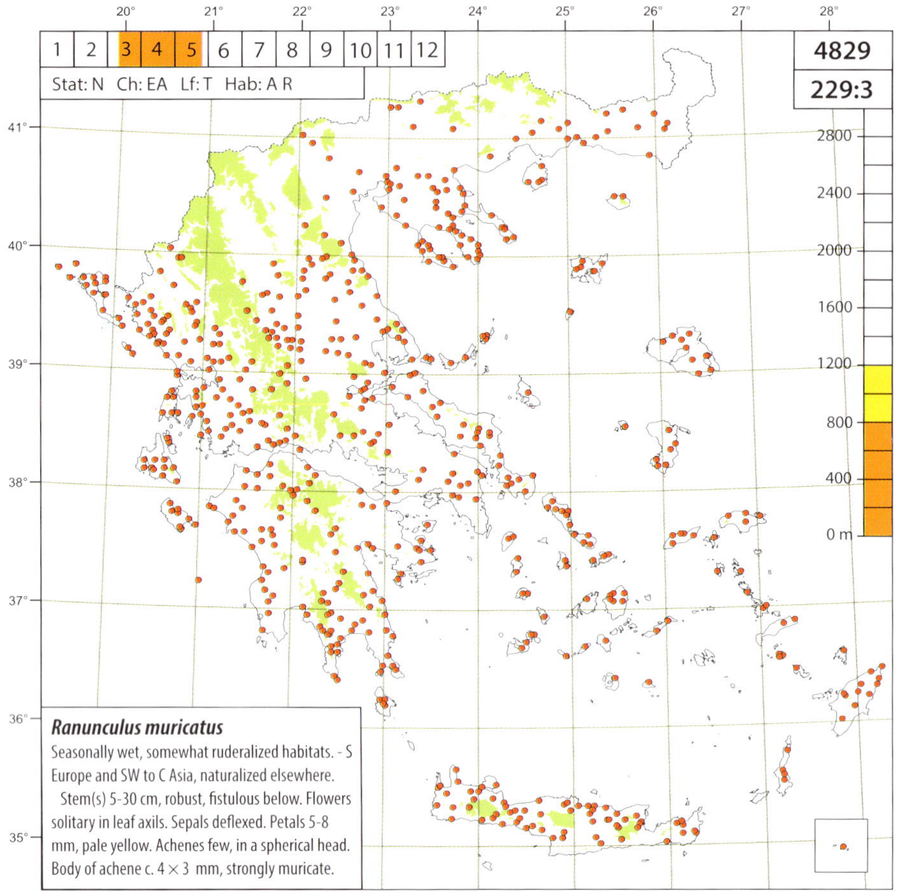

Ranunculus muricatus
Seasonally wet, somewhat ruderalized habitats. - S Europe and SW to C Asia, naturalized elsewhere.
Stem(s) 5-30 cm, robust, fistulous below. Flowers solitary in leaf axils. Sepals deflexed. Petals 5-8 mm, pale yellow. Achenes few, in a spherical head. Body of achene c. 4 × 3 mm, strongly muricate.

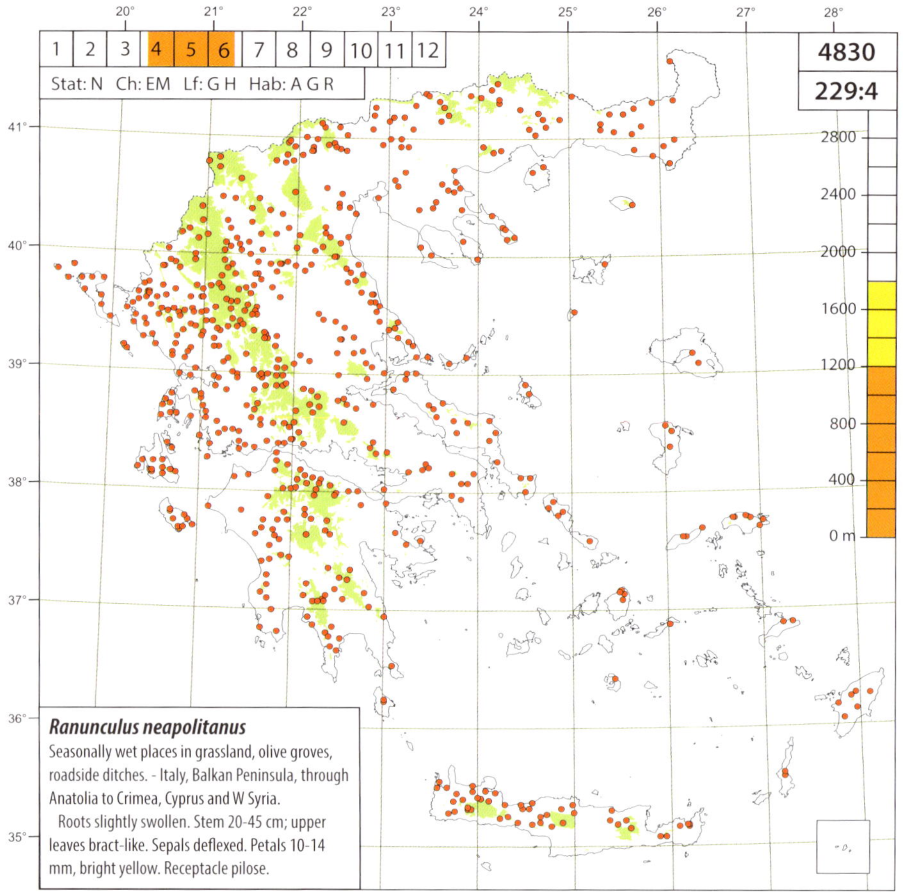

Ranunculus neapolitanus
Seasonally wet places in grassland, olive groves, roadside ditches. - Italy, Balkan Peninsula, through Anatolia to Crimea, Cyprus and W Syria.
Roots slightly swollen. Stem 20-45 cm; upper leaves bract-like. Sepals deflexed. Petals 10-14 mm, bright yellow. Receptacle pilose.

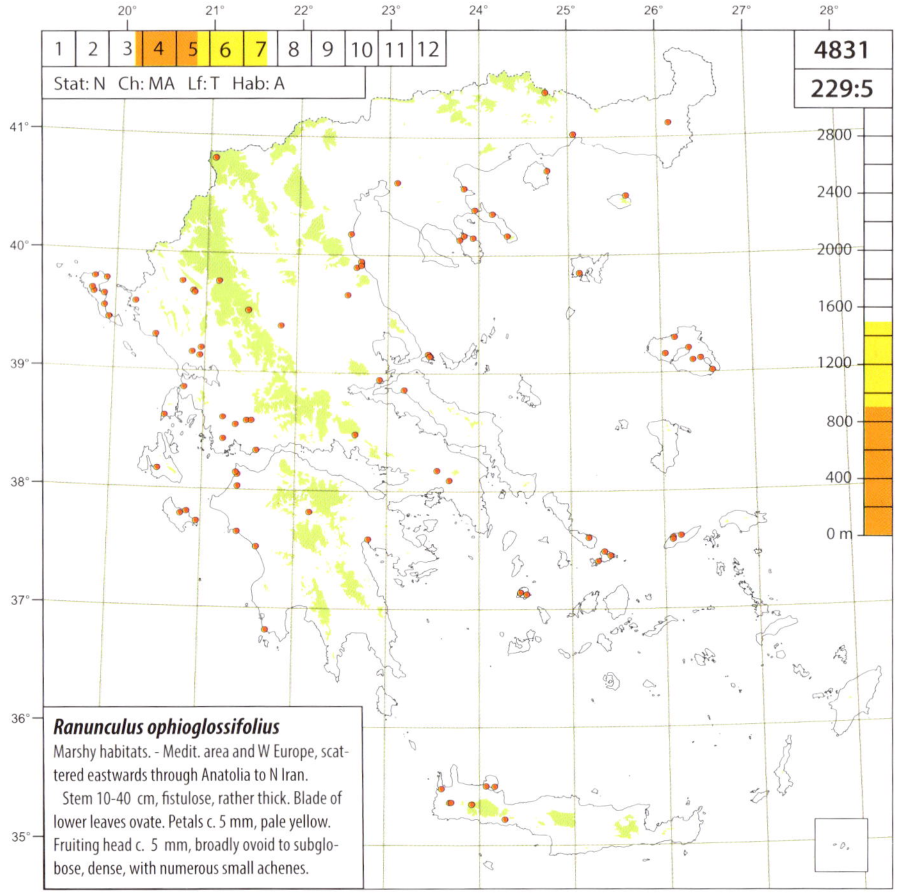

Ranunculus ophioglossifolius
Marshy habitats. - Medit. area and W Europe, scattered eastwards through Anatolia to N Iran.
Stem 10-40 cm, fistulose, rather thick. Blade of lower leaves ovate. Petals c. 5 mm, pale yellow. Fruiting head c. 5 mm, broadly ovoid to subglobose, dense, with numerous small achenes.

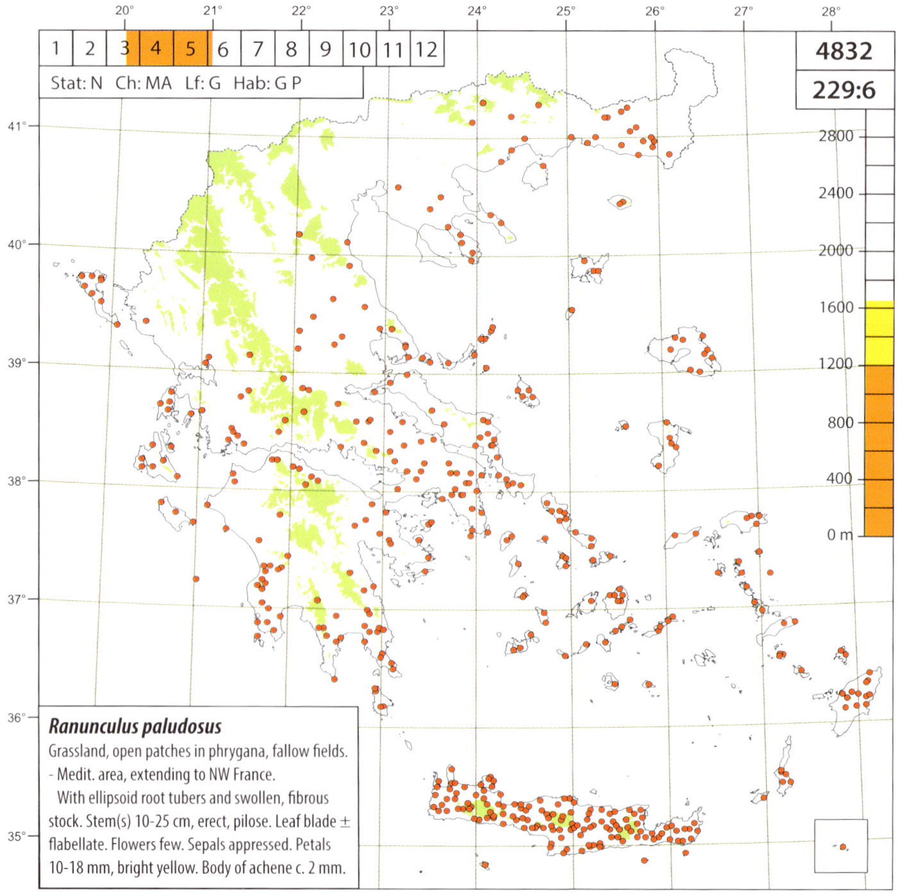

Ranunculus paludosus
Grassland, open patches in phrygana, fallow fields. - Medit. area, extending to NW France.
With ellipsoid root tubers and swollen, fibrous stock. Stem(s) 10-25 cm, erect, pilose. Leaf blade ± flabellate. Flowers few. Sepals appressed. Petals 10-18 mm, bright yellow. Body of achene c. 2 mm.

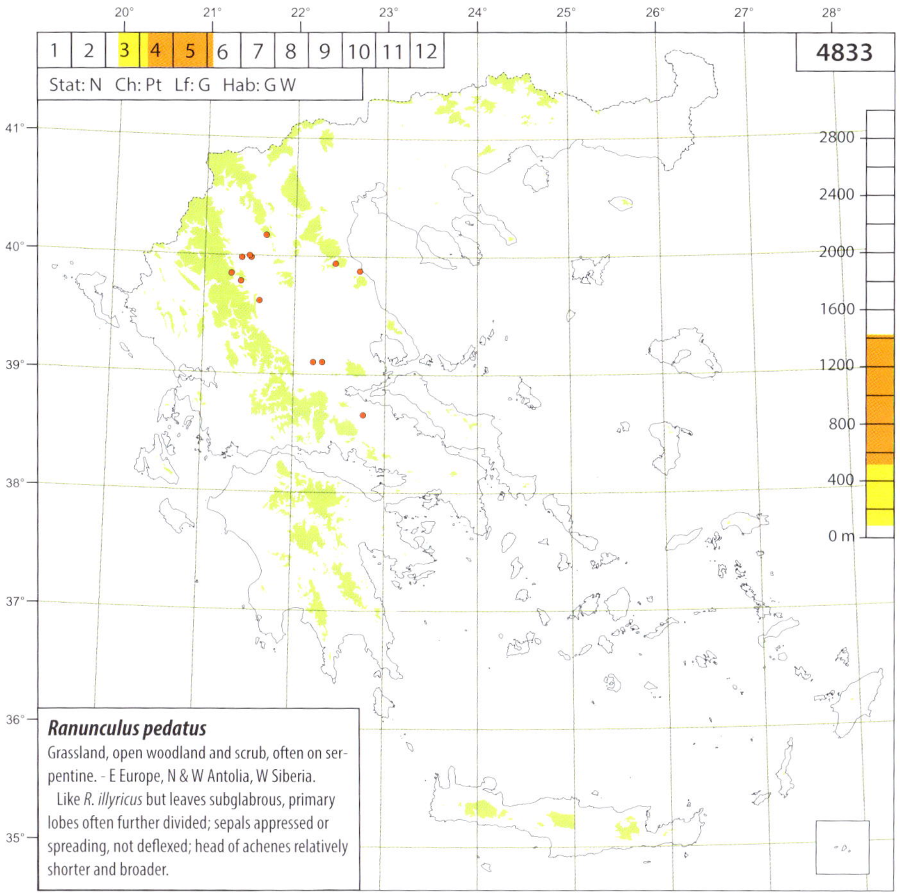

Ranunculus pedatus

Grassland, open woodland and scrub, often on serpentine. - E Europe, N & W Antolia, W Siberia.

Like *R. illyricus* but leaves subglabrous, primary lobes often further divided; sepals appressed or spreading, not deflexed; head of achenes relatively shorter and broader.

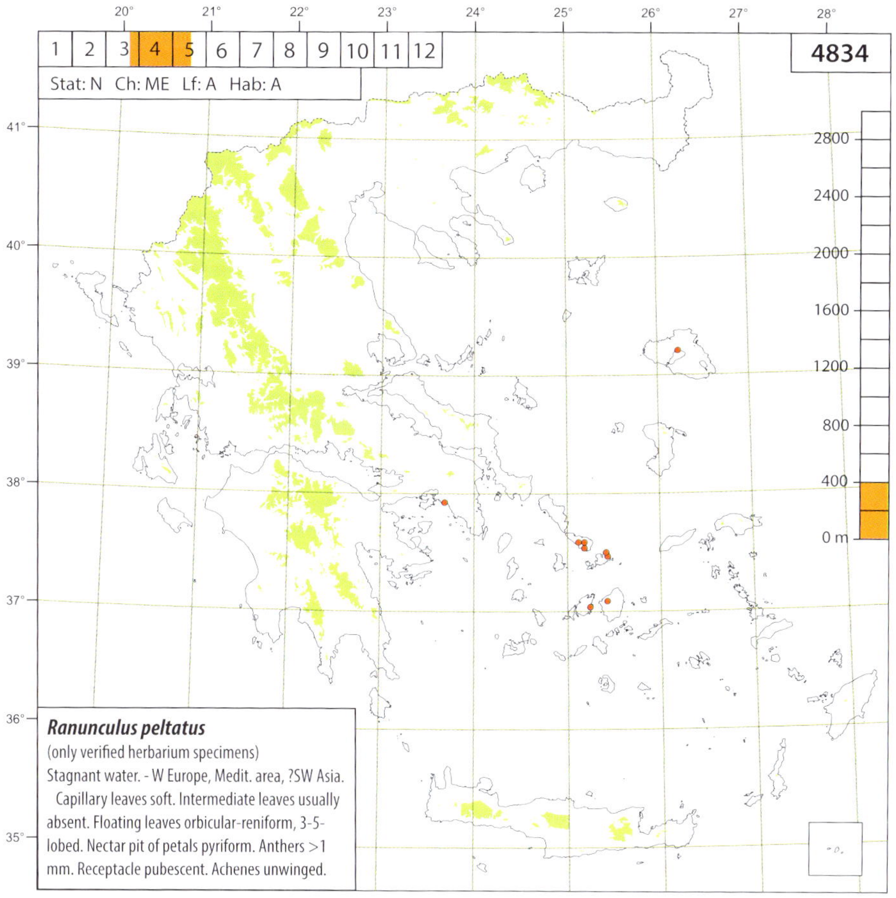

Ranunculus peltatus

(only verified herbarium specimens)

Stagnant water. - W Europe, Medit. area, ?SW Asia.

Capillary leaves soft. Intermediate leaves usually absent. Floating leaves orbicular-reniform, 3-5-lobed. Nectar pit of petals pyriform. Anthers >1 mm. Receptacle pubescent. Achenes unwinged.

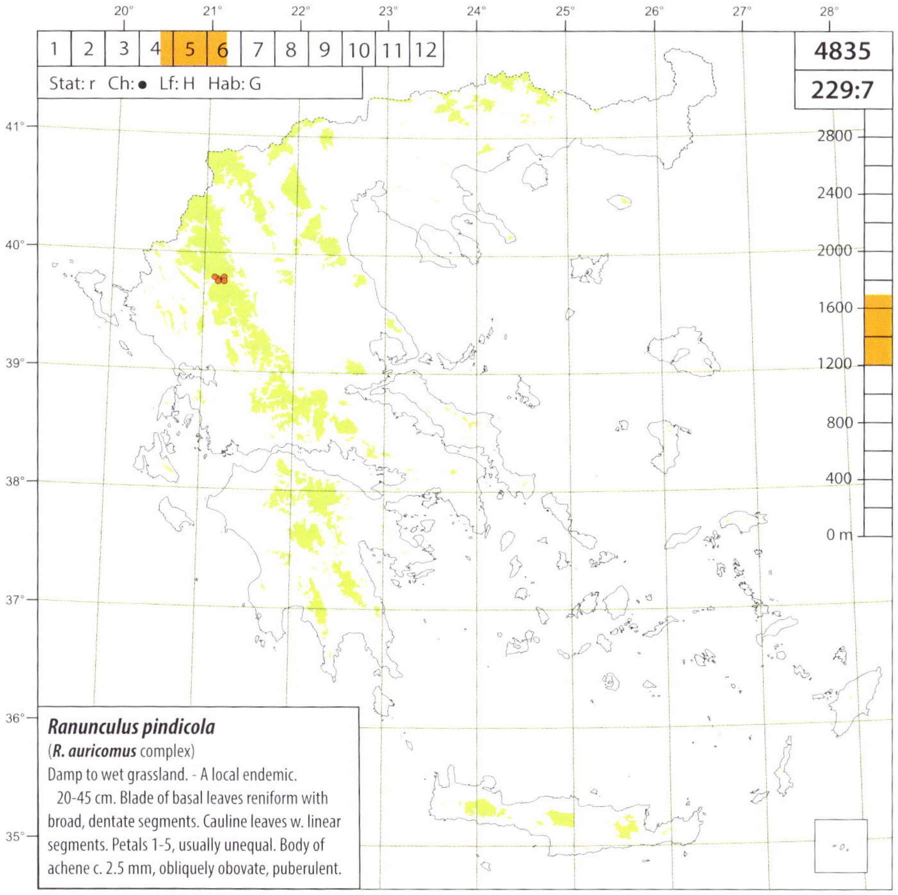

Ranunculus pindicola

(***R. auricomus*** complex)

Damp to wet grassland. - A local endemic.

20-45 cm. Blade of basal leaves reniform with broad, dentate segments. Cauline leaves w. linear segments. Petals 1-5, usually unequal. Body of achene c. 2.5 mm, obliquely obovate, puberulent.

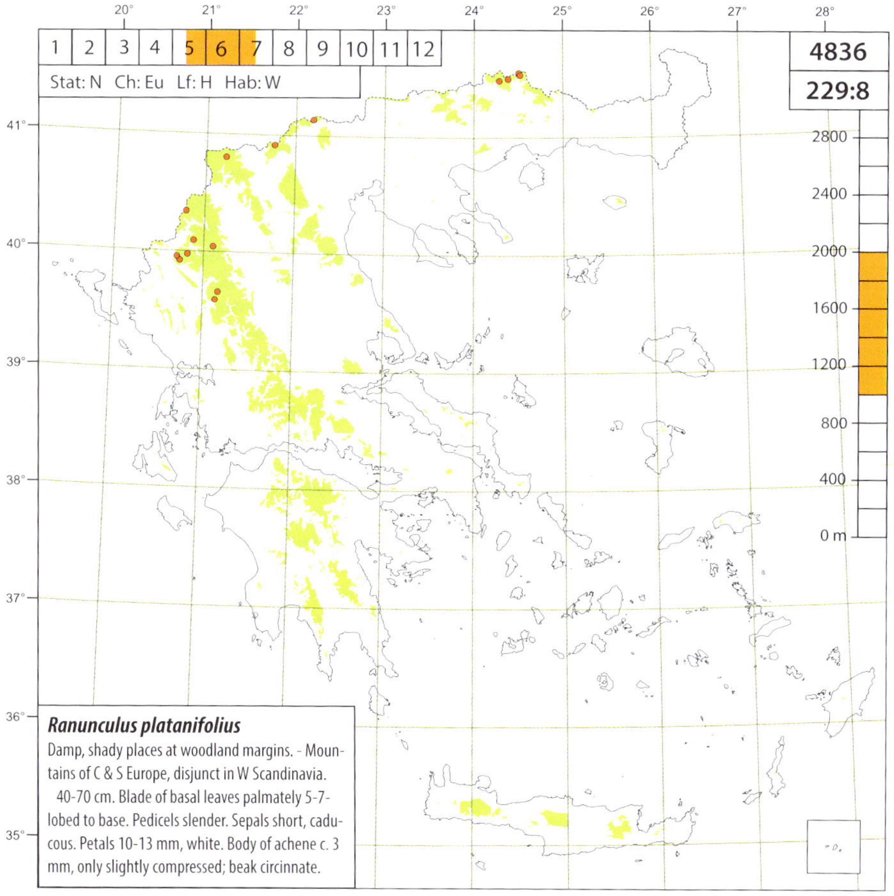

Ranunculus platanifolius

Damp, shady places at woodland margins. - Mountains of C & S Europe, disjunct in W Scandinavia.

40-70 cm. Blade of basal leaves palmately 5-7-lobed to base. Pedicels slender. Sepals short, caducous. Petals 10-13 mm, white. Body of achene c. 3 mm, only slightly compressed; beak circinnate.

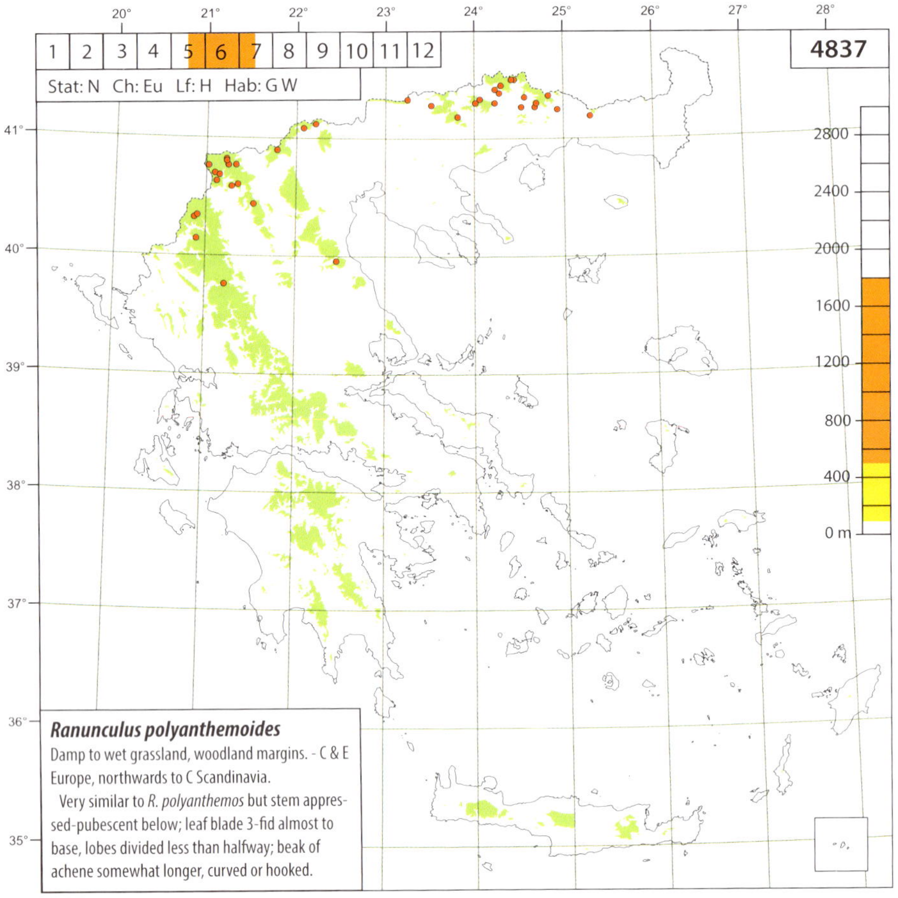

Ranunculus polyanthemoides
Damp to wet grassland, woodland margins. - C & E Europe, northwards to C Scandinavia.
Very similar to *R. polyanthemos* but stem appressed-pubescent below; leaf blade 3-fid almost to base, lobes divided less than halfway; beak of achene somewhat longer, curved or hooked.

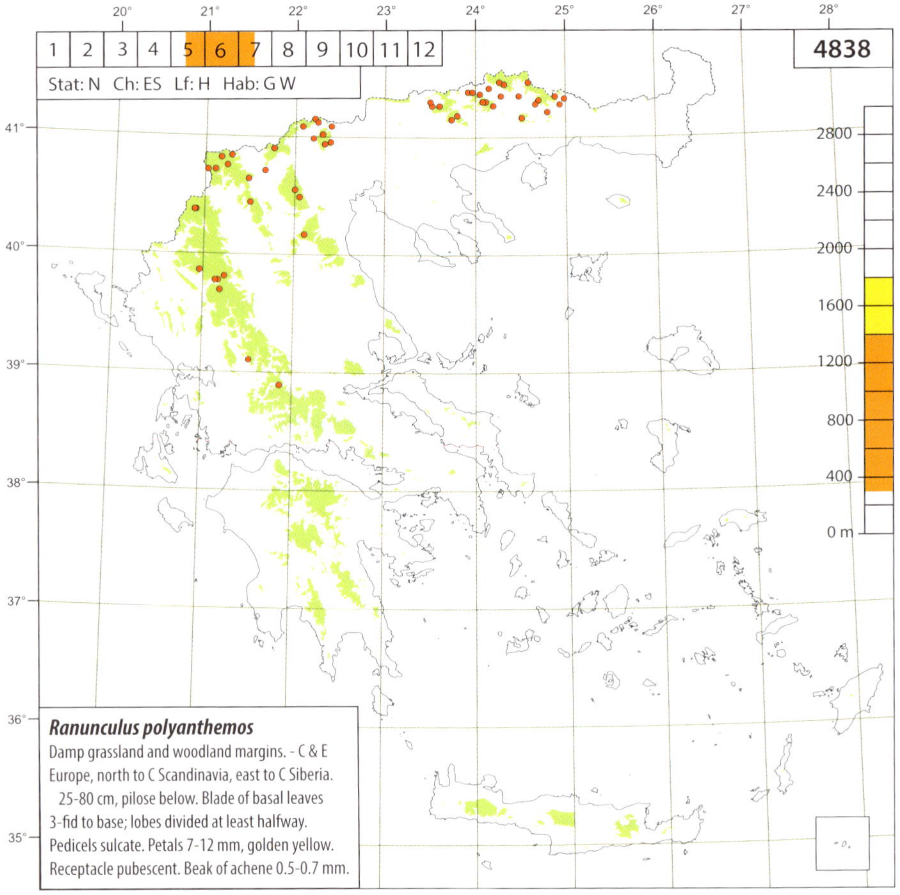

Ranunculus polyanthemos
Damp grassland and woodland margins. - C & E Europe, north to C Scandinavia, east to C Siberia.
25-80 cm, pilose below. Blade of basal leaves 3-fid to base; lobes divided at least halfway. Pedicels sulcate. Petals 7-12 mm, golden yellow. Receptacle pubescent. Beak of achene 0.5-0.7 mm.

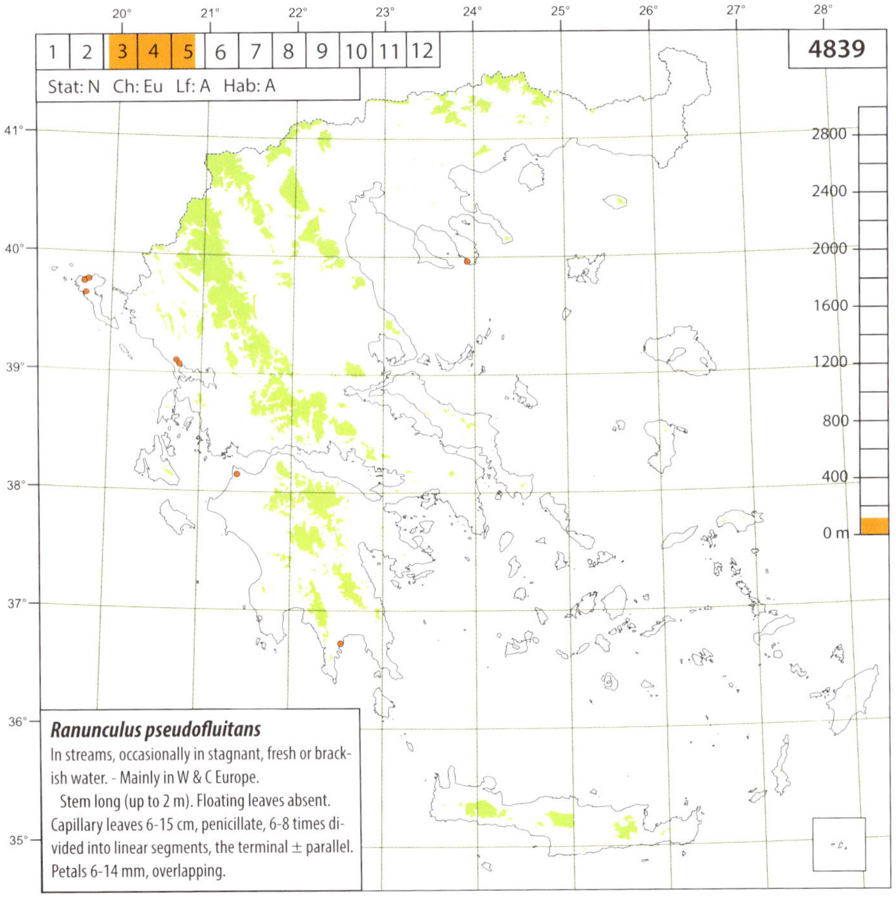

Ranunculus pseudofluitans
In streams, occasionally in stagnant, fresh or brackish water. - Mainly in W & C Europe.
Stem long (up to 2 m). Floating leaves absent. Capillary leaves 6-15 cm, penicillate, 6-8 times divided into linear segments, the terminal ± parallel. Petals 6-14 mm, overlapping.

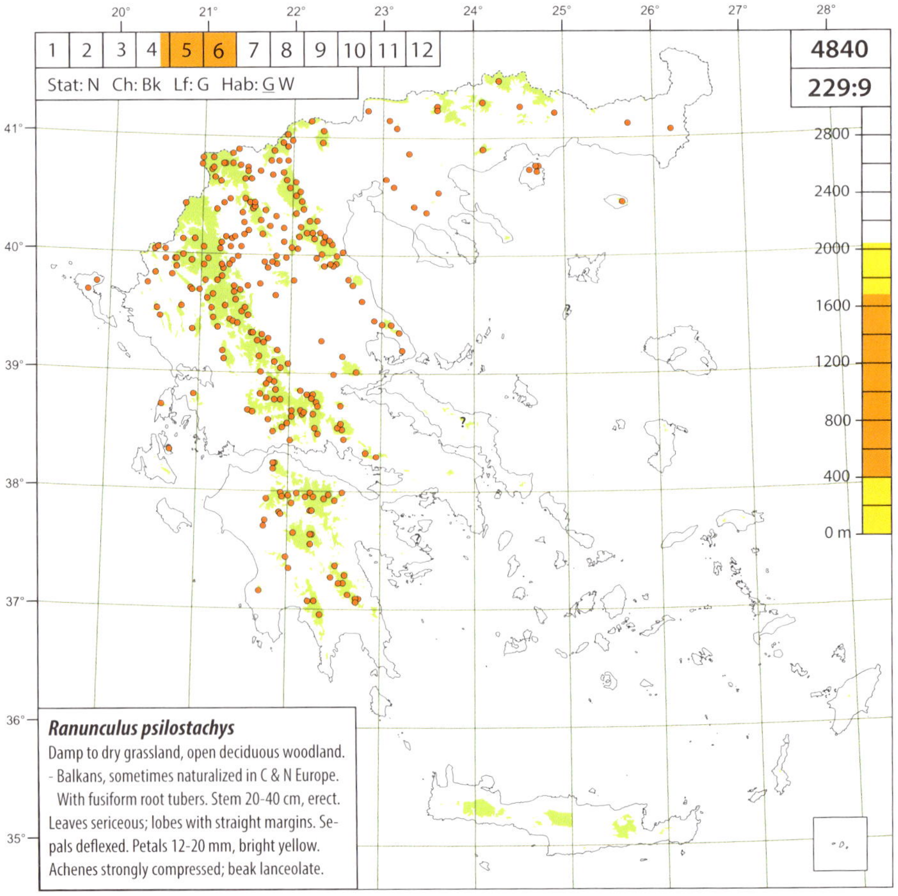

Ranunculus psilostachys
Damp to dry grassland, open deciduous woodland. - Balkans, sometimes naturalized in C & N Europe.
With fusiform root tubers. Stem 20-40 cm, erect. Leaves sericeous; lobes with straight margins. Sepals deflexed. Petals 12-20 mm, bright yellow. Achenes strongly compressed; beak lanceolate.

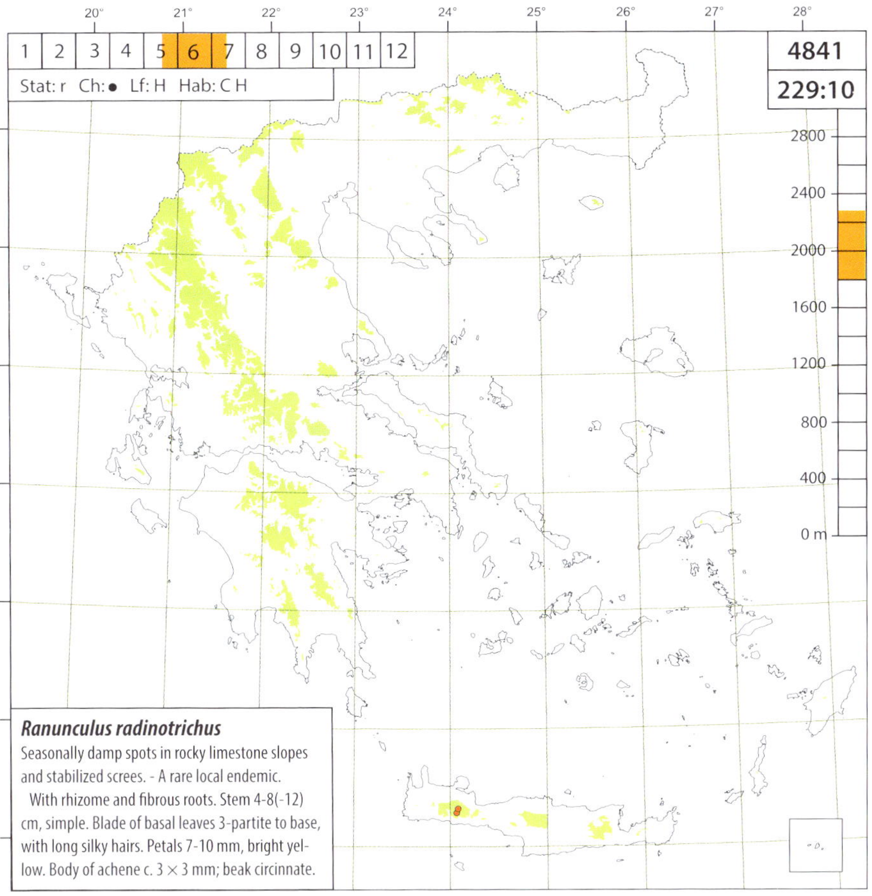

Ranunculus radinotrichus

Seasonally damp spots in rocky limestone slopes and stabilized screes. - A rare local endemic.

With rhizome and fibrous roots. Stem 4-8(-12) cm, simple. Blade of basal leaves 3-partite to base, with long silky hairs. Petals 7-10 mm, bright yellow. Body of achene c. 3 × 3 mm; beak circinnate.

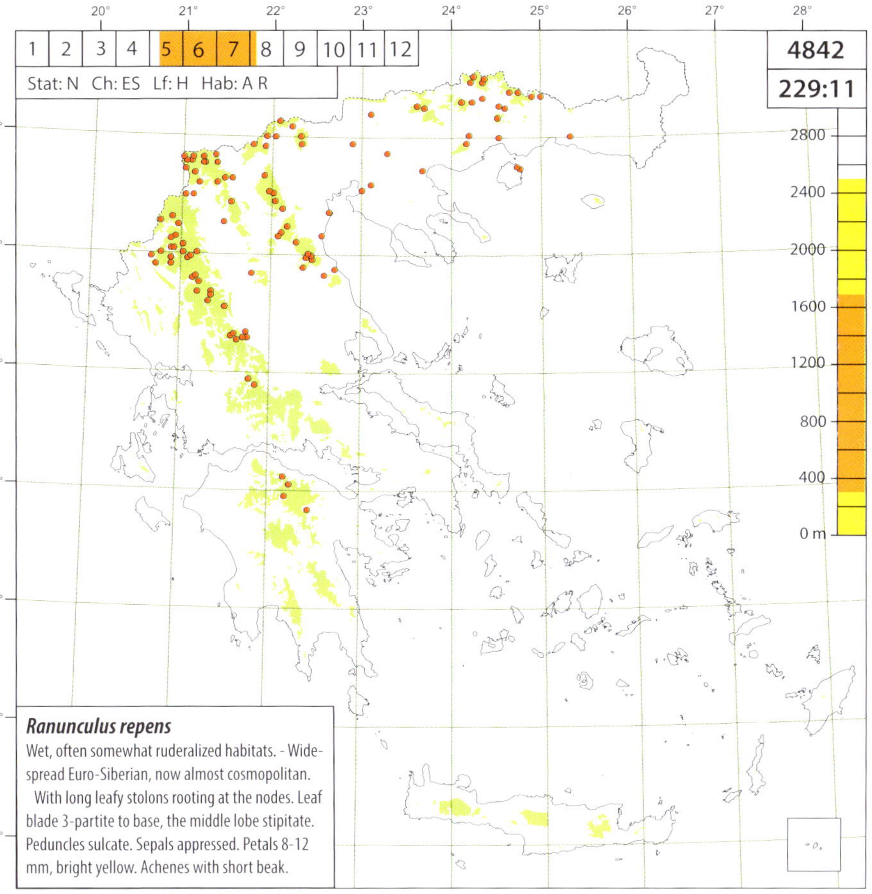

Ranunculus repens

Wet, often somewhat ruderalized habitats. - Widespread Euro-Siberian, now almost cosmopolitan.

With long leafy stolons rooting at the nodes. Leaf blade 3-partite to base, the middle lobe stipitate. Peduncles sulcate. Sepals appressed. Petals 8-12 mm, bright yellow. Achenes with short beak.

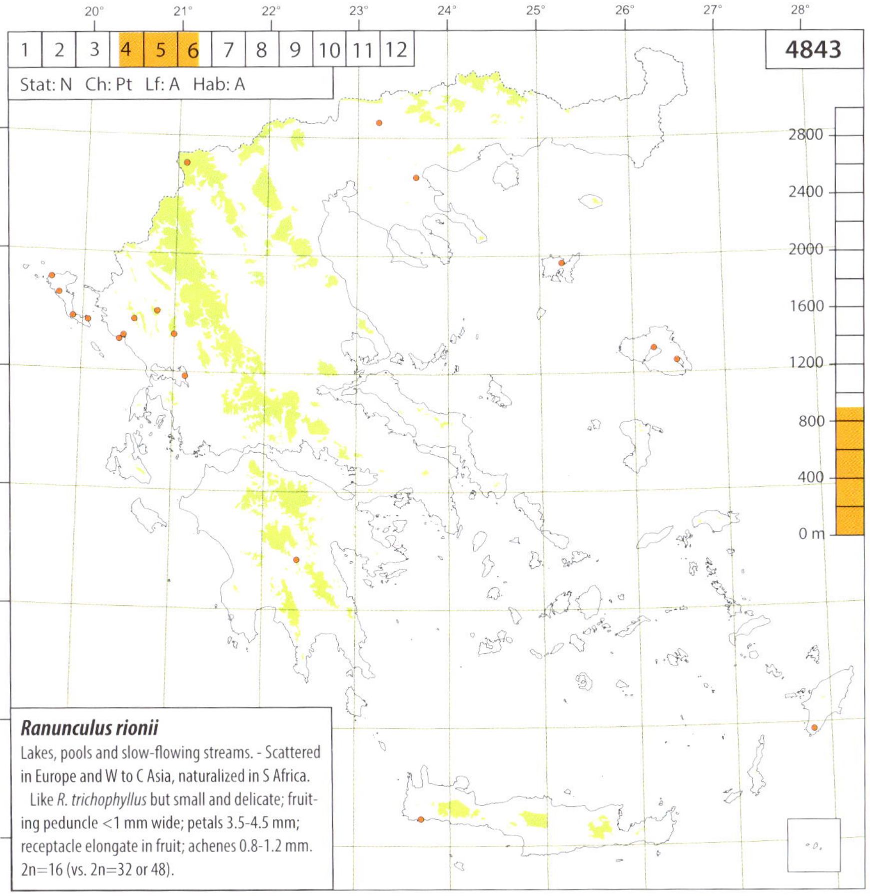

Ranunculus rionii

Lakes, pools and slow-flowing streams. - Scattered in Europe and W to C Asia, naturalized in S Africa.

Like *R. trichophyllus* but small and delicate; fruiting peduncle <1 mm wide; petals 3.5-4.5 mm; receptacle elongate in fruit; achenes 0.8-1.2 mm. 2n=16 (vs. 2n=32 or 48).

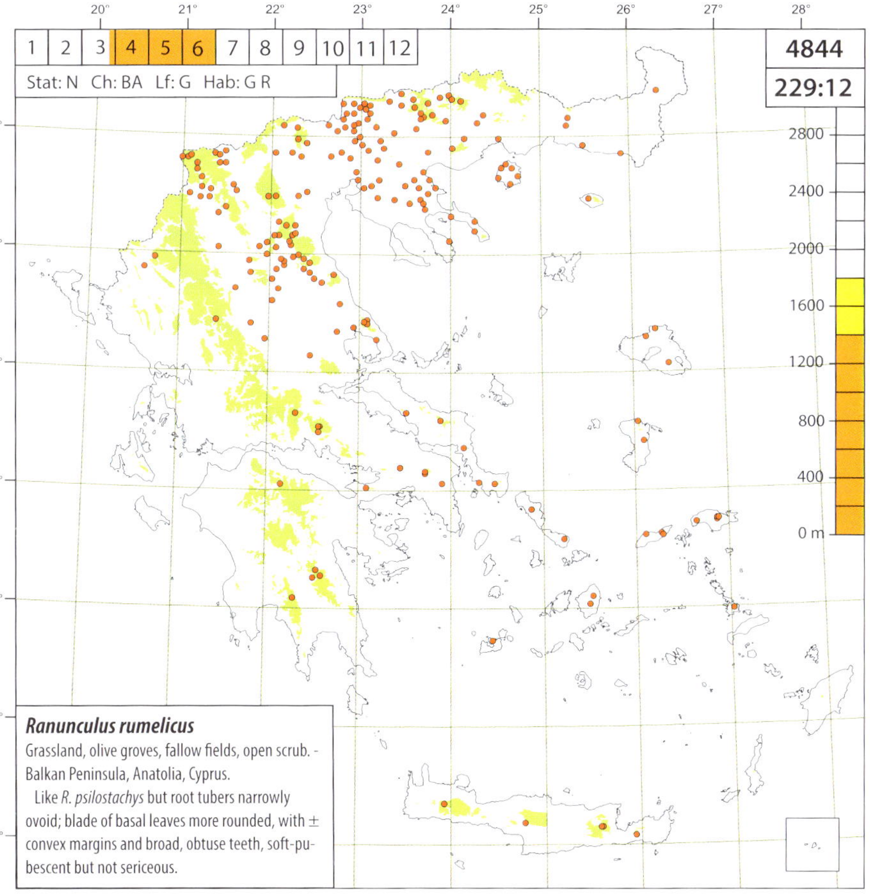

Ranunculus rumelicus

Grassland, olive groves, fallow fields, open scrub. - Balkan Peninsula, Anatolia, Cyprus.

Like *R. psilostachys* but root tubers narrowly ovoid; blade of basal leaves more rounded, with ± convex margins and broad, obtuse teeth, soft-pubescent but not sericeous.

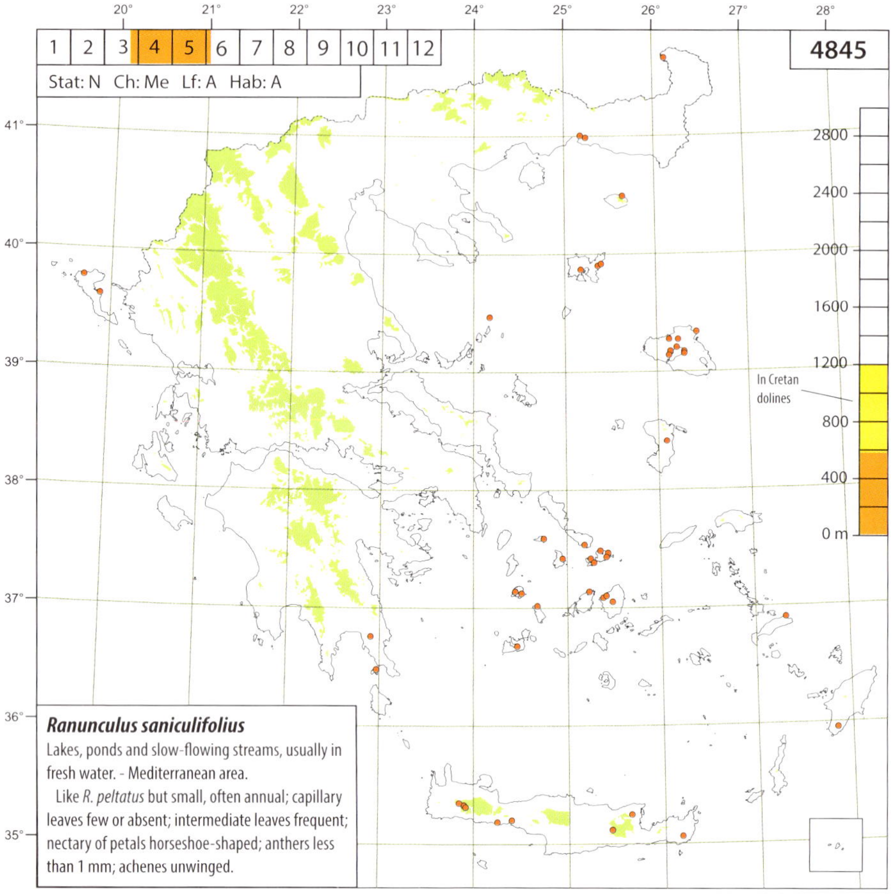

Ranunculus saniculifolius

Lakes, ponds and slow-flowing streams, usually in fresh water. - Mediterranean area.

Like *R. peltatus* but small, often annual; capillary leaves few or absent; intermediate leaves frequent; nectary of petals horseshoe-shaped; anthers less than 1 mm; achenes unwinged.

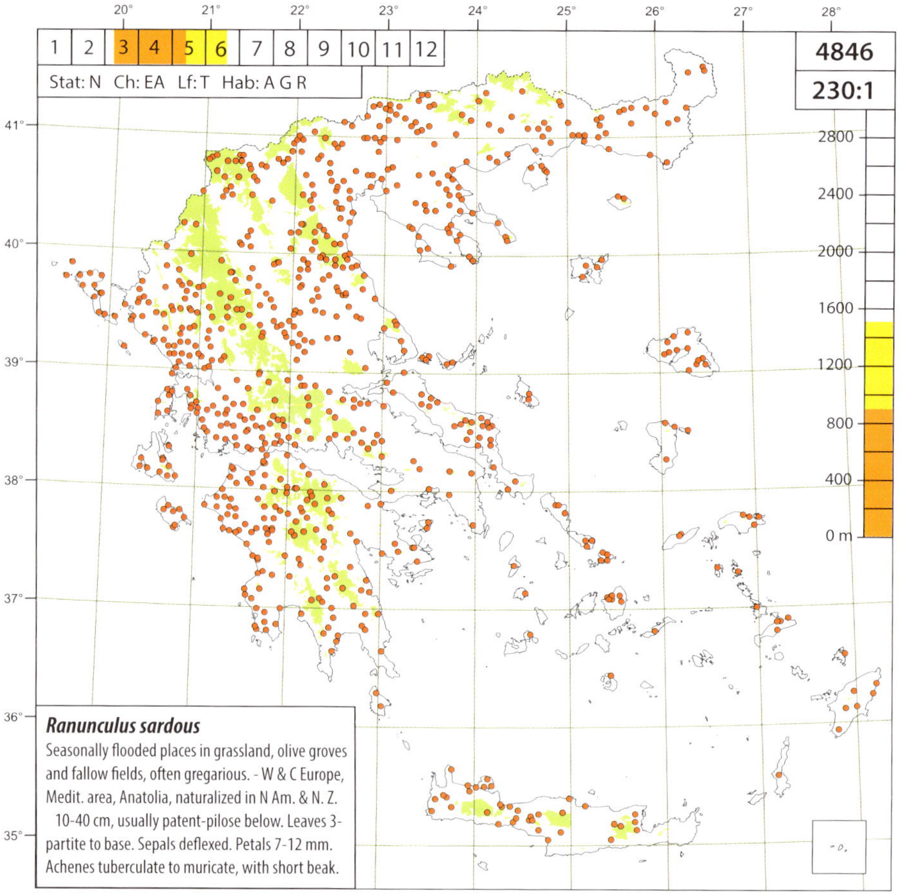

Ranunculus sardous

Seasonally flooded places in grassland, olive groves and fallow fields, often gregarious. - W & C Europe, Medit. area, Anatolia, naturalized in N Am. & N. Z.

10-40 cm, usually patent-pilose below. Leaves 3-partite to base. Sepals deflexed. Petals 7-12 mm. Achenes tuberculate to muricate, with short beak.

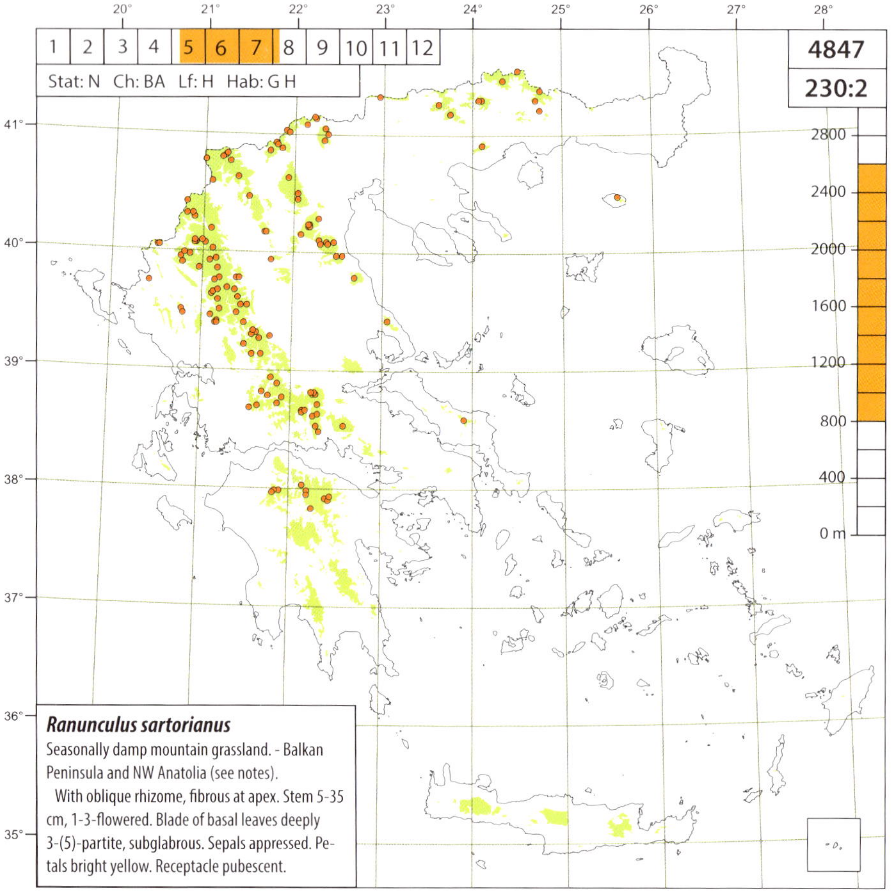

Ranunculus sartorianus

Seasonally damp mountain grassland. - Balkan Peninsula and NW Anatolia (see notes).

With oblique rhizome, fibrous at apex. Stem 5-35 cm, 1-3-flowered. Blade of basal leaves deeply 3-(5)-partite, subglabrous. Sepals appressed. Petals bright yellow. Receptacle pubescent.

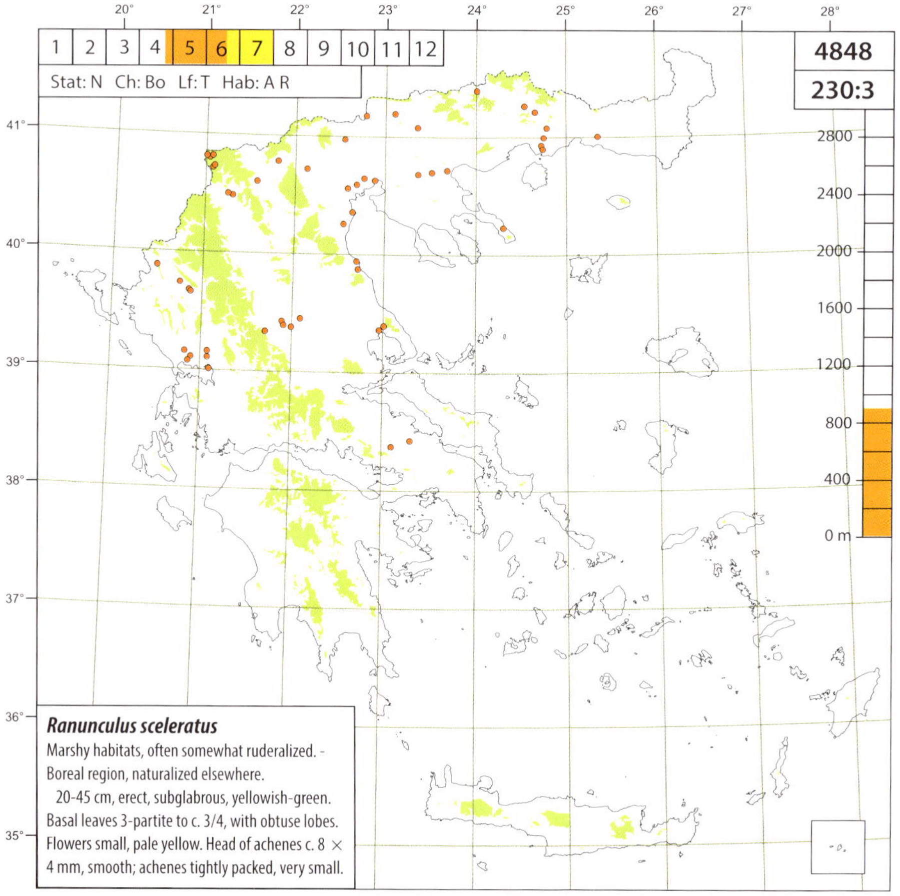

Ranunculus sceleratus

Marshy habitats, often somewhat ruderalized. - Boreal region, naturalized elsewhere.

20-45 cm, erect, subglabrous, yellowish-green. Basal leaves 3-partite to c. 3/4, with obtuse lobes. Flowers small, pale yellow. Head of achenes c. 8 × 4 mm, smooth; achenes tightly packed, very small.

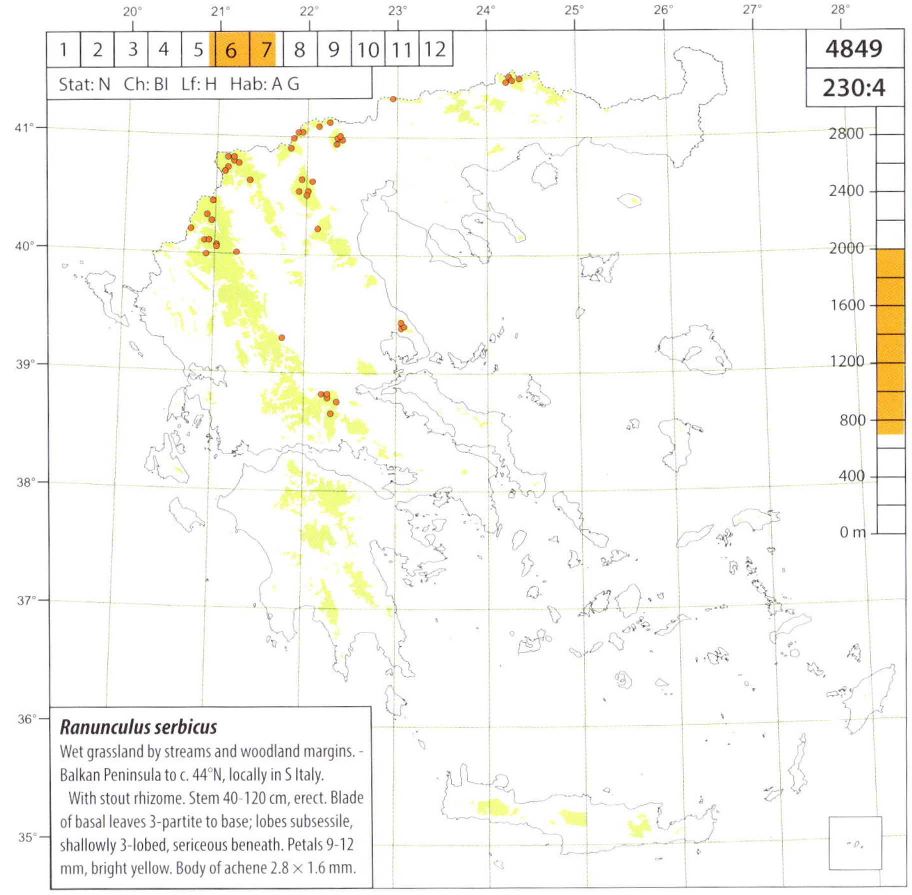

Ranunculus serbicus

Wet grassland by streams and woodland margins. - Balkan Peninsula to c. 44°N, locally in S Italy.

With stout rhizome. Stem 40-120 cm, erect. Blade of basal leaves 3-partite to base; lobes subsessile, shallowly 3-lobed, sericeous beneath. Petals 9-12 mm, bright yellow. Body of achene 2.8 × 1.6 mm.

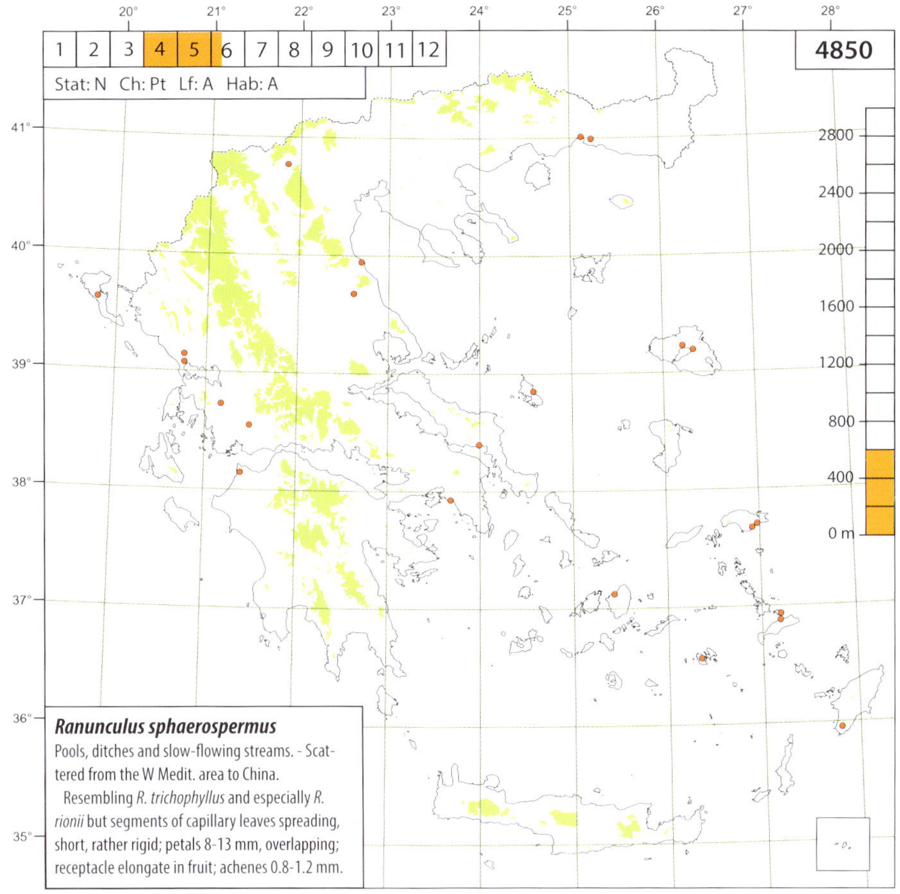

Ranunculus sphaerospermus

Pools, ditches and slow-flowing streams. - Scattered from the W Medit. area to China.

Resembling *R. trichophyllus* and especially *R. rionii* but segments of capillary leaves spreading, short, rather rigid; petals 8-13 mm, overlapping; receptacle elongate in fruit; achenes 0.8-1.2 mm.

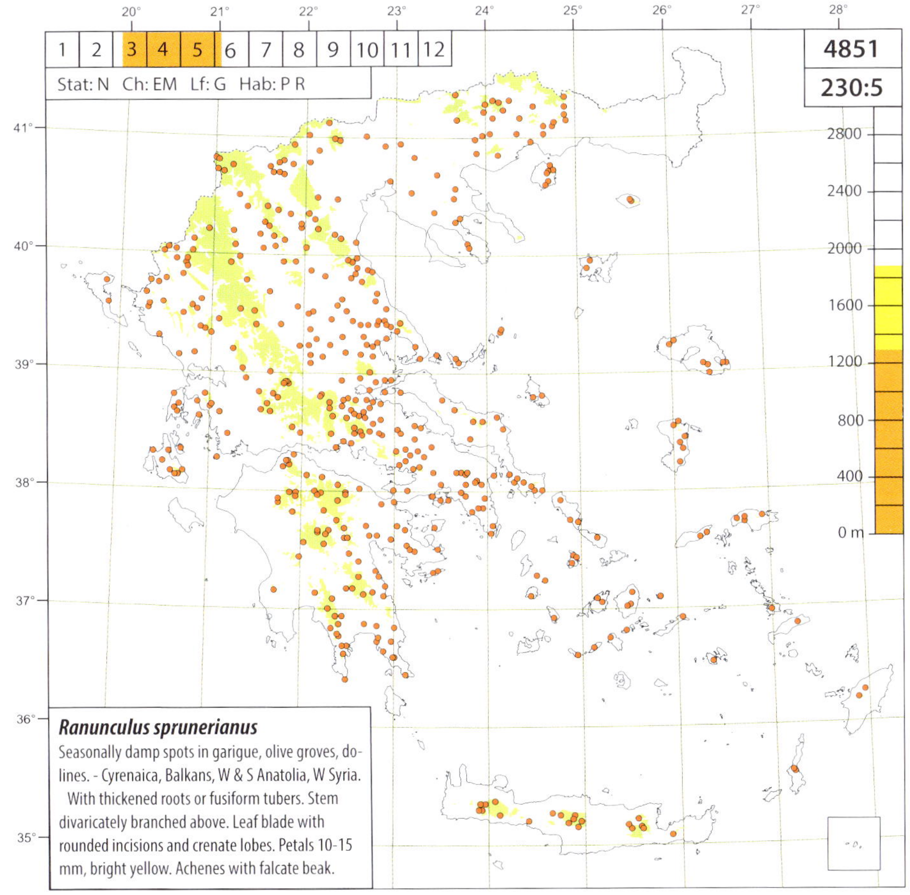

Ranunculus sprunerianus

Seasonally damp spots in garigue, olive groves, dolines. - Cyrenaica, Balkans, W & S Anatolia, W Syria.

With thickened roots or fusiform tubers. Stem divaricately branched above. Leaf blade with rounded incisions and crenate lobes. Petals 10-15 mm, bright yellow. Achenes with falcate beak.

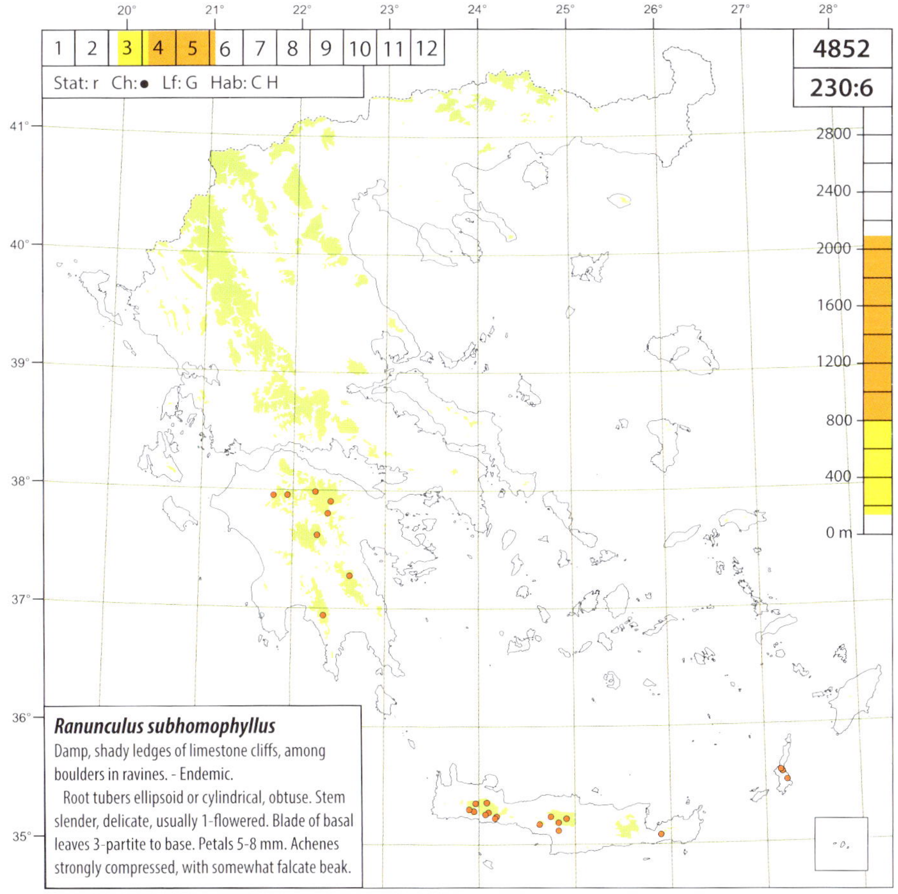

Ranunculus subhomophyllus

Damp, shady ledges of limestone cliffs, among boulders in ravines. - Endemic.

Root tubers ellipsoid or cylindrical, obtuse. Stem slender, delicate, usually 1-flowered. Blade of basal leaves 3-partite to base. Petals 5-8 mm. Achenes strongly compressed, with somewhat falcate beak.

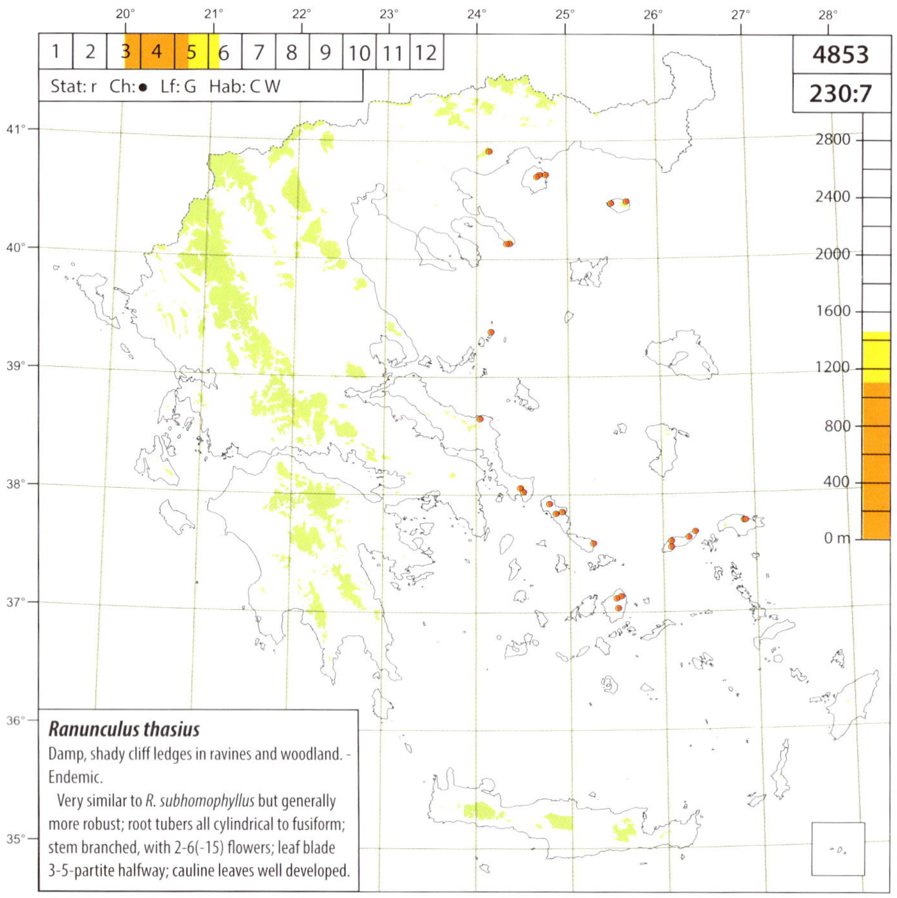

Ranunculus thasius

Damp, shady cliff ledges in ravines and woodland. - Endemic.

Very similar to *R. subhomophyllus* but generally more robust; root tubers all cylindrical to fusiform; stem branched, with 2-6(-15) flowers; leaf blade 3-5-partite halfway; cauline leaves well developed.

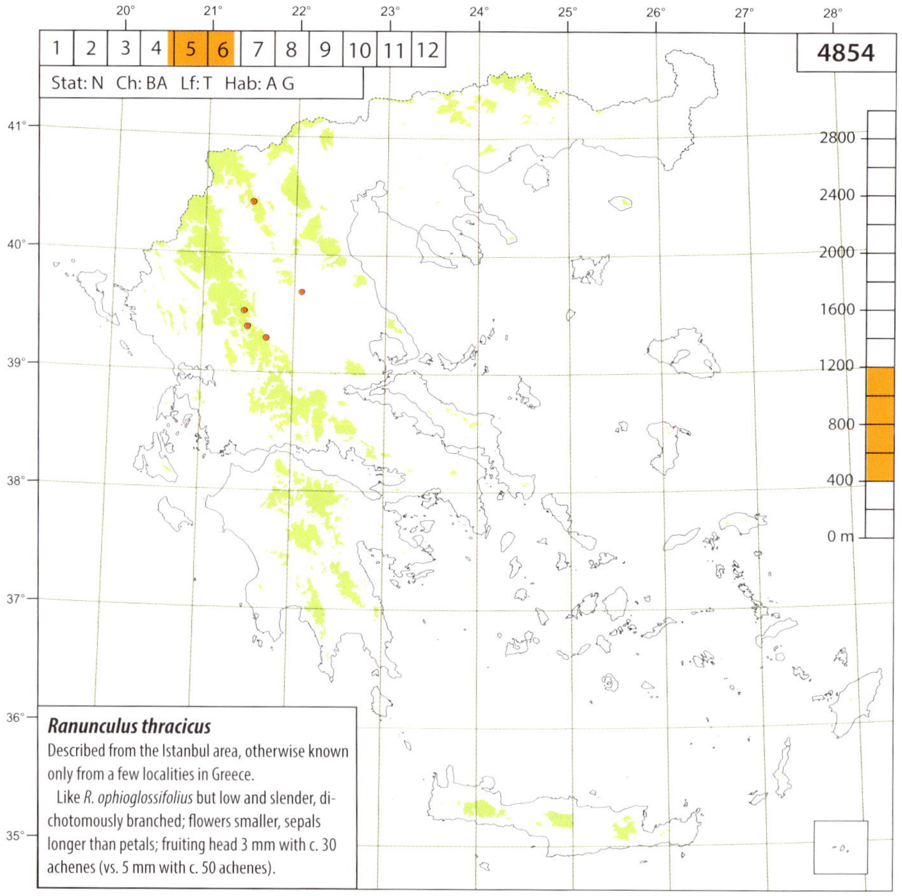

Ranunculus thracicus

Described from the Istanbul area, otherwise known only from a few localities in Greece.

Like *R. ophioglossifolius* but low and slender, dichotomously branched; flowers smaller, sepals longer than petals; fruiting head 3 mm with c. 30 achenes (vs. 5 mm with c. 50 achenes).

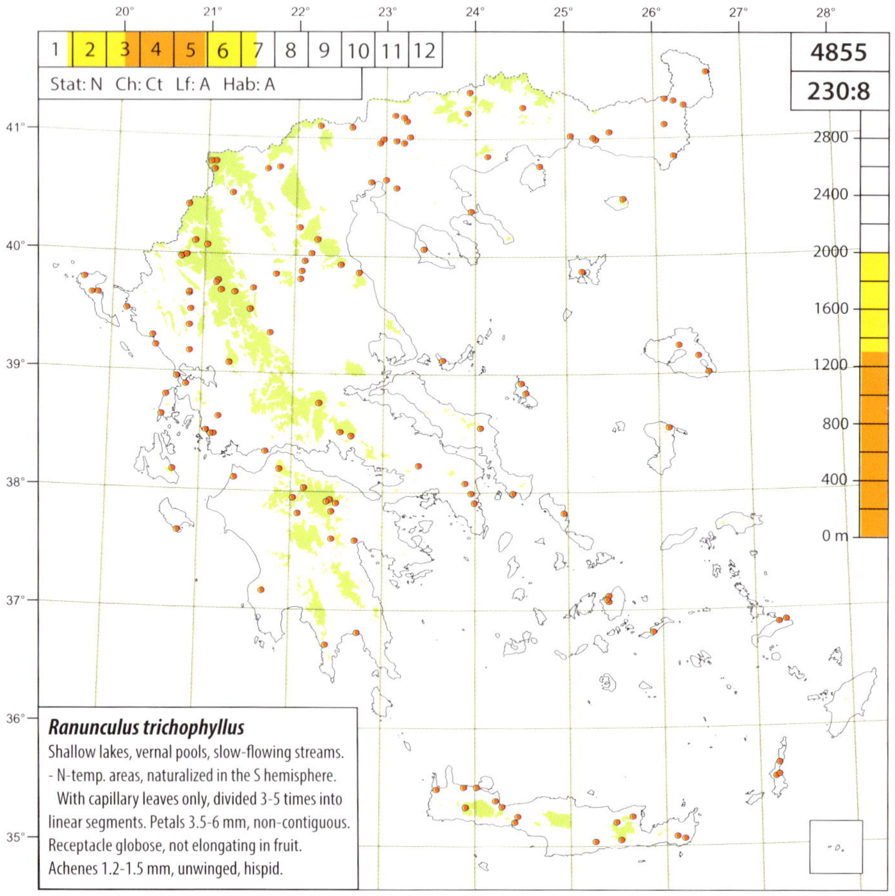

Ranunculus trichophyllus

Shallow lakes, vernal pools, slow-flowing streams. - N-temp. areas, naturalized in the S hemisphere.

With capillary leaves only, divided 3-5 times into linear segments. Petals 3.5-6 mm, non-contiguous. Receptacle globose, not elongating in fruit. Achenes 1.2-1.5 mm, unwinged, hispid.

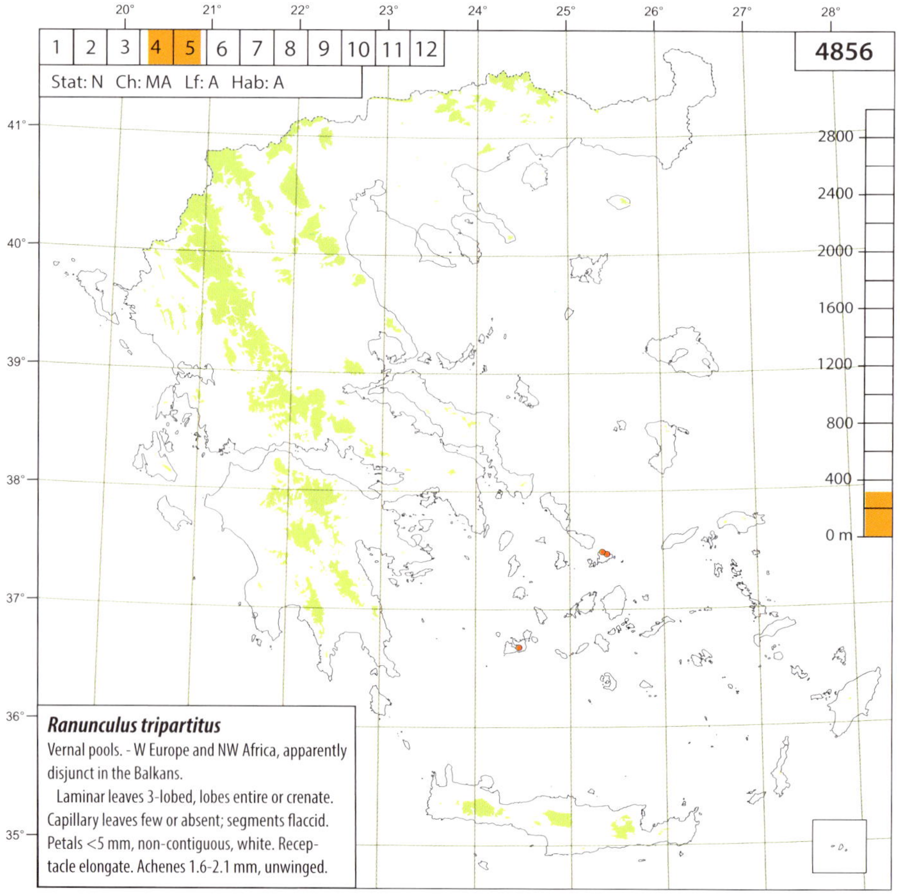

Ranunculus tripartitus

Vernal pools. - W Europe and NW Africa, apparently disjunct in the Balkans.

Laminar leaves 3-lobed, lobes entire or crenate. Capillary leaves few or absent; segments flaccid. Petals <5 mm, non-contiguous, white. Receptacle elongate. Achenes 1.6-2.1 mm, unwinged.

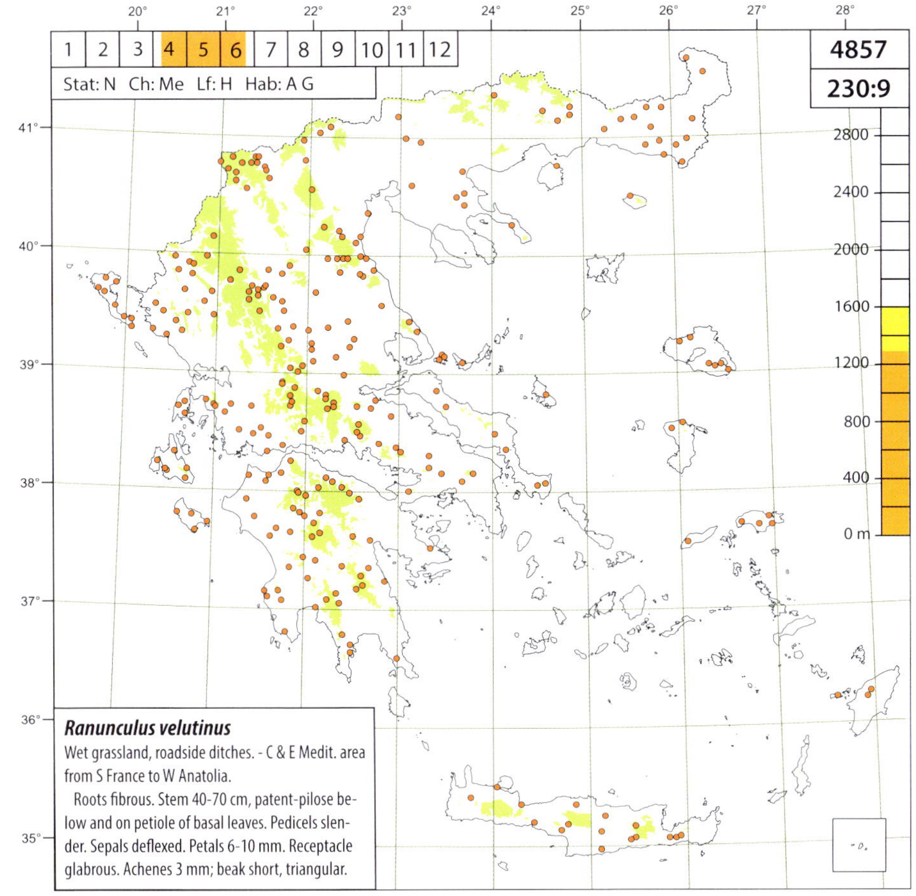

Ranunculus velutinus

Wet grassland, roadside ditches. - C & E Medit. area from S France to W Anatolia.

Roots fibrous. Stem 40-70 cm, patent-pilose below and on petiole of basal leaves. Pedicels slender. Sepals deflexed. Petals 6-10 mm. Receptacle glabrous. Achenes 3 mm; beak short, triangular.

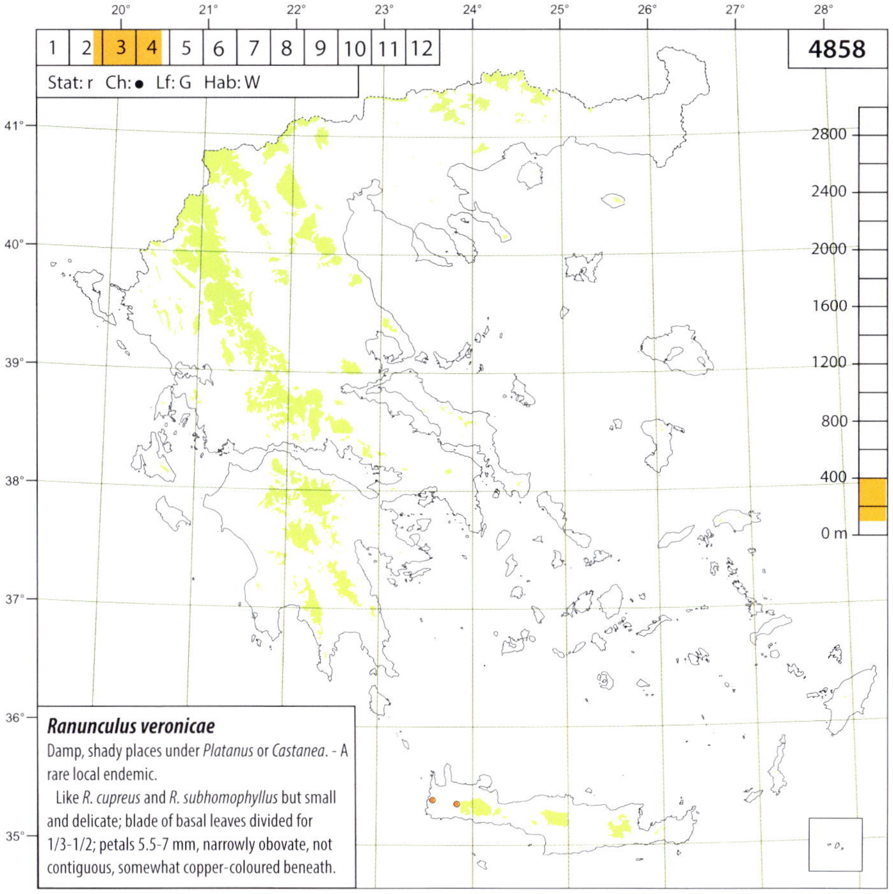

Ranunculus veronicae

Damp, shady places under *Platanus* or *Castanea*. - A rare local endemic.

Like *R. cupreus* and *R. subhomophyllus* but small and delicate; blade of basal leaves divided for 1/3-1/2; petals 5.5-7 mm, narrowly obovate, not contiguous, somewhat copper-coloured beneath.

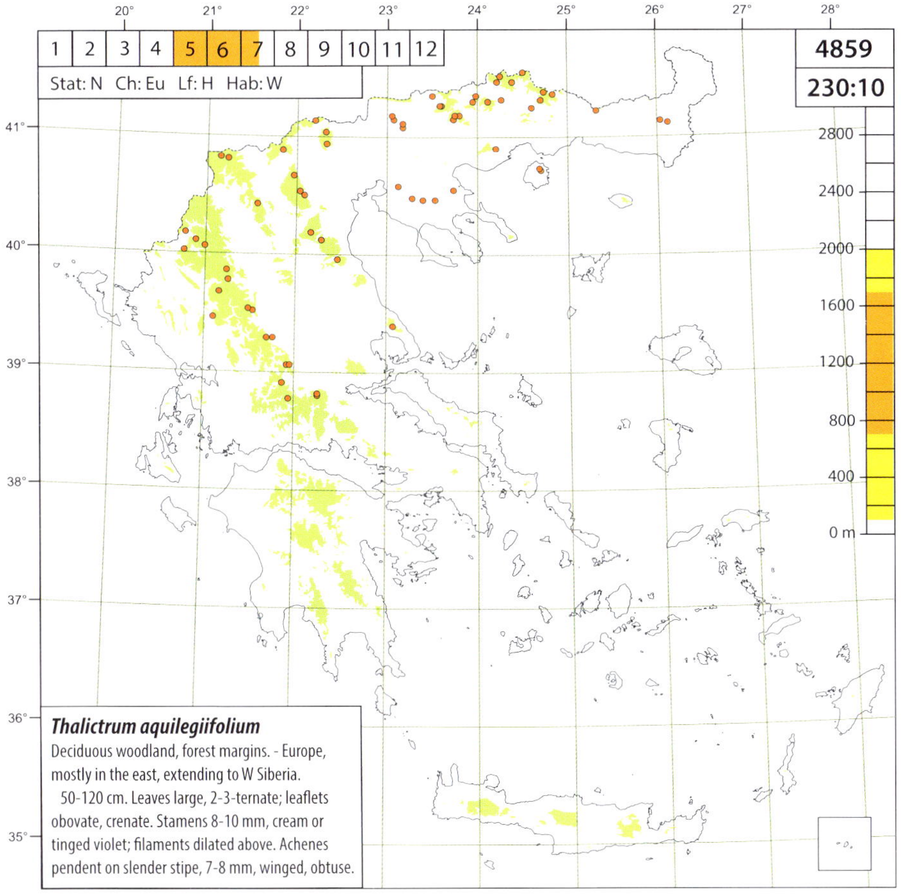

Thalictrum aquilegiifolium

Deciduous woodland, forest margins. - Europe, mostly in the east, extending to W Siberia.

50-120 cm. Leaves large, 2-3-ternate; leaflets obovate, crenate. Stamens 8-10 mm, cream or tinged violet; filaments dilated above. Achenes pendent on slender stipe, 7-8 mm, winged, obtuse.

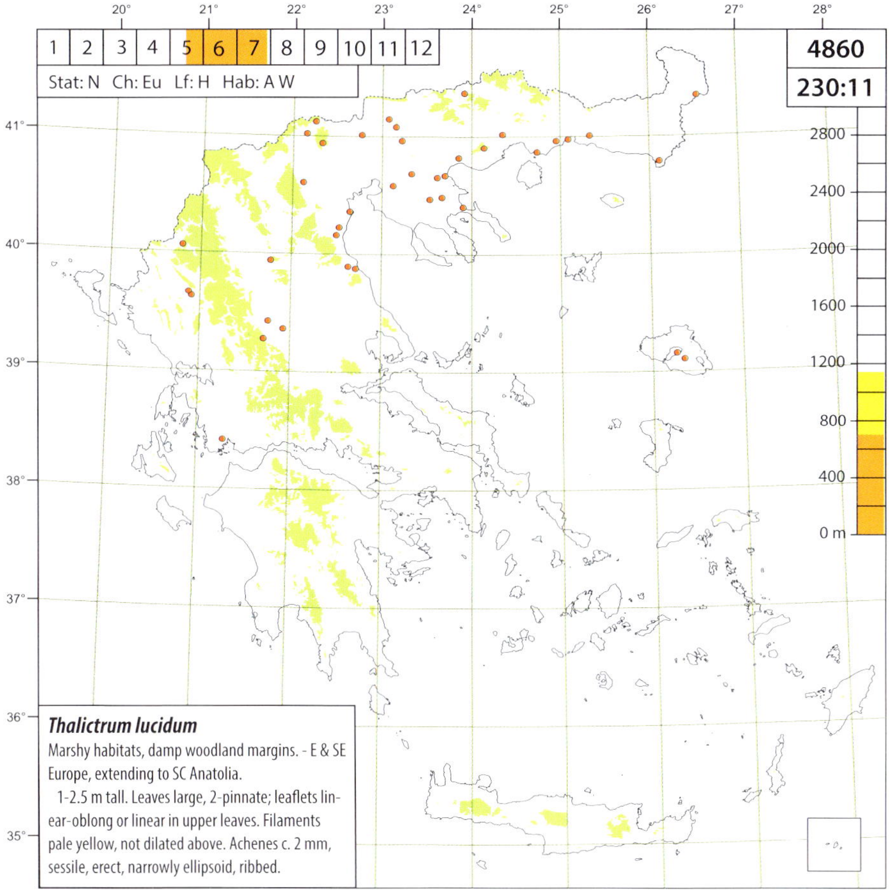

Thalictrum lucidum

Marshy habitats, damp woodland margins. - E & SE Europe, extending to SC Anatolia.

1-2.5 m tall. Leaves large, 2-pinnate; leaflets linear-oblong or linear in upper leaves. Filaments pale yellow, not dilated above. Achenes c. 2 mm, sessile, erect, narrowly ellipsoid, ribbed.

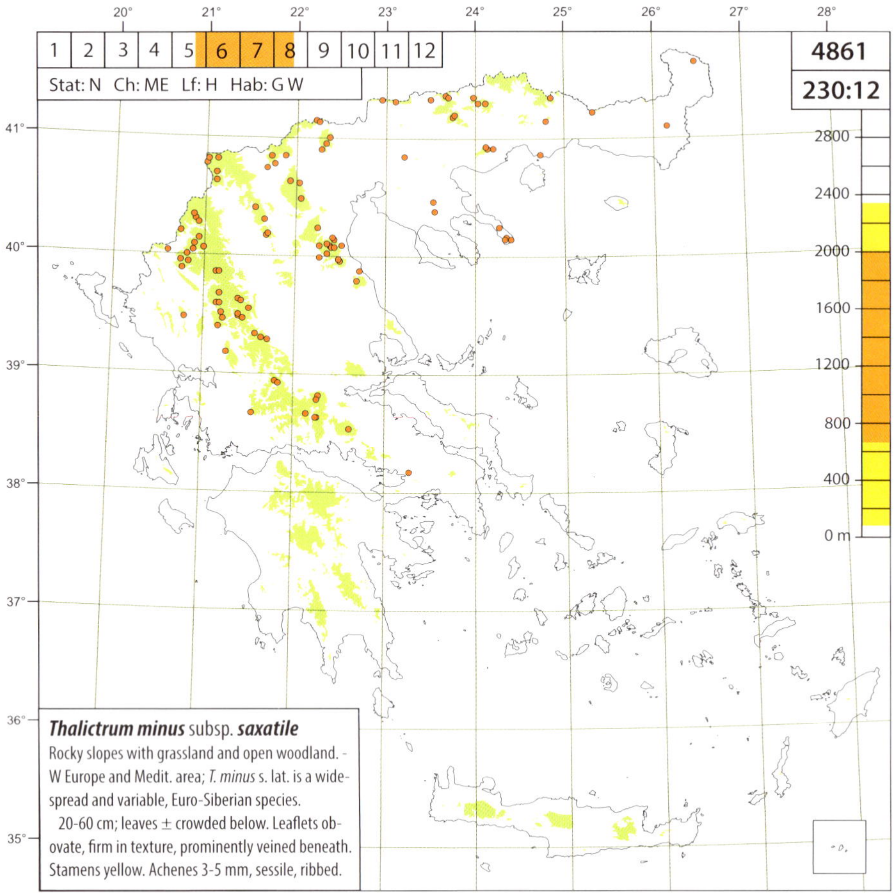

Thalictrum minus **subsp. *saxatile***
Rocky slopes with grassland and open woodland. - W Europe and Medit. area; *T. minus* s. lat. is a widespread and variable, Euro-Siberian species.
20-60 cm; leaves ± crowded below. Leaflets obovate, firm in texture, prominently veined beneath. Stamens yellow. Achenes 3-5 mm, sessile, ribbed.

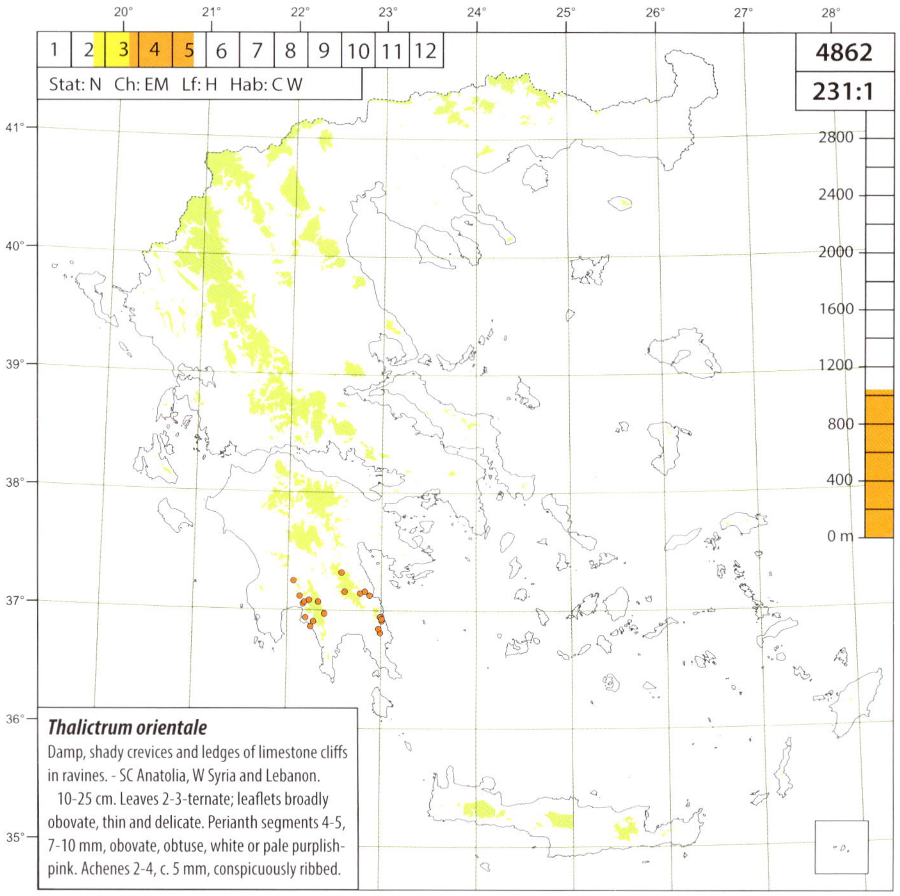

Thalictrum orientale
Damp, shady crevices and ledges of limestone cliffs in ravines. - SC Anatolia, W Syria and Lebanon.
10-25 cm. Leaves 2-3-ternate; leaflets broadly obovate, thin and delicate. Perianth segments 4-5, 7-10 mm, obovate, obtuse, white or pale purplish-pink. Achenes 2-4, c. 5 mm, conspicuously ribbed.

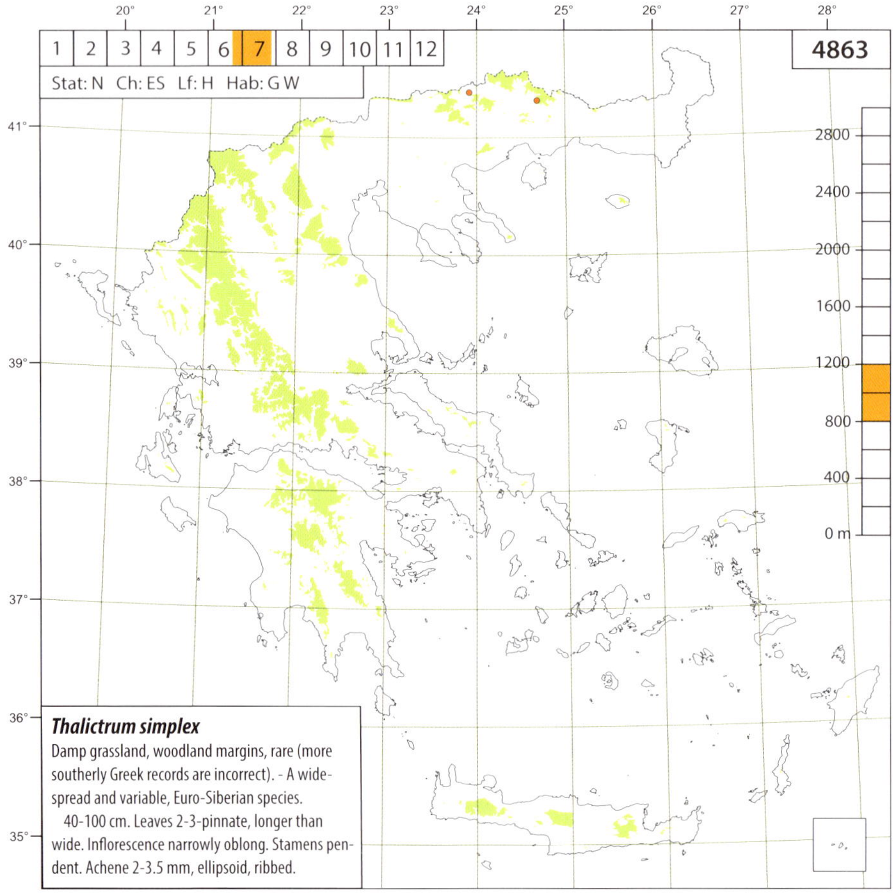

Thalictrum simplex
Damp grassland, woodland margins, rare (more southerly Greek records are incorrect). - A widespread and variable, Euro-Siberian species.
40-100 cm. Leaves 2-3-pinnate, longer than wide. Inflorescence narrowly oblong. Stamens pendent. Achene 2-3.5 mm, ellipsoid, ribbed.

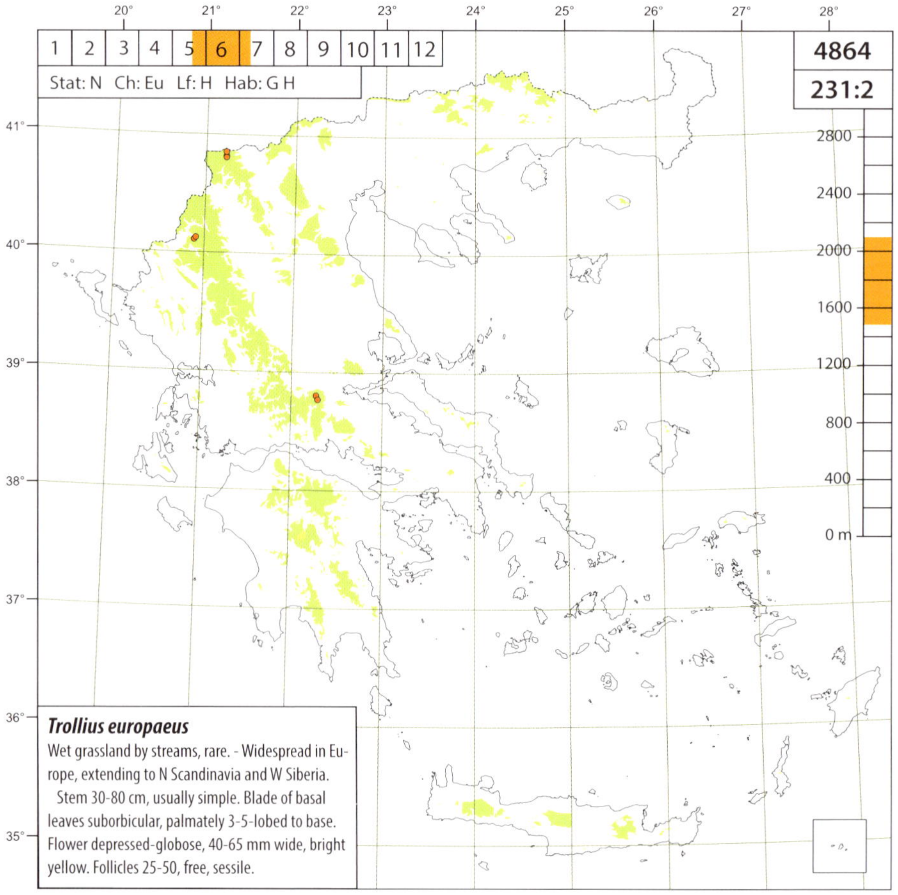

Trollius europaeus
Wet grassland by streams, rare. - Widespread in Europe, extending to N Scandinavia and W Siberia.
Stem 30-80 cm, usually simple. Blade of basal leaves suborbicular, palmately 3-5-lobed to base. Flower depressed-globose, 40-65 mm wide, bright yellow. Follicles 25-50, free, sessile.

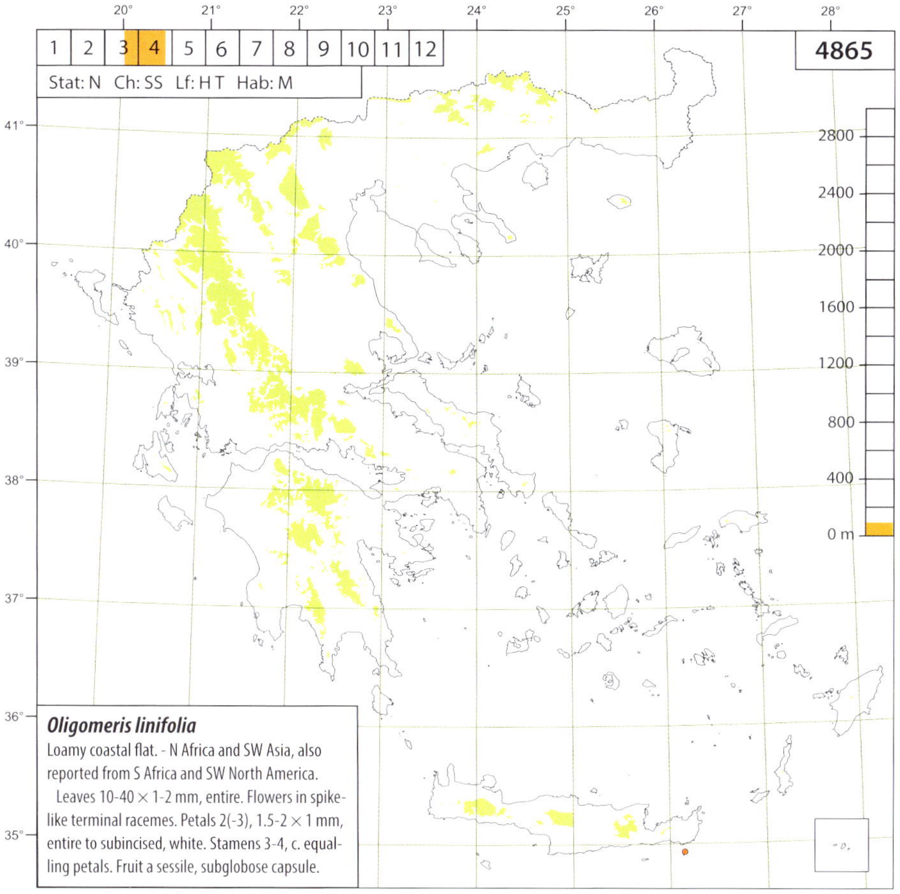

Oligomeris linifolia
Loamy coastal flat. - N Africa and SW Asia, also reported from S Africa and SW North America.
Leaves 10-40 × 1-2 mm, entire. Flowers in spike-like terminal racemes. Petals 2(-3), 1.5-2 × 1 mm, entire to subincised, white. Stamens 3-4, c. equalling petals. Fruit a sessile, subglobose capsule.

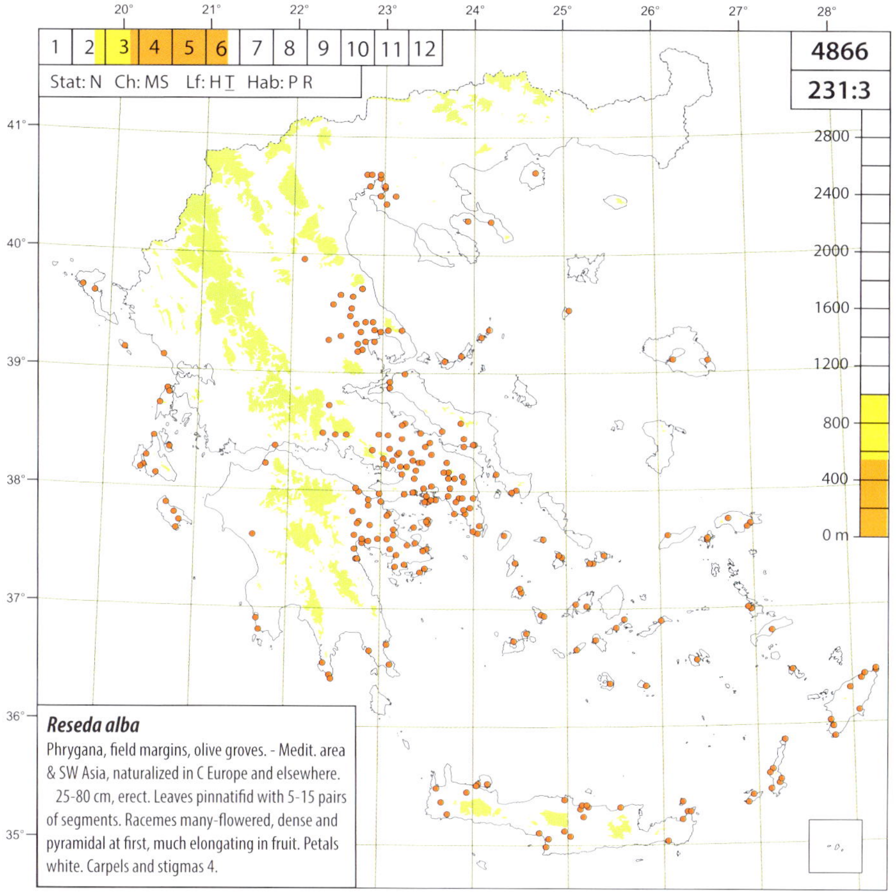

Reseda alba
Phrygana, field margins, olive groves. - Medit. area & SW Asia, naturalized in C Europe and elsewhere.
25-80 cm, erect. Leaves pinnatifid with 5-15 pairs of segments. Racemes many-flowered, dense and pyramidal at first, much elongating in fruit. Petals white. Carpels and stigmas 4.

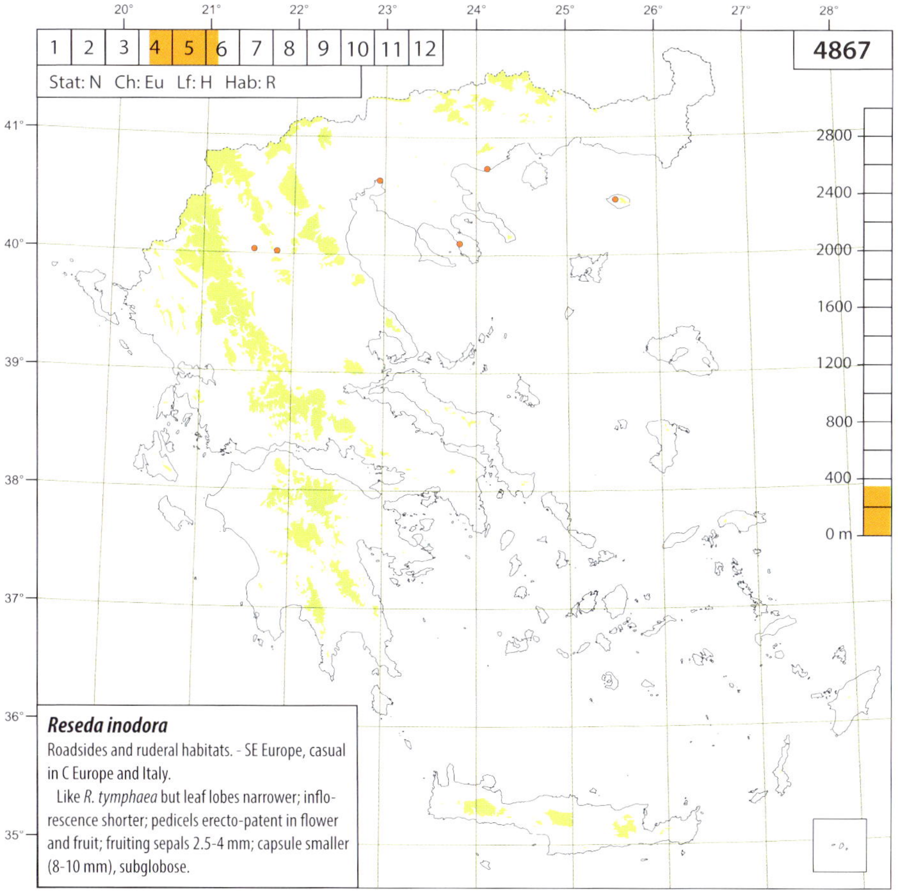

Reseda inodora
Roadsides and ruderal habitats. - SE Europe, casual in C Europe and Italy.
Like *R. tymphaea* but leaf lobes narrower; inflorescence shorter; pedicels erecto-patent in flower and fruit; fruiting sepals 2.5-4 mm; capsule smaller (8-10 mm), subglobose.

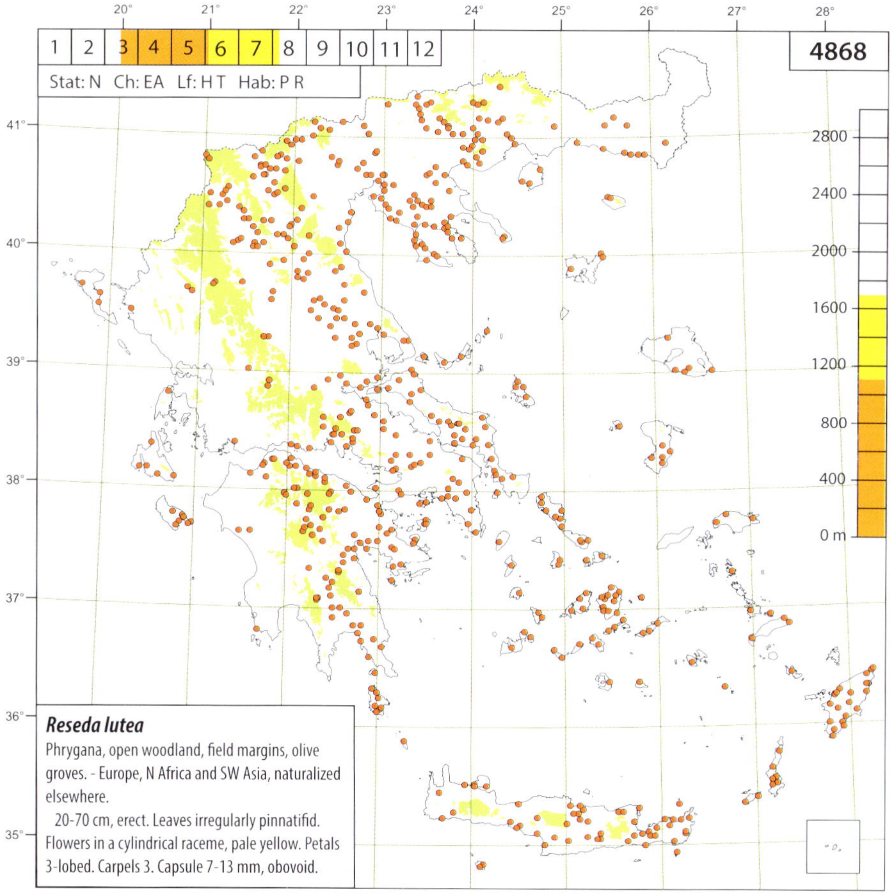

Reseda lutea
Phrygana, open woodland, field margins, olive groves. - Europe, N Africa and SW Asia, naturalized elsewhere.
20-70 cm, erect. Leaves irregularly pinnatifid. Flowers in a cylindrical raceme, pale yellow. Petals 3-lobed. Carpels 3. Capsule 7-13 mm, obovoid.

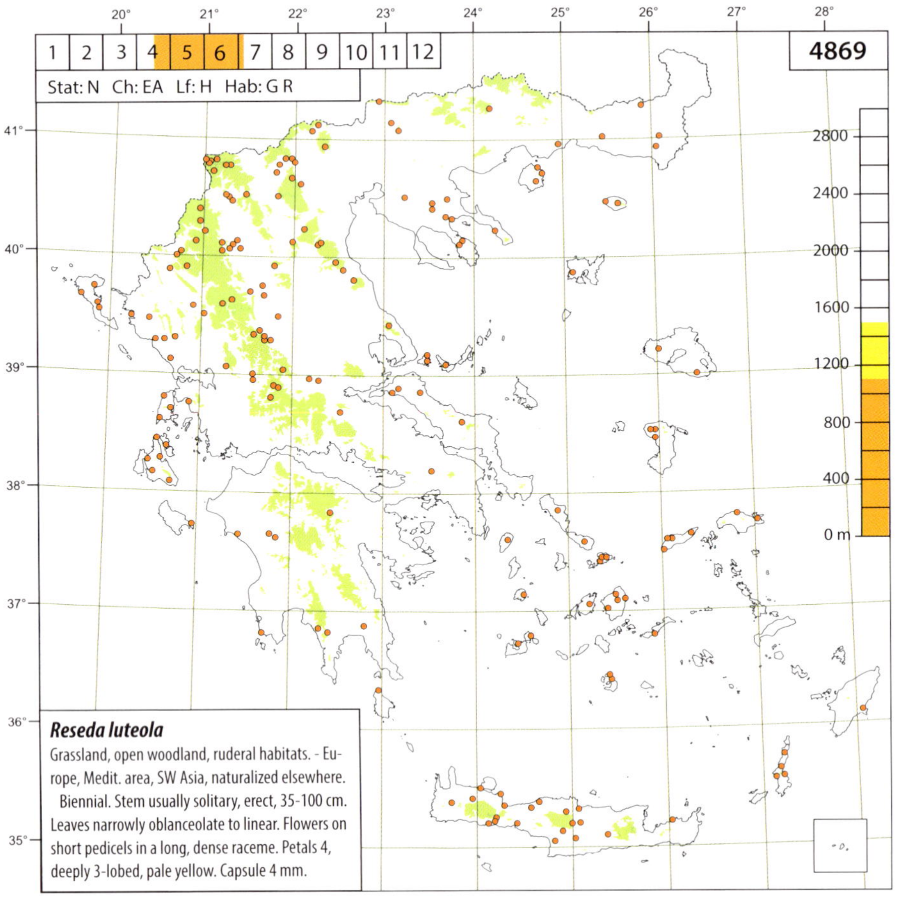

Reseda luteola
Grassland, open woodland, ruderal habitats. - Europe, Medit. area, SW Asia, naturalized elsewhere.
Biennial. Stem usually solitary, erect, 35-100 cm. Leaves narrowly oblanceolate to linear. Flowers on short pedicels in a long, dense raceme. Petals 4, deeply 3-lobed, pale yellow. Capsule 4 mm.

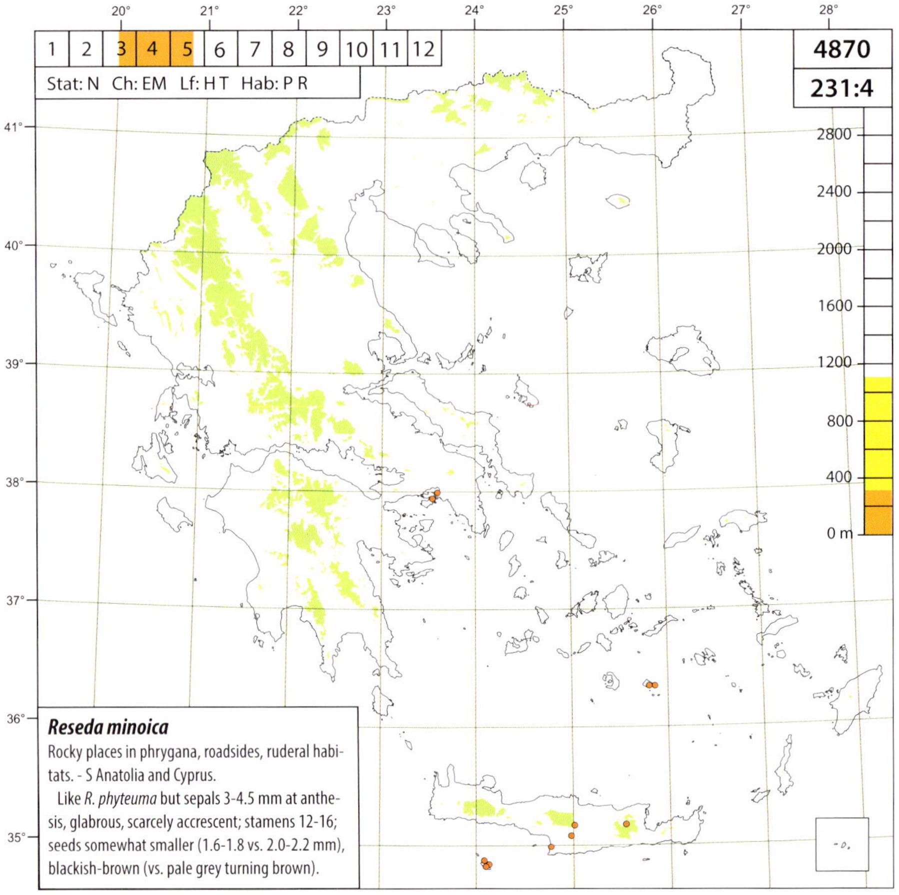

Reseda minoica
Rocky places in phrygana, roadsides, ruderal habitats. - S Anatolia and Cyprus.
Like *R. phyteuma* but sepals 3-4.5 mm at anthesis, glabrous, scarcely accrescent; stamens 12-16; seeds somewhat smaller (1.6-1.8 vs. 2.0-2.2 mm), blackish-brown (vs. pale grey turning brown).

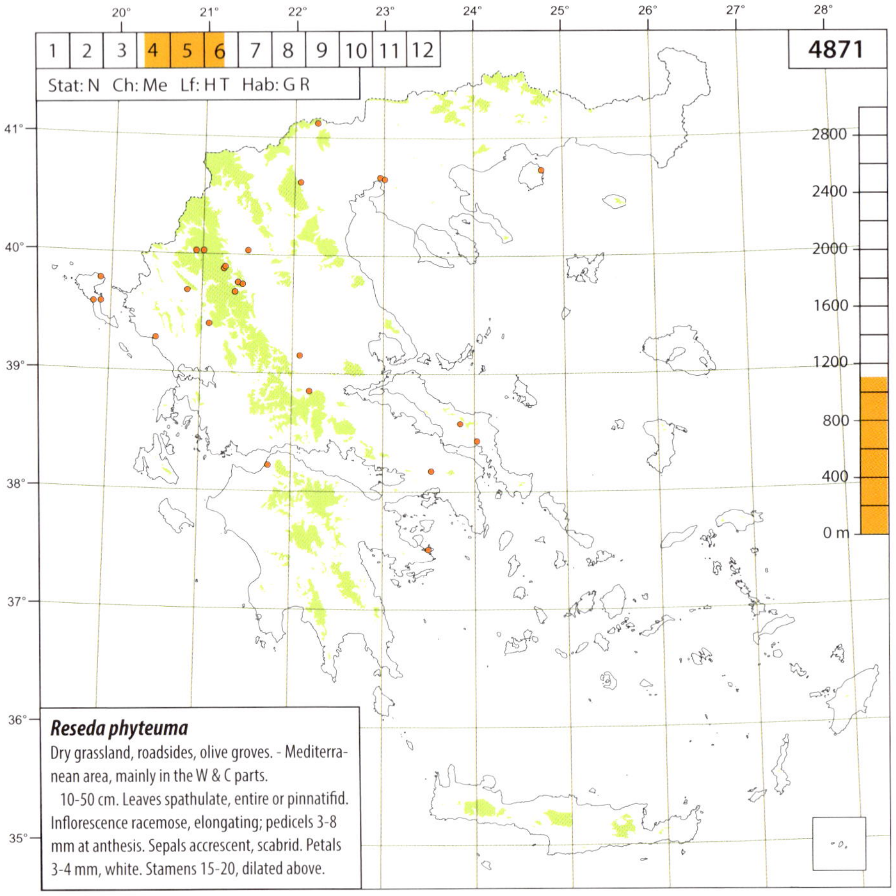

Reseda phyteuma
Dry grassland, roadsides, olive groves. - Mediterranean area, mainly in the W & C parts.
10-50 cm. Leaves spathulate, entire or pinnatifid. Inflorescence racemose, elongating; pedicels 3-8 mm at anthesis. Sepals accrescent, scabrid. Petals 3-4 mm, white. Stamens 15-20, dilated above.

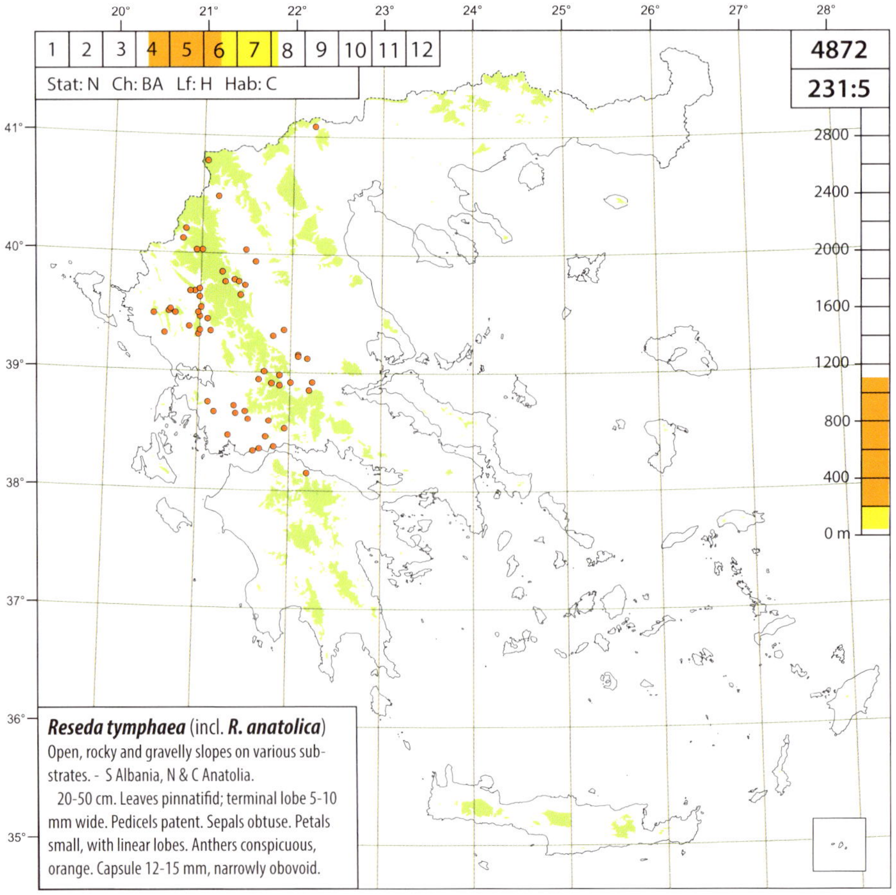

Reseda tymphaea (incl. ***R. anatolica***)
Open, rocky and gravelly slopes on various substrates. - S Albania, N & C Anatolia.
20-50 cm. Leaves pinnatifid; terminal lobe 5-10 mm wide. Pedicels patent. Sepals obtuse. Petals small, with linear lobes. Anthers conspicuous, orange. Capsule 12-15 mm, narrowly obovoid.

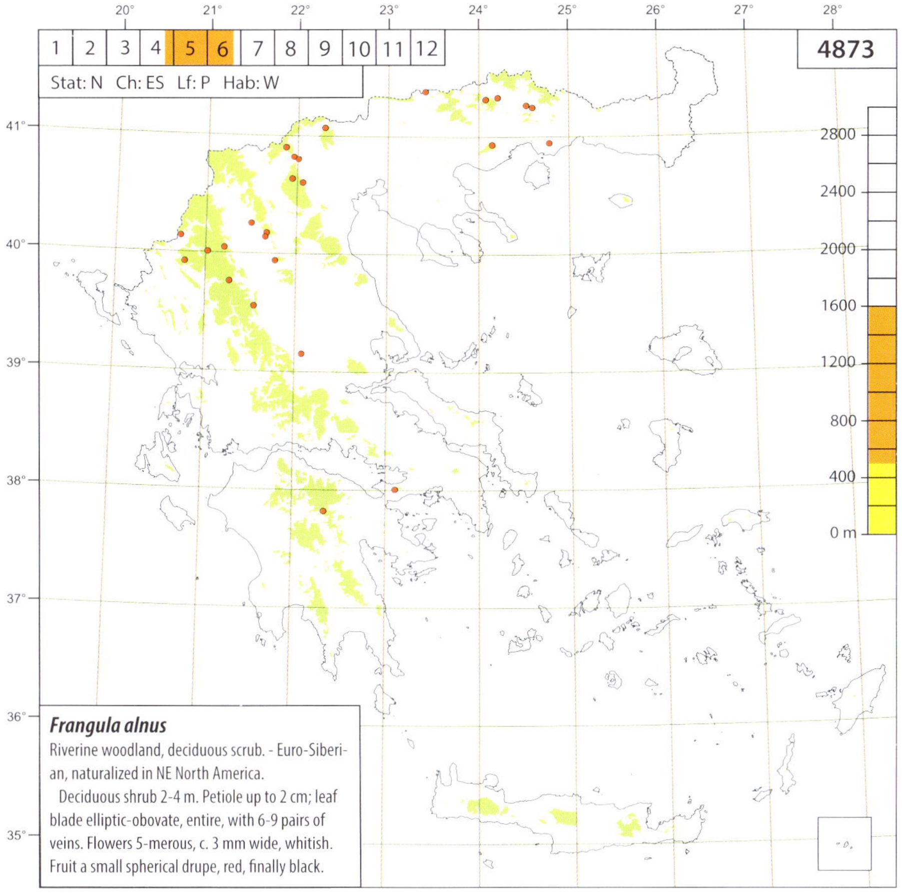

Frangula alnus
Riverine woodland, deciduous scrub. - Euro-Siberian, naturalized in NE North America.
Deciduous shrub 2-4 m. Petiole up to 2 cm; leaf blade elliptic-obovate, entire, with 6-9 pairs of veins. Flowers 5-merous, c. 3 mm wide, whitish. Fruit a small spherical drupe, red, finally black.

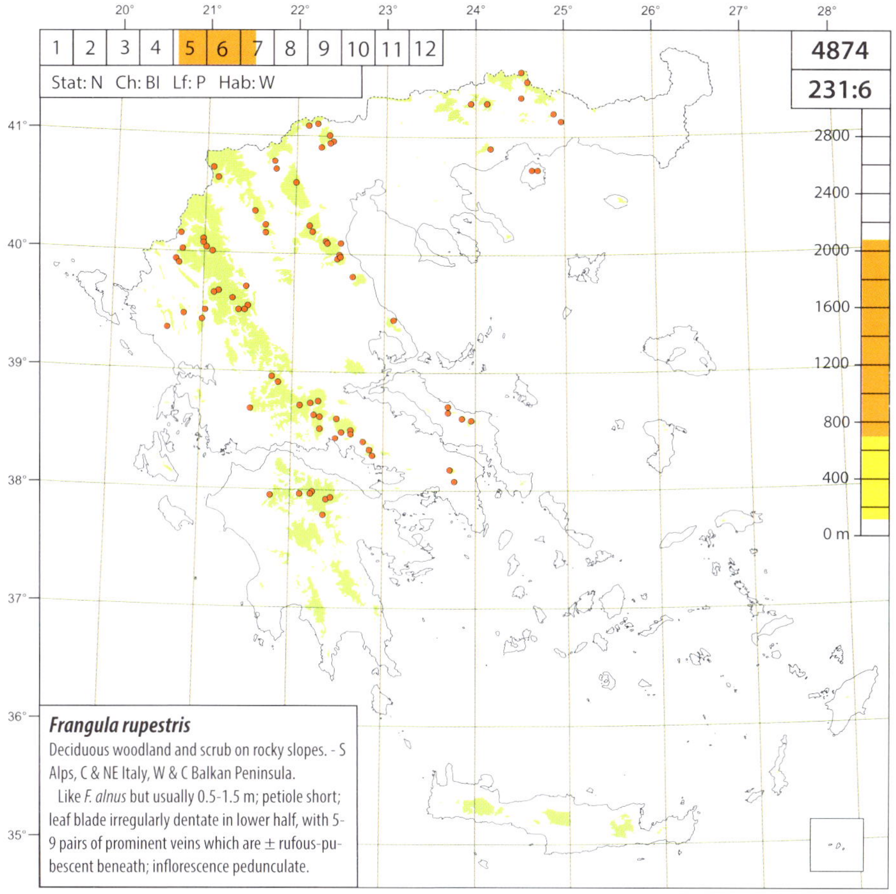

Frangula rupestris
Deciduous woodland and scrub on rocky slopes. - S Alps, C & NE Italy, W & C Balkan Peninsula.
Like *F. alnus* but usually 0.5-1.5 m; petiole short; leaf blade irregularly dentate in lower half, with 5-9 pairs of prominent veins which are ± rufous-pubescent beneath; inflorescence pedunculate.

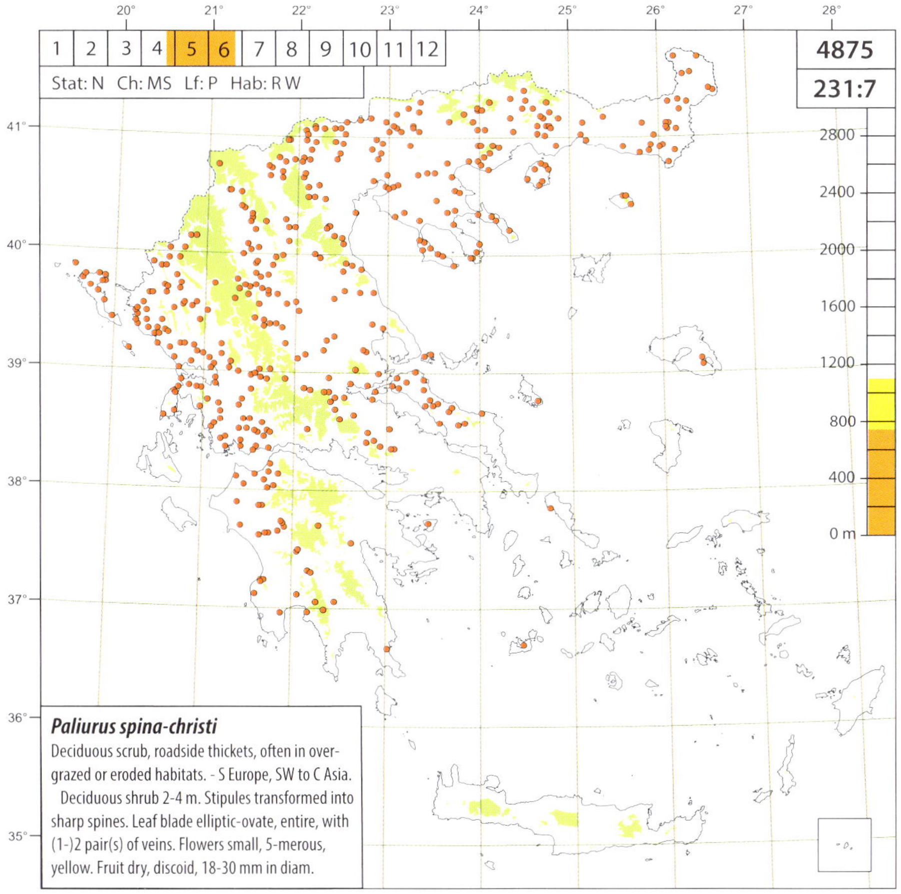

Paliurus spina-christi
Deciduous scrub, roadside thickets, often in over-grazed or eroded habitats. - S Europe, SW to C Asia.
Deciduous shrub 2-4 m. Stipules transformed into sharp spines. Leaf blade elliptic-ovate, entire, with (1-)2 pair(s) of veins. Flowers small, 5-merous, yellow. Fruit dry, discoid, 18-30 mm in diam.

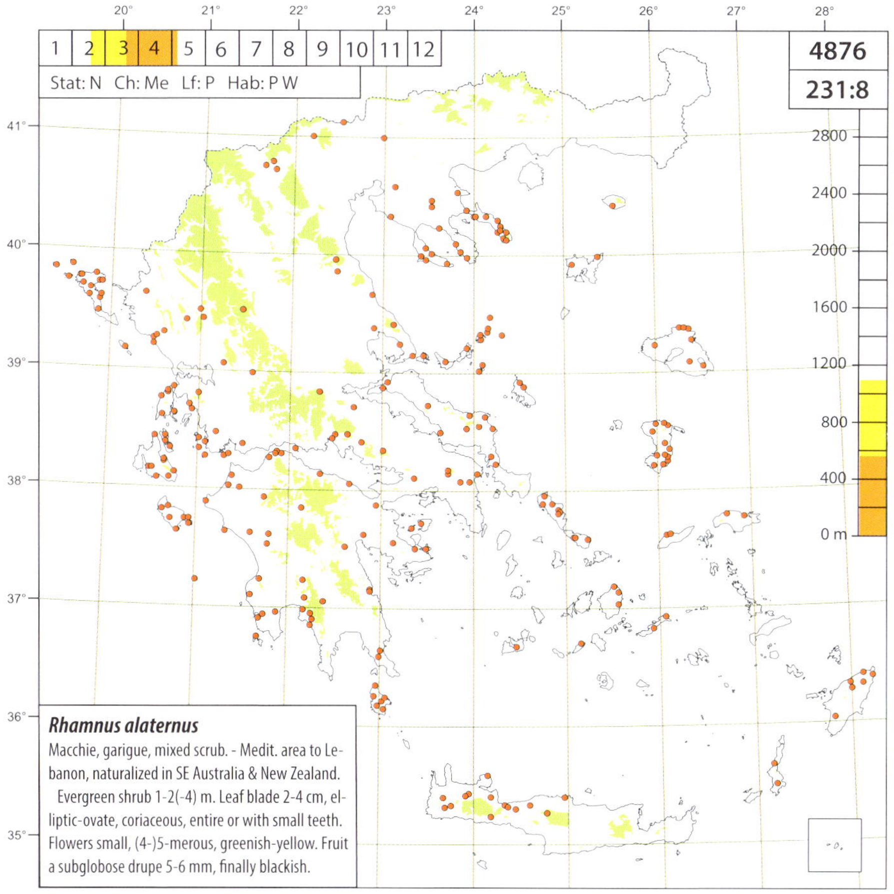

Rhamnus alaternus
Macchie, garigue, mixed scrub. - Medit. area to Lebanon, naturalized in SE Australia & New Zealand.
Evergreen shrub 1-2(-4) m. Leaf blade 2-4 cm, elliptic-ovate, coriaceous, entire or with small teeth. Flowers small, (4-)5-merous, greenish-yellow. Fruit a subglobose drupe 5-6 mm, finally blackish.

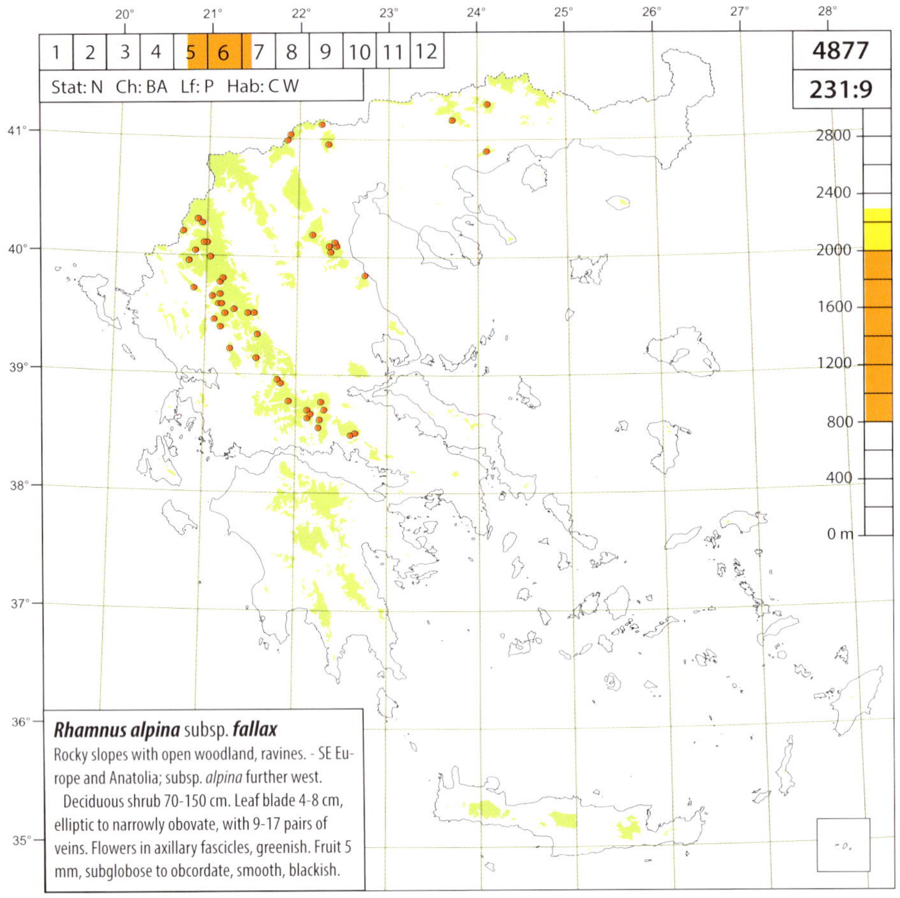

Rhamnus alpina* subsp. *fallax
Rocky slopes with open woodland, ravines. - SE Europe and Anatolia; subsp. *alpina* further west.
Deciduous shrub 70-150 cm. Leaf blade 4-8 cm, elliptic to narrowly obovate, with 9-17 pairs of veins. Flowers in axillary fascicles, greenish. Fruit 5 mm, subglobose to obcordate, smooth, blackish.

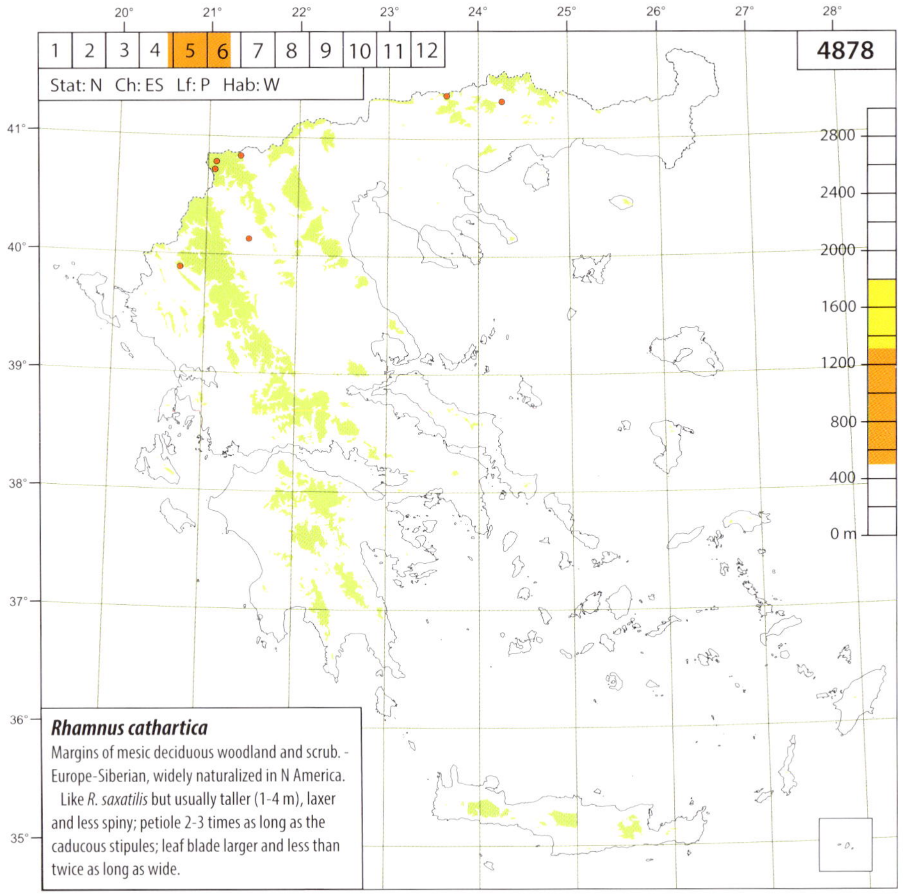

Rhamnus cathartica
Margins of mesic deciduous woodland and scrub. - Europe-Siberian, widely naturalized in N America.
Like *R. saxatilis* but usually taller (1-4 m), laxer and less spiny; petiole 2-3 times as long as the caducous stipules; leaf blade larger and less than twice as long as wide.

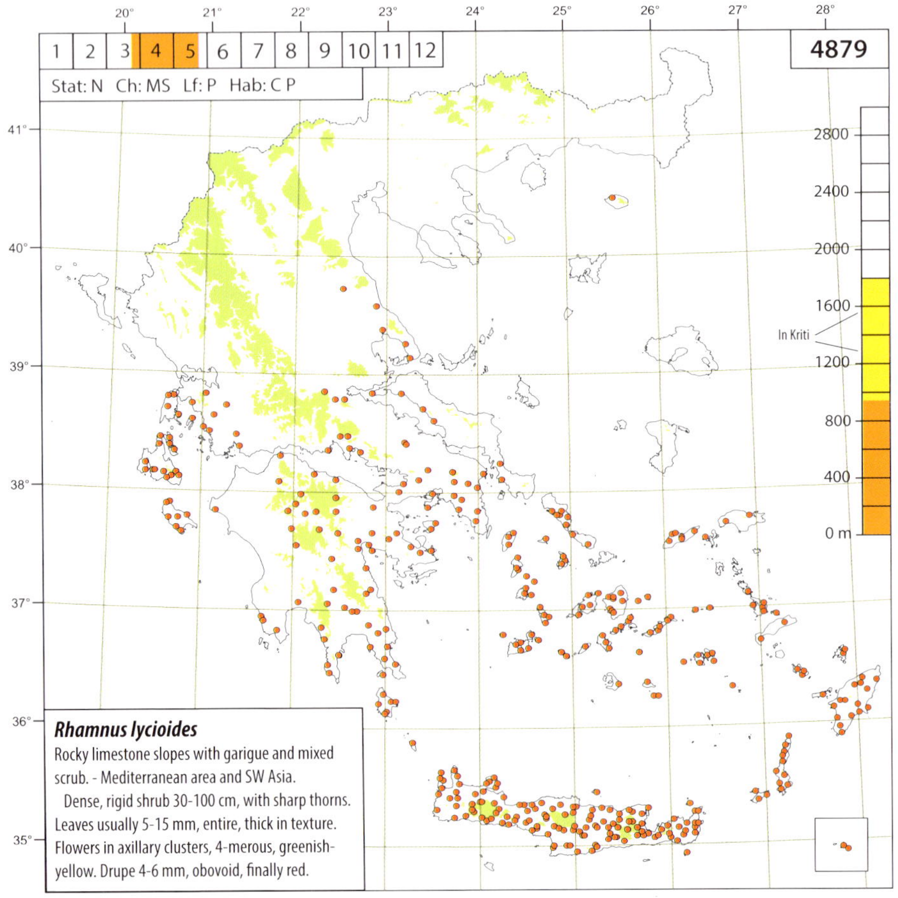

Rhamnus lycioides
Rocky limestone slopes with garigue and mixed scrub. - Mediterranean area and SW Asia.
Dense, rigid shrub 30-100 cm, with sharp thorns. Leaves usually 5-15 mm, entire, thick in texture. Flowers in axillary clusters, 4-merous, greenish-yellow. Drupe 4-6 mm, obovoid, finally red.

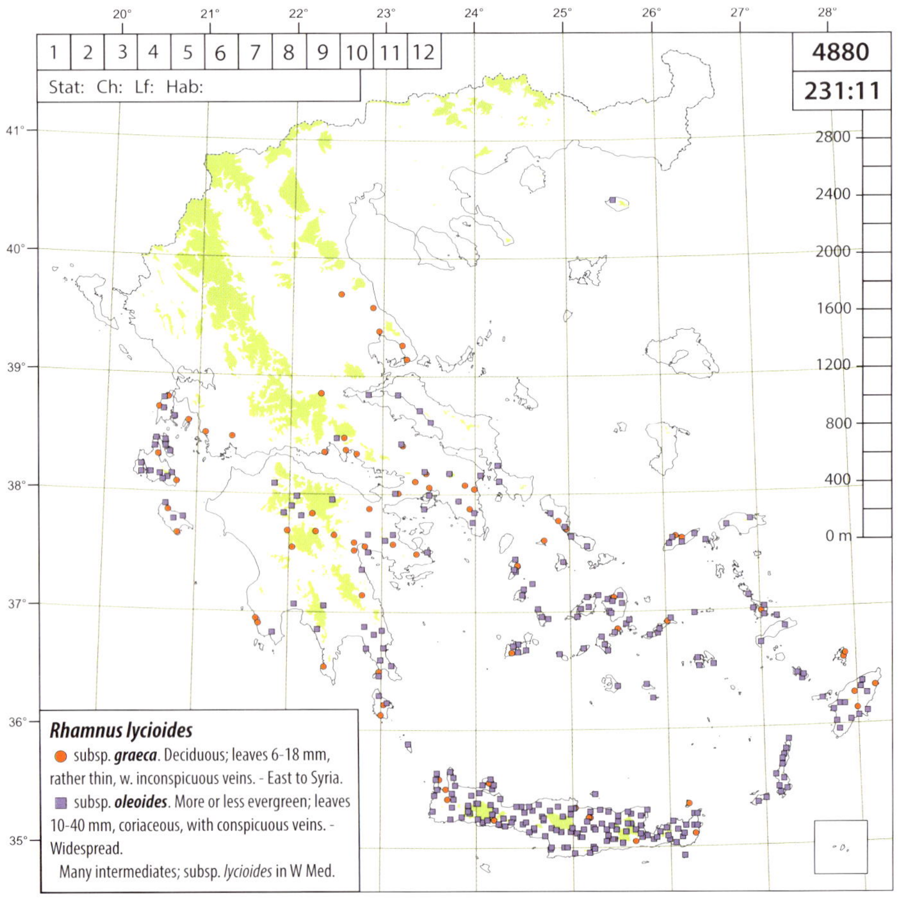

Rhamnus lycioides
● subsp. ***graeca***. Deciduous; leaves 6-18 mm, rather thin, w. inconspicuous veins. - East to Syria.
■ subsp. ***oleoides***. More or less evergreen; leaves 10-40 mm, coriaceous, with conspicuous veins. - Widespread.
Many intermediates; subsp. *lycioides* in W Med.

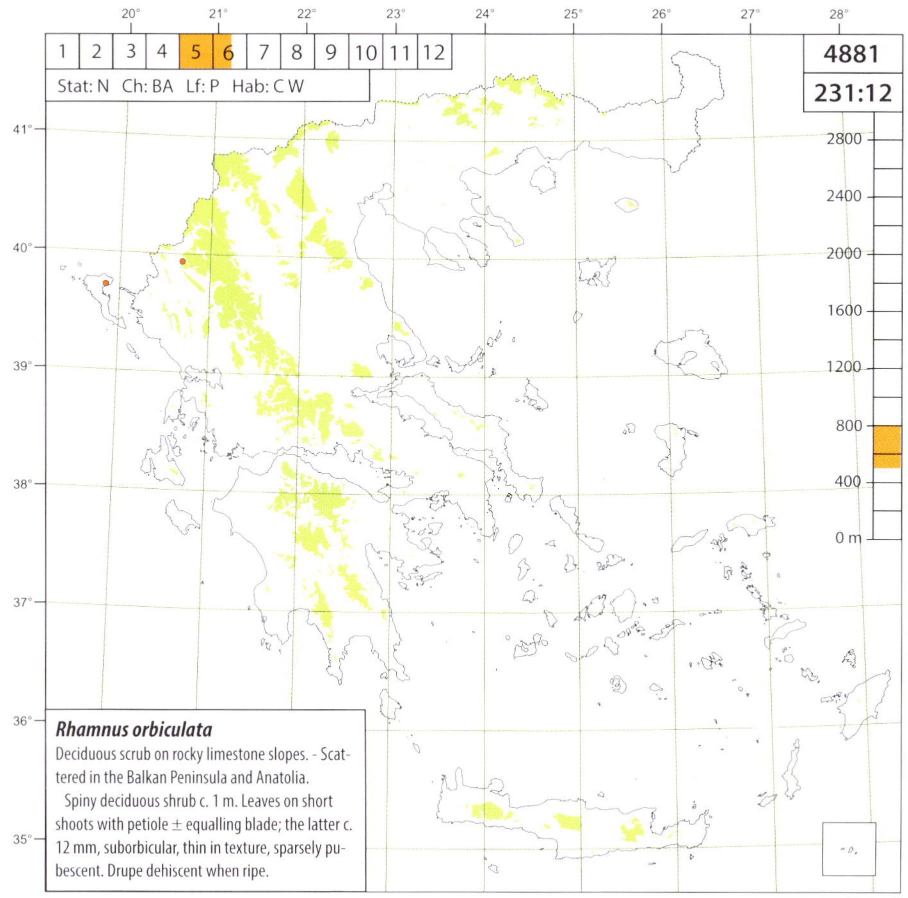

Rhamnus orbiculata
Deciduous scrub on rocky limestone slopes. - Scattered in the Balkan Peninsula and Anatolia.
Spiny deciduous shrub c. 1 m. Leaves on short shoots with petiole ± equalling blade; the latter c. 12 mm, suborbicular, thin in texture, sparsely pubescent. Drupe dehiscent when ripe.

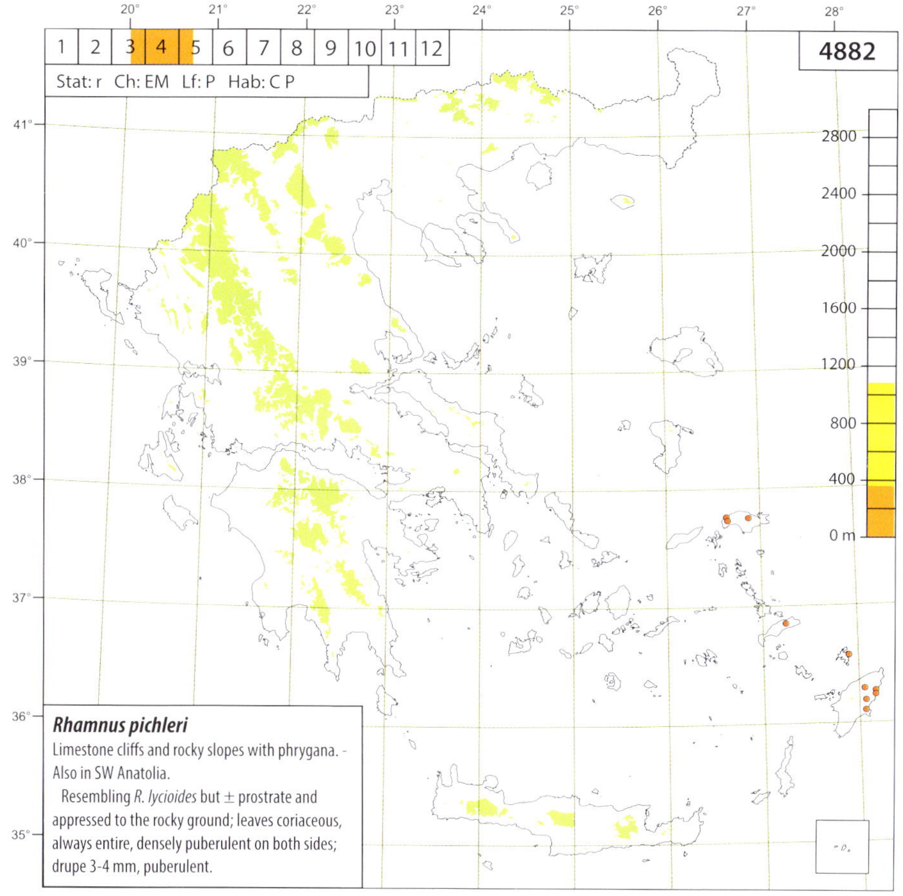

Rhamnus pichleri
Limestone cliffs and rocky slopes with phrygana. - Also in SW Anatolia.
Resembling *R. lycioides* but ± prostrate and appressed to the rocky ground; leaves coriaceous, always entire, densely puberulent on both sides; drupe 3-4 mm, puberulent.

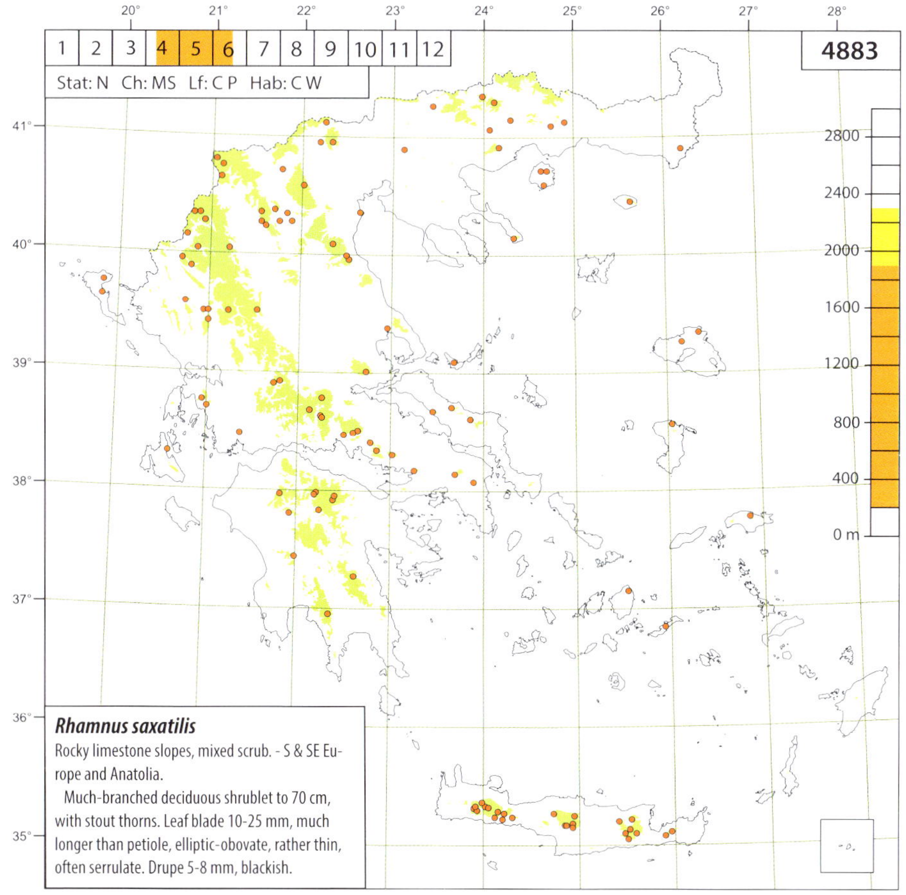

Rhamnus saxatilis
Rocky limestone slopes, mixed scrub. - S & SE Europe and Anatolia.
Much-branched deciduous shrublet to 70 cm, with stout thorns. Leaf blade 10-25 mm, much longer than petiole, elliptic-obovate, rather thin, often serrulate. Drupe 5-8 mm, blackish.

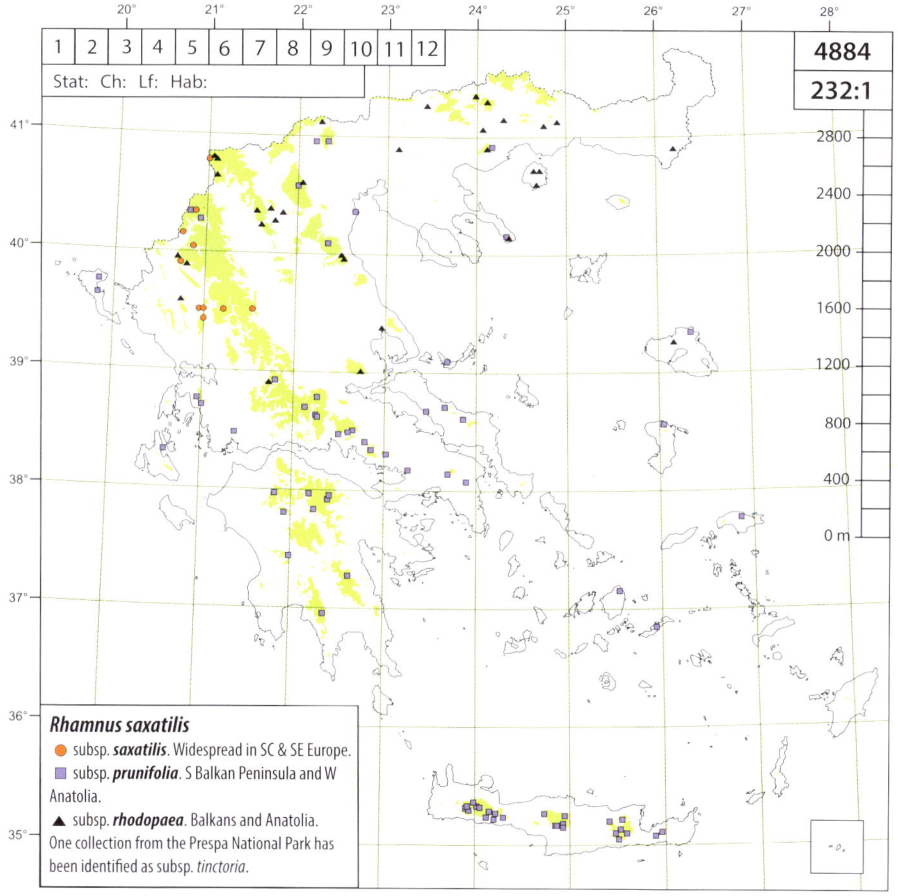

Rhamnus saxatilis
● subsp. ***saxatilis***. Widespread in SC & SE Europe.
■ subsp. ***prunifolia***. S Balkan Peninsula and W Anatolia.
▲ subsp. ***rhodopaea***. Balkans and Anatolia.
One collection from the Prespa National Park has been identified as subsp. *tinctoria*.

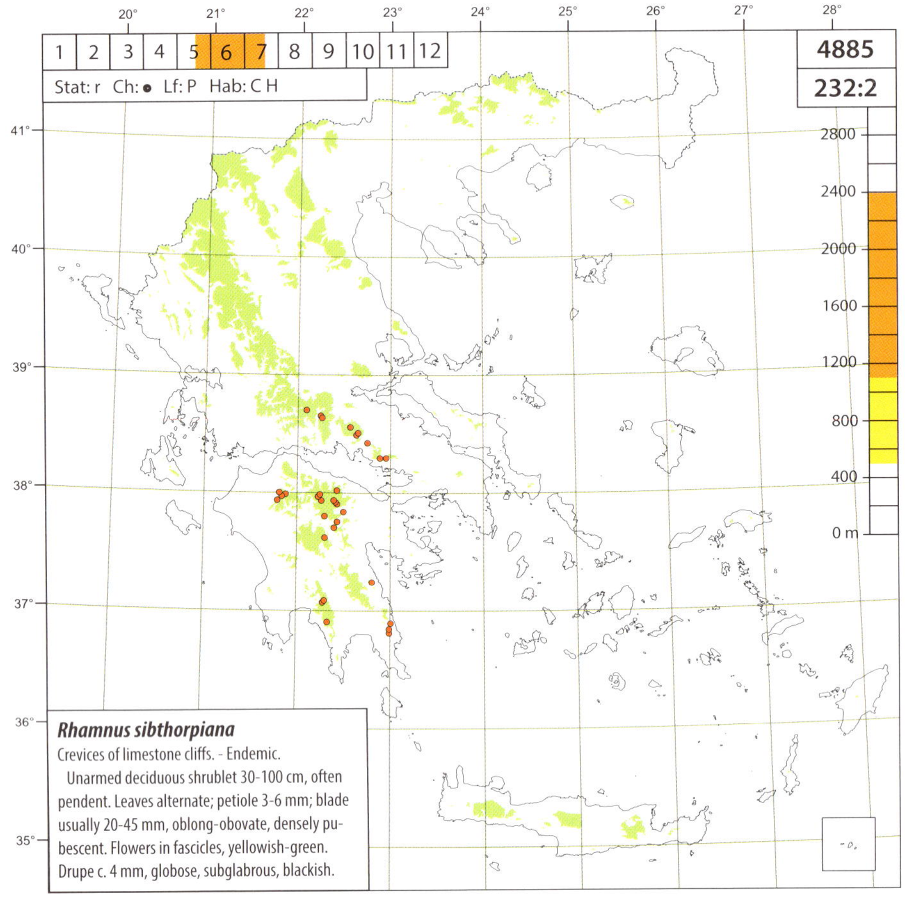

Rhamnus sibthorpiana
Crevices of limestone cliffs. - Endemic.
Unarmed deciduous shrublet 30-100 cm, often pendent. Leaves alternate; petiole 3-6 mm; blade usually 20-45 mm, oblong-obovate, densely pubescent. Flowers in fascicles, yellowish-green. Drupe c. 4 mm, globose, subglabrous, blackish.

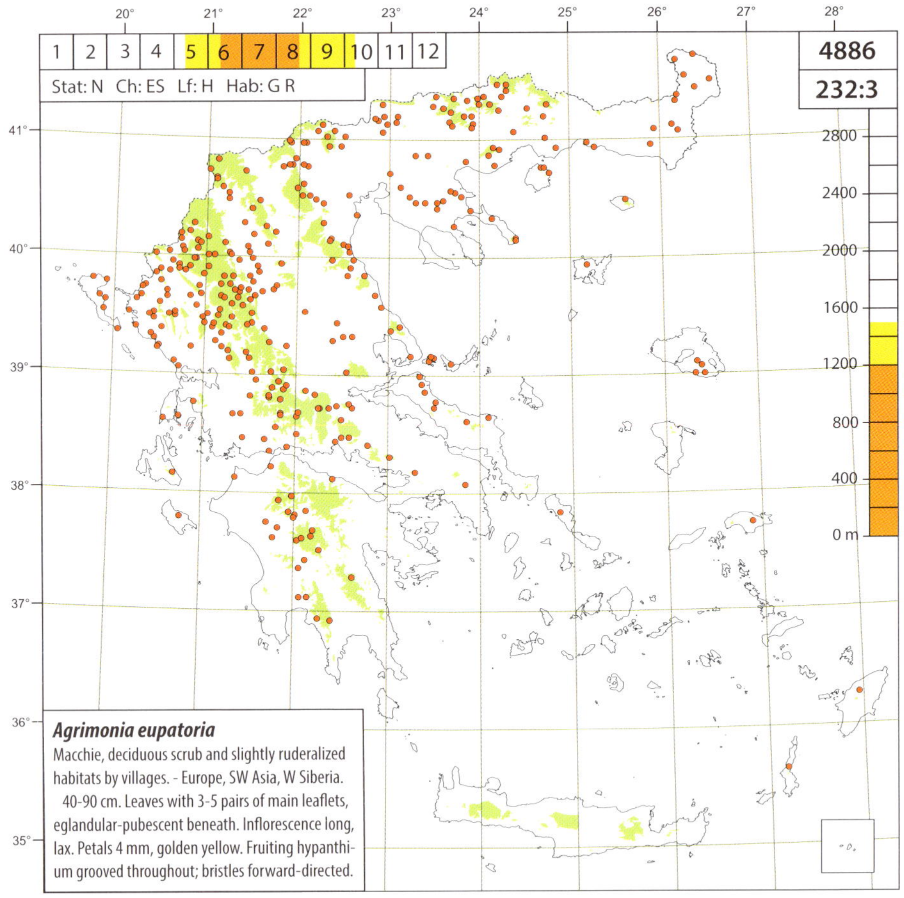

Agrimonia eupatoria
Macchie, deciduous scrub and slightly ruderalized habitats by villages. - Europe, SW Asia, W Siberia.
40-90 cm. Leaves with 3-5 pairs of main leaflets, eglandular-pubescent beneath. Inflorescence long, lax. Petals 4 mm, golden yellow. Fruiting hypanthium grooved throughout; bristles forward-directed.

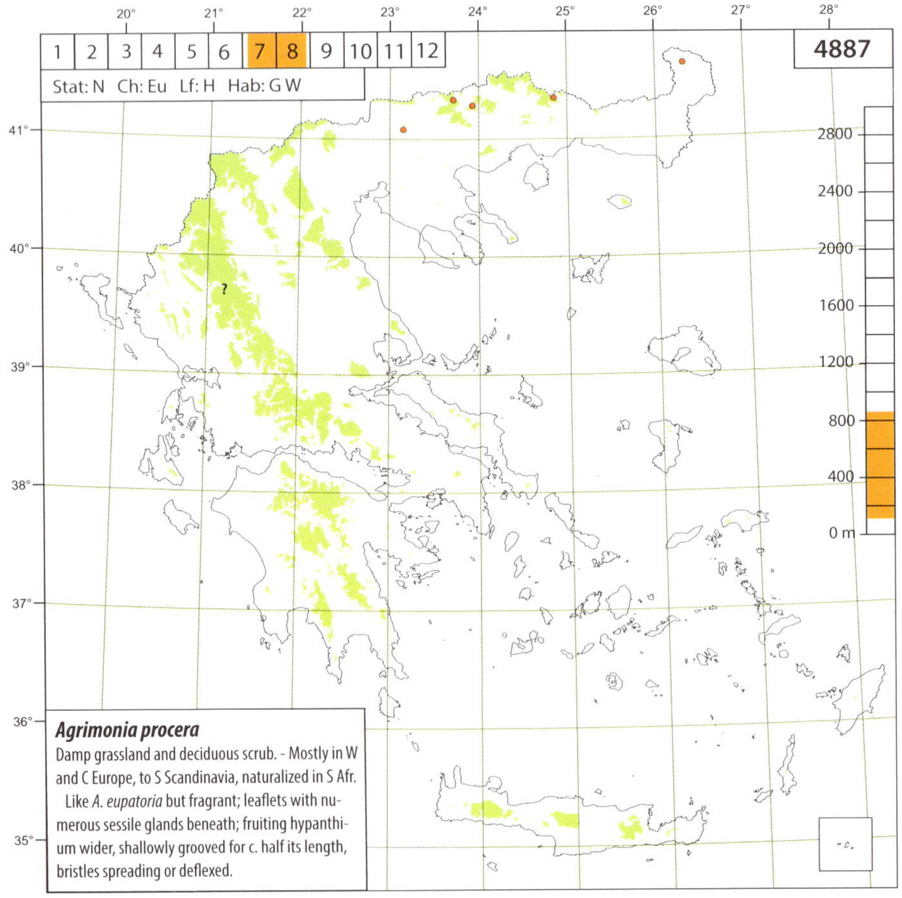

Agrimonia procera
Damp grassland and deciduous scrub. - Mostly in W and C Europe, to S Scandinavia, naturalized in S Afr.
Like *A. eupatoria* but fragrant; leaflets with numerous sessile glands beneath; fruiting hypanthium wider, shallowly grooved for c. half its length, bristles spreading or deflexed.

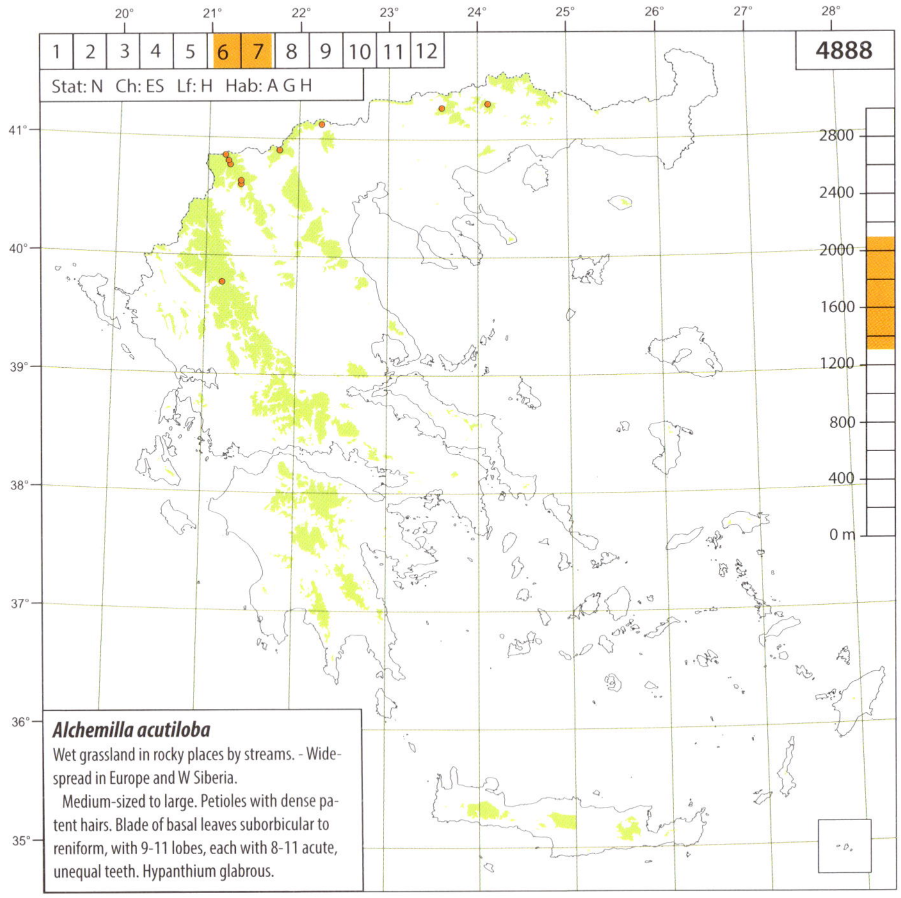

Alchemilla acutiloba
Wet grassland in rocky places by streams. - Widespread in Europe and W Siberia.
Medium-sized to large. Petioles with dense patent hairs. Blade of basal leaves suborbicular to reniform, with 9-11 lobes, each with 8-11 acute, unequal teeth. Hypanthium glabrous.

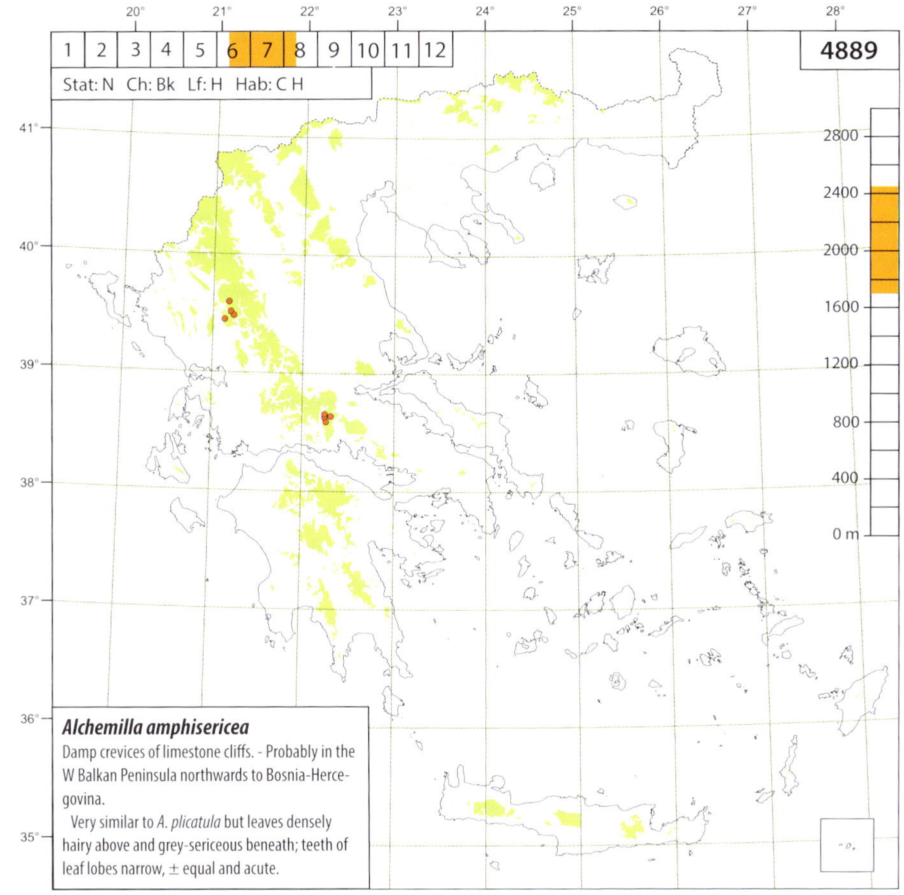

Alchemilla amphisericea

Damp crevices of limestone cliffs. - Probably in the W Balkan Peninsula northwards to Bosnia-Hercegovina.

Very similar to *A. plicatula* but leaves densely hairy above and grey-sericeous beneath; teeth of leaf lobes narrow, ± equal and acute.

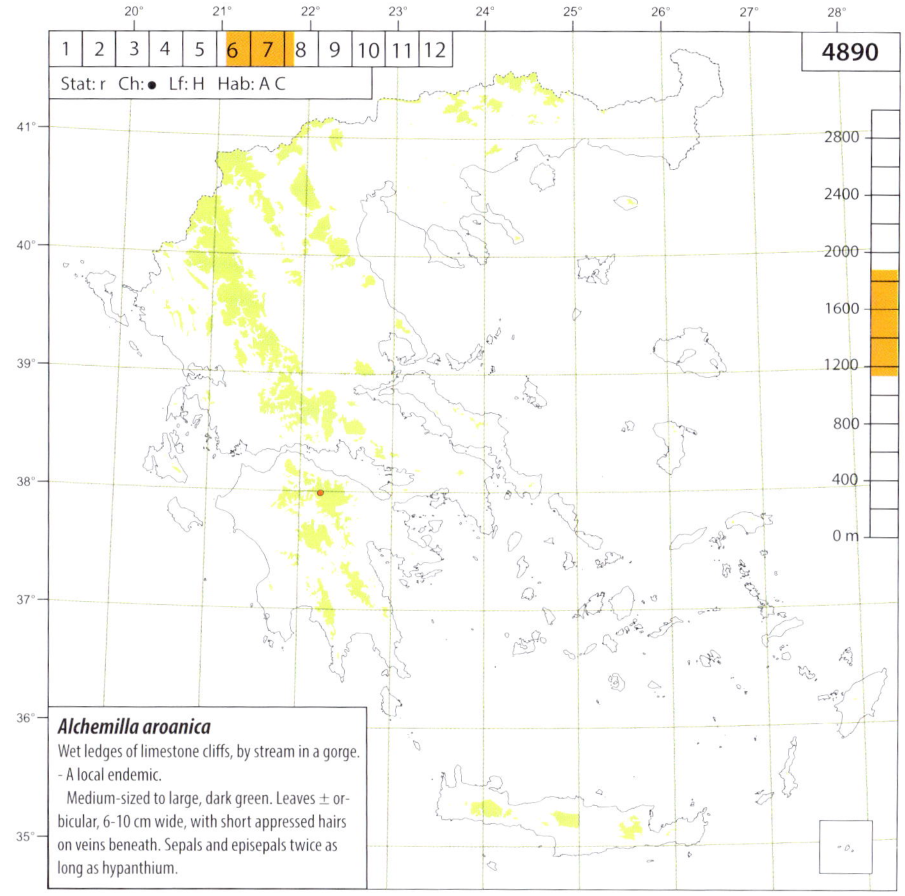

Alchemilla aroanica

Wet ledges of limestone cliffs, by stream in a gorge. - A local endemic.

Medium-sized to large, dark green. Leaves ± orbicular, 6-10 cm wide, with short appressed hairs on veins beneath. Sepals and episepals twice as long as hypanthium.

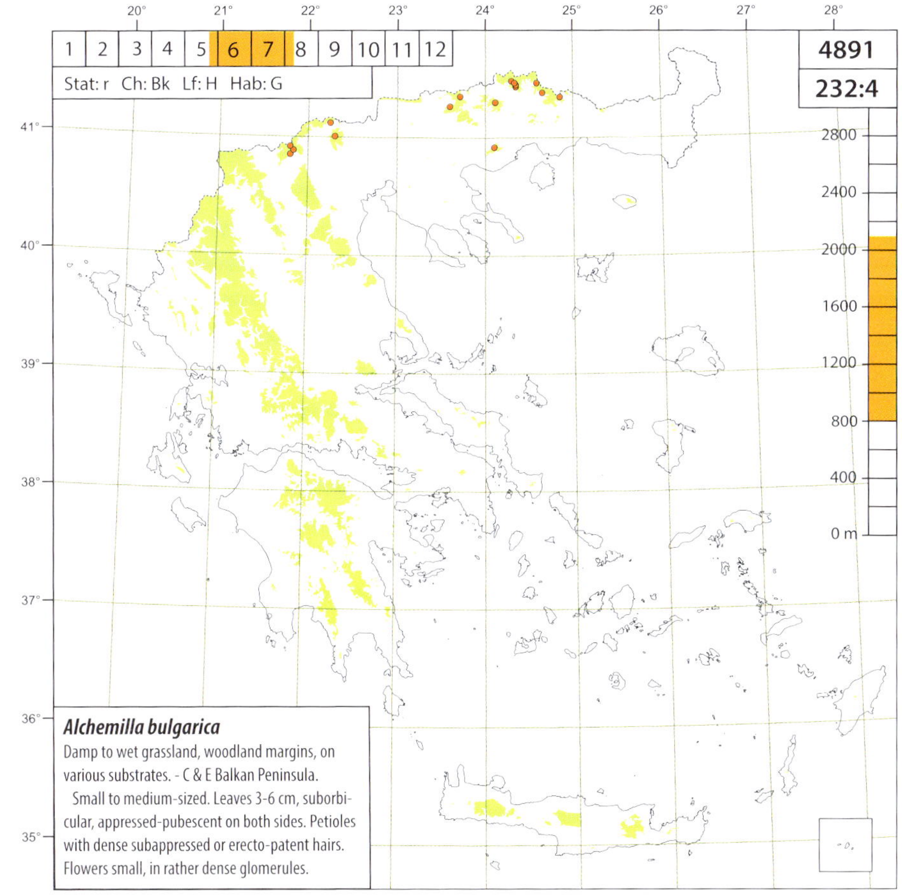

Alchemilla bulgarica

Damp to wet grassland, woodland margins, on various substrates. - C & E Balkan Peninsula.

Small to medium-sized. Leaves 3-6 cm, suborbicular, appressed-pubescent on both sides. Petioles with dense subappressed or erecto-patent hairs. Flowers small, in rather dense glomerules.

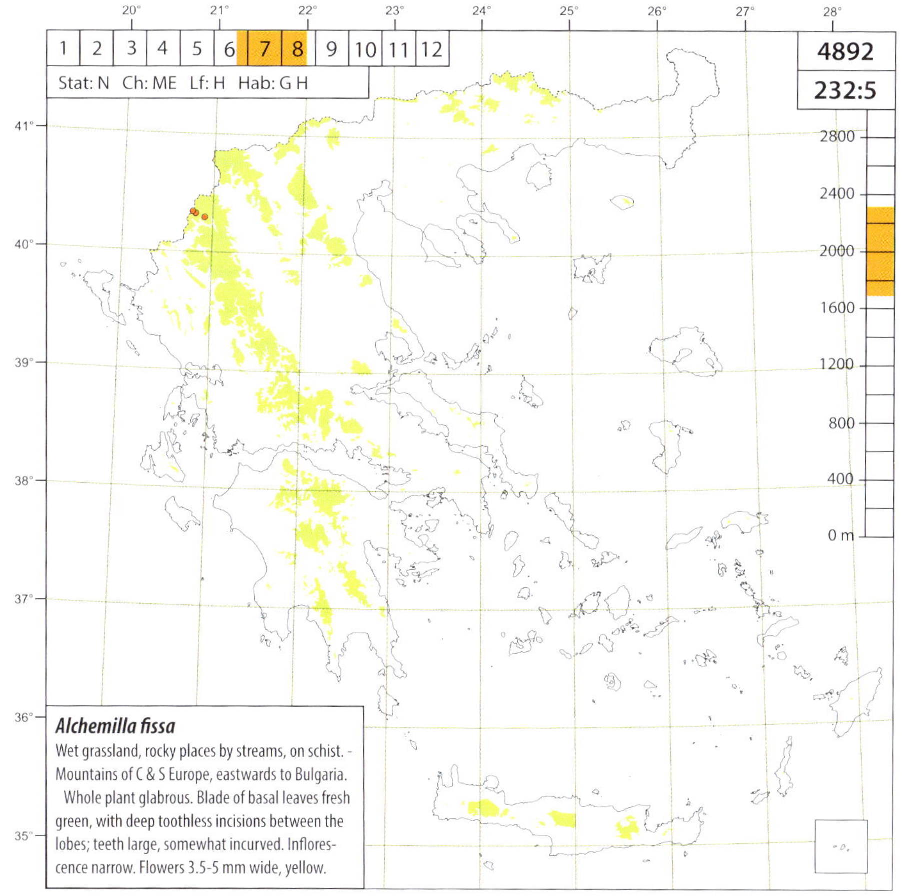

Alchemilla fissa

Wet grassland, rocky places by streams, on schist. - Mountains of C & S Europe, eastwards to Bulgaria.

Whole plant glabrous. Blade of basal leaves fresh green, with deep toothless incisions between the lobes; teeth large, somewhat incurved. Inflorescence narrow. Flowers 3.5-5 mm wide, yellow.

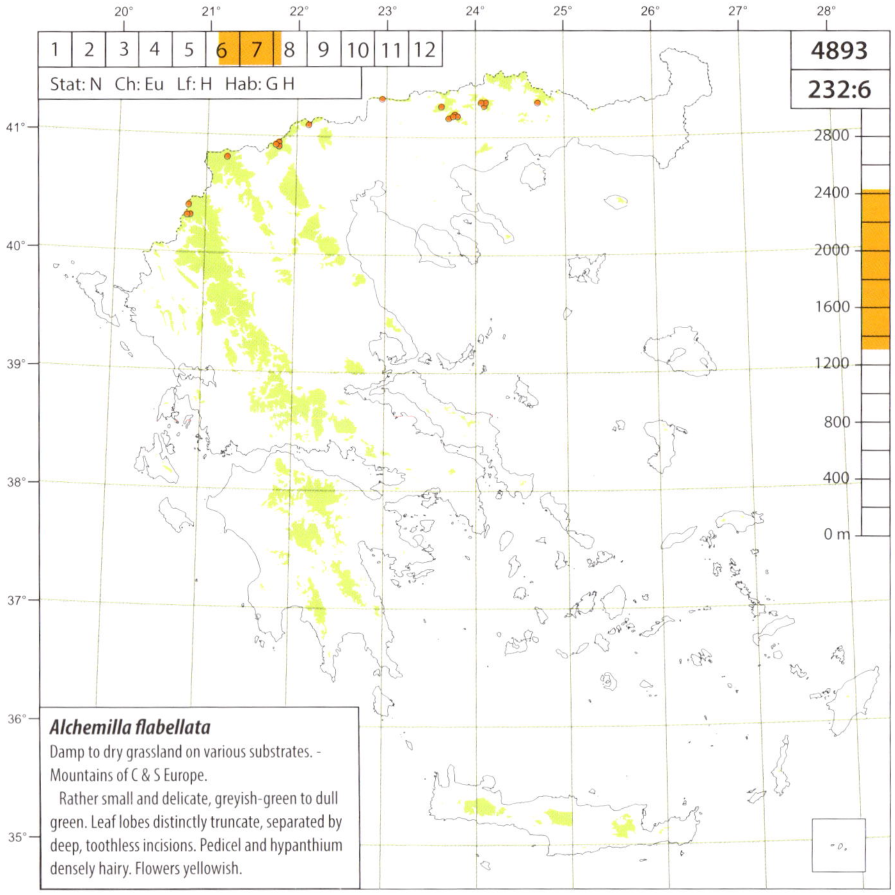

Alchemilla flabellata

Damp to dry grassland on various substrates. - Mountains of C & S Europe.

Rather small and delicate, greyish-green to dull green. Leaf lobes distinctly truncate, separated by deep, toothless incisions. Pedicel and hypanthium densely hairy. Flowers yellowish.

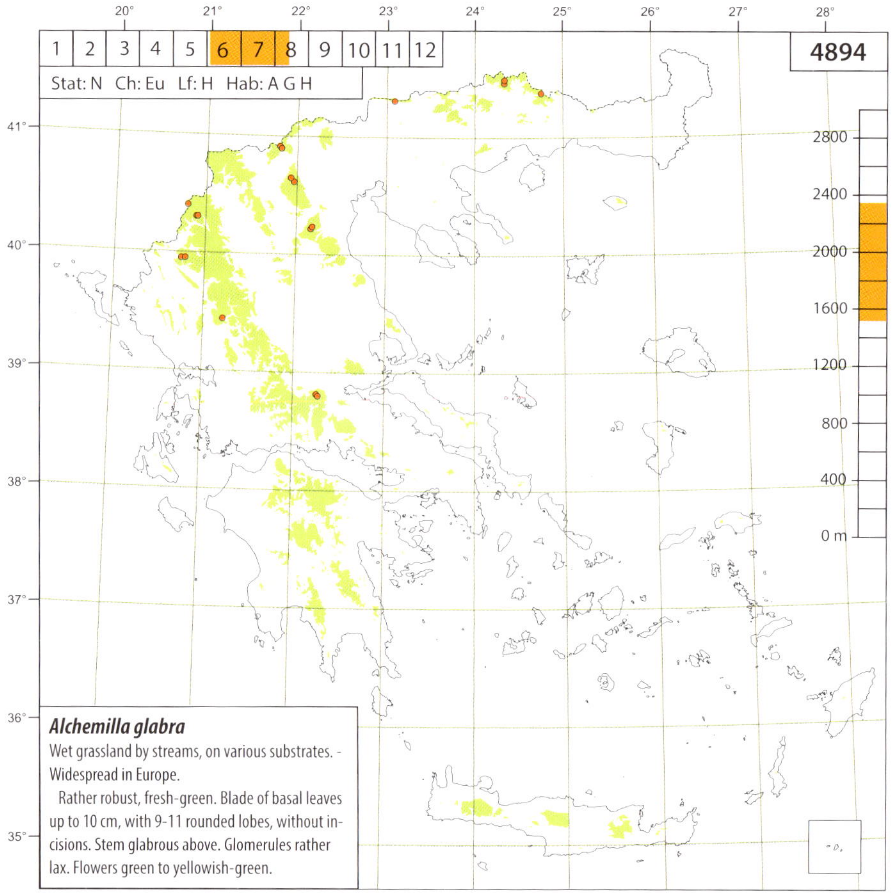

Alchemilla glabra

Wet grassland by streams, on various substrates. - Widespread in Europe.

Rather robust, fresh-green. Blade of basal leaves up to 10 cm, with 9-11 rounded lobes, without incisions. Stem glabrous above. Glomerules rather lax. Flowers green to yellowish-green.

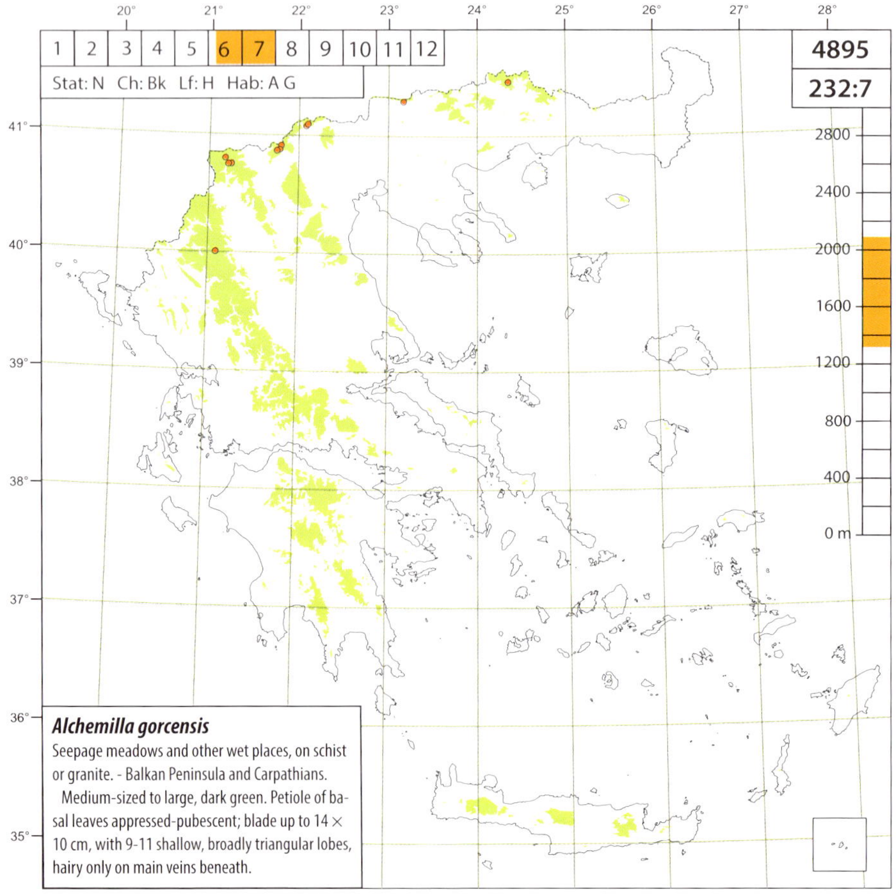

Alchemilla gorcensis

Seepage meadows and other wet places, on schist or granite. - Balkan Peninsula and Carpathians.

Medium-sized to large, dark green. Petiole of basal leaves appressed-pubescent; blade up to 14 × 10 cm, with 9-11 shallow, broadly triangular lobes, hairy only on main veins beneath.

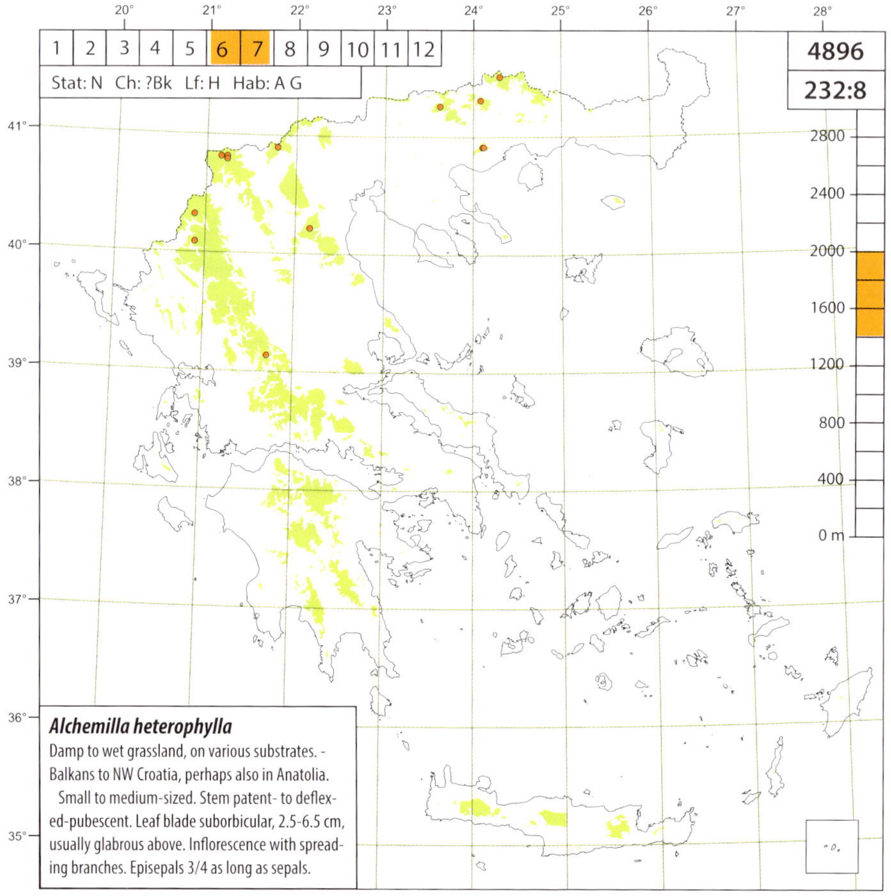

Alchemilla heterophylla

Damp to wet grassland, on various substrates. - Balkans to NW Croatia, perhaps also in Anatolia.

Small to medium-sized. Stem patent- to deflexed-pubescent. Leaf blade suborbicular, 2.5-6.5 cm, usually glabrous above. Inflorescence with spreading branches. Episepals 3/4 as long as sepals.

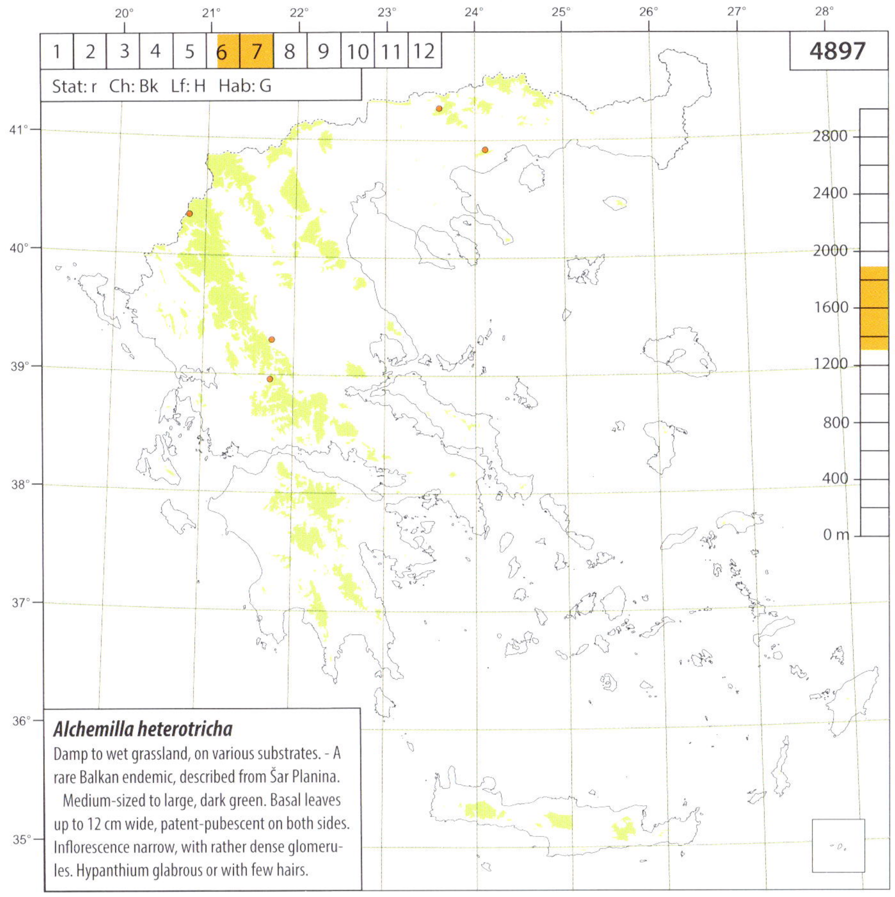

Alchemilla heterotricha

Damp to wet grassland, on various substrates. - A rare Balkan endemic, described from Šar Planina.

Medium-sized to large, dark green. Basal leaves up to 12 cm wide, patent-pubescent on both sides. Inflorescence narrow, with rather dense glomerules. Hypanthium glabrous or with few hairs.

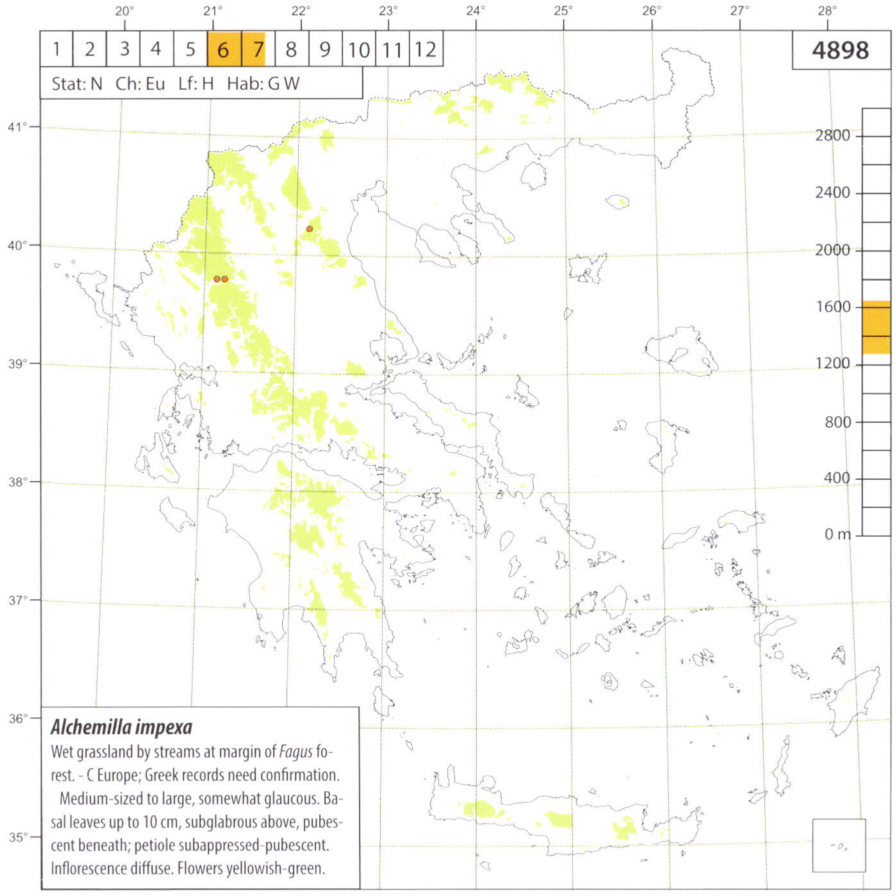

Alchemilla impexa

Wet grassland by streams at margin of *Fagus* forest. - C Europe; Greek records need confirmation.

Medium-sized to large, somewhat glaucous. Basal leaves up to 10 cm, subglabrous above, pubescent beneath; petiole subappressed-pubescent. Inflorescence diffuse. Flowers yellowish-green.

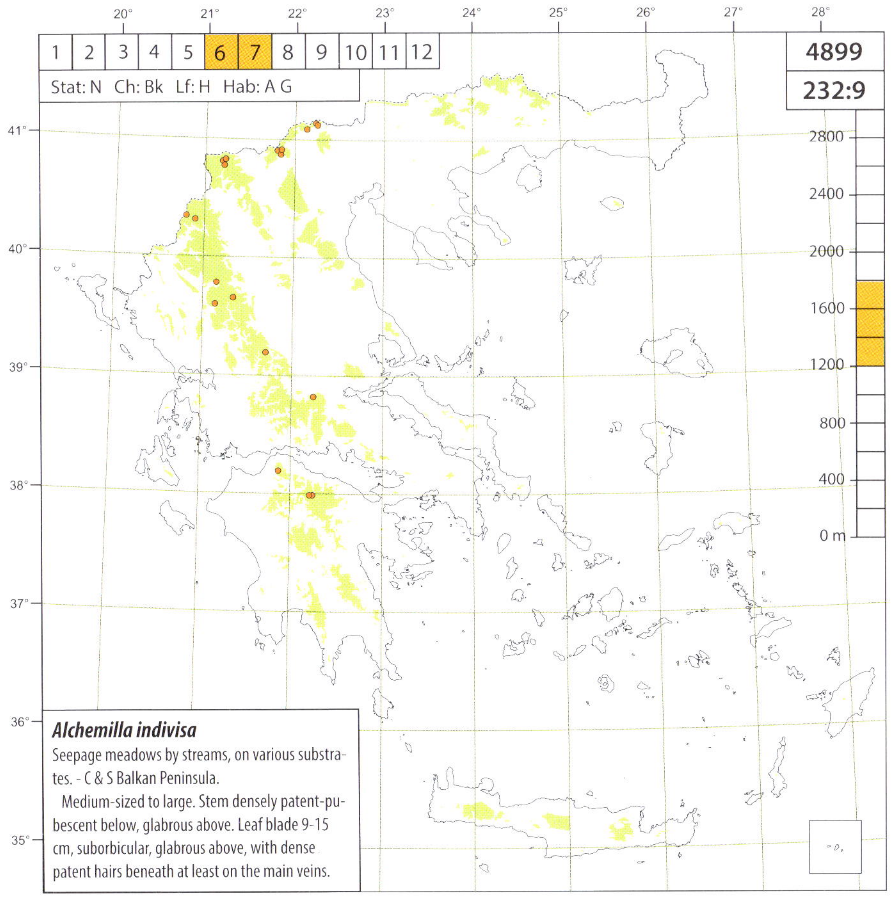

Alchemilla indivisa

Seepage meadows by streams, on various substrates. - C & S Balkan Peninsula.

Medium-sized to large. Stem densely patent-pubescent below, glabrous above. Leaf blade 9-15 cm, suborbicular, glabrous above, with dense patent hairs beneath at least on the main veins.

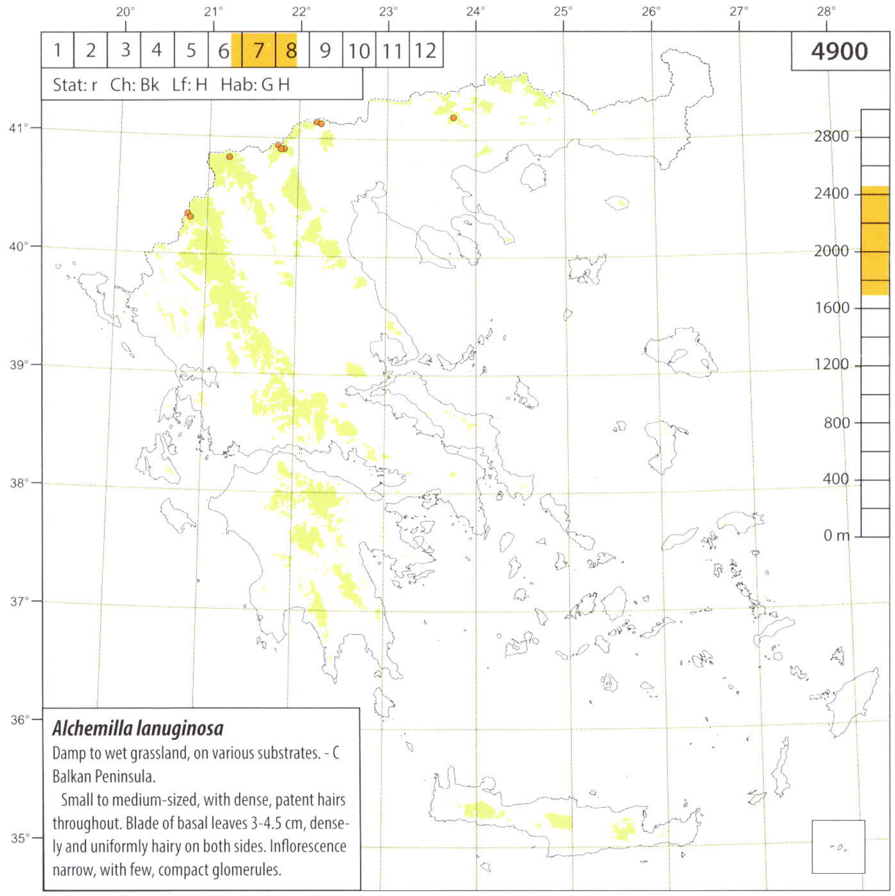

Alchemilla lanuginosa

Damp to wet grassland, on various substrates. - C Balkan Peninsula.

Small to medium-sized, with dense, patent hairs throughout. Blade of basal leaves 3-4.5 cm, densely and uniformly hairy on both sides. Inflorescence narrow, with few, compact glomerules.

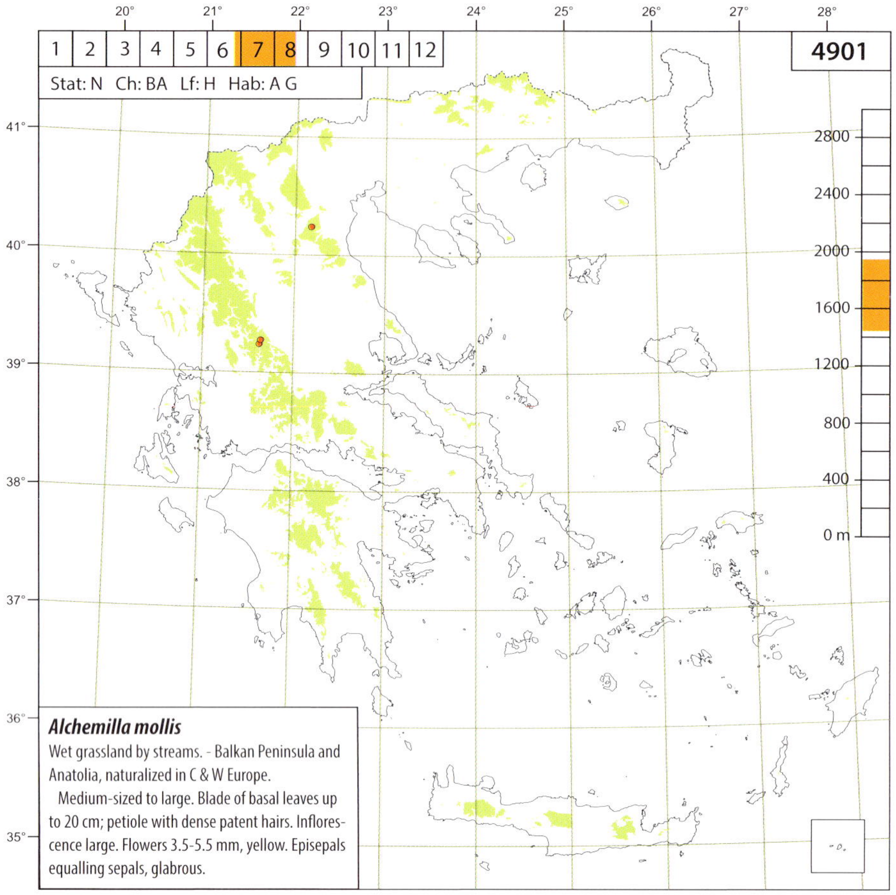

Alchemilla mollis

Wet grassland by streams. - Balkan Peninsula and Anatolia, naturalized in C & W Europe.

Medium-sized to large. Blade of basal leaves up to 20 cm; petiole with dense patent hairs. Inflorescence large. Flowers 3.5-5.5 mm, yellow. Episepals equalling sepals, glabrous.

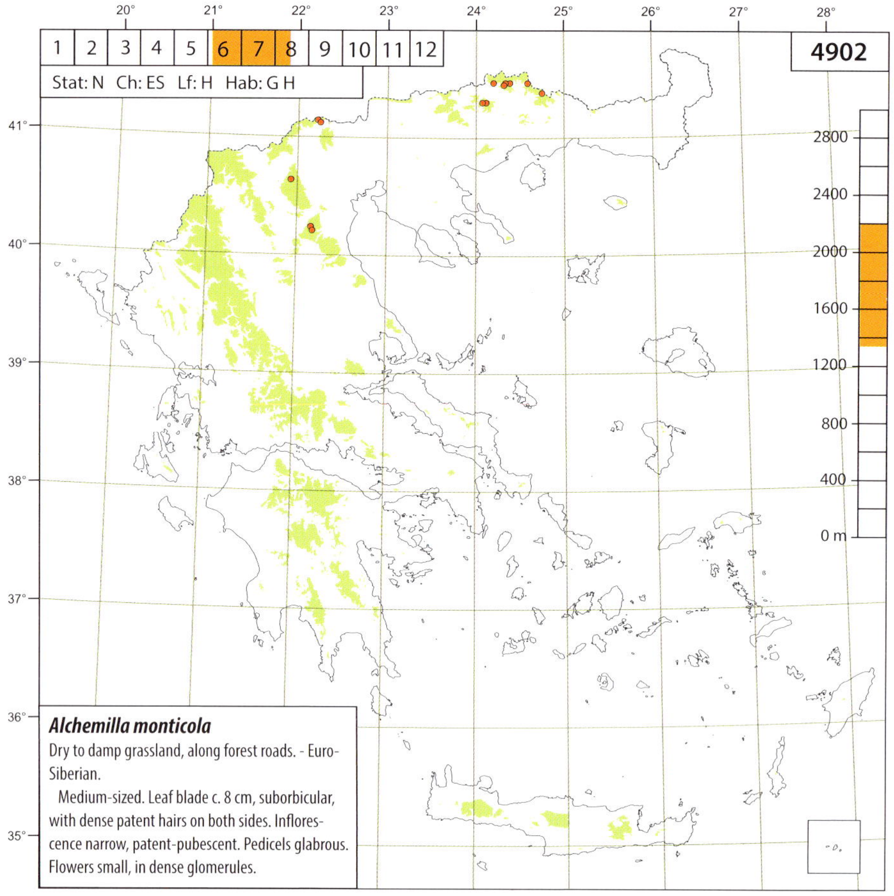

Alchemilla monticola

Dry to damp grassland, along forest roads. - Euro-Siberian.

Medium-sized. Leaf blade c. 8 cm, suborbicular, with dense patent hairs on both sides. Inflorescence narrow, patent-pubescent. Pedicels glabrous. Flowers small, in dense glomerules.

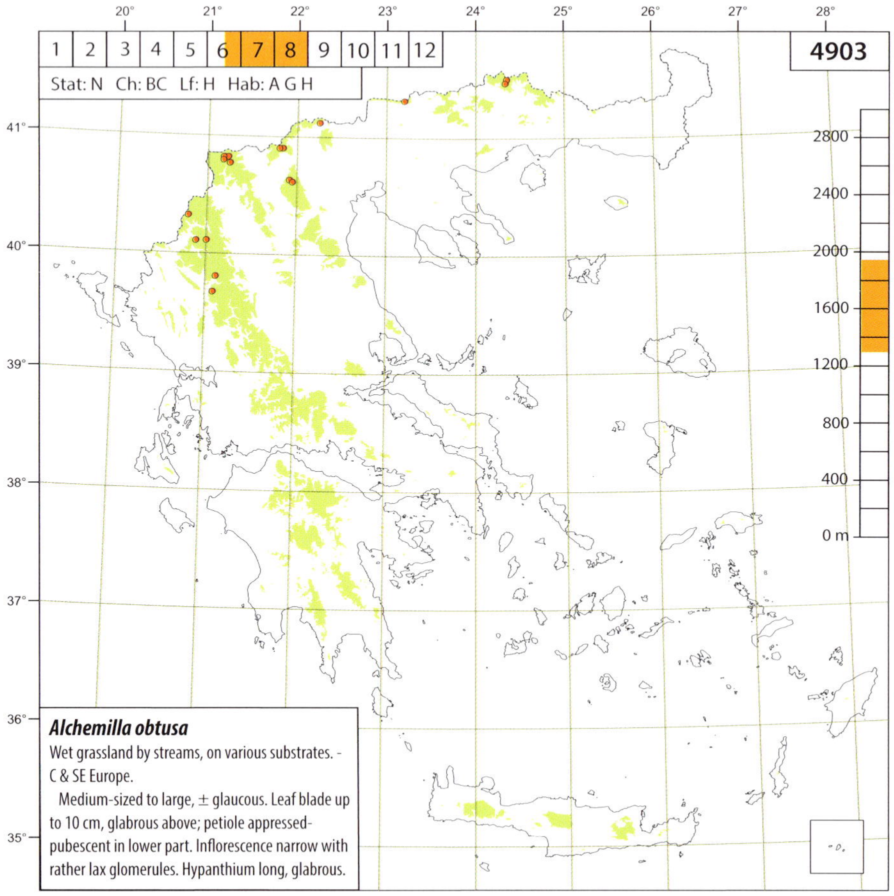

Alchemilla obtusa

Wet grassland by streams, on various substrates. - C & SE Europe.

Medium-sized to large, ± glaucous. Leaf blade up to 10 cm, glabrous above; petiole appressed-pubescent in lower part. Inflorescence narrow with rather lax glomerules. Hypanthium long, glabrous.

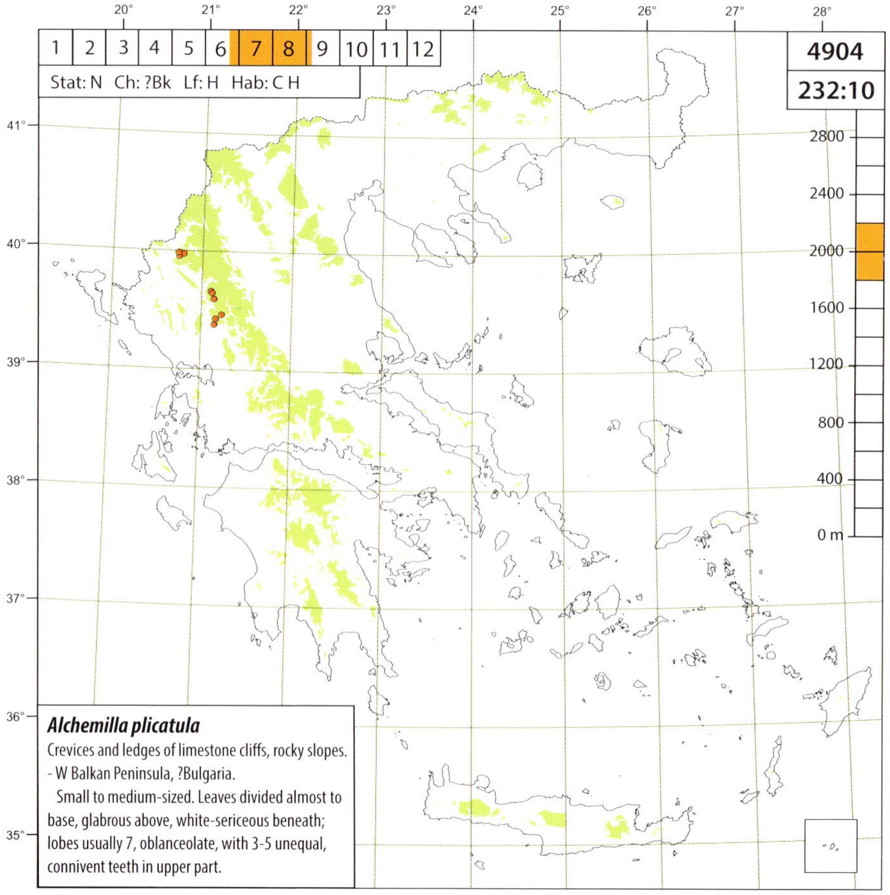

Alchemilla plicatula

Crevices and ledges of limestone cliffs, rocky slopes. - W Balkan Peninsula, ?Bulgaria.

Small to medium-sized. Leaves divided almost to base, glabrous above, white-sericeous beneath; lobes usually 7, oblanceolate, with 3-5 unequal, connivent teeth in upper part.

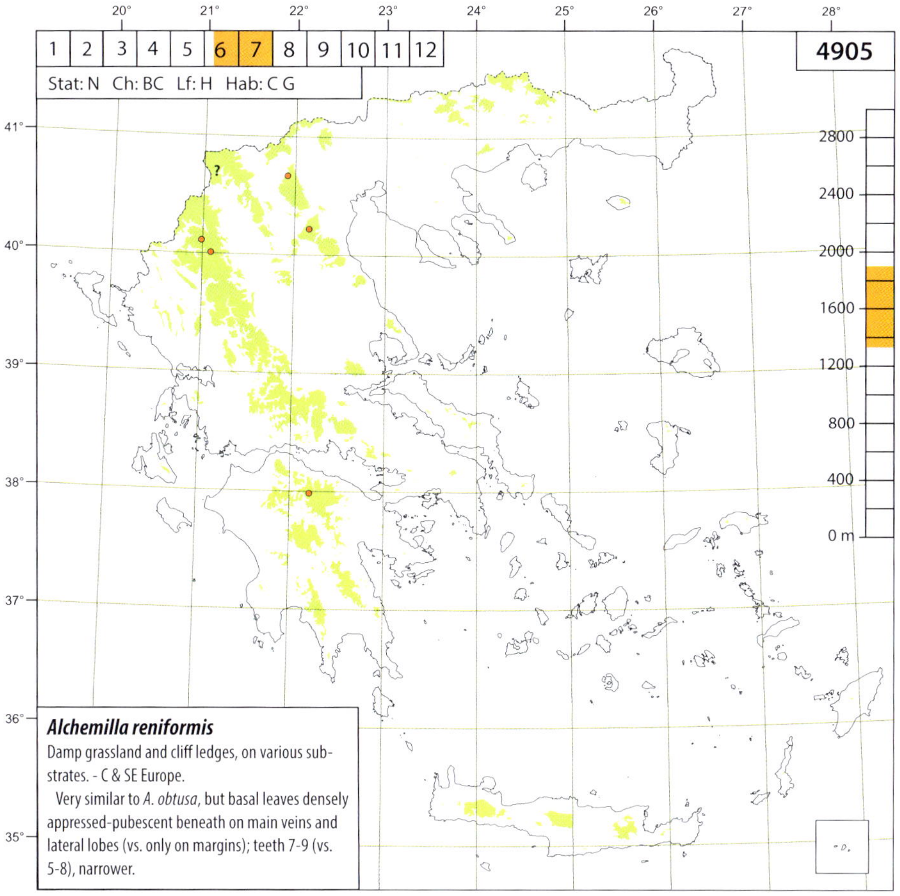

Alchemilla reniformis

Damp grassland and cliff ledges, on various substrates. - C & SE Europe.

Very similar to *A. obtusa*, but basal leaves densely appressed-pubescent beneath on main veins and lateral lobes (vs. only on margins); teeth 7-9 (vs. 5-8), narrower.

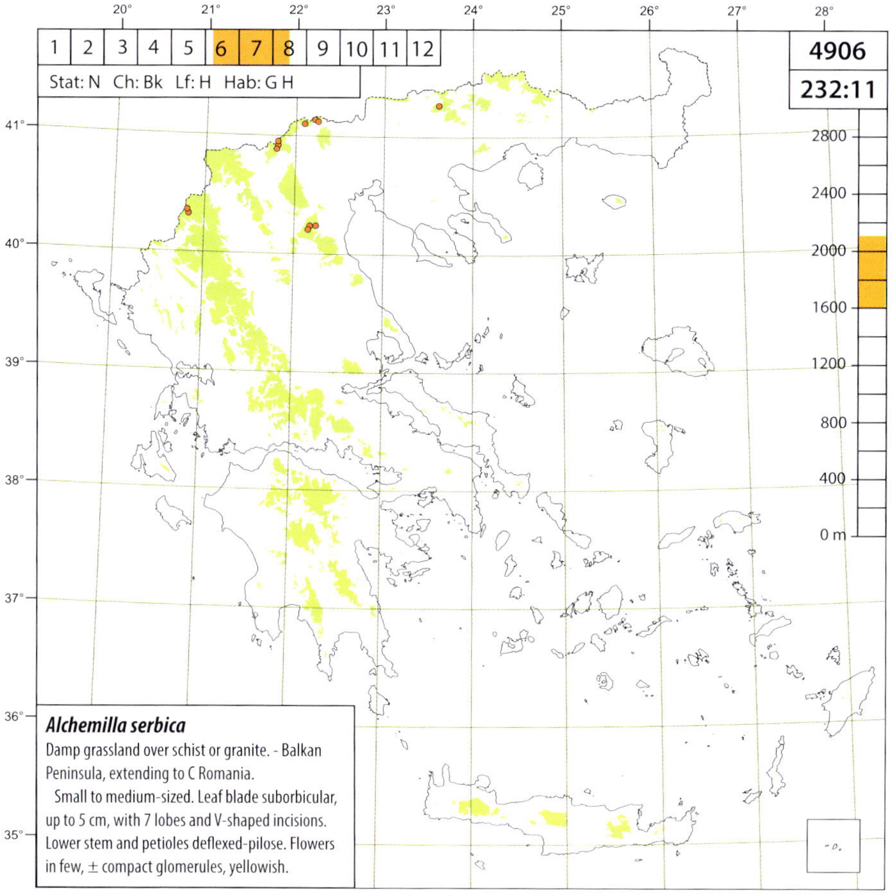

Alchemilla serbica

Damp grassland over schist or granite. - Balkan Peninsula, extending to C Romania.

Small to medium-sized. Leaf blade suborbicular, up to 5 cm, with 7 lobes and V-shaped incisions. Lower stem and petioles deflexed-pilose. Flowers in few, ± compact glomerules, yellowish.

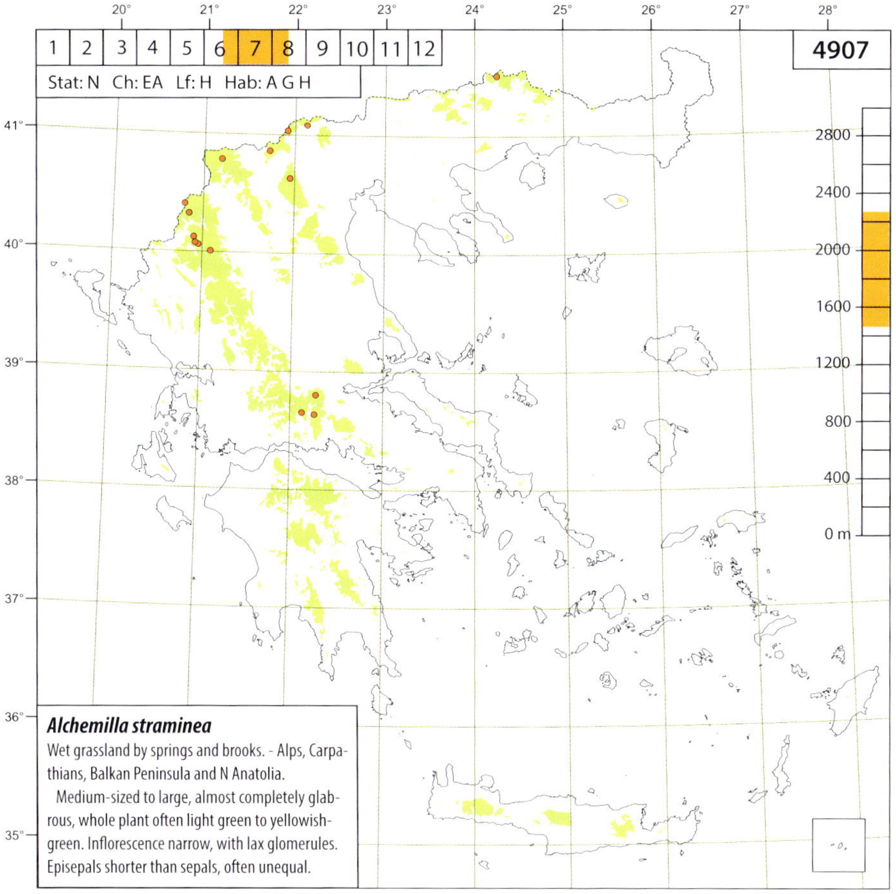

Alchemilla straminea

Wet grassland by springs and brooks. - Alps, Carpathians, Balkan Peninsula and N Anatolia.

Medium-sized to large, almost completely glabrous, whole plant often light green to yellowish-green. Inflorescence narrow, with lax glomerules. Episepals shorter than sepals, often unequal.

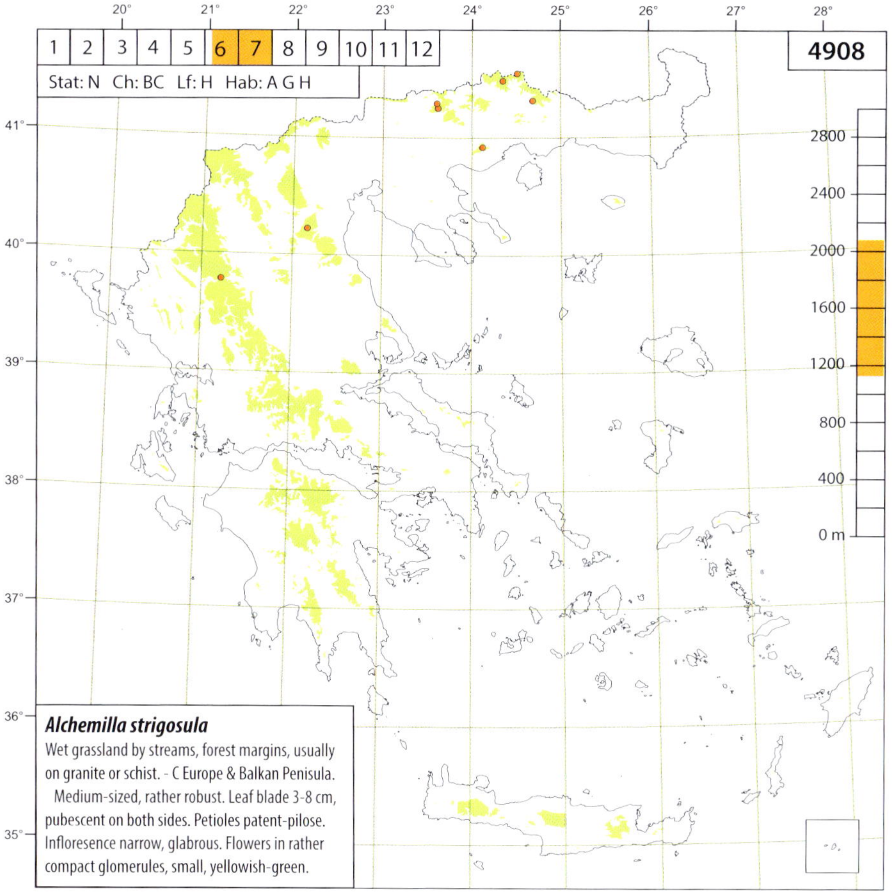

Alchemilla strigosula

Wet grassland by streams, forest margins, usually on granite or schist. - C Europe & Balkan Penisula.

Medium-sized, rather robust. Leaf blade 3-8 cm, pubescent on both sides. Petioles patent-pilose. Infloresence narrow, glabrous. Flowers in rather compact glomerules, small, yellowish-green.

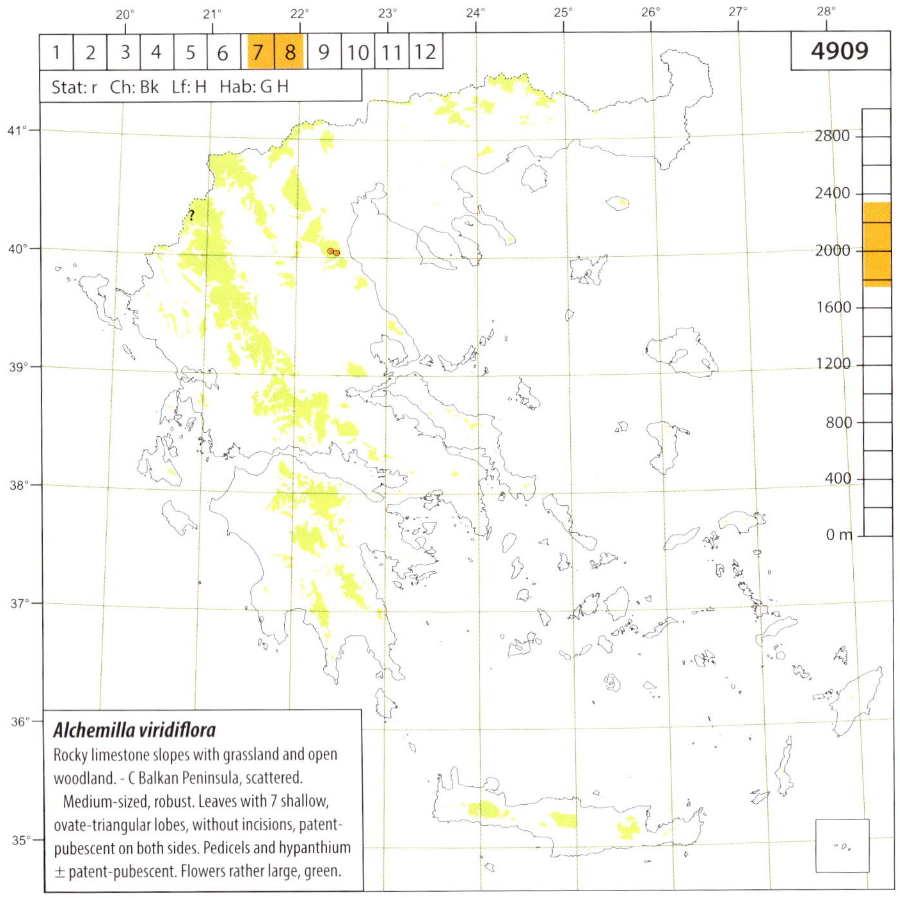

Alchemilla viridiflora

Rocky limestone slopes with grassland and open woodland. - C Balkan Peninsula, scattered.

Medium-sized, robust. Leaves with 7 shallow, ovate-triangular lobes, without incisions, patent-pubescent on both sides. Pedicels and hypanthium ± patent-pubescent. Flowers rather large, green.

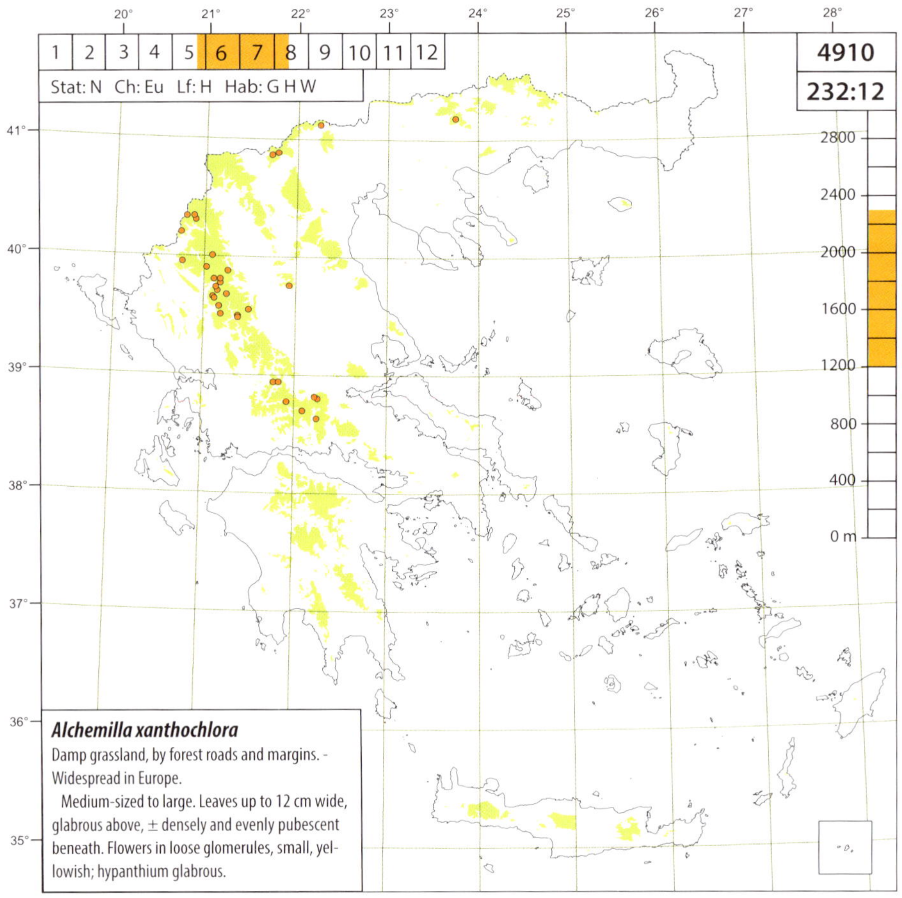

Alchemilla xanthochlora

Damp grassland, by forest roads and margins. - Widespread in Europe.

Medium-sized to large. Leaves up to 12 cm wide, glabrous above, ± densely and evenly pubescent beneath. Flowers in loose glomerules, small, yellowish; hypanthium glabrous.

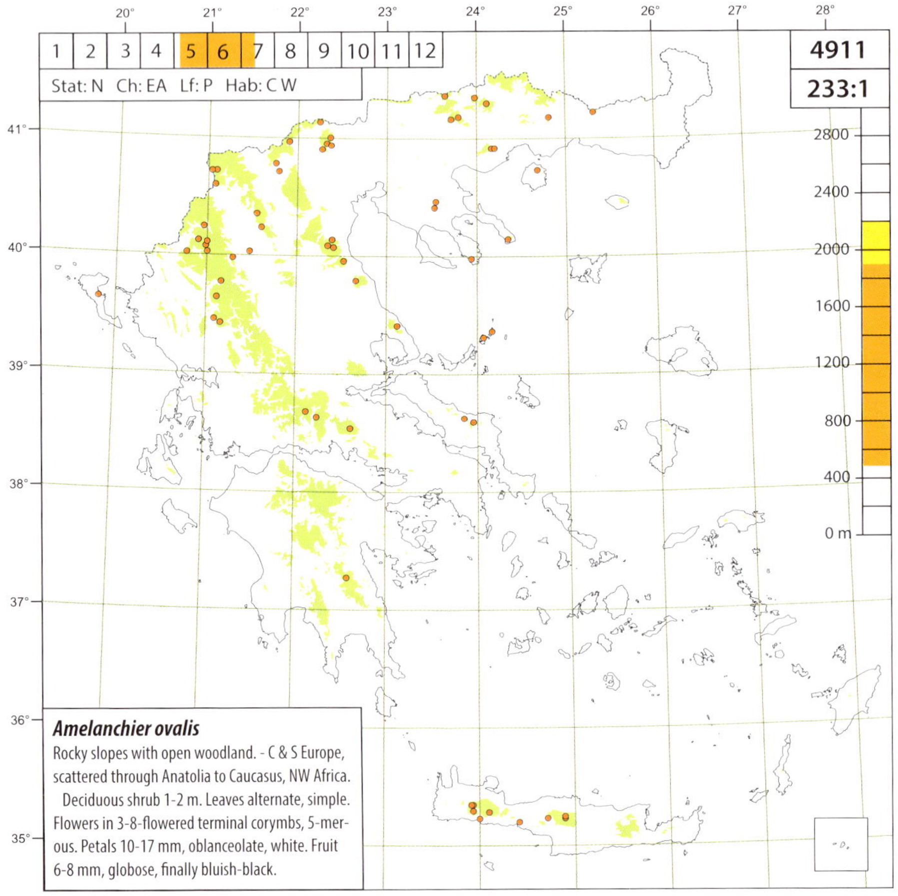

Amelanchier ovalis

Rocky slopes with open woodland. - C & S Europe, scattered through Anatolia to Caucasus, NW Africa.

Deciduous shrub 1-2 m. Leaves alternate, simple. Flowers in 3-8-flowered terminal corymbs, 5-merous. Petals 10-17 mm, oblanceolate, white. Fruit 6-8 mm, globose, finally bluish-black.

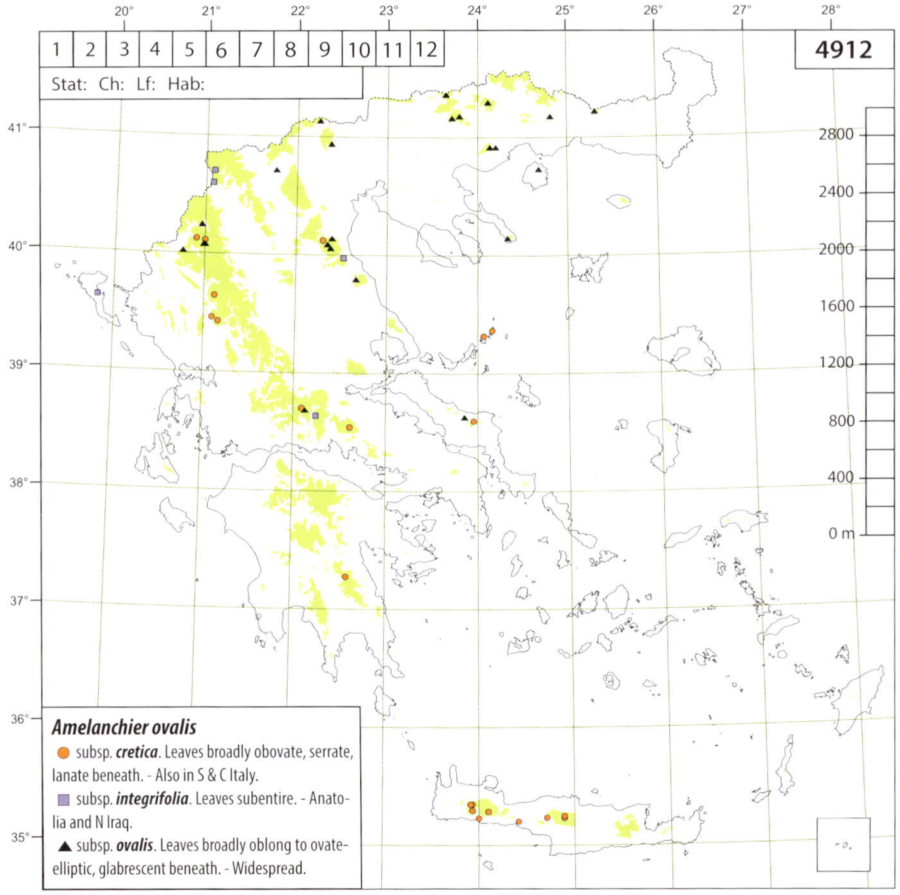

Amelanchier ovalis

● subsp. ***cretica***. Leaves broadly obovate, serrate, lanate beneath. - Also in S & C Italy.

■ subsp. ***integrifolia***. Leaves subentire. - Anatolia and N Iraq.

▲ subsp. ***ovalis***. Leaves broadly oblong to ovate-elliptic, glabrescent beneath. - Widespread.

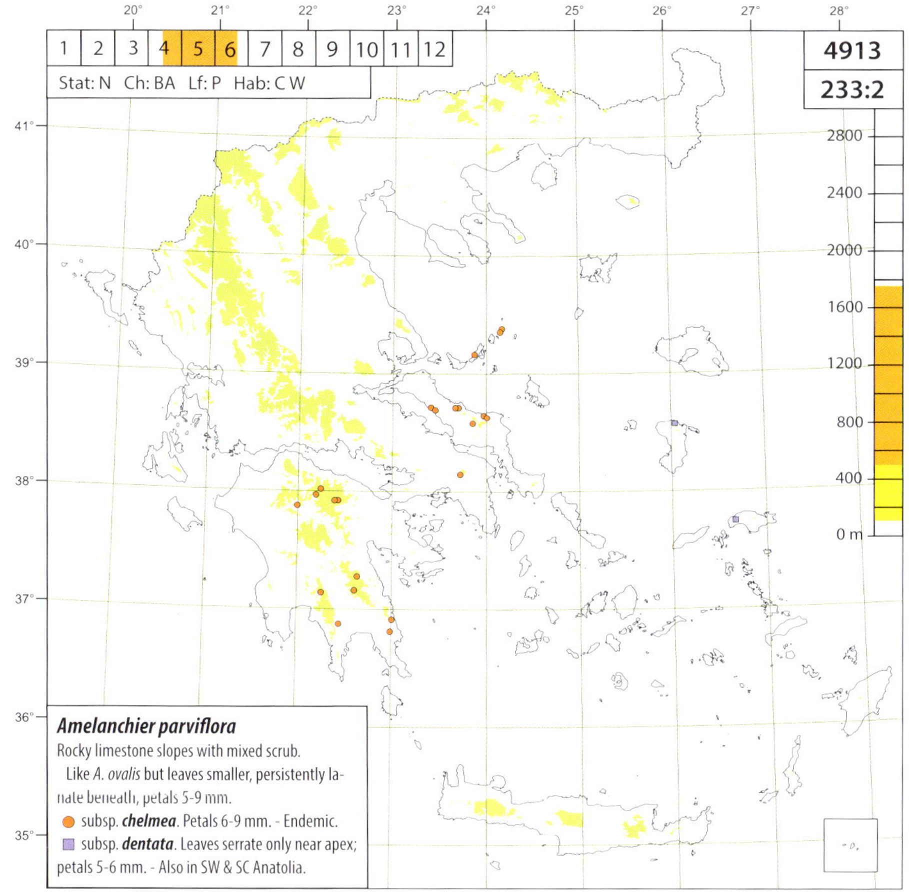

Amelanchier parviflora
Rocky limestone slopes with mixed scrub.
Like *A. ovalis* but leaves smaller, persistently lanate beneath, petals 5-9 mm.
● subsp. ***chelmea***. Petals 6-9 mm. - Endemic.
■ subsp. ***dentata***. Leaves serrate only near apex; petals 5-6 mm. - Also in SW & SC Anatolia.

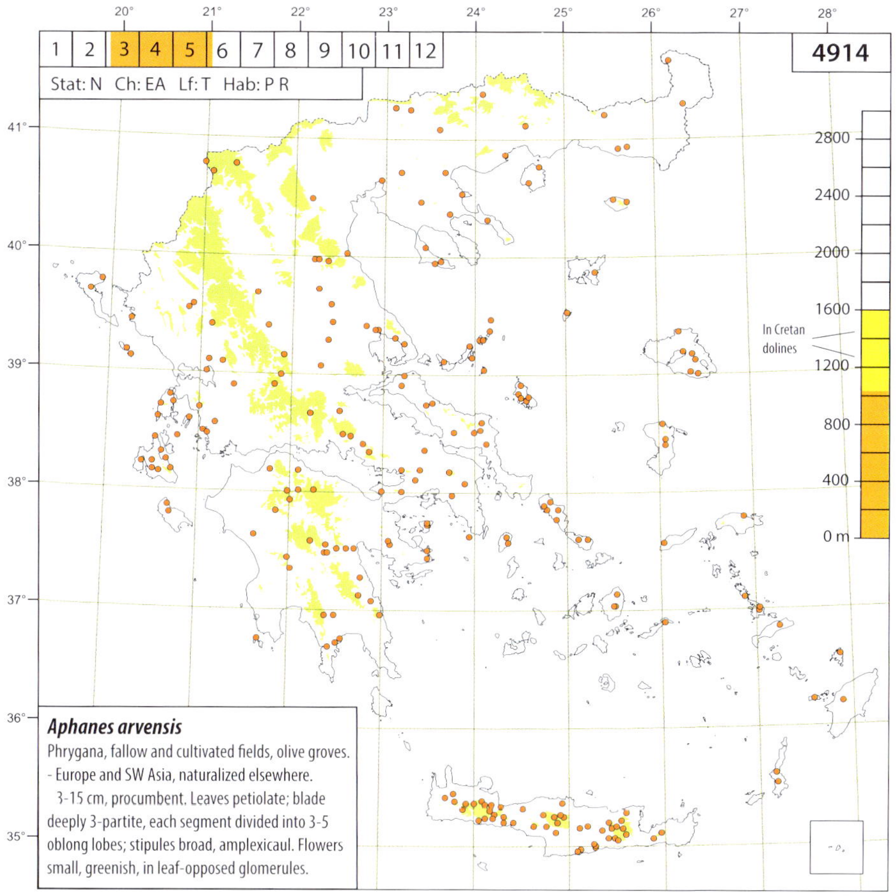

Aphanes arvensis
Phrygana, fallow and cultivated fields, olive groves. - Europe and SW Asia, naturalized elsewhere.
3-15 cm, procumbent. Leaves petiolate; blade deeply 3-partite, each segment divided into 3-5 oblong lobes; stipules broad, amplexicaul. Flowers small, greenish, in leaf-opposed glomerules.

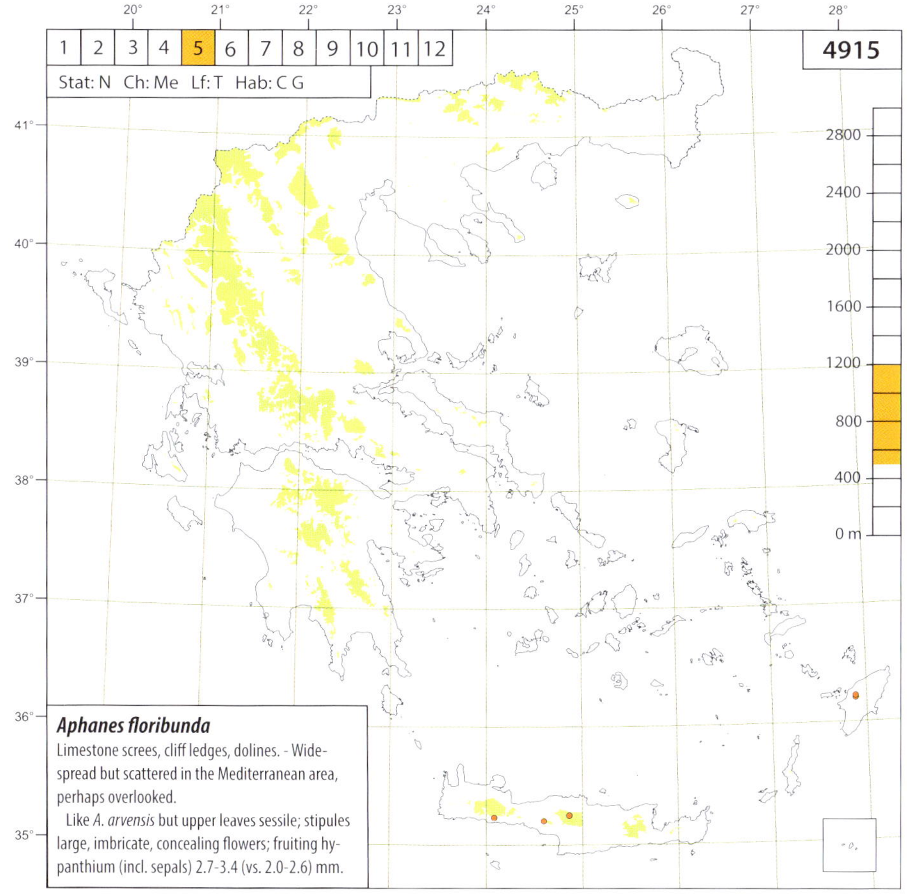

Aphanes floribunda
Limestone screes, cliff ledges, dolines. - Widespread but scattered in the Mediterranean area, perhaps overlooked.
Like *A. arvensis* but upper leaves sessile; stipules large, imbricate, concealing flowers; fruiting hypanthium (incl. sepals) 2.7-3.4 (vs. 2.0-2.6) mm.

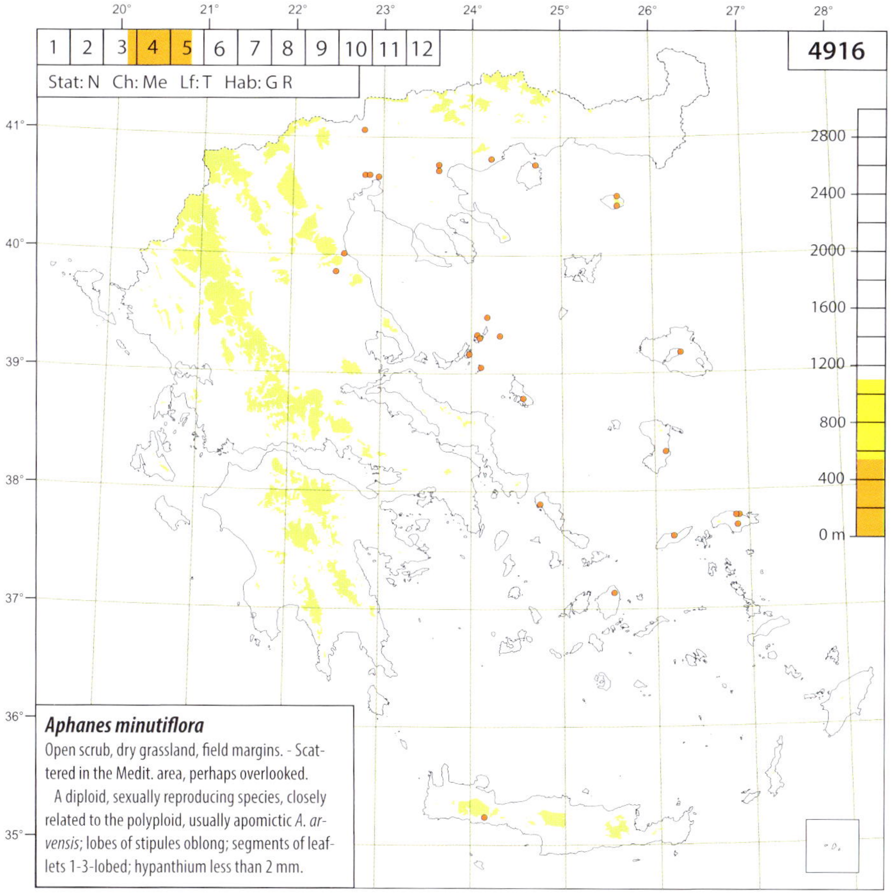

Aphanes minutiflora
Open scrub, dry grassland, field margins. - Scattered in the Medit. area, perhaps overlooked.
A diploid, sexually reproducing species, closely related to the polyploid, usually apomictic *A. arvensis*; lobes of stipules oblong; segments of leaflets 1-3-lobed; hypanthium less than 2 mm.

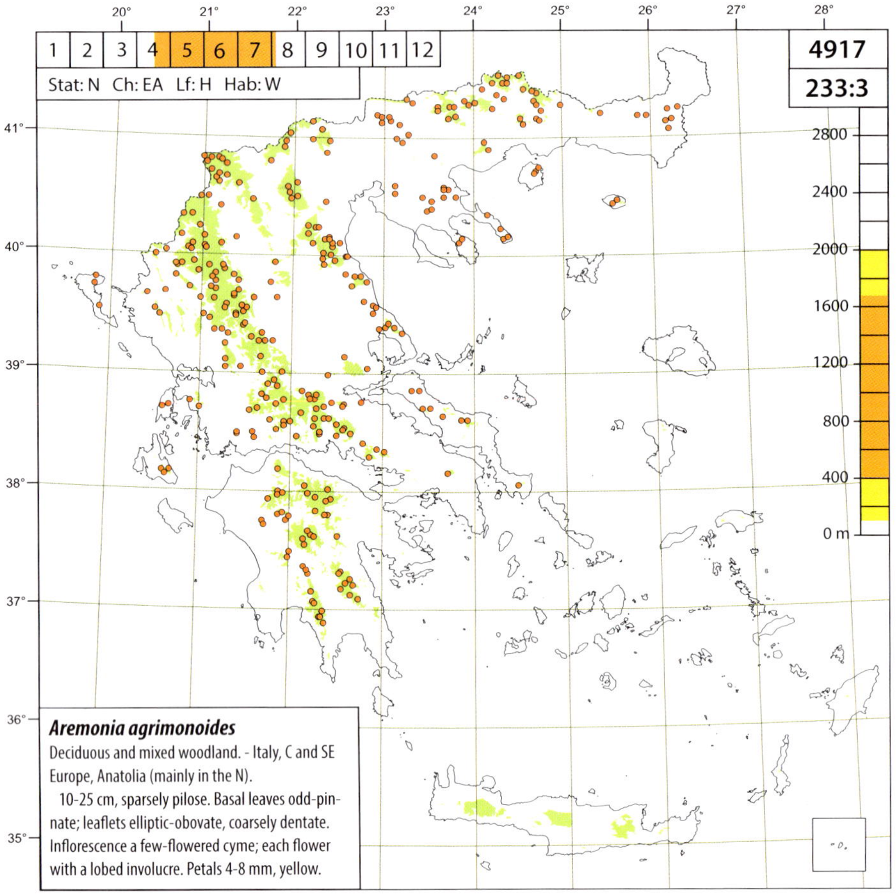

Aremonia agrimonoides
Deciduous and mixed woodland. - Italy, C and SE Europe, Anatolia (mainly in the N).
10-25 cm, sparsely pilose. Basal leaves odd-pinnate; leaflets elliptic-obovate, coarsely dentate. Inflorescence a few-flowered cyme; each flower with a lobed involucre. Petals 4-8 mm, yellow.

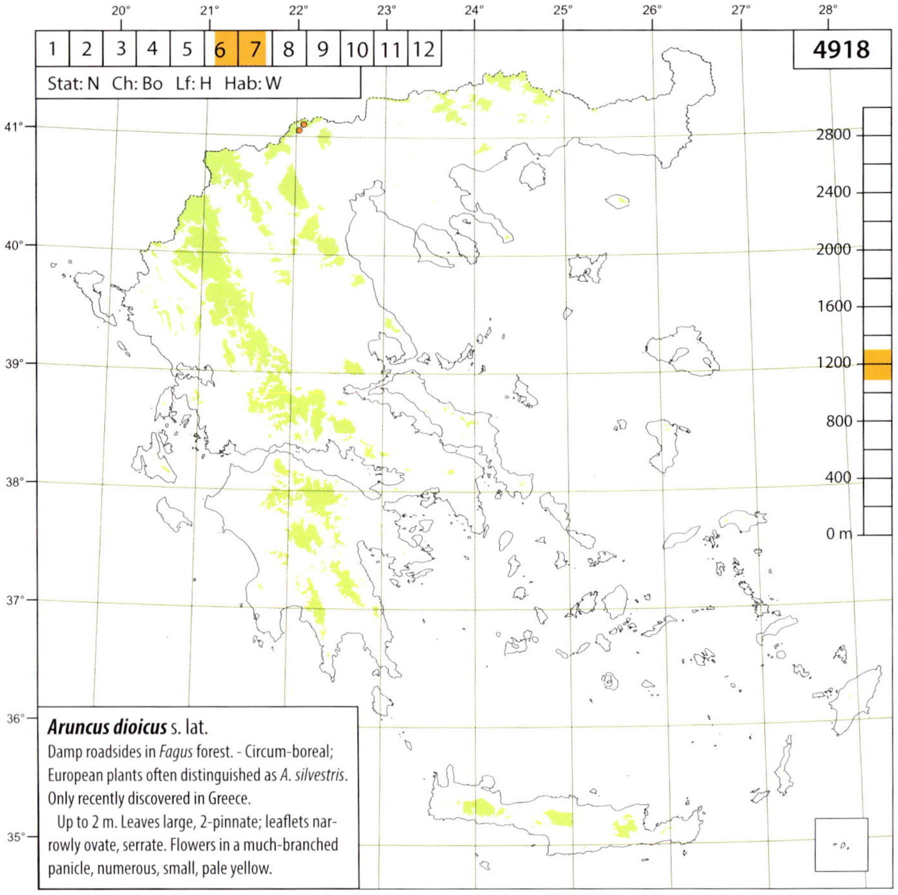

Aruncus dioicus s. lat.
Damp roadsides in *Fagus* forest. - Circum-boreal; European plants often distinguished as *A. silvestris*. Only recently discovered in Greece.
Up to 2 m. Leaves large, 2-pinnate; leaflets narrowly ovate, serrate. Flowers in a much-branched panicle, numerous, small, pale yellow.

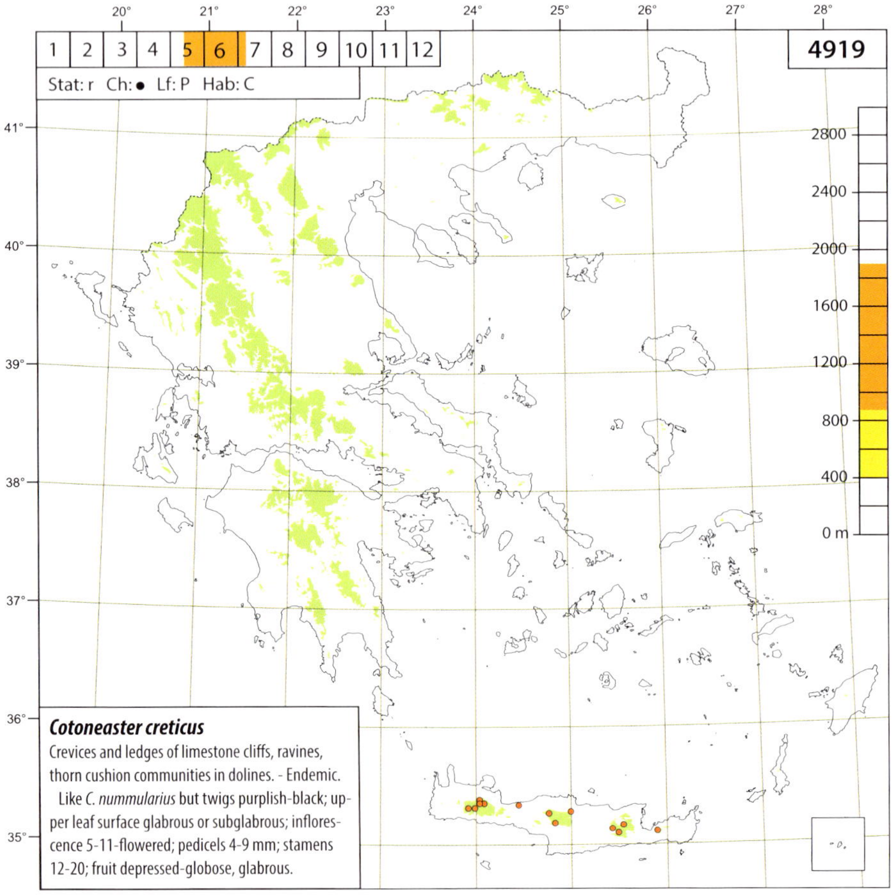

Cotoneaster creticus
Crevices and ledges of limestone cliffs, ravines, thorn cushion communities in dolines. - Endemic.
Like *C. nummularius* but twigs purplish-black; upper leaf surface glabrous or subglabrous; inflorescence 5-11-flowered; pedicels 4-9 mm; stamens 12-20; fruit depressed-globose, glabrous.

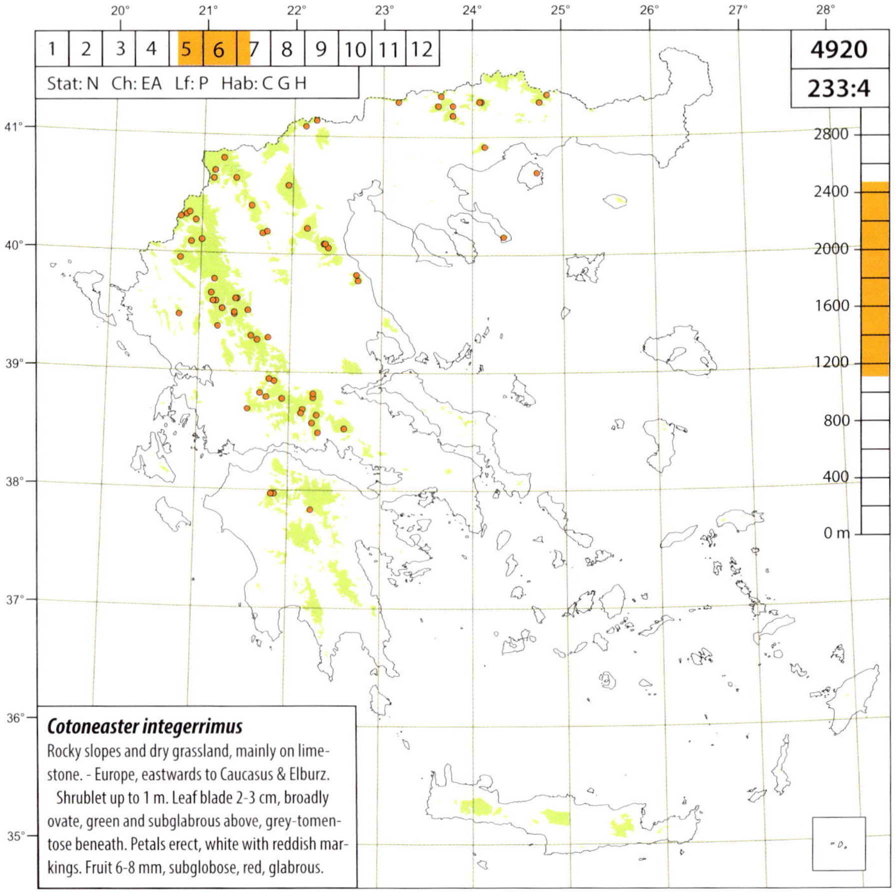

Cotoneaster integerrimus
Rocky slopes and dry grassland, mainly on limestone. - Europe, eastwards to Caucasus & Elburz.
Shrublet up to 1 m. Leaf blade 2-3 cm, broadly ovate, green and subglabrous above, grey-tomentose beneath. Petals erect, white with reddish markings. Fruit 6-8 mm, subglobose, red, glabrous.

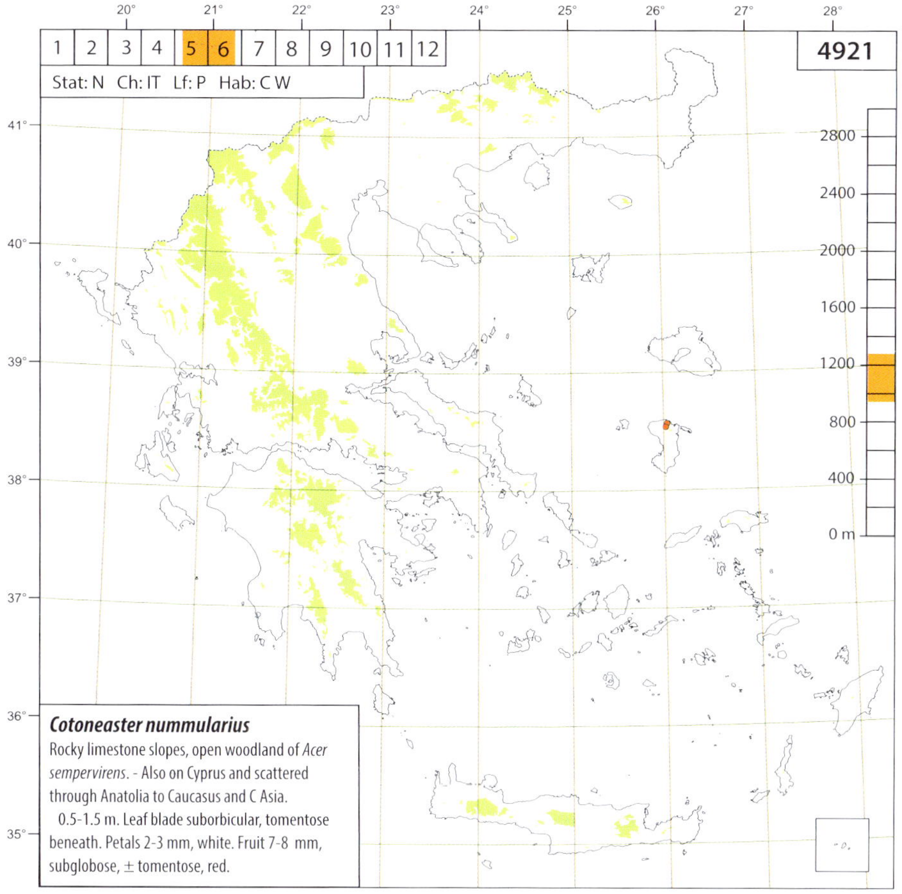

Cotoneaster nummularius

Rocky limestone slopes, open woodland of *Acer sempervirens*. - Also on Cyprus and scattered through Anatolia to Caucasus and C Asia.

0.5-1.5 m. Leaf blade suborbicular, tomentose beneath. Petals 2-3 mm, white. Fruit 7-8 mm, subglobose, ± tomentose, red.

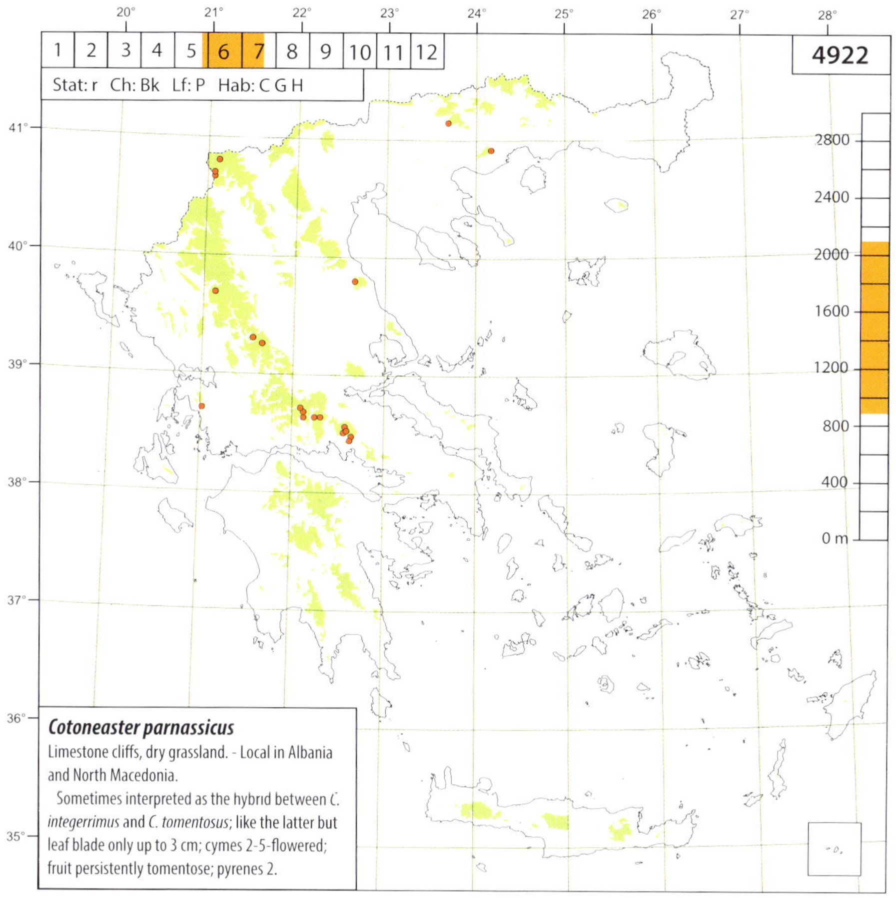

Cotoneaster parnassicus

Limestone cliffs, dry grassland. - Local in Albania and North Macedonia.

Sometimes interpreted as the hybrid between *C. integerrimus* and *C. tomentosus*; like the latter but leaf blade only up to 3 cm; cymes 2-5-flowered; fruit persistently tomentose; pyrenes 2.

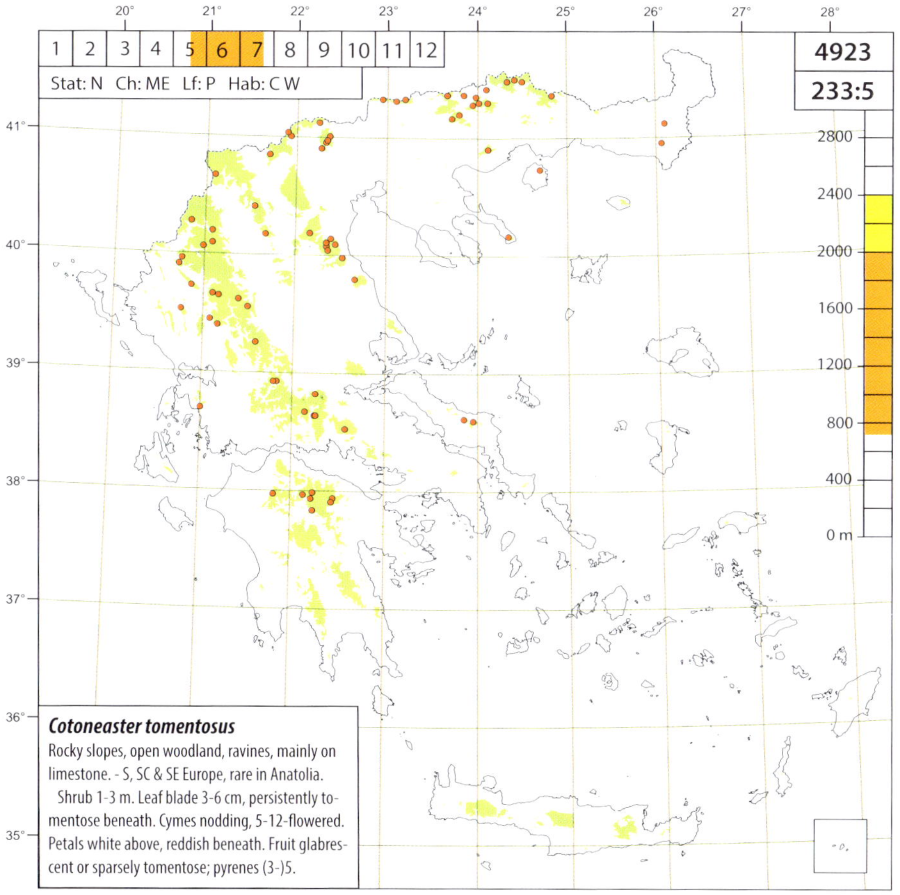

Cotoneaster tomentosus

Rocky slopes, open woodland, ravines, mainly on limestone. - S, SC & SE Europe, rare in Anatolia.

Shrub 1-3 m. Leaf blade 3-6 cm, persistently tomentose beneath. Cymes nodding, 5-12-flowered. Petals white above, reddish beneath. Fruit glabrescent or sparsely tomentose; pyrenes (3-)5.

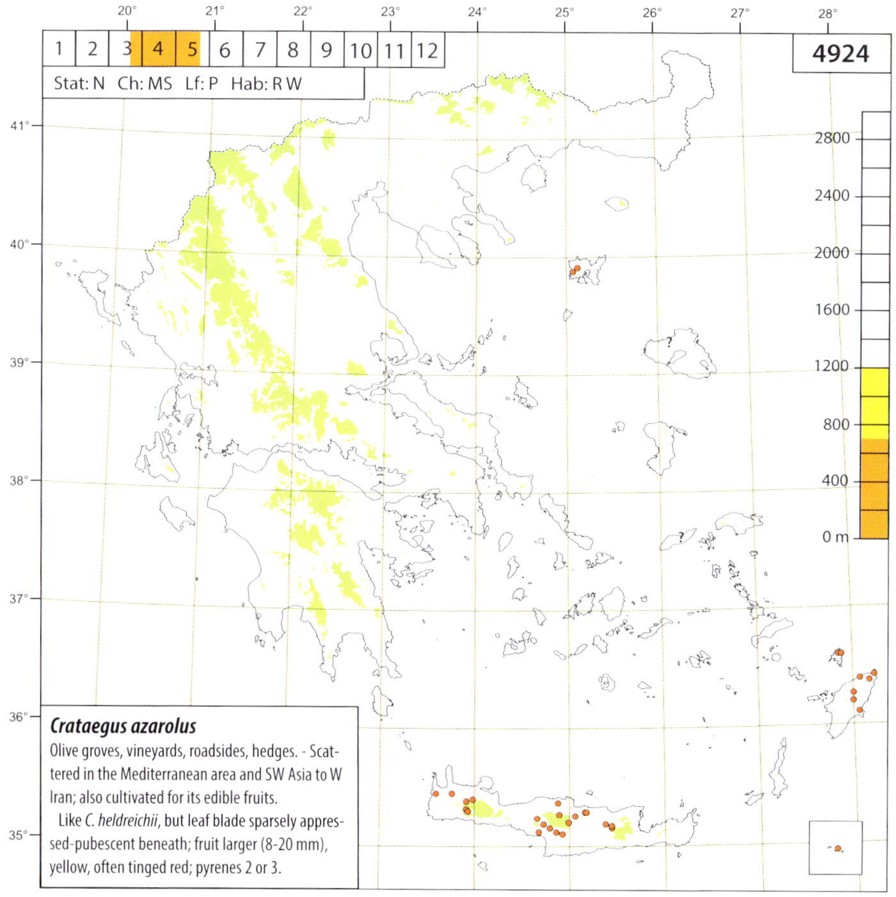

Crataegus azarolus

Olive groves, vineyards, roadsides, hedges. - Scattered in the Mediterranean area and SW Asia to W Iran; also cultivated for its edible fruits.

Like *C. heldreichii*, but leaf blade sparsely appressed-pubescent beneath; fruit larger (8-20 mm), yellow, often tinged red; pyrenes 2 or 3.

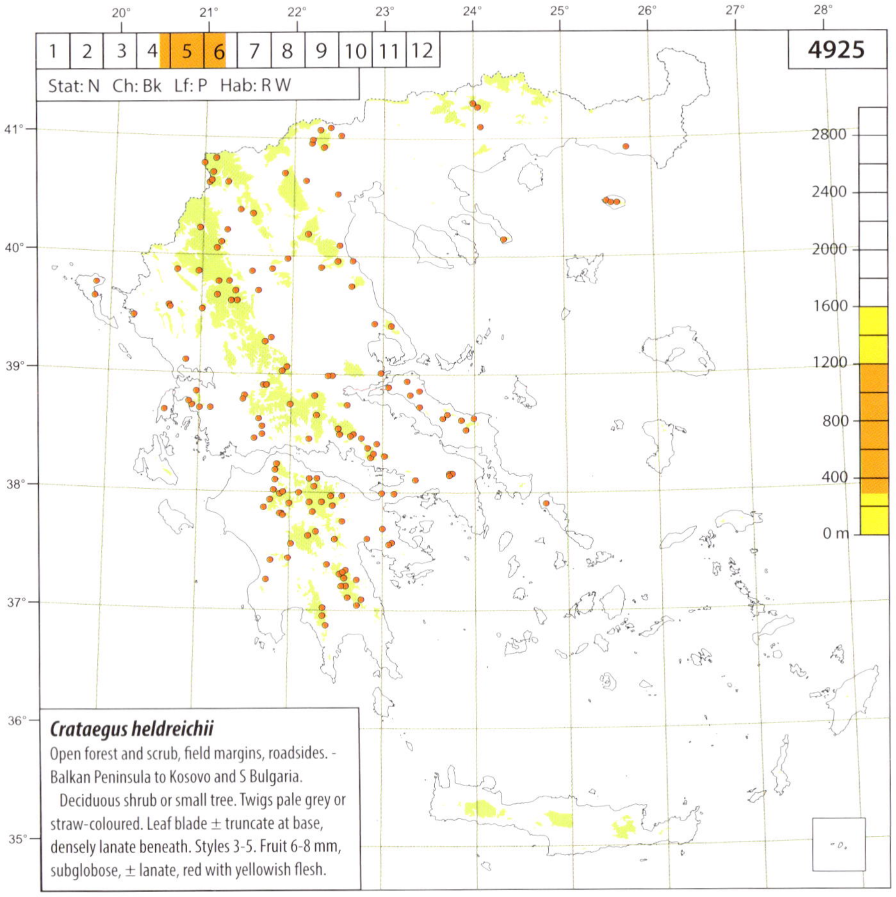

Crataegus heldreichii

Open forest and scrub, field margins, roadsides. - Balkan Peninsula to Kosovo and S Bulgaria.

Deciduous shrub or small tree. Twigs pale grey or straw-coloured. Leaf blade ± truncate at base, densely lanate beneath. Styles 3-5. Fruit 6-8 mm, subglobose, ± lanate, red with yellowish flesh.

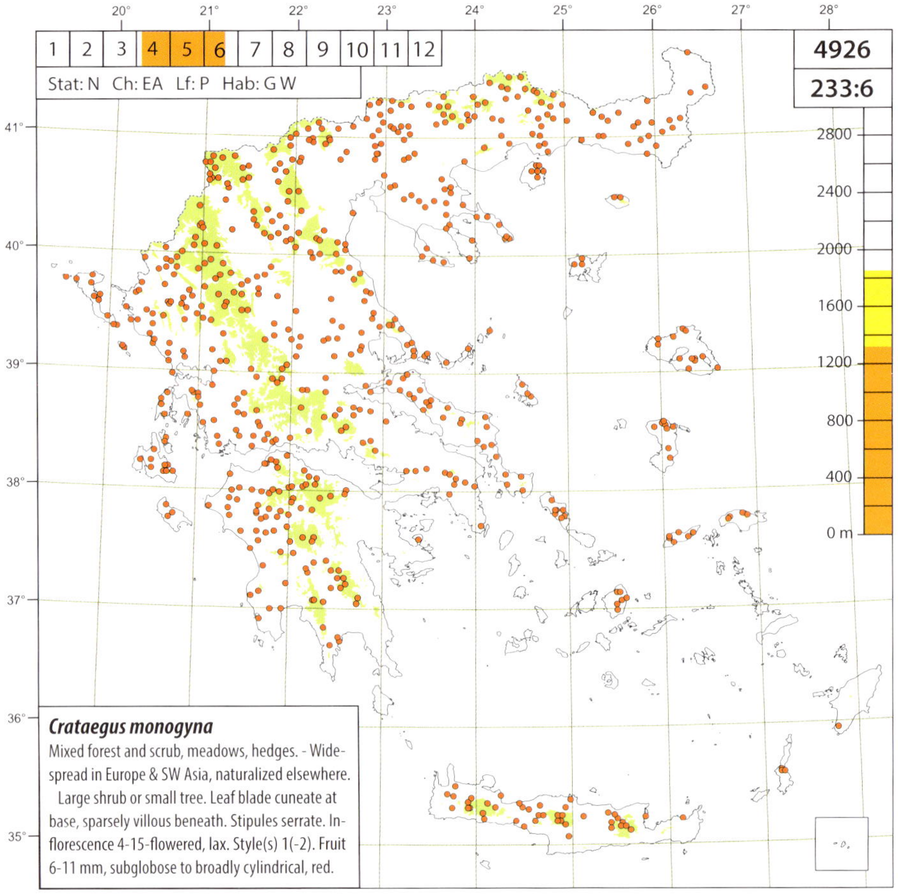

Crataegus monogyna

Mixed forest and scrub, meadows, hedges. - Widespread in Europe & SW Asia, naturalized elsewhere.

Large shrub or small tree. Leaf blade cuneate at base, sparsely villous beneath. Stipules serrate. Inflorescence 4-15-flowered, lax. Style(s) 1(-2). Fruit 6-11 mm, subglobose to broadly cylindrical, red.

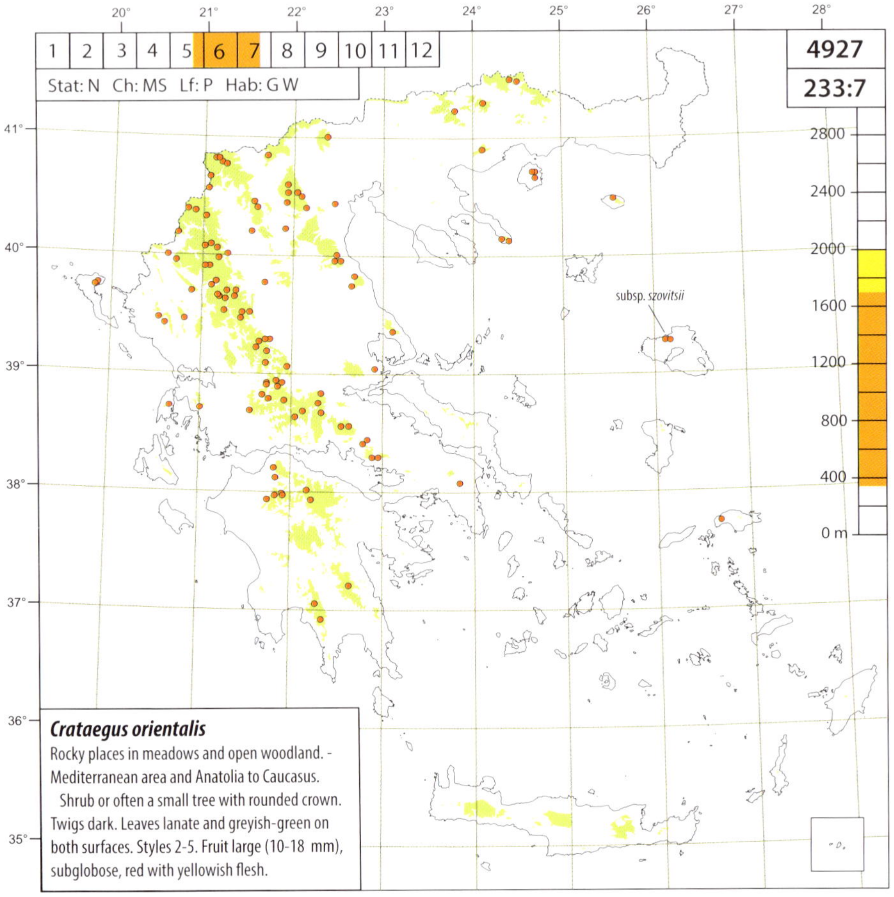

Crataegus orientalis

Rocky places in meadows and open woodland. - Mediterranean area and Anatolia to Caucasus.

Shrub or often a small tree with rounded crown. Twigs dark. Leaves lanate and greyish-green on both surfaces. Styles 2-5. Fruit large (10-18 mm), subglobose, red with yellowish flesh.

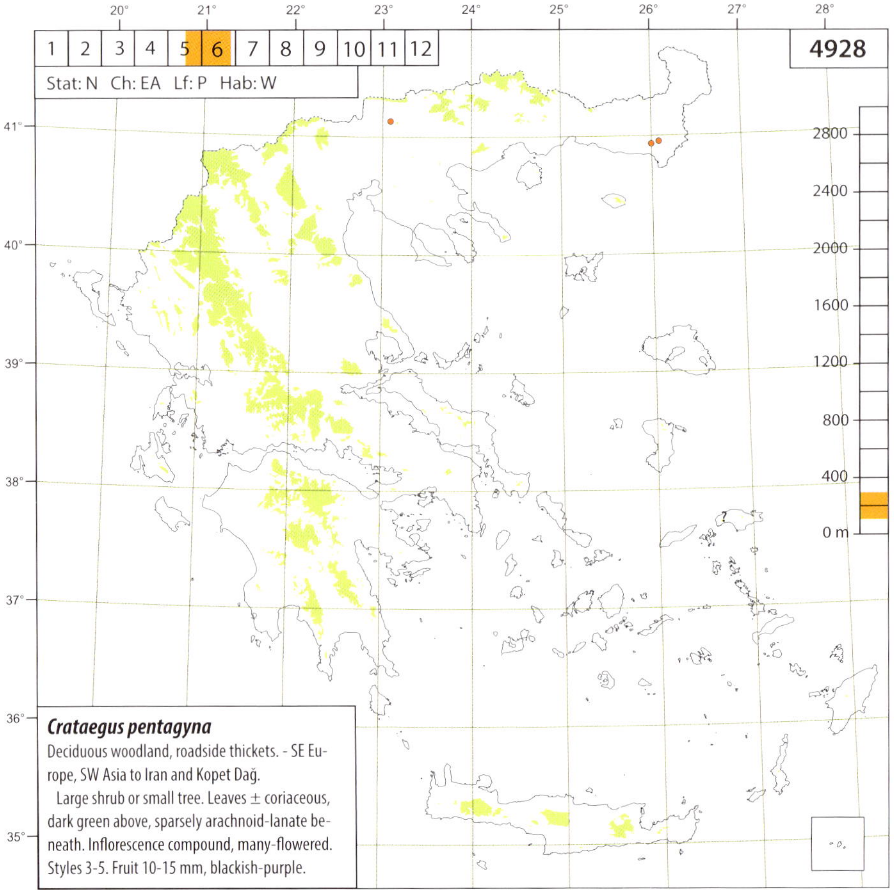

Crataegus pentagyna

Deciduous woodland, roadside thickets. - SE Europe, SW Asia to Iran and Kopet Dağ.

Large shrub or small tree. Leaves ± coriaceous, dark green above, sparsely arachnoid-lanate beneath. Inflorescence compound, many-flowered. Styles 3-5. Fruit 10-15 mm, blackish-purple.

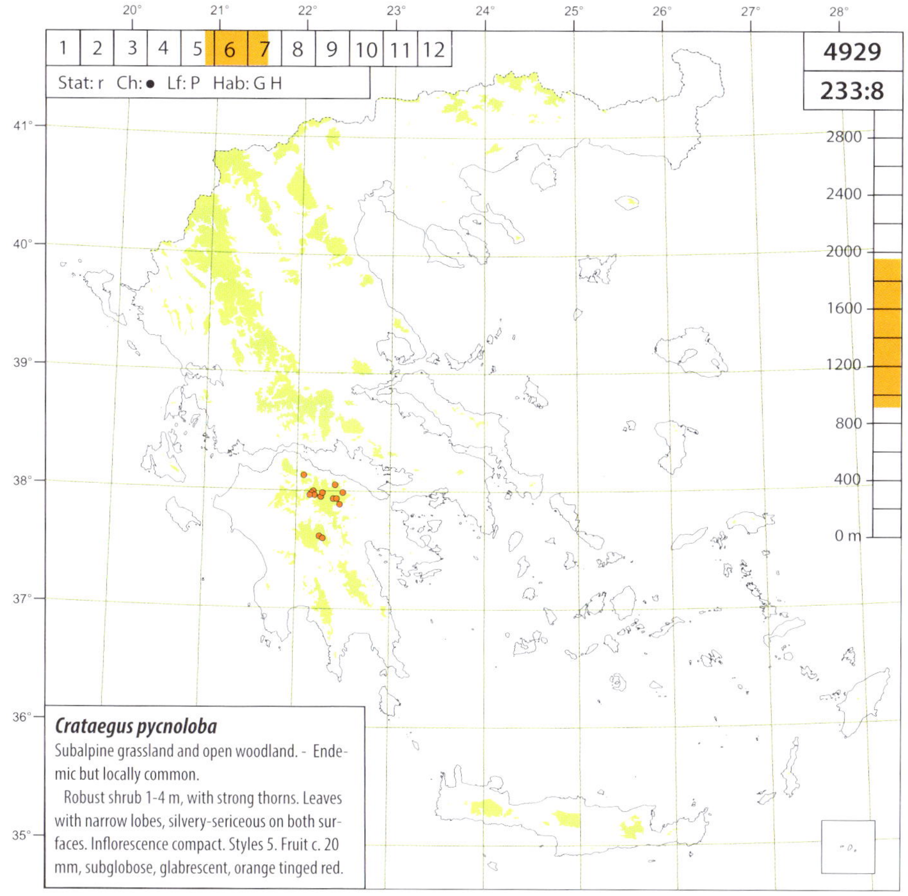

Crataegus pycnoloba

Subalpine grassland and open woodland. - Endemic but locally common.

Robust shrub 1-4 m, with strong thorns. Leaves with narrow lobes, silvery-sericeous on both surfaces. Inflorescence compact. Styles 5. Fruit c. 20 mm, subglobose, glabrescent, orange tinged red.

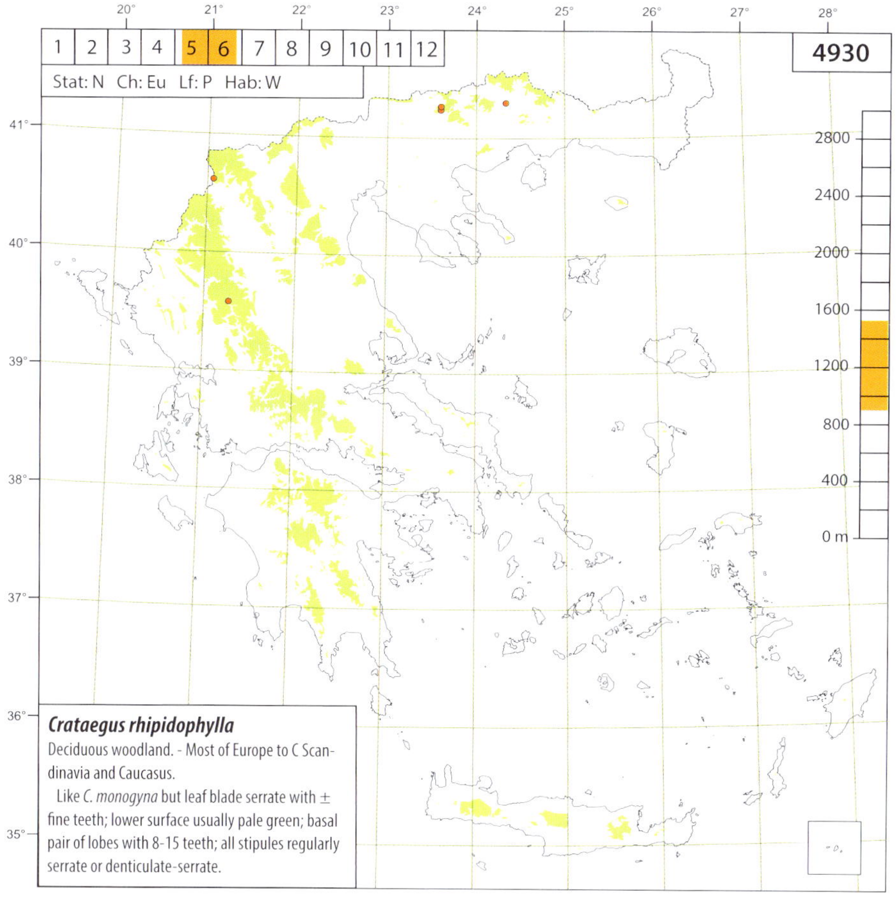

Crataegus rhipidophylla

Deciduous woodland. - Most of Europe to C Scandinavia and Caucasus.

Like *C. monogyna* but leaf blade serrate with ± fine teeth; lower surface usually pale green; basal pair of lobes with 8-15 teeth; all stipules regularly serrate or denticulate-serrate.

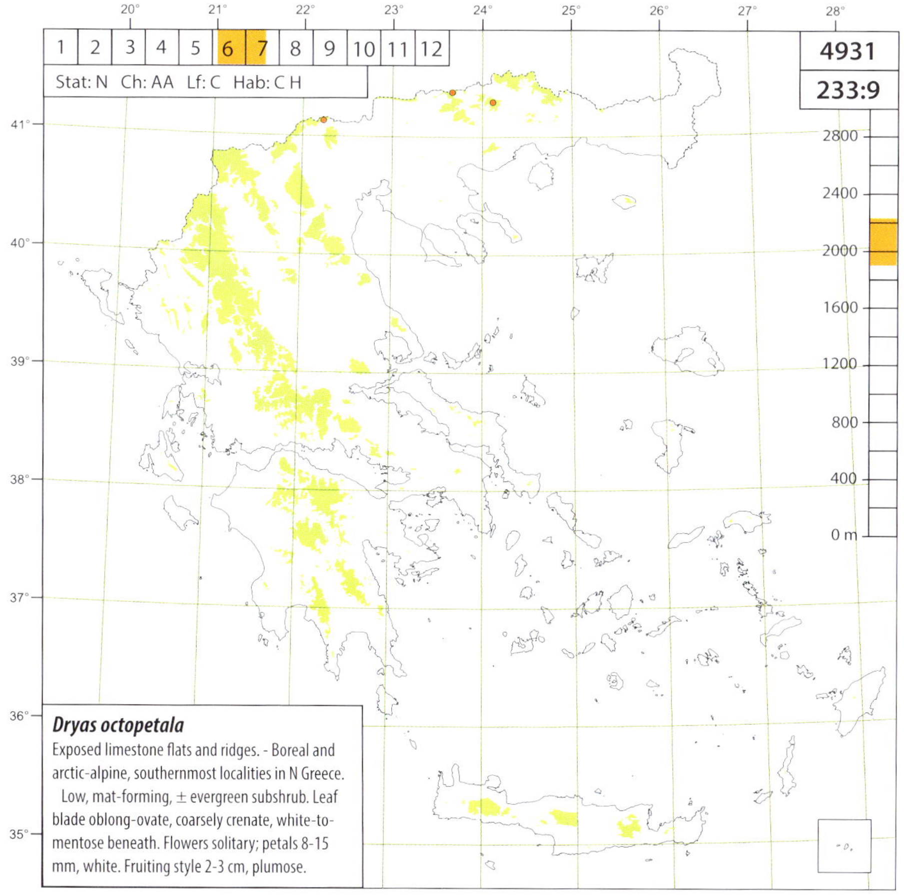

Dryas octopetala

Exposed limestone flats and ridges. - Boreal and arctic-alpine, southernmost localities in N Greece.

Low, mat-forming, ± evergreen subshrub. Leaf blade oblong-ovate, coarsely crenate, white-tomentose beneath. Flowers solitary; petals 8-15 mm, white. Fruiting style 2-3 cm, plumose.

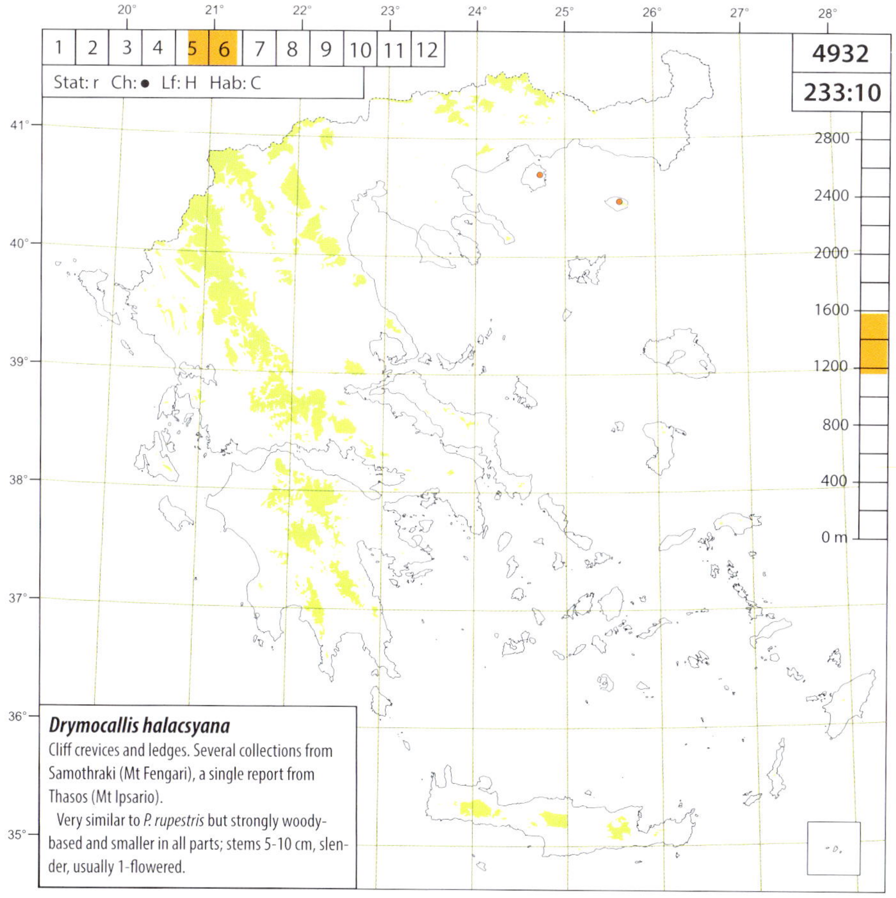

Drymocallis halacsyana

Cliff crevices and ledges. Several collections from Samothraki (Mt Fengari), a single report from Thasos (Mt Ipsario).

Very similar to *P. rupestris* but strongly woody-based and smaller in all parts; stems 5-10 cm, slender, usually 1-flowered.

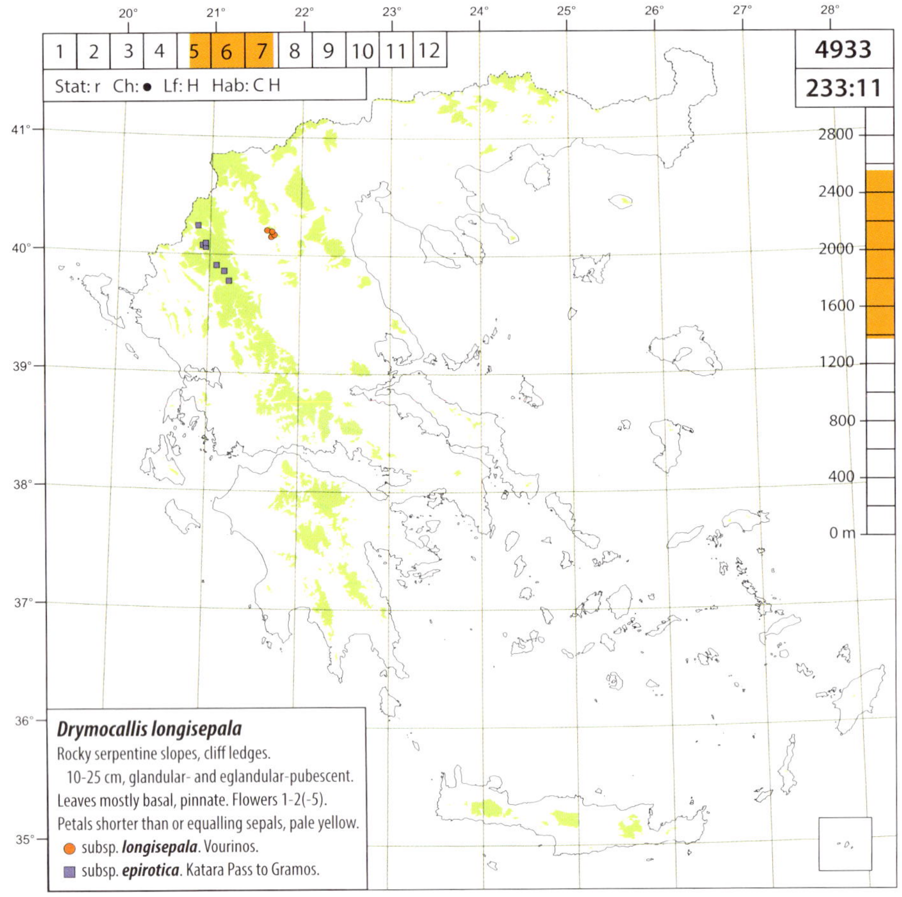

Drymocallis longisepala
Rocky serpentine slopes, cliff ledges.
10-25 cm, glandular- and eglandular-pubescent. Leaves mostly basal, pinnate. Flowers 1-2(-5). Petals shorter than or equalling sepals, pale yellow.
● subsp. ***longisepala***. Vourinos.
■ subsp. ***epirotica***. Katara Pass to Gramos.

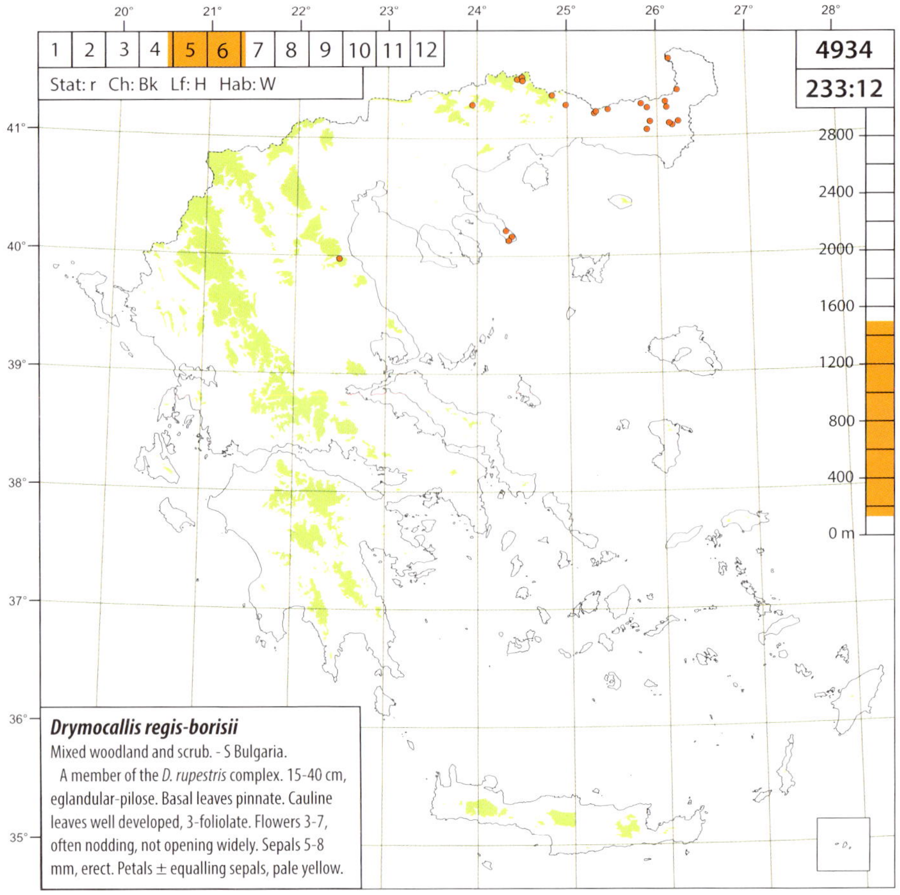

Drymocallis regis-borisii
Mixed woodland and scrub. - S Bulgaria.
A member of the *D. rupestris* complex. 15-40 cm, eglandular-pilose. Basal leaves pinnate. Cauline leaves well developed, 3-foliolate. Flowers 3-7, often nodding, not opening widely. Sepals 5-8 mm, erect. Petals ± equalling sepals, pale yellow.

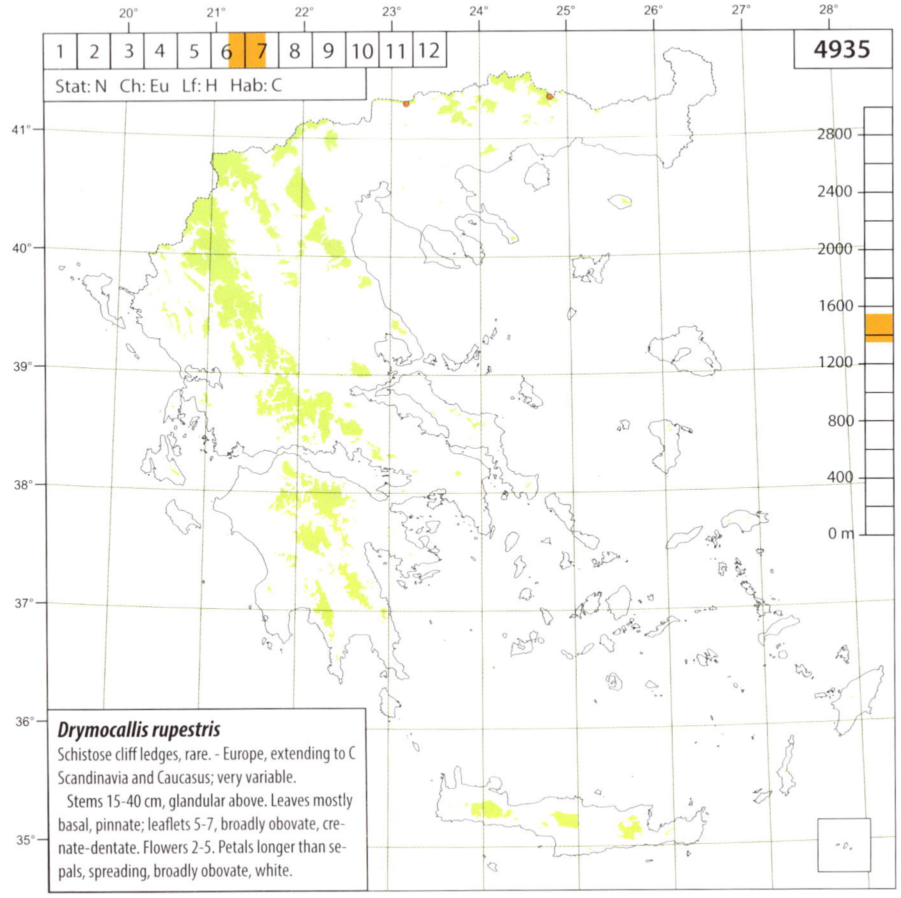

Drymocallis rupestris
Schistose cliff ledges, rare. - Europe, extending to C Scandinavia and Caucasus; very variable.
Stems 15-40 cm, glandular above. Leaves mostly basal, pinnate; leaflets 5-7, broadly obovate, crenate-dentate. Flowers 2-5. Petals longer than sepals, spreading, broadly obovate, white.

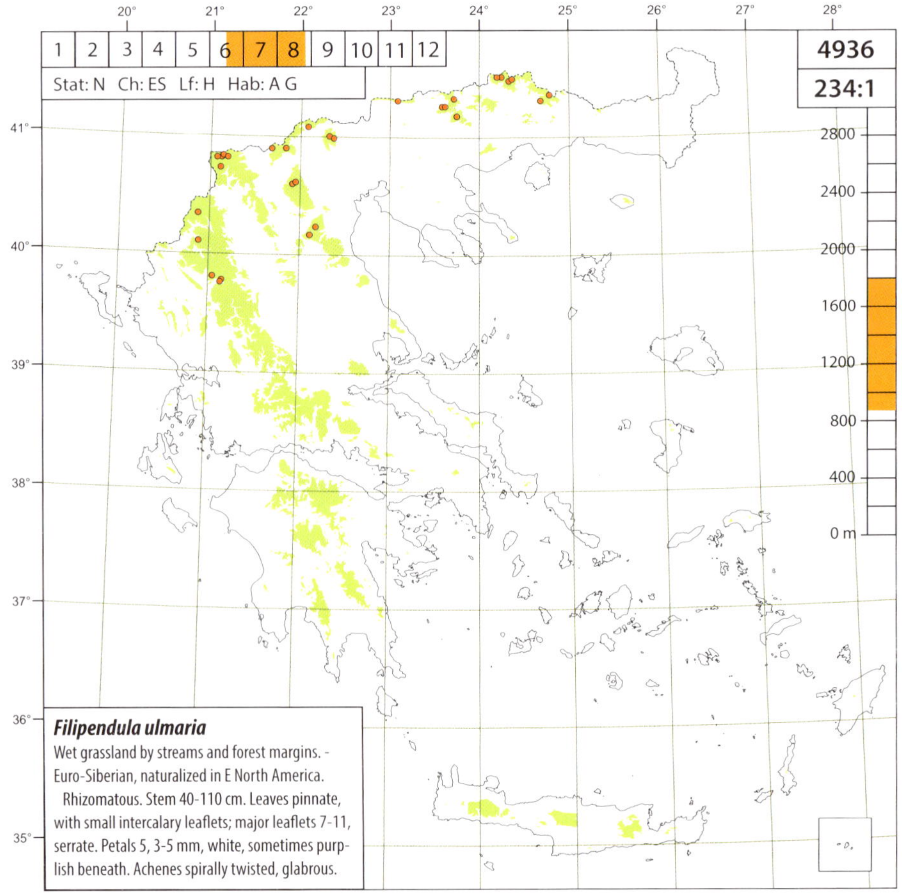

Filipendula ulmaria
Wet grassland by streams and forest margins. - Euro-Siberian, naturalized in E North America.
Rhizomatous. Stem 40-110 cm. Leaves pinnate, with small intercalary leaflets; major leaflets 7-11, serrate. Petals 5, 3-5 mm, white, sometimes purplish beneath. Achenes spirally twisted, glabrous.

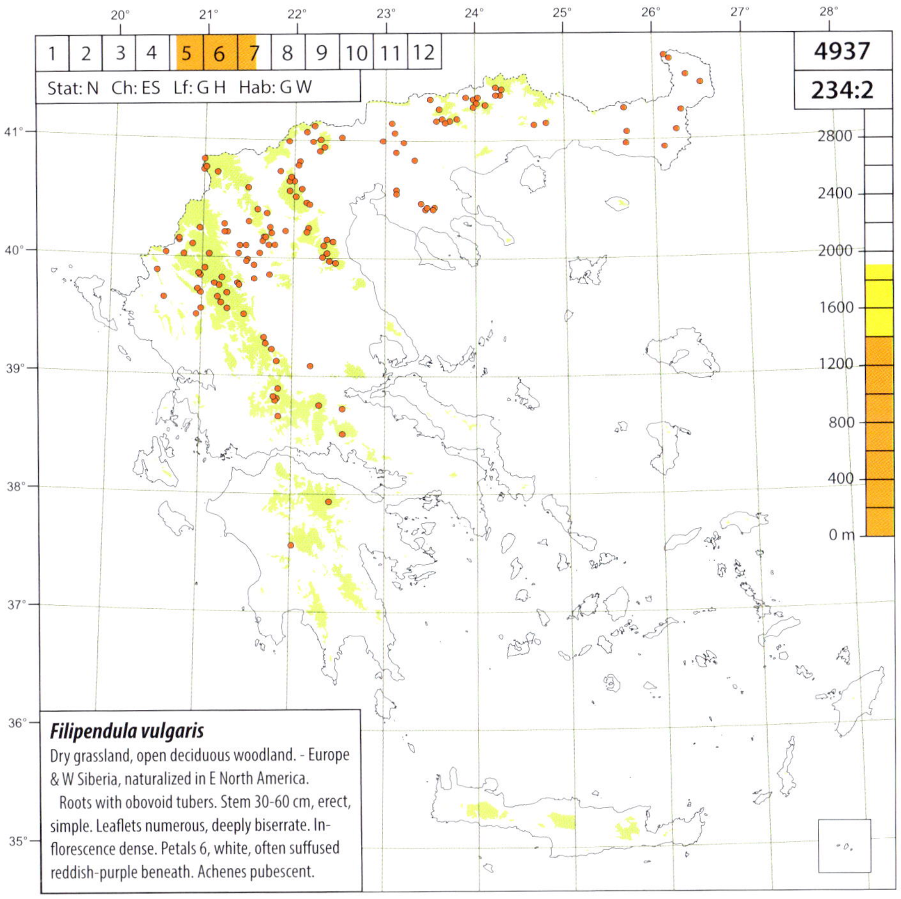

Filipendula vulgaris

Dry grassland, open deciduous woodland. - Europe & W Siberia, naturalized in E North America.

Roots with obovoid tubers. Stem 30-60 cm, erect, simple. Leaflets numerous, deeply biserrate. Inflorescence dense. Petals 6, white, often suffused reddish-purple beneath. Achenes pubescent.

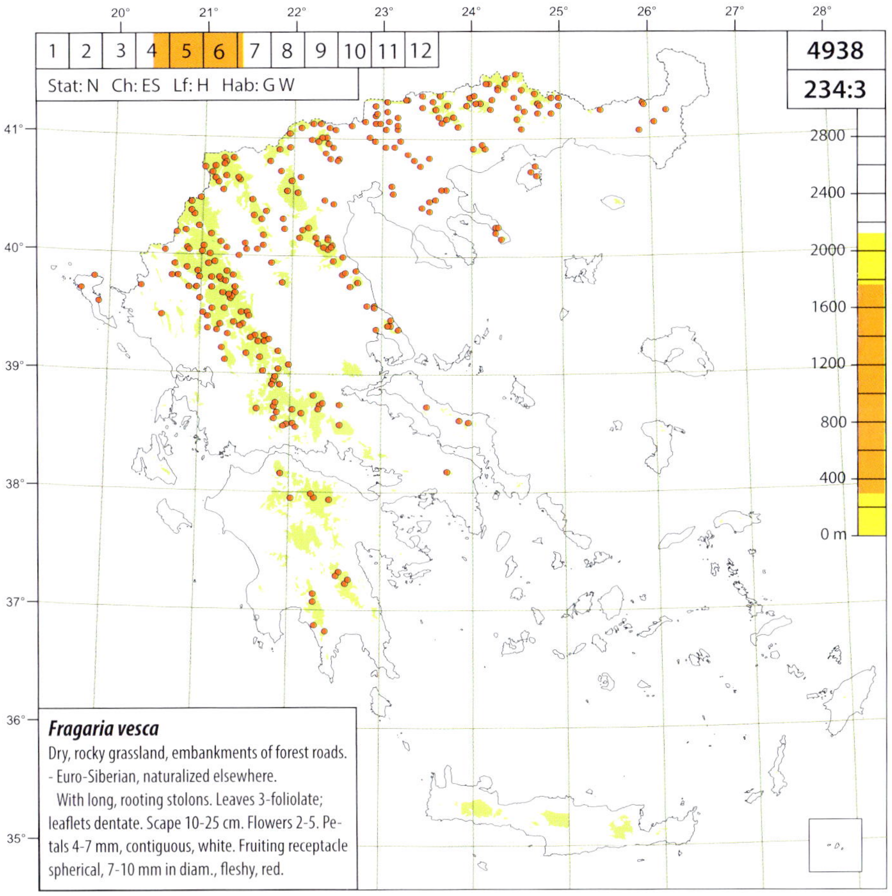

Fragaria vesca

Dry, rocky grassland, embankments of forest roads. - Euro-Siberian, naturalized elsewhere.

With long, rooting stolons. Leaves 3-foliolate; leaflets dentate. Scape 10-25 cm. Flowers 2-5. Petals 4-7 mm, contiguous, white. Fruiting receptacle spherical, 7-10 mm in diam., fleshy, red.

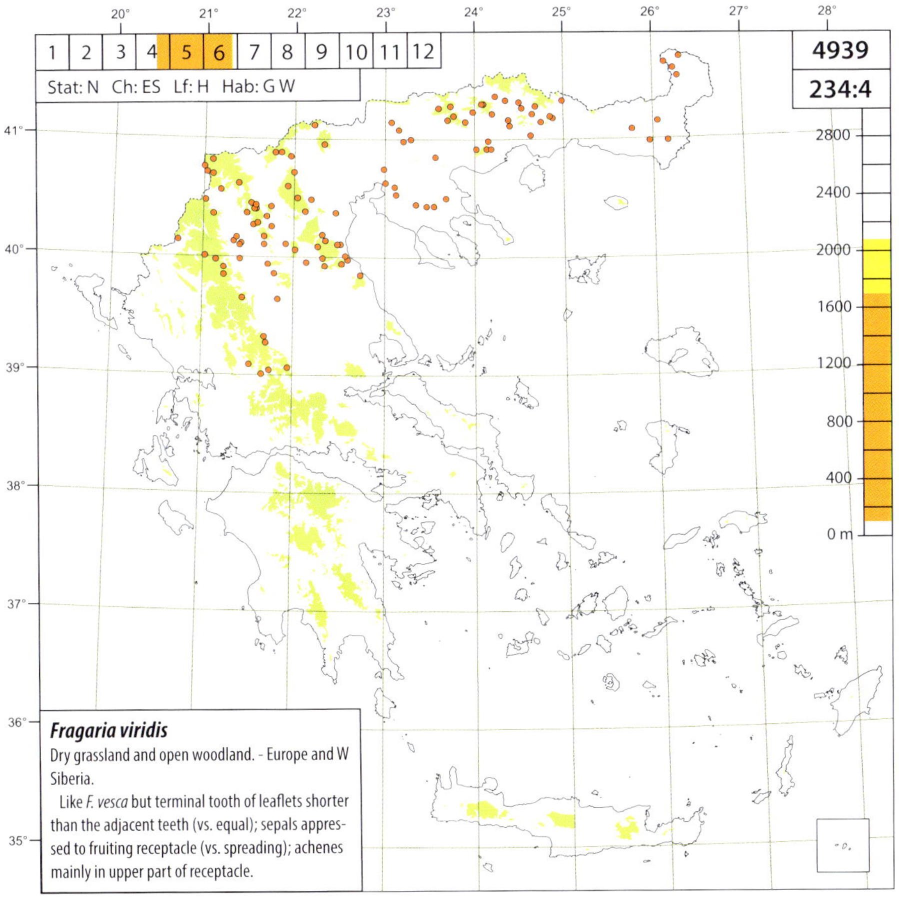

Fragaria viridis

Dry grassland and open woodland. - Europe and W Siberia.

Like *F. vesca* but terminal tooth of leaflets shorter than the adjacent teeth (vs. equal); sepals appressed to fruiting receptacle (vs. spreading); achenes mainly in upper part of receptacle.

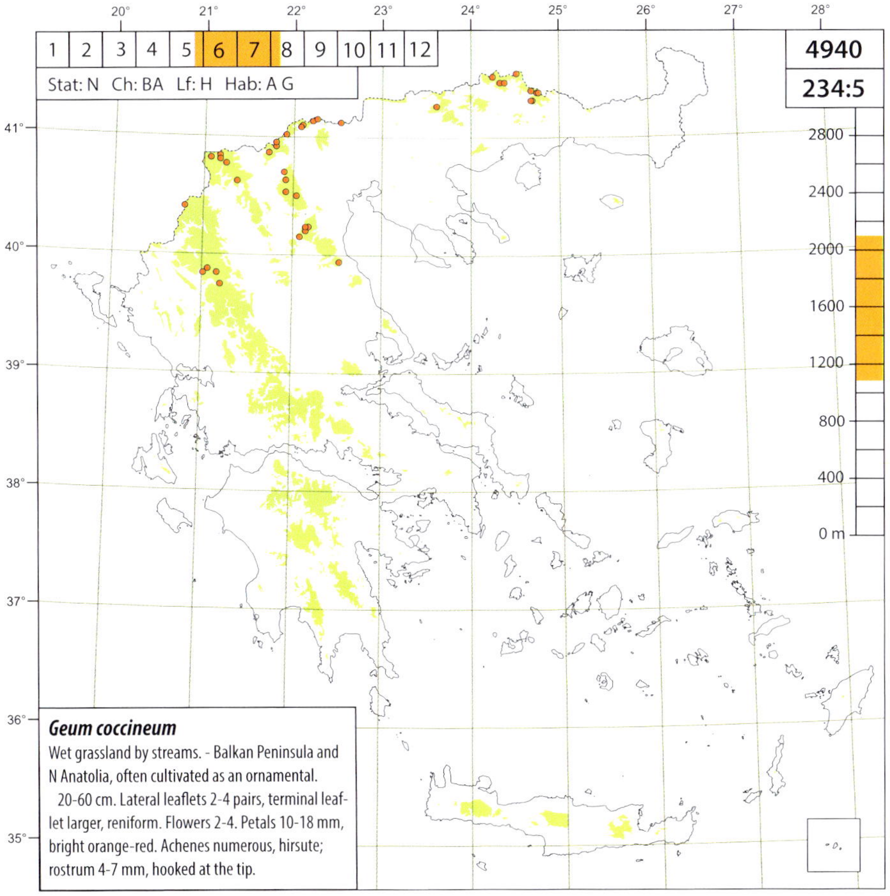

Geum coccineum

Wet grassland by streams. - Balkan Peninsula and N Anatolia, often cultivated as an ornamental.

20-60 cm. Lateral leaflets 2-4 pairs, terminal leaflet larger, reniform. Flowers 2-4. Petals 10-18 mm, bright orange-red. Achenes numerous, hirsute; rostrum 4-7 mm, hooked at the tip.

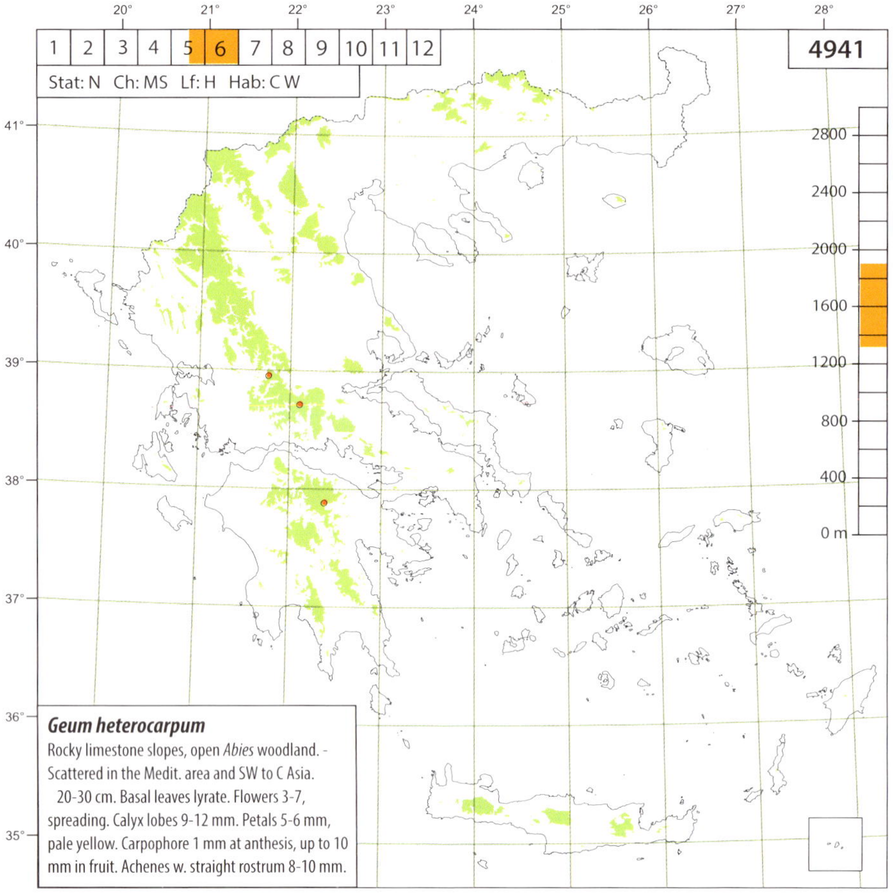

Geum heterocarpum

Rocky limestone slopes, open *Abies* woodland. - Scattered in the Medit. area and SW to C Asia.

20-30 cm. Basal leaves lyrate. Flowers 3-7, spreading. Calyx lobes 9-12 mm. Petals 5-6 mm, pale yellow. Carpophore 1 mm at anthesis, up to 10 mm in fruit. Achenes w. straight rostrum 8-10 mm.

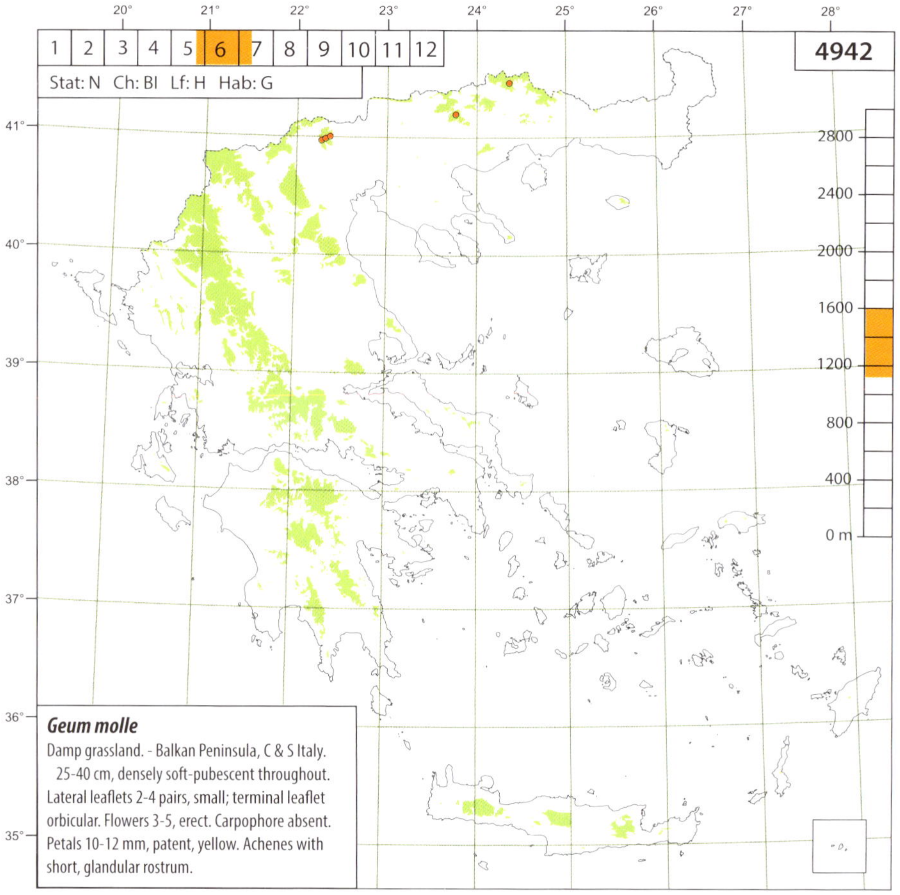

Geum molle

Damp grassland. - Balkan Peninsula, C & S Italy.

25-40 cm, densely soft-pubescent throughout. Lateral leaflets 2-4 pairs, small; terminal leaflet orbicular. Flowers 3-5, erect. Carpophore absent. Petals 10-12 mm, patent, yellow. Achenes with short, glandular rostrum.

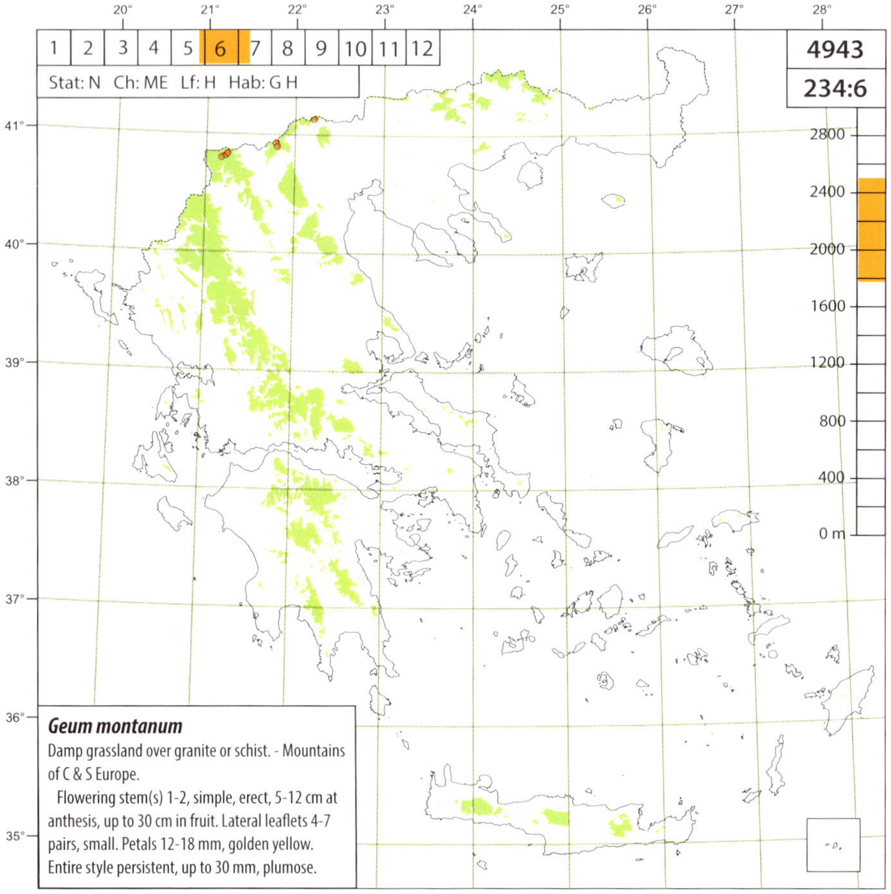

Geum montanum

Damp grassland over granite or schist. - Mountains of C & S Europe.

Flowering stem(s) 1-2, simple, erect, 5-12 cm at anthesis, up to 30 cm in fruit. Lateral leaflets 4-7 pairs, small. Petals 12-18 mm, golden yellow. Entire style persistent, up to 30 mm, plumose.

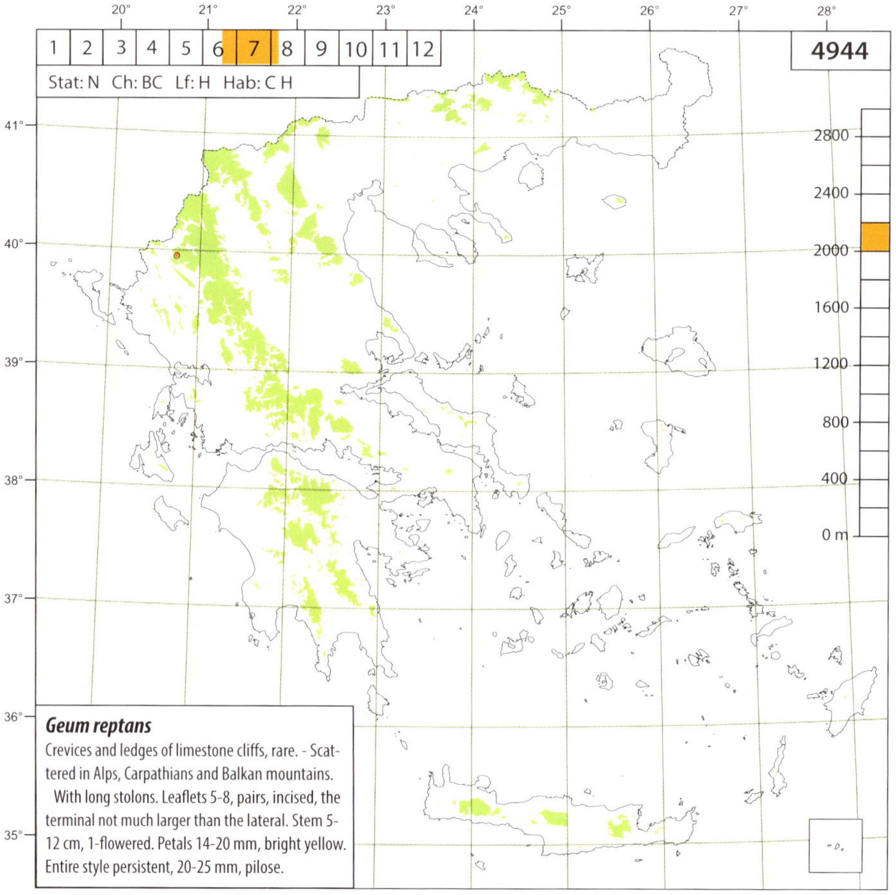

Geum reptans

Crevices and ledges of limestone cliffs, rare. - Scattered in Alps, Carpathians and Balkan mountains.

With long stolons. Leaflets 5-8, pairs, incised, the terminal not much larger than the lateral. Stem 5-12 cm, 1-flowered. Petals 14-20 mm, bright yellow. Entire style persistent, 20-25 mm, pilose.

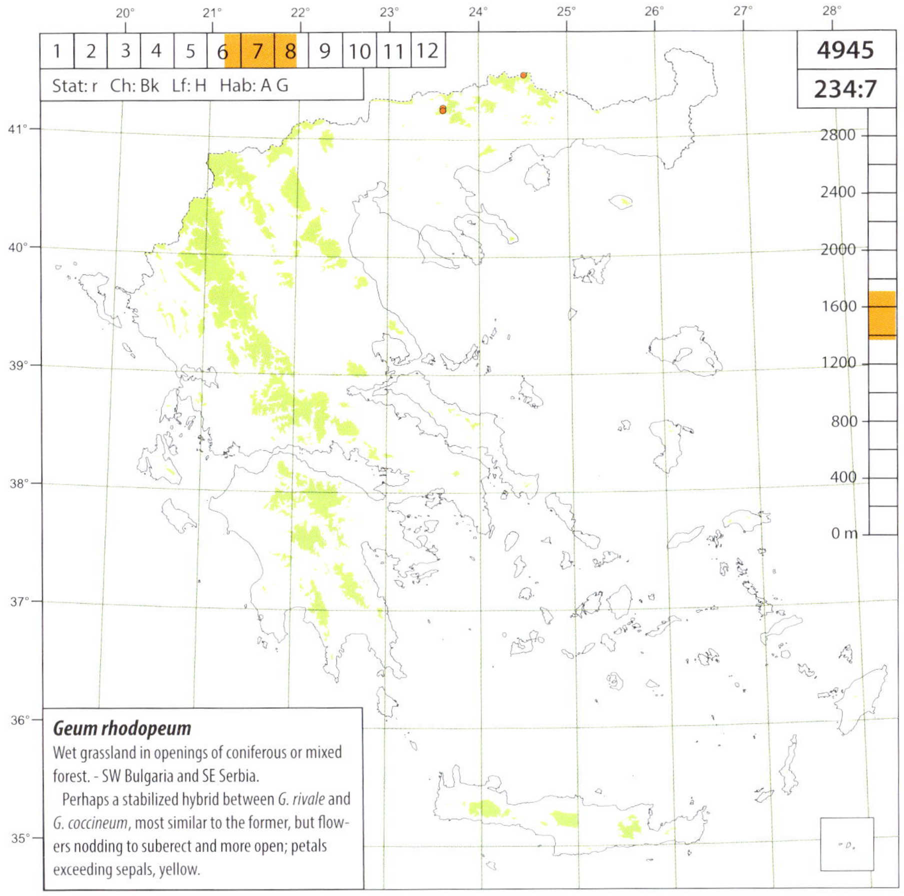

Geum rhodopeum

Wet grassland in openings of coniferous or mixed forest. - SW Bulgaria and SE Serbia.

Perhaps a stabilized hybrid between *G. rivale* and *G. coccineum*, most similar to the former, but flowers nodding to suberect and more open; petals exceeding sepals, yellow.

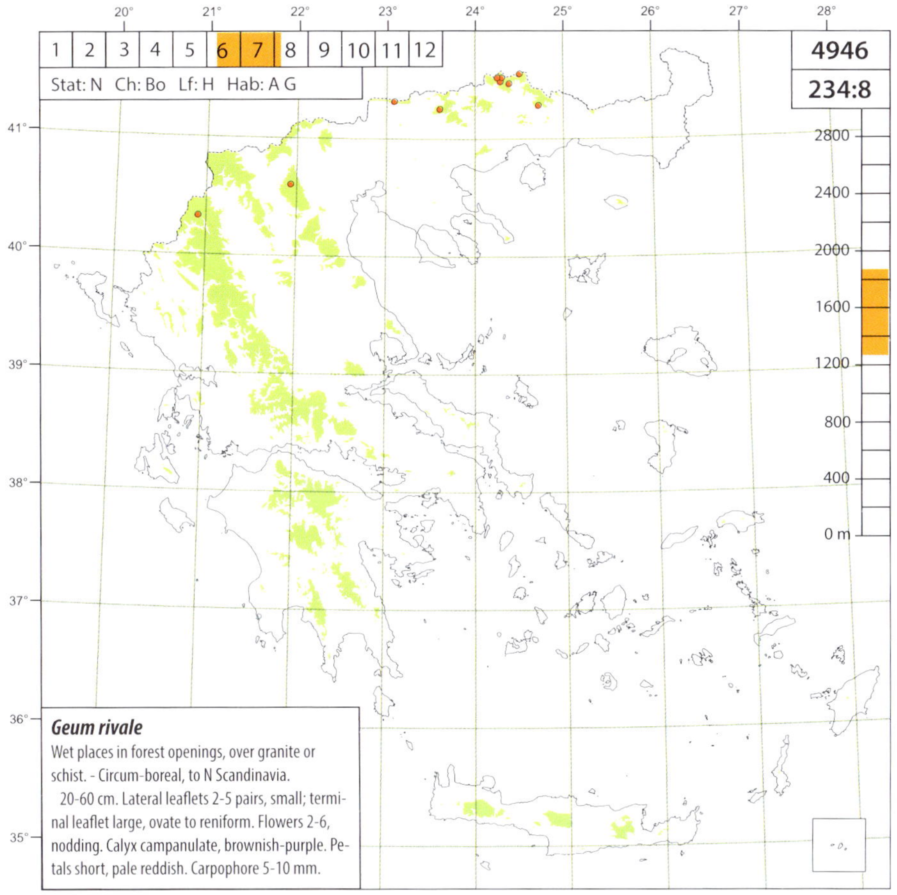

Geum rivale

Wet places in forest openings, over granite or schist. - Circum-boreal, to N Scandinavia.

20-60 cm. Lateral leaflets 2-5 pairs, small; terminal leaflet large, ovate to reniform. Flowers 2-6, nodding. Calyx campanulate, brownish-purple. Petals short, pale reddish. Carpophore 5-10 mm.

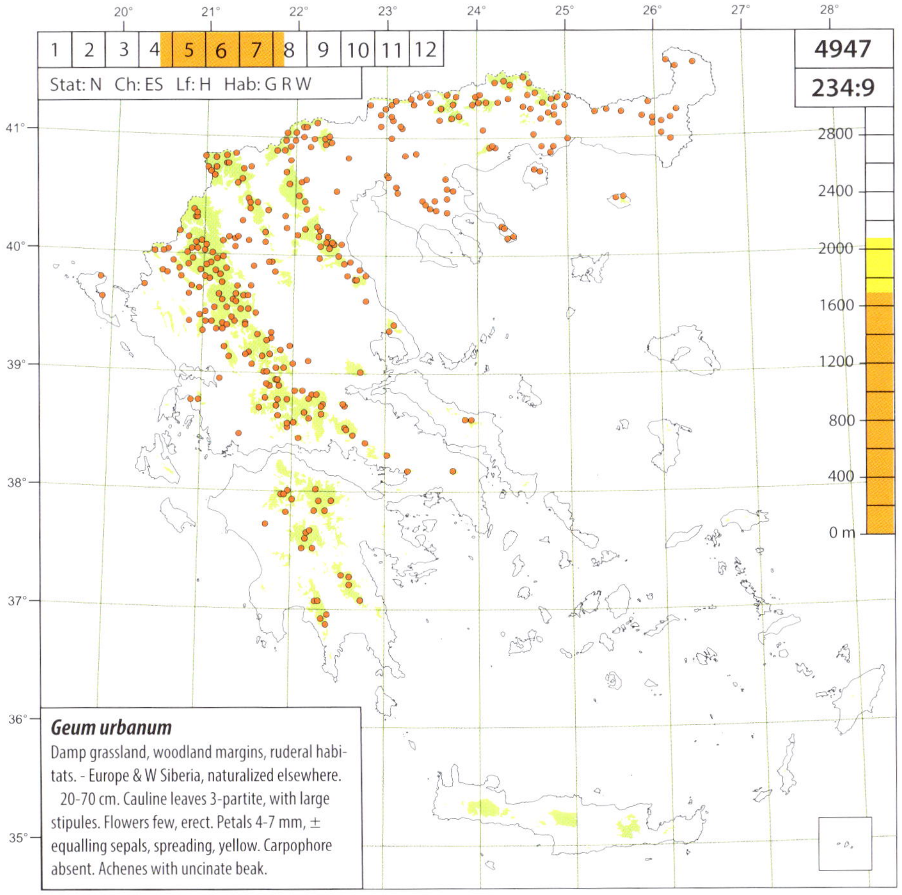

Geum urbanum

Damp grassland, woodland margins, ruderal habitats. - Europe & W Siberia, naturalized elsewhere.

20-70 cm. Cauline leaves 3-partite, with large stipules. Flowers few, erect. Petals 4-7 mm, ± equalling sepals, spreading, yellow. Carpophore absent. Achenes with uncinate beak.

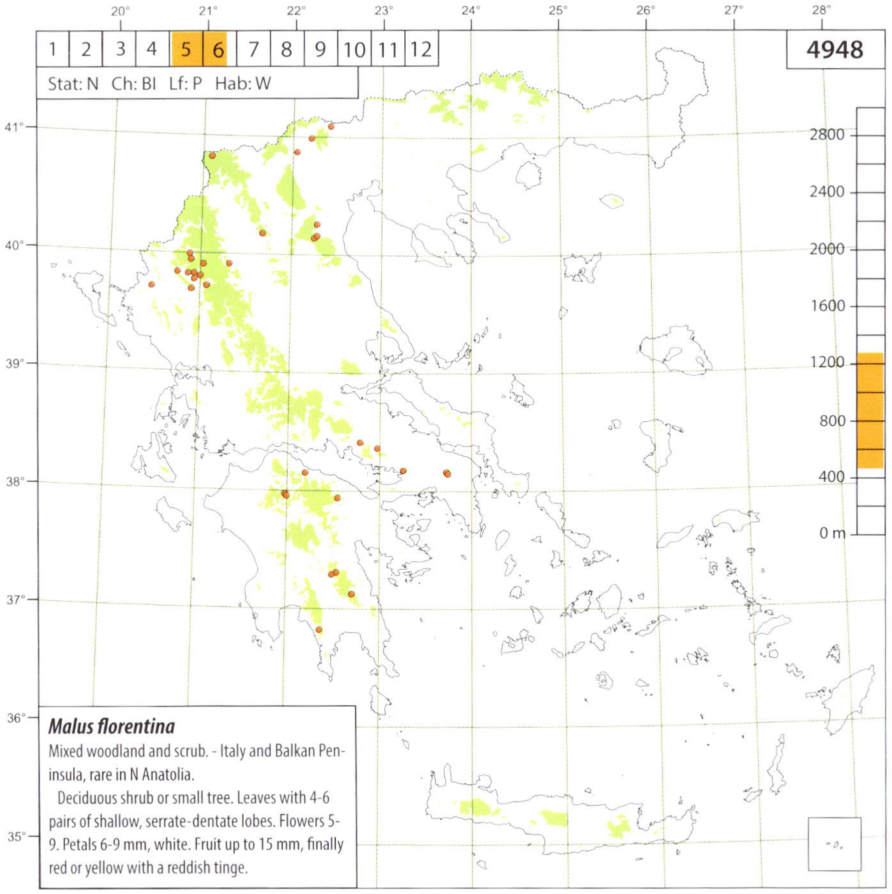

Malus florentina

Mixed woodland and scrub. - Italy and Balkan Peninsula, rare in N Anatolia.

Deciduous shrub or small tree. Leaves with 4-6 pairs of shallow, serrate-dentate lobes. Flowers 5-9. Petals 6-9 mm, white. Fruit up to 15 mm, finally red or yellow with a reddish tinge.

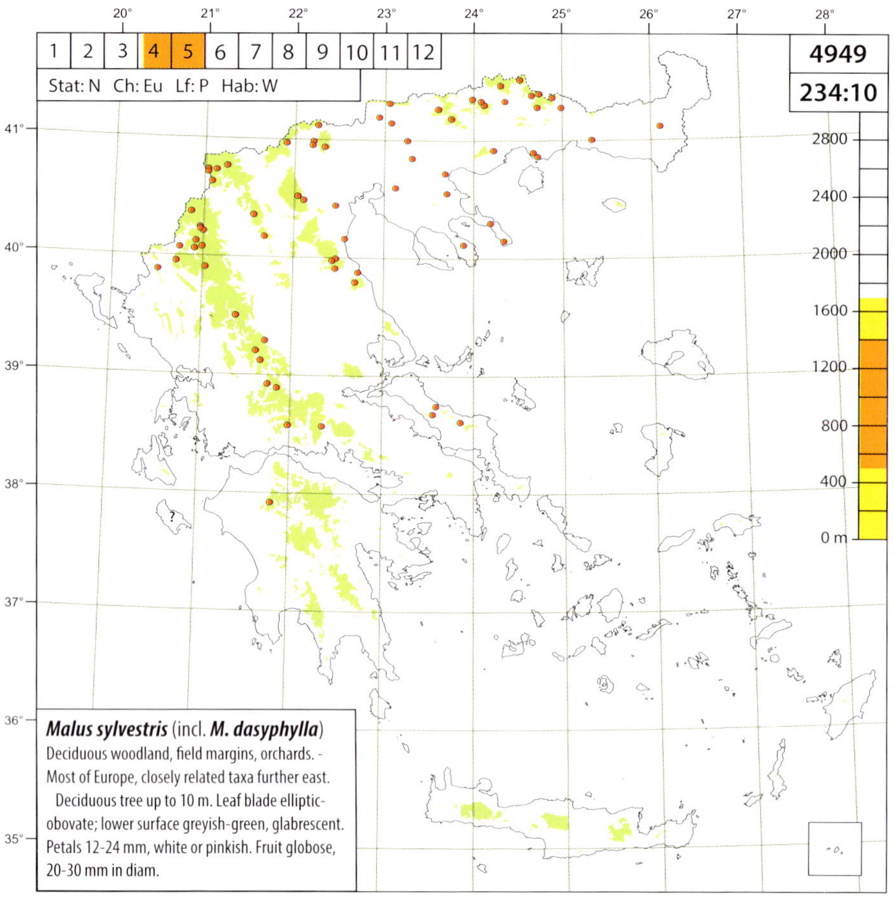

Malus sylvestris (incl. ***M. dasyphylla***)
Deciduous woodland, field margins, orchards. - Most of Europe, closely related taxa further east.
Deciduous tree up to 10 m. Leaf blade elliptic-obovate; lower surface greyish-green, glabrescent. Petals 12-24 mm, white or pinkish. Fruit globose, 20-30 mm in diam.

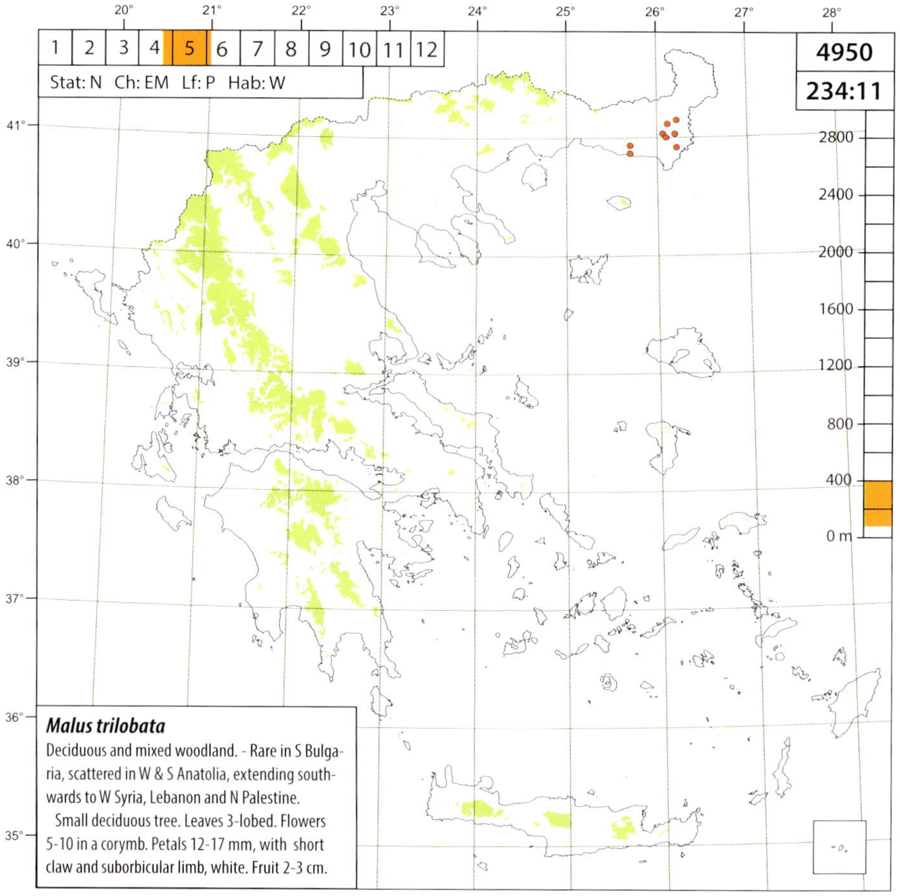

Malus trilobata
Deciduous and mixed woodland. - Rare in S Bulgaria, scattered in W & S Anatolia, extending southwards to W Syria, Lebanon and N Palestine.
Small deciduous tree. Leaves 3-lobed. Flowers 5-10 in a corymb. Petals 12-17 mm, with short claw and suborbicular limb, white. Fruit 2-3 cm.

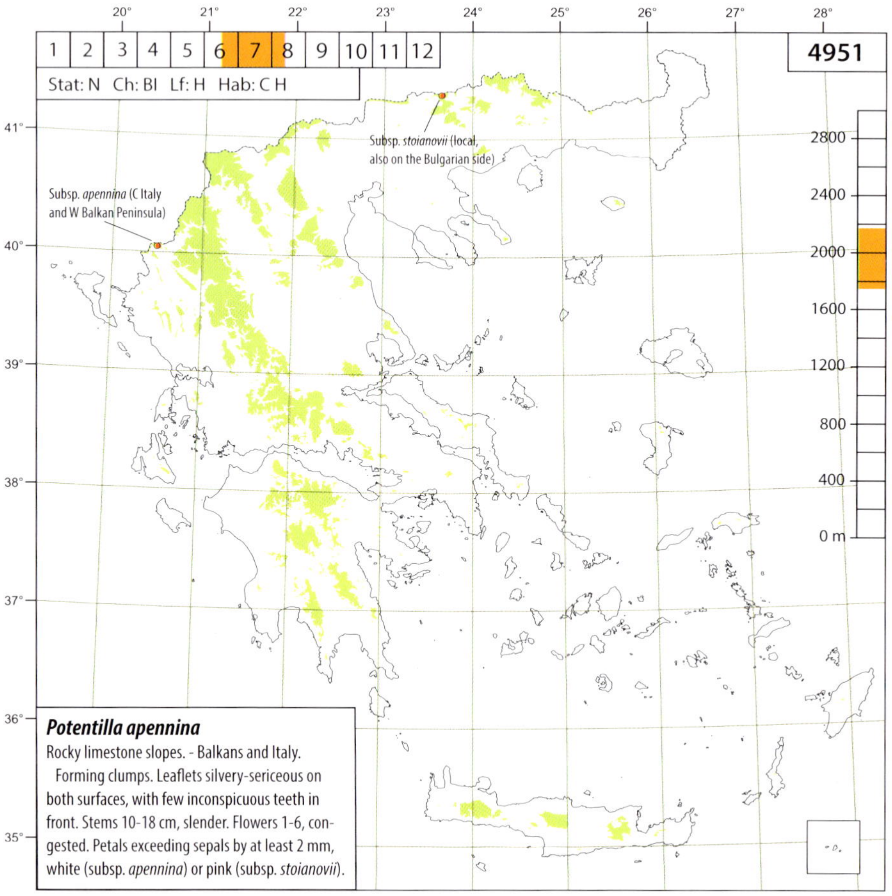

Potentilla apennina
Rocky limestone slopes. - Balkans and Italy.
Forming clumps. Leaflets silvery-sericeous on both surfaces, with few inconspicuous teeth in front. Stems 10-18 cm, slender. Flowers 1-6, congested. Petals exceeding sepals by at least 2 mm, white (subsp. *apennina*) or pink (subsp. *stoianovii*).

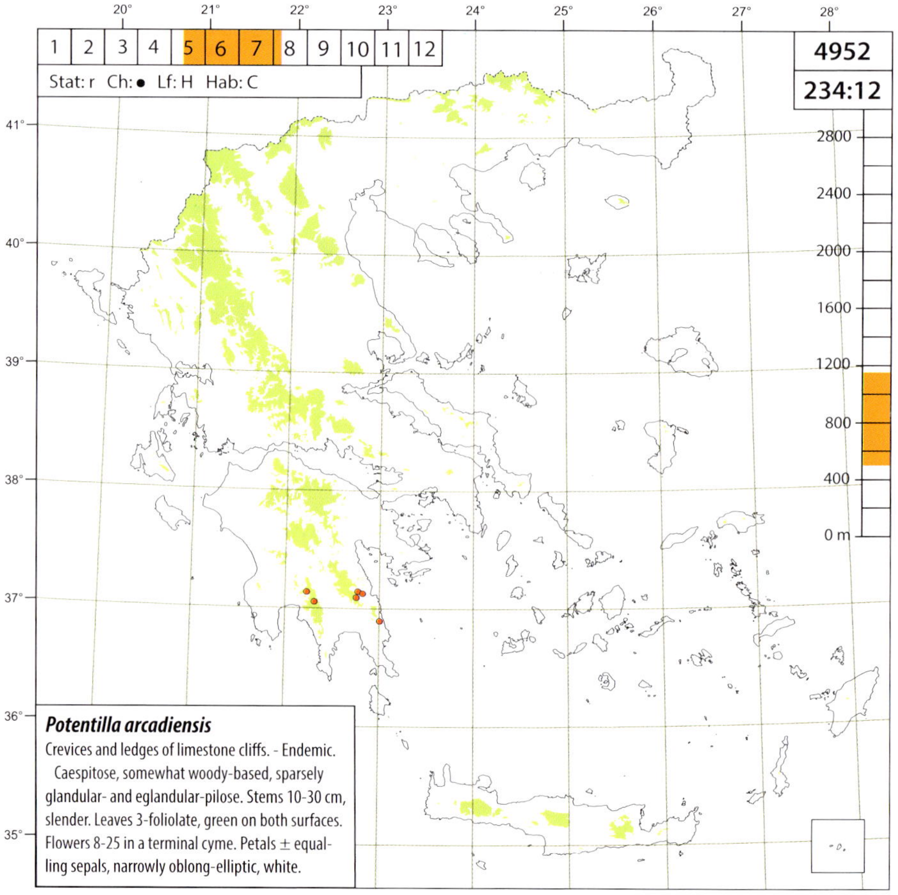

Potentilla arcadiensis
Crevices and ledges of limestone cliffs. - Endemic.
Caespitose, somewhat woody-based, sparsely glandular- and eglandular-pilose. Stems 10-30 cm, slender. Leaves 3-foliolate, green on both surfaces. Flowers 8-25 in a terminal cyme. Petals ± equalling sepals, narrowly oblong-elliptic, white.

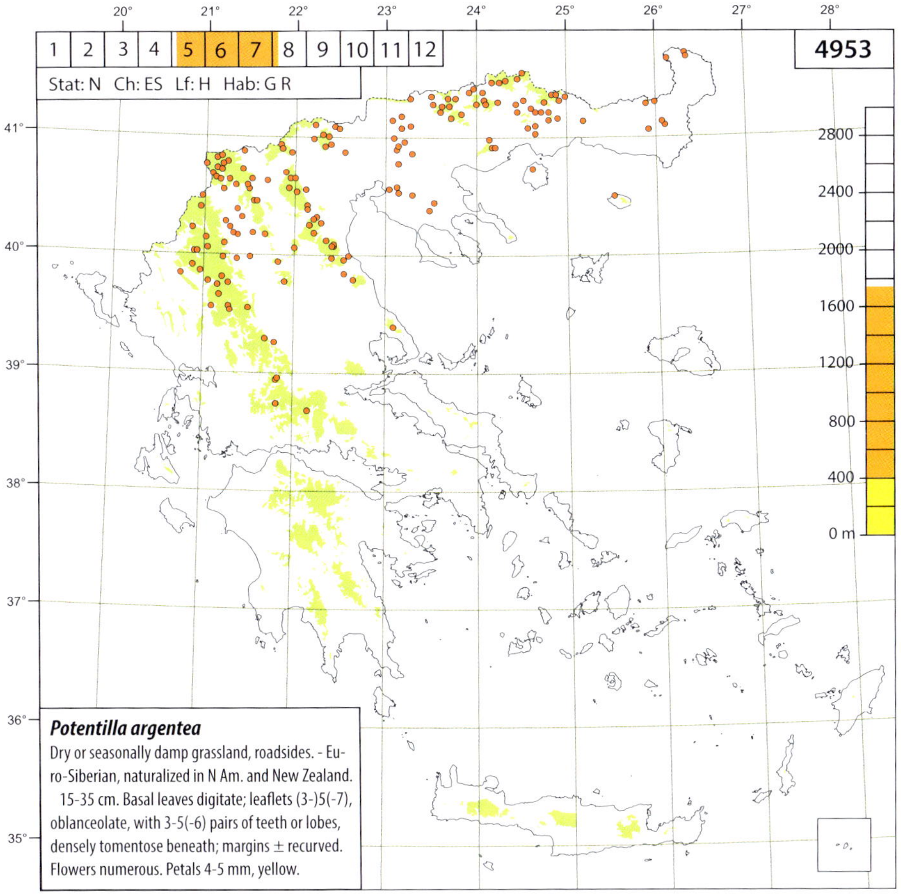

Potentilla argentea

Dry or seasonally damp grassland, roadsides. - Euro-Siberian, naturalized in N Am. and New Zealand.

15-35 cm. Basal leaves digitate; leaflets (3-)5(-7), oblanceolate, with 3-5(-6) pairs of teeth or lobes, densely tomentose beneath; margins ± recurved. Flowers numerous. Petals 4-5 mm, yellow.

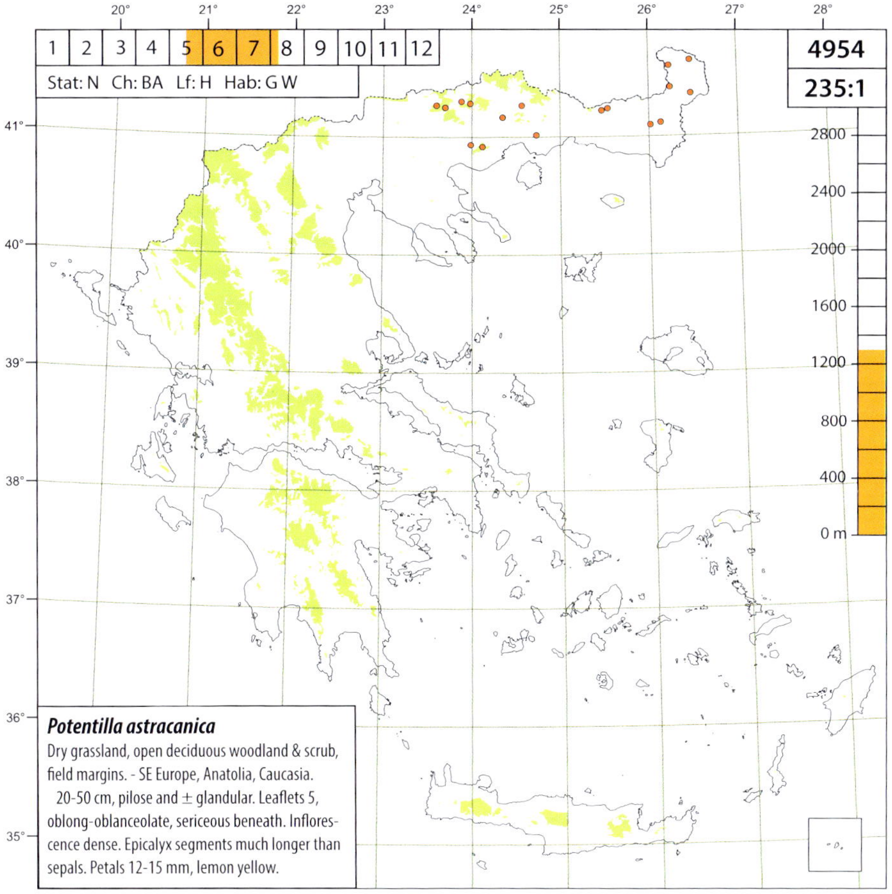

Potentilla astracanica

Dry grassland, open deciduous woodland & scrub, field margins. - SE Europe, Anatolia, Caucasia.

20-50 cm, pilose and ± glandular. Leaflets 5, oblong-oblanceolate, sericeous beneath. Inflorescence dense. Epicalyx segments much longer than sepals. Petals 12-15 mm, lemon yellow.

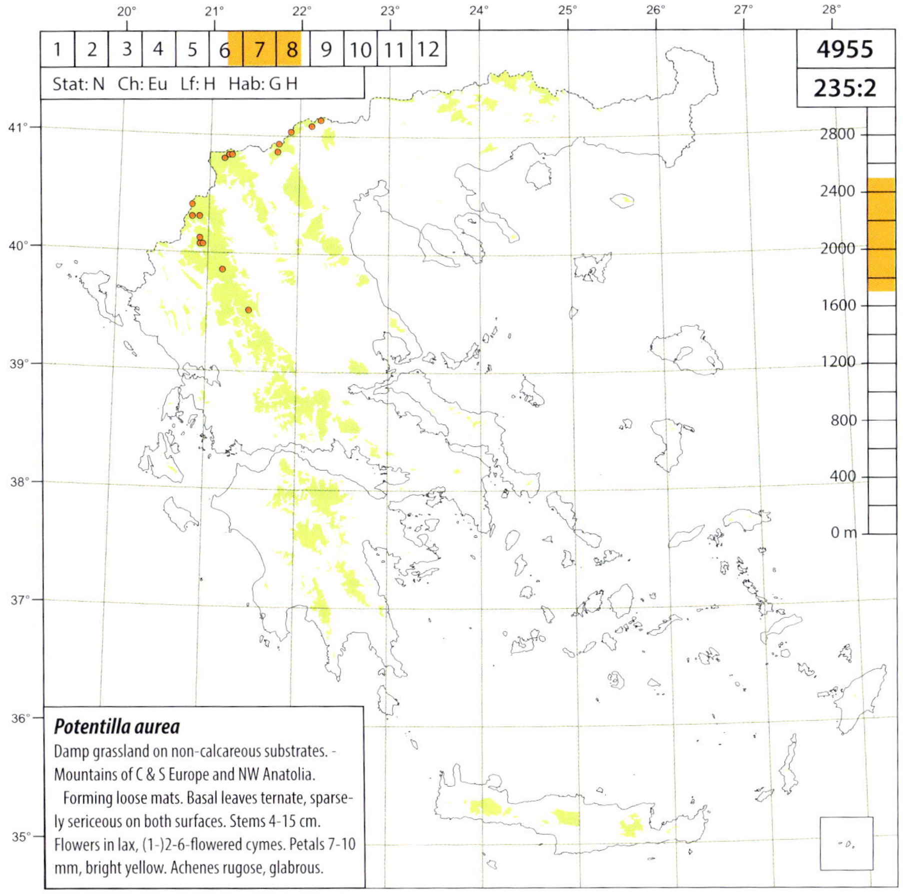

Potentilla aurea

Damp grassland on non-calcareous substrates. - Mountains of C & S Europe and NW Anatolia.

Forming loose mats. Basal leaves ternate, sparsely sericeous on both surfaces. Stems 4-15 cm. Flowers in lax, (1-)2-6-flowered cymes. Petals 7-10 mm, bright yellow. Achenes rugose, glabrous.

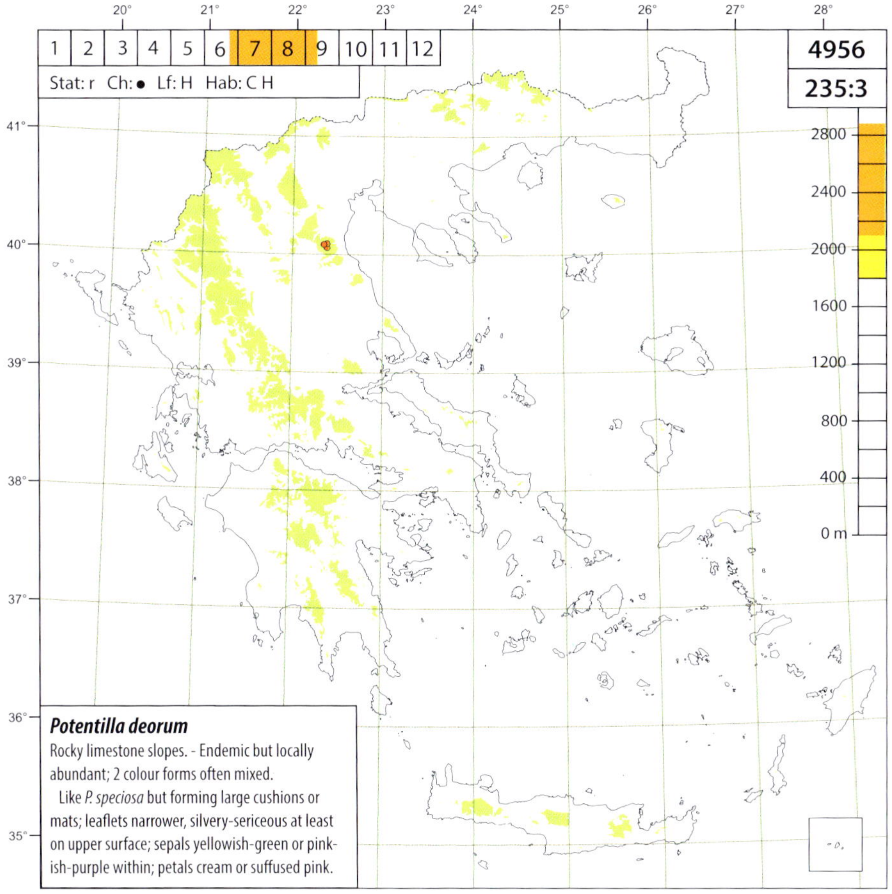

Potentilla deorum

Rocky limestone slopes. - Endemic but locally abundant; 2 colour forms often mixed.

Like *P. speciosa* but forming large cushions or mats; leaflets narrower, silvery-sericeous at least on upper surface; sepals yellowish-green or pinkish-purple within; petals cream or suffused pink.

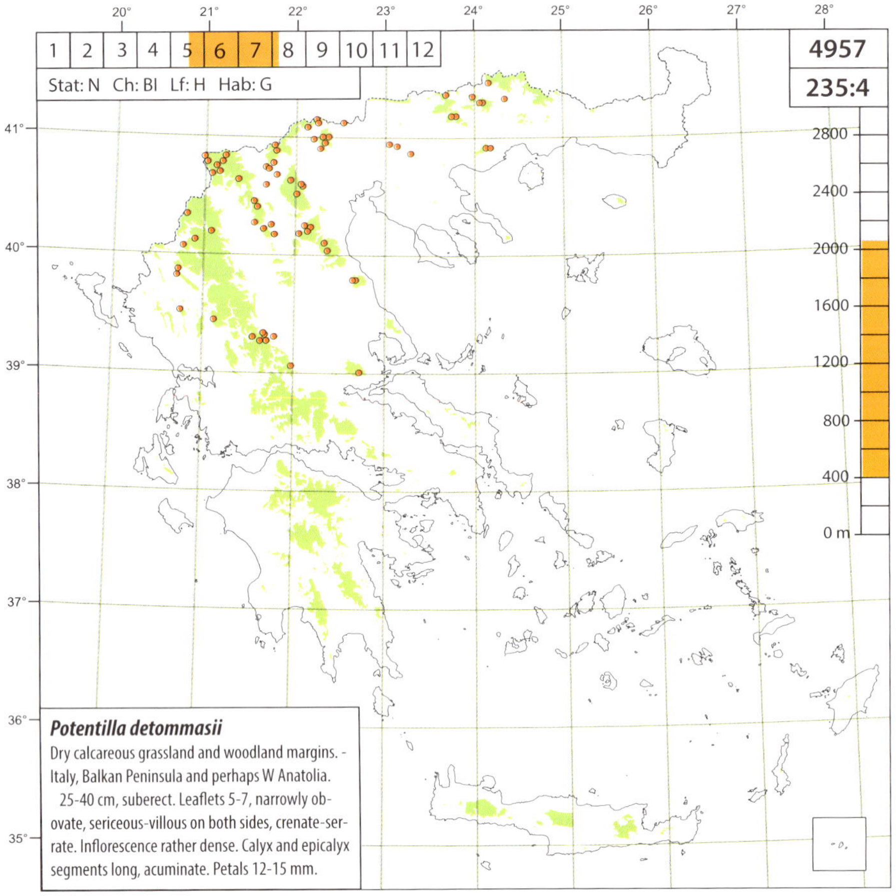

Potentilla detommasii

Dry calcareous grassland and woodland margins. - Italy, Balkan Peninsula and perhaps W Anatolia.

25-40 cm, suberect. Leaflets 5-7, narrowly obovate, sericeous-villous on both sides, crenate-serrate. Inflorescence rather dense. Calyx and epicalyx segments long, acuminate. Petals 12-15 mm.

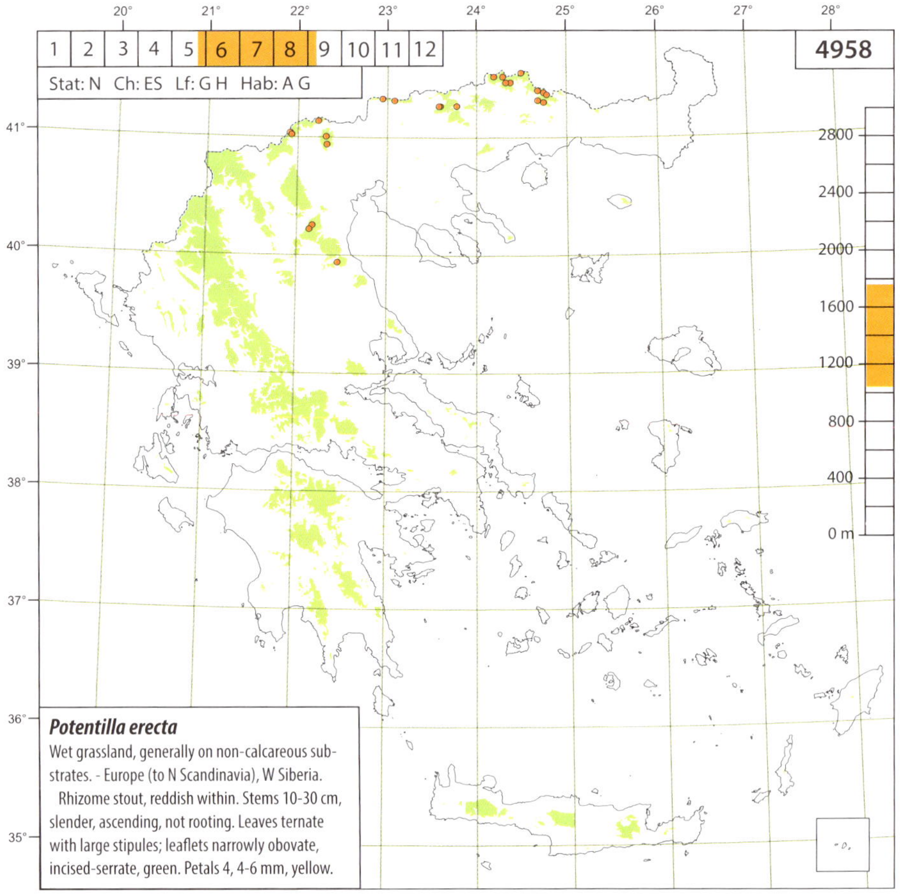

Potentilla erecta

Wet grassland, generally on non-calcareous substrates. - Europe (to N Scandinavia), W Siberia.

Rhizome stout, reddish within. Stems 10-30 cm, slender, ascending, not rooting. Leaves ternate with large stipules; leaflets narrowly obovate, incised-serrate, green. Petals 4, 4-6 mm, yellow.

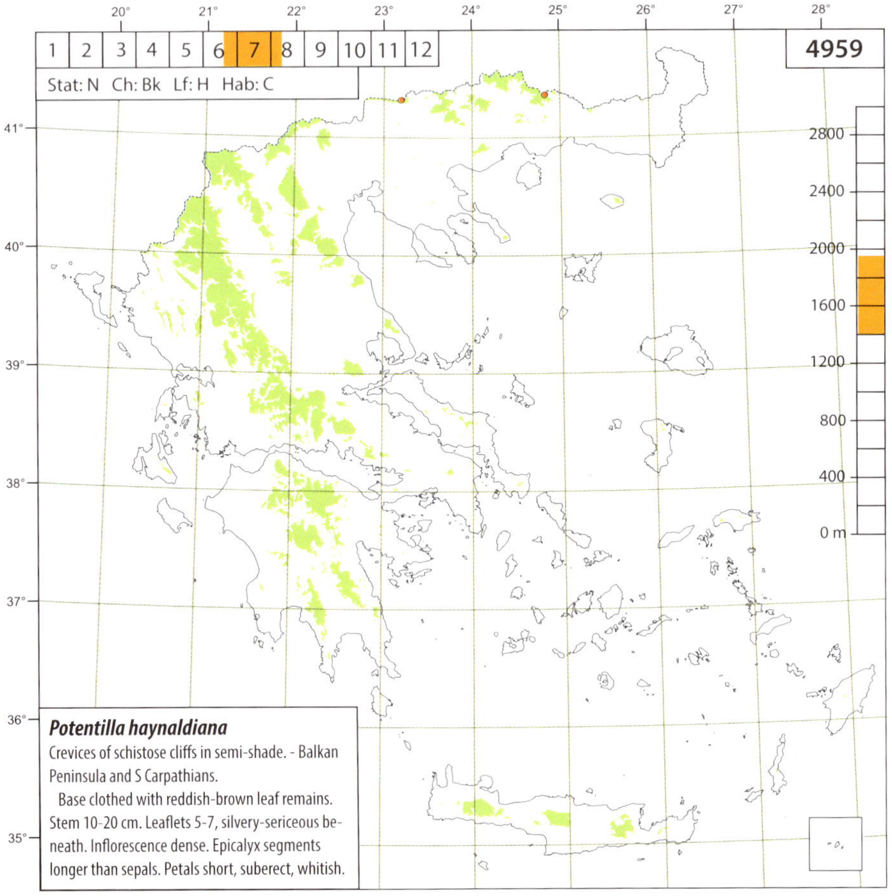

Potentilla haynaldiana

Crevices of schistose cliffs in semi-shade. - Balkan Peninsula and S Carpathians.

Base clothed with reddish-brown leaf remains. Stem 10-20 cm. Leaflets 5-7, silvery-sericeous beneath. Inflorescence dense. Epicalyx segments longer than sepals. Petals short, suberect, whitish.

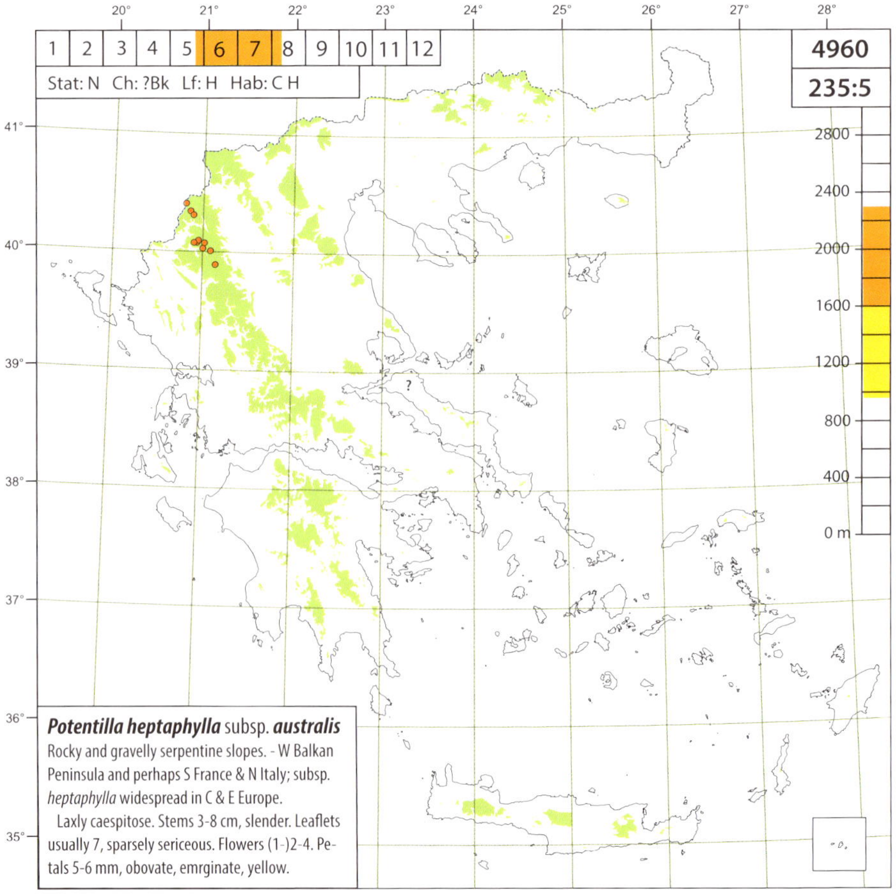

Potentilla heptaphylla subsp. ***australis***

Rocky and gravelly serpentine slopes. - W Balkan Peninsula and perhaps S France & N Italy; subsp. *heptaphylla* widespread in C & E Europe.

Laxly caespitose. Stems 3-8 cm, slender. Leaflets usually 7, sparsely sericeous. Flowers (1-)2-4. Petals 5-6 mm, obovate, emrginate, yellow.

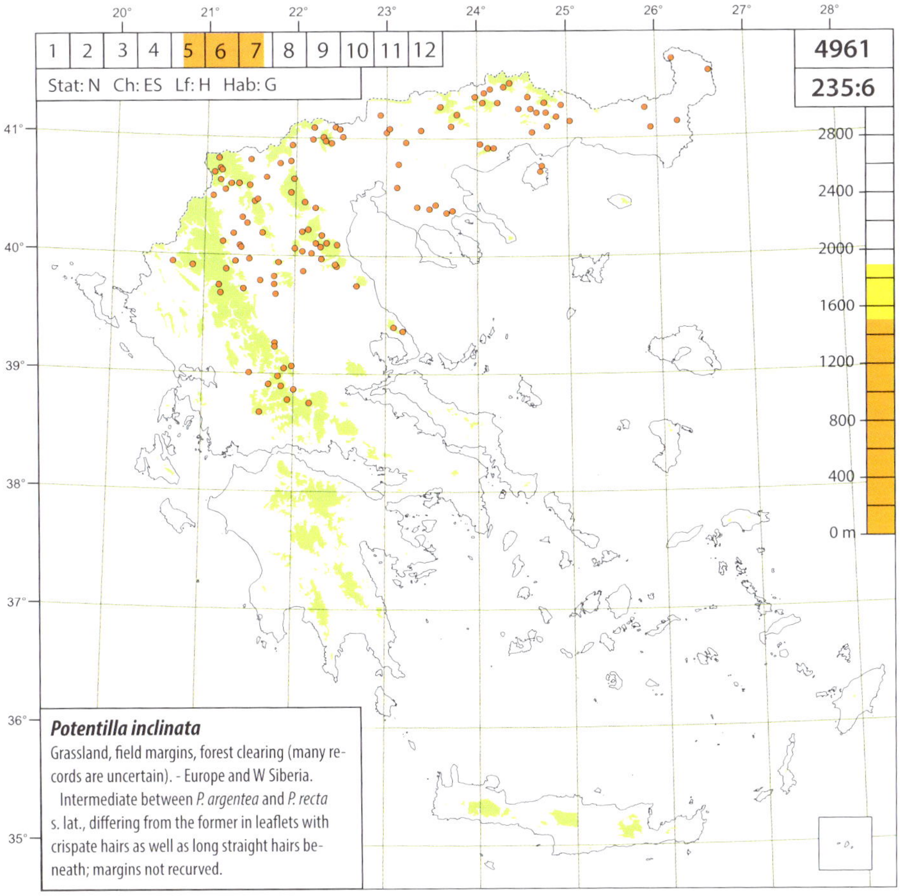

Potentilla inclinata

Grassland, field margins, forest clearing (many records are uncertain). - Europe and W Siberia.

Intermediate between *P. argentea* and *P. recta* s. lat., differing from the former in leaflets with crispate hairs as well as long straight hairs beneath; margins not recurved.

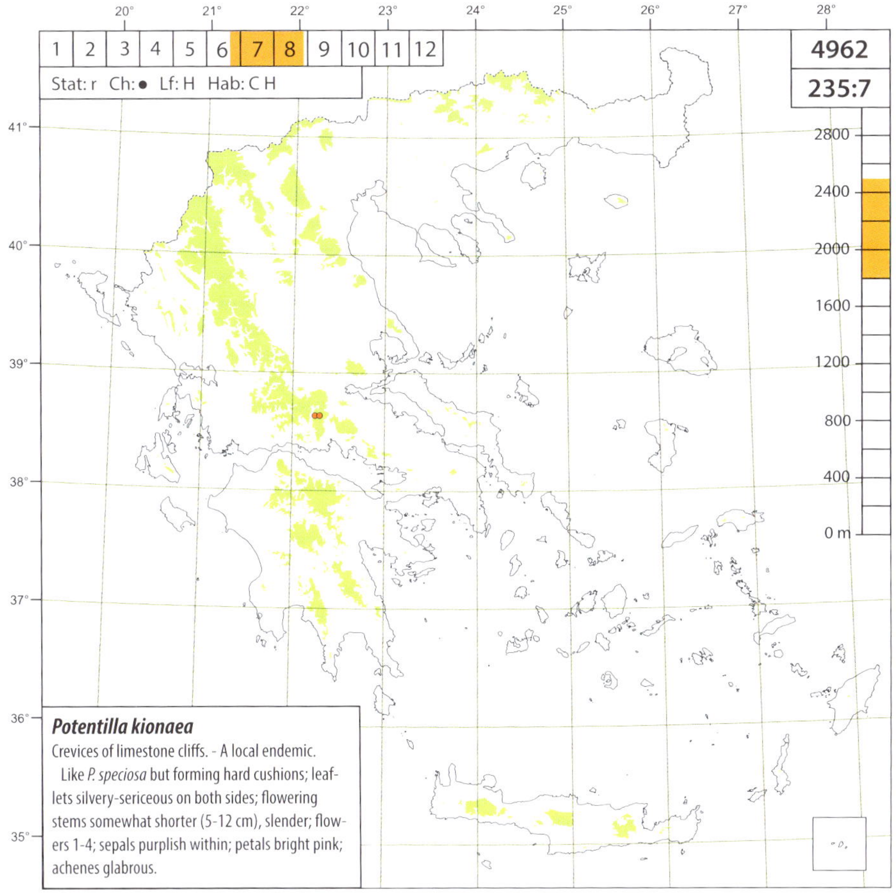

Potentilla kionaea

Crevices of limestone cliffs. - A local endemic.

Like *P. speciosa* but forming hard cushions; leaflets silvery-sericeous on both sides; flowering stems somewhat shorter (5-12 cm), slender; flowers 1-4; sepals purplish within; petals bright pink; achenes glabrous.

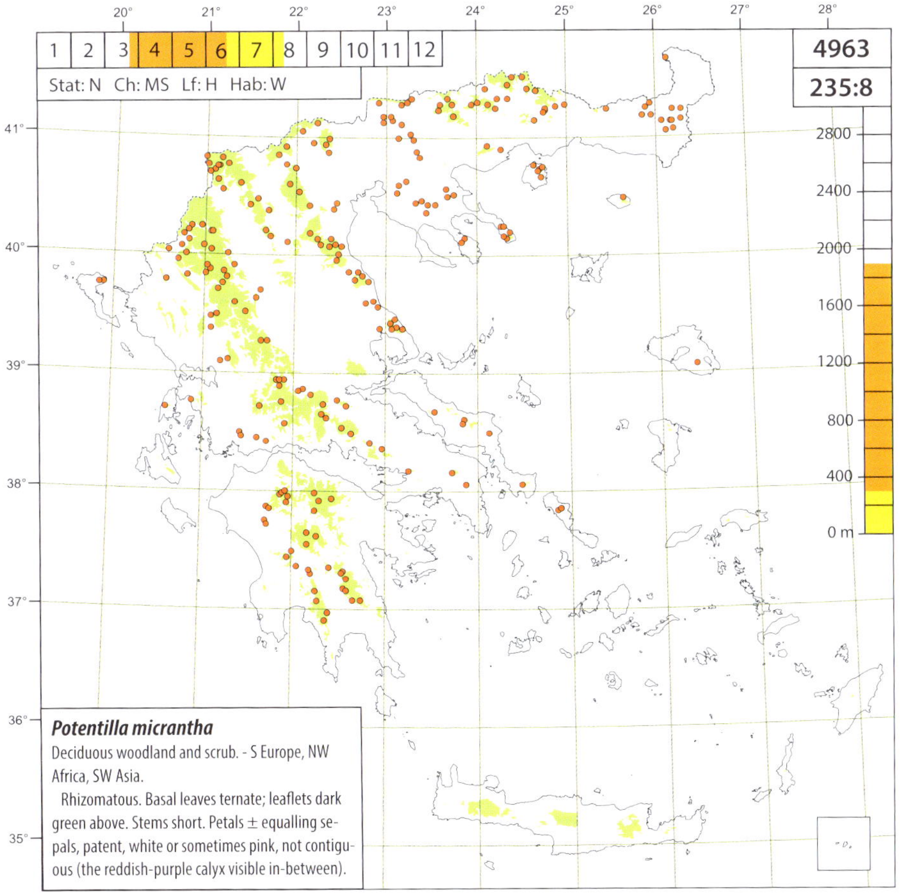

Potentilla micrantha

Deciduous woodland and scrub. - S Europe, NW Africa, SW Asia.

Rhizomatous. Basal leaves ternate; leaflets dark green above. Stems short. Petals ± equalling sepals, patent, white or sometimes pink, not contiguous (the reddish-purple calyx visible in-between).

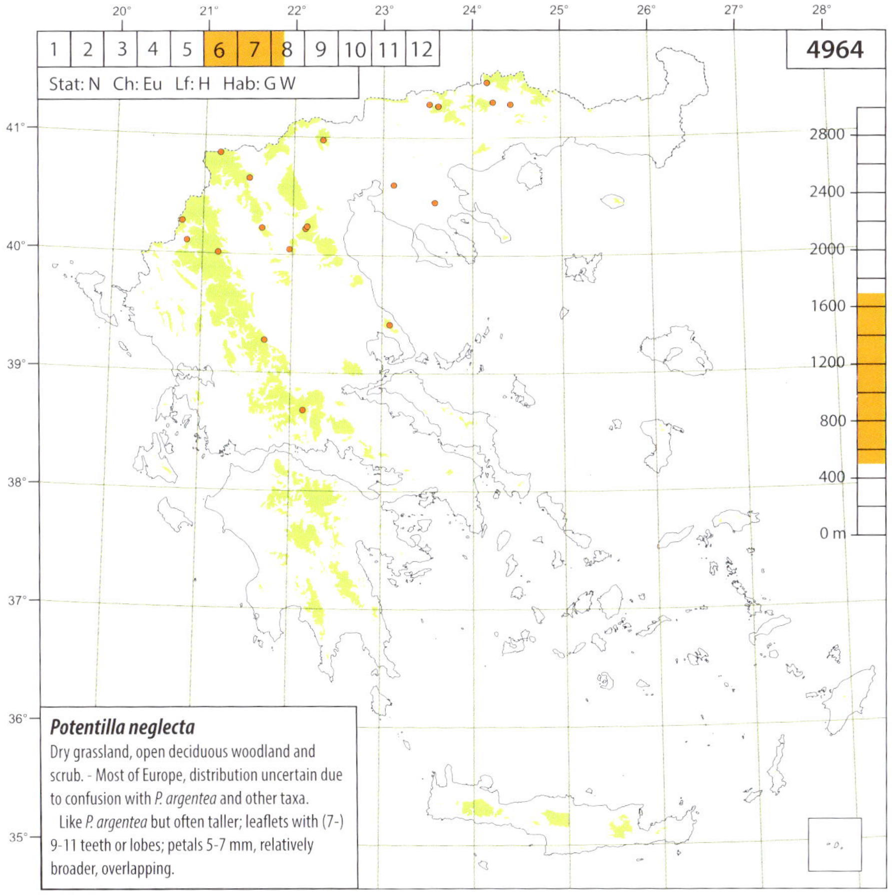

Potentilla neglecta

Dry grassland, open deciduous woodland and scrub. - Most of Europe, distribution uncertain due to confusion with *P. argentea* and other taxa.

Like *P. argentea* but often taller; leaflets with (7-) 9-11 teeth or lobes; petals 5-7 mm, relatively broader, overlapping.

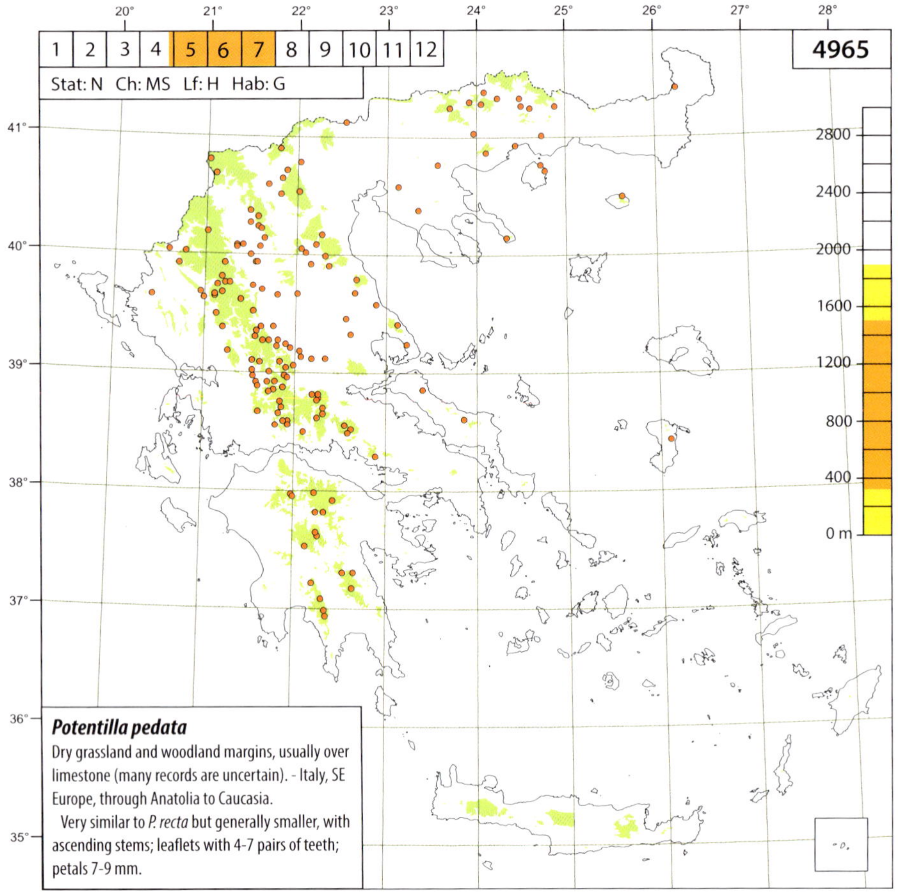

Potentilla pedata
Dry grassland and woodland margins, usually over limestone (many records are uncertain). - Italy, SE Europe, through Anatolia to Caucasia.
Very similar to *P. recta* but generally smaller, with ascending stems; leaflets with 4-7 pairs of teeth; petals 7-9 mm.

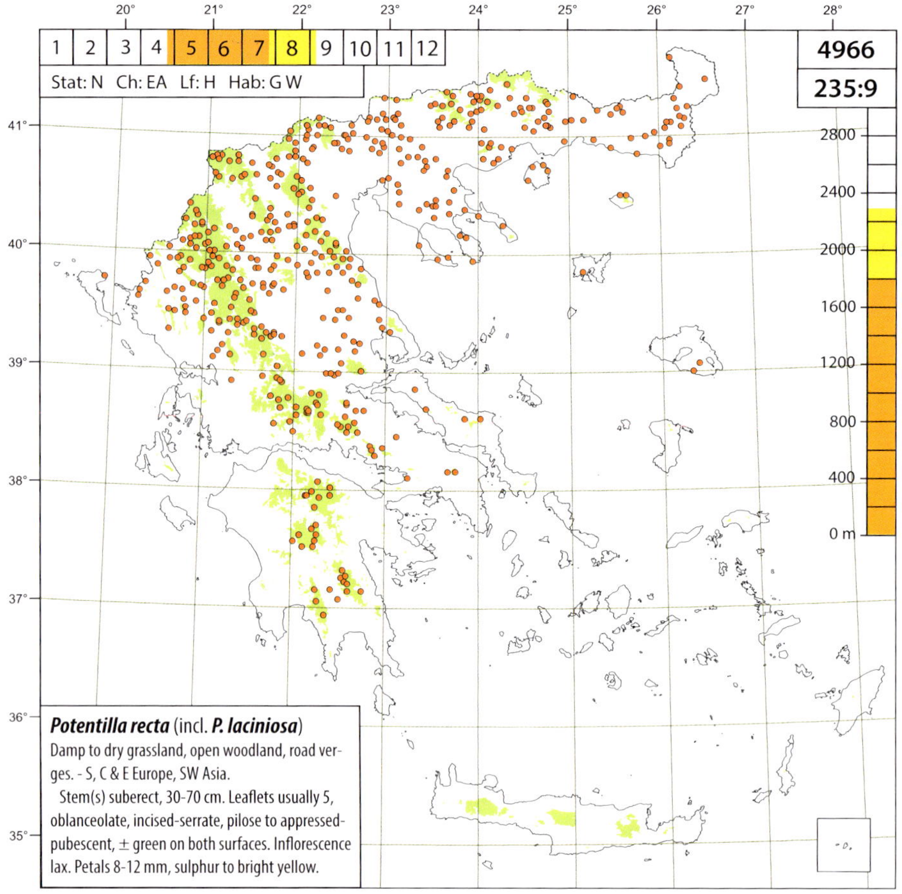

Potentilla recta (incl. ***P. laciniosa***)
Damp to dry grassland, open woodland, road verges. - S, C & E Europe, SW Asia.
Stem(s) suberect, 30-70 cm. Leaflets usually 5, oblanceolate, incised-serrate, pilose to appressed-pubescent, ± green on both surfaces. Inflorescence lax. Petals 8-12 mm, sulphur to bright yellow.

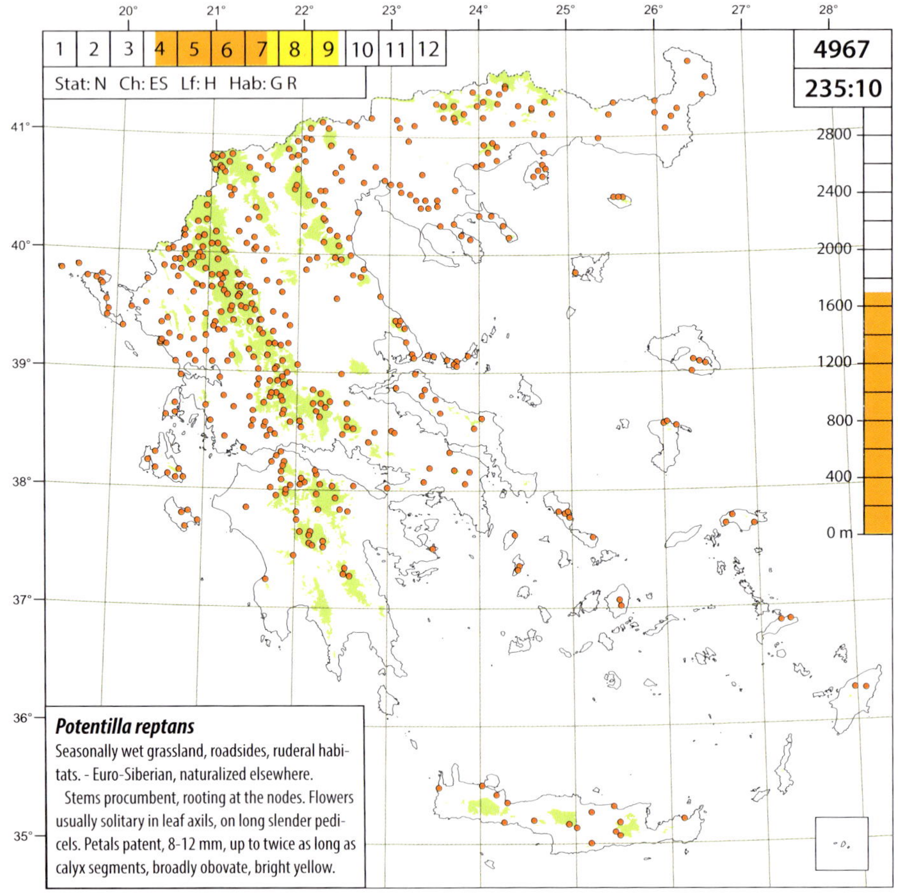

Potentilla reptans
Seasonally wet grassland, roadsides, ruderal habitats. - Euro-Siberian, naturalized elsewhere.
Stems procumbent, rooting at the nodes. Flowers usually solitary in leaf axils, on long slender pedicels. Petals patent, 8-12 mm, up to twice as long as calyx segments, broadly obovate, bright yellow.

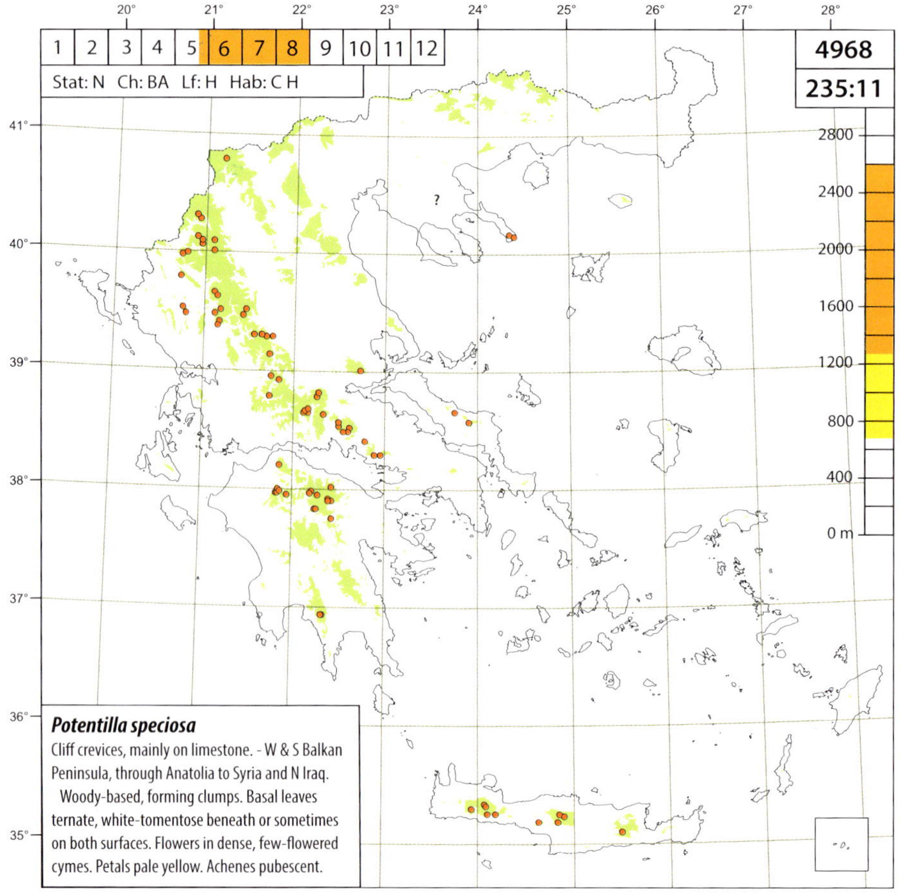

Potentilla speciosa
Cliff crevices, mainly on limestone. - W & S Balkan Peninsula, through Anatolia to Syria and N Iraq.
Woody-based, forming clumps. Basal leaves ternate, white-tomentose beneath or sometimes on both surfaces. Flowers in dense, few-flowered cymes. Petals pale yellow. Achenes pubescent.

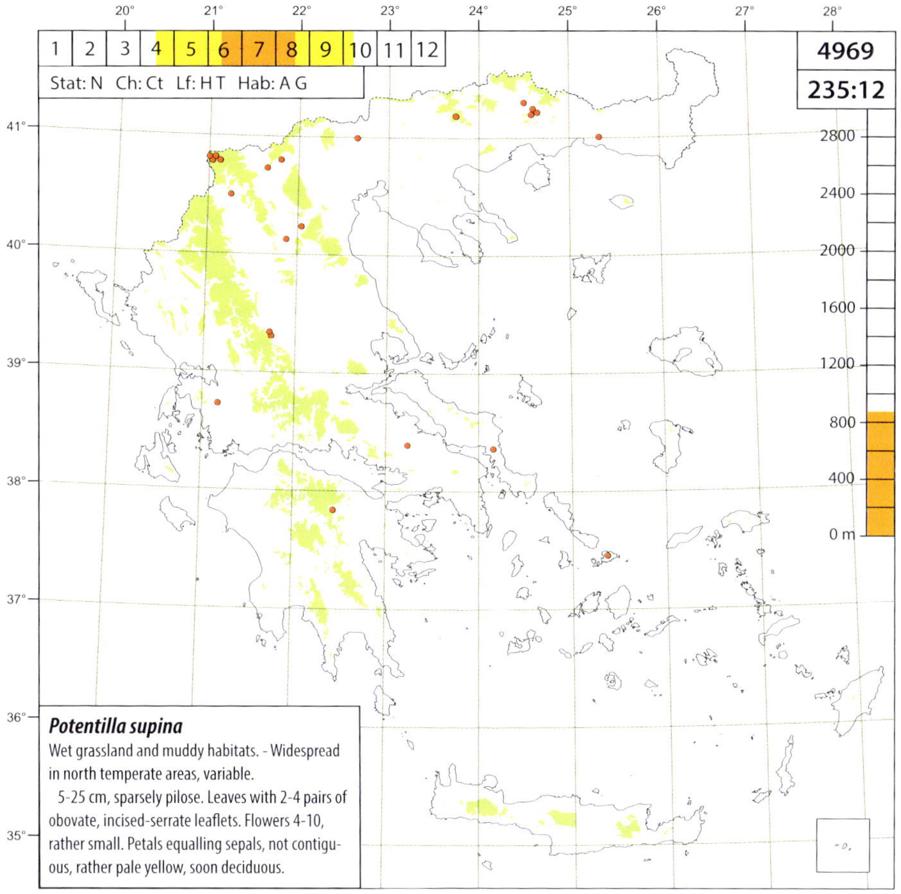

Potentilla supina
Wet grassland and muddy habitats. - Widespread in north temperate areas, variable.
5-25 cm, sparsely pilose. Leaves with 2-4 pairs of obovate, incised-serrate leaflets. Flowers 4-10, rather small. Petals equalling sepals, not contiguous, rather pale yellow, soon deciduous.

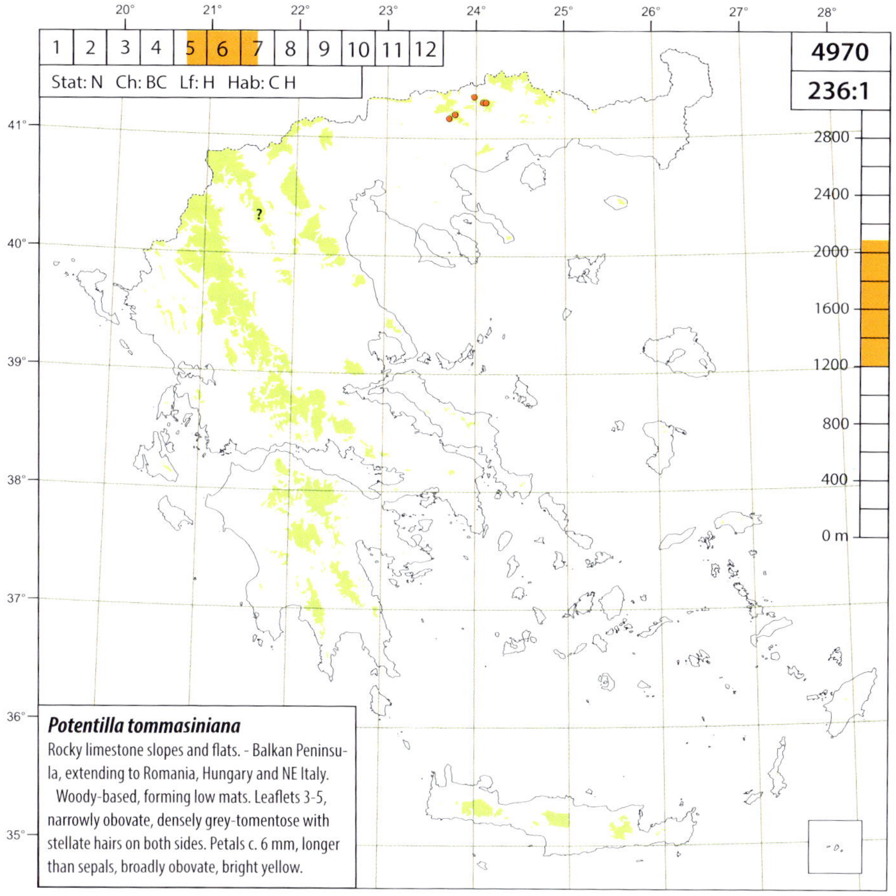

Potentilla tommasiniana
Rocky limestone slopes and flats. - Balkan Peninsula, extending to Romania, Hungary and NE Italy.
Woody-based, forming low mats. Leaflets 3-5, narrowly obovate, densely grey-tomentose with stellate hairs on both sides. Petals c. 6 mm, longer than sepals, broadly obovate, bright yellow.

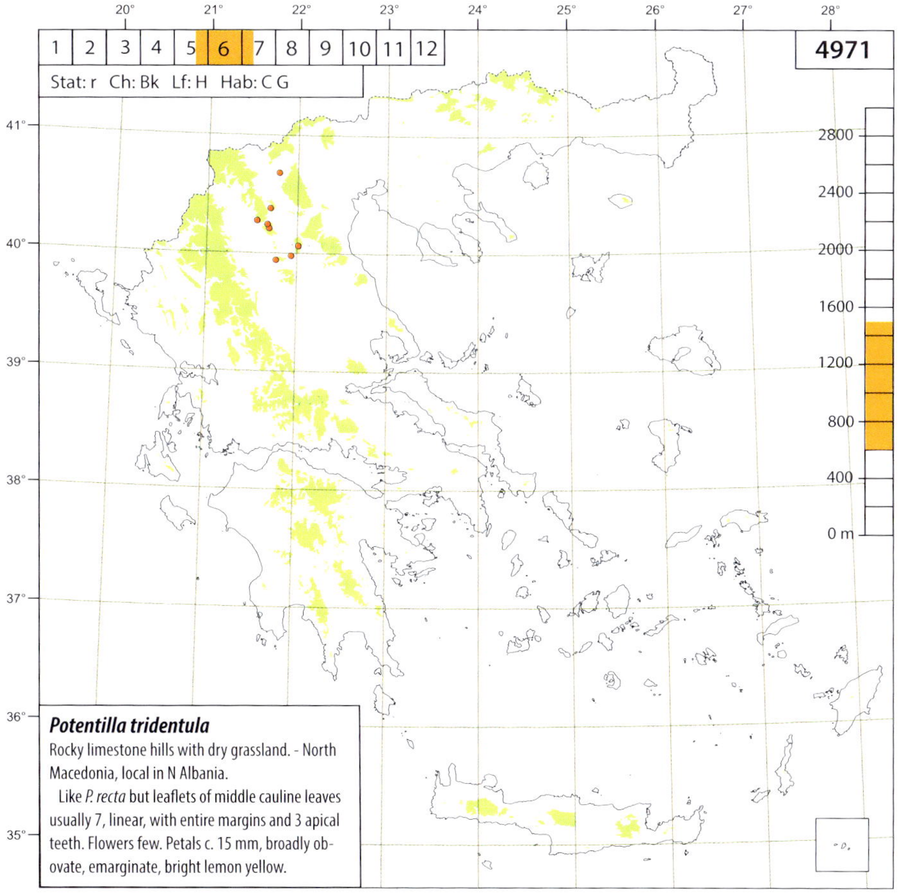

Potentilla tridentula
Rocky limestone hills with dry grassland. - North Macedonia, local in N Albania.
Like *P. recta* but leaflets of middle cauline leaves usually 7, linear, with entire margins and 3 apical teeth. Flowers few. Petals c. 15 mm, broadly obovate, emarginate, bright lemon yellow.

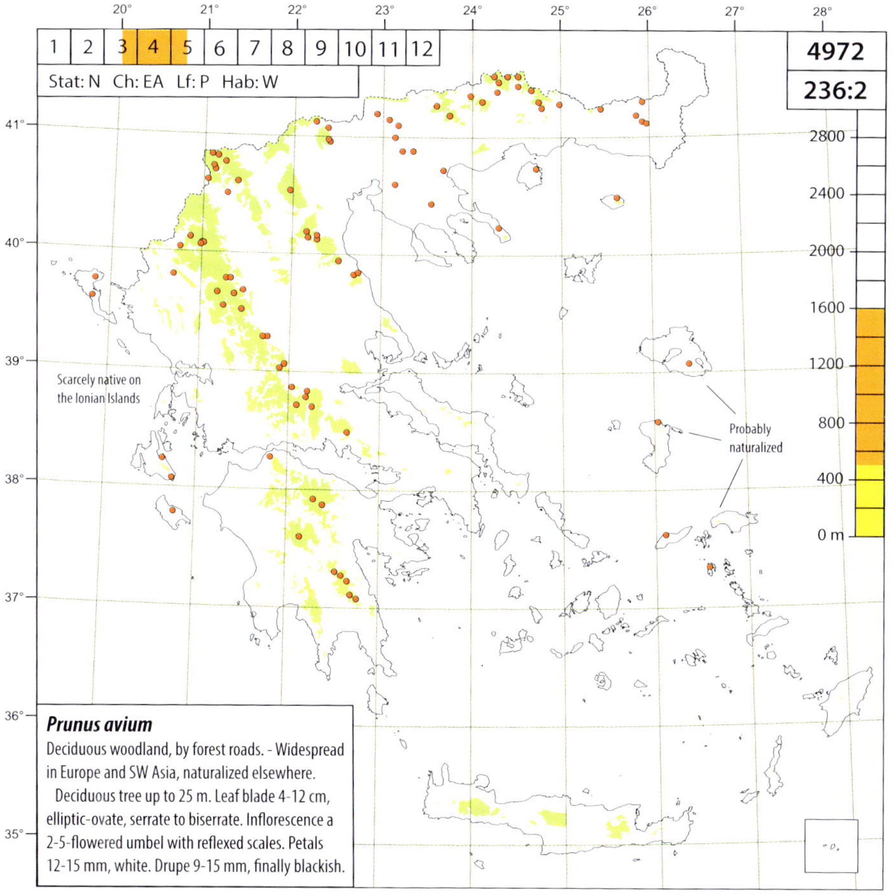

Prunus avium
Deciduous woodland, by forest roads. - Widespread in Europe and SW Asia, naturalized elsewhere.
Deciduous tree up to 25 m. Leaf blade 4-12 cm, elliptic-ovate, serrate to biserrate. Inflorescence a 2-5-flowered umbel with reflexed scales. Petals 12-15 mm, white. Drupe 9-15 mm, finally blackish.

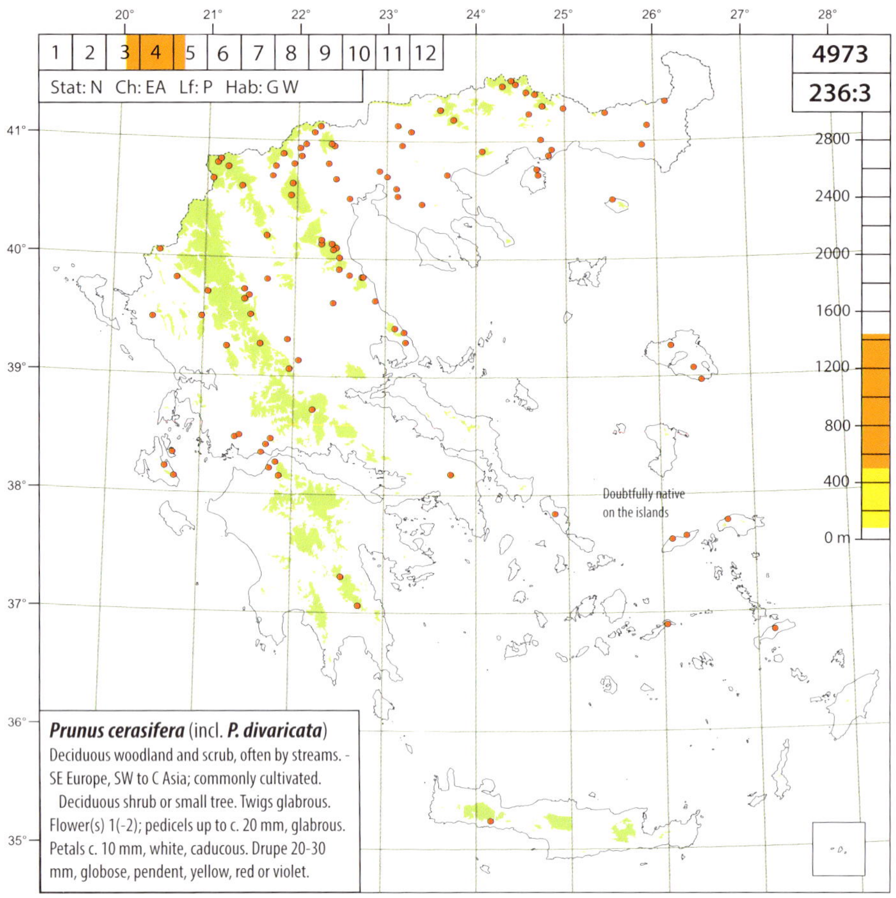

Prunus cerasifera (incl. ***P. divaricata***)
Deciduous woodland and scrub, often by streams. - SE Europe, SW to C Asia; commonly cultivated.
Deciduous shrub or small tree. Twigs glabrous. Flower(s) 1(-2); pedicels up to c. 20 mm, glabrous. Petals c. 10 mm, white, caducous. Drupe 20-30 mm, globose, pendent, yellow, red or violet.

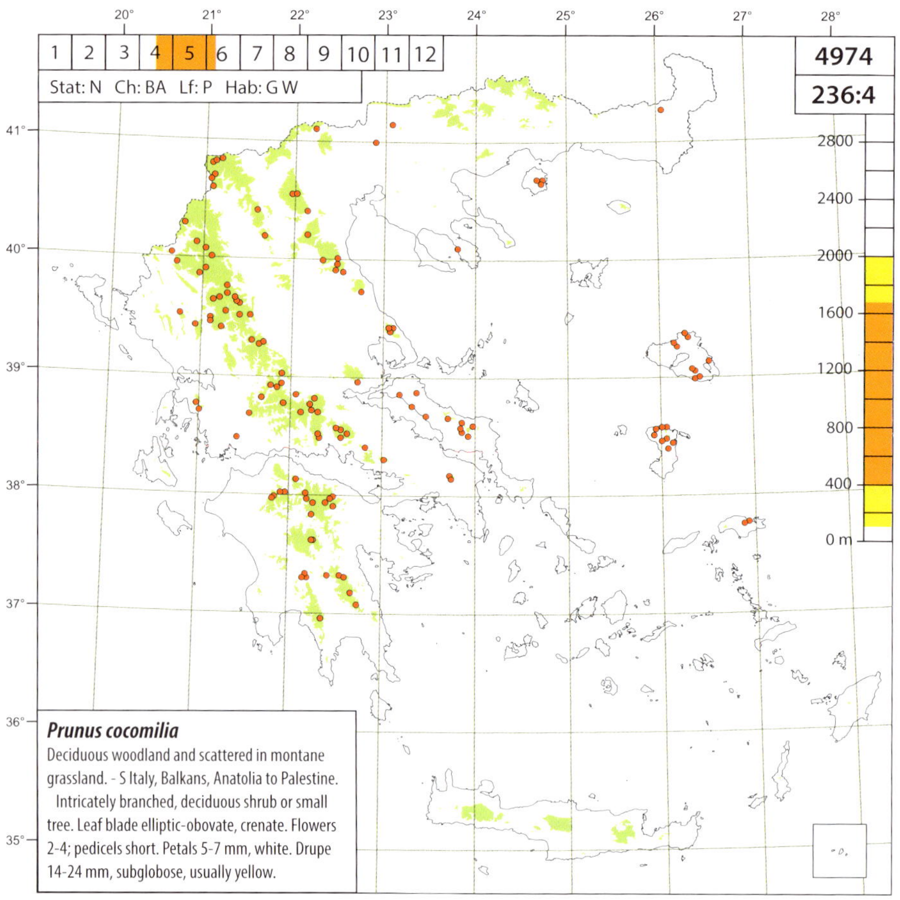

Prunus cocomilia
Deciduous woodland and scattered in montane grassland. - S Italy, Balkans, Anatolia to Palestine.
Intricately branched, deciduous shrub or small tree. Leaf blade elliptic-obovate, crenate. Flowers 2-4; pedicels short. Petals 5-7 mm, white. Drupe 14-24 mm, subglobose, usually yellow.

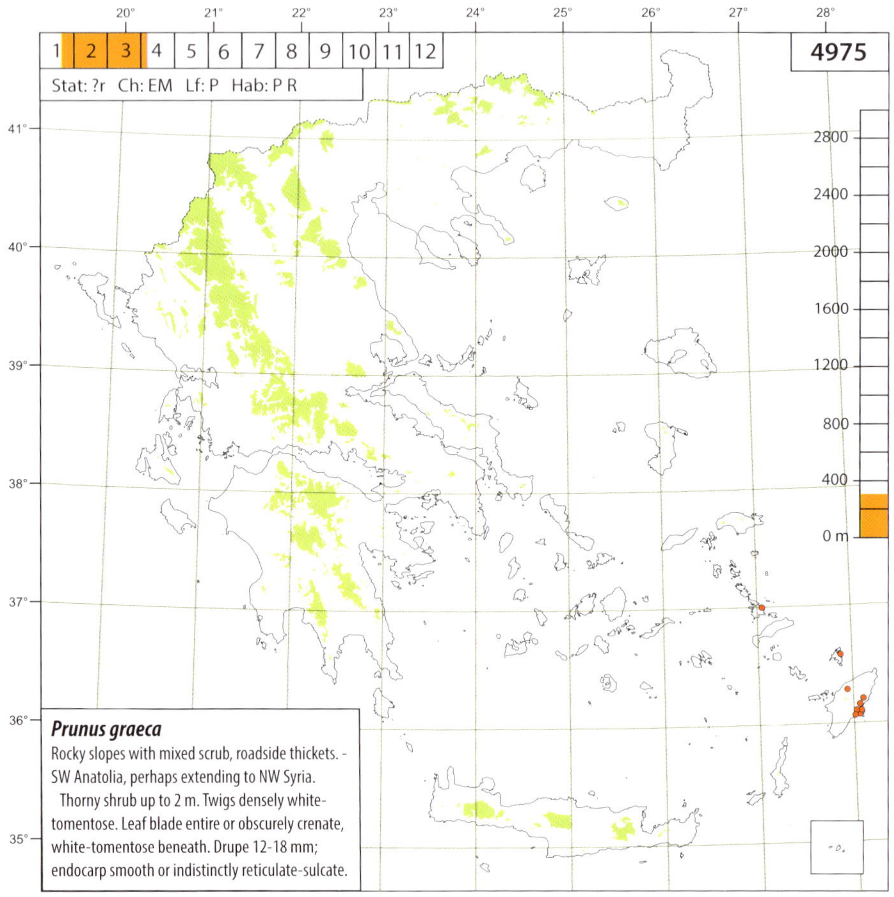

Prunus graeca
Rocky slopes with mixed scrub, roadside thickets. - SW Anatolia, perhaps extending to NW Syria.
Thorny shrub up to 2 m. Twigs densely white-tomentose. Leaf blade entire or obscurely crenate, white-tomentose beneath. Drupe 12-18 mm; endocarp smooth or indistinctly reticulate-sulcate.

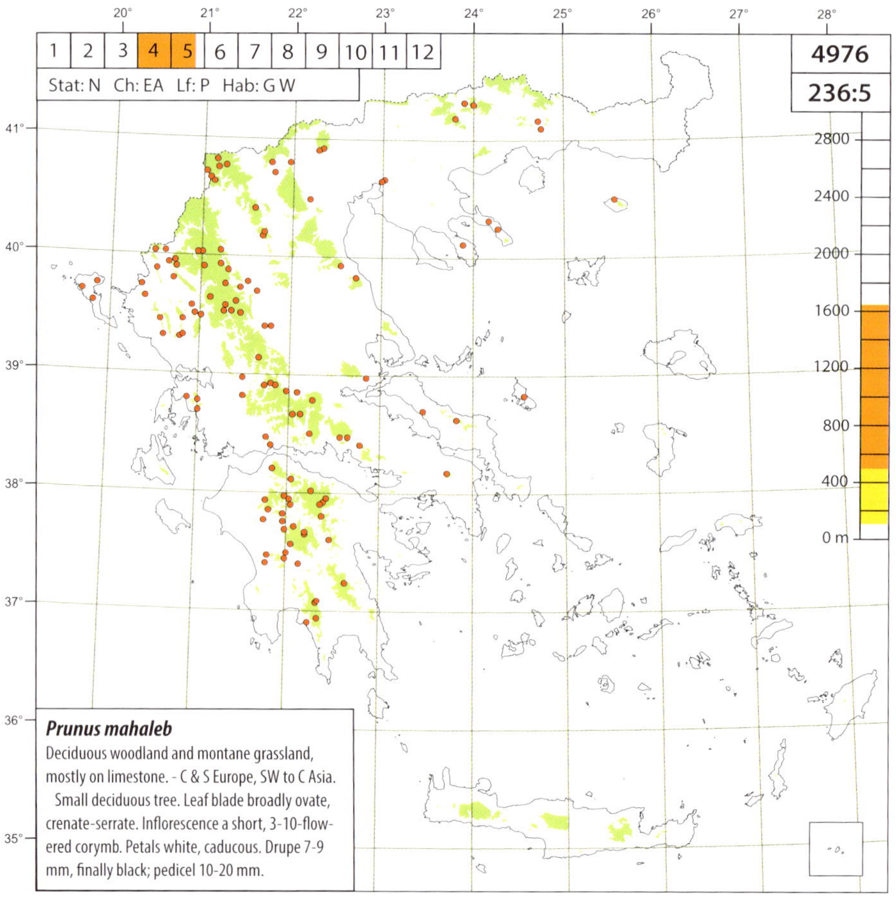

Prunus mahaleb
Deciduous woodland and montane grassland, mostly on limestone. - C & S Europe, SW to C Asia.
Small deciduous tree. Leaf blade broadly ovate, crenate-serrate. Inflorescence a short, 3-10-flowered corymb. Petals white, caducous. Drupe 7-9 mm, finally black; pedicel 10-20 mm.

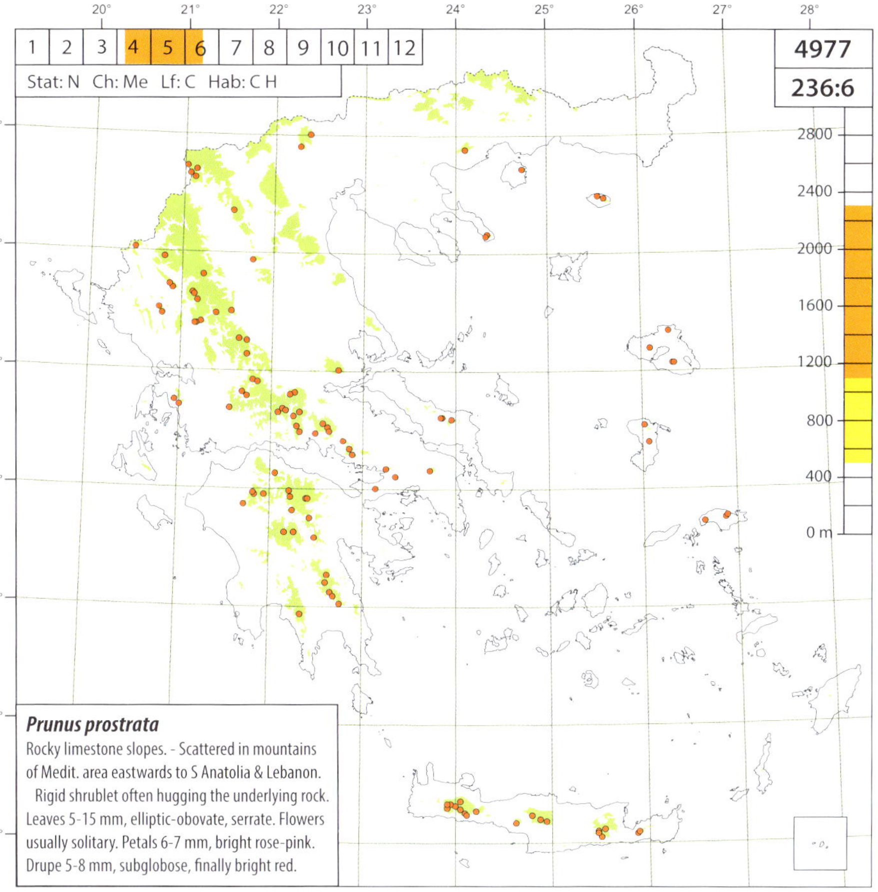

Prunus prostrata

Rocky limestone slopes. - Scattered in mountains of Medit. area eastwards to S Anatolia & Lebanon.

Rigid shrublet often hugging the underlying rock. Leaves 5-15 mm, elliptic-obovate, serrate. Flowers usually solitary. Petals 6-7 mm, bright rose-pink. Drupe 5-8 mm, subglobose, finally bright red.

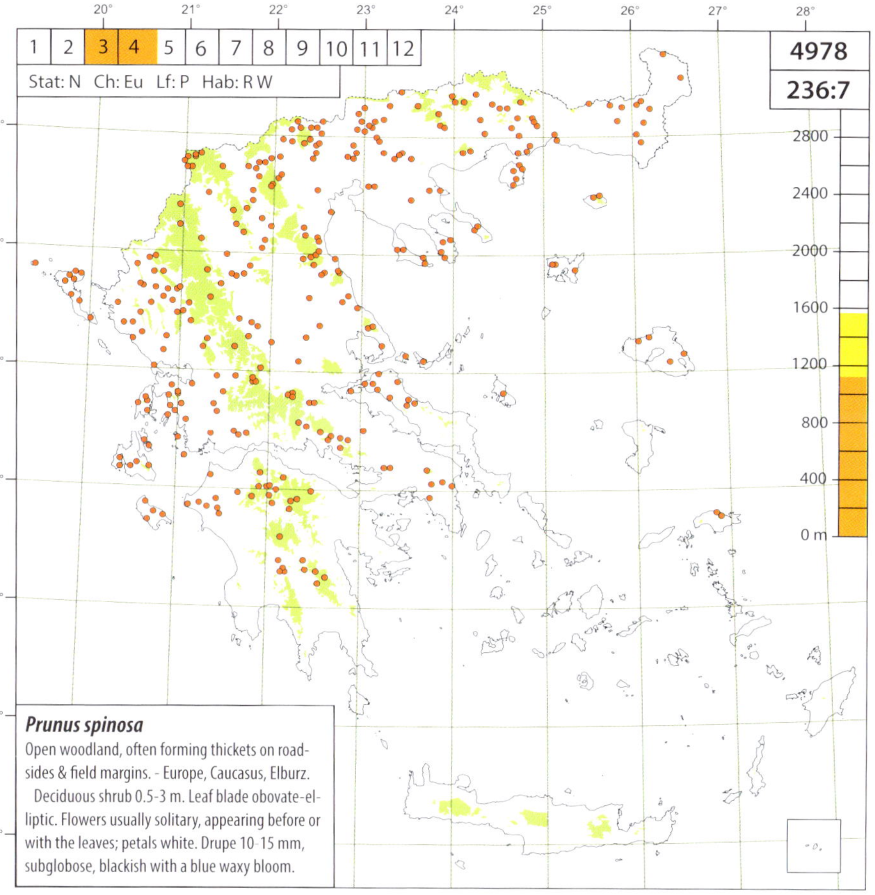

Prunus spinosa

Open woodland, often forming thickets on roadsides & field margins. - Europe, Caucasus, Elburz.

Deciduous shrub 0.5-3 m. Leaf blade obovate-elliptic. Flowers usually solitary, appearing before or with the leaves; petals white. Drupe 10-15 mm, subglobose, blackish with a blue waxy bloom.

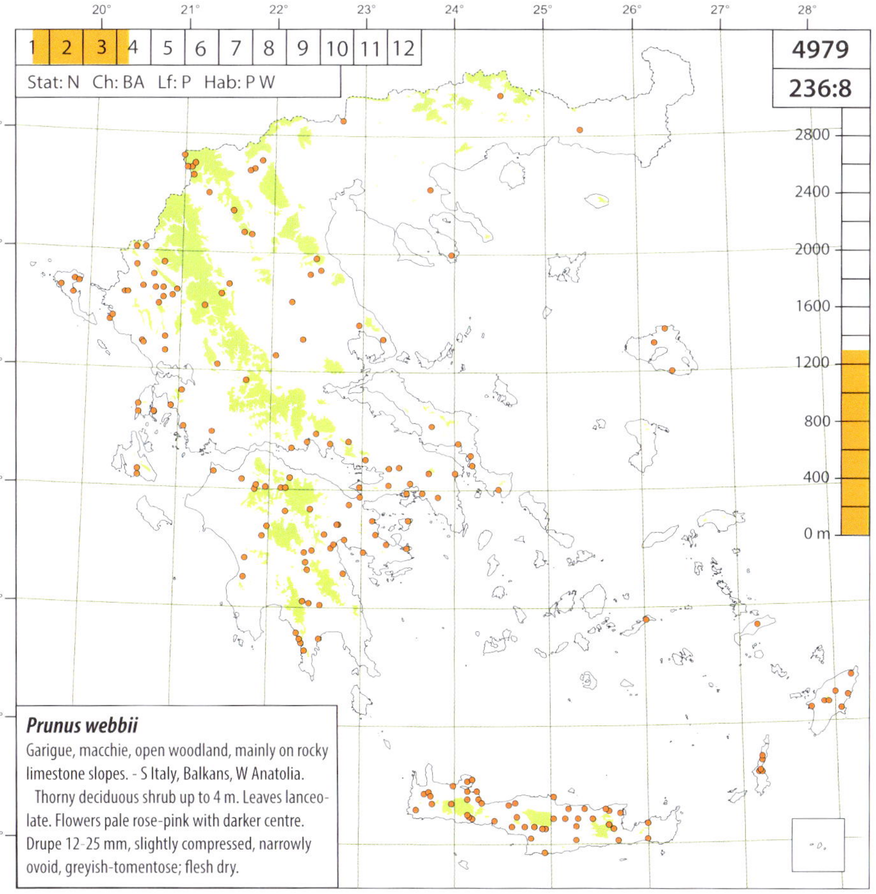

Prunus webbii

Garigue, macchie, open woodland, mainly on rocky limestone slopes. - S Italy, Balkans, W Anatolia.

Thorny deciduous shrub up to 4 m. Leaves lanceolate. Flowers pale rose-pink with darker centre. Drupe 12-25 mm, slightly compressed, narrowly ovoid, greyish-tomentose; flesh dry.

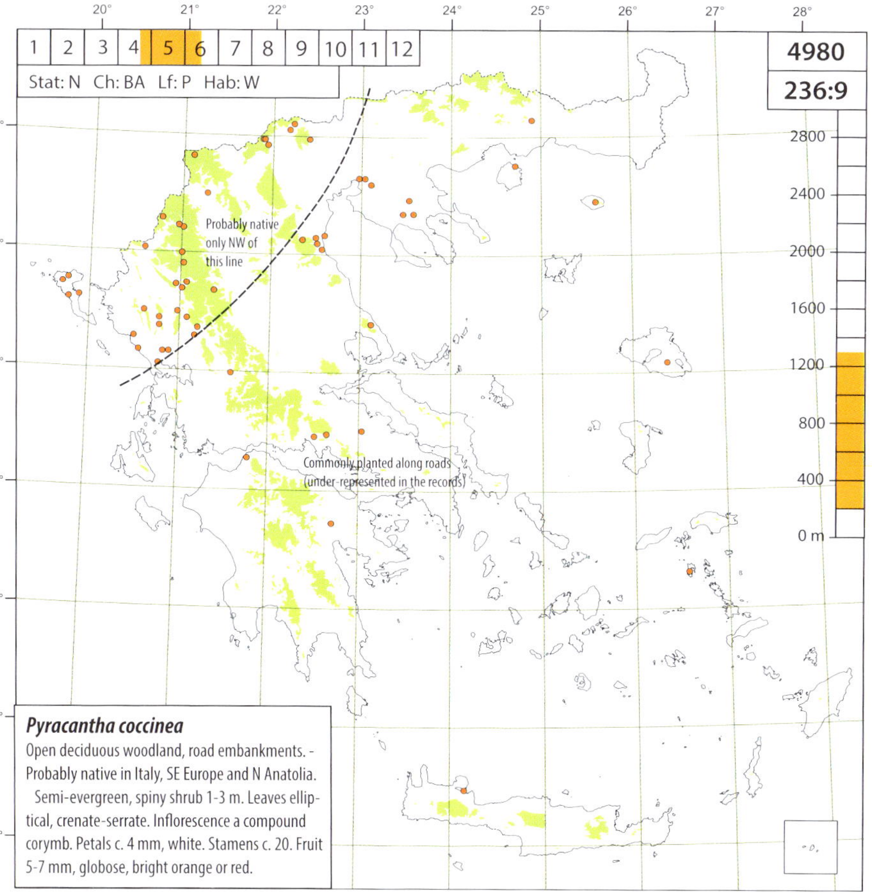

Pyracantha coccinea

Open deciduous woodland, road embankments. - Probably native in Italy, SE Europe and N Anatolia.

Semi-evergreen, spiny shrub 1-3 m. Leaves elliptical, crenate-serrate. Inflorescence a compound corymb. Petals c. 4 mm, white. Stamens c. 20. Fruit 5-7 mm, globose, bright orange or red.

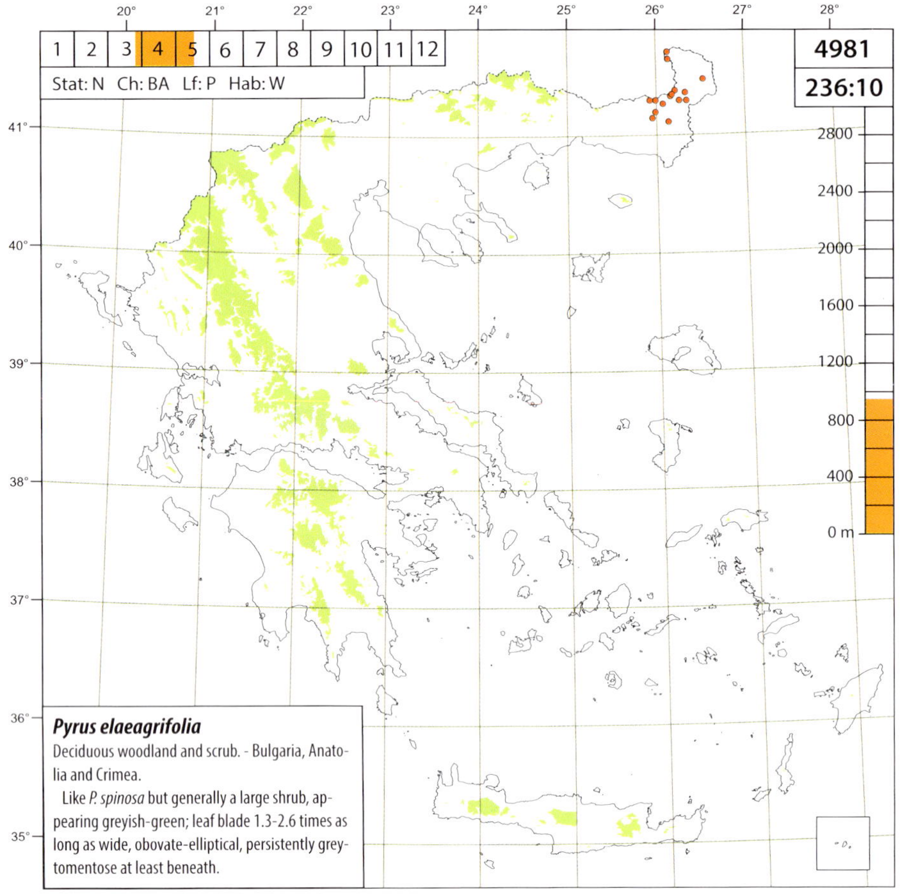

Pyrus elaeagrifolia

Deciduous woodland and scrub. - Bulgaria, Anatolia and Crimea.

Like *P. spinosa* but generally a large shrub, appearing greyish-green; leaf blade 1.3-2.6 times as long as wide, obovate-elliptical, persistently grey-tomentose at least beneath.

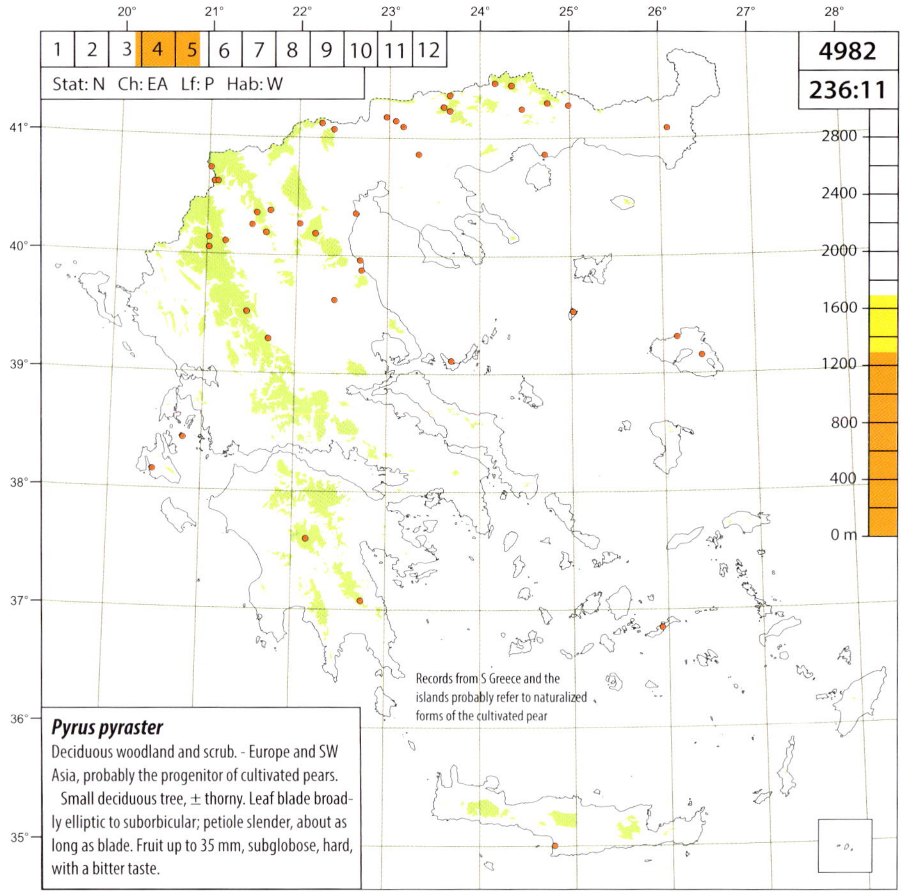

Pyrus pyraster

Deciduous woodland and scrub. - Europe and SW Asia, probably the progenitor of cultivated pears.

Small deciduous tree, ± thorny. Leaf blade broadly elliptic to suborbicular; petiole slender, about as long as blade. Fruit up to 35 mm, subglobose, hard, with a bitter taste.

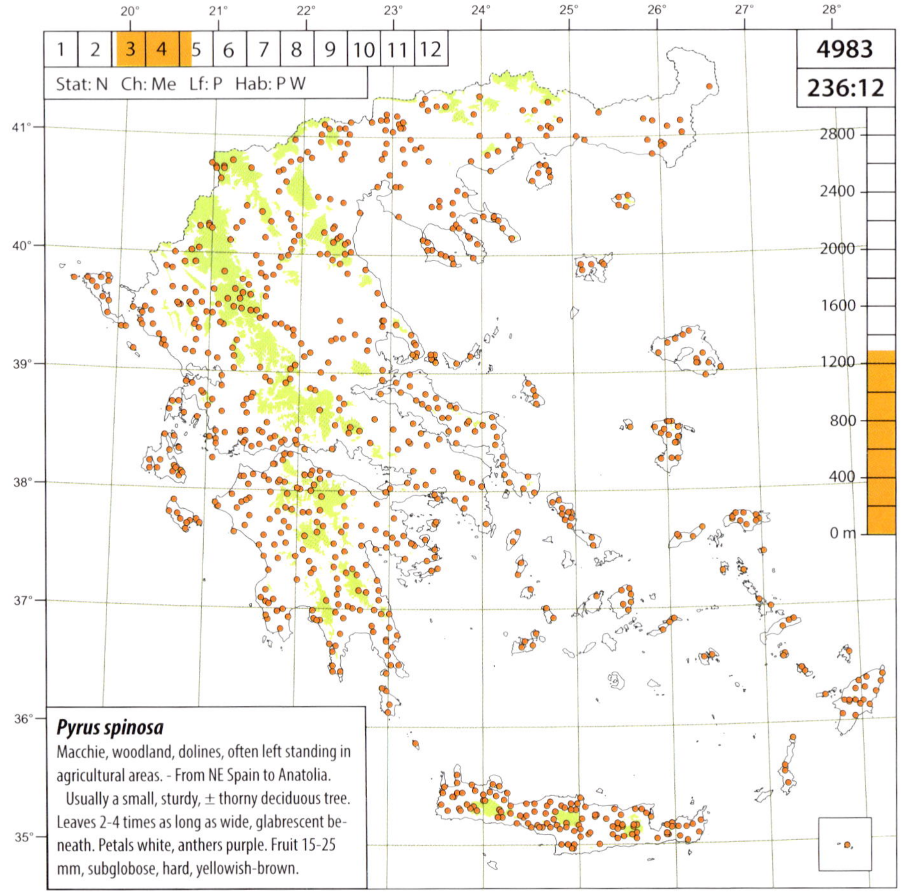

Pyrus spinosa

Macchie, woodland, dolines, often left standing in agricultural areas. - From NE Spain to Anatolia.

Usually a small, sturdy, ± thorny deciduous tree. Leaves 2-4 times as long as wide, glabrescent beneath. Petals white, anthers purple. Fruit 15-25 mm, subglobose, hard, yellowish-brown.

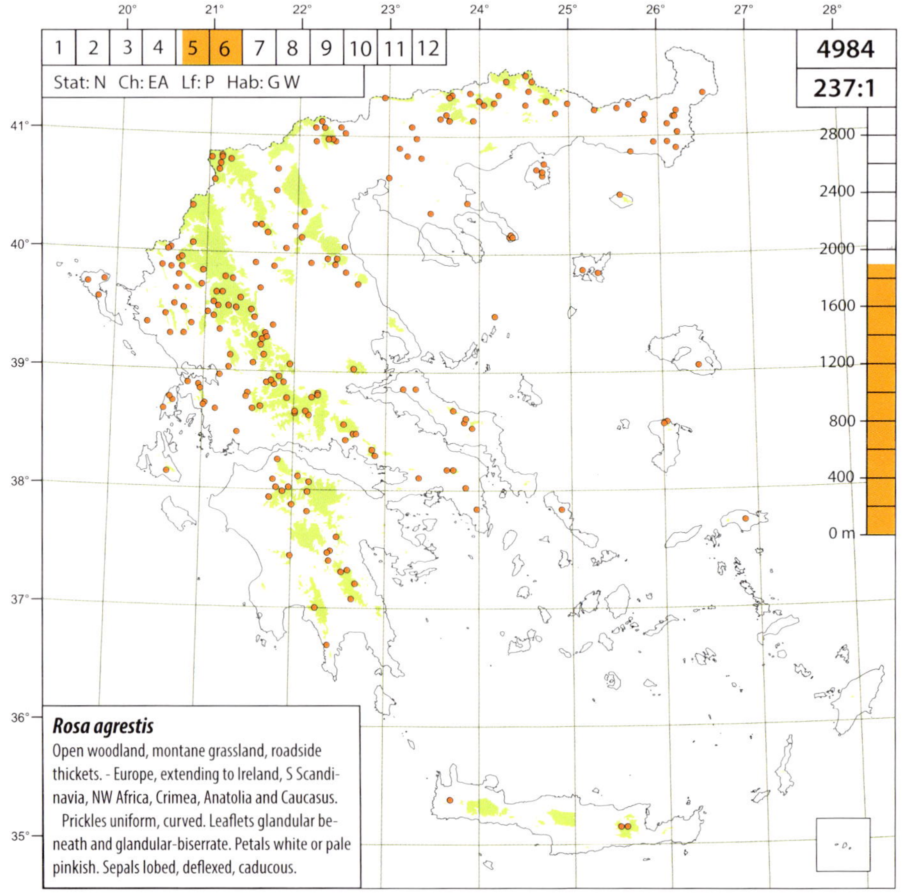

Rosa agrestis

Open woodland, montane grassland, roadside thickets. - Europe, extending to Ireland, S Scandinavia, NW Africa, Crimea, Anatolia and Caucasus.

Prickles uniform, curved. Leaflets glandular beneath and glandular-biserrate. Petals white or pale pinkish. Sepals lobed, deflexed, caducous.

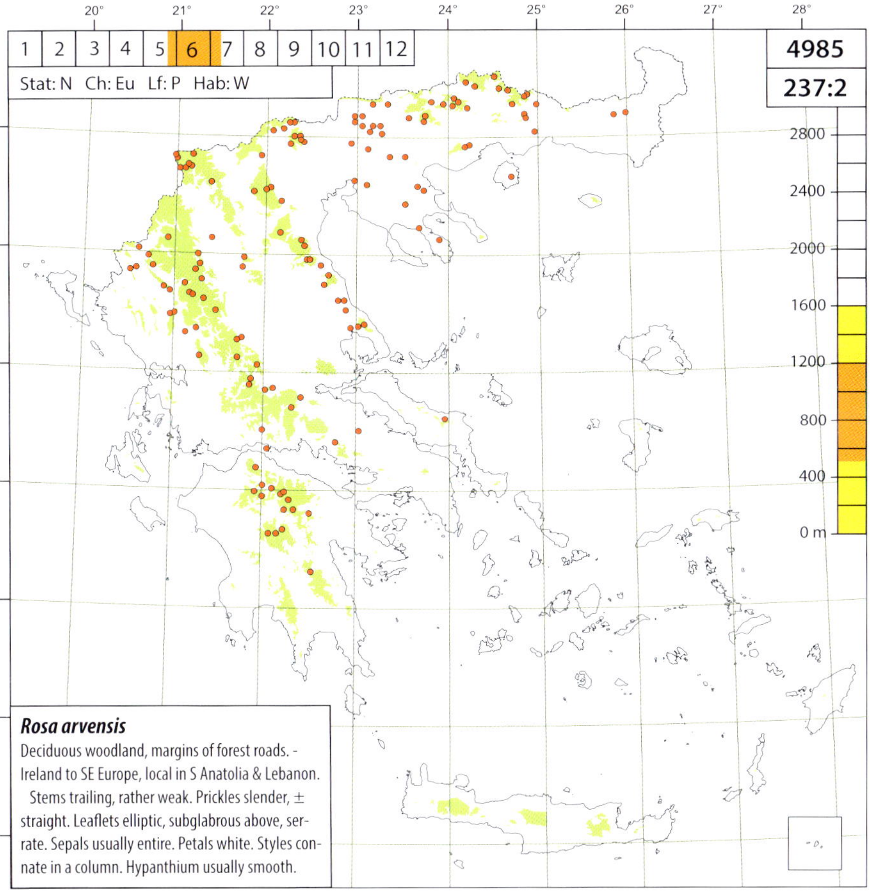

Rosa arvensis

Deciduous woodland, margins of forest roads. - Ireland to SE Europe, local in S Anatolia & Lebanon.

Stems trailing, rather weak. Prickles slender, ± straight. Leaflets elliptic, subglabrous above, serrate. Sepals usually entire. Petals white. Styles connate in a column. Hypanthium usually smooth.

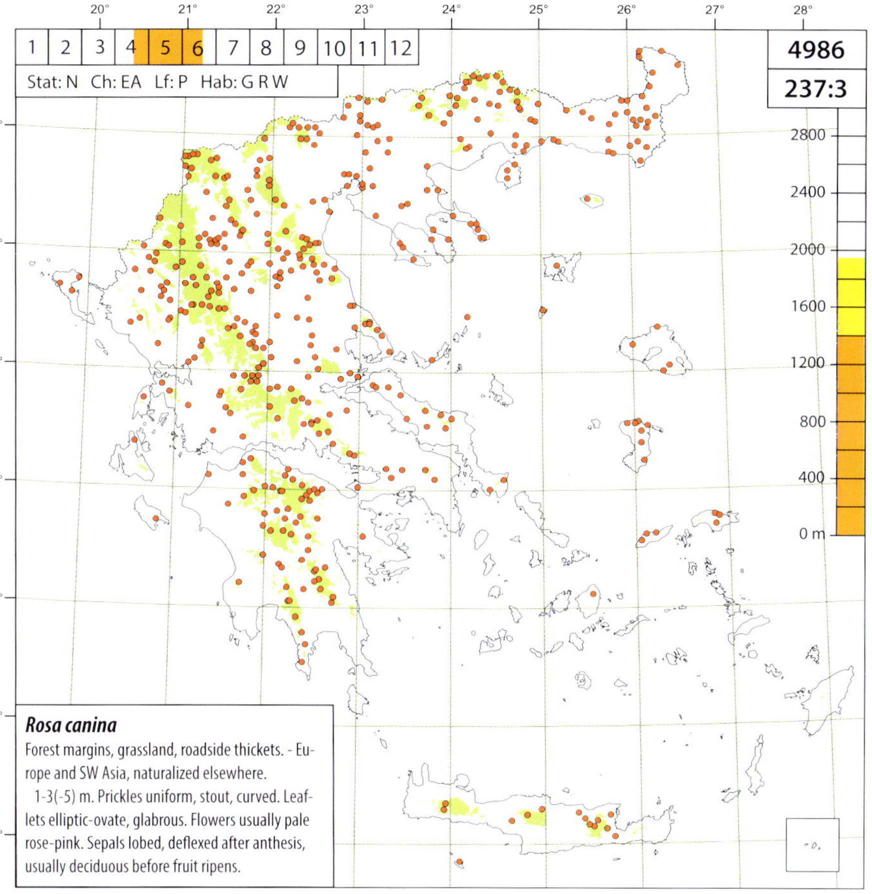

Rosa canina

Forest margins, grassland, roadside thickets. - Europe and SW Asia, naturalized elsewhere.

1-3(-5) m. Prickles uniform, stout, curved. Leaflets elliptic-ovate, glabrous. Flowers usually pale rose-pink. Sepals lobed, deflexed after anthesis, usually deciduous before fruit ripens.

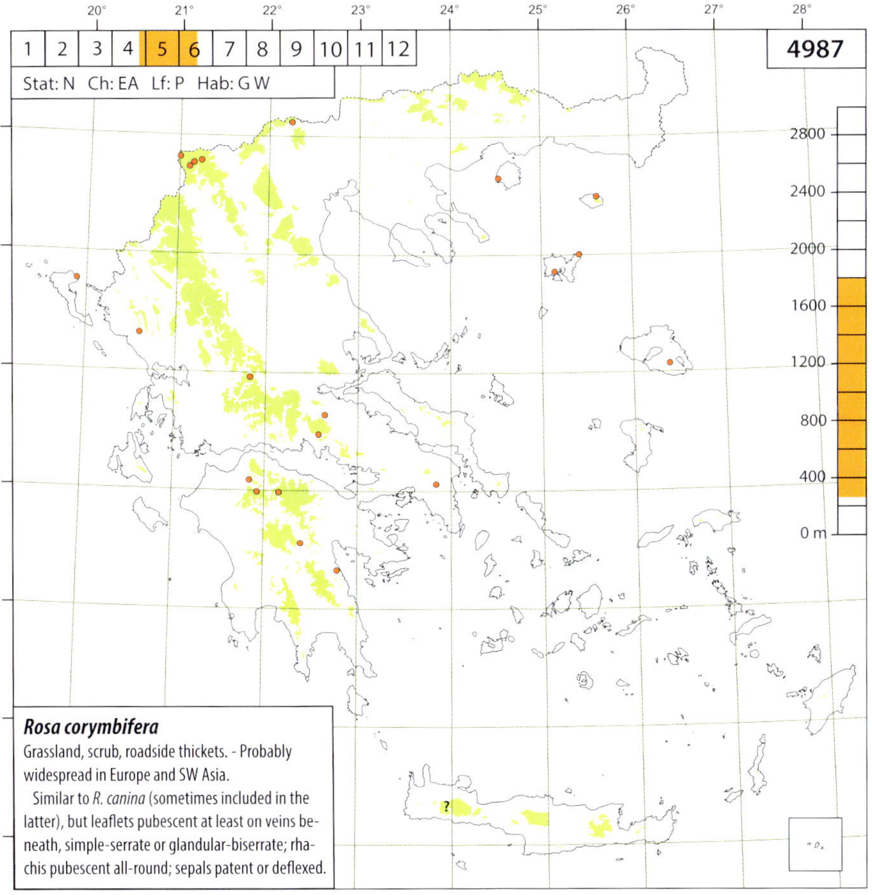

Rosa corymbifera

Grassland, scrub, roadside thickets. - Probably widespread in Europe and SW Asia.

Similar to *R. canina* (sometimes included in the latter), but leaflets pubescent at least on veins beneath, simple-serrate or glandular-biserrate; rhachis pubescent all-round; sepals patent or deflexed.

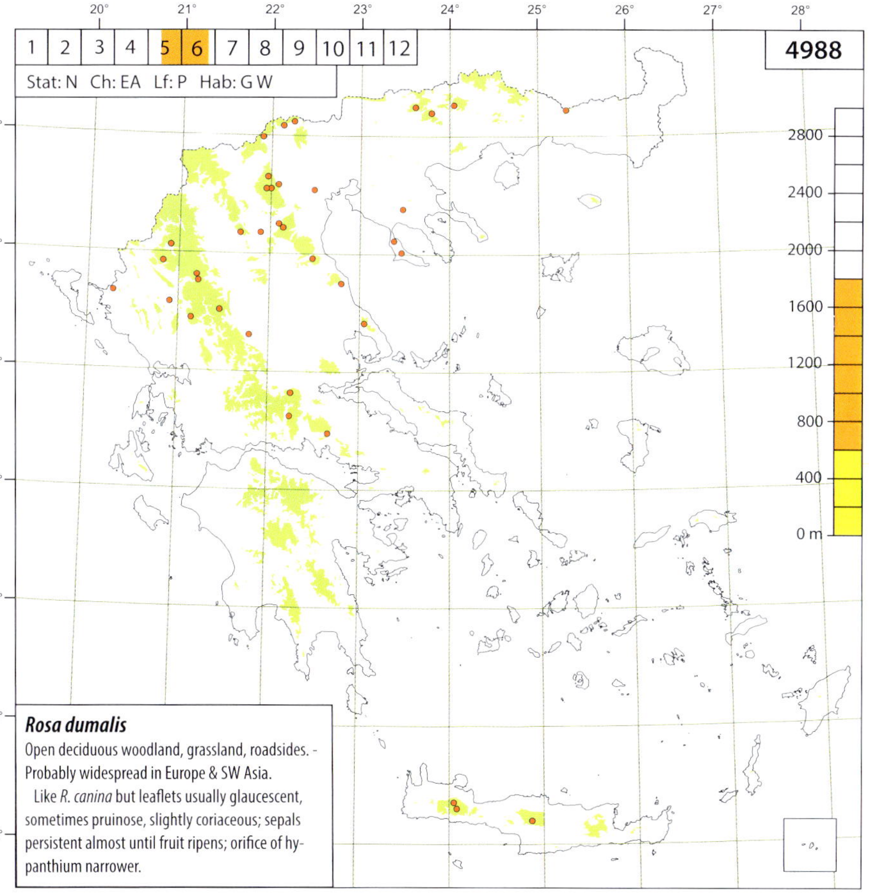

Rosa dumalis

Open deciduous woodland, grassland, roadsides. - Probably widespread in Europe & SW Asia.

Like *R. canina* but leaflets usually glaucescent, sometimes pruinose, slightly coriaceous; sepals persistent almost until fruit ripens; orifice of hypanthium narrower.

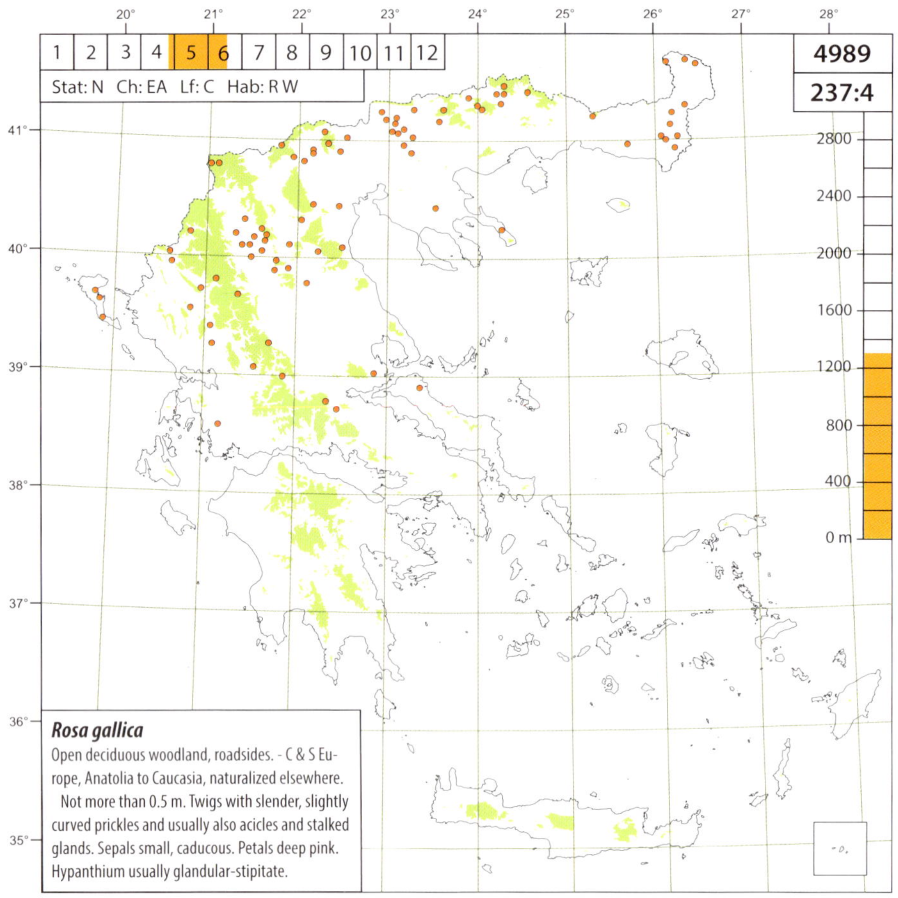

Rosa gallica
Open deciduous woodland, roadsides. - C & S Europe, Anatolia to Caucasia, naturalized elsewhere.
Not more than 0.5 m. Twigs with slender, slightly curved prickles and usually also acicles and stalked glands. Sepals small, caducous. Petals deep pink. Hypanthium usually glandular-stipitate.

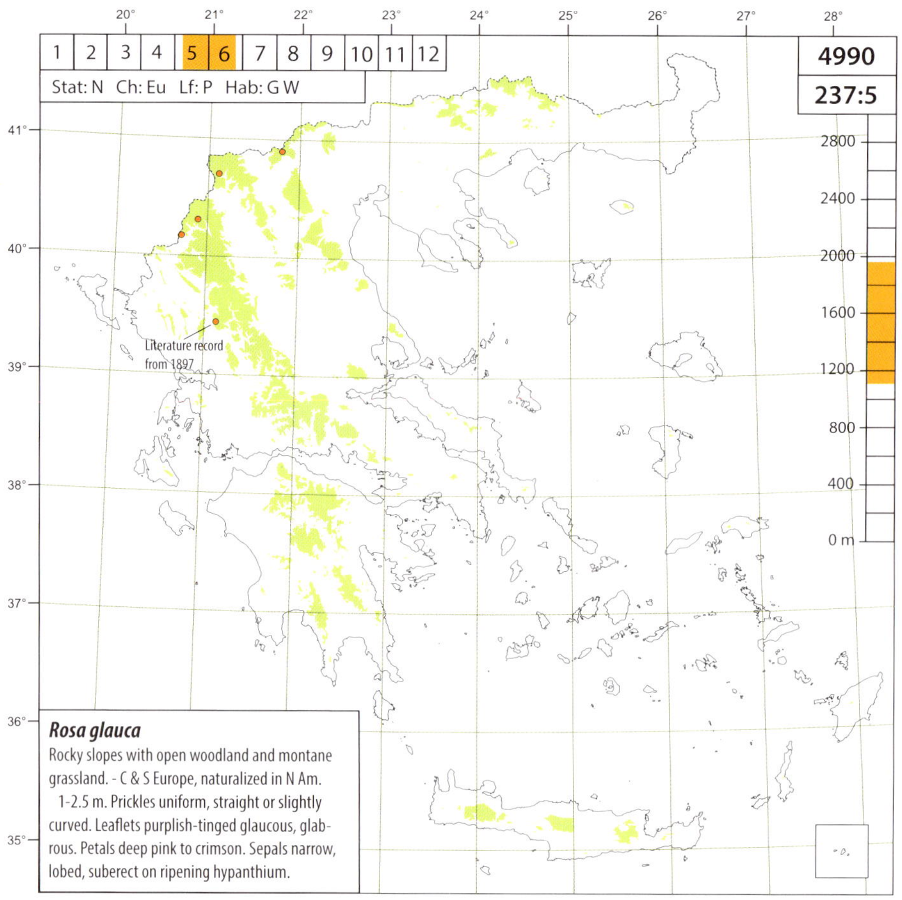

Rosa glauca
Rocky slopes with open woodland and montane grassland. - C & S Europe, naturalized in N Am.
1-2.5 m. Prickles uniform, straight or slightly curved. Leaflets purplish-tinged glaucous, glabrous. Petals deep pink to crimson. Sepals narrow, lobed, suberect on ripening hypanthium.

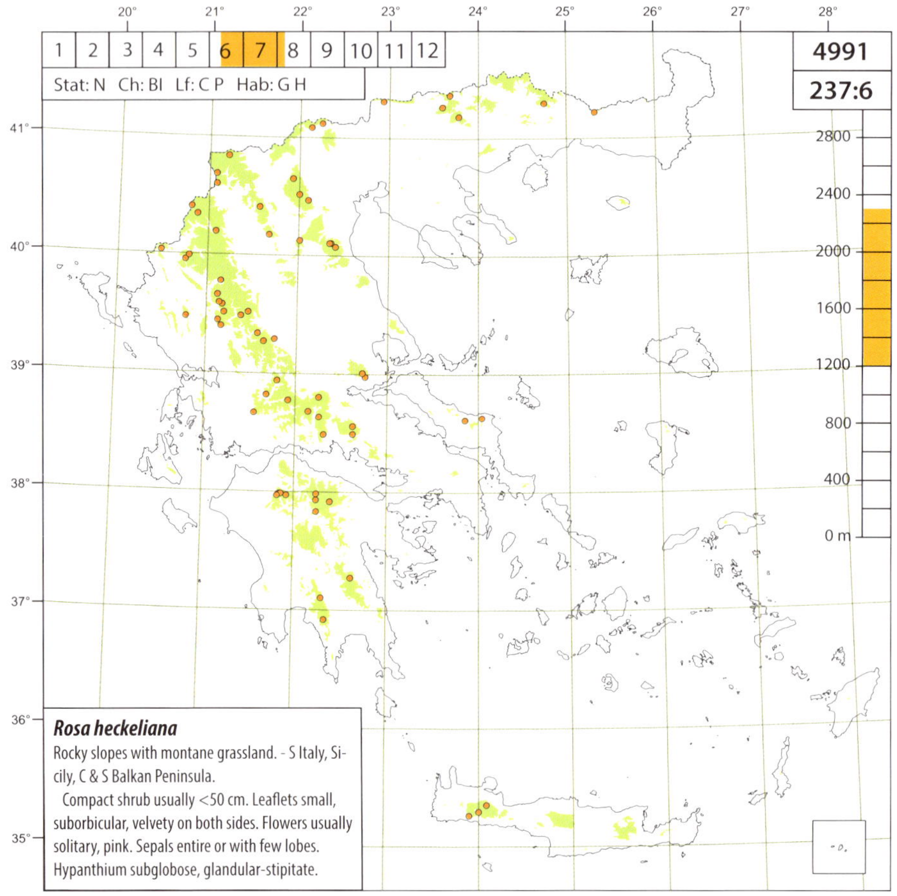

Rosa heckeliana
Rocky slopes with montane grassland. - S Italy, Sicily, C & S Balkan Peninsula.
Compact shrub usually <50 cm. Leaflets small, suborbicular, velvety on both sides. Flowers usually solitary, pink. Sepals entire or with few lobes. Hypanthium subglobose, glandular-stipitate.

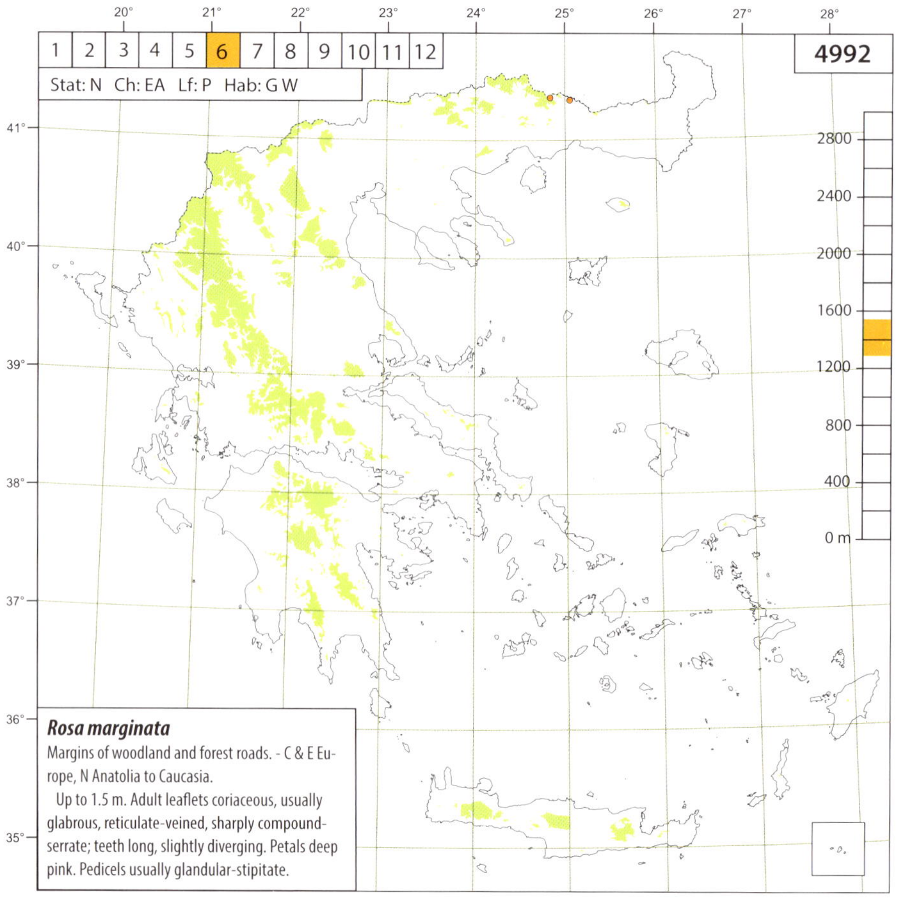

Rosa marginata
Margins of woodland and forest roads. - C & E Europe, N Anatolia to Caucasia.
Up to 1.5 m. Adult leaflets coriaceous, usually glabrous, reticulate-veined, sharply compound-serrate; teeth long, slightly diverging. Petals deep pink. Pedicels usually glandular-stipitate.

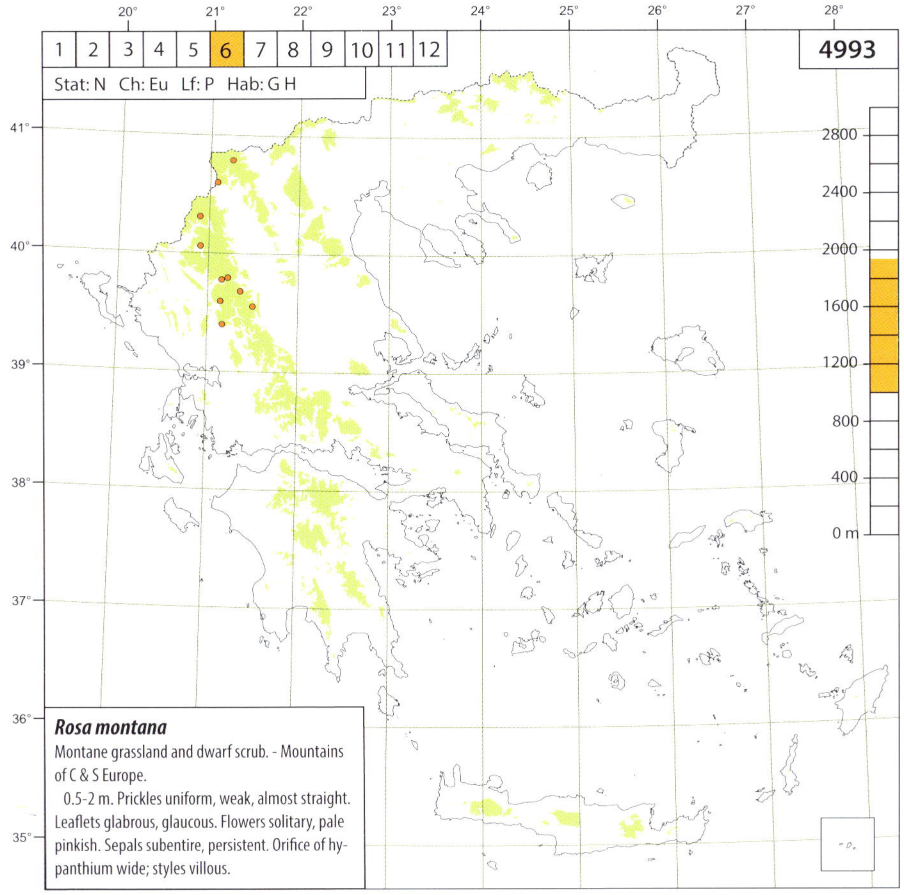

Rosa montana

Montane grassland and dwarf scrub. - Mountains of C & S Europe.

0.5-2 m. Prickles uniform, weak, almost straight. Leaflets glabrous, glaucous. Flowers solitary, pale pinkish. Sepals subentire, persistent. Orifice of hypanthium wide; styles villous.

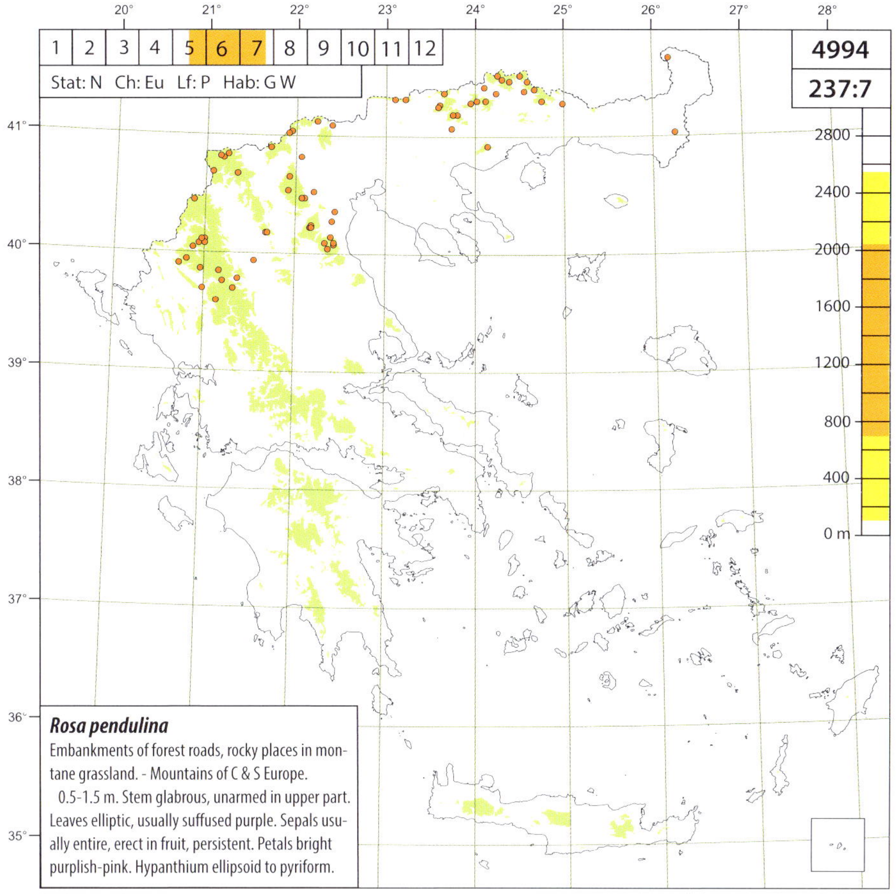

Rosa pendulina

Embankments of forest roads, rocky places in montane grassland. - Mountains of C & S Europe.

0.5-1.5 m. Stem glabrous, unarmed in upper part. Leaves elliptic, usually suffused purple. Sepals usually entire, erect in fruit, persistent. Petals bright purplish-pink. Hypanthium ellipsoid to pyriform.

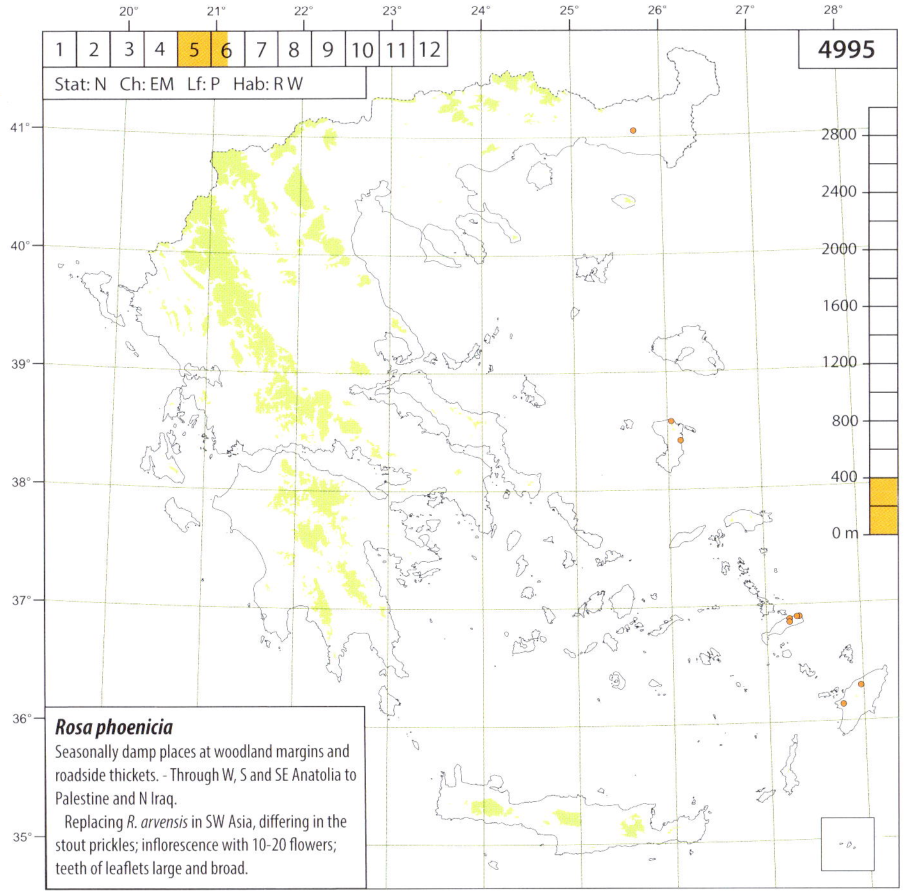

Rosa phoenicia

Seasonally damp places at woodland margins and roadside thickets. - Through W, S and SE Anatolia to Palestine and N Iraq.

Replacing *R. arvensis* in SW Asia, differing in the stout prickles; inflorescence with 10-20 flowers; teeth of leaflets large and broad.

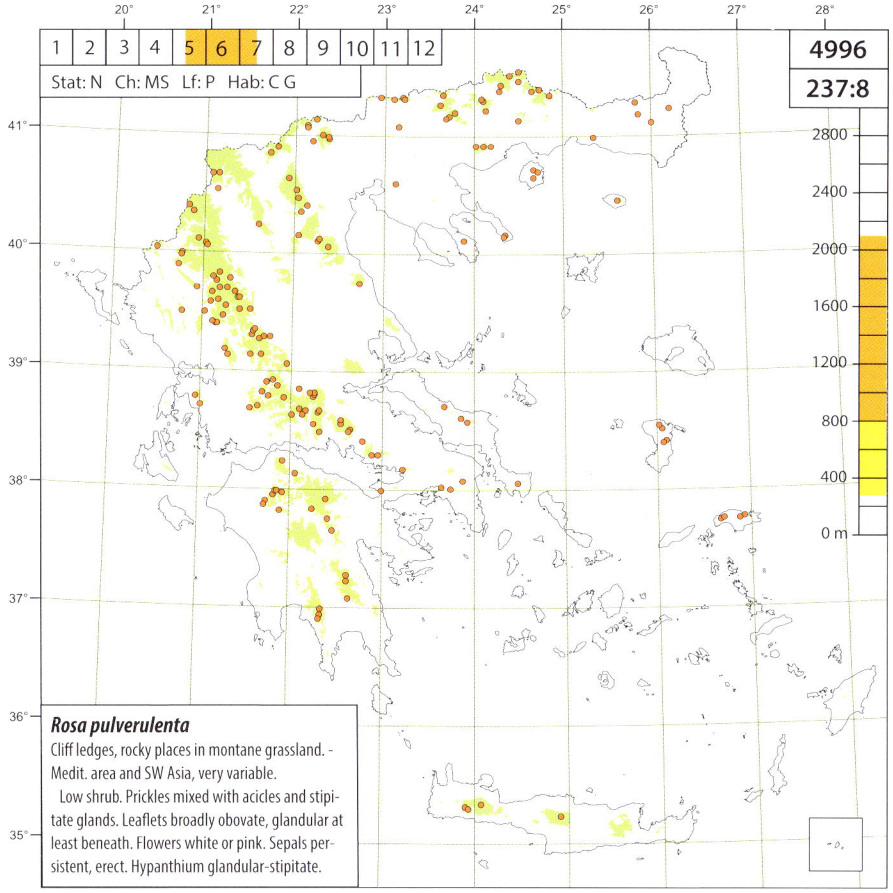

Rosa pulverulenta

Cliff ledges, rocky places in montane grassland. - Medit. area and SW Asia, very variable.

Low shrub. Prickles mixed with acicles and stipitate glands. Leaflets broadly obovate, glandular at least beneath. Flowers white or pink. Sepals persistent, erect. Hypanthium glandular-stipitate.

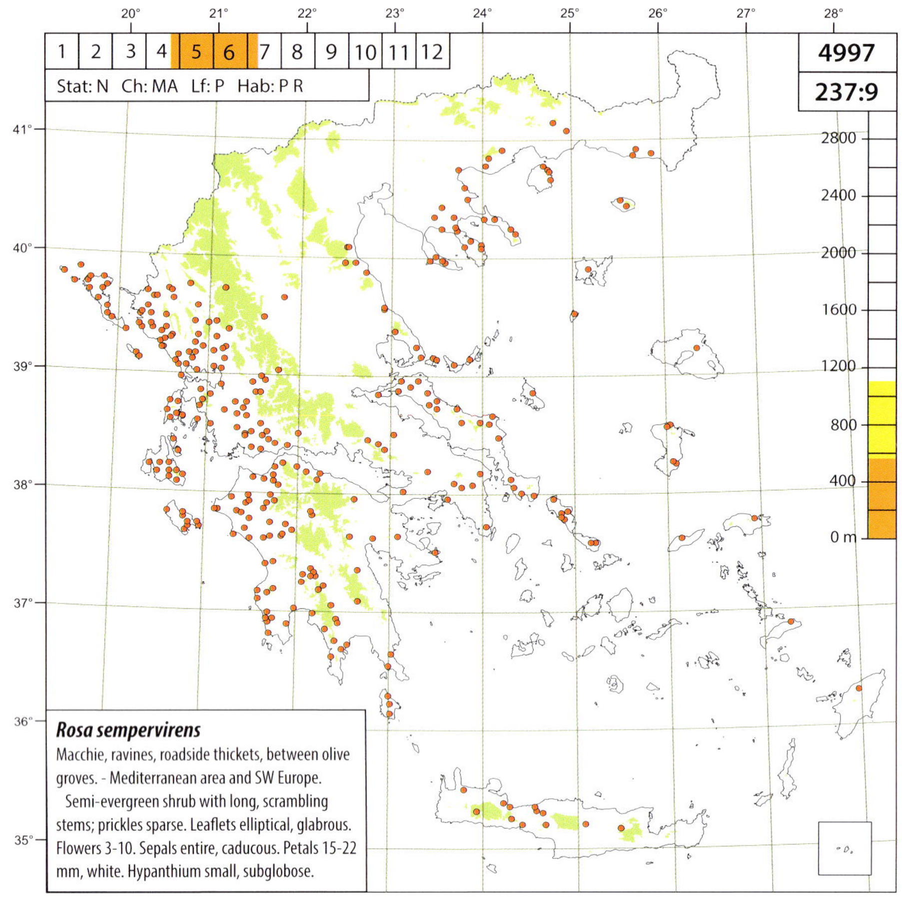

Rosa sempervirens

Macchie, ravines, roadside thickets, between olive groves. - Mediterranean area and SW Europe.

Semi-evergreen shrub with long, scrambling stems; prickles sparse. Leaflets elliptical, glabrous. Flowers 3-10. Sepals entire, caducous. Petals 15-22 mm, white. Hypanthium small, subglobose.

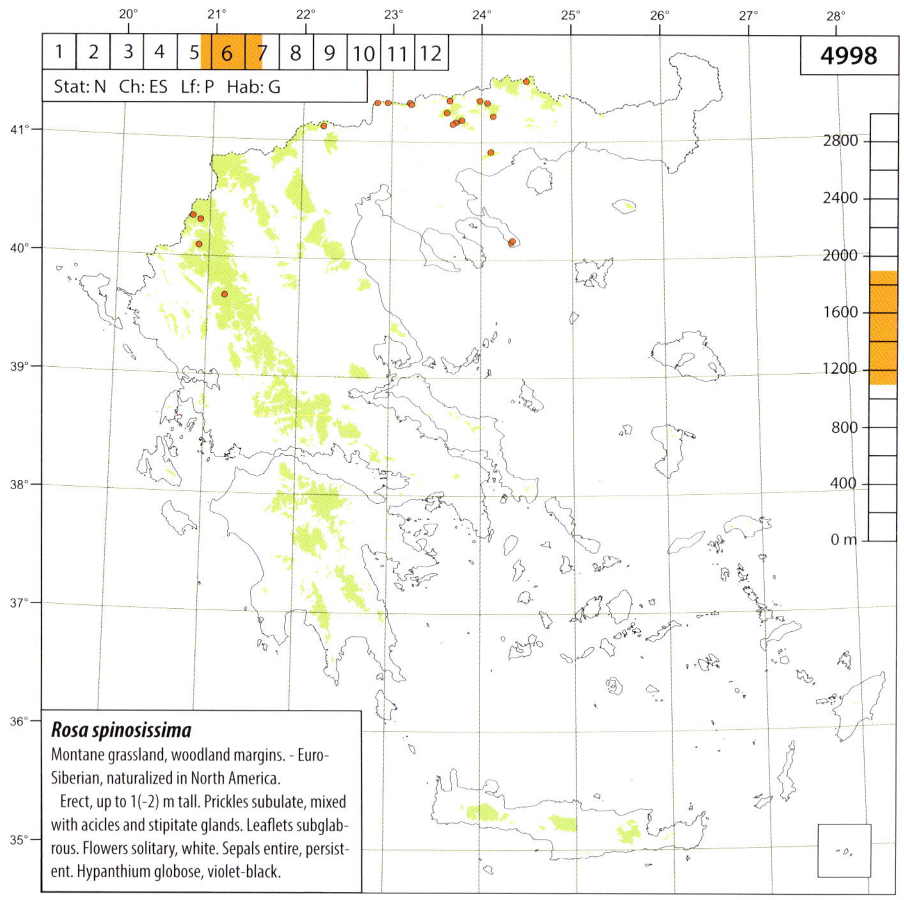

Rosa spinosissima

Montane grassland, woodland margins. - Euro-Siberian, naturalized in North America.

Erect, up to 1(-2) m tall. Prickles subulate, mixed with acicles and stipitate glands. Leaflets subglabrous. Flowers solitary, white. Sepals entire, persistent. Hypanthium globose, violet-black.

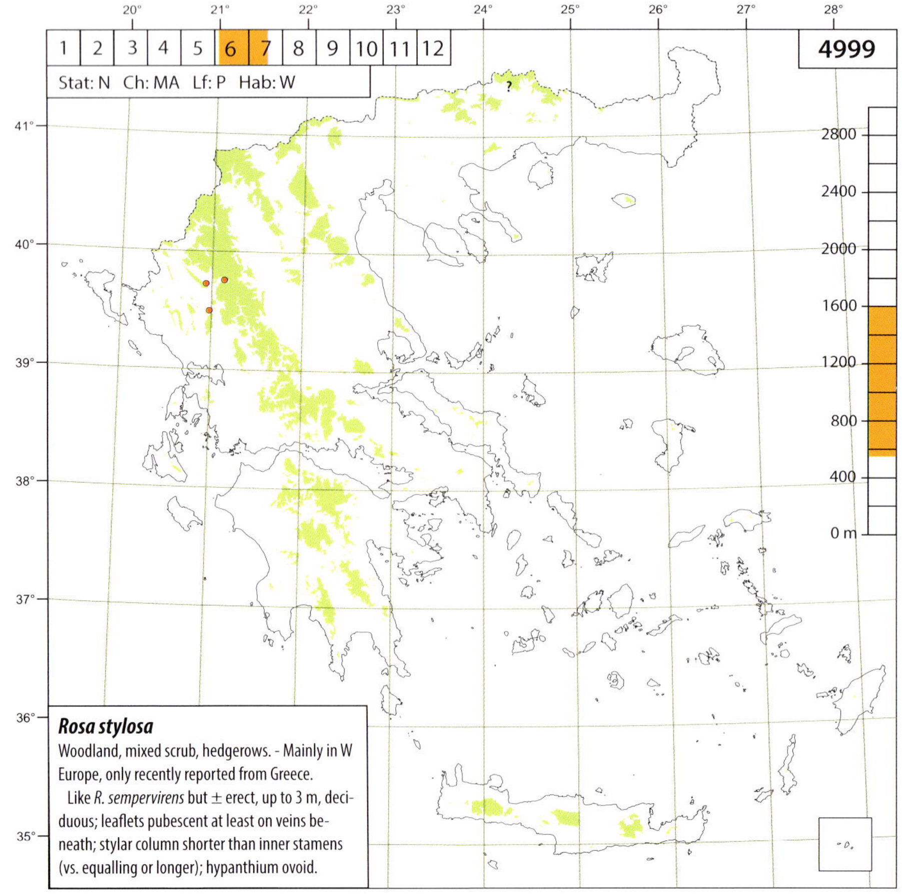

Rosa stylosa

Woodland, mixed scrub, hedgerows. - Mainly in W Europe, only recently reported from Greece.

Like *R. sempervirens* but ± erect, up to 3 m, deciduous; leaflets pubescent at least on veins beneath; stylar column shorter than inner stamens (vs. equalling or longer); hypanthium ovoid.

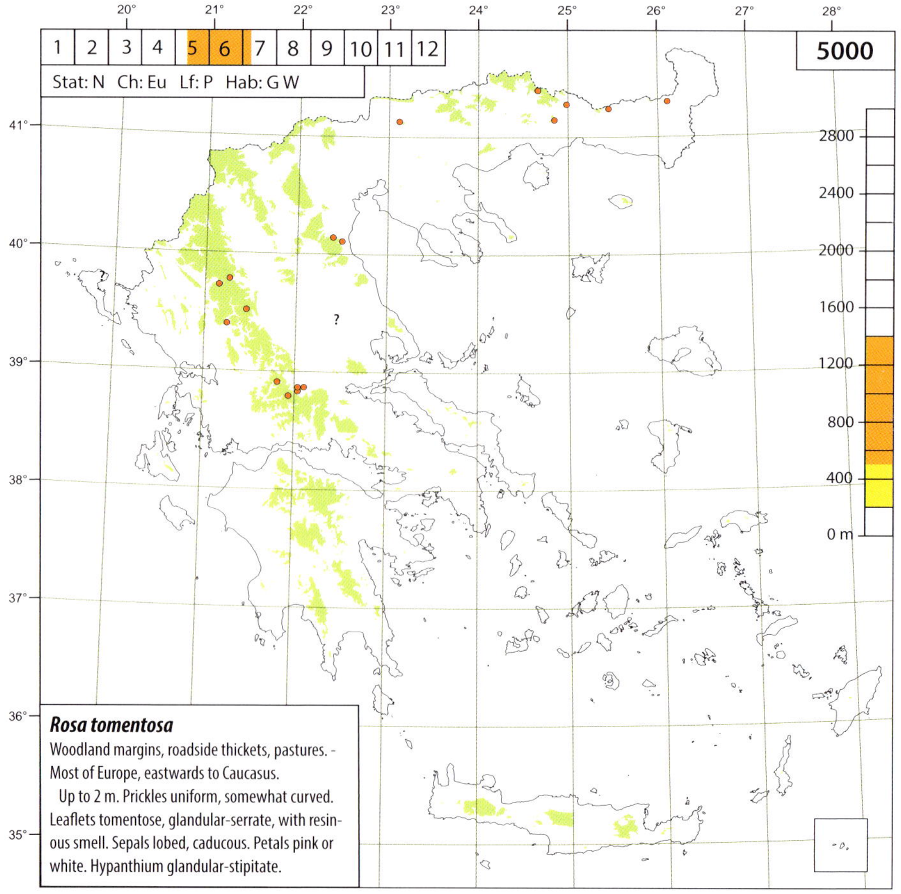

Rosa tomentosa

Woodland margins, roadside thickets, pastures. - Most of Europe, eastwards to Caucasus.

Up to 2 m. Prickles uniform, somewhat curved. Leaflets tomentose, glandular-serrate, with resinous smell. Sepals lobed, caducous. Petals pink or white. Hypanthium glandular-stipitate.

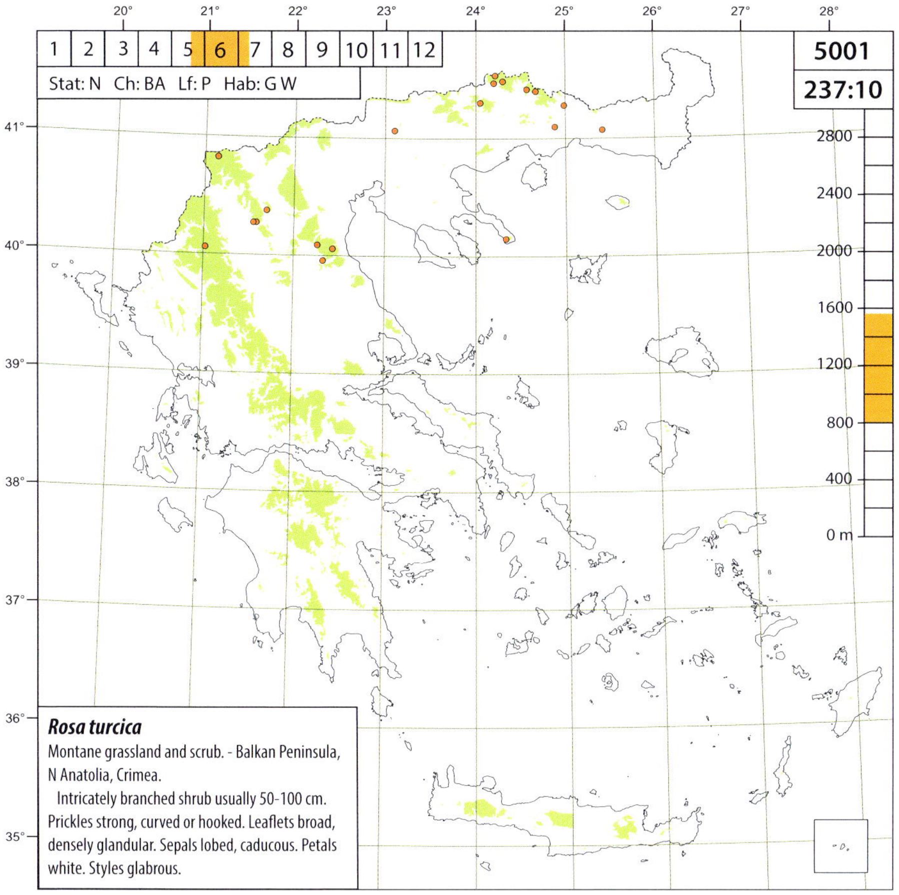

Rosa turcica
Montane grassland and scrub. - Balkan Peninsula, N Anatolia, Crimea.
Intricately branched shrub usually 50-100 cm. Prickles strong, curved or hooked. Leaflets broad, densely glandular. Sepals lobed, caducous. Petals white. Styles glabrous.

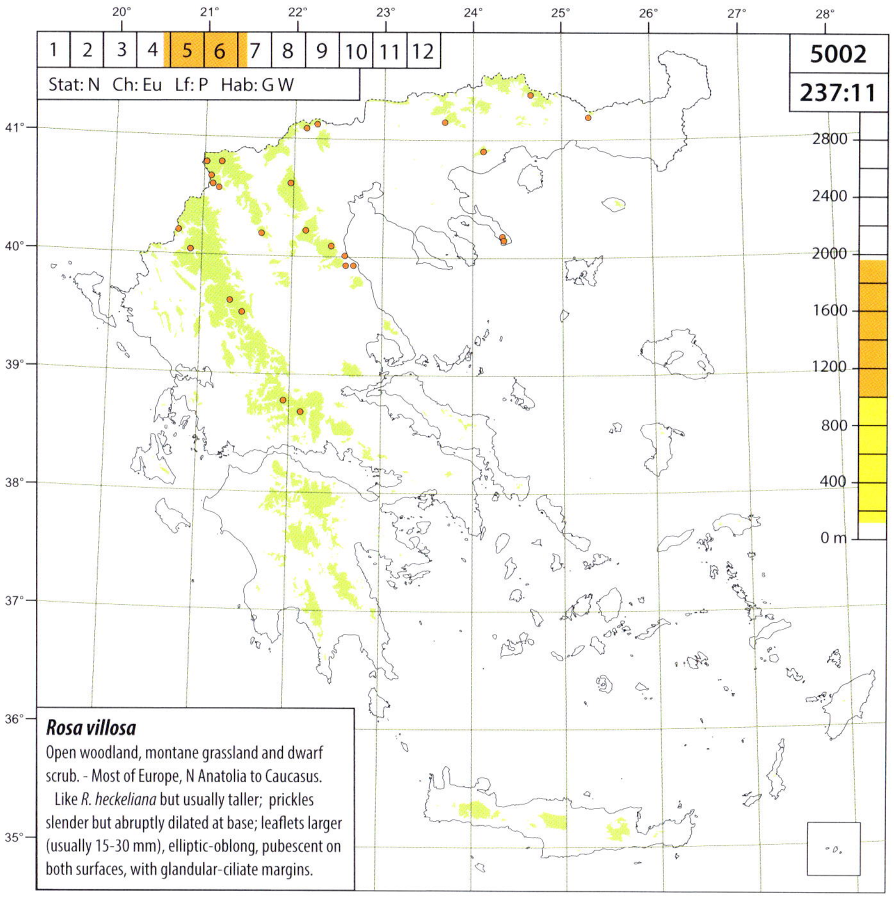

Rosa villosa
Open woodland, montane grassland and dwarf scrub. - Most of Europe, N Anatolia to Caucasus.
Like *R. heckeliana* but usually taller; prickles slender but abruptly dilated at base; leaflets larger (usually 15-30 mm), elliptic-oblong, pubescent on both surfaces, with glandular-ciliate margins.

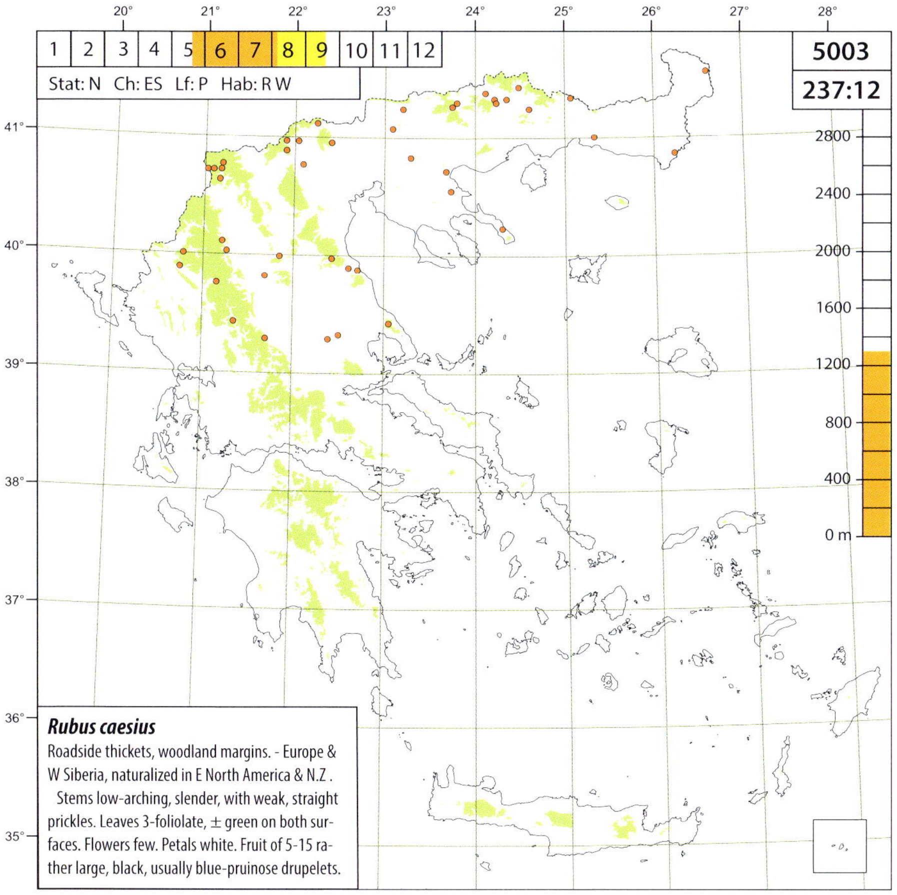

Rubus caesius
Roadside thickets, woodland margins. - Europe & W Siberia, naturalized in E North America & N.Z.
Stems low-arching, slender, with weak, straight prickles. Leaves 3-foliolate, ± green on both surfaces. Flowers few. Petals white. Fruit of 5-15 rather large, black, usually blue-pruinose drupelets.

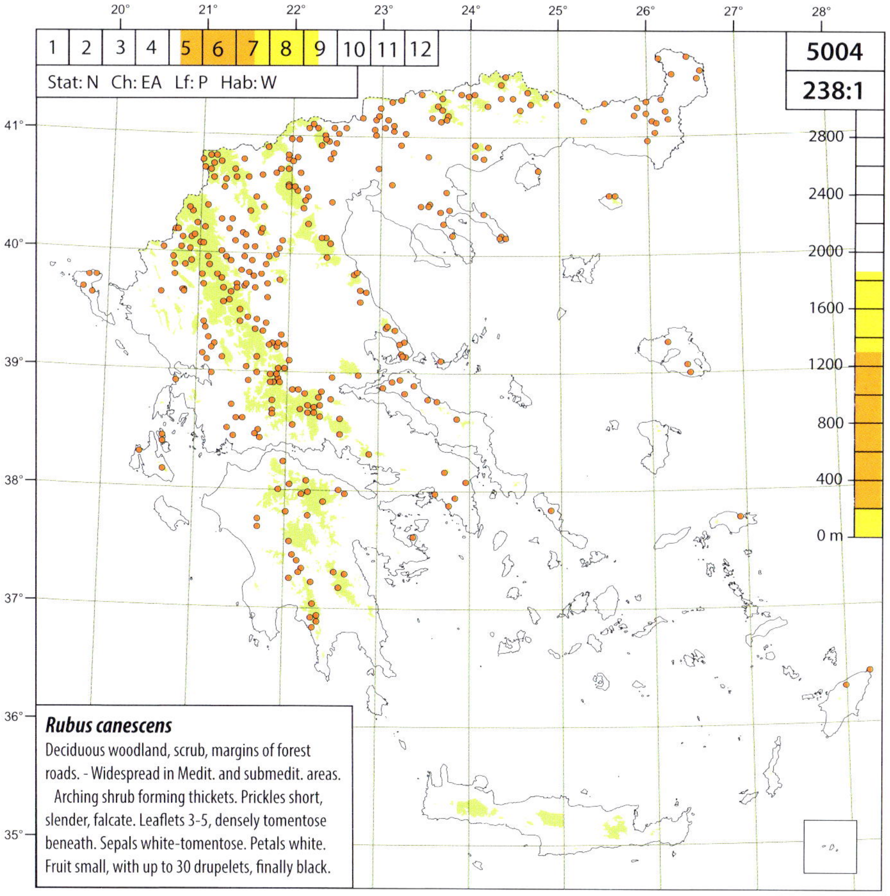

Rubus canescens
Deciduous woodland, scrub, margins of forest roads. - Widespread in Medit. and submedit. areas.
Arching shrub forming thickets. Prickles short, slender, falcate. Leaflets 3-5, densely tomentose beneath. Sepals white-tomentose. Petals white. Fruit small, with up to 30 drupelets, finally black.

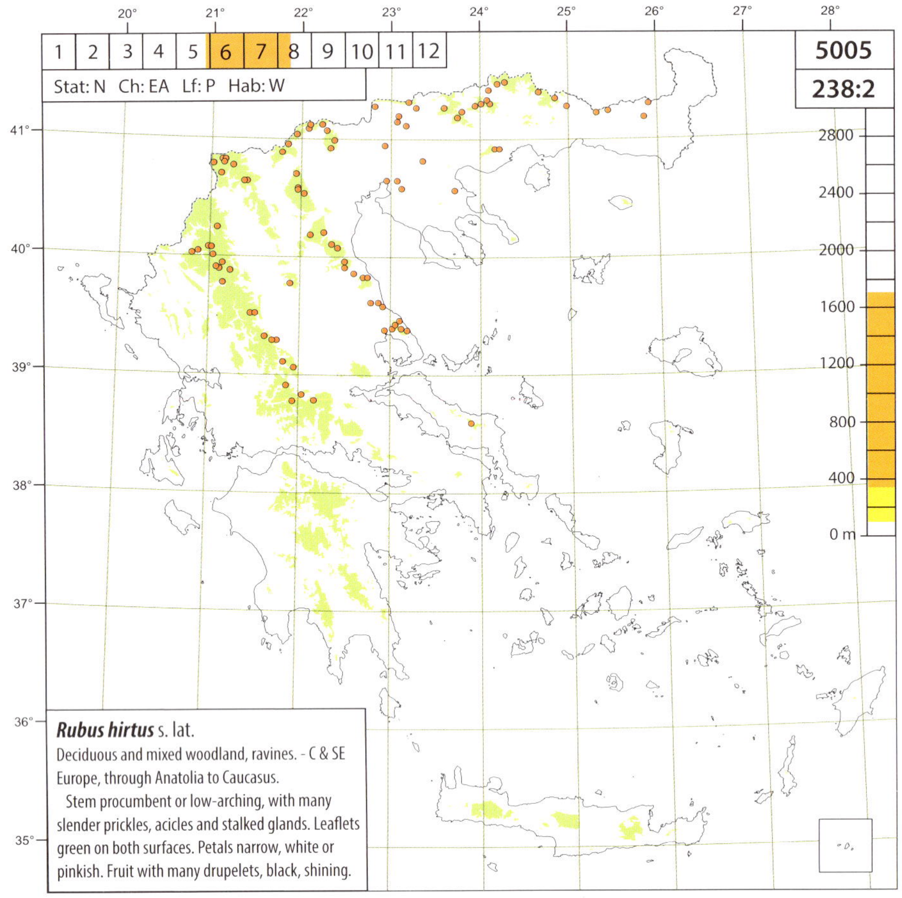

Rubus hirtus s. lat.
Deciduous and mixed woodland, ravines. - C & SE Europe, through Anatolia to Caucasus.
Stem procumbent or low-arching, with many slender prickles, acicles and stalked glands. Leaflets green on both surfaces. Petals narrow, white or pinkish. Fruit with many drupelets, black, shining.

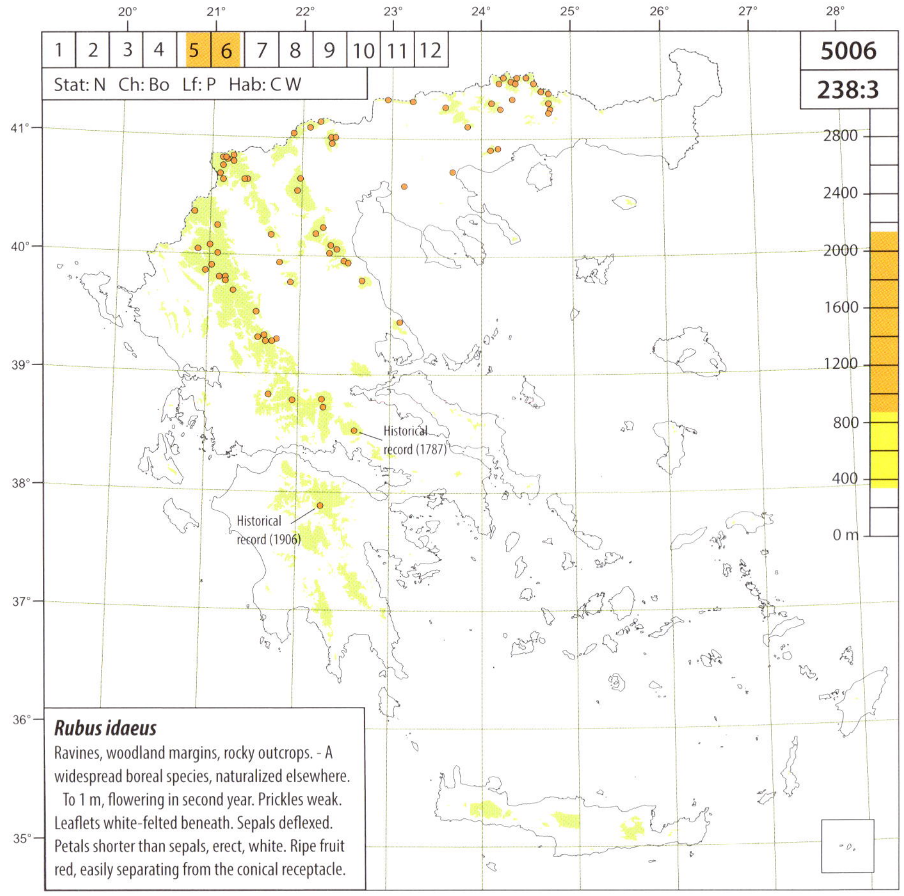

Rubus idaeus
Ravines, woodland margins, rocky outcrops. - A widespread boreal species, naturalized elsewhere.
To 1 m, flowering in second year. Prickles weak. Leaflets white-felted beneath. Sepals deflexed. Petals shorter than sepals, erect, white. Ripe fruit red, easily separating from the conical receptacle.

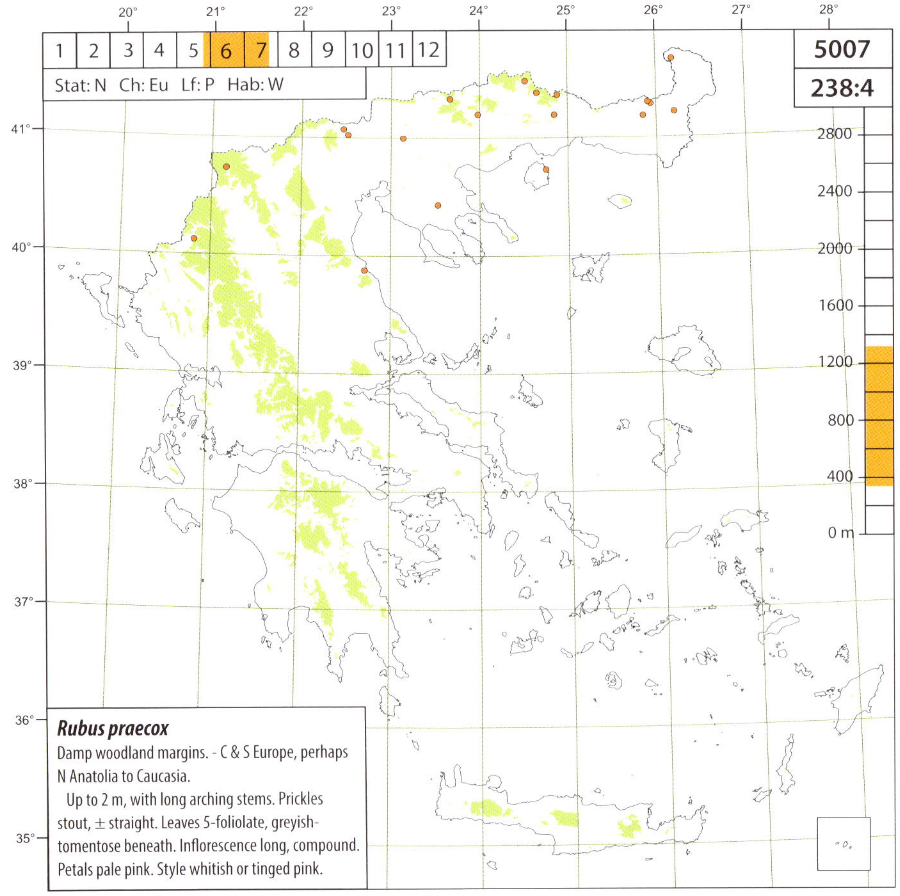

Rubus praecox
Damp woodland margins. - C & S Europe, perhaps N Anatolia to Caucasia.
Up to 2 m, with long arching stems. Prickles stout, ± straight. Leaves 5-foliolate, greyish-tomentose beneath. Inflorescence long, compound. Petals pale pink. Style whitish or tinged pink.

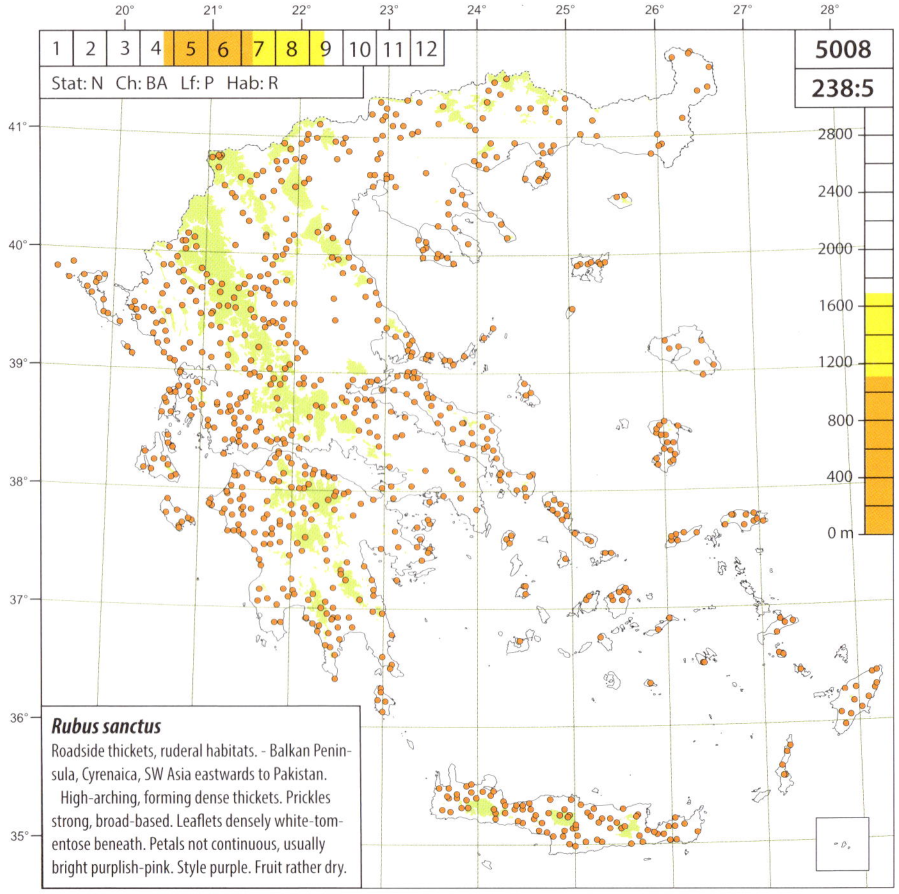

Rubus sanctus
Roadside thickets, ruderal habitats. - Balkan Peninsula, Cyrenaica, SW Asia eastwards to Pakistan.
High-arching, forming dense thickets. Prickles strong, broad-based. Leaflets densely white-tomentose beneath. Petals not continuous, usually bright purplish-pink. Style purple. Fruit rather dry.

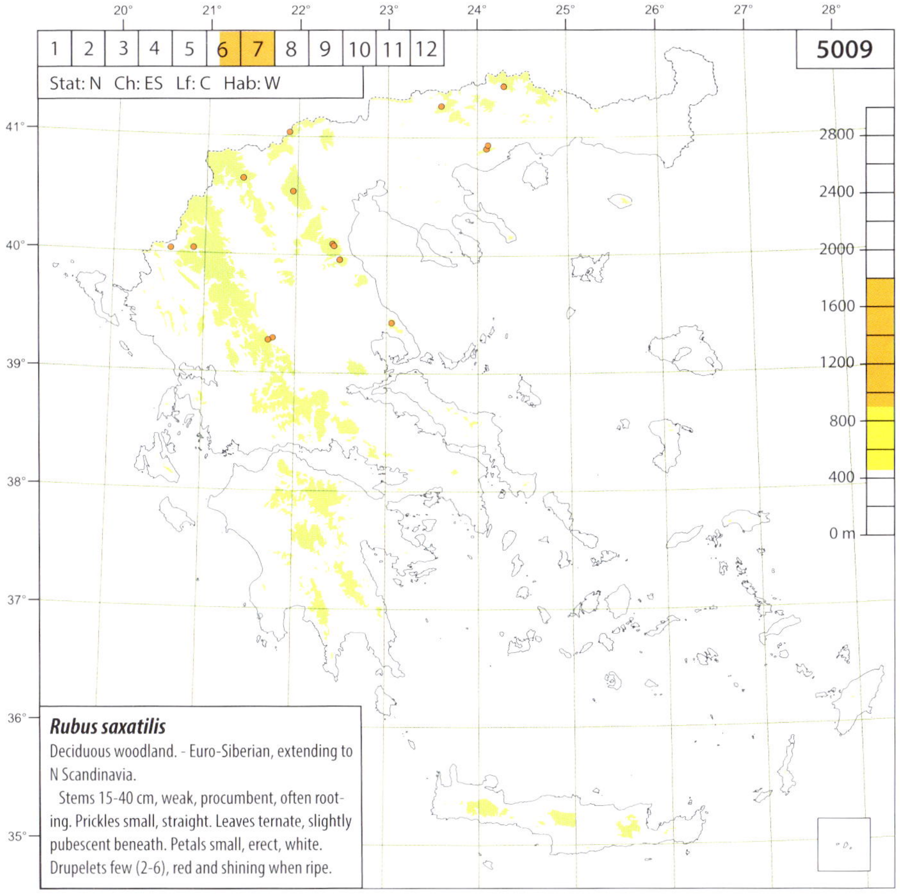

Rubus saxatilis
Deciduous woodland. - Euro-Siberian, extending to N Scandinavia.
Stems 15-40 cm, weak, procumbent, often rooting. Prickles small, straight. Leaves ternate, slightly pubescent beneath. Petals small, erect, white. Drupelets few (2-6), red and shining when ripe.

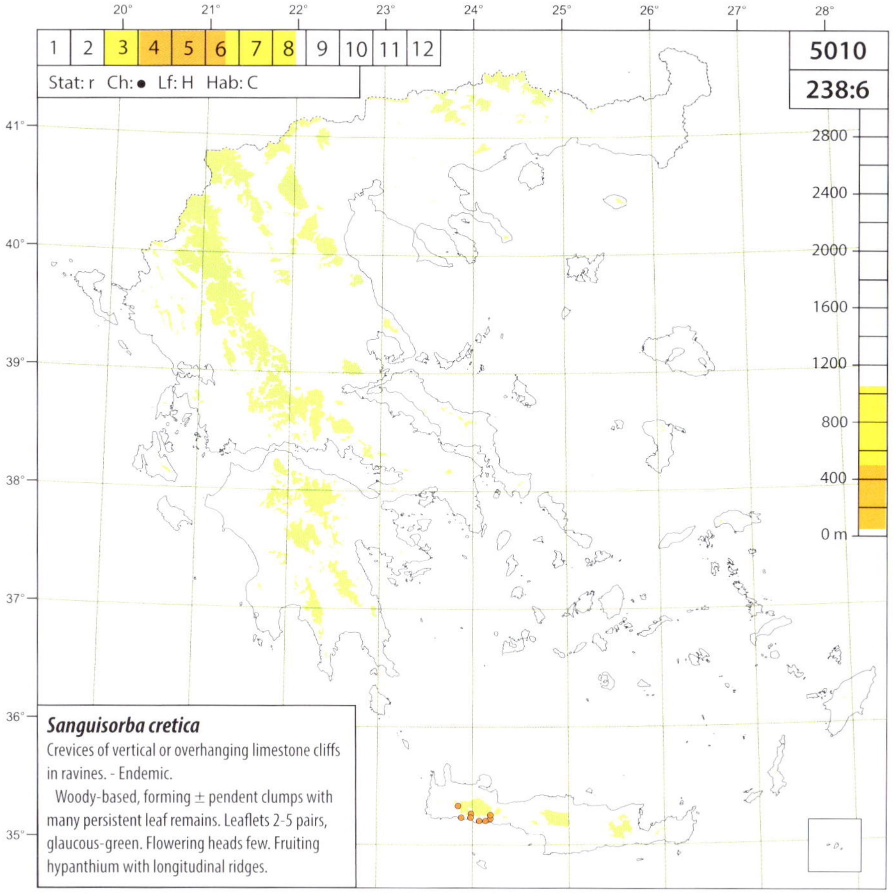

Sanguisorba cretica
Crevices of vertical or overhanging limestone cliffs in ravines. - Endemic.
Woody-based, forming ± pendent clumps with many persistent leaf remains. Leaflets 2-5 pairs, glaucous-green. Flowering heads few. Fruiting hypanthium with longitudinal ridges.

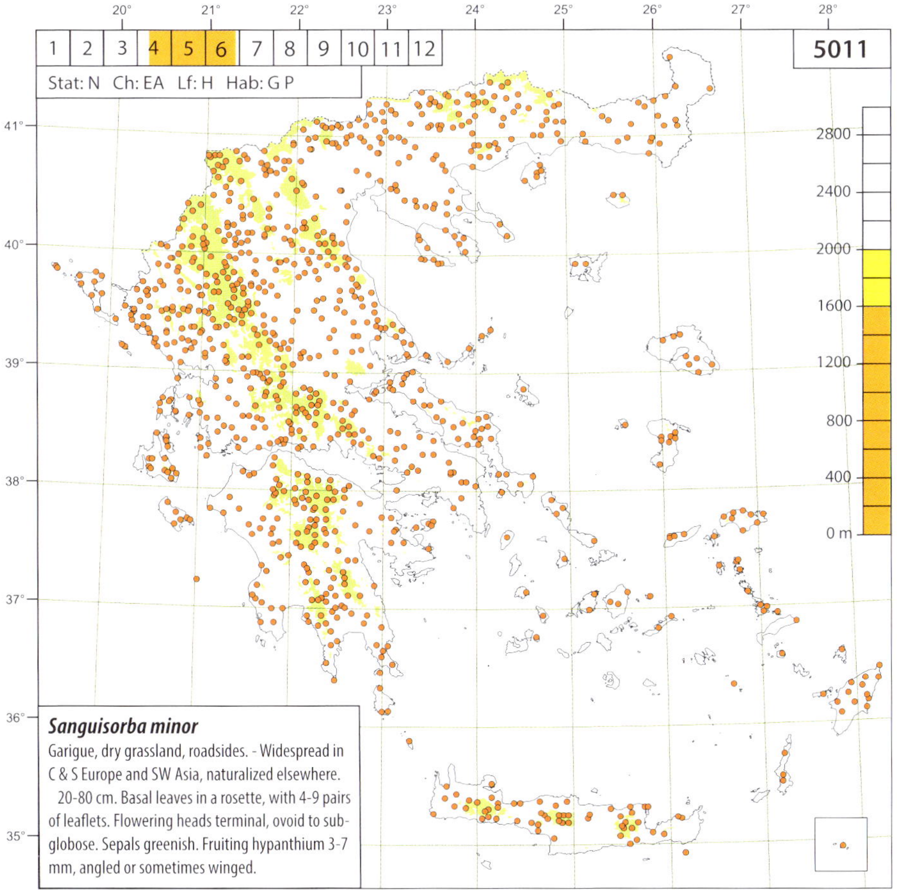

Sanguisorba minor
Garigue, dry grassland, roadsides. - Widespread in C & S Europe and SW Asia, naturalized elsewhere.
20-80 cm. Basal leaves in a rosette, with 4-9 pairs of leaflets. Flowering heads terminal, ovoid to subglobose. Sepals greenish. Fruiting hypanthium 3-7 mm, angled or sometimes winged.

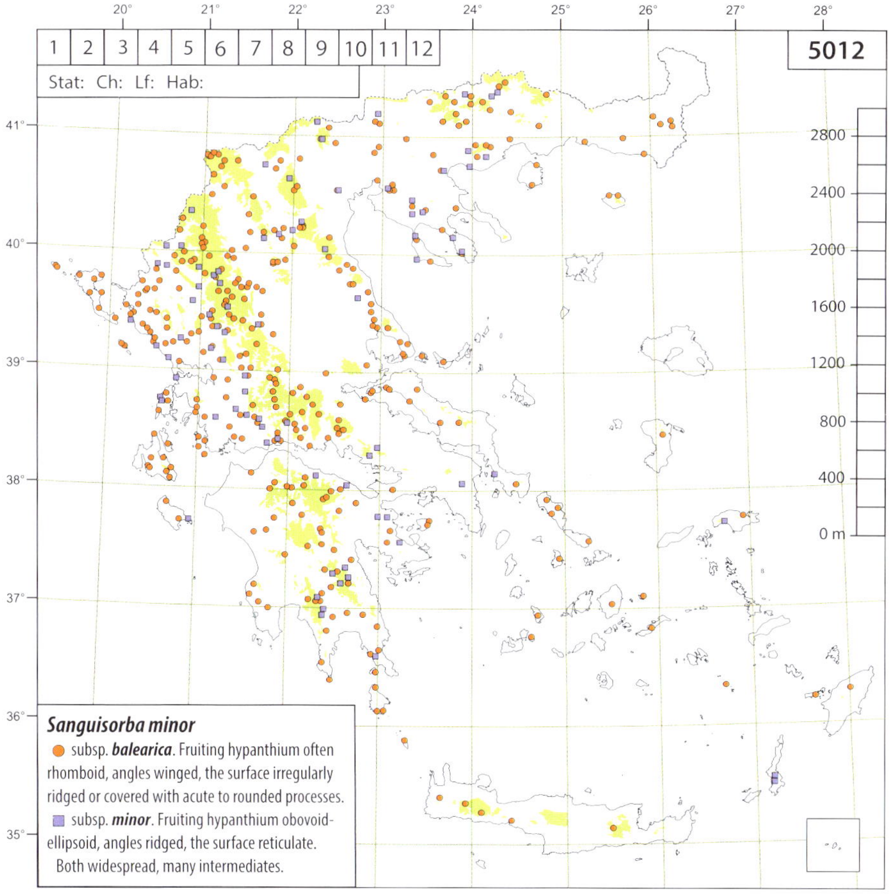

Sanguisorba minor
● subsp. ***balearica***. Fruiting hypanthium often rhomboid, angles winged, the surface irregularly ridged or covered with acute to rounded processes.
■ subsp. ***minor***. Fruiting hypanthium obovoid-ellipsoid, angles ridged, the surface reticulate.
Both widespread, many intermediates.

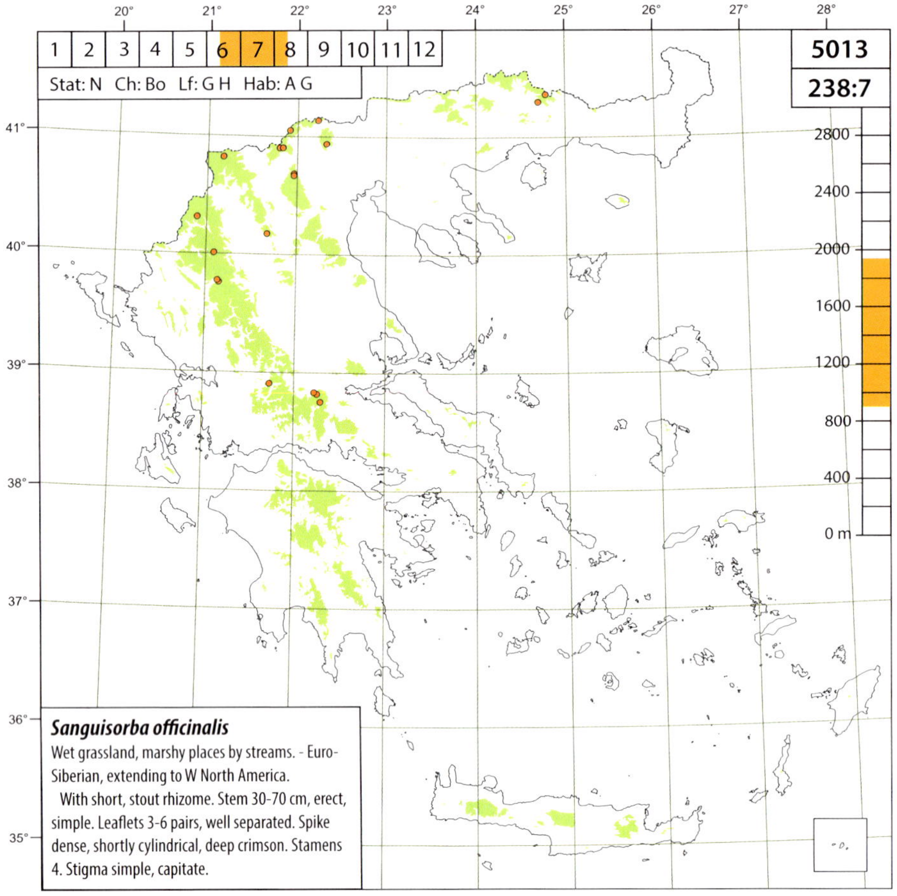

Sanguisorba officinalis

Wet grassland, marshy places by streams. - Euro-Siberian, extending to W North America.

With short, stout rhizome. Stem 30-70 cm, erect, simple. Leaflets 3-6 pairs, well separated. Spike dense, shortly cylindrical, deep crimson. Stamens 4. Stigma simple, capitate.

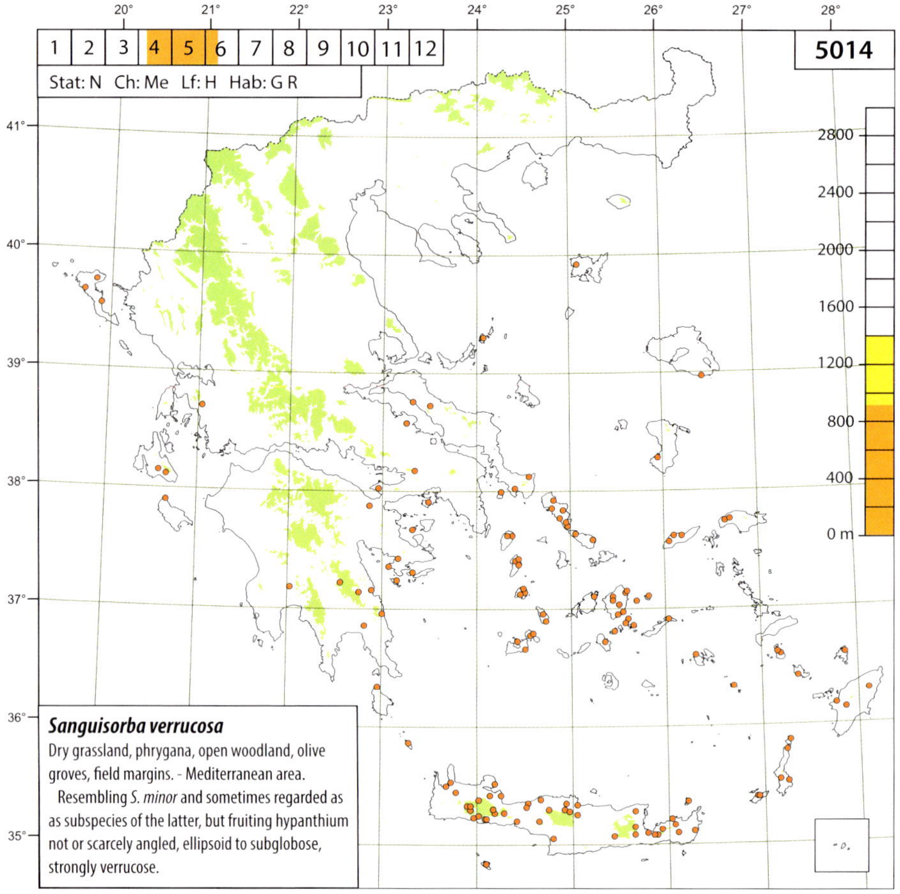

Sanguisorba verrucosa

Dry grassland, phrygana, open woodland, olive groves, field margins. - Mediterranean area.

Resembling *S. minor* and sometimes regarded as as subspecies of the latter, but fruiting hypanthium not or scarcely angled, ellipsoid to subglobose, strongly verrucose.

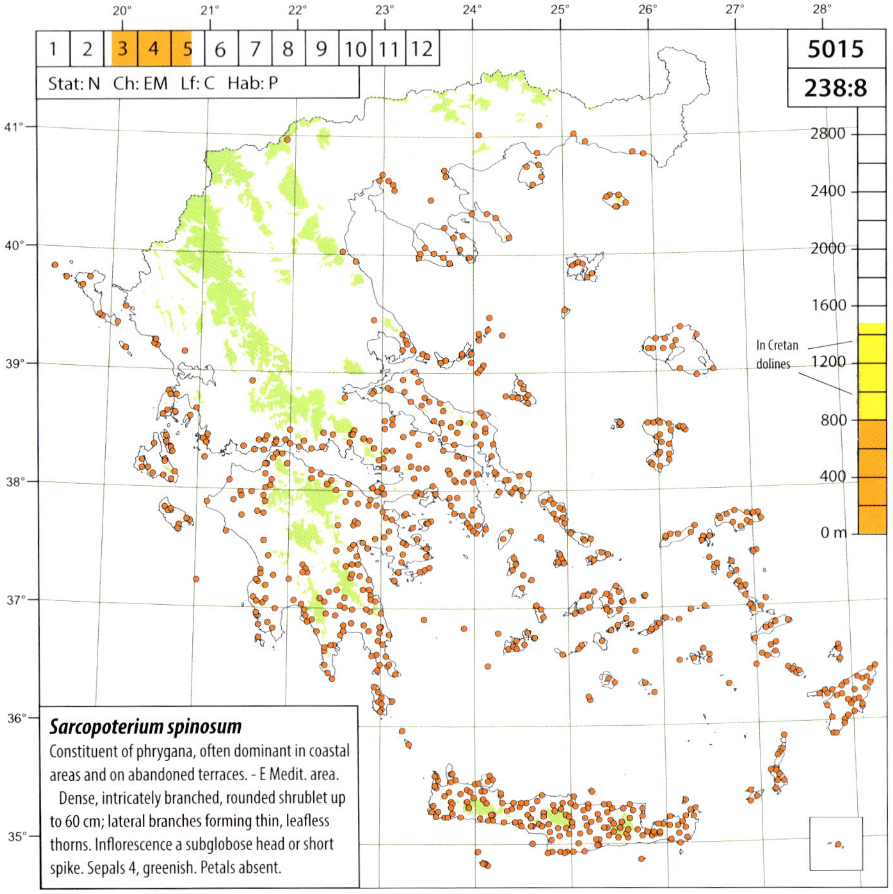

Sarcopoterium spinosum

Constituent of phrygana, often dominant in coastal areas and on abandoned terraces. - E Medit. area.

Dense, intricately branched, rounded shrublet up to 60 cm; lateral branches forming thin, leafless thorns. Inflorescence a subglobose head or short spike. Sepals 4, greenish. Petals absent.

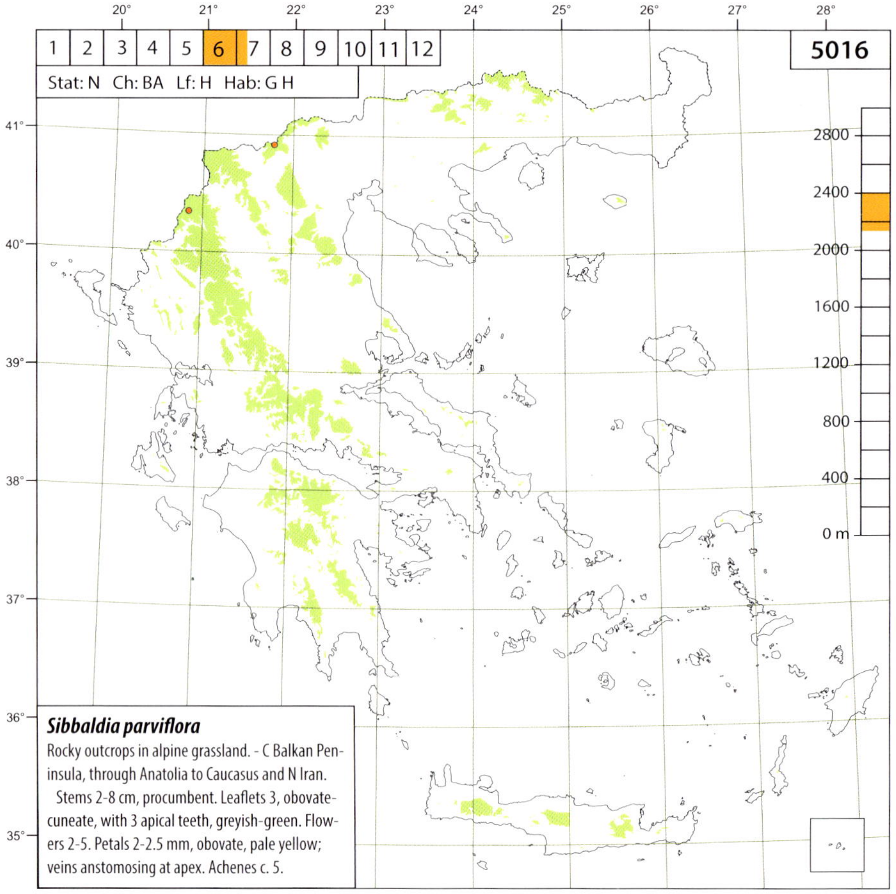

Sibbaldia parviflora

Rocky outcrops in alpine grassland. - C Balkan Peninsula, through Anatolia to Caucasus and N Iran.

Stems 2-8 cm, procumbent. Leaflets 3, obovate-cuneate, with 3 apical teeth, greyish-green. Flowers 2-5. Petals 2-2.5 mm, obovate, pale yellow; veins anstomosing at apex. Achenes c. 5.

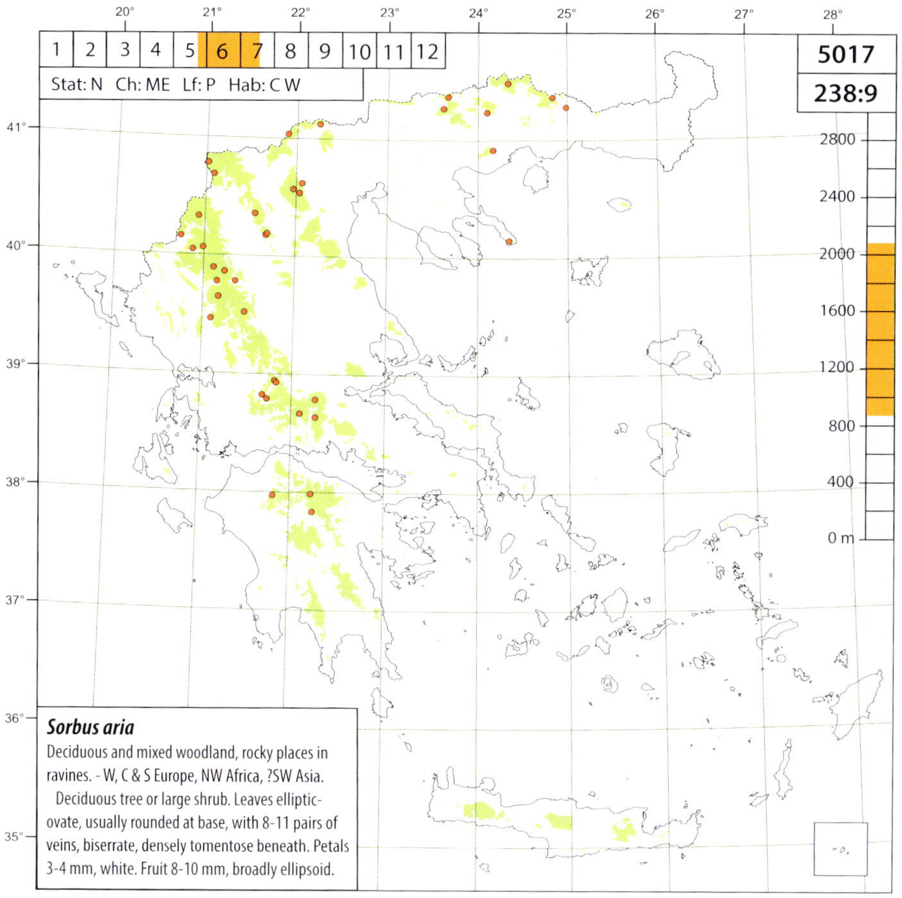

Sorbus aria
Deciduous and mixed woodland, rocky places in ravines. - W, C & S Europe, NW Africa, ?SW Asia.
Deciduous tree or large shrub. Leaves elliptic-ovate, usually rounded at base, with 8-11 pairs of veins, biserrate, densely tomentose beneath. Petals 3-4 mm, white. Fruit 8-10 mm, broadly ellipsoid.

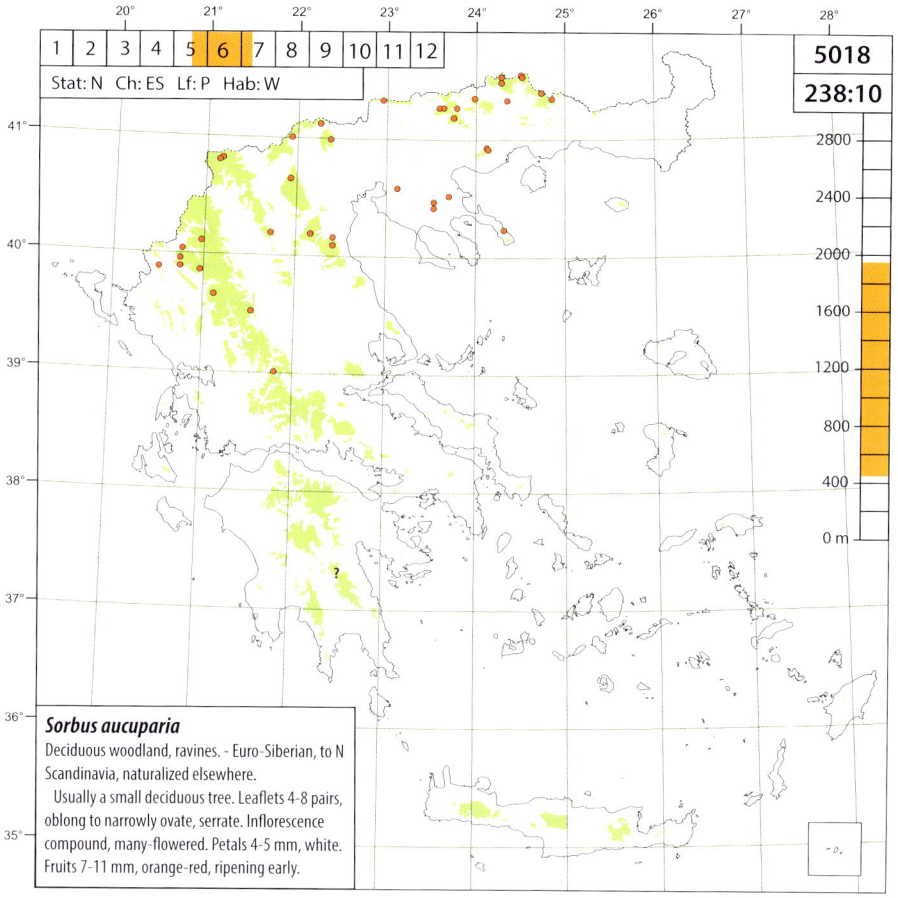

Sorbus aucuparia
Deciduous woodland, ravines. - Euro-Siberian, to N Scandinavia, naturalized elsewhere.
Usually a small deciduous tree. Leaflets 4-8 pairs, oblong to narrowly ovate, serrate. Inflorescence compound, many-flowered. Petals 4-5 mm, white. Fruits 7-11 mm, orange-red, ripening early.

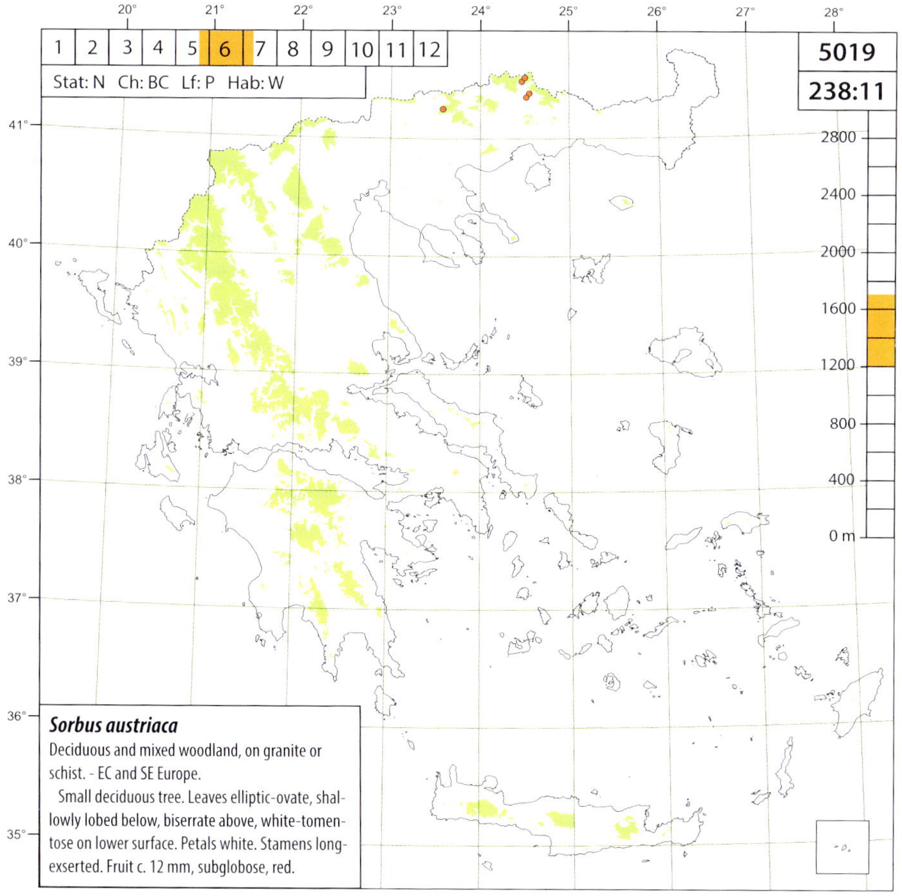

Sorbus austriaca
Deciduous and mixed woodland, on granite or schist. - EC and SE Europe.
Small deciduous tree. Leaves elliptic-ovate, shallowly lobed below, biserrate above, white-tomentose on lower surface. Petals white. Stamens long-exserted. Fruit c. 12 mm, subglobose, red.

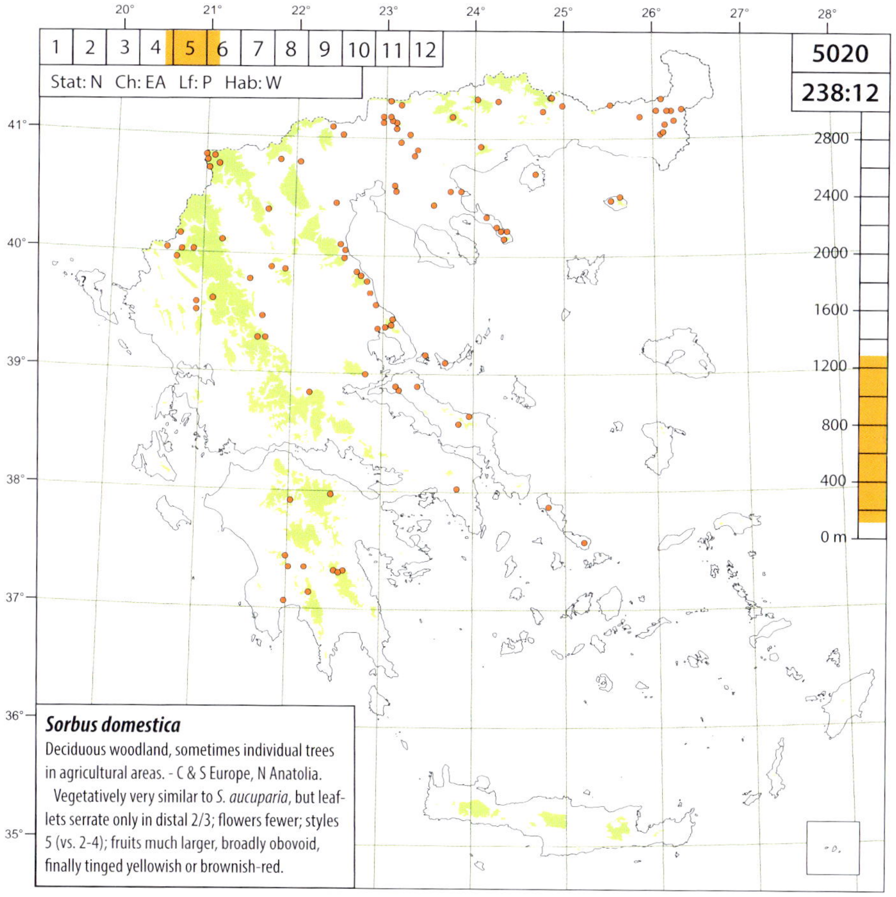

Sorbus domestica
Deciduous woodland, sometimes individual trees in agricultural areas. - C & S Europe, N Anatolia.
Vegetatively very similar to *S. aucuparia*, but leaflets serrate only in distal 2/3; flowers fewer; styles 5 (vs. 2-4); fruits much larger, broadly obovoid, finally tinged yellowish or brownish-red.

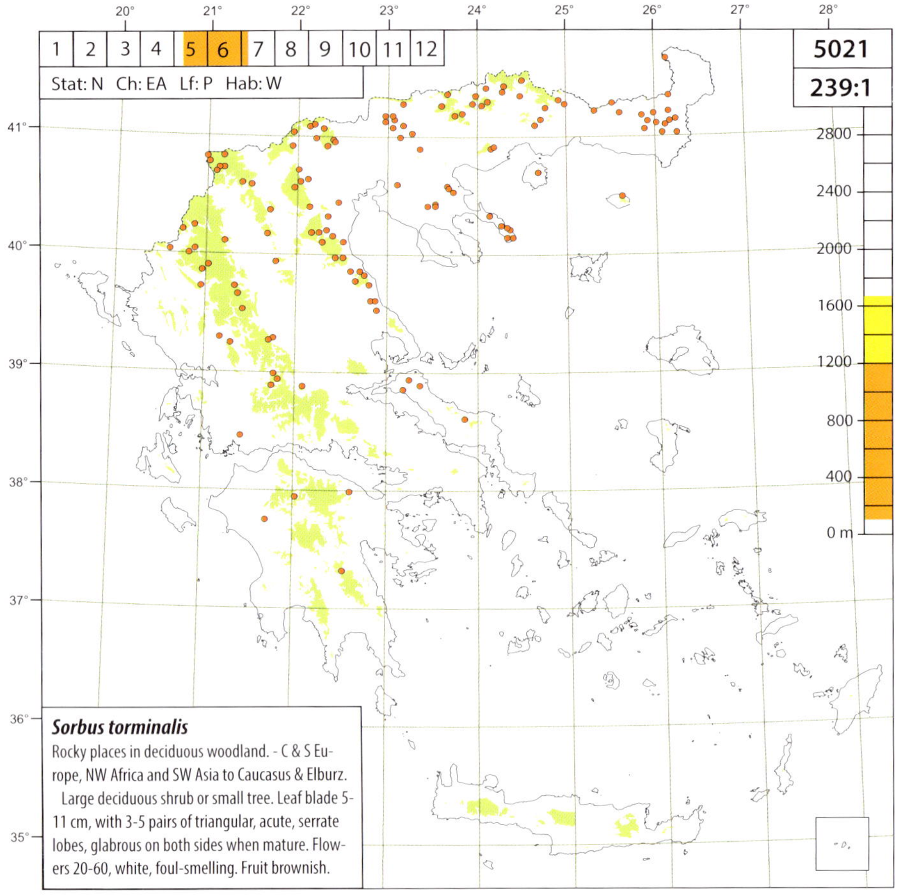

Sorbus torminalis
Rocky places in deciduous woodland. - C & S Europe, NW Africa and SW Asia to Caucasus & Elburz.
Large deciduous shrub or small tree. Leaf blade 5-11 cm, with 3-5 pairs of triangular, acute, serrate lobes, glabrous on both sides when mature. Flowers 20-60, white, foul-smelling. Fruit brownish.

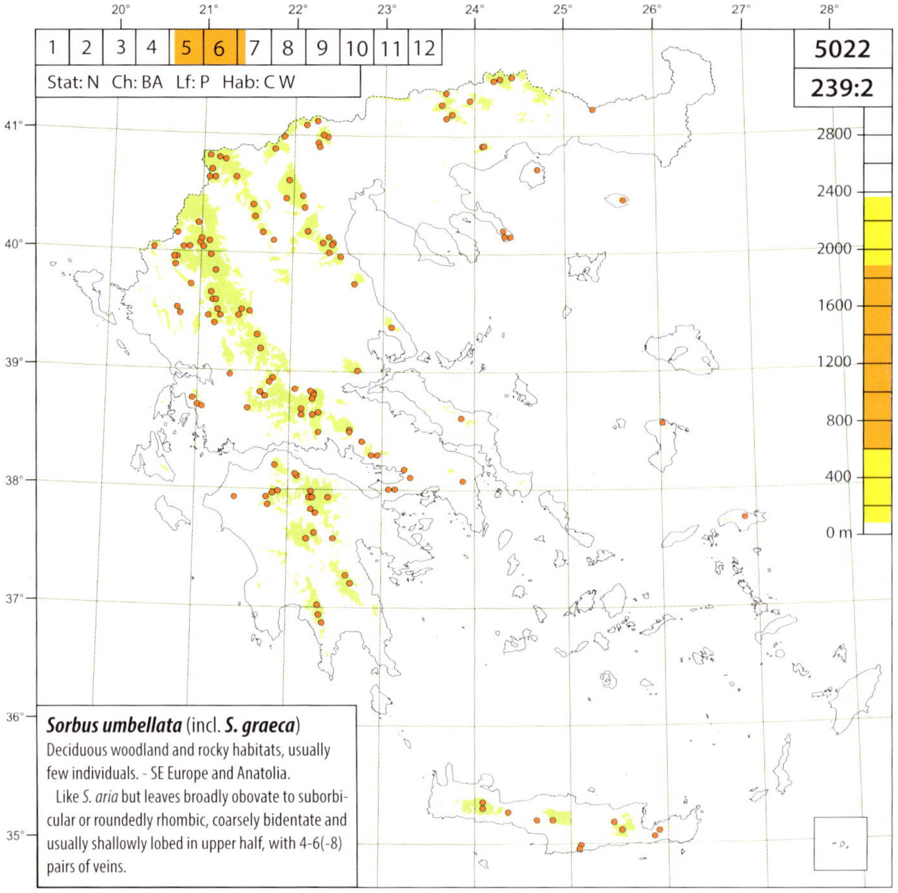

Sorbus umbellata (incl. ***S. graeca***)
Deciduous woodland and rocky habitats, usually few individuals. - SE Europe and Anatolia.
Like *S. aria* but leaves broadly obovate to suborbicular or roundedly rhombic, coarsely bidentate and usually shallowly lobed in upper half, with 4-6(-8) pairs of veins.

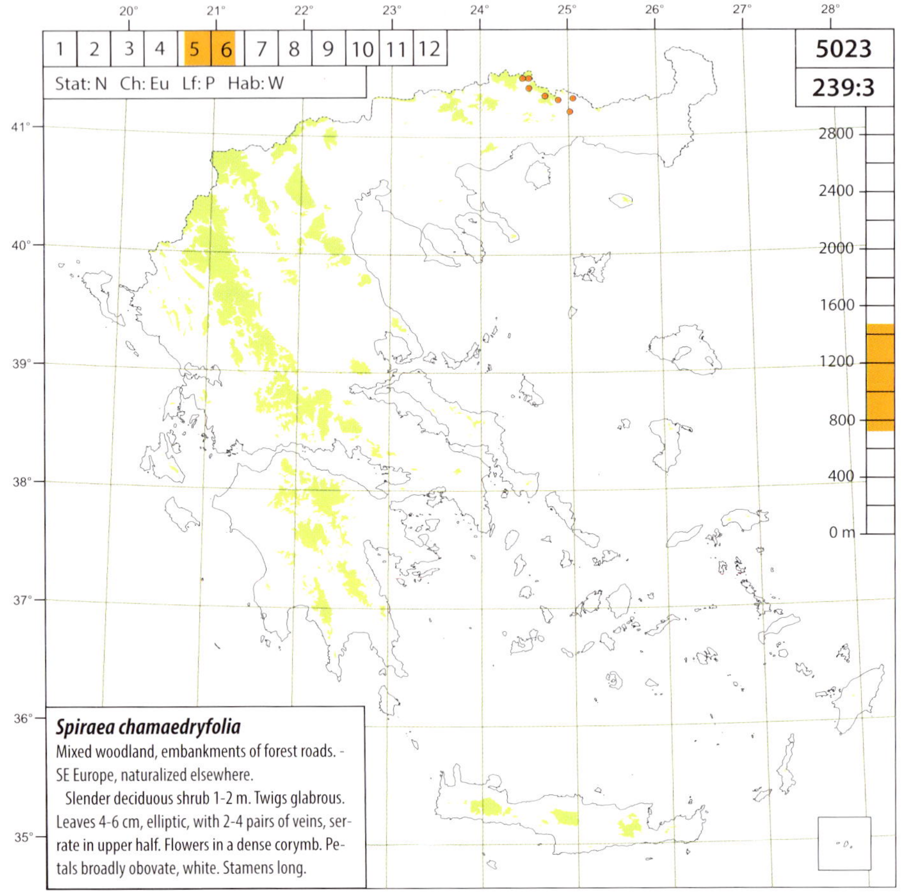

Spiraea chamaedryfolia
Mixed woodland, embankments of forest roads. - SE Europe, naturalized elsewhere.
Slender deciduous shrub 1-2 m. Twigs glabrous. Leaves 4-6 cm, elliptic, with 2-4 pairs of veins, serrate in upper half. Flowers in a dense corymb. Petals broadly obovate, white. Stamens long.

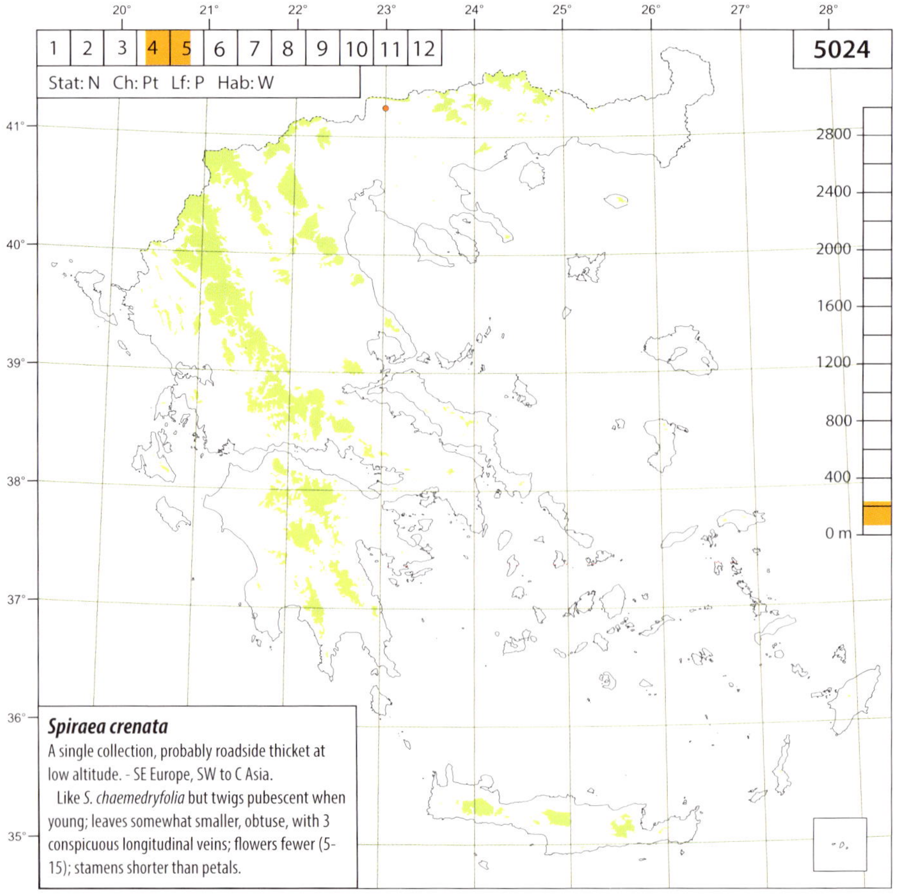

Spiraea crenata
A single collection, probably roadside thicket at low altitude. - SE Europe, SW to C Asia.
Like *S. chaemedryfolia* but twigs pubescent when young; leaves somewhat smaller, obtuse, with 3 conspicuous longitudinal veins; flowers fewer (5-15); stamens shorter than petals.

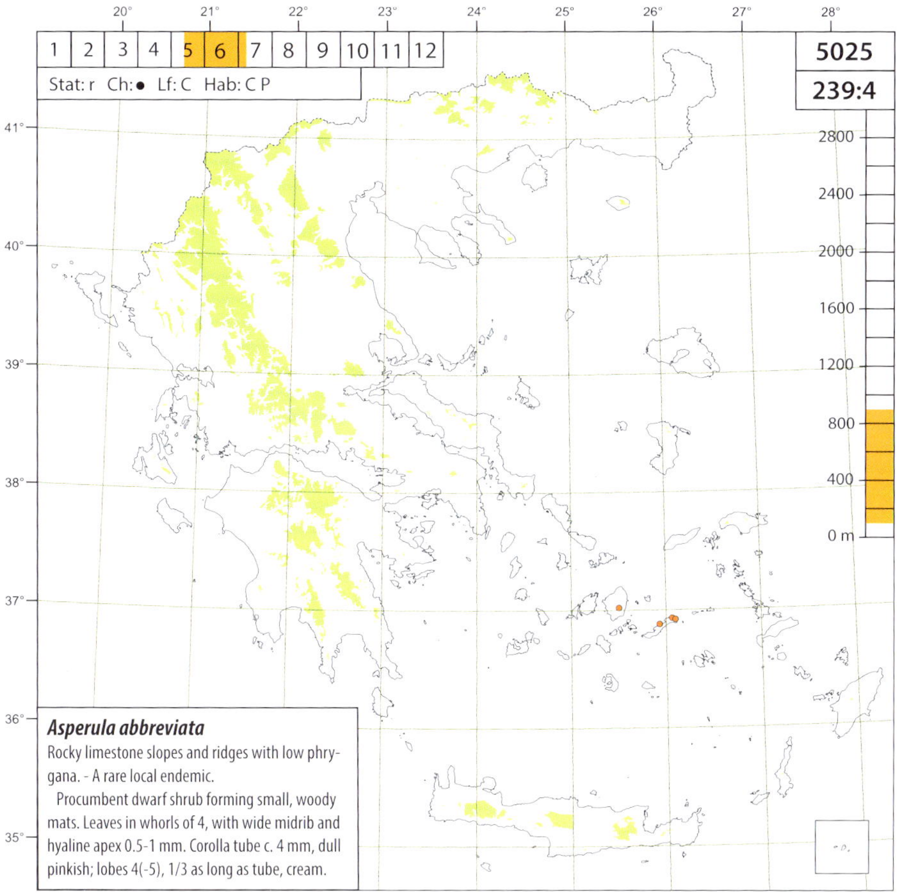

Asperula abbreviata
Rocky limestone slopes and ridges with low phrygana. - A rare local endemic.
Procumbent dwarf shrub forming small, woody mats. Leaves in whorls of 4, with wide midrib and hyaline apex 0.5-1 mm. Corolla tube c. 4 mm, dull pinkish; lobes 4(-5), 1/3 as long as tube, cream.

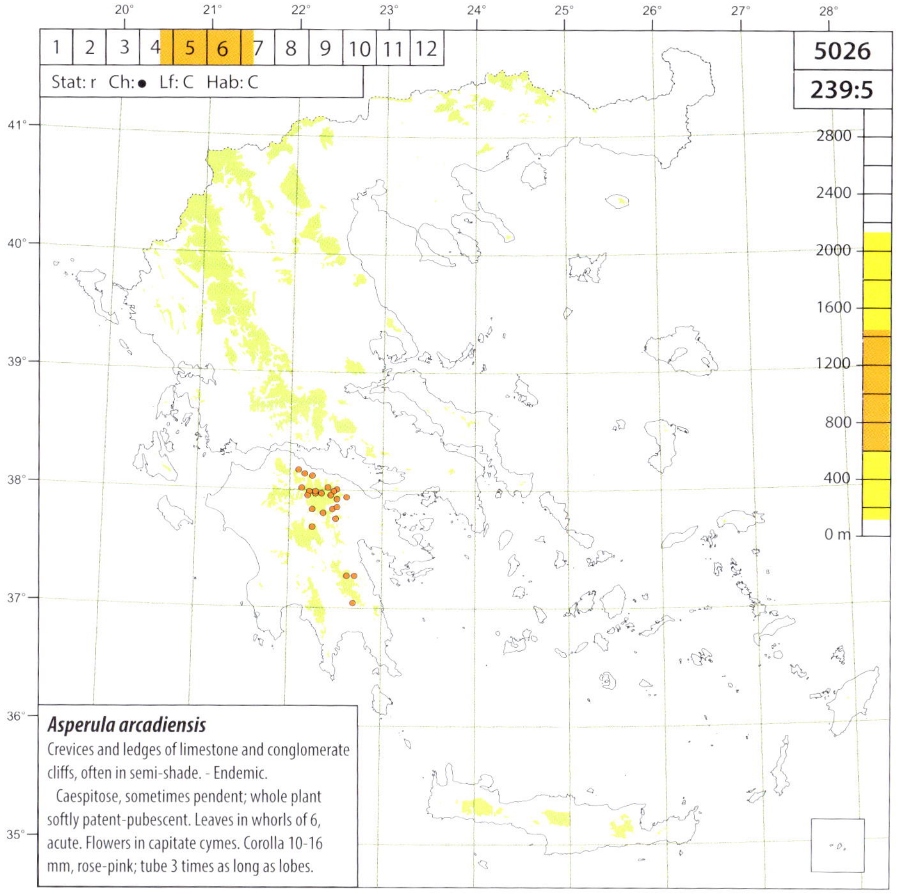

Asperula arcadiensis
Crevices and ledges of limestone and conglomerate cliffs, often in semi-shade. - Endemic.
Caespitose, sometimes pendent; whole plant softly patent-pubescent. Leaves in whorls of 6, acute. Flowers in capitate cymes. Corolla 10-16 mm, rose-pink; tube 3 times as long as lobes.

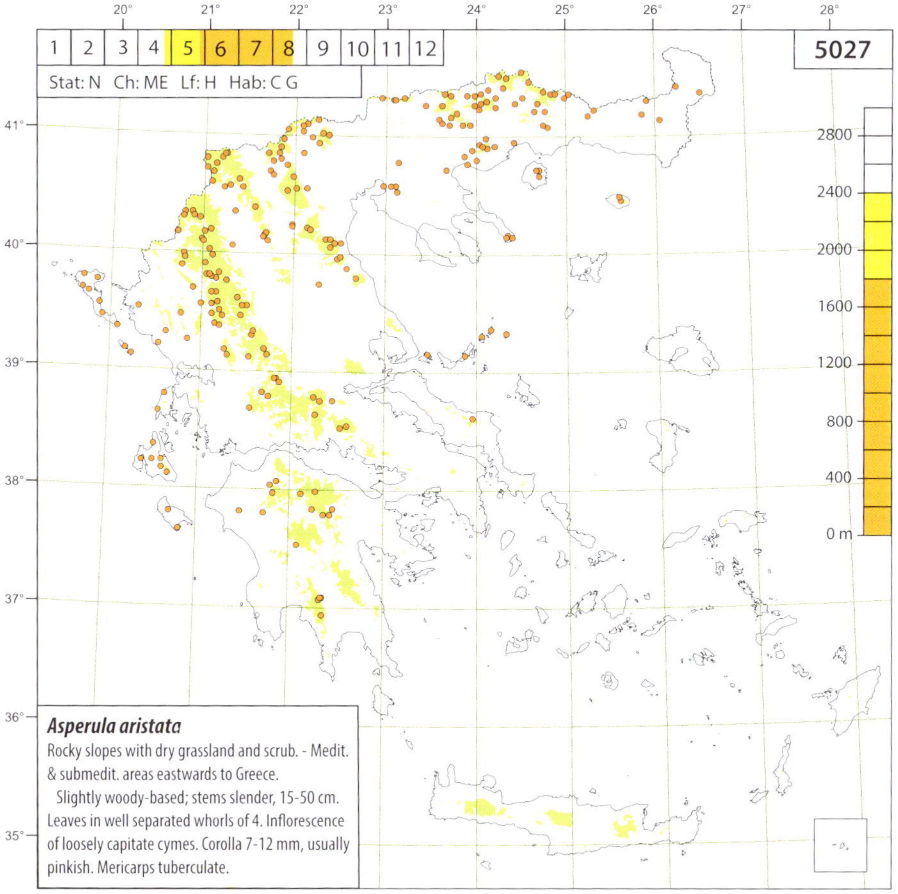

Asperula aristata
Rocky slopes with dry grassland and scrub. - Medit. & submedit. areas eastwards to Greece.
Slightly woody-based; stems slender, 15-50 cm. Leaves in well separated whorls of 4. Inflorescence of loosely capitate cymes. Corolla 7-12 mm, usually pinkish. Mericarps tuberculate.

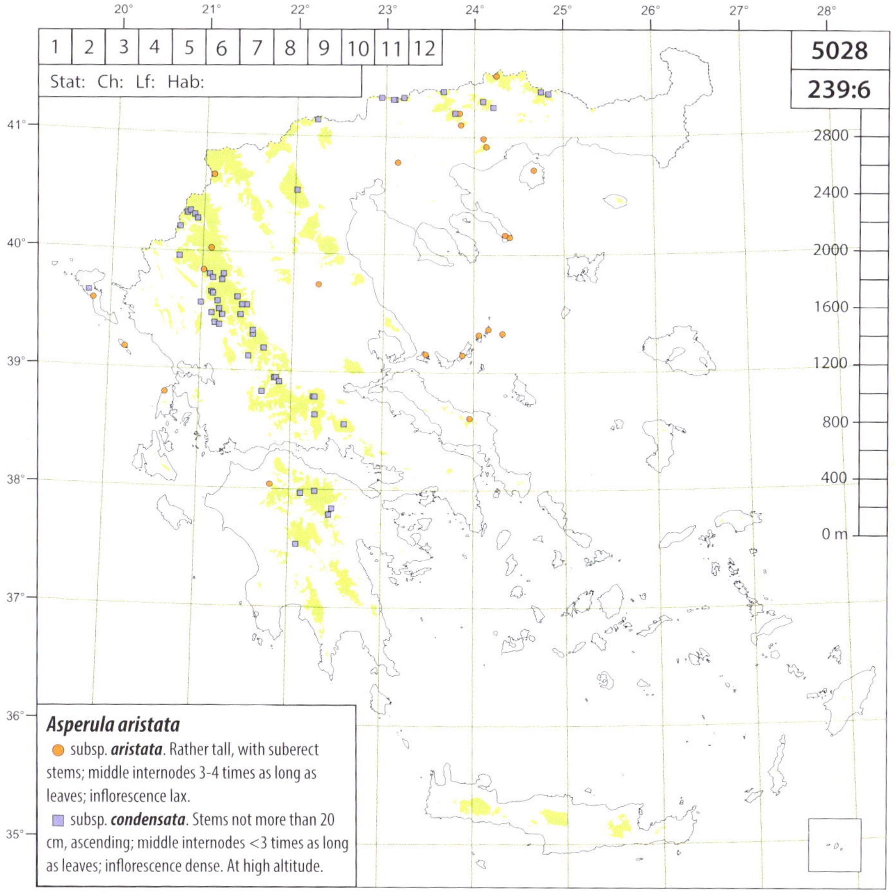

Asperula aristata
● subsp. ***aristata***. Rather tall, with suberect stems; middle internodes 3-4 times as long as leaves; inflorescence lax.
■ subsp. ***condensata***. Stems not more than 20 cm, ascending; middle internodes <3 times as long as leaves; inflorescence dense. At high altitude.

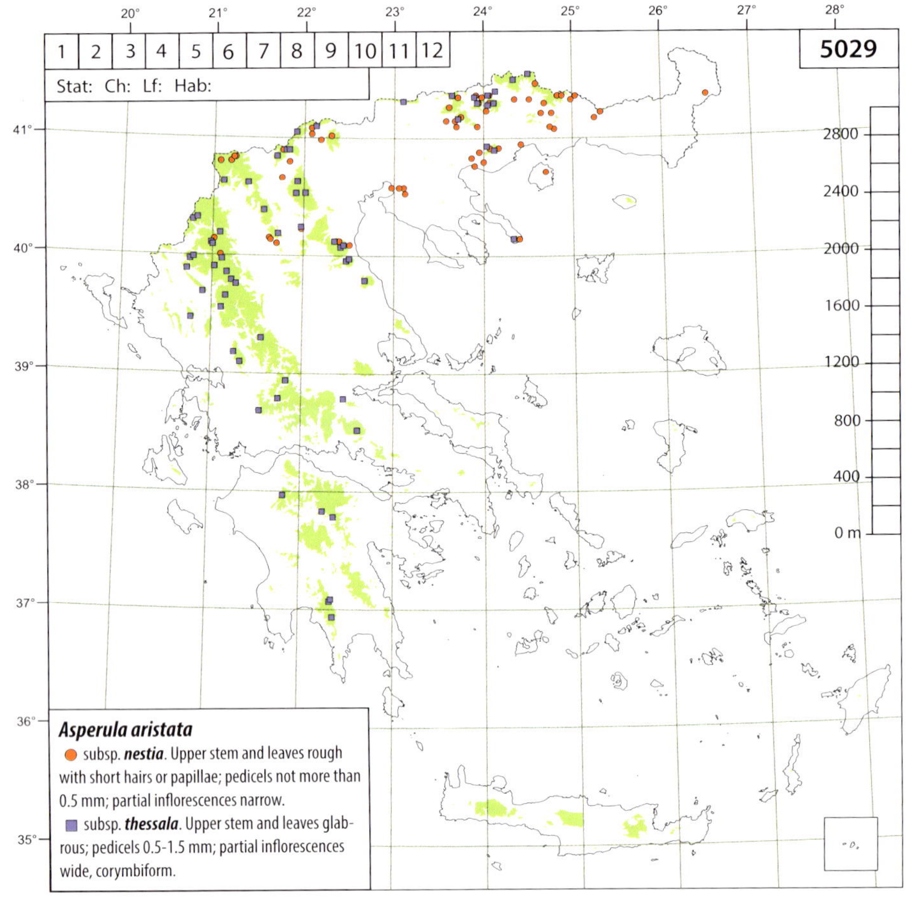

Asperula aristata
● subsp. ***nestia***. Upper stem and leaves rough with short hairs or papillae; pedicels not more than 0.5 mm; partial inflorescences narrow.
■ subsp. ***thessala***. Upper stem and leaves glabrous; pedicels 0.5-1.5 mm; partial inflorescences wide, corymbiform.

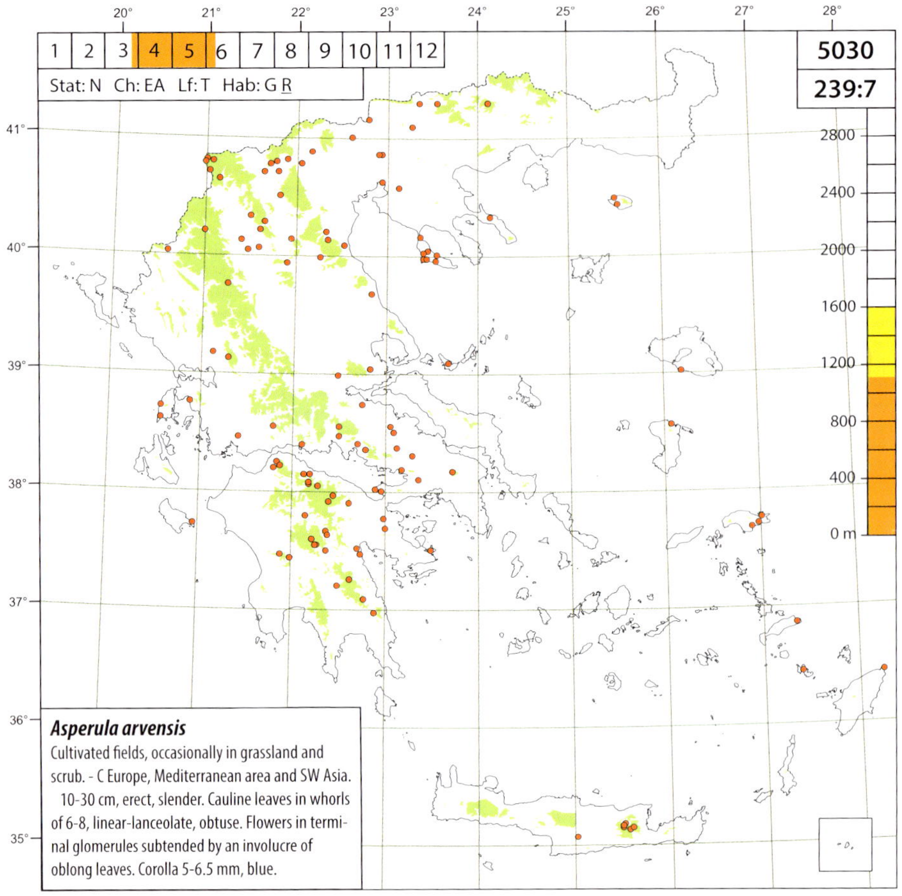

Asperula arvensis
Cultivated fields, occasionally in grassland and scrub. - C Europe, Mediterranean area and SW Asia.
10-30 cm, erect, slender. Cauline leaves in whorls of 6-8, linear-lanceolate, obtuse. Flowers in terminal glomerules subtended by an involucre of oblong leaves. Corolla 5-6.5 mm, blue.

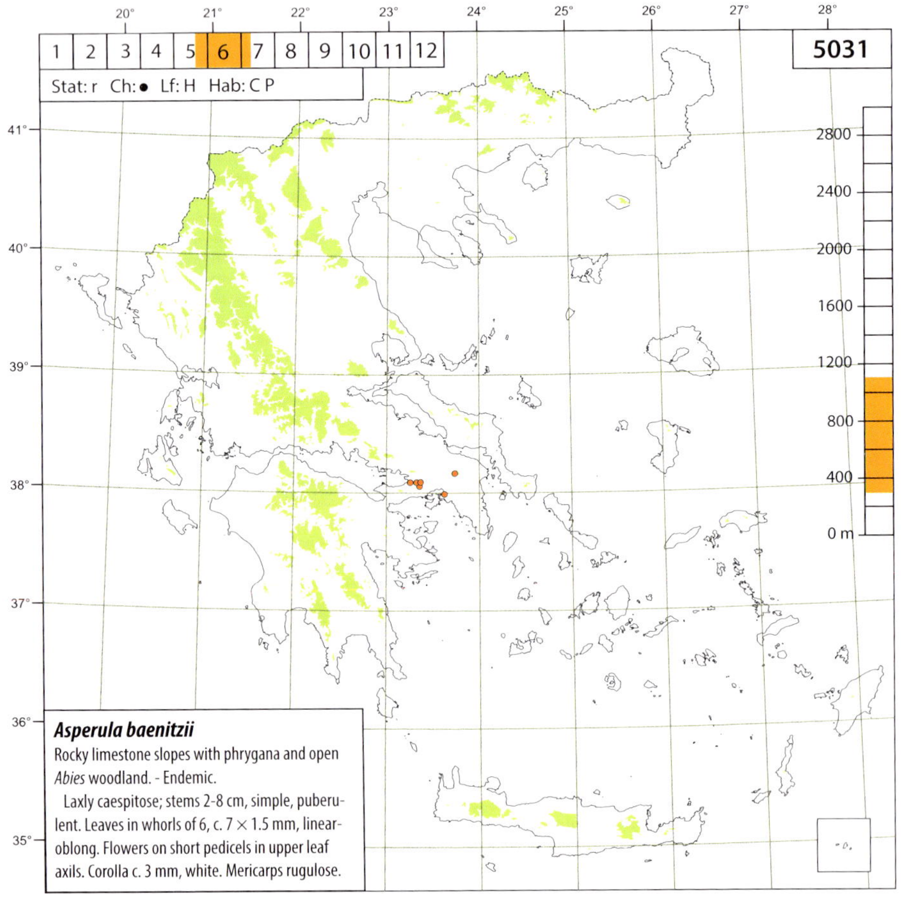

Asperula baenitzii
Rocky limestone slopes with phrygana and open *Abies* woodland. - Endemic.
Laxly caespitose; stems 2-8 cm, simple, puberulent. Leaves in whorls of 6, c. 7 × 1.5 mm, linear-oblong. Flowers on short pedicels in upper leaf axils. Corolla c. 3 mm, white. Mericarps rugulose.

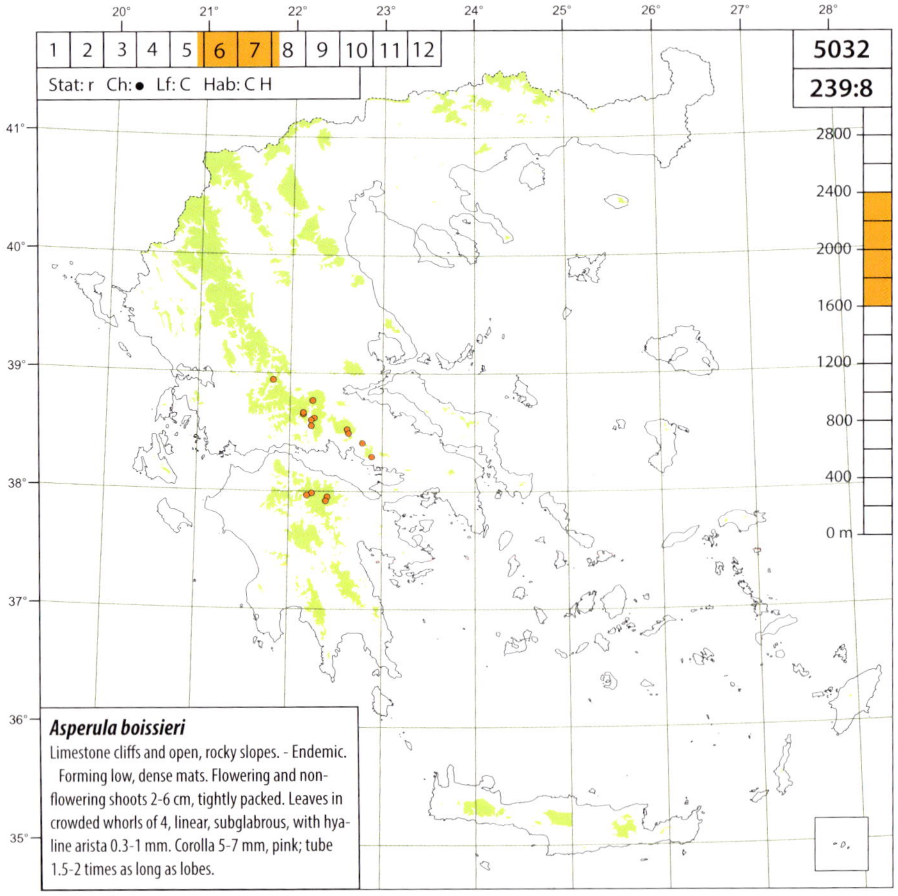

Asperula boissieri
Limestone cliffs and open, rocky slopes. - Endemic.
Forming low, dense mats. Flowering and non-flowering shoots 2-6 cm, tightly packed. Leaves in crowded whorls of 4, linear, subglabrous, with hyaline arista 0.3-1 mm. Corolla 5-7 mm, pink; tube 1.5-2 times as long as lobes.

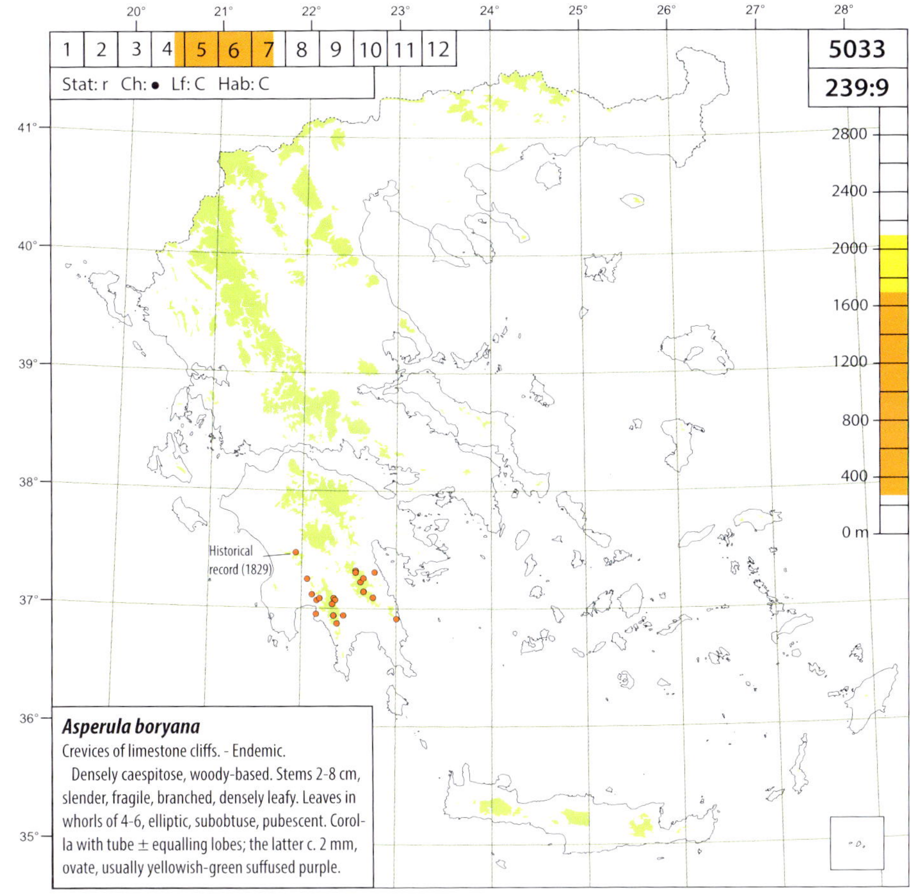

Asperula boryana

Crevices of limestone cliffs. - Endemic.

Densely caespitose, woody-based. Stems 2-8 cm, slender, fragile, branched, densely leafy. Leaves in whorls of 4-6, elliptic, subobtuse, pubescent. Corolla with tube ± equalling lobes; the latter c. 2 mm, ovate, usually yellowish-green suffused purple.

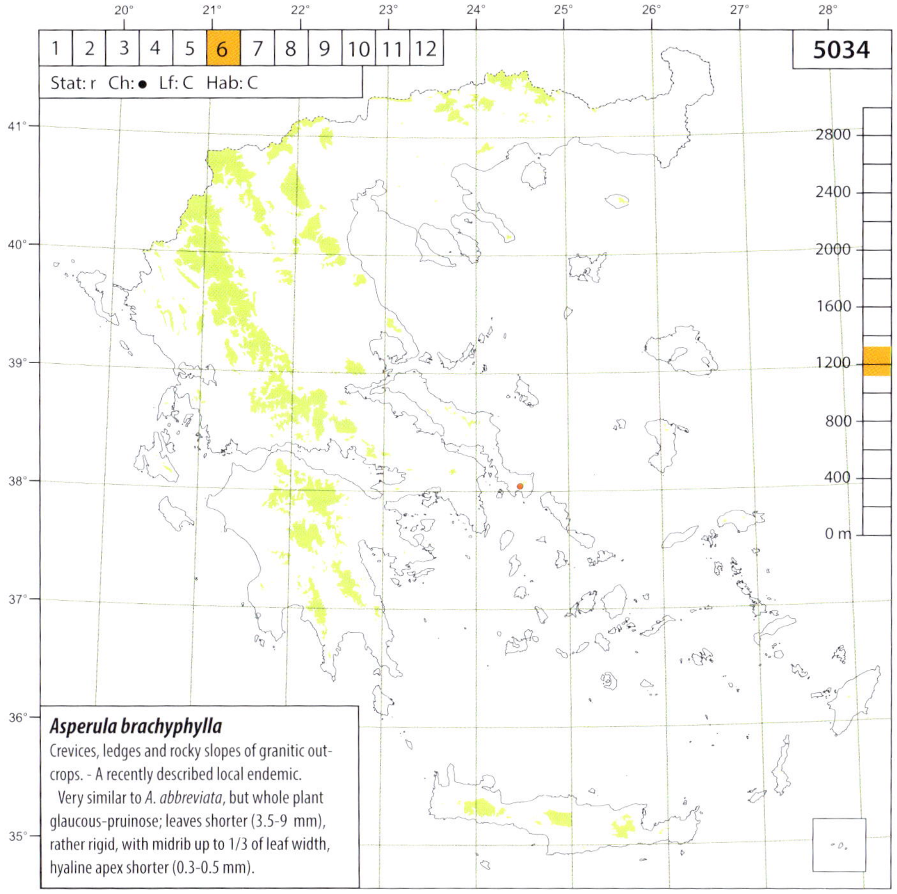

Asperula brachyphylla

Crevices, ledges and rocky slopes of granitic outcrops. - A recently described local endemic.

Very similar to *A. abbreviata*, but whole plant glaucous-pruinose; leaves shorter (3.5-9 mm), rather rigid, with midrib up to 1/3 of leaf width, hyaline apex shorter (0.3-0.5 mm).

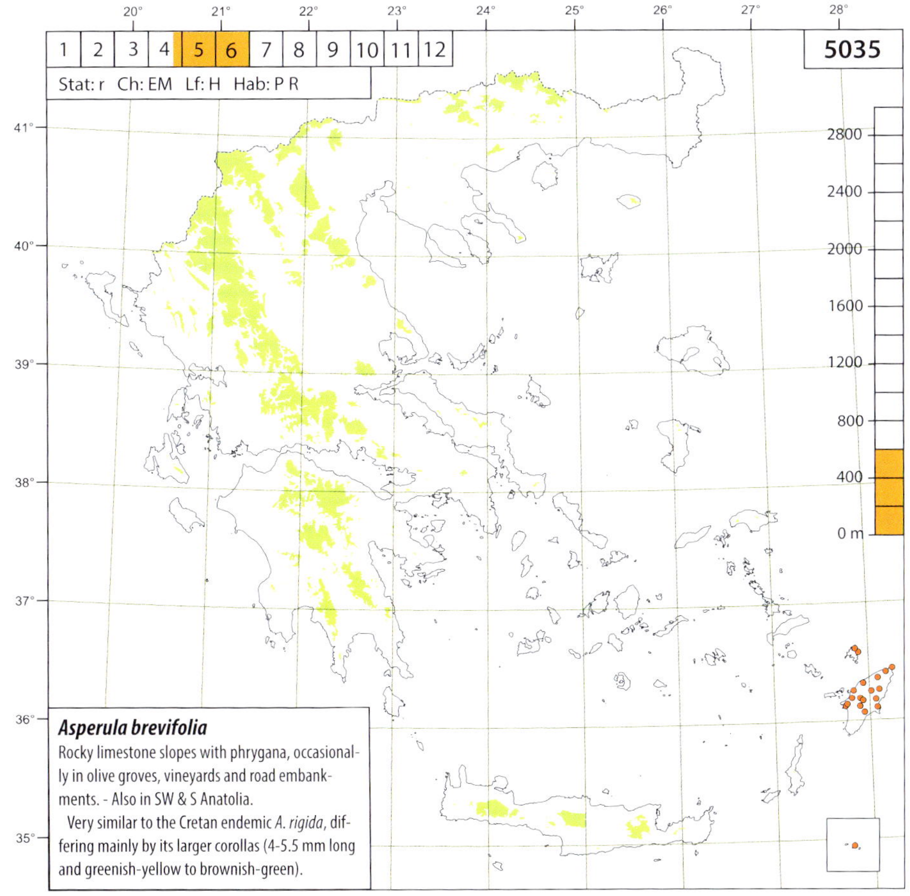

Asperula brevifolia

Rocky limestone slopes with phrygana, occasionally in olive groves, vineyards and road embankments. - Also in SW & S Anatolia.

Very similar to the Cretan endemic *A. rigida*, differing mainly by its larger corollas (4-5.5 mm long and greenish-yellow to brownish-green).

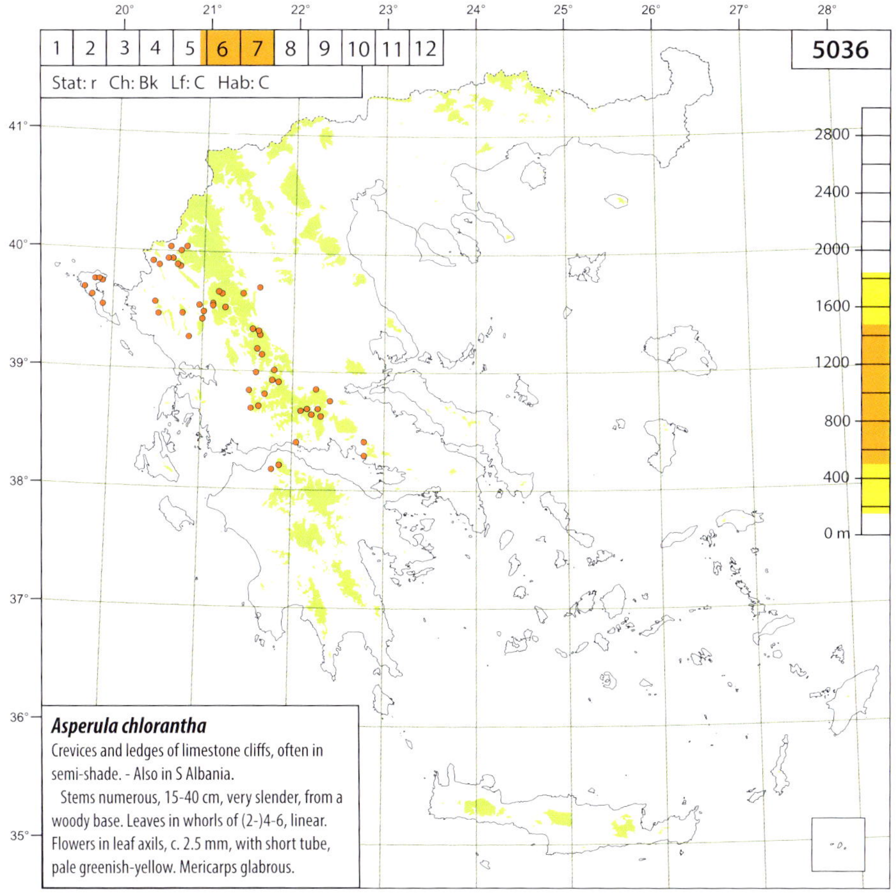

Asperula chlorantha

Crevices and ledges of limestone cliffs, often in semi-shade. - Also in S Albania.

Stems numerous, 15-40 cm, very slender, from a woody base. Leaves in whorls of (2-)4-6, linear. Flowers in leaf axils, c. 2.5 mm, with short tube, pale greenish-yellow. Mericarps glabrous.

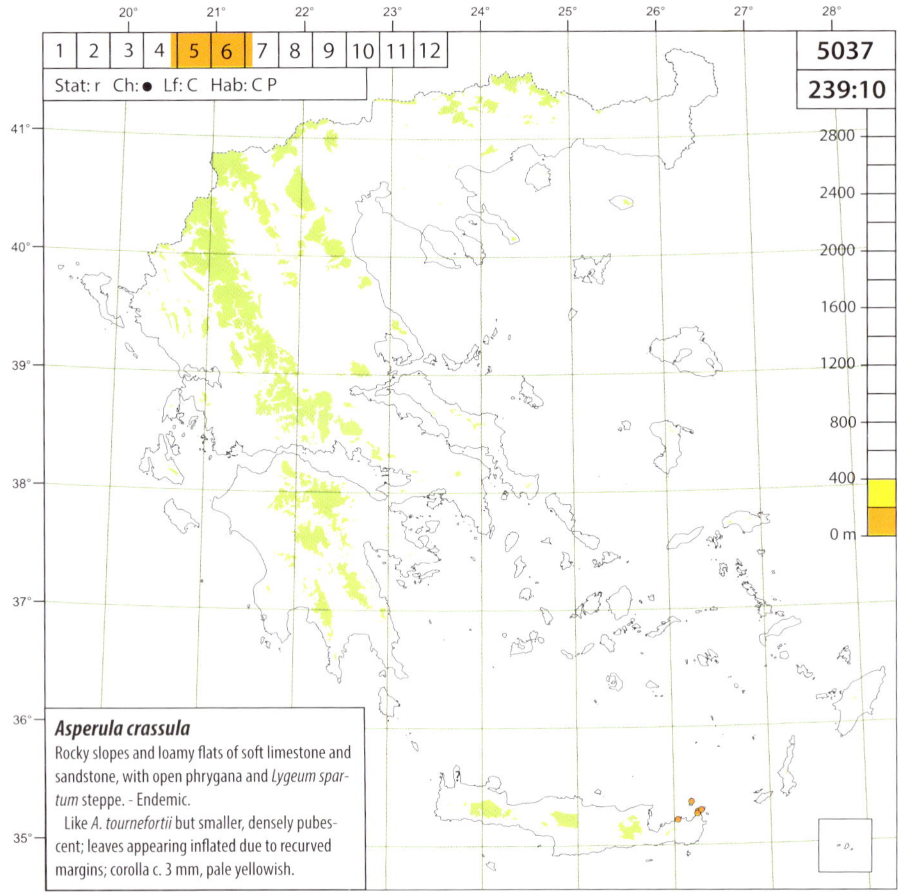

Asperula crassula
Rocky slopes and loamy flats of soft limestone and sandstone, with open phrygana and *Lygeum spartum* steppe. - Endemic.
Like *A. tournefortii* but smaller, densely pubescent; leaves appearing inflated due to recurved margins; corolla c. 3 mm, pale yellowish.

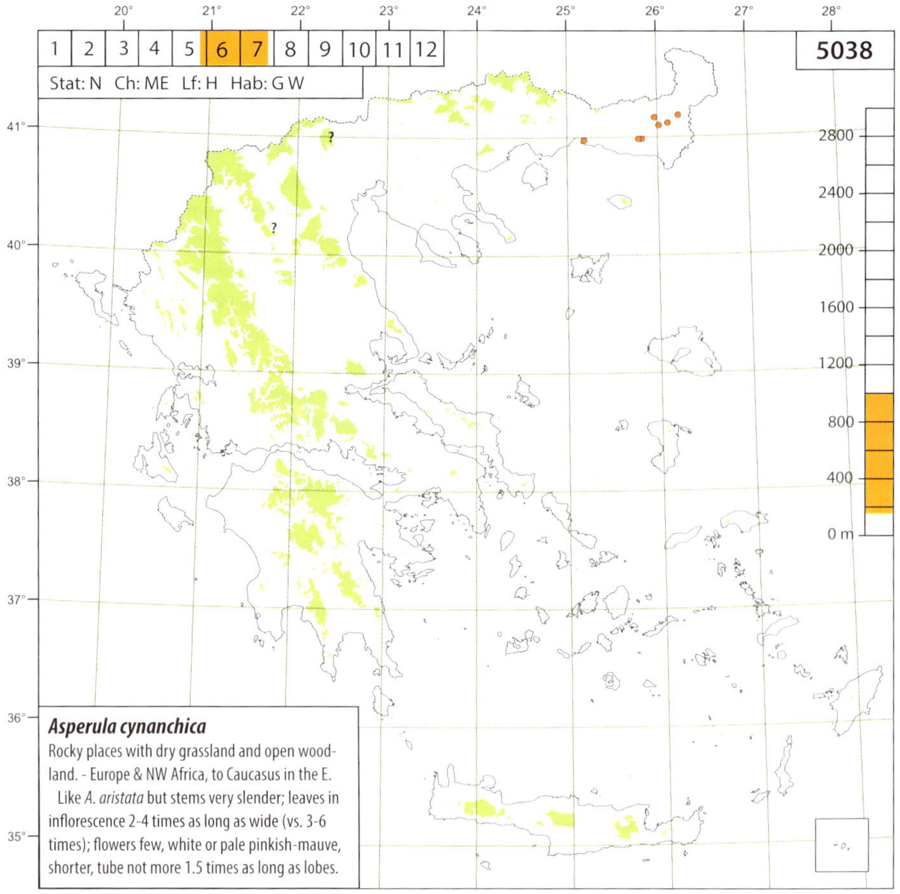

Asperula cynanchica
Rocky places with dry grassland and open woodland. - Europe & NW Africa, to Caucasus in the E.
Like *A. aristata* but stems very slender; leaves in inflorescence 2-4 times as long as wide (vs. 3-6 times); flowers few, white or pale pinkish-mauve, shorter, tube not more 1.5 times as long as lobes.

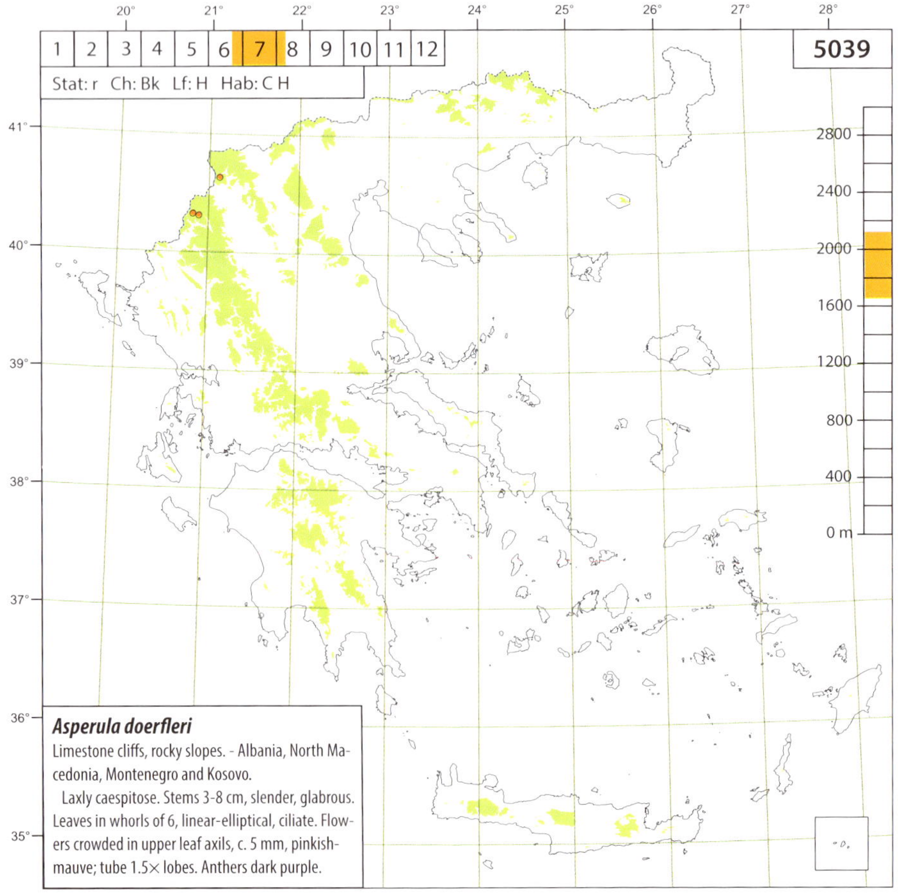

Asperula doerfleri
Limestone cliffs, rocky slopes. - Albania, North Macedonia, Montenegro and Kosovo.
Laxly caespitose. Stems 3-8 cm, slender, glabrous. Leaves in whorls of 6, linear-elliptical, ciliate. Flowers crowded in upper leaf axils, c. 5 mm, pinkish-mauve; tube 1.5× lobes. Anthers dark purple.

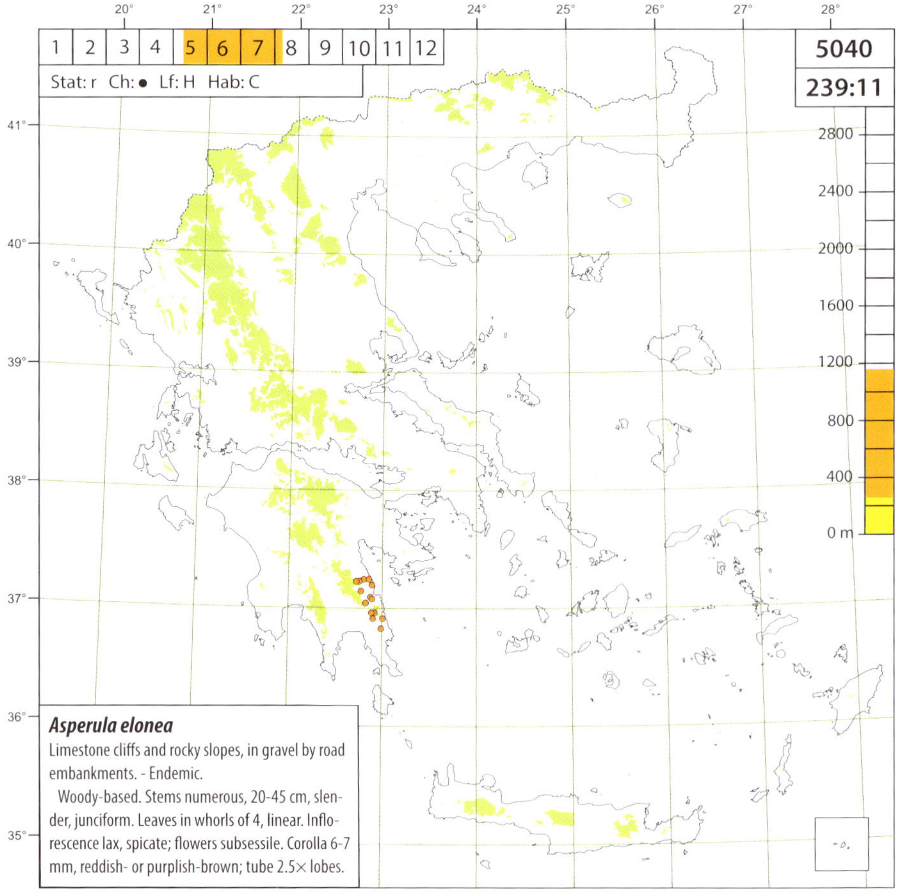

Asperula elonea
Limestone cliffs and rocky slopes, in gravel by road embankments. - Endemic.
Woody-based. Stems numerous, 20-45 cm, slender, junciform. Leaves in whorls of 4, linear. Inflorescence lax, spicate; flowers subsessile. Corolla 6-7 mm, reddish- or purplish-brown; tube 2.5× lobes.

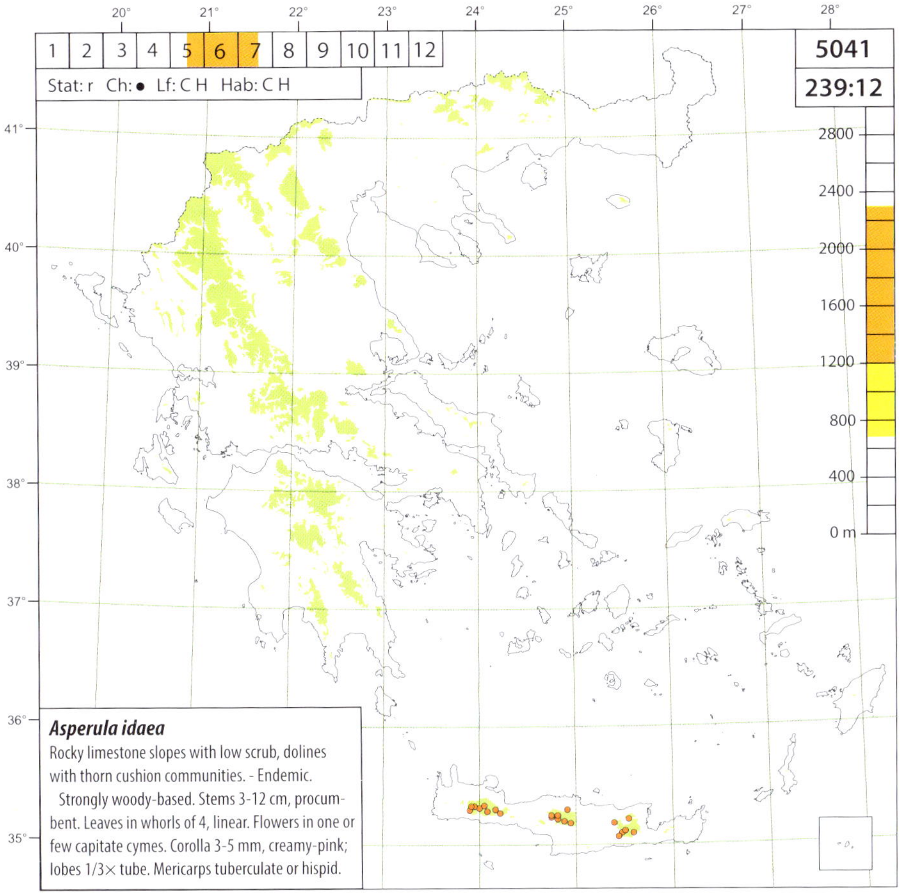

Asperula idaea
Rocky limestone slopes with low scrub, dolines with thorn cushion communities. - Endemic.
Strongly woody-based. Stems 3-12 cm, procumbent. Leaves in whorls of 4, linear. Flowers in one or few capitate cymes. Corolla 3-5 mm, creamy-pink; lobes 1/3× tube. Mericarps tuberculate or hispid.

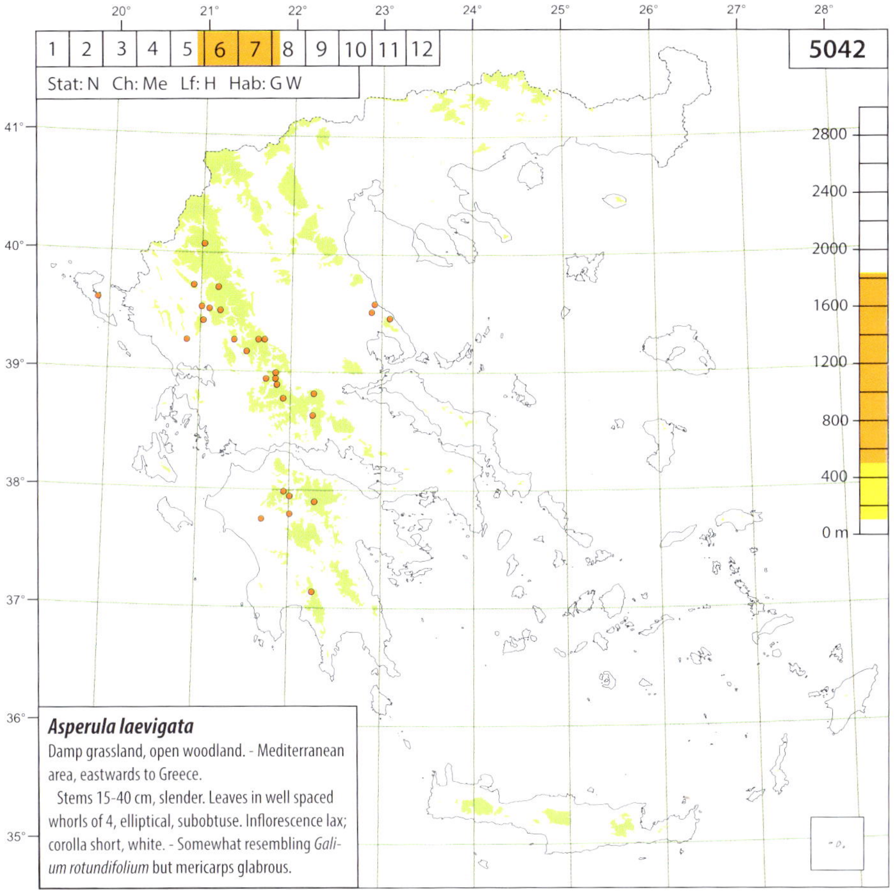

Asperula laevigata
Damp grassland, open woodland. - Mediterranean area, eastwards to Greece.
Stems 15-40 cm, slender. Leaves in well spaced whorls of 4, elliptical, subobtuse. Inflorescence lax; corolla short, white. - Somewhat resembling *Galium rotundifolium* but mericarps glabrous.

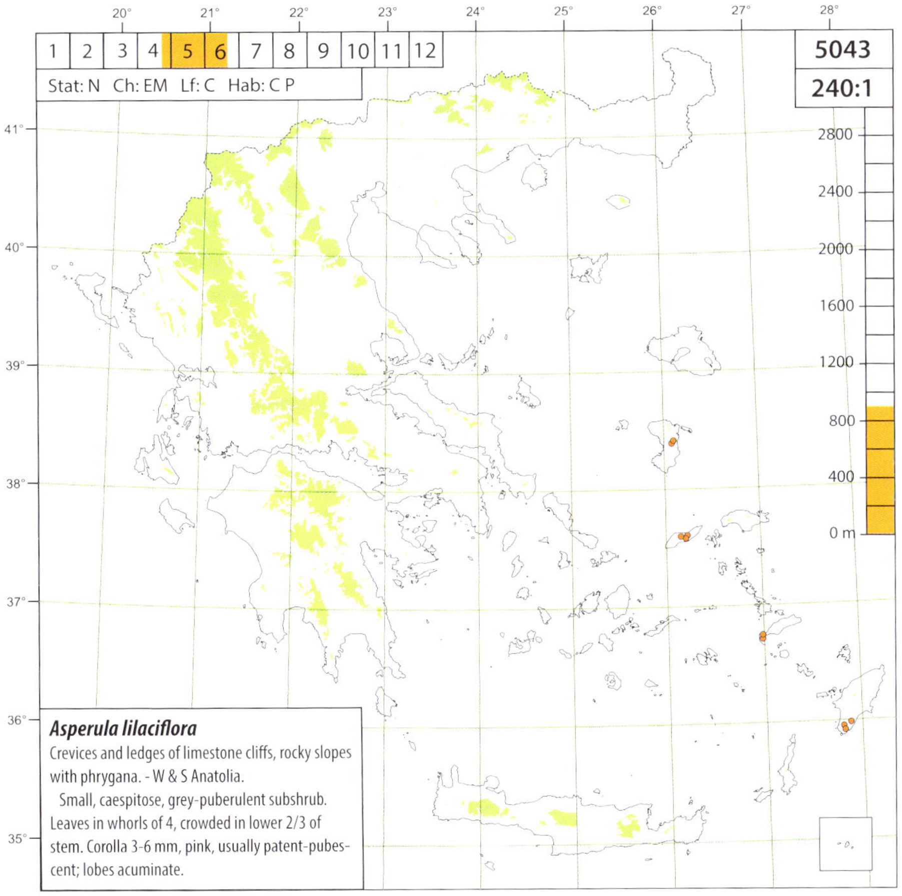

Asperula lilaciflora
Crevices and ledges of limestone cliffs, rocky slopes with phrygana. - W & S Anatolia.
Small, caespitose, grey-puberulent subshrub. Leaves in whorls of 4, crowded in lower 2/3 of stem. Corolla 3-6 mm, pink, usually patent-pubescent; lobes acuminate.

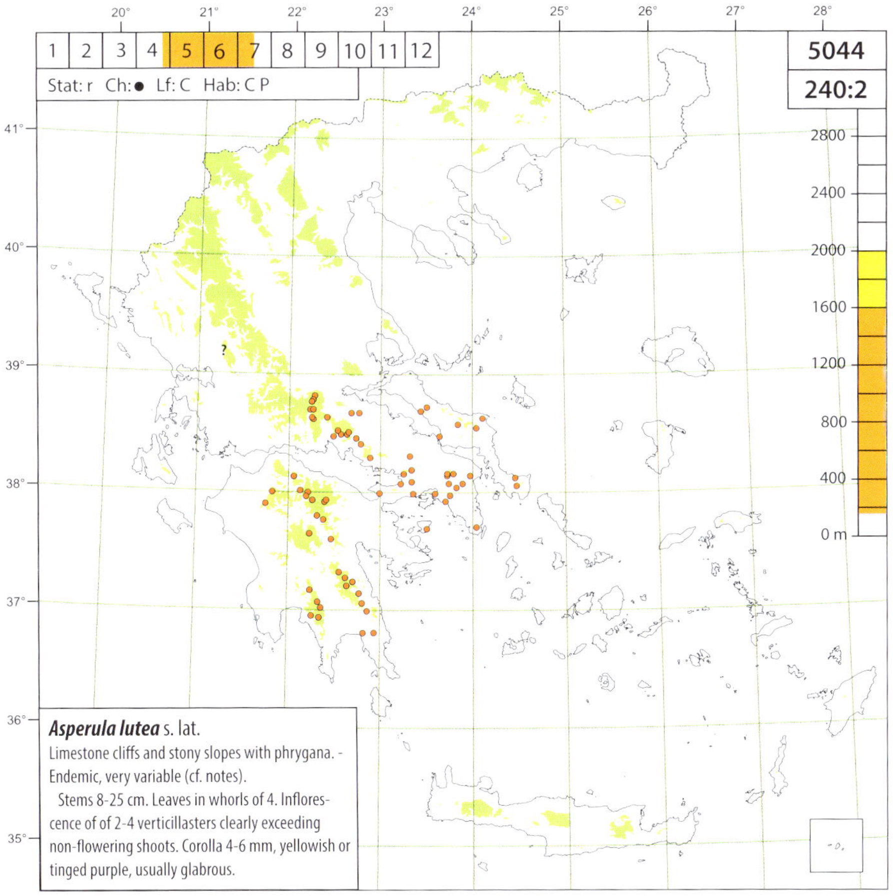

Asperula lutea s. lat.
Limestone cliffs and stony slopes with phrygana. - Endemic, very variable (cf. notes).
Stems 8-25 cm. Leaves in whorls of 4. Inflorescence of of 2-4 verticillasters clearly exceeding non-flowering shoots. Corolla 4-6 mm, yellowish or tinged purple, usually glabrous.

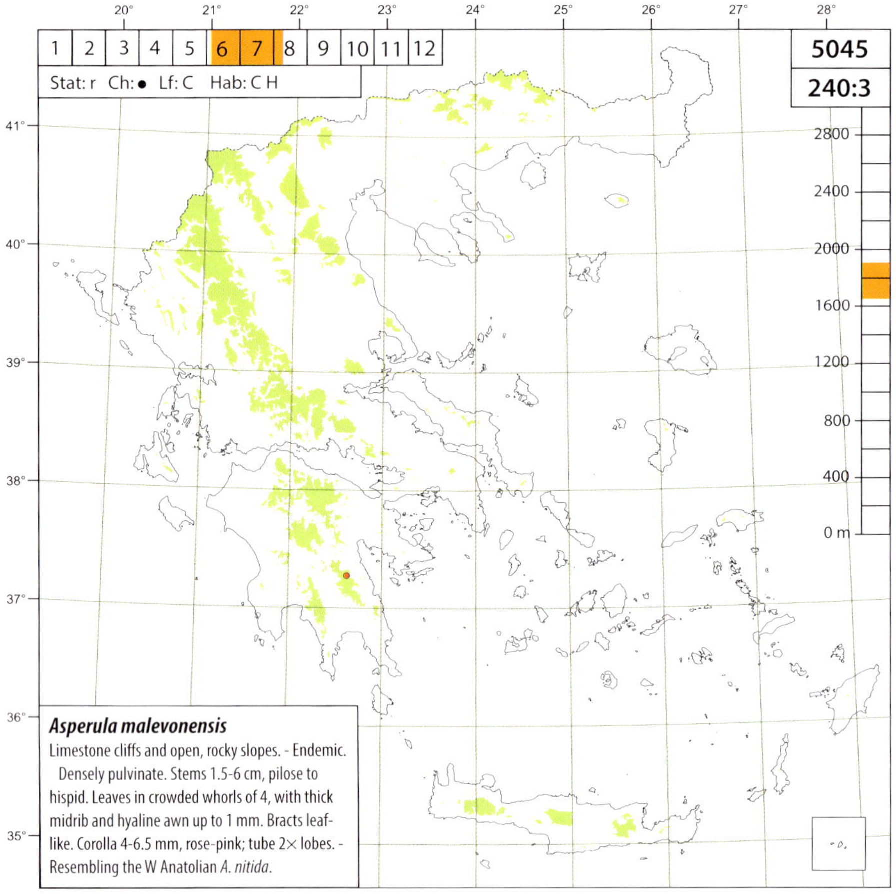

Asperula malevonensis
Limestone cliffs and open, rocky slopes. - Endemic.
Densely pulvinate. Stems 1.5-6 cm, pilose to hispid. Leaves in crowded whorls of 4, with thick midrib and hyaline awn up to 1 mm. Bracts leaf-like. Corolla 4-6.5 mm, rose-pink; tube 2× lobes. - Resembling the W Anatolian *A. nitida*.

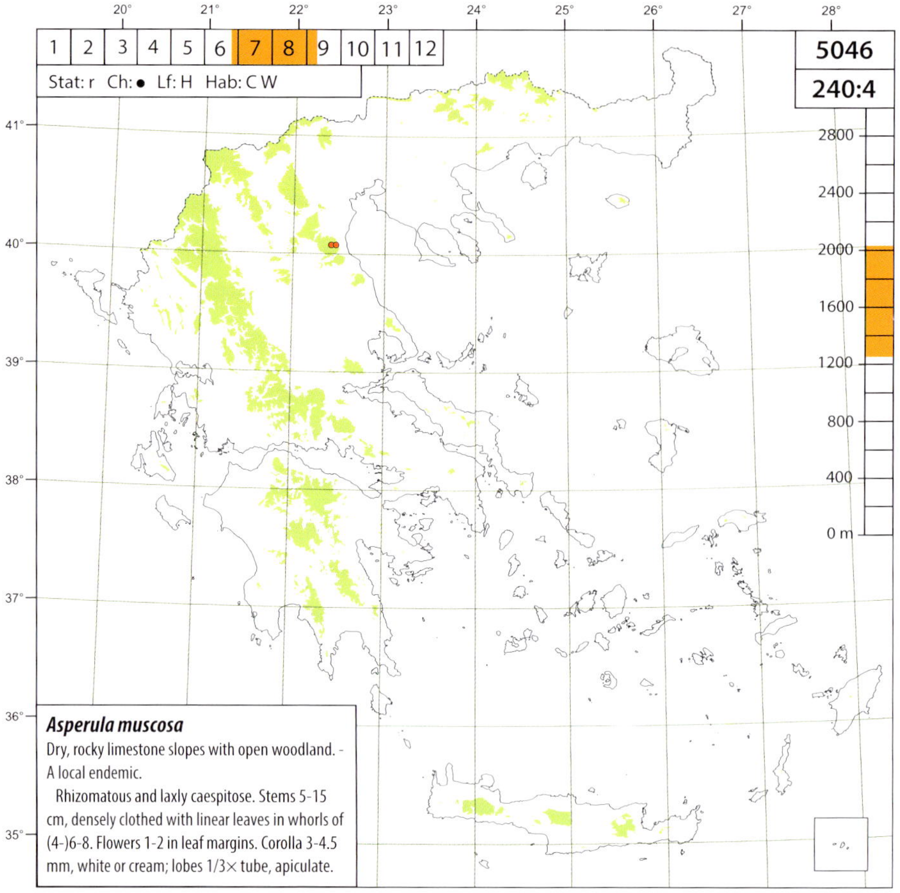

Asperula muscosa
Dry, rocky limestone slopes with open woodland. - A local endemic.
Rhizomatous and laxly caespitose. Stems 5-15 cm, densely clothed with linear leaves in whorls of (4-)6-8. Flowers 1-2 in leaf margins. Corolla 3-4.5 mm, white or cream; lobes 1/3× tube, apiculate.

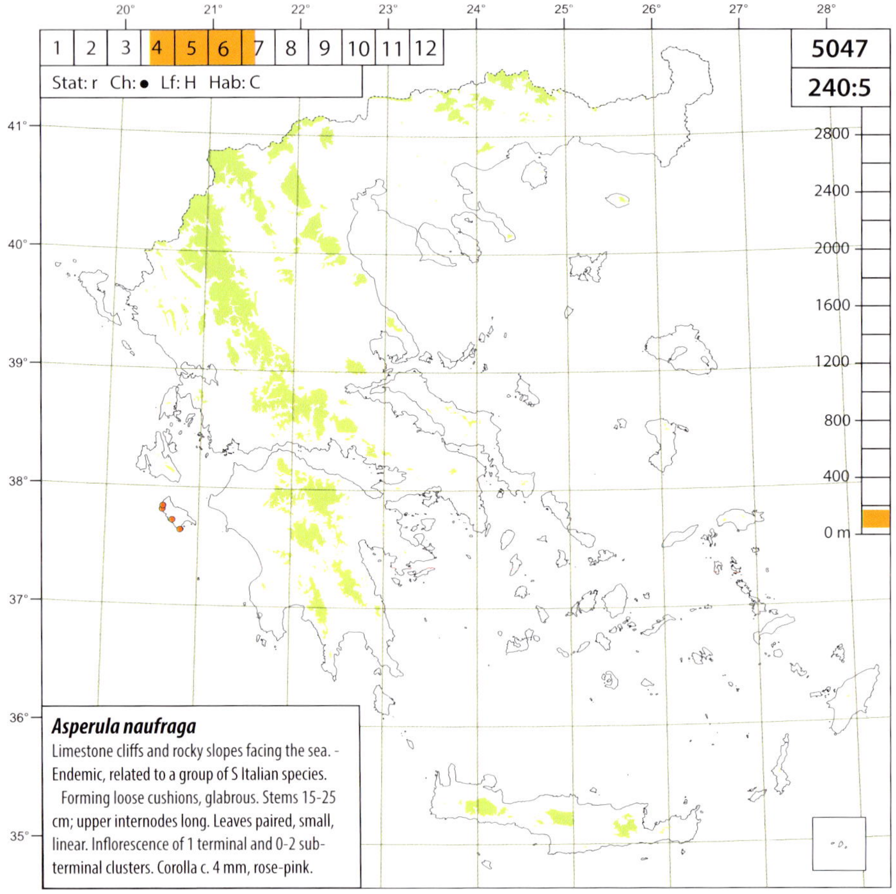

Asperula naufraga
Limestone cliffs and rocky slopes facing the sea. - Endemic, related to a group of S Italian species.
Forming loose cushions, glabrous. Stems 15-25 cm; upper internodes long. Leaves paired, small, linear. Inflorescence of 1 terminal and 0-2 sub-terminal clusters. Corolla c. 4 mm, rose-pink.

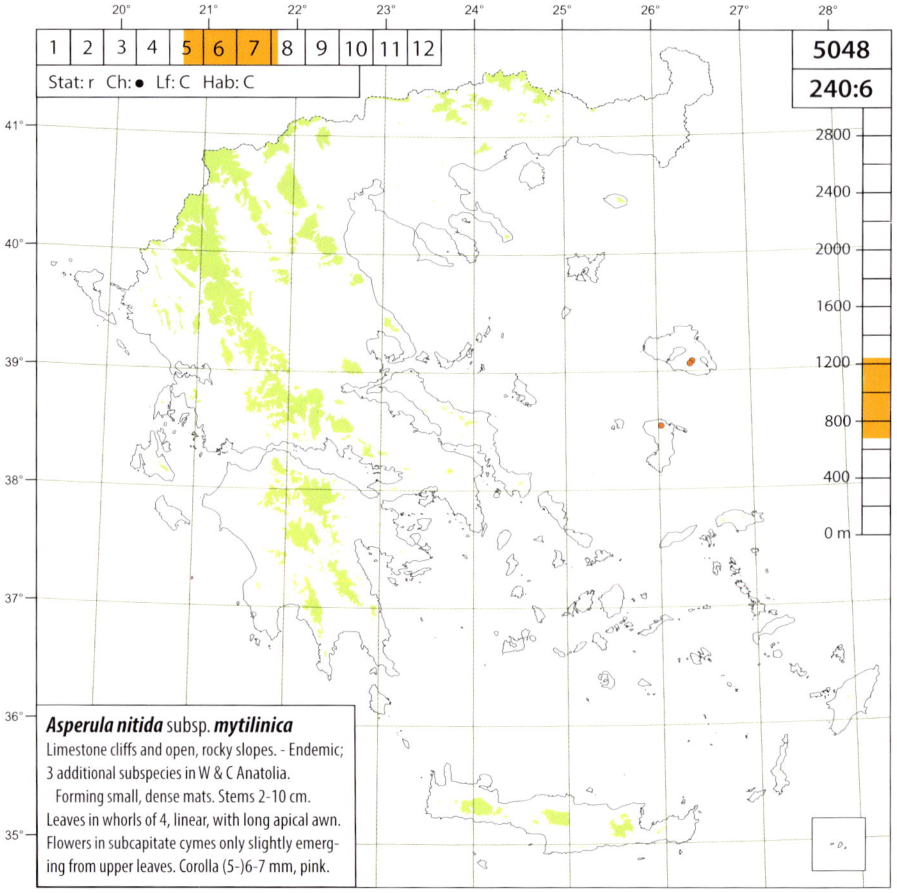

Asperula nitida* subsp. *mytilinica
Limestone cliffs and open, rocky slopes. - Endemic; 3 additional subspecies in W & C Anatolia.
Forming small, dense mats. Stems 2-10 cm. Leaves in whorls of 4, linear, with long apical awn. Flowers in subcapitate cymes only slightly emerging from upper leaves. Corolla (5-)6-7 mm, pink.

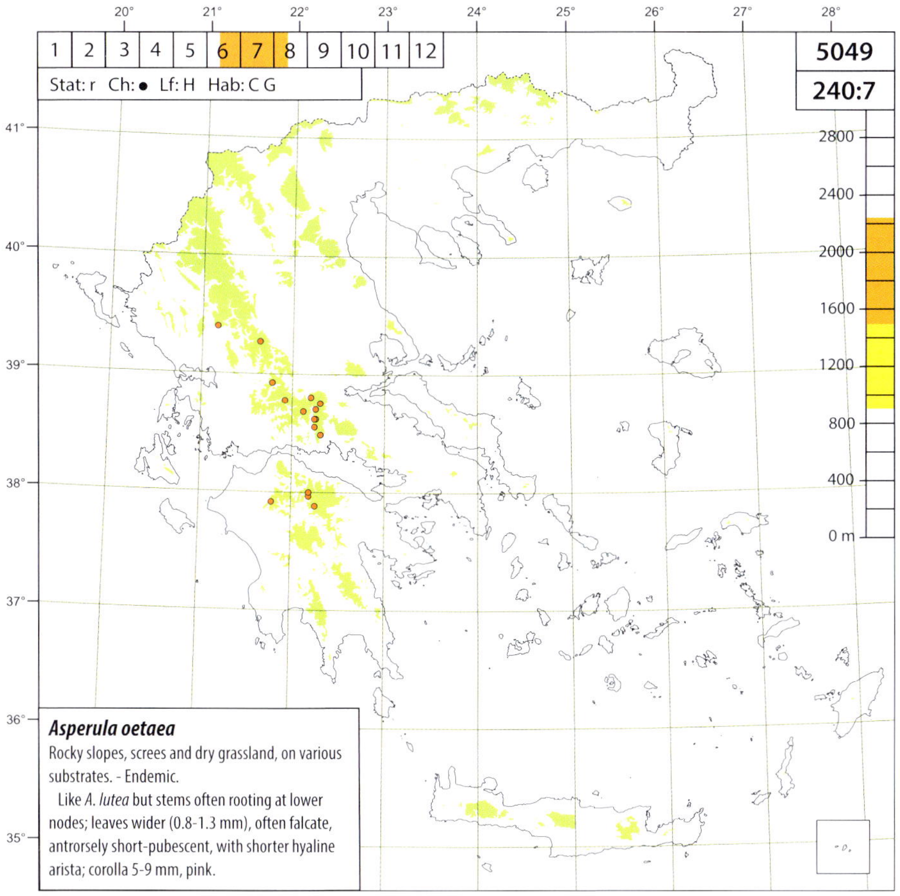

Asperula oetaea
Rocky slopes, screes and dry grassland, on various substrates. - Endemic.
Like *A. lutea* but stems often rooting at lower nodes; leaves wider (0.8-1.3 mm), often falcate, antrorsely short-pubescent, with shorter hyaline arista; corolla 5-9 mm, pink.

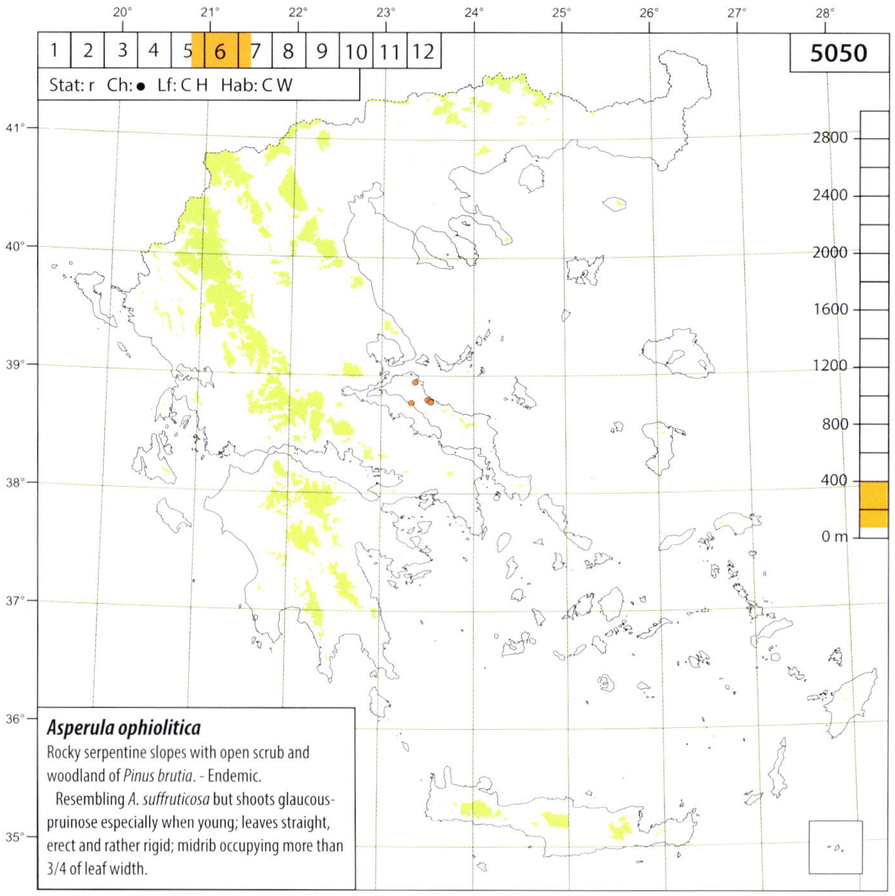

Asperula ophiolitica
Rocky serpentine slopes with open scrub and woodland of *Pinus brutia*. - Endemic.
Resembling *A. suffruticosa* but shoots glaucous-pruinose especially when young; leaves straight, erect and rather rigid; midrib occupying more than 3/4 of leaf width.

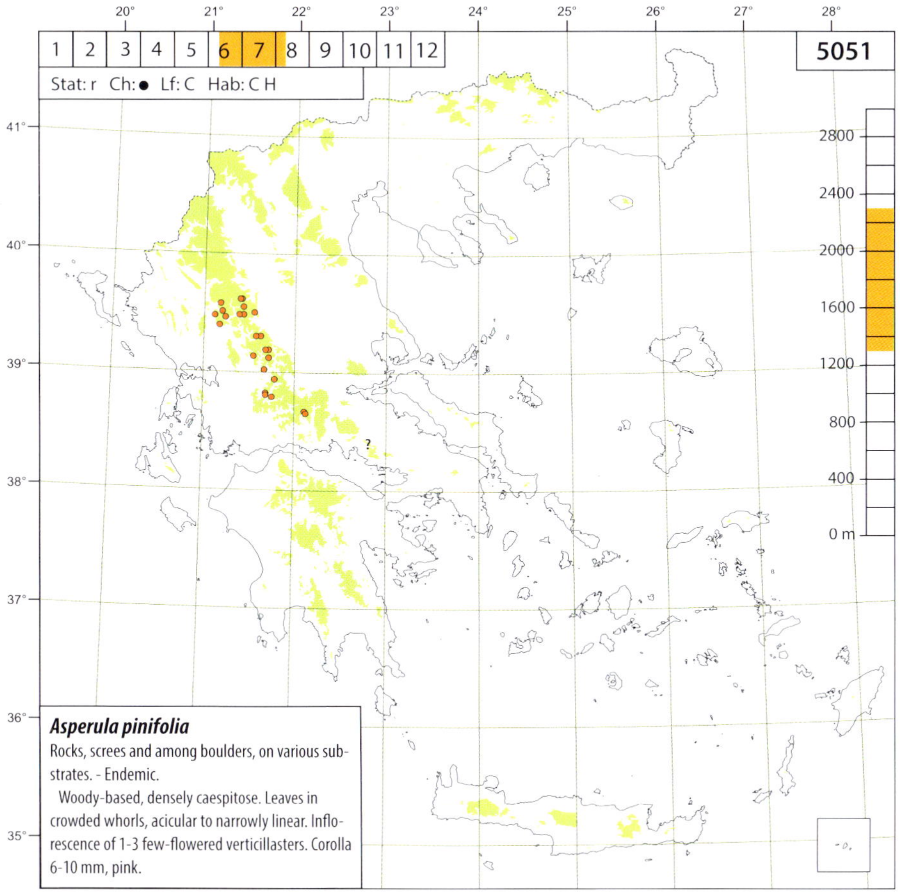

Asperula pinifolia
Rocks, screes and among boulders, on various substrates. - Endemic.
Woody-based, densely caespitose. Leaves in crowded whorls, acicular to narrowly linear. Inflorescence of 1-3 few-flowered verticillasters. Corolla 6-10 mm, pink.

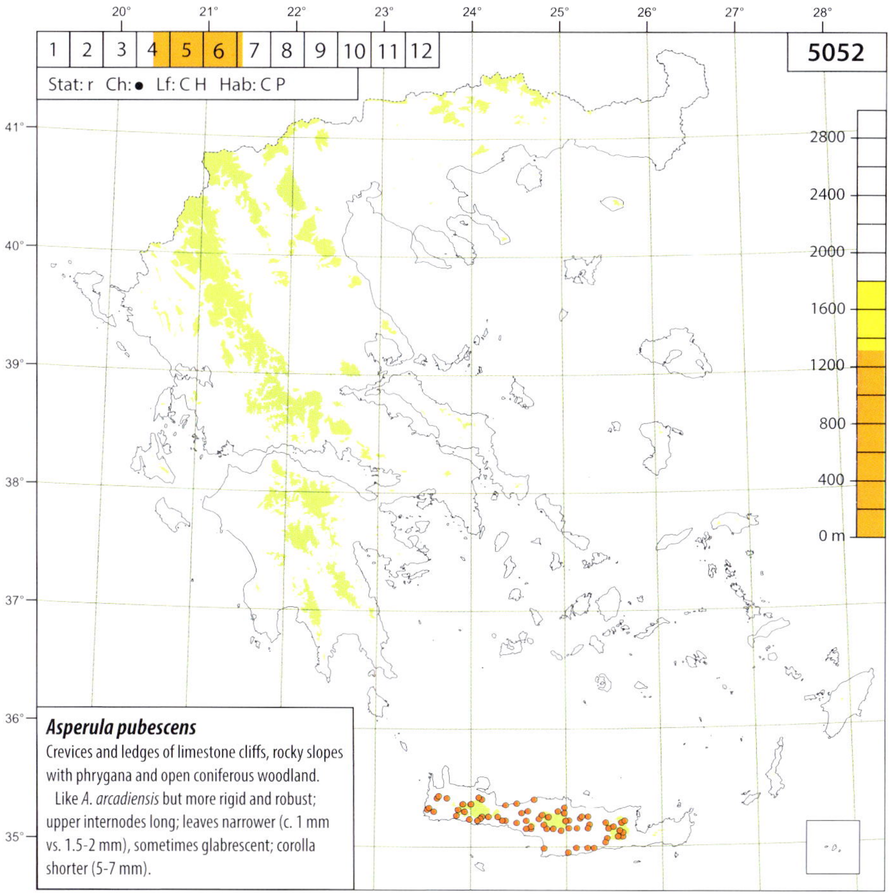

Asperula pubescens
Crevices and ledges of limestone cliffs, rocky slopes with phrygana and open coniferous woodland.
Like *A. arcadiensis* but more rigid and robust; upper internodes long; leaves narrower (c. 1 mm vs. 1.5-2 mm), sometimes glabrescent; corolla shorter (5-7 mm).

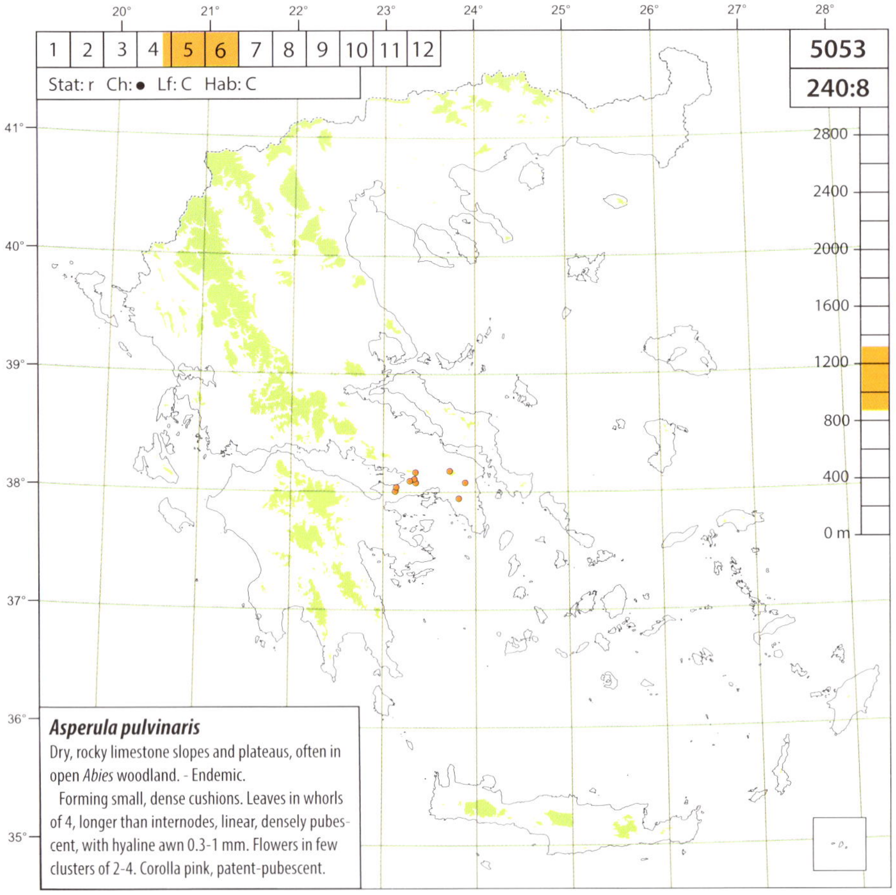

Asperula pulvinaris

Dry, rocky limestone slopes and plateaus, often in open *Abies* woodland. - Endemic.

Forming small, dense cushions. Leaves in whorls of 4, longer than internodes, linear, densely pubescent, with hyaline awn 0.3-1 mm. Flowers in few clusters of 2-4. Corolla pink, patent-pubescent.

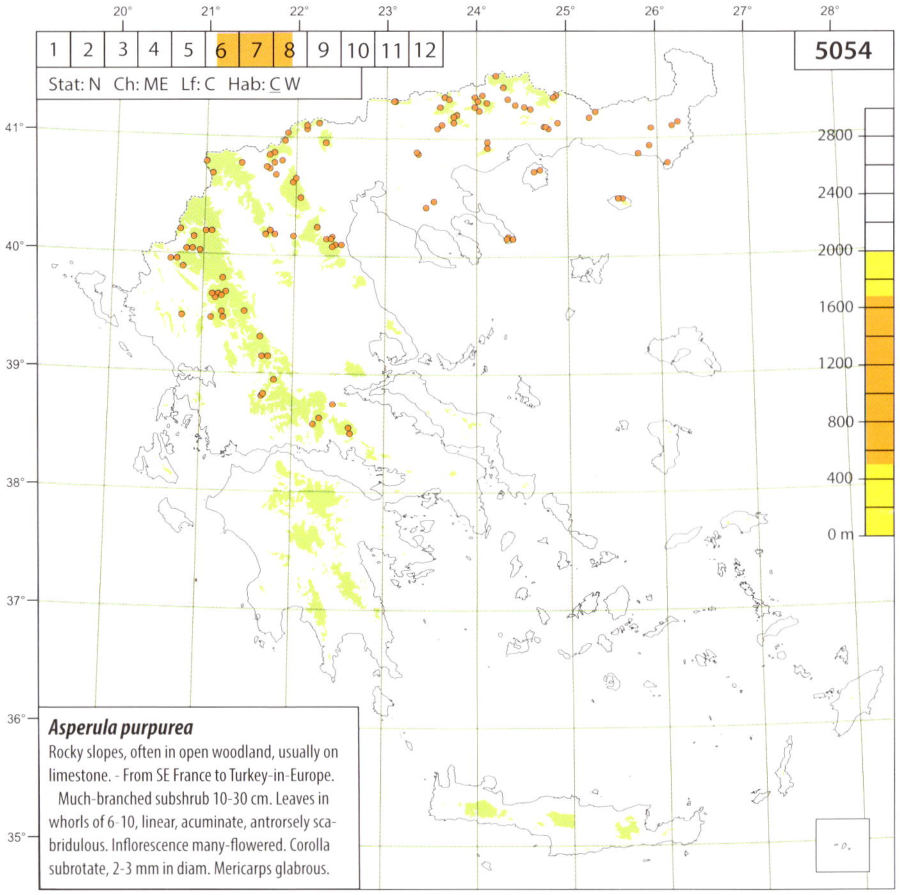

Asperula purpurea

Rocky slopes, often in open woodland, usually on limestone. - From SE France to Turkey-in-Europe.

Much-branched subshrub 10-30 cm. Leaves in whorls of 6-10, linear, acuminate, antrorsely scabridulous. Inflorescence many-flowered. Corolla subrotate, 2-3 mm in diam. Mericarps glabrous.

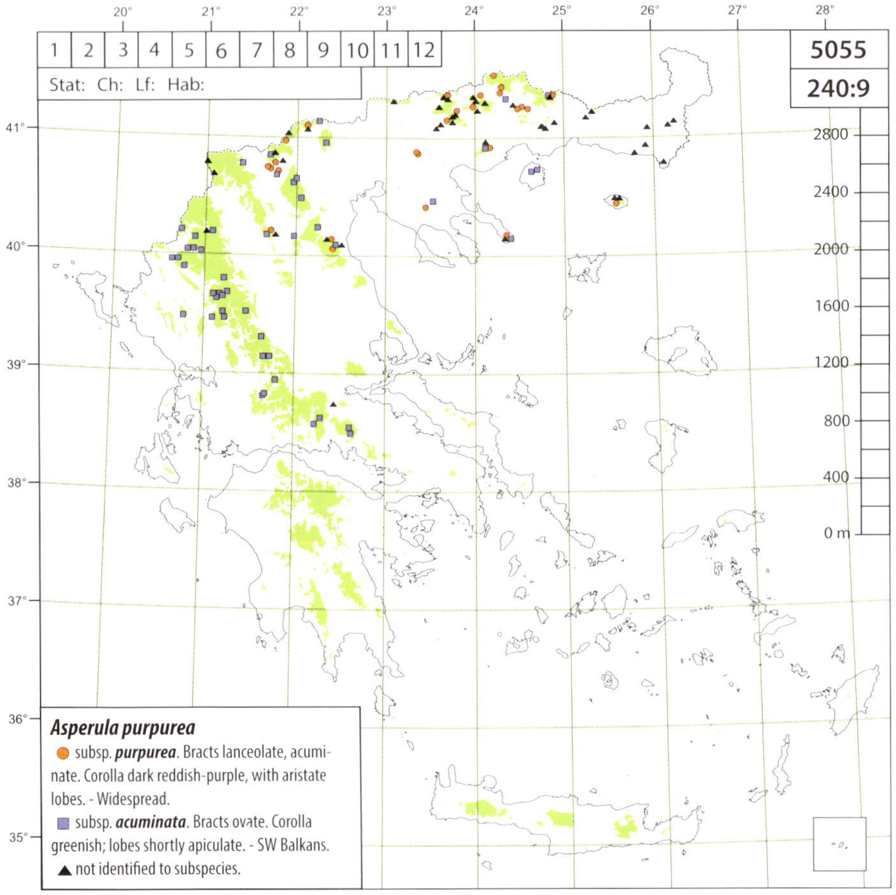

Asperula purpurea

● subsp. ***purpurea***. Bracts lanceolate, acuminate. Corolla dark reddish-purple, with aristate lobes. - Widespread.

■ subsp. ***acuminata***. Bracts ovate. Corolla greenish; lobes shortly apiculate. - SW Balkans.

▲ not identified to subspecies.

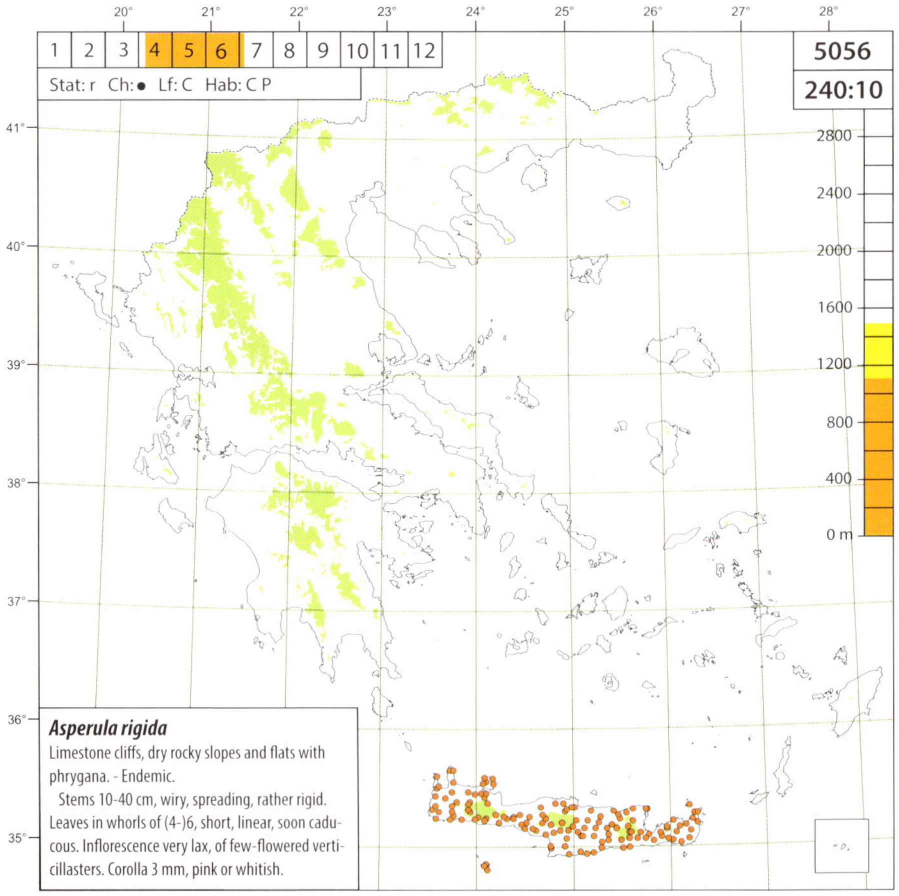

Asperula rigida

Limestone cliffs, dry rocky slopes and flats with phrygana. - Endemic.

Stems 10-40 cm, wiry, spreading, rather rigid. Leaves in whorls of (4-)6, short, linear, soon caducous. Inflorescence very lax, of few-flowered verticillasters. Corolla 3 mm, pink or whitish.

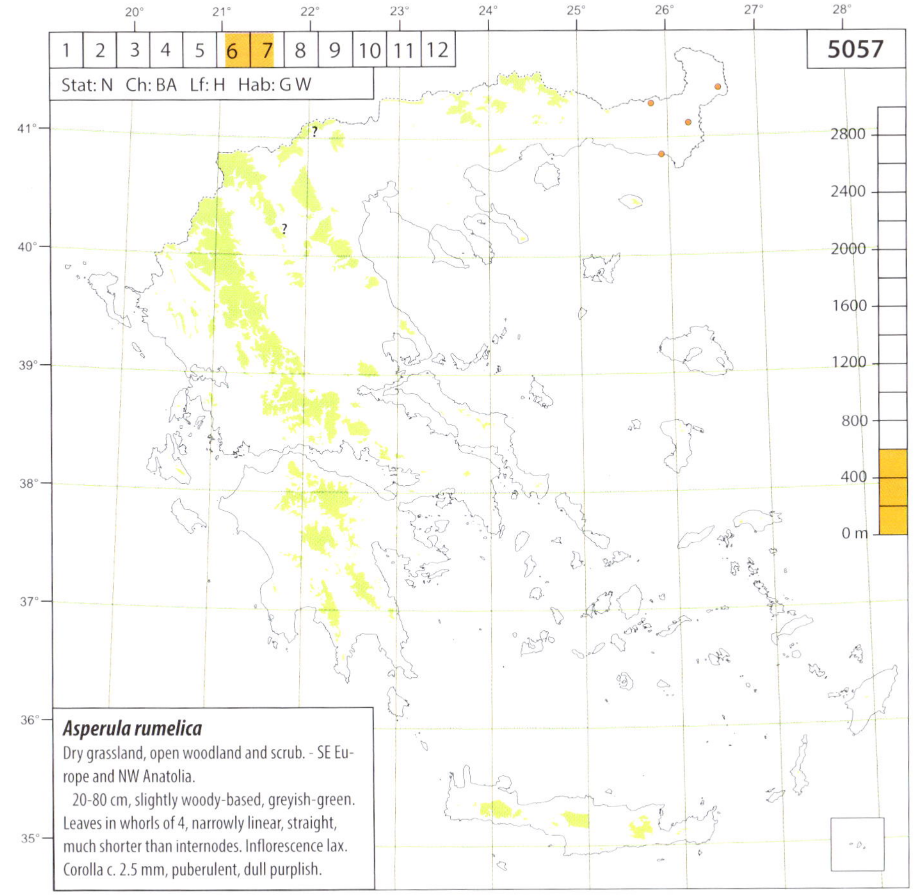

Asperula rumelica
Dry grassland, open woodland and scrub. - SE Europe and NW Anatolia.
20-80 cm, slightly woody-based, greyish-green. Leaves in whorls of 4, narrowly linear, straight, much shorter than internodes. Inflorescence lax. Corolla c. 2.5 mm, puberulent, dull purplish.

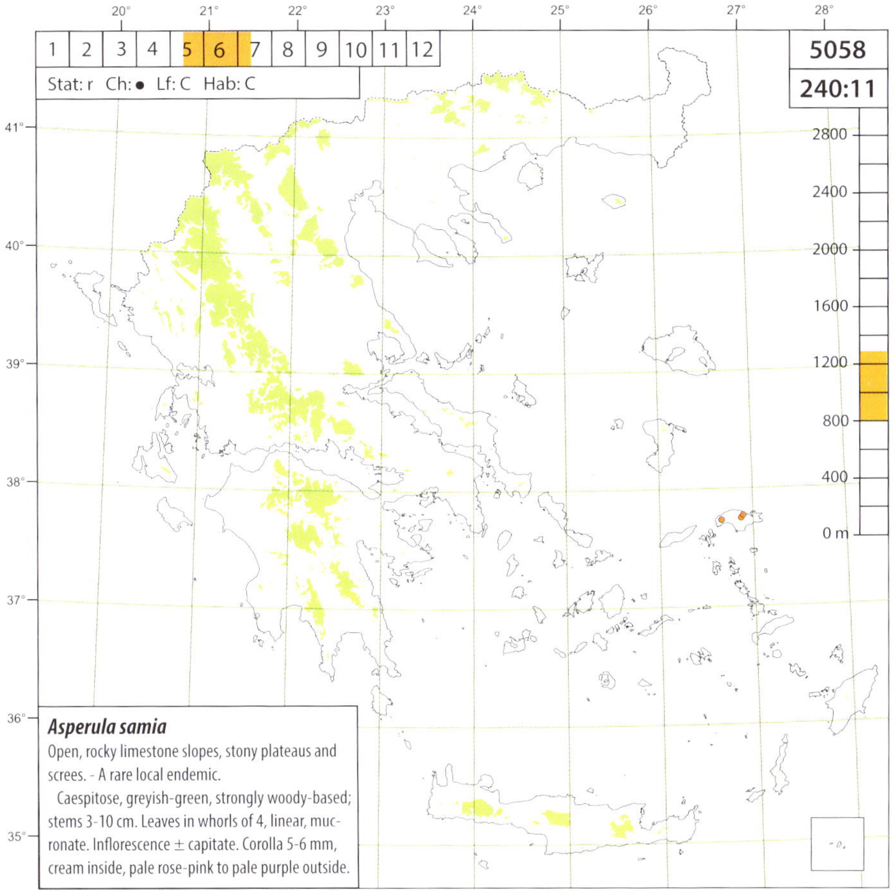

Asperula samia
Open, rocky limestone slopes, stony plateaus and screes. - A rare local endemic.
Caespitose, greyish-green, strongly woody-based; stems 3-10 cm. Leaves in whorls of 4, linear, mucronate. Inflorescence ± capitate. Corolla 5-6 mm, cream inside, pale rose-pink to pale purple outside.

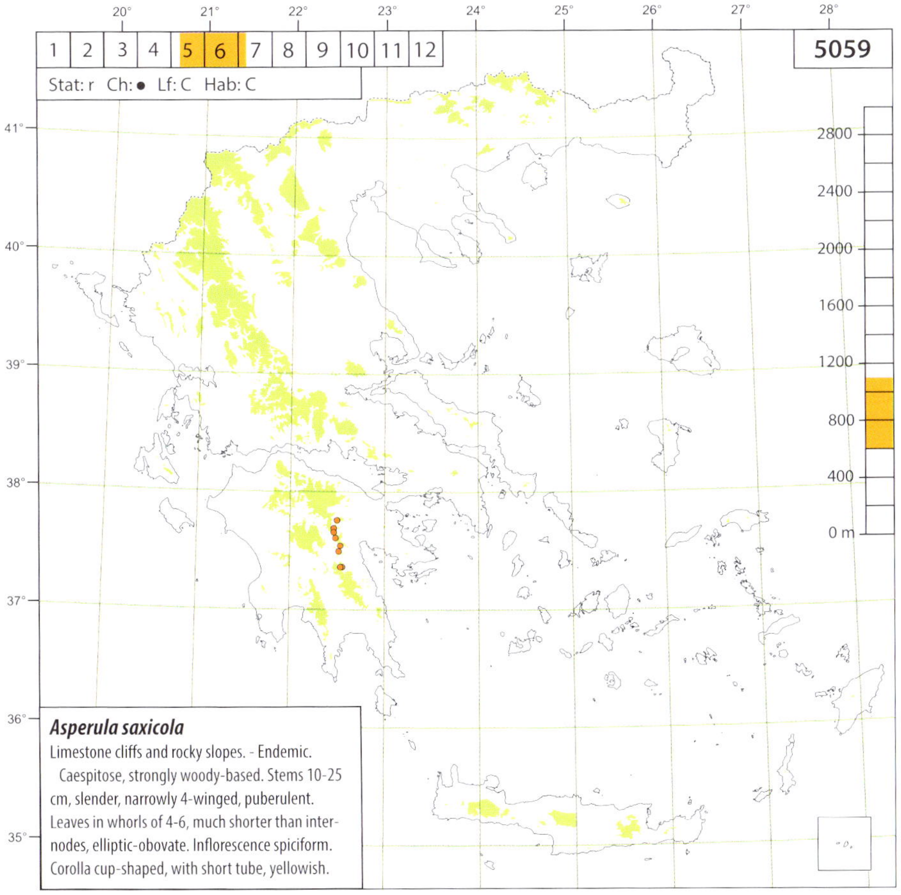

Asperula saxicola
Limestone cliffs and rocky slopes. - Endemic.
Caespitose, strongly woody-based. Stems 10-25 cm, slender, narrowly 4-winged, puberulent. Leaves in whorls of 4-6, much shorter than internodes, elliptic-obovate. Inflorescence spiciform. Corolla cup-shaped, with short tube, yellowish.

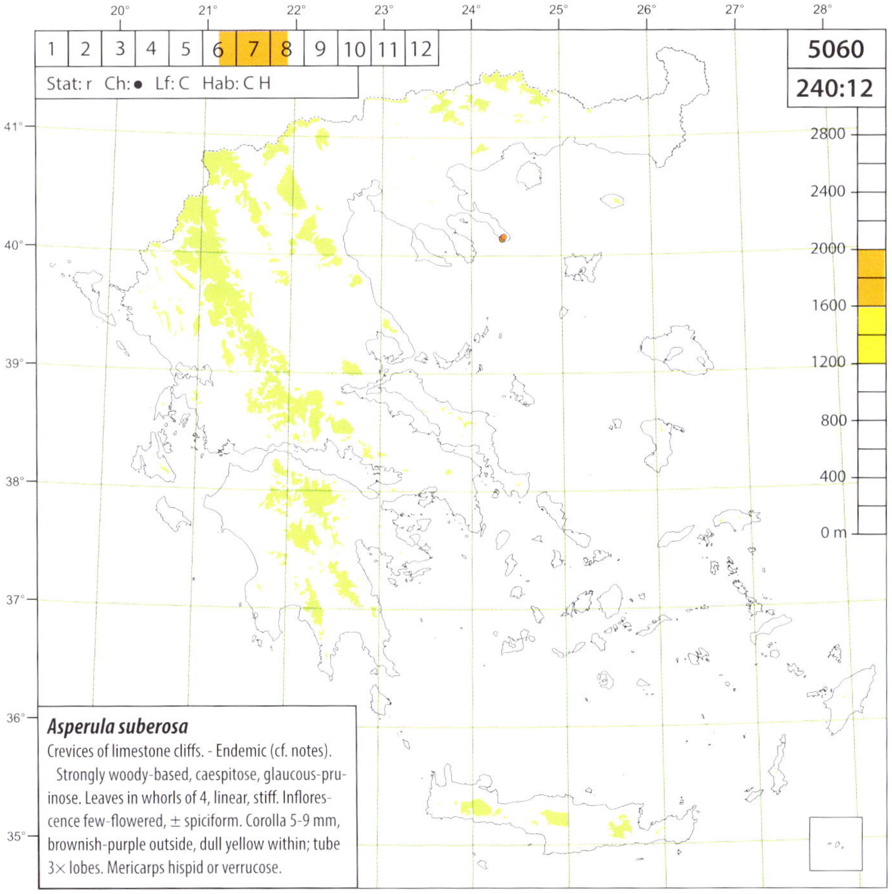

Asperula suberosa
Crevices of limestone cliffs. - Endemic (cf. notes).
Strongly woody-based, caespitose, glaucous-pruinose. Leaves in whorls of 4, linear, stiff. Inflorescence few-flowered, ± spiciform. Corolla 5-9 mm, brownish-purple outside, dull yellow within; tube 3× lobes. Mericarps hispid or verrucose.

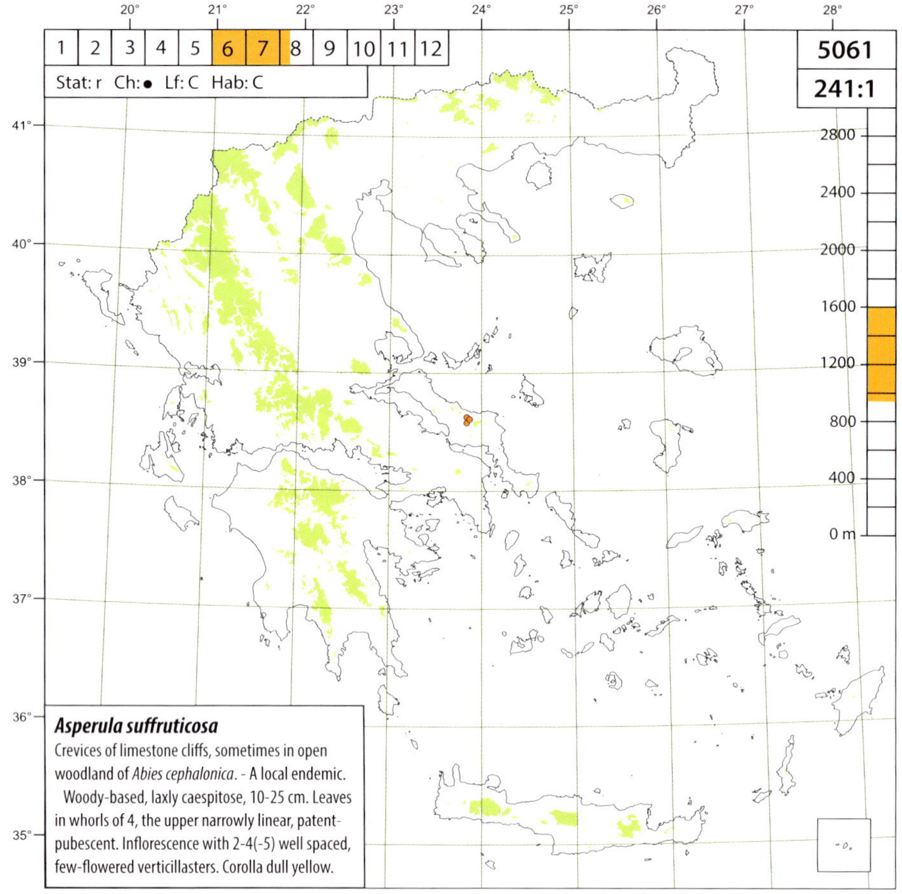

Asperula suffruticosa
Crevices of limestone cliffs, sometimes in open woodland of *Abies cephalonica*. - A local endemic.
Woody-based, laxly caespitose, 10-25 cm. Leaves in whorls of 4, the upper narrowly linear, patent-pubescent. Inflorescence with 2-4(-5) well spaced, few-flowered verticillasters. Corolla dull yellow.

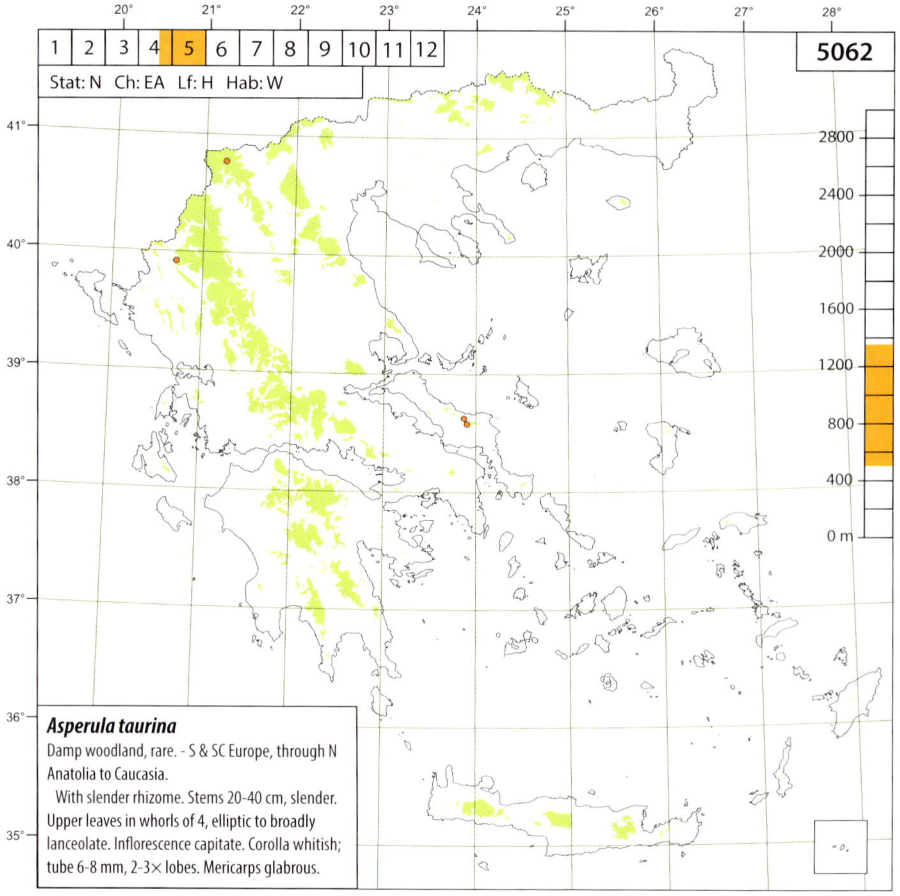

Asperula taurina
Damp woodland, rare. - S & SC Europe, through N Anatolia to Caucasia.
With slender rhizome. Stems 20-40 cm, slender. Upper leaves in whorls of 4, elliptic to broadly lanceolate. Inflorescence capitate. Corolla whitish; tube 6-8 mm, 2-3× lobes. Mericarps glabrous.

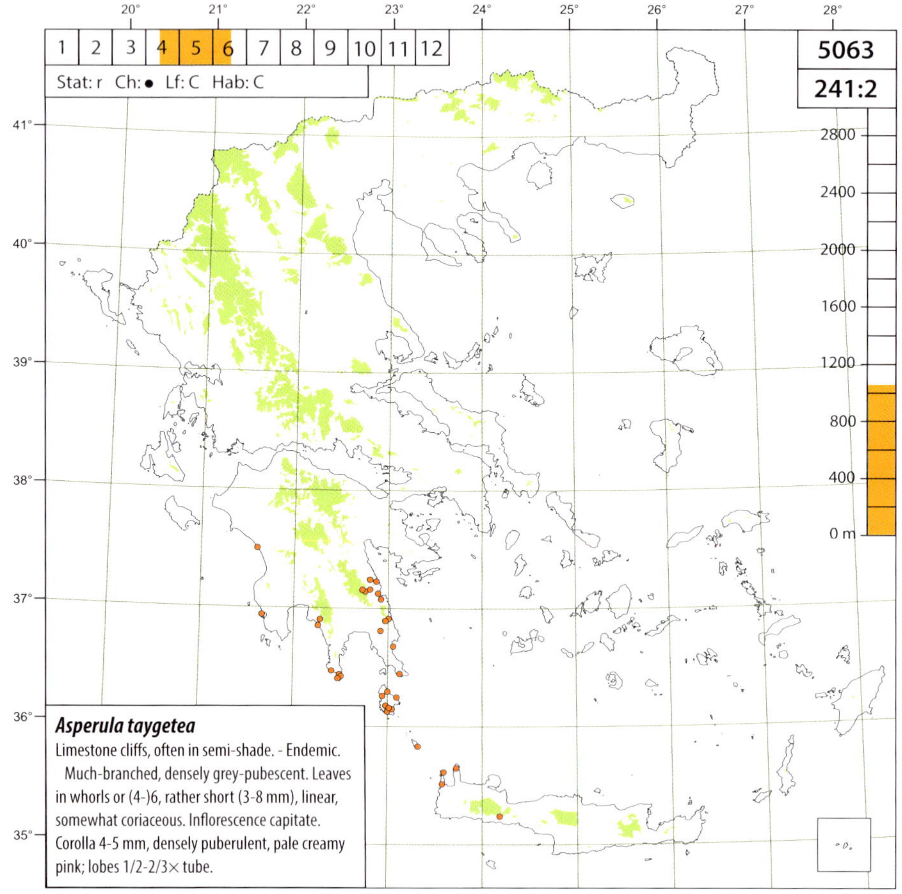

Asperula taygetea
Limestone cliffs, often in semi-shade. - Endemic.
Much-branched, densely grey-pubescent. Leaves in whorls or (4-)6, rather short (3-8 mm), linear, somewhat coriaceous. Inflorescence capitate. Corolla 4-5 mm, densely puberulent, pale creamy pink; lobes 1/2-2/3× tube.

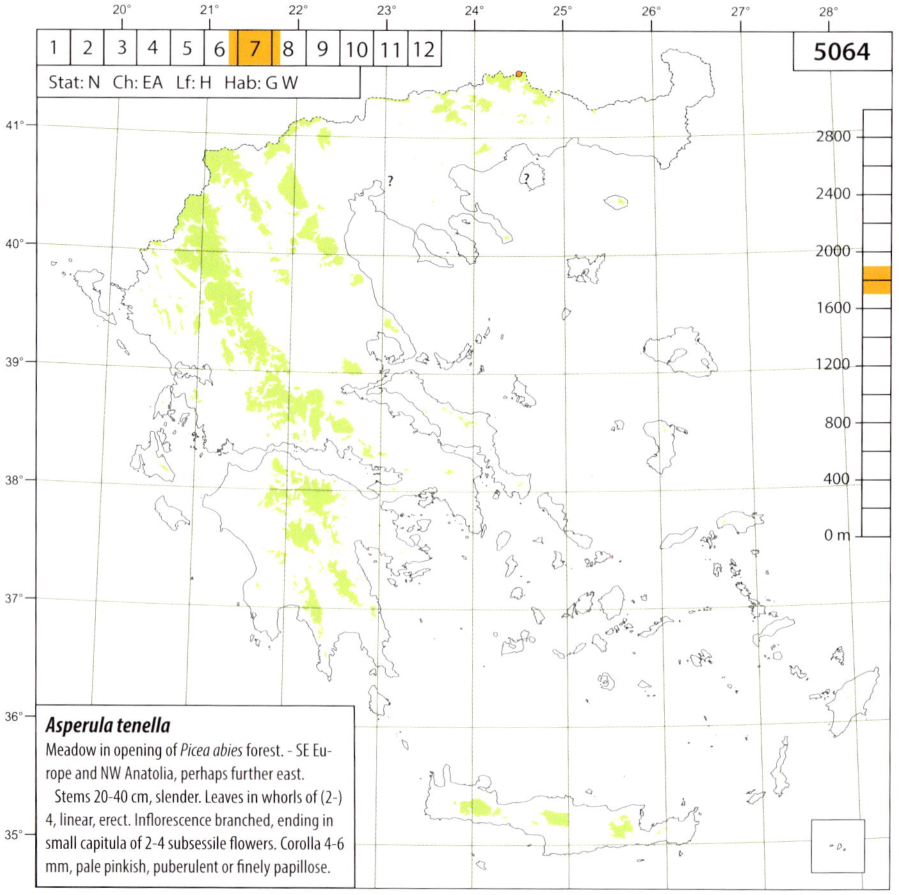

Asperula tenella
Meadow in opening of *Picea abies* forest. - SE Europe and NW Anatolia, perhaps further east.
Stems 20-40 cm, slender. Leaves in whorls of (2-)4, linear, erect. Inflorescence branched, ending in small capitula of 2-4 subsessile flowers. Corolla 4-6 mm, pale pinkish, puberulent or finely papillose.

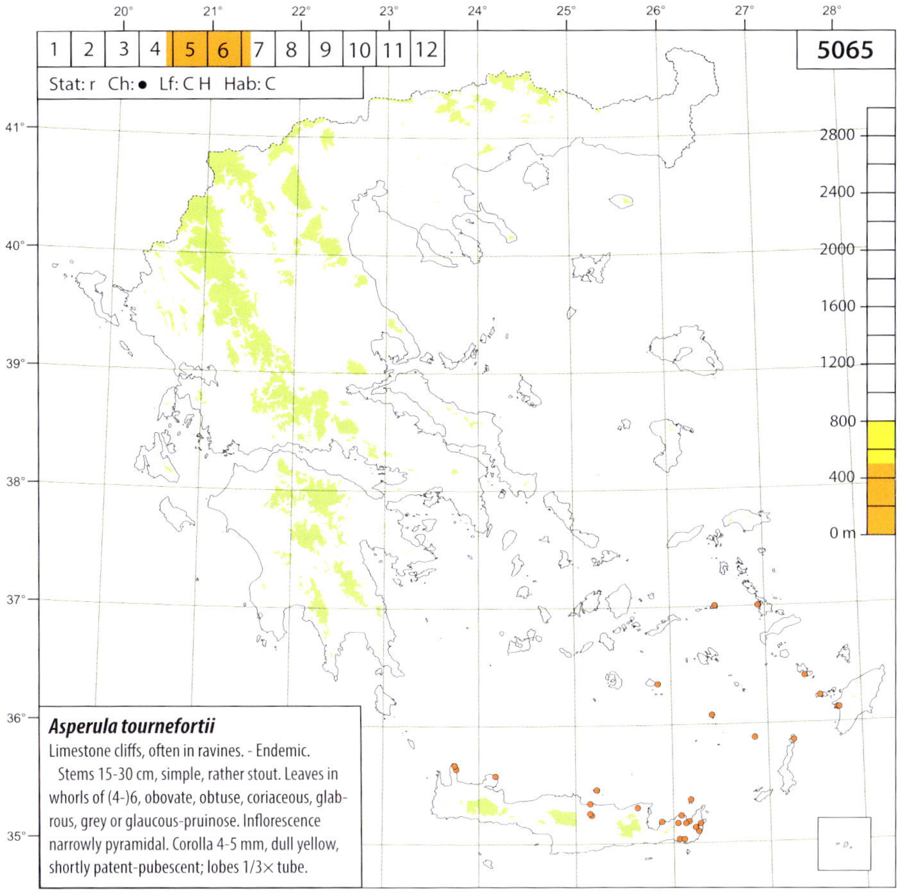

Asperula tournefortii

Limestone cliffs, often in ravines. - Endemic.

Stems 15-30 cm, simple, rather stout. Leaves in whorls of (4-)6, obovate, obtuse, coriaceous, glabrous, grey or glaucous-pruinose. Inflorescence narrowly pyramidal. Corolla 4-5 mm, dull yellow, shortly patent-pubescent; lobes 1/3× tube.

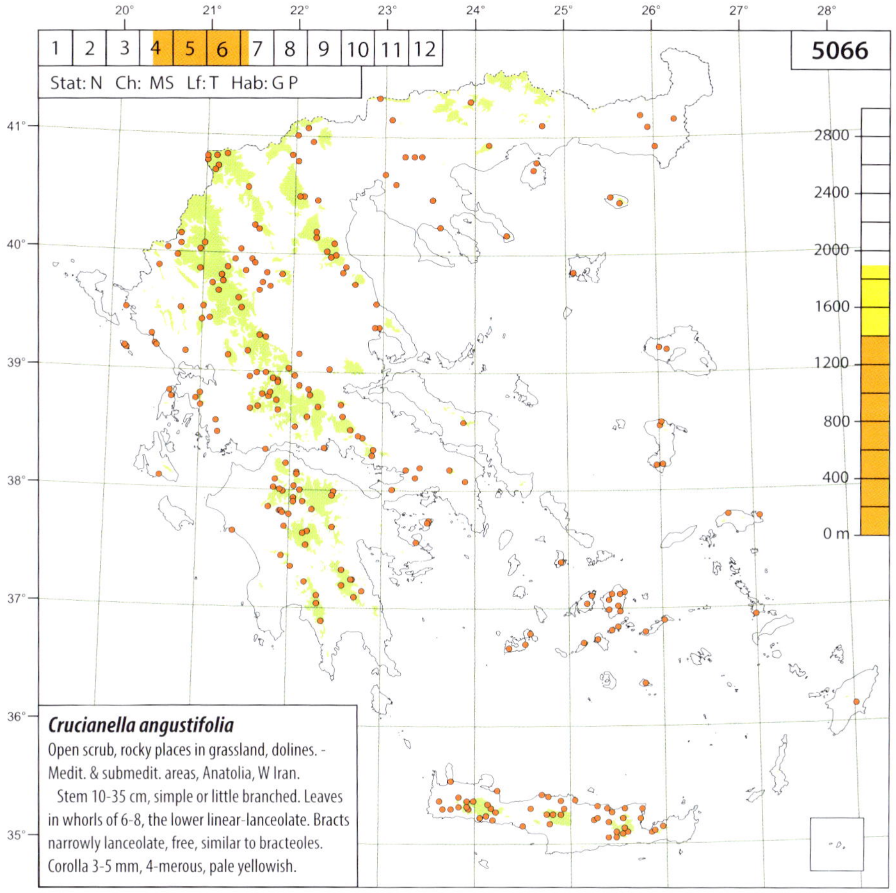

Crucianella angustifolia

Open scrub, rocky places in grassland, dolines. - Medit. & submedit. areas, Anatolia, W Iran.

Stem 10-35 cm, simple or little branched. Leaves in whorls of 6-8, the lower linear-lanceolate. Bracts narrowly lanceolate, free, similar to bracteoles. Corolla 3-5 mm, 4-merous, pale yellowish.

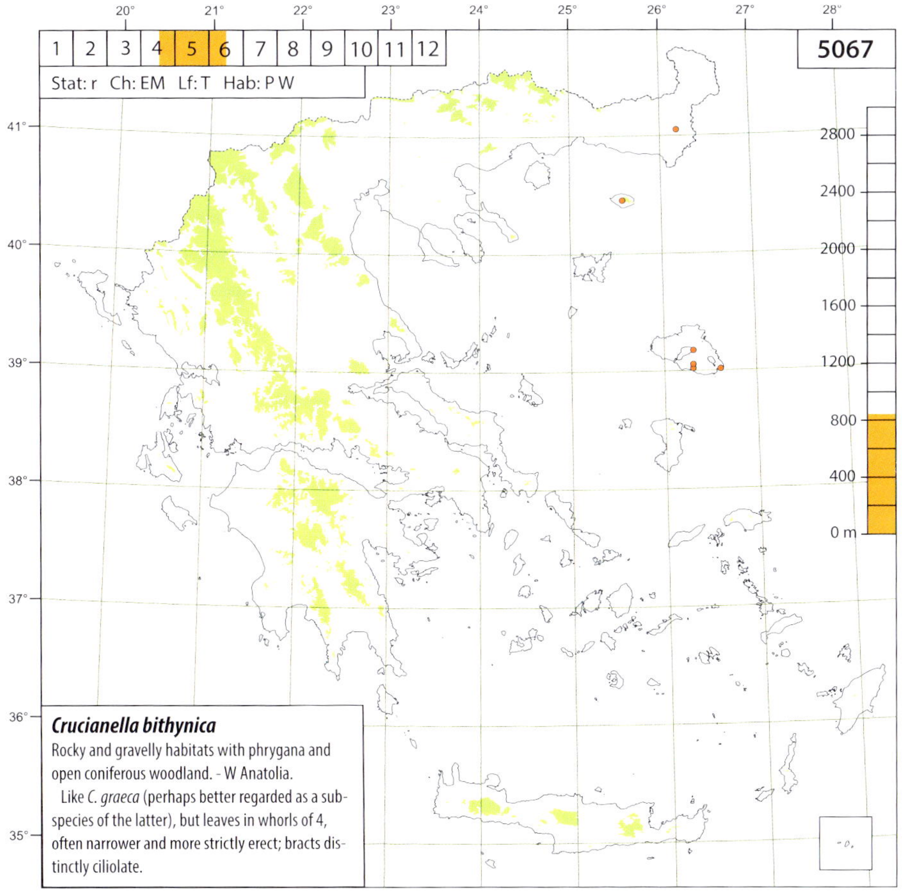

Crucianella bithynica

Rocky and gravelly habitats with phrygana and open coniferous woodland. - W Anatolia.

Like *C. graeca* (perhaps better regarded as a subspecies of the latter), but leaves in whorls of 4, often narrower and more strictly erect; bracts distinctly ciliolate.

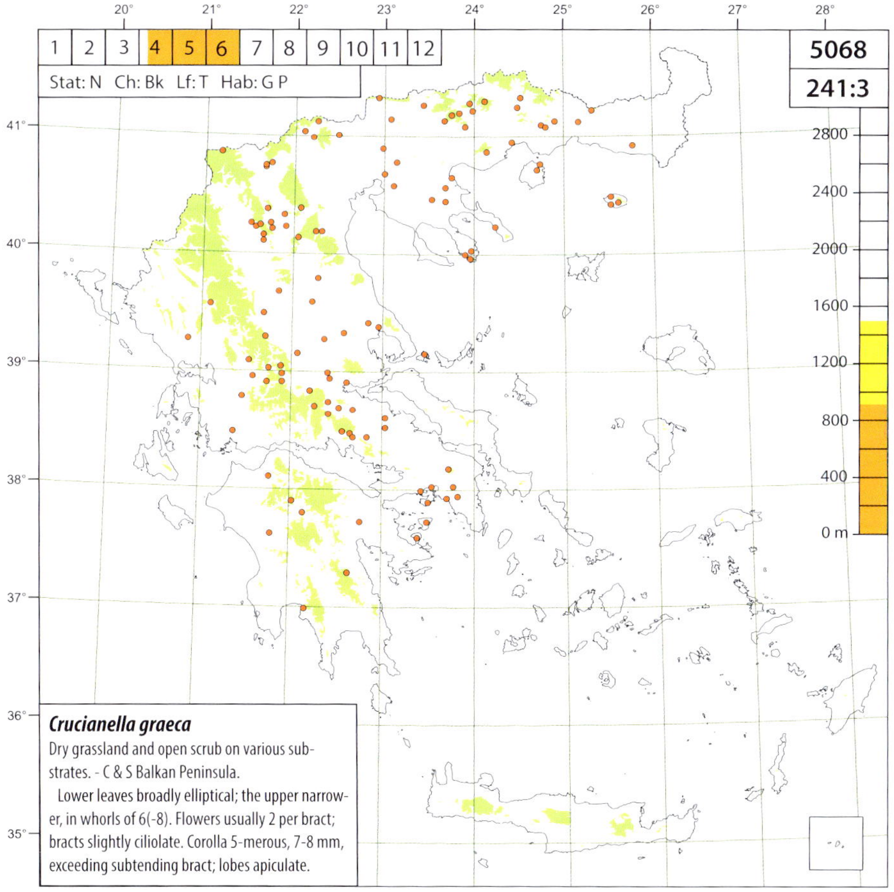

Crucianella graeca

Dry grassland and open scrub on various substrates. - C & S Balkan Peninsula.

Lower leaves broadly elliptical; the upper narrower, in whorls of 6(-8). Flowers usually 2 per bract; bracts slightly ciliolate. Corolla 5-merous, 7-8 mm, exceeding subtending bract; lobes apiculate.

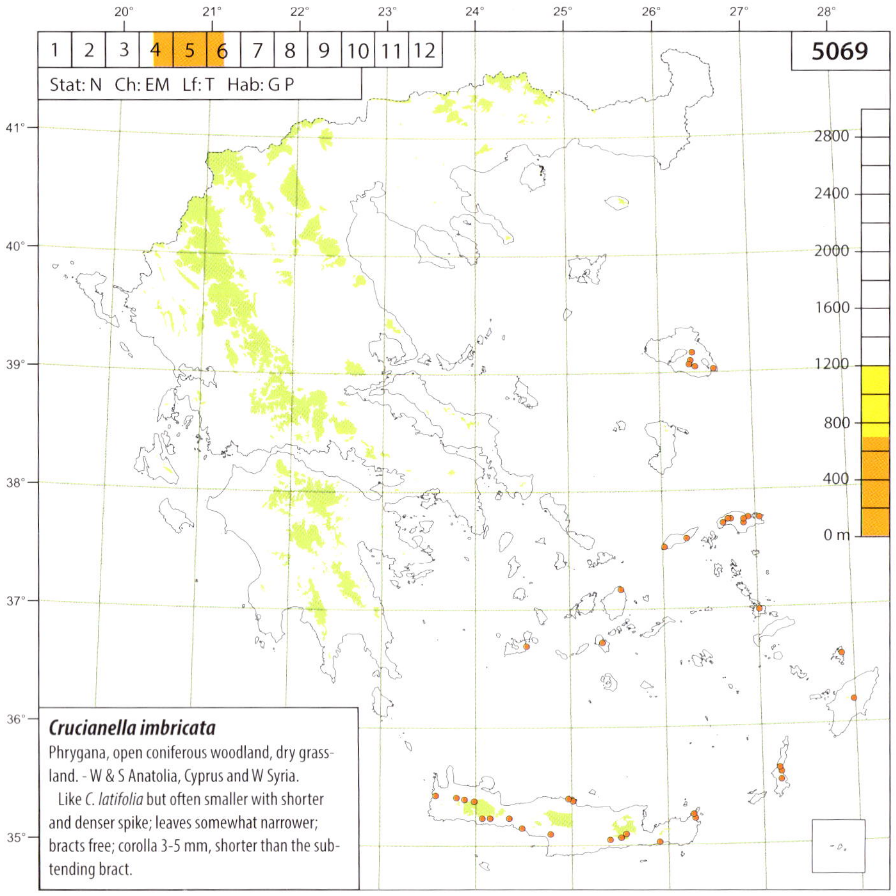

Crucianella imbricata

Phrygana, open coniferous woodland, dry grassland. - W & S Anatolia, Cyprus and W Syria.

Like *C. latifolia* but often smaller with shorter and denser spike; leaves somewhat narrower; bracts free; corolla 3-5 mm, shorter than the subtending bract.

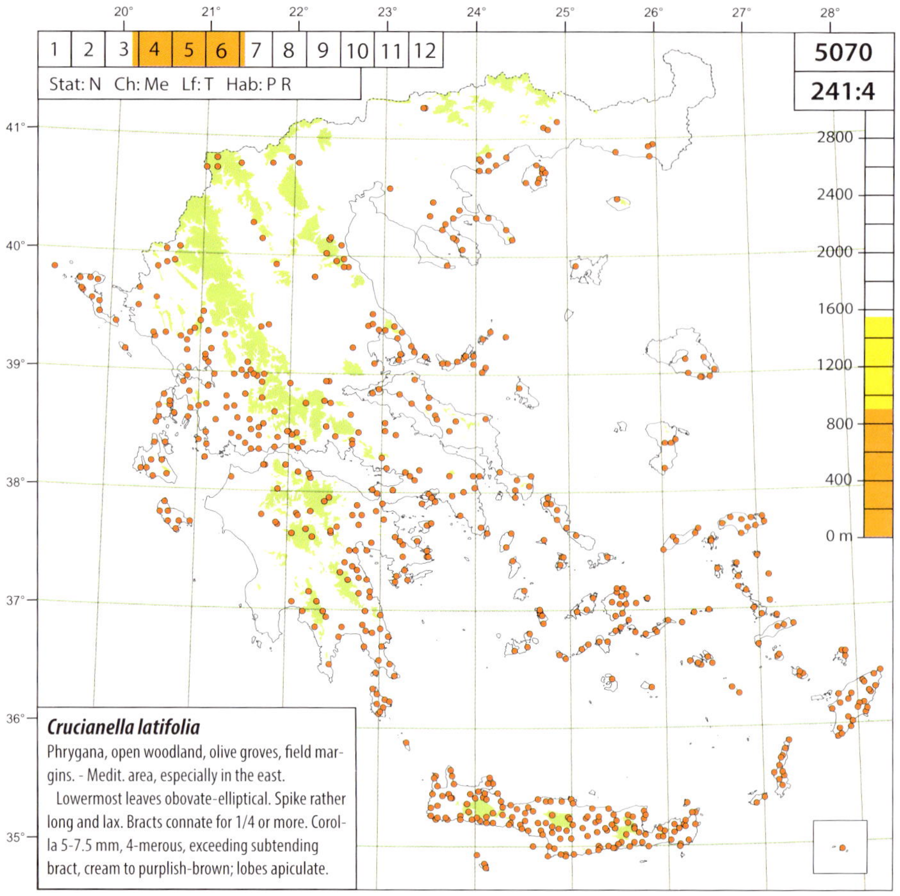

Crucianella latifolia

Phrygana, open woodland, olive groves, field margins. - Medit. area, especially in the east.

Lowermost leaves obovate-elliptical. Spike rather long and lax. Bracts connate for 1/4 or more. Corolla 5-7.5 mm, 4-merous, exceeding subtending bract, cream to purplish-brown; lobes apiculate.

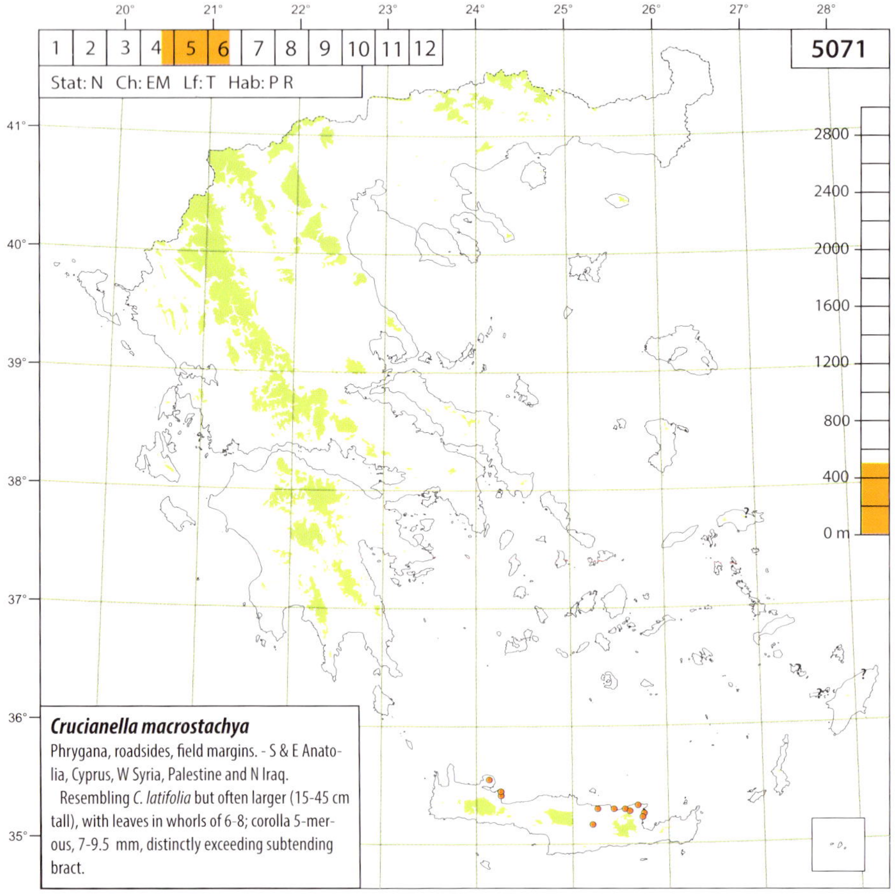

Crucianella macrostachya

Phrygana, roadsides, field margins. - S & E Anatolia, Cyprus, W Syria, Palestine and N Iraq.

Resembling *C. latifolia* but often larger (15-45 cm tall), with leaves in whorls of 6-8; corolla 5-merous, 7-9.5 mm, distinctly exceeding subtending bract.

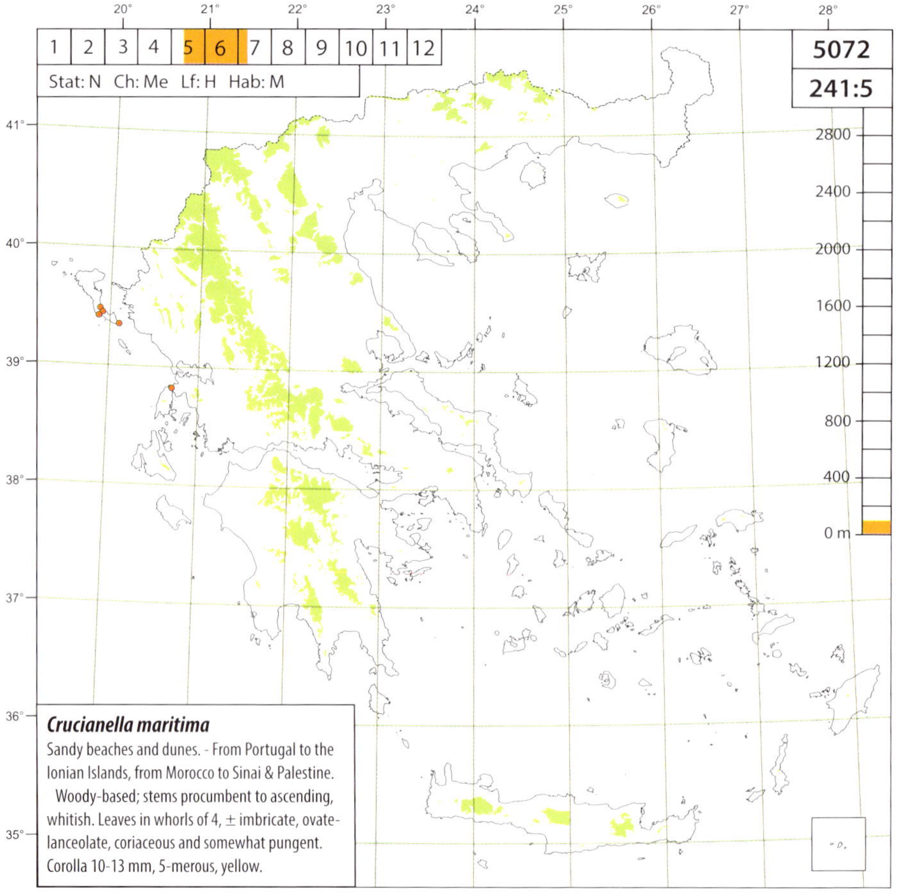

Crucianella maritima

Sandy beaches and dunes. - From Portugal to the Ionian Islands, from Morocco to Sinai & Palestine.

Woody-based; stems procumbent to ascending, whitish. Leaves in whorls of 4, ± imbricate, ovate-lanceolate, coriaceous and somewhat pungent. Corolla 10-13 mm, 5-merous, yellow.

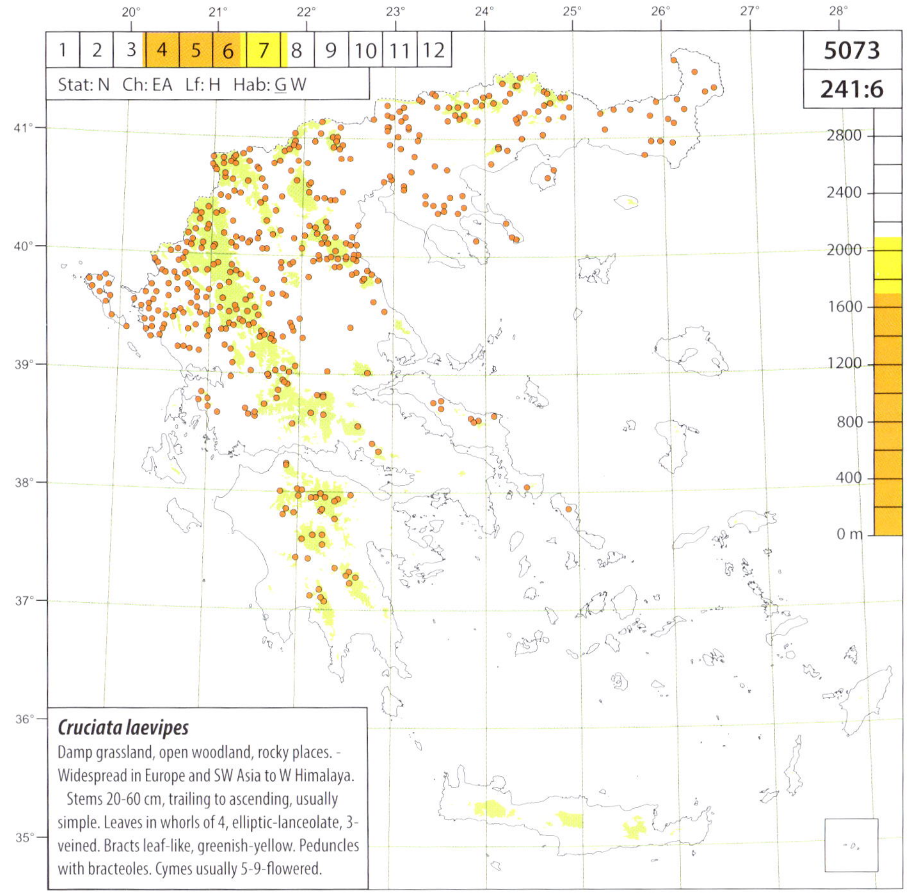

Cruciata laevipes
Damp grassland, open woodland, rocky places. - Widespread in Europe and SW Asia to W Himalaya.
Stems 20-60 cm, trailing to ascending, usually simple. Leaves in whorls of 4, elliptic-lanceolate, 3-veined. Bracts leaf-like, greenish-yellow. Peduncles with bracteoles. Cymes usually 5-9-flowered.

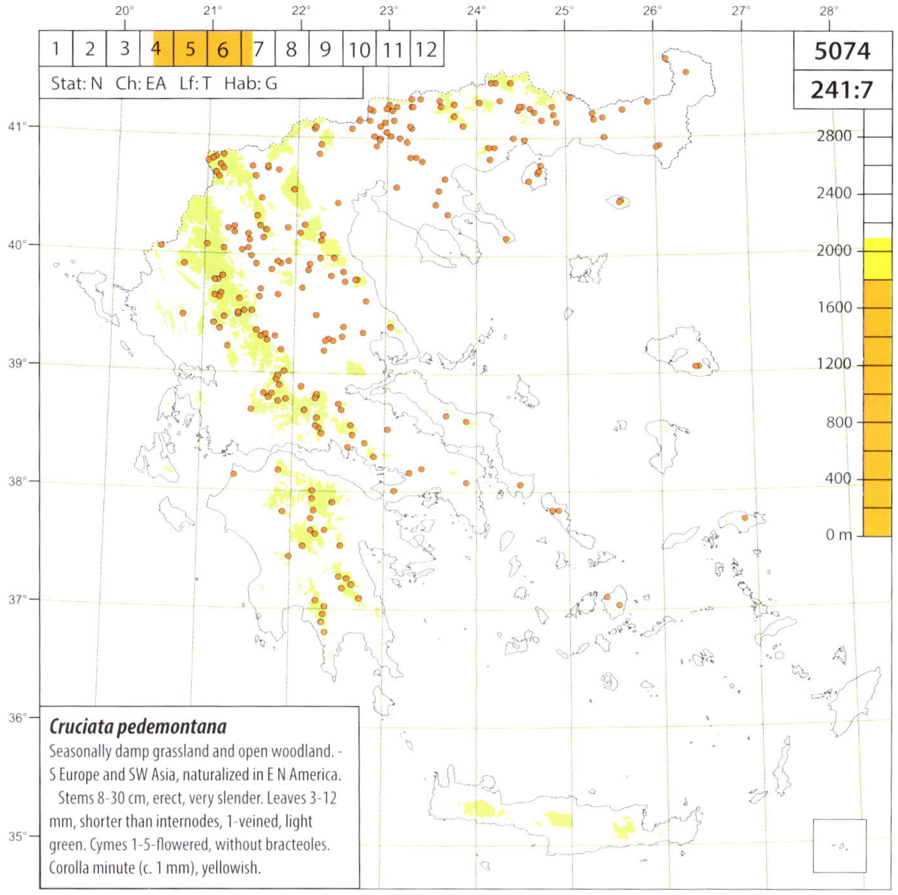

Cruciata pedemontana
Seasonally damp grassland and open woodland. - S Europe and SW Asia, naturalized in E N America.
Stems 8-30 cm, erect, very slender. Leaves 3-12 mm, shorter than internodes, 1-veined, light green. Cymes 1-5-flowered, without bracteoles. Corolla minute (c. 1 mm), yellowish.

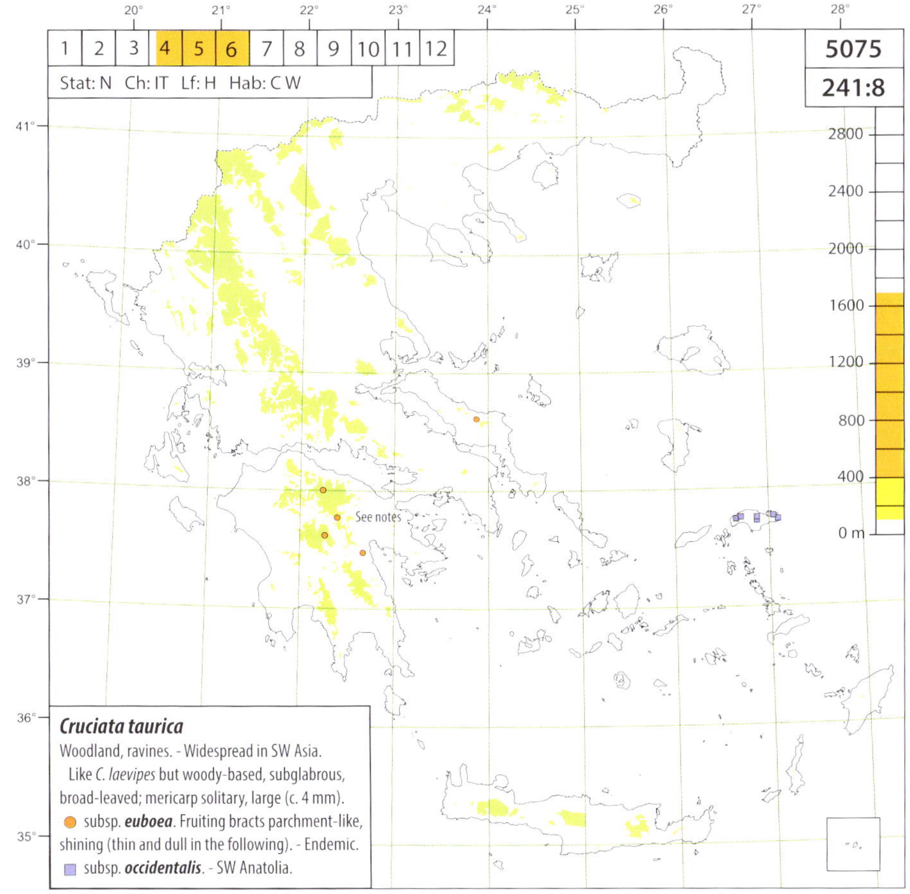

Cruciata taurica
Woodland, ravines. - Widespread in SW Asia.
Like *C. laevipes* but woody-based, subglabrous, broad-leaved; mericarp solitary, large (c. 4 mm).
● subsp. ***euboea***. Fruiting bracts parchment-like, shining (thin and dull in the following). - Endemic.
■ subsp. ***occidentalis***. - SW Anatolia.

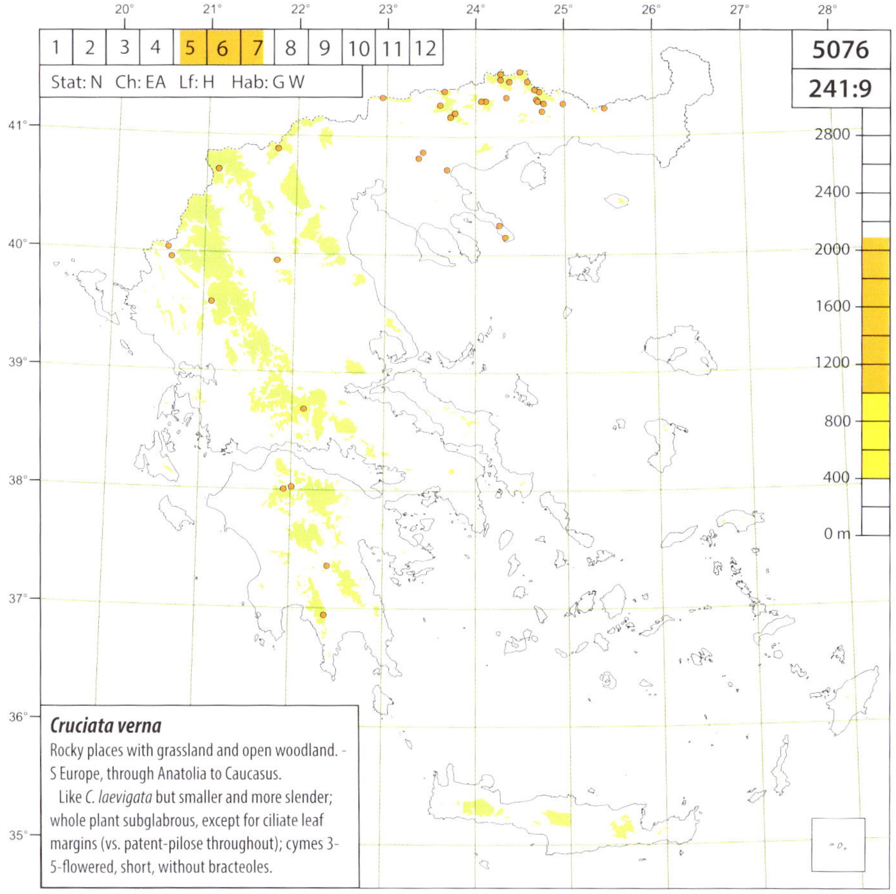

Cruciata verna
Rocky places with grassland and open woodland. - S Europe, through Anatolia to Caucasus.
Like *C. laevigata* but smaller and more slender; whole plant subglabrous, except for ciliate leaf margins (vs. patent-pilose throughout); cymes 3-5-flowered, short, without bracteoles.

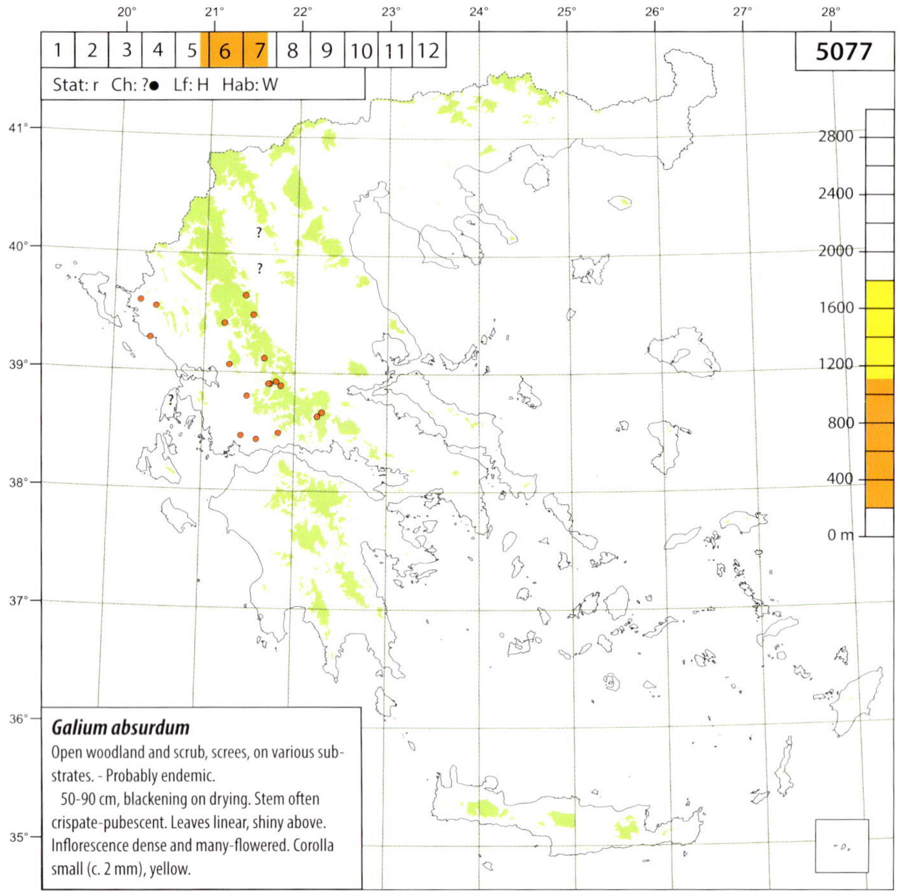

Galium absurdum

Open woodland and scrub, screes, on various substrates. - Probably endemic.

50-90 cm, blackening on drying. Stem often crispate-pubescent. Leaves linear, shiny above. Inflorescence dense and many-flowered. Corolla small (c. 2 mm), yellow.

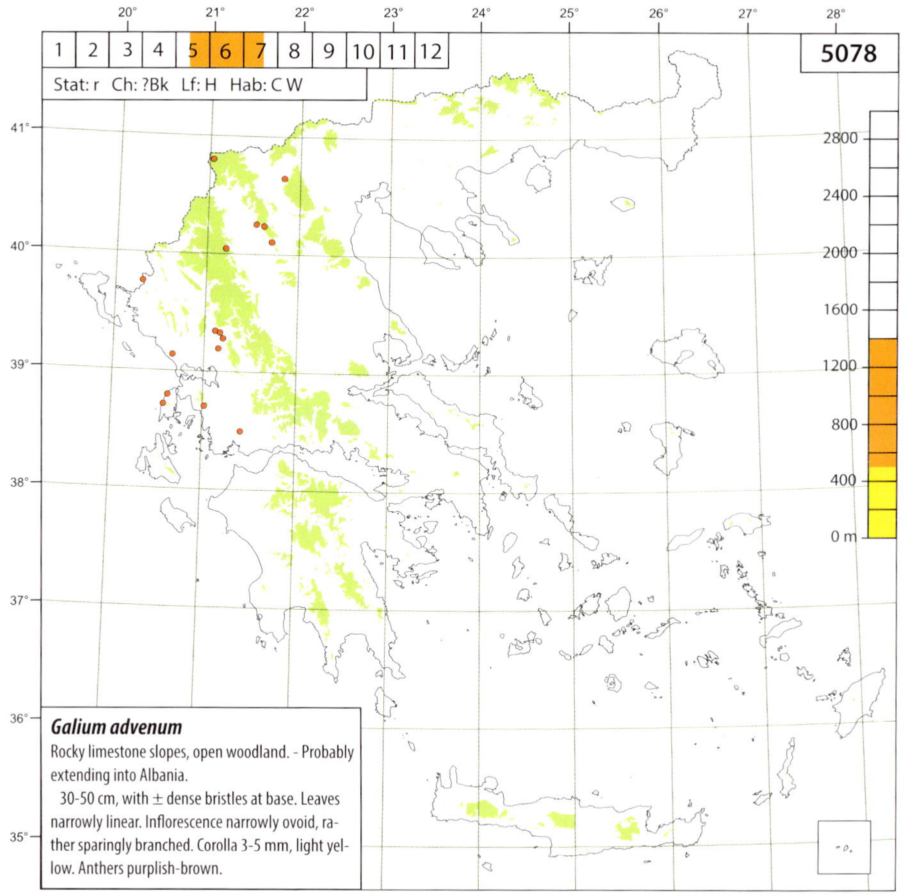

Galium advenum

Rocky limestone slopes, open woodland. - Probably extending into Albania.

30-50 cm, with ± dense bristles at base. Leaves narrowly linear. Inflorescence narrowly ovoid, rather sparingly branched. Corolla 3-5 mm, light yellow. Anthers purplish-brown.

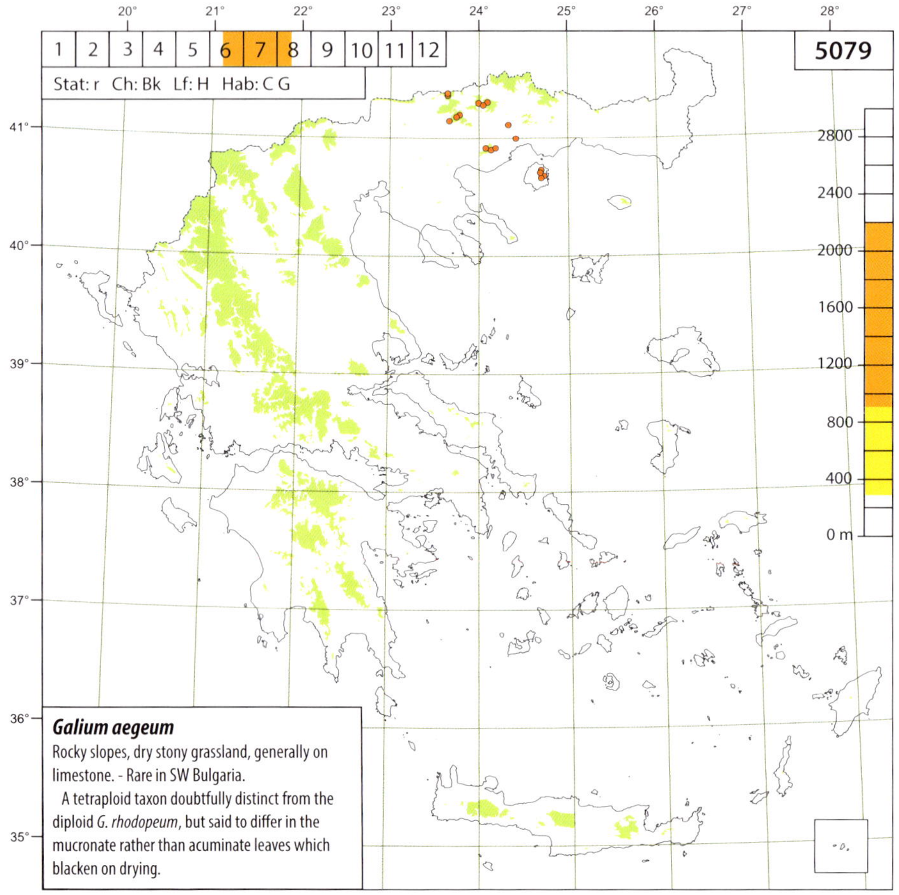

Galium aegeum

Rocky slopes, dry stony grassland, generally on limestone. - Rare in SW Bulgaria.

A tetraploid taxon doubtfully distinct from the diploid *G. rhodopeum*, but said to differ in the mucronate rather than acuminate leaves which blacken on drying.

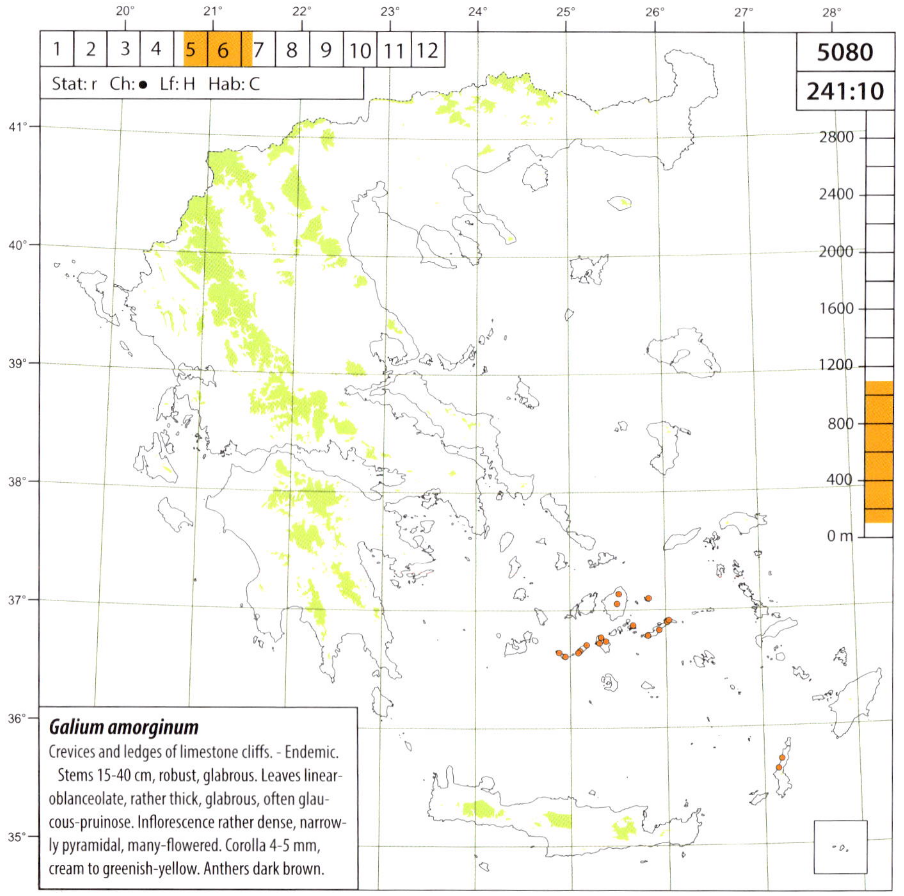

Galium amorginum

Crevices and ledges of limestone cliffs. - Endemic.

Stems 15-40 cm, robust, glabrous. Leaves linear-oblanceolate, rather thick, glabrous, often glaucous-pruinose. Inflorescence rather dense, narrowly pyramidal, many-flowered. Corolla 4-5 mm, cream to greenish-yellow. Anthers dark brown.

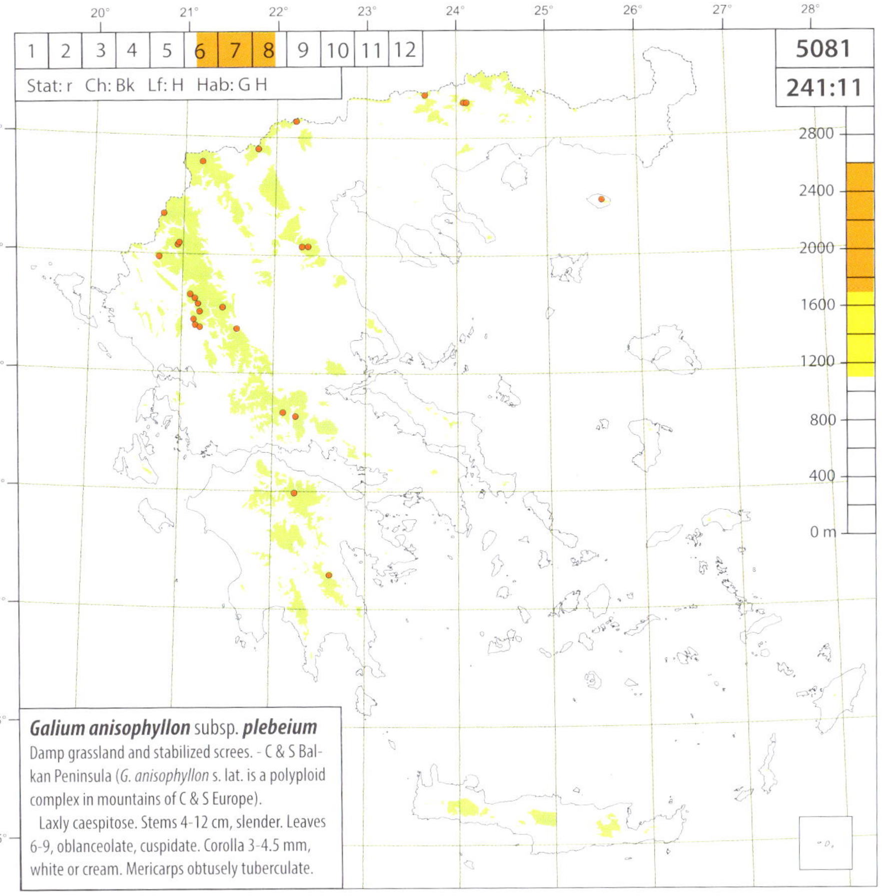

Galium anisophyllon subsp. ***plebeium***
Damp grassland and stabilized screes. - C & S Balkan Peninsula (*G. anisophyllon* s. lat. is a polyploid complex in mountains of C & S Europe).
Laxly caespitose. Stems 4-12 cm, slender. Leaves 6-9, oblanceolate, cuspidate. Corolla 3-4.5 mm, white or cream. Mericarps obtusely tuberculate.

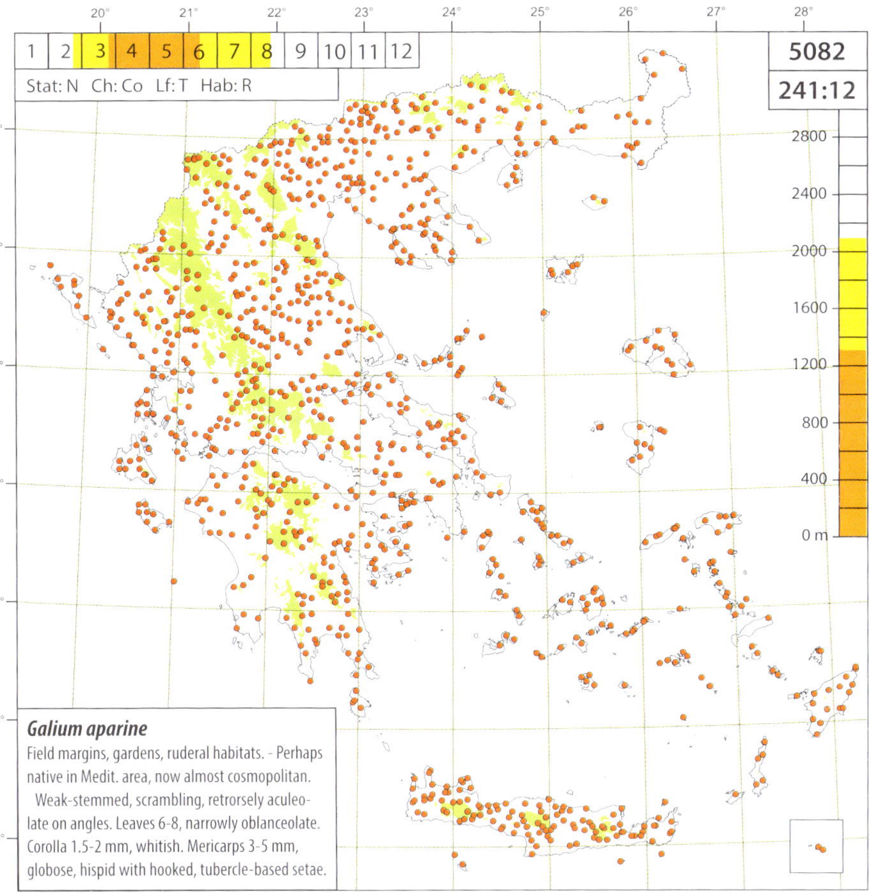

Galium aparine
Field margins, gardens, ruderal habitats. - Perhaps native in Medit. area, now almost cosmopolitan.
Weak-stemmed, scrambling, retrorsely aculeolate on angles. Leaves 6-8, narrowly oblanceolate. Corolla 1.5-2 mm, whitish. Mericarps 3-5 mm, globose, hispid with hooked, tubercle-based setae.

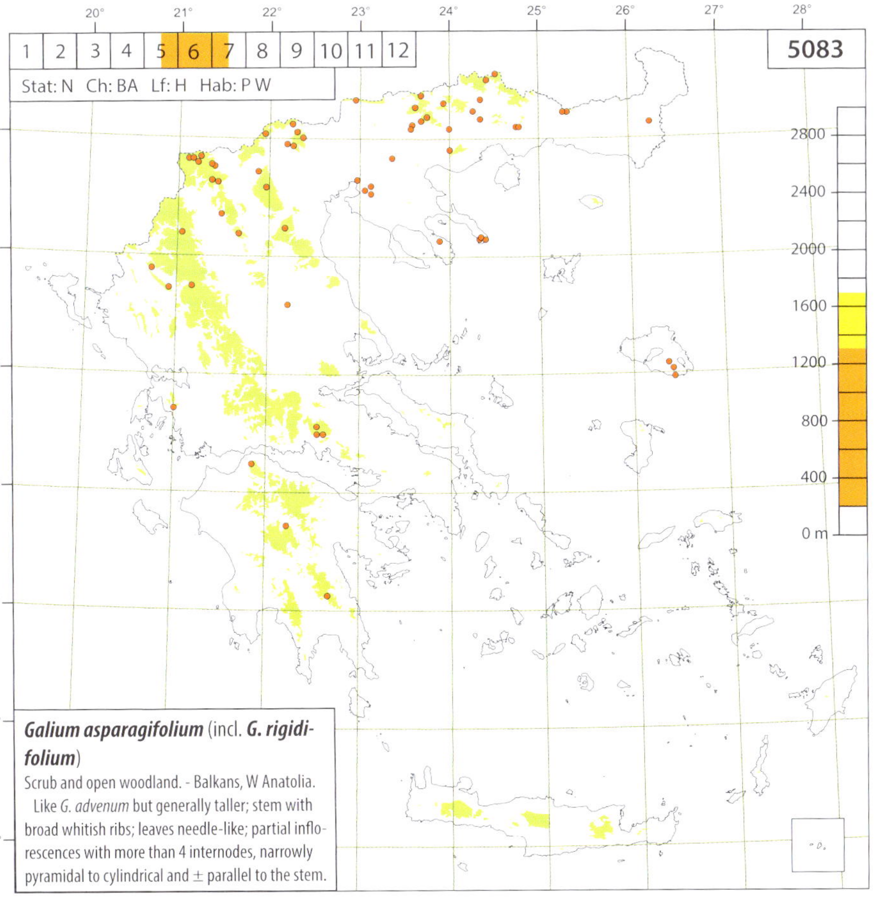

Galium asparagifolium (incl. ***G. rigidifolium***)
Scrub and open woodland. - Balkans, W Anatolia.
Like *G. advenum* but generally taller; stem with broad whitish ribs; leaves needle-like; partial inflorescences with more than 4 internodes, narrowly pyramidal to cylindrical and ± parallel to the stem.

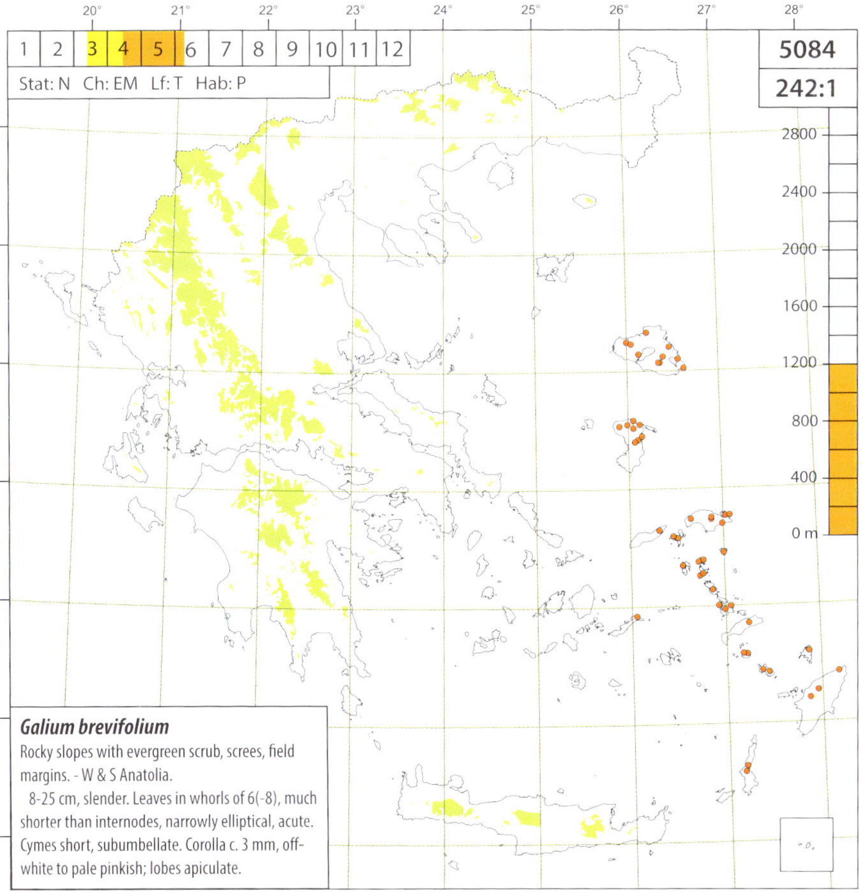

Galium brevifolium
Rocky slopes with evergreen scrub, screes, field margins. - W & S Anatolia.
8-25 cm, slender. Leaves in whorls of 6(-8), much shorter than internodes, narrowly elliptical, acute. Cymes short, subumbellate. Corolla c. 3 mm, off-white to pale pinkish; lobes apiculate.

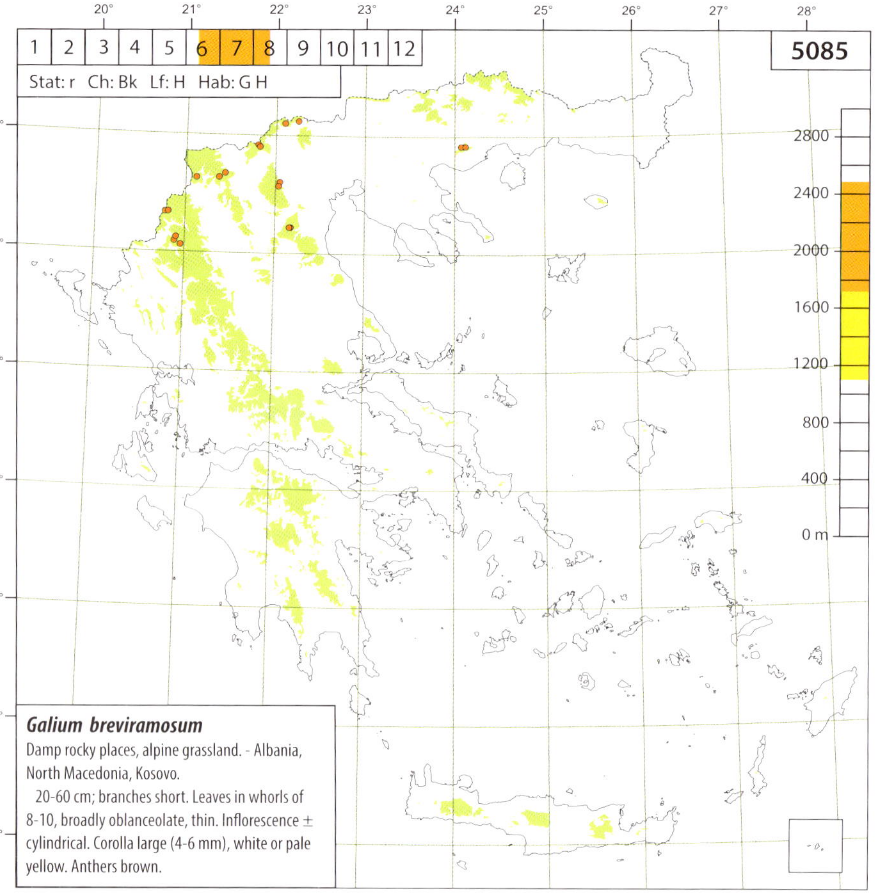

Galium breviramosum

Damp rocky places, alpine grassland. - Albania, North Macedonia, Kosovo.

20-60 cm; branches short. Leaves in whorls of 8-10, broadly oblanceolate, thin. Inflorescence ± cylindrical. Corolla large (4-6 mm), white or pale yellow. Anthers brown.

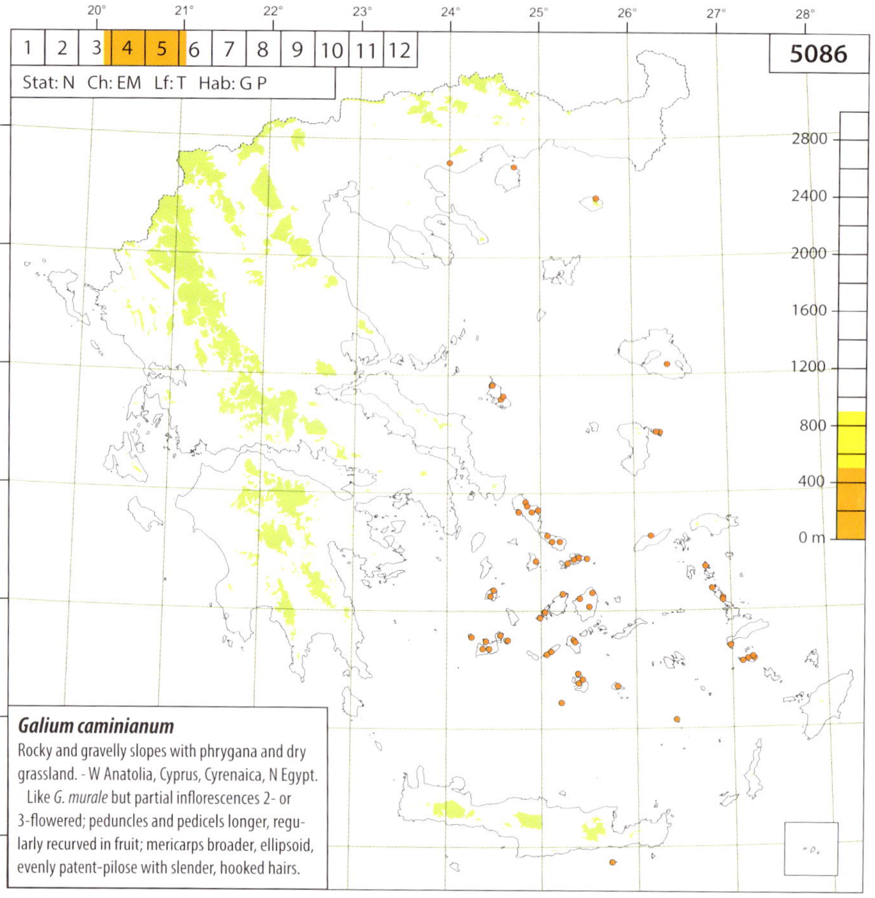

Galium caminianum

Rocky and gravelly slopes with phrygana and dry grassland. - W Anatolia, Cyprus, Cyrenaica, N Egypt.

Like *G. murale* but partial inflorescences 2- or 3-flowered; peduncles and pedicels longer, regularly recurved in fruit; mericarps broader, ellipsoid, evenly patent-pilose with slender, hooked hairs.

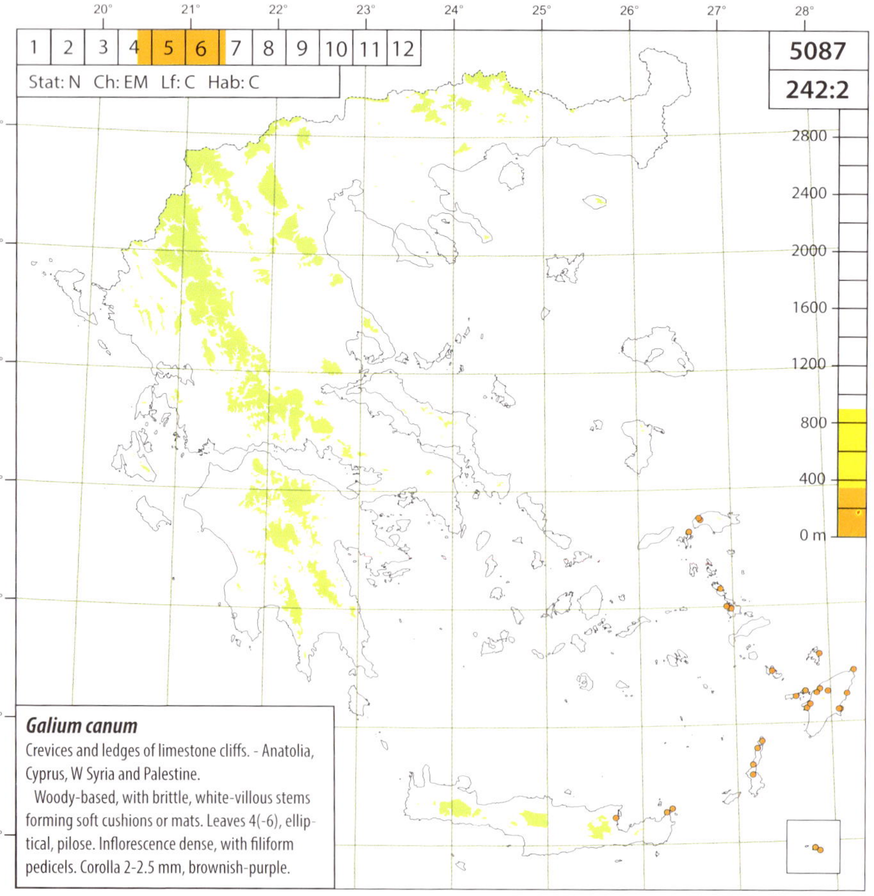

Galium canum

Crevices and ledges of limestone cliffs. - Anatolia, Cyprus, W Syria and Palestine.

Woody-based, with brittle, white-villous stems forming soft cushions or mats. Leaves 4(-6), elliptical, pilose. Inflorescence dense, with filiform pedicels. Corolla 2-2.5 mm, brownish-purple.

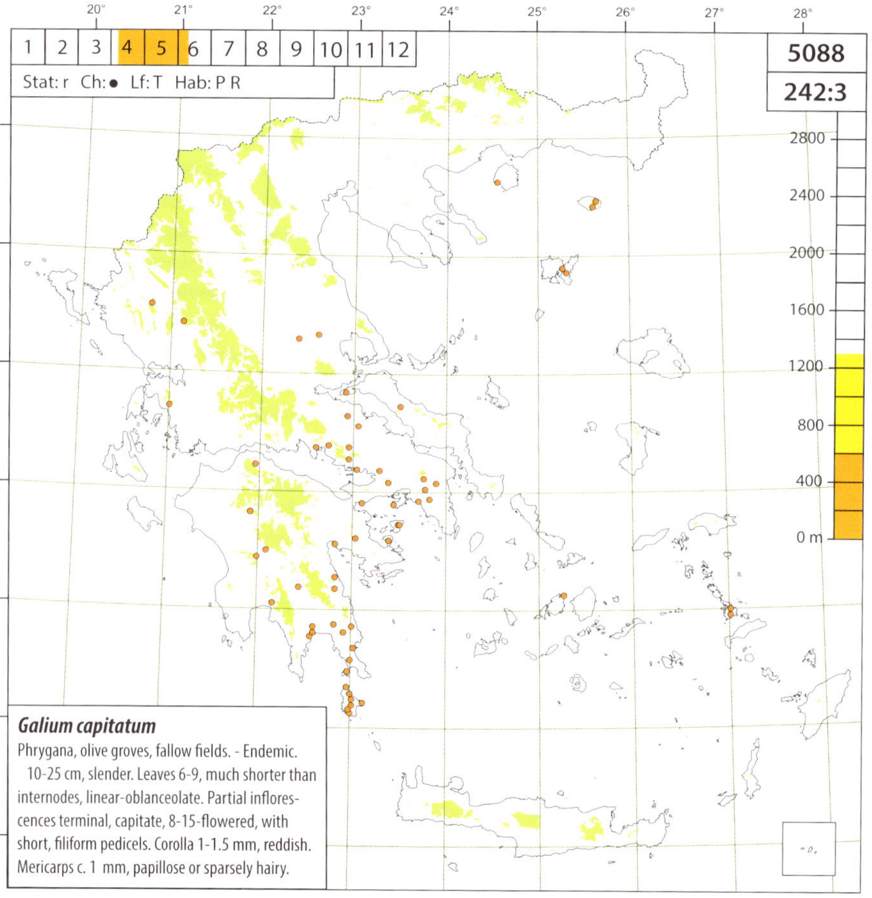

Galium capitatum

Phrygana, olive groves, fallow fields. - Endemic.

10-25 cm, slender. Leaves 6-9, much shorter than internodes, linear-oblanceolate. Partial inflorescences terminal, capitate, 8-15-flowered, with short, filiform pedicels. Corolla 1-1.5 mm, reddish. Mericarps c. 1 mm, papillose or sparsely hairy.

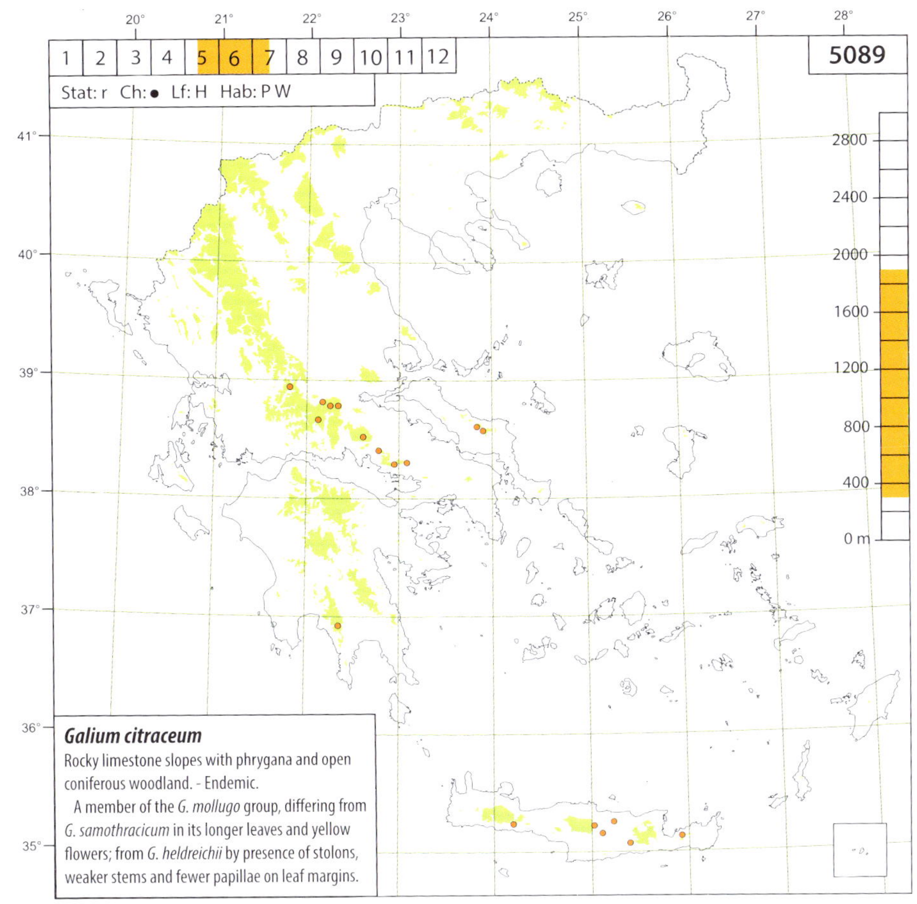

Galium citraceum

Rocky limestone slopes with phrygana and open coniferous woodland. - Endemic.

A member of the *G. mollugo* group, differing from *G. samothracicum* in its longer leaves and yellow flowers; from *G. heldreichii* by presence of stolons, weaker stems and fewer papillae on leaf margins.

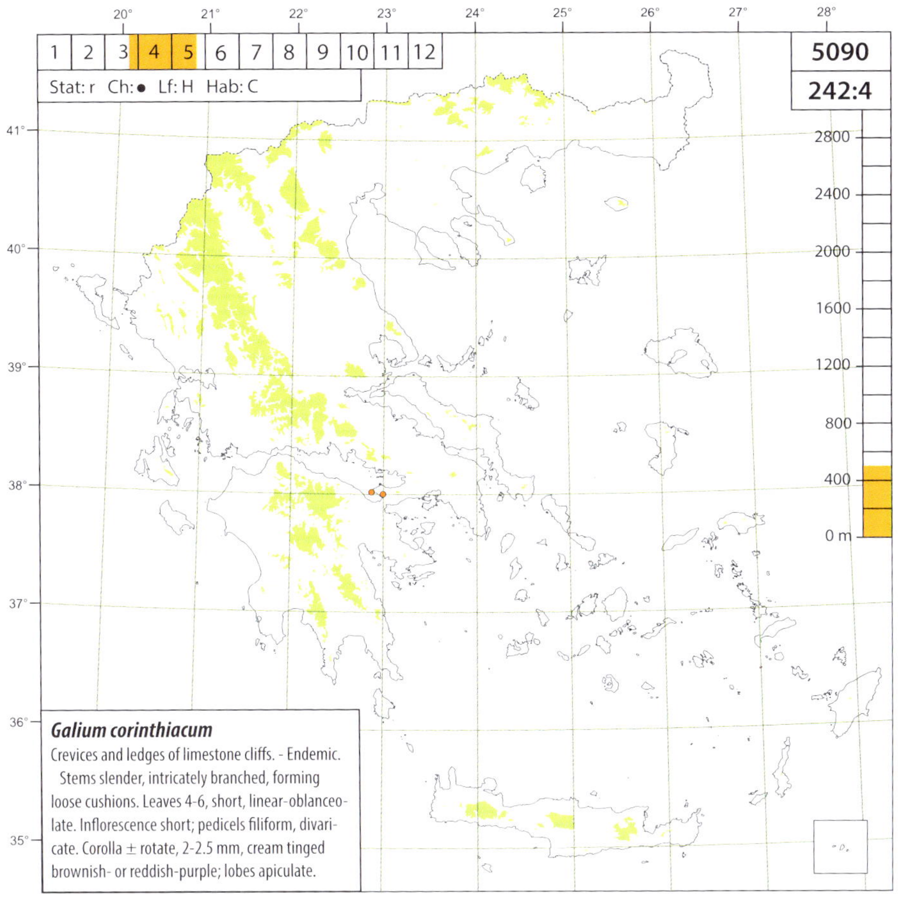

Galium corinthiacum

Crevices and ledges of limestone cliffs. - Endemic.

Stems slender, intricately branched, forming loose cushions. Leaves 4-6, short, linear-oblanceolate. Inflorescence short; pedicels filiform, divaricate. Corolla ± rotate, 2-2.5 mm, cream tinged brownish- or reddish-purple; lobes apiculate.

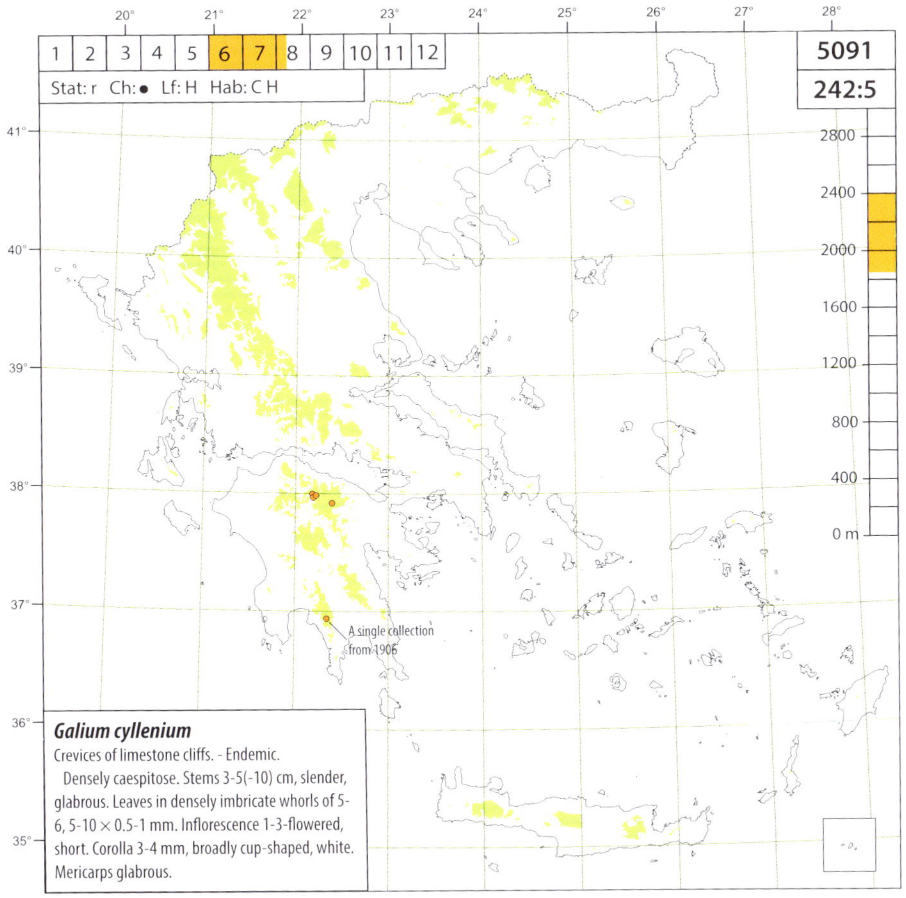

Galium cyllenium

Crevices of limestone cliffs. - Endemic.

Densely caespitose. Stems 3-5(-10) cm, slender, glabrous. Leaves in densely imbricate whorls of 5-6, 5-10 × 0.5-1 mm. Inflorescence 1-3-flowered, short. Corolla 3-4 mm, broadly cup-shaped, white. Mericarps glabrous.

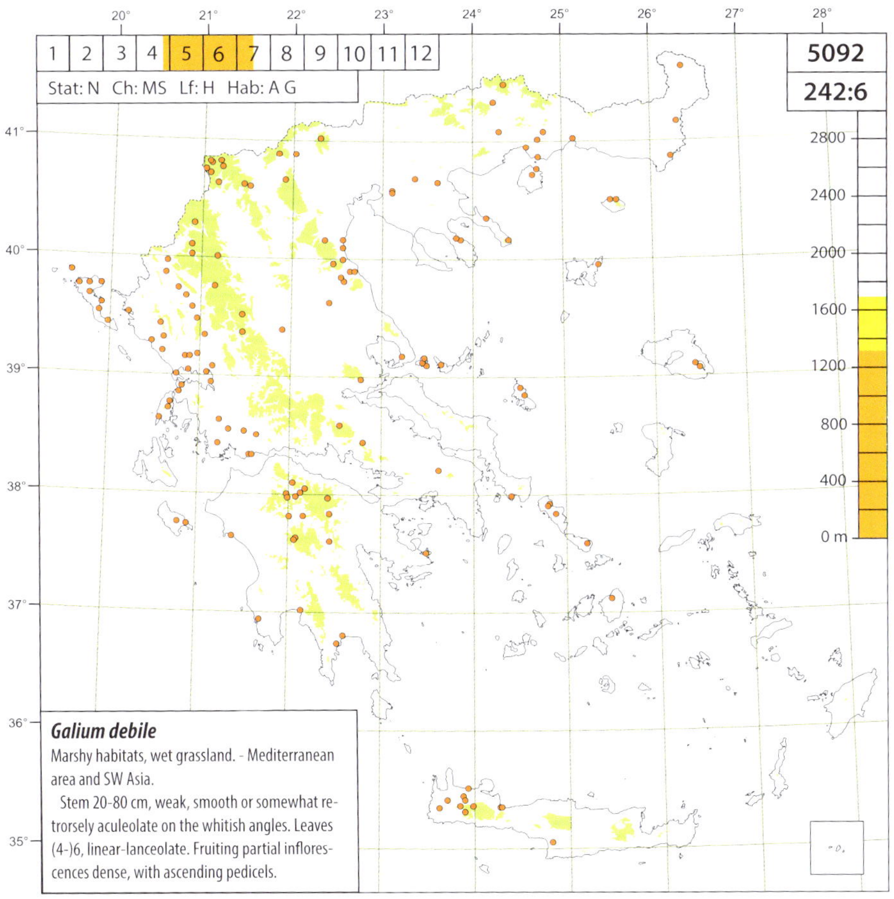

Galium debile

Marshy habitats, wet grassland. - Mediterranean area and SW Asia.

Stem 20-80 cm, weak, smooth or somewhat retrorsely aculeolate on the whitish angles. Leaves (4-)6, linear-lanceolate. Fruiting partial inflorescences dense, with ascending pedicels.

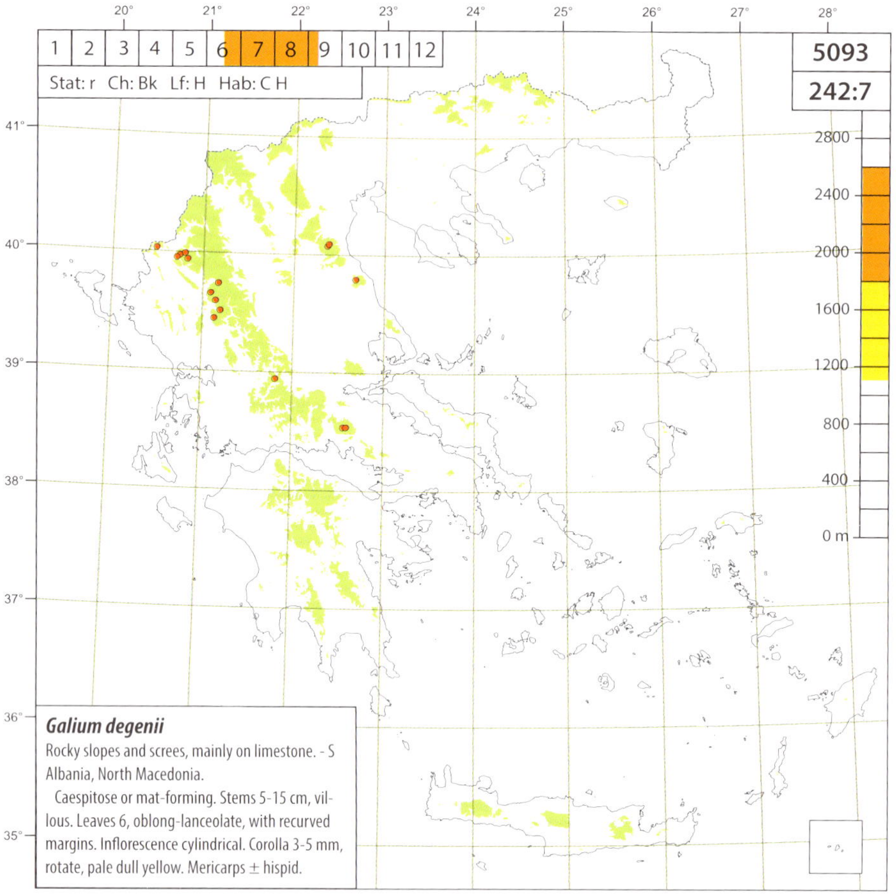

Galium degenii

Rocky slopes and screes, mainly on limestone. - S Albania, North Macedonia.

Caespitose or mat-forming. Stems 5-15 cm, villous. Leaves 6, oblong-lanceolate, with recurved margins. Inflorescence cylindrical. Corolla 3-5 mm, rotate, pale dull yellow. Mericarps ± hispid.

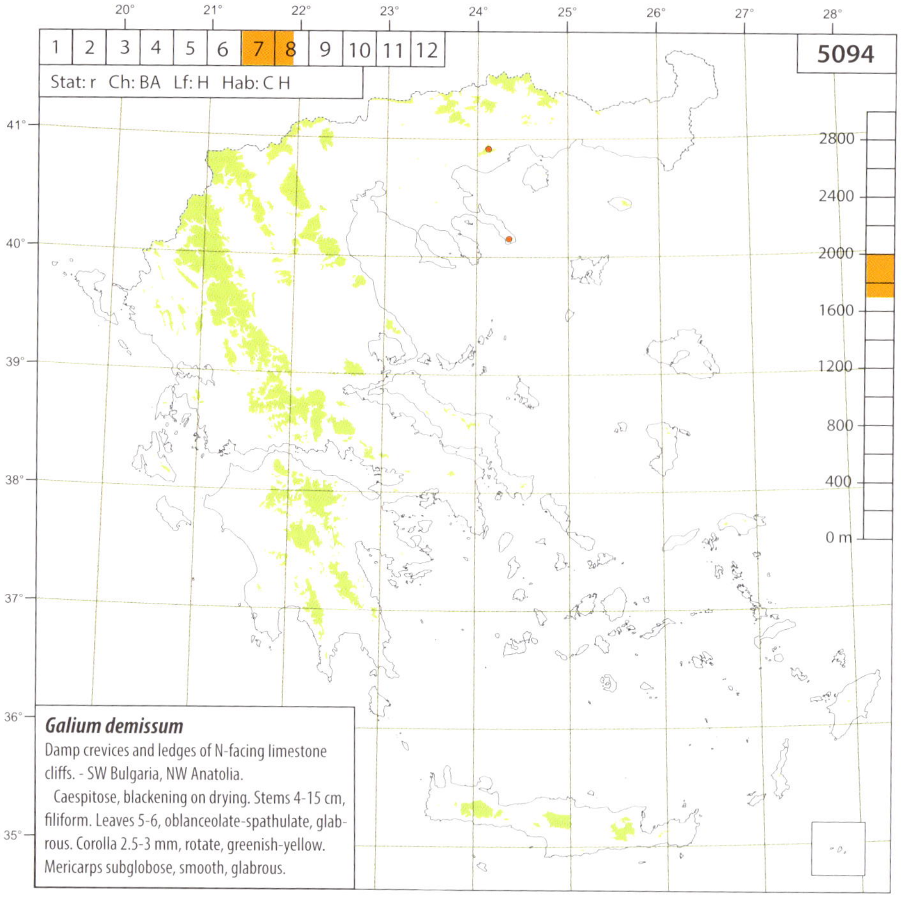

Galium demissum

Damp crevices and ledges of N-facing limestone cliffs. - SW Bulgaria, NW Anatolia.

Caespitose, blackening on drying. Stems 4-15 cm, filiform. Leaves 5-6, oblanceolate-spathulate, glabrous. Corolla 2.5-3 mm, rotate, greenish-yellow. Mericarps subglobose, smooth, glabrous.

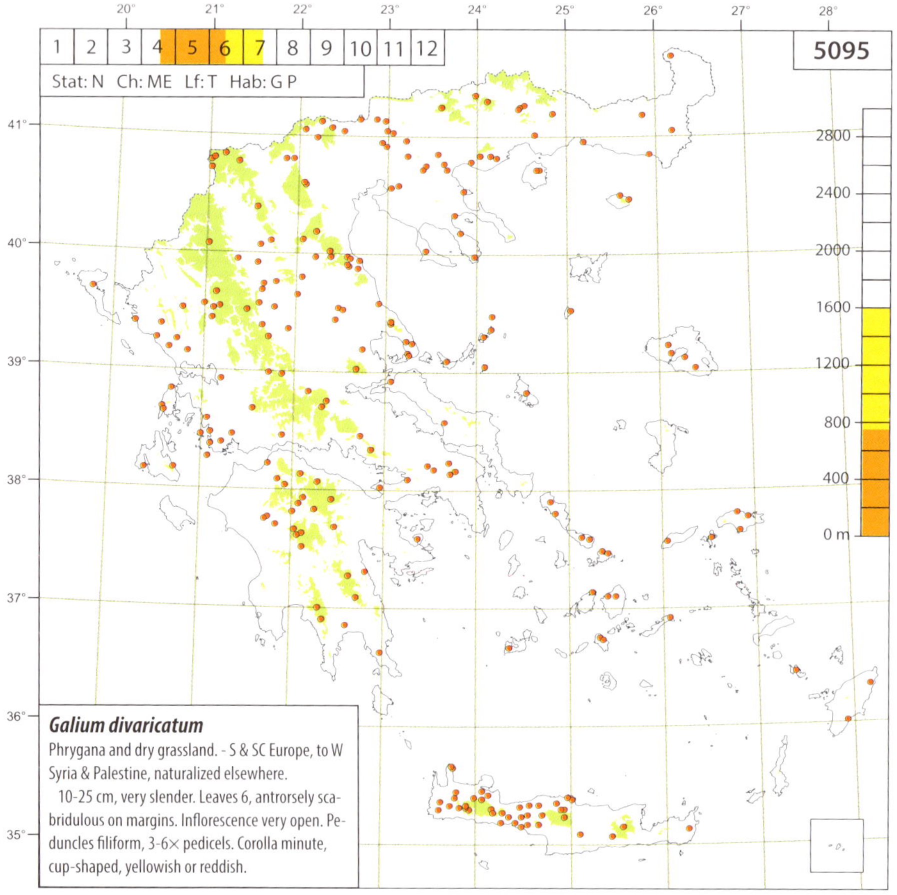

Galium divaricatum

Phrygana and dry grassland. - S & SC Europe, to W Syria & Palestine, naturalized elsewhere.

10-25 cm, very slender. Leaves 6, antrorsely scabridulous on margins. Inflorescence very open. Peduncles filiform, 3-6× pedicels. Corolla minute, cup-shaped, yellowish or reddish.

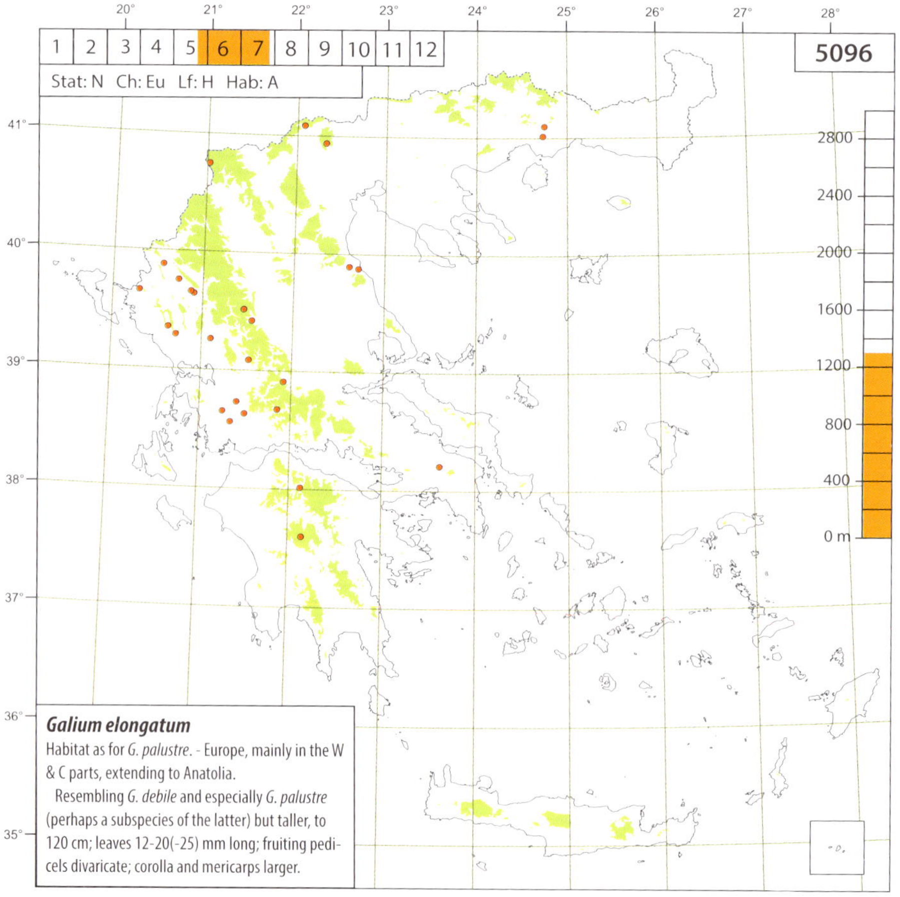

Galium elongatum

Habitat as for *G. palustre*. - Europe, mainly in the W & C parts, extending to Anatolia.

Resembling *G. debile* and especially *G. palustre* (perhaps a subspecies of the latter) but taller, to 120 cm; leaves 12-20(-25) mm long; fruiting pedicels divaricate; corolla and mericarps larger.

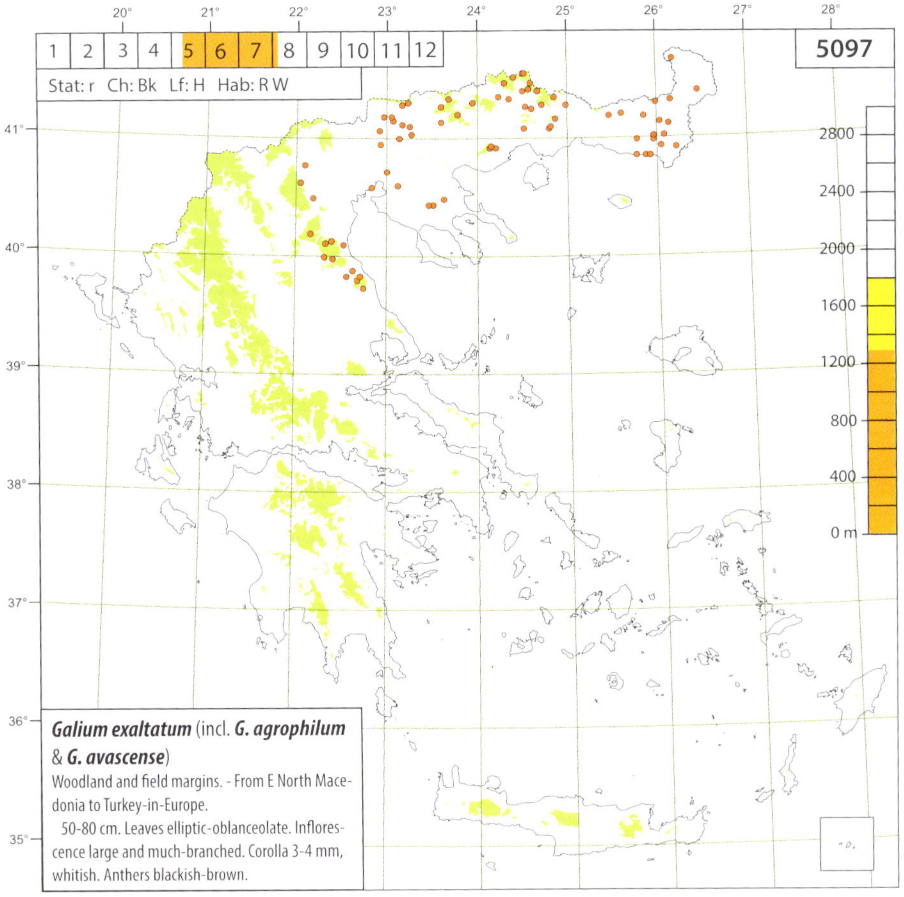

Galium exaltatum (incl. ***G. agrophilum*** & ***G. avascense***)
Woodland and field margins. - From E North Macedonia to Turkey-in-Europe.
50-80 cm. Leaves elliptic-oblanceolate. Inflorescence large and much-branched. Corolla 3-4 mm, whitish. Anthers blackish-brown.

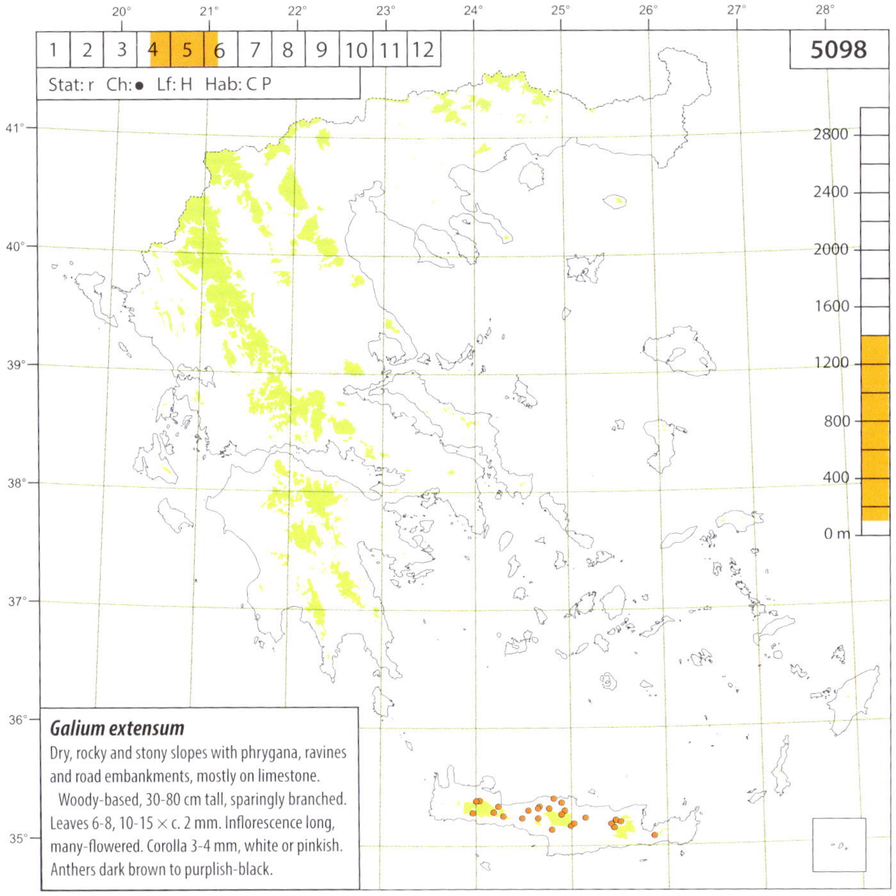

Galium extensum
Dry, rocky and stony slopes with phrygana, ravines and road embankments, mostly on limestone.
Woody-based, 30-80 cm tall, sparingly branched. Leaves 6-8, 10-15 × c. 2 mm. Inflorescence long, many-flowered. Corolla 3-4 mm, white or pinkish. Anthers dark brown to purplish-black.

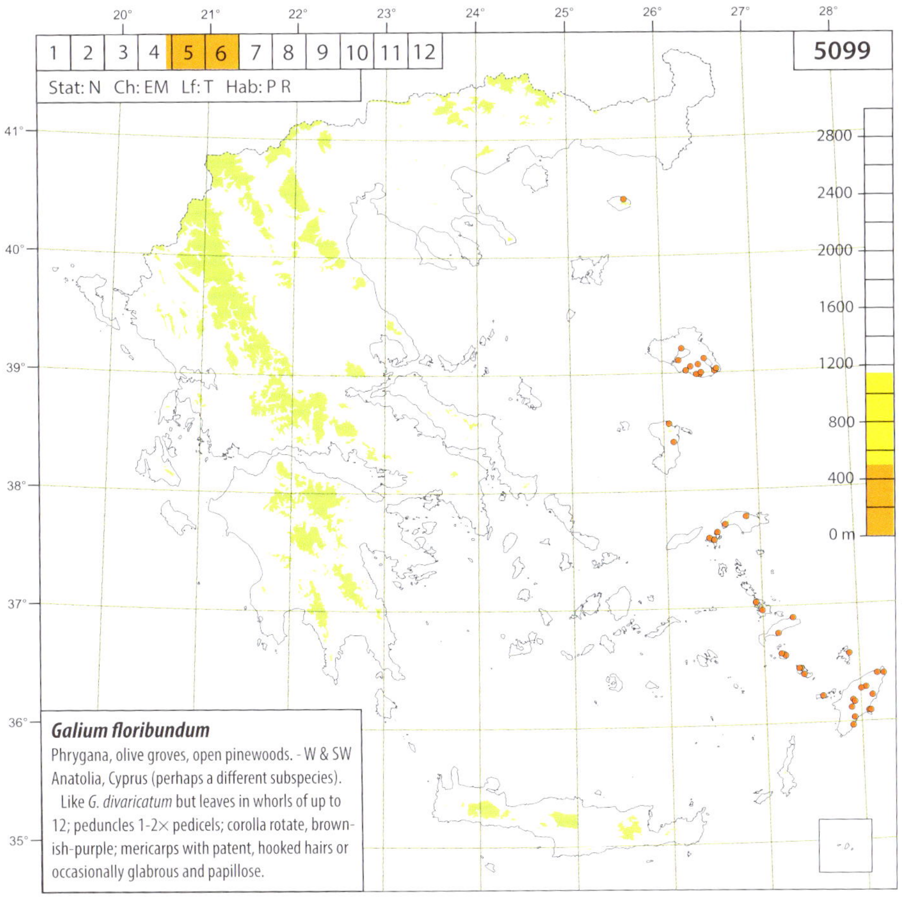

Galium floribundum
Phrygana, olive groves, open pinewoods. - W & SW Anatolia, Cyprus (perhaps a different subspecies).
Like *G. divaricatum* but leaves in whorls of up to 12; peduncles 1-2× pedicels; corolla rotate, brownish-purple; mericarps with patent, hooked hairs or occasionally glabrous and papillose.

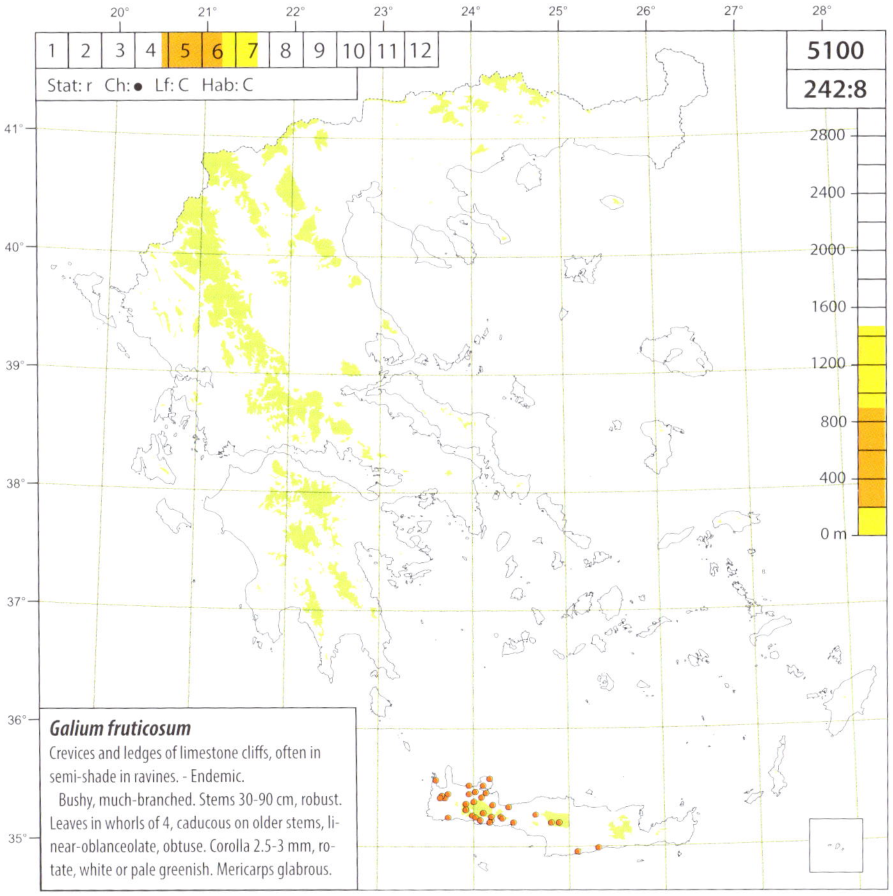

Galium fruticosum
Crevices and ledges of limestone cliffs, often in semi-shade in ravines. - Endemic.
Bushy, much-branched. Stems 30-90 cm, robust. Leaves in whorls of 4, caducous on older stems, linear-oblanceolate, obtuse. Corolla 2.5-3 mm, rotate, white or pale greenish. Mericarps glabrous.

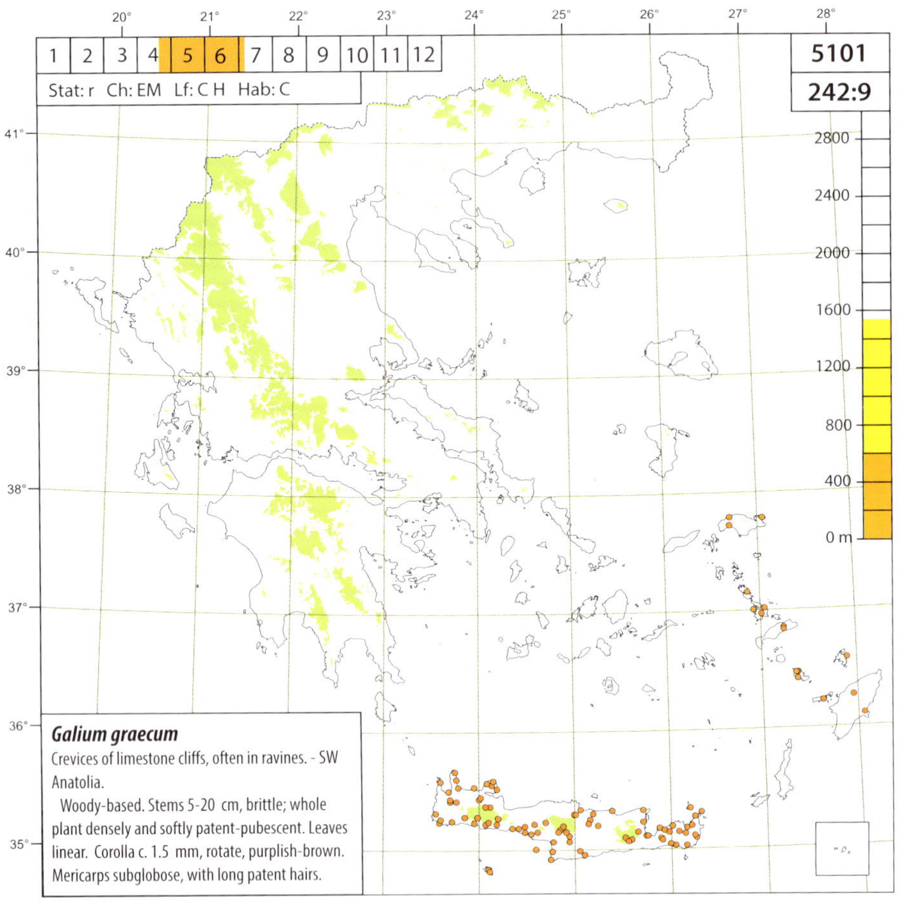

Galium graecum
Crevices of limestone cliffs, often in ravines. - SW Anatolia.
Woody-based. Stems 5-20 cm, brittle; whole plant densely and softly patent-pubescent. Leaves linear. Corolla c. 1.5 mm, rotate, purplish-brown. Mericarps subglobose, with long patent hairs.

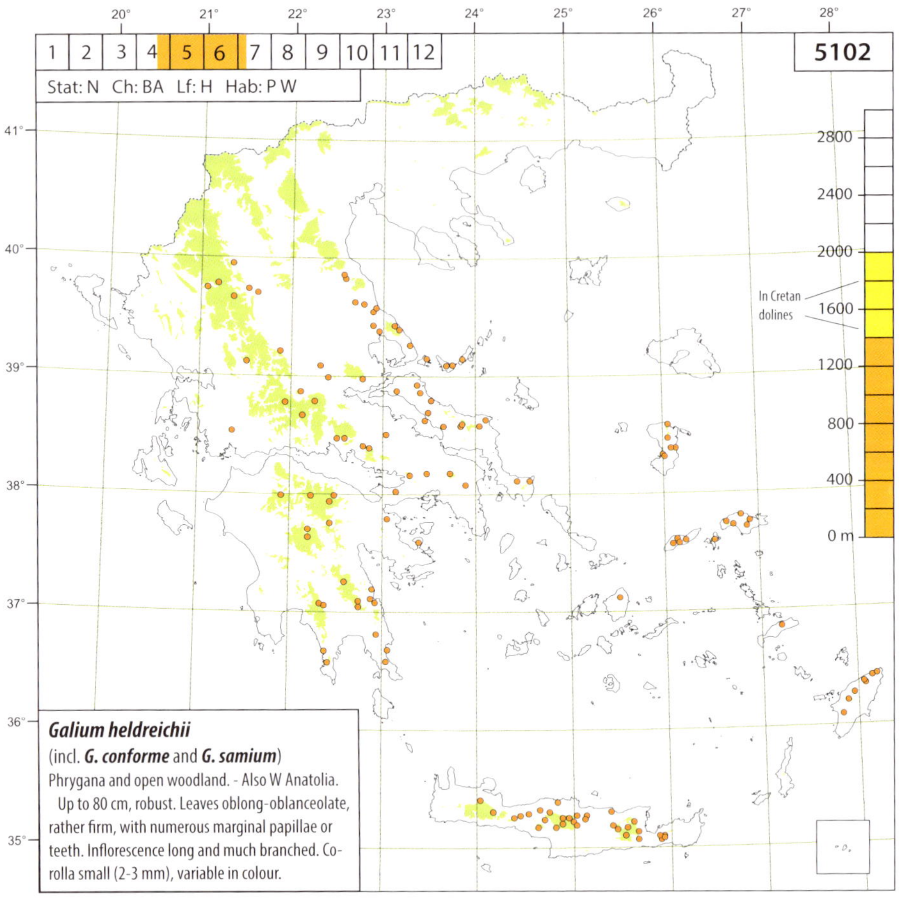

Galium heldreichii
(incl. ***G. conforme*** and ***G. samium***)
Phrygana and open woodland. - Also W Anatolia.
Up to 80 cm, robust. Leaves oblong-oblanceolate, rather firm, with numerous marginal papillae or teeth. Inflorescence long and much branched. Corolla small (2-3 mm), variable in colour.

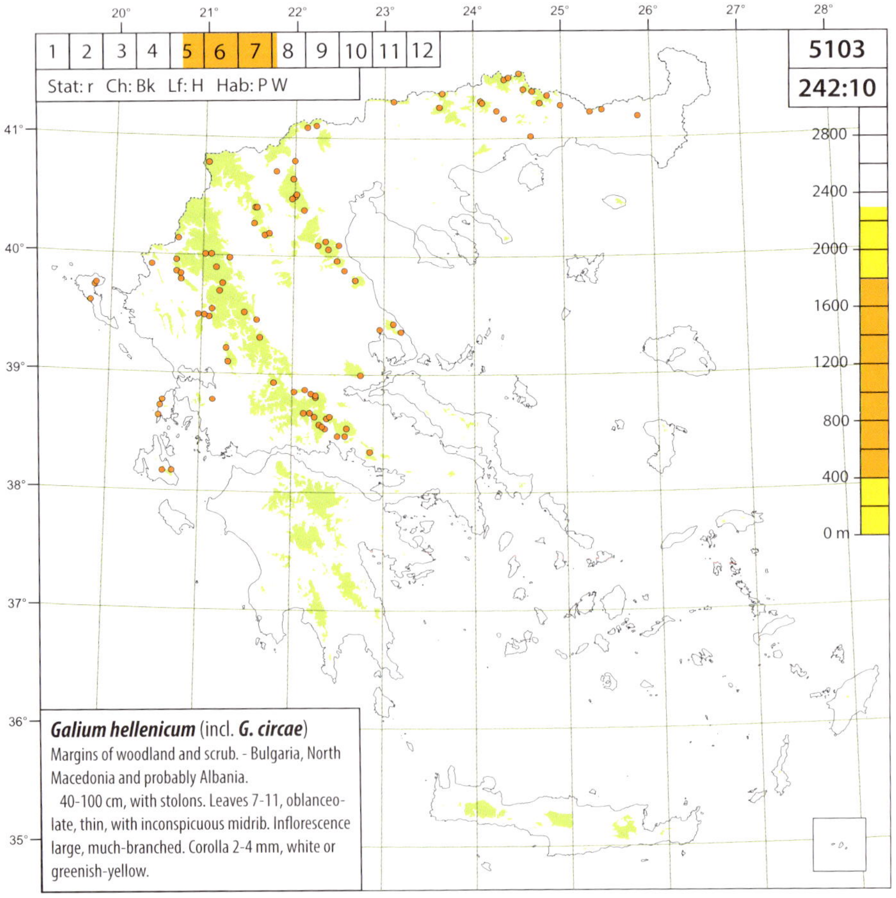

Galium hellenicum (incl. ***G. circae***)
Margins of woodland and scrub. - Bulgaria, North Macedonia and probably Albania.
40-100 cm, with stolons. Leaves 7-11, oblanceolate, thin, with inconspicuous midrib. Inflorescence large, much-branched. Corolla 2-4 mm, white or greenish-yellow.

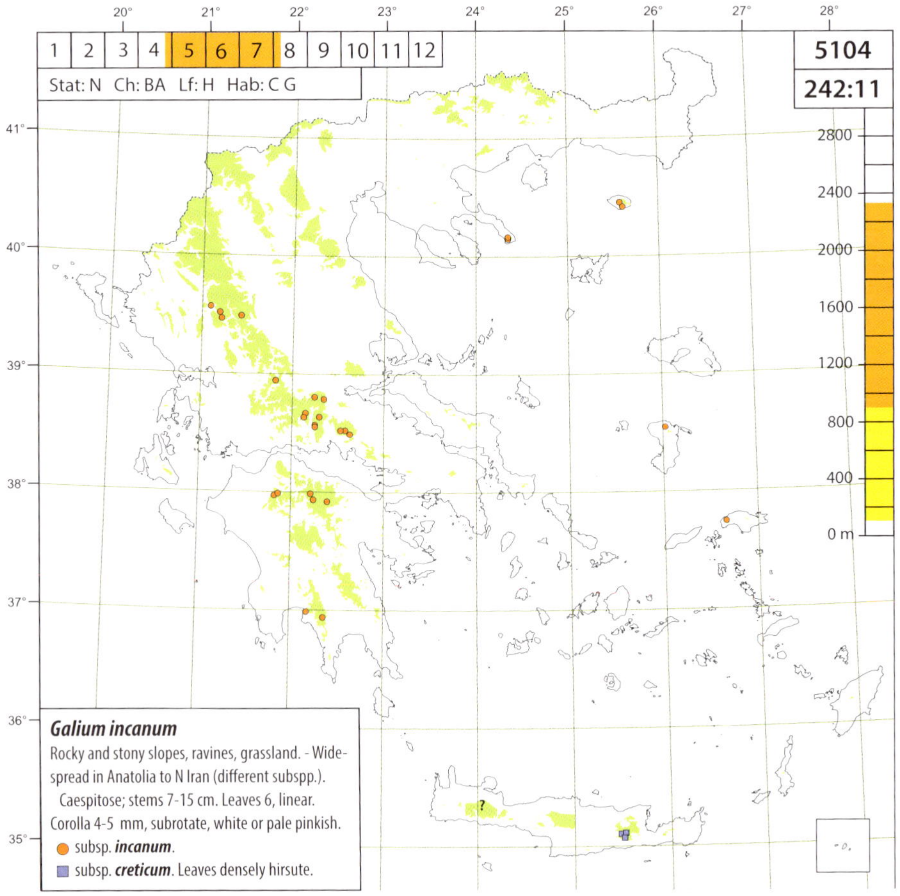

Galium incanum
Rocky and stony slopes, ravines, grassland. - Widespread in Anatolia to N Iran (different subspp.).
Caespitose; stems 7-15 cm. Leaves 6, linear. Corolla 4-5 mm, subrotate, white or pale pinkish.
● subsp. ***incanum***.
■ subsp. ***creticum***. Leaves densely hirsute.

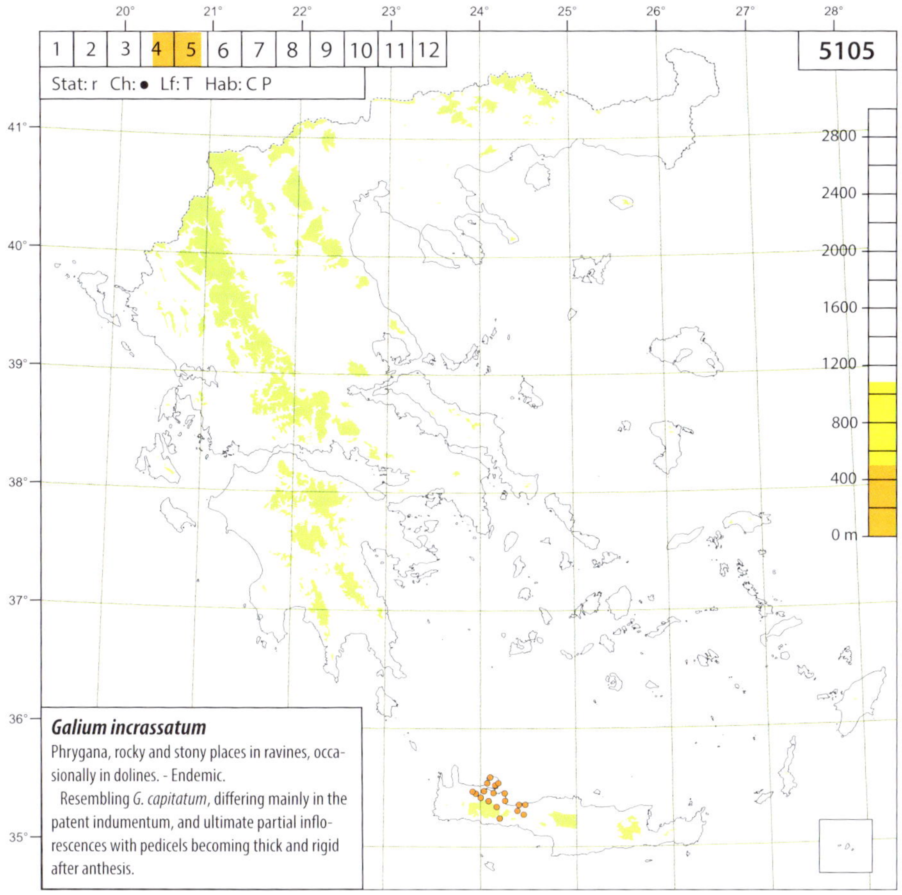

Galium incrassatum
Phrygana, rocky and stony places in ravines, occasionally in dolines. - Endemic.
Resembling *G. capitatum*, differing mainly in the patent indumentum, and ultimate partial inflorescences with pedicels becoming thick and rigid after anthesis.

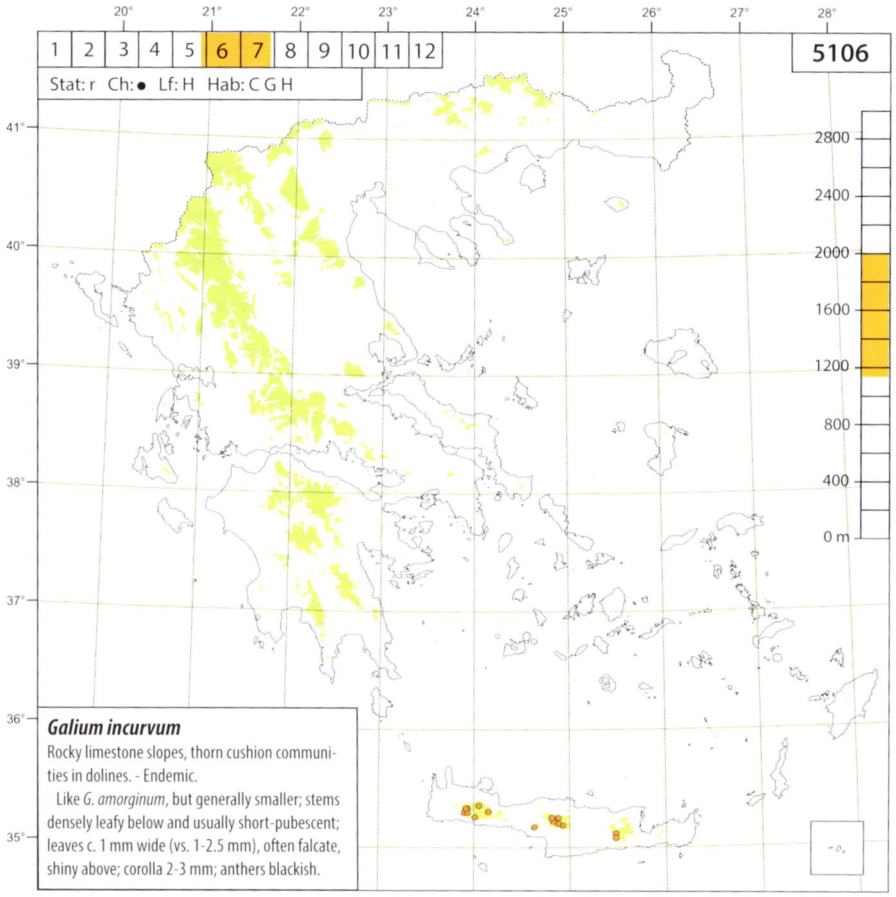

Galium incurvum
Rocky limestone slopes, thorn cushion communities in dolines. - Endemic.
Like *G. amorginum*, but generally smaller; stems densely leafy below and usually short-pubescent; leaves c. 1 mm wide (vs. 1-2.5 mm), often falcate, shiny above; corolla 2-3 mm; anthers blackish.

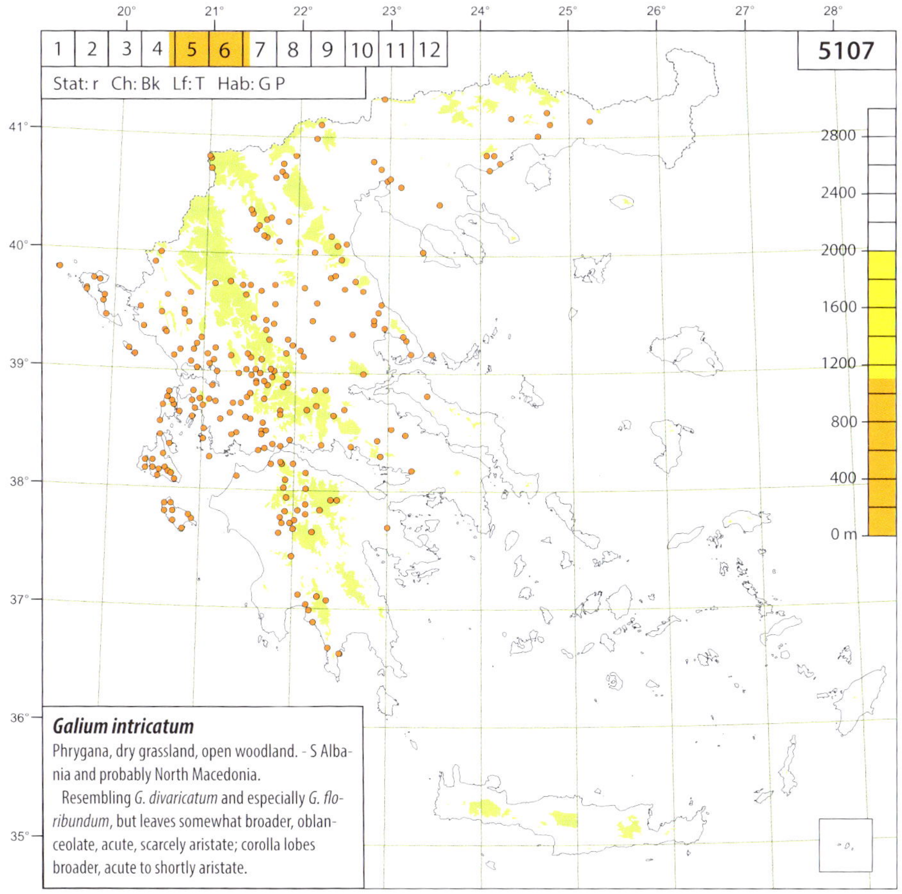

Galium intricatum
Phrygana, dry grassland, open woodland. - S Albania and probably North Macedonia.
Resembling *G. divaricatum* and especially *G. floribundum*, but leaves somewhat broader, oblanceolate, acute, scarcely aristate; corolla lobes broader, acute to shortly aristate.

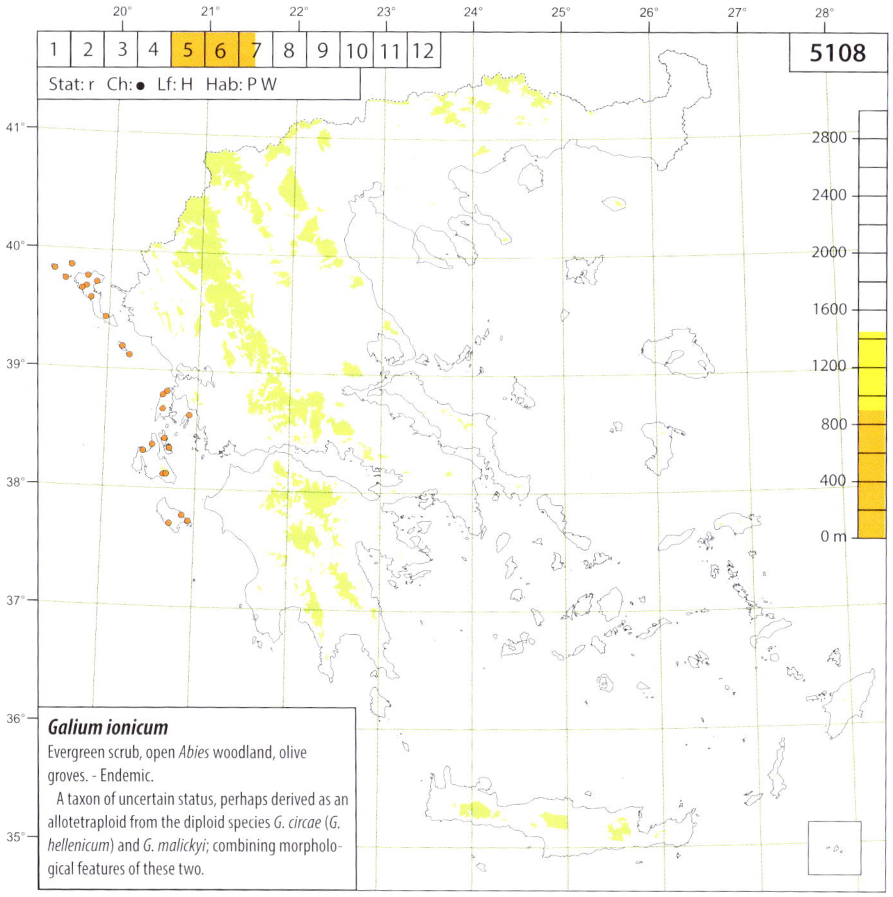

Galium ionicum
Evergreen scrub, open *Abies* woodland, olive groves. - Endemic.
A taxon of uncertain status, perhaps derived as an allotetraploid from the diploid species *G. circae* (*G. hellenicum*) and *G. malickyi*; combining morphological features of these two.

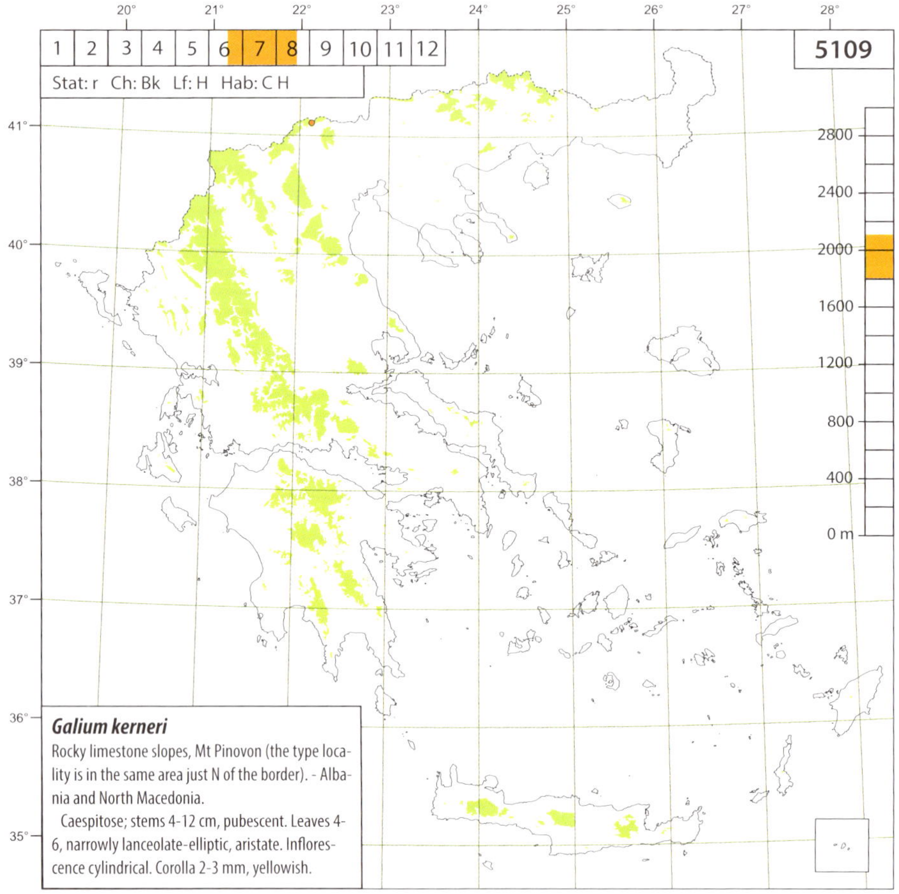

Galium kerneri
Rocky limestone slopes, Mt Pinovon (the type locality is in the same area just N of the border). - Albania and North Macedonia.
Caespitose; stems 4-12 cm, pubescent. Leaves 4-6, narrowly lanceolate-elliptic, aristate. Inflorescence cylindrical. Corolla 2-3 mm, yellowish.

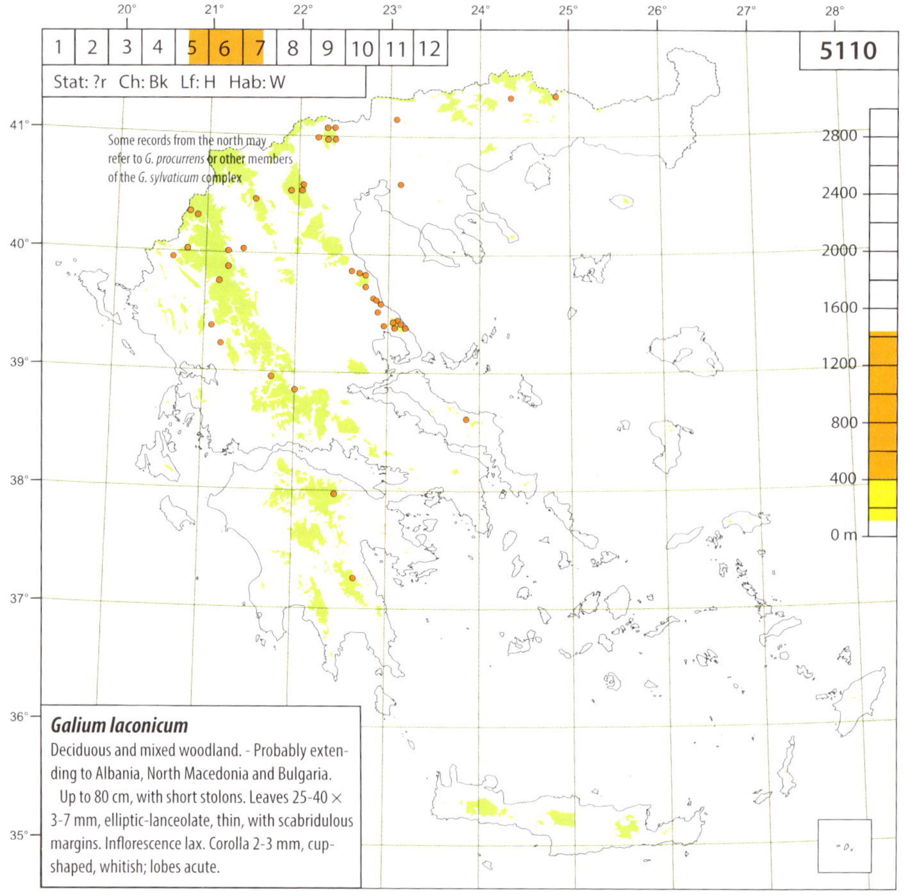

Galium laconicum
Deciduous and mixed woodland. - Probably extending to Albania, North Macedonia and Bulgaria.
Up to 80 cm, with short stolons. Leaves 25-40 × 3-7 mm, elliptic-lanceolate, thin, with scabridulous margins. Inflorescence lax. Corolla 2-3 mm, cup-shaped, whitish; lobes acute.

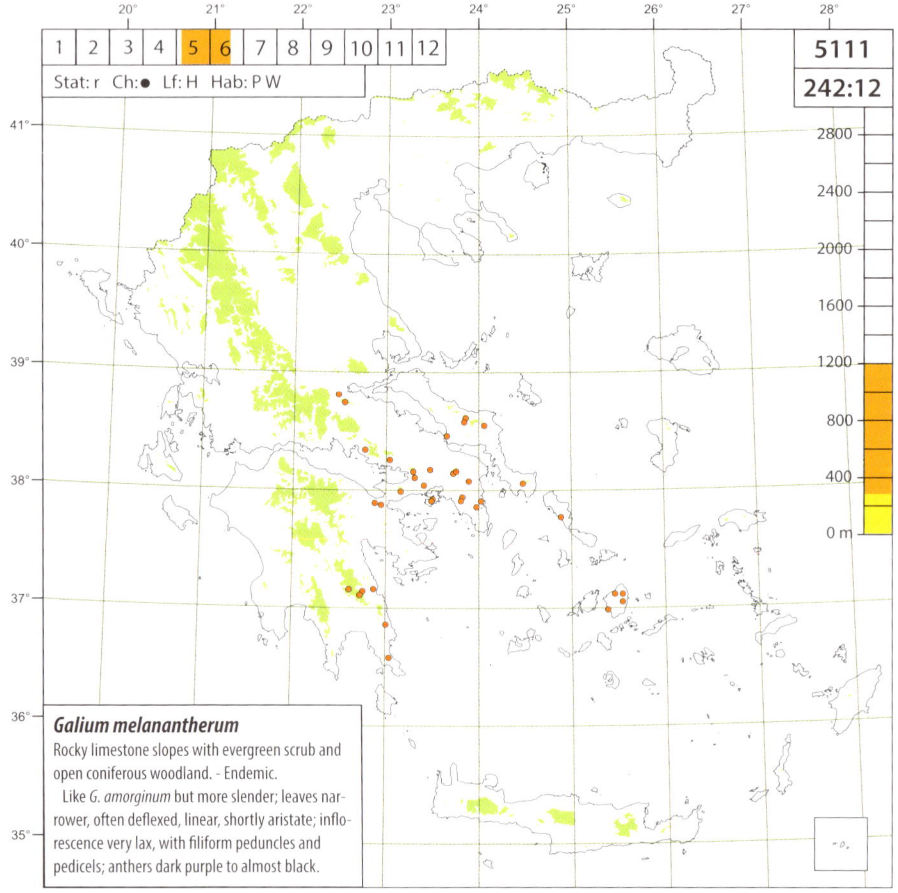

Galium melanantherum
Rocky limestone slopes with evergreen scrub and open coniferous woodland. - Endemic.
Like *G. amorginum* but more slender; leaves narrower, often deflexed, linear, shortly aristate; inflorescence very lax, with filiform peduncles and pedicels; anthers dark purple to almost black.

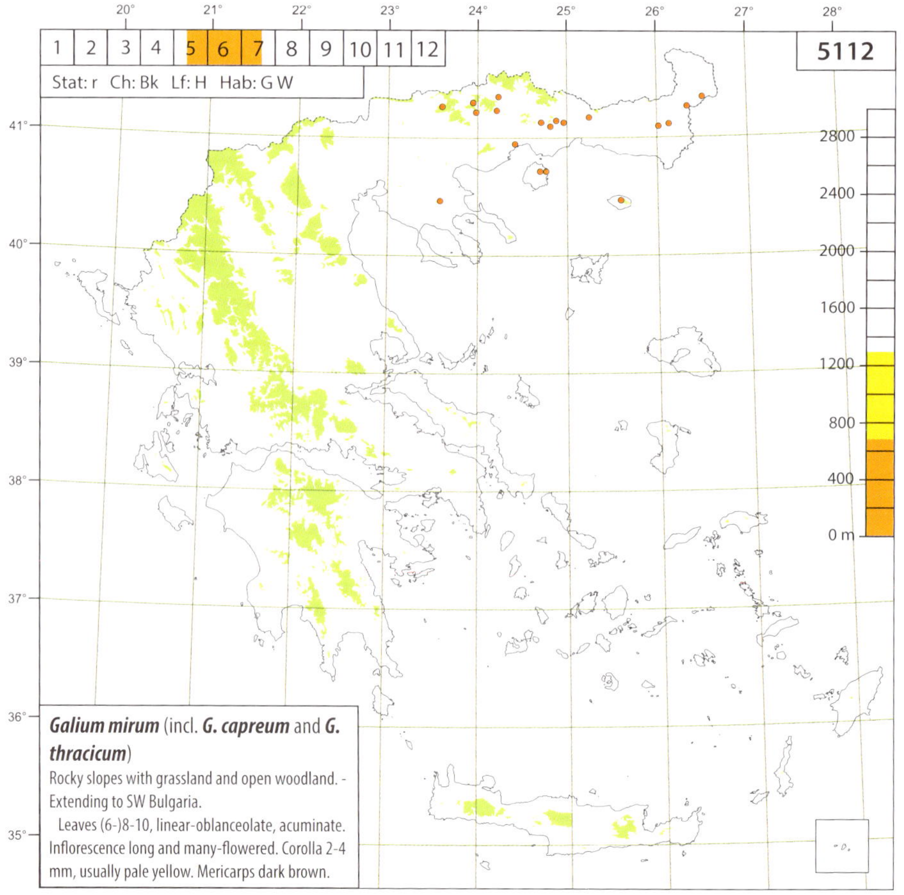

Galium mirum (incl. ***G. capreum*** and ***G. thracicum***)
Rocky slopes with grassland and open woodland. - Extending to SW Bulgaria.
Leaves (6-)8-10, linear-oblanceolate, acuminate. Inflorescence long and many-flowered. Corolla 2-4 mm, usually pale yellow. Mericarps dark brown.

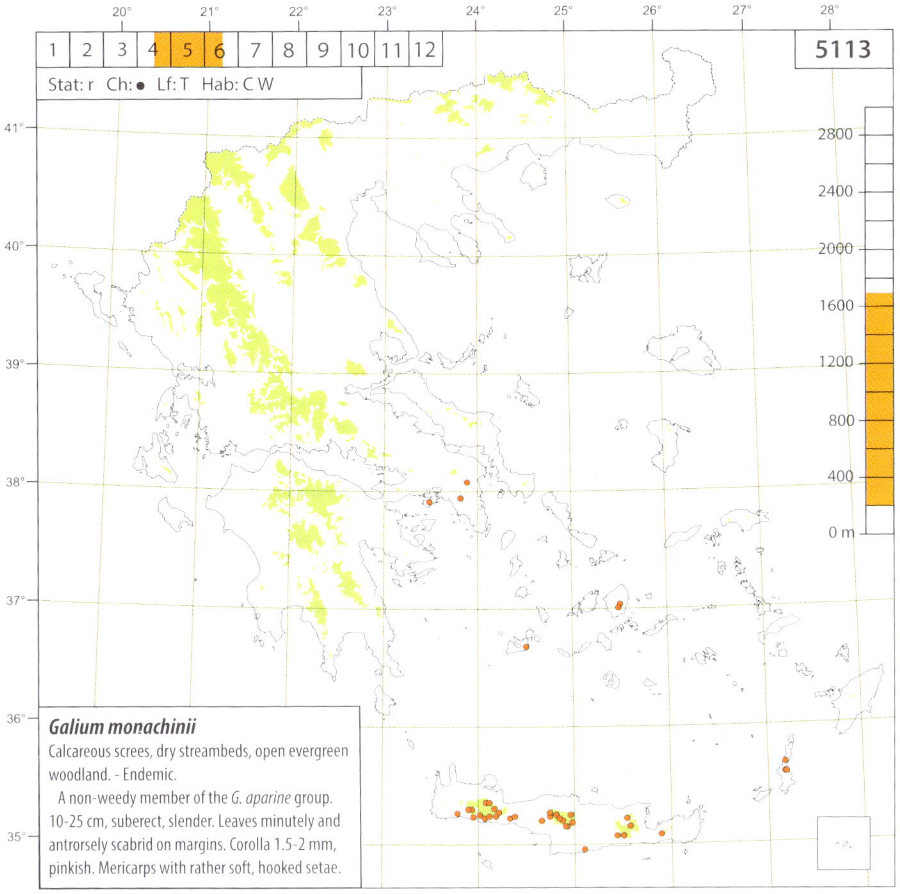

Galium monachinii
Calcareous screes, dry streambeds, open evergreen woodland. - Endemic.
A non-weedy member of the *G. aparine* group. 10-25 cm, suberect, slender. Leaves minutely and antrorsely scabrid on margins. Corolla 1.5-2 mm, pinkish. Mericarps with rather soft, hooked setae.

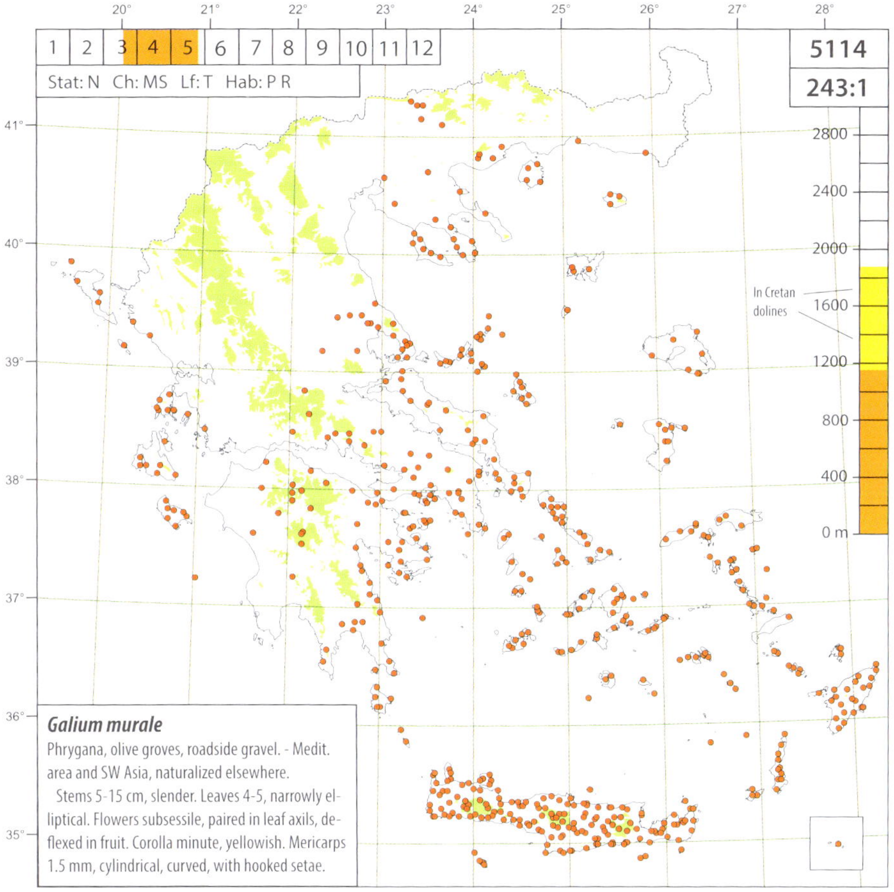

Galium murale
Phrygana, olive groves, roadside gravel. - Medit. area and SW Asia, naturalized elsewhere.
Stems 5-15 cm, slender. Leaves 4-5, narrowly elliptical. Flowers subsessile, paired in leaf axils, deflexed in fruit. Corolla minute, yellowish. Mericarps 1.5 mm, cylindrical, curved, with hooked setae.

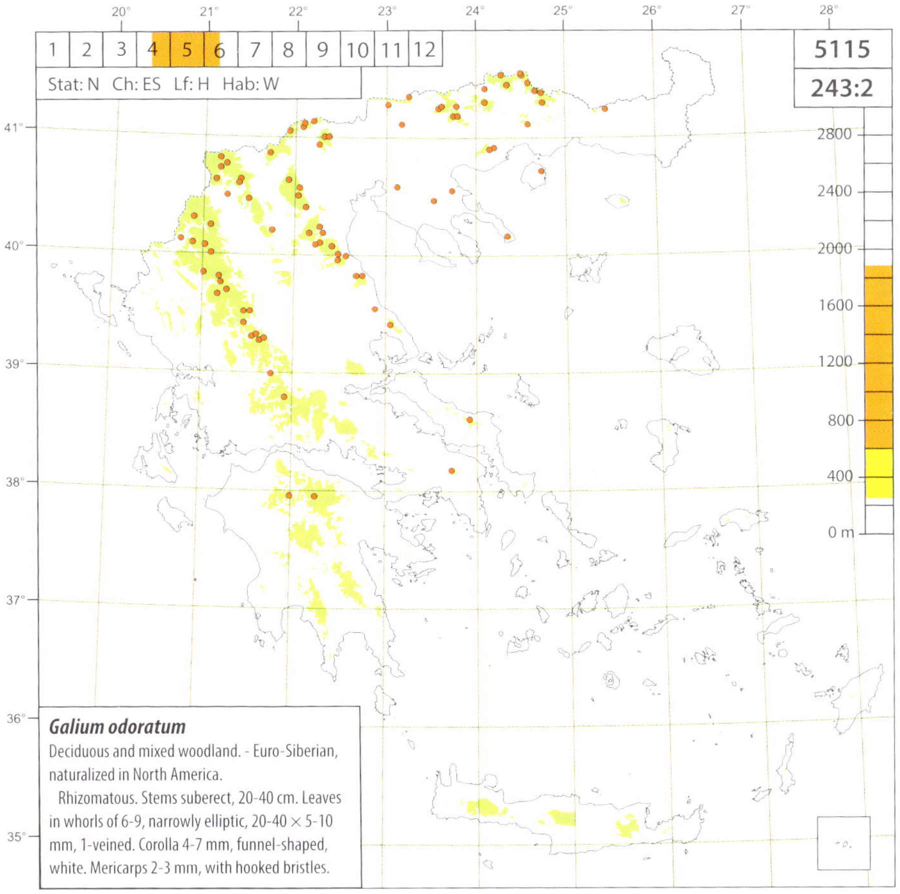

Galium odoratum
Deciduous and mixed woodland. - Euro-Siberian, naturalized in North America.
Rhizomatous. Stems suberect, 20-40 cm. Leaves in whorls of 6-9, narrowly elliptic, 20-40 × 5-10 mm, 1-veined. Corolla 4-7 mm, funnel-shaped, white. Mericarps 2-3 mm, with hooked bristles.

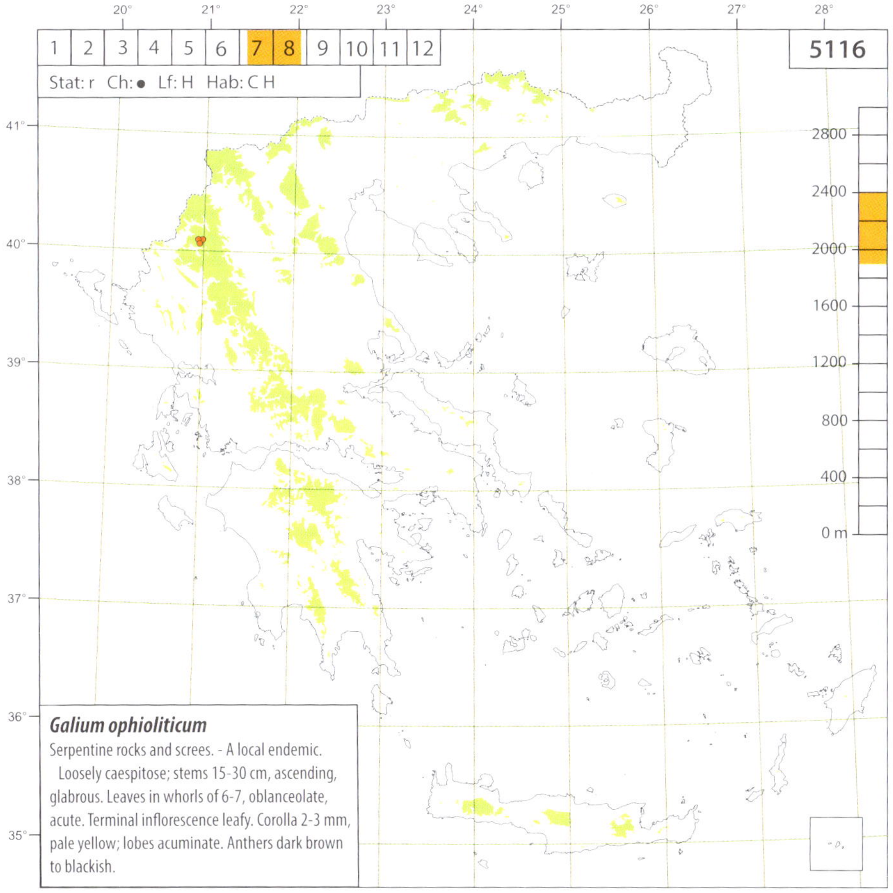

Galium ophioliticum
Serpentine rocks and screes. - A local endemic.
Loosely caespitose; stems 15-30 cm, ascending, glabrous. Leaves in whorls of 6-7, oblanceolate, acute. Terminal inflorescence leafy. Corolla 2-3 mm, pale yellow; lobes acuminate. Anthers dark brown to blackish.

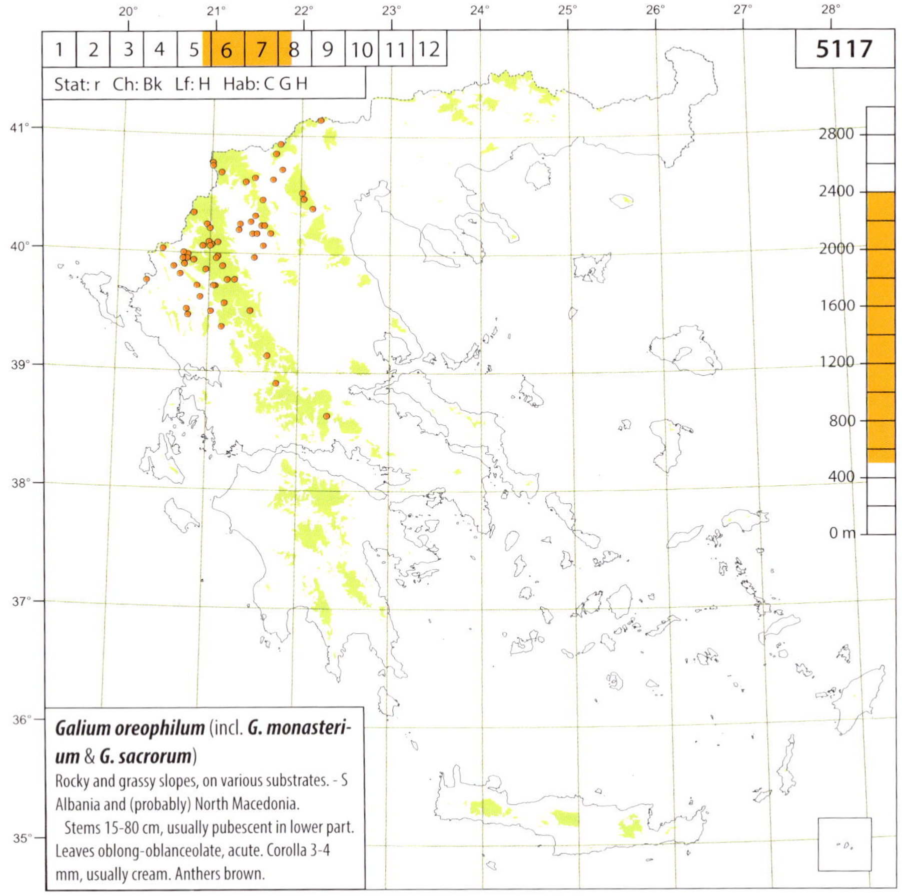

Galium oreophilum (incl. ***G. monasterium*** & ***G. sacrorum***)
Rocky and grassy slopes, on various substrates. - S Albania and (probably) North Macedonia.
Stems 15-80 cm, usually pubescent in lower part. Leaves oblong-oblanceolate, acute. Corolla 3-4 mm, usually cream. Anthers brown.

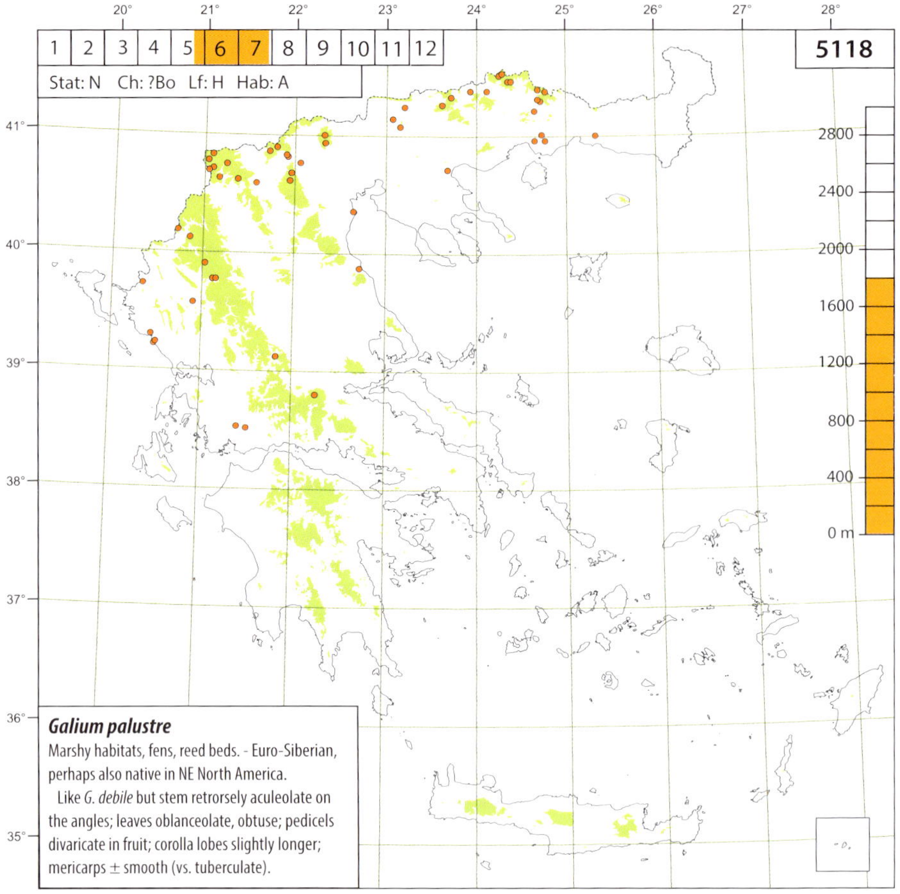

Galium palustre
Marshy habitats, fens, reed beds. - Euro-Siberian, perhaps also native in NE North America.
Like *G. debile* but stem retrorsely aculeolate on the angles; leaves oblanceolate, obtuse; pedicels divaricate in fruit; corolla lobes slightly longer; mericarps ± smooth (vs. tuberculate).

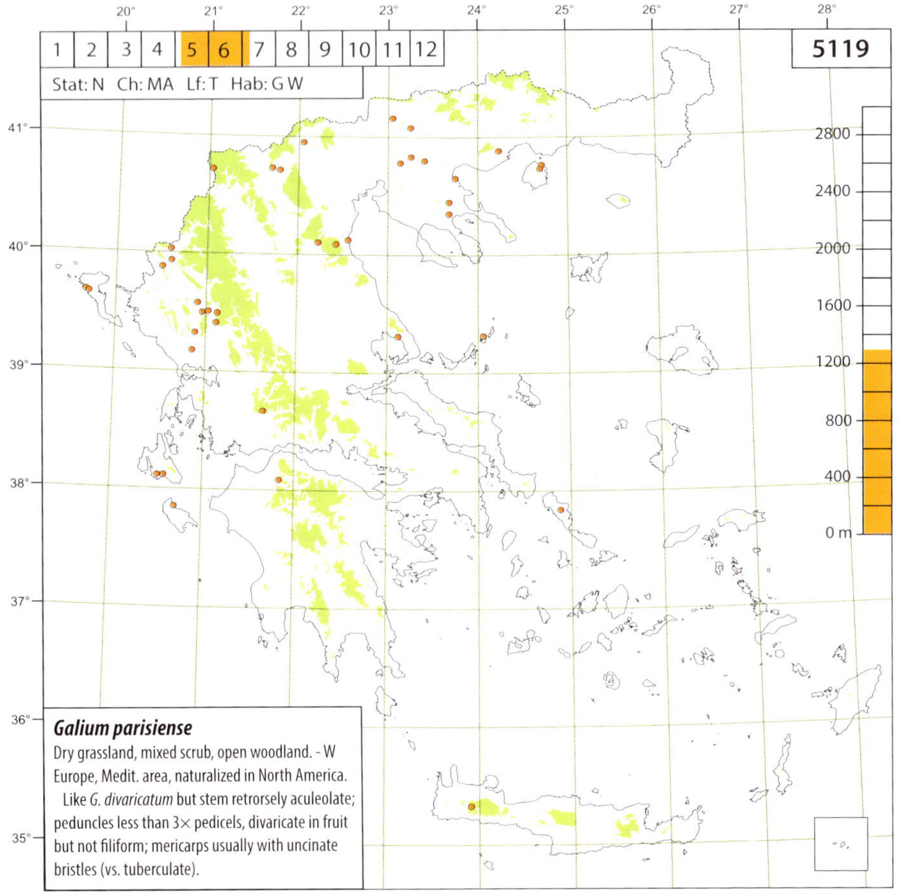

Galium parisiense
Dry grassland, mixed scrub, open woodland. - W Europe, Medit. area, naturalized in North America.
Like *G. divaricatum* but stem retrorsely aculeolate; peduncles less than 3× pedicels, divaricate in fruit but not filiform; mericarps usually with uncinate bristles (vs. tuberculate).

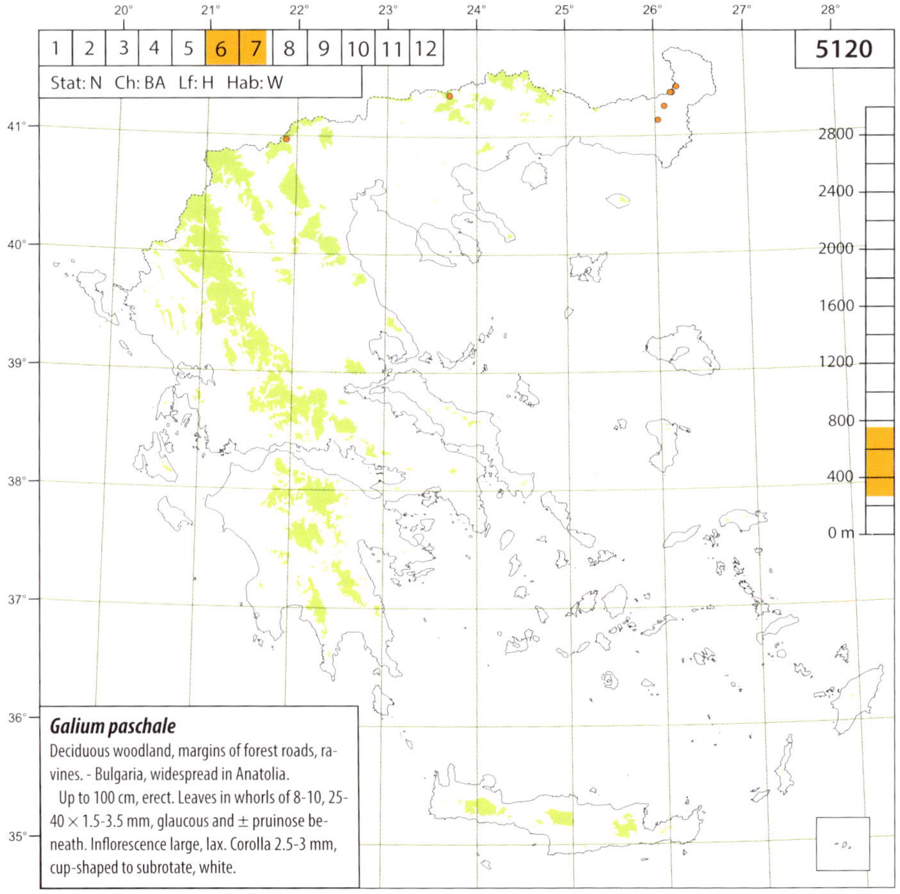

Galium paschale
Deciduous woodland, margins of forest roads, ravines. - Bulgaria, widespread in Anatolia.
Up to 100 cm, erect. Leaves in whorls of 8-10, 25-40 × 1.5-3.5 mm, glaucous and ± pruinose beneath. Inflorescence large, lax. Corolla 2.5-3 mm, cup-shaped to subrotate, white.

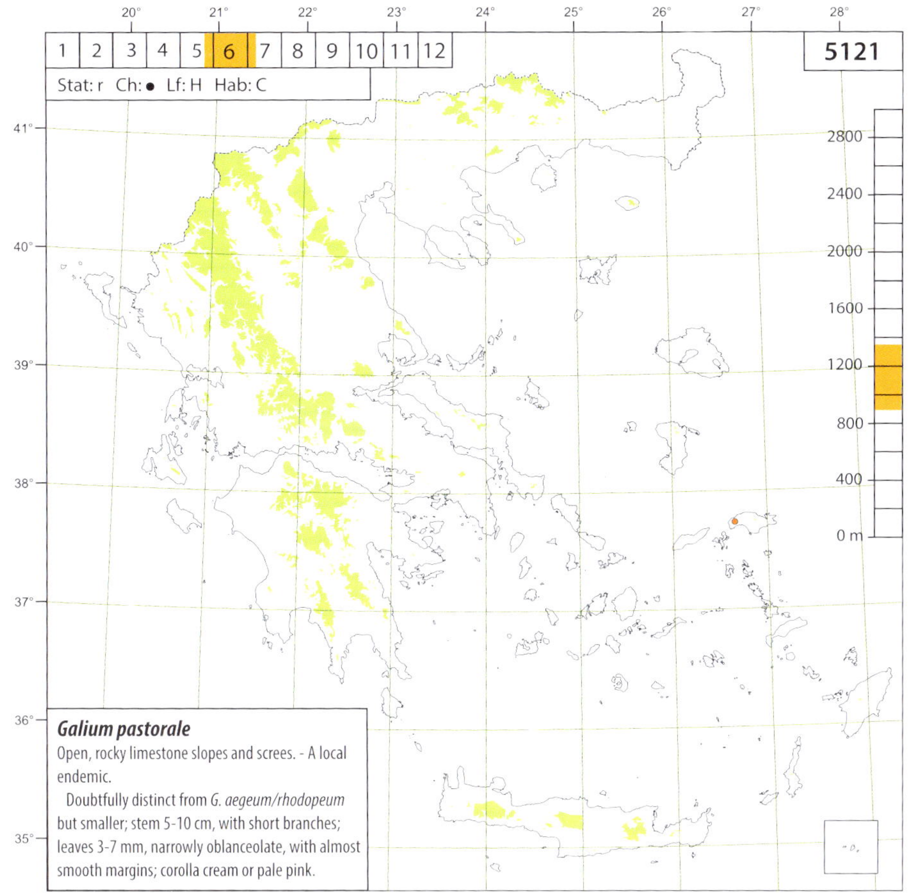

Galium pastorale
Open, rocky limestone slopes and screes. - A local endemic.
Doubtfully distinct from *G. aegeum/rhodopeum* but smaller; stem 5-10 cm, with short branches; leaves 3-7 mm, narrowly oblanceolate, with almost smooth margins; corolla cream or pale pink.

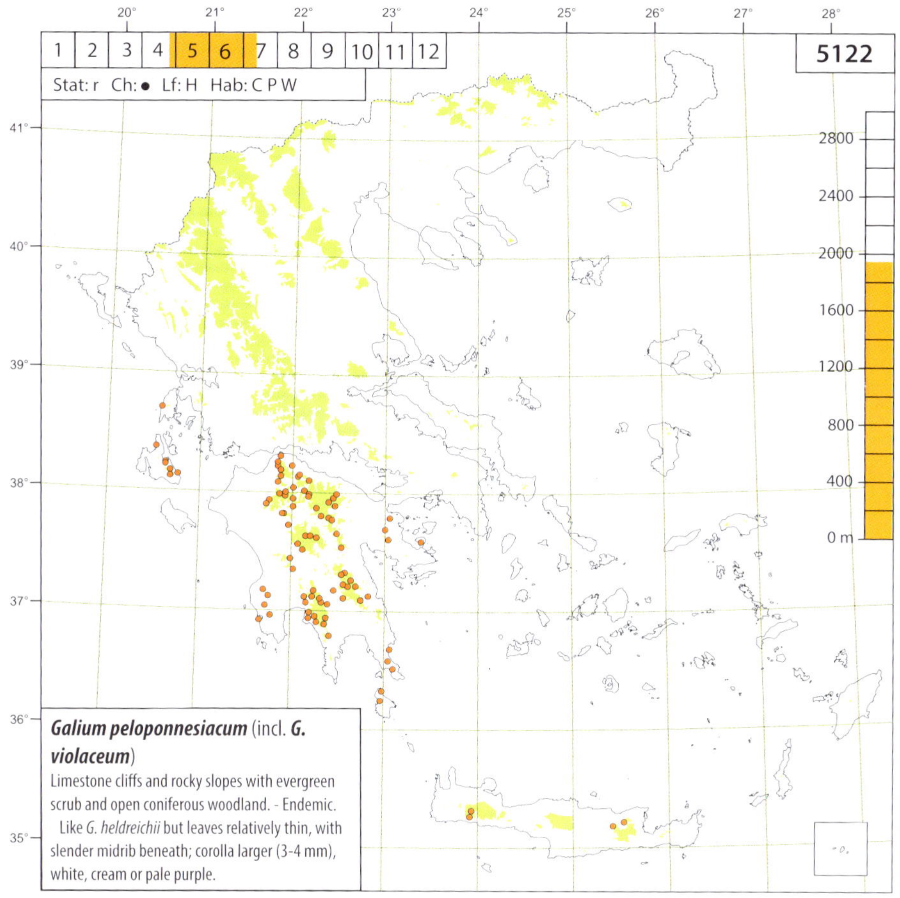

Galium peloponnesiacum (incl. ***G. violaceum***)
Limestone cliffs and rocky slopes with evergreen scrub and open coniferous woodland. - Endemic.
Like *G. heldreichii* but leaves relatively thin, with slender midrib beneath; corolla larger (3-4 mm), white, cream or pale purple.

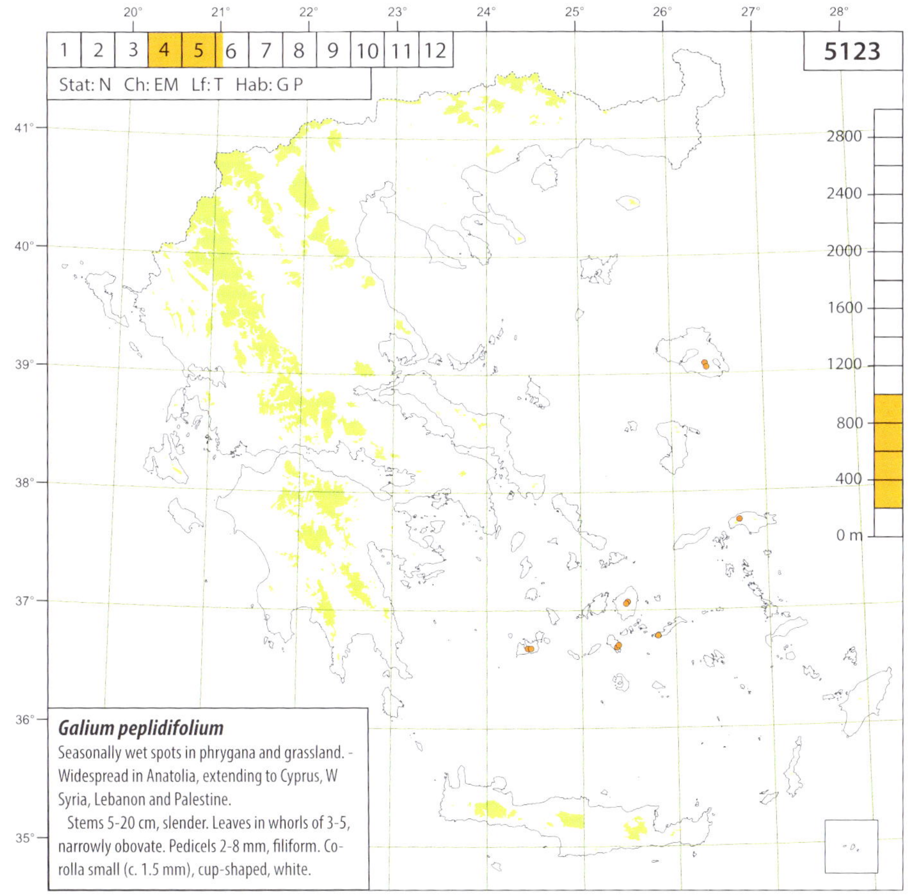

Galium peplidifolium
Seasonally wet spots in phrygana and grassland. - Widespread in Anatolia, extending to Cyprus, W Syria, Lebanon and Palestine.
Stems 5-20 cm, slender. Leaves in whorls of 3-5, narrowly obovate. Pedicels 2-8 mm, filiform. Corolla small (c. 1.5 mm), cup-shaped, white.

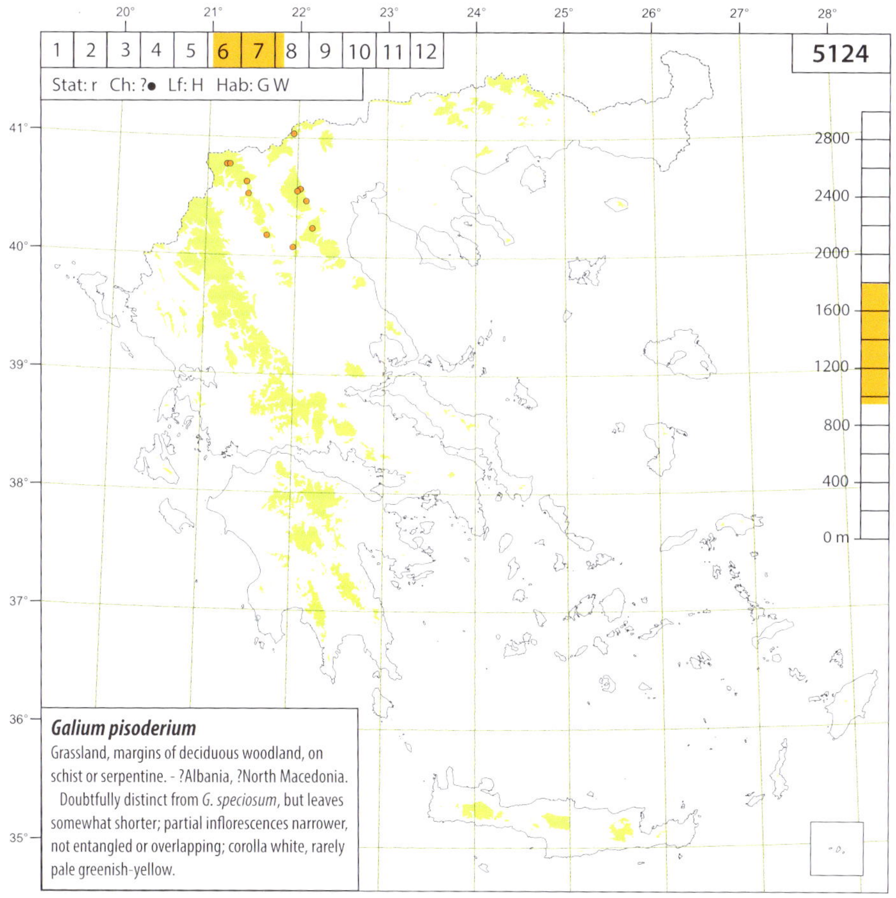

Galium pisoderium
Grassland, margins of deciduous woodland, on schist or serpentine. - ?Albania, ?North Macedonia.
Doubtfully distinct from *G. speciosum*, but leaves somewhat shorter; partial inflorescences narrower, not entangled or overlapping; corolla white, rarely pale greenish-yellow.

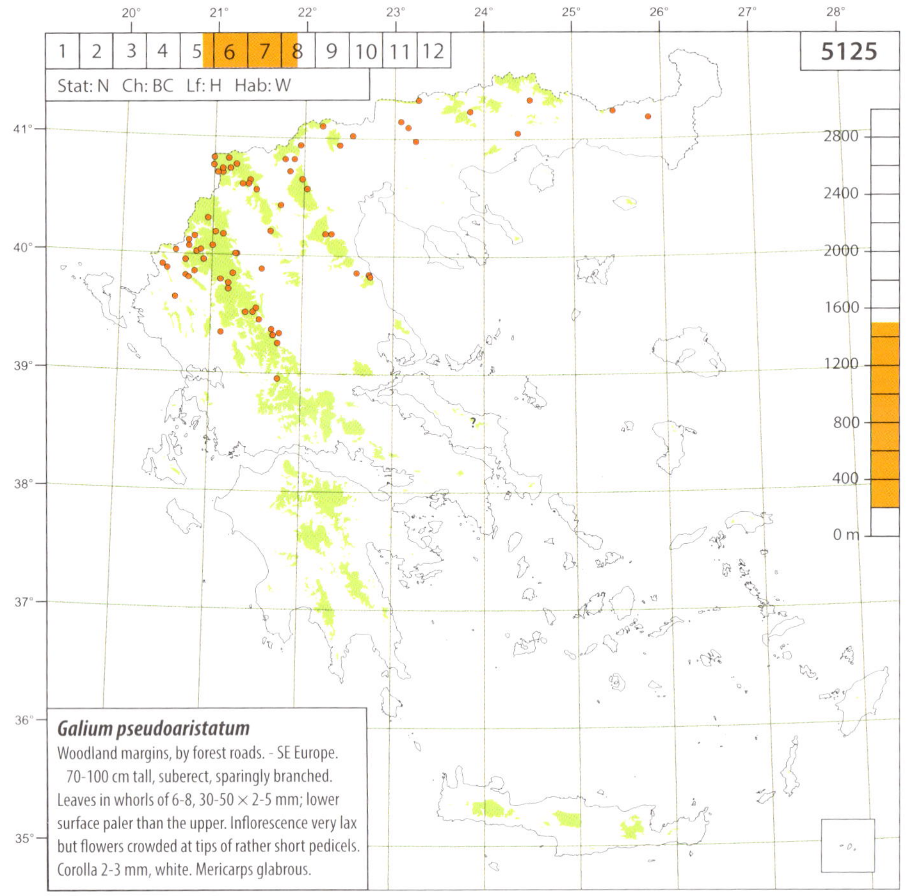

Galium pseudoaristatum

Woodland margins, by forest roads. - SE Europe.

70-100 cm tall, suberect, sparingly branched. Leaves in whorls of 6-8, 30-50 × 2-5 mm; lower surface paler than the upper. Inflorescence very lax but flowers crowded at tips of rather short pedicels. Corolla 2-3 mm, white. Mericarps glabrous.

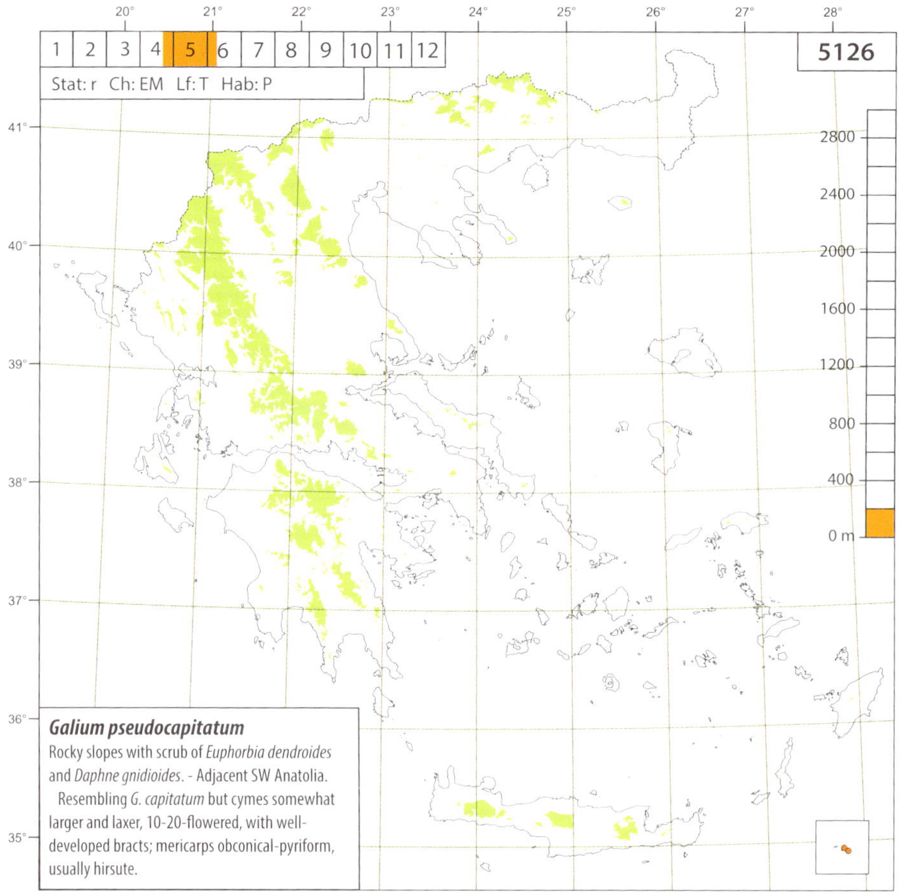

Galium pseudocapitatum

Rocky slopes with scrub of *Euphorbia dendroides* and *Daphne gnidioides*. - Adjacent SW Anatolia.

Resembling *G. capitatum* but cymes somewhat larger and laxer, 10-20-flowered, with well-developed bracts; mericarps obconical-pyriform, usually hirsute.

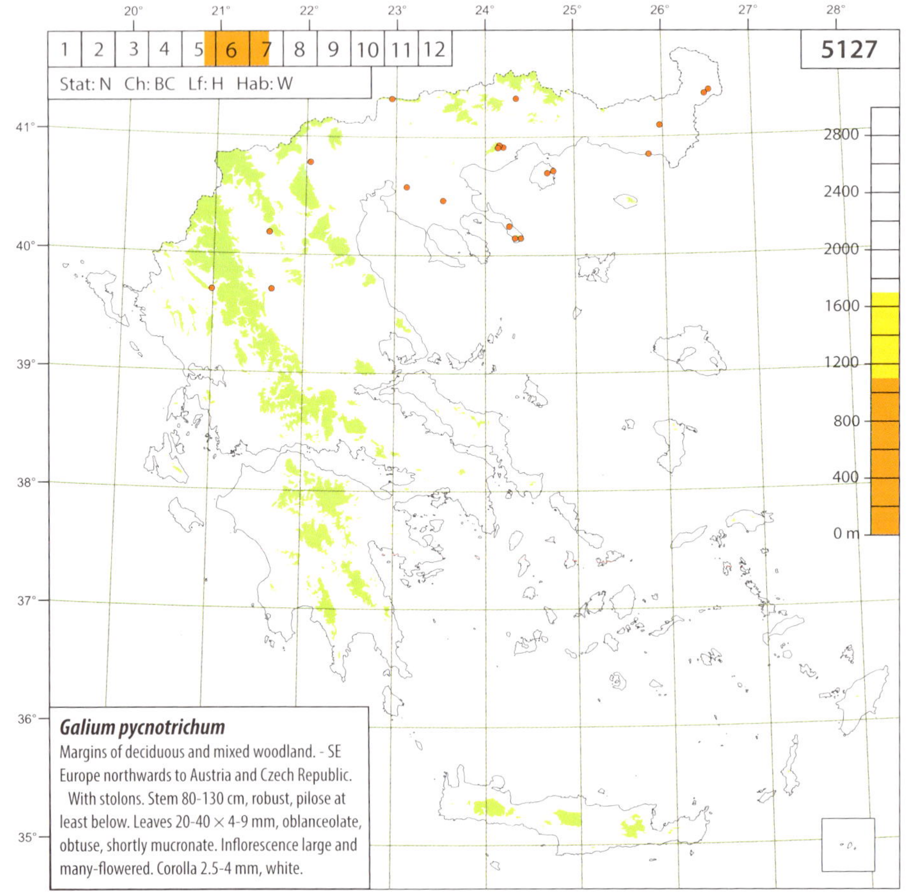

Galium pycnotrichum

Margins of deciduous and mixed woodland. - SE Europe northwards to Austria and Czech Republic.

With stolons. Stem 80-130 cm, robust, pilose at least below. Leaves 20-40 × 4-9 mm, oblanceolate, obtuse, shortly mucronate. Inflorescence large and many-flowered. Corolla 2.5-4 mm, white.

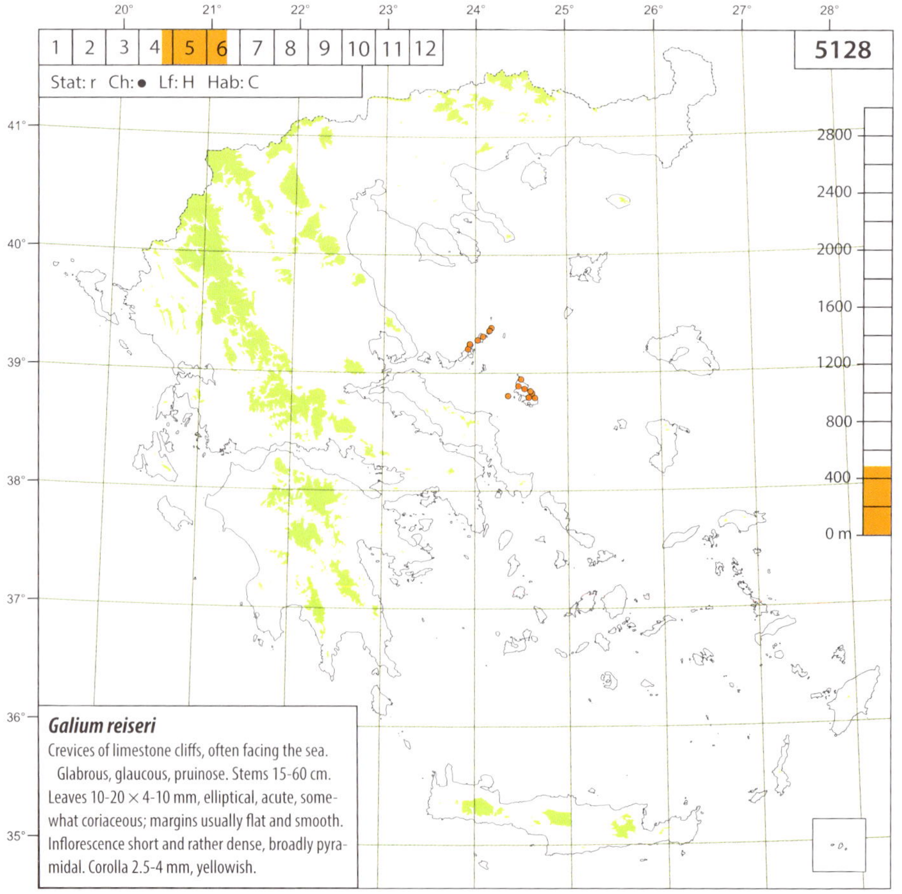

Galium reiseri

Crevices of limestone cliffs, often facing the sea.

Glabrous, glaucous, pruinose. Stems 15-60 cm. Leaves 10-20 × 4-10 mm, elliptical, acute, somewhat coriaceous; margins usually flat and smooth. Inflorescence short and rather dense, broadly pyramidal. Corolla 2.5-4 mm, yellowish.

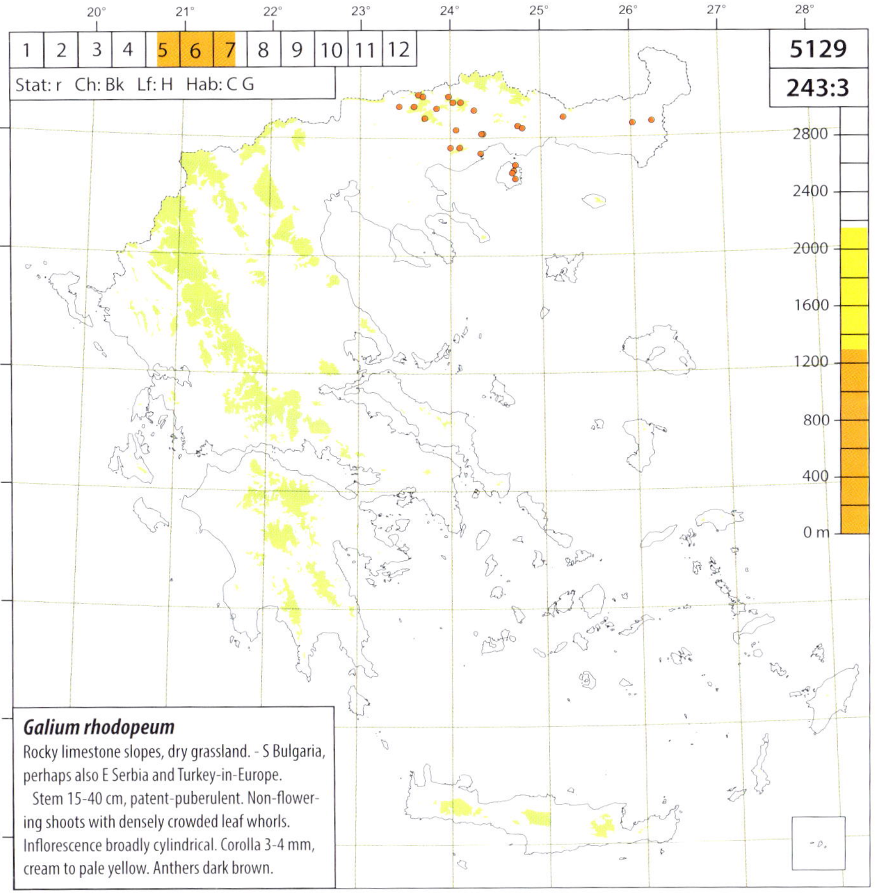

Galium rhodopeum

Rocky limestone slopes, dry grassland. - S Bulgaria, perhaps also E Serbia and Turkey-in-Europe.

Stem 15-40 cm, patent-puberulent. Non-flowering shoots with densely crowded leaf whorls. Inflorescence broadly cylindrical. Corolla 3-4 mm, cream to pale yellow. Anthers dark brown.

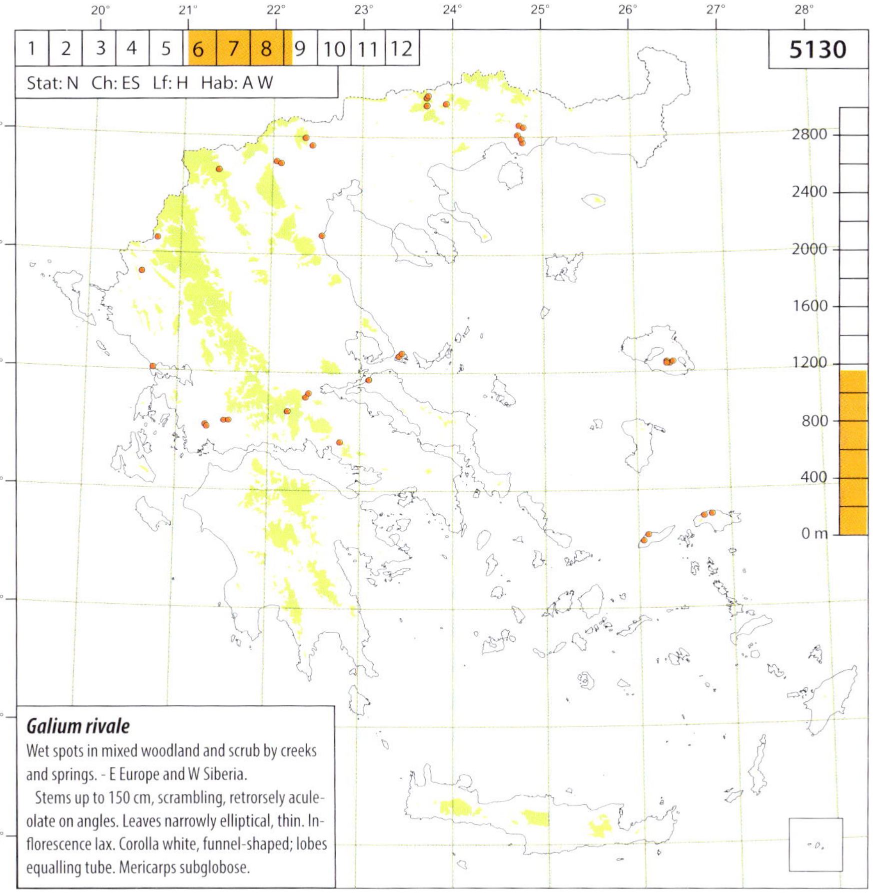

Galium rivale

Wet spots in mixed woodland and scrub by creeks and springs. - E Europe and W Siberia.

Stems up to 150 cm, scrambling, retrorsely aculeolate on angles. Leaves narrowly elliptical, thin. Inflorescence lax. Corolla white, funnel-shaped; lobes equalling tube. Mericarps subglobose.

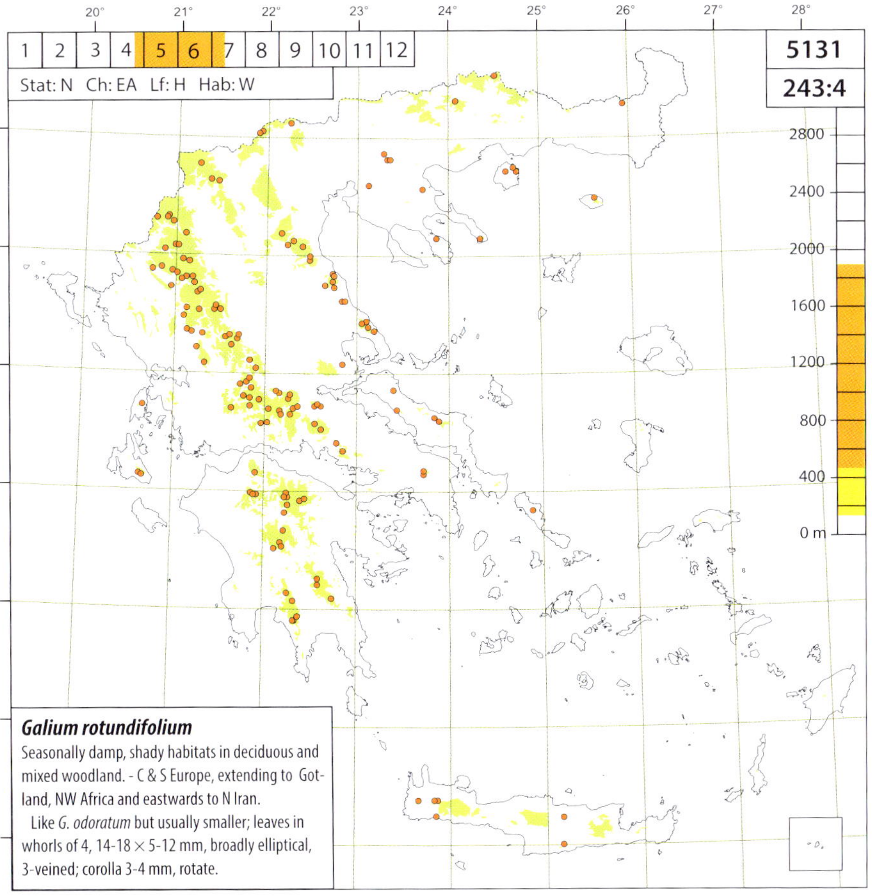

Galium rotundifolium

Seasonally damp, shady habitats in deciduous and mixed woodland. - C & S Europe, extending to Gotland, NW Africa and eastwards to N Iran.

Like *G. odoratum* but usually smaller; leaves in whorls of 4, 14-18 × 5-12 mm, broadly elliptical, 3-veined; corolla 3-4 mm, rotate.

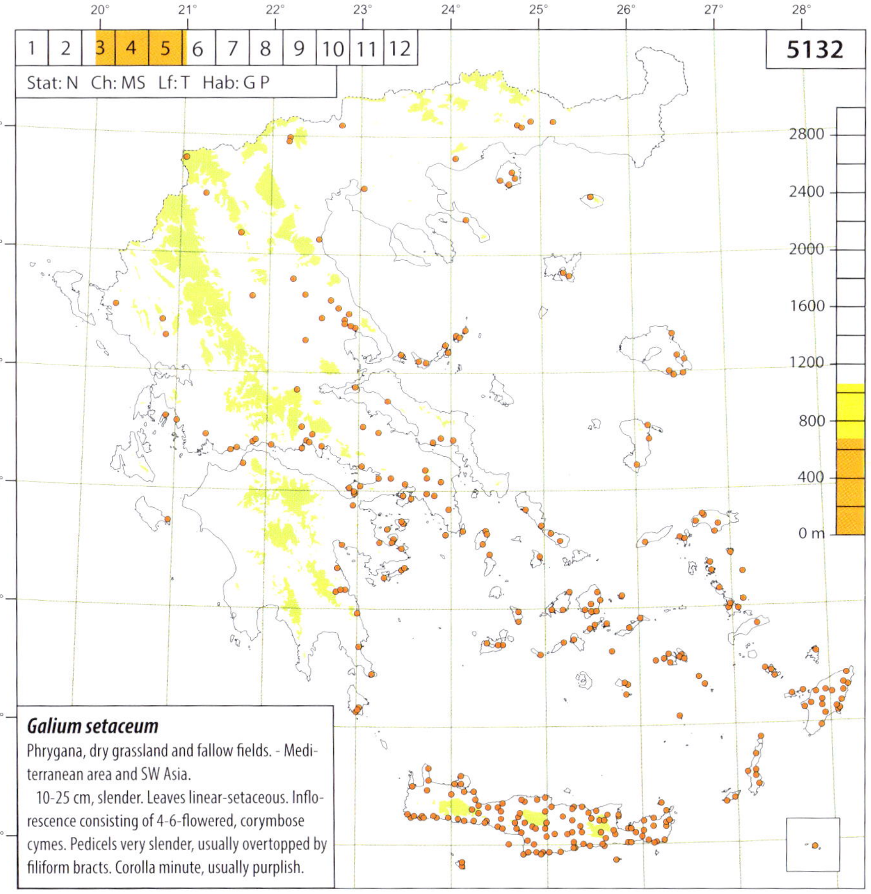

Galium setaceum

Phrygana, dry grassland and fallow fields. - Mediterranean area and SW Asia.

10-25 cm, slender. Leaves linear-setaceous. Inflorescence consisting of 4-6-flowered, corymbose cymes. Pedicels very slender, usually overtopped by filiform bracts. Corolla minute, usually purplish.

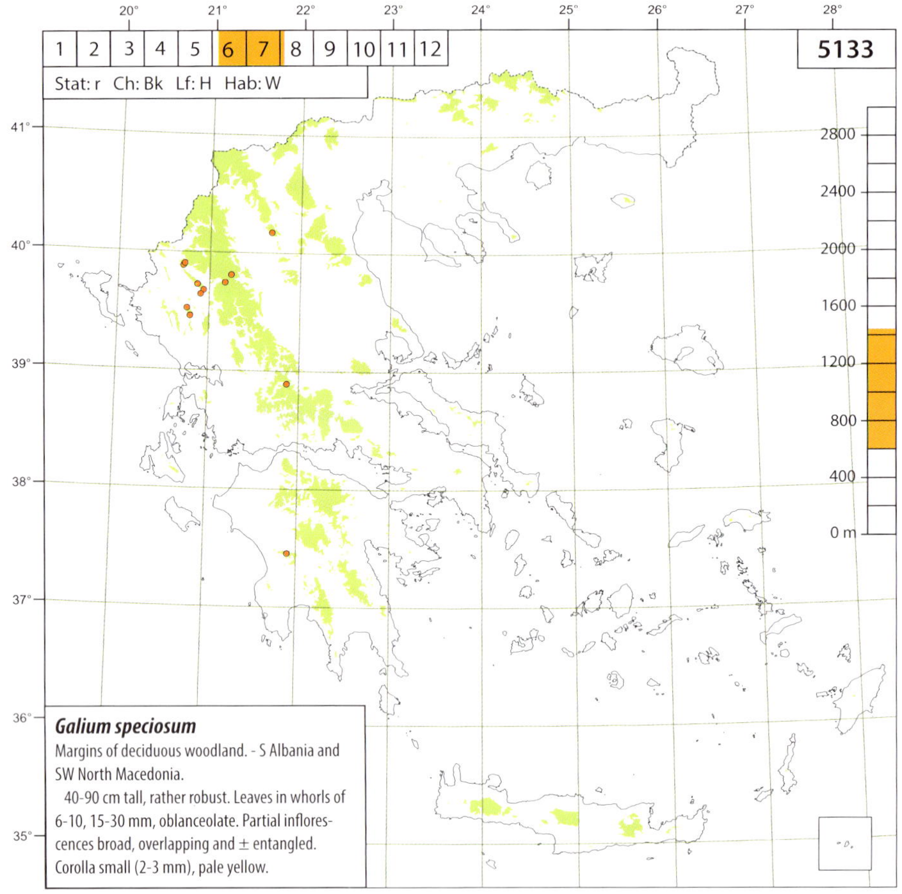

Galium speciosum
Margins of deciduous woodland. - S Albania and SW North Macedonia.
40-90 cm tall, rather robust. Leaves in whorls of 6-10, 15-30 mm, oblanceolate. Partial inflorescences broad, overlapping and ± entangled. Corolla small (2-3 mm), pale yellow.

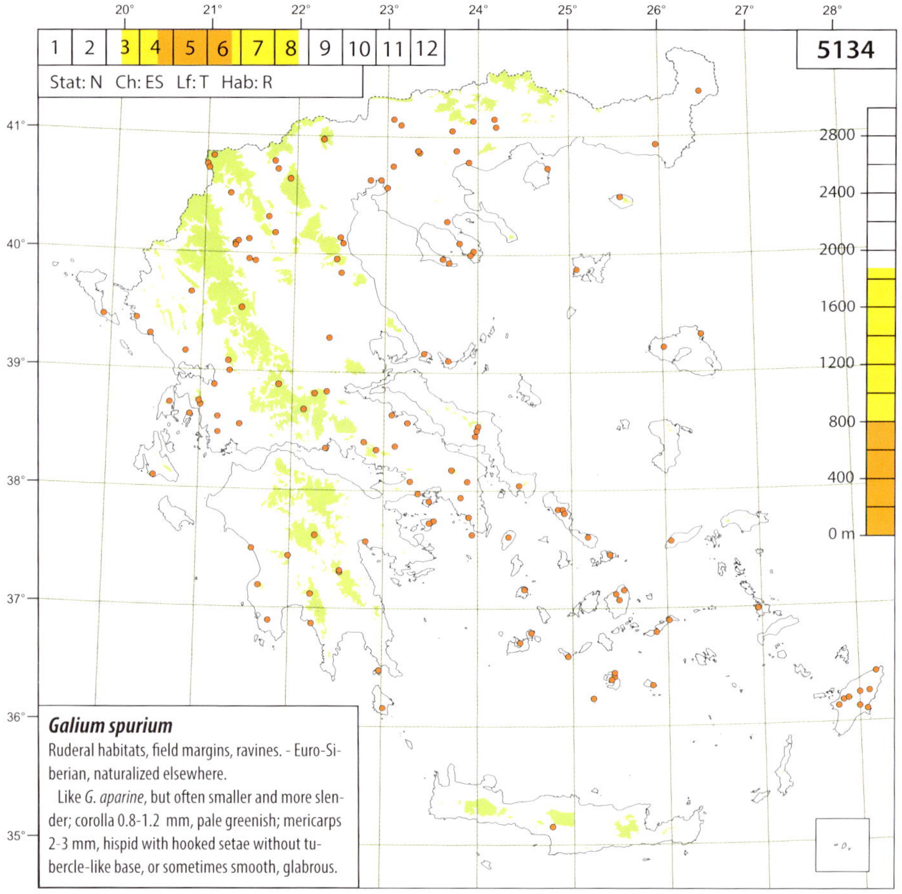

Galium spurium
Ruderal habitats, field margins, ravines. - Euro-Siberian, naturalized elsewhere.
Like *G. aparine*, but often smaller and more slender; corolla 0.8-1.2 mm, pale greenish; mericarps 2-3 mm, hispid with hooked setae without tubercle-like base, or sometimes smooth, glabrous.

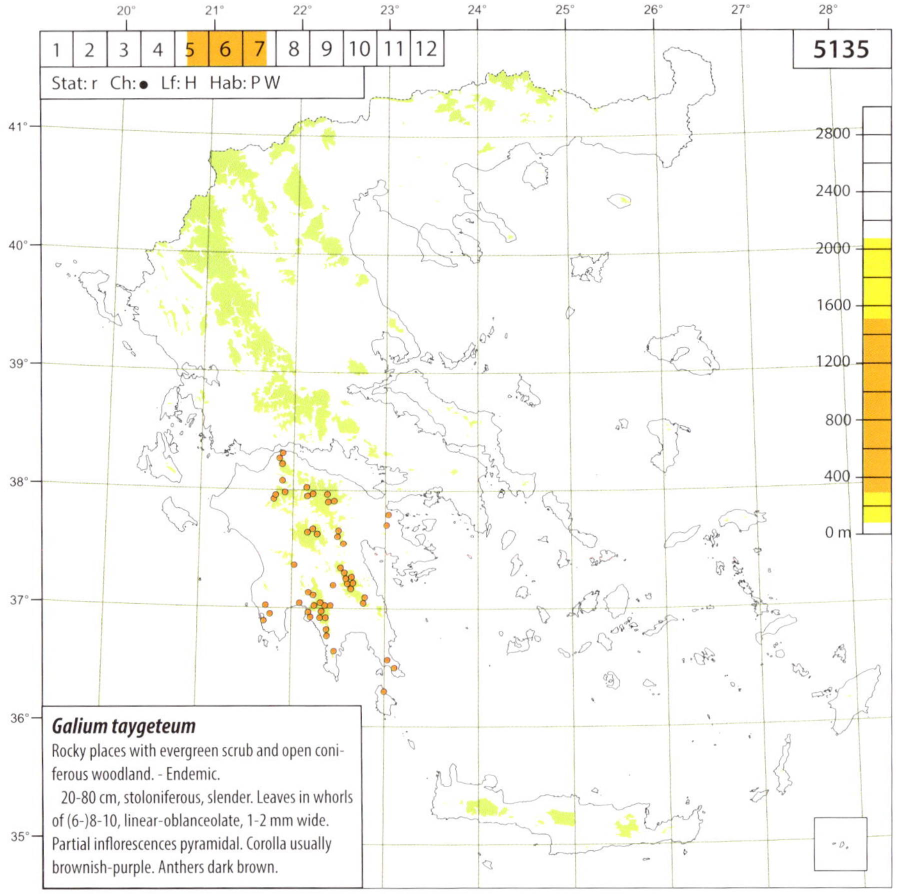

Galium taygeteum
Rocky places with evergreen scrub and open coniferous woodland. - Endemic.
20-80 cm, stoloniferous, slender. Leaves in whorls of (6-)8-10, linear-oblanceolate, 1-2 mm wide. Partial inflorescences pyramidal. Corolla usually brownish-purple. Anthers dark brown.

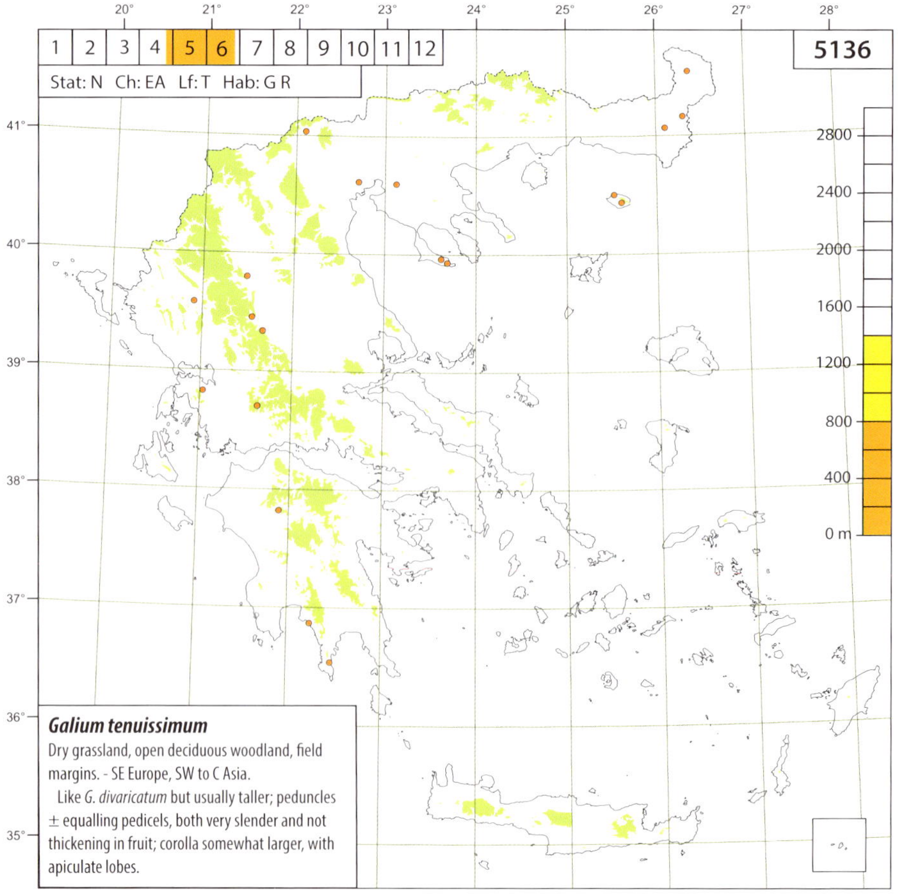

Galium tenuissimum
Dry grassland, open deciduous woodland, field margins. - SE Europe, SW to C Asia.
Like *G. divaricatum* but usually taller; peduncles ± equalling pedicels, both very slender and not thickening in fruit; corolla somewhat larger, with apiculate lobes.

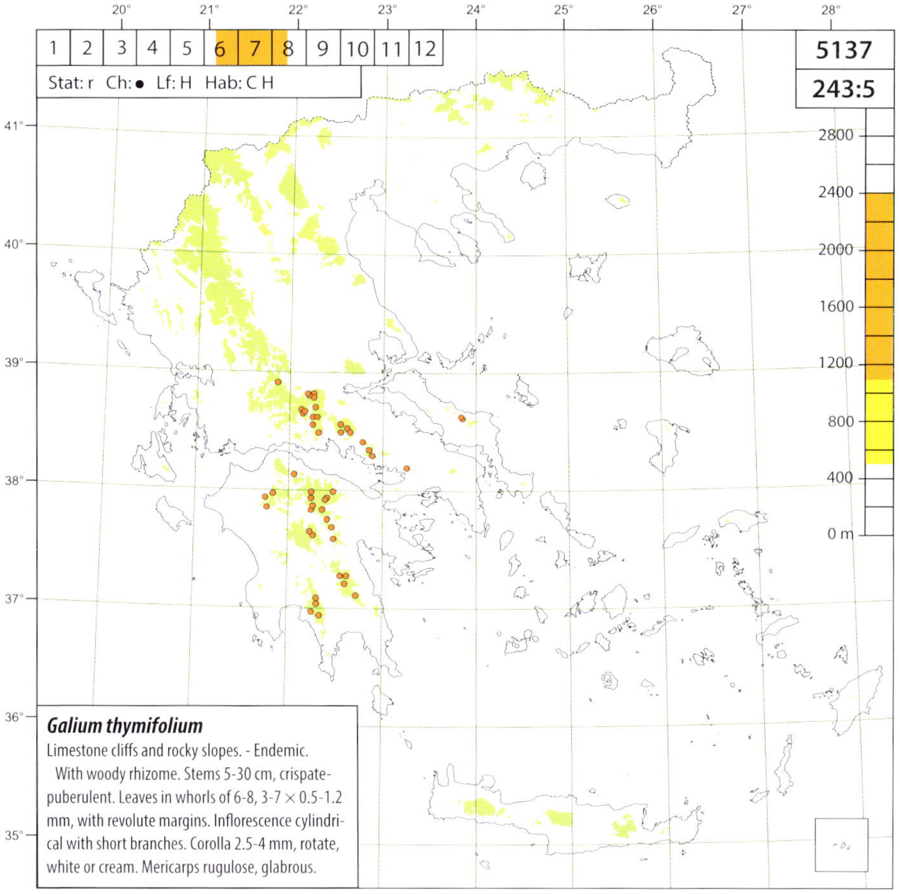

Galium thymifolium

Limestone cliffs and rocky slopes. - Endemic.

With woody rhizome. Stems 5-30 cm, crispate-puberulent. Leaves in whorls of 6-8, 3-7 × 0.5-1.2 mm, with revolute margins. Inflorescence cylindrical with short branches. Corolla 2.5-4 mm, rotate, white or cream. Mericarps rugulose, glabrous.

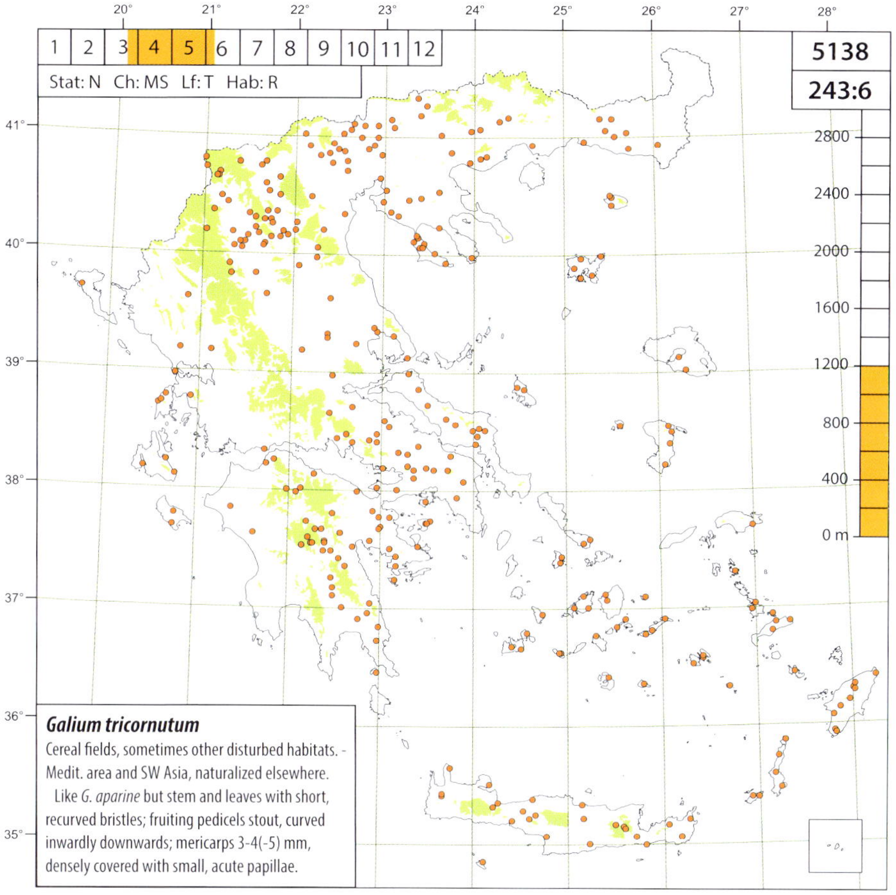

Galium tricornutum

Cereal fields, sometimes other disturbed habitats. - Medit. area and SW Asia, naturalized elsewhere.

Like *G. aparine* but stem and leaves with short, recurved bristles; fruiting pedicels stout, curved inwardly downwards; mericarps 3-4(-5) mm, densely covered with small, acute papillae.

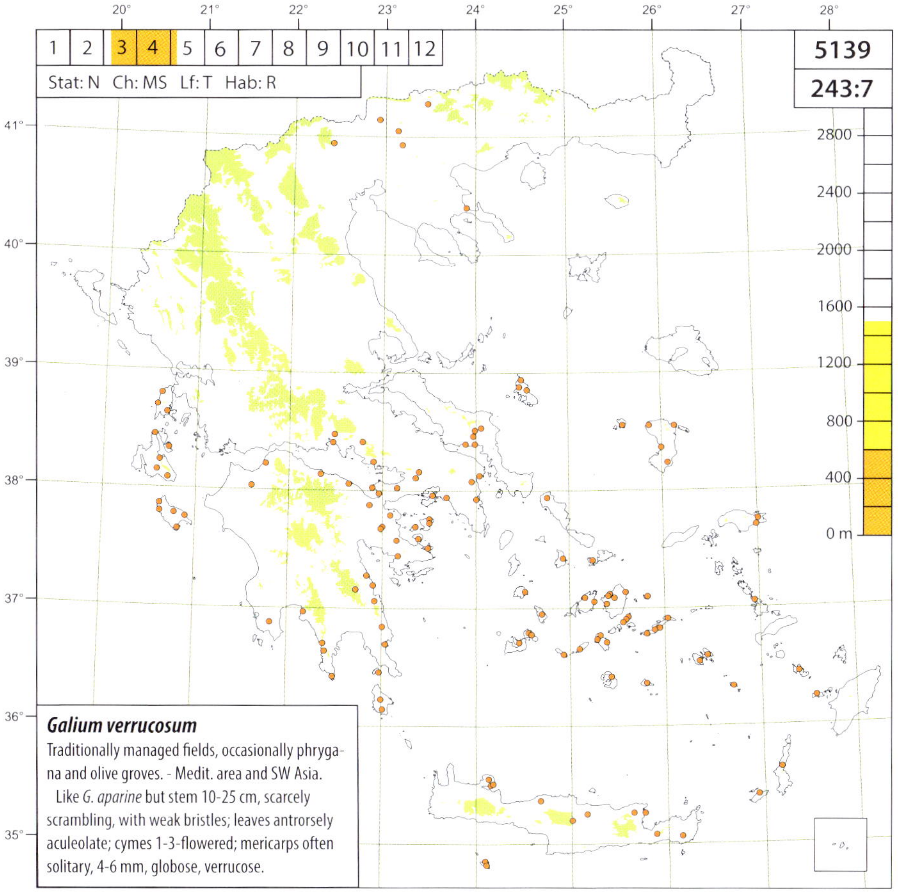

Galium verrucosum

Traditionally managed fields, occasionally phrygana and olive groves. - Medit. area and SW Asia.

Like *G. aparine* but stem 10-25 cm, scarcely scrambling, with weak bristles; leaves antrorsely aculeolate; cymes 1-3-flowered; mericarps often solitary, 4-6 mm, globose, verrucose.

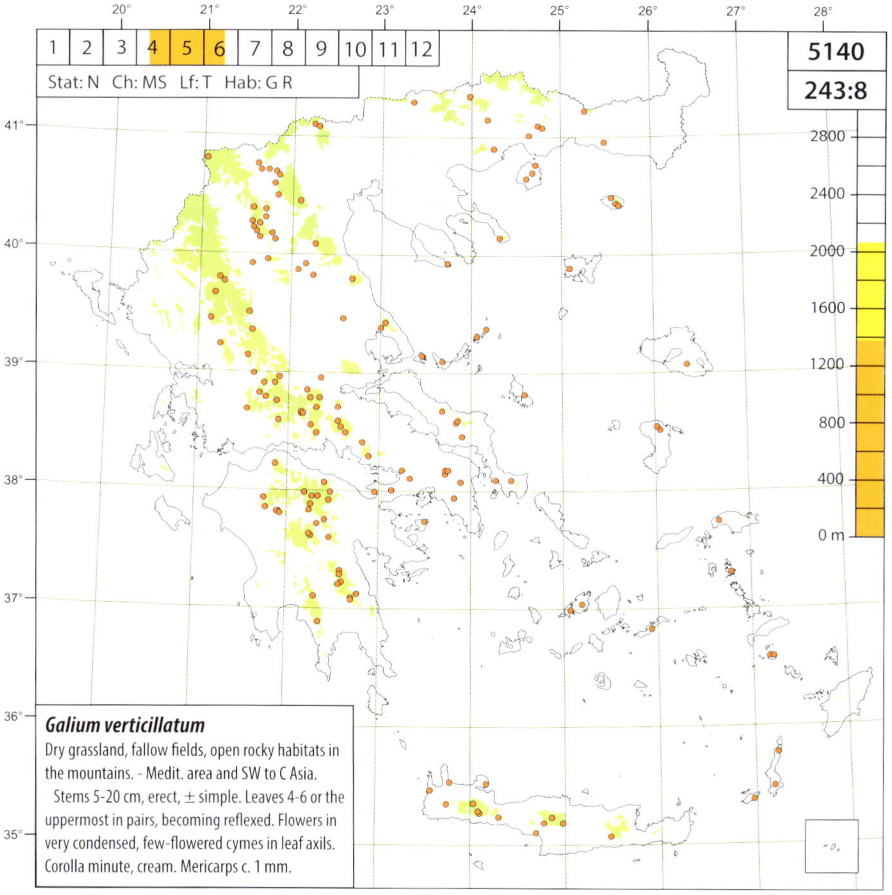

Galium verticillatum

Dry grassland, fallow fields, open rocky habitats in the mountains. - Medit. area and SW to C Asia.

Stems 5-20 cm, erect, ± simple. Leaves 4-6 or the uppermost in pairs, becoming reflexed. Flowers in very condensed, few-flowered cymes in leaf axils. Corolla minute, cream. Mericarps c. 1 mm.

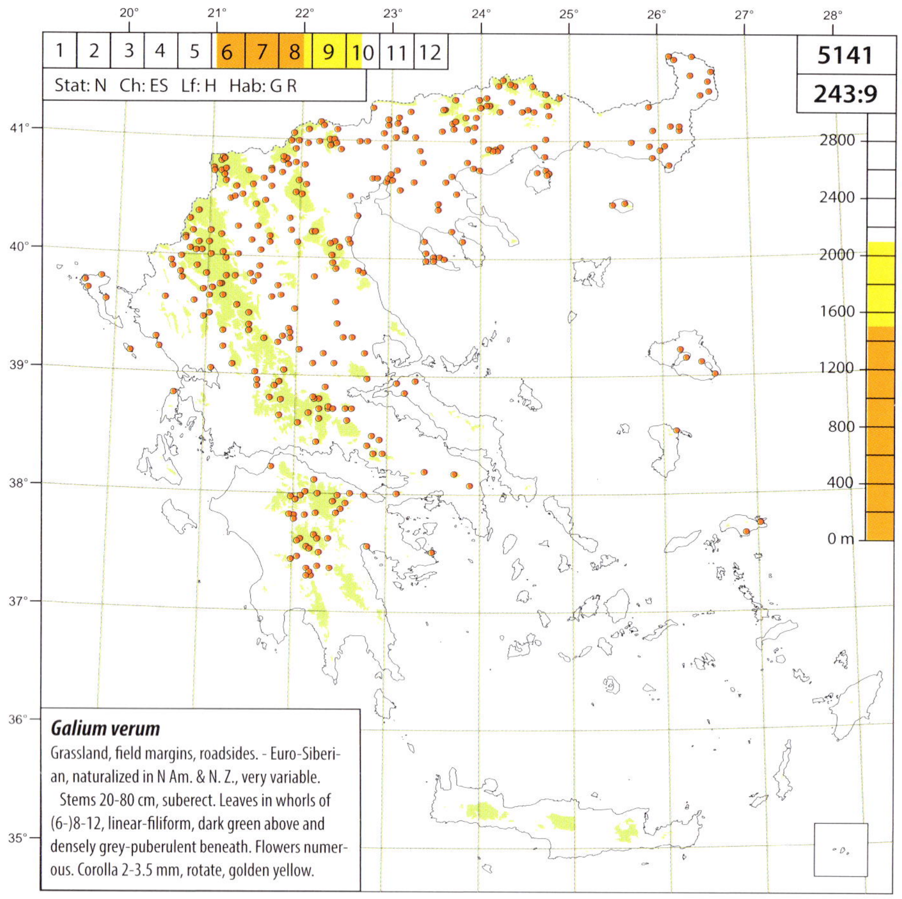

Galium verum

Grassland, field margins, roadsides. - Euro-Siberian, naturalized in N Am. & N. Z., very variable.
Stems 20-80 cm, suberect. Leaves in whorls of (6-)8-12, linear-filiform, dark green above and densely grey-puberulent beneath. Flowers numerous. Corolla 2-3.5 mm, rotate, golden yellow.

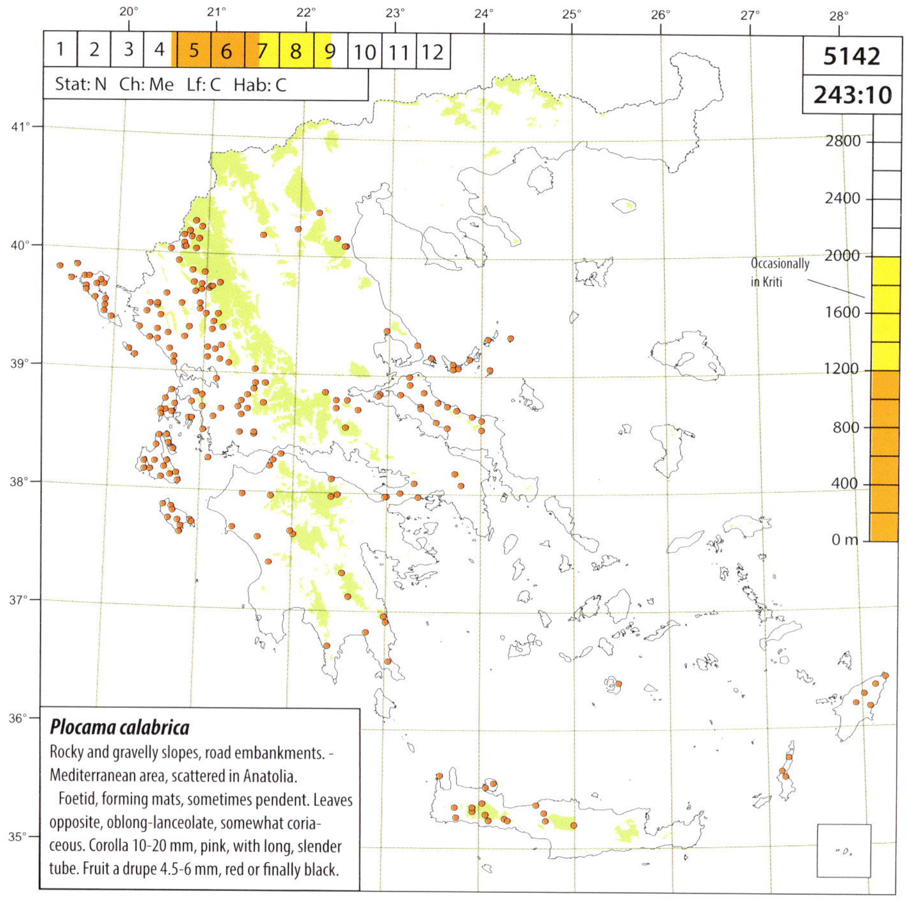

Plocama calabrica

Rocky and gravelly slopes, road embankments. - Mediterranean area, scattered in Anatolia.
Foetid, forming mats, sometimes pendent. Leaves opposite, oblong-lanceolate, somewhat coriaceous. Corolla 10-20 mm, pink, with long, slender tube. Fruit a drupe 4.5-6 mm, red or finally black.

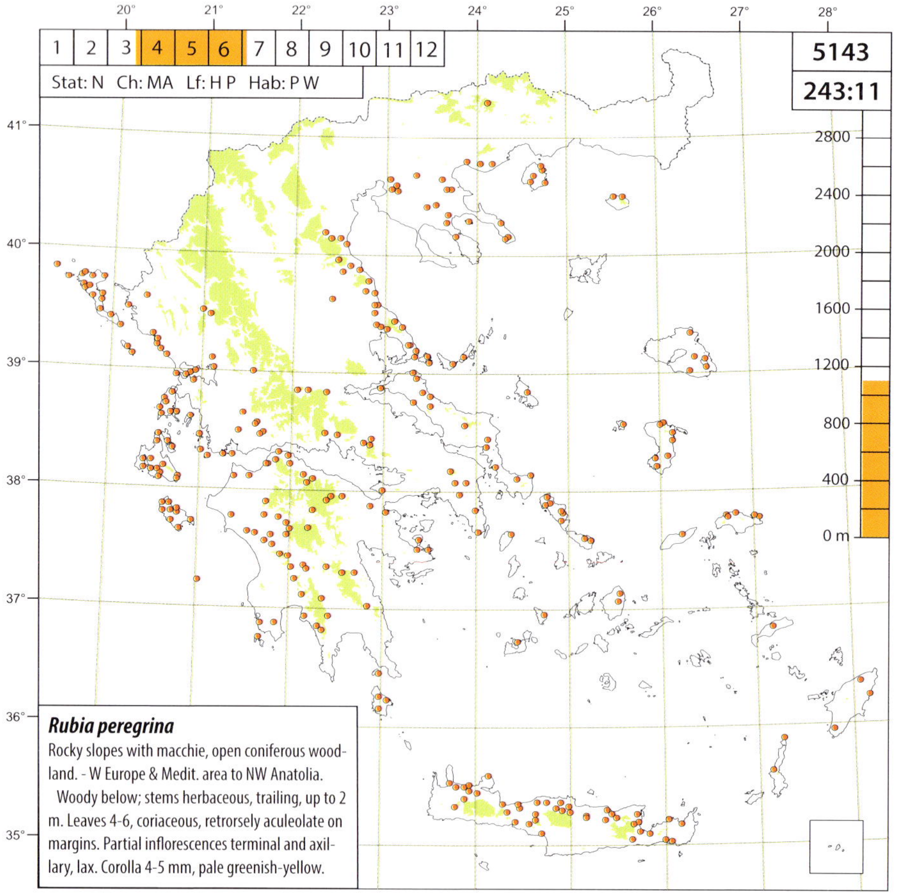

Rubia peregrina

Rocky slopes with macchie, open coniferous woodland. - W Europe & Medit. area to NW Anatolia.
Woody below; stems herbaceous, trailing, up to 2 m. Leaves 4-6, coriaceous, retrorsely aculeolate on margins. Partial inflorescences terminal and axillary, lax. Corolla 4-5 mm, pale greenish-yellow.

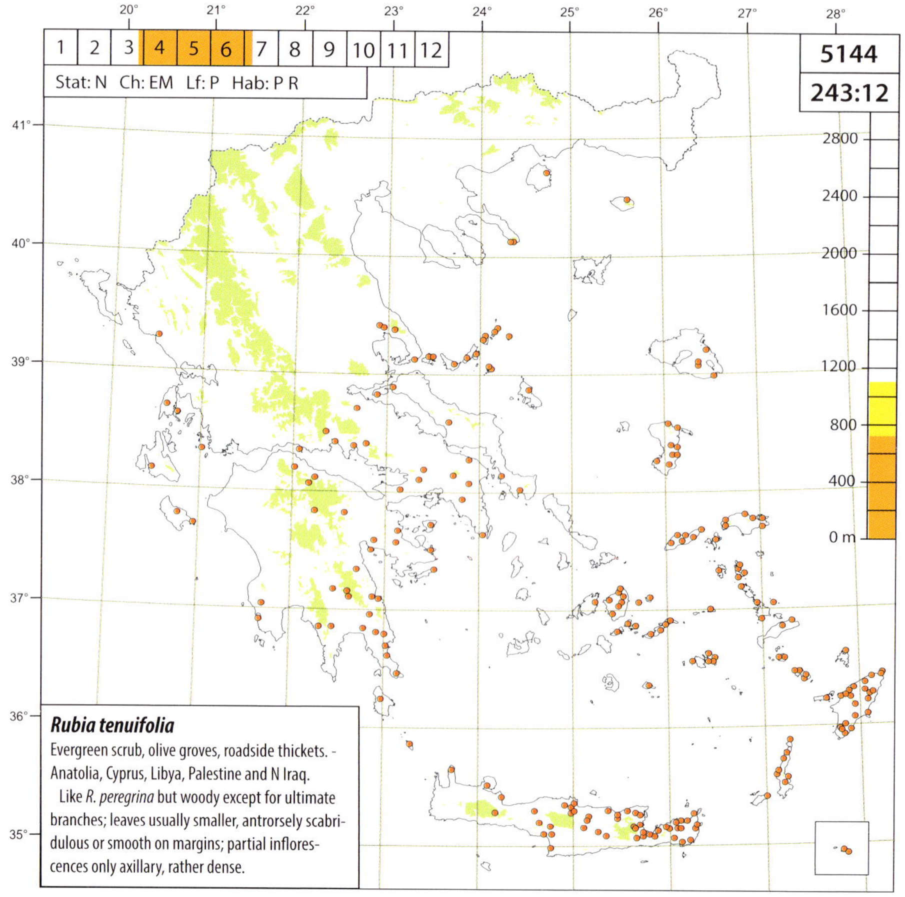

Rubia tenuifolia

Evergreen scrub, olive groves, roadside thickets. - Anatolia, Cyprus, Libya, Palestine and N Iraq.
Like *R. peregrina* but woody except for ultimate branches; leaves usually smaller, antrorsely scabridulous or smooth on margins; partial inflorescences only axillary, rather dense.

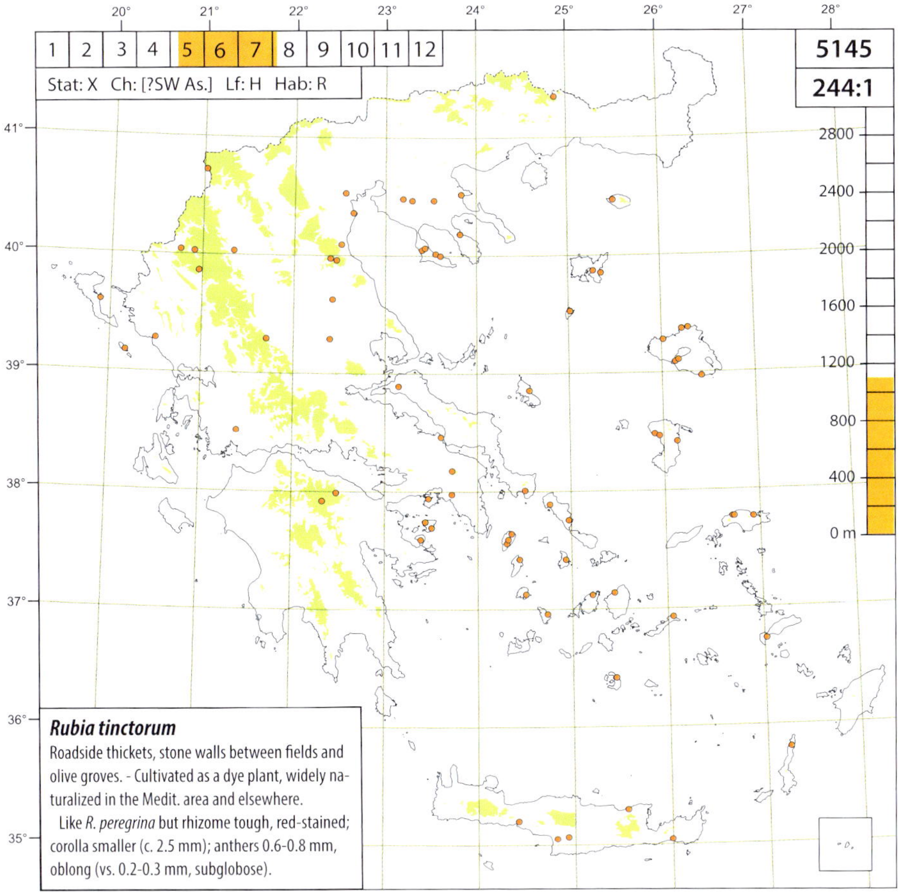

Rubia tinctorum

Roadside thickets, stone walls between fields and olive groves. - Cultivated as a dye plant, widely naturalized in the Medit. area and elsewhere.

Like *R. peregrina* but rhizome tough, red-stained; corolla smaller (c. 2.5 mm); anthers 0.6-0.8 mm, oblong (vs. 0.2-0.3 mm, subglobose).

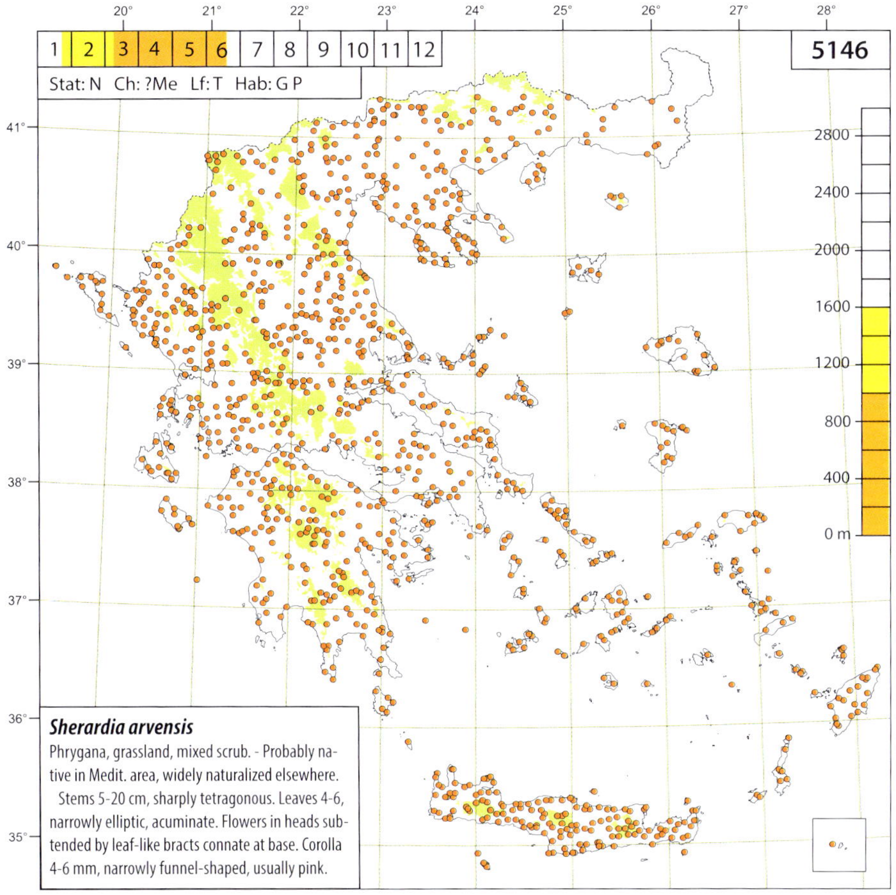

Sherardia arvensis

Phrygana, grassland, mixed scrub. - Probably native in Medit. area, widely naturalized elsewhere.

Stems 5-20 cm, sharply tetragonous. Leaves 4-6, narrowly elliptic, acuminate. Flowers in heads subtended by leaf-like bracts connate at base. Corolla 4-6 mm, narrowly funnel-shaped, usually pink.

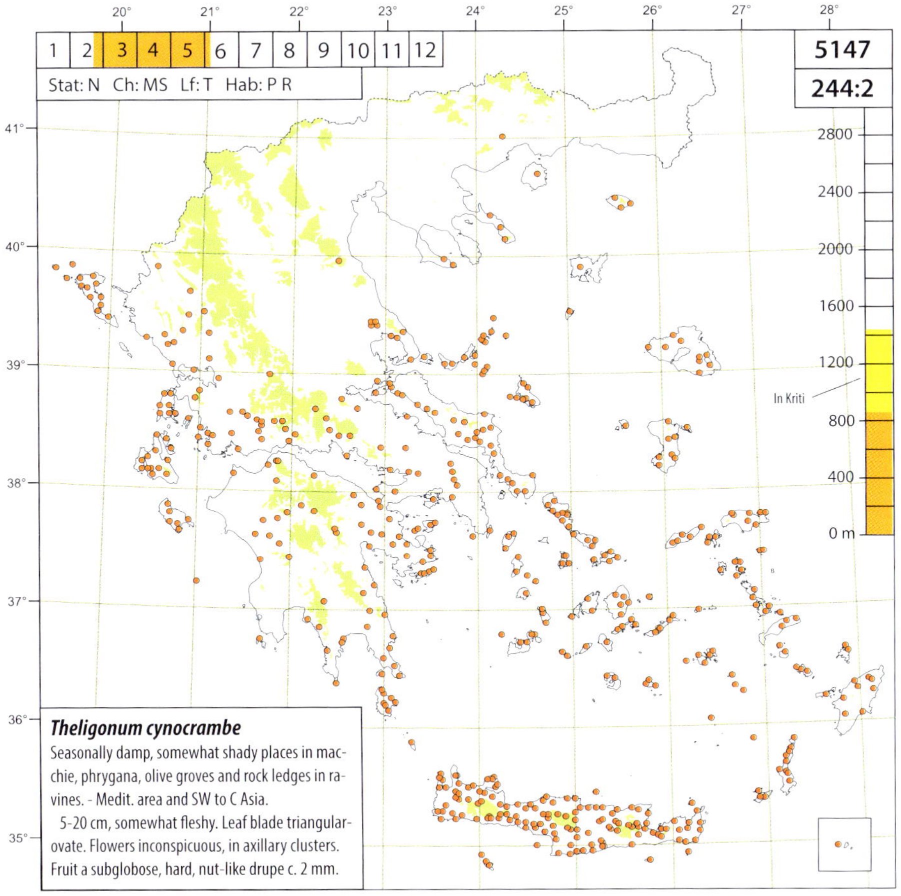

Theligonum cynocrambe

Seasonally damp, somewhat shady places in macchie, phrygana, olive groves and rock ledges in ravines. - Medit. area and SW to C Asia.

5-20 cm, somewhat fleshy. Leaf blade triangular-ovate. Flowers inconspicuous, in axillary clusters. Fruit a subglobose, hard, nut-like drupe c. 2 mm.

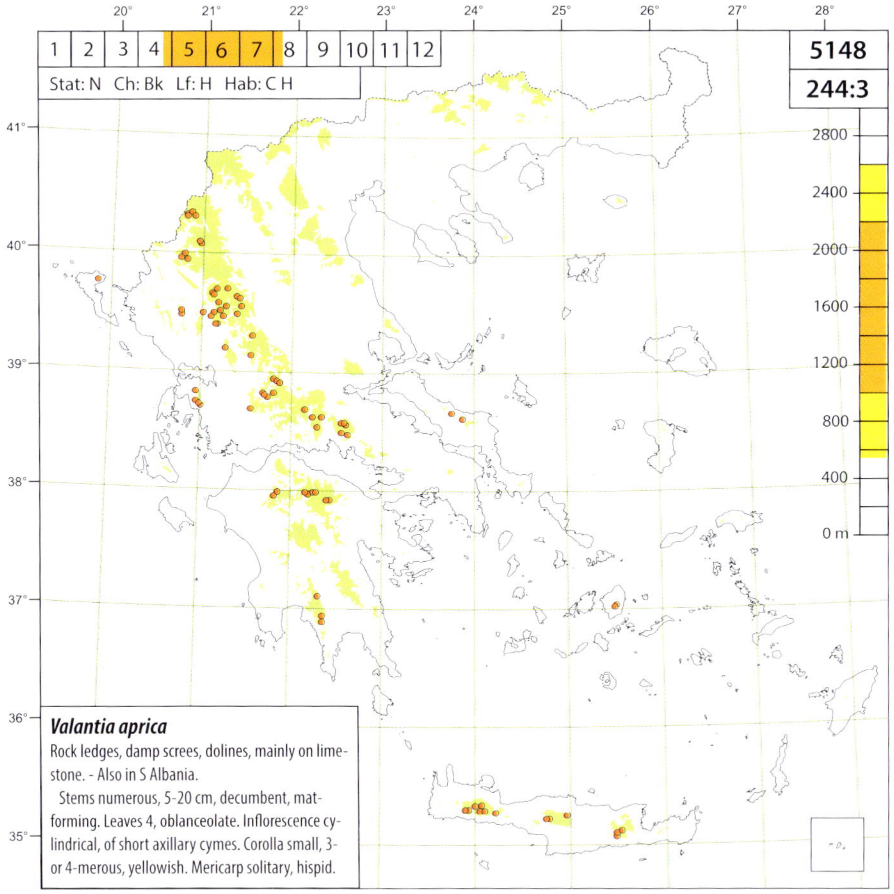

Valantia aprica

Rock ledges, damp screes, dolines, mainly on limestone. - Also in S Albania.

Stems numerous, 5-20 cm, decumbent, mat-forming. Leaves 4, oblanceolate. Inflorescence cylindrical, of short axillary cymes. Corolla small, 3- or 4-merous, yellowish. Mericarp solitary, hispid.

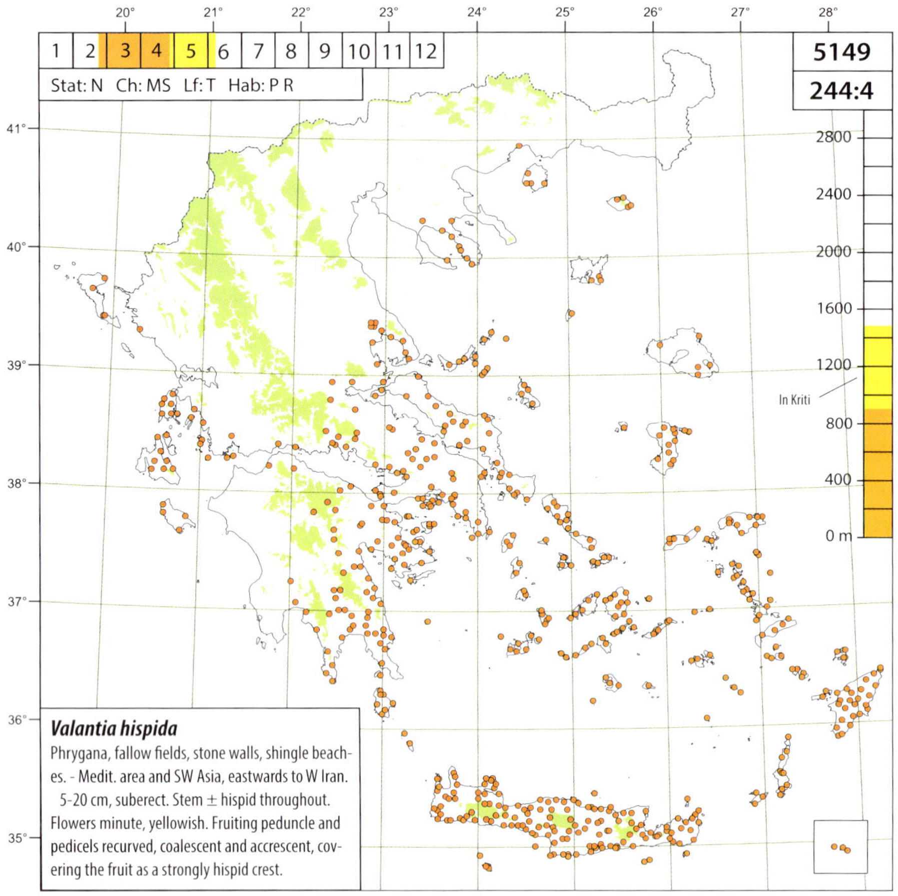

Valantia hispida

Phrygana, fallow fields, stone walls, shingle beaches. - Medit. area and SW Asia, eastwards to W Iran.

5-20 cm, suberect. Stem ± hispid throughout. Flowers minute, yellowish. Fruiting peduncle and pedicels recurved, coalescent and accrescent, covering the fruit as a strongly hispid crest.

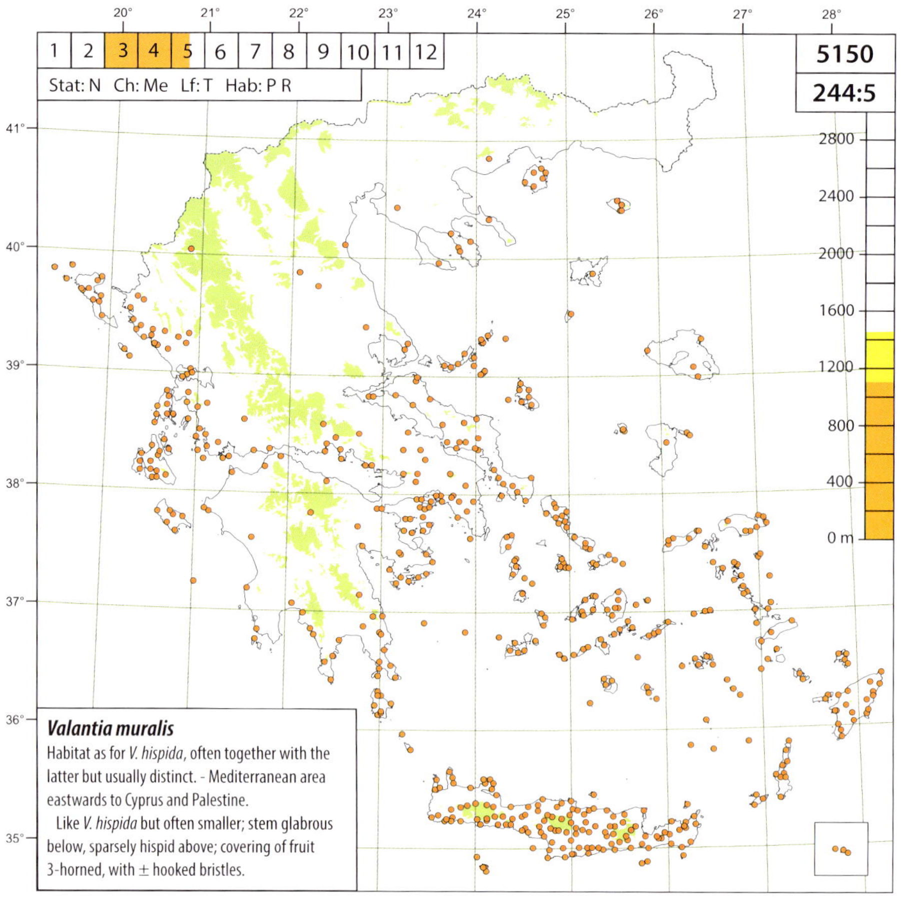

Valantia muralis

Habitat as for *V. hispida*, often together with the latter but usually distinct. - Mediterranean area eastwards to Cyprus and Palestine.

Like *V. hispida* but often smaller; stem glabrous below, sparsely hispid above; covering of fruit 3-horned, with ± hooked bristles.

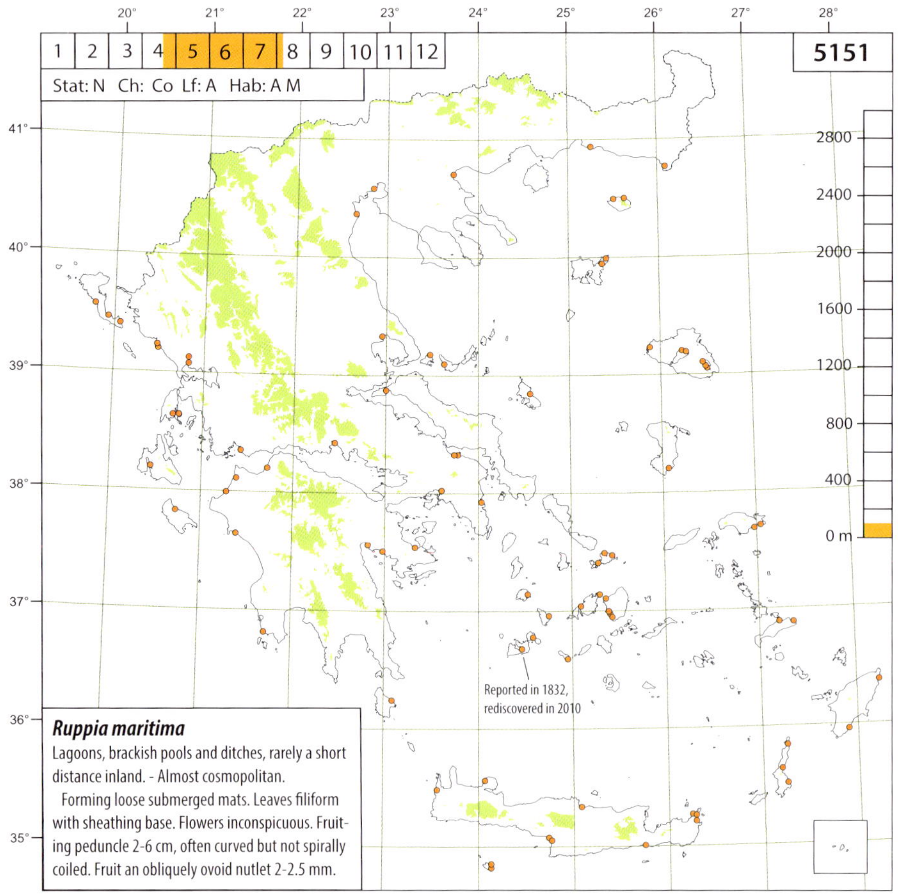

Ruppia maritima

Lagoons, brackish pools and ditches, rarely a short distance inland. - Almost cosmopolitan.

Forming loose submerged mats. Leaves filiform with sheathing base. Flowers inconspicuous. Fruiting peduncle 2-6 cm, often curved but not spirally coiled. Fruit an obliquely ovoid nutlet 2-2.5 mm.

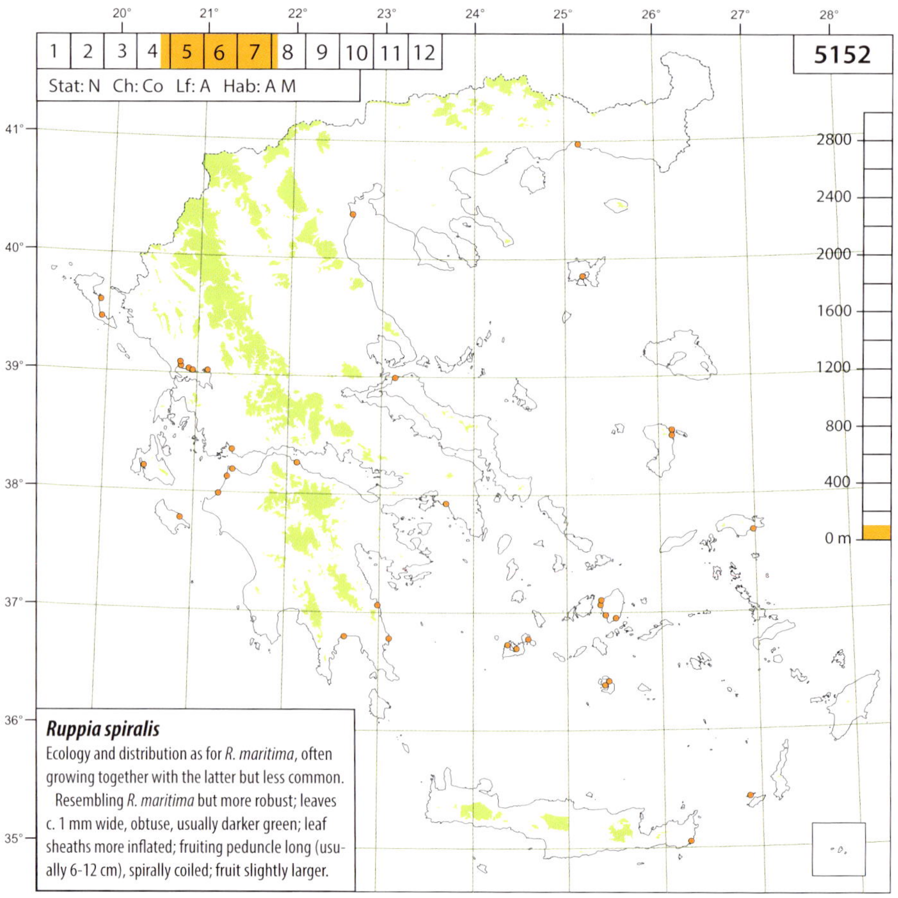

Ruppia spiralis

Ecology and distribution as for *R. maritima*, often growing together with the latter but less common.

Resembling *R. maritima* but more robust; leaves c. 1 mm wide, obtuse, usually darker green; leaf sheaths more inflated; fruiting peduncle long (usually 6-12 cm), spirally coiled; fruit slightly larger.

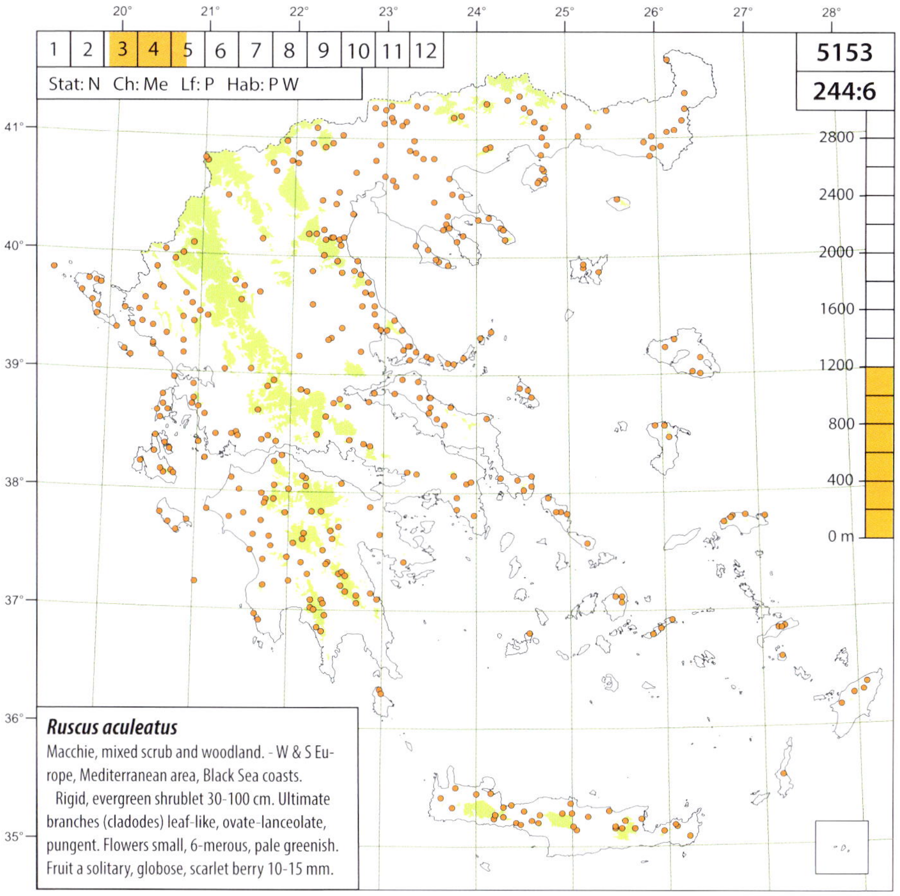

Ruscus aculeatus
Macchie, mixed scrub and woodland. - W & S Europe, Mediterranean area, Black Sea coasts.
Rigid, evergreen shrublet 30-100 cm. Ultimate branches (cladodes) leaf-like, ovate-lanceolate, pungent. Flowers small, 6-merous, pale greenish. Fruit a solitary, globose, scarlet berry 10-15 mm.

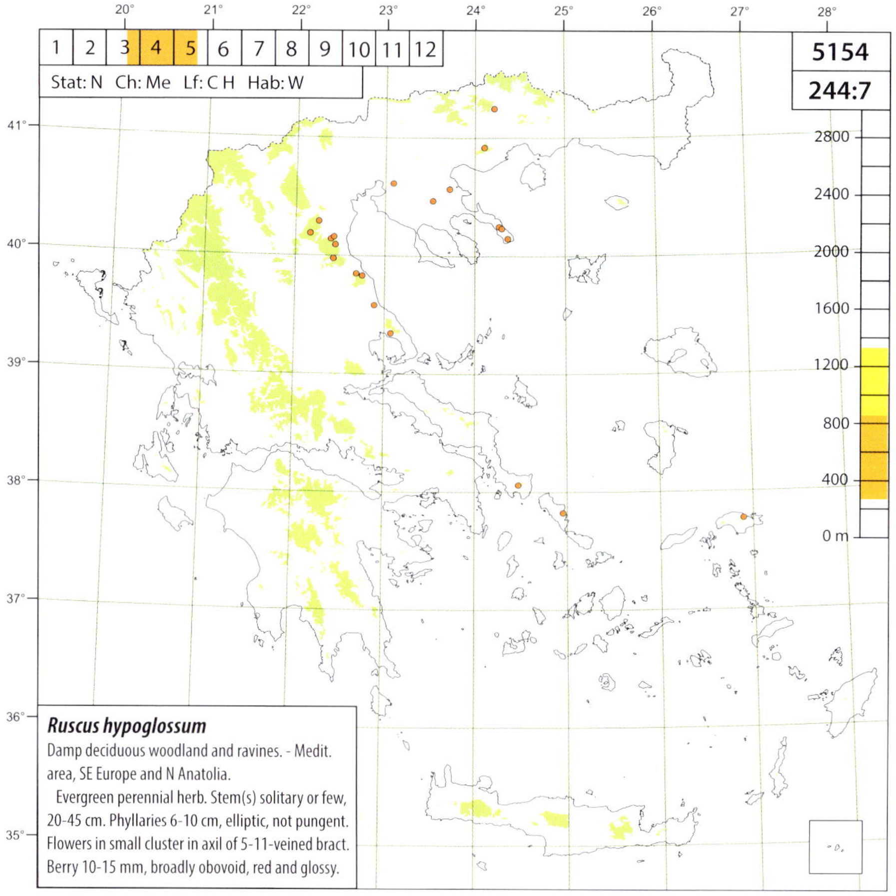

Ruscus hypoglossum
Damp deciduous woodland and ravines. - Medit. area, SE Europe and N Anatolia.
Evergreen perennial herb. Stem(s) solitary or few, 20-45 cm. Phyllaries 6-10 cm, elliptic, not pungent. Flowers in small cluster in axil of 5-11-veined bract. Berry 10-15 mm, broadly obovoid, red and glossy.

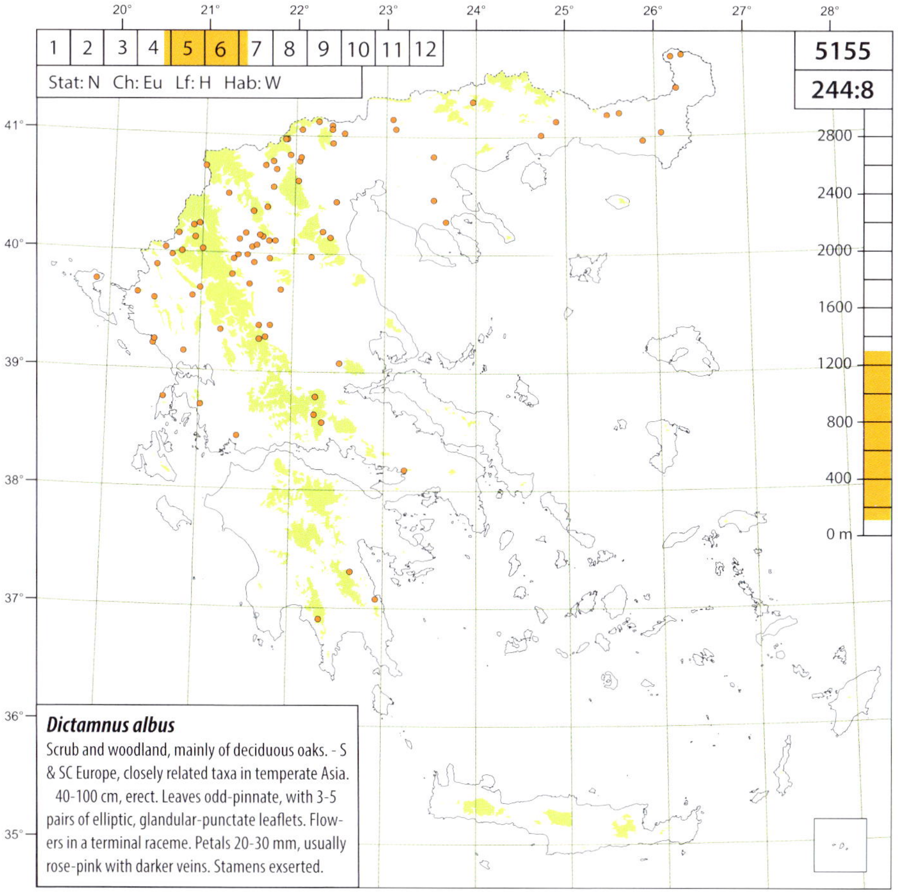

Dictamnus albus
Scrub and woodland, mainly of deciduous oaks. - S & SC Europe, closely related taxa in temperate Asia.
40-100 cm, erect. Leaves odd-pinnate, with 3-5 pairs of elliptic, glandular-punctate leaflets. Flowers in a terminal raceme. Petals 20-30 mm, usually rose-pink with darker veins. Stamens exserted.

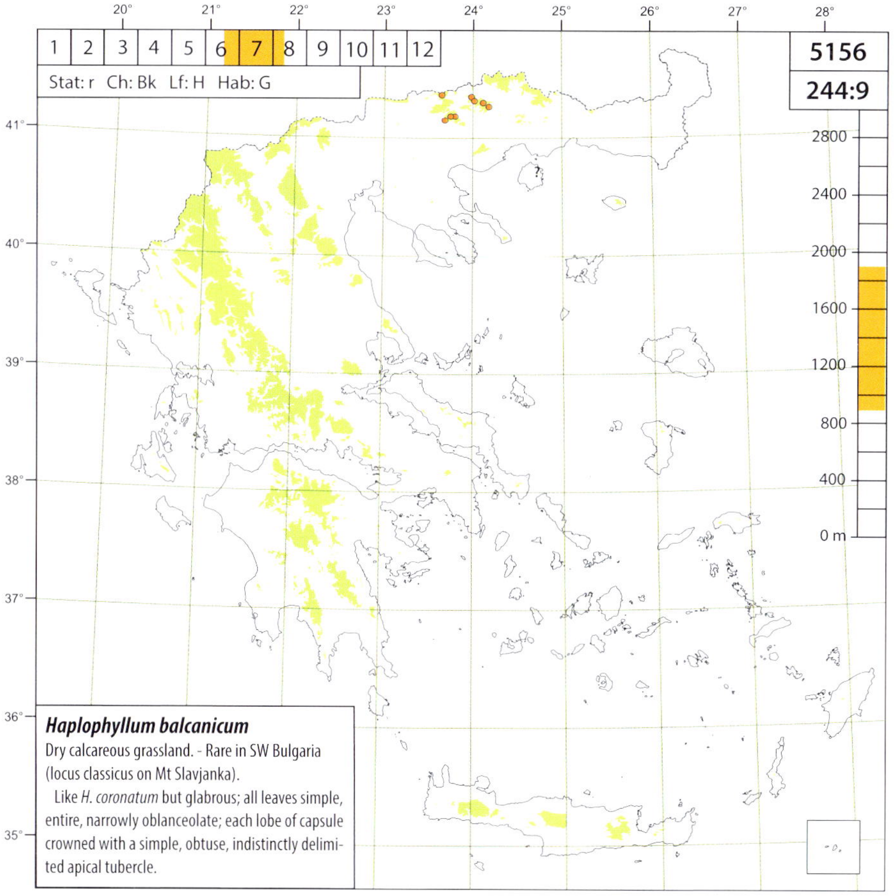

Haplophyllum balcanicum
Dry calcareous grassland. - Rare in SW Bulgaria (locus classicus on Mt Slavjanka).
Like *H. coronatum* but glabrous; all leaves simple, entire, narrowly oblanceolate; each lobe of capsule crowned with a simple, obtuse, indistinctly delimited apical tubercle.

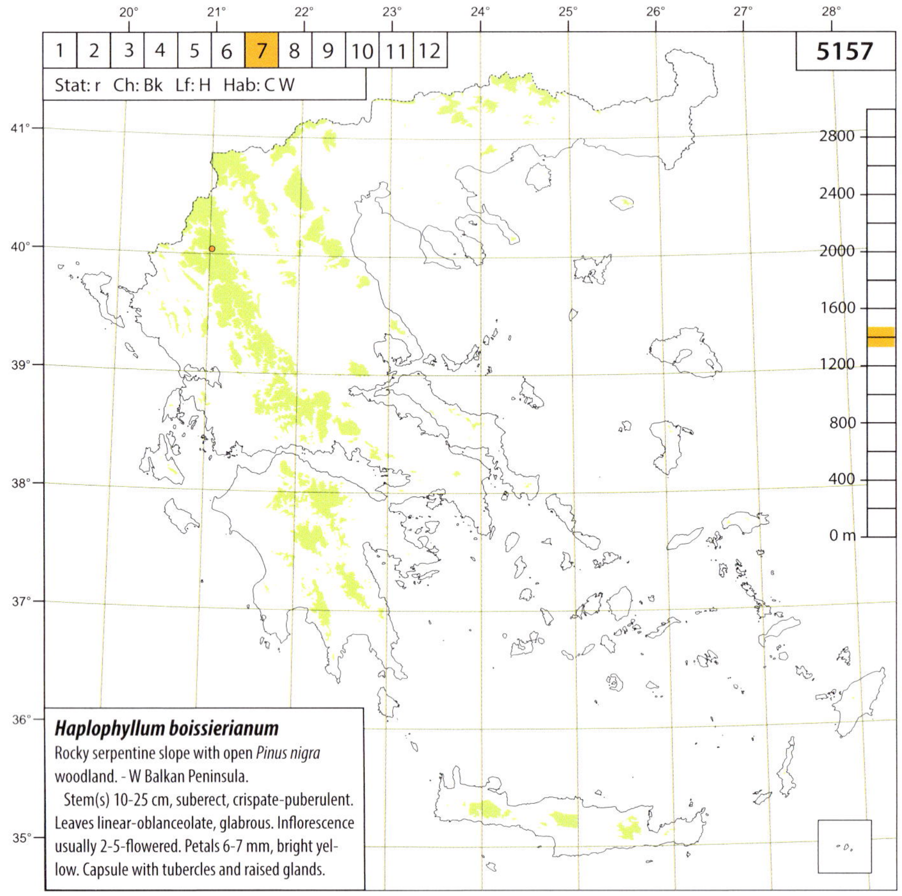

Haplophyllum boissierianum

Rocky serpentine slope with open *Pinus nigra* woodland. - W Balkan Peninsula.

Stem(s) 10-25 cm, suberect, crispate-puberulent. Leaves linear-oblanceolate, glabrous. Inflorescence usually 2-5-flowered. Petals 6-7 mm, bright yellow. Capsule with tubercles and raised glands.

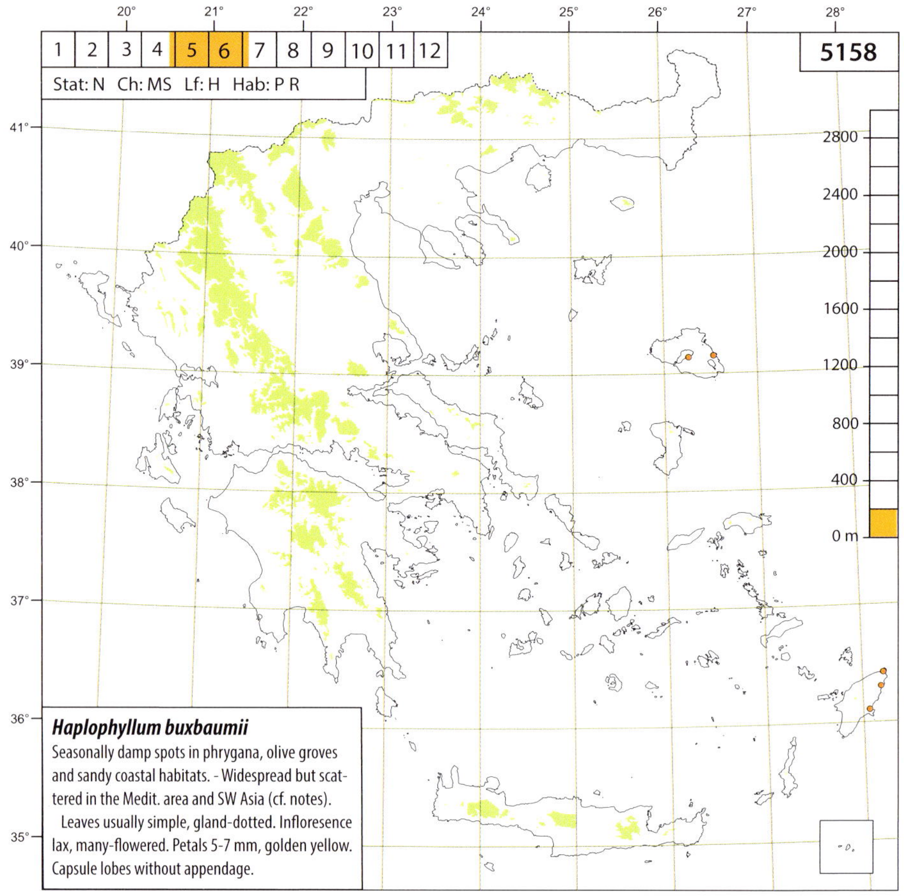

Haplophyllum buxbaumii

Seasonally damp spots in phrygana, olive groves and sandy coastal habitats. - Widespread but scattered in the Medit. area and SW Asia (cf. notes).

Leaves usually simple, gland-dotted. Inflorescence lax, many-flowered. Petals 5-7 mm, golden yellow. Capsule lobes without appendage.

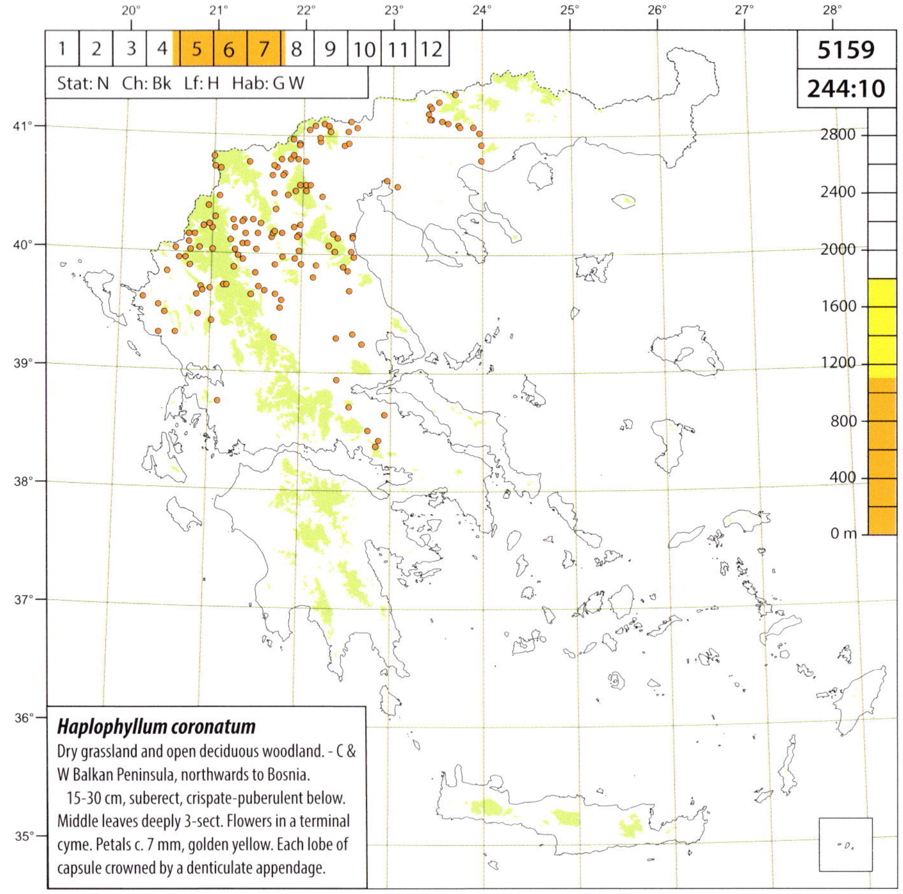

Haplophyllum coronatum

Dry grassland and open deciduous woodland. - C & W Balkan Peninsula, northwards to Bosnia.

15-30 cm, suberect, crispate-puberulent below. Middle leaves deeply 3-sect. Flowers in a terminal cyme. Petals c. 7 mm, golden yellow. Each lobe of capsule crowned by a denticulate appendage.

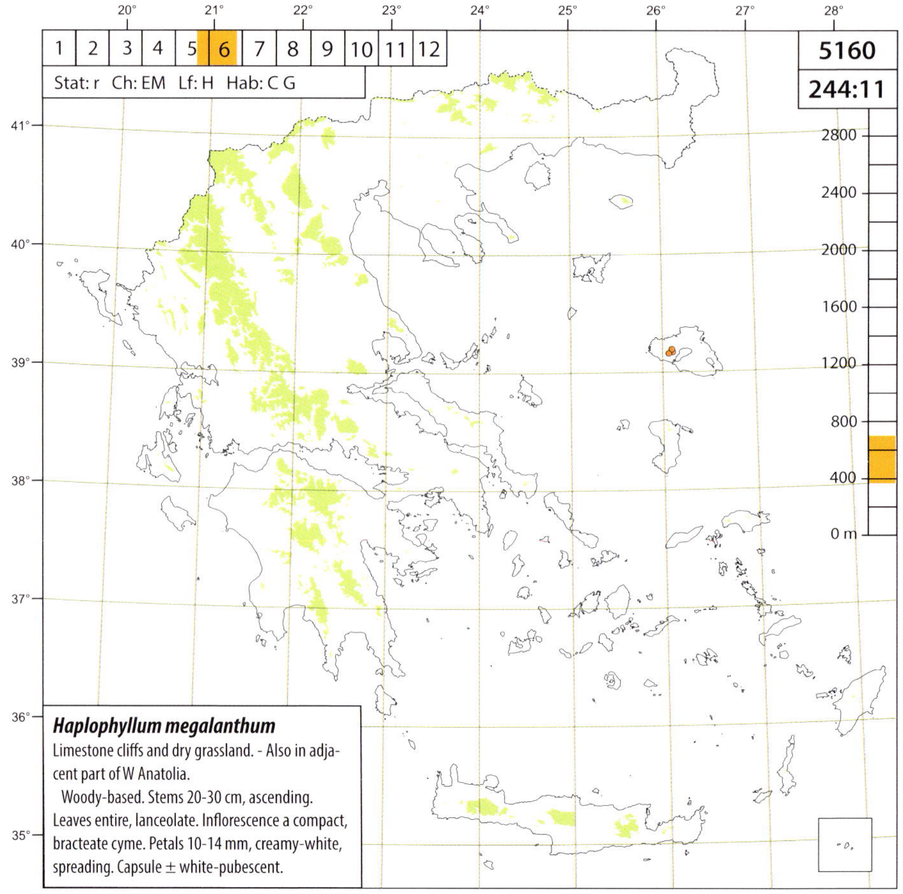

Haplophyllum megalanthum

Limestone cliffs and dry grassland. - Also in adjacent part of W Anatolia.

Woody-based. Stems 20-30 cm, ascending. Leaves entire, lanceolate. Inflorescence a compact, bracteate cyme. Petals 10-14 mm, creamy-white, spreading. Capsule ± white-pubescent.

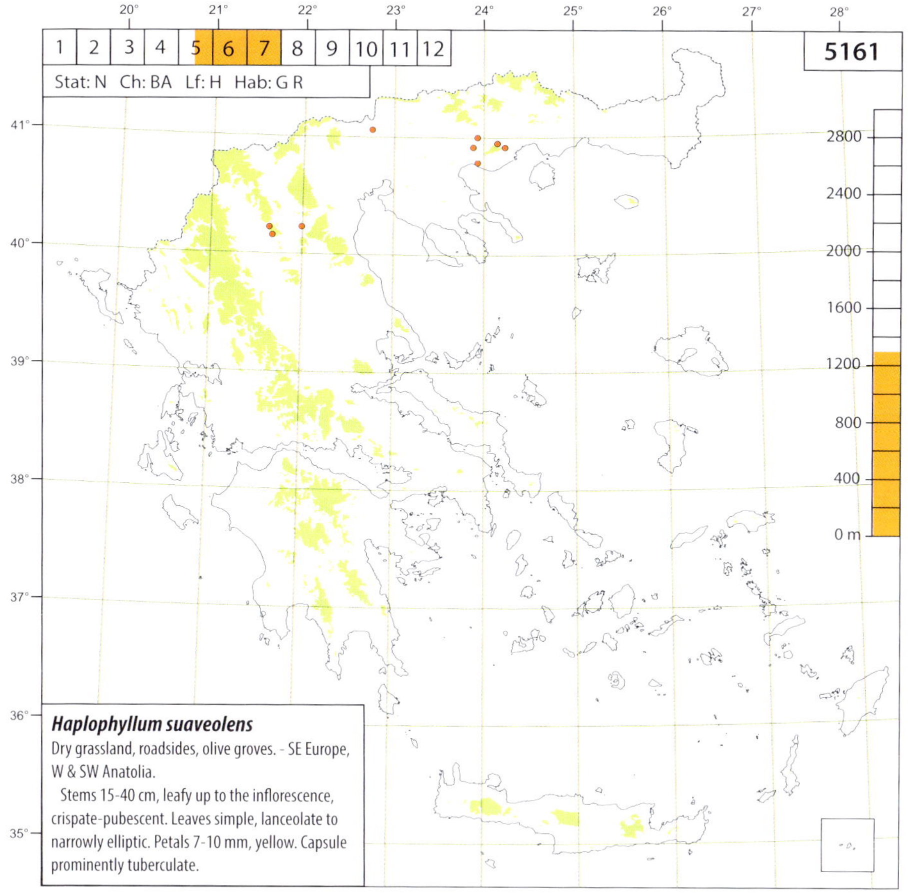

Haplophyllum suaveolens

Dry grassland, roadsides, olive groves. - SE Europe, W & SW Anatolia.

Stems 15-40 cm, leafy up to the inflorescence, crispate-pubescent. Leaves simple, lanceolate to narrowly elliptic. Petals 7-10 mm, yellow. Capsule prominently tuberculate.

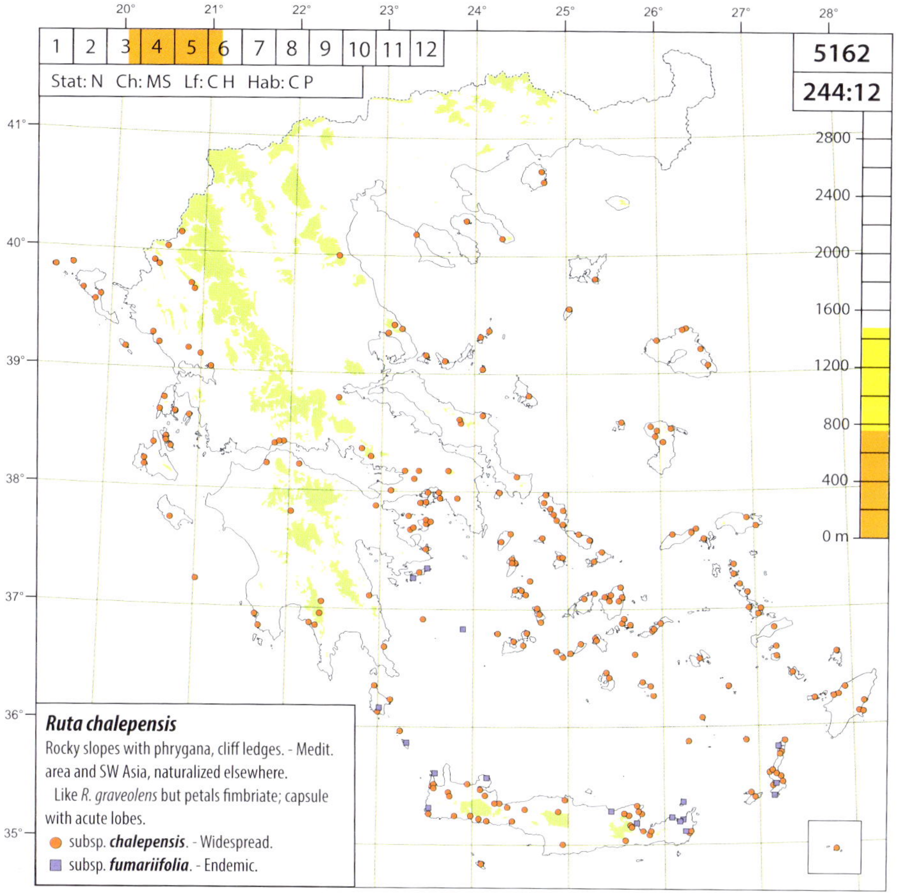

Ruta chalepensis

Rocky slopes with phrygana, cliff ledges. - Medit. area and SW Asia, naturalized elsewhere.

Like *R. graveolens* but petals fimbriate; capsule with acute lobes.

● subsp. ***chalepensis***. - Widespread.

■ subsp. ***fumariifolia***. - Endemic.

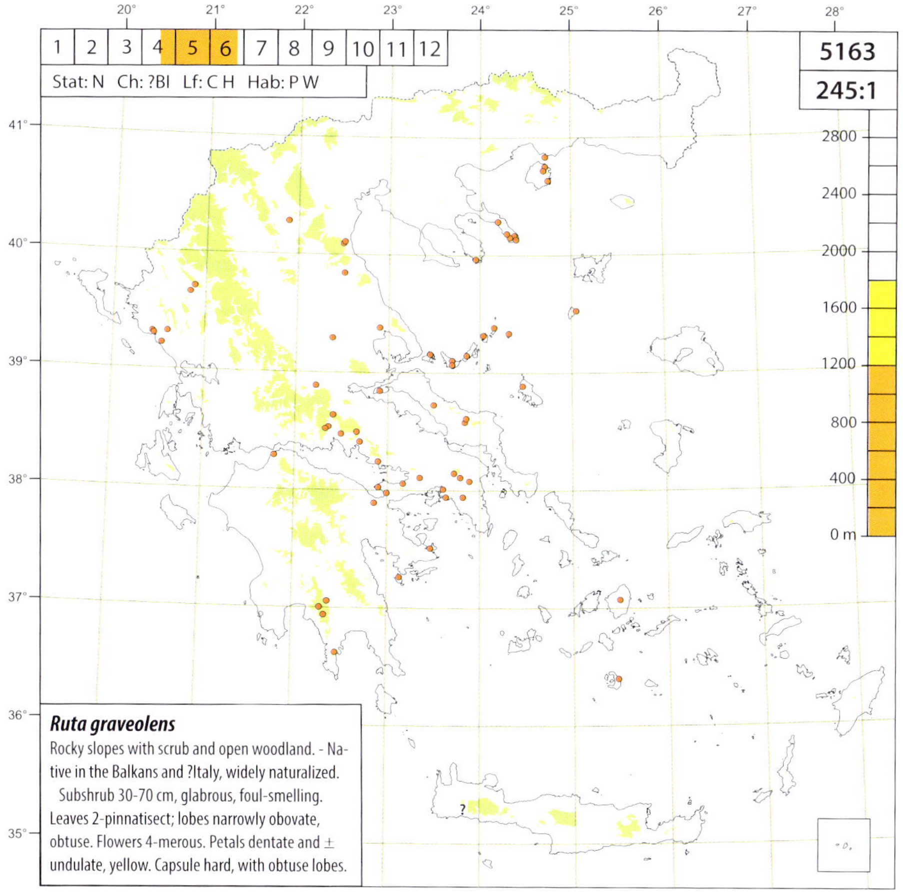

Ruta graveolens

Rocky slopes with scrub and open woodland. - Native in the Balkans and ?Italy, widely naturalized.

Subshrub 30-70 cm, glabrous, foul-smelling. Leaves 2-pinnatisect; lobes narrowly obovate, obtuse. Flowers 4-merous. Petals dentate and ± undulate, yellow. Capsule hard, with obtuse lobes.

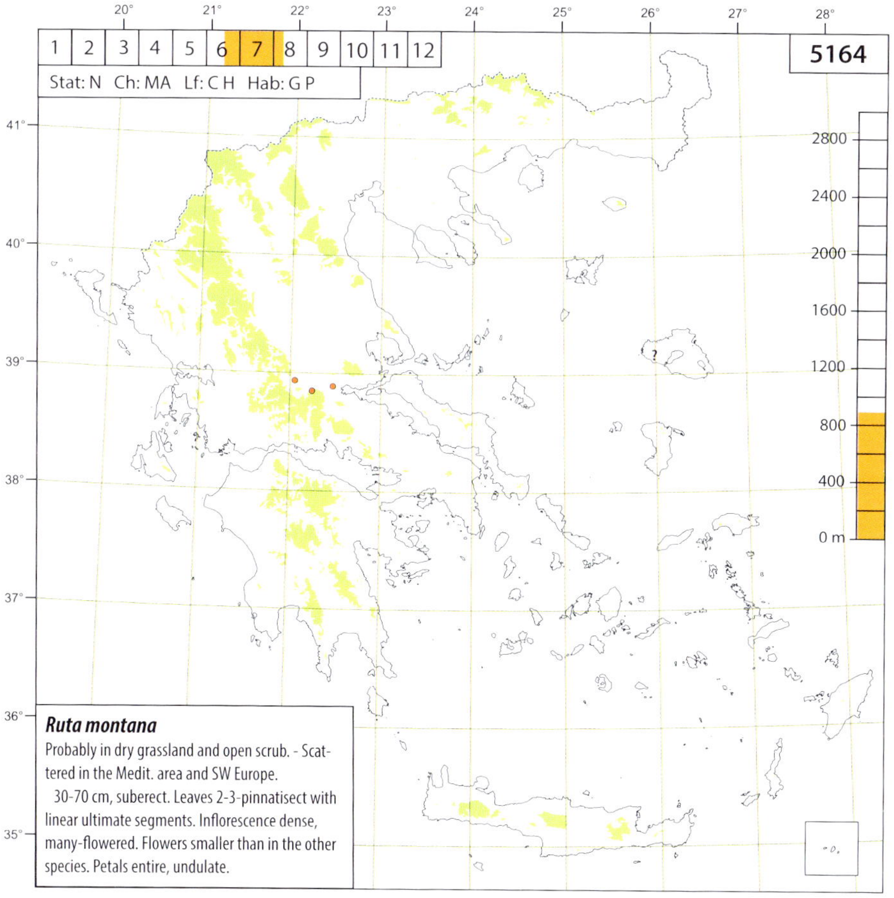

Ruta montana

Probably in dry grassland and open scrub. - Scattered in the Medit. area and SW Europe.

30-70 cm, suberect. Leaves 2-3-pinnatisect with linear ultimate segments. Inflorescence dense, many-flowered. Flowers smaller than in the other species. Petals entire, undulate.

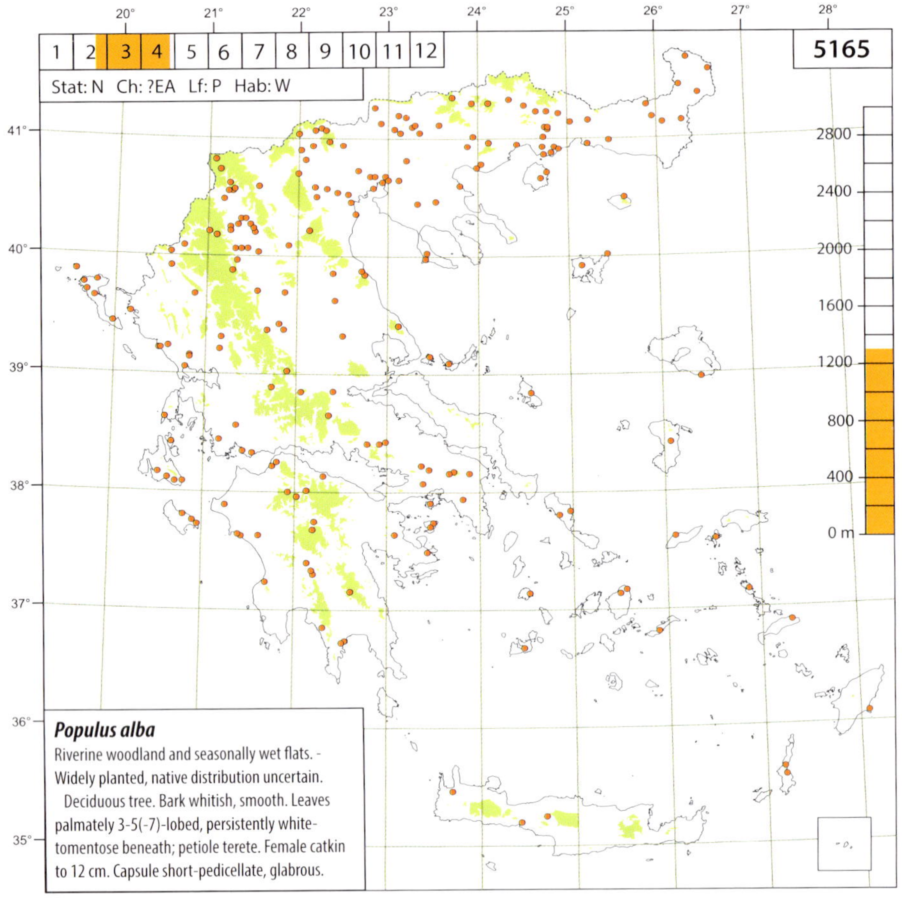

Populus alba

Riverine woodland and seasonally wet flats. - Widely planted, native distribution uncertain.

Deciduous tree. Bark whitish, smooth. Leaves palmately 3-5(-7)-lobed, persistently white-tomentose beneath; petiole terete. Female catkin to 12 cm. Capsule short-pedicellate, glabrous.

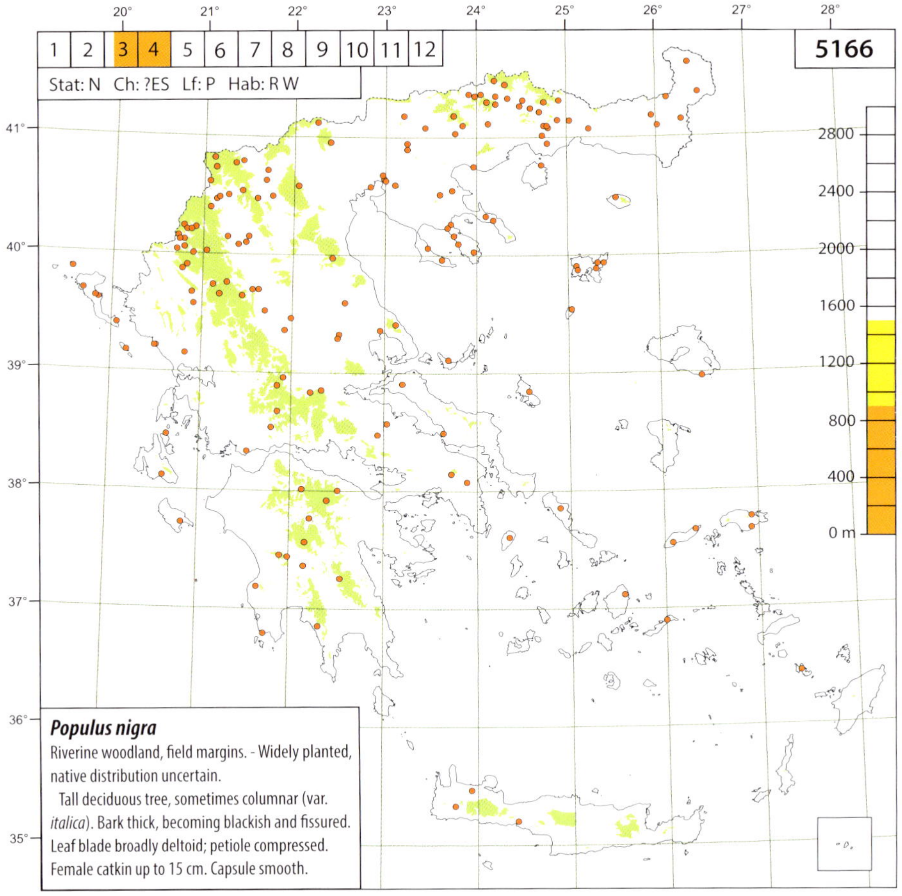

Populus nigra

Riverine woodland, field margins. - Widely planted, native distribution uncertain.

Tall deciduous tree, sometimes columnar (var. *italica*). Bark thick, becoming blackish and fissured. Leaf blade broadly deltoid; petiole compressed. Female catkin up to 15 cm. Capsule smooth.

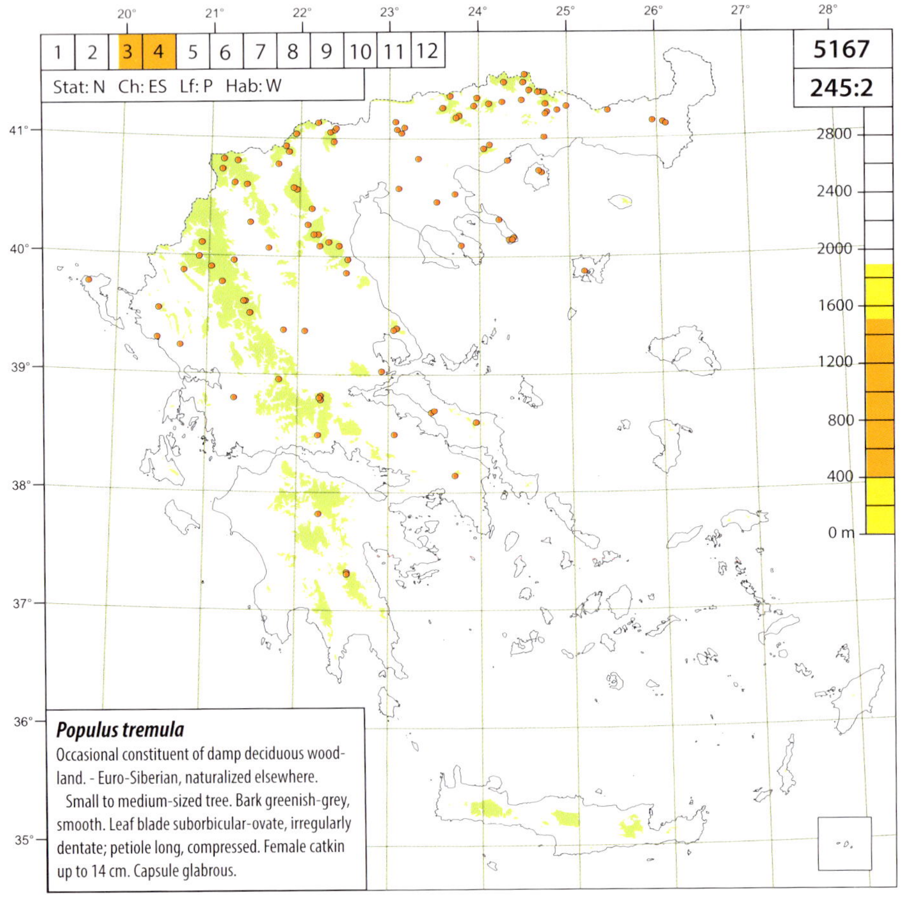

Populus tremula

Occasional constituent of damp deciduous woodland. - Euro-Siberian, naturalized elsewhere.

Small to medium-sized tree. Bark greenish-grey, smooth. Leaf blade suborbicular-ovate, irregularly dentate; petiole long, compressed. Female catkin up to 14 cm. Capsule glabrous.

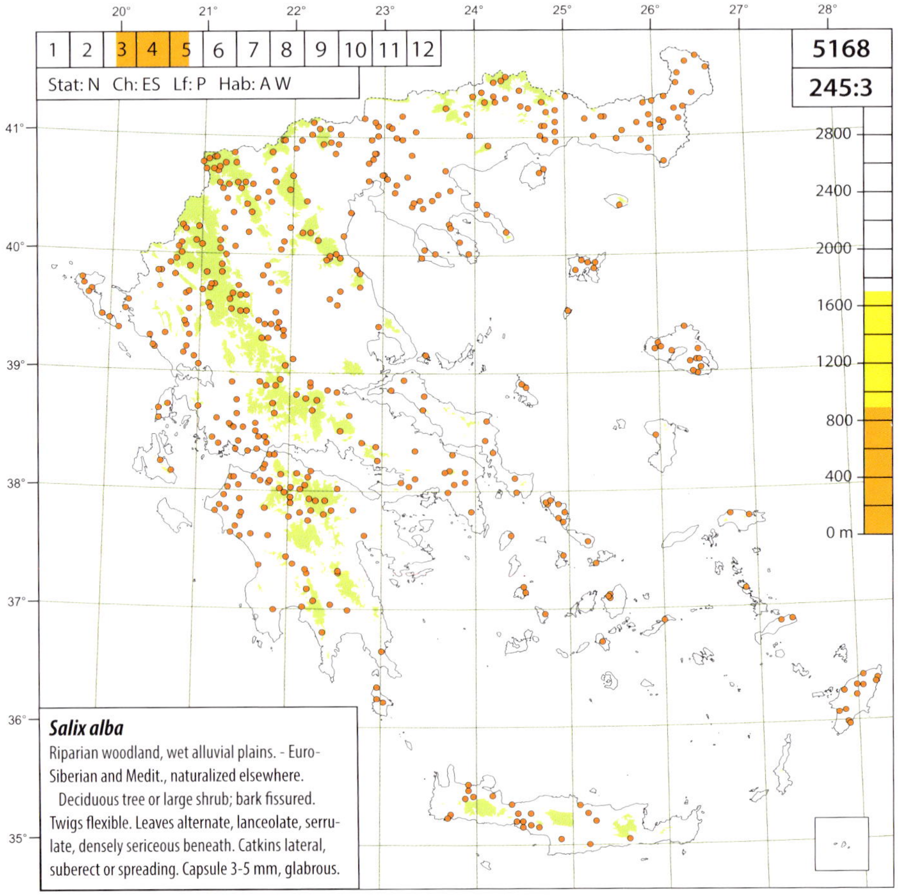

Salix alba

Riparian woodland, wet alluvial plains. - Euro-Siberian and Medit., naturalized elsewhere.

Deciduous tree or large shrub; bark fissured. Twigs flexible. Leaves alternate, lanceolate, serrulate, densely sericeous beneath. Catkins lateral, suberect or spreading. Capsule 3-5 mm, glabrous.

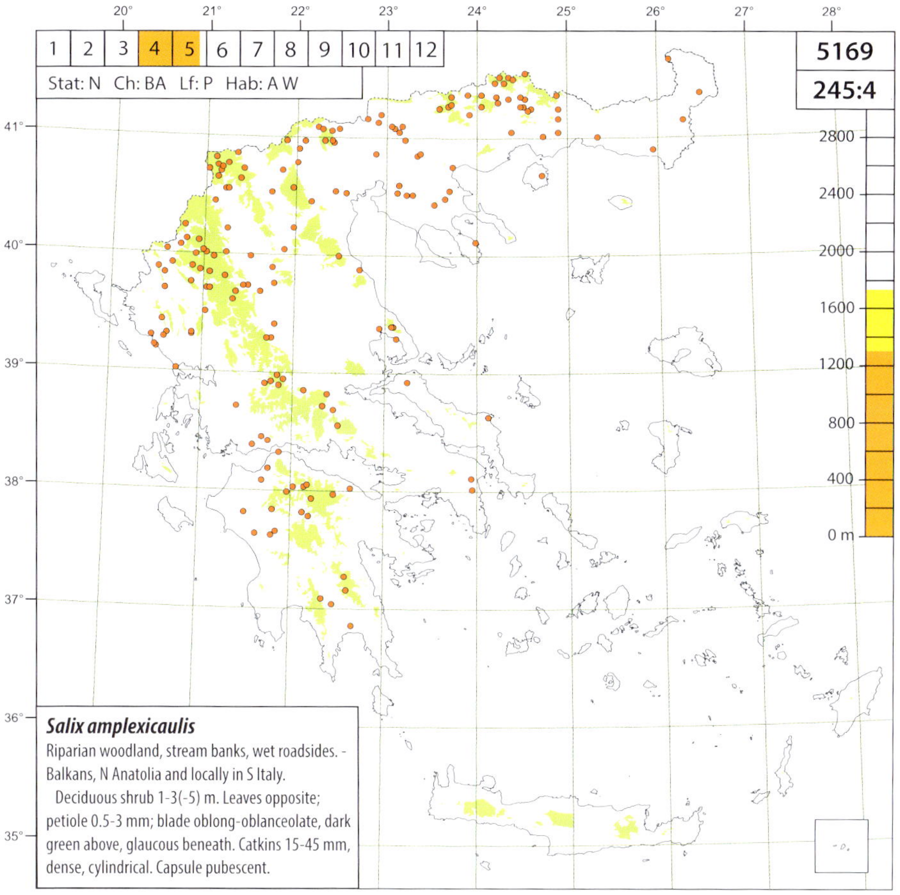

Salix amplexicaulis
Riparian woodland, stream banks, wet roadsides. - Balkans, N Anatolia and locally in S Italy.
Deciduous shrub 1-3(-5) m. Leaves opposite; petiole 0.5-3 mm; blade oblong-oblanceolate, dark green above, glaucous beneath. Catkins 15-45 mm, dense, cylindrical. Capsule pubescent.

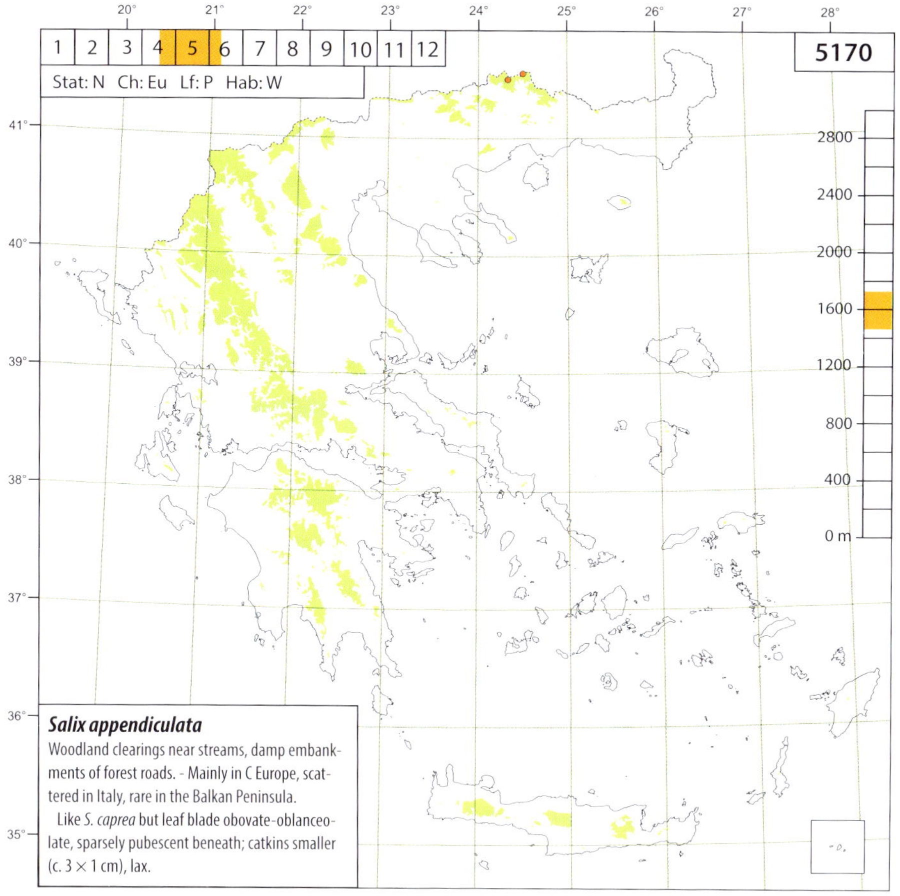

Salix appendiculata
Woodland clearings near streams, damp embankments of forest roads. - Mainly in C Europe, scattered in Italy, rare in the Balkan Peninsula.
Like *S. caprea* but leaf blade obovate-oblanceolate, sparsely pubescent beneath; catkins smaller (c. 3 × 1 cm), lax.

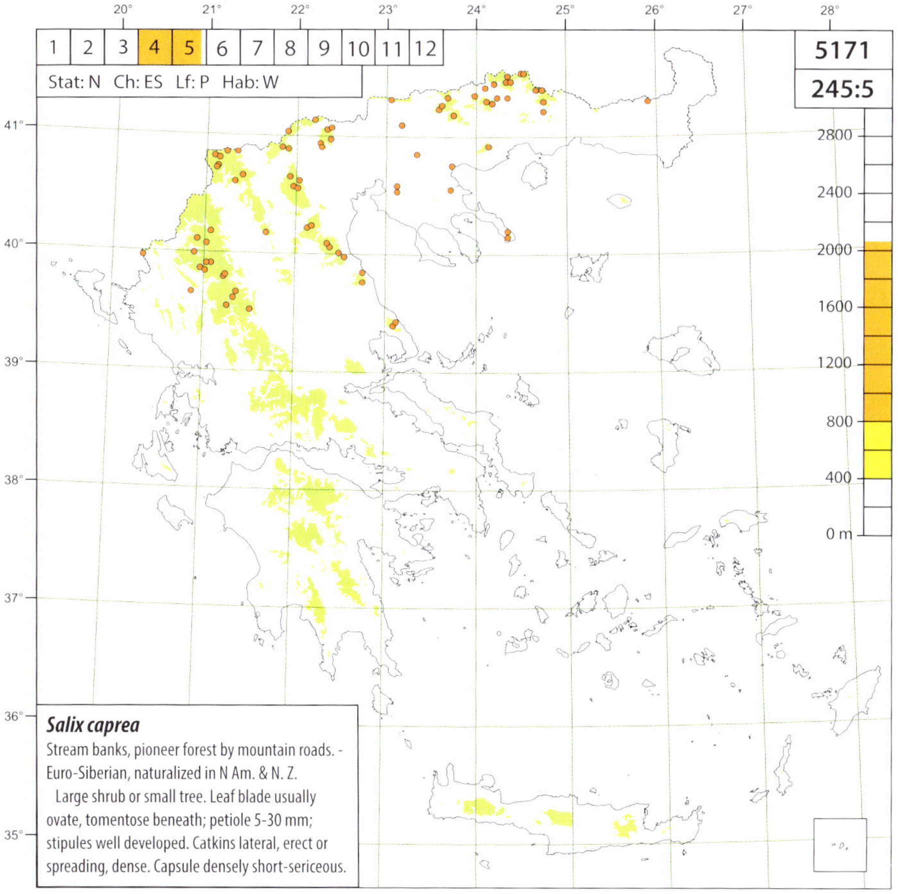

Salix caprea
Stream banks, pioneer forest by mountain roads. - Euro-Siberian, naturalized in N Am. & N. Z.
Large shrub or small tree. Leaf blade usually ovate, tomentose beneath; petiole 5-30 mm; stipules well developed. Catkins lateral, erect or spreading, dense. Capsule densely short-sericeous.

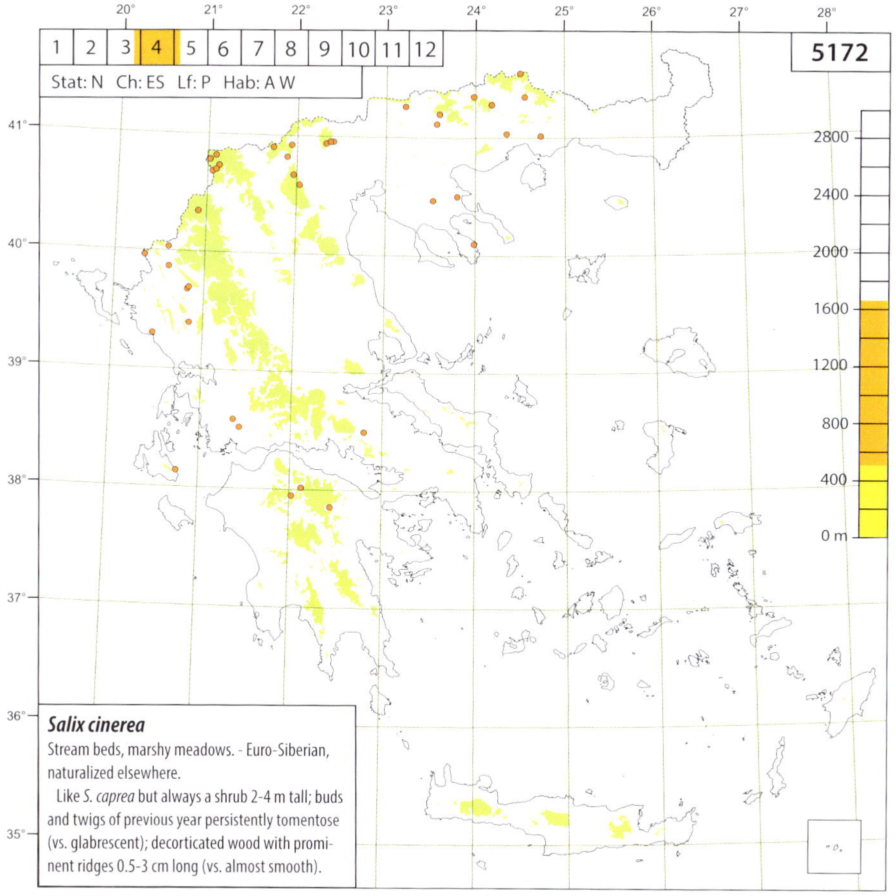

Salix cinerea
Stream beds, marshy meadows. - Euro-Siberian, naturalized elsewhere.
Like *S. caprea* but always a shrub 2-4 m tall; buds and twigs of previous year persistently tomentose (vs. glabrescent); decorticated wood with prominent ridges 0.5-3 cm long (vs. almost smooth).

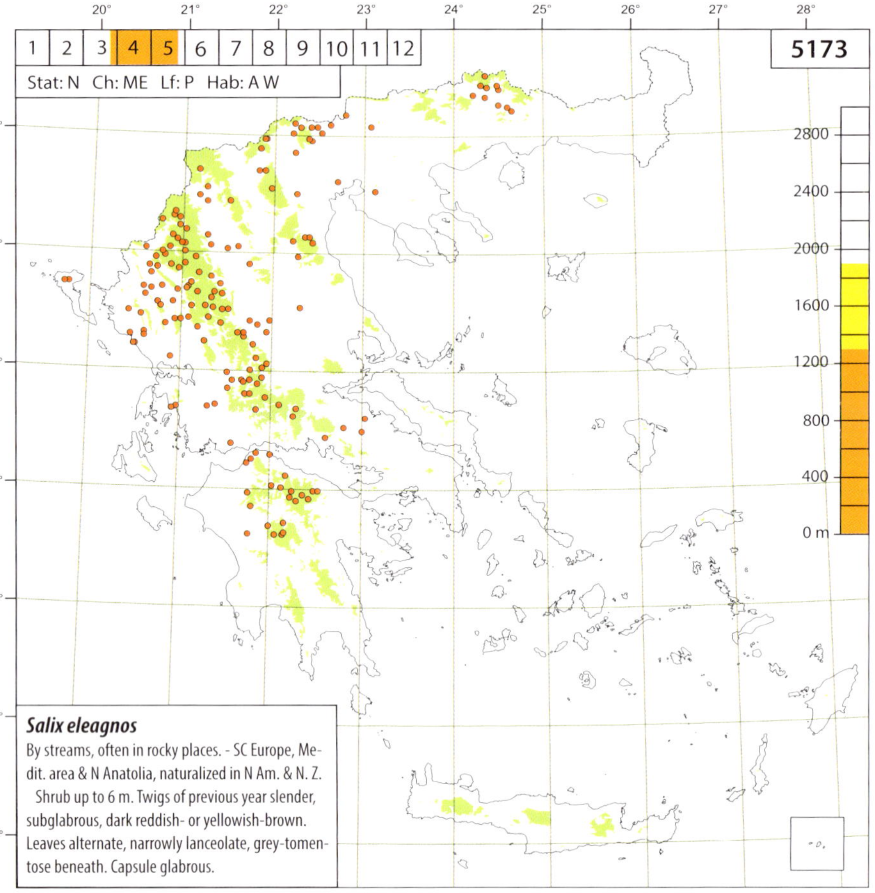

Salix eleagnos
By streams, often in rocky places. - SC Europe, Medit. area & N Anatolia, naturalized in N Am. & N. Z.
Shrub up to 6 m. Twigs of previous year slender, subglabrous, dark reddish- or yellowish-brown. Leaves alternate, narrowly lanceolate, grey-tomentose beneath. Capsule glabrous.

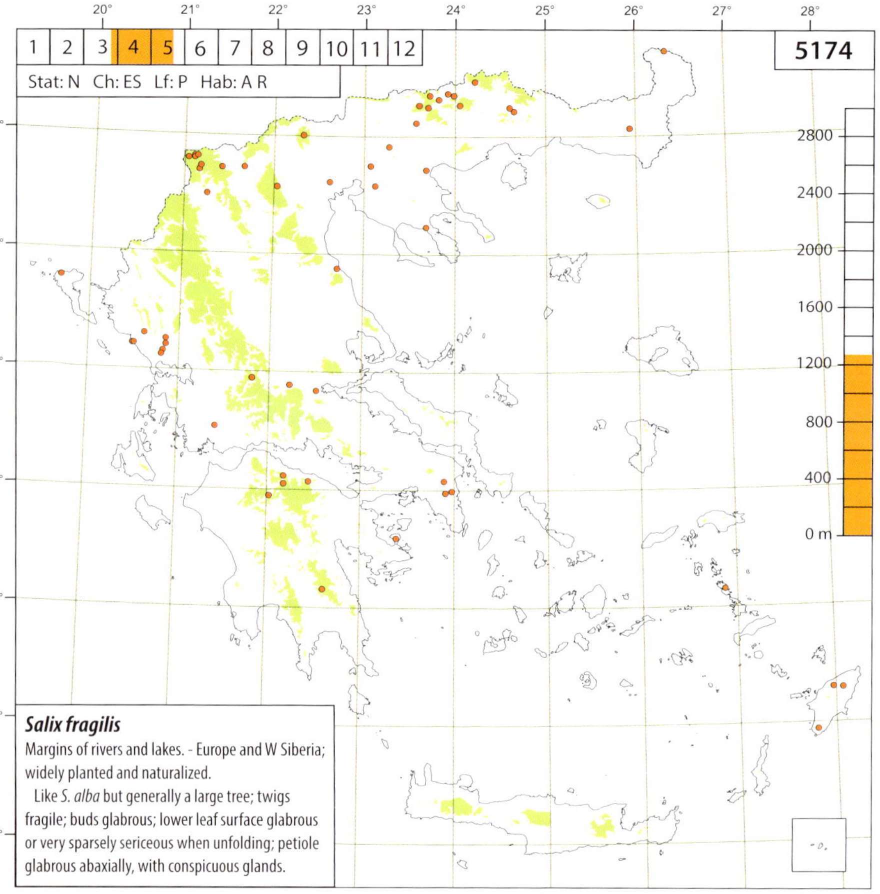

Salix fragilis
Margins of rivers and lakes. - Europe and W Siberia; widely planted and naturalized.
Like *S. alba* but generally a large tree; twigs fragile; buds glabrous; lower leaf surface glabrous or very sparsely sericeous when unfolding; petiole glabrous abaxially, with conspicuous glands.

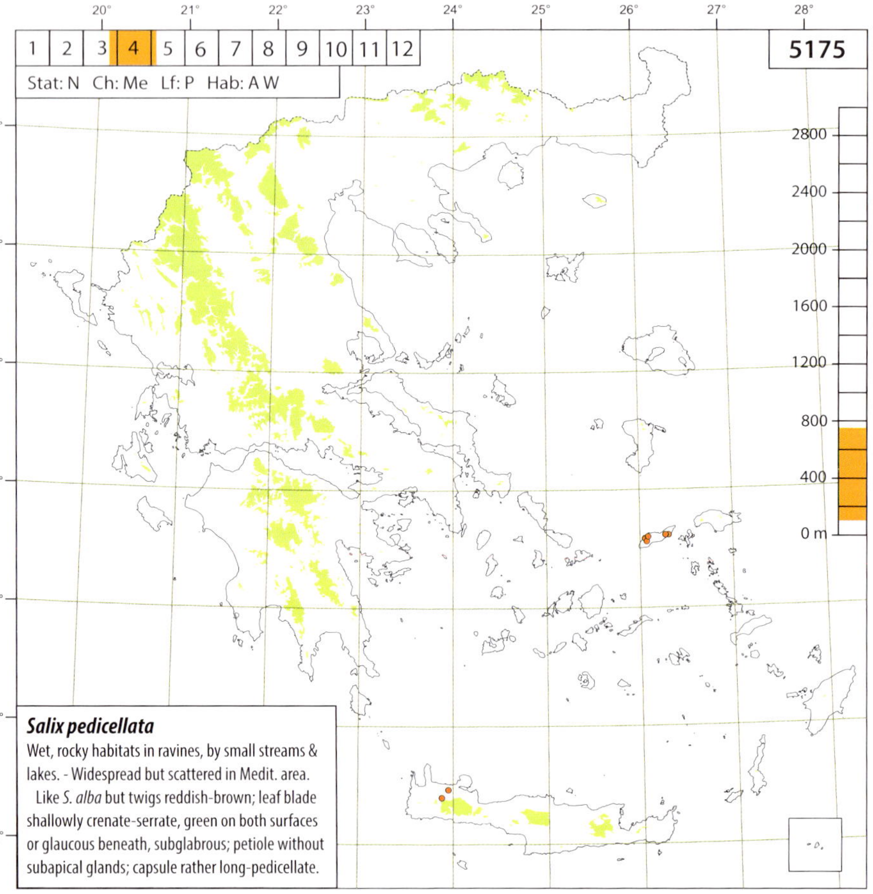

Salix pedicellata
Wet, rocky habitats in ravines, by small streams & lakes. - Widespread but scattered in Medit. area.
Like *S. alba* but twigs reddish-brown; leaf blade shallowly crenate-serrate, green on both surfaces or glaucous beneath, subglabrous; petiole without subapical glands; capsule rather long-pedicellate.

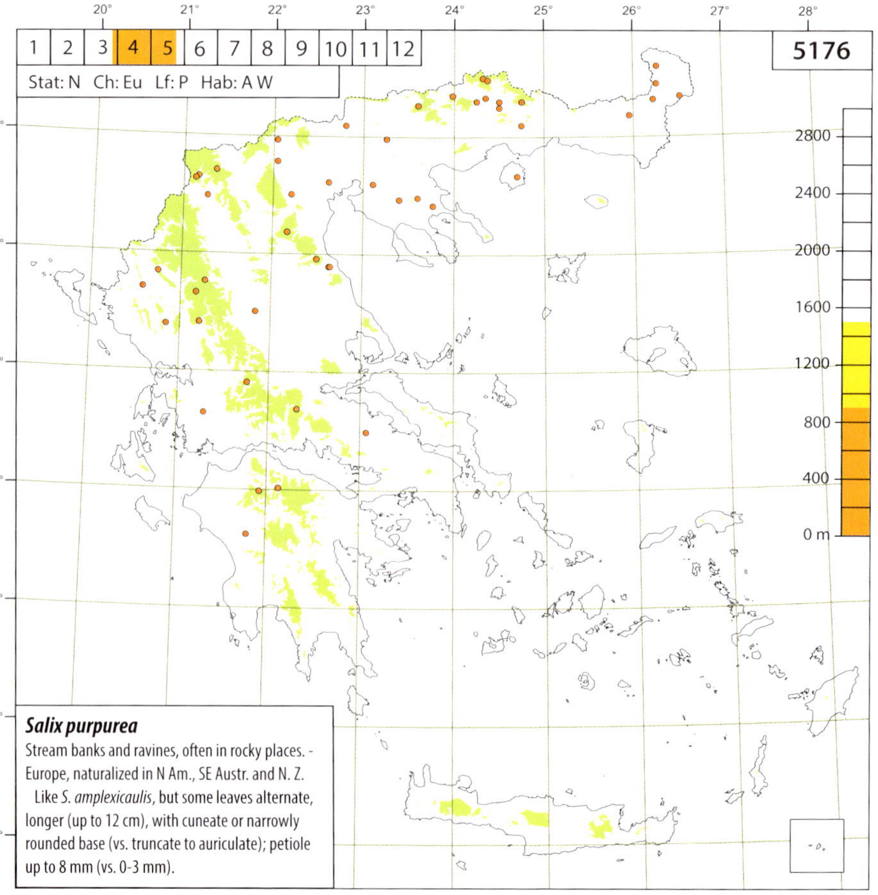

Salix purpurea
Stream banks and ravines, often in rocky places. - Europe, naturalized in N Am., SE Austr. and N. Z.
Like *S. amplexicaulis*, but some leaves alternate, longer (up to 12 cm), with cuneate or narrowly rounded base (vs. truncate to auriculate); petiole up to 8 mm (vs. 0-3 mm).

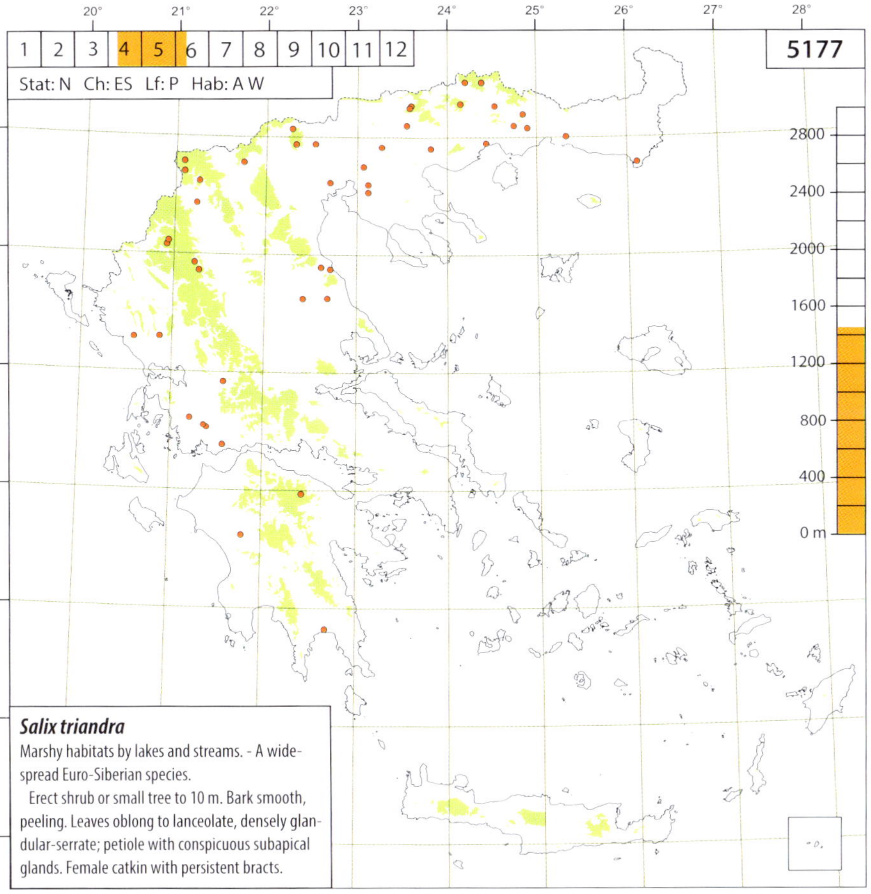

Salix triandra

Marshy habitats by lakes and streams. - A widespread Euro-Siberian species.

Erect shrub or small tree to 10 m. Bark smooth, peeling. Leaves oblong to lanceolate, densely glandular-serrate; petiole with conspicuous subapical glands. Female catkin with persistent bracts.

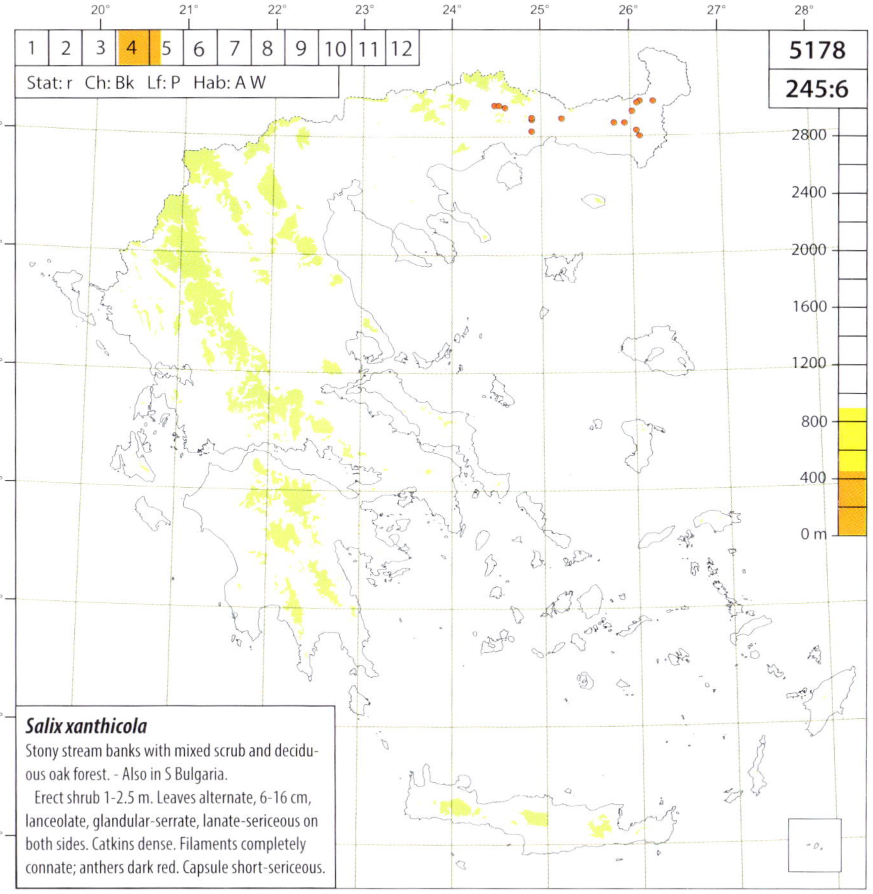

Salix xanthicola

Stony stream banks with mixed scrub and deciduous oak forest. - Also in S Bulgaria.

Erect shrub 1-2.5 m. Leaves alternate, 6-16 cm, lanceolate, glandular-serrate, lanate-sericeous on both sides. Catkins dense. Filaments completely connate; anthers dark red. Capsule short-sericeous.

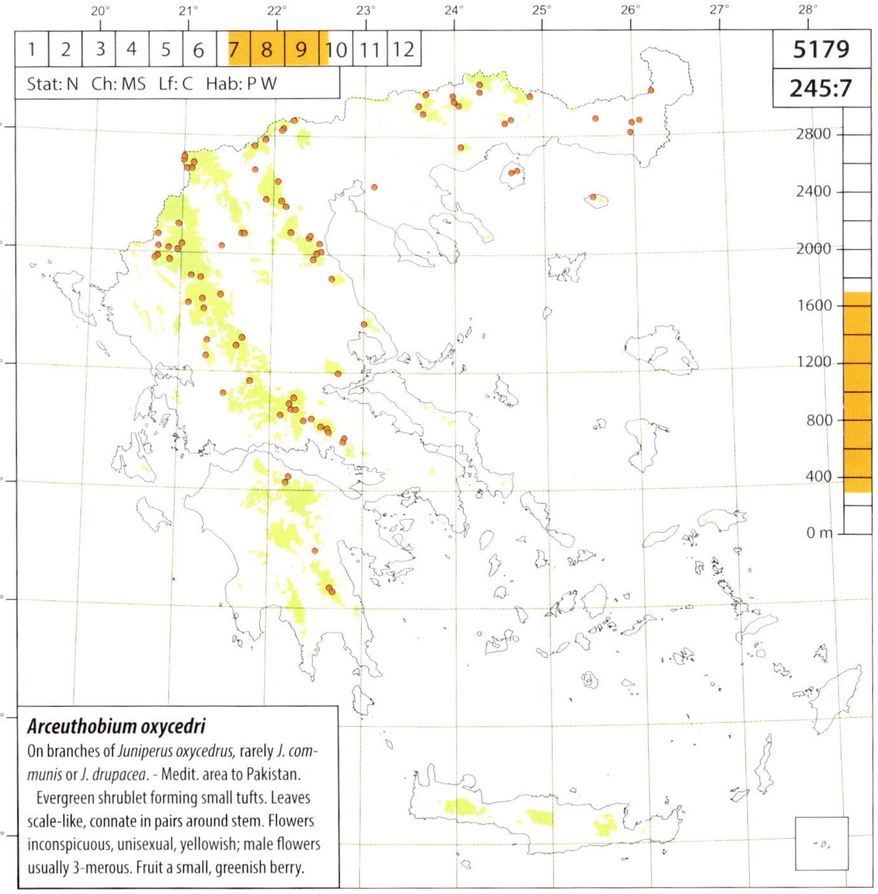

Arceuthobium oxycedri

On branches of *Juniperus oxycedrus*, rarely *J. communis* or *J. drupacea*. - Medit. area to Pakistan.

Evergreen shrublet forming small tufts. Leaves scale-like, connate in pairs around stem. Flowers inconspicuous, unisexual, yellowish; male flowers usually 3-merous. Fruit a small, greenish berry.

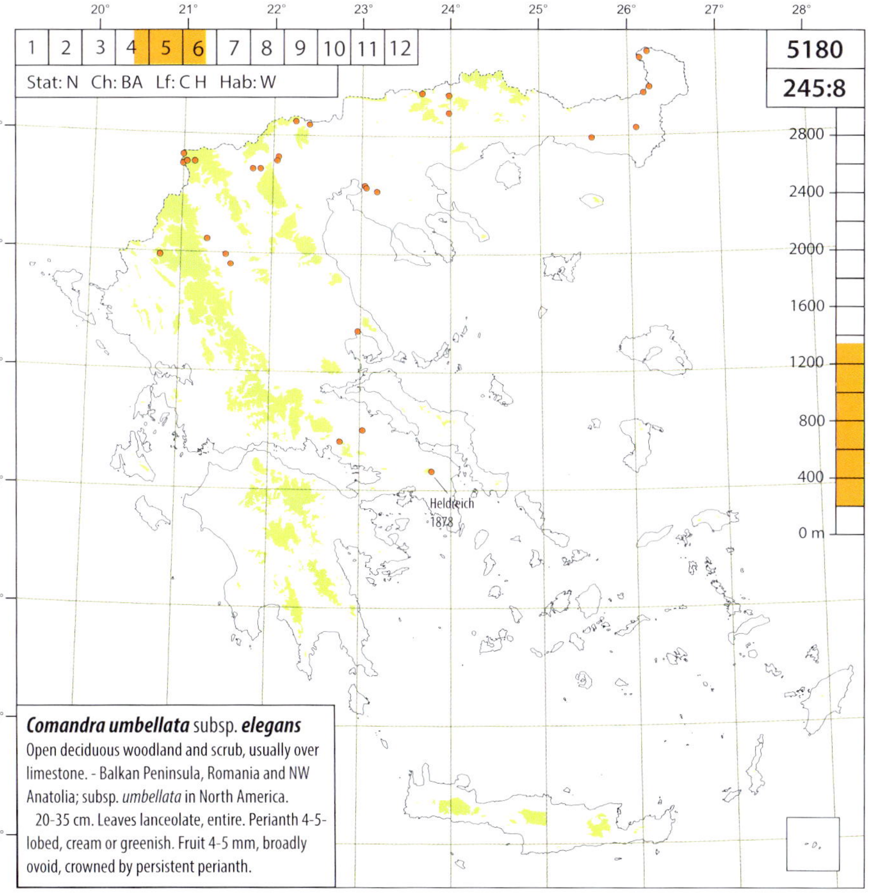

Comandra umbellata subsp. *elegans*

Open deciduous woodland and scrub, usually over limestone. - Balkan Peninsula, Romania and NW Anatolia; subsp. *umbellata* in North America.

20-35 cm. Leaves lanceolate, entire. Perianth 4-5-lobed, cream or greenish. Fruit 4-5 mm, broadly ovoid, crowned by persistent perianth.

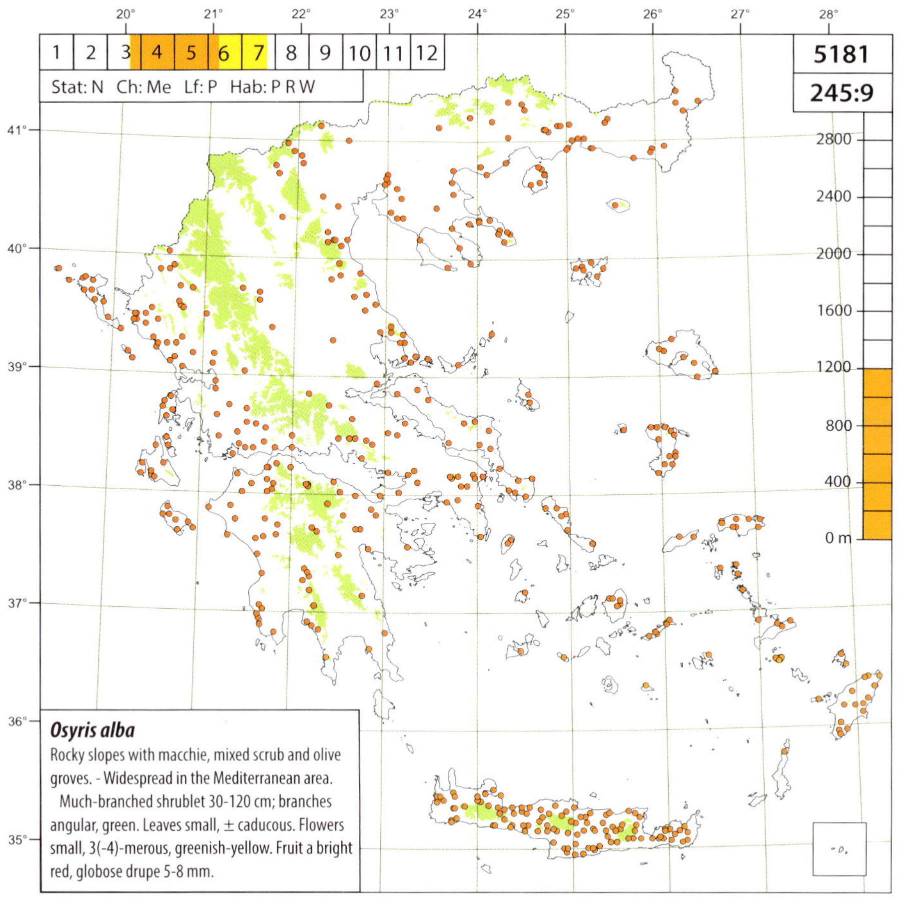

Osyris alba
Rocky slopes with macchie, mixed scrub and olive groves. - Widespread in the Mediterranean area.
Much-branched shrublet 30-120 cm; branches angular, green. Leaves small, ± caducous. Flowers small, 3(-4)-merous, greenish-yellow. Fruit a bright red, globose drupe 5-8 mm.

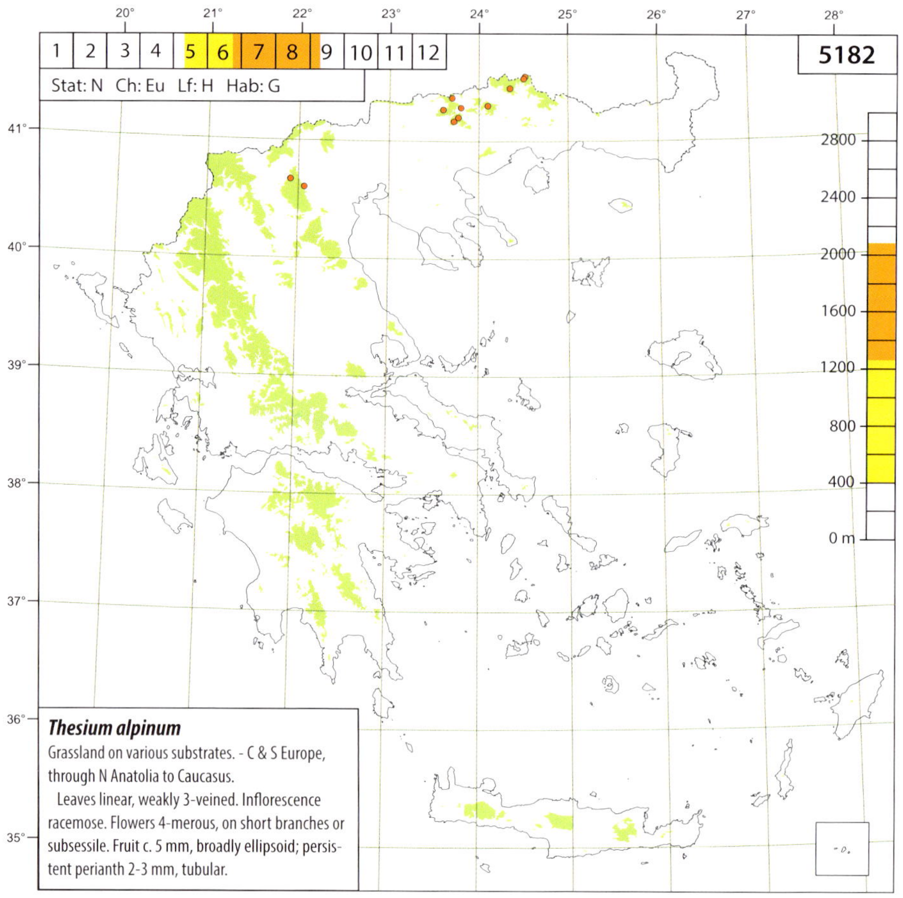

Thesium alpinum
Grassland on various substrates. - C & S Europe, through N Anatolia to Caucasus.
Leaves linear, weakly 3-veined. Inflorescence racemose. Flowers 4-merous, on short branches or subsessile. Fruit c. 5 mm, broadly ellipsoid; persistent perianth 2-3 mm, tubular.

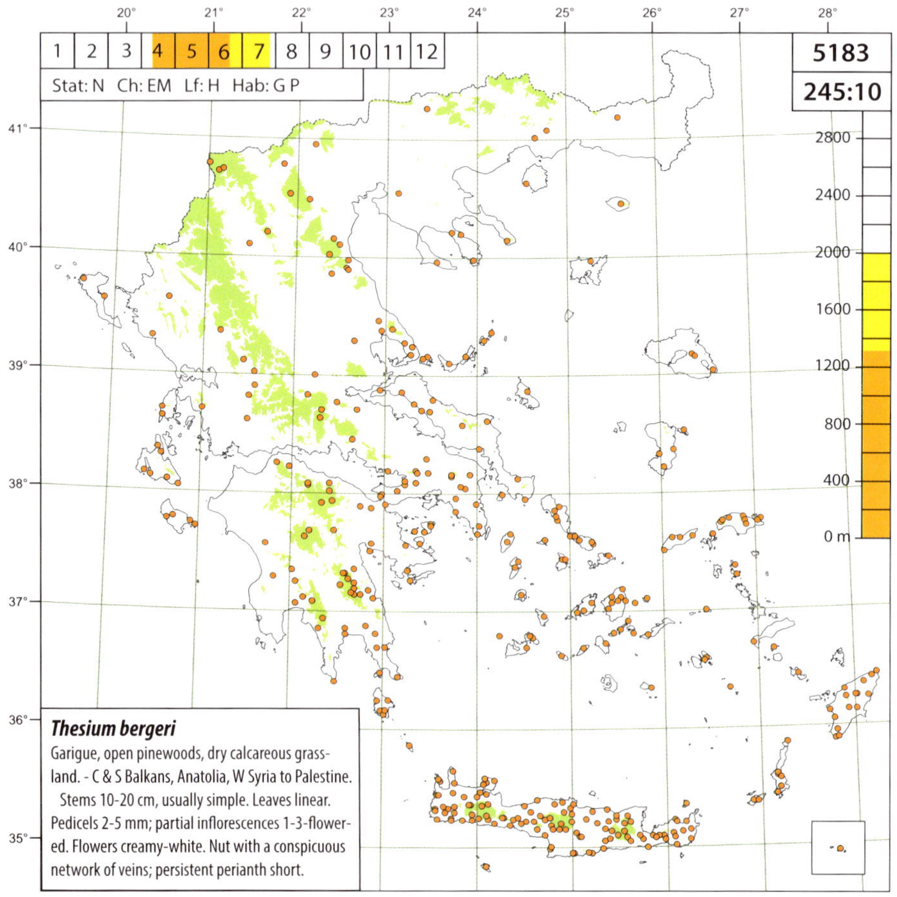

Thesium bergeri
Garigue, open pinewoods, dry calcareous grassland. - C & S Balkans, Anatolia, W Syria to Palestine.
Stems 10-20 cm, usually simple. Leaves linear. Pedicels 2-5 mm; partial inflorescences 1-3-flowered. Flowers creamy-white. Nut with a conspicuous network of veins; persistent perianth short.

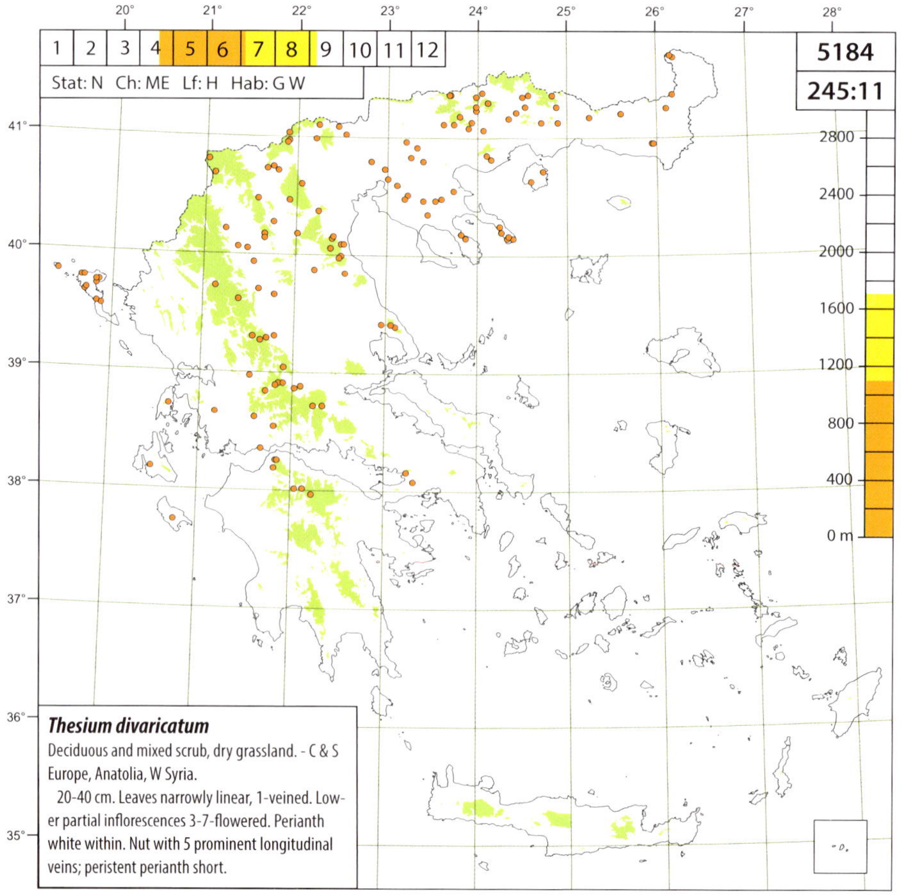

Thesium divaricatum
Deciduous and mixed scrub, dry grassland. - C & S Europe, Anatolia, W Syria.
20-40 cm. Leaves narrowly linear, 1-veined. Lower partial inflorescences 3-7-flowered. Perianth white within. Nut with 5 prominent longitudinal veins; peristent perianth short.

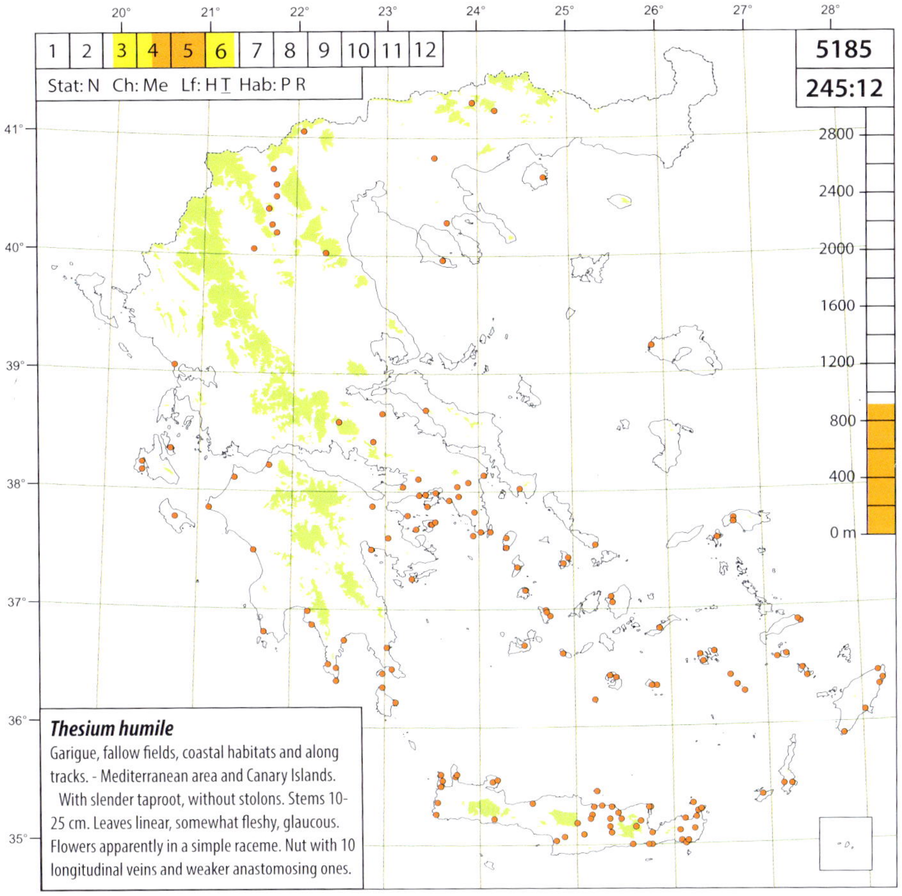

Thesium humile
Garigue, fallow fields, coastal habitats and along tracks. - Mediterranean area and Canary Islands.
With slender taproot, without stolons. Stems 10-25 cm. Leaves linear, somewhat fleshy, glaucous. Flowers apparently in a simple raceme. Nut with 10 longitudinal veins and weaker anastomosing ones.

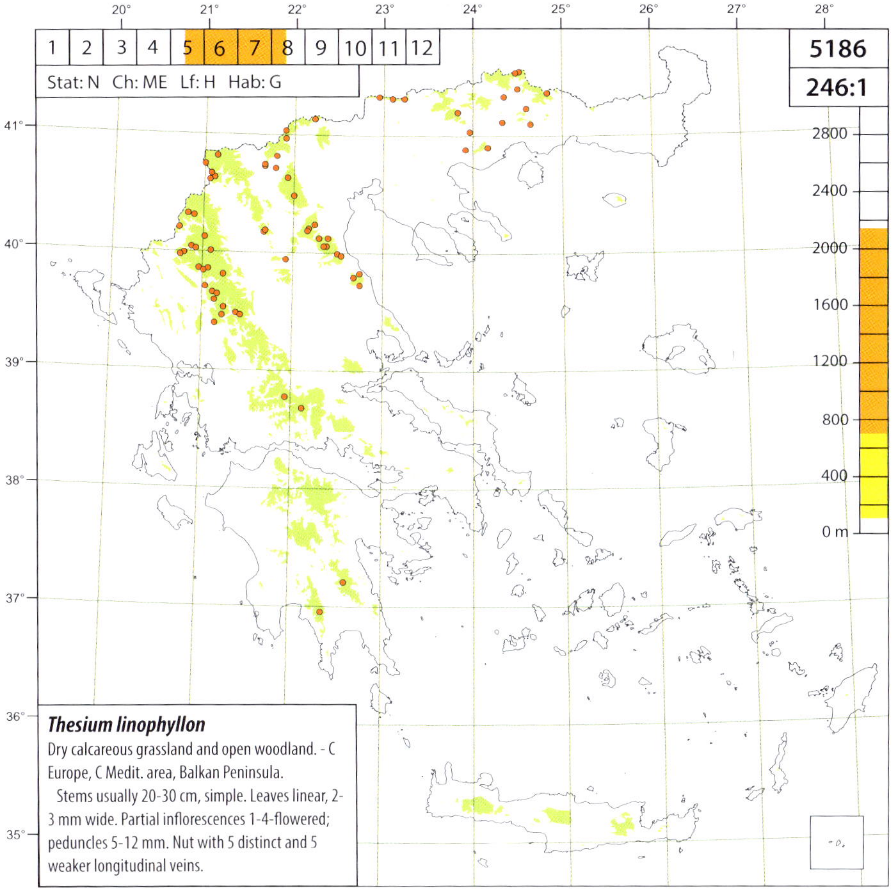

Thesium linophyllon
Dry calcareous grassland and open woodland. - C Europe, C Medit. area, Balkan Peninsula.
Stems usually 20-30 cm, simple. Leaves linear, 2-3 mm wide. Partial inflorescences 1-4-flowered; peduncles 5-12 mm. Nut with 5 distinct and 5 weaker longitudinal veins.

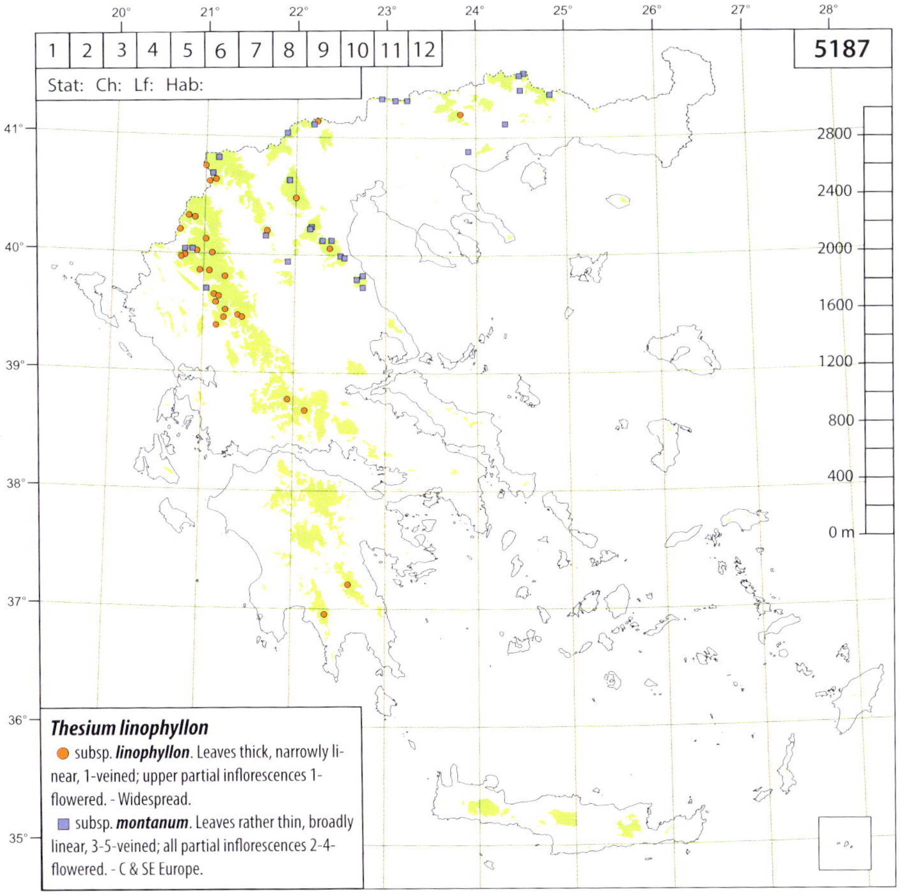

Thesium linophyllon
● subsp. ***linophyllon***. Leaves thick, narrowly linear, 1-veined; upper partial inflorescences 1-flowered. - Widespread.
■ subsp. ***montanum***. Leaves rather thin, broadly linear, 3-5-veined; all partial inflorescences 2-4-flowered. - C & SE Europe.

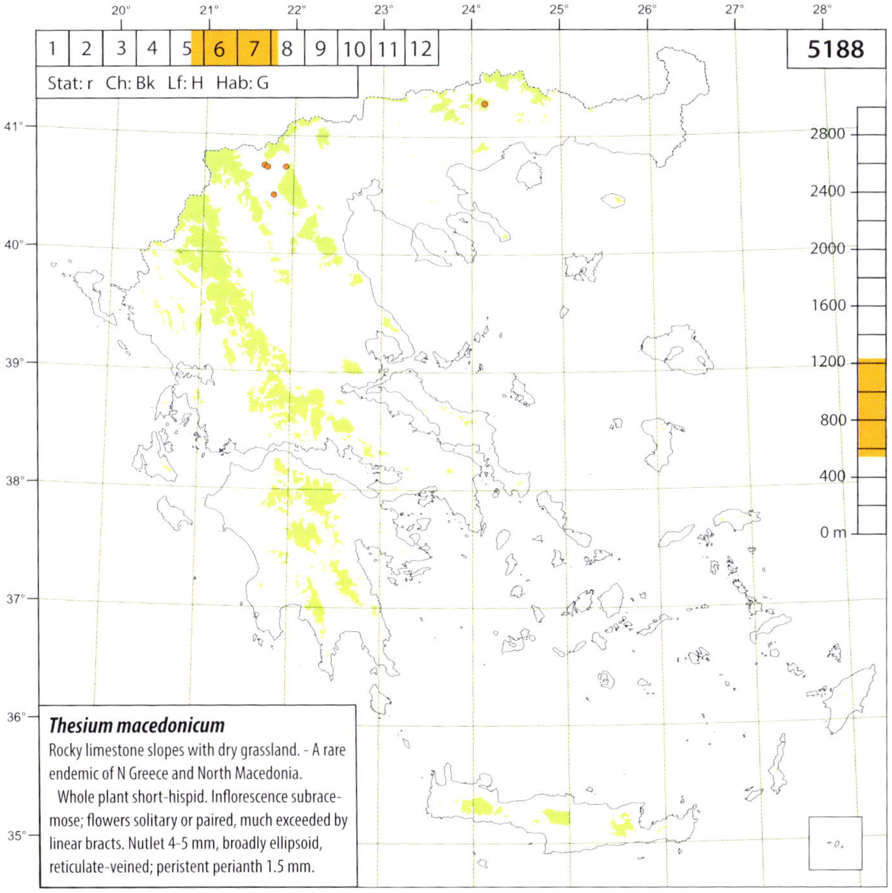

Thesium macedonicum
Rocky limestone slopes with dry grassland. - A rare endemic of N Greece and North Macedonia.
Whole plant short-hispid. Inflorescence subracemose; flowers solitary or paired, much exceeded by linear bracts. Nutlet 4-5 mm, broadly ellipsoid, reticulate-veined; persistent perianth 1.5 mm.

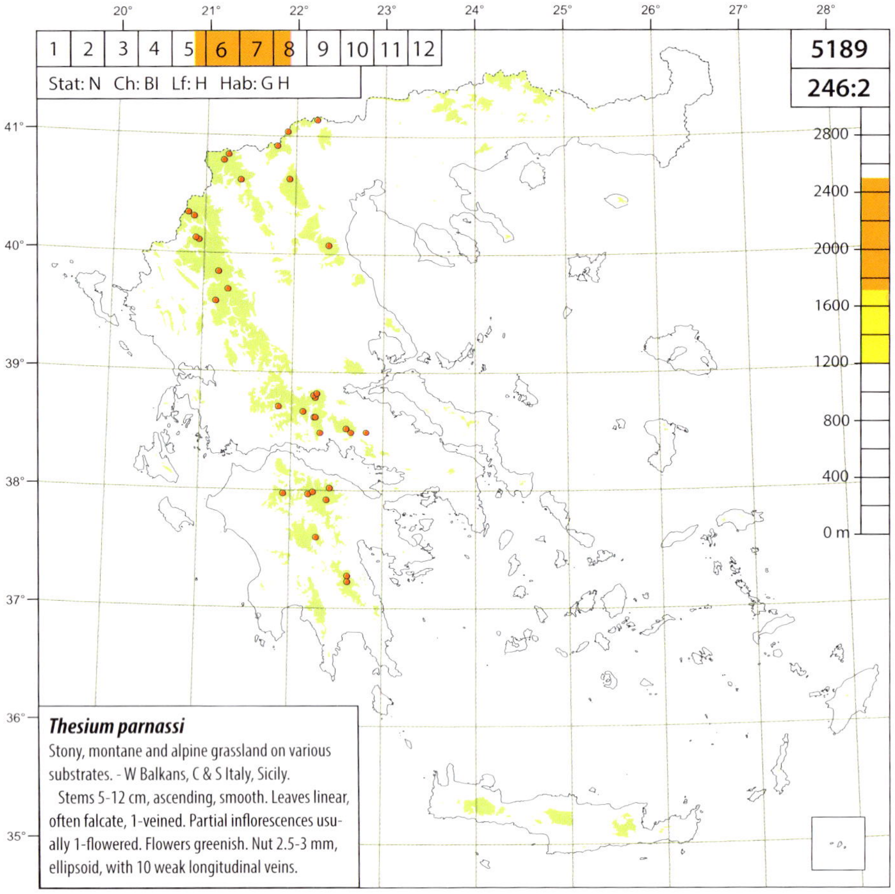

Thesium parnassi

Stony, montane and alpine grassland on various substrates. - W Balkans, C & S Italy, Sicily.

Stems 5-12 cm, ascending, smooth. Leaves linear, often falcate, 1-veined. Partial inflorescences usually 1-flowered. Flowers greenish. Nut 2.5-3 mm, ellipsoid, with 10 weak longitudinal veins.

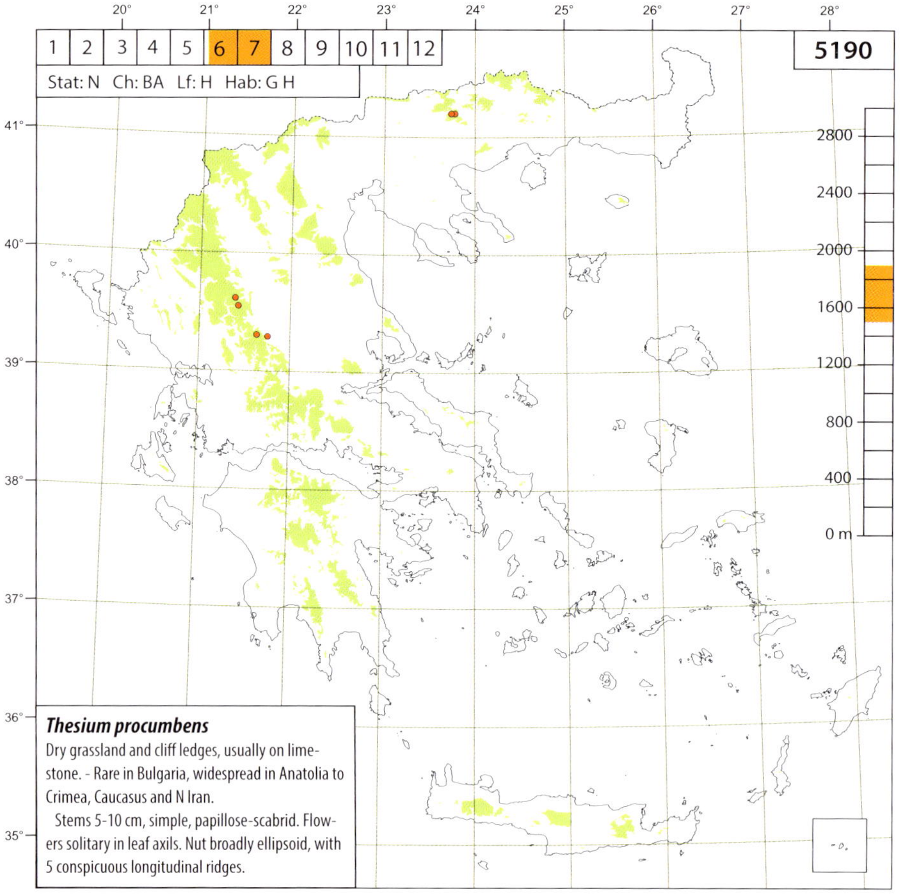

Thesium procumbens

Dry grassland and cliff ledges, usually on limestone. - Rare in Bulgaria, widespread in Anatolia to Crimea, Caucasus and N Iran.

Stems 5-10 cm, simple, papillose-scabrid. Flowers solitary in leaf axils. Nut broadly ellipsoid, with 5 conspicuous longitudinal ridges.

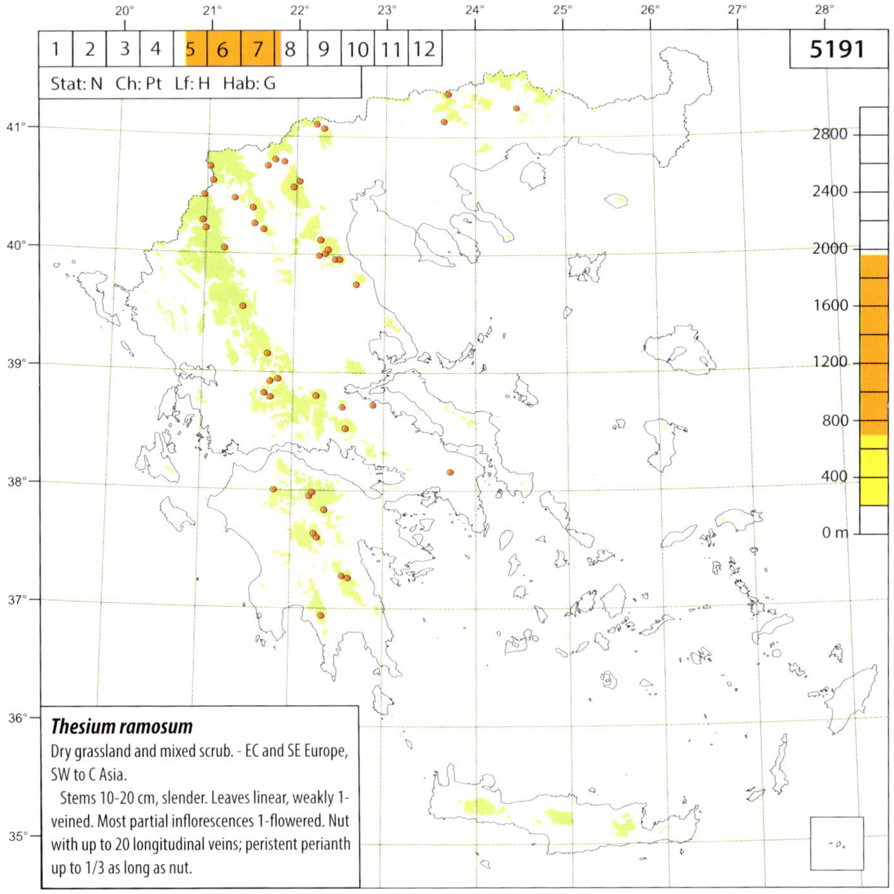

Thesium ramosum

Dry grassland and mixed scrub. - EC and SE Europe, SW to C Asia.

Stems 10-20 cm, slender. Leaves linear, weakly 1-veined. Most partial inflorescences 1-flowered. Nut with up to 20 longitudinal veins; persistent perianth up to 1/3 as long as nut.

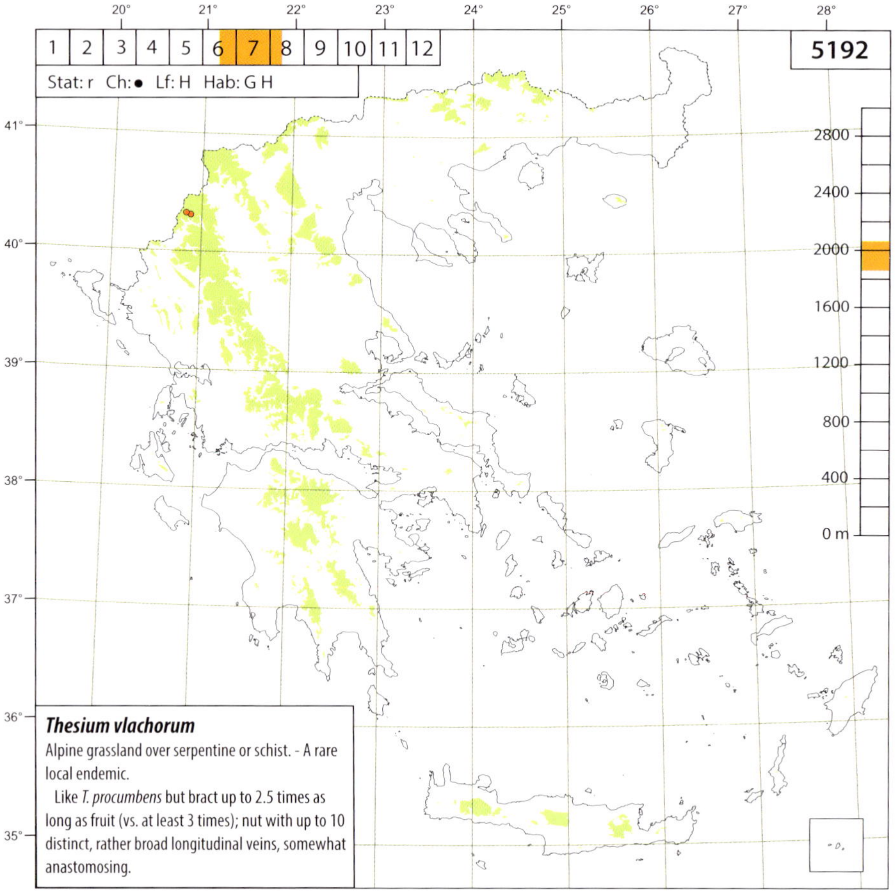

Thesium vlachorum

Alpine grassland over serpentine or schist. - A rare local endemic.

Like *T. procumbens* but bract up to 2.5 times as long as fruit (vs. at least 3 times); nut with up to 10 distinct, rather broad longitudinal veins, somewhat anastomosing.

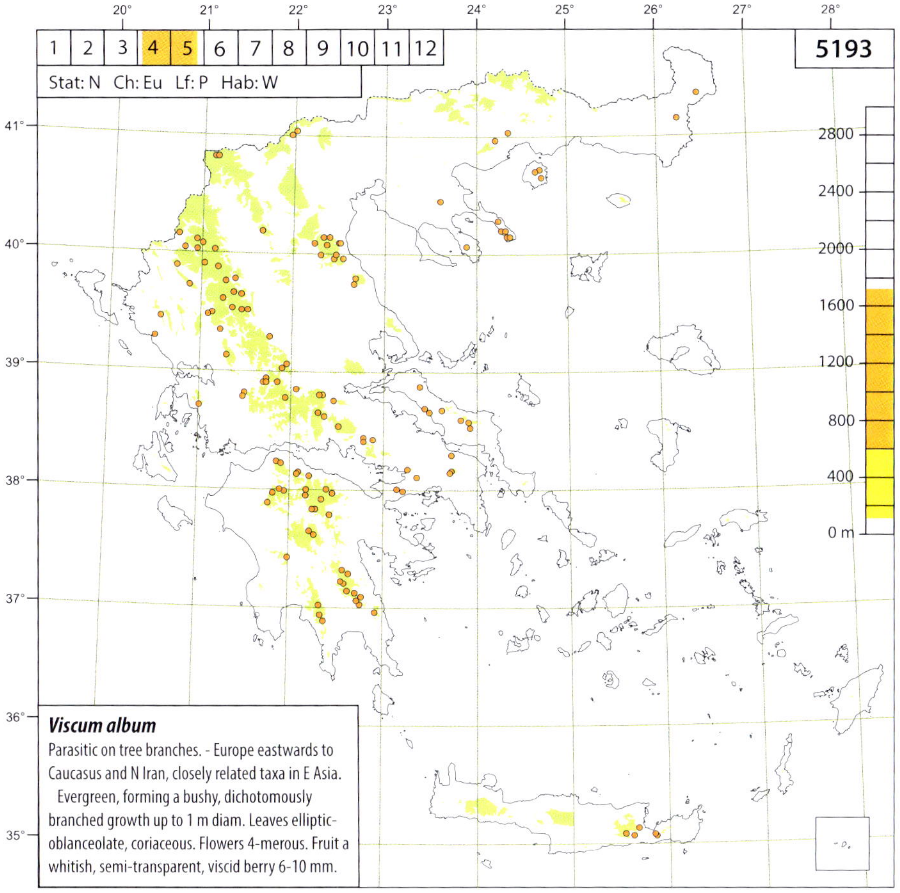

Viscum album

Parasitic on tree branches. - Europe eastwards to Caucasus and N Iran, closely related taxa in E Asia.

Evergreen, forming a bushy, dichotomously branched growth up to 1 m diam. Leaves elliptic-oblanceolate, coriaceous. Flowers 4-merous. Fruit a whitish, semi-transparent, viscid berry 6-10 mm.

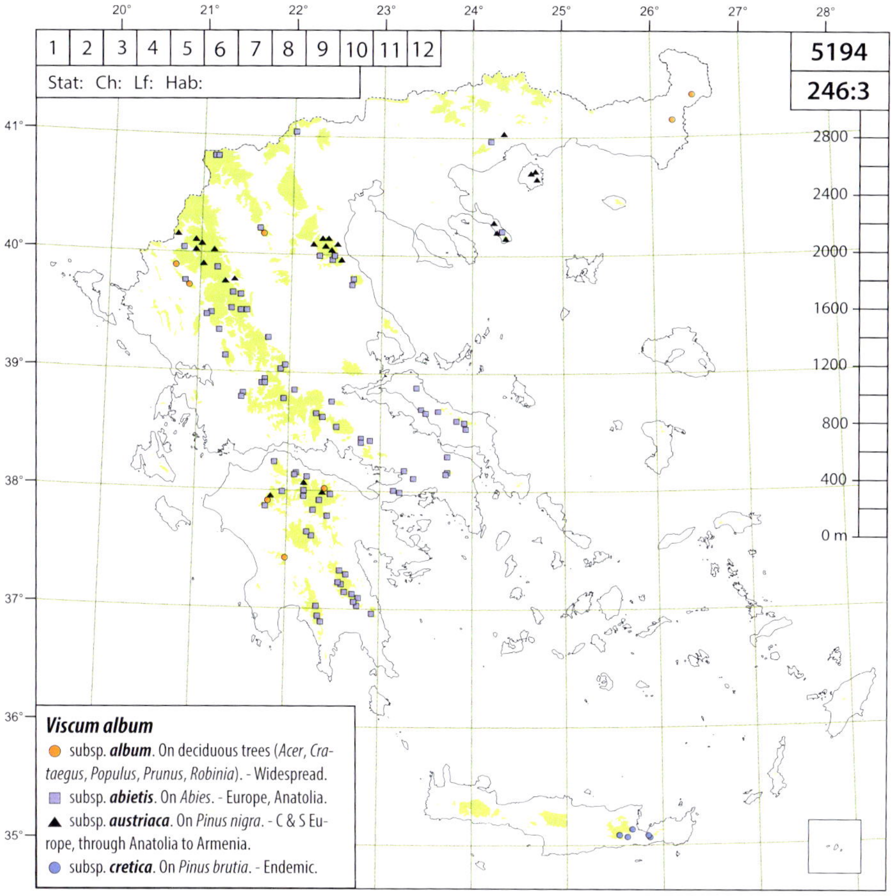

Viscum album

- subsp. ***album***. On deciduous trees (*Acer, Crataegus, Populus, Prunus, Robinia*). - Widespread.
- subsp. ***abietis***. On *Abies*. - Europe, Anatolia.
- subsp. ***austriaca***. On *Pinus nigra*. - C & S Europe, through Anatolia to Armenia.
- subsp. ***cretica***. On *Pinus brutia*. - Endemic.

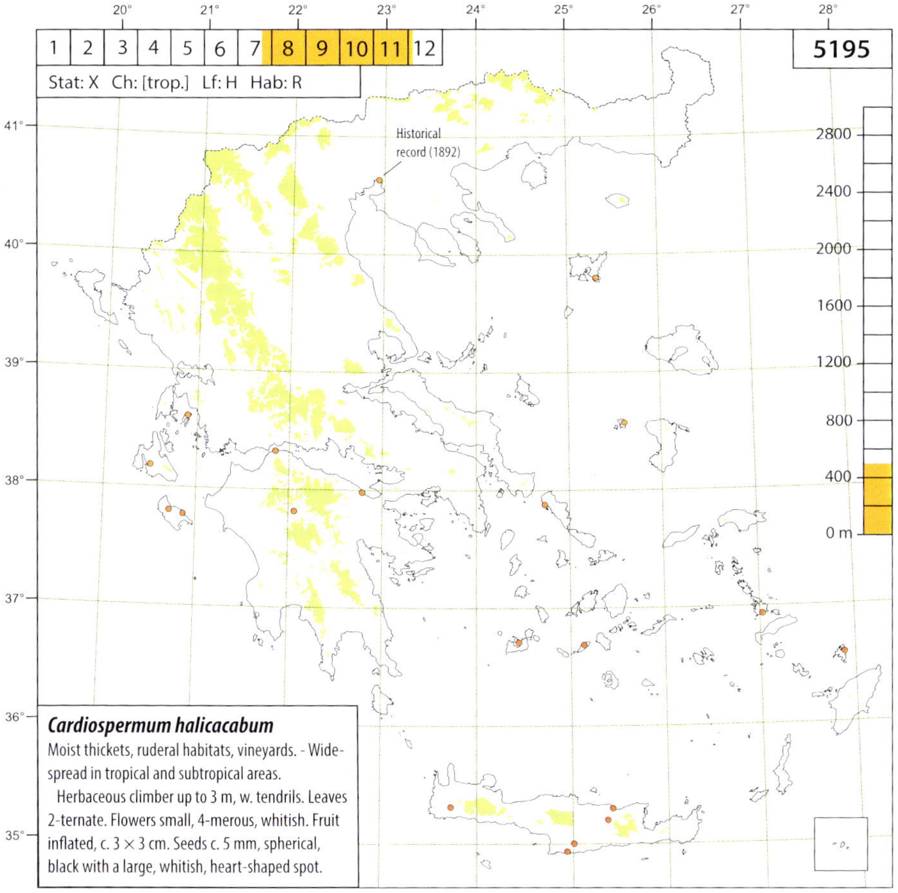

Cardiospermum halicacabum

Moist thickets, ruderal habitats, vineyards. - Widespread in tropical and subtropical areas.

Herbaceous climber up to 3 m, w. tendrils. Leaves 2-ternate. Flowers small, 4-merous, whitish. Fruit inflated, c. 3 × 3 cm. Seeds c. 5 mm, spherical, black with a large, whitish, heart-shaped spot.

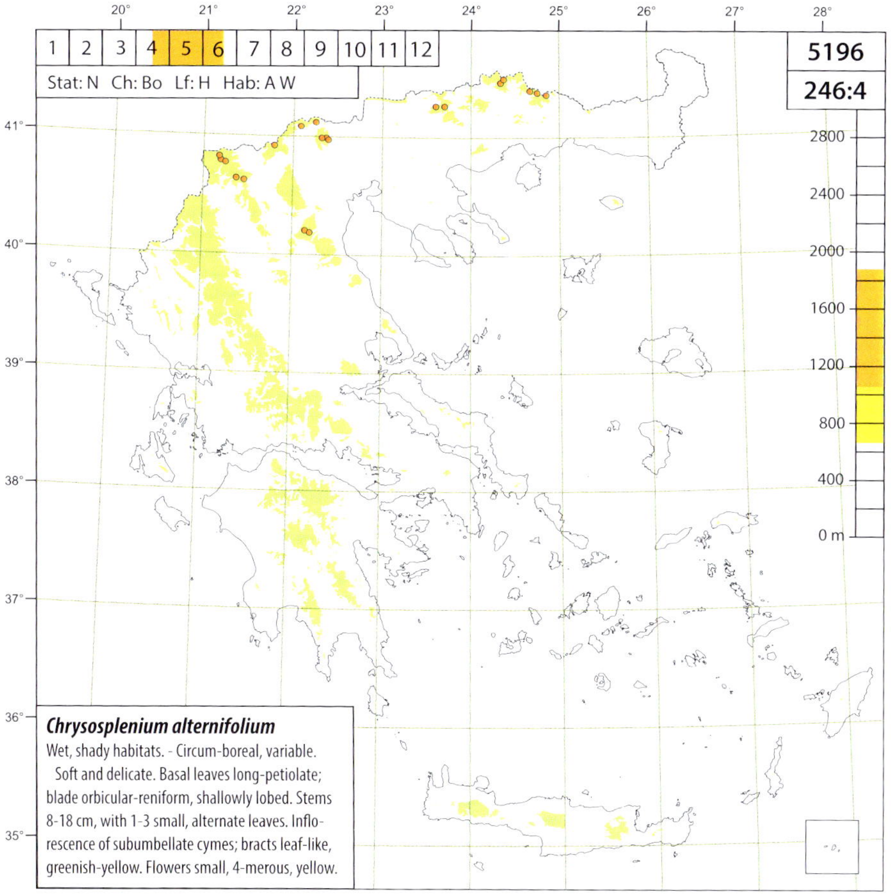

Chrysosplenium alternifolium

Wet, shady habitats. - Circum-boreal, variable.

Soft and delicate. Basal leaves long-petiolate; blade orbicular-reniform, shallowly lobed. Stems 8-18 cm, with 1-3 small, alternate leaves. Inflorescence of subumbellate cymes; bracts leaf-like, greenish-yellow. Flowers small, 4-merous, yellow.

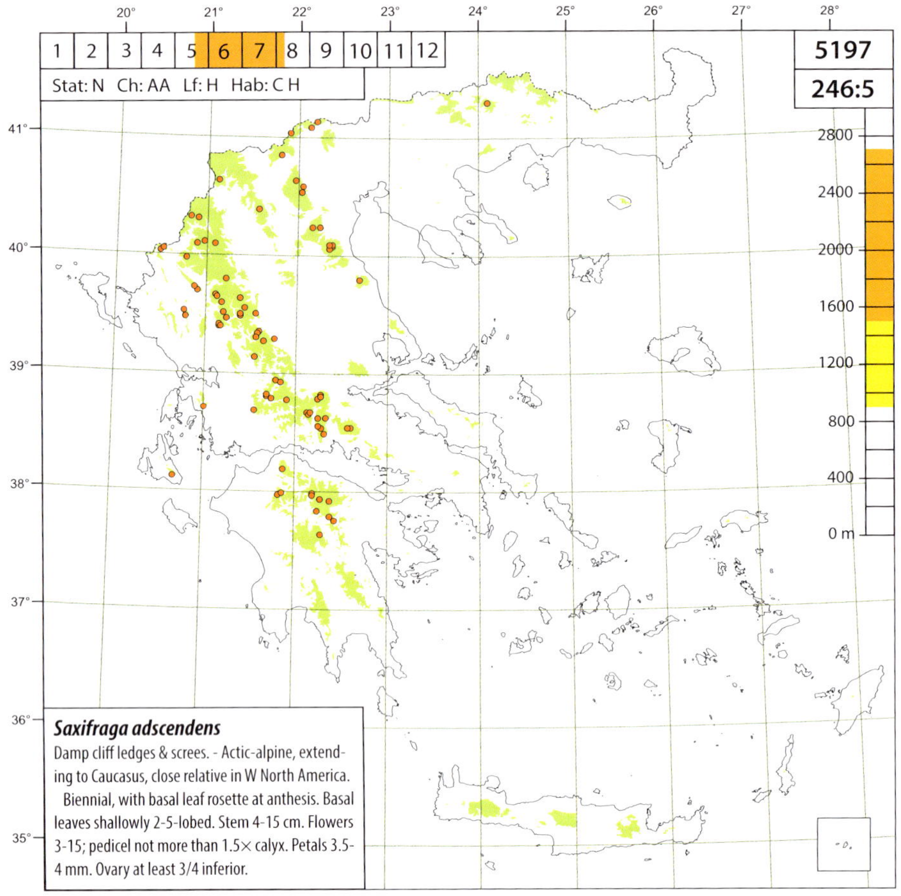

Saxifraga adscendens

Damp cliff ledges & screes. - Actic-alpine, extending to Caucasus, close relative in W North America.

Biennial, with basal leaf rosette at anthesis. Basal leaves shallowly 2-5-lobed. Stem 4-15 cm. Flowers 3-15; pedicel not more than 1.5× calyx. Petals 3.5-4 mm. Ovary at least 3/4 inferior.

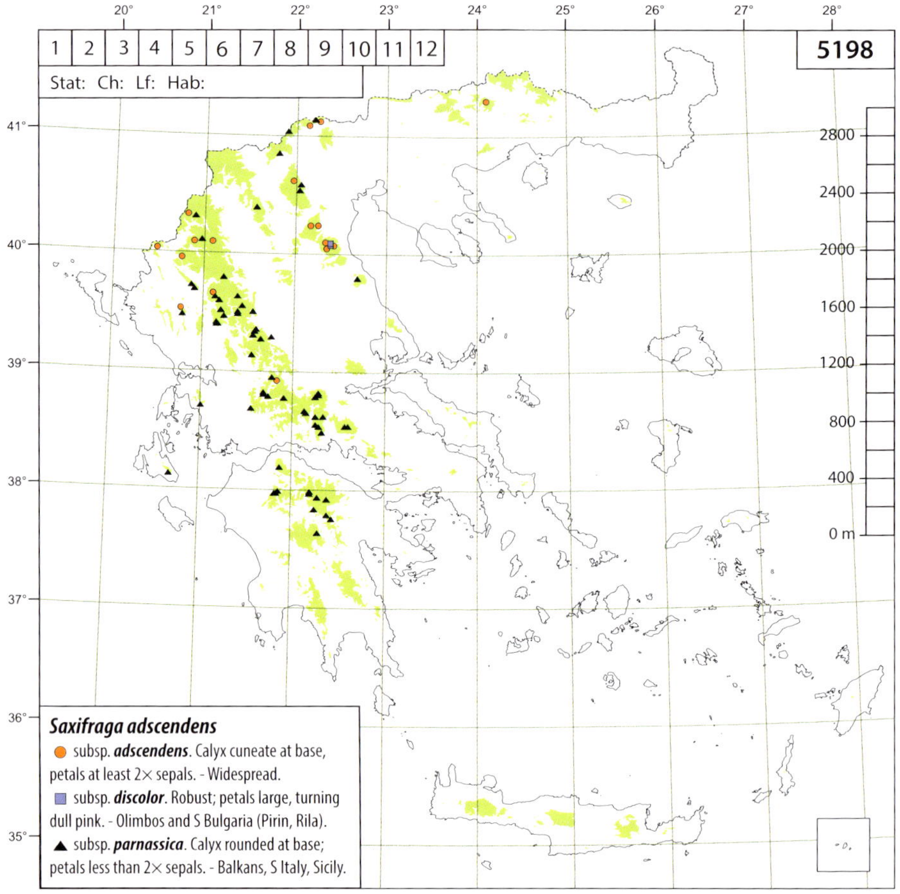

Saxifraga adscendens

● subsp. ***adscendens***. Calyx cuneate at base, petals at least 2× sepals. - Widespread.

■ subsp. ***discolor***. Robust; petals large, turning dull pink. - Olimbos and S Bulgaria (Pirin, Rila).

▲ subsp. ***parnassica***. Calyx rounded at base; petals less than 2× sepals. - Balkans, S Italy, Sicily.

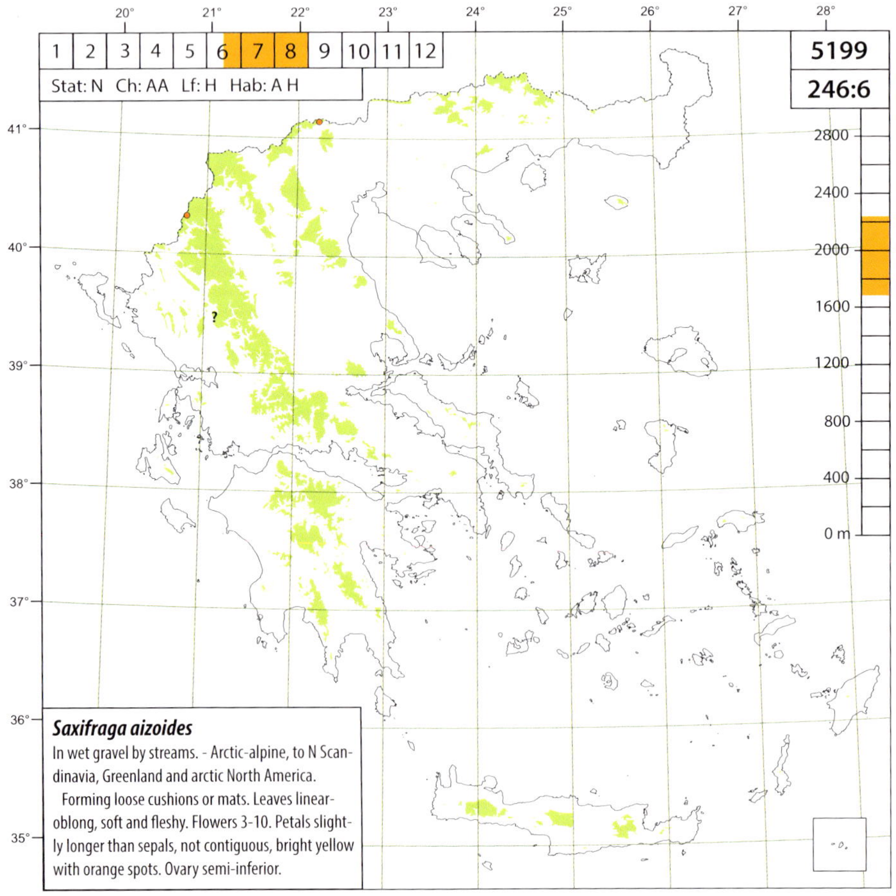

Saxifraga aizoides

In wet gravel by streams. - Arctic-alpine, to N Scandinavia, Greenland and arctic North America.

Forming loose cushions or mats. Leaves linear-oblong, soft and fleshy. Flowers 3-10. Petals slightly longer than sepals, not contiguous, bright yellow with orange spots. Ovary semi-inferior.

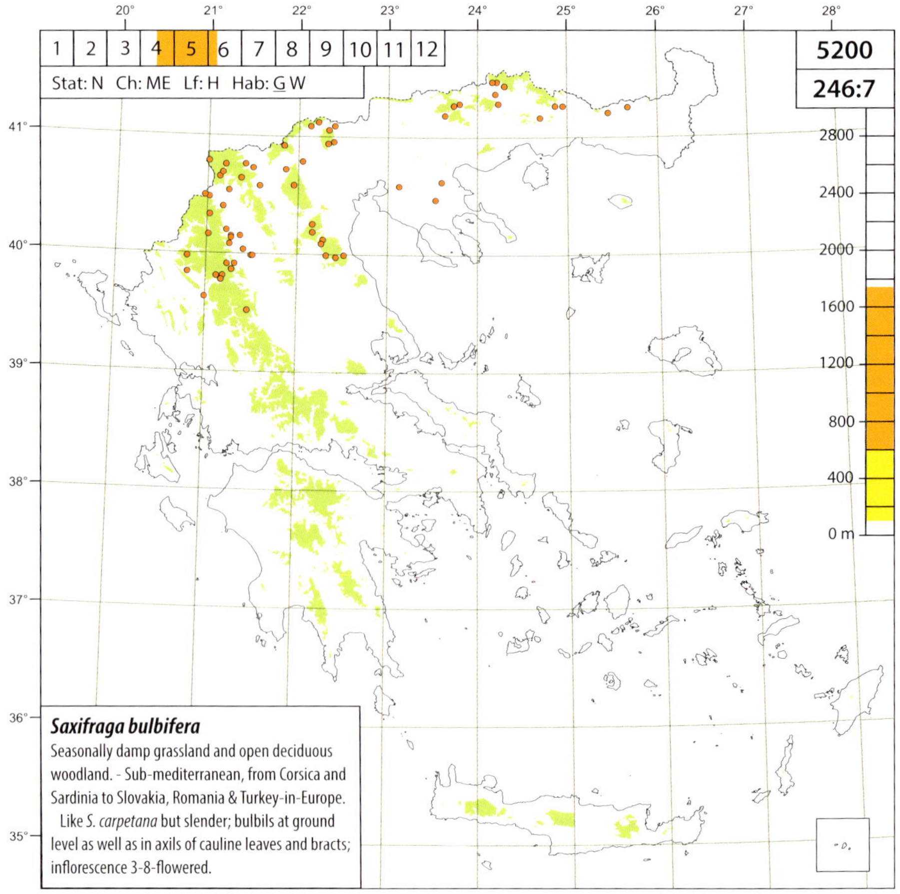

Saxifraga bulbifera

Seasonally damp grassland and open deciduous woodland. - Sub-mediterranean, from Corsica and Sardinia to Slovakia, Romania & Turkey-in-Europe.

Like *S. carpetana* but slender; bulbils at ground level as well as in axils of cauline leaves and bracts; inflorescence 3-8-flowered.

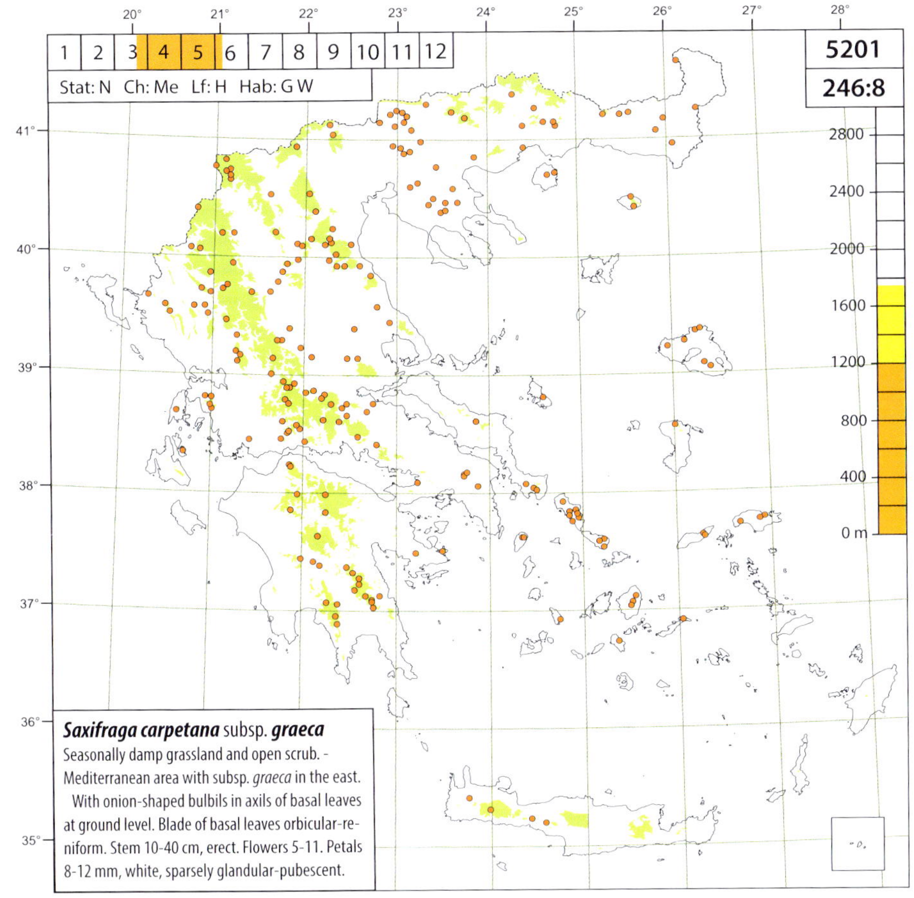

Saxifraga carpetana subsp. ***graeca***
Seasonally damp grassland and open scrub. - Mediterranean area with subsp. *graeca* in the east.
With onion-shaped bulbils in axils of basal leaves at ground level. Blade of basal leaves orbicular-reniform. Stem 10-40 cm, erect. Flowers 5-11. Petals 8-12 mm, white, sparsely glandular-pubescent.

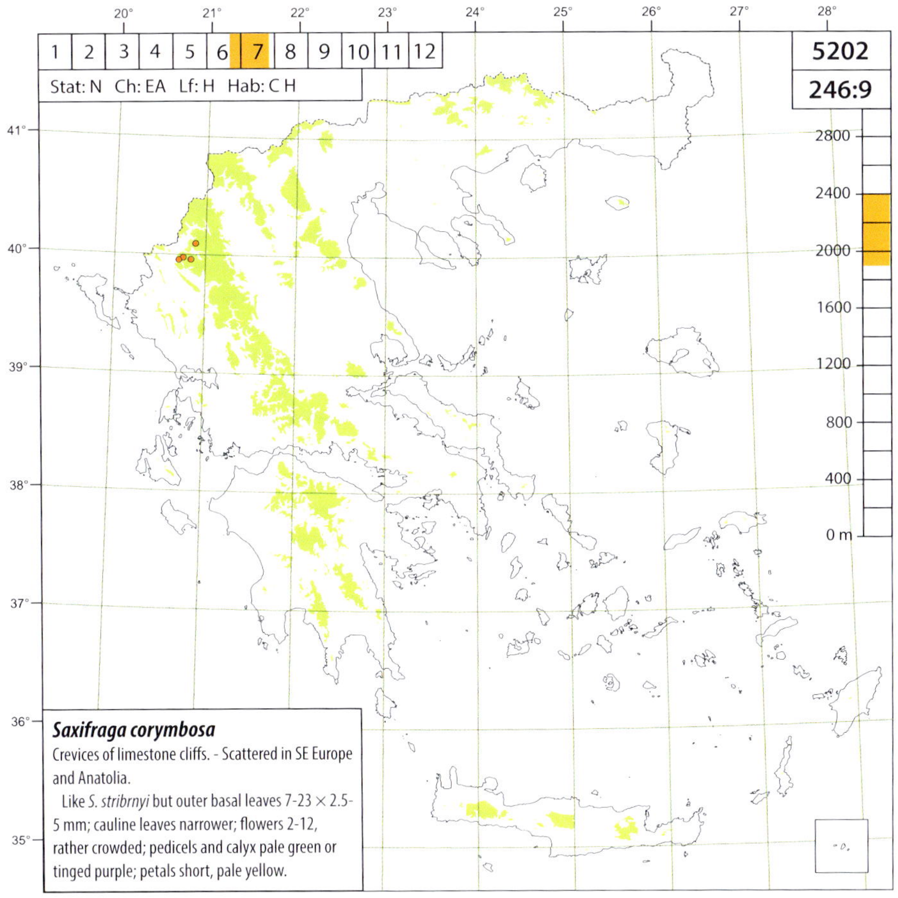

Saxifraga corymbosa
Crevices of limestone cliffs. - Scattered in SE Europe and Anatolia.
Like *S. stribrnyi* but outer basal leaves 7-23 × 2.5-5 mm; cauline leaves narrower; flowers 2-12, rather crowded; pedicels and calyx pale green or tinged purple; petals short, pale yellow.

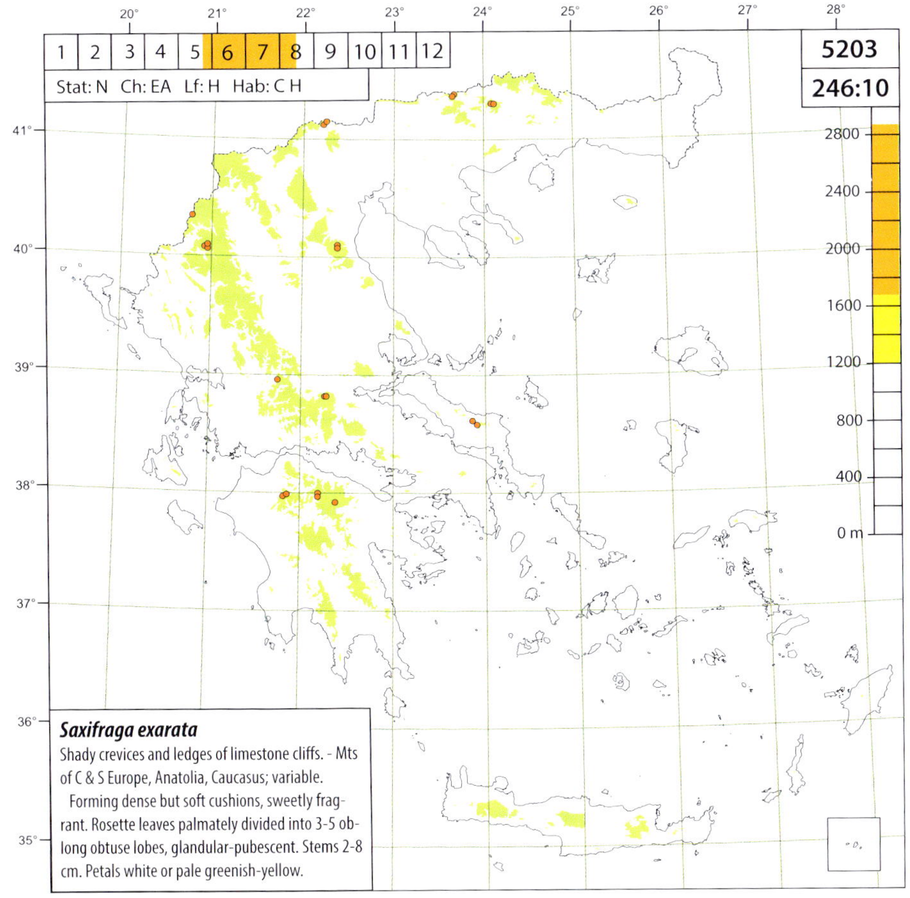

Saxifraga exarata
Shady crevices and ledges of limestone cliffs. - Mts of C & S Europe, Anatolia, Caucasus; variable.
Forming dense but soft cushions, sweetly fragrant. Rosette leaves palmately divided into 3-5 oblong obtuse lobes, glandular-pubescent. Stems 2-8 cm. Petals white or pale greenish-yellow.

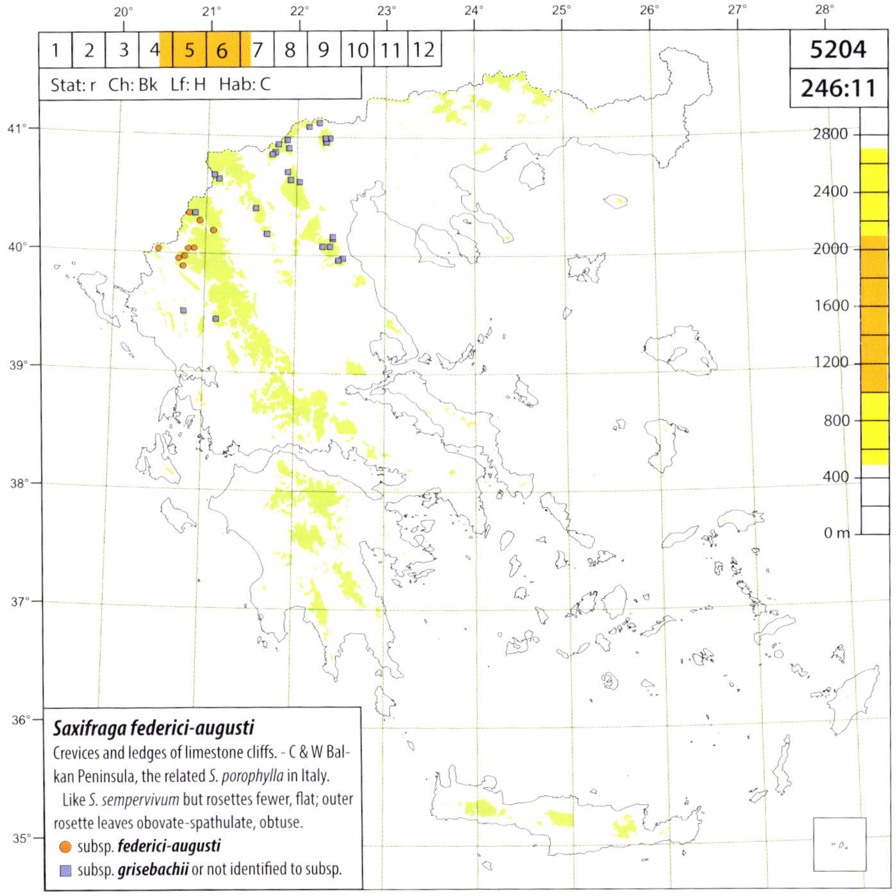

Saxifraga federici-augusti
Crevices and ledges of limestone cliffs. - C & W Balkan Peninsula, the related *S. porophylla* in Italy.
Like *S. sempervivum* but rosettes fewer, flat; outer rosette leaves obovate-spathulate, obtuse.
● subsp. ***federici-augusti***
■ subsp. ***grisebachii*** or not identified to subsp.

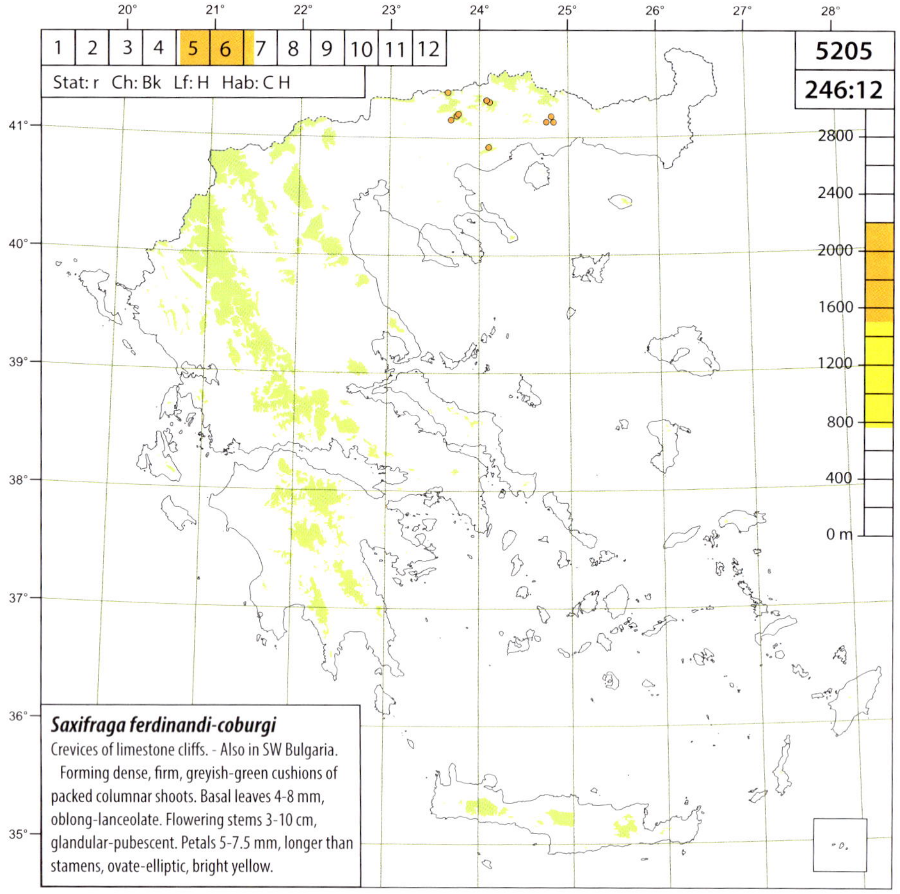

Saxifraga ferdinandi-coburgi
Crevices of limestone cliffs. - Also in SW Bulgaria.
Forming dense, firm, greyish-green cushions of packed columnar shoots. Basal leaves 4-8 mm, oblong-lanceolate. Flowering stems 3-10 cm, glandular-pubescent. Petals 5-7.5 mm, longer than stamens, ovate-elliptic, bright yellow.

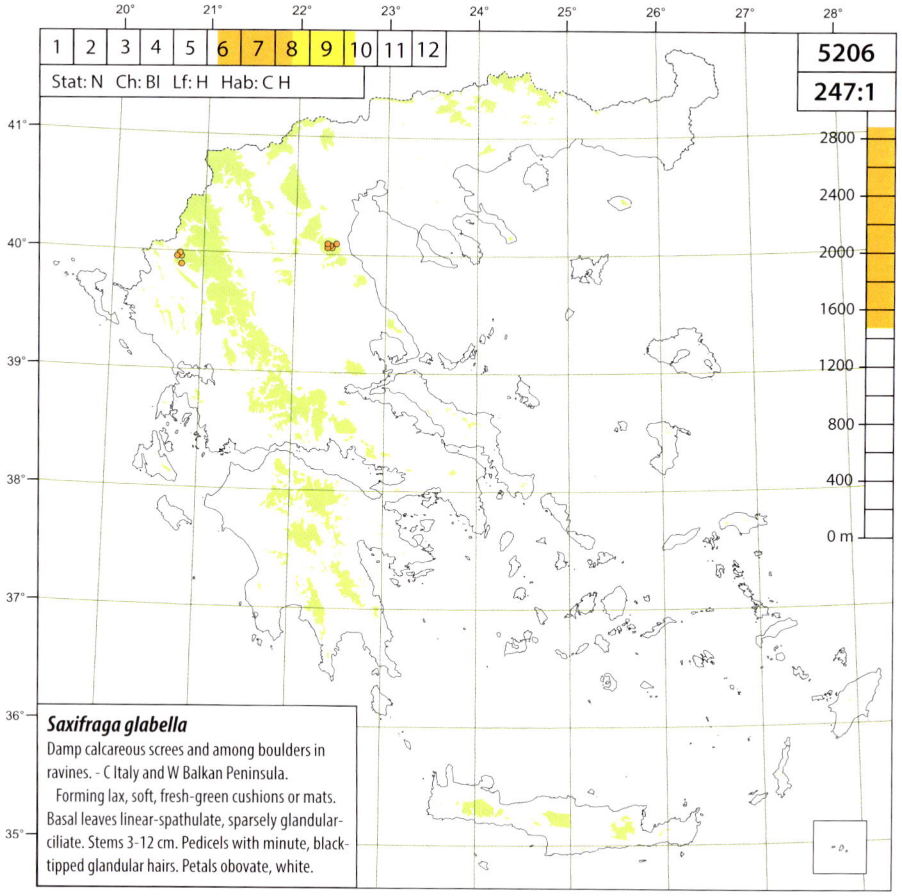

Saxifraga glabella
Damp calcareous screes and among boulders in ravines. - C Italy and W Balkan Peninsula.
Forming lax, soft, fresh-green cushions or mats. Basal leaves linear-spathulate, sparsely glandular-ciliate. Stems 3-12 cm. Pedicels with minute, black-tipped glandular hairs. Petals obovate, white.

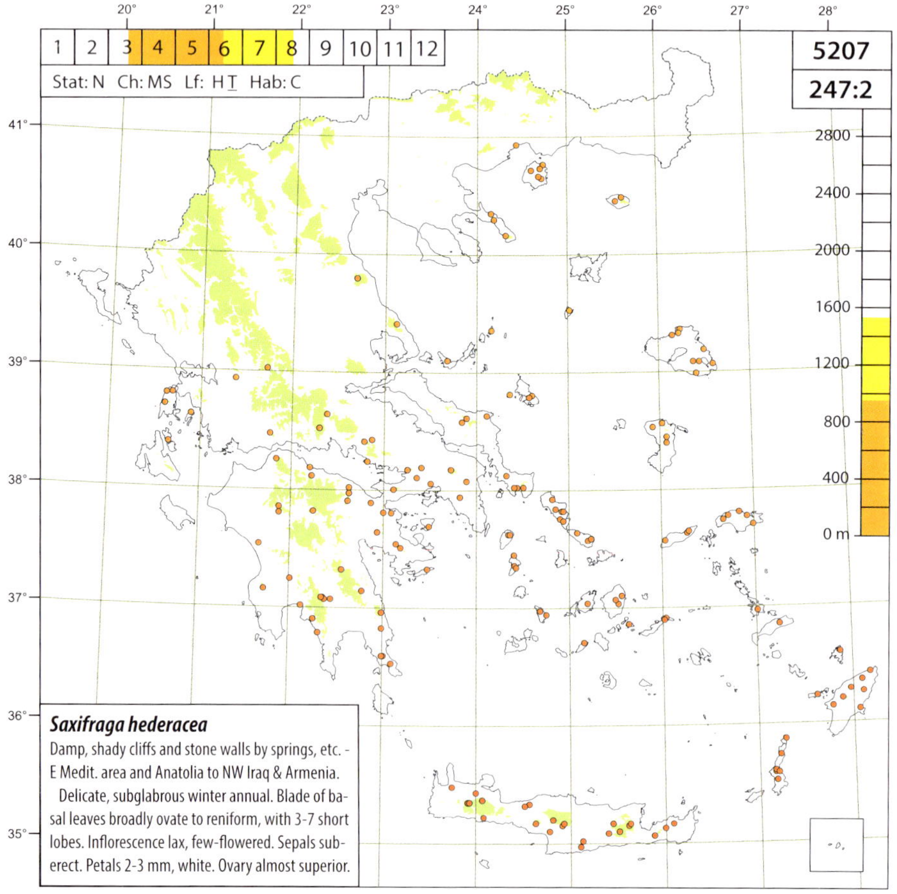

Saxifraga hederacea
Damp, shady cliffs and stone walls by springs, etc. - E Medit. area and Anatolia to NW Iraq & Armenia.
Delicate, subglabrous winter annual. Blade of basal leaves broadly ovate to reniform, with 3-7 short lobes. Inflorescence lax, few-flowered. Sepals suberect. Petals 2-3 mm, white. Ovary almost superior.

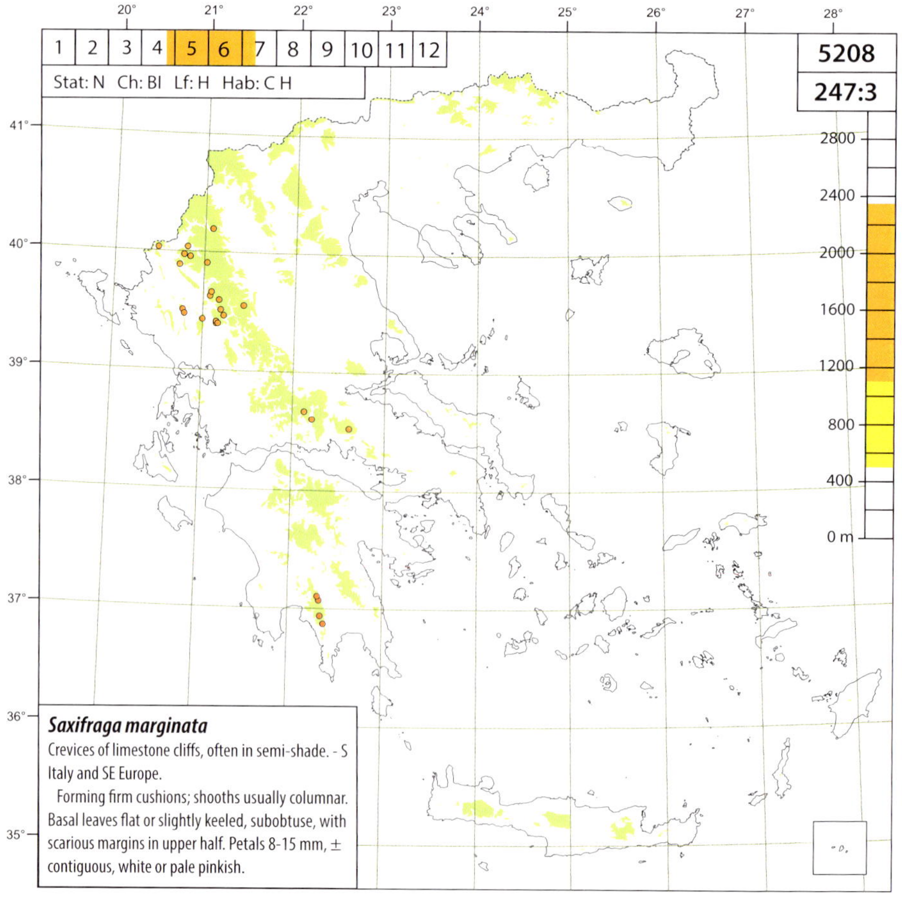

Saxifraga marginata
Crevices of limestone cliffs, often in semi-shade. - S Italy and SE Europe.
Forming firm cushions; shooths usually columnar. Basal leaves flat or slightly keeled, subobtuse, with scarious margins in upper half. Petals 8-15 mm, ± contiguous, white or pale pinkish.

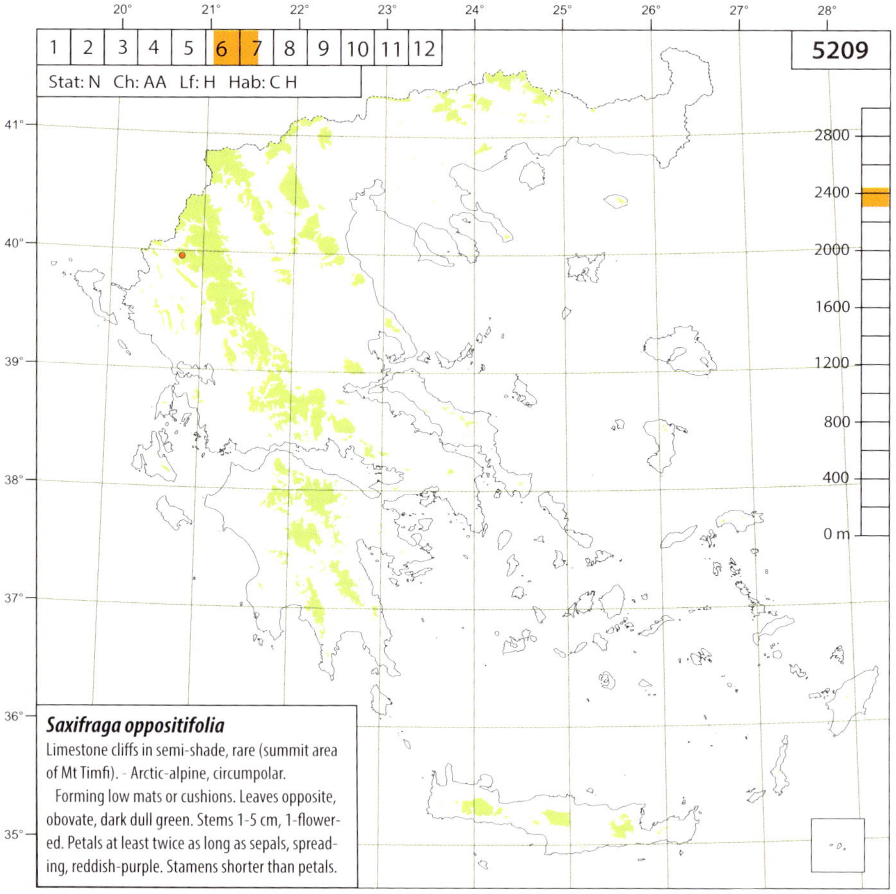

Saxifraga oppositifolia

Limestone cliffs in semi-shade, rare (summit area of Mt Timfi). - Arctic-alpine, circumpolar.

Forming low mats or cushions. Leaves opposite, obovate, dark dull green. Stems 1-5 cm, 1-flowered. Petals at least twice as long as sepals, spreading, reddish-purple. Stamens shorter than petals.

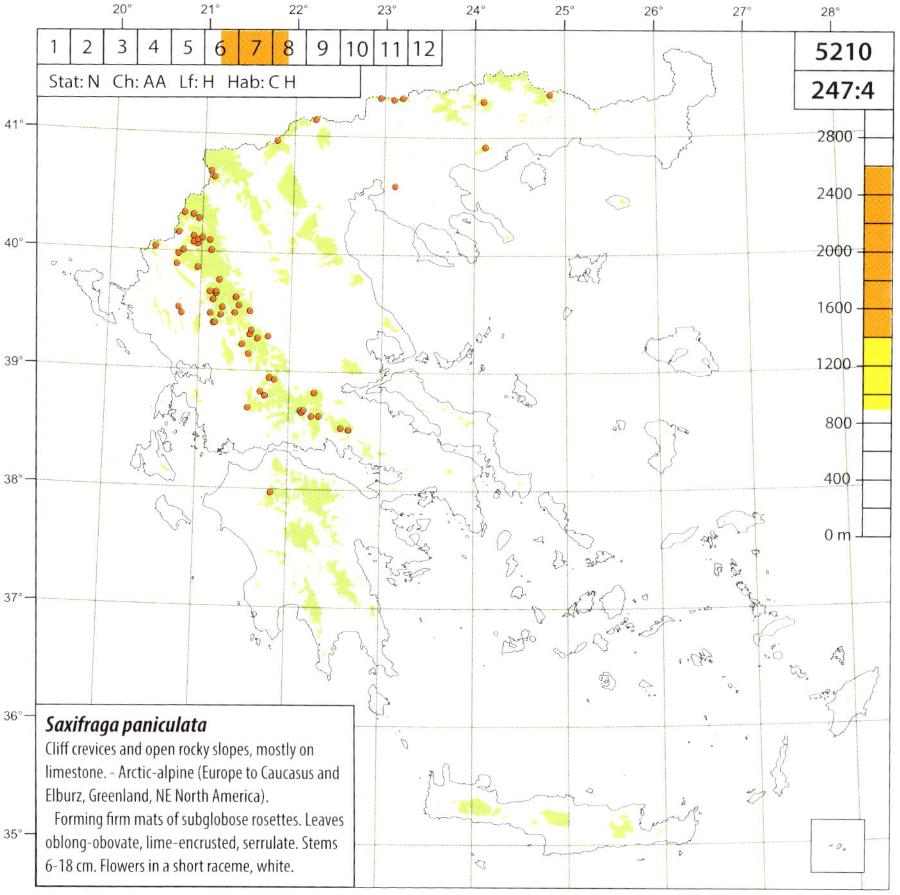

Saxifraga paniculata

Cliff crevices and open rocky slopes, mostly on limestone. - Arctic-alpine (Europe to Caucasus and Elburz, Greenland, NE North America).

Forming firm mats of subglobose rosettes. Leaves oblong-obovate, lime-encrusted, serrulate. Stems 6-18 cm. Flowers in a short raceme, white.

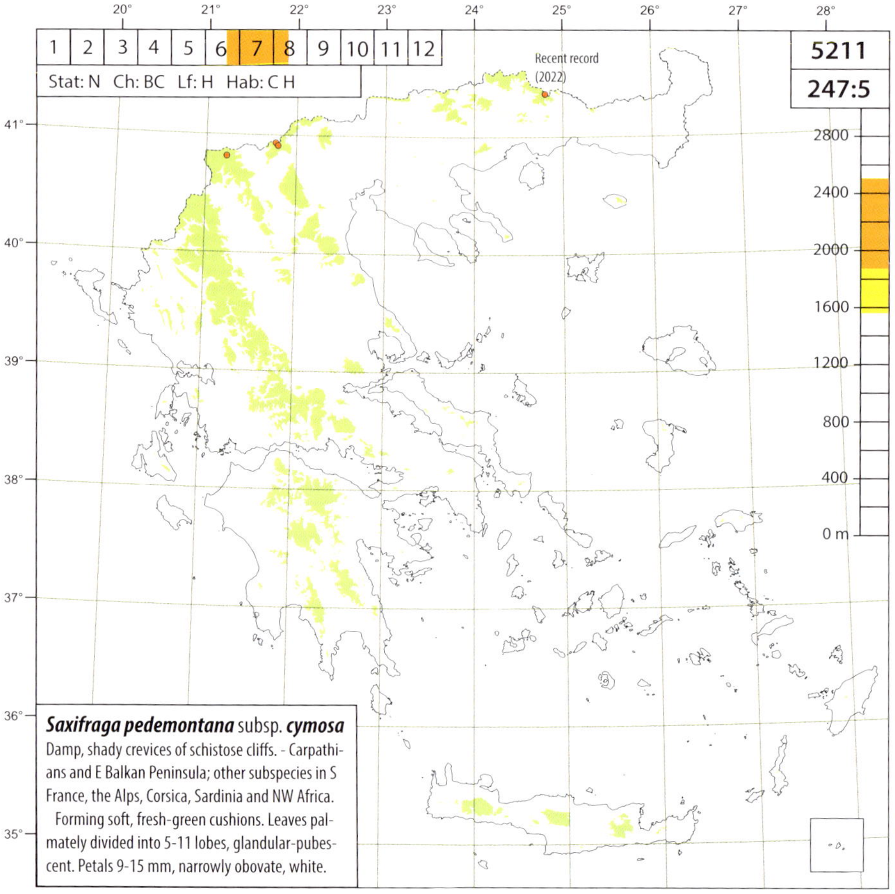

Saxifraga pedemontana subsp. ***cymosa***

Damp, shady crevices of schistose cliffs. - Carpathians and E Balkan Peninsula; other subspecies in S France, the Alps, Corsica, Sardinia and NW Africa.

Forming soft, fresh-green cushions. Leaves palmately divided into 5-11 lobes, glandular-pubescent. Petals 9-15 mm, narrowly obovate, white.

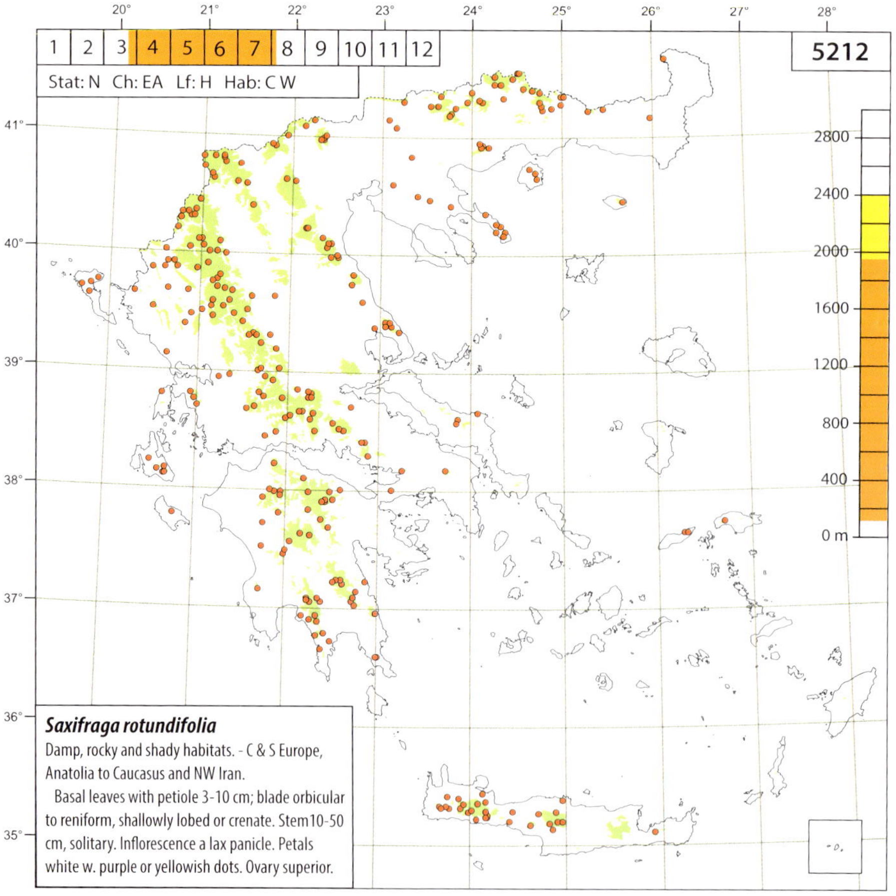

Saxifraga rotundifolia

Damp, rocky and shady habitats. - C & S Europe, Anatolia to Caucasus and NW Iran.

Basal leaves with petiole 3-10 cm; blade orbicular to reniform, shallowly lobed or crenate. Stem 10-50 cm, solitary. Inflorescence a lax panicle. Petals white w. purple or yellowish dots. Ovary superior.

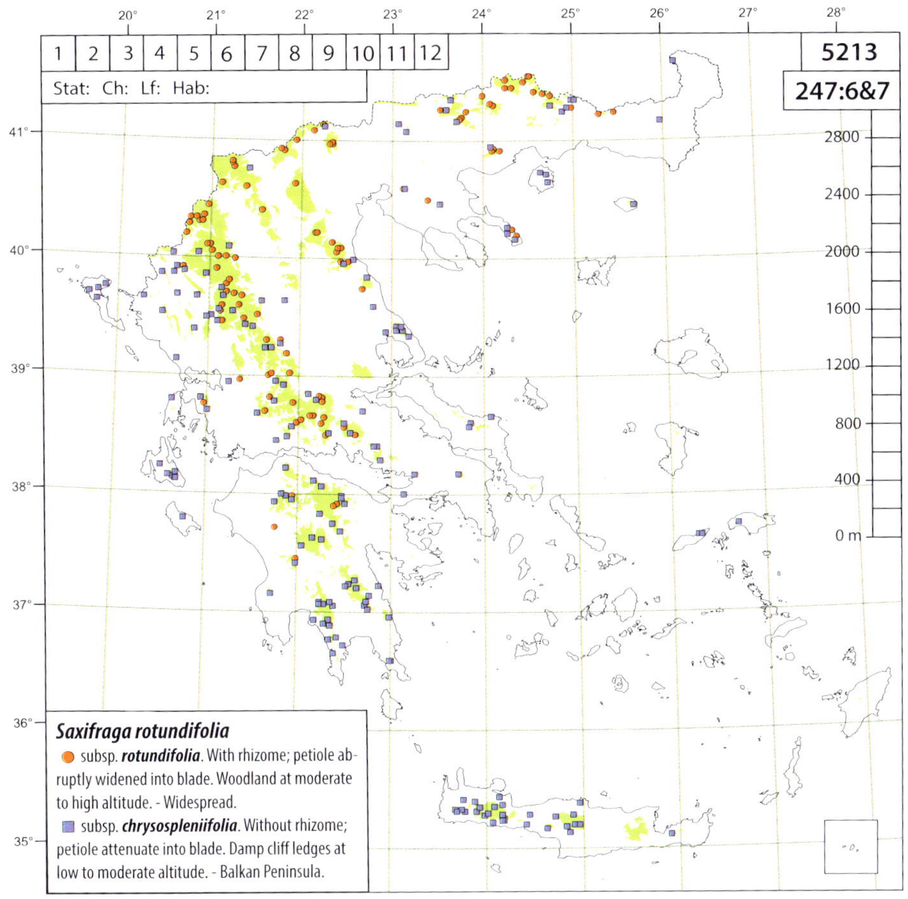

Saxifraga rotundifolia
subsp. ***rotundifolia***. With rhizome; petiole abruptly widened into blade. Woodland at moderate to high altitude. - Widespread.
subsp. ***chrysospleniifolia***. Without rhizome; petiole attenuate into blade. Damp cliff ledges at low to moderate altitude. - Balkan Peninsula.

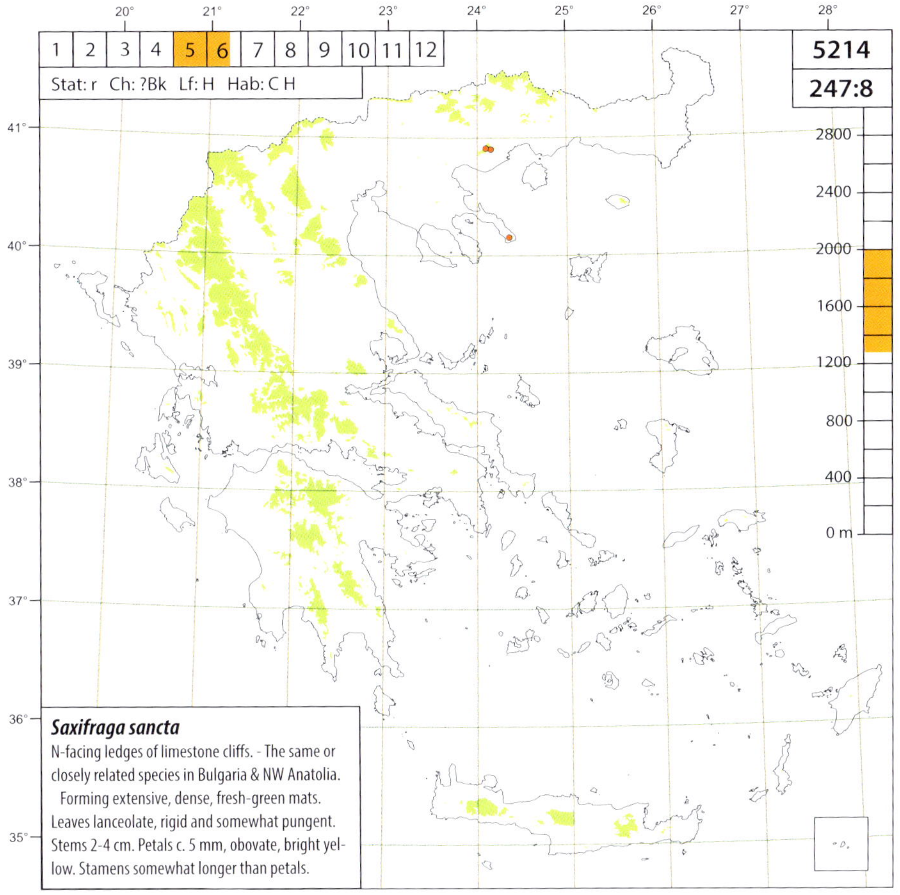

Saxifraga sancta
N-facing ledges of limestone cliffs. - The same or closely related species in Bulgaria & NW Anatolia.
Forming extensive, dense, fresh-green mats. Leaves lanceolate, rigid and somewhat pungent. Stems 2-4 cm. Petals c. 5 mm, obovate, bright yellow. Stamens somewhat longer than petals.

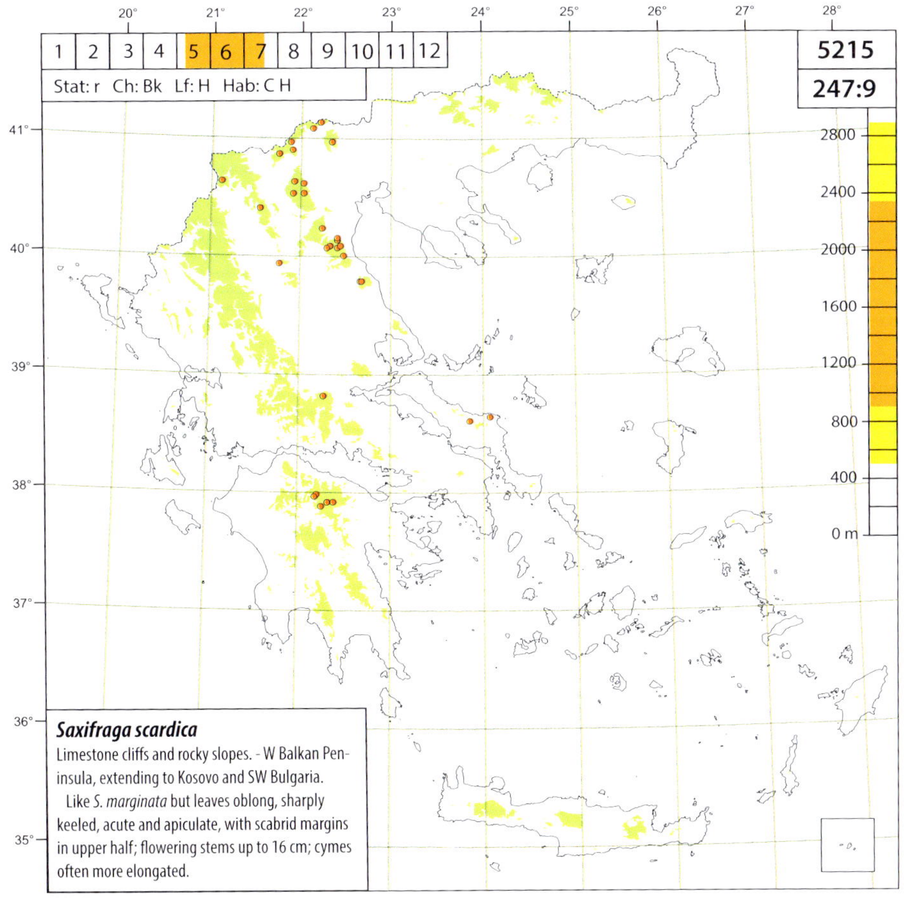

Saxifraga scardica
Limestone cliffs and rocky slopes. - W Balkan Peninsula, extending to Kosovo and SW Bulgaria.
Like *S. marginata* but leaves oblong, sharply keeled, acute and apiculate, with scabrid margins in upper half; flowering stems up to 16 cm; cymes often more elongated.

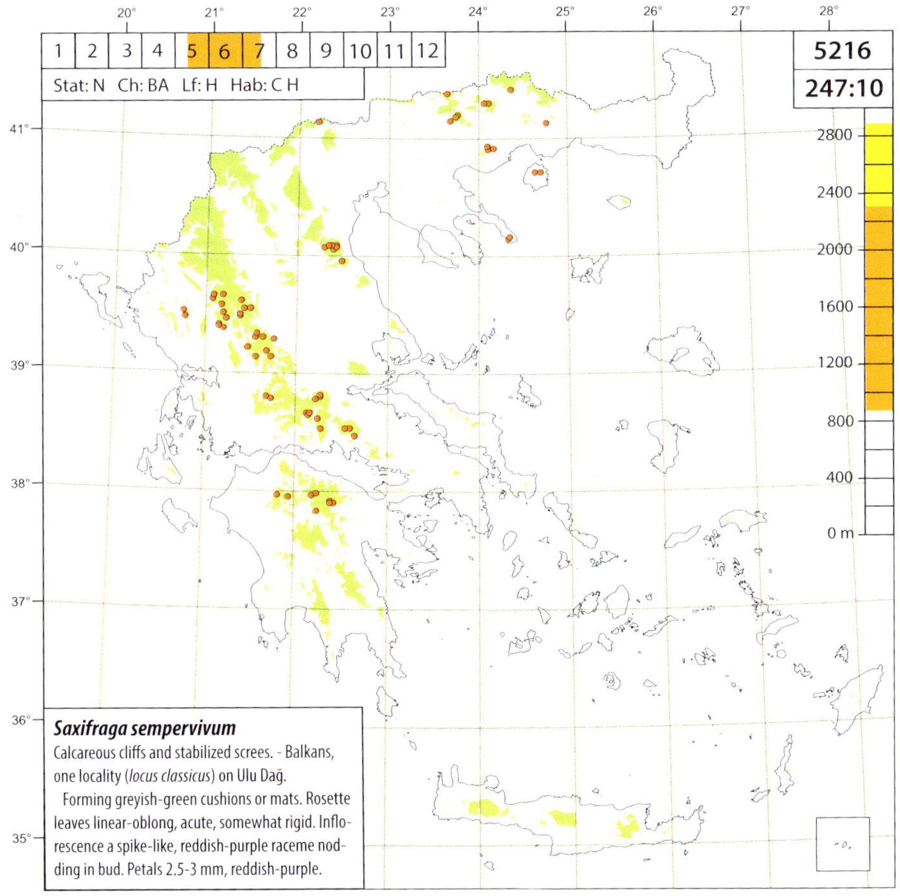

Saxifraga sempervivum
Calcareous cliffs and stabilized screes. - Balkans, one locality (*locus classicus*) on Ulu Dağ.
Forming greyish-green cushions or mats. Rosette leaves linear-oblong, acute, somewhat rigid. Inflorescence a spike-like, reddish-purple raceme nodding in bud. Petals 2.5-3 mm, reddish-purple.

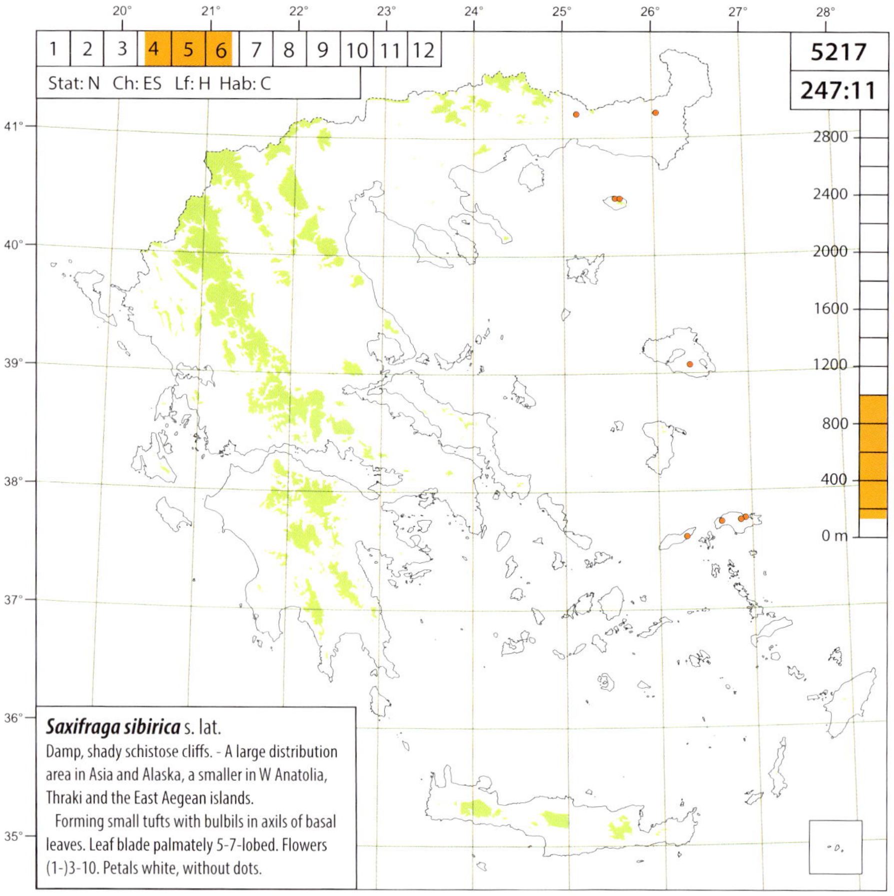

Saxifraga sibirica s. lat.
Damp, shady schistose cliffs. - A large distribution area in Asia and Alaska, a smaller in W Anatolia, Thraki and the East Aegean islands.
Forming small tufts with bulbils in axils of basal leaves. Leaf blade palmately 5-7-lobed. Flowers (1-)3-10. Petals white, without dots.

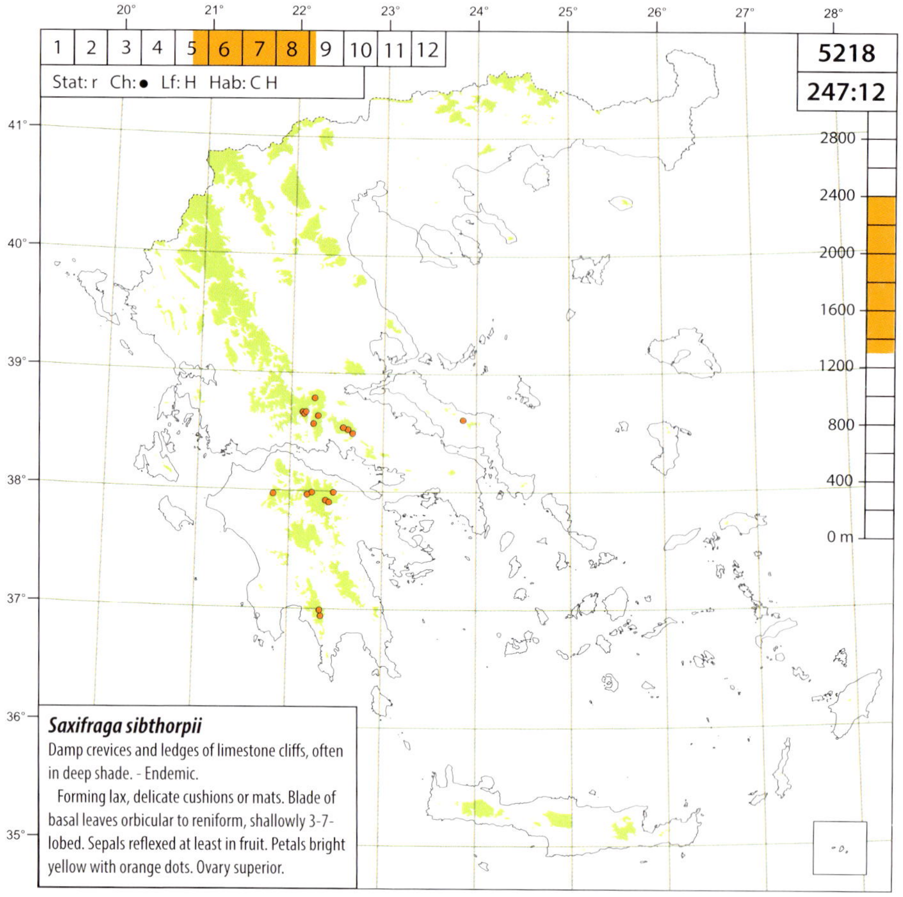

Saxifraga sibthorpii
Damp crevices and ledges of limestone cliffs, often in deep shade. - Endemic.
Forming lax, delicate cushions or mats. Blade of basal leaves orbicular to reniform, shallowly 3-7-lobed. Sepals reflexed at least in fruit. Petals bright yellow with orange dots. Ovary superior.

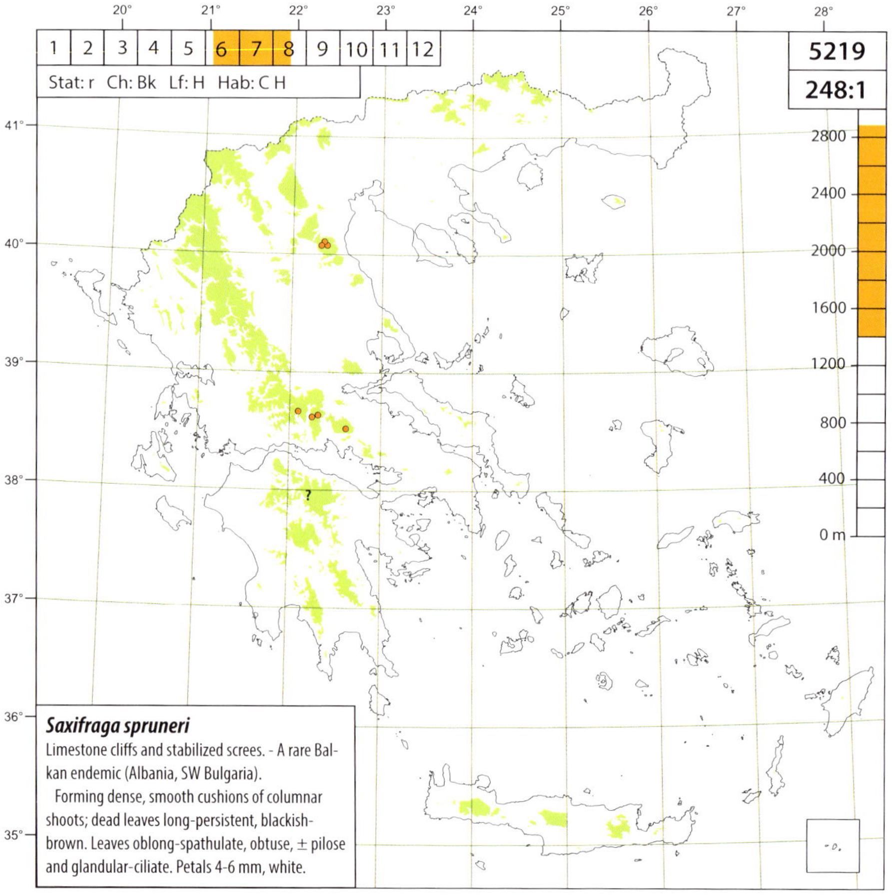

Saxifraga spruneri
Limestone cliffs and stabilized screes. - A rare Balkan endemic (Albania, SW Bulgaria).
Forming dense, smooth cushions of columnar shoots; dead leaves long-persistent, blackish-brown. Leaves oblong-spathulate, obtuse, ± pilose and glandular-ciliate. Petals 4-6 mm, white.

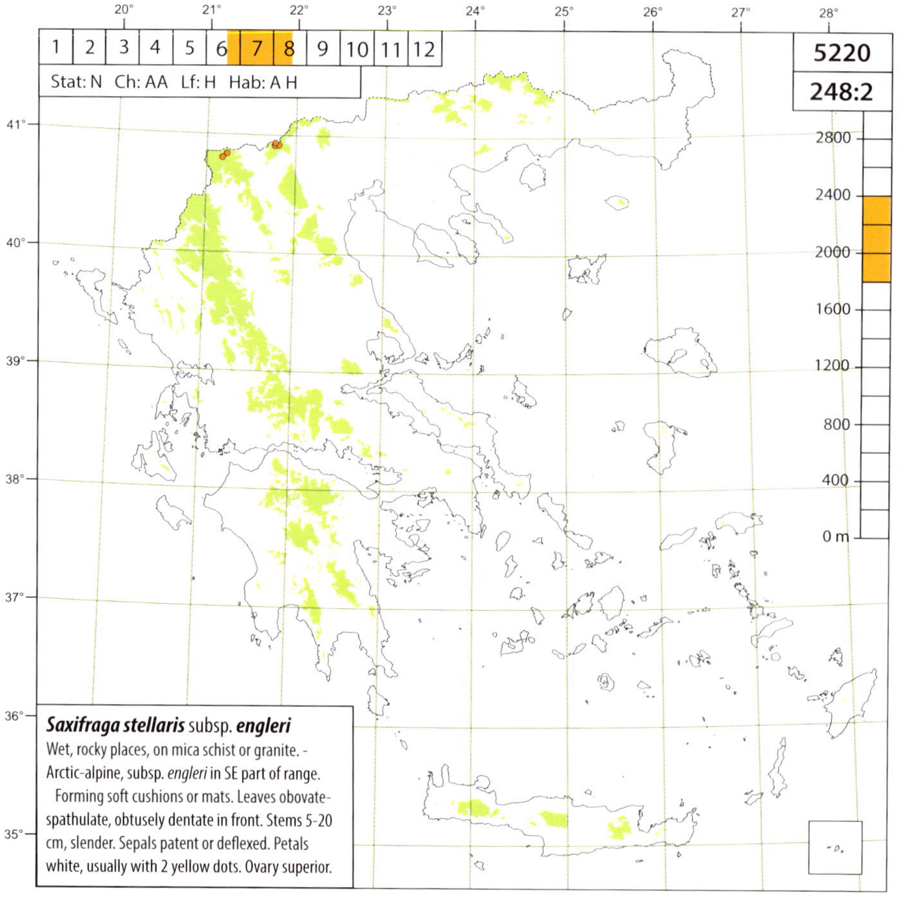

Saxifraga stellaris subsp. _engleri_
Wet, rocky places, on mica schist or granite. - Arctic-alpine, subsp. _engleri_ in SE part of range.
Forming soft cushions or mats. Leaves obovate-spathulate, obtusely dentate in front. Stems 5-20 cm, slender. Sepals patent or deflexed. Petals white, usually with 2 yellow dots. Ovary superior.

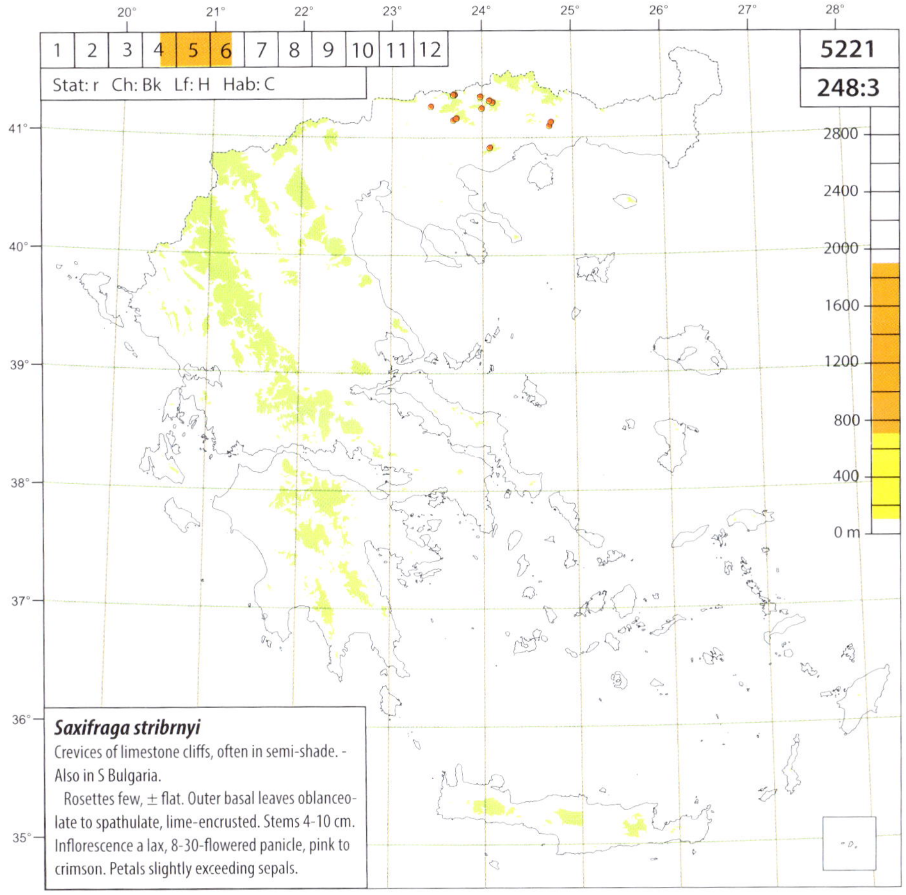

Saxifraga stribrnyi
Crevices of limestone cliffs, often in semi-shade. - Also in S Bulgaria.
Rosettes few, ± flat. Outer basal leaves oblanceolate to spathulate, lime-encrusted. Stems 4-10 cm. Inflorescence a lax, 8-30-flowered panicle, pink to crimson. Petals slightly exceeding sepals.

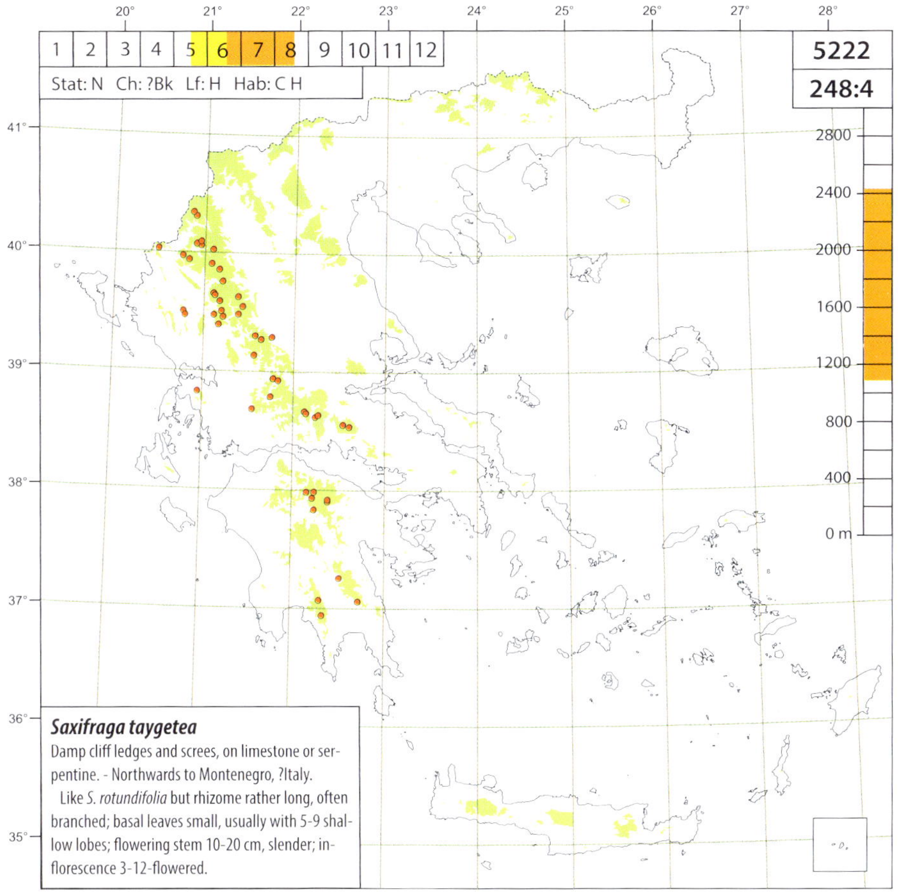

Saxifraga taygetea
Damp cliff ledges and screes, on limestone or serpentine. - Northwards to Montenegro, ?Italy.
Like *S. rotundifolia* but rhizome rather long, often branched; basal leaves small, usually with 5-9 shallow lobes; flowering stem 10-20 cm, slender; inflorescence 3-12-flowered.

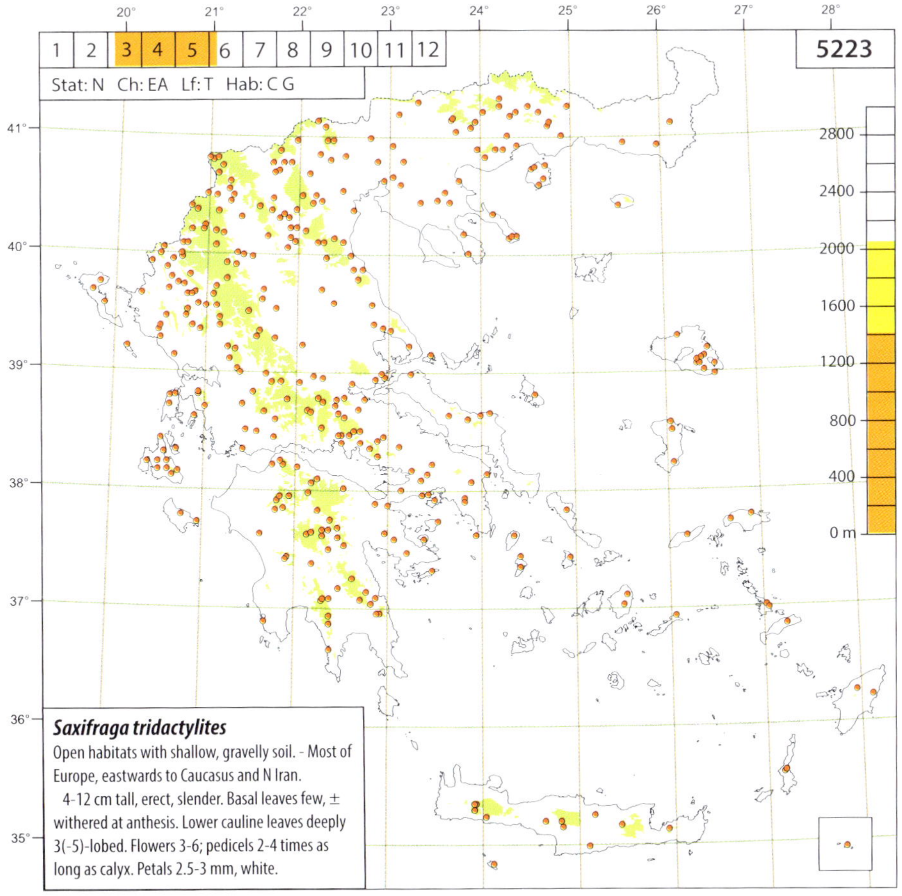

Saxifraga tridactylites
Open habitats with shallow, gravelly soil. - Most of Europe, eastwards to Caucasus and N Iran.
4-12 cm tall, erect, slender. Basal leaves few, ± withered at anthesis. Lower cauline leaves deeply 3(-5)-lobed. Flowers 3-6; pedicels 2-4 times as long as calyx. Petals 2.5-3 mm, white.

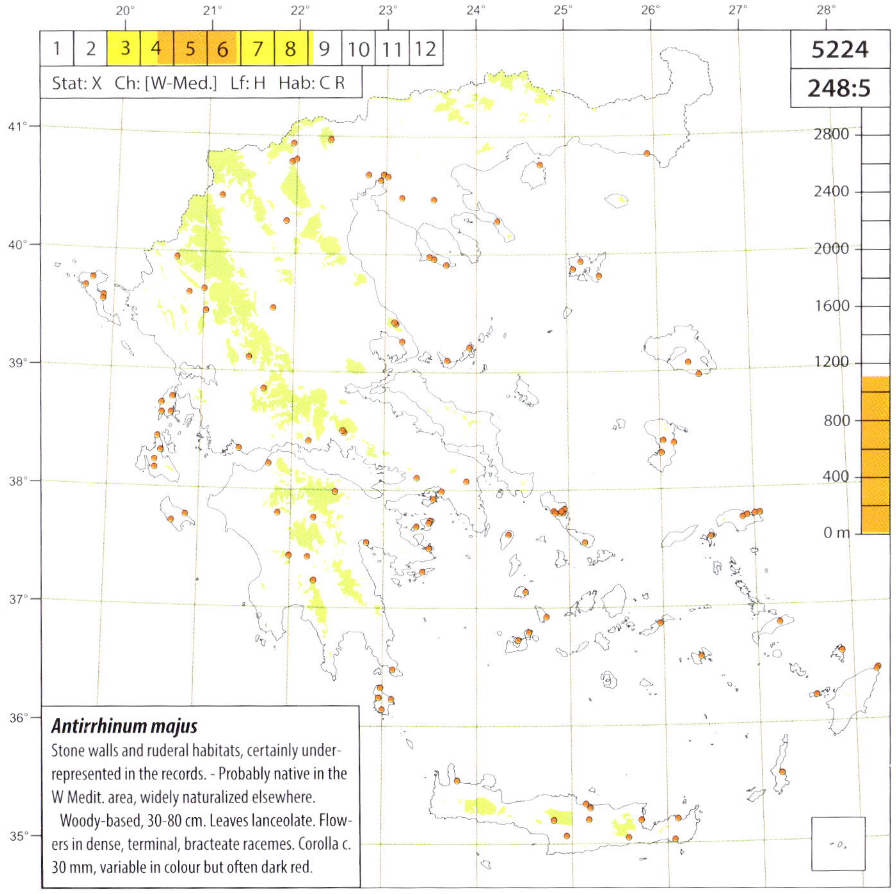

Antirrhinum majus
Stone walls and ruderal habitats, certainly under-represented in the records. - Probably native in the W Medit. area, widely naturalized elsewhere.
Woody-based, 30-80 cm. Leaves lanceolate. Flowers in dense, terminal, bracteate racemes. Corolla c. 30 mm, variable in colour but often dark red.

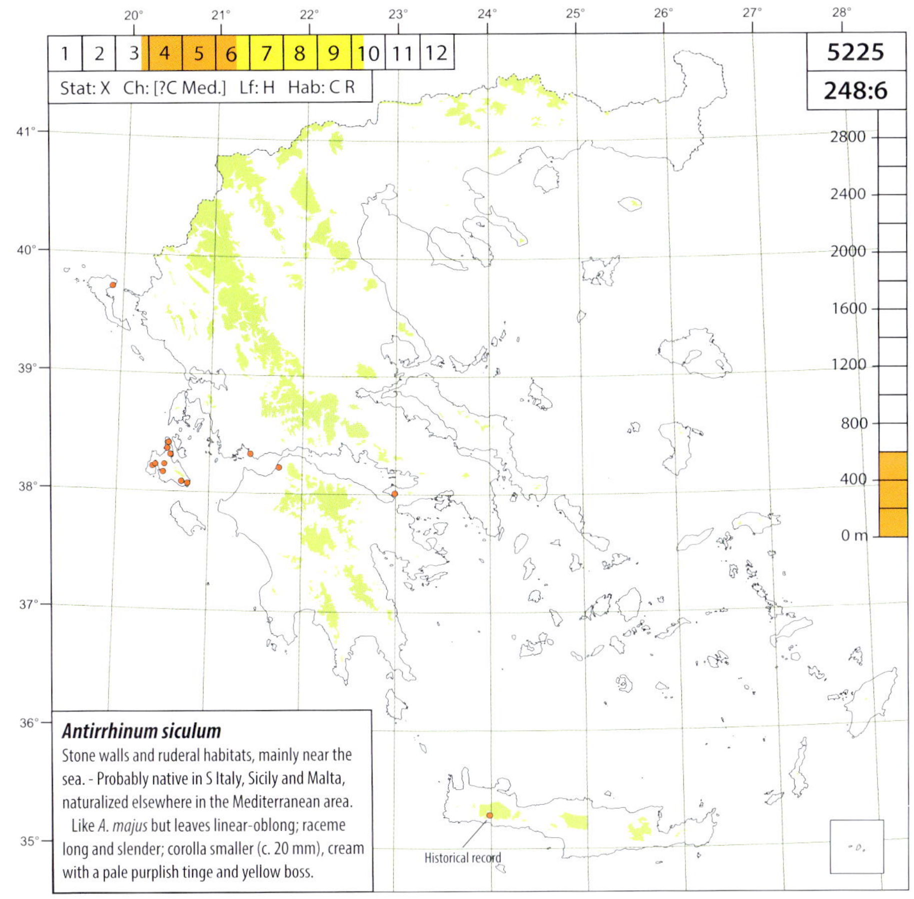

Antirrhinum siculum

Stone walls and ruderal habitats, mainly near the sea. - Probably native in S Italy, Sicily and Malta, naturalized elsewhere in the Mediterranean area.

Like *A. majus* but leaves linear-oblong; raceme long and slender; corolla smaller (c. 20 mm), cream with a pale purplish tinge and yellow boss.

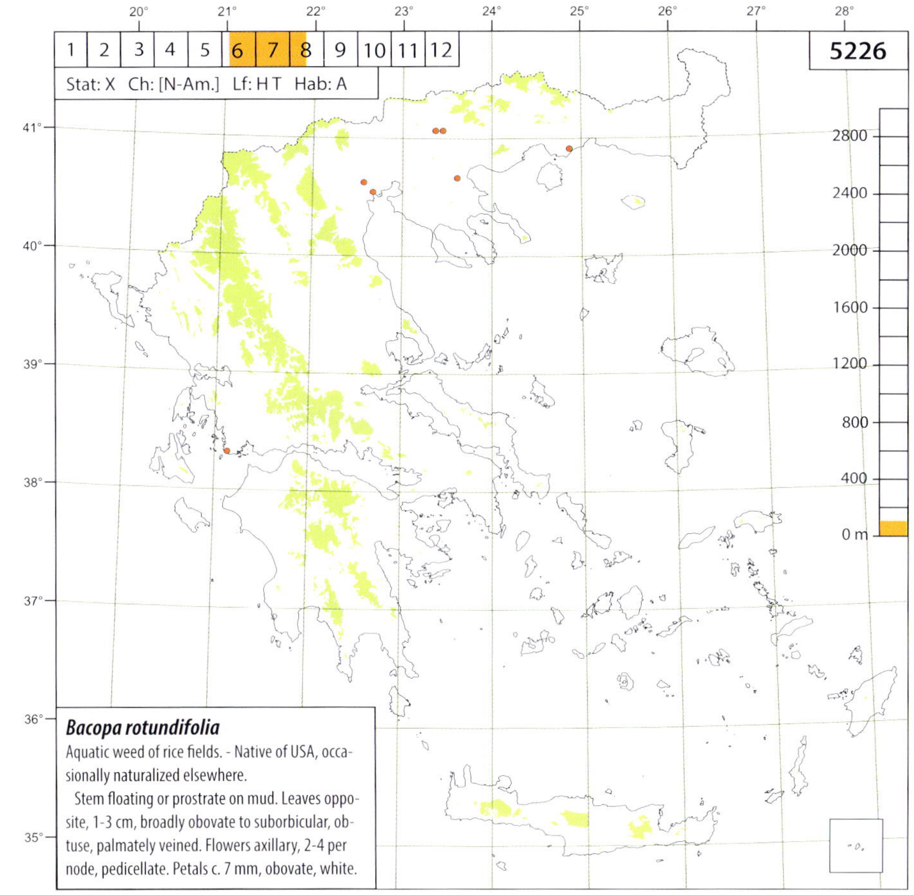

Bacopa rotundifolia

Aquatic weed of rice fields. - Native of USA, occasionally naturalized elsewhere.

Stem floating or prostrate on mud. Leaves opposite, 1-3 cm, broadly obovate to suborbicular, obtuse, palmately veined. Flowers axillary, 2-4 per node, pedicellate. Petals c. 7 mm, obovate, white.

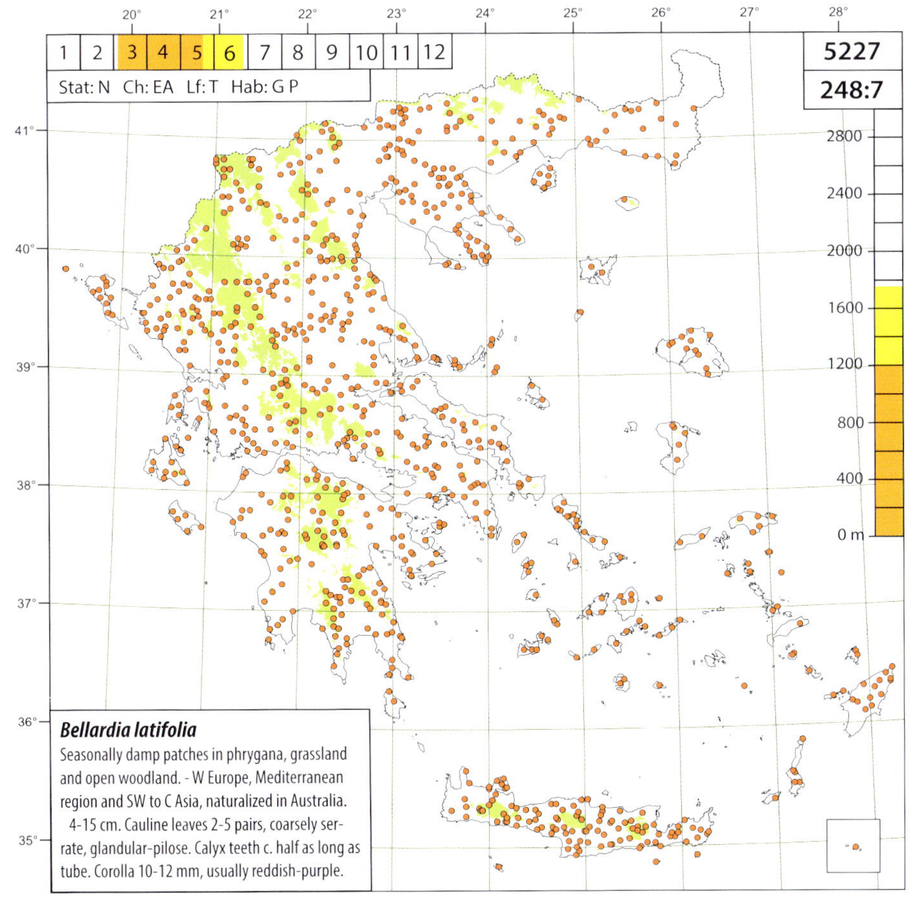

Bellardia latifolia

Seasonally damp patches in phrygana, grassland and open woodland. - W Europe, Mediterranean region and SW to C Asia, naturalized in Australia.

4-15 cm. Cauline leaves 2-5 pairs, coarsely serrate, glandular-pilose. Calyx teeth c. half as long as tube. Corolla 10-12 mm, usually reddish-purple.

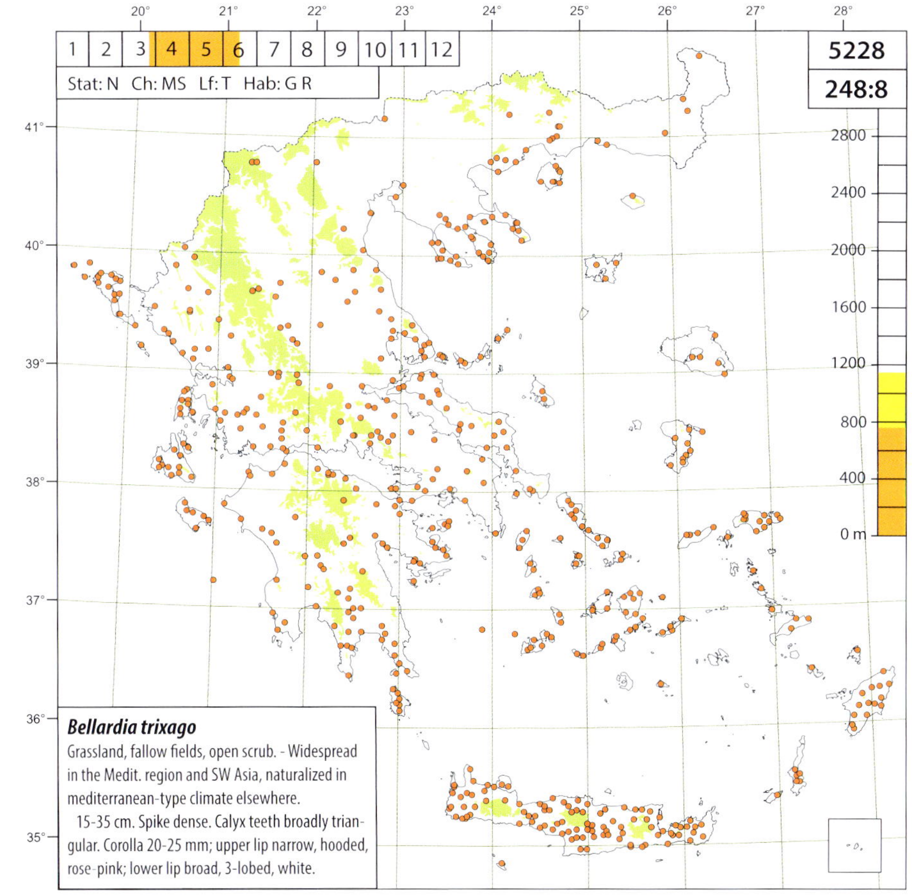

Bellardia trixago

Grassland, fallow fields, open scrub. - Widespread in the Medit. region and SW Asia, naturalized in mediterranean-type climate elsewhere.

15-35 cm. Spike dense. Calyx teeth broadly triangular. Corolla 20-25 mm; upper lip narrow, hooded, rose-pink; lower lip broad, 3-lobed, white.

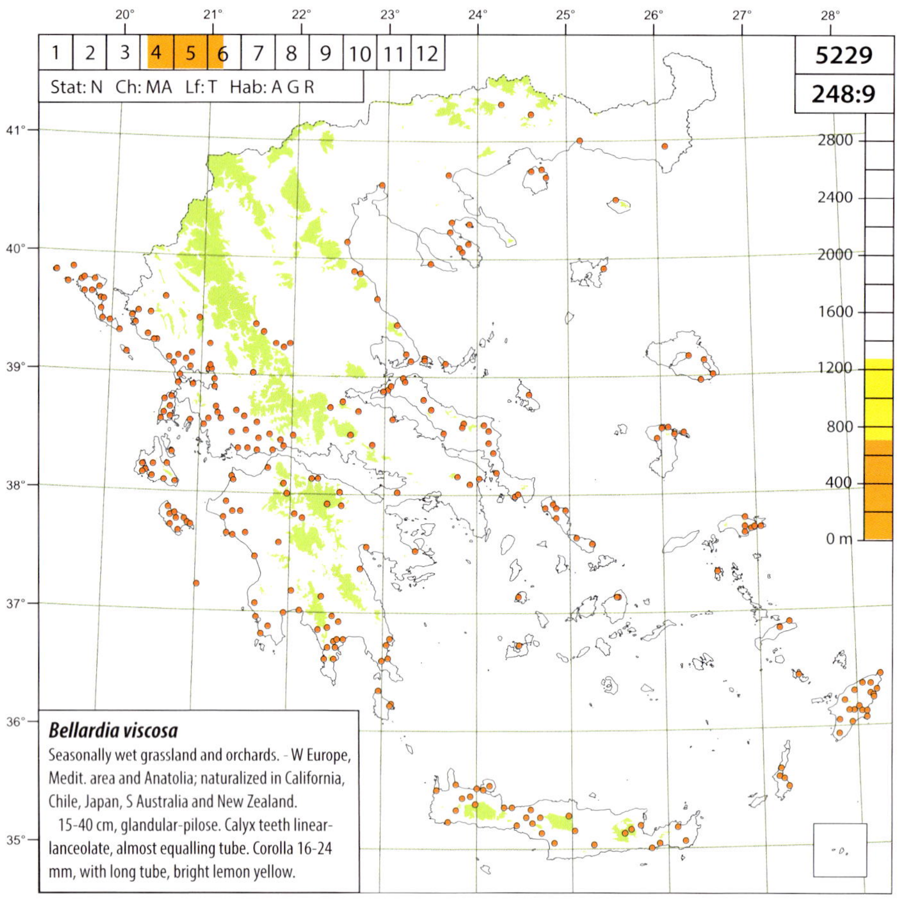

Bellardia viscosa
Seasonally wet grassland and orchards. - W Europe, Medit. area and Anatolia; naturalized in California, Chile, Japan, S Australia and New Zealand.
15-40 cm, glandular-pilose. Calyx teeth linear-lanceolate, almost equalling tube. Corolla 16-24 mm, with long tube, bright lemon yellow.

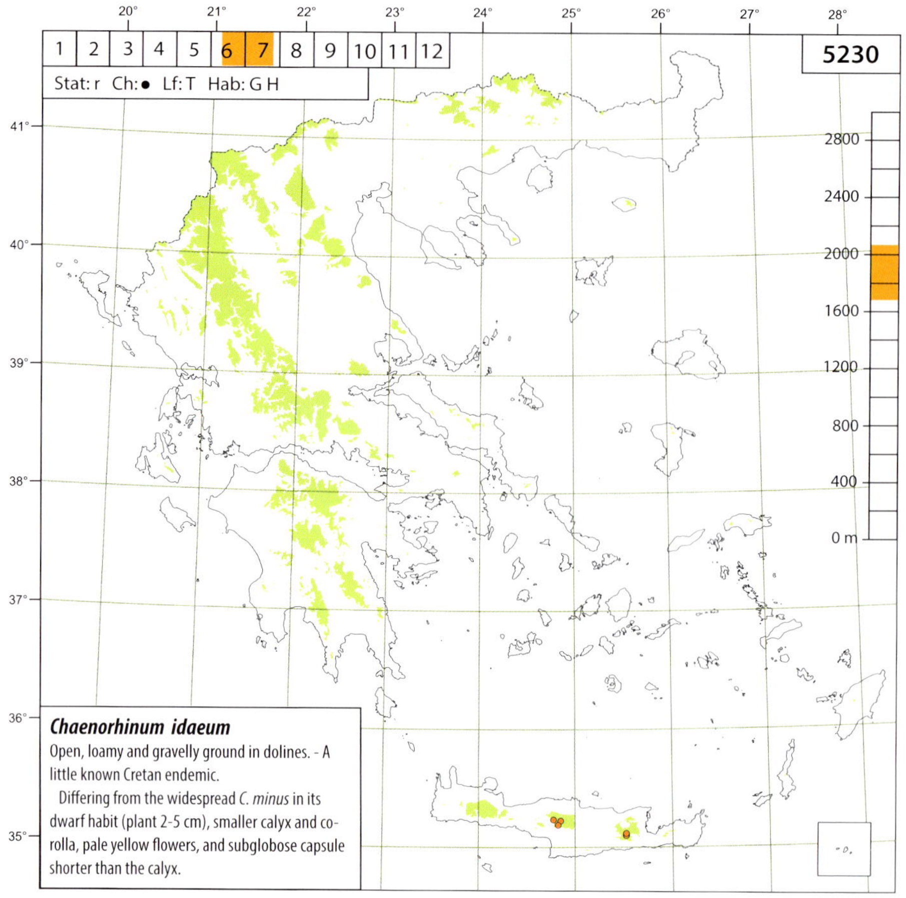

Chaenorhinum idaeum
Open, loamy and gravelly ground in dolines. - A little known Cretan endemic.
Differing from the widespread *C. minus* in its dwarf habit (plant 2-5 cm), smaller calyx and corolla, pale yellow flowers, and subglobose capsule shorter than the calyx.

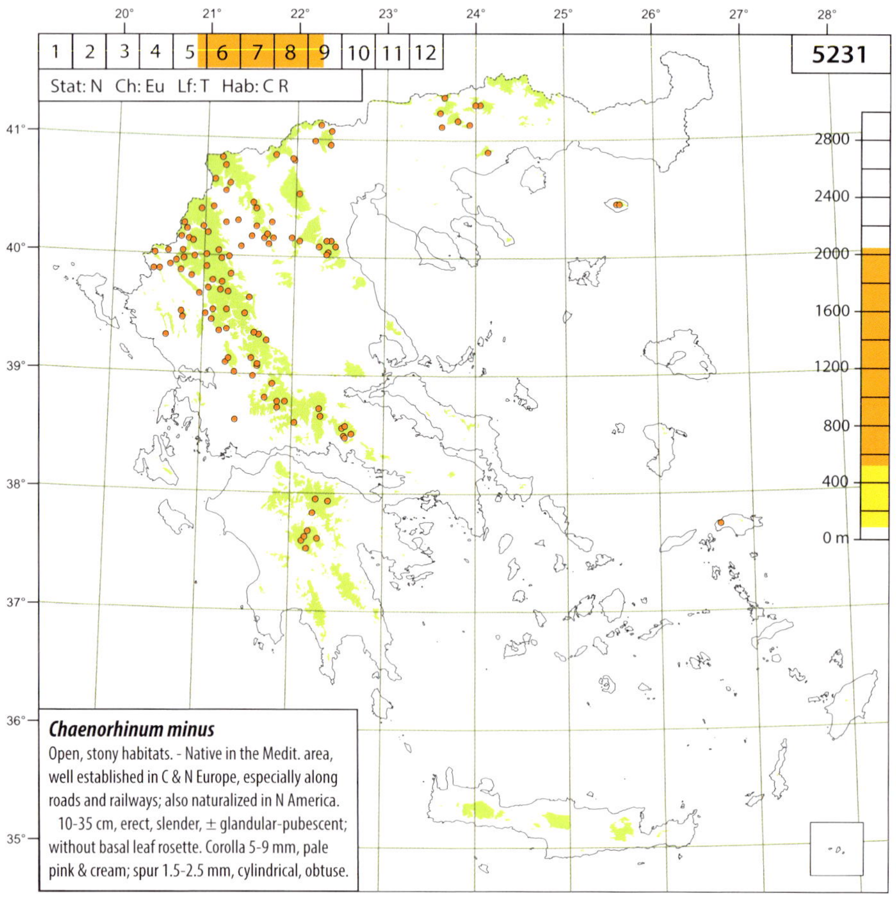

Chaenorhinum minus
Open, stony habitats. - Native in the Medit. area, well established in C & N Europe, especially along roads and railways; also naturalized in N America.
10-35 cm, erect, slender, ± glandular-pubescent; without basal leaf rosette. Corolla 5-9 mm, pale pink & cream; spur 1.5-2.5 mm, cylindrical, obtuse.

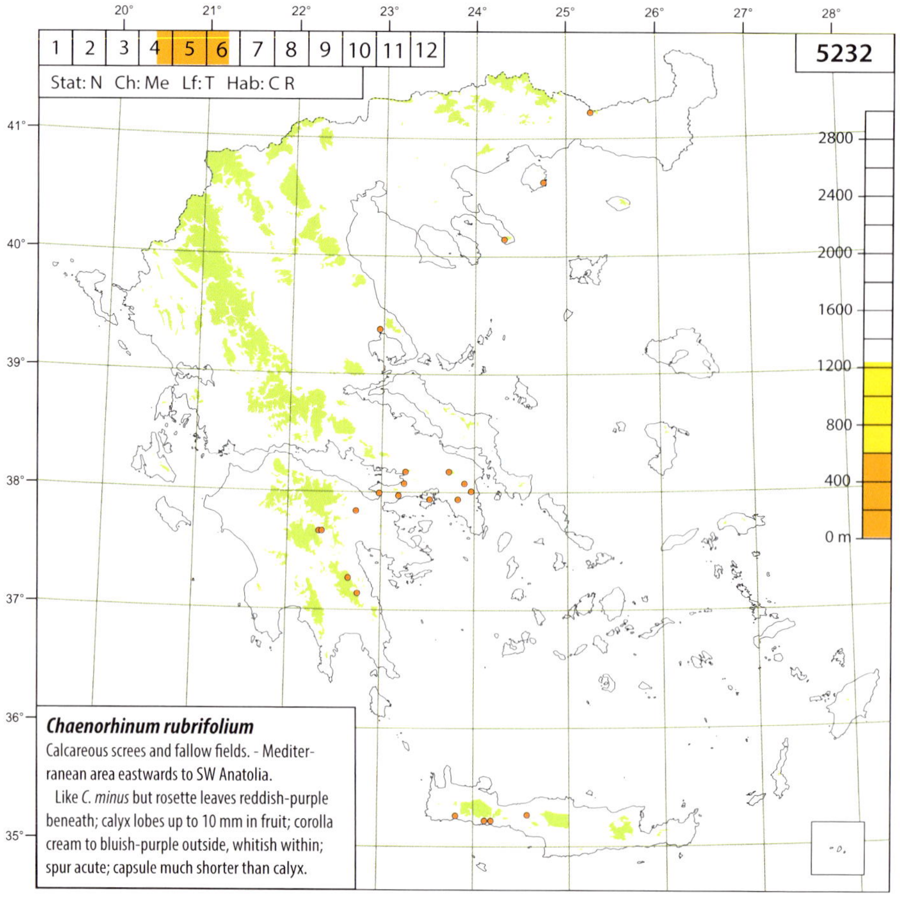

Chaenorhinum rubrifolium
Calcareous screes and fallow fields. - Mediterranean area eastwards to SW Anatolia.
Like *C. minus* but rosette leaves reddish-purple beneath; calyx lobes up to 10 mm in fruit; corolla cream to bluish-purple outside, whitish within; spur acute; capsule much shorter than calyx.

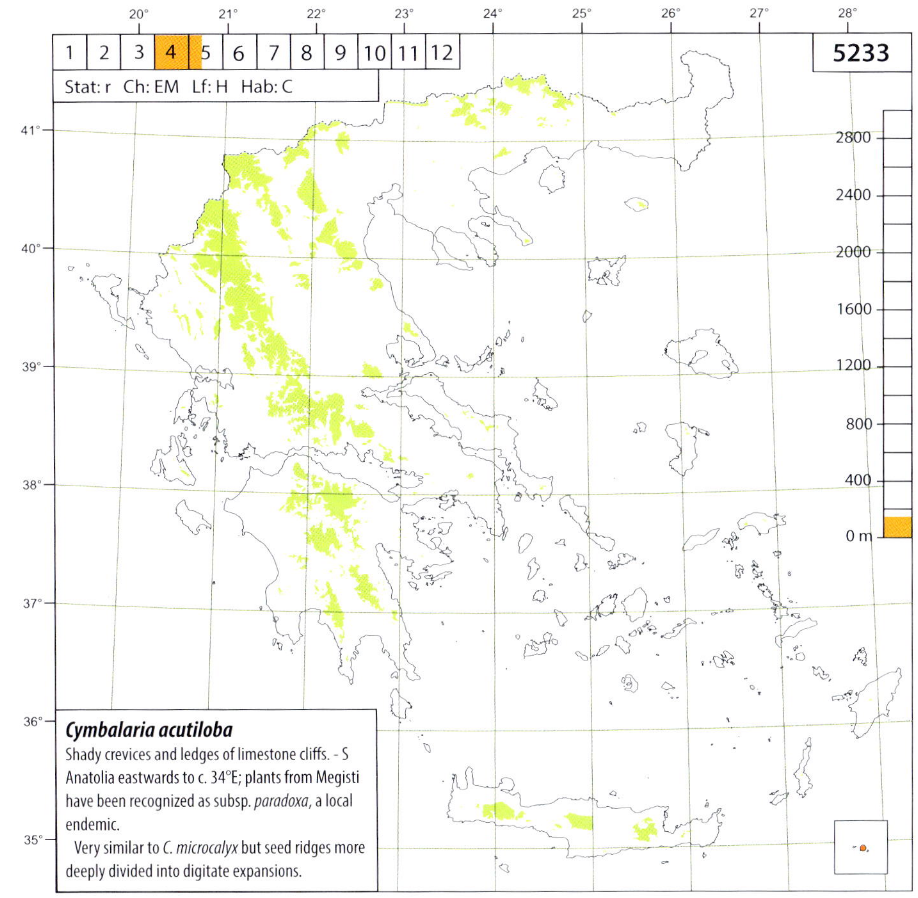

Cymbalaria acutiloba

Shady crevices and ledges of limestone cliffs. - S Anatolia eastwards to c. 34°E; plants from Megisti have been recognized as subsp. *paradoxa*, a local endemic.

Very similar to *C. microcalyx* but seed ridges more deeply divided into digitate expansions.

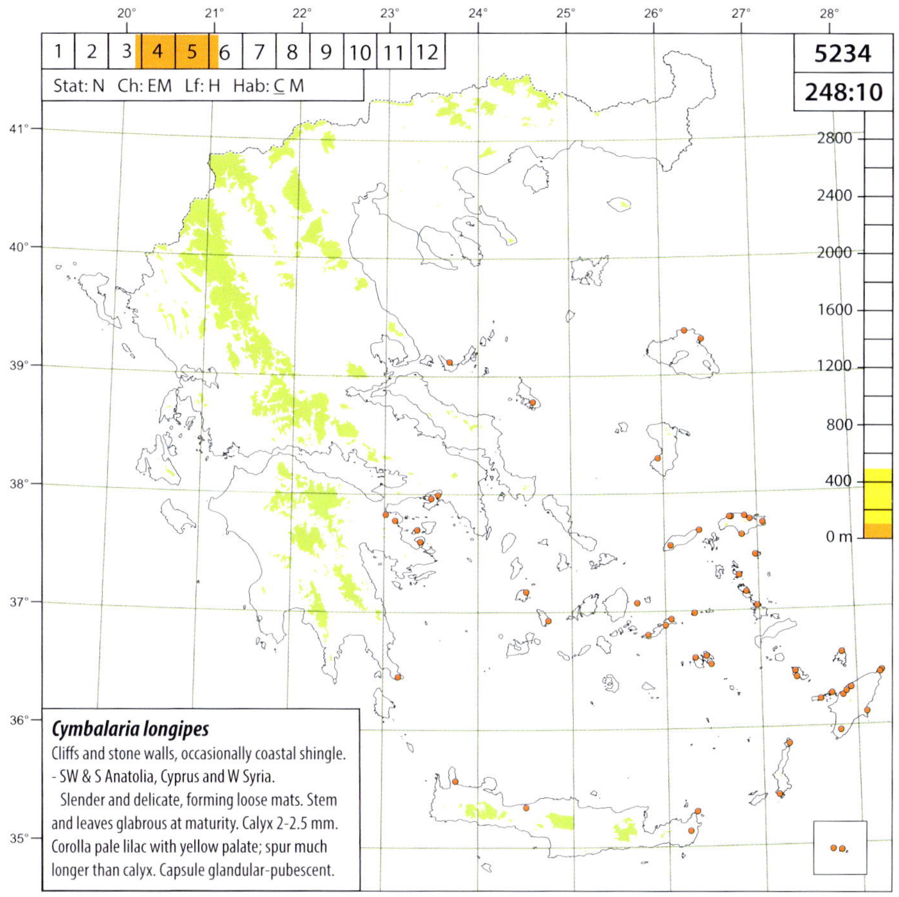

Cymbalaria longipes

Cliffs and stone walls, occasionally coastal shingle. - SW & S Anatolia, Cyprus and W Syria.

Slender and delicate, forming loose mats. Stem and leaves glabrous at maturity. Calyx 2-2.5 mm. Corolla pale lilac with yellow palate; spur much longer than calyx. Capsule glandular-pubescent.

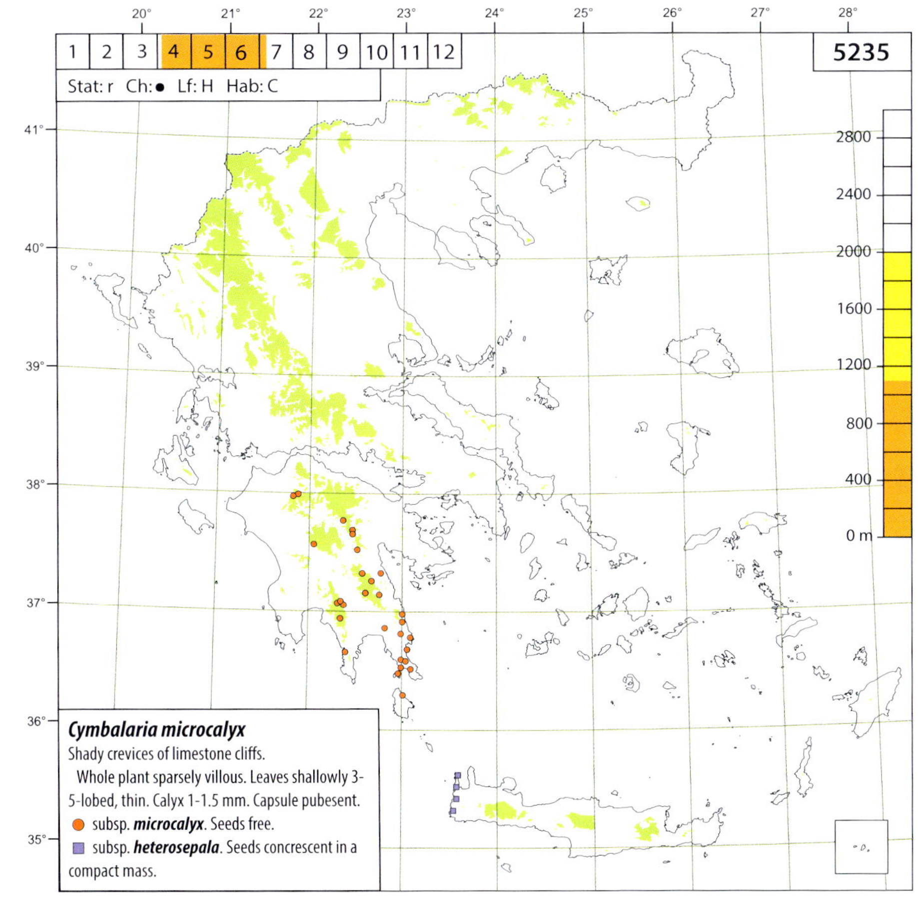

Cymbalaria microcalyx

Shady crevices of limestone cliffs.

Whole plant sparsely villous. Leaves shallowly 3-5-lobed, thin. Calyx 1-1.5 mm. Capsule pubescent.

- ● subsp. ***microcalyx***. Seeds free.
- ■ subsp. ***heterosepala***. Seeds concrescent in a compact mass.

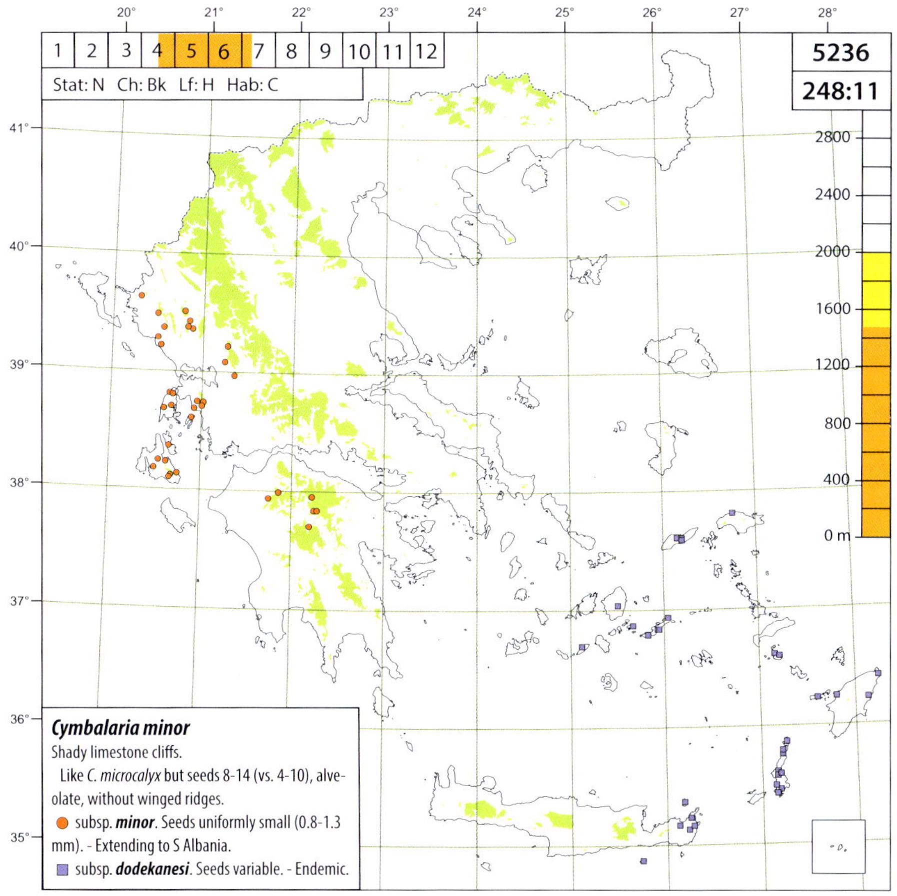

Cymbalaria minor

Shady limestone cliffs.

Like *C. microcalyx* but seeds 8-14 (vs. 4-10), alveolate, without winged ridges.

- ● subsp. ***minor***. Seeds uniformly small (0.8-1.3 mm). - Extending to S Albania.
- ■ subsp. ***dodekanesi***. Seeds variable. - Endemic.

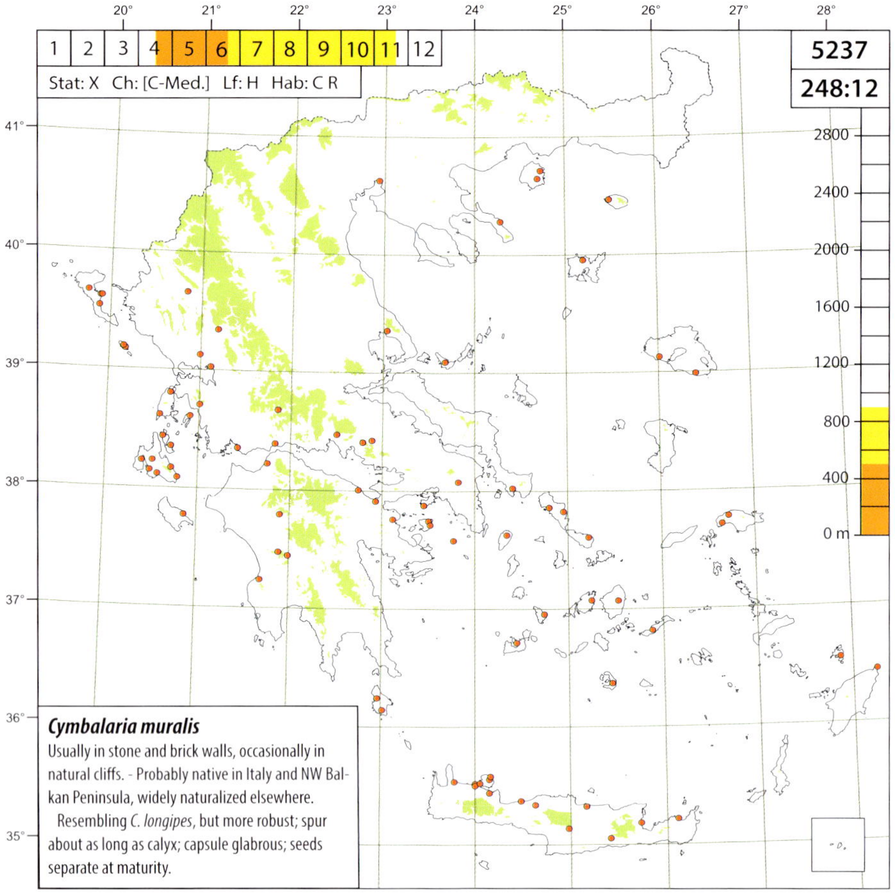

Cymbalaria muralis
Usually in stone and brick walls, occasionally in natural cliffs. - Probably native in Italy and NW Balkan Peninsula, widely naturalized elsewhere.
Resembling *C. longipes*, but more robust; spur about as long as calyx; capsule glabrous; seeds separate at maturity.

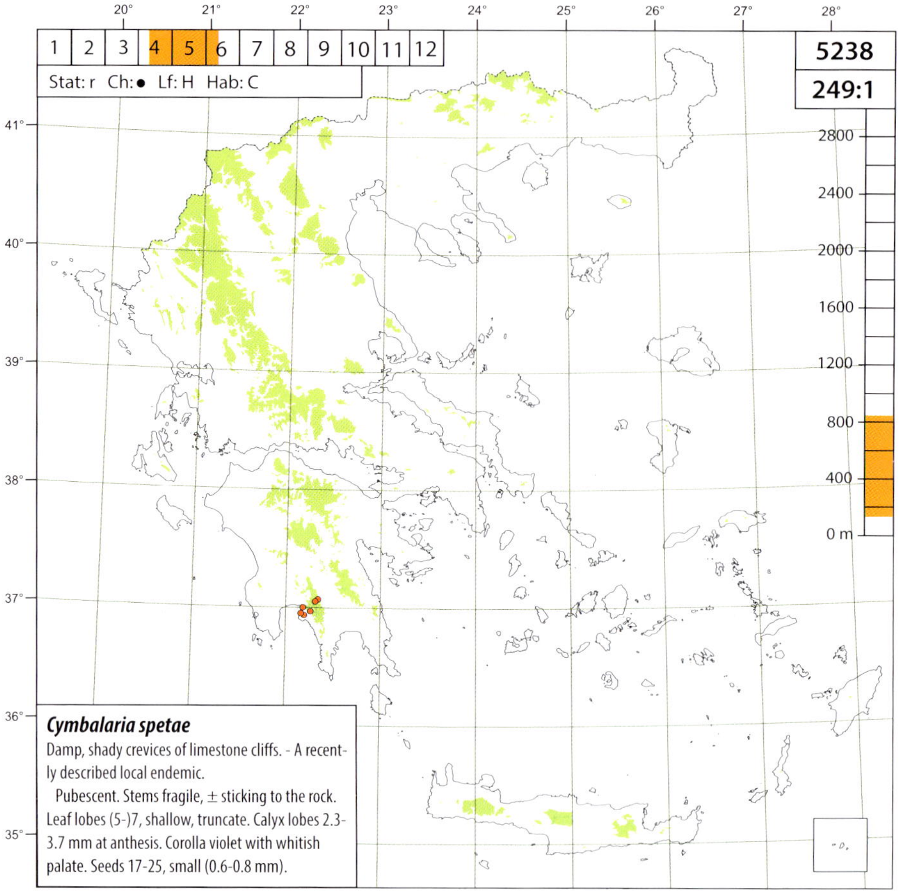

Cymbalaria spetae
Damp, shady crevices of limestone cliffs. - A recently described local endemic.
Pubescent. Stems fragile, ± sticking to the rock. Leaf lobes (5-)7, shallow, truncate. Calyx lobes 2.3-3.7 mm at anthesis. Corolla violet with whitish palate. Seeds 17-25, small (0.6-0.8 mm).

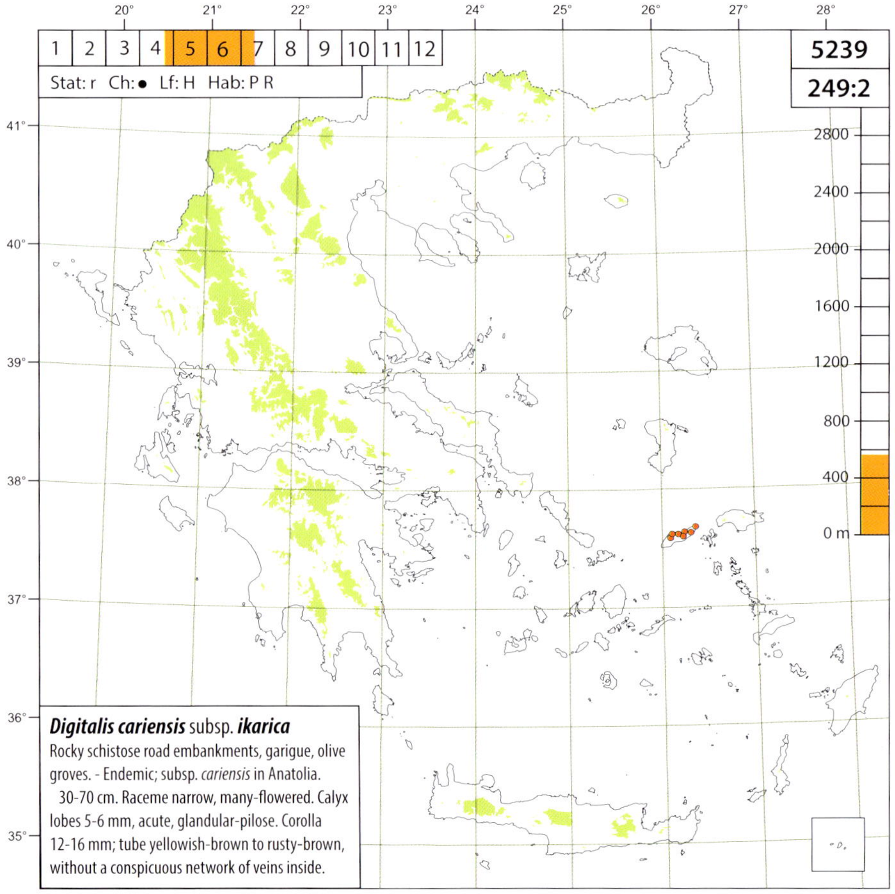

Digitalis cariensis* subsp. *ikarica
Rocky schistose road embankments, garigue, olive groves. - Endemic; subsp. *cariensis* in Anatolia.
30-70 cm. Raceme narrow, many-flowered. Calyx lobes 5-6 mm, acute, glandular-pilose. Corolla 12-16 mm; tube yellowish-brown to rusty-brown, without a conspicuous network of veins inside.

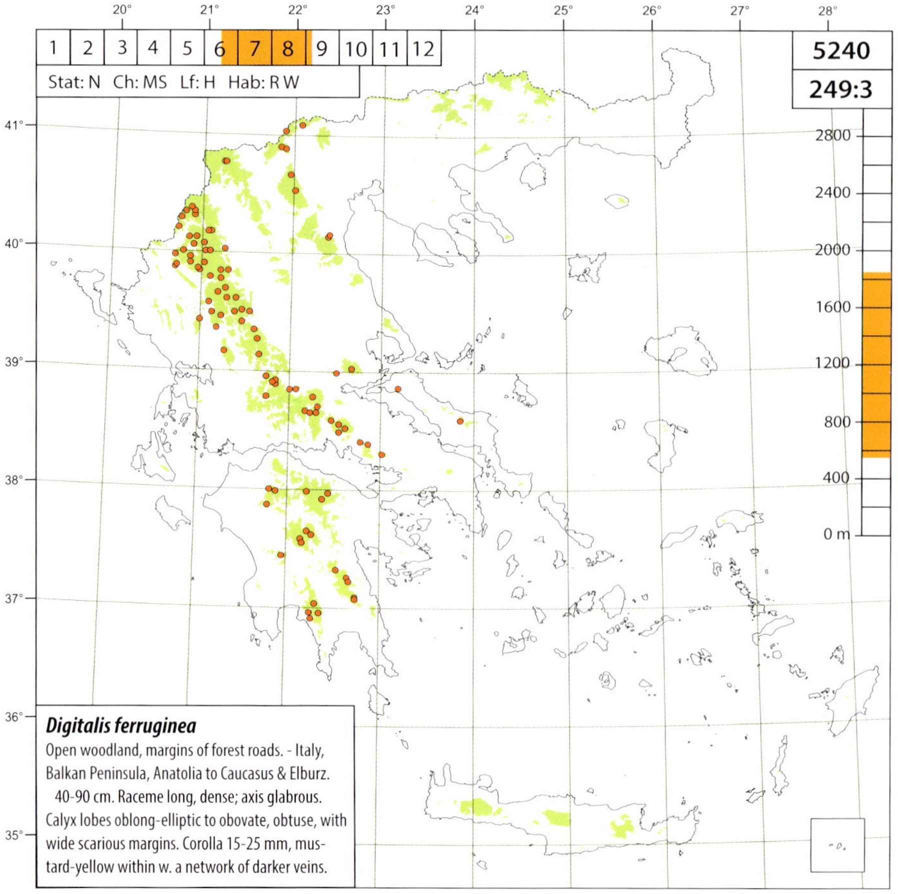

Digitalis ferruginea
Open woodland, margins of forest roads. - Italy, Balkan Peninsula, Anatolia to Caucasus & Elburz.
40-90 cm. Raceme long, dense; axis glabrous. Calyx lobes oblong-elliptic to obovate, obtuse, with wide scarious margins. Corolla 15-25 mm, mustard-yellow within w. a network of darker veins.

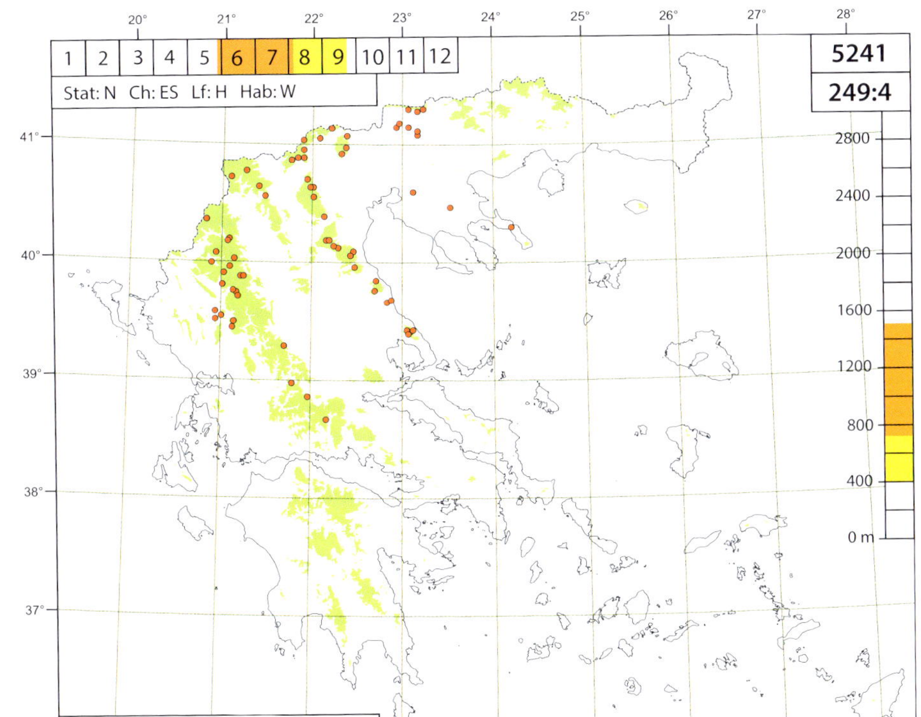

Digitalis grandiflora
Open woodland, *Pteridium* thickets, by forest roads. - C & E Europe to W Siberia, naturalized in E N Am.
40-100 cm. Cauline leaves 5-12 cm, lanceolate. Flowers few, in a ± unilateral raceme. Calyx lobes without scarious margins. Corolla 3-5 cm, rather pale yellow, indistinctly 2-lipped.

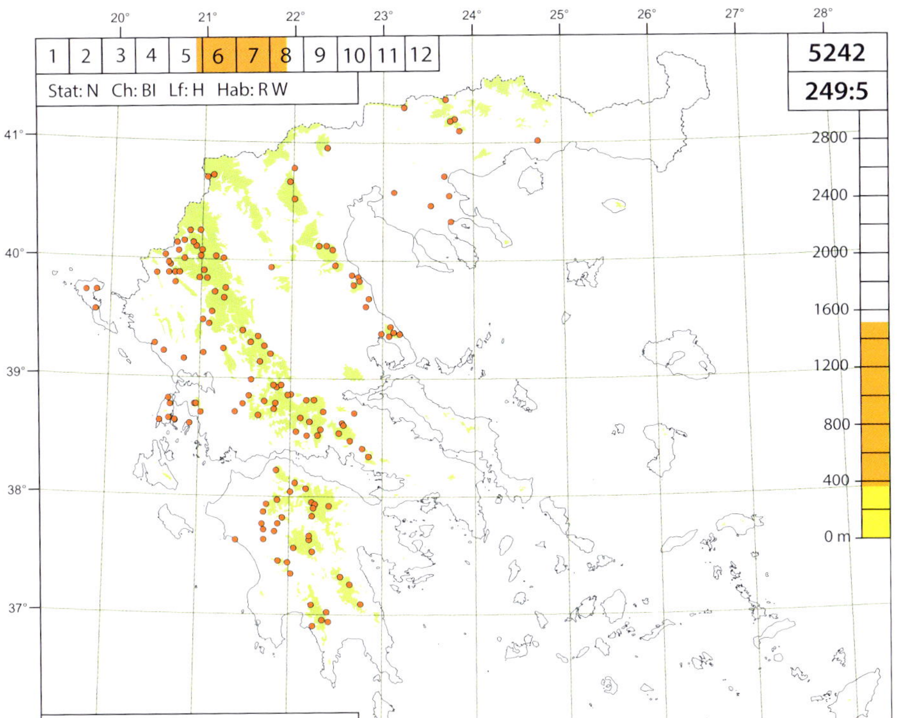

Digitalis laevigata
Open woodland, ravines, road embankments. - Balkan Peninsula (mainly in the west) and NE Italy.
Like *D. ferruginea* but raceme laxer; calyx lobes acute, without or with very narrow scarious margins; corolla 18-30 mm, inside of lower lip glabrous or sparsely (vs. densely) bearded.

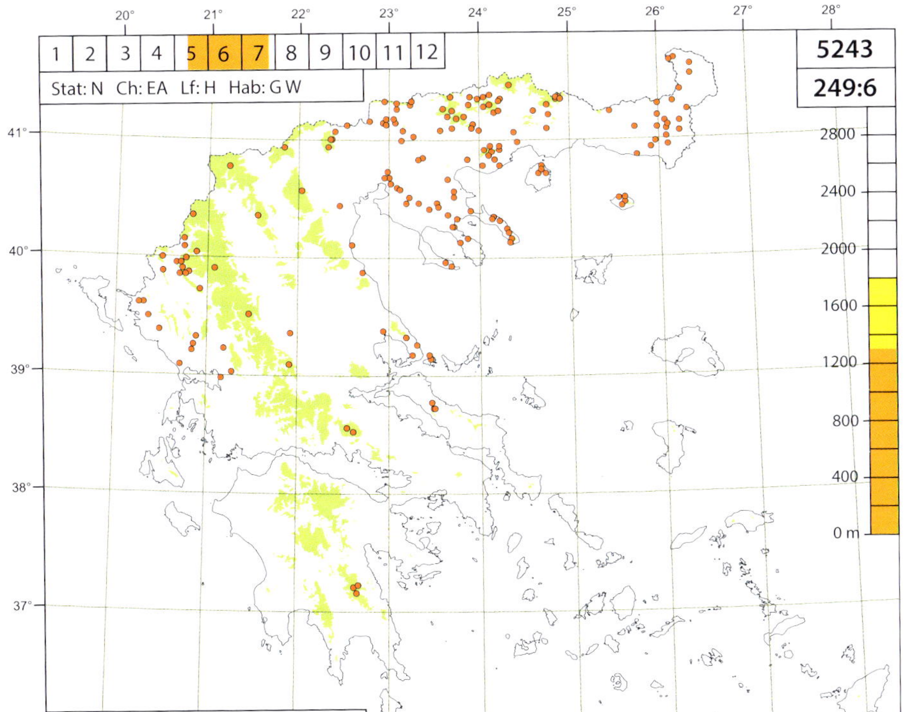

Digitalis lanata
Grassland, mixed scrub, woodland margins. - SE Europe and NW Anatolia, naturalized in N America.
30-100 cm. Raceme long, dense; axis lanate and glandular-villous. Calyx lobes acute, without scarious margins. Corolla with network of rusty-brown veins inside; lower lip white.

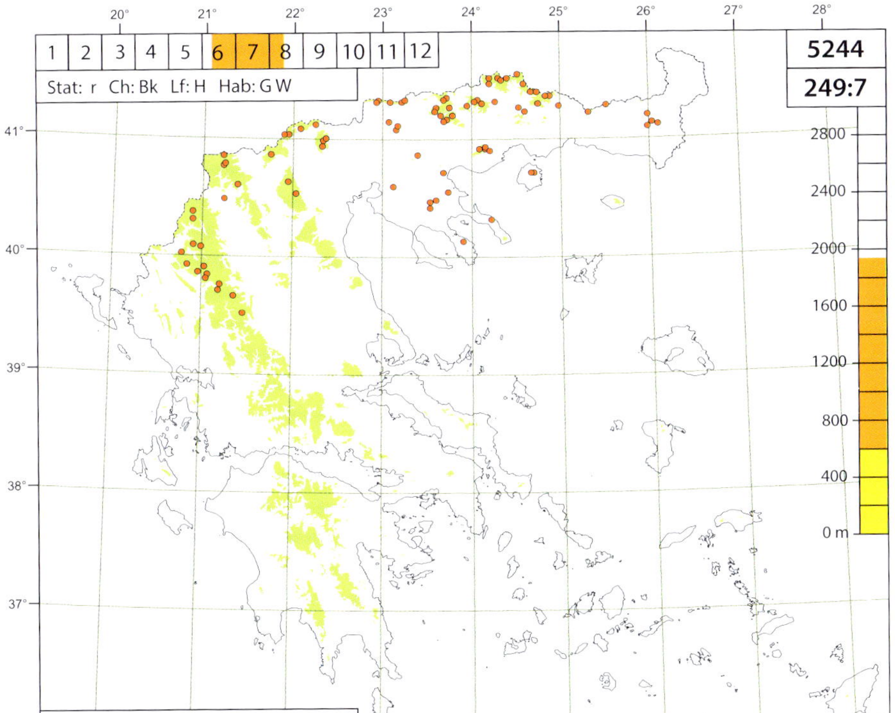

Digitalis viridiflora
Woodland margins and montane grassland. - C & E Balkan Peninsula to Turkey-in-Europe.
30-80 cm; stem sparsely lanate. Raceme short, dense, ± unilateral. Corolla 11-17 mm, greenish-white or cream, veined and marked brownish-orange in front; lobes short.

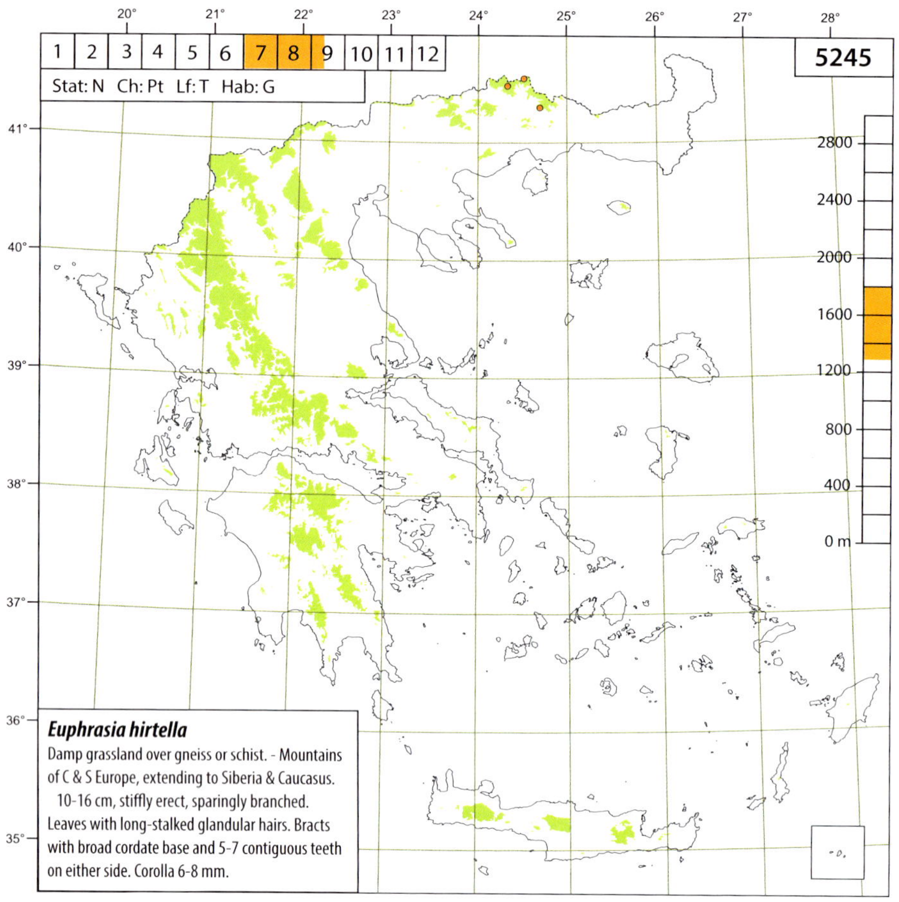

Euphrasia hirtella

Damp grassland over gneiss or schist. - Mountains of C & S Europe, extending to Siberia & Caucasus.
10-16 cm, stiffly erect, sparingly branched. Leaves with long-stalked glandular hairs. Bracts with broad cordate base and 5-7 contiguous teeth on either side. Corolla 6-8 mm.

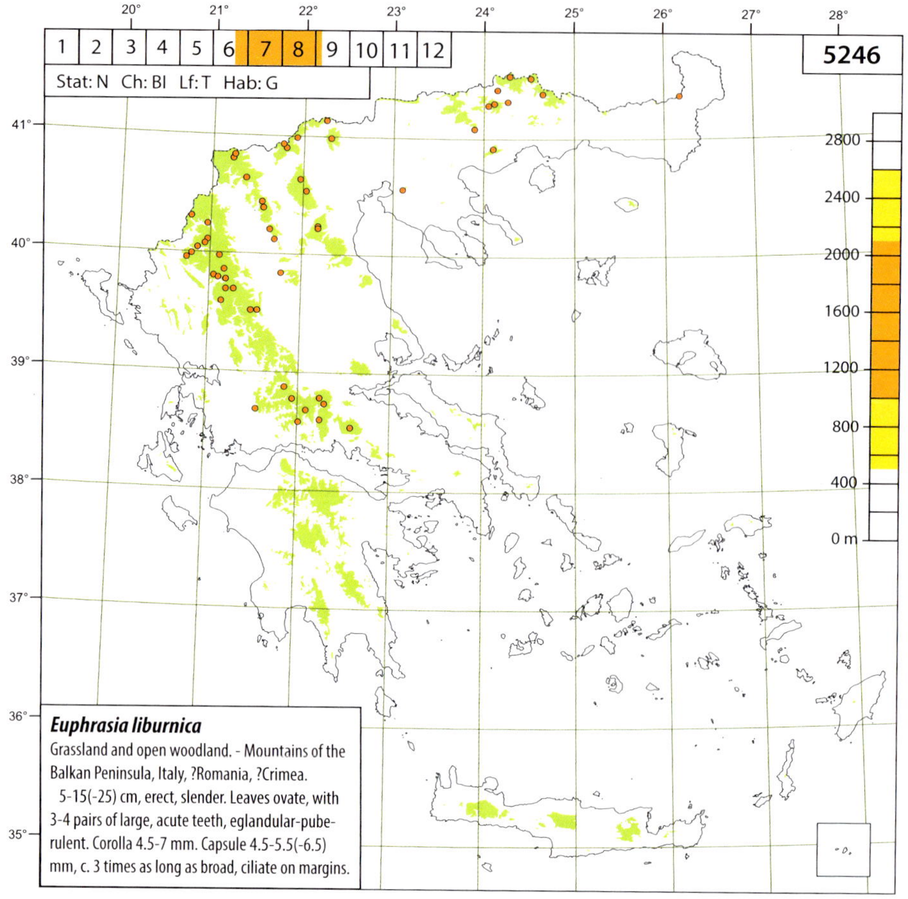

Euphrasia liburnica

Grassland and open woodland. - Mountains of the Balkan Peninsula, Italy, ?Romania, ?Crimea.
5-15(-25) cm, erect, slender. Leaves ovate, with 3-4 pairs of large, acute teeth, eglandular-pubescent. Corolla 4.5-7 mm. Capsule 4.5-5.5(-6.5) mm, c. 3 times as long as broad, ciliate on margins.

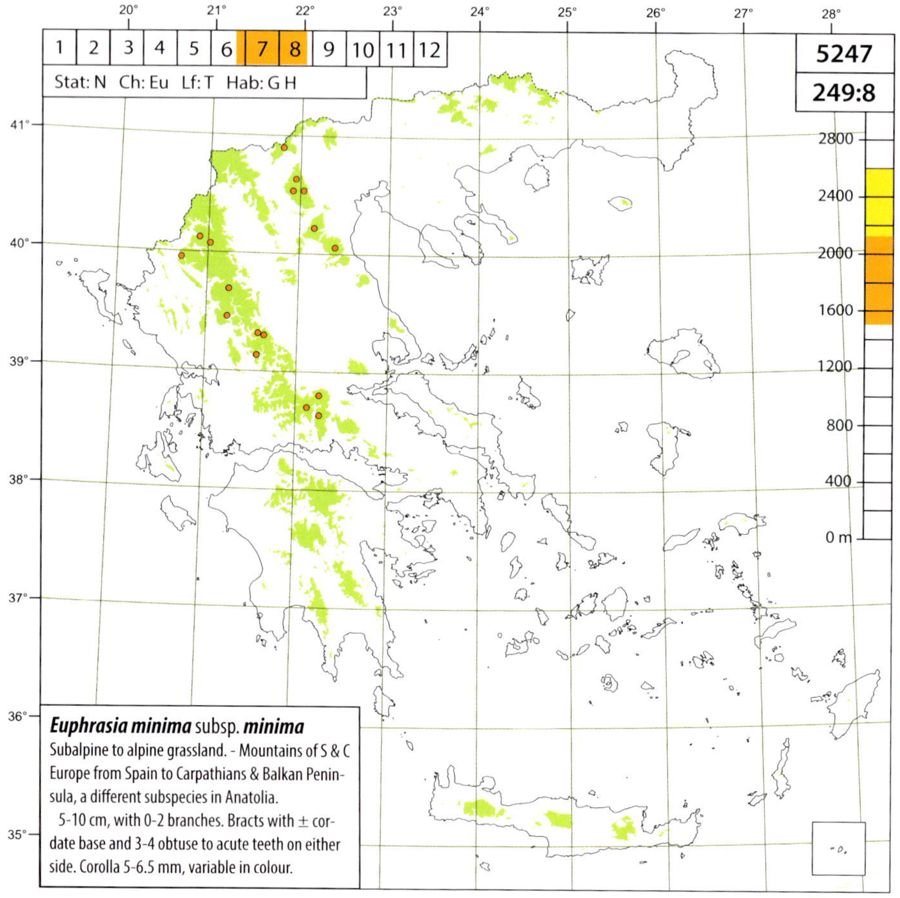

Euphrasia minima subsp. *minima*

Subalpine to alpine grassland. - Mountains of S & C Europe from Spain to Carpathians & Balkan Peninsula, a different subspecies in Anatolia.
5-10 cm, with 0-2 branches. Bracts with ± cordate base and 3-4 obtuse to acute teeth on either side. Corolla 5-6.5 mm, variable in colour.

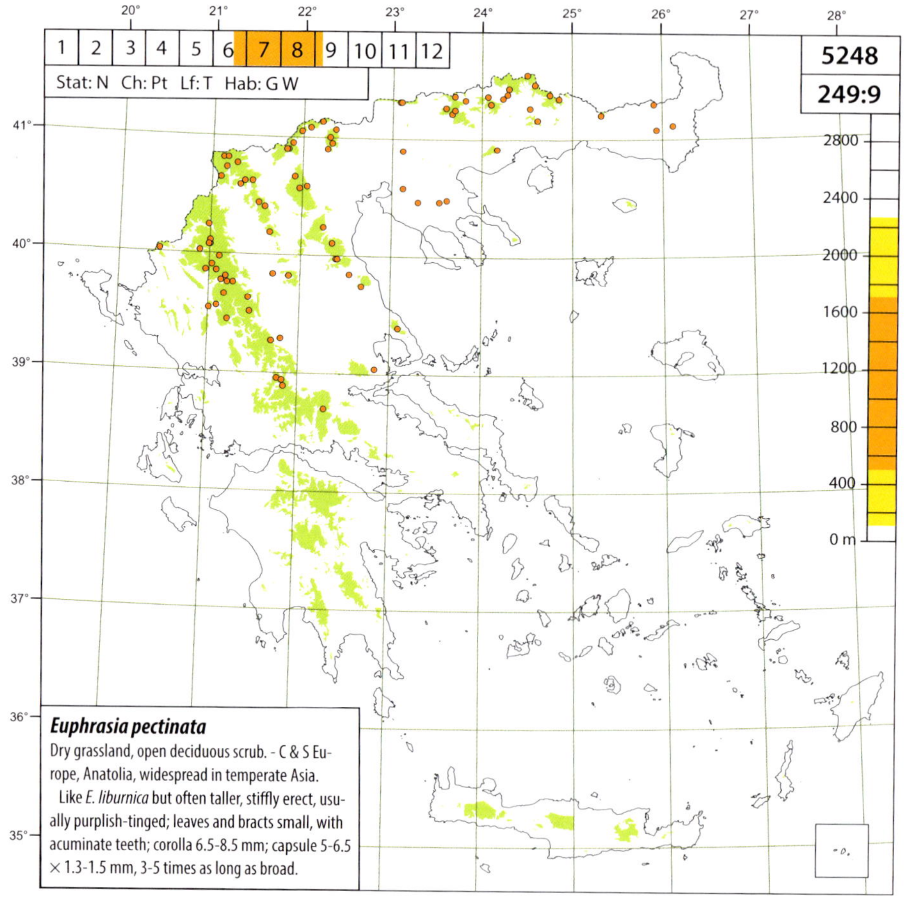

Euphrasia pectinata

Dry grassland, open deciduous scrub. - C & S Europe, Anatolia, widespread in temperate Asia.
Like *E. liburnica* but often taller, stiffly erect, usually purplish-tinged; leaves and bracts small, with acuminate teeth; corolla 6.5-8.5 mm; capsule 5-6.5 × 1.3-1.5 mm, 3-5 times as long as broad.

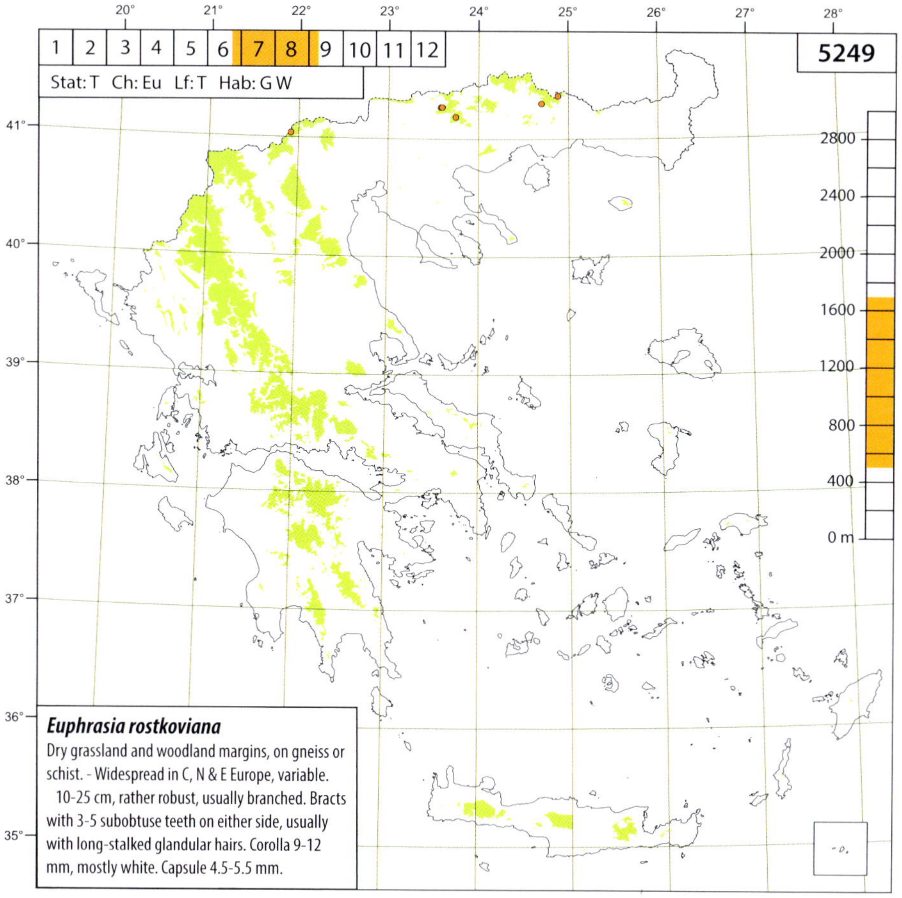

Euphrasia rostkoviana

Dry grassland and woodland margins, on gneiss or schist. - Widespread in C, N & E Europe, variable.

10-25 cm, rather robust, usually branched. Bracts with 3-5 subobtuse teeth on either side, usually with long-stalked glandular hairs. Corolla 9-12 mm, mostly white. Capsule 4.5-5.5 mm.

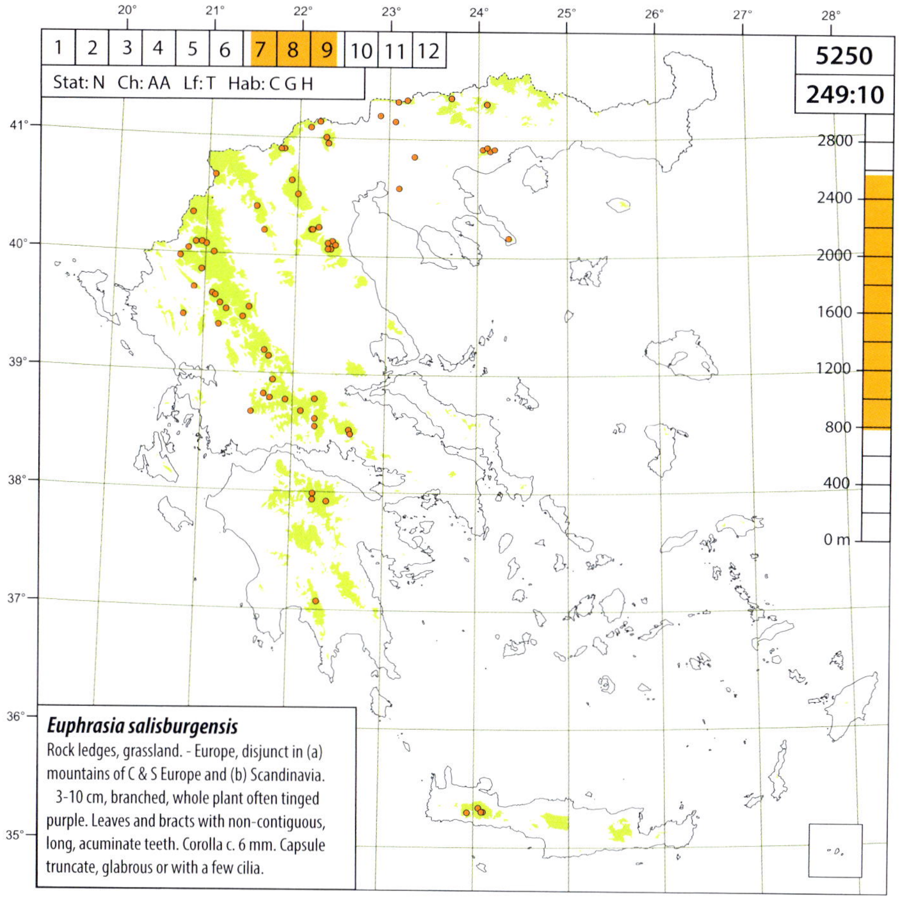

Euphrasia salisburgensis

Rock ledges, grassland. - Europe, disjunct in (a) mountains of C & S Europe and (b) Scandinavia.

3-10 cm, branched, whole plant often tinged purple. Leaves and bracts with non-contiguous, long, acuminate teeth. Corolla c. 6 mm. Capsule truncate, glabrous or with a few cilia.

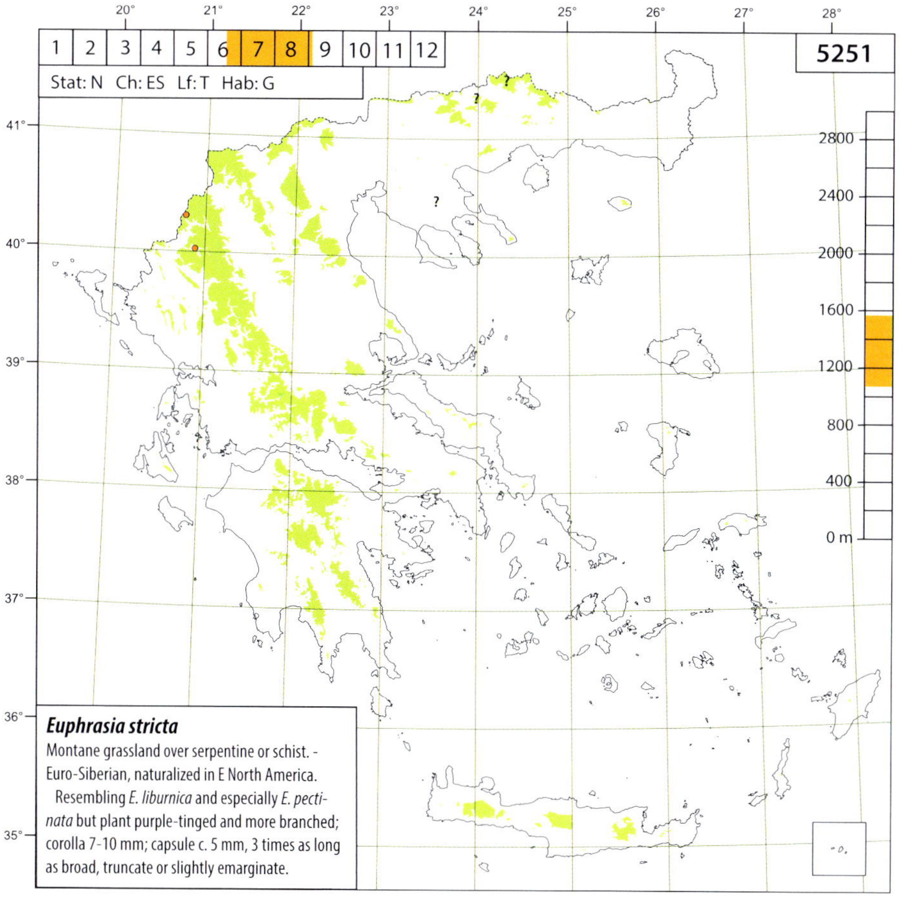

Euphrasia stricta

Montane grassland over serpentine or schist. - Euro-Siberian, naturalized in E North America.

Resembling *E. liburnica* and especially *E. pectinata* but plant purple-tinged and more branched; corolla 7-10 mm; capsule c. 5 mm, 3 times as long as broad, truncate or slightly emarginate.

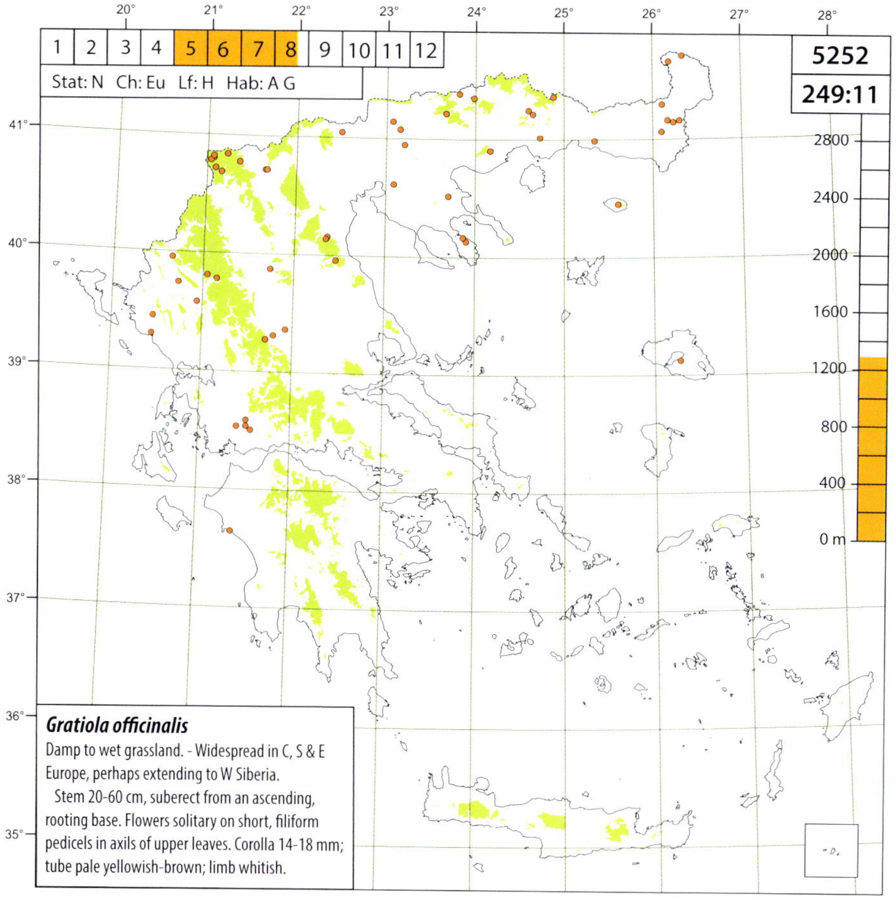

Gratiola officinalis

Damp to wet grassland. - Widespread in C, S & E Europe, perhaps extending to W Siberia.

Stem 20-60 cm, suberect from an ascending, rooting base. Flowers solitary on short, filiform pedicels in axils of upper leaves. Corolla 14-18 mm; tube pale yellowish-brown; limb whitish.

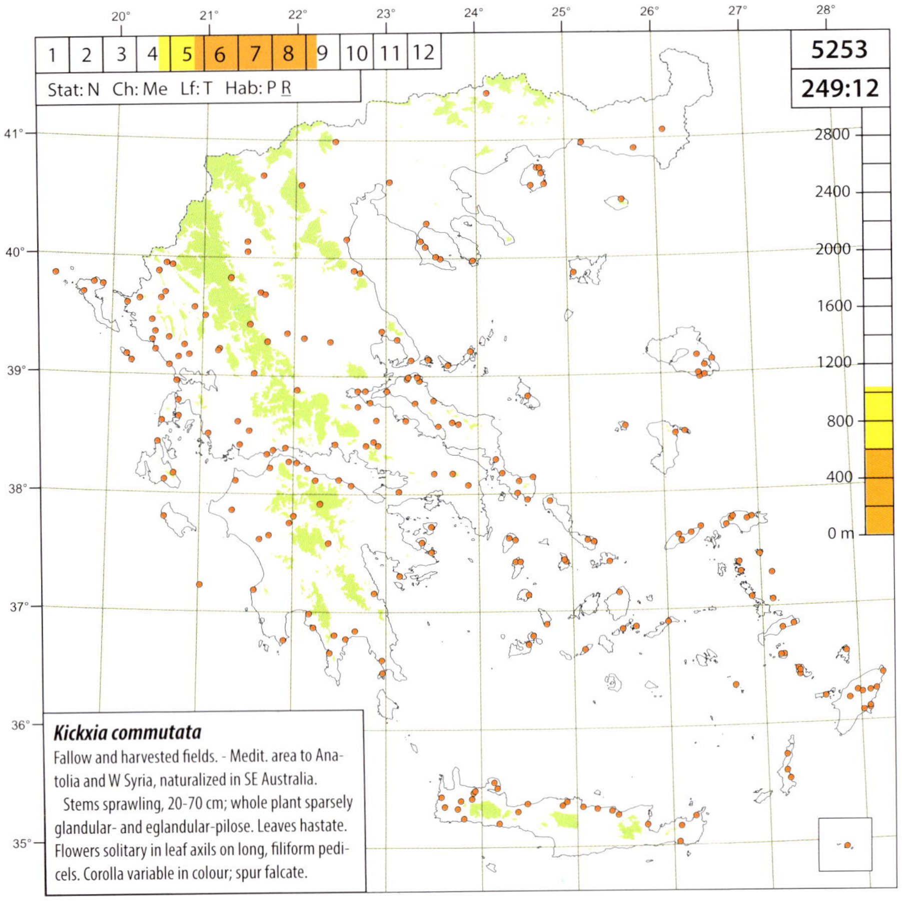

Kickxia commutata
Fallow and harvested fields. - Medit. area to Anatolia and W Syria, naturalized in SE Australia.
Stems sprawling, 20-70 cm; whole plant sparsely glandular- and eglandular-pilose. Leaves hastate. Flowers solitary in leaf axils on long, filiform pedicels. Corolla variable in colour; spur falcate.

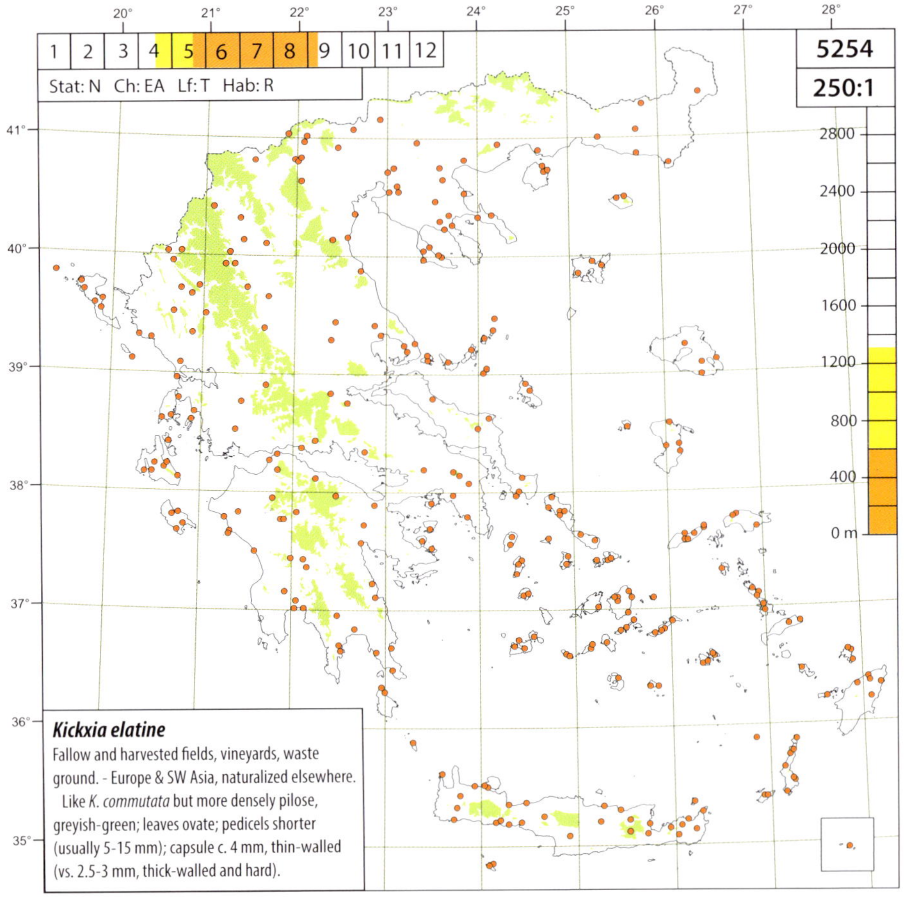

Kickxia elatine
Fallow and harvested fields, vineyards, waste ground. - Europe & SW Asia, naturalized elsewhere.
Like *K. commutata* but more densely pilose, greyish-green; leaves ovate; pedicels shorter (usually 5-15 mm); capsule c. 4 mm, thin-walled (vs. 2.5-3 mm, thick-walled and hard).

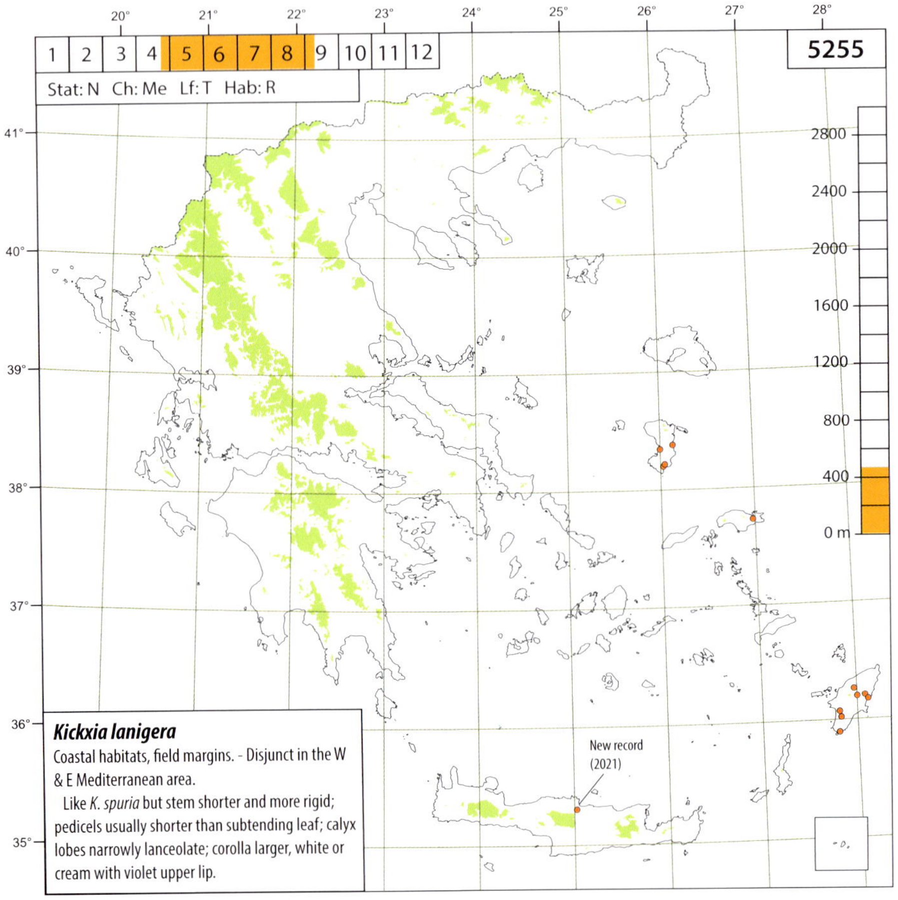

Kickxia lanigera
Coastal habitats, field margins. - Disjunct in the W & E Mediterranean area.
Like *K. spuria* but stem shorter and more rigid; pedicels usually shorter than subtending leaf; calyx lobes narrowly lanceolate; corolla larger, white or cream with violet upper lip.

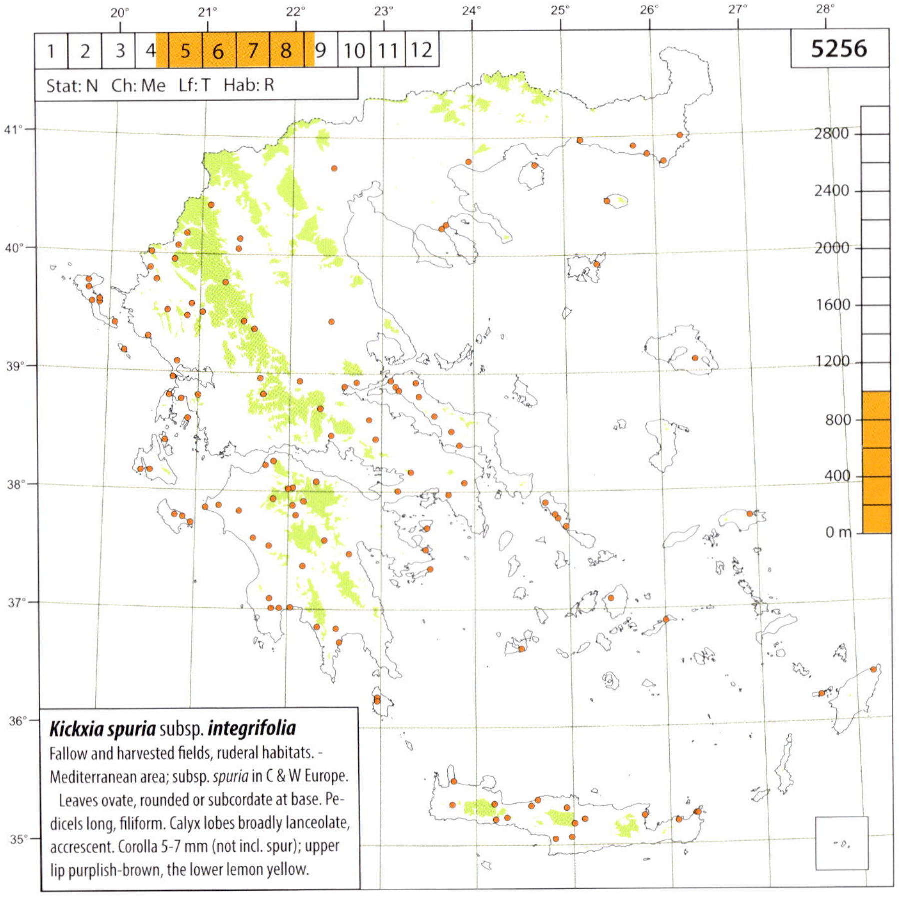

Kickxia spuria* subsp. *integrifolia
Fallow and harvested fields, ruderal habitats. - Mediterranean area; subsp. *spuria* in C & W Europe.
Leaves ovate, rounded or subcordate at base. Pedicels long, filiform. Calyx lobes broadly lanceolate, accrescent. Corolla 5-7 mm (not incl. spur); upper lip purplish-brown, the lower lemon yellow.

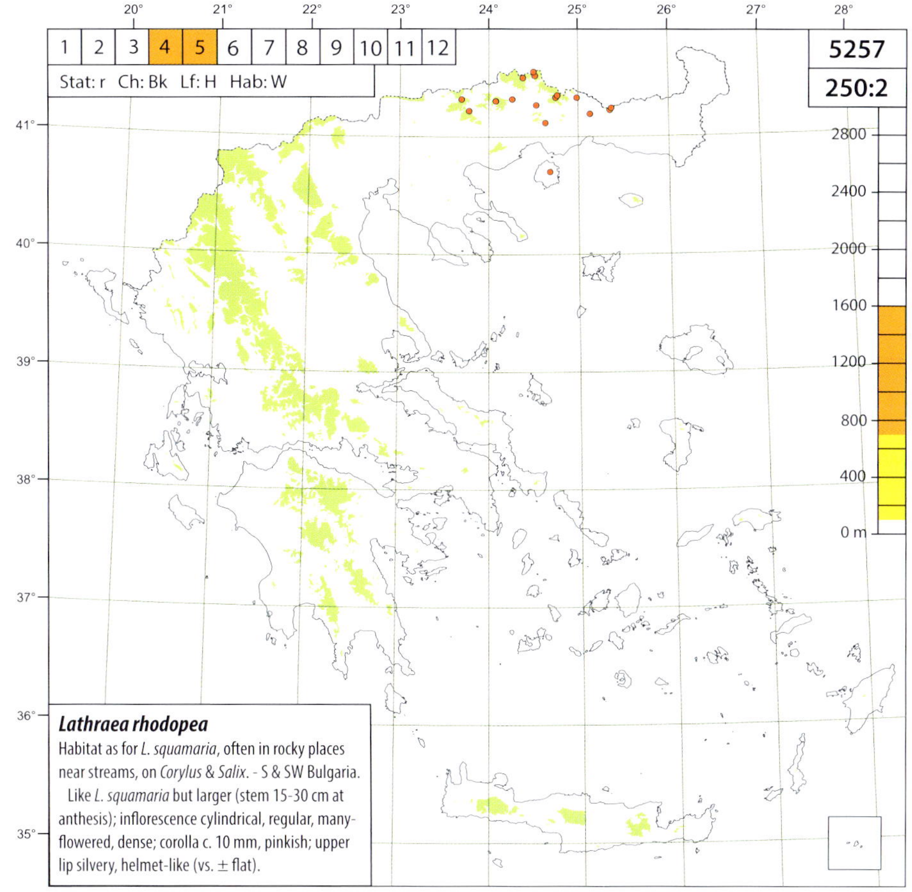

Lathraea rhodopea

Habitat as for *L. squamaria*, often in rocky places near streams, on *Corylus* & *Salix*. - S & SW Bulgaria.

Like *L. squamaria* but larger (stem 15-30 cm at anthesis); inflorescence cylindrical, regular, many-flowered, dense; corolla c. 10 mm, pinkish; upper lip silvery, helmet-like (vs. ± flat).

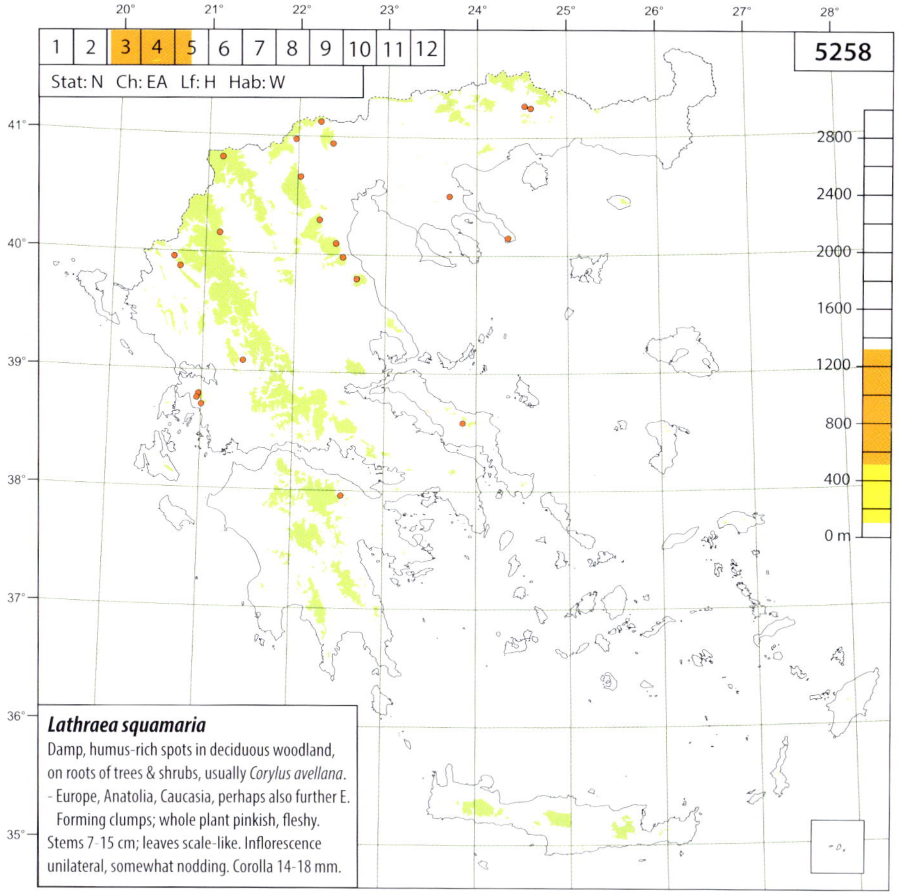

Lathraea squamaria

Damp, humus-rich spots in deciduous woodland, on roots of trees & shrubs, usually *Corylus avellana*. - Europe, Anatolia, Caucasia, perhaps also further E.

Forming clumps; whole plant pinkish, fleshy. Stems 7-15 cm; leaves scale-like. Inflorescence unilateral, somewhat nodding. Corolla 14-18 mm.

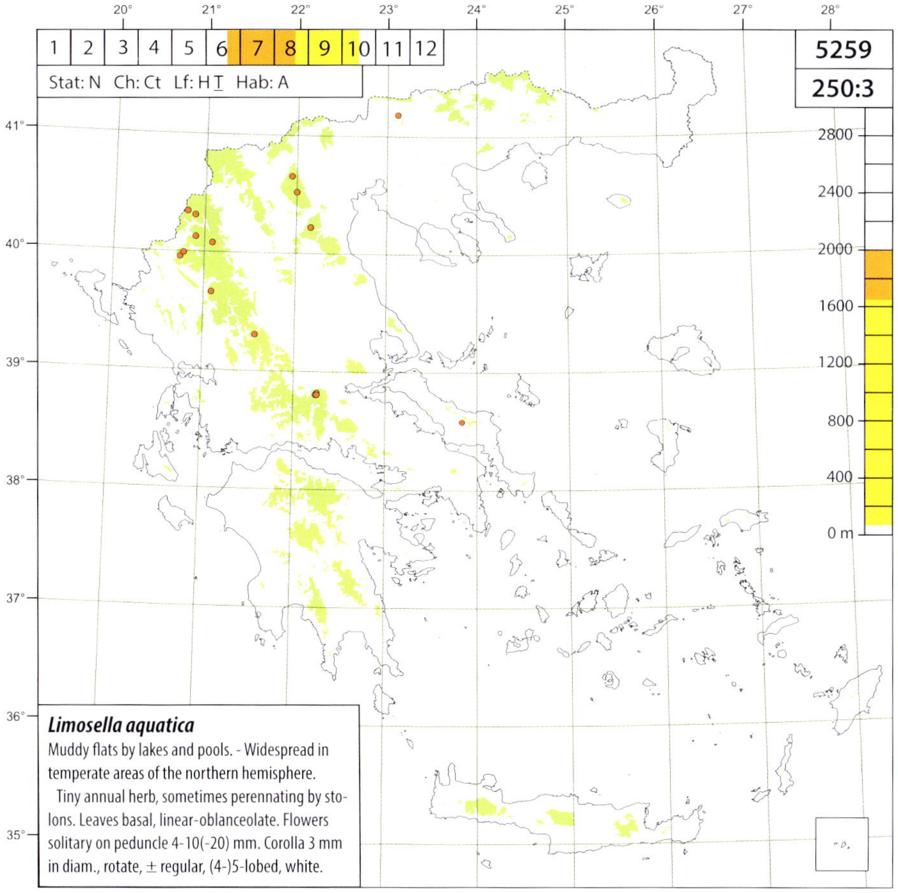

Limosella aquatica

Muddy flats by lakes and pools. - Widespread in temperate areas of the northern hemisphere.

Tiny annual herb, sometimes perennating by stolons. Leaves basal, linear-oblanceolate. Flowers solitary on peduncle 4-10(-20) mm. Corolla 3 mm in diam., rotate, ± regular, (4-)5-lobed, white.

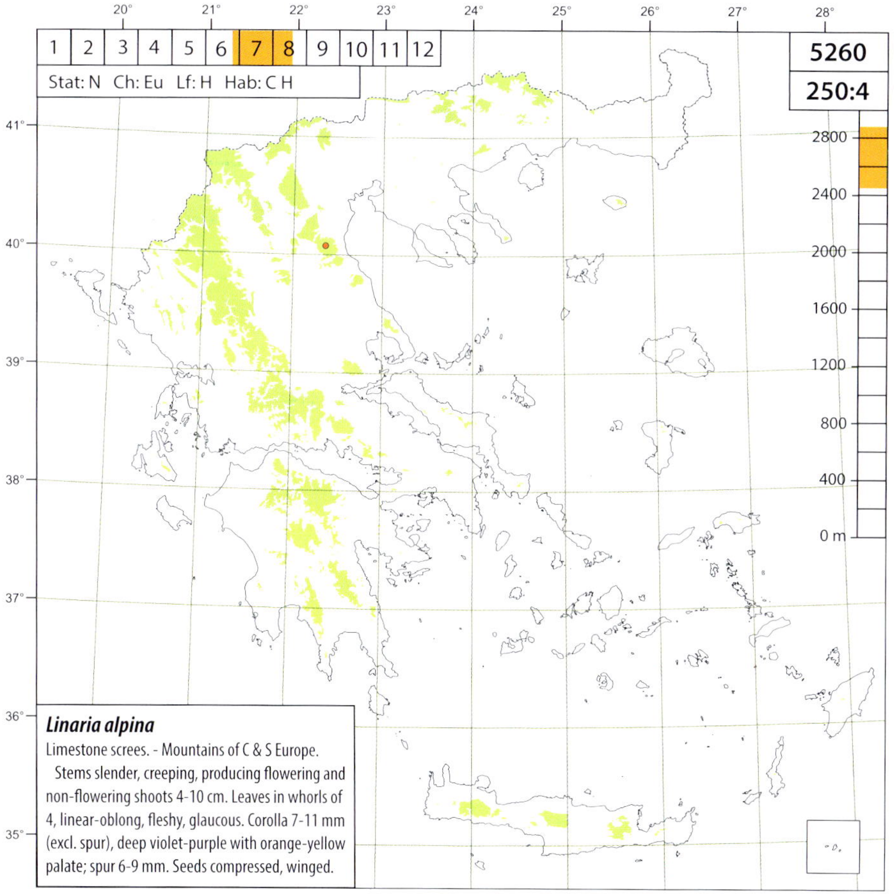

Linaria alpina

Limestone screes. - Mountains of C & S Europe.

Stems slender, creeping, producing flowering and non-flowering shoots 4-10 cm. Leaves in whorls of 4, linear-oblong, fleshy, glaucous. Corolla 7-11 mm (excl. spur), deep violet-purple with orange-yellow palate; spur 6-9 mm. Seeds compressed, winged.

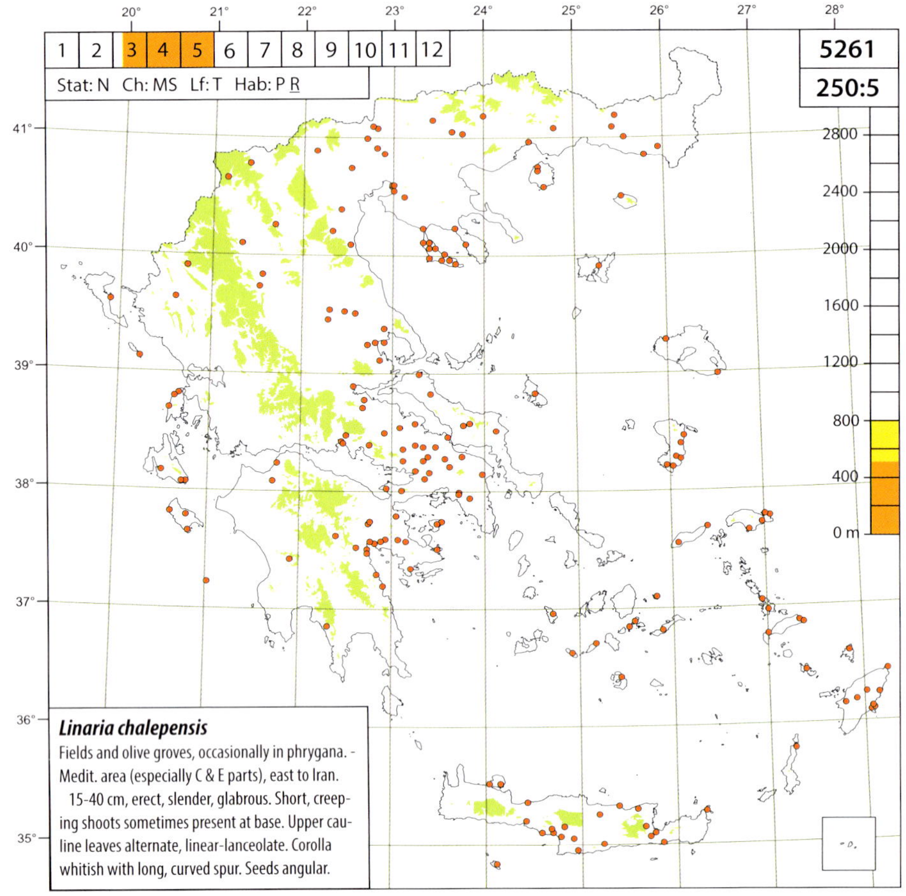

Linaria chalepensis

Fields and olive groves, occasionally in phrygana. - Medit. area (especially C & E parts), east to Iran.

15-40 cm, erect, slender, glabrous. Short, creeping shoots sometimes present at base. Upper cauline leaves alternate, linear-lanceolate. Corolla whitish with long, curved spur. Seeds angular.

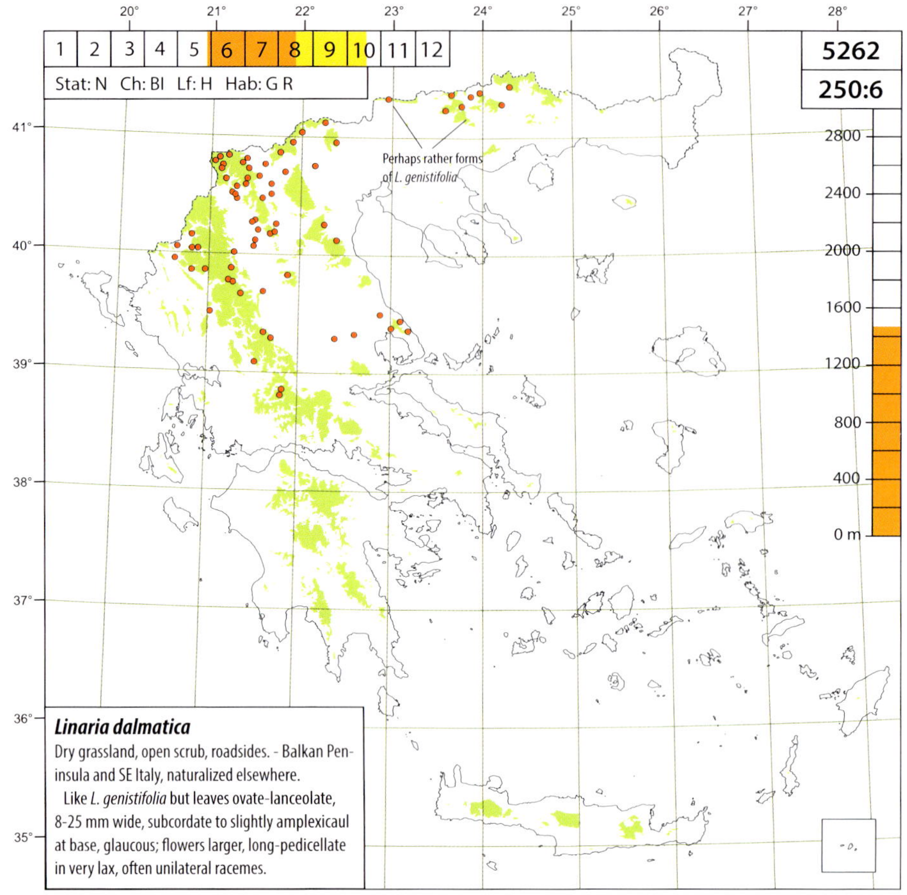

Linaria dalmatica

Dry grassland, open scrub, roadsides. - Balkan Peninsula and SE Italy, naturalized elsewhere.

Like *L. genistifolia* but leaves ovate-lanceolate, 8-25 mm wide, subcordate to slightly amplexicaul at base, glaucous; flowers larger, long-pedicellate in very lax, often unilateral racemes.

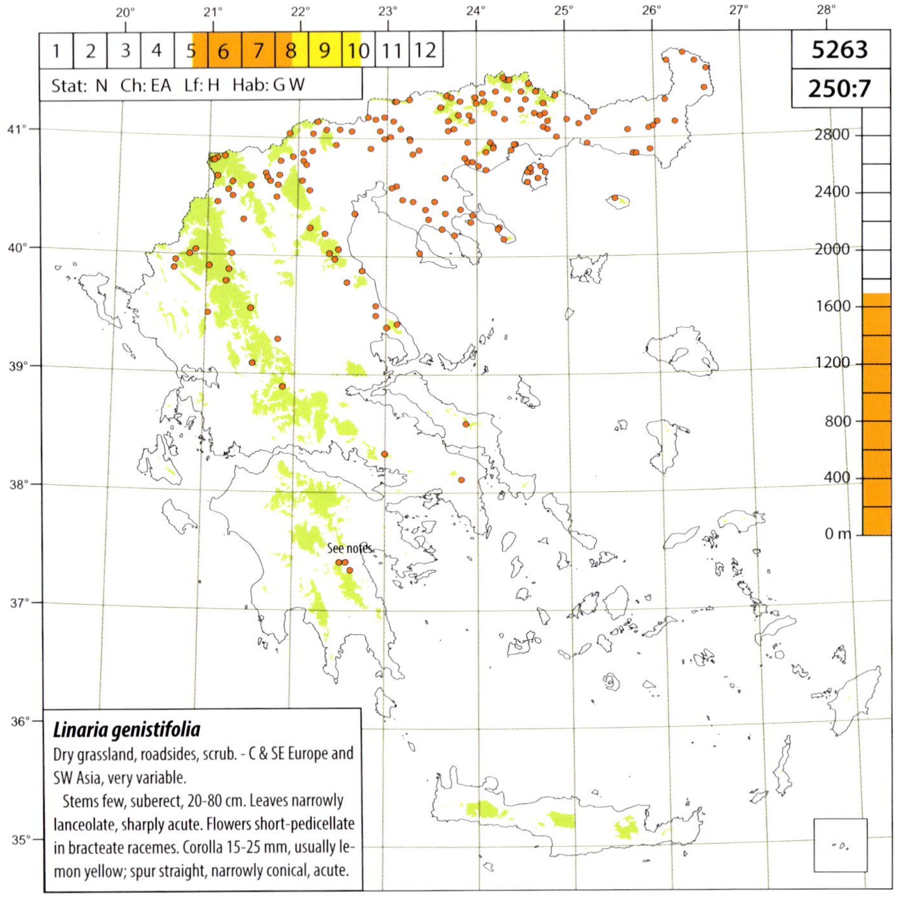

Linaria genistifolia

Dry grassland, roadsides, scrub. - C & SE Europe and SW Asia, very variable.

Stems few, suberect, 20-80 cm. Leaves narrowly lanceolate, sharply acute. Flowers short-pedicellate in bracteate racemes. Corolla 15-25 mm, usually lemon yellow; spur straight, narrowly conical, acute.

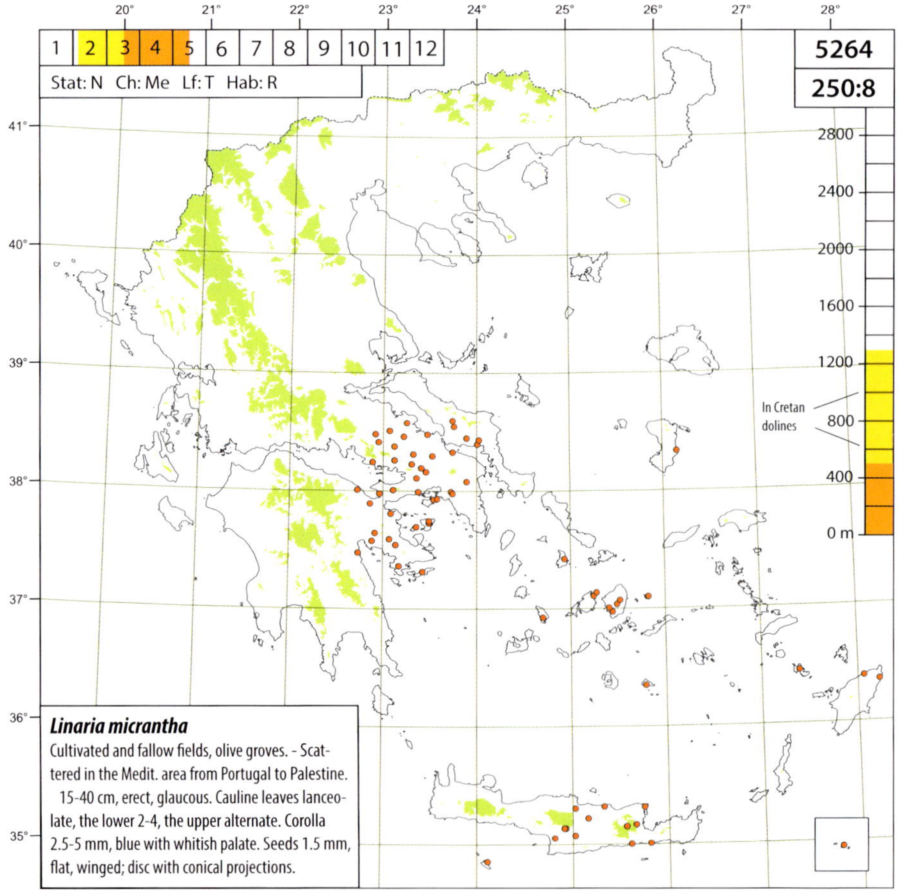

Linaria micrantha

Cultivated and fallow fields, olive groves. - Scattered in the Medit. area from Portugal to Palestine.

15-40 cm, erect, glaucous. Cauline leaves lanceolate, the lower 2-4, the upper alternate. Corolla 2.5-5 mm, blue with whitish palate. Seeds 1.5 mm, flat, winged; disc with conical projections.

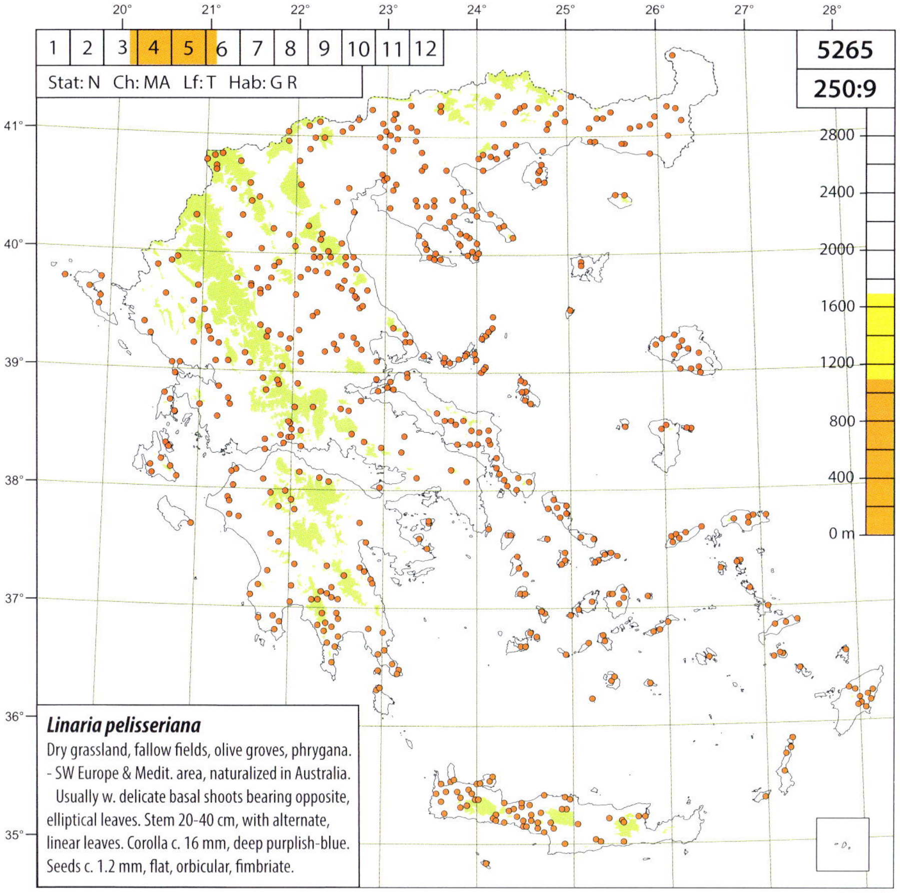

Linaria pelisseriana
Dry grassland, fallow fields, olive groves, phrygana. - SW Europe & Medit. area, naturalized in Australia.
Usually w. delicate basal shoots bearing opposite, elliptical leaves. Stem 20-40 cm, with alternate, linear leaves. Corolla c. 16 mm, deep purplish-blue. Seeds c. 1.2 mm, flat, orbicular, fimbriate.

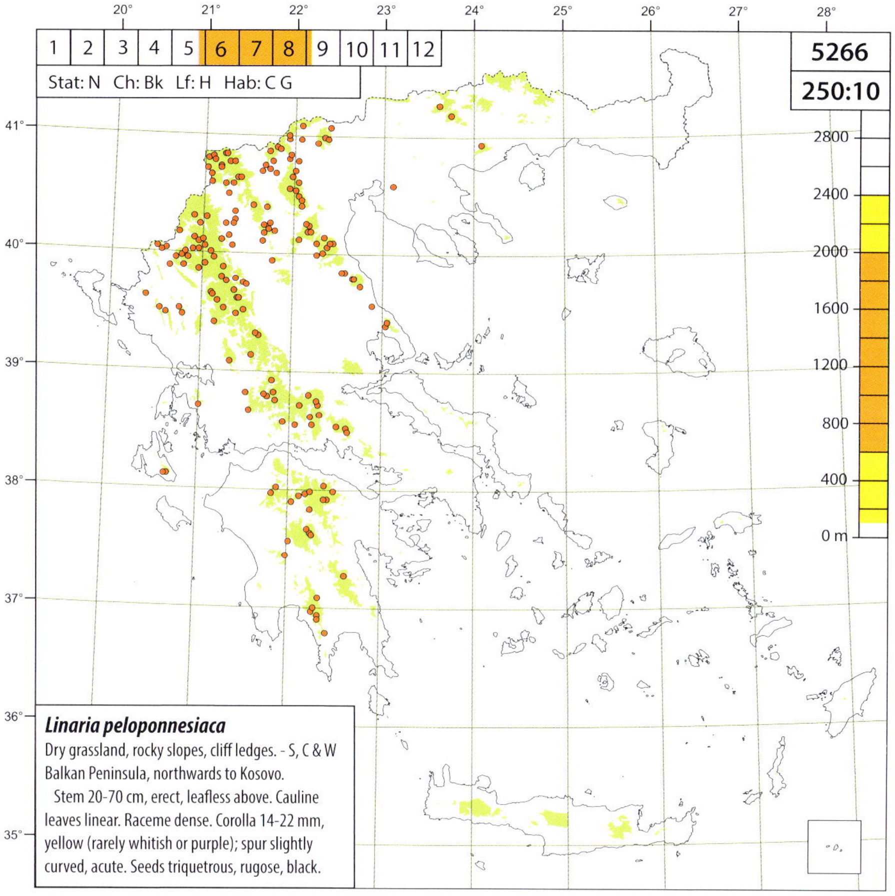

Linaria peloponnesiaca
Dry grassland, rocky slopes, cliff ledges. - S, C & W Balkan Peninsula, northwards to Kosovo.
Stem 20-70 cm, erect, leafless above. Cauline leaves linear. Raceme dense. Corolla 14-22 mm, yellow (rarely whitish or purple); spur slightly curved, acute. Seeds triquetrous, rugose, black.

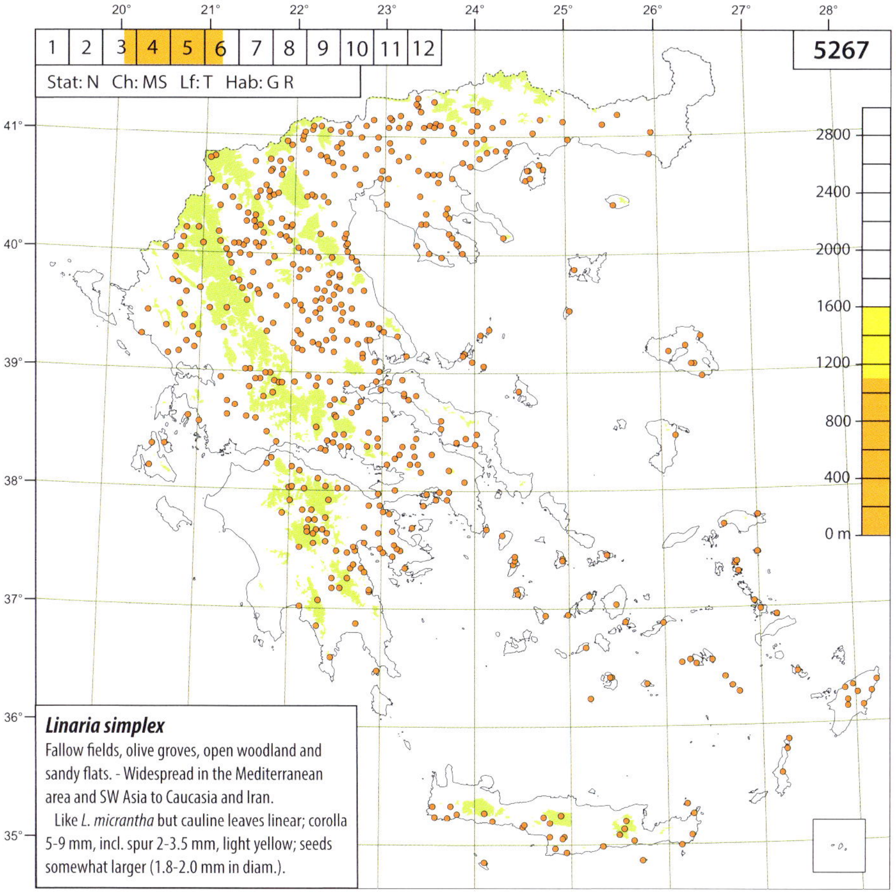

Linaria simplex
Fallow fields, olive groves, open woodland and sandy flats. - Widespread in the Mediterranean area and SW Asia to Caucasia and Iran.
Like *L. micrantha* but cauline leaves linear; corolla 5-9 mm, incl. spur 2-3.5 mm, light yellow; seeds somewhat larger (1.8-2.0 mm in diam.).

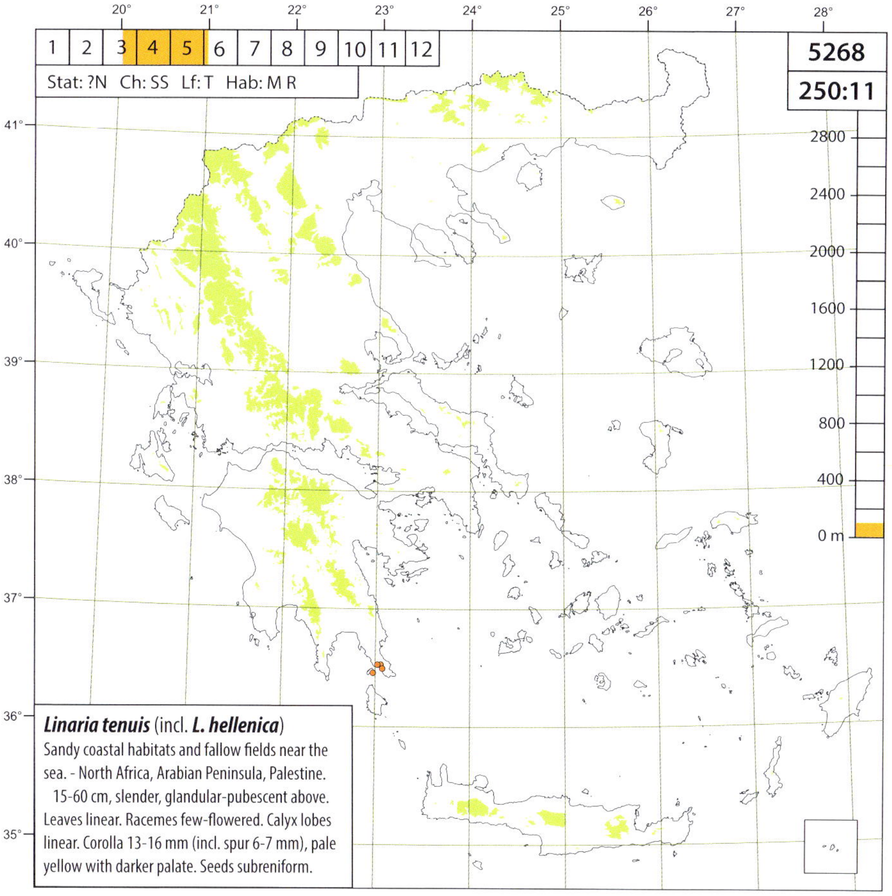

Linaria tenuis (incl. ***L. hellenica***)
Sandy coastal habitats and fallow fields near the sea. - North Africa, Arabian Peninsula, Palestine.
15-60 cm, slender, glandular-pubescent above. Leaves linear. Racemes few-flowered. Calyx lobes linear. Corolla 13-16 mm (incl. spur 6-7 mm), pale yellow with darker palate. Seeds subreniform.

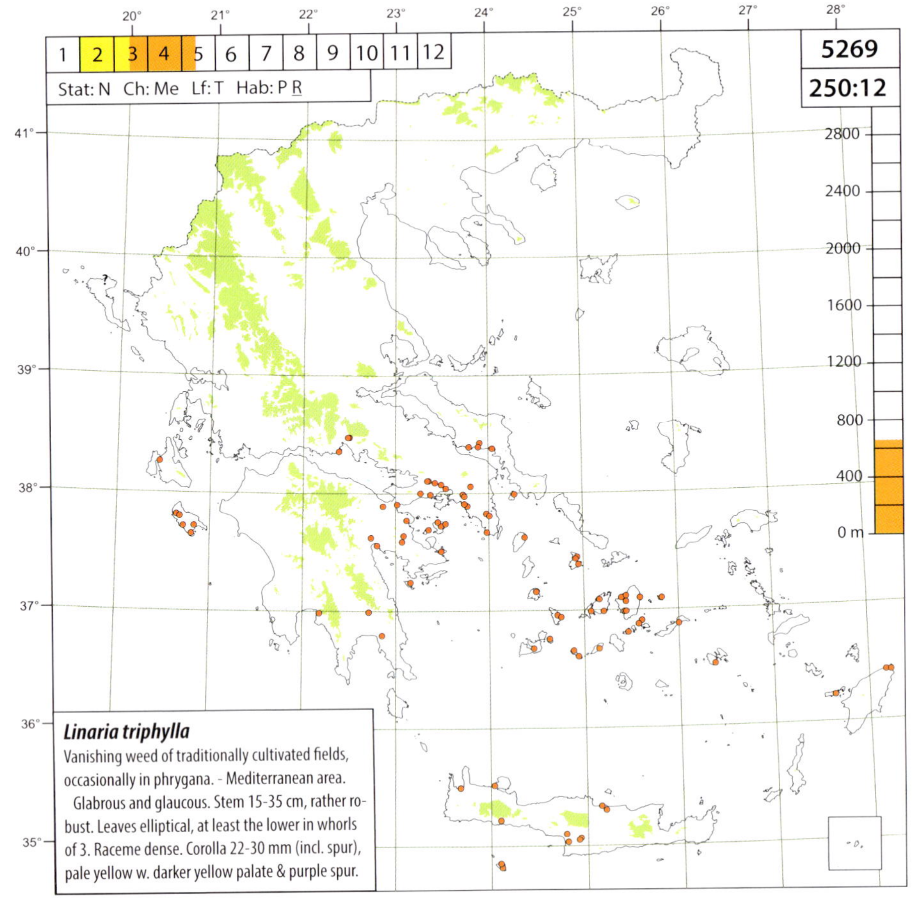

Linaria triphylla
Vanishing weed of traditionally cultivated fields, occasionally in phrygana. - Mediterranean area.
Glabrous and glaucous. Stem 15-35 cm, rather robust. Leaves elliptical, at least the lower in whorls of 3. Raceme dense. Corolla 22-30 mm (incl. spur), pale yellow w. darker yellow palate & purple spur.

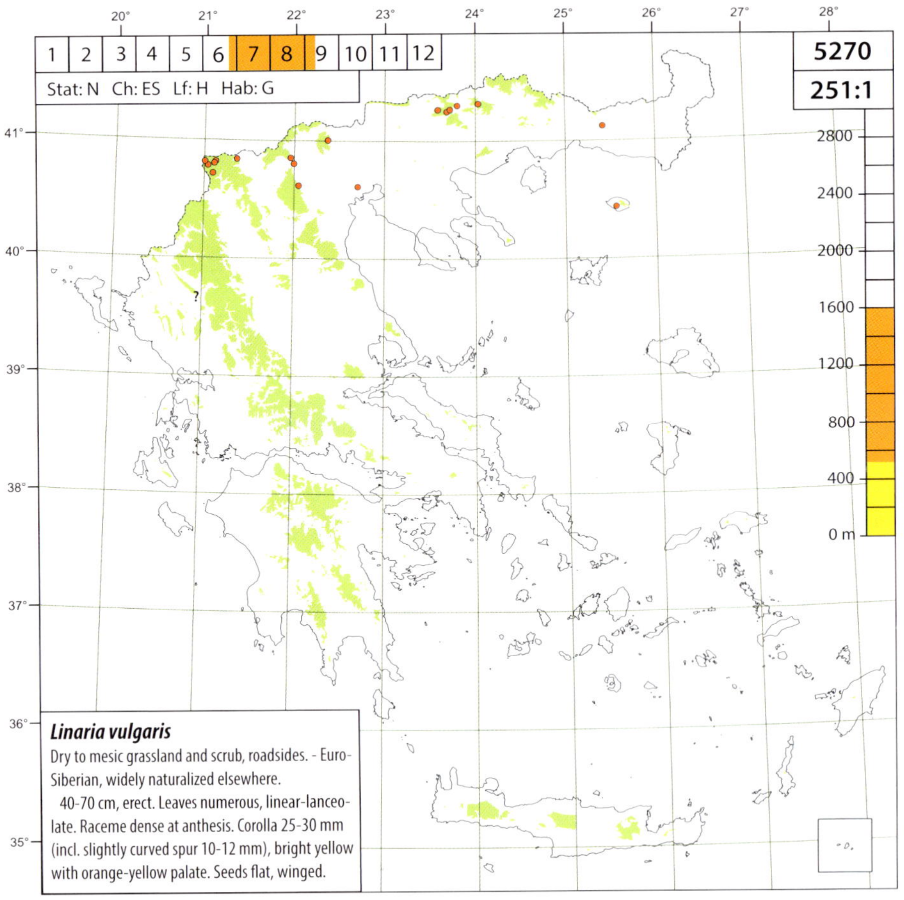

Linaria vulgaris
Dry to mesic grassland and scrub, roadsides. - Euro-Siberian, widely naturalized elsewhere.
40-70 cm, erect. Leaves numerous, linear-lanceolate. Raceme dense at anthesis. Corolla 25-30 mm (incl. slightly curved spur 10-12 mm), bright yellow with orange-yellow palate. Seeds flat, winged.

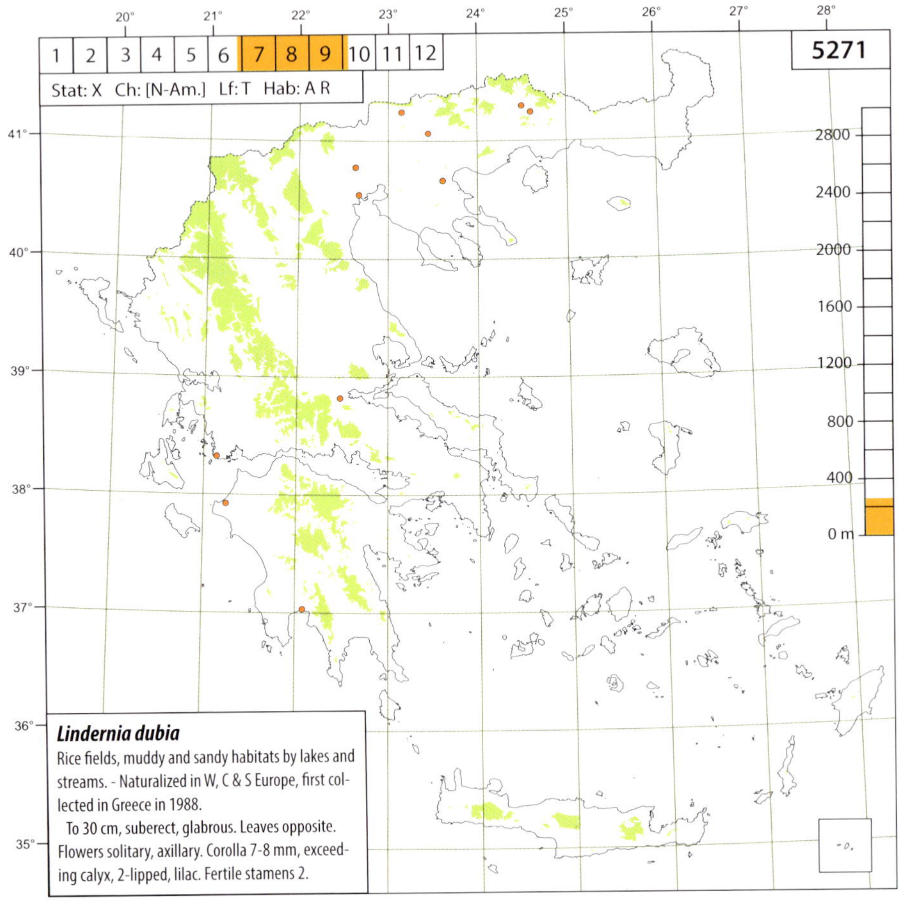

Lindernia dubia
Rice fields, muddy and sandy habitats by lakes and streams. - Naturalized in W, C & S Europe, first collected in Greece in 1988.
To 30 cm, suberect, glabrous. Leaves opposite. Flowers solitary, axillary. Corolla 7-8 mm, exceeding calyx, 2-lipped, lilac. Fertile stamens 2.

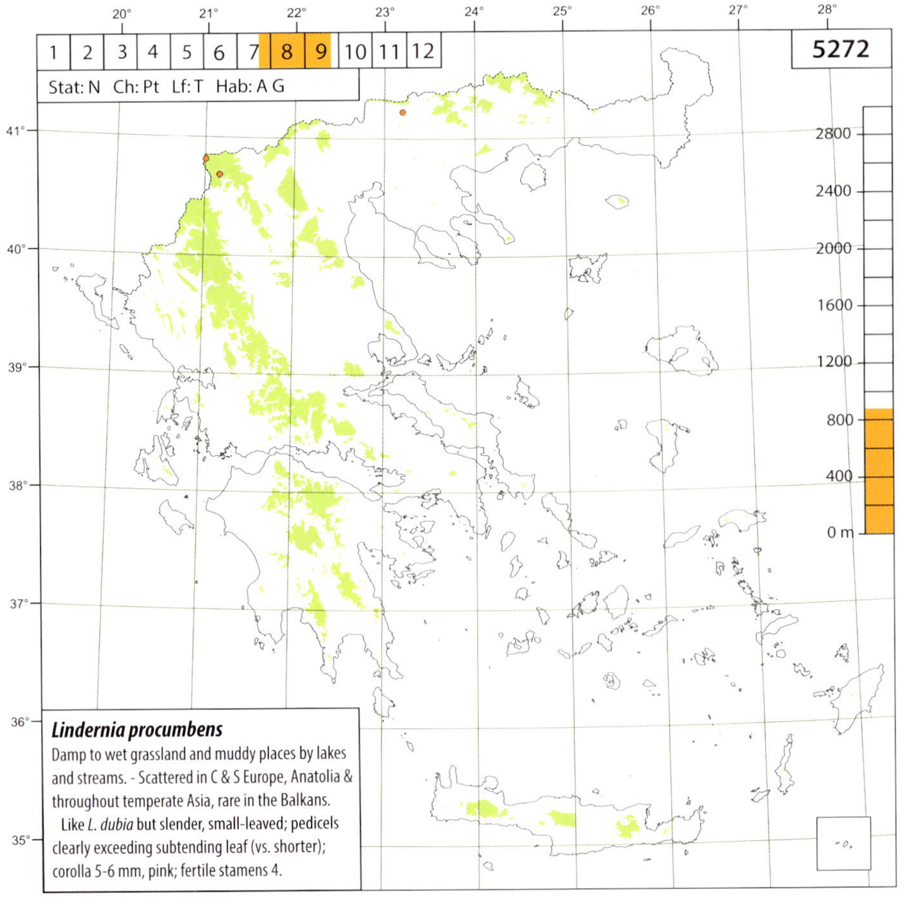

Lindernia procumbens
Damp to wet grassland and muddy places by lakes and streams. - Scattered in C & S Europe, Anatolia & throughout temperate Asia, rare in the Balkans.
Like *L. dubia* but slender, small-leaved; pedicels clearly exceeding subtending leaf (vs. shorter); corolla 5-6 mm, pink; fertile stamens 4.

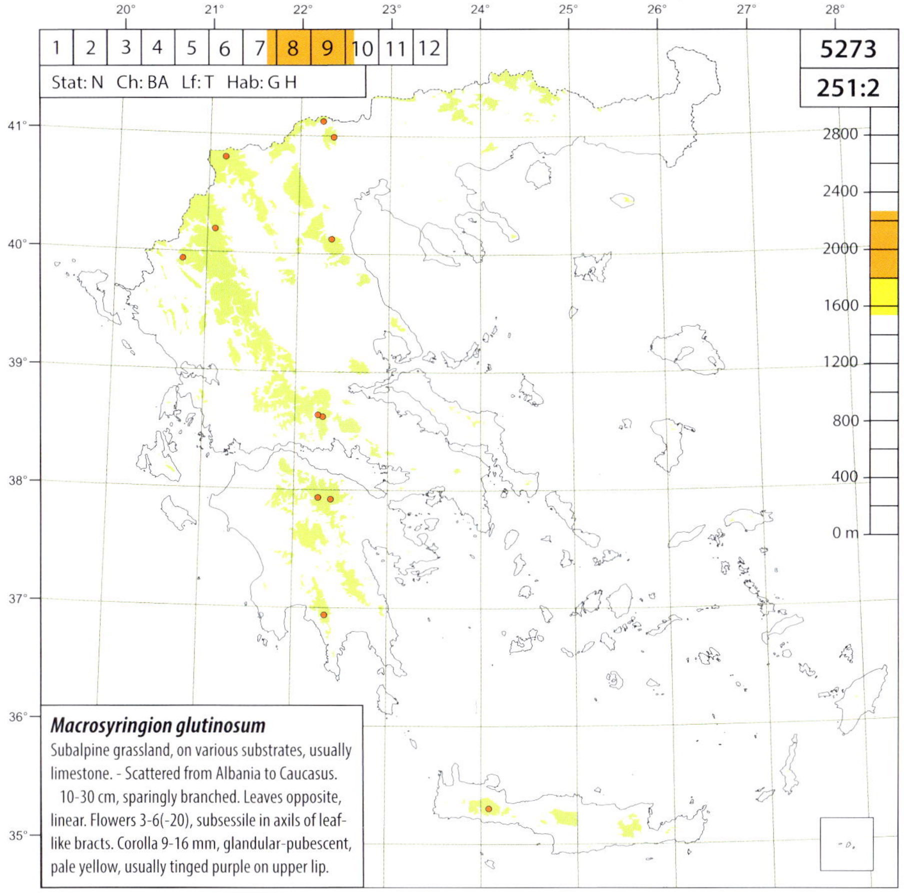

Macrosyringion glutinosum
Subalpine grassland, on various substrates, usually limestone. - Scattered from Albania to Caucasus.

10-30 cm, sparingly branched. Leaves opposite, linear. Flowers 3-6(-20), subsessile in axils of leaf-like bracts. Corolla 9-16 mm, glandular-pubescent, pale yellow, usually tinged purple on upper lip.

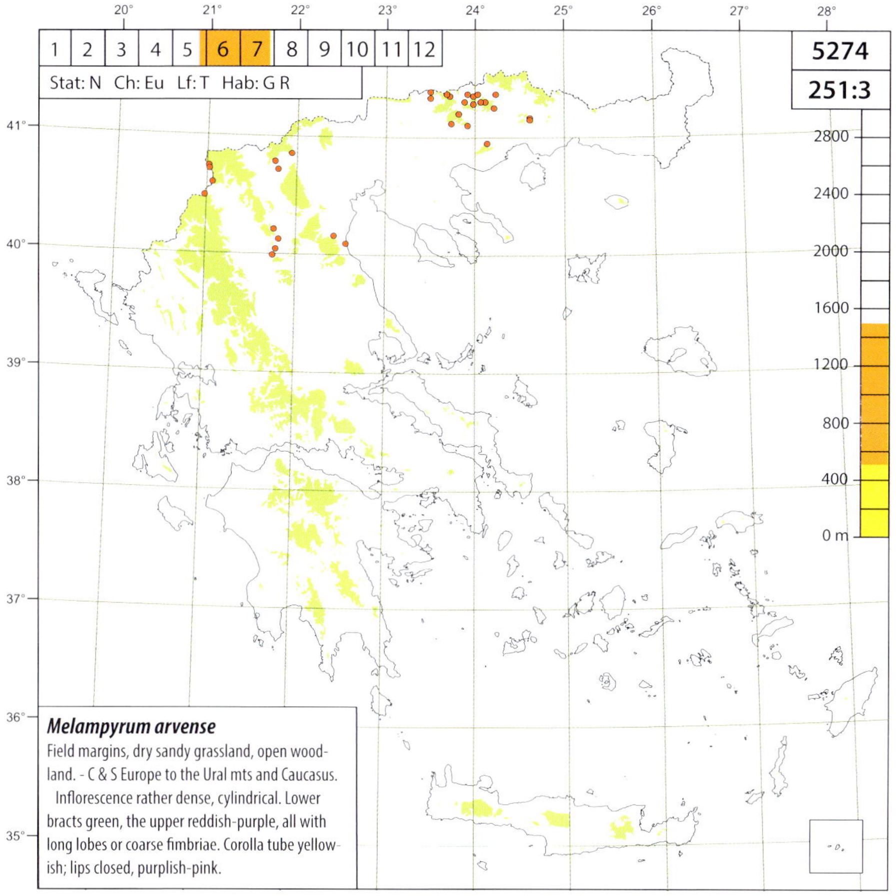

Melampyrum arvense
Field margins, dry sandy grassland, open woodland. - C & S Europe to the Ural mts and Caucasus.

Inflorescence rather dense, cylindrical. Lower bracts green, the upper reddish-purple, all with long lobes or coarse fimbriae. Corolla tube yellowish; lips closed, purplish-pink.

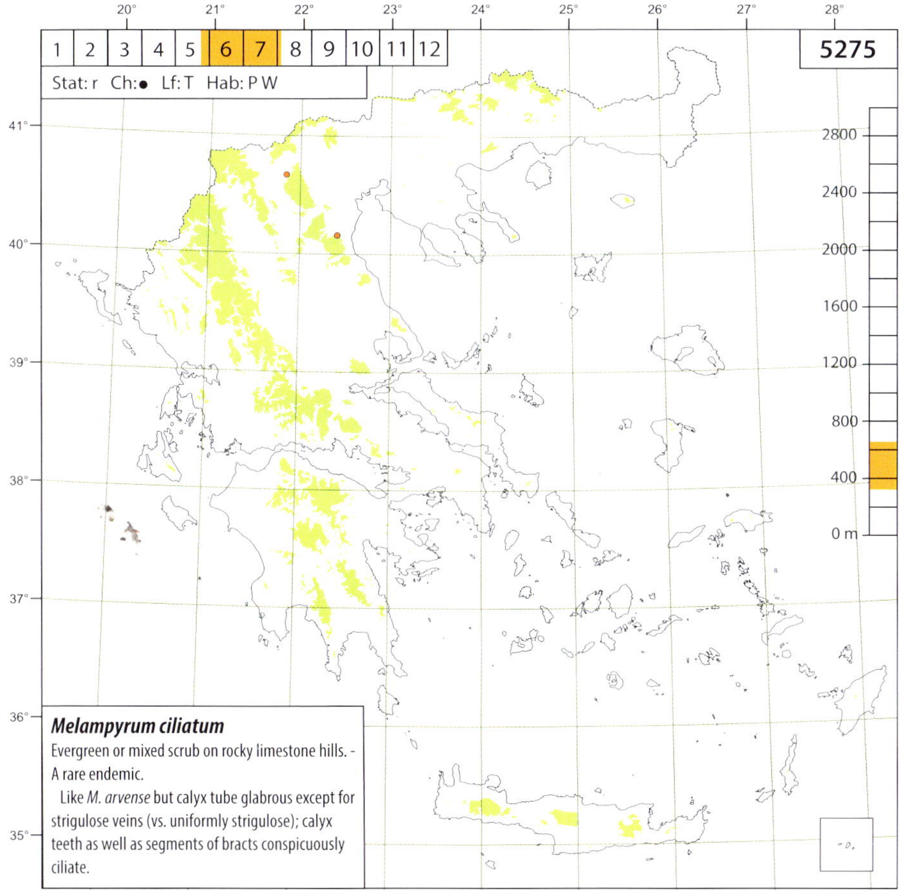

Melampyrum ciliatum
Evergreen or mixed scrub on rocky limestone hills. - A rare endemic.

Like *M. arvense* but calyx tube glabrous except for strigulose veins (vs. uniformly strigulose); calyx teeth as well as segments of bracts conspicuously ciliate.

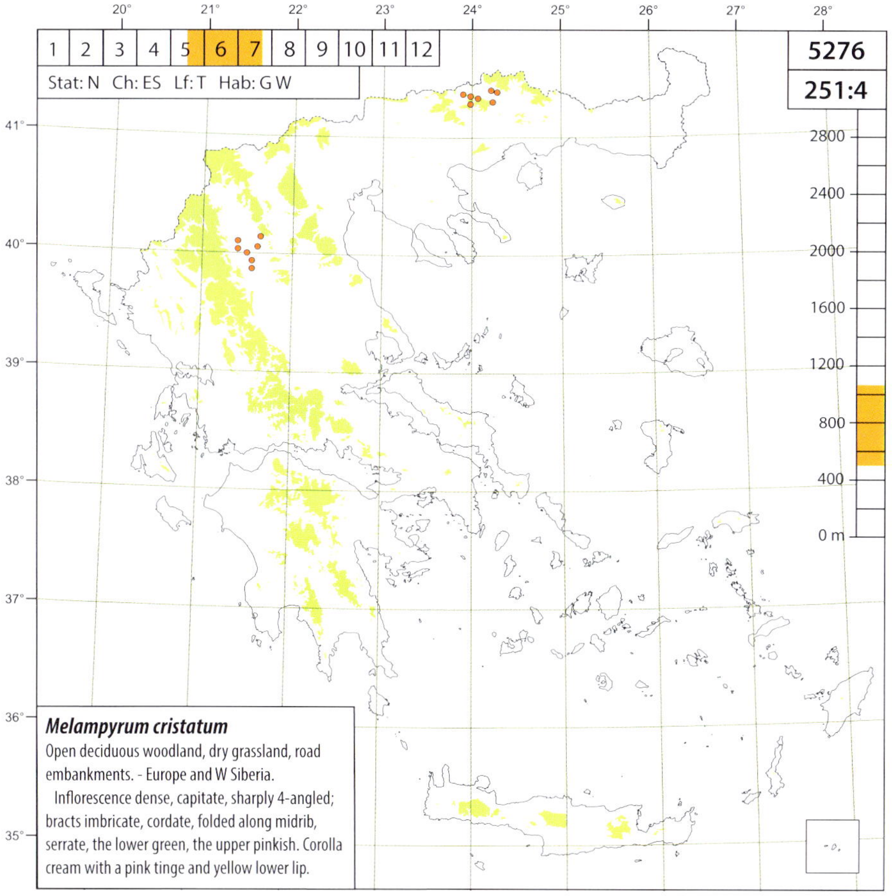

Melampyrum cristatum
Open deciduous woodland, dry grassland, road embankments. - Europe and W Siberia.

Inflorescence dense, capitate, sharply 4-angled; bracts imbricate, cordate, folded along midrib, serrate, the lower green, the upper pinkish. Corolla cream with a pink tinge and yellow lower lip.

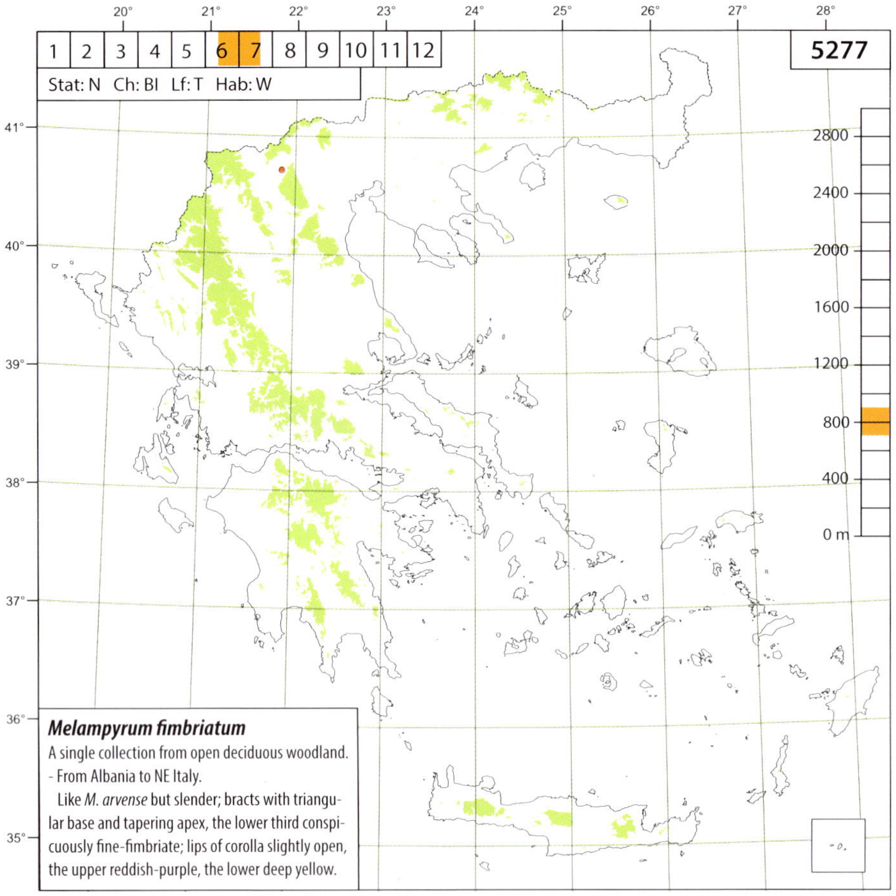

Melampyrum fimbriatum

A single collection from open deciduous woodland. - From Albania to NE Italy.

Like *M. arvense* but slender; bracts with triangular base and tapering apex, the lower third conspicuously fine-fimbriate; lips of corolla slightly open, the upper reddish-purple, the lower deep yellow.

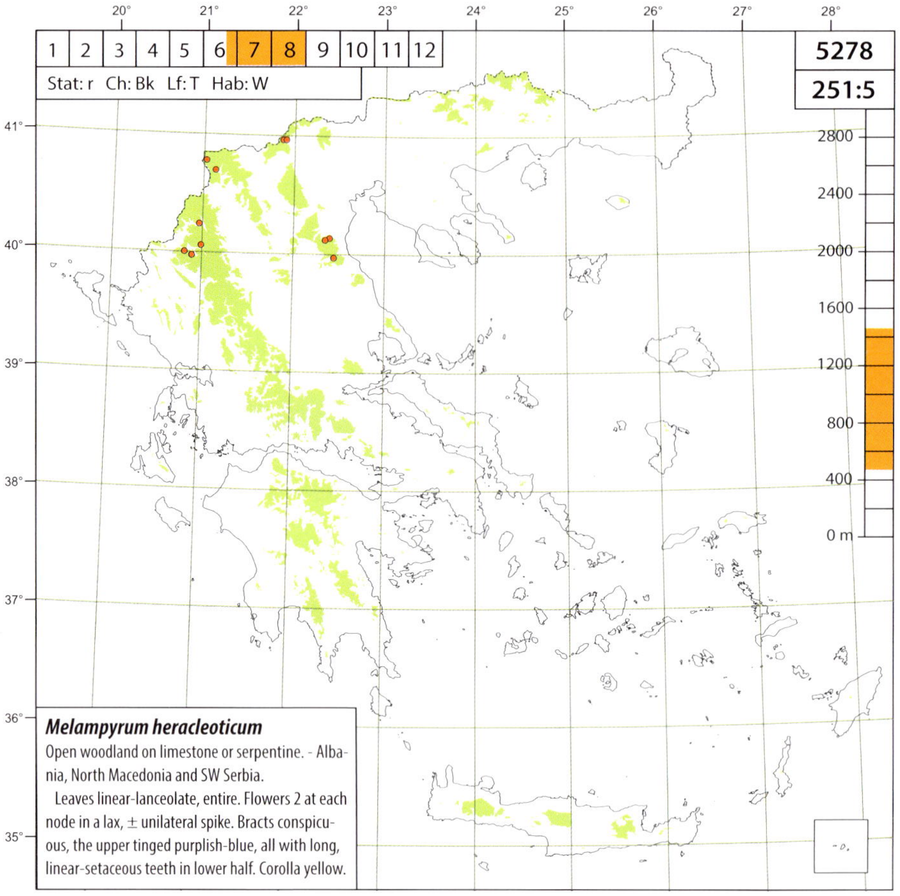

Melampyrum heracleoticum

Open woodland on limestone or serpentine. - Albania, North Macedonia and SW Serbia.

Leaves linear-lanceolate, entire. Flowers 2 at each node in a lax, ± unilateral spike. Bracts conspicuous, the upper tinged purplish-blue, all with long, linear-setaceous teeth in lower half. Corolla yellow.

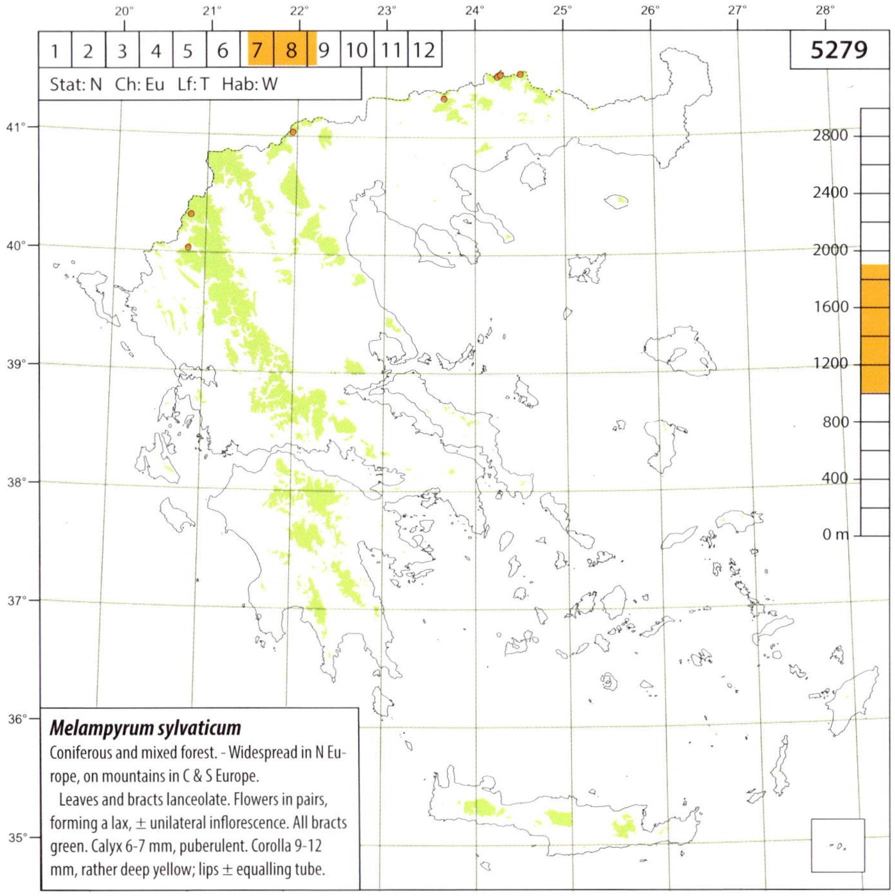

Melampyrum sylvaticum

Coniferous and mixed forest. - Widespread in N Europe, on mountains in C & S Europe.

Leaves and bracts lanceolate. Flowers in pairs, forming a lax, ± unilateral inflorescence. All bracts green. Calyx 6-7 mm, puberulent. Corolla 9-12 mm, rather deep yellow; lips ± equalling tube.

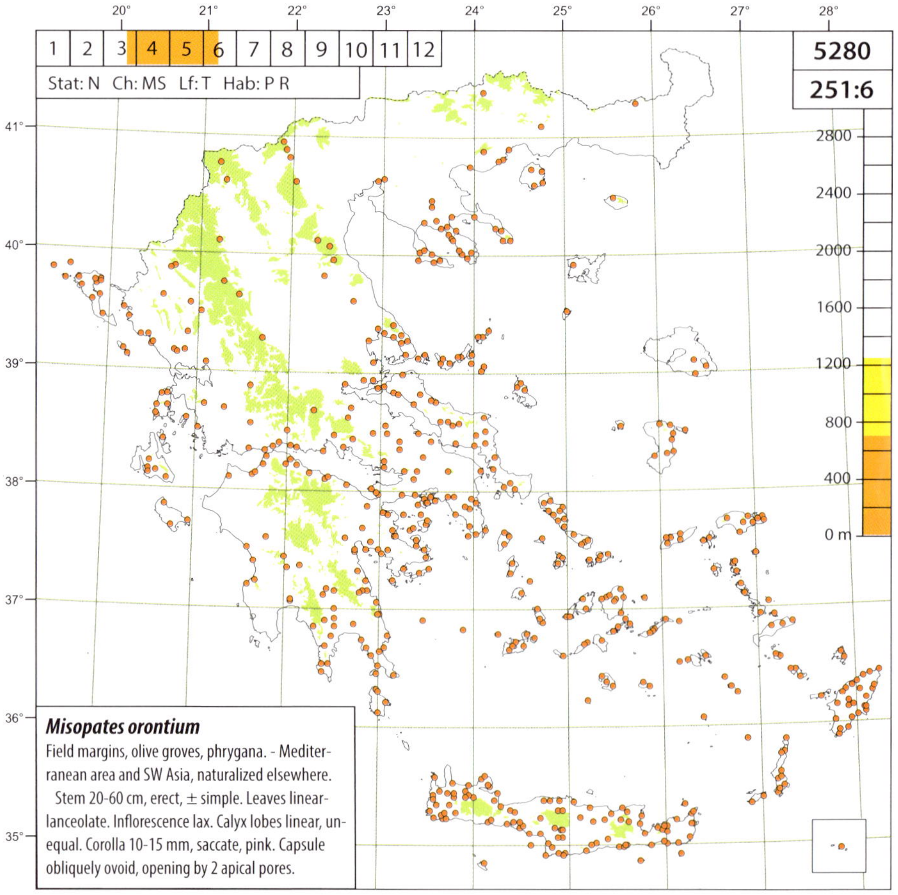

Misopates orontium

Field margins, olive groves, phrygana. - Mediterranean area and SW Asia, naturalized elsewhere.

Stem 20-60 cm, erect, ± simple. Leaves linear-lanceolate. Inflorescence lax. Calyx lobes linear, unequal. Corolla 10-15 mm, saccate, pink. Capsule obliquely ovoid, opening by 2 apical pores.

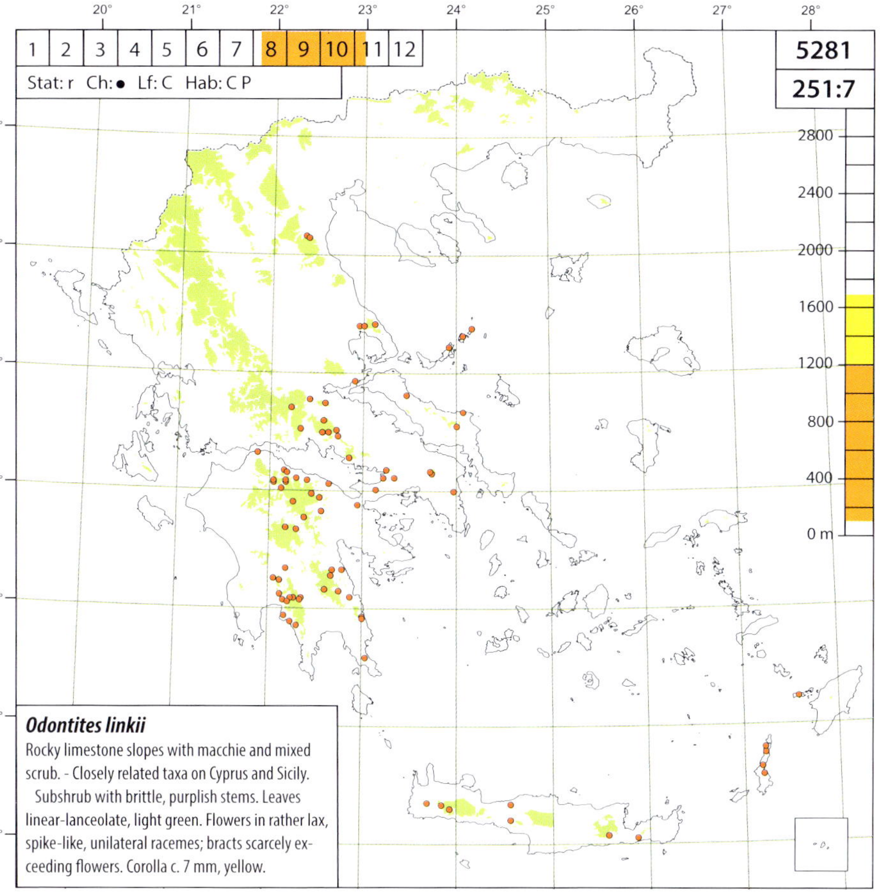

Odontites linkii
Rocky limestone slopes with macchie and mixed scrub. - Closely related taxa on Cyprus and Sicily.
Subshrub with brittle, purplish stems. Leaves linear-lanceolate, light green. Flowers in rather lax, spike-like, unilateral racemes; bracts scarcely exceeding flowers. Corolla c. 7 mm, yellow.

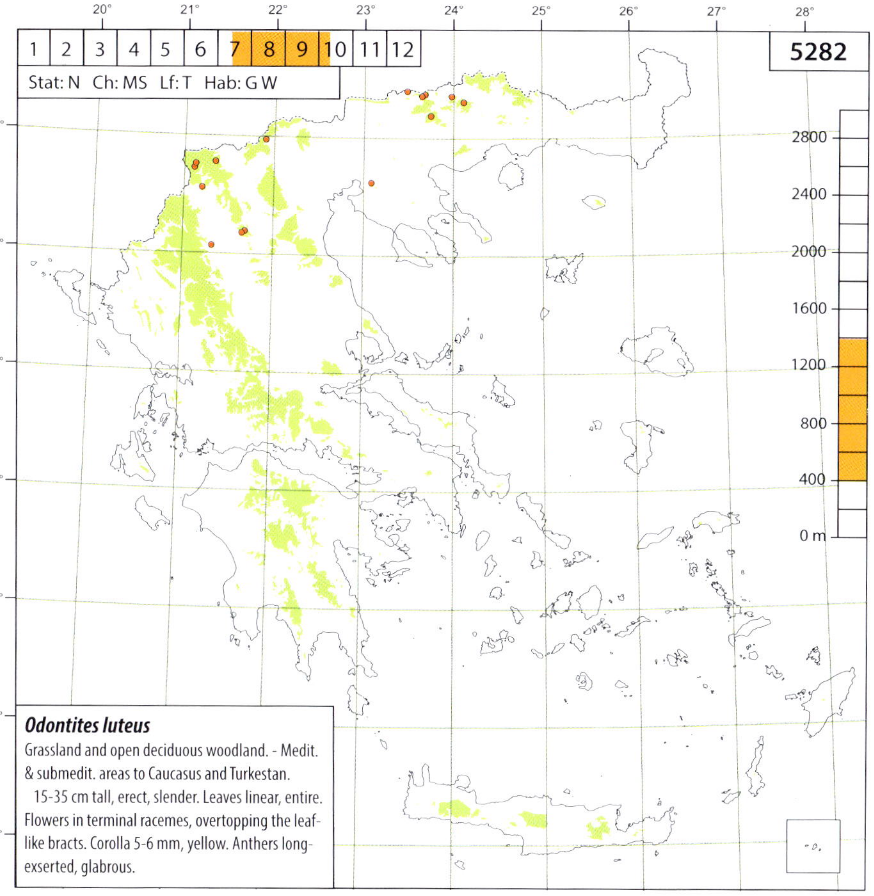

Odontites luteus
Grassland and open deciduous woodland. - Medit. & submedit. areas to Caucasus and Turkestan.
15-35 cm tall, erect, slender. Leaves linear, entire. Flowers in terminal racemes, overtopping the leaf-like bracts. Corolla 5-6 mm, yellow. Anthers long-exserted, glabrous.

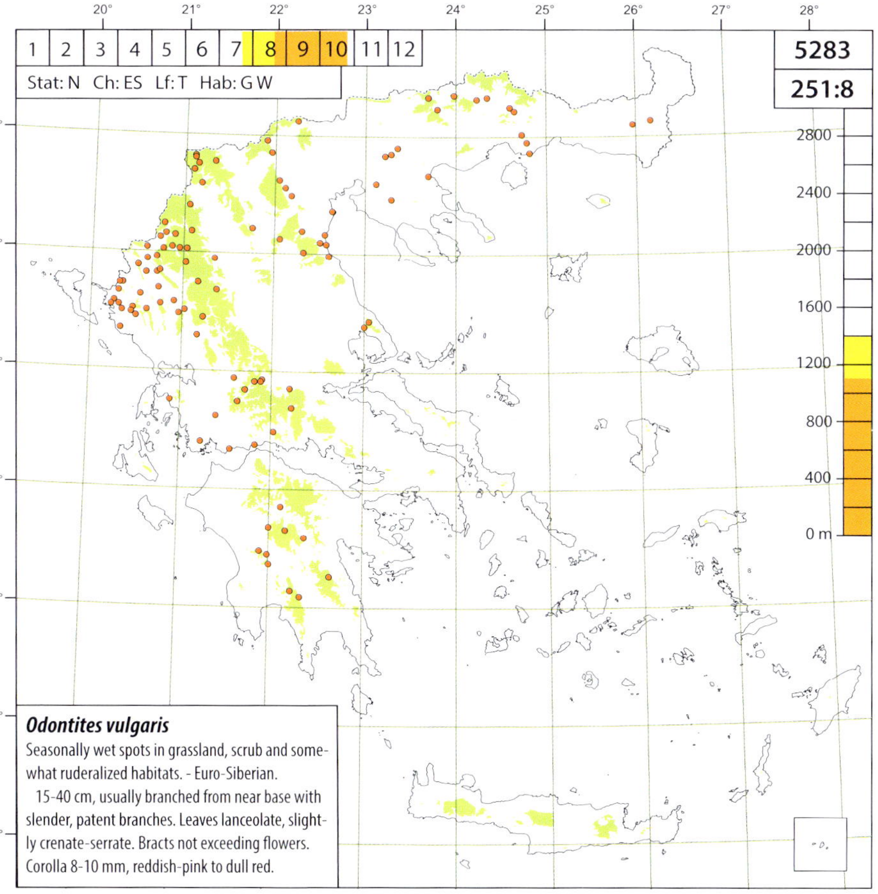

Odontites vulgaris
Seasonally wet spots in grassland, scrub and somewhat ruderalized habitats. - Euro-Siberian.
15-40 cm, usually branched from near base with slender, patent branches. Leaves lanceolate, slightly crenate-serrate. Bracts not exceeding flowers. Corolla 8-10 mm, reddish-pink to dull red.

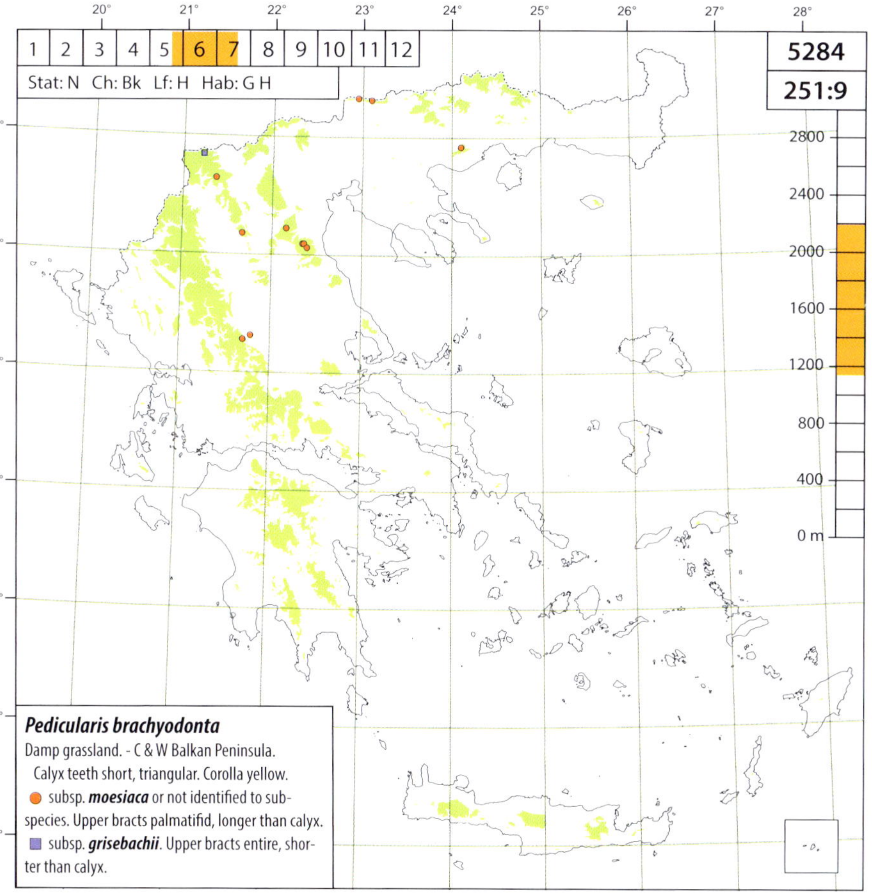

Pedicularis brachyodonta
Damp grassland. - C & W Balkan Peninsula.
Calyx teeth short, triangular. Corolla yellow.
● subsp. ***moesiaca*** or not identified to subspecies. Upper bracts palmatifid, longer than calyx.
■ subsp. ***grisebachii***. Upper bracts entire, shorter than calyx.

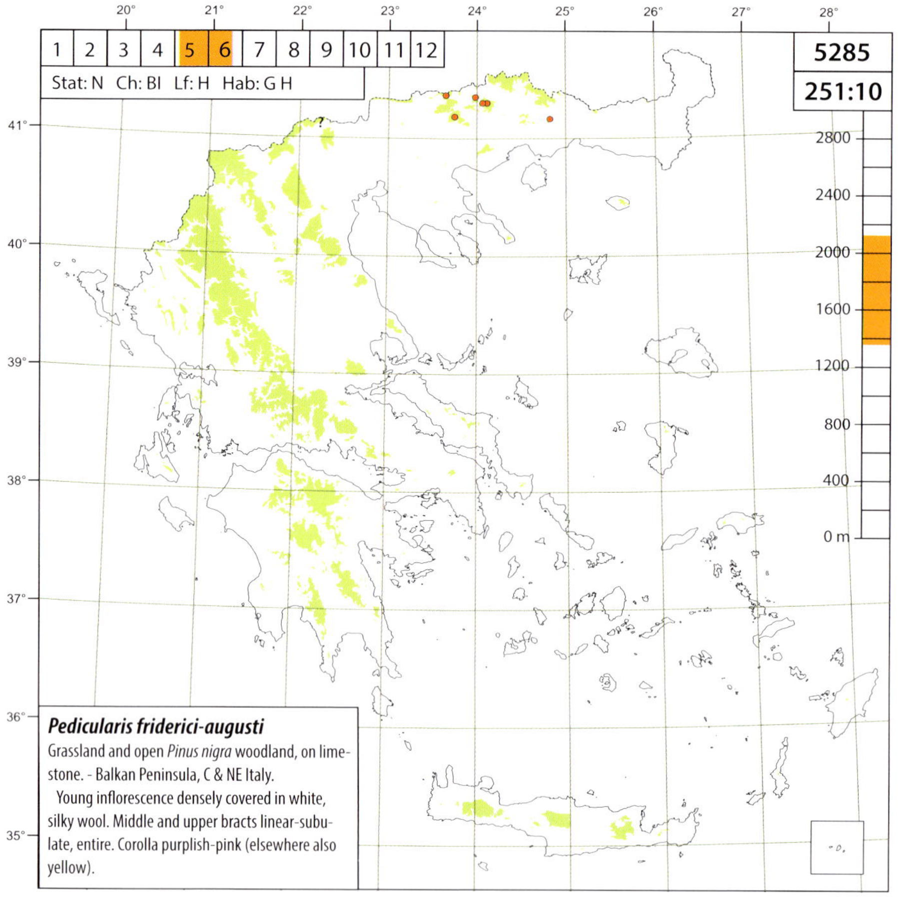

Pedicularis friderici-augusti
Grassland and open *Pinus nigra* woodland, on limestone. - Balkan Peninsula, C & NE Italy.
Young inflorescence densely covered in white, silky wool. Middle and upper bracts linear-subulate, entire. Corolla purplish-pink (elsewhere also yellow).

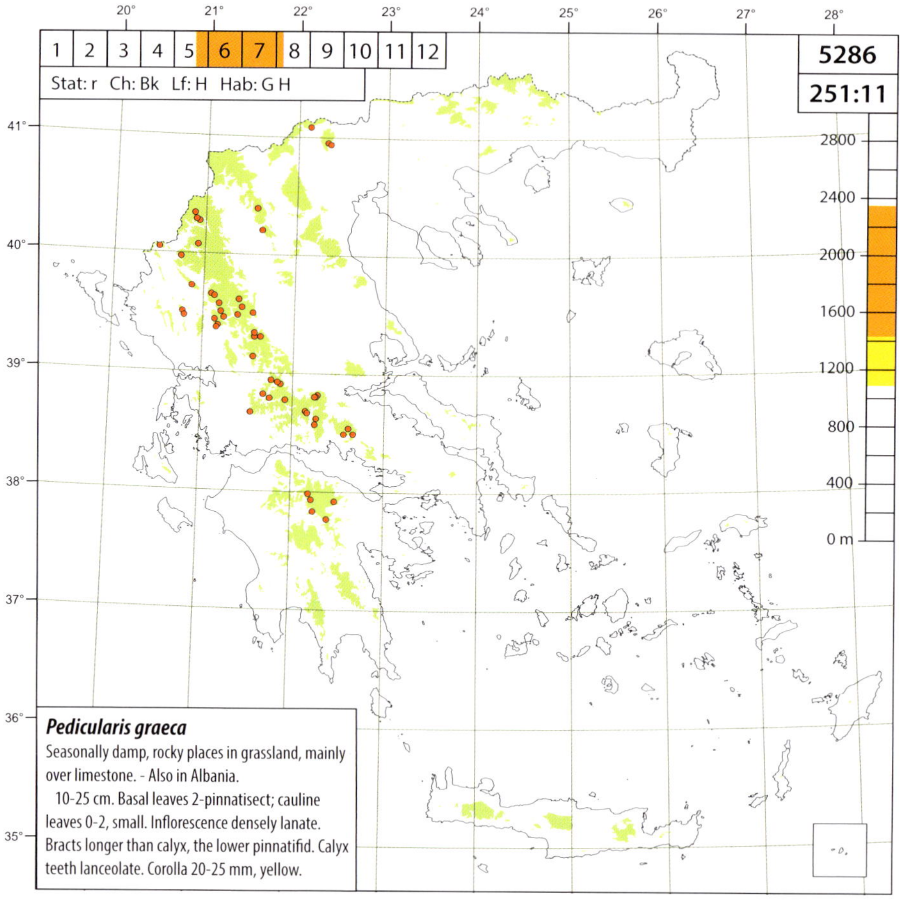

Pedicularis graeca
Seasonally damp, rocky places in grassland, mainly over limestone. - Also in Albania.
10-25 cm. Basal leaves 2-pinnatisect; cauline leaves 0-2, small. Inflorescence densely lanate. Bracts longer than calyx, the lower pinnatifid. Calyx teeth lanceolate. Corolla 20-25 mm, yellow.

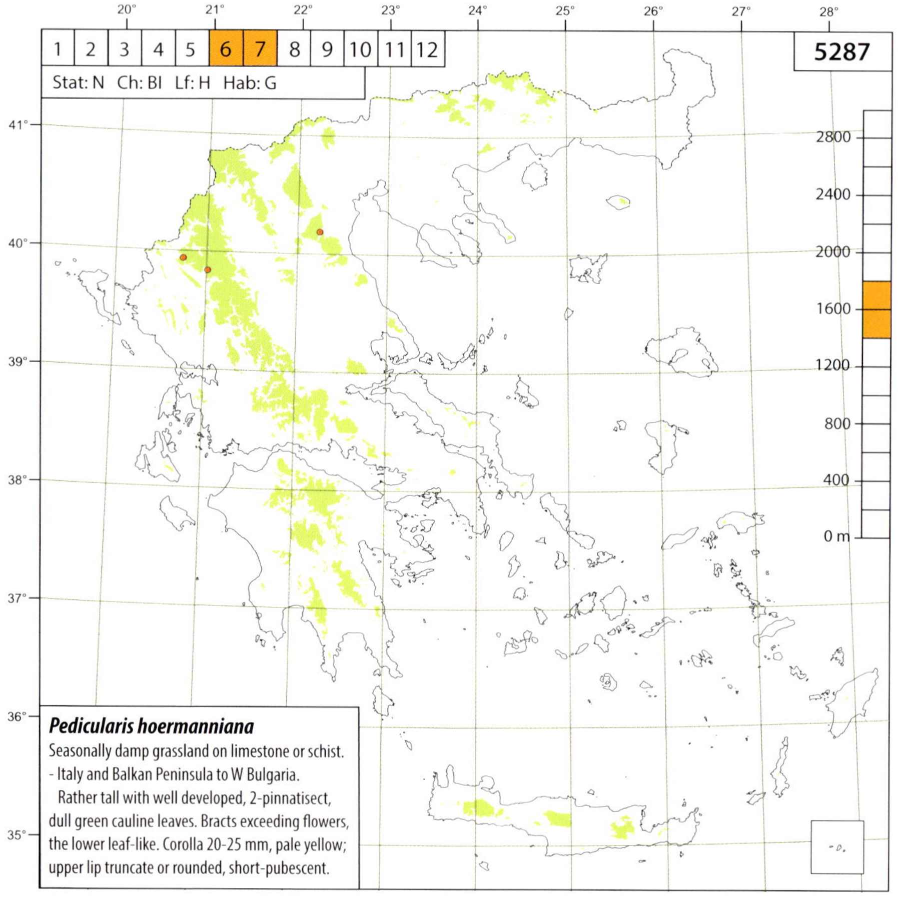

Pedicularis hoermanniana
Seasonally damp grassland on limestone or schist. - Italy and Balkan Peninsula to W Bulgaria.
Rather tall with well developed, 2-pinnatisect, dull green cauline leaves. Bracts exceeding flowers, the lower leaf-like. Corolla 20-25 mm, pale yellow; upper lip truncate or rounded, short-pubescent.

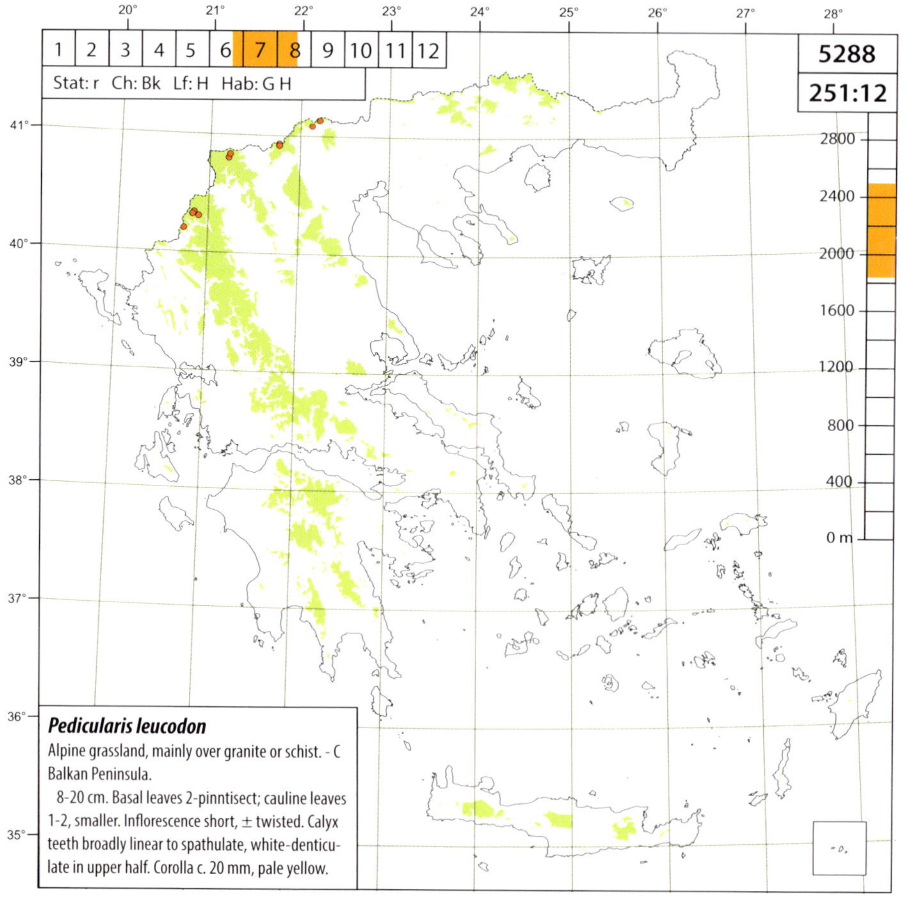

Pedicularis leucodon
Alpine grassland, mainly over granite or schist. - C Balkan Peninsula.
8-20 cm. Basal leaves 2-pinntisect; cauline leaves 1-2, smaller. Inflorescence short, ± twisted. Calyx teeth broadly linear to spathulate, white-denticulate in upper half. Corolla c. 20 mm, pale yellow.

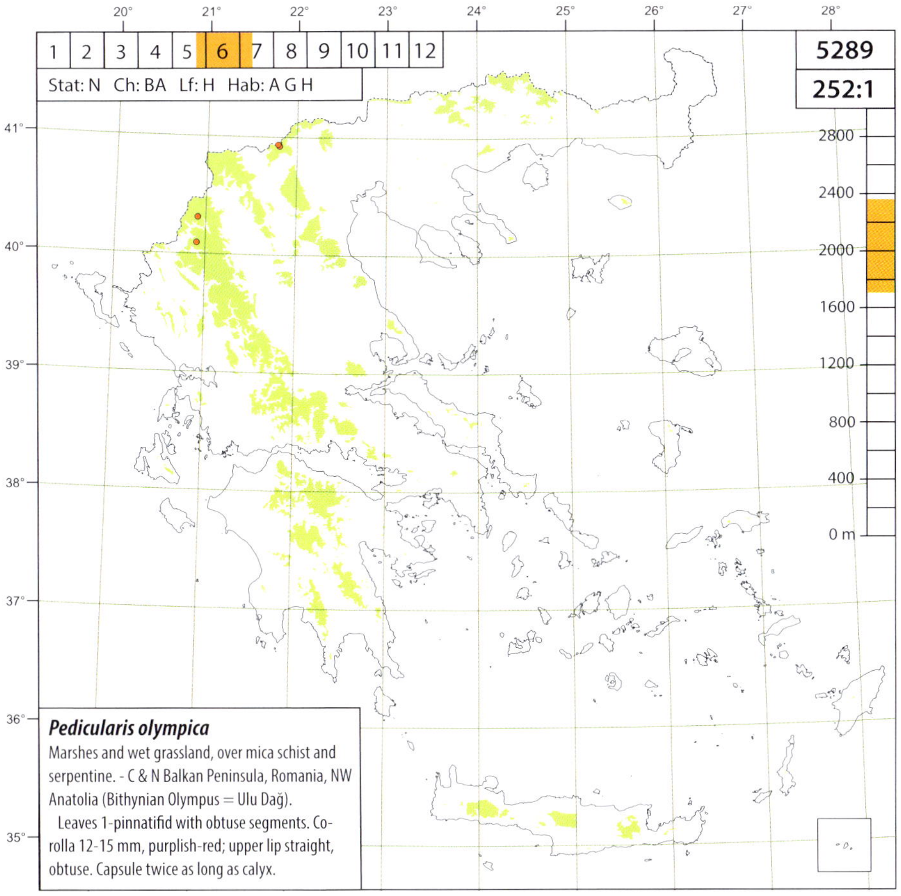

Pedicularis olympica

Marshes and wet grassland, over mica schist and serpentine. - C & N Balkan Peninsula, Romania, NW Anatolia (Bithynian Olympus = Ulu Dağ).

Leaves 1-pinnatifid with obtuse segments. Corolla 12-15 mm, purplish-red; upper lip straight, obtuse. Capsule twice as long as calyx.

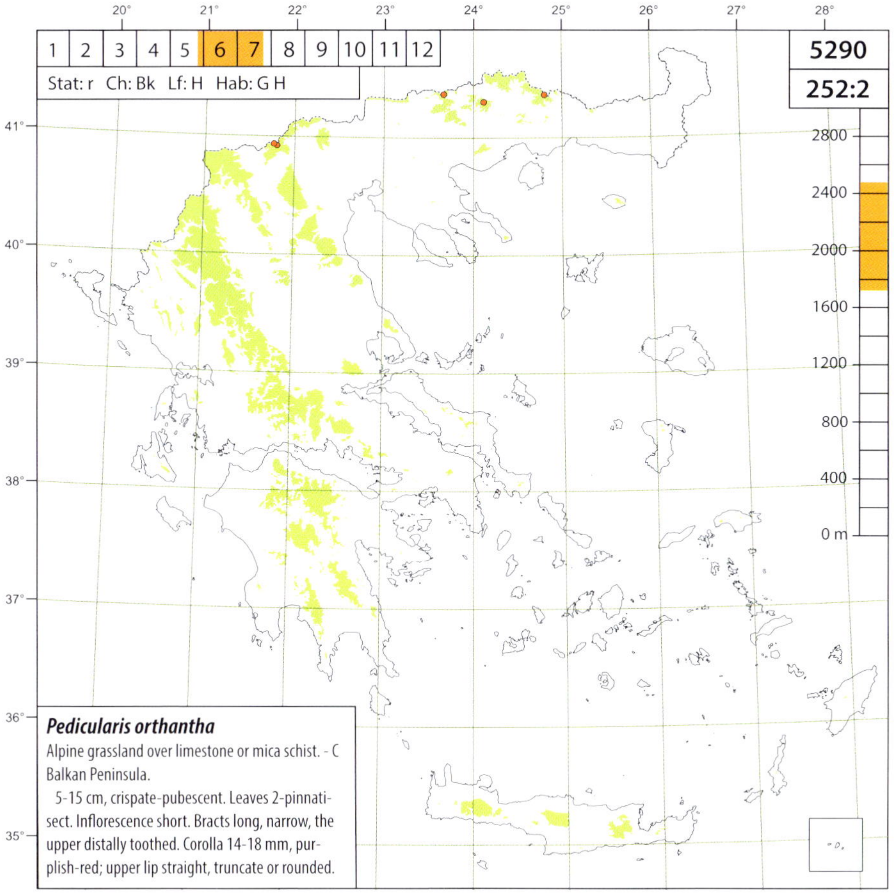

Pedicularis orthantha

Alpine grassland over limestone or mica schist. - C Balkan Peninsula.

5-15 cm, crispate-pubescent. Leaves 2-pinnatisect. Inflorescence short. Bracts long, narrow, the upper distally toothed. Corolla 14-18 mm, purplish-red; upper lip straight, truncate or rounded.

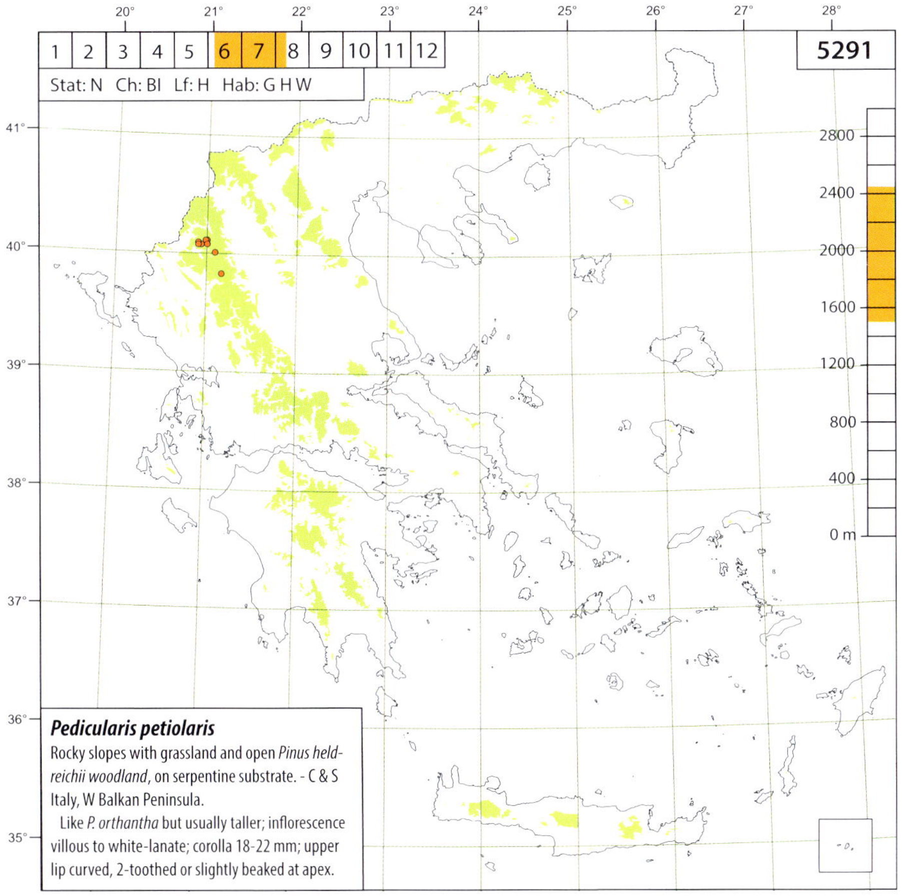

Pedicularis petiolaris

Rocky slopes with grassland and open *Pinus heldreichii woodland*, on serpentine substrate. - C & S Italy, W Balkan Peninsula.

Like *P. orthantha* but usually taller; inflorescence villous to white-lanate; corolla 18-22 mm; upper lip curved, 2-toothed or slightly beaked at apex.

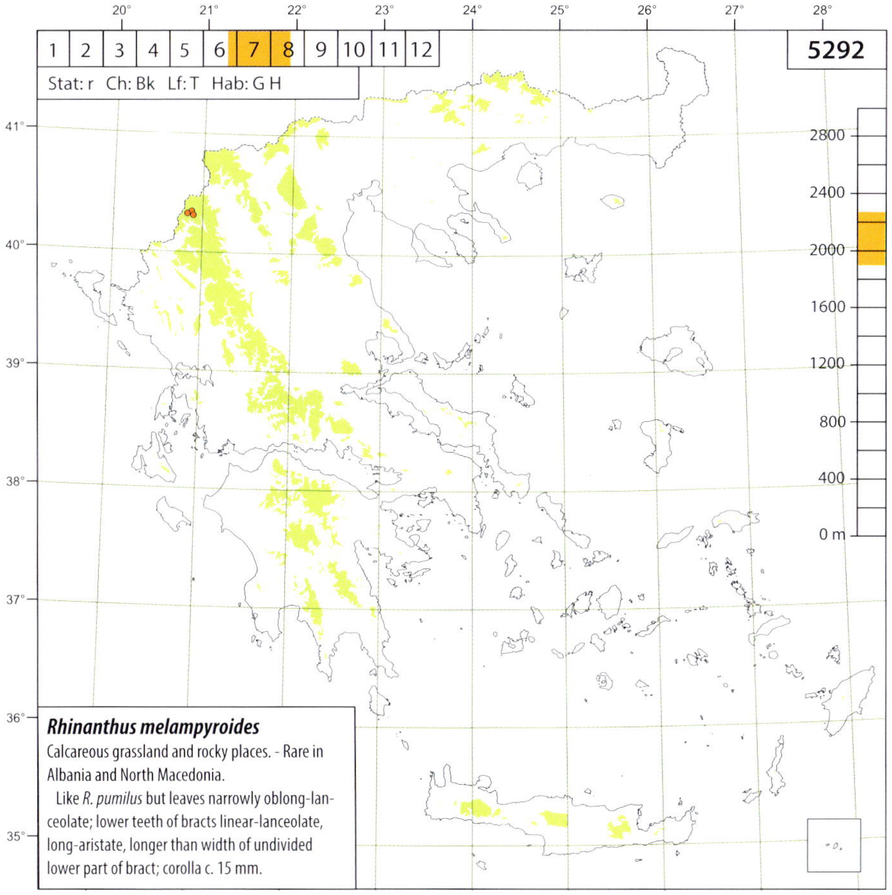

Rhinanthus melampyroides

Calcareous grassland and rocky places. - Rare in Albania and North Macedonia.

Like *R. pumilus* but leaves narrowly oblong-lanceolate; lower teeth of bracts linear-lanceolate, long-aristate, longer than width of undivided lower part of bract; corolla c. 15 mm.

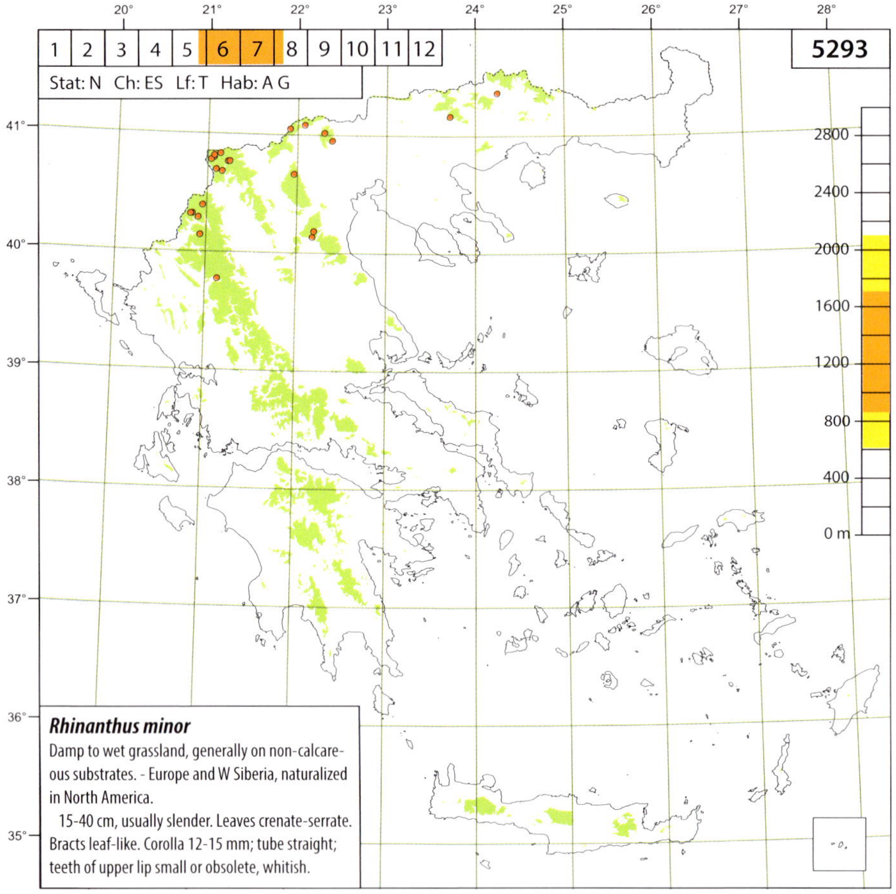

Rhinanthus minor

Damp to wet grassland, generally on non-calcareous substrates. - Europe and W Siberia, naturalized in North America.

15-40 cm, usually slender. Leaves crenate-serrate. Bracts leaf-like. Corolla 12-15 mm; tube straight; teeth of upper lip small or obsolete, whitish.

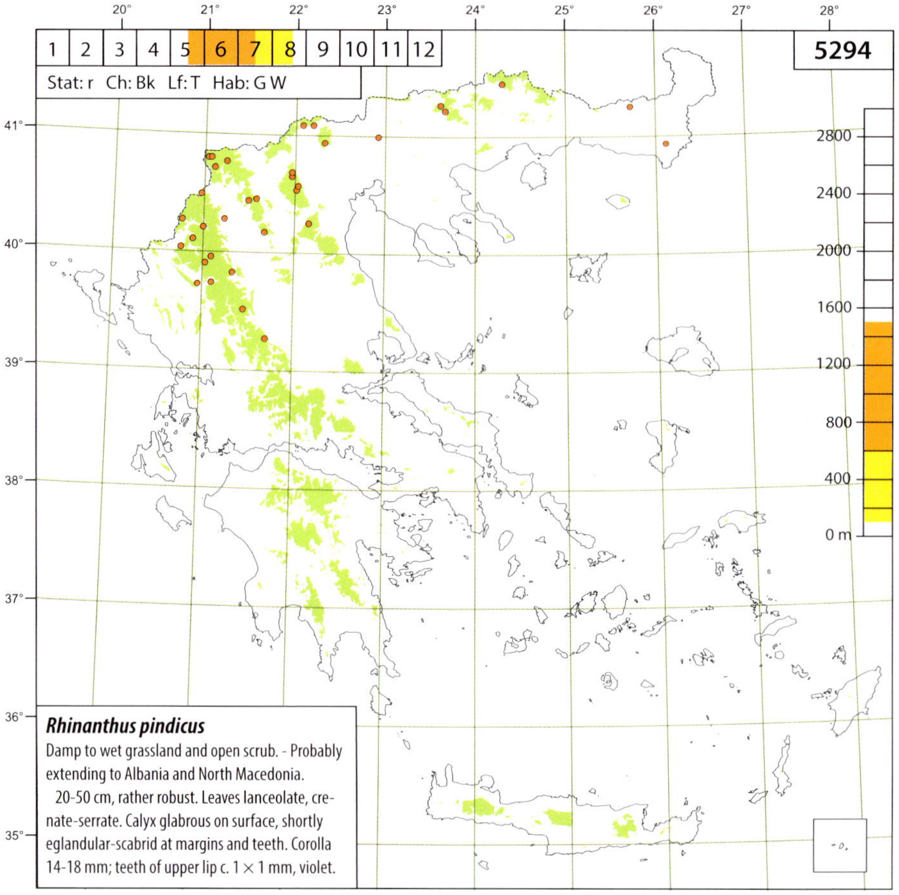

Rhinanthus pindicus

Damp to wet grassland and open scrub. - Probably extending to Albania and North Macedonia.

20-50 cm, rather robust. Leaves lanceolate, crenate-serrate. Calyx glabrous on surface, shortly eglandular-scabrid at margins and teeth. Corolla 14-18 mm; teeth of upper lip c. 1 × 1 mm, violet.

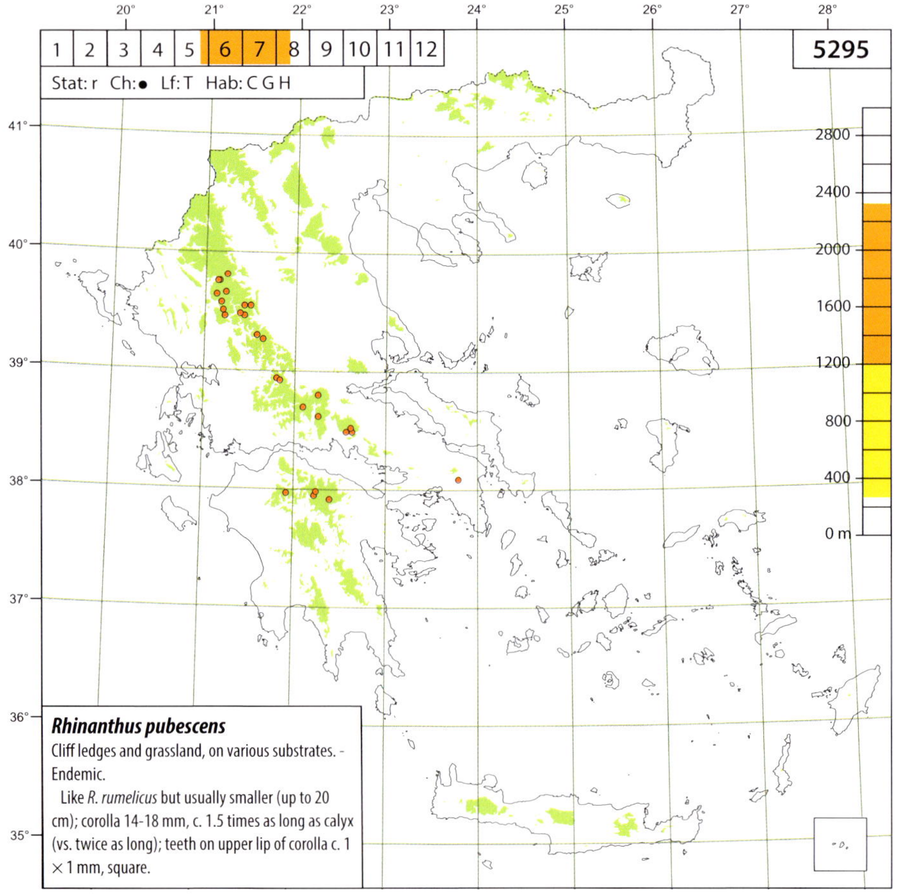

Rhinanthus pubescens

Cliff ledges and grassland, on various substrates. - Endemic.

Like *R. rumelicus* but usually smaller (up to 20 cm); corolla 14-18 mm, c. 1.5 times as long as calyx (vs. twice as long); teeth on upper lip of corolla c. 1 × 1 mm, square.

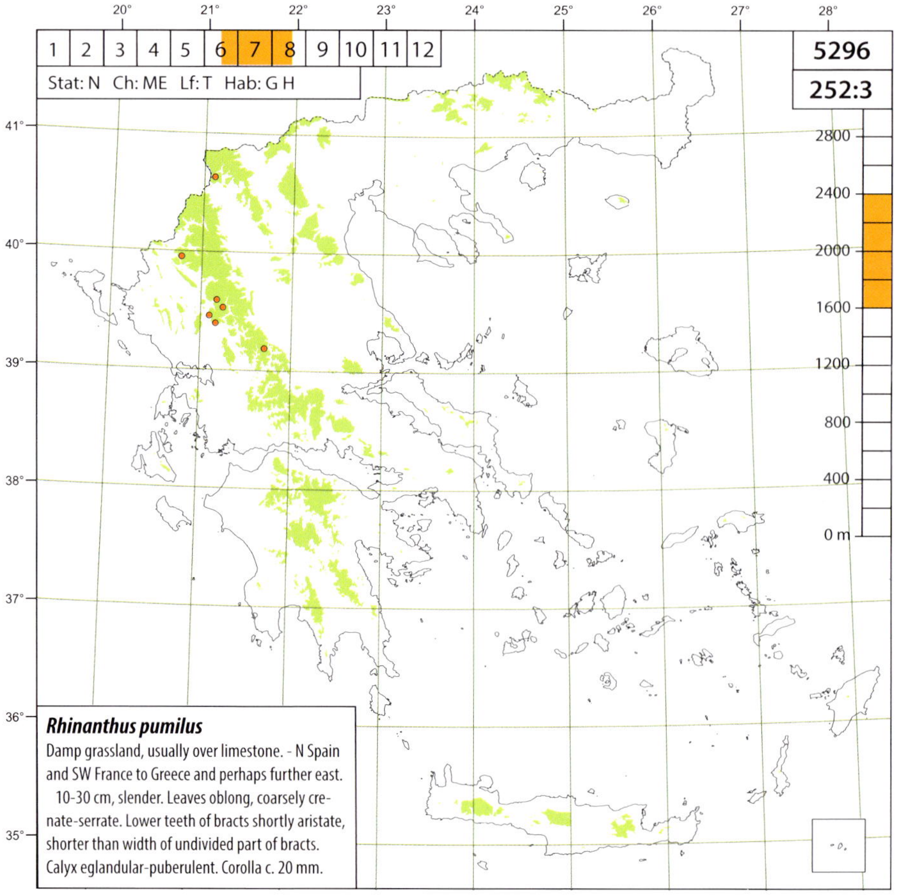

Rhinanthus pumilus

Damp grassland, usually over limestone. - N Spain and SW France to Greece and perhaps further east.

10-30 cm, slender. Leaves oblong, coarsely crenate-serrate. Lower teeth of bracts shortly aristate, shorter than width of undivided part of bracts. Calyx eglandular-puberulent. Corolla c. 20 mm.

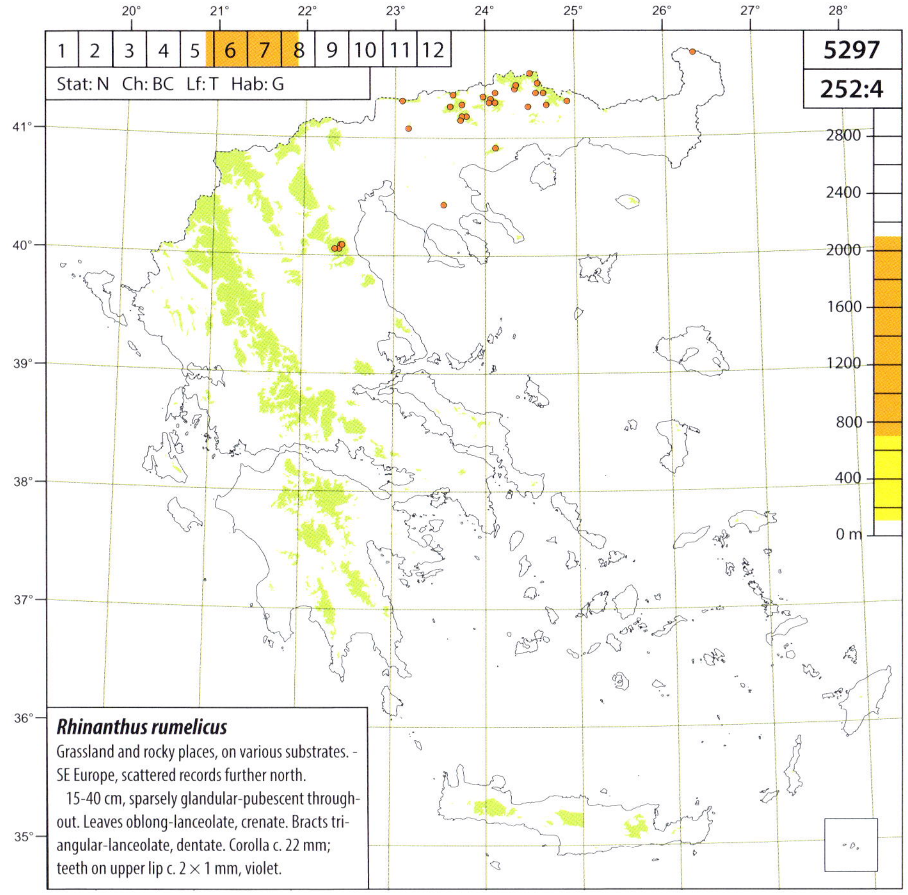

Rhinanthus rumelicus

Grassland and rocky places, on various substrates. - SE Europe, scattered records further north.

15-40 cm, sparsely glandular-pubescent throughout. Leaves oblong-lanceolate, crenate. Bracts triangular-lanceolate, dentate. Corolla c. 22 mm; teeth on upper lip c. 2 × 1 mm, violet.

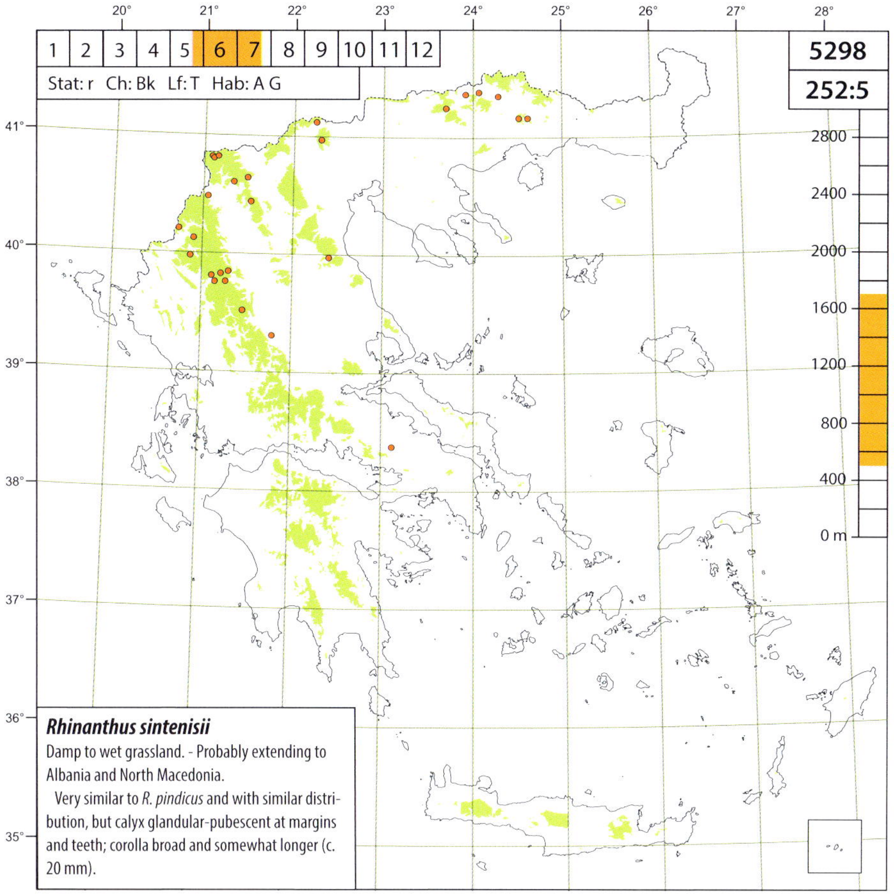

Rhinanthus sintenisii

Damp to wet grassland. - Probably extending to Albania and North Macedonia.

Very similar to *R. pindicus* and with similar distribution, but calyx glandular-pubescent at margins and teeth; corolla broad and somewhat longer (c. 20 mm).

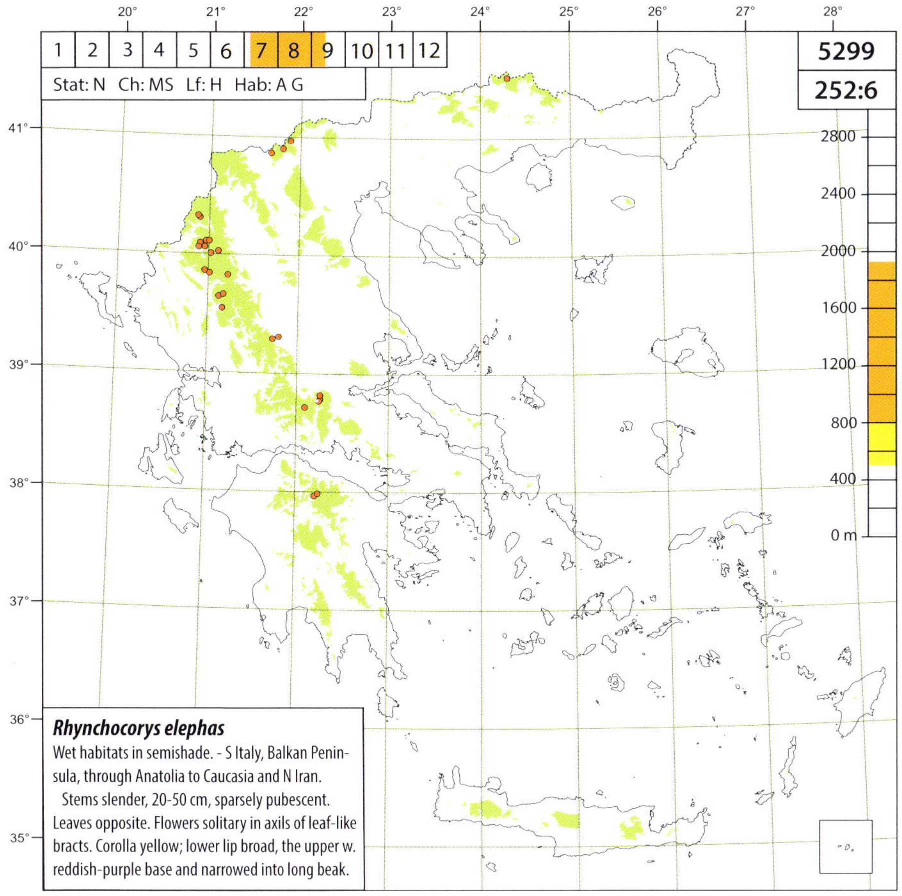

Rhynchocorys elephas

Wet habitats in semishade. - S Italy, Balkan Peninsula, through Anatolia to Caucasia and N Iran.

Stems slender, 20-50 cm, sparsely pubescent. Leaves opposite. Flowers solitary in axils of leaf-like bracts. Corolla yellow; lower lip broad, the upper w. reddish-purple base and narrowed into long beak.

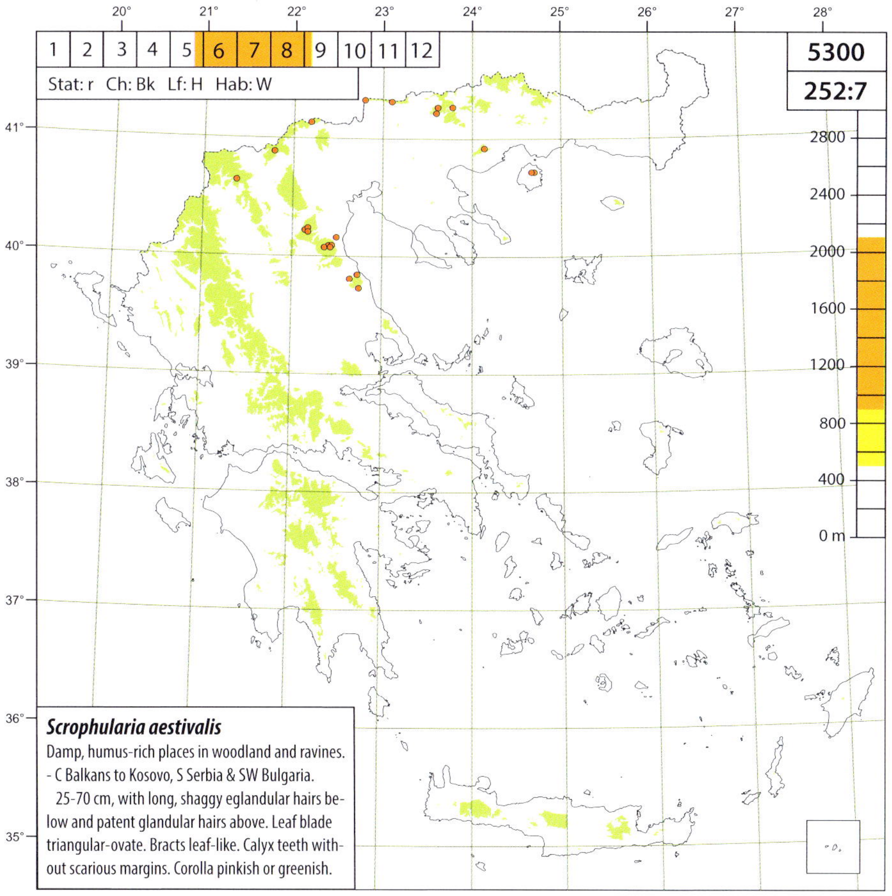

Scrophularia aestivalis

Damp, humus-rich places in woodland and ravines. - C Balkans to Kosovo, S Serbia & SW Bulgaria.

25-70 cm, with long, shaggy eglandular hairs below and patent glandular hairs above. Leaf blade triangular-ovate. Bracts leaf-like. Calyx teeth without scarious margins. Corolla pinkish or greenish.

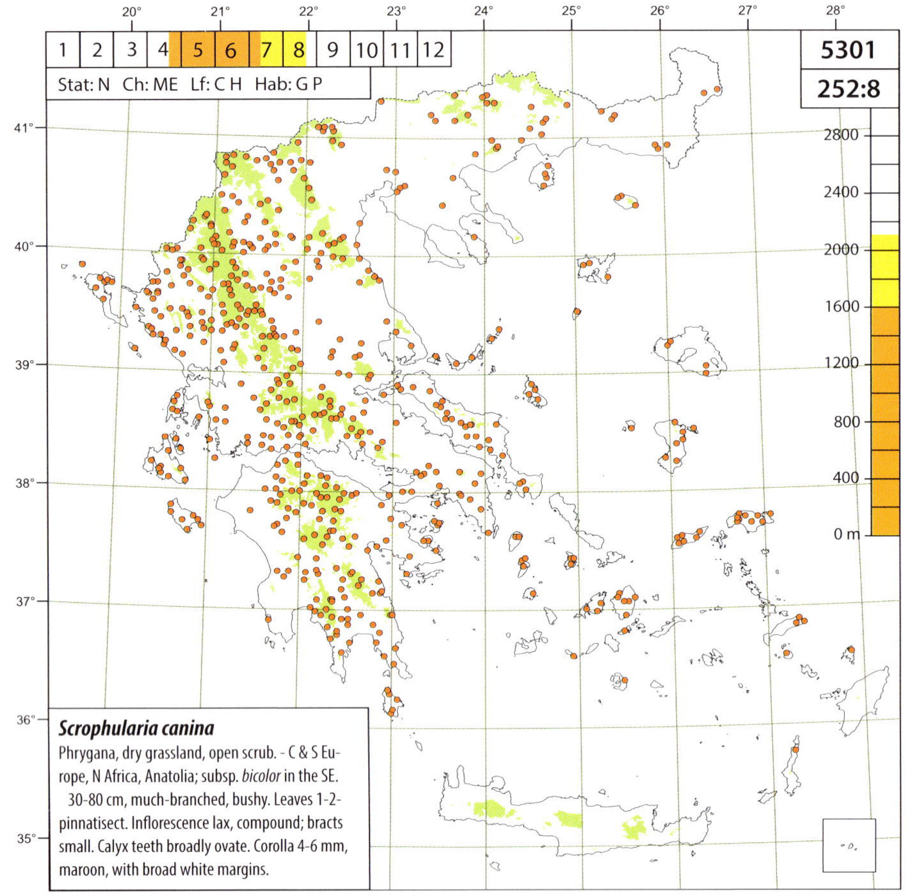

Scrophularia canina

Phrygana, dry grassland, open scrub. - C & S Europe, N Africa, Anatolia; subsp. *bicolor* in the SE.

30-80 cm, much-branched, bushy. Leaves 1-2-pinnatisect. Inflorescence lax, compound; bracts small. Calyx teeth broadly ovate. Corolla 4-6 mm, maroon, with broad white margins.

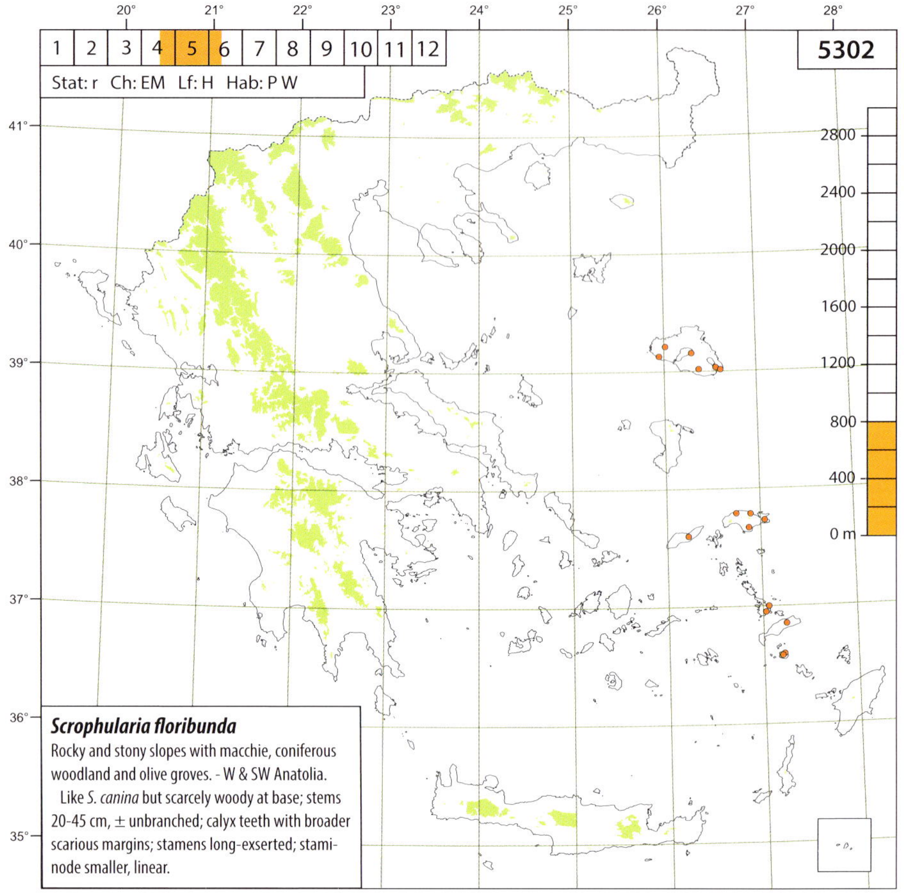

Scrophularia floribunda

Rocky and stony slopes with macchie, coniferous woodland and olive groves. - W & SW Anatolia.

Like *S. canina* but scarcely woody at base; stems 20-45 cm, ± unbranched; calyx teeth with broader scarious margins; stamens long-exserted; staminode smaller, linear.

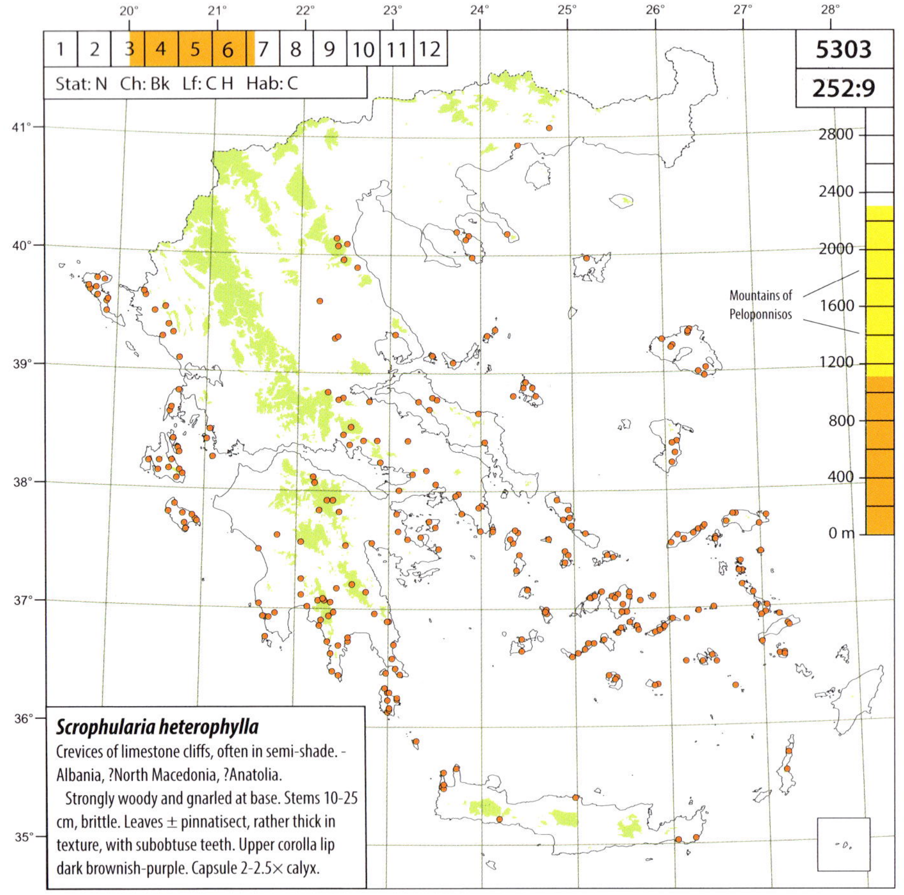

Scrophularia heterophylla

Crevices of limestone cliffs, often in semi-shade. - Albania, ?North Macedonia, ?Anatolia.

Strongly woody and gnarled at base. Stems 10-25 cm, brittle. Leaves ± pinnatisect, rather thick in texture, with subobtuse teeth. Upper corolla lip dark brownish-purple. Capsule 2-2.5× calyx.

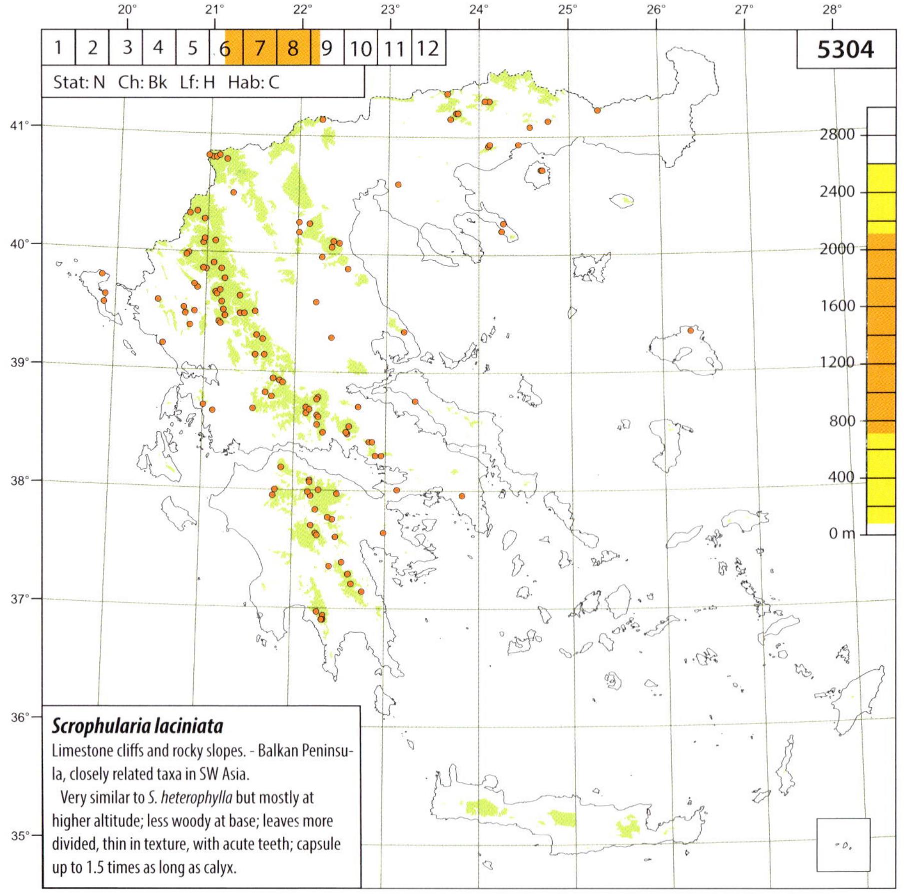

Scrophularia laciniata

Limestone cliffs and rocky slopes. - Balkan Peninsula, closely related taxa in SW Asia.

Very similar to *S. heterophylla* but mostly at higher altitude; less woody at base; leaves more divided, thin in texture, with acute teeth; capsule up to 1.5 times as long as calyx.

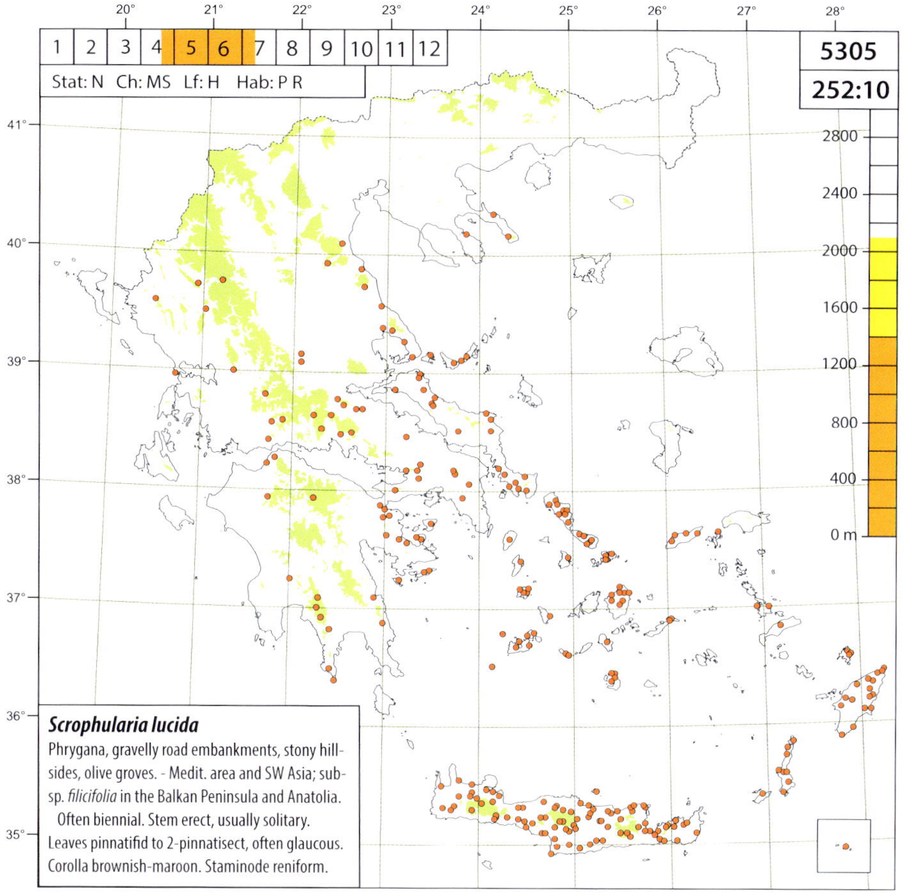

Scrophularia lucida

Phrygana, gravelly road embankments, stony hill-sides, olive groves. - Medit. area and SW Asia; subsp. *filicifolia* in the Balkan Peninsula and Anatolia.

Often biennial. Stem erect, usually solitary. Leaves pinnatifid to 2-pinnatisect, often glaucous. Corolla brownish-maroon. Staminode reniform.

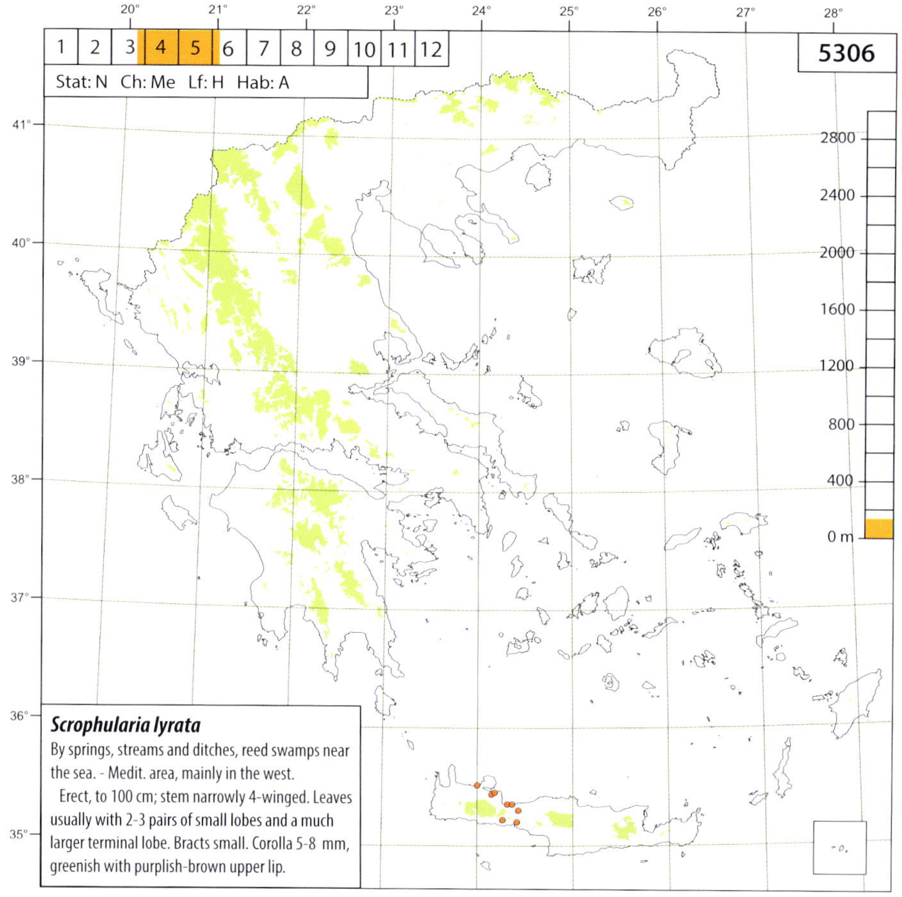

Scrophularia lyrata

By springs, streams and ditches, reed swamps near the sea. - Medit. area, mainly in the west.

Erect, to 100 cm; stem narrowly 4-winged. Leaves usually with 2-3 pairs of small lobes and a much larger terminal lobe. Bracts small. Corolla 5-8 mm, greenish with purplish-brown upper lip.

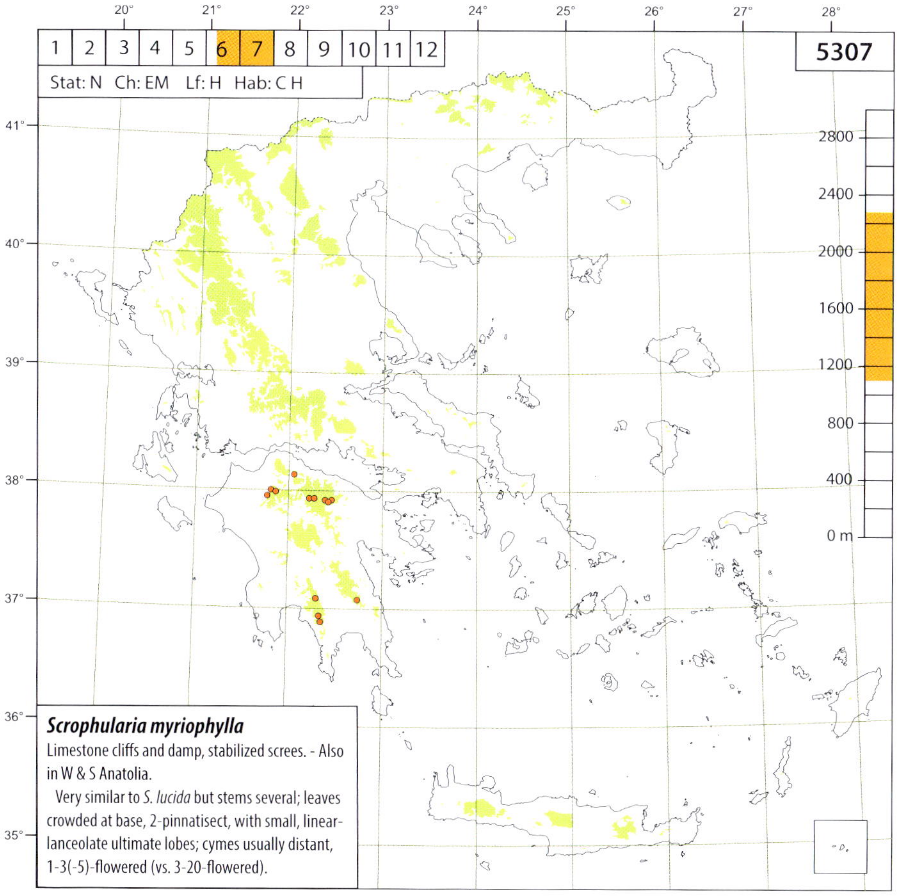

Scrophularia myriophylla

Limestone cliffs and damp, stabilized screes. - Also in W & S Anatolia.

Very similar to *S. lucida* but stems several; leaves crowded at base, 2-pinnatisect, with small, linear-lanceolate ultimate lobes; cymes usually distant, 1-3(-5)-flowered (vs. 3-20-flowered).

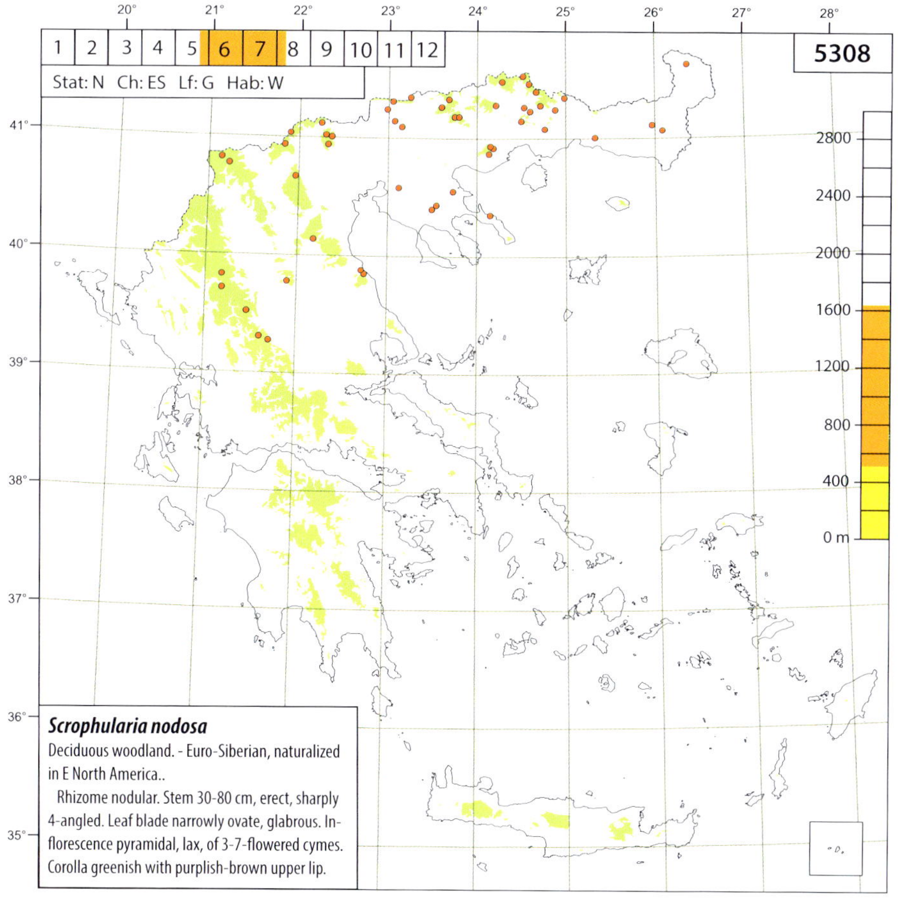

Scrophularia nodosa

Deciduous woodland. - Euro-Siberian, naturalized in E North America..

Rhizome nodular. Stem 30-80 cm, erect, sharply 4-angled. Leaf blade narrowly ovate, glabrous. Inflorescence pyramidal, lax, of 3-7-flowered cymes. Corolla greenish with purplish-brown upper lip.

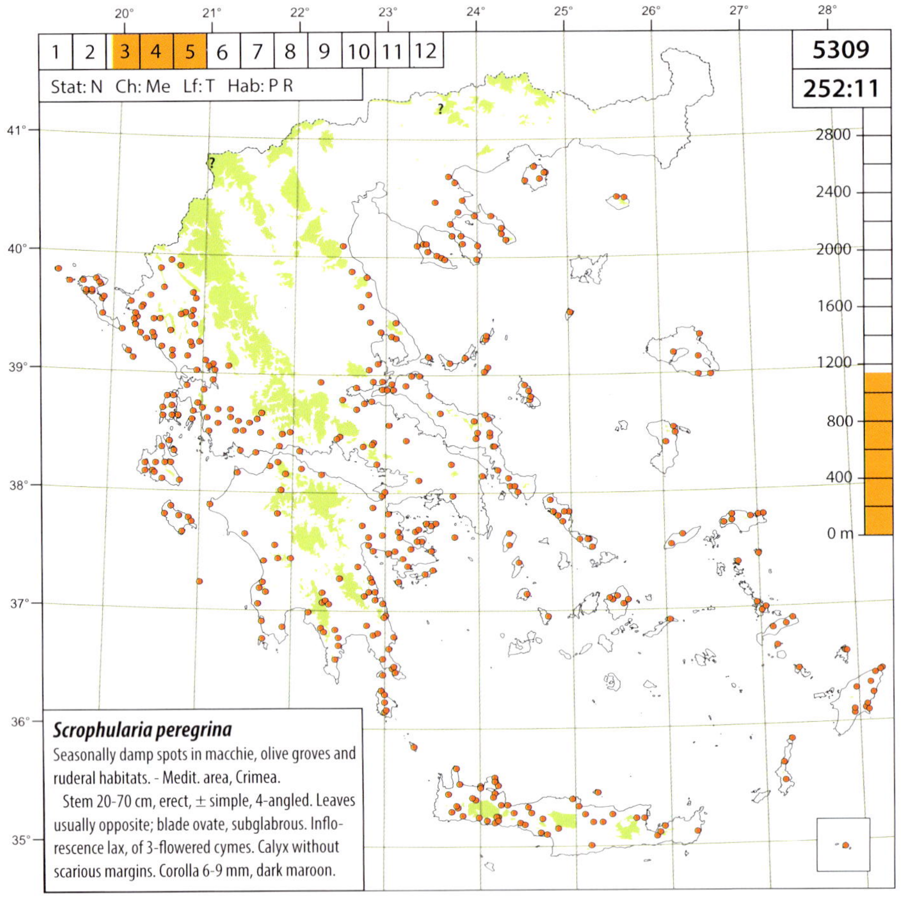

Scrophularia peregrina

Seasonally damp spots in macchie, olive groves and ruderal habitats. - Medit. area, Crimea.

Stem 20-70 cm, erect, ± simple, 4-angled. Leaves usually opposite; blade ovate, subglabrous. Inflorescence lax, of 3-flowered cymes. Calyx without scarious margins. Corolla 6-9 mm, dark maroon.

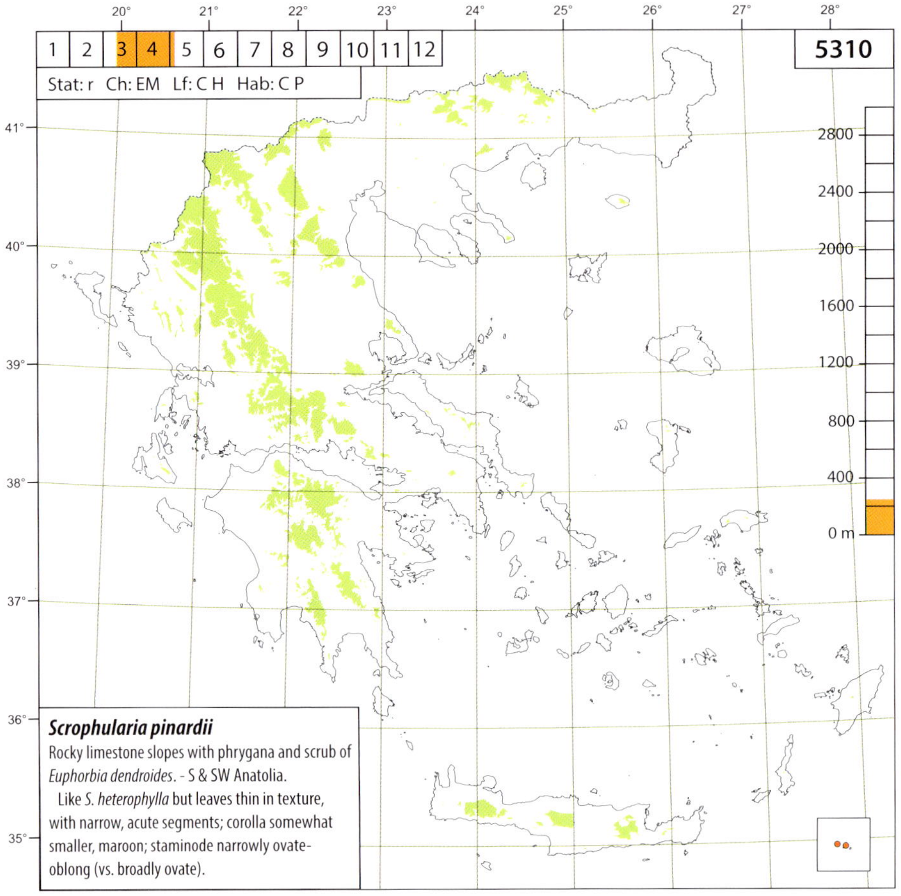

Scrophularia pinardii

Rocky limestone slopes with phrygana and scrub of *Euphorbia dendroides*. - S & SW Anatolia.

Like *S. heterophylla* but leaves thin in texture, with narrow, acute segments; corolla somewhat smaller, maroon; staminode narrowly ovate-oblong (vs. broadly ovate).

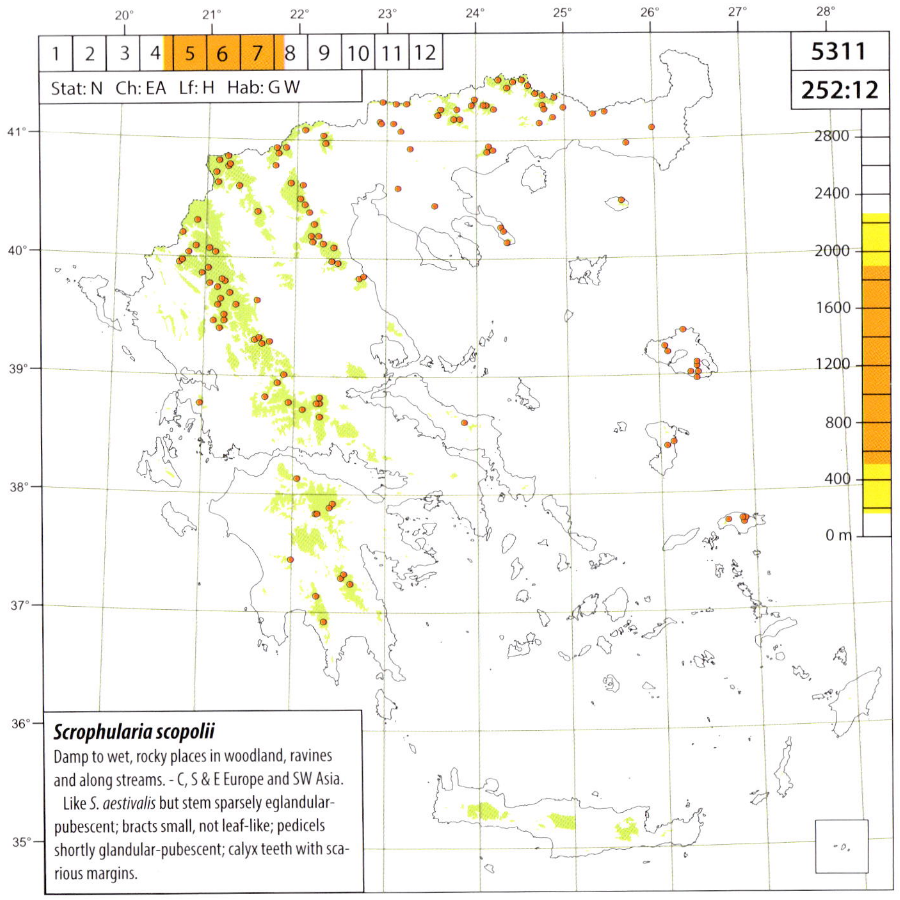

Scrophularia scopolii

Damp to wet, rocky places in woodland, ravines and along streams. - C, S & E Europe and SW Asia.

Like *S. aestivalis* but stem sparsely eglandular-pubescent; bracts small, not leaf-like; pedicels shortly glandular-pubescent; calyx teeth with scarious margins.

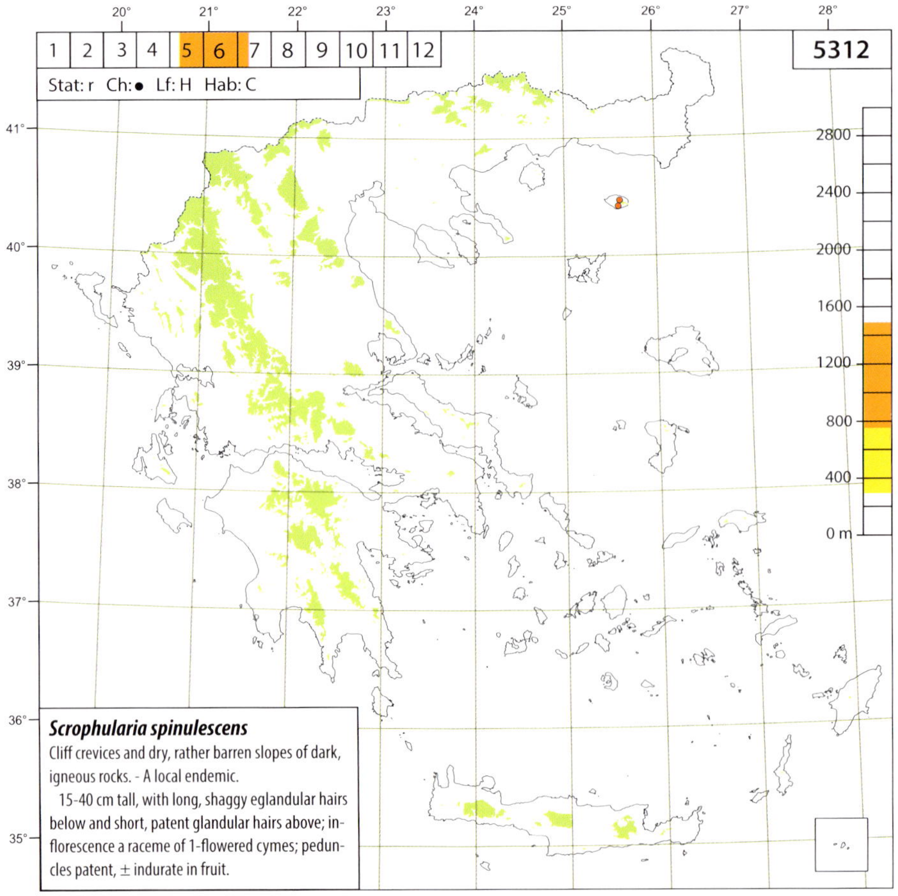

Scrophularia spinulescens

Cliff crevices and dry, rather barren slopes of dark, igneous rocks. - A local endemic.

15-40 cm tall, with long, shaggy eglandular hairs below and short, patent glandular hairs above; inflorescence a raceme of 1-flowered cymes; peduncles patent, ± indurate in fruit.

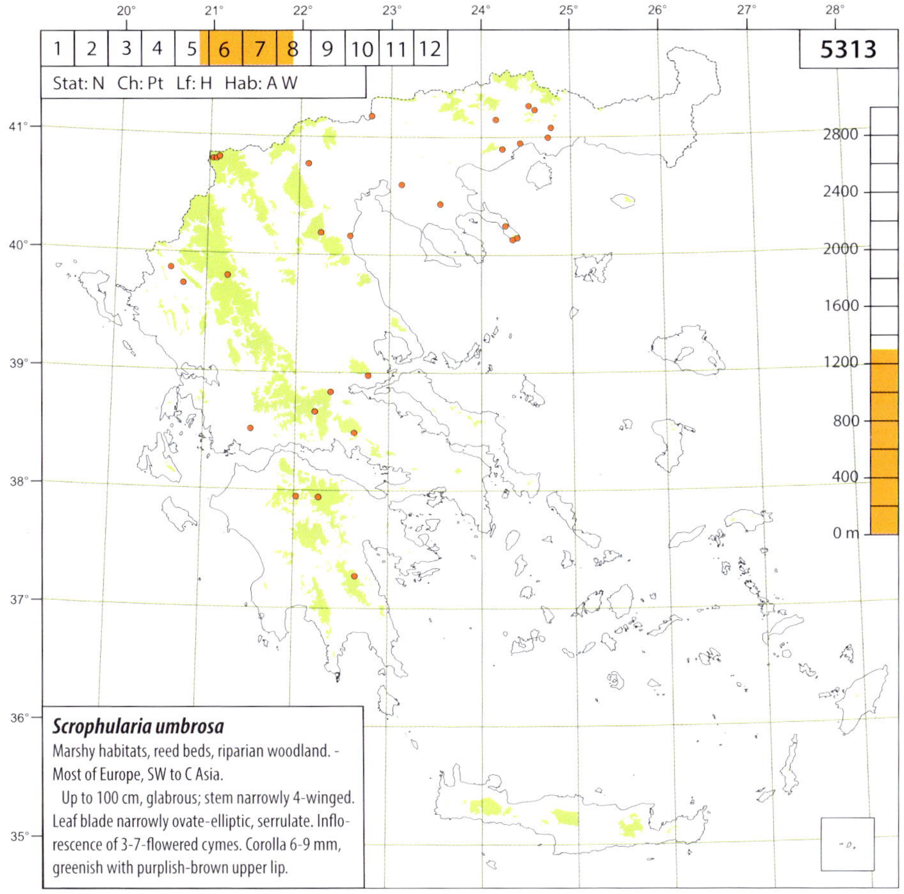

Scrophularia umbrosa
Marshy habitats, reed beds, riparian woodland. - Most of Europe, SW to C Asia.
Up to 100 cm, glabrous; stem narrowly 4-winged. Leaf blade narrowly ovate-elliptic, serrulate. Inflorescence of 3-7-flowered cymes. Corolla 6-9 mm, greenish with purplish-brown upper lip.

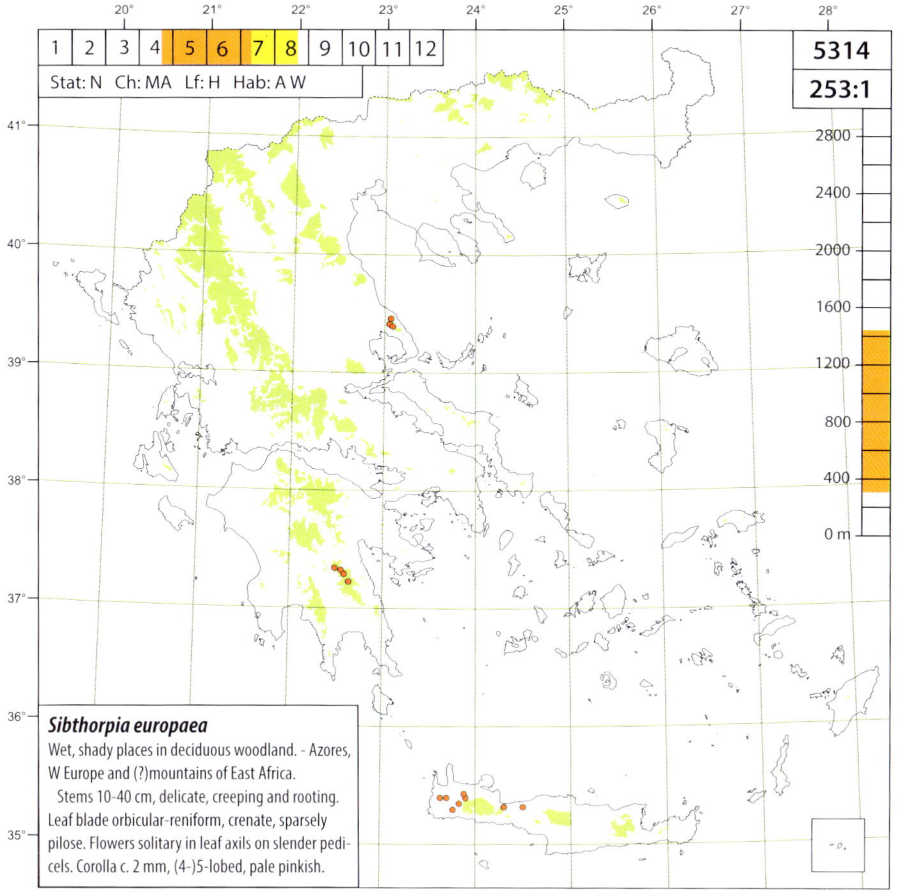

Sibthorpia europaea
Wet, shady places in deciduous woodland. - Azores, W Europe and (?)mountains of East Africa.
Stems 10-40 cm, delicate, creeping and rooting. Leaf blade orbicular-reniform, crenate, sparsely pilose. Flowers solitary in leaf axils on slender pedicels. Corolla c. 2 mm, (4-)5-lobed, pale pinkish.

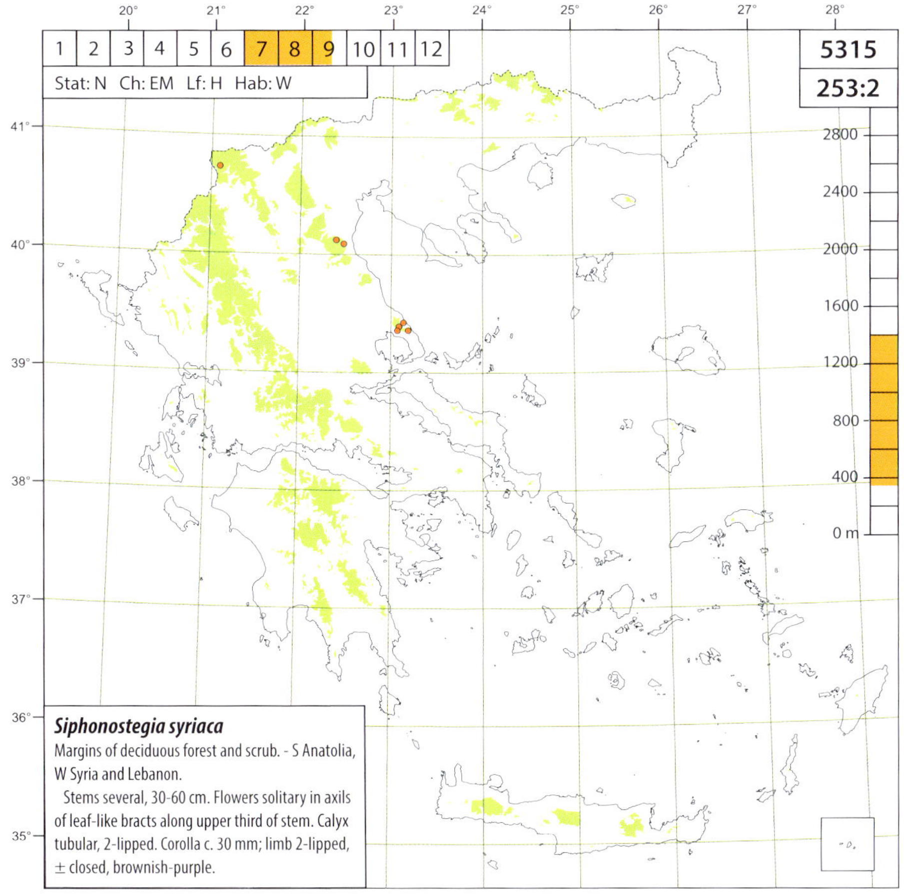

Siphonostegia syriaca
Margins of deciduous forest and scrub. - S Anatolia, W Syria and Lebanon.
Stems several, 30-60 cm. Flowers solitary in axils of leaf-like bracts along upper third of stem. Calyx tubular, 2-lipped. Corolla c. 30 mm; limb 2-lipped, ± closed, brownish-purple.

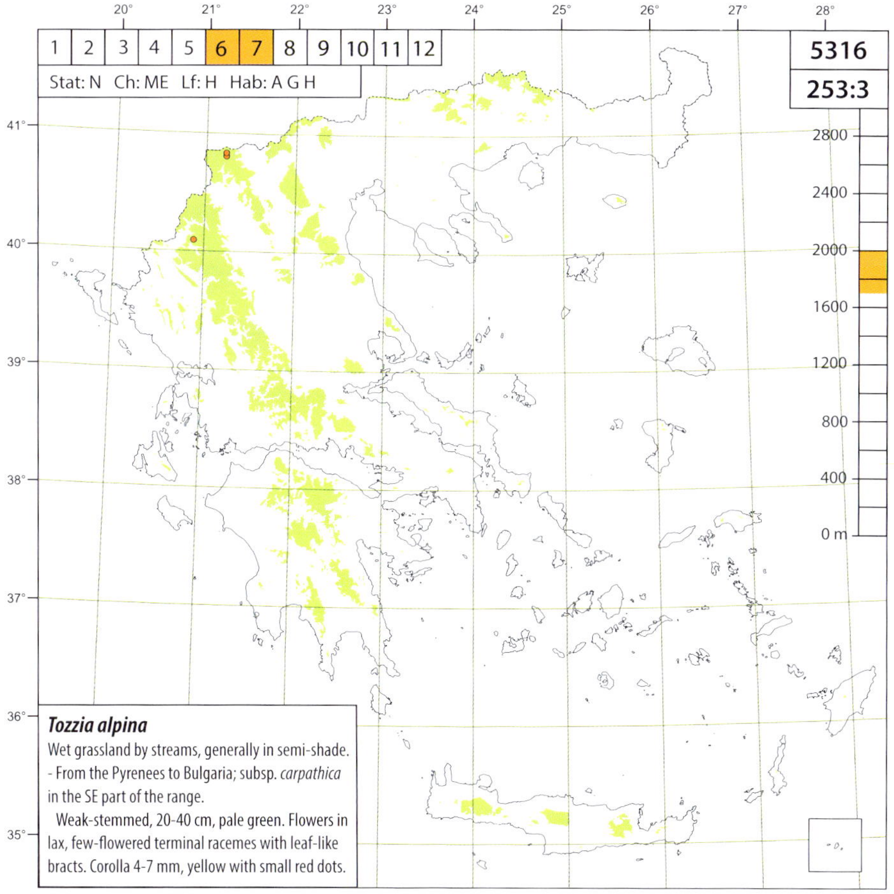

Tozzia alpina
Wet grassland by streams, generally in semi-shade. - From the Pyrenees to Bulgaria; subsp. *carpathica* in the SE part of the range.
Weak-stemmed, 20-40 cm, pale green. Flowers in lax, few-flowered terminal racemes with leaf-like bracts. Corolla 4-7 mm, yellow with small red dots.

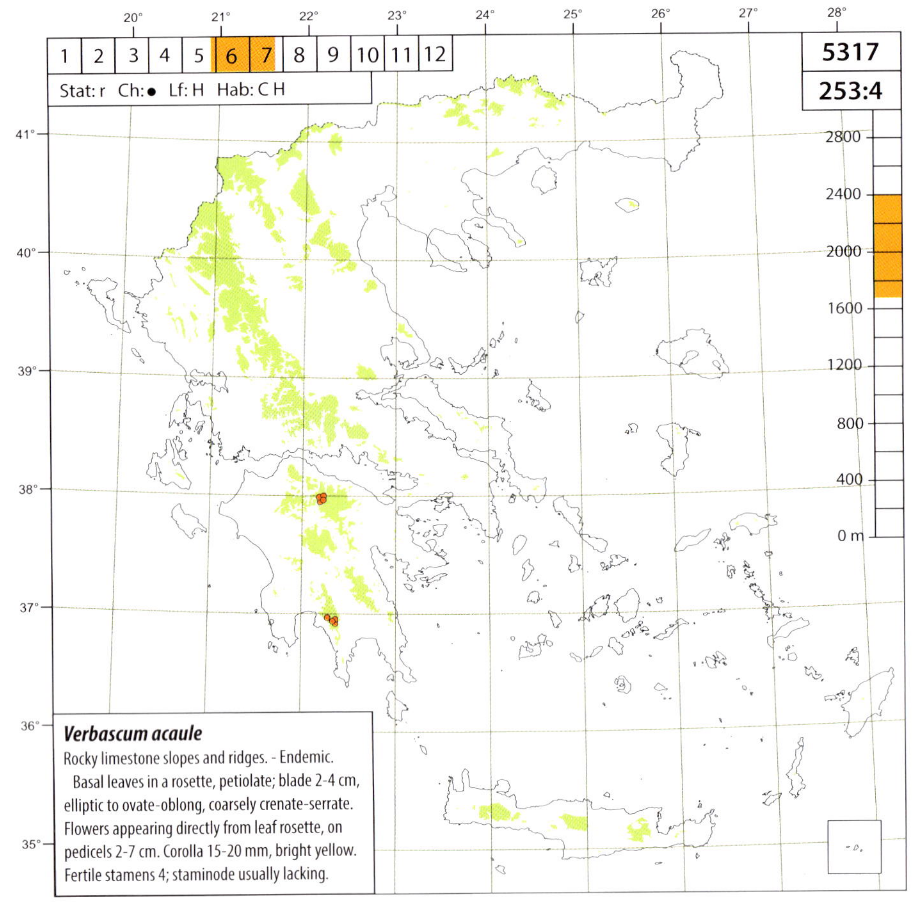

Verbascum acaule

Rocky limestone slopes and ridges. - Endemic.

Basal leaves in a rosette, petiolate; blade 2-4 cm, elliptic to ovate-oblong, coarsely crenate-serrate. Flowers appearing directly from leaf rosette, on pedicels 2-7 cm. Corolla 15-20 mm, bright yellow. Fertile stamens 4; staminode usually lacking.

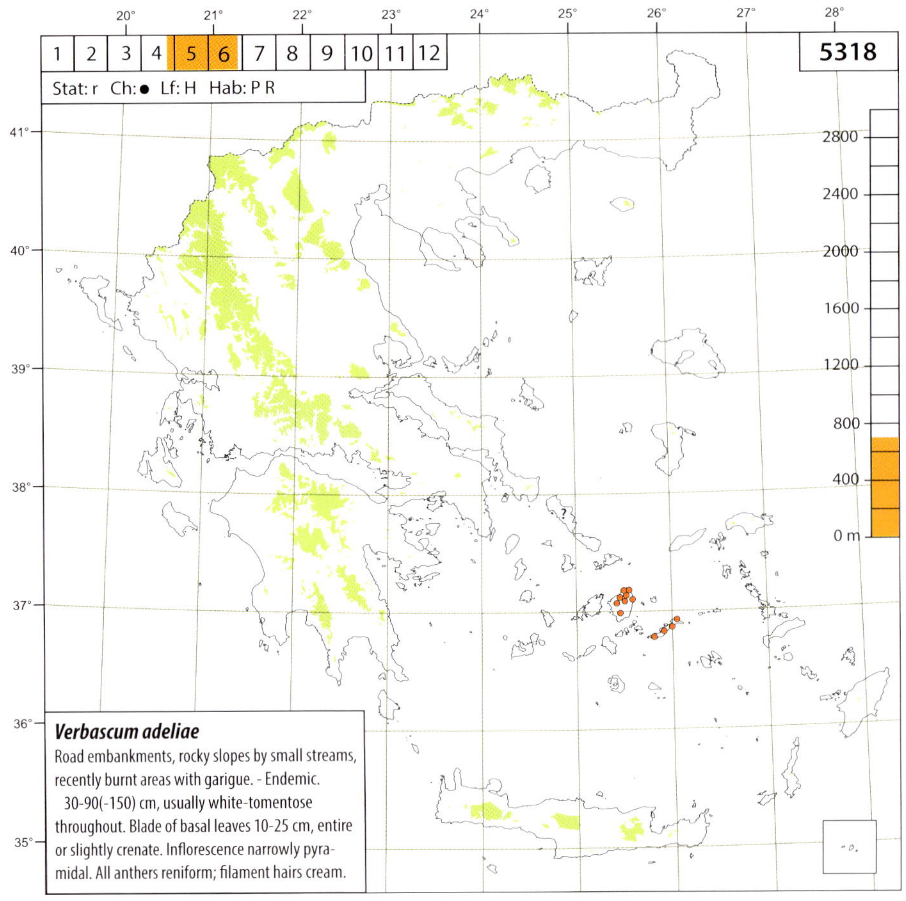

Verbascum adeliae

Road embankments, rocky slopes by small streams, recently burnt areas with garigue. - Endemic.

30-90(-150) cm, usually white-tomentose throughout. Blade of basal leaves 10-25 cm, entire or slightly crenate. Inflorescence narrowly pyramidal. All anthers reniform; filament hairs cream.

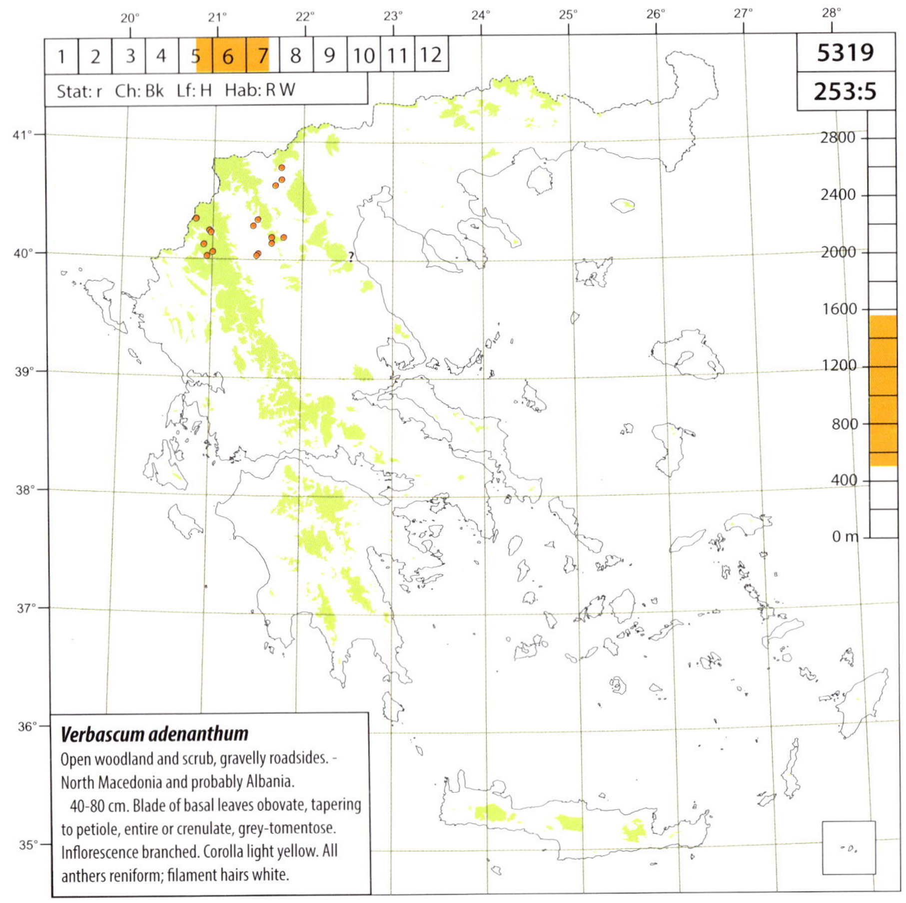

Verbascum adenanthum

Open woodland and scrub, gravelly roadsides. - North Macedonia and probably Albania.

40-80 cm. Blade of basal leaves obovate, tapering to petiole, entire or crenulate, grey-tomentose. Inflorescence branched. Corolla light yellow. All anthers reniform; filament hairs white.

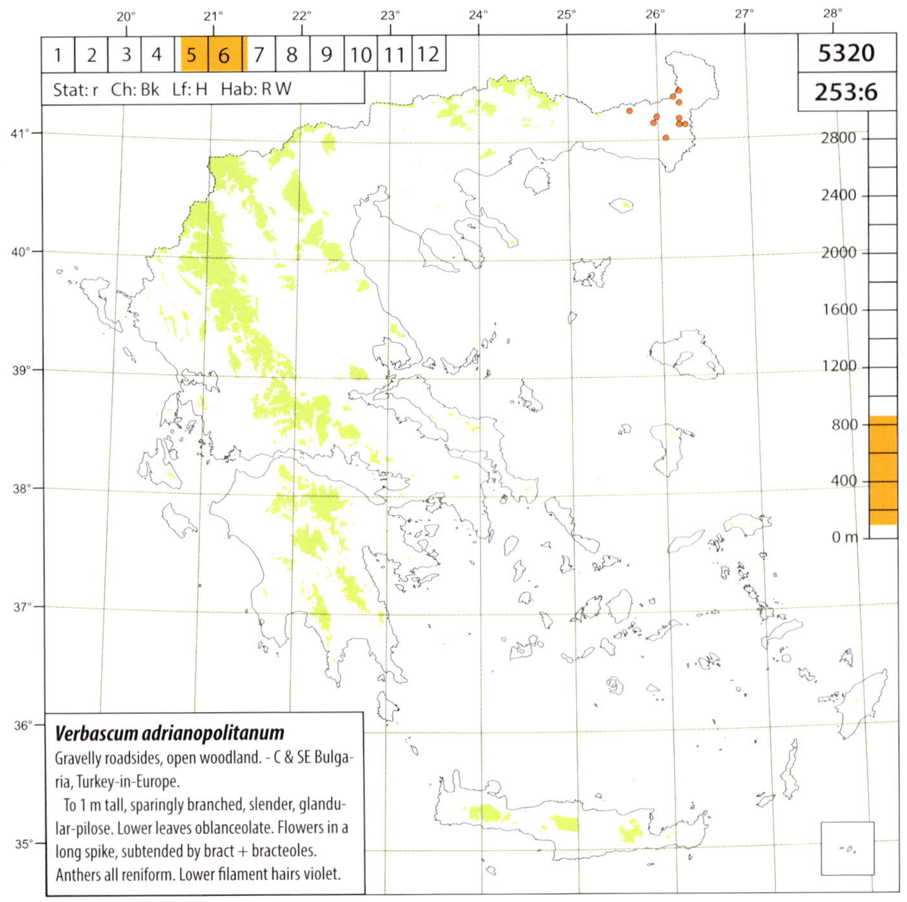

Verbascum adrianopolitanum

Gravelly roadsides, open woodland. - C & SE Bulgaria, Turkey-in-Europe.

To 1 m tall, sparingly branched, slender, glandular-pilose. Lower leaves oblanceolate. Flowers in a long spike, subtended by bract + bracteoles. Anthers all reniform. Lower filament hairs violet.

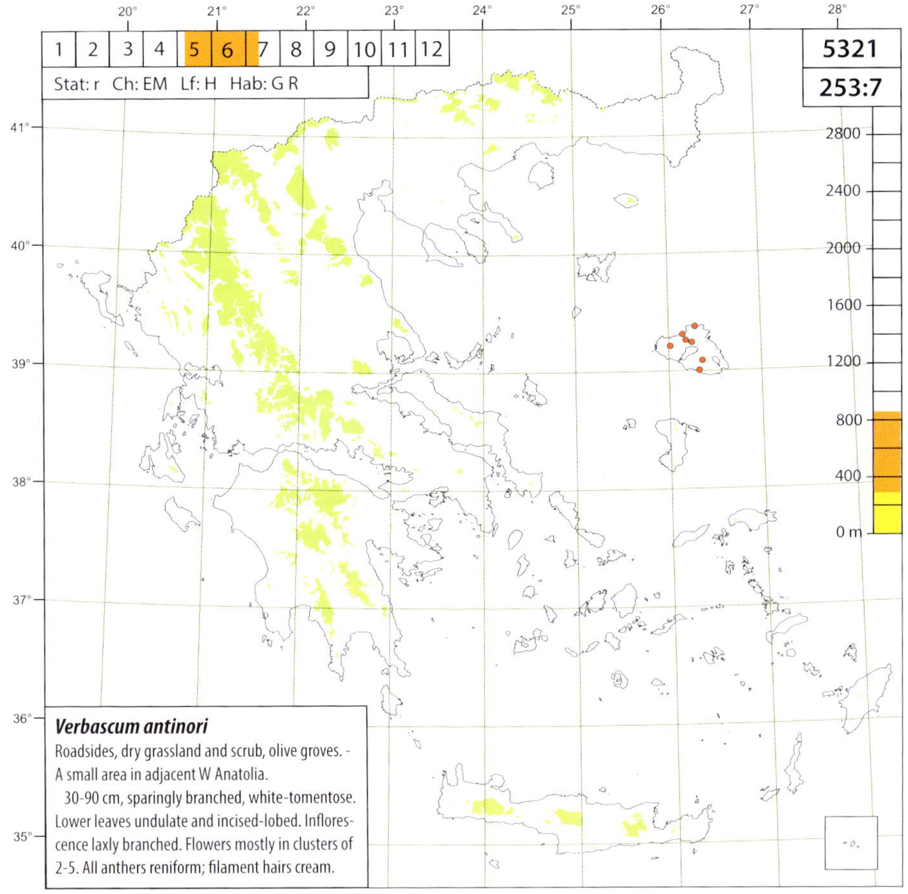

Verbascum antinori

Roadsides, dry grassland and scrub, olive groves. - A small area in adjacent W Anatolia.

30-90 cm, sparingly branched, white-tomentose. Lower leaves undulate and incised-lobed. Inflorescence laxly branched. Flowers mostly in clusters of 2-5. All anthers reniform; filament hairs cream.

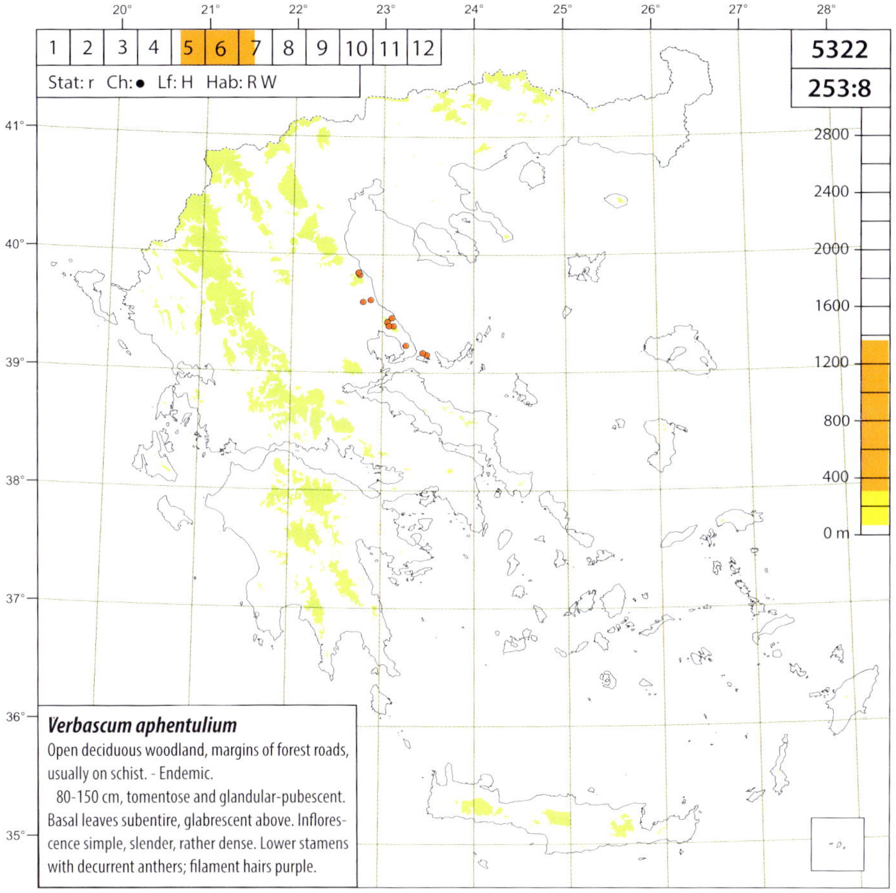

Verbascum aphentulium

Open deciduous woodland, margins of forest roads, usually on schist. - Endemic.

80-150 cm, tomentose and glandular-pubescent. Basal leaves subentire, glabrescent above. Inflorescence simple, slender, rather dense. Lower stamens with decurrent anthers; filament hairs purple.

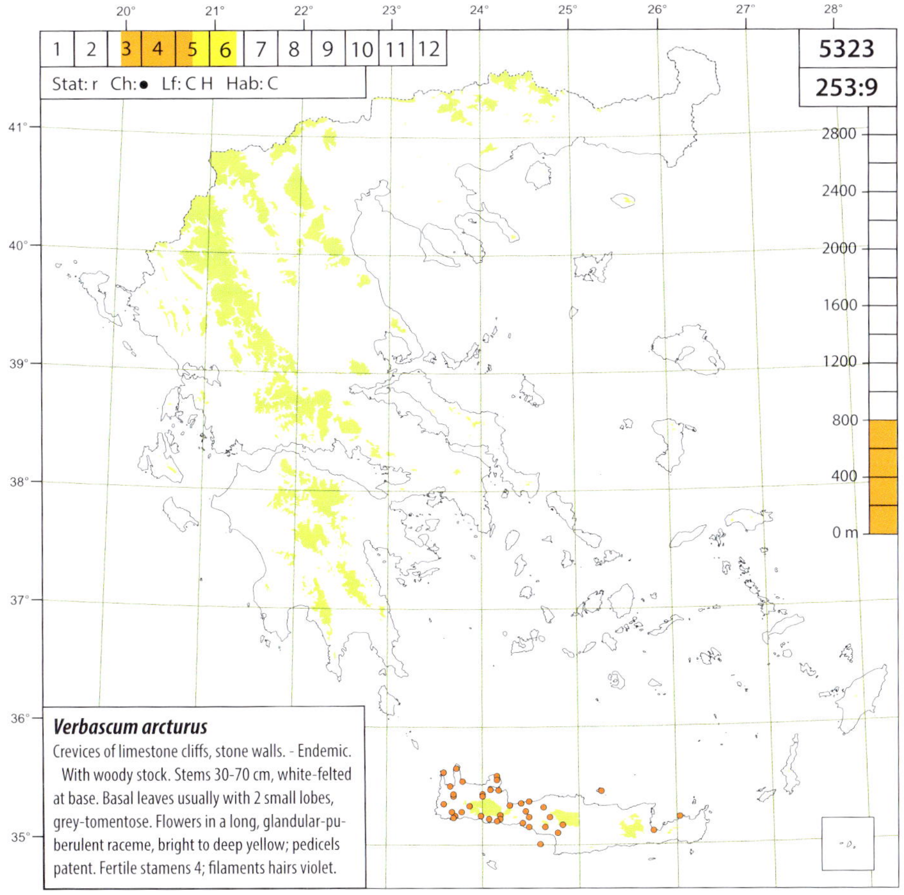

Verbascum arcturus

Crevices of limestone cliffs, stone walls. - Endemic.

With woody stock. Stems 30-70 cm, white-felted at base. Basal leaves usually with 2 small lobes, grey-tomentose. Flowers in a long, glandular-puberulent raceme, bright to deep yellow; pedicels patent. Fertile stamens 4; filaments hairs violet.

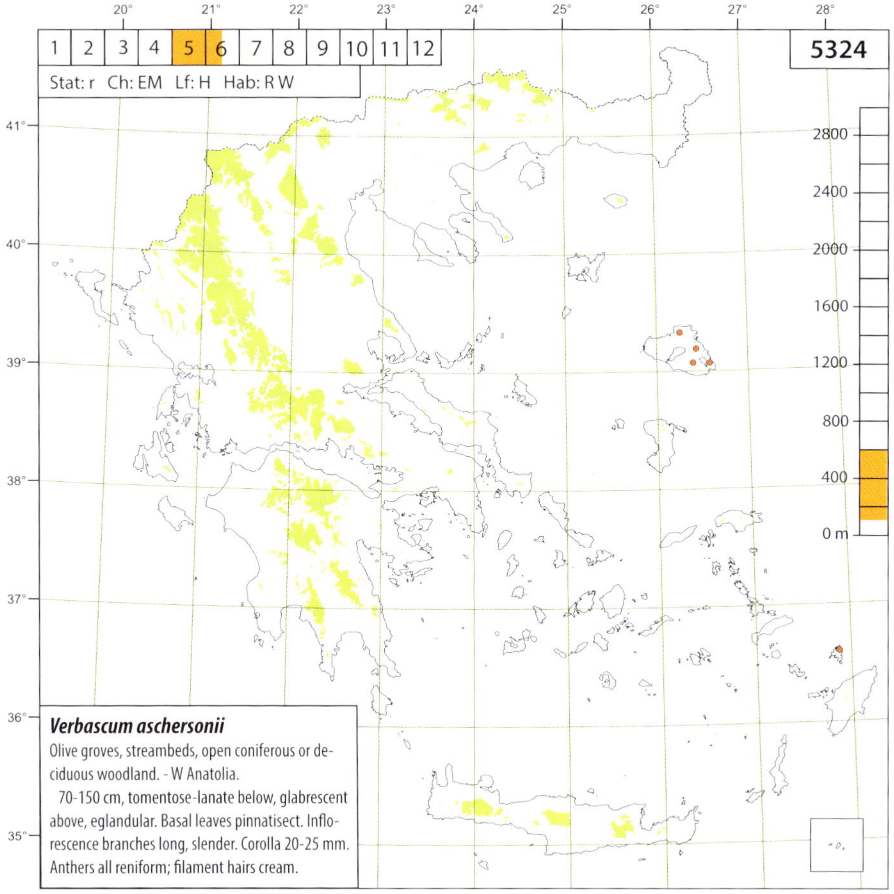

Verbascum aschersonii

Olive groves, streambeds, open coniferous or deciduous woodland. - W Anatolia.

70-150 cm, tomentose-lanate below, glabrescent above, eglandular. Basal leaves pinnatisect. Inflorescence branches long, slender. Corolla 20-25 mm. Anthers all reniform; filament hairs cream.

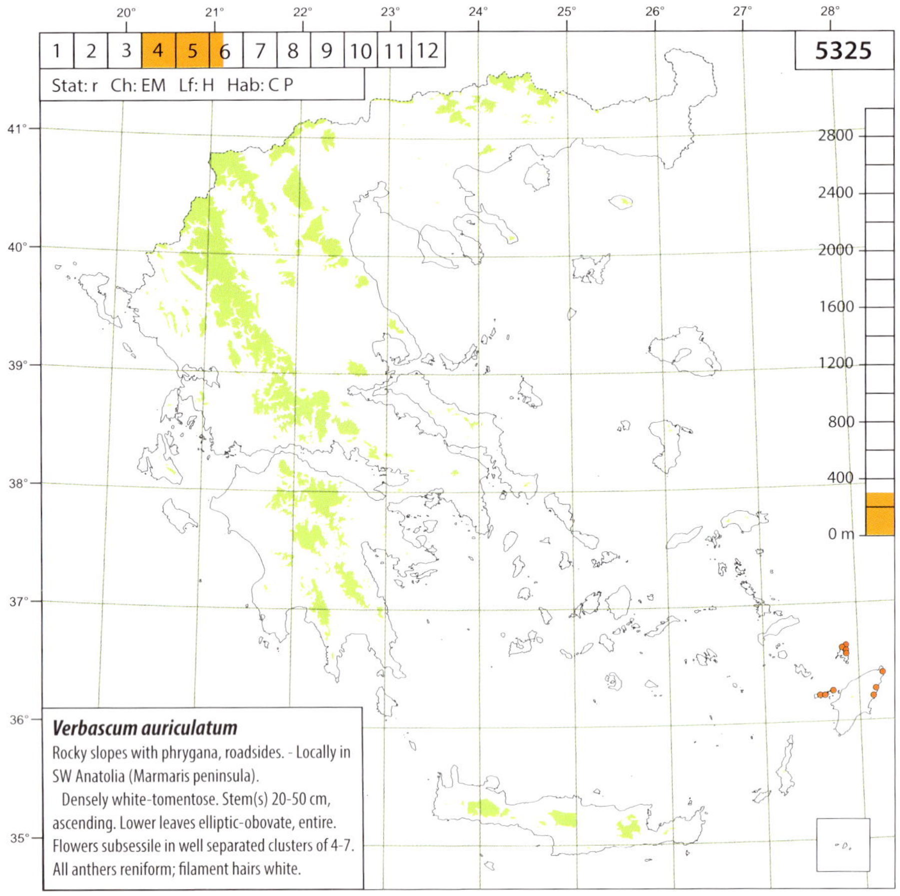

Verbascum auriculatum
Rocky slopes with phrygana, roadsides. - Locally in SW Anatolia (Marmaris peninsula).
Densely white-tomentose. Stem(s) 20-50 cm, ascending. Lower leaves elliptic-obovate, entire. Flowers subsessile in well separated clusters of 4-7. All anthers reniform; filament hairs white.

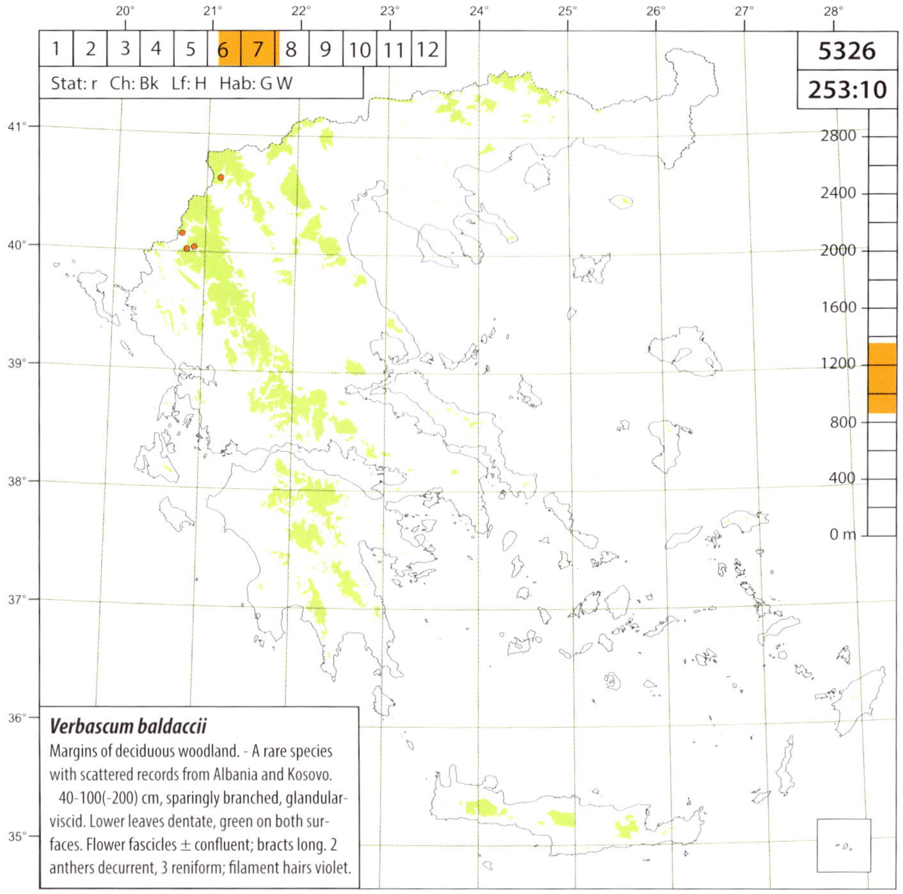

Verbascum baldaccii
Margins of deciduous woodland. - A rare species with scattered records from Albania and Kosovo.
40-100(-200) cm, sparingly branched, glandular-viscid. Lower leaves dentate, green on both surfaces. Flower fascicles ± confluent; bracts long. 2 anthers decurrent, 3 reniform; filament hairs violet.

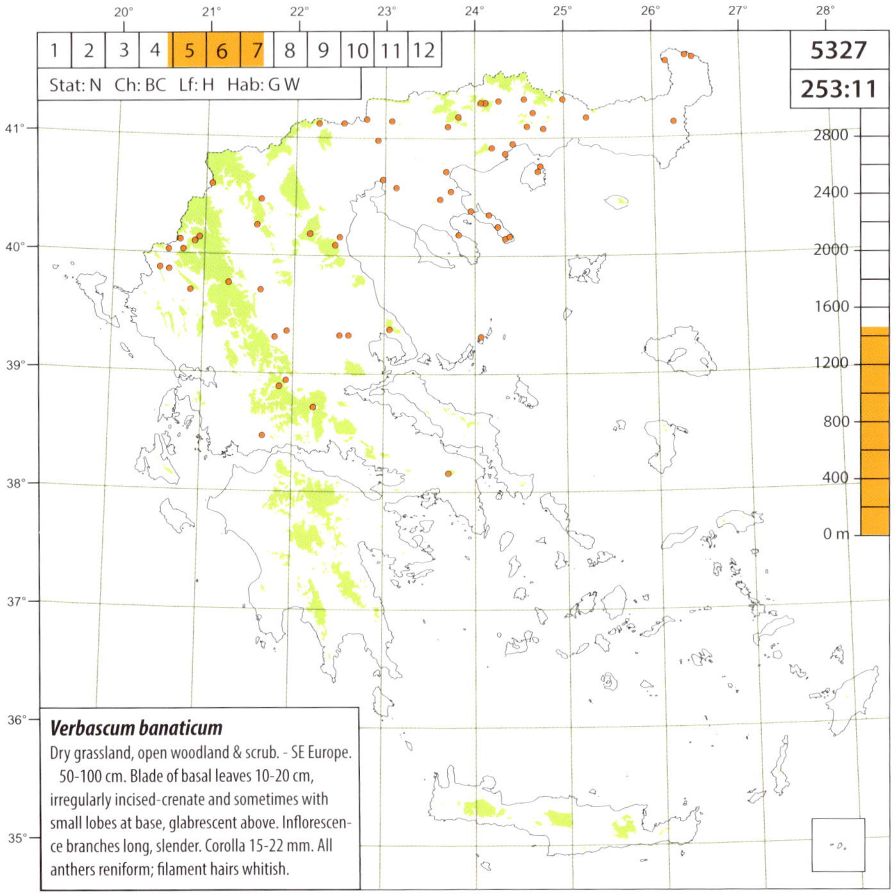

Verbascum banaticum
Dry grassland, open woodland & scrub. - SE Europe.
50-100 cm. Blade of basal leaves 10-20 cm, irregularly incised-crenate and sometimes with small lobes at base, glabrescent above. Inflorescence branches long, slender. Corolla 15-22 mm. All anthers reniform; filament hairs whitish.

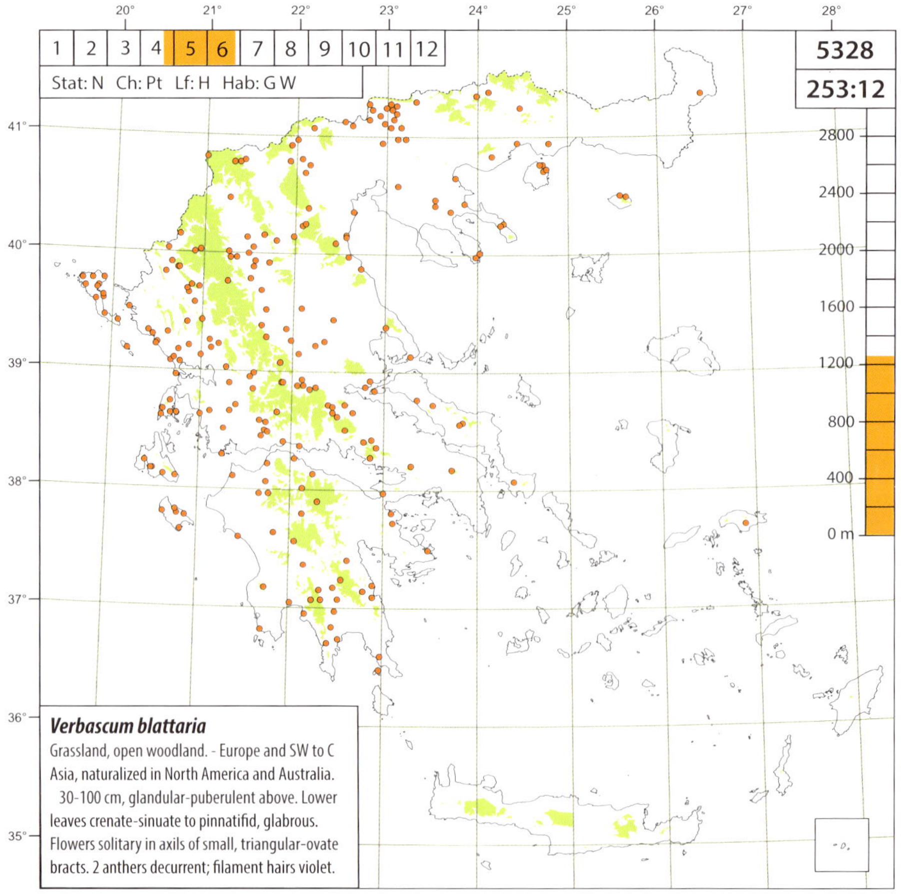

Verbascum blattaria
Grassland, open woodland. - Europe and SW to C Asia, naturalized in North America and Australia.
30-100 cm, glandular-puberulent above. Lower leaves crenate-sinuate to pinnatifid, glabrous. Flowers solitary in axils of small, triangular-ovate bracts. 2 anthers decurrent; filament hairs violet.

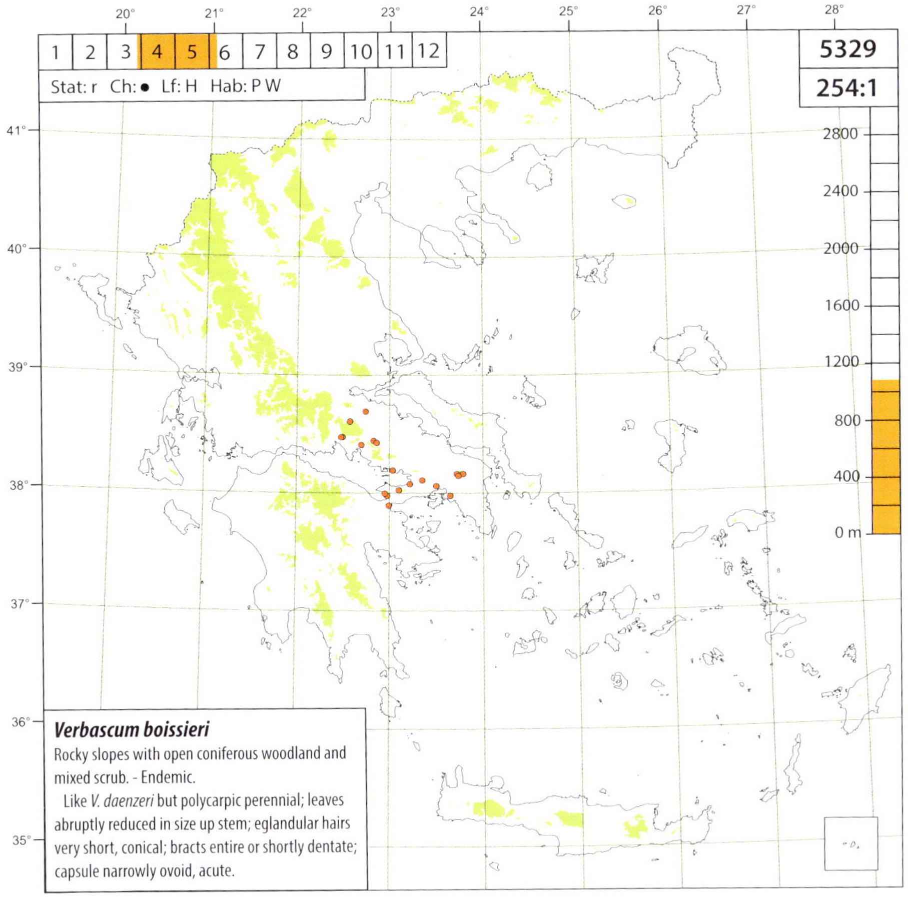

Verbascum boissieri

Rocky slopes with open coniferous woodland and mixed scrub. - Endemic.

Like *V. daenzeri* but polycarpic perennial; leaves abruptly reduced in size up stem; eglandular hairs very short, conical; bracts entire or shortly dentate; capsule narrowly ovoid, acute.

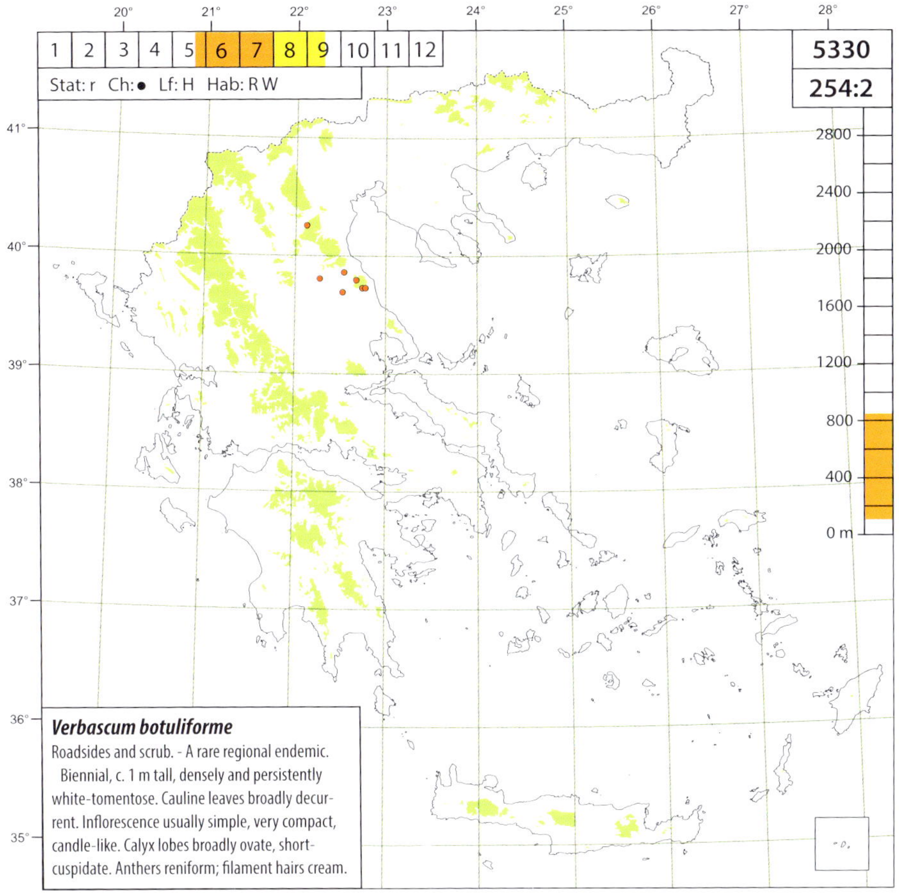

Verbascum botuliforme

Roadsides and scrub. - A rare regional endemic.

Biennial, c. 1 m tall, densely and persistently white-tomentose. Cauline leaves broadly decurrent. Inflorescence usually simple, very compact, candle-like. Calyx lobes broadly ovate, short-cuspidate. Anthers reniform; filament hairs cream.

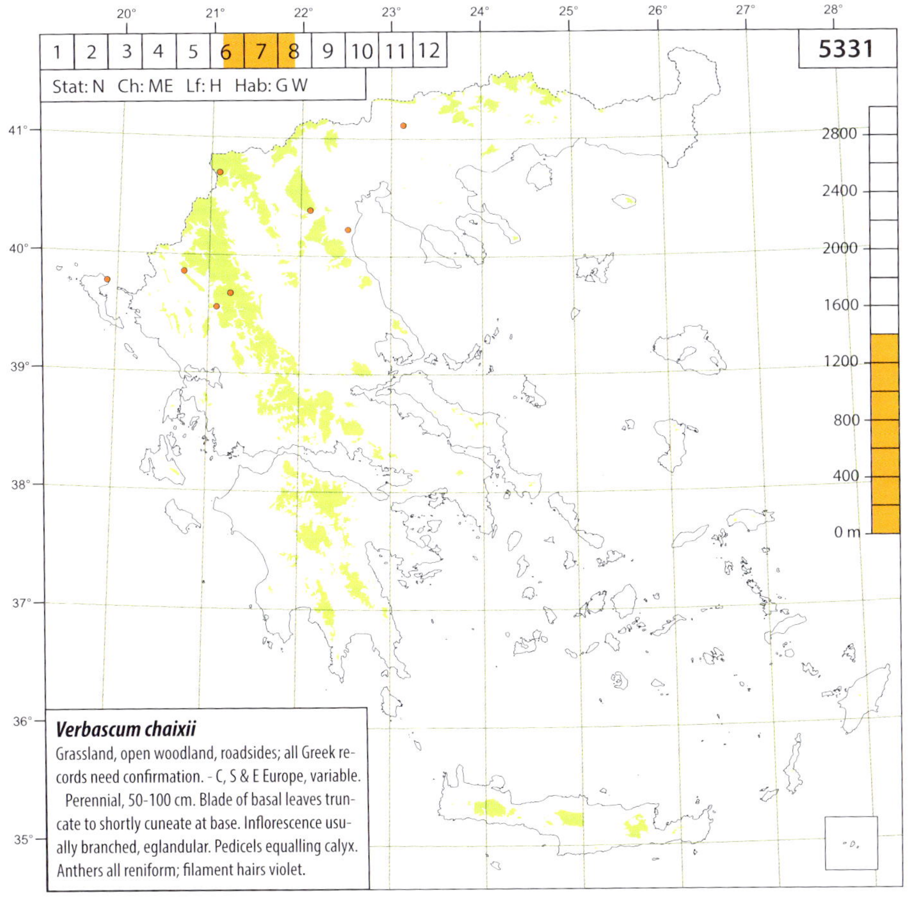

Verbascum chaixii

Grassland, open woodland, roadsides; all Greek records need confirmation. - C, S & E Europe, variable.

Perennial, 50-100 cm. Blade of basal leaves truncate to shortly cuneate at base. Inflorescence usually branched, eglandular. Pedicels equalling calyx. Anthers all reniform; filament hairs violet.

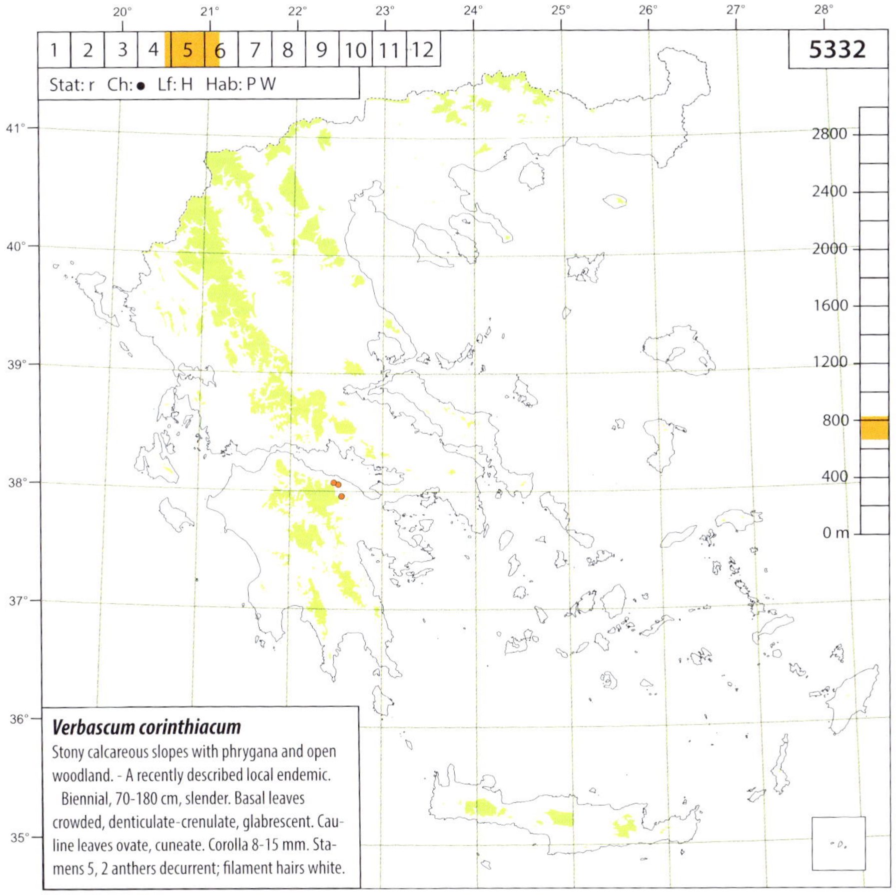

Verbascum corinthiacum

Stony calcareous slopes with phrygana and open woodland. - A recently described local endemic.

Biennial, 70-180 cm, slender. Basal leaves crowded, denticulate-crenulate, glabrescent. Cauline leaves ovate, cuneate. Corolla 8-15 mm. Stamens 5, 2 anthers decurrent; filament hairs white.

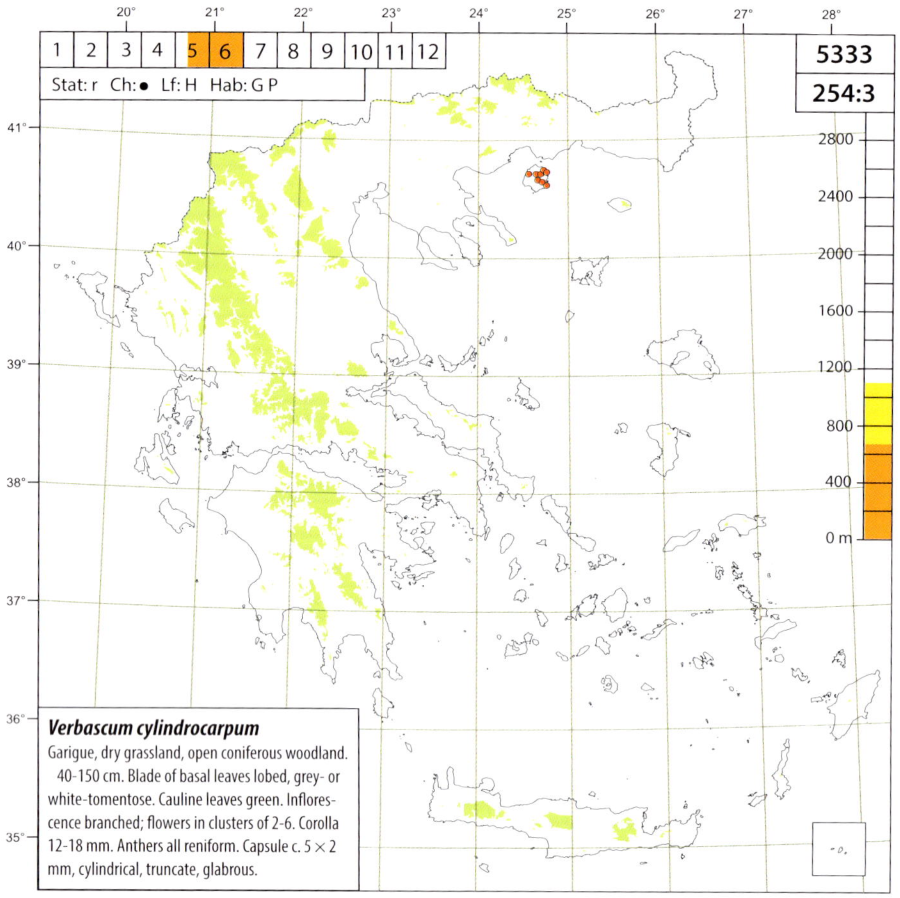

Verbascum cylindrocarpum
Garigue, dry grassland, open coniferous woodland.
40-150 cm. Blade of basal leaves lobed, grey- or white-tomentose. Cauline leaves green. Inflorescence branched; flowers in clusters of 2-6. Corolla 12-18 mm. Anthers all reniform. Capsule c. 5 × 2 mm, cylindrical, truncate, glabrous.

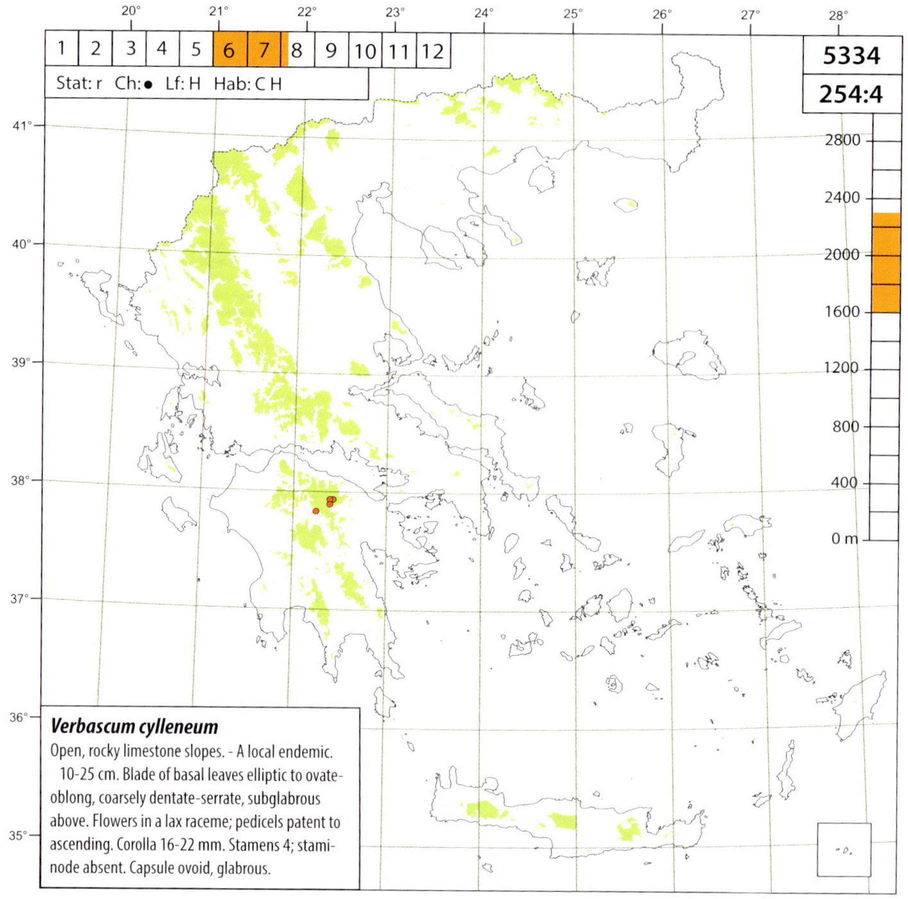

Verbascum cylleneum
Open, rocky limestone slopes. - A local endemic.
10-25 cm. Blade of basal leaves elliptic to ovate-oblong, coarsely dentate-serrate, subglabrous above. Flowers in a lax raceme; pedicels patent to ascending. Corolla 16-22 mm. Stamens 4; staminode absent. Capsule ovoid, glabrous.

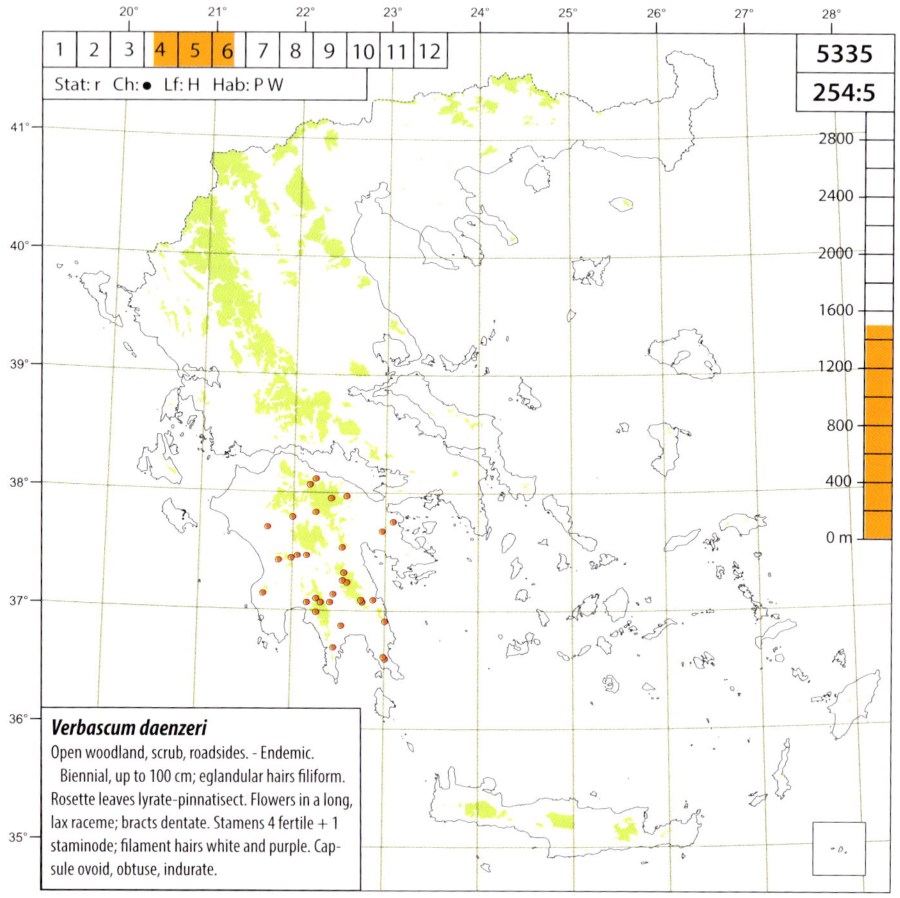

Verbascum daenzeri
Open woodland, scrub, roadsides. - Endemic.
Biennial, up to 100 cm; eglandular hairs filiform. Rosette leaves lyrate-pinnatisect. Flowers in a long, lax raceme; bracts dentate. Stamens 4 fertile + 1 staminode; filament hairs white and purple. Capsule ovoid, obtuse, indurate.

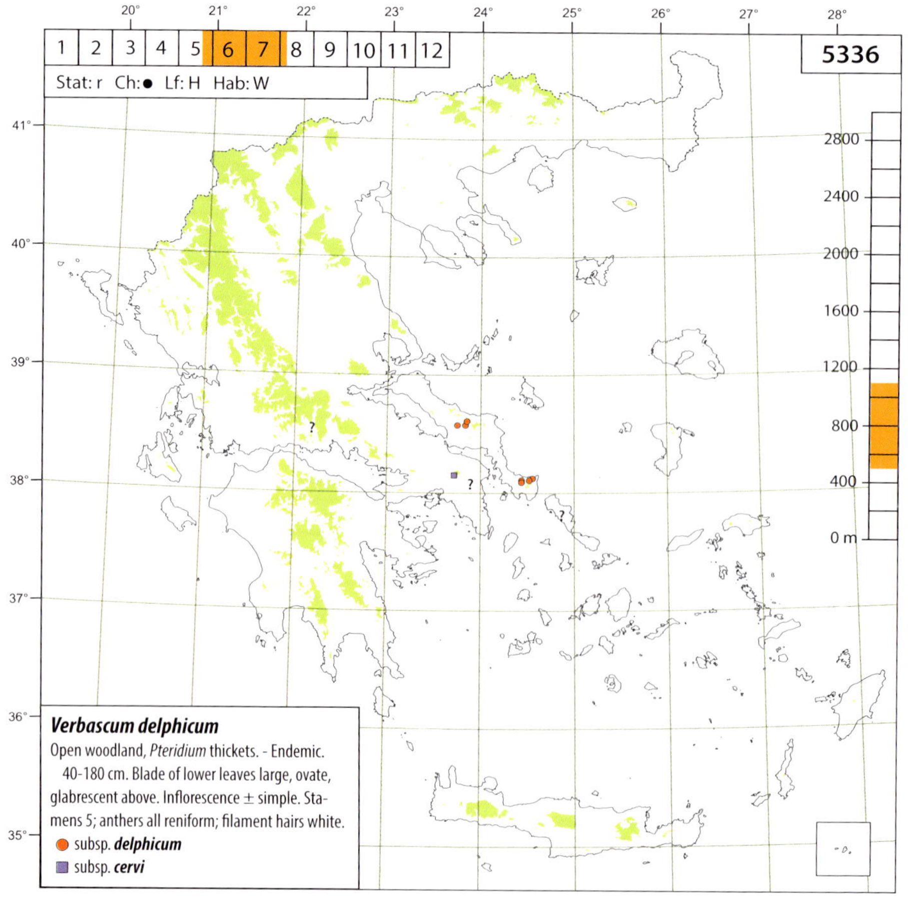

Verbascum delphicum
Open woodland, *Pteridium* thickets. - Endemic.
40-180 cm. Blade of lower leaves large, ovate, glabrescent above. Inflorescence ± simple. Stamens 5; anthers all reniform; filament hairs white.
● subsp. ***delphicum***
■ subsp. ***cervi***

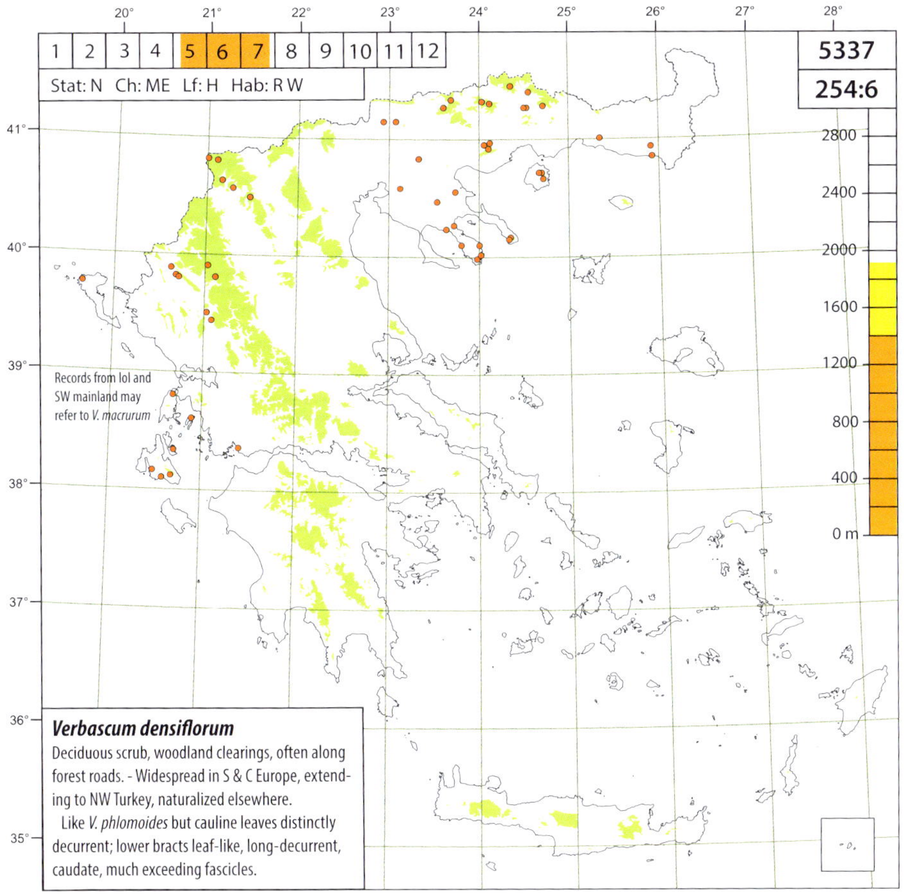

Verbascum densiflorum

Deciduous scrub, woodland clearings, often along forest roads. - Widespread in S & C Europe, extending to NW Turkey, naturalized elsewhere.

Like *V. phlomoides* but cauline leaves distinctly decurrent; lower bracts leaf-like, long-decurrent, caudate, much exceeding fascicles.

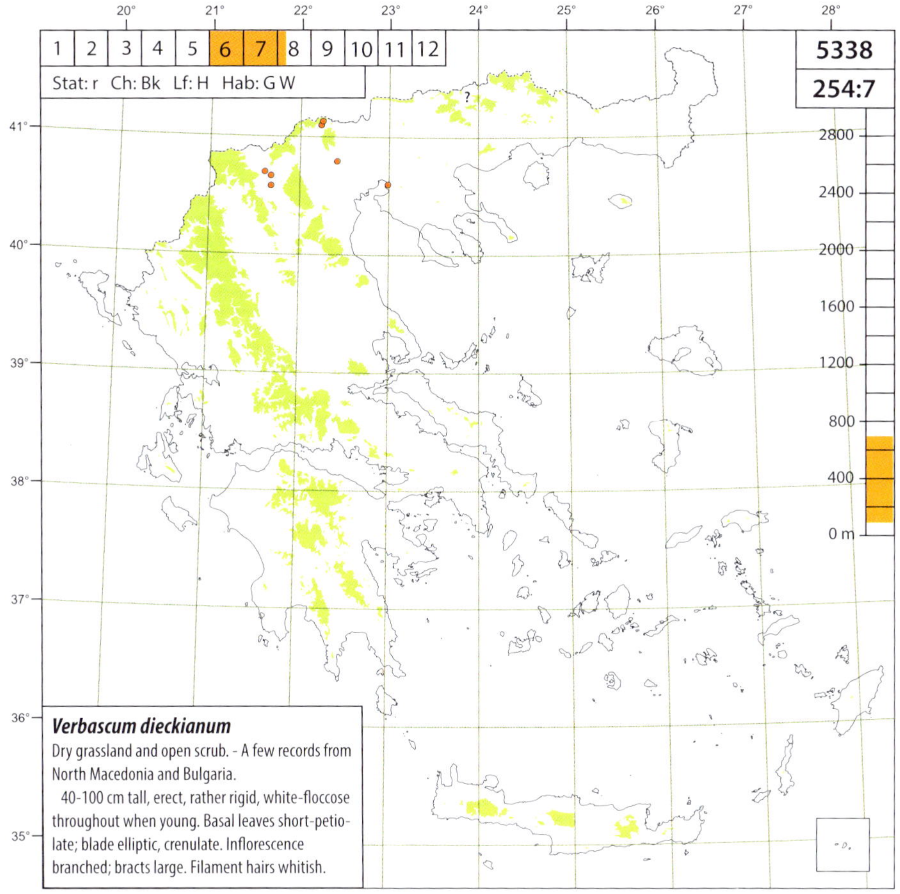

Verbascum dieckianum

Dry grassland and open scrub. - A few records from North Macedonia and Bulgaria.

40-100 cm tall, erect, rather rigid, white-floccose throughout when young. Basal leaves short-petiolate; blade elliptic, crenulate. Inflorescence branched; bracts large. Filament hairs whitish.

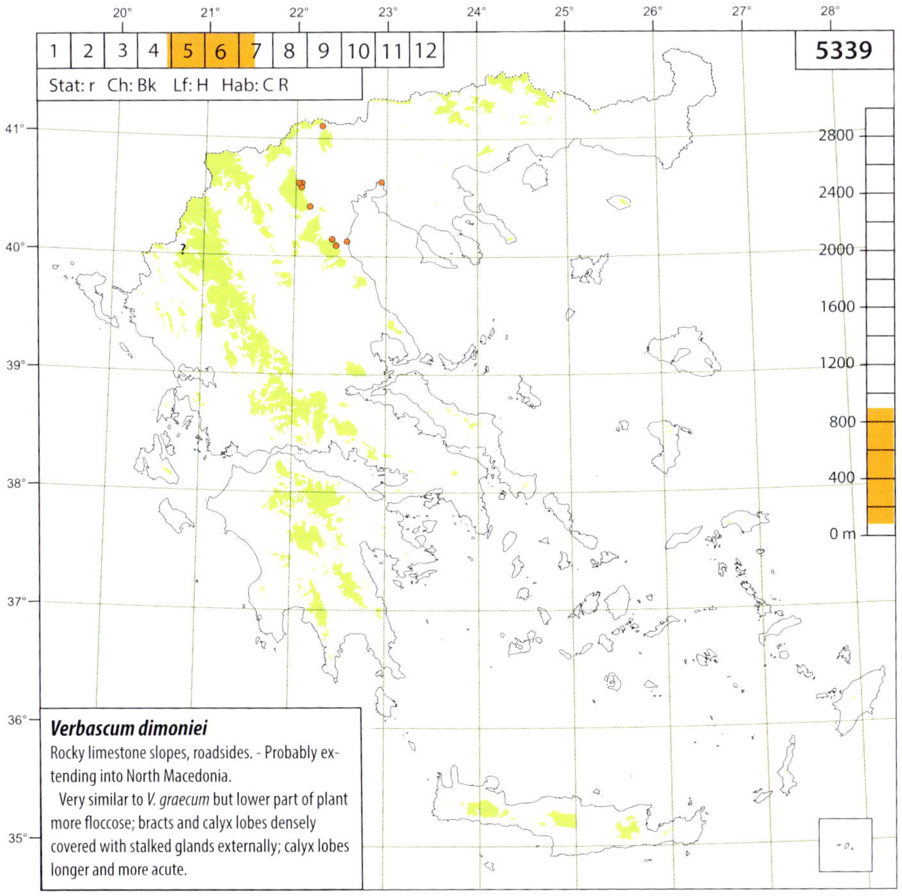

Verbascum dimoniei

Rocky limestone slopes, roadsides. - Probably extending into North Macedonia.

Very similar to *V. graecum* but lower part of plant more floccose; bracts and calyx lobes densely covered with stalked glands externally; calyx lobes longer and more acute.

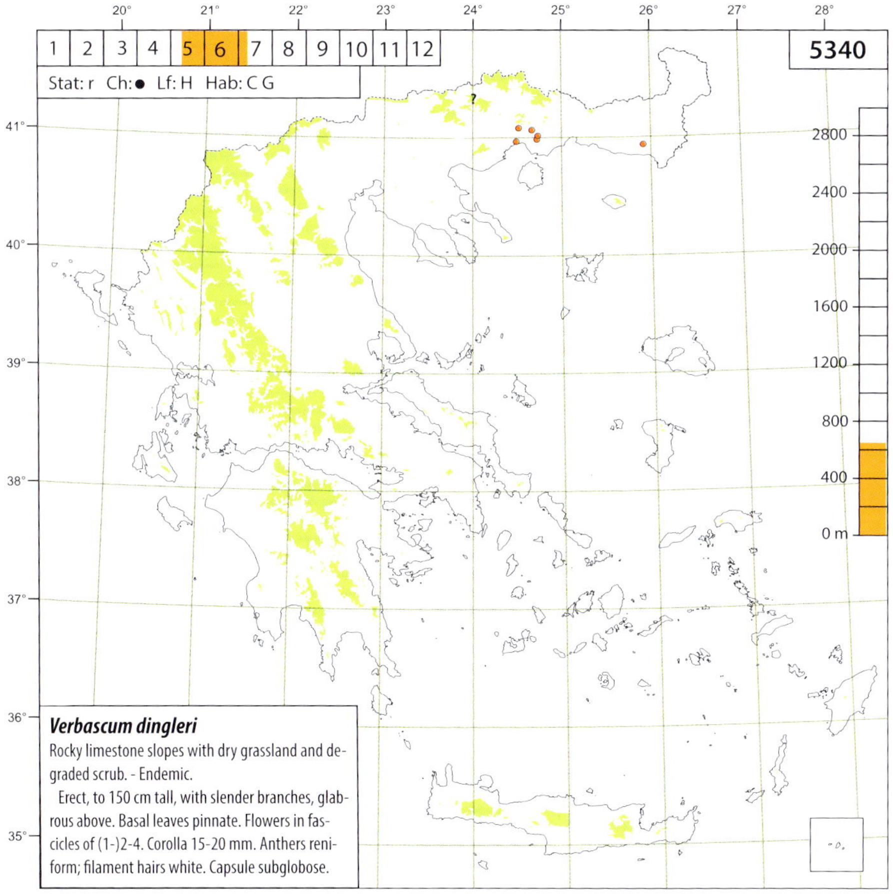

Verbascum dingleri

Rocky limestone slopes with dry grassland and degraded scrub. - Endemic.

Erect, to 150 cm tall, with slender branches, glabrous above. Basal leaves pinnate. Flowers in fascicles of (1-)2-4. Corolla 15-20 mm. Anthers reniform; filament hairs white. Capsule subglobose.

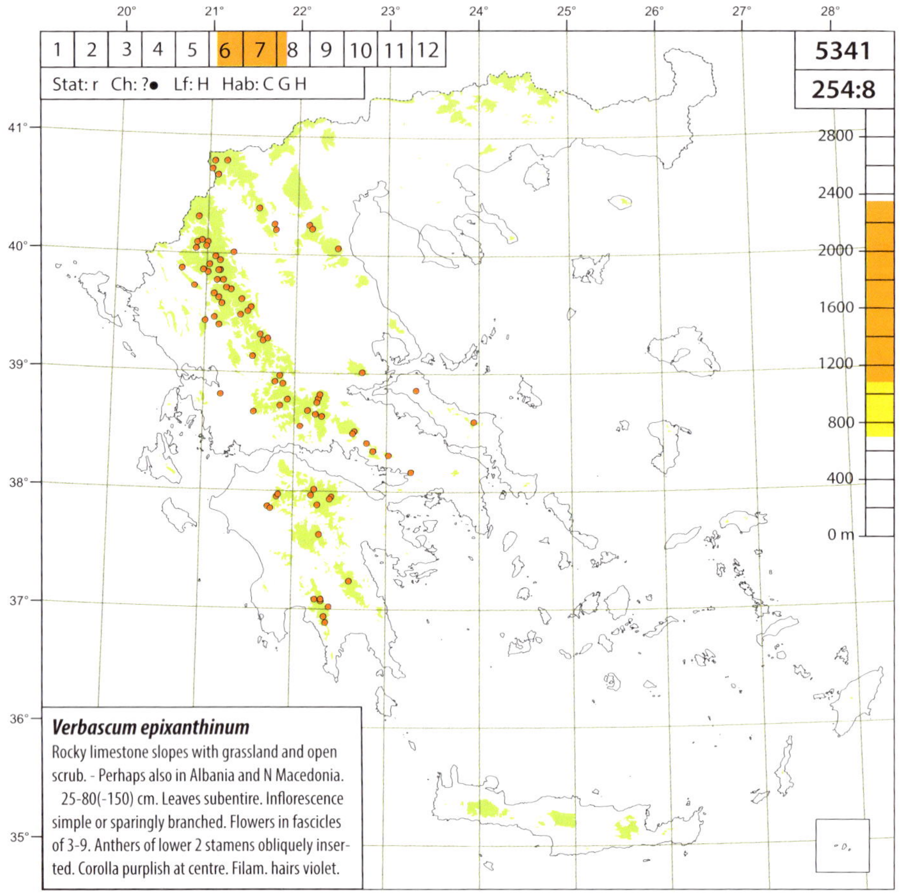

Verbascum epixanthinum

Rocky limestone slopes with grassland and open scrub. - Perhaps also in Albania and N Macedonia.

25-80(-150) cm. Leaves subentire. Inflorescence simple or sparingly branched. Flowers in fascicles of 3-9. Anthers of lower 2 stamens obliquely inserted. Corolla purplish at centre. Filam. hairs violet.

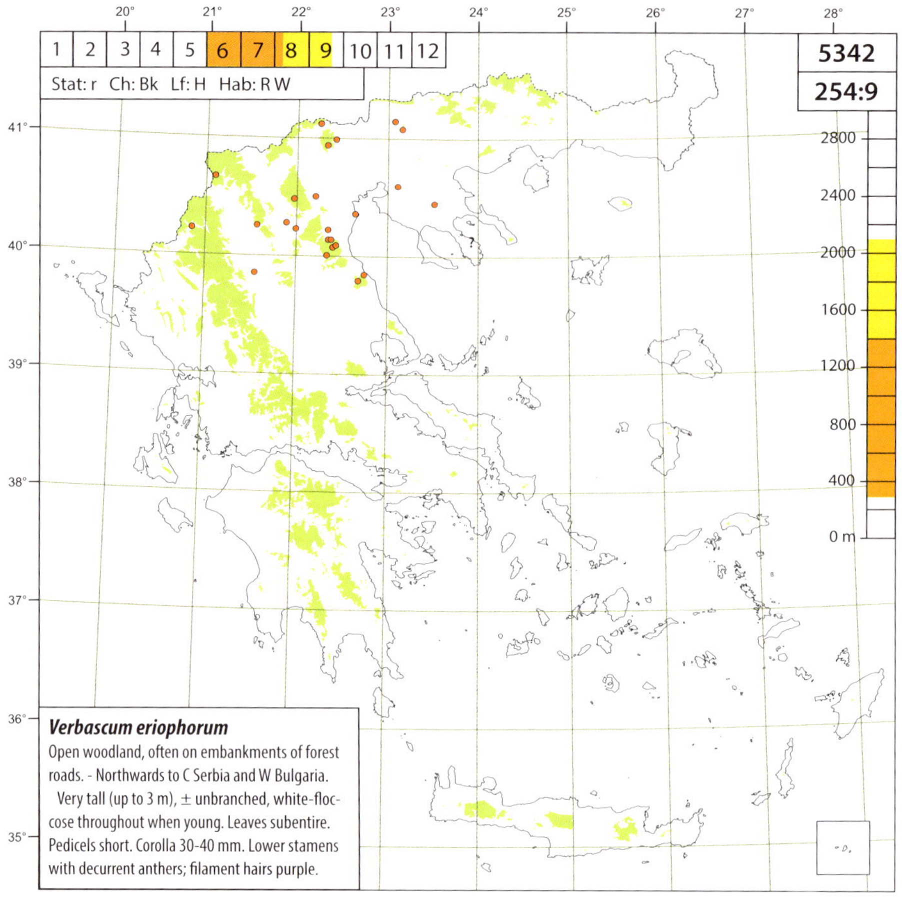

Verbascum eriophorum

Open woodland, often on embankments of forest roads. - Northwards to C Serbia and W Bulgaria.

Very tall (up to 3 m), ± unbranched, white-floccose throughout when young. Leaves subentire. Pedicels short. Corolla 30-40 mm. Lower stamens with decurrent anthers; filament hairs purple.

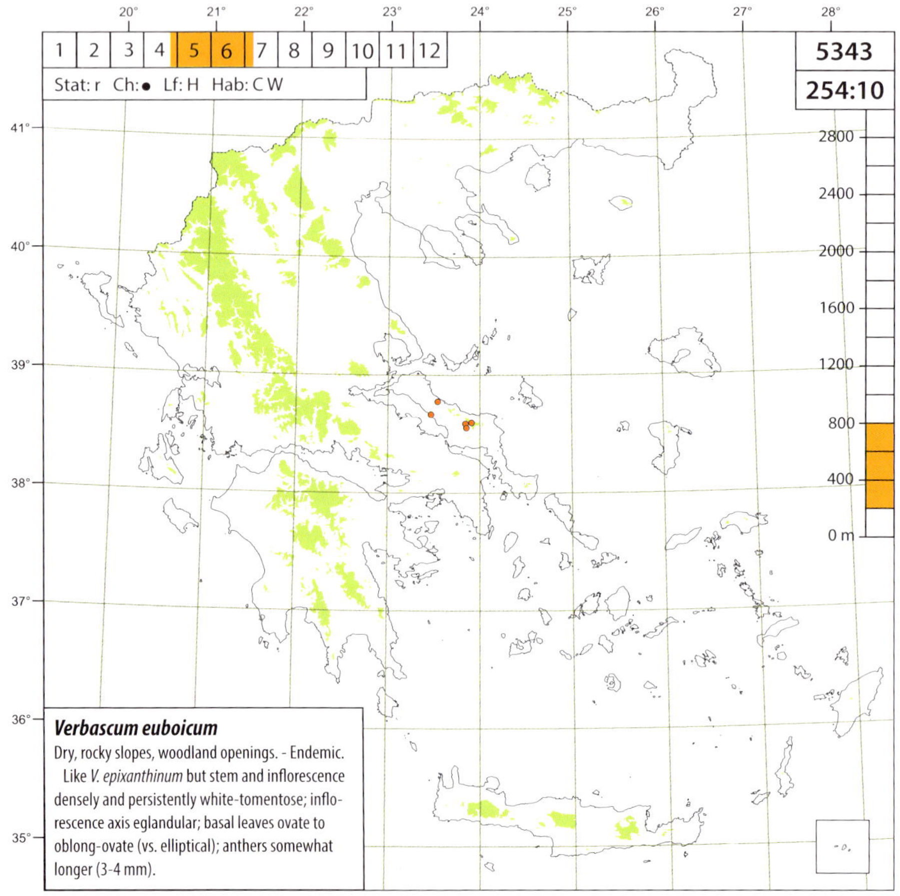

Verbascum euboicum

Dry, rocky slopes, woodland openings. - Endemic.

Like *V. epixanthinum* but stem and inflorescence densely and persistently white-tomentose; inflorescence axis eglandular; basal leaves ovate to oblong-ovate (vs. elliptical); anthers somewhat longer (3-4 mm).

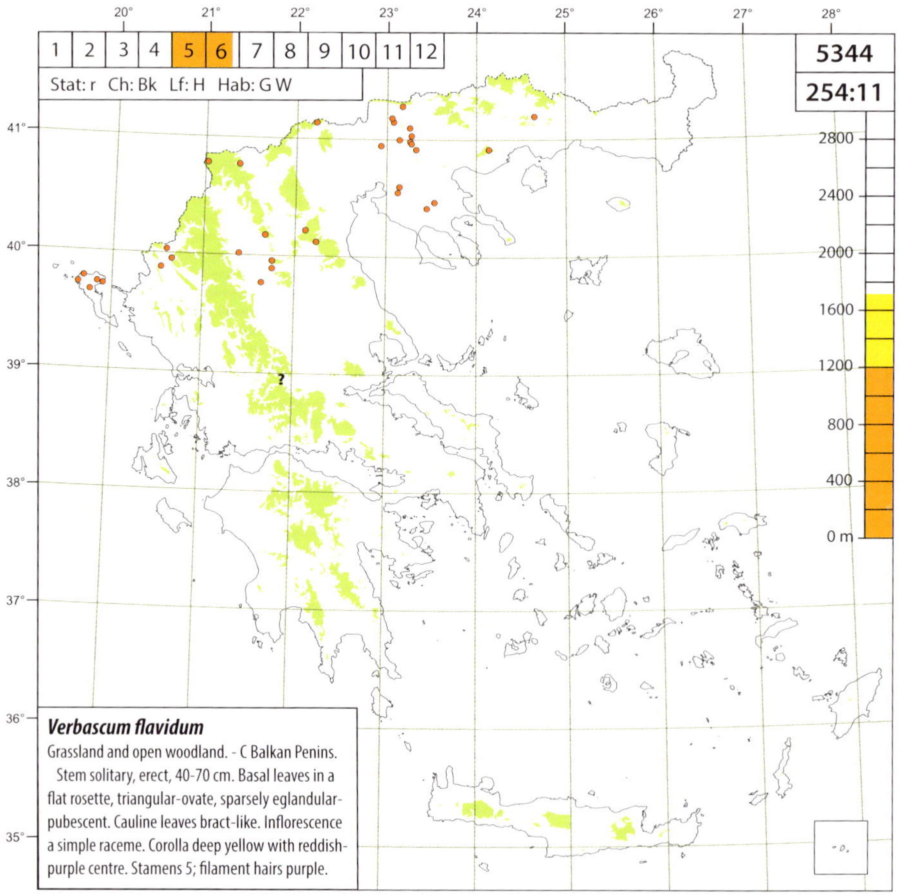

Verbascum flavidum

Grassland and open woodland. - C Balkan Penins.

Stem solitary, erect, 40-70 cm. Basal leaves in a flat rosette, triangular-ovate, sparsely eglandular-pubescent. Cauline leaves bract-like. Inflorescence a simple raceme. Corolla deep yellow with reddish-purple centre. Stamens 5; filament hairs purple.

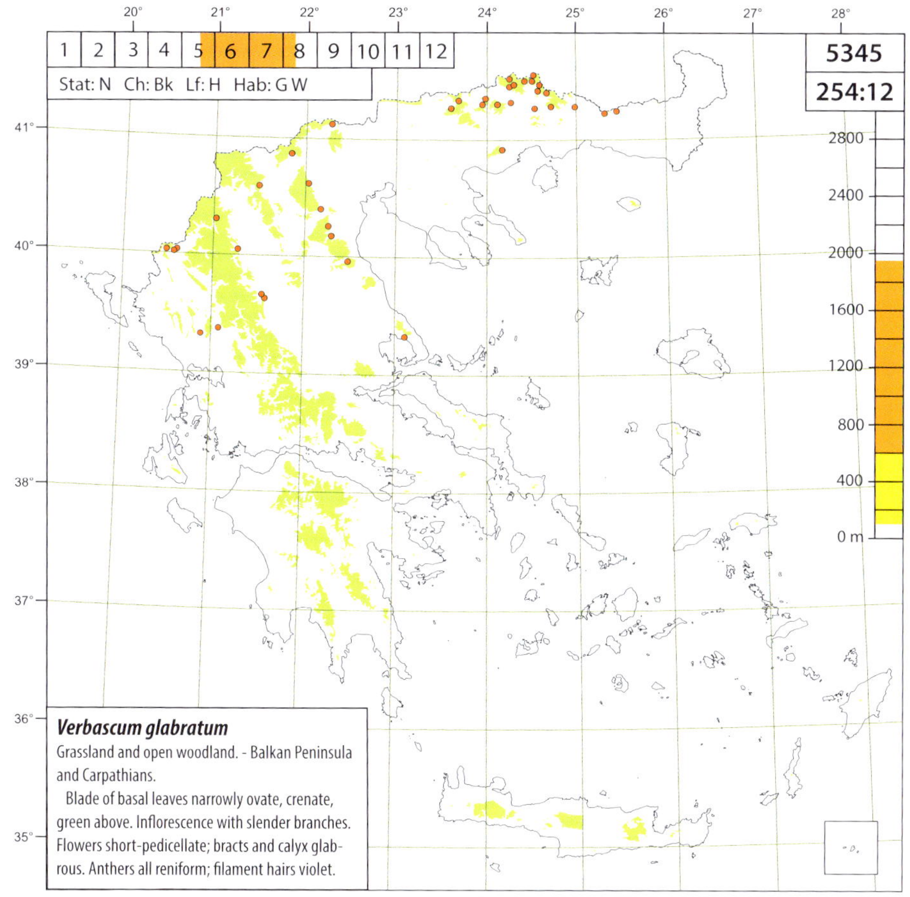

Verbascum glabratum
Grassland and open woodland. - Balkan Peninsula and Carpathians.
Blade of basal leaves narrowly ovate, crenate, green above. Inflorescence with slender branches. Flowers short-pedicellate; bracts and calyx glabrous. Anthers all reniform; filament hairs violet.

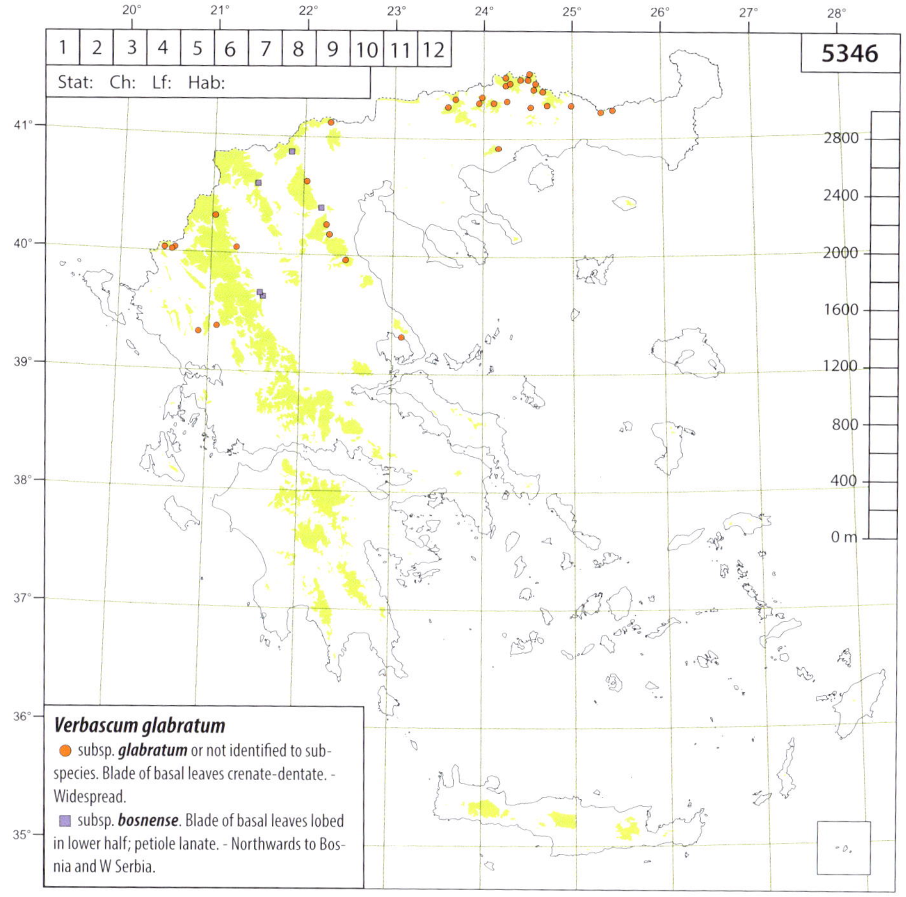

Verbascum glabratum
● subsp. ***glabratum*** or not identified to subspecies. Blade of basal leaves crenate-dentate. - Widespread.
■ subsp. ***bosnense***. Blade of basal leaves lobed in lower half; petiole lanate. - Northwards to Bosnia and W Serbia.

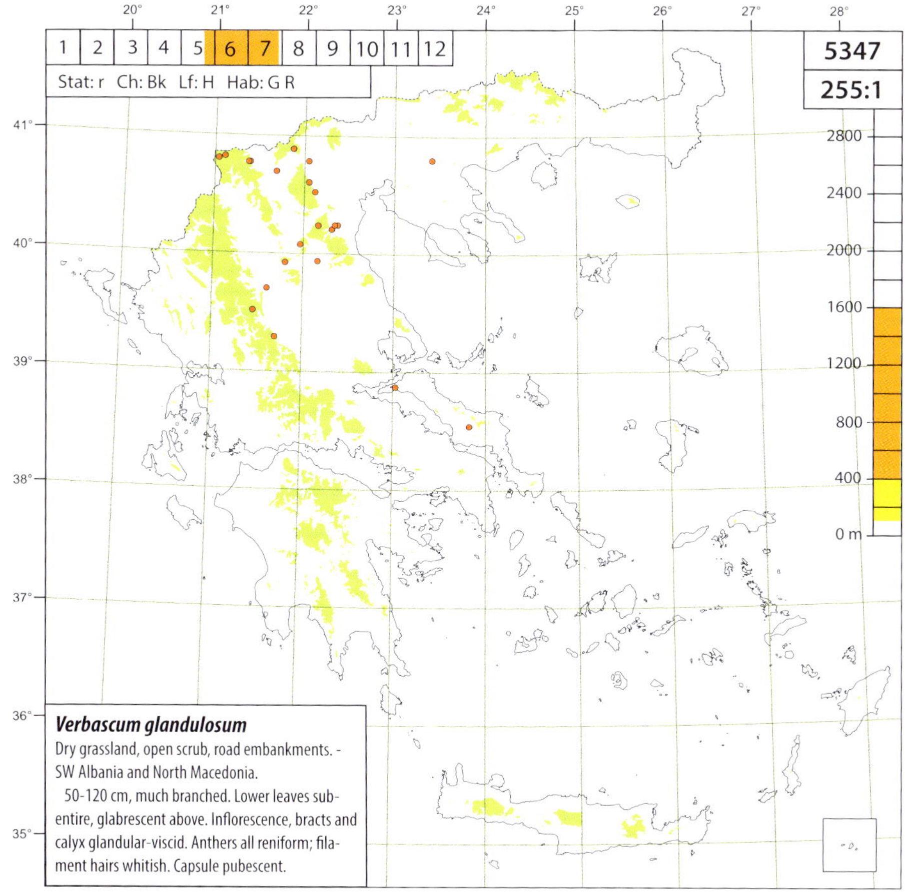

Verbascum glandulosum
Dry grassland, open scrub, road embankments. - SW Albania and North Macedonia.
50-120 cm, much branched. Lower leaves subentire, glabrescent above. Inflorescence, bracts and calyx glandular-viscid. Anthers all reniform; filament hairs whitish. Capsule pubescent.

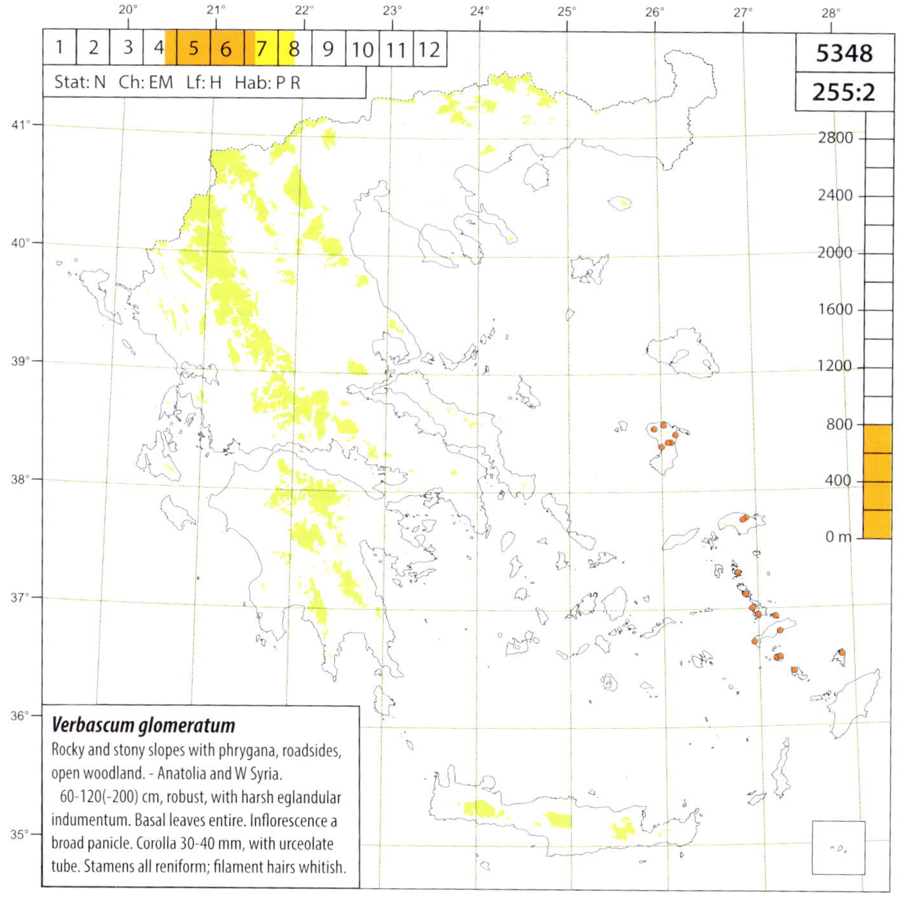

Verbascum glomeratum
Rocky and stony slopes with phrygana, roadsides, open woodland. - Anatolia and W Syria.
60-120(-200) cm, robust, with harsh eglandular indumentum. Basal leaves entire. Inflorescence a broad panicle. Corolla 30-40 mm, with urceolate tube. Stamens all reniform; filament hairs whitish.

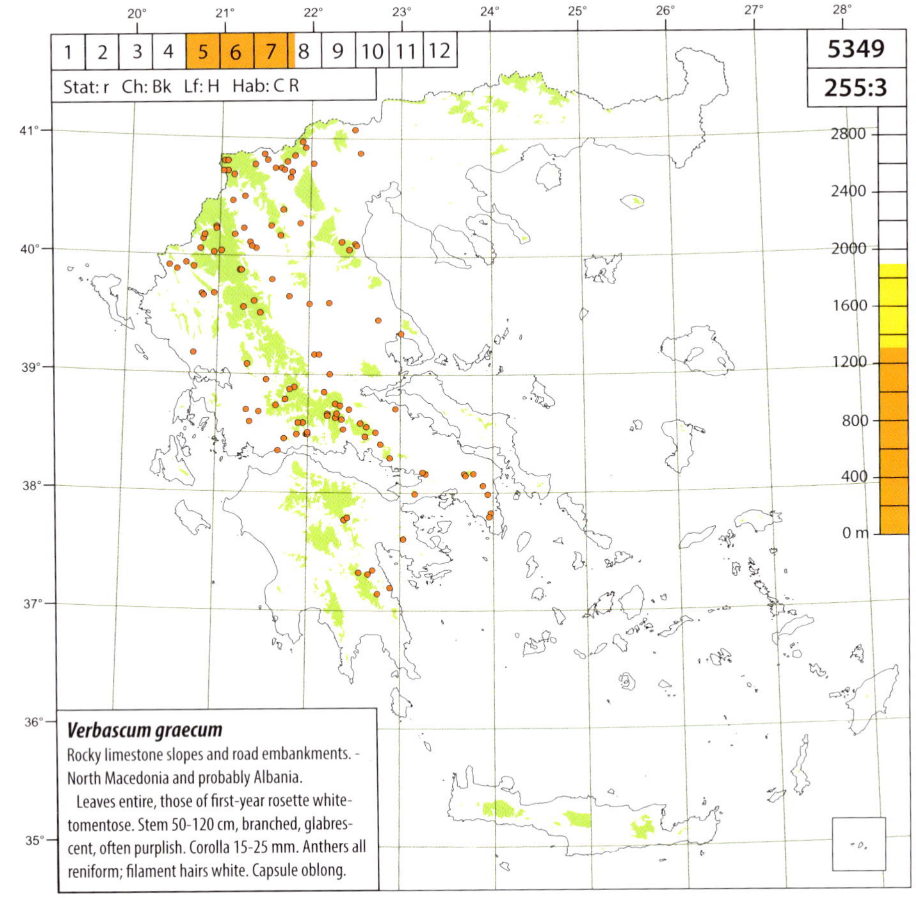

Verbascum graecum

Rocky limestone slopes and road embankments. - North Macedonia and probably Albania.

Leaves entire, those of first-year rosette white-tomentose. Stem 50-120 cm, branched, glabrescent, often purplish. Corolla 15-25 mm. Anthers all reniform; filament hairs white. Capsule oblong.

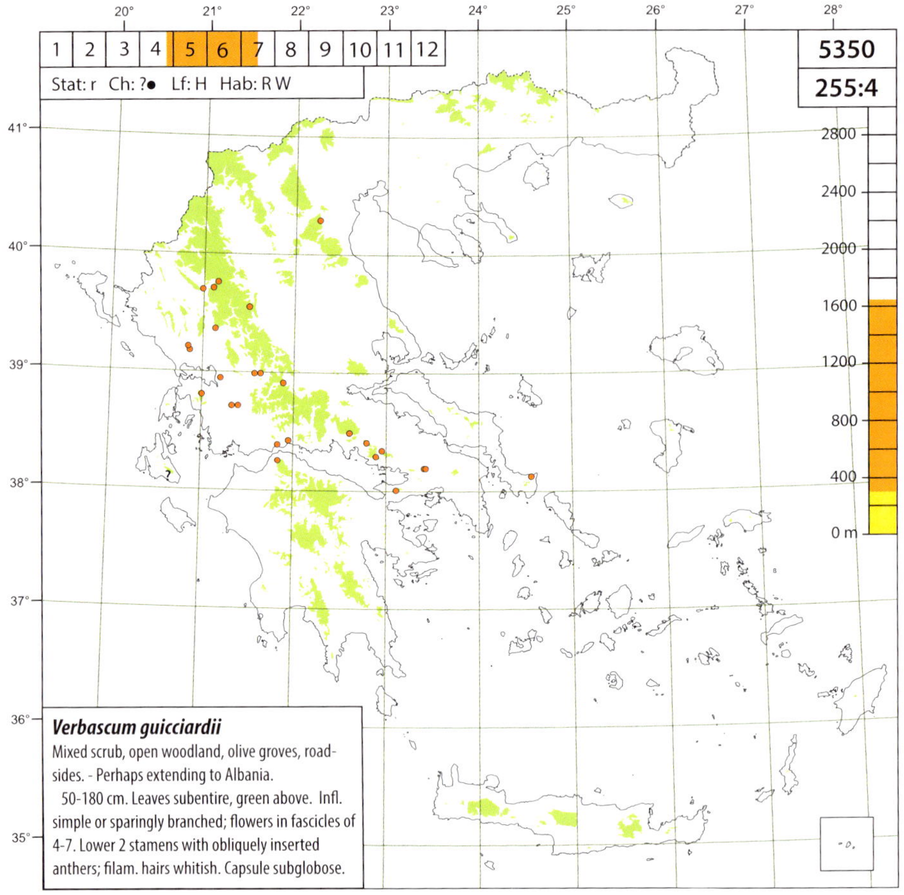

Verbascum guicciardii

Mixed scrub, open woodland, olive groves, roadsides. - Perhaps extending to Albania.

50-180 cm. Leaves subentire, green above. Infl. simple or sparingly branched; flowers in fascicles of 4-7. Lower 2 stamens with obliquely inserted anthers; filam. hairs whitish. Capsule subglobose.

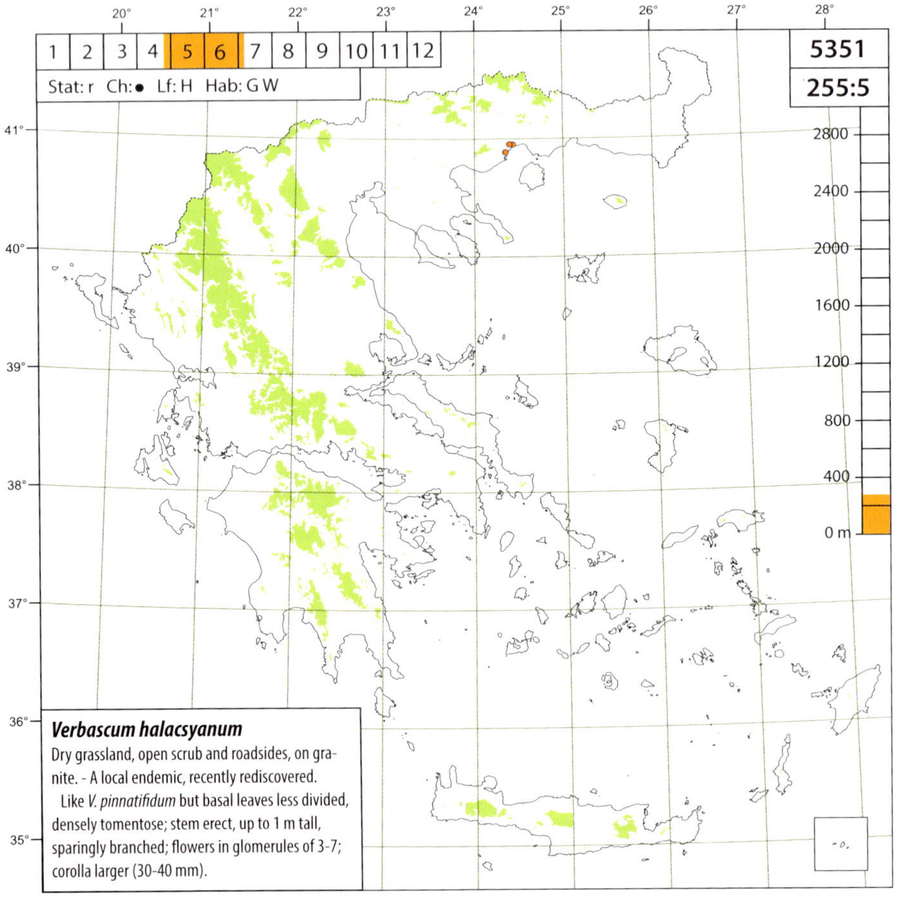

Verbascum halacsyanum

Dry grassland, open scrub and roadsides, on granite. - A local endemic, recently rediscovered.

Like *V. pinnatifidum* but basal leaves less divided, densely tomentose; stem erect, up to 1 m tall, sparingly branched; flowers in glomerules of 3-7; corolla larger (30-40 mm).

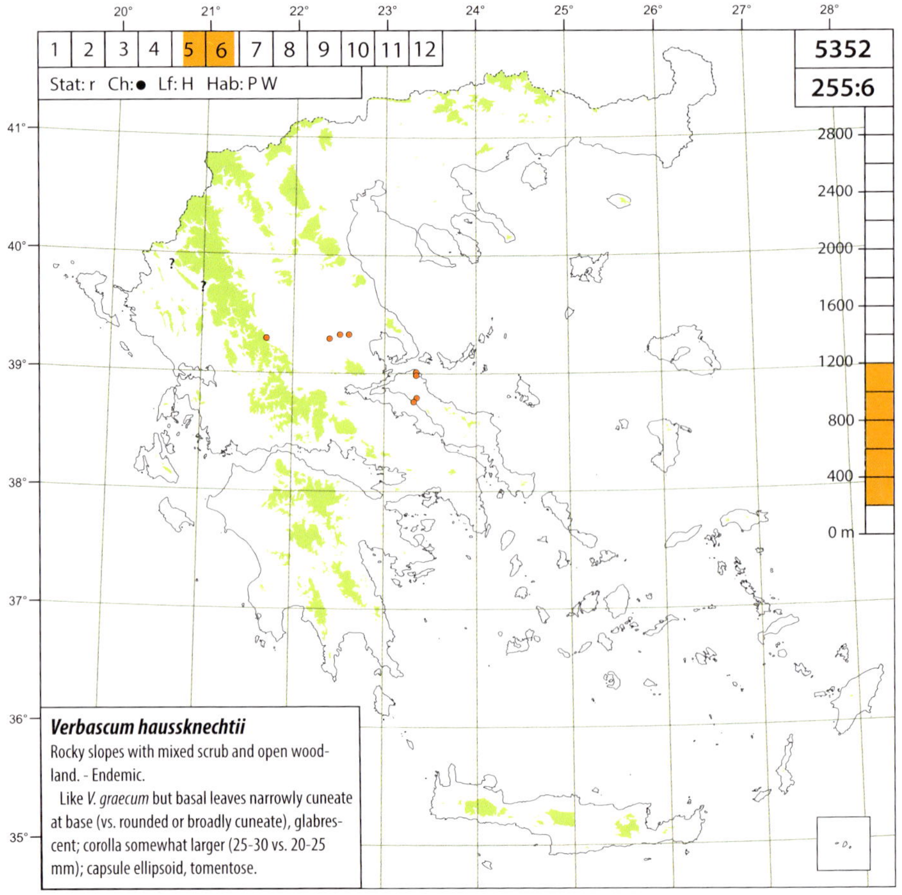

Verbascum haussknechtii

Rocky slopes with mixed scrub and open woodland. - Endemic.

Like *V. graecum* but basal leaves narrowly cuneate at base (vs. rounded or broadly cuneate), glabrescent; corolla somewhat larger (25-30 vs. 20-25 mm); capsule ellipsoid, tomentose.

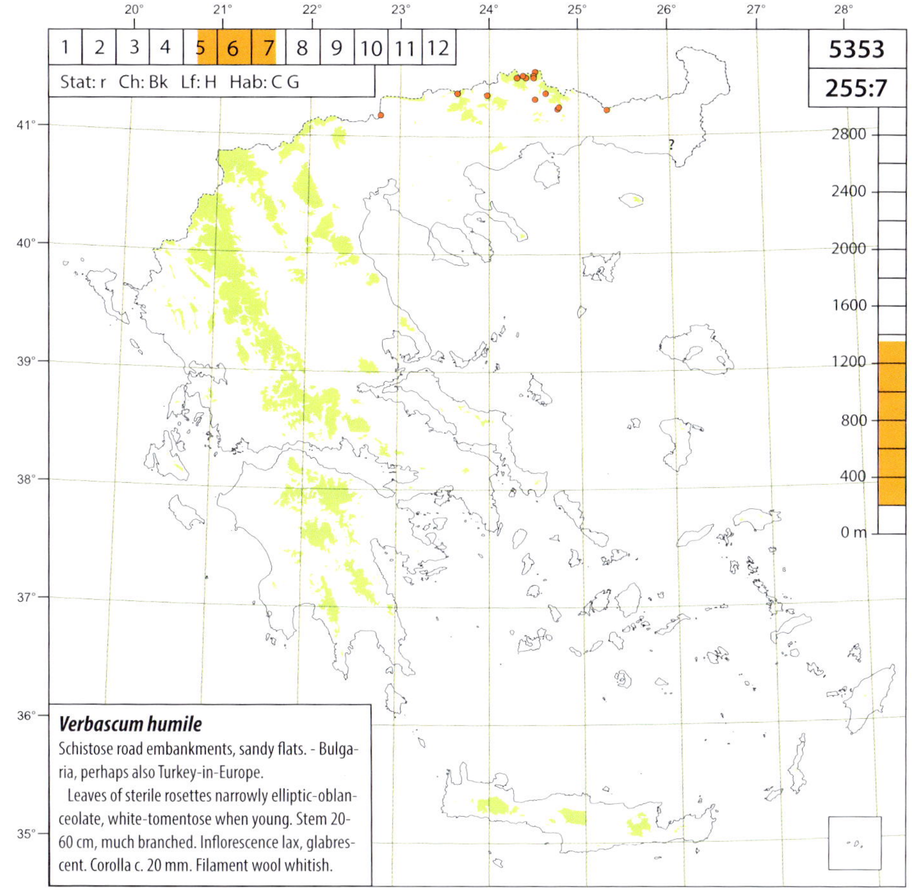

Verbascum humile

Schistose road embankments, sandy flats. - Bulgaria, perhaps also Turkey-in-Europe.

Leaves of sterile rosettes narrowly elliptic-oblanceolate, white-tomentose when young. Stem 20-60 cm, much branched. Inflorescence lax, glabrescent. Corolla c. 20 mm. Filament wool whitish.

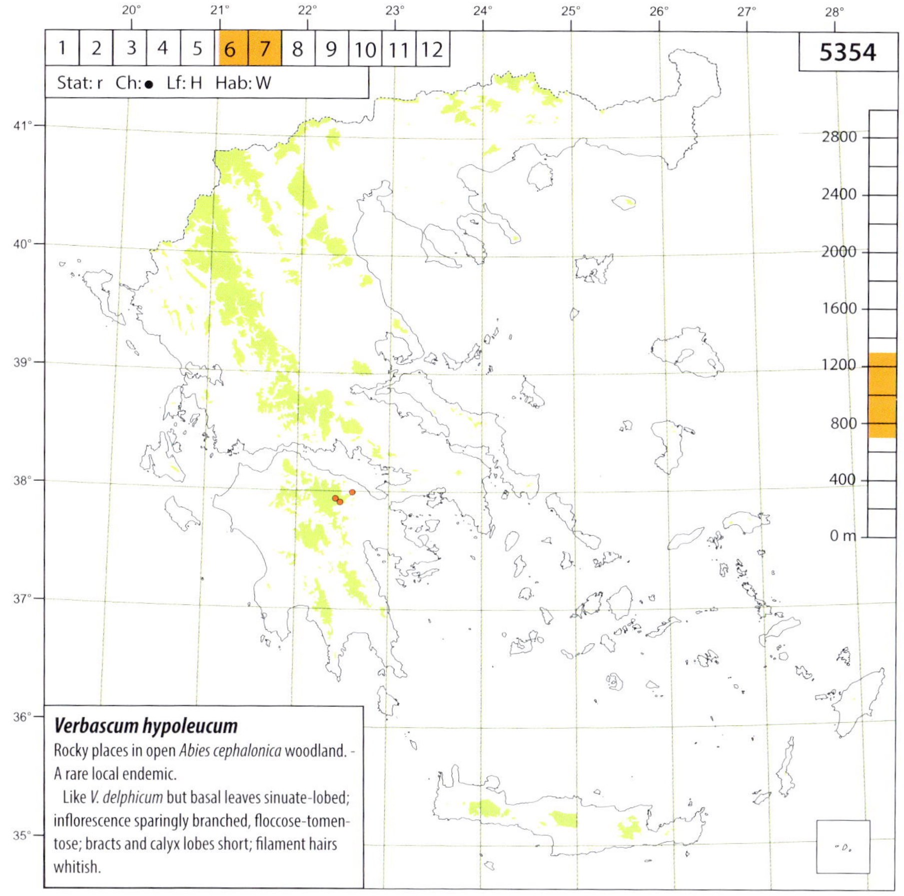

Verbascum hypoleucum

Rocky places in open *Abies cephalonica* woodland. - A rare local endemic.

Like *V. delphicum* but basal leaves sinuate-lobed; inflorescence sparingly branched, floccose-tomentose; bracts and calyx lobes short; filament hairs whitish.

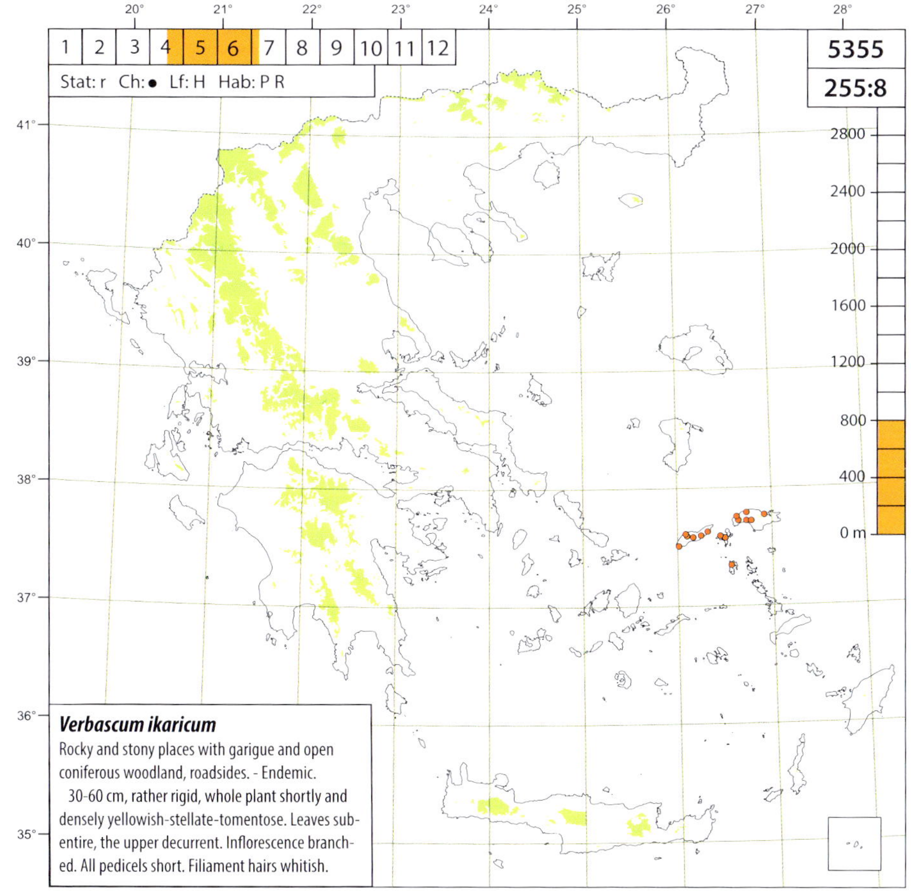

Verbascum ikaricum

Rocky and stony places with garigue and open coniferous woodland, roadsides. - Endemic.

30-60 cm, rather rigid, whole plant shortly and densely yellowish-stellate-tomentose. Leaves subentire, the upper decurrent. Inflorescence branched. All pedicels short. Filiament hairs whitish.

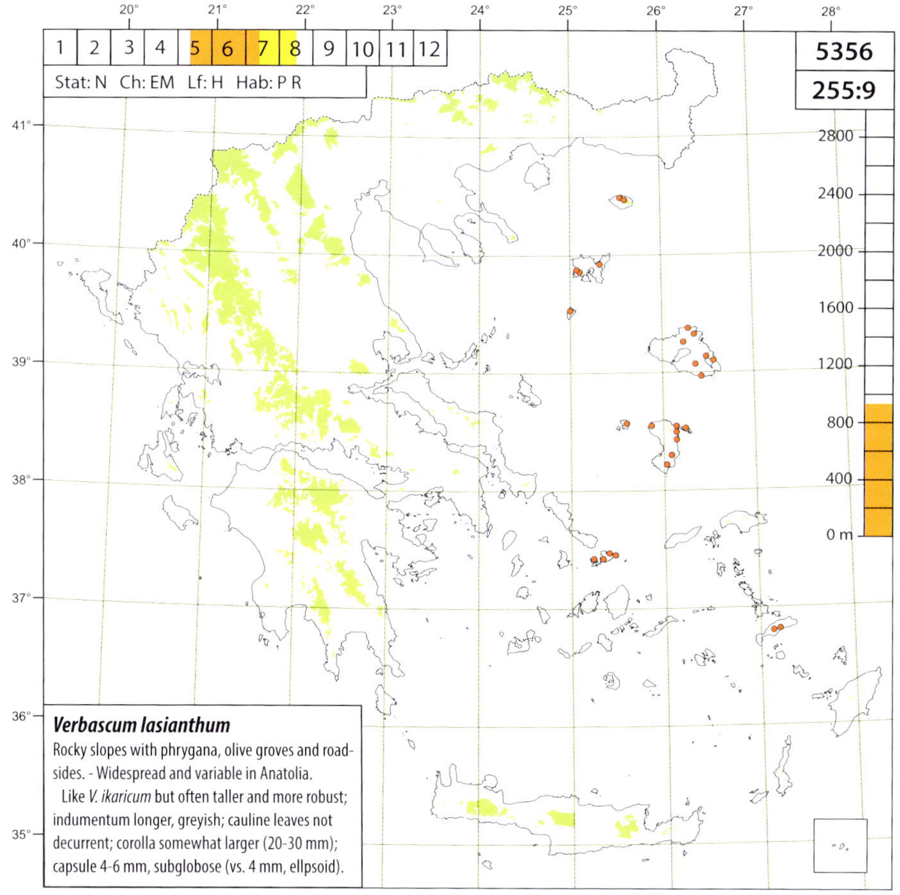

Verbascum lasianthum

Rocky slopes with phrygana, olive groves and roadsides. - Widespread and variable in Anatolia.

Like *V. ikaricum* but often taller and more robust; indumentum longer, greyish; cauline leaves not decurrent; corolla somewhat larger (20-30 mm); capsule 4-6 mm, subglobose (vs. 4 mm, ellpsoid).

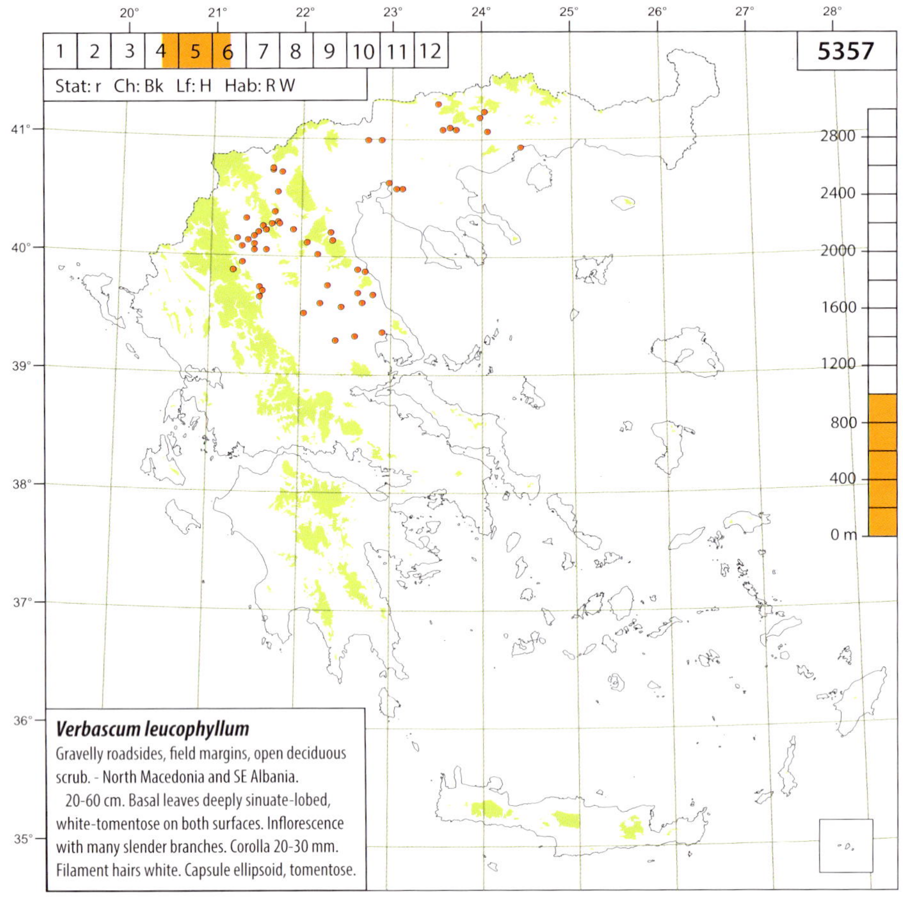

Verbascum leucophyllum

Gravelly roadsides, field margins, open deciduous scrub. - North Macedonia and SE Albania.

20-60 cm. Basal leaves deeply sinuate-lobed, white-tomentose on both surfaces. Inflorescence with many slender branches. Corolla 20-30 mm. Filament hairs white. Capsule ellipsoid, tomentose.

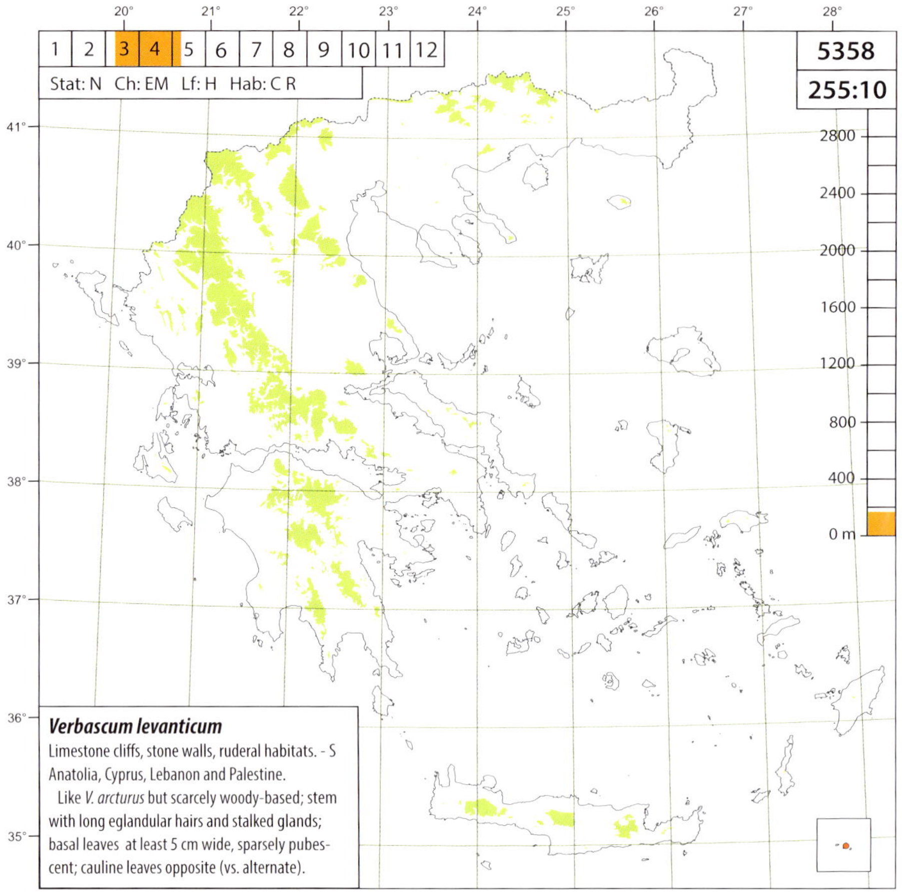

Verbascum levanticum

Limestone cliffs, stone walls, ruderal habitats. - S Anatolia, Cyprus, Lebanon and Palestine.

Like *V. arcturus* but scarcely woody-based; stem with long eglandular hairs and stalked glands; basal leaves at least 5 cm wide, sparsely pubescent; cauline leaves opposite (vs. alternate).

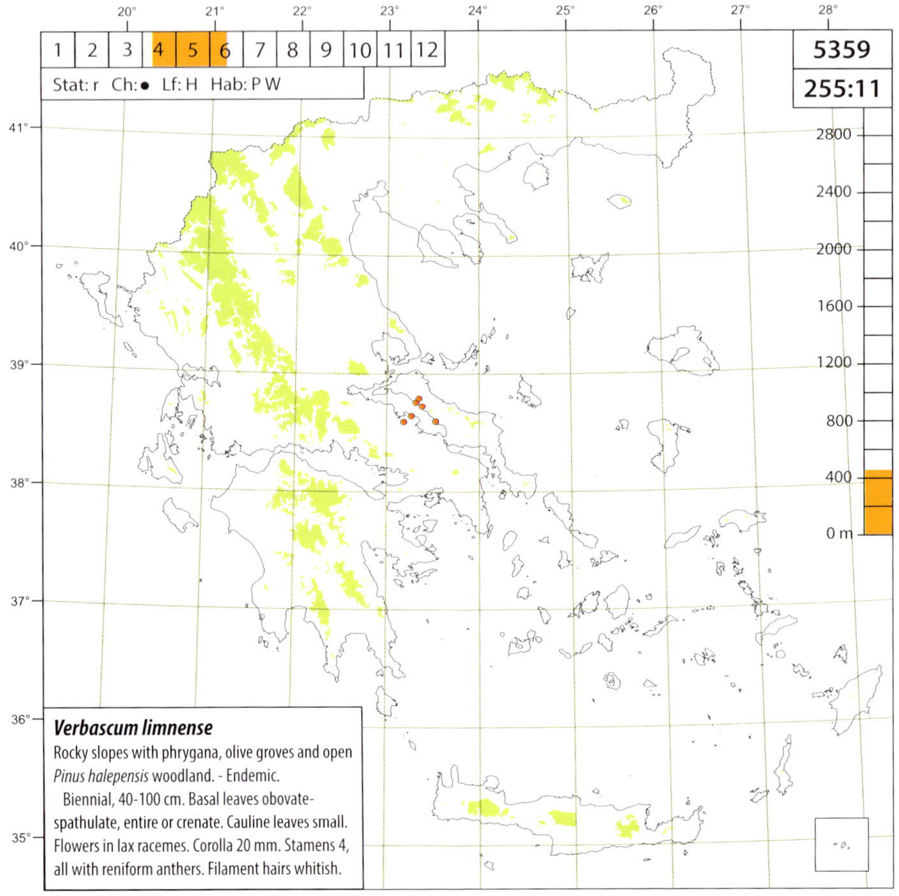

Verbascum limnense

Rocky slopes with phrygana, olive groves and open *Pinus halepensis* woodland. - Endemic.

Biennial, 40-100 cm. Basal leaves obovate-spathulate, entire or crenate. Cauline leaves small. Flowers in lax racemes. Corolla 20 mm. Stamens 4, all with reniform anthers. Filament hairs whitish.

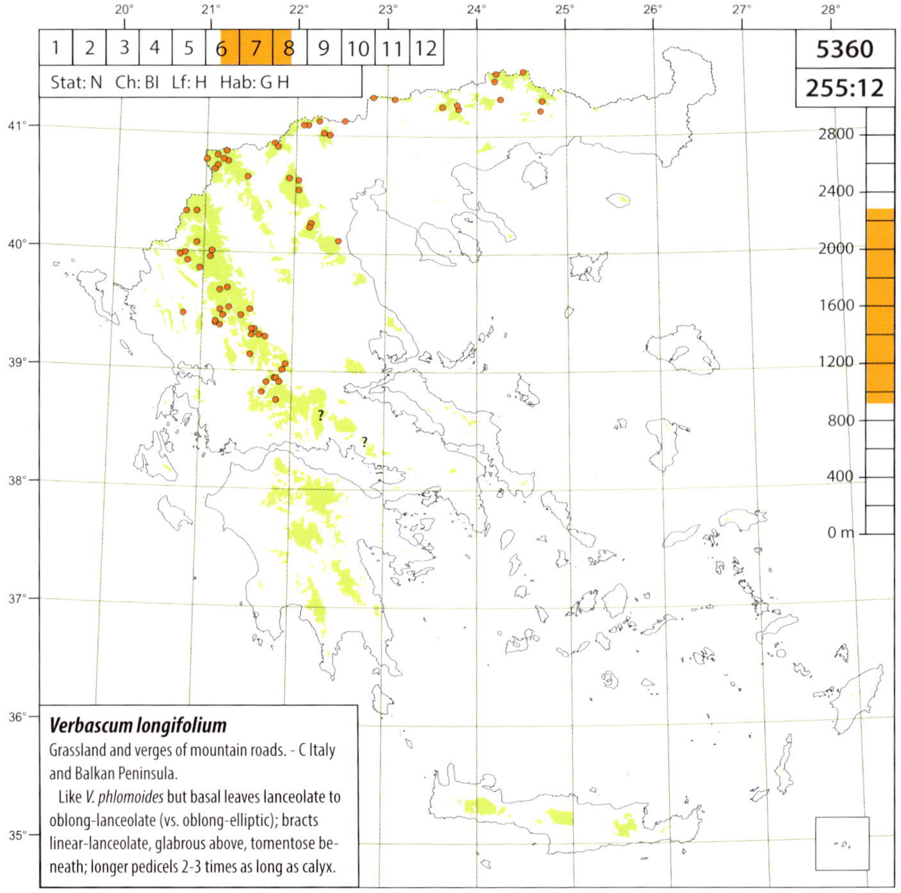

Verbascum longifolium

Grassland and verges of mountain roads. - C Italy and Balkan Peninsula.

Like *V. phlomoides* but basal leaves lanceolate to oblong-lanceolate (vs. oblong-elliptic); bracts linear-lanceolate, glabrous above, tomentose beneath; longer pedicels 2-3 times as long as calyx.

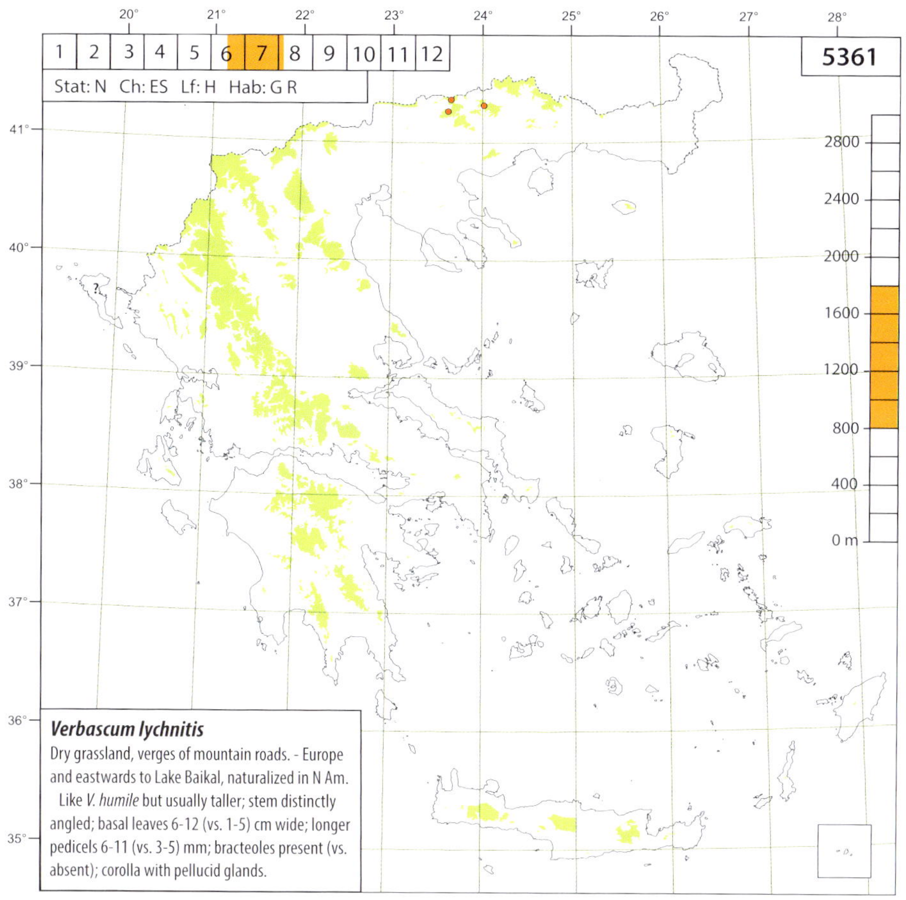

Verbascum lychnitis

Dry grassland, verges of mountain roads. - Europe and eastwards to Lake Baikal, naturalized in N Am.

Like *V. humile* but usually taller; stem distinctly angled; basal leaves 6-12 (vs. 1-5) cm wide; longer pedicels 6-11 (vs. 3-5) mm; bracteoles present (vs. absent); corolla with pellucid glands.

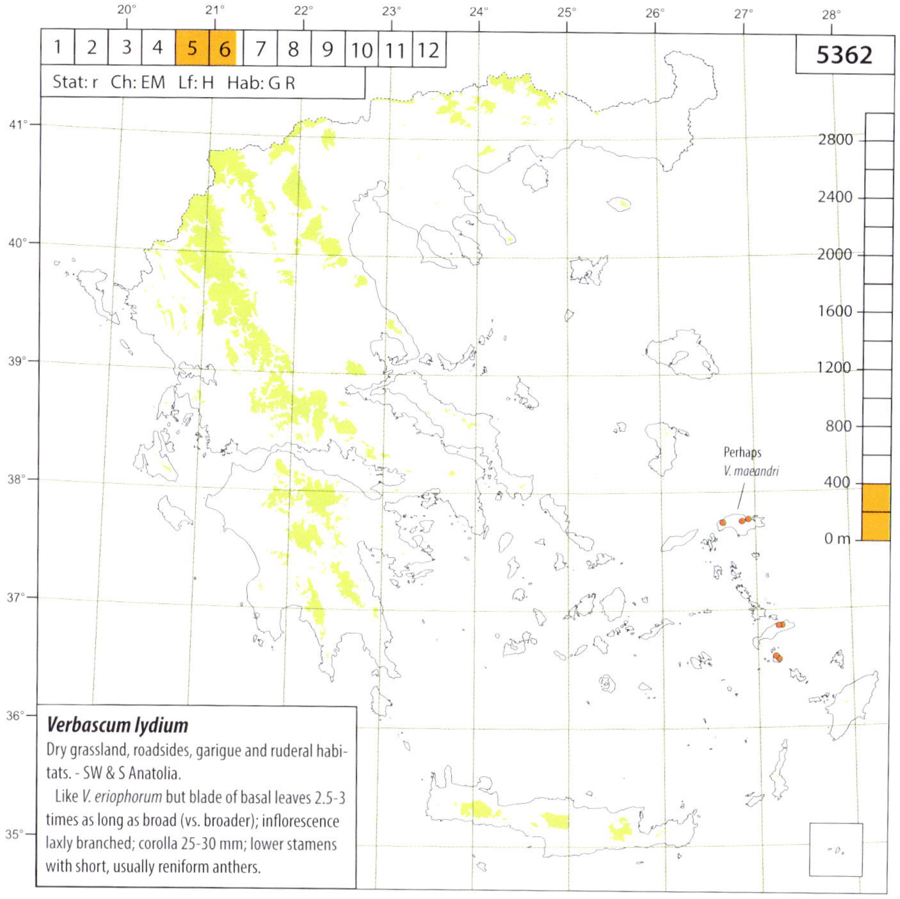

Verbascum lydium

Dry grassland, roadsides, garigue and ruderal habitats. - SW & S Anatolia.

Like *V. eriophorum* but blade of basal leaves 2.5-3 times as long as broad (vs. broader); inflorescence laxly branched; corolla 25-30 mm; lower stamens with short, usually reniform anthers.

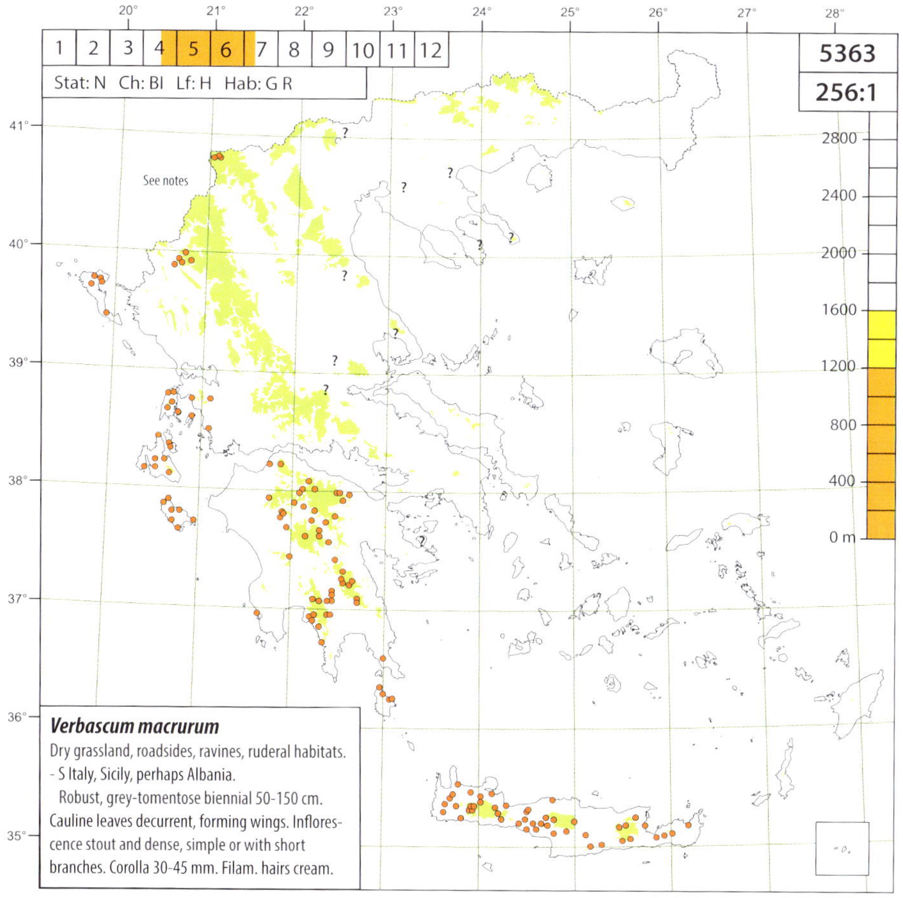

Verbascum macrurum

Dry grassland, roadsides, ravines, ruderal habitats. - S Italy, Sicily, perhaps Albania.

Robust, grey-tomentose biennial 50-150 cm. Cauline leaves decurrent, forming wings. Inflorescence stout and dense, simple or with short branches. Corolla 30-45 mm. Filam. hairs cream.

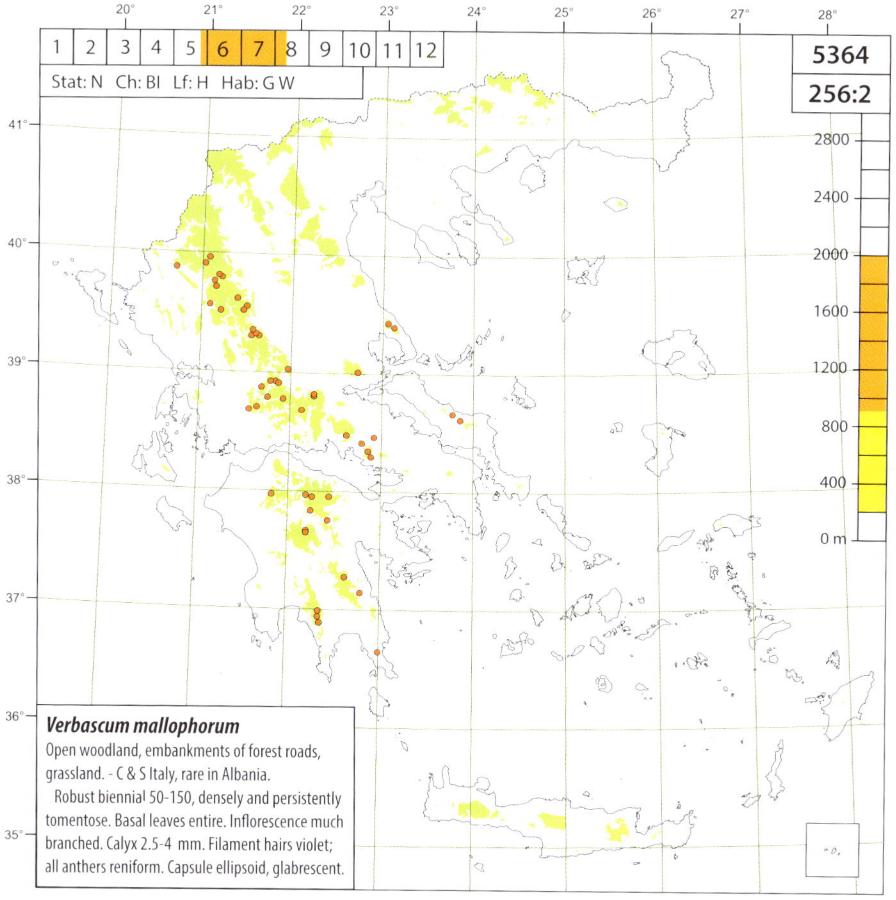

Verbascum mallophorum

Open woodland, embankments of forest roads, grassland. - C & S Italy, rare in Albania.

Robust biennial 50-150, densely and persistently tomentose. Basal leaves entire. Inflorescence much branched. Calyx 2.5-4 mm. Filament hairs violet; all anthers reniform. Capsule ellipsoid, glabrescent.

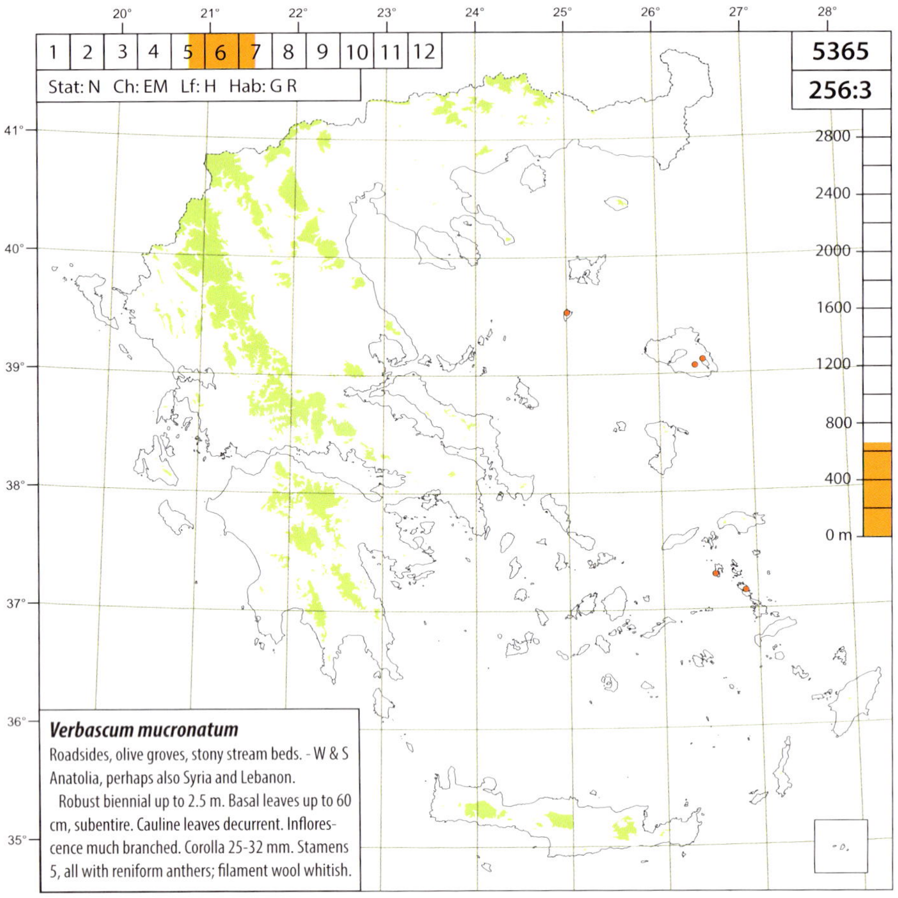

Verbascum mucronatum
Roadsides, olive groves, stony stream beds. - W & S Anatolia, perhaps also Syria and Lebanon.
Robust biennial up to 2.5 m. Basal leaves up to 60 cm, subentire. Cauline leaves decurrent. Inflorescence much branched. Corolla 25-32 mm. Stamens 5, all with reniform anthers; filament wool whitish.

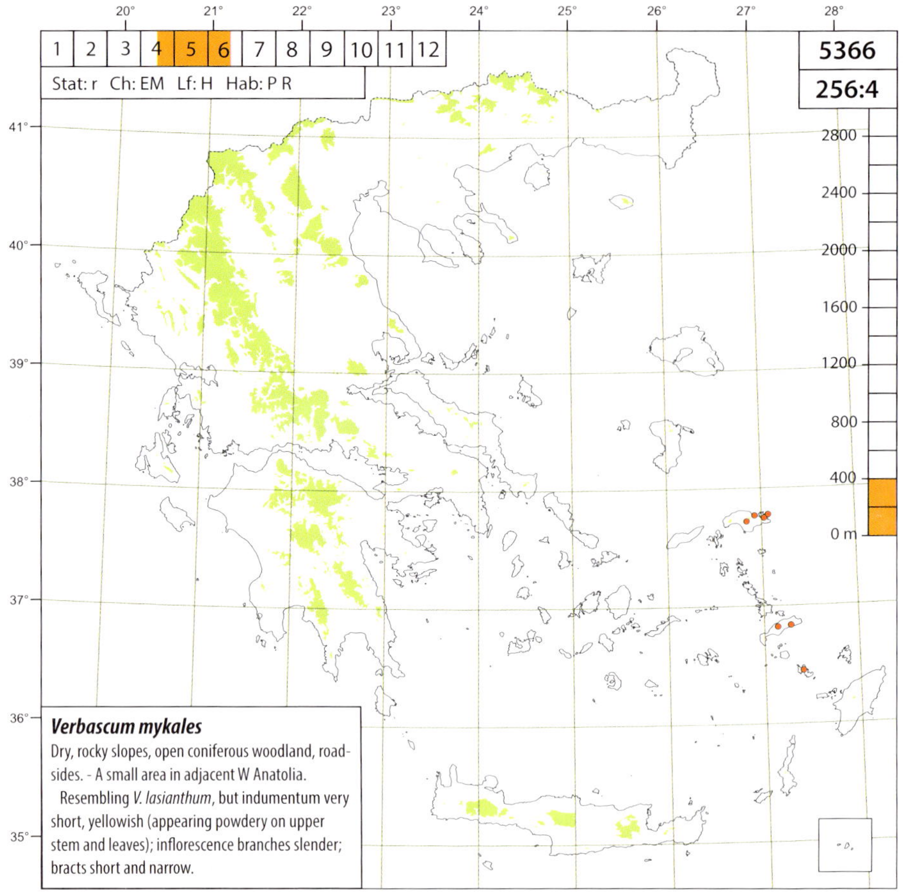

Verbascum mykales
Dry, rocky slopes, open coniferous woodland, roadsides. - A small area in adjacent W Anatolia.
Resembling *V. lasianthum*, but indumentum very short, yellowish (appearing powdery on upper stem and leaves); inflorescence branches slender; bracts short and narrow.

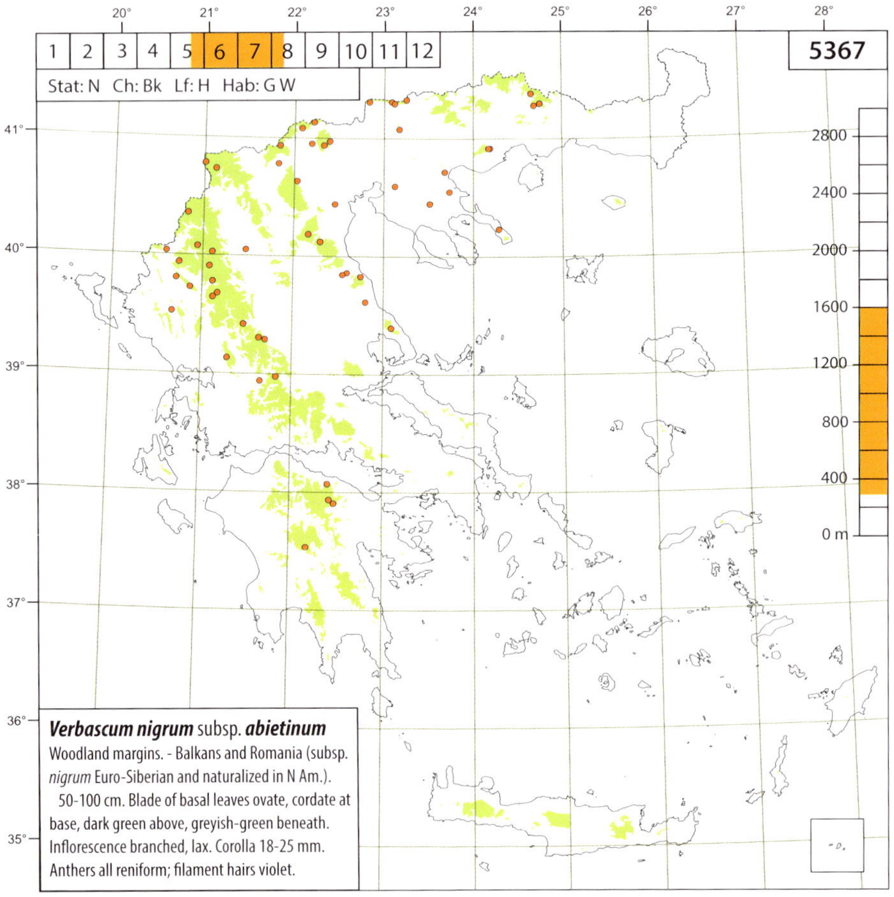

Verbascum nigrum* subsp. *abietinum
Woodland margins. - Balkans and Romania (subsp. *nigrum* Euro-Siberian and naturalized in N Am.).
50-100 cm. Blade of basal leaves ovate, cordate at base, dark green above, greyish-green beneath. Inflorescence branched, lax. Corolla 18-25 mm. Anthers all reniform; filament hairs violet.

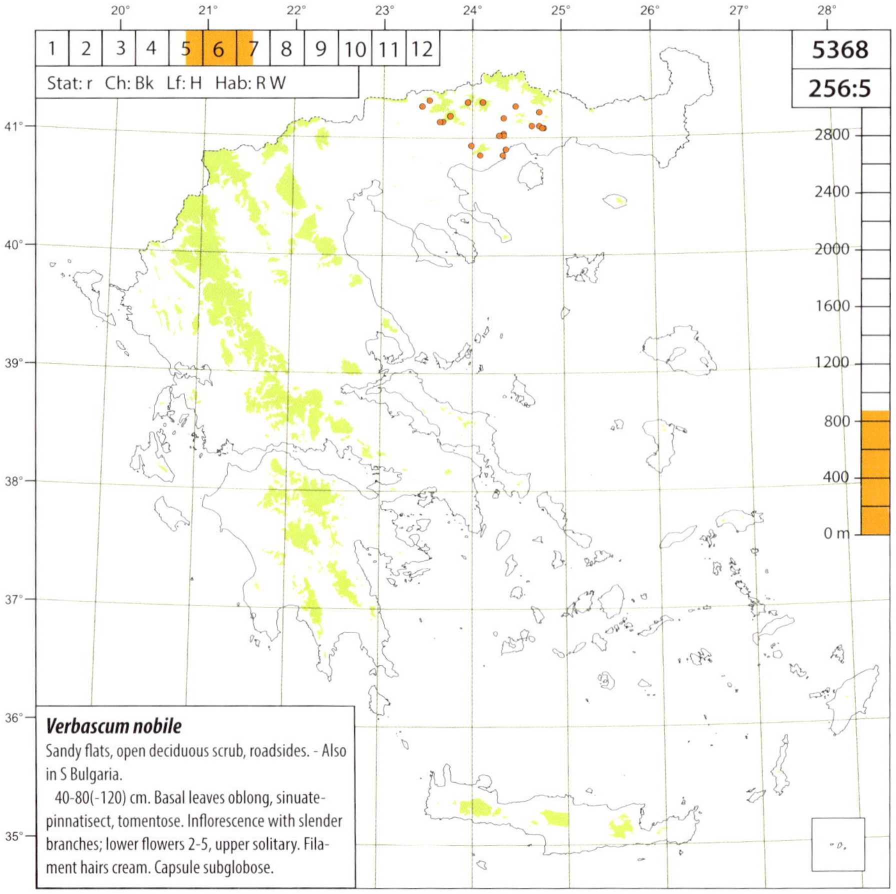

Verbascum nobile
Sandy flats, open deciduous scrub, roadsides. - Also in S Bulgaria.
40-80(-120) cm. Basal leaves oblong, sinuate-pinnatisect, tomentose. Inflorescence with slender branches; lower flowers 2-5, upper solitary. Filament hairs cream. Capsule subglobose.

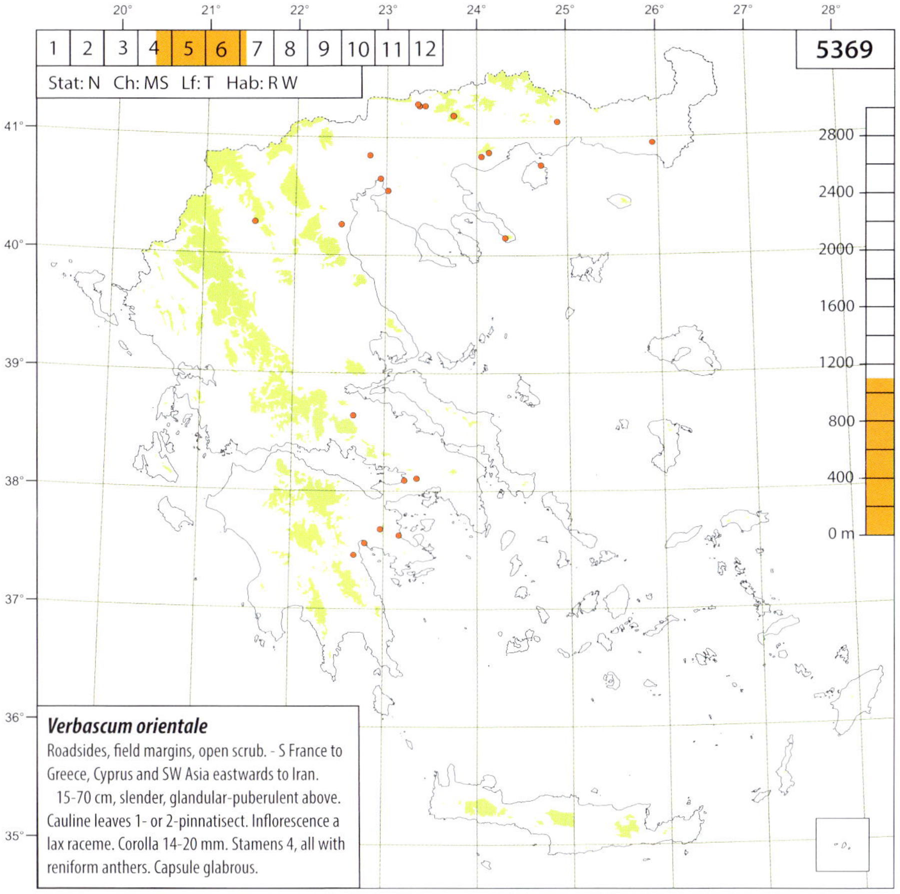

Verbascum orientale
Roadsides, field margins, open scrub. - S France to Greece, Cyprus and SW Asia eastwards to Iran.
15-70 cm, slender, glandular-puberulent above. Cauline leaves 1- or 2-pinnatisect. Inflorescence a lax raceme. Corolla 14-20 mm. Stamens 4, all with reniform anthers. Capsule glabrous.

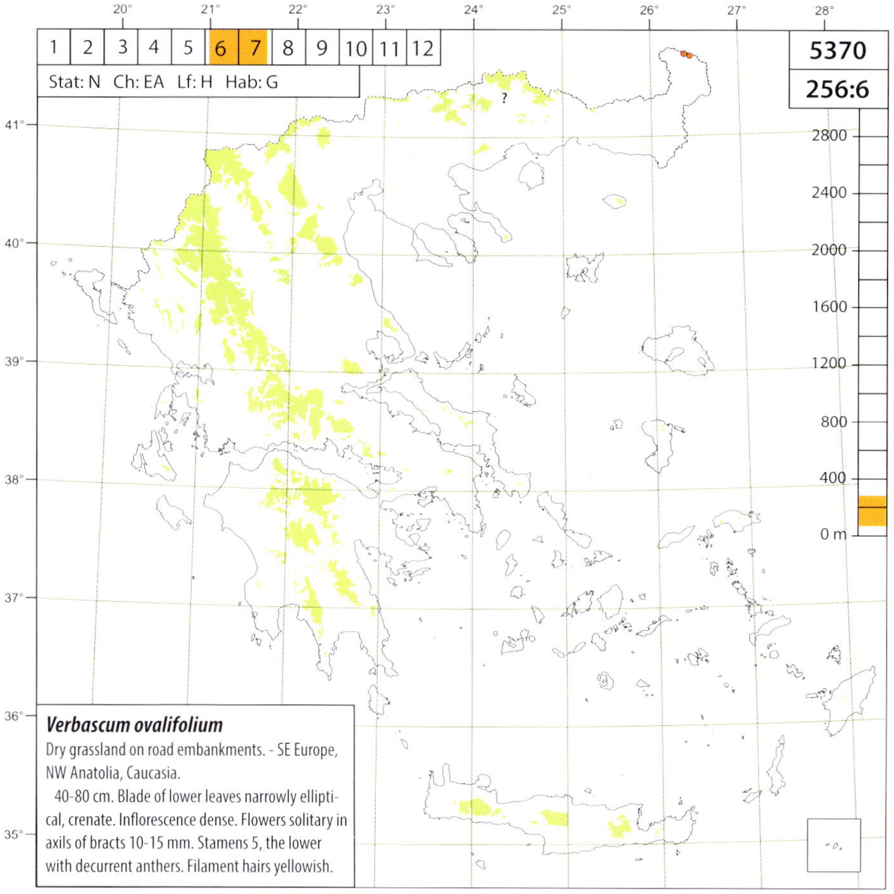

Verbascum ovalifolium
Dry grassland on road embankments. - SE Europe, NW Anatolia, Caucasia.
40-80 cm. Blade of lower leaves narrowly elliptical, crenate. Inflorescence dense. Flowers solitary in axils of bracts 10-15 mm. Stamens 5, the lower with decurrent anthers. Filament hairs yellowish.

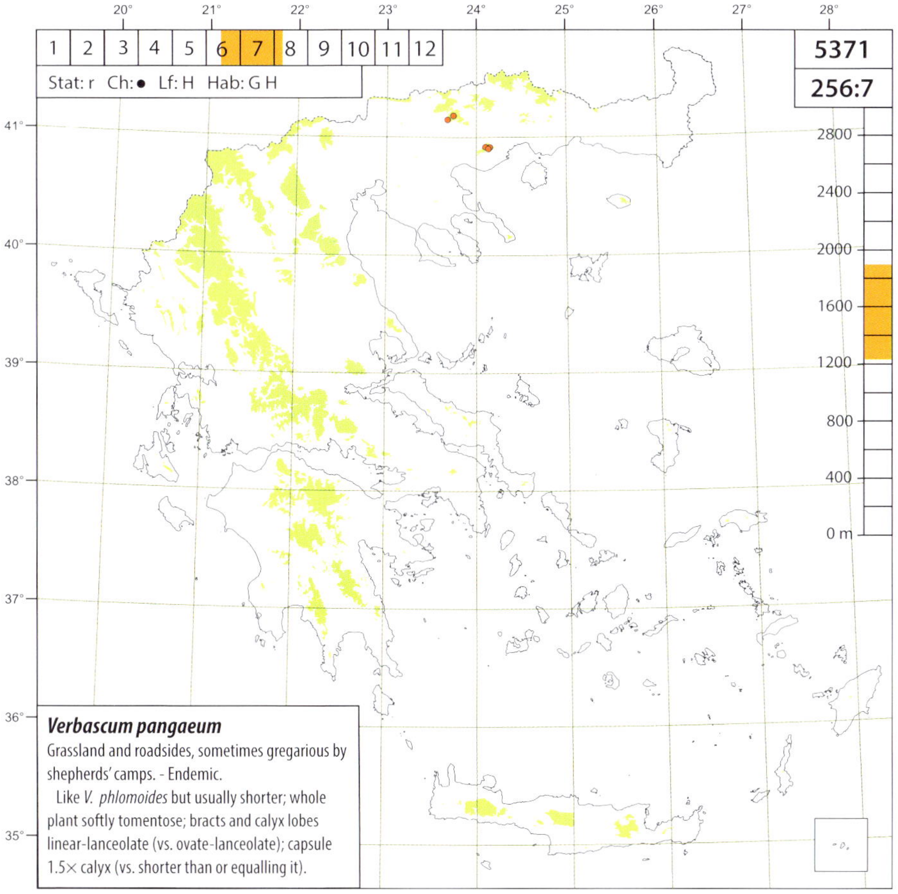

Verbascum pangaeum
Grassland and roadsides, sometimes gregarious by shepherds' camps. - Endemic.
Like *V. phlomoides* but usually shorter; whole plant softly tomentose; bracts and calyx lobes linear-lanceolate (vs. ovate-lanceolate); capsule 1.5× calyx (vs. shorter than or equalling it).

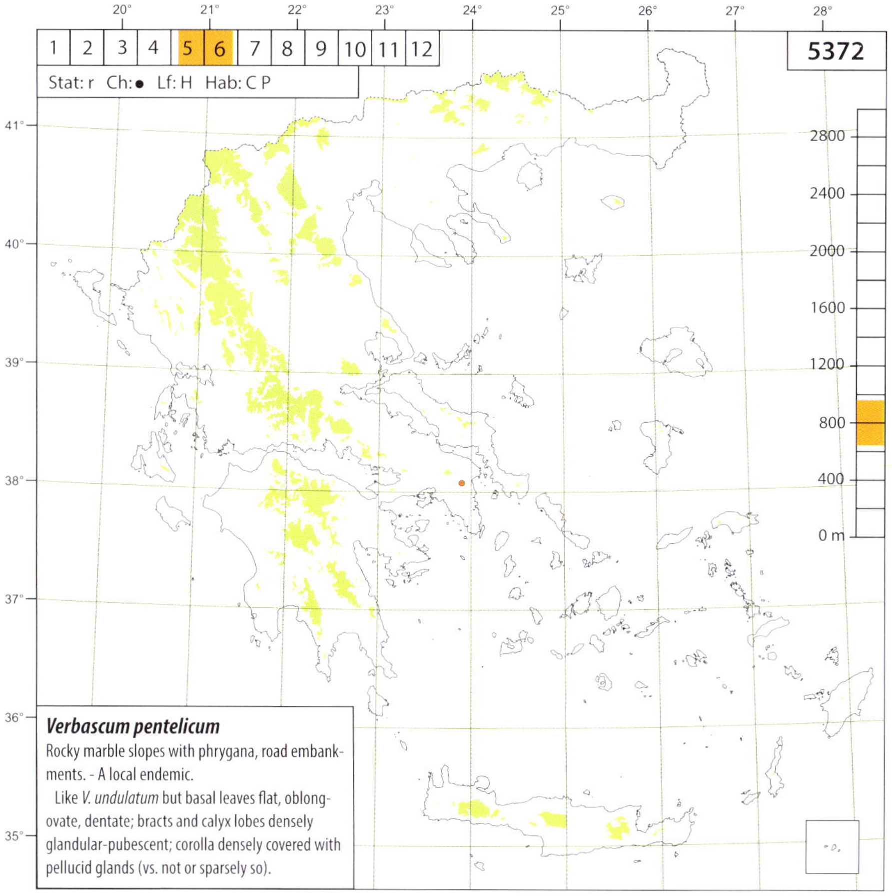

Verbascum pentelicum
Rocky marble slopes with phrygana, road embankments. - A local endemic.
Like *V. undulatum* but basal leaves flat, oblong-ovate, dentate; bracts and calyx lobes densely glandular-pubescent; corolla densely covered with pellucid glands (vs. not or sparsely so).

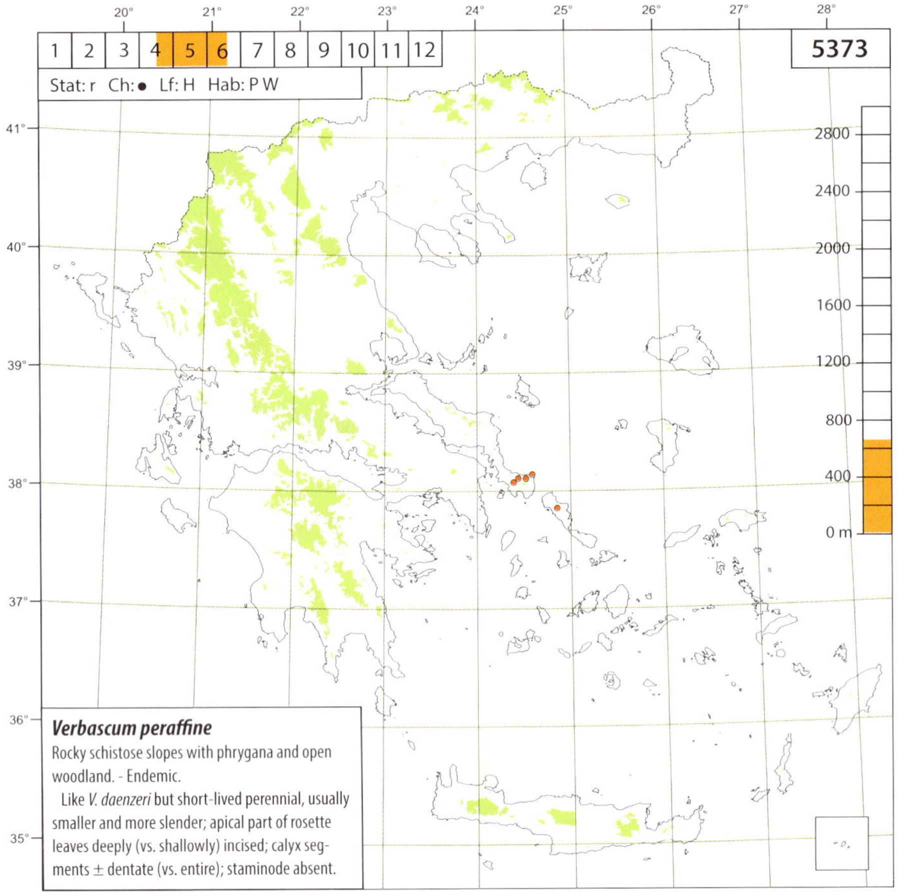

Verbascum peraffine

Rocky schistose slopes with phrygana and open woodland. - Endemic.

Like *V. daenzeri* but short-lived perennial, usually smaller and more slender; apical part of rosette leaves deeply (vs. shallowly) incised; calyx segments ± dentate (vs. entire); staminode absent.

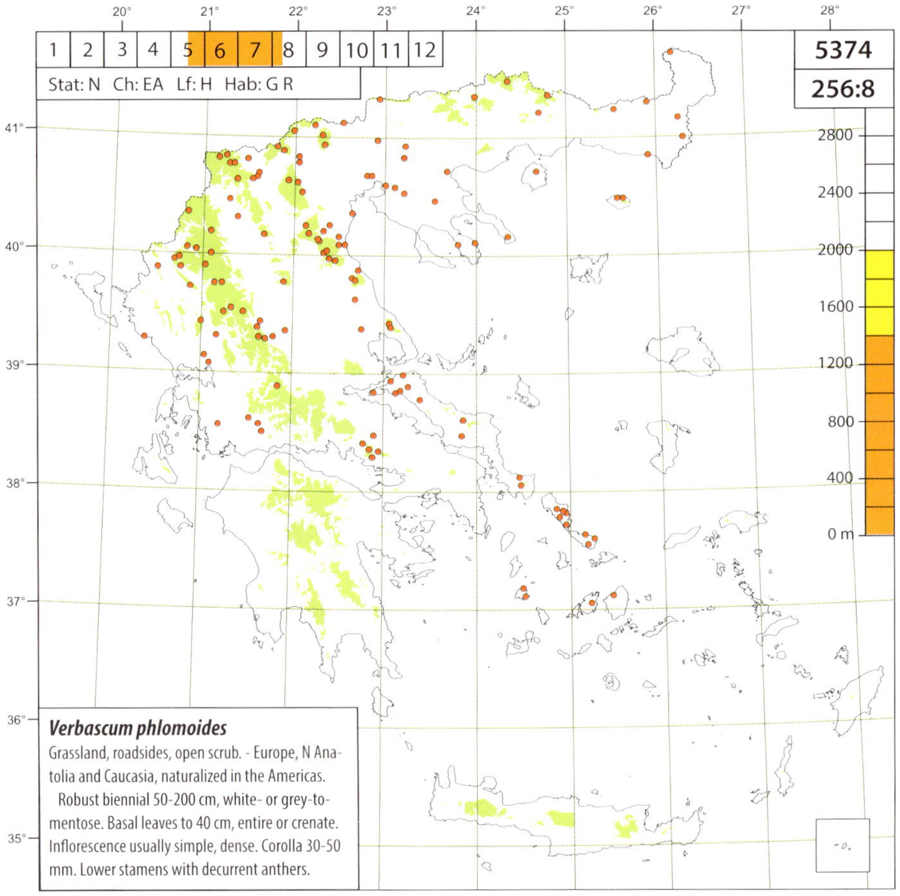

Verbascum phlomoides

Grassland, roadsides, open scrub. - Europe, N Anatolia and Caucasia, naturalized in the Americas.

Robust biennial 50-200 cm, white- or grey-tomentose. Basal leaves to 40 cm, entire or crenate. Inflorescence usually simple, dense. Corolla 30-50 mm. Lower stamens with decurrent anthers.

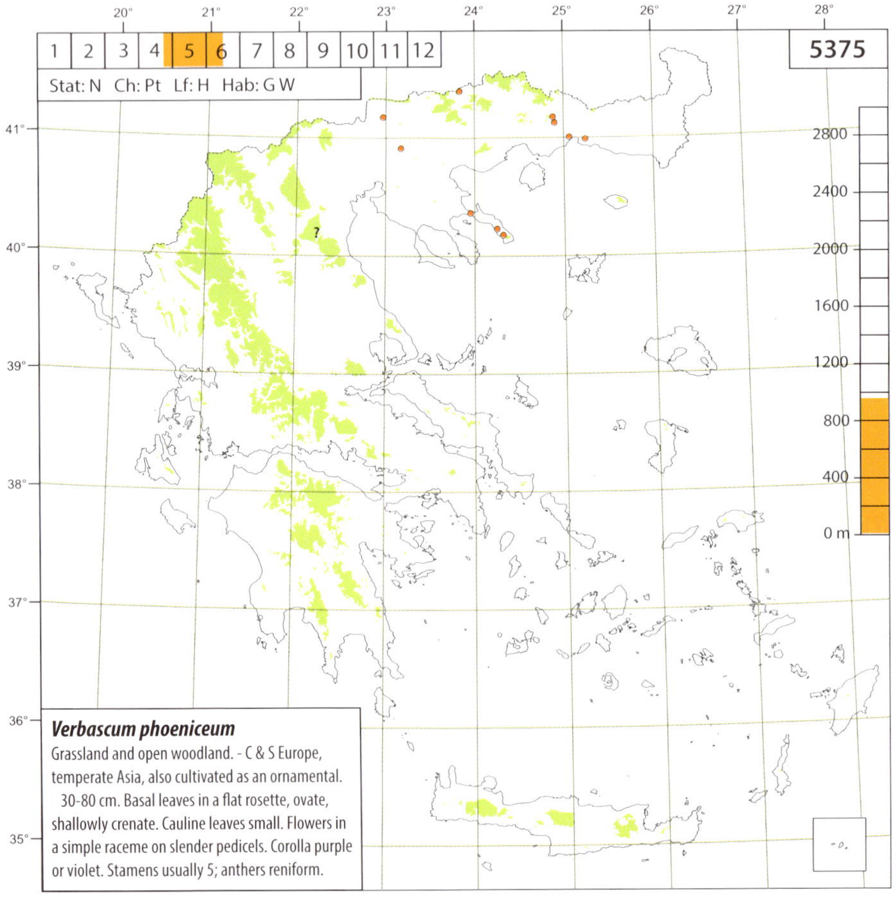

Verbascum phoeniceum

Grassland and open woodland. - C & S Europe, temperate Asia, also cultivated as an ornamental.

30-80 cm. Basal leaves in a flat rosette, ovate, shallowly crenate. Cauline leaves small. Flowers in a simple raceme on slender pedicels. Corolla purple or violet. Stamens usually 5; anthers reniform.

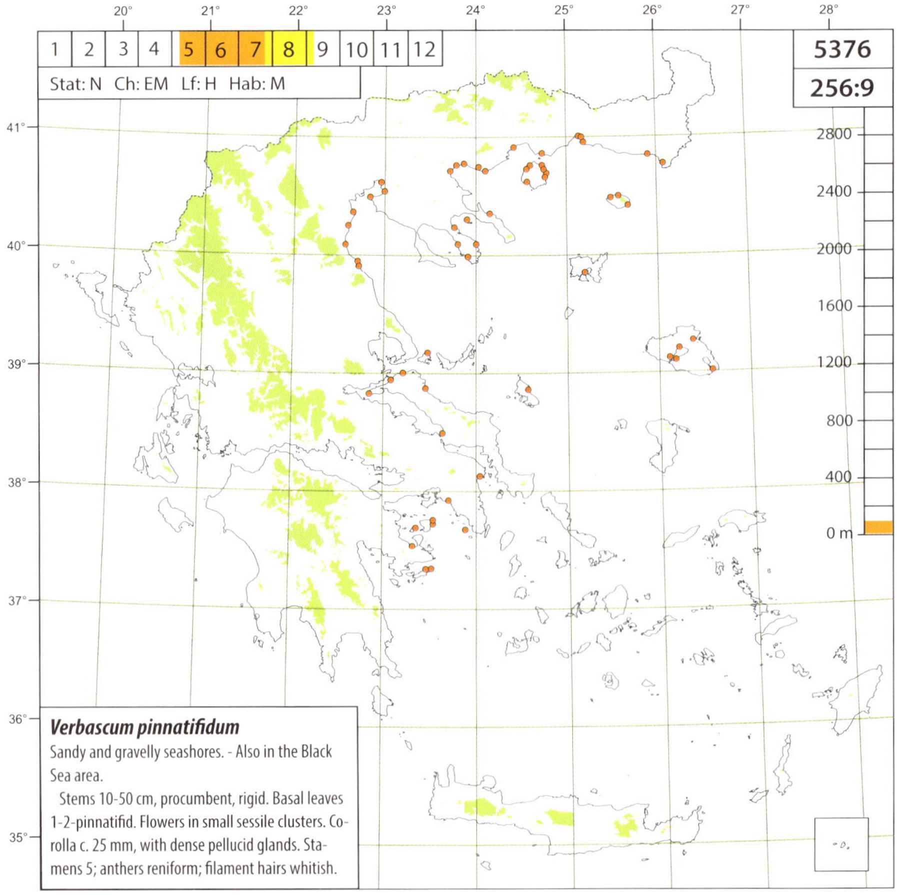

Verbascum pinnatifidum

Sandy and gravelly seashores. - Also in the Black Sea area.

Stems 10-50 cm, procumbent, rigid. Basal leaves 1-2-pinnatifid. Flowers in small sessile clusters. Corolla c. 25 mm, with dense pellucid glands. Stamens 5; anthers reniform; filament hairs whitish.

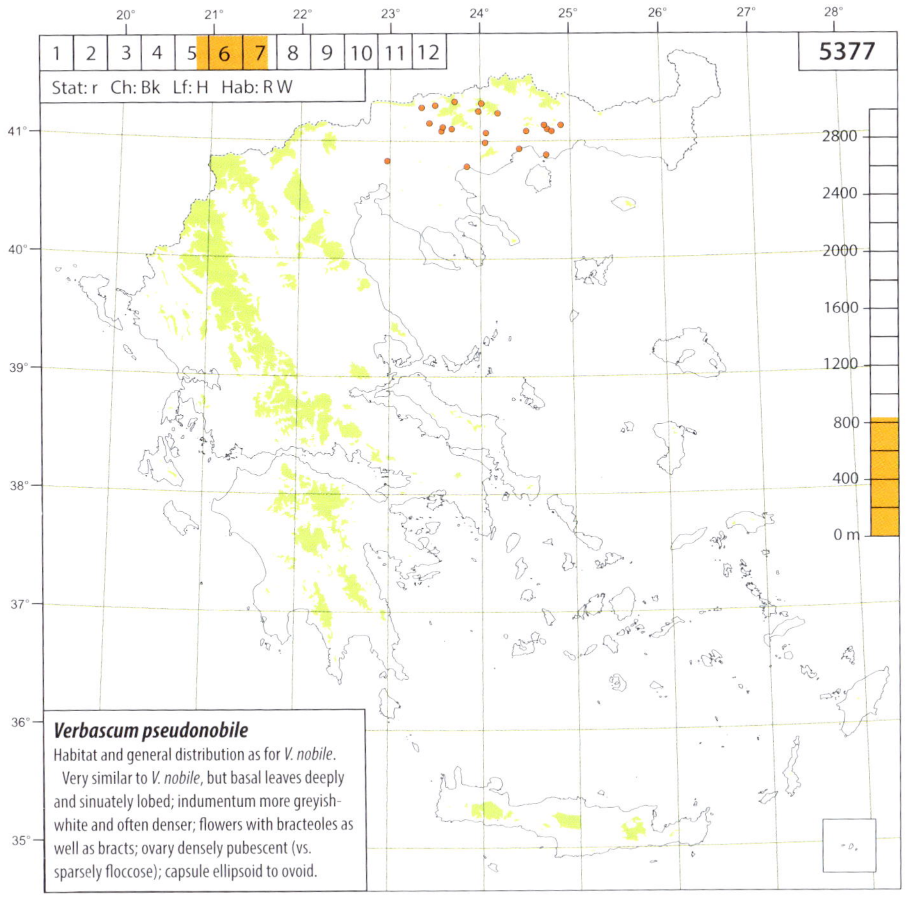

Verbascum pseudonobile
Habitat and general distribution as for *V. nobile*.
Very similar to *V. nobile*, but basal leaves deeply and sinuately lobed; indumentum more greyish-white and often denser; flowers with bracteoles as well as bracts; ovary densely pubescent (vs. sparsely floccose); capsule ellipsoid to ovoid.

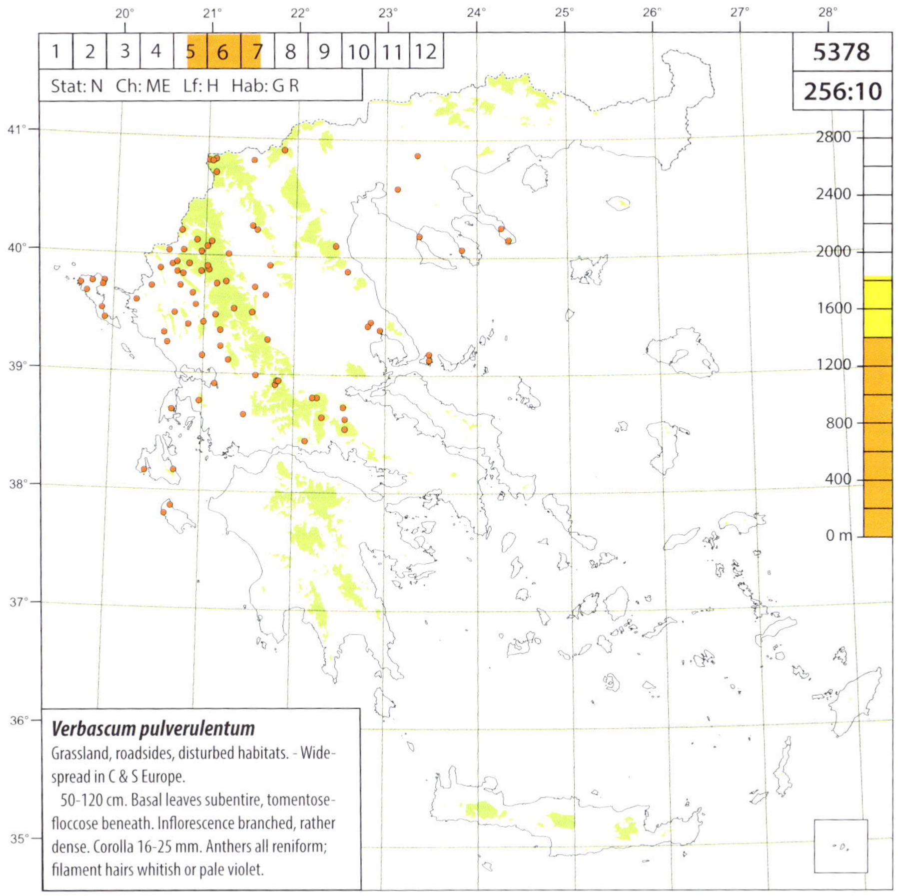

Verbascum pulverulentum
Grassland, roadsides, disturbed habitats. - Widespread in C & S Europe.
50-120 cm. Basal leaves subentire, tomentose-floccose beneath. Inflorescence branched, rather dense. Corolla 16-25 mm. Anthers all reniform; filament hairs whitish or pale violet.

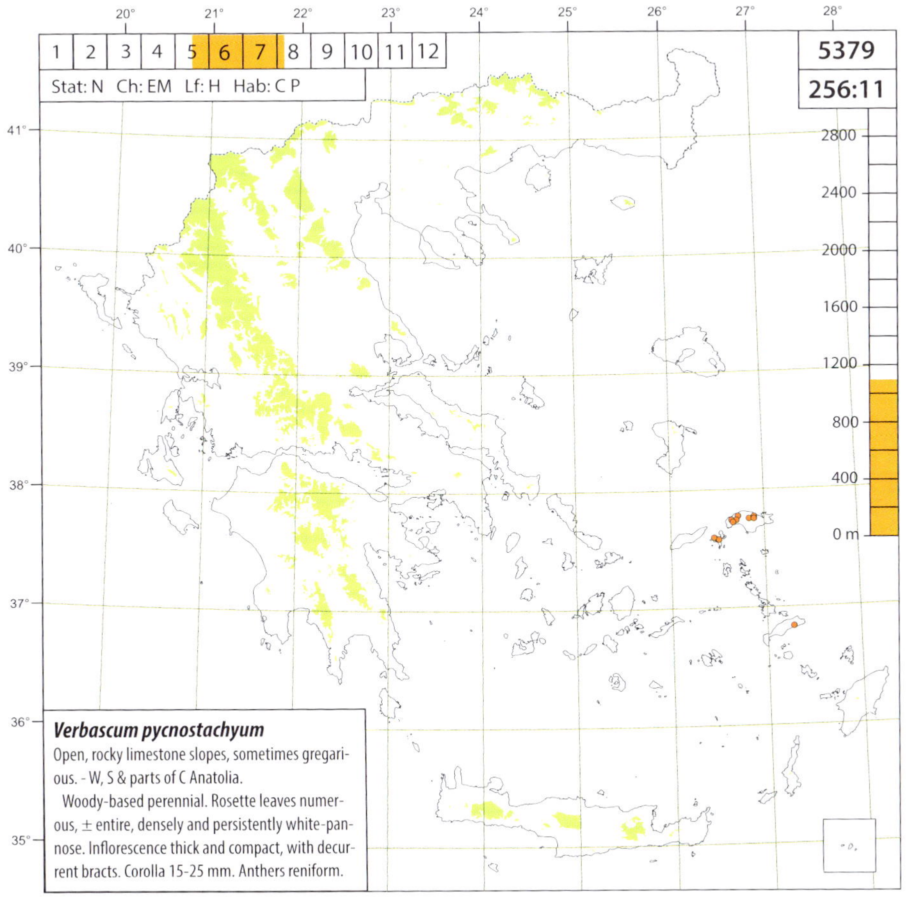

Verbascum pycnostachyum
Open, rocky limestone slopes, sometimes gregarious. - W, S & parts of C Anatolia.
Woody-based perennial. Rosette leaves numerous, ± entire, densely and persistently white-pannose. Inflorescence thick and compact, with decurrent bracts. Corolla 15-25 mm. Anthers reniform.

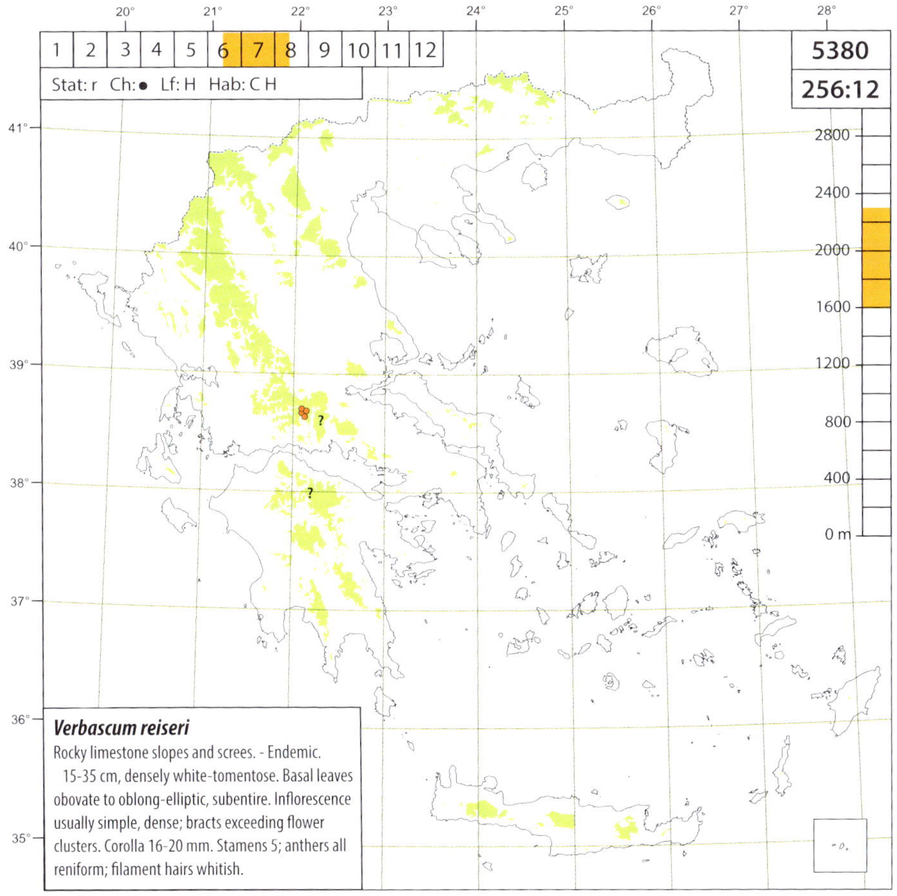

Verbascum reiseri
Rocky limestone slopes and screes. - Endemic.
15-35 cm, densely white-tomentose. Basal leaves obovate to oblong-elliptic, subentire. Inflorescence usually simple, dense; bracts exceeding flower clusters. Corolla 16-20 mm. Stamens 5; anthers all reniform; filament hairs whitish.

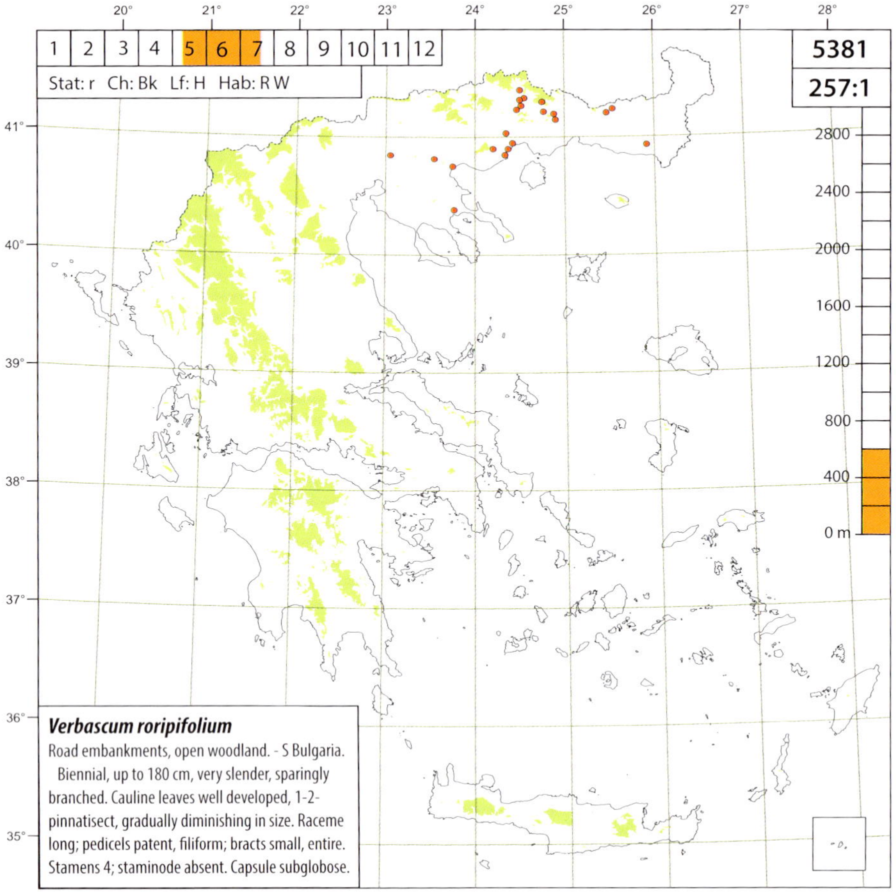

Verbascum roripifolium
Road embankments, open woodland. - S Bulgaria.
Biennial, up to 180 cm, very slender, sparingly branched. Cauline leaves well developed, 1-2-pinnatisect, gradually diminishing in size. Raceme long; pedicels patent, filiform; bracts small, entire. Stamens 4; staminode absent. Capsule subglobose.

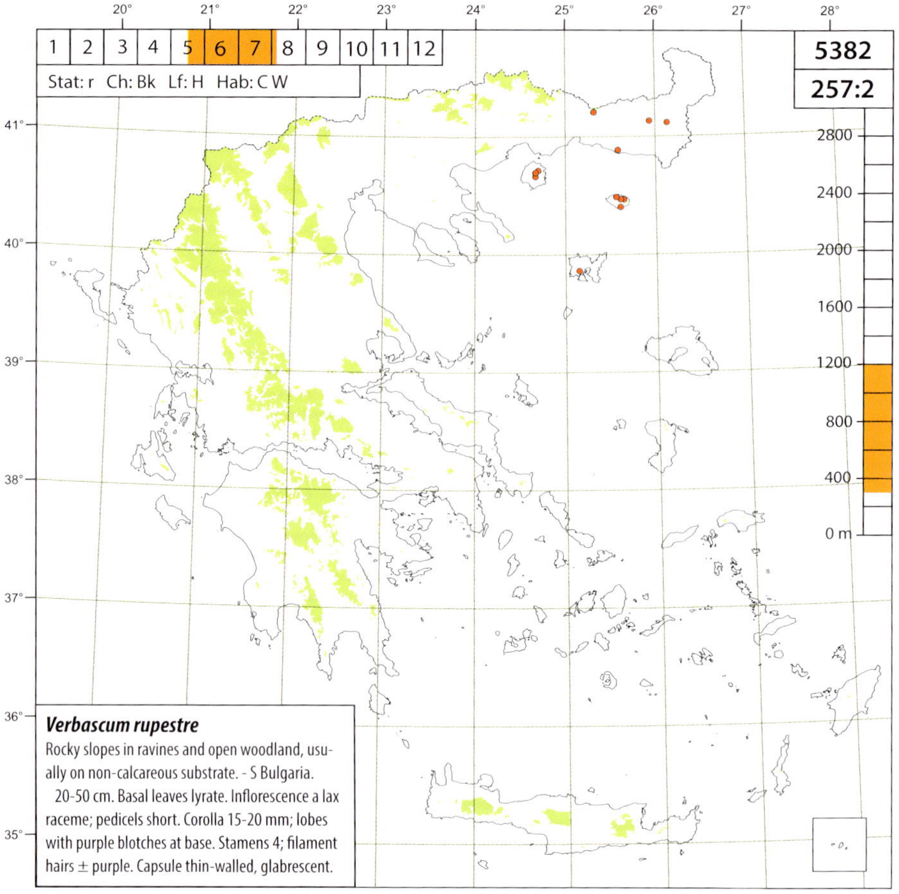

Verbascum rupestre
Rocky slopes in ravines and open woodland, usually on non-calcareous substrate. - S Bulgaria.
20-50 cm. Basal leaves lyrate. Inflorescence a lax raceme; pedicels short. Corolla 15-20 mm; lobes with purple blotches at base. Stamens 4; filament hairs ± purple. Capsule thin-walled, glabrescent.

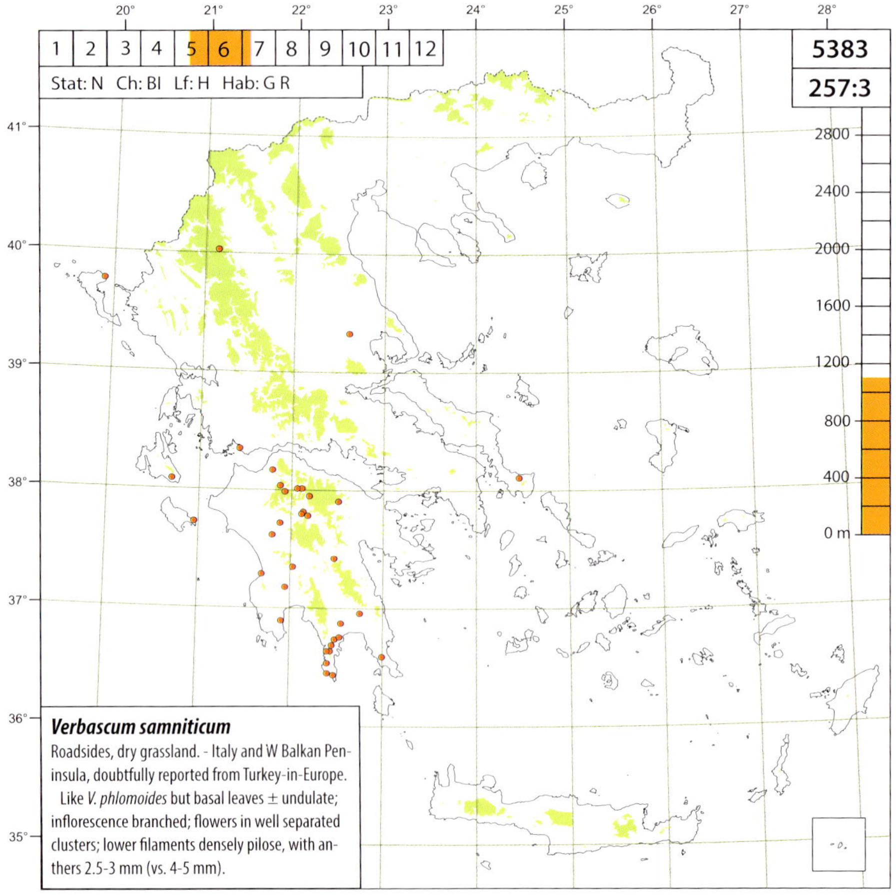

Verbascum samniticum
Roadsides, dry grassland. - Italy and W Balkan Peninsula, doubtfully reported from Turkey-in-Europe.
Like *V. phlomoides* but basal leaves ± undulate; inflorescence branched; flowers in well separated clusters; lower filaments densely pilose, with anthers 2.5-3 mm (vs. 4-5 mm).

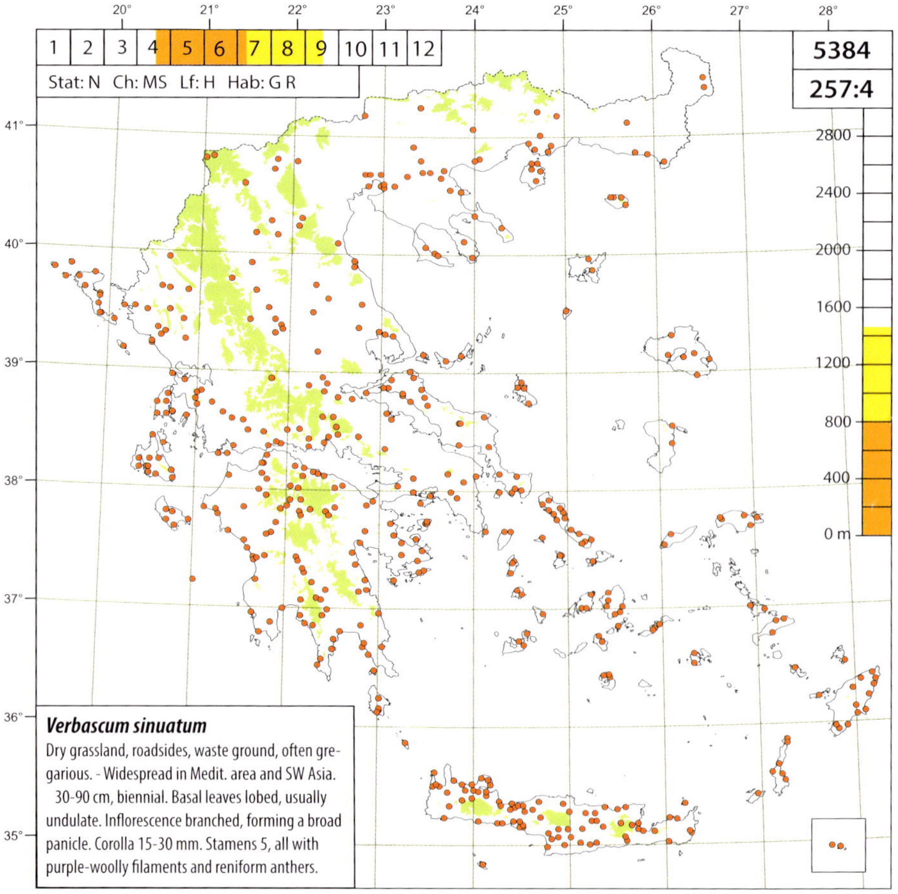

Verbascum sinuatum
Dry grassland, roadsides, waste ground, often gregarious. - Widespread in Medit. area and SW Asia.
30-90 cm, biennial. Basal leaves lobed, usually undulate. Inflorescence branched, forming a broad panicle. Corolla 15-30 mm. Stamens 5, all with purple-woolly filaments and reniform anthers.

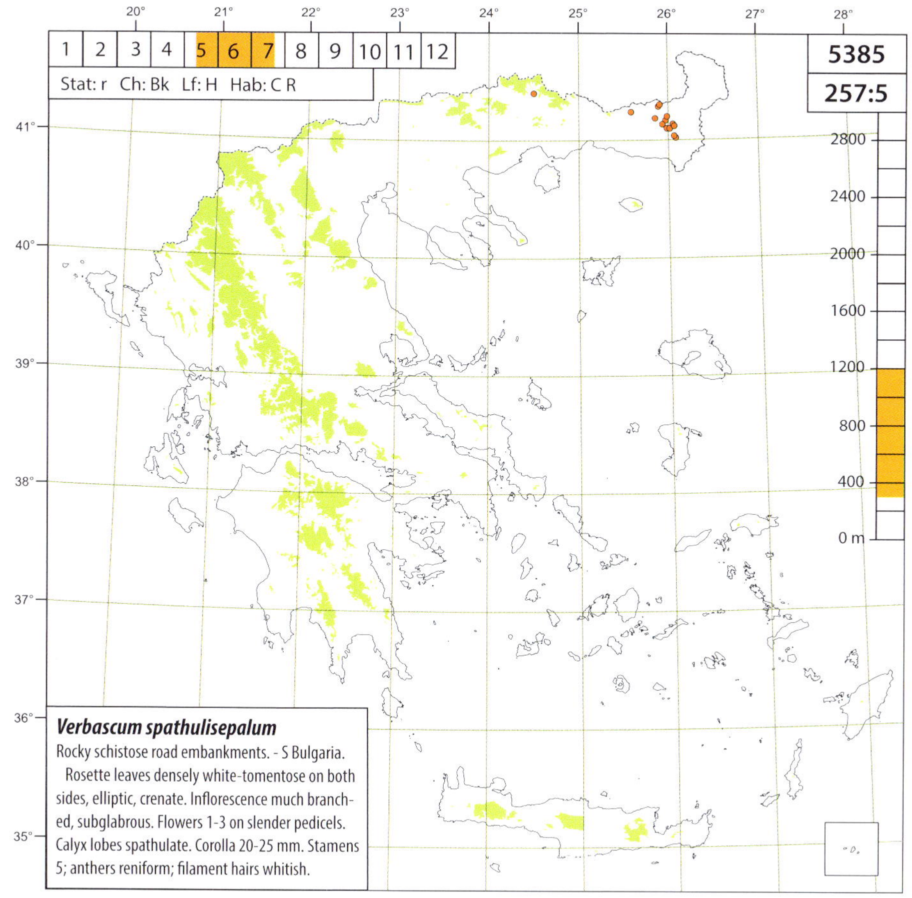

Verbascum spathulisepalum

Rocky schistose road embankments. - S Bulgaria.

Rosette leaves densely white-tomentose on both sides, elliptic, crenate. Inflorescence much branched, subglabrous. Flowers 1-3 on slender pedicels. Calyx lobes spathulate. Corolla 20-25 mm. Stamens 5; anthers reniform; filament hairs whitish.

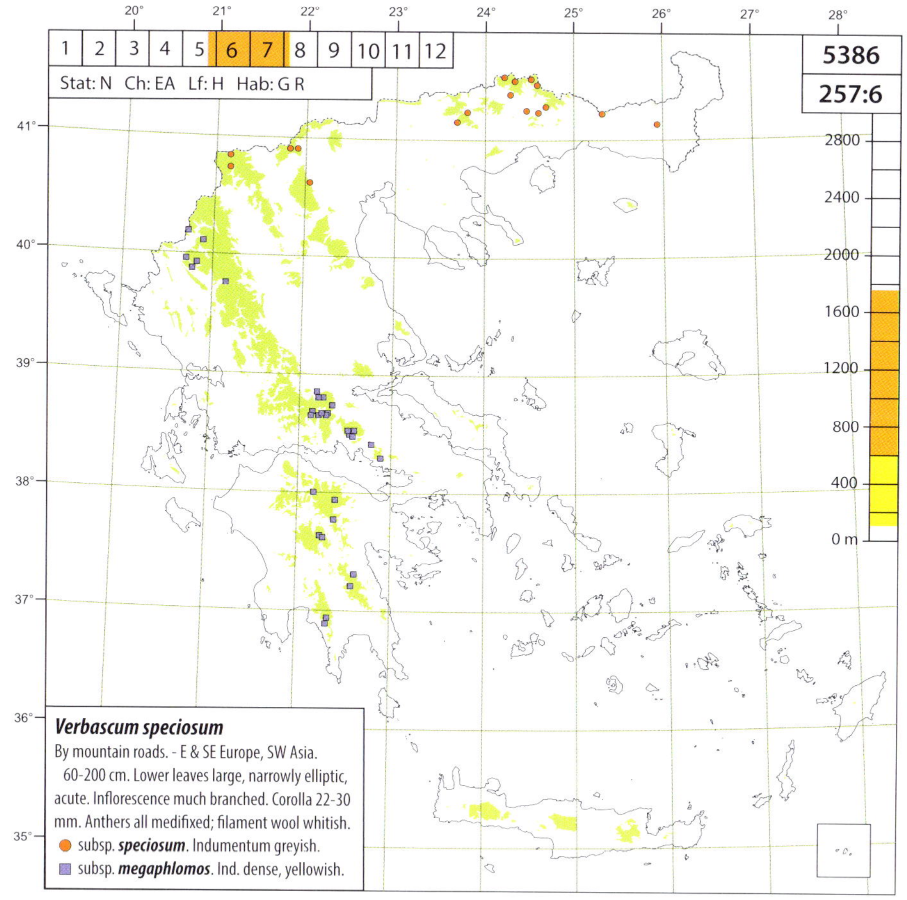

Verbascum speciosum

By mountain roads. - E & SE Europe, SW Asia.

60-200 cm. Lower leaves large, narrowly elliptic, acute. Inflorescence much branched. Corolla 22-30 mm. Anthers all medifixed; filament wool whitish.

● subsp. ***speciosum***. Indumentum greyish.

■ subsp. ***megaphlomos***. Ind. dense, yellowish.

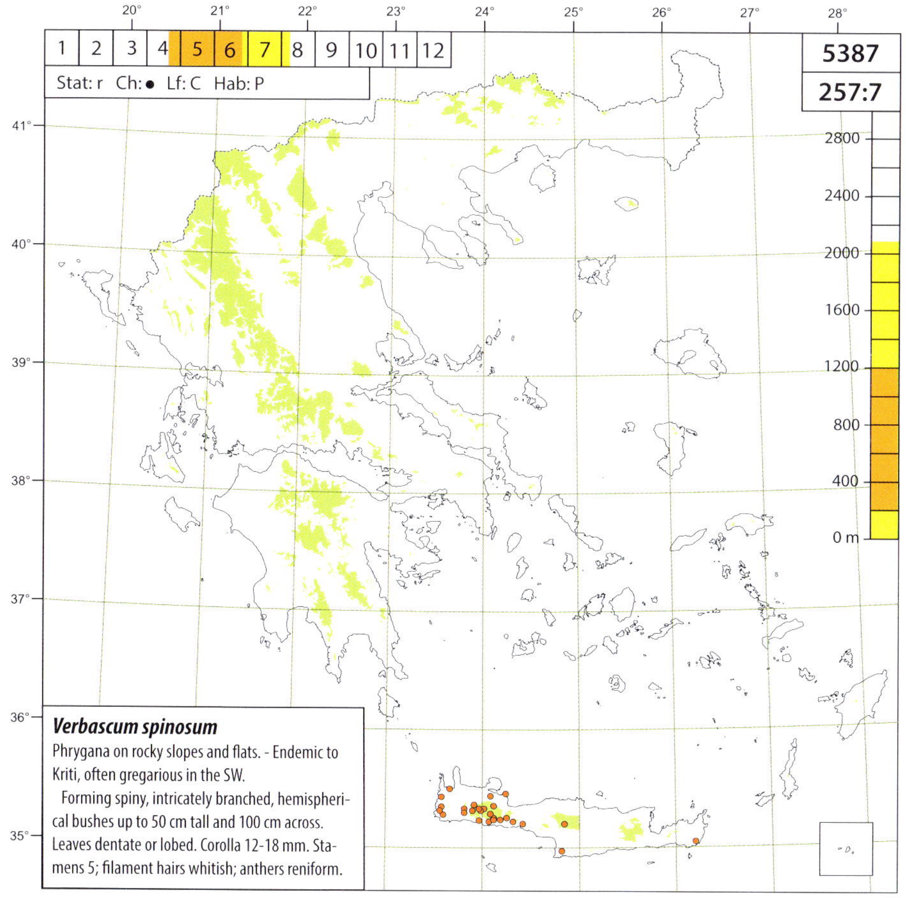

Verbascum spinosum

Phrygana on rocky slopes and flats. - Endemic to Kriti, often gregarious in the SW.

Forming spiny, intricately branched, hemispherical bushes up to 50 cm tall and 100 cm across. Leaves dentate or lobed. Corolla 12-18 mm. Stamens 5; filament hairs whitish; anthers reniform.

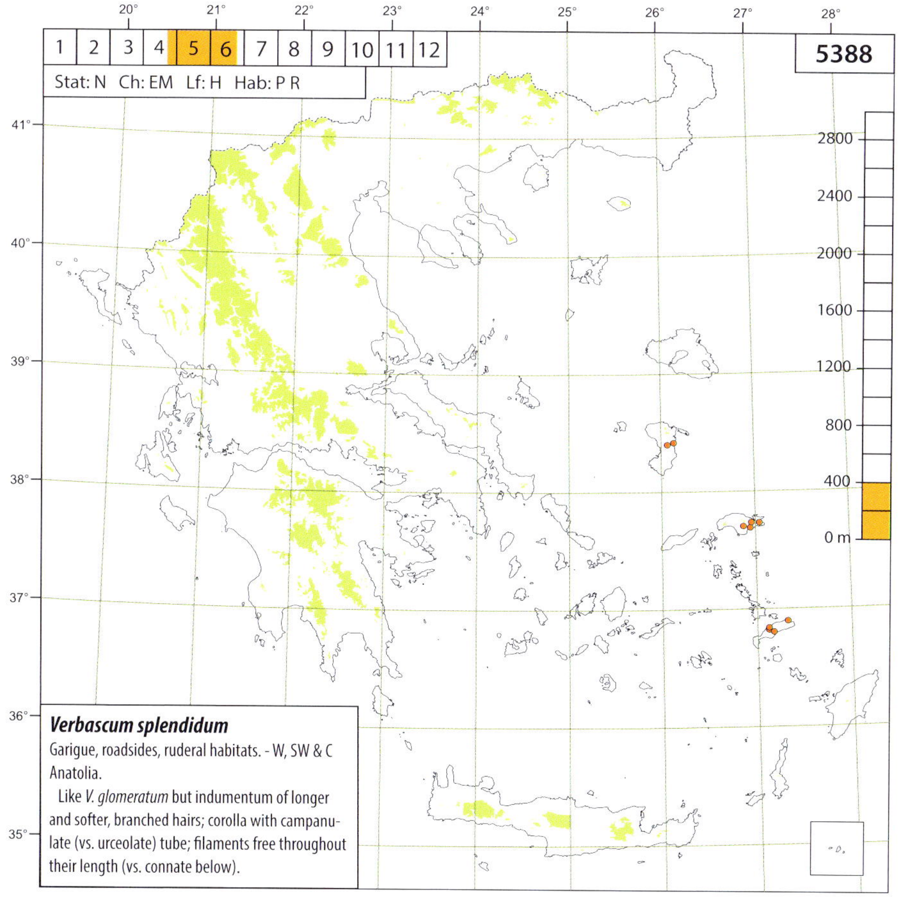

Verbascum splendidum

Garigue, roadsides, ruderal habitats. - W, SW & C Anatolia.

Like *V. glomeratum* but indumentum of longer and softer, branched hairs; corolla with campanulate (vs. urceolate) tube; filaments free throughout their length (vs. connate below).

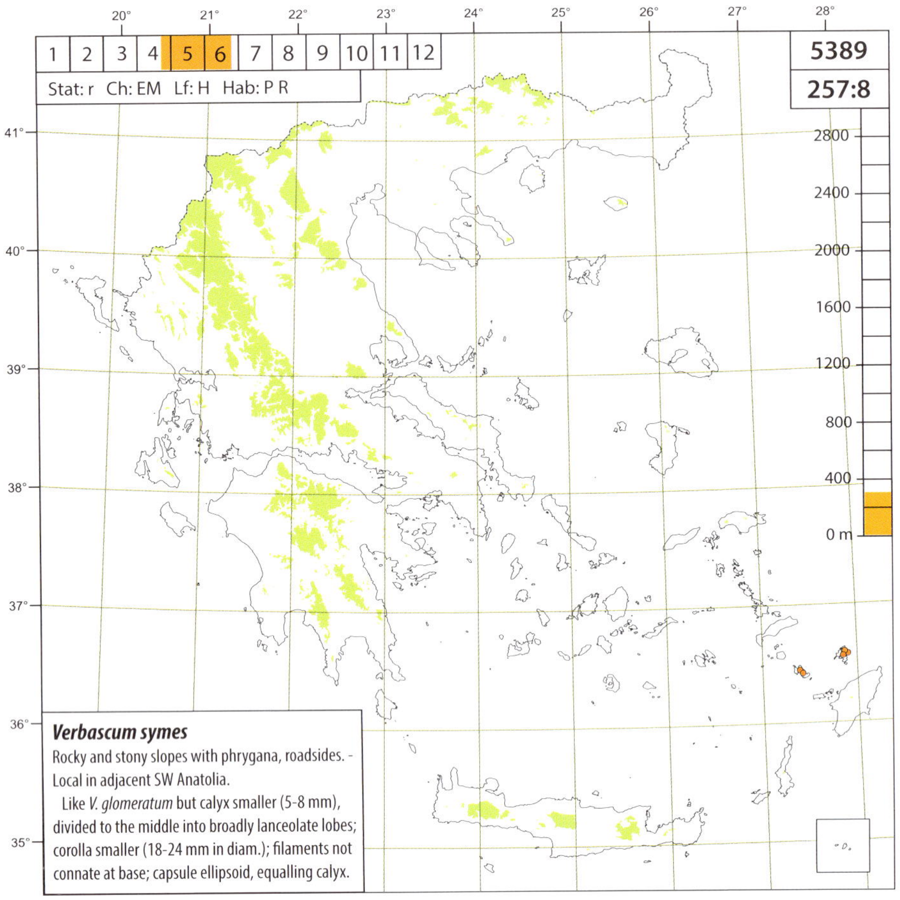

Verbascum symes
Rocky and stony slopes with phrygana, roadsides. - Local in adjacent SW Anatolia.
Like *V. glomeratum* but calyx smaller (5-8 mm), divided to the middle into broadly lanceolate lobes; corolla smaller (18-24 mm in diam.); filaments not connate at base; capsule ellipsoid, equalling calyx.

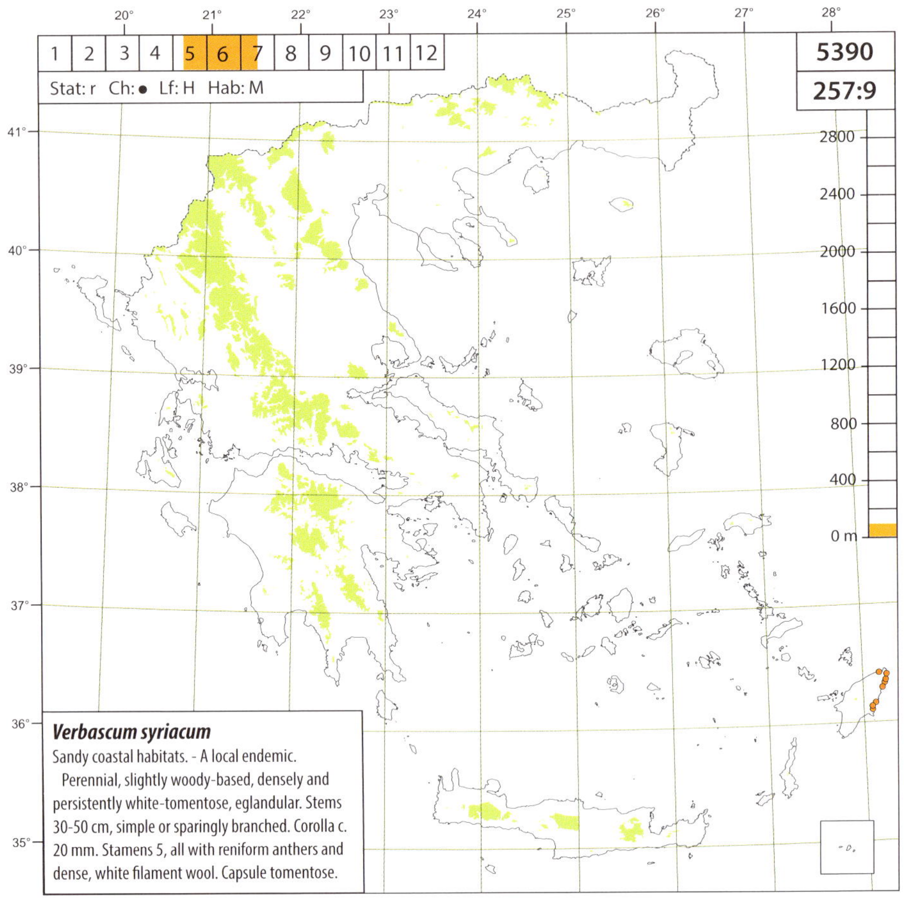

Verbascum syriacum
Sandy coastal habitats. - A local endemic.
Perennial, slightly woody-based, densely and persistently white-tomentose, eglandular. Stems 30-50 cm, simple or sparingly branched. Corolla c. 20 mm. Stamens 5, all with reniform anthers and dense, white filament wool. Capsule tomentose.

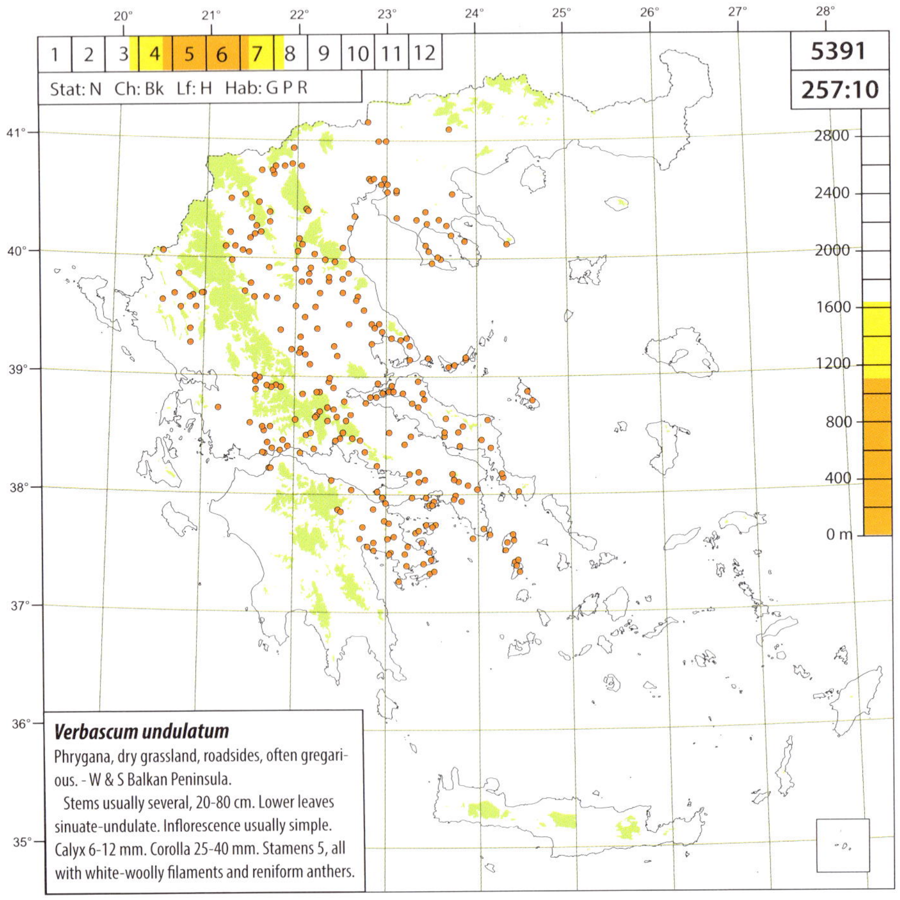

Verbascum undulatum
Phrygana, dry grassland, roadsides, often gregarious. - W & S Balkan Peninsula.
Stems usually several, 20-80 cm. Lower leaves sinuate-undulate. Inflorescence usually simple. Calyx 6-12 mm. Corolla 25-40 mm. Stamens 5, all with white-woolly filaments and reniform anthers.

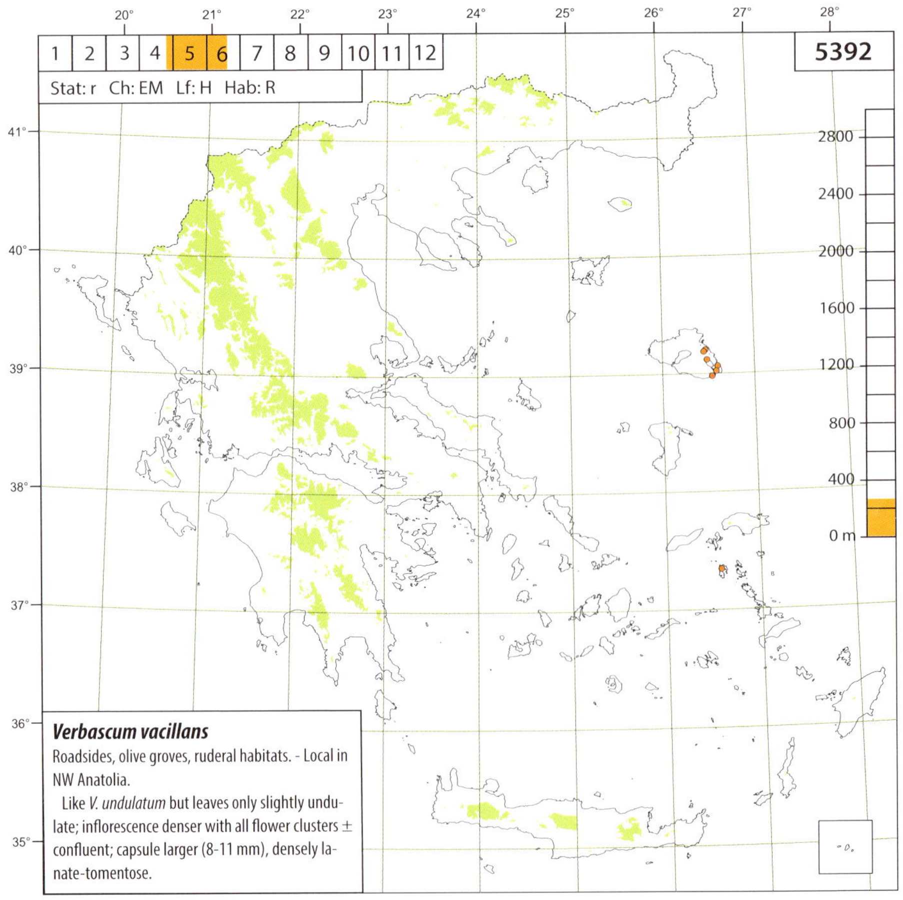

Verbascum vacillans
Roadsides, olive groves, ruderal habitats. - Local in NW Anatolia.
Like *V. undulatum* but leaves only slightly undulate; inflorescence denser with all flower clusters ± confluent; capsule larger (8-11 mm), densely lanate-tomentose.

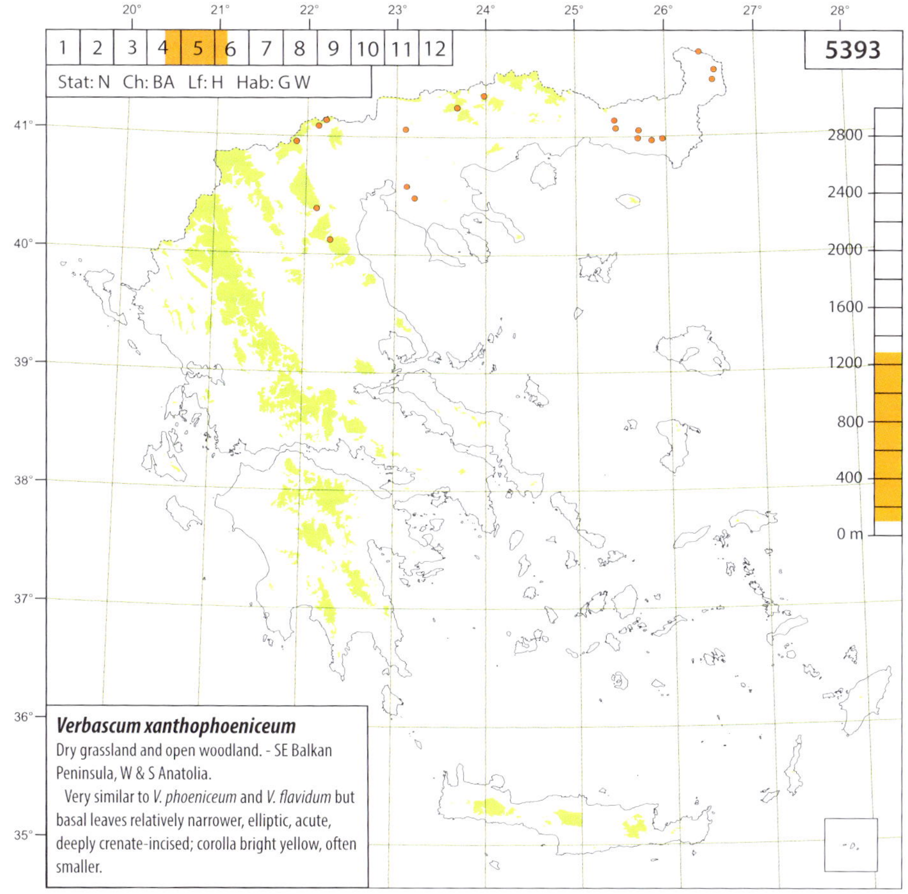

Verbascum xanthophoeniceum

Dry grassland and open woodland. - SE Balkan Peninsula, W & S Anatolia.

Very similar to *V. phoeniceum* and *V. flavidum* but basal leaves relatively narrower, elliptic, acute, deeply crenate-incised; corolla bright yellow, often smaller.

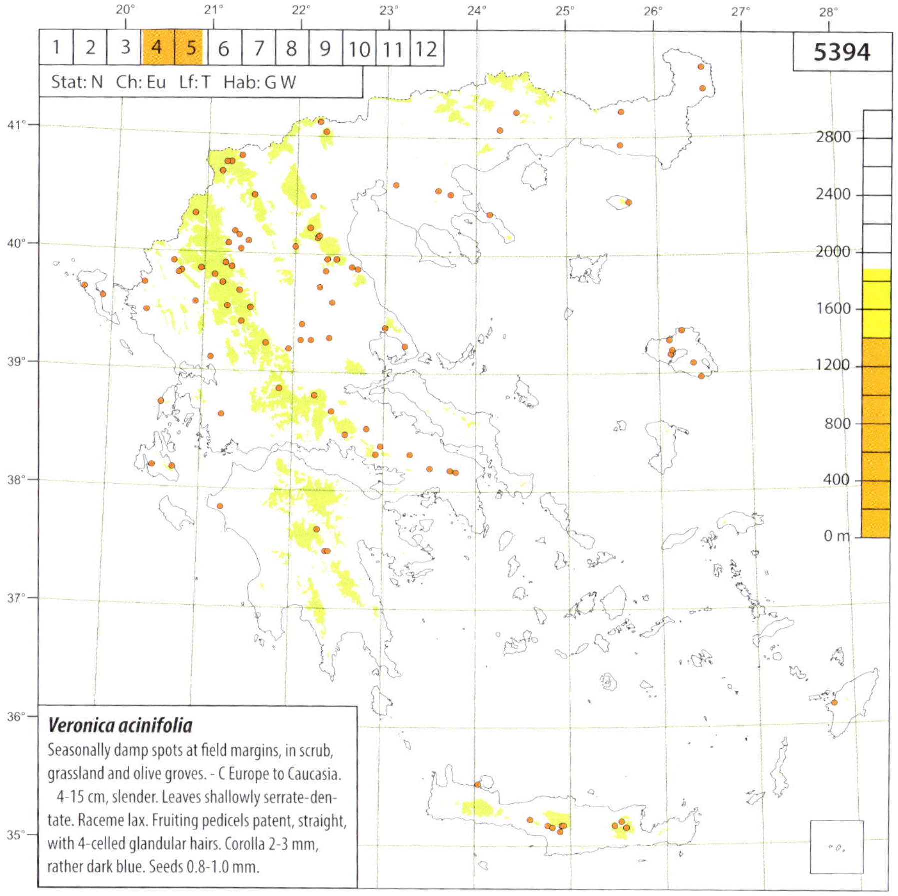

Veronica acinifolia

Seasonally damp spots at field margins, in scrub, grassland and olive groves. - C Europe to Caucasia.

4-15 cm, slender. Leaves shallowly serrate-dentate. Raceme lax. Fruiting pedicels patent, straight, with 4-celled glandular hairs. Corolla 2-3 mm, rather dark blue. Seeds 0.8-1.0 mm.

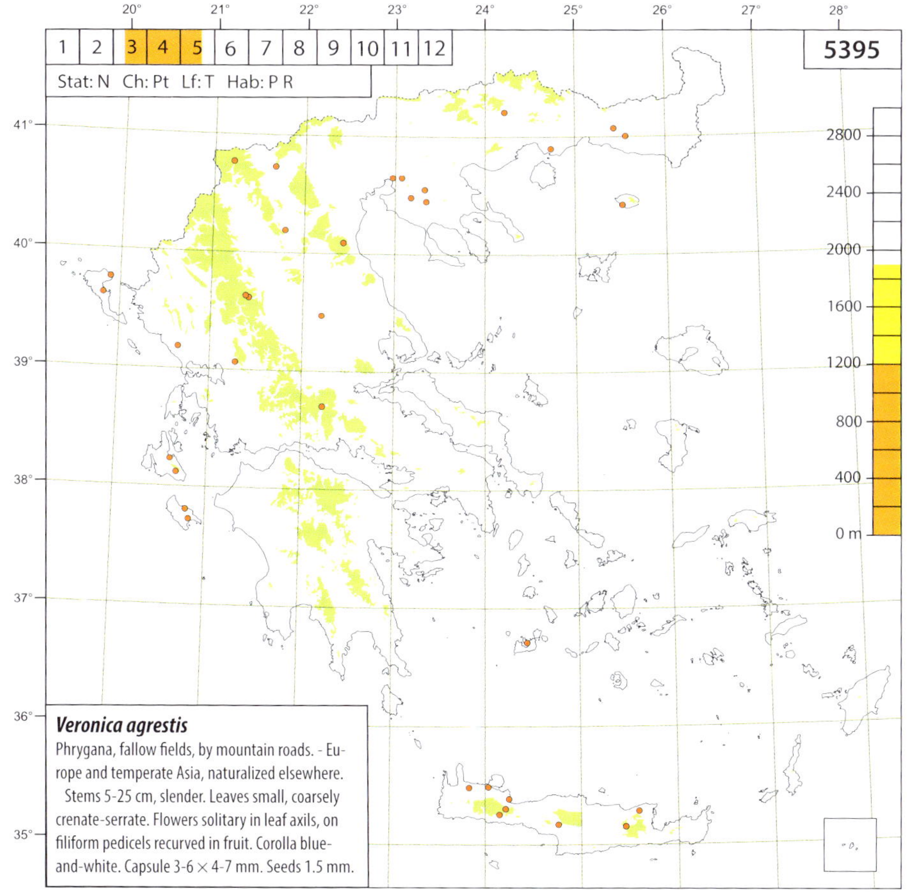

Veronica agrestis

Phrygana, fallow fields, by mountain roads. - Europe and temperate Asia, naturalized elsewhere.

Stems 5-25 cm, slender. Leaves small, coarsely crenate-serrate. Flowers solitary in leaf axils, on filiform pedicels recurved in fruit. Corolla blue-and-white. Capsule 3-6 × 4-7 mm. Seeds 1.5 mm.

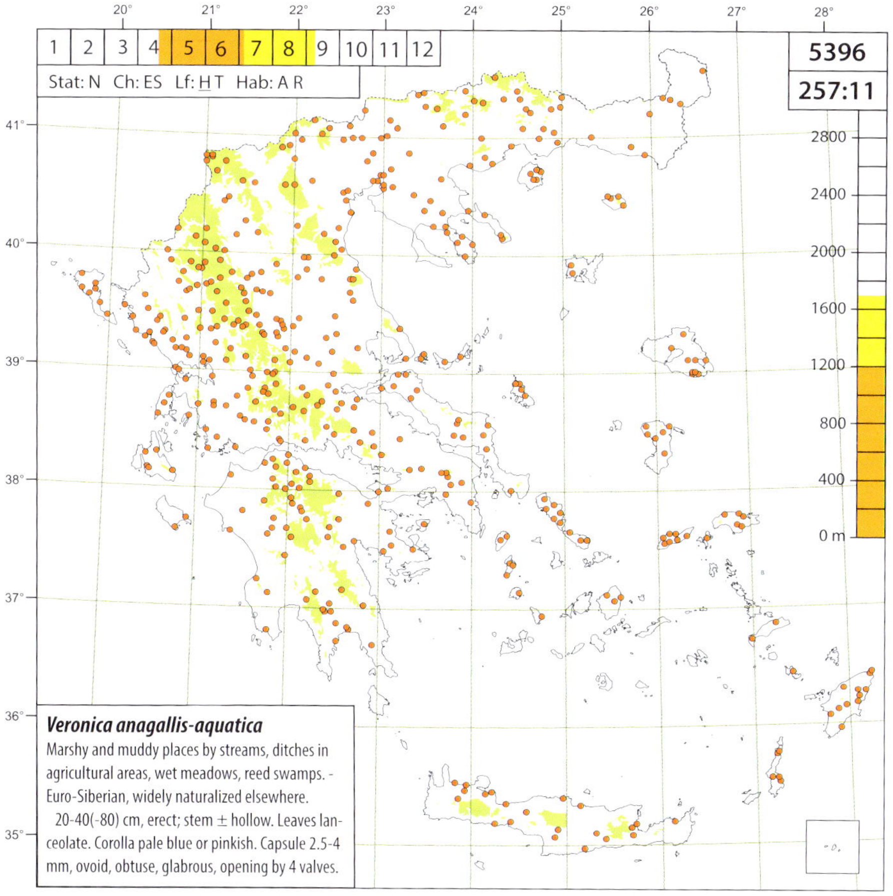

Veronica anagallis-aquatica

Marshy and muddy places by streams, ditches in agricultural areas, wet meadows, reed swamps. - Euro-Siberian, widely naturalized elsewhere.

20-40(-80) cm, erect; stem ± hollow. Leaves lanceolate. Corolla pale blue or pinkish. Capsule 2.5-4 mm, ovoid, obtuse, glabrous, opening by 4 valves.

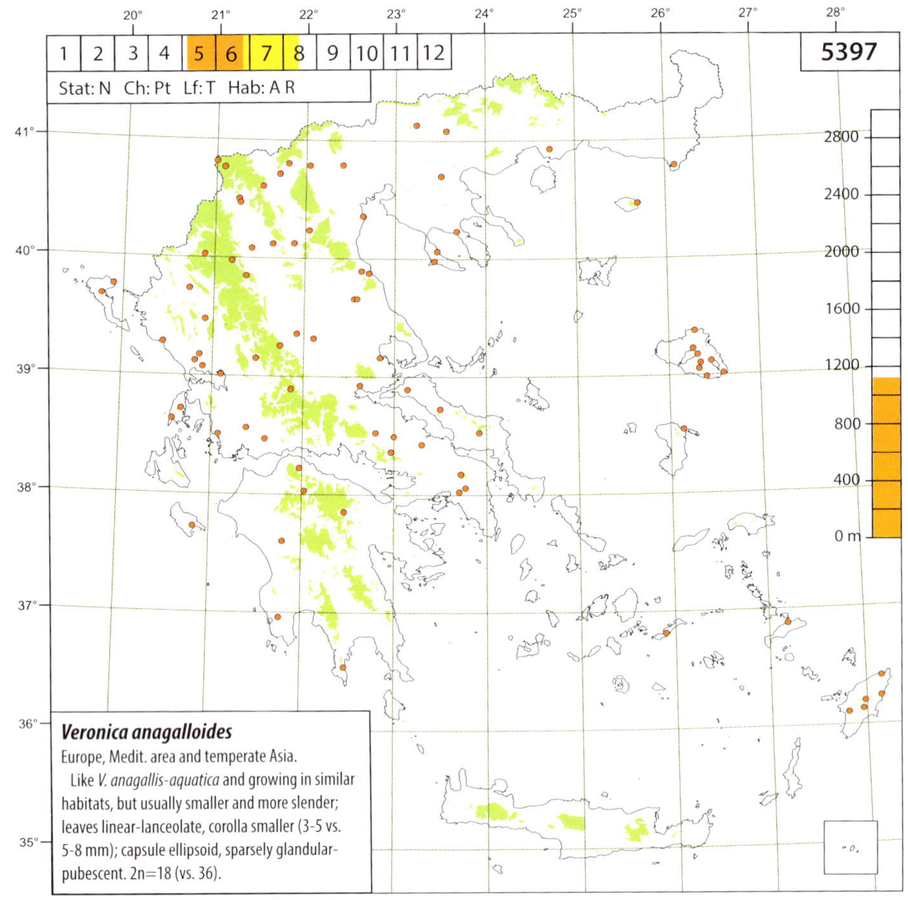

Veronica anagalloides

Europe, Medit. area and temperate Asia.

Like *V. anagallis-aquatica* and growing in similar habitats, but usually smaller and more slender; leaves linear-lanceolate, corolla smaller (3-5 vs. 5-8 mm); capsule ellipsoid, sparsely glandular-pubescent. 2n=18 (vs. 36).

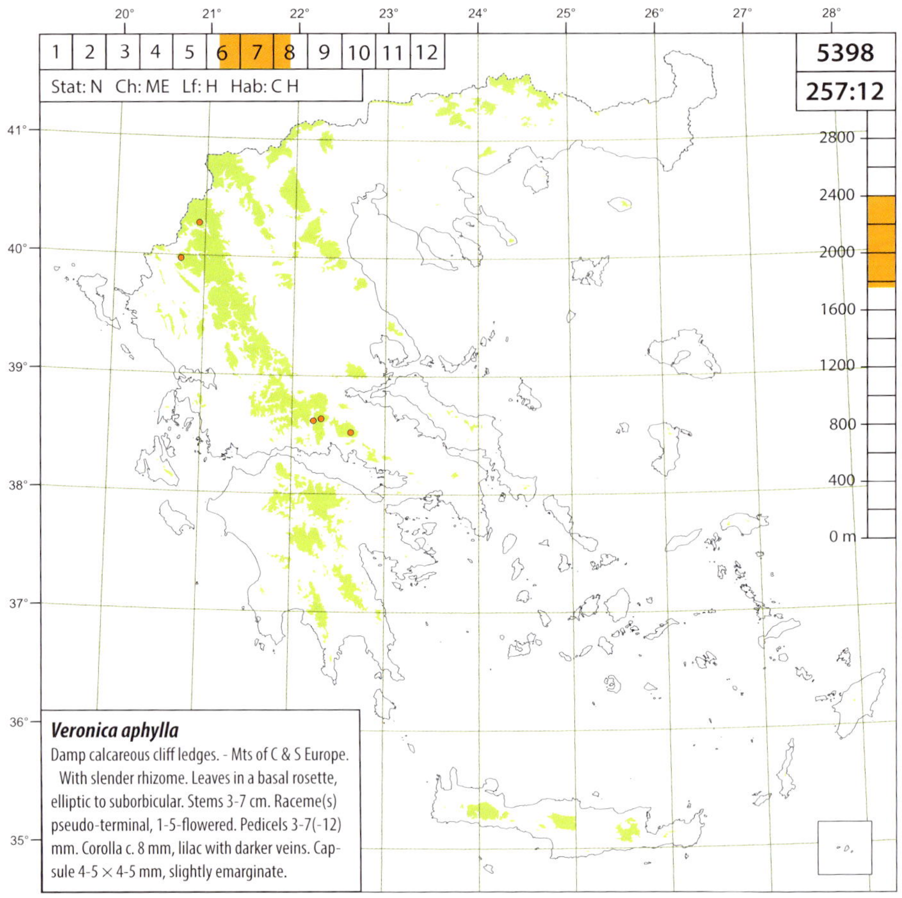

Veronica aphylla

Damp calcareous cliff ledges. - Mts of C & S Europe.

With slender rhizome. Leaves in a basal rosette, elliptic to suborbicular. Stems 3-7 cm. Raceme(s) pseudo-terminal, 1-5-flowered. Pedicels 3-7(-12) mm. Corolla c. 8 mm, lilac with darker veins. Capsule 4-5 × 4-5 mm, slightly emarginate.

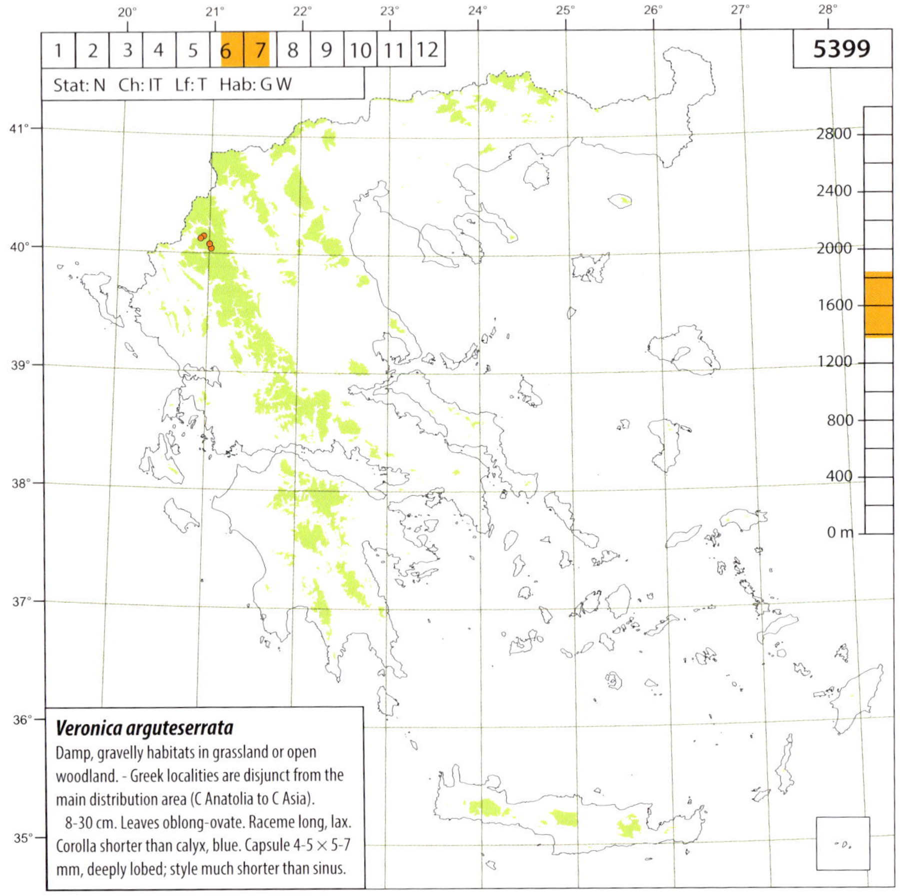

Veronica arguteserrata

Damp, gravelly habitats in grassland or open woodland. - Greek localities are disjunct from the main distribution area (C Anatolia to C Asia).

8-30 cm. Leaves oblong-ovate. Raceme long, lax. Corolla shorter than calyx, blue. Capsule 4-5 × 5-7 mm, deeply lobed; style much shorter than sinus.

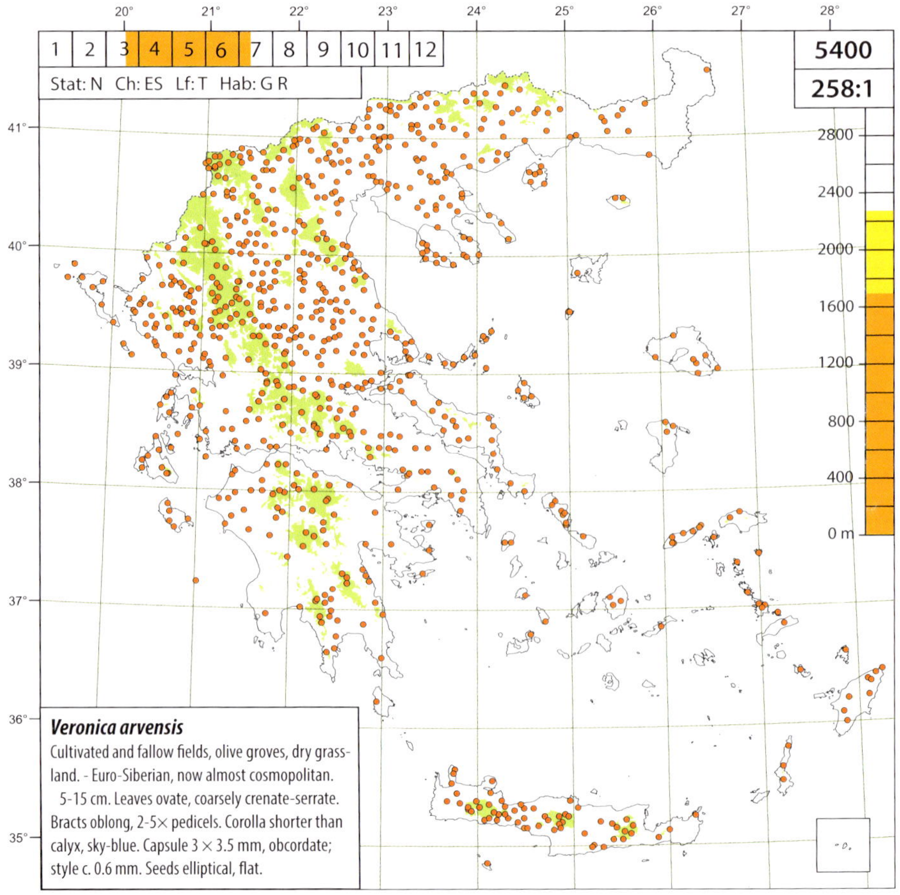

Veronica arvensis

Cultivated and fallow fields, olive groves, dry grassland. - Euro-Siberian, now almost cosmopolitan.

5-15 cm. Leaves ovate, coarsely crenate-serrate. Bracts oblong, 2-5× pedicels. Corolla shorter than calyx, sky-blue. Capsule 3 × 3.5 mm, obcordate; style c. 0.6 mm. Seeds elliptical, flat.

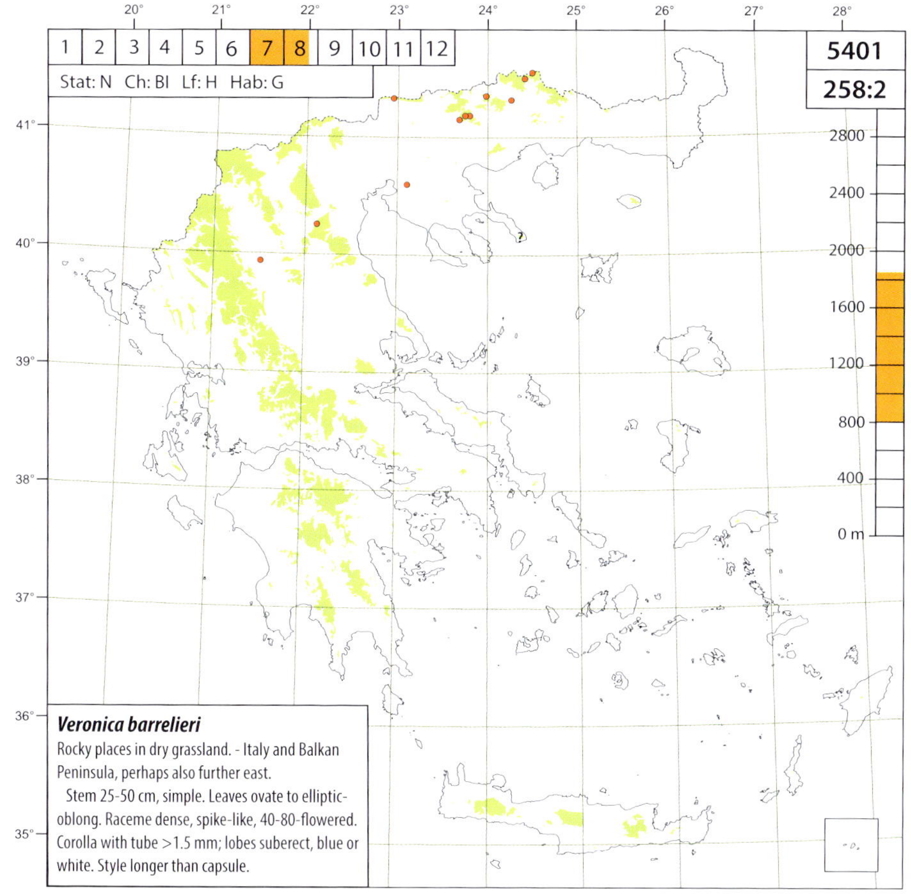

Veronica barrelieri

Rocky places in dry grassland. - Italy and Balkan Peninsula, perhaps also further east.

Stem 25-50 cm, simple. Leaves ovate to elliptic-oblong. Raceme dense, spike-like, 40-80-flowered. Corolla with tube >1.5 mm; lobes suberect, blue or white. Style longer than capsule.

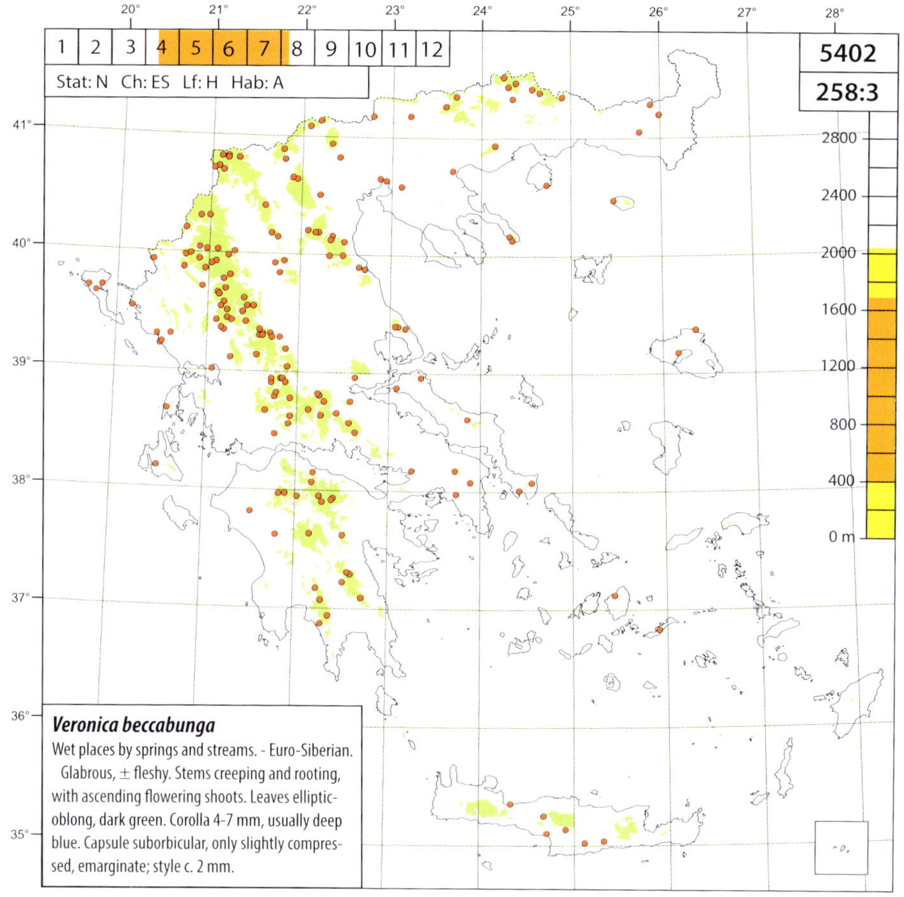

Veronica beccabunga

Wet places by springs and streams. - Euro-Siberian.

Glabrous, ± fleshy. Stems creeping and rooting, with ascending flowering shoots. Leaves elliptic-oblong, dark green. Corolla 4-7 mm, usually deep blue. Capsule suborbicular, only slightly compressed, emarginate; style c. 2 mm.

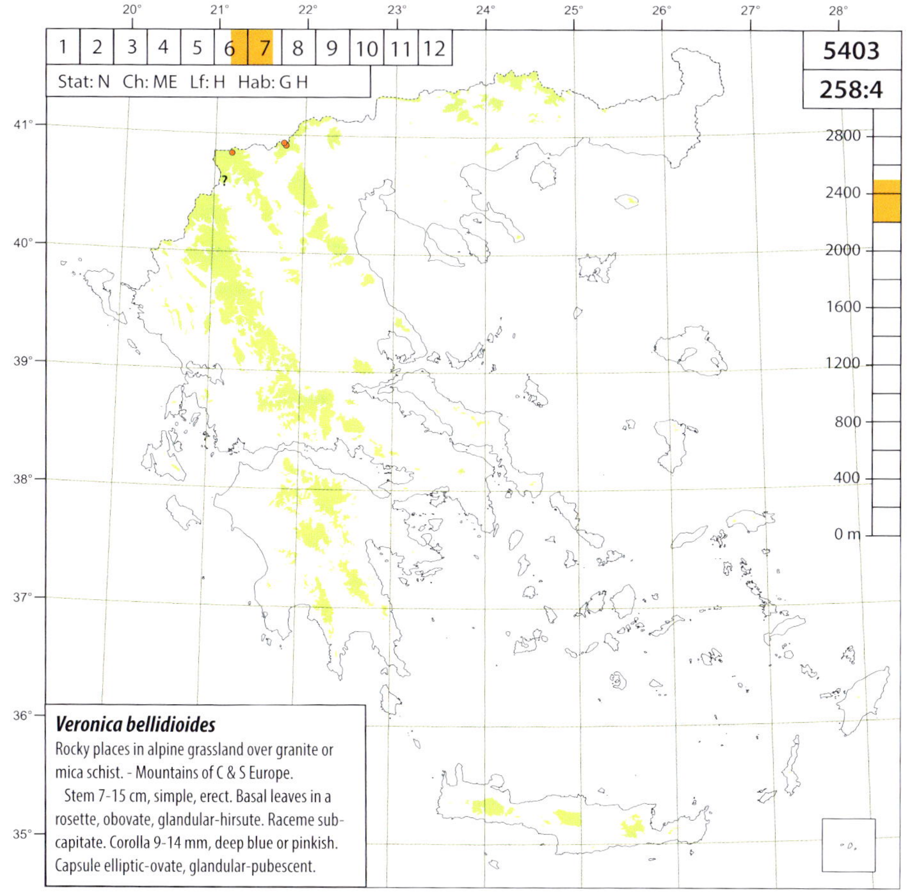

Veronica bellidioides

Rocky places in alpine grassland over granite or mica schist. - Mountains of C & S Europe.

Stem 7-15 cm, simple, erect. Basal leaves in a rosette, obovate, glandular-hirsute. Raceme subcapitate. Corolla 9-14 mm, deep blue or pinkish. Capsule elliptic-ovate, glandular-pubescent.

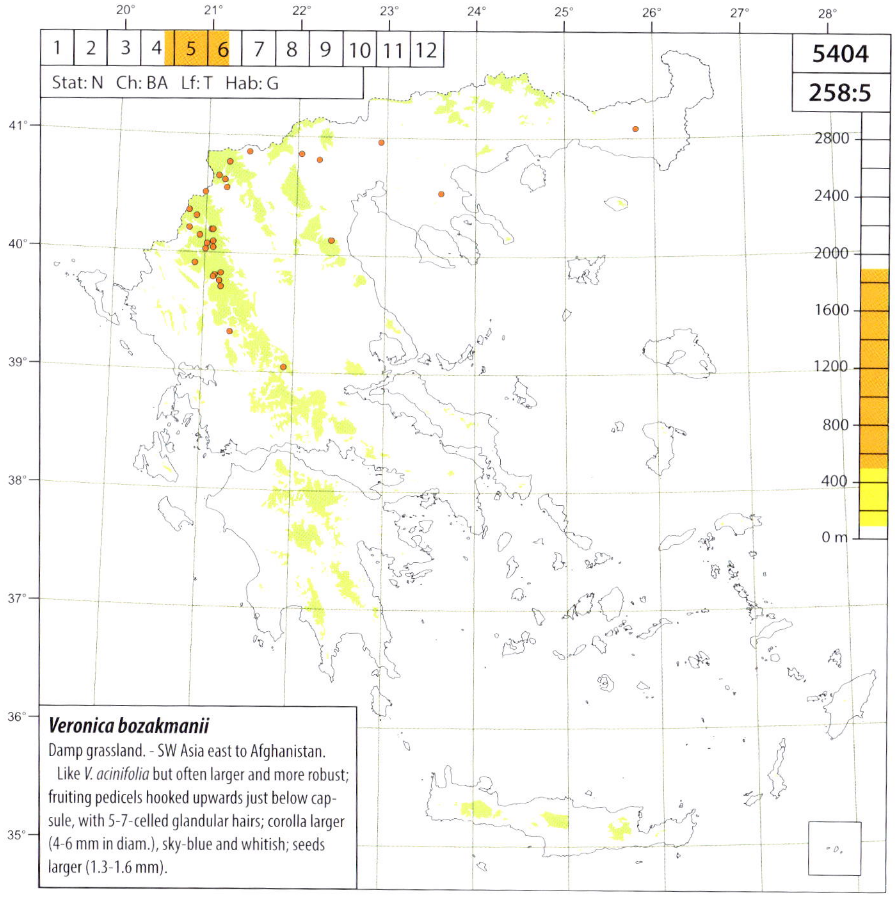

Veronica bozakmanii

Damp grassland. - SW Asia east to Afghanistan.

Like *V. acinifolia* but often larger and more robust; fruiting pedicels hooked upwards just below capsule, with 5-7-celled glandular hairs; corolla larger (4-6 mm in diam.), sky-blue and whitish; seeds larger (1.3-1.6 mm).

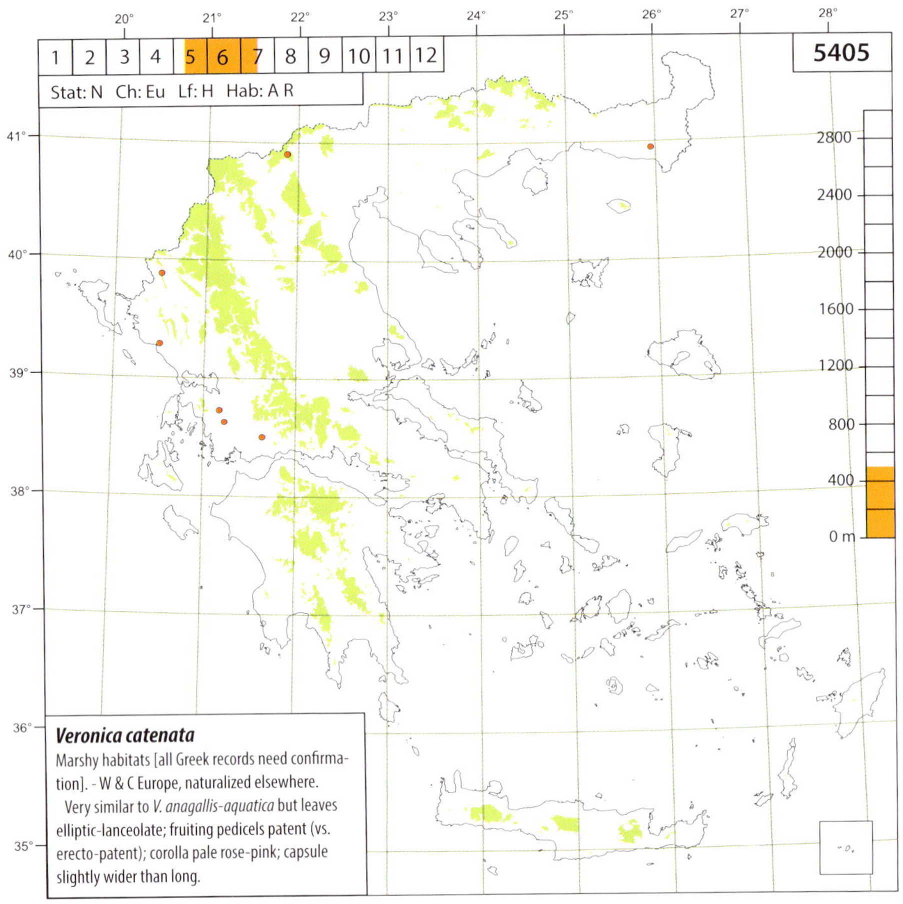

Veronica catenata
Marshy habitats [all Greek records need confirmation]. - W & C Europe, naturalized elsewhere.
Very similar to *V. anagallis-aquatica* but leaves elliptic-lanceolate; fruiting pedicels patent (vs. erecto-patent); corolla pale rose-pink; capsule slightly wider than long.

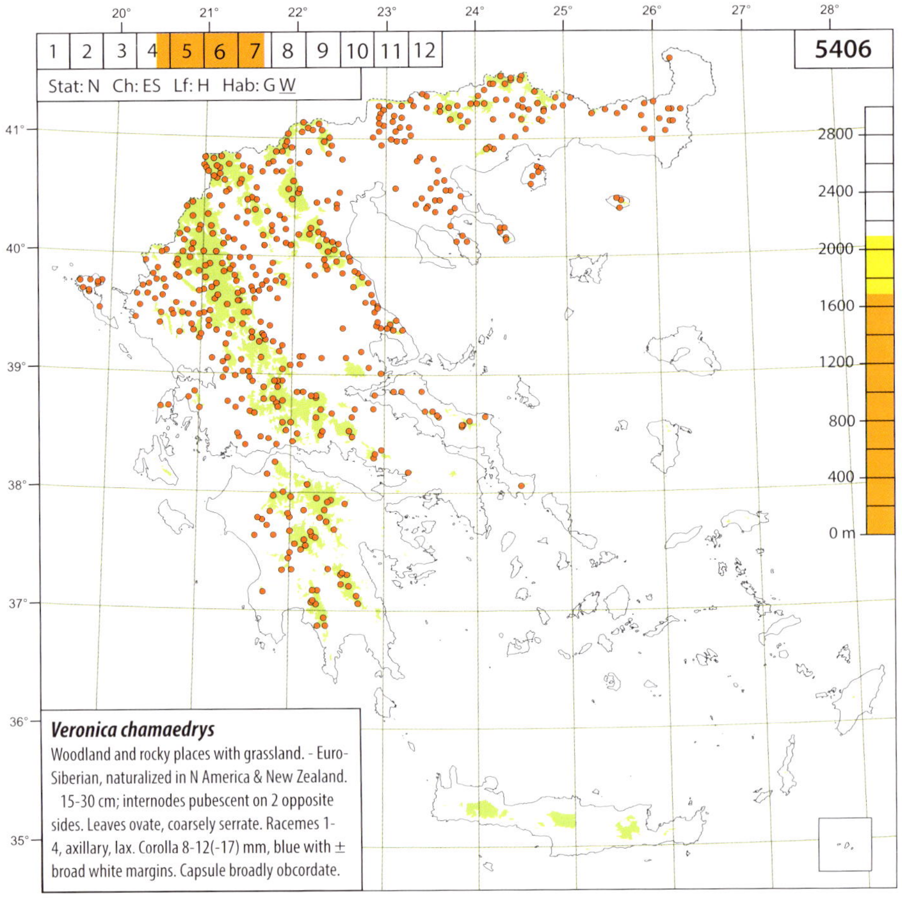

Veronica chamaedrys
Woodland and rocky places with grassland. - Euro-Siberian, naturalized in N America & New Zealand.
15-30 cm; internodes pubescent on 2 opposite sides. Leaves ovate, coarsely serrate. Racemes 1-4, axillary, lax. Corolla 8-12(-17) mm, blue with ± broad white margins. Capsule broadly obcordate.

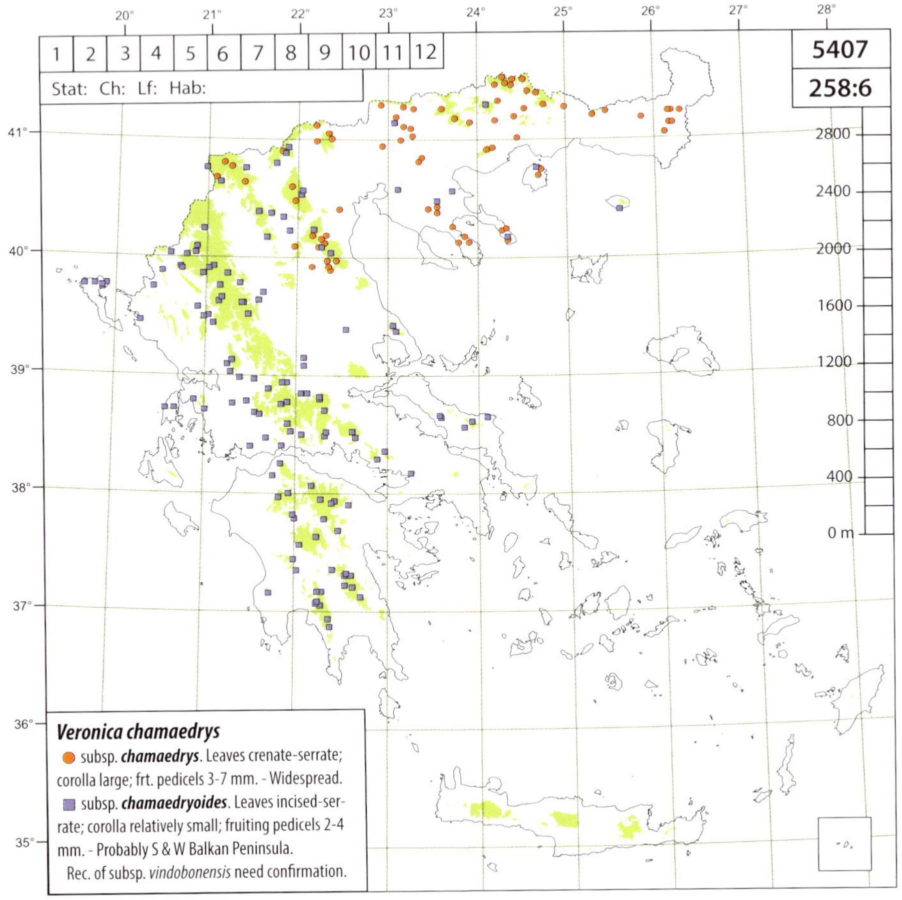

Veronica chamaedrys
● subsp. ***chamaedrys***. Leaves crenate-serrate; corolla large; frt. pedicels 3-7 mm. - Widespread.
■ subsp. ***chamaedryoides***. Leaves incised-serrate; corolla relatively small; fruiting pedicels 2-4 mm. - Probably S & W Balkan Peninsula.
Rec. of subsp. *vindobonensis* need confirmation.

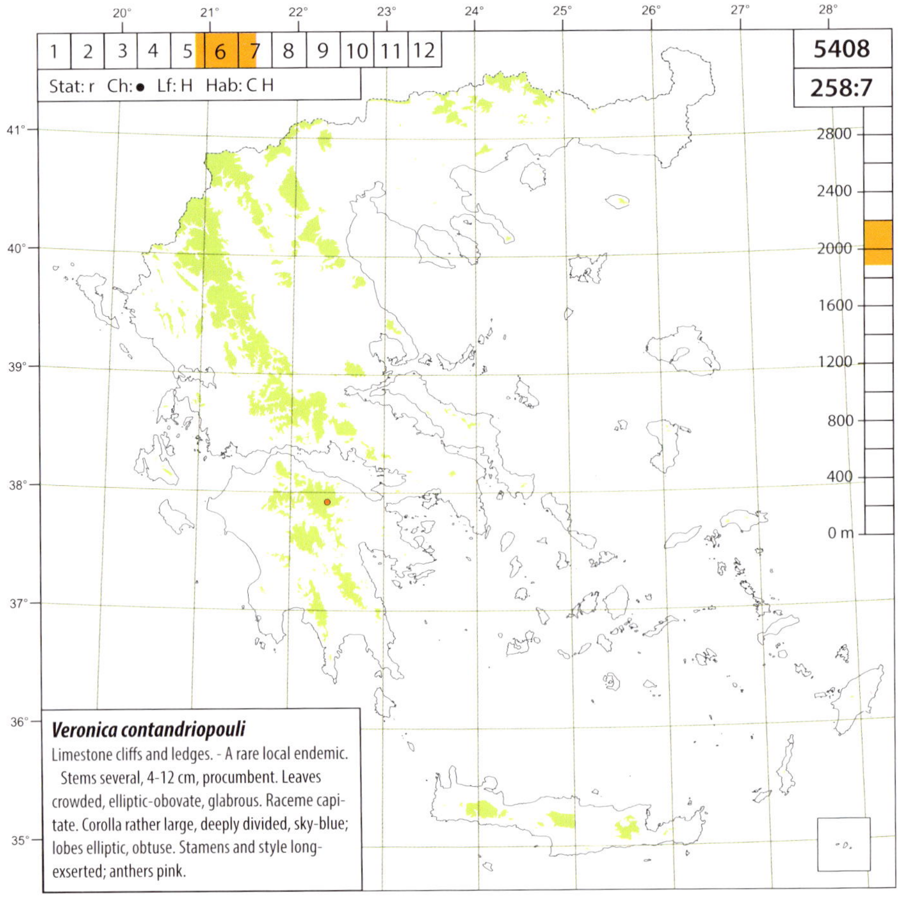

Veronica contandriopouli
Limestone cliffs and ledges. - A rare local endemic.
Stems several, 4-12 cm, procumbent. Leaves crowded, elliptic-obovate, glabrous. Raceme capitate. Corolla rather large, deeply divided, sky-blue; lobes elliptic, obtuse. Stamens and style long-exserted; anthers pink.

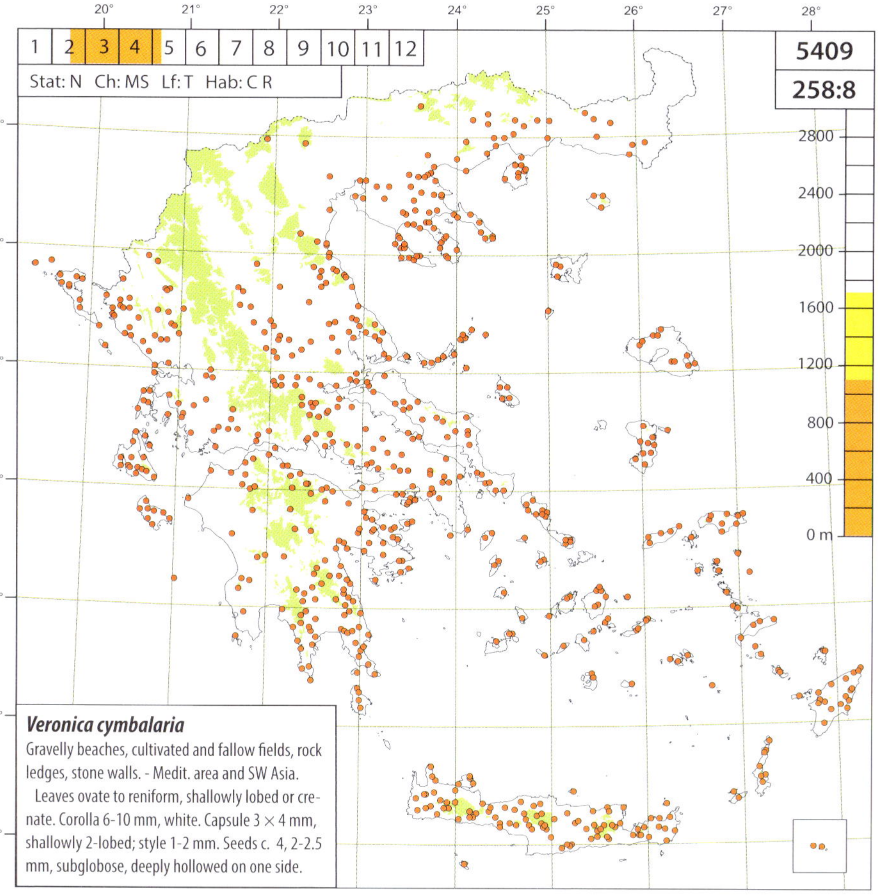

Veronica cymbalaria

Gravelly beaches, cultivated and fallow fields, rock ledges, stone walls. - Medit. area and SW Asia.

Leaves ovate to reniform, shallowly lobed or crenate. Corolla 6-10 mm, white. Capsule 3 × 4 mm, shallowly 2-lobed; style 1-2 mm. Seeds c. 4, 2-2.5 mm, subglobose, deeply hollowed on one side.

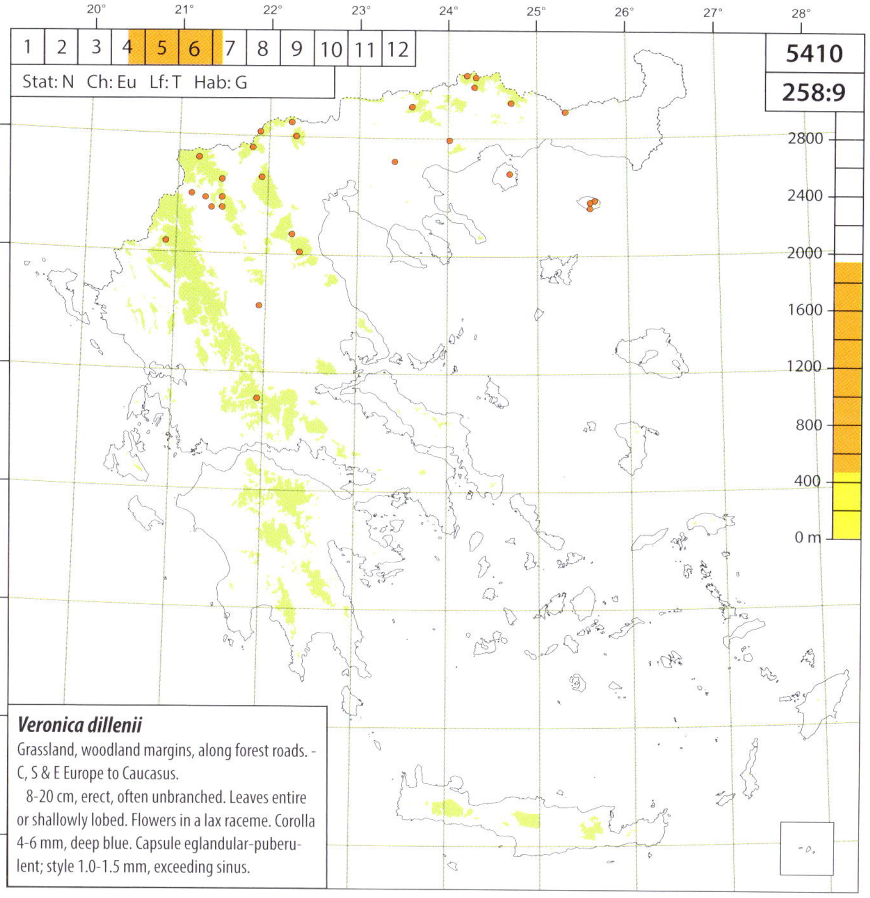

Veronica dillenii

Grassland, woodland margins, along forest roads. - C, S & E Europe to Caucasus.

8-20 cm, erect, often unbranched. Leaves entire or shallowly lobed. Flowers in a lax raceme. Corolla 4-6 mm, deep blue. Capsule eglandular-puberulent; style 1.0-1.5 mm, exceeding sinus.

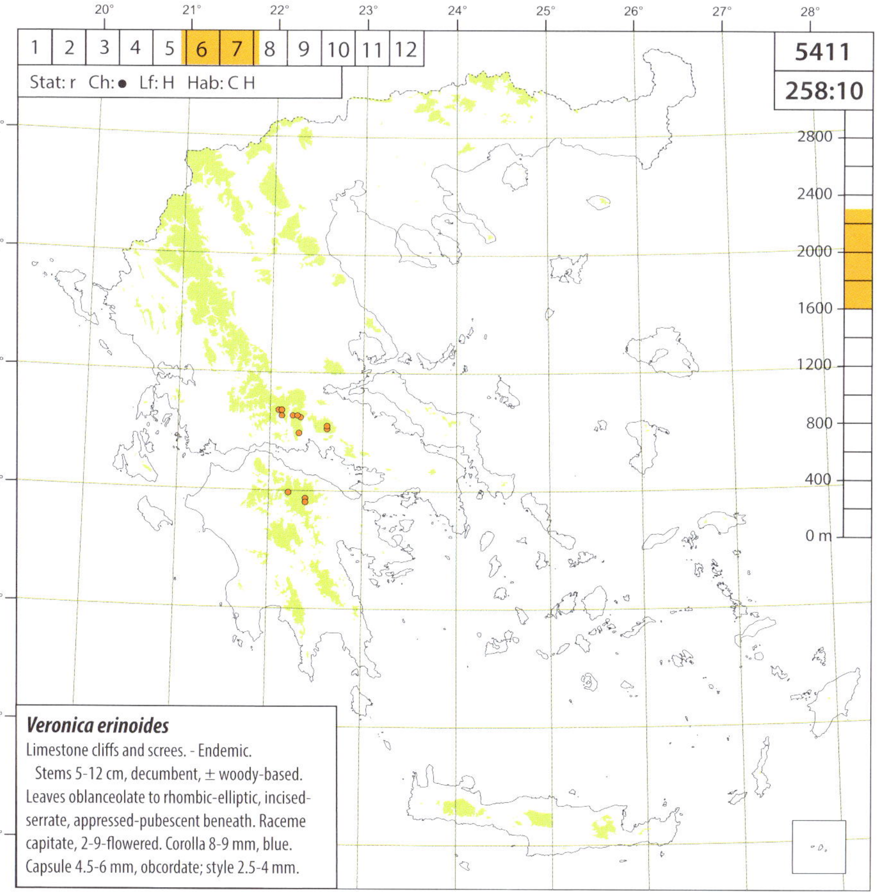

Veronica erinoides

Limestone cliffs and screes. - Endemic.

Stems 5-12 cm, decumbent, ± woody-based. Leaves oblanceolate to rhombic-elliptic, incised-serrate, appressed-pubescent beneath. Raceme capitate, 2-9-flowered. Corolla 8-9 mm, blue. Capsule 4.5-6 mm, obcordate; style 2.5-4 mm.

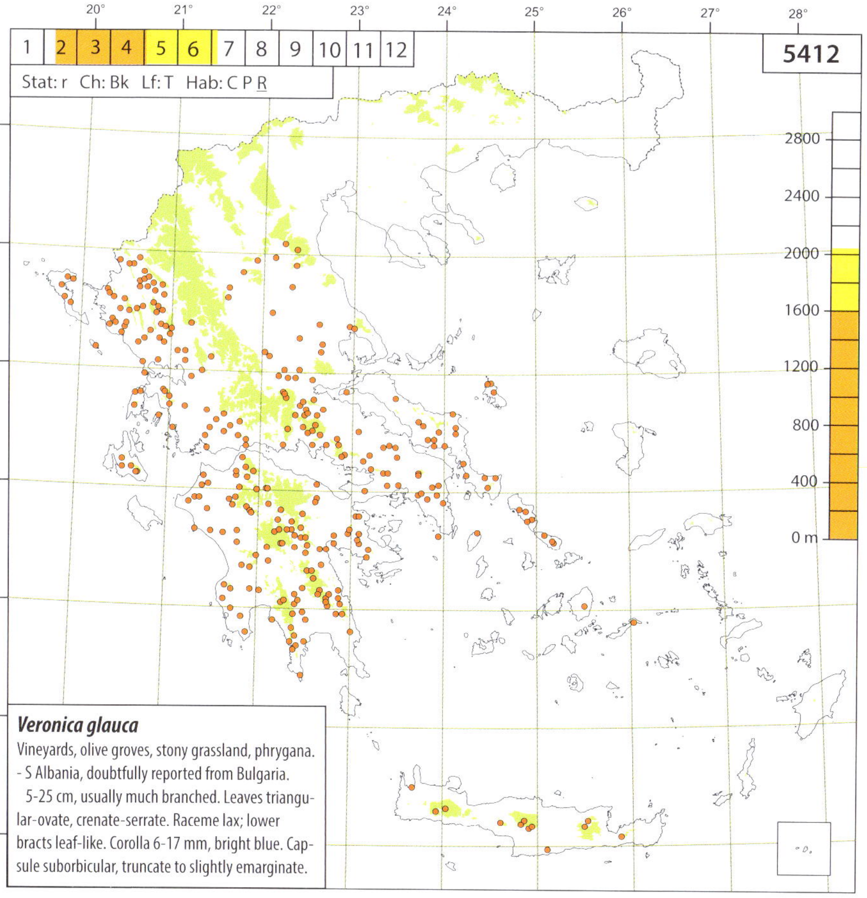

Veronica glauca

Vineyards, olive groves, stony grassland, phrygana. - S Albania, doubtfully reported from Bulgaria.

5-25 cm, usually much branched. Leaves triangular-ovate, crenate-serrate. Raceme lax; lower bracts leaf-like. Corolla 6-17 mm, bright blue. Capsule suborbicular, truncate to slightly emarginate.

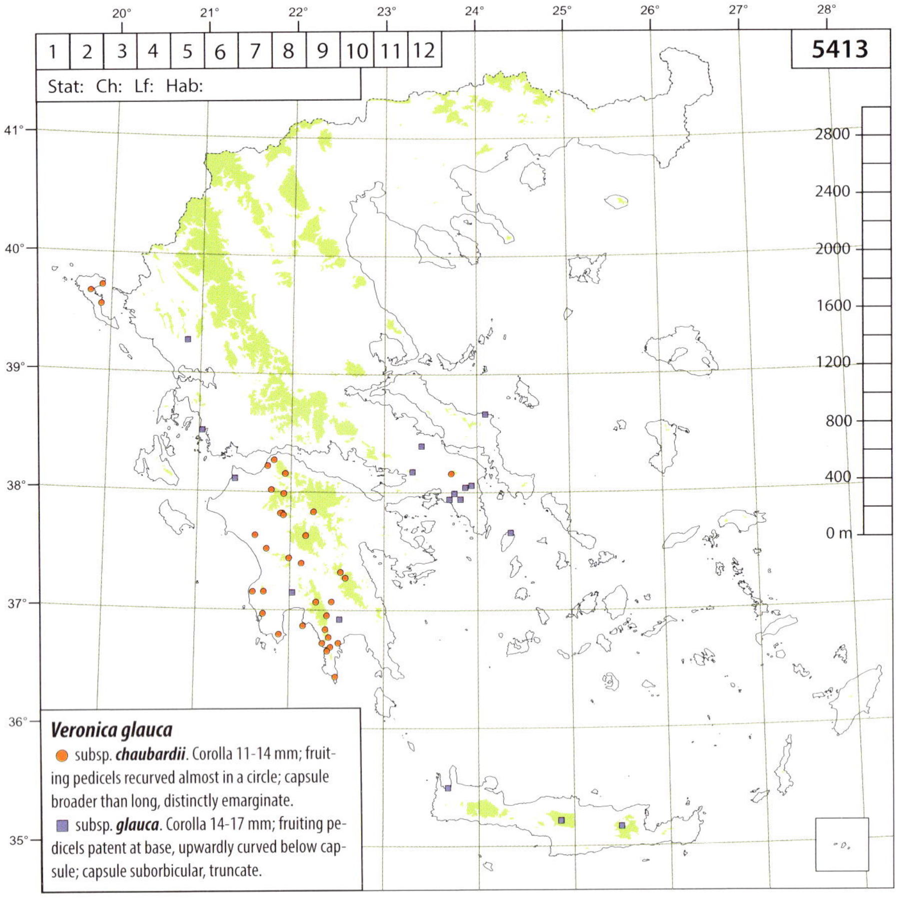

Veronica glauca

● subsp. **chaubardii**. Corolla 11-14 mm; fruiting pedicels recurved almost in a circle; capsule broader than long, distinctly emarginate.

■ subsp. **glauca**. Corolla 14-17 mm; fruiting pedicels patent at base, upwardly curved below capsule; capsule suborbicular, truncate.

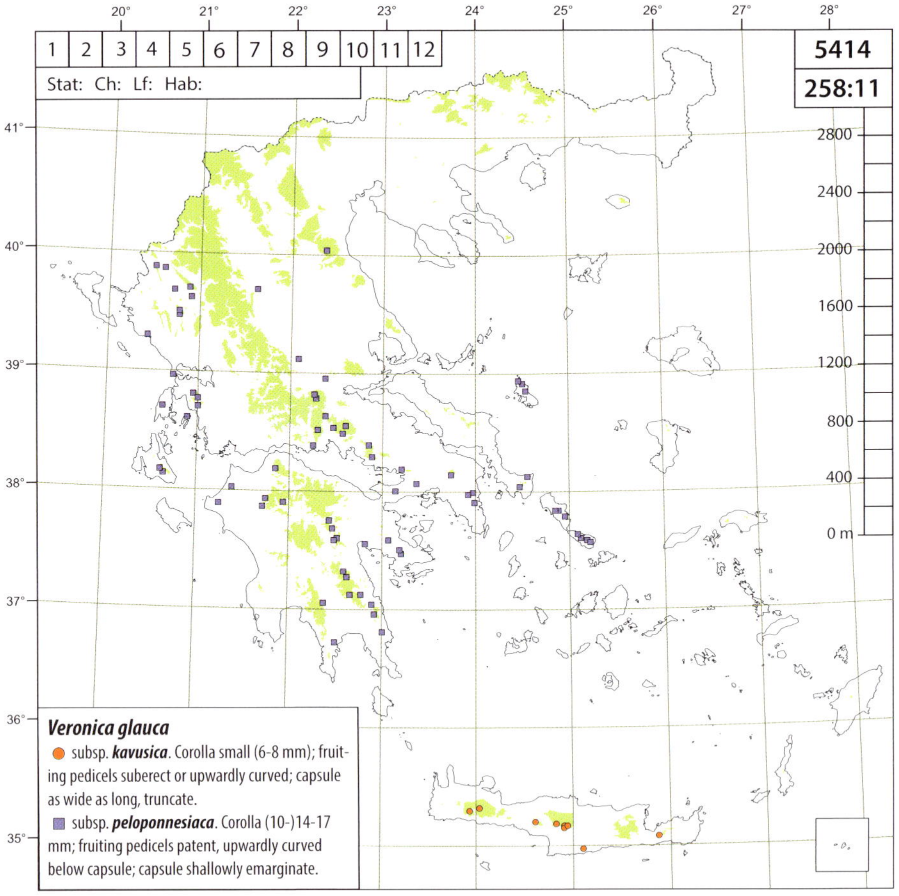

Veronica glauca

● subsp. **kavusica**. Corolla small (6-8 mm); fruiting pedicels suberect or upwardly curved; capsule as wide as long, truncate.

■ subsp. **peloponnesiaca**. Corolla (10-)14-17 mm; fruiting pedicels patent, upwardly curved below capsule; capsule shallowly emarginate.

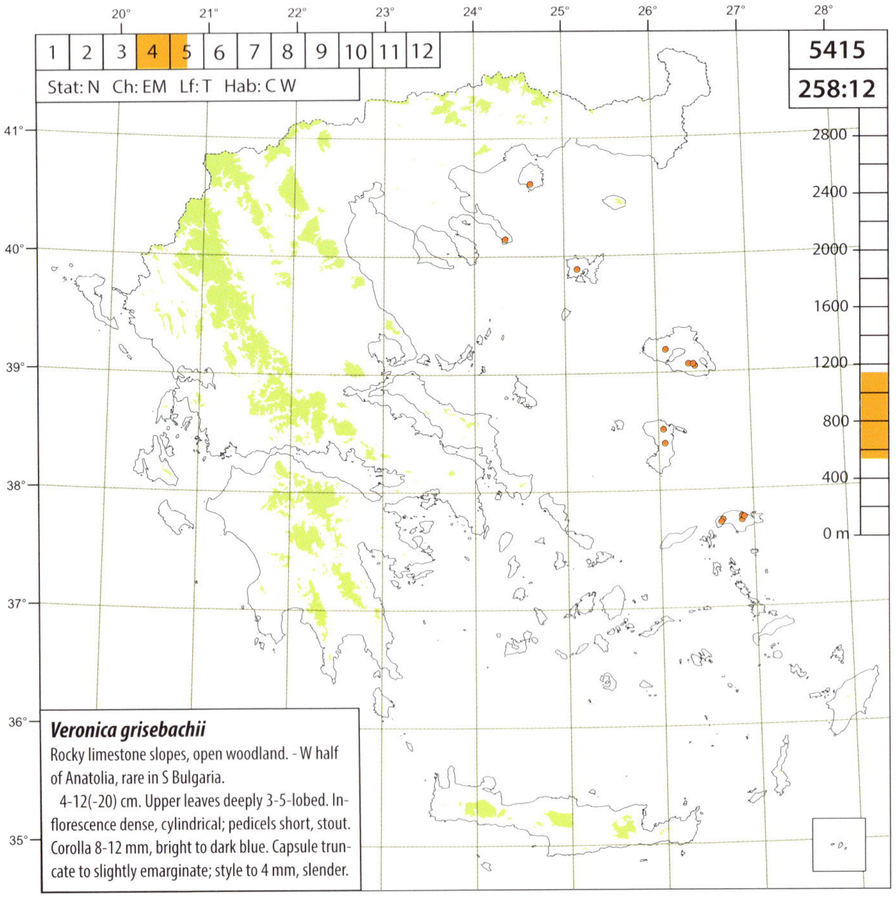

Veronica grisebachii

Rocky limestone slopes, open woodland. - W half of Anatolia, rare in S Bulgaria.

4-12(-20) cm. Upper leaves deeply 3-5-lobed. Inflorescence dense, cylindrical; pedicels short, stout. Corolla 8-12 mm, bright to dark blue. Capsule truncate to slightly emarginate; style to 4 mm, slender.

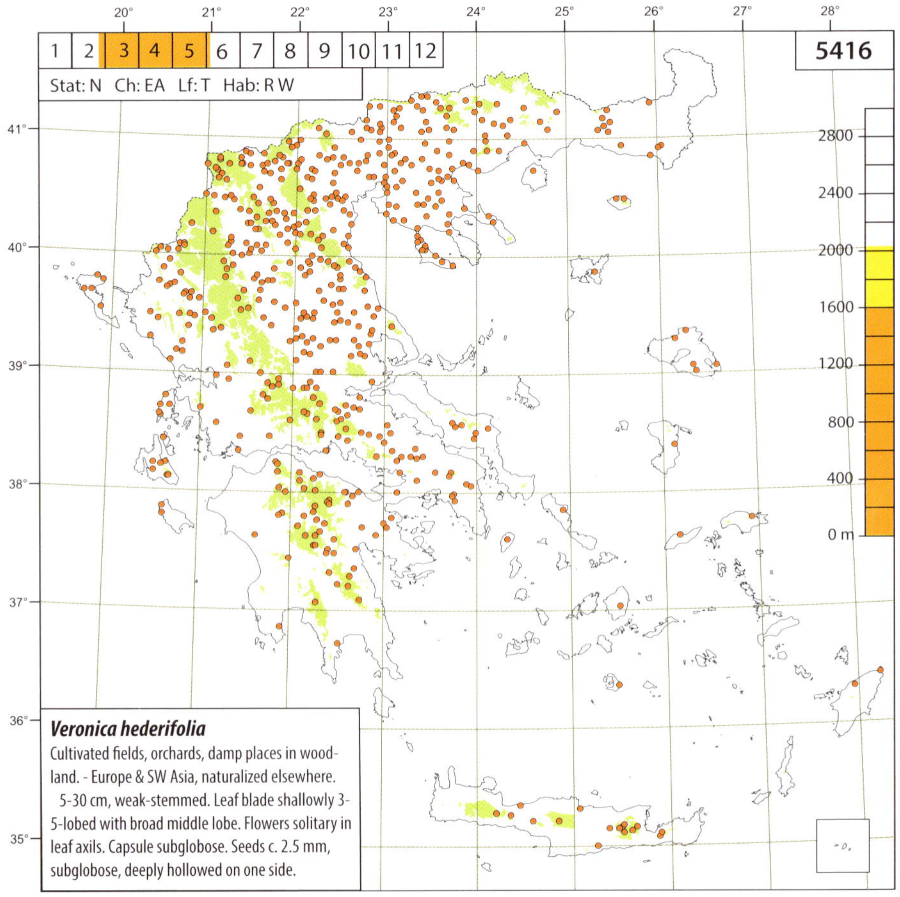

Veronica hederifolia

Cultivated fields, orchards, damp places in woodland. - Europe & SW Asia, naturalized elsewhere.

5-30 cm, weak-stemmed. Leaf blade shallowly 3-5-lobed with broad middle lobe. Flowers solitary in leaf axils. Capsule subglobose. Seeds c. 2.5 mm, subglobose, deeply hollowed on one side.

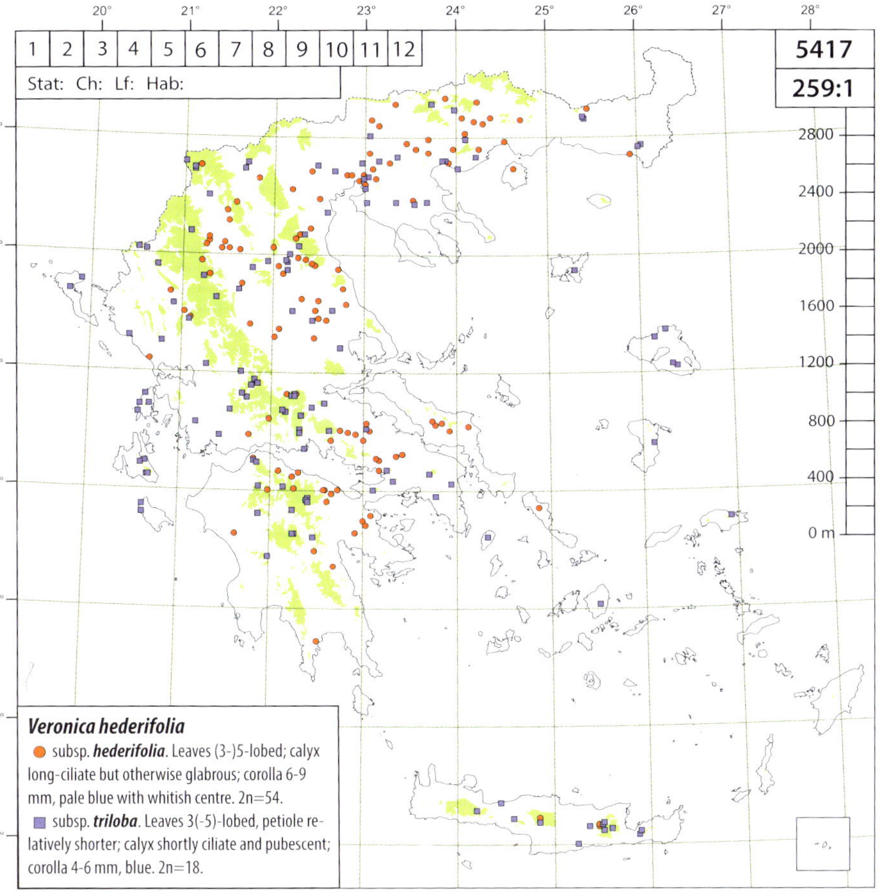

Veronica hederifolia

● subsp. ***hederifolia***. Leaves (3-)5-lobed; calyx long-ciliate but otherwise glabrous; corolla 6-9 mm, pale blue with whitish centre. 2n=54.

■ subsp. ***triloba***. Leaves 3(-5)-lobed, petiole relatively shorter; calyx shortly ciliate and pubescent; corolla 4-6 mm, blue. 2n=18.

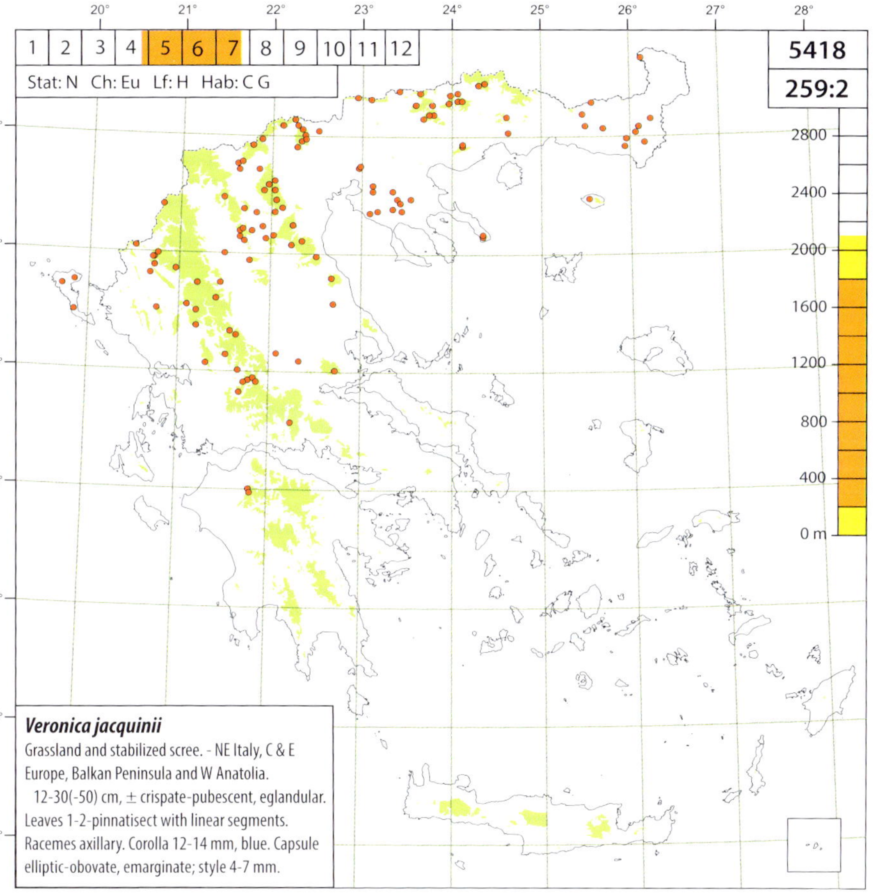

Veronica jacquinii

Grassland and stabilized scree. - NE Italy, C & E Europe, Balkan Peninsula and W Anatolia.

12-30(-50) cm, ± crispate-pubescent, eglandular. Leaves 1-2-pinnatisect with linear segments. Racemes axillary. Corolla 12-14 mm, blue. Capsule elliptic-obovate, emarginate; style 4-7 mm.

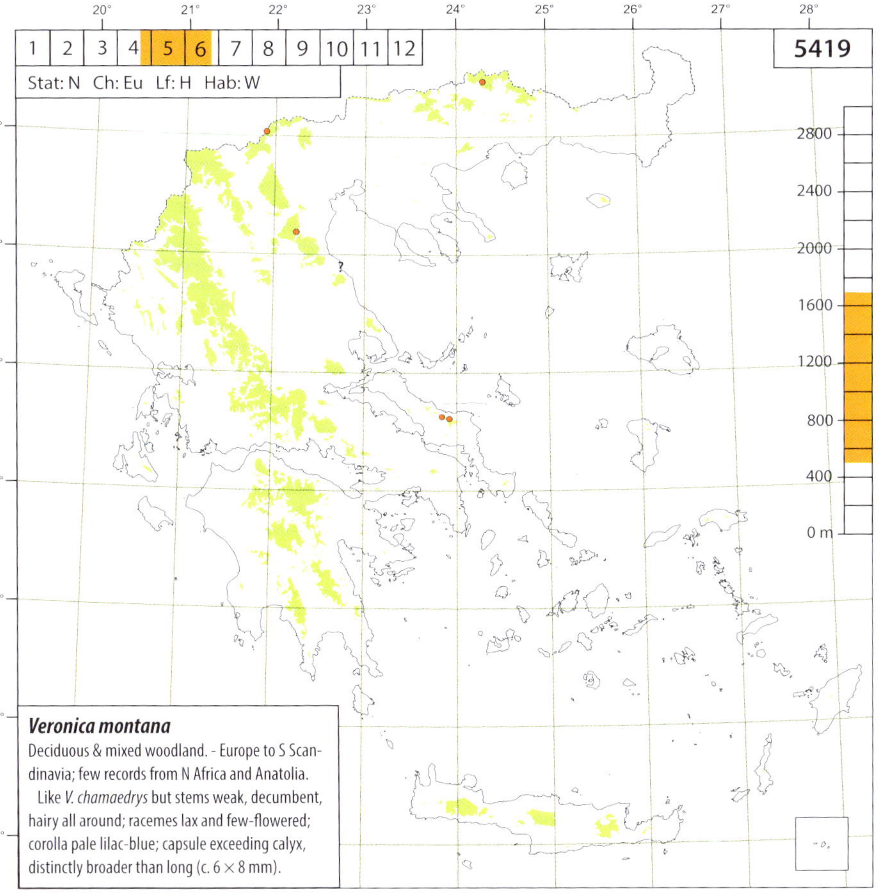

Veronica montana

Deciduous & mixed woodland. - Europe to S Scandinavia; few records from N Africa and Anatolia.

Like *V. chamaedrys* but stems weak, decumbent, hairy all around; racemes lax and few-flowered; corolla pale lilac-blue; capsule exceeding calyx, distinctly broader than long (c. 6 × 8 mm).

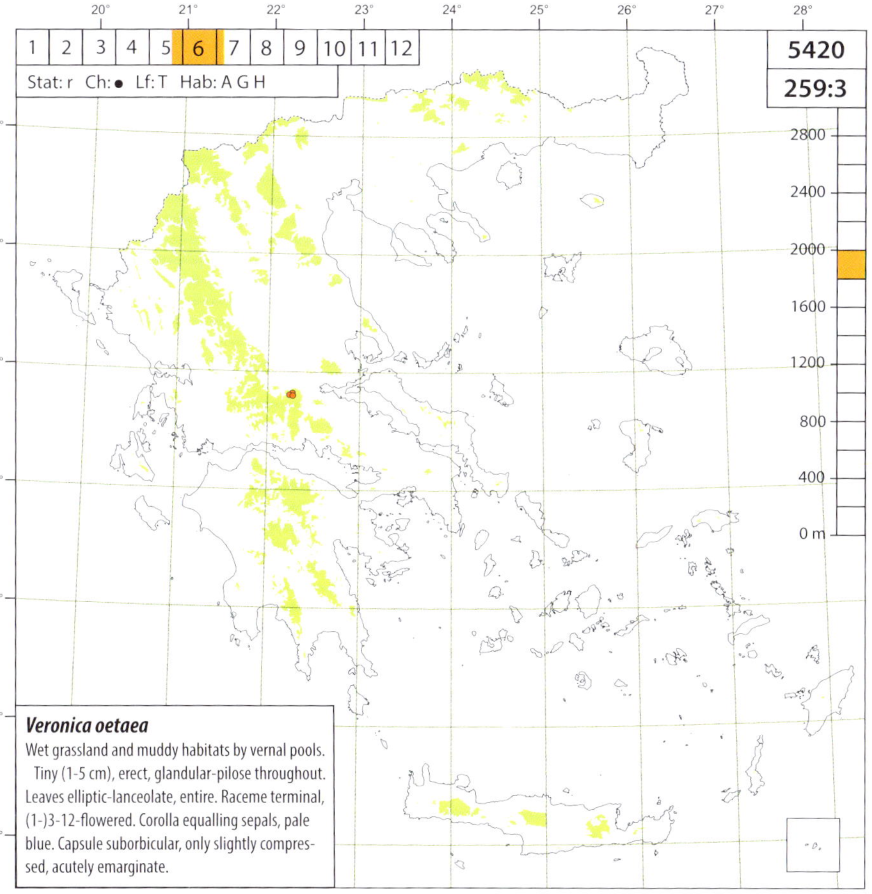

Veronica oetaea

Wet grassland and muddy habitats by vernal pools.

Tiny (1-5 cm), erect, glandular-pilose throughout. Leaves elliptic-lanceolate, entire. Raceme terminal, (1-)3-12-flowered. Corolla equalling sepals, pale blue. Capsule suborbicular, only slightly compressed, acutely emarginate.

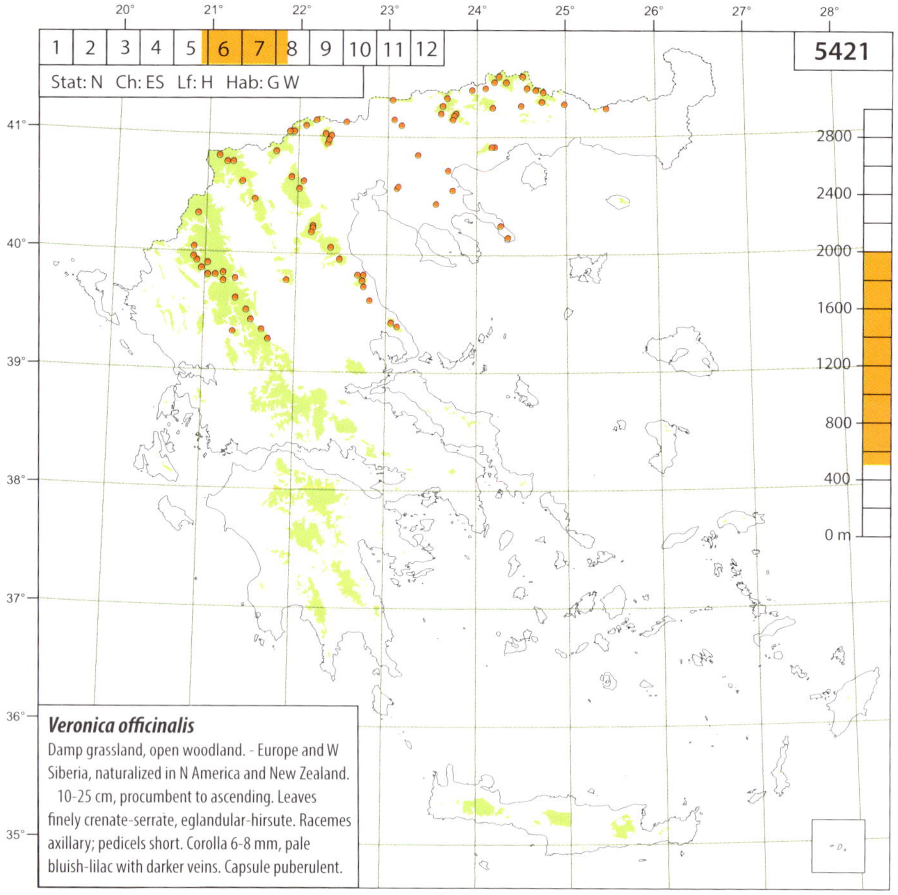

Veronica officinalis
Damp grassland, open woodland. - Europe and W Siberia, naturalized in N America and New Zealand.
10-25 cm, procumbent to ascending. Leaves finely crenate-serrate, eglandular-hirsute. Racemes axillary; pedicels short. Corolla 6-8 mm, pale bluish-lilac with darker veins. Capsule puberulent.

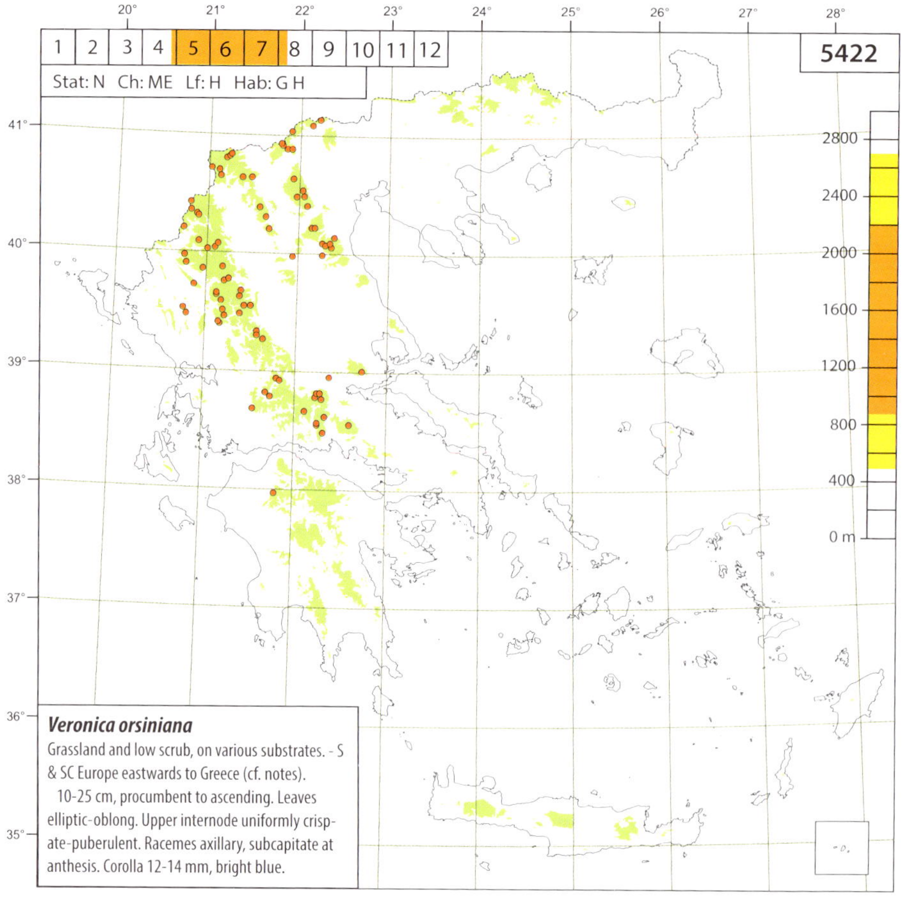

Veronica orsiniana
Grassland and low scrub, on various substrates. - S & SC Europe eastwards to Greece (cf. notes).
10-25 cm, procumbent to ascending. Leaves elliptic-oblong. Upper internode uniformly crisp-ate-puberulent. Racemes axillary, subcapitate at anthesis. Corolla 12-14 mm, bright blue.

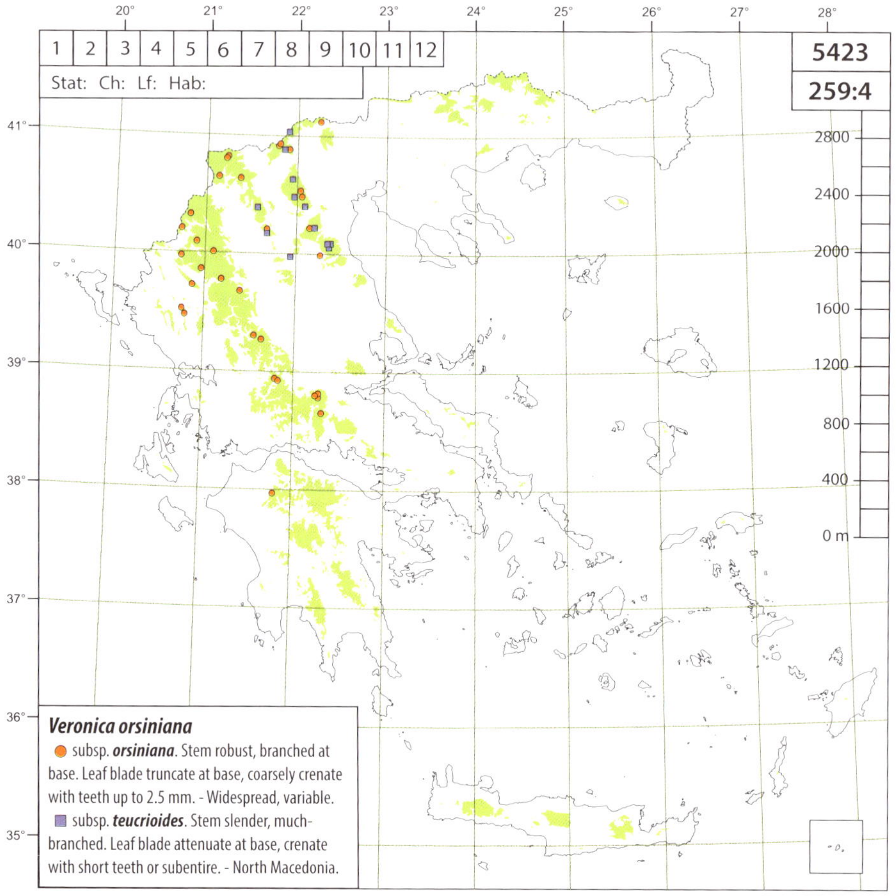

Veronica orsiniana
● subsp. ***orsiniana***. Stem robust, branched at base. Leaf blade truncate at base, coarsely crenate with teeth up to 2.5 mm. - Widespread, variable.
■ subsp. ***teucrioides***. Stem slender, much-branched. Leaf blade attenuate at base, crenate with short teeth or subentire. - North Macedonia.

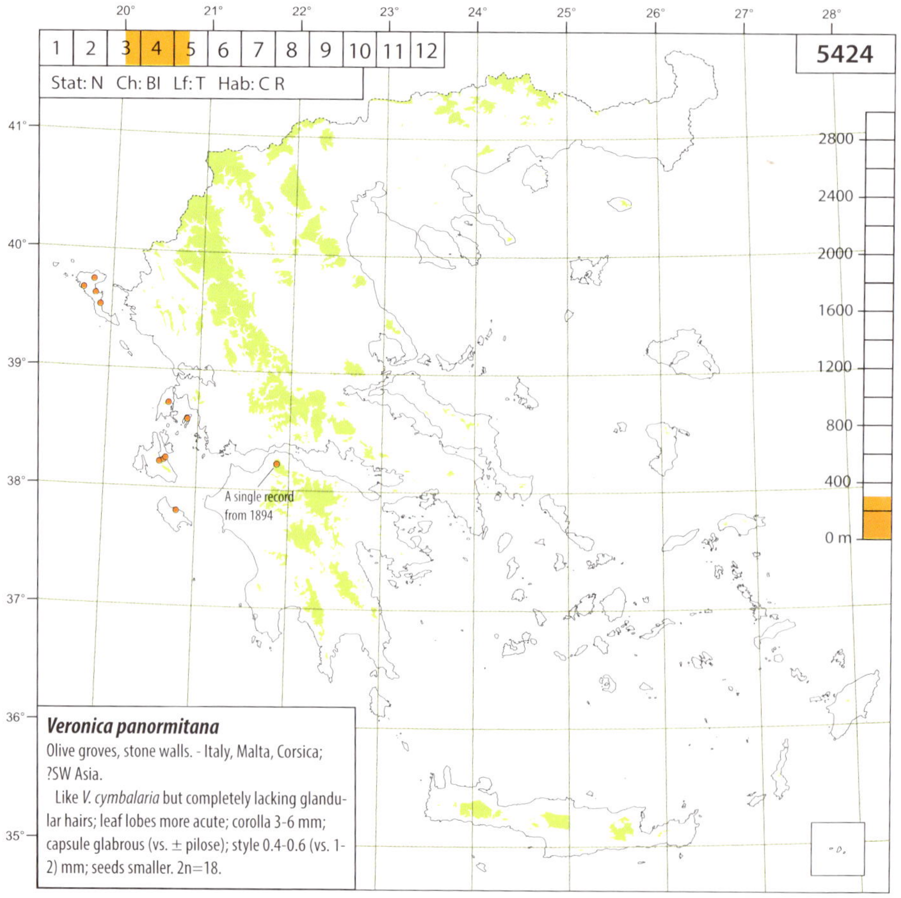

Veronica panormitana
Olive groves, stone walls. - Italy, Malta, Corsica; ?SW Asia.
Like *V. cymbalaria* but completely lacking glandular hairs; leaf lobes more acute; corolla 3-6 mm; capsule glabrous (vs. ± pilose); style 0.4-0.6 (vs. 1-2) mm; seeds smaller. 2n=18.

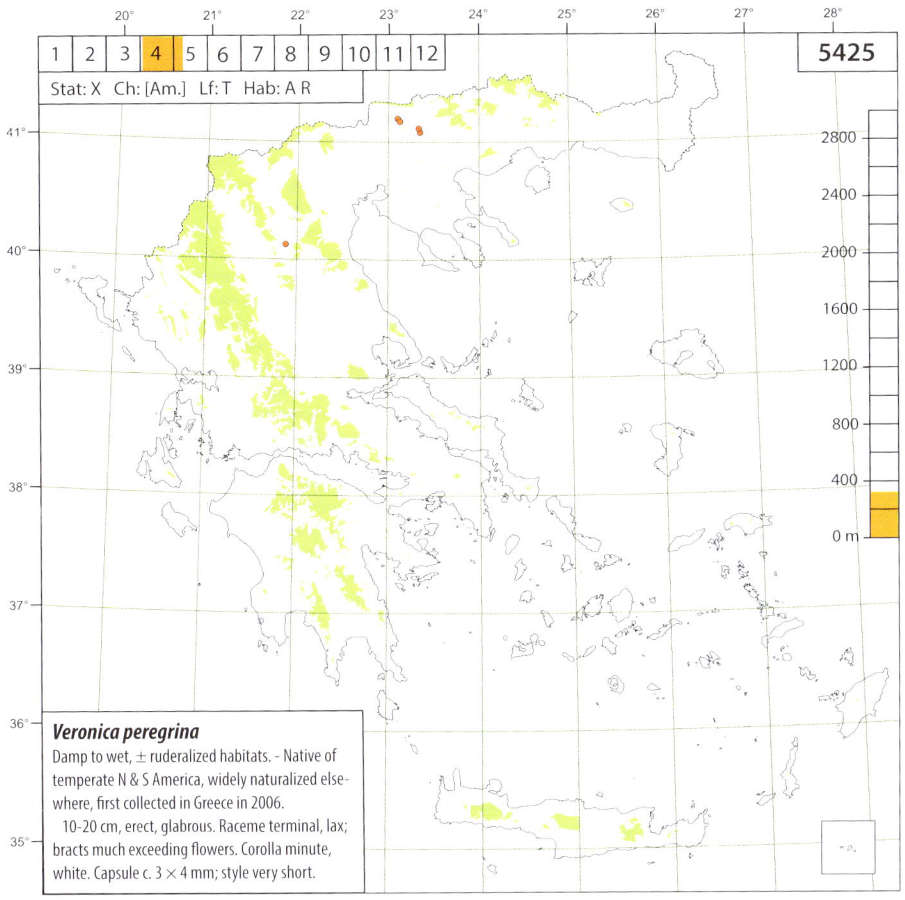

Veronica peregrina

Damp to wet, ± ruderalized habitats. - Native of temperate N & S America, widely naturalized elsewhere, first collected in Greece in 2006.

10-20 cm, erect, glabrous. Raceme terminal, lax; bracts much exceeding flowers. Corolla minute, white. Capsule c. 3 × 4 mm; style very short.

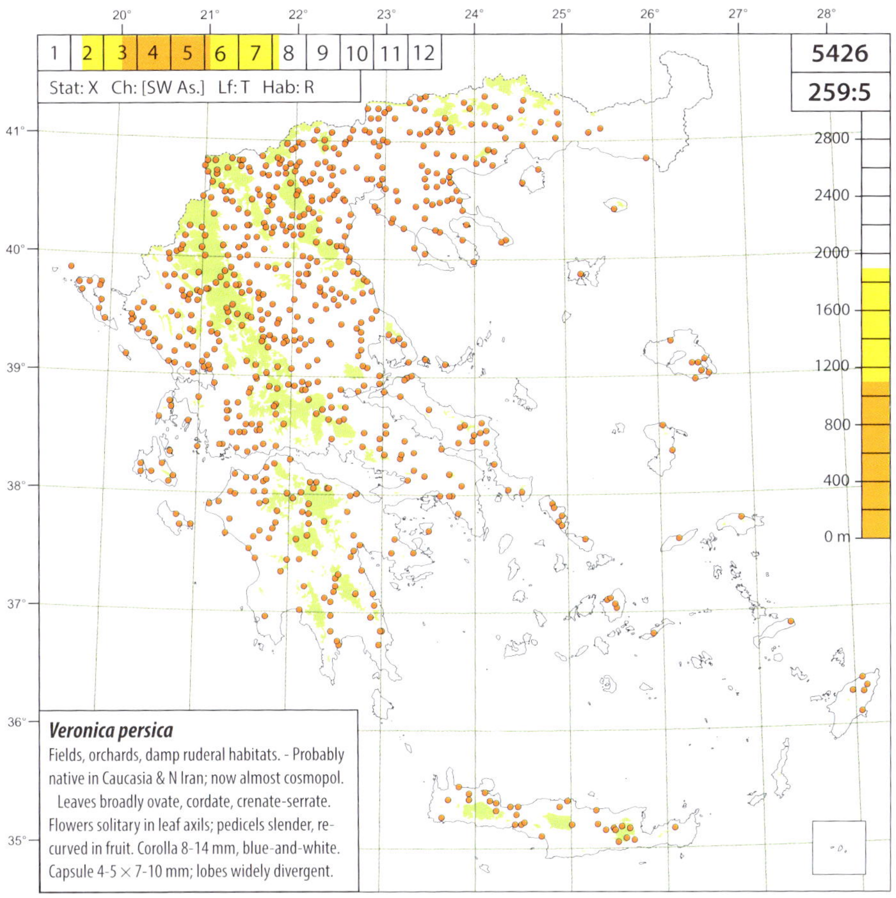

Veronica persica

Fields, orchards, damp ruderal habitats. - Probably native in Caucasia & N Iran; now almost cosmopol.

Leaves broadly ovate, cordate, crenate-serrate. Flowers solitary in leaf axils; pedicels slender, recurved in fruit. Corolla 8-14 mm, blue-and-white. Capsule 4-5 × 7-10 mm; lobes widely divergent.

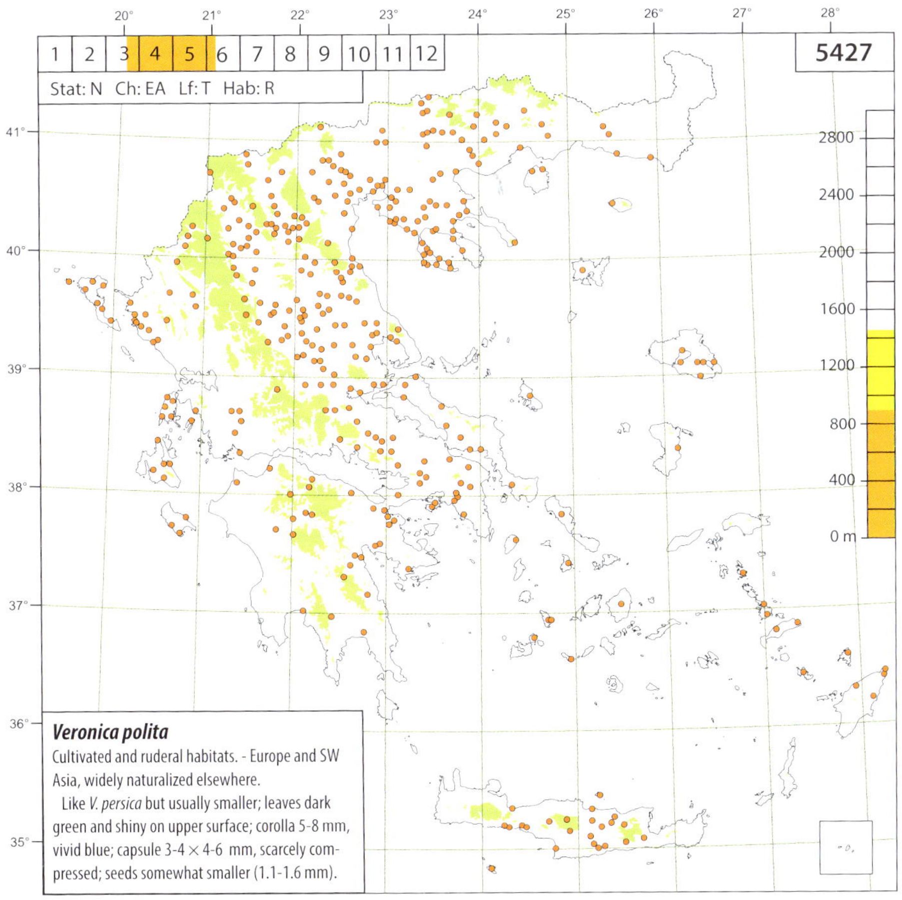

Veronica polita

Cultivated and ruderal habitats. - Europe and SW Asia, widely naturalized elsewhere.

Like *V. persica* but usually smaller; leaves dark green and shiny on upper surface; corolla 5-8 mm, vivid blue; capsule 3-4 × 4-6 mm, scarcely compressed; seeds somewhat smaller (1.1-1.6 mm).

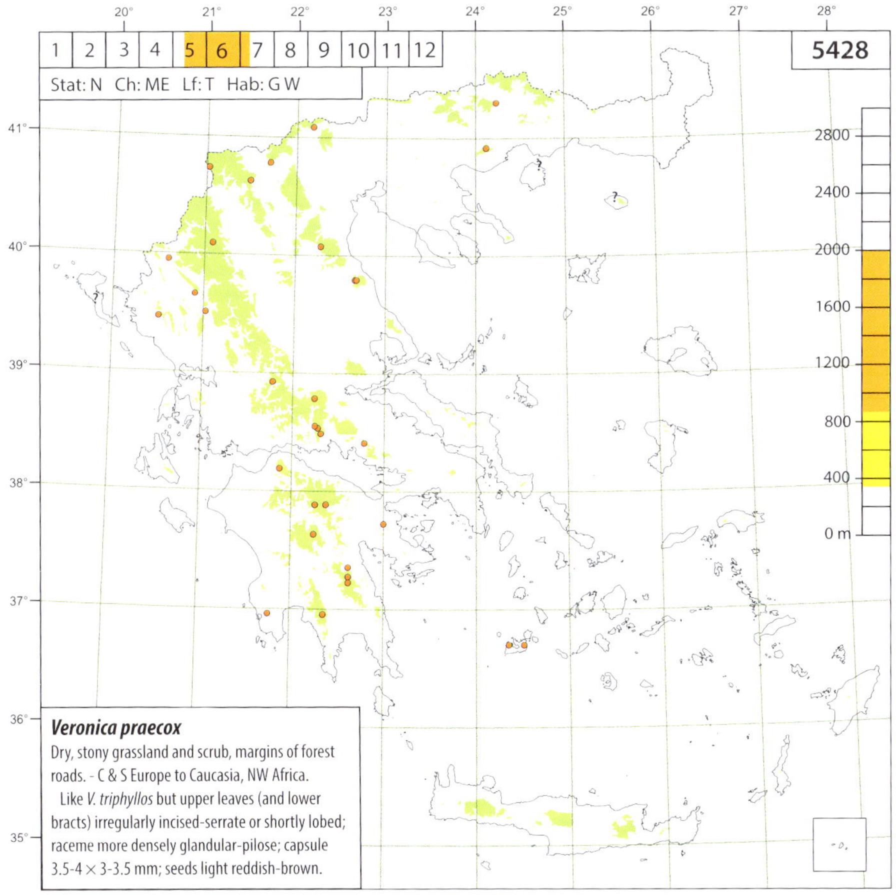

Veronica praecox

Dry, stony grassland and scrub, margins of forest roads. - C & S Europe to Caucasia, NW Africa.

Like *V. triphyllos* but upper leaves (and lower bracts) irregularly incised-serrate or shortly lobed; raceme more densely glandular-pilose; capsule 3.5-4 × 3-3.5 mm; seeds light reddish-brown.

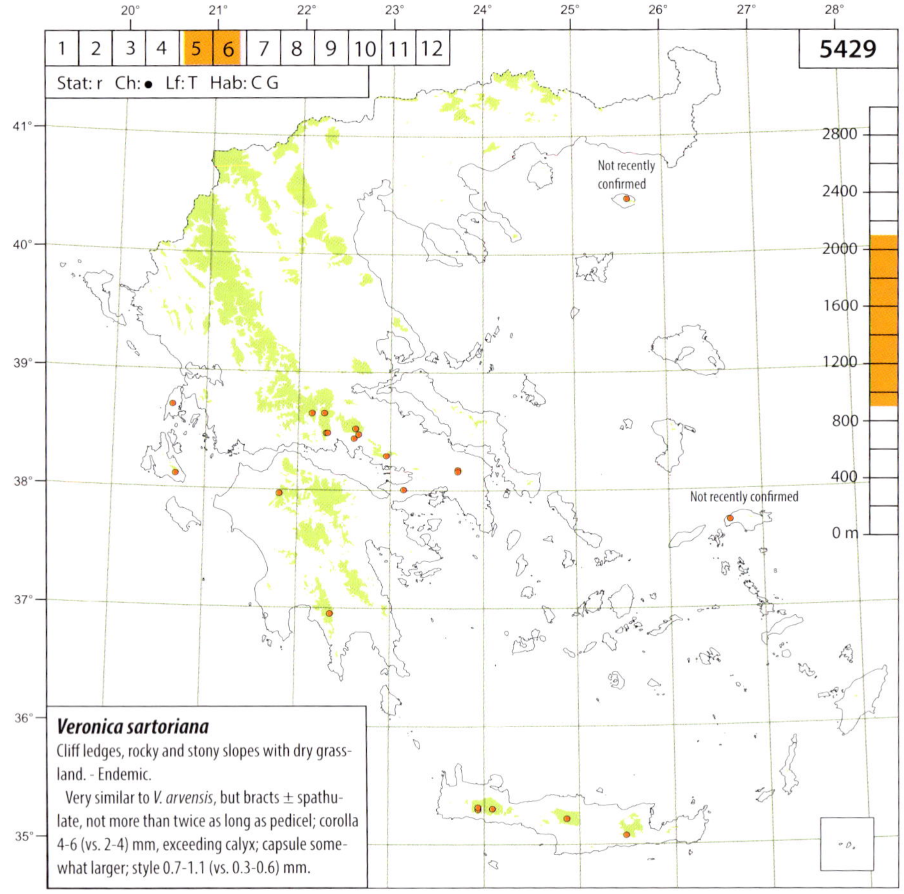

Veronica sartoriana
Cliff ledges, rocky and stony slopes with dry grassland. - Endemic.
Very similar to *V. arvensis*, but bracts ± spathulate, not more than twice as long as pedicel; corolla 4-6 (vs. 2-4) mm, exceeding calyx; capsule somewhat larger; style 0.7-1.1 (vs. 0.3-0.6) mm.

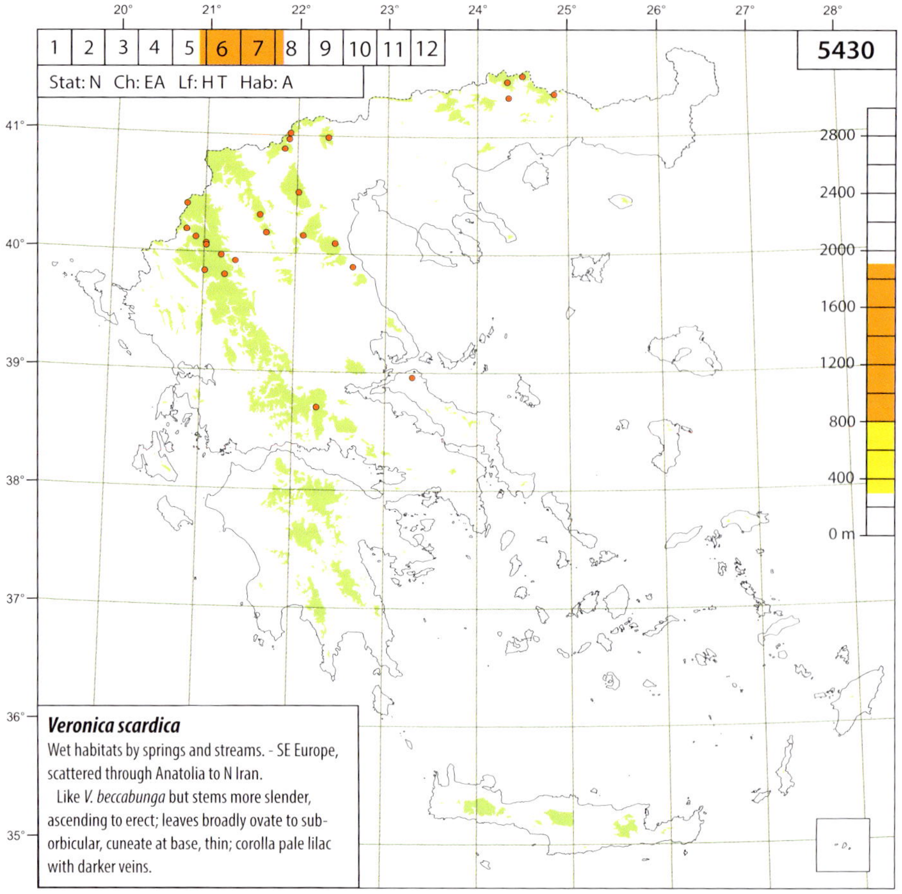

Veronica scardica
Wet habitats by springs and streams. - SE Europe, scattered through Anatolia to N Iran.
Like *V. beccabunga* but stems more slender, ascending to erect; leaves broadly ovate to suborbicular, cuneate at base, thin; corolla pale lilac with darker veins.

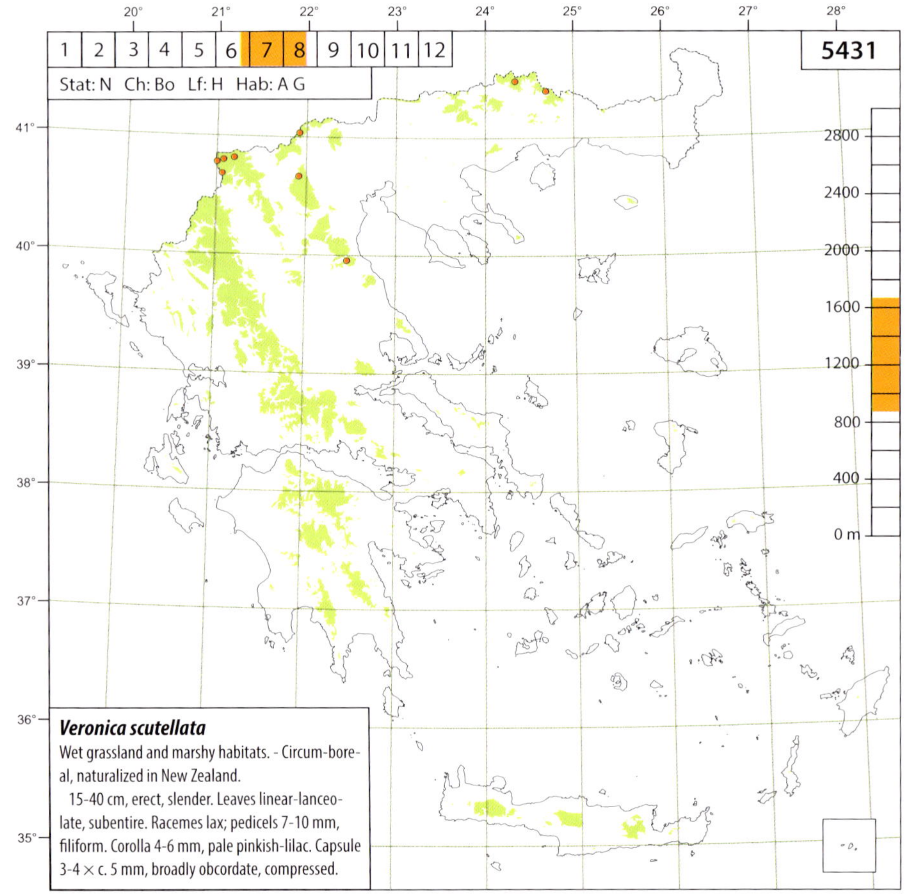

Veronica scutellata
Wet grassland and marshy habitats. - Circum-boreal, naturalized in New Zealand.
15-40 cm, erect, slender. Leaves linear-lanceolate, subentire. Racemes lax; pedicels 7-10 mm, filiform. Corolla 4-6 mm, pale pinkish-lilac. Capsule 3-4 × c. 5 mm, broadly obcordate, compressed.

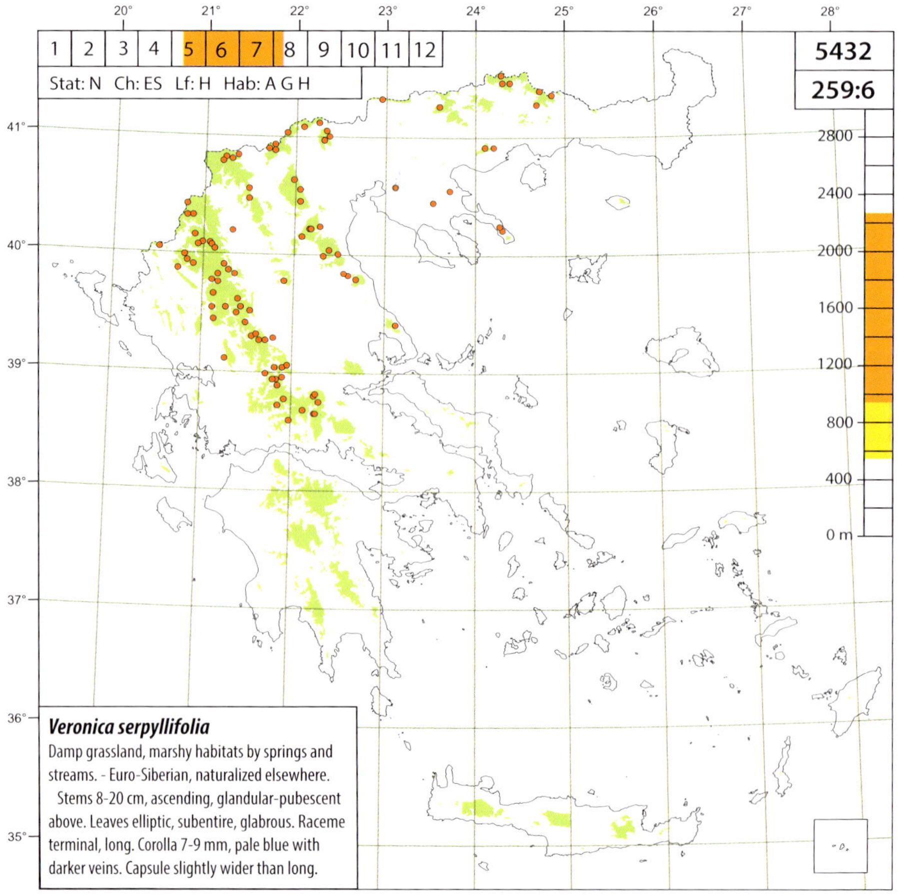

Veronica serpyllifolia
Damp grassland, marshy habitats by springs and streams. - Euro-Siberian, naturalized elsewhere.
Stems 8-20 cm, ascending, glandular-pubescent above. Leaves elliptic, subentire, glabrous. Raceme terminal, long. Corolla 7-9 mm, pale blue with darker veins. Capsule slightly wider than long.

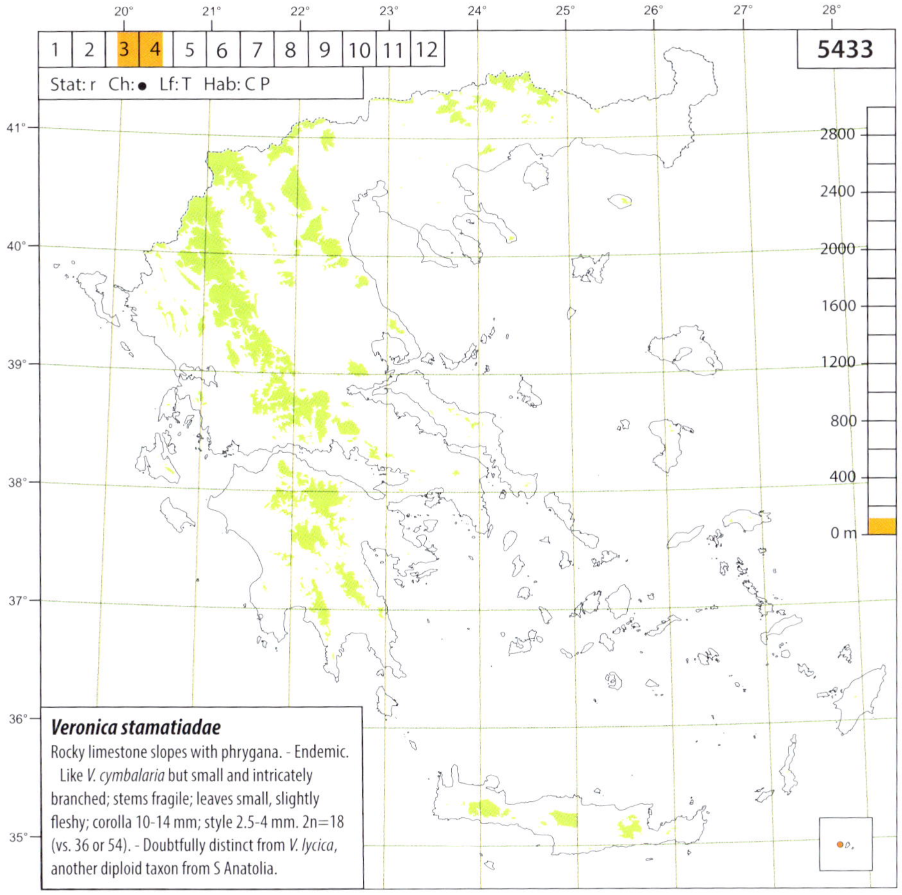

Veronica stamatiadae

Rocky limestone slopes with phrygana. - Endemic.

Like *V. cymbalaria* but small and intricately branched; stems fragile; leaves small, slightly fleshy; corolla 10-14 mm; style 2.5-4 mm. 2n=18 (vs. 36 or 54). - Doubtfully distinct from *V. lycica*, another diploid taxon from S Anatolia.

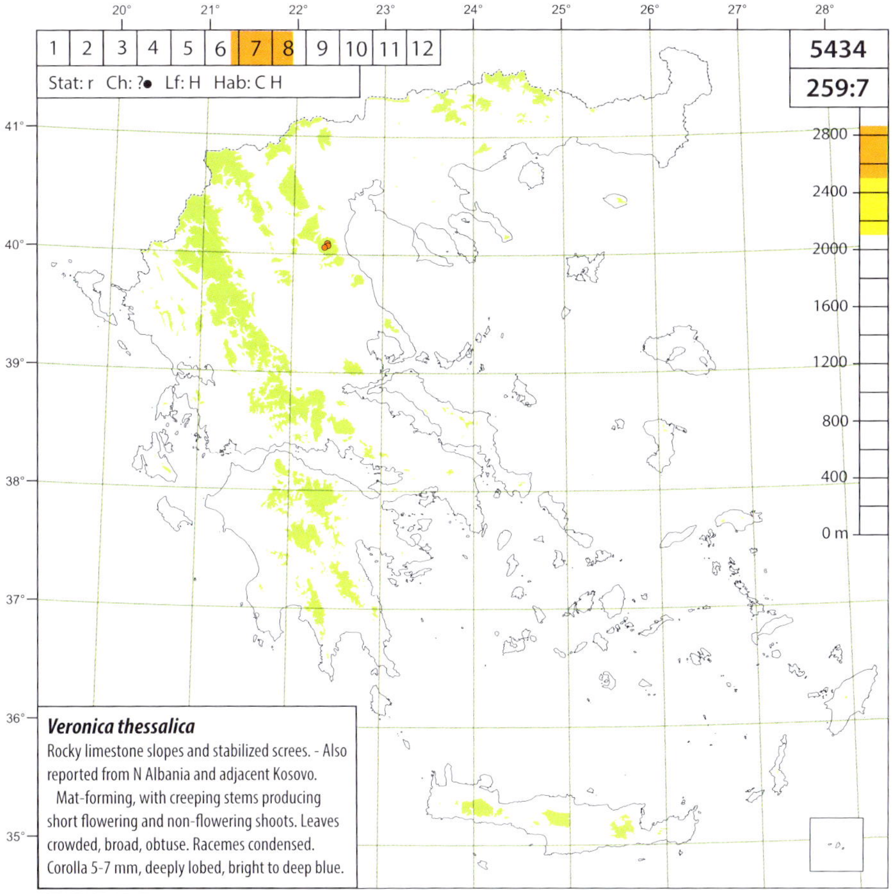

Veronica thessalica

Rocky limestone slopes and stabilized screes. - Also reported from N Albania and adjacent Kosovo.

Mat-forming, with creeping stems producing short flowering and non-flowering shoots. Leaves crowded, broad, obtuse. Racemes condensed. Corolla 5-7 mm, deeply lobed, bright to deep blue.

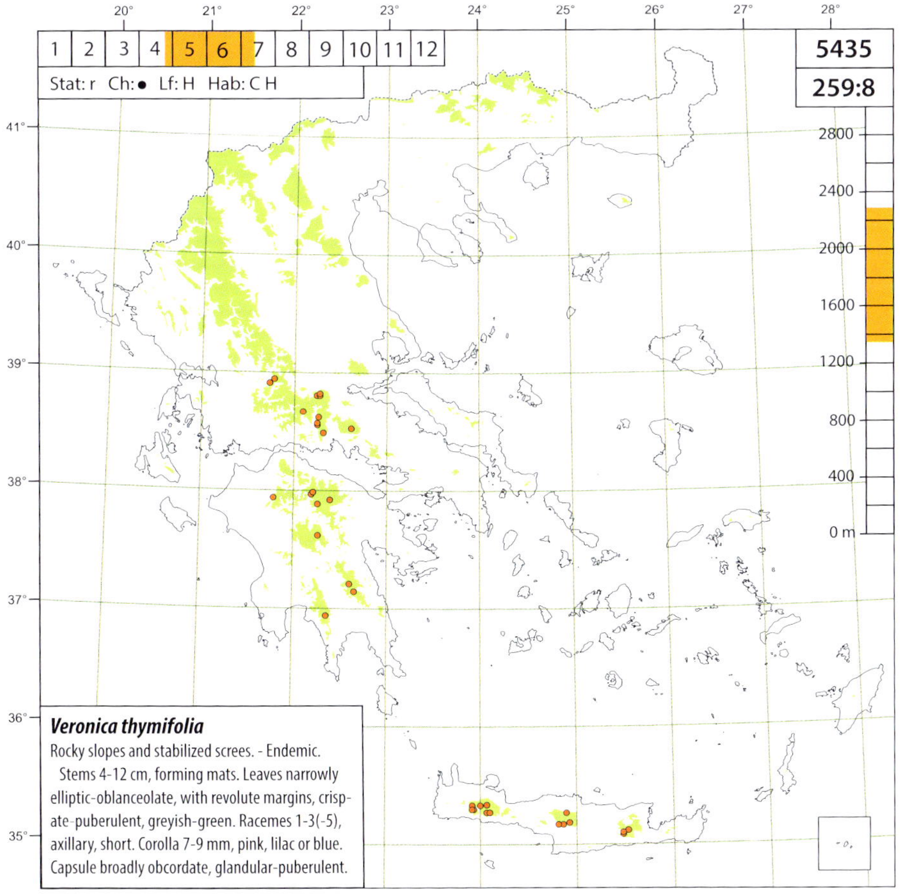

Veronica thymifolia

Rocky slopes and stabilized screes. - Endemic.

Stems 4-12 cm, forming mats. Leaves narrowly elliptic-oblanceolate, with revolute margins, crispate-puberulent, greyish-green. Racemes 1-3(-5), axillary, short. Corolla 7-9 mm, pink, lilac or blue. Capsule broadly obcordate, glandular-puberulent.

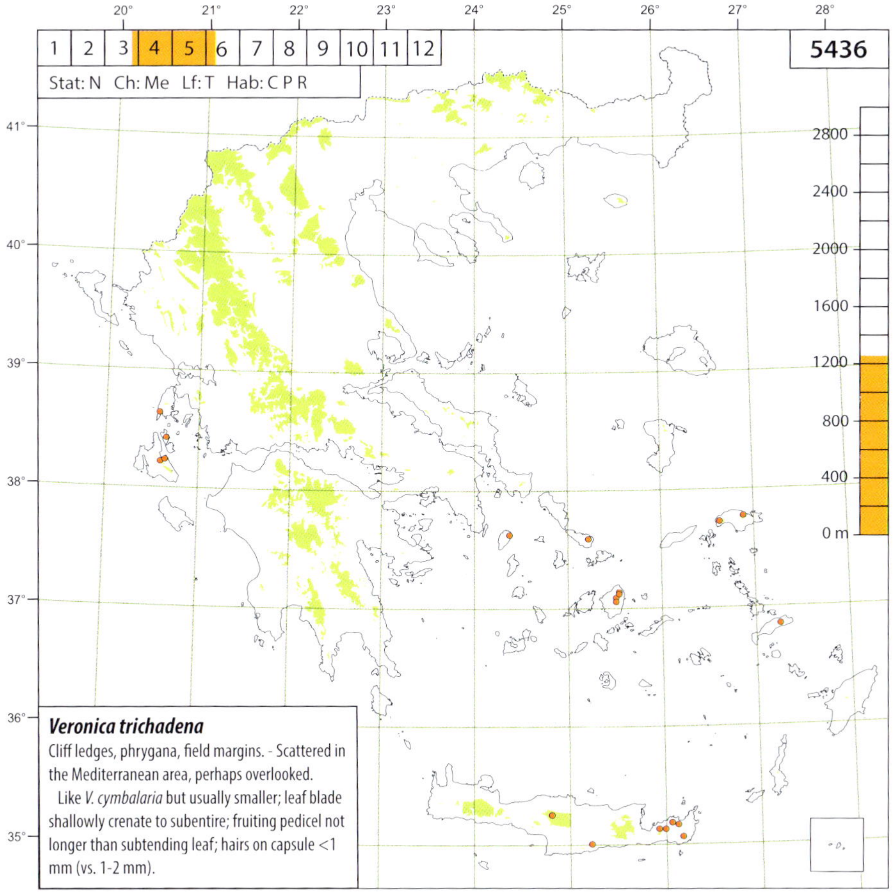

Veronica trichadena

Cliff ledges, phrygana, field margins. - Scattered in the Mediterranean area, perhaps overlooked.

Like *V. cymbalaria* but usually smaller; leaf blade shallowly crenate to subentire; fruiting pedicel not longer than subtending leaf; hairs on capsule <1 mm (vs. 1-2 mm).

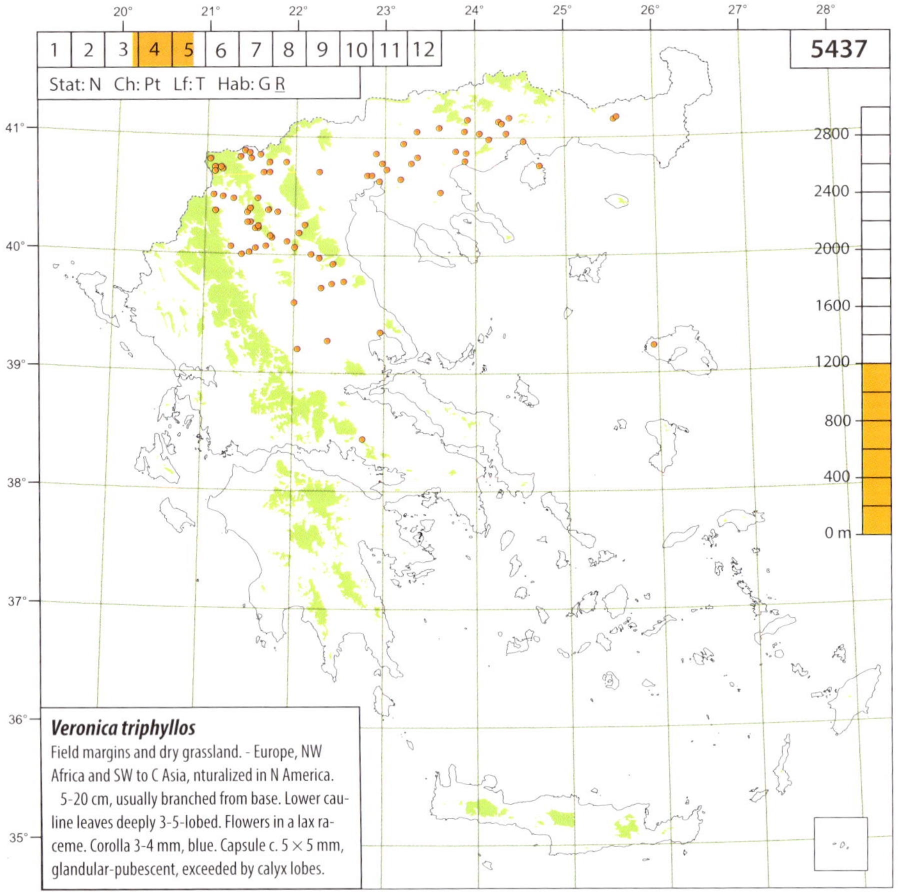

Veronica triphyllos
Field margins and dry grassland. - Europe, NW Africa and SW to C Asia, nturalized in N America.
5-20 cm, usually branched from base. Lower cauline leaves deeply 3-5-lobed. Flowers in a lax raceme. Corolla 3-4 mm, blue. Capsule c. 5 × 5 mm, glandular-pubescent, exceeded by calyx lobes.

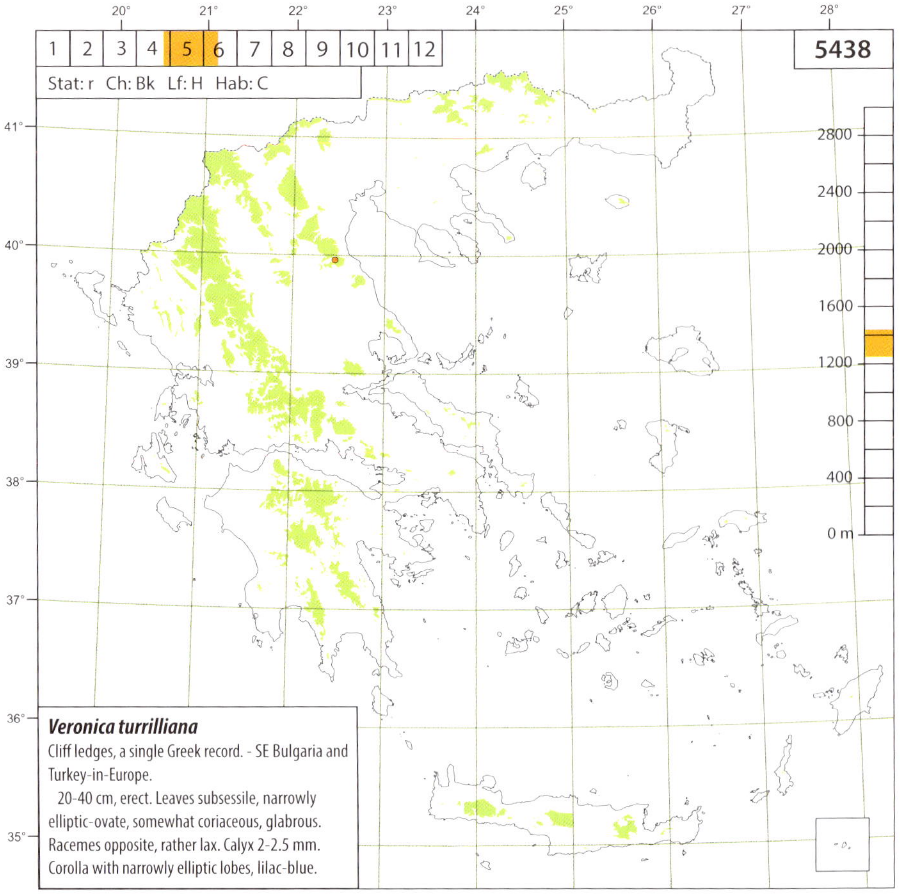

Veronica turrilliana
Cliff ledges, a single Greek record. - SE Bulgaria and Turkey-in-Europe.
20-40 cm, erect. Leaves subsessile, narrowly elliptic-ovate, somewhat coriaceous, glabrous. Racemes opposite, rather lax. Calyx 2-2.5 mm. Corolla with narrowly elliptic lobes, lilac-blue.

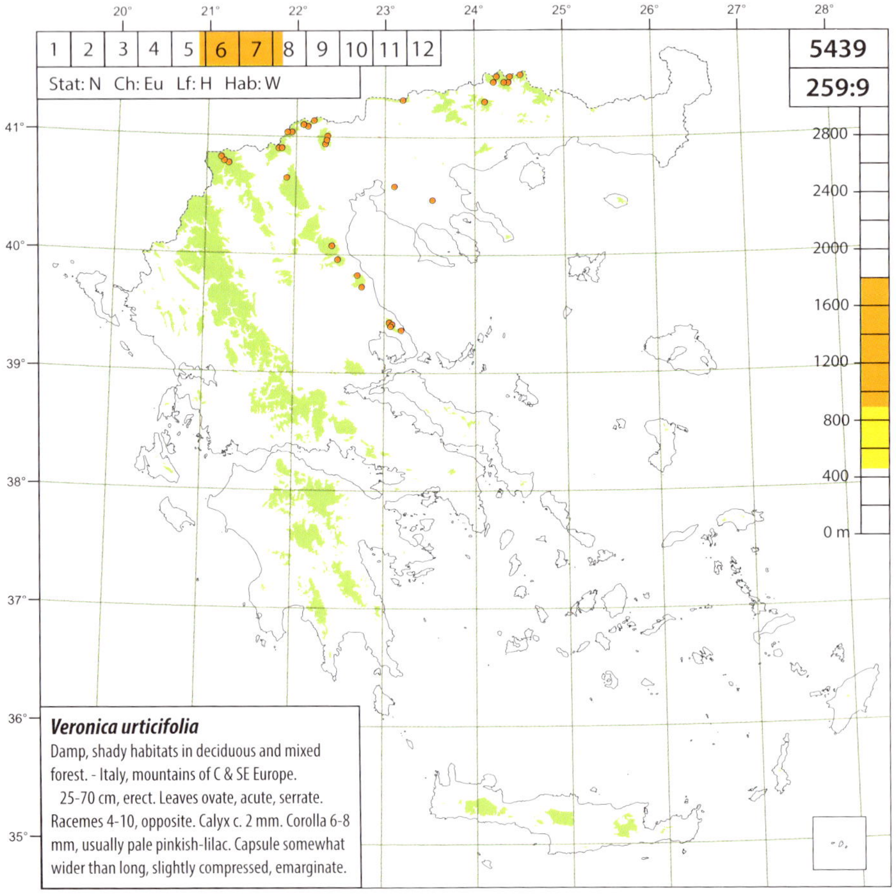

Veronica urticifolia
Damp, shady habitats in deciduous and mixed forest. - Italy, mountains of C & SE Europe.
25-70 cm, erect. Leaves ovate, acute, serrate. Racemes 4-10, opposite. Calyx c. 2 mm. Corolla 6-8 mm, usually pale pinkish-lilac. Capsule somewhat wider than long, slightly compressed, emarginate.

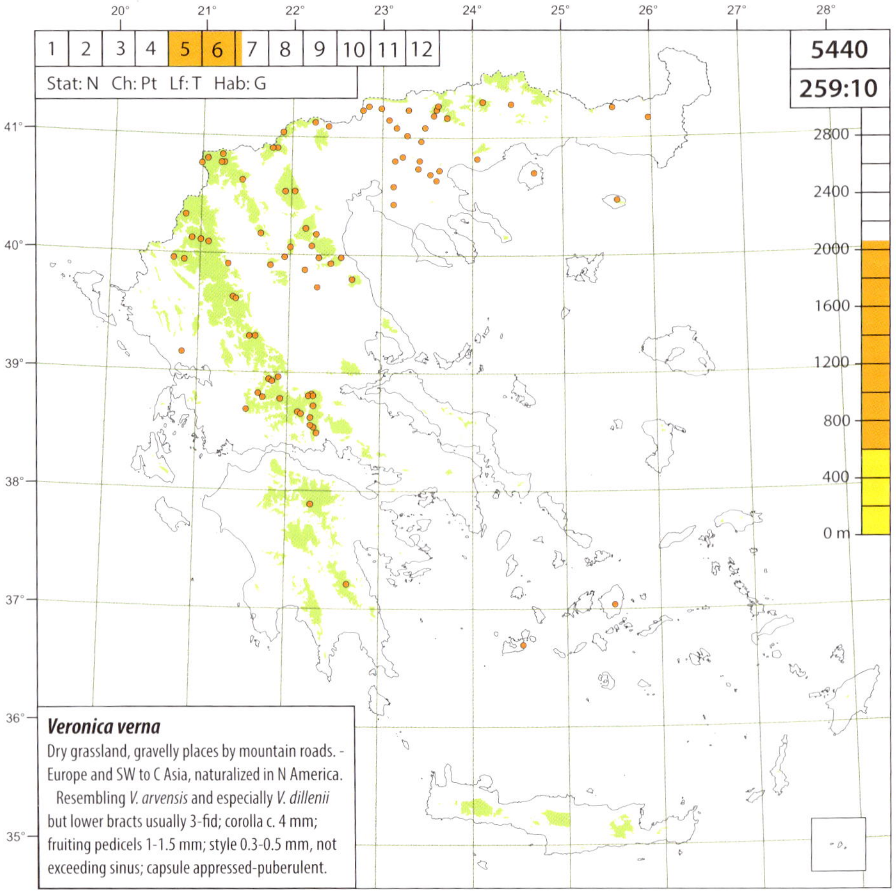

Veronica verna
Dry grassland, gravelly places by mountain roads. - Europe and SW to C Asia, naturalized in N America.
Resembling *V. arvensis* and especially *V. dillenii* but lower bracts usually 3-fid; corolla c. 4 mm; fruiting pedicels 1-1.5 mm; style 0.3-0.5 mm, not exceeding sinus; capsule appressed-puberulent.

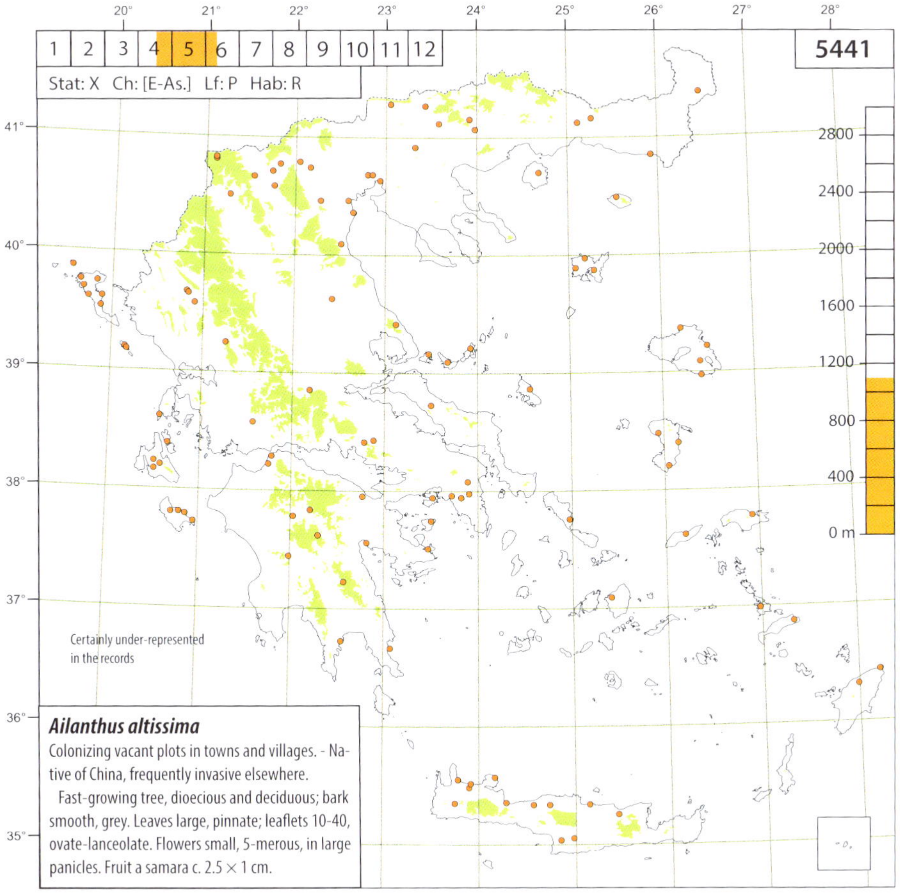

Ailanthus altissima

Colonizing vacant plots in towns and villages. - Native of China, frequently invasive elsewhere.

Fast-growing tree, dioecious and deciduous; bark smooth, grey. Leaves large, pinnate; leaflets 10-40, ovate-lanceolate. Flowers small, 5-merous, in large panicles. Fruit a samara c. 2.5 × 1 cm.

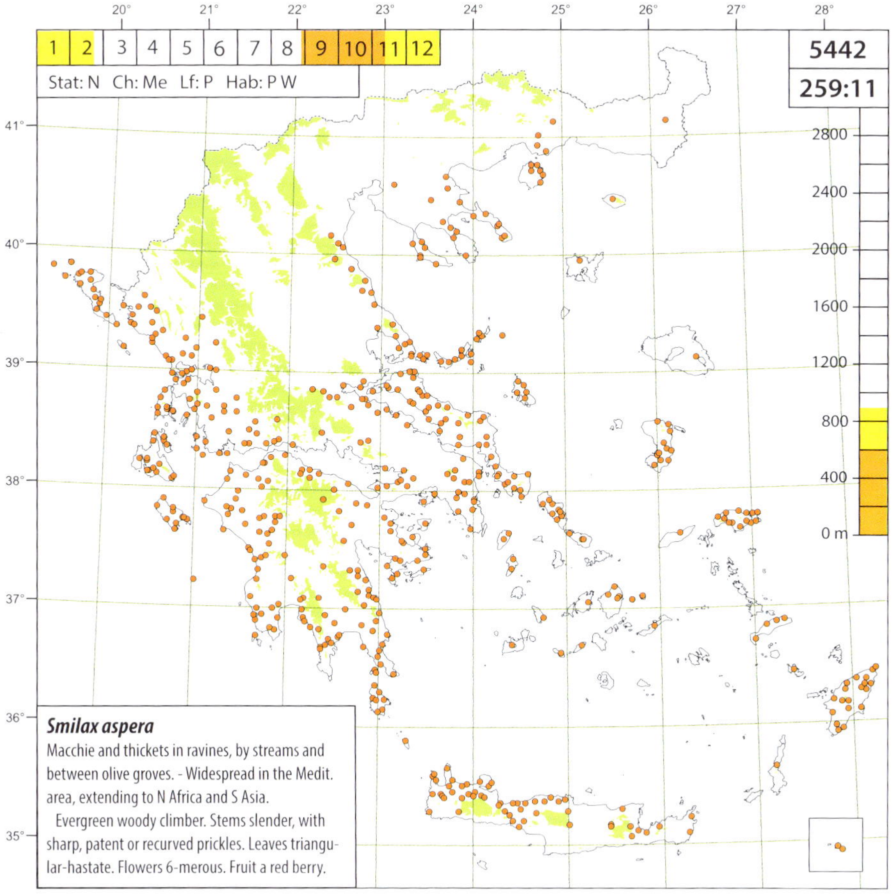

Smilax aspera

Macchie and thickets in ravines, by streams and between olive groves. - Widespread in the Medit. area, extending to N Africa and S Asia.

Evergreen woody climber. Stems slender, with sharp, patent or recurved prickles. Leaves triangular-hastate. Flowers 6-merous. Fruit a red berry.

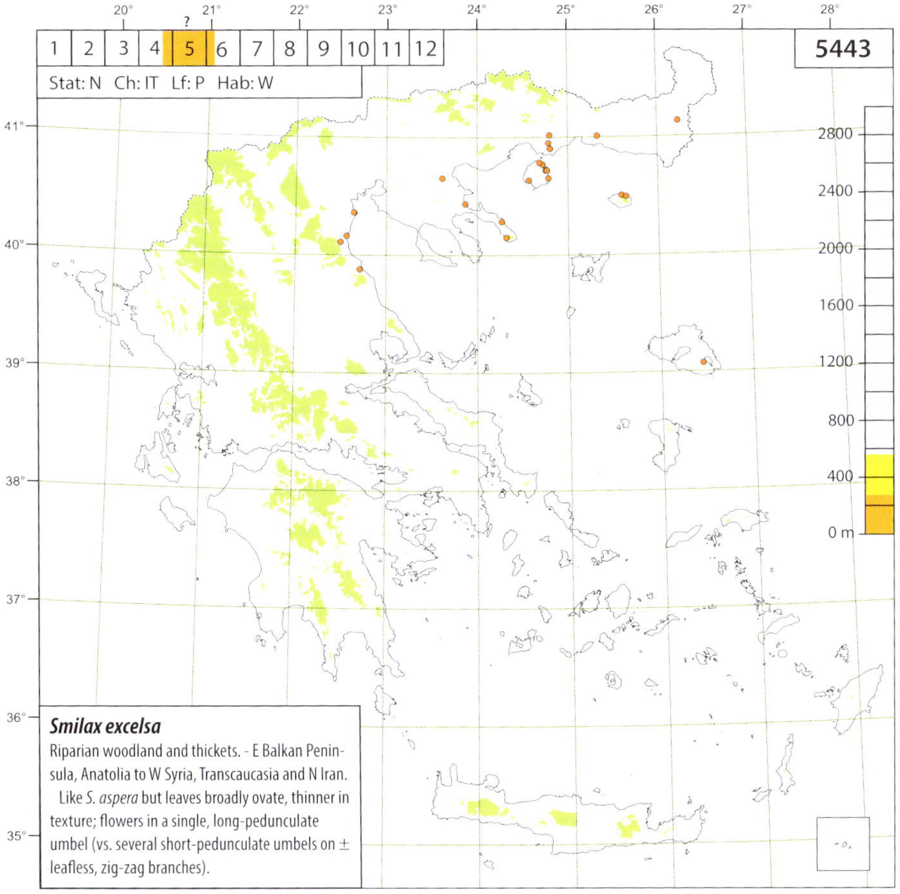

Smilax excelsa

Riparian woodland and thickets. - E Balkan Peninsula, Anatolia to W Syria, Transcaucasia and N Iran.

Like *S. aspera* but leaves broadly ovate, thinner in texture; flowers in a single, long-pedunculate umbel (vs. several short-pedunculate umbels on ± leafless, zig-zag branches).

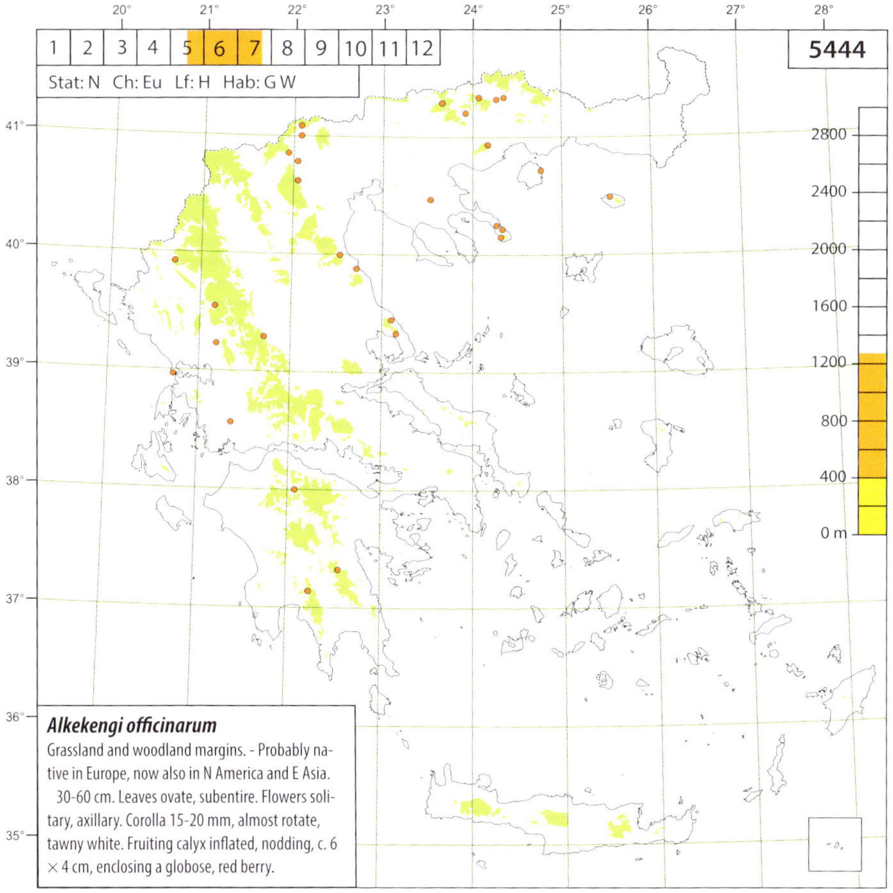

Alkekengi officinarum

Grassland and woodland margins. - Probably native in Europe, now also in N America and E Asia.

30-60 cm. Leaves ovate, subentire. Flowers solitary, axillary. Corolla 15-20 mm, almost rotate, tawny white. Fruiting calyx inflated, nodding, c. 6 × 4 cm, enclosing a globose, red berry.

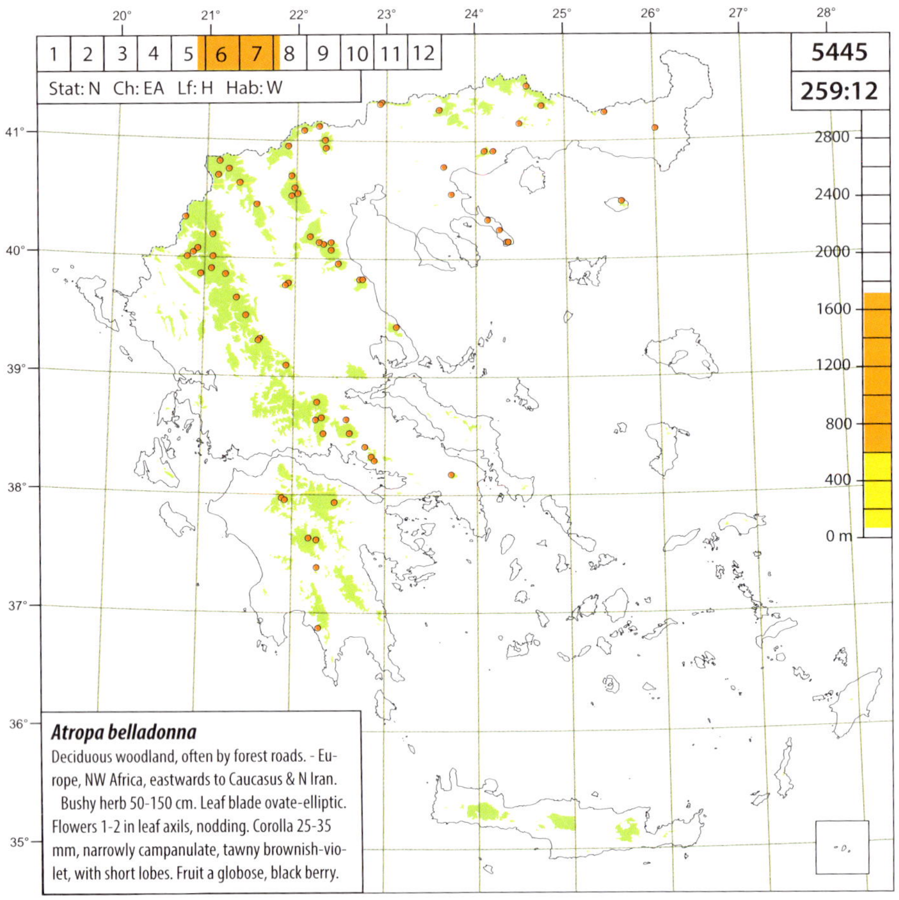

Atropa belladonna

Deciduous woodland, often by forest roads. - Europe, NW Africa, eastwards to Caucasus & N Iran.

Bushy herb 50-150 cm. Leaf blade ovate-elliptic. Flowers 1-2 in leaf axils, nodding. Corolla 25-35 mm, narrowly campanulate, tawny brownish-violet, with short lobes. Fruit a globose, black berry.

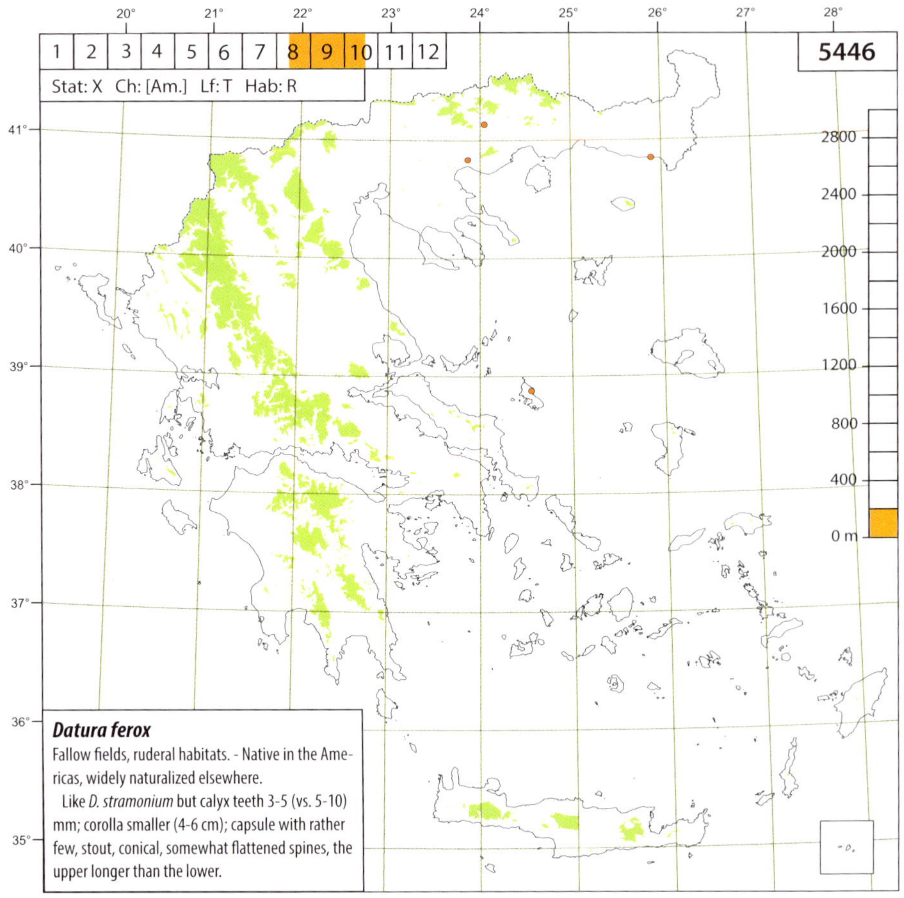

Datura ferox

Fallow fields, ruderal habitats. - Native in the Americas, widely naturalized elsewhere.

Like *D. stramonium* but calyx teeth 3-5 (vs. 5-10) mm; corolla smaller (4-6 cm); capsule with rather few, stout, conical, somewhat flattened spines, the upper longer than the lower.

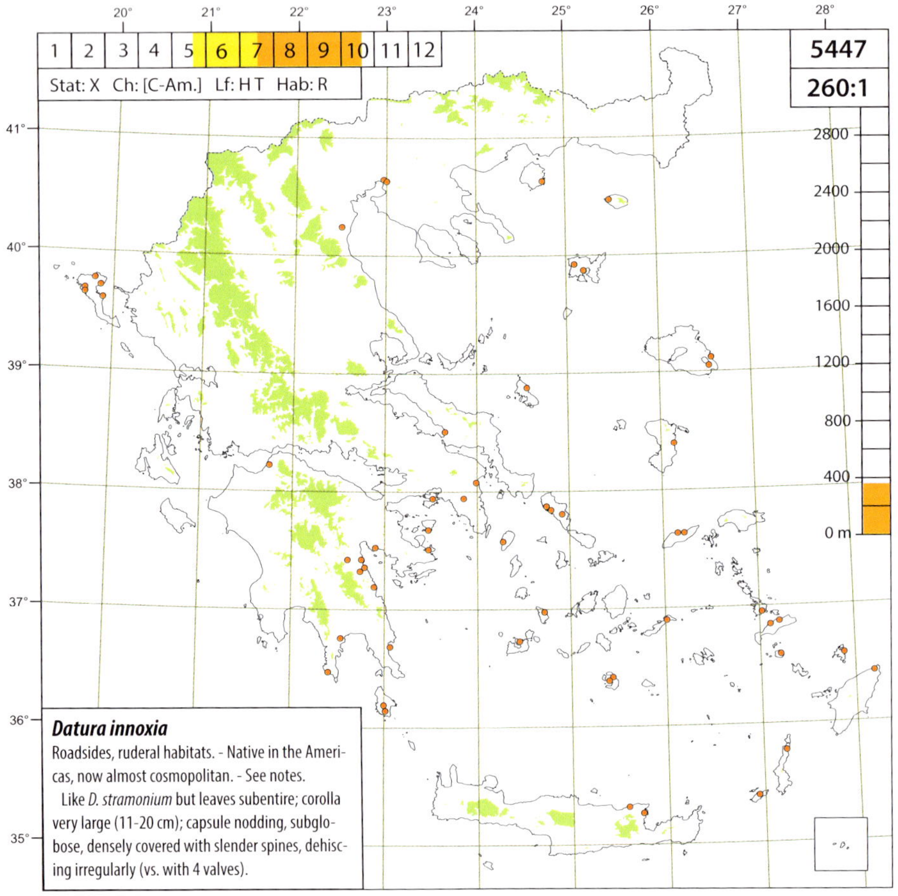

Datura innoxia

Roadsides, ruderal habitats. - Native in the Americas, now almost cosmopolitan. - See notes.

Like *D. stramonium* but leaves subentire; corolla very large (11-20 cm); capsule nodding, subglobose, densely covered with slender spines, dehiscing irregularly (vs. with 4 valves).

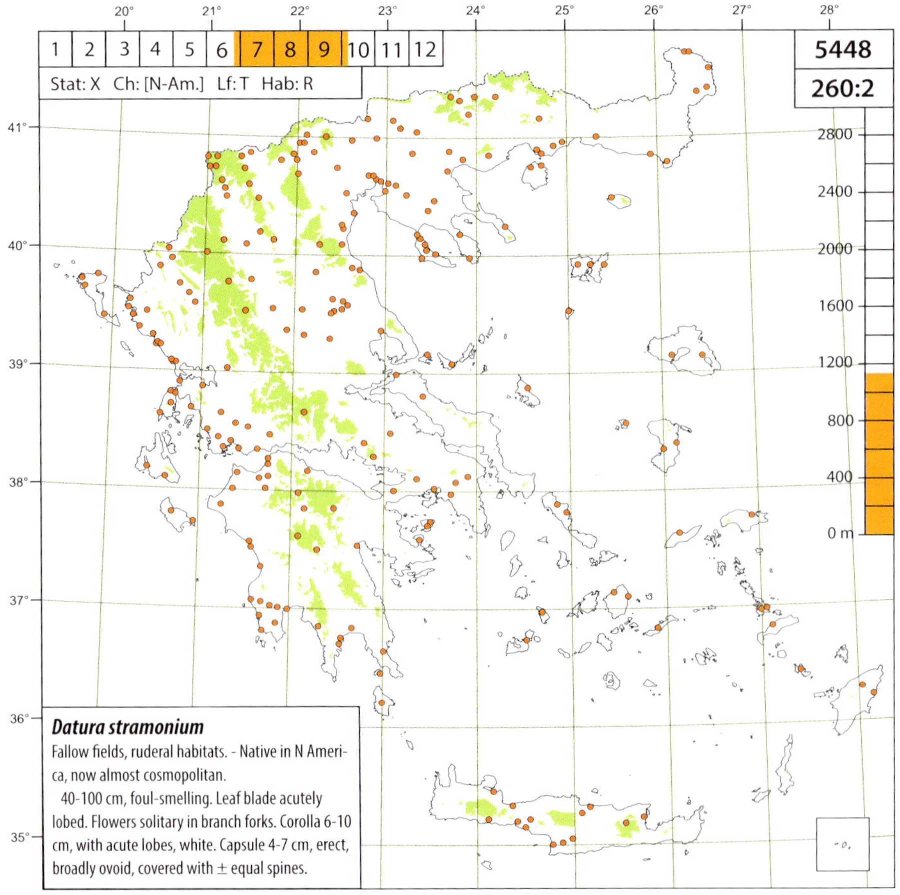

Datura stramonium

Fallow fields, ruderal habitats. - Native in N America, now almost cosmopolitan.

40-100 cm, foul-smelling. Leaf blade acutely lobed. Flowers solitary in branch forks. Corolla 6-10 cm, with acute lobes, white. Capsule 4-7 cm, erect, broadly ovoid, covered with ± equal spines.

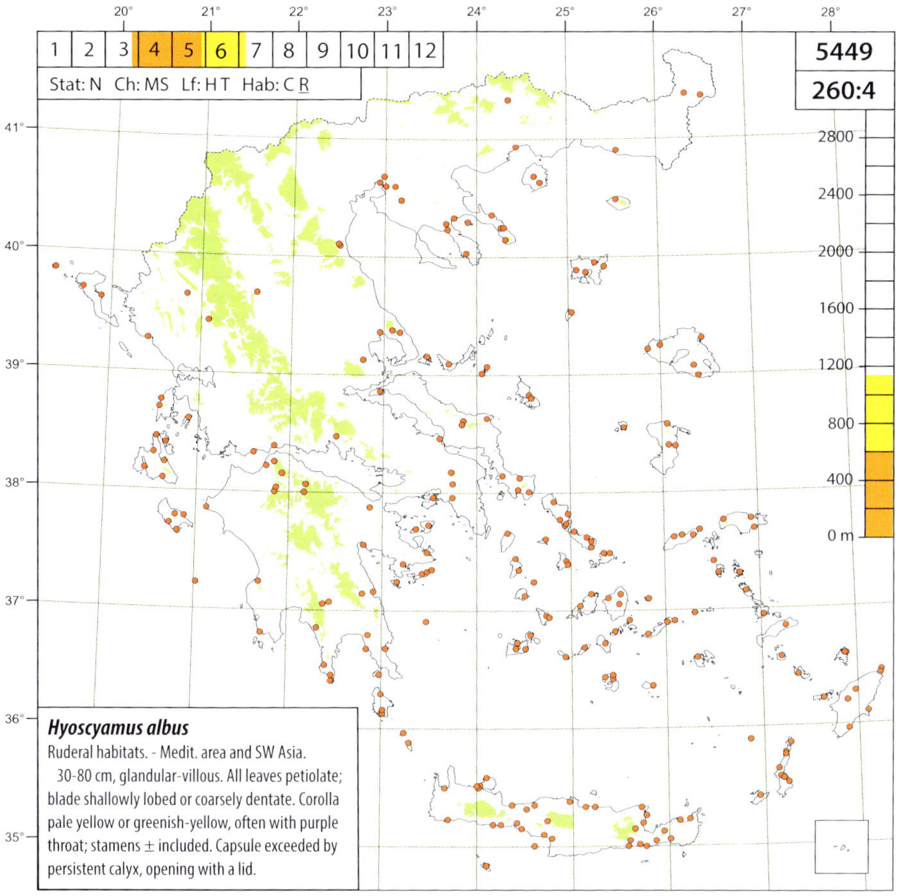

Hyoscyamus albus
Ruderal habitats. - Medit. area and SW Asia.
30-80 cm, glandular-villous. All leaves petiolate; blade shallowly lobed or coarsely dentate. Corolla pale yellow or greenish-yellow, often with purple throat; stamens ± included. Capsule exceeded by persistent calyx, opening with a lid.

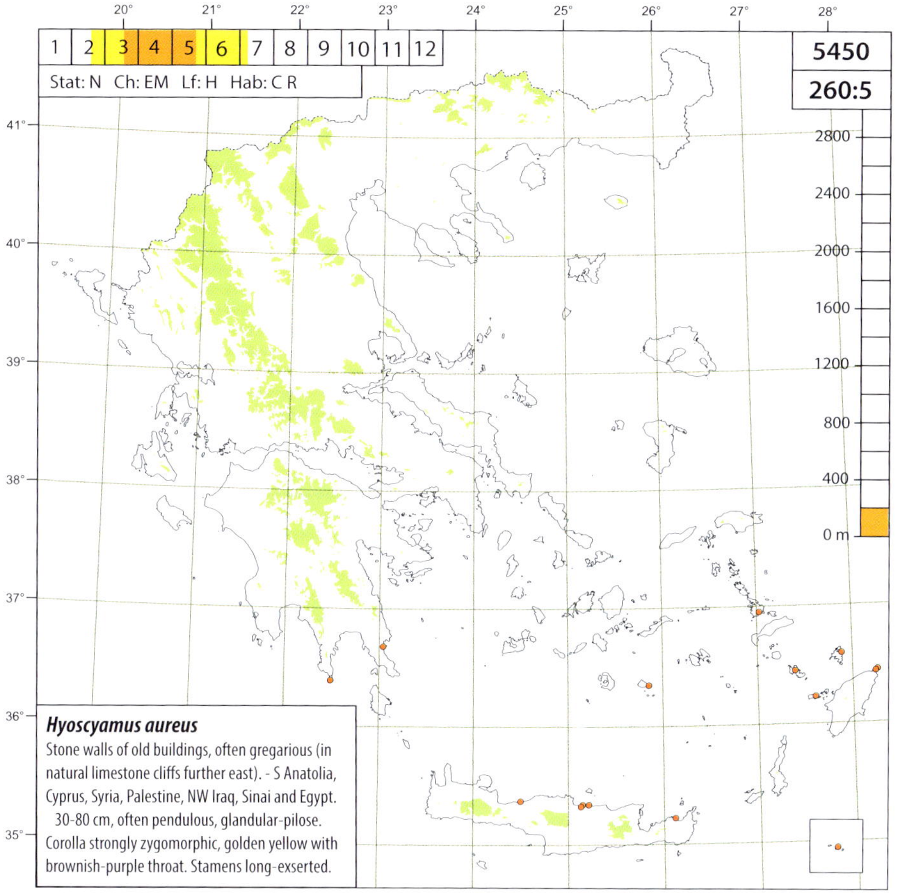

Hyoscyamus aureus
Stone walls of old buildings, often gregarious (in natural limestone cliffs further east). - S Anatolia, Cyprus, Syria, Palestine, NW Iraq, Sinai and Egypt.
30-80 cm, often pendulous, glandular-pilose. Corolla strongly zygomorphic, golden yellow with brownish-purple throat. Stamens long-exserted.

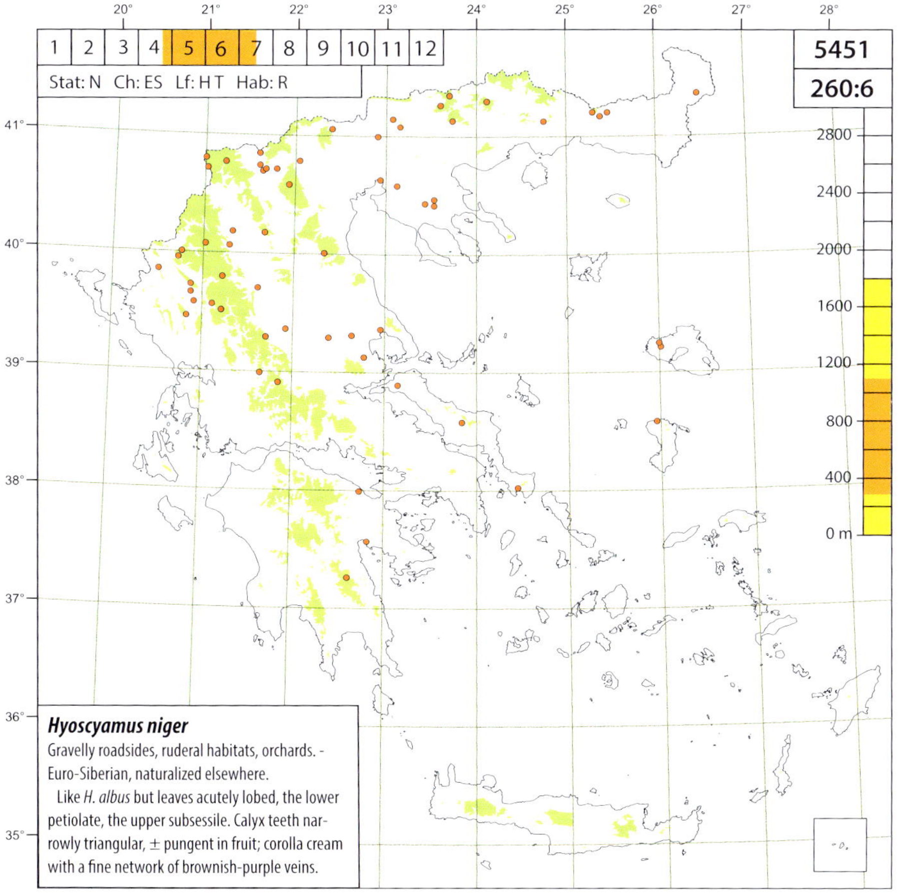

Hyoscyamus niger
Gravelly roadsides, ruderal habitats, orchards. - Euro-Siberian, naturalized elsewhere.
Like *H. albus* but leaves acutely lobed, the lower petiolate, the upper subsessile. Calyx teeth narrowly triangular, ± pungent in fruit; corolla cream with a fine network of brownish-purple veins.

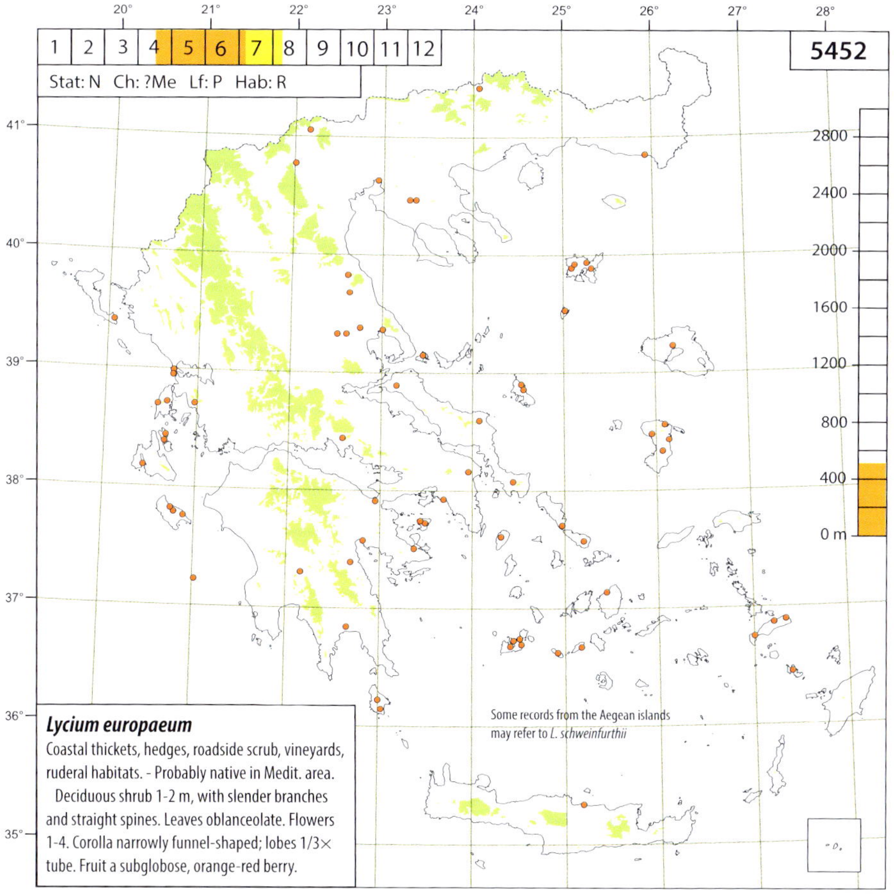

Lycium europaeum
Coastal thickets, hedges, roadside scrub, vineyards, ruderal habitats. - Probably native in Medit. area.
Deciduous shrub 1-2 m, with slender branches and straight spines. Leaves oblanceolate. Flowers 1-4. Corolla narrowly funnel-shaped; lobes 1/3× tube. Fruit a subglobose, orange-red berry.

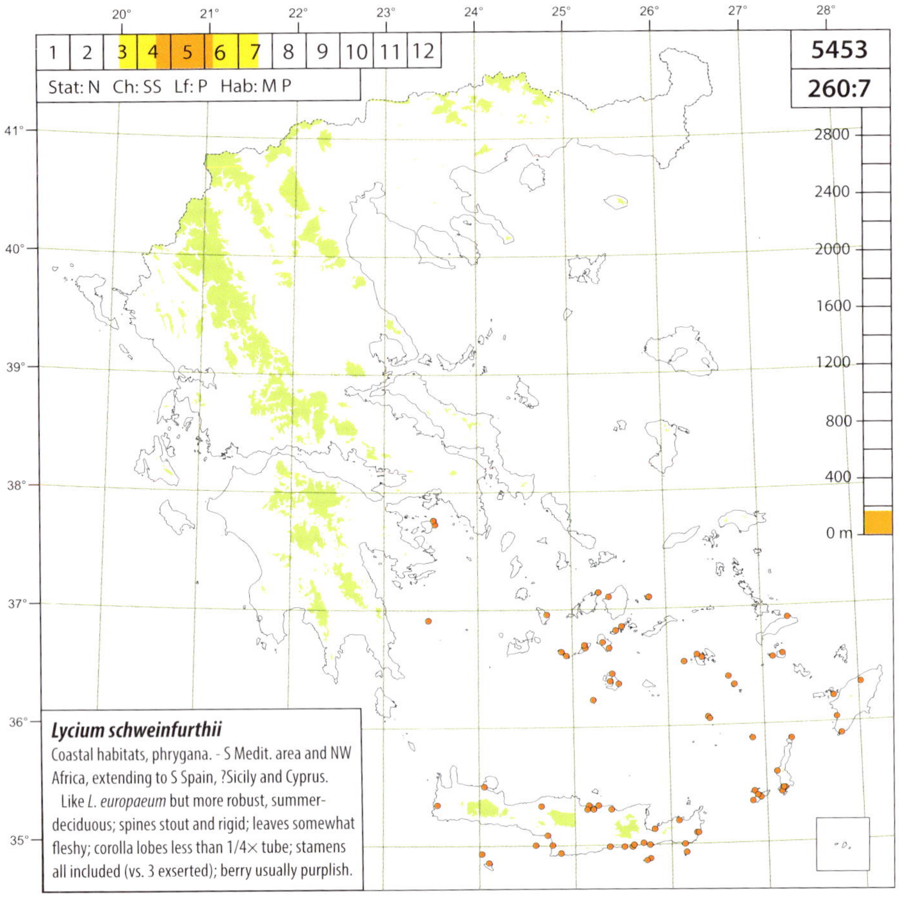

Lycium schweinfurthii
Coastal habitats, phrygana. - S Medit. area and NW Africa, extending to S Spain, ?Sicily and Cyprus.
Like *L. europaeum* but more robust, summer-deciduous; spines stout and rigid; leaves somewhat fleshy; corolla lobes less than 1/4× tube; stamens all included (vs. 3 exserted); berry usually purplish.

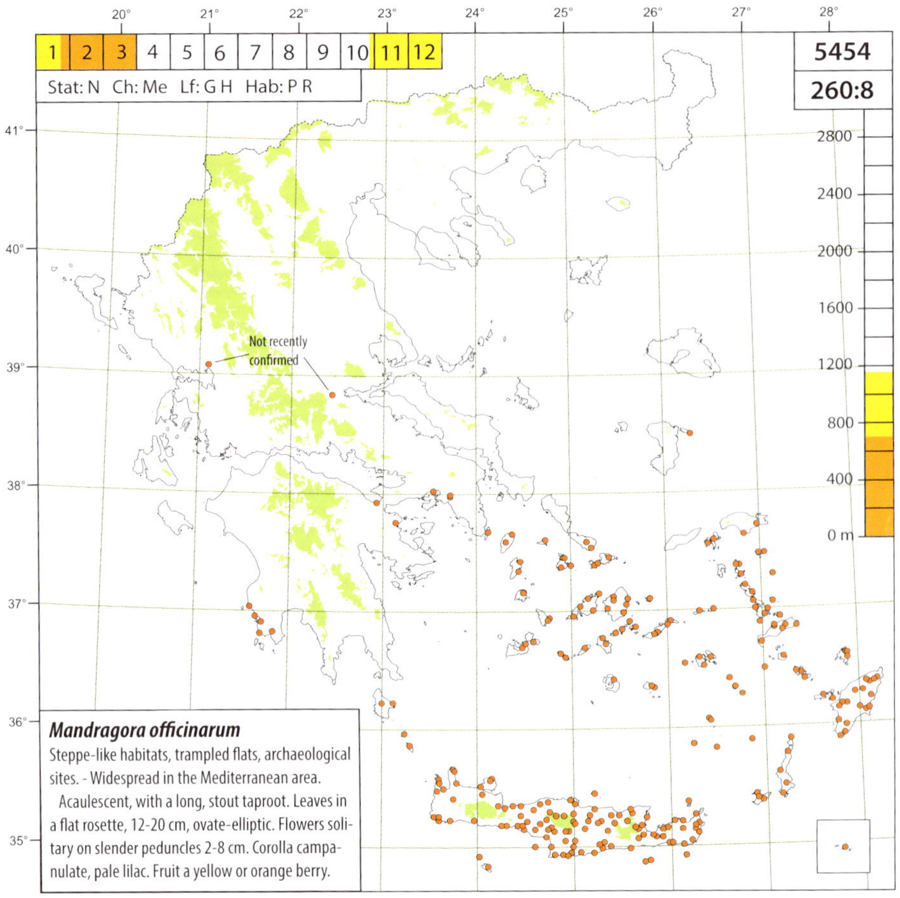

Mandragora officinarum
Steppe-like habitats, trampled flats, archaeological sites. - Widespread in the Mediterranean area.
Acaulescent, with a long, stout taproot. Leaves in a flat rosette, 12-20 cm, ovate-elliptic. Flowers solitary on slender peduncles 2-8 cm. Corolla campanulate, pale lilac. Fruit a yellow or orange berry.

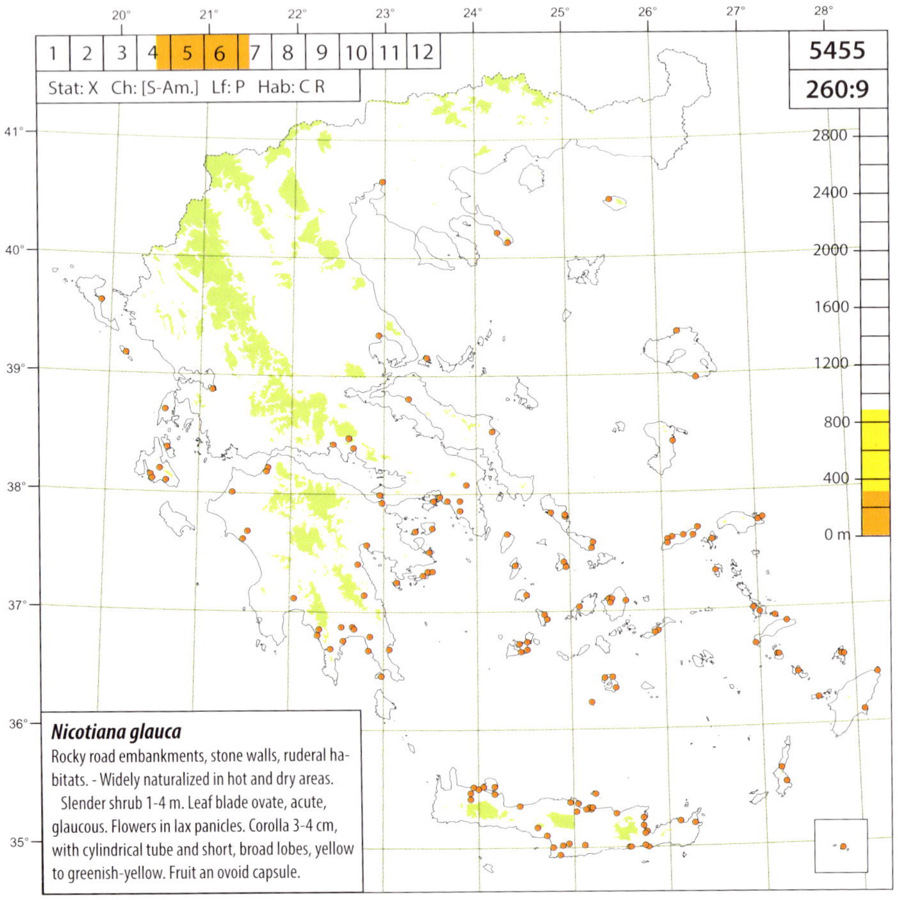

Nicotiana glauca
Rocky road embankments, stone walls, ruderal habitats. - Widely naturalized in hot and dry areas.
Slender shrub 1-4 m. Leaf blade ovate, acute, glaucous. Flowers in lax panicles. Corolla 3-4 cm, with cylindrical tube and short, broad lobes, yellow to greenish-yellow. Fruit an ovoid capsule.

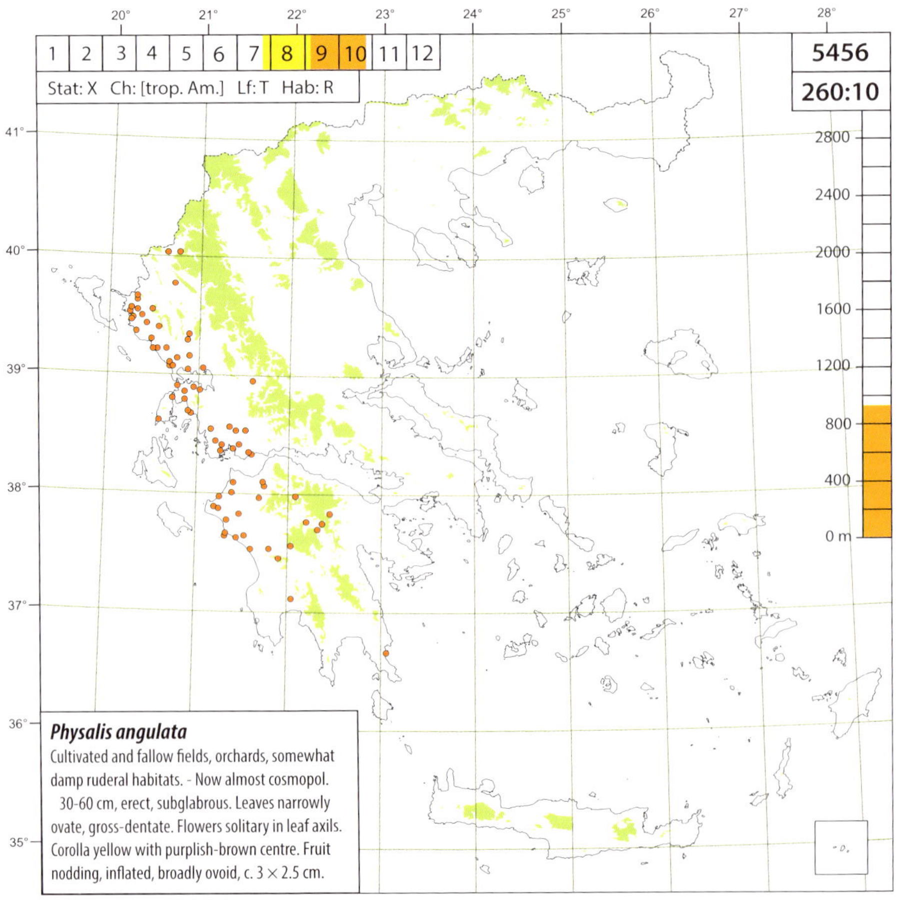

Physalis angulata
Cultivated and fallow fields, orchards, somewhat damp ruderal habitats. - Now almost cosmopol.
30-60 cm, erect, subglabrous. Leaves narrowly ovate, gross-dentate. Flowers solitary in leaf axils. Corolla yellow with purplish-brown centre. Fruit nodding, inflated, broadly ovoid, c. 3 × 2.5 cm.

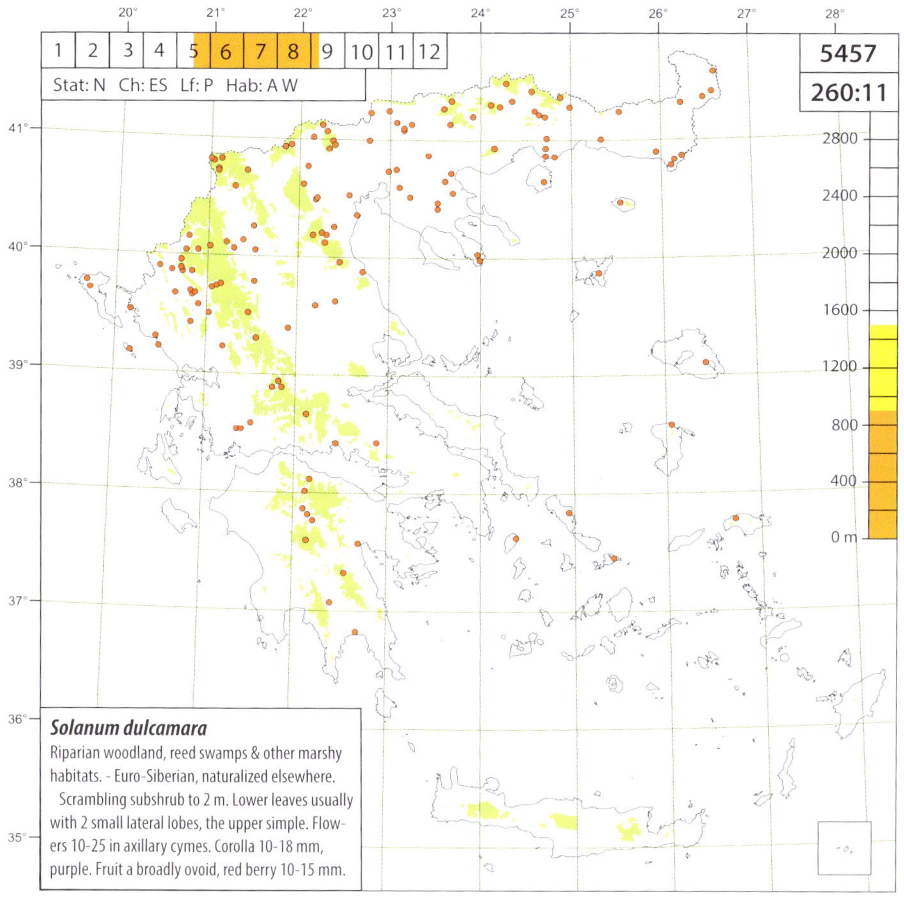

Solanum dulcamara

Riparian woodland, reed swamps & other marshy habitats. - Euro-Siberian, naturalized elsewhere.

Scrambling subshrub to 2 m. Lower leaves usually with 2 small lateral lobes, the upper simple. Flowers 10-25 in axillary cymes. Corolla 10-18 mm, purple. Fruit a broadly ovoid, red berry 10-15 mm.

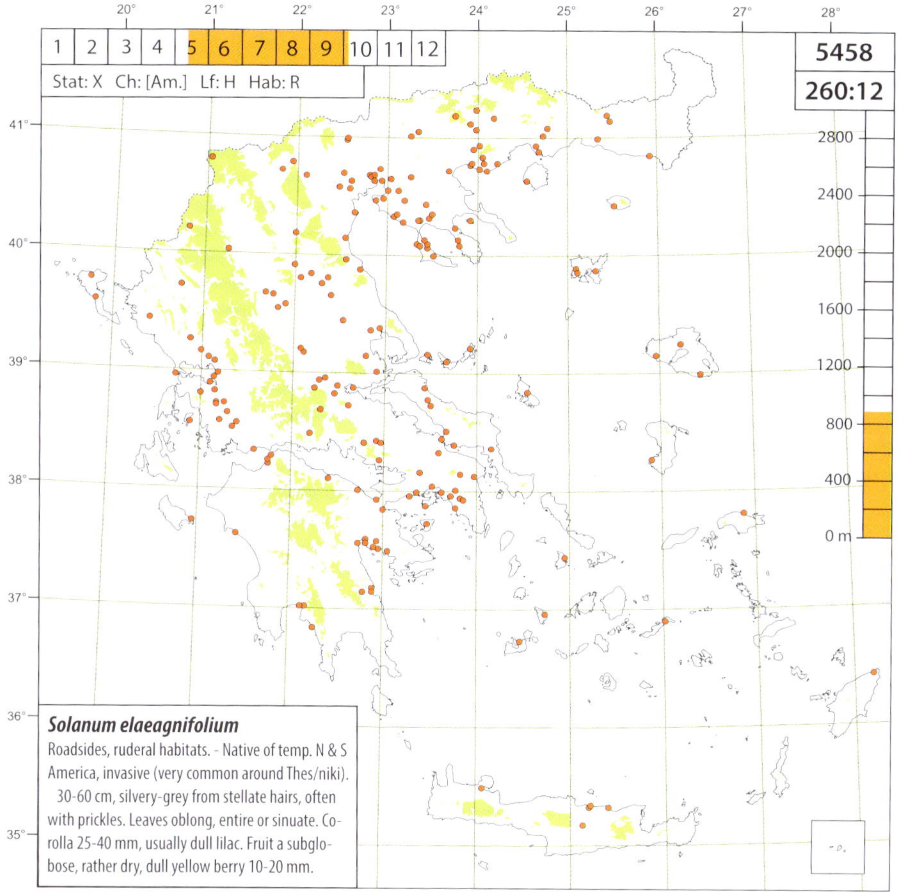

Solanum elaeagnifolium

Roadsides, ruderal habitats. - Native of temp. N & S America, invasive (very common around Thes/niki).

30-60 cm, silvery-grey from stellate hairs, often with prickles. Leaves oblong, entire or sinuate. Corolla 25-40 mm, usually dull lilac. Fruit a subglobose, rather dry, dull yellow berry 10-20 mm.

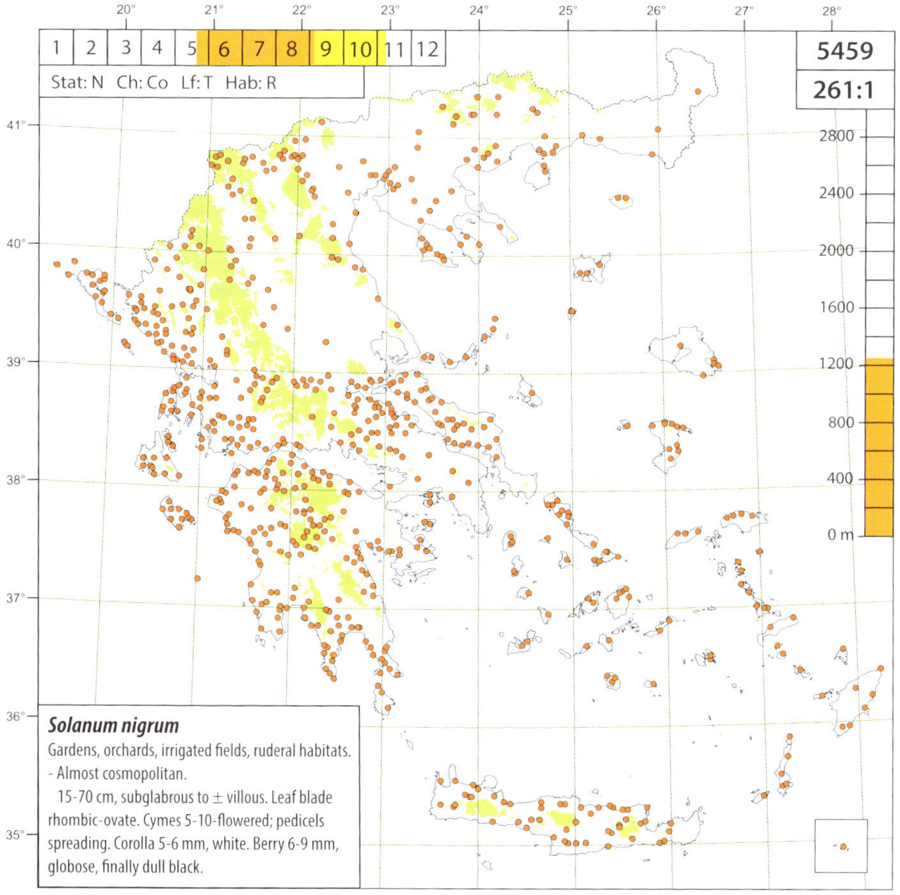

Solanum nigrum

Gardens, orchards, irrigated fields, ruderal habitats. - Almost cosmopolitan.

15-70 cm, subglabrous to ± villous. Leaf blade rhombic-ovate. Cymes 5-10-flowered; pedicels spreading. Corolla 5-6 mm, white. Berry 6-9 mm, globose, finally dull black.

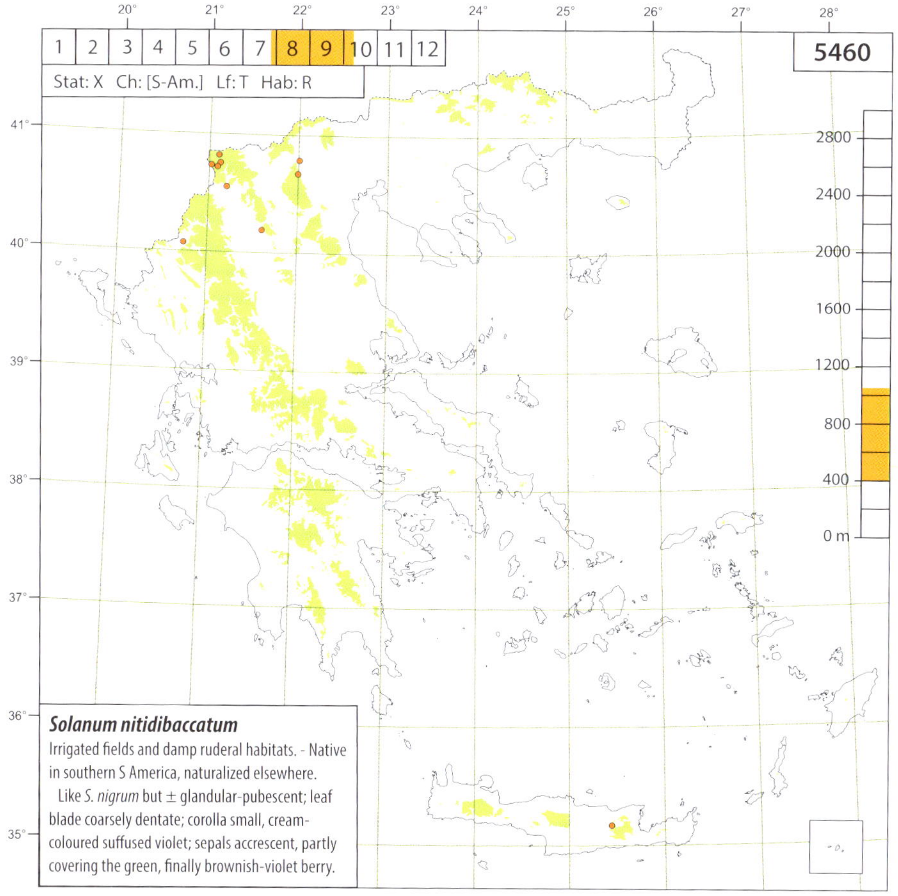

Solanum nitidibaccatum

Irrigated fields and damp ruderal habitats. - Native in southern S America, naturalized elsewhere.

Like *S. nigrum* but ± glandular-pubescent; leaf blade coarsely dentate; corolla small, cream-coloured suffused violet; sepals accrescent, partly covering the green, finally brownish-violet berry.

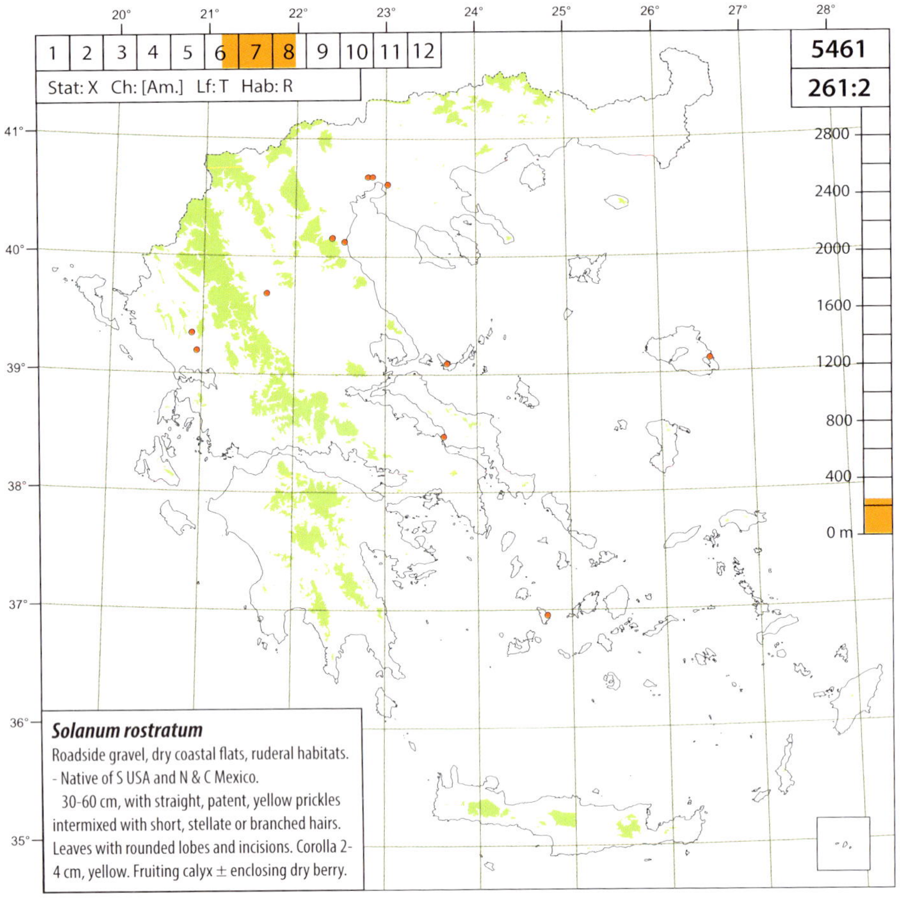

Solanum rostratum
Roadside gravel, dry coastal flats, ruderal habitats. - Native of S USA and N & C Mexico.
30-60 cm, with straight, patent, yellow prickles intermixed with short, stellate or branched hairs. Leaves with rounded lobes and incisions. Corolla 2-4 cm, yellow. Fruiting calyx ± enclosing dry berry.

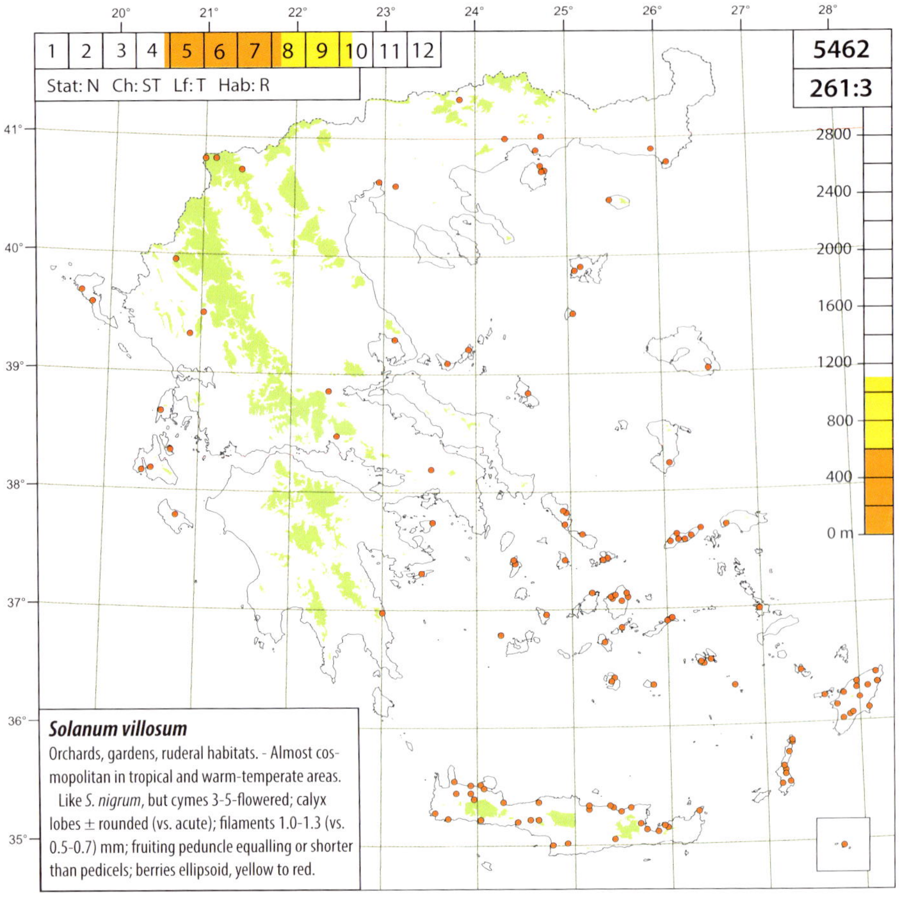

Solanum villosum
Orchards, gardens, ruderal habitats. - Almost cosmopolitan in tropical and warm-temperate areas.
Like *S. nigrum*, but cymes 3-5-flowered; calyx lobes ± rounded (vs. acute); filaments 1.0-1.3 (vs. 0.5-0.7) mm; fruiting peduncle equalling or shorter than pedicels; berries ellipsoid, yellow to red.

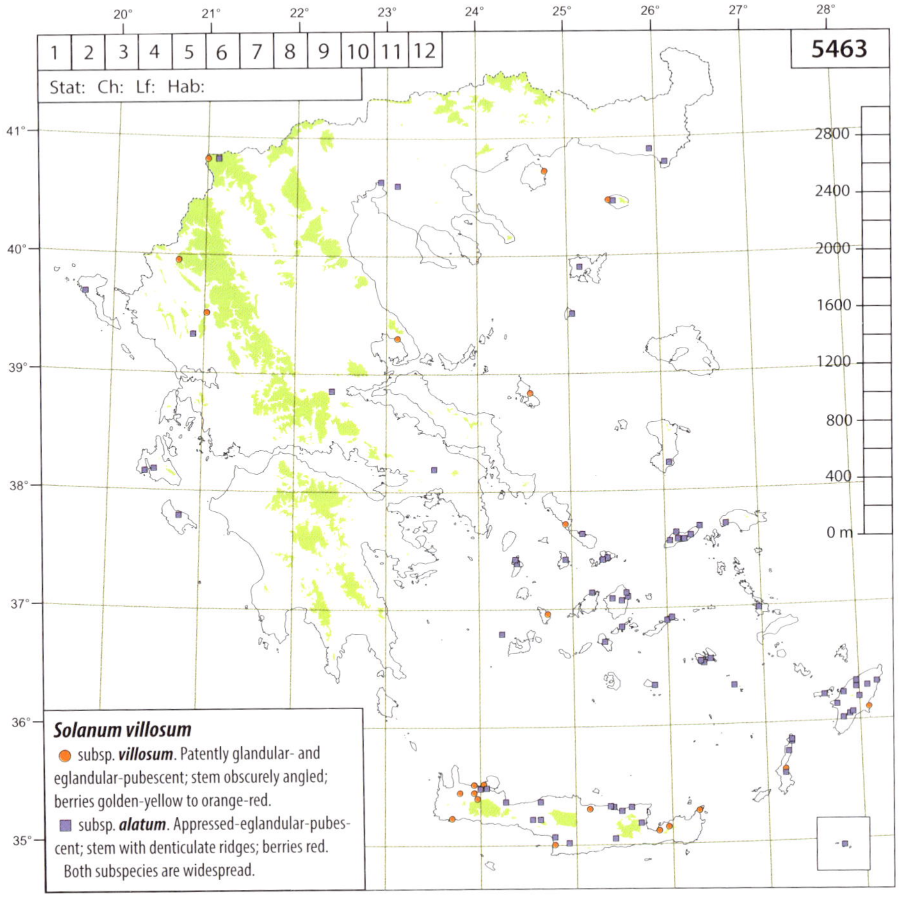

Solanum villosum
● subsp. ***villosum***. Patently glandular- and eglandular-pubescent; stem obscurely angled; berries golden-yellow to orange-red.
■ subsp. ***alatum***. Appressed-eglandular-pubescent; stem with denticulate ridges; berries red.
Both subspecies are widespread.

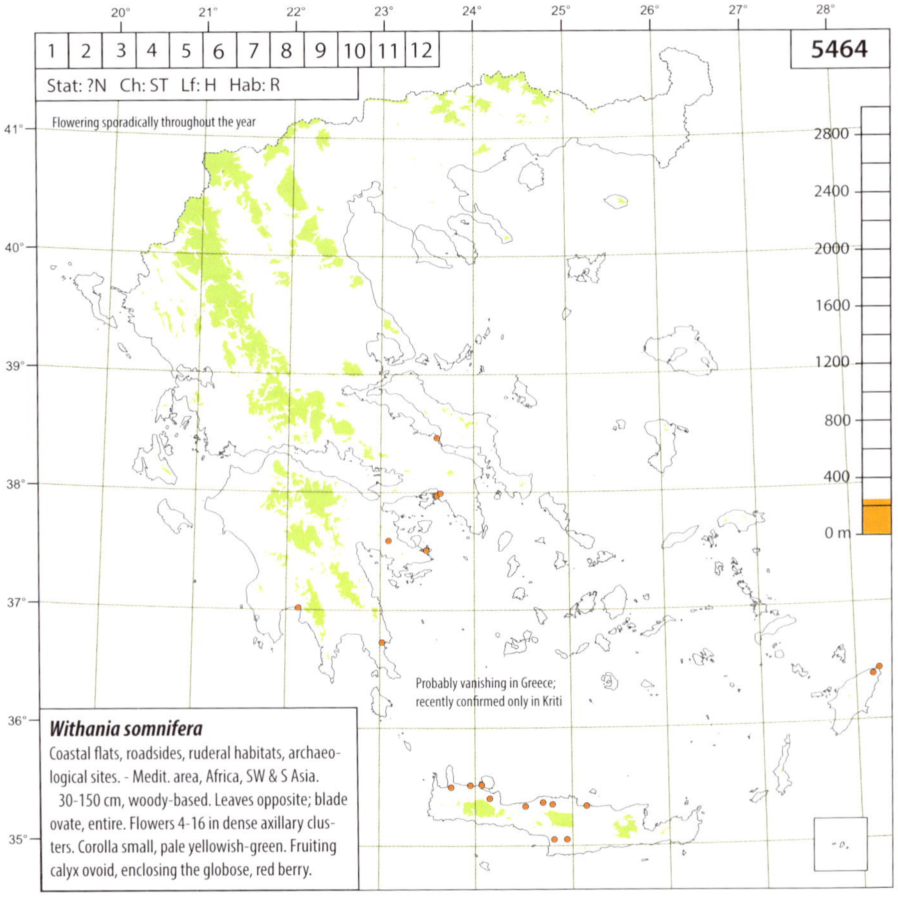

Withania somnifera
Coastal flats, roadsides, ruderal habitats, archaeological sites. - Medit. area, Africa, SW & S Asia.
30-150 cm, woody-based. Leaves opposite; blade ovate, entire. Flowers 4-16 in dense axillary clusters. Corolla small, pale yellowish-green. Fruiting calyx ovoid, enclosing the globose, red berry.

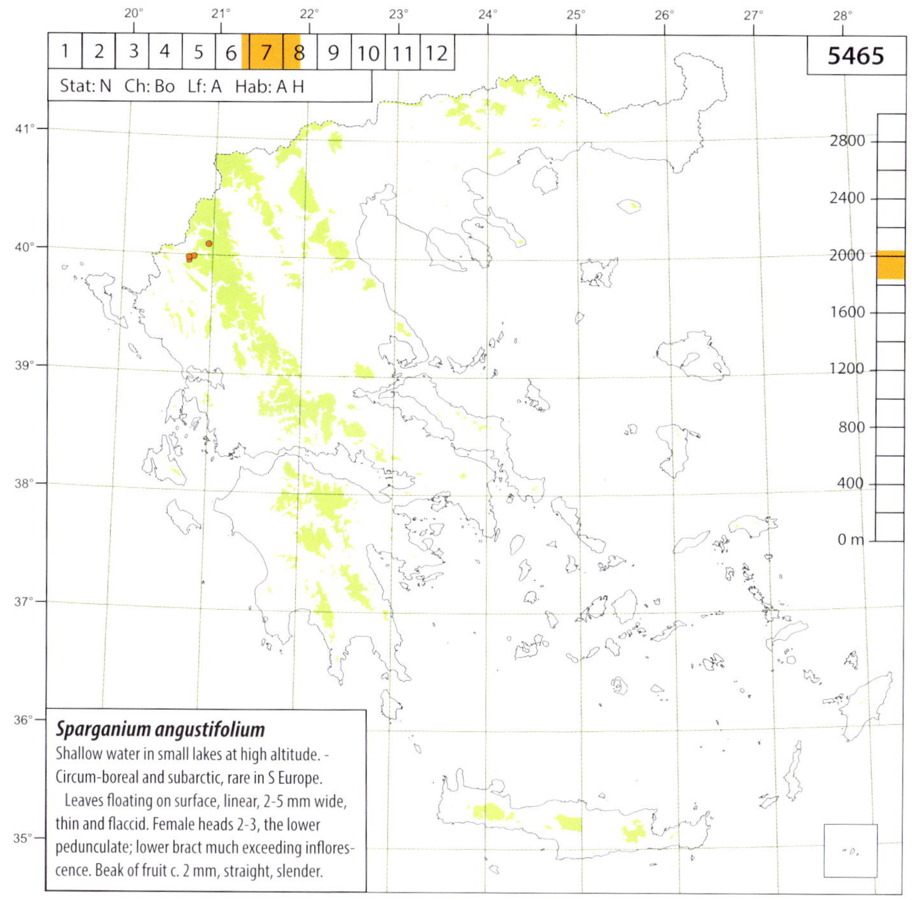

Sparganium angustifolium
Shallow water in small lakes at high altitude. - Circum-boreal and subarctic, rare in S Europe.
Leaves floating on surface, linear, 2-5 mm wide, thin and flaccid. Female heads 2-3, the lower pedunculate; lower bract much exceeding inflorescence. Beak of fruit c. 2 mm, straight, slender.

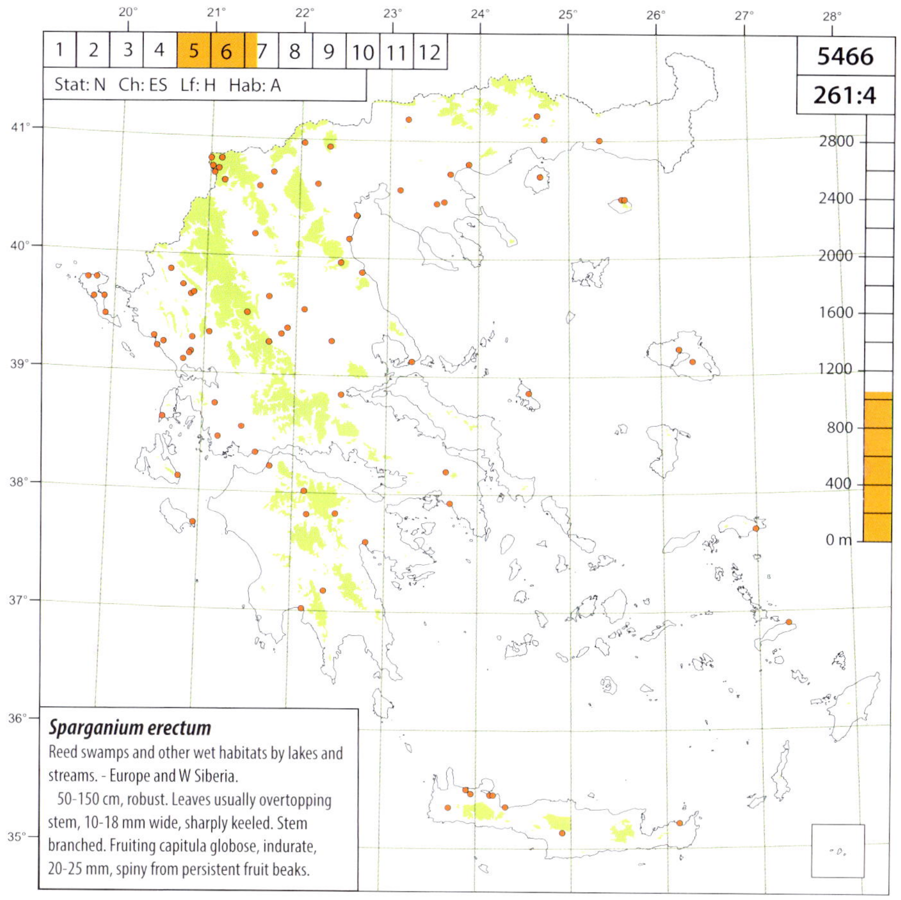

Sparganium erectum
Reed swamps and other wet habitats by lakes and streams. - Europe and W Siberia.
50-150 cm, robust. Leaves usually overtopping stem, 10-18 mm wide, sharply keeled. Stem branched. Fruiting capitula globose, indurate, 20-25 mm, spiny from persistent fruit beaks.

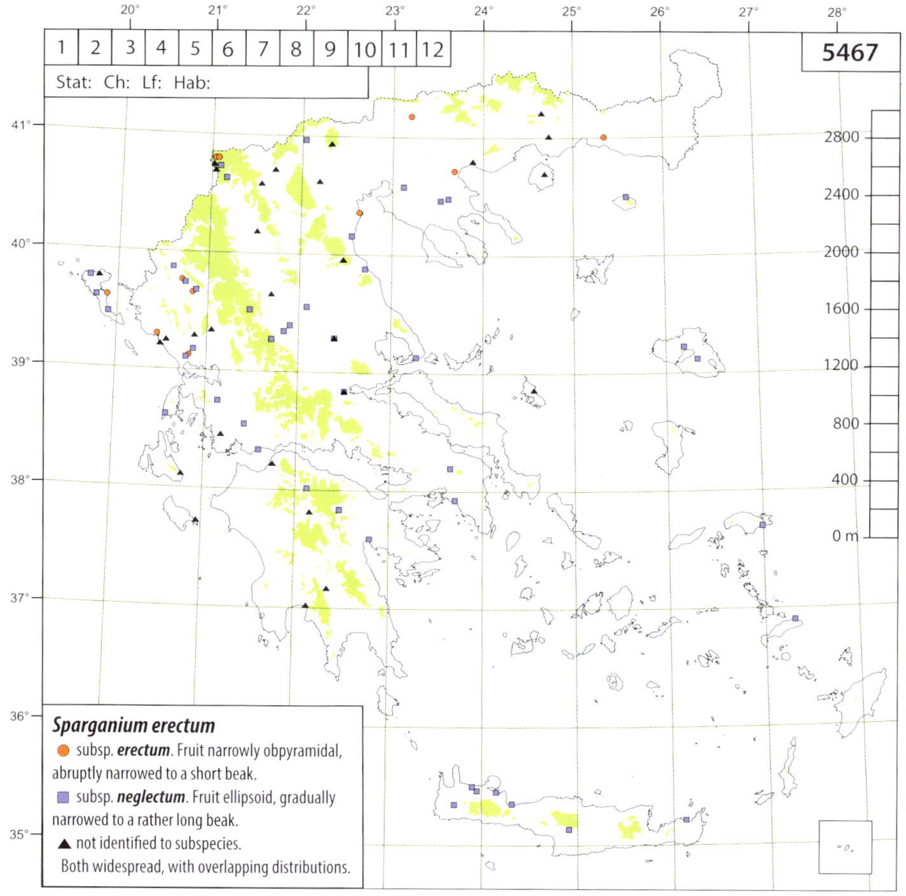

Sparganium erectum
● subsp. ***erectum***. Fruit narrowly obpyramidal, abruptly narrowed to a short beak.
■ subsp. ***neglectum***. Fruit ellipsoid, gradually narrowed to a rather long beak.
▲ not identified to subspecies.
Both widespread, with overlapping distributions.

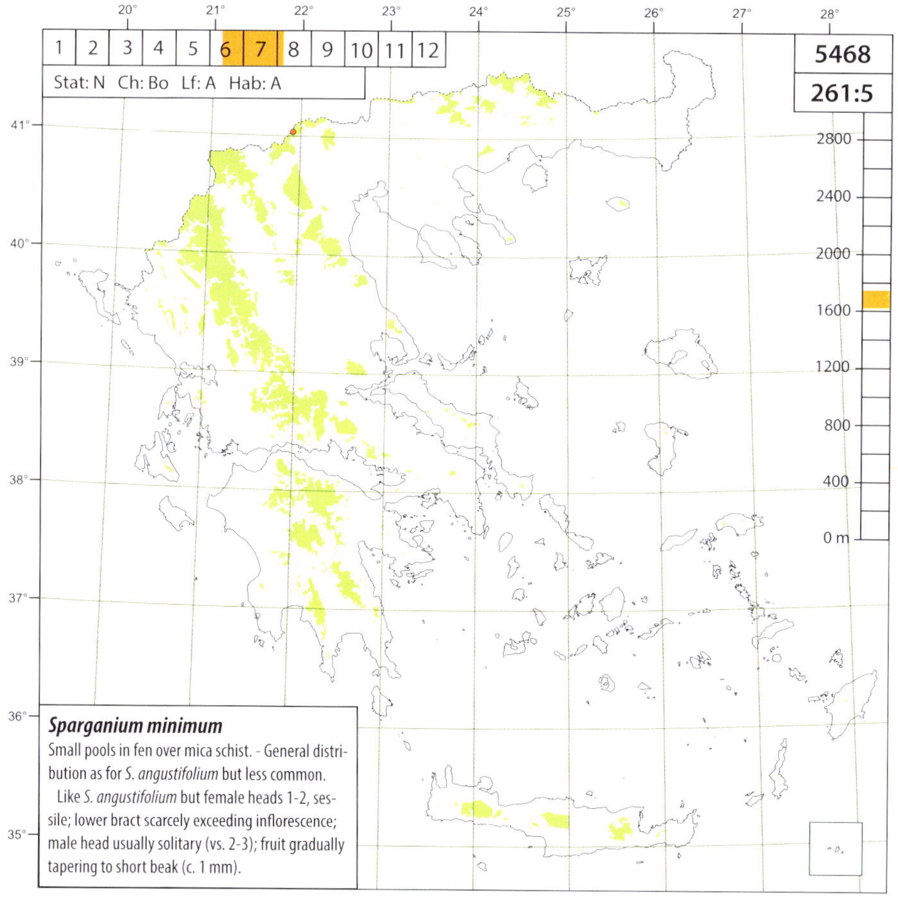

Sparganium minimum
Small pools in fen over mica schist. - General distribution as for *S. angustifolium* but less common.
Like *S. angustifolium* but female heads 1-2, sessile; lower bract scarcely exceeding inflorescence; male head usually solitary (vs. 2-3); fruit gradually tapering to short beak (c. 1 mm).

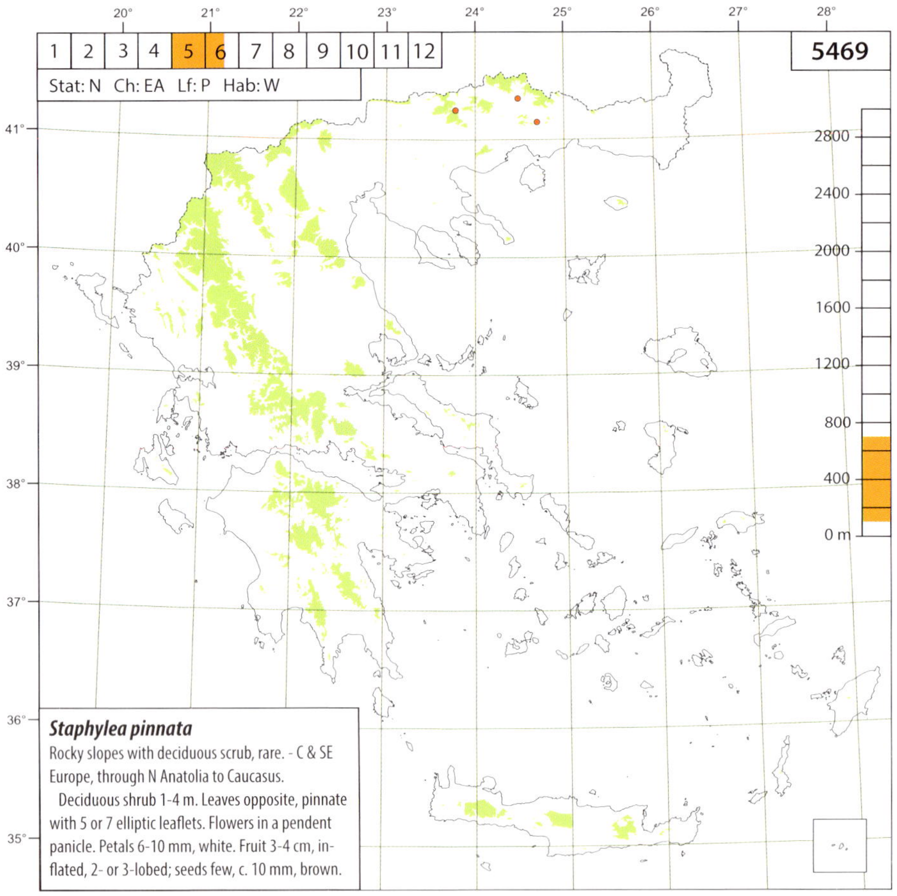

Staphylea pinnata

Rocky slopes with deciduous scrub, rare. - C & SE Europe, through N Anatolia to Caucasus.

Deciduous shrub 1-4 m. Leaves opposite, pinnate with 5 or 7 elliptic leaflets. Flowers in a pendent panicle. Petals 6-10 mm, white. Fruit 3-4 cm, inflated, 2- or 3-lobed; seeds few, c. 10 mm, brown.

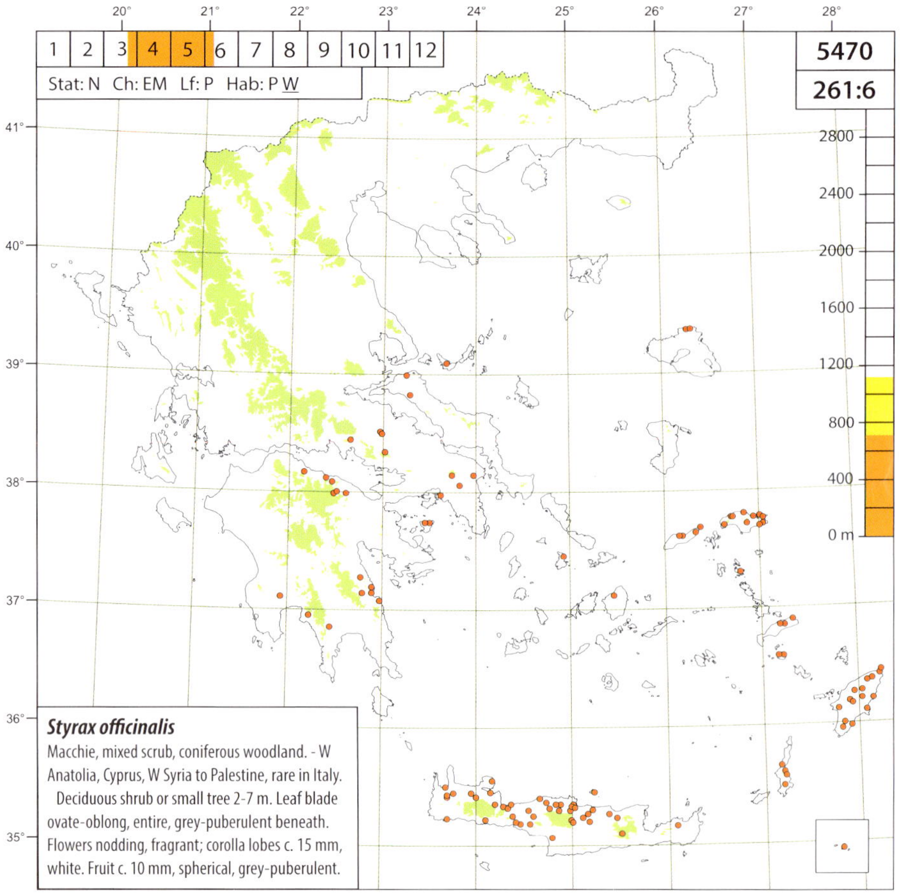

Styrax officinalis

Macchie, mixed scrub, coniferous woodland. - W Anatolia, Cyprus, W Syria to Palestine, rare in Italy.

Deciduous shrub or small tree 2-7 m. Leaf blade ovate-oblong, entire, grey-puberulent beneath. Flowers nodding, fragrant; corolla lobes c. 15 mm, white. Fruit c. 10 mm, spherical, grey-puberulent.

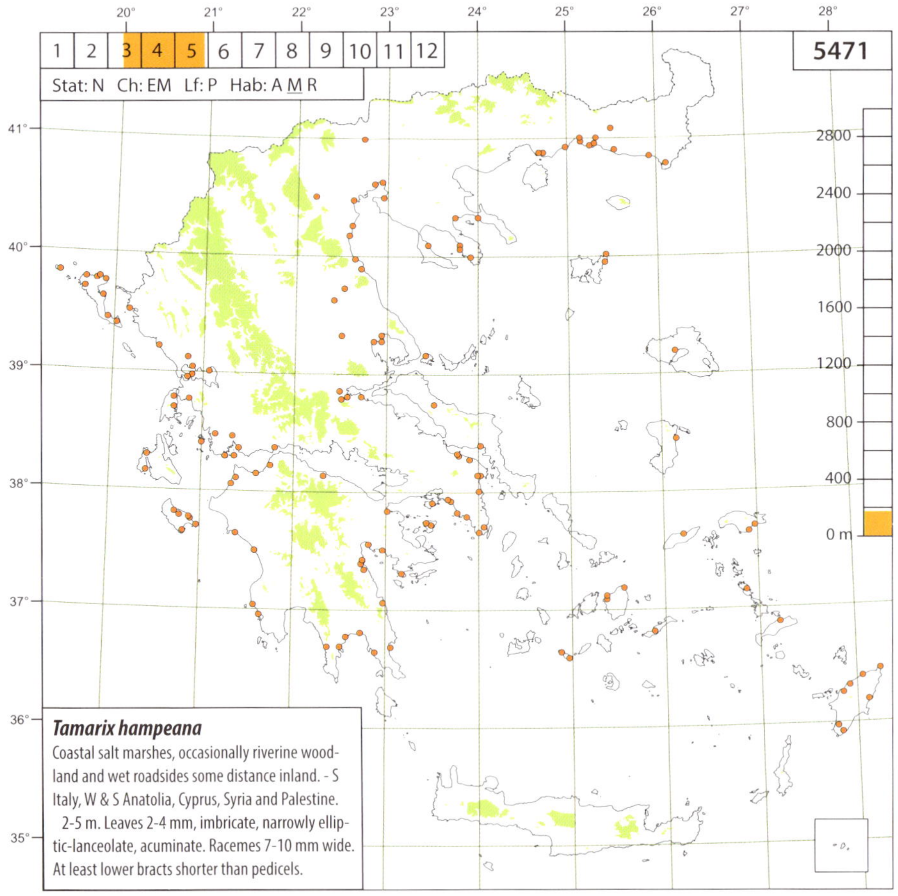

Tamarix hampeana

Coastal salt marshes, occasionally riverine woodland and wet roadsides some distance inland. - S Italy, W & S Anatolia, Cyprus, Syria and Palestine.

2-5 m. Leaves 2-4 mm, imbricate, narrowly elliptic-lanceolate, acuminate. Racemes 7-10 mm wide. At least lower bracts shorter than pedicels.

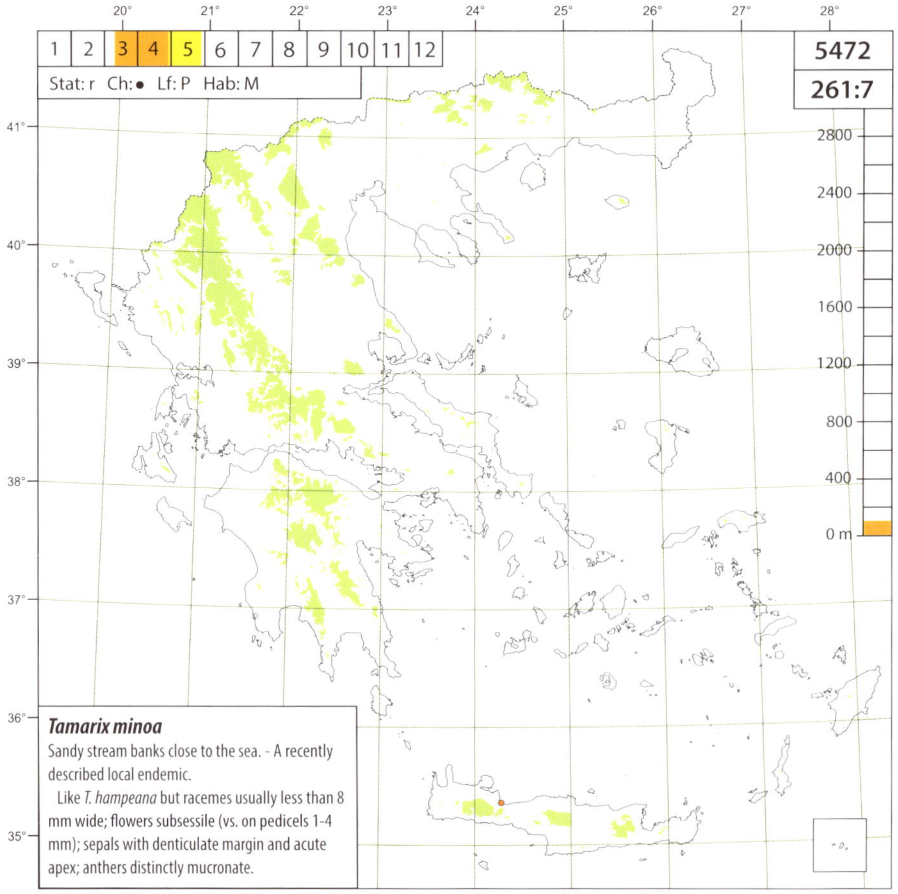

Tamarix minoa

Sandy stream banks close to the sea. - A recently described local endemic.

Like *T. hampeana* but racemes usually less than 8 mm wide; flowers subsessile (vs. on pedicels 1-4 mm); sepals with denticulate margin and acute apex; anthers distinctly mucronate.

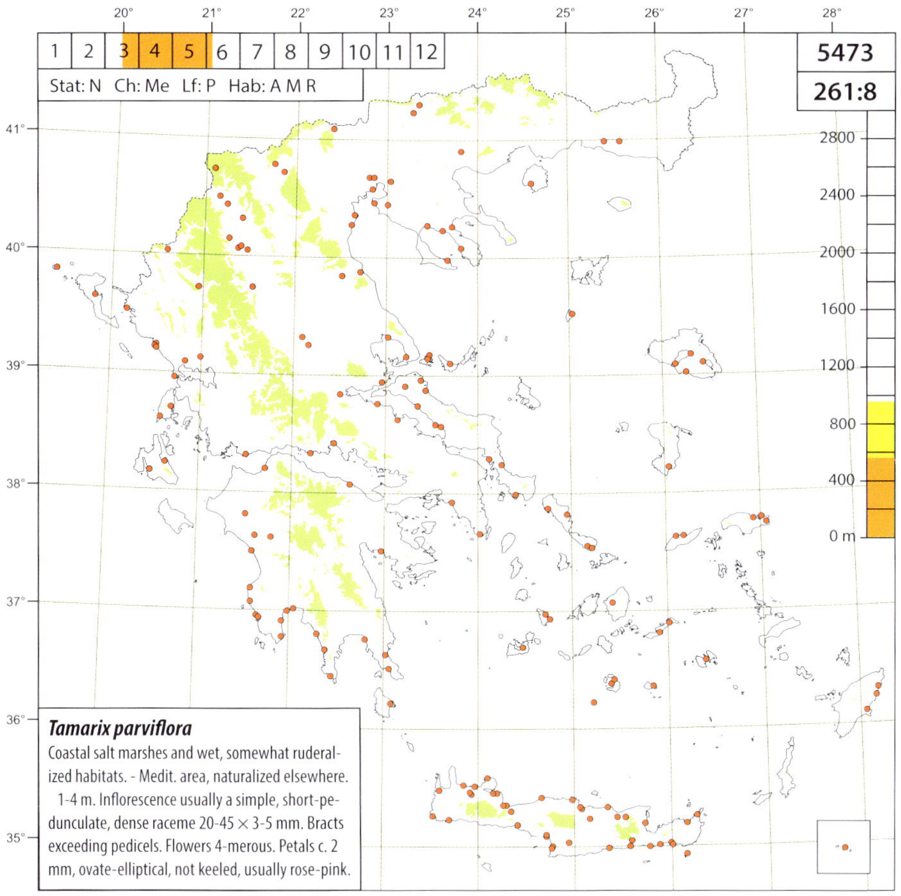

Tamarix parviflora

Coastal salt marshes and wet, somewhat ruderalized habitats. - Medit. area, naturalized elsewhere.

1-4 m. Inflorescence usually a simple, short-pedunculate, dense raceme 20-45 × 3-5 mm. Bracts exceeding pedicels. Flowers 4-merous. Petals c. 2 mm, ovate-elliptical, not keeled, usually rose-pink.

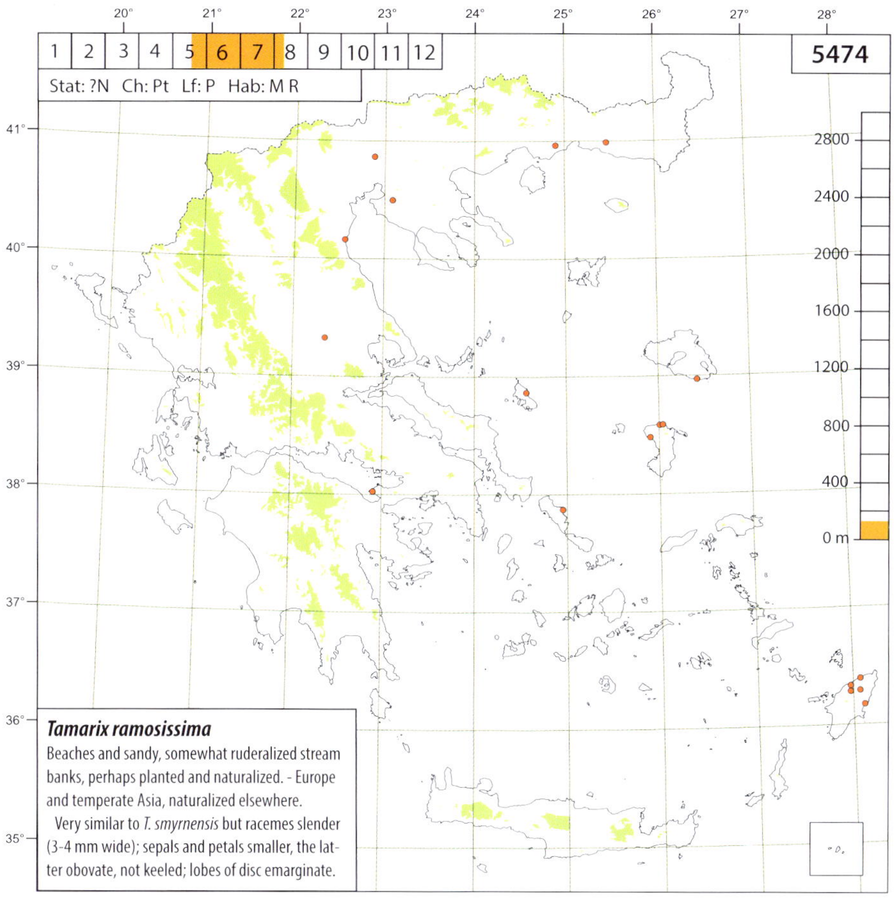

Tamarix ramosissima

Beaches and sandy, somewhat ruderalized stream banks, perhaps planted and naturalized. - Europe and temperate Asia, naturalized elsewhere.

Very similar to *T. smyrnensis* but racemes slender (3-4 mm wide); sepals and petals smaller, the latter obovate, not keeled; lobes of disc emarginate.

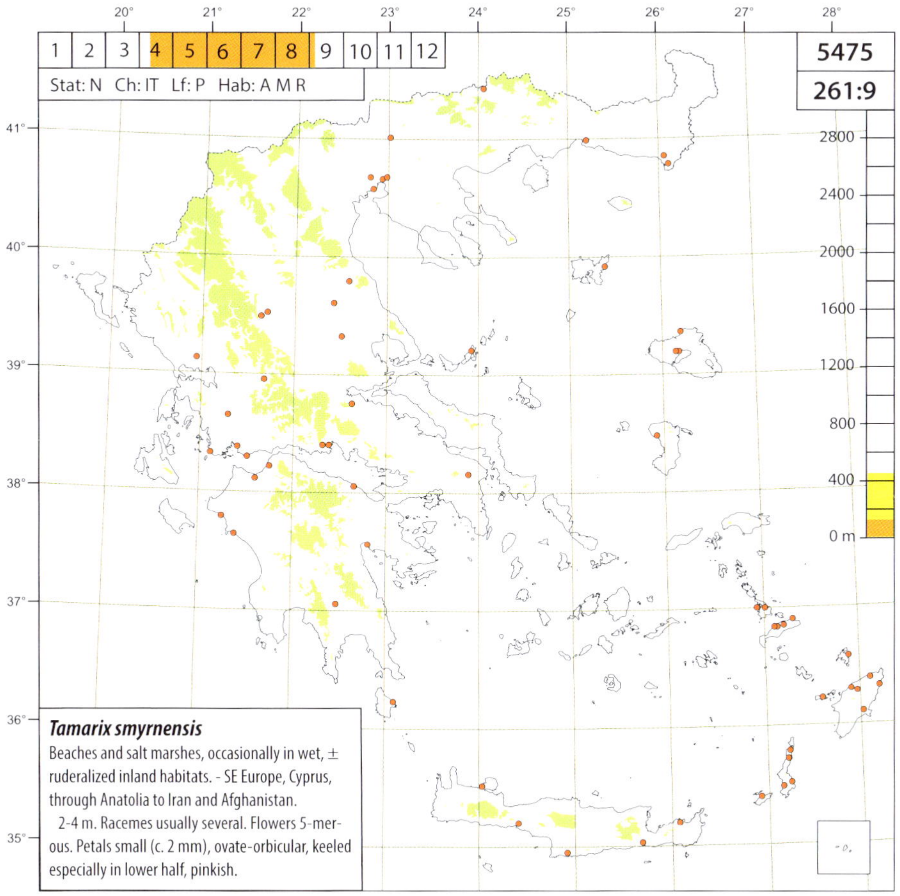

Tamarix smyrnensis

Beaches and salt marshes, occasionally in wet, ± ruderalized inland habitats. - SE Europe, Cyprus, through Anatolia to Iran and Afghanistan.

2-4 m. Racemes usually several. Flowers 5-merous. Petals small (c. 2 mm), ovate-orbicular, keeled especially in lower half, pinkish.

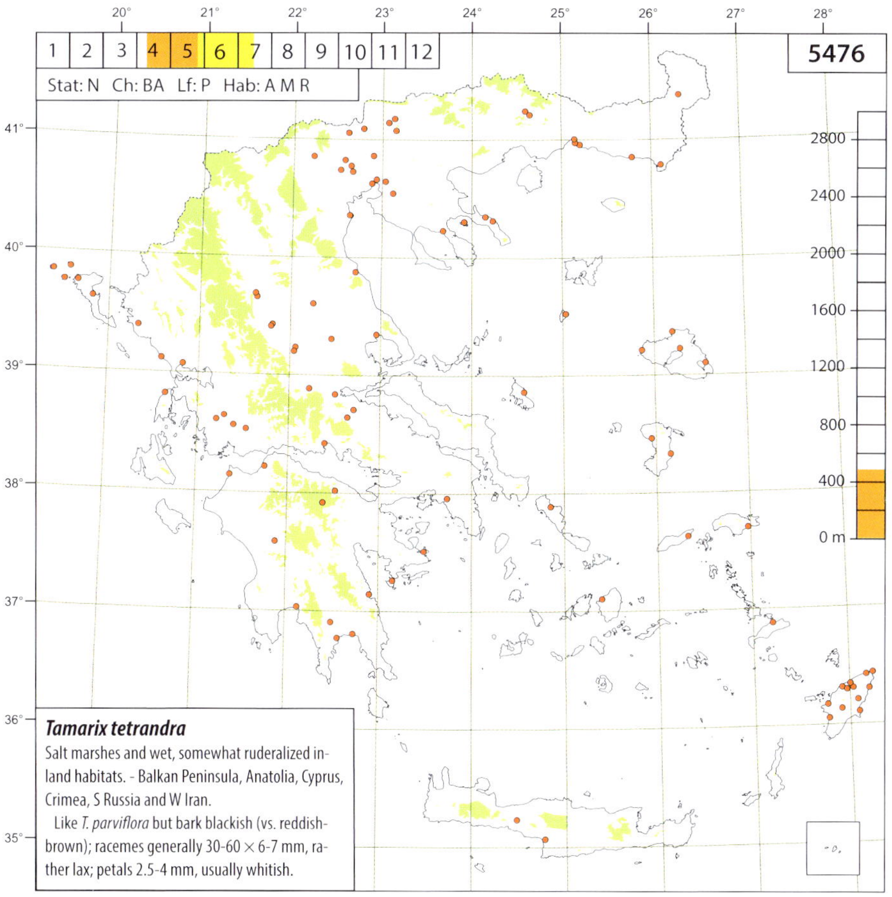

Tamarix tetrandra

Salt marshes and wet, somewhat ruderalized inland habitats. - Balkan Peninsula, Anatolia, Cyprus, Crimea, S Russia and W Iran.

Like *T. parviflora* but bark blackish (vs. reddish-brown); racemes generally 30-60 × 6-7 mm, rather lax; petals 2.5-4 mm, usually whitish.

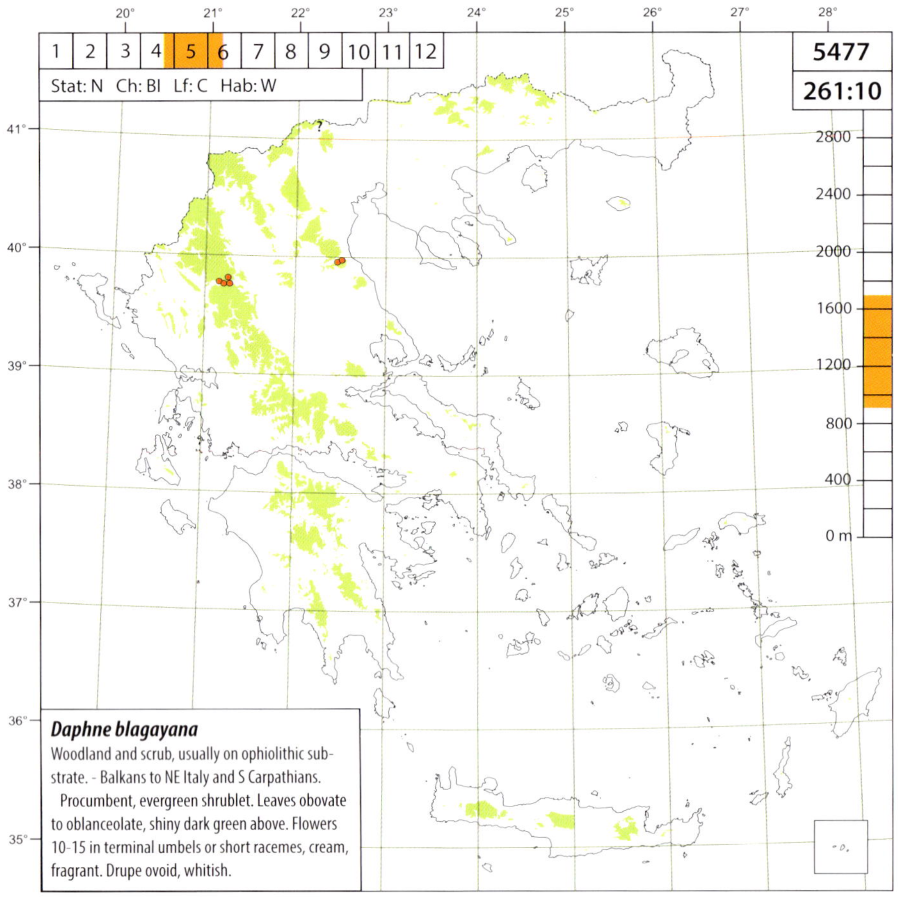

Daphne blagayana
Woodland and scrub, usually on ophiolithic substrate. - Balkans to NE Italy and S Carpathians.
Procumbent, evergreen shrublet. Leaves obovate to oblanceolate, shiny dark green above. Flowers 10-15 in terminal umbels or short racemes, cream, fragrant. Drupe ovoid, whitish.

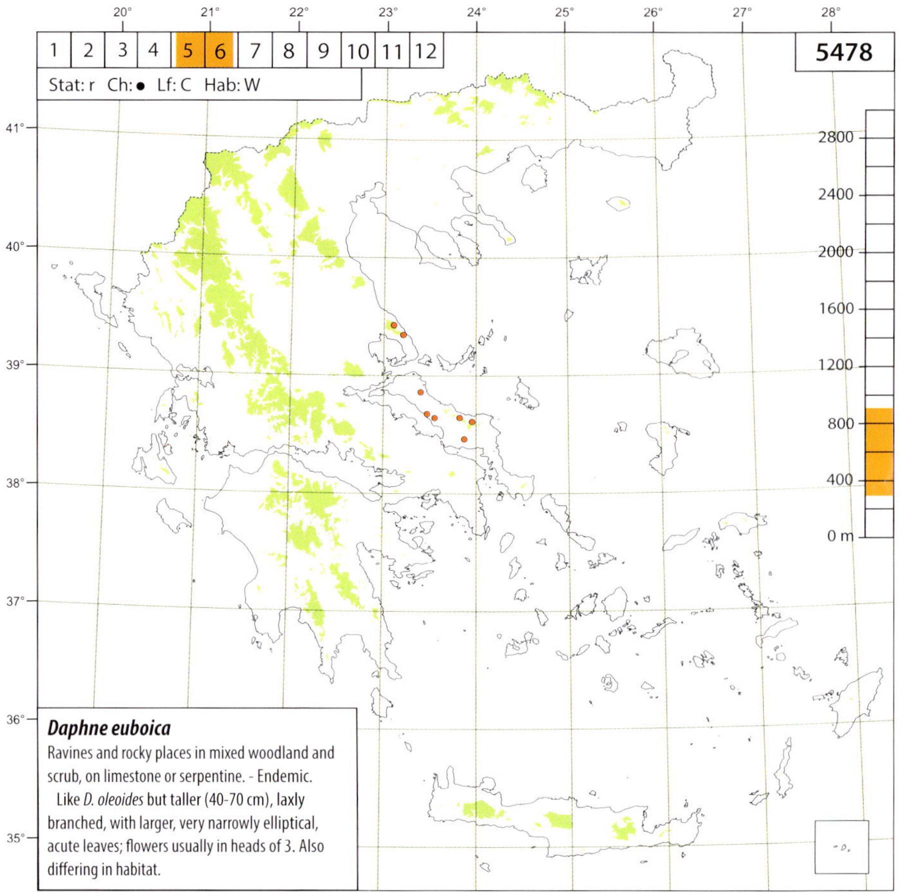

Daphne euboica
Ravines and rocky places in mixed woodland and scrub, on limestone or serpentine. - Endemic.
Like *D. oleoides* but taller (40-70 cm), laxly branched, with larger, very narrowly elliptical, acute leaves; flowers usually in heads of 3. Also differing in habitat.

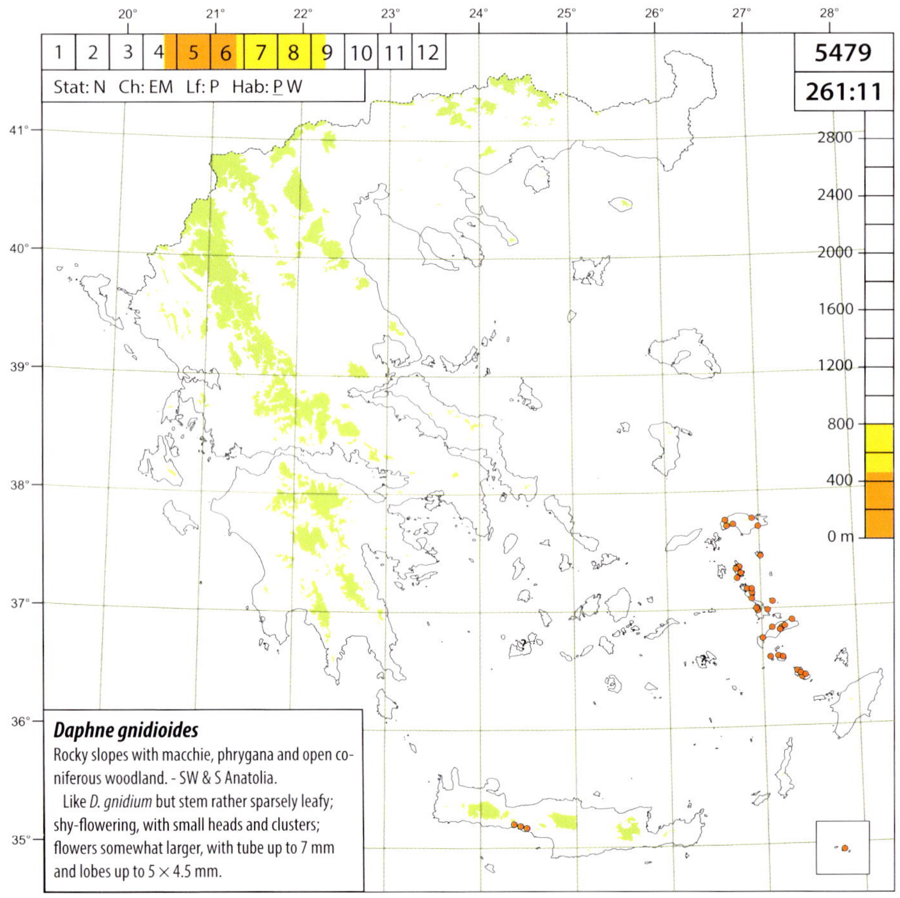

Daphne gnidioides
Rocky slopes with macchie, phrygana and open coniferous woodland. - SW & S Anatolia.
Like *D. gnidium* but stem rather sparsely leafy; shy-flowering, with small heads and clusters; flowers somewhat larger, with tube up to 7 mm and lobes up to 5 × 4.5 mm.

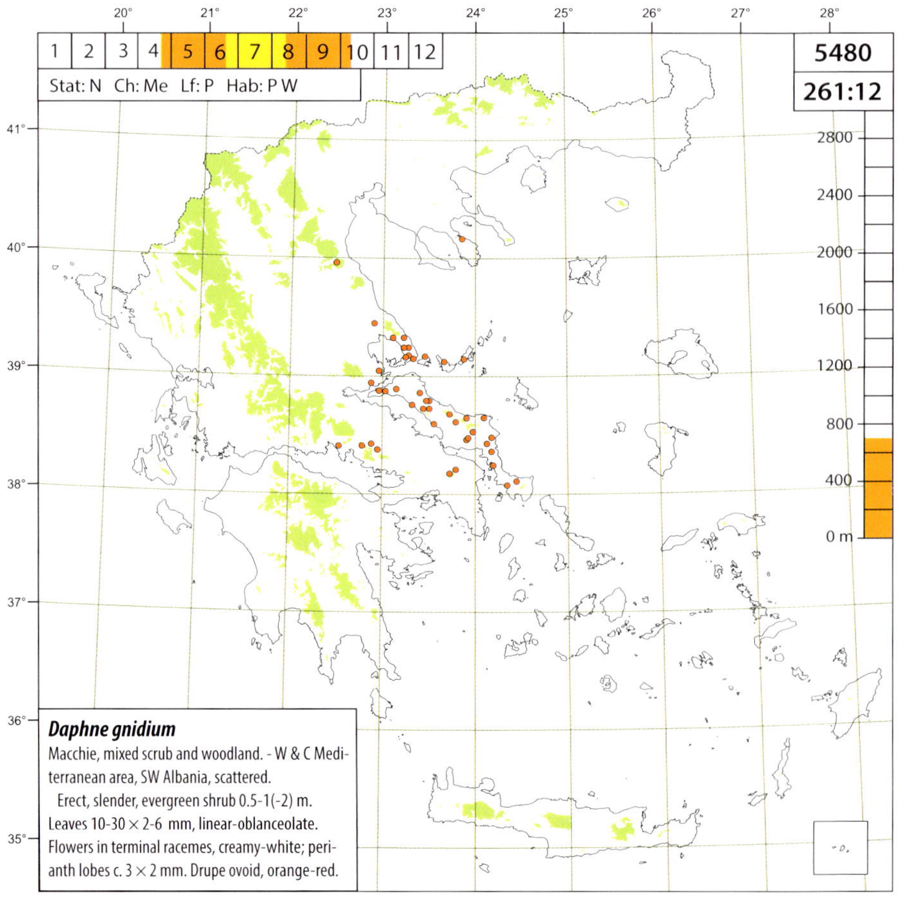

Daphne gnidium
Macchie, mixed scrub and woodland. - W & C Mediterranean area, SW Albania, scattered.
Erect, slender, evergreen shrub 0.5-1(-2) m. Leaves 10-30 × 2-6 mm, linear-oblanceolate. Flowers in terminal racemes, creamy-white; perianth lobes c. 3 × 2 mm. Drupe ovoid, orange-red.

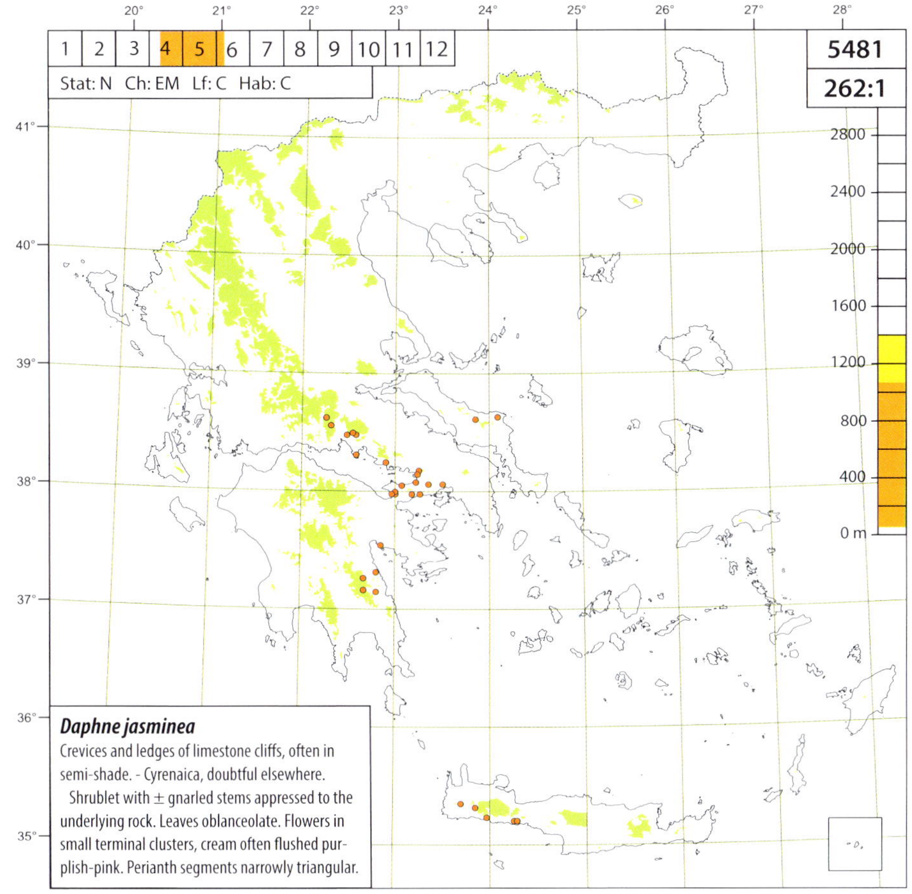

Daphne jasminea
Crevices and ledges of limestone cliffs, often in semi-shade. - Cyrenaica, doubtful elsewhere.
Shrublet with ± gnarled stems appressed to the underlying rock. Leaves oblanceolate. Flowers in small terminal clusters, cream often flushed purplish-pink. Perianth segments narrowly triangular.

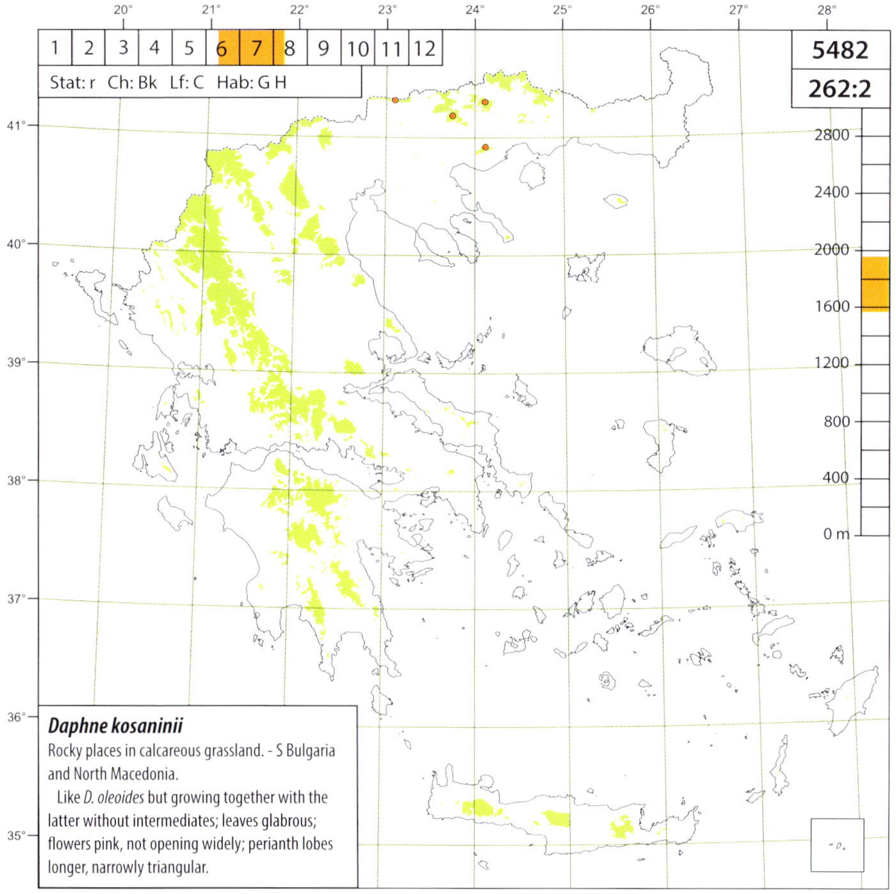

Daphne kosaninii
Rocky places in calcareous grassland. - S Bulgaria and North Macedonia.
Like *D. oleoides* but growing together with the latter without intermediates; leaves glabrous; flowers pink, not opening widely; perianth lobes longer, narrowly triangular.

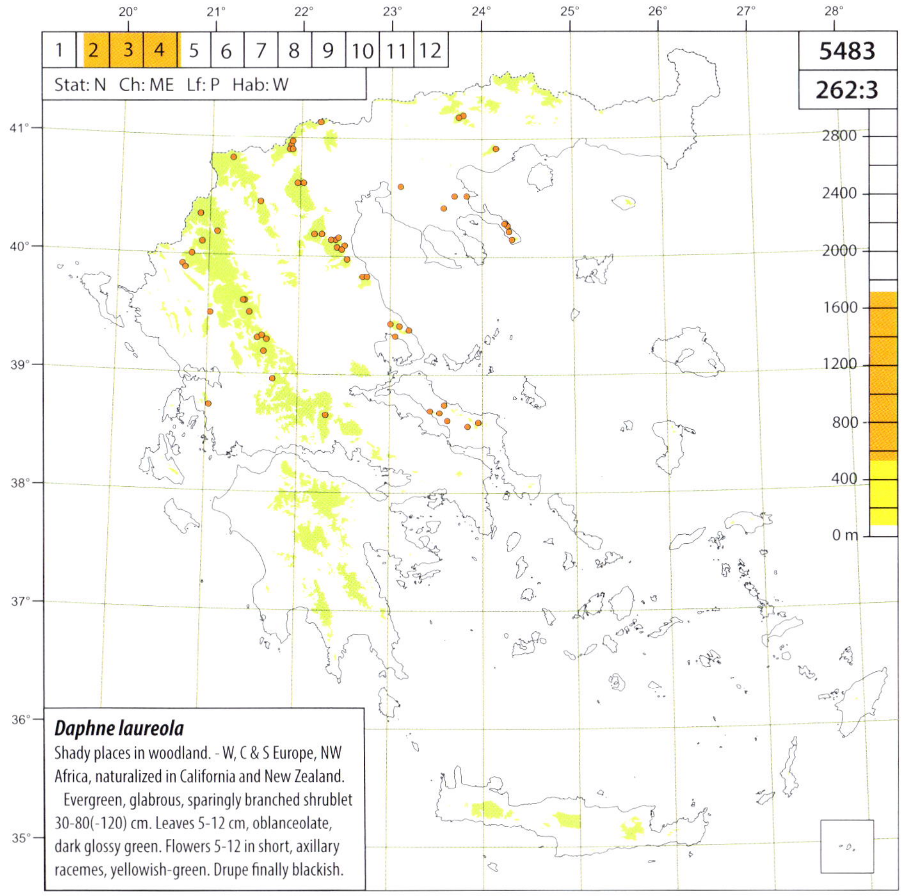

Daphne laureola
Shady places in woodland. - W, C & S Europe, NW Africa, naturalized in California and New Zealand.
Evergreen, glabrous, sparingly branched shrublet 30-80(-120) cm. Leaves 5-12 cm, oblanceolate, dark glossy green. Flowers 5-12 in short, axillary racemes, yellowish-green. Drupe finally blackish.

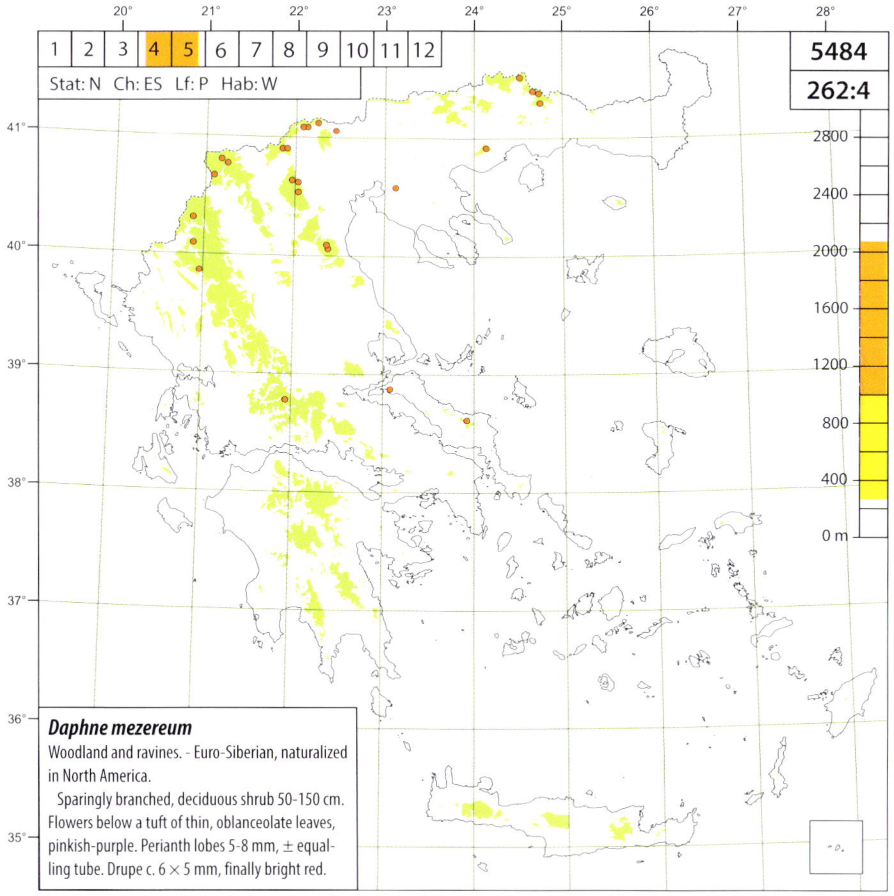

Daphne mezereum
Woodland and ravines. - Euro-Siberian, naturalized in North America.
Sparingly branched, deciduous shrub 50-150 cm. Flowers below a tuft of thin, oblanceolate leaves, pinkish-purple. Perianth lobes 5-8 mm, ± equalling tube. Drupe c. 6 × 5 mm, finally bright red.

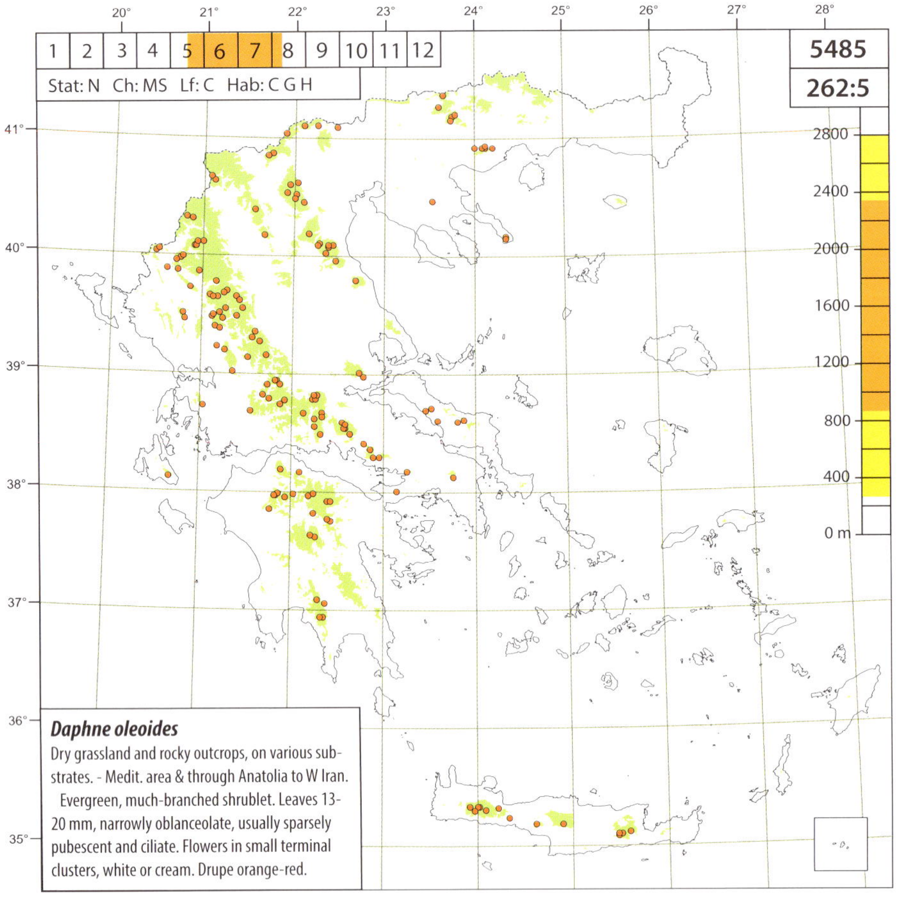

Daphne oleoides
Dry grassland and rocky outcrops, on various substrates. - Medit. area & through Anatolia to W Iran.
Evergreen, much-branched shrublet. Leaves 13-20 mm, narrowly oblanceolate, usually sparsely pubescent and ciliate. Flowers in small terminal clusters, white or cream. Drupe orange-red.

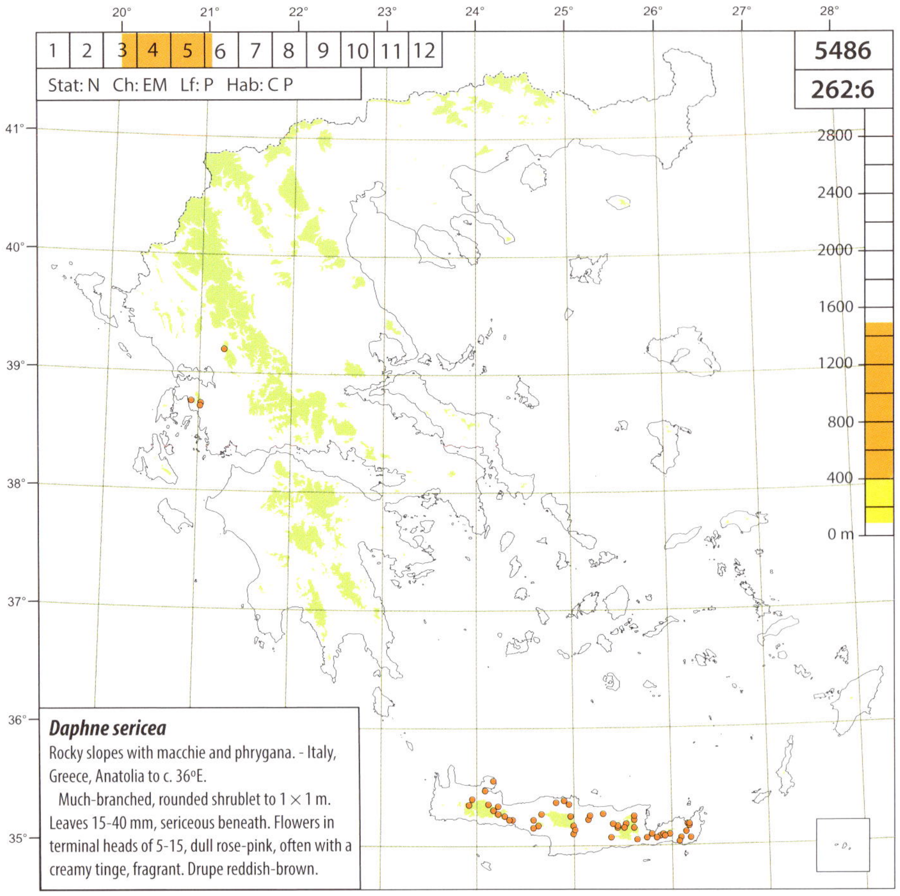

Daphne sericea
Rocky slopes with macchie and phrygana. - Italy, Greece, Anatolia to c. 36°E.
Much-branched, rounded shrublet to 1 × 1 m. Leaves 15-40 mm, sericeous beneath. Flowers in terminal heads of 5-15, dull rose-pink, often with a creamy tinge, fragrant. Drupe reddish-brown.

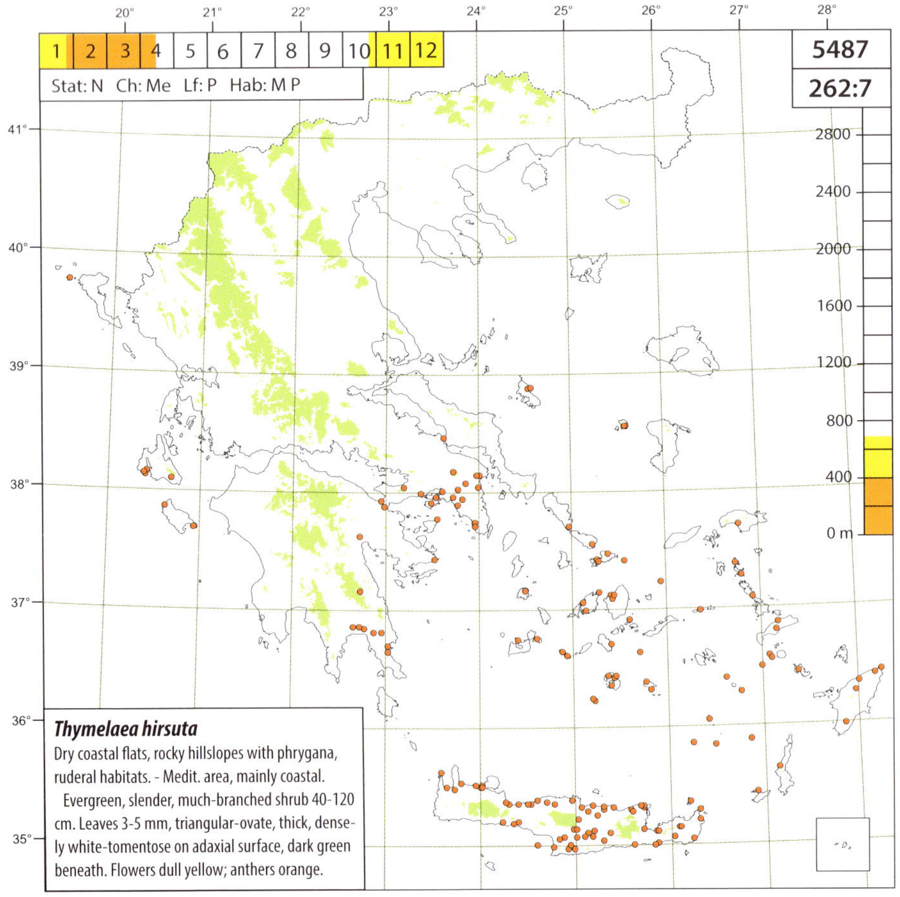

Thymelaea hirsuta
Dry coastal flats, rocky hillslopes with phrygana, ruderal habitats. - Medit. area, mainly coastal.
Evergreen, slender, much-branched shrub 40-120 cm. Leaves 3-5 mm, triangular-ovate, thick, densely white-tomentose on adaxial surface, dark green beneath. Flowers dull yellow; anthers orange.

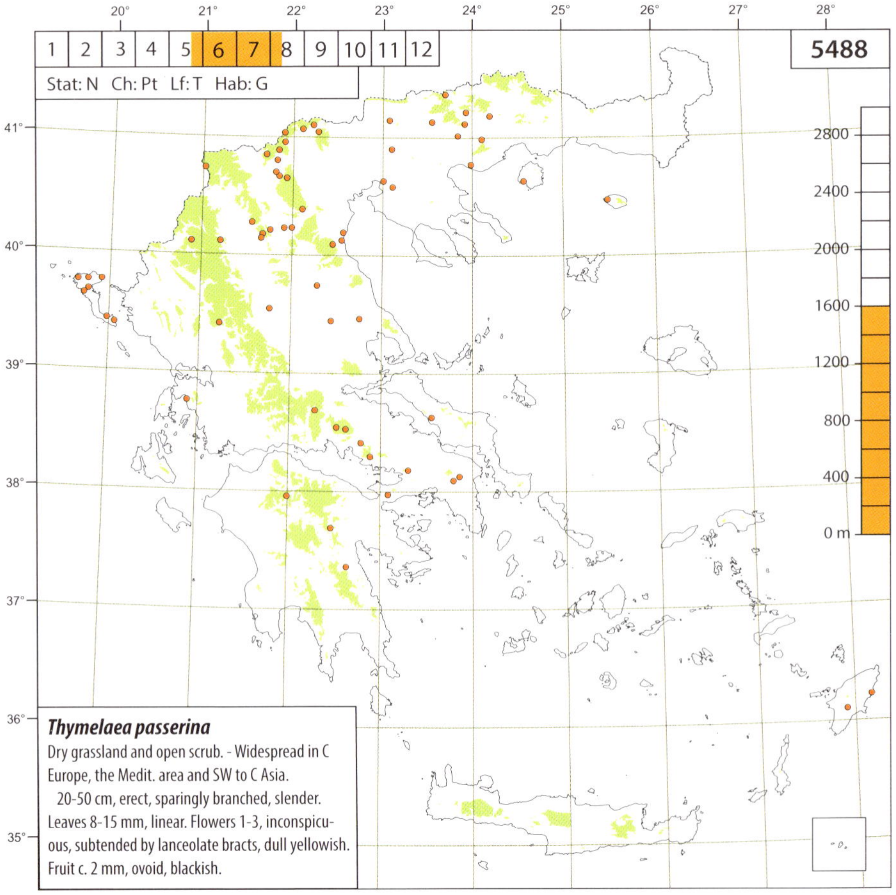

Thymelaea passerina
Dry grassland and open scrub. - Widespread in C Europe, the Medit. area and SW to C Asia.
20-50 cm, erect, sparingly branched, slender. Leaves 8-15 mm, linear. Flowers 1-3, inconspicuous, subtended by lanceolate bracts, dull yellowish. Fruit c. 2 mm, ovoid, blackish.

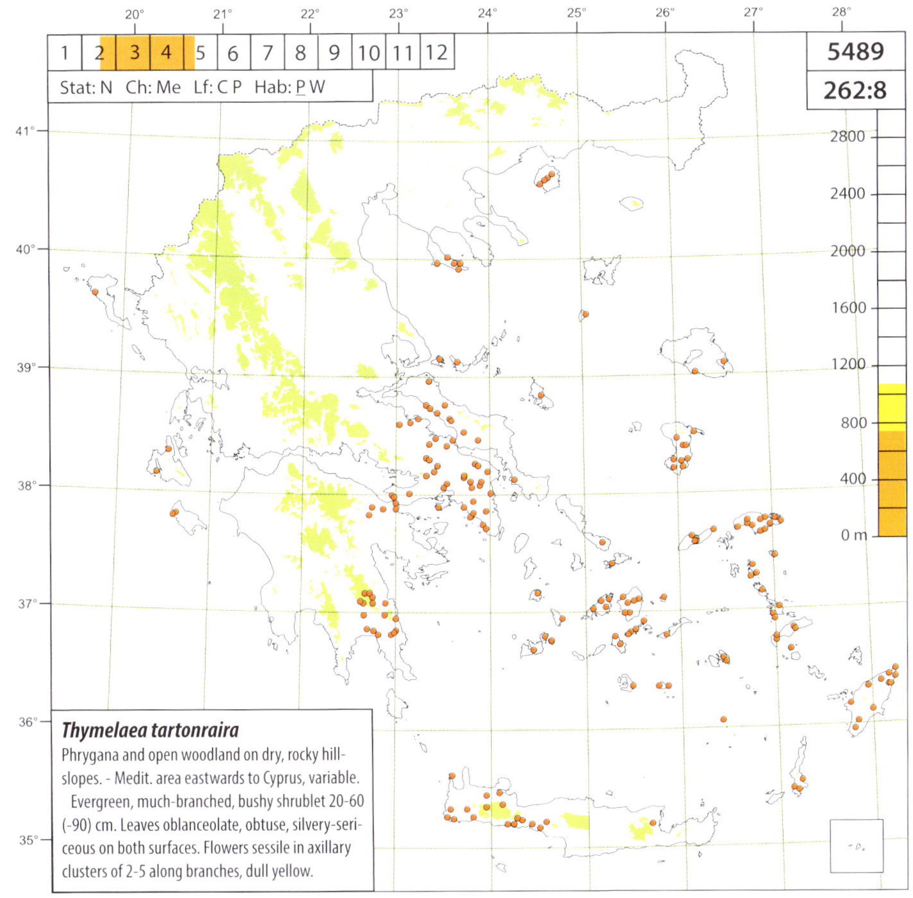

Thymelaea tartonraira
Phrygana and open woodland on dry, rocky hill-slopes. - Medit. area eastwards to Cyprus, variable.
Evergreen, much-branched, bushy shrublet 20-60 (-90) cm. Leaves oblanceolate, obtuse, silvery-sericeous on both surfaces. Flowers sessile in axillary clusters of 2-5 along branches, dull yellow.

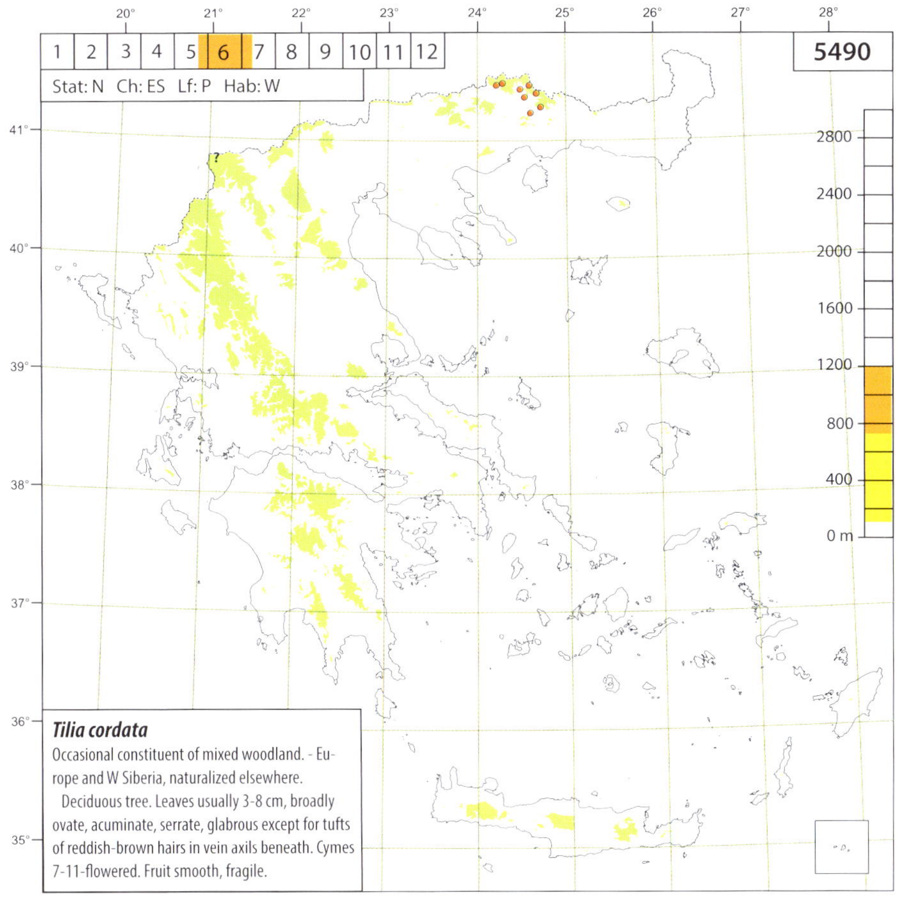

Tilia cordata
Occasional constituent of mixed woodland. - Europe and W Siberia, naturalized elsewhere.
Deciduous tree. Leaves usually 3-8 cm, broadly ovate, acuminate, serrate, glabrous except for tufts of reddish-brown hairs in vein axils beneath. Cymes 7-11-flowered. Fruit smooth, fragile.

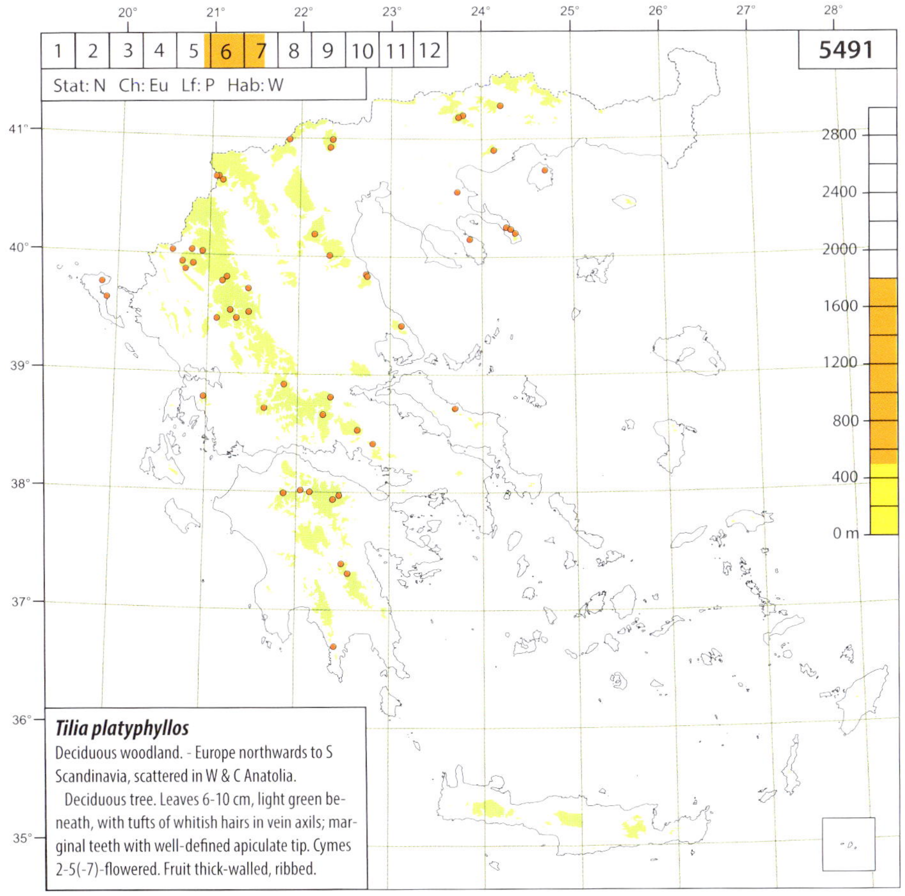

Tilia platyphyllos
Deciduous woodland. - Europe northwards to S Scandinavia, scattered in W & C Anatolia.
Deciduous tree. Leaves 6-10 cm, light green beneath, with tufts of whitish hairs in vein axils; marginal teeth with well-defined apiculate tip. Cymes 2-5(-7)-flowered. Fruit thick-walled, ribbed.

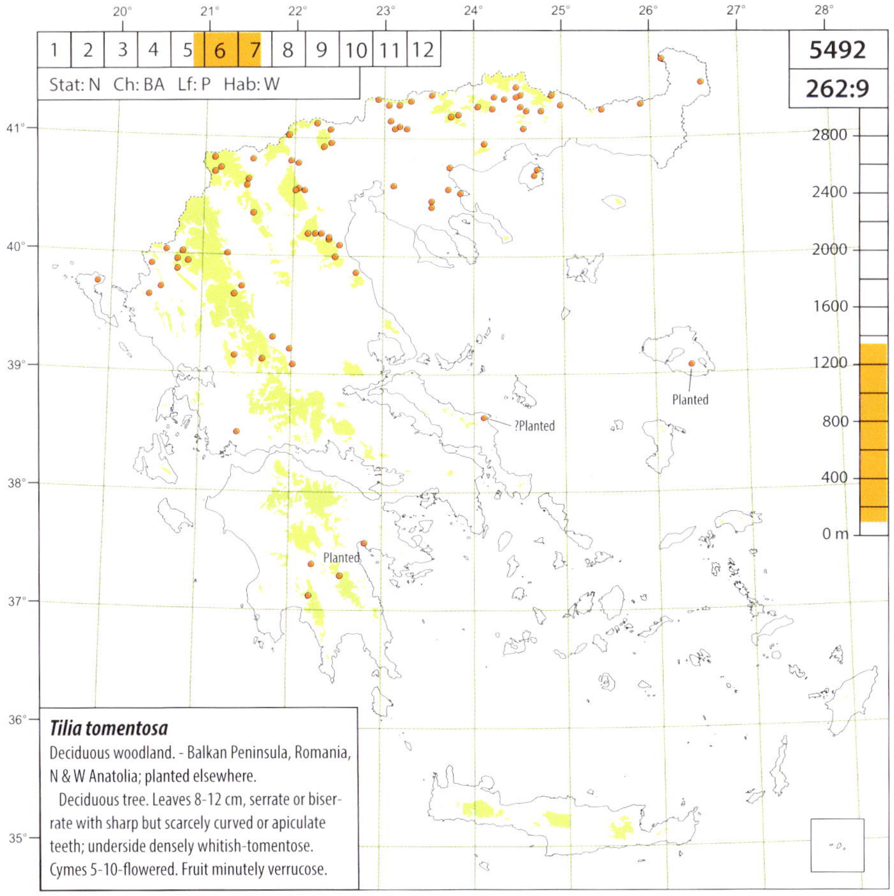

Tilia tomentosa
Deciduous woodland. - Balkan Peninsula, Romania, N & W Anatolia; planted elsewhere.
Deciduous tree. Leaves 8-12 cm, serrate or biserrate with sharp but scarcely curved or apiculate teeth; underside densely whitish-tomentose. Cymes 5-10-flowered. Fruit minutely verrucose.

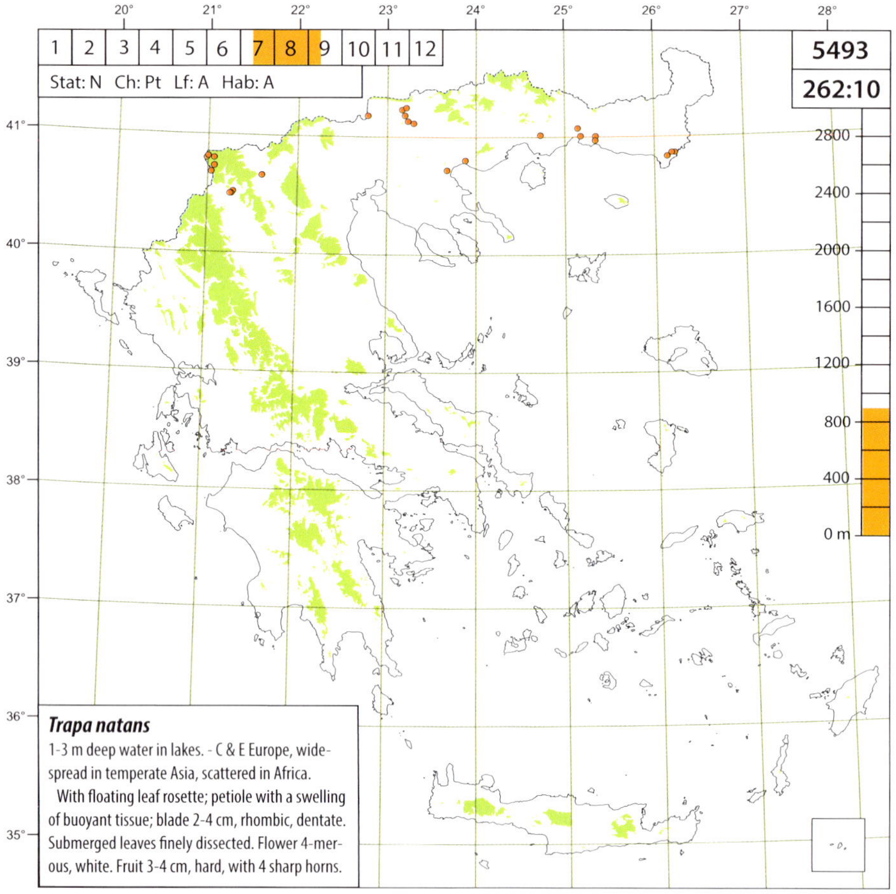

Trapa natans
1-3 m deep water in lakes. - C & E Europe, widespread in temperate Asia, scattered in Africa.
With floating leaf rosette; petiole with a swelling of buoyant tissue; blade 2-4 cm, rhombic, dentate. Submerged leaves finely dissected. Flower 4-merous, white. Fruit 3-4 cm, hard, with 4 sharp horns.

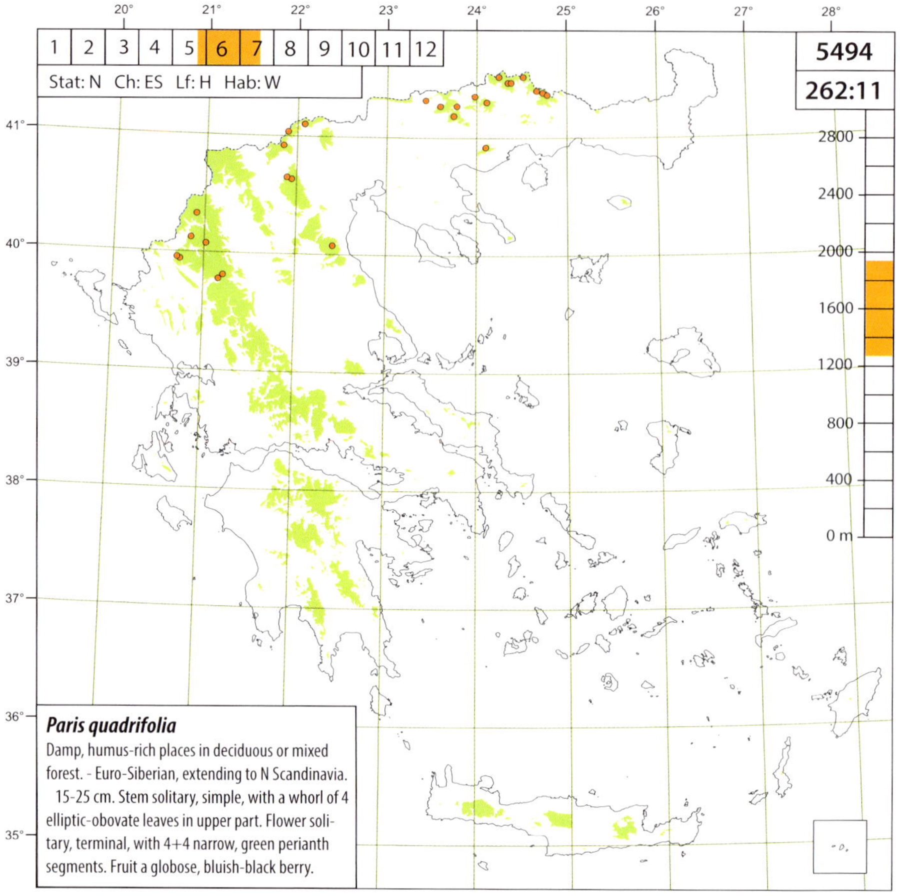

Paris quadrifolia
Damp, humus-rich places in deciduous or mixed forest. - Euro-Siberian, extending to N Scandinavia.
15-25 cm. Stem solitary, simple, with a whorl of 4 elliptic-obovate leaves in upper part. Flower solitary, terminal, with 4+4 narrow, green perianth segments. Fruit a globose, bluish-black berry.

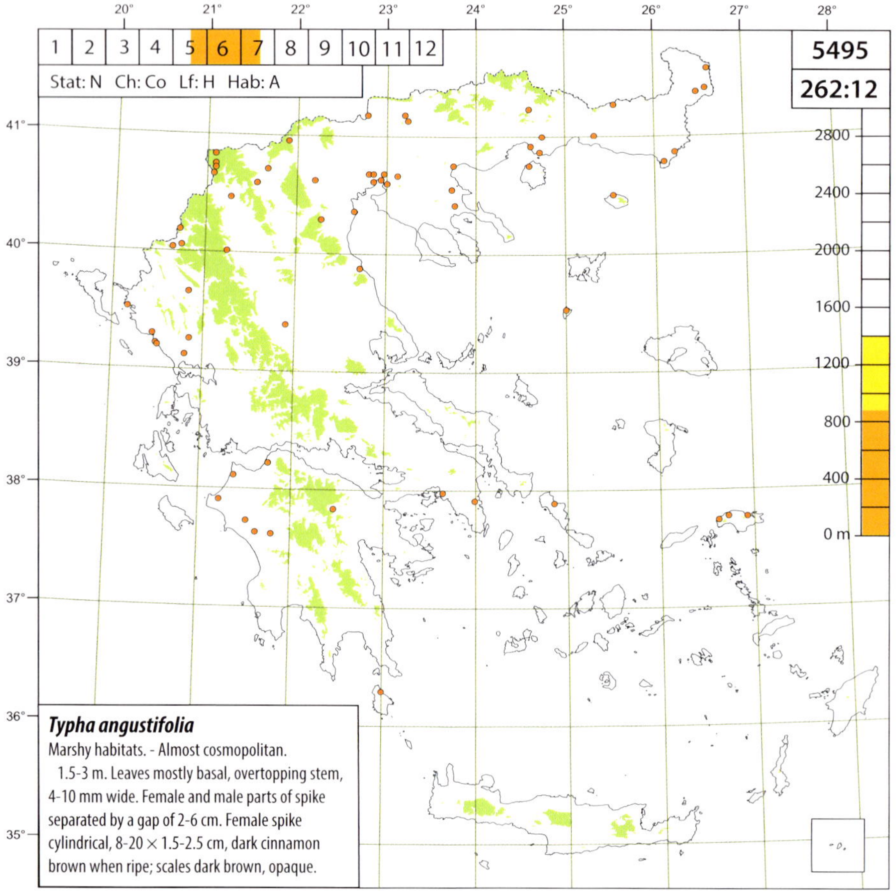

Typha angustifolia
Marshy habitats. - Almost cosmopolitan.
1.5-3 m. Leaves mostly basal, overtopping stem, 4-10 mm wide. Female and male parts of spike separated by a gap of 2-6 cm. Female spike cylindrical, 8-20 × 1.5-2.5 cm, dark cinnamon brown when ripe; scales dark brown, opaque.

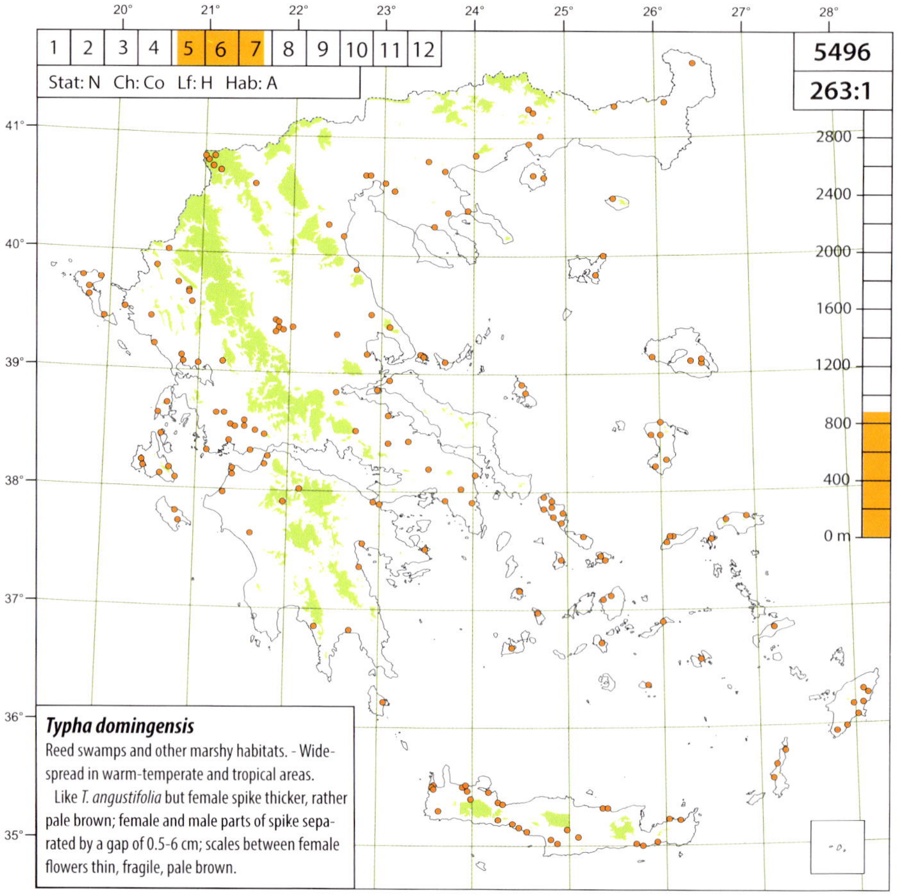

Typha domingensis
Reed swamps and other marshy habitats. - Widespread in warm-temperate and tropical areas.
Like *T. angustifolia* but female spike thicker, rather pale brown; female and male parts of spike separated by a gap of 0.5-6 cm; scales between female flowers thin, fragile, pale brown.

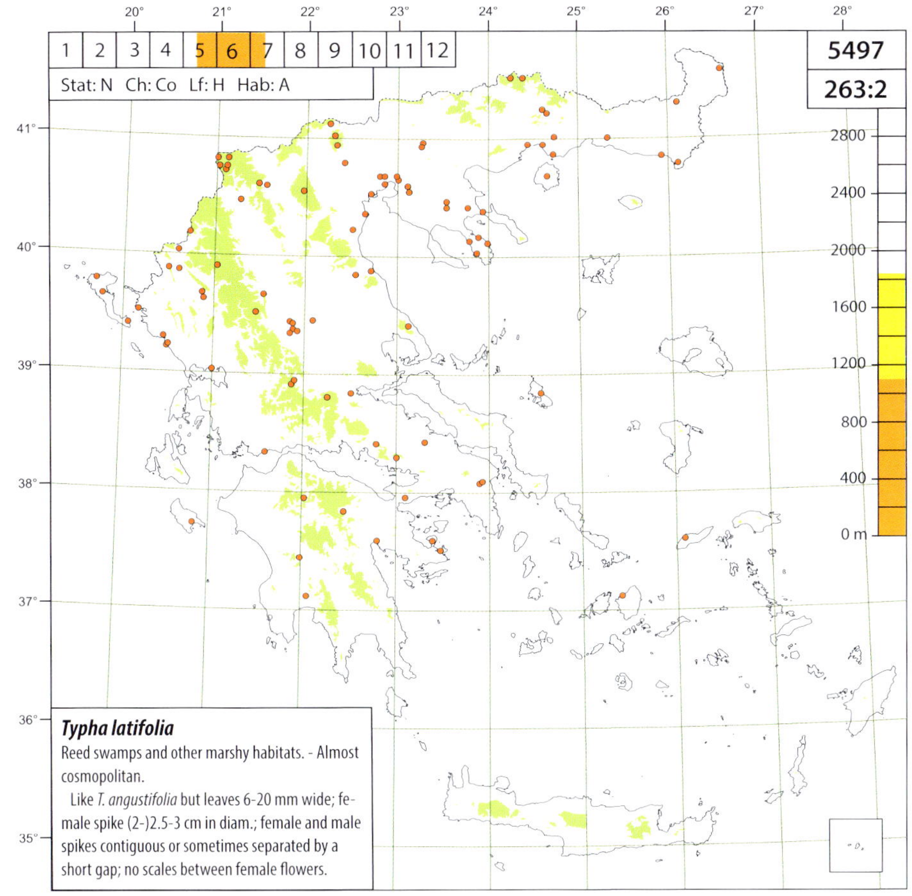

Typha latifolia

Reed swamps and other marshy habitats. - Almost cosmopolitan.

Like *T. angustifolia* but leaves 6-20 mm wide; female spike (2-)2.5-3 cm in diam.; female and male spikes contiguous or sometimes separated by a short gap; no scales between female flowers.

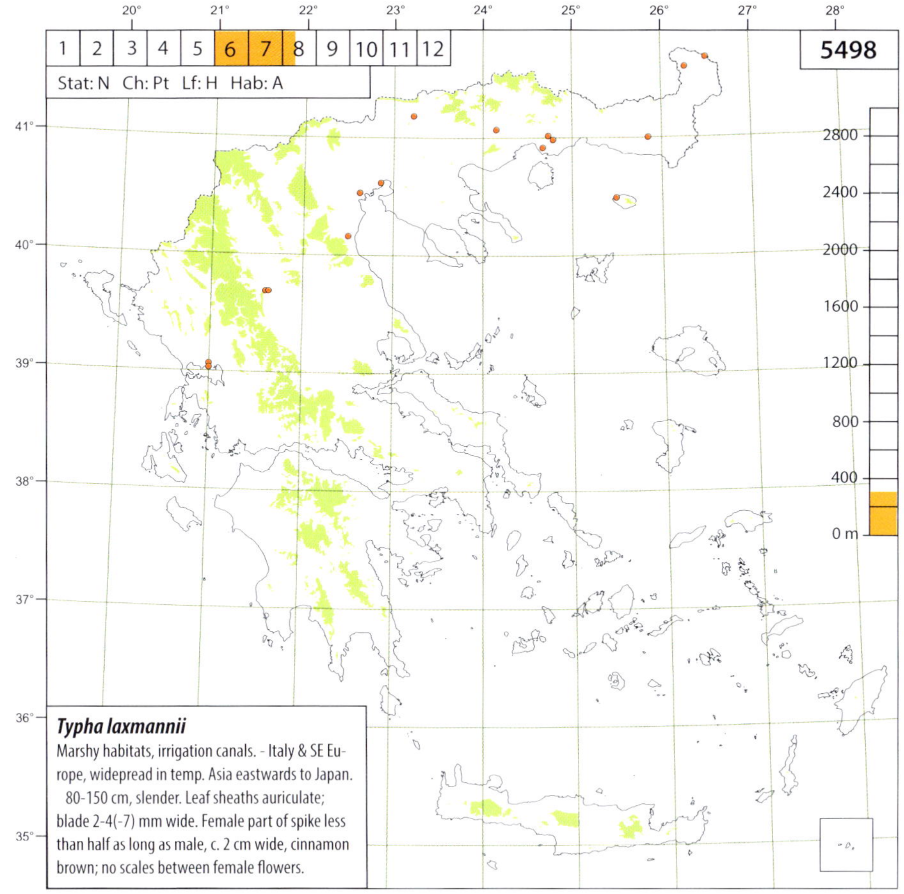

Typha laxmannii

Marshy habitats, irrigation canals. - Italy & SE Europe, widespread in temp. Asia eastwards to Japan.

80-150 cm, slender. Leaf sheaths auriculate; blade 2-4(-7) mm wide. Female part of spike less than half as long as male, c. 2 cm wide, cinnamon brown; no scales between female flowers.

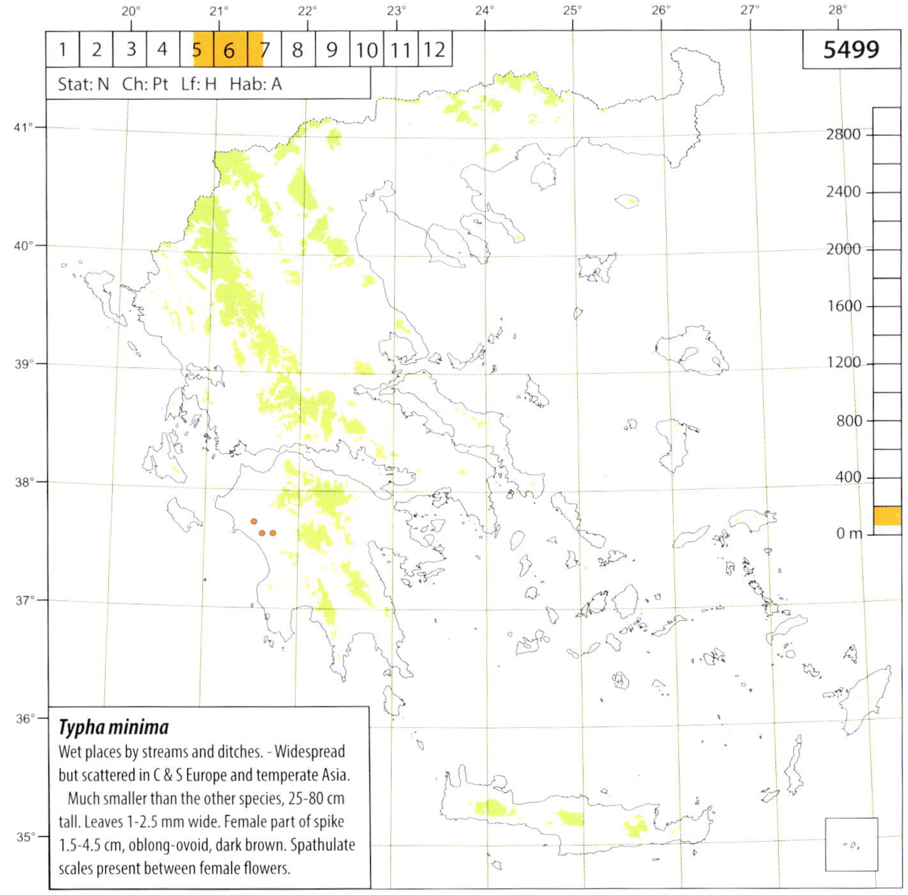

Typha minima

Wet places by streams and ditches. - Widespread but scattered in C & S Europe and temperate Asia.

Much smaller than the other species, 25-80 cm tall. Leaves 1-2.5 mm wide. Female part of spike 1.5-4.5 cm, oblong-ovoid, dark brown. Spathulate scales present between female flowers.

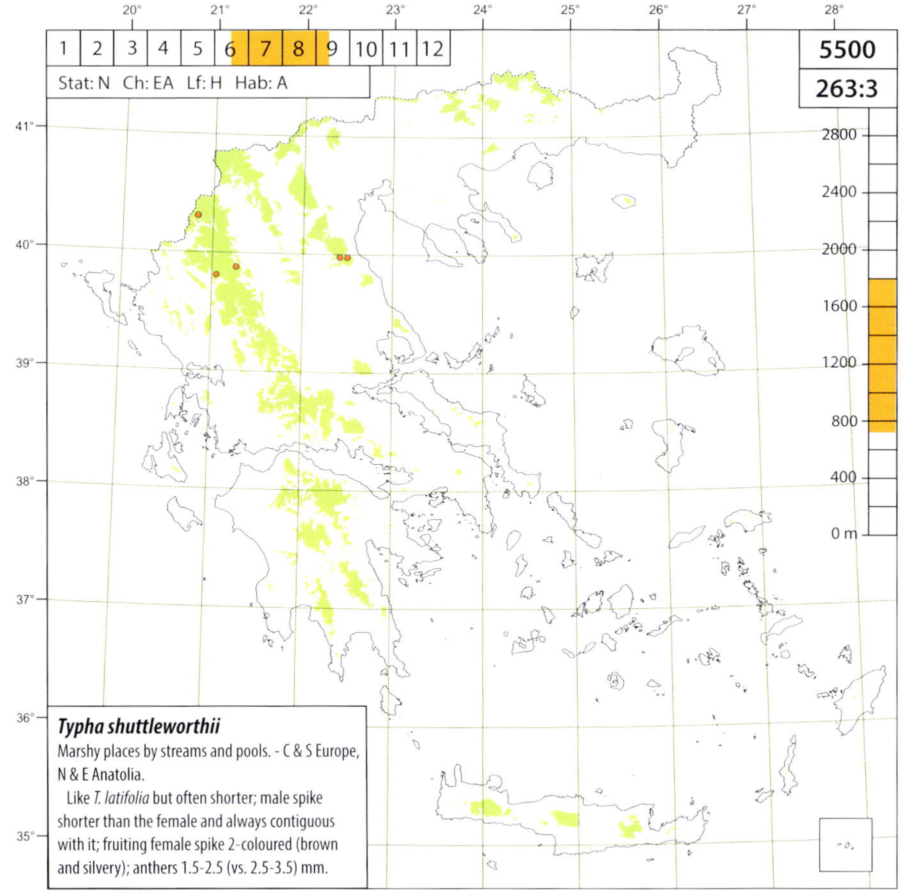

Typha shuttleworthii

Marshy places by streams and pools. - C & S Europe, N & E Anatolia.

Like *T. latifolia* but often shorter; male spike shorter than the female and always contiguous with it; fruiting female spike 2-coloured (brown and silvery); anthers 1.5-2.5 (vs. 2.5-3.5) mm.

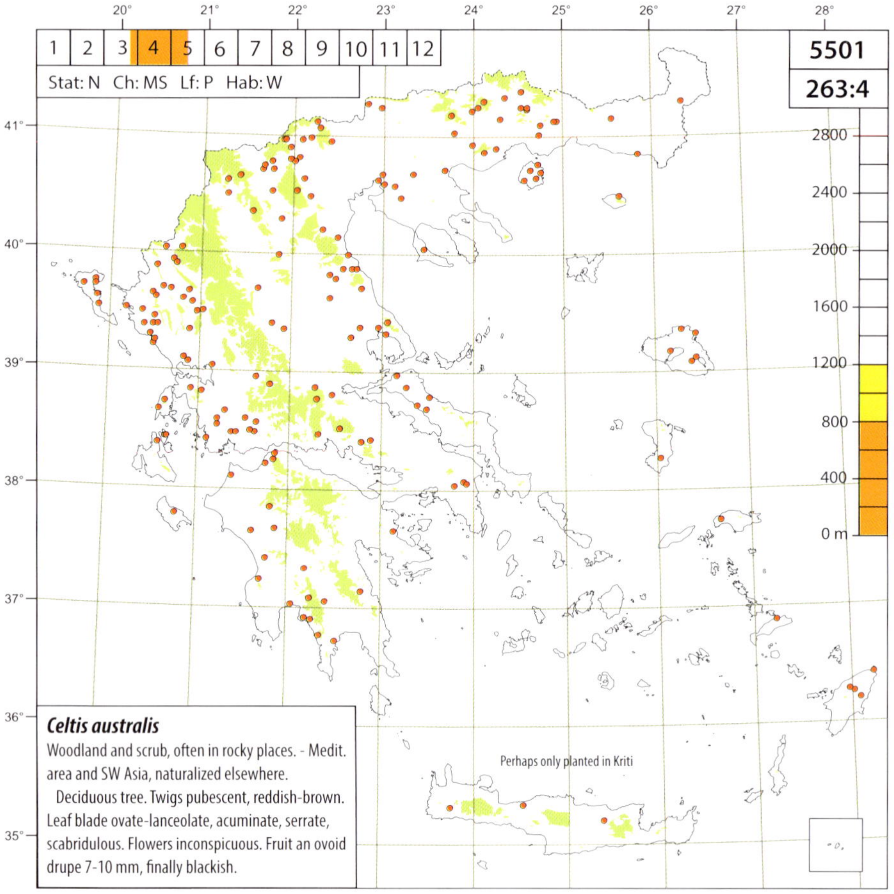

Celtis australis

Woodland and scrub, often in rocky places. - Medit. area and SW Asia, naturalized elsewhere.

Deciduous tree. Twigs pubescent, reddish-brown. Leaf blade ovate-lanceolate, acuminate, serrate, scabridulous. Flowers inconspicuous. Fruit an ovoid drupe 7-10 mm, finally blackish.

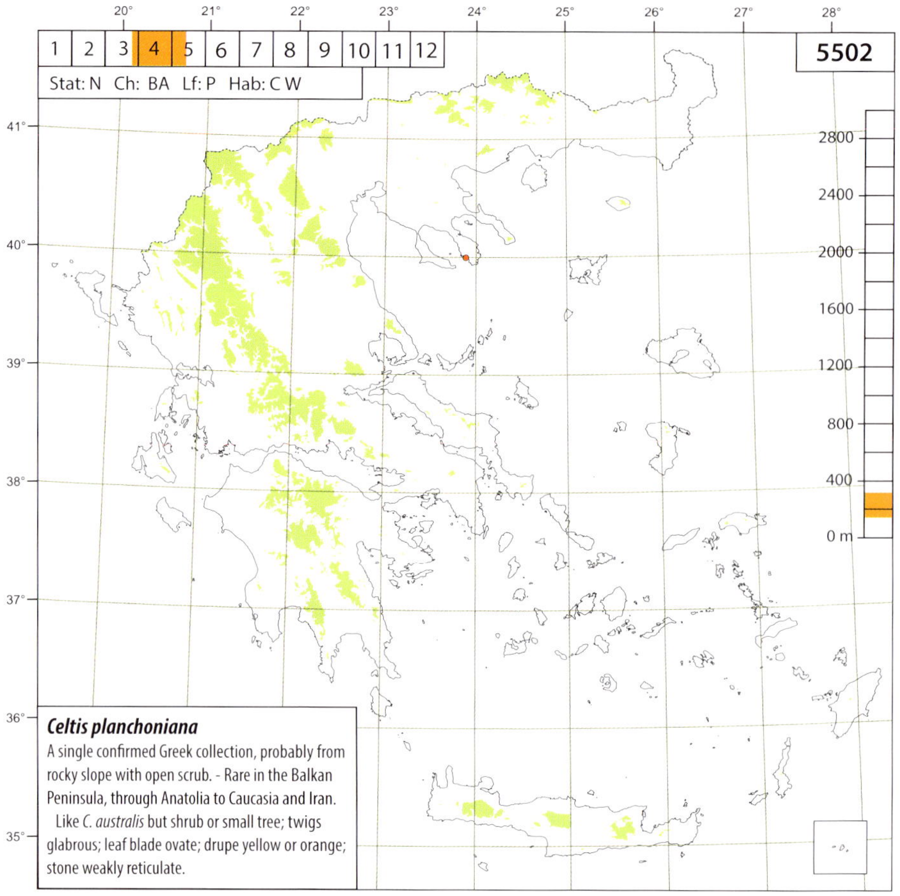

Celtis planchoniana

A single confirmed Greek collection, probably from rocky slope with open scrub. - Rare in the Balkan Peninsula, through Anatolia to Caucasia and Iran.

Like *C. australis* but shrub or small tree; twigs glabrous; leaf blade ovate; drupe yellow or orange; stone weakly reticulate.

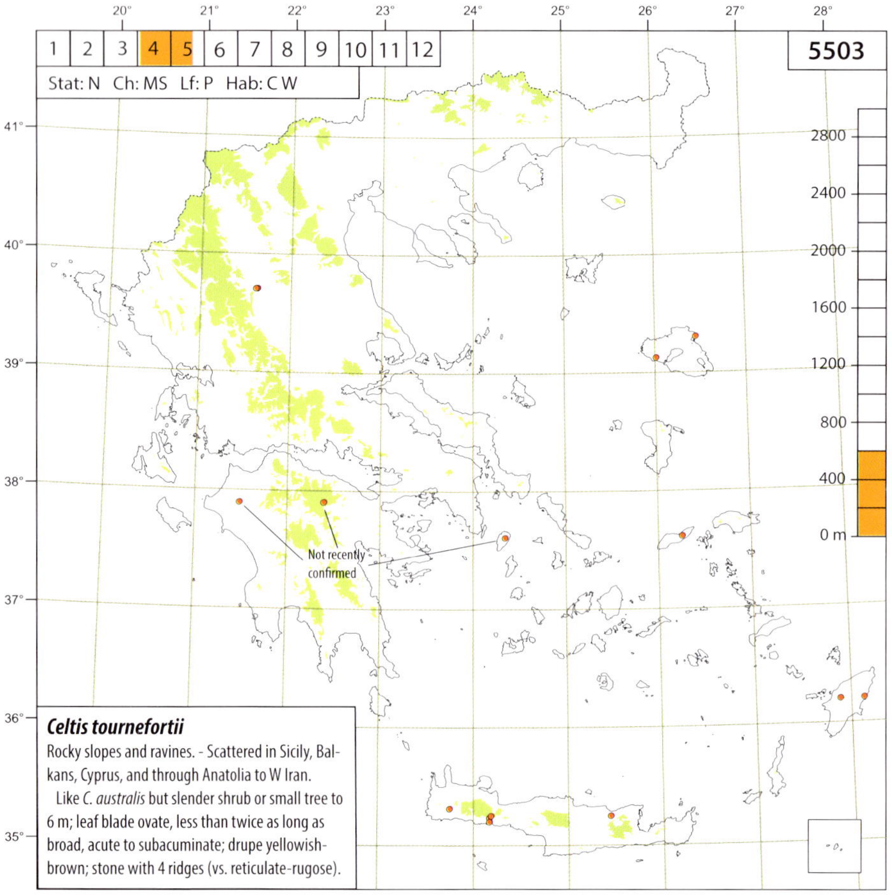

Celtis tournefortii

Rocky slopes and ravines. - Scattered in Sicily, Balkans, Cyprus, and through Anatolia to W Iran.

Like *C. australis* but slender shrub or small tree to 6 m; leaf blade ovate, less than twice as long as broad, acute to subacuminate; drupe yellowish-brown; stone with 4 ridges (vs. reticulate-rugose).

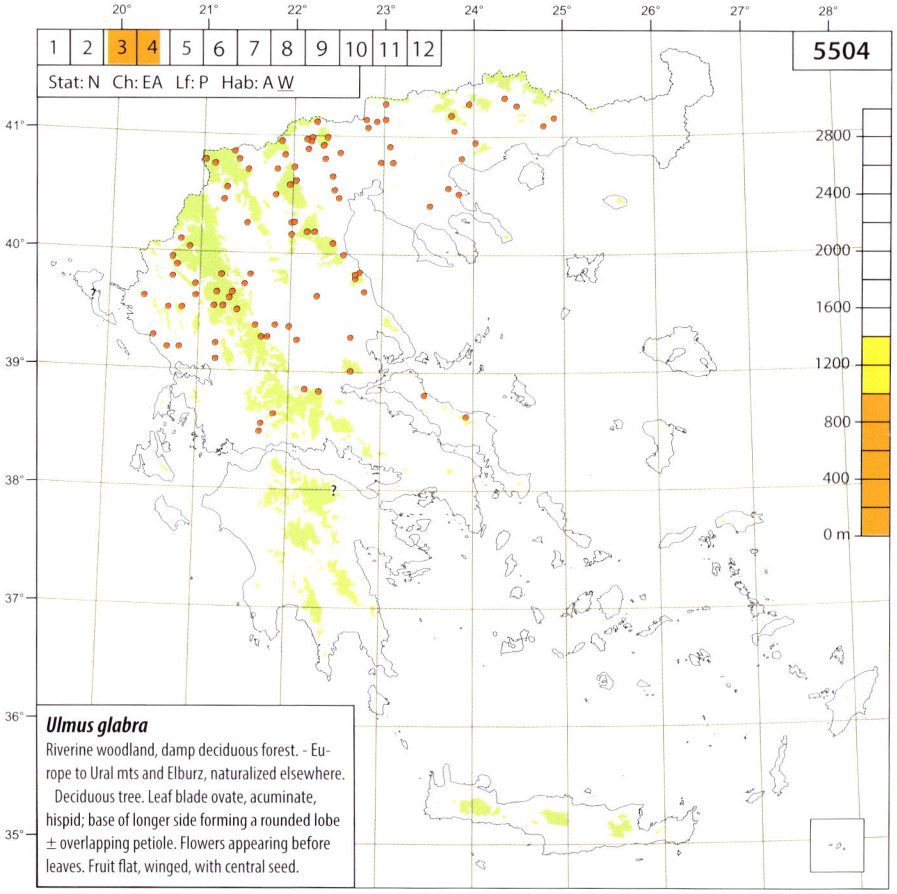

Ulmus glabra

Riverine woodland, damp deciduous forest. - Europe to Ural mts and Elburz, naturalized elsewhere.

Deciduous tree. Leaf blade ovate, acuminate, hispid; base of longer side forming a rounded lobe ± overlapping petiole. Flowers appearing before leaves. Fruit flat, winged, with central seed.

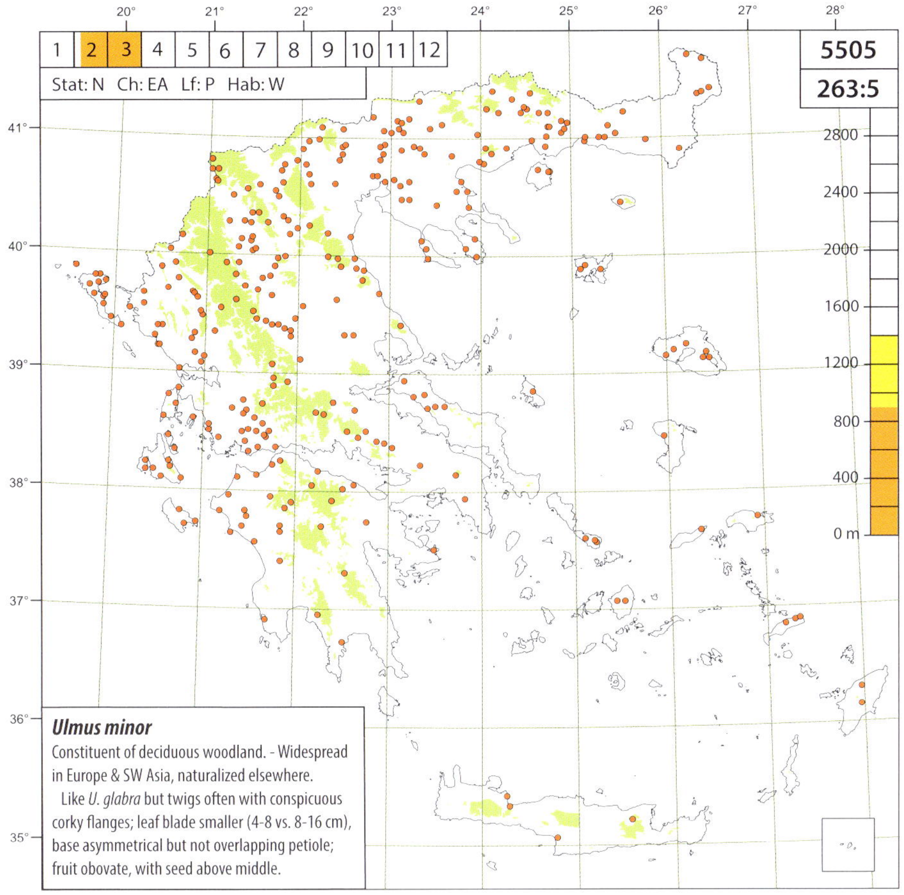

Ulmus minor

Constituent of deciduous woodland. - Widespread in Europe & SW Asia, naturalized elsewhere.

Like *U. glabra* but twigs often with conspicuous corky flanges; leaf blade smaller (4-8 vs. 8-16 cm), base asymmetrical but not overlapping petiole; fruit obovate, with seed above middle.

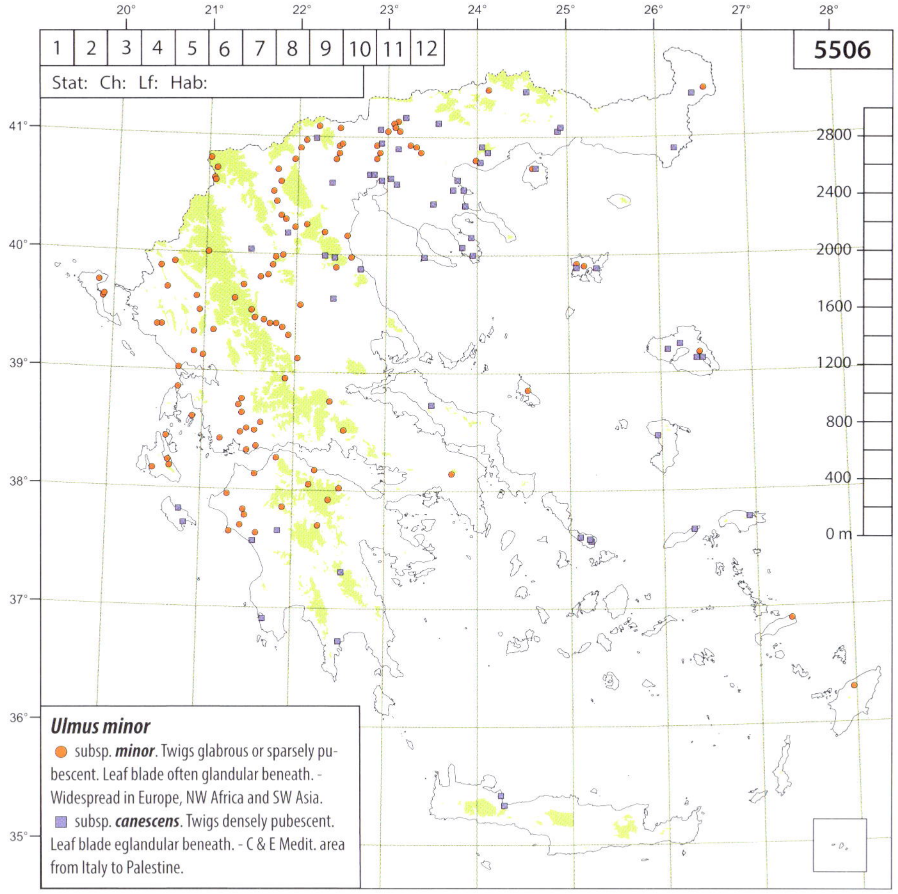

Ulmus minor

● subsp. ***minor***. Twigs glabrous or sparsely pubescent. Leaf blade often glandular beneath. - Widespread in Europe, NW Africa and SW Asia.

■ subsp. ***canescens***. Twigs densely pubescent. Leaf blade eglandular beneath. - C & E Medit. area from Italy to Palestine.

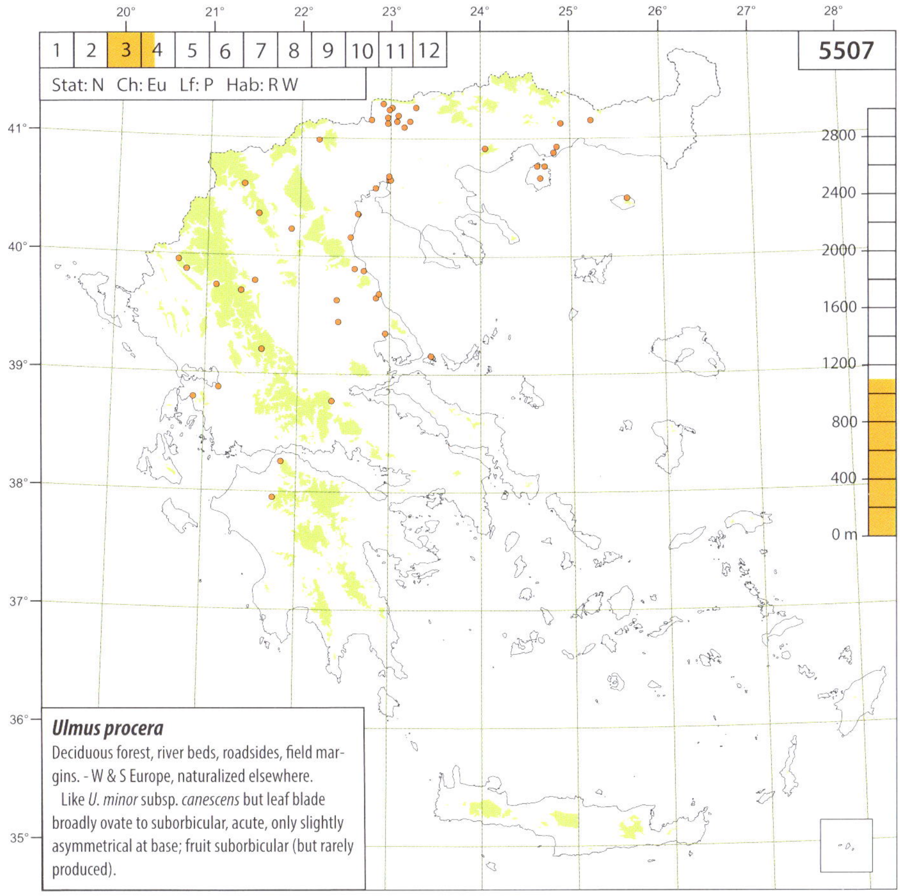

Ulmus procera

Deciduous forest, river beds, roadsides, field margins. - W & S Europe, naturalized elsewhere.

Like *U. minor* subsp. *canescens* but leaf blade broadly ovate to suborbicular, acute, only slightly asymmetrical at base; fruit suborbicular (but rarely produced).

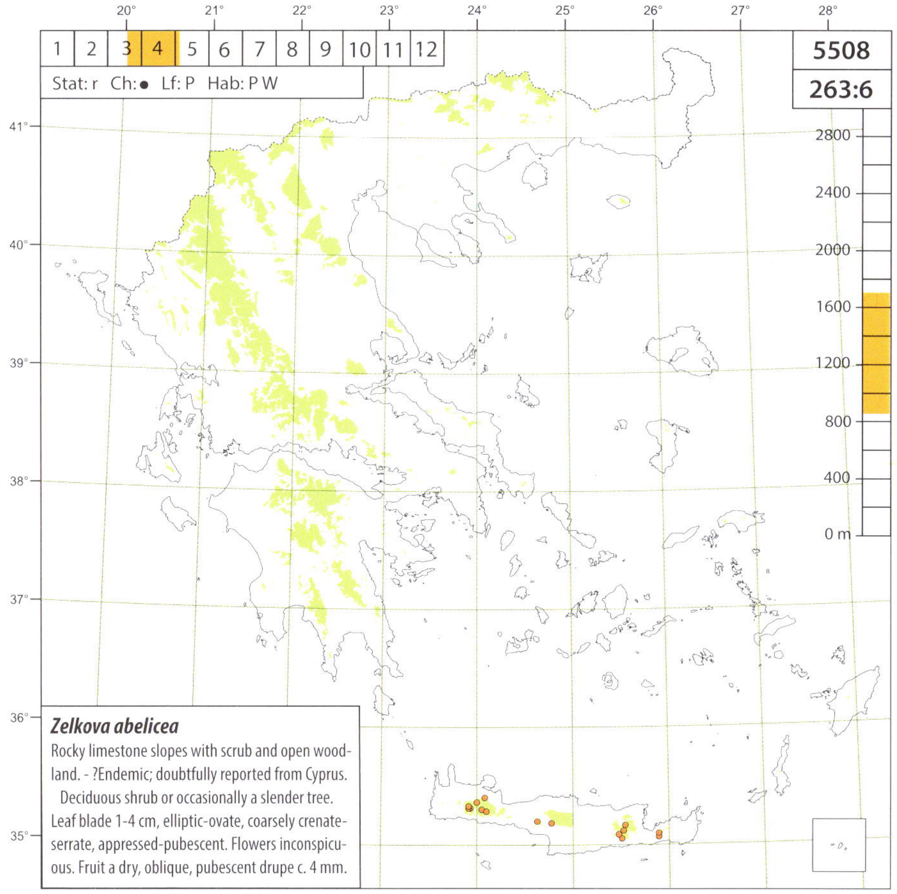

Zelkova abelicea

Rocky limestone slopes with scrub and open woodland. - ?Endemic; doubtfully reported from Cyprus.

Deciduous shrub or occasionally a slender tree. Leaf blade 1-4 cm, elliptic-ovate, coarsely crenate-serrate, appressed-pubescent. Flowers inconspicuous. Fruit a dry, oblique, pubescent drupe c. 4 mm.

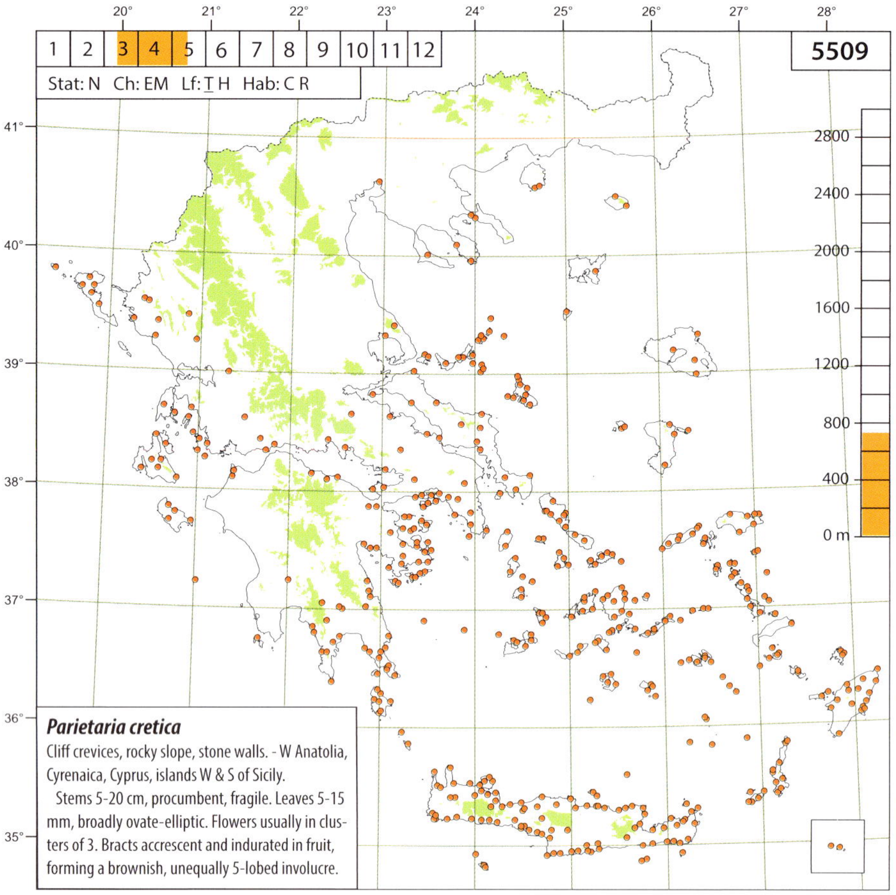

Parietaria cretica
Cliff crevices, rocky slope, stone walls. - W Anatolia, Cyrenaica, Cyprus, islands W & S of Sicily.
Stems 5-20 cm, procumbent, fragile. Leaves 5-15 mm, broadly ovate-elliptic. Flowers usually in clusters of 3. Bracts accrescent and indurated in fruit, forming a brownish, unequally 5-lobed involucre.

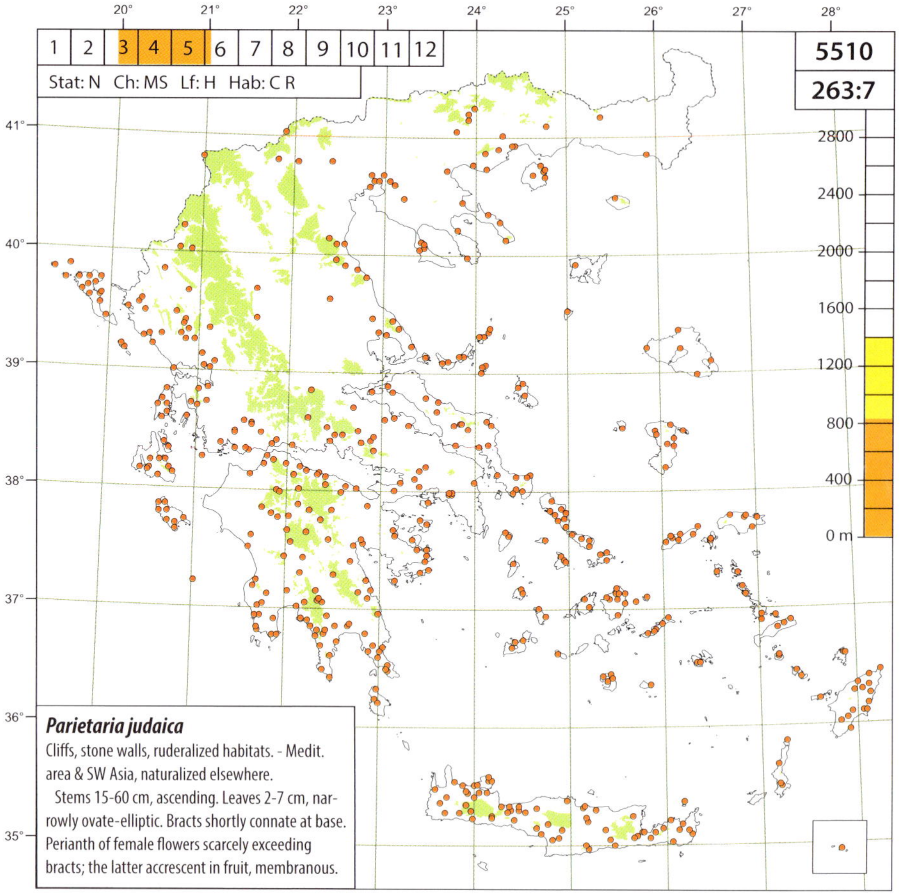

Parietaria judaica
Cliffs, stone walls, ruderalized habitats. - Medit. area & SW Asia, naturalized elsewhere.
Stems 15-60 cm, ascending. Leaves 2-7 cm, narrowly ovate-elliptic. Bracts shortly connate at base. Perianth of female flowers scarcely exceeding bracts; the latter accrescent in fruit, membranous.

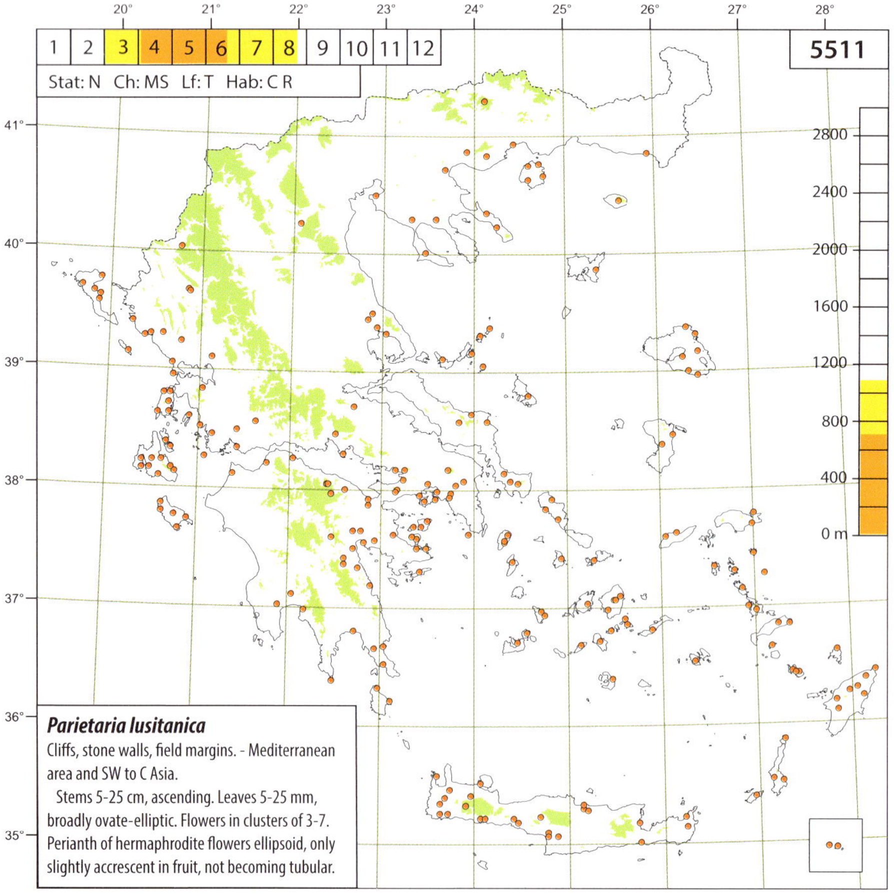

Parietaria lusitanica
Cliffs, stone walls, field margins. - Mediterranean area and SW to C Asia.
Stems 5-25 cm, ascending. Leaves 5-25 mm, broadly ovate-elliptic. Flowers in clusters of 3-7. Perianth of hermaphrodite flowers ellipsoid, only slightly accrescent in fruit, not becoming tubular.

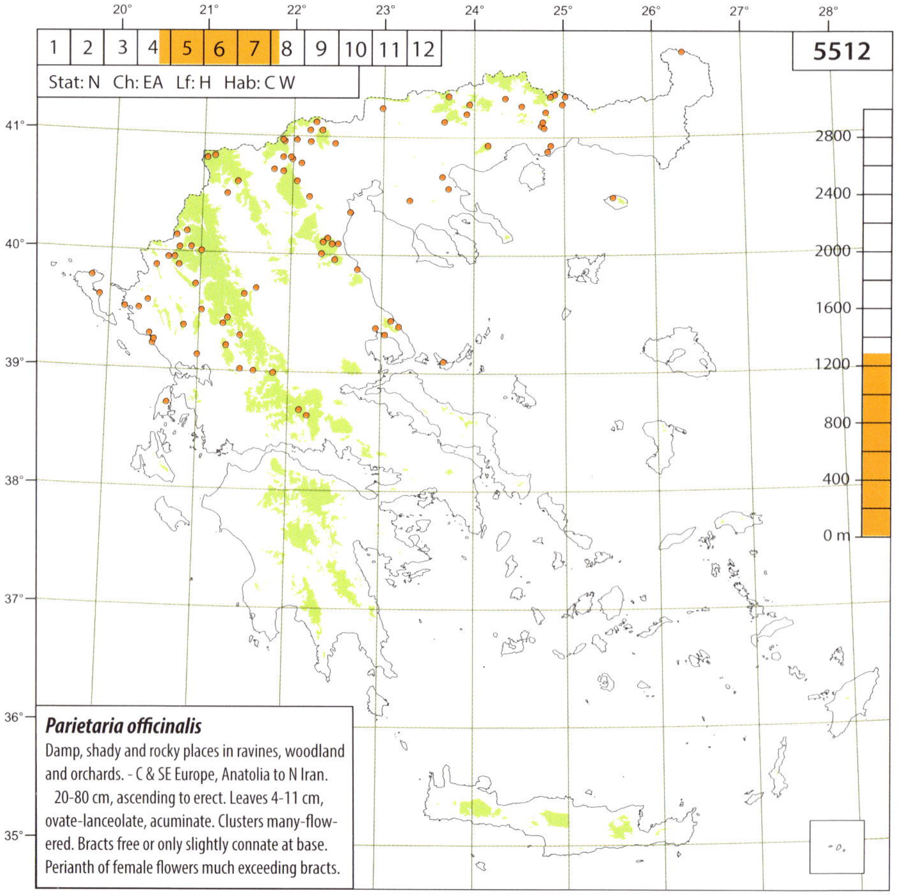

Parietaria officinalis
Damp, shady and rocky places in ravines, woodland and orchards. - C & SE Europe, Anatolia to N Iran.
20-80 cm, ascending to erect. Leaves 4-11 cm, ovate-lanceolate, acuminate. Clusters many-flowered. Bracts free or only slightly connate at base. Perianth of female flowers much exceeding bracts.

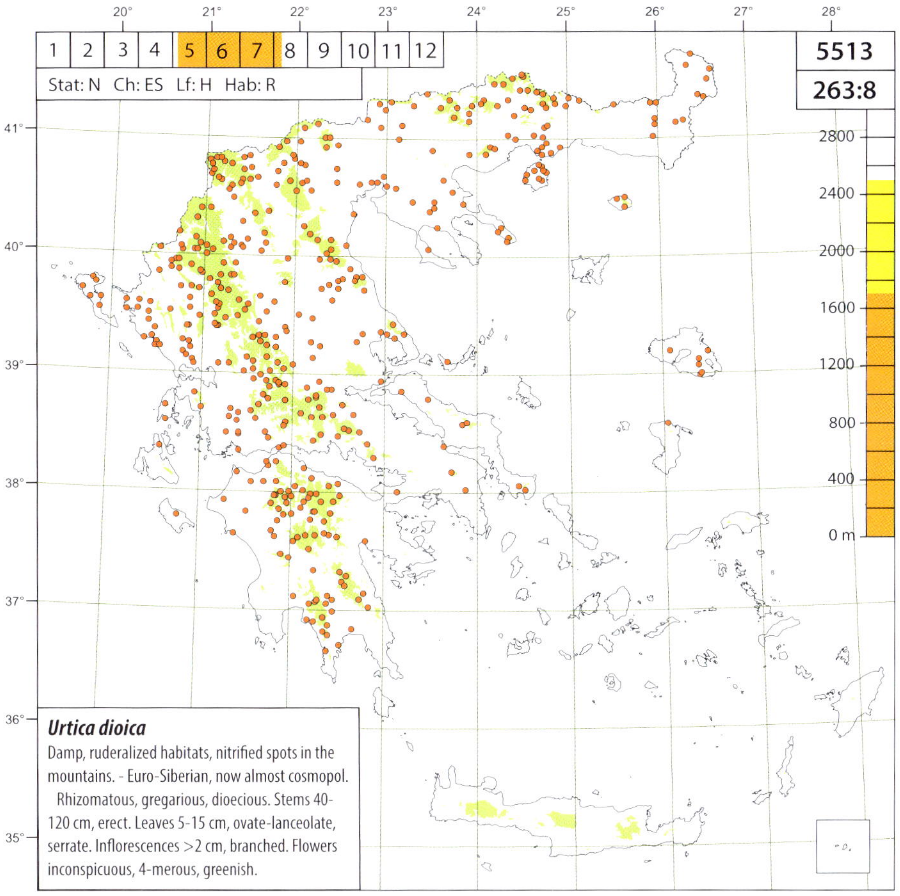

Urtica dioica

Damp, ruderalized habitats, nitrified spots in the mountains. - Euro-Siberian, now almost cosmopol.

Rhizomatous, gregarious, dioecious. Stems 40-120 cm, erect. Leaves 5-15 cm, ovate-lanceolate, serrate. Inflorescences >2 cm, branched. Flowers inconspicuous, 4-merous, greenish.

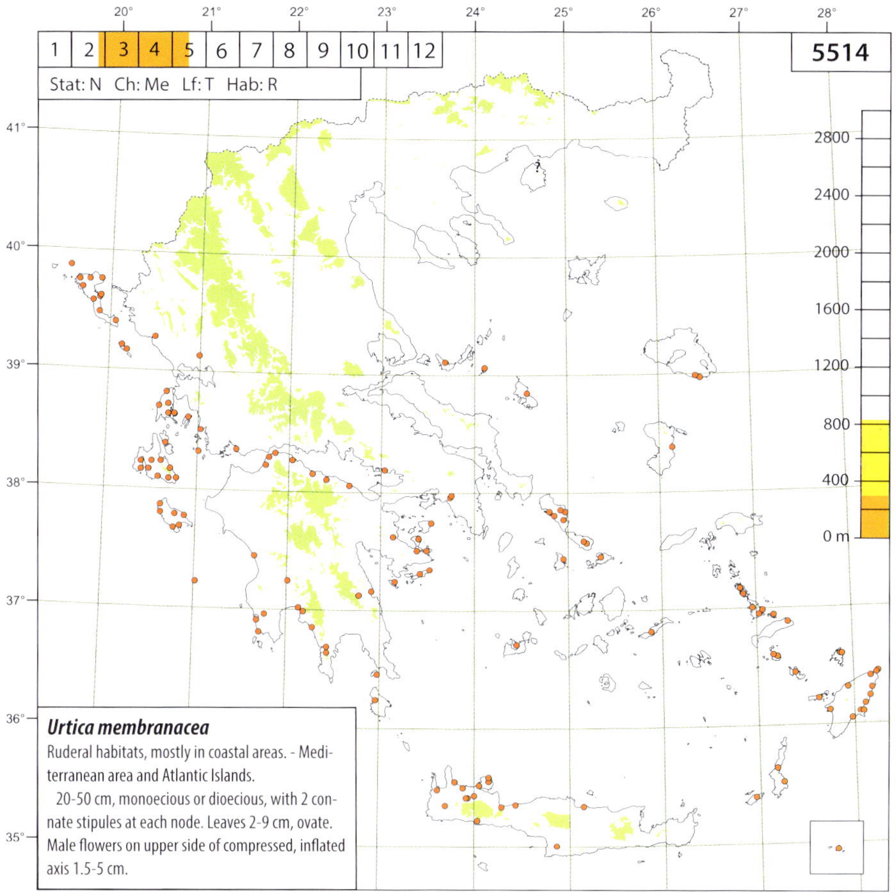

Urtica membranacea

Ruderal habitats, mostly in coastal areas. - Mediterranean area and Atlantic Islands.

20-50 cm, monoecious or dioecious, with 2 connate stipules at each node. Leaves 2-9 cm, ovate. Male flowers on upper side of compressed, inflated axis 1.5-5 cm.

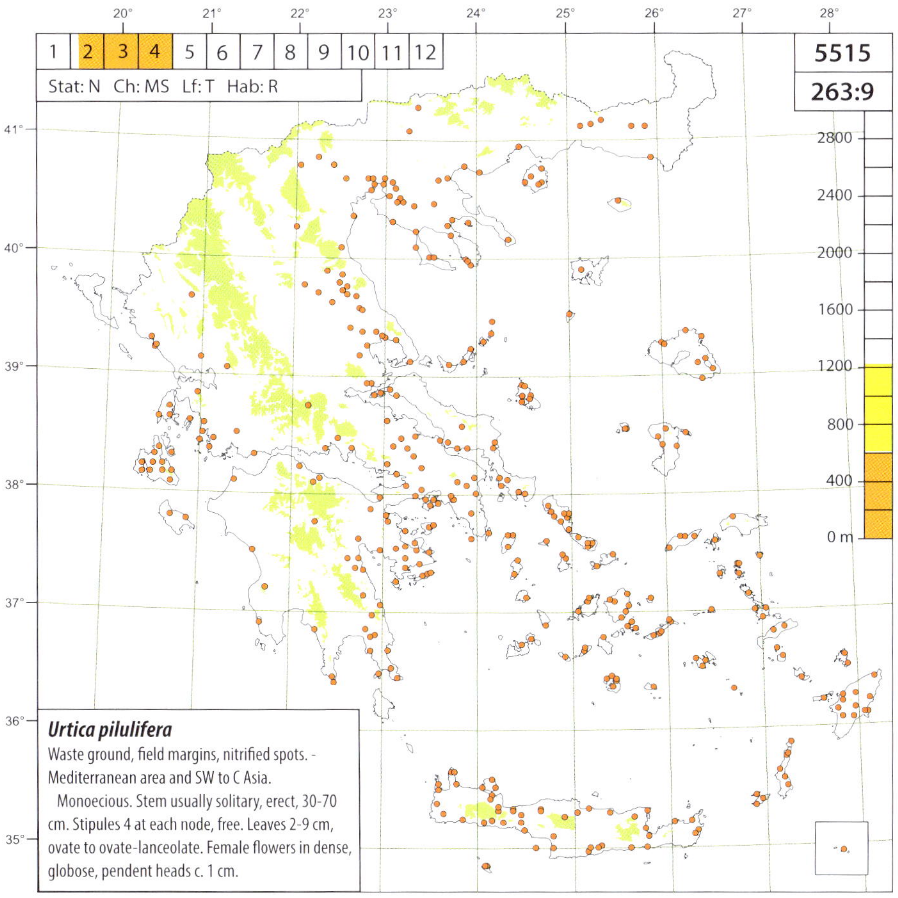

Urtica pilulifera

Waste ground, field margins, nitrified spots. - Mediterranean area and SW to C Asia.

Monoecious. Stem usually solitary, erect, 30-70 cm. Stipules 4 at each node, free. Leaves 2-9 cm, ovate to ovate-lanceolate. Female flowers in dense, globose, pendent heads c. 1 cm.

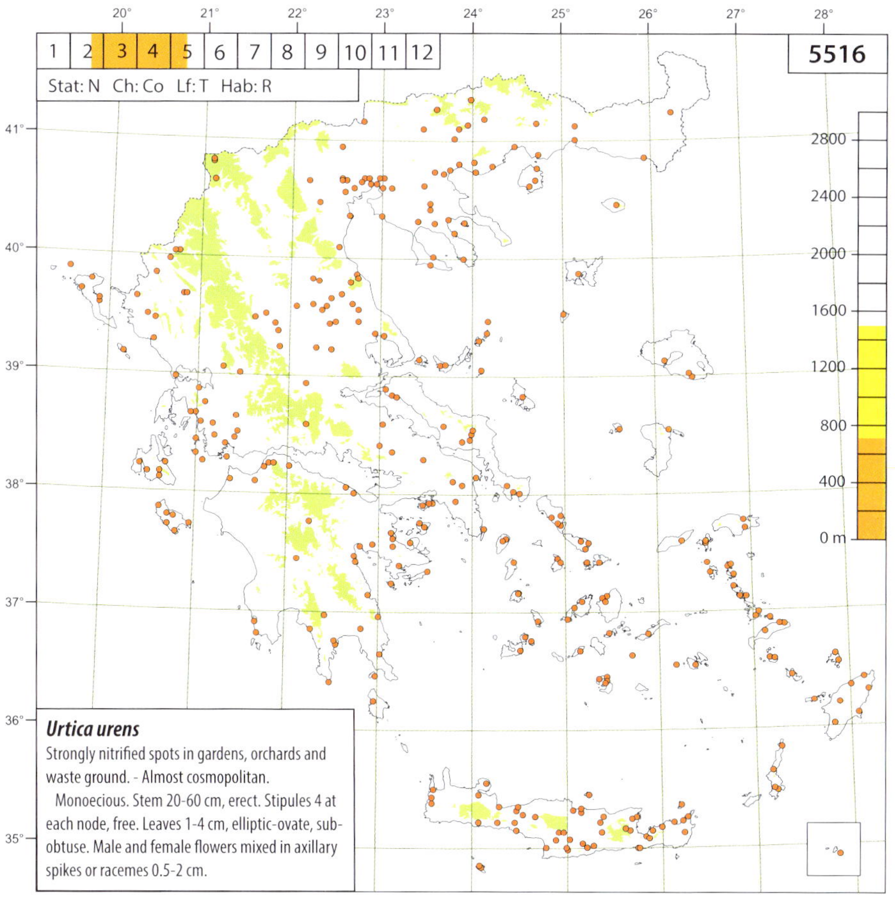

Urtica urens

Strongly nitrified spots in gardens, orchards and waste ground. - Almost cosmopolitan.

Monoecious. Stem 20-60 cm, erect. Stipules 4 at each node, free. Leaves 1-4 cm, elliptic-ovate, subobtuse. Male and female flowers mixed in axillary spikes or racemes 0.5-2 cm.

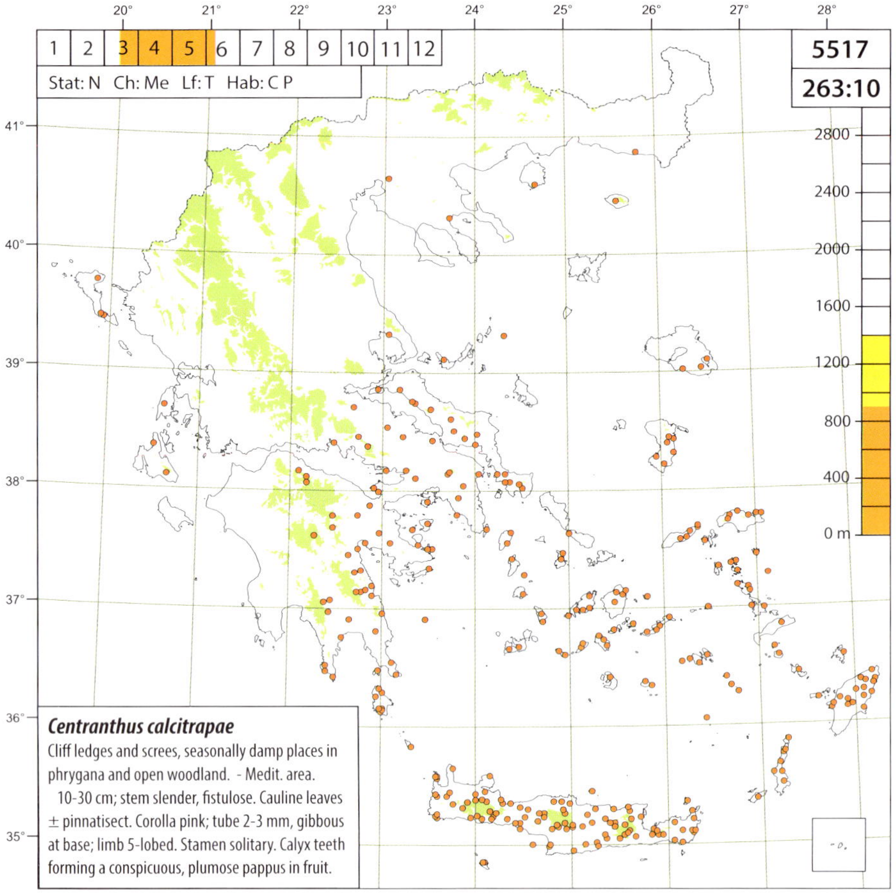

Centranthus calcitrapae
Cliff ledges and screes, seasonally damp places in phrygana and open woodland. - Medit. area.
10-30 cm; stem slender, fistulose. Cauline leaves ± pinnatisect. Corolla pink; tube 2-3 mm, gibbous at base; limb 5-lobed. Stamen solitary. Calyx teeth forming a conspicuous, plumose pappus in fruit.

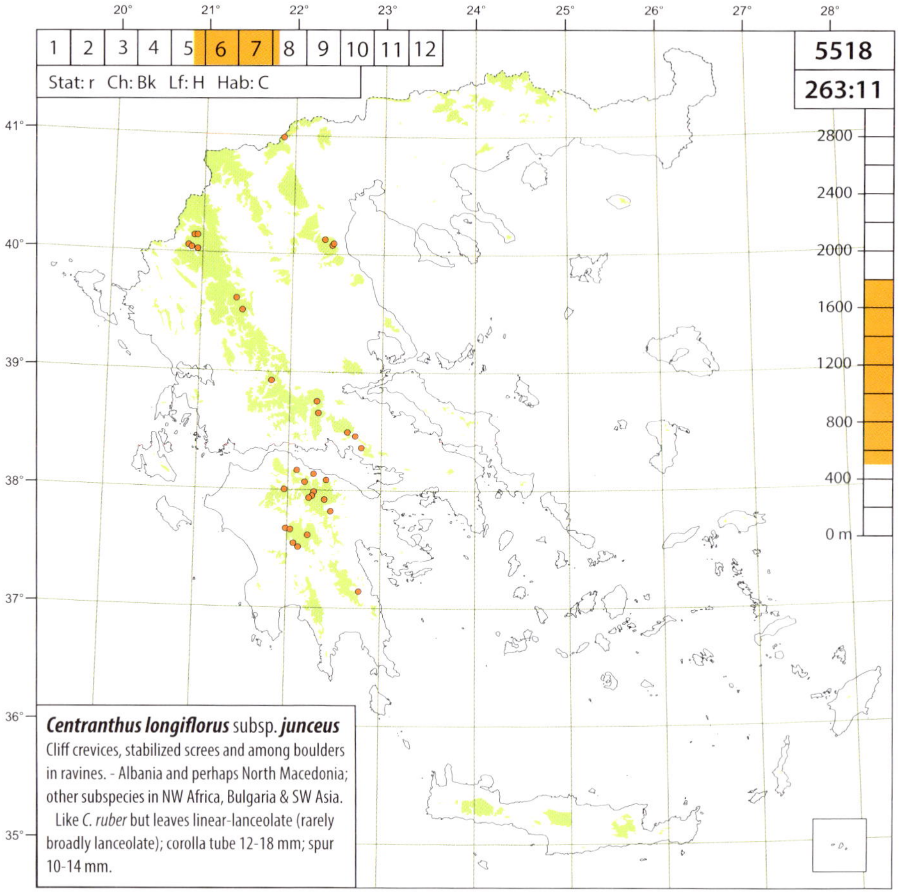

Centranthus longiflorus subsp. ***junceus***
Cliff crevices, stabilized screes and among boulders in ravines. - Albania and perhaps North Macedonia; other subspecies in NW Africa, Bulgaria & SW Asia.
Like *C. ruber* but leaves linear-lanceolate (rarely broadly lanceolate); corolla tube 12-18 mm; spur 10-14 mm.

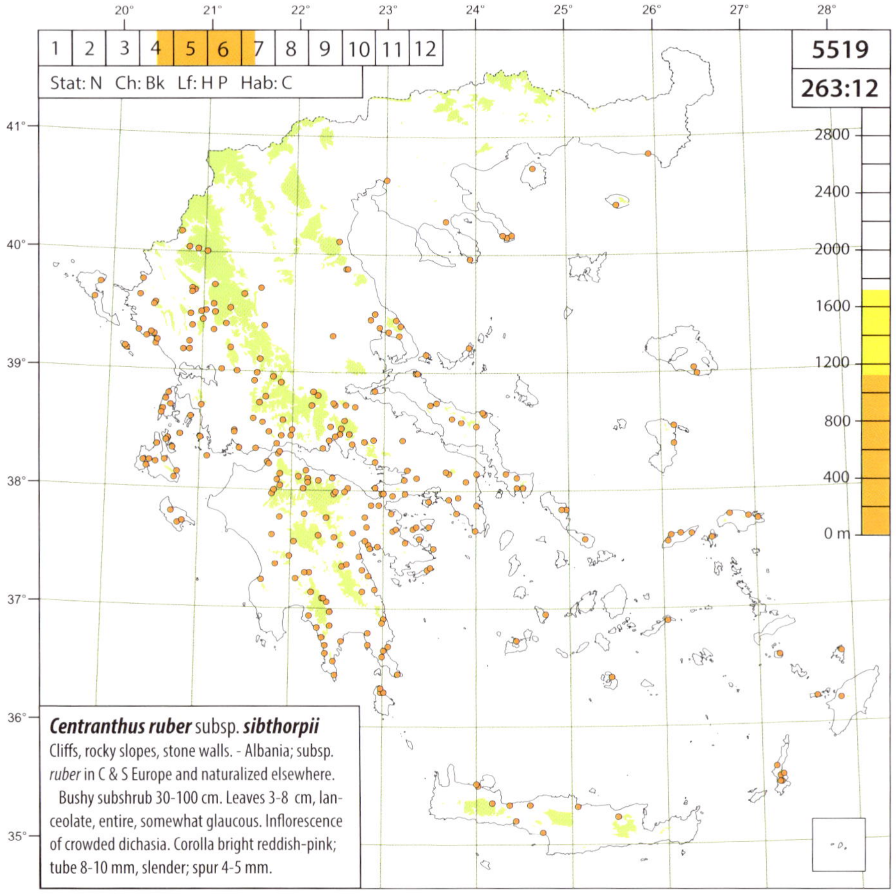

Centranthus ruber subsp. ***sibthorpii***
Cliffs, rocky slopes, stone walls. - Albania; subsp. *ruber* in C & S Europe and naturalized elsewhere.
Bushy subshrub 30-100 cm. Leaves 3-8 cm, lanceolate, entire, somewhat glaucous. Inflorescence of crowded dichasia. Corolla bright reddish-pink; tube 8-10 mm, slender; spur 4-5 mm.

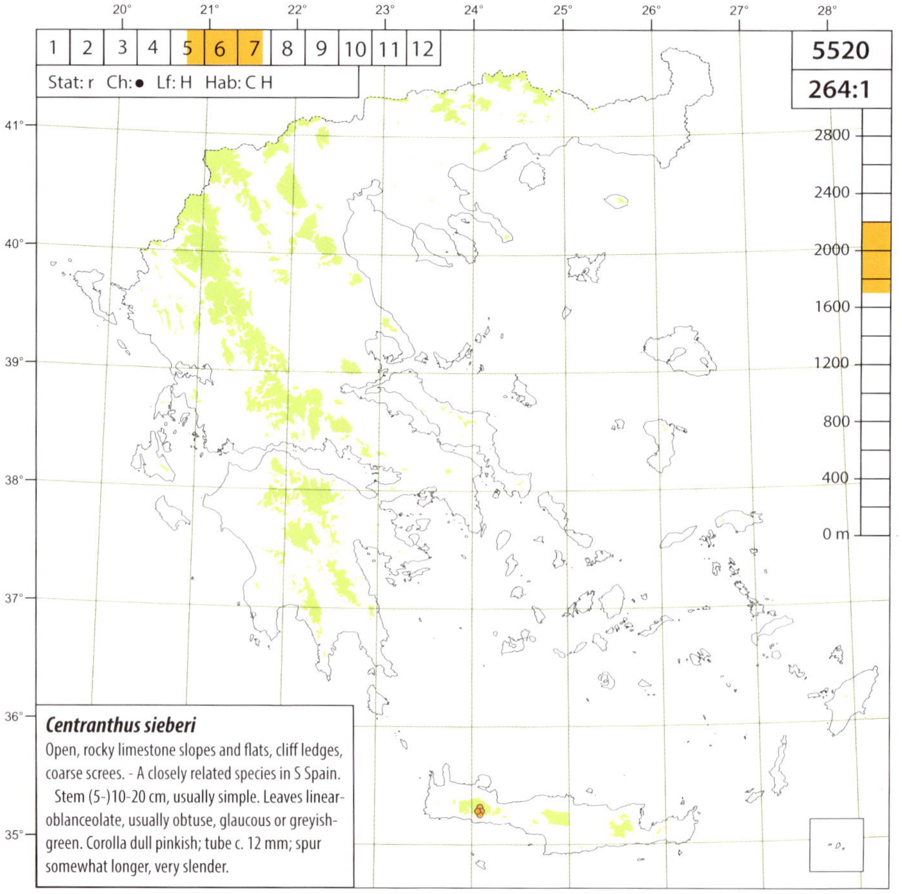

Centranthus sieberi
Open, rocky limestone slopes and flats, cliff ledges, coarse screes. - A closely related species in S Spain.
Stem (5-)10-20 cm, usually simple. Leaves linear-oblanceolate, usually obtuse, glaucous or greyish-green. Corolla dull pinkish; tube c. 12 mm; spur somewhat longer, very slender.

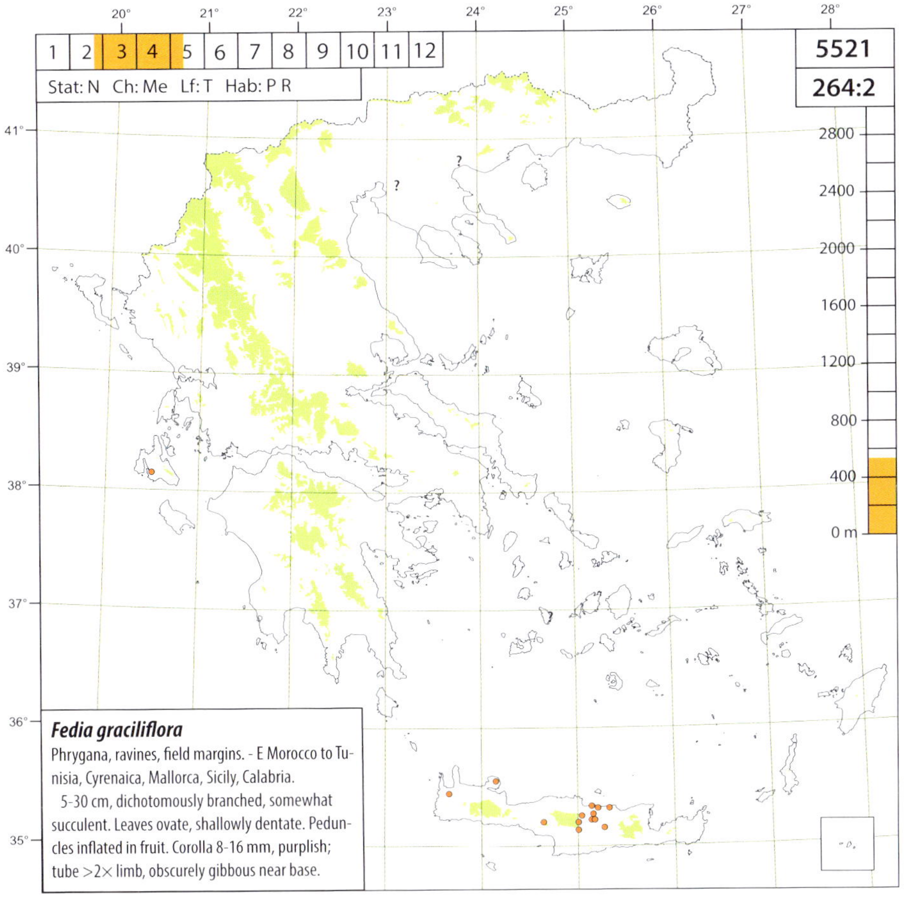

Fedia graciliflora

Phrygana, ravines, field margins. - E Morocco to Tunisia, Cyrenaica, Mallorca, Sicily, Calabria.

5-30 cm, dichotomously branched, somewhat succulent. Leaves ovate, shallowly dentate. Peduncles inflated in fruit. Corolla 8-16 mm, purplish; tube >2× limb, obscurely gibbous near base.

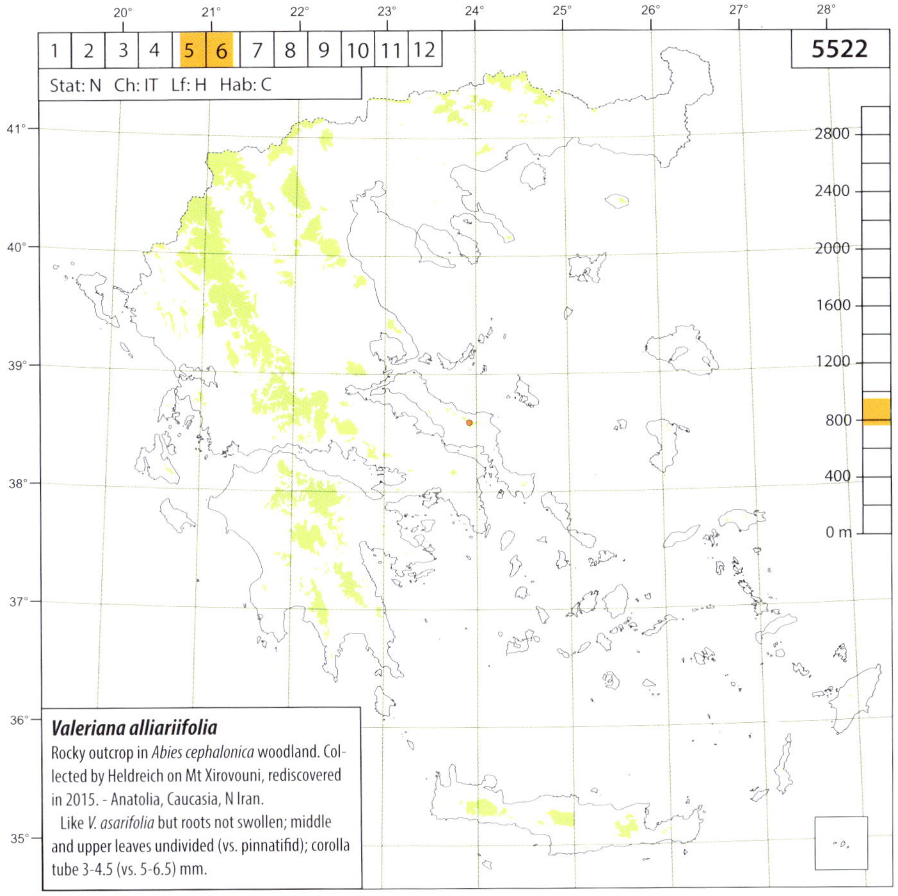

Valeriana alliariifolia

Rocky outcrop in *Abies cephalonica* woodland. Collected by Heldreich on Mt Xirovouni, rediscovered in 2015. - Anatolia, Caucasia, N Iran.

Like *V. asarifolia* but roots not swollen; middle and upper leaves undivided (vs. pinnatifid); corolla tube 3-4.5 (vs. 5-6.5) mm.

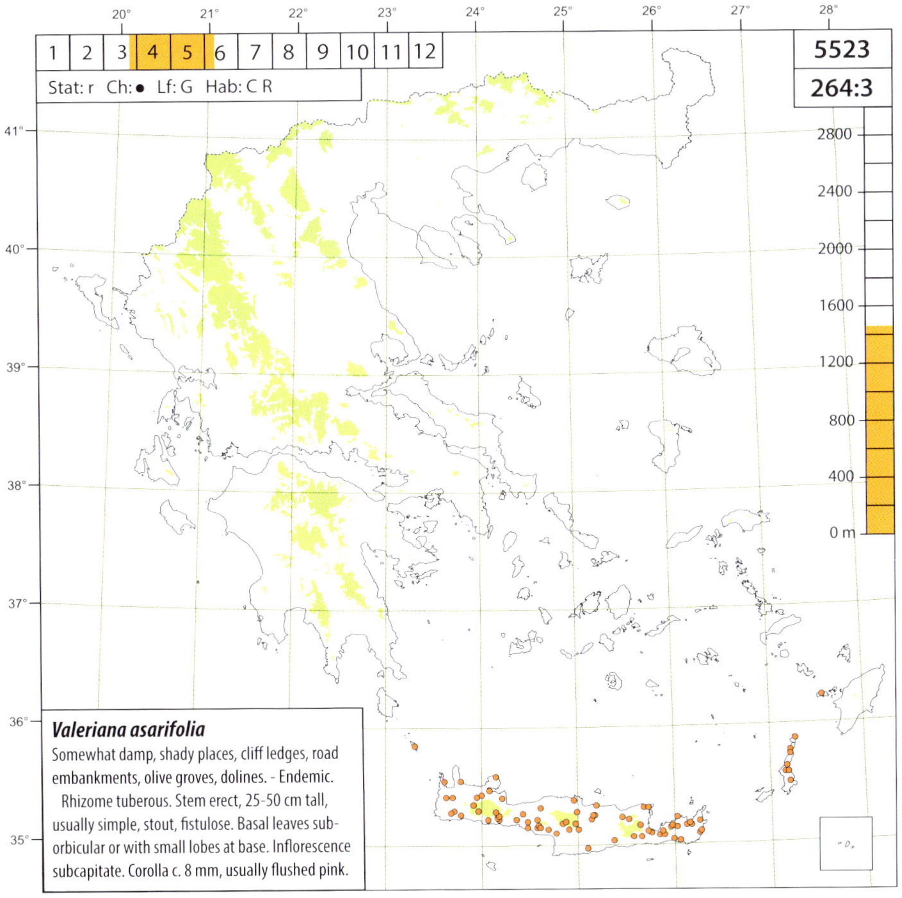

Valeriana asarifolia

Somewhat damp, shady places, cliff ledges, road embankments, olive groves, dolines. - Endemic.

Rhizome tuberous. Stem erect, 25-50 cm tall, usually simple, stout, fistulose. Basal leaves suborbicular or with small lobes at base. Inflorescence subcapitate. Corolla c. 8 mm, usually flushed pink.

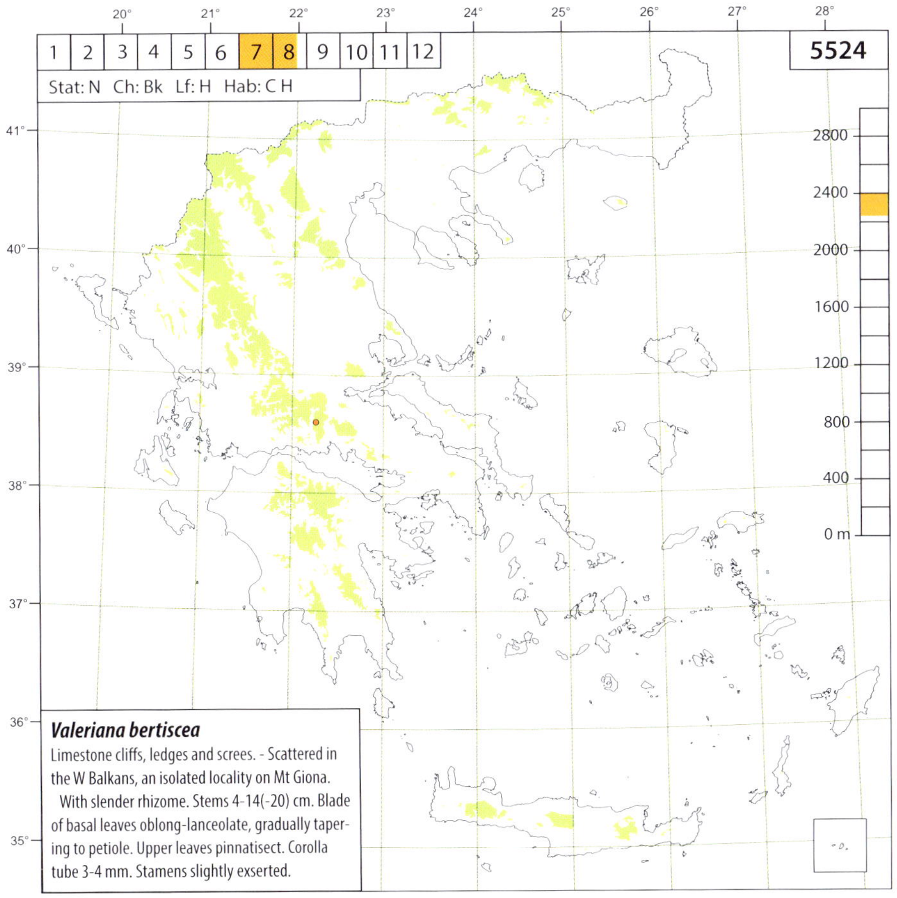

Valeriana bertiscea

Limestone cliffs, ledges and screes. - Scattered in the W Balkans, an isolated locality on Mt Giona.

With slender rhizome. Stems 4-14(-20) cm. Blade of basal leaves oblong-lanceolate, gradually tapering to petiole. Upper leaves pinnatisect. Corolla tube 3-4 mm. Stamens slightly exserted.

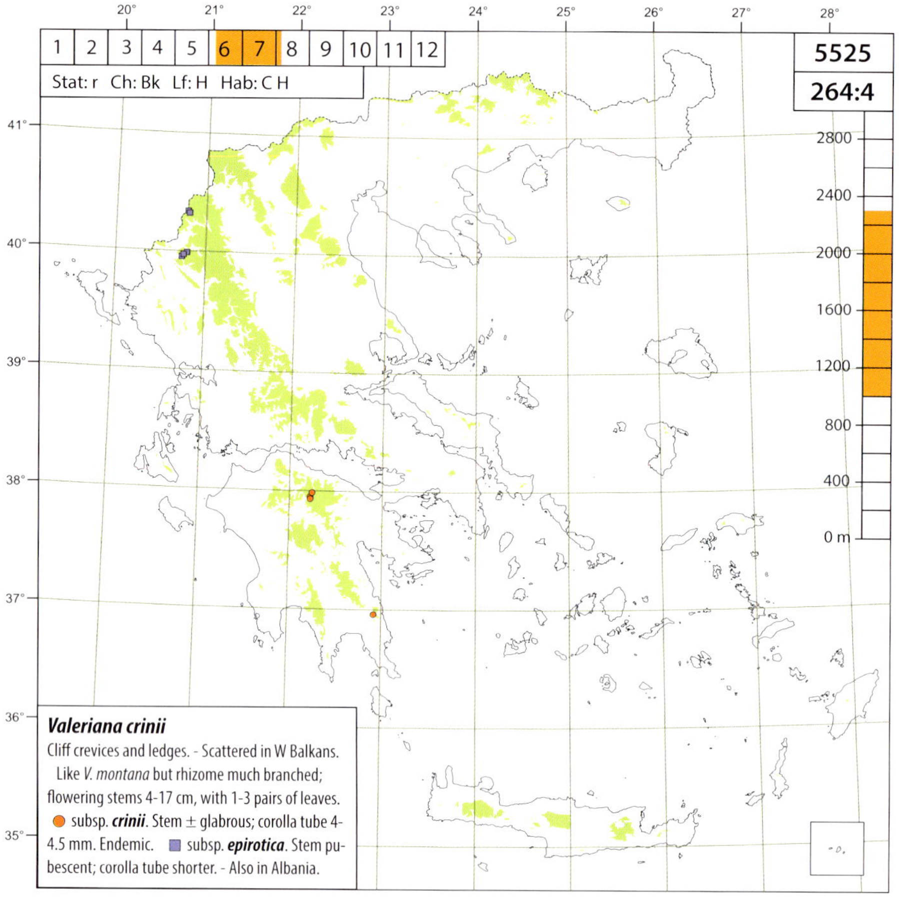

Valeriana crinii

Cliff crevices and ledges. - Scattered in W Balkans.
Like *V. montana* but rhizome much branched; flowering stems 4-17 cm, with 1-3 pairs of leaves. ● subsp. ***crinii***. Stem ± glabrous; corolla tube 4-4.5 mm. Endemic. ■ subsp. ***epirotica***. Stem pubescent; corolla tube shorter. - Also in Albania.

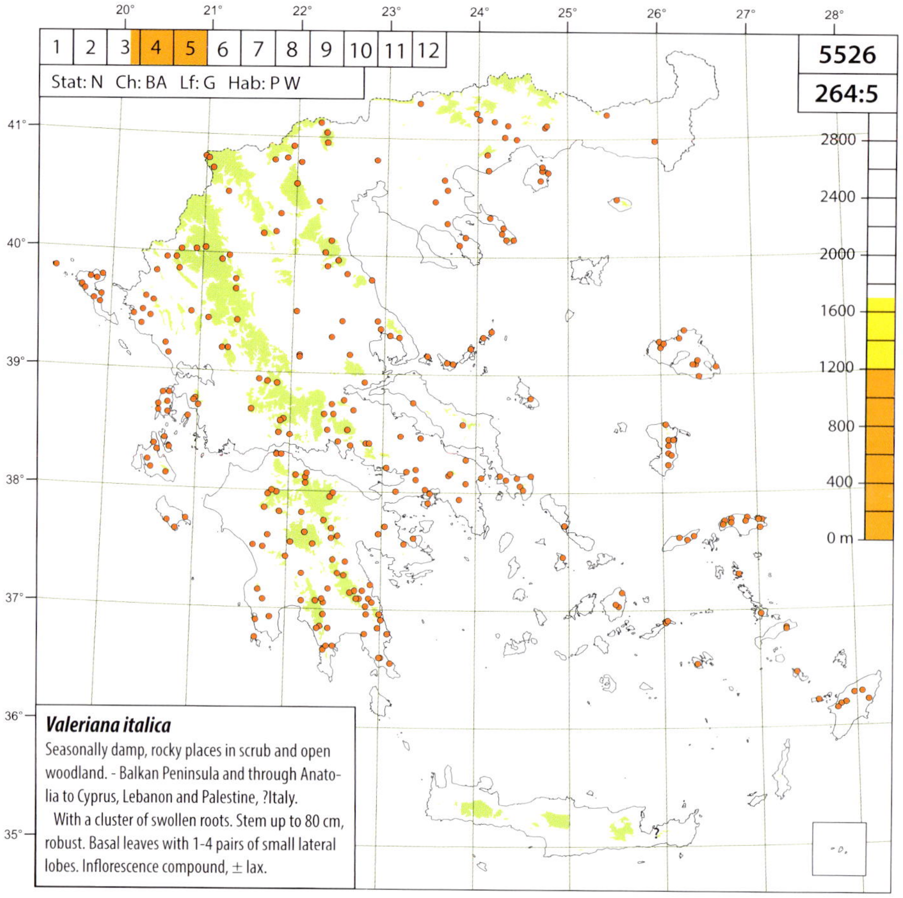

Valeriana italica

Seasonally damp, rocky places in scrub and open woodland. - Balkan Peninsula and through Anatolia to Cyprus, Lebanon and Palestine, ?Italy.
With a cluster of swollen roots. Stem up to 80 cm, robust. Basal leaves with 1-4 pairs of small lateral lobes. Inflorescence compound, ± lax.

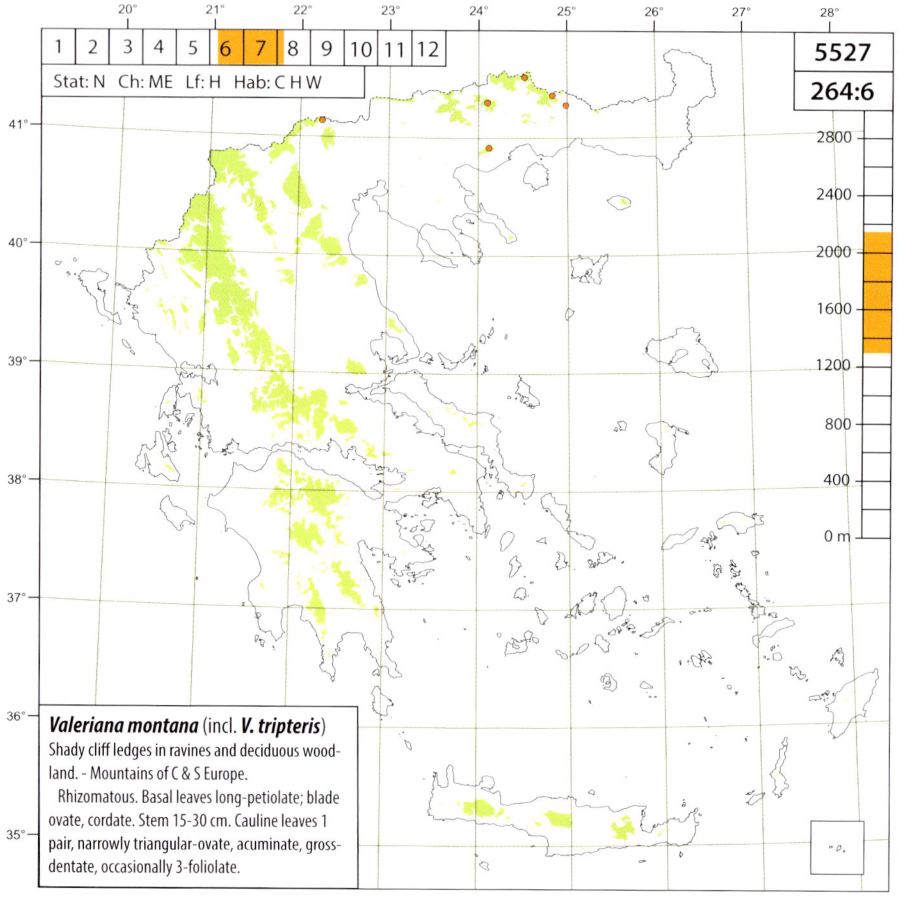

Valeriana montana (incl. *V. tripteris*)

Shady cliff ledges in ravines and deciduous woodland. - Mountains of C & S Europe.
Rhizomatous. Basal leaves long-petiolate; blade ovate, cordate. Stem 15-30 cm. Cauline leaves 1 pair, narrowly triangular-ovate, acuminate, grossdentate, occasionally 3-foliolate.

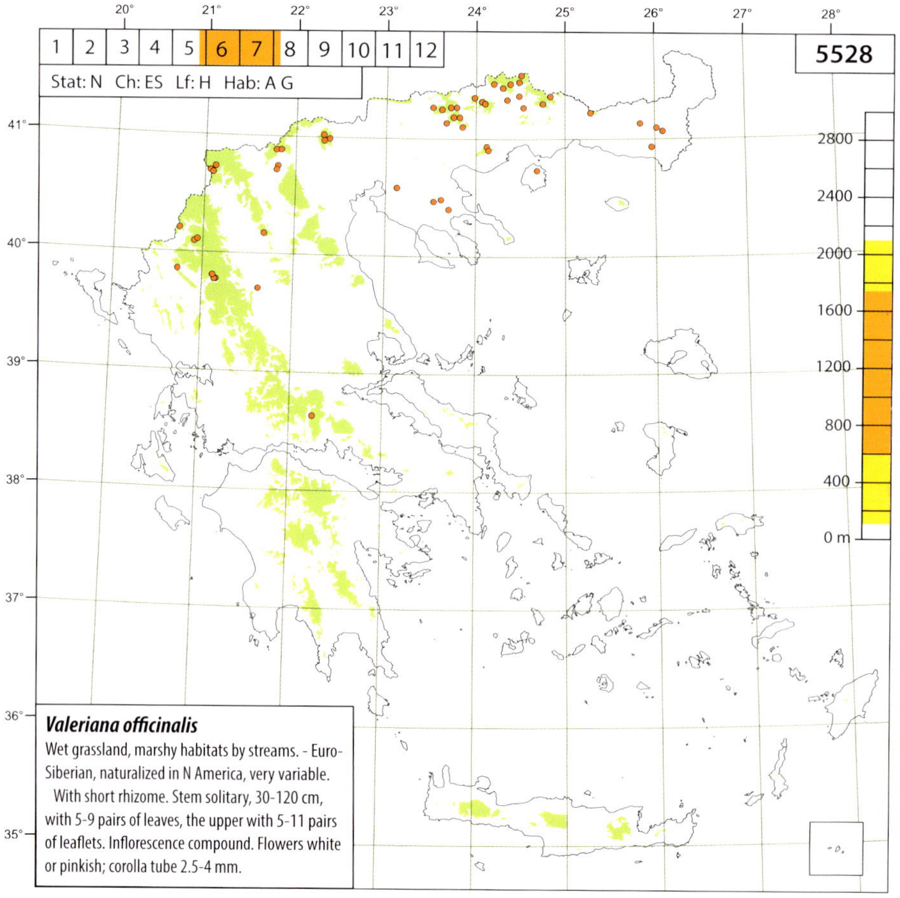

Valeriana officinalis

Wet grassland, marshy habitats by streams. - Euro-Siberian, naturalized in N America, very variable.
With short rhizome. Stem solitary, 30-120 cm, with 5-9 pairs of leaves, the upper with 5-11 pairs of leaflets. Inflorescence compound. Flowers white or pinkish; corolla tube 2.5-4 mm.

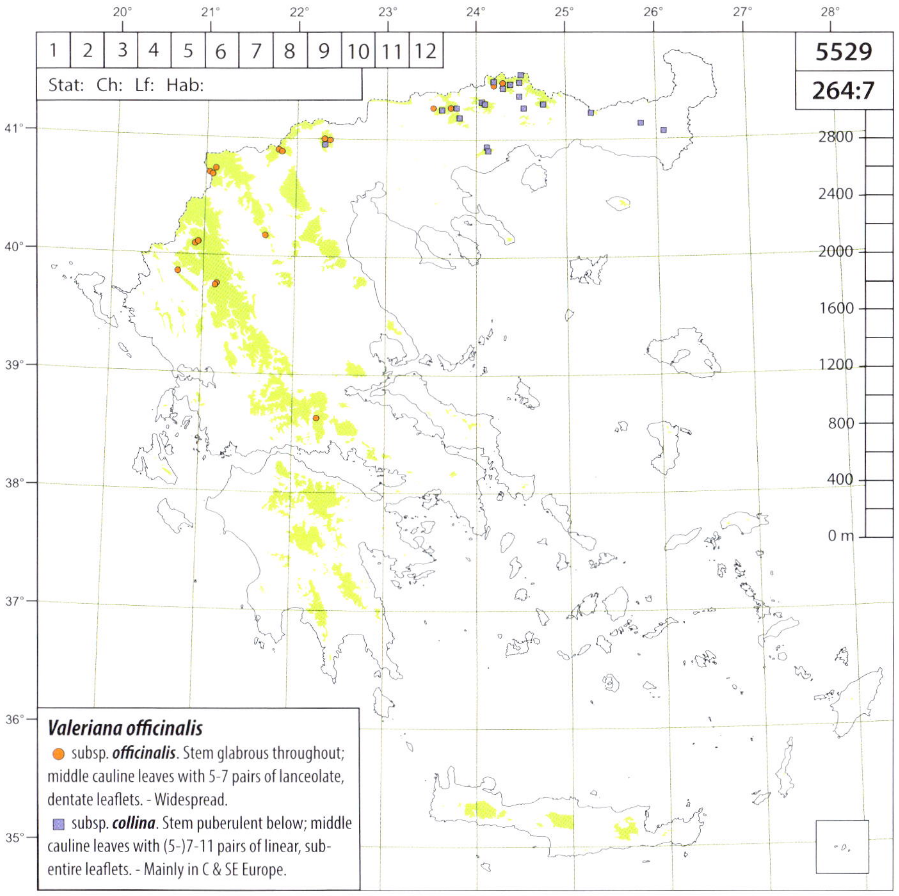

Valeriana officinalis

subsp. ***officinalis***. Stem glabrous throughout; middle cauline leaves with 5-7 pairs of lanceolate, dentate leaflets. - Widespread.

subsp. ***collina***. Stem puberulent below; middle cauline leaves with (5-)7-11 pairs of linear, sub-entire leaflets. - Mainly in C & SE Europe.

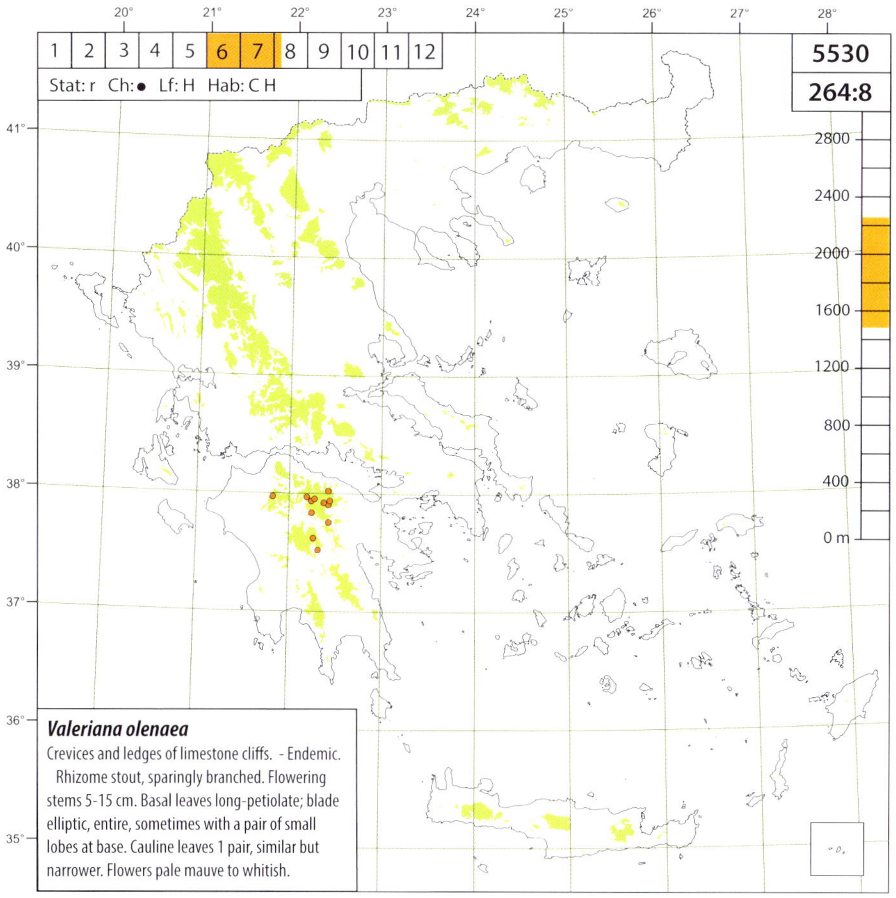

Valeriana olenaea

Crevices and ledges of limestone cliffs. - Endemic.

Rhizome stout, sparingly branched. Flowering stems 5-15 cm. Basal leaves long-petiolate; blade elliptic, entire, sometimes with a pair of small lobes at base. Cauline leaves 1 pair, similar but narrower. Flowers pale mauve to whitish.

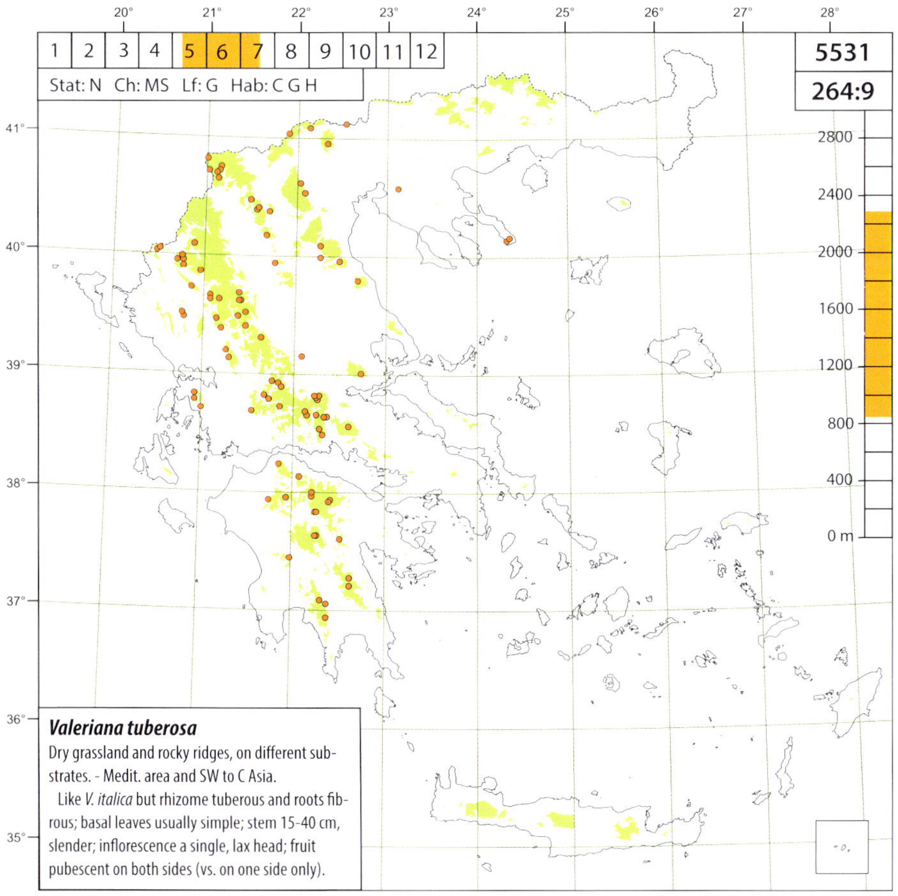

Valeriana tuberosa

Dry grassland and rocky ridges, on different substrates. - Medit. area and SW to C Asia.

Like *V. italica* but rhizome tuberous and roots fibrous; basal leaves usually simple; stem 15-40 cm, slender; inflorescence a single, lax head; fruit pubescent on both sides (vs. on one side only).

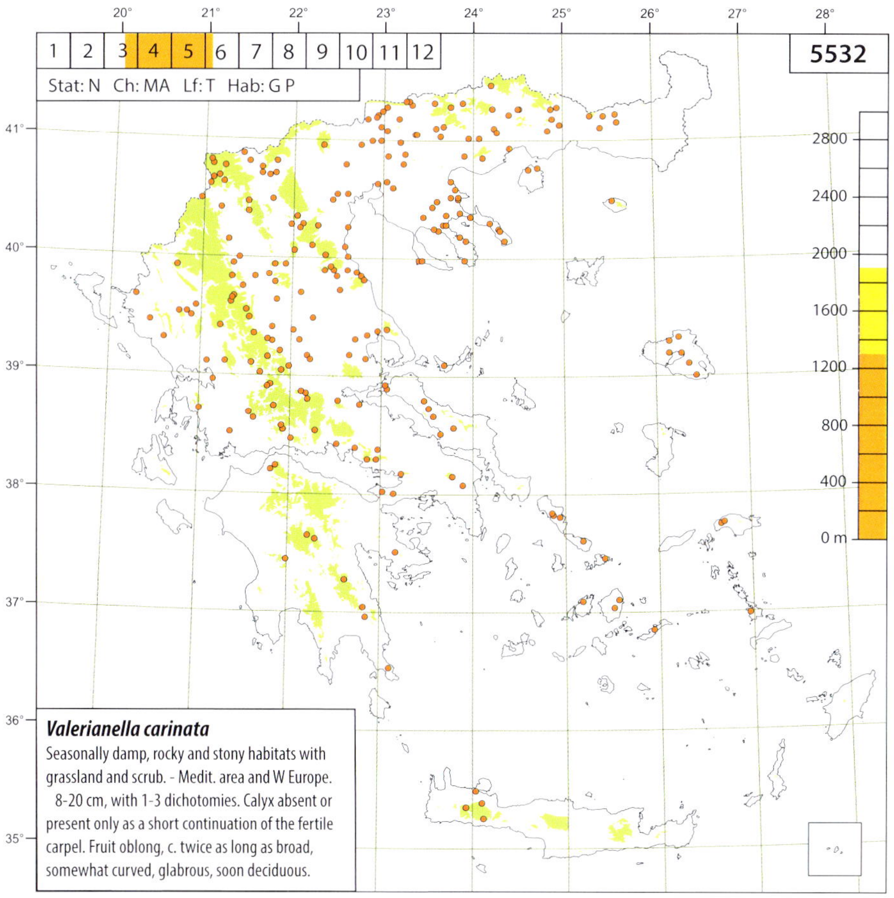

Valerianella carinata

Seasonally damp, rocky and stony habitats with grassland and scrub. - Medit. area and W Europe.

8-20 cm, with 1-3 dichotomies. Calyx absent or present only as a short continuation of the fertile carpel. Fruit oblong, c. twice as long as broad, somewhat curved, glabrous, soon deciduous.

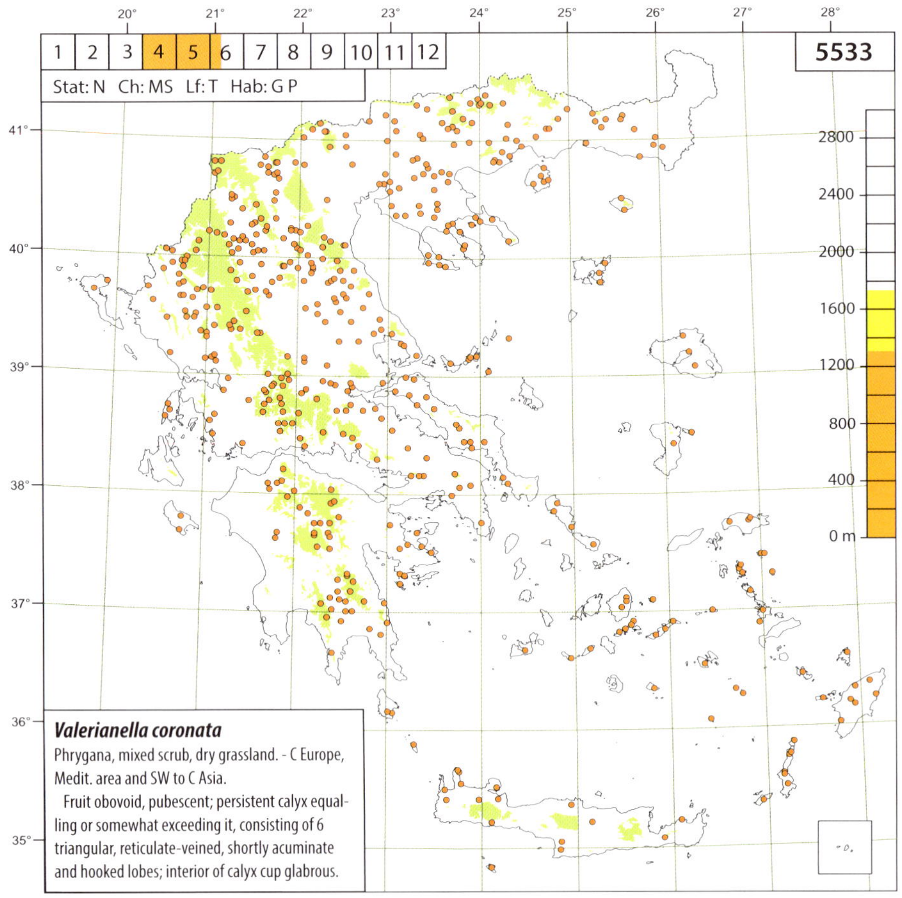

Valerianella coronata

Phrygana, mixed scrub, dry grassland. - C Europe, Medit. area and SW to C Asia.

Fruit obovoid, pubescent; persistent calyx equalling or somewhat exceeding it, consisting of 6 triangular, reticulate-veined, shortly acuminate and hooked lobes; interior of calyx cup glabrous.

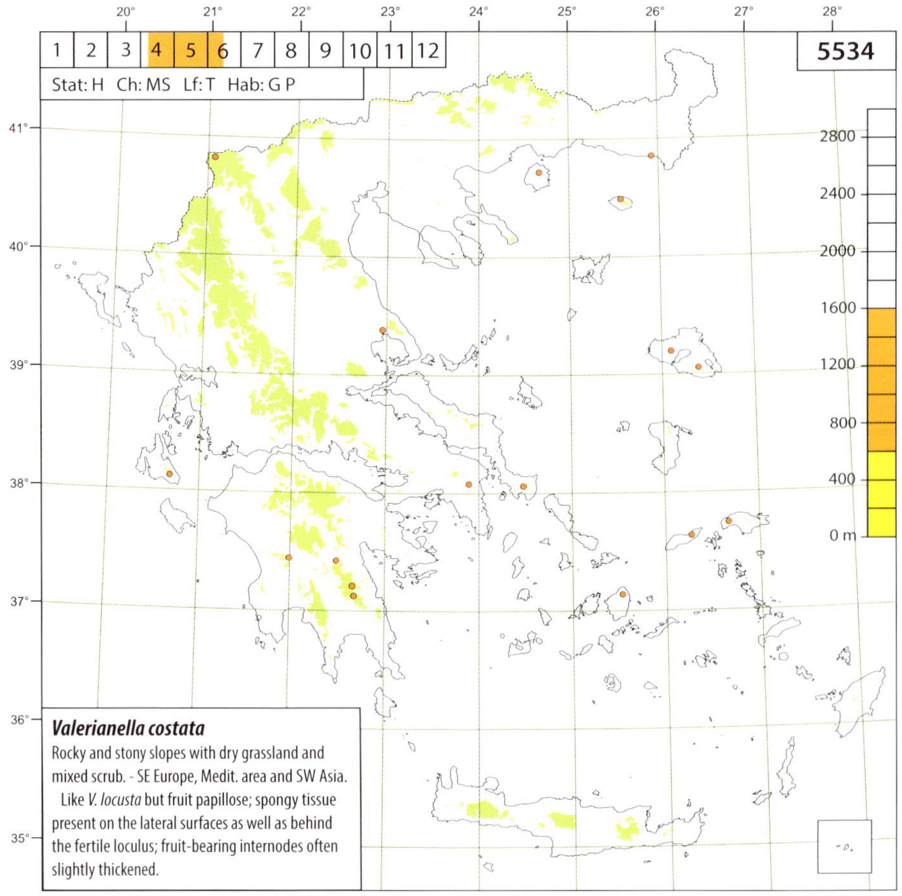

Valerianella costata

Rocky and stony slopes with dry grassland and mixed scrub. - SE Europe, Medit. area and SW Asia.

Like *V. locusta* but fruit papillose; spongy tissue present on the lateral surfaces as well as behind the fertile loculus; fruit-bearing internodes often slightly thickened.

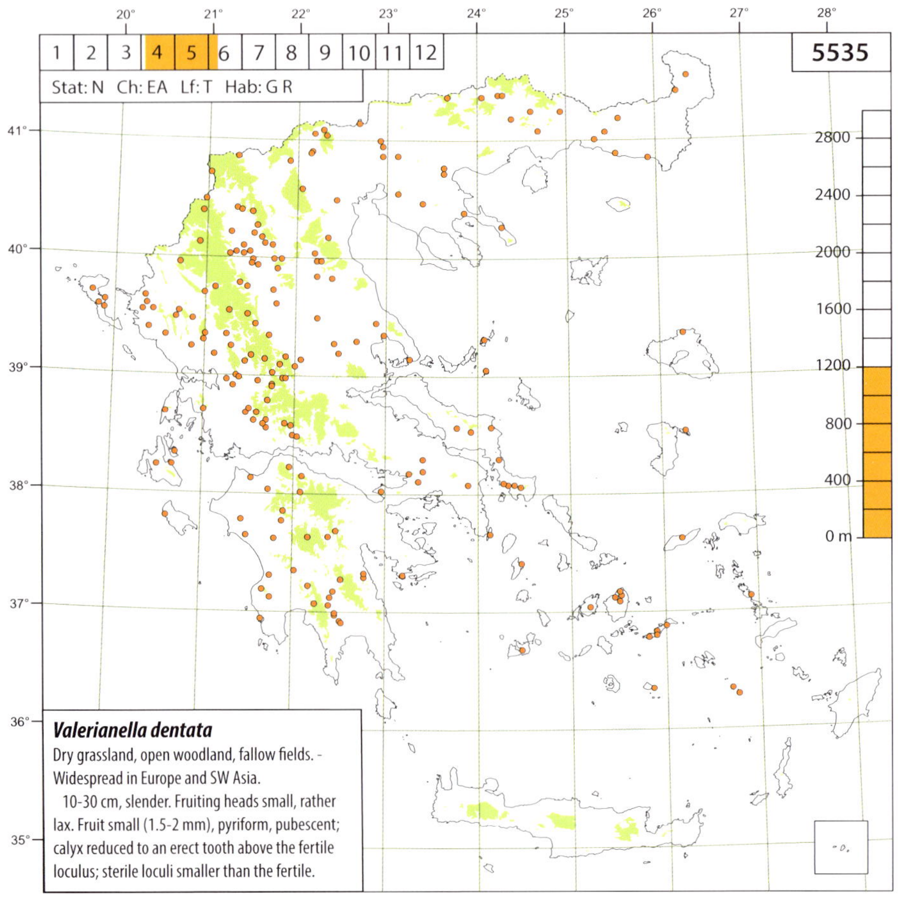

Valerianella dentata

Dry grassland, open woodland, fallow fields. - Widespread in Europe and SW Asia.

10-30 cm, slender. Fruiting heads small, rather lax. Fruit small (1.5-2 mm), pyriform, pubescent; calyx reduced to an erect tooth above the fertile loculus; sterile loculi smaller than the fertile.

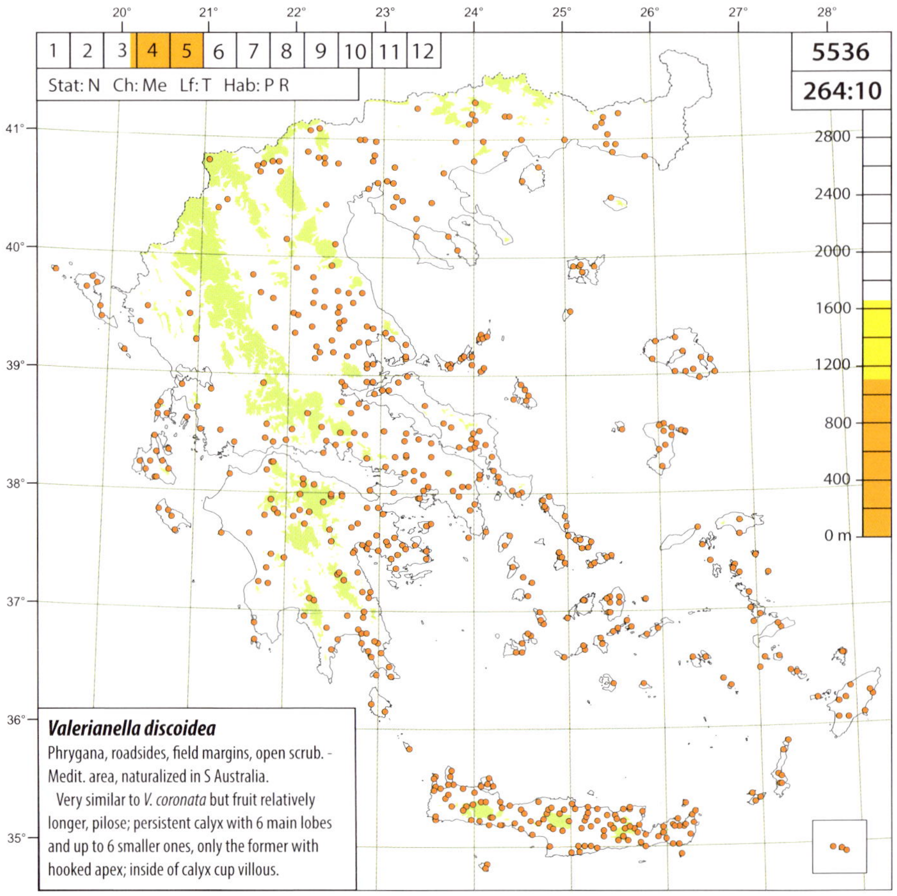

Valerianella discoidea

Phrygana, roadsides, field margins, open scrub. - Medit. area, naturalized in S Australia.

Very similar to *V. coronata* but fruit relatively longer, pilose; persistent calyx with 6 main lobes and up to 6 smaller ones, only the former with hooked apex; inside of calyx cup villous.

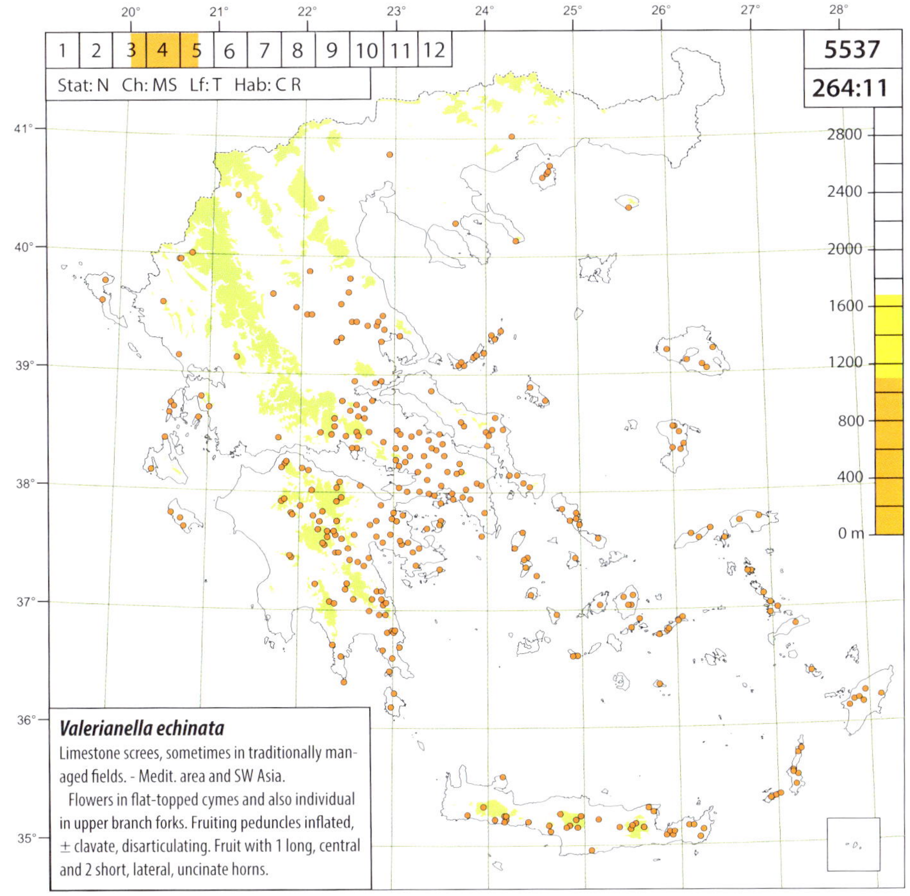

Valerianella echinata
Limestone screes, sometimes in traditionally managed fields. - Medit. area and SW Asia.
Flowers in flat-topped cymes and also individual in upper branch forks. Fruiting peduncles inflated, ± clavate, disarticulating. Fruit with 1 long, central and 2 short, lateral, uncinate horns.

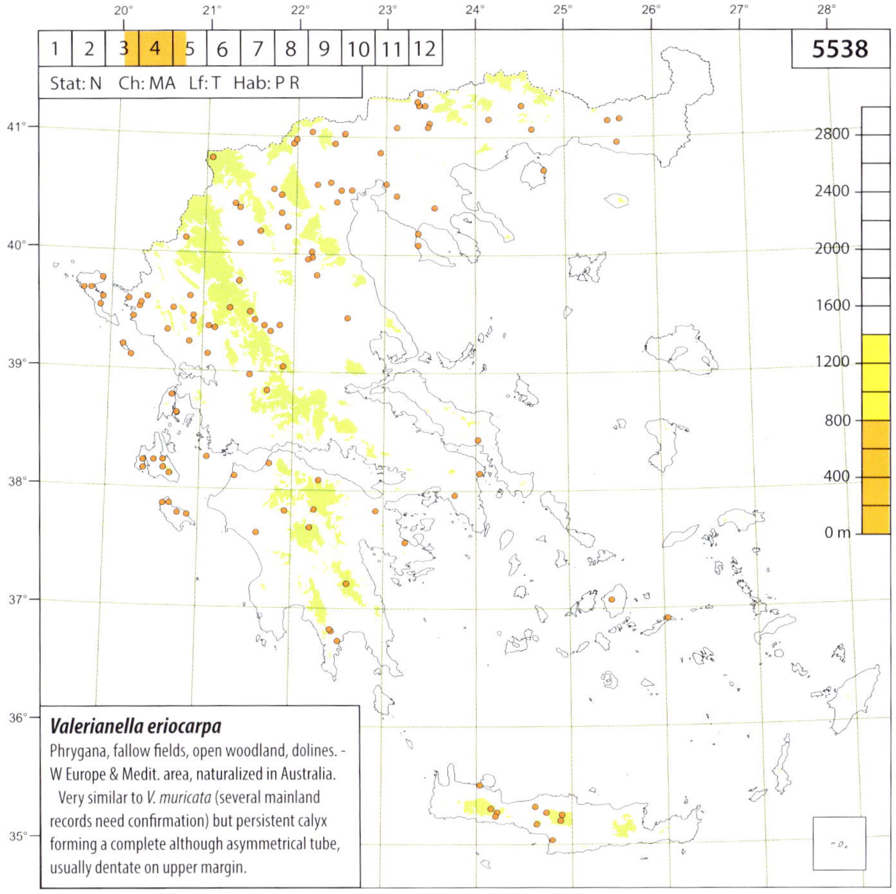

Valerianella eriocarpa
Phrygana, fallow fields, open woodland, dolines. - W Europe & Medit. area, naturalized in Australia.
Very similar to *V. muricata* (several mainland records need confirmation) but persistent calyx forming a complete although asymmetrical tube, usually dentate on upper margin.

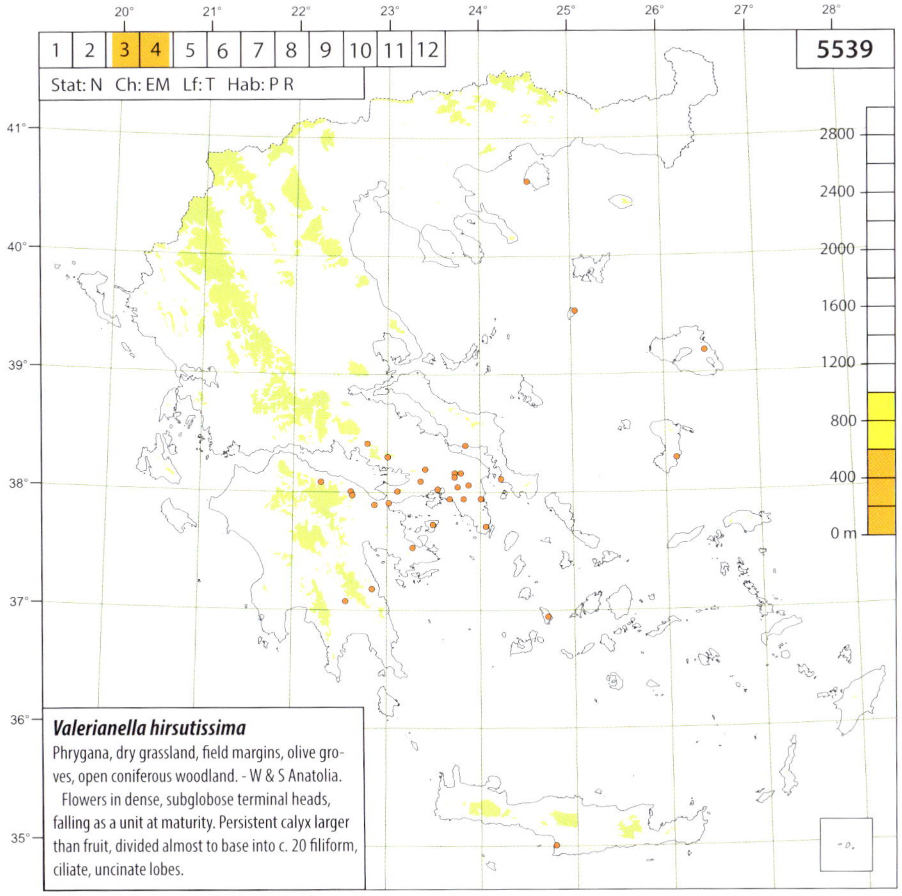

Valerianella hirsutissima
Phrygana, dry grassland, field margins, olive groves, open coniferous woodland. - W & S Anatolia.
Flowers in dense, subglobose terminal heads, falling as a unit at maturity. Persistent calyx larger than fruit, divided almost to base into c. 20 filiform, ciliate, uncinate lobes.

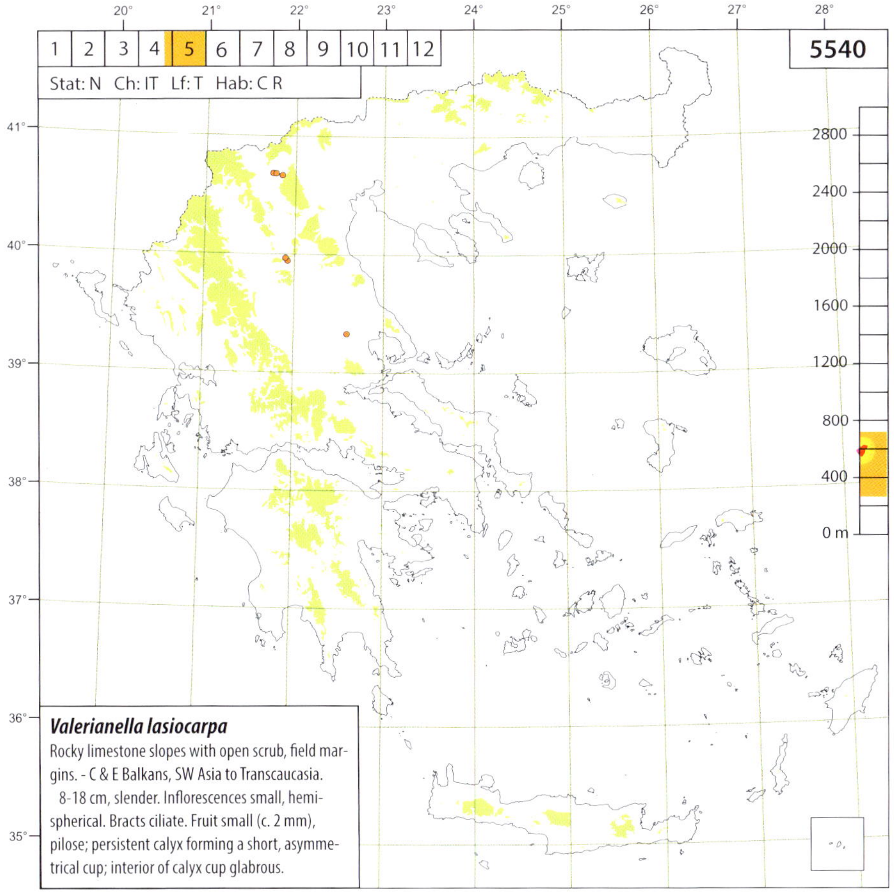

Valerianella lasiocarpa
Rocky limestone slopes with open scrub, field margins. - C & E Balkans, SW Asia to Transcaucasia.
8-18 cm, slender. Inflorescences small, hemispherical. Bracts ciliate. Fruit small (c. 2 mm), pilose; persistent calyx forming a short, asymmetrical cup; interior of calyx cup glabrous.

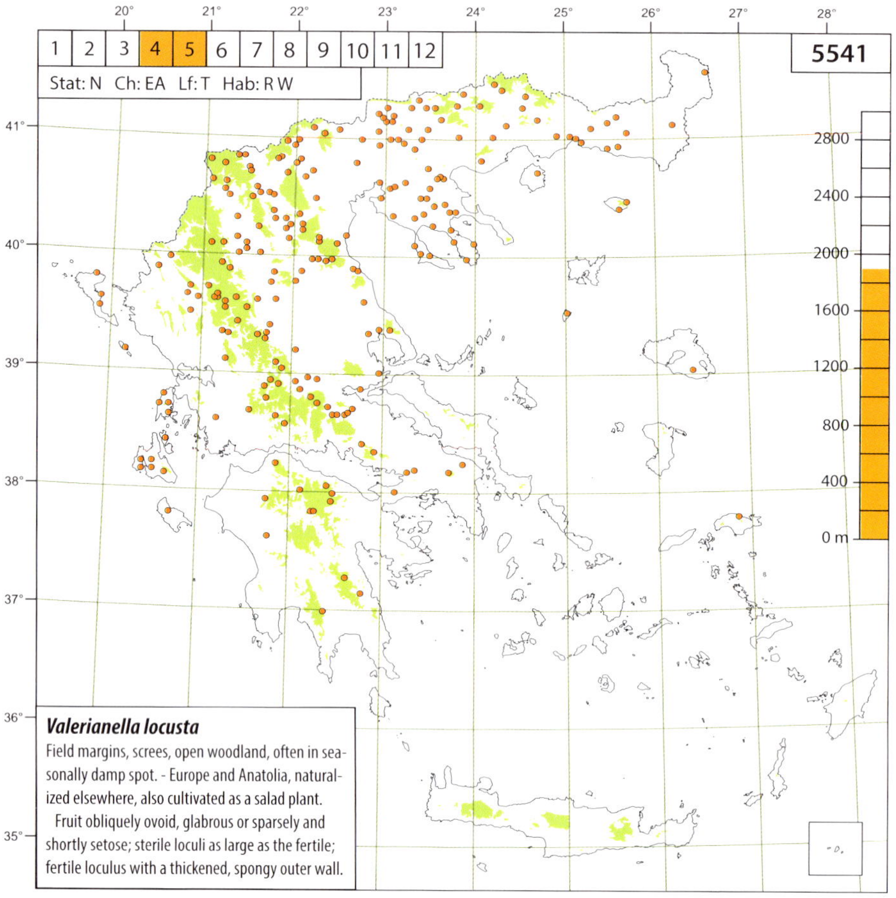

Valerianella locusta

Field margins, screes, open woodland, often in seasonally damp spot. - Europe and Anatolia, naturalized elsewhere, also cultivated as a salad plant.

Fruit obliquely ovoid, glabrous or sparsely and shortly setose; sterile loculi as large as the fertile; fertile loculus with a thickened, spongy outer wall.

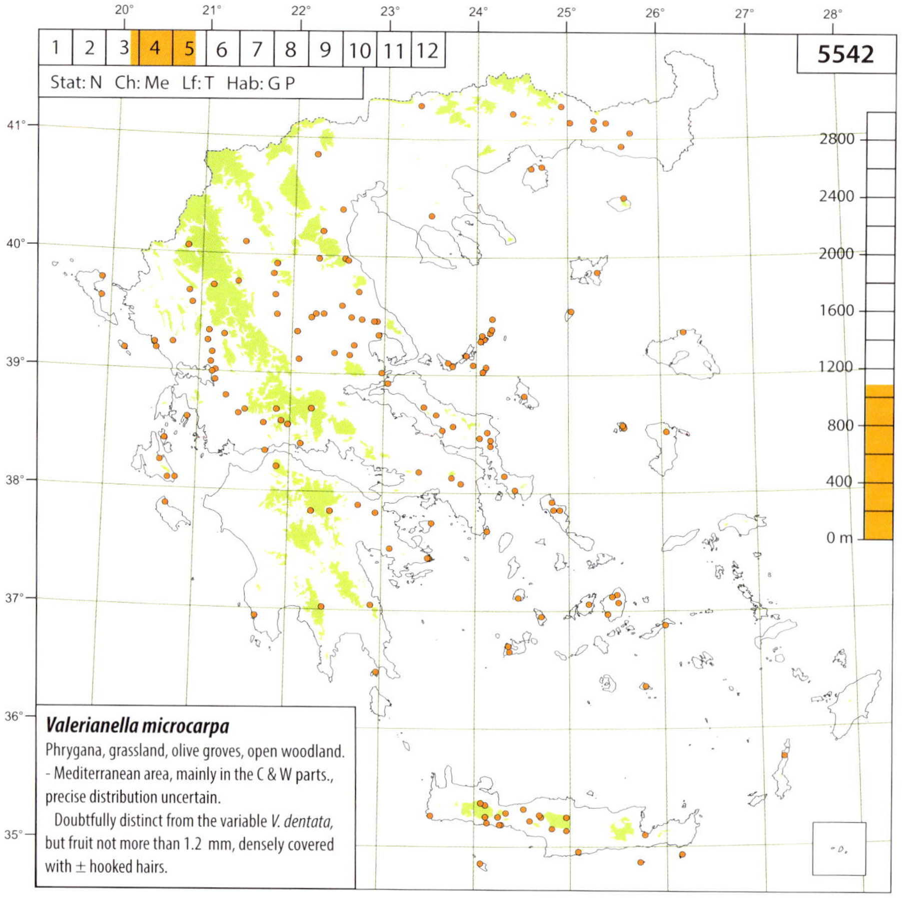

Valerianella microcarpa

Phrygana, grassland, olive groves, open woodland. - Mediterranean area, mainly in the C & W parts., precise distribution uncertain.

Doubtfully distinct from the variable *V. dentata*, but fruit not more than 1.2 mm, densely covered with ± hooked hairs.

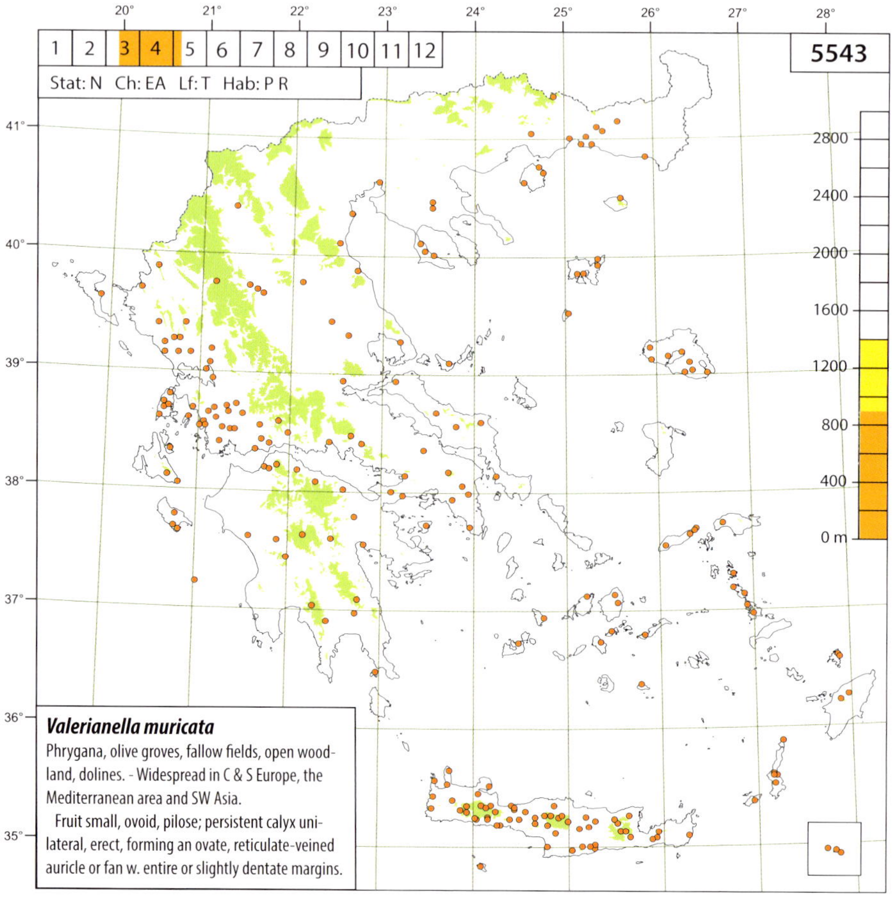

Valerianella muricata

Phrygana, olive groves, fallow fields, open woodland, dolines. - Widespread in C & S Europe, the Mediterranean area and SW Asia.

Fruit small, ovoid, pilose; persistent calyx unilateral, erect, forming an ovate, reticulate-veined auricle or fan w. entire or slightly dentate margins.

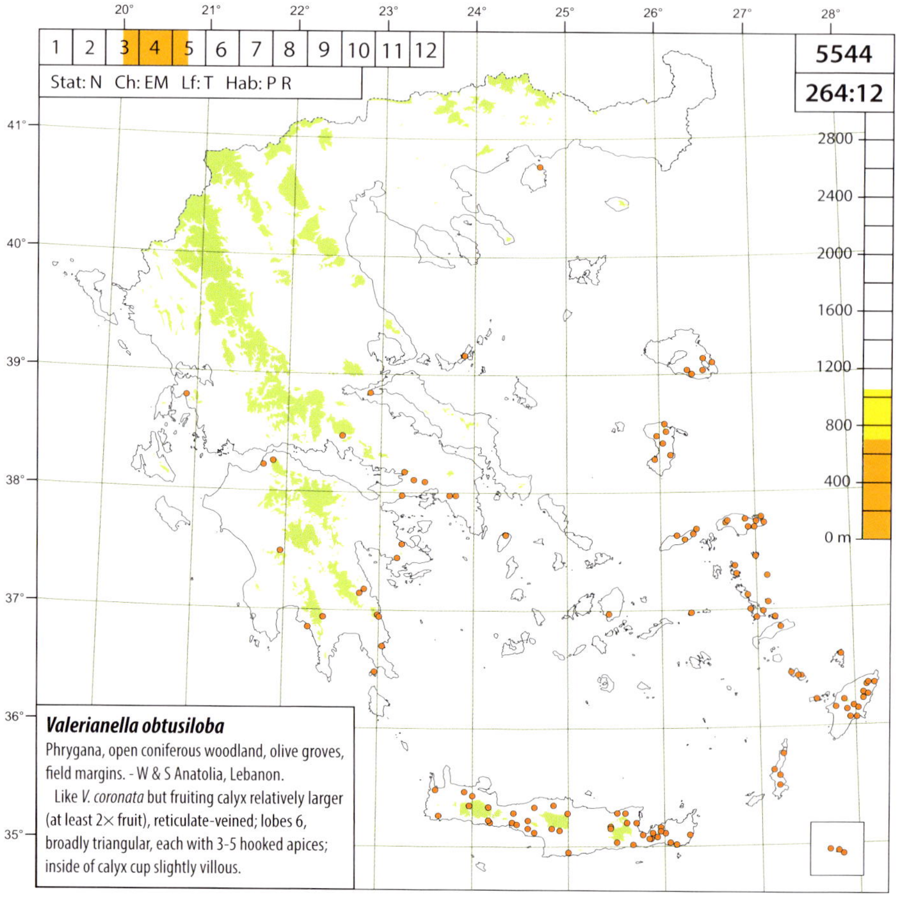

Valerianella obtusiloba

Phrygana, open coniferous woodland, olive groves, field margins. - W & S Anatolia, Lebanon.

Like *V. coronata* but fruiting calyx relatively larger (at least 2× fruit), reticulate-veined; lobes 6, broadly triangular, each with 3-5 hooked apices; inside of calyx cup slightly villous.

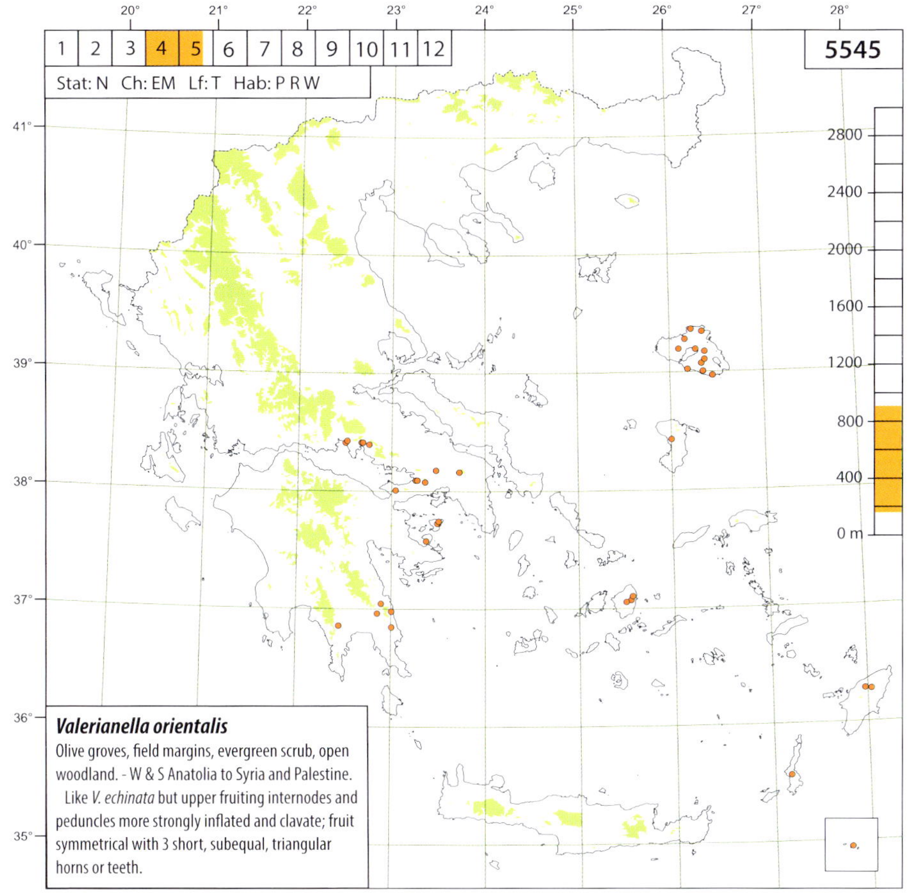

Valerianella orientalis
Olive groves, field margins, evergreen scrub, open woodland. - W & S Anatolia to Syria and Palestine.
Like *V. echinata* but upper fruiting internodes and peduncles more strongly inflated and clavate; fruit symmetrical with 3 short, subequal, triangular horns or teeth.

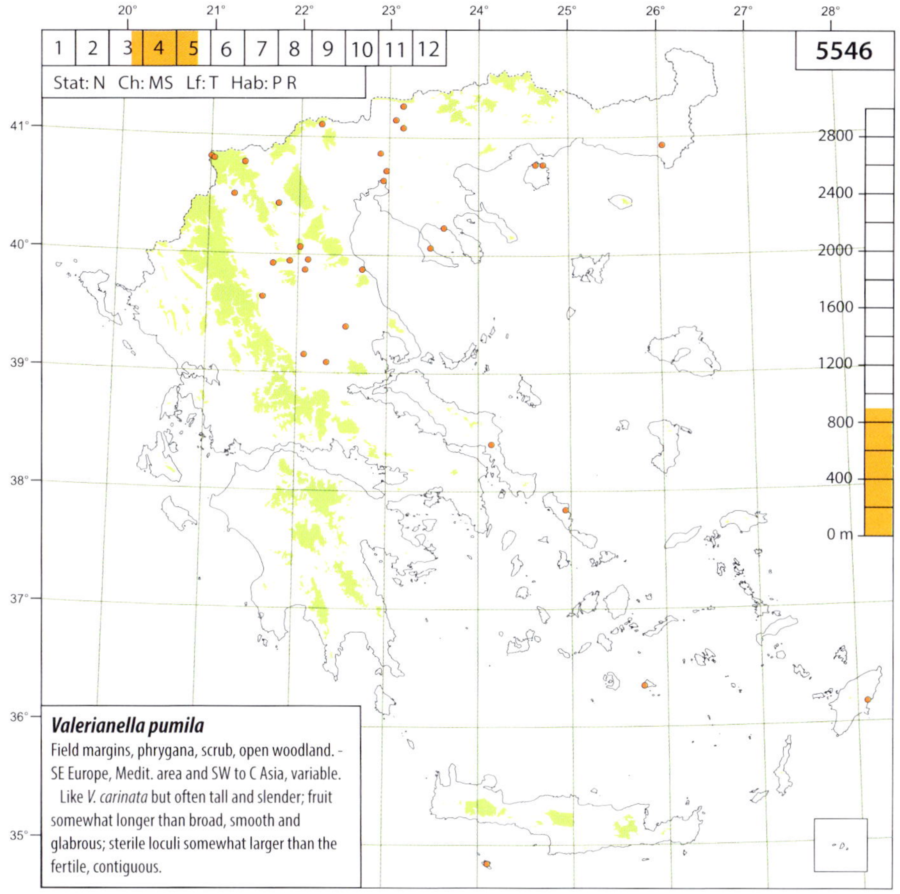

Valerianella pumila
Field margins, phrygana, scrub, open woodland. - SE Europe, Medit. area and SW to C Asia, variable.
Like *V. carinata* but often tall and slender; fruit somewhat longer than broad, smooth and glabrous; sterile loculi somewhat larger than the fertile, contiguous.

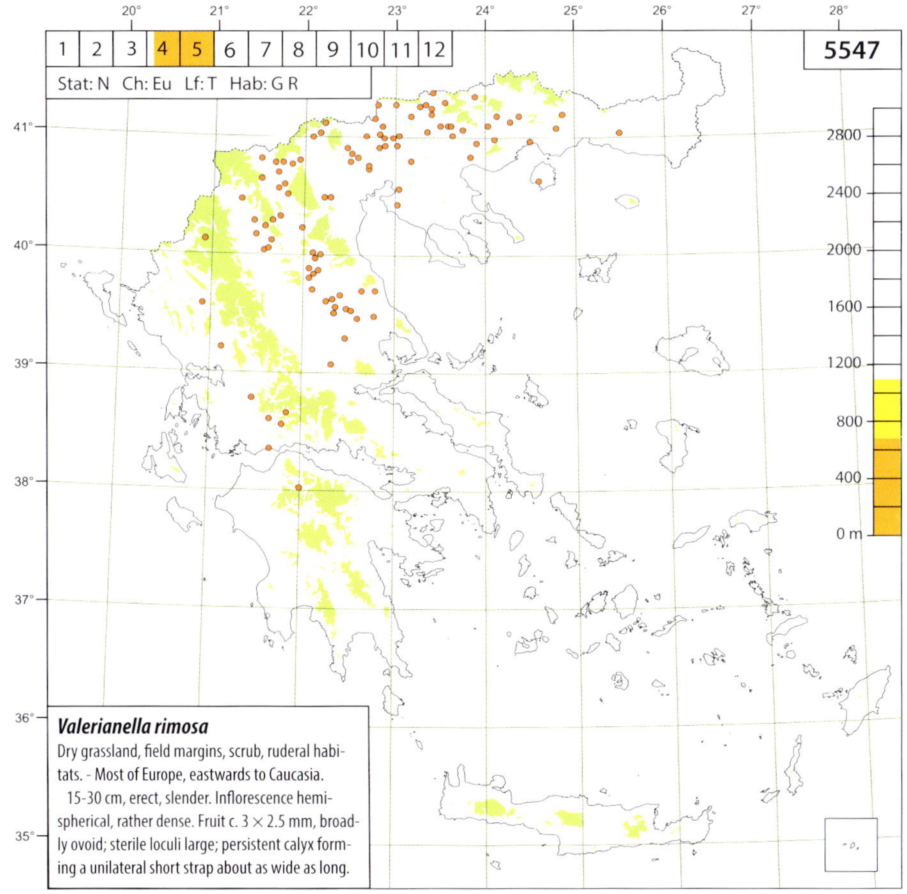

Valerianella rimosa
Dry grassland, field margins, scrub, ruderal habitats. - Most of Europe, eastwards to Caucasia.
15-30 cm, erect, slender. Inflorescence hemispherical, rather dense. Fruit c. 3 × 2.5 mm, broadly ovoid; sterile loculi large; persistent calyx forming a unilateral short strap about as wide as long.

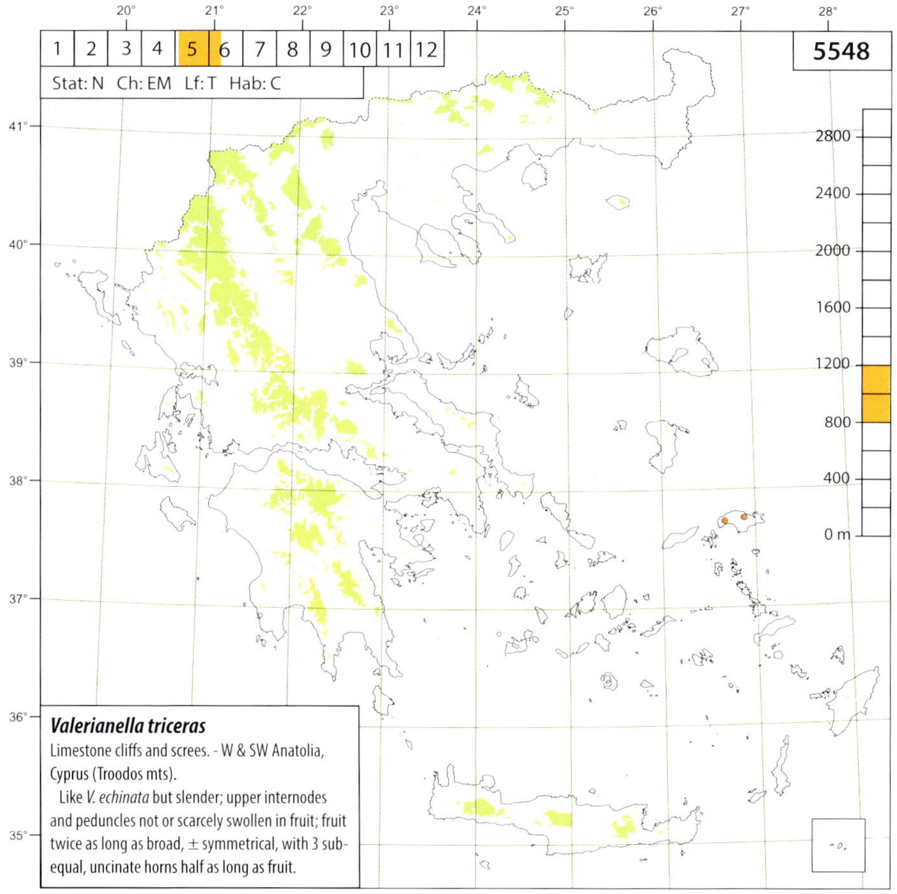

Valerianella triceras
Limestone cliffs and screes. - W & SW Anatolia, Cyprus (Troodos mts).
Like *V. echinata* but slender; upper internodes and peduncles not or scarcely swollen in fruit; fruit twice as long as broad, ± symmetrical, with 3 subequal, uncinate horns half as long as fruit.

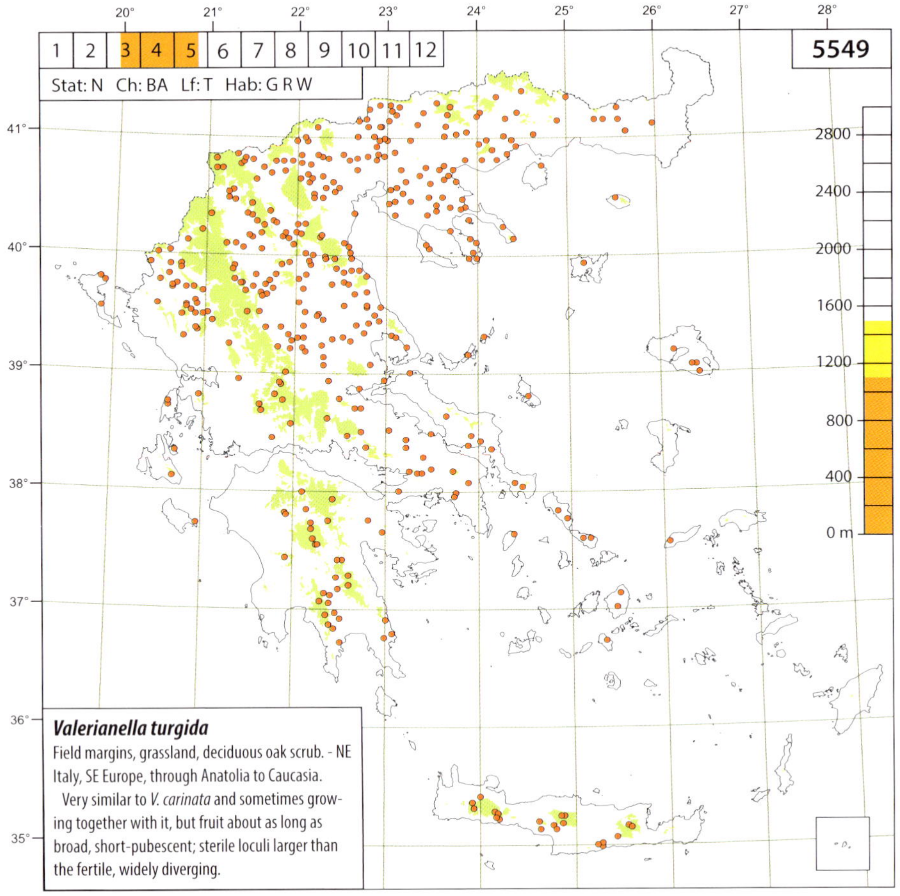

Valerianella turgida
Field margins, grassland, deciduous oak scrub. - NE Italy, SE Europe, through Anatolia to Caucasia.
Very similar to *V. carinata* and sometimes growing together with it, but fruit about as long as broad, short-pubescent; sterile loculi larger than the fertile, widely diverging.

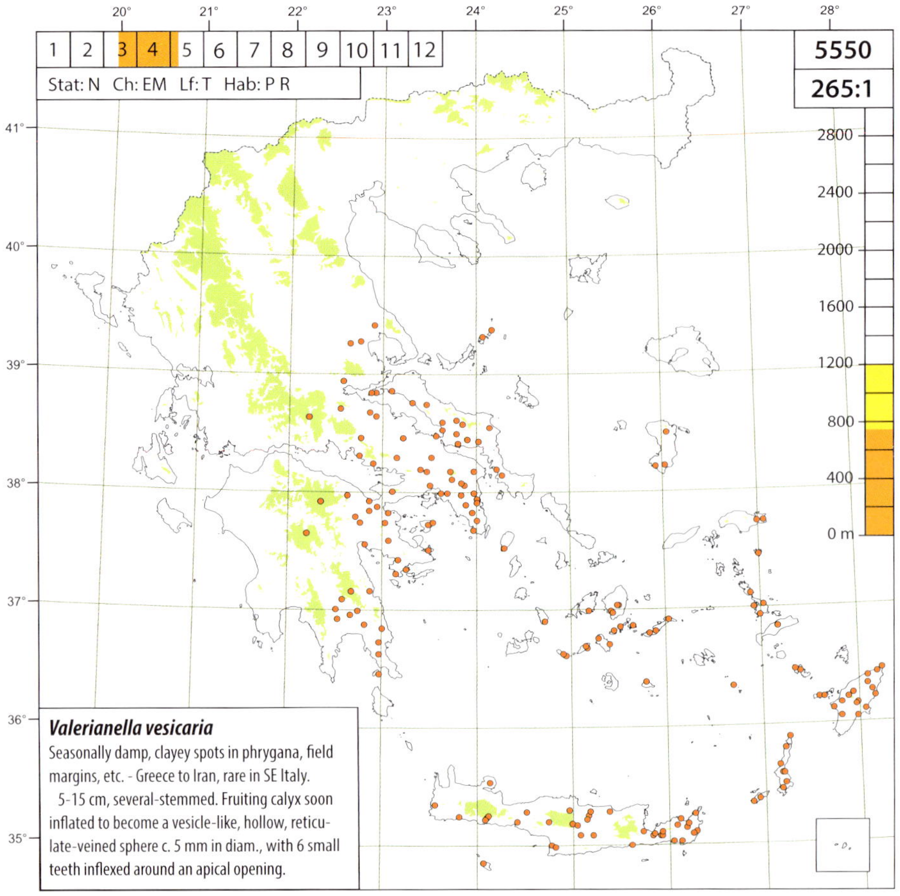

Valerianella vesicaria
Seasonally damp, clayey spots in phrygana, field margins, etc. - Greece to Iran, rare in SE Italy.
5-15 cm, several-stemmed. Fruiting calyx soon inflated to become a vesicle-like, hollow, reticulate-veined sphere c. 5 mm in diam., with 6 small teeth inflexed around an apical opening.

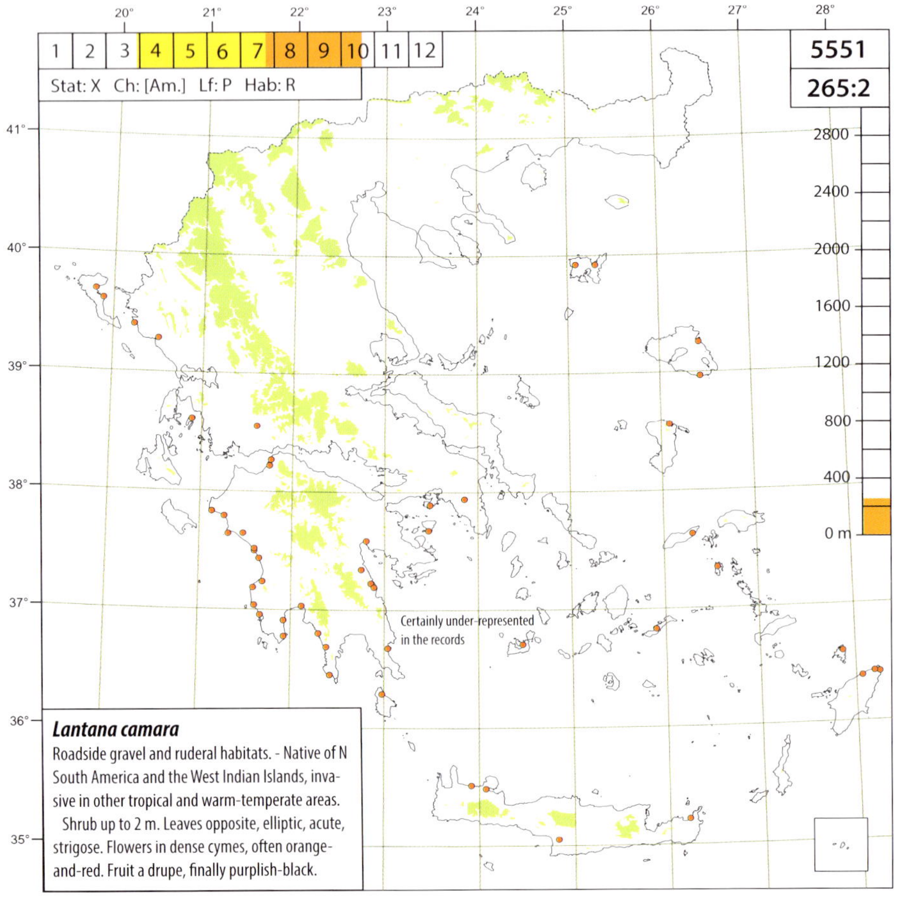

Lantana camara
Roadside gravel and ruderal habitats. - Native of N South America and the West Indian Islands, invasive in other tropical and warm-temperate areas.
Shrub up to 2 m. Leaves opposite, elliptic, acute, strigose. Flowers in dense cymes, often orange-and-red. Fruit a drupe, finally purplish-black.

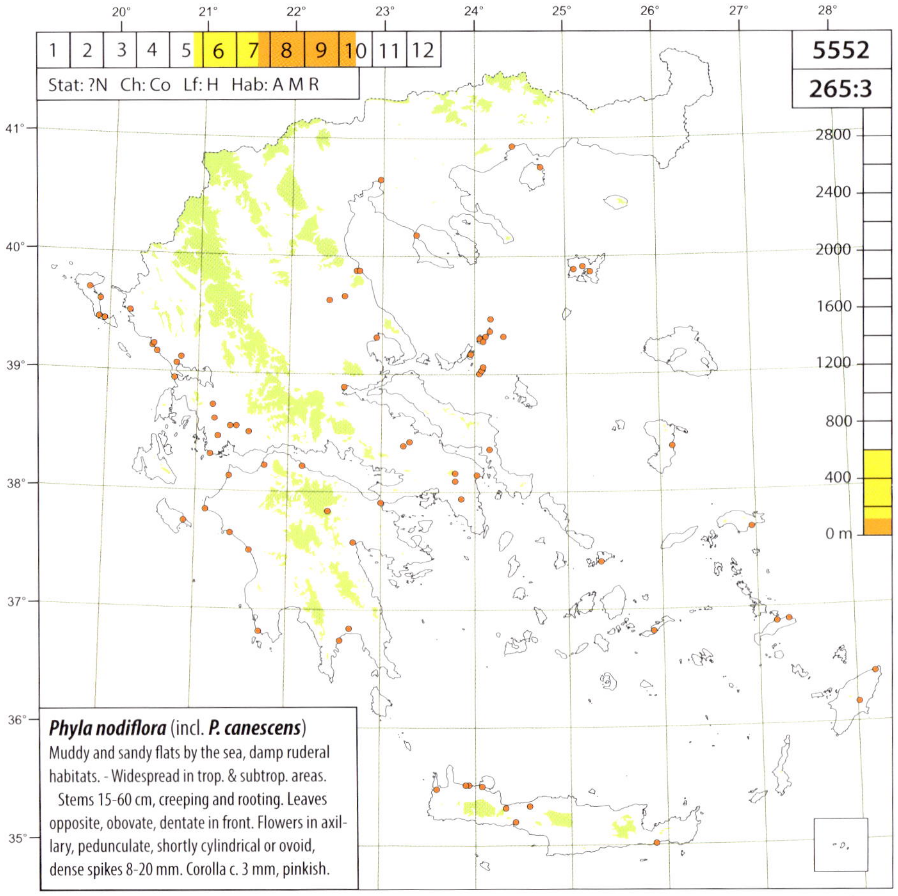

Phyla nodiflora (incl. ***P. canescens***)
Muddy and sandy flats by the sea, damp ruderal habitats. - Widespread in trop. & subtrop. areas.
Stems 15-60 cm, creeping and rooting. Leaves opposite, obovate, dentate in front. Flowers in axillary, pedunculate, shortly cylindrical or ovoid, dense spikes 8-20 mm. Corolla c. 3 mm, pinkish.

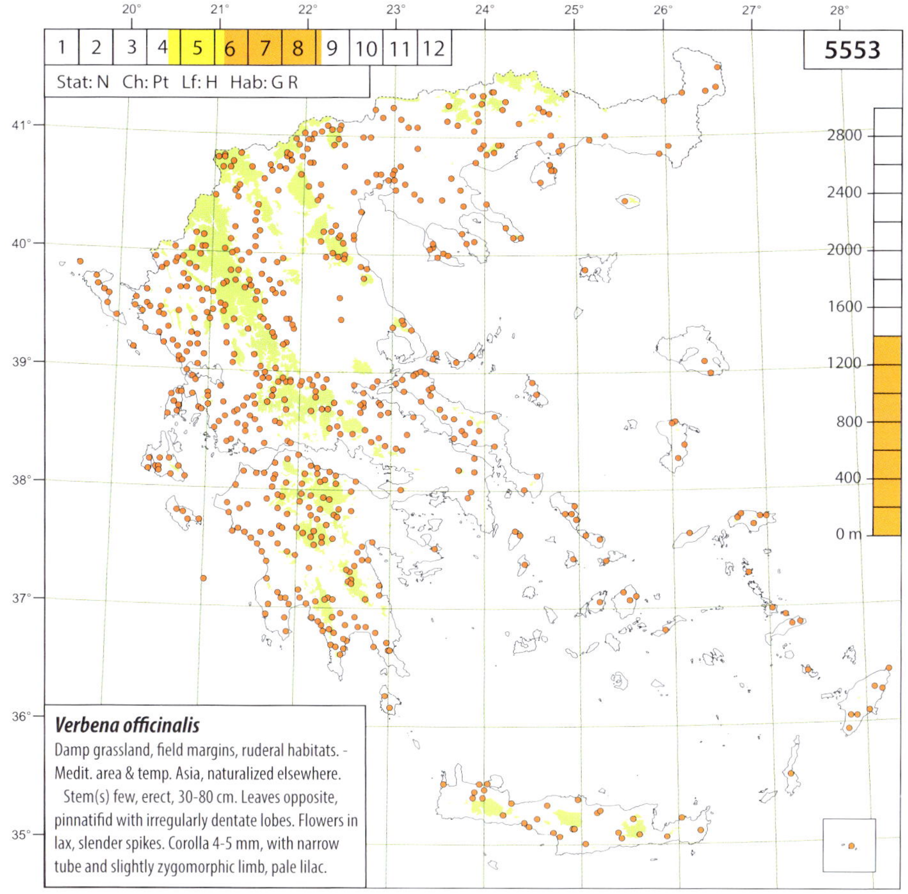

Verbena officinalis
Damp grassland, field margins, ruderal habitats. - Medit. area & temp. Asia, naturalized elsewhere.
Stem(s) few, erect, 30-80 cm. Leaves opposite, pinnatifid with irregularly dentate lobes. Flowers in lax, slender spikes. Corolla 4-5 mm, with narrow tube and slightly zygomorphic limb, pale lilac.

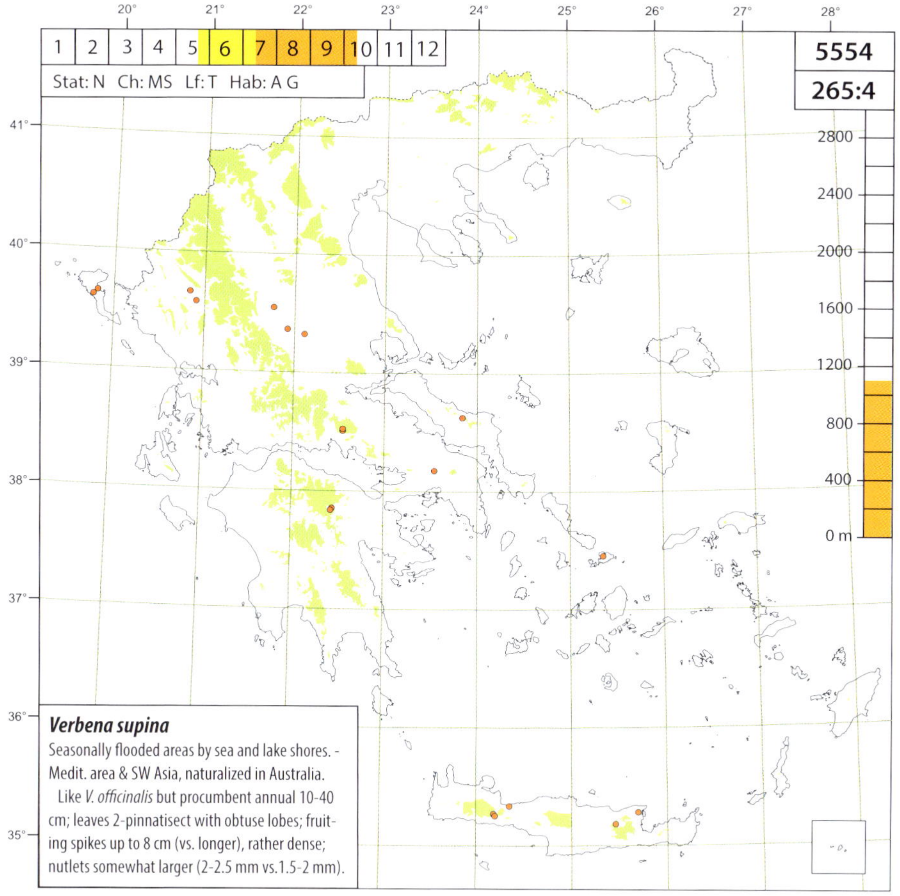

Verbena supina
Seasonally flooded areas by sea and lake shores. - Medit. area & SW Asia, naturalized in Australia.
Like *V. officinalis* but procumbent annual 10-40 cm; leaves 2-pinnatisect with obtuse lobes; fruiting spikes up to 8 cm (vs. longer), rather dense; nutlets somewhat larger (2-2.5 mm vs. 1.5-2 mm).

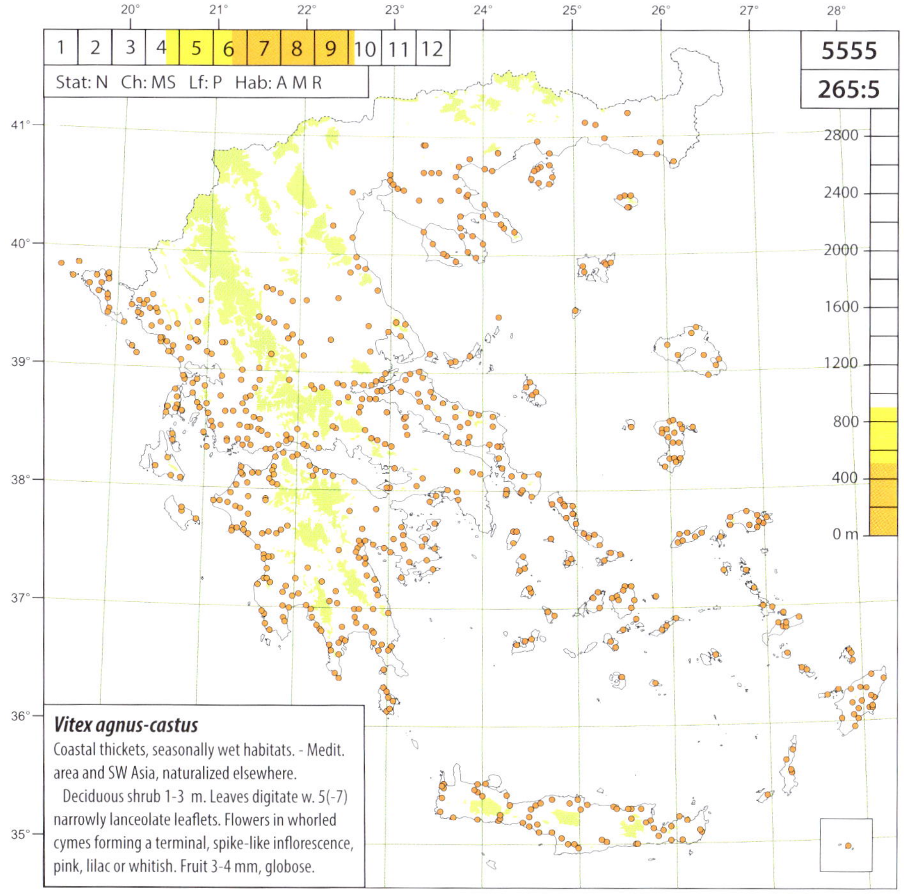

Vitex agnus-castus
Coastal thickets, seasonally wet habitats. - Medit. area and SW Asia, naturalized elsewhere.
Deciduous shrub 1-3 m. Leaves digitate w. 5(-7) narrowly lanceolate leaflets. Flowers in whorled cymes forming a terminal, spike-like inflorescence, pink, lilac or whitish. Fruit 3-4 mm, globose.

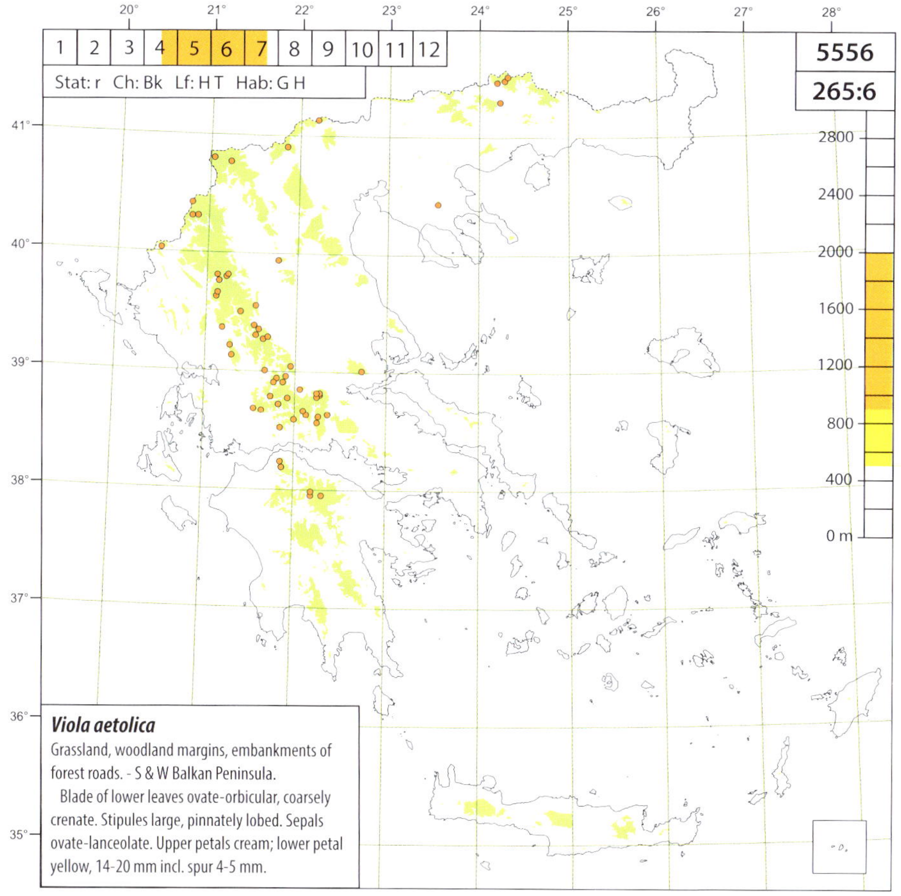

Viola aetolica
Grassland, woodland margins, embankments of forest roads. - S & W Balkan Peninsula.
Blade of lower leaves ovate-orbicular, coarsely crenate. Stipules large, pinnately lobed. Sepals ovate-lanceolate. Upper petals cream; lower petal yellow, 14-20 mm incl. spur 4-5 mm.

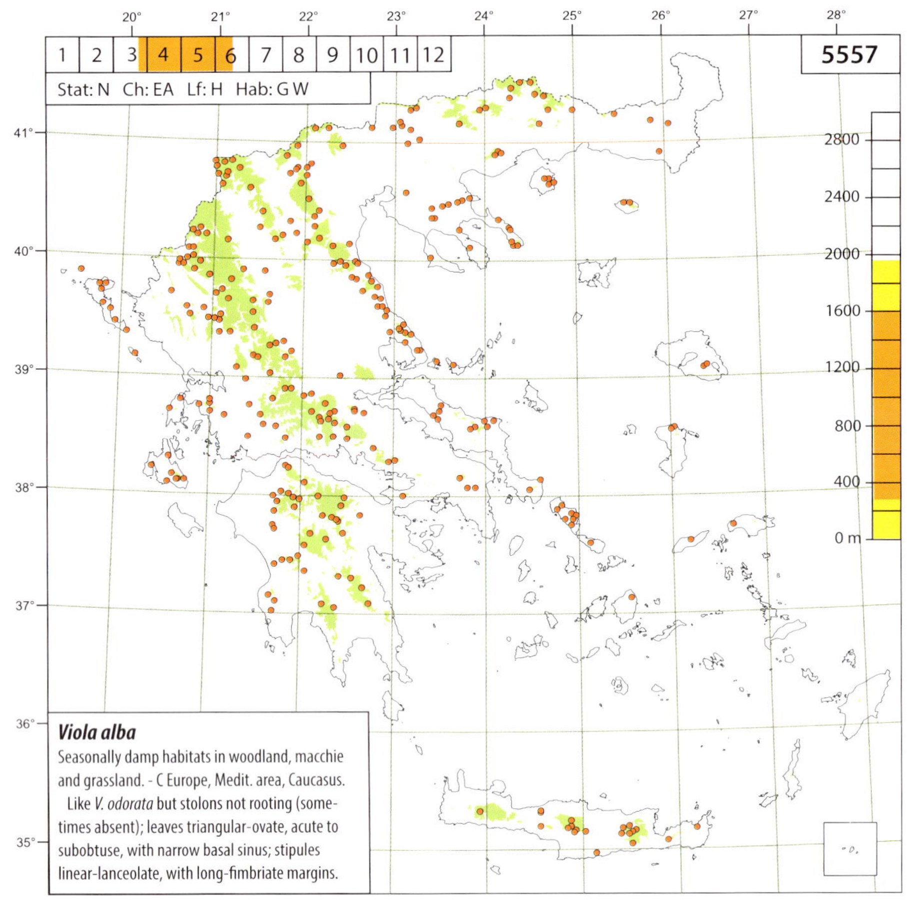

Viola alba

Seasonally damp habitats in woodland, macchie and grassland. - C Europe, Medit. area, Caucasus.

Like *V. odorata* but stolons not rooting (sometimes absent); leaves triangular-ovate, acute to subobtuse, with narrow basal sinus; stipules linear-lanceolate, with long-fimbriate margins.

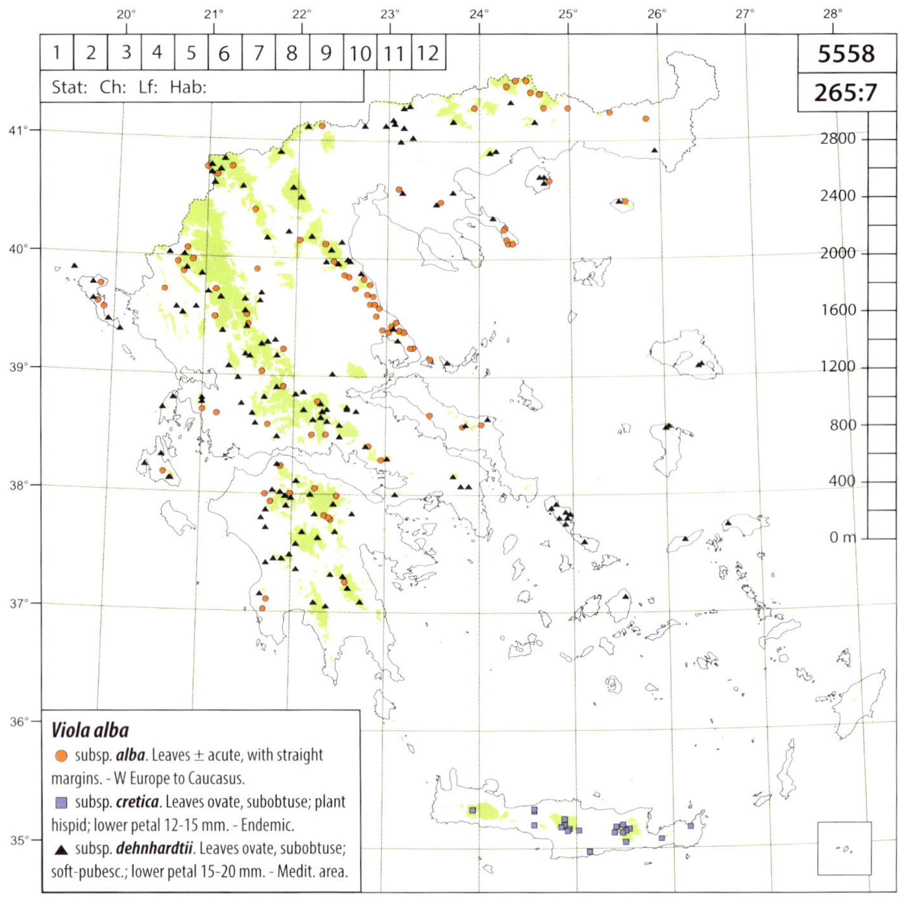

Viola alba

● subsp. ***alba***. Leaves ± acute, with straight margins. - W Europe to Caucasus.

■ subsp. ***cretica***. Leaves ovate, subobtuse; plant hispid; lower petal 12-15 mm. - Endemic.

▲ subsp. ***dehnhardtii***. Leaves ovate, subobtuse; soft-pubesc.; lower petal 15-20 mm. - Medit. area.

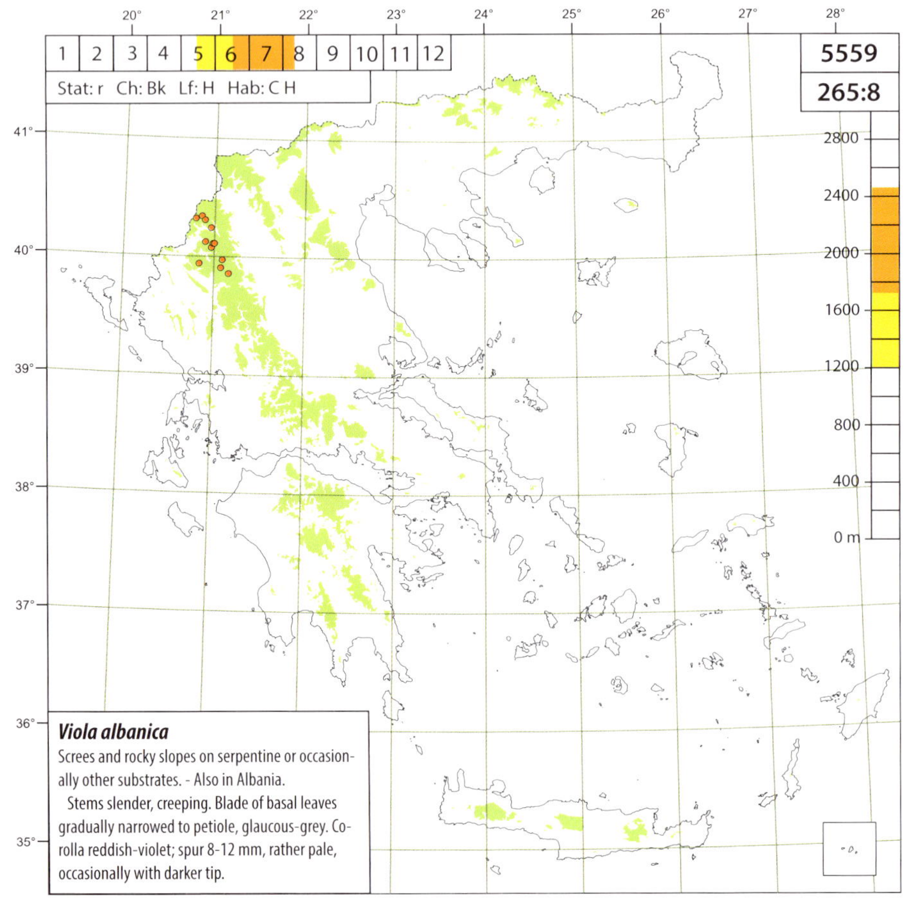

Viola albanica

Screes and rocky slopes on serpentine or occasionally other substrates. - Also in Albania.

Stems slender, creeping. Blade of basal leaves gradually narrowed to petiole, glaucous-grey. Corolla reddish-violet; spur 8-12 mm, rather pale, occasionally with darker tip.

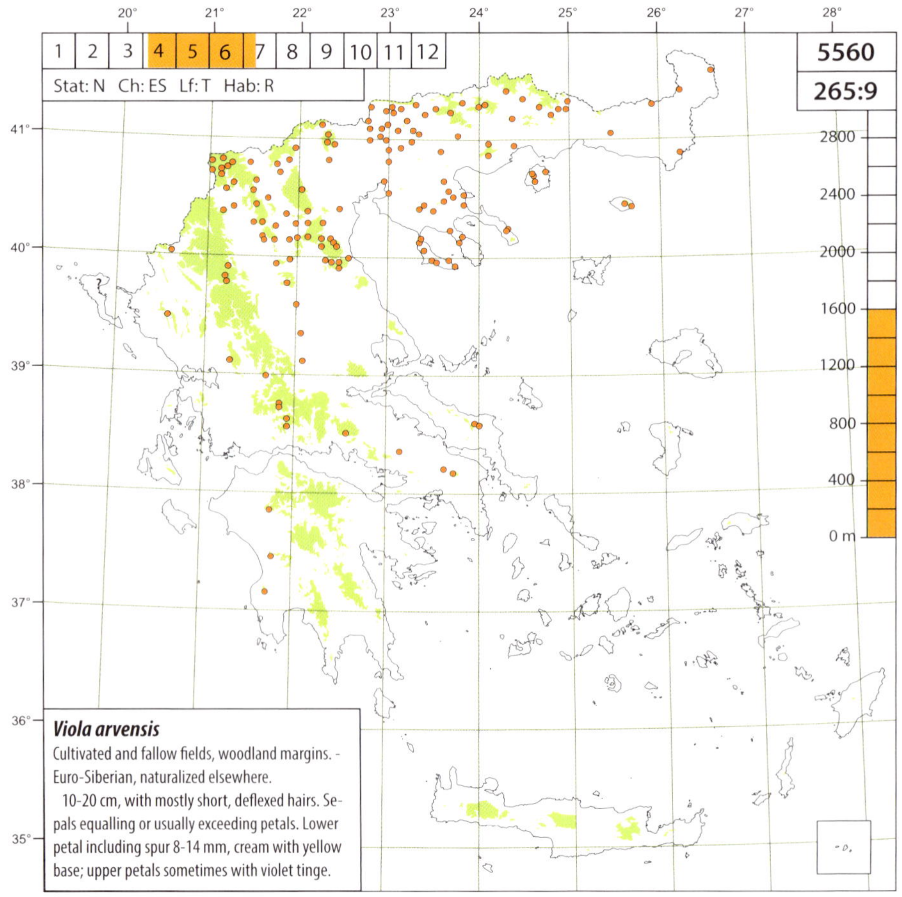

Viola arvensis

Cultivated and fallow fields, woodland margins. - Euro-Siberian, naturalized elsewhere.

10-20 cm, with mostly short, deflexed hairs. Sepals equalling or usually exceeding petals. Lower petal including spur 8-14 mm, cream with yellow base; upper petals sometimes with violet tinge.

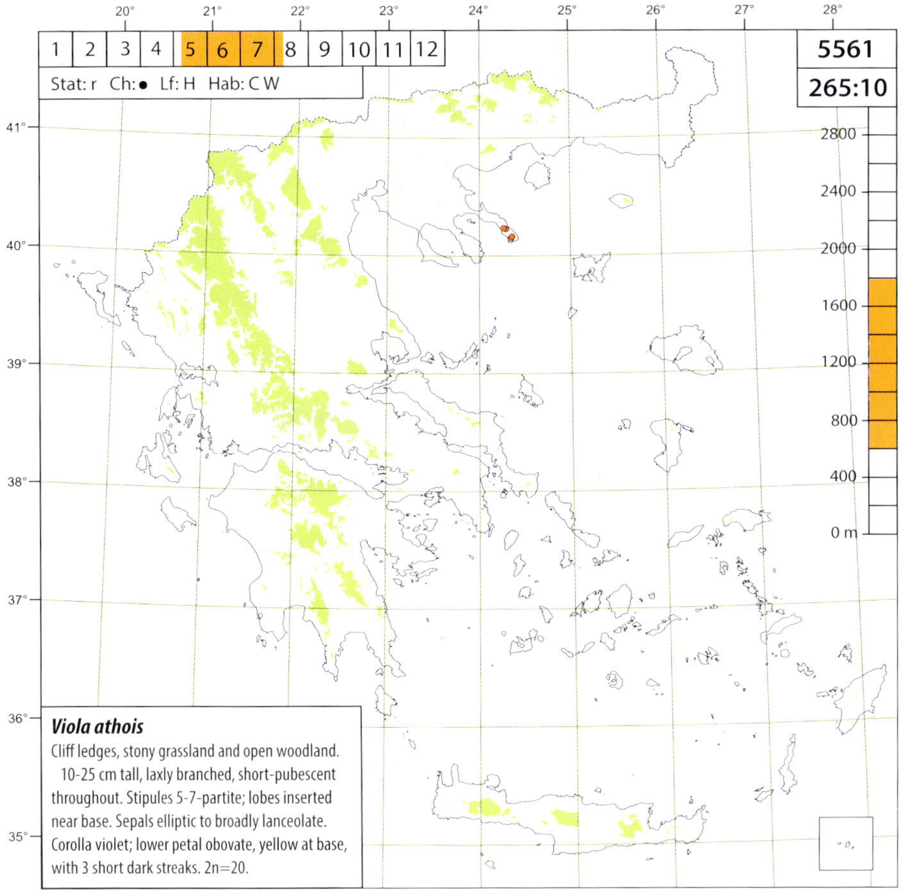

Viola athois

Cliff ledges, stony grassland and open woodland.

10-25 cm tall, laxly branched, short-pubescent throughout. Stipules 5-7-partite; lobes inserted near base. Sepals elliptic to broadly lanceolate. Corolla violet; lower petal obovate, yellow at base, with 3 short dark streaks. 2n=20.

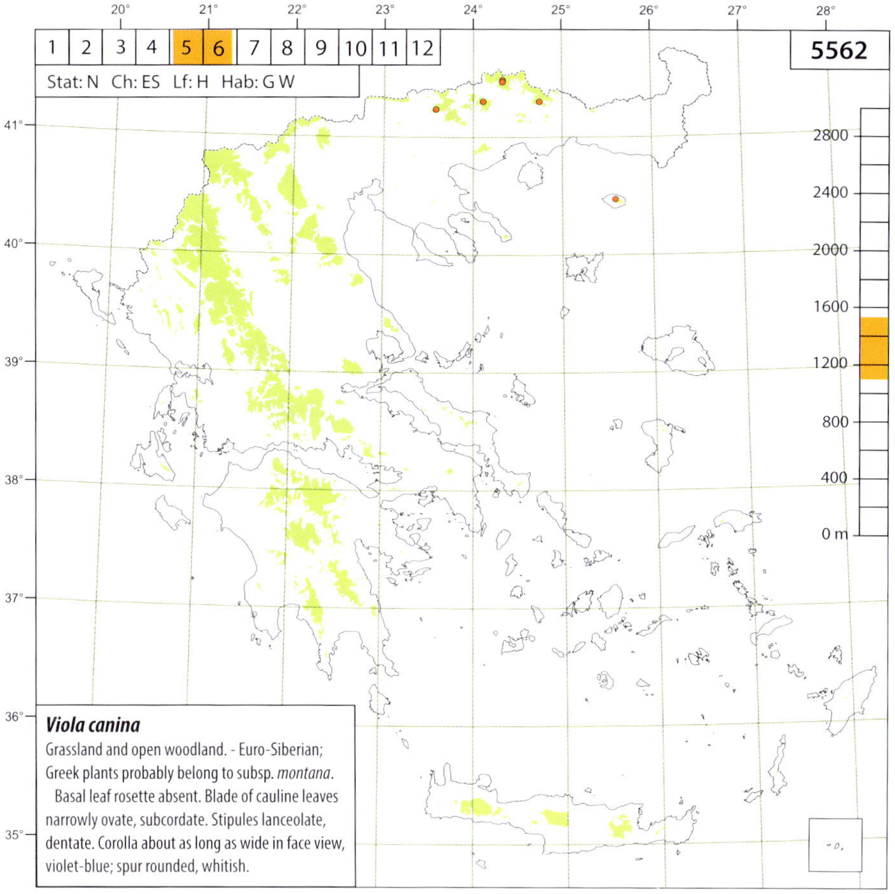

Viola canina

Grassland and open woodland. - Euro-Siberian; Greek plants probably belong to subsp. *montana*.

Basal leaf rosette absent. Blade of cauline leaves narrowly ovate, subcordate. Stipules lanceolate, dentate. Corolla about as long as wide in face view, violet-blue; spur rounded, whitish.

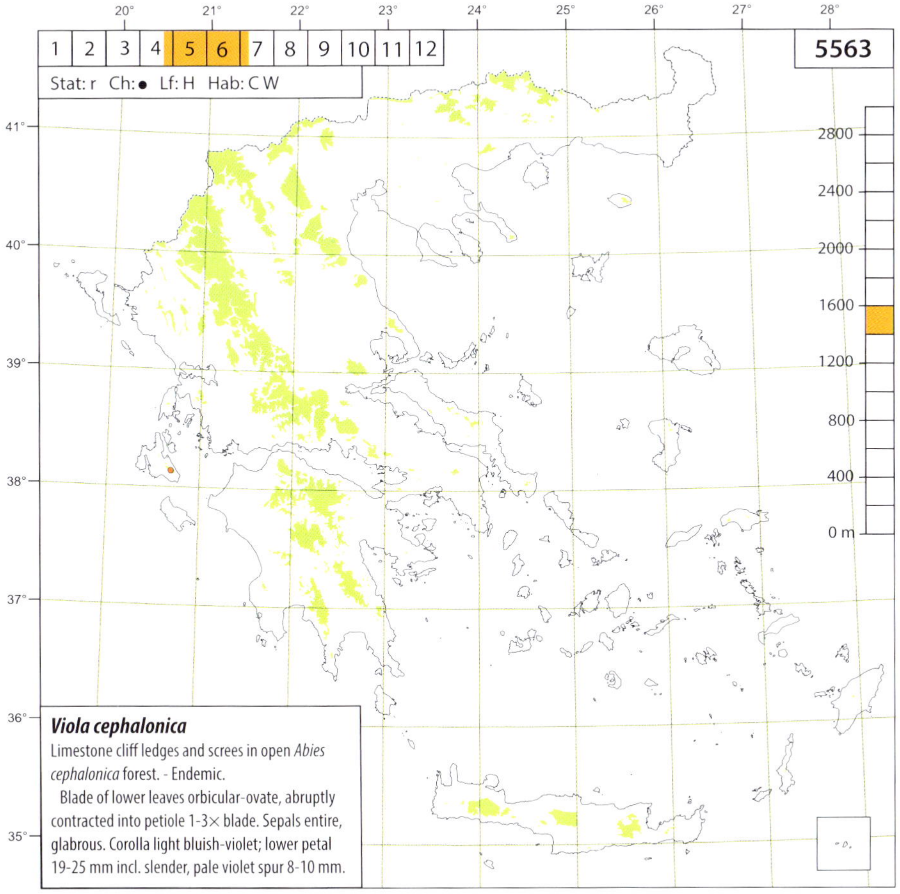

Viola cephalonica

Limestone cliff ledges and screes in open *Abies cephalonica* forest. - Endemic.

Blade of lower leaves orbicular-ovate, abruptly contracted into petiole 1-3× blade. Sepals entire, glabrous. Corolla light bluish-violet; lower petal 19-25 mm incl. slender, pale violet spur 8-10 mm.

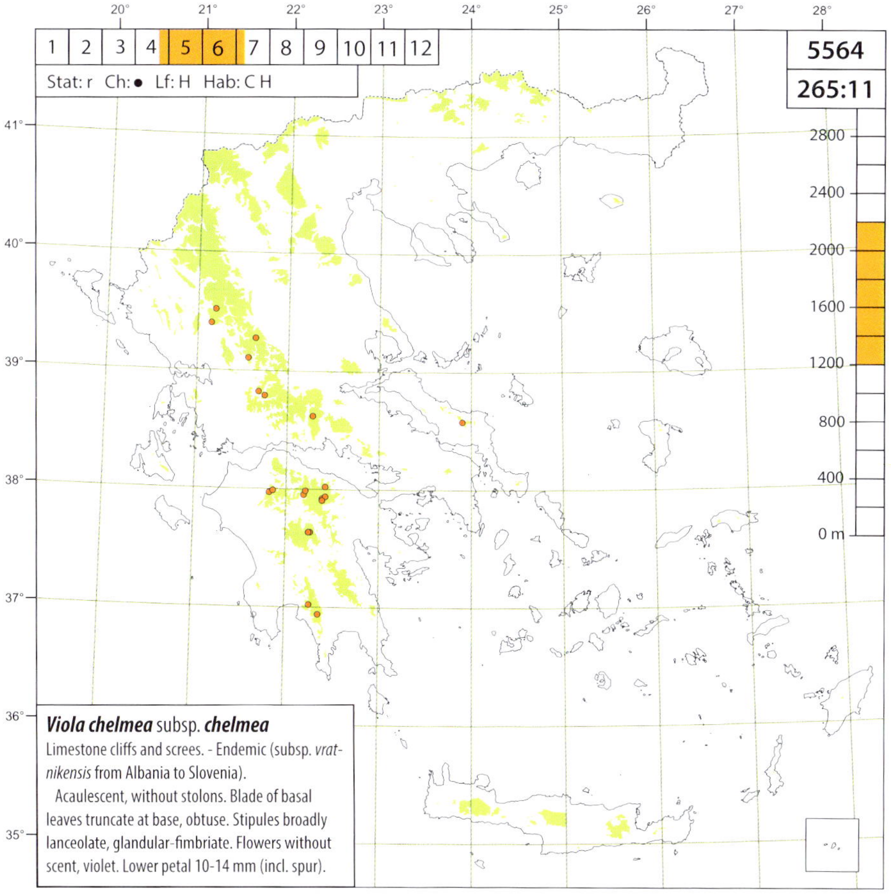

Viola chelmea* subsp. *chelmea

Limestone cliffs and screes. - Endemic (subsp. *vratnikensis* from Albania to Slovenia).

Acaulescent, without stolons. Blade of basal leaves truncate at base, obtuse. Stipules broadly lanceolate, glandular-fimbriate. Flowers without scent, violet. Lower petal 10-14 mm (incl. spur).

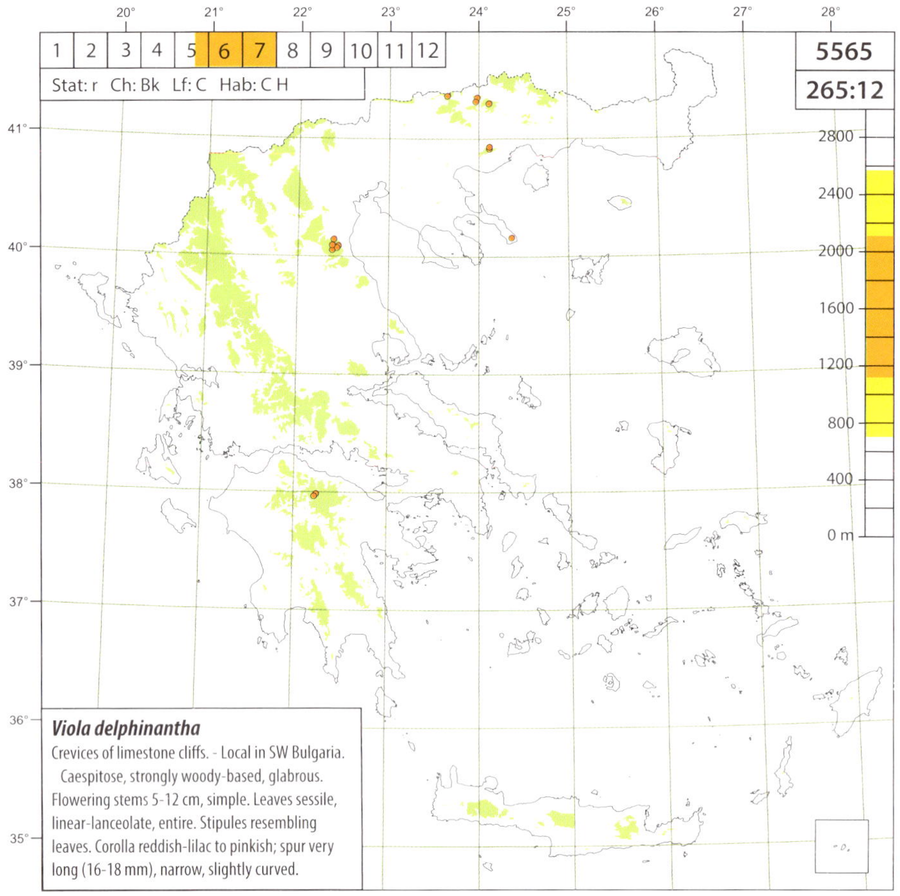

Viola delphinantha

Crevices of limestone cliffs. - Local in SW Bulgaria.

Caespitose, strongly woody-based, glabrous. Flowering stems 5-12 cm, simple. Leaves sessile, linear-lanceolate, entire. Stipules resembling leaves. Corolla reddish-lilac to pinkish; spur very long (16-18 mm), narrow, slightly curved.

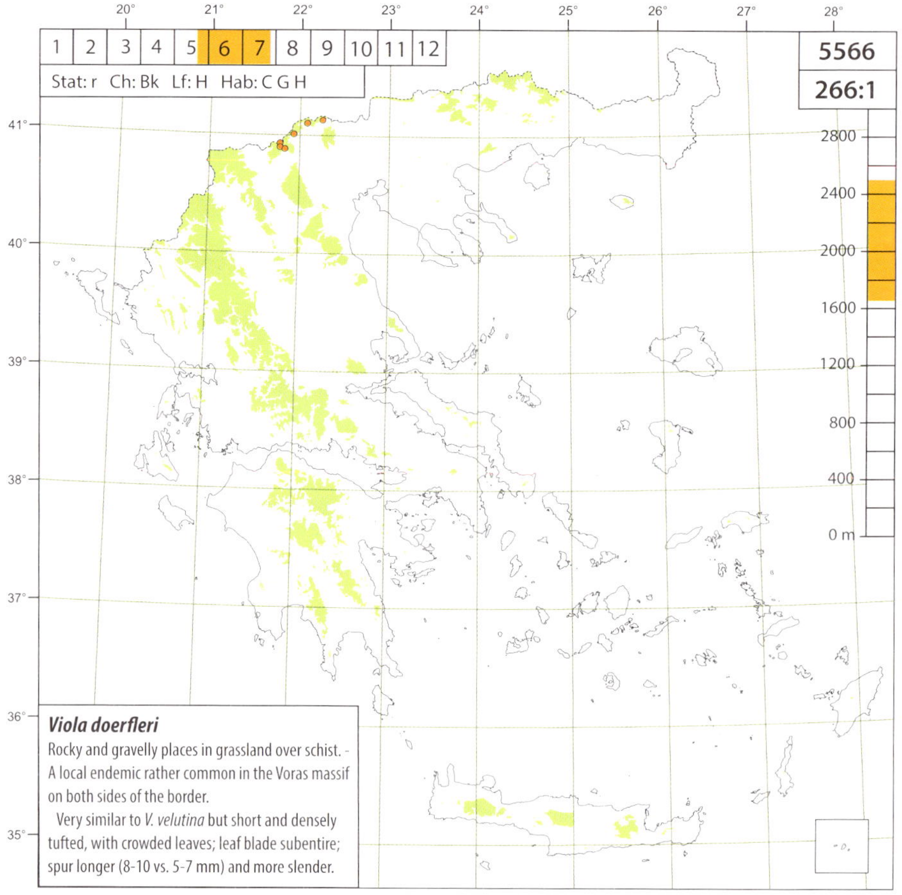

Viola doerfleri

Rocky and gravelly places in grassland over schist. - A local endemic rather common in the Voras massif on both sides of the border.

Very similar to *V. velutina* but short and densely tufted, with crowded leaves; leaf blade subentire; spur longer (8-10 vs. 5-7 mm) and more slender.

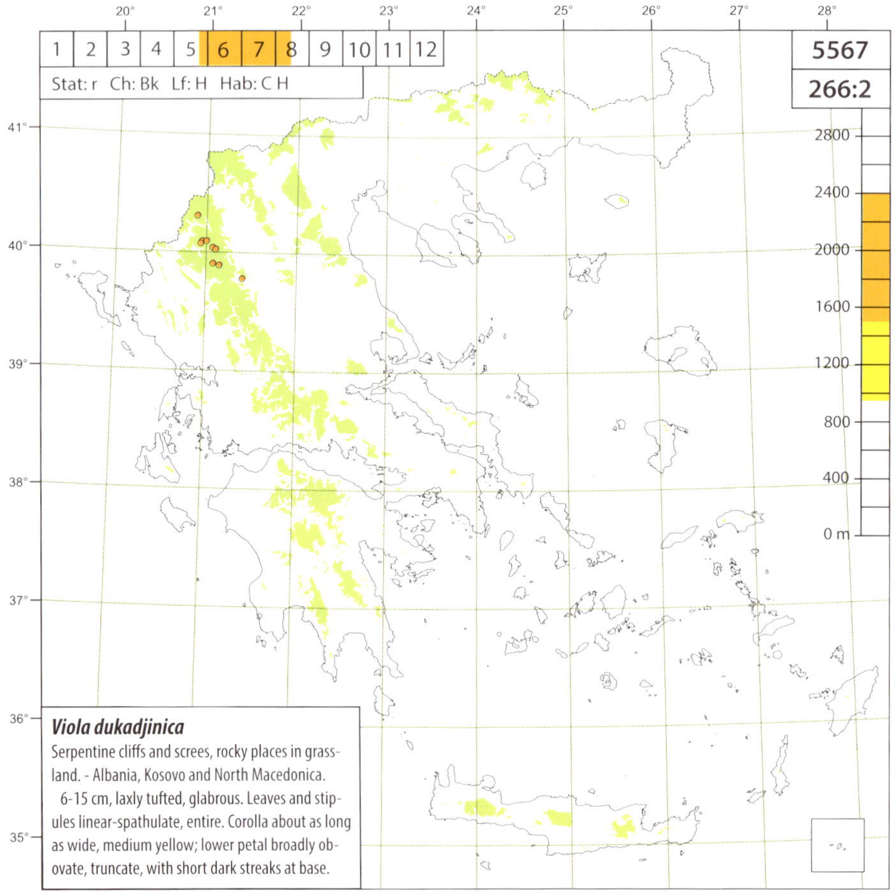

Viola dukadjinica

Serpentine cliffs and screes, rocky places in grassland. - Albania, Kosovo and North Macedonica.

6-15 cm, laxly tufted, glabrous. Leaves and stipules linear-spathulate, entire. Corolla about as long as wide, medium yellow; lower petal broadly obovate, truncate, with short dark streaks at base.

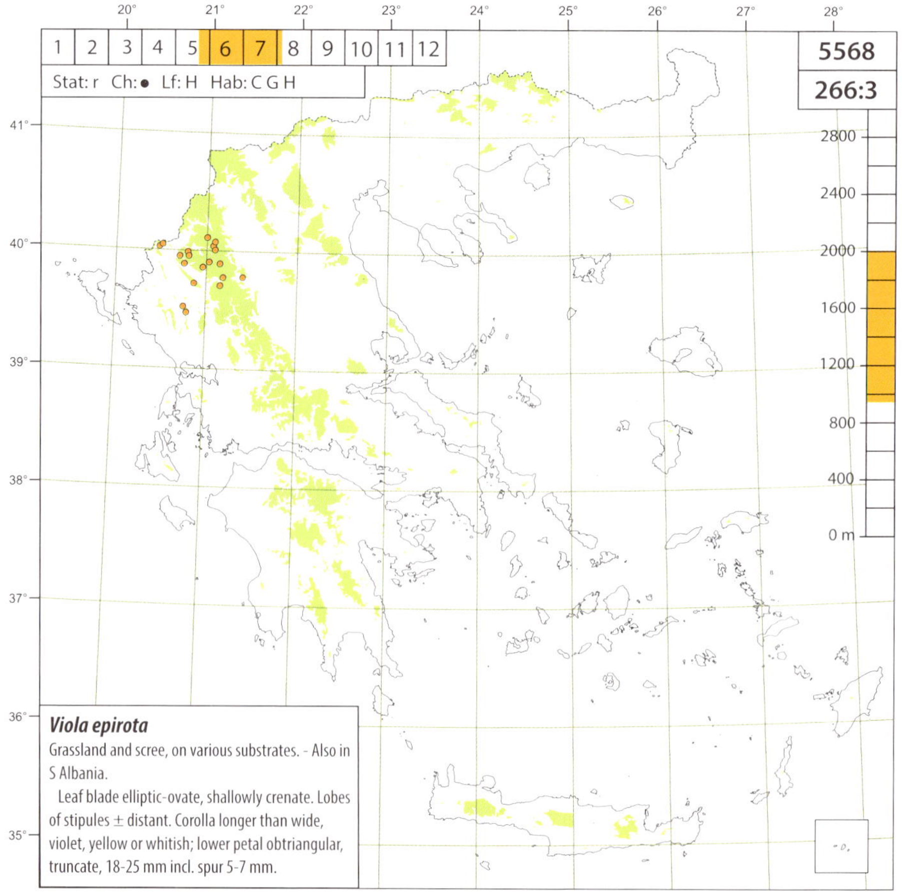

Viola epirota

Grassland and scree, on various substrates. - Also in S Albania.

Leaf blade elliptic-ovate, shallowly crenate. Lobes of stipules ± distant. Corolla longer than wide, violet, yellow or whitish; lower petal obtriangular, truncate, 18-25 mm incl. spur 5-7 mm.

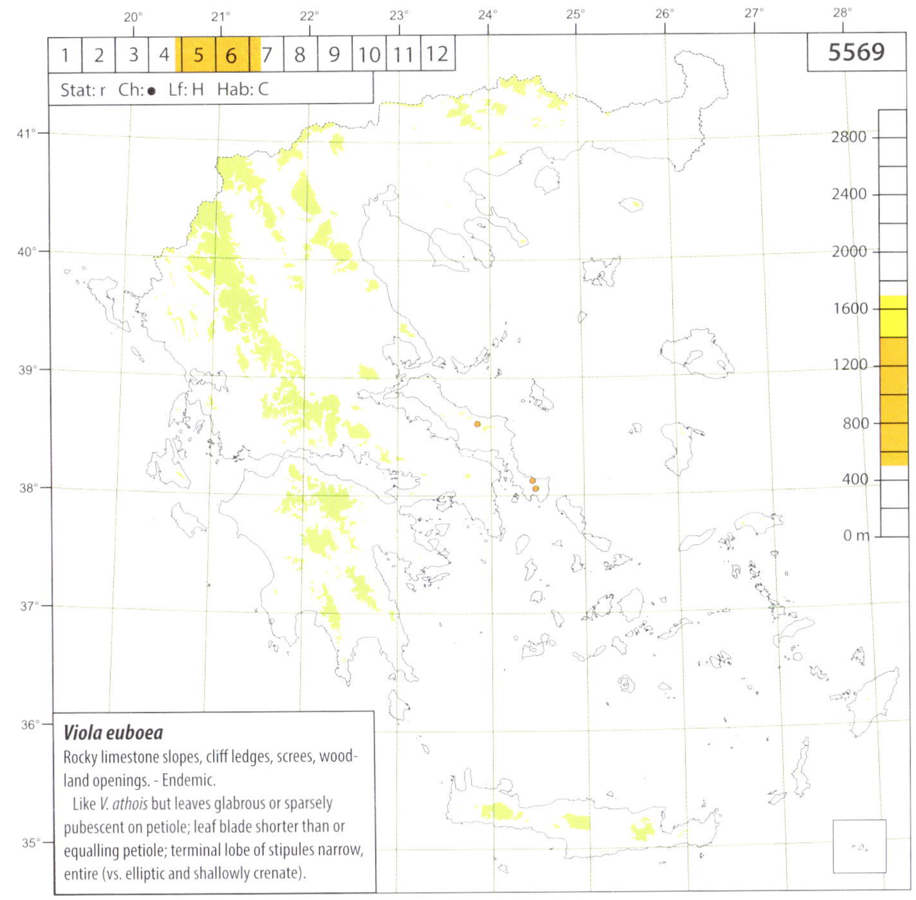

Viola euboea

Rocky limestone slopes, cliff ledges, screes, woodland openings. - Endemic.

Like *V. athois* but leaves glabrous or sparsely pubescent on petiole; leaf blade shorter than or equalling petiole; terminal lobe of stipules narrow, entire (vs. elliptic and shallowly crenate).

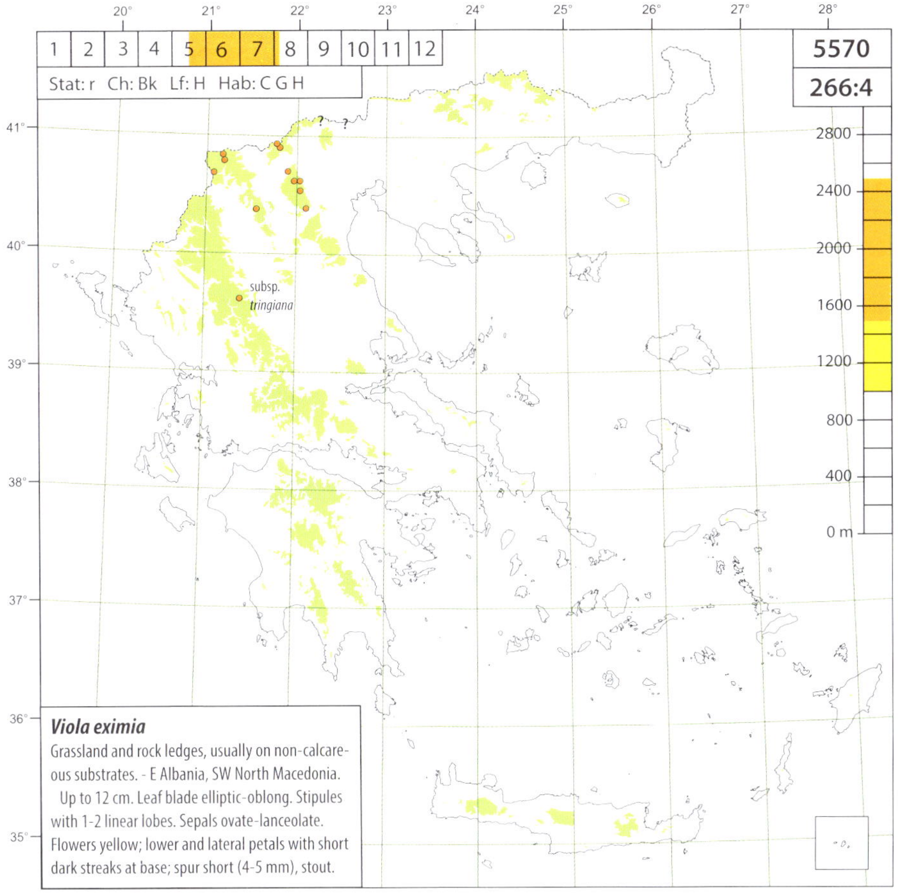

Viola eximia

Grassland and rock ledges, usually on non-calcareous substrates. - E Albania, SW North Macedonia.

Up to 12 cm. Leaf blade elliptic-oblong. Stipules with 1-2 linear lobes. Sepals ovate-lanceolate. Flowers yellow; lower and lateral petals with short dark streaks at base; spur short (4-5 mm), stout.

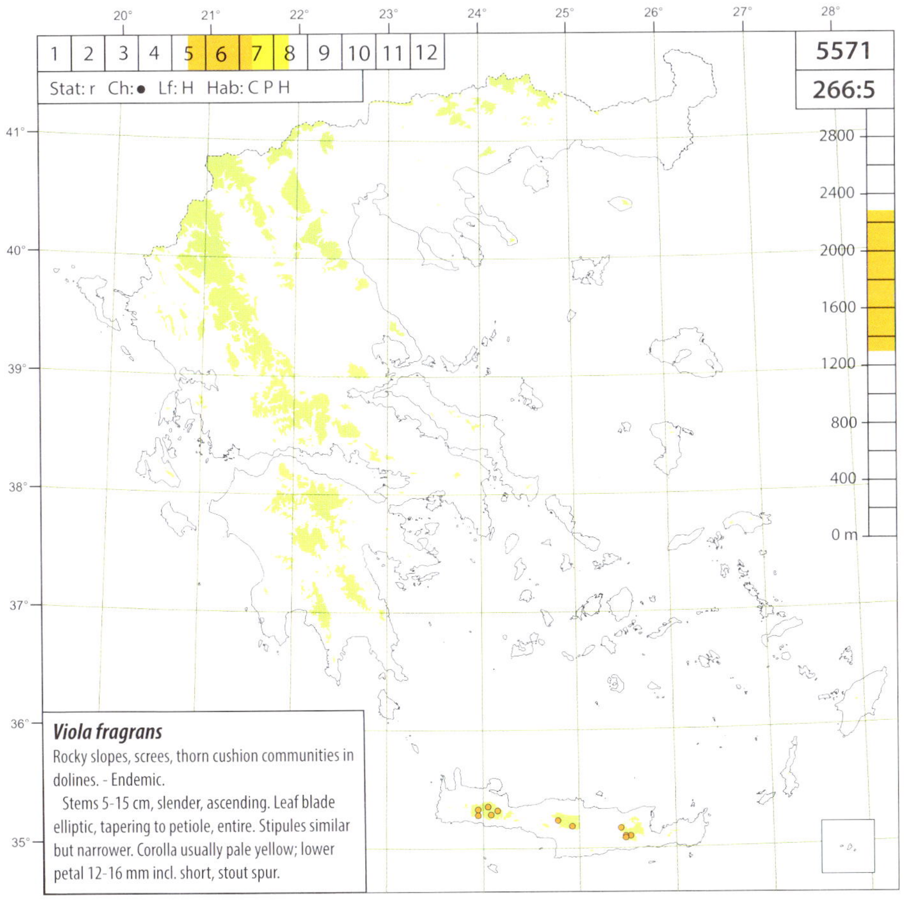

Viola fragrans

Rocky slopes, screes, thorn cushion communities in dolines. - Endemic.

Stems 5-15 cm, slender, ascending. Leaf blade elliptic, tapering to petiole, entire. Stipules similar but narrower. Corolla usually pale yellow; lower petal 12-16 mm incl. short, stout spur.

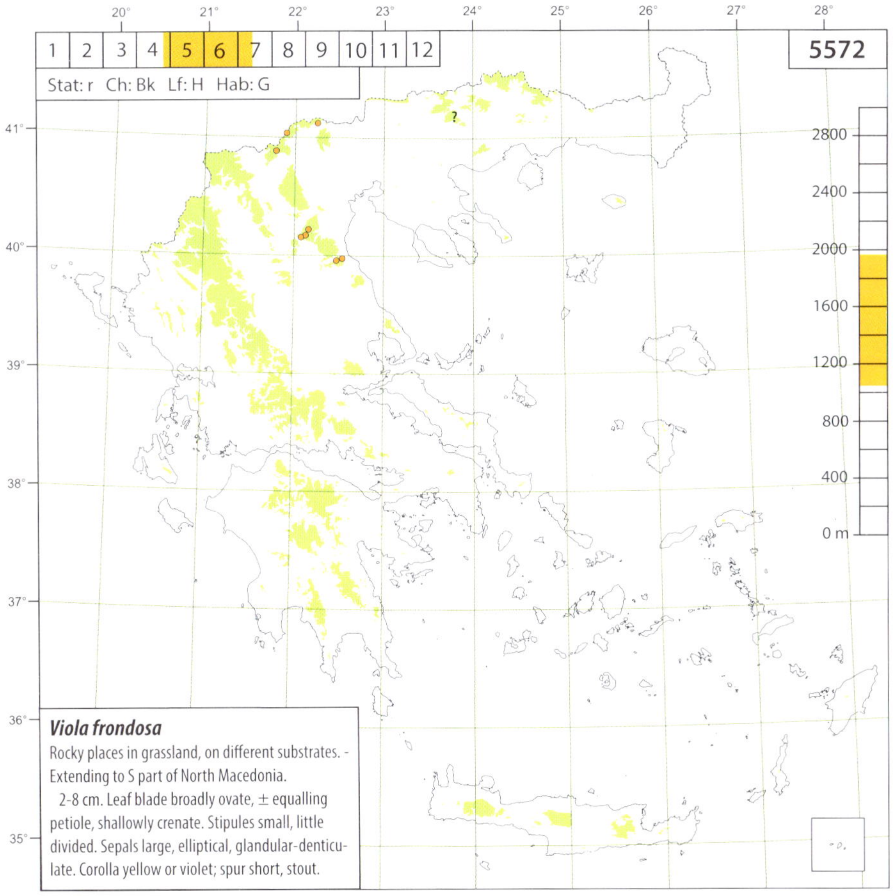

Viola frondosa

Rocky places in grassland, on different substrates. - Extending to S part of North Macedonia.

2-8 cm. Leaf blade broadly ovate, ± equalling petiole, shallowly crenate. Stipules small, little divided. Sepals large, elliptical, glandular-denticulate. Corolla yellow or violet; spur short, stout.

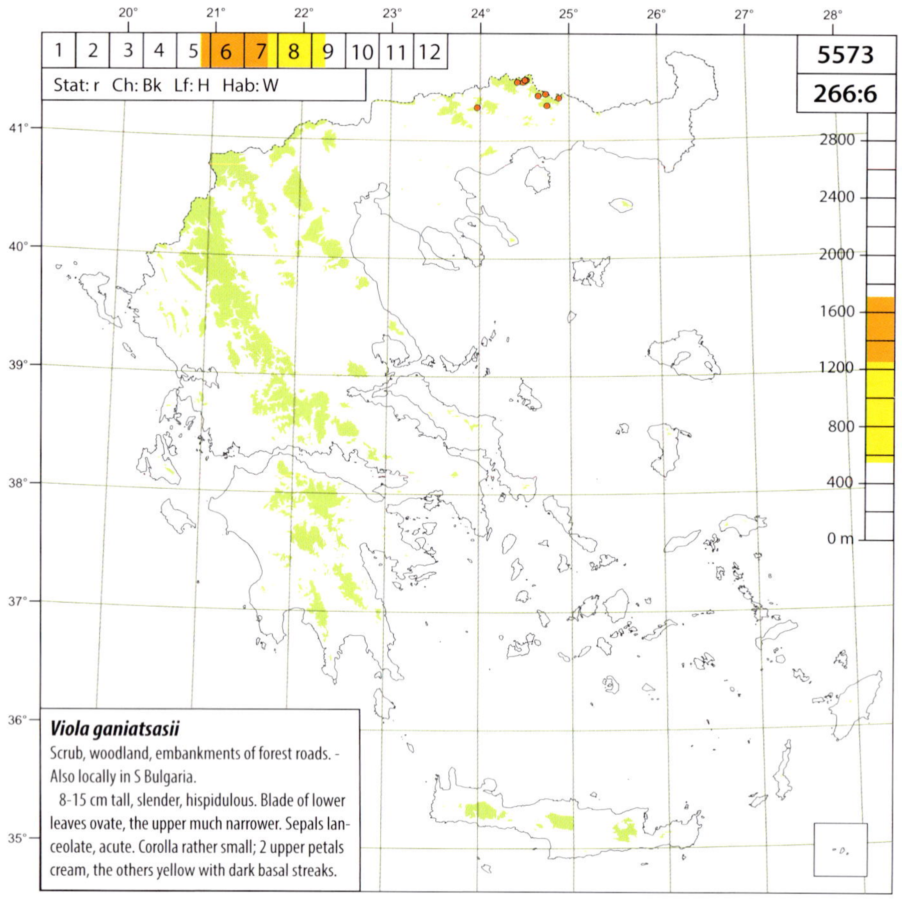

Viola ganiatsasii

Scrub, woodland, embankments of forest roads. - Also locally in S Bulgaria.

8-15 cm tall, slender, hispidulous. Blade of lower leaves ovate, the upper much narrower. Sepals lanceolate, acute. Corolla rather small; 2 upper petals cream, the others yellow with dark basal streaks.

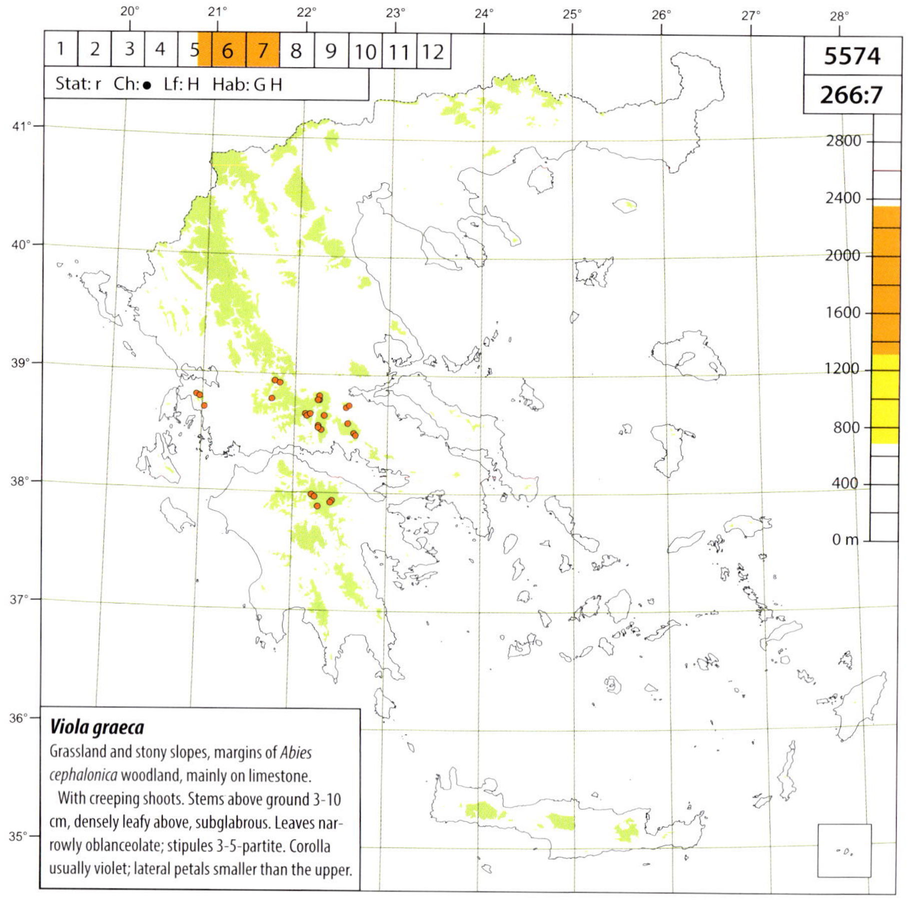

Viola graeca

Grassland and stony slopes, margins of *Abies cephalonica* woodland, mainly on limestone.

With creeping shoots. Stems above ground 3-10 cm, densely leafy above, subglabrous. Leaves narrowly oblanceolate; stipules 3-5-partite. Corolla usually violet; lateral petals smaller than the upper.

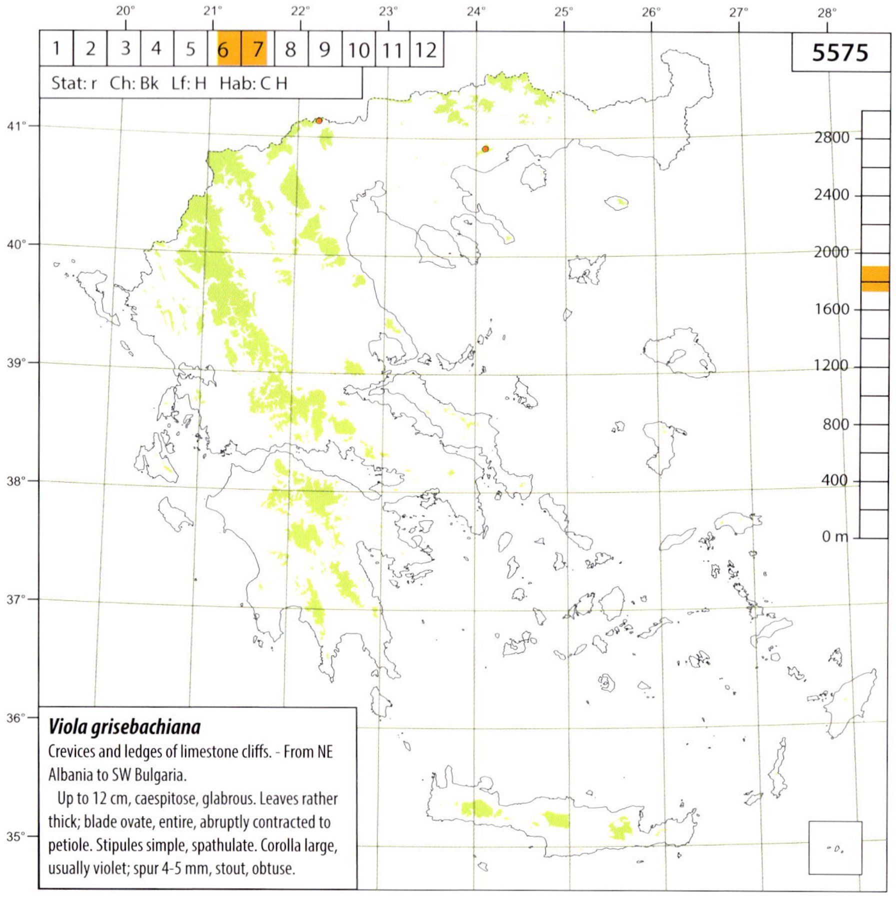

Viola grisebachiana

Crevices and ledges of limestone cliffs. - From NE Albania to SW Bulgaria.

Up to 12 cm, caespitose, glabrous. Leaves rather thick; blade ovate, entire, abruptly contracted to petiole. Stipules simple, spathulate. Corolla large, usually violet; spur 4-5 mm, stout, obtuse.

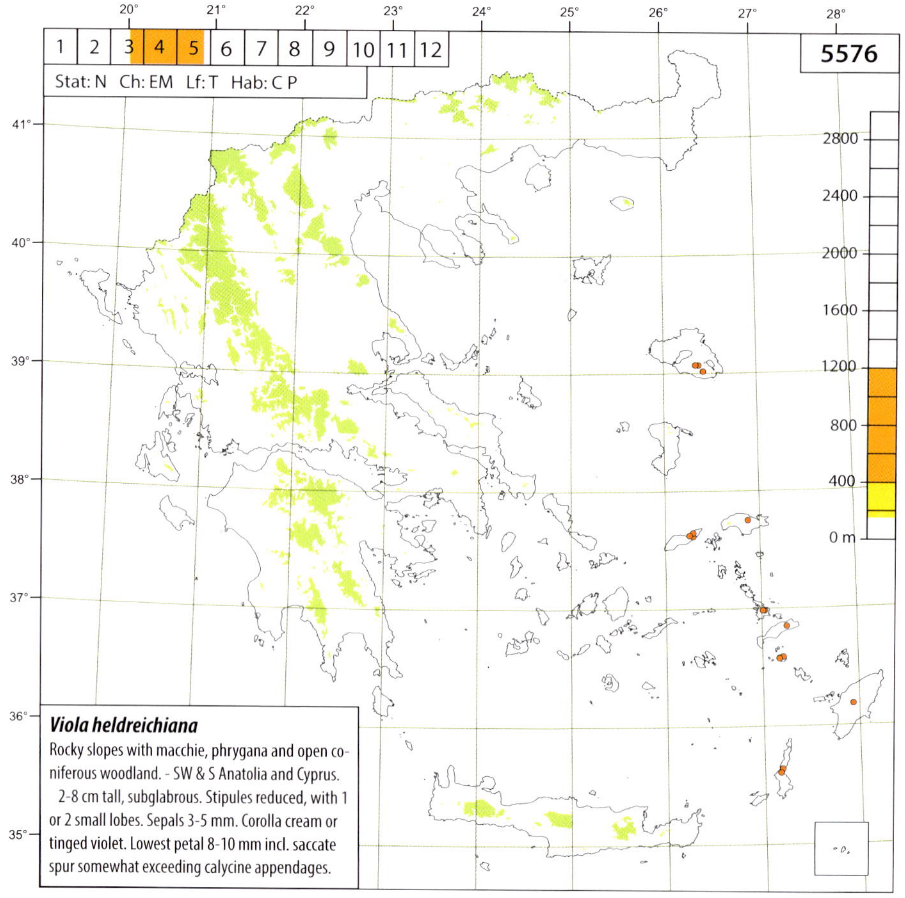

Viola heldreichiana

Rocky slopes with macchie, phrygana and open coniferous woodland. - SW & S Anatolia and Cyprus.

2-8 cm tall, subglabrous. Stipules reduced, with 1 or 2 small lobes. Sepals 3-5 mm. Corolla cream or tinged violet. Lowest petal 8-10 mm incl. saccate spur somewhat exceeding calycine appendages.

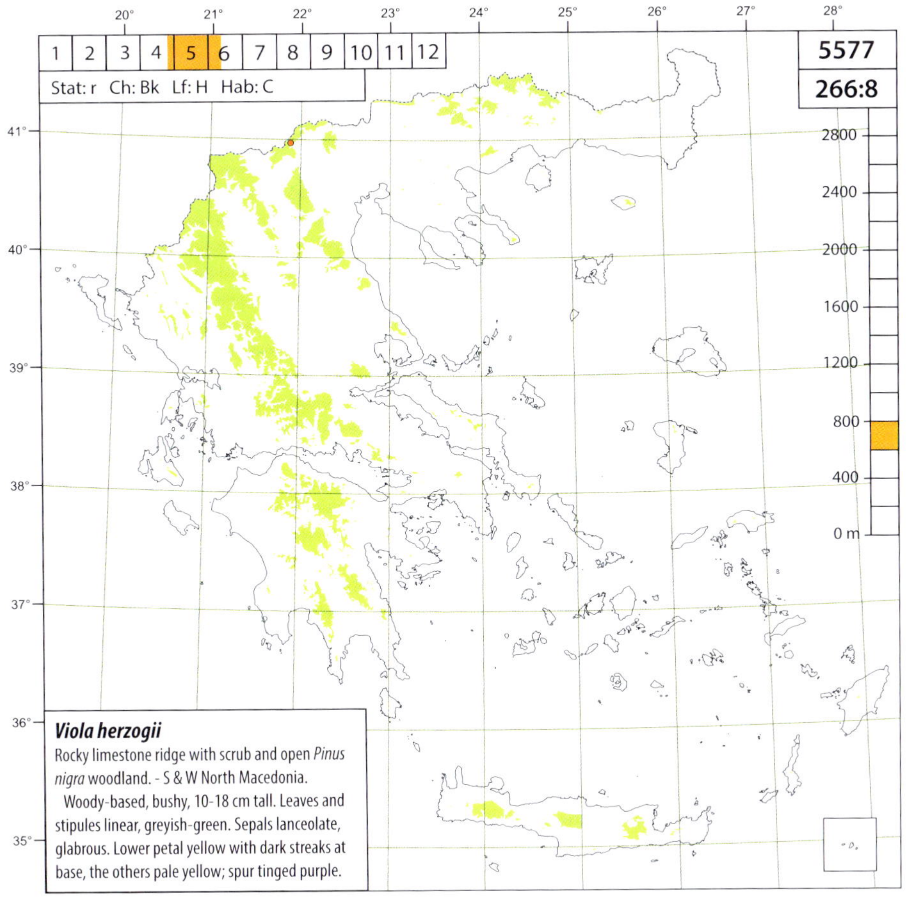

Viola herzogii
Rocky limestone ridge with scrub and open *Pinus nigra* woodland. - S & W North Macedonia.
Woody-based, bushy, 10-18 cm tall. Leaves and stipules linear, greyish-green. Sepals lanceolate, glabrous. Lower petal yellow with dark streaks at base, the others pale yellow; spur tinged purple.

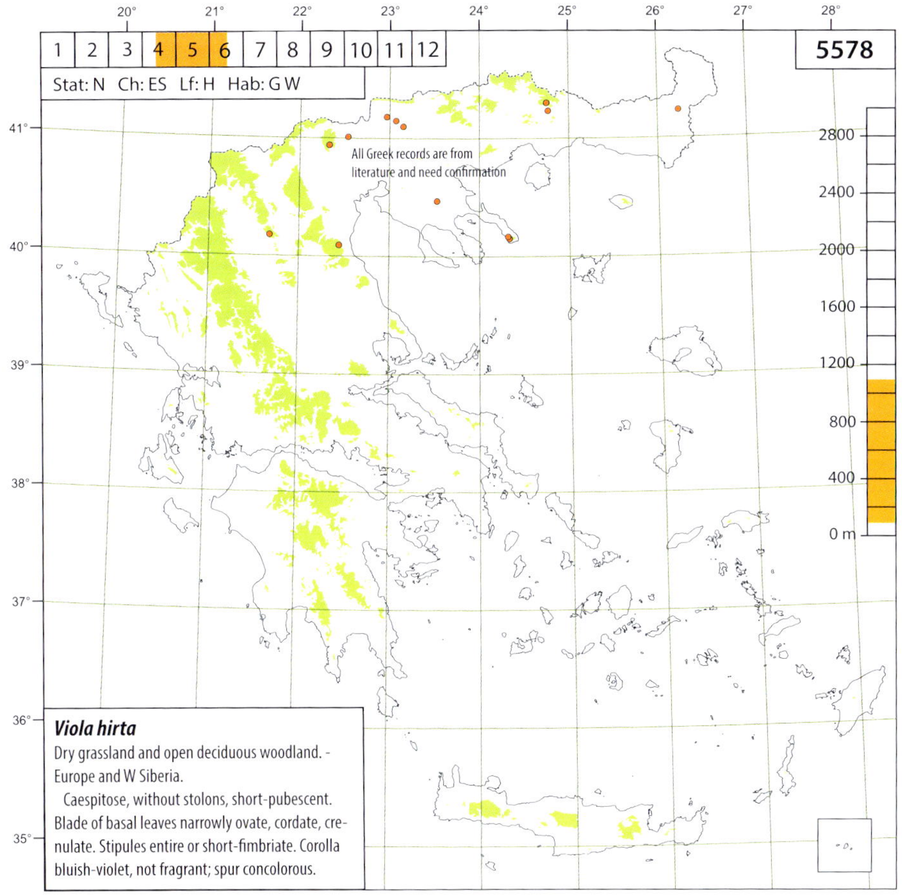

Viola hirta
Dry grassland and open deciduous woodland. - Europe and W Siberia.
Caespitose, without stolons, short-pubescent. Blade of basal leaves narrowly ovate, cordate, crenulate. Stipules entire or short-fimbriate. Corolla bluish-violet, not fragrant; spur concolorous.

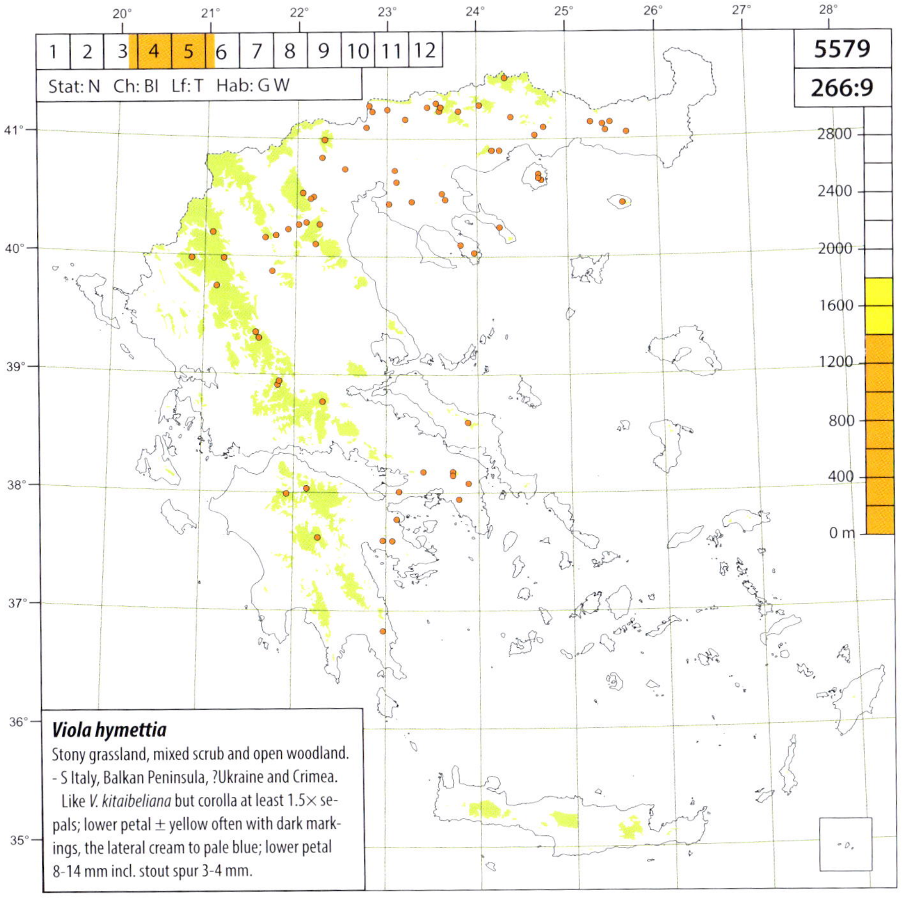

Viola hymettia
Stony grassland, mixed scrub and open woodland. - S Italy, Balkan Peninsula, ?Ukraine and Crimea.
Like *V. kitaibeliana* but corolla at least 1.5× sepals; lower petal ± yellow often with dark markings, the lateral cream to pale blue; lower petal 8-14 mm incl. stout spur 3-4 mm.

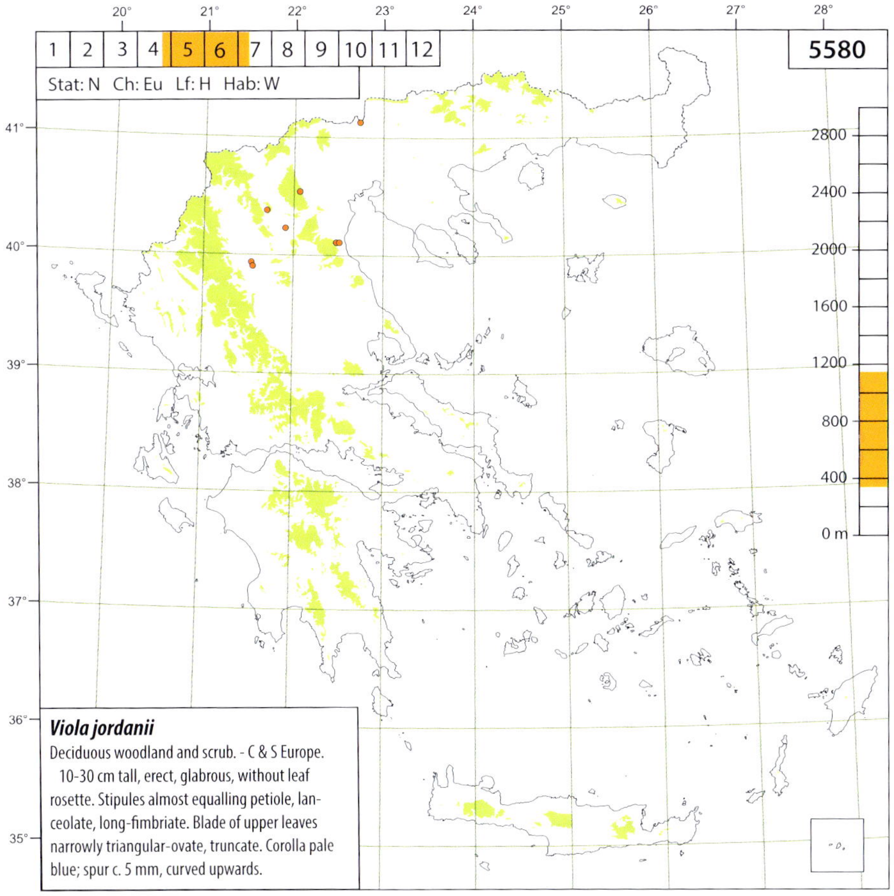

Viola jordanii
Deciduous woodland and scrub. - C & S Europe.
10-30 cm tall, erect, glabrous, without leaf rosette. Stipules almost equalling petiole, lanceolate, long-fimbriate. Blade of upper leaves narrowly triangular-ovate, truncate. Corolla pale blue; spur c. 5 mm, curved upwards.

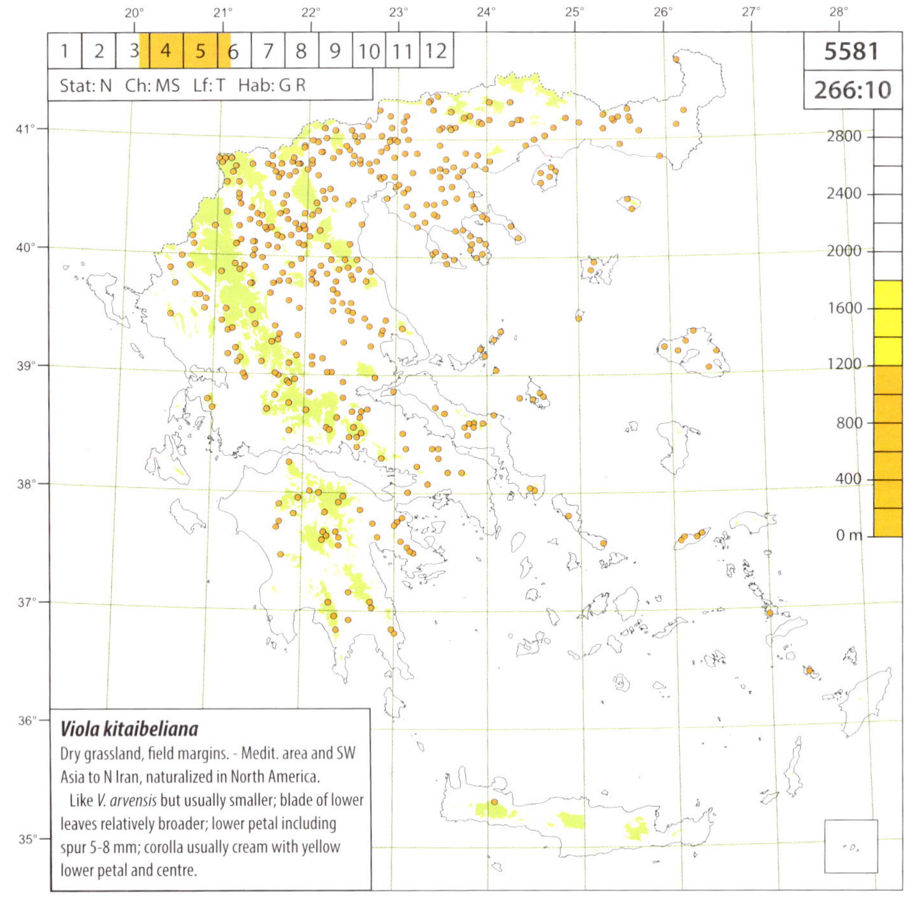

Viola kitaibeliana

Dry grassland, field margins. - Medit. area and SW Asia to N Iran, naturalized in North America.

Like *V. arvensis* but usually smaller; blade of lower leaves relatively broader; lower petal including spur 5-8 mm; corolla usually cream with yellow lower petal and centre.

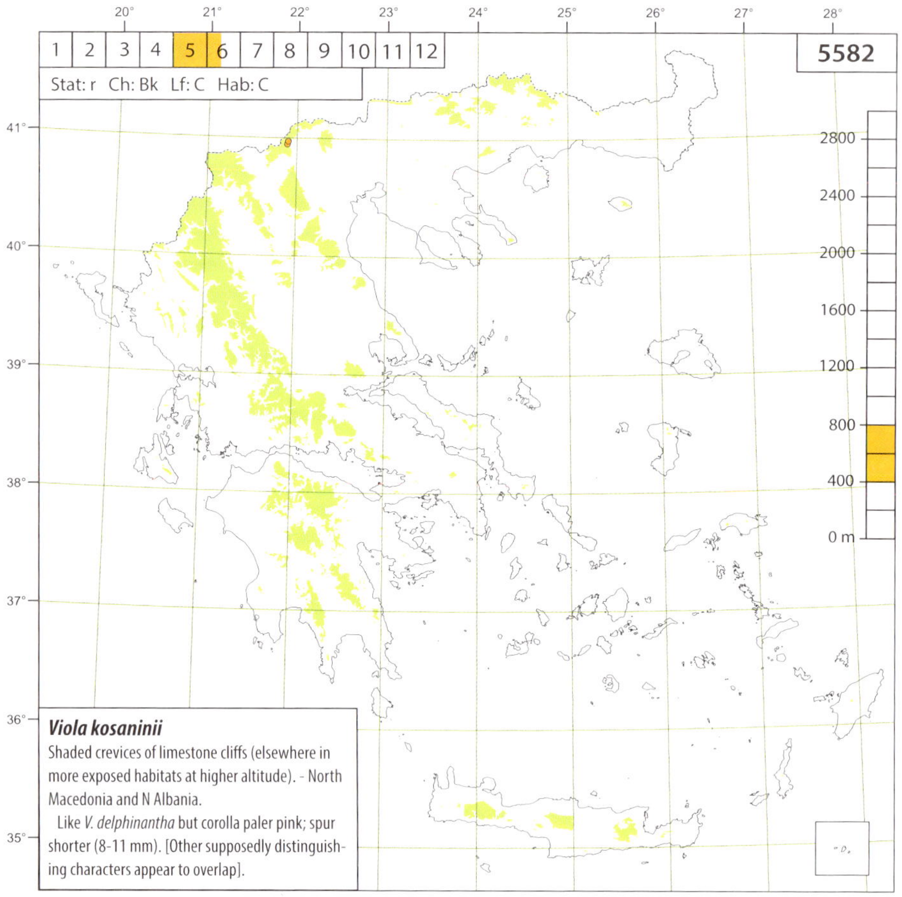

Viola kosaninii

Shaded crevices of limestone cliffs (elsewhere in more exposed habitats at higher altitude). - North Macedonia and N Albania.

Like *V. delphinantha* but corolla paler pink; spur shorter (8-11 mm). [Other supposedly distinguishing characters appear to overlap].

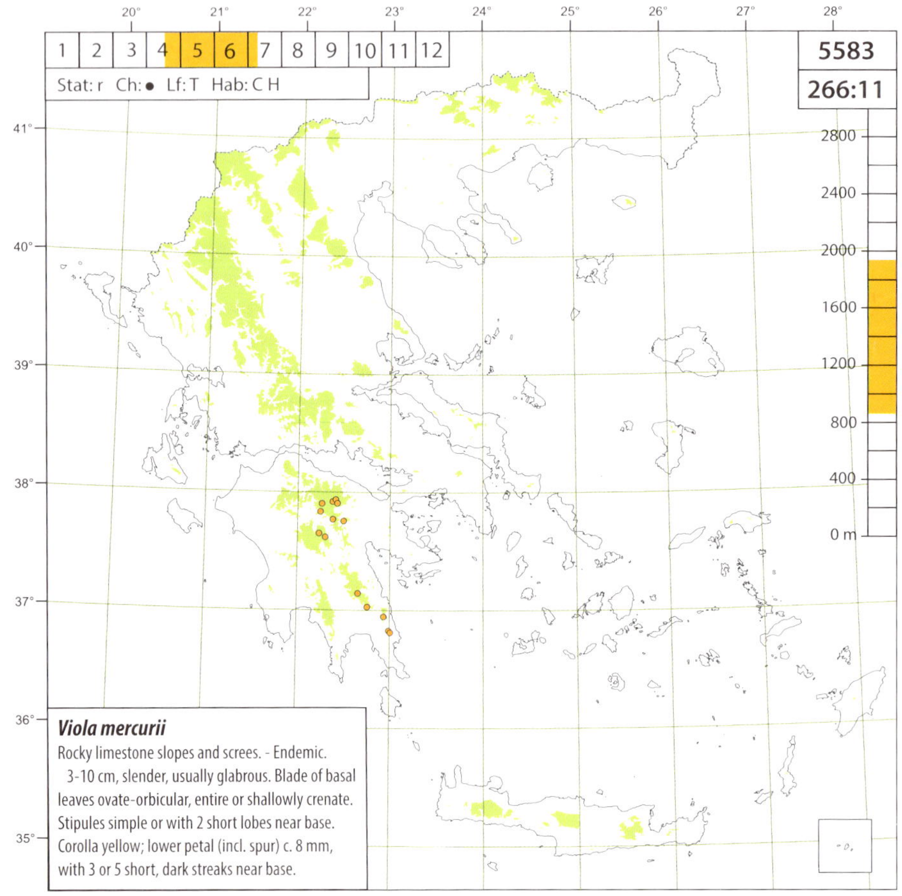

Viola mercurii

Rocky limestone slopes and screes. - Endemic.

3-10 cm, slender, usually glabrous. Blade of basal leaves ovate-orbicular, entire or shallowly crenate. Stipules simple or with 2 short lobes near base. Corolla yellow; lower petal (incl. spur) c. 8 mm, with 3 or 5 short, dark streaks near base.

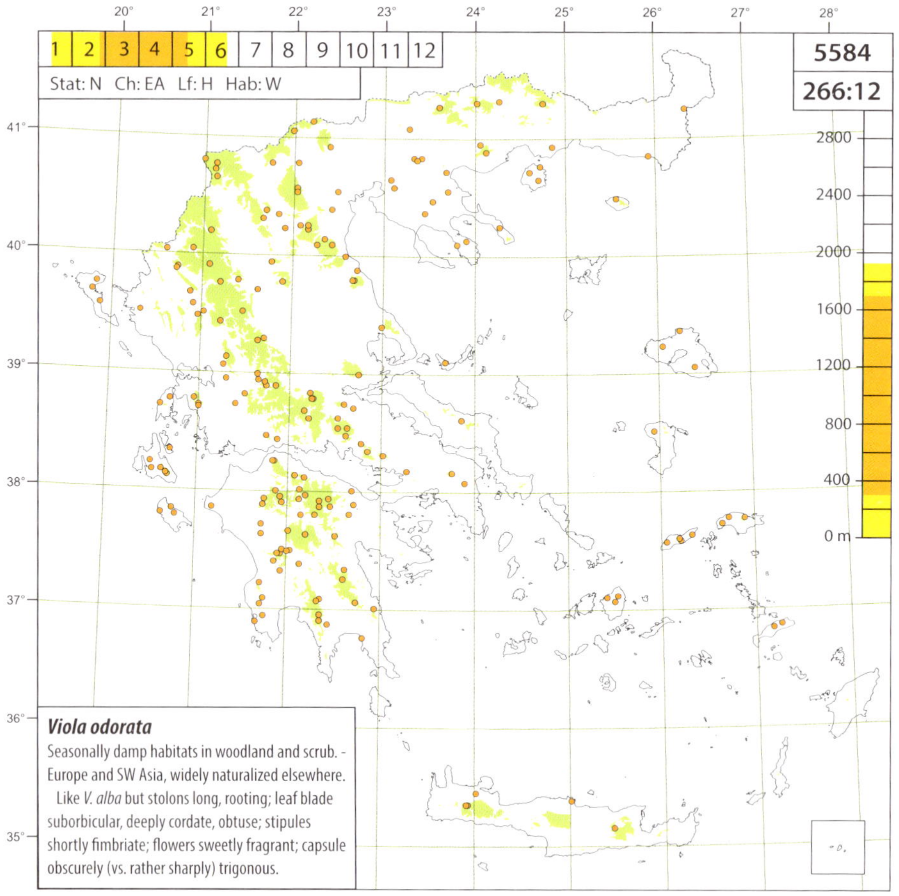

Viola odorata

Seasonally damp habitats in woodland and scrub. - Europe and SW Asia, widely naturalized elsewhere.

Like *V. alba* but stolons long, rooting; leaf blade suborbicular, deeply cordate, obtuse; stipules shortly fimbriate; flowers sweetly fragrant; capsule obscurely (vs. rather sharply) trigonous.

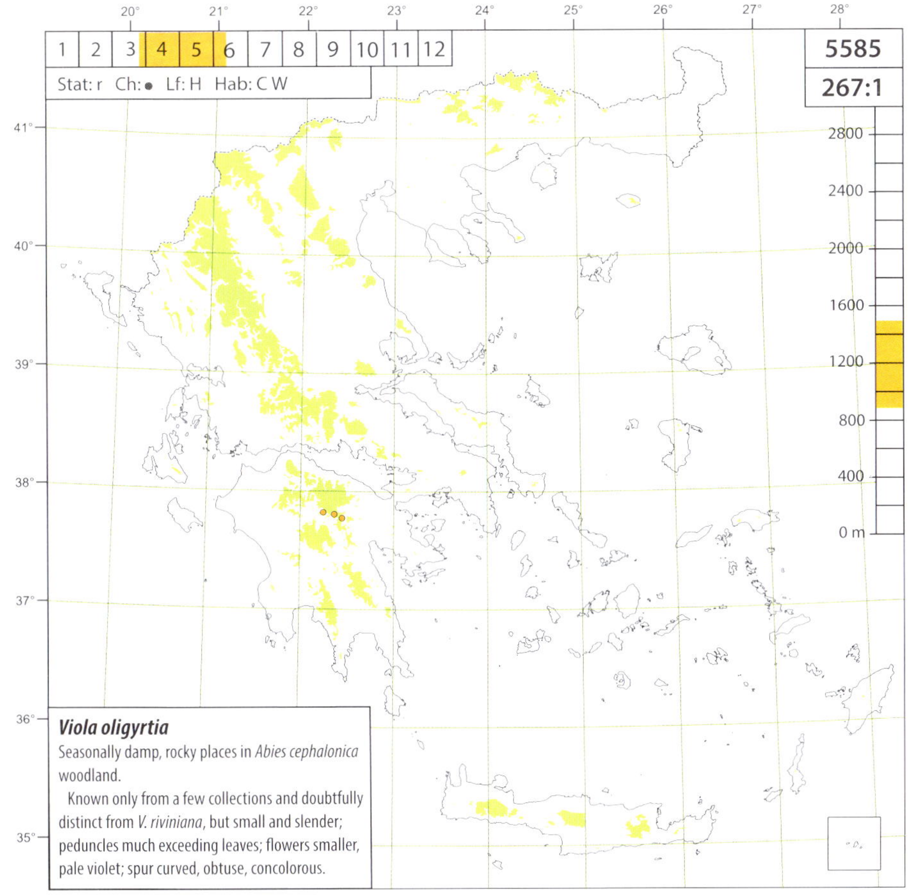

Viola oligyrtia
Seasonally damp, rocky places in *Abies cephalonica* woodland.
Known only from a few collections and doubtfully distinct from *V. riviniana*, but small and slender; peduncles much exceeding leaves; flowers smaller, pale violet; spur curved, obtuse, concolorous.

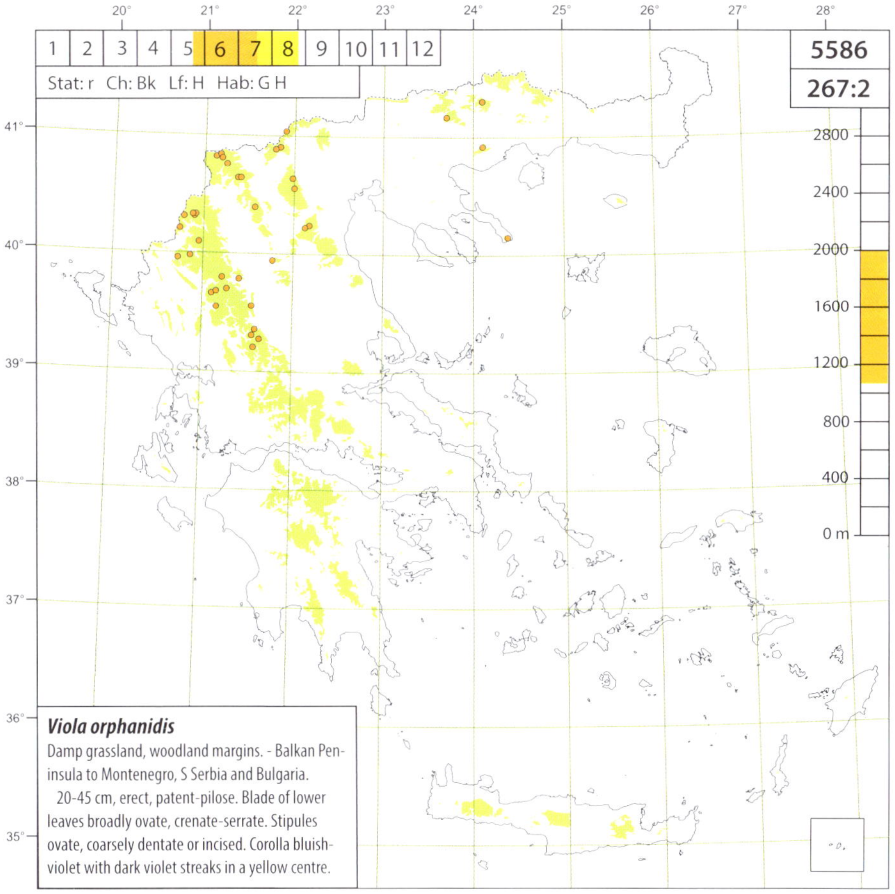

Viola orphanidis
Damp grassland, woodland margins. - Balkan Peninsula to Montenegro, S Serbia and Bulgaria.
20-45 cm, erect, patent-pilose. Blade of lower leaves broadly ovate, crenate-serrate. Stipules ovate, coarsely dentate or incised. Corolla bluish-violet with dark violet streaks in a yellow centre.

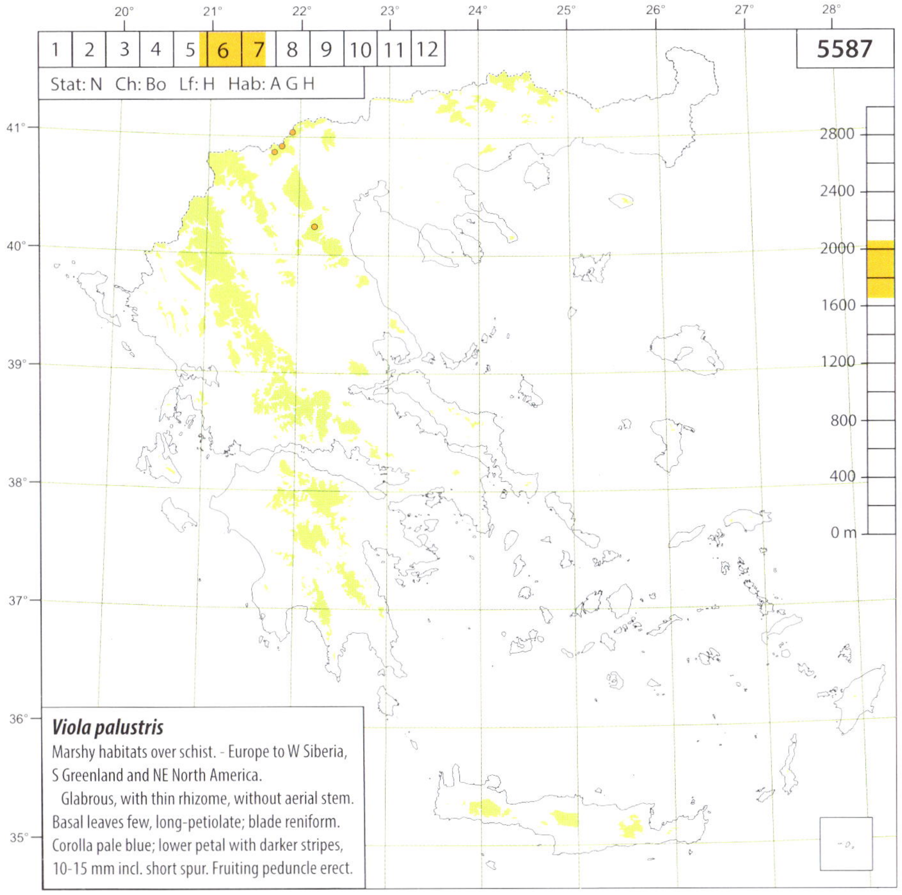

Viola palustris
Marshy habitats over schist. - Europe to W Siberia, S Greenland and NE North America.
Glabrous, with thin rhizome, without aerial stem. Basal leaves few, long-petiolate; blade reniform. Corolla pale blue; lower petal with darker stripes, 10-15 mm incl. short spur. Fruiting peduncle erect.

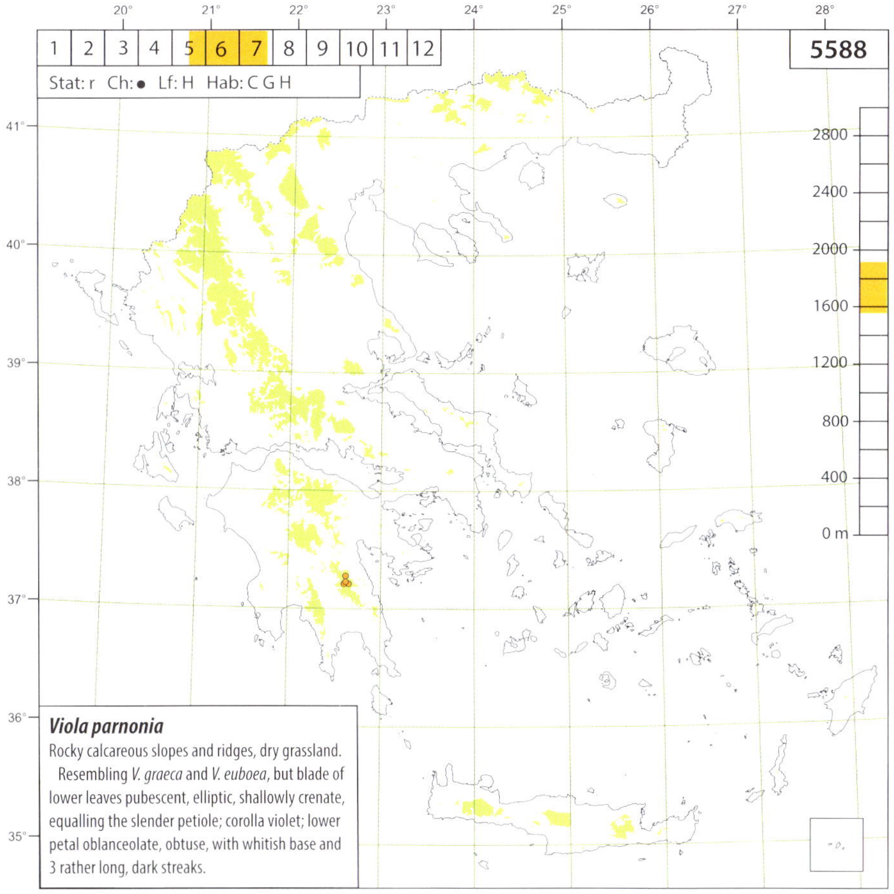

Viola parnonia
Rocky calcareous slopes and ridges, dry grassland.
Resembling *V. graeca* and *V. euboea*, but blade of lower leaves pubescent, elliptic, shallowly crenate, equalling the slender petiole; corolla violet; lower petal oblanceolate, obtuse, with whitish base and 3 rather long, dark streaks.

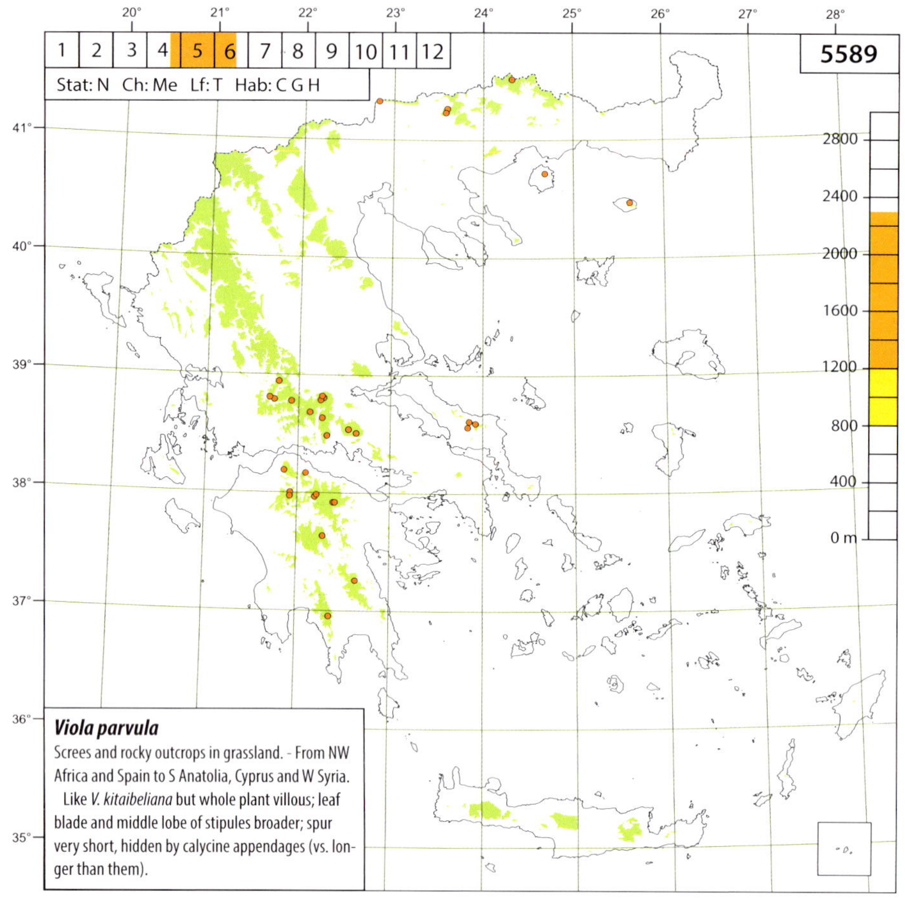

Viola parvula

Screes and rocky outcrops in grassland. - From NW Africa and Spain to S Anatolia, Cyprus and W Syria.

Like *V. kitaibeliana* but whole plant villous; leaf blade and middle lobe of stipules broader; spur very short, hidden by calycine appendages (vs. longer than them).

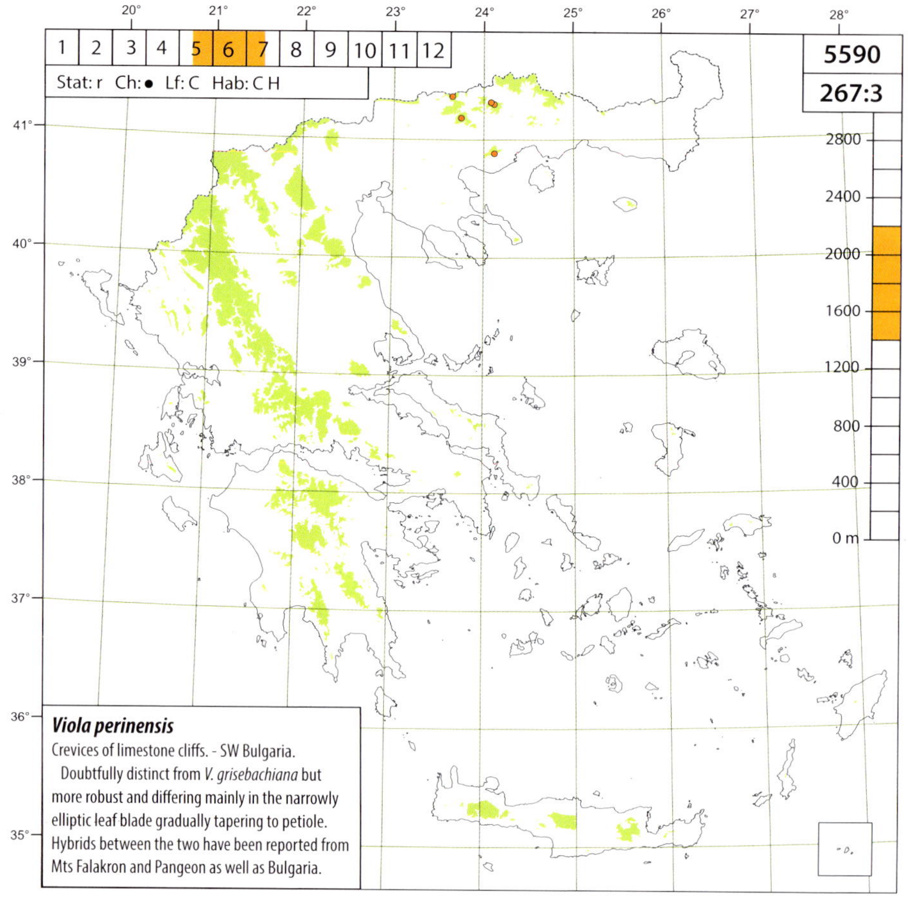

Viola perinensis

Crevices of limestone cliffs. - SW Bulgaria.

Doubtfully distinct from *V. grisebachiana* but more robust and differing mainly in the narrowly elliptic leaf blade gradually tapering to petiole. Hybrids between the two have been reported from Mts Falakron and Pangeon as well as Bulgaria.

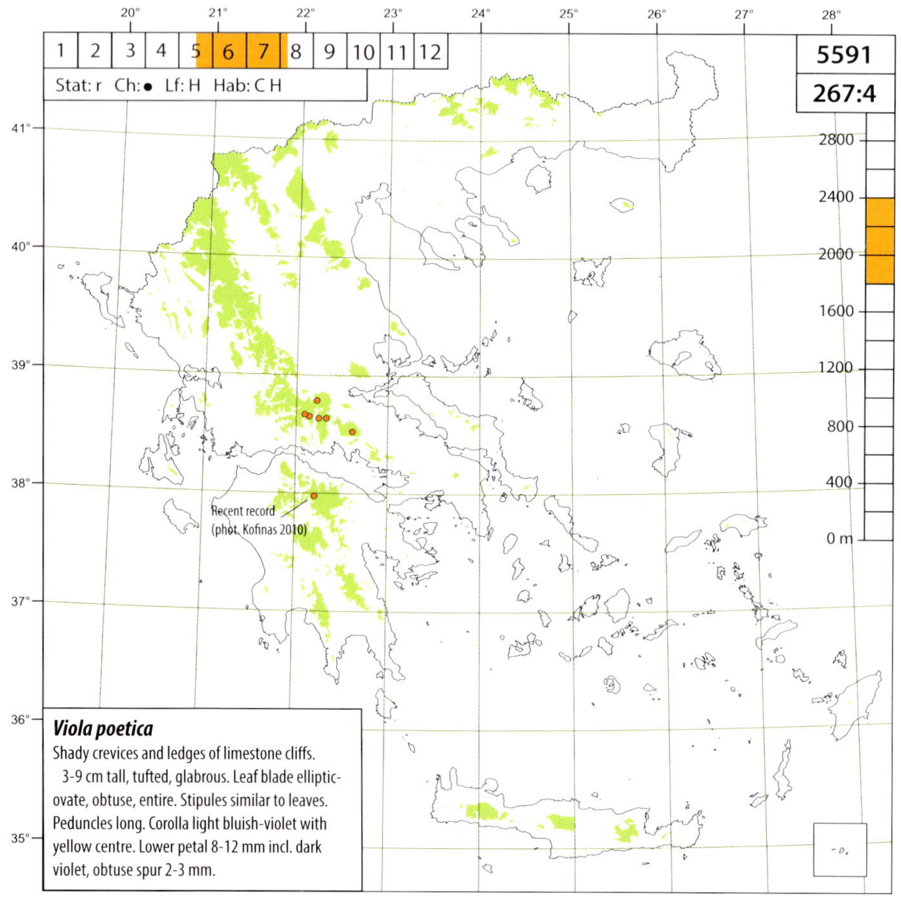

Viola poetica

Shady crevices and ledges of limestone cliffs.

3-9 cm tall, tufted, glabrous. Leaf blade elliptic-ovate, obtuse, entire. Stipules similar to leaves. Peduncles long. Corolla light bluish-violet with yellow centre. Lower petal 8-12 mm incl. dark violet, obtuse spur 2-3 mm.

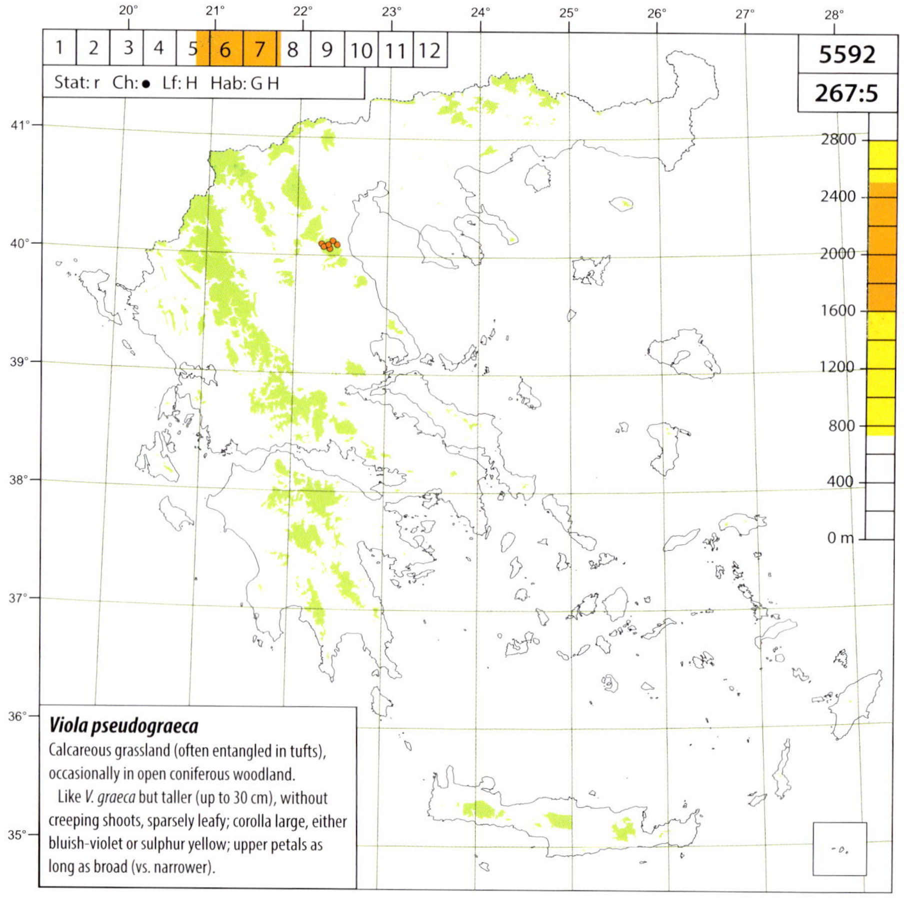

Viola pseudograeca

Calcareous grassland (often entangled in tufts), occasionally in open coniferous woodland.

Like *V. graeca* but taller (up to 30 cm), without creeping shoots, sparsely leafy; corolla large, either bluish-violet or sulphur yellow; upper petals as long as broad (vs. narrower).

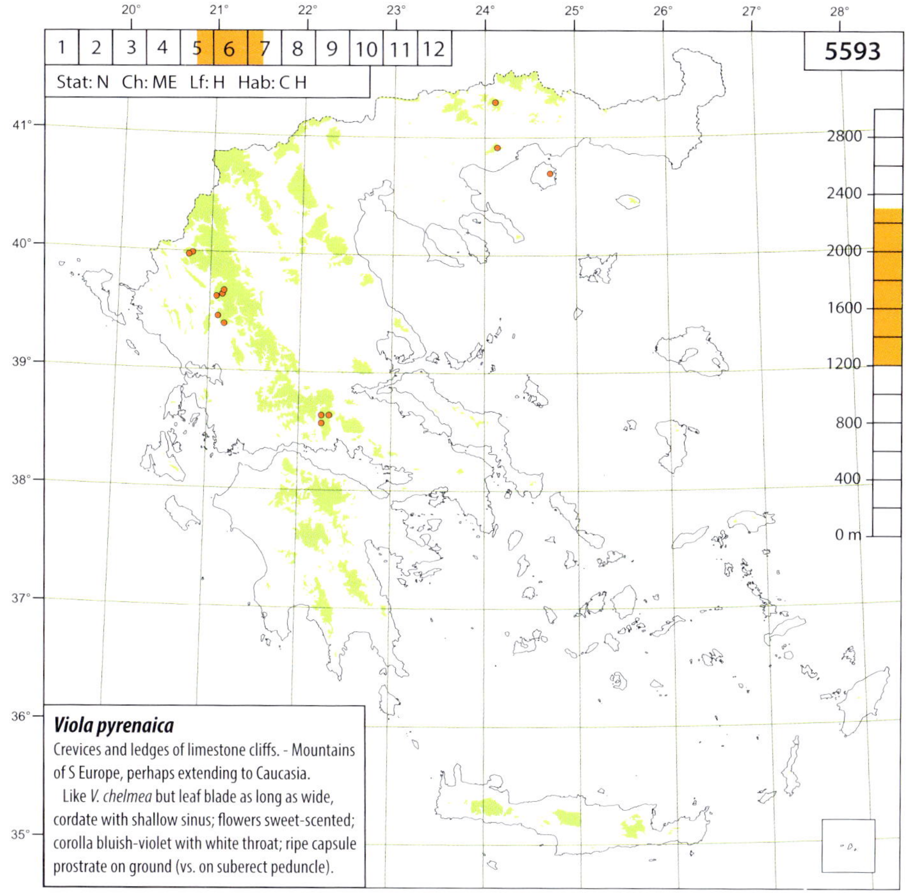

Viola pyrenaica
Crevices and ledges of limestone cliffs. - Mountains of S Europe, perhaps extending to Caucasia.
Like *V. chelmea* but leaf blade as long as wide, cordate with shallow sinus; flowers sweet-scented; corolla bluish-violet with white throat; ripe capsule prostrate on ground (vs. on suberect peduncle).

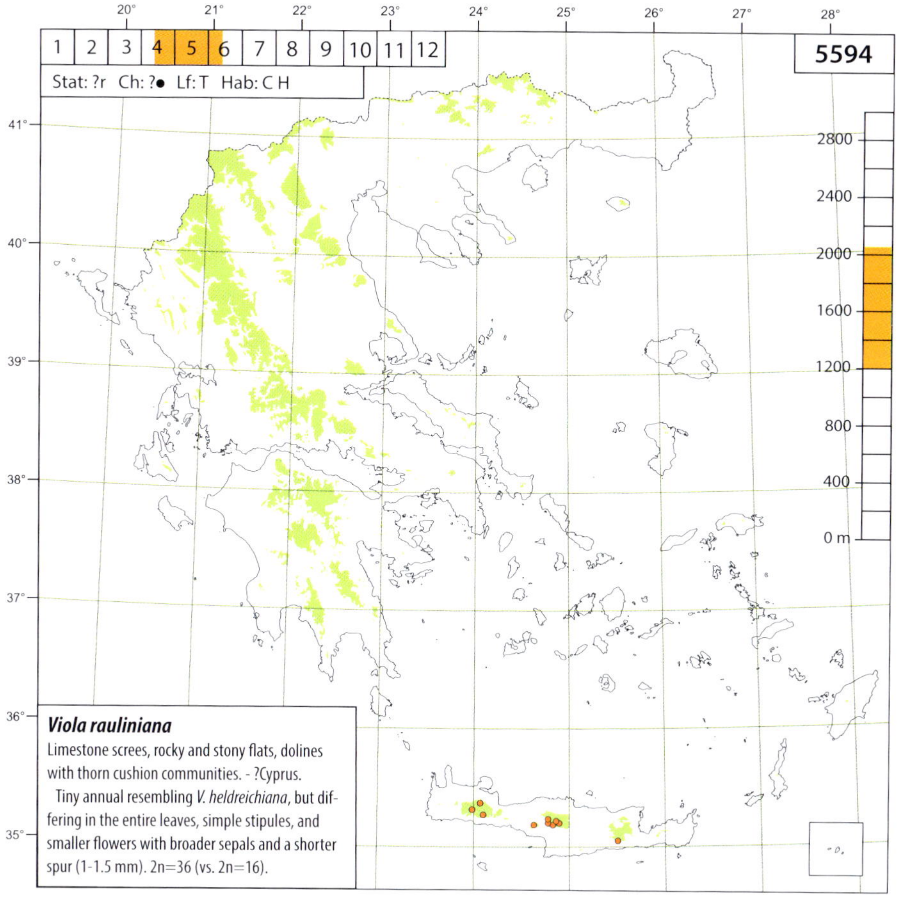

Viola rauliniana
Limestone screes, rocky and stony flats, dolines with thorn cushion communities. - ?Cyprus.
Tiny annual resembling *V. heldreichiana*, but differing in the entire leaves, simple stipules, and smaller flowers with broader sepals and a shorter spur (1-1.5 mm). 2n=36 (vs. 2n=16).

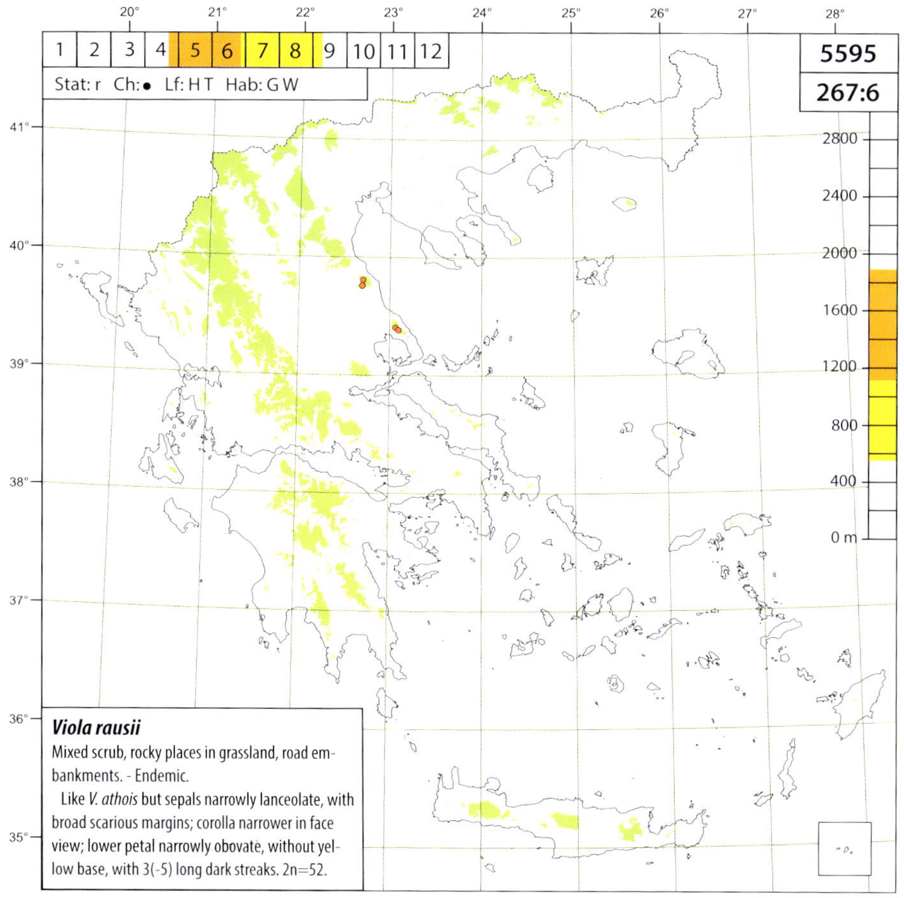

Viola rausii
Mixed scrub, rocky places in grassland, road embankments. - Endemic.
Like *V. athois* but sepals narrowly lanceolate, with broad scarious margins; corolla narrower in face view; lower petal narrowly obovate, without yellow base, with 3(-5) long dark streaks. 2n=52.

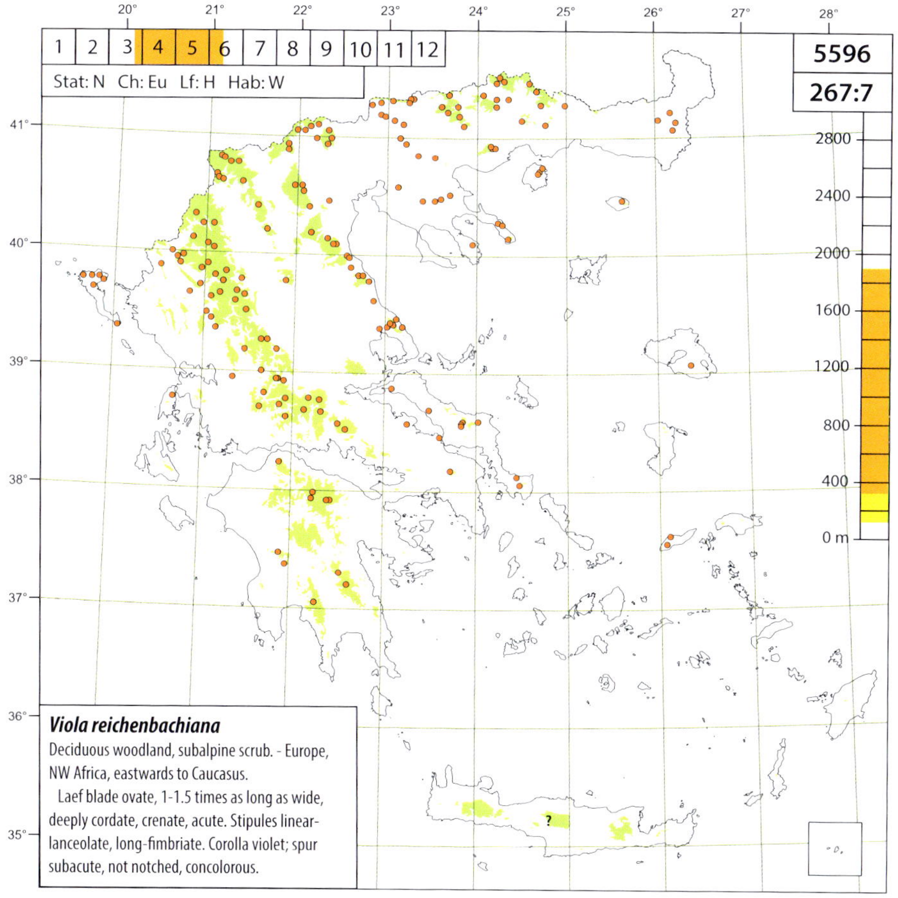

Viola reichenbachiana
Deciduous woodland, subalpine scrub. - Europe, NW Africa, eastwards to Caucasus.
Laef blade ovate, 1-1.5 times as long as wide, deeply cordate, crenate, acute. Stipules linear-lanceolate, long-fimbriate. Corolla violet; spur subacute, not notched, concolorous.

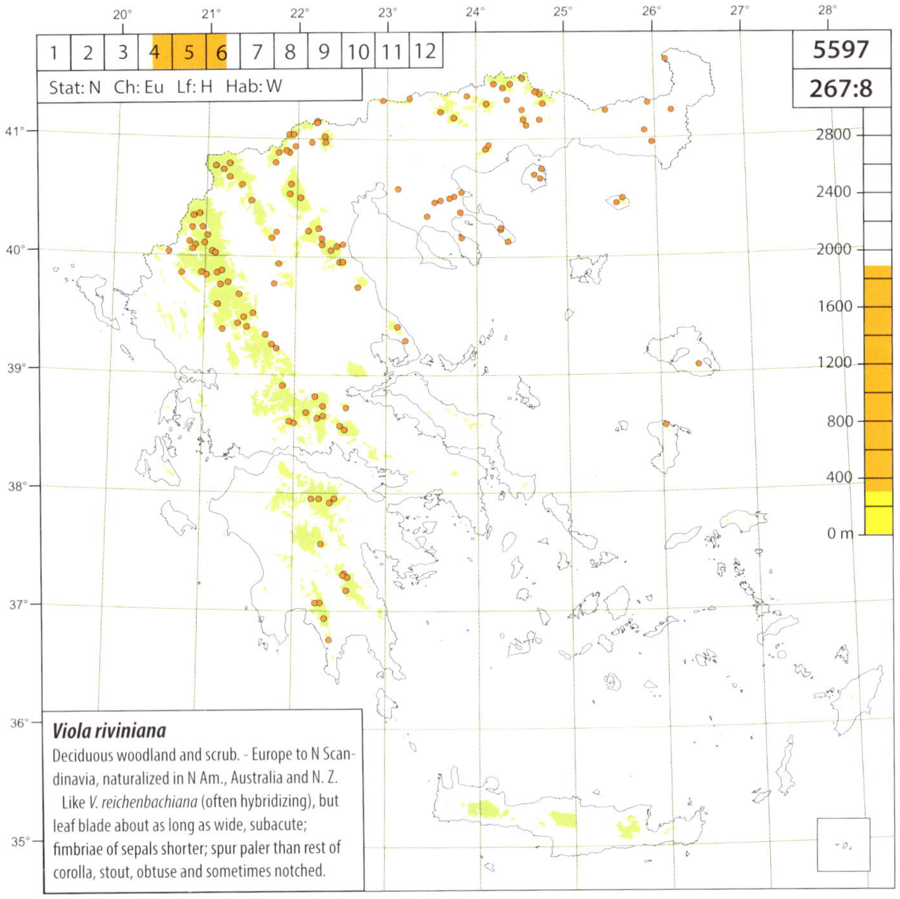

Viola riviniana

Deciduous woodland and scrub. - Europe to N Scandinavia, naturalized in N Am., Australia and N. Z.

Like *V. reichenbachiana* (often hybridizing), but leaf blade about as long as wide, subacute; fimbriae of sepals shorter; spur paler than rest of corolla, stout, obtuse and sometimes notched.

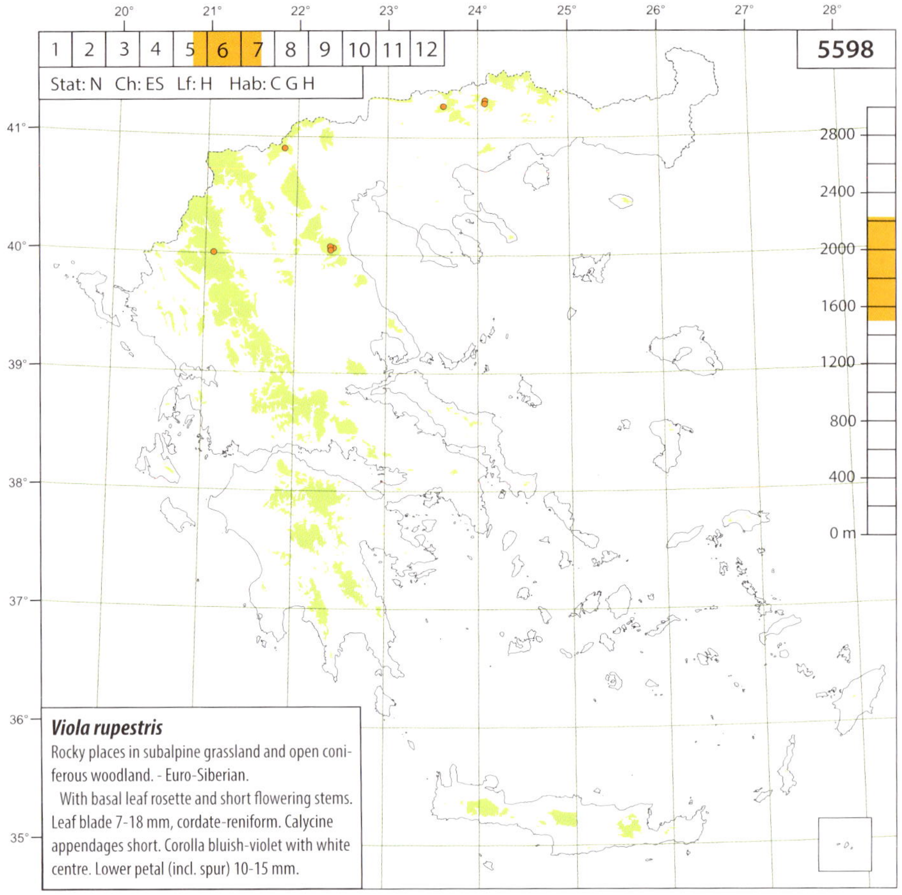

Viola rupestris

Rocky places in subalpine grassland and open coniferous woodland. - Euro-Siberian.

With basal leaf rosette and short flowering stems. Leaf blade 7-18 mm, cordate-reniform. Calycine appendages short. Corolla bluish-violet with white centre. Lower petal (incl. spur) 10-15 mm.

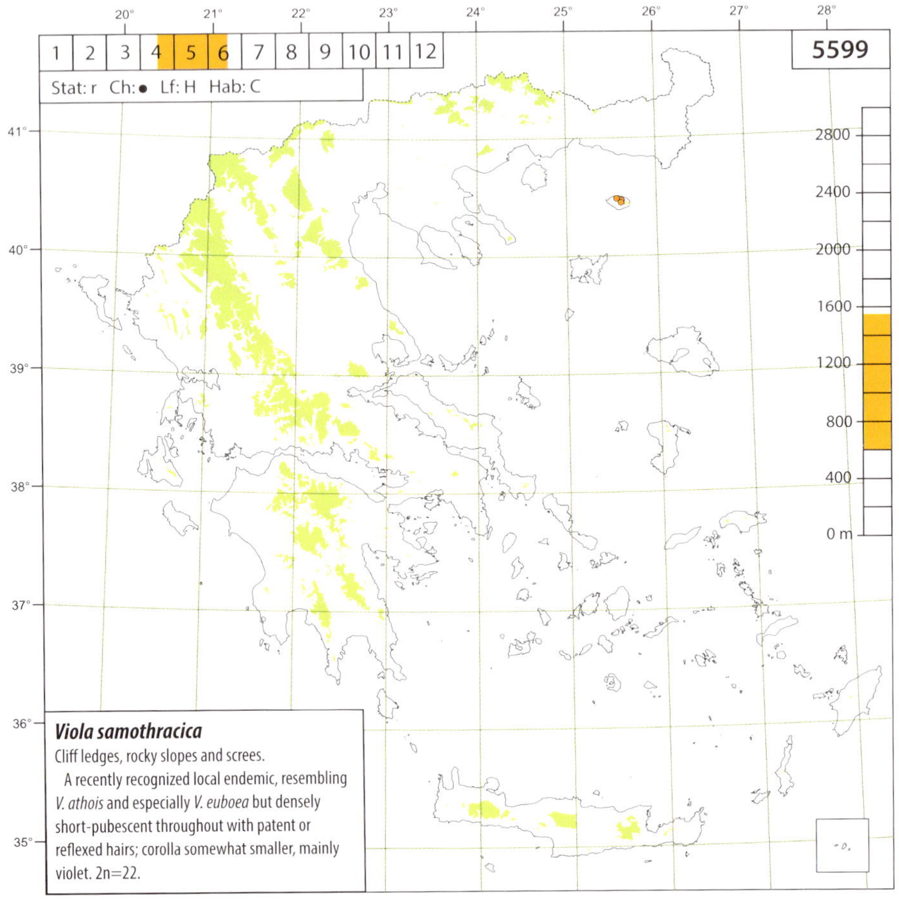

Viola samothracica

Cliff ledges, rocky slopes and screes.

A recently recognized local endemic, resembling *V. athois* and especially *V. euboea* but densely short-pubescent throughout with patent or reflexed hairs; corolla somewhat smaller, mainly violet. 2n=22.

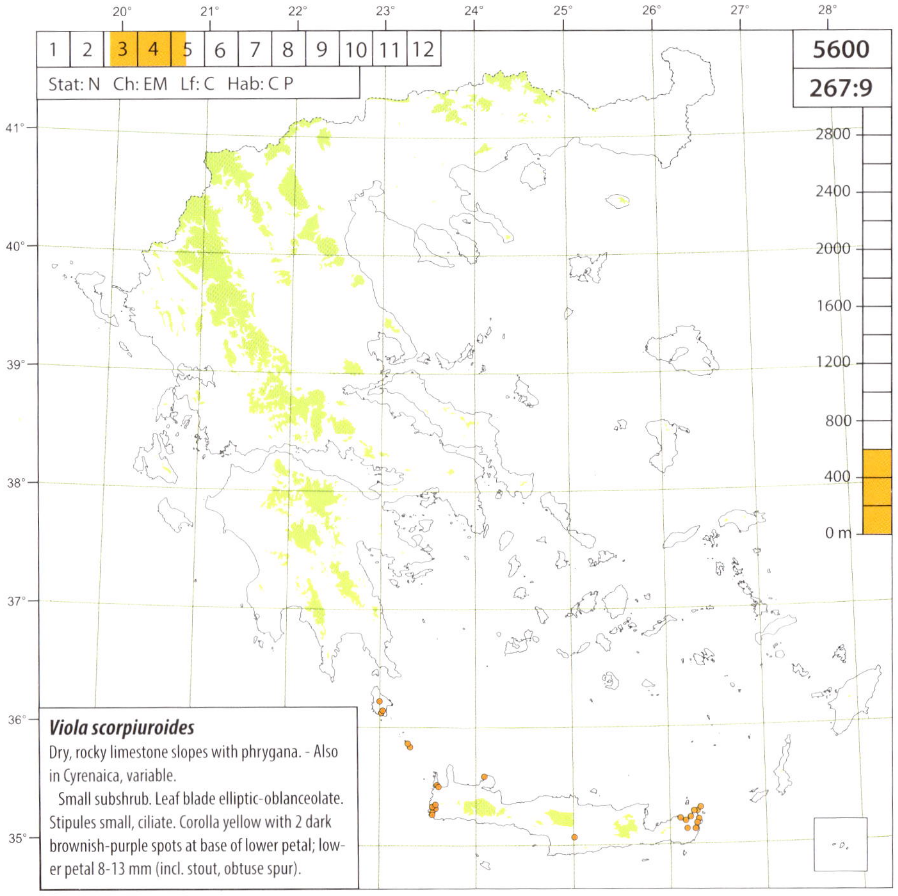

Viola scorpiuroides

Dry, rocky limestone slopes with phrygana. - Also in Cyrenaica, variable.

Small subshrub. Leaf blade elliptic-oblanceolate. Stipules small, ciliate. Corolla yellow with 2 dark brownish-purple spots at base of lower petal; lower petal 8-13 mm (incl. stout, obtuse spur).

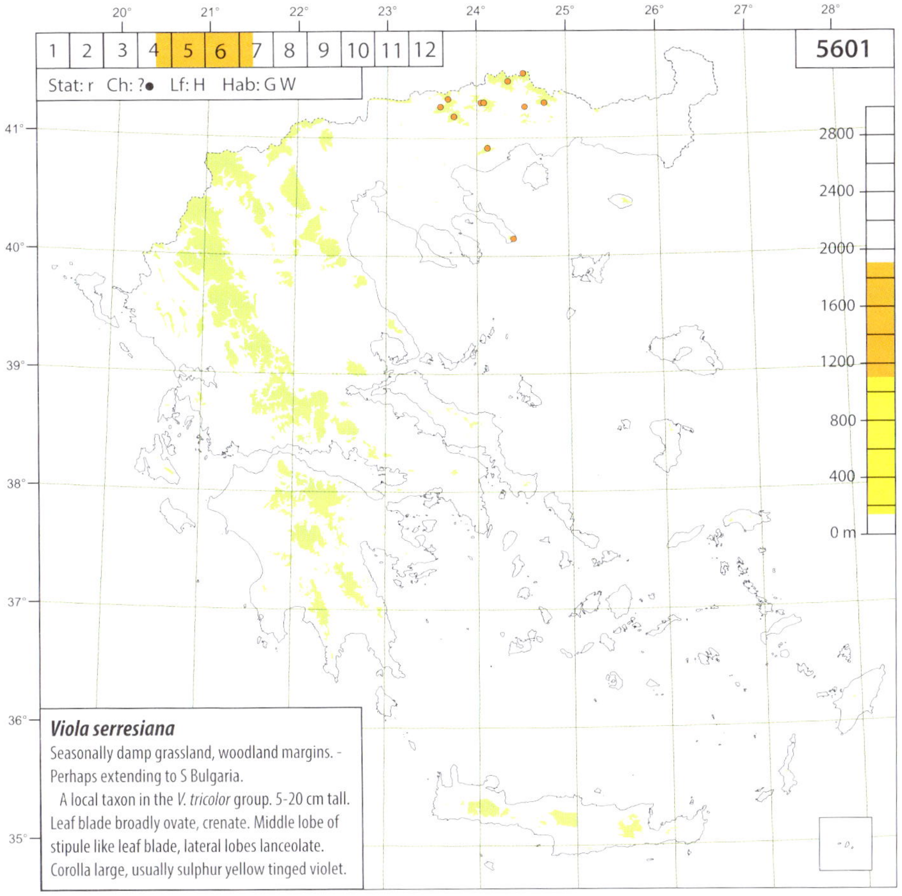

Viola serresiana

Seasonally damp grassland, woodland margins. - Perhaps extending to S Bulgaria.

A local taxon in the *V. tricolor* group. 5-20 cm tall. Leaf blade broadly ovate, crenate. Middle lobe of stipule like leaf blade, lateral lobes lanceolate. Corolla large, usually sulphur yellow tinged violet.

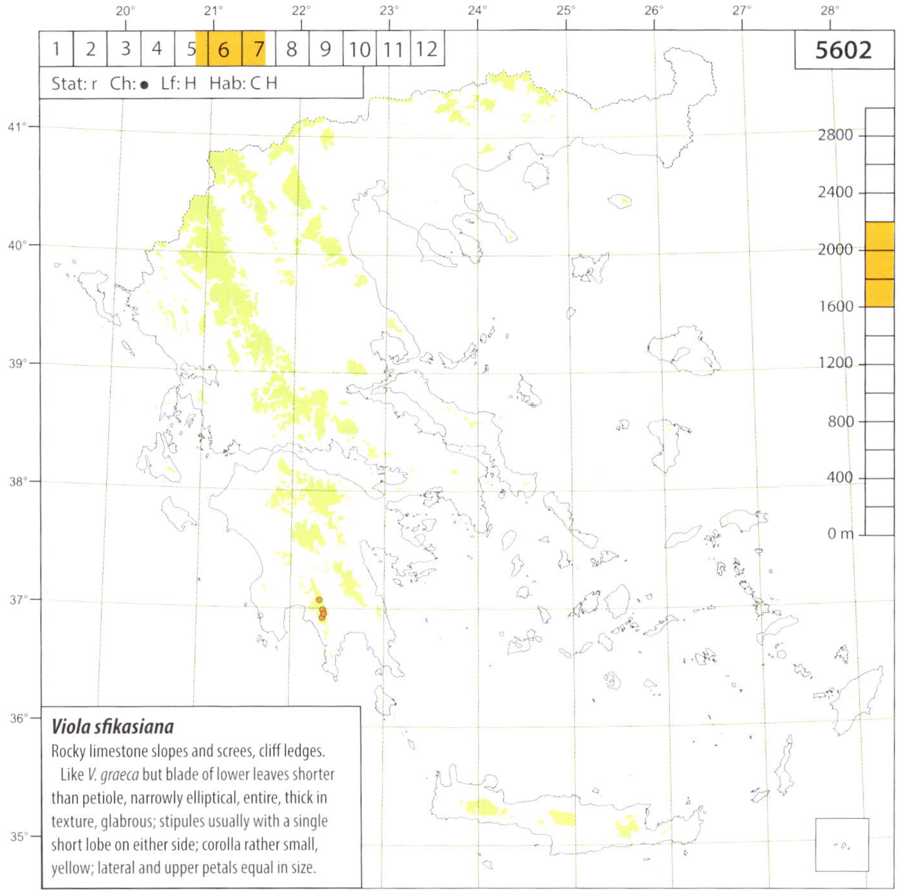

Viola sfikasiana

Rocky limestone slopes and screes, cliff ledges.

Like *V. graeca* but blade of lower leaves shorter than petiole, narrowly elliptical, entire, thick in texture, glabrous; stipules usually with a single short lobe on either side; corolla rather small, yellow; lateral and upper petals equal in size.

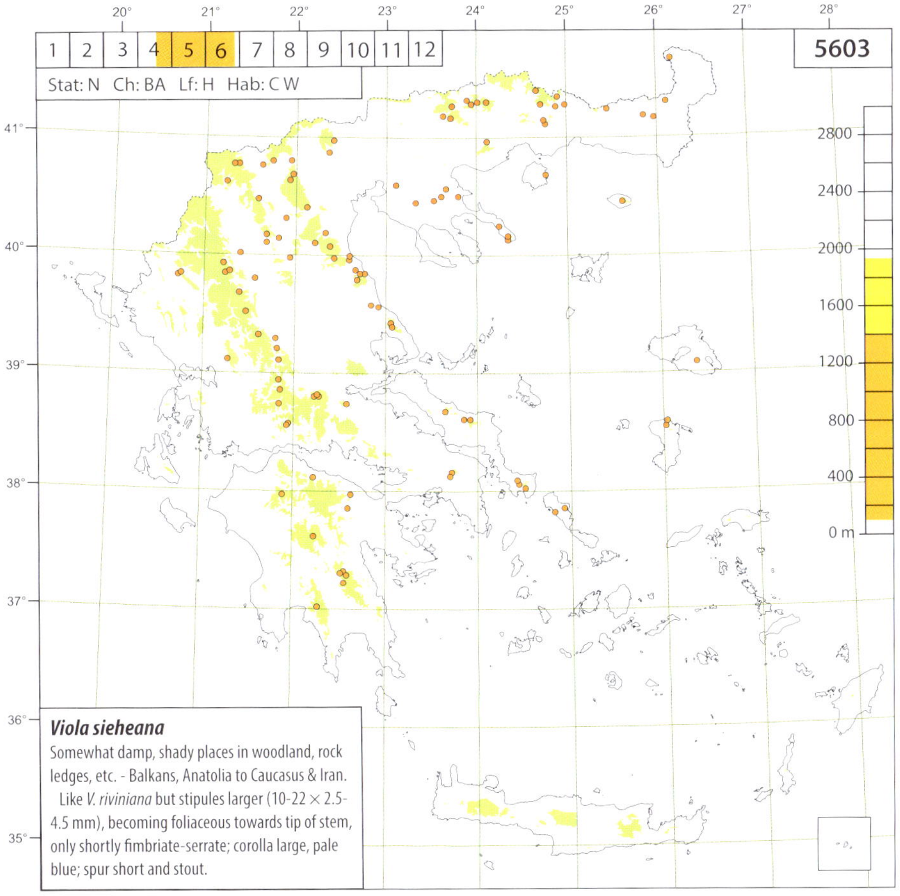

Viola sieheana

Somewhat damp, shady places in woodland, rock ledges, etc. - Balkans, Anatolia to Caucasus & Iran.

Like *V. riviniana* but stipules larger (10-22 × 2.5-4.5 mm), becoming foliaceous towards tip of stem, only shortly fimbriate-serrate; corolla large, pale blue; spur short and stout.

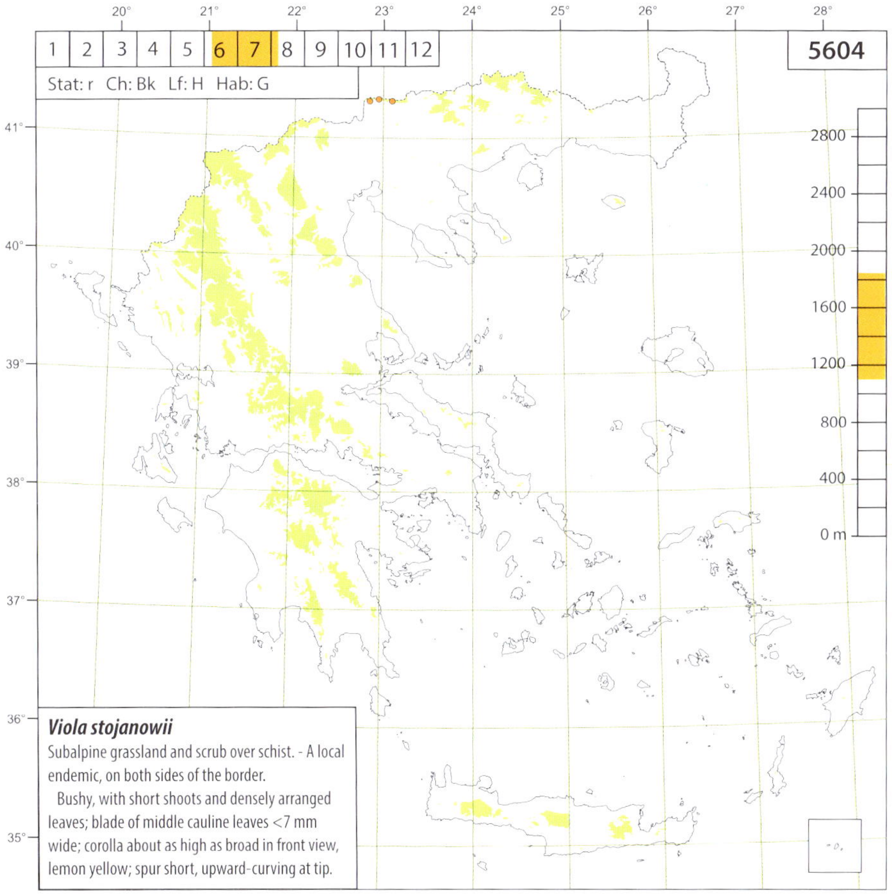

Viola stojanowii

Subalpine grassland and scrub over schist. - A local endemic, on both sides of the border.

Bushy, with short shoots and densely arranged leaves; blade of middle cauline leaves <7 mm wide; corolla about as high as broad in front view, lemon yellow; spur short, upward-curving at tip.

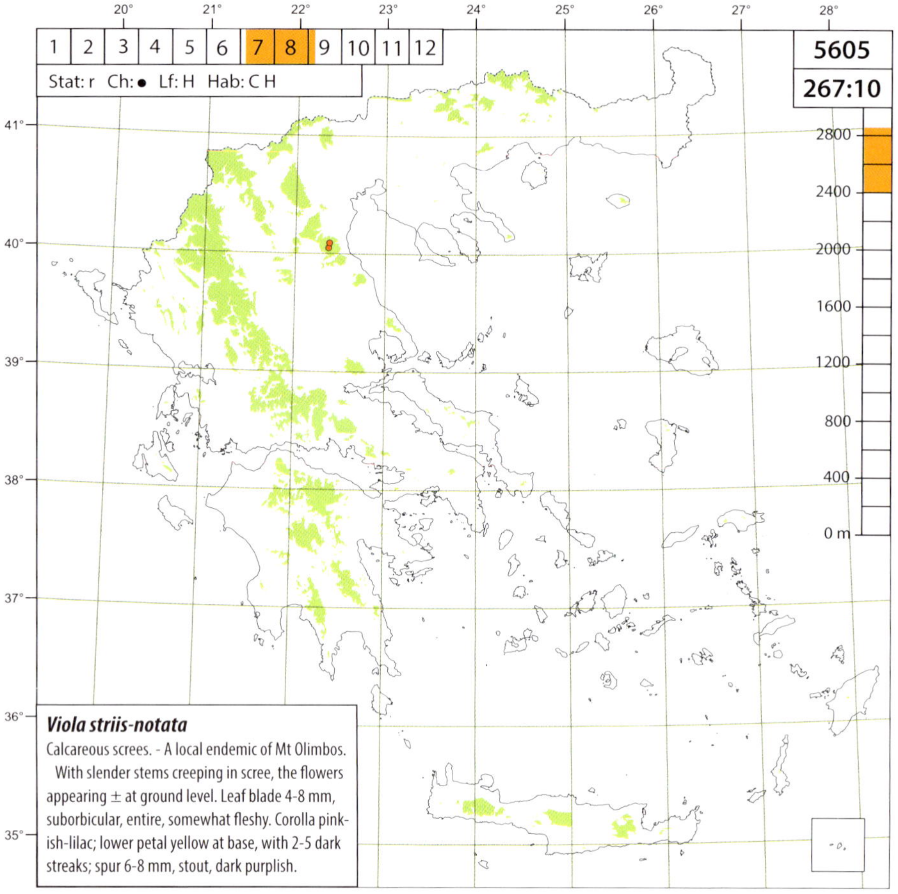

Viola striis-notata
Calcareous screes. - A local endemic of Mt Olimbos.
With slender stems creeping in scree, the flowers appearing ± at ground level. Leaf blade 4-8 mm, suborbicular, entire, somewhat fleshy. Corolla pink-ish-lilac; lower petal yellow at base, with 2-5 dark streaks; spur 6-8 mm, stout, dark purplish.

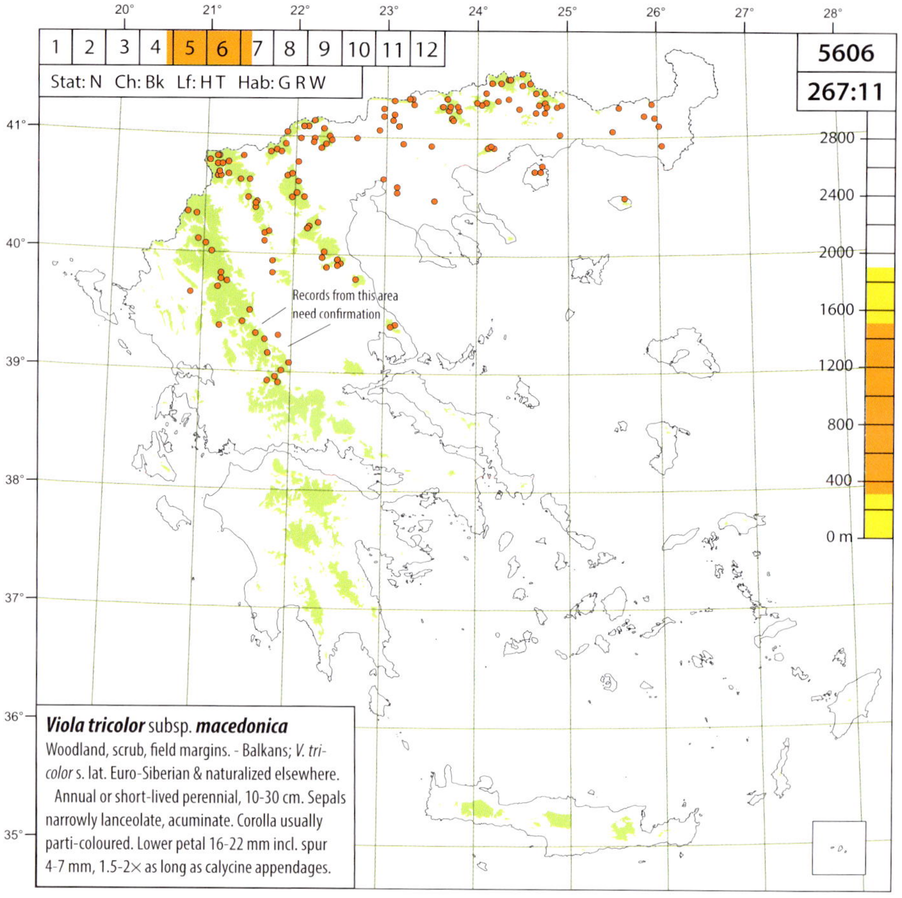

Viola tricolor* subsp. *macedonica
Woodland, scrub, field margins. - Balkans; *V. tricolor* s. lat. Euro-Siberian & naturalized elsewhere.
Annual or short-lived perennial, 10-30 cm. Sepals narrowly lanceolate, acuminate. Corolla usually parti-coloured. Lower petal 16-22 mm incl. spur 4-7 mm, 1.5-2× as long as calycine appendages.

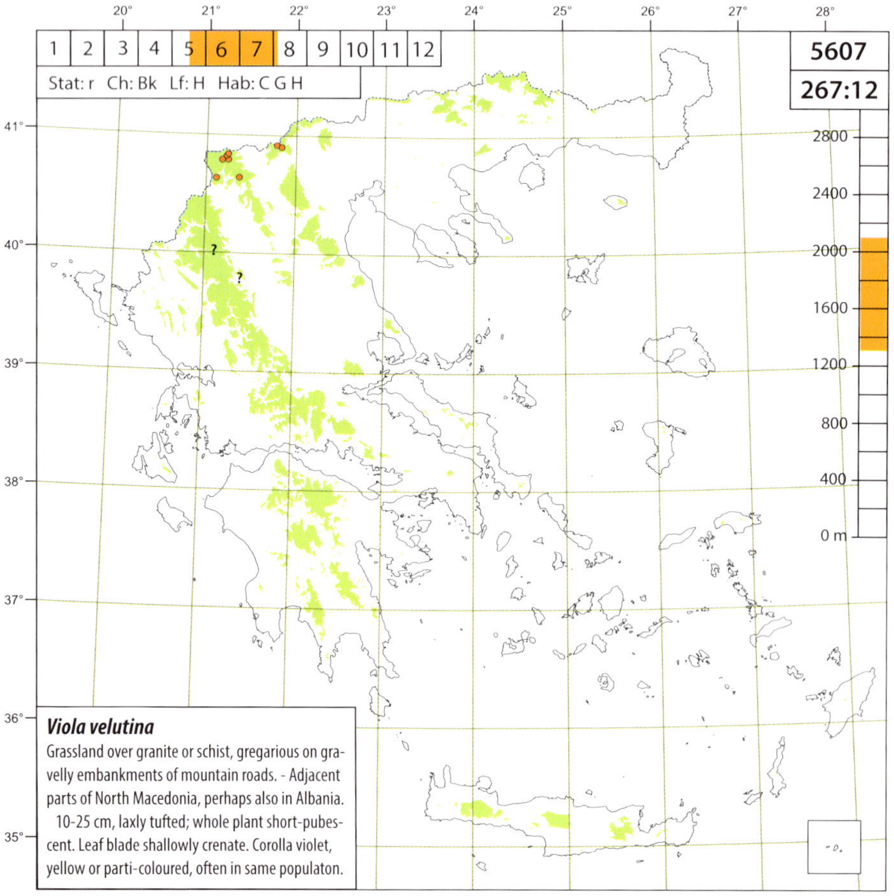

Viola velutina
Grassland over granite or schist, gregarious on gravelly embankments of mountain roads. - Adjacent parts of North Macedonia, perhaps also in Albania.
10-25 cm, laxly tufted; whole plant short-pubescent. Leaf blade shallowly crenate. Corolla violet, yellow or parti-coloured, often in same populaton.

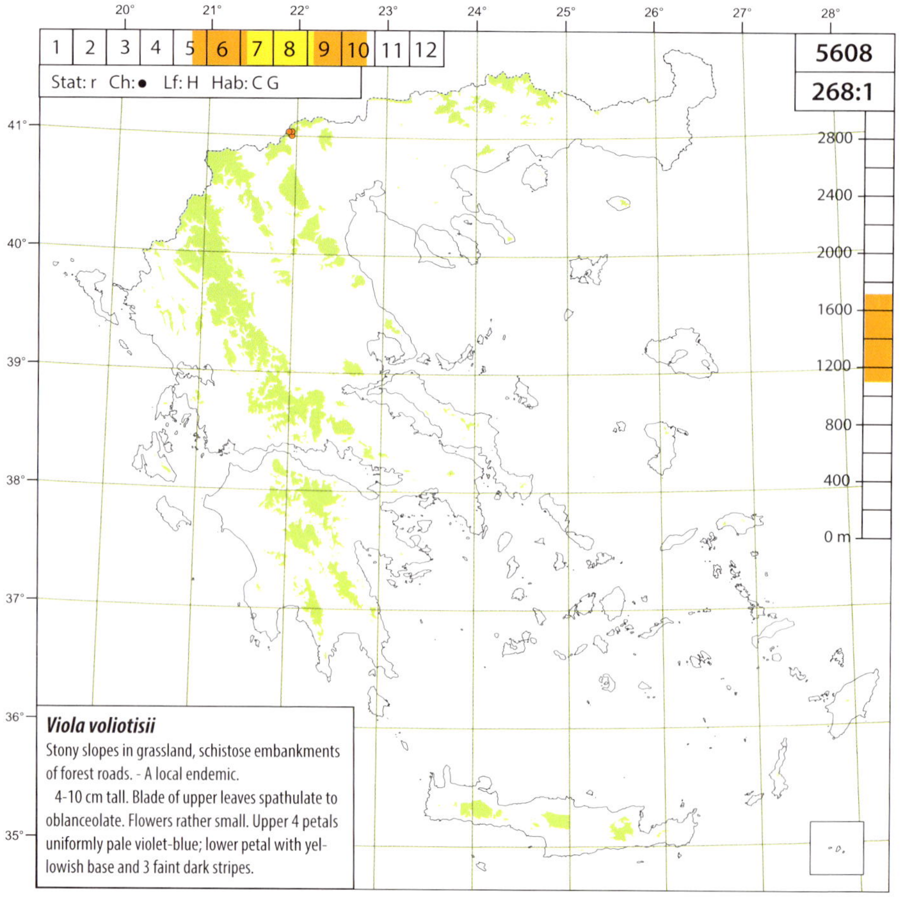

Viola voliotisii
Stony slopes in grassland, schistose embankments of forest roads. - A local endemic.
4-10 cm tall. Blade of upper leaves spathulate to oblanceolate. Flowers rather small. Upper 4 petals uniformly pale violet-blue; lower petal with yellowish base and 3 faint dark stripes.

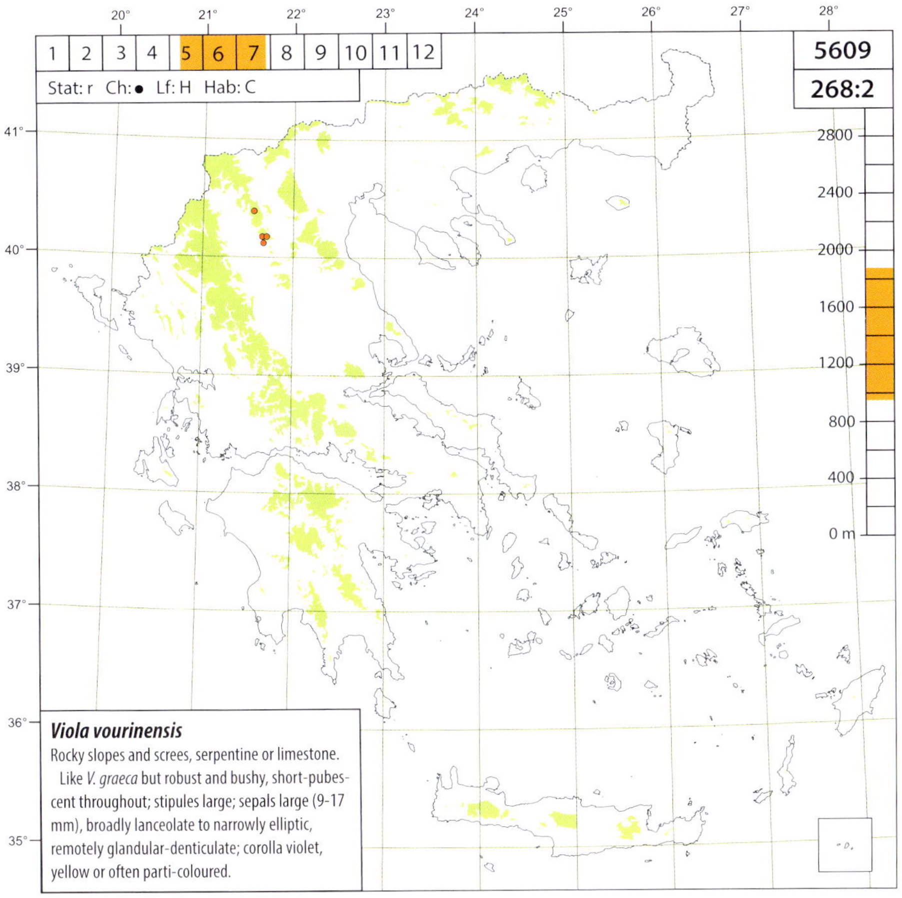

Viola vourinensis
Rocky slopes and screes, serpentine or limestone.
Like *V. graeca* but robust and bushy, short-pubescent throughout; stipules large; sepals large (9-17 mm), broadly lanceolate to narrowly elliptic, remotely glandular-denticulate; corolla violet, yellow or often parti-coloured.

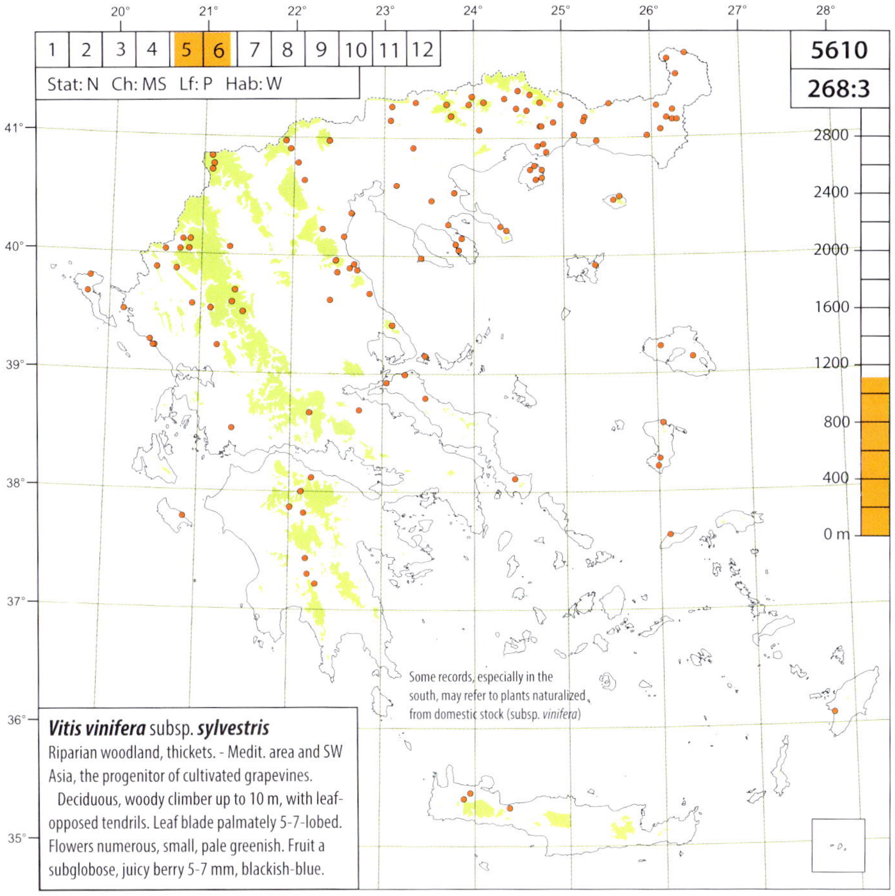

Vitis vinifera subsp. ***sylvestris***
Riparian woodland, thickets. - Medit. area and SW Asia, the progenitor of cultivated grapevines.
Deciduous, woody climber up to 10 m, with leaf-opposed tendrils. Leaf blade palmately 5-7-lobed. Flowers numerous, small, pale greenish. Fruit a subglobose, juicy berry 5-7 mm, blackish-blue.

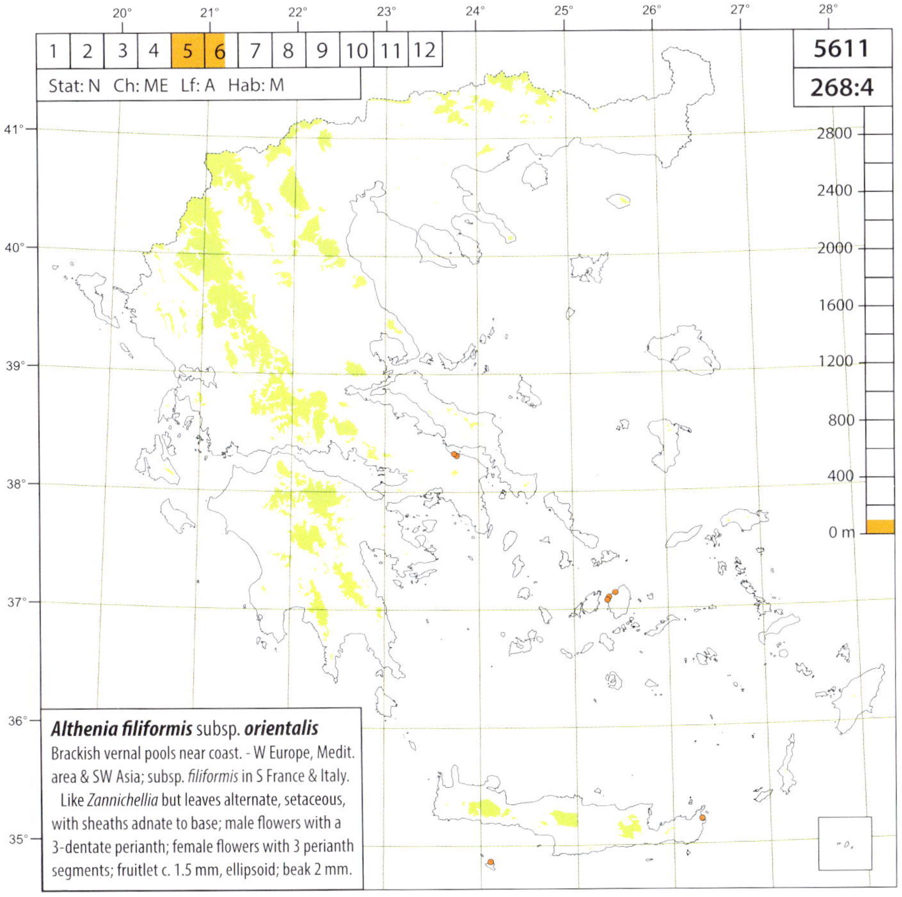

Althenia filiformis subsp. ***orientalis***
Brackish vernal pools near coast. - W Europe, Medit. area & SW Asia; subsp. *filiformis* in S France & Italy.
Like *Zannichellia* but leaves alternate, setaceous, with sheaths adnate to base; male flowers with a 3-dentate perianth; female flowers with 3 perianth segments; fruitlet c. 1.5 mm, ellipsoid; beak 2 mm.

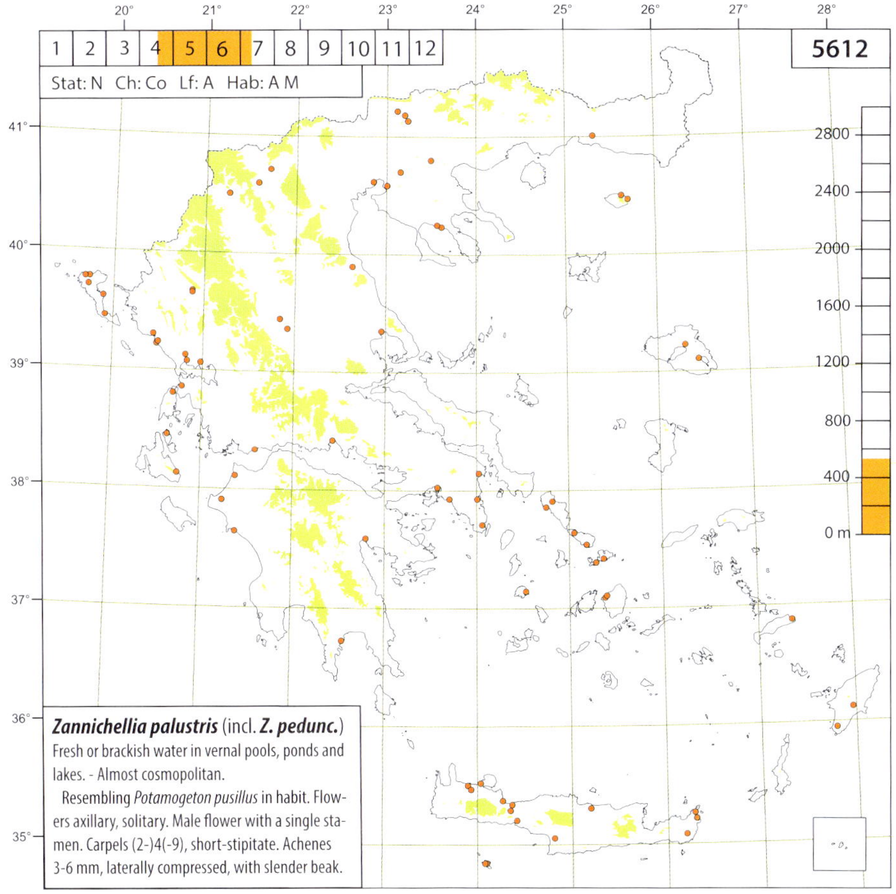

Zannichellia palustris (incl. ***Z. pedunc.***)
Fresh or brackish water in vernal pools, ponds and lakes. - Almost cosmopolitan.
Resembling *Potamogeton pusillus* in habit. Flowers axillary, solitary. Male flower with a single stamen. Carpels (2-)4(-9), short-stipitate. Achenes 3-6 mm, laterally compressed, with slender beak.

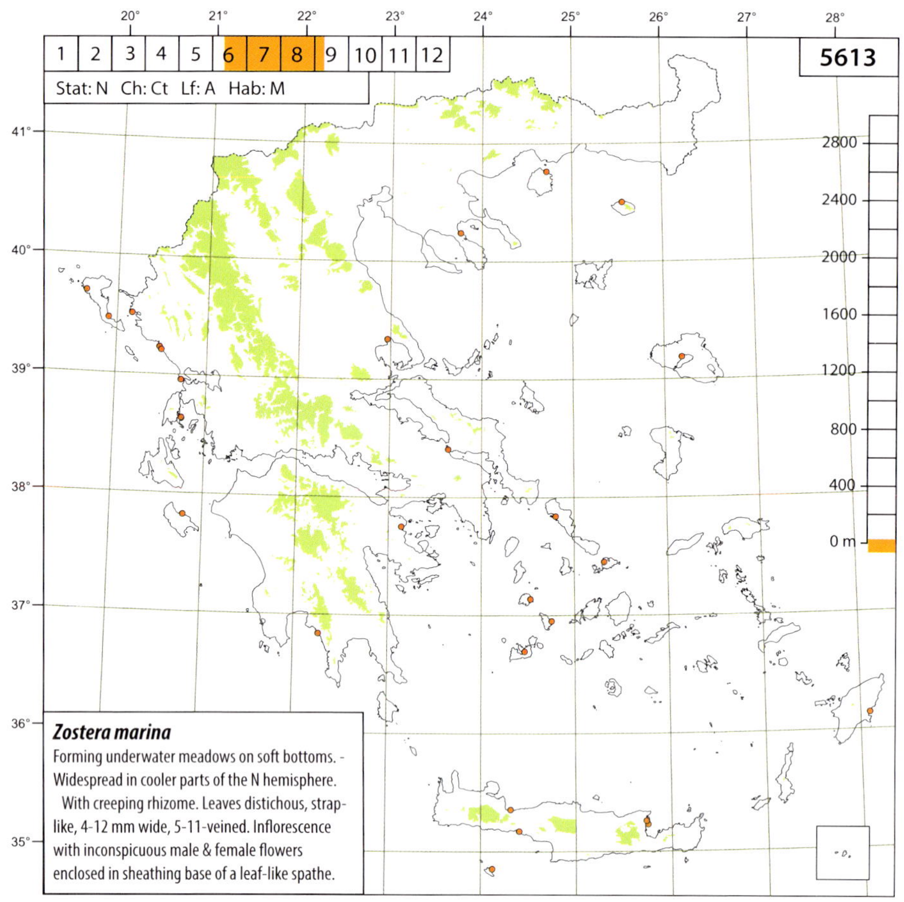

Zostera marina

Forming underwater meadows on soft bottoms. - Widespread in cooler parts of the N hemisphere.

With creeping rhizome. Leaves distichous, strap-like, 4-12 mm wide, 5-11-veined. Inflorescence with inconspicuous male & female flowers enclosed in sheathing base of a leaf-like spathe.

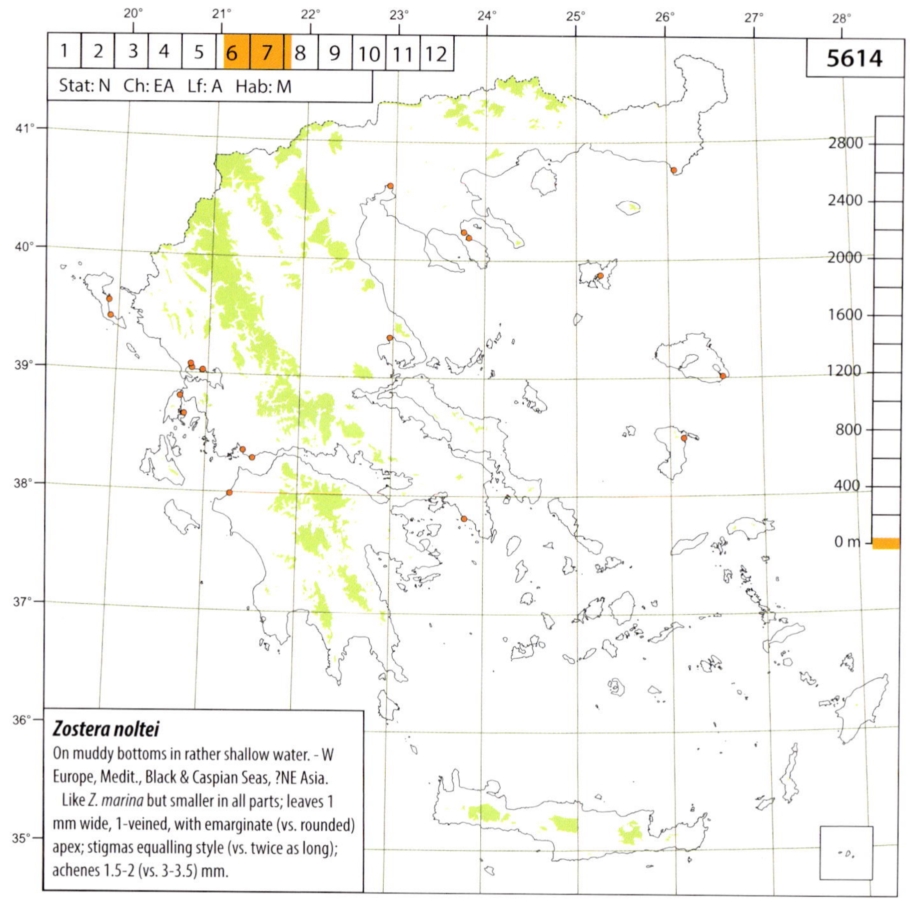

Zostera noltei

On muddy bottoms in rather shallow water. - W Europe, Medit., Black & Caspian Seas, ?NE Asia.

Like *Z. marina* but smaller in all parts; leaves 1 mm wide, 1-veined, with emarginate (vs. rounded) apex; stigmas equalling style (vs. twice as long); achenes 1.5-2 (vs. 3-3.5) mm.

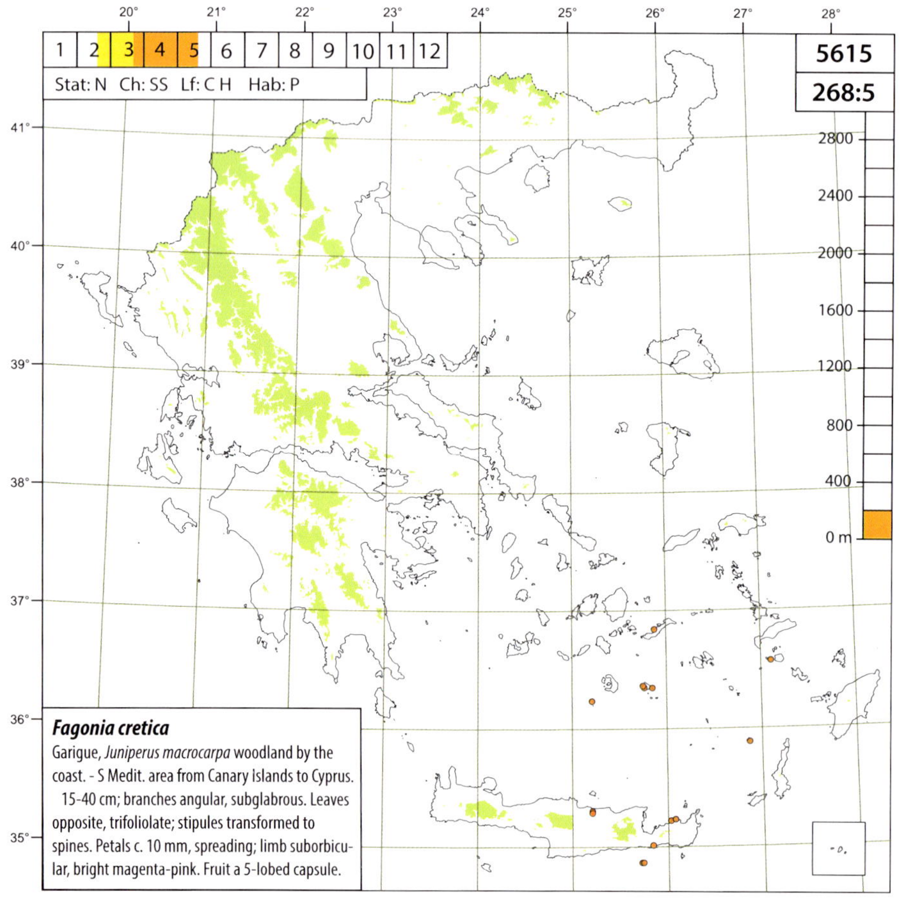

Fagonia cretica

Garigue, *Juniperus macrocarpa* woodland by the coast. - S Medit. area from Canary Islands to Cyprus.

15-40 cm; branches angular, subglabrous. Leaves opposite, trifoliolate; stipules transformed to spines. Petals c. 10 mm, spreading; limb suborbicular, bright magenta-pink. Fruit a 5-lobed capsule.

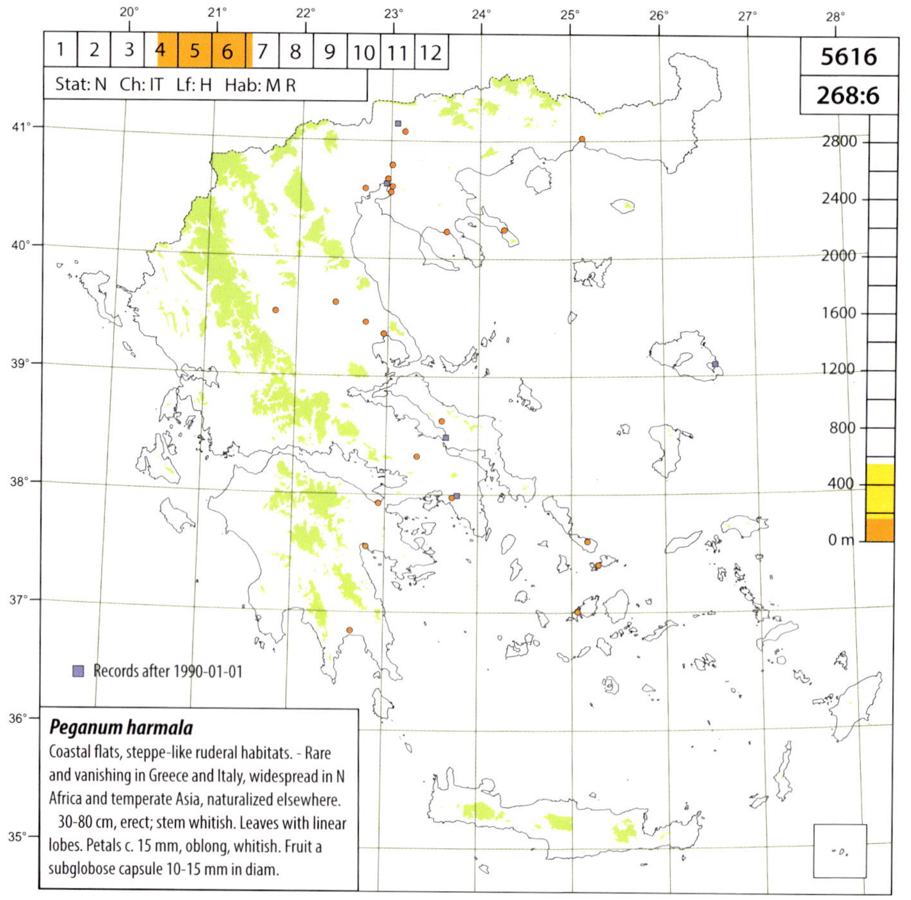

Peganum harmala

Coastal flats, steppe-like ruderal habitats. - Rare and vanishing in Greece and Italy, widespread in N Africa and temperate Asia, naturalized elsewhere.

30-80 cm, erect; stem whitish. Leaves with linear lobes. Petals c. 15 mm, oblong, whitish. Fruit a subglobose capsule 10-15 mm in diam.

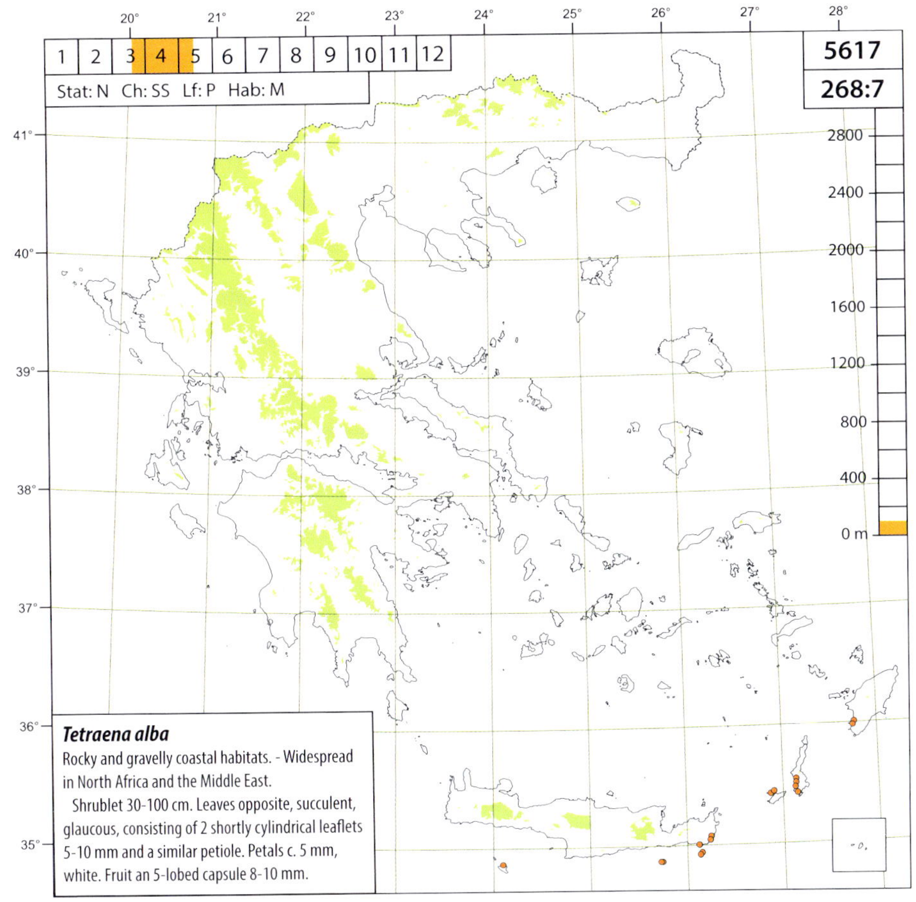

Tetraena alba

Rocky and gravelly coastal habitats. - Widespread in North Africa and the Middle East.

Shrublet 30-100 cm. Leaves opposite, succulent, glaucous, consisting of 2 shortly cylindrical leaflets 5-10 mm and a similar petiole. Petals c. 5 mm, white. Fruit an 5-lobed capsule 8-10 mm.

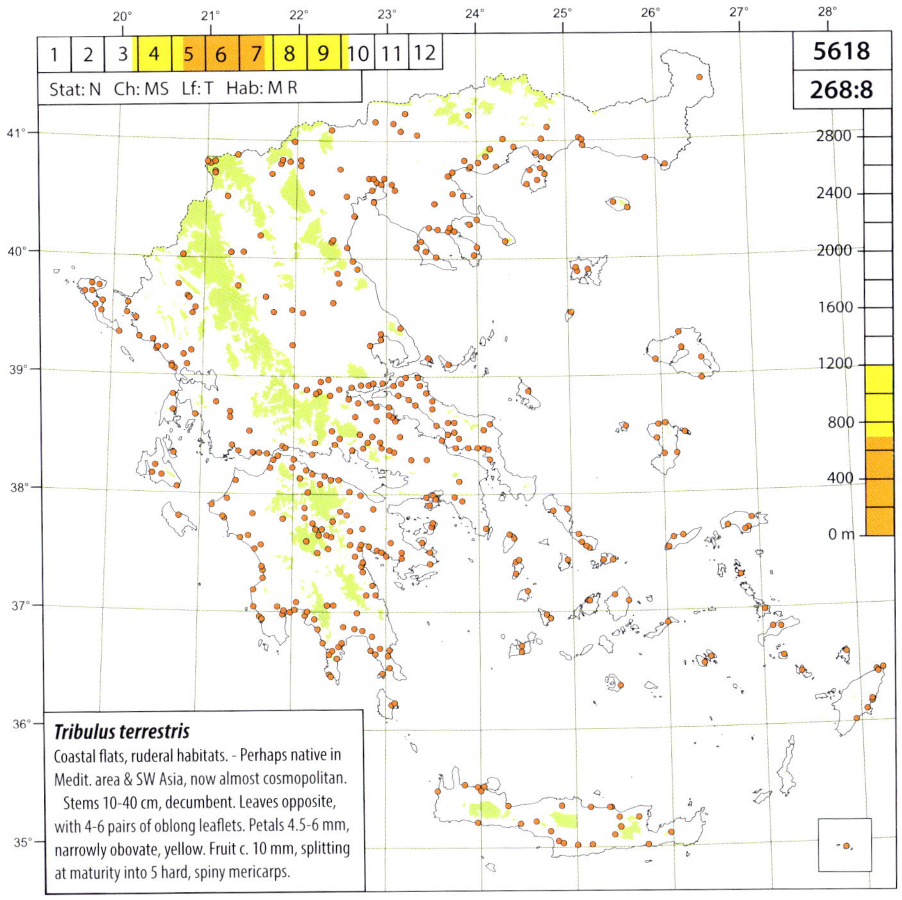

Tribulus terrestris

Coastal flats, ruderal habitats. - Perhaps native in Medit. area & SW Asia, now almost cosmopolitan.

Stems 10-40 cm, decumbent. Leaves opposite, with 4-6 pairs of oblong leaflets. Petals 4.5-6 mm, narrowly obovate, yellow. Fruit c. 10 mm, splitting at maturity into 5 hard, spiny mericarps.

Atlas
of the
Hellenic Flora

III

Arne Strid

Atlas of the Hellenic Flora

III

Published with support from the
Aage V. Jensen Charity Foundation

Broken Hill Publishers Ltd

ISBN: 978-9925-35-027-8

16 Zenas Kanther Princess De Tyras Str.
1065 Nicosia, CYPRUS
e-mail: info@brokenhill.com.cy
website: http://www.brokenhill.com.cy

EXCLUDED SPECIES AND TAXONOMIC COMMENTS

Notes on taxonomy, nomenclature, distribution, etc., provided in "Mountain Flora of Greece" (Strid 1986, Strid & Kit Tan 1991), "Flora Hellenica" (Strid & Kit Tan 1997, 2003), the re-issue of "Flora Graeca Sibthorpiana" (Strid & Strid 2009-2013), "Vascular Plants of Greece" (Dimopoulos & al. 2013, 2016), "Atlas of the Aegean Flora" (Strid 2016a&b) and "Flora and vegetation of the Prespa National Park" (Strid, Bergmeier & Fotiadis 2020) are not repeated here, unless there is additional information or other special reasons.

Acacia karroo Hayne, a native of S Africa, is occasionally planted along roads, e.g. on Rodos, but is scarcely naturalized. It is a large shrub with powerful thorns, bipinnate leaves and small, globose, bright yellow inflorescences.

Acacia saligna (Labill.) H. L. Wendl. (syn. *A. cyanophylla* Lindl.), a native of S & SW Australia, is frequently planted in coastal areas, being reported from Kriti, Kithira, Rodos and Evvia as well as Peloponnisos and the mainland, but is scarcely spreading into the natural vegetation. In other parts of the world with a Mediterranean-type climate, notably the SW Cape of South Africa, it is a troublesome invasive weed. It is an unarmed large shrub with oblanceolate phyllodia (flattened, leaf-like petioles) and inflorescences as in *A. karroo*.

A few other *Acacia* species are more rarely planted (see Arabatzis 2001: 22-27 and Fl. Cyprus 1: 596-603, 1977).

Acanthus caroli-alexandri Hausskn. is an enigmatic taxon based on a collection by Haussknecht 1885 from southern Pindos ("circa monasterium Korona, in nemorosis quercinis"). It resembles *A. spinosus* but is slender, with conspicuously narrow basal leaves and a relatively short inflorescence. Similar plants have been re-collected a few times in this area. Taxonomically it may be best considered to fall within the range of variation of *A. spinosus*. There is also some resemblance to the recently described *A. greuterianus*, which is a low plant with slender, long-creeping rhizome, soft, simply lobed basal leaves and a short inflorescence with usually pale greenish bracts. It is possible that *A. spinosus* and *A. greuterianus* hybridize, and that *A. caroli-alexandri* as well as occasional individuals of *A. greuterianus* with purplish bracts are derivatives of such hybridization.

Achillea filipendulina Lam. is cultivated as an ornamental and has been reported as an established alien in an abandoned cemetery in Kesari village (nom. Korinthias); photo V. Christodoulou, 2020 (see Zarkos & al. 2020: 568).

Achillea ptarmica L. is a widespread and variable, Euro-Siberian species. A record from Mt Vrondous (Goulimis 1956a: 11) has not been confirmed and is probably incorrect.

Achillea umbellata Sm. The umbel is occasionally reduced to a single capitulum, and such plants may look strikingly different. They have been described as *A. umbellata* var. *monocephala* Heldr. ex Heimerl, but are scattered throughout the range of the species and merge into the typical variety. Franzén (in Mount. Fl. Greece 2: 435-436) did not recognize them taxonomically.

Achyranthes sicula (L.) All. (*A. aspera* var. *sicula* L.) was illustrated on Plate 244 of Flora Graeca (1819), probably based on material from *locus classicus* in Sicily. Boissier (1879: 995) reported it from Rodos, based on a collection by Ehrenberg in 1820. There is no subsequent confirmation from Greek territory. It is a mainly African weed extending to the W Mediterranean area and the Canary Islands.

Acinos. Based on flimsy evidence, Melnikov (2016) reduced *Acinos* to a subgenus of *Ziziphora* and made a large number of new combinations.

Acinos alpinus (L.) Moench. The maps are preliminary. Lowland records by Willing (B) have been disregarded. As far as they have been possible to check most of them represent other taxa, usually *A. suaveolens*.

Acis ionica Bareka, Kamari & Phitos. See *Leucojum ionicum*.

Aconitum pentheri Hayek has been considered a subspecies of *A. burnatii* Gayer, but Rottensteiner (2018) divided the latter into four allopatric species with *A. pentheri* being restricted to the W & C Balkan Peninsula (described from "monte Žlep prope Ipek" = Peč, now in Montenegro). *A. humbertii* Beauverd, described from Mt Kajmakčalan, is a synonym of *A. pentheri*. *A. burnatii* s. str. is restricted to S France, and there are two additional, closely related species in the W Carpathians and S Spain (Sierra Nevada), respectively.

Aconitum variegatum subsp. ***judenbergense*** (Rchb.) Gayer was reported by Goulimis (1956a: 16) from "Papalazos", an unidentified locality in the area of Mt Voras. It is now considered a synonym of *A. variegatum* subsp. *nasutum* (Rchb.) Götz, and has never been confirmed on Greek territory. *A. variegatum* L. is a variable species occurring from NE Italy and the Balkan Peninsula to Caucasus.

Acorus calamus L. was listed for "Gr" in Euro+Med PlantBase, but there seem to be no confirmed records. The species is probably of North American origin but widely naturalized in Europe, including C & E Bulgaria. Occurrence in Greece is by no means unthinkable.

Acuston lunarioides (Willd.) Raf. was previously known as *Fibigia lunarioides* (Willd.) Sweet or *Alyssum lunarioides* Willd. It is a distinct species collected already by Tournefort and subsequently by Sibthorp (cf. comments in Strid 2016, 1: 164). According to Al-Shehbaz & Mabberley (2017) the monotypic genus *Acuston* comprises *A. perenne* (Mill.) Mabb. & Al-Shehbaz subsp. *perenne* in the Aegean area and *A. perenne* subsp. *obovatum* (Boiss. & Kotschy) Mabb. & Al-Shehbaz in Syria. Subsequently (German & Al-Shehbaz 2020) it has been proposed to reject the name *Lunaria perennis* Mill. to preserve the commonly used epithet *lunarioides*.

Adenocarpus complicatus (L.) J. Gay belongs to a species complex with centre of diversity in the Iberian Peninsula (cf. Cubas & al. 2010). *A. graecus* Griseb., described from the Athos peninsula, is sometimes regarded as a separate species, distributed in Greece, W & S Anatolia, Syria and Lebanon. Euro+Med PlantBase recognized *A. complicatus* subsp. *complicatus* and *A. complicatus* subsp. *commutatus* (Guss.) Cout. in Greece. There is no reason to suspect that there is more than one taxon in the country.

Adenostyles. The description of *A. orientalis* Boiss. was based on collections by Grisebach (1839) and Orphanides (1862) on Mt Pelister just N of the Greek border. It has subsequently been recombined as *A. alliariae* subsp. *orientalis* (Boiss.) Greuter. In its typical form this taxon is restricted to the Greek-Albanian-Macedonian border area, but plants from Montenegro, known as *A. kerneri* Simonk. or *A. alliariae* var. *kerneri* (Simonk.) K. Malý, are scarcely taxonomically different (cf. Wagenitz 1983a, Dillenberger & Kadereit 2012).

Adonis aestivalis L. is a vanishing weed of cereal fields. Earlier records have been referred to subsp. *squarrosa* (Steven) Nyman, but subsp. *aestivalis* was reported from localities in NE Greece by Meyer & Willing (2010: 198). The differences seem rather obscure.

Aegilops biuncialis subsp. ***archipelagica*** (Eig) Raus (syn. *A. biuncialis* var. *archipelagica* Eig; *Triticum macrochaetum* subsp. *archipelagicum* (Eig) Greuter) was described from Karpathos and has been recorded from a few islands in the C & S Aegean area. Its status is somewhat uncertain and it appears to co-exist with subsp. *biuncialis*.

Aegilops comosa Sm. subsp. ***comosa*** and subsp. ***heldreichii*** (Boiss.) Eig (syn. *Triticum heldreichii* (Boiss.) K. Richt.) differ in rather vague, quantitative characters and have completely overlapping distributions. The monographer Slageren (1994: 184) listed the latter as *A. comosa* var. *subventricosa* Boiss. Both were described from the foothills of Mt Parnitha, Attiki.

Aegilops contracta (Eig) H. Scholz is very close to *A. neglecta* Req. ex Bertol., but said to differ by its more contracted spikes, constantly 2-awned lower and upper glumes of the lower fertile spikelet, with one awn 1.5-2.5 mm wide at the base, and tricuspidate to triaristate lemmas. It has been reported from Kriti, Karpathos and Rodos, but the records need confirmation, and its taxonomic status seems dubious.

Aegilops ventricosa Tausch is a W Mediterranean species doubtfully recorded from the Ionian Islands. "In Europe the natural distribution area of this species does not extend further east than to Italy and the Kvarner islands (in N Africa it reaches from Morocco to Egypt); see map in Slageren (1994: 396). The only record for Corfu (Borkowsky 1994: 146) refers to a ruderal habitat, that for Cephalonia (lacking comments on status) was found in a phytosociological table (Bolòs & al. 1996: 107) relating to therophytic grassland" [cited from Flora Ionica website].

Aeonium arboreum (L.) Webb & Berthel., a native of Gran Canaria, has been cultivated as an ornamental in the Mediterranean area for a long time, and was illustrated on Plate 473 of *Flora Graeca* (1825), based on material collected in Cyprus, "in a Turkish village called Iktinia [now Ktima], upon rubbish ground on the outskirts of the town, 11th May 1787". In Greece it is frequently cultivated in hot and dry places, in pots or old feta (cheese) and olive-oil tins outside houses, etc., and is occasionally semi-naturalized on walls and in ruderal habitats.

Aeschynomene indica L. and ***A. rudis*** Benth. have been reported from the edge of rice fields in the Serres province (Kit Tan, based on observations and photos by I. Tsialtis). Both are widespread in the tropics and may be on the verge of becoming locally naturalized in Greece.

Aesculus hippocastanum L. is rather widespread in the Pindos, but usually in small populations (often only a few trees in ravines). At the time of Ottoman rule it was established as a decorative street tree in Konstantinopolis, and from there introduced into C Europe in the 17th century (cf. Lack 2000b, 2001).

Aethionema rhodopaeum D. K. Pavlova is a rare serpentinophyte related to *A. saxatile* and described from the eastern Rodopi mountains in Bulgaria. It has subsequently been recombined as *A. saxatile* subsp. *rhodopaeum* (D. K. Pavlova) Constantin., Kougioum. & Kalpoutz. and has been reported also from the Greek side of the border. For a key to the Greek subspecies of *A. saxatile*, see Doumas, Goula & Constantinidis (2022: 16).

Agrimonia eupatoria L. All Greek records are likely to represent subsp. *eupatoria*. The poorly defined subsp. *grandis* (C. A. Mey.) Bornm. has been reported from some localities in the north.

Agrostemma githago subsp. ***thessala*** (Bornm.) Greuter was originally described as *A. gracile* var. *thessalum* Bornm., based on a collection by Haussknecht (1885) from the vicinity of Farsala in the Thessalian plain. There are a few additional collections from semi-natural habitats in this area, but the last is from 1896 and this taxon is now likely to be extinct. It may have been one of the wild ancestors of the widespread cereal weed *A. githago* s. str. The closely related *A. gracile* Boiss. was described from Manisa Dağ near Izmir in W Anatolia, and has been recollected fairly recently in dry grassland in this area (e.g., Strid & al. 23204, 5 July 1984; B, EGE, G, UPA, herb. Strid). It is still scattered in Anatolia and perhaps further east.

Agrostis castellana Boiss. & Reut. was listed for Peloponnisos by Mermygkas, Zikos & Constantinidis (2021: supplementary file). The record is feasible, but no details are available.

Agrostis lachnantha Nees, a xenophyte from Yemen, E & S Africa, was reported from a ruderal habitat on Samothraki by Biel & Scholz (2008: 469). It is not clear to what extent it was naturalized.

Agrostis stolonifera subsp. ***scabriglumis*** (Boiss. & Reut.) Maire has been reported from the island of Samothraki (Biel & Kit Tan 2010a: 146). It was first described, as *A. scabriglumis* Boiss. & Reut., from river courses in Sierra Nevada, S Spain, and has subsequently been recorded from scattered localities in the Mediterranean area. Flora Europaea (5: 234, 1980) listed it merely as a synonym of *A. stolonifera* L.

Agrostis vinealis Schreb. A few collections from Mt Pieria and Mt Rodopi have been identified as *A. vinealis*, but differences between this and *A. canina* are not clear.

Aira caryophyllea L. The few Greek records are those listed by Frey (1997a: 260). No specimens clearly referable to *A. caryophyllea* have been seen by the author, but some may be hiding among those identified as *A. elegans* Willd. or *A. cupaniana* Guss.

Aira elegans subsp. ***ambigua*** (Asch.) Holub (with bifid lemmas, both bearing a dorsal awn) is an aptly named variant of dubious taxonomic value, scattered throughout the eastern range of the species.

Aira provincialis Jord. was listed for "?Cr" in Flora Europaea 5: 228 (1980). The source for the record is obscure; the species is likely to be endemic to SE France and Corsica.

Aira tenorei Guss. was listed for "Gr" in Flora Europaea 5: 228 (1980). The record may go back to reports by Haussknecht (1900: 43) of *Avena corymbosa* Nyman (a synonym of *Aira tenorei*) from Akrokorinth (N Peloponnisos) and S Pindos. *A. tenorei* is a species of the W & C Mediterranean area. The Greek records are dubious and do not appear to have been confirmed.

Ajuga iva (L.) Schreb. occurs in pink- and yellow-flowered forms and flowers both in the spring and autumn. Further studies are needed to determine whether it should be taxonomically subdivided.

Ajuga orientalis subsp. ***aenesia*** (Heldr.) Phitos & Damboldt has been described from Kefallinia. There are no essential differences between this and other populations of *A. orientalis* L.

Ajuga salicifolia (L.) Schreb. Greek specimens belong to the rather poorly defined subsp. *bassarabica* (Săvul. & Zahar.) P. W. Ball. This and the nominate subspecies show wide geographical overlap in Anatolia (cf. Fl. Turkey 7: 46, 1982).

Albizzia julibrissin Durazz., a native of China and perhaps N Iran, is frequently planted as an ornamental but scarcely naturalized. It is a flat-topped tree with bipinnate leaves and spherical flower heads with long, showy, rose-pink staminal filaments.

Alcea apterocarpa (Tchich.) Boiss. was listed from the East Aegean Islands in Euro+Med PlantBase, probably based on records by Rechinger (1944a: 370) from Rodos which must be considered as doubtful. Uzunhisarcıklı & Vural (2012: 613) recorded this species only from Turkey, Egypt and Iraq.

Alcea lavateriflora (DC.) Boiss. was described from Lebanon and has been doubtfully recorded from Turkey and Greece (cf. Uzunhisarcıklı & Vural (2012: 628).

Alcea rosea L. (hollyhock) was described from cultivated material. It may be an allotetraploid derived from *A. setosa* (Boiss.) Alef. and *A. biennis* Winterl. Following Uzunhisarcıklı & Vural (2012), *A. pallida* (Willd.) Waldst. & Kit. is regarded as a synonym of *A. biennis*. According to the same authors, *A. cretica* (Weinm.) Greuter is a synonym of *A. rosea*. Taxonomy and distribution of all Greek *Alcea* is provisional. The genus is taxonomically difficult with centre of diversity in the Flora Iranica area.

Alchemilla. The account of this critical genus is largely based on an unpublished typescript for Flora Hellenica by H. Kalheber & W. Lippert (2004).

Alchemilla alpina complex. A member of this group growing in limestone mountains of NW Greece (notably Mt Timfi) has been variously named *A. montenegrina* Gand., *A. plicatula* Gand. and *A. velebitica* (Janch.) Degen. Taxonomy and nomenclature are unsettled. The two first-mentioned were both described by Gandoger in the same publication (1883: 11), *A. montenegrina* from "Montenegro, in alpibus" and *A. plicatula* from Croatia ("in alpibus regionis adriaticae australis") as well as Austria. *A velebitica* has a convoluted nomenclature, but was recorded from several areas in the western Balkans (cf. Janchen 1919: 177). *A. montenegrina* has subsequently been considered endemic to Montenegro, whereas the more widespread taxon is probably to be named *A. plicatula* (cf. Nikolić 2020, 3: 483). The identity of *A. velebitica* is uncertain, but it is likely to be a synonym of *A. plicatula*.

Alchemilla exigua Buser ex Paulin was mapped from one locality in NE Greece by Kurtto, Fröhner & Lampinen (2007: 62), but no specified record has been seen.

Alchemilla fallax Buser, a species occurring in mountains from the Pyrenees to Serbia, has been reported once from Mt Tsoumerka (Kakarditsa), based on a collection by Halácsy, 9.7. 1893 (WU). Modern confirmation is needed.

Alchemilla glaucescens Wallr. was included in the Greek checklist (Dimopoulos & al. 2013: 135), probably based on Willing 9414p.p. from Mt Vermion which had been registered under that name. The specimen needs critical examination.

Alchemilla micans Buser is a widespread European species. There is one historical collection "in subalpinus Tymphresti" (Fraas s.n., c. 1840; M); in MFG 1: 425 (1986) this was listed under *A. gracilis* Opiz.. A single modern collection (Snogerup 15968 from nomos Florinis; LD, M) has been identified as *A. micans*, but critical revision is needed.

Alchemilla mollis (Buser) Rothm. After the preparation of the maps this species has been reported from the Rodopi mountains by Charalampidou, Eleftheriadou & Theodoropoulos in Fl. Medit. 32: 89 (2022).

Alchemilla peristerica Pawł. was mapped in Kurtto, Fröhner & Lampinen (2007: 157) from a single locality on the border of NW Greece and North Macedonia. However, the *locus classicus* is just north of the border and there appears to be no record from Greek territory.

Alhagi. Kleinsteuber, Ristow & Hassler (2016: 431) listed specimens from Rodos as *A. maurorum* Medik. Greek plants represent the western end of a species complex which is important in steppe vegetation in SW Asia. There is unlikely to be more than one taxon in Greece, and in this Atlas all Greek plants have provisionally been referred to *A. graecorum* Boiss. It is vanishing on the Greek mainland and the Kikladian islands. Bory & Chaubard (1832-33: 314) observed: "Nous n'avons d'abord recontré cette belle et singulère espèce qu'à Syra, où elle abondait, surtout au sud de la ville, et commencait à fleurir vers la mi-Août". Landerer (1856e: 189) noted "... findet sich im Hafen von Phalerno, auch auf Syra, und besonders auf der Insel Tinos fand ich sie häufig". There appears to be no modern confirmation from Siros. On Tinos it was collected in 1990 and on Amorgos in 1994. There is also a recent observation from Piraeus.

Alkanna areolata Boiss. Reports from Samos need confirmation (cf. Strid 2016a: 136).

Alkanna chrysanthiana Kit Tan was described from the Mani Peninsula and said to differ from *A. sartoriana* Boiss. & Heldr. in the sinuate-dentate leaves and white corolla. Plants from the Mani Peninsula were illustrated by Marchant (2018a: 205). There are scarcely any differences between these and plants collected and photographed by Strid 4.4. 2004 near Nafplion (*locus classicus* for *A. sartoriana*). In both the corolla is white with yellow to orange centre, and the leaves vary from entire to somewhat sinuate-dentate. See also Kit Tan & Marchant (2021).

Alkanna pindicola Hausskn. was listed for Peloponnisos by Mermygkas, Zikos & Constantinidis (2021: supplementary file). The record is feasible, but no details are available.

Alkanna sandwithii Rech. f. was described from S Albania and has been reported from NC Greece by Babalonas (1983: 18, 1989: 391). Most probably the Greek records refer to *A. pindicola*.

Alkanna scardica Griseb. A report of this species was probably based on Sintenis 640 (LD) from "Chaliki: in valle Negerli" [nom. Trikalon], which has turned out to be *A. noneiformis*.

Alkanna sfikasiana Kit Tan, Vold & Strid in Fisis [Journal of the Hellenic Society for the Protection of Nature] 109: 43 (2005) must be replaced by *A. versicolor* Orph. The latter name appeared in a little known publication, *Bulletin du congrès international de botanique et d'horticulture de St. Pétersbourg le 6/18 le 8/20 et le 10/22 mai 1869* [p. 138, publ. 1870], and was not cited in Euro+Med PlantBase. In Flora Orientalis 4: 231 (1879) it was listed as *A. graeca* var. *versicolor*. The type of *A. versicolor* is Orphanides 858 (LD, WU-Hal), "in monte Malevo Laconiae", dated 1877 but in fact collected in 1851. In the original publication the locality was given as "près du village Hajos Joannis", with the note: "elle brillait par ses fleurs de différentes couleurs; car les unes étaient jaunes, les autres d'un beau bleu, et les plus jeunes d'une vive couleur de rose". The description and the specimens leave little doubt that *A. versicolor* and *A. sfikasiana* are conspecific. [Kit Tan begs to differ, believing that Orphanides' *A. versicolor* is a form of *A. graeca* subsp. *baeotica*. Detailed examination of the type specimens and the range of variation is needed to settle the matter].

Allium achaium Boiss. & Orph. was based on a collection by Orphanides from Mt Klokos (N Peloponnisos), but the type matches the previously described *A. frigidum* Boiss. & Heldr. Tzanoudakis, Tsakiri & Raus (2019) coined the new name *A. oreohellenicum* for plants previously known as *A. achaium*. "*Allium achaium*" sensu Bogdanović & al. (2011a) is a member of the completely different *A. flavum/stamineum* group.

Allium apergii Trigas, Iatroú & Tzanoud. is known only from the type collection from Mt Ochi, S Evvia (see Trigas, Iatroú & Tzanoudakis 2010: 226). It is a member of the *A. paniculatum* group.

Allium balcanicum Brullo, Pavone & Salmeri. This name has been given to Balkan populations of *A. cupanii*. They are probably referable to subsp. *cupanii*.

Allium carlstroemi Cattaneo is a name recently applied to plants from the island of Simi (see Cattaneo 2020) here referred to *A. stamineum* Boiss. The description was based on two collections (probably a single population). The rather extensive Greek collections of *A. stamineum* (64 records in the database) were not examined. *A. carlstroemi* and *A. stamineum* were said to differ, e.g., in flowering time: April-May for the former and June-July for the latter. In fact, the two collections of *A. carlstroemi* are from 27 April and 8 May, whereas flowering plants identified as *A. stamineum* have been collected from early May to early August. The group is in need of critical revision throughout its range.

Allium cephalonicum Brullo, Pavone & Salmeri. The moderately variable *Allium cupanii* was fragmented by Brullo, Pavone & Salmeri (2015) into 12(!) poorly defined species, including 10 described by themselves. *A. cephalonicum*, based on a single collection from Mt Enos, Kefallinia, is probably a synonym of *A. cupanii* subsp. *cupanii*.

Allium circinatum Sieber has been divided into subsp. *circinatum* (in Kriti) and subsp. *peloponnesiacum* Tzanoud. (in Peloponnisos). They differ only vaguely in quantitative characters, the latter being somewhat larger in all parts.

Allium cirrhosum Vand. was listed as an accepted species in Dimopoulos & al. (2013: 45), but is apparently a synonym of *Allium carinatum* subsp. *pulchellum* (G. Don) Bonnier & Layens.

Allium cremnophilum Brullo, Pavone & Salmeri was based on a single collection from Mt Ipsario, Thasos, which evidently belongs to *A. cupanii* subsp. *hirtovaginatum*. Five previous collections of the latter from Mt Ipsario were ignored.

Allium cupanii Raf. was named after Francesco Cupani (1657-1710), an Italian monk and naturalist. The original spelling of the epithet is *cupani*, but this should be corrected in accordance with Art. 60.8 of the Code of Nomenclature. The same applies to *Colchicum cupanii* Guss. and *Melica cupanii* Guss. [Thanks are due to Nick Turland for putting this straight].

Allium cylleneum Brullo, Pavone & Salmeri is a name recently given to a single population (or individual?) of *A. cupanii* subsp. *hirtovaginatum* from Mt Killini. Two previous collections of the latter from Mt Killini were not mentioned.

Allium euboicum Rech. f. was described from serpentine substrate in N Evvia and has subsequently been reported also elsewhere in the area. It has been recombined as *A. paniculatum* subsp. *euboicum* (Rech. f.) Stearn.

Allium exile Boiss. & Orph., described from Chios, is probably conspecific with *A. sipyleum* Boiss., described from Manisa Dağ ("Montis Sipyli") in W Anatolia just opposite Chios.

Allium goulimyi Tzanoud. was based on a single collection from Mt Vermion. It seems to differ from *A. bornmuelleri* Hayek only in vague, quantitative characters, and the drawing accompanying the description is a perfect match for the latter. Mt Vermion is in-between known localities for *A. bornmuelleri*.

Allium goumenissanum Ioannidis & Tzanoud., a new, autumn flowering species of sect. *Codonoprasum*, has just been described from nomos Kilkis, N Greece – see Phytotaxa 554(1): 59-69 (2022). It is related to *A. stamatiadae*, *A. orestis* and *A. panormitisi*.

Allium guicciardii Heldr. was described from Mt Parnassos, and has later been considered a variety or subspecies of *A. flavum* L. It can scarcely be distinguished from other high-altitude forms of *A. flavum* subsp. *flavum*.

Allium hirtovaginatum subsp. ***samium*** Brullo, Pavone & Salmeri is a name recently given to a single population (or individual?) of *A. cupanii* from Mt Kerkis, Samos.

Allium hirtovaginum P. Candargy is a dubious species of sect. *Codonoprasum* described from the island of Lesvos and subsequently reported from scattered localities in Greece and W Anatolia. The type material has been lost and the original description is inconclusive. Brullo & Salmeri (2021) described 13(!) new species related to *A. hirtovaginum* from the Aegean area and SW Anatolia, all based on a single or few of their own collections whilst ignoring the much more extensive collections by others. The taxonomic value, if any, of these "species" is difficult to assess. The ones recorded from Greek territory are: *A. abanticum* (Evvia), *A. hippocraticum* (Kos), *A. pignattii* (Samos), *A. pythagoricum* (Samos) and *A. velutinum* (Monemvassia).

Allium iatrouinum Trigas is known only from the type collection from Mt Ochi, S Evvia (see Trigas, Kalpoutzakis & Constantinidis 2017: 180). It is a member of the *A. cupanii* complex.

Allium integerrimum Zahar. is a dubious species described from Thessalia (E foothills of Mt Olimbos and Meteora) and Lesvos (Kalloni). It is scarcely distinct from the variable *A. sphaerocephalon* L.

Allium karistanum Brullo, Pavone & Salmeri was based on a single collection from southern Evvia. It is scarcely different from *A. cupanii*. *A.* cf. *karistanum* has subsequently been reported also from the island of Skiros (Trigas & al. 2021: 325).

Allium karvounis Brullo, Giusso & Musarella is a name recently given to a single population (or individual?) of *A. stamineum* from Mt Karvouni (Ambelos), Samos.

Allium lefkadensis Brullo, Giusso & Musarella is a name recently given to a single population (or individual?) of *A. stamineum* from Mt Elati, Lefkas.

Allium makrianum C. Brullo & al. was based on a single collection from the island of Chios. It does not seem specifically distinct from *A. paniculatum*.

Allium neapolitanum Cirillo. In February 2022 Christian Gilli discovered a small population in a pine plantation on the Lassi peninsula north of Argostoli (Kefallinia), but regarded it as doubtfully native. A previous record from the island of Ithaki (Markantonatou 1991) has been considered dubious and was not included on the Flora Ionica website. The record on the mainland just east of Kerkira is based on Willing 101339 ("SO Filate, Krautfluren an Felsanriss unter Buschwald"), which has been seen and verified by the present author at Herbarium B in 2018.

Allium nigrum L. Records from the Kiklades are quite recent. It was reported by Biel & Kit Tan (2017b: 286) from Amorgos and subsequently by Scharlau from Naxos (photo 2022-04-27).

Allium orosamium Brullo, Giusso & Musarella is a name recently given to a single population (or individual?) of *A. stamineum* from Mt Kerkis, Samos.

Allium phitosianum Brullo & al. was based on a single collection from Mt Timfristos. It is apparently a high-altitude form of *A. paniculatum*.

Allium phthioticum Boiss. & Heldr. Reports from Italy apparently refer to the similar *A. permixtum* Guss. It should be critically examined whether these two are indeed specifically distinct. In the recent Flora d'Italia website they are listed as synonymous.

Allium pictistamineum O. Schwarz. A report of this species from Samos, cited in Flora of Turkey (8: 159, 1984) is plausible but needs confirmation. *A. pictistamineum* is otherwise known from a few localities in W Anatolia, and the record from Samos may refer to the closely related *A. candargyi*.

Allium rhodiacum Brullo, Pavone & Salmeri, described one year later than *A. chalkii* Tzanoud. & Kollmann, is scarcely distinct from the latter.

Allium rotundum L. Taxonomy follows Bothmer (1972), as modified by Stearn (1978a: 178-180) who divided *A. scorodoprasum* L. into four subspecies. All Greek plants are probable referable to subsp. *rotundum* (L.) Stearn. In the Greek checklist (Dimopoulos & al. 2013: 45) *A. rotundum* was recognized as a separate species.

Allium rupicola Boiss. ex Mouterde has been reported from the island of Rodos based on a single collection (Profitis Ilias above Salakos) by Bourgeau in 1870. The specimen was seen by Kollmann (cf. Fl. Turkey 8: 162, 1984) and the record is likely to be correct. However, there is no modern confirmation (Kleinsteuber, pers. comm. 2020), and the species must now be regarded as extinct in Greece. It occurs in S Anatolia and Lebanon.

Allium sphaerocephalon subsp. ***aegaeum*** (Heldr. & Halácsy) Karavok. & Tzanoud. (*A. aegaeum* Heldr. & Halácsy) was described from Naxos based on a collection by Leonis ("in arenosis maritimis prope urbem"), and has subsequently been reported from scattered localities elsewhere in Greece. It is doubtfully distinct from *A. sphaerocephalon* s. str.

Allium suaveolens Jacq. is a species of C Europe and N Italy with a few records also from the W Balkan Peninsula. It has been collected at high altitude in southern Pindos (see Aldén 1976: 316 as well as Meyer 2011: 11). Mermygas, Zikos & Constantinidis (2021: supplementary file) listed it from Peloponnisos, but no details are available. A recent collection from Mt Tzoumerka may also belong to this species (Trigas, in litt. 2022-02-10). Greek specimens should be critically compared with C European material.

Allium topalianum was reported as a new species from the islands of Evvia and Andros by Trigas & Bareka (2019). It does not appear to have been formally described.

Allium tzanoudakisianum Brullo, Pavone & Salmeri has been described from the island of Naxos. It falls within the range of variation of *A. cupanii* subsp. *hirtovaginatum*.

Aloe vera (L.) Burm. f., a native of NE Africa and parts of the Arabian Peninsula, is frequently cultivated as an ornamental in hot and dry parts of Greece. An excellent illustration appeared on Plate 341 of *Flora Graeca* (1823), probably based on material from Cyprus. The species may persist for some time in ruderal habitats but is scarcely naturalized.

Alopecurus bulbosus Gouan. A record from Strimonikos kolpos in nomos Thessalonikis (Eleftheriadou & Theodoropoulos 2003: 303) is feasible but needs confirmation. The species occurs mainly in W Europe and the W & C Mediterranean region, but has also been reported from the Istanbul area (Fl. Turkey 8: 375, 1985).

Alopecurus davisii Bor. Rechinger 4112 (1934-04-15, LD) from Mt Kerki, Samos, was listed as *A. vaginatus* (Willd.) Boiss. by Davis (1985: 378), but as *A. davisii* by Doğan (1999: 261). The type of the latter is Davis 1688 (E, K), collected in the same locality 1940-05-02. The first collection from Mt Kerki is probably Forsyth Major 532 (1887-05-06), listed by Stefani, Forsyth Major & Barbey (1892: 64) as *A. vaginatus*. *A. davisii* is restricted to Mt Kerki and Nif Dağ in the Izmir province, whereas *A. vaginatus* is a more widespread, Irano-Turanian species (described from Crimea). *A. lanatus* Sm., described from Ulu Dağ, is closely related.

Alopecurus myosuroides Huds. Most of the Greek plants represent subsp. *myosuroides*, but some from the East Aegean Islands may belong to subsp. *tonsus* (C. I. Blanche ex Boiss.) M. Doğan in which the awn is non-exserted or absent (cf. Doğan 1999: 249-250).

Alopecurus setarioides Gren. This species was described from S France; records from the Balkan Peninsula and W Anatolia need revision. In Greece there is one record from Kriti, cited by Böhling & Scholz (2003: 19), and one from Lesvos (Bazos 2124, B; det. Scholz). The identity of these plants is uncertain, and they may represent a form of *A. rendlei* Eig.

Aloysia citriodora Palau, a xenophyte of southern South American origin, has been reported from a ruderal habitat in the town of Nafplion (NE Peloponnisos) – see Raus (2007: 211).

Alyssum. See also *Odontarrhena*.

Alyssum bertolonii Desv. was listed for "Gr" in Fl. Eur. ed. 2, 1: 367 (1993). See, however, comments in Fl. Hellen. 2: 224 (2003).

Alyssum cuneifolium Ten. Plants from serpentine screes on Mt Smolikas as well as Evvia and Attiki have tentatively been referred to this species (not mapped), but may represent a form of *A. montanum* subsp. *repens* which has subsequently been included in *A. spruneri* (see below). *A. cuneifolium* was described from S Italy.

Alyssum densistellatum T. R. Dudley from serpentine areas of northern Evvia was tentatively accepted as a separate species by Strid (2016, 1: 151), but included in *A. spruneri* by Španiel, Zozomová-Lihová & Marhold (2017: 1437).

Alyssum doerfleri Degen. For variation in this species and the related *A. taygeteum* Heldr., see Greuter (1974a). The type of *A. doerfleri* is Dörfler, it. turc. sec. 1893 no. 48 (B, G, W, WU) from "Kossov, pr. Zborsko", a locality in North Macedonia near the Greek border.

Alyssum idaeum Boiss. & Heldr. was reported from south central Anatolia by Orcan & Binzet (2006). The article is confused and should be disregarded.

Alyssum montanum-repens complex. Numerous taxa have been described (cf. Hartvig in Mount. Fl. Greece 1: 286-291, 1986), but Španiel, Zozomová-Lihová & Marhold (2017) included all Greek material in *A. spruneri* Jord. & Fourr. This presumably includes Greek records of *A. scardicum* Wettst.

Alyssum pulvinare Velen. was included in Mountain Flora of Greece (1: 285, 1986), based on a single collection from Mt Iti which differs slightly from the Bulgarian type. No further information has become available.

Alyssum stribrnyi Velen. was listed for Greece in the Euro+Med PlantBase, but no Greek records have been located. A single dot in NW Greece in Jalas, Suominen & Lampinen (1996: 46) actually refers to a locality just N of the border.

Alyssum trichostachyum Rupr. Greek collections previously referred to this species were included in *A. spruneri* by Španiel, Zozomová-Lihová & Marhold (2017).

Amaranthus emarginatus subsp. ***pseudogracilis*** (Thell.) Hügin was reported from sandy banks of a freshwater stream by the harbour of Georgioupoli in W Kriti by Raus & Svensson (2017: 89) and has subsequently been collected in suburban Athens by Polymenakos (2020-07-26, no. 1059, ATH). Other Greek records are assumed to represent subsp. *emarginatus*.

Amaranthus graecizans L. Two subspecies, subsp. *graecizans* and subsp. *sylvestris* (Villiers) Brenan, have been reported in Greece and elsewhere in Europe, the latter being the most common.

Amaranthus watsonii Standl. has been reported once, from the vicinity of Githion in S Peloponnisos (Raus 2006b: 720), and may be on the verge of becoming naturalized. A native of SW North America, it is a dioecious, erect, glandular-pubescent annual with long-petiolate leaves. The fruit dehisces transversely to reveal a smooth, reddish-black seed.

Amaryllis belladonna L. has been reported from the Ionian Islands and Andros, but only as a garden escape.

Ammannia senegalensis Lam. has been reported once in Greece, in S Peloponnisos, between Kalamata and Messini, SE of the airport, weed in rice fields, associated with *Ammannia coccinea*, *Bergia capensis*, *Cyperus difformis*, *Lindernia dubia*, etc. (Raabe s.n., 23.9. 2002; B, MSTR, herb. Raabe) – see Raus & Raabe (2002b: 202). It differs from *A. baccifera* and *A. coccinea* chiefly by its effuse, clearly pedunculate inflorescences. The degree of naturalisation in Greece is uncertain.

Ammannia verticillata (Ard.) Lam. A report by Lavrentiades (1973) from rice fields at the Gallikos river mouth (W of Thessaloniki) is assumed to represent *A. baccifera* L.

Ammophila arenaria (L.) Link. It has recently been proposed to transfer this species to the genus *Calamagrostis* (see Banfi, Galasso & Bertolucci 2018).

Ampelodesmos mauritanicus (Poir.) T. Durand & Schinz is a species of the W & C Mediterranean area. It was recorded from Zakinthos already by Boissier (Fl. Orient. 5:

563, 1881) and has been recently confirmed. The single record from Kos (the only one in the Flora of Turkey area) is far out of range, and the specimen cited (Rechinger 7896) should be critically re-examined.

Amphoricarpos autariatus Blešić & E. Mayer. According to the very narrow species concept of Blešić & Mayer (1967) *A. neumayeri* Vis. (= *A. neumayerianus* (Vis.) Greuter) is endemic in a small coastal area in Dalmatia. Greek material would be referable to *A. autariatus* sp. nov. subsp. *bertisceus* subsp. nov. Differences are obscure, and probably all Balkan populations belong to a single species, *A. neumayerianus*, with some regional variation. See also map in Niketić & al. (2014: 62) and analysis by Caković & al. (2015). The widely used name *A. neumayeri* Vis. is illegitimate and has been replaced by *A. neumayerianus* (Vis.) Greuter.

Anacamptis coriophora (L.) R. M. Bateman, Pridgeon & M. W. Chase. All plants in southern Greece (S of c. 39°30'N) belong to the Mediterranean subsp. *fragrans* (Pollini) R. M. Bateman, Pridgeon & M. W. Chase. In the N & NE mainland this co-exists with the C European subsp. *coriophora*. Their distributions overlap completely in this area, and they are often difficult to distinguish. Thus only a collective map can be presented.

Anacamptis papilionacea (L.) R. M. Bateman, Pridgeon & M. W. Chase. The species has been excessively subdivided. Subsp. *thaliae* Kreutz, J. Essink & L. Essink from Rodos and adjacent SW Anatolia is probably a synonym of subsp. *aegaea* (P. Delforge) J. Lewis & Kreutz.

Anacyclus radiatus Loisel. has been listed for Greece, e.g., in Flora Europaea 4: 168 (1976), perhaps based on historical records from Kriti and Peloponnisos. The latter are likely to refer to *A. clavatus* (Desf.) Pers., and so is a recent record from the Thessaloniki area (Krigas & Kokkini 2004: 88). *A. radiatus* is a species of the C & W Mediterranean area.

Anagallis. Manns & Anderberg (2009) proposed an extended definition of the genus *Lysimachia*, to include *Asterolinon* and *Anagallis*. In this Atlas the two last-mentioned are retained in their traditional sense.

Anchusa arvensis (L.) M. Bieb. (*Lycopsis arvensis* L.) was listed for "Gr" in Flora Europaea 3: 108 (1972), but there appear to be no confirmed records (cf. Dimopoulos & al. 2013: 261).

Anchusa leucantha Selvi & Bigazzi (2003) is a later homonym for *A. leucantha* Willd. ex DC. (1846). The former name has subsequently been recombined as a subspecies of *A. officinalis* (see Kit Tan & al. 2013) and is provisionally retained as such although it may well deserve species rank.

Anchusa pusilla Guşul. and ***A. thessala*** Boiss. & Spuner. Several collections from NE Greece registered as *A. thessala* may in fact represent *A. pusilla*. Most of the specimens are young, and well developed fruits seem to provide the only safe clue.

Androsace maxima L. was listed for Peloponnisos by Mermygkas, Zikos & Constantinidis (2021: supplementary file), probably in error. Confirmed Greek records are all in the north.

Andryala dentata Sm. was described from the island of Milos and illustrated on Plate 831 of Flora Graeca (1837). It has often been regarded as a synonym of *A. integrifolia* L. (described from France and Sicily), but Polymenakos & Kit Tan (2017c: 429-430) kept them separate, reporting the former from Schinias, Attiki. Further studies are needed to determine whether differences in leaf shape and other characters are consistent. There is also a recent report of *A. dentata* from Rodos (Ristow, in litt.). If the two are maintained as separate species, all Greek material is likely to belong to *A. dentata*.

Andrzeiowskia cardamine Rchb. was listed for "Gr" in Fl. Eur. ed. 2, 1: 378 (1993), and the map in Atlas Fl. Eur. 11: 112 (1996) indicates a locality in the north-east. The record is feasible but needs confirmation (cf. also Fl. Hellen. 2: 297, 2003).

Anemone narcissiflora L. and ***A. sylvestris*** L. have both been reported by Zaganiaris (1940a: 41) from Karlik Dağ (Papikio) in NE Greece. Although Zaganiaris is notoriously unreliable, these records are feasible since both species have been confirmed to occur north of the border in Bulgaria.

Anthemis abrotanifolia (Willd.) Guss. was listed for Lefka Ori by Franzén in Mount. Fl. Greece 2: 427 (1991). The source for this record is not known, and there is no subsequent confirmation; it must thus be regarded as dubious.

Anthemis amblyolepis Eig, accepted as a separate species in the Greek checklist (Dimopoulos & al. 2013: 53), is scarcely distinct from *A. palaestina* (Reuter ex Kotschy) Boiss. It has subsequently been recombined as *A. palaestina* subsp. *amblyolepis* (Eig) Feinbrun, and it is not clear whether both taxa occur in Greece.

Anthemis chia L. After the preparation of the maps this species has been reported from the Rodopi mountains by Charalampidou, Eleftheriadou & Theodoropoulos in Fl. Medit. 32: 84 (2022). The record is out of range and probably incorrect.

Anthemis macedonica Boiss. & Orph. Dimopoulos & al. (2013: 53) recognized subsp. *macedonica* and subsp. *thracica* (Griseb.) Oberpr. & Greuter. The latter was based on *A. montana* var. *thracica* Griseb., which was described from Turkey-in-Europe and may rather be synonymous with *A. tenuiloba* (DC.) R. Fern., a taxon in the *A. cretica* complex – see Strid (2000b: 271).

Anthemis melampodina Delile, a desert species of NE Africa and Syria, was reported from Attiki by Heldreich (1877b: 500). No details are available and there is no subsequent confirmation.

Anthemis melia Biel & Kit Tan has recently been described from the island of Milos (see Biel & Kit Tan 2021c: 241-243). It differs from *A. auriculata* Boiss., which also occurs on Milos, by its smaller stature, tomentose-pubescent indumentum, smaller capitula with short ligules, involucral bracts with distinct brown margins, and much shorter corona. From *A. tomentosa* L. which so far has not been found on Milos, it differs by its much smaller capitula, shorter fertile ligules and brown (not scarious) involucral bract margins. *A. melia* was reported from several localities, growing in phrygana from sea level to 720 m and flowering from late April to mid-May. It should be looked for elsewhere in the Kiklades.

Anthemis parnesia Boiss. & Heldr. was described from Mt Parnitha and is known from a few, mostly historical, collections in NE Peloponnisos and SE Sterea Ellas. It differs from *A. auriculata* Boiss. only in vague quantitative characters and is probably conspecific.

Anthemis peregrina L. Populations of *A. tomentosa* from W Greece have been referred to *A. peregrina*, but there are no morphological differences. Occasional inland populations from Attiki and Peloponnisis have been called *A. peregrina* var. *heracleotica* Boiss. & Heldr. or *A. guicciardii* Heldr. & Sartori.

Anthemis pindicola Halásy was described from Mts Karava and Voutsikaki in Southern Pindos. It is probably synonymous with *A. cretica* subsp. *columnae* (Ten.) Franzén.

Anthemis rigida Heldr. is common on coastal flats in the C & S Aegean area and ascends to high altitude in Cretan dolines. It usually has discoid capitula and is rather uniform. Ligulate forms occur locally on some islands and are mutually different. They were subject to a detailed study by Goula & Constantinidis (2021) who recognized four ligulate subspecies in addition to the widespread, eligulate subsp.

rigida. An additional ligulate form on the island of Amorgos was described a few months later as subsp. *amorgina* Biel & Kit Tan; it occurs in the C & N parts whereas subsp. *rigida* is widespread on the island.

Anthemis segetalis Ten. All Greek records of this species need confirmation. At least some of them refer to *A. austriaca* Jacq. or *A. coelopoda* Boiss.

Anthemis tomentosa L. See under *A. peregrina* L.

Anthericum ramosum L., a species of C & E Europe, has recently been reported from Mt Orvilos (Doumas, Goula & Constantinidis 2022: 3-4). It differs from *A. liliago* in the more branched inflorescence, smaller flowers and shorter, obtuse capsule. However, there are at least three previous collections from Mt Orvilos identified as *A. liliago*, viz., Stamatiadou 20376 (ATH, C), Strid & al. 19335 (herb. Strid) and Strid & al. 27297 ((ATH, G, LD). My collection no. 19335 does not differ from other Greek populations of *A. liliago*, which frequently have a branched inflorescence. The report of *A. ramosum* on Mt Orvilos thus needs confirmation. [Thanks are due to Kit Tan for drawing my attention to this].

Anthoxanthum odoratum L. At low altitude it may be difficult to determine whether the plants are annual or perennial. Thus the difference between *A. odoratum* and *A. ovatum* is sometimes obscure. Both diploid and tetraploid plants of *A. odoratum* are widespread. High-altitude diploids from N & C Europe have been distinguished as *A. alpinum* Á. Löve & D. Löve. Such plants occur on several Greek mountains from the N borders to Mt Psiloritis (Kriti), but can scarcely be separated morphologically.

Anthoxanthum pauciflorum Adamović is a taxon of uncertain status, described from mountains in North Macedonia: "In summis alpinis montium Babuna-Planina et Kajmakčalan (Nidže-Planina)". The latter locality is on the Greek border. It probably falls within the range of variation of *A. odoratum*, although several differences were listed in the original description. There are subsequent collections of *A. odoratum* from high altitude on the Greek side of Kajmakčalan. In the Greek checklist (Dimopoulos & al. 2013: 122) *A. pauciflorum* was listed as an accepted species and recorded also from KK (Kriti-Karpathos). Occurrence in the latter area is highly unlikely.

Anthriscus nitidus (Wahlenb.) Hazsl. has been reported from one locality in nom. Kastorias, based on Hügin 6953 (B). The specimen could not be located at B in March 2016, and the record needs confirmation. *A. nitidus* is a member of the widespread and variable *A. sylvestris* group. It is mainly found in mountains of C & E Europe; Reduron (2007a: 379) recorded it southwards to Bulgaria and S Italy.

Anthyllis hermanniae L. According to Brullo & Giusso del Galdo (2006b) subsp. *hermanniae* is restricted to SW Albania, Greece and coastal areas of W Anatolia. Different subspecies occur in the C Mediterranean area, and a closely related species, *A. hystrix* (Willk. ex Barceló) Cardona, is endemic to Minorca (Balearic Islands).

Anthyllis montana L. is very variable, and several subspecies and varieties have been recognized. All Greek plants are probably referable to subsp. *jacquinii* (A. Kern.) Hayek which occurs in the SE part of the range of the species.

Anthyllis vulneraria L. Subdivision and data on morphology, ecology and distribution of the subspecies follow unpublished work by Bergmeier. For nomenclatural reasons subsp. *alpicola* (Brügger) Guterm. replaces subsp. *alpestris* Asch. & Graebn. (see Gutermann 2006: 72). Subsp. *hispidissima* (Sagorski) Cullen, a taxon described from C Anatolia, has been reported also from Greece, but these records are likely to refer to subsp. *rubriflora*.

Aphanes australis Rydb. was described from North America. The few and scattered Greek records probably refer to *A. minutiflora*.

Apium nodiflorum (L.) Lag. is now usually referred to the genus *Helosciadium*, as *H. nodiflorum* (L.) W. D. J. Koch. In the photo appendix it appears under both names (as Plate 15:2 and 19:9), and this inconsistency was discovered too late to correct.

Aquilegia. Plants from N Albania illustrated in Vangjeli (2017: 273) resemble *A. ottonis* subsp. *amaliae*, but detailed comparison is needed to establish their identity.

Aquilegia aurea Janka. This distinctive, yellow-flowered species was reported from Mt Kerkini (Beles) by Zaganiaris (1940a: 42). The record has never been confirmed but is feasible since the species occurs in SW Bulgaria as well as North Macedonia.

Aquilegia dumeticola Jord. One collection from Mt Gramos (Strid & Vassiliadis 55376, 2004-06-28; G, herb. Strid and photos), originally called *A. nigricans*, was revised by Nardi (2015: 269) as *A. dumeticola*. The latter is a member of the *A. vulgaris* complex, described from Corsica and widespread in Italy. It has also been reported from Montenegro and Bosnia-Herzegovina, but the identity of Balkan material is uncertain, and the Gramos collection is provisionally retained under *A. nigricans*.

Aquilegia meridionalis (Quézel & Contandr.) E. Nardi. An *Aquilegia* was collected in 1857 "ad rupes in faucibus umbrosis m. Parnassi l. d. Dypotamo" (Guicciardi 566; FI, GBOIS, K, WUHAL). It does not appear to have been confirmed on Mt Parnassos. The same taxon was collected on Mt Giona by Halácsy in 1888 (see Halácsy 1888: 748) and subsequently by Maire & Petitmengin in 1904 (see Maire & Petitmengin 1907: 7). Quézel & Contandriopoulos (1965a: 110) described plants from Mt Giona as a new species (or subspecies?) with the epithet *australis* (or *meridionalis*?): ".... nous amène à considérer les échantillons du Giona comme une sous-espèce particulière, la ssp. *australis* nov. sp.: *Aquilegia amaliae* Heldr. ssp. *meridionalis* nov. sp. .....". Nardi (2014: 105) designated Maire & Petitmengin 308 (FI) from "rochers calcaires humides exposés au nord-est dans les gorges au-dessus de Segditsa [Prosilion], près de la fontain de Platylithos" as neotype for *A. meridionalis* (Quézel & Contandr.) E. Nardi. Although further studies of variation are needed, the plants from Parnassos and Giona do not seem distinct from *A. ottonis* subsp. *ottonis* (see Strid in Mount. Fl. Greece 1: 73, 2003).

Arabidopsis halleri (L.) O'Kane & Al-Shehbaz has been reported once from the Rodopi Mountains by Eleftheriadou (1992: 43), sub nom. *Cardaminopsis halleri* (L.) Hayek. It is widespread from C Europe through temperate Asia (Al-Shehbaz & O'Kane 2002: 13-15); the Greek record needs confirmation.

Arabis hirsuta (L.) Scop. was listed for Peloponnisos by Mermygkas, Zikos & Constantinidis (2021: supplementary file). The record is feasible, but no details are available. After the preparation of the maps it has also been reported from the Rodopi mountains by Charalampidou, Eleftheriadou & Theodoropoulos in Fl. Medit. 32: 84 (2022).

Arabis nemorensis (Hoffm.) W. D. J. Koch. A report from NW Greece has turned out to be incorrect (see Bergmeier & Hand 2009: 338).

Arabis nova Vill. Greek records probably refer to *A. auriculata* Lam.

Arabis rosea DC. was listed for Greece in Euro+Med PlantBase, but is probably endemic to Sicily and S Italy.

Araujia sericifera Brot. is a woody climber native to S America. It is cultivated as an ornamental and naturalized in SW Europe. In Greece it has been reported from ruderal habitats in W Kriti, N Peloponnisos and the Preveza area, and may be on the verge of becoming established.

Aremonia agrimonoides (L.) DC. Both subsp. *agrimonoides* and subsp. *pouzarii* Skalický have been reported from northern and central Greece. The latter has more numerous leaflets and somewhat larger petals. Their distributions overlap completely and there are many intermediates.

Arenaria oxypetala Sm. is a species of uncertain status, see comments in Strid & Strid (2011: 74).

Arenaria phitosiana Greuter & Burdet was based on a single collection from the island of Gioura (N Sporades). There are no significant differences between this and *A. filicaulis* subsp. *graeca*.

Aristida adscensionis L. was first reported from Kriti by Rechinger (1943b: 178): "Gattung und Art sind neu für Kreta, die Ägäis und die Balkan-Halbinsel überhaupt". Greek plants have subsequently been referred to *A. caerulescens* Desf. Although perennial, the latter is doubtfully distinct from the pantropical weed *A. adscensionis*; it has been recombined as *A. adscensionis* subsp. *caerulescens* (Desf.) Bourreil & Trouin. Thiv, Reyes-Betancort & Fragman-Sapir (2019) concluded that the two cannot be separated as species or subspecies, and that *A. caerulescens* is merely a synonym of *A. adscensionis*. The unusual species epithet is derived from Ascension Island in the South Atlantic, where it was collected in 1751 by the Swedish priest and naturalist Peter Osbeck en route to Canton.

Aristolochia elongata (Duch.) E. Nardi is an illegitimate name and must be replaced by *A. nardiana* I. M. Turner (Ecology and Evolution. 2022;12:e8765. https://doi.org/10.1002/ece3.8765). [Thanks are due to Th. Raus for drawing my attention to this recent publication].

Armeria. All Greek material is in need of critical revision. *A. canescens* (described from Dalmatia) is taken in a wide sense, including *A. rumelica* Boiss. (described from "Rumelia", probably = S Bulgaria), *A. adamovicii* Halácsy (from Mt Peristeri in SW North Macedonia), *A. orphanidis* Boiss. (from Mt Chelmos) and *A. thessala* Boiss. & Heldr. (from Mt Olimbos). The provisionally recognized local taxa (*A. cariensis*, *A. icarica*, *A. johnsenii*, *A. sancta* and *A. undulata*) are all related to *A. canescens*.

Armeria cariensis Boiss., a species described from SW Anatolia, was tentatively reported from a small island off Chios (Fl. Turkey 7: 503, 1982) and subsequently from some other small East Aegean islands (Panitsa 1997: 76, Panitsa & Tzanoudakis 2010: 281). It is a taxon of uncertain status in the *A. canescens* complex, said to differ by the narrowly linear inner leaves (less than 1 mm wide), and short-pedicellate flowers.

Armeria undulata (Bory & Chaub.) Boiss. is a taxon of uncertain status, related to *A. canescens*. It was described from Mt Likeo and a locality near Megalopolis in Peloponnisos. An illustration in Bory & Chaubard (1832-33: Plate 10) shows a plant with conspicuously undulate margins of the outer leaves. There are scattered records from Peloponnisos and the Greek mainland as well as the islands of Lefkas and Limnos, most of them historical literature reports in need of confirmation. The few modern collections identified as *A. undulata* have broad outer leaves distinctly contrasting with the narrowly linear inner ones, but without the undulate margins of the type specimen.

Artemisia annua L. was included in the Greek checklist (Dimopoulos & al. 2013: 53), but the background for this record is obscure. In Schuler (2004: 152) there is an unspecified record of *Artemisia annua* listed under species from northern Greece with "Soziologisches Verhalten indifferent bzw. ungeklärt", but this may have been adopted from the maps of Meusel & Jäger (1992) or Meusel et al. (1965, 1978) [A. Schuler in litt., 2019-05-26]. The species is known from Bulgaria and occurrence in NE Greece is feasible, but so far unconfirmed.

Arthrocnemum macrostachyum (Moric.) K. Koch. The species usually known under this name has recently been transferred to the new genus *Arthrocaulon*, comprising only the widespread *A. macrostachyum* (Moric.) Piirainen & G. Kadereit and one local species from the Cape Verde Islands (Piirainen, Liebisch & Kadereit 2017). *Arthrocnemum glaucum* (Delile) Ung.-Sternb. (a name used, e.g., in Fl. Turkey 2: 320, 1967), is apparently a synonym of *A. macrostachyum*.

Arundo. Recent taxonomic and nomenclatural history of this genus has been rather turbulent. The present account follows Hardion & al. (2012a, 2012b, 2014). Three species are recognized to occur in Greece: *A. donax* L. (now almost cosmopolitan), *A. plinii* Turra (syn. *A. collina* Ten., *A. hellenica* Danin; Italy and Balkan Peninsula) and *A. micrantha* Lam. (syn. *A. mauritanica* Desf., *A. mediterranea* Danin; circum-Mediterranean).

Asparagus aphyllus L. All Greek records are likely to represent the E Mediterranean subsp. *orientalis* (Baker) P. H. Davis, although difference between this and the W Mediterranean subsp. *aphyllus* are not very clear.

Asparagus plumosus Baker, a native of South Africa, is sometimes cultivated as an ornamental and has been reported to be locally naturalized in coastal pine forest near Xilokastro (N Peloponnisos), fide Kit Tan based on Sfikas 14077.

Asparagus tenuifolius Lam. was listed for Peloponnisos by Mermygkas, Zikos & Constantinidis (2021: supplementary file), probably in error. The rather few verified Greek records are all in the north.

Asperula. Phylogenetic speculations based on molecular evidence have recently led to the proposal to raise the sections *Cynanchica*, *Thliphthisa* and *Hexaphylla* to the level of genus, thus making room for a large number of new combinations (Del Guacchio & Caputo 2020). Evidence for this seems, at best, to be flimsy.

Asperula incana Sm., which was listed as an accepted species in Euro+Med Plant Base, is a synonym of *A. pubescens* (Willd) Ehrend. & Schönb.-Tem.

Asperula involucrata Wahlenb. was listed for "Gr" in Flora Europaea 4: 12 (1976), perhaps going back to the imprecise "Thra. The." in Hayek's Prodromus (2: 444, 1930). No specimens have been seen, but occurrence in Greece is feasible since the species is known to occur in SE Bulgaria.

Asperula lilaciflora Boiss. Several subspecies have been recognized. Subsp. *coa* (Rech. f.) Ehrend., subsp. *phrygia* (Bornm.) Schönb.-Tem. and subsp. *runemarkii* Ehrend. & Schönb.-Tem. have been reported from the East Aegean Islands, but the differences are vague and scarcely consistent. *A. icarica* Ehrend. & Schönb.-Tem. is a synonym of *A. lilaciflora* subsp. *runemarkii*.

Asperula lutea Sm. is a variable species in need of further study. Five subspecies have been recognized in Greece, but the differences are not clear and thus only a collective map is provided. See also comments in Greuter (2012: 107) and Strid (2016, 1: 514).

Asperula suberosa Sm. Plants from high altitude on Mt Pirin in SW Bulgaria have been called *A. aristata* var. *pirinica* (Stoj. & Acht.) Ančev, and subsequently referred to *A. suberosa* (Peev & al. 2015: 193). Detailed comparison is needed, but the population from Mt Pirin is unlikely to be conspecific with *A. suberosa* from Mt Athos.

Asperula taurina L. This species is scattered in S Europe and through N Anatolia to Caucasia. Zografidis (2017) found a small population in damp *Abies* woodland on Mt Xirovouni, Evvia, confirming an unpublished record by Dinter (no. 6109, coll. 5.5. 1999) from the same area. A record by Goulimis (1960a: 20) from the Vikos gorge had previously been regarded as dubious, but Zografidis found two specimens at ATH: Goulimis no. 20277 from the Vikos gorge and no. 20278 from Pisoderion (nom. Florinis). Limited evidence suggests that Goulimis' collections represent subsp. *leucanthera* (Beck) Hayek, whereas specimens from Mt Xirovouni match subsp. *taurina*.

Asperula tenella Heuff. ex Degen. The inclusion of this species is based on a single collection from Rodopi Mts: Strid & al. 44487 (G, LD, herb. Strid). Literature records from low altitude in the Thessaloniki area (Krigas & Kokkini 2005: 66) and the island of Thasos (Konstantinou & Babalonas 1992: tab. 1) are probably incorrect.

Asperula tenuifolia Boiss. In Fl. Turkey 7: 749 (1982) this species was reported from Mt Kerkis, Samos, based on Rechinger 2038 and Runemark & Nordenstam 16971. The specimens have not been seen and may represent the subsequently described *A. samia*.

Asplenium adulterinum Milde was reported from serpentine rocks on Mt Smolikas by Quézel & Contandriopoulos (1965b: 53). The record has not been confirmed and may refer to a form of *A. trichomanes*. *A. adulterinum* is scattered in C & N Europe.

Asplenium ×aprutianum Lovis, a hybrid between *A. lepidum* and *A. trichomanes*, has been recorded once from Lefka Ori (Zaffran 1990: 66).

Asplenium ceterach subsp. ***bivalens*** (D. E. Mey.) Greuter & Burdet and subsp. ***mediterraneum*** Pinter were both listed for "Gr" in Euro+Med PlantBase. There are no clear morphological differences between these and the widespread subsp. *ceterach*.

Asplenium fontanum (L.) Bernh. Records of this species from Rodos and Kriti are almost certainly incorrect (cf. Finkl 1962: 116 and Greuter 1975g: 154).

Asplenium ×khaniense Jermy & Brownsey, a hybrid between *A. creticum* and *A. trichomanes* subsp. *quadrivalens*, has been described from Lefka Ori (Brownsey & Jermy 1975).

Asplenium trichomanes subsp. ***pachyrachis*** (H. Christ) Lovis & Reichst. was reported from Lefka Ori by Lovis & Reichstein (1985: 192). Two other subspecies have also been reported from Lefka Ori, and variation in this area needs further study. The recently recognized subsp. *hastatum* (H. Christ.) S. Jess. has been reported from one locality in Peloponnisos (see Jessen 1995: 114).

Aster amellus L., a widespread European species extending to the Caucasus and Ural mountains, has recently been reported from deciduous oak woodland at c. 750 m in the Greek part of Rodopi (Doumas, Goula & Constantinidis 2022: 8-9). Historical records from Attiki are probably incorrect.

Asteriscus maritimus (L.) Less. has been reported from Greece in several recent works including Flora Europaea (4: 139, 1976). The records all appear to go back to Halácsy (1902a: 17) who had not seen any specimens but cited older literature. There is no recent confirmation and the records are likely to be incorrect.

Asterolinon. Manns & Anderberg (2009) proposed an extended definition of the genus *Lysimachia*, to include *Asterolinon* and *Anagallis*. In this Atlas the two last-mentioned are retained in their traditional sense.

Astracantha Podlech. A large group of astragali, mainly spiny species of SW Asia, has long been known as "tragacantha". Podlech (1983) typified *Tragacantha* Mill. (1754) as to become a synomym of *Astragalus* L. (1753), a nomenclatural trick that made room for coining the new name *Astracantha* and making hundreds of new combinations. It has subsequently been recognized that *Astragalus* s. str. and *Tragacantha/Astracantha* can not be maintained as separate genera due to the occurrence of species with intermediate features.

Astragalus angustifolius Lam. is a widespread and moderately variable species which was excessively fragmented by Brullo, Giusso del Galdo & Musarella (2012).

Astragalus apollineus Boiss. & Heldr. was listed for Peloponnisos by Mermygkas, Zikos & Constantinidis (2021: supplementary file), but no details are available.

Astragalus baldaccii Degen was described from Mt Dousko (Njemerska) on the Greek/Albanian border. It was retained as a separate species by Podlech & Zarre (2013: 2054), but there appear to be no consistent differences between this and *A. lacteus*. A related species, *A. gracaninii* Micevski, has been described from Galičica planina and the upper Vardar valley in North Macedonia (Micevski 1971); it differs in the smaller, rose-purple corolla. Closely related to the latter is *A. wilmottianus* Stoj., a rare regional endemic of W Bulgaria (cf. Velčev 1984: 192).

Astragalus cephalonicus C. Presl, described from the island of Kefallinia, was accepted as *A. sempervirens* subsp. *cephalonicus* (C. Presl) Asch. & Graebn. in the Greek checklist, but merged into *A. sempervirens* without taxonomic rank by Podlech & Zarre (2013: 1372).

Astragalus corinthiacus Brullo, Giusso & Musarella is a name recently given to a single population (or individual?) of *A. sempervirens* from Mt Parnassos. In the Flora Hellenica Database are 12 previous records of *A. sempervirens* from Mt Parnassos and 27 from elsewhere in Sterea Ellas. None of them were considered.

Astragalus ictericus Dingler was based on a collection by Dingler, 3 May 1876 (M) from [NE Greece] "in declivitate meridionali Rhodopes prope pagum Koslukŏi inter virgulta alt. 1200 ped." [locality not precisely identified]. It is likely to be a synonym of *A. angustiflorus* subsp. *anatolicus*, although retained by Podlech & Zarre (2013: 558) as a separate species known only from the type collection. *A. maronensis* Dingler, based on another collection from the same area, "in fruticetis declivitatis septentrionalis promontorii montis Geicklü-Dagh inter Macri et Maronia Thraciae meridionalis [W of Alexandroupolis] alt. 1000 ped.", (Dingler s.n., 9 May 1876; M) is also likely to be a synonym of *A. angustiflorus* subsp. *anatolicus*.

Astragalus idaeus Bunge. For comments on this rare species, see Atlas of the Aegean Flora 1: 274 (2016). Subsequently it has been rediscovered also on Mt Psiloritis by Johan Grudemo, 5.7. 2016. Photos of an individual in late flowering stage were sent to Nick Turland (B) who informed Dionysios Vassiliades. The latter returned to the locality 20.11. 2016 and found plants still in recognizable condition. They have longer peduncles but otherwise appear to match plants from Mt Dikti. Flowering plants were photographed by Grudemo in the same locality, late May 2017. An old and previously uncertain record from Mt Psiloritis has thus been confirmed.

Astragalus maniaticus Kit Tan & Strid, described from S Peloponnisos, was retained as a separate species by Podlech & Zarre (2013: 964). With more material from the area now available it is evident that it represents somewhat depauperate, annual specimens of *A. suberosus* subsp. *haarbachii*. Plants with relatively narrow legumes occur throughout the distribution range of the latter. Such plants have been described as *A. argolicus* Hausskn. (type: "M. Palamidi supra Naupliam", 2 May 1885, Haussknecht; K, JE, LE, W, WU), and subsequently recombined as *A. suberosus* var. *argolicus* (Hausskn.) Podl.

Astragalus mesopterus Griseb. has been recorded from NE Greece and S Bulgaria and was accepted as a separate species by Podlech & Zarre (2013: 1774). It seems doubtfully distinct from *A. onobrychis*.

Astragalus sesameus L. is a species of the W & C Mediterranean area. Records from Greece are likely to refer to the similar *A. sinaicus*, with the possible exception of a specimen from Kerkira (Schratt & Gutermann 19863a) which is in early flowering stage. *A. sesameus* has subsequently been reported from SW Albania (Barina 2017: 241) and could well occur in Kerkira. Fruiting material is needed for positive identification.

Astragalus tymphresteus Boiss. & Spruner was based on a collection by Spruner, 1842 (G-BOIS, M, P) "in summo monte Velugo (Tymphresto veterum)". It is here considered conspecific with *A. sirinicus* Ten., described from the Apennines, "in montes Sirini, prope Lago negro", Tenore (BR, K, RE, W).

Asyneuma canescens (Waldst. & Kit.) Griseb. & Schenk was listed for Peloponnisos by Mermygkas, Zikos & Constantinidis (2021: supplementary file). The record is feasible, but no details are available.

Asymeuma limonifolium (L.) Janchen. The monographer Damboldt (1970: 68) aptly observed that this is "eine sehr variable, formenreiche Art, deren zahlreiche Standortsmodifikationen fast alle als eigene Arten beschrieben worden sind", but finally gave taxonomic recognition only to subsp. *limonifolium* and the Anatolian subsp. *pestalozzae* (Boiss.) Damboldt. In Greece this species has an unusually wide altitudinal range, in the Olimbos area from sea level to 2600 m; the extremes are strikingly different but there are more or less gradual transitions.

Atriplex littoralis L. All Greek records of this species (Tocl & Rohlena 1902: 2, Jalas & Suominen 1980: 40, Biel & Kit Tan 2007a: 109) are likely to refer to *A. patula*.

Atriplex oblongifolia Waldst. & Kit. Greek records of this species need confirmation. It was mapped from the Greek mainland in Atlas Fl. Eur. 5: 36 (1980), but this was apparently based on historical records by Formánek which were later revised by Vandas (1909: 511-512) as *A. rosea* and *A. patula*. There is a single, more recent report from the island of Kalimnos (Zervou, Raus & Yannitsaros 2009: 168) which is feasible although rather out of range.

Atriplex sagittata Borkh. There are recent collections from coastal areas of Andros and Karpathos as well as a historical record from the Thessaloniki area. Kit Tan (in Fl. Hellen. 1: 123, 1997) regarded it as "a rare casual".

Aubrieta. The account of this genus is based on Phitos (in Flora Hellenica 2: 192-195, 2003). Some of the more distinct local or regional variants were cherry-picked as separate species, whereas *A. deltoidea* is a heterogeneous conglomerate of all that remains. The genus is in need of critical revision. In cultivation even distinctly different species (such as *A. erubescens* and *A. thessala*) hybridize freely. The report of *A. scyria* Halácsy from Mt Giona is quite recent (see Kit Tan & Kofinas 2021: 268-269). Apparently both *A. deltoidea* and *A. scyria* occur in this area.

Aurinia petraea (Ard.) Schur. Greek records of this species probably refer to *A. corymbosa* Griseb.

Aurinia sinuata (L.) Griseb. has been reported from NW Greece, e.g., by Pavlides (1985: 194). A specimen in the Pavlides herbarium is in poor condition but may represent *A. saxatilis* subsp. *orientalis*. No confirmed Greek records have been located, but the species occurs in Italy and the W Balkan Peninsula and could well extend to Greece.

Austrocylindropuntia subulata (Mühlenpf.) Backeb., a native of the Peruvian Andes, has recently been reported from scattered localities in Peloponnisos and on the islands of Milos, Amorgos and Inousse. It may be on the verge of becoming naturalized.

Avena byzantina K. Koch is a species of cultivated oats adapted to hot and dry climate. It differs from the common cultivated oats (*A. sativa* L.) in the somewhat larger spikelets (27-30 mm) with rachilla segments breaking at their base and falling attached to upper floret, and awn lacking a distinct column. There are scattered reports from ruderal habitats in Greece.

Avena sterilis subsp. ***atherantha*** (C. Presl) H. Scholz and subsp. ***trichophylla*** (K. Koch) Malzev are dubious taxa listed for Greece in Euro+Med PlantBase.

Avena strigosa Schreb. Literature records from NE Greece (Kruse, Monser & Pausch 1995: 40) and Peloponnisos (Dinter 2004 and Mermygkas in litt.) need confirmation.

Azolla. Greek material was first reported as *A. caroliniana* Willd. (Raus 1982b: 183), but has subsequently been revised as *A. filiculoides* Lam.

Bacopa rotundifolia (Michx.) Wettst., a native of the Americas, has been reported from a few localities in NE Greece and one in Spain (see Raus 2003). The first Greek collection appears to be Strid & al. 27365, 25.6. 1988 (ATH, G, GB, herb. Strid, herb. Kit Tan) gathered "along main road Serre-Thessaloniki, just E of Strimonas river bridge". There is also one subsequent record from the Acheloos river estuaries in Etolias-Akarnanias (Koumpli-Sovantzi 2008: 322). The somewhat similar *B. monnieri* (L.) Wettst. (with spathulate, 1-veined leaves and solitary flowers) occurs in the SE Mediterranean area (Nile delta to upper Jordan valley) as well as NW Spain; it has recently been reported also from Hungary (Király, Steták & Bányász 2007) and should be looked for in Greece.

Ballota nigra L. Taxonomy at subspecies level is preliminary, and there are many intermediates between the four subspecies recognized here. Many Greek records have been referred to subsp. *uncinata* (Fiori & Bég.) Patzak, but according to Euro+Med PlantBase this is a synonym of subsp. *ruderalis* (Sw.) Briq. Subsp. *foetida* (Vis.) Hayek is apparently a synonym of subsp. *meridionalis* (Bég.) Bég. *B. nigra* subsp. *anomala* Greuter was based on a single collection from S Peloponnisos. It differs from subsp. *ruderalis* only in the presence of small supplementary calyx teeth, a character of dubious taxonomic value.

Barbarea balcana Pančić has been reported from the Prespa National Park by Pavlides (1985: 197), but specimens in the Pavlides herbarium are of uncertain status and confirmation is needed.

Barbarea conferta Boiss. & Heldr. was based on Heldreich 3646 (B, TAU, WU), collected "in montis Kyllenes Achaiae regione media, prope pagum Trikala. Apano-Markala". There is no subsequent confirmation on Mt Killini nor elsewhere, and the status of this taxon is uncertain. It may be a form of *B. vulgaris* which has been recorded from neighboring localities in N Peloponnisos.

Basella rubra L. has been reported from Peloponnisos and Kefallinia, growing in flower-beds and ruderal habitats and scarcely naturalized. It is a pantropical vine sometimes cultivated as a vegetable.

Bassia hirsuta (L.) Kuntze has been transferred to the monotypic genus *Spirobassia* (see Kadereit & Freitag 2011: 71).

Bassia hyssopifolia (Pall.) Kuntze has been reported once from a ruderal habitat in SE Kriti (Deschâtres & Greuter 2001: 320). It is native in E Europe and temperate Asia; the record from Kriti probably represents a casual introduction.

Bassia prostrata (L.) Beck (*Kochia prostrata* (L.) Schrad.) has been reported once from the Evros river delta (Babalonas 1981: 256). It is widespread in S Europe and temperate Asia. The Greek record is feasible but needs confirmation.

Bellevalia edirnensis Özhatay & B. Mathew (1991) was described from a locality in Turkey-in-Europe close to the Greek border. It was subsequently reported as new for Greece by Bareka, Phitos & Kamari (2008: 738), based on a collection by Bareka (2004) from the north-eastern corner of Greece. In fact it had been collected in this area already by Raus & Schiers (no. 18022, 7.5. 1991; B) and by Strid & Lassen (no. 50744, 10.5. 2000; G, LD).

Bellevalia sarmatica (Pall.) Miscz. was tentatively reported by Bazos (2005: 326) from the island of Lesvos. It is a later synonym for *B. speciosa* Woronow ex Grossh. which has subsequently been confirmed from Lesvos (see Meyer & Bazos 2020: 307-308). A historical report of *B. ciliata* (Cirillo) Nees from a locality on Lesvos (Candargy 1898) probably refers to *B. speciosa* as well.

Bellevalia spicata (Raf.) Boiss. was listed as a separate species in Euro+Med PlantBase, but is a synonym of *B. hyacinthoides* (Bertol.) K. Perss. & Wendelbo.

Bergia capensis L., a semi-aquatic weed of paleotropical origin, has been collected once in a rice field in S Peloponnisos (Raabe s.n., 3 Oct 2001). The nearest native occurrence is in Egypt.

Beta vulgaris L. Several inland collections by Willing have been named subsp. *macrocarpa* (Guss.) Thell. by Th. Raus. Most of the specimens are too young for positive identification and have provisionally been mapped only as *B. vulgaris* s. lat.

Biarum tenuifolium (L.) Schott. Taxonomy at subspecies level is provisional. The main flowering period is in the late summer and autumn, but some populations flower as early as May. It remains to be established to what extent flowering time is correlated with morphological variation.

Bidens albus (L.) DC., a native of tropical and subtropical areas in the Americas, has recently been found in ruderal habitats on the islands of Rodos (Kleinsteuber, Ristow & Hassler 2016: 154) and Karpathos (Strid 60375; UPA, herb. Strid), and may be on the verge of becoming naturalized. It resembles *B. pilosus* but has 4-7 showy white ligules.

Bistorta vivipara (L.) Delarbre (*Polygonum viviparum* L). There are no confirmed Greek records, but this species may well occur on granitic or schistose mountains in northernmost Greece as it is known from SW Bulgaria.

Blackstonia acuminata (W. D. J. Koch & Ziz) Domin is closely related to *B. perfoliata* and sometimes regarded as a subspecies of the latter. Zeltner (1991) recognized *B. acuminata* subsp. *acuminata* (apparently widespread in Greece) and subsp. *aestiva* (K. Malý) Zeltner (co-existing with subsp. *acuminata* in Rodos and possibly elsewhere). Morphological differences are not clear, but limited evidence suggests that subsp. *acuminata* is tetraploid with 2n=40 and subsp. *aestiva* diploid with 2n=20.

Blitum capitatum L. has recently been collected on a sandy beach at Igoumenitsa (Dimitrov s.n., 2013-08-04, SOM 171836). It resembles *B. virgatum* L., but has less toothed, broadly triangular leaves and fewer and larger cymes in the inflorescence. *B. capitatum* is native in North America and has been reported as naturalized or casual in several European countries; it may be on the verge of becoming established also in Greece.

Blysmus compressus (L.) Link. After the preparation of the maps this species has been reported from the Rodopi mountains by Charalampidou, Eleftheriadou & Theodoropoulos in Fl. Medit. 32: 90 (2022).

Bolanthus graecus (Schreb.) Barkoudah was listed for Peloponnisos by Mermygkas, Zikos & Constantinidis (2021: supplementary file). The record is feasible, but no details are available.

Bolanthus intermedius Phitos was based on a single collection from low altitude in northern Evvia. Further gatherings show that there are no consistent differences between this and *B. thymifolius*.

Bolboschoenus. Taxonomy follows Hroudová & al. (2007). *B. glaucus* (Lam.) S. G. Sm. is often difficult to distinguish from *B. maritimus* (L.) Palla, and in older literature the two have usually not been separated. Some mapped records of the latter may thus refer to the former.

Bornmuellera baldaccii subsp. ***rechingeri*** Greuter was based on a single collection differing only vaguely in leaf indumentum. It is scarcely worthy of taxonomic recognition at any level, so unfortunately this attractive name (commemorating three great botanists) will have to disappear into synonymy.

Bornmuellera emarginata (Boiss.) Rešetnik (2014) is a recombination of *Leptoplax emarginata* (Boiss.) O. E. Schulz. Although the latter hybridizes with both Greek species of *Bornmuellera*, the name is retained here because of the distinct morphological differences (cf. Hartvig in Mount. Fl. Greece 1: 275, 1986).

Bothriochloa pertusa (L.) A. Camus was recorded from a roadside in S Peloponnisos by Ristow (2013: 246). It occurs from North Africa to India with its only previous European localities in Sicily. The status of the Greek population, whether native or a casual introduction, is not clear. *B. pertusa* differs from the common *B. ischaemum*, e.g., by having a conspicious depression (pit) in the lower glume.

Bouteloua dactyloides (Nutt.) Columbus, a North American species ("buffalo grass"), has been reported once from a lawn in Iraklion, Kriti (Böhling & Scholz 2003: 30).

Brachypodium distachyon (L.) P. Beauv. is a polyploid complex comprising *B. distachyon* s. str. (2n=10), *B. stacei* Catalán (2n=20) and their derived allopolyploid *B. hybridum* Catalán (2n=30) – see Catalán & al. (2012, 2016). All three are widespread in the Mediterranean area, and the last-mentioned is also naturalized in other regions with a suitable climate (California, South Africa, Australia, etc.). All occur in Greece, but in practice it is not possible to identify herbarium specimens with any degree of confidence. Thus only a collective map is presented.

Brachypodium phoenicoides (L.) Roem. & Schult. is a species of the W & C Mediterranean area. Greek records (e.g., Hansen 1982: 34 from Kerkira, Phitos & Damboldt 1985: 157 from Kefallinia, Dimopoulos 1993: 109 from Mt Killini) refer to *B. retusum* (Pers.) P. Beauv. or *B. rupestre* (Host) Roem. & Schult.

Brachypodium sylvaticum (Huds.) P. Beauv. Three subspecies have been reported in Greece, viz. subsp. *sylvaticum*, subsp. *creticum* H. Scholz & Greuter and subsp. *glaucovirens* Murb., the first-mentioned being the most common. Subsp. *creticum* differs in quantitative characters (fewer spikelets, awn equalling lemma, etc.); it is restricted to high altitude in W Kriti and merges into subsp. *sylvaticum* at lower altitude. Subsp. *glaucovirens*, described from Herzegovina, has variously been regarded as a subspecies or variety of *B. sylvaticum* or as a separate species, *B. glaucovirens* (Murb.) Sagorski. There are scattered records from the Aegean area, including Samothraki (Biel & Kit Tan 2007a: 111) and Kriti (Bergmeier 2011: 173). The monographer Schippmann (1991: 165) reduced it to a synonym of *B. sylvaticum* subsp. *sylvaticum*.

Brassica cadmea Heldr. ex O. E. Schultz was based on Heldreich 821 (B, GB, K, LD, WU), collected 23 Apr. 1885 "in collibus argillosis prope Thebas". The plant illustrated in Flora Graeca 7: tab. 648 (1830) under the name *Brassica fruticulosa* Cyr. may be conspecific (the true *B. fruticulosa* Cyr. in a different, west Mediterranean species). The locality given in Flora Graeca is merely "in collibus Graeciae", but the plant may have been gathered and illustrated on Sibthorp's excursion to Mt Parnassos in June 1787. In a letter to Turrill dated 31.5. 1932 (in Kew archives) Pinatzi mentions having re-collected it, presumably in the same year, "près de la ville de Thebes, sur une colle aride". It does not appear to have been seen since.

Brassica napus L. is probably derived from hybrids between *B. oleracea* and *B. rapa*. Various forms are cultivated for their oil-producing seeds or swollen, edible roots. In Greece and elsewhere in Europe it is occasionally naturalized in ruderal habitats.

Brassica oleracea L. is probably native in coastal areas in W Europe. Different forms are widely cultivated and occasionally naturalized in ruderal habitats.

Brimeura fastigiata (Viv.) Chouard has been reported from Mt Taigetos; see Damboldt & Melzheimer (1976) for description and chromosome number (2n=28). The record is very doubtful. According to Almeida da Silva, Sáez & Rosselló (2001) *B. fastigiata* occurs in Corsica, Sardinia and the Balearic islands. The report by Damboldt & Melzheimer may have resulted from a confusion of labels with plants from Corsica cultivated in the same greenhouse. The genus *Brimeura* has 3 species, all in N Spain, the Balearic Island, Corsica and Sardinia.

Briza elatior Sm. was described from Mt Athos and has subsequently been recombined as *B. media* subsp. *elatior* (Sm.) Rohlena. There are no obvious differences between this and other forms of *B. media* (cf. Essi & al. 2017).

Bromopsis pindica (Hausskn.) Holub (= *Bromus pindicus* Hausskn.) was listed as an accepted species in Euro+Med PlantBase. It was described from "schistosis humidiusculis jugi Zygos [Lakmon] P[indi] T[ymphaei]", and is probably a synonym of *Bromus riparius* Rehmann.

Bromus arvensis subsp. ***parviflorus*** (Desf.) H. Scholz is an ill-defined taxon originally described (as *B. parviflorus* Desf.) from plants cultivated in the Royal Botanic Garden in Paris. It has subsequently been reported from widely scattered localities in NE Greece, Croatia and S Sweden (Raab-Straube & Scholz (2013).

Bromus catharticus Vahl (syn. *Ceratochloa cathartica* (Vahl) Herter) is an American weed that has been reported from a few localities in W Kriti, along streets and at margins of olive groves (see Böhling & Scholz 2003: 37, Turland & al. 2004: 399). It may be on the verge of becoming naturalized.

Bromus commutatus Schhrad. Most, perhaps all, Greek collections belong to subsp. *commutatus*. *B. commutatus* subsp. *decipiens* (Bomble & H. Scholz) H. Scholz, originally described from Germany as *B. secalinus* subsp. *decipiens* Bomble & H. Scholz, is a segetal weed which may occur in NW Greece. *B. commutatus* subsp. *neglectus* (Parl.) P. M. Sm. has been doubtfully recorded from the Thessalian plain and Mt Iti.

Bromus elidis H. Scholz was described from a somewhat ruderalized coastal habitat in NW Peloponnisos. It has not been reported elsewhere and is scarcely distinct from *B. racemosus*.

Bromus inermis Leyss. was listed for Peloponnisos by Mermygkas, Zikos & Constantinidis (2021: supplementary file), but no details are available. There are very few confirmed Greek records of this species, all in the central and northern mainland.

Bromus intermedius subsp. ***optimae*** (H. Scholz) H. Scholz is a dubious taxon described from Cyprus and subsequently reported also from the East Aegean Islands (cf. Greuter & Raus 2002: 205).

Bromus japonicus Thunb. Scholz (2005: 61-62) reported subsp. *anatolicus* (Boiss. & Heldr.) Pénzes from a few localities in Greece. Differences between this and subsp. *japonicus* are obscure.

Bromus kunkelii (H. Scholz) H. Scholz was described from the Canary Islands as *B. madritensis* subsp. *kunkelii* H. Scholz, and has subsequently been reported from a single coastal locality on the northern side of the Gulf of Kornithos (Constantinidis & Scholz 2006: 727). It is doubtfully distinct from *B. rubens* L.

Bromus madritensis subsp. ***madritensis*** and subsp. ***haussknechtii*** (Boiss.) H. Scholz have been recognized in Greece, but the morphological differences are obscure and there is no clear geographical separation. Most of the Greek records have not been identified to subspecies.

Bromus parvispiculatus H. Scholz has been reported from scattered localities on the Greek mainland and Peloponnisos. It appears to fall within the range of variation of *B. hordeaceus* L., representing individuals with small spikelets.

Bromus racemosus L. A few, scattered records of subsp. *lusitanicus* (Sales & P. M. Sm.) H. Scholz & Spalton need confirmation.

Bromus regnii H. Scholz was described from Cyprus and appears doubtfully distinct from *B. japonicus*. It has subsequently been reported from Lesvos (Scholz & Bazos 2004: 78) and from one locality in North Central Greece (Scholz 2008b: 470).

Bromus squarrosus subsp. ***consimilis*** H. Scholz is a highly dubious taxon, described from Samothraki on the basis of a single, fragmentary collection.

Buddleja asiatica Lour., ***B. davidii*** Franch. and ***B. madagascariensis*** Lam. have been recorded in ruderal habitats and along roads. All are doubtfully naturalized.

Bufonia euboica Phitos & Kamari was described from northern Evvia. It differs from *B. stricta* only in vague, quantitative characters, probably related to its low-altitude habitat. At the other extreme is a high-altitude ecotype from Mt Psiloritis (Kriti) which has been called *B. stricta* subsp. *cecconiana* (Bald.) Rech. f. The species has a wide altitudinal range in Kriti, and there are rather gradual transitions between subsp. *stricta* and subsp. *cecconiana*.

Bufonia tenuifolia L. There are several historical records of this species; those that have been possible to check represent *B. paniculata* Dubois, and *B. tenuifolia* may be restricted to the W Mediterranean area. *B. parviflora* Griseb., described from the Thessaloniki area, is apparently a synonym of *B. paniculata*. A report of *B. paniculata* from Rodos (Constantinidis 2004c: 72) refers to the subsequently described *B. ophiolitica* Ristow, Jos. Krause & Rätzel.

Buglossoides arvensis complex. The maps are provisional. Only a fraction of the vast collections by Willing at B have been critically examined. In particular, *B. incrassata* subsp. *splitgerberi* (Guss.) E. Zippel & Selvi is likely to be more widespread than indicated by the records. According to Zippel & Raus (2020) the latter occurs in much of C & E Europe, extending to Ukraine, Anatolia and Armenia.

Buglossoides glandulosa (Velen.) R. Fern. is a taxon of uncertain status in the *B. arvensis* complex, differing mainly by its glandular and strigose indumentum. It was described as *Lithospermum glandulosum* Velen., based on a collection by Střibrný from the vicinity of Sadovo in Bulgaria, and has subsequently been found also elsewhere in Bulgaria as well as Romania (cf. Kiryakov & Petrova 2004). In Greece it has been reported as a weed in a silvicultural nursery near Thessaloniki (Greuter & Raus 2004: 72).

Bunias orientalis L. was listed for "Gr" in Fl. Eur. ed. 2, 1: 325 (1993), but no specimens or specified records have been seen.

Bunium divaricatum Bertol. was reported from Akarnanika Ori by Maire & Petitmengin (1908: 97), observing "Cette plante est nouvelle pour la flore grecque. Son aire géographique s'étend de l'Istrie jusqu'a l'Acarnanie; elle est particulièrement fréquente en Dalmatie". The name is probably a synonym for *B. alpinum* subsp. *montanum* (W. D. J. Koch) P. W. Ball, which has subsequently been reported from Mt Parnassos by Garnweidner (1986: 124) and Mt Paikon by Athanasiadis & Drossos (1990: 110). All records need confirmation.

Bupleurum veronense Turra was listed for Peloponnisos by Mermygkas, Zikos & Constantinidis (2021: supplementary file), probably in error. Confirmed Greek records are all in the north-west.

Calamintha menthifolia Host (*C. sylvatica* Bromf.) and ***C. nepeta*** (L.) Savi are both variable and sometimes difficult to distinguish. Their synonymy is very convoluted. Three subspecies of *C. menthifolia* have been recognized in Greece, but differences are obscure. Most, perhaps all, Greek records of *C. nepeta* are referable to subsp. *glandulosa* (Req.) P. W. Ball.

Calamintha vardarensis Šilić was described from North Macedonia and has been reported also in Albania, Greece and Bulgaria. It is probably not specifically distinct from *C. nepeta* (L.) Savi.

Calendula officinalis L. is widely cultivated as an ornamental and may persist for a short time in ruderal habitats but is scarcely naturalized. It is a robust annual or short-lived perennial with capitula much larger than those of *C. arvensis*.

Callitriche hamulata Kütz. ex W. D. J. Koch has been reported from the islands of Mikonos and Samothraki. It is not clearly different from *C. brutia*, and Lansdown (2008: 132) reduced it to a variety of the latter.

Callitriche obtusangula Le Gall has been reported "in ulig. maritim. inter Naupliam et Tirynth" (Haussknecht 1894: 103), as well as from nomos Pellis (Papastergiadou & Babalonas 1993a: 140) and by Lake Kerkini (Papastergiadou & Babalonas 1993b: 194); all Greek records need confirmation. The species is widespread in C & W Europe. Lansdown (2008: 102) did not map any localities in the Balkans, but there is a subsequent record from NW Albania (Barina 2017: 340).

Calluna vulgaris (L.) Hull, a widespread and gregarious European species, was reported from Karlik Dağ (Papikio) in NE Greece by Zaganiaris (1940a: 85). There is no subsequent confirmation, but a mistake seems unlikely. The nearest locality is in SE Bulgaria, and it has recently been reported also from Turkey-in-Europe.

Camelina sativa (L.) Crantz. Greek records, as far as they have been possible to check, represent *C. microcarpa* Andrz. ex DC.

Campanula sect. ***Quinqueloculares***. Taxonomy in this group of rock-dwelling campanulas is provisional. It basically follows Phitos (1964) where too many ill-defined species and subspecies have been recognized. A detailed study by Liveri (2021) provided valuable data on morphology, palynology and karyology, but little new with respect to taxonomy and distribution.

Campanula abietina Griseb. & Schenk, described from Siebenbürgen, was reported from Greece in Euro+Med PlantBase. It is doubtfully distinct from the widespread *C. patula*.

Campanula andrewsii A. DC. Plants from Monemvassia (SE Peloponnisos), described as subsp. *hirsutula* Phitos, are scarcely distinct enough to merit taxonomic status.

Campanula asperuloides (Boiss. & Orph.) Engl. A shade form from rather low altitude in the Langada gorge (E foothills of Mt Taigetos) has been described as *Trachelium taygeteum* Quézel & Contandr. Greuter & Burdet recombined it as *Campanula asperuloides* subsp. *taygetea*, evidently without seeing any specimens. It has subsequently been recollected in the Langada gorge (e.g., Tzanoudakis & al. 4355; UPA, herb. Strid). There are no essential differences between these plants and other populations of *C. asperuloides*.

Campanula boreosporadum Raus is *nomen novum* for a taxon originally described as *Symphyandra sporadum* Halácsy (1895), based on a collection by Reiser from the rocky island of Gioura in the Northern Sporades. When transferred to *Campanula* the latter would be a homonym for *C. sporadum* Feer (1890). It belongs to a small group of three disjunct taxa (*C. cretica* (A. DC.) D. Dietr., *C. samothracica* (Degen) Greuter & Burdet and *C. boreosporadum*) optionally regarded as species or subspecies.

Campanula calaminthifolia Lam. was listed for the island of Ikaria in Fl. Turkey 6: 34 (1978), based on collections by Guiol. In the Flora Hellenica Database are no other collections by Guiol from Ikaria, and the record seems dubious.

Campanula chalcidica (Buser) Engl. was recognized as a Greek endemic in Euro+Med PlantBase. It was originally described as *Diosphaera chalcidica* Buser, based on a collection by Janka (1871) from Mt Athos. There are several subsequent collections from this locality, e.g., Strid & al. 15919 (C, G). They are referable to *C. jacquinii* (Sieber) A. DC. subsp. *rumeliana* (Hampe) Kit Tan.

Campanula jacquinii subsp. *jacquinii* and subsp. *rumeliana* are rather distinct and optionally regarded as separate species.

Campanula dichotoma L. is unlikely to occur in Greece. A record from the Edessa area (Goulimis 1960a: 23) probably refers to *C. scutellata* Griseb. 18th century records from elsewhere in Greece, e.g., Dumont d'Urville (1822: 24) from Kos and Friedrichsthal (1838: 277) from Poros are likely to refer to other annual species.

Campanula dimitrii Strid, stat. & nom. nov. = *C. carpatha* var. *oreophila* Phitos is a local endemic of cliff ledges in the summit area of Mt Kali Limni, Karpathos. It differs in several charcters from *C. carpatha* Halácsy which is rather widespead at low altitude in the Karpathos island group, and since there appear to be no intermediates it is reasonable to classifiy the small, white-flowered mountain plant from Kali Limni as a separate species (see Strid in Phytol. Balcan. 27: 390-391, 2022).

Campanula epigaea Janka ex Degen was based on syntypes from Bulgarian Rodopi (leg. Janka, 1871), Mt Pelister in SW North Macedonia (leg. Orphanides, 1862) and the island of Samothraki (leg. Degen, 1890). It has subsequently been reduced to *C. patula* subsp. *epigaea* (Janka ex Degen) Hayek, but was retained as a separate species in the Bulgarian Flora (11: 124, 2012). It may constitute an endemic Balkan taxon, but its relations to the widespread and variable *C. patula* are not clear.

Campanula garganica Ten. is a variable, amphi-adriatic species. Plants from the Ionian Islands have long been recognized as subsp. *cephallenica* (Feer) Hayek. A local population from Akarnanika Ori has been described as *C. acarnanica* Damboldt, and subsequently reduced to *C. garganica* subsp. *acarnanica* (Damboldt) Damboldt. It does not appear to be consistently different from subsp. *cephallenica*. The group has recently been subject to excessive splitting by Bogdanović & al. (2014) who recognized 11(!) species.

Campanula kamariana Kyriak., Liveri & Phitos. On 11th September 1997 Strid & Kit Tan observed non-flowering plants of an unidentified *Campanula* on the stone walls of Kelefa castle in the Mani peninsula. It appeared very distinctive by its much divided, subglabrous leaves. The species was first collected here some eight years later by D. Lakušić. Photographs were sent to the present author who could confirm his opinion that they represented an undescribed species. Publication was delayed awaiting molecular data for the new species and its presumed relatives to establish its phylogenetic position. A seed collection was made on 23rd October 2015 (Strid 58170), and the species was established in cultivation. By early 2017 word on the whereabouts of the undescribed species had spread to Georgia Kamari who sent a student to collect it. At record-breaking speed (less than four months after the "discovery") it was published in a friendly journal under the name *Campanula kamariana* by her husband Dimitrios Phitos and two students – without mentioning the real discoverer. Any qualified and unbiased reviewer would have alerted the editor to the circumstances, thus preventing such unfortunate publication (cf. also under *Leucojum ionicum*).

Campanula kastellorizana Carlström was based on a single collection (Greuter 11829) from the island of Ro in the Kastellorizo group. It was described as being densely hispid, with lax, corymbose inflorescence and short and broad corolla tube. A subsequent collection from the adjacent island of Megisti (Strid 60661) matches *C. kastellorizana* in inflorescence and corolla characters but is hispid only on the calyx teeth. Other collections from Megisti (e.g., Stamatiadou 16506), reported by Greuter (1979d: 589) as *C. drabifolia* Sm., are likely to represent the same taxon. Further studies are needed to determine whether *C. kastellorizana* is consistently different from *C. rhodensis* A. DC.

Campanula macrostachya Willd. After the preparation of the maps this species has been reported from the Rodopi mountains by Charalampidou, Eleftheriadou & Theodoropoulos in Fl. Medit. 32: 85 (2022).

Campanula nisyria Papatsou & Phitos was described from the East Aegean island of Nisiros. *C. hagielia* and *C. lyrata* are not clearly distinguishable in this area, and *C. nisyria* is one of the intermediate forms. It is probably best considered to fall within the range of variation of *C. lyrata* (in spite of claims to the opposite by Phitos 2021).

Campanula podocarpa Boiss., a species of S Anatolia, was listed in Flora of Turkey (6: 54, 1978) from the East Aegean Islands, based on Runemark & Nordenstam 16744 (LD) collected on a small island near Rodos. This specimen has subsequently been revised by Carlström as *C. rhodensis* A. DC.

Campanula rotundifolia L. Taken in the broadest sense this occurs throughout the boreal zone, extending into the arctic. Several regional taxa have been described from the Balkans, including *C. velebitica* Borbás, *C. albanica* Witasek, *C. rotundifolia* subsp. *sancta* Hayek, *C. rotundifolia* subsp. *hellenica* Hayek, and *C. pindicola* Aldén. No modern, critical revision is available, and therefore only a collective map can be presented.

Campanula versicolor Andrews. According to a recent taxonomic revision (Janković & al. 2019) all Greek populations belong to subsp. *versicolor*. In the C Balkan Peninsula (NE Albania to W Bulgaria) this is replaced by subsp. *korabensis* (F. K. Mey.) I. Janković & D. Lakušić, and in SE Italy by subsp. *tenorei* (Moretti) I. Janković & D. Lakušić. Related species occur in the W Balkan Peninsula.

Cannabis sativa L. is a tall, much-branched, dioecious annual with leaves palmately 5-9-lobed into narrowly lanceolate, serrate segments, and inconspicuous, greenish flowers. It is probably native in SW Asia and has been cultivated since ancient times for its fibres (hemp) and narcotic resin (*marijuana, hashish, dagga*). In Greece it is found occasionally in ruderal habitats but is scarcely naturalized.

Capnophyllum peregrinum (L.) Lange = *Krubera peregrina* (L.) Hoffm. was reported from ?Gr in Flora Europaea (2: 358, 1968). Occurrence in Greece is feasible, but no specimens or precise records have been seen.

Capsella rubella Reut. Diploid, Mediterranean plants in the *C. bursa-pastoris* complex have traditionally been called *C. rubella*. They have small, often reddish-tinged petals and small siliculae with slightly concave lateral margins. However, Svensson (1984) demonstrated that, at least in Greece, there is no correlation between chromosome number and morphology. All records of *C. rubella* have thus been mapped as *C. bursa-pastoris* (L.) Medik.

Cardamine acris/amara/barbareoides: Karol Marhold and Judita Zozomova have kindly examined a specimen list extracted from the Flora Hellenica Database. The maps are based on their revisions, as well as Šlenker & al. (2021).

Cardamine flexuosa With. was listed for Greece in Euro+Med PlantBase, but Greek specimens identified as *C. flexuosa* appear to be large plants of *C. hirsuta* L. *C. flexuosa* is widespread in Europe, and it is by no means unthinkable that it could occur in the northern Greek mainland.

Cardamine graeca L. Plants with conspicuously papillose-setose ovary and siliqua have been called var. *eriocarpa* (DC.) Fritsch, but this character seems to vary even within populations.

Cardamine impatiens L. subsp. ***impatiens*** and subsp. ***pectinata*** (DC.) Stoj. & Stef. have largely overlapping distribution areas and may be better treated as separate species (cf. Kučera, Lihová & Marhold 2006). However, many of the Greek records have been identified only as *C. impatiens* s. lat., and subspecies rank for the two taxa is therefore provisionally retained in the map.

Cardamine occulta Hornem. is an annual of East Asiatic origin, recently introduced into C Europe where it is found mainly as a weed in plant nurseries. Ardenghi, Cauzzi & Galasso (2015: 121) reported it from a flower bed in the city of Iraklion, Kriti (sub nom. *C. hamiltonii* G. Don). Until further notice it must be regarded as doubtfully naturalized in Greece. See Marhold & al. (2016) and Šlenker & al. (2018a, 2018b). There is also a recent record from Cyprus (Hand 2020: 3).

Cardamine resedifolia L. No Greek specimens have been seen. Goulimis (1960a: 16) reported this species from Mt Vourinos, but his collection (at ATH) represents *C. plumieri* Vill. Euro+Med PlantBase listed *C. resedifolia* from Greece, referring to "Marhold, pers. comm., 2003", but no details are available.

Cardopatium boryi Spach is an enigmatic species based on a collection by Bory de St Vincent in 1829 from "Laconia", where no *Cardopatium* has been confirmed (except for a recent record from Monemvassia). It has been considered a synonym of *C. corymbosum* (L.) Pers., but the long original description indicates several differences. It does not appear to have been recollected.

Carduus hamulosus Ehrh. has been recorded from several localities in mainland Greece as well as the islands of Thasos and Samothraki. Most of the records are historical and all need confirmation. It is otherwise a species of E Europe and N Anatolia.

Carduus nutans L. subsp. *leiophyllus* (Petrović) Stoj. & Stef., subsp. *nutans*, subsp. *scabrisquamus* Arènes and subsp. *taygeteus* (Boiss. & Heldr.) Hayek have been recorded from Greece, but the differences are not clear and only a collective map is presented.

Carduus tmoleus Boiss. was described from W Anatolia ("Lydia Tmolus") and is widespread in the Greek mountains where it is often gregarious by shepherds' camps. Persistently grey-tomentose plants from Mt Parnonas (SE Peloponnisos) have been described as *C. cronius* Boiss. & Heldr., but are probably best regarded as a variety of *C. tmoleus* (cf. Mount. Fl. Greece 2: 480, 1991).

Carex agastachys Ehrh. ex L. f. One specimen from northern Pindos, originally identified as *C. pendula*, has been revised as the closely related *C. agastachys* (E. Ljungstrand, Oct. 2017). According to Jiménez-Mejías & al. (2017), *C. pendula* Huds. s. lat. comprises two largely allopatric species, *C. pendula* s. str. in W Europe and the Mediterranean area, and *C. agastachys* from C & E Europe to Caucasus and N Iran. For differences, see also Mígues, Martín-Bravo & Jiménez-Mejías (2018). Critical examination of all *C. pendula* specimens from N Greece is needed. One collection from the Athos peninsula (Sutorý, 2019-06-15, BRNM 816448) has subsequently been reported as *C. agastachys*; *C. pendula* s. str. was reported from the same area. *C. agastachys* has recently been recombined as *C. pendula* subsp. *agastachys* (L. f.) Ljungstrand.

Carex appropinquata Schumach. There are several records of this species from mountains of C and N Greece, but all are likely to refer to slender forms of *C. paniculata* L. (cf. Hartvig in Mount Fl. Greece 2: 845, 1991). Parent (2005: 229) reported true *C. appropinquata* from Mt Tzena which is feasible but needs confirmation. There is also a recent report from Bulgaria (Hájek, Hájková & Apostolova 2006: 367-369).

Carex atrata L. Specimens from Mt Vardousia were confirmed by Sebastian Gebauer (e-mail 2018-09-13) to represent *C. atrata* s. str., the southernmost and rather isolated locality for this arctic-alpine species. It was collected here already by Heldreich (see Halácsy 1904a: 321). A record of *C. atrata* from "mt Dokimi in Pindo" (Formánek) is dubious; it was not mentioned by Vandas (1909).

Carex bulgarica (Domin) Lazare. Balkan plants previously referred to *C. sempervirens* Vill. were recognized as a separate species, *C. bulgarica*, by Lazare (1984, 1988). However, *C. bulgarica* (Domin) Lazare is a later homonym for *C. bulgarica* Velen. (1889), the latter being validly published as an alternative name for *C. digitata* var. *bulgarica* Velen. The Balkan taxon in the *C. sempervirens* complex has been renamed *C. lazarei* Jac. Koopman & al. (see Koopman & al. 2019).

Carex curvula All. Doubtfully present; see Mount Fl. Greece 2: 864 (1991).

Carex demissia Hornem. has been reported from the island of Samothraki (Biel & Kit Tan 2008e: 436). These specimens have long-pedunculate male spikes and small utricles (3-4 mm) with relatively long beak, thus matching *C. demissa* rather than the closely related *C. viridula* Michx. (as described and illustrated in Central European and British Floras). They belong to a taxonomically difficult group and their status is unsettled.

Carex egorovae A. M. Molina, Acedo & Llamas is a recently described species reported from localities in SE Europe (including the island of Thasos) and through Anatolia to N Iran. It seems doubtfully distinct from *C. divulsa* Stokes.

Carex elongata L. Flora Europaea (5: 301, 1980) reported this species from "?Gr". No substantiating specimens or records have been found. Koopman (2022: 129) did not map it in the Balkans.

Carex hostiana DC. Records from Mt Iti (Maire & Petitmengin 1908: 214, sub nom. *C. hornschuchiana* Hoppe) and from localities in NE Greece (Lavrentiades 1956: 76) need confirmation. It is known from Albania and Bulgaria (Koopman 2022: 160).

Carex kurdica Kük. ex Hand.-Mazz. Plants from Mt Varnous mapped as *C. acuta* may represent a westerly extension of the closely related *C. kurdica* (cf. Jiménez-Mejías & al. 2014: 71). It was first collected in the Pisoderi area by Alston & Sandwith 1932-06-12 (no. 809, K). However, the taxonomy is still unsettled (Jiménez-Mejías, e-mail 2018-07-16).

Carex laevigata Sm. Reports of this species from Greece probably refer to *C. distans* L.

Carex lepidocarpa Tausch. Records of this species from Greece refer to the subsequently described *C. castroviejoi* Luceño & Jim. Mejías.

Carex limosa L. The only Greek locality for this boreal species is a wet depression in the Voras massif called Dobro Polje very close to the border with North Macedonia where it was first collected by Mrkvička in 1916 (see Velenovský 1922: 32). The species is considered endangered in Bulgaria where it occurs in a few localities in the south and west (Peev & al. 2015: 429). It is also rare and threatened in the Vlasina plateau in SE Serbia (see Ranđelović 2010: 4). In the Dobro Polje locality it grows together with *C. lasiocarpa* Ehrh., another boreal species very rare in the Balkans.

Carex melanostachya M. Bieb. ex Willd. As for *C. elongata*.

Carex montana L. As for *C. curvula*.

Carex mucronata All. As for *C. curvula*.

Carex muricata complex. *C. muricata* L. s. str. is probably absent from Greece. For nomenclatural reasons the taxon previously known as *C. divulsa* subsp. *leersii* (Kneuck.) W. Koch has been renamed *C. polyphylla* Kir. & Ker. (see Buttler 2006). It is widespread in Greece and usually easily recognized on the distinctly separated lower spikes; all Greek material previously identified as *C. divulsa* is probably referable to this taxon. *C. spicata* Huds., with denser inflorescence and larger, spreading utricles, is also widespread. Scattered records of *C. pairae* F. W. Schultz all need confirmation.

Carex ornithopoda Willd. A single record of this species from the Nestos river gorge (Kirchhoff & Petermann 1992b: 62) may refer to *C. digitata*, but occurrence of *C. ornithopoda* in N Greece is feasible since there are localities in Albania, North Macedonia and Bulgaria.

Carex otomana A. M. Molina, Acedo & Llamas is a recently described and rather dubious species in the *C. muricata* complex, reported to occur from the Balkans to C Asia. In Greece there is a record from Mt Vrondous.

Carex pallidula Harmaja. Koopman, Więcław & Wilhelm (2016) reported this species from several European countries from Sweden and Finland to NE Greece. It has usually been included in *C. digitata*, and its taxonomic status remains unsettled.

Carex sempervirens Vill. See *C. bulgarica*.

Carex viridula Michx. Taxonomy and nomenclature are unsettled. Greek plants may be referable to *C. oederi* Retz.

Carex vulpina L. There are several historical records from Greece, but all are likely to represent misidentified material of *C. otrubae* Podp.

Carthamus lanatus L. Bergmeier & al. (2021 suppl. 1 p. 4) observed: "The plants [from the island of Limnos] have floral characters of subsp. *lanatus* (deep yellow flowers, rather short involucral bracts) and vegetative characters of subsp. *baeticus* (Boiss. & Reuter) Nyman (stems without glands, rather long leaf segments). Hanelt (1963) already mentioned transitional forms from the North Aegean".

Carthamus tinctorius L. is sometimes cultivated, but scarcely naturalized. It is not known in the wild state, but presumably originated in SW Asia. It has rather large, ovoid capitula (involucre 15-20 mm diam.) with orange-yellow florets and achenes without pappus. As a crop it is cultivated for its oil (extracted from the achenes) and red dye. The dried florets are sometimes sold as fake saffron.

Catapodium balearicum (Willk.) H. Scholz has been reported from few and widely scattered localities in the Mediterranean region, within the range of *C. marinum* (L.) C. E. Hubb. and often together with it. It has been tentatively accepted and mapped, but varietal rank is probably more appropriate.

Catapodium borgesii H. Scholz is a highly dubious taxon, based on a single specimen from Mt Dikti, Kriti, cultivated in Berlin. Judging from the description and illustration provided it is probably a form of *C. rigidum* (L.) C. E. Hubb.

Catapodium hemipoa (Spreng.) M. Laínz. A few records from sandy coastal habitats in Peloponnisos, Kriti, Rodos, Chios, Thasos and S Anatolia have been referred to this species which is optionally classified as *C. rigidum* subsp. *hemipoa* (Spreng.) Stace. It is a chiefly W Mediterranean taxon, rare in the east. Greek specimens need further study, and the distribution is imperfectly known. *C. hemipoa* subsp. *occidentale* (Paunero) H. Scholz & S. Scholz (= *C. zwierleinii* (Lojac.) Brullo) was recorded from KK and EAe in the Greek checklist (Dimopoulos & al. 2013: 124), but the identity of these records is uncertain.

Caucalis platycarpos L. was listed for Peloponnisos by Mermygkas, Zikos & Constantinidis (2021: supplementary file), probably in error. Confirmed Greek records are all in the north.

Cenchrus clandestinus (Hochst. ex Chiov.) Morrone (*Penniseum clandestinum* Hochst. ex Chiov.), a native of tropical East Africa, has been reported from irrigated lawns in Kriti, Rodos and the Ionian Islands as well as the Athens and Thessaloniki areas. It does not appear to be spreading into natural vegetation. Concerning taxonomy and nomenclature, see Thulin & Phillips (2015).

Cenchrus echinatus L. was reported by Verloove (2012b: 293) as new for Greece, without locality details; the record probably refers to a collection by Jagel (1991) from a coastal habitat on the island of Elafonisos S of Peloponnisos. The species is probably native in S USA, C & S America and the West Indies, and widely naturalized in tropical and subtropical areas mainly in the S hemisphere. In the Mediterranean area are scattered records from Spain to Egypt. It resembles *C. longispinus*, but the bur has numerous flexible, retrorsely barbellate outer bristles, and the inner bristles are ± arranged in a single whorl.

Cenchrus orientalis (Rich.) Morrone. No specimens have been seen, and the published records are only tentatively accepted as *C. orientalis*, a native of NE Africa and the Irano-Turanian region (Lebanon through Anatolia to Caucasia and India). It was first reported, as *Pennisetum orientale* Rich., from Evvia ("neu für Europa") by Rechinger (1961b: 452). It is locally common in serpentine scree in the area of Pili and Mantoudi; for ecology and phytosociology, see Pohl & Lenki (2004). Recently it has also been reported in a similar habitat on the adjacent mainland (Polymenakos & Kit Tan 2020b: 202). The Greek localities represent a considerable disjunction, and it should be critically examined whether these plants are indeed conspecific with those from Anatolia and elsewhere. The distribution and the specialized habitat on Evvia makes one suspect they could represent a separate, indigenous species.

Centaurea alba complex. Taxonomy and maps are provisional. Representatives of this group in the Iberian Peninsula were taxonomically revised by López & Devesa (2011), but no similar study is available for Greek and Balkan material.

Centaurea amplifolia Boiss. & Heldr. is a rare local endemic probably restricted to the Flambouritsa valley on Mt Killini, where it was first collected by Orphanides in 1851 (no. 125; C, G, GB, K, LD, WU) and by Heldreich a few years later. It was not rediscovered until 1993. Similar plants have been reported from SE Bulgaria and NW Turkey; the latter were subsequently described as a separate species, *C. wagenitziana* Bancheva & Kit Tan (see Kit Tan & al. 2009a). According to Bancheva & Dalcheva (2018: 115) it is now extinct in Turkey and known only from two small populations in Bulgaria. Another six or seven related species occur in Anatolia (Eren 2007, Doğu, Bağcı & Dinç 2009). This rather distinctive group of tall, large-leaved perennials is optionally referred to a separate genus, *Rhaponticoides*.

Centaurea armoraciifolia Sm. is a taxon of uncertain status. It was recorded from "Peloponnesi montibus", but the description is fragmentary and there are no subsequent reports. Euro+Med PlantBase listed it as a synonym of *C. ragusina* subsp. *lungensis* (Ginz.) Hayek which seems unlikely.

Centaurea atropurpurea Olivier. In spite of the distinct morphological differences this species hybridizes with *C. raphanina* Sm. *C. halacsyi* Dörfler and *C. nigrotriangulata* Rech. f. (both described from Amorgos) are apparently such hybrids. Strid 57757 (Herb. Strid and photo) from Naxos is also this hybrid. It was growing at the base of cliffs together with the putative parent, being intermediate in stature and flower colour.

Centaurea biokovensis Teyber was listed for "Cr" in Euro+Med PlantBase, referring to Hayek's Prodromus 2: 770 (1931). However, Hayek reported this species only from Dalmatia (Mt Biokovo) where it is likely to be endemic.

Centaurea charrelii Halácsy & Dörfl. is a rare Greek endemic belonging to the otherwise Irano-Turanian sect. *Cynaroides* (cf. Negaresh & Rahiminejad 2018: 32-36). It was discovered near Edessa by Louis Charrel, an amateur botanist in Thessaloniki who also called himself by the Turkish name Abdurrahman Nadji (type: Charrel s.n., 1893-07-18; B, K, JE, W). There is no confirmation from this area. The more southerly records are all fairly recent (Kokmotos & Georgiadis 2001: 212 & 2005: 413, Constantinidis & al. 2002: 114, Kokmotos 2008: 67).

Centaurea cylindrocephala Bornm., also known as *C. biebersteinii* subsp. *cylindrocephala* (Bornm.) Dostál is a member of sect. *Acrolophus* of uncertain taxonomic status, described from the Skopje area and reported also from a few localities in N Greece. It is an erect, short-lived perennial with numerous small capitula; involucre c. 8 × 4 mm, narrowly ovoid; phyllary appendage indistinctly demarcated, pale brown, with 3-4 pairs of fimbriae and a short apical mucro; florets pinkish.

Centaurea drakiensis Freyn & Sint. was described from the Volos area, and has been recombined as *C. thessala* subsp. *drakiensis* (Freyn & Sint.) T. Georgiadis. Differences between *C. thessala* subsp. *thessala* and subsp. *drakiensis* are obscure and they have overlapping distributions.

Centaurea epirota Halácsy is very close to *C. pindicola* (Griseb.) Boiss. Some plants from serpentine substrate on Mt Vourinos with pinkish-lilac florets are intermediate between the two. It should not be confused with the similarly named *C. epirotica* (Bald.) Halácsy (= *C. deusta* var. *epirotica* Bald.), a member of the *C. alba* complex.

Centaurea graeca subsp. ***ceccariniana*** (Boiss. & Heldr.) Dostál (*C. ceccariniana* Boiss. & Heldr.) was described from Mt Parnassos, based on a collection by Heldreich. It differs from *C. graeca* s. str. only in the long and stout spine of the phyllary appendages. This character is variable even within populations, and *C. ceccariniana* is scarcely worthy of taxonomic recognition (cf. Wagenitz & Gamal-Eldin 1985: 112-113).

Centaurea grbavacensis (Rohlena) Stoj. & Acht. The tongue-twisting specific epithet was derived from the town of Grbavac near Prilep in North Macedonia (not from the town with the same name in Croatia), where it was first collected by Vandas in 1923 (type at PRC). Plants from Mt Tzena cultivated by the present author in the Copenhagen Botanic Garden around 1980 hybridized spontaneously with the rather different *C. graeca* Griseb. from the foothills of Mt Olimbos.

Centaurea hyalolepis Boiss., was included in the Greek checklist (Dimopoulos & al. 2013: 55) based on historical records, and was also listed from Peloponnisos by Mermygas, Zikos & Constantinidis (2021: supplementary file). The species was described from Syria and is unlikely to occur in Greece.

Centaurea iberica Spreng. The epecific epithet is derived from Iberia, an ancient kingdom in present-day Eastern Georgia. Greek material has been provisionally referred to subsp. *holzmanniana* (Boiss.) Dostál, but differences are not clear between this and the widespread subsp. *iberica*.

Centaurea incompleta Halácsy was based on Sintenis & Bornmüller 1332, 21.8. 1891 (LD, WU) from the eastern foothills of Mt Olimbos, "welche ich mit keiner bekannten Art vereinigen kann, deren Beschreibung aber der im sehr vorgerückten Stadium befindlichen Exemplare wegen, eine unvollständige bleiben muss" (Halácsy 1898a: 583). For photos of this rare species, see Basdra & Kipopoulos (2018: 26-27).

Centaurea jurineifolia Boiss. from Bulgaria and Romania and ***C. peudobovina*** Hayek from Bulgaria are sometimes regarded as subspecies of *C. rutifolia* Sm. The last-mentioned was originally recorded "circa Byzantium", but the type specimen illustrated in Flora Graeca 10: tab. 916 (1840) was undoubtedly collected in the Athos peninsula in August, 1787. *C. rutifolia* s. str. is confined to rocky coastal habitats in the Athos peninsula. Plants from Bulgaria and Romania are likely to be specifically different.

Centaurea lactiflora Halácsy was described from "Konisko in Thessalien", based on a single collection by Formánek. It has been subsequently reported from a few additional localities in NC and NE Greece, and was accepted in the Greek checklist (Dimopoulos & al. 2013: 56). However, it is likely to be a synonym of *C. tymphaea* Hausskn. (cf. Georgiadis 1980: 55).

Centaurea laureotica Halácsy was listed for Peloponnisos by Mermygkas, Zikos & Constantinidis (2021: supplementary file), but no details are available.

Centaurea napulifera group. According to Skokanová & al. (2019) all Greek plants previous referred to *C. napulifera* Rochel and related taxa such as *C. tuberosa* Vis., *C. orbelica* Velen. and *C. velenovskyi* Adamović (*C. napulifera* subsp. *velenovskyi* (Adamović) Wagenitz & Gamal-Eldin) belong to two new species: *Cyanus austrobalcanicus* Skokanová and *Cyanus vorasanus* Skokanová. For further notes on synonymy and distribution, as well as recombinations of the two new species under *Centaurea*, see Raus & Strid (2020).

Centaurea prespana Rech. f. (1975), described from the Prespa National Park (NW Greece) is a synonym of *C. soskae* Košanin (1926a), described from the vicinity of Lake Ohrid in North Macedonia. The latter name appeared in a little known journal overlooked by Rechinger. *C. soskae* belongs to a small group of closely related species in the border area of Greece, Albania and North Macedonia (see Wagenitz & al. 2018).

Centaurea promota (E. Gamal-Eldin & Wagenitz) Strid was originally described as a subspecies of *C. nervosa* Willd. and appeared as such in the Greek checklist (Dimopoulos & al. 2013: 56). However, it differs rather distinctly in several morphological features and has subsequently been raised to species rank (Strid 2019c: 222).

Centaurea pseudobovina Hayek, see under *C. jurineifolia*.

Centaurea rufidula Bornm. is a poorly known species described from the vicinity of Lake Doirani in North Macedonia near the Greek border. Pavlides (1997a: 36) reported it from the Prespa National Park in NW Greece, probably in error.

Centaurea sanctae-annae J. Wagner, described from the Athos peninsula, is probably a hybrid between *C. diffusa* Lam. and *C. huljakii* J. Wagner.

Centaurea saxatilis K. Koch was recombined, on flimsy evidence, as *C. raphanina* subsp. *saxatilis* (K. Koch) Greuter. There is no reason to suspect that it is taxonomically different from *C. raphanina* subsp. *raphanina*.

Centaurea scabiosa L. subsp. ***fritschii*** (Hayek) Hayek was reported from the Rodopi mountains by Wagenitz & Gamal-Eldin (1985: 105), based on Strid & al. 19572. Subsp. *fritschii* is a SE European geographical race of the widespread *C. scabiosa*, differing in details of the phyllary appendages. Presumably all Greek records of *C. scabiosa* are referable to this subspecies.

Centaurea sicula L., a species of the W & C Mediterranean area, was reported from 3 localities in disturbed agricultural habitats in the Argolis prefecture (NE Peloponnisos) by Kalpoutzakis, Raabe & Constantinidis (2017). It resembles *C. solstitialis* L. and *C. melitensis* L. No details are available, and it is not clear whether the plants were native or represent a recent introduction.

Centaurea solstitialis subsp. ***erythracantha*** (Halácsy) Dostál (*C. erythracantha* Halácsy) was based on a collection by Adamović from the eastern foothills of Mt Olimbos. It is scarcely different from *C. solstitialis* s. str.

Centaurea sphaerocephala L. Flora Europaea (4: 283, 1976) listed this species as doubtfully present in Greece, based on imprecise records in earlier literature. It is probably restricted to the W Mediterranean region, and no evidence has appeared to support its presence in Greece.

Centaurea spinosa L. For notes on variation, see Aegean Atlas 1: 95-96 (2016). *C. cycladum* Heldr. (described from Mikonos), *C. tragacanthioides* Rech. f. (described from Mt Kerki, Samos) and *C. spinosa* subsp. *tomentosa* (Halácsy) Hayek (described from Attiki) do not appear to differ consistently from *C. spinosa* s. str.

Centaurea stereophylla Besser was listed for "Gr" in Flora Europaea (4: 268, 1976), probably based on the imprecise "Ma" and "Thra" in Hayek's Prodromus (2: 744, 1931). The species is related to *C. scabiosa* L. and distributed from Bulgaria to Ukraine. No Greek material has been located, but it is not unthinkable that it could occur in the north-east.

Centaurea stoebe subsp. ***serbica*** (Prodan) Ochsmann was included in the Greek checklist (Dimopoulos & al. 2013: 57), possibly based on a note in Strid & Kit Tan (1991: 496, in discussion under *C. affinis*). There appear to be no confirmed records from Greece.

Centaurea thasia Hayek may be conspecific with *C. ipsaria* Stoj. & Kitan., and is an older name than the latter. However, *C. thasia* must be rejected as a *nomen dubium* due to fragmentary description and lack of type specimen.

Centaurea triumfettii All. Recent studies (e.g., Skokanová & al. 2009, Olšavská & al. 2011) indicate that this species is confined to the Alps of SE France and NW Italy. Greek plants previously referred to *C. triumfettii* belong to *C. axillaris* Willd. (pending conservation of that name) and if treated at subspecies level they should be called *C. triumfettii* subsp. *axillaris* (Willd.) Stef. & T. Georgiev.

Centaurea tymphaea subsp. ***brevispina*** (Hausskn.) Dostál was based on a collection by Haussknecht from Mt Pilon, and has subsequently been reported also elsewhere in NC and NE Greece. Differences between this and *C. tymphaea* s. str. are scarcely consistent.

Centaurea urvillei DC. is variable, especially in Anatolia. Most of the Greek specimens belong to subsp. *urvillei* (based on a collection by Dumont d'Urville from Samos), but subsp. *stepposa* Wagenitz (described from C4 Konya) and subsp. *armata* Wagenitz (described from C5 Içel) have also been reported. Some collections from Kos and Lesvos have been tentatively referred to subsp. *stepposa*. At least one collection from Rodos (Bourgeau 106; E, G, GOET, LD) has been identified by Wagenitz as subsp. *armata*, and the latter has also been reported from Kalimnos (Zervou 2011: 83).

Centaurea vandasii Velen., a rare Balkan endemic from the border area of Bulgaria, North Macedonia and Greece, has recently been found on the Greek side of Mt Beles (Kit Tan & Kofinas unpubl.).

Centaurea vermia Rech. f. was based on two collections from Mt Vermion (type: Rechinger 8927; G, K, LD). It is characterized by greyish-green, little divided leaves with broad segments (the upper ± entire and oblanceolate) and relatively large, dark brown phyllary appendages. It seems doubtfully distinct from the widespread and variable *C. affinis* Friv.

Centaurea wettsteinii Degen & Dörfl. is a rare local endemic of the border area between Greece and North Macedonia. Greuter (in Greuter & Raab-Straube 2007: 141-142) concluded that the type, collected by Dörfler in 1893, is from a locality just inside present Greek territory, and reported a collection of his own "in latere boreo-occident. montis Pinovo (Kozuf)". Recently it has been found also on the adjacent Mt Tzena (Kit Tan & Kofinas, unpubl.).

Centaurium erythraea Rafn. At least three subspecies occur in Greece, viz. subsp. *erythraea*, subsp. *rhodense* (Boiss. & Reut.) Melderis and subsp. *rumelicum* (Velen.) Melderis, perhaps also subsp. *turcicum* (Velen.) Melderis. They are poorly characterized and their distributions completely overlap. Thus only a collective map can be presented. A local population from the island of Kithira has been described as *C. limoniiforme* Greuter.

Centaurium subspicatum (Velen.) Ronniger is a dubious taxon probably conspecific with *Schenkia spicata* (L.) G. Mans. It was described from Bulgaria and, according to Fl. Eur. 3: 59 (1972), has been reported also from Kriti.

Centaurium tenuiflorum (Hoffmanns. & Link) Fritsch. Both the tetraploid subsp. *tenuiflorum* and the diploid subsp. *acutiflorum* (Schott) Zeltner have been reported from Greece, but the morphological differences are obscure. Intermediates between *C. pulchellum* (Sw.) Druce and C. *tenuifolium* are occasionally found.

Centranthus macrosiphon Boiss. is native to S & SE Spain. It is frequently cultivated as an ornamental and has been reported as semi-naturalized in Kithira (Greuter & Rechinger 1967: 130, Yannitsaros 1998: 89), Salamis and Athens area (Vallianatou 2005: 257, Constantinidis & Yannitsaros 1993: 281), Mt Pateras (Constantinidis

1997b: 263), W Peloponnisos (Giannopoulos, Kit Tan & Vold 2016b: 268) and Kerkira (Flora Ionica website). It is an annual, somewhat reminiscent of *C. calcitrapae* (L.) Dufr., but with a thick, fistulose stem and a dense terminal inflorescence with pink to red flowers; the corolla tube is longer (6-8 mm, with slender spur c. 1 mm).

Centranthus ruber group. Taxonomic relations between *C. ruber* (L.) DC. s. lat. and *C. longiflorus* Steven s. lat. are not clear. The latter is usually recognized on the linear-lanceolate leaves, lax inflorescences and long flowers. Most Greek forms of *C. ruber* have broadly lanceolate, entire leaves and belong to subsp. *sibthorpii* (Boiss.) Hayek. *C. ruber* subsp. *ruber* (with broader, ± dentate leaves) is cultivated as an ornamental, but probably not native in Greece.

Cephalaria glaberrima Contandr. & Quézel was based on a single, rather depauperate collection from Mt Vardousia. Its status in uncertain.

Cephalaria leucantha (L.) Roem. & Schult. was listed for Peloponnisos by Mermygkas, Zikos & Constantinidis (2021: supplementary file), but no details are available.

Cephalaria pastricensis Dörfl. & Hayek, a species described from the border of Albania and Kosovo, was reported from a serpentine locality in the Smolikas area ("1 km westlich des Distrato-Dorfes") by Babalonas (1983: 22). According to Constantinidis & Phitos (in Greuter & Raus 2004: 74) Babalonas' collection as well a later one made at the same locality are likely to represent the subsequently described *C. fanourii* Perdetz. & Kit Tan.

Cephalaria syriaca (L.) Roem. & Schult. is a widespread, Irano-Turanian species occurring from Cyprus and C Anatolia to Afghanistan, Pakistan and Turkestan. It has been reported as a casual in scattered localities from Bulgaria to Spain, and there is one historical record from Greece ("Salonicae haud rara", Charrel 1892: 339).

Cephalaria transsylvanica (L.) Roem & Schult. was quite recently found by a roadside in Naxos (Astrid Scharlau, photo 2022-05-17).

Cephalaria uralensis (Murray) Roem. & Schult. was reported from "?Gr" in Fl. Eur. (4: 57, 1976), perhaps based on the imprecise "Ma" in Hayek (1930: 496). It is not unthinkable that it could occur in northern Greece, but no specimens or precise records have been seen.

Cerastium arvense L. Greek records (cf. Fl. Eur. and Euro+Med PlantBase) are probably incorrect. Reports by Pavlides (1985: 167) from the Prespa area refer to *C. fontanum* subsp. *vulgare* (Hartm.) Greuter & Burdet. Other literature records may refer to *C. decalvans* Schlosser & Vuk.

Cerastium brachypetalum subsp. ***tauricum*** (Spreng.) Murb. was listed for Greece in Euro+Med PlantBase, but no specimens have been seen and the source is obscure.

Cerastium decalvans Schloss. & Vuk. Most Greek plants were previously referred to subsp. *decalvans*, but according to Niketić & al. (2021) this in confined to the NW Balkan Peninsula, and Greek plants identified as such should be called subsp. *leontopodium* (Stoj. & Stef.) Niketić. In addition, subsp. *glutinosum* (Halácsy) Niketić is endemic to the island of Thasos, and subsp. *orbelicum* (Velen.) Stoj. & Stef. occurs locally along the northern borders. Subsp. *leontopodium* and subsp. *orbelicum* coexist on Mt Voras (Kajmakčalan) on the border of Greece and North Macedonia, the former occurring on marble and the latter on micaceous schist.

Cerastium ramosissimum Boiss. was described from Sierra Nevada in southern Spain. As currently understood, it occurs in two disjunct areas: (1) the Iberian Peninsula, and (2) the East Mediterranean area, perhaps extending to Ukraine. It belongs to a species complex which was briefly discussed by Strid in Fl. Hellen. 1: 213-214 (1997). The related *C. bulgaricum* Uechtr. was here believed to be a habitat modification, but with more material (including living plants) now available it appears to be a fairly distinct species.

Cerastium rectum subsp. ***petricola*** (Pančić) Greuter is a high-altitude ecotype. There are gradual transitions between this and the nominate subspecies.

Ceratocephala orthoceras DC. (= *C. testiculata* auct.) was listed for Greece in Euro+Med PlantBase, but there appear to be no confirmed records. The species occurs in E Europe southwards to Bulgaria, and is widespread in SW Asia. It could well occur in NE Greece. It differs from *C. falcata* (L.) Pers. in the smaller flowers and achenes, the latter with a ± straight beak.

Ceratonia siliqua L. has been cultivated in the Mediterranean area since ancient times (cf. Zohary 2002); its native distribution may be in the SE Mediterranean area and the Arabian Peninsula (see Browicz 1983a: 21-22 and references given here). It Greece it appears native in hot and dry areas in the south.

Cestrum parqui L'Hér. was listed by Bornmüller (1928c: 300) from "Patras, in Hecken verwildert". It is native to Chile and cultivated as an ornamental in Europe. It is a medium-sized shrub with lanceolate leaves and clusters of tubular, greenish-yellow, fragrant flowers 20-25 mm. The fruit is a blackish berry.

Chaenorhinum minus (L.) Lange. Speta (1980a) transferred this species to the poorly defined genus *Microrrhinum* and divided it into a number of highly dubious taxa. *M. praetermissum* (Delastre) Speta was reported from scattered localities on the Greek mainland and *M. janchenii* Speta from Mt Peristeri. Both are likely to be forms of *C. minus*. *C. litorale* (Willd.) R. M. Fritsch was incorrectly included in the Greek checklist (Dimopoulos & al. 2013: 144). It has not been reported from Greece, but occurs in coastal habitats in the NW Balkan Peninsula and is better regarded as a subspecies of *C. minus*.

Chaenorhinum origanifolium (L.) Fourr. Speta (1980a: 24) reported this species from Mt Trimfristos, based on a single collection by Spruner, 1842 (WU-HAL), labeled "Velugo in Aetolien". It is otherwise a species of the W Mediterranean area, and the Greek record is highly dubious.

Chamaecytisus. Several new taxa have recently been described from Albania (see Pifkó & Barina 2016, Pifkó 2018). Critical re-examination of the Greek material is needed. In the recent Italian checklist (Bartolucci & al. 2018) *Chamaecytisus* was merged with *Cytisus*.

Chamaecytisus frivaldszkyanus (Degen) Kuzmanov. Barina (2017: 243) mapped several records from SE Albania, and the species may well occur in adjacent parts of Greece.

Chamaecytisus jankae (Valen.) Rothm. There is an unspecified report from Greece by Cristofolini (1991: 210). This species otherwise occurs in Serbia, Bulgaria and perhaps Albania. No Greek specimens have been seen by the present author.

Chamaecytisus spinescens (C. Presl) Rothm. Both subsp. *spinescens* and subsp. *creticus* (Boiss. & Heldr.) K. I. Chr. were listed for Peloponnisos by Dimopoulos & al. (2013: 93), but the former is probably restricted to Sicily.

Chasmanthe floribunda (Salisb.) N. E. Br., a native of the Cape of South Africa, is locally naturalized in the W Mediterranean area, and has recently been reported from ruderal habitats on the island of Rodos (Rätzel, Kalaentzis & Ristow 2021: 151).

Cheilanthes. The generic name *Allosoros* Bernh. has been adopted in the Greek checklist (Dimopoulos & al. 2013: 39) and some other recent publications. However, it has recently been rejected under Art. 56 of the Code of Nomenclature (https://doi.org/10.1002/tax.12674, see p. 217, no. 2509). The correct name for the cheilanthoid fern genus that occurs in Greece, to which the name *Allosorus* was wrongly applied, is either *Oeosporangium* Vis. or *Cheilanthes* Sw., depending on how narrowly or broadly one circumscribes the genus. [Thanks are due to Nick Turland for this information].

Cheilanthes pteridioides (Reichard) C. Chr. Greek records are likely to refer to *C. acrostica* (Balb.) Tod.

Chenopodiastrum hybridum (L.) S. Fuentes, Uotila & Borsch was listed for Peloponnisos by Mermygkas, Zikos & Constantinidis (2021: supplementary file). Verified Greek records are all in the north.

Chenopodium album complex. The maps for *C. album* L., *C. opulifolium* W. D. J. Koch & Ziz, *C. striatiforme* Murr and *C. betaceum* Andrz. (*C. strictum* auct. non Roth) are largely based on the extensive collections from the mainland and Peloponnisos by Willing (B) which have been revised by Th. Raus. Most of the specimens from the Aegean islands (by other collectors) have not been critically examined. Consequently, some records from the latter area mapped as *C. album* may represent other members of the complex. A record of *C. probstii* Aellen from the island of Lesvos (https://herbarium.univie.ac.at/database/search.php) may belong to *C. strictum*/*betaceum*. *C. suecicum* Murr, another member of this complex, was reported from Limnos by Bergmeier & al. (2021 suppl. 1 p. 10). It is native in C & N Europe and naturalized or casual elsewhere.

Following Mosyakin (2017b) the species usually called *C. strictum* by European authors has been renamed *C. betaceum* Andrz. The latter name was accepted in the Italian checklist (Bartolucci & al. 2018: 243).

Chondrilla acantholepis Boiss. is a taxon of uncertain status, probably a mere form of *C. juncea* L. with setose midvein of the phyllaries.

Chrozophora obliqua (Vahl) A. Juss. ex Spreng. and ***C. tinctoria*** (L.) A. Juss. Only a collective map is presented. For taxonomic comments, see Aegean Atlas 1: 263 (2016).

Chrysojasminum. Banfi (2014) proposed to raise *Jasminum* sect. *Alternifolia* to the rank of genus, with *Jasminum fruticans* L. becoming *Chrysojasminum fruticans* (L.) Banfi.

Cirsium epiroticum Petr. was based on a single specimen (Rechinger 18556) from Mt Tsouka Rossa in northern Pindos, "auf Serpentin an feuchten Orten". It is of uncertain status and may be a form of *C. mairei* Halácsy.

Cirsium erisethales (Jacq.) Scop., was listed for "Gr" in Fl. Eur. 4: 238 (1976), but no specimens or specified records have been seen.

Cirsium heldreichii subsp. ***euboicum*** Petr. was described from Mt Kandili (N Evvia). The type seems conspecific with *C. eriophorum* (L.) Scop. – see *http://jacq.nhm-wien.ac.at/djatoka/jacq-viewer/viewer.html?rft_id=w_19590002848&identifiers=w_19590002848.*

Cirsium ligulare subsp. ***albanum*** Wettst. was described from Uesküb [Skopje] in North Macedonia and it is possible that all Greek records should be referred to this subspecies. The type of *C. ligulare* Boiss. is from Kalofer, a town in Bulgaria. Greek plants have also been called *C. armatum* Velen. or *C. ligulare* subsp. *armatum* (Velen.) Petr. The complex is in need of critical study.

Cirsium pindicolum Hausskn. is a taxon of uncertain status, described from Mt Baba (Tringia) in southern Pindos. The type specimen (at JE) was annotated by F. Petrak in 1910 as a hybrid between *C. siculum* Spreng. (= *C. creticum* (Lam.) d'Urv.) and *C. tymphaeum* Hausskn.

Cirsium steirolepis Petr. The type is Sintenis, Iter Trojanum 616 (31.7. 1883; BM, JE) from Mt Ida (= Kaz Dağı) in NW Anatolia – not Mt Ida (= Psiloritis) in Kriti.

Cistanche phelypaea (L.) Cout. According to Moreno Moral, Sánchez Pedraja & Piwowarczyk (2018), *C. phelypaea* s. str. is restricted to SW Spain and Portugal. The widespread taxon in the South Mediterranean area (from SE Spain eastwards, and presumably also in coastal areas of west tropical Africa) is *C. phelypaea* subsp. *lutea* (Desf.) Fern. Casas & M. Laínz, which is sometimes regarded as a separate species under the name *C. lutea* (Desf.) Hoffmanns. & Link. The latter is more robust with dense inflorescence and relatively longer and narrower bracts.

Citrullus colocynthis (L.) Schrad. has frequently been reported from ruderal and coastal habitats in southern Greece, but is scarcely native or naturalized.

Citrullus lanatus (Thunb.) Matsum. & Nakai (watermelon) is frequently cultivated and may persist for some time in ruderal habitats.

Cleistogenes Keng has been replaced by *Kengia* Packer, based on nomenclatural technicalities which were disputed by Bor in Fl. Iranica 70: 441-442 (1970). Following, e.g., the online Italian checklist (consulted 2021-09-16) the name *Cleistogenes serotina* (L.) Keng is preferred here.

Clematis orientalis L. Historical reports of this species from the East Aegean Islands probably refer to *C. cirrhosa* L.

Cleome aurea Čelak., described from the Athos peninsula, was included in *C. ornithopodioides* L. by Roalson (2021). As currently defined the latter species occurs in the Balkan Peninsula, Cyprus and throughout Anatolia.

Cleome iberica DC. This widespread species was described from Caucasia, the epithet alluding to the ancient kingdom of Iberia in present-day Georgia. Plants from serpentine substrate on Rodos and adjacent parts of SW Anatolia are somewhat deviating and may represent an undescribed, local taxon. They were not mentioned by Roalson (2021).

Clinopodium dalmaticum (Benth.) Bräuchler & Heubl and the related ***C. taygeteum*** (P. H. Davis) Bräuchler & Heubl were originally described in *Micromeria*, but Bräuchler, Meimberg & Heubl (2006) have shown that they are better placed in *Clinopodium*. For variation in *C. dalmaticum*, see Raus & Strid (2019).

Clinopodium vulgare L. Both subsp. *orientale* Bothmer and subsp. *vulgare* are common in Greece. The former has calyx 9.5-12 mm, with lower teeth 1.5 times as long as the upper, whereas the latter has calyx 7-9.5 mm, with lower teeth twice as long as the upper. The former occurs mainly in the Balkan Peninsula and Anatolia, whereas the latter is widespread. There are many intermediates and scarcely any geographical separation, although all specimens from Kriti and Rodos appear to represent subsp. *orientale*.

Clypeola jonthlaspi L. is often divided into subsp. *jonthlaspi* (silicula 3-3.5 mm; hairs on disc numerous, long, tapering to sharp point) and subsp. *microcarpa* (Moris) Arcang. (silicula somewhat smaller; hairs on disc few, short and rounded). Only a fraction of the Greek records have been identified to subspecies. The two show no difference in distribution and their taxonomic status remains unsettled.

Coeloglossum viride (L.) Hartm. Molecular evidence (Bateman & Rudall 2018) suggests inclusion of this distinctive species in the genus *Dactylorhiza*, as *D. viridis* (L.) R. M. Bateman, Pridgeon & M. W Chase. In view of the distinct morphological differences it is maintained here as *Coeloglossum*.

Coix lacryma-jobi L. This peculiar species, a native of tropical Asia and Africa, is sometimes cultivated and may be locally naturalized in Kriti. Damanakis (1984) reported it from an irrigation ditch in a *Citrus* orchard near Rethimno, where according to the local farmer it had existed for 40 years. Čelakovsky (1887a: 546) reported it from the monastery of Chilandar in the Athos peninsula, observing "Wird in Gärten an feuchten Orten gebaut der zu Rosenkränzen benützten Kugeln wegen".

Colchicum baytopiorum C. D. Brickell was reported from Rodos by Carlström (1987: 121), based on a collection by H. Wollin 1974-04-08, cultivated in Göteborg Botanical Garden and seen by K. Persson. According to Persson (e-mail 2011-11-14) collections of *C. baytopiorum* from SW Anatolia were cultivated at the same time, and there is a possibility that labels were mixed up. The record from Rodos thus needs confirmation.

Colchicum hungaricum Janka. Greek records of this species refer to *C. doerfleri* Halácsy.

Colchicum macedonicum Košanin has been reported from Mt Dudica in the Kozuf area just north of the Greek border (41°08´N, 22°09´E) and may well occur also on the Greek side, but so far there are confirmed records only from North Macedonia and Albania.

Colocasia esculenta (L.) Schott, a tropical crop plant of SE Asian origin, is said to be locally established from cultivation in wet areas of S Ikaria.

Colutea cilicica Boiss. & Balansa, a species described from SW Anatolia, was reported from Southern Pindos by Browicz (1963a: 49), based on historical collections from 1885 and 1896. Critical re-examination of the specimens is needed; they are likely to represent a form of *C. arborescens* L.

Comandra umbellata (L.) Nutt. is widespread in North America where it has been divided into at least four subspecies. In addition, a taxon originally described from "Bannatus" [W Romania] as *Thesium elegans* Spreng. has been recombined as *Comandra umbellata* subsp. *elegans* (Spreng.) Piehl. This is a most unusual disjunction, and it should be critically examined whether plants from Romania and the Balkans are indeed conspecific with American material.

Commelina communis L., a native of East Asia, may be on the verge of becoming naturalized in Greece. It has been reported from the shores of lake Mikri Prespa in the NW as well as urban habitats in Ioannina and Mesolongion, and most recently on the island of Limnos (Bergmeier & al. 2021 suppl. 1 p. 11).

Conringia austriaca (Jacq.) Sweet. Records from the Prespa area (Pavlides 1985: 197) probably refer to *C. orientalis* (L.) Dumort.

Conringia planisiliqua Fisch. & C. A. Mey. has recently been transferred to a monotypic genus, as *Iljinskaea planisiliqua* (Fisch. & C. A. Mey.) Al-Shehbaz, Özüdoğru & D. A. German, comb. nov. (see Al-Shehbaz, Özüdoğru & German 2021). *Iljinskaea* is distinguished from *Conringia* by having strongly latiseptate (vs. 4- or 8-angled, terete or terete-latiseptate) fruit, nearly veinless (vs. 1- or 3-veined) valves, presence (vs. absence) of median nectaries, purple- to brownish-veined (vs. yellowish-veined) petals, and subconduplicate (vs. incumbent) cotyledons.

Consolida phrygia (Boiss.) Soó. All Greek plants are probable referable to the rather poorly defined subsp. *thessalonica* (Soó) P. H. Davis (type: Sintenis 519, LD, from Kalambaka). A report of subsp. *phrygia* from Rodos (cf. Fl. Turkey 1: 126, 1965) has not been confirmed.

Consolida samia P. H. Davis was based on a single collection (Runemark & Snogerup 19592, LD) from calcareous scree in the SW foothills of Mt Kerkis, Samos, where it is undoubtedly very rare and has subsequently been looked for in vain by others, including the present author. Minareci, Altan & Aktan (2011) reported it from a single locality on the adjacent Anatolian mainland (Soma Kocasivri hill in the Manisa area), a record that needs confirmation. It is related to *C. hellespontica* (Boiss.) Chater.

Convolvulus althaeoides aggr. *C. althaeoides* L. and *C. elegantissimus* Mill. can be reasonably well distinguished in the Aegean area and were retained as separate species in the Aegean Atlas (Strid 2016). In the mainland they merge into each other, with the latter being the more common (including most of the numerous Willing collections from the mainland, Evvia and Peloponnisos). In view of the many intermediates, subspecies or varietal rank seems more appropriate, and only a collective map can be provided.

Convolvulus oleifolius Desr. A robust, broad-leaved form, var. *scopulorum* Rech. f., is mainly found in rocky places on small islands. It approaches the Cretan endemic *C. argyrothamnos* Greuter.

Convolvulus sabatius Viv. belongs to a species complex of the W Mediterranean area and is also cultivated as an ornamental (cf. Carine & Robba 2010, Wood & al. 2015: 142-143). It was described from Italy where subsp. *sabatius* is apparently confined to the area of Cabo di Noli in Liguria, whereas subsp. *mauritanicus* (Boiss.) Murb. is fairly widespread in Morocco and Algeria. The species has been reported to be locally established in stone walls and ruderal habitats in the town of Corfu and recently also in nomos Ilias.

Convolvulus tricolor L. is a species of the W & C Mediterranean area, also widely cultivated as an ornamental. It resembles *C. pentapetaloides* L. but has larger, 3-coloured corolla. A few records from the Thessaloniki and Kilkis areas (NE Greece) may refer to casual occurrences.

Coris monspeliensis L. Identification of the Ionian (and Greek) populations of *Coris monspeliensis* with subsp. *syrtica* (otherwise restricted to N Africa) as assumed in the "Med-Checklist" (Greuter & al. 1989: 375) is questionable; according to "Vascular Plants of Greece" subsp. *monspeliensis* is accepted as the only representative in Greece (Dimopoulos & al. 2013: 131). Due to the glandular indumentum Browicz & Zieliński (1989: 83) had considered the Zakynthos populations belonging to "var. *syrtica*" (based on material from Tunisia). The leaves of our plants, however, are glabrescent to nearly glabrous, and the uppermost ones are not entire (as described by Murbeck 1898: 2; cf. Masclans 1968: 757) but distinctly spinose-dentate like in subsp. *fontqueri* (Masclans 1968, Gómez in Castroviejo & al. 1997) that has shorter calyx teeth; apart from different branching pattern. The indumentum character (adult leaves become almost completely glabrous) and stem-leaf dentation are not sufficiently correlated to justify seven varieties (subsumed as three subspecies) as proposed by Masclans (1968) for the West Mediterranean region. Consequently we refrain for the present from subdividing the species. [Cited from Flora Ionica website, 2016-11-12].

Plants from maritime sand near Valona in Albania have been described as *C. monspeliensis* var. *annua* Halácsy & Bald. This taxon has subsequently been reported also from Italy, as *C. monspeliensis* subsp. *annua* (Halácsy & Bald.) Arrigoni.

Coronilla glauca/valentina. These two taxa are often regarded as subspecies of *C. valentina* L., but were accepted at species level by the monographer Schmidt (1979a). They co-exist on the Ionian Islands, apparently without intermediates. At species level the name *C. argentea* L. (1753) replaces *C. glauca* L. (1755).

Cortaderia selloana (Schult. & Schult. f.) Asch. & Graebn., a large tufted grass native in Argentina, is cultivated as an ornamental and doubtfully naturalized, e.g., in the Preveza area (Chitos 2009: 106) and the town of Rodos (Galanos 2015: 275). Some previous records represent misidentified material of *Saccharum ravennae* (L.) Murray.

Corydalis integra/wettsteinii. *C. integra* Barbey & Fors.-Major was described from Mt Kerkis, Samos; the description is accompanied by an excellent analytical line drawing (Stefani, Forsyth Major & Barbey 1892: tab. 1). In these plants the flowers are uniformly rose-pink as shown by photos from *locus classicus* sent by S. Tsiftsis, 2021-04-16. Plants from Mt Athos with inner petals tipped dark purple have been described as *C. wettsteinii* Adamović (1906a); they are also larger with a more elongate inflorescence. Similar plants occur on Mt Chortiatis. The two were merged by Lidén in Flora Hellenica 2: 105 (2003), but seem to be reasonably distinct.

Corydalis solida (L.) Clairv. Most of the Greek plants belong to subsp. *incisa* Lidén which tends to grow in open habitats and is relatively robust with elongate inflorescence and deeply divided bracts. Such plants have formerly been called *C. densiflora* J. & C. Presl, but the latter is now considered endemic to S Italy and Sicily. Some specimens from woodland habitats in the north are small and slender, with short, few-flowered inflorescence and entire or apically dentate bracts. They appear to match *C. solida* subsp. *solida* or subsp. *slivenensis* (Velen.) Hayek, but also approach the N & C European *C. intermedia* (L.) Mérat. Spyros Tsiftsis (e-mail 2021-04-02) found such plants on Mt Menikion together with *C. solida* subsp. *incisa*; they have also been recorded from the vicinity of the Katara Pass, Mt Tzena, and Mt Vrondous. Further studies are needed on variation in this group throughout the Balkan Peninsula.

Corylus maxima Mill. and ***C. pontica*** K. Koch have been reported from a few localities in Greece but almost certainly planted. They are native in N Anatolia and Caucasia.

Cotoneaster creticus J. Fryer & B. Hylmö. Kurtto, Sennikov & Lampinen (2013: 71) interpreted this recently described Cretan endemic as a synonym of the widespread *C. nummularius* Fisch. & C. A. Mey.

Crepis atheniensis Babc. is a taxon of dubious status known only from one 19th century collection from the vicinity of Athens. It is probably a synonym of *C. setosa* Haller f.

Crepis aurea subsp. ***glabrescens*** (Caruel) Arcang. *Crepis aurea* (L.) Cass. has been divided into three, poorly defined subspecies. Plants from the Balkan Peninsula have often been referred to subsp. *glabrescens*.

Crepis baldaccii Halácsy was described from Mt Tomor in Albania. Plants from Mt Timfi differing in leaf characters have been described as subsp. *carpini* Greuter but this merely represents individual variation. More typical individuals as well as intermediates occur in the same locality.

Crepis bithynica Boiss. was reported from Mt Olimbos by Halácsy (1902a: 222). It has never been confirmed here, nor elsewhere in Greece. This species was described from Ulu Dağ (Bithynian Olimbos) and is rare in NW & N Anatolia. In the Balkans it has been reported from Bosnia-Herzegovina and SW Bulgaria.

Crepis bursifolia L. has been collected once in a park in Thessaloniki (Snogerup 9831, LD). It is native in Italy and naturalized in France, Spain and California. The Greek collection may represent a casual introduction.

Crepis capillaris (L.) Wallr. Scattered reports of this species (e.g., Pavlides 1982: 121 from Vertiskos, Spanou & al. 2006: 239 from nomos Ilias and Sfikas 2008b: 11 from Mt Pendeli) are probably incorrect. It is widespread in C & W Europe and could perhaps occur in N Greece.

Crepis neglecta L. s. lat. Taxonomy basically follows Sell in Fl. Eur. 4: 355-356 (1976). The excessive splitting by Kamari (1976) does not seem to be based on scientific evidence.

Crepis paludosa (L.) Moench, a widespread European species, has been reported once from Samothraki (Biel & Kit Tan (2013b: 271). Confirmation would be welcome.

Crepis smyrnaea DC. ex Froel. was listed by Halácsy (1902a: 221), based on a collection by Haussknecht from Mt Pendeli, Attiki. It is unlikely to occur in this area, and the gathering may represent *C. fraasii* Sch. Bip. *C. smyrnaea* was described from the vicinity of Smyrna [Izmir] and is known from some other localities in W Anatolia. It should be looked for on the East Aegean Islands.

Crepis vesicaria L. One collection from low altitude on Samothraki has been identified as subsp. *taraxacifolia* (Thuill.) Thell., a taxon of uncertain status.

Crepis viscidula Froel. Subsp. *viscidula* and subsp. *geracioides* (Hausskn.) Kamari have been recognized in Greece, but the supposedly distinguishing characters occur in different combinations and there is no geographical separation.

Crocus. Taxonomy basically follows Mathew (1982a). The otherwise excellent monograph by Rukšāns (2017b) uses a narrower species concept, and modifications have been made to match the principles adopted elsewhere in this Atlas.

Crocus biflorus subsp. ***crewei*** (Hook. f.) B. Mathew (*C. crewei* Hook. f.). This name has sometimes been applied to Greek members of the *C. biflorus* complex. Indeed, the type of *C. crewei* was supposedly from the island of Siros, but this is evidently a mistake. As currently understood this taxon is restricted to a small area in Anatolia (near Denizli). The *C. biflorus* complex has been subject to much splitting, both in Turkey and Greece; several recently described local endemics from the East Aegean islands are of uncertain taxonomic status.

Crocus candidus E. D. Clarke was listed for the East Aegean Islands in Euro+Med PlantBase, but no specified record or specimen has been seen. According to Rukšāns (2017b: 153-154) it is endemic to NW Anatolia.

Crocus cartwrightianus Herb. A record from nomos Ilias in NW Peloponnisos goes back to Maire & Petitmengin (1908: 205), "Forêt de Manoladha, alluvion siliceuse", sub nom. *C. sativus* var. *cartwrightianus* (Herb.) Maw. It has never been confirmed in this area, and the record may refer to *C. hadriaticus*. [Thanks are due to Kit Tan for this observation].

Crocus chrysanthus subsp. ***multifolius*** Papan. & Zacharof was described from north central Greece and said to differ from subsp. *chrysanthus* in the 5-6 long leaves and the chromosome number 2n=8 (vs. 2n=12-20). Further studies are needed on its possible taxonomic status.

Crocus goulimyi subsp. ***leucanthus*** B. Mathew is a name given to a white-flowered form of the species. Such individuals are occasionally found amongst the usual ones with lilac-blue flowers, and scarcely deserve taxonomic rank.

Crocus hadriaticus subsp. ***parnonicus*** B. Mathew and subsp. ***parnassicus*** (B. Mathew) B. Mathew were described from Mt Parnonas and Mt Parnassos, respectively. They were later considered by Mathew himself (in litt.) to represent local populations not worthy of taxonomic status.

Crocus macedonicus Rukšān was described from N Greece and North Macedonia. There are scarcely any morphological differences between this and *C. pallasii* Goldb., although limited evidence suggests that plants from Mt Imittos and Samos (*C. pallasii* s. str.) have 2n=14 whereas plants from Mt Vertiskos (*C. macedonicus*) have 2n=16 – see Karamplianis, Tsiftsis & Constantinidis (2013: 60).

Crocus nivalis Bory & Chaub. For comments on variation and taxonomy, see Aegean Atlas 1: 347 (2016). As mentioned here, an unconfirmed record from Samos was based on corms collected by Rechinger and cultivated in Vienna (see Rechinger 1944: 738). Recently, similar plants have been collected on Ikaria by Arnis Seisums and cultivated by Janis Rukšāns, flowering in March 2022. They match *C. nivalis* s. lat., more specifically the taxon called *C. sublimis* Herb. or *C. sieberi* subsp. *sublimis* (Herb.) B. Mathew. [Thanks are due to Janis Rukšāns for sending photos].

Crocus olivieri J. Gay. Both subsp. *olivieri* and subsp. *balansae* (Baker) B. Mathew have been recorded from Chios. There appear to be no consistent differences.

Crocus orphei Karabl. & Constantin. is a recently described species currently known only from Mt Falakron and closely related to *C. reticulatus* Steven ex Adams. The latter was described from Northern Caucasus and is scattered on the north side of the Black Sea to Moldova in the west. Harpke & al. (2014) made a detailed study of the *C. reticulatus* group, dividing it into nine species. Plant on Mt Falakron were first collected by N. Jacobsen in 1998 and cultivated in the Göteborg Botanic Garden (sub. nom. *C. reticulatus*). The type of *C. orphei* is a subsequent collection from the same locality by Karablianis & Tsiftsis, 30.3. 2010 (ATHU).

Crocus pallasii Goldb. Local populations from the islands of Chios and Kos have recently been described as *C. homeri* Rukšāns and *C. samarasii* Rukšāns, respectively (see Rukšāns 2021).

Crocus pumilus (Rukšāns) Rukšāns is a name given to Cretan plants of *C. laevigatus* Bory & Chaub. As pointed out already by Mathew (1982: 114) these tend to be small, but overlap in measures is too great for separate taxa to be recognized.

Crocus sativus L., the saffron crocus, is a triploid species not known in the wild, presumably originating in Greece from the diploid *C. cartwrightianus* Herb. and spreading already in antiquity. In Greece it is traditionally cultivated by the aptly named village of Krokos near Kozani, but most of the commercial production is now in Spain, Iran and Kashmir.

Cruciata taurica (Willd.) Ehrend. When reporting this species from a new locality on Mt Iti, Trigas & al. (2021: 322) observed that these plants as well as others previously gathered in Peloponnisos differ from the type of subsp. *euboea* (Ehrend.) Ehrend. from Mt Dirfis, Evvia. In the map they have provisionally been merged with the latter, but may in fact represent an undescribed taxon.

Crypsis. Phylogenetic speculations based on molecular evidence have led to the recent proposal to reduce *Crypsis* to a section of *Sporobolus* (see Peterson & al. 2014).

Crypsis acuminata Trin. has been reported once, from an alluvial plan in Rodos ("Laerma - Apollona, frühjahrsnasse, sandig-kiesige Alluvionen", Böhling 8841, det. Scholz; B). It resembles *C. alopecuroides* (Piller & Mitterp.) Schrad., and the record needs confirmation.

Cucurbita pepo L. (pumpkin, squash) is commonly cultivated and may persist for some time in ruderal habitats.

Cupressus sempervirens L. The narrow, columnar form is commonly cultivated throughout the Mediterranean area. Native plants (generally with spreading branches) are distributed from Kriti through S Anatolia to Lebanon, with disjunct ocurrences in Cyrenaica and N Iran (cf. Browicz 1982b: 70).

Cuscuta approximata Bab. Both subsp. *approximata* and subsp. *macranthera* (Boiss.) Feinbrun & Greuter have been reported from Greece, but the differences are obscure. The latter was originally described as C. *macranthera* Boiss., based on a collection by Heldreich from Mt Taigetos.

Cuscuta palaestina Boiss. All Greek records are likely to refer to subsp. *palaestina*. The otherwise Anatolian subsp. *balansae* (Yunck.) Plitmann has been reported twice from the Aegean islands, but probably in error.

Cuscuta suaveolens Ser., a native of South America, is widely naturalized in temperate areas elsewhere. Two collections from northern Greece (Greuter 15408 from Thasos and Snogerup 21609 from the Ioannina area) have been registered as *C. suaveolens*, but the specimens have not been seen by the present author and confirmation is needed.

Cyclamen coum Mill. Although the epithet appears to mean "from Kos" the species has never been found on that island nor elsewhere on Greek territory. It is widespread in Anatolia, extending to the Black Sea region of Bulgaria and Romania in the west and to Lebanon and Palestine in the south. The etymology is uncertain (cf. Mathew 2013: 55-56).

Cyclamen hederifolium Sol. ex Aiton is very variable, especially in leaf characters. Some plants from Peloponnisos have been referred to subsp. *crassifolium* (Hildebr.) Culham, Denney & P. Moore. As implied by the epithet it tends to have thicker leaves, but the difference is not clear and most specimens have not been identified to subspecies. Limited evidence suggests that subsp. *crassifolium* is tetraploid with 2n=68 whereas subsp. *hederifolium* is (at least in most cases) diploid with 2n=34.

Cydonia oblonga Mill., a native of SW & C Asia, is commonly planted for its edible fruits (quince, κυδώνι) and occasionally naturalized near villages. It is a large shrub or small tree with entire leaves and large, solitary, white or pinkish flowers. The fruit is apple- or pear-like, 5-12 cm in diam., ± tomentose when young, glabrescent and yellow when ripe, turning rose-pink when boiled.

Cymbalaria. Taxonomy follows Carnicero (2017). In a more recently published article (Carnicero & al. 2021), *C. microcalyx* subsp. *paradoxa* Greuter has been raised to species level as *C. paradoxa* (Greuter) Carcinero, and *C. minor* subsp. *dodekanesi* (Greuter) Carcinero & al. has become *C. acutiloba* subsp. *dodekanesi* (Greuter) Speta.

Cymbalaria microcalyx subsp. ***alba*** (Voliotis) Kit Tan & Iatroú is a form with whitish corolla, described from the eastern foothills of Mt Taigetos and subsequently reported also from other localities in Peloponnisos. Its status is uncertain, and following Carnicero (2017: 191) it has been merged with subsp. *microcalyx*.

Cynara cyrenaica Maire & Weiller in known only from few populations in Cyrenaica and E Kriti. Plants from Cyprus, originally reported as *C. cyrenaica*, have recently been described as an endemic species, *C. makrisii* Hand & Hadjik.

Cynara scolymus L. (artichoke) is widely cultivated but scarcely naturalized. Its wild progenitor is probably *C. cardunculus* L.

Cynodon transvaalensis Burtt-Davy is widespread in tropical areas and commonly cultivated as a lawn grass. It has been reported as a casual escape in Attiki (Damanakis & Scholz 1990: 420). It resembles *C. dactylon* (L.) Pers., but has filiform rather than flat leaves and usually 2 shorter spikes.

Cynoglossum, etc. Delimitation of genera in the *Cynoglosseae* (*Cynoglossum* L., *Mattia* Schult., *Mattiastrum* Brand, *Paracaryum* (DC.) Boiss., *Rindera* Pall., *Solenanthus* Ledeb., etc.) is occasionally problematic. Finding that in this particular case lumping would give room for a larger number of new combinations than splitting, Greuter (1981b: 32-36) merged them all into *Cynoglossum*. In this Atlas the approach of Hilger & al. (2005) is followed, with the Greek genera *Cynoglossum*, *Paracaryum*, *Rindera* and *Solenathus* maintained as separate.

Cynoglossum hungaricum Simonk. As used by Greek authors this name apparently refers to *C. pustulatum* subsp. *parvifolium* (Vis.) Sutorý (see Sutorý 2000: 230-231).

Cynoglossum nebrodense Guss. Historical records from Mt Dirfis (Evvia) and mountains of Peloponnisos are likely to refer to *C. columnae* Ten.

Cynoglottis (Guşul.) Vural & Kit Tan is a small segregate of *Anchusa*, differing in details of flowers and pollen. *C. barrelieri* (All.) Vural & Kit Tan was based on *Buglossum barrelieri* All. from Italy. *Anchusa serpentinicola* Rech. f. was described as a serpentine endemic from Mt Vourinos (NC Greece) and has later been recombined as *Cynoglottis barrelieri* subsp. *serpentinicola* (Rech. f.) Vural & Kit Tan. Most of the Greek records belong to this taxon which is in fact more widespread on limestone (*Anchusa phocidica* Gustavsson, described from Mt Giona, is apparently a synonym). Records from low altitude in the NE are probably referable to subsp. *barrelieri*, but differences are quantitative and not very clear. The species has a wide altitudinal and ecological range, and further studies are needed. Selvi & al. (2004) maintained *Cynoglottis* as a separate genus.

Cyperus alopecuroides Rottb. has been reported as naturalized in the harbour area of the town of Rodos (Kleinsteuber, in litt., based on a collection by Ristow in 2003).

Cyperus alternifolius subsp. ***textilis*** (Thunb.) F. Verloove, a native of South Africa naturalized in the Azores, was reported by Verloove (2014: 196) from the island of Kerkira, based on a single collection from 1931. There appears to be no recent confirmation, and it may have been casual.

Cyperus glomeratus L. was listed for Greece in Fl. Eur. (5: 286, 1980), but the source for this record is not known. Subsequent reports from the Evros river mouth (Babalonas 1981: 258) and the Preveza area (Chitos 2009: 99) are feasible but need confirmation. Verloove (2014) did not list any specimens from Greece.

Cyperus involucratus Rottb. (*C. alternifolius* auct.) is cultivated as an ornamental and may be occasionally naturalized, e.g. on Rodos and in the city of Patras. It is a caespitose, leafless perennial with an umbrella-like inflorescence of long, flat bracts and 20-25 short rays.

Cyperus michelianus (L.) Delile. Both subsp. *michelianus* and subsp. *pygmaeus* (Rottb.) Asch. & Graebn. occur in Greece, sometimes in mixed populations. Normally they are readily distinguished by the glumes being distichous in the latter and 3-ranked in the former. Other supposedly distinguishing characters occur in different combinations. Further studies are needed to establish whether the two are consistently separate.

Cyperus serotinus Rottb. Greek records need confirmation; see comments in Atlas of the Aegean Flora 1: 254 (2016).

Cyperus squarrosus L. was listed for the Cretan area in Euro+Med PlantBase, referring to Verloove (2014), but this is apparently a mistake. Only specimens from Italy were mentioned in this publication.

Cypripedium calceolus L. This distinctive, Euro-Siberian orchid was reported from Mt Timfristos by Halácsy (1904a: 152), based on an observation (or collection?) by Spruner c. 1840. No herbarium material has been located and the record has never been confirmed. The species was reported from the Bulgarian side of Rodopi by Perry & Perry (2014) and subsequently by Popatanasov (2015: 203-204), and should be searched for on the Greek side.

Cystopteris alpina (Lam.) Desv. has been reported from scattered mountain localities in Greece, including Olimbos (Stojanov & Jordanov 1938: 159), Valtou Ori (Giannakos 2007: 68), Samothraki (Biel & Kit Tan 2008c: 293) and Parnassos (Polymenakos & Kit Tan 2014a: 112). It is difficult to distinguish from some forms of the variable *C. fragilis* (L.) Bernh., and all records need confirmation. *C. alpina* is a taxon of uncertain status and distribution, differing from *C. fragilis* mainly in the parallel-sided, emarginate ultimate leaf segments.

Cytisus striatus (Hill) Rothm. is a species of SW Europe and NW Africa, differing from *C. scoparius* (L.) Link, e.g., in having appressed-hairy calyces and densely pilose legumes. It was found near the town of Ouranopolis in NE Greece over a distance of 1 km along the road and did not appear to be planted, but is probably naturalized from former cultivation nearby (Christensen & Strid 2010).

Dactylis glomerata L. Whilst the extremes are strikingly different, the subspecies differ only in quantitative characters and merge into each other in zones of contact. Subsp. *hispanica* (Roth) Nyman (syn. *D. hispanica* Roth) is Mediterranean and submediterranean, and the most common form of the species in Greece; it merges into subsp. *glomerata* in the north. Subsp. *hackelii* (Asch. & Graebn.) Cif. & Giacom. (syn. subsp. *marina* (Borrill) Greuter; *D. marina* Borrill) is perhaps only a maritime ecotype of subsp. *hispanica*. Similarly, subsp. *rigida* (Boiss. & Heldr.) Hayek (syn. *D. rigida* Boiss. & Heldr.), which forms dense, hard cushions, may be regarded as a rather extreme ecotype adapted to the dry and harsh conditions of the Cretan limestone mountains.

Dactyloctenium aegyptium (L.) Willd. According to Flora of Turkey (9: 578, 1985) this species is naturalized also in Kriti, which seems feasible. However, no specimens or specified records have neen seen.

Dactylorhiza majalis (Rchb.) P. F. Hunt & Summerh. Based on molecular data from Hedrén & al. (2007), Kühn, Pedersen & Cribb (2019: 151 ff.) recombined *D. cordigera* (Fr.) Soó, *D. kalopissii* E. Nelson, *D. macedonica* J. Hölz. & Künkele, *D. pindica* B. & E. Willing and *D. pythagorae* Gölz & H. R. Reinhard as subspecies of *D. majalis*, a very variable species of C & N Europe.

Dalbergia rugosa Hepper, a tree of West Africa, was listed for Greece in Euro+Med PlantBase, referring to Flora Europaea, but this is apparently a mistake as the species was not mentioned in the latter.

Daphne jasminea Sm. Most of the Greek populations have been referred to subsp. *jarmilae* Halda, but this is doubtfully distinct from the nominate subspecies (cf. Aegean Atlas 1: 547, 2016).

Daphne sojakii Halda is a name given to pink-flowered plants from Mt Vermion resembling *D. oleoides* Schreb. They may be referable to *D. kosaninii* (Stoj.) Stoj.

Dasypyrum. The combination *D. breviaristatum* (H. Lindb.) Fred. in Nordic J. Bot. 11: 140 (1991) is illegitimate, and the correct name for the species is *D. hordeaceum* P. Candargy.

Datura wrightii Regel, a neophyte of North American origin, has recently been reported, as new for Greece, from the island of Limnos (Bergmeier 2021b: 364). It is similar to *D. innoxia* Mill., differing in the indumentum consisting of very short, curved, eglandular hairs (vs. somewhat longer, straight, glandular hairs in *D. innoxia*). Previous Greek collections identified as *D. innoxia* should be critically re-examined. Of the two specimens under that name in my private herbarium, one (Kit Tan & Strid 18710 from nomos Arkadias) matches *D. innoxia* and the other (Strid 60583 from the road between Githion and Areopolis) *D. wrightii*. Photos of fruiting plants from the village of Vathia in the Mani Peninsula (Strid, 2016-11-05) also match *D. wrightii*.

Daucus bicolor Sm. The identity of the flowering specimen illustrated in Flora Graeca is not clear (cf. Strid & Strid 2010: 140-141). Reduron (2007b: 964) used the name *D. bicolor* for the taxon here called *D. broteri* Ten., but the former may rather be a synonym of *D. guttatus* Sm. Fruiting material is needed for positive identification of these species.

Daucus carota L. is a widespread and very variable species or species complex. For an extensive account of variation and taxonomy, especially in France, see Reduron (2007b: 968 ff.). No detailed study of Greek material is available, and only a collective map can be presented.

Dianthus arpadianus Ade & Bornm. was described from Mt Fengari, Samothraki, and named after the Hungarian botanist Árpád von Degen (1866-1934) who is best known for his *Flora Velebitica* and was also the first to collect plants on Samothraki (in 1890). It has also been recorded from Kaz Dağ (Mt Ida) in NW Anatolia and, somewhat surprisingly, collected at low altitude on the Greek mainland opposite Samothraki (Rechinger 5949, 1934-05-28; G, LD). In the latter locality it was rediscovered and photographed by K. Goula, 2018-06-02. It should be critically examined whether the lowland plants are indeed conspecific with those from Mt Fengari.

Dianthus bessarabicus (Kleopow) Klokov, a species of Romania and Ukraine, was listed for Greece in Euro+Med PlantBase. The source is obscure, and no Greek specimens have been seen.

Dianthus brachyzonus Borbás & Formánek was reported from NW Greece in Fl. Eur. ed. 2, 1: 237, 1993). According to Euro+Med PlantBase it is a synonym of *D. cruentus* Griseb.

Dianthus campestris M. Bieb. As for *D. bessarabicus* above.

Dianthus capitatus DC. As for *D. bessarabicus* above.

Dianthus carthusianorum L. Greek records probably refer to other taxa in the *D. cruentus* complex.

Dianthus leptopetalus Willd., described from Caucasus, might occur in NE Greece, but records (e.g., Turrill 1918: 269) are more likely to refer to *D. monadelphus* subsp. *pallens* (Sm.) Greuter & Burdet.

Dianthus microlepis Boiss. occurs in SW Bulgaria. An unconfirmed record from Mt Beles just across the Greek border (Zaganiaris 1940a: 51) may refer to *D. simulans* Stoj. & Stef.

Dianthus moesiacus Vis. & Pančić. Greek records are probably incorrect (cf. Fl. Hellen. 1: 372, 1997).

Dianthus muglensis Hamzaoğlu & Koç. See Strid (2021: 265-266). Subsequently Manousakis has observed *D. muglensis* in a second locality on Rodos, on Mt Koutsoutis NW of Archangelos (e-mail 2021-11-09). *D. cyprius* A. K. Jacks. & Turrill, an endemic of northern Cyprus, has similar but darker petals.

Dianthus pallidiflorus Ser. was listed for Greece in Euro+Med PlantBase. The source is obscure and no Greek specimens have been seen.

Dianthus pelviformis Heuff. As for *D. moesiacus* above.

Dianthus quadrangulus Velen., described from Bulgaria, was listed for Greece in Euro+Med PlantBase. Greek records may refer to *D. strymonis* Rech. f. or other members of the *D. cruentus* complex.

Dianthus roseoluteus Velen. As for *D. moesiacus* above.

Dianthus serratifolius subsp. ***abbreviatus*** (Halácsy) Strid. With more specimens now available there appear to be gradual transitions between subsp. *serratifolius* and subsp. *abbreviatus*.

Dianthus stenopetalus Griseb. After the preparation of the maps this species has been reported from the Rodopi mountains by Charalampidou, Eleftheriadou & Theodoropoulos in Fl. Medit. 32: 85 (2022).

Dianthus strictus Banks & Sol. Greek plants have been referred to subsp. *multipunctatus* (Ser.) Greuter & Burdet. The latter was described from the island of Kos; it overlaps with subsp. *strictus* in distribution and is doubtfully worthy of taxonomic status.

Dianthus sylvestris Wulfen is a variable species of C & S Europe. Most of the Greek collections are referable to subsp. *sylvestris*, but plants from the Ionian Islands are slightly deviating (cf. Fl. Hellen. 1: 346, 1997) and have been called *D. arrostii* C. Presl. In Flora Ionica (https://floraionica.univie.ac.at/) plants from Kerkira and Paxi were referred to *D. sylvestris* subsp. *longicaulis* (Ten.) Greuter & Burdet. The latter was described, as *D. longicaulis* Ten., from S Italy, and recombined by Greuter & Burdet (in Greuter & Raus 1982b: 187) who presented no evidence for this course of action and apparently had not seen the types nor any other specimens. The group is in need of critical revision.

Dianthus tristis Velen., described from Bulgaria, is a taxon of uncertain status more or less intermediate between *D. cruentus* and *D. stenopetalus* Griseb. Specimens from Mt Varnous may belong here.

Dichanthium annulatum (Forssk.) Stapf is native in tropical Asia and parts of Africa, and has been introduced elsewhere as a forage grass. There is a single record from a garden in Iraklion, Kriti (Damanakis 7034, B; det. Scholz).

Digitalis laevigata Waldst. & Kit. Almost all Greek records refer to subsp. *graeca* (Ivanina) Werner. In the far north this merges into subsp. *laevigata* (described from Croatia). The two differ only vaguely in quantitative characters (mainly size of corolla).

Digitalis leucophaea Sm. was described from the Athos peninsula and illustrated on Plate 607 of Flora Graeca (1831). It was said to differ from *D. lanata* Ehrh. by its slender inflorescence with smaller flowers and relatively longer bracts. It can scarcely be maintained even as a subspecies since many intermediates occur on the Athos peninsula and elsewhere in N & NE Greece.

Digitalis macedonica Heywood was based on a single collection from Mt Smolikas (Balls & Gourlay B3529, K). It has never been confirmed and may be a hybrid between *D. laevigata* and *D. viridiflora* Lindl., both of which occur in the area.

Digitaria aegyptiaca (Retz.) Willd. was listed for "Gr" in Euro+Med PlantBase, but the source for this record is obscure. No Greek specimens or specified references have been seen.

Digitaria ciliaris (Retz.) Koeler is probably native in tropical Asia and widely introduced in other tropical and warm-temperate areas. It is established as a weed of rice fields in Italy, but Greek specimens identified as *D. ciliaris* are from other ruderal habitats and do not appear to differ consistently from *D. sanguinalis* (L.) Scop.

Dinacrusa Krebs (1994b) was established to accommodate a few species previously referred to *Althaea* or *Malva*, including *Malva aegyptia* L., *M. cretica* Cav. and *M. setigera* Schimp. & Spenn. This split has had few followers.

Diospyros lotus L. A record by Halácsy (1902a: 286) from "Phthiotis: pr. Hypata ad radices mt. Aeta (Heldr.)", cited by Regel (1951: 45) is likely to refer to cultivated plants. It is a small evergreen tree with spherical, yellow berries 1-2 cm, probably native in SE Asia and sometimes cultivated in the Middle East and the Mediterranean area.

Diplachne fusca (L.) P. Beauv. (*Leptochloa fusca* (L.) Kunth) has been reported as a weed of rice fields in western Peloponnisos and the Nestos river area in the north-east (Greuter & Raus 2002: 205), and subsequently from the delta of Axios-Loudias-Aliakmon (Hanlidou, Koutouvela & Tsoulkanaki 2019: 133). These are the first Greek records of a tufted perennial grass that is widely distributed in tropical and warm-temperate regions from Africa to India and through SE Asia to Australia. It may be on the verge of becoming established in Greece as well as in Cyprus.

Diplotaxis erucoides (L.) DC. is sometimes regarded as native only in the W Mediterranean area but, if so, it is long since established in Greece where it was reported already by Friedrichsthal (1838: 267) from "sandiges Ackerland gegen den Hafen Phalerus". There are several recent records from NE Peloponnisos (cf. Zarkos & al. 2016c: 463-464). It may have been overlooked because of its mainly winter-flowering habit. In was collected and photographed by the present author in December 2016 in olive groves in nomos Korinthias, where it did not appear to be a recent introduction. In 2020 and 2021 it was found also in Kriti (see http://cretanflora.com).

Dittrichia orientalis Brullo & De Marco. The widespread *Inula* (*Dittrichia*) *viscosa* was subdivided by Brullo & De Marco (2001), with most Greek plants being referred to a new species, *Dittrichia orientalis*. There appears to be little or no evidence for this. Some Greek plants have also been called *D. viscosa* subsp. *angustifolia* (Bég.) Greuter.

Doronicum caucasicum/orientale. Hind (2019) reported that "critical examination and reading of the works in which the protologues were published has led to the conclusion that *D. caucasicum* M. Bieb. was published earlier than *D. orientale* Hoffm., but it is highly likely that the two names were based on the same Adams collection that Adams had annotated as *Doronicum orientale*". Consequently, the name *D. caucasicum* will have to replace *D. orientale* which has been used in most of the relevant literature in recent decades.

Dorycnium strictum (Fisch. & C. A. Mey.) Lassen (*Lotus strictus* Fisch. & C. A. Mey.) is scattered through Anatolia to C Asia, with outpost localities in SE Bulgaria. Fl. Eur. (2: 176, 1968) listed it for Greece, perhaps based on the imprecise "Thra" in Hayek (1926a: 882). It is not unthinkable that *D. strictum* could occur in NE Greece, but the record may also refer to a glabrous form of *D. hirsutum* (L.) Ser. (see Greuter & Raus 1986b: 109).

Draba heterocoma Fenzl belongs to the complex of *D. bruniifolia* Stev. which is widespread and very variable in Anatolia. Plants from Samos have been referred to subsp. *archipelagi* (O. E. Schulz) Buttler which also occurs in adjacent parts of W Anatolia. A record of *D. heterocoma* from Karpathos goes back to a collection by Pichler (1876) which has not been traced and may have been erroneously cited; there is no subsequent confirmation.

Draba parnassica Boiss. & Heldr. Reports from SW Albania (Vangjeli 2017: 352) need confirmation.

Drimia maritima (L.) Stearn, s. lat. For comments on taxonomy, see Aegean Atlas (1: 332, 2016). Martínez-Azorín, Crespo & Alonso-Vargas have quite recently (Medit. Bot. 43, e78272, 2022) transferred this species complex to *Squilla* Steinh., considering *Charybdis* Speta an illegitimate name. There is presumably only one taxon in Greece, and in this Atlas it is provisionally retained as *Drimia maritima*. Adopting a narrow species concept it would be referable to *Drimia aphylla* (Forssk.) J. C. Manning & Goldblatt, a taxon of the E Mediterranean area.

Dryopteris carthusiana (Vill.) H. P. Fuchs has been reported from Mt Vrondous (Greuter 1977b: 6), Mt Voras (Humbert & Topali 129, specimen at G) and Mt Rodopi (Eleftheriadou 1992: 28); a collection from Samothraki was tentatively identified as *D. carthusiana* by Biel & Kit Tan (2014a: 61). All records need confirmation and may refer to *D. expansa* (C. Presl) Fraser-Jenk. & Jermy or *D. dilatata* (Hoffm.) A. Gray. *D. carthusiana* is rare and scattered in North Macedonia (Micevski 1985c: 71) and Albania (Barina 2017: 51).

Dysphania ambrosioides (L.) Mosyakin & Clemants. Willing 250628 & 251299 (B) from the Florina area and Willing 252768 (B) from the Naoussa area have been annotated by P. Uotila (2020) as "*D. ambrosioides* ad *D. chilensis* (Schrad.) Mosyakin & Clemants vergens".

Dysphania anthelmintica (L.) Mosyakin & Clemants, a native of North America, has been reported from Rodos (Kleinsteuber & al. 2016: 369-370). It is closely related to *D. ambrosioides* and possibly conspecific (cf. comments in Aegean Atlas 1: 229, 2016). Willing 253352 (B) from north central Greece ("Imathias, Trilofos. Krautfluren im Ort") and Willing 248307 (B) from the Grevena area, originally reported as *D. anthelmintica*, appear to represent *D. ambrosioides* (Th. Raus, in litt., 2020-03-20). Possible occurrence of the former in Greece needs confirmation. Uotila & al. (2021: 542) stated: "*Dysphania ambrosioides* and *D. anthelmintica* have been cultivated as medicinal plants, but many of the plants cultivated under the name *Chenopodium anthelminticum* L. or *Dysphania anthelmintica* seem to be misidentified and belong to *D. ambrosioides*".

Dysphania carinata (R. Br.) Mosyakin & Clemants. A single collection from nomos Pellis (Willing 252492, B) originally reported as *D. antihelmintica*, has been revised as *D. carinata* by P. Uotila, 2020. It is presumably not established.

Echinochloa crus-galli (L.) P. Beauv. Awned and unawned forms may occur in the same area. Most of the Greek collections are probably referable to subsp. *crus-galli*. Subsp. *hispidula* (Retz.) Honda has been reported from rice fields in the north, first by Raus & Scholz in Greuter & Raus (2002: 205), and subsp. *spiralis* (Vasinger) Tzvelev has been reported from Kriti (Böhling & Scholz 2003: 44) and SW mainland Greece (Zotos, Raus & Dimopoulos 2006: 736).

Echinochloa oryzoides (Ard.) Fritsch was described and illustrated already by Arduino (1764: 16 tab. 5, as *Panicum oryzoides*) from plants he had grown in Italy from seeds received mixed up among rice samples. It has only recently been reported from Greece.

Echinops ritro L. subsp. ***ritro***, subsp. ***ruthenicus*** (Boiss. & Heldr.) Kožuharov and subsp. ***sartorianus*** (M. Bieb.) Nyman have been recognized in Greece, but the morphological differences are obscure and their distributions overlap completely. Plants from Kerkira have recently been referred to *E. siculus* Strobl (Conti, Reich & Gutermann 2020: 96) which was previous considered endemic to S Italy and Sicily. The latter is in some respects intermediate between *E. ritro* and *E. sphaerocephalus*, and is here classified as *E. ritro* subsp. *siculus* (Strobl) Greuter.

Echinops sphaerocephalus L. Three subspecies have been recognized in Greece, based mainly on the amount of glandular pubescence on upper part of stem and phyllaries. Most of the Greek collections belong to subsp. *albidus* (Boiss. & Spruner) Maire & Petitm. (with eglandular stem and glabrous outer phyllaries), but there are some reports also of subsp. *sphaerocephalus* (with glandular upper stem and some short glandular hairs on outer phyllaries). Plants from high altitude in Peloponnisos match subsp. *taygeteus* (Boiss. & Heldr.) Maire & Petitm. (with densely glandular-pubescent upper stem and outer phyllaries). Further studies are needed to determine the taxonomic value, if any, of this variation.

Echinops spinosissimus Turra. Almost all records from the Aegean area can be referred to subsp. *spinosissimus*, but some islands populations merge into the poorly defined subsp. *bithynicus* (Boiss.) Greuter. Plants from the Ionian Islands have been identified as subsp. *neumayeri* (Vis.) Kožuharov, a taxon previously recorded from Croatia and Albania (see Conti, Reich & Gutermann 2020: 100).

Echium italicum L. Large, much-branched individuals have been called subsp. *biebersteinii* (Lacaita) Greuter & Burdet, but there is continuous variation.

Echium maculatum L. is the oldest name for a species previously known as *E. russicum* J. F. Gmel. It is rare in Greece and was first collected on Mt Falakron by Stamatiadou (1981-07-01, no. 22355, ATH). For details on plants from this area, see Asenov (2015b: 368-369) as well as Simoglou, Dordas & Koutsos (2015: 397-398). There is a recent record also from nomos Kilkis (Ioannidis, Doulkeridou & Strid 2021b: 260). The species is considered vulnerable in Bulgaria where it is widespread but scattered (Peev & al. 2015: 672). Hilger & Böhle (2000) proposed to transfer it to the monotypic genus *Pontechium*; this split has had few if any followers.

Echium sabulicola Pomel was reported from a coastal locality in western Greece (map in Klotz 1962c: 1090). It is otherwise a W & C Mediterranean species extending to Calabria and Cyrenaica. Occurrence in W Greece is feasible but needs confirmation. A collection from the small Kikladian island of Kavouras near Andiparos (Runemark & Bentzer 30822, LD) may represent a casual introduction.

Echium spurium Lojac. is a taxon of uncertain status tentatively reported from Sicily and Greece by Klotz (1962c: 1098). It was not mentioned by Pignatti (2018), and listed as a synonym of *E. parviflorum* Moench in the *Flora d'Italia* website.

Edraianthus graminifolius (L.) A. DC. is a variable species. Greek populations belong to the nominate subspecies whereas subsp. *siculus* (Strobl) Greuter & Burdet is confined to southern Italy and Sicily. Two taxa described at species rank within this complex have been reported to occur in Greece, viz. *E. australis* (Wettst.) Lakušić (StE, SPi, NPi, NC, NE, widespread from Parnassos to Pangeon) and *E. horvatii* Lakušić (NC: Mt Boutsi). Morphological differences are minute, and in terms of molecular phylogeny both are nested within *E. graminifolius* (Sefanović & al. 2008: 468-469, figs. 7 A & 7C). Several other taxa occur in the C & W Balkan Peninsula. Euro+Med PlantBase listed *E. tenuifolius* (Waldst. & Kit.) A. DC. from Greece, but this is confined to the W Balkans from Montenegro to Istria. *Halacsyella parnassica* (Boiss. & Spruner) Janch., here referred to a monotypic genus, is optionally included in *Edraianthus*.

Elaeagnus commutata Bernh. ex Rydb., a native of North America, has been collected once in N Peloponnisos ("4 km from Valimi to Akrata", 28.5. 2021, Kit Tan & G. Vold 33190; ATH, C, UPA). It is not clear to what extent it was naturalized, but the species has been reported as an established alien in Italy and some other European countries.

Elatine. In addition to *E. alsinastrum* L. (with whorled leaves and almost straight seeds), Takács & al. (2018) regognized five species in the Mediterranean area, all with opposite leaves and more or less curved seeds. They constitute a polyploid complex with chromosome numbers from 2n=18 to 2n=54. Greek plants previously called *E. macropoda* Guss. were referred to *E. campylosperma* Seub. and are provisionally mapped as such in this Atlas. According to Popiela & Łysko (2010: 83) *E. macropoda* is widespread in the Mediterranean area; it was not reported from Greece but from neighbouring countries including S Italy, W Anatolia and Cyprus. Only a fraction of the Greek specimens have been critically examined. Bazos (pers. comm., 2021) found that the majority of plants from Lesvos have seeds matching those of *E. macropoda* rather than *E. campylosperma*, and it is possible that both species occur in Greece.

Eleocharis geniculata (L.) Roem. & Schult. was listed for IoI by Dimopoulos & al. (2013: 89), based on a single collection from Kerkira kept in a private herbarium. It may refer to *E. caduca* (Delile) Schult.

Eleocharis mitracarpa Steud. appears as an accepted species in Euro+Med PlantBase as well as Flora Europaea and Flora of Turkey, although Strandhede's monograph (1966) concluded that it falls within the range of variation of *E. palustris* (L.) Roem. & Schult. subsp. *palustris*. Plants matching *E. mitracarpa* occur from Anatolia to Pakistan, and have also been reported from the island of Kos (Fl. Turkey 9: 52, 1985) as well as the western part of the Greek mainland, viz. Lake Ioanninon (Sarika-Hartzinikolau 1999: 150) and the Louros river floodplain in nomos Prevezis (Sarika, Dimopoulos & Yannitsaros 2005: 77).

Eleocharis palustris var. ***repens*** Maire & Petitm. was described from wet grassland at 1120-1400 m on Mt Parnassos and Mt Killini: *Luteo-virens, rhizomate longe repente, caulibus floriferis prostratis vel apice ascendentibus*. Such plants were observed by Hartvig & Strid 13 June 2021 on Mt Parnassos close to the original locality of Maire & Petitmengin. A distinctive feature not mentioned by M & P appears to be the flattened culms. These populations probably deserve higher taxonomic status than variety. [Thanks are due to Per Hartvig for these observations].

Eleocharis parvula (Roem. & Schult.) Bluff & al., a tiny, widespread wetland species, has been reported once in Greece, as a rice field weed near the Sperchios river in nomos Fthiotidos (Greuter & Raus 2002: 204). It is doubtfully native, but should be looked for in similar habitats elsewhere.

Eleocharis quinqueflora (Hartmann) O. Schwarz. After the preparation of the maps this species has been reported from the Rodopi mountains by Charalampidou, Eleftheriadou & Theodoropoulos in Fl. Medit. 32: 90 (2022).

Elytrigia. Banfi (2018) transferred all European species of *Elytrigia* to the genus *Thinopyrum*. Evidence for this seems, at best, to be flimsy.

Elytrigia campestris (Godr. & Gren.) Kerguélen is a taxon of uncertain status, occurring mainly in W Europe. There are credible reports from grassland and woodland margins on Mt Chortiatis (Karagiannakidou & Raus 1996: 542) and Mt Gerania (Constantinidis 1997b: 281) as well as dubious ones from a few coastal localities.

Elytrigia juncea (L.) Nevski and related species. See comments in Strid (2016a: 452).

Elytrigia lazica subsp. ***divaricata*** (Boiss. & Balansa) Valdés & H. Scholz (*Elymus lazicus* subsp. *divaricatus* (Boiss. & Balansa) Melderis) has been collected only once in Europe, on Mt Pinovon in north central Greece, rocky limestone outcrops in *Fagus* woodland, 1400-1700 m, 16.8. 1979, Strid & al. 16552 (B, G, herb. Strid). 2n=14 has been reported in the Greek collection (Strid 1986c: 902). The species is widespread and variable in Anatolia.

Elytrigia strigosa (M. Bieb.) Nevski has been reported once in Greece, from a serpentine area in the foothills of Mt Voras (Babalonas & Moustakas 1986). It is otherwise endemic to mountains of Crimea, and the Greek record needs confirmation.

Empetrum nigrum L., a widespread boreal and subarctic species, was reported from Karlik Dağ (Papikio) in NE Greece by Zaganiaris (1940a: 85). There is no subsequent confirmation, but a mistake seems unlikely. The nearest localities are in SW Bulgaria (Mts Rila and Pirin).

Epilobium adenocaulon Hausskn. is a species of North American origin, widely naturalized in C & N Europe. It has been reported once in Greece, in a flood plain habitat by the Nestos river near Paranestion (Schuler & Snogerup 2006: 724). In Vascular Plants of Greece (Dimopoulos & al. 2013) it was merged with *E. ciliatum* Raf., another alien of North American origin. The two are usually regarded as distinct in Central and North European Floras. They differ from all native European species in the seeds which have papillose ribs and a conspicuous neck.

Epilobium alpestre (Jacq.) Krock. has recently been reported from Mt Varnous in nomos Florinis (Giannakis & al. 2021: 363), growing by a stream at c. 1780 m with *Doronicum austriacum, Geum coccineum, Cirsium appendiculatum*, etc. It is previously known to occur in the Pyrenees, Alps, Carpathians, N Balkan Peninsula, NE Anatolia and Caucasus. According to Nikolić (2020, 3: 189) it extends to Croatia, Bosnia-Herzegovina and Albania. It is a distinctive species characterized by solitary, erect stem 20-80 cm, narrowly ovate, finely serrate leaves in whorls of 3(-4), rather large, pink flowers and capitate stigma.

Epilobium collinum Vill. was listed for Peloponnisos by Mermygkas, Zikos & Constantinidis (2021: supplementary file), but no details are available.

Epilobium obscurum Schreb. As for *E. collinum*.

Epipactis spp. Taxonomy follows Tsiftsis & Antonopoulos (2017), although some of the recently described species are dubiously distinct (cf. also Kühn, Pedersen & Cribb 2019).

Epipactis cretica Kalop. & Robatsch. The name needs to be validated as no type specimen was cited with the description.

Equisetum fluviatile L. After the preparation of the maps this species has been reported from the Rodopi mountains by Charalampidou, Eleftheriadou & Theodoropoulos in Fl. Medit. 32: 83 (2022).

Eragrostis barrelieri Daveau has been reported as a weed in the Floriculture Experimental Station of Attiki (Damanakis & Scholz 1990: 421), and subsequently from the island of Limnos (Bergmeier & al. 2021: suppl. 1 p. 24). The species is of uncertain origin and now has an almost worldwide distribution, including much of the Mediterranean area. It may be on the verge of becoming naturalized in Greece.

Eragrostis curvula (Schrad.) Nees has been reported from lawns and ruderal habitats in the cities of Athens and Patras (Damanakis & Scholz 1990: 421, Chronopoulos & Christodoulakis 1996: 172). It is probably native of southern Africa and widely introduced elsewhere but scarcely established in the Mediterranean area.

Eragrostis leptocarpa Benth., an alien of Australasian origin, has been reported once from a ruderal habitat in the town of Corfu (Scholz 2000a: 242).

Eragrostis minor Host subsp. ***angustata*** H. Scholz & Raus has been reported from a few, scattered localities, and subsp. *roborovskii* (Tzvelev) H. Scholz once from the Kavala area. Their status is uncertain, and most of the Greek specimens have not been identified to subspecies.

Eragrostis multiglandulosa H. Scholz was based on a single, fragmentary collection from Samothraki. It appears to differ from *E. cilianensis* (All.) Janch. only in the presence of glands on the leaves and lemmas. Further material is needed to establish whether it is of any taxonomic significance.

Eragrostis neomexicana Vasey ex L. H. Dewey has been reported once, from a ruderal habitat on the island of Thasos (Biel & Kit Tan 2017a: 124). It is native from SW USA to N South America, and a rare alien in Europe.

Eragrostis pectinacea (Michx.) Nees is widespread as native in the Americas and may be on the verge of becoming established as a weed in the Mediterranean area. In Greece there are reports from cultivated fields in northern Peloponnisos (Greuter & Raus 2004: 78) and Attiki (Giannopolitis 675, H).

Eragrostis pilosa (L.) P. Beauv. belongs to a complex of five closely related species, some of which are regionally invasive in Europe (see Wróbel & al. 2021).

Erigeron epiroticus (Vierh.) Halácsy. Plants from Mt Taigetos (leg. Kofinas), originally registered as *E. epiroticus*, differ in their densely hairy leaves and are of uncertain identity (Kit Tan, in litt. 2021-08-04).

Erigeron uniflorus subsp. ***parnassensis*** A. G. Game ex M. J. Y. Foley was based on a single specimen from the summit area of Mt Parnassos; it may be a form of *E. glabratus* Bluff & Fingerh.

Eriobotrya japonica (Thunb.) Lindl. (English "loquat" from Cantonese "lo guat", Greek μούσμουλο), a native of SC China, is cultivated for its edible fruits and frequently seen at the outskirts of villages, but scarcely spreading into natural vegetation. It is a small, evergreen tree with simple, dark green leaves 10-25 cm long. The rather inconspicuous flowers appear in winter, and the fruit ripens already in April-May; it is broadly ellipsoid, 3-5 cm long and brownish-yellow, with 1-3 large seeds.

Eriochloa contracta Hitchc., a native of C USA ("prairie cupgrass") has been been reported from an irrigated lawn in N Kriti (Bergmeier 2007: 441). It is scarcely naturalized.

Erodium absinthoides Willd. All Greek material is likely to belong to subsp. *guicciardii* (Heldr.) Maire & Petitm. which occurs also in SE Albania and North Macedonia. In the foothills of Mt Siniatsiko it grows together with *E. hartvigianum* Strid & Kit Tan (at *locus clasicus* for the latter), and occasional hybrids occur (Strid 56135; G, herb. Strid).

Erodium acaule (L.) Bech. & Thell. There are scattered literature records almost throughout Greece, but all need confirmation and may refer to large-flowered, ± acaulescent forms of *E. cicutarium* (L.) L'Hér.

Erodium aethiopicum (Lam.) Brumh. & Thell. Records from the islands of Gavdos and Chrisi S of Kriti (Bergmeier 2011: 171) as well as a record from a small island near Egina (Vallianatou 2005: 183) need confirmation. They may refer to large-flowered forms of *E. cicutarium* (L.) L'Hér.

Erysimum cheiri (L.) Crantz is cultivated as an ornamental and occasionally naturalized on stone walls near settlements and in coastal areas. It differs from the native species of sect. *Cheiranthus* in the conspicuously fragrant flowers with glandular, usually purplish sepals and thickened style.

Erysimum crassistylum C. Presl was listed for Peloponnisos by Mermygkas, Zikos & Constantinidis (2021: supplementary file). The record is feasible, but no details are available.

Erysimum creticum/raulinii. As pointed out by Polatschek (1973: 116) descriptions of these two species were based on mixed gatherings. Although the type of *E. creticum* Boiss. & Heldr. is labeled "In saxosis Contopoura eparch. Apocorono, Apr. 1846" [i.e., in western Kriti] it was most probably collected in the Sitia area of eastern Kriti, whereas the type of *E. raulinii* Boiss. is indeed from "rupibus et saxosis Contopoura".

Erysimum diffusum Ehrh. There are several records of this species from N Greece, but all are dubious, most of them probably referring to *E. crassistylum* C. Presl.

Erysimum graecum Boiss. & Heldr. Reverchon 217, 12.4. 1884 (BP, BRNM, LD, W, WU-HAL) was reported from the Topalia area in W Kriti by Polatschek (1973: 112), apparently in error. The record was omitted in Flora Hellenica.

Erysimum kuemmerlei Jáv. was listed for Greece in Flora Europaea (ed. 2 1: 334, 1993) as well as in Euro+Med PlantBase. Occurrence in northern Greece is feasible, but no specified Greek record has been seen, and the species was not included by Polatschek & Snogerup (in Flora Hellenica 2, 2003).

Erysimum moesiacum Velen. was reported from NC in Dimopoulos & al. (2013: 72). It is apparently a synonym of *E. crassistylum* C. Presl.

Eucalyptus camaldulensis Dehnh. is commonly cultivated at low altitude in S Greece, but is scarcely naturalized unlike, e.g., in Portugal, where it is invasive. It is native along river courses and floodplains in much of Australia (known as Red River Gum) and is one of the most widely planted eucalypts elsewhere. It is a fast-growing, evergreen tree with smooth bark shed in long ribbons, and falcate leaves. Oddly, the species was described from trees cultivated in the vicinity of the Camaldoli monastery near Naples, hence the name.

A few other *Eucalyptus* species are occasionally cultivated in Greece (see Arabatzis 2001: 208-213 as well as Fl. Cyprus 1: 660-664, 1977).

Euphorbia agraria M. Bieb., see *E. euboea* Halácsy.

Euphorbia anacampseros Boiss. was listed for Greece in Euro+Med PlantBase, but the source for the record is obscure and no specimens have been seen.

Euphorbia anthula Lavrent. & Papan., listed as an accepted species in Euro+Med PlantBase, is a synonym of *E. apios* L.

Euphorbia arguta Banks & Sol. Flora of Turkey (7: 599, 1982) listed this species for the East Aegean islands of Chios and Kos, based on historical collections, but there is no recent confirmation. The species is otherwise known from Cyprus and S Anatolia to Palestine, Egypt and Cyrenaica. It was listed for Peloponnisos by Mermygkas, Zikos & Constantinidis (2021: supplementary file), probably in error.

Euphorbia barrelieri. According to Frajman & Schönswetter (2017) all Greek material previously referred to *E. barrelieri* and *E. baselicis* belongs to *E. thessala* (Formánek) Degen & Dörfl. *E. barrelieri* Savi (syn. *E. baselicis* Ten.) is endemic to Italy; four other taxa in the *E. barrelieri* group occur in the W Balkan Peninsula. The type of *E. baselicis* subsp. *thessala* Formánek (Verh. Naturf. Vereins Brünn 34: 282, 1896) is Formánek s.n., 7.9.1895 (BRNM), "Habitat in petrosis prope statmum Patsios in O. th.". Concerning distribution in North Macedonia, see Teofilovski (2011: 115).

Euphorbia brittingeri Opiz ex Samp. was described from Portugal. Greek records of this species refer to *E. montenegrina* (Bald.) K. Malý. For taxonomic accounts of this complex, see Cresti & al. (2019) and Caković & al. (2021).

Euphorbia carniolica Jacq., a species of the Alps and Carpathians, has recently been reported as new for Greece based on collections from nomos Xanthis (for details, see Doumas, Goula & Constantinidis 2022: 27-28).

Euphorbia chamaesyce L. Greek material has been divided into subsp. *chamaesyce* and subsp. *massiliensis* (DC.) Thell. Both are widespread with no difference in distribution pattern. They are distinguished mainly on the leaf margin being entire to crenulate in the former and serrate in the latter. This character is variable, and according to Pahlevani & Riina (2011: 307) there is no basis for retaining the two as separate taxa.

Euphorbia cybirensis Boiss. was described from W Anatolia and has generally been regarded as a synonym of *E. valerianifolia* Lam., but may deserve subspecies rank. Small and slender plants matching *E. cybirensis* as well as intermediates to *E. valerianifolia* s. str., occur from Kalimnos to Rodos.

Euphorbia davidii Subils has been reported from a cotton field in Evvia (Bouchet in an e-mail 2021-01-16). It is native to SW & C North America, and it is not clear to what extent it was naturalized in the Greek locality.

Euphorbia dulcis L. is a species of W & C Europe which has been listed for Greece, probably going back to a report by Grisebach (1843: 137) "in herbosis apricis m. Nidgé [Piperitsa]". There are subsequent reports of *E. dulcis* also from Mt Cholomondas (Voliotis 1967: 79), Mt Athos (Zoller, Geissler & Athanasiadis 1977: tab. 4) and the island of Thasos (Düll, unpubl.). Voliotis' specimen from Mt Cholomondas belongs to *E. oblongata* Griseb., and the other reports are likely to represent the latter as well.

Euphorbia esula subsp. ***tommasiniana*** (Bertol.) Kuzmanov. The name is provisional. According to Reichert, Gregor & Meierott (2018), the correct name for this taxon is *E. saratoi* Ard. at species level and *E. esula* subsp. *saratoi* (Ard.) P. Fourn. at subspecies level, whereas *E. tommasiniana* Bertol. s. str. (Bertoloni 1842: 78-79) is endemic to the karstic area around Trieste in NE Italy and W Slovenia. *E. esula* s. lat. is a widespread and variable Euro-Siberian species which is also considered invasive in North America; the latter records are frequently referred to *E. virgata* Waldst. & Kit.

Euphorbia euboea Halácsy is an enigmatic species known only from a single collection (Pichler s.n., June 1876; WU) at Steni in the foothills of Mt Dirfis, Evvia. The type consists of two flowering stems without basal parts. It has been reduced to *E. agraria* var. *euboea* (Halácsy) Hayek, but may in fact be a form of *E. platyphyllos* L. Fruiting material is needed to determine its status.

Euphorbia flavicoma DC. Greek reports of this species have been referred to *E. verrucosa* L. (see Cresti & al. 2019) and subsequently to *E. montenegrina* (Bald.) K. Malý (Caković & al. 2021).

Euphorbia fragifera Jan has been recorded from the Ionian Islands and elsewhere, but probably in error. No specimens have been seen and it was not included in the Flora Ionica website.

Euphorbia glabriflora Vis. has generally been regarded as an endemic of the C & S Balkan Peninsula, but Stevanoski & al. (2020) have demonstrated that it merges into the amphi-Adriatic *E. spinosa* L. in Bosnia-Herzegovina, N Albania and Montenegro. As currently understood, *E. spinosa* subsp. *spinosa* occurs in S France (Provence), Italy and the W Balkan Peninsula southwards to N Albania, whereas *E. spinosa* subsp. *glabriflora* (Vis.) Frajman is distributed from C Greece to S Serbia.

Euphorbia heterophylla L., a native of C & N America, has been reported from ruderal habitats in Peloponnisos (Thornberg & Kit Tan 2017: 137), Samos (Trigas, in litt.) and Rodos (Manousakis, in litt.), but is doubtfully naturalized.

Euphorbia hirta L. was listed for the East Aegean Islands in Euro+Med PlantBase, but the source for the record is obscure and no specimens have been seen.

Euphorbia humifusa Willd. has been reported from a ruderal habitat on the island of Thasos (Biel & Kit Tan 2018c: 403) and subsequently from an apparently natural habitat on Mt Zas, Naxos (Biel & Kit Tan 2020d: 541). It may be on the verge of becoming established. It resembles *E. chamaesyce* L. but differs, e.g., in the smooth seeds.

Euphorbia hyssopifolia L. has recently been recorded from a ruderal habitats in N Peloponnisos (Zarkos & al. 2015c: 391). It is evidently a printing error for *E. hypericifolia* L. as shown by a photo in the publication.

Euphorbia lingulata Heuff. Greek records of this species apparently refer to *E. epithymoides* L.

Euphorbia lucida Waldst. & Kit. As for *E. anacampseros*.

Euphorbia myrsinites L. s. str. is probably amphi-Adriatic. Plants occurring from E Greece to Caucasia may be referable to the closely related *E. pontica* Prokh. (Frajman unpubl.).

Euphorbia seguieriana subsp. ***seguieriana*** and subsp. ***niciciana*** **(**Borbás ex Novák) Rech. f. have traditionally been recognized. Rechinger (in Fl. Iranica 6: 43, 1964) concluded that the two could not be reliably separated in this area. Frajman & al. (2019) retained the two at species level and reported only *E. niciciana* in Greece. *E. seguieriana* Neck. s. str. was reported to be widespread from Spain to Caucasus, but scarcely extending south of the Belgrade area in the Balkans.

Euphorbia verrucosa L. In a recent revision (Caković & al. 2021) populations from the south-eastern part of the range of *E. verrucosa* s. lat. have been separated as *E. montenegrina* (Bald.) K. Malý, whereas *E. verrucosa* s. str. occurs from the Pyrenees to the NW Balkan Peninsula.

Euphorbia villosa Willd. For nomenclatural reasons this name has to be replaced by *E. illirica* Lam. The species is widespread in EC & SE Europe and frequently subdivided. A recent study (Frajman, Graniszewska & Schönswetter 2016) concluded that it is best retained as a single, variable species.

Euphorbia zahnii Heldr. was described from the western foothills of Mt Taigetos, and has subsequently been regarded as a synonym of *E. valerianifolia* Lam. (= *E. akenocarpa* auct., non Guss.), but differs in some aspects. It does not appear to have been re-collected, and further studies are needed.

Euphrasia stricta J. F. Lehm. The map is based on two collections from Mt Gramos (Karl s.n., 28.6. 2006, W) and Mt Smolikas (Karl s.n., 25.6. 2006, W), identified by E. Vitek (2007). Previous literature records from N Greece (Voliotis 1967: 103 from Mt Cholomonda, Kitanov 1943: 268 from the Falakron area, and Eleftheridou 1992: 78 from Rodopi) are probably incorrect. *E. stricta* is a variable species of C & N Europe, recorded southwards to Montenegro and Albania. Further studies are needed of Balkan plants identified as *E. stricta*, *E. pectinata* and *E. liburnica*.

Fagus sylvatica L. Intermediates between subsp. *sylvatica* and subsp. *orientalis* (Lipsky) Greuter & Burdet occur in NE Greece. They have been called *F. moesiaca* (K. Malý) Czecz.

Fallopia aubertii (L. Henry) Holub (*F. baldschuanica* (Regel) Holub) is a native of temperate Asia (China, Russia, Kazakhstan, etc.), widely cultivated an an ornamental and occasionally naturalized in ravines, roadside thickets, etc. It is a fast-growing, slender, much-branched woody climber with drooping clusters of small, white flowers.

Fallopia japonica (Houtt.) Ronse Decr. (*Reynoutria japonica* Houtt.), a native of NE Asia, is invasive in several parts of Europe and North America. In Greece it has been reported locally in the Prespa area. It is a large perennial herb with a powerful rhizome and large, ovate leaves.

Fedia. Xena de Enrech & Mathez (1990) divided *F. cornucopiae* s. lat. into three species: *F. cornucopiae* (L.) Gaertn. s. str. (S Spain, S Portugal and N Morocco), *F. pallescens* (Maire) Mathez (Morocco) and *F. graciliflora* Fisch. & C. A. Mey. (widespread but scattered in the Mediterranean region from E Morocco to Cyrenaica and Kriti). The plant illustrated on Plate 32 of Flora Graeca (1806), supposedly from "Byzantium", may have been collected on Sicily, where *Fedia* is common. Plants from Sicily and Greece are referable to *F. graciliflora* subsp. *graciliflora*. A record from Kefallinia is based on a verified specimen (Phitos & Kamari 19416; UPA, herb. Strid). A record from Sifnos (Kalheber, in litt., 2011) as well as old literature records from Kerkira (Mazziari 1834-35: 452) and NE Greece (Jeanpert 1919: 517, Turrill 1920c: 186) need confirmation.

Ferulago. There are different opinions on the nomenclature of *F. campestris* and *F. galbanifera*. Lyskov (2020) has shown that the name *Ferulago galbanifera* (Mill.) W. D. J. Koch has priority over *F. campestris* (Besser) Grecescu, although Reduron (2007c: 1295) argued that the former is a superfluous name.

Ferulago asparagifolia Boiss. was incorrectly listed for the Cretan area in Flora Europaea 2: 359 (1968); cf. Dimopoulos & al. (2013: 274).

Ferulago nodosa (L.) Boiss. Greek material belongs to subsp. *nodosa*. An endemic taxon on Sicily was originally described as *F. geniculata* Guss. and subsequently recombined as *F. nodosa* subsp. *geniculata* (Guss.) Troia & Raimondo. For nomenclatural reasons the latter name replaces *F. nodosa* subsp. *rigida* Troia & Raimondo (cf. Peruzzi & al. 2015: 80).

Festuca apennina De Not is an Italian and Central European species in the *F. pratensis* complex, reported also from southern Pindos (Dimopoulos & al. 2013: 125). The identity of the Greek specimens is uncertain.

Festuca graeca (Hack.) Markgr.-Dann., see *F. varia*.

Festuca grandiaristata Markgr.-Dann. is a dubious species described from Mt Cholomondas. It is probably not distinct from *F. hirtovaginata* (Acht.) Markgr.-Dann.

Festuca hercegovinica Markgr.-Dann. has been reported from Mt Varnous by Markgraf-Dannenberg (1978: 323), based on Alston & Sandwith 323, 6.6. 1932 (K). It is doubtfully distinct from *F. koritnicensis* Hayek & J. Vetter which has subsequently been collected on this mountain (e.g., Strid & al. 18110, 26.6. 1981; C, G, M, UPA) and is locally dominant in rather dry, subalpine grassland over granite.

Festuca kozanensis Foggi & Joch. Müll. is nomen novum for *F. rechingeri* Markgr.-Dann., non E. B. Alexeev, a member of the *F. varia* complex occurring in north central Greece.

Festuca macedonica J. Vetter was described from Mt Pangeon and has been reported also from a few other limestone mountains in northern Greece. It is scarcely distinct from *F. koritnicensis* Hayek & J. Vetter.

Festuca nigrescens Lam. is a member of the *F. rubra* complex. It is of uncertain status and has provisionally been included in *F. rubra*.

Festuca ovina group. Members of this group (a polyploid complex of densely caespitose plants with setaceous leaves) occur throughout the temperate regions of the northern hemisphere. In Greece they are important constituents of montane to alpine grassland on various substrates. Taxonomic treatment varies from the broad view of the monographer Hackel (1882), who accepted only one species, with a complicated hierarchy of infraspecific units, to the extreme splitting of Markgraf-Dannenberg (1976 and in Flora Europaea 5, 1980) who recognized c. 40 species in Europe alone. An attempt has been made here to merge Greek material into recognizable groups, but the taxonomy is still provisional. Taxonomy relies heavily on leaf anatomical characters, especially the amount and distribution of sclerenchyma tissue. Cultivation trials in preparation for the Mountain Flora of Greece (2: 755-762, 1991) have shown that these features are, at least in some cases, subject to considerable environmental modification. The grass specialist N. L. Bor (Fl. Iranica 70: 74 ff., 1970) counted no less than 519(!) described taxa under *F. ovina* s. lat., and expressed considerable doubt as to the taxonomic usefulness of leaf anatomical characters, finally reverting to recognizing only *F. ovina* L., s. lat.

Festuca oviniformis J. Vetter was based on Rechinger 9515 (G, K, LD, S) from the Greek part of eastern Rodopi: "supra pagum Iasmos (Jasi-Köj), reg. querc. substr. silic.". It is doubtfully distinct from *F. hirtovaginata* (Acht.) Markgr.-Dann.

Festuca penzesii (Acht.) Markgr.-Dann. is a member of the *F. varia* complex reported from Mt Falakron in NE Greece and Mt Ali Botuš in SW Bulgaria (Markgraf-Dannenberg 1976: 166).

Festuca pindica (Markgr.-Dann.) Markgr.-Dann. is another local taxon of dubious status in the *F. varia* complex, mainly found on serpentine in NW Greece (Markgraf-Dannenberg, op. cit.).

Festuca pluriflora Schult. has been recombined as *F. pratensis* subsp. *pluriflora* (Schult.) Zangheri, and is scarcely distinct from *F. arundinacea* Schreb. This group is optionally referred to the genus *Schedonorus*.

Festuca rivularis Boiss. was reported from Mt Vrondous by Scholz (1993: 114), based on Willing 7031, 22.6. 1989 (B). This species was described from Sierra Nevada in S Spain, and occurrence in Greece seems unlikely. Subsequent reports from Mt Vermion (Chochliouros 2005: 124) also need confirmation.

Festuca taurica (Hack.) Trautv. is a taxon of SE Europe, doubtfully distinct from *F. valesiaca* Schleich. ex Gaudin. It has been reported from a few localities in the Nestos river gorge (Rechinger 1939a: 548, Markgraf-Dannenberg 1976: 120, Voliotis 1984a: 163).

Festuca thracica (Acht.) Markgr.-Dann. has been reported from limestone mountains in northernmost Greece (Markgraf-Dannenberg 1976: 101, Chasapis & al. 2019: 207) as well as SW Bulgaria (Peev & al. 2015: 494). It is doubtfully distinct from *F. hirtovaginata* (Acht.) Markgr.-Dann.

Festuca trichophylla subsp. ***asperifolia*** (St-Yves) Al-Bermani was listed in Dimopoulos & al. (2013: 125), but the source is obscure and the records are probably referable to *F. rubra* s. lat.

Festuca ustulata (St.-Yves) Markgr.-Dann. was reported from Mt Psiloritis by Böhling & Scholz (2003: 48), based on Risse 652 (B). This taxon was described from S. Anatolia and Lebanon, as *F. indigesta* f. *ustulata* St.-Yves. The Cretan record is likely to represent *F. pseudosupina* J. Vetter.

Festuca valida subsp. ***leilaensis*** Markgr.-Dann. was described from Mt Vrondous, based on Rechinger 10821 (G, LD, M, W) which had previously been identified by Vetter as *F. varia* var. *genuina*.

Festuca varia Haenke was described from the E Alps. Taken in a wide sense it occurs in most of C & S Europe and through Anatolia to Caucasus. Greek material was divided into a number of poorly defined species and subspecies by Markgraf-Dannenberg (1976 and in Flora Europaea 5: 136-137, 1980). For a brief discussion of the variation, see Strid in Mount. Fl. Greece 2: 753-754 (1991).

Festuca violacea subsp. ***handelii*** Markgr.-Dann. is a highly dubious taxon based on a single individual in a mixed collection (Handel-Mazetti s.n., 1927-07-17, W) from Mt Olimbos.

Fibigia clypeata var. ***eriocarpa*** (DC.) E. Fourn. is sometimes regarded as a separate species or subspecies. It differs from *F. clypeata s. str.* in a single character (siliculae with long, simple or forked hairs in addition to dense coverage of short stellate hairs), and there is no geographical separation.

Fibigia lunarioides (Willd.) Sweet. See *Acuston lunarioides*.

Filipendula ulmaria (L.) Maxim. Most of the Greek plants have leaflets glabrous above, sparsely pubescent beneath and green on both surfaces. They would thus belong to var. *denudata* (J. Presl & C. Presl) Maxim., but the value of the infraspecific taxa is dubious (cf. Kurtto, Lampinen & Junikka 2004: 35-37).

Foeniculum piperitum (Ucria) C. Presl, described from Sicily, has generally been regarded as a subspecies or synonym of *F. vulgare* Mill., but was restored to species rank by Ilardi & Troia (2021). Its general distribution is not clear. From Greece are many records both of *F. vulgare* subsp. *vulgare* and *F. vulgare* subsp. *piperitum* (Ucria) Bég. According to Rechinger (in Fl. Iranica 162: 347, 1987) the distinguishing features are "nicht scharf ausgeprägt bezw. nicht korreliert".

Frankenia corymbosa Desf. is a mainly North African species, in Greece previously known only from small islands south of Kriti. However, Kit Tan, Polymenakos & Pantakis (2021: 272-273) reported it also from rocky coastal habitats on small islands off eastern Attiki. For an account of its morphology and distribution, see this article.

Fraxinus angustifolia Vahl. Two collections from Lesvos have been identified as subsp. *syriaca* (Boiss.) Yalt., and variation in this area needs further study. According to Green (1985) the correct name for *F. angustifolia* Vahl is *F. rotundifolia* Mill., but this was not accepted in Euro+Med PlantBase.

Fraxinus excelsior L. Almost all Greek reports of this widespread European species are literature records which need confirmation. It is sometimes difficult to distinguish from *F. angustifolia* subsp. *oxycarpa*, especially when sterile. In Albania *F. excelsior* is sporadic in the mountains, whereas lowland records probably refer to *F. angustifolia* (Barina 2017: 318). The latter has been divided into three ill-defined subspecies; most, perhaps all, Greek specimens are referable to subsp. *oxycarpa* (Willd.) Franco & Rocha Afonso.

Fraxinus pallisae Wilmott. Flora Europaea (3: 54, 1972) gives the specific epithet as *pallisiae*, but the spelling in the original publication is *pallisæ*.

Fritillaria conica Boiss. The type is a collection by Bory (probably in March or early April 1829) from "Peloponneso prope Navarin". The first to observe a yellow-flowered *Fritillaria* in this area was Hawkins c. 1794. Florae Graecae Prodromus (1: 229, 1806) states under *Fritillaria sibthorpiana*: "Prope Cressam, hodiè Porto Cavalieri, Asiae minoris, primùm invenit Sibthorp; et posteà in monte humile saxoso prope Pilon Peloponnesi, hodiè Navarin. D. Hawkins". The former refers to *Fritillaria sibthorpiana* (Sm.) Baker (illustrated on Plate 330 of Flora Graeca), the latter most probably to *F. conica*.

Fritillaria graeca Boiss. & Spruner. In central mainland Greece the southerly subsp. *graeca* merges into subsp. *thessala* (Boiss.) Rix (syn. *F. thessala* (Boiss.) Kamari), which extends to the Ionian Islands, Albania and North Macedonia. Intermediates (in N Peloponnisos and Sterea Ellas) have been called *F. mutabilis* Kamari. *F. ionica* Halácsy, described from Kerkira, is a synonym of *F. graeca* subsp. *thessala*. Some plants from Mt Parnitha and mountains of Peloponnisos with bell-shaped flowers and acute, somewhat outward-curving tepals have been called *F. guicciardii* Heldr. & Sartori. They are of uncertain status but probably best regarded as a variety of *F. graeca* subsp. *graeca* (cf. Zaharof 1986a).

Fritillaria macedonica Bornm. This rare species, which was recently discovered in a single locality on Mt Olimbos by Thomas Giannakis, appeared on a North Macedonian postage stamp in 2008. See also photos in Basdra & Kipopoulos (2018: 144-145), sub nom. *Fritillaria* sp.

Fritillaria pelinaea Kamari, described from Mt Pelineo on N Chios and said to grow together with *F. carica* Rix, merely represents individual variation within the latter. For details on the Turkish distribution of *F. carica*, see Tekşen & Aytaç (2011: 466). A. record of *F. forbesii* Baker from Samos refers to *F. carica*. A collection from Chios listed by Rechinger (1944a: 721) as *F. pinardii* Boiss. probably represents *F. carica*.

Fritillaria phitosia Kamari, Zahos & Siagou (2017: 230) was described from the Magnisia peninsula just north of Evvia. There are no differences between this and other populations of *F. euboeica* Rix.

Fritillaria rixii Zaharof, described from Evvia, falls within the range of variation of *F. euboeica*.

Fritillaria spetsiotica Kamari. In its typical form *F. rhodocanakis* Orph. ex Baker is restricted to the island of Idra (where it is fairly common) plus a few adjacent localities. Similar plants from the Argolis peninsula have been described as *F. rhodocanakis* subsp. *argolica* Zaharof, as a separate species (*F. spetsiotica* Kamari) or regarded as hybrids between *F. spetsiotica*, *F. rhodocanakis* and *F. graeca* Boiss. & Spruner. Hybrid status is highly unlikely since these plants show no reduction in fertility. *F. davisii* Turrill from the Mani Peninsula and "*F. rhodocanakis* subsp. *argolica*" from the vicinity of Didima in Argolida were cultivated side by side in the Göteborg Botanic Garden and turned out to be indistinguishable. Already in the original description of *F. davisii*, Turrill (1942: 239) observed: "The affinity of *F. Davisii* is with *F. rhodocanakis*. It differs in the broader lowest leaves and in the shorter style with relatively longer arms". *F. rhodocanakis* subsp. *argolica* (*F. spetsiotica*) has thus been merged with *F. davisii*.

Fritillaria theophrasti Kamari & Phitos is a name given to a local population of *F. pontica* Wahlenb. from the island of Lesvos.

Fumana. There are several taxonomic problems in this small genus:

F. aciphylla Boiss. Plants growing on serpentine outcrops at low and moderate altitude in C and NE Greece have been tentatively referred to this species, which was originally described from calcareous mountains in C Anatolia [B5 and C6 in Flora of Turkey]. It has recently been reported from E Rodopi in Bulgaria (Kunev 2020); these plants appear to match those from adjacent Greek localities. It is possible that the Balkan plants represent an undescribed species.

Two collections from the Stix ravine (Mt Chelmos), viz. Stamatiadou 3773 (ATH) and Strid 44108 (G, herb. Strid) have been labeled *Fumana chelmea*, sp. nov. ined. by J. Güemes, 2001. The name does not appear to have been published, and further studies are needed. One historical record of *F. procumbens* (Dunal) Gren. & Godr. from this locality (Maire & Petitmengin 1908: 37) is likely to refer to the same taxon, and so may other high-altitude records of *F. procumbens* from Chelmos and Taigetos.

F. laevis (Cav.) Pau is probably under-represented in the records, having often been regarded as a synonym or variety of *F. thymifolia* (L.) Webb. In a floristic report from the island of Limnos, Bergmeier & al. (2021 suppl. 1 p. 11) observed: "*Fumana laevis* and *F. thymifolia* have not been distinguished in the region (and the former disregarded by Strid 2016). The distribution of *F. laevis* in Greece and the Aegean is therefore unknown. The two species are variable but well differentiated, and the Limnian plants can be identified by the differences in petal, leaf and internode size and leaf hairiness (see Güemes & Molero 1993 and Raus & Kleinsteuber 2016). Mixed stands with both species occur." [The citation Güemes & Molero 1993 may in fact refer to Güemes & Molero 2006, Fl. Iberica 3: 422-436. Raus & Kleinsteuber 2016 refers to Flora von Rhodos und Chalki 1: 373-379].

F. paphlagonica subsp. ***alpina*** (Janch.) Greuter is provisionally classified as a Cretan endemic. It was originally described as *F. vulgaris* f. *alpina* Maire ex Janch. and reported from Lefka Ori, Chelmos and Athos.

F. scoparia Pomel. Records include *F. ericoides* auct., non (Cav.) Gand. As currently interpreted the latter is restricted to the W Mediterranean area.

Fumaria flabellata Gasp. was listed for Peloponnisos by Mermygkas, Zikos & Constantinidis (2021: supplementary file), but no details are available.

Fumaria judaica Boiss. Two collections from Kefallinia have been identified as subsp. *insignis* (Pugsley) Lidén. The latter apparently co-exists with subsp. *judaica*, and further studies of variation in this area are needed.

Fumaria rostellata Knaf seems to be much more common and widespread in Greece than indicated by Lidén in his account for Flora Hellenica (2: 110 and map 919, 2003). At Herbarium B is a large set of specimens collected by Willing. Only some of them have been critically examined by the present author, but these appear to be a good match for collections confirmed by Lidén.

Fumaria schleicheri Soy.-Will. Greek records probably refer to *F. vaillantii* Loisel.

Gagea fragifera Vill. ex Ehr. Bayer & G. López. The nomenclature is convoluted, and the earliest available name at species level may be *G. liotardii* (Sternb.) Schult. & Schult. f.

Gagea parnassica J.-M. Tison, ined. Plants belonging to the *G. pratensis/pusilla* group are scattered in the mountains from N Peloponnisos to the northern borders and have also been reported from the island of Samothraki. They have provisionally been called *G. parnassica*, but further studies of their status and affinities are needed.

Gagea ramulosa A. Terracc. was listed as an accepted species in the Greek checklist (Dimopoulos & al. 2013: 112), but appears to be a synonym of *G. dubia* A. Terracc.

Gagea reticulata (Pall.) Schult. & Schult. f. As used by authors on the Greek flora this name probably refers to *G. rigida* Boiss. & Spruner.

Gagea rhodiaca A. Terracc., described from Rodos, is probably a synonym of *G. rigida* Boiss. & Spruner. Plants collected on Mt Ataviros (just W of summit, 1000-1100 m, Strid 56072, 2006-05-10) and cultivated in the Göteborg Botanic Garden match the latter in all essestial characters.

Gagea saxatilis (Mert. & W. D. J. Koch) Schult. & Schult. f. A record from Samothraki (Biel & Kit Tan 2007a: 110) probably refers to *G. bohemica* (Zauschn.) Schult. & Schult. f.

Gagea tymphaea J.-M. Tison, ined. A few recent collections from calcareous grassland at high altitude on Mt Timfi and Mt Douskon have been identified as an undescribed species related to *G. pratensis* and perhaps restricted to this area. Further studies and description will follow.

Galanthus gracilis **Čelak.** Greek records (e.g., from Samos) probably refer to *G. elwesii* Hook. f. See Davis (199: 140-147) for relationships between these two species.

Galanthus reginae-olgae Orph. was described from the foothills of Mt Taigetos. Plants from this area flower in October/November without or with poorly developed leaves. In northern Peloponnisos and western mainland Greece are plants flowering in December to March, with leaves typically 3-8 cm at anthesis; they have been named *G. reginae-olgae* subsp. *vernalis* Kamari. There are rather gradual transitions. Plants from Kerkira have been described as *G. nivalis* f. *corcyrensis* Beck and subsequently raised to species level as *G. corcyrensis* (Beck) Stern. They are autumn-flowering with short leaves, thus more or less matching *G. reginae-olgae* s. str. The scarcity of morphological characters makes any taxonomic subdivision arbitrary. In cultivation there is much individual variation in flowering time and degree of leaf development even among plants from the same locality.

Galeopsis pubescens Besser was reported by Voliotis (1981a: 118) from Mt Tzena and the Athos peninsula, the latter referring to a record by Ganiatsas (1963). It is a species of C Europe extending southwards to N Albania. Occurrence in Greece seems unlikely.

Galeopsis segetum Neck. was included in the Greek checklist (Dimopoulos & al. 2013: 106), based on Snogerup 15201 (LD) from Mt Paikon. However, this specimen represents *G. ladanum* L. (revised by A. Strid, 2020-11-03, based on photos sent from Herbarium LD, courtesy P. Frödén & A. Thell). Literature records of *G. segetum* from N Greece (Reif & Löblich-Ille 1999: 99 and Manolaki, Tsakiri & Papastergiadou 2011: 866) are also likely to refer to *G. ladanum*.

Galium agrophilum Krendl is a dubious species reported from some localities in NE Greece. *G. exaltatum* Krendl and *G. agrophilum* were described in the same publication (Krendl 1988). The author appears himself to have been in doubt. One collection (Greuter 11362; ATH, G, W) was cited as *G. exaltatum* on p. 47 and as *G. agrophilum* on p. 53.

Galium album L. was included as an accepted species in the Greek checklist (Dimopoulos & al. 2013: 138), but scattered records are likely to refer to members of the *G. mollugo* group such as *G. hellenicum* Krendl and *G. pycnotrichum* (Heinr. Braun) Borbás.

Galium avascense Krendl is known only from a few collections in NE Greece (nom Evrou). It may be a narrow-leaved form of *G. exaltatum* Krendl.

Galium brevifolium Sm. Most Greek records are likely to represent subsp. *insulare* Ehrend. & Schönb.-Tem., but one collection from Rodos (Carlström 4842; LD) and one from Samos (Strid 54750; G, herb. Strid) have divaricate, scarcely thickened fruiting pedicels and appear to match subsp. *brevifolium*. Both subspecies are scattered in W & SW Anatolia. Subsp. *insulare* was based on the same type (a collection from Chios by Orphanides) as *G. caudatum* var. *chium* Boiss., later recombined as *G. brevifolium* var. *chium* (Boiss.) Rech. f. Subsp. *brevifolium* was illustrated on Plate 135 of Flora Graeca (1813), based on material from SW Anatolia.

Galium canum DC. Greek plants belong to the rather vaguely defined subsp. *ovatum* Ehrend., which also occurs in SW & S Anatolia. Two additional subspecies have been recognized in Anatolia, Cyprus, W Syria and Palestine.

Galium capreum Krendl was described from Mt Fengari, Samothraki. It appears to be a small form of *G. thracicum* Krendl which occurs at somewhat lower altitude on the same island. The latter in turn is scarcely distinguishable from *G. mirum* Rech. f.

Galium corrudifolium Vill. was listed for "Gr" in Fl. Eur. 4: 25 (1976), perhaps referring to a report of *G. lucidum* subsp. *corrudifolium* (Vill.) Hayek from Mt Chortiatis by Regel (1942a: 127). The latter probably refers to *G. asparagifolium* Boiss. & Heldr.

Galium decorum Krendl was based on a single collection from Mt Ipsario, Thasos. It is scarcely distinct from *G. aegaeum/rhodopeum*.

Galium dumosum Boiss. A single collection from Ro in the Kastellorizo island group (Greuter 11820) was listed by Ehrendorfer & Schönbeck-Temesy (in Fl. Turkey 7: 827, 1982) as *G. dumosum*, "approaching *G. graecum*". Subsequent collections from the adjacent island of Megisti (Strid 57925 and 60648) are a good match for *G. dumosum* as described in Fl. Turkey (loc. cit.). The latter is a species of SW & S Anatolia more or less intermediate between *G. canum* Req. ex DC. and *G. graecum* L. Further studies are needed to establish its taxonomic status.

Galium fruticosum Willd. was based on a collection by Tournefort from Kriti. It has been reported also from Spain, Morocco and Algeria, but these plants are likely to refer to the similar *G. ephedroides* Willk.

Galium graecum L. Densely hairy plants from low altitude in E Kriti have been described as subsp. *pseudocanum* Ehrend. Such plants have subsequently been found also on Samos (Dinter in Greuter & Raus 2009: 339-340; Strid 58056, G, herb. Strid). They seem to merge into *G. graecum* s. str.

Galium humifusum M. Bieb. was described from Caucasus and Crimea. It is widespread in S Russia and Ukraine, extending to N & E Anatolia and possibly NE Bulgaria. The record for "Gr" in Fl. Eur. 4: 22 (1976) may have been based on the imprecise "Ma" in Hayek (1930: 447). The species is unlikely to occur in Greece (or North Macedonia). Greek specimens labelled *G. humifusum* have all turned out to be misidentified.

Galium macedonicum Krendl has been reported once from Greece, based on Rechinger 45364 (W) from Mt Vrondous. It is otherwise known from North Macedonia, S Serbia and SW Bulgaria, and is doubtfully distinct from *G. asparagifolium/rigidifolium*.

Galium malickyi Krendl is a diploid species known only from Mt Pantokrator, Kerkira. It is doubtfully distinct from *G. hellenicum* Krendl.

Galium mixtum Krendl was described from Kerkira where intermediates to *G. malickyi* Krendl and *G. circae* Krendl were also said to occur. Neither of these three "species" seem distinguishable from *G. hellenicum*.

Galium mollugo group. Identifying Greek taxa in this group was moderately difficult until Krendl (1988) made it virtually impossible by recognizing 41 species, most of them new and often with obscure morphological differences (or none at all). Use of this publication is further complicated by poor editing; for instance, every single page reference in the Index is incorrect. The publication was dated 1986-7, but in fact appeared in 1988. An attempt has been made to merge Krendl's species into recognizable groups, but several of the maps remain highly provisional.

Galium nigricans Boiss. was listed in Dimopoulos & al. (2013: 139), based on a single record from the island of Thasos by Düll (in litt., 2011-10-28) which needs confirmation. It is a species of E Anatolia and Iran, unlikely to occur in Greece.

Galium octonarium (Klokov) Pobed., a species of E Europe and SW Asia, was reported for "?Gr" in Fl. Eur. 4: 27 (1976). No records or specimens have been seen.

Galium oreophilum Krendl, ***G. monasterium*** Krendl and ***G. sacrorum*** Krendl were all described from Mt Timfi. There seem to be no consistent morphological differences, and they have been merged under the first-mentioned name.

Galium palustre L. was listed for Peloponnisos by Mermygkas, Zikos & Constantinidis (2021: supplementary file), but the record may refer to the similar *G. debile* Desv.

Galium procurrens Ehrend. from the C Balkan Peninsula is very similar to *G. laconicum* Boiss. & Heldr. Some records from N Greece mapped as the latter may in fact belong to *G. procurrens* or other members of the *G. sylvaticum* complex.

Galium recurvum DC. was listed for "AE(G)", "Cr" and "Gr" in Euro+Med PlantBase, but this name is a synonym of *G. caminianum* Schult.

G. micranthum was described by Dumont d'Urville (1822: 272) based on his collection "ad marginem Vulcani insulae novae Camini [Nea Kaimeni] prope Thera". This name is andedated by *G. micranthum* Pursh (1813), a North American species. *G. caminianum* Schult. in Roem. & Schult., Mantissa 3: 186, was published in 1827, based on the same type as *G. micranthum* d'Urv. Up to now the species has generally been known by the younger name *G. recurvum* Req. ex DC., Prodr. 4: 609 (1830), also based on Dumont d'Urville's collection from Nea Kaimeni. Occurrence has been confirmed by at least three independent collections from this small island (Runemark & Bentzer in 1967, Raus in 1984 and Bowen in 1993).

Galium samothracicum Rech. f. (*G. insulare* Krendl) is a species of doubtful status, described from Samothraki and recorded by Krendl also from several localities on Kriti. It is closely related to *G. citraceum* Boiss. and *G. heldreichii* Halácsy, but said to be relatively small and slender, with oblanceolate, antrorsely scabridulous leaves, few-flowered, ± patent partial inflorescences, and rather large, white to greenish-yellow flowers.

Galium scabrifolium (Boiss.) Hausskn. was listed for "Gr" in Euro+Med PlantBase. It is probably a synonym of *G. asparagifolium* Boiss. & Heldr.

Galium setaceum Lam. has been divided into the western subsp. *setaceum* (described from Spain) and the eastern subsp. *decaisnei* (Boiss.) Ehrend. (described from Iran). Most of the Greek specimens have not been identified to subspecies, but limited evidence suggests that Aegean records refer to subsp. *decaisnei* whereas intermediates occur on the mainland.

Galium thasium Stoj. & Kitan. was described from the island of Thasos ("in fissuris rupium sub cacuminem Prophitis Ilias, ad cca. 900 m s. m. Planta florifera lecta 14 Julio a. 1943"). The species was not mentioned by Krendl (1988) nor in the Euro+Med PlantBase or Flora Europaea. It may be conspecific with *G. pycnotrichum* (Heinr. Braun) Borbás.

Galium thracicum Krendl. See under *G. capreum*.

Galium uliginosum L., a widespread Euro-Siberian species, has recently been reported from the Greek side of the Rodopi mountains (Charalampidou, Eleftheriadou & Theodoropoulos 2021: 364). There are several previous records of the similar *G. palustre* L. from the same area.

Galium violaceum Krendl has been recombined as *G. taygeteum* var. *violaceum* (Krendl) Greuter. "*Galium taygeteum* and *G. violaceum*, described simultaneously by Krendl (1987) [actually 1988] cannot in my opinion be upheld even as subspecies. The former is the mountain ecotype of the latter, but ..... they grade into each other over a wide belt of medium altitude" (Greuter 2012: 81). Morphological differences between the two are indeed vague and inconsistent, but several chromosome counts by Krendl (op. cit.) indicate 2n=22 in *G. violaceum* and 2n=44 in *G. taygeteum*.

Gastridium phleoides (Nees & Meyen) C. E. Hubb. was described from Chile, presumably based on introduced plants of Mediterranean stock. It cannot be consistently distinguished from *G. ventricosum* (Gouan) Schinz & Thell. Scholz (1986b) cited a few Greek specimens that could clearly be identified as one or the other, but ignored the majority than could not. Scattered records of *G. scabrum* C. Presl (Yannitsaros 1969: 127 from Kithira, Pavlides 1976a: 139 from the Sithonia peninsula) are dubious. The latter was described from Sicily and is now mostly classified as a synonym of *G. ventricosum*. Scoppola (2019), however, adopted a narrow species concept, recognizing *G. phleoides*, *G. scabrum* and *G. ventricosum* as separate taxa.

Gaudiniopsis macra (M. Bieb.) Eig was described from Caucasus and is widespread and variable in the Irano-Turanian region. Greek specimens may belong to subsp. *antalyensis* M. Doğan, which is otherwise recorded from SW Anatolia.

Genista acanthoclada DC. Following Vierhapper (1919a: 157-166) some specimens from Karpathos, Rodos, Tilos, Kastellorizo and adjacent parts of Anatolia have been referred to subsp. *echinus* (Spach) Vierh. The differences are not clear, and the taxon is not mentioned in subsequent Floras (e.g., Flora of Turkey and Kleinsteuber & al. 2016).

Genista aetnensis (Biv.) DC., an endemic of Sardinia and Sicily, has been found planted by a dam in nomos Ilias (Kit Tan & al. 31192, 18 Dec 2011, flowering and fruiting). It may be on the verge of becoming naturalized.

Genista melia Boiss. was based on a collection by Fauché (1840), supposedly from the island of Milos. It was never confirmed and long remained enigmatic. However, the type specimen (at G) turned out to represent the W Mediterranean *G. scorpius* (L.) DC. It is not clear whether the specimen had been erroneously labelled or whether *G. scorpius* once occurred on Milos; if so, it is likely to have been a casual introduction (see Kit Tan & Zieliński (2017: 311).

Genista millii group. The taxonomy is highly provisional and modern revision is needed. In Greece this group comprises *G. halacsyi* Heldr., *G. millii* Heldr. ex Boiss., *G. parnassica* Halácsy and *G. subsericans* (Bornm.) Rech. f. Greek specimens identified as *G. albida* Willd. are probably referable to *G. millii*.

Genista sagittalis L. The population on Mt Parnonas is disjunct. The species was collected here already by Orphanides, and there are several recent confirmations. Some plants (e.g., Strid 60752; UPA, herb. Strid) are atypical, being pilose throughout and with only narrowly winged stems, perhaps indicating introgression from *G. halacsyi* Heldr.

Genista sericea Wulfen. All Greek records need critical revision. At least some of them are likely to represent subsp. *rigida* (Pamp.) Feoli Chiapella & F. Conti.

Gentiana cruciata L. Both subsp. *cruciata* and subsp. *phlogifolia* (Schott & Kotschy) Tutin have been reported from northern Greece, but the differences are not clear (cf. Hartvig in Strid & Kit Tan 1991: 5-6). Limited evidence suggests that subsp. *phlogifolia* may be restricted to the Carpathians.

Gentianella austriaca (A. Kern. & Jos. Kern.) Holub was reported from Mt Falakron by Akeroyd & Preston (1981b: 287); the collection was tentatively referred to var. *neilreichii* (Dörfler & Wettst.) Hayek. A report of *G. lutescens* (Velen.) Holub from Mt Falakron (Sfikas 1983c: 57) may refer to the same taxon. No specimens have been seen, and it remains to be established whether these plants are different from *G. bulgarica* (Velen.) Holub.

Geocaryum divaricatum (Boiss. & Orph.) Engstrand. For comments on this presumably extinct species, see Kit Tan, Zarkos & Mermygkas (2022: 403-406).

Geranium lanuginosum Lam. After the preparation of the maps this species has been reported from the Rodopi mountains by Charalampidou, Eleftheriadou & Theodoropoulos in Fl. Medit. 32: 86 (2022).

Geranium macrorrhizum group. Wagensommer & Venanzoni (2021) recognized *G. macrorhizum* L. in Italy, the SE Alps and SE Europe (a record from Crimea may refer to non-native plants), *G. dalmaticum* (Beck) Rech. f. in the W Balkan Peninsula and *G. kikianum* Kit Tan & Vold in Mt Taigetos, as well as *G. lucarinii* sp. nov., a local endemic of C Italy.

Geranium nodosum L. was listed for Greece in Euro+Med PlantBase, citing "personal communication (Valcarcel & Aedo, 2003)". This is probably a mistake going back to *G. nodosum* var. *freyeri* (Griseb.) Nyman. Grisebach's *G. freyeri* was based on material from Mt Cholomonda and the Athos Peninsula in NE Greece. It has subsequently been confirmed in both localities and is clearly a synonym of *G. versicolor* L. For *G. nodosum* Linnaeus originally recorded "Habitat in Delphinatu" [Dauphiné, S France]. Both species have been reported from Italy.

Geranium pratense L. was listed for Greece in Fl. Eur. (2: 195, 1968) as well as in Euro+Med PlantBase. The record was probably based on a report by Goulimis (1956a: 13) from Mt Falakron, Chionotripa. Goulimis' plant is very likely to represent *G. sylvaticum* L. which has subsequently been collected several times in this locality, whereas *G. pratense* has never been confirmed in Greece.

Geranium subcaulescens L'Hér. ex DC. A local population on Mt Kitheronas (e.g., Strid 58162; G, herb. Strid) has white or pale pink petals, but matches *G. subcaulescens* in calyx indumentum and other characters. Plants from Turkey with similarly pale petals have been recognized as *G. subcaulescens* var. *subacutum* Boiss. (syn. *G. subcaulescens* var. *leucophaeum* Hausskn. ex Bornm.) and later raised to species level as *G. subacutum* (Boiss.) Aedo. The latter is widespread in Turkey (with syntypes from Cappadocia, the Pontic mountains and Armenia), and detailed comparison is needed to establish whether plants from Mt Kitheronas are conspecific (cf. also Robinson & Kit Tan 2018). *G. subcaulescens* was described from Mt Parnassos and is probably a Balkan endemic; see, however, notes in Fl. Turkey 10: 105 (1988).

Geranium tuberosum L. The record from Milos is very recent (Biel & Kit Tan in Phytol. Balcan. 28: 149-150, 2022) and represents a significant extension of the range.

Gladiolus atroviolaceus Boiss. was listed for Greece in Euro+Med PlantBase, but the basis for the record is obscure and no specimens have been seen. It is an Irano-Turanian species unlikely to grow wild in Greece.

Gladiolus communis L. has recently been reported from SE Peloponnisos where it is apparently well established in olive groves (Kit Tan & Kofinas 2021: 267-268). Previous Greek records of *G. communis* have been interpreted as referring to *G. italicus* Mill. which is widespread in arable habitats. The latter is smaller in all parts, with fewer flowers and unwinged seeds.

Gladiolus glaucus Heldr. ex Halácsy, described from the western foothills of Mt Taigetos, is probably a small form of *G. illyricus*.

Gladiolus kotschyanus Boiss. was included in the Greek checklist (Dimopoulos & al. 2013: 104), based on a few collections from the area of the Katara pass which probably represent *G. imbricatus* L. *G. kotschyanus* was described from SE Anatolia and is unlikely to occur in Greece.

Gladiolus tristis subsp. ***spiralis*** (Pers.) Maire & Weiller, a taxon from South Africa, has been reported to be locally naturalized in NW Peloponnisos (Bareka, Constantinidis & Kamari 1998: 305, Giannopoulos & Kit Tan 2021: 134-135).

Glaucium serpieri Heldr. was described from Lavrion (SE Attiki) and said to differ from the widespread *G. flavum* Crantz in several characters, including 2-pinnatisect lower leaves, crisp white hairs on various parts of plant, and petals with a dark blotch at base – cf. Aegean Atlas 1: 424 (2016). On 4th April 2017 I had the opportunity to visit this locality and found the supposedly distinguishing characters occurring in different combinations. *G. serpieri* thus appears to represent individual variation within *G. flavum*.

Glechoma hirsuta Waldst. & Kit. was listed for Peloponnisos by Mermygkas, Zikos & Constantinidis (2021: supplementary file), but no details are available. Verified Greek records are all in the north, and occurrence in Peloponnisos seems unlikely.

Gleditsia triacanthos L., a deciduous tree native to SE North America, is sometimes cultivated and has been reported a few times from roadsides. It is naturalized elsewhere in C & S Europe and may become so also in Greece.

Glyceria spicata (Biv.) Guss. is a species of uncertain status, scattered in the Mediterranean area and recorded from a few Greek localities (W foothills of Taigetos, Lesvos, and the Pisoderi area in the NW). It resembles *G. fluitans* (L.) R. Br. but many spikelets are subsessile, giving the inflorescence a different appearance. The glumes are rounded or truncate at apex and only weakly veined (vs. subacute and strongly veined).

Glycyrrhiza macedonica Orph. & Boiss. was based on Orphanides 427 (ATHU, G) collected 1862-08-13 "entre la ville de Genitzà [Giannitsa] et celle de Vadena [Edessa], près du village Dort-Armut [perhaps Kria Vrisi]". It is probably a synonym of *G. echinata* L. although accepted as a separate species by Rechinger in Flora Iranica 157: 169 (1984).

Godetia amoena G. Don, a native of W North America, is occasionally naturalized in Europe. From Greece is a single collection, Bazos 1170 (ATHU, LD) from a damp ruderal habitat in Kalloni Bay on the island of Lesvos.

Goniolimon collinum (Griseb.) Boiss. and ***G. incanum*** (L.) Hepper both appear as accepted species in Euro+Med PlantBase. As pointed out by Hepper (Fl. Turkey 10: 212, 1988), Linnaeus described *Statice incana* from plants cultivated in Uppsala raised from seeds sent to him by Forsskål. These were assumed to come from Egypt, but had in fact been collected on the island of Imbros (Gökçeada) in the NE Aegean area, where the Danish expedition called en route to *Arabia Felix*. Unaware of this, Grisebach described *Statice collina* from the same general area (cf. Strid 2000b: 302). The two are clearly conspecific. The eipthet *incana* is somewhat misleading since a distinctive feature of this species is the glabrous bracts and calyx [type specimen at LINN kindly checked by M. Spencer, 2015].

Goniolimon dalmaticum (C. Presl) Rchb. f. has been reported from scattered localities in the Balkan Peninsula, including northern Greece, but most of these records refer to *G. tataricum*. According to Buzurović & al. (2016) *G. dalmaticum* is endemic to Croatia where it grows on saline ground near the sea.

Goniolimon tataricum (L.) Boiss. was divided into four subspecies by Buzurović & al. (2020), including subsp. *graecum* Buzurović, which was said to be endemic to a small area near the town of Serres in NE Greece. Differences are rather obscure, and most of the Balkan populations were not assigned to subspecies.

Groenlandia, a monotypic genus of Potamogetonaceae, was not named after Greenland (where it does not occur), but after Johannes or Jean Groenland (1824-1891), a German botanist, horticulturist and microscopist. The generic name *Groenlandia* was created by J. E. Gay in 1854, but the nomenclaturally valid combination *Groenlandia densa* was made by J. P. Fourreau in 1869 to accommodate *Potamogeton densus* L., Sp. Pl.: 126 (1753; "habitat in Gallia").

Gymnadenia rhellicani (Teppner & E. Klein) Teppner & E. Klein was listed for Peloponnisos by Mermygkas, Zikos & Constantinidis (2021: supplementary file), probably in error.

Gypsophila macedonica Vandas, described from a single locality near Bitola in SW North Macedonia, was mapped to occur also in NC Greece in Atlas Florae Europaeae 7: 133 (1986), based on a report by Goulimis (1960a: 14) from the vicinity of Verria. The Greek record needs confirmation and may refer to *G. muralis*.

Hablitzia tamnoides M. Bieb., a species from the Caucasus, has been collected once on the island of Chios (cf. Fl. Hellen. 1: 110, 1997) where it may have been a casual introduction. It is a scrambling and twining plant with a superficial resemblance to *Tamus communis* L.

Halimium voldii Kit Tan, Perdetz. & Raus. In Flora of Turkey (1: 522, 1965) is a record from the island of Rodos (sub nom. *H. umbellatum*), based on a Heldreich collection at K. Heldreich is known to have visited Rodos briefly in March 1845 en route to Anatolia and collected a few specimens in coastal habitats near the town. *H. voldii/umbellatum* is very unlikely to occur in this locality. Heldreich collected extensively on Mt Taigetos in 1844, and a mixup of labels seems probable.

Haplophyllum boissierianum Vis. & Pančić is a serpentine endemic, known from a single collection in Greece: Mt Smolikas, between Samarina and Armata, rocky slope in open *Pinus nigra* woodland, 1400 m (Hartvig & al. 5848, cf. Townsend 1986: 324). For total distribution and notes, see Stevanović & al. (2014: 262-264). *H. albanicum* (Bald.) Bornm. is related but not conspecific (see Townsend 1986: 318 ff.).

Haplophyllum buxbaumii (Poir.) G. Don. *Ruta buxbaumii* Poir. was described from Tunisia, and it should be critically examined whether it is indeed conspecific with *R. spatulata* as assumed by Townsend (1986: 205). The latter was first recorded in Florae Graecae Prodromus (1: 273, 1809) under the name *R. linifolia* L., and was later described as a new species, *R. spatulata* Sm., in Flora Graeca (4: 63, tab. 370, 1824). Sibthorp's specimen was probably collected in Cyprus in the spring of 1787. Plants from E Anatolia, Iraq and W Iran have been recognized as *H. buxbaumii* subsp. *mesopotamicum* (Boiss.) C. C. Towns.

Haplophyllum patavinum (L.) G. Don occurs in Italy and the W Balkan Peninsula. A record from Mt Vourinos by Goulimis (1960a: 20; ATH), belongs to *H. coronatum* Griseb. (see Townsend 1986: 270).

Hedera helix L. Yellow-fruited plants from Italy and Greece have been regarded as a separate variety (*H. helix* var. *poetarum* Nicotra) or even species (*H. poetarum* Bertol. nom. illeg.), but Raus (2019) has shown that they merely represent a mutant form with a deficiency of anthocyanine making the leaves appear brighter green and the fruits yellowish-orange. Such plants are cultivated and occasionally naturalized.

Hedysarum pestalozzae Boiss. was reported from the vicinity of Megara by Polymenakos & Constantinidis in a poster at the congress of the Hellenic Botanical Society in Chania 14.-17.8. 2017. The species was described from SW Anatolia ("in Lyciâ propè Elmalu"); it is widespread but scattered in Turkey.

Helianthus annuus L. (sunflower), a species of North American origin, is commonly cultivated and occasionally found in ruderal habitats, but is not persistent.

Helichrysum amorginum Boiss. & Orph. The records from Anafi and Folegandros are recent ones (see Trigas & al. 2021: 325-326). Plants from Anafi are somewhat deviating whereas those from Folegandros match specimens from *locus classicus* on Amorgos.

Helichrysum italicum (Roth) G. Don. Galbany-Casals, Sáez & Benedí (2006) recognized subsp. *italicum*, subsp. *microphyllum* (Willd.) Nyman and subsp. *siculum* (Jord. & Fourr.) Galbany, L. Sáez & Benedí, the two first-mentioned occurring in Greece. The Cretan populations were referred to subsp. *microphyllum* which has a disjunct distribution on some of the major Mediterranean islands from Mallorca to Cyprus. Many plants from the Aegean islands (north of Kriti) were classified as intermediate between subsp. *italicum* and subsp. *microphyllum*, whereas plants from Kerkira belong to subsp. *italicum* which has an amphi-Adriatic distribution.

Helichrysum stoechas (L.) Moench. Two subspecies have often been recognized, subsp. *stoechas* in the W Mediterranean area and subsp. *barrelieri* (Ten.) Nyman in the east, but after analysis of variation in several morphological characters, Galbany-Casals, Sáez & Benedí (2006) concluded that there are no significant differences.

Helosciadium repens (Jacq.) W. D. J. Koch. Haristos 1161 (ATH) from nomos Pellis, near the village of Rizari, has been identified as *H. repens* but is not typical and needs critical revision. A subsequent report from nomos Prevezis, by Acheron river (Manolaki, Tsakiri & Papastergiagou 2011: 868) also needs confirmation. *H. repens* resembles *H. nodiflorum* (L.) W. D. J. Koch, but has decumbent stems rooting at most nodes, as well as broader, incised-serrate leaf segments and longer peduncles. The species is scattered in C & S Europe and NW Africa.

Hepatica nobilis Schreb. The inclusion of this unmistakable species is based on a single field note by Kit Tan & G. Vold from Mt Voras, 12.5. 1993. It is apparently rare and local in this area, but occurs in several localities in Albania and North Macedonia. In Flora Graeca 6: Plate 513 (1825) is an illustration of a flowering plant of *Hepatica nobilis* said to originate from Mt Athos. The locality is undoubtedly incorrect. Sibthorp & Bauer visited Athos in early August, 1787, when this species would long have been over. The illustration is likely to have been prepared by Bauer when he and Sibthorp made their way overland from Vienna to Florence in the spring of 1786. Somewhat mysteriously, there is a subsequent report from Mt Athos by Zoller, Geissler & Athanasiadis (1977: tab. 4), "unterhalb der Panagia, 1100 m". There are no suitable habitats in this frequently visited locality (along the track to the summit) where the species has been searched for in vain by others, including the present author.

Heracleum humile Sm. The only Greek collection is one by Pichler in 1873 "in lapidosis montis Atho". Pichler is known to have visited Mt Athos in 1873 collecting, e.g., the local endemic *Aubrieta erubescens* Griseb. Nevertheless there is a possibility that the specimen is from the Ulu Dağ (locus classicus for *Heracleum humile*).

Heracleum sphondylium s. lat. For an account of this complex, especially in France, see Reduron (2007c: 1393 ff.). The three Greek taxa, here recognized as subspecies of *H. sphondylium* L., are rather distinct and optionally regarded as separate species. Literature reports of subsp. *sphondylium* from N Greece (Athanasiadis & Drossos 1990: 111 from Mt Paikon, Drossos 1992: 112 from Lake Mitrikou, Pavlides 1982: 147 from Mt Vertiskos and Pavlides 1985: 259 from the Prespa National Park) may refer to subsp. *ternatum* (Velen.) Briq.

Herniaria olympica J. Gay was described from the Bithynian Olympus (Ulu Dağ). Greek records need confirmation (cf. Fl. Hellen 1: 232, 1997).

Hesperis balansae E. Fourn. has been divided into subsp. *balansae* (described from Manisa Dağ, W Anatolia) and subsp. *mytilensis* F. Dvořák (described from Lesvos). All Greek material probably belongs to the latter, although the differences are rather obscure.

Hesperis matronalis L. Plants from Rodopi (Arkoudorema, c. 1200 m) have recently been reported as *H. matronalis* subsp. *matronalis* (Doumas, Goula & Constantinidis 2022: 17-19). Further studies are needed on the distinction between *H. matronalis* and *H. laciniata* in NE Greece.

Hesperis sylvestris Crantz was listed for "NC" by Dimopoulos & al. (2013: 72), probably based on a record from Mt Voras by Voliotis (1981d: 254). It has recently been confirmed to occur in NE Greece (Ioannidis, Doulkeridou & Strid 2020b: 198-199).

Hibiscus syriacus L. is a late-flowering ornamental shrub native to China and Korea (the specific epithet is a geographical mistake by Linnaeus). It has been recorded from scattered localities in Greece, collected already by Sintenis & Bornmüller, 16.6. 1891 (LD; "Peninsula Hagion Oros. Karyes, cult. in hort.") and reported by Charrel (1892: 272), "Aufugus Salonicae". It may persist for some time in ruderal habitats but is scarcely naturalized.

Hieracium breazense Zahn was reported from Mt Vermion by Dunkel & Gottschlich (2021: 73). It was previously regarded as endemic to Romania.

Hieracium chalasmenoense Zahn is known only from Orphanides' type collection (Taigetos, Chalasmeno Vouno). It is apparently very rare, and modern collections are needed to ascertain its status. The eipthet was first published as *chalasinense*, but corrected by Dunkel & Gottschlich (2021: 82).

Hieracium gracilifurcum Zahn was reported from Mt Olimbos by Ascherson & al. (1936: 106), based on a collection by Orphanides (G-BOIS); the specimen has a note by Buttler: "similar to *leithneri* but totally different indumentum", which fits the description. Further material is needed to ascertain its status.

Hieracium guthnikianum Hegetschw. There is a single, recent collection (Gottschlich & Dunkel 70230) from Mt Beles near the Bulgarian border.

Hieracium juranomorphum Zahn was reported from Mt Gramos by Dunkel & Gottschlich (2021: 73). It was described from North Caucasus and has previously been reported also from Bulgaria.

Hieracium maculatum Schrank. There is a single, recent collection from Mt Cholomondas (Theodoropoulos 066, det. Gottschlich).

Hieracium noeanum Zahn (= *H. brevifolium* subsp. *noeanum* (Zahn) Greuter) was reported from Mt Timfi, as new for Greece, by Authier (2021: 96).

Hieracium oroglaucum O. Behr, E. Behr & Zahn was reported from Mt Voras by Dunkel & Gottschlich (2021: 73). It was previously known only from Albania.

Hieracium ossaeum Zahn was listed in Euro+Med PlantBase from "?Bu Gr". In spite of its name, apparently alluding to Mt Ossa, it was in fact described (by Zahn in Vandas, Reliquiae Formánekianae 1909: 364) from "Visko Tepe, Kara Baïr ad Ichtiman et Sakardža in Bg [Bulgaria]". There appear to be no records from Greece.

Hieracium pelagae subsp. ***denticulatum*** Gottschl. & Dunkel. There are a few recent collections from nomos Xanthis.

Hieracium penzesii F. Kováts & Zahn ex Pénzes can be regarded as a forgotten species. Described just before the World War II in a new but little-known journal (Pénzes 1939), it was never adopted in Bulgarian Floras. Only Greuter listed it recently (Greuter & Raab-Straube 2007). According to the estimation of *H. jankae* as "*sparsum* > *pannosum*" by Zahn, Greuter transferred it to *H. jankae* in the rank of subspecies. However, *H. jankae* is densely villous in all parts and possesses subplumose hairs. Contrary to *H. jankae*, *H. penzesii* is only covered with (not subplumose, only distinctly dentate) hairs on both surfaces of the leaves. The upper part, including the synflorescence, is almost glabrous and resembles completely *H. sparsum* (isotype: B-101024849! represents a plant that is not completely developed; a better one is illustrated in Fig. XVII in Pénzes 1939). Therefore, the taxonomical circumscription of *H. jankae* does not correspond at all to the plants that were found. Summarily, *H. penzesii* is set apart from *H. jankae*, and its ranking with species seems to be more appropriate. [Cited from Gottschlich & Dunkel 2018a: 221].

Hieracium peristericum Zahn was based on a collection by Formánek from "Berg Peristeri im Pindus". There is no confirmation from Greek territory, and it is possible that Formánek's specimen was in fact collected on Mt Varnous (Pelister or Peristeri) just north of the present border between NW Greece and North Macedonia.

Hieracium petraeum Bluff & Fingerh. was reported from "Gr" in Fl. Eur. 4: 396 (1976). The record may refer to *H. petraeum* Friv. [non Bluff & Fingerh.] which was described from "Rumelia" and is likely to be a synonym of *H. rhodopeum* Griseb.

Hieracium phocaicum Zahn. The type is Heldreich 725pp (WU!), which is part of a collection of 3 plants. The two others were determined by Zahn as *H. sartorianum* Boiss. & Heldr. The status of *H. phocaicum* is uncertain.

Hieracium psaridianum Zahn was accidentally left out of the maps. There are credible records from the mountains of Peloponnisos (Taigetos, Chelmos and Erimanthos). Two collections from Mt Koronas, Naxos, also appear to match this species.

Hieracium pseudobifidum Schur, a member of sect. *Transsilvanica*, was reported from a locality in nomos Dramas by Dunkel & Gottschlich (2021: 73). It is widespread in the Balkans, extending to E Austria.

Hieracium pseudolympicum Rech. f. & Zahn was recognized at species level in the Greek checklist (Dimopoulos & al. 2013: 60), but has subsequently been recombined as *H. sparsum* subsp. *pseudolympicum* (Rech. f. & Zahn) Gottschl. & Dunkel.

Hieracium rotundatum Kit ex Schult., a member of sect. *Transsilvanica*, was reported from a locality in nomos Dramas by Dunkel & Gottschlich (2021: 73). It is a Balkan species, more recently reported also from SE Germany and the Czech Republic. Greek records of *H. transsylvanicum* Heuff. (Dimopoulos & al. 2013: 61) have not been confirmed and may refer to *H. rotundatum* or *H. pseudobifidum*.

Hieracium schmidtii subsp. ***lasiophyllum*** (W. D. J. Koch) Zahn was reported from Mt Timfi, as new for Greece, by Authier (2021: 91).

Himantoglossum caprinum group. Plants from SE Europe have traditionally been called *H. caprinum* (M. Bieb.) Spreng. Taken in a narrow sense this is restricted to the Crimea, and plants from elsewhere have been renamed *H. jankae* Somlyay, Kreutz & Óvári (see Molnár & al. 2012). The latter is close to *H. calcaratum* (Beck) Schltr., and Lorenz (2019) referred Greek plants to *H. calcaratum* subsp. *rumelicum* (H. Baumann & R. Lorenz) R. Lorenz. The name *H. jankae* is provisionally retained in this Atlas. Following Tsiftsis & Antonopoulos (2017: 496 ff.) Cretan plants called *H. samarisense* C. Alibertis & A. Alibertis or *H. affine* subsp. *samarisense* (C. Alibertis & A. Alibertis) H. Baumann & R. Lorenz have been included in *H. jankae*. Plants from Lesvos called *H. affine* (Boiss.) Schltr. or *H. montis-tauri* Kreutz & W. Lüders are scarcely distinguishable from other Greek populations of *H. jankae*.

Hippocrepis comosa/glauca. Records from Attiki and Evvia are only tentatively referred to *H. glauca* Ten., a species described from S Italy and apparently scattered in the W & C Mediterranean area. Records of *H. glauca* from N Greece are almost certainly incorrect and likely to refer to forms of *H. comosa* L. The group is in need of detailed study.

Holcus annuus Salzm. ex C. A. Mey. According to Bergmeier (in litt., 2021) at least some Greek plants belong to subsp. *annuus* and not to subsp. *setiglumis* (Boiss. & Reut.) M. Seq. & Castrov. as assumed by Strid (2016, 1: 457).

Hordeum murinum subsp. ***setariurum*** H. Scholz & Raus was based on a single collection (Raus & Schiers 19587, B) from the Pinios river valley in nomos Lasisis.

Hordeum vulgare subsp. ***agriocrithon*** (Åberg) Á. Löve & D. Löve was listed for KK and EAe in Euro+Med PlantBase. *H. agriocrithon* Åberg is a wild form of six-rowed barley with brittle rhachis, originally described from Tibet (cf. Bothmer & al. 1991: 23). As applied to Greek material the name probably refers to *H. vulgare* subsp. *spontaneum* (K. Koch) Thell.

Hyacinthus orientalis L. has been reported from the islands of Chios, Samos and Nisiros as well as a few localities on the mainland, but probably only as an escape from cultivation. It is native in S Anatolia, Syria, Lebanon and Iraq, and is the wild progenitor of the commonly cultivated hyacinths.

Hydrocotyle. This genus of small, semi-aquatic plants, traditionally included in Apiaceae, is now considered to be better placed in Araliaceae.

Hylotelephium telephium (L.) H. Ohba is variable and sometimes subdivided. If so, all Greek plants are probably referable to *H. maximum* (L.) Holub.

Hymenonema graecum (L.) DC. There is a single, historical record from "Byzantium" (Istanbul) which has not been confirmed and seems unlikely for phytogeographical reasons. According to Fl. Turkey (5: 626 (1975) the specimen was annotated by Boissier, but there may have been a mixup of labels.

Hyoseris radiata L. All Greek records of this species are probably referable to *H. lucida* L.

Hypecoum procumbens L. Three rather poorly characterized subspecies have been recognized in Greece: subsp. *procumbens* (widespread), subsp. *fragrantissimum* Å. Dahl on the islands of Mikonos and Rhinia, and subsp. *atropunctatum* Å. Dahl on the East Aegean Islands and adjacent parts of W Anatolia.

Hypericum athoum Boiss. & Orph. "J'ai trouvé cette belle espèce de la section *Euhypericum* dans la même localité du mont Athos, avec le *Silene Orphanidis*, un peu au dessous du sommet, où il n'est pas abondant, et où il forme des touffes dans les fentes des rochers" (Orphanides 1870b: 136). It has never been found again on Mt Athos where it is certainly very rare.

Hypericum atomarium Boiss., ***H. cuisinii*** Barbey and ***H. icaricum*** Kit Tan. Taxonomy of this species complex is unsettled. *H. atomarium* was described from "montes Sipyli" [Manisa Dağ] in W Anatolia (opposite Chios) and is fairly widespread in W & SW Anatolia. As interpreted here, *H. cuisinii* is endemic to the Karpathos island group, although plants from the Aydin province in W Anatolia (opposite Samos) have recently been described as *H. cuisinii* subsp. *anatolicum* Dirmenci & N. Robson (see Dirmenci & Robson 2019). Adopting a narrow species concept, Kit Tan (in Kit Tan & Strid 2018: 167) referred plants from Ikaria, Samos and Chios to *H. icaricum* sp. nov. In the map all East Aegean records from Rodos in the south to Lesvos in the north have provisionally been classified as *H. atomarium*.

Hypericum dubium Leers was accepted in Euro+Med PlantBase and listed for several European countries including Greece. According to Robson (2002: 73) it is a synonym of *H. maculatum* subsp. *obtusiusculum* (Tourlet) Hayek, a taxon of C & W Europe unlikely to occur in Greece.

Hypericum empetrifolium Willd. Throughout most of its range *H. empetrifolium* is rather uniform, being an erect shrublet 30-60 cm tall with elongate, 8-40-flowered inflorescence raised well above the leaves. In Kriti there is pronounced ecotypic differentiation along an altitudinal gradient with the tall subsp. *empetrifolium* in the lowlands, the low, dense and few-flowered subsp. *oliganthum* (Rech. f.) I. Hagemann at intermediate altitudes and the even more extreme (dense, prostrate and rooting) subsp. *tortuosum* (Rech. f.) I. Hagemann at 1200-1900 m. Although the extremes are strikingly different and retain their habit in cultivation, there are more or less gradual transitions.

Hypericum hircinum L. is a widespread and variable species, also cultivated as an ornamental and naturalized in W Europe. Subsp. *albimontanum* (Greuter) N. Robson is a regional taxon from Greece and Cyprus, while subsp. *majus* (Aiton) N. Robson is more widespread in the C & E Mediterranean area and naturalized elsewhere. The former is usually recognized on the relatively broad leaves with crispate-undulate margins, but intermediates occur, e.g., in E Kriti. The plant illustrated on Plate 773 of Flora Graeca (1833) was apparently collected in W Kriti and represents subsp. *albimontanum*.

Hypericum hyssopifolium Chaix. A record from Lesvos by Candargy cited by Rechinger (1944a: 264) has not been confirmed and its status is uncertain. *H. hyssopifolium* is a variable species of Anatolia, with the closely related *H. tymphresteum* Boiss. & Spruner in a few localities on the Greek mainland and in Peloponnisos.

Hypericum maculatum Crantz. All Greek material belongs to subsp. *immaculatum* (Murb.) A. Fröhl. which is restricted to a rather small area from the eastern Rodopi to Montenegro and Bosnia, perhaps with outlying localities in the Carpathians.

Hypericum montanum L. is mainly a C European species. Literature records from N Greece (Turrill 1918: 273 from "Krusa Balkan", Ganiatsas 1963: 102 from the Athos Peninsula and Leibfritz & Schuler 1997: 130 from the Nestos river valley) all need confirmation.

Hypericum perfoliatum subsp. ***phitosianum*** Greuter & Karl. Robson (2010a: 62-63) pointed out that some populations from the Kiklades are ± intermediate between the widespread *Hypericum perfoliatum* L. and the Cretan endemic *H. trichocaulon* Boiss. & Heldr. Having looked further into the matter Trigas (2018) described them as *H. cycladicum*. One year later Greuter & Karl (2019) described the same taxon as *H. perfoliatum* subsp. *phitosianum*, without citing the previous description of *H. cycladicum* (although the latter was listed in IPNI and other databases, and correspondence shows that Greuter was well aware of it). By means of a nomenclatural trick he can thus claim priority at subspecies level. For occurrence of this plant on Mikonos, see Polymenakos & Kit Tan (2019: 109).

Hypericum perforatum L. Most of the Greek records are referable to subsp. *veronense* (Schrank) Ces. (syn. var. *angustifolium* DC.). This occurs in S Europe and SW Asia and merges into the widespread subsp. *perforatum* in N Greece.

Hypericum polyphyllum subsp. ***subcordatum*** N. Robson & Hub.-Mor. was reported from Ikaria by Christodoulakis (1996a: 76). The record is likely to refer to *H. olympicum* L.

Hypericum richeri Vill. Rechinger 10853 (BM, G, LD, W) from Mt Vrondous is the type of *H. aucheri* var. *punctato-fimbriatum* Rech. f. (see Rechinger 1939a: 463). This collection was identified by Robson in 1967 as *H. richeri* subsp. *grisebachii* (Boiss.) Nyman, and is the source of the record of this subspecies from Greece in Fl. Eur. 2: 267 (1968). Later Robson (2010a: 77) corrected himself, including the Rechinger collection in *H. barbatum* Jacq.

Hypericum rochelii Griseb. & Schenk. Some Greek collections have been identified as subsp. *pseudotenellum* (Vandas) D. Jord. and others as subsp. *rochelii*. There appear to be no clear differences. *H. pseudotenellum* Vandas was based on a collection by Formánek a. 1900 from Mt Alibotus, probably on the Bulgarian side.

Hypericum rumeliacum Boiss. Many records are likely to be incorrect. Only those based on verified herbarium specimens have been mapped.

Hypericum spruneri Boiss. Only records based on verified herbarium specimens have been mapped. Many others, especially from Peloponnisos and the far north-east need confirmation.

Iberis amara L. was listed for "Gr" in Euro+Med PlantBase referring to Atlas Florae Europaeae 11 (1996) where, however, there is no Greek record. It is a species of SW Europe, very unlikely to occur in Greece.

Iberis aurosica Chaix. Euro+Med PlantBase listed *I. epirota* Halácsy as a synonym of *I. aurosica*. The latter was described from Dauphiné and is otherwise restricted to the border area of NW Italy and SE France. The two are not likely to be conspecific.

Iberis pinnata L. As for *I. amara* above.

Iberis spathulata DC. Greek records are likely to refer to *I. carnosa* Willd. As currently understood *I. spathulata* is endemic to the Pyrenees.

Ibicella lutea (Lindl.) Van Eselt., a native of E Brazil and N Argentina, was reported by Yannitsaros & Bazos (2006) from ruderal habitats in Lesvos, where it is apparently well established. It is a glandular-viscid annual with large, orbicular leaves and bright yellow flowers in dense racemes. The mature fruit is very characteristic, being woody and consisting of 2 echinate follicles 6-7 cm, each produced into a slender, hooked beak c. 10 cm, hence the names devil's claw, unicorn plant or proboscis flower. Plants from Lesvos were first reported as *Proboscidea lousianica* (see under the latter). Recently *Ibicella lutea* has been found also by the airport on Limnos (Dimitrakopoulos, e-mail 2021-12-18).

Inula subfloccosa Rech. f. is a taxon of uncertain status, known from a few populations in southern Evvia. It is perhaps best regarded as another subspecies of *I. verbascifolia* (Willd.) Hausskn.

Ipomoea batatas (L.) Lam. (sweet potato), a native of S & C America, is cultivated and may persist for some time in ruderal habitats but is scarcely naturalized.

Iris albicans Lange (*I. florentina* auct.) is cultivated in gardens, cemeteries, etc., and occasionally naturalized along roads and in ruderal habitats. It is probably native in the E Mediterranean area, but taxonomic relations between this and *I. germanica* are not clear (cf. Mathew 1981: 21-22).

Iris albida Davidov was described from "aquis et paludosis circa urbem Dede-Aghač [Alexandroupolis]". It is a synonym of *I. orientalis* Mill. (see Boltenkov & Güner 2021).

Iris attica Boiss. & Heldr. Morphological differences between this species and the more northerly *I. pumila* L. are not clear, but limited evidence suggests that the former is diploid and the latter tetraploid. Plants from northern Greece should be critically studied.

Iris germanica L. This species is certainly native in NW Greece where it is found on cliff ledges, in rocky calcareous grassland and open deciduous woodland in natural settings far from habitations, sometimes ascending to 2100 m. More southerly records, including all from the islands, are likely to refer to naturalized plants. The general distribution is uncertain due to widespread cultivation. In the Prespa area white-flowered plants occur in mixed populations with the more common blue ones.

Iris unguicularis Poir. Davis & Jury (1990) classified plants from the Greek mainland as subsp. *carica* var. *angustifolia* (Boiss. & Heldr.) A. P. Davis & Jury and those from the East Aegean islands and SW Anatolia as subsp. *carica* var. *carica*. The former are optionally called *I. unguicularis* subsp. *angustifolia* (Boiss. & Heldr.) Greuter.

Isatis corymbosa Boiss. (syntype from "Thessaliá, circà Larissam") was listed as a Greek endemic in Euro+Med PlantBase as well as Atlas Florae Europaeae (10: 45, 1994), but is probably a synonym of *I. tomentella* Boiss. & Balansa.

Isoetes echinospora Durieu was listed for "Pe" in Dimopoulos & al. (2013: 38), but the source for this record is obscure. It is a species of W & N Europe, unlikely to occur in Greece.

Isoetes gymnocarpa (Gennari) A. Braun and ***I. todaroana*** Troia & Raimondo cannot be distinguished from *I. histrix* Bory and have been merged with the last-mentioned.

Isoetes haussknechtii Troia & Greuter was described from Arkadia and differs from *I. heldreichii* Wettst. only in being terrestrial with shorter and usually more numerous leaves. The two are likely to represent terrestrial and aquatic forms of the same species. The type of *I. heldreichii* is Heldreich s.n., 1885-07-10 (B, LD, WU) gathered "prope Karditza, submersa in palude prope Palaeokastron". Haussknecht s.n. (B, BR, JE, LD, WUHAL) gathered at "Planities Thessaliae: in ulig. pr. Palaeokastron" is probably from the same locality but dated 1885-06-15.

A village now called Mitropolis had the name "Paliokastro" during Ottoman rule. Perhaps the small artificial lake SW of Mitropolis (39°19'45"N, 21°49'15"E) was once a suitable *Isoetes* habitat and could be *locus classicus* for *Isoetes heldreichii*. There are many other watercourses around this village. [R. Jahn, in litt. 2014-02-02]. Troia & Greuter (2015) assumed that *I. heldreichii* is now extinct.

Isoetes pirinica D. F. Brunton & D. Ivanova has recently been described from high-altitude lakes on Mt Pirin in SW Bulgaria (see Brunton, Ivanova & Sokoloff 2021). It is related to the North European *I. lacustris* L., and should be looked for on the Greek side of the border.

Isoetes sicula Tod. has been reported from a single locality in Peloponnisos (Greuter (2012: 24). The record is likely to refer to *I. durieui*.

Isoetes velata A. Braun was listed from Kik and KK in Dimopoulos & al. (2013: 38). *I. longissima* Bory (replacing the illegitimate name *I. velata* A. Braun) was reported from the Omalos plain in W Kriti by Bergmeier (2011: 171). This as well as other Greek records of *I. velata* are likely to refer to *I. phrygia* (Boiss.) Hausskn. (cf. Troìa & Greuter 2015: 399-400).

Isopyrum aquilegioides L. Records by Chaubard & Bory (1838: 35) from localities in Peloponnisos ("en plusieurs endroits du Magne; aux environs de Kalamata, etc.") undoubtedly refer to the subsequently described *Thalictrum orientale* Boiss. *Isopyrum aquilegioides* is a synonym of *I. thalictroides* L. which has not been recorded in Greece but is fairly widespread in Bulgaria. It could be expected also on the Greek side of the border in the Rodopi Mountains.

Jacobaea erucifolia (L.) G. Gaertn., B. Mey. & Scherb. (*Senecio erucifolius* L.). All Greek records of this species are likely to refer to *Jacobaea vulgaris* Gaertn. (= *Senecio jacobaea* L.).

Jacobaea othonnae (M. Bieb.) C. A. Mey. was described from Georgia, as *Senecio othonnae* M. Bieb. (1808). Similar plants from the Athos peninsula were recorded already by Sibthorp (1787), subsequently described as *Cineraria anomala* Sm. in Florae Graecae Prodromus 2: 179 (1813) and illustrated under that name in Flora Graeca 9: Plate 872 (1837). There are several subsequent confirmations from the Athos peninsula and a few recent collections from elsewhere in northern Greece. Sutorý (2021: 65) made the nomenclatural recombination *Jacobaea othonnae* subsp. *anomala* (Sm.) Sutorý for the Athos plant, the diagnostic character being the woolly-tomentose synflorescence and involucre (well visible in my collection no 56856 from Kria Nera in the SE foothills of Mt Athos). It is not clear whether other Balkan populations (not examined by Sutorý) belong to the same subspecies. Plants recently reported from Achladovouno in nomos Xanthis (Doumas, Goula & Constantinidis 2022: 11) match subsp. *anomala*. *Jacobaea othonnae* has a large distribution area from the Balkan Peninsula to Caucasus as well as mountains of N Iran, Iraq and Syria. Plants from the far east of this range were described as subsp. *adamsii* Sutorý, being characterized by short-pubescent synflorescence and involucre. The taxonomic value, if any, of the indumentum characters can be assessed only after analysis of variation throughout the range of the species.

Jacobaea pancicii (Degen) Vladimirov & Raab-Straube. A record of this species from Mt Falakron by Goulimis (1956a: 12, sub nom. *Senecio pancicii* Degen) needs confirmation. It is otherwise restricted to Bulgaria and Serbia.

Jacobaea subalpina (W. D. J. Koch) Pelser & Veldkamp was listed for Peloponnisos by Mermygkas, Zikos & Constantinidis (2021: supplementary file), probably in error. Verified Greek records are all in wet habitats in the north.

Jasminum officinale L. is native from Caucasus to W China. It is a climbing, semi-evergreen shrub with clusters of white, fragrant flowers. In Greece it is cultivated as an ornamental and has been reported to be more or less naturalized in river beds, e.g., on the islands of Chios and Lesvos.

Johrenia thessala Bornm. is only known from the type collection, Sintenis 528 (LD) from the Kalambaka area. It may represent a form of *Pimpinella tragium* (fide Constantinidis, in litt. 2011).

Juglans regia L. The walnut is scattered throughout Greece but generally close to villages. It is native in mountains of C & SW Asia and may have been brought to Greece already in antiquity (Theophrastos called it "Persian nut"). It is widely cultivated for its valuable hardwood, edible nuts and also as an ornamental park tree.

Juncus alpinus Vill. This name, which was used in Flora Europaea, Mountain Flora of Greece and many other publications, has to be replaced by *J. alpinoarticulatus* Chaix (see Hämet-Ahti 1980).

Juncus fontanesii J. Gay. Both subsp. *fontanesii* and subsp. *pyramidatus* (Laharpe) Snogerup have been reported from Greece, and the differences are not always clear. Intermediates are also said to occur in C & S Italy.

Juncus ranarius Songeon & E. P. Perrier. Few and scattered records from Greece (e.g., Rechinger 1939a: 542 from Porto Lagos in nomos Xanthis and Raus 1999: 414 from Kerkira) probably refer to forms of *J. bufonius* L.

Juncus sorrentinii Parl. was listed for Greece in Euro+Med PlantBase, but the source is obscure and the record is likely to refer to *J. bufonius* L.

Juncus sphaerocarpus Nees. After the preparation of the maps this species has been reported from the Rodopi mountains by Charalampidou, Eleftheriadou & Theodoropoulos in Fl. Medit. 32: 91 (2022).

Juniperus excelsa M. Bieb. Farjon (in Fl. Turkey 11: 9-10, 2000) recognized subsp. *excelsa* and subsp. *polycarpos* (K. Koch) Takht. Both are widespread, and all European plants belong to subsp. *excelsa* which was described from the Crimea.

Juniperus turbinata Guss., described from Sicily, was long regarded as a synonym of *J. phoenicea* L. but subsequently resurrected as an independent species (see Adams 2013, Mazur & al. 2016, Ferrer Gallego & al. 2017). More recently (Pavon, Véla & Médail 2020) *J. phoenicea* has been divided into three subspecies: subsp. *phoenicea* in E Spain and S France (plus very locally in Italy), subsp. *turbinata* (Guss.) Nyman (circum-Mediterranean) and subsp. *canariensis* (Guyot) D. Pavon, E. Véla & F. Médail in the Canary Islands. All Greek plants are thus referred to *J. phoenicea* subsp. *turbinata*.

Jurinea consanguinea DC. was listed for Peloponnisos by Mermygkas, Zikos & Constantinidis (2021: supplementary file), but no details are available and the record may refer to *J. mollis* s. lat.

Jurinea kilaea Azn. (*J. albicaulis* Bunge subsp. *kilaea* (Azn.) Kožuharov) was described from the vicinity of Constantinople and has also been reported from coastal sand in E Bulgaria. Euro-Med PlantBase listed it for "Gr"; it may occur in the North Aegean area but there appear to be no confirmed records.

Jurinea polycephala Formánek is a taxon of uncertain status, perhaps an unusual form of *J. mollis* with branched stem. It was described from Demirkapu (Demir Kapija), a town in North Macedonia some 30 km from the Greek border. Euro+Med PlantBase listed it from Greece, probably based on the imprecise "Ma" in Hayek (1931: 700). There appear to be no specified records from Greece.

Justicia adhatoda L. is native from Afghanistan to Vietnam and frequently cultivated elsewhere as an ornamental. There are scattered records in the Mediterranean area including at least one (Kosonen s.n., LD) from Kriti, nomos Lasithiou, Xerocampos.

Kengia, see *Cleistogenes*.

Kickxia commutata (Rchb.) Fritsch. The vast majority of Greek records represent subsp. *graeca* (Bory & Chaub.) R. Fern. which was described from the Mani Peninsula in S Peloponnisos and illustrated in Bory & Chaubard (1832-33: Plate XXI). Subsp. *graeca* occurs in the S & C Balkan Peninsula, W & S Anatolia, Cyprus and Lebanon/Palestine. There are credible reports of the widespread subsp. *commutata* from the Ionian Islands and the far west of the mainland whereas records from elsewhere need confirmation. The latter differs in the obtuse leaves, relatively shorter pedicels and larger corolla.

Kickxia elatine (L.) Dumort. Almost all Greek records refer to the Mediterranean subsp. *crinita* (Mabille) Greuter, but the widespread subsp. *elatine* may occur in the far north.They differ in quantitative characters and there are many intermediates.

Knautia albanica Briq. is a poorly known Balkan endemic. It was reported from Alibotuš (Orvilos) on the Greek/Bulgarian border by Drenowski (1934a: 4). There is no subsequent confirmation.

Koeleria macrantha group. Taxonomy and nomenclature are very unsettled for the Greek taxa, and only a collective map can be presented. *K. eriostachya* Pančić, *K. lobata* (M. Bieb.) Roem. & Schult., *K. macrantha* (Ledeb.) Schult., *K. nitidula* Velen., *K. pyramidata* (Lam.) P. Beauv. and *K. splendens* C. Presl have been reported from Greece. The Italian members of the group were revised by Brullo, Giusso del Galdo & Minissale (2009).

Koelreuteria paniculata Laxmann, a native of China, has been cultivated as an ornamental street tree in Greece since the 19th century. Yannitsaros (2004a) reported is as subspontaneous in parks in Athens.

Laburnum anagyroides Medik. Reports of this species from Southern Pindos (e.g., Sfikas & Sfikas 1995: 2) refer to the similar *L. alpinum* (Mill.) Bercht. & J. Presl., see Kit Tan & al. (2015a: 382-383).

Lactuca aculeata Boiss. & Kotschy is a recent addition to the Greek (and European) flora – see Bergmeier & Meyer (2021). Resembling the widespread *L. serriola* L. it may have been previously overlooked. Further records could well be hiding among specimens identified as the latter, especially from the East Aegean Islands.

Lactuca alpina (L.) A. Gray was listed for Peloponnisos by Mermygkas, Zikos & Constantinidis (2021: supplementary file), certainly in error. The only confirmed Greek records are in the Rodopi mountains.

Lactuca quercina L. was listed for "Gr" in Fl. Eur. 4: 330 (1976), probably based on Halácsy (1902a: 209) who reported it from a locality in Epirus, sub nom. *L. chaixi* Vill., citing a collection by Formánek (not seen by Halácsy). According to Vandas (1909: 342) there is no extant herbarium material, and the record must be regarded as dubious. The nearest localities are in northern Albania. A recent report of *L. quercina* from SE Peloponnisos (Koutsotheodoris 2008: 21) is almost certainly incorrect.

Lactuca virosa L. was listed for Peloponnisos by Mermygkas, Zikos & Constantinidis (2021: supplementary file), probably in error. Verified Greek records are all in the north.

Lagurus ovatus subsp. ***nanus*** (Guss.) Messeri and subsp. ***vestitus*** (Messeri) Brullo were based on few specimens from widely scattered localities in the Mediterranean region. If taxonomically recognizable at all, they should be treated as varieties (as originally described).

Lamium album L. was listed for "Gr" in Fl. Eur. 3: 148 (1972) and has subsequently been recorded by Borkowsky (1994: 95) from Kerkira and by Voliotis (1976a: 50) from Mt Vrondous, probably in error. Several specimens from Pe and StE at B named *L. album* are misidentified Willing collections of either *L. moschatum* Mill. or *L. bifidum* Cirillo (rev. Th. Raus).

Lamium bifidum subsp. ***albimontanum*** Rech. f., described from Kriti and recognized in Euro+Med PlantBase, is probably a synonym of *L. purpureum* L. (see Mennema 1989: 125).

Lamium cariense R. R. Mill, otherwise a species of W Anatolia, occurs also in Greece according to Fl. Turkey 7: 130 (1982). It is doubtfully distinct from *L. garganicum* subsp. *striatum* (Sm.) Hayek of which there are several records from the East Aegean Islands.

Lamium hybridum Vill. Greek records (e.g., Halácsy 1902a: 513, Minoglou 2000: 67) probably refer to *L. purpureum* L.

Lampranthus roseus (Willd.) Schwantes, an ornamental of South African origin, has been reported from beach and waste ground on the island of Amorgos (Biel & Kit Tan 2015a: 56). It was recorded from Naxos (Biel & Kit Tan 2020d: 541), but later revised as *Malephora purpureocrocea* (pers. comm., Kit Tan 2021). It has also been reported from Thira (Raus, in litt. as *Lampranthus* sp., but probably referring to *Malephora*).

Lantana achyranthifolia Desf. has been reported as a casual in the island of Milos – see Biel & Kit Tan (2019c: 300). It is smaller than *L. camara* L., with few-flowered corymbs and whitish corollas.

Lappula barbata (M. Bieb.) Gürke. A few specimens from north central Greece have dense, appressed, greyish indumentum matching that of *L. barbata*, but are nevertheless likely to belong to *L. squarrosa* (Retz.) Dumort.

Lappula patula (Lehm.) Gürke has been reported from the area of Kozani and Kastoria in north central Greece, but specimens have turned out to represent *L. squarrosa*. The latter is probably the only *Lappula* species in Greece.

Laserpitium pesudomeum Boiss., a Greek endemic described from Mt Killini, is morphologically distinct, and recent molecular studies have confirmed its isolated taxonomic position. It has recently been transferred to a separate, monotypic genus, as *Laserocarpum pseudomeum* (Boiss.) Spalik & al. (2019).

Laserpitium siler L. It has recently been proposed to transfer the two taxa here mapped as *L. siler* subsp. *garganicum* (Ten.) Arcang. and subsp. *laeve* (Halácsy) Hartvig to the genus *Siler*, as *S. zernyi* (Hayek) Thell. subsp. *zernyi* and subsp. *laeve* (Halácsy) Niketić & al., respectively – see Conti & al. (2021). Both are Balkan endemics extending northwards to Serbia/Kosovo. A third subspecies, subsp. *ochridanum* (Micevski) Niketić & al., was recognized on Mt Galičica in SW North Macedonia and the Albanian part of the Prespa National Park. *S. zernyi* is a member of the *S. montanum* complex which, according to Conti & al. (op. cit.), comprises 10 species and subspecies in the Alps, Italy and Balkan Peninsula.

Lathraea squamaria L. In the southernmost locality (Korinthia, near Agia Triada) the host plant was *Platanus orientalis* (Zarkos & al. 2014a: 132). A record from Mt Athos has not been confirmed since first reported by Grisebach (1844: 60, "sparsim in quercetis m. Athonis"), but is undoubtedly correct. [After the preparation of the maps this species has been recorded even further south, near the village of Abelonia in nomos Messinias at 37°26'N, 21°57'E – Giannopoulos 2022-04-06].

Lathyrus angulatus L. is a species of the W & C Mediterranean area. Literature records from Greece, e.g., Voliotis (1967: 90) from Mt Cholomondas, Haeupler (1991: 237) from NE Peloponnisos, and Krigas & Kokkini (2005: 57) from Thessaloniki, probably refer to *L. sphaericus* Retz.

Lathyrus aureus (Steven) D. Brândză. Records from Mt Pinovo (Voliotis 1987b: 188) as well as a locality NE of Thessaloniki (Turrill 1918: 284) are feasible but need confirmation. The species in widespread in Caucasia and Anatolia, extending to Romania and Bulgaria.

Lathyrus hallersteinii Baumg. was listed for Greece in Fl. Eur. (2: 140, 1968), but the source for this record is obscure and no specimens have been seen. The species is closely related to *L. pratensis* L.

Lathyrus odoratus L., a native of S Italy & Sicily, is widely cultivated as an ornamental and has been reported a few times from ruderal habitats in Greece, but is scarcely naturalized.

Lathyrus pannonicus (Jacq.) Garcke is a variable species or species complex occurring from Spain to C & SE Russia. Greek plants resemble specimens from N Italy and the W Balkan Peninsula (southwards to Albania) which have been referred to subsp. *varius* (C. Koch) P. W. Ball, but differ in the non-ciliate calyx lobes.

Lathyrus stenophyllus Boiss. & Heldr. was described from the Alanya area in S Anatolia. Records from Rodos and Kriti (cf. Fl. Turkey 3: 361, 1970) probably refer to forms of *L. sativus* L.

Lathyrus vernus (L.) Bernh. has been reported several times from Greece, but no verified specimens have been seen and most of the records certainly refer to *L. venetus* (Mill.) Wohlf. which largely replaces *L. vernus* in SE Europe. Both species may occur in northernmost Greece; ripe legumes are needed for positive identification.

Lavandula angustifolia Mill., a native of the W & C Mediterranean area is cultivated for ornament and perfume, but scarcely naturalized. It is a slender, much-branched subshrub with linear, entire leaves and a lax spike without enlarged upper bacts.

Lavandula dentata L., a native of the W Mediterranean area, has been reported from a few localities in Greece, e.g., on Amorgos (Biel & Kit Tan 2019b: 102), probably as a casual. It resembles *L. angustifolia* but has crenate-dentate to pectinate leaves, somewhat denser spike and smaller corollas.

Legousia pentagonia (L.) Druce. According to a recent revision (Wahlsteen & Tyler 2019) this is better treated as a geographical race, *L. speculum-veneris* subsp. *pentagonia* (L.) Wahlsteen, replacing the nominate subspecies in the SE part of the range, with occasional intermediates in the Aegean area.

Legousia scabra (Lowe) Gamisans was listed as an accepted species in Euro+Med PlantBase, but according to Wahlsteen & Tyler (2019: 32) it is a synonym of *L. falcata* (Ten.) Fritsch ex Janch.

Lemna aequinoctialis Welw. has been reported once, in rice fields in the Acheloos river estuaries (Koumpli-Sovantzi 2008: 321). It resembles *L. minor* but has ovate-lanceolate fronds, laterally winged root sheath and sharply pointed root cap. The species is widespread in tropical and subtropical areas with the nearest localities in the Nile delta. It may be naturalizing in Greece.

Lens culinaris Medik. (*Ervum lens* L., lentil) is one of the oldest crop plants of the Near East, probably originating in the Fertile Crescent and cultivated since Neolithic times. In Greece it may persist for some time in ruderal habitats, but is not fully naturalized. The origin is uncertain, but it may have been derived from *L. orientalis* (Boiss.) Schmalh. (which is sometimes regarded as a subspecies of *L. culinaris*). The cultivated plants are erect and more robust, with more numerous leaflets, lanceolate stipules and indehiscent legumes.

Leontodon crispus subsp. ***asper*** (Waldst. & Kit.) Rohlena is sometimes regarded as a separate species, and the correct name is then *L. biscutellifolius* DC.

Leontodon crispus subsp. ***rossianus*** (Degen & Lengyel) Hayek, a taxon of the NW Balkan Peninsula, has been reported from Mt Vourinos (Hagemann & al., in litt.), probably in error.

Leontodon taraxacoides (Vill.) Mérat. This name has often been used for plants from the Ionian Islands and elsewhere in Europe, but according to Euro+Med PlantBase it is synonymous with *L. saxatilis* Lam. Plants from Kerkira have been referred to *L. saxatilis* subsp. *rothii* Maire which may be synonymous with *L. longirostris* (Finch & P. D. Sell) Talavera, recently reported also from the island of Limnos (see Bergmeier 2021a: 356-357). The record of *L. saxatilis* from Milos is very recent (Biel & Kit Tan in Phytol. Balcan. 28: 147-148, 2022) and represents a significant extension of the range.

Leopoldia maritima (Desf.) Parl. (*Muscari maritimum* Desf.), otherwise a North African species, was listed for the Cretan area in Euro+Med PlantBase. No details are known, and the record needs confirmation.

Leopoldia neumannii Böhnert & Lobin (2017), described from the Vikos gorge (Epirus) is probably a form of *Muscari comosum* (L.) Mill. with poorly developed sterile flowers.

Lepidium densiflorum Schrad. Willing 86225 from the vicinity of Aridea in north central Greece was originally called *L. ruderale* L. but has subsequently been revised as *L. densiflorum*, a species of North American origin widely naturalized in C Europe. It may be on the verge of becoming established in Greece.

Lepidium sativum L. is probably indigenous in SW Asia and Egypt. It is frequently grown to seedling stage as a salad plant or used in bird seed, occurring as a casual in ruderal habitats but scarcely naturalized.

Leptoplax emarginata (Boiss.) O. E. Schulz. See *Bornmuellera*.

Lesquereuxia syriaca Boiss. & Reut. According to McNeal & al. (2013) the monotypic genus *Lesquereuxia* Boiss. & Reut. (1853) is nested within *Siphonostegia* Benth. (1835). The correct name is thus *Siphonostegia syriaca* (Boiss. & Reut.) Boiss.

Leucanthemum vulgare complex. Most of the Greek plants are relatively tall and occur in damp to wet habitats at moderate altitude, flowering from late May to mid-July. They are diploid and probably referable to *L. praecox* (Horvatić) Villard. Plants from somewhat drier grassland at high altitude on Mts Smolikas, Gramos and perhaps elsewhere in the Pindos are short with linear, subentire cauline leaves; they are late-flowering (end of July and August) and probably referable to *L. pallens* (Perreym.) DC. Both are often regarded as subspecies of *L. vulgare* Lam.

Leucojum ionicum Kit Tan, Mullaj, Sfikas & Strid was described in 2004. Two years later, Bareka, Kamari & Phitos (2006) described *Acis ionica*, pretending not to realize that the two are conspecific, although identical in morphology, ecology, distribution, flowering time and chromosome number. Everything in the latter article, except the name, has been copied from the former. *Acis* is variously regarded as a subgenus/section of *Leucojum* or as a separate genus. If referred to *Acis* the species will have to be named *A. orientalis* Strid (see Strid 2019b).

In this case, Art. 53.2 of the Code of Nomenclature applies: "When two or more names of genera or species based on different types are so similar that they are likely to be confused (because they are applied to related taxa or for any other reason) they are to be treated as homonyms". When Bareka, Kamari & Phitos published *Acis ionica* and explicitly referenced *Leucojum ionicum* as a different species based on a different type, they knew perfectly well that the latter would be transferred to *Acis* along with some other former *Leucojum* species, and so the epithet "*ionica*" for what they claimed was a different species was deliberately chosen to produce confusion.

Leymus arenarius (L.) Hochst. was included in the Greek checklist (Dimopoulos & al. 2013: 126), based on Strid & Kit Tan 31348 (ATH, G, herb. Strid) from Porto Lagos in nomos Rodopis. This collection has subsequently been revised as *L. racemosus* subsp. *sabulosus* (M. Bieb.) Tzvelev.

Lilium albanicum Griseb. In Greek plants the flowers are always yellow. Further north are yellow-, orange- and red-flowered individuals sometimes in mixed populations.

Lilium candidum L. "Die Lilie war seit dem grauesten Alterthum Sinnbild des Unschuld und der Sittsamkeit, sie entstand aus der Milch der Hera" (Landerer 1856c: 59). For comments on distribution and variation, see Aegean Atlas 1: 388-389 (2016).

Lilium martagon L. Greek plants belong to a variety with villous buds and rather dark purple flowers. In C European plants the flowers are generally paler.

Limodorum trabutianum Batt. is a peloric form of *L. abortivum* (L.) Sw. scattered throughout the distribution range of the latter from NW Africa and Portugal eastwards at least to Greece. In Greece it has been reported from S Peloponnisos (Ellenbast 2008) and the Prespa area (Rohmer 2018b, mixed with *L. abortivum*). Plants from the Dadia area in NE Greece have been recognized as *L. trabutianum* subsp. *thracum* Presser; see also Tsiftsis & Antonopoulos (2017: 154-155), Kati & al. (2000: 274) and Wucherpfennig (2019: 322-325). Taxonomically it is probably best regarded as *L. abortivum* var. *trabutianum* (Batt.) Raynaud.

Limonium. Brullo & Erben (2016) described 39 new species from the Aegean area, most of them based on one or a few of their own collections, whereas the much larger sets of specimens from major herbaria such as ATH (c. 210 nos), C (c. 130 nos) and LD (c. 910 nos) were ignored. Most of their "species" are local populations or slightly deviating individuals. Especially in the taxonomic neighbourhoods of *L. roridum* and *L. sitiacum* several new species were based on single collections (or individuals?) and show only vague statistical differences (or none at all). An attempt has been made here to merge them into recognizable groups, but taxonomy and distribution remain highly provisional. See also comments in Aegean Atlas (1: 430-435, 2016).

Limonium albomarginatum Brullo is a species of uncertain status based on a single collection from the southern Mani Peninsula (near Gerolimin). It is very similar to *L. doerfleri* (Halácsy) Rech. f.

Limonium archaeothirae Erben & Brullo is a dubious species recently described from the island of Thira (Santorin). It is scarcely distinct from *L. ocymifolium* (Poir.) Kuntze.

Limonium astypaleanum Erben & Brullo was based on a single collection from the island of Astipalea. It may belong to *L. roridum* (Sm.) Brullo & Guarino which has been collected several times on that island by Runemark & al. (LD).

Limonium athinense Erben & Brullo is a dubious species based on a single collection from Voula (near Athens). It seems closely related to *L. roridum*, and further material is needed to verify its status.

Limonium aucheri (Girard) Greuter & Burdet and ***L. ocymifolium*** (Poir.) Kuntze were regarded as "distinct species" by Brullo & Erben (2016: 159), but careful examination of their descriptions reveals that even the best "distinguishing" characters overlap to a considerable degree. The species have therefore been merged under the latter name.

Limonium avei (De Not.) Brullo & Erben was reported from the island of Gavdos S of Kriti by Brullo (1988: 17). The record almost certainly refers to *L. echioides* (L.) Mill.

Limonium cancellatum (Bertol.) Kuntze has been reported from the Ionian islands of Lefkas (Maire & Petitmengin 1908: 186) and Kerkira (Borkowsky 1994: 108). As currently understood *L. cancellatum* is restricted to Croatia. Records from Kerkira are likely to refer to the subsequently described *L. arcuatum* R. Artelari.

Limonium cephalonicum R. Artelari. There are no clear differences between *L. cephalonicum* (described in 1984) and *L. damboldtianum* Phitos & R. Artelari (described in 1981). Both are endemic to the island of Kefallinia, and intermediates occur near the town of Assos.

Limonium chersonesum Erben & Brullo, ***L. ikaricum*** Erben & Brullo and ***L. parosicum*** Erben & Brullo, all based on single collections (or individuals?), appear to fall within the range of variation of *L. roridum* (Sm.) Brullo & Guarino.

Limonium chrisianum Brullo & Guarino was based on a single collection and appears to differ from *L. creticum* R. Artelari only in the slightly smaller leaves. It is not likely to be specifically distinct.

Limonium compactum Erben & Brullo was described from the island of Skiros and differs from *L. narbonense* Mill. only in the more densely arranged spikelets. The two overlap completely in distribution and ecology, and the rank of variety may be appropriate.

Limonium contractum Erben & Brullo and ***L. pusillum*** Erben & Brullo were both based on collections from Astipalea. There appear to be no consistent differences between these and *L. palmare* (Sm.) Rech. f. which has been collected several times on this island by Runemark & al. (specimens at LD).

Limonium crateriforme Erben & Brullo, described from Karpathos and eastern Kriti, was said to be close to *L. proliferum* (d'Urv.) Erben & Brullo. Neither can be clearly separated from *L. roridum* (Sm.) Brullo & Guarino.

Limonium dolihiense Erben & Brullo was based a single collection and is a very dubious species. According to the descriptions for *L. dolihiense* and *L. roridum* all morphological characters and measurements overlap, the only possible difference being in the rosette leaves (15-30 × 2-4 mm and narrowly oblanceolate in *L. dolihiense*, 10-30 × 4-8 mm and spathulate in *L. roridum*).

Limonium fragile Erben & Brullo is a dubious species reported from a few localities in the South Aegean area, including a recent report from the island of Kasos (Cattaneo & Grano 2020: 542). It seems to differ from *L. roridum* only in the slightly more erect branches and somewhat smaller spikelets and flowers. 2n=43 has been reported in both taxa.

Limonium grabusae Erben & Brullo, and ***L. sougiae*** Erben & Brullo, were both described from single populations from W Kriti. They are unlikely to be specifically distinct from *L. sieberi* (Boiss.) Kuntze.

Limonium graecum (Poir.) Rech. f. Brullo & Erben (2016: 130) regarded *L. graecum* as a "taxonomically quite isolated species". Careful examination of the long descriptions of *L. graecum* and *L. palmare* (Sm.) Rech. f. (op. cit.: p. 129 and p. 114, respectively) reveal only one difference: "Stems verrucate to rugose" in the former and "more or less tuberculate" in the latter – scarcely sufficient to separate species. When the two are merged, *L. palmare* is the older name.

Limonium helenae Erben & Brullo is a dubious species, described from Chios. One of the collections cited (Snogerup 10804, B) has been identified by Artelari as *L. roridum*, which seems feasible.

Limonium isidorum Erben & Brullo is a dubious species, described from Samos. It is likely to fall within the range of variation of *L. ocymifolium* (Poir.) Kuntze.

Limonium ithacense Artelari appears to fall within the range of variation of *L. saracinatum* R. Artelari. In fact, already in the original description intermediates were reported in northern Kefallinia.

Limonium kirikosicum Erben & Brullo was based on a single collection from Ikaria. It is scarcely different from *L. virgatum* (Willd.) Fourr.

Limonium korakonisicum R. Artelari & Valli is an apomictic polyploid (2n=51), recently described from a single collection from the island of Zakinthos. It is doubtfully distinct from *L. sieberi* (Boiss.) Kuntze.

Limonium ladikanum Erben & Brullo was listed by Dimopoulos & al. (2013: 120), but this is apparently an unpublished manuscript name. It does not appear in Erben & Brullo (2016), nor in IPNI.

Limonium meandrinum Erben & Brullo was described from the island of Karpathos and said to resemble *L. astypaleanum* Erben & Brullo and *L. kirikosicum* Erben & Brullo. All are likely to represent local populations of *L. roridum*.

Limonium microcycladicum Erben & Brullo is a dubious species known from two collections. According to the description it is characterized by relatively broad, glaucous leaves and long spikes. *L. roridum* occurs in the same area and the differences seem obscure.

Limonium minoicum Erben & Brullo is a dubious species known from a small area in S Kriti and differing from *L. creticum* R. Artelari only in the sometimes 3-veined leaves and marginally shorter spikes.

Limonium monolithicum Erben & Brullo was described as endemic to a small area in W Rodos and is scarcely different from *L. sitiacum* Rech. f., showing only vague statistical differences in the dimensions of leaves, bracts and calyces. Plants at LD collected elsewhere in Rodos as well as Simi were identified as *L. sitiacum* by Runemark and evidently belong to *L. monolithicum* as defined by Erben & Brullo. Runemark counted 2n=43 in one of the collections from Simi.

Limonium oligotrichum Brullo & Erben was described from Karpathos. It is scarcely distinguishable from *L. amopicum* Erben & Brullo, and both are doubtfully distinct from *L. sitiacum* Rech. f.

Limonium pagasaeum Erben & Brullo is a name given to an octoploid population from eastern Thessaly. Morphologically it falls within the range of variation of *L. narbonense* Mill.

Limonium phitosianum R. Artelari, described from the island of Zakinthos, is scarcely different from *L. damboldtianum* Phitos & R. Artelari (from Kefallinia).

Limonium pigadiense (Rech. f.) Rech. f. was described from Karpathos. According to Artelari & Georgiou (2003: 488) it does not differ consistently from *L. ocymifolium* (Poir.) Kuntze.

Limonium proliferum (d'Urv.) Erben & Brullo is very similar to *L. roridum* and likely to be a variety or habitat modification of the latter. In the long descriptions provided by Erben & Brullo (2016: 136-137 and 48), all morphological characters are identical except for the fact that *L. proliferum* tends to have more proliferating stems. Boissier (1879: 862) treated this taxon as *Statice rorida* var. *prolifera* which seems appropriate.

Limonium recticaule Erben & Brullo, described from northern Kriti, is doubtfully distinct from *L. roridum*.

Limonium rigidum Alf. Mayer was based on a single collection very close to the type locality for *L. creticum* R. Artelari. The two are likely to represent individual variation within one species.

Limonium runemarkii Rech. f. is a taxon of uncertain status, described from S Evvia. It has been regarded as synonymous with either *L. ocymifolium* (Poir.) Kuntze or *L. sieberi* (Boiss.) Kuntze.

Limonium samium Erben & Brullo is dubious species described from NW Samos. It was said to differ from *L. vanandense* Erben & Brullo in the longer spikelets, tuberculate bracts and more densely hairy calyces. Both are doubtfully distinct from *L. sitiacum* Rech. f.

Limonium schinousae Erben & Brullo and ***L. thirae*** Erben & Brullo are dubious species based on very limited material from the Kikladian islands of Schinousa and Thira (Santorin), respectively. They are likely to fall within the range of variation of *L. roridum*.

Limonium sirinicum Erben & Brullo was based on a single collection from the island of Sikinos. It is scarcely different from *L. doerfleri* (Halácsy) Rech. f. which has been collected several times on that island by Runemark & al. (specimens at LD).

Limonium spreitzenhoferi Erben & Brullo is a local endemic recently described from the island of Kithira. It is doubtfully distinct from *L. sieberi* (Boiss.) Kuntze.

Limonium stenotatum (Rech. f.) Erben & Brullo appears to fall within the range of variation of *L. roridum*.

Limonium taenari Erben & Brullo was described from a single locality at the southern tip of the Mani Peninsula. It was said to be close to *L. proliferum* which in turn is scarcely distinguishable from *L. roridum*.

Limonium vanandense Erben & Brullo is a dubious species described from N Karpathos and said to differ from *L. sitiacum* Rech. f. only in the slightly shorter spikelets (7.5-8.2 mm vs. 8.5-10.0 mm) and calyces (5.5-6.2 mm vs. 6.3-7.3 mm).

Limonium vravronense Erben & Brullo is a dubious species from the eastern coast of Attiki. It is scarcely different from *L. atticum* Erben & Brullo.

Limonium xerocamposicum Erben & Brullo is a dubious species known only from the type collection. It differs from *L. roridum* only in being somewhat smaller in all parts, with narrower leaves and fewer sterile branches. It is likely to represent a depauperate form of *L. roridum*, modified to the hot and dry climate of SE Kriti.

Limonium xiliense Erben & Brullo is a species of uncertain status, described from SE Peloponnisos. It appears very close to *L. elaphonisicum* Alf. Mayer.

Linaria albifrons (Sm.) Spreng. was recorded "in insula Rhodô", as *Antirrhinum albifrons* Sm., in Fl. Graec. Prodr. 1: 432 (1809) and illustrated in Fl. Graeca 6: tab. 588 (1826). It has never been confirmed in Rodos, and Sibthorp's specimen may have been collected in Cyprus. The species occurs from Tunisia to Palestine, Cyprus, and probably further east.

Linaria angustissima (Loisel.) Borbás. Records from the Nestos river gorge (Voliotis 1984a: 162), the Prespa National Park (Pavlides 1985: 251) and Mt Menikion (Karagiannakidou & Kokkini 1987: 282) all need confirmation. The species is similar to *L. vulgaris* Mill. but has somewhat smaller, lemon yellow flowers with relatively longer spur. It is distributed in mountains of south central Europe at least to Croatia in the SE, and it is not unthinkable that it could occur in northern Greece.

Linaria arvensis (L.) Desf. was listed for "Gr" in Fl. Eur. 3: 236 (1972) perhaps going back to a report by Maire & Petitmengin (1908: 162) of "*L. arvensis* var. *flaviflora* Boiss." from Mt Killini. The record is likely to refer to *L. simplex* which has subsequently been collected in this area.

Linaria concolor Griseb. This name has been interpreted differently and was listed as an accepted species in Euro+Med PlantBase. After examining the type, Niketić & Tomović (2008: 622) concluded that it is a synonym of *L. genistifolia* subsp. *sofiana* (Velen.) Chater & D. A. Webb, a taxon occurring in Bulgaria and NE Greece.

Linaria genistifolia (L.) Mill. A tall and large-flowered plant was collected by

Orphanides (no 711, 4.6. 1856; WU) in the foothills of Mt Parnonas (E Peloponnisos). It was listed by Halácsy (1902a: 409) as *L. dalmatica* (L.) Mill. In 2020 it was recollected in the same area by Strid (no. 60599, 6.10. 2020; B, UPA, herb. Strid) and subsequently by Kit Tan & Vold (no. 33155, 31.10. 2020; ATH, C, herb. Kit). Specimens of the two last-mentioned collections have both flowers and ripe capsules from earlier flowering in the summer. They do not match *L. dalmatica* of northern Greece and have provisionally been classified as a form of the variable *L. genistifolia* (L.) Mill., perhaps an undescribed subspecies. Recent collections of *L. genistifolia* by Polymenakos & Pantavos from the foothills of Mt Elikonas (no. 1021, ATH), the lower slopes of Mt Dirfis above Steni (no. 867, ATH), and Kifissia, Attiki (no. 3, ATH) should be critically examined as they may represent the same, southerly taxon (cf. Strid, Kit Tan & Kofinas 2021). For a review of the *L. genistifolia* complex, see Niketić & Tomović (2008).

Linaria reflexa (L.) Desf. was reported by Halácsy (1902a: 411), and the record was repeated by Hayek (1929: 138). The original source is Sibthorp & Smith (1809: 433), "in agro Argolico", which is most probably a mistake. *L. reflexa* is an Italian species unlikely to occur in Greece.

Linum caespitosum Sm. and ***L. doerfleri*** Rech. f. For comments, see Aegean Atlas 1: 392 (2016).

Linum euboeum Bornm. is a taxon of uncertain status, reported from a few localities in C & N Evvia. It may be conspecific with *L. elegans* Spruner ex Boiss.

Linum flavum complex. This taxonomically difficult group is widespread from C & SE Europe to Caucasus, and numerous taxa have been described, especially from the Balkans. For a brief discussion, see Hartvig in Strid (1986: 556-558). Greek plants have provisionally been referred to *L. flavum* subsp. *albanicum* (Janch.) Hartvig. *Linum tauricum* Willd. was recognized as a separate species in Fl. Eur. (2: 207, 1968) as well as Euro+Med PlantBase, but does not appear to be clearly distinguishable from *L. flavum*.

Linum goulimyi Rech. f. is a member of the *L. flavum* complex of uncertain status, known only from the type collection from *Pinus halepensis* woodland in northern Evvia (Rechinger 16564; G, K, S).

Linum hirsutum L. is represented in Greece by subsp. *hirsutum* and subsp. *spathulatum* (Halácsy & Bald.) Hayek. They have completely overlapping distribution areas, and the differentiation is altitudinal rather than geographical, the latter occurring at higher altitude. They differ in quantitative characters and merge into each other at intermediate levels.

Linum olympicum Boiss. was described from the Bithynian Olympus (Ulu Dağ, NW Anatolia). *L. olympicum* subsp. *athoum* Hartvig was based on a single, historical collection (Aucher-Éloy a. 1837, G), supposedly from Mt Athos. The species has never been confirmed here and is likely to be restricted to Anatolia.

Linum perenne L. Plants from limestone mountains of NE Greece, here classified as *L. perenne* subsp. *alpinum* (Jacq.) Stoj. & Stef., are optionally regarded as a separate species, *L. alpinum* Jacq. Reports of the latter from NE Europe (cf. Euro+Med PlantBase) are likely to refer to a different taxon.

Linum phitosianum Christod. & Iatroú is merely a white-flowered form of *L. hellenicum* Iatroú. Apart from the difference in petal colour they are identical in every respect – see descriptions and illustrations in Kit Tan & Iatroú (2001: 198-200). Petals sometimes vary from white to purplish-mauve in the same population (e.g., Strid 60005; UPA, herb. Strid and photos).

Linum punctatum C. Presl. Taken in a wide sense this species occurs in three disjunct areas, with subsp. *punctatum* in Sicily, subsp. *pycnophyllum* (Boiss. & Heldr.) Gustavsson in Greece and S Albania, and subsp. *kurdicum* (P. H. Davis) Greuter & Burdet in SE Anatolia.

Linum strictum L. Most of the Greek plants belong to subsp. *spicatum* (Pers.) Nyman (with flowers crowded on very short branches, forming a spike-like inflorescence), but differences between this and subsp. *strictum* are not clear and varietal rank may be more appropriate.

Linum uninerve (Rchb.) Jáv. was listed as an accepted species in Euro+Med PlantBase, but is probably a synonym of *L. flavum* L.

Linum usitatissimum L. (flax) is only known as a cultivated plant. It was probably derived by selection from *L. bienne* Mill. in the Fertile Crescent during Paleolithic times. In Greece there are scattered records from the mainland and Ionian Islands where it may persist for a short time in ruderal habitats but is scarcely naturalized.

Liquidambar orientalis Mill. Records from Cyprus apparently refer to cultivated and naturalized trees (cf. Browicz 1982b: 42, Tsintides, Hadjikyriakou & Christodoulou 2002: 173).

Lithospermum goulandriorum Rech. f. was described in 1971 based on a collection from Mt Timfi, and has since led a peripatetic existence, being transferred to the genera *Aegonychon*, *Buglossoides* and most recently to *Glandora* (Cecchi & al. 2014). A large-flowered population from Mt Koziakas has been described as *L. goulandriorum* subsp. *thessalicum* Aldén, but falls within the range of variation of the nominate subspecies.

Lolium remotum Schrank was formerly a weed of flax fields in most of Europe. Terrell (1968: 40) cited it from Kriti, without precise locality; there appears to be confirmation. A report from the mouth of the Evros river in the far north-east (Babalonas 1981: 261) is dubious. Sands (1980: 31) listed it from the Ionian island of Paxi, but occurrence here seems unlikely and was not cited on the Flora Ionica website.

Lolium scholtzii Greuter was based on a single collection from the foothills of Mt Taigetos. It is scarcely different from *L. rigidum* Gaudin.

Lomelosia calocephala (Boiss.) Greuter & Burdet was reported from NC Greece by Strid (1998: 167, sub nom. *Scabiosa calocephala* Boiss.), but the specimens have later been revised as the closely related *L. rotata* (M. Bieb.) Greuter & Burdet. *L. calocephala* is an Irano-Turanian species probably absent from Greece.

Lomelosia cretica (L.) Greuter & Burdet. In spite of its name this species is restricted to Sicily, S Italy and the Balearic Islands. It is closely related to *L. albocincta*, *L. hymettia*, *L. minoana* and *L. variifolia*. For an account of this group, see Greuter (1967b: 237-243).

Lomelosia prolifera (L.) Greuter & Burdet was reported from Rodos by Carlström (1987: 87), based on a collection by Finkl (1944). It is otherwise known from SE Anatolia, Cyprus and Lebanon, resembling *L. divaricata* (Jacq.) Greuter & Burdet but white-flowered. Occurrence on Rodos has not been confirmed and seems dubious.

Lomelosia rutifolia (Vahl) Avino & P. Caputo was reported from coastal sand in W Peloponnisos by Georgiadis (1999: 54), sub nom. *Pycnocomon rutifolium* (Vahl) Hoffmanns. & Link. It is a W Mediterranean species doubtfully native in Italy and Greece.

Lonicera caprifolium L. Several literature records from Chalkidiki, Sterea Ellas and the Ionian Islands have been disregarded. They may refer to *L. etrusca* Santi, *L. implexa* Aiton or to cultivated material of *L. caprifolium*.

Lonicera japonica Thunb., a woody climber of East Asiatic origin, is cultivated as an ornamental and has been reported from ruderal habitats in several towns and villages, but is doubtfully naturalized.

Lonicera periclymenum L. is a species of W and C Europe. Scattered records from the Greek mainland, Peloponnisos and Kerkira are all likely to be incorrect, mostly referring to *L. etrusca*.

Lotus aduncus (Griseb.) Nyman. Grisebach spent only a few hours on the island of Thasos on 2nd June 1839 en route from Enos to Athos, but managed to collect several interesting plants including a red-flowered legume ("in pratis maritimis ins. Tassos pr. Casavi [Skala Prinos] sparse"), which was subsequently described as *Tetragonolobus aduncus* Griseb. It was accepted as an endemic species by Drossos (1995b) as well as in Euro+Med PlantBase, but is apparently a colour form of *Lotus gebelia* Vent.

Lotus aegaeus (Griseb.) Boiss. The type specimen was collected by Grisebach "in litori arenoso Thraciae pr. Makri, oppidum insula Samothrake oppositum" one or two days before that of *L. aduncus*. It is a conspicuous plant with large, yellow flowers, fairly common in NE Greece and extending through Anatolia to Syria and N Iraq. However, it is scarcely distinct from the variable *L. gebelia*.

Lotus biflorus Desr., see *Tetragonolobus biflorus* (Desr.) Ser.

Lotus hispidus DC. Records from the East Aegean islands of Kalimnos (Zervou, Raus & Yannitsaros 2009: 170) and Ikaria (Christodoulakis 1996a: 78) apparently refer to *L. angustissimus* L. and *L. subbiflorus* Lag., respectively.

Lotus longisiliquosus R. Roem. Euro+Med PlantBase uses the name form *longesiliquosus*, but the spelling in the original publication is *longisiliquosus*.

Lotus stenodon (Boiss. & Heldr.) Heldr., described from alpine levels on Chelmos and Parnassos, was listed as an accepted species in Fl. Eur. (2: 174, 1968) as well as Euro+Med PlantBase, but appears to fall within the range of variation of *L. corniculatus* L.

Lotus strictus Fisch. & C. A. Mey. See *Dorycnium strictum* (Fisch. & C. A. Mey.) Lassen.

Ludwigia grandiflora (Michx.) Greuter & Burdet has been reported from a single locality by Lake Lysimachia in Sterea Ellas where it was growing together with *L. peploides* (Kunth) P. H. Raven (Kit Tan & al. 2009c: 285). For distinguishing features, see this publication.

Lupinus albus L. Almost all Greek records refer to the native subsp. *graecus* (Boiss. & Spruner) Franco & P. Silva, which was described from the foothills of Mt Taigetos. Subsp. *albus* is cultivated and occasionally naturalized in ruderal habitats. It has white flowers and larger, uniformly pale brown seeds.

Lupinus hispanicus Boiss. & Reuter. Unspecified reports of this species from Greece (e.g., Fl. Turkey 3: 40, 1970) probably refer to *L. gredensis* Gand.

Lupinus micranthus Guss. For nomenclatural reasons this name, which has been widely used, must be replaced by *L. gussoneanus* J. Agardh (see Soldano 1994a).

Lutzia cretica (L.) Greuter & Burdet. Following Al-Shehbaz (2012: 933) *Lutzia* has been reduced to a synonym of *Alyssoides*.

Luzula forsteri (Sm.) DC. Some collections from northern Greece have been identified as subsp. *rhizomata* (Ebinger) Z. Kaplan, a taxon originally described in 1964 from the Ilgaz province in C Anatolia (but not recognized in the Flora of Turkey 9, 1985). Differences are obscure and there appear to be many intermediates.

Luzula luzuloides (Lam.) Dandy & Wilmott. Both subsp. *luzuloides* and subsp. *rubella* (Mert. & W. D. J. Koch) Holub (= subsp. *cuprina* (Asch. & Graebn.) Chrtek & Křísa) have been reported from Greece. They have completely overlapping distributions and the differences are not clear.

Luzula multiflora (Ehrh.) Lej. is a widespread and variable species which has been differently subdivided. Some plants from N Greece have been referred to *L. fallax* Kirschner, *L. taurica* (V. I. Krecz.) Novikov, *L. multiflora* subsp. *multiflora* and *L. multiflora* subsp. *snogerupii* Kirschner. The taxonomy is unsettled and only a collective map can be presented.

Luzula spicata (L.) DC. is a widespread species extending to N Scandinavia and Greenland. It is variable especially in the southern part of its range, with four subspecies recognized in Italy. Greek plants have been called *L. spicata* subsp. *pindica* (Hausskn.) Gamisan, *L. italica* Parl. and *L. bulgarica* Chrtek & Křísa, but the taxonomy is unsettled.

Lycium barbarum L., a native of China, has been reported from scattered localities almost throughout Greece. It is cultivated as an ornamental and may persist for some time in ruderal habitats, but is scarcely naturalized. It differs from the native *L. europaeum* L. and *L. schweinfurthii* Dammer by having leaves broadest at or below the middle, corolla lobes ± equalling tube, corolla throat densely hairy, and filaments with a dense tuft of hairs at base.

Lycium chinense Mill., also a native of China, is similar to *L. barbarum* but more broad-leaved; it is an occasional escape from cultivation.

Lycium intricatum Boiss. is a W Mediterranean species. Greek records refer to *L. schweinfurthii* Dammer.

Lygos, see *Retama*.

Malabaila involucrata Boiss. & Spruner. A collection from Rodos (Rechinger 7177, LD) has been revised as *M. aurea* (Sm.) Boiss. (Carlström 1987: 84). Reports of *M. involucrata* from adjacent areas in SW Anatolia should be critically re-examined. Engstrand (1970a: 388) reported 2n=22 in a collection of *M. involucrata* from Naxos (Runemark & Snogerup 9141, LD); in Fl. Turkey 10: 417 (1988) the locality was incorrectly given as Ikaria.

Malcolmia macrocalyx subsp. ***scyria*** (Rech. f.) P. W. Ball supposedly differs from the nominate subspecies in the single character of having racemes bracteate almost to the apex (vs. mostly at the base). There seems to be continuous variation in this feature.

Malcolmia orsiniana (Ten.) Ten. Greek material has been reported as subsp. *angulifolia* (Boiss. & Orph.) Stork and subsp. *serbica* (Pančić) Greuter & Burdet, but as pointed out by Stork (1972a) as well as in Flora Hellenica (2: 164-165, 2003) there are intermediates in different geographical areas. All Greek plants are better included in subsp. *angulifolia* which occurs also in Albania, North Macedonia and Serbia whereas subsp. *orsiniana* is a taxon of the NW Balkan Peninsula and C Italy.

Malcolmia ramosissima (Desf.) Thell. was listed for Greece in Fl. Eur. (ed. 2, 1: 338, 1993), but the source is not clear. Reports from maritime habitats in NE Greece (Drossos 1992: 106) and the island of Kerkira (Borkowsky 1994: 62) evidently refer to *M. nana* (DC.) Boiss. *M. ramosissima* occurs in the W & C Mediterranean area.

Malus domestica (Suckow) Borkh., the cultivated apple, is probably the result of complex hybridisation between several ancestors including *M. sylvestris* (L.) Mill. For references, see Kurtto, Sennikov & Lampinen (2013: 35-40). It usually differs from *M. sylvestris* in the larger fruits and leaves persistently tomentose beneath.

Malva pusilla Sm. has been reported from scattered localities almost throughout Greece, but all need confirmation and may refer to the similar *M. parviflora* L. All

Willing specimens at B reported as *M. pusilla* (nos. 284340, 285279, 288590, 288687 and 289341) have been revised as *M. parviflora* (A.S., 2021-09-20). *M. pusilla* is mainly a Euro-Siberian species, rare in the Mediterranean area but naturalized in North America, etc.

Malva thuringiaca (L.) Vis. has been divided into subsp. *thuringiaca* and subsp. *ambigua* (DC.) Valdés, based on quantitative differences in leaves and inflorescences. The former is widespread whereas the latter has been recorded from Italy and parts of the Balkan Peninsula. In Greece their distributions overlap completely and there are some intermediates.

Malva verticillata L. Records of this species from the Thessaloniki area, Mt Paikon and the island of Kimolos all need confirmation. It is a native of E Asia, cultivated as a salad plant in C & S Europe and occasionally naturalized. It resembles *M. neglecta* Wallr. but is generally larger, with shorter petals and strongly accrescent calyx.

Mandragora officinarum L. is variable in morphology and flowering time, and has often been subdivided. However, a recent generic monograph (Ungricht, Knapp & Press 1998) recognized only *M. officinarum* in the Mediterranean area, with two additional species further east.

Marrubium alternidens Rech. f. was based on numerous specimens from Iran and Afghanistan, and is apparently conspecific with *M. anisodon* K. Koch, described from "Derbend am kaspischen Meere". The distinguishing feature for both is the fact that the calyx has 5 long and 5 short teeth as opposed to 5 long, equal teeth in the similar *M. vulgare* L. Greek material is variable in this respect. Of the numerous Willing specimens at B identified as *M. alternidens* most have 5 equal teeth, thus matching *M. vulgare*, but in some the teeth are unequal to various extent. There appear to be no consistent differences in indumentum or other features, nor in distribution, and all specimens must be considered to fall within the range of variation of *M. vulgare*. *M. alternidens/anisodon* should be excluded from the Greek (and European) flora; it occurs from eastern Anatolia to Kashmir and central Asia.

Marrubium friwaldskyanum Boiss. Historical records from W & NW Greece (Halácsy 1894a: 252, Maire & Petitmengin 1908: 174, Vandas 1909: 474) are likely to refer to *M. velutinum* Sm. *M. friwaldskyanum* was described from "Rumelia" and is considered endemic to a small area in SC Bulgaria. It is not unthinkable that it could occur in Greece, but then probably in the NE. It resembles *M. velutinum* but the leaves are dark green and only sparsely pubescent above, and the corollas white rather than pale yellow.

Marrubium pestalozzae Boiss. Records from NW Greece (Chitos 2009: 75 and Bouharmont unpubl.) probably refer to *M. velutinum*. *M. pestalozzae* was described from SW Anatolia, and its status is uncertain. Cullen (in Fl. Turkey 7: 169-170, 1982) regarded it as synonymous with *M. parviflorum* Fisch. & C. A. Mey.

Marsilea aegyptiaca Willd. was described from Egypt and is scattered in Algeria, Tunesia, tropical Africa, India and C Asia. The total distribution is uncertain. There are reports also from Spain and eastern Europe (Moldavia and Volga delta). The Spanish record probably refers to *M. batardae* Launert, and those from E Europe need confirmation. The single Greek locality on the island of Elafonisos ("Saisonaler Tümpel bei Leptos Kavos") was first reported by Jagel (1992: 23) and is very isolated.

Matthiola euboea Boiss. was based on Aucher-Éloy no. 89 (G, P), supposedly collected on Evvia in 1837. However, the specimen is most probably from Syria (see Al-Shehbaz & Barriera 2019: 20), and the species does not occur in Greece.

Matthiola fruticulosa s. lat. As currently understood *M. fruticulosa* subsp. *valesiaca* (Gaudin) P. W. Ball occurs in two widely disjunct areas: (1) in NW Italy and S Switzerland, and (2) in N Greece and elsewhere in the central Balkan Peninsula. Taxonomy and nomenclature are convoluted, but the Balkan plants may well deserve species rank, and then *M. thessala* Boiss. & Orph. appears to be the earliest name.

Medicago arborea subsp. ***strasseri*** (Greuter & al.) Sobr.-Vesp. & Ceresuela, described from north central Kriti, differs from the nominate subspecies only in the slightly smaller and more curved fruits and may represent native plants from Kriti. Plants with larger and less curved fruits (subsp. *arborea*) also occur on the island, and are perhaps naturalized from cultivation. Intermediates are found, e.g., on Karpathos.

Medicago bonarotiana Arcang. (1876) was described from the vicinity of Florence in N Italy, "in località molto prossime alla villa dell'illustre Michelangelo Buonarroti". It was later recombined, by the author himself, as a variety of *M. blancheana* Boiss. (1856) which had been described from cereal fields near Saida in Syria [now Lebanon]. The East Aegean records are referable to this variety, which is characterized by the absence of spines on the legume; it has also been reported from NW Turkey and (probably in error) from Spain. The taxonomic value, if any, of *M. bonarotiana* is not clear. Euro+Med PlantBase as well as the Flora d'Italia website list it as a mere synonym of *M. blancheana*. It does not appear to have been confirmed in Italy and may have beeen a casual introduction. *M. blancheana* s. str. occurs in Anatolia and southwards to Palestine.

Melampyrum barbatum Waldst. & Kit. ex Willd. was listed for ?Gr in Fl. Eur. (3: 254, 1972), but the source for this record is obscure, and the species is unlikely to occur in Greece. The nearest localities seem to be unconfirmed reports from N Albania (Barina 2017: 331).

Melampyrum fimbriatum Vandas. In Herbarium B are several collections by Willing from NC & NE Greece identified as *M. fimbriatum*, but only one of them (no. 21539 from Mt Vermion) matches authentic *M. fimbriatum*, represented at B by Vandas s.n., VII 1891, collected at *locus classicus* in Bosnia-Herzegovina ("in dumetis montis Gliova prope Trebinje") two years after the description of the species. All other Willing collections named *M. fimbriatum* represent forms of *M. arvense* L. For a photo of the rare *M. fimbriatum*, see http://flora.nhm-wien.ac.at/Seiten-Arten/Melampyrum-fimbriatum.htm.

Melia azedarach L., a native of SE Asia, is commonly planted as a street tree, but is scarcely naturalized in Greece, although it is reported to be invasive in the S United States (South Carolina and Georgia). It is a small tree with compound leaves and lilac or purple flowers in large panicles. The fruit is a subglobose, yellow drupe 10-20 mm in diam.

Melica ciliata L. was divided by Hempel (2012) into four poorly defined subspecies, which all meet in Greece, viz. subsp. *ciliata*, subsp. *glauca* (F. W. Schultz) K. Richt., subsp. *magnolii* (Gren. & Godr.) K. Richt. and subsp. *taurica* (K. Koch) Tzvelev. Most of the Greek specimens are difficult or impossible to identify to subspecies.

Melica cretica Boiss. & Heldr. is scarely distinct from *M. ciliata* and probably best regarded as an ecotype of the latter restricted to moderate and high altitude in Kriti. According to Hempel (2012), *M. cretica* occurs in the Balkan Peninsula and much of Anatolia. It was divided into five very poorly defined subspecies, three of which were reported from Greece (subsp. *cretica*, subsp. *greuteri* W. Hempel and subsp. *monticola* (Prokudin) W. Hempel), as well as several varieties and formae.

Melica cupanii Guss. A record from Mt Boumistos (Akarnanika Ori) by Vlachos, Christodoulakis & Kamari (2003: 433) needs confirmation. The species is otherwise restricted to Italy, Sicily and Tunisia.

Melilotus altissimus Thuill. Several of the localities are based on literature and need confirmation. The species may be more common than indicated by the records, having not always been distinguished from *M. officinalis* (L.) Lam.

Melilotus polonicus (L.) Pall. has been reported once from nomos Prevezis (Willing 104054, det. P. Lassen), perhaps as a casual. It resembles *M. officinalis* but has shorter racemes with only 4-9 flowers, and larger, lanceolate to oblong-rhomboid legume. It is a species of E Europe and WC Asia.

Melilotus wolgicus Poir. has been reported once from a ruderal habitat in nomos Prevezis (Raus, Willing & Yena 2010: 197). It is native in steppe, usually on saline soil, from Ukraine and SE Russia to C Asia. It resembles *M. albus* Medik. but has longer pedicels and smaller corollas.

Mentha arvensis L. has been reported from W Greece ("Champs cultivés dans la plaine de Mytikas") by Maire & Petitmengin (1908: 183). The record may refer to *M. pulegium* L.

Mentha pulegium L. Kokkini (1983) recognized subsp. *cephalonica* (Heinr. Braun) Kokkini, subsp. *erinoides* (Heldr.) Kokkini and subsp. *pulegioides* (Sieber) Kokkini in Greece, but the differences are obscure and only a collective map can be presented.

Mespilus germanica L. (*Crataegus germanica* (L.) Kuntze) is native from SE Bulgaria through N Anatolia to Caucasus and Elburz (cf. Browicz 1982b: 48-49, 146); it has long been cultivated and naturalized in Europe. In Greece there are scattered records from roadsides and ruderal habitats. One record from the far north-east (mixed deciduous woodland 7 km W of Pentalofos, Strid & Hartvig, obs. 2018-05-11) may refer to a native occurrence. *M. germanica* is a deciduous shrub with obovate leaves, solitary white flowers and a pyriform, brownish fruit.

Micromeria acropolitana Halácsy was first discovered by Maire & Petitmengin on 30 August 1906 at one of the most iconic landmarks in the world: "Athènes, rocailles calcaires sur l'Arcopole". It remained elusive for a century until rediscovered by Gregory Tsounis and Lambros Tsounis (see Tsounis, Tsounis & Kit Tan 2009; Kit Tan, Tsounis & Tsounis 2010; Šiljak-Yakovlev & al. 2011). It appears most closely related to *M. microphylla* (d'Urv.) Benth., which was described from Malta and has been incorrectly reported from Kriti and Karpathos.

Micromeria cremnophila Boiss. & Heldr. Hartvig (in Mount. Fl. Greece 2: 134-135, 1991) considered this species to be primarily a mountain plant. With more material now available the majority of collections are from altitudes between c. 400 and 1400 m. Subsp. *cremnophila* is endemic to the SW Balkan Peninsula. Two additional subspecies have been described from S & E Anatolia.

Micromeria cristata (Hampe) Griseb. was described from "Rumelia", based on a collection by Frivaldszky. Greek plants belong to subsp. *cristata*. The species is very variable further east, with five subspecies being recognized in Fl. Turkey 7: 345-346 (1982).

Micromeria hispida Boiss. & Heldr. ex Benth. The type is a collection by Heldreich, 21 April 1846, "in rupestribus pr. Kainurio Chorio (Eparchiae Mirabello) Cretae" (G-BOIS, BP, GOET, P, TCD, WU-HAL). Bräuchler, Ryding & Heubl (2008: 387) considered it conspecific with *M. microphylla* (d'Urv.) Benth. which is scarcely correct (see above). In the "Cretanflora" website *M. hispida* was listed also from Lefka Ori, but no specified record from this area has been seen by the present author. A related species, *M. aybalae* H. Duman & Dirmenci, has recently been described from the Muğla province in SW Anatolia (see Duman & Dirmenci 2017).

Microthlaspi, one of 12 segregates of *Thlaspi* proposed by Meyer (1973), was recognized by Dimopoulos & al. (2013: 73), with 2 species in Greece. It is now generally regarded as a synonym of *Noccaea* (see Al-Shehbaz 2014, Özüdoğru & al. 2019). Plants from Rodos have been described as *Microthlaspi mediterraneo-orientale* Tahir Ali & Thines (in Taxon 65: 94, 2016); there seems to be little evidence for this and they are here classified as *Noccaea natolica* (Boiss.) Al-Shehbaz.

Milium vernale subsp. ***montianum*** (Parl.) K. Richt. was described from Sicily. Greek records, e.g., from Kriti (Egli 1993b: 260) and Kefallinia (Phitos & Damboldt 1985: 161) need confirmation. Authentic material of subsp. *montianum* from Sicily seems rather different from any Greek specimens.

Mimulus guttatus DC. (*Erythranthe guttata* (DC.) G. L. Nesom), a native of W North America, is cultivated as an ornamental and widely naturalized, including in much of Europe. In Greece it has been collected only once (Strid, 1974) by a stream in the Enipevs valley in the eastern foothills of Mt Olimbos. It can be expected elsewhere in N Greece.

Mimulus moschatus Douglas ex Lindl. (*Erythranthe moschata* (Douglas ex Lindl.) G. L. Nesom) was found in a wet spot in a ravine at Dafni (Athos peninsula) by P. Hartvig & A. Strid on 28th July, 2022. It is a native of North America, previously reported as naturalized in scattered localities in W & C Europe. [The very last amendment to this Atlas, added in proof 2022-08-06].

Minuartia is a genus of c. 150 species, many in the Mediterranean area. In spite of several attempts there is no satisfactory taxonomic treatment, and several species groups are in need of critical revision. The account by Kamari in Fl. Hellen. 1: 170-191 (1997) added to the confusion. Recent phylogenetic speculations (Dillenberger & Kadereit 2014) have led to proposed fragmentation of *Minuartia* into 11(!) genera.

Minuartia bosniaca (Beck) K. Malý was listed for Greece in Euro+Med PlantBase, possibly based on a report by Quézel & Contandriopoulos (1968: 20) from Mt Falakron. No specimens have been seen, and the occurrence needs confirmation.

Minuartia erythrosepala (Boiss.) Hand.-Mazz. Greek records of this species are probably referable to *M. anatolica* (Boiss.) Woronow.

Minuartia glomerata (M. Bieb.) Degen. Plants from C Italy previously known *M. trichocalycina* (Ten. & Guss.) Grande have been recombined as *M. glomerata* subsp. *trichocalycina* (Ten. & Guss.) Conti (see Conti 1997). Reports of *M. glomerata* from the Iberian Peninsula are probably incorrect.

Minuartia graminifolia (Ard.) Jáv. was divided into five subspecies by Conti (2003). Plants from the single Greek locality belong to subsp. *brachypetala* Kamari.

Minuartia hirsuta (M. Bieb.) Hand.-Mazz. was described from the Crimea and has been divided into three poorly characterized subspecies. Most Greek plants are probably referable to subsp. *falcata* (Griseb.) Mattf., based on *Alsine falcata* Griseb. from S Bulgaria, but intermediates between this and subsp. *hirsuta* have also been recorded. Intermediates between *M. hirsuta* subsp. *falcata* and *M. eurytanica* (Boiss. & Heldr.) Hand.-Mazz. occur in the Pindos.

Minuartia hybrida (Vill.) Schischk. The map includes records identified as *M. lydia* (Boiss.) Bornm.

Minuartia kamariana Greuter was based on a single collection from SE Peloponnisos. It is a synonym of *M. attica* (Boiss. & Spruner) Vierh.

Minuartia kitanovii Panov is a dubious species based on a single collection from Mt Falakron (cf. Jalas & Suominen 1983: 46).

Minuartia lydia (Boiss.) Bornm. Greek material called *M. lydia* (Kamari in Fl. Hellen. 1: 175-176) is probably referable to *M. hybrida* and falls entirely within the geographical range of the latter. The true *M. lydia* is likely to be confined to Anatolia and Lebanon and is doubtfully distinct from *M. mesogitana*.

Minuartia mesogitana (Boiss.) Hand.-Mazz. Greek material has been divided into three subspecies, of which subsp. *kotschyana* (Boiss.) McNeill and subsp. *mesogitana* are scarcely distinguishable. The third, subsp. *velenovskyi* (Rohlena) McNeill, appears more distinct and is sometimes regarded as a separate species under the name *M. velenovskyi* (Rohlena) Hayek or *Alsine orphanidis* Boiss. The taxonomy is, however, highly preliminary.

Minuartia pichleri (Boiss.) Maire & Petitm. The type is from Mt Killini where it was collected by Pichler in 1876 ("in fiss. rup. mtis Kyllene"), but has not been confirmed. Dimopoulos and Zarkos are not aware of any recent records from this mountain (pers. comm., 2018-12-11). Pichler collected in the same year on Mt Taigetos, and a mix-up of labels may have occurred. *M. favargeri* Iatroú & T. Georgiadis, described from the vicinity of Moni Elonis (S foothills Mt Parnonas) does not appear to differ consistently from *M. pichleri*.

Minuartia recurva (All.) Schinz & Thell. Densely caespitose, alpine forms have been called subsp. *juressi* (Schltdl.) Mattf. or subsp. *condensata* (C. Presl) Greuter & Burdet, but there are gradual transitions between these and subsp. *recurva*.

Minuartia rumelica Panov was listed for Greece in Euro+Med PlantBase. It is a species of uncertain status, described from the vicinity of Plovdiv in S Bulgaria. No Greek records have been located.

Minuartia stellata (E. D. Clarke) Maire & Petitm. Records from S Albania (Barina 2017: 199, Vangjeli 2017: 159) are likely to refer to *M. pseudosaxifraga* (Mattf.) Greuter & Burdet.

Molinia caerulea (L.) Moench. Taxonomy at subspecies level is not clear. Most of the Greek records refer to subsp. *arundinacea* (Schrank.) K. Richt. which occurs in wet grassland and woodland by streams over limestone or serpentine. A few records from base-poor fens in northernmost Greece (Voras, Rodopi) appear to refer to subsp. *caerulea*. See also comments in Strid (2016, 1: 461-462).

Molopospermum peloponnesiacum (L.) W. D. J. Koch was based on *Ligusticum peloponnesiacum* L., Sp. Pl.: 250 (1753). The epithet goes back to Mattioli, a 16th century author. In spite of its name this species has never been found in Greece, but is restricted to the W & C Mediterranean area and SC Europe.

Moluccella spinosa L. An excellent illustration of this distinctive species appeared in Flora Graeca 6: Plate 567 (1827), the specimen allegedly collected "in monte Parnasso". It was reported from Kriti already by Sieber in 1817 (see Raulin 1869: 533), but this record was questioned by Rechinger (1944a: 513) and omitted in Med-Checklist; however, it was rediscovered in this area by Bergmeier in 2009. The species is apparently rare and vanishing throughout the Mediterranean area, but has recently been reported from localities in SE Peloponnisos (see Kit Tan, Kofinas & Strid 2020: 560-561). Strangely, in the protologue (Species Plantarum 2: 587, 1753) Linnaeus recorded "Habitat in Moluccis" (referring to the Moluccan or Maluku Islands in Indonesia), apparently based on a misinterpretation of the text in Dodones, Stirpium Historiae Pemptades (1583: 92-93).

Monotropa hypopitys L. Glabrous plants (var. *glabra* Roth) are scattered throughout the Greek range of the species. In central Europe the latter are said to differ also in the relatively broader capsule and diploid chromosome number (2n=16 vs. 2n=48), and are sometimes referred to a separate species, *M. hypophegea* Wallr. Elsewhere the differences seem to be less clear-cut.

Moraea mediterranea Goldblatt. Records from the islands of Kefallinia and Lefkas (Bareka & al. 2018: 91) are likely to refer to small forms of *M. sisyrinchium* (L.) Ker Gawl. The generic name is not derived from Morea (the old name for Peloponnisos), but from Johan Moraeus (1672-1742), a physician in Falun, Sweden, and father-in-law of Linnaeus.

Morus alba L. and ***M. nigra*** L. (white and black mulberry, respectively), both of Asiatic origin, are commonly cultivated for their edible fruits and may persist for some time along roads and in ravines. They are deciduous trees with undivided, ovate leaves, and male or female flowers in pendulous catkins. The fruit is a juicy, broadly cylindrical syncarp 2-2.5(-4) cm. According to one theory, the old name for Peloponnisos, Morea, was derived from μουριά, the Greek name for the mulberry tree. Mulberry leaves, particularly those of *M. alba*, are the sole food source for the silkworm. A modest silk industry flourished in NE Greece in the 19th and early 20th centuries, and there is a small silk museum in Soufli.

Muscari anatolicum Cowley & Özhatay was described from several localities in south central Anatolia (C3 to C6). In the Göteborg Botanic Garden a collection from Mt Klokos, N Peloponnisos (Strid 41151) was cultivated side by side with one from Anatolia (between Akseki to Beysehir, JP 87-77), and the two turned out to be indistinguishable – a disjunction reminiscent of *Juniperus drupacea* or *Biebersteinia orphanidis*. There may well be additional localities for *Muscari anatolicum* in Peloponnisos, but it is easily confused with the widespread *M. neglectum* Guss. ex Ten.

Muscari armeniacum/neglectum. Without notes on flower colour, it may be almost impossible to distinguish herbarium material of *M. armeniacum* Leichtlin ex Baker and *M. neglectum* Guss. ex Ten. The former is less common in Greece, and verified records are mostly in the north-east. At Herbarium B are vast quantities of Willing specimens identified as *M. armeniacum*; they have not been mapped and most of them certainly represent *M. neglectum*. *M. armeniacum* was listed for Peloponnisos by Mermygkas, Zikos & Constantinidis (2021: supplementary file), but no details are available and the record seems dubious.

Muscari heldreichii Boiss. was described from Mt Parnassos and has been reported also from Giona and Vardousia. It is scarcely distinguishable from *M. botryoides* (L.) Mill. and must be regarded as a high-altitude form of the latter.

Muscari kerkis Karlén was described from Mt Kerkis, Samos, and has recently been reported also from the Aydın and Denizli provinces on the opposite Anatolian mainland (Eker & Armağan 2020). Similar plants from Mt Karvouni, Samos, here provisionally classified as *M. neglectum*, are of uncertain status and are currently being investigated. They approach *M. kerkis* and differ from *M. pulchellum/neglectum* in the seeds which are c. 1.5 mm, somewhat compressed, dull black and reticulate-pitted (vs. c. 2 mm, almost spherical, shiny black and smooth in the latter). The seeds of *M. kerkis* are not known. [Thanks are due to Kit Tan for these observations].

Muscari weissii Freyn. Taxonomy of this and related taxa is somewhat unsettled. Two similar species, *M. elmasii* Yıldırım and *M. muglaensis* Eker, H. Duman & Yıldırım, have recently been described from SW Anatolia.

Myosotis margaritae Štěpánková was described from Bulgaria, and a single Greek collection was also cited (Rechinger 3184 from Mt Voras). The species is of uncertain status and seems very close to *M. sicula* Guss.

Myosotis paucipilosa (Grau) Ristow & Hand. Following Ristow & Hand (in Hand 2015: 251) this is regarded as a separate species rather than a subspecies of *M. refracta* as originally described. It is nested within the distribution area of the latter and differs in some rather distinct morphological characters as well as basic chromosome number (2n=20 vs. 2n=44). Being a small and inconspicuous plant it may well be overlooked and more common than indicated by the few records.

Myosotis ramosissima subsp. ***gracillima*** (Loscos & J. Prado) Rivas Mart. is a taxon of dubious status listed for KK by Dimopoulos & al. (2013: 69) and subsequently reported also from Lesvos (EAe) and Mt Giona (StE).

Myosotis refracta subsp. ***aegagrophila*** Greuter & Grau, described from Lefka Ori, is probably a habitat modification of subsp. *refracta*.

Myosotis secunda A. Murray was listed for Greece in Euro+Med PlantBase. The source for the record is not clear. It is a western European species unlikely to occur in Greece.

Myosotis sylvatica subsp. ***subarvensis*** Grau. The status of this taxon remains unsettled (cf. Mount. Fl. Greece 2: 53-54, 1991).

Myosurus heldreichii Heldr. ex H. Lév. and ***M. minimus*** L. The maps are provisional, and all Greek material needs critical examination. *M. heldreichii* was based on specimens collected by Heldreich in 1901 on the island of Dilos and posthumously published by Léveillé in 1902 (cf. Strid in Fl. Hellen. 2: 70, 2003). Limited evidence suggests that this is a lowland plant whereas the widespread *M. minimus* is found in Greece by vernal pools at rather high altitude. In *M. heldreichii* the scape is very short and swollen, but also in Greek plants identified as *M. minimus* the scape is generally shorter than in C European material of the latter. Plants from the shallow lake of Limni Dasiou at c. 1500 m on Mt Killini first reported as *M. sessilis* M. Watson (Zarkos & al. 2017b: 324) were subsequently revised as *M. minimus* (Zarkos & al. 2018: 290), and the latter approach is followed here.

Myrtus communis L. Plants with whitish fruits are scattered throughout the distribution range of the species, usually growing together with the normal variety with blackish-blue fruits. Such plants were known already to Linnaeus and named var. *tarentina* L. In Greece they have been reported at least from SE Peloponnisos (several records incl. Strid & al. 55414; G, LD), E Kriti (Strid 58120; G, herb. Strid) and Lesvos (Hansen & Nielsen 8802; B, C). There are scarcely any other consistent differences, and there is likely to be a simple genetic background for this conspicuous variation (cf. Kit Tan, Kofinas & Zieliński 2021: 145-146).

Najas gracillima (Engelm.) Magnus, ***N. graminea*** Delile and ***N. orientalis*** Triest & Uotila are alien weeds of rice fields, each reported from a few localities in Greece, the two first-mentioned in the north-east and the last-mentioned in Peloponnisos (see Raus & Raabe 2002a: 296). Their degree of naturalization is not clear.

Najas marina L. A report of subsp. *armata* (H. Lindb.) Horn from the Propokos lagoon in nomos Achaias (Christia & al. 2011: 175) needs confirmation. All other Greek records are likely to represent subsp. *marina*.

Narcissus poeticus L. Subsp. *poeticus* and subsp *radiiflorus* (Salisb.) Baker differ in shape of perianth segments, but there are many intermediates and no geographical separation. In large populations in the vicinity of the Katara Pass, individuals matching one or the other subspecies (as well as intermediates) can be found side by side.

Narcissus papyraceus Ker Gawl., a species of the W Mediterranean area, has been reported from the Ionian Islands and recently also from Rodos, but is doubtfully naturalized.

Narcissus tazetta L. All native Greek plants are likely to belong to subsp. *tazetta*. Scattered reports of subsp. *aureus* (Jord. & Fourr.) Baker and subsp. *italicus* (Ker Gawl.) Baker refer to cultivated and doubtfully naturalized plants.

Nasturtium microphyllum Rchb. was listed for Greece in Euro+Med PlantBase, probably in error. It is a taxon of W & C Europe, sometimes included in *N. officinale* W. T. Aiton.

Nepeta argolica complex. *N. argolica* Bory & Chaub. (1832-33) has priority over *N. sibthorpii* Benth. (1834). *N. sibthorpii* Benth. ssp. *tumeniana* T. Dirmenci has recently been described from Kaz Dağ in NW Anatolia; its relations to the Greek taxa are not clear.

Nepeta spruneri Boiss. Willing 280813, 281696 and 281730 (B) from the foothills of Mt Falakron have been reported as *N. spruneri* which is otherwise restricted to the range of mountains from Sterea Ellas to S Albania. Plants photographed in this area (rocky calcareous grassland by road from Volakas to the ski centre, 1400 m) by the present author, 2018-07-02, are somewhat deviating and may represent an undescribed taxon.

Nepeta ucranica L. A record from Sidironeri (nom. Dramas) by Goulimis (1956a: 13) is probably incorrect.

Neurotropis platycarpa (Fisch. & C. A. Mey.) F. K. Mey. was recognized by Dimopoulos & al. (2013: 74). It is now classified as *Noccaea platycarpa* (Fisch. & C. A. Mey.) Al-Shehbaz (see Al-Shehbaz 2014, Özüdoğru & al. 2019).

Nicandra physalodes (L.) Gaertn., a native of Peru, is cultivated as an ornamental and is casual or semi-naturalized in ruderal habitats in the Mediterranean region and Europe. In Greece it seems to have been first collected on the island of Tinos (Runemark & Engstrand 36417, 25.5. 1968; LD); it has subsequently been reported also from Kriti, Lesvos, Skiros, Kerkira and Milos, and may be on the verge of becoming established. It is somewhat reminiscent of *Alkekengi officinarum* Moench, but has a larger calyx, which is angular at anthesis and divided almost to the base; the corolla is also larger and violet-blue.

Nigella arvensis complex. For comments and references, see Aegean Atlas (1: 491-493, 2016). The group was taxonomically revised by Strid (1970a), based on extensive crossing experiments combined with analyses of morphological variation, breeding systems and distribution patterns. This study was subsequently repeated with molecular methods by Bittkau & Comes (2005, 2009) which resulted in several new terms and acronyms but little or nothing with respect to taxonomy, biogeography and evolution. In a recent taxonomic revision (Dönmez, Uğurlu Aydin & Oybak Dönmez 2021) the authors have completely failed to understand variation and evolution in this group as demonstrated, for instance, by the fact that the two self-pollinating species *N. doerfleri* Vierh. and *N. stricta* Strid were simply listed as synonyms of the outbreeding *N. arvensis* L., although morphologically distinct, reproductively isolated and biogeographically well characterized.

Nigella sativa L. is a "homeless" species probably originating in Anatolia where it is widely cultivated for its aromatic seeds. It was also formerly cultivated in Kriti, the East Aegean Islands and perhaps elsewhere in Greece, but does not appear to have become naturalized.

Nitrosalsola aegaea (Rech. f.) T. A. Theodorova is a new combination recently proposed for the species here called *Caroxylon aegaeum* (Rech. f.) Akhani & Roalson (see Theodorova 2015).

Noccaea boeotica F. K. Mey. was described from Mt Elikonas and has been reported from scattered localities almost throughout the Greek mainland. These plants can scarcely be distinguished from *N. ochroleuca* (Boiss. & Heldr.) F. K. Mey.

Noccaea viridisepala (Podp.) F. K. Mey. is a Balkan endemic related to *N. kovatsii* (Heuff.) F. K. Mey. It has been recorded from mountains in north central Greece, and it is possible that records from Pieria and Vermion mapped as *N. kovatsii* are in fact referable to *N. viridisepala*.

Nonea pulla (L.) DC. All Greek records of this species are likely to refer to *N. atra* Griseb.

Ocimum basilicum L. (basil), a species of tropical origin, is commonly cultivated as a pot herb and may persist for some time in ruderal habitats, but is scarcely naturalized. The word basil is derived from the Greek βασιλεύς, king, as it was said to have grown at the spot where St Helena, mother of the emperor Constantine, discovered the original cross of Christ.

Odontarrhena C. A. Mey ex Ledeb. has often been regarded as a section of *Alyssum*, but recent studies (e.g., Španiel & al. 2015, Cecchi & al. 2018) have resurrected it to separate generic status. It comprises c. 87 species mainly distributed in the Euro-Mediterranean and Irano-Turanian regions. They are perennials, usually with compound subumbellate racemes, and seeds solitary in each loculus. Several species are gregarious in serpentine areas.

Odontarrhena decipiens (Nyár.) L. Cecchi & Selvi. Baldacci 206 (BP, K, WU, WU-HAL) from Mt Smolikas is the type collection for both *Alyssum smolikanum* Nyár. and *A. decipiens* Nyár. Whilst there are several subsequent collections of *A. smolikanum* (*Odontarrhena smolikana*) from Mt Smolikas, there is no confirmation of *A. decipiens* (*Odontarrhena decipiens*). The latter may in fact be endemic to Albania, and a confusion of labels can be suspected.

Odontarrhena sibirica (Willd.) Španiel & al. is interpreted in a broad sense and should probably be subdivided. Plant from maritime sand in Rodos may be referable to *O. borzeana* (Nyár.) D. A. German. The latter was reported also from N Greece in Fl. Turkey 1: 401 (1965), but the basis for this is not clear. Plants from serpentine substrate in the Dadia area, nomos Evrou (e.g., Strid & al. 53079; ATH, G, herb. Strid) are somewhat deviating and may represent an undescribed taxon.

Odontites vernus (Bellardi) Dumort. A single collection from the Rodopi (Strid & al. 13808; B, G, herb. Strid) has been identified by Th. Raus as representing the C European *O. vernus* subsp. *vernus*. All other records appear to refer to *O. vulgaris* Moench, which is also known as *O. vernus* subsp. *serotinus* (Dumort.) Corb. The two are sometimes difficult to distinguish. A single collection from the Volos area (EC Greece) has been listed as *O. vulgaris* subsp. *siculus* (Guss.) Bolliger (see Bolliger 1996: 113), a taxon otherwise reported from scattered localities in Sicily and Albania.

Oenanthe banatica Heuff. Greek records need confirmation. They include a report from the Pisoderi area by Quézel & Contandriopoulos (1968: 29). A collection from the vicinity of Dimarion in nomos Xanthis (Strid & al. 13461; ATH, C, G) distributed as *O. banatica* is likely to represent *O. lachenalii* C. C. Gmel.

Oenanthe peucedanifolia Pollich. After the preparation of the maps this species has been reported from the Rodopi mountains by Charalampidou, Eleftheriadou & Theodoropoulos in Fl. Medit. 32: 83 (2022).

Oenanthe tenuifolia Boiss. & Orph. was described from Mt Chortiatis. However, it is a later homonym for *O. tenuifolia* Thunb., a species from South Africa. The Greek plant has been renamed *O. montis-korthiati* Soldano.

Oenanthe tricholoba Greuter, recently described from Peloponnisos, appears to be a form of *O. pimpinelloides* L. with unusually narrow ultimate leaf lobes.

Oenothera. Some species are known only from one or two records in ruderal habitats and are doubtfully naturalized. They include at least the following:

Oenothera fallax Renner. Recorded from Samothraki (Biel 09.128 and 11.244). Perhaps originating in Europe through hybridization between *Oe. biennis* L. and *Oe. glazioviana* Micheli.

Oenothera laciniata Hill. A single collection from *Populus* plantation SW of Rodopoli in nomos Serron (Willing 155170, B; see Raus 2006b: 725). The species is native in Central and Atlantic North America.

Oenothera lindheimeri (Engelm. & A. Gray) W. L. Wagner & Hoch. Collected in Attiki (Raabe s.n., ATHU, B, MSTR; see Raabe & Raus 2016: 430) and subsequently reported from the islands of Rodos (Galanos 2020: 545) and Thasos (Biel & Kit Tan 2021a: 130). Native in Louisiana and Texas; cultivated as an ornamental, with scattered reports of naturalized plants in Portugal, Spain and Belgium. This species may be on the verge of becoming established in Greece.

Oenothera parviflora L. was listed for NC and NE in Dimopoulos & al. (2013: 115), but the specimens have later been revised as *Oe. indecora* Cambess.

Oenothera stricta Ledeb. ex Link has been reported once from the island of Egina (coll. Issigoni, comm. Kit Tan 2011). It is native of Chile and widely naturalized elsewhere.

Oenothera suaveolens Pers. has been reported from Samothraki (Biel 11.255, priv. herb.) and floodplain of the lower Nestos river (Schuler 00/288, B). It is of uncertain status and perhaps synonymous with *Oe. biennis* L.

Oenothera speciosa Nutt. is native in prairies of C & S USA and Mexico. It is commonly cultivated as an ornamental and has become popular in Greece in recent decades, occasionally persisting for some time in ruderal habitats. It is a somewhat bushy perennial with stems up to 40 cm. The flowers are large and showy, pink with a greenish-yellow centre.

Oloptum M. Röser & H. R. Hamasha has recently been proposed as a new genus to accommodate *Piptatherum miliaceum* (L.) Coss. and its subspecies *thomasii* (Duby) Freitag (Hamasha, Hagen & Röser 2014). The two differ in a single morphological character and have completely overlapping distributions; they are consequently better regarded as varieties.

Onobrychis arenaria subsp. ***cana*** (Boiss.) Hayek was listed for the East Aegean Islands in Euro+Med PlantBase, but records are likely to refer to *O. armena* Boiss. & A. Huet or *O. arenaria* subsp. *lasiostachya* (Boiss.) Hayek.

Onobrychis citrina Kit Tan, Stevanović & G. Vold, ***O. citrinoides*** Kit Tan and ***O. lassenii*** Kit Tan & G. Vold are recently described local endemics from N Greece in the *O. alba* group. See Phytol. Balcan. 22(2): 282 (2016), 26(3): 560 (2020) and 27(3): 395-400 (2022) for details.

Onobrychis degenii Dörfl. was described from North Macedonia not far from the Greek border ("In rupibus arsenicis prope Allchar", 30.6. 1893, Dörfler 149; B, JE). Literature records from N Greece (Goulimis 1956a: 16 from Mt Voras and Mt Vermio; Voliotis 1983c: 164 from Mt Tzena) need confirmation. *O. degenii* is a member of the *O. alba* group with narrowly elliptical, acute leaflets, dense grey-villous indumentum and larger legumes (7-10 mm) with longer spines – cf. Kit Tan & Vold in Phytol. Balcan. 27(3): 399 (2022).

Onobrychis hypargyrea Boiss. was collected on Rodos by Fiori (1923) and subsequently by Goulimis (1948), but is now likely to be extinct in its only Greek locality although fairly widespread in the W half of Anatolia. It belongs to sect. *Hymenobrychis* which is represented by several species in the Irano-Turanian region and by three closely related local endemics in Europe: *O. megalophylla* Griseb. in North Macedonia, *O. aliacmonia* Rech. f. in NC Greece and *O. peloponnesiaca* (Iatroú & Kit Tan) Iatroú & Kit Tan in S Peloponnisos.

Onobrychis oxyodonta Boiss. was reported from the Prespa National Park by Pavlides (1985: 220), but a specimen in the Pavlides herbarium at SPP (Ag. Germanos) belongs to *O. alba* (Waldst. & Kit.) Desv. *O. oxyodonta* was described from C Anatolia and is unlikely to occur in Europe.

Ononis adenotricha Boiss. was listed for Peloponnisos by Mermygkas, Zikos & Constantinidis (2021: supplementary file), probably in error. There are only a few verified Greek records, all in the north-east.

Ononis arvensis L. Literature records from N Greece probably refer to forms of *O. spinosa* L.

Ononis biflora Desf. was described from NW Africa. It occurs in the W Mediterranean area and has also been reported in E Anatolia, Palestine, etc. It remains to be established whether plants from these widely disjunct areas are indeed conspecific. Samuelsson (1934: 15) reported *O. biflora* from Delphi, "inter ruinas specimen unicum inventum", specifically stating that it was "eine für die Flora Griechenlands neue Art". It has not been confirmed in Greece and may have been casual. Willing 31858 (B) from the SW part of the mainland ("2.0 km NNW Vathilopoulo") was originally reported as *O. biflora*, but has subsequently been revised by Constantinidis as *O. ornithopodioides* L.

Ononis natrix L. The *O. natrix* complex is widespread in the Mediterranean area. Most of the Greek records are referable to *O. ramosissima* Desf.

Ononis spinosa L. No less than six subspecies have been reported from Greece, but morphological differences as well as distribution patterns are not clear, and thus only a collective map can be presented.

Ononis talaverae Devesa & G. López was listed for Peloponnisos by Mermygas, Zikos & Constantinidis (2021: supplementary file), but no details are available. The record may refer to *O. ramosissima* Desf. (which was not listed).

Onopordum illyricum subsp. ***cardunculus*** (Boiss.) Franco. Most of the Greek records are likely to refer to this subspecies which, however, is poorly defined.

Onopordum parnassicum Boiss. & Heldr. was described from the lower slopes of Mt Parnassos ("In reg. abietina m. Parnassi prope Gourna"). It was later recombined as *O. acanthium* subsp. *parnassicum* (Boiss. & Heldr.) Nyman. Further studies are needed to determine its taxonomic value, if any.

Onopordum rhodense Rech. f. was described from Rodos, near Salakos. According to Kleinsteuber, Ristow & Hassler (2016: 209) it is a synonym of *O. boissierianum* Raab-Straube & Greuter (nom. nov. for *O. corymbosum* Boiss., non Willk., a species described from SW Anatolia).

Onosma. A comprehensive nomenclatural synopsis has recently been published by Cecchi & Hilger (2021).

Onosma arenaria Waldst. & Kit. is widespread in C & E Europe, and was reported from two localities in Southern Pindos by Chitos (2009: 73). The Greek records probably refer to a form of *O. heterophylla* Griseb.

Onosma bulgarica D. K. Pavlova (2009) was described from serpentine substrate in S Bulgaria, but the name is a later homonym for *O. bulgarica* Velen. (1890). It was renamed *O. pavlovae* Petrova & Kit Tan (2009), but does not appear to be taxonomically different from *O. kittanae* Strid, described in 2003 from serpentine substrate in NE Greece (see Teppner & Karl 2017).

Onosma epirotica Teppner. This species was based on a single collection from the Aoos river gorge near Konitsa. Apart from a small and inconsistent difference in corolla colour, it is identical with *O. heterophylla* Griseb. and cannot be maintained as a separate taxon (cf. Authier 2000c: 159-160).

Onosma erecta subsp. ***malickyi*** Teppner was described from mountains of Peloponnisos. It overlaps with the nominate subspecies morphologically as well as geographically, and cannot be maintained as a separate taxon.

Onosma helvetica (A. DC.) Boiss. Plants discovered by Halácsy in 1893 in the Tsoumerka massif in Southern Pindos were identified as *O. helvetica*, but later described by Hayek (1928a: 88) as *O. halacsyi*. Teppner recollected it in the same area in 1980 (see Teppner 1991: 279-281). Balkan plants must apparently be named *O. pseudoarenaria* Schur, whereas the true *O. helvetica* is a species of the Alps.

Onosma kaheirei Teppner. Some plants from Ikaria key out as intermediate between *O. kaheirei* and *O. aucheriana* DC.

Onosma montana Sm. may be at least partly conspecific with the taxon here called *O. aucheriana* DC., but must be regarded as a *nomen confusum* since two taxa were mixed on the type sheet (which now appears to be lost) and the localities indicated in the Florae Graecae Prodromus (Kriti and Peloponnisos) are undoubtedly incorrect. Plants from E Greece have also been called *O. rigida* Ledeb. The latter was described from Crimea and is probably different from *O. aucheriana* (described from Ulu Dağ).

Onosma pseudoeuboica Teppner & Karl differs from *O. euboica* Rech. f. only in vague quantitative characters and appears to be a local population rather than a separate taxon.

Onosma taurica Pallas ex Willd. (*O. cinerea* Schreb.). Historical reports from Greece need confirmation. No specimens or specified records have been seen.

Onosma viridis (Borbás) Jáv. was listed for Greece in Euro+Med PlantBase. The source of the record is obscure, and it is likely to refer to a form of *O. heterophylla* Griseb.

Ophrys spp. No other genus in the Greek flora has suffered from taxonomic inflation to the same extent as *Ophrys*. In recent decades hundreds of geographical races, local populations and deviating individuals have been named according to the principle "if two plants are different they represent different species". Reasonable species and subspecies concepts were restored by Pedersen & Faurholdt (2007) as well as Kühn, Pedersen & Cribb (2019) which are followed here, with minor amendments. Extensive synonymy is also provided in these publication.

Ophrys aegilica D. Engelen is a name recently given to a local population of *O. sphegodes* subsp. *mammosa* (Desf.) Soó ex E. Nelson from the island of Andikithira (Engelen 2020).

Ophrys cretica (Vierh.) E. Nelson. As currently defined, *O. cretica* comprises subsp. *cretica* and subsp. *karpathensis* E. Nelson (syn. subsp. *ariadnae* (Paulus) H. Kretzschmar), with largely overlapping distributions; they differ rather vaguely in the shape of the stigmatic cavity the lateral lobes of the labellum. Pedersen & Faurholdt (2007: 202 ff.) merged *O. cretica* with the conspicuously large-flowered *O. kotschyi* H. Fleischm. & Soó from Cyprus, which may be better retained as a separate, endemic species.

Ophys fuciflora/holoserica. Different interpretations of nomenclatural technicalities have led to the intermittant use of the names *O. fuciflora* (F.W. Schmidt) Moench and *O. holoserica* (Burm. f.) Greuter for the same taxon. The Atlas follows Pedersen & Faurholdt (2007) and Kühn, Pedersen & Cribb (2019) using the former name, whereas Dimopoulos & al. (2016) adopt the latter.

Ophrys scolopax subsp. ***phrygia*** (H. Fleischm. & Bornm.) H. Æ. Pedersen & P. J. Cribb has recently been recognized as a taxon of NE Greece (Chios, Lesvos and Evros) and S Anatolia (see Kühn, Pedersen & Cribb 2019: 304).

Opuntia spp. In addition to the widespread *O. ficus-indica* (L.) Mill. the following species have recently been reported in Greece:

O. humifusa Raf.: Islands of Zakinthos, Kefallinia, Poros and Thasos, Lake Vouliagmeni (W of Loutraki) and foothills of Mt Falakron.

O. microdasys (Lehm.) Pfeiff.: Island of Salamis and Limnos.

O. robusta H. L. Wendl. ex Pfeiff.: Cape Sounion.

O. rufida Engelm.: Island of Milos.

O. stricta Haw.: Chalkidiki.

O. vulgaris Mill.: Attiki and island of Milos.

All are cultivated and doubtfully naturalized.

Orchis anatolica Boiss. was listed for Peloponnisos by Mermygkas, Zikos & Constantinidis (2021: supplementary file), but no details are available.

Orchis pinetorum Boiss. & Kotschy. This name has frequently been applied to slender plants with unspotted leaves and lax inflorescence, tending to grow in open coniferous woodland. According to Tsiftsis & Antonopoulos (2017: 449) they fall within the range of variation of *O. mascula* (L.) L.

Origanum* ×*intercedens Rech. f. is the hybrid between *O. onites* L. and *O. vulgare* subsp. *hirtum* (Link) A. Terracc. It has been reported from localities in north central Kriti.

Origanum ×lirium Heldr. ex Halácsy is probably the hybrid between *O. scabrum* Boiss. & Heldr. and *O. vulgare* subsp. *hirtum* (Link) A. Terracc. It is known from a few collections on rocky limestone slopes at 500-1300 m on Mt Dirfis and Mt Xerovouni (Evvia) as well as Taigetos and Parnonas. Kalpoutzakis & Constantinidis (2005: 123-124) did not observe independent reproduction of *O.× lirium* and concluded that it depends exclusively on the co-existence of the parents. For notes on *O. scabrum*, see Aegean Atlas (1: 365, 2016).

Origanum majorana L. is native to Cyprus and adjacent parts of S Anatolia. It is widely cultivated as a medicinal and culinary herb (sweet majoram). A collection from Naxos (Runemark 811, LD) probably represents a garden escape.

Orlaya daucoides Greuter. The nomenclature of this species is very convoluted. It has been named *O. kochii* Heywood (1960) and later *O. daucoides* Greuter (1967), but according to a recent analysis (Reduron 2008a: 1918-1919) the correct name is *O. platycarpos* W. D. J. Koch.

Ornithogalum. Taxonomy and distribution is unsettled for many groups in this difficult genus. The extensive collections by Willing at B need critical examination and have for the major part not been mapped.

Ornithogalum alatum Turrill was described from nomos Xanthis in NE Greece. It has been interpreted as a synonym of *O. wiedemannii* Boiss., although listed as an accepted species in Euro+Med PlantBase, recorded from Greece, Bulgaria and Turkey-in-Europe.

Ornithogalum amphibolum Zahar. A record from Mt Paikon (Athanasiadis & Drossos 1990: 93) is probably incorrect. The species was described from Romania and has been doubtfully reported also from Bulgaria.

Ornithogalum collinum subsp. ***rhodium*** Speta is taxon of uncertain status, based on cultivated material from a single locality in northern Rodos.

Ornithogalum graecum Zahar. The type of this species, a collection from Mt Killini, has been identified by W.T. Stearn (1989) as *O. exscapum* Ten.

Ornithogalum gussonei Ten. The map is highly preliminary. *O. gussonei* is often confused with *O. exscapum* Ten. and *O. kochii* Parl. Records of *O. collinum* Guss. in older literature frequently refer to *O. gussonei*.

Ornithogalum orthophyllum Ten. Greek records of this species (Voliotis 1983c: 169 from Mt Tzena, Pavlides 1985: 283 from Mt Varnous) probably refer to *O. kochii* Parl.

Ornithogalum pascheanum Speta was described from NW Anatolia (shores of Lake Bolu). It was included in the Greek checklist (Dimopoulos & al. 2013: 102) based on Snogerup 11030 (LD) from Chios identified as *O. pascheanum* but probably belonging to *O. pumilum* Zahar.

Ornithogalum pluttulum Speta was described from Turkey-in-Europe and reported also from NE Greece (Chalkidiki) and SE North Macedonia. It is probably a synonym of *O. refractum* Kit. ex Schltdl.

Ornithogalum sibthorpii Greuter, *nom. ign.* has been reported from several localities in Greece. The name was coined to replace *O. nanum* Sm. (1809), non Thunb. (1794). *O. nanum* Sm. was based on a very young specimen illustrated on Plate 333 of Flora Graeca (1823). The plate matches 3-4 species of *Ornithogalum*, and the associated description and habitat reference ("in Arcadiâ, et prope Abydum") are inconclusive. The identity of *O. sibthorpii/nanum* is therefore anybody's guess and the names should not be used.

Ornithogalum sigmoideum Freyn & Sint. was described from NE Antolia (A7 Gümüşane) and has been interpreted as conspecific with *O. sibthorpii/nanum* (Fl. Turkey 8: 239, 1984).

Ornithogalum umbellatum/divergens. Taxonomy and nomenclature is unsettled (cf. Dimopoulos & al. 2016: 343). High-altitude records from the north have been omitted; many of them are likely to refer to *O. kochii* Parl.

Orobanche artemisiae-campestris Gaudin (1829) was described from Switzerland and is mainly a species of C Europe. *O. loricata* Rchb. (1829) is a synonym published slightly later. Greek records of *O. artemisiae-campestris* (e.g., Dimopoulos & al. 2013: 118) need confirmation and may refer to *O. picridis* W. D. J. Koch. The two have frequently been regarded as synonymous, but are retained as separate species in the Orobanchaceae online index.

Orobanche ballotae A. Pujadas has recently been reported from one locality in NE Peloponnisos where it was parasitic on *Lamium garganicum*; see Rätzel, Ristow & Uhlich (2019: 435-436). It is otherwise known only from Spain, but may be more widespread.

Orobanche canescens C. Presl. Scattered reports of this species all need confirmation and may refer to *O. pubescens* d'Urv. *O. canescens* is mainly an Italian species, but records from Kefallinia (Flora Ionica website) and Kerkira (map in Domina, Uhlich & Barone 2022: 94) are feasible.

Orobanche cernua Loefl. is a widespread species described from Spain. Reports from NW Greece and from a few Aegean islands need confirmation.

Orobanche litorea Guss. was described from Sicily and has been reported from scattered localities in Portugal, Tunisia, S France, Sardinia and mainland Italy (see Rätzel & Uhlich 2018). It was tentatively included in the Aegean Atlas (Strid 2016, 1: 420, sub nom. *O. fuliginosa* Reut. ex Jord.), based on a few records from the Aegean area which have subsequently been questioned. At least some records of *O. grisebachii* Reut. from coastal areas in Kriti and Karpathos are likely to represent *O. litorea*, and the latter may also occur in coastal sand dunes on Kefallinia. Scattered Greek records of *O. fuliginosa* are all dubious, For the time being it is not possible to produce a map with any degree of confidence.

Orobanche rapum-genistae Thuill. was reported by Lafranchis & Sfikas (2009, 2: 144-145). The record refers to a locality in the Rodopi Mts, along the path from Frakto forest station to the Virgin Forest, above Stenorema, c. 1300 m, parasitic on *Genista* sp. (probably *G. carinalis*), phot. Lafranchis, 28.07. 2008. The record is likely to be correct (fide H. Uhlich, in litt. 13.11. 2020). This is the only locality in Greece; confirmation and herbarium specimens would thus be welcome. *O. rapum-genistae* is widespread in W Europe and the Mediterranean area (cf. Foley 2000c).

Otospermum glabrum (Lag.) Willk. was reported by Vogt & Willing (2015: 452) from a locality in SW Peloponnisos, growing in ruderal vegetation in a road construction area, perhaps as a casual introduction. It is native in the south of the Iberian Peninsula and in the Maghreb countries (Morocco, Algeria and Tunisia).

Oxalis corymbosa DC. and ***O. debilis*** Kunth. There are scattered records of these species from ruderal habitats throughout Greece, but the few specimens seen do not have the mass of sessile bulblets characteristic for *O. corymbosa/debilis* which is occasionally naturalized in W Europe. It is possible that all Greek records refer to *O. articulata* Savigny.

Oxalis exilis A. Cunn., a native of Australia and New Zealand, has been collected once, in a ruderal habitat on the island of Andros (Snogerup 11722, LD). It looks rather like a small form of *O. corniculata* L. and is occasionally naturalized in W Europe, doubtfully so in Greece.

Oxytropis purpurea (Bald.) Markgr. (1927) was based on *O. pilosa* var. *purpurea* Bald. The type of the latter is Baldacci 272, collected in 1897 in northern Albania ("in saxosis subalp. m. Šeint distr. Oroši"). Guiseppi 47 (K), collected 26.6. 1929 on Mt Olimbos, was described as *O. olympica* by Turrill (1930a: 122). Having discovered that this name is a later homonym for *O. olympica* St John (from Mt Olympus, Wash., USA!) Turrill (1935a: 54) changed the name to *O. thessala*. However, the Greek and Albanian plants are clearly conspecific, and the oldest name at species level is *O. purpurea*.

Paeonia mascula subsp. ***russoi*** (Biv.) Cullen & Heywood. Hong (2010) distinguished plants from Sicily and Calabria as *P. mascula* subsp. *russoi* and similar plants from Corsica, Sardinia, the Ionian islands and the adjacent mainland of western Greece as *P. corsica* Sieber ex Tausch. The distribution is odd, and I fail to see any significant differences between plants from Sicily (called *P. mascula* subsp. *russoi* by Hong) and those from Kefallinia and Lefkas (called *P. corsica*). Greek plants have provisionally been mapped under the latter name.

Paeonia saueri D. Y. Hong, X. Q. Wang & D. M. Zhang was tentatively listed as an accepted species by Dimopoulos & al. (2013: 119). It was described from Mt Pangeon and is scarcely different from *P. peregrina* Mill. (see Stearn, Davis & Kit Tan 2012).

Panicum dichotomiflorum Michx., a native of North America, has been reported once from an irrigated field in Kriti (Böhling & Scholz 2003: 60). It resembles *P. capillare* L., but is generally larger, with glabrous leaf sheaths and lower glume only 1/4 as long as floret.

Panicum hygrocharis Steud., a species of E & S Africa, was listed for KK in Dimopoulos & al. (2013: 127), probably in error. No record from the Cretan area has been located, and the species was not included in Euro+Med PlantBase.

Papaver. Banfi & al. (2022) have recently proposed to transfer the species of sect. *Argemonidium* – in Greece *P. apulum* Ten., *P. argemone* L., *P. davisii* (Kadereit) M. V. Agab., *P. nigrotinctum* Fedde and *P. virchowii* Asch. & Sint. ex Boiss. – to the genus *Roemeria*.

Papaver argemone L. There are several Greek records of the European *P. argemone* s. str., but those seen by the author represent *P. nigrotinctum* Fedde. The two are not always distinct, and further studies are needed of Balkan populations (cf. Kadereit in Strid & Kit Tan 2003: 91).

Papaver argemone subsp. ***davisii*** Kadereit was described from W Anatolia. It was subsequently raised to species level, as *P. davisii* (Kadereit) M. V. Agab., and has been reported from one locality in Peloponnisos and one in Sterea Ellas. Greek material needs confirmation and may rather be referable to *P. nigrotinctum*.

Papaver confine Jord. or *P. dubium* subsp. *confine* (Jord.) Hörandl is a tetraploid taxon of uncertain status in the *P. dubium* complex. It has been reported from scattered localities in the Aegean area and mountains of Sterea Ellas.

Papaver guerlekense Stapf occurs in S Anatolia, Cyprus and the Aegean area. It was recognized as a separate species in Euro+Med PlantBase, but Kadereit (in Strid & Kit Tan 2003: 88) concluded that it cannot be unambiguously separated from *P. rhoeas* L.

Paramoltkia doerfleri (Wettst.) Greuter & Burdet (*Moltkia doerfleri* Wettst.) was reported from the Konitsa area by Snogerup & Snogerup (2006: 721) and this record was repeated by Phitos, Constantinidis & Kamari (2009b: 249-250). However, the only specimen (Snogerup & Snogerup 21294, LD) has subsequently been revised as *Aegonychon purpurocaeruleum* (L.) Holub (see Selvi 2011: 312). *Paramoltkia doerfleri* is not a member of the Greek flora but remains, at least for the time being, an endemic of N Albania and SW Serbia.

Parapholis pycnantha (Hack.) C. E. Hubb. is mainly a W Mediterranean species. There are two Greek records: Hansen (1982: 34) from Kerkira and Zervou, Raus & Yannitsaros (2009: 175) from Kalimnos. Both are feasible but need confirmation.

Parapholis strigosa (Dumort.) C. E. Hubb. was reported fom Kerkira by Borkowsky (1994). The record is dubious and was not accepted on the Flora Ionica website. A subsequent record from Limnos has turned out to be incorrect. *P. strigosa* occurs along the coasts of W Europe, extending locally to Italy (see map in Runemark 1962a: 12).

Parkinsonia aculeata L., a native of SW USA and Mexico, is cultivated as an ornamental and occasionally found on sand dunes, along roads, etc., but doubtfully naturalized in Greece although invasive in dry parts of Africa and Australia. It is a large shrub with slender, thorny twigs, linear leaves with numerous small leaflets and showy, yellow flowers.

Paronychia carica Chaudhri was reported from Chios in Fl. Turkey 2: 261 (1967), based on a single, historical collection. It is otherwise a species of W & S Anatolia; occurrence on Chios or elsewhere in the East Aegean area is feasible but needs confirmation.

Paronychia cephalotes subsp. ***thracica*** Chaudhri has been described from NE Greece. There appear to be no consistent differences between this and *P. cephalotes* s. str.

Paronychia sintenisii Chaudhri was listed also for Greece in Euro+Med PlantBase, but the evidence for this is obscure. According to Chaudhri (1968: 236) it is a local endemic of Mt Ida (Kaz Dağ) in NW Anatolia.

Paspalum notatum Flüggé was reported from a ruderal coastal habitat in Kerkira by Scholz (2002: 206). The species is native in Mexico, the West Indies and S America, and the degree of naturalization in Kerkira is uncertain.

Pastinaca argyrophylla Delip. was described in 1990 from Mt Slavjanka (Alibotuš) in SW Bulgaria and should be looked for on the Greek side of the border. It is related to the W Anatolian *P. glandulosa* Boiss. & Hausskn.

Pedicularis friderici-augusti Tomm. was described from Monte Slavnik in Slovenia and has subsequently been reported from a few localities in NE and C Italy. These plants have pale yellow corollas, and it remains to be established whether the plants from NE Greece with purplish-pink corollas here referred to *P. friderici-augusti* are indeed conspecific.

Pedicularis hoermanniana K. Malý is rare in Greece and was not included in the Mountain Flora (Raus in Strid & Kit Tan 1991) as a report by Goulimis (1960a: 21) was considered dubious. However, it has subsequently been confirmed to occur in Greece (Authier 2000d) and North Macedonia (Teofilovski 2018). The total distribution comprises C & NE Italy, Slovenia and much of the Balkan Peninsula southwards to Stara Planina in Bulgaria and Mts Timfi and Pieria in Greece.

Pedicularis verticillata L. was reported from Mt Aftia (Grevenon/Ioanninon) by Goulimis (1960a: 21), probably in error.

Peltaria angustifolia DC. has been collected once on the island of Rodos: "In fissuris rupium calc. montis Hag. Elias prope Archangelos, 400 m", Rechinger 8434, 25 Jun 1935 (BM, G). It has not been confirmed in the subsequent thorough floristic studies of this island and must be assumed extinct in Greece. The species is widespread in S Anatolia, extending to Syria and N Iran.

Pentanema. Phylogenetic speculations based on molecular evidence (Gutiérrez-Larruscain & al. 2018) has led to the proposed transfer of several species of *Inula* to the genus *Pentanema*.

Perovskia abrotanoides Kar. is native in SW to C Asia and cultivated as an ornamental elsewhere. There are several reports of casual occurrences in C Europe and one recent collection from Greece, island of Amorgos (Biel & Kit Tan 2016a: 97).

Persicaria bistorta (L.) Samp. After the preparation of the maps this species has been reported from the Rodopi mountains by Charalampidou, Eleftheriadou & Theodoropoulos in Fl. Medit. 32: 88 (2022).

Persicaria capitata (D. Don) H. Gross, a native of C Asia, is locally naturalized in W & S Europe, and has been reported from ruderal habitats in the cities of Patras and Ioannina. It is a procumbent perennial with flowers in dense heads 5-10 mm.

Persicaria decipiens (R. Br.) K. L. Wilson is a name used in recent literature for a species long known as *Persicaria salicifolia* (Willd.) Asenov or *Polygonum salicifolium* Willd. *Persicaria decipiens* is native to Australia where it was collected at Botany Bay already by Banks and Solander in 1772. The type of *Polygonum decipiens* R. Br. is a specimen at BM collected in almost the same locality ("Port Jackson") by Robert Brown in 1804 and described in 1810. *Polygonum salicifolium* Willd. was described in 1809 from the Canary Islands and is widespread in Europe, much of Africa and SW Asia. When making the recombination *Persicaria decipiens*, Wilson (1988: 178) remarked "...it is very close to *Persicaria salicifolia* (Brouss. ex Willd.) Asenov (incl. *Polygonum serrulatum* Lag.) of Europe, Africa and South-West Asia, and study of more material may show that they are synonymous". Although *Polygonum salicifolium* Willd. is one year older than *Polygonum decipiens* R. Br., the former epithet is not available when Willdenow's species is transferred to *Persicaria*, since there is already *Persicaria salicifolia* Gray 1821 (presumably a synonym of *Persicaria maculosa* Gray). It seems a far-fetched conclusion that the European and native Australian plants should be conspecific, and further studies are needed on the correct name for the former.

Persicaria dubia (Stein) Fourr. Euro+Med PlantBase uses this name for the species here called *Persicaria mitis* (Schrank) Asenov, claiming that the latter name is illegitimate.

Persicaria lapathifolia subsp. ***brittingeri*** (Opiz) Soják is a taxon of uncertain status. A report by Snogerup (in Fl. Hellen. 1: 89, 1997) was based on a single collection from the island of Lesvos where it was said to grow together with subsp. *lapathifolia*. A subsequent report from Chalkidiki (Düll, in litt., 2011) needs confirmation.

Persicaria orientalis (L.) Spach, probably a native of S & SE Asia, is cultivated as an ornamental and has been reported from scattered Greek localities as casual in ruderal habitats. It is erect to scandent and up to 1.5 m tall, densely short-pubescent throughout, with narrowly ovate leaves and nodding, pink spikes.

Persicaria senegalensis (Meisn.) Soják, a native of Africa and Australia, is cultivated as an ornamental and has been reported as naturalized or casual in wet ruderal habitats in Kriti and Rodos. It is a robust rhizomatous perennial up to 2 m tall with leaves grey-tomentose beneath, and pink spikes.

Persicaria vivipara (L.) Ronse Decr. A map in Stevanović & al. (2009: 225) appears to show dots on the Greek/Macedonian and Greek/Bulgarian borders including at least one inside Greece. It is not unthinkable that this arctic-alpine species could extend to Greece, but no specimens or specified reports have been seen.

Petasites pyrenaicus (L.) G. López (syn. *P. fragrans* (Vill.) C. Presl) has been collected once in a ruderal habitat on the island of Kerkira. It has scapes appearing at the same time as the basal leaves and with only 2-7 scale-leaves. The flowers are whitish-pink and fragrant. The species is native in W Europe and often cultivated as an ornamental; it is not clear whether the specimens from Kerkira were naturalized.

Petrorhagia fasciculata (Margot & Reut.) P. W. Ball & Heywood was listed for Peloponnisos by Mermygkas, Zikos & Constantinidis (2021: supplementary file). The record is feasible, but no details are available.

Petrosimonia oppositifolia (Pall.) Litv. was listed for Greece in Euro+Med PlantBase, probably based on the report of *P. crassifolia* (Pall.) Bunge (a synonym of *P. oppositifolia*) from "Thra" in Hayek (1924b: 159). There is no precise locality or confirmation. Atlas Fl. Eur. (5: 89, 1980) mapped *P. oppositifolia* from E Romania, Crimea and eastwards, with a single locality also in SW Albania; the latter was not confirmed by Barina (2017).

Petteria ramentacea (Sieber) C. Presl was reported from the Preveza area by Chitos & Thanopoulos (1998: 168), and the record was repeated by Phitos, Constantinidis & Kamari (2009b: 251-252). This species is native in mountain scrub in Albania and Croatia, and is also cultivated as an ornamental. The Greek record is somewhat out of range geographically as well as ecologically, and it may represent a casual introduction.

Peucedanum longifolium Waldst. & Kit. Frey (1989) reduced *P. longifolium* to one of four subspecies of *P. officinale* L. Subsp. *officinale* and subsp. *longifolium* (Waldst. & Kit.) Frey appear to meet in N Greece, and some records are uncertain as to subspecies.

Phagnalon methanaeum Hausskn. Taken in a wide sense *P. saxatile* (L.) Cass. occurs in most of the W & C Mediterranean area eastwards to the Kiklades. Greek plants have been described as *P. methanaeum*, but the differences are not clear.

Phagnalon rupestre (L.) DC. is represented by subsp. *rupestre* and subsp. *graecum* (Boiss. & Heldr.) Batt. which have frequently been regarded as different species. However, the distinguishing features (mainly shape of the outer phyllaries) are subtle and inconsistent, and in Greece their distribution areas overlap completely.

Phalaris aquatica L. Greek plants have also been reported under the names *P. bulbosa* L. and *P. tuberosa* L. Concerning the nomenclatural tangle surrounding these three names, see Bor in Fl. Iranica 70: 348-349 (1970).

Phalaris elongata Braun-Blanq. was described from NW Africa and has subsequently been reported also from Portugal and Italy. A single Greek collection from Mt Pastra, Viotias (Constantinidis s.n., 1995-05-20; B, UPA) has been identified as *P. elongata* by H. Scholz. The taxonomic status is uncertain. *P. elongata* appears as an accepted species in Euro+Med PlantBase and in the recent Italian Flora (Pignatti 2017a: 516), but is often regarded as a synonym of *P. aquatica* L.

Phalaris truncata Bertol. was listed for "Gr" in Euro+Med PlantBase, but no details are available. The only Greek record appears to be a dubious literature report from Strimonikos kolpos east of Thessaloniki (Eleftheriadou & Theodoropoulos 2003: 304).

Phaseolus vulgaris L. (common bean), a native of the Americas, is widely cultivated in many different varieties. It may persist for some time in ruderal habitats but is scarcely naturalized.

Phelipanche mutelii group. Following Atlas of the Aegean Flora (1: 422, 2016), *P. nana* (Reut.) Soják is treated as a variety of *P. mutelii* (F. W. Schultz) Pomel, whereas the closely related *P. ramosa* (L.) Pomel (*Orobanche ramosa* L.) is considered absent from Greece.

Phelipanche oxyloba (Reut.) Soják has been reported from scattered localities throughout Greece. Rätzel, Ristow & Uhlich (2017: 649) recombined it as *P. mutelii* var. *oxyloba* (Reut.) Rätzel & Uhlich.

Phelipanche schultzii (Mutel) Pomel (*Orobanche schultzii* Mutel). There are several records of this species from scattered localities throughout Greece, but according to Sánchez Pedraja & al. (http://www.farmaliergaes.com/Otrospdf/publica/Orobanchaceae%20Index.htm) all need confirmation and may refer to related species such as *P. libanotica* Boiss. and *P. gussoneana* (Lojac) Domina & al.

Phelipanche schultzioides M. J. Y. Foley was described from the foothills of Mt Chelmos, where it was parasitic on *Urtica dioica*. According to Rätzel & al. (2018: 209-210) it is conspecific with *P. gussoneana*.

Phelypaea boissieri (Reuter) Stapf. The only Greek locality for this rare and spectacular species is in the Prespa National Park. According to Cullen (2010) there are three species of *Phelypaea*: *P. boissieri* (NW Greece, North Macedonia, S & E Turkey and adjacent NW Iran), *P. coccinea* Poir. (Caucasus, Crimea and perhaps N Iran) and *P. tournefortii* Desf. (E Turkey and adjacent Georgia and Armenia). Barina (2017: 336) mapped *P. boissieri* in 3 localities in S Albania where the host was said to be *Centaurea graeca*. The species has recently been recorded also from N Iraq, growing on *Cousinia hausskneehtii* (Al-Asidi & Al-Mayah 2017).

Philadelphus coronarius L., a deciduous shrub with white, fragrant flowers, is probably native in C & N Italy and Austria, and has long been cultivated as an ornamental. There are few and scattered records from Greece and it is scarcely naturalized.

Phleum paniculatum Huds. was listed for Greece in Fl. Eur. (5: 240, 1980), but no specimens or precise records have been seen.

Phleum rhaeticum (Humphries) Rauschert was listed for Greece in Euro+Med PlantBase. It was originally described from the Alps, as *P. alpinum* subsp. *rhaeticum* Humphries. Its relations to *P. alpinum* s. str. are not clear, and it is unlikely to occur in Greece.

Phoenix theophrasti Greuter is close to the date palm, *P. dactylifera* L., but recent molecular studies (Boydak & al. 2021) appear to support its status as a separate species.

Phyla canescens (Kunth) Greene was recognized at species level in the Greek checklist (Dimopoulos & al. 2103: 144) as well as Euro+Med PlantBase, but in a recent revision it has been reduced to *P. nodiflora* var. *minor* (Hook.) N. O'Leary & Múlgura, comb. nov. (see O'Leary & Múlgura 2012: 589). *P. nodiflora* has an almost worldwide distribution in tropical and warm-temperate areas whereas the other four species in the genus are exclusively American.

Phyllolepidum. *Ptilotrichum cyclocarpum* Boiss., now recombined as *Phyllolepidium cyclocarpum* (Boiss.) L. Cecchi, was described from Kurdistan, based on a collection by Aucher. It has been reported from scattered localities in mountains of E & C Anatolia, and it remains to be established whether plants from the Balkans are indeed conspecific. For distribution in North Macedonia, see Teofilovski (2011).

Physalis angulata L. Naturalized Mediterranean plants have also been called *P. ixocarpa* Brot. ex Hornem. and *P. philadelphica* Lam. The differences are not clear, but there appears to be only one species in Greece. Plants of this group are commonly cultivated (known as tomatillo or Mexican husk tomato). The somewhat outlying locality in SE Peloponnisos is based on Stamatiadou 4360 (ATH) with the note "Probably formerly cultivated for its edible fruits".

Phyteuma pseudorbiculare Pant. was described from alpine meadows on Mt Durmitor (Montenegro). In the original description *P. orbiculare* var. *cordatum* Griseb. (described from Mt Kobelitza on the border of Kosovo and North Macedonia) was listed as a synonym. The differences between *P. orbiculare* L. and *P. pseudorbiculare* are mainly in leaf shape. In the only confirmed Greek locality on Mt Gramos plants are variable in this respect; in some the blade of the basal leaves is broadly ovate and cordate at base (as in *P. pseudorbiculare*), in others it is elliptic-oblong and rounded at base (as in *P. orbiculare*). Both species have been recorded from Albania, but further studies are needed to establish whether they are consistently different. The *P. orbiculare* complex comprises several closely related taxa in mountains of C & S Europe.

Picris hieracioides L. Most, perhaps all, Greek records refer to subsp. *spinulosa* (Guss.) Arcang. Subsp. *hieracioides* may also occur in the north, but taxonomy at subspecies level is unsettled.

Picris hispidissima (Bartl.) W. D. J. Koch was listed for "?Gr" in Fl. Eur. (4: 317, 1976) as well as in Euro+Med PlantBase, possibly based on a report from Mt Vermion by Goulimis (1960a: 15). The record is dubious and there is no recent confirmation.

Pilosella alpicola (Steud. & Hochst.) F. W. Schultz & Sch. Bip. was listed for NC by Dimopoulos & al. (2013: 63). As currently understood the *P. alpicola* group is a polyploid complex with four vicariant species, two of which occur in northern Greece: *P. rhodopea* (Griseb.) Szeląg and *P. serbica* (F. W. Schultz & Sch. Bip.) Szeląg (see Szeląg 2008, Šingliarová, Zozomová-Lihova & Mráz 2019).

Pilosella balansae (Boiss.) S. Bräut. & Greuter was listed for NC by Dimopoulos & al. (2013: 63). The record probably goes back to Adamović 428, 27.07. 1905 (WU) from "regione media mt. Olympi Thessaliae" which had been seen by Zahn and erroneously included in *Hieracium macrotrichum* subsp. *balansae* Boiss. The specimen has later been identified as *Pilosella leontocephala* (Halácsy) Vladimirov, Coşkunçelebi & Kit Tan, but belongs to *P. macrotricha* (Boiss.) F. W. Schultz & Sch. Bip. (fide Gottschlich).

Pilosella bonaquae (Buttler & Lippert) S. Bräut. & Greuter was based on a single collection from Mt Varnous (Kalo Nero) in NW Greece. It has subsequently been interpreted as a hybrid derivative of *P. cymosa* (L.) F. W. Schultz & Sch. Bip. under the name of *P. tubulata* (Vollm.) Soják.

Pilosella budensis (Borbás) Soják has been reported once from Mt Chortiatis (Karagiannakidou & Raus 1996: 506), based on a collection by Orphanides. The latter is the type of *Hieracium hortatschicum* Zahn which was listed, probably in error, as a synonym of *Pilosella budensis* in Euro+Med PlantBase. The identity of Orphanides' collection is uncertain, and there appears to be no modern confirmation. *Hieracium budense* Borbás was described from Hungary and is unlikely to occur in Greece.

Pilosella halacsyi (Heldr. ex Halácsy) Soják was described from Mt Parnitha (Attiki) and has recently been reported also from localities in NW Greece. It has been interpreted as the hybrid *P. cymosa* subsp. *heldreichiana* (Nägeli & Peter) Gottschl. × *P. testimonialis* (Nägeli ex J. Hofm.) Gottschl.

Pilosella leucopsilon (Arv.-Touv.) Gottschl. For nomenclatural reasons the name has changed to *P. testimonialis* (Nägeli ex J. Hofm.) Gottschl. (see Gottschlich 2020: 137).

Pilosella sphaerocephala (Rchb.) F. W. Schultz & Sch. Bip. was listed for Greece in Euro+Med PlantBase, erroneously referring to Dimopoulos & al. (2013). It is a central European species unlikely to occur in Greece.

Pimpinella saxifraga L. was listed for Peloponnisos by Mermygkas, Zikos & Constantinidis (2021: supplementary file), probably in error. Verified Greek records are all in the far north.

Pinguicula megaspilea Boiss. & Heldr. Fleischmann (2015) resurrected *P. megaspilaea* as a species separate from *P. hirtiflora* Ten., citing material from *locus classicus* by the monastery of Megaspileo N of Mt Chelmos as well as Parnassos,

Giona and locally in SW Anatolia. However, the distinguishing characters indicated (shape of summer leaves and relative length of spur) fall within the range of variation of *P. crystallina* subsp. *hirtiflora* (Ten.) Strid, and there is no justification for retaining *P. megaspilaea* as a separate taxon. *P. louisii* Markgr., described from Albania and reported also from a few localities in N Greece, is scarcely different from *P. crystallina* subsp. *hirtiflora*. A report of *P. hirtiflora* var. *louisii* (Markgraf) Ernst from "Philippos, près de Drama; rochers calcaires" (Contandriopoulos & Quézel 1974: 28) is highly dubious and has been omitted (there are no suitable localities in this area). As currently defined, *P. crystallina* subsp. *hirtiflora* occurs in the W Balkan Peninsula from Taigetos to the vicinity of Split, in a few localities in S Italy and locally naturalized by the Italian/French border. Subsp. *crystallina* occurs in Cyprus and Anatolia (mainly in the south).

Pinus mugo Turra. Records from Vitsi (Zagariaris 1940a: 137) and Orvilos (Sfikas 1984a: 13) are feasible but need confirmation. This species forms *Krummholz* at subalpine level in the Alps and Carpathians, extending southwards to Bulgaria and North Macedonia.

Pistacia vera L. (pistachio nut), a native of C Asia, is sometimes cultivated in S Greece, first on the island of Egina (hence the Greek name φυστίκι Αιγίνης), but scarcely naturalized. It is a small, deciduous tree with larger leaves and much larger fruits than in the native species. It is commercially cultivated notably in Iran.

Plantago altissima L. There are scattered records of this species almost throughout Greece, but all are dubious and may refer to large forms of *P. lanceolata* L., although Willing 10716 (B) from the foothills of Mt Menikion and Willing 102401 (B) from nomos Thesprotias approach typical *P. altissima*.

Plantago bellardii All. subsp. ***bellardii*** and subsp. ***deflexa*** (Pilg.) Rech. f. are both widespread in Greece. The former has scapes equalling or exceeding the leaves and remaining ascending to erect in fruit. The latter has short scapes which become arcuate-recurved in fruit. There is no geographical separation, and varietal rank seems more appropriate.

Plantago coronopus group. Taxonomy and nomenclature remain unsettled. There are two entities in Greece: diploids called *P. coronopus* L. subsp. *coronopus* and tetraploids usually called *P. weldenii* Rchb. or *P. coronopus* subsp. *weldenii* (Rchb.) Arcang., the latter being more common. The name *P. commutata* Guss. has intermittantly been used for both taxa. For lectotypification of the names *P. weldenii* and *P. commutata*, see Lazzaro & al. (2010). In this Atlas the names *P. coronopus* subsp. *coronopus* and *P. coronopus* subsp. *weldenii* are provisionally retained, although the two entities may well merit species rank. Only records specifically identified as *P. coronopus* subsp. *coronopus* have been mapped as such. All other records of the group have been assumed to represent subsp. *weldenii*.

Plantago gentianoides Sm. was listed for Greece in Fl. Eur. (4: 42, 1976), but the background for this record is obscure. One collection from the vicinity of the Katara Pass (Snogerup 20717, LD), identified as *P. gentianoides*, is probably a form of *P. media*. *P. gentianoides* was described from Ulu Dağ and has been reported also from the C Balkan Peninsula and Romania.

Plantago indica L. This name has been proposed for rejection (Doweld & Shipunov 2017). At the time of writing there is still no decision by the nomenclature committee, and the name is thus provisionally retained. Greek records of *P. arenaria* Waldst. & Kit. are presumably referable to the taxon here called *P. indica*.

Plantago subulata L. Greek records refer to *P. holosteum* Scop.

Platanthera chlorantha (Custer) Rchb. Plants from Lesvos tend to have small flowers. They have been referred to *P. holmboei* Lindb. f. (described from Cyprus), and subsequently described as an endemic species, *P. lesbiaca* Devillers-Tersch. & al. Kühn, Pedersen & Cribb (2019: 103) reported *P. holmboei* only from Cyprus, S Anatolia and coastal areas of Syria-Palestine.

Platanthera fornicata (Bab.) Buttler. Greek plants formerly identified as *P. bifolia* (L.) Rich. were referred to the recently recognized *P. fornicata* by Buttler & Raus (2017: 91-92), but retained as *P. bifolia* by Tsiftsis & Antonopoulos (2017: 196-199). *P. fornicata* has been further divided into subsp. *fornicata* and subsp. *atropatanica* (B. Baumann & al.) Buttler. All differences are vague and quantitative. For that reason Greek plants are provisionally retained under *P. bifolia*. Cf. also Durka & al. (2017), Kühn, Pedersen & Cribb (2019: 100) and Wucherpfennig (2021). According to the last-mentioned reference, *P. fornicata* ("kleine *bifolia*") has been confirmed only from England, Scandinavia, Benelux countries and Germany.

Plumbago auriculata Lam., a native of South Africa, is cultivated as an ornamental and has been reported from ruderal habitats and streambeds in Kriti, Lesvos and Amorgos, but is doubtfully naturalized. It resembles *P. europaea* L., but has larger, pale blue flowers.

Poa bivonae Parl. was listed for "Gr" in Euro+Med PlantBase, probably referring to a report by Rechinger (1939a: 545) which was based on his own collection no. 9441 (G, W) from the Nestos river gorge. This collection was identified by Scholz (1986a: 397) as *P. sinaica* Steud. (new for Greece). It is scarcely distinct from *P. bulbosa* L. *P. bivonae* was described from Sicily, and was accepted as an endemic species by Pignatti (2017a: 529).

Poa cephalonica H. Scholz is a taxon of somewhat uncertain status, originally described as endemic to the *Abies* zone on Mt Enos, Kefallinia, and subsequently reported from several localities on the Ionian Islands as well as on the adjacent mainland. Further studies are needed to establish whether it is consistently different from *P. bulbosa*.

Poa chaixii Vill. was listed for "Gr" in Fl. Eur. 5: 163 (1980). This was based on a single specimen (Strid 1208) from nomos Kozanis which has subsequently been revised as *P. hybrida* Gaudin (see Scholz 1986a: 394).

Poa maroccana Nannf. has been reported from scattered localities in Greece and elsewhere in the Mediterranean area. It is very similar to *P. infirma* Kunth and may be an autotetraploid variant of the latter, differing mainly in the larger anthers (0.7-0.9 mm).

Poa perligularis H. Scholz (nom. nov. for *P. bulbosa* subsp. *perligulata* H. Scholz) is a dubious taxon in the *P. bulbosa* complex, reported from scattered localities in the Mediterranean area including Kefallinia.

Poa sinaica subsp. ***graeca*** H. Scholz was based on Strid 17143 (C) from SE Peloponnisos. Strid 14181 from Attiki (Shineas near Marathon) was cited as a paratype. Both appear to represent *P. pelasgis* H. Scholz.

Poa stiriaca Dörfl. was reported as new for Greece by Dimitrov & Vutov (2015: 461), based on their own collection from serpentine cliffs near Metsovon. *P. stiriaca* was described from Austria and has also been reported from elsewhere in east central Europe. It is unlikely to occur in Greece.

Poa timoleontis Heldr. ex Boiss. A large set of specimens collected by Willing (B) and reported as *P. timoleontis* has been excluded from the map. They need critical examination, and many of them are likely to represent other taxa in the *P. bulbosa* group.

Poa trichophylla/trichopoda. *Poa trichopoda* Heldr. & Sartori in Boiss., Diagn. ser. 2, 4: 136 (1859) was described from Mt Parnassos, based on collections by Sartori (1846) and Heldreich (1852). In Fl. Orient. 6: 604 (1884) Boissier used the name *Poa* *trichophylla* Boiss. for the same species, citing "*Poa trichopoda* errore typogr." as a synonym. The Greek checklist and some other modern publications use the name *Poa trichopoda*, but Boissier's intention was apparently to call the species *Poa trichophylla*. The Code (Art. 60.1) explicitly allows correction of typographical errors, and in this case it is obvious because the author wrote that he was correcting his own error. [Thanks are due to Nick Turland for advice on this matter].

Poa versicolor Besser was listed for "Gr" in Fl. Eur. (5: 165, 1980). The record is likely to refer to *P. sterilis* M. Bieb.

Poacynum. Eurasian species previously included in *Apocynum* or *Trachomitum* were referred to the genus *Poacynum* by Mavrodiev, Laktionov & Alexeev (2015). They include a species previously called *Trachomitum sarmatiense* Woodson which has its main distribution in southern Russia and Asia Minor, but is also known from the island of Limnos (see Aegean Atlas 1: 66, 2016). It has recently been reported also from a sandy beach in nomos Xanthis (Doumas, Goula & Constantinidis 2022: 5-6, sub nom. *Trachomitum venetum* (L.) Woodson).

Polycarpon alsinifolium (Biv.) DC. was included as an accepted species in the Greek checklist (Dimopoulos & al. 2013: 81), but is probably a maritime ecotype of *P. tetraphyllum* (L.) L. The latter has often been divided into *P. tetraphyllum* s. str. (with leaves usually in whorls of 4, and 3-5 stamens) and *P. diphyllum* Cav. (with opposite leves and 1-3 stamens), but there are many intermediates and no geographical separation.

Polycnemum heuffelii A. F. Láng was mapped for C Greece in Atlas Fl. Eur. 5: 11 (1980), probably based on historical records by Haussknecht (1898a: 57). According to Kit Tan (in Fl. Hellen. 1: 110, 1997) they may represent misidentified material of *P. arvense* L. The genus *Polycnemum* is sometimes included in Amaranthaceae rather than Chenopodiaceae.

Polygala acarnanica (Chodat) Kožuharov & Petrova. Chodat (1893: 436) described *P. major* f. *acarnanica* "in clivibus Acarnaniae Graeciae". It was raised to species level by Kožuharov & Petrova in Flora na Bulgarija 7: 201 (1979) who recognized both *P. major* Jacq. and *P. acarnanica* in Bulgaria (the latter on Mt Pirin). However, they seem doubtfully distinct.

Polygala anatolica Boiss. & Heldr. Literature records from mountains of northern Greece (Athanasiadis & Drossos 1990: 98, Eleftheriadou 1992: 58, Sfikas 1998b: 5) probably refer to *P. major*.

Polygala comosa Schkuhr. Greek plants may belong to subsp. *alibotusensis* Degen & Dren., a little-known taxon described from the Bulgarian side of Mt Orvilos (Ali Botuš). Mermygkas, Zikos & Constantinidis (2021: supplementary file) listed *P. comosa* from Peloponnisos, probably in error. All verified Greek records are north of 40°N.

Polygala nicaeensis W. D. J. Koch. Most of the Greek records refer to subsp. *mediterranea* Chodat. The pubescent subsp. *tomentella* (Boiss.) Chodat occurs particularly in Attiki and Evvia, and there are also scattered records of subsp. *carniolica* (A. Kern.) P. Graebn. In view of the many uncertainties only a collective map is provided.

Polygala supina Schreb. At least some of the Greek records represent subsp. *rhodopaea* (Velen.) McNeill, but the subspecies are poorly defined.

Polygala vulgaris L. There are several records from mountainous areas throughout the mainland and northern Peloponnisos. All are likely to refer to forms of *P. nicaeensis*.

Polygonatum latifolium Desf. This name, which was used by Dimopoulos & al. (2013: 86) for a species closely related to *P. odoratum*, is a synonym of *P. hirtum* (Bosc ex Poir.) Pursh. *Convallaria pruinosa* Halácsy, based on Sintenis & Bornmüller 1496 (LD, WU) from the eastern foothills of Mt Olimbos, is also a synonym of *P. hirtum*.

Polygonum praelongum Coode & Cullen has been reported once from Kastellorizo (Megisti), based on an immature specimen. It is otherwise known only from the type collection in S Anatolia (C3 Antalya). Several related species have been described from Anatolia (cf. Leblebici, Duman & Aytaç 1993), and the group is in need of revision.

Polypogon maritimus Willd. Recent Floras have often recognized subsp. *maritimus* and subsp. *subspathaceus* (Req.) Bonnier & Layens. Scholz (1991c: 139) treated them as different species, whereas Meikle (1985: 1784) "would, at most, regard *P. subspathaceus* as a form of *P. maritimus*". The morphological differences are very small and their distributions overlap completely. Consequently, Meikle's view appears more appropriate.

Polypogon viridis (Gouan) Breistr. Euro+Med PlantBase uses this name based on *Agrostis viridis* Gouan, but there appear to be good reasons to regard it as a *nomen dubium* (cf. Meikle 1985: 1785-1786). If so, the correct name for the species will be *Polypogon semiverticillatus* (Forssk.) Hyl.

Pontechium maculatum (L.) Böhle & Hilger. Phylogenetic speculations based on molecular evidence have led to the proposed transfer of *Echium maculatum* L. to the monotypic genus *Pontechium* (Hilger & Böhle 2000). Morphologically there is little or no evidence for such a split. The species has also been known as *Echium rubrum* Jacq. (non Forssk.) or *E. russicum* J. F. Gmel.

Populus ×canadensis Moench, generally interpreted as the hybrid between *P. deltoides* Marshall (a North American species) and *P. nigra* L., is sometimes planted. There are scattered records from the mainland and Peloponnisos.

Populus ×canescens (Aiton) Sm., the hybrid between *P. alba* L. and *P. tremula* L., has been reported from riparian woodland in the Prespa area, probably resulting from natural hybridization. Elsewhere in Greece it is commonly planted; there are scattered records almost throughout the country.

Populus nigra L. Both subsp. *nigra* and subsp. *neapolitana* (Ten.) Maire have been reported in Greece, but most of the records have not been identified to subspecies. The former is widespread and has glabrous twigs, leaves and inflorescence axis. The latter occurs in the Mediterranean area and SW Asia and is more or less pubescent.

Populus thevestina Dode, a native of C Asia, is planted on Lesvos and probably elsewhere. It resembles the columnar form of *P. nigra*, but has whitish, rather smooth bark and shallowly serrate or subentire (vs. glandular-crenate) leaves.

Portulaca oleracea L. s. lat. is a polyploid complex with an almost worldwide distribution. Based on minute differences in seed size and testa characters several "species" have been described (cf. Danin, Domina & Raimondo 2008, Danin 2011), two or three of which sometimes grow together in the same population. Recent studies involving flow cytometric, chromosomal and morphometric analyses (Walter, Vekslyarska & Dobeš 2015) have shown this species concept to be untenable. The cultivated *P. grandiflora* Hook. has been reported as a garden escape in a few localities.

Potamogeton acutifolius Link was reported from Greece by Hayek (1933: 13) as well as in Fl. Eur. (5: 10, 1980), perhaps based on Haussknecht (1900: 23), "in rivulis Th[essaliae] pr. Trikkalam". There is no modern confirmation. One collection from a dam in the Arachoos river by the hydroelectric station of Louros (Strid 29760; ATH, G, LD, herb. Strid) has been distributed as *P. cf. acutifolius*. The specimens are very young, but Z. Kaplan (in litt., Febr. 2021) identified them as either *P. pusillus* or *P.*

berchtoldii. *P. acutifolius* is unlikely to occur in Greece. According to information kindly supplied by Kaplan there are very few confirmed records from S Europe (Italy, Croatia and Serbia).

Potamogeton alpinus Balb., a widespread boreal species resembling *P. gramineus* L., was listed for "Gr" in Fl. Eur. 5: 9 (1980), probably based on a record from the Ionian Islands in Hayek's Prodromus (3: 11, 1933). It was not reported in the online Flora Ionica (https://floraionica.univie.ac.at/). A subsequent record from Lake Ioannina (Chitos 2009: 97) needs confirmation.

Potamogeton compressus L. A single record by Haeupler (1991: 213) from Peloponnisos ("Tal des Ladonas südlich von Toumbitsi") needs confirmation. *P. compressus* resembles *P. berchtoldii* Fieber but is larger in all parts with compressed, narrowly winged stem, and leaves 2-4 mm wide. Flora Europaea (5: 10, 1980) reported "an isolated record in NW Greece", but the evidence for this is not clear. Hayek (1933: 13) listed the species for "Jon", but it was not included in the online Flora Ionica (https://floraionica.univie.ac.at/). *P. compressus* is a widespread boreal species, and occurrence in Greece seems rather unlikely.

Potamogeton obtusifolius Mert. & W. D. J. Koch was reported by Halácsy (1890: 41), based on Sintenis no. 1869 (LD, WU) from [nomos Pierias] "in stagnant. inter Katarina et Spigi". This record may be the basis for "?Gr" in Fl. Eur. (5: 10, 1980). Sintenis' specimen was revised as *P. berchtoldii* by Zdeněk Kaplan in 2004. *P. obtusifolius* is a widespread boreal species unlikely to occur in Greece.

Potamogeton polygonifolius Pourr. has been reported from a few, scattered localities in Greece but probably in error. A record from Kerkira goes back to Stephanides (1948) and has not been confirmed. Strid & al. 28103 ("SE of the village of Pirgion NE of Parga, shallow lake and damp meadows, 16.4. 1989"; G, LD, herb. Strid) has been distributed as *P. polygonifolius*, but Uwe Raabe and Zdeněk Kaplan have independently (February 2021) identified it as *P. natans* L. based on scanned image of a herbarium specimen. Sarika-Hatzinikolaou 859 (ATHU) from Kalodiki wetland in the same area needs confirmation. *P. polygonifolius* is a species of W & N Europe unlikely to occur in Greece.

Potamogeton schweinfurthii A. Benn., a widespread African species, was first reported from Greece by Kaplan (2005: 424), based on Greuter & Charpin 13279 (20.4. 1976) from a water reservoir in eastern Kriti. According to Lastrucci, Frignani & Kaplan (2010) it occurs also in Kefallinia, Malta, Corsica, Sardinia and C & SE Italy. There are recent collections also from the Aegean islands of Milos and Limnos.

Potentilla aurea L. All Greek specimens are likely to belong to subsp. *chrysocraspeda* (Lehm.) Nyman, which occurs in the Balkan Peninsula, Carpathians and NW Anatolia. Subsp. *aurea* occurs from the Pyrenees to SE Europe; distribution areas overlap in the Balkans and Carpathians.

Potentilla calabra Ten. is a member of the *P. argentea* complex probably restricted to S Italy. Greek records refer to other members of the complex.

Potentilla cinerea Chaix ex Vill. was listed for "Gr" in Fl. Eur. 2: 45 (1968), but no specimens or specified records have been seen.

Potentilla greuteriana Kyriak., Kamari, Kofinas & Phitos is a name given to a local population of *P. arcadiensis* Iatroú from Mt Taigetos. It differs vaguely in quantitative characters, presumably as a result of growing in a more exposed habitat at relatively high altitude. The type of *P. arcadiensis* in from shady, somewhat overhanging limestone cliffs in the foothills of the neighboring Mt Parnonas.

Potentilla macedonica Micevski has been reported once from the Prespa National Park. It is one of several dubious new species in the *P. recta* complex described from North Macedonia (cf. Micevski 1998).

Potentilla pindicola Hausskn. is a taxon of uncertain status, probably not distinct from *P. inclinata* Vill. It has been reported from scattered localities in N Greece, North Macedonia, Bulgaria and Crimea.

Potentilla speciosa Willd. Based on achene characters Soják (1993) divided this species into three subspecies. Most, perhaps all, Greek plants belong to subsp. *illyrica* Soják.

Potentilla supina L. Almost all Greek specimens belong to subsp. *supina*, but subsp. *paradoxa* (Nutt.) Soják has been reported from a few localities in N & C Greece. The former has flowers ± deflexed after anthesis and achenes almost smooth, whereas the latter has flowers suberect after anthesis and achenes with whitish lamellae. Soják (1987) recognized four subspecies of *P. supina* in Eurasia and North America, but very little detail is given in this article and taxonomy at subspecies level is preliminary. Subsp. *paradoxa* (*P. paradoxa* Nutt.) was described from North America and is unlikely to occur in Greece.

Potentilla velenovskyi Hayek is a member of the *P. recta* complex of uncertain status, probably endemic to the C Balkan Peninsula and reported once from the Prespa National Park. It is rather tall with usually 7 leaflets which are glabrous above and have a mixture of long eglandular hairs and shorter glandular hairs on the lower surface.

Primula veris L. Both subsp. *veris*, subsp. *columnae* (Ten.) Lüdi and subsp. *suaveolens* (Bertol.) Gutermann & Ehrend. have been reported from Greece, but the differences are obscure and Greek plants are rather uniform.

Primulaceae. In recent decades this family has been variously divided and reshuffled, sometimes with rather absurd results. Rather than trying to keep up with shifting fashion I have used a traditional, broad circumscription.

Proboscidea lousianica (Mill.) Thell. was reported from Lesvos by Hecht-Markou (1999), but these plants have later been revised as *Ibicella lutea* (Lindl.) Van Eselt. (Yannitsaros & Bazos 2001, 2006). The true *Proboscidea lousianica* was reported from NE Peloponnisos ("SSW of Arachneo alongside road to Gatzia, scattered individuals in cereal fields") by Raus & Raabe (2016: 429-430). The latter is a native of SW United States and Mexico; it is introduced or casual in Australia and South Africa, and has also been reported from scattered localities in the Mediterranean area and S Russia. It may be on the verge of becoming established in Greece.

Prunella grandiflora (L.) Scholler. There are a few literature records from northern Greece: Kitanov (1943: 263) from "Schilka-Gebirge" [Falakron area], Politis (1953: 34) from "Taxiarchis, au lieu dit Itamos" [Chalkidiki], Voliotis (1967: 84) from Mt Cholomondas, and Athanasiadis & Drossos (1990: 84) from Mt Paikon. Two collections by Willing (nos. 283996 and 284088) registered as *P. grandiflora* could not be located at B, 2021-09-22. Whilst it is not unthinkable that this C European species could reach northern Greece, the records so far are likely to refer to forms of *P. vulgaris*.

Prunus cerasus L. (sour cherry) is presumably native in N Anatolia and Caucasia. It is widely cultivated in Greece and elsewhere in Europe and is occasionally long-persistent from abandoned cultivation rather than naturalized. It differs from *P. avium* (L.) L. in the following characters: Petiole up to 2 cm; inflorescence with a few leafy bracts at base; scales of involucre erect; drupe ± acid-tasting. *Prunus cerasus* is a tetraploid, perhaps derived from hybridization between *P. avium* and *P. fruticosa* Pall., a species of E Europe and SW Asia.

Prunus cocomilia Ten. After the preparation of the maps this species has been reported from the Rodopi mountains by Charalampidou, Eleftheriadou & Theodoropoulos in Fl. Medit. 32: 89 (2022).

Prunus discolor (Spach) C. K. Schneid., listed for Greece in Euro+Med PlantBase, is apparently a synonym of *P. graeca* Steud. (cf. Eisenman 2015: 188).

Prunus domestica L. (plum) is commonly cultivated. It differs from *P. cerasifera* Ehrh. in the usually pubescent twigs and leaves, and larger, pendent fruits which are usually broadly ellipsoid. Naturalized forms with ± thorny twigs are often called subsp. *insititia* (L.) Bonnier & Layens.

Prunus dulcis (Mill.) D. A. Webb (almond), probably originating in Anatolia and Iran, is widely cultivated and sometimes naturalized. It differs from the native *P. webbii* (Spach) Vierh. by being usually taller and unarmed, with leaf blade 10-40 mm wide and drupe at least 30 mm long.

Prunus ×rhodia (Browicz) Eisenman has been described from the island of Rodos as a hybrid between *P. dulcis* and *P. graeca*. It is more or less intermediate between the parents.

Prunus spinosa L. Almost all Greek plants have densely and persistently pubescent leaves, petioles and young twigs, thus matching subsp. *dasyphylla* (Schur) Domin, a somewhat poorly defined taxon which appears to represent the species in the eastern part of its range.

Pseudorchis albida (L.) Á. & D. Löve. The single Greek population probably belongs to the East European subsp. *tricuspis* (Beck) E. Klein (cf. Tsiftsis & Antonopoulos 2011: 798-799).

Pteris cretica L. In spite of its name and the fact that it has been listed in reference works such as Flora Europaea and Euro+Med PlantBase, this species has never been found in Greece. The reference to Crete in the original description of Linnaeus goes back to misinterpretation of older literature. From the island of Poros or the adjacent coast of NE Peloponnisos is a single report by Friedrichthal (1838: 275). It has never been confirmed and is most probably incorrect.

Puccinellia bilykiana Klokov was reported from a ruderal habitat (bottom of mine pit) in nomos Kozanis by Brofas, Trigas & Scholz (2010: 201). It is a halophyte native in an area from Moldova to Ciscaucasia; the Greek locality may represent a casual introduction by migratory birds. The Greek plants are doubtfully distinct from *P. festuciformis* (Host) Parl.

Puccinellia fasciculata (Torr.) E. P. Bicknell has been reported once (Ljungstrand & Strid 2008: 471) from a saltmarsh near Vonitsa in W Greece. Otherwise a species of W Europe and the W & C Mediterranean area (eastwards to Croatia), it is very similar to *P. distans* (L.) Parl., but differs in having some short panicle branches with spikelets at base (vs. all long, with spikelets only in distal half); the lemma of *P. fasciculata* is more acute with midvein reaching apex.

Pulicaria clausonis Pomel was listed for "Gr" in Euro+Med PlantBase, referring to Halácsy (1912) where, however, this species is not mentioned. It was described from Algeria and is unlikely to occur in Greece.

Pulicaria dysenterica (L.) Bernh. Greek material may belong to subsp. *uliginosa* Nyman (cf. Fl. Cyprus 2: 894 (1985).

Pulmonaria obscura Dumort. was reported from Mt Vrondous by Goulimis (1956a: 9), based on his collection 21.6. 1955. The record was cited by Voliotis (1976a: 47). D. Mermygkas has kindly sent a scan of the specimen at ATH, which most probably represents *P. rubra* Schott.

Pycnocomon rutifolium (Vahl) Hoffmanns. & Link. See *Lomelosia rutifolia*.

Pycreus flavidus (Retz.) T. Koyama was listed for Peloponnisos by Mermygkas, Zikos & Constantinidis (2021: supplementary file), but no details are available.

Pyrola rotundifolia L. Reports from the Rodopi mountains (Eleftheriadou 1992: 65) and the island of Thasos (Stojanov & Kitanov 1946: 129) need confirmation. *P. rotundifolia* resembles *P. chlorantha* Sw. and *P. media* Sw. but has more numerous flowers (6-15) with broadly campanulate, white corolla and long, curved style.

Pyrus communis L., the cultivated pear, is probably derived by selection from *P. pyraster* (L.) Burgsd. It has non-thorny twigs, relatively shorter petiole and usually larger, pyriform, fleshy fruit. *P. pyraster* is often regarded as a subspecies or variety of *P. communis*.

Quercus aucheri Jaub. & Spach. Browicz (1988: 21) refers to a record from the island of Gioura in the Northern Sporades. No details are available, and it is likely to refer to misidentified material of *Q. coccifera*.

Quercus congesta C. Presl was listed for Greece and the Cretan area in Euro+Med PlantBase, probably in error. No specimens or records have been located, and the species is likely to be endemic to SW Italy, Sicily and Sardinia.

Quercus infectoria Olivier. Subsp. *infectoria* has been reported from Rodos, Samos, Chios and Lesvos, and subsp. *veneris* (A. Kern.) Meikle (syn. *Q. boissieri* Reut.) from Rodos and Thasos. They differ only vaguely in leaf size, etc.

Ranunculus sect. ***Batrachium***. Taxonomy follows Wiegleb, Bobrov & Zalewska-Gałosz (2017), with minor deviations.

Ranunculus aquatilis L. has been reported from a single locality in E Kriti (see comments in Strid 2016, 1: 494).

Ranunculus auricomus group. Only one member of this complex has been reported from Greece and it is known only from the vicinity of the Katara pass where it is locally common. In Flora Hellenica (Strid & Kit Tan 2003: 58) it was provisionally listed as *R. auricomus* L., but has subsequently been described as *R. pindicola* Dunkel, a local endemic.

Ranunculus crenatus L. has been collected only once in Greece, on Mt Gramos (Strid & al. 55320, 2004-06-28; ATH, B, G, LD, herb. Strid). This collection is referable to *R. crenatus* s. str. which occurs in mountains of the Balkan Peninsula, Carpathians and locally in the eastern Alps, generally on siliceous substrate. A related species, *R. bertisceus* Kuzmanović & al., has recently been described from limestone mountains in the Albanian/Montenegrinian border area (Kuzmanović & al. 2021). Similar plants from C & S Italy were referred to *R. magellensis* Ten.

Ranunculus cytheraeus (Halácsy) Baldini was described from the island of Kithira, based on a collection by Leonis. It has never been confirmed on this island, but has subsequently been reported, e.g., from Kriti and Cyprus. There seem to be no consistent differences between this and *R. bullatus* L.

Ranunculus ficaria L. (*Ficaria verna* Huds.). At least three subspecies occur in Greece, but morphological variation and distribution patterns need further study, and only a collective map can be presented. Subsp. *chrysocephalus* P. D. Sell occurs in Kriti and probably elsewhere in the Aegean area, but the difference between this and the widespread subsp. *ficariiformis* Rouy & Foucard is not clear. Subsp. *calthifolius* (Rchb.) Arcang. occurs in the C & N parts of the mainland, usually at high altitude, and subsp. *ficaria* may also be present in this area (cf. Veldkamp 2015).

Ranunculus lingua L. This widespread European species was reported from coastal localities in Peloponnisos by Bory & Chaubard (1832-33: 155). Its habitat – lowland swamps with fresh or brackish water – was certainly scarce already at that time, and most of the sites have since fallen victim to development. In 2002 it was rediscovered by Uwe Raabe in a single locality in NE Peloponnisos ("wetland between Mili and Nea Chios") which may be identical with "marais des moulins d'Argos" where it had been reported by Bory & Chaubard (loc. cit.). This small population may already have gone extinct.

Ranunculus millefolius Banks & Sol. was listed for Greece in Euro+Med PlantBase, referring to Hayek's Prodromus where, however, this species is not mentioned. It has an Irano-Turanian distribution and is unlikely to occur in Greece.

Ranunculus oreophilus auct., see *R. sartorianus*.

Ranunculus parviflorus L. Records from Thasos (Stojanov & Kitanov 1945: 302) and the Ionian island of Paxi (Sands 1980: 31) as well as unpublished records from Sterea Ellas probably refer to *R. chius* DC. which has sometimes been regarded as a subspecies of *R. parviflorus*.

Ranunculus sartorianus Boiss. & Heldr. All Greek populations of the *R. montanus* complex are provisionally called *R. sartorianus* Boiss. & Heldr., but they probably represent more than one taxon. Some plants from non-calcareous mountains in northernmost Greece have been referred to *R. oreophilus* subsp. *balcanicus* Micevski. They are comparatively tall and have achenes with very short, dark beak. Several other taxa have been reported from North Macedonia (see, e.g., Micevski 1983d). *R. oreophilus* M. Bieb. was described from the Crimea and is unlikely to be conspecific with any Greek plant. Reports of *R. carinthiacus* Hoppe from Greece are probably incorrect. All Balkan members of the group are in need of critical revision.

Raparia is one of the many segregates of *Thlaspi* by F. K. Meyer (1973). Al-Shehbaz (2014) and Özüdoğru & al. (2019) listed it as a synonym of *Noccaea*. Plants from Ikaria have been described as *Raparia bulbosa* subsp. *aegaea* F. K. Mey. They cannot be distinguished from *Noccaea bulbosa* s. str.

Raphanus sativus L. (radish) was included in the Greek checklist (Dimopoulos & al. 2013: 74). It is commonly cultivated and may persist for a short time in ruderal habitats. It differs from *R. raphanistrum* L. in the following characters: Taproot swollen, red; leaves somewhat fleshy, ± glabrous; fruit 6-20 mm in diam., spongy, not or only slightly constricted between the 4-10 seeds.

Reseda alba subsp. ***hookeri*** (Guss.) Arcang. occurs in the W & C Mediterranean area. Historical records from Attiki and N Peloponnisos (Haussknecht 1894: 41) have not been confirmed.

Reseda arabica Boiss. A single record from eastern Kriti is probably incorrect (see Snogerup in Strid & Kit Tan 2003: 302).

Reseda odorata L. is a cultivated species of uncertain origin. Greek records, as far as they have been possible to check, represent *R. minoica* Martín-Bravo & Jim. Mejías and *R. phyteuma* L. (see Aegean Atlas 1: 501, 2016).

Reseda orientalis (Müll. Arg.) Boiss. According to Snogerup (in Fl. Hellen 2: 301, 2003) reports from Greece are incorrect.

Retama monosperma (L.) Boiss., a native of Spain and Portugal, is found occasionally in ruderal habitats, persisting from cultivation for ornament but doubtfully naturalized. A similar species of W Mediterranean origin, *R. raetam* (Forssk.) Webb, has been reported to invade semi-natural habitats in Attiki (Bergmeier, Dimopoulos & Zogaris 2010: 197).

Rhagadiolus edulis/stellatus. For notes on variation, see Aegean Atlas 1: 125 (2016). According to Feinbrun-Dothan (in Fl. Palaestina 3: 413, 1978) *R. edulis* Gaertn. prefers more mesic habitats, but there seems to be no such difference in Greece where the two varieties often grow in mixed populations.

Rhamnus pumila Turra. A single collection from Mt Tzena (Strid & al. 16678; G, herb. Strid) matches this species, which has been recombined as *Oreoherzogia pumila* (Turra) W. Vent (see Hauenschild & al. 2016: 73). A literature report from the neighbouring Mt Voras (Babalonas 1983: 23) may also belong here.

Rhazya thracica Davidov was reported from NE Greece (between Porto Lagos and Xanthi) by Zahariadis (1973: 175). It is probably conspecific with *Amsonia orientalis* Decne., described from the Bursa area in NW Anatolia.

Rhinanthus angustifolius C. C. Gmel. was reported from the vicinity of Pisoderi (Mt Varnous) by Quézel & Contandriopoulos (1968: 31). The record probably refers to *R. pindicus* (Sterneck) Soó.

Rhinanthus asperulus (Murb.) Soó was listed for Greece in Euro+Med PlantBase, referring to Flora Europaea. The record is likely to refer to *R. pumilus* (Sterneck) Pau.

Rhinanthus illyricus (Beck & Sterneck) Soó was described from a single locality in Bosnia and has been doubtfully reported from Greece. No matching specimens have been seen.

Rhinanthus mediterraneus (Sterneck) Adamović. For nomenclatural reasons this name is replaced by *R. pumilus* (Sterneck) Pau.

Rhinanthus ovifugus Chabert was reported from Mt Vourinos by Quézel & Contandriopoulos (1968: 30). The record probably refers to *R. pindicus* (Sterneck) Soó.

Rhinanthus wagneri Degen was described from the Balkan mountains in Bulgaria. It occurs northwards to Hungary, and has recently been reported also from North Macedonia (Teofilovski 2011: 117). A few collections from NE Greece, viz. Strid & al. 16260 (B, G, herb. Strid) and Strid & al. 52876 (ATH, G, LD, herb. Strid) are of uncertain status, being more or less intermediate between *R. wagneri* and *R. rumelicus* Velen. (see also Raus in Strid & Kit Tan 1991: 252).

Rhynchocorys elephas (L.) Griseb. Burbidge & Richardson (1970) recognized subsp. *elephas* in S Italy, Greece and Bulgaria, and four additional subspecies further east. Subsp. *elephas* has subsequently been reported also from the Prilep area in North Macedonia (Dimitrov in Greuter & Raus 2001: 324).

Ribes multiflorum Kit. ex Roem. & Schult. Plants collected in the Stix ravine (Mt Chelmos) by Orphanides (no. 213, 17/29 Jun 1852) were distributed under the unpublished manuscript name *R. roeseri* Orph., and are represented at least in BM, S, W and WU. The population was rediscovered in 1988 (Strid & al. 27380; E, G, UPA, herb. Strid). These plants deviate in some respects, but detailed studies are needed to establish whether they represent a separate, endemic taxon. On Mt Parnassos the species has not been confirmed since first collected by Heldreich, 1852-08-15, (W; fruiting specimen labelled "In monte Parnasso. reg. abietina, Hag. Nikolaos").

Ribes odoratum H. L. Wendl., a xenophyte of North American origin, has been reported from a ruderal habitat in a village in nomos Ioanninon (Raus 2007: 210). It is doubtfully naturalized.

Rindera graeca (A. DC.) Boiss. & Heldr. In 1860 Heldreich described plants from Mt Dirfis, Evvia, as *Mattia schmidtii* (see Strid 2016a: 147). Plants from the same locality were subsequently described by Orphanides (1870b: 138) as *Rindera graeca* var. *foliolata* Orph., alluding to its numerous leaves. These plants are scarcely taxonomically distinct from other populations of *R. graeca*.

Romulea columnae subsp. ***rollii*** (Parl.) Marais was listed for "Cr" and "Gr" in Euro+Med PlantBase, but the source is murky. It was described from Italy, as *R. rollii* Parl., and no evidence has been found to support that it should occur in Greece.

Rorippa lippizensis (Wulfen) Rchb. Records from Greece and Bulgaria refer to *R. thracica* (Griseb.) Fritsch. As currently defined *R. lippizensis* is restricted to the W Balkan Peninsula from N Albania to Istria (cf. Ančev & Tomšovic 1999).

Rosa caryophyllacea Besser was listed for "Gr" in Fl. Eur. 2: 32 (1968), but no specimens or specified records have been seen.

Rosa gallica L. was listed for Peloponnisos by Mermygkas, Zikos & Constantinidis (2021: supplementary file), but no details are available.

Rosa micrantha Sm. There are several literature records of this species, but all are likely to refer to *R. agrestis* Savi.

Rosa nitidula Besser and ***R. squarrosa*** (Rau) Boreau, which were listed for "Gr" in Fl. Eur. 2: 29 (1968) have been provisionally merged with *R. canina* L.

Rosa orientalis Dupont ex Ser. Greek records apparently refer to *R. heckeliana* Tratt.

Rostraria hadjikyriakou Christodoulou & Hand, a local endemic resembling the widespread *R. cristata* (L.) Tzvelev, has recently been described from Cyprus (Hand & al. 2021). It should be looked for in the East Aegean area.

Rostraria hispida (Savi) Doğan, ***R. obtusiflora*** (Boiss.) Holub and ***R. smyrnacea*** (Trin.) H. Scholz may all be under-represented in the records due to confusion with the much more common *R. cristata* (L.) Tzvelev.

Rostraria pubescens (Lam.) Trin. There appear to be no confirmed Greek records of this species, which is mainly found in the W & C Mediterranean region. Reports from Imeri Gramvousa off NW Kriti (Christodoulakis, Economidou & Georgiadis 1991: 62) and the East Aegean island of Nisiros (Burton 1991: 35) are dubious.

Rubia tenuifolia d'Urv. Both subsp. *tenuifolia* and subsp. *brachypoda* (Boiss.) Ehrend. & Schönb.-Tem. have been reported in the Aegean area, but the differences are obscure.

Rubus is poorly known in northern Greece and in need of critical study. There are likely to be several taxa not yet identified.

Rubus caesius L. Many of the Greek specimens are atypical, having rather well developed prickles on the stem, and black fruits with little or no blue bloom. The species was listed for Peloponnisos by Mermygkas, Zikos & Constantinidis (2021: supplementary file), probably in error. All verified Greek records are from the northern and central mainland.

Rubus ulmifolius Schott. This name has frequently been applied to the common *Rubus* of roadside thickets at low altitude. Taken in a narrow sense *R. ulmifolius* is probably restricted to W Europe and the W & C Mediterranean area, with similar plants in the E Mediterranean area belonging to *R. sanctus* Schreb. The two are doubtfully distinct and perhaps better regarded as geographical races. Other names are also involved, including *R. creticus* L. which is arguably not validly published. For a discussion of the convoluted nomenclature, see Ferrer-Gallego & Van de Beek (2021).

Rumex bucephalophorus L. Scattered collections from the Ionian and Aegean islands have been referred to subsp. *gallicus* (Steinh.) Rech. f., a taxon otherwise confined to the W & C Mediterranean area. It lacks geocarpic flowers and otherwise combines characters of subsp. *aegaeus* Rech. f. and subsp. *bucephalophorus*.

Rumex obtusifolius L. Four different subspecies have been recognized in Greece, viz. subsp. *obtusifolius*, subsp. *subalpinus* (Schur) Rech. f., subsp. *sylvestris* (Wallr.) Čelak. and subsp. *transiens* (Simonk.) Rech. f. They are difficult to distinguish and their distribution areas overlap completely. Only a collective map is provided.

Rumex patientia L. Most, perhaps all, Greek collections represent subsp. *orientalis* (Bernh.) Danser. Records from Evvia and Peloponnisos are feasible, but the specimens are too young for positive identification.

Ruppia cirrhosa (Petagna) Grande. This name was used in the Greek checklist (Dimopoulos & al. 2013: 139) as well as Euro+Med PlantBase. Ito & al. (2017a) have subsequently shown that *R. spiralis* L. ex Dumort. is the correct name for this species which differs from *R. maritima* L. in the long and coiled peduncles.

Ruscus hypophyllum L., a W Mediterranean species, is sometimes cultivated as an ornamental and may be locally naturalized, e.g., in Kriti, Rodos and Kerkira. It resembles *R. hypoglossum* L., but has smaller, ± scarious bracts with 1-3 veins.

Ruta chalepensis L. There are gradual transitions between subsp. *chalepensis* and subsp. *fumariifolia* (Boiss. & Heldr.) Nyman (syn. *R. fumariifolia* Boiss. & Heldr.), and the latter is at least partly a modification to hot and dry habitats at low altitude in the southern part of the range of the species. It was described from Cape Sidero in E Kriti, based on a collection by Heldreich.

Saccharum spontaneum L. has been reported from Kriti and Kos as well as a few localities in W Peloponnisos and most recently from the island of Limnos (Ristow, Krause & Bergmeier 2021: 361). It is cultivated as an ornamental and doubtfully naturalized. It differs from *S. ravennae* (L.) Murray in being rhizomatous, with longer silky hairs from the callus, and lemma without awn.

Sagina subulata (Sw.) C. Presl. For nomenclatural reasons the name has changed to *C. alexandrae* Iamonico (see Iamonico 2016: 164-165).

Salicornia europaea complex. According to Kadereit & al. (2012), *S. europaea* L. is a species of the Atlantic coasts of W Europe. In Greece it is replaced by two species with the rather misleading names *S. perennans* Willd. and *S. procumbens* Sm. The holdings at B have been revised by Th. Raus, but many other records have been identified only as *S. europaea* s. lat. Thus three maps are provided, one for *S. perennans* and one for *S. procumbens* as well as a collective map for the *S. europaea* complex.

Salix aegaea Cambria & al. is a name recently given to a local population of *S. pedicellata* Desf. from Ikaria (Cambria & al. 2020).

Salix aegyptiaca L. was reported from NE Greece ("near Komotini") in Fl. Eur. ed. 2, 1: 60 (1993). The record refers to the subsequently described *S. xanthicola* K. I. Chr.

Salix ×ardana Ziel. & Pancheva (= *S. alba* L. × *S. xanthicola* K. I. Chr.) was described from Bulgarian Rodopi and may well occur also on the Greek side.

Salix aurita L. was listed for "Gr" in Euro+Med PlantBase, but no specimens have been seen, and literature records from N Greece (Voliotis 1967: 101, Pavlides 1985: 248) probably refer to misidentified material of *S. caprea* L. or *S. cinerea* L.

Salix babylonica L., a native of China, is sometimes planted in parks, by dams, etc. It is a medium-sized to large tree resembling *S. alba* L., but with very slender, pendent twigs ("weeping willow") and narrower leaves which soon become sparsely sericeous or glabrous on both sides.

Salix euxina I. V. Belyaeva is a recently described species related to *S. alba* L. and *S. fragilis* L. It is widespread in Europe and Anatolia, and may also occur in Greece. Herbarium specimens have so far mostly been identified as *S. fragilis*.

Salix kaptarae Cambria, C. Brullo & Brullo is a name recently given to a local population of *S. pedicellata* Desf. from W Kriti (Cambria, C. Brullo & S. Brullo 2019).

Salix pentandra L. A report by Lavrentiades (1956: 56) from the vicinity of Kavalla in NE Greece is probably incorrect.

Salix ×rubens Schrank (= *S. alba* L. × *S. fragilis* L.) is widely planted and sometimes naturalized. There are scattered records almost throughout Greece.

Salix* ×*velchevii Ziel. & Pancheva (= *S. amplexicaulis* Bory & Chaub. × *S. xanthicola* K. I. Chr.) was described from Bulgarian Rodopi and has subsequently been reported also south of the border (Kailis & Eleftheriadou 2011: 280).

Salsola kali complex. Taxonomy follows Rilke (1999b) and Mosyakin (2017a). *S. kali* L. s. str. is a coastal, Atlantic-Baltic taxon not occurring in Greece. *S. tragus* L. occurs mainly in ruderal, inland habitats. All coastal records previously called *S. kali* s. lat. have been referred to S. *squarrosa* Steven ex Moq., an eastern Mediterranean taxon described from the island of Naxos. The latter corresponds to what Rilke (1999b: 133) called *S. tragus* subsp. *pontica* (Pall.) Rilke. Some coastal records may actually represent *S. tragus*, but detailed examination of fruiting material is needed for positive identification. *Kali dodecanesicum* C. Brullo & al., described from the islands of Rodos and Kos, is a synonym of *Salsola squarrosa*.

Salsola soda L. Following Mosyakin (2017a) the coastal species generally known as *Salsola soda* is now named *Soda inermis* Fourr. The latter name has also been accepted in the recent Italian checklist (Bartolucci & al. 2018).

Salvia candidissima Vahl was described from Armenia. Hedge (in Fl. Turkey 7: 447-448) recognized subsp. *candidissima* in C & E Anatolia and subsp. *occidentalis* Hedge in W Anatolia and the Balkan Peninsula. In Greece the latter occurs mainly on screes of serpentine and flysch in the north-west. However, it was collected much further south, near the Isthmus of Korinthos, already by Mill in 1855 (see Halácsy 1902a: 487 and Pearce 2006: 160). This record was considered dubious, but in 2018 the species was recollected in the same area by Willing (no. 286884, B; conf. A.S. 2021-09-22).

Salvia forskahlei L. was reported from "Agion Pnevma" in nomos Dramas by Goulimis (1956a: 14). The specimen at ATH was revised as *S. teddii* Turrill by Hedge in 1969. However, *S. forskahlei* is known from SE Bulgaria and could well occur in adjacent parts of Greece. The original spelling of the epithet in Linnaeus's Mantissa (1767: 26) is *Forskåhlea*.

Salvia microphylla Kunth has been collected once in W Peloponnisos, near Taxiarches (Kit Tan & al. 31119). It was apparently cultivated as a hedge plant and scarcely naturalized. It is native to SE USA and Mexico.

Salvia nemorosa L. Both subsp. *nemorosa* and subsp. *tesquicola* (Klokov & Pobed.) Soó (= var. *pseudosylvestris* (Stapf) Bornm.) have been reported from Greece, but the differences are obscure.

Salvia nutans L. has been reported from the foothills of Mt Vermion by Grisebach (1844: 113) and from Granitis in nomos Dramas by Goulimis (1960a: 17). Goulimis' specimen at ATH was revised as *S. teddii* Turrill by I. C. Hedge, 1969.

Salvia pinnata L. was listed for the East Aegean Islands in Euro+Med PlantBase, probably based on a record by Candargy (1897) cited in Flora of Turkey 7: 418 (1982). There is no subsequent confirmation (I. Bazos, e-mail 2020-11-01), and the record must be regarded as dubious.

Sanguisorba officinalis L. was listed for Peloponnisos by Mermygkas, Zikos & Constantinidis (2021: supplementary file), but no details are available.

Sanguisorba rupicola (Boiss. & Reut.) A. Braun & C. D. Bouché was listed from NC in the Greek checklist (Dimopoulos & al. 2013: 137), but this was apparently based on a misidentified specimen. *S. rupicola* occurs in the W Mediterranean area eastwards to Sicily.

Santolina chamaecyparissius L. has been reported from a few localities in the Aegean area, including Kriti, Milos, Leros, Ikaria and Thasos, but probably only as a garden escape. It belongs to a species complex native in the W & C Mediterranean region. It is a bushy perennial with narrow, pectinate, grey leaves, and solitary, discoid, yellow capitula on long peduncles.

Saponaria calabrica Guss. Local populations from the islands of Kefallinia and Elafonisos have been described as *S. aenesia* Heldr. and *S. jagelii* Phitos & Greuter, respectively. The latter is a fairly distinct coastal ecotype, recently reported also from the island of Limnos (Bergmeier & al. 2021: suppl. 1 p. 9).

Saponaria haussknechtii Simmler was based on a collection by Haussknecht from "Pindus tymphaeus, in summo monte Zygus (Lakmon veterum) supra Metsovo, substratu silico-serpentino". On the type specimen at Herbarium K, Turrill has made the following note dated 30.12. 1922: "While I agree with Simmler that the Thessalian plant is distinct from *S. depressa* of N. Afr. & Sicily I think there is only one species in Thessaly. I cannot find sufficient morphological characters to separate *S. haussknechtii* specifically from *S. intermedia* and regard the former as a habitat form of the latter". Turrill's view agrees with my own experience and is followed here. *S. stranjensis* D. Jord. from the E Balkan Peninsula and *S. sicula* Raf. (= *S. depressa* Biv., nom. illeg.) from Sicily, Sardinia and NW Africa are closely related to *S. intermedia* Simmler. *S. stranjensis* has only recently been reported from Greece; for details, see Doumas, Goula & Constantinidis (2022: 20-21).

Sarcocornia A. J. Scott was retained as a separate genus in Fl. Eur. and Fl. Hellen., but re-united with *Salicornia* by Piirainen, Liebisch & Kadereit (2017). In Greece it comprises two woody-based perennials, *Salicornia fruticosa* (L.) L. and *S. perennis* Mill.

Satureja hellenica Heldr. ex Halácsy is a taxon in the *S. montana* complex of uncertain status, closely related to *S. parnassica* Heldr. & Sartori ex Boiss. and mapped under the latter. It occurs at low altitude (100-1000 m) and has relatively large (c. 10 mm), narrowly elliptical leaves. The corolla is whitish flushed or dotted pink.

Saxifraga rotundifolia L. Some plants from wet habitats at high altitude in NW Greece match *S. heucherifolia* Griseb. & Schenk, having long rhizome and petals with fairly large and distinct purple dots. They have here been merged into subsp. *rotundifolia*, but may merit subspecies rank.

Scabiosa atropurpurea L. (*Sixalix atropurpurea* (L.) Greuter & Burdet). Most of the Greek plants have been referred to subsp. *maritima* (L.) Arcang. and some to subsp. *chaniotica* (Rech. f.) Rech. f., but according to Devesa (in Fl. Iberica 15: 332-334, 2007) subspecific taxa are poorly defined.

Scabiosa cinerea Lam. has recently been collected in dry calcicolous grassland on Mt Triklario in NW Greece (Bergmeier 18-475). It is a member of the *S. ochroleuca* group distinguished on the densely stellate-pubescent leaves. It is scattered in S Europe from the Pyrenees to the W Balkan Peninsula.

Scaligeria napiformis (Spreng.) Grande. A related, endemic species, *S. alziarii* Hand, Hadjik. & Zetzsche has recently been described from Cyprus. It grows together with *S. napiformis* and differs mainly in being perennial with a long, branched, horizontal rhizome. Similar plants should be looked for in Greece.

Schedonorus plurifolius (Schult.) Bergmeier & H. Scholz has been reported from wetland habitats in Kriti (and Sicily). It differs from the widespread European meadow grass *Schedonorus pratensis* (Huds.) P. Beauv. (*Festuca pratensis* Huds.) by its less robust habit and somewhat smaller spikelets with awned lemmas. Its taxonomic status is uncertain, and it has been included in the map for *Festuca pratensis* s. lat.

Schinus molle L., a native of the Peruvian Andes, is planted along streets and occasionally spreading into vacant plots, etc. It is a slender, evergreen tree with drooping branches. The leaves are odd-pinnate with 9-20 pairs of narrow leaflets. The fruits ("pink peppercorns") are used as a spice.

Schismus barbatus (L.) Thell. has been reported from the Messara plain in SC Kriti, based on Hügin 6914 (B). In Rechinger (1944a: 777) there is a record of *S. arabicus* Nees from the same area. An old record of *S. calycinus* (Loefl.) K. Koch from Thessaloniki (Charrel 1892: 411) may also refer to *S. arabicus*. *S. barbatus* is a mainly N African species, closely related to *S. arabicus*, but differing in details of lemma and palea. Its status in Greece is uncertain.

Scilla autumnalis L., s. lat. is a polyploid complex (2n = 12, 14, 26, 28, 42, etc.) in need of critical study throughout its range. Speta (2000b, sub nom. *Prospero*) named six new species from Kriti (*P. battagliae*, *P. depressum*, *P. hierapytnense*, *P. idaeum*, *P. minimum* and *P. rhadamanthi*), all based on single collections and with fragmentary descriptions. Several others have been described from elsewhere. Only *P. talosii* Tzanoud. & Kypr. (*S. talosii* (Tzanoud. & Kypr.) Speta from the island of Dia (N of Kriti) is reasonably distinct, with very broad leaves and a high chromosome number (2n = c.150).

Adopting similar hyperinflation at the generic level, Speta (1998a) recognized 34(!) genera within *Scilla* s. lat., most of them new.

Scilla bifolia L. s. lat. is a polyploid complex with 2n = 18, 36 and 54. It was divided by Speta (1981, 1991a, 1998a, etc.) into a number of taxa none of which can be clearly distinguished morphologically. Binomials applied to Greek material include *S. andria* Speta, *S. cydonia* Speta, *S. longistylosa* Speta, *S. nivalis* Boiss., *S. pneumonanthe* Speta, *S. reuteri* Speta, *S. subnivalis* (Halácsy) Speta and *S. voethorum* Speta.

Scilla hyacinthoides L. is probably native from W Anatolia to Palestine (cf. Speta 1981: 182-188). It is cultivated and occasionally naturalized elsewhere in the Mediterranean area. Oddly, Greek records include one from an islet NW of Karpathos (Runemark & Persson s.n., 1966-04-07; plants cultivated in Lund, herbarium specimen at LD). Recent reports from Lachania and Monte Smith on Rodos (Manousakis, in litt. 20.1. 2021) could perhaps represent native occurrences.

Scilla nana (Schult. & Schult. f.) Speta has sometimes been divided into *S. nana* s. str. (on Lefka Ori) and *S. albescens* Speta (on other Cretan mountains), but plants from Lefka Ori and Dikti cultivated side by side in the Göteborg Botanic Garden turned out to be indistinguishable.

Scilla peruviana L., a W Mediterranean species, is sometimes cultivated as an ornamental. It has been reported from Kerkira and Kefallinia (https://floraionica.univie.ac.at/), probably as a casual escape.

Scirpoides holoschoenus (L.) Soják. According to Desfayes (2004b) all Greek material belongs to subsp. *australis* (Murray) Soják, which occurs from Italy to Turkestan and India. Subsp. *holoschoenus* occurs in the W Mediterranean area, and there is a third subspecies in South Africa.

Scorzonera L. Recent phylogenetic speculations, based on molecular studies, have led to proposed fragmentation of *Scorzonera* into seven genera (cf. Zaika & al. 2020, Bartolucci, Galasso & Conti 2020).

Scorzonera austriaca Willd. was long known only from a single Greek collection from the northern foothills of Mt Olimbos (Strid & al. 1976-06-10, no. 11340; ATH, LD, herb. Strid). A record from Mt Smolikas is of uncertain status and may refer to a different taxon. Plants matching the Olimbos specimens were recently reported from Mt Achladovouno in nomos Xanthis, sub nom. *Takhtajaniantha austriaca* (Willd.) Zaika, Sukhor. & N. Kilian (Doumas, Goula & Constantinidis 2022: 12).

Scorzonera judaica Eig. Böhling (2001: 322) reported this species from Rodos, but it was not recognized by Kleinsteuber, Ristow & Hassler (2016: 221) who listed Rodos populations under *S. sublanata* Lipsch. Manousakis (in litt., 2021) also reported *S. judaica* from Rodos (Mt Attaviros) as well as the island of Simi. In Fl. Turkey 5: 652 (1975) *S. judaica* appeared under the younger synonym *S. pseudolanata* Grossh. Euro+Med PlantBase listed *S. judaica* as a synonym of *S. psychrophila* Boiss. & Hausskn. which appears unlikely; the latter was described from alpine level in Iran. The group is in need of detailed study.

Scorzonera parviflora Jacq. was reported by Haussknecht (1895: 55) from nomos Karditsis ("In nemorosis oropedii Neuropolis"). There are modern records from nomos Evrou in the far north-east (Korakis & al. 2006: 20). The latter are feasible, but no specimens have been seen, and the occurrence of this species in Greece needs confirmation.

Scorzonera psychrophila Boiss. & Hausskn. See under *S. judaica*.

Scorzonera serpentinica Rech. f. has been collected several times in open pinewoods on ophiolithic substrate in N Evvia and has recently been reported also from a similar habitat on Mt Gerania (Constantinidis 6681, UPA). It is very close to *S. crocifolia* Sm. (differing, e.g., in the less prominent leaf sheaths) and probably not specifically distinct.

Scrophularia canina L. subsp. ***canina*** occurs mainly in C & S Europe including S Russia, and subsp. *bicolor* (Sm.) Greuter occurs in the Balkans and Anatolia, but the precise distribution of the subspecies is uncertain. In Greece subsp. *canina* may be restricted to mountains of the C & NW mainland, preferring serpentine substrate (cf. Raus in Mount. Fl. Greece 2: 198-199, 1991), but there are many intermediates. Records of *S. canina* from Kriti are probably incorrect (Raus, in litt., 2014).

Scrophularia lucida L. is a variable species (or species complex) of the Mediterranean area and SW Asia. No less than four plants illustrated under different names on Plates 598, 599, 600 and 601 (1827 & 1831) of Flora Graeca are now all believed to represent forms of *S. lucida* s. lat.; cf. Boissier (1879: 403-404), Lall & Mill in Fl. Turkey 6: 635-636 (1978) and Raus in Mount. Fl. Greece 2: 191-192 (1991). Most, perhaps all, Greek records refer to subsp. *filicifolia* (Mill.) Rech. f.

Scrophularia lyrata Willd. The Cretan plant was collected already by Heldreich in 1846 ("in fossis subhumidis circa hortos ad Cydoniam [Chania]") and named *S. auriculata* in his herbarium. It was subsequently described as *S. cretica* Boiss. & Heldr. and is now believed to be conspecific with *S. lyrata* Willd., a species described from Portugal. The name *S. auriculata* L. has been conserved (see Taxon 53: 821, 2004) to permit its continued use for a mainly W European taxon with 2n = 84, whereas *S. lyrata* is a closely related, mainly Mediterranean taxon with 2n = 58.

Scrophularia myriophylla Boiss. & Heldr. was described from [Turkey C4 Antalya] "montis Gheidagh Tauri Isaurici", based on a collection by Heldreich. It is known from scattered localities in W & S Anatolia (cf. Uzunhisarcıklı, Güner & Ekici 2018:176). Further studies are needed to establish whether Greek and Turkish plants are indeed conspecific.

Scrophularia nodosa L. was listed for Peloponnisos by Mermygkas, Zikos & Constantinidis (2021: supplementary file), but no details are available.

Scrophularia pindicola Hausskn. was listed as an accepted species and a Greek endemic in Euro+Med PlantBase, but Raus (in Strid & Kit Tan 1991: 198), who had examined the type specimen, regarded it as a synonym of *S. canina* subsp. *bicolor* (Sm.) Greuter.

Scrophulariaceae. Following shifting, ephemeral fashion in recent decades this family has been divided and the fragments have been variously reshuffled, sometimes with absurd results. Like, for instance, Nikolić in the Croatian Flora (2020), I have reverted to a broad circumscription of the family, matching that of Flora Europaea (1972), Flora of Turkey (1978) and Flora of Cyprus (1985).

Scutellaria pinnatifida A. Ham. was listed for the East Aegean Islands in Euro+Med PlantBase. The record probably refers to *S. orientalis* L., which occurs on Mt Kerkis and Mt Karvouni, Samos.

Scutellaria rubicunda Hornem. There are many Greek records of this species, but according to Bothmer (1987) all refer to forms of *S. rupestris* Boiss. & Heldr. *S. rubicunda* s. str. is endemic to Sicily.

Secale strictum (C. Presl) C. Presl. Taxonomy at subspecies level is unsettled. Plants from N Peloponnisos have been reported as subsp. *anatolicum* (Boiss.) K. Hammer (Ristow & Scholz 2013: 162); most other Greek collections are referable to subsp. *strictum*. Several other subspecies occur elsewhere. On the mainland *S. strictum* is often found along mountain roads, but it has recently been reported from low altitude on the island of Limnos (Bergmeier & al. 2021: suppl. 1 p. 28).

Sedum eriocarpum Sm. Eight local and regional subspecies have been recognized in Greece (see 't Hart in Flora Hellenica 2: 324-325, 2003, as well as Polymenakos & Kit Tan 2016a: 101).

Sedum rupestre L. There are several Greek records of this C European species, but all are likely to refer to *S. ochroleucum* Chaix.

Sedum sexangulare L. is known from two localities in the Pindos, but perhaps as a garden escape. Other Greek records refer to *S. apoleipon* 't Hart. *S. sexangulare* is a C European species occurring southwards to N Albania.

Sempervivum ciliosum Craib. In Greek material there is complete overlap between plants identified as *S. ciliosum* and *S. ruthenicum* Schnittsp. & C. B. Lehm. The supposedly distinguishing characters (length of stolons, shape of rosettes, length of cilia of rosette leaves, density of inflorescence, colour of petals and filaments, etc.) occur in different combinations, and there is no geographical or ecological separation. *S. ciliosum* var. *galicicum* A. C. Sm. was based on a single collection from Mt Galičica in North Macedonia not far from the Greek border, and has subsequently been raised to species level as *S. galicicum* (A. C. Sm.) Micevski. Several other poorly defined species have been described.

Sempervivum octopodes Turrill. Sterile plants (photo U. Bergmeier) from granitic rocks at 2200 m on Mt Varnous may belong to this species which is closely related to *S. ciliosum/ruthenicum*. It would be the only Greek record for *S. octopodes* which was described from Mt Pelister, just N of the border; the latter is characterized by long stolons, smaller petals (c. 8 × 1.5 mm) which are pale yellow tinged purple at base, and pale purple filaments.

Senecio angulatus L. f. is a native of South Africa and a more or less established alien in Australia, New Zealand and the W Mediterranean area. In Greece it has recently been reported from the SE Aegean islands of Rodos, Simi and Tilos, currently as a garden escape but perhaps on the verge of becoming established.

Senecio doronicum (L.) L. According to Calvo, Álvarez & Aedo (2015) Greek plants previously referred to this species belong to *S. transylvanicus* Boiss., which was described from the vicinity of Kronstadt in Siebenbürgen [now Brasov in Romania] and has also been recorded from Montenegro, SW Bulgaria and N Greece, whereas *S. doronicum* occurs from the Cantabrian Mountains to the northern Dinaric Alps. For the convoluted nomenclature of *S. transylvanicus*, see Calvo, Álvarez & Aedo (op. cit. p. 37). The only confirmed Greek localities are at high altitude on Mt Smolikas. A collection by Orphanides (1862) from Mt Persisteri (Varnous) most probably refers to a locality just north of the present border. An unconfirmed record of "*S. doronicum*" from Mt Voras (Goulimis 1956a: 12) may also refer to *S. transylvanicus*.

Senecio hercynicus Herborg. Some Greek records have been referred to subsp. *dalmaticus* (Griseb.) Greuter or var. *expansus* (Boiss. & Heldr.) Herborg. The taxonomy is unsettled.

Senecio inaequidens DC., a xenophyte of South African origin, was collected at a locality in nomos Kastorias in 2014 (see Willing & Raus 2015). It is well established in C Europe and has recently reached Bulgaria (Vladimirov & Petrova 2009). Further possible expansion in Greece should be monitored. It is a slender, erect perennial with linear-lanceolate, ± amplexicaul leaves.

Senecio longipedunculatus Halácsy is perhaps conspecific with *S. macedonicus* Griseb. "Unfortunately, *S. longipedunculatus* is only known from the original material [from Mt Parnassos] with immature achenes that lack an essential character to compare it with *S. castagneanus*. Consequently, we prefer to include *S. longipedunculatus* under 'Doubtful or excluded names'. More collections and further investigation is needed to elucidate its taxonomic position" (Calvo, Álvarez & Aedo 2015: 50).

Senecio nemorensis L. Both subsp. *bulgaricus* (Velen.) Greuter and subsp. *jacquinianus* (Rchb.) Čelak. were included in the Greek checklist (Dimopoulos & al. 2013: 64), but all Greek material is likely to belong to the latter.

Senecio tamoides DC., a xenophyte of South African origin, has been reported as a garden escape on the islands of Amorgos (Biel & Kit Tan 2015a: 57), Naxos (Biel & Kit Tan 2020d: 541) and Milos (Biel & Kit Tan 2021c: 244).

Serapias bergonii E. G. Camus. Kühn, Pedersen & Cribb (2019: 349) treated this taxon as *S. vomeracea* subsp. *laxiflora* (Soó) Gölz & H. R. Reinhard.

Serapias cordigera L. Plants from Kriti and some other Aegean islands have been referred to subsp. *cretica* B. & H. Baumann, but the differences are obscure.

Serapias neglecta De Not. The taxonomy is unsettled. Kühn, Pedersen & Cribb (2019: 346 ff.) regarded *S. neglecta* as endemic to the French and Italian Riviera, Corsica and Sardina; *S. neglecta* subsp. *ionica* (H. Baumann & Künkele) H. Baumann & R. Lorenz was listed as a synonym of *S. orientalis*.

Serapias orientalis (Greuter) H. Baumann & Künkele has been divided into subsp. *orientalis* and subsp. *carica* H. Baumann & Künkele. The morphological differences are obscure and their geographical ranges completely overlap.

Serratula tinctoria L. Maire & Petitmengin (1908: 128) reported this species from the vicinity of Malakasi in Northern Pindos, observing: "Nous n'avont malheureusement rapporté aucun spécimen de cette plants, que nous ne savions pas inconnue en Grèce à ce moment, mais nous est suffisamment familière pour que nous l'ayons notée avec certitude ...". The occurrence of *Serratula tinctoria* in Greece is feasible but needs confirmation.

Sesamum indicum L. is one of the oldest oilseed crops, first domesticated on the Indian subcontinent perhaps 5,000 years ago and now widely cultivated in tropical and subtropical countries.There are scattered records from Greece, the first already by Clementi (1855: 67) "in Attica, ad Kiephyssum". Modern records are available from Kriti, the Thessaloniki area and some Kikladian islands, but the species is scarcely naturalized.

Seseli annuum L. was reported by Quézel & Contandriopoulos (1968: 29) from "forêts de hêtres et de chênes sur le Vermion". The authors are notoriously unreliable, but the fact that the species is known from Bulgaria and has recently been recorded also from North Macedonia (Teofilovski 2017: 100) renders some credibility to their report.

Seseli crithmifolium (DC.) Boiss. "Tournefort a trouvé cette plante rare sur un bloc de marbre inhabité au milieu de la mer, non loin de l'île d'Amorgo, et nommé Nicouria. Moi, j'ai trouvé cette plante sur les rochers maritimes de l'île de Pholegandros" (Orphanides 1870b: 131). The species was recollected on Nikouria by Runemark & Bentzer (30.6. 1967, no. 30586, LD). There are several subsequent records from Folegandros.

Seseli halkensis Cattaneo, Kit Tan & Biel was described from the East Aegean islands of Chalki and Tilos, growing together with *S. crithmifolium*. It appears to represent individual variation within the latter.

Seseli tortuosum L. Lyskov & al. (2018) divided *S. tortuosum* s. lat. into *S. tortuosum* L. in the W & C Mediterranean area and N Africa, and *S. arenarium* M. Bieb. in SE Europe and SW Asia. Greek material of *S. arenarium* is not homogeneous and should perhaps be subdivided. It occurs both in maritime sand in the north-east and in rocky inland habitats from C Peloponnisos to Northern Pindos. In spite of the very different habitats the morphological differences are not clear. Plants from steppic habitats in E Bulgaria and E Romania have recently been named *S. besserianum* Stoyanov & Ostr., nom. nov. (= *S. peucedanifolium* Besser, nom. illeg.), see Stoyanov & al. (2020), but the populations from coastal habitats in NE Greece are scarcely conspecific with the latter.

Sesleria alba/anatolica. *S. alba* Sm. was described from the Belgrade forest near Istanbul and illustrated on Plate 72 of Flora Graeca (1806). *S. anatolica* Deyl was described from W Anatolia and Lebanon. Although the monographer Deyl (1946) regarded them as separate, the latter has subsequently been reduced to a synonym of the former (cf. Kit Tan in Fl. Turkey 9: 525, 1985 and Di Pietro & al. 2017: 102).

Sesleria coerulans Friv. was listed for Peloponnisos by Mermygkas, Zikos & Constantinidis (2021: supplementary file), probably in error. Confirmed Greek records are all near the northern borders.

Sesleria korabensis (Kümmerle & Jáv.) Deyl was described from the border area of Albania and North Macedonia. Greek records probably refer to *S. tenerrima* (Fritsch) Hayek. The two are doubtfully distinct (cf. comments in Mount. Fl. Greece 2: 779, 1991).

Sesleria latifolia (Adamović) Degen (*S. heuflerana* var. *latifolia* Adamović) was described from Serbia and has been lectotypified with a collection from Suva Planina (see Di Pietro & al. 2017: 104). Greek records are likely to refer to *S. robusta* Schott, Nyman & Kotschy.

Sesleria rigida subsp. ***achtarovii*** (Deyl) Deyl has been recorded from NE Greece, e.g., in Mount. Fl. Greece 2: 780 (1991). Kuzmanović & al. (2013a) restored it to species rank (as originally described) and regarded *S. rigida* Rchb. as an endemic of the Romanian Carpathians.

Sesleria robusta Schott, Nyman & Kotschy was listed for Peloponnisos by Mermygkas, Zikos & Constantinidis (2021: supplementary file), probably in error. All confirmed Greek records are in the northern and central mainland.

Setaria adhaerens (Forssk.) Chiov. and ***S. verticillata*** (L.) P. Beauv. have generally been considered different species in European Floras, but Morrone & al. (2014: 128 ff.) found that the supposedly distinguishing characters show continuous variation, and merged them under the latter name. Crespo & al. (2020: 144) typified *S. adhaerens* with a Forsskål specimen from Yemen and maintained that it is different from *S. verticillata*. Since they have been kept separate in the Flora Hellenica Database, two maps are presented.

Setaria faberi R. A. W. Herrm., a xenophyte of E Asiatic origin resembling *S. viridis* (L.) P. Beauv., was reported by Bergmeier (2007: 442) from irrigated gardens in W Kriti. In the monograph by Morrone & al. (2014: 44-46) it was recorded only from China, Korea and Japan. In Greece it is probably not yet established.

Setaria italica (L.) P. Beauv. is grown as a crop in southern Europe and in tropical and subtropical Asia. It is almost certainly derived from *S. viridis* (L.) P. Beauv. (cf. Morrone & al. 2014: 68). Euro+Med PlantBase reported it as casual or naturalized in Greece, but no specimens or specified records have been seen.

Setaria verticilliformis Dumort. has been reported from Lesvos (e.g., Nielsen 10823; B, G, UPA; sub nom. *S. decipiens* Schimp.) and subsequently from Milos (Biel & Kit Tan 2019c: 301) and Naxos (Biel & Kit Tan 2020d: 542), and also listed for Peloponnisos by Mermygkas, Zikos & Constantinidis (2021: supplementary file). It has been considered to fall within the range of variation of *S. verticillata* (L.) P. Beauv., but was accepted as a separate species by Morrone & al. (2014). *S. verticilliformis* was described from Luxembourg already in 1827. It is probably a European species but widely naturalized in other continents. Its status in Greece is uncertain.

Sida spinosa L., a widespread tropical/subtropical weed, has been observed by Uwe Raabe at the edge of a maize field in Peloponnisos. It may be on the verge of becoming established. There are rather few and scattered records elsewhere in Europe.

Sideritis. It has recently been proposed to merge the genus *Sideritis* into *Stachys* (see Bartolucci, Peruzzi & Soldano 2014, Bartolucci & al. 2018), thus making room for many new combinations.

Silene cephallenia Heldr. Plants from the mainland have been recognized as subsp. *epirotica* Melzh. They are scarcely different from *S. cephallenia* s. str.

Silene crassipes Fenzl has been reported from Greece, based on a single collection (Stamatiadou 5489, ATH) which, however, matches *S. muscipula* L. *Silene crassipes* is probably restricted to south central Anatolia, Syria and Palestine, whereas *S. muscipula* is (or was) widespread in the Mediterranean area. Somewhat mysteriously, Flora Europaea (ed. 2, 1: 214, 1993, in a note under *S. linicola*) reported *S. crassipes* from "Thrace". This was probably taken from the imprecise "Thra" in Hayek's Prodromus. The latter may refer either to NE Greece or S Bulgaria. The species is not included in the Bulgarian Flora, and there seem to be no records from NE Greece; the report is thus unconfirmed and probably incorrect.

Silene falcata Sm. A single, historical record from Mt Athos is almost certainly incorrect – see comments in Strid & Strid (2011: 72).

Silene fetlerii D. Pavlova, previously known only from a small area in S Bulgaria, has recently been reported from a locality in nomos Xanthis, on serpentine substrate (Doumas, Goula & Constantinidis 2022: 22-23).

Silene gemmata Meikle, previously believed to be endemic to Cyprus, was reported from the summit area of Mt Attaviros, Rodos, by Kleinsteuber, Ristow & Hassler (2016: 346).

Silene gigantea L. Subdivision of this variable species follows Du Pasquier, Naciri & Jeanmonod (2015) and Du Pasquier, Jeanmonod & Yamama (2017).

Silene guicciardii Boiss. & Heldr. is a local endemic first collected by Samaritani & Guicciardi in 1857 in the southern foothills of Mt Parnassos. It was not rediscovered until 2015, presumably in the same locality. After a careful study, Trigas, Kokkoris & Kougioumoutzis (2017) reduced it to a subspecies of *S. marschallii* C. A. Mey, which is widespread in SW Asia.

Silene heldreichii Boiss. According to Iamonico (2018: 122) *S. remotiflora* Vis. (described from Dalmatia) is a synonym of *S. heldreichii* Boiss. (described from S Anatolia). Greuter (in Fl. Hellen. 1: 310, 1997) regarded the two as separate species, and the latter opinion is followed here.

Silene horvatii Micevski was described from Mt Kajmakčalan just N of the Greek border. It is of uncertain status and perhaps conspecific with *S. schwarzenbergeri* Halácsy.

Silene integripetala Bory & Chaub. Local populations from the island of Elafonisos, NW Kriti and C Kriti have been described as subsp. *elaphonisica* Oxelman, subsp. *lidenii* Oxelman and subsp. *greuteri* (Phitos) Akeroyd, respectively.

Silene multicaulis Guss. A compact, broad-leaved form occurring locally at high altitude in Peloponnisos and Sterea Ellas has been included in subsp. *multicaulis* (cf. Zarkos & al. 2020: 568-569).

Silene nemoralis Waldst. & Kit. Literature records from N Greece, sub nom. *S. italica* subsp. *nemoralis* (Waldst. & Kit.) Nyman, need confirmation. They may refer to *S. italica* (L.) Pers. subsp. *italica*.

Silene nutans L. was listed for Greece in Fl. Eur. (ed. 2, 1: 199, 1993), but the source for this record is not clear. Reports from the Vikos gorge on Mt Timfi (Chitos 2009: 69) and the Vouraikos gorge in N Peloponnisos (Koutsopoulos & Sarlis 2003: 309) are probably incorrect. It is not unthinkable that this species could occur in the far north of Greece as there are confirmed reports from Albania, North Macedonia and Bulgaria.

Silene pendula L. is an annual with fairly large, pink flowers, native in S Italy and cultivated elsewhere as an ornamental. It has been reported from the Thessaloniki area (Krigas & Kokkini 2004: 87), but is doubtfully naturalized.

Silene pinetorum Boiss. & Heldr. has been divided into the easterly subsp. *pinetorum* and the westerly subsp. *sphaciotica* Oxelman & Greuter, but the differences are obscure.

Silene pseudobehen Boiss. (1843) was based on a collection by Aucher from "Asiâ minore", differing from *S. behen* L. in the small size, asymmetrical inflorescence, larger petals and longer stamens. There are a few reports from the East Aegean Islands as well as Santorin. Although retained as a separate species by Greuter in Fl. Hellen. 1: 306 (1997) it is doubtfully distinct, and probably best treated as a variety of *S. behen*. Indeed, Boissier later (1867: 584) changed his mind and reduced it to *S. behen* var. *minor* Boiss.

Silene pusilla Waldst. & Kit. Taxonomy of the Greek representatives in the *Silene* (*Heliosperma*) *pusilla* complex is provisional, but they seem to fall into two distinct groups: (1) many-stemmed, caespitose, pink-flowered plants growing in wet, non-calcareous habitats, here called *S. pusilla* subsp. *albanica* (K. Malý) Greuter & Burdet, and (2) few-stemmed, laxly caespitose, white-flowered plants growing in shady crevices of calcareous cliffs. The latter comprise the taxa called *S. pusilla* subsp. *chromodonta* (Boiss. & Reut.) Greuter (Mt Olimbos), *S. pusilla* subsp. *tymphaea* Greuter (high altitude on Mt Timfi), and *S. intonsa* Greuter & Melzh. (Aoos river gorge in the foothills of Mt Timfi). The two first-mentioned are very similar and have here been merged under the name of *S. chromodonta* Boiss. & Reut., whereas *S. intonsa* is provisionally retained as a separate species. According to Stevanović & al. (2014: 254-255) *S. pusilla* subsp. *tymphaea* is a synonym of *Heliosperma pusillum* (Waldst. & Kit.) Rchb. subsp. *monachorum* (Vis. & Pančić) Niketić & Stevan.

Silene retzdorffiana (K. Malý) H. Neumayer was listed for Greece in Euro+Med PlantBase. The record probably goes back to a report by Quézel & Contandriopoulos (1965b: 55) which in fact refers to *S. intonsa* Greuter & Melzh.

Silene salamandra Pamp. was listed from "Tu" in Euro+Med PlantBase, referring to Flora of Turkey vol. 10 (1988: 80). However, in this publication the species is recorded only from Rodos. It is by no means unthinkable that it could occur also in adjacent parts of SW Anatolia, but for the time being it remains a local endemic of the island of Rodos.

Silene thymifolia Sm. Greek records (cf. Fl. Turkey 2: 210, 1967) are probably incorrect. The species is related to *S. supina* (M. Bieb.) and *S. spergulifolia* (Willd.) M. Bieb., and probably restricted to coastal sand in the Istanbul area – see comments in Strid & Strid (2011: 22).

Silene uniflora Roth. Greek records certainly refer to forms of *S. vulgaris* L.

Silene viscaria (L.) Jessen, also known as *Lychnis viscaria* L. or *Viscaria vulgaris* Bernh., is widespread in Europe. Most Greek literature records are likely to refer to the closely related *S. atropurpurea* (Griseb.) Greuter & Burdet, which replaces *S. viscaria* from the C Balkan Peninsula southwards, but true *S. viscaria* may occur in the far NE of Greece.

Silene vulgaris subsp. ***bosniaca*** (Beck) Greuter & al. is apparently a synonym of subsp. *commutata* (Guss.) Hayek.

Siler, see *Laserpitium*.

Sinapis alba L. Almost all Greek records refer to subsp. *mairei* (H. Lindb.) Maire, which is the wild form of the species. The cultivated subsp. *alba* is occasionally found in ruderal habitats.

Siphonostylis. *Iris unguicularis* Poir. and a few related taxa were transferred to the new genus *Siphonostylis* by Schulze (1965). This split has had few followers. Subdivision of the variable *I. unguicularis* is according to Davis & Jury (1991).

Sisymbrium altissimum L. was listed for Peloponnisos by Mermygkas, Zikos & Constantinidis (2021: supplementary file), probably in error. Verified Greek records are all in the north.

Solanum chenopodioides Lam. was reported from a ruderal habitat on the island of Thasos by Biel & Kit Tan (2017a: 123). It is a native of temperate S America and widely naturalized elsewhere, resembling *S. nigrum* L. but with strongly deflexed fruting pedicels (bent downwards at junction with the stem) and ± appressed fruiting calyx lobes (see also Särkinen & al. 2018: 65 ff.). Additional Greek records may be hiding among specimens identified as *S. nigrum*.

Solanum elaeagnifolium Cav., the "Texas tomato", first appeared in the Thessaloniki area at the time of the German occupation of World War II, hence being nicknamed "Germans". In this case, however, the Germans are likely to have been innocent, the prime suspect being the American Farm School in Thessaloniki. It is one of the most invasive species in Greece, often abundant in ruderal habitats and flowering throughout the summer.

Solanum linnaeanum Hepper & P.-M. L. Jaeger (*S. sodomaeum* L.). Rechinger 19104 (herb. Strid) from the vicinity of Chalkis, Evvia, originally registered as *S. sodomaeum*, has been revised as *S. rostratum* Dunal, and Snogerup & Snogerup 19899 (LD) from the vicinity of Mesolongion (labelled *S. sodomeum*) has turned out to be *S. elaeagnifolium* Cav. A record of *S. sodomaeum* "in hortulanus ad Prevesa" (Baldacci 1894b: 99) is feasible but has not been subsequently confirmed. The illustration of *S. sodomaeum* in Flora Graeca 3: tab. 235 (1819) was apparently based on material from Sicily. The species occurs in the W & C Mediterranean area.

Solanum nigrum L. Pubescent specimens have been recognized as subsp. *schultesii* (Opiz) Wessely. The latter has been recorded almost throughout Greece, but according to Särkinen & al. (2018) there is continuous variation and no infraspecific taxa can be distinguished. *S. decipiens* Opiz was listed as an accepted species in Dimopoulos & al. (2013: 142), but is now regarded as a synonym of *S. nigrum* subsp. *schultesii*.

Solanum nitidibaccatum Bitter is a diploid species native to temperate S America and naturalized in N America, Europe, Australia and New Zealand. In Greece it may be more common than indicated by the records, having perhaps been confused with *S. nigrum* L. *S. physalifolium* Rusby is a synonym of *S. nitidibaccatum*.

Solanum pseudocapsicum L., a native of South America, is cultivated as an ornamental mainly for its globose, shiny, orange-red berries 10-15 mm in diam. It has been reported from ruderal habitats in Peloponnisos, NW Greece, Limnos and Lesvos, but is scarcely naturalized.

Solenopsis minuta subsp. ***annua*** Greuter & al. has recently been reported also from Cyprus and simultaneously raised to species rank, as *S. annua* (Greuter & al.) Hand & Christodoulou, comb. nov. (see Christodoulou, Hand & Iosif 2020: 5).

Solenopsis sp. On the Flora Ionica website (https://floraionica.univie.ac) *Solenopsis notabilis* was listed from the island of Kefallinia. This name does not appear in IPNI, nor in the recent revision of *Solenopsis* (Crespo, Serra & Juan 1998), and the identity of the record is uncertain.

Sonchus palustris L. has been reported from coastal localities in the north-east (Drossos 1992: 105) as well as the delta of the Kalamas river and the Acheron river area in the west (Georgiadis & al. 1994: 180 and Manolaki, Tsakiri & Papastergiadou 2011: 865, respectively). The records may refer to coastal forms of *S. arvensis* L. sometimes distinguished as subsp. *uliginosus* (M. Bieb.) Nyman. *S. palustris* is widespread in C Europe and W Asia; it is considered endangered in Bulgaria where it is scattered in the east (Peev & al. 2015: 614). Mermygkas, Zikos & Constantinidis (2021: supplementary file) reported *S. arvensis* subsp. *uliginosus* from Peloponnisos, probably in error; the rather few verified Greek records are all in the north.

Sorbus baldaccii (C. K. Schneid.) Zinserl. is a taxon of uncertain status, originally described as a variety of *S. aria* (L.) Crantz "aus den Gebirgen Albaniens". Kurtto, Sennikov & Lampinen (2018: 51) mapped it from several localities in the W Balkan Peninsula from Istria to NW Greece, but it was not recognized from Albania by Barina (2017: 414), nor from Croatia by Niketić (2020, 3: 531). From Greece is a single collection from Mt Vourinos (Greuter 16306; B, UPA, herb. Strid) identified as *S. umbellata* subsp. *baldaccii* (C. K. Schneid.) K. I. Chr. – see Christensen in Phitos & al. (1996: 476-477). There is also a report of *S. baldaccii* from N Anatolia (Fl. Turkey 10: 137, 1988). It seems doubtfully distinct from *S. umbellata* (Desf.) Fritsch.

Sorbus borbasii Jáv. is a taxon of uncertain status. Leaf shape suggests that it may be a hybrid between a member of subgen. *Aria* (*S. aria* or *S. umbellata*) and one of subgen. *Cormus* (probably *S. aucuparia*). There are few and scattered records from NE Greece, W Bulgaria and SW Romania. "*S. borbasii* is one of several dozen similar apomictic 'microspecies' of hybrid origin, and recognizing it on the basis of individual specimens is hardly justified" (cited from Kit Tan, Kofinas & Zieliński (2021: 146-147).

Sorbus chamaemespilus (L.) Crantz was reported from the Athos peninsula (above Agios Pavlos) by Grisebach (1843: 93) and from *Fagus* forest on the eastern side of Mt Olimbos by Stojanov & Jordanov (1938: 187). There are no subsequent confirmations from these rather well explored areas, but occurrence in Greece is feasible since the species occurs in C & S Europe with confirmed records from North Macedonia and SW Bulgaria at the far south-east of its range.

Sorbus graeca (Spach) Lodd. ex S. Schauer has been variously treated as a separate species or as a subspecies/variety of *S. aria* (L.) Crantz or *S. umbellata* (Desf.) Fritsch. It is scarcely distinguishable from the latter.

Sorghum bicolor (L.) Moench is widely cultivated in tropical and subtropical areas, probably originating in Africa. In Greece there are few and scattered records from ruderal habitats, and it is scarcely naturalized.

Sorghum drummondii (Steud.) Millsp. & Chase has been reported, as *S. sudanense* (Piper) Stapf, from ruderal habitats in Rodos (Böhling & Scholz 2003: 77) and Naxos (Böhling, pers. comm. 2011). It is of tropical African origin and occasionally cultivated as a fodder crop in SE Europe. It is doubtfully naturalized in Greece.

Spartina maritima (Curtis) Fernald has been reported from lake Mitrikou in nomos Rodopis by Drossos (1992: 115) and from Mamousia in nomos Achaias by Verroios & Georgiadis (2011: 614). Both need confirmation. *S. maritima* is a W European species unlikely to occur in Greece.

Spartina versicolor E. Fabre was included in the Greek checklist (Dimopoulos & al. 2013: 129), based on Schuler 98/320 from the Nestos river delta which had been registered under that name in the Flora Hellenica Database. The specimen has subsequently been revised as *Phacelurus digitatus* (Sm.) Griseb. (Raus, 2021-03-08). *S. versicolor* occurs in the W & C Mediterranean area with its easternmost locality in sand dunes near Vlora in Albania (Barina 2017: 378, Raus & Schuler 2021: 150).

Spartium junceum L. Most of the Greek records certainly refer to subspontaneous occurrences. It is commonly planted along roads and spreads for some distance into the natural vegetation, sometimes in large quantity. However, it is also a constituent of natural scrub far from roads and habitations, e.g., in the Athos peninsula.

Spergularia lycia P. Monnier & Quézel is a dubious species, described from Sandras Dağ and Ak Dağ in SW Anatolia. Zaffran (1990: 96) reported it from a doline in Lefka Ori, but this record is likely to refer to a form of *S. rubra* (L.) J. Presl & C. Presl.

Spinacia oleracea L. is cultivated as a vegetable (spinach) and may persist for some time in ruderal habitats, but is scarcely naturalized.

Sporobolus virginicus (L.) Kunth is a native of southern USA, widely naturalized in tropical and subtropical regions elsewhere. Records from the Mediterranean area, including Greece, probably refer to *S. pungens* (Schreb.) Kunth. The Italian online checklist (consulted 2019-09-16) regards the two as synonymous, with priority given to *S. virginicus*.

Stachys atherocalyx K. Koch was reported from the Nestos river gorge by Goulimis (1956a: 14) and has been confirmed in this area by Strid & al. (1997-05-21, no. 42502, C) and Dinter (2011: 58). These specimens match Bulgarian plants identified as *S. atherocalyx* (cf. illustration in Jordanov & al. 1989: 389). However, this species was described from the Caucasus, and it remains to be established whether Balkan plants are indeed conspecific. *S. atherocalyx* has also been reported from the island of Thasos (Biel & Kit Tan 2018c: 403), but a photo in this publication shows a broader-leaved plant with triangular calyx teeth and white flowers – perhaps a form of *S. recta* L.

Stachys cretica L. is extremely variable and the subdivision is provisional. A new subspecies, subsp. ***gulendamiana*** Akçiçek, has recently been described from maritime sand in W Anatolia, see Akçiçek & Güner in Phytotaxa 539(3): 257-264 (2022). It should be looked for on the East Aegean islands.

Stachys iva Griseb. According to Micevski (1969b) this species is endemic to a small area near Prilep in the SW part of North Macedonia. Other records previously identified as *S. iva*, including all from Greece, were referred to the closely related *S. horvaticii* Micevski. Krestovskaya (2017) listed the latter as a synonym of the former, and divided *S. iva* into three rather poorly defined varieties. The latter approach is followed here.

Stachys menthifolia Vis. was listed for Peloponnisos by Mermygkas, Zikos & Constantinidis (2021: supplementary file), probably in error. Verified Greek records are all in the north-west.

Stachys pangaea Phitos was accepted as an endemic species in Mountain Flora of Greece (Baden in Strid & Kit Tan 1991: 103-104). However, with more material now available it is evident that it falls within the range of variation of *S. plumosa* Griseb.

Stachys plumosa Griseb. Some records from N Greece refer to subsp. *freynii* (Hausskn.) Hayek, a taxon of uncertain status originally described (as *S. freynii* Hausskn.) from Mt Voutsikaki, "in subalpinis P[indi] D[olopici] in saxosis Gionscala [Kazarma] supra Sermeniko".

Stachys recta L., s. lat. Variation in this species is complex and imperfectly understood. For the time being only a preliminary, collective map can be presented. Some of the more distinct entities are subsp. *olympica* Stoj. & Jordanov from Mt Olimbos and subsp. *baldaccii* (K. Malý) Hayek from the Pindos (often on serpentine), but especially at lower altitude the variation is confusing. See also Lenherr (1983), Chrtek (1992a) and Hartvig in Strid & Kit Tan (1991: 105-107).

Stachys spinosa L. was listed for Peloponnisos by Mermygkas, Zikos & Constantinidis (2021: supplementary file), but no details are available. So far the species has been considered endemic to Kriti and Karpathos with an outlying locality on Anafi.

Stachys thirkei K. Koch is a species of somewhat uncertain status, belonging to the complex of *S. germanica* L. and *S. cretica* L. It was described from the vicinity of Bursa and has subsequently been reported from scattered localities in W Anatolia, the Balkan Peninsula and Italy. It is closely related to *S. cretica* and to *S. byzantina* K. Koch, another Anatolian species, but was accepted as separate in the recent Greek and Italian checklists.

Stachys tymphaea Hausskn. was listed for Peloponnisos by Mermygkas, Zikos & Constantinidis (2021: supplementary file). The record is feasible but no details are available.

Staphisagria. *Delphinium staphisagria* L. is rather different from other members of this genus and is optionally named *Staphisagria macrosperma* Spach.

Stellaria alsine Grimm. After the preparation of the maps this species has been reported from the Rodopi mountains by Charalampidou, Eleftheriadou & Theodoropoulos in Fl. Medit. 32: 85 (2022).

Stellaria media group. The account is provisional as only a small part of the vast holdings in the major herbaria (notably at B) has been critically revised. Matters have been complicated by the recent description of a new species, *S. ruderalis* (Lepší & al. 2019), which is most similar to *S. media* (L.) Vill. s. str.; it was recorded mainly from C Europe but is probably much more widespread. It appears to be a well characterized evolutionary unit, but herbarium specimens can be identified only after careful examination of several morphological characters, with ripe seeds proving the only safe clue. Many, perhaps most, Greek records mapped as *S. media* are likely to refer to *S. ruderalis*. The "lesser chick weed" was called *S. apetala* Ucria in the Greek checklist (Dimopoulos & al. 2013: 83), but the correct name is apparently *S. pallida* (Dumort.) Piré.

Stenotaphrum secundatum (Walter) Kuntze has been collected in a few localities in Greece, including the islands of Kerkira, Kefallinia, Zakinthos and Skiros, growing in irrigated lawns and scarcely spreading beyond such habitats. It is a tropical species, reported to be locally naturalized in SW Europe. It is a stoloniferous perennial with short spikes in which the small, coriaceous spikelets are partly sunk into cavities on one side of a broad, corky rhachis.

Sternbergia minoica Ravenna was based on a single collection from Kriti cultivated in Santiago, Chile. It is almost certainly conspecific with *S. lutea* (L.) Ker. Gawl. ex Spreng.

Sternbergia vernalis (Mill.) Gorer & J. H. Harvey is a spring-flowering species similar to *S. lutea*. It has been reported from a single locality in E Kriti where it was probably introduced and not fully naturalized.

Stipa balcanica (Martinovský) Kožuharov, see *S. pennata* complex.

Stipa capensis Thunb. The Mediterranean plant was generally known as *S. tortilis* Desf. until found to be conspecific with *S. capensis*. The latter is fairly widespread in the Western Cape of South Africa northwards to Namibia, and was first described by Thunberg in 1794. Meikle (1985: 1792) failed to find any consistent differences between Mediterranean and South African plants. The latter may well have been naturalized from Mediterranean stock already at the time of Thunberg.

Stipa endotricha Martinovský, see *S. pennata* complex.

Stipa eriocaulis Borbás, see *S. pennata* complex.

Stipa isoldae H. Scholz (1989b) was based on a single collection from serpentine substrate on Mt Vourinos. There appear to be no differences between this and *S. rechingeri* Martinovský (1972) which was likewise described from serpentine mountains in NW Greece, but not mentioned in the description of *S. isoldae*. Both can probably be included in *S. pennata* s. lat.

Stipa monticola H. Scholz, see *S. pennata* complex.

Stipa neesiana Trin. & Rupr. = *Nassella neesiana* (Trin. & Rupr.) Barkworth has been reported once (Greuter & Raus 2004: 78) from a lawn at the University of Thessaloniki campus. It is native in temperate South America and sometimes naturalized elsewhere (including some records from C & W Europe).

Stipa novakii Martinovský, see *S. pennata* complex.

Stipa pennata complex. Adopting an extremely narrow species concept, Martinovský (1967a, 1967b, 1970, 1972, 1975, etc.) recognized numerous local and regional species in E & SE Europe, including at least four in Greece. According to the much wider species concept of Freitag (1985) all Greek material is referable to *S. pennata* subsp. *pulcherrima* (K. Koch) Freitag, with the possible exception of some plants from granitic and schistose mountains in N Greece which have been referred to *S. joannis* subsp. *balcanica* Martinovský. According to Gonzales, Aedo & García (2013: 361) the latter is also a synonym of *S. pennata* L. See further comments in Mountain Flora of Greece 2: 828-829 (1991).

Stipa rechingeri Martinovský, see *S. pennata* complex.

Stipa thessala Hausskn. is a taxon of uncertain status. In the original description Haussknecht cited two of his own collections, one "in schistosis supra Malakasi" and the other "in saxosis supra Pharsalum". Both specimens are at JE. According to Scholz (1984a) the former is *S. capillata* L., whereas the latter differs is indumentum characters and was selected as lectotype for *S. thessala*. The differences are not convincing, and *S. thessala* is likely to fall within the range of variation of the widespread *S. capillata*.

Stipa tirsa Steven is widespread from C Europe to Kazakhstan. Gonzalo, Aedo & García (2013: 359) cited a single collection from Greece, viz. Lippert 20891 (M) from Mt Vourinos. *S. tirsa* differs from members of the *S. pennata* complex by the very short ligule (0.1-0.4 mm) of the basal leaves which are conspicuously narrow (c. 0.5 mm diam.) with sparse, stiff hairs on the abaxial surface and a long setaceous apex.

Stratiotes aloides L. was reported by Dimitrov & Raus (2001: 326), based on a historical collection (Petvov, 9.9. 1908, SO) from the vicinity of Enidze [Giannitsa] in the Vardar [Axios] river plain. No suitable habitats remain in this area, and the species must be assumed extinct in Greece. It is considered critically endangered in Bulgaria where it is known from a few localities in the north (Peev & al. 2015: 327).

Symphytum ottomanum Friv. was listed for Peloponnisos by Mermygkas, Zikos & Constantinidis (2021: supplementary file). The record is feasible, but no details are available.

Taeniatherum caput-medusae (L.) Nevski. Taken in a wide sense this species occurs throughout the Mediterranean region and Anatolia eastwards to Afghanistan and C Asia, with subsp. *caput-medusae* in Spain and S France, subsp. *asperum* (Simonk.) Melderis (syn. *T. asperum* (Simonk.) Nevski) throughout the range of the species and subsp. *crinitum* (Schreb.) Melderis (syn. *T. crinitum* (Schreb.) Nevski) from the Balkans eastwards. The subspecies differ vaguely in quantitative characters and overlap for most of their range. In Greece it is often difficult to distinguish between subsp. *asperum* and subsp. *crinitum* (cf. Frederiksen 1986).

Tagetes patula L., a native of C America, has been reported a few times from ruderal habitats, e.g., Akeroyd & Preston (1987: 358) from the Thessaloniki area, Panitsa & al. (2021: 51) from NE Peloponnisos, and collections by Hansen & Nielsen (C) from the islands of Kos and Lesvos. It is cultivated as an ornamental and doubtfully naturalized.

Tamarix arborea (Ehrenb.) Bunge, a N African species, is sometimes cultivated on beaches and has been reported from the islands of Siros and Milos. It resembles *T. nilotica*, but has racemes up to 6 mm broad, filaments arising from the top of the discal lobes, and the base of the filaments dilated.

Tamarix dalmatica Baum was listed for "Gr" in Fl. Eur. 2: 293 (1968), but no specimens have been seen and the occurrence needs confirmation. It resembles *T. hampeana* Boiss. & Heldr. but has longer bracts and sepals.

Tamarix nilotica (Ehrenb.) Bunge is a species of N Africa and the Saharo-Arabian region, doubtfully native in the Aegean area but sometimes cultivated on beaches (reported from the islands of Kriti, Rodos, Samos, Patmos, Chios, Lesvos and Zakinthos). It can be recognized on the following combination of features: Usually a small tree 3-6 m; leaves broad-based, ± amplexicaul; inflorescence compound, paniculate, borne on branches of current year; racemes c. 4 mm broad; flowers 5-merous, white or pale pinkish; petals deciduous; filaments arising between the sinuses of the disc; base of filaments not dilated. Autumn-flowering.

Tanacetum balsamita L. (*Balsamita major* Desf.) has been collected in ruderal habitats in the village of Chania on Mt Pilio (Stamatiadou 14021; ATH, herb. Strid) and in the village of Sarandapicho in nomos Korinthias (Kit Tan & al. 32025; herb. Kit). It is probably native in SW Asia and has long been cultivated in Europe as an ornamental and medicinal herb. The Greek populations are doubtfully naturalized.

Tanacetum corymbosum (L.) Sch. Bip. A few records from N Greece have been referred to subsp. *cinereum* (Griseb.) Grierson which differs from the nominate subspecies mainly in being more densely grey-pubescent. It may be a geographical race of SE Europe and W Anatolia.

Tanacetum vulgare L. was reported from Kriti at http://www.cretanflora.com/tanacetum_vulgare.html, but the photo shows a different, albeit unidentified, species.

Taraxacum. The account of this taxonomically difficult genus is highly provisional. The Greek flora comprises both diploid, sexually reproducing species (mainly in the mountains) and polyploid apomictic taxa (mainly in ruderal habitats). Only some of the better known species have been mapped, whereas others (often based on single collections) are briefly listed below. Vast amounts of Greek *taraxaca* await critical revision in the major herbaria.

Taraxacum abrotanum Kirschner, Sonck & Štěpánek was described in 1993, based on a collection by Sonck from the vicinity of the Katara Pass, but later (Kirschner & Štěpánek 1998a: 169) listed as a synonym of *T. apiculatiforme* Soest.

Taraxacum aganippeum Sonck was described from a wet habitat near Metsovon. Kirschner & Štěpánek (1998a: 219) listed it as a synonym of *T. scaturiginosum* G. E. Haglund.

Taraxacum albomarginatum A. J. Richards (1991) is a later homonym of *T. albomarginatum* Kitam. (1935). The former name has been replaced by *T. voricola* Štěpánek & al.

Taraxacum aldenii A. J. Richards. Aldén no. 3613 (LD) from a snowbed meadow on Mt Kakarditsa has been annotated "type of *Taraxacum aldenii*" by A. J. Richards. It was included in the Greek checklist (Dimopoulos & al. 2013: 65), but is apparently a manuscript name without any published description.

Taraxacum apiculatoides Małecka. Willing 81658 (B) from deciduous oak woodland in the Kalambaka area has been identified as *T. apiculatoides* by J. Štěpánek. This species is otherwise known from North Macedonia and Bulgaria.

Taraxacum aznavourii Soest. A single collection (Strid 881) from calcicolous grassland in the summit area of Mt Pangeo has been identified as *T. aznavourii*, a species described from European Turkey (cf. Richards in Mount. Fl. Greece 2: 566-567, 1991).

Taraxacum balcanicum Rech. f. was described from alpine grassland at 2200 m on Mt Voras (Kajmakčalan). The name was overlooked in Mount. Fl. Greece (1991), but *T. graecofontanum* A. J. Richards & Sonck sp. nov. was reported from the same locality, and it remains to be established whether the two are conspecific. If so, *T. balcanicum* is the older name.

Taraxacum bithynicum DC. is a sexual diploid (2n=16) described from the Bithynian Olympus (Ulu Dağ). It is probably endemic in that area; records from moderate altitude on Mts Dikti, Lefka Ori and Taigetos need confirmation.

Taraxacum buttleri Soest was described from Elmalı Dağ in SW Anatolia. Two collections from calcicolous grassland on Mt Dirfis, Evvia (Strid nos 24627 & 24643) have been referred to this species by A. J. Richards.

Taraxacum capricum Soest, otherwise known from Italy and S France, has been reported from Samothraki (Biel & Kit Tan 2013a: 134) and Thasos (Biel & Kit Tan 2017b: 285). The subsequently described *T. copidophylloides* A. J. Richards is apparently close to *T. capricum*.

Taraxacum copidophyllum Dahlst. was listed for Greece in Euro+Med PlantBase, referring to the Mountain Flora of Greece, vol. 2 (Strid & Kit Tan 1991). This is apparently a mistake since the species is not mentioned in the Mountain Flora.

Taraxacum cylleneum Fürnkrantz, a distinct diploid species, is believed to be endemic to limestone mountains of N Peloponnisos. However, Meyer (2011: 172) reported it also from serpentine in the Korça area of Albania, claiming that plants from serpentine substrate on Smolikas, Milea and in the Metsovon area (NW Greece) belong to this species as well. Given the geographical and ecological differences this seems unlikely. The Albanian record was not cited by Kirschner & al. (2021), who also regarded records from the Pindos as probably incorrect.

Taraxacum declivicola Kirschner, Sonck & Štěpánek was described from cultivated plants raised from seeds collected near the Katara Pass. It is related to the widespread C European *T. subalpinum* Hudziok, but differs in having long, non-imbricate outer phyllaries with a broad margin. For description and illustration, see also Kirschner & Štěpánek (1998a: 168-169). No subsequent collections have been reported.

Taraxacum decrepitum Kirschner & Štěpánek was described from the western foothills of Mt Taigetos. It resembles *T. scaturiginosum* G. E. Haglund but differs in the patent, ± linear-triangular leaf lobes, outer bracts usually arcuate or recurved with broad membraneous margins, and achenes with cone 1.2-1.8 mm long (Kirschner & Štěpánek 1998a: 225).

Taraxacum deorum A. J. Richards is known from a single, high-altitude collection on Mt Olimbos. It resembles *T. graecofontanum* A. J. Richards & Sonck and *T. voricola* Štěpánek (*T. albomarginatum* auct.) which both occur on Olimbos, differing from the former in the well demarcated, blackish-purple apex of the inner phyllaries and from the latter in the straw-coloured (vs. reddish) achenes.

Taraxacum divinum Sonck, based on a single collection from the eastern foothills of Mt Olimbos, was listed as a Greek endemic in the Greek checklist (Dimopoulos & al. 2013: 66) as well as the Euro+Med PlantBase. According to Štěpánek & Kirschner (2014: 140) it is a synonym of *T. gracilens* Dahlst.

Taraxacum edessicum Sonck was based on a single collection from the railway station of Edessa (north central Greece). According to Štěpánek & Kirschner (2018: 366) it is conspecific with *T. salonikiense* Sonck which had been described in the same article with the holotype from the city of Thessaloniki. *T. aestuans* Sonck, described from Albania (Korça), was also listed by Štěpánek & Kirschner (loc. cit.) as a synonym of *T. salonikiense*.

Taraxacum fibratum Sonck was based on two collections from the foothills of Mt Olimbos and Mt Pilion. Also specimens from Mt Vermion and elsewhere have subsequently been referred to *T. fibratum*. According to Štěpánek & Kirschner (2014: 142) it is a synonym of *T. herae* Sonck.

Taraxacum gionense A. J. Richards, described from Mt Giona, has been synonymized with *T. botanicorum* Sonck (cf. Štěpánek & Kirschner 2018: 368).

Taxacacum hemicyclum G. E. Haglund. A single collection from nomos Karditsis (Willing 94821, B) has been identified by J. Štěpánek as *T. hemicyclum*, otherwise a species of C & N Europe.

Taraxacum hoppeanum auct. This name has been applied to plants from the Athos peninsula as well as the Thessaloniki area (Turrill 1937c: 248-249). Their identity is uncertain.

Taracaxum insolitum Kirschner, Sonck & Štěpánek was based on a collection by Sonck from the vicinity of the Katara Pass. There appear to be no other records in Greece, but it has subsequently been reported also from Bulgaria.

Taraxacum janchenii Kirschner & Štěpánek was reported from NE Greece in the Greek checklist (Dimopoulos & al. 2013: 66), but the source for this record is obscure. In the original description it was reported from a rather wide area from Dalmatia to C Romania.

Taraxacum kalambakae Sonck was based on a single collection from the town of Kalambaka ("behind the Café Edelweiss").

Taraxacum lacistophylloides Sonck was described from the railway station of Kalambaka. The name was later changed to *T. vexatum* Sonck. nom. nov. (Sonck 1986: 167), the epithet *lacistophylloides* having been used for another species by Dahlstedt in 1933. According to Štěpánek & Kirschner (2014: 140) *T. vexatum* is a synonym of *T. epirense* Soest.

Taraxacum magnesicum Sonck, described from "Magnesia, roadside at Zagora", was listed as a Greek endemic in the Greek checklist (Dimopoulos & al. 2013: 66) as well as in the Euro+Med PlantBase. However, Štěpánek & Kirschner (2014: 140) reduced it to a synonym of *T. fragosum* Sonck.

Taraxacum molybdocephalum Sonck (described from several localities in the Greek mainland and Peloponnisos), was listed as a synonym of *T. dialeptum* Sonck (described from the eastern foothills of Mt Olimbos) by Štěpánek & Kirschner (2014: 139).

Taraxacum multisinuatum Kirschner, Sonck & Štěpánek was based on a single collection from the vicinity of the Katara Pass. There appear to be no subsequent records from elsewhere (see also Kirschner & Štěpánek (1998a).

Taraxacum nanulum Sonck was based on a single collection from the southern foothills of Mt Parnassos. According to Štěpánek & Kirschner (2014: 142) it is a synonym of *T. herae* Sonck.

Taraxacum nervatum A. J. Richards. Although included in the Greek checklist (Dimopoulos & al. 2013: 66) this is apparently a manuscript name that was never published.

Taraxacum noterophilum Kirschner, Sonck & Štěpánek. As for *T. multisinuatum*.

Taraxacum olympicola Sonck has been reported from the eastern foothills of Mt Olimbos as well as the Edessa railway station. According to Štěpánek & Kirschner (2014: 142) it is a synonym of *T. herae* Sonck.

Taraxacum phitosii Soest was based on a single collection from Mt Pilio.

Taraxacum poliochlorum Dahlst. is a species of uncertain status, based on a single collection from the southern foothills of Mt Parnassos. There appear to be no other Greek records, but it has been reported from widely scattered localities in Anatolia (Fl. Turkey 5: 897, 1975).

Taraxacum pseudobrachyglossum Soest was described from Turkey-in-Europe and may be endemic in this area. A record from Lesvos (Hansen & Nielsen 6559, C) needs confirmation.

Taraxacum radinum Sonck, described from two localities in Thessaly, was listed as endemic in the Greek checklist (Dimopoulos & al. 2013: 66) as well as in the Euro+Med PlantBase. However, Štěpánek & Kirschner (2014: 140) reduced it to a synonym of *T. fragosum* Sonck.

Taraxacum refectum Sonck was based on a single collection from the eastern foothills of Mt Olimbos. There appear to be no other Greek records, but it has subsequently been reported from North Macedonia and Bulgaria (see Kirschner & Štěpánek 1998a: 204-205).

Taraxacum rigidifolium Sonck was described from two adjacent localities in nomos Ioanninon. There appear to be no subsequent records.

Taraxacum scanicum Dahlst. was described from S Sweden and is widespread in C & N Europe. One collection from Mt Gramos (Willing 11344D, B) has been identified as *T. scanicum* by Štěpánek.

Taraxacum stenospermum Sennen. This species was included in the Greek checklist (Dimopoulos & al. 2013: 66), based on a single collection from the eastern foothills of Mt Olimbos (Strid 8165, herb. Strid) identified as such by A. J. Richards. It is otherwise native to the Iberian Peninsula, and occurrence in Greece seems unlikely.

Taraxacum sublimiforme Sonck was based on two collections from the same locality in the city of Ioannina ("near the old castle").

Taraxacum submicrocranum Sonck was collected together with *T. sublimiforme* by the old castle of Ioannina and at a second locality on the island in Lake Ioannina. There appear to be no other records, and the status of both species is uncertain.

Taraxacum terenodes Sonck was described from the eastern foothills of Mt Olimbos and has subsequently been reported also from Mt Vermion (Chochliouros & Georgiadis 1997b: 105, Chochliouros 2005: 81). According to Štěpánek & Kirschner (2014: 142) is it a synonym of *T. herae* Sonck.

Taraxacum thessalicum Soest is a species of uncertain status, apparently known only from the type collection.

Taraxacum trigonense Sonck was based on a single collection from the village of Trigona in nomos Trikalon ("on a moist loamy slope on the roadside").

Taraxacum vexatum Sonck. See under *T. lacistophylloides*.

Taraxacum viale Sonck was based on a single collection from nomos Karditsis. According to Štěpánek & Kirschner (2018: 366) it is conspecific with *T. amborum* G. E. Haglund.

Taraxacum xanthiense Soest was based on collections by Tedd in the town of Xanthi. It has not been confirmed elsewhere except for a dubious record from Thessaloniki (Krigas & Kokkini 2005: 46).

Taraxacum zagorae Sonck was based on a single collection from the village of Zagora (nomos Magnisias).

Teloxys aristata (L.) Moq. (*Chenopodium aristatum* L.) was listed for the Cretan area in Euro+Med PlantBase, referring to Hayek (1924b: 144). This is apparently a mistake as there is no such record in Hayek.

Tephroseris. Greek material has often been referred to the widespread *T. integrifolia* (L.) Holub, with subsp. *integrifolia* (with pubescent achenes) occurring in the north and north-east, and subsp. *aucheri* (DC.) B. Nord. (with subglabrous achenes) throughout the Greek range except in the north-east. All Greek specimens are now included in *T. papposa* (Rchb.) Schur (see Kadereit, Laux & Dillenberger 2021).

Tetragonia tetragonoides (Pall.) Kuntze (New Zealand spinach) is cultivated as a leafy vegetable and may persist for some time in ruderal habitats but is scarcely naturalized. It is a trailing plant with thick, triangular leaves covered by small papillae.

Tetragonolobus biflorus (Desr.) Ser. was listed for "Gr" in Fl. Eur. 2: 177 (1968), probably based on Baldacci (1894b: 94) "in herbidis Gajduronissi sinu ambracico (Epirus)", sub nom. *Lotus biflorus* Desv. [sic!]. The record was cited by Halácsy (1900: 416) who, however, had not seen any specimen. There is no modern confirmation, and the species may be restricted to S Italy, Sicily and NW Africa.

Teucrium chamaedrys L. Several infraspecific taxa have been described (cf. Rechinger 1941a). Of those reported from Greece only subsp. *olympicum* Rech. f. is reasonably distinct, having dense indumentum, broad, deeply incised-serrate leaves and dark corollas. It is restricted to the forest zone of mts Olimbos and Ossa.

Teucrium creticum L. The specific epithet represents one of the geographical mistakes by Linnaeus. The species is endemic to Cyprus and coastal areas from S Anatolia to Palestine (see map in Browicz 1996a: 35).

Teucrium divaricatum Heldr. Three rather poorly defined subspecies are recognized in Greece. A fourth, subsp. *villosum* (Čelak.) Rech. f., has been reported from Kriti and the East Aegean Islands; it does not appear to differ significantly from subsp. *divaricatum*.

Teucrium flavum L. Based on indumentum characters, Rechinger (1941a: 393 ff.) recognized subsp. *glaucum* (Jord. & Fourr.) Ronniger, subsp. *gymnocalyx* Rech. f. and subsp. *hellenicum* Rech. f. in Greece. They are poorly characterized and have overlapping distributions.

Teucrium fruticans L., a native of the W & C Mediterranean area, is commonly cultivated as an ornamental and may be on the verge of becoming naturalized in some localities in S Greece, e.g., near Sitia in E Kriti (Strid obs., 2022-04-10). It is a much-branched, grey- to white-tomentose shrub with large, pale lilac-blue flowers.

Teucrium parnassicum (Čelak.) Wettst. has generally been considered a synonym of *T. montanum* L. subsp. *montanum*, but the group is currently being studied by D. Lakušić, and preliminary results indicate that it may indeed be a separate, endemic species. Details will be given in a forthcoming publication. *T. montanum* var. *parnassicum* Čelak. was based on Guicciardi 393 ("In m. Parnassi, reg. inferiori supra Rachova"). It was raised to species level by Wettstein (1889: 376), who considered plants from Anatolia to represent the same taxon ("Auf felsigen Stellen des Aglassan-Dagh. Stimmt vollkommen überein mit den von Guiccardi auf dem Parnasse gesammelten Exemplaren, auf Grund deren Čelakovský sein *T. Parnassicum* aufstellte, das sowohl von *T. Pannonicum* A. Kerner als von *T. montanum* L. gut zu unterscheiden ist"). Plants from Samos resemble *T. parnassicum* in their densely grey-pubescent leaves; they have here been provisionally included in *T. montanum* s. lat.

Teucrium polium f. ***dumulosum*** Rech. f., a local form from Kriti, has been recombined as *T. dumulosum* (Rech. f.) Brullo & Guarino (Brullo & Guarino 2000: 275). No evidence was presented for the proposed change of rank.

Teucrium spinosum L. was reported by Candargy (1898) from Lesvos, but there is no recent confirmation. The species is widespread in Anatolia, extending to W Syria and Palestine, and was illustrated on Plate 539 of Flora Graeca (1825), based on material collected "in arvis inter Smyrnam [Izmir] et Bursam".

Thalictrum flavum L. Records from NE Greece (Vandas 1909: 4 from Mt Chortiatis, Turrill 1918: 261 from Krusa Balkan NNE of Thessaloniki, Voliotis 1967: 97 from Mt Cholomondas and Babalonas 1981: 266 from the Evros delta) probably all refer to *T. lucidum* L.

Thalictrum foetidum L. was listed for "Gr" in Fl. Eur. ed. 2, 1: 291 (1993), but records probably refer to *T. minus* subsp. *saxatile* DC. ex Ces.

Thesium arvense Horv. was included as an accepted species in the Greek checklist (Dimopoulos & al. 2013: 140). However, the name is illegitimate and has been replaced by *T. ramosum* Hayne.

Thesium brachyphyllum Boiss. is closely related to *T. procumbens* C. A. Mey. and sometimes regarded as conspecific, although kept separate by Romo, Didukh & Boratyński (2004). It is assumed that all Greek material belongs to the latter, but further studies of variation are needed.

Thymelaea gussonei Boreau is doubtfully distinct from *T. passerina* (L.) Coss. & Germ. In the former the stem is hairy at least above (vs. glabrous) and the flowers are mixed male, female and hermaphrodite (vs. entirely or predominantly hermaphrodite). *T. gussonei* is widespread in the Mediterranean area and has been reported from Kerkira (Gutermann & Kit Tan 2006). One collection from Mt Parnassos (Greuter 5978) has been registered as *T. gussonei*, whereas others have reported *T. passerina* from the same area. *T. passerina* subsp. *pubescens* (Guss.) Meikle is a synonym of *T. gussonei* Boreau. For comments on variation, see Fl. Cyprus 2: 1427 (1985).

Thymelaea tartonraira (L.) All. Almost all Greek plants belong to subsp. *argentea* (Sm.) Holmboe which occurs in Greece, W & S Anatolia and Cyprus (cf. Kit Tan 1980b). There are credible reports of the W & C Mediterranean subsp. *tartonraira* only from the Ionian Islands (Kerkira, Kefallinia and Zakinthos).

Thymus atticus Čelak. was described from mountains of Attiki (Imittos and Pendeli) where it is fairly rare, being more common in NE Greece. These plants have linear leaves (c. 1 mm wide) and almost white flowers. Plants from the western part of the range, from north central Greece southwards to northern Peloponnisos, have oblanceolate leaves and pinkish flowers. They appear to constitute a geographical race worthy of subspecies rank.

Thymus aznavourii Velen., a species described from the Istanbul area and previously believed to be endemic to Turkey-in-Europe, was reported from SE Bulgaria and NE Greece by Stoyanov & Marinov (2021). The Greek record (from the Orestiada area) was based on a historical literature report (Stojanov 1921c); it is plausible but needs confirmation. *T. aznavourii* differs from *T. zygioides* Griseb. in the non-lignified stems, linear leaves, greenish calyx and whitish corolla (with purple spots).

Thymus ciliatopubescens (Halácsy) Halácsy, described from Mt Giona, was listed as an accepted species in Euro+Med PlantBase, but is probably a synonym of *T. boissieri* Halácsy.

Thymus dolopicus Formánek was based on a rather fragmentary collection from "m. Ghavallu [Voutsikaki] in montibus Agrapha Thessaliae". Although accepted by Baden in Mount. Fl. Greece 2: 148 (1991) it is of dubious status and the name has been applied to rather different-looking specimens. It may be a form of *T. longicaulis* C. Presl.

Thymus grisebachii Ronniger was listed as an accepted species in Euro+Med PlantBase, but has been reduced to *T. sibthorpii* var. *grisebachii* (Ronniger) Jalas & Baden.

Thymus heterotrichus Griseb., described from Mt Athos, was listed as an accepted species in Euro+Med PlantBase, but may fall within the range of variation of *T. sibthorpii* Benth. (cf. Lacaita 1918: 125).

Thymus holosericeus Čelak. The name of this distinct endemic of the Ionian islands alludes to the short and dense pubescence. "Die ausgezeichnetste Character dieser Pflanze besteht in der feinen und dichten, äusserst kurzhaarigen, vollkommen sammerartigen Bekleidung aller grünen Theile, des Stengels, der Blätter, Deckblätter und Kelche" (Čelakovský 1883b: 166). This character is obviously variable. In one collection from Kefallinia (Damboldt 24/94/77, B) the leaves are glabrous except for a few long marginal cilia in the lower part, whereas bracts and calyces are more or less velutinous.

Thymus laconicus Jalas. This rare local endemic was first described by Turrill as *T. pastoralis*, the epithet alluding to a local shepherd who had collected material for Goulimis. However, *T. pastoralis* Turrill is a later homonym of *T. pastoralis* Iljin, and the Peloponnesian plant was renamed *T. laconicus*. The original description and illustration in Turrill (1956b) is somewhat misleading as it was stated to be a prostrate subshrub, whereas in fact it is a much-branched, strongly woody shrublet superficially resembling *Thymbra capitata*. It appears to be threatened by fragmentation of habitat due to recent planting of new olive groves on former phrygana (Strid obs., 26.5. 2021). An excellent illustration is found in Kit Tan w/ Iatroú (2001: 317).

Thymus leucospermus Hartvig was listed for Peloponnisos by Mermygkas, Zikos & Constantinidis (2021: supplementary file), probably in error.

Thymus leucotrichus Halácsy is variable, especially in indumentum characters. Plants from NE Greece have been recognized as subsp. *neiceffii* (Degen & Urum.) Jalas. The identity of similar plants from S Anatolia and Cyprus is not clear (cf. Fl. Turkey 7: 356-357, 1982).

Thymus longidentatus (Degen & Urum.) Ronniger was listed for "?Gr" in Fl. Eur. 3: 180 (1972). It was described from S Bulgaria and could well occur in NE Greece, but is doubtfully distinct from *T. zygioides* Griseb.

Thymus pulegioides L. is a widespread species of Europe and W Siberia. One collection from Mt Chelidona has been identified as *T. pulegioides* by Jalas (1990), and there are a few additional literature reports from N Peloponnisos to the NW border area. The occurrence of this species in Greece needs confirmation.

Thymus sibthorpii Benth. was listed for Peloponnisos by Mermygkas, Zikos & Constantinidis (2021: supplementary file), but no details are available.

Tilia cordata Mill. Bergmeier collected a specimen from a single, mature tree in the Prespa National Park (NW of Lemos, 2018-06-08, no. 18-181). It had first been reported from this area by Pavlides (1985: 256) between Vrondero and Agathoto and between Oxya and Sfikas. The latter reports are most likely erroneous for *T. platyphyllos* Scop. which occurs at the indicated sites. *Tilia cordata* was listed for Peloponnisos by Mermygkas, Zikos & Constantinidis (2021: supplementary file), probably in error; the few confirmed Greek records are all in the far north.

Tilia platyphyllos Scop. Most of the Greek specimens belong to subsp. *corinthiaca* (Bosc ex K. Koch) Pigott which is restricted to Greece, S Albania and NW Anatolia (cf. Pigott 2012: 102), but at least some trees from the Prespa National Park match subsp. *pseudorubra* C. K. Schneid. which is widespread in the N Mediterranean area from N Greece to S France.

Tilia rubra DC. was listed for "Gr" in Fl. Eur. 2: 248 (1968), but according to Pigott (2012: 90) it is a synonym of *T. platyphyllos* Scop.

Tilia ×vulgaris Hayne. Scattered Greek records (e.g., Regel 1952: 40 from Mt Iti, Pavlides 1985: 256 from Mt Triklario, Voliotis 1981b: 259 from nomos Pellis and Dimopoulos 1993: 103 from Mt Killini) all need confirmation. *T. vulgaris* is generally interpreted as the hybrid between *T. cordata* Mill. and *T. platyphyllos* Scop. In Greece the former is probably restricted to the Rodopi area.

Tordylium byzantinum (Azn.) Hayek was listed for "Cr" in Fl. Eur. 2: 367 (1968); occurrence in this area seems unlikely. Rechinger (1939a: 484) reported it from two localities in NE Greece. The corresponding specimens (Rechinger 8963e and 9010) have not been seen, but occurrence in NE Greece is feasible since the species was described from the Istanbul area. It has subsequently been interpreted as a synonym of *T. trachycarpum* (Boiss.) Al-Eisawi.

Torilis arvensis subsp. ***recta*** Jury appears to fall within the range of variation of subsp. *arvensis*.

Torilis pseudonodosa Bianca and ***T. webbii*** Jury are names given to occasional variants of *T. nodosa* (L.) Gaertn. with both mericarps equally spiny. They grow in amongst the usual heterocarpic variant and there are no other consistent differences.

Trachelium caeruleum L. subsp. ***caeruleum*** was reported from Naxos by Constantinidis & Gavalas (2017). It is a native of the Iberian Peninsula and NW Africa, previously reported as naturalized in France, Italy, U.K., Belgium, etc. The Naxos population has persisted for at least 5 years (2012-2016), but does not appear to be spreading. There are previous reports from Kerkira (Borkowsky 1997), but here also the species is doubtfully naturalized (cf. Flora Ionica website).

Trachomitum. See *Poacynum*.

Trachystemon orientalis (L.) G. Don was reported from Mt Timfi by Goulimis (1960a: 23). D. Mermygkas has kindly sent a scan of the specimen at ATH; it is in poor conditions but probably represents *Symphytum ottomanum* Friv. Other Greek records of *Trachystemon orientalis* may refer to *Symphytum creticum* (Willd.) Runemark ex Greuter & Rech. f. (cf. Dimopoulos & al. 2013: 300).

Tradescantia fluminensis Vell., a native of South America has been reported from damp, ruderal habitats in several localities including the islands of Kriti, Kerkira and

Samothraki. It may be on the verge of becoming naturalized in Greece. It has recently been reported also from Cyprus (Spitale & Papatheodoulo 2019).

Tragopogon hayekii (Soó) I. Richardson has been reported once, from [nom. Kilkis] "Salzwiesen des Pikrolimni-Sees" (Babalonas & Papastergiadou 1990: 118). It is closely related to *T. pratensis* L. and the record needs confirmation.

Tragopogon orientalis L. was included as an accepted species in the Greek checklist (Dimopoulos & al. 2013: 66). It cannot be clearly distinguished from *T. pratensis*, and is often classified as *T. pratensis* subsp. *orientalis* (L.) Čelak., gradually replacing the nominate subspecies in the SE part of the range. Many specimens by Willing (B) identified as *T. pratensis* have been omitted from the map as they need critical revision.

Tragopogon pichleri Boiss. is a little known, yellow-flowered species described from W Anatolia (C2 Bursa) and reported from Chios by Meikle (1954: 149) and Snogerup & al. (2001: 64), based on Pauli s.n. and Rechinger 5400. The record is feasible but needs confirmation.

Trifolium alexandrinum L. is a cultivated fodder plant ("Egyptian clover") related to *T. constantinopolitanum* Ser. and probably originating in the Eastern Mediterranean area. It was reported from SW Peloponnisos already by Bory & Chaubard (1832-33: 218) and there are subsequent records from W Kriti as well as NW mainland Greece. All are likely to refer to casual escapes.

Trifolium aureum Pollich was listed for Peloponnisos by Mermygkas, Zikos & Constantinidis (2021: supplementary file). The southernmost Greek record so far is on Mt Iti, and occurrence in Peloponnisos seems rather unlikely.

Trifolium brutium Ten. Records from Peloponnisos (e.g., Baliousis 2013: 32) are probably referable to *T. aurantiacum* Boiss. & Spruner.

Trifolium clusii Gren. & Godr. has been reported from the Ionian Islands (cf. Flora Ionica website). It is a taxon of uncertain status, resembling *T. resupinatum* L. and *T. tomentosum* L.

Trifolium congestum Link was listed for the East Aegean Islands in Euro+Med PlantBase. It is probably a synonym of *T. fragiferum* L.

Trifolium glanduliferum Boiss. is represented by two varieties or subspecies:

subsp. *glanduliferum* (Samos and Chios): Stipules, leaflets and calyx teeth glandular-pubescent; heads subtended by a row of connate involucral bracts.

subsp. *nervulosum* (Boiss. & Heldr.) Zohary (Lesvos and NE mainland): Stipules, leaflets and calyx teeth glabrous; such bracts rudimentary.

In Greece the two are rather distinct, but according to Zohary & Heller (1984: 269) intermediates occur elsewhere.

Trifolium hybridum L. Most Greek plants are probably referable to subsp. *elegans* (Savi) Aschers. & Graebn. which cannot be clearly separated from subsp. *anatolicum* (Boiss.) Hossain. Subsp. *hybridum* is the cultivated plant of C & N Europe.

Trifolium incarnatum L. The taxonomy of this and related species is unsettled (cf. Aegean Atlas 1: 311, 2016). Subsp. *incarnatum* is a red-flowered, cultivated plant; a recent record from W Kriti (http://www.cretanflora.com/trifolium_incarnatum.html) probably refers to a casual occurrence. Specimens from N Greece apparently belong to subsp. *molinerii* (Hornem.) Syme. Specimens from S Peloponnisos, which are often gregarious at roadsides, have large, pale yellow corollas and probably belong to an undescribed taxon which has been called *T. incarnatum* subsp. *luteum* Lassen, ined. They approach *T. xanthinum* Freyn, a species of W Greece which has sometimes been regarded as a mere variety of the widespread *T. stellatum* L. (although normally distinct).

Trifolium medium L. Most of the Greek plants match subsp. *balcanicum* Velen. (with patent-pubescent stem and upper calyx teeth longer than tube) which replaces the widespread subsp. *medium* (with appressed-pubescent stem and upper calyx teeth shorter than or equaling tube) in the C & S Balkan Peninsula. Intermediates occur in N Greece, e.g., on Mt Tzena and Mt Pangeon.

Trifolium michaelis Greuter was based on a collection from the NW foothills of Mt Taigetos. It appears to be a synonym of *T. pallidum* Waldst. & Kit.

Trifolium mutabile Port. was listed for Greece in Euro+Med PlantBase, but is doubtfully distinct from *T. vesiculosum* Savi. Some specimens from NE Greece, here included in *T. vesiculosum*, have fruiting calyx only slightly inflated and without transverse veins, thus keying out as *T. mutabile*.

Trifolium noricum Wulfen. Most of the Greek plants are mat-forming with small leaflets and pink corollas, thus matching *T. ottonis* Boiss., which was described from Mt Timfristos and has subsequently been considered a geographical race of *T. noricum*. Plants from Mt Timfi match *T. noricum* s. str., being taller with large leaflets and white corollas (cf. Mount. Fl. Greece 1: 511-512, 1986). *T. ottonis* appears to be worthy at least of subspecies rank.

Trifolium obscurum Savi is mainly a species of the W Mediterranean area, fairly rare in Italy and scattered further east. There are only two confirmed records in Greece, one from the island of Samos based on a collection from 1993 (see Lassen 1999: 61) and a more recent one from the foothills of Mt Kitheronas (Polymenakos 546, ATH). Greek records of *T. obscurum* published on the GBIF website must be treated with considerable suspicion; apparently there are no details and no vouchers.

Trifolium pachycalyx Zohary was reported from the island of Limnos by Bergmeier, Krause & Ristow (2021: 362). This is apparently the first record of this species since it was originally described in 1969 based on historical collections from NW Turkey, on both sides of the Bosporus. Being small and easily overlooked it may be more widely distributed in the NE Aegean area.

Trifolium pannonicum Jacq. Literature records from Kerkira (Hansen 1982: 32), nomos Artis (Chitos 2009: 60) and Mt Iti (Karetsos 2002: 102) all need confirmation; at least some of them refer to *T. ochroleucon* Huds. A report from the Prespa National Park (Pavlides 1985: 222) turned out to represent *T. vesiculosum* Savi. *T. pannonicum* is widespread in Europe and Anatolia, and could well occur in Greece.

Trifolium resupinatum L. Most of the Greek records belong to the native subsp. *resupinatum*. The cultivated and naturalized subsp. *suaveolens* (Willd.) Ponert (tall, with fistulose stem and large, pinkish-purple corollas) has been reported from a few localities in W Greece, including the Ionian Islands.

Trifolium setiferum Boiss., described from W Anatolia, was listed as an accepted species in Euro+Med PlantBase and reported from Italy and the Balkan Peninsula, including Greece. It is probably a synonym of *T. vesiculosum* Savi.

Trifolium velenovskyi Vandas has been reported from several mountains in northern Greece (Rodopi, Timfi, Tzena, Varnous, Voras, Vrondous), but the records are likely to refer to *T. aureum* Pollich.

Trigonella spp. A large number of *Melilotus* species have recently been recombined under *Trigonella* (see references at dryades.units.it/floritaly/). Evidence for this seems, at best, to be flimsy.

Trigonella caerulea (L.) Ser. Most of the Greek plants belong to the native subsp. *procumbens* (Besser) Thell. Subsp. *caerulea*, which is cultivated for fodder, has been collected once in the Alexandroupolis area (Sintenis & Bornmüller 247, LD), but the differences are not clear.

Trigonella esculenta Willd. was listed as an accepted species in Euro+Med PlantBase. The original description was based on cultivated material probably of Indian origin. Greek records are likely to refer to *T. corniculata* (L.) L.

Trigonella sprunneriana Boiss. The original spelling of the epithet is *sprunneriana*, which is optionally regarded as an orthographic error for *spruneriana*. The species was named after Wilhelm von Spruner (1805-1874), a German botanist who collected in Sterea Ellas and Peloponnisos c. 1837-1844 and whose name is occasionally spelt "Sprunner".

Trigonella strangulata Boiss. was reported from the southern foothills of Mt Parnassos by Greuter (1975b: 324). No material has been seen and there appears to be no subsequent confirmation. *T. strangulata* is a widespread Irano-Turanian species not otherwise known in Europe.

Trigonella velutina Boiss. was reported in the same area as the previous and in the same publication (Greuter 1975b: 325). No material has been seen, and it remains to be established whether it is distinct from *T. sprunneriana* Boiss. Like *T. strangulata*, *T. velutina* is a widespread Irano-Turanian species not otherwise known in Europe.

Trinia dalechampii (Ten.) Janchen. Greek records of this species probably refer to *T. frigida* (Boiss. & Heldr.) Drude and perhaps also to *T. glauca* (L.) Dumort. subsp. *glauca*. *T. dalechampii* was described from Italy and has also been recorded in the western Balkan Peninsula southwards to Albania.

Tripleurospermum caucasicum (Willd.) Hayek was described from Caucasus where it occurs at high altitude. Plants from Albania and NW Greece referred to this species may well be taxonomically different. Populations throughout the wide geographical range should be carefully compared.

Tripleurospermum disciforme (C. A. Mey.) Sch. Bip. has recently been reported from a locality in the north-eastern corner of Greece not far from the Turkish city of Edirne (Dimitrov 2021: 357). It is widespread in Anatolia and the Flora Iranica area.

Tripleurospermum rosellum (Boiss. & Orph.) Hayek was based on a collection by Orphanides from Mt Parnonas (SE Peloponnisos) and is likely to be endemic in this area. Plants from Lesvos tentatively referred to *T. rosellum* by Strid (2016a: 133), following Fl. Turkey 5: 300 (1975), match specimens from nom. Evrou identified as *T. oreades* (Boiss.) Rech. f. *T. lesbiacum* (Cand.) Rech. f. from Lesvos is probably a synonym of *T. oreades*, although the type specimen is lost.

Tripleurosperum tempskyanum (Freyn & Sint.) Hayek is generally believed to be a serpentine endemic of NW Greece. However, it was reported from Ulu Dağ in NW Anatolia by Inceer & Haryilioglu (2010: 153). A photo in Inceer (2021a: 349) is indeed similar to *T. tempskyanum*, but such a disjunction is biogeographically unlikely. Careful comparison between Greek and Turkish material is needed to establish if there are any differences.

Trisetum flavescens (L.) P. Beauv. is a widespread and polymorphic species. Subsp. *splendens* (C. Presl.) Arcang. and subsp. *tenue* (Formánek) Strid have been recognized in Greece (cf. Mount. Fl. Greece 2: 806-808, 1991), but Barberá, Romero-Zarco & Aedo (2018: 368 ff.) merged both into subsp. *flavescens*.

Trisetum rechingeri Chrtek from Mt Dirfis, Evvia, is a highly dubious species. The description was based on a single sheet (Heldreich 3522, 7 Aug 1858; P), originally labelled *T. flavescens* var. *splendens* and differing from *T. laconicum* Boiss. & Orph. from Mt Parnonas (SE Peloponnisos) only in vague, quantitative characters. According to Chrtek (1966b: 93), the same collection number at herbarium W is a mixture of *T. laconicum* and *T. rechingeri*: "In diesem Falle, meiner Meinung nach, handelt es sich wahrscheinlich um eine Verwechslung, die vielleicht bei der Manipulation mit den Herbarbelegen entstand". It seems far more likely that "*T. rechingeri*" represents individual variation within *T. laconicum*, a taxon closely related to *T. flavescens*. No *Trisetum* appears to have been re-collected on Mt Dirfis. Barberá, Romero-Zarco & Aedo (2018: 387) also concluded that *T. rechingeri* is a synonym of *T. laconicum*.

Triticum. In spite of much work on the genetics and evolution of the cultivated wheats, taxonomy of the wild relatives is still unsettled (cf. Caligari & Brandham 2001). *T. monococcum* subsp. *aegilopoides* (Link) Thell. (syn. *Crithodium aegilopoides* Link; *T. boeoticum* Boiss.) has been reported from a few localities in Sterea Ellas, Peloponnisos and the Aegean area, but it is not clear if and how this differs from naturalized plants of cultivated *T. monococcum* L. It was illustrated by Link (1834: tab. 3, sub nom. *C. aegilopoides*), based on material collected between Nafplio and Korinthos. There are scattered reports of *T. aestivum* L., *T. durum* Desf., *T. monococcum* L. and *T. turgidum* L. from ruderal habitats throughout Greece, all probably referring to casual escapes from cultivation.

Tropaeolum majus L., a native of South America commonly cultivated as an ornamental, was recorded from KK in the Greek checklist (Dimopoulos & al. 2013: 143), but probably only as a casual escape.

Tuberaria lipopetala. In *T. guttata* s. lat. populations or individuals with rudimentary petals are scattered throughout the range. From NW Africa such plants have been described as *Helianthemum guttatum* subsp. *lipopetalum* Murb. and subsequently recombined as *Tuberaria lipopetala* (Murb.) Greuter & Burdet. This name has been applied also to Greek plants. If the latter are to be given taxonomic status at all, they are unlikely to represent the same taxon as the type of *Helianthemum guttatum* subsp. *lipopetalum*.

Tulipa celsiana DC. was listed as an accepted species in Euro+Med PlantBase, but is apparently a synonym of *T. australis* Link.

Tulipa clusiana DC. A single, historical record from Peloponnisos is apparently incorrect (Raus 2021: 152).

Tulipa doerfleri Gand. The habitat makes it reasonable to suspect that this species respresents an ancient introduction of Anatolian stock, but no studies are available as to its origin.

Tulipa goulimyi Sealey & Turrill. According to photos by N. Turland plants from the outlying locality in western Kriti (Falassarna) have distinctly narrower tepals than those from Peloponnisos and Kithira.

Tulipa hageri Heldr. (1874: 98) was described from Mt Parnitha, "Nordseite des Gebirges, auf einer mit Getreide bebauten Hochebene, Megalo Chorafi genannt, unterhalb der Ruinen der alten attischen Grenzfeste Dekeleia". A good colour illustration appeared with the original description. *T. orphanidea* Boiss. ex Heldr. was reported to occur in the same locality, to be more common and distinctly different. *T. hageri* has remained somewhat enigmatic, but is provisionally retained as a separate species. Plants collected by Polymenakos 3.4. 2018 in a cereal field in eparchia Thivon are a close match for the original illustration of *T. hageri* (see Polymenakos & Kit Tan 2020a: 142).

Tulipa praecox Ten. (1811). Due to nomenclatural uncertainties this name was provisionally retained in the Aegean Atlas (Strid 2016a: 391) for a tulip that is locally abundant in olive groves and cultivated fields on central Chios, although it is a later homonym of *T. praecox* Cav. (1802). The latter was described from material cultivated in the Madrid Botanical Garden and is apparently a synonym of *T. clusiana* DC. The next name available for the Chios tulip is probably *T. raddii* Reboul (1822), although the identity of the latter should be critically examined (according to Everett 2013: 165 it is a synonym of *T. agenensis* DC. in Redouté). *T. praecox* Ten. was said to grow at margins of cultivated fields on the island of Capri. The illustration in

Tenore's Flora Napolitana shows a broad-leaved tulip likely to be conspecific with the one on Chios. Conserving the widely used name *T. praecox* Ten. would appear to be the best solution.

Tulipa scardica Bornm. has been reported twice from foothills of Southern Pindos, based on Sintenis 1555 (LD) and Stamatiadou 21536 (ATH). These specimens are scarcely disinguishable from *T. undulatifolia* Boiss. *T. scardica* was described from the foothills of Šar Planina in North Macedonia; according to Mayer & Micevski (1970) it is endemic to the middle and upper Vardar valley. It resembles *T. undulatifolia* but has shorter and broader perianth segments.

Ulmus laevis Pall. Christensen (in Fl. Hellen. 1: 52, 1997) listed records from Evvia as doubtful. Reports from riverine woodland in the Tembi valley (Hayek 1928b: 310), the Pinios delta (Eleftheriadou & al. 1995: 225) and the Nestos river gorge (Kirchhoff & Petermann 1992b: 66, Schuler 2000: 422) are plausible, but no specimens have been seen and confirmation is needed. *U. laevis* is recognized on the following combination of characters: Bud scales bicoloured, distinctly darker towards apex; pedicels much longer than flowers; fruits ciliate, pendent on long pedicels; seed central.

Urtica dioica group. A form with small leaves and few stinging hairs has been collected by the Erythropotamos river WSW of Ellinochori in the far NE of the Greek mainland (Strid & Hartvig 59559; UPA, herb. Strid). It is probably referable to *U. galeopsifolia* Wierzb. ex Opiz, but further studies of variation are needed.

Utricularia gibba L. has been reported once, from the Kalodiki wetland area (Sarika-Hatzinikolaou, Koumpli-Sovantzi & Yannitsaros 1996: 228). The record is feasible but needs confirmation since both *U. australis* R. Br. and *U. vulgaris* L. have also been reported from this locality. *U. gibba* differs by the narrower upper lip of the corolla. It is widespread in tropical and warm-temperate areas; in Europe it is otherwise known with certainty only from Portugal and SW Spain.

Vaccinium uliginosum L. Greek specimens appear to match subsp. *microphyllum* (Lange) Tolm., a high-altitude form of uncertain distribution and status.

Valeriana alliariifolia Vahl was reported from Mt Xirovouni (Evvia) by Halácsy (1901: 744), based on Heldreich, Herb. Norm. 798. The record was repeated in Fl. Eur. 4: 53 (1976), as *V. alliariifolia* Adams. *V. alliariifolia* Vahl, Enum. Pl. 2: 11 (1805) was described from Cappadocia (C Anatolia), whereas *V. alliariifolia* Adams in Beitr. Naturk. (Weber & Mohr) 1: 44 (1805) was described from Caucasus. Richardson in Flora of Turkey 4: 552 (1972) regarded the two as conspecific, with priority given to *V. alliariifolia* Adams which was reported to be widespread in Anatolia, Caucasia and N Iran. The record from Mt Xirovouni (the only one in Europe) has been considered dubious, and the species was omitted in the Greek checklist (Dimopoulos & al. 2013). However, it was confirmed in 2015 (Kalpoutzakis 4718; ACA, ATHU – see Trigas & al. 2021: 328), probably very close to the original locality of Heldreich.

Valeriana bertiscea Pančić is a rare species described from Montenegro (see Vukojičić & al. 2021: 7) and subsequently reported from scattered localities in the W Balkan Peninsula. *V. heideri* Halácsy was based on a collection by Halácsy (12.7. 1888, WU-Hal) from Mt Giona where it has subsequently been confirmed. Franzén (in Mount. Fl. Greece 2: 346, 1991) interpreted *V. heideri* as a synonym of *V. bertiscea*, and this opinion is followed here.

Valeriana italica. Application of the names *V. italica* Lam. and *V. dioscoridis* Sm. is uncertain – see comments in Strid (2016a: 553).

Valeriana montana group. *V. montana* L. and *V. tripteris* L. are difficult to distinguish and have overlapping distributions in C & S Europe. Intermediates and hybrids have been reported (cf. Weberling in Hegi, Flora von Mittelleuropa VI/2: p. 158, 1970; Parent 2005: 219). 2n=16 and 2n=32 have been counted in both species. Most of the Greek plants match *V. montana* but those from Mt Pangeon as well as some from Mt Falakron have 3-foliolate cauline leaves and have been referred to *V. tripteris*. There is unlikely to be more than one taxon in Greece, and this has provisionally been called *V. montana*.

Valeriana phitosiana Quézel & Contandr. was described from schistose cliffs by Drakolimni on Mt Timfi. There are several collections of *V. crinii* subsp. *epirotica* (Phitos) Franzén from the same locality, and *V. phitosiana* almost certainly represents a deviating individual of the latter.

Valeriana pratensis subsp. ***angustifolia*** (Soó) Kirschner, Buttler & Hand. Greek records probably refer to *V. officinalis* subsp. *collina* (Wallr.) Nyman.

Valeriana stolonifera Czern. was listed for "Gr" in Euro+Med PlantBase. The record is likely to refer to a form of *V. officinalis* L.

Valerianella. The extensive collections by Willing (B) have, for the main part, been mapped as reported at www.willing-botanik.de, but critical examination of herbarium specimens is needed in some cases.

Veratrum lobelianum Bernh. is a widespread, Euro-Siberian species sometimes regarded as a subspecies or variety of the white-flowered *V. album* L. from mountains of C & S Europe. However, Euro+Med PlantBase uses the name *V. lobelianum* for a Russian-Transcaucasian taxon. The complex is in need of taxonomic revision (cf. Szeliga, Ciura & Tyrka 2017).

Verbascum auriculatum Sm. has long been regarded as a synonym of *V. mucronatum* Lam., but Zografidis (2018) demonstrated that the two are quite distinct. The type of *V. auriculatum*, illustrated on Plate 225 of Flora Graeca (1819), was probably collected and sketched in northern Rodos on 28 May, 1787. *V. propontideum* Murb. is a synonym of *V. auriculatum*. The former epithet was derived from Propontis, an ancient name for the Sea of Marmara, since the type specimen was believed to originate from Chalki (Turkish Heybeliada), one of the Princes' Islands near Istanbul, whereas in fact it had been collected on the island with the same name just W of Rodos.

Verbascum chaixii Vill. is a widespread and variable species of C & S Europe. The rather few Greek records are mostly literature reports in need of confirmation. Most of them refer to subsp. *austriacum* (Roem. & Schult.) Hayek, but subsp. *chaixii* has been reported from the vicinity of Kavala by Dimitrov (2013: 160). However, the latter was revised as *V. sinuatum* L. by A. Zografidis (April 2021), based on a scan of the specimen.

Verbascum delphicum Boiss. & Heldr. Zografidis (2016) made a detailed study of this rare species, describing plants from the isolated locality on Mt Parnitha as subsp. *cervi* Zograf. Populations from southern Evvia also differ somewhat from those at *locus classicus* (on Mt Dirfis) and were recognized as var. *filictorum* Zograf. Literature records from the island of Andros (Malakates 1933b: 94) and a locality in nom. Fokidos (Aplada 2013: 254) are feasible but need confirmation.

Verbascum densiflorum Bertol. An unspecified record from Peloponnisos (Mermygkas, Zikos & Constantinidis 2021: supplementary file) may refer to *V. macrurum* Ten.

Verbascum dimoniei Velen. is a little known species occurring in a small area in north central Greece and perhaps extending into adjacent parts of North Macedonia. The type is Dimonie, Plantae Macedonicae no. 327 (PRC) from "Fetitza prope Njausta [Naoussa]" in the eastern foothills of Mt Vermion where it has subsequently been confirmed. It is closely related to *V. graecum* Heldr. & Sartori ex Boiss. and was only tentatively accepted as a different species by Murbeck (1933: 331).

Verbascum dingleri Mattf. & Stef. is a rare regional endemic. The description was based on a specimen collected "auf Kalkfelsen bei Derwent unweit Dedeagatsch [Alexandroupolis]". Derwent was the Turkish name for the village of Avas or Avantas c. 10 km N of Alexandroupolis [information courtesy Th. Raus, January 2021]. Murbeck (1933: 481) had examined the type in the Berlin Herbarium, but this was probably destroyed in 1943. There are only a few subsequent collections, all from the Kavala area, and no specimens at B.

Verbascum foetidum Boiss. & Heldr. is a species of uncertain status combining some features of *V. epixanthinum* Boiss. & Heldr. and *V. eriophorum* Godr. It was based on Heldreich 2737, 6.8. 1852 (B, G-BOIS, TAU, W) from Mt Parnassos. Murbeck (1933: 171), who had examined the type specimen, "halte es für wahrscheinlich, dass *V. foetidum* ein südlicher, vielleicht am Parnassus endemischer Paralleltypus zu *V. eriophorum* ist". Zografidis found it in a different locality on Mt Parnassos as well as Vardousia. Preliminary studies indicate that it may well be a separate species endemic in this area.

Verbascum friedrichsthalianum Kuntze was listed as a Greek endemic in Euro+Med PlantBase. It is apparently a synonym of *V. daenzeri* (Fauché & Chaub.) Kuntze (see Zografidis & Strid 2017: 66).

Verbascum glandulosum Delile was listed as an accepted species in Euro+Med PlantBase as well as the Greek checklist (Dimopoulos & al. 2013: 141), but replaced with *V. gloeotrichum* Hausskn. & Heldr. in the Aegean Atlas (Strid 2016a: 537) on the grounds that *V. glandulosum* Delile (1849) is antedated by *V. glandulosum* Thome ex Roem. & Schult. (1819) – see also Murbeck (1933: 300). However, the last-mentioned name was listed only in synonymy and thus cannot invalidate Delile's *V. glandulosum*.

Verbascum halacsyanum Sint. & Bornm. ex Halácsy. The description of this local endemic was based on Sintenis & Bornmüller 140 (B, JE, LD, WU-HAL), collected near Kavala on 10th May, 1891, although it had in fact been gathered here already by Friedrichsthal c. 1835. It has subsequently been confirmed in this area by Rechinger (in 1936), Willing (in 2009), Zografidis (in 2016) and Strid (in 2018).

Verbascum hypoleucum Boiss. & Heldr. is a rare local endemic long known only from Heldreich's type collection "in faucibus umbrosis reg. abietinae inter Kyllenes et Gymno-vouno montes versus Phlampouritza" (10.7. 1848; ?B, G). Murbeck (1933: 434-435) provisionally retained it as a separate species, noting that it is closely related to *V. abietinum* Borbás (*V. nigrum* subsp. *abietinum* (Borbás) I. K. Ferguson). The plant found by Zarkos and Christodoulou at "Mt Killini, Flabouritsa gorge, along rocky footpath from Manna to Markou Lakka, 1160 m, 37°57'N, 22°28'E, 16.06.2012", and reported as *V. banaticum* Schrad. in Phytol. Balcan. 18: 228-229 (2012) is in fact *V. hypoleucum* (A. Zografidis, e-mail 18.10. 2019). A second locality was found in 2019 by Kostas Polymenakos "in the aesthetic Dryodasos Mongostou (foothills of Mt Kyllene)".

Verbascum lanatum Schrad. A single record from the Anthero area (nom. Dramas) by Tsiripidis & Athanasiadis (2003: 294) needs confirmation. The species is otherwise known from N Italy and EC Europe southwards to N Bulgaria.

Verbascum limnense Fraas, a member of subg. *Celsia*, hybridizes with *V. haussknechtii* Heldr. ex Hausskn. in northern Evvia (e.g., Strid 60769, UPA). The occurrence of such hybrids corroborates the fact that *Verbascum* s. str. and *Celsia* are not tenable as separate genera.

Verbascum longifolium Ten. was listed for Peloponnisos by Mermygkas, Zikos & Constantinidis (2021: supplementary file), but no details are available.

Verbascum lychnitis L. is widespread in Europe. Records from the island of Kerkira (Hansen 1982: 44, Borkowsky 1994: 121, Dinter 1997: 38) need confirmation and were not included in the Flora Ionica website. Historical reports from NE Greece (Rechinger 1939a: 491, Kitanov 1943: 266, Goulimis 1960a: 28) were believed to be incorrect. However, specimens at ATH collected by Goulimis on Mt Vrondous 21 June 1955 (beech forest to rocks of Ali Baba, 1500-1700 m), have been confirmed by A. Zografidis (pers. comm. 2021-06-07) to represent *V. lychnitis*. The locality is nearly identical with that reported by Rechinger (loc. cit.). Strid & al. 19365, 23 Jul 1981 (C, C, G, UPA) from Mt Orvilos was distributed as *V. humile* Janka, tentatively identified by Raus (1988) as the hybrid *V. humile* × *V. speciosum*, and revised by Zografidis (2021) as *V. lychnitis*. This renders credibility also to the report by Kitanov (loc. cit.) from Mt Falakron.

Verbascum macrurum Ten. Further studies of variation are needed. Records from northern Greece may represent *V. niveum* subsp. *pannosiforme* (Stoj.) Murb.

Verbascum orphanideum Murb. is an enigmatic species. The description was apparently based on a single individual, Orphanides s.n., 27.7. 1850 (?B), labelled merely "Laconia". It is likely to have been collected on the northern or upper part of Mt Parnonas where there are other collections by Orphanides the same day (e.g., the similar *V. megaphlomos*, Orphanides 2993; B, UPS, WU). "Der wichtigste Unterschied ist an den zwei vorderen Staubblättern vorhanden, deren Filament bei der hier beschiebenen Art im oberen Drittel oder Viertel nackt ist und deren Konnektiv gleichfalls vollkommen ohne Papillen ist; bei *V. megaphlomos* dagegen ist das Filament der genannten Staubblätter immer bis zur Anthere hinauf dicht papillös, und das dazugehörige Konnektiv ist auf seiner Innenseite gleichfalls mit einem Büschel von langen Papillen versehen" (Murbeck 1933: 313). *V. megaphlomos* Boiss. & Heldr. is now regarded as a subspecies of *V. speciosum* Schrad. Further studies are needed to determine the taxonomic value, if any, of *V. orphanideum*.

Verbascum ovalifolium Sims was not included in the Greek checklist (Dimopoulos & al. 2013), as previous reports (Rechinger 1963, Ferguson in Fl. Eur. 3: 208, 1972) were considered dubious. A small population of c. 30 plants was discovered by Zografidis, 14.7. 2016, near Dilofos in the north-easternmost part of the Greek mainland (see Zografidis 2017: 64). Hybrids with *V. banaticum* Schrad. were observed c. 5 km SE of the Dilofos site. *V. ovalifolium* was collected in the same area by Strid, 9.6. 2017 (no. 58944). Both subsp. *ovalifolium* and subsp. *thracicum* (Velen.) Murb. have been reported from Bulgaria and W Turkey (Dane & Yılmaz 2005), but there appear to be no consistent differences.

Verbascum phlomoides L. was listed for Peloponnisos by Mermygkas, Zikos & Constantinidis (2021: supplementary file), but no details are available.

Verbascum phoeniceum group. The taxonomy of *V. phoeniceum* L., *V. flavidum* (Boiss.) Freyn & Bornm. and *V. xanthophoeniceum* Griseb. is unsettled. *V. phoeniceum* is widespread in C & E Europe and SW to C Asia (and frequently cultivated as an ornamental); it is characterized by deep reddish-purple to violet corolla, an unusual feature in the genus. Such plants are scattered in NE Greece and have been photographed by the present author in the Athos peninsula and the provinces of Xanthis and Rodopis. In the same area and further to the west are plants with bright to deep yellow corolla (usually with reddish-purple centre). They have been registered as *V. flavidum*, but scarcely differ in any other characters and are perhaps better regarded as *V. phoeniceum* var. *flavidum* (as originally described by Boissier). *Celsia rechingeri* Murb., described from Kerkira, probably belongs here (although lacking the fifth stamen). *V. xanthophoeniceum*, which was described from NW Turkey (Çanakkale), is yellow-flowered and appears to differ also in leaf characters. Photos from Mt Timfristos show purple- and yellow-flowered plants as well as intermediates. For comments on this species group, see Murbeck (1925: 106-107), Bornmüller (1928d: 48), Murbeck (1933: 582 ff.), Huber-Morath in Flora of Turkey 6: 495-497 (1978) and Meyer (2011: 132). *V. phoeniceum* was listed for Peloponnisos by Mermygkas, Zikos & Constantinidis (2021: supplementary file), probably in error; all confirmed Greek records are in the north.

Verbascum salicifolium Zograf. & al. has just been described from a single population in the vicinity of Lake Doirani, nomos Kilkis (see Zografidis & al. in Phytotaxa 552(3): 182-190, 2022). It resembles *V. adenanthum* but has oblanceolate to narrowly elliptic, ± acute basal leaves, leaf-like bracts subtending the lower branches of the inflorescence, and corolla with obovate lower lobes.

Verbascum syriacum Schrad. See comments in Strid (2016, 1: 541). Records from Syria, Lebanon and Iraq are probably incorrect.

Verbascum undulatum Lam. The illustration of "Verbascum Graecum, fruticosum, folio sinuato candidissimo" in Tournefort, Rel. Voy. Levant., ed. Paris, 1: t. facing p. 335 (1717) was designated as lectotype of *V. undulatum* Lam. by Zografidis (2019a: 244). A collection from Kea (Zografidis 429; B, UPA) was designated as an epitype. The name *V. undulatum* is thus preserved in its traditional sense. *V. plicatum* Sm. is a synonym of *V. undulatum*.

Verbena aristigera S. Moore has been collected in a coastal marsh on the island of Skiathos (see Biel & Kit Tan 2006a: 110). It is a perennial herb with dissected leaves and spikes of mauve flowers, native in South America and naturalized in SE North America, Australia, etc. The degree of naturalization in Greece is not clear.

Veronica acinifolia L. and ***V. bozakmanii*** M. A. Fisch. are provisionally retained as separate species, although the morphological differences are not as clearcut as claimed in the original description of the latter. Several chromosome counts indicate 2n=14 in *V. acinifolia* and 2n=28 in *V. bozakmanii*, but Sanchez Agudo & al. (in Taxon 60: 1795, 2011) found both numbers in Spanish material matching *V. acinifolia*.

Veronica alpina L. was listed for "Gr" in Euro+Med PlantBase, but no specimens or records have been found. The southernmost occurrences for this arctic-alpine species are in NW North Macedonia and SW Bulgaria.

Veronica aznavourii Dörfler (*V. pontica* Velen.) was listed for "Gr" in Fl. Eur. (3: 249, 1972), perhaps based on the imprecise "Thra" in Hayek's Prodromus (2: 163, 1929). It is otherwise restricted to E Bulgaria and NW to C Anatolia, and occurrence in NE Greece is feasible.

Veronica contandriopouli Quézel. This species was originally reported from Mt Killini, "sur les falaises calcaires verticales exposees au Nord en compagnie de *Valeriana olenaea*, *Minuartia stellata*, *Carum rupestre*, *Achillea umbellata* et *Hieracium sartorianum*" (Quézel & Contandriopoulos 1967). The description was somewhat contradictory, and it was listed as a dubious species by Fischer in Mount. Fl. Greece 2: 216 (1991) who had not seen any specimens. It was rediscovered in 2004 by Spyros Apostolopoulos and Georgios Zarkos and is indeed a distinct local endemic, very rare and known only from a few small populations (cf. Zarkos, Christodoulou & Kit Tan 2013b).

Veronica hederifolia L. and ***V. triloba*** (Opiz) Opiz. See comments in Strid (2016a: 563).

Veronica jacquinii Baumg. is optionally regarded as a subspecies of *V. austriaca* L. (Rojas-Andrés & Martínez-Ortega 2016: 10 ff.).

Veronica kindlii Adamović was based on a single collection from Mt Voras on the border between Greece and North Macedonia ("in summis alpinis m. Nidže-Planina"). It is probably a form of *V. orsiniana* Ten. which is rather common on this mountain and extends to the summit area. According to Rojas-Andrés & Martínez-Ortega (2016: 29-33) populations identified as *V. orsiniana* from the Balkan Peninsula constitute an independent evolutionary lineage for which the earliest available name is *V. kindlii*. The name *V. orsiniana* was reserved for plants occurring in NE Spain, S France and Italy.

Veronica praecox All. The records from Milos are very recent (Biel & Kit Tan in Phytol. Balcan. 28: 150, 2022) and represents a significant extension of the range.

Veronica serpyllifolia L. Differences between subsp. *serpyllifolia* and subsp. *humifusa* (Dicks.) Syme (syn. *V. balcanica* Velen.) are rather obscure. Most Greek specimens belong to the latter, but the former may occur at low altitude. Both are widespread and the differentiation is altitudinal rather than geographical.

Veronica teucrioides Boiss. & Heldr., here classified as a subspecies of *V. orsiniana*, may be better regarded as a separate species restricted to a rather small area in N Greece and North Macedonia (Rojas-Andrés & Martínez-Ortega 2016: 75-78).

Veronica thessalica Benth. is a characteristic species of stabilized calcareous screes in the summit area of Mt Olimbos and has been reported also from N Albania and adjacent Kosovo. Detailed studies are needed to determine whether the latter plants are indeed conspecific with those from Olimbos or perhaps rather referable to *V. saturejoides* Vis. Plants from Olimbos have also been called *V. saturejoides* var. *olympica* Stoj. & Jordanov.

Veronica turrilliana Stoj. & Stef. The inclusion of this species is based on a report by Bergmeier (1988) from Kato Olimbos. It is otherwise known only from a small area in SE Bulgaria and Turkey-in-Europe.

Vicia cracca L. Only verified herbarium specimens have been mapped. Numerous records, especially from low altitude in the south, are likely to refer to forms of *V. villosa*. Most Greek specimens of *V. cracca* have rather stiff, erect stems, thus matching subsp. *incana* (Gouan) Rouy which occurs mainly in mountains of SE Europe and Anatolia. A few specimens from the far north appear to belong to subsp. *cracca* (with weak, flexuous stems). *V. cracca* s. lat. is a widespread and extremely variable, circum-boreal species.

Vicia davisii Greuter was based on a single collection from an islet NE of Megisti (Kastellorizo). It is probably conspecific with *V. cassia* Boiss.

Vicia dumetorum L. was listed for Greece in Fl. Eur. (2: 132, 1968) as well as Euro+Med PlantBase, possibly based on the imprecise "Thra" in Hayek's Prodromus. It is a Central European species extending as far as Bulgaria in the SE. No Greek material has been seen, but occurrence in the far north is feasible.

Vicia ervilia (L.) Willd. Many of the Greek records are old and may refer to plants naturalized from former cultivation for fodder. However, recent low-altitude records from agricultural or ruderal habitats are available from Lefkas, Kriti, Kasos, Karpathos and Rodos. In addition, there are at least three collections from limestone scree at rather high altitude, viz., Orphanides 3345 (ATHU) from Mt Parnonas, Willing 13821 (B) from Mt Timfristos and Strid 57446 (ATH, G, herb. Strid) from Mt Klokos. The latter may represent a native form of the species; such plants are known also from Anatolia and further east.

Vicia faba L. (broad bean) is a cultivated plant of uncertain origin. It is a robust annual with 2- or 3-paired leaves without tendril, whitish corolla with blackish-purple wings, and large legumes with compressed seeds 20-30 mm wide. It is widely cultivated in Greece and may persist for a short time in ruderal habitats, but is scarcely naturalized. *V. faba* is probably related to the native *V. narbonensis* L., which is less robust and has much smaller seeds.

Vicia galilaea Plitm. & Zohary is a member of the *V. narbonensis* complex, described from Palestine. It is widespread in Anatolia and has been reported also from the island of Lesvos. Differences between this and *V. narbonensis* s. str. are not clear and there are many intermediates (cf. Fl. Turkey 3: 324, 1970).

Vicia incana Gouan. See *V. cracca*.

Vicia johannis Tamamsch. is a member of the *V. narbonensis* complex, described from Azerbaijan. Records from northern mainland Greece as well as the island of Kos need confirmation.

Vicia lutea L. Most of the Greek specimens have rather abundant, stiff, tubercle-based hairs on the upper internodes. Such plants have been called *V. lutea* var. *hirta* (DC.) Loisel. or subsp. *vestita* (Boiss.) Rouy. They co-exist with the less common subsp. *lutea* which has few, short, slender and ± appressed hairs without tuberculate base.

Vicia monantha Retz. There is only one credible report of this species in Greece, from the vicinity of Kattavia in S Rodos (Böhling & Bässler in Greuter & Raus 1999: 62, see also Kleinsteuber & al. 2016: 541), and even for this the precise identity remains uncertain. Several previous literature records from various Greek localities are dubious and probably refer to different species. No specimens have been seen by the present author, nor by Lassen in his draft account of *Vicia* for Flora Hellenica.

Vicia pannonica L. Almost all Greek records represent the widespread subsp. *striata* (M. Bieb.) Nyman, which has tawny purplish flowers and is mainly found in cultivated fields (occasionally in semi-natural habitats). Subsp. *pannonica*, with pale yellow flowers, has been recorded from damp scrub in a few localities in northern Greece.

Vicia sativa L. is a very variable species (or species complex), widespread in the Euro-Siberian region and naturalized in temperate areas almost worldwide. Flowering and fruiting material as well as underground parts of the plant are generally needed for positive identification of the subspecies. Many herbarium specimens and literature records remain identified only as *V. sativa* s. lat.

Vicia sicula (Rafin.) Guss. Turrill (1918: 284) reported *Orobus atropurpureus* Desf. from a locality 8-10 km N of Thessaloniki, remarking: "The distribution of this plant is peculiar. It is not uncommon north of Thessaloniki". It is now considered a synonym of *Vicia sicula*, a species occurring in S Italy, Sicily and NW Africa. There is no subsequent confirmation from Greece, and the specimen (Turrill 39, K) should be critically re-examined.

Vinca minor L. All Greek records of this species are likely to refer to *V. herbacea* Waldst. & Kit.

Vincetoxicum funebre Boiss. & Kotschy. This species has been reported from calcareous screes on Mts Timfristos and Iti (Sterea Ellas), but specimens are likely to fall within the range of variation of *V. fuscatum* (Hornem.) Rchb. f. *V. funebre* was described from N Iran and is otherwise known from NE Anatolia and Caucasia.

Viola allchariensis Beck was based on Dörfler 73 (M, WU, herb. Beck) from Alšar, a locality in the Voras area just north of the Greek border, and has subsequently been recollected here, but is not known elsewhere. Voliotis (1981b: 259) reported it from the Greek side of the Voras massif, but according to Erben this record as well as others from Greece "haben sich als falsch erwiesen".

Viola arborescens L. is a W Mediterranean species. In 1818 Sieber found a woody-based, yellow-flowered violet at Cape Sidero in easternmost Kriti and observed ".... ich bemerkte hier an einigen Felsen das baumförmige Veilchen (*Viola arborea*) .... Desfontaines, ein französischer Botaniker, hatte es in der Barbarei [NW Africa] entdeckt". Greuter & Rechinger (1967: 58-59) discussed taxonomy of the *V. arborescens* group. Greek plants are now referred to *V. scorpiuroides* Coss. See also under *V. methodiana*.

Viola brachyphylla W. Becker is a yellow-flowered perennial of uncertain status. The description was based on a single collection (Mrkvička 74, 30.5. 1916; PRC) from Dobro Pojle, a doline at c. 1700 m in the Voras massif close to the North Macedonian border. Erben (1985: 471) was of the opinion that it may be a hybrid involving *V. eximia* Formánek. A subsequent collection from the same locality (Strid & al. 30061, 2.6. 1989; ATH, C, G, LD, UPA) was identified as *V. eximia* × *V. frondosa* by A. Livaniou-Tiniakou, 1995. Records of *V. brachyphylla* from Mt Vourinos (Babalonas 1983: 24 & 1989: 398) are probably incorrect.

Viola calcarata L. was listed for "?Gr" in Fl. Eur. 2: 277 (1968), but no specimens or specified records have been seen. It occurs in the Alps and mountains of the W Balkan Peninsula southwards to N Albania. Occurrence in Greece seems unlikely.

Viola dirphya Tiniakou was described as a local endemic of Mt Dirfis, Evvia, growing in somewhat damp, rocky woodland habitats at 900-1300 m. Morphologically it can scarcely be separated from *V. sieheana* W. Becker, which occurs in the same area.

Viola elatior Fries was listed for "Gr" in Fl. Eur. 2: 275 (1968), but the basis for this record is not clear. *V. elatior* is a tall, narrow-leaved plant of wet habitats, widespread from C Europe to W Siberia. It is unlikely to occur in Greece.

Viola graeca (W. Becker) Halácsy. Raus (in Mount. Fl. Greece 1: 628, 1986) discussed the variation but refrained from subdividing the species. Erben (1985: 477 ff.) divided it into *V. graeca* s. str. (lectotype from Mt Parnassos) and the local endemics *V. pseudograeca* (Mt Olimbos), *V. sfikasiana* (Mt Taigetos) and *V. vourinensis* (Mts Vourinos and Siniatsikon). The latter approach is followed here. Erben (1985: 497) observed that plants from Mt Parnonas (SE Peloponnisos) are somewhat different, but did not recognize them as a separate taxon. With more material available, Kit Tan, Sfikas & Vold (1997) described them as *V. parnonia*. The group is similarly variable in Italy, with several local taxa and several chromosome numbers.

Viola jordanii Hanry. Greek plants are referred to this species following Tiniakou (1996), although in some respects they are more similar to *V. persicifolia* Schreb. (*V. stagnina* Kit.), and the taxonomy of the group is not clear. The Greek plants typically grow in rather dry deciduous oak woodland and scrub.

Viola methodiana Coustur. & Gand. was described from Cape Sidero in easternmost Kriti. These plants differ from *V. scorpiuroides* Coss. from W Kriti and Kithira in the strongly woody-based, compact habit, rather broad leaves and small flowers; they may deserve subspecies rank under the latter.

Viola orbelica Pančić was reported by Sfikas (1998b: 5) from the area of Mt Chaidou (Koula) in the eastern Rodopi. It is a member of the *V. tricolor* complex otherwise endemic to Pirin and Rila mountains in Bulgaria. The Greek record needs confirmation.

Viola phitosiana Erben is a taxon of uncertain status combining features of *V. hymettia* Boiss. & Heldr. and *V. kitaibeliana* Schult. and perhaps representing hybrids between these two.

Viola rauliniana Erben. Very similar plants occur on Cyprus, and it remains to be established whether these are conspecific with the ones from Kriti. *V. rauliniana* was listed also for Anatolia in Euro+Med Checklist, but the source for this record is obscure.

Viola stojanowii W. Becker. Some of the plants referred to this species by Raus in Mount. Fl. Greece (1: 622, 1986) have subsequently been described as *V. ganiatsasii* Erben. As currently understood *V. stojanowii* is a local endemic of Mt Beles (Belasica Planina) on both sides of the Greek/Bulgarian border.

Viola suavis M. Bieb. Rechinger 3772 (LD) from Samos was tentatively listed as *V. pontica* W. Becker (a synonym of *V. suavis*) by Rechinger (1944a: 255). It needs confirmation and may in fact refer to a form of *V. odorata* L. The same applies to records of *V. suavis* from Samothraki (Biel & Kit Tan 2011a: 132).

Viola thasia W. Becker was described as a local endemic of the island of Thasos (type: Sintenis & Bornmüller 348; G-BOIS, JE, LD, WU). According to Raus (Mount. Fl. Greece 1: 633, 1986) there are no morphological differences between plants from Thasos and those from the adjacent mainland here called *V. tricolor* subsp. *macedonica* (Boiss. & Heldr.) A. Schmidt.

Vitis labrusca L., a native of E North America, was reported by Ardenghi & Galasso (2013: 245) as a casual in the Nafpaktos area. The hybrid *V. riparia* Michx. × *V. rupestris* Scheele (two other North American species) was found in the same area. The hybrid is widely used in the Mediterranean region as a phylloxera-free rootstock for grafting cultivated vines. *V. rupestris* has also been collected by Snogerup (no. 8579, 23.6. 1991; LD) in the town of Chios, "around the castle ruins".

Vulpia is close to *Festuca* and optionally included in the latter, as in *Portale della Flora d'Italia* (consulted online 2021-09-16).

Vulpia membranacea (L.) Dumort. is a W Mediterranean species. Greek records (e.g., Haussknecht 1900: 52 from Attiki, Gandoger 1920b: 282 from Siros, Hansen, Snogerup & Rungby 1966: 369 from Rodos, Cotton & Stace 1976: 242 from Kriti and Mikonos) probably refer to *V. fasciculata* (Forssk.) R. M. Fritsch.

Willemetia stipitata (Jacq.) Dalla Torre. Greek plants belong to subsp. *albanica* (Kümmerle & Jáv.) Kirschnerová, which was described from Mt Korab in Albania and has been recorded also from North Macedonia, Montenegro and SW & C Serbia.

Xanthium strumarium L. has been variously subdivided with Greek plants referred to *X. strumarium* subsp. *strumarium*, *X. strumarium* subsp. *brasilicum* (Vell.) O. Bolòs & Vigo, *X. strumarium* subsp. *cavanillesii* (Schouw) D. Löve & P. Dansereau, *X. orientale* subsp. *italicum* (Moretti) Greuter and *X. pungens* Wallr. The taxonomy is unsettled and only few records have been precisely identified; thus only a collective map can be presented.

Zannichellia pedunculata Rchb. (syn. *Z. palustris* var. *pedicellata* Wahlenb. & Rosén) differs from *Z. palustris* L. only vaguely in quantitative characters such as length of peduncle and dimensions of fruits. There appear to be no differences in ecology or distribution.

Ziziphus jujuba Mill. (*Z. zizyphus* (L.) Meikle) is a small deciduous tree probably native in Caucasus and C Asia. It is widely cultivated for its edible fruits and casual in the Mediterranean area. There are scattered records throughout Greece, but it does not appear to be naturalized.

Ziziphus lotus (L.) Lam. is a spiny shrub or small tree of North Africa, also cultivated for its edible fruits. There are few and scattered records from Greece, probably referring to subspontaneous occurrences.

FLORA HELLENICA BIBLIOGRAPHY

listing 16,078 items relevant to the taxonomy and distribution of vascular plants of Greece (continuation of Strid 1996a and 2006a)

For principles, see the previous editions

AAS, G. & AL. 1994: Morphology, isozyme variation, cytology, and reproduction of hybrids between *Sorbus aria* (L.) Crantz and *S. torminalis* (L.) Crantz. - *Bot. Helv.* **104(2)**: 195-214.

ABBOTT, R.J. & AL. 2000: Hybrid origin of the Oxford ragwort, *Senecio squalidus* L. - *Watsonia* **23(1)**: 123-138.

ABBOTT, R.J. & AL. 2002: Hybrid origin of the Oxford ragwort, *Senecio squalidus* L.: morphological and allozyme differences between *S. squalidus* and *S. rupestris* Waldst. & Kit. - *Watsonia* **24(1)**: 17-29.

ABDALLAH, M.S. & DE WIT, H.C.D. 1978: The Resedaceae. A taxonomical revision of the family (final instalment). - *Meded. Landbouwhogeschool* **78(14)**: 95-416 + 91 plates.

ABDEDDAIM-BOUGHANMI, K. & AL. 2019: A single species, two basic chromosomal numbers: case of *Lygeum spartum* (Poaceae). - *Pl. Biosyst.* **153(6)**: 775-783.

ABDELGUERFI, A. & GUITTONNEAU, G.-G. 1979: [Chromosome numbers in *Medicago*]. In Löve, Á. (ed.): IOPB chromosome number reports LXIV. - *Taxon* **28(4)**: 392-393.

ABDELGUERFI, A. & GUITTONNEAU, G.-G. 1989: Morphologie, écologie et répartition du *Medicago heterocarpa* Spach (= *M. lesinii* Small) en Algérie, espèce méditerranéenne du complexe *M. murex* Willd. - *Candollea* **44(2)**: 423-433.

ABRAHAM, A. 1987: The northernmost and related Romuleas. - *Feddes Repert.* **98(11-12)**: 599-616.

ABRAMOVA, L.I. 1975: On the taxonomical structure of the genus *Polygonatum* Mill. [In Russian with English summary]. - *Bot. Zhurn. (Moscow & Leningrad)* **60(4)**: 490-497.

ABU SBAIH, H.A. & JURY, S.L. 1994: Seed micromorphology and taxonomy in *Orobanche* (Orobanchaceae). - *Fl. Medit.* **4**: 41-48.

ABUHADRA, M.N. 2000: Taxonomic studies on the *Arenaria serpyllifolia* group (Caryophyllaceae). - *Fl. Medit.* **10**: 185-190.

ACEDO, C. & LLAMAS, F. 1991: Revision del género *Dactylis* L. (Poaceae) en el N.O. de la Península Ibérica. - *Bull. Soc. Bot. France, Lett. Bot.* **138(4-5)**: 329-338.

ACEDO, C. & LLAMAS, F. 1994: *Bromus alopecuros* a new record for the Iberian Peninsula, with morphological, chorological and nomenclatural observations on the *B. lanceolatus* group. - *Fl. Medit.* **4**: 203-212.

ACEDO, C. & LLAMAS, F. 1999: The genus *Bromus* L. (Poaceae) in the Iberian Peninsula. - *Phanerog. Monogr.* **22**: ix + 293 pp.

ACHEROY, M. 1987: Morphological and genetical variations between and within populations of *Petrorhagia prolifera* (Caryophyllaceae). - *Bull. Soc. Roy. Bot. Belgique* **120(1)**: 72-80.

ACHTAROV, B. 1938: Floristisches Material aus den Pirin- und Rila-Gebirgen, mit kritischen Bemerkungen. - *Izv. Tsarsk. Prir. Inst. Sofiya* **11**: 70-80.

ACHTAROV, B. 1939: Phylogenetische und systematische Studien über die *Poa*-Arten Bulgariens. - *Izv. Bulg. Bot. Druzh.* **8**: 120-187.

ACHTAROV, B. 1951: Kritische Studien über die Gattung *Koeleria* Pers. in Bulgarien [In Bulgarian with Russian and German summaries]. - *Izv. Bot. Inst. (Sofia)* **2**: 209-232.

ACHTAROV, B. 1953: Die Gattung *Festuca* L. (Schwingel) in Bulgarien [In Bulgarian with German summary]. - *Izv. Bot. Inst. (Sofia)* **3**: 3-89.

ACHTAROV, B. 1957: Die Gattung *Carex* L. (Segge) in Bulgarien [In Bulgarian with German summary]. - Sofia: Bulgarian Academy of Sciences, 179 pp.

ACHTAROV, B. & KELLERER, J. 1939: Einige seltene Orchideen-Arten, die Seine Majestät König Ferdinand I von Bulgarien auf der Insel Rhodos gesammelt hat. - *Izv. Tsarsk. Prir. Inst. Sofiya* **12**: 206-208.

ACHTAROV, B. & LINDTNER, V. 1940: Beitrag zur Kenntnis der *Dianthus*- und *Centaurea*-Arten Mazedoniens und Albaniens. (Ergebnisse zweier von Vojtel Lindtner, Belgrad, in den Jahren 1938 und 1939 unternommenen Reisen). - *Izv. Tsarsk. Prir. Inst. Sofiya* **13**: 195-200.

ACKERFIELD, J. & WEN, J. 2002: A morphometric analysis of *Hedera* (the ivy genus, Araliaceae) and its taxonomic implications. - *Adansonia* **24(2)**: 197-212.

ACKERMANN, M. & ACKERMANN, M. 1986: *Ophrys vernixia* Brot.: Neu für Kreta. - *Mitt. Arbeitskreis. Heimische Orchid. Baden-Württemberg* **18(1)**: 151-158.

ACKERS, G. 2007: The fine ferns of Flora Graeca. - *Pteridologist* **4(6)**: 203-208.

ADAMAKOPOULOS, T., SFIKAS, G. & CHATZIRVASANIS, V. 1988: Axiologisi ke diachirisi tou ikosistimatos tis Gionas [In Greek]. - Athine: 94 pp.

ADAME, P. & TALAVERA, S. 1980: Números cromosómicos para la flora española. 130-131. - *Lagascalia* **9(2)**: 256-257.

ADAMIDIS, G.C. & AL. 2014: The effect of harsh abiotic conditions on the diversity of serpentine plant communities on Lesbos, an eastern Mediterranean island. - *Pl. Ecol. Divers.* **7(3)**: 433-444.

ADAMOVIĆ, L. 1902: Die Šibljak-Formation, ein wenig bekanntes Buschwerk der Balkanländer. - *Bot. Jahrb. Syst.* **31(1-2)**: 1-29.

ADAMOVIĆ, L. 1904: Beiträge zur Flora von Macedonien und Altserbien. - *Denkschr. Kaiserl. Akad. Wiss., Wien. Math.-Naturwiss. Kl.* **74(1)**: 115-151 + 5 plates.

ADAMOVIĆ, L. 1905a: Plantae macedonicae novae. - *Oesterr. Bot. Z.* **55(5-6)**: 178-181, 235-238.

ADAMOVIĆ, L. 1905b: Die Entwicklung der Balkanflora seit der Tertiärzeit. - *Bot. Jahrb. Syst.* **36 Beibl. 81**: 62-76.

ADAMOVIĆ, L. 1905c: Bericht über die mit Unterstützung der kaiserl. Akademie unternommene botanische Forschungsreise in den Balkanländern. - *Oesterr. Bot. Z.* **55(12)**: 493-495.

ADAMOVIĆ, L. 1906a: *Corydalis Wettsteinii*. Eine neue *Corydalis*-Art der Balkanhalbinsel. - *Oesterr. Bot. Z.* **56(5-6)**: 174-176.

ADAMOVIĆ, L. 1906b: Die Panzerföhre im Pindusgebiete. - *Oesterr. Bot. Z.* **56(12)**: 487.

ADAMOVIĆ, L. 1906c: Über eine bisher nicht unterschiedene Vegetationsformation der Balkanhalbinsel, die Pseudomacchie. - *Verh. K.K. Zool.-Bot. Ges. Wien* **56**: 355-360.

ADAMOVIĆ, L. 1907a: Die pflanzengeographische Stellung und Gliederung der Balkanhalbinsel. - *Denkschr. Kaiserl. Akad. Wiss., Wien. Math.-Naturwiss. Kl.* **80**: 405-495 + 3 maps.

ADAMOVIĆ, L. 1907b: *Thymus Plasonii* Adamov., eine gelblichblühende, neue *Thymus*-Art der Balkanhalbinsel. - *Oesterr. Bot. Z.* **57(5)**: 200-201.

ADAMOVIĆ, L. 1908: Die Rosskastanie im Balkan. - *Bot. Jahrb. Syst.* **41 Beibl. 94**: 1-9 + 2 plates.

ADAMOVIĆ, L. 1909a: Die Vegetationsverhältnisse der Balkanländer (Mösische Länder), umfassend Serbien, Altserbien, Bulgarien, Ostrumelien, Nordthrakien und Nordmazedonien. - Leipzig: Engelmann, xvi + 567 pp. + 1 map [repr. 1976 by A.R. Gantner, Vaduz].

ADAMOVIĆ, L. 1909b: Verbreitung der Holzgewächse in Bulgarien und Ostrumelien. - *Denkschr. Kaiserl. Akad. Wiss., Wien. Math.-Naturwiss. Kl.* **84**: 625-639 + 1 map.

ADAMOVIĆ, L. 1910: Vegetationsbilder aus Bosnien und der Herzegowina. In Karsten, G. & Schenk, H. (eds): Vegetationsbilder, 8. Reihe, Heft 4. - Jena: Gustav Fischer Verlag, Tafel 19-24 + Text.

ADAMOVIĆ, L. 1911: Bericht über die im Jahre 1910 ... unternommene botanische Forschungsreise durch Montenegro, Albanien, Altserbien, Mazedonien, Epirus, Thessalien und Nordgriechenland. - *Oesterr. Bot. Z.* **61(6)**: 244-245.

ADAMOWSKI, W. 2009: *Impatiens balfourii* as an emerging invader in Europe. In: Pyšek P. & Pergl J. (eds): Biological invasions: towards a synthesis. - *NeoBiota* **8**: 183-194.

ADAMS, D.E., PERKINS, W.E. & ESTES, J.R. 1981: Pollination systems in *Paspalum dilatatum* Poir. (Poaceae): an example of insect pollination in a temperate grass. - *Amer. J. Bot.* **68(3)**: 389-394.

ADAMS, J.E. 1940: A systematic study of the genus *Arctostaphylos* Adans. - *J. Elisha Mitchell Sci. Soc.* **56(1)**: 1-62.

ADAMS, R.P. 2004: *Juniperus deltoides*, a new species, and nomenclatural notes on *J. polycarpos* and *J. turcomanica* (Cupressaceae). - *Phytologia* **86(3)**: 47-51.

ADAMS, R.P. 2008: Junipers of the world: The genus *Juniperus*, 2nd edition. - Vancouver: Trafford Publishing Co.: 402 pp. [3rd ed., 2011; 4th ed. 2014].

ADAMS, R.P. 2014: Morphological comparison and key to *Juniperus deltoides* and *J. oxycedrus*. - *Phytologia* **96(2)**: 58-62.

ADAMS, R.P. & AL. 1999: The leaf essential oils and taxonomy of *Juniperus oxycedrus* L. subsp. *oxycedrus*, subsp. *badia* (H. Gay) Desbeaux, and subsp. *macrocarpa* (Sibth. & Sm.) Ball. - *J. Essential Oil. Res.* **11(2)**: 167-172.

ADAMS, R.P. & AL. 2005: Cryptic speciation between *Juniperus deltoides* and *Juniperus oxycedrus* (Cupressaceae) in the Mediterranean. - *Biochem. Syst. Ecol.* **33(8)**: 771-787.

ADAMS, R.P. & AL. 2013: Analysis of *Juniperus phoenicea* from throughout its range in the Mediterranean using DNA sequence data from nrDNA and petN-psbM: The case for recognition of *J. turbinata* Guss. - *Phytologia* **95**: 202-209.

ADE, A. 1934: *Dianthus Arpadianus* Ade et Bornmüller, spec. nov. - *Repert. Spec. Nov. Regni Veg.* **36**: 385-388.

ADE, A. & RECHINGER, K.H. 1938: Samothrake. - *Repert. Spec. Nov. Regni Veg. Beih.* **100**: 106-146 + 4 plates.

ADEMA, F. & RAUSCHERT, S. 1979: (464) Proposal to conserve 8485 *Asperula* L. (1753) with ... type species, *A. arvensis* L., against *Asperula* L. (1753) with the lectotype *A. odorata* L. - *Taxon* **28(4)**: 422-423. [approved, cf. Taxon 31(3): 543-544, 1982].

ADIGÜZEL, N. & REEVES, R.D. 2002: A new nickel-accumulating species of *Alyssum* (Cruciferae) from western Turkey. - *Edinburgh J. Bot.* **59(2)**: 215-219.

ADOLPHI, K. 1982: (665) Proposal to conserve *Glechoma* L. against *Glecoma* L. - *Taxon* **31(1)**: 118. [approved, cf. Taxon 33(2): 301, 1984].

AEDO, C. 1996: Revision of *Geranium* subgenus *Erodioidea* (Geraniaceae). - *Monogr. Syst. Bot. Missouri Bot. Gard.* **49**: 104 pp.

AEDO, C. 2003: Taxonomic revision of *Geranium* sect. *Trygonium* (Geraniaceae). - *Bot. Zhurn. (Moscow & St. Petersburg)* **88(4)**: 124-131.

AEDO, C. 2012: Geranium taxonomic information system [continuously updated]. - http://www.geranium.es.

AEDO, C. 2017: Taxonomic revision of *Geranium* sect. *Ruberta* and *Unguiculata* (Geraniaceae). - *Ann. Missouri Bot. Gard.* **102(3)**: 409-465.

AEDO, C. & AL. 2005: Taxonomic revision of *Geranium* sect. *Dissecta* (Geraniaceae). - *Syst. Bot.* **30 (3)**: 533-558.

AEDO, C. & AL. 2007: Taxonomic revision of *Geranium* subsect *Mediterranea*. - *Syst. Bot.* **32(1)**: 93-128.

AEDO, C., ALDASORO, J.J. & NAVARRO, C. 1998: Revision of *Geranium* sections *Batrachioidea* and *Divaricata* (Geraniaceae). - *Ann. Missouri Bot. Gard.* **85(4)**: 594-630.

AEDO, C. & ESTRELLA, M. DE LA 2006: Taxonomic revision of *Geranium* subsect *Tuberosa* (Boiss.) Yeo (Geraniaceae). - *Israel J. Pl. Sci.* **54(1)**: 19-54.

AEDO, C. & FUMEAUX, N. 2012: (2047) Proposal to conserve the name *Geranium pyrenaicum* (Geraniaceae) with a conserved type. - *Taxon* **61(1)**: 256. [approved, cf. Taxon 62(6): 1315, 2013].

AEDO, C. & MUÑOZ GARMENDIA, F. 1996: Some notes on the sectional nomenclature of *Geranium* (Geraniaceae). - *Taxon* **45(1)**: 104-106.

AEDO, C., MUÑOZ GARMENDIA, F. & PANDO, F. 1998: World checklist of *Geranium* L. (Geraniaceae). - *Anales Jard. Bot. Madrid* **56(2)**: 211-252.

AELLEN, P. 1938a: Die orientalischen *Beta*-Arten. - *Ber. Schweiz. Bot. Ges.* 48: 470-484.

AELLEN, P. 1938b: Beitrag zur Kenntnis von Spinacia L. - *Ber. Schweiz. Bot. Ges.* **48**: 485-490.

AELLEN, P. 1938c: *Halimione* Aellen, eine rehabilitierte Chenopodiaceen-Gattung. - *Verh. Naturf. Ges. Basel* **49**: 118-130.

AELLEN, P. 1939: Die *Atriplex*-Arten des Orients. - *Bot. Jahrb. Syst.* **70(1)**: 1-66 + 3 plates.

AELLEN, P. 1967: New Chenopodiaceae from Turkey. - *Notes Roy. Bot. Gard. Edinburgh* **28(1)**: 29-34.

AESCHIMANN, D. 1984a: Etude biosystématique du *Silene vulgaris* s.l. (Caryophyllaceae) dans le domaine alpin. Morphologie de la graine. - *Candollea* **39(1)**: 135-149.

AESCHIMANN, D. 1984b: Etude biosystématique du *Silene vulgaris* s.l. (Caryophyllaceae) dans le domaine alpin. Typification du subsp. *marginata* (Kit.) Hayek. - *Candollea* **39(2)**: 395-397.

AESCHIMANN, D. 1984c: Etude biosystématique du *Silene vulgaris* s.l. (Caryophyllaceae) dans le domaine alpin. Traitement numérique des populations des Alpes et de quelques chaînes voisines. - *Candollea* **39(2)**: 399-415.

AESCHIMANN, D. 1984d: Etude biosystématique du *Silene vulgaris* s.l. (Caryophyllaceae) dans le domaine alpin. Bibliographie. - *Candollea* **39(2)**: 417-422.

AESCHIMANN, D. 1985a: Etude biosystématique du *Silene vulgaris* s.l. dans le domaine alpin. Cartes de distribution. - *Candollea* **40(1)**: 57-65.

AESCHIMANN, D. 1985b: Etude biosystématique du *Silene vulgaris* s.l. dans le domaine alpin. Essai d'interprétation évolutif et propositions taxonomiques. - *Candollea* **40(1)**: 67-98.

AESCHIMANN, D. & BOCQUET, G. 1980: Les types biologiques du *Silene vulgaris* s.l. (Caryophyllaceae). - *Candollea* **35(2)**: 451-495.

AESCHIMANN, D. & BOCQUET, G. 1983: Étude biosystématique du *Silene vulgaris* s.l. (Caryophyllaceae) dans le domain alpin. Notes nomenclaturales. - *Candollea* **38(1)**: 203-209.

AFORDAKOS, G. 1999: To *Lilium candidum* L. stin Kriti [In Greek]. - *Fisis [Journal of the Hellenic Society for the Protection of Nature]* **84**: 17-18.

AFZAL-RAFII, Z. 1972: Contribution à l'étude cytotaxinomique des *Salvia* de Turquie, II. - *Bull. Soc. Bot. France* **119(3-4)**: 167-175.

AFZAL-RAFII, Z. 1975: Recherches sur le genre *Salvia* en région méditerranéenne & au Proche-Orient. Considérations cytotaxonomique, palynologique et phylogénetique. - Marseille: Univ. d'Aix- Marseille III, Thèse, 225 pp.

AFZAL-RAFII, Z. 1976: Etude cytotaxonomique et phylogénétique de quelques *Salvia* de la région méditerranéenne: Groupe du *Salvia officinalis* L. - *Bull. Soc. Bot. France* **123(9)**: 515-531.

AFZAL-RAFII, Z. 1977: Contribution à l'étude cytotaxonomique et phylogénique de *Salvia verticillata* et des espèces qui lui sont apparentées. - *Acta Ecol. Iran.* **2**: 80-94.

AFZAL-RAFII, Z. 1979: Contribution à l'étude cytotaxonomique du groupe *Salvia verbenaca* L. - *Bull. Soc. Bot. France, Lett. Bot.* **126(1)**: 79-86.

AFZAL-RAFII, Z. 1980: [Chromosome numbers in *Salvia*]. In Löve, Á. (ed.): Chromosome number reports LXVII. - *Taxon* **29(2-3)**: 365-366.

AFZAL, Z. 1971: Contribution à l'étude cytotaxonomique des *Salvia* de Turquie. - *Bull. Soc. Bot. France* **118(1-2)**: 69-75.

AGABABIAN, M. 1997: *Centaurea* subg. *Centaurea* (Compositae): Delimitation and distribution of sections and subsections. - *Lagascalia* **19(1-2)**: 889-901.

AGAPOVA, N.D. 1976a: [Chromosome numbers in *Ornithogalum*]. In Löve, Á. (ed.): IOPB chromosome number reports LIV. - *Taxon* **25(5-6)**: 635-636.

AGAPOVA, N.D. 1976b: The caryotypical analysis of *Ornithogalum gussonei* Ten. (Liliaceae) from the European part of the U.S.S.R. [In Russian]. - *Bot. Zhurn. (Moscow & Leningrad)* **61(12)**: 1730-1740.

AGAPOVA, N.D. 1980: Cytosystematic study of the European representatives of the genus *Ornithogalum* (Liliaceae) of the flora of the U.S.S.R. II. Subgenus *Ornithogalum* [In Russian with English summary]. - *Bot. Zhurn. (Moscow & Leningrad)* **65(6)**: 783-794.

AGAPOVA, N.D. 1981: A comparative karyological study of some *Ranunculus* species from the European part of the U.S.S.R. [In Russian with English summary]. - *Bot. Zhurn. (Moscow & Leningrad)* **66(4)**: 483-492.

AGARDH, J.G. 1835: Synopsis generis Lupini. - Lund: C.F. Berling, xiv + 43 pp. + 2 plates.

AGARWAL, K. & GUPTA, P.K. 1983a: Cytological studies in the genus *Trigonella* Linn. - *Cytologia* **48(4)**: 771-779.

AGARWAL, K. & GUPTA, P.K. 1983b: Cytological studies in the genus *Medicago* Linn. - *Cytologia* **48(4)**: 781-793.

AGAYEV, Y.M. 2002: New features in karyotype structure and origin of saffron, *Crocus sativus* L. - *Cytologia* **67**: 245-252.

AGERER-KIRCHHOFF, C. 1976: Revision von *Astragalus* L. sect. *Astragalus* (Leguminosae). – *Boissiera* **25**: 1-197.

AGHABABIAN, M. & RAIMONDO, F.M. 2001: Typification of the name *Papaver hybridum* L. (Papaveraceae). - *Taxon* **60(5)**: 1475-1476.

AGHABABYAN, M. & AL. 2007: Typification of Sicilian *Helichrysum* (Compositae) revisited. - *Taxon* **56 (4)**: 1285-1288.

AGHABABYAN, M.V. 2011: A revision of *Papaver* sect. *Argemonidium* Spach (Papaveraceae). - *Takhtajania* **1**: 38-42.

AGNEW, A.D.Q. 1968: The interspecific relationships of *Juncus effusus* and *J. conglomeratus* in Britain. - *Watsonia* **6(6)**: 377-388.

AGOSTINI, R. 1960: *Aster squamatus* Hieron. in Sicilia e stato attuale della sua espansione in Italia. - *Boll. Ist. Bot. Univ. Catania, ser. 2* **3**: 131-136.

AGOSTINI, R. 1962: Notizie su una nuova stazione e sull'areale di *Coris monspeliensis* L. - *Delpinoa, n.s.* **3**: 153-181.

AGUINAGALDE, I., GÓMEZ-CAMPO, C. & SÁNCHEZ-YÉLAMO, M.D. 1992: A chemotaxonomic survey of wild relatives of *Brassica oleracea* L. - *Bot. J. Linn. Soc.* **109(1)**: 57-67.

AGUINAGALDE, I., SÁNCHEZ-YÉLAMO, M.D. & SHAGARODSKY, T. 1991: Phytochemical diversity in *Brassica cretica* Lam. - *Bot. Chron. (Patras)* **10**: 667-672.

AHLBURG, M.S. 1993: Hellebores: Christmas rose, Lenten rose. - London: Batsford, 128 pp. + 4 colour plates [first published in German in 1989].

AHMAD, M. & AL. 2021: Spatial and ecological drivers of genetic structure in Greek populations of *Alkanna tinctoria* (Boraginaceae), a polyploid medicinal herb. - *Front. Pl. Sci.* **12**: doi: 10.3389/fpls.2021.706574.

AHMED, M. 1968: Some aspects of the autecology of *Ranunculus arvensis*. - *Ege Üniv. Fen Fak. Ilmi Raporlar Ser.* **62**: 1-19 + 1 plate.

AHMED, M. 1969: Some autecological studies of *Ranunculus muricatus* L. - *Ege Üniv. Fen Fak. Ilmi Raporlar Ser.* **65**: 1-14 + 1 plate.

AHMED, M. & VARDAR, Y. 1973: Distribution and plasticity of *Myrtus communis*. - *Phyton (Horn)* **15(1-2)**: 145-150.

AHRENDT, L.W.A. 1961: *Berberis* and *Mahonia*. A taxonomic revision. - *J. Linn. Soc., Bot.* **57(369)**: 1-410.

AICHELE, D. & SCHWEGLER, H.-W. 1957: Die Taxonomie der Gattung *Pulsatilla*. - *Feddes Repert. Spec. Nov. Regni Veg.* **60(1-3)**: 1-230 + 32 plates.

AIELLO, M., BRULLO, S. & PICCIONE, V. 1980: Numerical analysis applied to the taxonomy of the genus *Valantia* L. (Rubiaceae). - *Anales Jard. Bot. Madrid* **37(2)**: 577-586.

AINSCOUGH, M.M., BARKER, C.M. & STACE, C.A. 1986: Natural hybrids between *Festuca* and species of *Vulpia* section *Vulpia*. - *Watsonia* **16(3)**: 143-151.

AINSWORTH, C.C., PARKER, J.S. & HORTON, D.M. 1983: Chromosome variation and evolution in *Scilla autumnalis*. In Brandham, P.E. & Bennett, M.D. (eds): Kew chromosome conference II, pp. 261-268. - London: George Allen & Unwin.

AINSWORTH, P. 1989: The genus *Zelkova*. - *Plantsman* **11(2)**: 80-86.

AKALIN, E. & ÖZHATAY, N. 2000: *Ferulago* species in western Turkey. In Özhatay, N. & al. (eds): Abstracts from the Second Balkan Botanical Congress, Istanbul, Turkey, May 14-18, 2000, p. 5.

AKALIN, E. & ÖZHATAY, N. 2001: *Ferulago* species in western Turkey. In Özhatay, N. (ed.): Proceedings of the Second Balkan Botanical Congress, 1: 77-86. - Istanbul.

AKALIN, E. & PIMENOV, M.G. 2004: *Ferulago trojana* (Umbelliferae), a new species from western Turkey. - *Bot. J. Linn. Soc.* **146(4)**: 499-504.

AKAN, H. & BALOS, M.M. 2008: Check-list of the genus *Biarum* Schott in the flora of Turkey, with a new record for Turkey: *Biarum syriacum* (Spreng.) H. Riedl. - *Turk. J. Bot.* **32(4)**: 305-310.

AKAN, H. & EKER, İ. 2005: Check-list of the genus *Colchicum* in the flora of Turkey. - *Turk. J. Bot.* **29 (4)**: 327-331.

AKAN, H., EKICI, M. & AYTAÇ, Z. 2020: The synopsis of the genus *Trigonella* L. (Fabaceae) in Turkey. - *Turk. J. Bot.* **44(6)**: 670-693.

AKÇIÇEK, E. 2010: A new subspecies of *Stachys cretica* (section *Eriostomum*, Lamiaceae) from Turkey. - *Turk. J. Bot.* **34(2)**: 131-136.

AKÇIÇEK, E., DIRMENÇI, T. & DÜNDAR, E. 2012: Taxonomical notes on *Stachys* sect. *Eriostomum* (Lamiaceae) in Turkey. - *Turk. J. Bot.* **36(3)**: 217-234.

AKÇIÇEK, E.2020: Taxonomic revision of *Stachys* sect. *Olisia* (Lamiaceae: Lamioideae) in Turkey. - *Phytotaxa* **449(2)**: 109-148.

ÅKERBERG, E. 1942: Cytogenetic studies in *Poa pratensis* and its hybrid with *Poa alpina*. - *Hereditas (Lund)* **28(1-2)**: 1-126.

AKEROYD, J.R. 1985: *Oxalis articulata* Savigny. In Greuter, W. & Raus, Th. (eds): Med-Checklist notulae, 11. - *Willdenowia* **15(1)**: 83.

AKEROYD, J.R. 1986a: *Capsella rubella* Reuter. In Greuter, W. & Raus, Th. (eds): Med-Checklist notulae, 12. - *Willdenowia* **15(2)**: 417.

AKEROYD, J.R. 1986b: *Astragalus physocalyx* Fischer. In Greuter, W. & Raus, Th. (eds): Med-Checklist notulae, 13. - *Willdenowia* **16(1)**: 108.

AKEROYD, J.R. 1986c: *Lupinus luteus* L. In Greuter, W. & Raus, Th. (eds): Med-Checklist notulae, 13. - *Willdenowia* **16(1)**: 112.

AKEROYD, J.R. 1987: Two overlooked species of *Polygonum* from SE Europe. In Chater, A.O. (ed.): Flora Europaea. Notulae systematicae ad Floram Europaeam spectantes. Series 2. No. 1. - *Bot. J. Linn. Soc.* **95(4)**: 251-257.

AKEROYD, J.R. 1988a: *Arenaria serpyllifolia* L. and related species. In Chater. A.O. (ed.): Flora Europaea: Notulae systematicae ad Floram Europaeam spectantes. Series 2. No. 2. - *Bot. J. Linn. Soc.* **97(4)**: 335-338.

AKEROYD, J.R. 1988b: Taxonomic and nomenclatural notes on *Polygonum* L. In Chater. A.O. (ed.): Flora Europaea: Notulae systematicae ad Floram Europaeam spectantes. Series 2. No. 2. - *Bot. J. Linn. Soc.* **97(4)**: 343-344.

AKEROYD, J.R. 1989a: *Arctotheca calendula* (L.) Levyns. In Greuter, W. & Raus, Th. (eds): Med- Checklist notulae, 15. - *Willdenowia* **19(1)**: 28.

AKEROYD, J.R. 1989b: *Ranunculus ficaria* subsp. *calthifolius* (Reichenb.) Arcangeli. In Greuter, W. & Raus, Th. (eds): Med-Checklist notulae, 15. - *Willdenowia* **19(1)**: 45.

AKEROYD, J.R. 1998: *Haberlea rhodopensis*: a classic Balkan endemic. - *New Plantsman* **5(1)**: 5-9.

AKEROYD, J.R. & BRIGGS, D. 1983a: Genecological studies of *Rumex crispus* L. I. Garden experiments using transplanted material. - *New Phytol.* **94(2)**: 309-323.

AKEROYD, J.R. & BRIGGS, D. 1983b: Genecological studies of *Rumex crispus* L. II. Variation in plants grown from wild-collected seed. - *New Phytol.* **94(2)**: 325-343.

AKEROYD, J.R. & JURY, S.L. 1991: Updating "Flora Europaea". - *Bot. Chron. (Patras)* **10**: 49-54.

AKEROYD, J.R., MELLORS, & PRESTON, C.D. 1976: [Unpublished flora list]. - Photocopy: 13 pp.

AKEROYD, J.R. & PRESTON, C.D. 1981a: Observations on two narrowly endemic plants, *Moehringia minutiflora* Bornm. and *Silene viscariopsis* Bornm., from Prilep, Yugoslavia. - *Biol. Conservation* **19**: 223-233.

AKEROYD, J.R. & PRESTON, C.D. 1981b: Floristic notes from Greek Macedonia. (Materials for the Mountain Flora of Greece, 11). - *Willdenowia* **11(2)**: 281-290.

AKEROYD, J.R. & PRESTON, C.D. 1987: Floristic notes from the Aegean region of Greece. - *Willdenowia* **16(2)**: 349-372.

AKEROYD, J.R. & PRESTON, C.D. 1990: Notes on some Aizoaceae naturalized in Europe. In Chater, A.O. (ed.): Flora Europaea: Notulae systematicae ad Floram Europaeam spectantes. Series 2. No. 3. - *Bot. J. Linn. Soc.* **103(3)**: 197-200.

AKEROYD, J.R. & STEARN, W.T. 1986: Joice Nan Kivell Loch (1893-1982). - *Ann. Mus. Goulandris* **7**: 17-24.

AKEROYD, J.R. & WALTERS, S.M. 1988: Taxonomic notes on *Silene* L. In Chater, A.O. (ed.): Flora Europaea: Notulae systematicae ad Floram Europaeam spectantes. Series 2. No. 2. - *Bot. J. Linn. Soc.* **97(4)**: 338-342.

AKHANI, H., EDWARDS, G. & ROALSON, E.H. 2007: Diversification of the Old World Salsoleae s.l. (Chenopodiaceae): Molecular phylogenetic analysis of nuclear and chloroplast data sets and a revised classification. - *Int. J. Plant Sci.* **168(6)**: 931-956.

AKHANI, H. & FÖRTHER, H. 1994: The genus *Heliotropium* L. (Boraginaceae) in Flora Iranica area. - *Sendtnera* **2**: 187-276.

AKHANI, H., GREUTER, W. & ROALSON, E.H. 2014: Notes on the typiification and nomenclature of *Salsola* and *Kali* (Chenopodiaceae). - *Taxon* **63(3)**: 647-650.

AKINCI, S. & ÖZHATAY, E. 1997: Taxonomic studies on *Linum tauricum* s.l. in European Turkey. In: Abstracts from the First Balkan Botanical Congress, Thessaloniki, September 19-22, 1997, p. 1. - Thessaloniki.

AKSOY, A. & AL. 2020: A phylogenetic analysis and biogeographical distribution of *Teucrium* Sect. *Teucrium* (Lamiaceae) and taxonomic notes for a new species from southwest Turkey. - *Turk. J. Bot.* **44(3)**: 322-337.

AKSOY, A., DIXON, J.M. & HALE, W.H.G. 1998: Biological Flora of the British Isles. *Capsella bursa-pastoris* (L.) Medikus (*Thlaspi bursa-pastoris* L., *Bursa bursa-pastoris* (L.) Shull, *Bursa pastoris* (L.) Weber). - *J. Ecol.* **86(1)**: 171-186.

AKSOY, A., HALE, W.H.G. & DIXON, J.M. 1996: Variation in leaf and capsule morphology in *Capsella bursa-pastoris* (L.) Medic. In Öztürk, M.A., Seçmen, Ö. & Görk, G. (eds): Plant life in southwest and central Asia, 1: pp. 110-121. - Izmir: Ege University Press.

AKSOY, A., HALE, W.H.G. & DIXON, J.M. 1999: Towards a simplified taxonomy of *Capsella bursa- pastoris* (L.) Medik. (Brassicaceae). - *Watsonia* **22(3)**: 243-250.

AKSOY, N. 2018: On Georges Vincent Aznavour, the last Ottoman plant collector and his herbarium held in Robert College (Istanbul), Turkey. In Aščerić-Todd, I. & al. (eds): Travellers in Ottoman lands. The Botanical Legacy, pp. 160-172. - Oxford: ASTENE and Archaeopress Publishing Ltd, 379 pp.

AL-ASADI, W.M. & AL-MAYAH, A.A. 2017: *Diphelypaea boissieri* (Reuter) Nicolson with a new host and distribution. - *Int. J. Conserv. Sci.* **6(8)**: 193-197.

AL-EISAWI, D. & JURY, S.L. 1988: A taxonomic revision of the genus *Tordylium* L. (Apiaceae). - *Bot. J. Linn. Soc.* **97(4)**: 357-403.

AL-EISAWI, D.M. 1982: List of Jordan vascular plants. - *Mitt. Bot. Staatssamml. München* **18**: 79-182.

AL-KELIDAR, R.K. & RICHARDS, A.J. 1981: Chromosomal indications of evolutionary trends in the genus *Delphinium* L. - *Cytologia* **46(3)**: 623-633.

AL-SAGHIR, M.G. & PORTER, D.M. 2012: Taxonomic revision of the genus *Pistacia* L. (Anacardiaceae). - *Amer. J. Pl. Sci.* **3(1)**: 12-32.

AL-SHEHBAZ, I.A. 2005: (1678) Proposal to conserve the name *Erucastrum* against *Kibera* and *Hirschfeldia* (Brassicaceae). - *Taxon* **54(1)**: 204-204. [approved, cf. Taxon 56(2): 590, 2007].

AL-SHEHBAZ, I.A. 2012: A generic and tribal synopsis of the Brassicaceae (Cruciferae). - *Taxon* **61(5)**: 931-954.

AL-SHEHBAZ, I.A. 2013: *Clypeola* is united with *Alyssum* (Brassicaceae). - *Harvard Pap. Bot.* **18(2)**: 125-128.

AL-SHEHBAZ, I.A. 2014: A synopsis of the genus *Noccaea* (Coluteocarpae, Brassicaceae). – *Harvard Pap. Bot.* **19(1)**: 25-51.

AL-SHEHBAZ, I.A. 2019: Discovery of the type collection of *Arabis nuda* Bél. (Brassicaceae). - *Adansonia* **41(1)**: 97-100.

AL-SHEHBAZ, I.A. & AL. 2014: Systematics, tribal placements, and synopses of the *Malcolmia* s.l. segregates (Brassicaceae). - *Harvard Pap. Bot.* **19(1)**: 53-71.

AL-SHEHBAZ, I.A. & BARRIWRA, G. 2019: Typification of Edmond Boissier's Cruciferae (Brassicaceae) names enumerated in Flora Orientalis. - *Boissiera* **72**: 1-192.

AL-SHEHBAZ, I.A., BEILSTEIN, M.A. & KELLOGG, E.A. 2006: Systematics and phylogeny of the Brassicaceae (Cruciferae): an overview. - *Pl. Syst. Evol.* **259(2)**: 89-120.

AL-SHEHBAZ, I.A. & KOCH, M. 2003: *Drabopsis* is united with *Draba* (Brassicaceae). - *Novon* **13(2)**: 172-173.

AL-SHEHBAZ, I.A. & MABBERLEY, D.J. 2017: What is Philip Miller's "Perennial Moonwort," *Lunaria perennis* (Brassicaceae/Cruciferae)? - *Novon* **25(4)**: 414-418.

AL-SHEHBAZ, I.A., MUMMENHOFF, K. & APPEL, O. 2002: *Cardaria, Coronopus* and *Stroganowia* are united with *Lepidium* (Brassicaceae). - *Novon* **12(1)**: 5-11.

AL-SHEHBAZ, I.A., MUTLU, B. & DÖNMEZ, A.A. 2007: The Brassicaceae (Cruciferae) of Turkey, updated. - *Turk. J. Bot.* **31(4)**: 327-336.

AL-SHEHBAZ, I.A. & O'KANE, S.L. JR. 2002: Taxonomy and phylogeny of *Arabidopsis* (Brassicaceae). - *The Arabidopsis Book* **2002**: 1-22.

AL-SHEHBAZ, I.A., ÖZÜDOĞRU, B. & GERMAN, D.A. 2021: *Iljinskaea* (Brassicaceae), a new genus based on *Conringia planisiliqua*. - *Phytotaxa* **500(2)**: 142-146.

AL-SHEHBAZ, I.A. & WARWICK, S.I. 2008: (1854) Proposal to conserve the name *Teesdalia* (Cruciferae). - *Taxon* **57(3)**: 1357-1358. [approved, cf. Taxon 60(1): 228, 2011].

AL-SHEHBAZ. I.A. & PRICE, R.A. 1998: Delimitation of the genus *Nasturtium* (Brassicaceae). - *Novon* **8**: 124-126.

ALARCÓN, M.L. & AEDO, C. 2002: Revisión taxonómica del género *Cephalanthera* (Orchidaceae) en la Península Ibérica e Islas Baleares. - *Anales Jard. Bot. Madrid* **59(2)**: 227-248.

ALBACH, D.C. 2002: Evolution and biogeography of southwestern Asian species of *Veronica*. In Koyuncu, M. (ed.): VIth Plant Life of Southwest Asia Symposium, Abstracts p. 37. - Van: Yüzüncü Yil University.

ALBACH, D.C. 2006: Evolution of *Veronica* (Plantaginaceae) on the Balkan Peninsula. - *Phytol. Balcan.* **12(2)**: 231-244.

ALBACH, D.C. 2008: Further arguments for the rejection of paraphyletic taxa: *Veronica* subgen. *Pseudolysimachium* (Plantaginaceae). - *Taxon* **57(1)**: 1-6.

ALBACH, D.C. & AL. 2008: Chromosome numbers in Veronicae (Plantaginaceae): review and several new counts. - *Ann. Missouri Bot. Gard.* **95(4)**: 543-566.

ALBACH, D.C. & AL. 2009: Phylogenetic analysis and differentiation of *Veronica* subgenus *Stenocarpon* in the Balkan Peninsula. - *Bot. J. Linn. Soc.* **159(4)**: 616-636.

ALBACH, D.C. & FISCHER, M.A. 2004: AFLP- and genome size analyses: contribution to the taxonomy of *Veronica* subg. *Pseudolysimachium* sect. *Pseudolysimachion* (Plantaginaceae), with a key to the European taxa. - *Phytol. Balcan.* **9(3)**: 401-424.

ALBACH, D.C., MARTÍNEZ-ORTEGA, M.M. & CHASE, M.W. 2004: *Veronica*: Parallel morphological evolution and phylogeography in the Mediterranean. - *Pl. Syst. Evol.* **246(3-4)**: 177-194.

ALBERGONI, F.G., BASSO, B. & TEDESCO, G. 1978: Considérations sur l'anatomie de *Posidonia oceanica* (Zosteraceae). - *Pl. Syst. Evol.* **130(3-4)**: 191-201.

ALBERS, F. 1972a: Cytosystematische Untersuchungen in der Subtribus Deschampsiineae Holub (Tribus Aveneae Nees). II. Die Gattungen *Vahlodea* Fr. und *Avenella* Koch. - *Ber. Deutsch. Bot. Ges.* **85(5-6)**: 279-285.

ALBERS, F. 1972b: Cytotaxonomie und B-Chromosomen bei *Deschampsia caespitosa* (L.) P.B. und verwandten Arten. - *Beitr. Biol. Pflanzen* **48(1)**: 1-62.

ALBERS, F. 1973: Cytosystematische Untersuchungen in der Subtribus Deschampsiineae Holub (Tribus Avenae Nees). I. Zwei Arten der Gattung *Corynephorus* P.B. - *Preslia* **45(1)**: 11-18.

ALBERS, F. 1974: Zwei Ausgangskaryotypen bei der Gräser-Gattung *Aira* L. - *Oesterr. Bot. Z.* **121(4)**: 251-254.

ALBERS, F. 1978: Karyologische und genomatische Veränderungen innerhalb der Gräser-Subtriben Aristaveninae und Airinae. - *Ber. Deutsch. Bot. Ges.* **91(2-3)**: 693-697.

ALBERS, F. 1979: Zwei neue Arten der Gattung *Aira* (Gramineae): *A. armoricana* und *A. byzantina*. - *Willdenowia* **9(2)**: 283-284.

ALBERS, F. 1980a: Karyogeobotanik der Gräser-Subtriben Aristaveninae und Airinae (Aveneae). - *Flora, Morphol. Geobot. Oekophysiol.* **169(2-3)**: 150-167.

ALBERS, F. 1980b: Systematik, Variation und Entwicklungstendenzen der Subtriben Aristaveninae und Airinae (Gramineae-Aveneae). - *Phyton (Horn)* **20(1-2)**: 95-116.

ALBERS, F. 1980c: Vergleichende Karyologie der Gräser-Subtriben Aristaveninae und Airinae (Poaceae-Aveneae). - *Pl. Syst. Evol.* **136(3-4)**: 137-167.

ALBRE, J., QUILICHINI, A. & GIBERNAU, M. 2003: Pollination ecology of *Arum italicum* (Araceae). - *Bot. J. Linn. Soc.* **141(2)**: 205-214.

ALCOBER, J.A. & AL. 1979: *Marsilea aegyptiaca* Willd. nueva cita para España. - *Anales Jard. Bot. Madrid* **36**: 55-59.

ALDASORO, J.J. & AL. 1998: The genus *Sorbus* (Maloideae, Rosaceae) in Europe and in North Africa: morphological analysis and systematics. - *Syst. Bot.* **23(2)**: 189-212.

ALDÉN, B. 1976: Floristic reports from the high mountains of Pindhos, Greece. - *Bot. Not.* **129(3)**: 297-321.

ALDÉN, B. 1982a: *Andrachne telephioides* subsp. *oreocretensis* (Euphorbiaceae), a new subspecies. (Materials for the Mountain Flora of Greece, 16). - *Willdenowia* **12(2)**: 211-212.

ALDÉN, B. 1982b: *Thesium vlachorum* sp. nov. (Materials for the Mountain Flora of Greece, 9). - *Nordic J. Bot.* **1(6)**: 709-710.

ALEX, J.F. 1962: The taxonomy, history, and distribution of *Linaria dalmatica*. - *Canad. J. Bot.* **40(2)**: 295-307 + 4 plates.

ALEXANDER, J.C.M. 1979: The Mediterranean species of *Senecio* sections *Senecio* and *Delphinifolius*. - *Notes Roy. Bot. Gard. Edinburgh* **37(3)**: 387-428.

ALEXEEV, E.B. 1975: Generis *Festuca* L. species angustifoliae partis europaeae URSS [In Russian]. - *Novosti Sist. Vysš. Rast.* **12**: 11-43.

ALEXIOU, S. 2013: The genus *Colchicum*. L. (Colchicaceae) in Greece. - *Parnassiana Arch.* **1**: 59-73.

ALEXIOU, S. 2014: New locality for *Bellevalia trifoliata* (Hyacinthaceae) from mainland Greece. - *Parnassiana Arch.* **2**: 3-5.

ALI, S.I. 1968: Typification of the genus *Medicago* and some of its Linnaean species. - *Taxon* **17(5)**: 540-541.

ALI, S.I. 1972: Typification of the genus *Medicago* and some of its Linnaean species - a reply. - *Taxon* **21(5-6)**: 665-668.

ALI, T. & AL. 2016: Morphology, phylogeny, and taxonomy of *Microthlaspi* (Brassicaceae: Coluteocarpeae) and related genera. - *Taxon* **65(1)**: 79-98.

ALIBERTIS, A. 1994: The gorge of Samaria and its plants. - Iraklion: Typokreta G. Kazanakis & Co S. A., 136 pp.

ALIBERTIS, A. 1997: Die Orchideen von Kreta und Karpathos. - Heraklion: Typokreta ABE, 159 pp.

ALIBERTIS, A. 2007: Healing, aromatic and edible plants of Crete. - Heraklion: Mystis, 351 pp.

ALIBERTIS, A. 2011: *Ophrys cretensis* subsp. *samica*, a new subspecies from Kephalonia (Greece). - *J. Eur. Orch.* **43(4)**: 753-758.

ALIBERTIS, A. 2015: The self-sown orchids of Greece. - Iraklion: Mystis Editions, 621 pp.

ALIBERTIS, A. & TRIANTAFYLLAKI, A. 2015: *Ophrys mavromata*, a new species from Cephalonia (Greece). - *J. Eur. Orch.* **47(2-4)**: 285-292.

ALIBERTIS, C. & ALIBERTIS, A. 1985: Wild orchids of Crete [also editions in German and French]. - Iraklion: Typokreta G. Kazanakis A.S., 37 pp. + 88 plates.

ALIBERTIS, C. & ALIBERTIS, A. 1988a: La Crète n'a pas fini de livrer ses secrets aux orchidophiles (1ère partie). - *Orchidophile (Ansières)* **20(86)**: 79-83.

ALIBERTIS, C. & ALIBERTIS, A. 1988b: Es gibt noch Hoffnung für das Kretische Waldvögelein *Cephalanthera cucullata* Boiss. & Heldr. - *Ber. Arbeitskreis. Heimische Orchid.* **5(1-2)**: 50-53.

ALIBERTIS, C. & ALIBERTIS, A. 1988c: Raynon d'espoir en ce qui concerne le céphalanthère de Crète (*Cephalanthera cucullata* Boiss. & Heldr.). - *Orchidophile (Ansières)* **20(81)**: 68-70.

ALIBERTIS, C. & ALIBERTIS, A. 1989a: La Crète n'a pas fini de livrer ses secrets aux orchidophiles (2ème partie). - *Orchidophile (Ansières)* **20(87)**: 108-112.

ALIBERTIS, C. & ALIBERTIS, A. 1989b: The wild orchids of Crete. Second revised and enlarged edition. - Iraklion: 176 pp. [also available in German and French].

ALIBERTIS, C., ALIBERTIS, A. & REINHARD, H.R. 1990: Untersuchungen am *Ophrys omegaifera*-Komplex Kretas. - *Mitt. Arbeitskreis Heimische Orchid. Baden-Württemberg* **22(1)**: 181-236.

ALISCIONI, S.S., OSPINA, J.C. & GOMIZ, N.E. 2016: Morphology and leaf anatomy of *Setaria* s. l. (Poaceae: Panicoideae: Paniceae) and its taxonomic significance. - *Pl. Syst. Evol.* **302(2)**: 173-185.

ALKIMOS, A. 1988a: Die wildwachsenden Orchideen Griechenlands: ihre Verbreitung und ihre Gefährdung. - *Orchidee (Hamburg)* **39(6)**: 230-239.

ALKIMOS, A. 1988b: Die Orchideen Griechenlands [In Greek with German summary]. - Athen: Giorgos Psichalos, 133 pp.

ALKIMOS, A. 1995: I orchidees tis Florinas sto chilos tis katastrofis [In Greek]. - *Fisis [Journal of the Hellenic Society for the Protection of Nature]* **71**: 15-17.

ALLAN, B. 2009: *Tulipa goulimyi* in Crete. - *Rock Gard.* **122**: 29.

ALLEN, K. & CHILTON, L. 1995: Plant list for Lindos & Pefkos, Rhodes. - Hunstanton, Marengo Press: 16 pp.

ALLIONI, C. 1785: Flora Pedemontana sive enumeratio methodica stirpium indigenarum Pedemontii. - Torino: J.M. Briolus, 3 vols. (344 + 366 + 92 pp.).

ALLKIN, R. & AL. 1983a: List of species and subspecies in the Vicieae, issue 2. In: Vicieae database project, Publ. no. 1, 20 pp. - Southampton.

ALLKIN, R. & AL. 1983b: Names and synonyms of species and subspecies in the Vicieae, issue 2. In: Vicieae database project, Publ. no. 2, 46 pp. - Southampton.

ALLKIN, R. & AL. 1983c: The geographical distribution of *Vicia*. In: Vicieae database project, Publ. no. 5, 35 pp. - Southampton.

ALLKIN, R. & AL. 1985: The geographical distribution of *Lathyrus*. In: Vicieae database project, Publ. no. 6, 35 pp. - Southampton.

ALLKIN, R. & AL. 1986: Names and synonyms of species and subspecies in the Vicieae, issue 3. In: Vicieae database project, Publ. no. 7, 75 pp. - Southampton.

ALMEIDA DA SILVA, R.M., SÁEZ, L. & ROSSELLÓ, J.A. 2001: Taxonomy of the genus *Brimeura* (Hyacinthaceae). - *Folia Geobot.* **36(2)**: 193-208.

ALMEIDA, J.L.F. DE 1948: Sobre a cariologia de *Ceratonia Siliqua* L. - *Agron. Lusit.* **10(4)**: 263-277 + 1 plate.

ALMEIDA, M.T. 1984: On some type-specimens of "Centaurea" in Willkomm's herbarium. 1. – *Webbia* **38**: 585-590.

ALMOND, M. 1988: A Balkan journey: part 1. - *Rock Gard.* **20(4)**: 459-466.

ALMOND, M. 2008: Mountain flower walks: the Greek mainland [Addendum to J. Richards]. – *Alpine Gardener* **76(3)**: 281-283.

ALMOND, M.J.B. 1999: Looking for *Fritillaria tuntasia*. - *Rock Gard.* **26(3)**: 195-197.

ALMQUIST, E. 1907: Studien über die *Capsella bursa-pastoris* (L.). - *Acta Horti Berg.* **4(6)**: 1-92.

ALMQUIST, E. 1923: Studien über *Capsella bursa-pastoris* (L.) II. - *Acta Horti Berg.* **7(2)**: 41-95.

ALONSO BEATO, M.T. & NAVARRO ANDRÉS, F. 1996: Números cromosómicos para la flora española. 774-779. - *Lagascalia* **18(2)**: 217-220.

ALPINAR, A. 1986: Some observations and findings on *Arum* L. (Araceae) species of West Turkey [In Turkish with English summary]. - *Türk Biyol. Derg.* **10(3)**: 240-253.

ALPINAR, A. 1987: [Chromosome numbers in *Arum* from Turkey]. In Löve, Á. (ed.): IOPB chromosome number reports XCIV. - *Taxon* **36(1)**: 285.

ALPINAR, K. 1998a: The presence of *Carduncellus caeruleus* (L.) C. Presl var. *incisus* DC. in Turkey [In Turkish with English abstract]. - *OT Sist. Bot. Derg.* **4(1)**: 43-48.

ALPINAR, K. 1998b: A new taxon for Turkey: *Galium recurvum* Req. ex DC. (Rubiaceae). - *Turk. J. Bot.* **22(4)**: 295-298.

ALSTON, A.H.G. & SANDWITH, N.Y. 1940: Results of two botanical expeditions to South Albania. - *J. Bot. (London)* **78**: 119-126, 147-151, 167-174, 193-199, 219-224, 232-246.

ALTAMURA, L., CIMA, F. & COLASANTE, M. 1991: Variabilità intraspecifica in *Papaver rhoeas* L. s.l. - *Giorn. Bot. Ital., n.s.* **125(3)**: 382.

ALTAMURA, L., COLASANTE, M. & D'AMATO, G. 1986: Numeri cromosomici per la Flora Italiana: 1022-1036. - *Inform. Bot. Ital.* **16(2-3)**: 261-270.

ALTAMURA, L. & RICCI, I. 1978: Cariologia e sistematica di *Hermodactylus tuberosus* Mill. - *Ann. Bot. (Rome)* **37(1)**: 1-13.

ÁLVAREZ FERNÁNDEZ, I. 2003: Systematics of Eurasian and North African *Doronicum* (Asteraceae: Senecioneae). - *Ann. Missouri Bot. Gard.* **90**: 319-389.

ALVES, M.C. & LEITÃO, M.T. 1976: Contribuição para o conhecimento citotaxonómico das Spermatophyta de Portugal. XIII. Geraniaceae. - *Bol. Soc. Brot., Ser. 2* **50**: 241-245.

ALYAFI, J. 1979: Approches systématique et écologique du genre *Pistacia* L. dans la région méditerranéenne. - Aix-Marseille: Thèse, vi + 123 pp.

ALZIAR, G. 1985: Contribution à l'histoire naturelle de l'île de Chypre - La flore. - *Biocosme Mésogéen* **2(1)**: 1-28.

ALZIAR, G. 1986: Contribution à l'histoire naturelle de l'île de Chypre - La flore. II. - *Biocosme Mésogéen* **3(2)**: 49-57.

ALZIAR, G. 1988a: Rectifications nomenclaturales et chorologiques dans les genres *Salvia* L. (Lamiaceae) et *Kicxia* Dumort. (Scrophulariaceae). - *Biocosme Mésogéen* **5(3-4)**: 85-86.

ALZIAR, G. 1988b: Catalogue synonymique des *Salvia* L. du monde (Lamiaceae). I. – *Biocosme Mésogéen* **5(3-4)**: 87-136.

ALZIAR, G. 1989a: Rectifications nomenclaturales dans le genre *Salvia*. Catalogue synonymique des *Salvia* du monde. II. - *Biocosme Mésogéen* **6(1-2)**: 79-115.

ALZIAR, G. 1989b: Corrections á apporter aux numéros 5(3-4) et 6(1-2). Concernent les genres: *Salvia*, *Lavandula*, *Teucrium* (Lamiaceae) et *Kickxia* (Scrophulariaceae). - *Biocosme Mésogéen* **6 (3)**: 161-162.

ALZIAR, G. 1989c: Catalogue synonymique des *Salvia* L. du monde (Lamiaceae). III. – *Biocosme Mésogéen* **6(4)**: 163-204.

ALZIAR, G. 1990: Catalogue synonymique des *Salvia* L. (Lamiaceae). IV. - *Biocosme Mésogéen* **7(1-2)**: 59-109.

ALZIAR, G. 1992: Catalogue synonymique des *Salvia* du monde (Lamiaceae). V. – *Biocosme Mésogéen* **9(2-3)**: 413-497.

ALZIAR, G. 1993: Catalogue synonymique des *Salvia* du monde (Lamiaceae). VI. – *Biocosme Mésogéen* **10(3-4)**: 33-117.

ALZIAR, G. & CAFFERTY, S. 1998: Typification of the names of several species of *Phlomis* of Linnaeus (Lamiaceae) [In French and English]. - *Biocosme Mésogéen* **14(4)**: 119-127.

AMADEI, L., GIORDANI, A. & TOMEI, P.E. 1984a: Numeri cromosomici per la Flora Italiana: 882-888. - *Inform. Bot. Ital.* **14(2-3)**: 243-247.

AMADEI, L., GIORDANI, A. & TOMEI, P.E. 1984b: Numeri cromosomici per la Flora Italiana: 941-947. - *Inform. Bot. Ital.* **15(1)**: 39-43.

AMARI, D. & AL. 2015: Nouvelles observations de *Malva arborea* (L.) Webb et Berthel. sur le littoral algerien. - *Acta Bot. Malac.* **40**: 213-216.

AMBASTHA, H.N.S. 1957: Cytological investigations in *Phalaris*. - *Genetica* **28**: 64-98.

AMICO, A. 1947: Osservazioni su *Sternbergia lutea* Ker-Gawl. - *Nuovo Giorn. Bot. Ital., n.s.* **54(3-4)**: 748-771.

AMIDŽIĆ, L., KRIVOŠEJ, Z. & STEVANOVIĆ, V. 1998: *Fritillaria macedonica* Bornm. - New species in Yugoslav Flora. - *Univ. Thought, Publ. Nat. Sci.* **5(2)**: 59-61.

AMIGO, J., BUJÁN, M. & ROMERO, I. 1991: Révision taxonomique du genre *Setaria* (Gramineae) dans la Péninsule Ibérique. - *Bull. Soc. Bot. France, Lett. Bot.* **138(2)**: 155-165.

AMIGUES, S. 1991: Note sur la statut de *Datura stramonium* L. en Europe. - *Naturalia Monspel., Sér. Bot.* **55**: 21-25.

AMIGUES, S. 2001: To kitrino Rododendro tis Lesvou to gnorize o Theofrastos [In Greek]. – *Fisis [Journal of the Hellenic Society for the Protection of Nature]* **93**: 16-17.

AMIGUES, S. & GREUTER, W. 2004: L'aconit du mont Saint-Jean: solution d'une énigme historique de botanique crétoise. - *Bot. Chron. (Patras)* **16**: 5-11.

AMMANN, K. 1974: Die mehrjährigen *Bromus*-Arten der Schweiz. - *Beiträge Kart. Schweiz. Fl.* **5**: 1-19.

AMMANN, K. 1981: Bestimmungsschwierigkeiten bei europäischen *Bromus*-Arten. - *Bot. Jahrb. Syst.* **102(1-4)**: 459-469.

AMMANN, K. 1990: Bestimmungsschlüssel der Umbelliferae Mitteleuropas ohne Fruchtmerkmale. Arten mit linealen bis lanzettlichen Blattzipfeln mit besonderer Berücksichtigung der Schweiz. - *Candollea* **45(2)**: 691-762.

AMMANN, K., KOŽUHAROV, S.I. & STRASSER, W. 1992: Exkursion vom 13. bis 24. Juli 1991 nach Westbulgarien. - *Mitth. Naturf. Ges. Bern, N.S.* **49**: 169-218.

AMORE, I., BEDINI, G. & GARBARI, F. 1999: Reports (1066-1070). In Kamari, G., Felber, F. & Garbari, F. (eds): Mediterranean chromosome number reports - 9. - *Fl. Medit.* **9**: 359-361.

AMOUTZOPOULOU-SCHINA, E. & HARITONIDIS, S. 1997: A study of a *Posidonia oceanica* (L.) Del. shallow meadow on the Pagassitikos Gulf, Aegean Sea. In: Abstracts from the First Balkan Botanical Congress, Thessaloniki, September 19-22, 1997, p. 2. – Thessaloniki.

AMVRAKIS, E. & AL. 2019: Flora diversity in olive orchards located in different agroecological zones. In Daskalakou, E.N. & al. (eds): 16th Conference of the Hellenic Botanical Society – Athens 10-13 October 2019. Program and book of abstracts, p. 11.

ANAGNOSTIDIS, K., ECONOMOU-AMILLI, A. & KOUMPLI-SOVANTZI, L. 1982: The aquatic fern *Azolla filiculoides* Lam., its *Anabaena* symbiont and periphyton in lake Trichonis, Greece. - *Bot. Chron. (Patras)* **2(2)**: 190-191.

ANAGNOSTIDIS, K. & PHITOS, D. 1975: Die Forschung der Flora und Vegetation Griechenlands und ihre Probleme. In Jordanov, D. & al. (eds): Problems of Balkan flora and vegetation, pp. 143-145. - Sofia: Publishing House of the Bulgarian Academy of Sciences.

ANAGNOSTOPOULOS, A. 1993: Karyotype variation within *Crepis fraasii* Schultz Bip. and *C. reuteriana* Boiss. in Greece. In Petrova, A.V. (ed.): Abstracts ... VII OPTIMA meeting in Borovetz, Bulgaria, 18-30 July 1993, p. 135.

ANAGNOSTOPOULOS, A. & ATHANASIOU, K. 1994: Registration of the rare, endemic and threatened plants of Zakinthos (Ionian Islands, Greece) [WWF project MR 4108]. - Patras: 40 pp.

ANAGNOSTOPOULOS, A. & KAMARI, G. 1992: Reports (46-50). In Kamari, G., Felber, F. & Garbari, F. (eds): Mediterranean chromosome number reports - 2. - *Fl. Medit.* **2**: 223-229.

ANAGNOU-VERONIKI, M. & AL. 2008: New records of plant pests and weeds in Greece. - *Hellen. Plant. Prot. J.* **1(2)**: 55-78.

ANČEV, M.E. 1971: On the taxonomy of *Galium rhodopeum* Vel. - *Dokl. Bulg. Akad. Nauk* **24(11)**: 1537-1540.

ANČEV, M.E. 1974a: [Chromosome numbers in *Galium*]. In Löve, Á. (ed.): IOPB chromosome number reports XLIV. - *Taxon* **23(2-3)**: 374-375.

ANČEV, M.E. 1974b: [Chromosome numbers from Bulgaria]. In Löve, Á. (ed.): IOPB chromosome number reports XLVI. - *Taxon* **23(5-6)**: 806.

ANČEV, M.E. 1975a: New taxa and new combinations in genus *Galium* L. - *Dokl. Bulg. Akad. Nauk* **28 (11)**: 1533-1535.

ANČEV, M.E. 1975b: [Chromosome numbers in *Galium* from Bulgaria]. In Löve, Á. (ed.): IOPB chromosome number reports XLVII. - *Taxon* **24(1)**: 146.

ANČEV, M.E. 1975c: [Chromosome numbers from Bulgaria]. In Löve, Á. (ed.): IOPB chromosome number reports XLIX. - *Taxon* **24(4)**: 514.

ANČEV, M.E. 1976a: Cytotaxonomic investigation of the genus *Cruciata* Mill. in Bulgaria [In Bulgarian with Russian and English summaries]. - *Fitologiya* **5**: 51-56.

ANČEV, M.E. 1976b: Karyologic study of Bulgarian Campanulaceae. - *Fitologiya* **5**: 66-74.

ANČEV, M.E. 1976c: [Chromosome numbers from Bulgaria]. In Löve, Á. (ed.): IOPB chromosome number reports LIII. - *Taxon* **25(4)**: 495-496.

ANČEV, M.E. 1978a: Taxonomic study of the genus *Galium* L. in Bulgaria [In Bulgarian with Russian and English summaries]. In Kožuharov, S.I. & Kuzmanov, B.A. (eds): Evolution of flowering plants ..., pp. 107-146. - Sofia: Publishing House of the Bulgarian Academy of Sciences.

ANČEV, M.E. 1978b: [Miscellaneous chromosome numbers from Bulgaria]. In Löve, Á. (ed.): IOPB chromosome number reports LXII. - *Taxon* **27(5-6)**: 532-533.

ANČEV, M.E. 1979: Chromosome numbers, karyotypes and dispersal mechanisms of Bulgarian *Crucianella* species. - *Fitologiya* **11**: 34-39.

ANČEV, M.E. 1981: [Chromosome numbers in Brassicaceae]. In Löve, Á. (ed.): Chromosome number reports LXIII. - *Taxon* **30(4)**: 855. [cf. also correction in *Taxon* 31(4): 777, 1982].

ANČEV, M.E. 1982a: Taxonomic study of genus *Galium* L. in Bulgaria. II. Karyological and pollen structural investigation [In Bulgarian with Russian and English summaries]. - *Fitologiya* **19**: 43-68.

ANČEV, M.E. 1982b: [Chromosome numbers from Bulgaria]. In Löve, Á. (ed.): Chromosome number reports LXXVI. - *Taxon* **31(3)**: 596.

ANČEV, M.E. 1982c: [Chromosome numbers from Bulgaria]. In Löve, Á. (ed.): Chromosome number reports LXXVII. - *Taxon* **31(4)**: 769.

ANČEV, M.E. 1983a: Karyology and reproductive characteristics of ... Cruciferae in Bulgaria [In Bulgarian with English abstract]. In Velčev, V.I. & al. (eds): Third national conference of botany, pp. 232-239. - Sofia: Publishing House of the Bulgarian Academy of Sciences.

ANČEV, M.E. 1983b: Cytotaxonomical study of *Campanula rotundifolia* group in Bulgaria [In Bulgarian with English abstract]. In Velčev, V.I. & al. (eds): Third national conference of botany, pp. 240-247. - Sofia: Publishing House of the Bulgarian Academy of Sciences.

ANČEV, M.E. 1984a: Natural hybrids of higher plants and their distribution in the Bulgarian flora [In Bulgarian with Russian and English summaries]. - *Fitologiya* **24**: 35-63.

ANČEV, M.E. 1984b: Ecological and karyol. characteristic of Bulgarian *Valerianella* species [In Bulgarian w. English abstr.]. In Velčev, V.I. & al. (eds): Savremenni teoretični i priložni aspecti ..., pp. 260-267. - Sofia: Publishing House of the Bulgarian Academy of Sciences.

ANČEV, M.E. 1984c: Karyological characteristics of *Alyssum umbellatum* Desv. and *Alyssum hirsutum* M.B. (Brassicaceae). In Drjanovska, O. & al. (eds): Proceedings of third nat. conf. on cytogenetics, pp. 428-431. - Sofia: Publishing House of the Bulgarian Academy of Sciences.

ANČEV, M.E. 1988: Karyologic study of Bulgarian Campanulaceae. In Velčev, V.I. & al. (eds): 100th anniversary of Academician Nikolaj A. Stojanov, pp. 125-137. - Sofia: Publishing House of the Bulgarian Academy of Sciences.

ANČEV, M.E. 1991: Genus *Alyssum* L. in Bulgarian flora [In Bulgarian with English summary]. In Kožuharov, S.I. & Kuzmanov, B.A. (eds): Evolution of flowering plants and florogenesis 2, pp. 85-118 + 16 plates. - Sofia: Publishing House of the Bulgarian Academy of Sciences.

ANČEV, M.E. 1993: Reports (217-227). In Kamari, G., Felber, F. & Garbari, F. (eds): Mediterranean chromosome number reports - 3. - *Fl. Medit.* **3**: 358-363.

ANČEV, M.E. 1995: Karyological variation and taxonomical notes on *Erysimum* L. (Brassicaceae) in Bulgarian flora. - *Giorn. Bot. Ital., n.s.* **129(1)**: 94-103.

ANČEV, M.E. 1996: Phytotaxonomic and phytogeographical studies in Bulgaria during the last decade (1983-1993). - *Bocconea* **5(1)**: 33-44.

ANČEV, M.E. 1997: Form-differentiation patterns and evolutionary trends in the family Brassicaceae (Cruciferae) in the Bulgarian flora. - *Phytol. Balcan.* **3(2-3)**: 65-74.

ANČEV, M.E. 1999: Prof. Dr. Stefan Kožuharov (1933-1997). - *Phytol. Balcan.* **5(2-3)**: 169-171.

ANČEV, M.E. 2000: The trichomes of *Alyssum*. - *Bot. Chron. (Patras)* **13**: 151-168.

ANČEV, M.E. 2006: Polyploidy and hybridogenesis in Bulgarian Brassicaceae - distribution and evolutionary role. In Peev, D.R. (ed.): Book of abstracts, IV Balkan botanical congress, Sofia, Bulgaria, 20-26 June 2006, p. 29.

ANČEV, M.E. 2007: Catalogue of the family Brassicaceae (Cruciferae) in the flora of Bulgaria. - *Phytol. Balcan.* **13(2)**: 153-178.

ANČEV, M.E. & DUDLEY, T.R. 1981: [Chromosome numbers in *Alyssum*]. In Löve, Á. (ed.): Chromosome number reports LXXIII. - *Taxon* **30(4)**: 856.

ANČEV, M.E. & GORANOVA, V.Y. 1997a: Reports (855-872). In Kamari, G., Felber, F. & Garbari, F. (eds): Mediterranean chromosome number reports - 7. - *Fl. Medit.* **7**: 246-258.

ANČEV, M.E. & GORANOVA, V.Y. 1997b: *Conringia planisiliqua* Fischer & C.A. Meyer (Brassicaceae) - a new species to the Bulgarian flora. - *Phytol. Balcan.* **3(1)**: 15-18.

ANČEV, M.E. & GORANOVA, V.Y. 1999: Reports (1031-1049). In Kamari, G., Felber, F. & Garbari, F. (eds): Mediterranean chromosome number reports - 9. - *Fl. Medit.* **9**: 331-339.

ANČEV, M.E. & GORANOVA, V.Y. 2003: Reports (1319-1327). In: Kamari, G., Blanché, C. & Garbari, F. (eds): Mediterranean chromosome number reports - 13. - *Fl. Medit.* **13**: 348-354.

ANČEV, M.E. & GORANOVA, V.Y. 2009: *Aubrieta* (Brassicaceae) in the Bulgarian flora. - *Phytol. Balcan.* **15(1)**: 43-50.

ANČEV, M.E., HARDALOVA, R.H. & TRIFONOVA, D.R. 1987: Cytotaxonomic study of the perennial species of *Erysimum* L. in Bulgaria [In Bulgarian with English summary]. In Kuzmanov, B. & al. (eds): Proceedings, fourth national conference on botany, pp. 24-31. - Sofia: Publishing House of the Bulgarian Academy of Sciences.

ANČEV, M.E. & KRENDL, F. 2011: *Galium* sect. *Leiogalium* (Rubiaceae) in the Bulgarian flora. - *Phytol. Balcan.* **17(3)**: 291-314.

ANČEV, M.E., MARHOLD, K. & GORANOVA, V.Y. 1997: Reports (873-877). In Kamari, G., Felber, F. & Garbari, F. (eds): Mediterranean chromosome number reports - 7. - *Fl. Medit.* **7**: 258-261.

ANČEV, M.E. & PENEVA, V.T. 1984: [Chromosome numbers from Bulgaria]. In Löve, Á. (ed.): Chromosome number reports LXXXII. - *Taxon* **33(1)**: 131-132.

ANČEV, M.E. & POLATSCHEK, A. 1998: Three new species of *Erysimum* L. (Brassicaceae) from Bulgarian flora. - *Ann. Naturhist. Mus. Wien, B* **100B**: 725-737.

ANČEV, M.E. & POLATSCHEK, A. 2003: *Erysimum bulgaricum* (Brassicaceae), a newly distinguished species for the Balkan Peninsula. - *Ann. Naturhist. Mus. Wien* **104B**: 691-698.

ANČEV, M.E. & POLATSCHEK, A. 2006: The genus *Erysimum* (Brassicaceae) in Bulgaria. - *Ann. Naturhist. Mus. Wien* **107B**: 227-273.

ANČEV, M.E. & TOMŠOVIC, P. 1999: The *Rorippa pyrenaica* group (Brassicaceae) in the Balkan Peninsula. - *Folia Geobot.* **34(2)**: 261-276.

ANČEV, M.E. & UZUNOV, D. 2001: *Alyssum orbelicum*, a new high-mountainous species of sect. *Odontarrhena* (Brassicaceae) from south-western Bulgaria. In Raimondo, F.M. (ed.): Abstracts from the X OPTIMA meeting, Palermo, 13-19 September 2001, p. 105.

ANČEV, M.E. & UZUNOV, D. 2002: *Alyssum orbelicum*, a new high-mountainous species of sect. *Odontarrhena* (Brassicaceae) from southwest Bulgaria. - *Phytol. Balcan.* **8(1)**: 25-30.

ANČEV, M.E. & AL. 2013: *Cardamine ×rhodopaea* (Brassicaceae), a triploid hybrid from the West Rhodope Mts: Morphology, distribution, relationships and origin. - *Phytol. Balcan.* **19(3)**: 328-333.

ANCHISI, E. & AL. 1987: Primule d'Europa. Primevères d'Europe [In Italian and French]. - Pavia: Gruppo Naturalistico Oltrepo Pavese, 136 pp.

ANDERBERG, A.A. 1991a: Taxonomy and phylogeny of the tribe Gnaphalieae (Asteraceae). - *Opera Bot.* **104**: 1-195.

ANDERBERG, A.A. 1991b: Taxonomy and phylogeny of the tribe Inuleae (Asteraceae). - *Pl. Syst. Evol.* **176(1-2)**: 75-123.

ANDERBERG, A.A. 1994: Phylogeny and subgeneric classification of *Cyclamen* L. (Primulaceae). - *Kew Bull.* **49(3)**: 455-467.

ANDERBERG, A.A., MANNS, U. & KÄLLERSJÖ, M. 2007: Phylogeny and floral evolution of the Lysimachieae (Ericales, Myrsinaceae): evdence from *ndhF* sequence data. - *Willdenowia* **37(2)**: 407-421.

ANDERBERG, A.A., SALTER, M.E. & NORDENSTAM, B. 1996: Type material of the Asteraceae in the Swedish Museum of Natural History (S). - *Compositae Newslett.* **28**: 1-172.

ANDERBERG, A.A., TRIFT, I. & KÄLLERSJÖ, M. 1998: On the systematic position of the genus *Coris* (Primulaceae). - *Nordic J. Bot.* **18(2)**: 203-207.

ANDERS-GASSER, I. 1988: Cytotaxonomische Untersuchungen an *Ranunculus ficaria* L. - *Mitth. Naturf. Ges. Bern, N.S.* **42**: 79-85.

ANDERSEN, M. & CHRISTENSEN, B. 1975: Kretas orchideer. - *Orchideen* **16(2)**: 13-18.

ANDERSON, B. & ANDERSON, E. 1989: The wild orchids of Samos. - *Mitt. Arbeitskreis Heimische Orchid. Baden-Württemberg* **21(4)**: 1136-1155.

ANDERSON, B. & ANDERSON, E. 1991: In search of *Ophrys aegaea*. - *Rock Gard.* **22(3)**: 321-324.

ANDERSON, D.E. 1961: Taxonomy and distribution of the genus *Phalaris*. - *Iowa State J. Sci.* **36(1)**: 1-96.

ANDERSON, E.B. 1971: *Iris unguicularis*. - *J. Roy. Hort. Soc.* **96(2)**: 82-83 + 1 plate.

ANDERSON, N.O. 1986: The distribution of the genus *Lilium* with reference to its evolution. – *Herbertia* **42**: 31-50.

ANDERSON, N.O. 1988: The distribution of the genus *Lilium* with reference to its evolution. - *Lily Yearb. N. Amer. Lily Soc.* **41**: 1-18 + 1 map.

ANDERSSON, I.A. & AL. 1983: A revision of the *Aethionema saxatile* complex (Brassicaceae). - *Willdenowia* **13(1)**: 3-42.

ANDERSSON, N.J. 1867: Monographia Salicum hucusque cognitarum. - *Kongl. Svenska Vetensk. Acad. Handl.* **6(1)**: i-iv, 1-180.

ANDONOVSKI, A. 1966: *Spiraea ulmifolia* Scop. (Syn.: *Sp. chamaedryfolia* L. p.p.) une nouvelle espèce dans le flore de la R.S. Macedonienne [In Macedonian with French summary]. - *Godišen Zborn. Zemjod.-Šumarsk. Fak. Univ. Skopje, Šumarstvo* **19**: 269-272.

ANDONOVSKI, A. 1972: *Alnus viridis* Lam. et DC. - nouvelle espèce dans le flore de la R.S. Macédoine [In Macedonian with French summary]. - *Godišen Zborn. Zemjod.-Šumarsk. Fak. Univ. Skopje, Šumarstvo* **24**: 33-36.

ANDREAS, CH.H. 1956: Notes on *Stellaria nemorum* L. subsp. *glochidisperma* Murbeck in the Netherlands. - *Acta Bot. Neerl.* **5(2)**: 145-156.

ANDREEV, N. 1979: [Miscellaneous chromosome numbers from Bulgaria]. In Löve, Á. (ed.): IOPB chromosome number reports LXIV. - *Taxon* **28(4)**: 397.

ANDREEV, N. 1981: [Chromosome numbers in Bulgarian plants]. In Löve, Á. (ed.): Chromosome number reports LXX. - *Taxon* **30(1)**: 74-75.

ANDREEV, N. 1982: [Chromosome numbers from Bulgaria]. In Löve, Á. (ed.): Chromosome number reports LXXVI. - *Taxon* **31(3)**: 575-576.

ANDRÉS-SÁNCHEZ, S. & AL. 2011a: A nomenclatural treatment for *Logfia* Cass. and *Filago* L. (Asteraceae) as newly circumscribed: Typification of several names. - *Taxon* **60(2)**: 572-576.

ANDRÉS-SÁNCHEZ, S. & AL. 2011b: (2008) Proposal to conserve the name *Filago arvensis* L. (Asteraceae) with a conserved type. - *Taxon* **60(2)**: 599-600. [not voted on, cf. Taxon 63(6): 1359, 2014].

ANDRÉS-SÁNCHEZ, S. & AL. 2011c: (2009) Proposal to conserve the name *Filago vulgaris* against *Gnaphalium germanicum* (*Filago germanica*) (Asteraceae). - *Taxon* **60(2)**: 600-602. [not approved, cf. *Taxon* 61(5): 1112, 2012].

ANDRÉS-SÁNCHEZ, S. & AL. 2021: (2839) Proposal to conserve the name *Triticum caninum* (*Elymus caninus*) (Poaceae) with a conserved type. - *Taxon* **70(5)**: 1138-1139.

ANDRÉS-SÁNCHEZ, S., MARTÍNEZ-ORTEGA, M.M. & RICO, E. 2013: Taxonomic revision of the genus *Logfia* (Asteraceae, Gnaphalieae) in the Mediterranean region. - *Anales Jard. Bot. Madrid* **70(1)**: 7-18.

ANDRÉS-SÁNCHEZ, S., MARTÍNEZ-ORTEGA, M.M. & RICO, E. 2014: Revisión taxonómica del género *Bombycilaena* (DC.) Smoljan. (Asteraceae). - *Candollea* **69(1)**: 55-62.

ANDRES, H. 1913: Studien zur speziellen Systematik der Pirolaceae. I. Revision der Sektion *Eu- Thelaia* H. Andr. - *Oesterr. Bot. Z.* **63(11)**: 445-450.

ANDRES, H. 1914a: Studien zur speziellen Systematik der Pirolaceae. Subsektion II. *Alefeldiana* H. Andr. - *Oesterr. Bot. Z.* **64**: 45-50, 232-254.

ANDRES, H. 1914b: Beiträge zur Kenntnis der Morphologie, Phytogeographie und allgemeinen Systematik der Pirolaceae. - *Verh. Bot. Vereins Prov. Brandenburg* **56**: 1-76.

ANDRES, H. 1923: Studien zur speziellen Systematik der Pirolaceae. IV. Revision der Gattung *Ramischia* Opiz. - *Repert. Spec. Nov. Regni Veg.* **19**: 209-224.

ANDRÉS, J. & AL. 1988: *Carex hallerana* Asso: problemas de identificación. - *Lagascalia* **15(Extra)**: 397-399.

ANDREW, S. 1985a: European orchid holidays. Greece - Part 1. - *Orchid Rev.* **93(1100)**: 194-198.

ANDREW, S. 1985b: European orchid holidays. Greece - Part 2. - *Orchid Rev.* **93(1101)**: 233-237.

ANDREW, S. 1985c: European orchid holidays. Crete - Part 1. - *Orchid Rev.* **93(1102)**: 263-266.

ANDREW, S. 1985d: European orchid holidays. Crete - Part 2. - *Orchid Rev.* **93(1103)**: 290-292.

ANDREWS, S. 1994: Tree of the year: *Zelkova*. - *Int. Dendrol. Soc. Year Book* **1993**: 11-30.

ANDRIKOU-CHARITIDOU, A. & CHANLIDOU, E. 2013: Survey of *Ramonda serbica* Pančić populations in Greece. In Gkelis, S. & al. (eds): 13th Panhellenic Scientific Conference, Hellenic Botanical Society, Thessaloniki 3-6 October. Program and book of abstracts, p. 43.

ANDROSHCHUK, A.F. & KOSTINENKO, L.D. 1981: Chromosome numbers of the genus *Achillea* L. Certain species cultivated in botanical gardens. - *Ukrajinsk Bot. Zhum. (Kiev)* **38(2)**: 53-57.

ANDRZEJEWSKA-GOLEC, E. 1992: A taxonomic study of *Plantago* subgenus *Psyllium* (Miller) Harms. - *Bot. J. Linn. Soc.* **108(1)**: 49-63.

ANEVA, I., DIMITROV, D.S. & VUTOV, V.M. 2015: Flora and vegetation of Slavyanka Mountain. - *Bulg. J. Agric. Sci.* **21(5)**: 926-934.

ANEVA, I. & ZHELEV, P. 2018: The ecological and floristic characteristics of populations of *Sideritis scardica* Griseb. in Olympus Mts., Greece. - *Ecol. Balkan.* **10(2)**: 93-99.

ANGELOV, G.B. 2003: Isozyme variation and genetic relationships among four Balkan endemics of the *Festuca ovina* group (Poaceae-Poeae). - *Phyton (Horn)* **43(2)**: 271-280.

ANGULO, M.D. & FIGUERAS, M.C. DE 1979: Números cromosómicos y sacos polínicos en especies del género *Trifolium*. - *Genét. Ibér.* **30-31**: 129-160.

ANGULO, M.D. & SÁNCHEZ DE RIVERA, A.M. 1975: Studies on *Trifolium subterraneum* ecotypes. - *Cytologia* **40(2)**: 415-423.

ANGULO, M.D. & SÁNCHEZ DE RIVERA, A.M. 1983: Karyologic studies on Spanish taxa of *Trifolium subterraneum* L. - *Cytologia* **48(2)**: 305-312.

ANGULO, M.D., SÁNCHEZ DE RIVERA, A.M. & GONZÁLEZ-BERNÁLDEZ, F. 1969: The chromosomes of *Trifolium subterraneum* L. - *Israel J. Bot.* **17(3)**: 155-161.

ANGULO, M.D., SÁNCHEZ DE RIVERA, A.M. & GONZÁLEZ-BERNÁLDEZ, F. 1971: Estudios cromosómicos en el género *Trifolium*. - V. - *Bol. Soc. Brot., Ser. 2* **45**: 253-267 + 4 plates.

ANONYMUS 1859: Über die botanischen Untersuchungen, welche in Griechenland seit Ankunft des Königs Otto I. gemacht worden sind (Übersetzung aus dem Moniteur Grec vom 27. Mai 1855). - *Flora* **42**: 483-495.

ANONYMUS 1889: Botanische Forschungsreisen [announcing travels of Christos Leonis to the Kiklades]. - *Oesterr. Bot. Z.* **39(4)**: 159.

ANONYMUS 1977: Activities of the Goulandris Natural History Museum 1964-1975. - *Ann. Mus. Goulandris* **3**: 9-18.

ANONYMUS 1982: IUCN Threatened Plants Committee Secretariat: The rare, threatened and endemic plants of Greece. - *Ann. Mus. Goulandris* **5**: 69-105.

ANONYMUS 1997: A guide to species Irises: Their identification and cultivation. Edited by the Species Group of the British Iris Society. - Cambridge: Cambridge Univ. Press, xvi + 371 pp.

ANONYMUS 2000: Background regarding *Sedum creticum* Presl. - *Sedum Soc. Newsletter* **52**: 32-34.

ANONYMUS 2003: [Photo of *Cyclamen hederifolium* subsp. *confusum*]. - *Alpine Gardener* **71(2)**: 195.

ANONYMUS 2004: I fotografisi enos endimikou fitou [In Greek]. - *Fisis [Journal of the Hellenic Society for the Protection of Nature]* **105**: 37-38.

ANONYMUS 2005a: BSBI database - Cytology. Available on-line: http://rbg-web2.rbge.org.uk/BSBI/.

ANONYMUS 2005b: w3TROPICOS - Index to plant chromosome numbers data base. Available on-line: http://mobot.mobot.org/W3T/Search/ipcn/html.

ANTOINE, F. 1864: *Pinus leucodermis* Ant. - *Oesterr. Bot. Z.* **14(12)**: 366-368.

ANTON, A.M. 1981: The genus *Tragus* (Gramineae). - *Kew Bull.* **36(1)**: 55-61.

ANTONIA, D. & KOKKINI, S. 1997: Distribution and morphological variation of *Satureja montana* group in East Macedonia and Thrace. In: Abstracts from the First Balkan Botanical Congress, Thessaloniki, September 19-22, 1997, p. 11. – Thessaloniki.

ANTONIS, A. 1997: Die Orchideen von Kreta und Karpathos. - Iraklion: W. Kuwidis & W. Manuras, 159 pp.

ANTONOPOULOS, Z. 2005: Avtofiis orchidees ton Kithiron [In Greek]. - *Fisis [Journal of the Hellenic Society for the Protection of Nature]* **111**: 8-10.

ANTONOPOULOS, Z. 2009a: Bee orchids of Greece. The genus *Ophrys*. - Rethymno: Mediterraneo Editions, 320 pp.

ANTONOPOULOS, Z. 2009b: I orchidees tou Olimbou. [In Greek]. - *Fisis [Journal of the Hellenic Society for the Protection of Nature]* **127**: 6-11.

ANTONOPOULOS, Z., BERGFELD, D. & TSIFTSIS, S. 2011: *Epipactis helleborine* subsp. *moratoria* Riech. & Zirnsack, a new subspecies for the flora of Greece. - *J. Eur. Orch.* **43(1)**: 85-98.

ANTONOPOULOS, Z., GAVALAS, G. & KREUTZ, K. 2011: The orchids of the Aegean island of Herakleia (Cyclades) and *Ophrys heracleotica* Gavalas, Kreutz & Z. Antonopoulos, a new *Ophrys* species. - *Ber. Arbeitskreis. Heimische Orchid.* **27(2)**: 266-281.

ANTONOPOULOS, Z. & TSIFTSIS, S. 2012: *Epipactis purpurata* Sm. and *Epipactis leptochila* (Godf.) Godf. subsp. *neglecta* Kümpel (Orchidaceae), two new *Epipactis* taxa for the flora of Greece. - *Ber. Arbeitskreis. Heimische Orchid.* **29(1)**: 81-99.

ANTONOPOULOS, Z. & TSIFTSIS, S. 2017: Atlas of the Greek orchids, vol. 2. The genus *Ophrys*. - Rethymno: Mediterraneo Editions, 503 pp.

ANTROPOV, V.I. & ANTROPOV, V.F. 1929: Rye in U.S.S.R. and in the adjoining countries [In Russian with English summary]. - *Trudy Prikl. Bot. Selekt. Suppl.* **36**: 1-366.

ANZALONE, B. 1987: Sistematica e corologia di *Pastinaca sativa* L. in Italia. - *Arch. Bot. Biogeogr. Ital.* **63(1-2)**: 1-21.

ANZALONE, B. & LATTANZI, E. 1988: Studio sistematico e corologico su alcuni *Seseli* della flora italiana. - *Acrch. Bot. Biogeogr. Ital.* **61(1-2)**: 55-83.

ANZALONE, B. & LATTANZI, E. 1989: Il *Lupinus graecus* Boiss. & Spruner in Italia e osservazioni su *Lupinus termis* Forsskal (1). - *Ann. Bot. (Rome)* **47 (Suppl.)**: 115-121.

ANZALONE, B., LATTANZI, E. & LEPORATTI, M.L. 1991: Il gruppo di *Ferula communis* L. (Umbelliferae) in Italia: ricerche sistematiche e corologiche. - *Arch. Bot. Biogeogr. Ital.* **67(3-4)**: 221-236.

APARICIO, A. & SILVESTRE, S. 1985: Números cromosómicos para la flora española. 422-434. - *Lagascalia* **13(2)**: 318-323.

APERGIS, M. 1997: Mia mistiriodis elliniki orchidea [In Greek]. - *Fisis [Journal of the Hellenic Society for the Protection of Nature]* **78**: 8-10.

APERGIS, M. 1998a: *Orthilia secunda* and *Neottia nidus-avis* on Mt. Taygetos. – *Anthophoros [privately distributed newsletter]* **3-1998**: 8.

APERGIS, M. 1998b: *Paeonia parnassica*. - *Fisis [Journal of the Hellenic Society for the Protection of Nature]* **81**: front cover.

APERGIS, M. 1998c: The fritillaries of Greece [In Greek with English summary]. - *Fisis [Journal of the Hellenic Society for the Protection of Nature]* **81**: 23-29, 44-46.

APERGIS, M. 1999: *Dactylorhiza sambucina*. - *Fisis [Journal of the Hellenic Society for the Protection of Nature]* **85**: front cover.

APERGIS, M. 2000: *Tulipa bakeri* on the Omalos plateau (Crete). - *Fisis [Journal of the Hellenic Society for the Protection of Nature]* **89**: front cover.

APERGIS, M. 2001: Peonies in Greece [In Greek with English summary]. - *Fisis [Journal of the Hellenic Society for the Protection of Nature]* **93**: 23-27.

APERGIS, M. 2002a: *Cyclamen peloponnesiacum*. - *Fisis [Journal of the Hellenic Society for the Protection of Nature]* **97**: front cover.

APERGIS, M. 2002b: Anixiatiki Kefallonia [In Greek]. - *Fisis [Journal of the Hellenic Society for the Protection of Nature]* **97**: 4-6.

APERGIS, M. 2003a: Taxidi sti Voria Ellada [In Greek]. - *Fisis [Journal of the Hellenic Society for the Protection of Nature]* **100**: 23-28.

APERGIS, M. 2003b: Fthinoporini Mani [In Greek]. - *Fisis [Journal of the Hellenic Society for the Protection of Nature]* **101**: 4-5.

APERGIS, M. 2003c: *Paeonia mascula* subsp. *russi* in Kefallinia [In Greek with English abstract]. - *Fisis [Journal of the Hellenic Society for the Protection of Nature]* **101**: 32.

APERGIS, M. 2003d: *Galanthus reginae-olgae* subsp. *vernalis*. Nea topothesia [In Greek and English]. - *Fisis [Journal of the Hellenic Society for the Protection of Nature]* **101**: 32.

APERGIS, M. 2003e: I Chios tin anixi [In Greek]. - *Fisis [Journal of the Hellenic Society for the Protection of Nature]* **102**: 23-26.

APERGIS, M. 2003f: Nea tis ellinikis chloridas [In Greek]. - *Fisis [Journal of the Hellenic Society for the Protection of Nature]* **102**: 29.

APERGIS, M. 2004a: [Prunus prostrata]. - *Fisis [Journal of the Hellenic Society for the Protection of Nature]* **105**: front cover.

APERGIS, M. 2004b: Kastoria - Grammos - Prespes 2003. - *Fisis [Journal of the Hellenic Society for the Protection of Nature]* **105**: 23-26.

APERGIS, M. 2004c: Naxos [In Greek]. - *Fisis [Journal of the Hellenic Society for the Protection of Nature]* **106**: 4-7.

APERGIS, M. 2005a: Kritis paralipomena [In Greek]. - *Fisis [Journal of the Hellenic Society for the Protection of Nature]* **108**: 23-26.

APERGIS, M. 2005b: Samos, Aprilios 2005 [In Greek]. - *Fisis [Journal of the Hellenic Society for the Protection of Nature]* **110**: 23-26.

APERGIS, M. 2006a: Ta fita andidroun stin klimatiki allagi [In Greek]. - *Fisis [Journal of the Hellenic Society for the Protection of Nature]* **112**: 23-26.

APERGIS, M. 2006b: Peonies stin Ochi [In Greek]. - *Fisis [Journal of the Hellenic Society for the Protection of Nature]* **114**: 18-19.

APERGIS, M. & APERGIS, M. 1999: *Ophrys* the trickster [In Greek with English summary]. – *Fisis [Journal of the Hellenic Society for the Protection of Nature]* **84**: 23-30, 44.

APERGIS, M. & APERGIS, M. 2006: Anazitontas tin elliniki fisi [In Greek]. - Privately published: 144 pp. (ISBN 960-631-664-5).

APERGIS, M. & SFIKAS, G. 2002a: Ta nea tis ellinikis chloridas [In Greek]. - *Fisis [Journal of the Hellenic Society for the Protection of Nature]* **97**: 41.

APERGIS, M. & SFIKAS, G. 2002b: Ta nea tis ellinikis chloridas ke panidas [In Greek]. - *Fisis [Journal of the Hellenic Society for the Protection of Nature]* **99**: 20-21.

APLADA, E. & AL. 2007: Phytogeography and ecological evaluation of the flora and vegetation of Mt Parnitha (Attica, Greece). - *Edinburgh J. Bot.* **64(2)**: 185-207.

APLADA, E., TINIAKOU, A. & GEORGIADIS, TH. 2012: The orchid flora of Mount Giona (Sterea Ellas, Greece). - *J. Eur. Orch.* **44(4)**: 765-792.

APLADA, I. 2013: Chlorida ke vlastisis ton ikosistimaton tou Oros Giona. Axiologisi - prostasia - diachirisi. [In Greek with English abstract]. - Patras: Ph.D. thesis, Univ. of Patras, 453 pp. + photos, tables and maps.

APOSTOLOS, G. 1997: Genetic studies on Greek mountain tea (*Sideritis* L.). - *Lamiales Newslett.* **5**: 8-9.

APOSTOLOVA, I. & DENČEV, C.M. 1993: Current state of *Centaurea imannuelis-loewii* [sic!] Degen in Bulgaria. In Petrova, A.V. (ed.): Abstracts from the VII OPTIMA meeting in Borovetz, Bulgaria, 18-30 July 1993, p. 120.

APOSTOLOVA, I. & DENCHEV, C. 1997: The currect status of *Centaurea immanuelis-loewii* (Compositae) in Bulgaria. - *Bocconea* **5**: 703-706.

APOSTOLOVA, I. & GANEVA, A. 2000: New data on *Edraianthus serbicus* (Kern.) Petrovic in Bulgaria. - *Phytol. Balcan.* **6(1)**: 65-71.

APOSTOLOVA, I. & PETROVA, A.S. 1997: New data on *Merendera sobolifera* C.A. Mey. - *Fl. Medit.* **7**: 185-196.

APPEL, O. & AL-SHEHBAZ, I.A. 1997: Generic limits and taxonomy of *Hornungia, Pritzelago,* and *Hymenolobus* (Brassicaceae). - *Novon* **7(4)**: 338-340.

APPLEQUIST, W.L. 2002: A reassessment of the nomenclature of *Matricaria* L. and *Tripleurospermum* Sch. Bip. (Asteraceae). - *Taxon* **51(4)**: 757-761.

APPLEQUIST, W.L. 2006: (1714-1715) Proposal to reject the names *Plantago psyllium* and *P. cynops* (Plantaginaceae). - *Taxon* **55(1)**: 235-236. [approved, cf. Taxon 58(1): 281, 2009].

AQUARO, G. & PERUZZI, L. 2006: Reports (1457-1461). In Kamari, G., Blanché, C. & Garbari, F. (eds): Mediterranean chromosome number reports - 16. - *Fl. Medit.* **16**: 386-396.

AQUARO, G., PERUZZI, L. & CESCA, G. 2004: Chromosome numbers of 20 flowering plants from ex-Yugoslav countries. In Vasić, O. (ed.): Abstracts, XI OPTIMA Meeting, Beograd 5-11.IX. 2004, p. 117.

AQUARO, G., PERUZZI, L. & CESCA, G. 2007: Chromosome numbers of 20 flowering plants from the ex-Yugoslav countries. - *Bocconea* **21**: 303-312.

ARABATZIS, TH. 1998: Thamni ke dendra stin Ellada. Tomos 1 [In Greek]. - Drama: Ikologiki Kinisi, xii + 292 pp.

ARABATZIS, TH. 2001: Thamni ke dendra stin Ellada. Tomos 2 [In Greek]. - Drama: Ikologiki Kinisi, xiv + 435 pp.

ARABATZIS, TH. & AL. 1997: Agrioloulouda tou Parthenou Dasous Dramas [In Greek]. - Drama: Ikologiki Kinisi, 76 pp.

ARABATZIS, TH., KARAGIANNAKIDOU, V. & TSIFTSIS, S. 2001: Contribution to the flora of Mount Menikion, NE Greece. In Özhatay, N. (ed.): Proceedings of the second Balkan Botanical Congress, 1: 87-90. - Istanbul.

ARABATZIS, TH., KOLATSIDAKIS, E. & TZANOUDAKIS, D. 1999: Contribution to the knowledge of flora of northern Greece and the widening of geographical distribution ... In Elefheriou, E.P. (ed.): Proceedings, 7th Panhellenic scientific conference, Alexandroupolis, pp. 327-331 [In Greek with English abstract]. - Thessaloniki: Univ. Studio Press.

ARABATZIS, TH. & PAPADIMITRIOU, S. 1996: Records of new taxa for the flora of Paiko [In Greek with English summary]. In: Proceedings of the 6th scientific conference of the Hellenic Botanical Society, pp. 296-300. - Paralimni, Cyprus.

ARABATZIS, TH. & VIDAKIS, K. 1996: *Spiraea chamaedryfolia* L., a new species for the Greek flora [In Greek with English summary]. In: Proceedings of the 6th scientific conference of the Hellenic Botanical Society, pp. 292-295. - Paralimni, Cyprus.

ARATA, M. 1944: Sulla cariologia del genere *Carlina* L. - *Nuovo Giorn. Bot. Ital., n.s.* **51(1-4)**: 39-44.

ARAUJO, E. & TALAVERA, S. 1981: Números cromosómicos para la flora española. 206-209. - *Lagascalia* **10(2)**: 233-235.

ARBEZ, M. 1969: Répartition, écologie et variabilité des sapins de Turquie du nord: *Abies nordmanniana* Spach, *Abies bornmuelleriana* Mattfeld, *Abies equi-trojani* Ascherson et Sintenis. - *Ann. Sci. Forest.* **26(2)**: 257-284.

ARCANGELI, G. 1882: Compendio della flora italiana ossia Manuale per la determinazione pelle piante che trovansi od inselvatichite nell'Italia e nelle isole adiacenti. - Torino: xx + 889 pp.

ARCANGELI, G. 1894: Compendio della flora italiana ossia Manuale per la determinazione pelle piante che trovansi od inselvatichite nell'Italia e nelle isole adiacenti. Ed. 2. - Torino & Roma: Ermanno Loescher, xix + 836 pp

ARCHIBALD, J.C. 1971: Plant portraits: *Ranunculus asiaticus* L. - *Quart. Bull. Alpine Gard. Soc. Gr. Brit.* **39(2)**: 120-123.

ARCUS, M. & DOROFTEI, E. 2001: *Opopanax chironium* (L.) Koch ssp. *bulgaricum* (Velen.) Beldie (Apiaceae) in the Romanian flora. In Özhatay, N. (ed.): Proceedings of the Second Balkan Botanical Congress, 1: 91-94. - Istanbul.

ARDENGHI, N.M.G. & AL. 2016: Chromosome numbers and karyomorphology of *Festuca circummediterranea* Patzke group (Poaceae, Loliinae). - *Phytotaxa* **263(3)**: 255-269.

ARDENGHI, N.M.G., CAUZZI, P. & GALASSO, G. 2015: *Cardamine hamiltonii* G. Don and *Oenothera speciosa* Nutt. In Raab-Straube, E. v. & Raus, Th. (eds): Euro+Med-Checklist notulae, 4. - *Willdenowia* **45(1)**: 121-123.

ARDENGHI, N.M.G. & GALASSO, G. 2013: *Vicia labrusca* L.). In Raab-Straube, E. v. & Raus, Th. (eds): Euro+Med-Checklist notulae, 2. - *Willdenowia* **43(2)**: 235.

ARDÉVOL GONZÁLEZ, J.F., BORGEN, L. & PÉREZ DE PAZ, P.L. 1993: Checklist of chromosome numbers counted in Canarian vascular plants. - *Sommerfeltia* **18**: 1-59.

ARDUINO, P. 1759: Animadversionum botanicarum specimen. - Padova: Conzatti, xxviii pp. + 12 plates.

ARDUINO, P. 1764: Animadversionum botanicarum specimen alterum. - Venetia: Typographia Sansoniana, xlii pp. + 20 plates.

ARDUINO, P. & AL. 1995: Genetic variation in *Orchis papilionacea* (Orchidaceae) from the Central Mediterranean region: taxonomic inferences at the intraspecific level. - *Pl. Syst. Evol.* **194(1-2)**: 9-23.

ARDUINO, P. & AL. 1996: Genetic variation and natural hybridization between *Orchis laxiflora* and *Orchis palustris* (Orchidaceae). - *Pl. Syst. Evol.* **202(1-2)**: 87-109.

ARENDS, J.C. & LAAN, F.M. VAN DER 1979: [Miscellaneous chromosome numbers]. In Löve, Á. (ed.): IOPB chromosome number reports LXV. - *Taxon* **28(5-6)**: 636-637.

ARÈNES, J. 1949: Contribution à l'étude du genre *Carduus*. - *Mém. Mus. Natl. Hist. Nat.* **24(4)**: 183-255.

ARÈNES, J. 1950: Monographie du genre *Arctium* L. - *Bull. Jard. Bot. État. Bruxelles* **20(1)**: 67-156

ARÈNES, J. 1957: Les centaurées de la sous-section *Jacea*. Systématique, chorologie et phylogénie. - *Bull. Jard. Bot. État. Bruxelles* **27(1)**: 143-157.

ARÈNES, J. 1959a: Sur les Centaurées de la sous-section "Phrygiae" Boiss. (in sect. *Jacea* Whlbg). - *Notul. Syst. (Paris)* **15(4)**: 376-390.

ARÈNES, J. 1959b: Sur la systématique des quelques "Carduus". - *Notul. Syst. (Paris)* **15(4)**: 390-410.

ARIANOUTSOU-FARAGGITAKI, M. & DIAMANTOPOULOS, J. 1985: Comparative phenology of five dominant plant species in maquis and phrygana ecosystems in Greece. - *Phyton (Horn)* **25(1)**: 77-85.

ARIANOUTSOU-FARAGGITAKI, M. & MARGARIS, N.S. 1982: Phryganic (east Mediterranean) ecosystems and fire. - *Ecol. Medit.* **8**: 473-480.

ARIANOUTSOU, M. & AL. 2010a: Comparing naturalized alien plants and recipient habitats across an east-west gradient in the Mediterranean Basin. - *J. Biogeogr.* **37(9)**: 1811-1823.

ARIANOUTSOU, M. & AL. 2010b: The alien flora of Greece: taxonomy, life traits and habitat preferences. - *Biol. Invasions* **12(10)**: 3525-3549.

ARISTA, M., ORTIZ, P.L. & TALAVERA, S. 1995: Reproductive isolation of two co-occurring subspecies of *Juniperus phoenicea*. In Valdés, B. & Silvestre, S. (eds): Abstracts. VIII OPTIMA meeting, Sevilla, Spain, 25 September - 1 October, 1995, p. 77.

ARISTA, M. & TALAVERA, S. 1990: Números cromosómicos para la flora española. 620-629. - *Lagascalia* **16(2)**: 323-328.

ARMAĞAN, M. 2021: *Allium muratozelii* (Amaryllidaceae), a new species from Turkey. - *Phytotaxa* **498(4)**: 255-264.

ARMSTRONG, J.V. 1997: A morphometric study of the woodrushes *Luzula campestris* (L.) DC., *L. multiflora* subsp. *multiflora* (Ehrh.) Lej. and *L. congesta* (Thuill.) Lej. (Juncaceae) in the British Isles. - *Watsonia* **21(3)**: 259-270.

ARMSTRONG, K.C. 1981: The evolution of *Bromus inermis* and related species of *Bromus* sect. *Pnigma*. - *Bot. Jahrb. Syst.* **102(1-4)**: 427-443.

ARNOLD, B. & JURY, S.L. 1996: Delimitation of *Pseudorlaya* taxa. - Lagascalia **18(2)**: 285-286.

ARNOLD, N. & AL. 1991: Short note on endemics of Cyprus. - Bot. Chron. (Patras) **10**: 521-526.

ARNOLD, R.E. 1942: *Fritillaria obiqua* and *F. tuntasiana*. - *Gard. Chron., ser 3* **111**: 76.

ARNOW, L.A. 1994: *Koeleria macrantha* and *K. pyramidata* (Poaceae): Nomenclatural problems and biological distinctions. - *Syst. Bot.* **19(1)**: 6-20.

ARONNE, G., WILCOCK, C.C. & PIZZOLONGO, P. 1993: Pollination biology and sexual differentiation of *Osyris alba* (Santalaceae) in the Mediterranean region. - *Pl. Syst. Evol.* **188(1-2)**: 1-16.

ARRIGO, N. & AL. 2010: Origin and expansion of the allotetraploid *Aegilops geniculata*, a wild relative of wheat. - *New Phytol.* **187(4)**: 1170-1180.

ARRIGONI, P.V. 1967: Ricerche sulla distribuzione del "Pinus halepensis" Mill. e del "Pinus pinea" L. in Sardegna. - *Webbia* **22(2)**: 405-418.

ARRIGONI, P.V. 1969: Ad floram italicam notulae taxonomicae et geobotanicae. 4. Nuove stazioni di "Cystopteris dickieana" Sim in Sardegna. - *Webbia* **24(1)**: 425-427.

ARRIGONI, P.V. 1974: Ricerche sulla querce caducifoglie italiana. 3. "Quercus frainetto" Ten. in Toscana. - *Webbia* **29(1)**: 87-104.

ARRIGONI, P.V. 1979: La piante endemiche della Sardegna: 40-53. - *Boll. Soc. Sarda Sci. Nat.* **18**: 223-295.

ARRIGONI, P.V. 1987: Inventario del genere *Hieracium* in Sardegna. - *Webbia* **41(1)**: 1-25. [Also distributed as *OPTIMA leaflet* no. 168].

ARRIGONI, P.V. 2014: Revisione tassonomica e corologica del genere *Polygala* in Italia. - *Inform. Bot. Ital.* **46(2)**: 235-263.

ARRIGONI, P.V. & DIANA CORRIAS, S. 1993: Contribution à la connaissance du genre *Limonium* en Corse. - *Candollea* **48(2)**: 631-677.

ARRIGONI, P.V., GIANNERINI, M. & MORI, B. 1981: Numeri cromosomici per la Flora Italiana: 714-721. - *Inform. Bot. Ital.* **12(2)**: 137-143.

ARRIGONI, P.V. & MORI, B. 1972: Numeri cromosomici per la Flora Italiana: 92-97. - *Inform. Bot. Ital.* **3 (3)**: 226-233.

ARRIGONI, P.V. & MORI, B. 1977: Numeri cromosomici per la Flora Italiana: 270-275. - *Inform. Bot. Ital.* **8(3)**: 269-276.

ARRIGONI, P.V. & MORI, B. 1981: Numeri cromosomici per la Flora Italiana: 722-727. - *Inform. Bot. Ital.* **12(2)**: 145-148.

ARTELARI, R. 1984a: Biosystematic study of the genus *Limonium* (Plumbaginaceae) in the Ionian islands [In Greek with English summary]. - Patras: Ph.D. thesis, Univ. of Patras, 200 pp.

ARTELARI, R. 1984b: The genus *Limonium* (Plumbaginaceae) in the Ionian area, Greece. - *Bot. Chron. (Patras)* **4(1-2)**: 17-39.

ARTELARI, R. 1984c: Two new species of *Limonium* (Plumbaginaceae) from Zakynthos island (Greece). - *Mitt. Bot. Staatssamml. München* **20**: 429-440.

ARTELARI, R. 1989a: Biosystematic study of the genus *Limonium* (Plumbaginaceae) in the Aegean area (Greece). I. Some *Limonium* species from the Kikladhes islands. - *Willdenowia* **18(2)**: 399-408.

ARTELARI, R. 1989b: *Limonium creticum* (Plumbaginaceae), a new species from Kriti island (Aegean sea) Greece. - *Candollea* **44(2)**: 415-421.

ARTELARI, R. 1989c: Biosystematic study of the genus *Limonium* (Plumbaginaceae) in the Aegean area (Greece). II. *Limonium hierapetrae* Rech. fil. from Kriti island. - *Webbia* **43(1)**: 33-40.

ARTELARI, R. 1992: Reports (51-55). In Kamari, G., Felber, F. & Garbari, F. (eds): Mediterranean chromosome number reports - 2. - *Fl. Medit.* **2**: 229-232.

ARTELARI, R. & CHRISTODOULAKIS, D. 1997: The genus *Limonium* (Plumbaginaceae) on the islands Samos, Ikaria and Fourni (E. Aegean, Greece). In: Abstracts from the First Balkan Botanical Congress, Thessaloniki, September 19-22, 1997, p. 3. – Thessaloniki.

ARTELARI, R. & ERBEN, M. 1986: *Limonium brevipetiolatum* - eine neue hexaploide Sippe aus Süd-Griechenland. - *Mitt. Bot. Staatssamml. München* **22**: 507-511.

ARTELARI, R. & GEORGIOU, O. 1999a: The genus *Limonium* (Plumbaginaceae) on the islands Kithira and Antikithira. In Elefheriou, E.P. (ed.): Proceedings, 7th Panhellenic scientific conference, Alexandroupolis, p. 252 [In English and Greek]. - Thessaloniki: Univ. Studio Press.

ARTELARI, R. & GEORGIOU, O. 1999b: Two new species of *Limonium* (Plumbaginaceae) from the island of Kithira (Greece). - *Bot. J. Linn. Soc.* **131(3)**: 399-415.

ARTELARI, R. & GEORGIOU, O. 2001: The genus *Limonium* in South Peloponnisos. In Kamari, G., Psaras, G. & Constantinidis, Th. (eds): Proceedings of the 8th scientific congress of the Greek Botanical Society, October 5-8, 2000, pp. 72-77 [In Greek with English abstract].

ARTELARI, R. & GEORGIOU, O. 2002: *Limonium runemarkii* Rech. fil. (Plumbaginaceae), from Evvia to Kriti. In Kamari, G. & al. (eds): Ellinini Votaniki Eteria. 9° Panelliniko Epistimoniko Sinedrio. Praktika, pp. 47-51 [In Greek with English abstract].

ARTELARI, R. & GEORGIOU, O. 2003: Biosystematic study of the genus *Limonium* (Plumbaginaceae) in the Aegean area, Greece. III. *Limonium* on the islands Kithira and Antikithira and the surrounding islets. - *Nordic J. Bot.* **22(4)**: 483-501.

ARTELARI, R. & KAMARI, G. 1986: A karyological study of ten *Limonium* species (Plumbaginaceae) endemic in the Ionian area, Greece. - *Willdenowia* **15(2)**: 497-513.

ARTELARI, R. & KAMARI, G. 1989: The genus *Sternbergia* in Greece: Taxonomy and karyology. In Phitos, D. (ed.): Abstracts. Sixth OPTIMA meeting, Delphi 10-16 Sept. 1989, p. 33.

ARTELARI, R. & KAMARI, G. 1991: The genus *Sternbergia* (Amaryllidaceae) in Greece: taxonomy and karyology. II. - *Bot. Chron. (Patras)* **10**: 239-251.

ARTELARI, R. & KAMARI, G. 1995: *Limonium kardamylii* (Plumbaginaceae), a new species from S Peloponnisos (Greece). - *Phyton (Horn)* **35(1)**: 131-137.

ARTELARI, R. & KAMARI, G. 2000: *Limonium messeniacum* (Plumbaginaceae), a new species from S. Peloponnisos (Greece). - *Bot. Chron. (Patras)* **13**: 45-49.

ARTIUSHENKO, Z.T. 1966: A critical review of the genus *Galanthus* L. [In Russian with English summary]. - *Bot. Zhurn. (Moscow & Leningrad)* **51(10)**: 1437-1451.

ARTIUSHENKO, Z.T. 1967: Taxonomy of the genus *Galanthus* L. [Translated from Bot. Žurn.]. - *Daffodil Tulip Year Book* **32**: 62-82.

ARTIUSHENKO, Z.T. 1975: *Galanthus* L. (Amaryllidaceae) in Greece. - *Ann. Mus. Goulandris* **2**: 9-21.

ARTS-DAMLER, TH. 1960: Cytogenetical studies on *Verbascum*-species and their hybrids. – *Genetica* **31**: 241-328.

ARYA, I.D., RAMA RAO, S. & RAINA, S.N. 1988: Cytomorphological studies of *Trigonella foenum- graecum*. Autotetraploids in three (C1, C2, C3) generations. - *Cytologia* **53(3)**: 525-534.

ARYAVAND, A. 1975: Contribution à l'étude cytotaxinomique de quelques crucifères de l'Iran et de la Turquie. - *Bull. Soc. Neuchâteloise Sci. Nat.* **98**: 43-58.

ARYAVAND, A. 1978: Contribution à l'étude cytotaxinomique des crucifères de l'Iran. II. - *Bull. Soc. Neuchâteloise Sci. Nat.* **101**: 95-106.

AŠČERIĆ-TODD, I. & AL. (EDS) 2018: Travellers in Ottoman lands. The Botanical Legacy. - Oxford: ASTENE and Archaeopress Publishing Ltd, 379 pp.

ASCHERSON, P. 1872: Kleine phytographische Bemerkungen. - *Bot. Zeitung (Berlin)* **30**: 629-632.

ASCHERSON, P. 1881: Rückreise von Alexandrien nach Berlin. - *Verh. Bot. Vereins Prov. Brandenburg* **22**: 50-57.

ASCHERSON, P. 1883: Geographische Verbreitung der bisher aus der Cyrenaica bekannten Pflanzen. - *Sitzungsber. Ges. Naturf. Freunde Berlin* **9**: 150-154.

ASCHERSON, P. 1906: Nachrichten über das Leben von Wilhelm von Spruner. - *Bull. Herb. Boissier, Ser. 2* **6(7)**: 582-584.

ASCHERSON, P. 1910: Die Heimat der Reseda (*Reseda odorata* L.). - *Verh. Bot. Vereins Prov. Brandenburg* **51**: 129-131.

ASCHERSON, P. & & AL. 1896-1939: Synopsis der mitteleuropäischen Flora. - Leipzig: Engelmann/Borntraeger, 12 vols. [for details, see TL-2 1: 74-77, 1976].

ASENOV, A. 2015a: Condition and threats to the populations of three Balkan endemic plants on Mt Falakron, North-Eastern Greece. In Bogdanović, S. & Jogan, N.: Book of abstracts - 6th Balkan Botanical Congress, p. 46. - Rijeka.

ASENOV, A. 2015b: Report 9. In Vladimirov, V., Dane, F. & Tan, Kit (eds): New floristic records in the Balkans: 28. - *Phytol. Balcan.* **21(3)**: 368-369.

ASENOV, A.I. & PAVLOVA, D.K. 2009: The high-altitude serpentine flora of Mt Belasitsa (Bulgaria). - *Phytol. Balcan.* **15(2)**: 191-198.

ASHTON, P.A. & ABBOTT, R.J. 1992: Isozyme evidence and the origin of *Senecio vulgaris* (Compositae). - *Pl. Syst. Evol.* **179(3-4)**: 167-174.

ASKER, S. 1970a: Apomixis and sexuality in the *Potentilla argentea* complex. I. Crosses with other s pecies. - *Hereditas (Lund)* **66(1)**: 127-144.

ASKER, S. 1970b: Apomixis and sexuality in the *Potentilla argentea* complex. II. Crosses within the complex. - *Hereditas (Lund)* **66(2)**: 189-204.

ASKER, S. 1985: Chromosome studies in *Potentilla*. - *Hereditas (Lund)* **102(2)**: 289-292.

ASPLAND, V. 1994: Expedition to Crete 1994. - *Cyclamen* **18(2)**: 42-44 + 2 plates.

ASSADI, M. & RUNEMARK, H. 1995: Hybridisation, genomic constitution and generic delimitation in *Elymus* s.l. (Poaceae: Triticeae). - *Pl. Syst. Evol.* **194(3-4)**: 189-205.

ASSYOV, B. & AL. (EDS) 2006: Conspectus of the Bulgarian vascular flora. Distribution maps and floristic elements. Third revised and enlarged edition [In Bulgarian and English]. - Sofia: Bulgarian Biodiversity Foundation, 452 pp.

ASSYOV, B. & AL. (EDS) 2012: Conspectus of the Bulgarian vascular flora. Distribution maps and floristic elements. 4th ed. - Sofia: Bulgarian Biodiversity Foundation, 488 pp.

ATCHLEY, S.C. 1938: Wild flowers of Attica. Illustrated with coloured plates by W.O. Everett. - Oxford: Clarendon Press, xix + 60 pp. + 21 pp. of colour plates.

ATHANASIADIS, N. 1975: Zur postglazialen Vegetationsentwicklung von Litochoro Katerinis und Pertouli Trikalon (Griechenland). - *Flora, Morphol. Geobot. Oekophysiol.* **164(1)**: 99-132.

ATHANASIADIS, N. 1977: *Gentianella bulgarica* (Velen.) J. Holub. Eine neue Art der griechischen Flora [In Greek with German summary]. - *Sci. Ann. Fac. Phys. Math. Aristot. Univ. Thessalonika* **1977**: 557-567.

ATHANASIADIS, N. & AL. 1993: Zur postglazialen Vegetationsentwicklung des Rhodopi-Gebirges (Elatias Dramas - Griechenland). - *Diss. Bot.* **196**: 427-437.

ATHANASIADIS, N. & AL. 2000: New localities of two rare species of the Greek flora. - *Bot. Chron. (Patras)* **13**: 271-272.

ATHANASIADIS, N. & DROSSOS, E.G. 1990: Flora and vegetation of Mount Païko [In Greek with English summary]. - *Sci. Ann. Dept. Forest. Nat. Environm. Aristot. Univ. Thessalonika* **LC/1(1)**: 35-149.

ATHANASIADIS, N. & DROSSOS, E.G. 1992: Leucojo-Fraxinetum parvifoliae Glavač 59 and Pruno- Fraxinetum Oberdorfer 53 of the Thessalian Pinios Delta (Greece). - *Bot. Helv.* **102(2)**: 159-170.

ATHANASIADIS, N. & ELEFTHERIADOU, E. 1989: *Alnus incana* (L.) Moench subsp. *incana*. A new species and subspecies for the Greek flora [In Greek with English summary]. - *Sci. Ann. Dept. Forest. Nat. Environm. Aristot. Univ. Thessalonika* **LB/2(19)**: 389-402.

ATHANASIADIS, N. & ELEFTHERIADOU, E. 1995: Flora of *Quercus trojana* Webb (*Quercus macedonica* DC.) forests in division of Kozani [In Greek]. - *Sci. Ann. Dept. Forest. Nat. Environm. Aristot. Univ. Thessalonika* **LH/1(37)**: 125-158.

ATHANASIADIS, N. & GERASIMIDIS, A. 1978: *Drosera rotundifolia* L., *Drosera intermedia* Hayne, zwei neue Arten der griechischen Flora. - *Sci. Ann. Fac. Agric. Forest. Univ. Thessalonika* **21(3)**: 65-82.

ATHANASIADIS, N. & GERASIMIDIS, A. 1981: *Ribes alpinum* L. a new species of Greek flora and a new range of appearance of *Ribes multiflorum* Kit. [In Greek with English summary]. - *Sci. Ann. Fac. Agric. Forest. Univ. Thessalonika* **KD-Annex(9)**: 285-300.

ATHANASIADIS, N. & GERASIMIDIS, A. 1985: Forest ecosystem of Strofilia, NW Peloponnesos, and its plantsociological units [In Greek with English summary]. - *Sci. Ann. Dept. Forest. Nat. Environm. Aristot. Univ. Thessalonika* **KH(10)**: 245-263 + 1 table.

ATHANASIADIS, N. & GERASIMIDIS, A. 1986: Zur postglazialen Vegetationsentwicklung von Boras-Gebirge (Almopia - Griechenland) [In Greek with German summary]. - *Sci. Ann. Dept. Forest. Nat. Environm. Aristot. Univ. Thessalonika* **KF(4)**: 211-249 + 1 table.

ATHANASIADIS, N. & GERASIMIDIS, A. 1987: Zur postglazialen Vegetationsentwicklung von Paikon-Gebirge (Nord Griechenland) [In Greek with German summary]. - *Sci. Ann. Dept. Forest. Nat. Environm. Aristot. Univ. Thessalonika* **L(11)**: 403-445 + 1 table.

ATHANASIADIS, N. & THEODOROPOULOS, K. 1990: Contribution to the knowledge of the Mount Cholomon flora [In Greek with English summary]. - *Sci. Ann. Dept. Forest. Nat. Environm. Aristot. Univ. Thessalonika* **LC/2(9)**: 167-211.

ATHANASIADIS, N., THEODOROPOULOS, K. & ELEFTHERIADOU, E. 1993: Contribution to the knowledge of the Mount Krusia flora [In Greek]. - *Sci. Ann. Dept. Forest. Nat. Environm. Aristot. Univ. Thessalonika* **36 (2)**: 255-290.

ATHANASIADIS, N., THEODOROPOULOS, K. & ELEFTHERIADOU, E. 2001: Phytosociological analysis of the Taxo-Fagetum Etter 1947 in the NE Chalkidiki. In Kamari, Psaras & Georgiadis (eds): Proceedings, 8th scientific congress, Greek Botanical Society, Patras October 5-8, 2000, pp. 42-50 [In Greek with English abstract].

ATHANASIOU, K. 1988: Some cytogeographical notes on the genus *Tulipa* L. (Liliaceae) in Greece. - *Bot. Chron. (Patras)* **8(1-2)**: 51-59.

ATHANASIOU, K. & KAMARI, G. 1992: Reports (61-63). In Kamari, G., Felber, F. & Garbari, F. (eds): Mediterranean chromosome number reports - 2. - *Fl. Medit.* **2**: 236-239.

ATHANASIOU, K. & KAMARI, G. 1993: Disploidy and karyotype differentiation in the genus *Biarum* Schott (Araceae) in Greece. In Petrova, A.V. (ed.): Abstracts from the VII OPTIMA meeting in Borovetz, Bulgaria, 18-30 July 1993, p. 26.

ATHANASIOU, K. & YANNITSAROS, A. 1995: Taxonomical revision of the genus *Biarum* Schott (Araceae) in Greece. In: Proc., 5th sci. conf., Hellenic Bot. Soc., Delphi 21-23 October 1994, pp. 273-276 [In Greek with English summary]. - Thessaloniki: Univ. Studio Press.

ATKINSON, M.D. 1992: Biological Flora of the British Isles. *Betula pendula* Roth (*B. verrucosa* Ehrh.) and *B. pubescens* Ehrh. - *J. Ecol.* **80(4)**: 837-870.

ATKINSON, M.D. & ATKINSON, E. 2002: Biological Flora of the British Isles. *Sambucus nigra* L. - *J. Ecol.* **90(5)**: 895-923.

ATSMON, D. & FEINBRUN, N. 1960: Chromosome counts in Israeli Cistaceae. - *Caryologia* **13(1)**: 240-246 + 1 plate.

AUBIN, P. 1984: *Garidella nigellastrum* L., plante enigmatique (Renonculacees). - *Bull. Mens. Soc. Linn. Soc. Bot. Lyon* **53(5)**: 185-186.

AUCHER-ÉLOY, P.M.R. 1843: Relations de voyages en Orient de 1830 à 1838. - Paris (Roret): Première partie pp. i-xxxi +1-352, seconde partie pp. i-iii + 353-775.

AUGUSTINOS, A. & AL. 2014: Genetic variation in three closely related *Minuartia* (Caryophyllaceae) species endemic to Greece: Implications for conservation management. - *Folia Geobot.* **49(4)**: 603-621.

AUQUIER, P. & DUVIGNEAUD, J. 1976: Sur la distinction des *Aristida adscensionis* L. et *A. caerulescens* Desf. - *Bull. Soc. Échange Pl. Vasc. Eur. Occid. Bassin Médit.* **16**: 133-138.

AURICH, CH. & PODLECH, D. 1989: Zur Gliederung von *Asteriscus spinosus* (L.) Schultz Bip. sensu lato (Compositae). - *Mitt. Bot. Staatssamml. München* **28**: 239-296.

AUSTIN, D.F. 2000: A revision of *Cressa* L. (Convolvulaceae). - *Bot. J. Linn. Soc.* **133(1)**: 27-39.

AUTHIER, P. 1988a: *Conringia planisiliqua* Fischer et C.A. Meyer (Cruciferae). Espèce irano-touranienne, nouvelle pour la flore grecque et européenne. - *Bull. Soc. Bot. France, Lett. Bot.* **135(2)**: 191-197.

AUTHIER, P. 1988b: Les campanulacées du Mont Timfi. - Paris: Privately printed, 70 pp.

AUTHIER, P. 1989: *Physalis ixocarpa* Brot. ex Hornem. (Solanaceae): une nouvelle espèce adventice pour la flore grecque. - *Bull. Soc. Bot. France, Lett. Bot.* **136(2)**: 159-163.

AUTHIER, P. 1991: Contribution à l'étude de la flore de la région des Monts Timfi (Epire, nord-ouest Grèce) (I). - *Bull. Soc. Bot. France, Lett. Bot.* **138(1)**: 71-78.

AUTHIER, P. 1992: Contribution à l'étude de la flore de la région des Monts Timfi (Epire, nord-ouest Grèce) (II). A propos du *Silene remotiflora* Vis. (Caryophyllaceae) en Grèce. - *Bull. Soc. Bot. France, Lett. Bot.* **139(3)**: 265-274.

AUTHIER, P. 1994a: Contributions à "Flora hellenica": la flore de la région des Monts Timfi (Epire, nord-ouest Grèce) (I). - *Acta Bot. Gallica* **140(5)**: 525-533.

AUTHIER, P. 1994b: Contributions à "Flora hellenica": la flore de la région des Monts Timfi (Parc national du Vikos-Aoos, Epire, nord-ouest Grèce) (II). - *Candollea* **49(2)**: 483-497.

AUTHIER, P. 1995: Catalogue commenté des plantes vasculaires de la région des Monts Timfi (Epire, nord-ouest Grèce) (Parc national du Vikos-Aoos et environs). - Paris: Privately printed, 143 pp.

AUTHIER, P. 1996: Catalogue commenté de la flore de la région des monts Timfi (parc national du Vikos-Aoos et environs - Epire - Nord-ouest Grèce). 1. Ranunculaceae-Fumariaceae. - *J. Bot. Soc. Bot. France* **6**: 57-77.

AUTHIER, P. 1997a: Flore rudérale de la ville de Konitsa (Grèce). - *J.A.T.B.A.* **39(2)**: 449-454.

AUTHIER, P. 1997b: *Carex phyllostachys* C.A. Meyer, une rare et intéressante espèce de la flore de Grèce. - *Lagascalia* **19(1-2)**: 927-936.

AUTHIER, P. 1998a: *Rindera graeca* (A. DC.) Boiss. & Heldr., une Boraginée endémique des montagnes de Grèce. - *Pl. Mont. Rocaille* **12(186)**: 369-370.

AUTHIER, P. 1998b: Catalogue commenté de la flore de la région des monts Timfi (parc national du Vikos-Aoos et environs - Epire - Nord-ouest Grèce). 2. Brassicaceae. - *J. Bot. Soc. Bot. France* **8**: 55-86.

AUTHIER, P. 2000a: Catalogue commenté de la flore de la région des monts Timfi (Parc national du Vikos-Aoos et environs, Epire, Nord-ouest Grèce). 2. Cyperaceae. - *Bot. Chron. (Patras)* **13**: 341-356.

AUTHIER, P. 2000b: Jany Renz 1907-1999 [obituary]. - *Orchidophile (Ansières)* **142**: 134.

AUTHIER, P. 2000c: Catalogue commenté de la flore de la région des monts Timfi (parc national du Vikos-Aoos et environs - Epire - Nord-ouest Grèce). 4. Boraginaceae. - *Candollea* **55(1)**: 153-178.

AUTHIER, P. 2000d: *Pedicularis hoermanniana* K. Maly (Scrophulariaceae), une "bonne" espèce de la flore de Grèce. - *Webbia* **55(1)**: 37-45.

AUTHIER, P. 2000e: Catalogue commenté de la flore de la région des monts Timfi (parc national du Vikos-Aoos et environs - Epire - Nord-ouest Grèce). 5 ["3"]. Fabaceae. - *J. Bot. Soc. Bot. France* **9**: 35-75.

AUTHIER, P. 2001a: Catalogue commenté de la flore de la région des monts Timfi (parc national du Vikos-Aoos et environs - Epire - Nord-ouest Grèce). 6. Verbenaceae à Solanaceae. - *J. Bot. Soc. Bot. France* **14**: 3-36.

AUTHIER, P. 2001b: Un *Ophrys* gréco-albanais: l'*Ophrys helenae* (1ière partie). - *Orchidophile (Ansières)* **148**: 171-178.

AUTHIER, P. 2001c: Un *Ophrys* gréco-albanais: l'*Ophrys helenae* (2e partie). - *Orchidophile (Ansières)* **149**: 209-216.

AUTHIER, P. 2002a: Catalogue commenté de la flore de la région des monts Timfi (parc national du Vikos-Aoos et environs - Epire - Nord-ouest Grèce). 7. Liliaceae. - *Candollea* **56(2)**: 209-244.

AUTHIER, P. 2002b: Catalogue commenté de la flore de la région des monts Timfi (parc national du Vikos-Aoos et environs - Epire - Nord-ouest Grèce). 8. Scrophulariaceae à Lentibulariaceae. - *J. Bot. Soc. Bot. France* **16**: 3-32.

AUTHIER, P. 2002c: Catalogue commenté de la flore de la région des monts Timfi (parc national du Vikos-Aoos et environs - Epire - Nord-ouest Grèce). 9. Apiaceae. - *Candollea* **57(2)**: 359-409.

AUTHIER, P. 2014: Reports 1-24. In Vladimirov, V. & Tan, Kit (eds): New floristic records in the Balkans: 24. - *Phytol. Balcan.* **20(1)**: 100-105.

AUTHIER, P. 2021: Catalogue commenté de la flore de la région des monts Timfi (Parc National du Vikos-Aoos et environs, Épire, nord-ouest Grèce). 13. Asteraceae 2ème partie et fin sous-famille des Liguliflorae (= Cichorioideae). - *Saussurea* **50**: 55-104.

AUTHIER, P. & AL. 2001: Herborisations dans le massif du Timfi (Épire, nord-ouest Grèce). Première partie. - *Pl. Mont. Rocaille* **13(200)**: 330-352 + front cover.

AUTHIER, P. & AL. 2002a: Herborisations dans le massif du Timfi (Epire, nord-ouest Grèce). Deuxième partie. - *Pl. Mont. Rocaille* **13(201)**: 378-390.

AUTHIER, P. & AL. 2002b: Herborisations dans le massif du Timfi (Épire, nord-ouest Grèce). 3. partie et fin. - *Pl. Mont. Rocaille* **13(202)**: 404-415.

AUTHIER, P. & COVILLOT, J. 2011: Catalogue actualisé des plantes de l'île de Rhodes (Grèce). - *Saussurea* **41**: 131-170.

AUTHIER, P. & GASNIER, D. 1995: Une orchidée bien connue, nouvelle pour la Grèce: *Goodyera repens* (L.) R. Br. - *Orchidophile (Ansières)* **26(119)**: 208-213.

AUTHIER, P. & TAN, KIT 2015: Report 10. In Vladimirov, V., Dane, F. & Tan, Kit (eds): New floristic records in the Balkans: 28. - *Phytol. Balcan.* **21(3)**: 269.

AVOLIO, M. 1996: *Euphorbia dendroides* L. - *Euphorbia J.* **10**: 54-59.

AVTZIS, N.D. & AL. 2007: A contribution to the natural distribution of *Aesculus hippocastanum* (Hippocastanaceae) in Greece. - *Phytol. Balcan.* **13(2)**: 183-187.

AWMACK, C.S. & LOCK, J.M. 2002: The genus *Alhagi* (Leguminosae: Papilionoideae) in the Middle East. - *Kew Bull.* **57(2)**: 435-443.

AXIOTIS, E., AXIOTIS, M. & TAN, KIT 2018: Reports 1-3. In Vladimirov, V., Aybeke, M. & Tan, Kit (eds): New floristic records in the Balkans: 37. - *Phytol. Balcan.* **24(3)**: 398.

AXIOTIS, E., AXIOTIS, M. & TAN, KIT 2020a: Report 1. In Vladimirov, V., Aybeke, M. & Tan, Kit (eds): New floristic records in the Balkans: 42. - *Phytol. Balcan.* **26(2)**: 188.

AXIOTIS, E., AXIOTIS, M. & TAN, KIT 2020b: Report 1. In Vladimirov, V., Aybeke, M. & Tan, Kit: New floristic records in the Balkans: 43. - *Phytol. Balcan.* **26(3)**: 538.

AXIOTIS, E., VERLOOVE, F. & TAN, KIT 2019: Report 1. In Vladimirov, V. & al. (eds): New floristic records in the Balkans: 38. - *Phytol. Balcan.* **25(1)**: 98.

AXIOTIS, M. & AXIOTIS, V. 2011: Fitá tis Elládas. I érevna sti Lésvo. Votanikí - Chrísis - Toxikótita. [In Greek]. - Mitilini: Entelechia, 717 pp.

AYBEKE, M. & DANE, F. 2017: *Onobrychis mehmetchiquii* (Fabaceae) sp. nov, a new species from European Turkey. - *Phytotaxa* **298(1)**: 96-100.

AYDIN, A. 1978: [Chromosome numbers in *Stachys* from the Istanbul area]. In Löve, Á. (ed.): IOPB chromosome number reports LXI. - *Taxon* **27(4)**: 375-376.

AYERS, T.J. 1991: The cultivated species of *Phyteuma* and *Asyneuma* (Campanulaceae). – *Baileya* **23(3)**: 126-138.

AYKURT, C. & AL. 2020: Taxonomic treatment and phylogenetic analysis of the family Potamogetonaceae in Turkey. - *Taxon* **69(6)**: 1172-1190.

AYKURT, C. & AL. 2021: *Omphalodes nedimeae* (Boraginaceae), a new species from the Mediterranean region of Turkey. - *Phytotaxa* **498(4)**: 242-254.

AYMERICH, P. & AL. 2012: *Potamogeton schweinfurthii* in the Iberian Peninsula. - *Anales Jard. Bot. Madrid* **69(2)**: 187-192.

AYMONIN, G.G. 1959: Essai d'une monographie écologique du *Daphne cneorum* L. - *Rev. Gén. Bot.* **66**: 281-328.

AYMONIN, G.G. 1974a: Polymorphisme chez le *Thymelaea Tartonraira* (L.) All. et position du "Passerina" *Thomasii* Duby de la Corse. - *Bull. Soc. Bot. France* **121, 95. Sess. extr.**: 41-43.

AYMONIN, G.G. 1974b: L'*Asplenium obovatum* Viv., plante de Corse. Observations sur la définition de l'espèce. - *Bull. Soc. Bot. France* **121, 95. Sess. extr.**: 61-65.

AYMONIN, G.G. 1976: Deux singulières chênaies en "zone bioclimatique méditerranéenne". - *Bull. Soc. Bot. France* - **123(1-2)**: 79-81. [Also distributed as *OPTIMA leaflet* no. 21].

AYMONIN, G.G. & CHARPIN, A. 1999: Bibliographie sélective des flores de France - nationales, régionales et départementales. - *J. Bot. Soc. Bot. France* **10**: 1-88.

AYTAÇ, Z., DUMAN, H. & EKICI, M. 2016: Two new species of *Achillea* L. (Asteraceae) from Turkey. - *Turk. J. Bot.* **40(4)**: 373-379.

AYTAÇ, Z. 1998: A new species of *Silene* (Caryophyllaceae) from Anatolia (Turkey). - *Thaiszia* **8(1)**: 7-11.

AYTAÇ, Z. 2000: The genus *Ebenus* L. (Leguminosae/Fabaceae) in Turkey. - *Karaca Arbor. Mag.* **5 (4)**: 145-171.

AYTAÇ, Z. & AKSOY, A. 2000a: A new *Sideritis* species (Labiatae) from Turkey. - *Fl. Medit.* **10**: 181-184.

AYTAÇ, Z. & AKSOY, A. 2000b: A new species of *Bornmuellera* Hausskn. (Brassicaceae) from south Anatolia, Turkey. - *Bot. J. Linn. Soc.* **134(3)**: 485-490.

AYTAÇ, Z. & MILL, R.R. 2005: Two new species of Boraginaceae (tribe Cynoglosseae) from Turkey. - *Edinburgh J. Bot.* **61(2-3)**: 109-118.

AYYAD, M.A. & HILMY, S. 1974: Morphological variation in *Asphodelus microcarpus*. - *Egypt. J. Bot.* **16 (1-3)**: 387-392.

AZER, S.A. 2013: Taxonomic revision of genus *Lemna* L. (Lemnaceae Gray) in Egypt. - *Ann. Agric. Sci. (Ain Shams Univ.)* **58(2)**: 257-263.

AZIZIAN, D. & MOORE, D.M. 1982: Morphological and palynological studies in *Phlomis* L., *Eremostachys* Bunge and *Paraphlomis* Prain (Labiatae). - *Bot. J. Linn. Soc.* **85(4)**: 225-248.

AZNAVOUR, G.V. 1897: Note sur la flore des environs de Constantinople. - *Bull. Soc. Bot. France* **44**: 164-177.

AZNAVOUR, G.V. 1899: Nouvelle contribution à la flore des environs de Constantinople. - *Bull. Soc. Bot. France* **46**: 135-153.

AZNAVOUR, G.V. 1902a: Plantes récueillies par M.F.X. Lobry dans l'île de Syra. - *Magyar Bot. Lapok* **1 (7)**: 193-196.

AZNAVOUR, G.V. 1902b: Enumération d'espèces nouvelles pour la flore de Constantinople, accompagnée de notes sur quelques plantes peu connues ou isuffisamment décrites qui se recontrent à l'état spontané aux environs de cette ville. - *Magyar Bot. Lapok* **1(10)**: 291-304.

AZNAVOUR, G.V. 1903: Enumération d'espèces nouvelles pour la flore de Constantinople ... [cont.]. - *Magyar Bot. Lapok* **2(5)**: 137-144.

AZNAVOUR, G.V. 1904: Enumération d'espèces nouvelles pour la flore de Constantinople ... [cont.]. - *Magyar Bot. Lapok* **3(1-2)**: 2-9.

AZNAVOUR, G.V. 1905: Enumération d'espèces nouvelles pour la flore de Constantinople ... [cont.]. - *Magyar Bot. Lapok* **4(6-7)**: 136-143.

AZNAVOUR, G.V. 1906: Enumération d'espèces nouvelles pour la flore de Constantinople ... [cont.]. - *Magyar Bot. Lapok* **5(5-7)**: 156-169.

AZNAVOUR, G.V. 1907: Deux plantes nouvelles de la flore constantinopolitaine. - *Magyar Bot. Lapok* **6 (1-4)**: 7-11.

AZNAVOUR, G.V. 1911: Énumération d'espèces nouvelles pour la flore de Constantinople ... [end]. - *Magyar Bot. Lapok* **10(1-3)**: 10-22.

AZNAVOUR, G.V. 1913: Nouveaux matériaux pour la flore de Constantinople. - *Magyar Bot. Lapok* **12 (6-7)**: 156-185.

BAATOUT, H. 1982: Analyse du polymorphisme dans le complexe *Hedysarum spinosissimum* L. - *Bull. Soc. Bot. France, Lett. Bot.* **129(2)**: 155-165.

BABALONAS, D. 1976: Über die Vegetation der östlichen Kavala-Küsten (Nordgriechenland). I. Salzbödenvegetation. - *Sci. Ann. Fac. Phys. Math. Aristot. Univ. Thessalonika* **16**: 295-307.

BABALONAS, D. 1977: Zwei in Griechenland eingeschleppte amerikanische Pflanzenarten: *Cenchrus pauciflorus* und *Heliotropium curassavicum*. - *Ann. Mus. Goulandris* **3**: 19-22.

BABALONAS, D. 1979: Pflanzensoziologisches Studium der Vegetation des Evros Deltas (Aisision Delta) [In Greek with German summary]. - *Sci. Ann. Fac. Phys. Math. Aristot. Univ. Thessalonika* **19**: 158 pp. + 2 fold-out tables.

BABALONAS, D. 1980: Vegetationseinheiten und Vegetationskartierung in dem Mündungsgebiet des Flusses Evros. - *Feddes Repert.* **91(9-10)**: 615-627.

BABALONAS, D. 1981: Floristischer Katalog des Mündungsgebiets des Evros. - *Candollea* **36(1)**: 251-269.

BABALONAS, D. 1983: Neue Fundorte für seltene Pflanzen in Nordgriechenland. - *Ann. Mus. Goulandris* **6**: 17-25.

BABALONAS, D. 1984a: Chorologie und Soziologie von *Leymus racemosus* subsp. *sabulosus* (Poaceae). - *Phyton (Horn)* **24(2)**: 165-171.

BABALONAS, D. 1984b: *Armeria maritima* subsp. *smolikana*, ein neues Taxon aus NW-Griechenland. - *Willdenowia* **14(1)**: 61-64.

BABALONAS, D. 1984c: Floristisches aus dem Vourinos-Gebirge (N-Griechenland). - *Willdenowia* **14 (1)**: 65-70.

BABALONAS, D. 1984d: Zur Kenntnis der Flora und Vegetation auf Serpentin-Standorten Nordgriechenlands. I. Serpentinvegetation im Voras-Gebirge. - *Feddes Repert.* **95(9-10)**: 687-697.

BABALONAS, D. 1988: Zur Kenntnis der Flora und Vegetation auf Serpentinstandorten Nordgriechenlands. II. Serpentinvegetation auf submontanen Stufen. - *Bot. Jahrb. Syst.* **110(2)**: 145-156.

BABALONAS, D. 1989: Beitrag zur Flora des serpentinischen Vourinos-Gebirges (Nordgriechenland). - *Willdenowia* **18(2)**: 387-398.

BABALONAS, D. 1995: Mavrovouni Krousion. Ikotouristikos odigos. - Kilkis: 239 pp.

BABALONAS, D., KARAGIANNAKIDOU, V. & DROSSOS, E.G. 1989: Contribution to the vegetation of the Vistonis lake [In Greek with English abstract]. - *Bios (Thessaloniki)* **1989**: 9-17.

BABALONAS, D., KARAGIANNAKIDOU, V. & KOKKINI, S. 1980: Untersuchungen über die Salzwiesen des ausgetrockneten Lantza-Salzsees (Nordgriechenland). - *Feddes Repert.* **91(1-2)**: 117-126.

BABALONAS, D. & KARATAGLIS, S. 1982: Some North Greece plant species in relation to the serpentine problem [In Greek with English abstract]. - *Bot. Chron. (Patras)* **2(2)**: 87-92.

BABALONAS, D., KARATAGLIS, S. & KABASSAKALIS, V. 1984: The ecology of plant populations growing on serpentine soils. III. Some plant species from North Greece in relation to the serpentine problem. - *Phyton (Horn)* **24(2)**: 225-238.

BABALONAS, D., KONSTANTINOS, M. & CHARALAMBIDIS, S. 1998: I chlorida tou Agiou Orous. - Thessaloniki: 155 pp.

BABALONAS, D. & MOUSTAKAS, M. 1986: *Agropyron strigosum* (Poaceae), a Pontic element found in Northern Greece. - *Willdenowia* **16(1)**: 69-71.

BABALONAS, D. & PAPANICOLAOU, K. 1984: A new species of *Iris* (Iridaceae) from NC Greece. - *Willdenowia* **14(1)**: 71-74.

BABALONAS, D. & PAPASTERGIADOU, E. 1989: The waterfern *Salvinia natans* (L.) All. in the Kerkini lake (North Greece). - *Arch. Hydrobiol.* **116**: 487-498.

BABALONAS, D. & PAPASTERGIADOU, E. 1990: Ein halophiles Gesellschaftsrelikt im griechischen Binnenland. - *Tuexenia* **10**: 115-121.

BABALONAS, D. & PAVLIDES, G. 1989: Contribution to the vegetation of aquatic plants of lake Mikri Prespa [In Greek with English abstract]. - *Bios (Thessaloniki)* **1989**: 19-29.

BABALONAS, D., SYKORA, K.V. & PAPASTERGIADOU, E. 1995: Review of plant communities from Greek dunes and salt-marshes, a preliminary summarizing list. - *Ann. Bot. (Rome)* **53**: 107-117.

BABCOCK, E.B. 1938: *Crepis foetida* and four closely related species. - *J. Bot. (London)* **76**: 202-211.

BABCOCK, E.B. 1947a: The genus *Crepis*. Part one. - *Univ. Calif. Publ. Bot.* **21**: 1-197 + 1 plate.

BABCOCK, E.B. 1947b: The genus Crepis. Part two. - *Univ. Calif. Publ. Bot.* **22**: 199-1030 + 36 plates.

BABCOCK, E.B. 1947c: Cytogenetics and speciation in *Crepis*. - *Advances Genet.* **1**: 69-93.

BABCOCK, E.B. 1949: Supplementary notes on *Crepis* I. - Some new or noteworthy species of the Mediterranean region. - *Mém. Soc. Hist. Nat. Afrique N., Hors Sér.* **2**: 9-20 + 2 plates.

BABCOCK, E.B. & CAMERON, D.R. 1934: Chromosomes and phylogeny in *Crepis*. II. The relationships of one hundred eight species. - *Univ. Calif. Publ. Agric. Sci.* **6(11)**: 287-324.

BABCOCK, E.B. & CAVE, M.S. 1938: A study of intra- and interspecific relations of *Crepis foetida* L. - *Z. Indukt. Abstammungs- Vererbungsl.* **75(1)**: 124-160.

BABIJ, V. & AL. 1997: A contribution to the taxonomy of *Stellaria nemorum* aggregat. In: Abstracts from the First Balkan Botanical Congress, Thessaloniki, September 19-22, 1997, p. 4. - Thessaloniki.

BABORKA, M. 1994: Bestäuber von Nigritellen sowie Beschreibung eines Bastards zwischen *Nigritella widderi* und *Gymnadenia*. - *J. Eur. Orch.* **26(1)**: 88-93.

BABORKA, M. 1998: Bestätigung von *Ophrys umbilicata* Desf. auf Rhodos sowie Neufund von

Ophrys cretica (Vierh.) E. Nelson im Nord-West-Teil der Insel. - *Orchidee (Hamburg)* **49(4)**: 183-184.

BABU, C.R. 1969: (301) Proposal to conserve the generic name sub 7305 *Micromeria* Benth. (1829) ... - *Taxon* **18(6)**: 733-734. [approved, cf. Taxon 21(4): 531, 1972].

BACCHETTA, G. & AL. 2010: Taxonomic revision of the *Dianthus sylvestris* group (Caryophyllaceae) in central-southern Italy, Sicily and Sardinia. - *Nordic J. Bot.* **28(2)**: 137-173.

BAČIČ, T., FRAJMAN, B. & KOCE, J.D. 2015: Diversification of *Luzula* sect. *Luzula* (Juncaceae) on the Balkan Peninsula - a cytogenetic approach. In Bogdanović, S. & Jogan, N.: Book of abstracts - 6th Balkan Botanical Congress, pp. 47-48. - Rijeka.

BAČIČ, T., FRAJMAN, B. & KOCE, J.D. 2015: Diversification of *Luzula* sect. *Luzula* (Juncaceae) on the Balkan Peninsula - a cytogenetic approach. - Folia Geobot. **51(1)**: 51-63.

BACKLUND, M. & THULIN, M. 2007: Revision of the Mediterranean species of *Plocama* (Rubiaceae). - *Taxon* **56(2)**: 516-520.

BACON, L. 1972a: Plant portraits: *Himantoglossum hircinum* (L.) Sprengel subsp. *caprinum* (Bieb.) K. Richter. - *Quart. Bull. Alpine Gard. Soc. Gr. Brit.* **40(2)**: 132, 135.

BACON, L. 1972b: Plant portraits: *Ophrys lutea* Cav. - *Quart. Bull. Alpine Gard. Soc. Gr. Brit.* **40(4)**: 295-296, 299.

BACON, L. 1973: Plant portraits: *Haplophyllum boissierianum* Vis. et Pančić, syn. *H. albanicum* (Bald.) Bornm. - *Quart. Bull. Alpine Gard. Soc. Gr. Brit.* **41(2)**: 160-162.

BACON, L. 1982: Plant portraits: *Campanula versicolor* Andrews. - *Quart. Bull. Alpine Gard. Soc. Gr. Brit.* **50(2)**: 144-147.

BACON, L. 1990: Erodiums for the rock garden. - *Quart. Bull. Alpine Gard. Soc. Gr. Brit.* **58(1)**: 65-83.

BADEN, C. 1984a: A floristic note from the E Aegean region. - *Willdenowia* **13(2)**: 321-322.

BADEN, C. 1984b: Chromosome numbers in some Greek angiosperms. (Materials for the Mountain Flora of Greece, 23). - *Willdenowia* **13(2)**: 335-336.

BADEN, C. 1984c: Chromosome numbers in the *Nepeta sibthorpii* group (Lamiaceae). (Materials for the Mountain Flora of Greece, 24). - *Willdenowia* **13(2)**: 337-340.

BADEN, C. 1985: Biosystematic studies in the *Nepeta sibthorpii* group. *N. heldreichii* included in *N. camphorata*. (Materials for the Mountain Flora of Greece, 26). - *Willdenowia* **14(2)**: 335-341.

BADEN, C. 1987: Biosystematic studies in the *Nepeta sibthorpii* group (Lamiaceae) in Greece. - *Opera Bot.* **93**: 1-54.

BADEN, C. & AL. 1989: Intergeneric hybridization between *Psathyrostachys* and *Hordeum*. - *Nordic J. Bot.* **9(4)**: 333-342.

BADER, B. & RAUS, TH. 1986: *Listera cordata* (L.) R. Br. In Greuter, W. & Raus, Th. (eds): Med- Checklist notulae, 13. - *Willdenowia* **16(1)**: 115.

BADR, A. & AL. 2002: Ancestors of white clover (*Trifolium repens* L.), as revealed by isozyme polymorphisms. - *Theor. Appl. Genet.* **106(1)**: 143-148.

BADR, A. & HAMOUD, M.A. 1985: The karyotypes of two species of *Asphodelus* L. Liliaceae and five species of *Erodium* L'Hérit. Geraniaceae. - *Egypt. J. Bot.* **28**: 145-148.

BADRÉ, F. & DESCHÂTRES, R. 1979: Les ptéridophytes de la France, liste commentée des espèces (taxinomie, cytologie, écologie et répartition générale). - *Candollea* **34(2)**: 379-457.

BADRÉ, F., DESCHÂTRES, R. & GAMISANS, J. 1986: Les ptéridophytes de la Corse. - *Bull. Mus. Natl. Hist. Nat., B, Adansonia* **8(4)**: 423-461.

BADRÉ, F., TRYON, A.F. & DESCHÂTRES, R. 1982: Les espèces du genre *Cheilanthes* Swartz (Pteridaceae, Pteridophyta) en France. - *Webbia* **36(1)**: 1-38.

BAENITZ, C. 1896: [Account of a botanical expedition to Kerkira]. - *Oesterr. Bot. Z.* **46(2)**: 79.

BAGELLA, S. & AL. 2014: Unraveling the taxonomy and nomenclature of the *Isoëtes histrix* Bory species complex (Isoetaceae, Lycopodiidae). - *Turk. J. Bot.* **39(2)**: 383-387.

BAHL, J.R. & TYAGI, B.R. 1989: Karyotype analysis of *Papaver dubium* L. - *Curr. Sci.* **58(24)**: 1384-1385.

BAILEY, R.H. 1992a: The society's expedition to Rhodes and Karpathos, 3-17 April 1991. - *Cyclamen* **15(2)**: 46-50.

BAILEY, R.H. 1992b: *Cyclamen graecum* and *C. persicum* on Rhodes and Karpathos. - *Cyclamen* **16(1)**: 13-17.

BAILEY, R.H. 1992c: *Cyclamen creticum* on Karpathos. - *Cyclamen* **16(2)**: 41-43 + 1 plate.

BAILLARGEON, G. 1988: (916) Proposal to conserve 2945 *Sinapis* (Brassicaceae) with *S. alba*, typ. cons. prop. - *Taxon* **37(4)**: 969-973. [not approved, cf. Taxon 42(3): 688, 1993].

BAILLON, H. 1866-95: Histoire des plantes. - Paris, London, Leipzig: 13 vols. [cf. TL-2 1: 97-99, 1976].

BAINBRIDGE, I. & BAINBRIDGE, C. 2002: *Aceras anthropophorum* (L.) Aiton f., P.C. - *Alpine Gardener* **70(4)**: 475-476.

BAJIĆ, D. 1963: Die Art der *Lippia nodiflora* Rich. Neu in der Flora Jugoslawiens [In Serbo-Croatian with German summary]. - *Rad. Šumarsk. Fak. Inst. Šumarstvo Sarajevu* **12(14)**: 209-211.

BAKALE, V.L. & SRINIVASU, T. 1988: Mitotic and karyological studies in *Xanthium strumarium* Linn. - *Advances Pl. Sci.* **1**: 201-207.

BAKER, E.G. 1890: Synopsis of genera and species of Malveae. - *J. Bot. (London)* **28**: 140-145, 207-213, 239-243, 339-343, 367-371.

BAKER, E.G. 1893: Synopsis of genera and species of Malveae [cont.]. - *J. Bot. (London)* **31**: 68-76, 212-217, 267-273, 334-338, 361-368.

BAKER, E.G. & SALMON, C.E. 1921: Some segregates of *Erodium cicutarium* L'Hérit. - *J. Bot. (London)* **58**: 121-127.

BAKER, G.P. 1929: Plant hunting in Crete, 1925-26-27. - *J. Roy. Hort. Soc.* **54(2)**: 387-409 + 12 plates.

BAKER, G.P. 1930: *Ramondia*, *Haberlea* and *Jankaea*. - *Quart. Bull. Alpine Gard. Soc. Gr. Brit.* **1(1)**: 14-17 + 3 plates.

BAKER, H.G. 1947: Biological Flora of the British Isles. *Melandrium* (Roehling em.) Fries. - *J. Ecol.* **35 (1-2)**: 271-292.

BAKER, H.G. 1955: *Geranium purpureum* Vill. and *G. robertianum* L. in the British flora. I. *Geranium purpureum*. - *Watsonia* **3(3)**: 160-167.

BAKER, H.G. 1956: *Geranium purpureum* Vill. and *G. robertianum* L. in the British flora. II. *Geranium robertianum*. - *Watsonia* **3(5)**: 270-279.

BAKER, H.G. 1957: Genecological studies in *Geranium* (section *Robertiana*). General considerations and the races of *G. purpureum* Vill. - *New Phytol.* **56(2)**: 172-192 + 2 plates.

BAKER, H.G. & BAKER, I. 1967: The cytotaxonomy of *Filipendula* (Rosaceae) and its implications. - *Amer. J. Bot.* **54(8)**: 1027-1034.

BAKER, J.G. 1873a: A review of the known species of *Crocus* I-V. - *Gard. Chron.* **1873**: 107-108, 179, 291-292, 434-436, 542-543.

BAKER, J.G. 1873b: A review of the known species of *Crocus* VI-IX. - *Gard. Chron.* **1873**: 609, 680, 1402-1403, 1431-1432.

BAKER, J.G. 1873c: A review of the known species of *Crocus* X-XII. - *Gard. Chron.* **1873**: 1466-1467, 1533, 1633.

BAKER, J.G. 1874: Revision of the genera and species of Tulipeae. - *J. Linn. Soc., Bot.* **14(76)**: 211-310.

BAKER, J.G. 1875a: *Crocus Boryi*. - *Bot. Mag.* **101**: tab. 6187 + 2 pp. text.

BAKER, J.G. 1875b: *Crocus veluchensis*. - *Bot. Mag.* **101**: tab. 6197 + 2 pp. text.

BAKER, J.G. 1875c: New garden plants. *Fritillaria (Monocodon) macrandra*, Baker, n. sp. - *Gard. Chron.*, n.s. **3**: 715-716.

BAKER, J.G. 1876a: A synopsis of the known species of *Iris*. - I-VI. - *Gard. Chron., n.s.* **5**: 526-527, 559, 623-624, 692, 723, 787-788.

BAKER, J.G. 1876b: A synopsis of the known species of *Iris*. - VII-IX. - *Gard. Chron., n.s.* **6**: 36-38, 143-144, 226, 323-324, 517-518.

BAKER, J.G. 1876c: A synopsis of the known species of *Iris*. - X-XIII. - *Gard. Chron., n.s.* **6**: 583-584, 614-615, 647-648, 708-710.

BAKER, J.G. 1876d: A synopsis of the known species of *Iris*. - XIV-XVI. - *Gard. Chron., n.s.* **6**: 740-741, 774-775, 806.

BAKER, J.G. 1876e: *Tulipa Hageri*. - *Bot. Mag.* **102**: tab. 6242 + 2 pp. text.
BAKER, J.G. 1877a: Systema Iridacearum. - *J. Linn. Soc., Bot.* **16(1)**: 61-180.
BAKER, J.G. 1877b: *Tulipa undulatifolia*. - *Bot. Mag.* **103**: tab. 6308 + 2 pp. text.
BAKER, J.G. 1877c: *Tulipa Orphanidea*. - *Bot. Mag.* **103**: tab. 6310 + 2 pp. text.
BAKER, J.G. 1878a: *Iris cretensis*. - *Bot. Mag.* **104**: tab. 6343 + 2 pp. text.
BAKER, J.G. 1878b: *Tulipa saxatilis*. - *Bot. Mag.* **104**: tab. 6374 + 2 pp. text.
BAKER, J.G. 1878c: Descriptions of new and little known Liliaceae. - *J. Bot. (London)* **16**: 321-326.
BAKER, J.G. 1879a: A synopsis of Colchicaceae and the aberrant tribes of Liliaceae. - *J. Linn. Soc., Bot.* **17**: 405-510.
BAKER, J.G. 1879b: *Colchicum montanum*. - *Bot. Mag.* **105**: tab. 6443 + 2 pp. text.
BAKER, J.G. 1879c: *Chionodoxa nana*. - *Bot. Mag.* **105**: tab. 6453 + 2 pp. text.
BAKER, J.G. 1888: Handbook of the Amaryllideae, including the Alstroemerieae and Agaveae. - London: George Bell & Sons, xii + 216 pp.
BAKER, J.G. 1891: *Colchicum sibthorpii*. - *Bot. Mag.* **117**: tab. 7181 + 2 pp. text.
BAKER. J.G. 1892: Handbook of the Irideae. - London: George Bell & Sons, xii + 247 pp.
BAKER, J.G. 1893: New or noteworthy plants. *Galanthus byzantinus* Baker s. sp. - *Gard. Chron., ser. 3* **13**: 226.
BAKŠAY, L. 1954: Chromosomenstudien an den ungarischen *Vicia*-Arten. - *Ann. Hist.-Nat. Mus. Natl. Hung. n.s.* **5**: 139-148.
BAKŠAY, L. 1956: Cytotaxonomical studies on the flora of Hungary. - *Ann. Hist.-Nat. Mus. Natl. Hung. n.s.* **7**: 321-334.
BAKŠAY, L. 1958: The chromosome numbers of Ponto-Mediterranean plant species. - *Ann. Hist.-Nat. Mus. Natl. Hung. n.s.* **9**: 121-125.
BALDACCI, A. 1893: Ricordi di un viaggio botanico fra Prevesa e Janina. - *Boll. Soc. Bot. Ital.* **1893**: 84-88.
BALDACCI, A. 1894a: Rivista critica della collezione botanica fatta nel 1892 in Albania. - *Malpighia* **8**: 69-87, 159-192, 278-301.
BALDACCI, A. 1894b: Contributo alla conoscenza della flora dalmata, montenegrina, albanese, epirota e greca. - *Nuovo Giorn. Bot. Ital., n.s.* **1(2)**: 90-103.
BALDACCI, A. 1894c: Monografia della sezione "Aizopsis DC." del genere *Draba* L. - *Nuovo Giorn. Bot. Ital., n.s.* **1(2)**: 103-121.
BALDACCI, A. 1895: Risultati botanici del viaggio compiuto in Creta nel 1893. - *Malpighia* **9**: 31-70, 251-279, 329-355.
BALDACCI, A. 1896a: Una corsa botanica nell'isola di Santa Maura (Leukas). - *Bull. Herb. Boissier* **4 (3)**: 201-206.
BALDACCI, A. 1896b: Rivista della collezione botanica fatta nel 1894 in Albania. - *Bull. Herb. Boissier* **4(9)**: 609-653.
BALDACCI, A. 1896c: Escursione botanica nell'Albania e nell'Epiro. - *Boll. Soc. Geogr. Ital. ser. 3* **9**: 295-297.
BALDACCI, A. 1897a: Die pflanzengeographische Karte von Mittel-Albanien und Epirus. - *Petermann's Mitth. Justus Pethers' Geogr. Anst.* **43(7-8)**: 12 pp. + 1 map.
BALDACCI, A. 1897b: Rivista della collezione botanica fatta nel 1895 in Albania. - *Nuovo Giorn. Bot. Ital., n.s.* **4(4)**: 386-419.
BALDACCI, A. 1898: Rivista della collezione botanica fatta nel 1895 in Albania (continuazione e fine). - *Nuovo Giorn. Bot. Ital., n.s.* **5(1)**: 5-44.
BALDACCI, A. 1899: Rivista della collezione botanica fatta nel 1896 in Albania. - *Nuovo Giorn. Bot. Ital., n.s.* **6(1-4)**: 5-37, 149-187, 334-356.
BALDACCI, A. 1900: Contributo alla conoscenza della flora del confine Montenegrino-Albanese. - *Mem. Reale Accad. Sci. Ist. Bologna, Cl. Sci. Fis., Ser. 5* **9**: 1-43.
BALDACCI, A. 1902: Rivista della collezione botanica fatta nel 1897 nell'Albania settentrionale. - *Mem. Reale Accad. Sci. Ist. Bologna, Cl. Sci. Fis., Ser. 5* **9**: 513-553.
BALDACCI, A. 1903: Itinerari fitogeografici del mio secondo viaggio in Creta (1899). - *Mem. Reale Accad. Sci. Ist. Bologna, Cl. Sci. Fis., Ser. 5* **10**: 253-274 + 1 map.
BALDACCI, A. 1904: Le esplorazione botaniche nell'isola di Creta nei secoli XVI et XVII. - *Atti Congr. Int. Sci. St. Roma* **10**: 81-88 + 1 map.
BALDACCI, A. 1906: Le relazioni fitogeografiche fra Creta e Karpathos. - *Mem. Reale Accad. Sci. Ist. Bologna, Cl. Sci. Fis., Ser. 6* **3**: 109-116.
BALDACCI, A. 1912: I rapporti fitogeografici fra l'isola di Creta e la Cirenaica. - *Boll. Reale Orto Bot. Palermo* **11(4)**: 193-201.
BALDACCI, A. 1917: Itinerari Albanesi (1892-1902) con uno sguardo generale all'Albania e alle sue comunicazioni stradali. - Roma: Presso la Reale Società Geografica Italiana, xxiii + 541 pp. + 6 maps.
BALDACCI, A. 1925: Le fonti della flora albanese. - *Mem. Reale Accad. Sci. Ist. Bologna, Cl. Sci. Fis., Ser. 8* **2**: 91-108.
BALDACCI, A. 1936: Sull'*Aesculus Hippocastanum* L. A proposito di un relitto dendroide eurasiatico della flora balcanica e specialmente illirico-ellenica. - *Mem. Reale Accad. Sci. Ist. Bologna, Cl. Sci. Fis., Ser. 9* **3**: 87-108 + 1 map.
BALDACCI, A. 1941: La *Pinus peuce* Gris. della penisola balcanica e la *Pinus excelsa* Wall. dell'Himalaya nel tempo e nello spazio. - *Mem. Reale Accad. Sci. Ist. Bologna, Cl. Sci. Fis., Ser. 9* **8**: 49-69 + 2 plates.
BALDACCI, A. 1945: Sugli endemismi albanesi nelle loro relazioni con la flora eurasiatica. - *Mem. Reale Accad. Sci. Ist. Bologna, Cl. Sci. Fis., Ser. 10* **2**: 14 pp.
BALDACCI, A. 1949: Sulla monografia del genere "Verbascum" del Murbeck e sull'appartenenza della priorità della scoperta del "Verbascum Baldaccii" Degen al botanico Vandas. - *Mem. Reale Accad. Sci. Ist. Bologna, Cl. Sci. Fis., Ser. 10* **6**: 8 pp.
BALDACCI, A. & BÉGUINOT, A. 1918: Contributo alla flora autumnale ed inverne dei dintorni di Vallona. - *Nuov. Giorn. Bot. Ital. , n.s.* **25**: 70-86.
BALDACCI, A. & SACCARDO, P.A. 1900: Onorio Belli e Prospero Alpino e la flora dell'isola di Creta. - *Malpighia* **14**: 140-163.
BALDINI, R.M. 1990: Numeri cromosomici per la Flora Italiana: 1164-1166. - *Inform. Bot. Ital.* **20(2-3)**: 624-626.
BALDINI, R.M. 1991: Numeri cromosomici per la Flora Italiana: 1231-1238. - *Inform. Bot. Ital.* **22(3)**: 227-236.
BALDINI, R.M. 1993a: Numeri cromosomici per la Flora Italiana: 1284-1289. - *Inform. Bot. Ital.* **24(3)**: 189-195.
BALDINI, R.M. 1993b: The genus *Phalaris* L. (Gramineae) in Italy. - *Webbia* **47(1)**: 1-53.
BALDINI, R.M. 1995: Revision of the genus *Phalaris* L. (Gramineae). - *Webbia* **49(2)**: 265-329.
BALDINI, R.M. 1996: Report (544). In Kamari, G., Felber, F. & Garbari, F. (eds): Mediterranean chromosome number reports - 5. - *Fl. Medit.* **5**: 346-350.
BALDINI, R.M. 1997: Reports (779-782). In Kamari, G., Felber, F. & Garbari, F. (eds): Mediterranean chromosome number reports - 7. - *Fl. Medit.* **7**: 198-204.
BALDINI, R.M. 1999: *Fumana scoparia* Pomel (Cistaceae): Considerations on a poorly known species for the Italian flora. - *Webbia* **54**: 73-84.
BALDINI, R.M. 2002: Typification in the genus *Rhamnus* L. section *Rhamnastrum* Rouy: *R. pumilus* Turra and *R. alpinus*. - *Taxon* **51(2)**: 377-380.
BALDINI, R.M. & FIORINI, G. 1986: Numeri cromosomici per la Flora Italiana: 1052. - *Inform. Bot. Ital.* **17(1-3)**: 102-103.
BALDINI, R.M. & JARVIS, C.E. 1991: Typification of some Linnaean names in *Phalaris* (Gramineae). - *Taxon* **40(3)**: 475-485.
BALDINI, R.M. & JARVIS, C.E. 2002: Typification of *Ranunculus bullatus* L. (Ranunculaceae). - *Bull. Nat. Hist. Mus. London, Bot.* **32(1)**: 7-11.
BALDWIN, J.T. JR. 1937: The cyto-taxonomy of the *Telephium* section of *Sedum*. - *Amer. J. Bot.* **24(3)**: 126-132.
BALDWIN, J.T. JR. 1939: Certain cytophyletic relations of Crassulaceae. - *Chron. Bot.* **5(4-6)**: 415-417.
BALI, TH., BAZOS, I. & CONSTANTINIDIS, TH. 2019: Flora of urban and suburban areas of Greece: a preliminary report on Chalandri, Attica. In Daskalakou, E.N. & al. (eds): 16th

Conference of the Hellenic Botanical Society – Athens 10-13 October 2019. Program and book of abstracts, p. 84.

BALIOUSIS, E. 2011: I chlorida ke i vlastisi tou Orous Pendeliki (Anatoliki Attiki). [In Greek]. - Athens: Ph.D. thesis, National and Kapodistrian University of Athens, 561 pp.

BALIOUSIS, E. 2013: Flora and vegetation of Mt Likeo (Peloponnisos, Greece). - *Fl. Medit.* **23**: 15-47.

BALIOUSIS, E. 2013: New floristic records from the island of Limnos. In Gkelis, S. & al. (eds): 13th Panhellenic Scientific Conference, Hellenic Botanical Society, Thessaloniki 3-6 October. Program and book of abstracts, p. 116.

BALIOUSIS, E. 2014: Recent data from the flora of the island of Limnos (NE Aegean, Greece): New alien invasive species affecting the agricultural economy of the island. - *Edinburgh J. Bot.* **71(2)**: 275-285.

BALIOUSIS, E. 2015: Flora and vegetation of the island of Kalamos (Ionian Sea, Greece) – floristic analysis and phytogeographical aspects. - *Hacquetia* **14(1)**: 307-318.

BALIOUSIS, E. 2016a: Flora and vegetation of Mt Aphrodisio (Peloponnisos, Greece). - *Fl. Medit.* **26**: 31-61.

BALIOUSIS, E. 2016b: New floristic records from the islands Psara and Antipsara (East Aegean, Greece). - *Bot. Chron. (Patras)* **21**: 5-16.

BALIOUSIS, E. 2018: The flora of Mt Imittos (Sterea Ellas, Greece): checklist, new records, analysis and phytogeographical aspects.- *Fl. Medit.* **28**: 99-110 + electronic suppl.

BALIOUSIS, E., SAMAROPOULOU, S. & BAREKA, P. 2021: Micromorphological study of some *Trifolium* taxa (Fabaceae) seeds from Greece and its systematic significance. - *Fl. Medit.* **31**: 247-259.

BALIOUSIS, E. & TAN, KIT 2012: Reports 18-43. In Vladimirov, V., Dane, F. & Tan, Kit (eds): New floristic records in the Balkans: 20. - *Phytol. Balcan.* **18(3)**: 336-339.

BALIOUSIS, E. & YANNITSAROS, A. 2010: Flora and vegetation of the island of Kalamos (Ionian Sea, Greece): A preliminary study. - *Bot. Chron. (Patras)* **20**: 5-23.

BALIOUSIS, E. & YANNITSAROS, A. 2011: Vascular plant diversity of Mt Pendelikon (Sterea Ellas, Greece): a recent inventory reflecting contemporary dynamics. - *Willdenowia* **41(1)**: 151-165.

BALIOUSIS, E. & YANNITSAROS, A. 2013: The rediscovery of the single-mountain endemic *Verbascum pentelicum* Murb. In Gkelis, S. & al. (eds): 13th Panhellenic Scientific Conference, Hellenic Botanical Society, Thessaloniki 3-6 October. Program and book of abstracts, p. 117.

BALL, P.W. 1961a: The taxonomic status of *Neslia paniculata* (L.) Desv. and *N. apiculata* Fisch., Mey. & Avé-Lall. In Heywood, V.H. (ed.): Notulae systematicae ad Floram Europaeam spectantes, no. 1. - *Feddes Repert. Spec. Nov. Regni Veg.* **64(1)**: 11-13.

BALL, P.W. 1961b: The identity of *Sisymbrium pyrenaicum* L. ex Loefl. and later homonyms. In Heywood, V.H. (ed.): Notulae systematicae ad Floram Europaeam spectantes, no. 1. – *Feddes Repert. Spec. Nov. Regni Veg.* **64(1)**: 14-17.

BALL, P.W. 1963a: A review of *Malcolmia maritima* and allied species. In Heywood, V.H. (ed.): Notulae systematicae ad Floram Europaeam spectantes, no. 2. - *Feddes Repert. Spec. Nov. Regni Veg.* **68(3)**: 179-186.

BALL, P.W. 1963b: [Note on *Matthiola longipetala*]. In Heywood, V.H. (ed.): Notulae systematicae ad Floram Europaeam spectantes, no. 2. - *Feddes Repert. Spec. Nov. Regni Veg.* **68(3)**: 194.

BALL, P.W. 1964a: A taxonomic review of *Salicornia* in Europe. In Heywood, V.H. (ed.): Notulae systematicae ad Floram Europaeam spectantes, no. 3. - *Feddes Repert. Spec. Nov. Regni Veg.* **69 (1)**: 1-8.

BALL, P.W. 1964b: A revision of *Cakile* in Europe. In Heywood, V.H. (ed.): Notulae systematicae ad Floram Europaeam spectantes, no. 3. - *Feddes Repert. Spec. Nov. Regni Veg.* **69(1)**: 35-40.

BALL, P.W. 1968a: A revision of the genus *Huetia* Boiss. (*Freyera* Reichenb.). In Heywood, V.H. (ed.): Notulae systematicae ad Floram Europaeam spectantes, no. 7. - *Feddes Repert.* **79(1-2)**: 3-18.

BALL, P.W. 1968b: [Notes on *Onobrychis* and *Vicia*]. In Heywood, V.H. (ed.): Notulae systematicae ad Floram Europaeam spectantes, no. 7. - *Feddes Repert.* **79(1-2)**: 41-45.

BALL, P.W. 1968c: [Notes on *Bunium*]. In Heywood, V.H. (ed.): Notulae systematicae ad Floram Europaeam spectantes, no. 7. - *Feddes Repert.* **79(1-2)**: 62-64.

BALL, P.W. 1968d: [New combinations in *Seseli*]. In Heywood, V.H. (ed.): Notulae systematicae ad Floram Europaeam spectantes, no. 7. - *Feddes Repert.* **79(1-2)**: 64.

BALL, P.W. 1971: The genus *Melittis* in Europe. In Heywood, V.H. (ed.): Notulae systematicae ad Floram Europaeam spectantes, no. 8. - *Bot. J. Linn. Soc.* **64(1)**: 69-71.

BALL, P.W. 1972a: Taxonomic and nomenclatural notes on European Labiatae (*Acinos* Miller, *Calamintha* Miller, *Lamium* L., *Satureja* L.). In Heywood, V.H. (ed.): Notulae system. ad Floram Europaeam spectantes, no. 13. - *Bot. J. Linn. Soc.* **65(4)**: 342-352.

BALL, P.W. 1972b: [Notes on *Onosma*]. In Heywood, V.H. (ed.): Notulae systematicae ad Floram Europaeam spectantes, no. 13. - *Bot. J. Linn. Soc.* **65(4)**: 353-354.

BALL, P.W. 1990: Notes on the genus *Erysimum* L. in Europe. In Chater, A.O. (ed.): Flora Europaea: Notulae systematicae ad Floram Europaeam spectantes. Series 2. No. 3. - *Bot. J. Linn. Soc.* **103(3)**: 200-213.

BALL, P.W. & HEYWOOD, V.H. 1962: The taxonomic separation of the cytological races of *Kohlrauschia prolifera* (L.) Kunth sensu lato. - *Watsonia* **5(3)**: 113-116.

BALL, P.W. & HEYWOOD, V.H. 1964: A revision of the genus *Petrorhagia*. - *Bull. Brit. Mus. (Nat. Hist.), Bot.* **3(4)**: 121-172 + 3 plates.

BALL, P.W. & TUTIN, T.G. 1959: Notes on annual species of *Salicornia* in Britain. - *Watsonia* **4(4)**: 193-205.

BALLE, S. & AL. 1960: *Loranthus*. - *Taxon* **9(7)**: 208-210.

BALLERO, M., SCRUGLI, A. & PIU, G. 1983: *Parietaria judaica* L. (Urticaceae) in Sardegna. Distribuzione, cariologia e considerazioni allergologiche. - *Atti Soc. Tosc. Sci. Nat. Pisa Mem., Ser. B* **89**: 47-55.

BALTISBERGER, M. 1981: Verwandtschaftsbeziehungen zwischen der Gruppe des *Ranunculus polyanthemos* L. und *R. repens* L. sowie Arten der Gruppen des *R. acris* L. und des *R. bulbosus* L. - *Bot. Helv.* **91**: 61-74.

BALTISBERGER, M. 1982: Die Artengruppe des *Ranunculus polyanthemos* L. in Europa. - *Ber. Schweiz. Bot. Ges.* **90(3-4)**: 143-188.

BALTISBERGER, M. 1983: Die *Ranunculus polyanthemos*-Gruppe in Bayern - Taxonomie und Anmerkungen zur Verbreitung. - *Ber. Bayer. Bot. Ges.* **54**: 107-115.

BALTISBERGER, M. 1984a: Neue Chromosomenzahlen aus der Gruppe der *Stachys recta* L. und anderen, verwandten Artengruppen. - *Ber. Geobot. Inst. E.T.H. Stiftung Rübel* **51**: 39-62.

BALTISBERGER, M. 1984b: Zytologische Untersuchungen an einigen Pflanzen aus Albanien. - *Ber. Geobot. Inst. E.T.H. Stiftung Rübel* **51**: 63-77.

BALTISBERGER, M. 1986: Biosystematic studies in *Ranunculus cacuminis* Strid. In: 5th OPTIMA meeting, Book of abstracts, p. 42. - Istanbul.

BALTISBERGER, M. 1988a: Additional notes on the flora of Mount Kajmakcalan (Greece). – *Willdenowia* **17(1-2)**: 33-36.

BALTISBERGER, M. 1988b: Chromosomenzahlen einiger Pflanzen aus Albanien. II. - *Ber. Geobot. Inst. E.T.H. Stiftung Rübel* **54**: 42-50.

BALTISBERGER, M. 1988c: Atlas Florae Europaeae notes. 8. New nomenclatural combinations within *Ranunculus polyanthemos*. - *Ann. Bot. Fenn.* **25(3)**: 293-294.

BALTISBERGER, M. 1989: Etudes biosystématiques du genre *Betonica*. Nombres chromosomiques. - *Biocosme Mésogéen* **6(1-2)**: 1-19.

BALTISBERGER, M. 1990a: Chromsosomenzahlen einiger Pflanzen aus Jugoslawien. - *Candollea* **45(2)**: 439-446.

BALTISBERGER, M. 1990b: Zytologische Untersuchungen an einigen Arten aus Italien. - *Arch. Bot. Biogeogr. Ital.* **66(3-4)**: 153-165.

BALTISBERGER, M. 1990c: Numeri cromosomici per la Flora Italiana: 1167-1184. - *Inform. Bot. Ital.* **20(2-3)**: 627-636.

BALTISBERGER, M. 1991a: Chromosomenzahlen einiger Labiaten aus Albanien. - *Ber. Geobot. Inst. E. T.H. Stiftung Rübel* **57**: 165-181.

BALTISBERGER, M. 1991b: Cytological investigations of some Greek plants. - *Fl. Medit.* **1**: 157-173.

BALTISBERGER, M. 1991c: Cytological investigations of some plants from Turkey. - *Willdenowia* **21(1-2)**: 225-232.

BALTISBERGER, M. 1991d: Numeri cromosomici per la Flora Italiana: 1208-1230. - *Inform. Bot. Ital.* **22(3)**: 216-226.

BALTISBERGER, M. 1992: Botanische Notizen und zytologische Untersuchungen an einigen Pflanzen (insbesondere aus den Gattungen *Ranunculus* und *Achillea*) aus dem albanisch-jugoslawischen Grenzgebiet (Korab, Sar Planina). - *Ber. Geobot. Inst. E.T.H. Stiftung Rübel* **58**: 192-211.

BALTISBERGER, M. 1993a: Biosystematic studies in *Ranunculus cacuminis* Strid. In: Demiriz, H. & Özhatay, N. (eds): Proceedings of the V OPTIMA meeting, Istanbul, 8-15 September, 1986, pp. 481-484. - Istanbul Üniversitesi Fen Fakültesi.

BALTISBERGER, M. 1993b: Cytologische Untersuchungen an Compositen aus Albanien. - *Candollea* **48(2)**: 437-448.

BALTISBERGER, M. 1993c: Two interesting chromosome numbers from the Balkans. - *Int. Organ. Pl. Biosyst. Newslett.* **20**: 12-15.

BALTISBERGER, M. 1994a: Chromosome numbers in some species from Greece. - *Bot. Chron. (Patras)* **11**: 15-29.

BALTISBERGER, M. 1994b: *Ranunculus cacuminis* and *R. crenatus*, representatives of the *R. alpestris*- group (Ranunculaceae) on the Balkan Peninsula. - *Pl. Syst. Evol.* **190(3-4)**: 231-244.

BALTISBERGER, M. 2002: Cytological investigations on some Albanian plant species. - *Candollea* **56(2)**: 245-259.

BALTISBERGER, M. 2006: Cytological investigations on Bulgarian phanerogams.- *Willdenowia* **36(Special issue)**: 205-216.

BALTISBERGER, M. 2015: [Chromosome numbers in Greek *Stachys*]. - In Marhold, K. & Breitwieser, I. (eds): IAPT/IOPB chromosome data 20. - *Taxon* **64(6)**: 1344-1345 + e-suppl.

BALTISBERGER, M. & AESCHIMANN, D. 1988: Die Chromosomenzahlen der *Silene*-Arten der Sektion *Inflatae* (Caryophyllaceae). - *Ber. Geobot. Inst. E.T.H. Stiftung Rübel* **54**: 51-60.

BALTISBERGER, M. & BALTISBERGER, E. 1995: Cytological data of Albanian plants. - *Candollea* **50(2)**: 457-493.

BALTISBERGER, M. & HESS, H.E. 1986: Zur Verbreitung von *Ranunculus polyanthemoides* Boreau und *R. nemorosus* DC. - *Veröff. Geobot. Inst. E.T.H. Stiftung Rübel Zürich* **87**: 79-90.

BALTISBERGER, M. & HUBER, W. 1987a: [Chromosome number of *Lamium moschatum*]. In: Löve, Á. (ed.): Chromosome number reports XCVI. - *Taxon* **36(3)**: 661.

BALTISBERGER, M. & HUBER, W. 1987b: Chromosome number report. - *Newslett. Int. Organ. Pl. Biosyst.* **9**: 4-5.

BALTISBERGER, M. & LENHERR, A. 1984a: Labiaten aus Albanien. - *Candollea* **39(2)**: 423-439.

BALTISBERGER, M. & LENHERR, A. 1984b: Neue Chromosomenzahlen aus der Gruppe der *Stachys recta* L. und anderen, verwandten Artengruppen. - *Ber. Geobot. Inst. E.T.H. Stiftung Rübel* **51**: 39-62.

BALTISBERGER, M. & LEUCHTMANN, A. 1991: Investigations on some Gramineae from Albania and Greece (Chromosome numbers and endophyte infection). - *Ber. Geobot. Inst. E.T.H. Stiftung Rübel* **57**: 182-192.

BALTISBERGER, M. & LIPPERT, W. 1987: Compositen aus Albanien. - *Candollea* **42(2)**: 679-691.

BALTISBERGER, M., MULLAJ, A. & TARTARI, V. 1993: Reports (185-203). In Kamari, G., Felber, F. & Garbari, F. (eds): Mediterranean chromosome number reports - 3. - *Fl. Medit.* **3**: 348-353.

BALTISBERGER, M. & MÜLLER, M. 1981: Vergleichende cytotaxonomische Untersuchungen an *Ranunculus seguieri* und der Artengruppe des *R. alpestris* (Ranunculaceae). - *Pl. Syst. Evol.* **138(1-2)**: 47-60.

BALTISBERGER, M. & VOELGER, M. 2006: Reports. In Marhold, K. (ed.): IAPT/IOPB chromosome data 1. - *Taxon* **55(2)**: 444 + electronic supplement.

BALTISBERGER, M. & WASER-WALTER, J. 2009: Kreuzbarkeit des invasiven *Geranium purpureum* mit dem einheimischen *Geranium robertianum* (Geraniaceae). - *Bot. Helv.* **119(1)**: 63-64.

BALTISBERGER, M. & WIDMER, A. 2005: Cytological investigations on some *Ranunculus*-species from Crete. - *Candollea* **60**: 335-344.

BALTISBERGER, M. & WIDMER, A. 2009: Karyological data of some angiosperms from Romania. - *Willdenowia* **39(2)**: 353-363.

BANACH-POGAN, E. 1955: Further cytological studies in *Cardamine* L. [In Polish]. - *Acta Soc. Bot. Poloniae* **24(2)**: 275-286.

BANACH-POGAN, E. 1956: Karyological studies in *Parnassia palustris* L. - *Acta Soc. Bot. Poloniae* **25 (4)**: 675-679.

BANCHEVA, S. & DELCHEVA, M. 2018: Investigation of the population status of the critically endangered plant species *Centaurea wagenitziana* (Asteraceae). - *Bot. Serbica* **142 (suppl. 1)**: 115.

BANCHEVA, S. & GORGOROV, R. 2010: Taxonomic revision and conservation status of *Centaurea davidovii* (sect. *Lepteranthus*, Asteraceae). - *Phytol. Balcan.* **16(2)**: 255-261.

BANCHEVA, S.T. 1998: Reports (970-976). In Kamari, G., Felber, F. & Garbari, F. (eds): Mediterranean chromosome number reports - 8. - *Fl. Medit.* **8**: 273-280.

BANCHEVA, S.T. 1999: Reports (1050-1054). In Kamari, G., Felber, F. & Garbari, F. (eds): Mediterranean chromosome number reports - 9. - *Fl. Medit.* **9**: 339-342.

BANCHEVA, S.T. 2006: The Balkan endemic *Colymbada finazzeri* (Centaureinae, Asteraceae) in the Bulgarian flora. - *Phytol. Balcan.* **12(2)**: 245-248.

BANCHEVA, S.T. & DELCHEVA, M. 2016: New locality for *Centaurea pichleri* (Asteraceae) in Bulgaria. - *Biol. Nyssana* **7(2)**: 87-90.

BANCHEVA, S.T. & DENCHEV, C.M. 2000a: Biosystematic study on *Centaurea parilica* in Bulgaria. In Özhatay, N. & al. (eds): Abstracts from the Second Balkan Botanical Congress, Istanbul, Turkey, May 14-18, 2000, p. 25.

BANCHEVA, S.T. & DENCHEV, C.M. 2000b: Occurrence and taxonomic investigation on *Centaurea pichleri* (Asteraceae) in Bulgaria. In Özhatay, N. & al. (eds): Abstracts from the Second Balkan Botanical Congress, Istanbul, Turkey, May 14-18, 2000, p. 26.

BANCHEVA, S.T. & DENCHEV, C.M. 2000c: Occurrence and taxonomic investigation of *Centaurea pichleri* Boiss. (Asteraceae) in Bulgaria. - *Phytol. Balcan.* **6(2-3)**: 167-175.

BANCHEVA, S.T. & RAIMONDO, F.M. 2001: Biosystematic studies on *Centaurea napulifera* Gr. (Compositae). In Raimondo, F.M. (ed.): Abstracts from the X OPTIMA meeting, Palermo, 13-19 September 2001, p. 106.

BANCHEVA, S.T. & RAIMONDO, F.M. 2003: Biosystematic studies of seven Balkan species from genus *Cyanus* (Compositae). - *Bocconea* **16(2)**: 507-527.

BANCHEVA, S.T. & RAIMONDO, F.M. 2013: A new *Centaurea* species (Asteraceae) from Mt Sakar, south-eastern Bulgaria. - *Pl. Biosyst.* **147(3)**: 800-805.

BANCHEVA, S.T. & STOYANOV, S. 2009: A new species of *Cyanus* (sect. *Napuliferae*, Compositae: Centaureinae) from SE Bulgaria. - *Novon* **19(4)**: 421-425.

BANFI, E. 1989: Osservazioni sulla specie italiane del genere *Setaria* P. Beauv. (Poaceae). – *Atti Soc. Ital. Sci. Nat. Mus. Civico Storia Nat. Milano* **130(13)**: 189-196.

BANFI, E. 2014: *Chrysojasminum*, a new genus for *Jasminum* sect. *Alternifolia* (Oleaceae, Jasmineae). - *Nat. Hist. Sci.* **1(1)**: 1-4.

BANFI, E. 2018: A survey of the *Elymus* L. s. l. species complex (Triticeae, Poaceae) in Italy: taxa and nothotaxa, new combinations and identification key. - *Nat. Hist. Sci.* **5(2)**: 57-64.

BANFI, E., GALASSO, G. & BARTOLUCCI, F. 2018: Nomenclatural novelties for the Euro+Med flora. - *Nat. Hist. Sci.* **5(1)**: 53-57.

BANFI, E. & AL. 2022: A new genus for *Papaver* sect. *Meconella* and new combinations in

Roemeria (Papaveraceae) in Europe and the Mediterranean area. - *Atti Soc. it. Sci. nat. Museo civ. Stor. nat. Milano* **9(1)**: 67-72.

BANGA, O. 1957: Origin of the European cultivated carrot. - *Euphytica* **6(1)**: 54-63.

BĂRA, I.I., CARAGHIN, M. & TRUȚĂ, E. 1988: Aspects de l'évolution du caryotype chez *Matricaria chamomilla* L. (x=9). - *Bot. Helv.* **98(1)**: 1-6.

BARANDA, J. 1983: Sobre la nomenclatura de *Rapistrum rugosum* (L.) All. subsp. *linnaeanum* (Cosson) Rouy & Fouc. - *Anales Jard. Bot. Madrid* **40(1)**: 278.

BARANOVA, M.V. 1971: Ad systematicam generis *Lilium* L. [In Russian; English translation of this volume by Amerind Publ. Co., New Delhi, 1978]. - *Novosti Sist. Vysš. Rast.* **8**: 89-95.

BARANOVA, M.V. & MORDAK, E.V. 2004: The genus *Prospero* (Hyacinthaceae) in the Eastern Europe and the Caucasus [In Russian]. - *Bot. Zhurn. (Moscow & St. Petersburg)* **89(7)**: 1165-1176.

BARBER, H.N. 1941: Spontaneous hybrids between *Sonchus asper* and *S. oleraceus*. - *Ann. Bot. (Oxford), n.s.* **5(18)**: 375-377.

BARBERÁ, P., ROMERO-ZARCO, C. & AEDO, C. 2018: Taxonomic revision of *Trisetum* sect. *Trisetum* (Poaceae: Pooideae: Aveninae) from Eurasia and North Africa. - *Ann. Missouri Bot. Gard.* **103(3)**: 350-392.

BARBERIS, G. 1988: *Cerastium utriense* sp. nov. (Caryophyllaceae) nell'Appennino ligure-piemontese. - *Webbia* **42(2)**: 153-160.

BARBERO, M. & AL. 1987: Impact of forest fires on structure and architecture of Mediterranean ecosystems. - *Ecol. Medit.* **13(4)**: 39-50.

BARBERO, M. & QUÉZEL, P. 1975: Les forêts de sapin sur le pourtour Méditerranéen. - *Anales Inst. Bot. Cavanilles* **32(2)**: 1245-1289.

BARBERO, M. & QUÉZEL, P. 1976: Les groupements forestiers de Grèce centro-méridionale. - *Ecol. Medit.* **2**: 3-86.

BARBERO, M. & QUÉZEL, P. 1980: La végétation forestière de Crète. - *Ecol. Medit.* **5**: 175-210.

BARBERO, M. & QUÉZEL, P. 1983: La végétation de la Grèce et l'action de l'homme. - *Méditerranée* **2**: 65-71.

BARBERO, M. & QUÉZEL, P. 1989: Contribution à l'étude phytosociologique des matorrals de Méditerranée orientale. - *Lazaroa* **11**: 37-60 + 4 fold-out tables.

BARBEY, C. & BARBEY, W. 1882: Herborisations au Levant. Égypte, Syrie et Méditerranée. - Lausanne: Georges Bridel, 187 pp. + 11 plates and 1 map.

BARBEY, W. 1884: Florae Sardoae compendium. Catalogue raisonné des végétaux observés dans l'île de Sardaigne. - Lausanne: Georges Bridel, 263 pp. + 7 plates.

BARBEY, W. 1885: *Epilobium* genus a cl. Ch. Cuisin illustratum. - Lausanne: Georges Bridel, 24 planches avec texte.

BARBEY, W. 1886: Additions à la flore de Carpathos et de Lycie. - *Bull. Soc. Vaud. Sci. Nat.* **21(93)**: 219-224.

BARBEY, W. 1898: *Sternbergia colchiciflora* W. et K. var. *aetnensis* Rouy. - *Bull. Herb. Boissier* **6(4)**: 344 + 1 plate.

BARBIER, E. & MATHEZ, J. 1973: Contribution à l'étude des Cynoglossées (Boraginacées): *Pardoglossum*, genre nouveau du Bassin méditerranéen occidental. - *Candollea* **28(2)**: 281-323 + 11 plates.

BARBO, M., CELA RENZONI, G. & POLDINI, L. 1998: Numeri cromosomici per la Flora Italiana: 1391-1395. - *Inform. Bot. Ital.* **29(2-3)**: 273-279.

BARBUJANI, G. & PIGLIUCCI, M. 1989: Geographical patterns of karyotype polymorphism in Italian populations of *Ornithogalum montanum* (Liliaceae). - *Heredity* **62(1)**: 67-75.

BARCLAY, C. 1968: Searching for the *Phoenix* in Crete. - *Gard. Chron., ser. 3* **164**: 15-17.

BARCLAY, C. 1975: A new locality of wild *Phoenix* in Crete. - *Ann. Mus. Goulandris* **2**: 23-29.

BARCLAY, C. 1986: Crete. Checklist of the vascular plants. - *Englera* **6**: xiii + 138 pp.

BAREKA, E.-P. 2000: Karyosystematic study of the genus *Leucojum* L. in Greece [In Greek with English summary]. - Patras: M.Sc. thesis, Univ. of Patras [n.v.].

BAREKA, E.-P. 2007: Viosistematiki meleti idon tou genous *Bellevalia* Lapeyr. (Hyacinthaceae). [In Greek with English summary]. - Patras: Ph.D. thesis, Univ. of Patras, 195 pp.

BAREKA, E.-P., CONSTANTINIDIS, TH. & KAMARI, G. 1998: Reports (1001-1008). In Kamari, G., Felber, F. & Garbari, F. (eds): Mediterranean chromosome number reports - 8. - *Fl. Medit.* **8**: 298-307.

BAREKA, E.-P. & KAMARI, G. 1999: Reports (1026-1030). In Kamari, G., Felber, F. & Garbari, F. (eds): Mediterranean chromosome number reports - 9. - *Fl. Medit.* **9**: 324-330.

BAREKA, E.-P. & KAMARI, G. 2001a: A cytogeographical study of the genus *Leucojum* L. (Amaryllidaceae) in Greece. In Raimondo, F.M. (ed.): Abstracts from the X OPTIMA meeting, Palermo, 13-19 September 2001, p. 107.

BAREKA, E.-P. & KAMARI, G. 2001b: Cytotaxonomic study of the genus *Leucojum* L. (Amaryllidaceae) in Greece. In Kamari, Psaras & Constantinidis (eds): Proceedings of the 8th scientific congress of the Greek Botanical Society, October 5-8, 2000, pp. 302-307 [In Greek with English abstract].

BAREKA, E.-P. & KAMARI, G. 2004: A karyosystematic study of the genus *Bellevalia* Layeyr. (Liliaceae) in Greece. In Vasić, O. (ed.): Abstracts, XI OPTIMA Meeting, Beograd 5-11.IX.2004, p. 24.

BAREKA, E.-P., KAMARI, G. & PHITOS, D. 2003: Cytogeographic study of the genus *Leucojum* (Amaryllidaceae) in Greece. - *Bocconea* **16(2)**: 529-536.

BAREKA, E.-P., KAMARI, G. & PHITOS, D. 2006: *Acis ionica* (Amaryllidaceae), a new species from the Ionian area (W Greece, S Albania). - *Willdenowia* **36(1)**: 357-366.

BAREKA, E.-P., KAPASA, M. & KAMARI, G. 2002: Karyological study of some plant taxa from Cephalonia island (Greece). In Kamari, G. & al. (eds): Ellinini Votaniki Eteria. 9° Panelliniko Epistimoniko Sinedrio. Praktika, pp. 198-205 [In Greek with English abstract].

BAREKA, E.-P., KOUTOULA, M. & KAMARI, G. 2000: Reports (1106-1109) In Kamari, G., Felber, F. & Garbari, F. (eds): Mediterranean chromosome number reports - 10. - *Fl. Medit.* **10**: 382-386.

BAREKA, E.-P., PHITOS, D. & KAMARI, G. 2008: A karyosystematic study of the genus *Bellevalia* Lapeyr. (Hyacinthaceae) in Greece. - *Bot. J. Linn. Soc.* **157(4)**: 723-739.

BAREKA, E.-P., ŠILJAK-YAKOVLEV, S. & KAMARI, G. 2012: Molecular cytogenetics of *Bellevalia* (Hyacinthaceae) species occurring in Greece. - *Pl. Syst. Evol.* **298(2)**: 421-430.

BAREKA, P. & AL. 2015: Karyomorphological study of some Cretan archeophytes. - *Fl. Medit.* **25 (Special Issue)**: 127-141.

BAREKA, P. & AL. 2018: Karyosystematic study of some taxa from the Ionian floristic region (Greece). I. - *Fl. Medit.* **28**: 85-97.

BAREKA, P., TURLAND, N.J. & KAMARI, G. 2015: *Bellevalia juliana* (Asparagaceae), a new hexaploid species from E Kriti (Greece). - *Pl. Biosyst.* **149(4)**: 703-709.

BARGHI, N., MUGNIER, C. & ŠILJAK-YAKOVLEV, S. 1989: Karyological study in some *Hypochoeris* spp. (Compositae) from Sicily. - *Pl. Syst. Evol.* **168(1-2)**: 49-57.

BARGHI, N. & ŠILJAK-YAKOVLEV, S. 1990: Karyological study in three species of *Glycyrrhiza* genus (*G. glabra*, *G. lepidota* and *G. echinata*). - *Caryologia* **43(3-4)**: 223-234.

BARI, E.A. 1973: Cytological studies in the genus *Silene* L. - *New Phytol.* **72(4)**: 833-838.

BARIGOZZI, C. (ED.) 1986: The origin and domestication of cultivated plants. - Amsterdam, etc.: Elsevier, vi + 218 pp.

BARINA, Z. 2015: A critical checklist of the Albanian vascular flora. In Bogdanović, S. & Jogan, N.: Book of abstracts - 6th Balkan Botanical Congress, p. 17. - Rijeka.

BARINA, Z. 2017: Revision of F. K. Meyer's herbarium from Albania. - *Phytotaxa* **331(2)**: 213-223.

BARINA, Z. & AL. 2015: Phylogenetic relationships among and morphological comparison of European taxa of *Gymnospermium* (Berberidaceae). In Bogdanović, S. & Jogan, N.: Book of abstracts - 6th Balkan Botanical Congress, pp. 49-50. - Rijeka.

BARINA, Z. & AL. 2017: Phylogenetic relationships, biogeography and taxonomic revision of European taxa of *Gymnospermium* (Berberidaceae). - *Bot. J. Linn. Soc.* **184(3)**: 298-311.

BARINA, Z. & AL. 2018: Checklist of vascular plants of Albania. - *Phytotaxa* **378(1)**: 1-339.

BARINA, Z. & AL. 2020: Erratum to Barina & al. (2018): Checklist of vascular plants of Albania. Phytotaxa 378(1): 1-339. - *Phytotaxa* **428(2)**: 162.

BARINA, Z. & PIFKÓ, D. 2008: Additions and amendments to the flora of Albania. - *Willdenowia* **38(2)**: 455-464.

BARINA, Z. & PIFKÓ, D. 2011a: Distribution of *Carex distachya* in Albania. - *Bot. Serbica* **35(2)**: 125-130.

BARINA, Z. & PIFKÓ, D. 2011b: Contributions to the flora of Albania, 2. - *Willdenowia* **41(1)**: 139-149.

BARINA, Z., PIFKÓ, D. & MESTERHÁZY, A. 2011: Contributions to the flora of Albania, 3. - *Willdenowia* **41(2)**: 329-339.

BARINA, Z., PIFKÓ, D. & RAKAJ, M. 2015a: Taxonomy of *Scilla albanica* Turrill (Asparagaceae). - *Folia Geobot.* **50(2)**: 151-159.

BARINA, Z., PIFKÓ, D. & RAKAJ, M. 2015b: Contributions to the flora of Albania, 5. - *Studia Bot. Hung.* **46(2)**: 119-140.

BARINA, Z., RAKAJ, M. & PIFKÓ, D. 2013: Contributions to the flora of Albania, 4. - *Willdenowia* **43(1)**: 165-184.

BARINA, Z. (ED.) 2017: Distribution atlas of vascular plants in Albania. - Budapest: Hungarian Natural History Museum, 492 pp.

BARKER, C.M. & STACE, C.A. 1986: Hybridization in the genera *Vulpia* and *Festuca* (Poaceae): meiotic behaviour of artificial hybrids. - *Nordic J. Bot.* **6(1)**: 1-10.

BARKOUDAH, Y.I. 1962: A revision of *Gypsophila*, *Bolanthus*, *Ankyropetalum* and *Phryna*. - *Wentia* **9**: 1-203.

BARKWORTH, M.E. & RABEI, S.H. 2019: (2728) Proposal to conserve the name *Datura innoxia* (Solanaceae) with that spelling. - *Taxon* **68(6)**: 1370-1371.

BARLING, D.M. 1959: Biological studies in *Poa angustifolia*. - *Watsonia* **4(4)**: 147-168.

BARLOW, P.W. & NEVIN, D. 1976: Quantitative karyology of some species of *Luzula*. - *Pl. Syst. Evol.* **125(2)**: 77-86.

BARNEBY, R.C. 1939: Plants from eastern Crete. - *New Fl. & Silva* **11(3)**: 202-210 + 1 plate.

BARNES, B. 1933: Notes on *Lathyrus aphaca* L. - *J. Bot. London* **71**: 10-15, 25-33.

BARNES, P. 1997: Some Albanian ferns. - *Pteridologist* **3(2)**: 99-102.

BARNES, P. & HODA, P. 2001: Plant exploration in Albania. A short history. - *Curtis's Bot. Mag., ser. 6* **18(3)**: 170-180 + 1 plate.

BARRA, A. & LÓPEZ GONZÁLEZ, G. 1984: Tipificación de los táxones del género *Narcissus* (Amaryllidaceae) descritos por Linneo. - *Anales Jard. Bot. Madrid* **40(2)**: 345-360.

BARRENSCHEEN, I. 1986: Die *Fragaria*-Arten Niedersachsens und ihre Bestimmung. - *Göttinger Florist. Rundbr.* **20(1)**: 1-13.

BARROW, S.C. 1998: A monograph of *Phoenix* L. (Palmae: Coryphoideae). - *Kew Bull.* **53(3)**: 513-575.

BARRY, R. & WADE, P.M. 1986: Biological Flora of the British Isles. *Callitriche truncata* Guss. - *J. Ecol.* **74(1)**: 289-294.

BARTHA, D. & AL. (EDS) 2015: Atlas Florae Hungaricae. Distribution maps of vascular plants of Hungary. [Text in Hungarian and English]. - Sopron: University of West Hungary Press, 329 pp.

BARTOLINI, G. 2002: The classification, origin, diffusion and history of the olive. - Rome: FAO, xi + 74 pp.

BARTOLO, G. & AL. 1977: Numeri cromosomici per la Flora Italiana: 315-328. - *Inform. Bot. Ital.* **9(1)**: 71-87.

BARTOLO, G. & AL. 1978: Numeri cromosomici per la Flora Italiana: 382-397. - *Inform. Bot. Ital.* **10(1)**: 64-80.

BARTOLO, G., BRULLO, S. & GIANGUZZI, L. 2004: Note tassonomiche su *Tillaea alata* Viv. (Crassulaceae), nuova per la flora italiana. - *Inform. Bot. Ital.* **36(2)**: 524-528.

BARTOLO, G., BRULLO, S. & PAVONE, P. 1979: Numeri cromosomici per la Flora Italiana: 484-493. - *Inform. Bot. Ital.* **10(2)**: 267-277.

BARTOLO, G., BRULLO, S. & PAVONE, P. 1980: Numeri cromosomici per la Flora Italiana: 617-631. - *Inform. Bot. Ital.* **11(2)**: 149-160.

BARTOLO, G., D'EMERICO, S. & PULVIRENTI, S. 2010: Studio cariologico di *Ophrys ferrum-equinum* Desfontaines e *Ophrys oestrifera* Steven della Grecia (Orchidaceae). - *J. Eur. Orch.* **42(3-4)**: 535-540.

BARTOLO, G. & PULVIRENTI, S. 1993a: *Serapias orientalis* subsp. *siciliensis* (Orchidaceae), a new subspecies from Sicily. - *Candollea* **48(1)**: 231-236.

BARTOLO, G. & PULVIRENTI, S. 1993b: *Limodorum brulloi* (Orchidaceae), a new species from Calabria (S Italy). - *Candollea* **48(2)**: 485-491.

BARTOLO, G. & PULVIRENTI, S. 2000: *Limodorum rubriflorum* (Orchidaceae): specie nuova del Mediterraneo orientale. [In Italian and English]. - *Caesiana* **12**: 1-12.

BARTOLOCCI, F. & GALASSO, G. 2016: *Stachys talbotii* Bartolucci & Galasso, nom. nov. (Lamiaceae). In Galasso, G. & al. (eds): Notulae to the Italian alien vascular flora: 2. - *Italian Botanist* **2**: 66-67.

BARTOLUCCI, F. & AL. 2018: An updated checklist of the vascular flora native to Italy. - *Pl. Biosyst.* **152 (2)**: 179-303.

BARTOLUCCI, F., CAPARELLI, K.F. & PERUZZI, L. 2009: A biometric study of *Fritillaria montana* Hoppe ex W.D.J. Koch s.l. (Liliaceae) shows a single polymorphic species, with no infraspecific taxa. - *Pl. Biosyst.* **143(3)**: 516-527.

BARTOLUCCI, F., GALASSO, G. & CONTI, F. 2020: Nomenclatural novelties and typification of names in *Scorzonera* sensu lato (Asteraceae, Cichorieae) for the Italian vascular flora. - *Phytotaxa* **437(5)**: 279-290.

BARTOLUCCI, F., PERUZZI, L. & SOLDANO, A. 2014: Notulae 2064-2069. In Bartolucci, F. & al. (eds): Notulae alla checklist della flora vascolare italiana: 16 (2027-2070). - *Inform. Bot. Ital.* **46(1)**: 81-83.

BARTOLUCCI, F. & WALTER, J. 2015: Typification of names in the genus *Thymus* (Lamiaceae). - *Phytotaxa* **221(2)**: 137-147.

BAŠAK, N. 1991: The genus *Fritillaria* (Liliaceae) in European Turkey. - *Bot. Chron. (Patras)* **10**: 841-844.

BAŠAK, N. 1993: A contribution to the fern flora of European Turkey. - *Istanbul Üniv. Eczac. Fak. Mecm.* **29(1)**: 49-52.

BAŠAK, N. & GÜLER, N. 2000: Karyological investigations and distribution of some rare and endemic plants of European Turkey. In Özhatay, N. & al. (eds): Abstracts from the Second Balkan Botanical Congress, Istanbul, Turkey, May 14-18, 2000, p. 99.

BAŠAK, N. & GÜLER, N. 2001: Karyological investigations and distribution of some rare and endemic plants of European Turkey. In Özhatay, N. (ed.): Proceedings of the Second Balkan Botanical Congress, 1: 435-438. - Istanbul.

BAŠAK, N. & ÖZHATAY, N. 1993: Cytotaxonomical notes on the genus *Tulipa* L. in European Turkey. In Petrova, A.V. (ed.): Abstracts from the VII OPTIMA meeting in Borovetz, Bulgaria, 18-30 July 1993, p. 157.

BAŠAK, N. & ÖZHATAY, N. 1997: Cytotaxonomic notes on the *Tulipa* species (Liliaceae) of European Turkey. - *Bocconea* **5(2)**: 727-731.

BAŠAK, N. & SAVAŞ, G. 2001: The genus *Trifolium* L. in Edirne (European Turkey). In Özhatay, N. (ed.): Proceedings of the Second Balkan Botanical Congress, 1: 121-128. - Istanbul.

BASDRA, E. & KIPOPOULOS, E. 2018: Flowers. Olympus terra & mare. [In Greek and English]. - Nicosia: Broken Hill Publishers Ltd., 207 pp.

BAŞER, K.H.C., SATIL, F. & TÜMEN, G. 2001: *Sideritis athoa* Papanikolau & Kokkini. - *Karaca Arbor. Mag.* **6(1)**: 35-38.

BASKIN, J.M. & BASKIN, C.C. 1973: Studies on the ecological life cycle of *Holosteum umbellatum*. - *Bull. Torrey Bot. Club* **100(2)**: 110-116.

BAŞLAR, S. 2000: An investigation on *Chrozophora tinctoria* (L.) Rafin. Distributed in West Anatolia. - *Türk. Bot. Derg.* **24(2)**: 103-112.

BASLER, A. 1972: Cytotaxonomische Untersuchungen an der Boraginaceen-Gattung *Symphytum* L. - *Bot. Jahrb. Syst.* **92(4)**: 508-553.

BÄSSLER, M. 1963: Zur Taxonomie der Gattung *Achillea*. I. Die Formenkreise um *A. nobilis* L. und *A. virescens* (Fenzl) Heimerl. - *Feddes Repert. Spec. Nov. Regni Veg.* **68(2)**: 139-162 + 1 plate.

BÄSSLER, M. 1966: Die Stellung des Subgenus *Orobus* (L.) Baker in der Gattung *Lathyrus* L. und seine systematische Gliederung. - *Feddes Repert.* **72(2)**: 69-97.

BÄSSLER, M. 1971: Beiträge zur Nomenklatur der Gattung *Lathyrus* L. - *Feddes Repert.* **82(6)**: 433-439.

BÄSSLER, M. 1973: Revision der eurasiatischen Arten von *Lathyrus* L. sect. *Orobus* (L.) Gren. et Godr. - *Feddes Repert.* **84(5-6)**: 329-447.

BÄSSLER, M. 1981: Revision von *Lathyrus* L. sect. *Lathyrostylis* (Griseb.) Bässler (Fabaceae). - *Feddes Repert.* **92(3)**: 179-254.

BÄSSLER, M. & BÖHLING, N. 1999: Notes on *Vicia tenuifolia*. In Greuter, W. & Raus, Th. (eds): Med- Checklist notulae, 18. - *Willdenowia* **29(1-2)**: 62.

BATCHELLER, F.N. 1982: The alpine gesneriads. - *Gloxinian* **32(2)**: 17-23.

BATE-SMITH, E.C. 1974: Chemotaxonomy of *Geranium*. - *Bot. J. Linn. Soc.* **67(4)**: 347-359.

BATEMAN, R.M. 2009: Evolutionary classification of European orchids: the crucial importance of maximising explicit evidence and minimising authoritarian speculation. - *J. Eur. Orch.* **41(2)**: 243-318.

BATEMAN, R.M. & AL. 2003: Molecular phylogenetics and evolution of Orchidinae and selected Habenariinae (Orchidaceae). - *Bot. J. Linn. Soc.* **142(1)**: 1-40.

BATEMAN, R.M., PRIDGEON, A.M. & CHASE, M.W. 1997: Phylogenetics of subtribe Orchidinae (Orchidoideae, Orchidaceae) based on nuclear ITS sequences. 2. Infrageneric relationships and reclassification to achieve monophyly of *Orchis* sensu stricto. - *Lindleyana* **12**: 113-141.

BATEMAN, R.M. & RUDALL, P.J. 2018: Clarified relationship between *Dactylorhiza viridis* and *Dactylorhiza iberica* renders obsolete the former genus *Coeloglossum* (Orchidaceae: Orchidinae). - *Kew Bull.* **73(1)**: 17 pp.

BATEMAN, R.M., RUDALL, P.J. & JAMES, K.E. 2006: Phylogenetic context, generic affinities and evolutionary origin of the enigmatic Balkan orchid *Gymnadenia frivaldii* Hampe ex Griseb. - *Taxon* **55(1)**: 107-118.

BATLLE, I. & TOUS, J. 1997: Cabob tree, *Ceratonia siliqua* L. - Rome: International Plant Genetic Resources Institute, 92 pp.

BATTAGLIA, E. 1949a: Diploidia e triploidia in *Sternbergia lutea* (L.) Ker.-Gawl. - *Caryologia* **1(2)**: 269-279 + 1 plate.

BATTAGLIA, E. 1949b: L'alterazione della meiosi nella riproduzione apomittica di *Chondrilla juncea* L. - *Caryologia* **2(1)**: 23-30.

BATTAGLIA, E. 1952: Filogenesi del cariotipo nel genere *Scilla*. II. Il cariotipo diploide di *Scilla autumnalis* L. - *Atti Soc. Tosc. Sci. Nat. Pisa Mem., Ser. B* **59**: 130-145.

BATTAGLIA, E. 1957a: *Urginea maritima* (L.) Baker: Biotipi 2n, 3n, 4n, 6n e loro distribuzione geografica. - *Caryologia* **9(2)**: 293-314 + 4 plates.

BATTAGLIA, E. 1957b: *Scilla autumnalis* L.: Biotipi 2n, 4n, 6n e loro distribuzione geografica. - *Caryologia* **10(1)**: 75-95 + 3 plates.

BATTAGLIA, E. 1957c: Ricerche citotassonomiche nel genere *Urginea*: *Urginea maritima* (L.) Baker e *Urginea maura* Maire. - *Caryologia* **10(2)**: 244-275 + 3 plates.

BATTAGLIA, E. 1963: Una mutazione con B-cromosomi, 2n=14+3B, in *Scilla autumnalis* L. (Liliaceae). - *Caryologia* **16(3)**: 609-618.

BATTAGLIA, E. 1964a: Un secondo caso di B-cromosomi (2n=14+6-8B) in *Scilla autumnalis* L. (Liliaceae) proveniente dalla Palestina. - *Caryologia* **17(1)**: 65-76.

BATTAGLIA, E. 1964b: Determinazione del cariotipo, dall'ovulo, in *Scilla autumnalis* L. (Liliaceae). - *Caryologia* **17(2)**: 417-425.

BATTAGLIA, E. 1964c: *Urginea maritima* (L.) Baker: Nuovi reperti di biotipi cariologici 2n, 3n, 4n, 6n. - *Caryologia* **17(3)**: 509-518.

BATTAGLIA, E. 1964d: *Scilla autumnalis* L.: Nuovi reperti di biotipi cariologici 2n, 4n, 6n. – *Caryologia* **17(3)**: 557-565.

BATTAGLIA, E., CESCA, G. & MAGGINI, F. 1969: Mutazioni cromosomiche in *Scilla peruviana* L. (Quarto contributo). - *Caryologia* **22(2)**: 177-185.

BAUDET, J.C. 1977: Origine et classification des espèces cultivées du genre *Phaceolus*. - *Bull. Soc. Roy. Bot. Belgique* **110(1-2)**: 65-76.

BAUDET, J.C. 1978: Prodrome d'une classification générique des Papilionaceae-Phaceoleae. - *Bull. Jard. Bot. Natl. Belg.* **48(1-2)**: 183-220.

BAUDIÈRE, A. & CAUWET, A.-M. 1968: Sur quelques plantes inédites, rares ou critiques de la flore des Pyrénées-Orientales et des Corbières Audoises. - *Naturalia Monspel., Sér. Bot.* **19**: 179-200.

BAUER, E.-M. & BERGMEIER, E. 2011: The mountain woodlands of western Crete - plant communities, forest goods, grazing impact and conservation. - *Phytocoenologia* **41(2)**: 73-105 + appendix.

BAUER, Z. 1964: Cytological and embryological studies in the genus *Inula* L. - *Acta Biol. Cracov., Ser. Bot.* **7(2)**: 117-130 + 2 plates.

BAUER, Z. 1968: Karyological studies in the genus *Euphorbia* L. I. *Euphorbia amygdaloides* L. - *Acta Biol. Cracov., Ser. Bot.* **11(1)**: 77-85 + 1 plate.

BAUER, Z. 1971: Further studies in the karyology of the genus *Euphorbia*. - *Acta Biol. Cracov., Ser. Bot.* **14(2)**: 159-178 + 2 plates.

BAUM, B.R. 1966: A monograph of the genus *Tamarix*. - Jerusalem: Ph.D. thesis, Hebrew Univ., iii + 193 pp.

BAUM, B.R. 1968a: A clarification of the generic limits of *Trigonella* and *Medicago*. - *Canad. J. Bot.* **46 (6)**: 741-749.

BAUM, B.R. 1968b: (233) Proposal to conserve the generic name 280. *Danthonia* (Gramineae). - *Taxon* **17(4)**: 444-446. [approved, cf. *Taxon* 18(4): 478-479, 1969].

BAUM, B.R. 1973: The genus *Danthoniastrum*, about its circumscription, past and present status, and some taxonomic principles. - *Oesterr. Bot. Z.* **122(1-2)**: 51-57.

BAUM, B.R. 1974: Typification of Linnaean species of oats, *Avena*. - *Taxon* **23(4)**: 579-583.

BAUM, B.R. 1977: Oats: wild and cultivated. A monograph of the genus *Avena* L. (Poaceae). - Ottawa: Canada Dept. of Agriculture, xv + 463 pp.

BAUM, B.R. 1978a: The genus *Tamarix*. - Jerusalem: Israel Academy of Sciences and Humanities, xii + 209 pp.

BAUM, B.R. 1978b: Taxonomy of the tribe Triticeae (Poaceae) using various numerical taxonomic techniques. II. Classification. - *Canad. J. Bot.* **56(1)**: 27-56.

BAUM, B.R. 1978c: Taxonomy of the tribe Triticeae (Poaceae) using various numerical taxonomic techniques. III. Synoptic key to genera and synopses. - *Canad. J. Bot.* **56(4)**: 374-385.

BAUM, B.R. 1982: The generic problem in the Triticeae: Numerical taxonomy and related concepts. In Estes, J.R., Tyrl, R.J. & Brunken, J.N. (eds): Grasses and grasslands. Systematics and ecology, pp. 109-143. - Norman: Univ. of Oklahoma Press.

BAUM, B.R. 1983: A phylogenetic analysis of the tribe Triticeae (Poaceae) based on morphological characters of the genera. - *Canad. J. Bot.* **61(2)**: 518-535.

BAUM, B.R. 1991: (995) Proposal to conserve the name *Avena fatua* L. for the common wild oats (Poaceae). - *Taxon* **40(1)**: 132-134. [not approved, cf. Taxon 43(1): 114-115, 1994].

BAUM, B.R. & AL. 1972: Notes on the habitat and distribution of *Avena* species in the Mediterranean and Middle East. - *Canad. J. Bot.* **50(6)**: 1385-1397.

BAUM, B.R. & BAILEY, L.G. 1984a: Taxonomic studies in wall barley (*Hordeum murinum*) and sea barley (*Hordeum marinum*). 1. Character investigation: assessment of new and traditional characters. - *Canad. J. Bot.* **62(4)**: 753-762.

BAUM, B.R. & BAILEY, L.G. 1984b: Taxonomic studies in wall barley (*Hordeum murinum* sensu lato) and sea barley (*Hordeum marinum* sensu lato). 2. Multivariate morphometrics. - *Canad. J. Bot.* **62 (12)**: 2754-2764.

BAUM, B.R. & BAILEY, L.G. 1985: Relationships between *Hordeum bulbosum* L. subsp. *bulbosum* and *H. bulbosum* subsp. *nodosum* comb. et stat. nov. - *Canad. J. Bot.* **63(4)**: 735-743.

BAUM, B.R. & JARVIS, C.E. 1985: The typification of three Linnaean species of *Hordeum* (Poaceae). - *Taxon* **34(3)**: 528-532.

BAUMANN, B. & BAUMANN, H. 1984: Die Orchideenflora der Ionischen Inseln Ithaki und Kefallinia. - *Mitt. Arbeitskreis Heimische Orchid. Baden-Württemberg* **16(1)**: 105-183.

BAUMANN, B. & BAUMANN, H. 1988: Ein Beitrag zur Kenntnis der Gattung *Epipactis* Zinn im Mittelmeergebiet. - *Mitt. Arbeitskreis Heimische Orchid. Baden-Württemberg* **20(1)**: 1-68. [Also distributed as *OPTIMA leaflet* no. 175].

BAUMANN, B. & BAUMANN, H. 1990a: Orchideenfunde von der ostägäischen Insel Ikaria (nördliche Sporaden, Griechenland). - *Mitt. Arbeitskreis Heimische Orchid. Baden-Württemberg* **22(4)**: 730-735.

BAUMANN, B. & BAUMANN, H. 1990b: Zur Höhenverbreitung der Gattung *Ophrys* L. - *Mitt. Arbeitskreis Heimische Orchid. Baden-Württemberg* **22(4)**: 818-829.

BAUMANN, B. & BAUMANN, H. 1991: Hybridogene Populationen zwischen *Orchis anatolia* Boiss. und *Orchis quadripunctata* Cyr. ex Ten. in der Ostmediterraneis. - *Mitt. Arbeitskreis Heimische Orchid. Baden-Württemberg* **23(1)**: 203-242.

BAUMANN, B. & BAUMANN, H. 1999: Ein Beitrag zur Kenntnis der *Serapias cordigera*-Gruppe. - *J. Eur. Orch.* **31(2)**: 495-521.

BAUMANN, B. & BAUMANN, H. 2001: Zur Kenntnis der Orchideenflora von Cyrenaika (Libyen). - *J. Eur. Orch.* **33(2)**: 691-725.

BAUMANN, B. & BAUMANN, H. 2007: *Ophrys holoserica* subsp. *cerigona* B. Baumann & H. Baumann, eine neue endemische Unterart der südgriechischen Insel Kithira. - *J. Eur. Orch.* **39(1)**: 105-122.

BAUMANN, H. 1969: *Spiranthes aestivalis* und *Spiranthes autumnalis (spiralis)*. - *Mitt. Arbeitskreis Heimische Orchid. Baden-Württemberg* **1**: 5-11.

BAUMANN, H. 1972: *Ophrys cretica* (Vierh.) Nelson ssp. *cretica* × *Ophrys sphecodes* Mill. ssp. *sphecodes*. - *Orchidee (Hamburg)* **23(5)**: 206-207.

BAUMANN, H. 1978a: *Himantoglossum adriaticum* spec. nov. - eine bislang übersehene Riemenzunge aus dem zentralen nördlichen Mittelmeergebiet. - *Orchidee (Hamburg)* **29(4)**: 165-172. [Also distributed as *OPTIMA leaflet* no. 71].

BAUMANN, H. 1978b: *Leucorchis* ×*vitosana* Baumann hybr. nat. nov. - der einzig mögliche infragenerische Bastard der Gattung *Leucorchis* E. Mey. - *Orchidee (Hamburg)* **29(5)**: 217-223.

BAUMANN, H. 1982: Die griechische Pflanzenwelt im Mythos, Kunst und Literatur. - München: Hirmer Verlag, 252 pp. [second edition 1986].

BAUMANN, H. 1983: Die balkanisch-orientalischen *Dactylorhiza*-Arten - ein Vergleich. - *Mitt. Arbeitskreis Heimische Orchid. Baden-Württemberg* **15(1)**: 43-108.

BAUMANN, H. 1986: Zur Polymorphie von *Orchis papilionacea* L. - *Jahresber. Naturwiss. Vereins Wuppertal* **39**: 87-97.

BAUMANN, H. 1990: Die "immergrüne" Platane von Gortyn (Kreta). - *Bot. Chron. (Patras)* **9(1-2)**: 13-18.

BAUMANN, H. 1991: La botanique dans l'antiquité grecque. - *Bot. Chron. (Patras)* **10**: 481-494.

BAUMANN, H. 1993: Greek wild flowers and plant lore in ancient Greece. Translated and augmented by W.T. Stearn and E.R. Stearn [German original first published in 1982]. - London: The Herbert Press, 252 pp.

BAUMANN, H. 1997: *Soldanella pelia*. A rare species of the endemic flora of Greece [In Greek with English summary]. - *Fisis [Journal of the Hellenic Society for the Protection of Nature]* **79**: 5-6, 44.

BAUMANN, H. 2006: Von Lilienblüten aus minoischer Sicht. - *Willdenowia* **36(1)**: 389-395.

BAUMANN, H. 2007: Flora Mythologica. Griechische Pflanzenwelt in der Antike [Vollständig überarbeitete Ausgabe]. - Zürich: Akanthus, 173 pp.

BAUMANN, H. & DAFNI, A. 1981a: Differenzierung und Arealform des *Ophrys omegaifera*-Komplexes im Mittelmeergebiet. - *Beih. Veröff. Naturschutz Landschaftspflege Baden-Württemberg* **19**: 129-153.

BAUMANN, H. & DAFNI, A. 1981b: *Orchis dinsmorei* (R. Schlechter) H. Baumann & A. Dafni comb. et stat. nov. - eine eigenständige Art aus dem *Orchis laxiflora* Komplex. - *Mitt. Arbeitskreis Heimische Orchid. Baden-Württemberg* **13(3)**: 311-336.

BAUMANN, H. & KÜNKELE, S. 1979: Das OPTIMA-Projekt zur Kartierung der mediterranen Orchideen. - *Mitt. Arbeitskreis Heimische Orchid. Baden-Württemberg* **11(1)**: 12-53.

BAUMANN, H. & KÜNKELE, S. 1980: Das OPTIMA-Projekt zur Kartierung der mediterranen Orchideen. - *Jahresber. Naturwiss. Vereins Wuppertal* **33**: 146-163.

BAUMANN, H. & KÜNKELE, S. 1981a: Beiträge zur Taxonomie orientalischer *Dactylorhiza*-Arten. - *Mitt. Arbeitskreis Heimische Orchid. Baden-Württemberg* **13(2)**: 220-266.

BAUMANN, H. & KÜNKELE, S. 1981b: *Ophrys umbilicata* Desf., eine gute aber falsch interpretierte *Ophrys*-Art aus dem östlichen Mittelmeergebiet. - *Mitt. Arbeitskreis Heimische Orchid. Baden- Württemberg* **13(3)**: 285-310.

BAUMANN, H. & KÜNKELE, S. 1981c: Beiträge zur Taxonomie europäischer Orchideenarten. - *Mitt. Arbeitskreis Heimische Orchid. Baden-Württemberg* **13(3)**: 337-373.

BAUMANN, H. & KÜNKELE, S. 1981d: Beiträge zur Nomenklatur und Verbreitung der mediterranen *Dactylorhiza*-Arten der Sektion *Sambucinae*. - *Mitt. Arbeitskreis Heimische Orchid. Baden-Württemberg* **13(4)**: 455-478.

BAUMANN, H. & KÜNKELE, S. 1982a: Die wildwachsenden Orchideen Europas. - Stuttgart: Kosmos, 432 pp.

BAUMANN, H. & KÜNKELE, S. 1982b: Beiträge zur Taxonomie von *Ophrys oestrifera* M.-Bieb. und *Ophrys scolopax* Cav. - *Mitt. Arbeitskreis Heimische Orchid. Baden-Württemberg* **14(2)**: 204-239.

BAUMANN, H. & KÜNKELE, S. 1983: Über einige von Linnaeus beschriebenen europäischen Orchideenarten mit besonderer Berücksichtigung der Gattung *Dactylorhiza* Necker ex Nevski. - *Mitt. Arbeitskreis Heimische Orchid. Baden-Württemberg* **15(4)**: 417-477.

BAUMANN, H. & KÜNKELE, S. 1984: Nochmals: *Orchis latifolia* L. (1753). - *Mitt. Arbeitskreis Heimische Orchid. Baden-Württemberg* **16(4)**: 598-632.

BAUMANN, H. & KÜNKELE, S. 1986: Die Gattung *Ophrys* L. - eine taxonomische Übersicht. - *Mitt. Arbeitskreis Heimische Orchid. Baden-Württemberg* **18(3)**: 305-688.

BAUMANN, H. & KÜNKELE, S. 1988a: Neue Beiträge zur Taxonomie europäischer und mediterraner Orchideen. - *Mitt. Arbeitskreis Heimische Orchid. Baden-Württemberg* **20(3)**: 610-651.

BAUMANN, H. & KÜNKELE, S. 1988b: Die Orchideen Europas. - Stuttgart: Kosmos Naturführer, 191 pp.

BAUMANN, H. & KÜNKELE, S. 1989: Die Gattung *Serapias* L. - eine taxonomische Übersicht. - *Mitt. Arbeitskreis Heimische Orchid. Baden-Württemberg* **21(3)**: 701-946.

BAUMANN, H. & KÜNKELE, S. 1994a: *Orchis prisca* Hautzinger - eine gefährdete und endemische Orchidee von Kreta. - *J. Eur. Orch.* **26(2)**: 147-167.

BAUMANN, H. & KÜNKELE, S. 1994b: *Ophrys kotschyi* H. Fleischm. & Soó - eine gefährdete und endemische Orchidee von Zypern. - *J. Eur. Orch.* **26(3-4)**: 317-364.

BAUMANN, H. & KÜNKELE, S. 1995: *Orchis prisca* Hautzinger (1. Nachtrag). - *J. Eur. Orch.* **27(1)**: 87-92.

BAUMANN, H., KÜNKELE, S. & LORENZ, R. 1989: Die nomenklatorischen Typen der von Linnaeus veröffentlichten Namen europäischer Orchideen. - *Mitt. Arbeitskreis Heimische Orchid. Baden- Württemberg* **21(3)**: 355-700.

BAUMANN, H., KÜNKELE, S. & LORENZ, R. 2004: *Ophrys speculum* Link, ein illegitimer Name. - *J. Eur. Orch.* **36(2)**: 561-583.

BAUMANN, H., KÜNKELE, S. & LORENZ, R. 2006: Orchideen Europas mit angrenzenden Gebieten. - Stuttgart: Verlag Eugen Ulmer, 333 pp.

BAUMANN, H. & LORENZ, R. 1988: Beiträge zur Kenntnis der Gattung *Epipactis* Zinn in Mittel- und Süditalien und der Verbreitung einiger in diesem Gebiet spät blühenden Orchideen. - *Mitt. Arbeitskreis Heimische Orchid. Baden-Württemberg* **20(3)**: 652-694.

BAUMANN, H. & LORENZ, R. 2005a: Beiträge zur Taxonomie europäischer und mediterraner Orchideen. - *J. Eur. Orch.* **37(3)**: 705-743.

BAUMANN, H. & LORENZ, R. 2005b: Beiträge zur Taxonomie europäischer und mediterraner Orchideen, Teil 2. - *J. Eur. Orch.* **37(4)**: 939-974.

BAUMANN, K. 1994: Die Zistrosenwürger, *Cytinus hypocistis* (L.) L. - *Palmengarten* **58(2)**: 154-157.

BAUMBERGER, H. 1971: Chromosomenzahlbestimmungen und Karyotypanalysen bei den Gattungen *Anemone*, *Hepatica* und *Pulsatilla*. - *Ber. Schweiz. Bot. Ges.* **80**: 17-95.

BAUMGARTNER, H., DAIBENZEIHER, A. & MARK, CH. 1999: Erstfund von *Ophrys vernixia* Brot. auf der Insel Keffalinia. - *J. Eur. Orch.* **31(3)**: 674-677.

BAUMGARTNER, J. 1907: Die ausdauernden Arten der Sectio *Eualyssum* aus der Gattung *Alyssum*. - *Beilage zum Jahresb. des n.-ö. Landes-Lehrerseminars in Wiener-Neustad* **34**: xiv + 1-35.

BAUMGARTNER, J. 1908: Die ausdauernden Arten der Sectio *Eualyssum* aus der Gattung *Alyssum*. II. Teil. - *Beilage zum Jahresb. des n.-ö. Landes-Lehrerseminars in Wiener-Neustad* **35**: 1-58.

BAUMGARTNER, J. 1909: Die ausdauernden Arten der Sectio *Eualyssum* aus der Gattung *Alyssum*. III. Teil. - *Beilage zum Jahresb. des n.-ö. Landes-Lehrerseminars in Wiener-Neustad* **36**: 1-32.

BAUMGARTNER, J. 1911: Die ausdauernden Arten der Sectio *Eualyssum* aus der Gattung *Alyssum*. IV. Teil. - *Jahresber. Kaiser Franz-Josefs Oberrealsch. Baden bei Wien* **48**: 1-18.

BAXTER, F. 1976: *Lilium chalcedonicum* and *L. albanicum*. - *Lilies* **1976**: 69-70.

BAXTER, F. 1977: Painting plants on Corfu. - *Garden (London)* **102(6)**: 259-263.

BAXTER, L. & BAXTER, F. 1973: Searching for lilies in Greece 1970-1972. - *Lilies* **1973**: 9-14.

BAXTER, L. & BAXTER, F. 1974: Some Fritillaries in the Pindus range. - *Lilies* **1973-74**: 80-84.

BAYER, E. & AL. 1987: Pflanzen des Mittelmeerraumes (Steinbachs Naturführer). - München: Mosaik Verlag, 288 pp.

BAYER, E. & LÓPEZ GONZÁLEZ, G. 1989: Nomenclatural notes on some names in *Gagea* Salisb. (Liliaceae). - *Taxon* **38(4)**: 643-645.

BAYER, E. & LÓPEZ GONZÁLEZ, G. 1994: Observaciones sobre el género *Deschampsia* P. Beauv. (Gramineae) en la Península Ibérica. - *Anales Jard. Bot. Madrid* **52(1)**: 53-65.

BAYER, M., KÜNKELE, S. & WILLING, E. 1978: Interimskarten zur Verbreitung der südgriechischen Orchideen. - *Mitt. Arbeitskreis Heimische Orchid. Baden-Württemberg* **10(3-4)**: 114-216.

BAYER, R.J. 2000: Observations on the morphology and geographic range of *Antennaria dioica* (L.) Gaertn. (Asteraceae: Gnaphalieae). - *Madroño* **46(4)**: 205-207.

BAYÓN, N.D. 2015: Revisión taxonómica de las especies monoicas de *Amaranthus* (Amaranthaceae): *Amaranthus* subg. *Amaranthus* y *Amaranthus* subg. *Albersia*. - *Ann. Missouri Bot. Gard.* **101**: 261-383.

BAYTOP, A. 1966: Une liste des Graminées de la flore d'Istanbul. - *Istanbul Üniv. Eczac. Fak. Mecm.* **2(1)**: 14-45.

BAYTOP, A. 1969a: The genus *Phalaris* in Turkey. - *Istanbul Üniv. Eczac. Fak. Mecm.* **5(1)**: 9-26.

BAYTOP, A. 1969b: The genus *Setaria* in Turkey. - *Istanbul Üniv. Eczac. Fak. Mecm.* **5(1)**: 34-45.

BAYTOP, A. 1969c: The grasses of European Turkey I: Arundineae [In Turkish with English summary]. - *Istanbul Üniv. Eczac. Fak. Mecm.* **5(3)**: 119-129.

BAYTOP, A. 1969d: The grasses of European Turkey II: Aeleuropodeae [In Turkish with English summary]. - *Istanbul Üniv. Eczac. Fak. Mecm.* **5(3)**: 175-177.

BAYTOP, A. 1971a: Étude systématique des plantes de la Turquie d'Europe et des environs d'Istanbul I. Apocynaceae [In Turkish with French summary]. - *Istanbul Üniv. Eczac. Fak. Mecm.* **7(1)**: 11-29.

BAYTOP, A. 1971b: Étude systématique des plantes de la Turquie d'Europe et des environs d'Istanbul II. Solanaceae [In Turkish with French summary]. - *Istanbul Üniv. Eczac. Fak. Mecm.* **7(2)**: 109-137.

BAYTOP, A. 1982: Deux nouvelles Acarées pour la Turquie. - *Istanbul Üniv. Eczac. Fak. Mecm.* **18**: 60-64.

BAYTOP, A. 1983: A new *Roemeria* from Turkey. - *Notes Roy. Bot. Gard. Edinburgh* **41(2)**: 281.

BAYTOP, A. 1989: The genus *Roemeria* in Turkey. In Tan, Kit (ed.): Plant taxonomy, phytogeography and related subjects. The Davis & Hedge festschrift, pp. 65-70. - Edinburgh: Edinburgh Univ. Press.

BAYTOP, A. 2010: Plant collectors in Anatolia (Turkey). - *Phytol. Balcan.* **16(2)**: 187-213.

BAYTOP, A. & DEMIRIZ, H. 1980: Rare plants and endemics in Turkey-in-Europe. - *Istanbul Üniv. Fen Fak. Mecm., B* **45**: 109-111.

BAYTOP, A. & ÖKTEM, F. 1970: The genus *Typha* in Turkey-in-Europe [In Turkish with English summary]. - *Istanbul Üniv. Eczac. Fak. Mecm.* **6(2)**: 53-64.

BAYTOP, A. & ÖZHATAY, N. 1975: The genus *Orobanche* in Turkey-in-Europe and around Istanbul [In Turkish with English summary]. - *Istanbul Üniv. Eczac. Fak. Mecm.* **11**: 65-81.

BAYTOP, A. & SARAÇOĞLU, M. 1982a: The annual species of *Papaver* in Turkey [In Turkish with English summary]. - *Istanbul Üniv. Eczac. Fak. Yayinl.* **34**: 28-32.

BAYTOP, A. & SARAÇOĞLU, M. 1982b: Systematic studies on annual *Papaver* species of Turkey. - *Istanbul Üniv. Eczac. Fak. Mecm.* **18**: 25-44.

BAYTOP, A. & TAN, KIT 2008: Theodor von Heldreich (1822-1902) and his Turkish collections. - *Turk. J. Bot.* **32(6)**: 471-479.

BAYTOP, T. 1975: The genus *Tulipa* in European Turkey. - *Istanbul Üniv. Eczac. Fak. Mecm.* **11(2)**: 167-170.

BAYTOP, T. 1976: The genus *Colchicum* in European Turkey. - *Istanbul Üniv. Eczac. Fak. Mecm.* **12 (1)**: 163-165.

BAYTOP, T. 1993: Quelques observations sur les *Colchicum* (Liliaceae) de la Turquie. In: Demiriz, H. & Özhatay, N. (eds): Proceedings of the V OPTIMA meeting, Istanbul, 8-15 September, 1986, pp. 273-277 + 3 plates. - Istanbul Üniversitesi Fen Fakültesi.

BAYTOP, T. & MATHEW, B.F. 1984: The bulbous plants of Turkey. An illustrated guide to the bulbous petaloid monocotyledons of Turkey. Amaryllidaceae. Iridaceae. Liliaceae. - London: B.T. Batsford Ltd, xii + 132 pp. + 32 plates.

BAYTOP, T., MATHEW, B.F. & BRIGHTON, C.A. 1975: Four new taxa in Turkish *Crocus* (Iridaceae). - *Kew Bull.* **30(2)**: 241-246.

BAZOS, I. 2005: Study of the flora and vegetation of Lesvos (East Aegean Islands, Greece). [In Greek]. - Ph.D. Thesis, University of Athens, 339 pp.

BAZOS, I. 2006: Chlorida ke vlastisis tis Lesvou [In Greek]. - *Fisis [Journal of the Hellenic Society for the Protection of Nature]* **114**: 8-10.

BAZOS, I. & AL. 2009: The alien vascular flora of Greece: Floristic analysis and chorology. - *Bocconea* **23**: 281-284.

BAZOS, I. & AL. 2017: Flora and phytogeography of Makronisos (Kiklades, Greece). In Dousis, M.A. & Thanos, K.A.: 15th conference of the Hellenic Botanical Society, book of abstracts: p. 104.

BAZOS, I. & AXIOTIS, M. 2015: *Teucrium flavum* subsp. *hellenicum*. In Raab-Straube, E. & Raus, Th. (eds): Euro+Med-Checklist notulae, 5. - *Willdenowia* **45(3)**: 455.

BAZOS, I., BOUTLAS, A. & CONSTANTINIDIS, TH. 2018: Filling the floristic gaps in the Aegean: The case of Dokos island (Mirtoön Archipelago, Greece). - *Bot. Serbica* **142 (suppl. 1)**: 92.

BAZOS, I. & CHIOTELIS, CH. 2019: *Hypopitys monotropa* subsp. *hypophegea*. In Raab-Straube, E. v. & Raus, Th. (eds): Euro+Med-Checklist notulae, 10. - *Willdenowia* **49(1)**: 102.

BAZOS, I. & SALATELLIS, D. 2015: *Teucrium lamiifolium* subsp. *lamiifolium*. In Raab-Straube, E. & Raus, Th. (eds): Euro+Med-Checklist notulae, 5. - *Willdenowia* **45(3)**: 455.

BAZOS, I., SNOGERUP, B. & SNOGERUP, S. 2007: *Amaranthus bouchonii*. In Greuter, W. & Raus, Th. (eds): Med-Checklist notulae, 25. - *Willdenowia* **37(1)**: 206.

BAZOS, I. & YANNITSAROS, A. 1992: A contribution to the flora and vegetation of the islets Tokmakia and Prasologos (Lesvos, Greece) [In Greek with English summary]. In: Proc. 14th Panhell. Biol. Conf., Nicosia-Cyprus, pp. 153-156.

BAZOS, I. & YANNITSAROS, A. 1995: Floristic investigation of some vernal pools of Lesvos (East Aegean, Greece). In: Praktika 16ou Sinedriou E.E.V.E., Volos, 5.-7.5. 1994, 2, pp. 7-9 [In Greek with English summary].

BAZOS, I. & YANNITSAROS, A. 1998: Rare plants from aquatic and semi-aquatic habitats of Lesvos island (East Aegean, Greece). In Moret, J. (ed.): Abstracts of the IX. OPTIMA meeting, Paris, 11-17 May 1998. - Paris: Mus. Nat. d'Hist. Naturelle.

BAZOS, I. & YANNITSAROS, A. 1999: Pteridophyte flora of Lesvos (East Aegean islands, Greece). - *Edinburgh J. Bot.* **56(3)**: 421-448.

BAZOS, I. & YANNITSAROS, A. 2000: The history of botanical investigations in Lesvos island (East Aegean, Greece). - *Biol. Gallo-Hellen., Suppl.* **26**: 55-68.

BAZOS, I. & YANNITSAROS, A. 2001: Rare plants from aquatic and semi-aquatic habitats of Lesvos island (East Aegean, Greece). - *Bocconea* **13**: 589-597.

BAZOS, I. & YANNITSAROS, A. 2004: Floristic reports from the island of Lesvos (Greece). I. Dicotyledons: Aceraceae to Guttiferae. - *Edinburgh J. Bot.* **61(1)**: 49-86.

BAZOS, I. & YANNITSAROS, A. 2005: East Mediterranean floristic element in Lesbos island. In Karamanos, A.J. & Thanos, C.A. (eds): Biodiversity and natural heritage in the Aegean. Proceedings of the conference "Theophrastus 2000", pp. 161-175 [In Greek with English summary]. - Athina: Afoi Fragoudi O.E.

BAZOS, I. & YANNITSAROS, A. 2009: *Taraxacum delphicum* Dahlst. In Greuter, W. & Raus, Th. (eds): Med-Checklist notulae, 28. - *Willdenowia* **39(2)**: 337.

BEAL, E.O. 1956: Taxonomic revision of the genus *Nuphar* Sm. of North America and Europe. - *J. Elisha Mitchell Sci. Soc.* **72(2)**: 319-346.

BEAN, W.J. 1907: *Bruckenthalia spiculifolia*. - *Bot. Mag.* **133**: tab. 8148 + 2 pp. text.

BEATOVIĆ, R. 2008: Rasprostranjenje endemicnih taksona roda *Crocus* i roda *Iris* na balkanskom poluostrvu [In Serbian]. - Beograd: Diploma thesis, Univ. of Beograd, 106 pp.

BEAUGÉ, A. 1974: *Chenopodium album* et espèces affines. Étude historique et statistique. - Paris: Société d'Edition d'Enseignement Supérieur, xx + 447 pp.

BEAUVERD, G. 1910: Contribution à l'étude de Composées. III. Le genre *Cicerbita*. - *Bull. Soc. Bot. Genève, Sér. 2* **2**: 99-144.

BEAUVERD, G. 1914a: Une collection de phanérogames du Mont-Athos (Nouvelle Grèce). - *Bull. Soc. Bot. Genève, Sér. 2* **6**: 102.

BEAUVERD, G. 1914b: Contribution à l'étude des Composées. - *Bull. Soc. Bot. Genève, Sér. 2* **6**: 142-156.

BEAUVERD, G. 1916: Monographie du genre *Melampyrum* L. - *Mém. Soc. Phys. Genève* **38(6)**: 291-657.

BEAUVERD, G. 1920: Un nouveau seneçon de la flore méditerranéenne. - *Bull. Soc. Bot. Genève, Sér. 2* **12**: 15.

BEAUVERD, G. 1925: Polymorphisme du *Listera ovata* (L.) R. Br. et clé analytique des espèces du genre *Listera*. - *Bull. Soc. Bot. Genève, Sér. 2* **17**: 338-343.

BEAUVERD, G. 1940: [Drawings of *Medicago* fruits]. - *Bull. Soc. Bot. Genève, Sér. 2* **31**: 12 plates + 1 p. text (at end of vol.).

BEAUVERD, G. 1941: Quelques nouveautés de la flore du Kaimaktchalan. - *Bull. Soc. Bot. Genève, Sér. 2* **32**: 186-188.

BEAUVERD, G. & TOPALI, S. 1936: Excursions botaniques en Grèce (Pélion, Eubée et Péloponèse). - *Bull. Soc. Bot. Genève, Sér. 2* **28**: 94-183 + 4 plates.

BEAUVERD, G. & TOPALI, S. 1937: Plantae novae graecae. - *Candollea* **7**: 255-278 + 12 plates.

BECHERER, A. 1928: Zur Nomenklatur der Gattung *Aremonia*. - *Magyar Bot. Lapok* **27 (1-12)**: 15-17.

BECHERER, A. 1929: Notes critiques sur le *Scirpus Holoschoenus* L. - *Candollea* **4**: 130-145.

BECHERER, A. 1938: Die *Aira*-Arten der Schweiz. - *Ber. Schweiz. Bot. Ges.* **48**: 9-21.

BECHERER, A. 1960: Zur Nomenklatur und geographischen Verbreitung von *Carex laevis* Kit. - *Ber. Schweiz. Bot. Ges.* **70**: 177-179.

BECHI, N. & GARBARI, F. 1992: Reports (121-124). In Kamari, G., Felber, F. & Garbari, F. (eds): Mediterranean chromosome number reports - 2. - *Fl. Medit.* **2**: 275-278.

BECHI, N. & GARBARI, F. 1994: Intraspecific variation and taxonomic aspects of some plants from the Apuan Alps (Tuscany, Italy). - *Fl. Medit.* **4**: 213-225.

BECK V. MANNAGETTA, G. 1882: Inulae Europae. Die europäischen *Inula*-Arten. - *Denkschr. Kaiserl. Akad. Wiss., Wien. Math.-Naturwiss. Kl.* **44**: 283-339 + 1 map.

BECK V. MANNAGETTA, G. 1886a: Flora von Südbosnien und der angrenzenden Hercegovina. I. Theil. - *Ann. K.K. Naturhist. Hofmus.* **1**: 271-325.

BECK V. MANNAGETTA, G. 1886b: Versuch einer Gliederung des Formenkreises der *Caltha palustris* L. - *Verh. K. K. Zool.-Bot. Ges. Wien* **36**: 347-352.

BECK V. MANNAGETTA, G. 1887a: Flora von Südbosnien und der angrenzenden Hercegovina. II. Theil. - *Ann. K.K. Naturhist. Hofmus.* **2**: 35-76.

BECK V. MANNAGETTA, G. 1887b: Flora von Südbosnien und der angrenzenden Hercegovina. III. Theil. - *Ann. K.K. Naturhist. Hofmus.* **2**: 81-184 + 6 plates.

BECK V. MANNAGETTA, G. 1890a: Monographie der Gattung *Orobanche*. - *Biblioth. Bot.* **4/19**: 275 pp. + 4 plates and 3 maps.

BECK V. MANNAGETTA, G. 1890b: Flora von Südbosnien und der angrenzenden Hercegovina. IV. Theil. - *Ann. K.K. Naturhist. Hofmus.* **4**: 339-372.

BECK V. MANNAGETTA, G. 1890c: Flora von Südbosnien und der angrenzenden Hercegovina. V. Theil. - *Ann. K.K. Naturhist. Hofmus.* **4**: 549-578.

BECK V. MANNAGETTA, G. 1891: Flora von Südbosnien und der angrenzenden Hercegovina. VI. Theil. - *Ann. K.K. Naturhist. Hofmus.* **6(3-4)**: 307-344 + 3 plates.

BECK V. MANNAGETTA, G. 1894a: *Viola (Melanium) allchariensis* G. Beck. sp. n. - *Jahreskat. Wiener Bot. Tauschvereins* **1894**: 6.

BECK V. MANNAGETTA, G. 1894b: *Viola (Melanium) arsenica* G. Beck. sp. n. - *Jahreskat. Wiener Bot. Tauschvereins* **1894**: 6.

BECK V. MANNAGETTA, G. 1894c: *Hesperis dinarica* G. Beck sp. n. - *Jahreskat. Wiener Bot. Tauschvereins* **1894**: 6.

BECK V. MANNAGETTA, G. 1894d: Die Schneeglöckchen. Eine monographische Skizze der Gattung *Galanthus*. - *Wiener Ill. Gart.-Zeitung* **19**: 45-58.

BECK V. MANNAGETTA, G. 1895: Flora von Südbosnien und der angrenzenden Hercegovina. VII. Theil. *Ann. K.K. Naturhist. Hofmus.* **10(2)**: 166-212.

BECK V. MANNAGETTA, G. 1896a: Flora von Südbosnien und der angrenzenden Hercegovina. VIII. Theil. - *Ann. K.K. Naturhist. Hofmus.* **11(1)**: 39-80.

BECK V. MANNAGETTA, G. 1896b: Die *Geum*-Arten der Balkanländer. - *Verh. K.K. Zool.-Bot. Ges. Wien* **45**: 101-104.

BECK V. MANNAGETTA, G. 1897: Die *Armeria*-Arten der Balkanhalbinsel. - *Verh. K.K. Zool.-Bot. Ges. Wien* **47**: 577-580.

BECK V. MANNAGETTA, G. 1898: Flora von Südbosnien und der angrenzenden Hercegovina. IX. Theil. - *Ann. K.K. Naturhist. Hofmus.* **13(1)**: 1-32.

BECK V. MANNAGETTA, G. 1901: Vie Vegetationsverhältnisse der illyrischen Länder. In Engler, A. & Drude, O.: Die Vegetation der Erde 4: xv + 534 pp. - Leipzig: Verlag von W. Engelmann.

BECK V. MANNAGETTA, G. 1903: Flora Bosne, Hercegovine i Novopazarskog sandžaka [In Serbo-Croatian]. - *Glasn. Zemaljsk. Muz. Bosni Hercegovini* **15(1)**: 1-230. [Concerning subsequent vols. and fascs., see also TL-2 1: 158-159, 1976].

BECK V. MANNAGETTA, G. 1904: Flora von Bosnien, der Herzegovina und des Sandžaks Novipazar. I Theil. - Wien: Carl Gerold's Sohn, 261 pp.

BECK V. MANNAGETTA, G. 1906a: Flora Bosne, Hercegovine i Novopazarskog sandžaka [In Serbo- Croatian] (cont.). - *Glasn. Zemaljsk. Muz. Bosni Hercegovini* **18(1)**: 69-82.

BECK V. MANNAGETTA, G. 1906b: Flora Bosne, Hercegovine i Novopazarskog sandžaka [In Serbo- Croatian] (cont.). - *Glasn. Zemaljsk. Muz. Bosni Hercegovini* **18(2)**: 137-150.

BECK V. MANNAGETTA, G. 1906c: Flora Bosne, Hercegovine i Novopazarskog sandžaka [In Serbo- Croatian] (cont.). - *Glasn. Zemaljsk. Muz. Bosni Hercegovini* **18(4)**: 469-496.

BECK V. MANNAGETTA, G. 1907: Flora Bosne, Hercegovine i Novopazarskog sandžaka [In Serbo- Croatian] (cont.). - *Glasn. Zemaljsk. Muz. Bosni Hercegovini* **19(1)**: 15-30.

BECK V. MANNAGETTA, G. 1916: Flora Bosne, Hercegovine i Novopazarskog sandžaka [In Serbo- Croatian] (cont.). - *Glasn. Zemaljsk. Muz. Bosni Hercegovini* **28(1-2)**: 41-168.

BECK V. MANNAGETTA, G. 1917: Flora Bosne, Hercegovine i Novopazarskog sandžaka [In Serbo- Croatian] (cont.). - *Glasn. Zemaljsk. Muz. Bosni Hercegovini* **28(3-4)**: 311-336.

BECK V. MANNAGETTA, G. 1919: Flora Bosne, Hercegovine i bivšeg sandžaka Novog Pazara [In Serbo-Croatian]. - *Glasn. Zemaljsk. Muz. Bosni Hercegovini* **30**: 177-218.

BECK V. MANNAGETTA, G. 1920: Flora Bosne, Hercegovine i bivšeg sandžaka Novog Pazara [In Serbo-Croatian] (cont.). - *Glasn. Zemaljsk. Muz. Bosni Hercegovini* **32**: 83-128.
BECK V. MANNAGETTA, G. 1922a: Flora Bosne, Hercegovine i bivšeg sandžaka Novog Pazara [In Serbo-Croatian] (cont.). - *Glasn. Zemaljsk. Muz. Bosni Hercegovini* **33**: 1-17.
BECK V. MANNAGETTA, G. 1922b: Orobancheae novae. - *Repert. Spec. Nov. Regni Veg.* **18**: 22-40.
BECK V. MANNAGETTA, G. 1923: Flora Bosne, Hercegovine i bivšeg sandžaka Novog Pazara [In Serbo-Croatian] (cont.). - *Glasn. Zemaljsk. Muz. Bosni Hercegovini* **35**: 49-74.
BECK V. MANNAGETTA, G. 1927: Flora Bosnae, Hercegovinae et regionis Novipazar. II. Choripetalae (finis). - Beograd-Sarajevo: 487 pp.
BECK V. MANNAGETTA, G. & SZYSZYŁOWICZ, I. 1889: Plantas a Dr. Ign. Szyszyłowicz in itinere per Cernagoram et in Albania adiacenti anno 1886 lectas. - *Rozpr. Spraw. Posiedzeń Wydz. Mat.-Przyr. Akad. Umiejetn.* **19**: 1-166.
BECKER, W. 1902: Ergebnisse einer Revision der Violae des Herbarium Barbey-Boissier. - *Bull. Herb. Boissier, Ser. 2* **2(10)**: 852-856.
BECKER, W. 1905: Die systematische Behandlung der Formenkreise der *Viola calcarata* und *lutea* (im weitesten Sinne genommen) auf Grundlage ihrer Entwicklungsgeschichte. - *Beih. Bot. Centralbl., Abt. 2* **18**: 347-393.
BECKER, W. 1906: Die systematische Behandlung der *Viola cenisia* (im weitesten Sinne genommen) auf Grundlage ihrer mutmasslichen Phylogenie. - *Beih. Bot. Centralbl., Abt. 2* **20**: 108-124.
BECKER, W. 1909: Violenstudien. I. - *Beih. Bot. Centralbl., Abt. 2* **26**: 1-44.
BECKER, W. 1910a: Violenstudien. II. - *Beih. Bot. Centralbl., Abt. 2* **26(3)**: 289-390.
BECKER, W. 1910b: Bearbeitung der *Anthyllis*-Sektion *Vulneraria* DC. - *Beih. Bot. Centralbl., Abt. 2* **27**: 256-287.
BECKER, W. 1912: Anthyllisstudien. - *Beih. Bot. Centralbl., Abt. 2* **29**: 16-40.
BECKER, W. 1921: Violae novae praecipue Asiaticae. - *Repert. Spec. Nov. Regni Veg.* **17**: 72-76.
BECKER, W. 1923a: Violae Asiaticae et Australenses. IV. - *Beih. Bot. Centralbl., Abt. 2* **40(1)**: 20-68.
BECKER, W. 1923b: Violae Asiaticae et Australenses. V. - *Beih. Bot. Centralbl., Abt. 2* **40(1)**: 69-118.
BECKER, W. 1923c: Violae Asiaticae et Australenses. VI. - *Beih. Bot. Centralbl., Abt. 2* **40(1)**: 119-171.
BECKER, W. 1924a: *Viola Stojanowii* sp. nov. - *Repert. Spec. Nov. Regni Veg.* **19**: 332-333.
BECKER, W. 1924b: *Viola brachyphylla* sp. nov. - *Repert. Spec. Nov. Regni Veg.* **20**: 73.
BECKER, W. 1926: *Viola dukadjinica* Becker & Kosanin sp. nov. und *Viola albanica* × *dukadjinica* (*V. Markgrafii* V. Bckr. hybr. nov.). - *Repert. Spec. Nov. Regni Veg.* **23**: 145-146.
BECKER, W. 1928: Violae novae balcanicae. - *Bull. Inst. Jard. Bot. Univ. Belgrade* **1(1)**: 33-35 + 1 plate.
BEDALOV, M. 1969: Chromosome number of *Biarum tenuifolium* (L.) Schott [In Serbo-Croatian with English summary]. - *Acta Bot. Croat.* **28**: 39-41 + 1 plate.
BEDALOV, M. 1972: Eine neue Chromosomenzahl für *Dracunculus vulgaris* Schott [In Serbo-Croatian with German summary]. - *Acta Bot. Croat.* **31**: 87-89 + 1 plate.
BEDALOV, M. 1973: [Chromosome numbers in Araceae from Jugoslavia]. In Löve, Á. (ed.): IOPB chromosome number reports XL. - *Taxon* **22(2-3)**: 285.
BEDALOV, M. 1975a: Taxonomic problems and distribution of the species *Arum nigrum* Schott in the Balkan flora. In Jordanov, D. & al. (eds): Problems of Balkan flora and vegetation, pp. 202-208. - Sofia: Publishing House of the Bulgarian Academy of Sciences.
BEDALOV, M. 1975b: Cytotaxonomical and phytogeographical investigation of the species *Arum italicum* Mill. in Jugoslavia. - *Acta Bot. Croat.* **34**: 143-150.
BEDALOV, M. 1976a: Cytotaxonomical investigation in the genus *Dracunculus* Adans. - *Biol. Écol. Médit.* **3(2)**: 41-43.
BEDALOV, M. 1976b: Diploid and triploid types of the genus *Sternbergia lutea* (L.) Ker-Gawler in Yugoslavia. - *Biol. Écol. Médit.* **3(2)**: 45-48.
BEDALOV, M. 1976c: Cytotaxonomical and phytogeographical investigation of the species *Arum alpinum* Schott et Kotschy in Jugoslavia [In Serbo-Croatian with English summary]. - *Glasn. Prir. Mus. Beogradu, Ser. B, Biol. Nauke* **31**: 111-118.
BEDALOV, M. 1977: Cytotaxonomical and phytogeographical investigation of the species *Arum maculatum* L. in Yugoslavia [In Serbo-Croatian with English summary]. - *Acta Bot. Croat.* **36**: 107-117.
BEDALOV, M. 1978: Sur quelques espèces diploïdes du genre *Arum* L. - *Bull. Soc. Neuchâteloise Sci. Nat.* **101**: 85-93.
BEDALOV, M. 1981: Cytotaxonomy of the genus *Arum* (Araceae) in the Balkans and the Aegean area. - *Bot. Jahrb. Syst.* **102(1-4)**: 183-200.
BEDALOV, M. 1984: A new pentaploid of the genus *Arum* (Araceae). - *Bot. Helv.* **94(2)**: 385-390.
BEDALOV, M. 1993: Considerazioni tassonomiche su *Arum alpinum* Schott et Kotschy, *Arum cylindraceum* Gasparr. e *Arum lucanum* Cavara et Grande. - *Webbia* **48**: 209-212.
BEDALOV, M. & AL. 1991: Comparison of *Arum nigrum* Schott and *A. apulum* (Carano) Bedalov. - *Bot. Chron. (Patras)* **10**: 861-868.
BEDALOV, M. & AL. 1998: [Chromosome numbers in *Arum*]. In Stace, C.A. (ed.): IOPB chromosome data, 13. - *Int. Organ. Pl. Biosyst. Newslett.* **29**: 17-18.
BEDALOV, M. & AL. 2002: [Chromosome numbers in Araceae]. In Stace, C.A. (ed.): IOPB chromosome data, 18. - *Int. Organ. Pl. Biosyst. Newslett.* **34**: 18-22.
BEDALOV, M. & DRENKOVSKI, R. 1993: The genus *Arum* in Macedonia. In Petrova, A.V. (ed.): Abstracts from the VII OPTIMA meeting in Borovetz, Bulgaria, 18-30 July 1993, p. 122.
BEDALOV, M. & FISCHER, M.A. 1995: *Arum alpinum* (Araceae) and its distribution in the Eastern Mediterranean. - *Phyton (Horn)* **35(1)**: 103-113.
BEDALOV, M. & GAŽI-BASKOVA, V. 1969: Die Verbreitung der Art *Convolvulus lineatus* L. in Jugoslawien [In Serbo-Croatian with German summary]. - *Acta Bot. Croat.* **28**: 417-418 + 1 plate.
BEDALOV, M. & HESSE, M. 1989: On some pollen types within *Arum* (Araceae). - *Pl. Syst. Evol.* **166(1-2)**: 41-44.
BEDALOV, M. & KÜPFER, PH. 2005: Studies on the genus *Arum* (Araceae). - *Bull. Soc. Neuchâteloise Sci. Nat.* **128**: 43-70.
BEDALOV, M. & SUŠNIK, F. 1971: Karyological studies in the genus *Sternbergia* L. in Yugoslavia. I. - *Caryologia* **23(4)**: 519-524.
BEDDOWS, A.R. 1959: Biological Flora of the British Isles. *Dactylis glomerata* L. - *J. Ecol.* **47(1)**: 22-239.
BEDDOWS, A.R. 1961: Biological Flora of the British Isles. *Holcus lanatus* L. - *J. Ecol.* **49(2)**: 421-430.
BEDDOWS, A.R. 1967. Biological Flora of the British Isles. *Lolium perenne* L. - *J. Ecol.* **55(2)**: 567-587.
BEDDOWS, A.R. 1971: The inter- and intraspecific relationships of *Holcus lanatus* L. and *H. mollis* L. sensu lato (Gramineae). - *Bot. J. Linn. Soc.* **64(2)**: 183-198.
BEDDOWS, A.R. 1973: Biological Flora of the British Isles. *Lolium multiflorum* Lam. (*L. perenne* L. ssp. *multiflorum* (Lam.) Husnot, *L. italicum* A. Braun). - *J. Ecol.* **61(2)**: 587-600.
BEETLE, A.A. 1949: Annotated list of original descriptions in *Scirpus*. - *Amer. Midl. Naturalist* **41(2)**: 453-493.
BEG, M.Z., KAHN, A.H. & AHMAD, M. 1989: Morphological diversification and genetic isolating mechanisms between some members of *Solanum nigrum* complex. - *Cytologia* **54(3)**: 419-424.
BEG, M.Z. & KHAN, A.H. 1988: The mechanism of speciation and evolutionary trend in *Solanum nigrum* complex. - *Cytologia* **53**: 297-306.
BEG, M.Z. & KHAN, A.H. 1989: Factors creating isolating barrier between species of *Solanum nigrum* complex. - *Cytologia* **54(3)**: 437-444.
BÉGUINOT, A. 1904: Materiali per una monografia del genere *Myosotis* L. - *Ann. Bot. (Rome)* **1(4)**: 275-295.
BÉGUINOT, A. 1905: Osservazioni floristiche e fitogeografiche sul gen. *Drypis* in Italia. - *Boll. Soc. Bot. Ital.* **1905**: 54-60.

BÉGUINOT, A. 1907: Revisione monografica del genere *Romulea* Maratti. - *Malpighia* **21**: 49-122, 364-478.

BÉGUINOT, A. 1908a: Revisione monografica del genere *Romulea* Maratti [cont.]. - *Malpighia* **22**: 377-469.

BÉGUINOT, A. 1908b: Il nanismo nel genere "Plantago" e le sue cause. - *Nuovo Giorn. Bot. Ital., n.s.* **15(2)**: 205-306.

BÉGUINOT, A. 1909: Revisione monografica del genere *Romulea* Maratti [cont.]. - *Malpighia* **23**: 55-117, 185-239, 257-296.

BÉGUINOT, A. 1910: Ricerche intorno al polimorfismo della *Stellaria media* (L.) Cir. in rapporto alle sue condizioni di esistenza. - *Nuovo Giorn. Bot. Ital., n.s.* **17**: 299-326, 348-390.

BÉGUINOT, A. 1914: Revisione monografica del genere *Enarthrocarpus* Labill. - *Nuovo Giorn. Bot. Ital., n.s.* **21(4)**: 361-397.

BÉGUINOT, A. 1920: Ricerche intorno al polimorfismo della *Stellaria media* (L.) Cir. in rapporto alle sue condizioni di esistenza [cont.]. - Padova: Fratelli Gallina, 147 pp. + 3 plates.

BÉGUINOT, A. 1921: Ricerche intorno al polimorfismo della *Stellaria media* (L.) Cir. in rapporto alle sue condizioni di esistenza [cont.]. - Ferrara: Industrie Grafiche Italiane, pp. 145-196 + 4 plates.

BÉGUINOT, A. 1925: Osservazioni sull'indigenato del "Platanus orientalis" L. nell'Italia del sud e nella Sicilia orientale. - *Arch. Bot. (Forlì)* **1**: 81-100.

BÉGUINOT, A. & ANDREUCCI, A. 1925: Piante della Macedonia (Reminiscenze di guerra). - *Arch. Bot. (Forlì)* **1**: 205-216.

BÉGUINOT, A. & VACCARI, A. 1912: Contribuzione alla flora di Rodi e di Stampalia. - *Atti Reale Ist. Veneto Sci. Lett. Arti* **72(2)**: 309-330.

BÉGUINOT, A. & VACCARI, A. 1914: Secondo contributo alla flora di Rodi. - *Mem. Reale Accad. Sci. Modena, Ser. 3* **12**: 267-279.

BEHR, O., BEHR, E. & ZAHN, K.H. 1939: Beiträge zur Kenntnis der Hieracien von Südserbien, Montenegro und Griechenland. - *Glasn. Skopsk. Naučn. Društva* **20(7)**: 121-129.

BEHRENDSEN, W. 1904: Zwei neue *Alectorolophus*-Formen. - *Allg. Bot. Z. Syst.* **10(1-2)**: 35-38.

BEIER, B.-A. 2005: A revision of the desert shrub *Fagonia* (Zygophyllaceae). - *Syst. Biodivers.* **3(3)**: 221-263.

BEIER, B.-A., CHASE, M.W. & THULIN, M. 2003: Phylogenetic relationships and taxonomy of subfamily Zygophylloideae (Zygophyllaceae) based on molecular and morphological data. - *Pl. Syst. Evol.* **240 (1-4)**: 11-39.

BEISSNER, L. 1887: Handbuch der Coniferen-Benennung. - Leipzig: Verlag von Hugo Voigt, iv + 90 pp.

BELL, S. 1990: Cultivation of tuberous European terrestrial orchids at Kew. - *Quart. Bull. Alpine Gard. Soc. Gr. Brit.* **58(2)**: 96-100.

BELLA, E. & AL. 2013: Plant hybridization in Greece: the case of fir (*Abies* sp.). In Gkelis, S. & al. (eds): 13th Panhellenic Scientific Conference, Hellenic Botanical Society, Thessaloniki 3-6 October. Program and book of abstracts, p. 120.

BELLOMARIA, B. & HRUŠKA, K. 1983: Numeri cromosomici per la Flora Italiana: 842-845. - *Inform. Bot. Ital.* **13(2-3)**: 176-178.

BELLOMARIA, B. & HRUŠKA, K. 1984: Numeri cromosomici per la Flora Italiana: 962-966. - *Inform. Bot. Ital.* **15(1)**: 53-57.

BELLUSCI, F. & AL. 2010: Exploring the history of *Serapias politisii* (Orchidaceae), a narrow endemic on the opposite coast of the Otranto channel: insights from molecular investigations. - *Bot. J. Linn. Soc.* **162(4)**: 572-580.

BELO-CORREIA, A.L. & COSTA, M.F. 1987: *Setaria verticillata* (L.) P. Beauv. e *Setaria adhaerens* (Forssk.) Chiov.: 1. - *Revista Biol. (Lisbon)* **13(1-4)**: 117-143.

BELO-CORREIA, A.L. & COSTA, M.F. 1989: *Setaria verticillata* (L.) P. Beauv. e *Setaria adhaerens* (Forssk.) Chiov. - 1. Aditamento. - *Bol. Soc. Brot., Ser. 2* **62**: 289-290.

BELYAEVA, I. V. 2009: Nomenclature of *Salix fragilis* L. and a new species, *S. euxina* (Salicaceae). - *Taxon* **58(4)**: 1344-1348.

BEN-ZE'EV, N. & ZOHARY, D. 1973: Species relationships in the genus *Pisum* L. - *Israel J. Bot.* **22(2)**: 73-91.

BENA, G. 2001: Molecular phylogeny supports the morphologically based taxonomic transfer of the "medicagoid" *Trigonella* species to the genus *Medicago*. - *Pl. Syst. Evol.* **229(3-4)**: 217-236.

BENBASAT, E. 1983: Micromorphological investigation ... amphidiploids between *Triticum* L. and *Secale* L. [In Bulgarian with English abstract]. In Velčev, V.I. & al. (eds): Third nat. conference of botany, pp. 190-198. - Sofia: Publishing House of the Bulgarian Academy of Sciences.

BENDIKSBY, M. & AL. 2011a: Molecular phylogeny and taxonomy of the genus *Lamium* L. (Lamiaceae): Disentangling origins of presumed allotetraploid. - *Taxon* **60(4)**: 986-1000.

BENDIKSBY, M. & AL. 2011b: Allopolyploid origin of the *Galeopsis* tetraploids - revisiting Müntzing's classical textbook example using molecular tools. - *New Phytol.* **191(4)**: 1150-1167.

BENEDÍ, C. 1991: Taxonomía de *Chaenorhinum rubrifolium* aggr. (Scrophulariaceae) en el área mediterránea occidental. - *Collect. Bot. (Barcelona)* **20**: 35-77.

BENEDÍ, C. 1998: Consideraciones sobre el género *Anthyllis* L. (Loteae-Leguminosae) y su tratamiento en Flora Iberica. - *Anales Jard. Bot. Madrid* **56(2)**: 279-303.

BENEDÍ, C. & ORELL, J.J. 1992a: Taxonomy of the genus *Chamaesyce* S.F. Gray (Euphorbiaceae) in the Iberian Peninsula and the Balearic Islands. - *Collect. Bot. (Barcelona)* **21**: 9-55.

BENEDÍ, C. & ORELL, J.J. 1992b: Reports (64-69). In Kamari, G., Felber, F. & Garbari, F.: Mediterranean chromosome number reports - 2. - *Fl. Medit.* **2**: 239-243.

BENEDÍ, C. & ORELL, J.J. 1993: *Euphorbia chamaesyce* L. dentro del genero *Chamaesyce* S.F. Gray (Euphorbiaceae). - *Collect. Bot. (Barcelona)* **22**: 148-151.

BENEDÍ, C. STRID, A. & VICENS, J. 1999: First results from the revision of *Euphorbia* s.l. (Euphorbiaceae) for Flora Hellenica. In Correia, A.I.D. & al. (eds): Livro de resumos. Guia de excursões. V jornadas de taxonomia botânica, 16-19/IX 1999, p. 57.

BENL, G. & MAYER, E. 1975: *Erythronium dens-canis* L., ein interessanter Neufund auf der Učka. - *Acta Bot. Croat.* **34**: 151-155.

BENNELL, A.P. & HU, C.-M. 1983: The pollen morphology and taxonomy of *Lysimachia*. - *Notes Roy. Bot. Gard. Edinburgh* **40(3)**: 425-458.

BENNERT, H.W. & AL. 1989: *Asplenium* ×*helii* Lusina, the valid name for the hybrids between *A. petrarchae* (Guérin) DC. and *A. trichomanes* L. (Aspleniaceae, Pteridophyta). II. Detailed description and illustrations. - *Webbia* **43(2)**: 311-337.

BENNERT, H.W. & FISCHER, G. 1993: Biosystematics and evolution of the *Asplenium trichomanes* complex. - *Webbia* **48**: 743-760.

BENNERT, H.W., JÄGER, W. & THEREN, G. 1982: Sporenmerkmale von Sippen des *Asplenium adiantum-nigrum*-Komplexes und ihre systematische Bedeutung. - *Ber. Deutsch. Bot. Ges.* **95(2)**: 297-312.

BENNETT, S.J. & MAXTED, N. 1997: An ecogeographical analysis of the *Vicia narbonensis* complex. - *Genet. Resources Crop Evol.* **44(5)**: 411-428.

BENNETT, S.T. & GRIMSHAW, J.M. 1991: Cytological studies in *Cyclamen* subg. *Cyclamen* (Primulaceae). - *Pl. Syst. Evol.* **176(3-4)**: 135-143.

BENNETT, S.T. & THOMAS, S.M. 1991: Karyological analysis and genome size in *Milium* (Gramineae) with special reference to polyploidy and chromosomal evolution. - *Genome* **34(6)**: 868-878.

BENOS, C. 1934: The distribution of *Betula pendula* and *Pinus peuce* in Greece [In Greek]. - *Dasiki Zoi* **24**: 1-12.

BENTHAM, G. 1832-36: Labiatarum genera et species: or, a description of the genera and species of plants of the order Labiatae. - London: J. Ridgeway and Sons, 8 fascs. [cf. TL-2 1: 174, 1976].

BENTHAM, G. & HOOKER, J.D. 1862-83: Genera plantarum ad exemplaria imprimis in herbariis kewensibus servata definita. - London: 3 vols. [cf. TL-2 1: 179-180, 1976].

BENTZER, B. 1969: Chromosome morphology in Aegean populations of *Leopoldia* Parl. (Liliaceae). - *Bot. Not.* **122(4)**: 457-480.

BENTZER, B. 1972a: Structural chromosome polymorphism in diploid *Leopoldia weissii* (Freyn) Freyn ex Heldr. (Liliaceae). - *Bot. Not.* **125(2)**: 180-185.

BENTZER, B. 1972b: Variation in the chromosome complement of *Leopoldia comosa* (L.) Parl. (Liliaceae) in the Aegean (Greece). - *Bot. Not.* **125(4)**: 406-418.

BENTZER, B. 1973: Taxonomy, variation and evolution in representatives of *Leopoldia* Parl. (Liliaceae) in the southern and central Aegean. - *Bot. Not.* **126(1)**: 69-132 + 1 plate and 1 map.

BENTZER, B. 1974: Karyotypes and meiosis in *Leopoldia* Parl. (Liliaceae) from the southern and central Aegean (Greece). - *Bot. Not.* **127(1)**: 69-86.

BENTZER, B., BOTHMER, R.V. & WENDELBO, P. 1974: Cytology and morphology of the genus *Hyacinthus* L. s. str. (Liliaceae). - *Bot. Not.* **127(3)**: 297-301.

BENTZER, B. & ELLMER, M. 1975: A case of stable chromosomal polymorphism in *Leopoldia comosa* (Liliaceae). - *Hereditas (Lund)* **81(2)**: 127-131.

BERCHEM, P. VAN 1904: Theodor von Heldreich [obituary]. *Mém. Soc. Phys. Hist. Nat. Genève* **34(4)**: 373-375.

BERG, CH. & AL. 1989: *Bellevalia ciliata* (Cyr.) Nees (Hyacinthaceae) neu für Bulgarien. – *Folia Geobot. Phytotax.* **24(3)**: 297-304.

BERG, E.V. 1853: Notiz über die Flora Graeca Sibthorpiana. - *St. Petersb. Zeitung* **248**: 4 pp.

BERGDOLT, E. 1932: Morphologische und physiologische Untersuchungen über *Viola*, zugleich ein Beitrag zur Lösung des Problems der Kleistogamie. - *Bot. Abh.* **20**: 1-120.

BERGER, A. 1924: A taxonomic review of the currants and gooseberries. - *Bull. New York Agric. Exp. Sta., Geneva* **109**: 3-118 + 8 plates.

BERGERON, M. 1978: Voyage en Grèce (Attique-Péloponèse 2-9 April 1977). Compte rendu du voyage des orchidophiles. - *Orchidophile (Ansières)* **9(33)**: 1091-1102.

BERGERON, M. 1980: Excursion de la section "Orchidées d'Europe" en 1979. - *Orchidophile (Ansières)* **11(43)**: 1632-1633.

BERGERON, M. 1981: Prospections faites en 1981 dans le Péloponèse et au nord du golf de Corinthe. In: 5e colloque d'automne. Orchidées d'Europe. Soc. franc. d'orchidophilie, pp. 5-10. - Paris [n.v.].

BERGFELD, D. & BERLINGHOF, N. 2008: Mitteilung über einen weiteren Fundort von *Ophrys pallidula* Paulus auf Ostkreta. - *J. Eur. Orch.* **40(3)**: 569-574.

BERGFELD, D. & SISCHKA, N. 2016: Bemerkungen zu *Neotinea lactea* und *Ophrys hebes* in Epirus (NW-Griechenland). - *J. Eur. Orch.* **48(2-4)**: 233-244.

BERGMAN, B., DRALEVA, S. & UZNOV, S. 2004: *Ophrys reinholdii* (Orchidaceae) - a new species for the Bulgarian flora. - *Phytol. Balcan.* **10(2-3)**: 175-177.

BERGMAN, B.T. 1950: Meiosis in two different clones of the apomictic *Chondrilla juncea*. - *Hereditas (Lund)* **36(3)**: 297-320.

BERGMAN, B.T. 1952a: Chromosome morphological studies in *Chondrilla juncea* and some remarks on the microsporogenesis. - *Hereditas (Lund)* **38(2)**: 128-130.

BERGMAN, B.T. 1952b: *Chondrilla chondrilloides*, a new sexual *Chondrilla* species. - *Hereditas (Lund)* **38(3)**: 367-369.

BERGMEIER, E. 1987: *Ranunculus flammula* L. In Greuter, W. & Raus, Th. (eds): Med-Checklist notulae, 14. - *Willdenowia* **16(2)**: 452.

BERGMEIER, E. 1988: Floristic notes on the Kato Olimbos area (NE Thessaly, Greece). – *Willdenowia* **17(1-2)**: 37-58.

BERGMEIER, E. 1989: Artemisietae-Gesellschaften in Bergdörfern Nordgriechenlands und Beobachtungen zur Apophytisierung. In Ubriszy Savoia, A. (ed.): Spontaneous vegetation in settlements (Proc. IAVS-Symposium 1988: Part 1). - *Braun-Blanquetia* **3**: 211-218.

BERGMEIER, E. 1990a: Wälder und Gebüsche des Niederen Olymp (Káto Olimbos, NO-Thessalien). Ein Beitrag zur systematischen und orographischen Vegetationsgliederung Griechenlands. - *Phytocoenologia* **18(2-3)**: 161-342.

BERGMEIER, E. 1990b: Spontanvegetation nordgriechischer Bergdörfer. - *Folia Geobot. Phytotax.* **25(1)**: 27-61.

BERGMEIER, E. 1995: Die Höhenstufung der Vegetation in Südwest-Kreta (Griechenland) entlang eines 2450 m-Transektes. - *Phytocoenologia* **25(3)**: 317-361.

BERGMEIER, E. 1996: Zur Bedeutung der Beweidung für die griechische Phrygana. - *Ber. Reinhard-Tüxen-Ges.* **8**: 221-236.

BERGMEIER, E. 1998a: Notes on *Aphanes minutiflora*. In Greuter, W. & Raus, Th. (eds): Med-Checklist notulae, 17. - *Willdenowia* **28(1-2)**: 163-174.

BERGMEIER, E. 1998b: Seasonal pools in the vegetation of Gavdos (Greece). In Moret, J. (ed.): Abstracts of the IX. OPTIMA meeting, Paris, 11-17 May 1998. - Paris: Mus. Nat. d'Hist. Naturelle.

BERGMEIER, E. 1998c: The phenological approach in Mediterranean landscape analysis. - *Phytocoenologia* **28(2)**: 145-156.

BERGMEIER, E. 1998d: Combined effects of fire and grazing on phrygana vegetation - a case study in SW Crete (Greece). - *Ecol. Medit.* **23(3-4)**: 1-10.

BERGMEIER, E. 1998e: Flowering intensity of phrygana plants after fencing. - *Israel J. Pl. Sci.* **46**: 41-46.

BERGMEIER, E. 1999: Gavdos - Europe's southernmost island between stability and change. - *Geoökodynamik* **30**: 87-96.

BERGMEIER, E. 2001: Seasonal pools in the flora of Gavdos (Greece) - in situ conservation required. - *Bocconea* **13**: 511-516.

BERGMEIER, E. 2002a: The vegetation of the high mountains of Crete - a revision and multivariate analysis. - *Phytocoenologia* **32(2)**: 205-249.

BERGMEIER, E. 2002b: Plant communities and habitat differentiation in the mediterranean coniferous woodlands of Mt. Parnon (Greece). - *Folia Geobot.* **37(3)**: 309-331.

BERGMEIER, E. 2005a: *Anthyllis vulneraria* subsp. *scardica* (Wettst.) Bergmeier, comb. nov. In Greuter, W. & Raus, Th. (eds): Med-Checklist notulae, 23. - *Willdenowia* **35(1)**: 59.

BERGMEIER, E. 2005b: Phytodiversity in deciduous oak forest vegetation in Greece - a comparative approach. - *Bot. Chron. (Patras)* **18(1)**: 37-49.

BERGMEIER, E. 2005c: Eine pflanzensoziologische Studie zu traditionell bewirtschafteten Getreideäckern auf Kreta. - *Hoppea* **66**: 351-375.

BERGMEIER, E. 2006: The diversity of segetal weeds in Crete (Greece) at species and community level. - *Ann. Bot. (Rome), ser. 2* **6**: 53-64.

BERGMEIER, E. 2007: *Buddleja madagascariensis, Dichondra micrantha, Chamaesyce* spp., *Cardiospermum halicacabum*, *Echinochloa oryzoides*, *Eriochloa contracta*, *Setaria faberi*, *Sphenopus divaricatus*, *Sporobolus indicus*. In Greuter, W. & Raus, Th. (eds): Med-Checklist notulae, 26. - *Willdenowia* **37(2)**: 436-442.

BERGMEIER, E. 2008a: *Ambrosia artemisiifolia* L., *Arabis nemorensis* (Wolff ex Hoffm.) W.D.J. Koch, *Hesperis matronalis* subsp. *cladotricha* (Borbás) Hayek. In Greuter, W. & Raus, Th. (eds): Med-Checklist notulae, 27. - *Willdenowia* **38(2)**: 466-467.

BERGMEIER, E. 2008b: Diversity of aestival plant communities of irrigated garden cropland in Cretan villages. - *Braunschweig. Geobot. Arb.* **9**: 65-80.

BERGMEIER, E. 2009: *Moluccella spinosa* L. In Greuter, W. & Raus, Th. (eds): Med-Checklist notulae, 28. - *Willdenowia* **39(2)**: 338.

BERGMEIER, E. 2010: *Filago wagenitziana* (Asteraceae, Gnaphalieae), a new species from western Crete, Greece. - *Willdenowia* **40(2)**: 183-188.

BERGMEIER, E. 2011: New floristic records, confirmations and other phytogeographical notes from Crete (Greece). - *Willdenowia* **41(1)**: 167-177.

BERGMEIER, E. 2013: *Papaver argemone* L. In Raab-Straube, E. v. & Raus, Th. (eds): Euro+Med- Checklist notulae, 2. - *Willdenowia* **43(2)**: 243-244.

BERGMEIER, E. 2014: [Note on *Dysphania multifida*, *Persicaria maculosa* and *Oligomeris linifolia*]. In Raab-Straube, E. v. & Raus, Th. (eds): Euro+Med-Checklist notulae, 3. - *Willdenowia* **44(2)**: 289-294.

BERGMEIER, E. 2021a: *Leontodon longirostris*. In Raus, Th. & Raab-Straube, E.v. (eds): Euro+Med-Checklist notulae, 14. - *Willdenowia* **51(3)**: 356-357.

BERGMEIER, E. 2021b: *Datura wrightii*. In Raus, Th. & Raab-Straube, E.v. (eds): Euro+Med-Checklist notulae, 14. - *Willdenowia* **51(3)**: 364.

BERGMEIER, E. & ABRAHAMCZYK, S. 2007: Ecology and distribution of the Aegean wetland endemics *Carex cretica* and *Lathyrus neurolobus*. - *Nova Hedwigia, Beiheft* **131**: 207-219.

BERGMEIER, E. & ABRAHAMCZYK, S. 2008: Current and historical diversity and new records of wetland plants in Crete, Greece. - *Willdenowia* **38(2)**: 433-453.

BERGMEIER, E. & AL. 1999: Notes on *Atriplex mollis* and *Frankenia corymbosa*. In Greuter, W. & Raus, Th. (eds): Med-Checklist notulae, 18. - *Willdenowia* **29(1-2)**: 52-55.

BERGMEIER, E. & AL. 2001: Flora and phytogeographical significance of the islands Chrisi, Koufonisi and nearby islets (S Aegean, Greece). - *Willdenowia* **31(2)**: 329-356.

BERGMEIER, E. & AL. 2004: Zonale sommergrüne Laubwälder der südlichen Balkanhalbinsel - eine Übersicht. - *Tuexenia* **24**: 89-111.

BERGMEIER, E. & AL. 2009: Plant communities on metalliferous soils in northern Greece. - *Phytocoenologia* **39(4)**: 411-438.

BERGMEIER, E. & AL. 2011: An inventory of the vascular plants and bryophytes of Gavdopoula island (S Aegean, Greece) and its phytogeographical significance. - *Willdenowia* **41(1)**: 179-190.

BERGMEIER, E. & AL. 2020: New additions to the flora of Prespa, Greece. - *Phytol. Balcan.* **26 (1)**: 103-130.

BERGMEIER, E. & AL. 2021: Phytodiversity of Limnos (North Aegean, Greece) - an update and evaluation. - *Fl. Medit.* **31**: 233-246 + suppl.

BERGMEIER, E. & DIMOPOULOS, P. 1999a: Classification of Greek *Fagus* woodlands: A preliminary survey. - *Ann. Bot. (Rome)* **57**: 91-104.

BERGMEIER, E. & DIMOPOULOS, P. 1999b: Notes on *Galium recurvum*. In Greuter, W. & Raus, Th. (eds): Med-Checklist notulae, 18. - *Willdenowia* **29(1-2)**: 62-63.

BERGMEIER, E. & DIMOPOULOS, P. 2001a: *Fagus sylvatica* forest vegetation in Greece: Syntaxonomy and gradient analysis. - *J. Veg. Sci.* **12**: 109-126.

BERGMEIER, E. & DIMOPOULOS, P. 2001b: Chances and limits of floristic island inventories - the Dionysades group (South Aegean, Greece) re-visited. - *Phyton (Horn)* **41(2)**: 277-293.

BERGMEIER, E. & DIMOPOULOS, P. 2003: The vegetation of islets in the Aegean and the relation between the occurrence of islet specialists, island size and grazing. - *Phytocoenologia* **33(2-3)**: 447-474.

BERGMEIER, E. & DIMOPOULOS, P. 2008: Identifying plant communities of thermophilous deciduous forest in Greece: Species composition, distribution, ecology and syntaxonomy. - *Pl. Biosyst.* **142(2)**: 228-254.

BERGMEIER, E. & DIMOPOULOS, P. 2010: *Rosa stylosa* Desv. In Greuter, W. & Raus, Th. (eds): Med- Checklist notulae, 29. - *Willdenowia* **40(2)**: 198.

BERGMEIER, E., DIMOPOULOS, P. & GEORGIOU, K. 1998: To notiotero nisi tis Evropis allazi [In Greek]. - *Ionos [Journal of the Hellenic Ornithological Society]* **7**: 27-29.

BERGMEIER, E., DIMOPOULOS, P. & ZOGARIS, S. 2010: *Retama raetam* (Forssk.) Webb. In Greuter, W. & Raus, Th. (eds): Med-Checklist notulae, 29. - *Willdenowia* **40(2)**: 197.

BERGMEIER, E., JAHN, R. & JAGEL, A. 1997: Flora and vegetation of Gávdos (Greece), the southernmost European island. I. Vascular flora and chorological relations. - *Candollea* **52(2)**: 305-358.

BERGMEIER, E., KRAUSE, J. & RISTOW, M. 2021: *Trifolium pachycalyx*. In Raus, Th. & Raab-Straube, E. v. (eds): Euro+Med-Checklist notulae, 14. - *Willdenowia* **51(3)**: 362.

BERGMEIER, E. & LANDOLT, E. 2009: *Lemna minuta* Kunth. In Greuter, W. & Raus, Th. (eds): Med- Checklist notulae, 28. - *Willdenowia* **39(2)**: 342.

BERGMEIER, E. & MATTHÄS, U. 1995: Additions to the flora of W Crete. - *Willdenowia* **25(1)**: 81-98.

BERGMEIER, E. & MATTHÄS, U. 1996a: Quantitative studies of phenology and early effects of non-grazing in Cretan phrygana vegetation. - *J. Veg. Sci.* **7**: 229-236.

BERGMEIER, E. & MATTHÄS, U. 1996b: Phänologisches Spektrum einer kretischen Phrygana. - *Tuexenia* **16**: 433-450.

BERGMEIER, E. & MEYER, S. 2018: Segetal plants of winter-annual crop fields in the Aegean islands – viewed in the contexts of landscape and traditional agricultural practice. - *Ber. d. Reinhold-Tüxen-Ges.* **30**: 73-84.

BERGMEIER, E. & MEYER, S. 2021: *Lactuca aculeata* (Asteraceae), a crop wild relative new to Europe. - *Fl. Medit.* **31**: 53-58.

BERGMEIER, E. & RAUS, TH. 2000: Verbreitung und Einnischung von Arten der Isoëto-Nanojuncetea in Griechenland. - *Mitt. Bad. Landesver. Naturkunde u. Naturschutz, N.F.* **17(2)**: 463-479.

BERGMEIER, E. & STRID, A. 2014: Regional diversity, population trends and threat assessment of the weeds of traditional agriculture in Greece. - *Bot. J. Linn. Soc.* **175(4)**: 607-623.

BERGMEIER, E. & STRID, A. 2019: *Centaurea devasiana* - a new species from Prespa, NW Greece. - *Phytol. Balcan.* **25(3)**: 281-286.

BERJANO, R., TALAVERA, M. & TALAVERA, S. 2014: El género *Urospermum* (Compositae) en el oeste de la Región Mediterránea. - *Acta Bot. Malac.* **39**: 117-128.

BERNAL, M. 1987: Notas sobre algunas especies del género *Dianthus* L. del ne de la Península Ibérica. - *Anales Jard. Bot. Madrid* **44(2)**: 568-571.

BERNAL, M. 1988: Acerca de la nomenclatura y tipificación de algunos táxones del género *Dianthus* L. (Caryophyllaceae). - *Anales Jard. Bot. Madrid* **45(1)**: 363-367.

BERNARD, Y. 1999: *Silene auriculata*. - *Pl. Mont. Rocaille* **12(191)**: 565-566.

BERNARD, Y. 2000: Prunus prostrata. - *Pl. Mont. Rocaille* **13(193)**: 9-11.

BERNARDI, L. 1979: Tentamen revisionis generis *Ferulago*. - *Boissiera* **30**: 1-182.

BERNARDO, L. & al. 2018: *Alyssum doerfleri* Degen (Brassicaceae), new for the Italian flora. - *Phytotaxa* **334(1)**: 10-18.

BERNHARD, T. 1931: Die Kiefern Kleinasiens. - *Mitt. Deutsch. Dendrol. Ges.* **43**: 29-50 + 5 plates and 1 map.

BERNHARDT, K.G. 1986: *Oxalis pes-caprae* L., ein anpassungsfähiger Neophyt in Sizilien. - *Bauhinia* **8 (3)**: 141-148.

BERNINI, A., PIAGGI, E. & POLANI, F. 2008: Gigli e tulipani d'Europa. - Pavia: Verba & Scripta s.a.s., 207 pp.

BERNSTRÖM, P. 1944: Two new hybrids in *Lamium*. - *Hereditas (Lund)* **30(1-2)**: 257-260.

BERNSTRÖM, P. 1946: Chromosome numbers in *Anemone nemorosa* and *A. ranunculoides*. - *Hereditas (Lund)* **32(3-4)**: 514-520.

BERNSTRÖM, P. 1952: Cytogenetic intraspecific studies in *Lamium*. I. - *Hereditas (Lund)* **38(2)**: 163-220.

BERNSTRÖM, P. 1953: Cytogenetic intraspecific studies in *Lamium*. II. - *Hereditas (Lund)* **39(3-4)**: 381-437.

BERNSTRÖM, P. 1955: Cytogenetic studies on relationships between annual species of *Lamium*. - *Hereditas (Lund)* **41(1-2)**: 1-122.

BERRY, M.I. & JACKSON, B.P. 1976: European mandrake (*Mandragora officinarum* and *M. autumnalis*). - *Pl. Med. (Stuttgart)* **30(3)**: 281-290.

BERTOLONI, A. 1833-57: Flora Italica sistens plantas in Italia et insulis circumstantibus sponte nascentes. - Bologna: 10 vols. [cf. TL-2 1: 203-205, 1976].

BERTON, A. 1974: Observations sur les formes du *Polypodium vulgare* L. - *Bull. Soc. Bot. France* **121, 95. Sess. extr.**: 45-53.

BEUERMANN, A. 1956: Die Waldverhältnisse im Peloponnes unter besonderer Berücksichtigung der Entwaldung und Aufforstung. - *Erdkunde* **10(2)**: 122-136.

BEURET, É. 1971: Répartition géographique de quelques *Arum* des groupes *Maculatum* L. et *Italicum* Mill. - *Bull. Soc. Neuchâteloise Sci. Nat.* **94**: 29-36.

BEURET, É. 1972: Présence d'un *Arum* diploïde en Italie. - *Bull. Soc. Neuchâteloise Sci. Nat.* **95**: 35-41 + 1 plate.

BEURET, É. 1977: Contribution à l'étude de le distribution géographique et de la physiologie de taxons affines di- et polyploides. - *Biblioth. Bot.* **32/133**: 80 pp.

BEURTON, CH. 1985: Zur Morphologie und Verbreitung von *Achillea collina* J. Becker ex Rchb. und *A. pannonica* Scheele (Compositae) in der DDR. I. *Achillea pannonica*. - *Gleditschia* **13(1)**: 113-125.

BEVAN, R. 1964: Two notable Greek endemics. - *J. Roy. Hort. Soc.* **89(1)**: 22 + 1 plate.

BEZZANT, L. 1990: Plant portraits: *Trachelium asperuloides*. - *Rock Gard.* **22(1)**: 52-53 + 1 plate.

BEZZANT, L. & GRAHAM, D. 1997: Plant portraits: *Dianthus superbus, Tulipa cretica*. - *Rock Gard.* **25 (2)**: 142, 144.
BHATTACHARJEE, R. 1974: Taxonomic studies in *Stachys* I: New species and infra-specific taxa from Turkey. - *Notes Roy. Bot. Gard. Edinburgh* **33(2)**: 275-292.
BHATTACHARJEE, R. 1980: Taxonomic studies in *Stachys*. II: A new infrageneric classification of *Stachys* L. - *Notes Roy. Bot. Gard. Edinburgh* **38(1)**: 65-96.
BHATTACHARYA, N.K. 1958: A comparative study on the cytology of a few species of two allied genera *Trigonella* and *Melilotus*. - *Caryologia* **11(2)**: 165-180 + 1 plate.
BHATTACHARYA, S. 1979: The somatic chromosomes in *Ajuga genevensis* L. (Labiatae). - *Curr. Sci.* **48 (19)**: 870-871.
BHATTACHARYA, S. 1983: Chromosome study of some members of the tribe Ajugoideae (Labiatae) with reference to their taxonomic elucidation. - *Rev. Roumaine Biol., Sér. Biol. Vég.* **28(1)**: 7-11.
BIANCO, P. & AL. 1988: Numeri cromosomici per la Flora Italiana: 1139-1155. - *Inform. Bot. Ital.* **19(3)**: 322-332.
BIANCO, P. & AL. 1989: Karyological studies of some taxa of the genus *Ophrys* (Orchidaceae) from Apulia (Italy). - *Caryologia* **42(1)**: 57-63.
BIANCO, P. & AL. 1991a: Considérations sur *Euphorbia dendroides* L., espèce sténo-méditerranéenne macaronésienne. - *Bot. Chron. (Patras)* **10**: 689-696.
BIANCO, P. & AL. 1991b: Polyploidy and aneuploidy in *Ophrys*, *Orchis*, and *Anacamptis* (Orchidaceae). - *Pl. Syst. Evol.* **178(3-4)**: 235-245.
BIANCO, P., D'EMERICO, S. & MEDAGLI, P. 1991: Numeri cromosomici per la Flora Italiana: 1239-1243. - *Inform. Bot. Ital.* **22(3)**: 237-239.
BIANCO, P., MEDAGLI, P. & D'EMERICO, S. 1988: Numeri cromosomici per la Flora Italiana: 1136-1138. - *Inform. Bot. Ital.* **19(3)**: 319-321.
BIANCO, P., MEDAGLI, P. & D'EMERICO, S. 1989: Nuovi dati distributivi e osservazioni morfologiche su *Aegilops uniaristata* Vis. (Gramineae), entità mediterraneo-orientale riaccertata per la flora italiana. - *Webbia* **43(1)**: 19-24.
BIANCO, P. & SCHIRONE, B. 1985: On *Quercus coccifera* L. s.l.: variation in reproductive phenology. - *Taxon* **34(3)**: 436-439.
BIEL, B. 1998a: Die Orchideenflora der Insel Lesvos (Griechenland). - *J. Eur. Orch.* **30(2)**: 251-443.
BIEL, B. 1998b: Die Orchideenflora der Insel Lesvos (Griechenland). Berichtigung zu *Jour. Eur. Orch.* 30(2): 251-443. 1998. - *J. Eur. Orch.* **30(3)**: 659.
BIEL, B. 1999a: Anmerkungen zur Taxonomie im *Ophrys scolopax*- und *Ophrys umbilicata*-Komplex am Beispiel der Insel Lesvos. - *Ber. Arbeitskreis. Heimische Orchid.* **16(1)**: 52-65.
BIEL, B. 1999b: Nachtrag zur Orchideeenflora von Lesvos (Griechenland). - *J. Eur. Orch.* **31(4)**: 852-876.
BIEL, B. 2000: Zur Orchideenflora der Insel Limnos. - *Ber. Arbeitskreis. Heimische Orchid.* **17(2)**: 51-78.
BIEL, B. 2001: Zur Orchideenflora der Inseln Santorin (Thira) und Anafi, südliche Kykladen, Griechenland. - *Ber. Arbeitskreis. Heimische Orchid.* **18(1)**: 87-127.
BIEL, B. 2002: Contributions to the flora of the Aegean islands of Lesvos and Limnos, Greece. - *Willdenowia* **32(2)**: 209-219.
BIEL, B. 2003: Eine neue intergenerische Hybride aus der Gattung *Orchis*. - *Ber. Arbeitskreis. Heimische Orchid.* **20(1)**: 73-74.
BIEL, B. 2004: Zur Orchideenflora den Inseln Samothraki (Griechenland) und Gökçeada (Türkei) - mit kurzem Ausflug auf das griechische Festland bei Alexandroupoli (Nomos Evros). - *Ber. Arbeitskreis. Heimische Orchid.* **20(2)**: 63-94.
BIEL, B. 2005a: Ergänzungen zur Orchideenflora der Nördlichen Sporaden (Griechenland) - Skiathos, Skopelos und Alonnisos. - *Ber. Arbeitskreis. Heimische Orchid.* **21(2)**: 4-79.
BIEL, B. 2005b: Contributions to the flora of the Aegean islands of Santorini and Anafi (Kiklades, Greece). - *Willdenowia* **35(1)**: 87-96.
BIEL, B. 2008: Ergänzungen zur Orchideenflora der Kykladen (Griechenland) - Kythnos, Serifos, Sifnos, Folegandros und Sikinos. - *Ber. Arbeitskreis. Heimische Orchid.* **25(1)**: 195-253.
BIEL, B., KRETZSCHMAR, G. & KRETZSCHMAR, H. 1998: Zur Orchideeenflora der Insel Skyros (Sporaden). - *Ber. Arbeitskreis. Heimische Orchid.* **15(1)**: 27-47.
BIEL, B. & RUDOLPH, F. 1992: Orchideenkartierung auf der Pelion-Halbinsel, Griechenland. - *Ber. Arbeitskreis. Heimische Orchid.* **9(2)**: 31-63.
BIEL, B. & SCHOLZ, H. 2006: *Avena barbata*. In Greuter, W. & Raus, Th. (eds): Med-Checklist notulae, 24. - *Willdenowia* **36(2)**: 728.
BIEL, B. & SCHOLZ, H. 2008: *Agrostis lachnantha* Nees. In Greuter, W. & Raus, Th. (eds): Med- Checklist notulae, 27. - *Willdenowia* **38(2)**: 469.
BIEL, B. & TAN, KIT 2006a: Reports 1-13. In Vladimirov, V., Tan, Kit & Stevanović, V. (eds): New floristic records in the Balkans: 1. - *Phytol. Balcan.* **12(1)**: 108-110.
BIEL, B. & TAN, KIT 2006b: Reports 1-10. In Vladimirov, V. & al. (eds): New floristic records in the Balkans: 2. - *Phytol. Balcan.* **12(2)**: 280-281.
BIEL, B. & TAN, KIT 2006c: Reports 13-70. In Vladimirov, V., Dihoru, G. & Tan, Kit (eds): New floristic records in the Balkans: 3. - *Phytol. Balcan.* **12(3)**: 415-419.
BIEL, B. & TAN, KIT 2007a: Reports 1-37. In Vladimirov, V. & al. (eds): New floristic records in the Balkans: 4. - *Phytol. Balcan.* **13(1)**: 108-111.
BIEL, B. & TAN, KIT 2007b: Reports 2-36. In Vladimirov, V. & al. (eds): New floristic records in the Balkans: 6. - *Phytol. Balcan.* **13(3)**: 434-437.
BIEL, B. & TAN, KIT 2008a: Reports 2-15. In Vladimirov, V., Dane, F. & Tan, Kit (eds): New floristic records in the Balkans: 7. - *Phytol. Balcan.* **14(1)**: 132-134.
BIEL, B. & TAN, KIT 2008b: Reports 16-60. In Vladimirov, V., Dane, F. & Tan, Kit (eds): New floristic records in the Balkans: 7. - *Phytol. Balcan.* **14(1)**: 134-139.
BIEL, B. & TAN, KIT 2008c: Reports 2-17. In Vladimirov, V., Dane, F. & Tan, Kit (eds): New floristic records in the Balkans: 8. - *Phytol. Balcan.* **14(2)**: 292-294.
BIEL, B. & TAN, KIT 2008d: Reports 18-36. In Vladimirov, V., Dane, F. & Tan, Kit (eds): New floristic records in the Balkans: 8. - *Phytol. Balcan.* **14(2)**: 294-296.
BIEL, B. & TAN, KIT 2008e: Reports 14-43. In Vladimirov, V., Dane, F. & Tan, Kit (eds): New floristic records in the Balkans: 9. - *Phytol. Balcan.* **14(3)**: 433-436.
BIEL, B. & TAN, KIT 2009a: Reports 9-16. In Vladimirov, V., Dane, F. & Tan, Kit (eds): New floristic records in the Balkans: 10. - *Phytol. Balcan.* **15(1)**: 117-120.
BIEL, B. & TAN, KIT 2009b: Reports 14-23. In Vladimirov, V., Dane, F. & Tan, Kit (eds): New floristic records in the Balkans: 11. - *Phytol. Balcan.* **15(2)**: 277-278.
BIEL, B. & TAN, KIT 2009c: Reports 1-36. In Vladimirov & al. (eds): New floristic records in the Balkans: 12. - *Phytol. Balcan.* **15(3)**: 432-437.
BIEL, B. & TAN, KIT 2010a: Reports 3-23. In Vladimirov, V., Dane, F. & Tan, Kit (eds): New floristic records in the Balkans: 13. - *Phytol. Balcan.* **16(1)**: 144-148.
BIEL, B. & TAN, KIT 2010b: A new species of *Campanula* (Campanulaceae) from the island of Samothraki, NE Greece. - *Phytol. Balcan.* **16(3)**: 351-353.
BIEL, B. & TAN, KIT 2010c: Reports 27-49. In Vladimirov, V. & al. (eds): New floristic records in the Balkans: 14. - *Phytol. Balcan.* **16(3)**: 418-420.
BIEL, B. & TAN, KIT 2011a: Reports 6-37. In Vladimirov, V. & al. (eds): New floristic records in the Balkans: 15. - *Phytol. Balcan.* **17(1)**: 131-134.
BIEL, B. & TAN, KIT 2011b: Nomenclatural note. - *Phytol. Balcan.* **17(2)**: 265.
BIEL, B. & TAN, KIT 2011c: Reports 1-19. In Vladimirov, V. & Tan, Kit (eds): New floristic records in the Balkans: 16. - *Phytol. Balcan.* **17(2)**: 248-250.
BIEL, B. & TAN, KIT 2011d: Reports 4-19. In Vladimirov, V., Dane, F & Tan, Kit (eds): New floristic records in the Balkans: 17. - *Phytol. Balcan.* **17(3)**: 362-365.
BIEL, B. & TAN, KIT 2012a: Reports 22-54. In Vladimirov, V. & al. (eds): New floristic records in the Balkans: 18. - *Phytol. Balcan.* **18(1)**: 72-76.
BIEL, B. & TAN, KIT 2012b: Reports 4-39. In Vladimirov, V., Dane, F. & Tan, Kit (eds): New floristic records in the Balkans: 19. - *Phytol. Balcan.* **18(2)**: 206-211.
BIEL, B. & TAN, KIT 2012c: Reports 44-64. In Vladimirov, V., Dane, F. & Tan, Kit (eds): New floristic records in the Balkans: 20. - *Phytol. Balcan.* **18(3)**: 339-342.

BIEL, B. & TAN, KIT 2013a: Reports 17-42. In Vladimirov, V., Dane, F. & Tan, Kit (eds): New floristic records in the Balkans: 21. - *Phytol. Balcan.* **19(1)**: 134-138.

BIEL, B. & TAN, KIT 2013b: Reports 16-24. In Vladimirov, V. & al. (eds): New floristic records in the Balkans: 22. - *Phytol. Balcan.* **19(2)**: 270-272.

BIEL, B. & TAN, KIT 2013c: Reports 16-24. In Vladimirov, V., Dane, F. & Tan, Kit (eds): New floristic records in the Balkans: 23. - *Phytol. Balcan.* **19(3)**: 375-378.

BIEL, B. & TAN, KIT 2014a: The flora of Samothraki. - Kifissia: Goulandris Nat. Hist. Museum, 244 pp.

BIEL, B. & TAN, KIT 2014b: Reports 25-41. In Vladimirov, V. & Tan, Kit (eds): New floristic records in the Balkans: 24. - *Phytol. Balcan.* **20(1)**: 105-107.

BIEL, B. & TAN, KIT 2014c: Reports 13-61. In Vladimirov, V. & al. (eds): New floristic records in the Balkans: 25. - *Phytol. Balcan.* **20(2-3)**: 185-189.

BIEL, B. & TAN, KIT 2014d: Reports 62-112. In Vladimirov, V. & al. (eds): New floristic records in the Balkans: 25. - *Phytol. Balcan.* **20(2-3)**: 189-195.

BIEL, B. & TAN, KIT 2015a: Reports 17-69. In Vladimirov, V., Dane, F. & Tan, Kit (eds): New floristic records in the Balkans: 26. - *Phytol. Balcan.* **21(1)**: 56-61.

BIEL, B. & TAN, KIT 2015b: Reports 70-187. In Vladimirov, V., Dane, F. & Tan, Kit (eds): New floristic records in the Balkans: 26. - *Phytol. Balcan.* **21(1)**: 62-72.

BIEL, B. & TAN, KIT 2015c: Reports 11-44. In Vladimirov, V. & al. (eds): New floristic records in the Balkans: 27. - *Phytol. Balcan.* **21(2)**: 191-195.

BIEL, B. & TAN, KIT 2015d: Reports 16-50. In Vladimirov, V., Dane, F. & Tan, Kit (eds): New floristic records in the Balkans: 28. - *Phytol. Balcan.* **21(3)**: 370-373.

BIEL, B. & TAN, KIT 2015e: Reports 51-91. In Vladimirov, V., Dane, F. & Tan, Kit (eds): New floristic records in the Balkans: 28. - *Phytol. Balcan.* **21(3)**: 373-376.

BIEL, B. & TAN, KIT 2016a: Reports 12-29. In Vladimirov, V. & al. (eds): New floristic records in the Balkans: 29. - *Phytol. Balcan.* **22(1)**: 95-98.

BIEL, B. & TAN, KIT 2016b: Reports 8-51. In Vladimirov, V. & al. (eds): New floristic records in the Balkans: 30. - *Phytol. Balcan.* **22(2)**: 261-265.

BIEL, B. & TAN, KIT 2016c: Reports 2-19. In Vladimirov, V. & Tan, Kit (eds): New floristic records in the Balkans: 31. - *Phytol. Balcan.* **22(3)**: 430-432.

BIEL, B. & TAN, KIT 2016d: Reports 20-72. In Vladimirov, V. & Tan, Kit (eds): New floristic records in the Balkans: 31. - *Phytol. Balcan.* **22(3)**: 432-437.

BIEL, B. & TAN, KIT 2017a: Reports 9-32. In Vladimirov, V., Aybeke, M. & Tan, Kit (eds): New floristic records in the Balkans: 32. - *Phytol. Balcan.* **23(1)**: 121-124.

BIEL, B. & TAN, KIT 2017b: Reports 14-49. In Vladimirov, V. & al. (eds): New floristic records in the Balkans: 33. - *Phytol. Balcan.* **23(2)**: 284-288.

BIEL, B. & TAN, KIT 2018a: Endemic plants of Greece. Amorgos. - Municipality of Amorgos: 73 pp.

BIEL, B. & TAN, KIT 2018b: Reports 11-22. In Vladimirov, V., Aybeke, M. & Tan, Kit (eds): New floristic records in the Balkans: 36. - *Phytol. Balcan.* **24(2)**: 265-267.

BIEL, B. & TAN, KIT 2018c: Reports 19-54. In Vladimirov, V., Aybeke, M. & Tan, Kit (eds): New floristic records in the Balkans: 37. - *Phytol. Balcan.* **24(3)**: 401-405.

BIEL, B. & TAN, KIT 2019a: Reports 8-13. In Vladimirov, V. & al. (eds): New floristic records in the Balkans: 38. - *Phytol. Balcan.* **25(1)**: 99-100.

BIEL, B. & TAN, KIT 2019b: Flora of Amorgos. - Kifissia: Goulandris Nat. Hist. Mus., 220 pp.

BIEL, B. & TAN, KIT 2019c: Reports 6-46. In Vladimirov, V., Aybeke, M. & Tan, Kit (eds): New floristic records in the Balkans: 40. - *Phytol. Balcan.* **25(3)**: 296-301.

BIEL, B. & TAN, KIT 2019d: Reports 47-72. In Vladimirov, V., Aybeke, M. & Tan, Kit (eds): New floristic records in the Balkans: 40. - *Phytol. Balcan.* **25(3)**: 301-303.

BIEL, B. & TAN, KIT 2020a: Reports 9-32. In Vladimirov, V., Aybeke, M. & Tan, Kit (eds): New floristic records in the Balkans: 41. - *Phytol. Balcan.* **26(1)**: 164-167.

BIEL, B. & TAN, KIT 2020b: Reports 12-15. In Vladimirov, V., Aybeke, M. & Tan, Kit (eds): New floristic records in the Balkans: 42. - *Phytol. Balcan.* **26(2)**: 190-192.

BIEL, B. & TAN, KIT 2020c: Reports 13-16. In Vladimirov, V., Aybeke, M. & Tan, Kit: New floristic records in the Balkans: 43. - *Phytol. Balcan.* **26(3)**: 539-540.

BIEL, B. & TAN, KIT 2020d: Reports 17-26. In Vladimirov, V., Aybeke, M. & Tan, Kit: New floristic records in the Balkans: 43. - *Phytol. Balcan.* **26(3)**: 540-542.

BIEL, B. & TAN, KIT 2021a: Reports 10-22. In Vladimirov, V., Aybeke, M. & Tan, Kit (eds): New floristic records in the Balkans: 44. - *Phytol. Balcan.* **27(1)**: 129-131.

BIEL, B. & TAN, KIT 2021b: Report 23. In Vladimirov, V., Aybeke, M. & Tan, Kit (eds): New floristic records in the Balkans: 44. - *Phytol. Balcan.* **27(1)**: 131-133.

BIEL, B. & TAN, KIT 2021c: Reports 11-38. In Vladimirov, V., Aybeke, M. & Tan, Kit (eds): New floristic records in the Balkans: 45. - *Phytol. Balcan.* **27(2)**: 241-246.

BIEL, B., TAN, KIT & TZANOUDAKIS, D. 2006: A new autumn-flowering species of *Allium* (Liliaceae) from the island of Sifnos (Cyclades, Greece). - *Willdenowia* **36(1)**: 367-372.

BIELBY, P. 1993: *C[yclamen] hederifolium* on Crete. - *Cyclamen* **17(1)**: 8-9.

BIGAZZI, M., DUMAN, H. & SELVI, F. 2003: *Anchusa limbata* Boiss. & Heldr. (Boraginaceae): contribution to the knowledge of an enigmatic species from SW Turkey. - *Candollea* **58(2)**: 339-349.

BIGAZZI, M., FIORINI, G. & SELVI, F. 2000: Reports (1193-1200) In Kamari, G., Felber, F. & Garbari, F. (eds): Mediterranean chromosome number reports - 10. - *Fl. Medit.* **10**: 407-414.

BIGAZZI, M., HILGER, H.H. & SELVI, F. 2002: Evidence from nuclear and chloroplast DNA for the placement of *Anchusa macedonica* in the genus *Gastrocotyle* (Boraginaceae). - *Webbia* **57(2)**: 173-180.

BIGAZZI, M., NARDI, E. & SELVI, F. 1997: *Anchusella*, a new genus of Boraginaceae from the Central-Eastern Mediterranean. - *Pl. Syst. Evol.* **205(3-4)**: 241-264.

BIGAZZI, M., NARDI, E. & SELVI, F. 2006: Palynological contribution to the systematics of *Rindera* and the allied genera *Paracaryum* and *Solenanthus* (Boraginaceae-Cynoglosseae). - *Willdenowia* **36(1)**: 37-46.

BIGAZZI, M. & SELVI, F. 1998: Pollen morphology in the Boragineae (Boraginaceae) in relation to the taxonomy of the tribe. - *Pl. Syst. Evol.* **213(1-2)**: 121-151.

BIGAZZI, M. & SELVI, F. 2000: *Anchusa samothracica* (Boraginaceae), a new species from the island of Samothraki, Greece. - *Nordic J. Bot.* **20(2)**: 141-148.

BIGAZZI, M. & SELVI, F. 2001: Karyotype morphology and cytogeography in *Brunnera* and *Cynoglottis* (Boraginaceae). - *Bot. J. Linn. Soc.* **136(4)**: 365-378.

BILLERI, G. 1954: Osservazioni sul cariogramma di *Allium moschatum* L. (Liliaceae). - *Caryologia* **6 (1)**: 45-51 + 1 plate.

BINGÖL, F. & TEMIZER, H. 1996: *Viburnum opulus* L. - *Karaca Arbor. Mag.* **3(3)**: 145-146.

BINI MALECI, L. 1976: Numeri cromosomici per la Flora Italiana: 196-198. - *Inform. Bot. Ital.* **7(3)**: 365-366.

BINI MALECI, L. & MORI, B. 1972: Numeri cromosomici per la Flora Italiana: 131-134. - *Inform. Bot. Ital.* **4(3)**: 224-227.

BIONDI, E. 1989: The vegetation of sedimentary low coasts in Corfu island. - *Colloq. Phytosoc.* **19**: 401-427.

BIONDI, E. & GÉHU, J.-M. 1987: A study of some phytocoenoses of *Euphorbia dendroides* L. found in Greece. - *Acta Bot. Croat.* **46**: 81-84.

BIR, S.S. 1971: Cytological observations on some Indian species of *Cystopteris* Bernh. - *Nucleus (Calcutta)* **14(1)**: 56-62.

BIRD, P.N. 1984: *Cyclamen* from Rhodes. - *Cyclamen Soc. J.* **8(1)**: 20-21.

BIRKINSHAW, C.R. 1994: Aspects of the ecology and conservation of *Damasonium alisma* Miller in Western Europe. - *Watsonia* **20(1)**: 33-39.

BIRKS, H.J.B. 1976: The distribution of European pteridophytes: a numerical analysis. - *New Phytol.* **77(1)**: 257-287.

BIRKS, J. 2003: [Photo of *Iris unguicularis* subsp. *cretensis*. In: Alpines and extremes]. – *Alpine Gardener* **71(1)**: 86-98.

BISBY, F.A. 1981: Tribe 32. Genisteae (Adans.) Benth. (1865). In Polhill, R.M. & Raven, P.H. (eds): Advances in legume systematics 1: pp. 409-425. - Kew: Royal Botanic Gardens.

BISBY, F.A. 1984: The Vicieae database project: Products and services. - *Webbia* **38**: 639-644.

BISBY, F.A. & NICHOLLS, K.W. 1977: Effects on varying character definitions on classification of Genisteae (Leguminosae). - *Bot. J. Linn. Soc.* **74(2)**: 97-121.
BISHOP, M. 2000: Some notable plants [of the mountains of northern Greece]. - *Quart. Bull. Alpine Gard. Soc. Gr. Brit.* **68(3)**: 397-406.
BITTKAU, C. & COMES, H.P. 2005: Evolutionary processes in a continental island system: molecular phylogeography of the Aegean *Nigella arvensis* alliance (Ranunculaceae) inferred from chloroplast DNA. - *Molec. Ecol.* **14(13)**: 4065-4083.
BITTKAU, CH. & COMES, H.-P. 2009: Molecular inference of a Late Pleistocene diversification shift in *Nigella* s.lat. (Ranunculaceae) resulting from increased speciation in the Aegean archipelago. - *J. Biogeogr.* **36(7)**: 1346-1360.
BIVONA-BERNARDI, A. DE 1813-19: Stirpium rariorum minusque cognitarum in Sicilia sponte provenientium descriptiones nonnullis iconibus auctae. - Palermo: Typis Regiis, 4 fascs [cf. TL-2 1: 223, 1976].
BIÇAKÇI, A. & MALYER, H. 2000: Morphology of *Echinophora* L. (Apiaceae) species in Turkey [In Turkish with English summary]. - *OT Sist. Bot. Derg.* **5(1)**: 1-30.
BJELČIĆ, Ž. 1960: Biljnogeografsko rasprostranjenje vrste *Gentiana crispata* Vis. [In Serbo-Croatian]. - *God. Biol. Inst. Univ. Sarajevu* **13(1-2)**: 3-9 + 1 map.
BJELČIĆ, Ž. 1967: Flora Bosnae et Hercegovinae. IV. Sympetalae. Pars 2 [continuation of the work of G. Beck v. Mannagetta & K. Malý]. - Sarajevo: Grafičko Preduzeće "Štampa", 111 pp.
BJELČIĆ, Ž. 1968: Über die endemische Art *Achillea abrotanoides* Vis. [In Serbo-Croatian with German summary]. - *Glasn. Zemaljsk. Muz. Sarajevu* **6**: 45-51.
BJELČIĆ, Ž. 1972: *Achillea* ×*albinea* Bjelčić et Malý (*A. abrotanoides* Vis. × *A. clavenae* L.). - *Wiss. Mitt. Bosnisch-Herzegowinischen Landesmus., C* **2**: 61-64.
BJELČIĆ, Ž. 1974: Flora Bosnae et Hercegovinae. IV. Sympetalae. Pars 3 [continuation of the work of G. Beck v. Mannagetta & K. Malý]. - Sarajevo: Štamparija Trebinje, 83 pp.
BJELČIĆ, Ž. 1983: Flora Bosnae et Hercegovinae. IV. Sympetalae. Pars 4 [continuation of the work of G. Beck v. Mannagetta & K. Malý]. - Sarajevo: Štamparija Trebinje, 188 pp.
BJELČIĆ, Ž. 1989: Vegetational geographical distribution of the *Cardamine acris* Griseb. species [In Serbo-Croatian with English summary]. - *Glasn. Zemaljsk. Muz. Sarajevu* **28**: 97-102.
BJELČIĆ, Ž. & MAYER, E. 1974: Zur Taxonomie des *Gentianella crispata*-Komplexes. - *Oesterr. Bot. Z.* **122(5)**: 353-358.
BJELČIĆ, Ž. & STEVANOVIĆ, V. 1986: *Phytolacca americana* L. in the flora and vegetation of Bosnia and Herzegovina [In Serbo-Croatian with English summary]. - *God. Biol. Inst. Univ. Sarajevu* **39**: 5-11.
BJELČIĆ, Ž., ŽIVKOVIĆ, D. & MEĐEDOVIĆ, S. 1982: The spread of the forms *Lamium galeobdolon* (L.) Nath. aggr. in the territory of Bosnia and Herzegovina and Montenegro [In Serbo-Croatian with English summary]. - *Glasn. Republ. Zavoda Zaštitu Prir. Prirodnjačke Zbirke Titogradu* **15**: 133-145.
BJÖRKMAN, S.O. 1951: Chromosome studies in *Agrotis*. (A preliminary report). - *Hereditas (Lund)* **37 (3)**: 465-468.
BJÖRKMAN, S.O. 1954: Chromosome studies in *Agrostis*. II. - *Hereditas (Lund)* **40(1-2)**: 254-258.
BJÖRKMAN, S.O. 1960: Studies in *Agrostis* and related genera. - *Symb. Bot. Upsal.* **17(1)**: 1-112 + 4 plates.
BJÖRKQVIST, I. 1961: Kromosomtalsförhållandena inom släktet *Alisma* [In Swedish with English summary]. - *Bot. Not.* **114(3)**: 281-299.
BJÖRKQVIST, I. 1967: Studies in *Alisma* L. I. Distribution, variation and germination. - *Opera Bot.* **17**: 1-128.
BJÖRKQVIST, I. 1968: Studies in *Alisma* L. II. Chromosome studies, crossing experiments and taxonomy. - *Opera Bot.* **19**: 1-138.
BJÖRKQVIST, I. & AL. 1969: Chromosome numbers in Iberian angiosperms. - *Bot. Not.* **122(2)**: 271-283.
BLACKBURN, K.B. 1933: On the relation between geographic races and polyploidy in *Silene ciliata* Pourr. - *Genetica* **15**: 49-66.
BLACKBURN, K.B. & BOULT, J.J. 1930: The status of the genus *Saponaria* and its near allies considered in the light of their cytology. - *Proc. Univ. Durham Philos. Soc.* **8(3)**: 260-266.
BLACKMORE, S. 1982: Palynology of subtribe Scorzonerinae (Compositae: Lactuceae) and its taxonomic significance. - *Grana* **21(3)**: 149-160.
BLAICH, G. 1991: Ergänzungen zur Kenntnis der Orchideenflora Westkretas. - *Mitt. Arbeitskreis Heimische Orchid. Baden-Württemberg* **23(3)**: 467-472.
BLAISE, S. 1965: Observations sur la caryosystématique et l'écologie de *Myosotis silvatica* (Ehrh.) Hoffm. sensu latissimo. - *Compt. Rend. Hebd. Séances Acad. Sci.* **260(2)**: 647-650.
BLAISE, S. 1966: Caryologie et morphologie comparée de *Myosotis stricta* Link et *M. ruscinonensis* Rouy. - *Compt. Rend. Hebd. Séances Acad. Sci., Sér. D* **262**: 103-106.
BLAISE, S. 1969: Considérations biosystématiques sur le groupe *Myosotis "discolor"*. - *Compt. Rend. Hebd. Séances Acad. Sci., Sér. D* **268(22)**: 2682-2685.
BLAISE, S. 1970a: Biosystématique de *Myosotis congesta* Shuttleworth ex Albert & Reynier. – *Compt. Rend. Hebd. Séances Acad. Sci., Sér. D* **270(3)**: 502-505.
BLAISE, S. 1970b: [Chromosome numbers in *Myosotis*]. In Löve, Á. (ed.): IOPB chromosome number reports XXV. - *Taxon* **19(1)**: 102.
BLAISE, S. 1972: Problèmes taxonomiques posés par l'homogénéité apparente du genre *Myosotis*. - *Candollea* **27(1)**: 65-81.
BLAISE, S. 1976: Étude de la répartition géographique d'un certain nombre de taxons eurasiatiques de *Myosotis*. - *Compte Rend. Séances Soc. Biogéogr.* **452-460**: 10-17.
BLAISE, S. 1981: [Chromosome numbers in *Myosotis*]. In Löve, Á. (ed.): Chromosome number reports LXX. - *Taxon* **30(1)**: 76.
BLAKELOCK, R.A. 1949: *Cytisus absinthoides*. - *Bot. Mag.* **166(4)**: tab. 87 + 4 pp. text.
BLAKELOCK, R.A. 1951: A synopsis of the genus *Euonymus*. - *Kew Bull.* **1951(2)**: 210-290.
BLAKELOCK, R.A. & TURRILL, W.B. 1939: On the flora of the Nearer East: XX. - *Bull. Misc. Inform. Kew* **1939**: 186-193.
BLAKESLEE, A.F. 1959: The genus *Datura* [published after the death of the author by A.G. Avery, S. Satina and J. Rietsema]. - *Chron. Bot.* **20**: i-xli + 1-289.
BLAMEY, M. & GREY-WILSON, CH. 1993: Mediterranean wild flowers. A complete guide to the flowers of the Mediterranean coasts and islands, native and introduced. - London: Harper Collins Publishers, 560 pp.
BLANCA LÓPEZ, G. 1981a: Origen, evolución y endemismo en la sección *Willkommia* G. Blanca (Gén. *Centaurea* L.). - *Anales Jard. Bot. Madrid* **37(2)**: 607-618.
BLANCA LÓPEZ, G. 1981b: Revisión del género *Centaurea* L. sect. *Willkommia* G. Blanca, nom. nov. - *Lagascalia* **10(2)**: 131-205.
BLANCA LÓPEZ, G. 1981c: Notas cariosistemáticas en el género *Centaurea* L. sect. *Willkommia* G. Blanca. II. Conclusiones. - *Anales Jard. Bot. Madrid* **38(1)**: 109-125.
BLANCA LÓPEZ, G. & DÍAZ DE LA GUARDIA, C. 1996: Sinopsis del género *Tragopogon* L. (Asteraceae) en la Península Ibérica. - *Anales Jard. Bot. Madrid* **54(1)**: 358-363.
BLANCHARD, J.W. 1990: *Narcissus*: a guide to wild daffodils. - St. John's Woking, Surrey: Alpine Garden Society, 203 pp.
BLANCHÉ, C. 1990: *Delphinium* L. subgen. *Delphinium*: origin and evolutionary trends. - *Collect. Bot. (Barcelona)* **19**: 75-95.
BLANCHÉ, C. 1991: Revisió biosistemàtica del gènere *Delphinium* L. a la Península Ibèrica i a les Illes Balears. - Barcelona: Institut d'Estudis Catalans, 290 pp.
BLANCHÉ, C. & AL. 1997: *Delphinium* L. subgen. *Delphinium* in the Iberian Peninsula and North Africa: A new taxonomic approach. - *Lagascalia* **19(1-2)**: 59-82.
BLANCHÉ, C. & MOLERO, J. 1983: *Delphinium bolosii* C. Blanché & J. Molero spec. nova - Etude de sa position systématique dans la série *Fissa* B. Pawł. - *Candollea* **38(2)**: 709-716.
BLANCHÉ, C., MOLERO, J. & SIMÓN, J. 1987: Données cytotaxonomiques sur les *Consolida* S.F. Gray (Ranunculaceae) en Méditerranée Occidentale. - *Saussurea* **18**: 1-10.
BLANCHÉ, C. & MORELO, J. 1993: Taxonomic and nomenclatural notes on *Delphinium* L. - *Bot. J. Linn. Soc.* **113(2)**: 125-133.

BLANCO, A. & PERRINO, P. 1974: Analisi citologica di alcune specie del genere *Vicia*. - *Giorn. Bot. Ital., n.s.* **108(3-4)**: 123-133.

BLANKENHORN, B. 1978: Pollenmorphologisch-systematische Untersuchungen an Valerianaceae. - *Bot. Jahrb. Syst.* **99(1)**: 108-138.

BLASDELL, R.F. 1963: A monographic study of the fern genus *Cystopteris*. - *Mem. Torrey Bot. Club* **21 (4)**: 1-102.

BLASI, C. & AL. 1990: Life forms variability of Mediterranan sclerophyllous forests. - *Vegetatio* **88(2)**: 93-102.

BLATT, H. & HENKE, E. 1985: *Orchis coriophora* L. × *Orchis lactea* Poir. Eine seltene Orchideen- Hybride aus Griechenland. - *Ber. Arbeitskreis. Heimische Orchid.* **2(1)**: 99, 106.

BLATT, H. & HERTEL, H. 1982: Beiträge zur Verbreitung der Orchideen in Kreta nebst einer Beschreibung des Bastardes *Ophrys bombyliflora* Link × *Ophrys spruneri* Nym. - *Mitt. Arbeitskreis Heimische Orchid. Hessen* **4(1-2)**: 1-40.

BLATT, H. & WIRTH, H.W. 1990: Anmerkungen zu "Die nomenklatorischen Typen der von Linnaeus veröffentlichten Namen europäischer Orchideen". I. *Ophrys fuciflora* versus *Ophrys holoserica*. - *Ber. Arbeitskreis. Heimische Orchid.* **7(1)**: 4-8.

BLAXLAND, K. 2004: A new species of *Viola* (Violaceae) from south-west Turkey. - *Bot. J. Linn. Soc.* **145(4)**: 505-509.

BLEČIĆ, V. 1967: Catalogus Flora Jugoslaviae. I/2. Gymnospermae. - Ljubljana: Acad. Sci. Art. Slovenica, 14 pp.

BLEČIĆ, V. & LAKUŠIĆ, R. 1969: Forests of *Pinion heldreichii* Horv. on Štitovo and Bjelasica in Montenegro [In Serbo-Croatian with English summary]. - *Glasn. Republ. Zavoda Zaštitu Prir. Prirodnjačke Zbirke Titogradu* **2**: 5-10 + 2 tables.

BLEČIĆ, V. & MAYER, E. 1967: Die europäischen Sippen der Gattung *Amphoricarpos* Visiani. - *Phyton (Horn)* **12(1-4)**: 150-158.

BLEČIĆ, V. & MAYER, E. 1974: Zur Kenntnis der balkanischen Taxa von *Lonicera* L. ser. *Alpigenae* Rehd. - *Feddes Repert.* **84(9-10)**: 647-653.

BLEČIĆ, V. & MAYER, E. 1981: Kritische Betrachtung über *Helleborus cyclophyllus* Boiss. in Jugoslawien. - *Glasn. Republ. Zavoda Zaštitu Prir. Prirodnjačke Zbirke Titogradu* **14**: 7-10.

BLEČIĆ, V. & PULEVIĆ, V. 1979: New data from the flora of Montenegro [In Serbo-Croatian with English summary]. - *Glasn. Republ. Zavoda Zaštitu Prir. Prirodnjačke Zbirke Titogradu* **12**: 189-193.

BLEČIĆ, V. & PULEVIĆ, V. 1983: Critical review of the *Amelanchier ovalis* in the flora of Yugoslavien [In Serbo-Croatian with English and Italian summaries]. - *Povrem. Izd. Muz. Grada Šibenika* **10**: 169-173.

BLIONIS, G.J., KOKKINI, S. & VOKOU, D. 1999: *Campanula* chasmophytes of Mt Olympos. In: Proceedings of 21st Panhellenic Conference of EEBE, Syros, May 28-31, 1999, pp. 191-192 [In Greek and English].

BLIONIS, G.J. & VOKOU, D. 2002: Structural and functional divergence of *Campanula spatulata* subspecies on Mt Olympos (Greece). - *Pl. Syst. Evol.* **232(1-2)**: 89-105.

BOBROV, A.E. 1967: The family Osmundaceae (R. Br.) Kaulf., its taxonomy and geography [In Russian with English summary]. - *Bot. Zhurn. (Moscow & Leningrad)* **52(11)**: 1600-1610.

BOCCHIERI, E., DE MARTIS, B. & ONNIS, A. 1979: "Solanum cornutum" Lam. e "Solanum elaeagnifolium" Cav., nuove avventizie per la Sardegna. - *Inform. Bot. Ital.* **10(2)**: 226-229.

BÖCHER, T.W. 1947: Cytogenetic and biological studies in *Geranium robertianum* L. - *Biol. Meddel. Kongel. Danske Vidensk. Selsk.* **20(8)**: 1-29.

BÖCHER, T.W. 1960: Experimental and cytological studies on plant species. V. The *Campanula rotundifolia* complex. - *Biol. Skr.* **11(4)**: 1-69 + 14 plates.

BÖCHER, T.W. 1964: Chromosome connections and abberations in the *Campanula persicifolia* group. - *Svensk Bot. Tidskr.* **58(1)**: 1-17.

BÖCHER, T.W. & LARSEN, K. 1957: Cytotaxonomical studies in the *Sanguisorba minor* complex. - *Bot. Tidsskr.* **53**: 284-290.

BÖCHER, T.W. & LARSEN, K. 1958: Geographical distribution of initiation of flowering, growth habit, and other characters in *Holcus lanatus* L. - *Bot. Not.* **111(1)**: 289-300.

BÖCHER, T.W., LARSEN, K. & RAHN, K. 1953: Experimental and cytological studies on plant species. I. *Kohlrauschia prolifera* and *Plantago coronopus*. - *Hereditas (Lund)* **39(1-2)**: 289-304.

BÖCHER, T.W., LARSEN, K. & RAHN, K. 1955a: Experimental and cytological studies on plant species. *Trifolium arvense* and some other pauciennial herbs. - *Biol. Skr.* **8(3)**: 1-31.

BÖCHER, T.W., LARSEN, K. & RAHN, K. 1955b: Experimental and cytological studies on plant species. *Plantago coronopus* and allied species. - *Hereditas (Lund)* **41(3-4)**: 423-453.

BÖCHER, T.W. & LEWIS, M.C. 1962: Experimental and cytological studies on plant species. VII. *Geranium sanguineum*. - *Biol. Skr.* **11(5)**: 1-25 + 2 plates.

BOCQUET, G. 1980: Crise de salinité messinienne et floristique Méditerranéenne. In: Colloque de la Fondation L. Emberger, Montpellier 9-10 avril 1980. - *Naturalia Monspel., Sér. Bot.* **No. hors série**: 21-31.

BOCQUET, G., WIDLER, B.E. & KIEFER, H. 1978: The Messinian model - A new outlook for the floristics and systematics of the Mediterranean area. - *Candollea* **33(2)**: 269-287.

BOCQUET, G. & ZERBST, K.-J. 1974: Beiträge zur Kenntnis der Gattung *Digitalis* L. (Scrophulariaceae). *Digitalis laevigata* Waldst. & Kit., *D. graeca* Ivanina et *D. graeca* var. *megalantha* Bocquet & Zerbst, var. nova. - *Candollea* **29(2)**: 251-266.

BOCZANTZEVA, V.V. 1973: *Thalictrum minus* L. s. lat. of the U.S.S.R. European part [In Russian]. - *Bot. Zhurn. (Moscow & Leningrad)* **58(11)**: 1641-1644.

BODEGOM, J. VAN 1973: Orchideeën in Z.O.-Europa en West-Turkije (2): Griekenland. - *Orchideeën* **35 (6)**: 132-136.

BODEGOM, J. VAN 1974: Orchideeën in Z.O.-Europa en West-Turkije (3): Kreta. - *Orchideeën* **36(1)**: 4-9.

BOGAERT, H. VAN 1982: Plantportret: *Origanum dictamnus* L. [In Dutch with English summary]. - *Bull. Arbor. Waasland* **5(2)**: 6-12.

BOGDANOVIĆ, S. & AL. 2011a: *Allium achaium* Boiss. (Alliaceae), a critical species of Greek flora. - *Candollea* **66(1)**: 57-64.

BOGDANOVIĆ, S. & AL. 2011b: *Allium cithaeronis* Bogdanović, C. Brullo, Brullo, Giusso, Musarella & Salmeri (Alliaceae), a new species from Greece. - *Candollea* **66(2)**: 377-382.

BOGDANOVIĆ, S. & AL. 2012: Taxonomic revision of the genus *Fumana* (Cistaceae) in Croatia. - *Pl. Biosyst.* **146 suppl.**: 69-85.

BOGDANOVIĆ, S. & AL. 2014: *Campanula skanderbegii*: Molecular and morphological evidence of a new *Campanula* species (Campanulaceae) endemic to Albania. - *Syst. Bot.* **39(4)**: 1250-1260.

BOGDANOVIĆ, S. & AL. 2015: *Aegilops uniaristata* Vis. (Poaceae): typification and occurrence in Croatia. - *Acta Bot. Croat.* **74(1)**: 165-172.

BOGDAVOVIĆ, S, & JOGAN, N. (EDS) 2015: Book of abstracts - 6th Balkan Botanical Congress. - Rijeka: 160 pp.

BOGIN, C. 1955: Revision of the genus *Sagittaria* (Alismataceae). - *Mem. New York Bot. Gard.* **9(2)**: 179-233.

BOGNER, J. 1979: A critical list of the aroid genera. - *Aroideana* **1(3)**: 63-73.

BOGNER, J. & BOYCE, P.C. 1989: A remarkable new *Biarum* (Araceae) from Turkey. - *Willdenowia* **18 (2)**: 409-417.

BOGNER, J. & NICOLSON, D.H. 1991: A revised classification of Araceae with dichotomous keys. - *Willdenowia* **21(1-2)**: 35-50.

BÖHLING, N. 1994: Studien zur landschaftsökologischen Raumgliederung auf der mediterranen Insel Naxos (Griechenland) unter besonderer Berücksichtigung von Zeigerpflanzen. - *Diss. Bot.* **230**: i-vii, 1-247 + 8 maps and charts.

BÖHLING, N. 1995: Zeigerwerte der Phanerogamen-Flora von Naxos (Griechenland). Ein Beitrag zur ökologischen Kennzeichnung der mediterranen Pflanzenwelt. - *Stuttgarter Beitr. Naturk., A* **533**: 1-75.

BÖHLING, N. 1997: Ergänzungen und Anmerkungen zur Flora der Insel Naxos (Kykladen, Griechenland): Zur botanischen Besonderheit und pflanzengeographischen Bedeutung des Engares-Tales. - *Stuttgarter Beitr. Naturk., A* **560**: 1-25.

BÖHLING, N. 1999: Notes on *Lycopus europaeus, Melilotus albus, Ononis mollis, Eucalyptus camaldulensis, Solanum nigrum* and *Solanum physalifolium*. In Greuter, W. & Raus, Th. (eds): Med- Checklist notulae, 18. - *Willdenowia* **29(1-2)**: 55-64.

BÖHLING, N. 2000a: *Myosotis litoralis, Pulicaria vulgaris, Melilotus officinalis, Aphanes floribunda, Cardiospermum halicacabum, Peucedanum chryseum, Isolepis setacea*. In Greuter, W. & Raus, Th. (eds): Med-Checklist notulae, 19. - *Willdenowia* **30(2)**: 231-242.

BÖHLING, N. 2000b: *Ranunculus veronicae* (Ranunculaceae), a new species from W Crete (Greece). - *Willdenowia* **30(2)**: 245-250.

BÖHLING, N. 2001: *Centaurea cyanus, Phagnalon saxatile, Scorzonera judaica, Juglans regia, Physalis angulata*. In Greuter, W. & Raus, Th. (eds): Med-Checklist notulae, 20. - *Willdenowia* **31(2)**: 321-325.

BÖHLING, N. 2004: *Robinia pseudoacacia* L. In Greuter, W. & Raus, Th. (eds): Med-Checklist notulae, 22. - *Willdenowia* **34(1)**: 76.

BÖHLING, N. 2006a: Report 14. In Vladimirov, V., Tan, Kit & Stevanović, V. (eds): New floristic records in the Balkans: 1. - *Phytol. Balcan.* **12(1)**: 111.

BÖHLING, N. 2006b: *Solanum pseudocapsicum*. In Greuter, W. & Raus, Th. (eds): Med-Checklist notulae, 24. - *Willdenowia* **36(2)**: 726.

BÖHLING, N. 2006c: Reports 20-27. In Vladimirov, V. & al. (eds): New floristic records in the Balkans: 2. - *Phytol. Balcan.* **12(2)**: 281-282.

BÖHLING, N. & AL. 2002: Notes on the Cretan mistletoe, *Viscum album* subsp. *creticum* subsp. nov. (Loranthaceae/Viscaceae). - *Israel J. Pl. Sci.* **50**: 77-84.

BÖHLING, N. & BÄSSLER, M. 1999: Notes on *Vicia monantha*. In Greuter, W. & Raus, Th. (eds): Med- Checklist notulae, 18. - *Willdenowia* **29(1-2)**: 62.

BÖHLING, N., GREUTER, W. & RAUS, TH. 2000: *Trifolium phitosianum* (Leguminosae), a new annual clover species from Greece. - *Bot. Chron. (Patras)* **13**: 37-44.

BÖHLING, N., GREUTER, W. & RAUS, TH. 2002: Zeigerwerte der Gefässpflanzen der Südägäis. - *Braun- Blanquetia* **32**: 109 pp.

BÖHLING, N. & RAUS, TH. 1999: Notes on *Stellaria media*. In Greuter, W. & Raus, Th. (eds): Med- Checklist notulae, 18. - *Willdenowia* **29(1-2)**: 52.

BÖHLING, N. & RAUS, TH. 2000: *Isoetes durieui* Bory, *Asclepias physocarpa* (E. Mey.) Schltr., *Lythrum tribracteatum* Spreng. In Greuter, W. & Raus, Th. (eds): Med-Checklist notulae, 19. - *Willdenowia* **30(2)**: 229-238.

BÖHLING, N. & SCHOLZ, H. 2003: The Gramineae (Poaceae) flora of the Southern Aegean islands (Greece). Checklist, new records, internal distribution. - *Ber. Inst. Landschafts-Pflanzenökologie Univ. Hohenheim Beih.* **16**: 1-88.

BÖHLING, N. & SCHOLZ, H. 2005: *Pennisetum clandestinum* Chiov. In Greuter, W. & Raus, Th. (eds): Med-Checklist notulae, 23. - *Willdenowia* **35(1)**: 63-64.

BÖHLING, N. & SCHOLZ, H. 2012: *Eragrostis minor* subsp. *roborovskii*. In Greuter, W. & Raus, Th. (eds): Med-Checklist notulae, 31. - *Willdenowia* **42(2)**: 293.

BÖHLING, N. & SNOGERUP, S. 2000: *Scleranthus polycarpos, Rumex obtusifolius* subsp. *subalpinus, Bupleurum euboeum*. In Greuter, W. & Raus, Th. (eds): Med-Checklist notulae, 19. - *Willdenowia* **30 (2)**: 232-240.

BÖHLING, N., TAN, KIT & NIELSEN, H. 2004: *Vicia galilaea* Plitm. & Zoh. In Greuter, W. & Raus, Th. (eds): Med-Checklist notulae, 22. - *Willdenowia* **34(1)**: 76.

BÖHME, B. 2001: Neues über das isolierte Vorkommen von *Rhinanthus rumelicus* Velen. (Drüsiger Klappertopf) bei Jena. - *Haussknechtia* **8**: 85-92.

BÖHNERT, T. & LOBIN, W. 2017: *Leopoldia neumannii* sp. nov. (Asparagaceae: Scilloideae): a new species of *Muscari* sensu lato from Greece. - *Willdenowia* **47(2)**: 179-185.

BÖHR, H.-J. 1989: Der Urwaldrest und der Beginn der forstlichen Nutzung in den Rhodopen. In Lienau, C. (ed.): "Europapark" in Nordost-Griechenland. - *Ber. Arbeitsgeb. Entwicklungsforsch. (Münster)* **15**: 37-40.

BOIE, K. 2004: Wiederentdeckung von *Ophrys bertolonii* auf Korfu. - *J. Eur. Orch.* **36(4)**: 1079-1080.

BOIE, K. 2008: *Ophrys delphinensis* auch im Süden des Peloponnes. - *Ber. Arbeitskreis. Heimische Orchid.* **25(2)**: 160-161.

BOISSIER, E. 1838: Elenchus plantarum novarum minusque cognitarum quas in itinere hispanico legit. - Genevae: Typographia Lador et Ramboz, iv + 94 pp.

BOISSIER, E. 1841: Plantae Aucherianae orientales enumeratae, cum novarum specierum descriptione. - *Ann. Sci. Nat. Bot., Sér. 2* **16**: 347-377.

BOISSIER, E. 1842: Plantae Aucherianae orientales enumeratae, cum novarum specierum descriptione (suite). - *Ann. Sci. Nat. Bot., Sér. 2* **17**: 45-90.

BOISSIER, E. 1842-59: Diagnoses plantarum orientalium novarum. Ser. 1, 13 fascs. in 2 vols. Ser. 2, 6 fascs. - Genève, etc.; fascimile ed. by A. Asher & Co. 1969 [cf. TL-2 1: 257-258, 1976].

BOISSIER, E. 1844a: Plantae Aucherianae adjunctis nonnullis e regionibus Mediterraneis et Orientalibus aliis cum novarum specierum descriptione. - *Ann. Sci. Nat. Bot., Sér. 3* **1**: 297-349.

BOISSIER, E. 1844b: Plantae Aucherianae adjunctis nonnullis e regionibus Mediterraneis et Orientalibus aliis cum novarum specierum descriptione. - *Ann. Sci. Nat. Bot., Sér. 3* **2**: 46-96.

BOISSIER, E. 1867: Flora Orientalis. Vol. 1 (Thalamiflorae). - Basel & Genève. (Reprinted 1975 by A. Asher & Co.), xxxiv + 1017 pp.

BOISSIER, E. 1872: Flora Orientalis. Vol. 2 (Calyciflorae). - Basel & Genève. (Reprinted 1975 by A. Asher & Co.), 1159 pp.

BOISSIER, E. 1875a: Flora Orientalis. Vol. 3 (Calyciflorae & Gamopetalae). - Basel & Genève. (Reprinted 1975 by A. Asher & Co.), 1033 pp.

BOISSIER, E. 1875b: Flora Orientalis. Vol. 4 (Coralliflorae & Monochlamydeae). - Basel & Genève. (Reprinted 1975 by A. Asher & Co.), pp. 1-280.

BOISSIER, E. 1879: Flora Orientalis. Vol. 4 (Coralliflorae & Monochlamydeae). - Basel & Genève. (Reprinted 1975 by A. Asher & Co.), pp. 281-1276.

BOISSIER, E. 1882: Flora Orientalis. Vol. 5 (Monocotyledoneae). - Basel & Genève. (Reprinted 1975 by A. Asher & Co.), pp. 1-428.

BOISSIER, E. 1884: Flora Orientalis. Vol. 5 (Monocotyledoneae). - Basel & Genève. (Reprinted 1975 by A. Asher & Co.), pp. 429-868.

BOISSIER, E. 1888: Flora Orientalis. Suppl. - Basel & Genève. (Reprinted 1975 by A. Asher & Co.), xxxiii + 466 pp. + 7 plates.

BOISSIEU, M. DE 1896: Quelques notes sur la flore d'Orient. - *Bull. Soc. Bot. France* **43**: 283-291.

BOKHARI, M.H. 1970: Materials for a Flora of Turkey: XXII. Plumbaginaceae. - *Notes Roy. Bot. Gard. Edinburgh* **30(2)**: 295-304.

BOKHARI, M.H. 1972a: Synopsis of Plumbaginaceae in Turkey. - *Notes Roy. Bot. Gard. Edinburgh* **32 (1)**: 57-77.

BOKHARI, M.H. 1972b: A brief review of stigma and pollen types in *Acantholimon* and *Limonium*. - *Notes Roy. Bot. Gard. Edinburgh* **32(1)**: 79-84.

BOKHARI, M.H. 1972c: Anatomical characters and their taxonomic importance in *Acantholimon*. - *Notes Roy. Bot. Gard. Edinburgh* **32(1)**: 85-92.

BOKHARI, M.H. 1973: Variation and taxonomic importance of anatomical characters in *Limonium*. - *Notes Roy. Bot. Gard. Edinburgh* **32(2)**: 275-290 + 2 plates.

BOKHARI, M.H. & HEDGE, I.C. 1975: Anatomical characters in *Capparis spinosa* and its allies. - *Notes Roy. Bot. Gard. Edinburgh* **34(2)**: 231-240.

BOLIN, J.F. & AL. 2008: The genus *Isoetes* (Isoetaceae, Lycophyta) in South-Western Asia. - *Turk. J. Bot.* **32(6)**: 447-457.

BOLLE, F. 1933: Eine Übersicht über die Gattung *Geum* L. und die ihr nahestehenden Gattungen. - *Repert. Spec. Nov. Regni Veg. Beih.* **72**: 1-119.

BOLLI, R. 1994: Revision of the genus *Sambucus*. - *Diss. Bot.* **223**: 1-227 + 29 plates.

BOLLIGER, M. 1982: Die Gattung *Pulmonaria* in Westeuropa. - Vaduz: J. Cramer, 215 pp.

BOLLIGER, M. 1985: Die Drüsenhaare der Gattung *Odontites* Ludwig (Scrophulariaceae) und ihre systematische Bedeutung. - *Bot. Jahrb. Syst.* **107(1-4)**: 153-175.

BOLLIGER, M. 1993: Systematik und Chorologie der Gattung *Odontites* Ludwig s.l. (Scrophulariaceae). - *Flora, Morphol. Geobot. Oekophysiol.* **188(4)**: 345-365.

BOLLIGER, M. 1996: Monographie der Gattung *Odontites* (Scrophulariaceae) sowie der verwandten Gattungen *Macrosyringion, Odontitella, Bornmuellerantha* und *Bartsiella*. - *Willdenowia* **26(1-2)**: 37-168.

BOLMGREN, K. & OXELMAN, B. 2004: Generic limits in *Rhamnus* L. s.l. (Rhamnaceae) inferred from nuclear and chloroplast DNA sequence phylogenies. - *Taxon* **53(2)**: 383-390.

BOLÒS, O. DE & AL. 1996: A survey on the vegetation of Cephalonia (Ionian islands). - *Phytocoenologia* **26(1)**: 81-123.

BOLÒS, O. DE & VIGO, J. 1974: Notes sobre taxonomia i nomenclatura de plantes, I. - *Butl. Inst. Catalana Hist. Nat., Secc. Bot.* **38**: 61-89.

BOLTENKOV, E.V. & GÜNER, A. 2021: Nomenclatural remarks on *Iris orientalis* (Iridaceae) and four related names including the Linnaean *I. ochroleuca*. - *Phytotaxa* **512(3)**: 205-212.

BOMBLE, W. & SCHOLZ, H. 1999: Eine neue Unterart des *Bromus secalinus* (Gramineae): ein sekundäres Unkraut. - *Feddes Repert.* **110(5-6)**: 425-438.

BONAFEDE, F. & VERIVAKI, A. 1988: Note botaniche sull'isola di Creta. - *Nat. Montagna* **35(4)**: 39-46.

BONAR, A. 1980: Wild flowers of Crete. - *Garden (London)* **105(9)**: 367-370.

BONATI, G. 1918: Le genre *Pedicularis* L. Morphologie, classification, distribution géographique, évolution et hybridation. - Nancy: Imprimerie Berger-Levrault, xiii + 168 pp.

BONAVENTURA, G. 1935: *Paspalum dilatatum* Poir. in Italia. - *Nuovo Giorn. Bot. Ital., n.s.* **41(4)**: 750-751.

BONAVENTURA, G. 1950: Diffusione di *Paspalum paspalodes* e *Paspalum dilatatum*. - *Nuovo Giorn. Bot. Ital., n.s.* **56(4)**: 648-650.

BONDEV, I. 1961: Neue und seltene Pflanzen für die Flora der Westrhodopen und des Mestatales (im Gebiet von Goce Delčev) [In Bulgarian with Russian and German summaries]. - *Izv. Bot. Inst. (Sofia)* **8**: 225-230.

BONDEV, I. 1964: *Paspalum distichum* L. im Strumatal [In Bulgarian with Russian and German summaries]. - *Izv. Bot. Inst. (Sofia)* **13**: 165-168.

BONDEV, I. 1991: The vegetation of Bulgaria. Map 1: 600.000 with explanatory text [In Bulgarian with English summary]. - Sofia: St. Kliment Ohridski University Press, 184 pp. + 1 fold-out map.

BONDEV, I. & KUZMANOV, B.A. 1963: Materials on Bulgaria's flora [In Bulgarian with Russian and English summaries]. - *Izv. Bot. Inst. (Sofia)* **11**: 153-154.

BONDEV, I., MARKOVA, M.L. & RADENKOVA, Y. 1967: Materials about the flora of the Rhodope and south- eastern Bulgaria [In Bulgarian with Russian and English summaries]. - *Izv. Bot. Inst. (Sofia)* **17**: 263-266.

BONNET, A.L.M. 1958: Étude caryologique de quelques *Thymus* méditerranéens. - *Naturalia Monspel., Sér. Bot.* **10**: 3-6.

BONNET, A.L.M. 1963a: Contribution à l'étude caryologique de *Clypeola jonthlaspi* L. - *Naturalia Monspel., Sér. Bot.* **15**: 29-40.

BONNET, A.L.M. 1963b: Contribution à l'étude caryologique du genre *Alyssum* L. (s. lat.). - *Naturalia Monspel., Sér. Bot.* **15**: 41-52.

BOOTH, T.A. & RICHARDS, A.J. 1976: Studies in the *Hordeum murinum* aggregate. 1. Morphology. - *Bot. J. Linn. Soc.* **72(1-4)**: 149-159.

BOOTT, F. 1858: Illustrations of the genus *Carex*. Vol. 1. - London: pp. i-xii + 1-74 pp. + plates 1-200.

BOOTT, F. 1860: Illustrations of the genus *Carex*. Vol. 2. - London: i-iv + 75-103 + plates 201-310.

BOOTT, F. 1862: Illustrations of the genus *Carex*. Vol. 3. - London: pp. i-iv + 105-126 + plates 311-411.

BOOTT, F. 1867: Illustrations of the genus *Carex*. Vol. 4. - London: pp. 127-233 + plates 412-460.

BOR, N.L. 1963: Three new species of Gramineae from the Middle East. - *Notes Roy. Bot. Gard. Edinburgh* **25(1)**: 63-65.

BOR, N.L. 1968: Two new grasses. - *Notes Roy. Bot. Gard. Edinburgh* **28(3)**: 299-300.

BOR, N.L. 1969: The strange case of the genus "Aeluropus" Trin. - *Webbia* **24(1)**: 401-409.

BORATYŃSKA, K. 1976: Distribution of *Populus tremula* L. in southwestern Asia. - *Arbor. Kórnickie* **21**: 9-14.

BORATYŃSKA, K. 1991: Chorology of the family Ulmaceae (sensu stricto). - *Arbor. Kórnickie* **34**: 3-29.

BORATYŃSKA, K. & AL. 1986: Trees and shrubs of Rodhos - a monographic study. - *Arbor. Kórnickie* **30**: 127-199.

BORATYŃSKA, K. & BORATYŃSKI, A. 1985: Distribution of *Liquidambar orientalis* Miller on Rodhos island. - *Arbor. Kórnickie* **29**: 3-11.

BORATYŃSKA, K. & DOLATOWSKI, J. 1986: Distribution of *Amygdalus graeca* Lindl. (Rosaceae) on Rhodes. - *Ann. Mus. Goulandris* **7**: 25-32.

BORATYŃSKI, A. & AL. 1991: Woody flora of Zakinthos island (Greece). - *Arbor. Kórnickie* **35**: 47-76.

BORATYŃSKI, A. & AL. 2014: The biogeography and genetic relationships of *Juniperus oxycedrus* and related taxa from the Mediterranean and Macaronesian regions. - *Bot. J. Linn. Soc.* **174(4)**: 637-653.

BORATYŃSKI, A. & BROWICZ, K. 1983: *Juniperus drupacea* Labill. in Greece. - *Arbor. Kórnickie* **27**: 3-16.

BORATYŃSKI, A. & BROWICZ, K. 1993: Contribution to the woody flora of Kefallinia (Ionian Islands, Greece). - *Arbor. Kórnickie* **38**: 7-15.

BORATYŃSKI, A. & BROWICZ, K. 1994: Trees and shrubs of Itháki (Ionian Islands, Greece). - *Arbor. Kórnickie* **39**: 9-30.

BORATYŃSKI, A. & BROWICZ, K. 1996: Trees and shrubs of the Ionian Islands (Greece). A preliminary list. - *Arbor. Kórnickie* **41**: 47-54.

BORATYŃSKI, A. & BROWICZ, K. 1997: Woody flora of Korfu (Ionian Islands, Greece). - *Arbor. Kórnickie* **42**: 65-100.

BORATYŃSKI, A., BROWICZ, K. & ZIELIŃSKI, J. 1984: Contribution to the woody flora of Greece. - *Arbor. Kórnickie* **28**: 7-68.

BORATYŃSKI, A., BROWICZ, K. & ZIELIŃSKI, J. 1987: Woody flora of Chios. - *Arbor. Kórnickie* **32**: 37-92.

BORATYŃSKI, A., BROWICZ, K. & ZIELIŃSKI, J. 1991: Woody flora of Euboea (Evvoia). - *Arbor. Kórnickie* **33**: 13-74.

BORATYŃSKI, A., BROWICZ, K. & ZIELIŃSKI, J. 1992: Chorology of trees and shrubs in Greece. - Poznań: Poznanska Drukarnia Naukowa, 286 pp.

BORATYŃSKI, A., TOMLIK, A. & ZIELIŃSKI, J. 1991: *Hypericum aegyptiacum* L. and *Dianthus fruticosus* L. on the island of Zakinthos (Greece). - *Arbor. Kórnickie* **34**: 55-63.

BORATYŃSKI, A. & ZIELIŃSKI, J. 1990: Some interesting woody species new for the island of Chios, Greece. - *Ann. Mus. Goulandris* **8**: 33-40.

BORATYŃSKI, A. & ZIELIŃSKI, J. 1991: New data to the distribution of *Scabiosa hymettia* Boiss. et Spruner. - *Arbor. Kórnickie* **34**: 65-69.

BORBÁS, V.V. 1880: A magyar birodalom vadon termö Roszsai monographiájának kisérlete. (Primitiae monographiae Rosarum imperii Hungarici) [In Hungarian]. - *Math. Term. Közlem.* **16**: 305-560.

BORBÁS, V.V. 1886: *Coronilla emeroides* Boiss. et Sprunn. - *Oesterr. Bot. Z.* **36(7)**: 230-232.

BORBÁS, V.V. 1889: Conspectus Ajugarum (e sectione *Bugulae*) novarum dubiarumque. - *Természetrajzi Füz.* **12**: 108-112.

BORBÁS, V.V. 1890: Közép-Europa, különözen magyarország kakakfüveinek ismertetése. (Symbolae Thymos Europae mediae, praecipue Hungariae cognoscendos) [In Hungarian]. - *Math. Term. Közlem.* **24**: 37-116.

BORBÁS, V.V. 1902: Species Hesperidium Hungariae atque Haemi. - *Magyar Bot. Lapok* **1**: 161-167, 196-205, 229-237, 261-272, 304-313, 344-349, 369-380.

BORBÁS, V.V. 1903: Species Hesperidium Hungariae atque Haemi [finis]. - *Magyar Bot. Lapok* **2(1-2)**: 13-23.

BORG, J. 1927: Descriptive Flora of the Maltese islands, including the ferns and flowering plants. - Malta: Govt. Printing Office, 846 pp. [reprinted 1976 by Otto Koeltz].

BORG, S.J. TER (ED.) 1986: Proceedings of a workshop on biology and control of *Orobanche*. - Wageningen: LH/VPO, vii + 206 pp.

BORGEN, L. 1975: Chromosome numbers of vascular plants from Macaronesia. - *Norweg. J. Bot.* **22(2)**: 71-76.

BORGEN, L. 1987: *Lobularia* (Cruciferae). A biosystematic study with special reference to the Macaronesian region. - *Opera Bot.* **91**: 1-96.

BORHIDI, A. 1957: *Centaurea achtarovii* Urumoff dans les Carpathes oientaux. Annotations sur le groupe *Centaurea triumfettii* All. - *Ann. Hist.-Nat. Mus. Natl. Hung. n.s.* **8**: 219-225.

BORHIDI, A. 1966: The variability range of *Paronychia cephalotes* (M.B.) Bess. and new data to the knowledge of south-European *Paronychia* species. - *Acta Bot. Acad. Sci. Hung.* **12(1-2)**: 33-40.

BORHIDI, A. 1968: Karyological studies on southeast European plant species, I. - *Acta Bot. Acad. Sci. Hung.* **14(3-4)**: 253-260.

BORHIDI, A. & ISÉPY, I. 1965: Taxa et combinationes novae generis *Potentilla* L. - *Acta Bot. Acad. Sci. Hung.* **11(3-4)**: 297-302.

BORHIDI, A. & PRISZTER, SZ. 1966: Eine neue *Cynanchum*-Art (*C. pannonicum* n.sp.) in Ungarn. - *Acta Bot. Acad. Sci. Hung.* **12(3-4)**: 241-254.

BORHIDI, A. & SIKURA, J.J. 1961: Observations on the *Paronychia* species of south-eastern Europe [In Russian with English summary]. - *Acta Bot. Acad. Sci. Hung.* **7(1-2)**: 1-5.

BORJA CARBONELL, J. 1965: Revisión de las especies españolas del género *Lythrum* L. - *Anales Inst. Bot. Cavanilles* **23**: 145-162 + 8 plates.

BORKOWSKY, O. 1994: Übersicht der Flora von Korfu. - *Braunschweig. Geobot. Arb.* **3**: 1-202.

BORKOWSKY, O. 1997: Vorläufige Sippenliste der Neufunde von Korfu/Kerkira 1993-1996. Letzte Fassung vom 15. 10. 96. - Pp. 84-86 in: Dinter I.: Insel Korfu (Ionische Inseln, Griechenland. Botanische Studienreise. - Lauffen: Privately distributed, 89 pp.

BORNMÜLLER, J. 1887: *Rhamnus orbiculata* Brnmllr. n. sp. - *Oesterr. Bot. Z.* **37(7)**: 225-227.

BORNMÜLLER, J. 1889: Ein Beitrag zur Eichenflora des südöstlichen Europa. - *Bot. Centralbl.* **37(5)**: 129-131.

BORNMÜLLER, J. 1894: Nachtrag zu "Florula insulae Thasos". - *Oesterr. Bot. Z.* **44(4-6)**: 124-128, 173-176, 212-216.

BORNMÜLLER, J. 1897: *Rhamnus orbiculata* Bornm. - *Bot. Centralbl.* **72(7)**: 225-229.

BORNMÜLLER, J. 1900: Ein neuer, bisher verkannter Bürger der europäischen Flora. - *Oesterr. Bot. Z.* **50(3)**: 90-93.

BORNMÜLLER, J. 1902a: Ueber die systematische Stellung von *Nigella elata* Boiss. - *Bull. Herb. Boissier, Ser. 2* **2(3)**: 329-332.

BORNMÜLLER, J. 1902b: Ueber die weitere Verbreitung von *Sideritis curvidens* Stapf. - *Mitth. Thüring. Bot. Vereins, N.F.* **16**: 122-123.

BORNMÜLLER, J. 1902c: *Viola Aetolica* Boiss. & Heldr. in Montenegro. - *Magyar Bot. Lapok* **1(11)**: 350.

BORNMÜLLER, J. 1905a: Kritische Bemerkungen über *Centaurea depressa* M.B. der europäischen Flora. - *Magyar Bot. Lapok* **4(8-11)**: 260-262.

BORNMÜLLER, J. 1905b: Novitiae florae orientalis. Series I (1-53). - *Mitt. Thüring. Bot. Vereins, N.F.* **20**: 1-51.

BORNMÜLLER, J. 1910: Über *Scabiosa Palaestina* L., neu für die Flora Europas. - *Magyar Bot. Lapok* **9 (5-9)**: 144-145.

BORNMÜLLER, J. 1911: *Verbascum lasianthum* Boiss., ein neuer Bürger der Flora Europas. - *Magyar Bot. Lapok* **10(4-7)**: 118-120.

BORNMÜLLER, J. 1912: Zur Nomenclatur von "Phleum exaratum". - *Magyar Bot. Lapok* **11(1-4)**: 18-20.

BORNMÜLLER, J. 1918a: Über den Formenkreis von *Cercis Siliquastrum* L. und *Cercis Griffithii* Boiss. - *Beih. Bot. Centralbl., Abt. 2* **36**: 1-14.

BORNMÜLLER, J. 1918b: *Echinops oxyodontus* Bornm. et Diels. Eine neue Art der Sektion *Ritro* aus Mazedonien. - *Magyar Bot. Lapok* **17(1-12)**: 42-44.

BORNMÜLLER, J. 1919: Zur Gattung *Moehringia*. - *Repert. Spec. Nov. Regni Veg.* **16**: 183-186.

BORNMÜLLER, J. 1921a: Über zwei neue Nelken aus dem Balkan. - *Repert. Spec. Nov. Regni Veg.* **17**: 38-41.

BORNMÜLLER, J. 1921b: Über eine neue *Solenanthus*-Art aus dem Balkan. - *Repert. Spec. Nov. Regni Veg.* **17**: 276-279.

BORNMÜLLER, J. 1921c: Ein Beitrag zur Kenntnis der Gattung *Asyneuma* Griseb. - *Beih. Bot. Centralbl., Abt. 2* **38**: 333-351.

BORNMÜLLER, J. 1921d: Über eine neue *Carum*-Art aus dem Balkan. - *Oesterr. Bot. Z.* **70(3-5)**: 101-102.

BORNMÜLLER, J. 1921e: Zur Gattung *Centaurea*. - *Beih. Bot. Centralbl., Abt. 2* **38**: 458-465.

BORNMÜLLER, J. 1921f: Zur Gattung *Ballota*. - *Repert. Spec. Nov. Regni Veg.* **17**: 282-286.

BORNMÜLLER, J. 1921g: *Centaurea cylindrocephala* Bornm. (spec. nov. sect. *Acrolophus-Paniculatae*). - *Repert. Spec. Nov. Regni Veg.* **17**: 453-455.

BORNMÜLLER, J. 1922a: *Pedicularis Ferdinandi* Bornm. spec. nov. (sect. *Bicuspidatae*) e flora macedonica. - *Notizbl. Bot. Gart. Berlin-Dahlem* **8(73)**: 213-218.

BORNMÜLLER, J. 1922b: Neues und Bemerkenswertes über *Verbascum*-Arten Mazedoniens. - *Repert. Spec. Nov. Regni Veg.* **18**: 133-141.

BORNMÜLLER, J. 1923a: Zur *Verbascum*-Flora Mazedoniens. - *Repert. Spec. Nov. Regni Veg.* **19**: 97-99.

BORNMÜLLER, J. 1923b: Zur Gattung *Centaurea* der mazedonischen und kleinasiatischen Flora. - *Repert. Spec. Nov. Regni Veg.* **19**: 99-104.

BORNMÜLLER, J. 1923c: Über eine neue *Aristolochia* aus Mazedonien. - *Repert. Spec. Nov. Regni Veg.* **19**: 195-197.

BORNMÜLLER, J. 1923d: Über eine neue Tulpe der Flora Mazedoniens. - *Repert. Spec. Nov. Regni Veg.* **19**: 199-200.

BORNMÜLLER, J. 1923e: *Fritillaria Macedonica* Bornm. (spec. nov.). - *Repert. Spec. Nov. Regni Veg.* **19**: 201-202.

BORNMÜLLER, J. 1924: *Polygonatum pruinosum* Boiss. und *Carex phyllostachys* C.A.M. in Europa. - *Repert. Spec. Nov. Regni Veg.* **19**: 321-326.

BORNMÜLLER, J. 1925a: Beiträge zur Flora Mazedoniens I. Sammlungen in den Kriegsjahren 1916-1918. - *Bot. Jahrb. Syst.* **59(2-5)**: 294-504 + 18 plates.

BORNMÜLLER, J. 1925b: *Hedysarum macedonicum* Bornm. - *Mitt. Thüring. Bot. Vereins, N.F.* **36**: 43-44.

BORNMÜLLER, J. 1925c: Über eine neue *Silene* der Flora Mazedoniens. - *Mitt. Thüring. Bot. Vereins, N.F.* **36**: 44-45.

BORNMÜLLER, J. 1925d: Was ist *Onobrychis megalophylla* Griseb.? - *Oesterr. Bot. Z.* **74(10)**: 237-244.

BORNMÜLLER, J. 1925e: Bemerkungen zu K. Weins Abhandlung "Was ist *Alyssum paniculatum* Desf.?". - *Oesterr. Bot. Z.* **74(11-12)**: 262-266.

BORNMÜLLER, J. 1926a: Beiträge zur Flora Mazedoniens II. Sammlungen in den Kriegsjahren 1916-1918. - *Bot. Jahrb. Syst.* **60 Beibl. 136**: 1-125.

BORNMÜLLER, J. 1926b: Bearbeitung der von H. Burgeff und Th. Herzog in den Kriegsjahren 1916-1918 in Mazedonien gesammelten Pflanzen. - *Allg. Bot. Z. Syst.* **32(1-12)**: 16-37.

BORNMÜLLER, J. 1927a: Bearbeitung der von H. Burgeff und Th. Herzog in den Kriegsjahren 1916-1918 in Mazedonien gesammelten Pflanzen [cont.]. - *Allg. Bot. Z. Syst.* **33(1-12)**: 25-38.

BORNMÜLLER, J. 1927b: *Viola cephalonica* Bornm. (sp. nov.). - *Mitt. Thüring. Bot. Vereins, N.F.* **37**: 50-52.

BORNMÜLLER, J. 1927c: Zur Kenntnis einiger orientalischen *Teucrium*-Arten der Section *Chamaedrys*. *Teucrium ixodes* Bornm. (spec. nov.) und die ihm nächst verwandten Arten. - *Mitt. Thüring. Bot. Vereins, N.F.* **37**: 55-61.

BORNMÜLLER, J. 1928a: Neue Arten und Formen aus Griechenland. - *Repert. Spec. Nov. Regni Veg.* **25**: 33-37.

BORNMÜLLER, J. 1928b: Kritische Bemerkungen über annuelle *Hippocrepis*-Arten. - *Notizbl. Bot. Gart. Berlin-Dahlem* **10(95)**: 421-451.

BORNMÜLLER, J. 1928c: Ergebnis einer botanischen Reise nach Griechenland im Jahre 1926 (Zante, Cephalonia, Achaia, Phokis, Aetolien). - *Repert. Spec. Nov. Regni Veg.* **25**: 161-203, 270-350.

BORNMÜLLER, J. 1928d: Beiträge zur Flora Mazedoniens III. Sammlungen in den Kriegsjahren 1916-1918. - *Bot. Jahrb. Syst.* **61 Beibl. 140**: 1-195.

BORNMÜLLER, J. 1929: Zur Kenntnis einiger orientalischer *Teucrium*-Arten der Section *Chamaedrys*. Der Formenkreis der *T. divaricatum* (Sieber exs.) Boiss. - *Mitt. Thüring. Bot. Vereins, N.F.* **38**: 26-32.

BORNMÜLLER, J. 1930: Kritische Bemerkungen über einige orientalische Arten der Gattung *Johrenia* nebst Beschreibung zweier neuer Typen. - *Repert. Spec. Nov. Regni Veg.* **28**: 33-53.

BORNMÜLLER, J. 1932a: Bearbeitung der von H. Burgeff und Th. Herzog in den Kriegsjahren 1916-1918 in Mazedonien gesammelten Pflanzen [cont.]. - *Repert. Spec. Nov. Regni Veg.* **30**: 337-362.

BORNMÜLLER, J. 1932b: Neue und kritische *Sideritis*-Arten (Sectio *Empedoclea*) der vorderasiatischen Flora. - *Magyar Bot. Lapok* **31(1-12)**: 127-143.

BORNMÜLLER, J. 1933a: Zur Flora von Montenegro, Albanien und Mazedonien. - *Magyar Bot. Lapok* **32(1-6)**: 109-142.

BORNMÜLLER, J. 1933b: Über *Euphorbia nuda* Velen. und *E. Soongarica* Boiss. (vera!). - *Bot. Jahrb. Syst.* **66(1)**: 113-118.

BORNMÜLLER, J. 1936-44: Symbolae ad floram anatolicam. - *Repert. Spec. Nov. Regni Veg. Beih.* **89**: 1-420.

BORNMÜLLER, J. 1936a: Über eine neue *Centaurea* der Sektion *Acrolophus-Pannosae* aus Epirus. - *Repert. Spec. Nov. Regni Veg.* **40**: 374-375.

BORNMÜLLER, J. 1936b: *Dianthus Behriorum* Bornm. (spec. nov.) aus der Flora von Mazedonien. - *Repert. Spec. Nov. Regni Veg.* **41**: 188.

BORNMÜLLER, J. 1937: Was ist *Viscaria atropurpurea* Griseb.? - *Repert. Spec. Nov. Regni Veg.* **42**: 142-147.

BORNMÜLLER, J. 1938: *Convolvulus* (sect. *Inermes*) *Sündermannii* Bornm. (spec. nov.). - *Repert. Spec. Nov. Regni Veg.* **43**: 152-153.

BORNMÜLLER, J. 1944: *Crocus athous* Bornm. (sp. n.). - *Mitt. Thüring. Bot. Vereins, N.F.* **51(2)**: 342-346.

BORODINA, A.E. 1979a: De generis *Rumex* L. speciebus in parte europaea URSS crescentibus. 2. Subgenera *Acetosa* (Mill.) Rech. f. et *Acetosella* (Meissn.) Rech. f. [In Russian]. - *Novosti Sist. Vysš. Rast.* **15**: 99-112.

BORODINA, A.E. 1979b: Main directions of evolution in the genus *Rumex* L. (Polygonaceae) and its system [In Russian with English summary]. - *Bot. Zhurn. (Moscow & Leningrad)* **64(4)**: 541-553.

BORÓN, P. & AL. 2020: Phylogeny of *Aconitum* subgenus *Aconitum* in Europe. - *Acta Soc. Bot. Poloniae* **89(3)**: 22 pp.

BORRE, A. VAN DEN & WATSON, L. 1994: The infrageneric classification of *Eragrostis* (Poaceae). - *Taxon* **43(3)**: 383-422.

BORRILL, M. 1956a: A biosystematic study of some *Glyceria* species in Britain. 1. Taxonomy. - *Watsonia* **3(6)**: 291-298.

BORRILL, M. 1956b: A biosystematic study of some *Glyceria* species in Britain. 2. Cytology. - *Watsonia* **3(6)**: 299-306.

BORRILL, M. 1958a: A biosystematic study of some *Glyceria* species in Britain. 3. Biometrical studies. - *Watsonia* **4(2)**: 77-88.

BORRILL, M. 1958b: A biosystematic study of some *Glyceria* species in Britain. 4. Breeding systems, fertility relationships and general discussion. - *Watsonia* **4(2)**: 89-100 + 1 plate.

BORRILL, M. 1961a: *Dactylis marina* Borrill, sp. nov., a natural group of related tetraploid forms. - *J. Linn. Soc., Bot.* **56(368)**: 431-439 + 3 plates.

BORRILL, M. 1961b: The pattern of morphological variation in diploid and tetraploid *Dactylis*. - *J. Linn. Soc., Bot.* **56(368)**: 441-452 + 1 plate.

BORRILL, M. 1961c: A morphological and genetical study of population differentiation in Cretan *Dactylis*. - *J. Linn. Soc., Bot.* **58(370)**: 87-93 + 1 plate.

BORRILL, M. 1963: Experimental studies of evolution in *Anthoxanthum* (Gramineae). - *Genetica* **34**: 183-210.

BORRILL, M. & CARROLL, C.P. 1969: A chromosome atlas of the genus *Dactylis* (part two). – *Cytologia* **34(1)**: 6-17 + 1 map.

BORRILL, M., KIRBY, M. & MORGAN, W.G. 1977: Studies in *Festuca*. 11. Interrelationships of some putative diploid ancestors of the polyploid broad-leaved fescues. - *New Phytol.* **78(3)**: 661-674.

BORRILL, M. & LINDNER, R. 1971: Diploid-tetraploid sympatry in *Dactylis* (Gramineae). - *New Phytol.* **70 (6)**: 1111-1124.

BORRILL, M., TYLER, B.F. & MORGAN, W.G. 1976: Studies in *Festuca* 7. Chromosome atlas (Part 2). An appraisal of chromosome race distribution and ecology, including *F. pratensis* var. *apennina* (De Not.) Hack - tetraploid. - *Cytologia* **41(2)**: 219-236.

BORTENSCHLAGER, S. 1967: Vorläufige Mitteilungen zur Pollenmorphologie in der Familie der Geraniaceen und ihre systematische Bedeutung. - *Grana* **7(2-3)**: 400-468.

BORY DE SAINT-VINCENT, J.B.G.M. & CHAUBARD, L.A. 1832-33: Expédition scientifique de Morée. Tome II - 2e partie. Botanique. - Paris: F.G. Levrault, 367 pp. + 38 pl. (publ. 1835-36) [cf. TL-2 1: 286, 1976].

BORZATTI DE LOEWENSTERN, A. & GARBARI, F. 2001: *Bellevalia dubia* (Guss.) Kunth ssp. *dubia* (Hyacinthaceae), an endemic unit of Sicilian flora. In Raimondo, F.M. (ed.): Abstracts from the X OPTIMA meeting, Palermo, 13-19 September 2001, p. 110.

BORZATTI DE LOEWENSTERN, A. & GARBARI, F. 2003a: *Bellevalia dubia* subsp. *boissieri* (Hyacinthaceae) in Italy. - *Webbia* **57(2)**: 181-186.

BORZATTI DE LOEWENSTERN, A. & GARBARI, F. 2003b: *Bellevalia dubia* subsp. *dubia* (Hyacinthaceae), an endemic unit for the Italian flora. - *Bocconea* **16(2)**: 543-548.

BOSABALIDIS, A.M. & KOKKINI, S. 1997: Infraspecific variation of leaf anatomy in *Origanum vulgare* grown wild in Greece. - *Bot. J. Linn. Soc.* **123(4)**: 353-362.

BOSBACH, K. & HURKA, H. 1981: Biosystematic studies on *Capsella bursa-pastoris* (Brassicaceae): enzyme polymorphism in natural populations. - *Pl. Syst. Evol.* **137(1-2)**: 73-94.

BOSCAIU, M. & AL. 1997: Números cromosomáticos de plantas occidentales, 751-776. - *Anales Jard. Bot. Madrid* **55(2)**: 430-431.

BOSCAIU, M. & AL. 1998: Números cromosomáticos de plantas occidentales, 786-808. - *Anales Jard. Bot. Madrid* **56(1)**: 119-120.

BOSCAIU, M., BACCHETTA, G. & GÜEMES, J. 2003: Morphological differentiation within the diploid cytotypes of *Urginea maritima* s.l. (Hyacinthaceae). - *Bocconea* **16(2)**: 549-555.

BOŞCAIU, M., MARHOLD, K. & EHRENDORFER, F. 1997: Karyological and nomenclatural notes on the *Cerastium banaticum* group (Caryophyllaceae). - *Fl. Medit.* **6**: 71-79.

BOSCH, J. 1992: Floral biology and pollinators of three co-occurring *Cistus* species (Cistaceae). - *Bot. J. Linn. Soc.* **109(1)**: 39-55.

BOSCH, M., SIMON, J. & BLANCHÉ, C. 2003: Reports (1305-1311). In Kamari, G., Blanché, C. & Garbari, F. (eds): Mediterranean chromosome number reports - 12. - *Fl. Medit.* **12**: 470-475.

BOSEMARK, O. 1956a: On accessory chromosomes in *Festuca pratensis*. III. Frequency and geographical distribution of plants with accessory chromosomes. - *Hereditas (Lund)* **42(1-2)**: 189-211.

BOSEMARK, O. 1956b: Cytogenetics of accessory chromosomes in *Phleum phleoides*. - *Hereditas (Lund)* **42(3-4)**: 443-466.

BOSEMARK, O. 1967: Edaphic factors and the geographical distribution of accessory chromosomes in *Phleum phleoides*. - *Hereditas (Lund)* **57(1-2)**: 239-262.

BOTEVA, D., GRIFFITHS, D.J. & DIMOPOULOS, P. 2004: Evaluation and mapping of the conservation significance of habitats using GIS: an example from Crete, Greece. - *J. Nature Conserv.* **12**: 237-250.

BOTEVA, D. & WOLF, W. 2000: *Cyperus michelianus* (L.) Link. In Greuter, W. & Raus, Th. (eds): Med- Checklist notulae, 19. - *Willdenowia* **30(2)**: 241.

BOTHMER, R.V. 1967: Intraspecific variation in *Clinopodium vulgare* L. (Labiatae). - *Bot. Not.* **120(2)**: 202-208.

BOTHMER, R.V. 1969: Studies in the Aegean flora. XIV. Studies in *Scutellaria* section *Vulgares* subsection *Peregrinae* from Greece and adjacent Turkey. - *Bot. Not.* **122(1)**: 38-56.

BOTHMER, R.V. 1970a: Studies in the Aegean flora. XV. Chromosome numbers in Labiatae. - *Bot. Not.* **123(1)**: 52-60.

BOTHMER, R.V. 1970b: Cytological studies in *Allium*. I. Chromosome numbers and morphology in *Allium* sect. *Allium* from Greece. - *Bot. Not.* **123(4)**: 518-550.

BOTHMER, R.V. 1972: Four species of *Allium* sect. *Allium* in Greece. - *Bot. Not.* **125(1)**: 62-76.

BOTHMER, R.V. 1974a: Studies in the Aegean flora. XXI. Biosystematic studies in the *Allium ampeloprasum* complex. - *Opera Bot.* **34**: 1-104 + 1 map.

BOTHMER, R.V. 1974b: Karyotype of *Allium chamaespathum* Boiss. - *Bot. Not.* **127(4)**: 546-547.

BOTHMER, R.V. 1975a: The *Allium ampeloprasum* complex on Crete. - *Mitt. Bot. Staatssamml. München* **12(1)**: 267-287. [Also distributed as *OPTIMA leaflet* no. 6].

BOTHMER, R.V. 1975b: Karyotype variation in *Allium bourgeaui*. - *Hereditas (Lund)* **79(1)**: 125-132.

BOTHMER, R.V. 1981: Differentiation patterns in the genus *Scutellaria* from the Balkan Peninsula. - *Bot. Jahrb. Syst.* **102(1-4)**: 271-283.

BOTHMER, R.V. 1982: Karyotype variation in *Allium commutatum* (Liliaceae s. lato). - *Pl. Syst. Evol.* **140(2-3)**: 179-189.

BOTHMER, R.V. 1985: Differentiation patterns in the *Scutellaria albida* group (Lamiaceae) in the Aegean area. - *Nordic J. Bot.* **5(5)**: 421-439.

BOTHMER, R.V. 1987: Differentiation patterns in the E. Mediterranean *Scutellaria rubicunda* group (Lamiaceae). - *Pl. Syst. Evol.* **155(1-4)**: 219-249.

BOTHMER, R.V. 1990: Evolutionary patterns in wild species of *Hordeum*. - *Sommerfeltia* **11**: 65-72.

BOTHMER, R.V. 1991: Interspecific hybridization in *Scutellaria* (Lamiaceae) from Greece and Sicily. - *Willdenowia* **20(1-2)**: 5-13. [Also distributed as *OPTIMA leaflet* no. 208].

BOTHMER, R.V., & AL. 1989: Variation and differentiation in *Hordeum marinum* (Poaceae). - *Nordic J. Bot.* **9(1)**: 1-10.

BOTHMER, R.V. & AL. 1991: An ecogeographical study of the genus *Hordeum* [Systematic and ecogeographic studies of crop genepools, 7]. - Rome: IBPGR, 126 pp. [2nd ed., 1995].

BOTHMER, R.V. & AL. 2014: Report 113. In Vladimirov, V. & al. (eds): New floristic records in the Balkans: 25. - *Phytol. Balcan.* **20(2-3)**: 279-280.

BOTHMER, R.V. & BENTZER, B. 1973: Chromosome morphology in *Bellevalia dubia* subsp. *boissieri* and *B. trifoliata* (Liliaceae) from Greece. - *Ann. Mus. Goulandris* **1**: 11-14.

BOTHMER, R.V. & JACOBSEN, N. 1980: A taxonomic revision of *Hordeum secalinum* and *H. capense*. - *Bot. Tidsskr.* **74(4)**: 223-235.

BOTHMER, R.V. & JACOBSEN, N. 1989: Intergeneric hybridization between *Hordeum* and *Hordelymus* (Poaceae). - *Nordic J. Bot.* **9(2)**: 113-117.

BOTHMER, R.V., JACOBSEN, N. & JØRGENSEN, R.B. 1981: Phylogeny and taxonomy in the genus *Hordeum*. In: Proceedings of the fourth international barley genetics symposium, pp. 13-21. - Edinburgh.

BOTHMER, R.V. & WENDELBO, P. 1981: Cytological and morphological variation in *Bellevalia*. - *Nordic J. Bot.* **1(1)**: 4-11.

BOTSCHANTZEVA, Z.P. 1982: Tulips. Taxonomy, morphology, cytology, phytogeography and physiology [Translated and edited by H.Q. Varekamp; originally published in Russian in 1962]. - Rotterdam: Balkema publishers, ix + 230 pp.

BOTTACCI, A. & AL. 1992: Numeri cromosomici per la Flora Italiana: 1267-1269. - *Inform. Bot. Ital.* **23 (2-3)**: 117-118.

BÖTTCHER, W. & JÄGER, E.J. 1984: Zur Interpretation der Verbreitung der Gattung *Sedum* L. s.l. (Crassulaceae) und ihrer Wuchsformtypen. - *Wiss. Z. Martin-Luther-Univ. Halle-Wittenberg, Math.- Naturwiss. Reihe* **33(1)**: 127-141.

BOTTEGA, S., GARBARI, F. & PERUZZI, L. 2001: Reports (1227-1231). In Kamari, G., Blanché, G. & Garbari, F. (eds): Mediterranean chromosome number reports. - *Fl. Medit.* **11**: 436-439.

BOTTEMA, S. 1974: Late Quaternary vegetation history of northwestern Greece. - Groningen: Ph.D. thesis, 190 pp. + maps and charts.

BOTTEMA, S. 1980a: On the history of the walnut (*Juglans regia* L.) in southeastern Europe. - *Acta Bot. Neerl.* **29(5-6)**: 343-349.

BOTTEMA, S. 1980b: Palynological investigations on Crete. - *Rev. Palaeobot. Palynol.* **31**: 193-217 + 1 chart.

BOTTEMA, S., ENTJES-NIEBORG, G. & ZEIST, W. VAN (EDS) 1990: Man's role in the shaping of the eastern Mediterranean landscape. - Rotterdam: Balkema publishers, x + 349 pp.

BOUDRIE, M. & LABATUT, A.J. 1990: *Anogramma leptophylla* (L.) Link en Dordogne, une découverte inattendue et intéressante. Sa distribution en France. - *Bull. Soc. Bot. Centre-Ouest, n.s.* **21**: 63-70.

BOUFFORD, D.E. 1983: The systematics and evolution of *Circaea* (Onagraceae). - *Ann. Missouri Bot. Gard.* **69(4)**: 804-994.

BOUFFORD, D.E. 2005: *Circaea lutetiana* sensu lato (Onagraceae) reconsidered. - *Harvard Pap. Bot.* **9(2)**: 255-256.

BOUHACHE, M. & TANJI, A. 1985: Evaluation du stock en semences de la morelle jaune (*Solanum elaeagnifolium* Cav.) dans le sol du Tadla (Maroc). - *Weed Res.* **25(1)**: 11-14.

BOUHARMONT, J. 1968: Les formes chromosomiques d'*Asplenium trichomanes* L. - *Bull. Jard. Bot. Natl. Belg.* **38(1)**: 103-114.

BOUHARMONT, J. 1972: Origine de la polyploïdie chez *Asplenium ruta-muraria* L. - *Bull. Jard. Bot. Natl. Belg.* **42(4)**: 375-383 + 3 plates.

BOUHARMONT, J. 1977: Patterns of chromosome evolution in Aspleniaceae. - *Nucleus (Calcutta)* **20(1-2)**: 65-70.

BOUHARMONT, J. 1980: Plantes récoltées en Crète avril 1980. [Unpublished flora list]. - Photocopy: 9 pp.

BOUHARMONT, J. 1981: Plantes récoltées en Grèce mai-juin 1981. [Unpublished flora list]. - Photocopy: 7 pp.

BOUHARMONT, J. 1983: Plantes récoltées en Grèce mai-juin 1983. [Unpublished flora list]. - Photocopy: 10 pp.

BOUHARMONT, J. 1985: Plantes récoltées en Grèce juillet 1985. [Unpublished flora list]. - Photocopy: 7 pp.

BOUHARMONT, J. 1986: Plantes récoltées en Créte mai 1986. [Unpublished flora list]. - Photocopy: 6 pp.

BOUHARMONT, J. 1987: Plants collected in Greece in 1983. Additions and corrections. [Unpublished flora list]. - Photocopy: 1 p.

BOULOS, L. 1960: Cytotaxonomic studies in the genus *Sonchus*. 2. The genus *Sonchus*, a general systematic treatment. - *Bot. Not.* **113(4)**: 400-420.

BOULOS, L. 1961: Cytotaxonomic studies in the genus *Sonchus*. 3. On the cytotaxonomy and distribution of *Sonchus arvensis* L. - *Bot. Not.* **114(1)**: 57-64.

BOULOS, L. 1972a: Révision systématique du genre *Sonchus* L. s.l. I. Introduction et classification. - *Bot. Not.* **125(4)**: 287-305.

BOULOS, L. 1972b: Our present knowledge on the flora and vegetation of Libya. Bibliography. - *Webbia* **26(2)**: 365-400.

BOULOS, L. 1973: Révision systématique du genre *Sonchus* L. s.l. IV. Sous-genre 1. *Sonchus*. - *Bot. Not.* **126(2)**: 155-196.

BOULOS, L. 1999: Flora of Egypt. Volume one. Azollaceae - Oxalidaceae. - Cairo: Al Hadara Publishing, xv + 419 pp.

BOULOS, L. 2000: Flora of Egypt. Volume two. Geraniaceae - Boraginaceae. - Cairo: Al Hadara Publishing, xvi + 352 pp.

BOULOS, L. 2003: Flora of Egypt. Volume three. Verbenaceae - Compositae. - Cairo: Al Hadara Publishing, xvi + 373 pp.

BOULOS, L. 2005: Flora of Egypt. Volume four. Monocotyledons (Alismataceae - Orchidaceae). - Cairo: Al Hadara Publishing, xvi + 617 pp.

BOURGUIGNON, V. 2010: *Sedum aetnense* Tineo ex Guss. (Crassulaceae), a new species record for the flora of Greece. - *Adansonia, sér. 3* **32(1)**: 121-123.

BOURNÉRIAS, J. & AL. 1991: Voyage d'étude de la S.F.O. en Crète, avril 1990. - *Orchidophile (Ansières)* **22(96)**: 75-82.

BOWDEN, W.M. 1959: The taxonomy and nomenclature of the wheats, barleys, and ryes and their wild relatives. - *Canad. J. Bot.* **37(4)**: 657-684.

BOWDEN, W.M. 1966: Chromosome numbers in seven genera of the tribe Triticeae. - *Canad. J. Genet. Cytol.* **8(1)**: 130-136.

BOWLES, E.A. 1918: Snowdrops. - *J. Roy. Hort. Soc.* **43(1)**: 26-36 + 16 plates.

BOWLES, E.A. 1923: Autumn crocuses. - *J. Roy. Hort. Soc.* **48(2-3)**: 161-167 + 4 plates.

BOWLES, E.A. 1934: A handbook of *Narcissus*. - London: Martin Hopkinson Ltd., 248 pp.

BOWLES, E.A. 1948: *Anemone hortensis* and *A. pavonina*. A history of confusion. - *J. Roy. Hort. Soc.* **73(3)**: 57-70 + 3 plates.

BOWLES, E.A. 1952: A handbook of *Crocus* and *Colchicum* for gardeners. Second, revised edition. With a chapter on changes of nomenclature by B.L. Burtt. - London: The Bodley Head, 222 pp. + 32 plates.

BOWLES, E.A. 1985: A handbook of *Narcissus*. - London: Waterstone & Co. Ltd, 248 pp. + 32 plates (first published in 1934).

BOWN, D. 1988: Aroids. Plants of the *Arum* family. - London, etc.: Century, 256 pp. + 24 plates.

BOX, E.O. 1987: Plant life forms and mediterranean environments. - *Ann. Bot. (Rome)* **45(2)**: 7-42.

BOYCE, P.C. 1986: Observations on Aroids. - *Quart. Bull. Alpine Gard. Soc. Gr. Brit.* **54(1)**: 35-42.

BOYCE, P.C. 1987a: A new species of *Arum* L. from Crete. - *Aroideana* **10(1)**: 6-8.

BOYCE, P.C. 1987b: A new subspecies of *Biarum davisii* Turrill from Turkey. - *Aroideana* **10(4)**: 14-15.

BOYCE, P.C. 1989: A new classification of *Arum* with keys to the infrageneric taxa. - *Kew Bull.* **44(3)**: 383-395.

BOYCE, P.C. 1990: Friar's cowls and mouse plants: the genus *Arisarum*. - *Bot. Mag. (Kew Mag.)* **7(1)**: tabs. 142-144 + pp. 14-21.

BOYCE, P.C. 1993: The genus *Arum*. - London: H.M.S.O. (Kew Mag. Monogr.), 196 pp. + 16 plates.

BOYCE, P.C. 1994: The genera *Dracunculus* and *Helicodiceros* (Araceae: Aroideae). - *Thaiszia* **4(2)**: 175-182.

BOYCE, P.C. 1995: The genus *Arum* (Araceae) in Greece and Cyprus. - *Ann. Mus. Goulandris* **9**: 27-38 [updated version on internet].

BOYCE, P.C. 2004: A review of *Biarum*: small aroids of the Mediterranean. - *Rock Gard. Quart.* **62(4)**: 244-253, 273.

BOYCE, P.C. 2006a: New observations on *Arum italicum*. - *Plantsman* **5(1)**: 36-39.

BOYCE, P.C. 2006b: A taxonomic revision of *Biarum* (Aracaeae). - *Aroideana* **29**: 2-36.

BOYCE, P.C. 2006c: *Arum* - a decade of change. - *Aroideana* **29**: 132-137.

BOYCE, P.C. 2008: A taxonomic revision of *Biarum* (Araceae). - *Curtis's Bot. Mag.* **25(1)**: 2-118.

BOYCE, P.C. & ATHANASIOU, K. 1991: A new subspecies of *Biarum tenuifolium* (Araceae) from Crete. - *Fl. Medit.* **1**: 5-13.

BOYD, J.W., MURRAY, D.S. & TYRL, R.J. 1984: Silverleaf nightshade, *Solanum elaeagnifolium*. Origin, distribution, and relation to man. - *Econ. Bot.* **38(2)**: 210-217.

BOYDAK, M. 1985: The distribution of *Phoenix theophrasti* in the Datça Peninsula, Turkey. - *Biol. Conservation* **32**: 129-135.

BOYDAK, M. 1986: A new natural distribution of *Phoenix theophrasti* in Kumluca-Karaöz, Turkey [In Turkish with English summary]. - *Istanbul Üniv. Orman Fak. Derg., A* **36(1)**: 1-13.

BOYDAK, M. 1987: A new occurrence of *Phoenix theophrasti* in Kumluca-Karaöz, Turkey. – *Principes* **31(2)**: 89-95.

BOYDAK, M. 2005: Plant diversity, *Phoenix theophrasti* and *Pinus brutia* in Turkey. In Karamanos, A.J. & Thanos, C.A. (eds): Biodiversity and natural heritage in the Aegean. Proceedings of the conference "Theophrastus 2000", pp. 251-277. - Athina: Afoi Fragoudi O.E.

BOYDAK, M. & AKA, M. 1983: *Phoenix theophrasti* Greuter and its natural distribution in Datça Peninsula of Turkey [In Turkish with English summary]. - *Istanbul Üniv. Orman Fak. Derg., A* **33(1)**: 73-92.

BOYDAK, M. & AL. 2021: ISSR genotyping of *Phoenix theophrasti* natural populations in Turkey and Crete (Greece) and *P. dactylifera*. - *Nordic J. Bot.* **39(10)**: e03104, doi: 10.1111/njb.03104.

BOYDAK, M. & BARROW, S.C. 1995: A new locality for *Phoenix* in Turkey: Gölköy-Bödrum. – *Principes* **39(3)**: 117-122.

BOŽA, P. 1990: Neue Taxa der Gattung *Glycyrrhiza* L. (Fabaceae) in Jugoslawien. - *Razprave IV. Razreda Sazu* **31(2)**: 23-36.

BOZIKA, P., TINIAKOU, A. & GEORGIADIS, TH. 2001: Comparative study of the flora and vegetation of the lakes Limnopoula Paramithias and Touba Ioanninon. In Kamari, Psaras & Constantinidis (eds): Proceedings of the 8th scientific congress of the Greek Botanical Society, October 5-8, 2000, pp. 318-326 [In Greek with English abstract].

BOZZINI, A. 1959: Revisione cito-sistematica del genere *Asparagus* L. I: Le specie di *Asparagus* della flora italiana e chiave analitica per la loro determinazione. - *Caryologia* **12(2)**: 199-264 + 3 plates.

BRADSHAW, A.D. 1958: Natural hybridization of *Agrostis tenuis* Sibth. and *A. stolonifera* L. - *New Phytol.* **57(1)**: 66-84.

BRAMWELL, D. (ED.) 1979: Plants and islands. - London, etc.: Academic Press, x + 459 pp.

BRANA, S., VUKOVIĆ, N. & KALIGARIČ, M. 2014: Least adder's-tongue (*Ophioglossum lusitanicum* L.) in Croatia - distribution, ecology and conservation. - *Acta Bot. Croat.* **73(2)**: 471-480.

BRAND, A. 1895a: Monographie der Gattung *Nigella*. - *Helios* **12(11-12)**: 171-174, 182-197.

BRAND, A. 1895b: Monographie der Gattung *Nigella* [cont.]. - *Helios* **13(1-3)**: 8-15, 22-28, 33-36.

BRAND, A. 1898: Monographie der Gattung *Lotus*. - *Bot. Jahrb. Syst.* **25(1-2)**: 166-232.

BRANDE, A. 1976: Zur Ausbreitungsgeschichte von *Xanthium* im südöstlichen Europa. - *Bot. Jahrb. Syst.* **95(3)**: 406-410.

BRANDENBURG, W.A., VOOREN, J.G. VAN DE & JARVIS, C.E. 2001: Lectotypification and description of *Clematis viticella* L. (Ranunculaceae). - *Bot. J. Linn. Soc.* **135(1)**: 13-23.

BRANDES, D. 1998: Vegetation der Strassenränder Korfus. In Brandes, D. (ed.): Vegetationsökologie von Habitatinseln und linearen Strukturen. Tagungsbericht des Braunschweiger Kolloquiums vom 22.-24. November 1996. - *Braunschweig. Geobot. Arb.* **5**: 247-262.

BRANDES, D. 2012: Beitrag zur Flora der Strassenränder von Rhodos. - Braunschweig: 14 pp.

BRANDES, R. 2007: Waldgrenzen griechischer Hochgebirge. - *Erlanger Geogr. Arb.* **Sonderband 36**: xiv + 361 pp.

BRANDT, J.-P. 1952: Contribution à la cytologie du genre *Veronica*. - *Bull. Soc. Neuchâteloise Sci. Nat.* **75**: 180-188 + 1 plate.

BRANDT, J.-P. 1953: Nouvelle contribution à la cytologie du genre *Veronica*. - *Bull. Soc. Neuchâteloise Sci. Nat.* **76**: 111-119.

BRANDT, J.-P. 1961: Cytotaxinomie et cytogéographie de *Veronica prostrata* L. - *Bull. Soc. Neuchâteloise Sci. Nat.* **84**: 35-88.

BRÄUCHLER, CH. 2018: And now for something completely different - new names in *Clinopodium* with comments on some types. - *Phytotaxa* **356(1)**: 71-80.

BRÄUCHLER, CH. 2018: Delimitation and revision of the genus *Thymbra* (Lamiaceae). - *Phytotaxa* **369 (1)**: 15-27.

BRÄUCHLER, CH. & AL. 2005: Polyphyly of the genus *Micromeria* (Lamiaceae) - evidence from cpDNA sequence data. - *Taxon* **54(3)**: 639-650.

BRÄUCHLER, CH., MEIMBERG, H. & HEUBL, G. 2006: New names in Old World *Clinopodium* - the transfer of the species of *Micromeria* sect. *Pseudomelissa* to *Clinopodium*. - *Taxon* **55(4)**: 977-981.

BRÄUCHLER, CH., RYDING, O. & HEUBL, G. 2008: The genus *Micromeria* (Lamiaceae), a synoptical update. - *Willdenowia* **38(2)**: 363-410.

BRAUN, H. 1886: Beiträge zur Kenntnis einiger Arten und Formen der Gattung *Rosa*. - *Verh. K. K. Zool.-Bot. Ges. Wien* **35**: 61-136.

BRÄUTIGAM, S. 1985: Beitrag zur Kenntnis der Gattung *Hieracium* L. in Bulgarien. - *Bot. Jahrb. Syst.* **107(1-4)**: 1-9.

BRÄUTIGAM, S. & GREUTER, W. 2007: A new treatment of *Pilosella* for the Euro-Mediterranean flora. - *Willdenowia* **37(1)**: 123-137.

BRÄUTIGAM, S. & KNAPP, H.D. 1976: Zur Verbreitung von *Hieracium*-Arten im hercynischen Gebiet. Teil 2. - *Feddes Repert.* **87(1-2)**: 31-48.

BREDEMANN, G. 1959: Die grosse Brennessel *Urtica dioica* L. Forschungen über ihren Anbau und Fasergewinnung (mit einem Anhang über ihre Nutzung für Arznei- und Futtermittel ... von K. Graber). - Berlin: Akademie-Verlag, viii + 137 pp.

BREISTROFFER, M. 1936: Révision systématique des variations du *Clypeola jonthlaspi* L. – *Candollea* **7**: 140-166.

BREISTROFFER, M. 1937: Sur la répartition géographique des diverses races du *Clypeola jonthlaspi* L. *Bull. Soc. Sci. Isère* **56**: 347-451.

BREISTROFFER, M. 1946: Nouvelles contributions à l'étude monographique du *Clypeola Jonthlaspi* L. - *Candollea* **10**: 241-280.

BREMER, K. 1993: New subtribes of the Lactuceae (Asteraceae). - *Novon* **3(4)**: 328-330.

BREMER, K. & HUMPHRIES, C.J. 1993: Generic monograph of Asteraceae-Anthemideae. - *Bull. Nat. Hist. Mus. London, Bot.* **23(2)**: 71-177.

BRENAN, J.P.M. 1961: *Amaranthus* in Britain. - *Watsonia* **4(6)**: 261-280.

BRENAN, J.P.M. & TOWNSEND, C.C. 1980: (540) Proposal to reject *Amaranthus blitum* L. (Amaranthaceae) under Art. 69 in favour of *A. lividus* L. - *Taxon* **29(5-6)**: 695-696. [not approved, cf. Taxon 33(2): 298, 1984].

BRETON-SINTÈS, S. 1974a: Étude biosystématique du genre *Knautia* (Dipsacaceae) dans le massif central français I. - Analyse morphologique, phytosociologique et cytogénétique des espèces. - *Ann. Sci. Nat., Bot. Biol. Vég.* **15(3)**: 197-254.

BRETON-SINTÈS, S. 1974b: Étude biosystématique du genre *Knautia* (Dipsacaceae) dans le massif central français II. - Analyse morphologique et cytogénétique d'hybrides expérimentaux. - *Ann. Sci. Nat., Bot. Biol. Vég.* **15(4)**: 277-319.

BRETON-SINTÈS, S. & CAUDERON, Y. 1970: Analyse morphologique et cytogénétique d'hybrides expérimentaux entre *Knautia* tétraploïdes. - *Bull. Soc. Bot. France* **117(5-6)**: 221-242.

BRETSCHNEIDER, J. 1962: Bestimmungsschlüssel der europäischen Sippen der Gattung *Glechoma* L. - *Drudea* **2(1-4)**: 69-71.

BRETT, O.E. 1952: Basic chromosome numbers in the genus *Cerastium*. - *Nature* **170**: 251-252.

BRETT, O.E. 1955: Cyto-taxonomy of the genus *Cerastium*. I. Cytology. - *New Phytol.* **54(2)**: 138-148 + 1 plate.

BRETZLER, E. 1899: Studien über die Gattung *Platanus* L. - *Nova Acta Acad. Caes. Leop.-Carol. German. Nat. Cur.* **77(2)**: 115-226 + 10 plates.

BREVIGLIERI, N. & BATTAGLIA, E. 1955: Ricerche cariologiche in *Olea europaea* L. - *Caryologia* **6(2-3)**: 271-283 + 2 plates.

BRICKELL, C.D. & MATHEW, B.F. 1973: Three recent discoveries in *Crocus*. - *J. Roy. Hort. Soc.* **98(8)**: 360-366 + 1 plate.

BRICKELL, C.D. & MATHEW, B.F. 1976: *Daphne*. The genus in the wild and in cultivation. - St. John's Woking, Surrey: Alpine Garden Society, 194 pp.

BRICKELL, C.D. & MCCLINTOCK, D. 1987: Proposal to conserve *Erica carnea* L. against *E. herbacea* L. (Ericaceae). - *Taxon* **36(2)**: 479-481.

BRIDSON, G.D.R. & SMITH, E.R. (EDS) 1991: B-P-H/S. Botanico-Periodicum-Huntianum/Supplementum. – Pittsburgh, Pa.: Hunt Institute for Botanical Documentation, 1068 pp.

BRIGGS, M. 1989: The fifth tulip in Crete. - *Quart. Bull. Alpine Gard. Soc. Gr. Brit.* **57(1)**: 41-47.

BRIGGS, M. 1990: First record of *Tulipa goulimyi* (Liliaceae) on Crete. - *Ann. Mus. Goulandris* **8**: 41-44.

BRIGHTON, C.A. 1976a: Cytological problems in the genus *Crocus* (Iridaceae). I. *Crocus vernus* aggregate. - *Kew Bull.* **31(1)**: 33-46 + 1 plate.

BRIGHTON, C.A. 1976b: Cytology of *Crocus olivieri* and its allies. - *Kew Bull.* **31(2)**: 209-217 + 1 plate.

BRIGHTON, C.A. 1976c: Chromosome variation in *Crocus*. In Jones, K. & Brandham, P.E. (eds): Current chromosome research, pp. 208-209. - Amsterdam, etc.: North-Holland.

BRIGHTON, C.A. 1977a: Cytological problems in the genus *Crocus* (Iridaceae): II. *Crocus cancellatus* aggregate. - *Kew Bull.* **32(1)**: 33-45 + 2 plates.

BRIGHTON, C.A. 1977b: Cytology of *Crocus sativus* and its allies (Iridaceae). - *Pl. Syst. Evol.* **128(3-4)**: 137-157.

BRIGHTON, C.A., MATHEW, B.F. & MARCHANT, C.J. 1974: Chromosome counts in the genus *Crocus* (Iridaceae). - *Kew Bull.* **28(3)**: 451-464.

BRIGHTON, C.A., MATHEW, B.F. & RUDALL, P. 1983: A detailed study of *Crocus speciosus* and its ally *C. pulchellus* (Iridaceae). - *Pl. Syst. Evol.* **142(3-4)**: 187-206.

BRIGHTON, C.A., SCARLETT, C.J. & MATHEW, B.F. 1980: Cytological studies and origins of some *Crocus* cultivars. In Brickell, C.D., Cuttler, D.F. & Gregory, M. (eds): Petaloid monocotyledons. Linnean Society symposium series no. 8, pp. 139-160 + 1 plate.

BRILLI-CATTARINI, A.J.B. & GUBELLINI, L. 1986: Una nuova specie di *Cardamine* (Cruciferae) della Peninsola Italiana e Sicilia. - *Webbia* **39(2)**: 397-407.

BRIQUET, J. 1889: Fragmenta Monographiae Labiatarum. I. Révision systématique des groupes spécifiques et subspécifiques dans le sous-genre *Menthastrum* du genre *Mentha*. - *Bull. Trav. Soc. Bot. Genève* **5**: 20-107.

BRIQUET, J. 1893: Monographie du genre *Galeopsis*. - *Mém. Couronnés Autres Mém. Acad. Roy. Sci. Belgique* **1893**: xii + 323 pp.

BRIQUET, J. 1894: Fragmenta Monographiae Labiatarum. Fascicule troisième. - *Bull. Herb. Boissier* **2 (12)**: 689-724.

BRIQUET, J. 1896: Fragmenta Monographiae Labiatarum [cont.]. - *Bull. Herb. Boissier* **4(10-12)**: 676-696, 762-808, 847-878.

BRIQUET, J. 1898: Fragmenta Monographiae Labiatarum. Fasc. V. Observations sur quelques Labiées intéressantes ou nouvelles principalement de l'herbier Delessert. - *Annuaire Conserv. Jard. Bot. Genève* **2**: 102-251.

BRIQUET, J. 1899: Notes sur quelques Buplèvres de l'Herbier de Linné. - *Bull. Soc. Bot. France* **46**: 289-291.

BRIQUET, J. 1910-55: Prodrome de la flore corse comprenant les résultats botaniques de six voyages exécutés en Corse sous de M. Émile Burnat [continued by R. de Litardière]. - Genève, etc.: 3 vols. [cf. TL-2 1: 329, 1976].

BRITTEN, E.J. 1963: Chromosome numbers in the genus *Trifolium*. - *Cytologia* **28(4)**: 428-449.

BRITTEN, J. 1893: Bibliographical notes. III. "Flora Corcirese". - *J. Bot. (London)* **31**: 555-357.

BRITTEN, J. 1906: Overlooked plants described by Schreber. - *J. Bot. (London)* **44**: 342-346.

BRITTON, D.M. 1951: Cytogenetic studies on the Boraginaceae. - *Brittonia* **7(4)**: 233-266.

BROCKMANN, I. & BOCQUET, G. 1978: Ökologische Einflüsse auf die Geschlechtsverteilung bei *Silene vulgaris* (Moench) Garcke (Caryophyllaceae). - *Ber. Deutsch. Bot. Ges.* **91(1)**: 217-230.

BRODY, T. & MENDLINGER, S. 1980: Species relationships and genetic variation in the diploid wheats (*Triticum, Aegilops*) as revealed by starch gel electrophoresis. - *Pl. Syst. Evol.* **136(3-4)**: 247-258.

BROFAS, G. & AL. 2001: The flora and vegetation of Gyali island, SE Aegean, Greece. – *Willdenowia* **31(1)**: 51-70.

BROFAS, G. & AL. 2007: Rehabilitation of disturbed areas by mining activities in high floristic diversity areas: the case of Mt Giona. In: 3rd International Conference on Sustainable Development Indicators in the Minerals Industry, June 2007, Milos island, Greece.

BROFAS, G. & KARETSOS, G. 1992: Simvoli sti dierevnisi tis ikologias tou Schinia-Marathona [In Greek with English summary]. - *Geôtechn. Epistimon. Themata* **3(1)**: 32-41.

BROOKS, R.R. 1987: Serpentine and its vegetation. A multidisciplinary approach. - Portland, Oregon: Dioscorides Press, vi + 454 pp.

BROOKS, R.R. & AL. 1979: Hyperaccumulation of nickel by *Alyssum* Linnaeus (Cruciferae). - *Proc. Roy. Soc. London, Ser. B, Biol. Sci.* **203**: 387-403.

BROUGH, M.A. 1953: *Jankaea Heldreichii* in cultivation. - *Quart. Bull. Alpine Gard. Soc. Gr. Brit.* **21(2)**: 102-111.

BROUSSALIS, P. 1977: The protection of the flora in Greece and its problems. - *Ann. Mus. Goulandris* **3**: 23-30.

BROUSSALIS, P. 1978a: Greek species of *Fritillaria* (Liliaceae). - *Ann. Mus. Goulandris* **4**: 13-18.

BROUSSALIS, P. 1978b: The genus *Paeonia* in Greece [In Greek with English summary]. – *Fisis [Journal of the Hellenic Society for the Protection of Nature]* **14**: 10-14, 38-39.

BROUSSALIS, P. 1987: The gesneriads of Greece [In Greek with English summary]. - *Fisis [Journal of the Hellenic Society for the Protection of Nature]* **38**: 7-9, 46.

BROUSSALIS, P. 1989a: *Comperia comperiana*, the elusive orchid [In Greek with English summary]. - *Fisis [Journal of the Hellenic Society for the Protection of Nature]* **44/45**: 4-6, 43-44.

BROUSSALIS, P. 1989b: *Posidonia oceanica* [In Greek with English summary]. - *Fisis [Journal of the Hellenic Society for the Protection of Nature]* **46/47**: 8-9, 37.

BROUSSALIS, P. 1989c: *Zelkova cretica* and its protection [In Greek with English summary]. - *Fisis [Journal of the Hellenic Society for the Protection of Nature]* **46/47**: 12, 36.

BROUSSALIS, P. 1991: The aroids (Araceae) in Greece [In Greek with English summary]. - *Fisis [Journal of the Hellenic Society for the Protection of Nature]* **51**: 12-16, 34-36.

BROUSSALIS, P. 1993: Aristolochias in Greece [In Greek with English abstract]. - *Fisis [Journal of the Hellenic Society for the Protection of Nature]* **60**: 13-16, 31.

BROUSSALIS, P. 1994a: *Gentiana* in Greece [In Greek with English summary]. - *Fisis [Journal of the Hellenic Society for the Protection of Nature]* **64**: 3-6, 46.

BROUSSALIS, P. 1994b: The genus *Narcissus* in Greece [In Greek with English summary]. - *Fisis [Journal of the Hellenic Society for the Protection of Nature]* **66**: 3-5, 44.

BROUSSALIS, P. 1995: The *Muscari* [In Greek with English summary]. - *Fisis [Journal of the Hellenic Society for the Protection of Nature]* **71**: 3-5, 45-46.

BROUSSALIS, P. 1996: The Greek flora news [In Greek with English summary]. - *Fisis [Journal of the Hellenic Society for the Protection of Nature]* **72**: 28-29, 45.

BROUSSALIS, P. 2001: Sta Idata tis Stigos [In Greek]. - *Fisis [Journal of the Hellenic Society for the Protection of Nature]* **93**: 4-7.

BROUSSALIS, P. 2002: Ta Gerania ke ta syngenis tous [In Greek]. - *Fisis [Journal of the Hellenic Society for the Protection of Nature]* **99**: 24-26.

BROUSSALIS, P. 2004: O Ixos [In Greek]. - *Fisis [Journal of the Hellenic Society for the Protection of Nature]* **106**: 10-11.

BROUSSALIS, P. 2006: Votanikes anamnisis [In Greek]. - *Fisis [Journal of the Hellenic Society for the Protection of Nature]* **114**: 4-5.

BROWICZ, K. 1963a: The genus *Colutea* L. A monograph. - *Monogr. Bot.* **14**: 3-136 + 28 plates.

BROWICZ, K. 1963b: Monografija rodzaju *Colutea* L. [In Polish]. - *Arbor. Kórnickie* **8**: 5-28.

BROWICZ, K. 1964: On the geographical distribution of *Platanus orientalis* L. in Bulgaria. - *Arbor. Kórnickie* **9**: 37-57.

BROWICZ, K. 1966: The genus *Periploca* L. A monograph. - *Arbor. Kórnickie* **11**: 5-104.

BROWICZ, K. 1967a: Genus *Cionura* Grisebach - distribution, history, and application. - *Arbor. Kórnickie* **12**: 5-32.

BROWICZ, K. 1967b: A supplement to the monograph of the genus *Colutea* L. - *Arbor. Kórnickie* **12**: 33-43.

BROWICZ, K. 1968: Distribution of woody Rosaceae in W. Asia. II. On the distribution of *Mespilus germanica* L. - *Arbor. Kórnickie* **13**: 27-36.

BROWICZ, K. 1969: Distribution of woody Rosaceae in W. Asia. III. *Erilobus trilobatus* (Poir.) Roem. - *Arbor. Kórnickie* **14**: 2-23.

BROWICZ, K. 1970a: Distribution of woody Rosaceae in W. Asia. VI. *Pyracantha coccinea* Roem. - *Arbor. Kórnickie* **15**: 17-27.

BROWICZ, K. 1970b: "Malus florentina" - its history, systematic position and geographical distribution. - *Fragm. Florist. Geobot.* **16(1)**: 61-83.

BROWICZ, K. 1971a: Distribution of woody Rosaceae in W Asia. VII. Genus *Amelanchier* Med. - *Arbor. Kórnickie* **16**: 5-26.

BROWICZ, K. 1971b: Verbreitungsareal und Veränderlichkeit von *Paliurus spina-christi* Mill. in Westasien. - *Ann. Naturhist. Mus. Wien* **75**: 27-32.

BROWICZ, K. 1972a: Materials for a Flora of Turkey XXVI: New taxa of woody Rosaceae. - *Notes Roy. Bot. Gard. Edinburgh* **31(2)**: 321-324.

BROWICZ, K. 1972b: Distribution of woody Rosaceae in W. Asia. IX. *Amygdalus orientalis* Duh. and closely related species. - *Arbor. Kórnickie* **17**: 5-18.

BROWICZ, K. 1973: Distribution of woody Rosaceae in W. Asia. XI. *Crataegus aronia* (L.) Bosc. ex DC. - *Arbor. Kórnickie* **18**: 5-21.

BROWICZ, K. 1974a: Distribution of woody Rosaceae in W. Asia. XIII. *Amygdalus webbii* Spach and closely related species. - *Arbor. Kórnickie* **19**: 5-21.

BROWICZ, K. 1974b: Distribution of woody Rosaceae in W. Asia. XIV. *Sorbus torminalis* (L.) Crantz. - *Arbor. Kórnickie* **19**: 23-30.

BROWICZ, K. 1975a: *Vincetoxicum canescens* (Willd.) Decne and *V. tmoleum* Boiss. – Systematic problems and geographical distribution. - *Fragm. Florist. Geobot.* **21(3)**: 261-271.

BROWICZ, K. 1975b: Distribution of woody Rosaceae in W. Asia. XV. *Cerasus prostrata* (Lab.) Ser. - *Arbor. Kórnickie* **20**: 23-35.

BROWICZ, K. 1976a: Genus *Bongardia* C.A. Mey. and its range. - *Fragm. Florist. Geobot.* **22(4)**: 435-444 + 2 plates.

BROWICZ, K. 1976b: The genus *Amygdalus* L. in Turkey. In: Kazdaği göknari ve Türkiye florasi uluslararasi simpozyomu bildirileri, pp. 239-255. - Istanbul Üniversitesi Orman Fakültesi Yayinlari.

BROWICZ, K. 1978a: Geographic distribution of some shrubs from the family Leguminosae in southwestern Asia. - *Arbor. Kórnickie* **23**: 5-30.

BROWICZ, K. 1978b: Distribution of woody Rosaceae in W. Asia. XVI. *Cerasus mahaleb* (L.) Miller. - *Arbor. Kórnickie* **23**: 31-40.

BROWICZ, K. 1979: Distribution of woody *Thymelaea* in the eastern Mediterranean region. - *Arbor. Kórnickie* **24**: 5-18.

BROWICZ, K. 1980: Distribution and variability of *Clematis cirrhosa* L. and *Myrtus communis* L. in the eastern Mediterranean. - *Arbor. Kórnickie* **25**: 23-36.

BROWICZ, K. 1982a: *Eriolobus trilobatus* (Poiret) M.J. Roemer in Greece. - *Ann. Mus. Goulandris* **5**: 23-31.

BROWICZ, K. 1982b: Chorology of trees and shrubs in south-west Asia and adjacent regions. 1. - Warszawa & Poznań: Polish Scientific Publishers, 172 pp

BROWICZ, K. 1982c: Distribution of species from the genus *Rhus* L. in the eastern Mediterranean region and in southwestern Asia. - *Arbor. Kórnickie* **26**: 3-14.

BROWICZ, K. 1983a: Chorology of trees and shrubs in south-west Asia and adjacent regions. 2. - Warszawa & Poznań: Polish Scientific Publishers, 86 pp.

BROWICZ, K. 1983b: ×*Malosorbus florentina* (Zuccagni) Browicz (Rosaceae) in Greece. - *Ann. Mus. Goulandris* **6**: 27-35.

BROWICZ, K. 1983c: *Bupleurum fruticosum* L. and *Lonicera implexa* Ait. in the eastern Mediterranean. - *Arbor. Kórnickie* **27**: 17-26.

BROWICZ, K. 1984a: Chorology of trees and shrubs in south-west Asia and adjacent regions. 3. - Warszawa & Poznań: Polish Scientific Publishers, 88 pp.

BROWICZ, K. 1984b: Chorology of trees and shrubs in south-west Asia and adjacent regions. 4. - Warszawa & Poznań: Polish Scientific Publishers, 79 pp.

BROWICZ, K. 1985: Bibliography of Polish botanical works in the field of plant systematics and geobotany for the years 1945-1980 dealing with flora of foreign countries. - *Arbor. Kórnickie* **29**: 197-218.

BROWICZ, K. 1986a: *Amygdalus ×rhodia* (Rosaceae), a new hybrid almond from Rhodes. - *Ann. Mus. Goulandris* **7**: 33-37.

BROWICZ, K. 1986b: The genus *Lithodora* Griseb. (Boraginaceae) in the eastern Mediterranean region. - *Ann. Mus. Goulandris* **7**: 39-48.

BROWICZ, K. 1986c: Chorology of trees and shrubs in south-west Asia and adjacent regions. 5. - Warszawa & Poznań: Polish Scientific Publishers, 88 pp.

BROWICZ, K. 1987: *Pistacia lentiscus* cv. *Chia* (Anacardiaceae) on Chios island. - *Pl. Syst. Evol.* **155 (1-4)**: 189-195.

BROWICZ, K. 1988: Chorology of trees and shrubs in south-west Asia and adjacent regions. 6. - Warszawa & Poznań: Polish Scientific Publishers, 86 pp.

BROWICZ, K. 1989a: On the occurrence of *Daphne gnidioides* Jaub. & Spach and *D. gnidium* in Greece. - *Bios (Thessaloniki)* **1989**: 31-39.

BROWICZ, K. 1989b: *Quercus aucheri* Jaub. et Spach and its range of occurrence. - *Arbor. Kórnickie* **31**: 3-11.

BROWICZ, K. 1990: Chorology of trees and shrubs in south-west Asia and adjacent regions. 7. - Warszawa & Poznań: Polish Scientific Publishers, 87 pp.

BROWICZ, K. 1991a: *Crataegus ×ruscinonensis* var. *aronioides* (Rosaceae), a new nothospecies and nothovariety for the flora of Greece. - *Willdenowia* **21(1-2)**: 113-117.

BROWICZ, K. 1991b: Chorology of trees and shrubs in south-west Asia and adjacent regions. 8. - Warszawa & Poznań: Polish Scientific Publishers, 86 pp.

BROWICZ, K. 1991c: On the geographic distribution of the genus *Thymbra* L. (Labiatae). - *Arbor. Kórnickie* **33**: 75-81.

BROWICZ, K. 1991d: Concept and chorology of the genera *Amygdalus* L. and *Louiseania* Carrière. - *Arbor. Kórnickie* **34**: 31-54.

BROWICZ, K. 1991e: Chorology of *Aristolochia sempervirens* L. (Aristolochiaceae). - *Arbor. Kórnickie* **35**: 77-81.

BROWICZ, K. 1991f: Trees and shrubs of Lemnos Is. (Greece). - *Arbor. Kórnickie* **36**: 5-25.

BROWICZ, K. 1991g: Woody species of *Euphorbia* in Turkey. - *Karaca Arbor. Mag.* **1(1)**: 13-19 + 1 plate.

BROWICZ, K. 1992a: A new species of *Amygdalus* from South Anatolia. - *Karaca Arbor. Mag.* **1(4)**: 125-126.

BROWICZ, K. 1992b: Chorology of trees and shrubs in south-west Asia and adjacent regions. 9. - Poznań/Daszewice: Sorus, 85 pp.

BROWICZ, K. 1993a: *Nicotiana glauca* and *Solanum elaeagnifolium* (Solanaceae) - two xenophytes from South America and the history of their spreading in the eastern Mediterranean. - *Fragm. Florist. Geobot. Suppl.* **2(1)**: 299-305.

BROWICZ, K. 1993b: Conspectus and chorology of the genus *Pyrus* L. - *Arbor. Kórnickie* **38**: 17-33.

BROWICZ, K. 1994a: Trees and shrubs of Kos island (Dodecanese, Greece). - *Arbor. Kórnickie* **39**: 31-55.

BROWICZ, K. 1994b: ×*Malosorbus florentina* (Zuccagni) Browicz in Turkey. - *Karaca Arbor. Mag.* **2(4)**: 159-164.

BROWICZ, K. 1995: Fifteen years of floristic-dendrological investigations in Greece [In Polish with English abstract]. - *Arbor. Kórnickie* **40**: 37-53.

BROWICZ, K. 1996a: Chorology of trees and shrubs in south-west Asia and adjacent regions. Supplement. - Poznań: Polish Academy of Sciences, 48 pp.

BROWICZ, K. 1996b: *Juniperus macrocarpa* Sibth. & Sm. in the area of the "Flora of Turkey and the East Aegean Islands". - *Karaca Arbor. Mag.* **3(3)**: 117-121.

BROWICZ, K. 1997a: Chorology of trees and shrubs in south-west Asia and adjacent regions. Phytogeographical analysis. - Poznań: Polish Academy of Sciences, 18 pp.

BROWICZ, K. 1997b: Woody flora of Melos and Kimolos (Cyclades, Greece). - *Arbor. Kórnickie* **42**: 45-63.

BROWICZ, K. 1997c: *Cichorium spinosum* L. in the eastern Mediterranean. - *Karaca Arbor. Mag.* **4(1)**: 19-21.

BROWICZ, K. 1998: Trees and shrubs of Poros (Saronic Gulf, Greece). - *Arbor. Kórnickie* **43**: 5-19.

BROWICZ, K. 1999a: Fifteen years of floristic and dendrological investigations in Greece. - *Ann. Mus. Goulandris* **10**: 29-44.

BROWICZ, K. 1999b: Chorological connections of woody Rosaceae between SE Europe and SW Asia. - *Acta Bot. Fenn.* **162**: 185-190.

BROWICZ, K. 2000: Trees and shrubs of the island of Skyros (Northern Sporades, Greece). - *Bot. Chron. (Patras)* **13**: 325-339.

BROWICZ, K. 2001a: Woody flora of Hydra (Argosaronic islands, Greece). - *Dendrobiology* **46**: 9-12.

BROWICZ, K. 2001b: Where is the place of origin of *Morus nigra* (Moraceae)? - *Fragm. Florist. Geobot.* **45(1-2)**: 273-280.

BROWICZ, K. & KARACA, H. 1993: *Eriolobus trilobatus* (Poiret) Roemer in Turkey. - *Karaca Arbor. Mag.* **2(2)**: 79-86.

BROWICZ, K. & ZIELIŃSKI, J. 1977: Two new taxa within the Ulmaceae family for the flora of Bulgaria and their geographical distribution. - *Fragm. Florist. Geobot.* **23(2)**: 141-150 + 1 plate.

BROWICZ, K. & ZIELIŃSKI, J. 1984: Chorology of trees and shrubs in south-west Asia and adjacent regions. 4. - Warszawa & Poznań: Polish Scientific Publishers, 80 pp.

BROWICZ, K. & ZIELIŃSKI, J. 1989a: On the geographic distribution of *Crataegus heldreichii* Boiss. - *Arbor. Kórnickie* **31**: 13-20.

BROWICZ, K. & ZIELIŃSKI, J. 1989b: *Coris monspeliensis* (Primulaceae) and *Laburnum alpinum* (Fabaceae), two species new for the flora of Greece. - *Willdenowia* **19(1)**: 83-86. [Also distributed as *OPTIMA leaflet* no. 188].

BROWICZ, K. & ZIELIŃSKI, J. 1996: Trees and shrubs of Ikaria Island (Greece). - *Arbor. Kórnickie* **41**: 15-45.

BROWICZ, K., ZIELIŃSKI, J. & TAN, KIT 2001: *Rumex vesicarius* (Polygonaceae) in the Eastern Mediterranean region. - *Polish Bot. J.* **46(1)**: 71-73.

BROWICZ, K. & ZOHARY, D. 1996: The genus *Amygdalus* L. (Rosaceae): Species relationships, distribution and evolution under domestication. - *Genet. Resources Crop Evol.* **43**: 229-247.

BROWN, R. 1986: Plant portraits: *Crocus sieberi*. - *Rock Gard.* **19(4)**: 383.

BROWN, W.L. 1941: The cytogenetics of *Poa pratensis*. - *Ann. Missouri Bot. Gard.* **28(4)**: 493-522 + 1 plate.

BROWNING, J. & GORDON-GRAY, K.D. 2000: Pattern of fruit morphology in *Bolboschoenus* (Cyperaceae) and their global distribution. - *S. African J. Bot.* **66(1)**: 63-71.

BROWNSEY, P.J. 1976a: A biosystematic investigation of the *Asplenium lepidum* complex. - *Bot. J. Linn. Soc.* **72(4)**: 235-267 + 1 plate.

BROWNSEY, P.J. 1976b: The origins of *Asplenium creticum* and *A. haussknechtii*. - *New Phytol.* **76(3)**: 523-542 + 2 plates.

BROWNSEY, P.J. & JERMY, A.C. 1973: A fern collecting expedition to Crete. - *Brit. Fern Gaz.* **10(6)**: 331-348.

BROWNSEY, P.J. & JERMY, A.C. 1975: *Asplenium ×khaniense*, a new wild hybrid from Crete. - *Candollea* **30(1)**: 21-28. [Also distributed as *OPTIMA leaflet* no. 3].

BRUCE, M.R. 1970: John Sibthorp [a lecture delivered at the School of Botany, Oxford (Research Seminar) on 29 May 1969]. - *Taxon* **19(3)**: 353-362.

BRUGGEN, H.W.E. VAN 1974: De orchideeën van Rodos. - *Orchideeën* **36(5)**: 134-138.

BRUHNS, T.B. 1981: Cytological identification of the genomes in pentaploid *Allium neapolitanum* using Giemsa C-banding. - *Pl. Syst. Evol.* **139(1-2)**: 1-10.

BRULLO, C. & AL. 2010: *Allium makrianum* (Alliaceae), a new autumnal species from Greece. - *Phyton (Horn)* **49(2)**: 267-278.

BRULLO, C. & AL. 2011: Taxonomical, morphological and geobotanical considerations on *Seseli tortuosum* L. (Apiaceae) from Sicily. - *Phyton (Horn)* **51(2)**: 201-210.

BRULLO, C. & AL. 2015: *Kali dodecanesicum* (Chenopodiaceae, Salsoloideae) a new species from Greece. - *Phytotaxa* **218(1)**: 61-68.

BRULLO, S. 1979: *Valantia calva*, a new species from Linosa, Sicily. - *Bot. Not.* **132(1)**: 61-64.

BRULLO, S. 1980a: *Valantia deltoidea* Brullo, sp. nov. from Sicily. - *Bot. Not.* **133(1)**: 63-66.

BRULLO, S. 1980b: Taxonomic and nomenclatural notes on the genus *Limonium* in Sicily. - *Bot. Not.* **133(3)**: 281-293.

BRULLO, S. 1988: Miscellaneous notes on the genus *Limonium* (Plumbaginaceae). - *Willdenowia* **17 (1-2)**: 11-18.

BRULLO, S. 1991: *Valantia eburnea* (Rubiaceae), a new species from Cyprus. - *Willdenowia* **20(1-2)**: 73-76. [Also distributed as *OPTIMA leaflet* no. 214].

BRULLO, S. 1992: *Limonium brutium*, a new species from S Italy. - *Fl. Medit.* **2**: 109-112.

BRULLO, S. 1998: Considerations phytogeographiques sur la Cyréniïque septentrionale. In Moret, J. (ed.): Abstracts of the IX. OPTIMA meeting, Paris, 11-17 May 1998, p. 21. - Paris: Mus. Nat. d'Hist. Naturelle.

BRULLO, S. 2009: Criteria and techniques in the taxonomic investigation of the Mediterranean flora. - *Bocconea* **23**: 7-18.

BRULLO, S. & AL. 1977a: Numeri cromosomici per la Flora Italiana: 283-298. - *Inform. Bot. Ital.* **9(1)**: 40-56.

BRULLO, S. & AL. 1977b: Numeri cromosomici per la Flora Italiana: 299-314. - *Inform. Bot. Ital.* **9(1)**: 57-71.

BRULLO, S. & AL. 1978a: Numeri cromosomici per la Flora Italiana: 366-374. - *Inform. Bot. Ital.* **10(1)**: 46-55.

BRULLO, S. & AL. 1978b: Numeri cromosomici per la Flora Italiana: 375-381. - *Inform. Bot. Ital.* **10(1)**: 57-63.

BRULLO, S. & AL. 1991: Numeri cromosomici per la Flora Italiana: 1251-1266. - *Inform. Bot. Ital.* **23(1)**: 39-47.

BRULLO, S. & AL. 1994: Cytotaxonomical revision of the *Allium obtusiflorum* group (Alliaceae). - *Fl. Medit.* **4**: 179-190.

BRULLO, S. & AL. 1995: Numeri cromosomici per la Flora Italiana: 1314-1334. - *Inform. Bot. Ital.* **26(2-3)**: 200-213.

BRULLO, S. & AL. 1996: Cytotaxonomic consideration of *Allium fuscum* Waldst. et Kit. (Liliaceae), a critical species of the European flora. - *Folia Geobot. Phytotax.* **31(4)**: 465-472.

BRULLO, S. & AL. 1997: Reports (885-898). In Kamari, G., Felber, F. & Garbari, F. (eds): Mediterranean chromosome number reports - 7. - *Fl. Medit.* **7**: 267-275.

BRULLO, S. & AL. 2001a: Cytotaxonomical remarks on *Allium pallens* L. and its relationship with *Allium convallarioides* Grossh. In Raimondo, F.M. (ed.): Abstracts from the X OPTIMA meeting, Palermo, 13-19 September 2001, p. 112.

BRULLO, S. & AL. 2001b: Cytotaxonomical notes on some rare endemic species of *Allium* (Alliaceae) from Greece. - *Caryologia* **54(1)**: 37-57.

BRULLO, S. & AL. 2003a: Three new species of *Allium* sect. *Codonoprasum* from Greece. - *Pl. Biosyst.* **137(2)**: 131-140.

BRULLO, S. & AL. 2003b: Cytotaxonomical remarks on *Allium pallens* and its relationships with *A. convallarioides* (Alliaceae). - *Bocconea* **16(2)**: 557-571.

BRULLO, S. & AL. 2004: Cytotaxonomic considerations on *Allium stamineum* Boiss. group (Alliaceae). In Vasić, O. (ed.): Abstracts, XI OPTIMA Meeting, Beograd 5-11.IX.2004, p. 122.

BRULLO, S. & AL. 2007: Cytotaxonomic considerations on *Allium stamineum* Boiss. group (Alliaceae). - *Bocconea* **21**: 325-343.

BRULLO, S. & AL. 2008: Taxonomic study on *Allium dentiferum* Webb & Berthel. (Alliaceae) and its relations with the allied species from the Mediterranean. - *Taxon* **57(1)**: 243-253.

BRULLO, S. & AL. 2019: *Poa jubata*, a rare Balkan species, first record for the Italian flora. - *Acta Bot. Croat.* **78(2)**: 147-154.

BRULLO, S. & AL. 2021: Taxonomical and chorological remarks on the Mediterranean *Poa maroccana* (Poaceae) and the first record in Italy from the Sicilian flora. - *Acta Bot. Croat.* **80(1)**: 63-73.

BRULLO, S., DE LEONARDIS, W. & PAVONE, P. 1982: Chromosome numbers of some Sicilian ferns. 1. - *Webbia* **35(2)**: 275-281.

BRULLO, S. & DE MARCO, G. 1999: Taxonomical revision of the genus *Dittrichia* (Asteraceae). In Correia, A.I.D. & al. (eds): Livro de resumos. Guia de excursões. V jornadas de taxonomia botânica, 16-19/IX 1999, p. 64.

BRULLO, S. & DE MARCO, G. 2001: Taxonomical revision of the genus *Dittrichia* (Asteraceae). - *Portug. Acta Biol.* **19(1-4)**: 341-354.

BRULLO, S., DE MARCO, G. & SIRACUSA, G. 1998: Taxonomical note on the genus *Adenocarpus* DC. (Leguminosae) in Italy. In Moret, J. (ed.): Abstracts of the IX. OPTIMA meeting, Paris, 11-17 May 1998. - Paris: Mus. Nat. d'Hist. Naturelle.

BRULLO, S., DE MARCO, G. & SIRACUSA, G. 2001: Taxonomical notes on the genus *Adenoocarpus* DC. (Leguminosae) in Italy. - *Bocconea* **13**: 425-436.

BRULLO, S. & ERBEN, M. 2016: The genus *Limonium* (Plumbaginaceae) in Greece. - *Phytotaxa* **240(1)**: 1-212.

BRULLO, S. & FURNARI, F. 1981: Phytogeographical considerations on the coastal vegetation of Cyrenaica. - *Anales Jard. Bot. Madrid* **37(2)**: 765-772.

BRULLO, S. & FURNARI, F. 1988: La vegetazione costiera della Cirenaica. - *Boll. Accad. Gioenia Sci. Nat. Catania* **21**: 37-117.

BRULLO, S. & GIUSSO DEL GALDO, G. 2001: *Astracantha dolinicola* (Fabaceae): a new species from Crete. - *Nordic J. Bot.* **21(5)**: 475-480.

BRULLO, S. & GIUSSO DEL GALDO, G. 2003: Taxonomical considerations of *Astragalus creticus* Lam. group (Fabaceae) from Crete. - *Israel J. Pl. Sci.* **51**: 307-313.

BRULLO, S. & GIUSSO DEL GALDO, G. 2006: Taxonomic remarks on the *Anthyllis hermanniae* L. (Fabaceae, Faboideae) species complex of the Mediterranean flora. - *Novon* **16(3)**: 304-314.

BRULLO, S., GIUSSO DEL GALDO, G. & MINISSALE, P. 2009: Taxonomic revision of the *Koeleria splendens* C. Presl group (Poaceae) in Italy based on morphological characters. - *Pl. Biosyst.* **143(1)**: 140-161.

BRULLO, S., GIUSSO DEL GALDO, G. & MUSARELLA, C.M. 2012: Taxonomic revision of *Astragalus angustifolius* group (Fabaceae). - *Bocconea* **24**: 19-52.

BRULLO, S., GIUSSO DEL GALDO, G. & TERRACI, M.C. 2008: *Allium aeginiense* Brullo, Giuso & Terraci (Alliaceae), a new species from Greece. - *Candollea* **63(2)**: 197-203.

BRULLO, S. & GIUSSO DEL GALDO, G.P. 2006: Taxonomic remarks on *Sesleria nitida* Ten. (Poaceae), an orophyte endemic to Sicily and the central-southern Apennines. - *Pl. Biosyst.* **140(1)**: 43-49.

BRULLO, S. & GUARINO, R. 1998: *Festuca humifusa* (Gramineae), a new species from Sicily. In Moret, J. (ed.): Abstracts of the IX. OPTIMA meeting, Paris, 11-17 May 1998. - Paris: Mus. Nat. d'Hist. Naturelle.

BRULLO, S. & GUARINO, R. 2000: Contribution to the knowledge of flora and vegetation of Khrisi islet (Crete, SE Mediterranean sea). - *Fl. Medit.* **10**: 265-282.

BRULLO, S. & GUARINO, R. 2001: *Festuca humifusa* (Gramineae), a new species from Sicily. - *Bocconea* **13**: 409-412.

BRULLO, S., GUARINO, R. & MINISSALE, P. 2000: Taxonomic and phytogeographical remarks on *Trifolium savianum* Guss., a mininterpreted species of the Italian flora. - *Bot. Jahrb. Syst.* **122(4)**: 469-480.

BRULLO, S. & GUGLIELMO, A. 2001: Considérations phytogéographiques sur la Cyrénaïque septentrionale. - *Bocconea* **13**: 209-222.
BRULLO, S., GUGLIELMO, A. & TERRASI, M.C. 1995: On *Allium pilosum* Sm. (Alliaceae), a rare Aegean endemic. - *Giorn. Bot. Ital., n.s.* **129(2)**: 34.
BRULLO, S., GUGLIELMO, A. & TERRASI, M.C. 1998: Notes on *Allium rhodopeum* (Alliaceae), a neglected species from the E Mediterranean area. - *Pl. Biosyst.* **132(1)**: 63-69.
BRULLO, S., MARCENÒ, C. & PAVONE, P. 1981: *Jasione sphaerocephala* sp. nov. (Campanulaceae) from Italy. - *Nordic J. Bot.* **1(2)**: 137-139.
BRULLO, S., MINISSALE, P. & TERRASI, M.C. 2003: *Elaeoselinum tunetanum* (Umbelliferae) a new species from Tunisia. - *Bocconea* **16(2)**: 573-578.
BRULLO, S. & MINISSALE, P. SAMPINATO, G. 1997: *Arrhenatherum nebrodense*, a new species from Sicily. - *Lagascalia* **19(1-2)**: 903-909.
BRULLO, S. & PAVONE, P. 1979: Numeri cromosomici per la Flora Italiana: 464-483. - *Inform. Bot. Ital.* **10(2)**: 248-265.
BRULLO, S. & PAVONE, P. 1983: *Allium greuteri* sp. nova (Liliaceae) from Cyrenaika and its relationships with the *Allium cupanii* group. - *Willdenowia* **13(1)**: 115-122.
BRULLO, S. & PAVONE, P. 1985: Taxonomic considerations on the genus *Desmazeria* (Gramineae) with description of a new species: *Desmazeria pignattii*. - *Willdenowia* **15(1)**: 99-106.
BRULLO, S., PAVONE, P. & SALMERI, C. 1991a: Cytotaxonomical notes on *Allium dentiferum* Webb & Berthelot, an unknown species of the Mediterranean flora. - *Bot. Chron. (Patras)* **10**: 785-796.
BRULLO, S., PAVONE, P. & SALMERI, C. 1991b: Cytotaxonomical notes on *Allium erythraeum* Griseb. from Greece and its relationships with *Allium obtusiflorum* DC. - *Giorn. Bot. Ital., n.s.* **125(3)**: 242.
BRULLO, S., PAVONE, P. & SALMERI, C. 1992: *Allium rhodiacum* (Liliaceae), a new species from Rhodos (Greece). - *Willdenowia* **22(1-2)**: 89-95.
BRULLO, S., PAVONE, P. & SALMERI, C. 1993a: Three new species of *Allium* from Cyprus. - *Candollea* **48(1)**: 279-290.
BRULLO, S., PAVONE, P. & SALMERI, C. 1993b: *Allium karistanum*, a new species from Evvoia (Greece). In Petrova, A.V. (ed.): Abstracts from the VII OPTIMA meeting in Borovetz, Bulgaria, 18-30 July 1993, p. 160.
BRULLO, S., PAVONE, P. & SALMERI, C. 1995a: Considerazioni citotassonomiche su *Allium pallens* L. (Alliaceae), specie critica del Mediterraneo. - *Inform. Bot. Ital.* **27(2-3)**: 309.
BRULLO, S., PAVONE, P. & SALMERI, C. 1995b: Considerazioni citotassonomiche e fitogeografiche su *Allium cupanii* Rafin. s.l., gruppo critico dell'area mediterranea. - *Giorn. Bot. Ital., n.s.* **129(1)**: 117-119.
BRULLO, S., PAVONE, P. & SALMERI, C. 1995c: Cytotaxonomic and phytogeographic considerations on *Allium meteoricum* (Alliaceae), rare endemic of the Greek flora. - *Giorn. Bot. Ital., n.s.* **129(2)**: 35.
BRULLO, S., PAVONE, P. & SALMERI, C. 1997: *Allium karistanum* (Liliaceae), a new species from Evvia (Greece). - *Bocconea* **5(2)**: 759-764.
BRULLO, S., PAVONE, P. & SALMERI, C. 1998a: *Allium brachyspathum* (Alliaceae), a new species from Karpathos (Aegean islands). In Moret, J. (ed.): Abstracts of the IX. OPTIMA meeting, Paris, 11-17 May 1998. - Paris: Mus. Nat. d'Hist. Naturelle.
BRULLO, S., PAVONE, P. & SALMERI, C. 1998b: Specie a fioritura tardiva del gruppo di *Allium paniculatum* nell'area mediterranea. - *Inform. Bot. Ital.* **29(2-3)**: 300-302.
BRULLO, S., PAVONE, P. & SALMERI, C. 1999: *Allium archeotrichon* (Alliaceae), a new species from Rhodos (Dodekannisos, Greece). - *Nordic J. Bot.* **19(1)**: 41-46.
BRULLO, S., PAVONE, P. & SALMERI, C. 2001: *Allium brachyspathum* (Alliaceae), a new species from the island of Karpathos (S Aegean area, Greece). - *Bocconea* **13**: 413-417.
BRULLO, S., PAVONE, P. & SALMERI, C. 2015: Biosystematic researches on *Allium cupani* group (Amaryllidaceae) in the Mediterranean area. - *Fl. Medit.* **25 (Special Issue)**: 209-244.
BRULLO, S., PAVONE, P. & SPAMPINATO, G. 1989: *Allium pentadactyli* (Liliaceae), a new species from S Italy. - *Willdenowia* **19(1)**: 115-120.
BRULLO, S., PAVONE, P. & TERRASI, M.C. 1979: Numeri cromosomici per la Flora Italiana: 494-505. - *Inform. Bot. Ital.* **10(2)**: 278-290.
BRULLO, S., PAVONE, P. & TERRASI, M.C. 1980: Numeri cromosomici per la Flora Italiana: 632-646. - *Inform. Bot. Ital.* **11(2)**: 161-171.
BRULLO, S., PAVONE, P. & TERRASI, M.C. 1985: Considerazioni cariologiche sul genere *Plantago* in Sicilia. - *Candollea* **40(1)**: 217-230.
BRULLO, S., PAVONE, P. & TERRASI, M.C. 1993: Karyological studies on *Plantago* in Cyrenaice. In: Demiriz, H. & Özhatay, N. (eds): Proceedings of the V OPTIMA meeting, Istanbul, 8-15 September, 1986, pp. 487-495. - Istanbul Üniversitesi Fen Fakültesi.
BRULLO, S., PAVONE, P. & TZANOUDAKIS, D. 1993: *Allium lagarophyllum* (Liliaceae), a new species from Greece. - *Willdenowia* **23(1-2)**: 107-111.
BRULLO, S., PAVONE, P. & ZIZZA, A. 1979: Numeri cromosomici per la Flora Italiana: 506-516. - *Inform. Bot. Ital.* **10(3)**: 397-405.
BRULLO, S., PAVONE, P. & ZIZZA, A. 1980: Numeri cromosomici per la Flora Italiana: 602-616. - *Inform. Bot. Ital.* **11(2)**: 139-148.
BRULLO, S. & SALMERI, C. 2021: Taxonomic investigation on *Allium hirtovaginum* group (Amaryllidaceae) from East Mediterranean area. *Fl. Medit.* **31 (special issue)**: 169-211.
BRULLO, S., SCELSI, F. & SIRACUSA, G. 1993: *Genista demarcoi*, a new species from Sicily. - *Fl. Medit.* **3**: 299-307.
BRULLO, S., SCELSI, F. & SPAMPINATO, G. 1996a: *Salix oropotamica* (Salicaceae), a new species from S. Italy. - *Fl. Medit.* **5**: 53-58.
BRULLO, S., SCELSI, F. & SPAMPINATO, G. 1996b: A new species of *Crepis* (Compositae) from Calabria (S. Italy). - *Fl. Medit.* **5**: 59-63.
BRULLO, S., SCELSI, F. & SPAMPINATO, G. 1997: A new species of *Armeria* (Plumbaginaceae) from S. Italy. - *Edinburgh J. Bot.* **54(1)**: 91-97.
BRULLO, S., SCELSI, F. & SPAMPINATO, G. 2001: *Salix ionica* (Salicaceae), a new species from S Italy. - *Bocconea* **13**: 419-423.
BRULLO, S. & SPAMPINATO, G. 1993: A new species of *Salix* (Salicaceae) from Calabria (S Italy). - *Candollea* **48(1)**: 291-295.
BRULLO, S. & SPAMPINATO, G. 2003: *Sedum annuum* subsp. *gussonei* Brullo & Spampinato subsp. nov. from S Italy. - *Bocconea* **16(2)**: 579-583.
BRULLO, S. & TZANOUDAKIS, D. 1989: *Allium maniaticum* (Liliaceae), a new species from S Greece. - *Willdenowia* **19(1)**: 111-114.
BRULLO, S. & TZANOUDAKIS, D. 1994: *Allium ionicum* (Liliceae), a new species from the Ionian islands (W Greece). - *Willdenowia* **24(1-2)**: 53-57.
BRUMHARD, P. 1905: Monographische Übersicht der Gattung *Erodium*. - *Arbeiten Königl. Bot. Gart. Breslau* **1905**: 59 pp. + 1 fold-out table.
BRUMMITT, R.K. 1968: [Notes on *Heracleum sphondylium*]. In Heywood, V.H. (ed.): Notulae systematicae ad Floram Europaeam spectantes, no. 7. - *Feddes Repert.* **79(1-2)**: 65.
BRUMMITT, R.K. 1971: The diagnostic characters of the European *Heliotropium* species. In Heywood, V.H. (ed.): Notulae systematicae ad Floram Europaeam spectantes, no. 8. - *Bot. J. Linn. Soc.* **64(1)**: 60-67. [Correction in *Bot. J. Linn. Soc.* 64(3): 275 (1971)].
BRUMMITT, R.K. & CHAUDHRI, M.N. 1967: Materials for a Flora of Turkey, XV: Illecebraceae. - *Notes Roy. Bot. Gard. Edinburgh* **28(1)**: 25-28.
BRUMMITT, R.K. & MEIKLE, R.D. 1993: The correct Latin names for the primrose and the oxlip, *Primula vulgaris* Hudson and *P. elatior* (L.) Hill. - *Watsonia* **19(3)**: 181-184.
BRUNDU, G. & AL. 2015: *Arctotheca calendula* (L.) Levyns: An emerging invasive species in Italy. - *Pl. Biosyst.* **149(6)**: 954-957.
BRUNNBAUER, W. 1989: Bibliographie zur Botanik von Sardinien (Italien) - I. - *Linzer Biol. Beitr.* **21(2)**: 405-488.
BRUNNBAUER, W. 1990: Bibliographie zur Botanik von Sardinien (Italien) - II. - *Linzer Biol. Beitr.* **22(2)**: 597-617.

BRUNNBAUER, W. 1991: Bibliographie zur Botanik von Sardinien (Italien) - III. - *Linzer Biol. Beitr.* **23(2)**: 593-613.
BRUNNBAUER, W. 1993: Bibliographie zur Botanik von Sardinien (Italien) - IV. - *Linzer Biol. Beitr.* **25 (2)**: 1015-1032.
BRUNNBAUER, W. 1994: Bibliographie zur Botanik von Sardinien (Italien) - V. - *Linzer Biol. Beitr.* **26(1)**: 539-578.
BRUNSBERG, K. 1976: Biosystematics in the *Lathyrus pratensis* complex. - Lund: Ph.D. thesis, Univ. of Lund, 120 + xvi pp.
BRUNSBERG, K. 1977: Biosystematics of the *Lathyrus pratensis* complex. - *Opera Bot.* **41**: 1-78.
BRUNTON, D.F., IVANOVA, D. & SOKOLOFF, P.C. 2021: Pirin Quillwort, *Isoetes pirinica* sp. nov., a new endemic lycophyte from Bulgaria. - *Fern Gaz.* **21(6)**: 240-252.
BRUS, R. & AL. 2011: Leaflet morphometric variation of service tree (*Sorbus domestrica* L.) in the Balkan Peninsula. - *Pl. Biosyst.* **145(2)**: 278-285.
BRÜTSCH, J.-P.J. 2000: Die Gattung *Nigritella* Rich. - *Bauhinia* **14**: 21-32.
BRUUN, H.G. 1932: Cytological studies in *Primula* with special reference to the relation between the karyology and taxonomy of the genus. - *Symb. Bot. Upsal.* **1**: 1-239.
BUCHEGGER, J. 1912: Beitrag zur Systematik von *Genista Hassertiana*, *G. holopetala* und *G. radiata*. - *Oesterr. Bot. Z.* **62**: 303-312, 368-376, 416-423, 458-465.
BUCHENAU, F. 1890: Monographia Juncacearum. - *Bot. Jahrb. Syst.* **12**: 1-495, post 621 + plates.
BUCHHEIM, G. & CONERT, H.J. 1969: Über *Danthonia alpina* Vest (Poaceae, Arundinoideae, Danthonieae). - *Senckenberg. Biol.* **50(5-6)**: 473-475.
BUCKNALL, C. 1913: A revision of the genus *Symphytum*, Tourn. - *J. Linn. Soc., Bot.* **41(284)**: 491-556.
BUIA, M.C. & MAZZELLA, L. 1991: Reproductive phenology of the Mediterranean seagrasses *Posidonia oceanica* (L.) Delile, *Cymodocea nodosa* (Ucria) Aschers., and *Zostera noltii* Hornem. - *Aquatic Bot.* **40(4)**: 343-362.
BUKLIEV, R. 1988: *Azolla filiculoides* Lam., a new alien species in the flora of Macedonia. - *Fragm. Balcan.* **14(2)**: 11-15.
BULIČ, Z. 1989: A contribution to the flora of the Tara river canyon's valley [In Serbo-Croatian with English summary]. - *Glasn. Republ. Zavoda Zaštitu Prir. Prirodnjačke Zbirke Titogradu* **22**: 113-138.
BULIŃSKA-RADOMSKA, Z. & LESTER, R.N. 1985: Relationships between five species of *Lolium* (Poaceae). - *Pl. Syst. Evol.* **148(3-4)**: 169-175.
BULIŃSKA-RADOMSKA, Z. & LESTER, R.N. 1986a: Relationships between seven species of *Vulpia* (Poaceae). - *Pl. Syst. Evol.* **153(1-2)**: 7-12.
BULIŃSKA-RADOMSKA, Z. & LESTER, R.N. 1986b: Species relationships in *Festuca* (sect. *Ovinae*, Poaceae). - *Pl. Syst. Evol.* **154(3-4)**: 175-182.
BULLITTA, S. & HAYWARD, M.D. 1996: Application of RAPD markers to a study of species relationships in the genus *Trifolium*. In Pickersgill, B. & Lock, J.M. (eds): Advances in legume systematics. Part 8, pp. 127-135. - Kew: Royal Botanic Gardens.
BULLOCK, A.A. 1967: *Vincetoxicum* again! - *Kew Bull.* **21(2)**: 351-352.
BULLOCK, A.A. 1969: On the typification of *Rhinanthus* L. - *Kew Bull.* **23(2)**: 265-266.
BUNGE, A.A. 1868: Generis Astragali species Gerontogeae. Pars prior. Claves diagnisticae. - *Mém. Acad. Imp. Sci. Saint Pétersbourg, Sér. 7* **11(16)**: 1-140.
BURBIDGE, F.W. & BAKER, J.G. 1875: The *Narcissus*: its history and culture, with coloured plates and descriptions of all known species and principal varieties. - London: L. Reeve & Co., xvi + 95 pp. + 48 plates.
BURBIDGE, R.B. & RICHARDSON, I.B.K. 1970: A revision of the genus *Rhynchocorys*. - *Notes Roy. Bot. Gard. Edinburgh* **30(1)**: 97-107.
BURDET, H.M. 1967: Contribution à l'étude caryologique des genres *Cardaminopsis*, *Turritis* et *Arabis* en Europe. - *Candollea* **22(1)**: 107-156.
BURDET, H.M. 1969b: *Arabis ciliata* Clairv. - *Candollea* **24(1)**: 139-143.
BURDET, H.M., CHARPIN, A. & JACQUEMOUD, F. 1979: Types nomenclaturaux des taxons décrits par Boissier, Leresche et Levier à la suite de leurs excursions en Espagne de 1878 et 1879. - *Mém. Soc. Bot. Genève* **1**: 63-82.
BURDET, H.M., CHARPIN, A. & JACQUEMOUD, F. 1981: Types nomenclaturaux des taxa ibériques décrits par Boissier ou Reuter. I. Gymnospermes à Graminées. - *Candollea* **36(2)**: 543-584.
BURDET, H.M., CHARPIN, A. & JACQUEMOUD, F. 1982a: Index Boissierianus Hispanicus. - *Candollea* **37(2)**: 397-427.
BURDET, H.M., CHARPIN, A. & JACQUEMOUD, F. 1982b: Types nomenclaturaux des taxa ibériques décrits par Boissier ou Reuter. II. Iridacées à Potamogetonacées. - *Candollea* **37(2)**: 381-395.
BURDET, H.M., CHARPIN, A. & JACQUEMOUD, F. 1982c: Thesaurus Boissierianus Hispanicus. - *Candollea* **37(2)**: 429-438.
BURDET, H.M., CHARPIN, A. & JACQUEMOUD, F. 1983a: Types nomenclaturaux des taxa ibériques décrits par Boissier ou Reuter. III. Aracées à Chenopodiacées. - *Candollea* **38(1)**: 401-441.
BURDET, H.M., CHARPIN, A. & JACQUEMOUD, F. 1983b: Types nomenclaturaux des taxa ibériques décrits par Boissier ou Reuter. IV. Cistacées à Composées. - *Candollea* **38(2)**: 751-802.
BURDET, H.M., CHARPIN, A. & JACQUEMOUD, F. 1984a: Types nomenclaturaux des taxa ibériques décrits par Boissier ou Reuter. V. Convolvulacées à Ericacées. - *Candollea* **39(1)**: 349-373.
BURDET, H.M., CHARPIN, A. & JACQUEMOUD, F. 1984b: Types nomenclaturaux des taxa iberiques décrits par Boissier ou Reuter. VI. Euphorbiacées à Guttifères. - *Candollea* **39(2)**: 771-789.
BURDET, H.M., CHARPIN, A. & JACQUEMOUD, F. 1987: Types nomenclaturaux des taxa ibériques décrits par Boissier ou Reuter. VII. Labiées. - *Candollea* **42(1)**: 111-127.
BURDET, H.M., CHARPIN, A. & JACQUEMOUD, F. 1988a: Types nomenclaturaux des taxa ibériques décrits par Boissier ou Reuter. VIII. Légumineuses à Oenothéracées. - *Candollea* **43(1)**: 293-329.
BURDET, H.M., CHARPIN, A. & JACQUEMOUD, F. 1988b: Types nomenclaturaux des taxa ibériques décrits par Boissier ou Reuter. IX. Orobanchacées á Polygonacées. - *Candollea* **43(2)**: 727-746.
BURDET, H.M., CHARPIN, A. & JACQUEMOUD, F. 1989: Types nomenclaturaux des taxa ibériques décrits par Boissier ou Reuter. X. Renonculacées à Rosacées. - *Candollea* **44(1)**: 43-58.
BURDET, H.M., CHARPIN, A. & JACQUEMOUD, F. 1990a: Types nomenclaturaux des taxa ibériques décrits par Boissier ou Reuter. XI. Rubiacées à Saxifragacées. - *Candollea* **45(1)**: 47-64.
BURDET, H.M., CHARPIN, A. & JACQUEMOUD, F. 1990b: Types nomenclaturaux des taxa ibériques décrits par Boissier ou Reuter. XII. Scrophulariacées. - *Candollea* **45(2)**: 609-625.
BURDET, H.M., CHARPIN, A. & JACQUEMOUD, F. 1991: Types nomenclaturaux des taxa ibériques décrits par Boissier ou Reuter. XIII. Solanacées à Violacées. - *Candollea* **46(2)**: 579-600.
BURDON, J.J. 1983: Biological Flora of the British Isles. *Trifolium repens* L. - *J. Ecol.* **71(1)**: 307-330.
BURES, P. & AL. 2004: Cytogeographical survey of *Eleocharis* subser. *Eleocharis* in Europe 1: *Eleocharis palustris*. - *Folia Geobot. Phytotax.* **39(3)**: 235-257.
BURGEL, J. 1990: Notes on the variability of *S[axifraga] marginata* in the mountains of Bosnia, Herzegovina, Montenegro and Macedonia. - *Saxifraga Group J.* **1**: 1-6.
BURGEL, J. 2002: A new subspecies of *Saxifraga scardica*. - *Saxifr. Mag.* **10**: 27-32.
BURKILL, I.H. 1937: The life-cycle of *Tamus communis* L. - *J. Bot. (London)* **75**: 1-12, 33-43, 65-74.
BURKILL, I.H. 1944: Biological Flora of the British Isles. *Tamus communis* L. - *J. Ecol.* **32(1)**: 121-129.

BURNAT, É. & GREMLI, A. 1887: Genre *Rosa* - Revision du groupe des *Orientales*. Étude sur les cinq espèces qui composent ce groupe dans le Flora Orientalis de Boissier. - Genève & Bâle: H. Georg, Libraire-Éditeur, vii + 90 pp. + 1 table.

BURRET, M. 1926: Beiträge zur Kenntnis der Tiliaceen. - *Notizbl. Bot. Gart. Berlin-Dahlem* **9(88)**: 592-880.

BURRI, CH. & AL. 2012: Zur Orchideenflora der Insel Kea (nordwestliche Kykladen, Griechenland). - *J. Eur. Orch.* **44(1)**: 83-116.

BURRI, CH. & BROGGI, M.F. 2011: Zur Orchideenflora der Insel Alonnisos (Nordsporaden, Griechenland). - *J. Eur. Orch.* **43(2)**: 378-400.

BURRI, CH. & BROGGI, M.F. 2013: Auf Orchideensuche am südlichsten Punkt Europas - die Insel Gavdos (Griechenland). - *J. Eur. Orch.* **45(1)**: 3-24.

BURRI, CH. & DELFORGE, P. 2013: Complément à la cartographie des orchidées de l'île d'Icaria (Îles égéennes orientales, Grèce). - *Naturalistes Belges* **94**: 107-114.

BURTON, R.E. 2006: A new subspecies of *Allium circinnatum* Sieber from S.W. Turkey. - *Ann. Mus. Goulandris* **11**: 81-84.

BURTON, R.E. 2009: Reports 24-30. In Vladimirov, V., Dane, F. & Tan, Kit (eds): New floristic records in the Balkans: 11. - *Phytol. Balcan.* **15(2)**: 278-179.

BURTON, R.M. 1989: *Radiola linoides* Roth. In Greuter, W. & Raus, Th. (eds): Med-Checklist notulae, 15. - *Willdenowia* **19(1)**: 35.

BURTON, R.M. 1991: A check-list and evaluation of the flora of Nisyros (Dodecanese, Greece). - *Willdenowia* **20(1-2)**: 15-38.

BURTON, R.M. 1996: A new thistle species from south-west Turkey. - *Karaca Arbor. Mag.* **3(3)**: 123-126.

BURTON, R.M. 1997a: Two new flowering plant species for the flora of Crete. - *Fl. Medit.* **6**: 69-70.

BURTON, R.M. 1997b: A new variety of *Dianthus elegans* d'Urv. - *Karaca Arbor. Mag.* **3(4)**: 183-186.

BURTON, R.M. 1999: Yet more new plants from the East Aegean island of Samos. - *Ann. Mus. Goulandris* **10**: 45-51.

BURTON, R.M. & TAN, KIT 2017: Reports 12-67. In Vladimirov, V. & al. (eds): New floristic records in the Balkans: 34. - *Phytol. Balcan.* **23(3)**: 415-419.

BURTON, R.M. & TAN, KIT 2018: Report 12. In Vladimirov, V., Aybeke, M. & Tan, Kit (eds): New floristic records in the Balkans: 35. - *Phytol. Balcan.* **24(1)**: 157.

BURTT, B.L. 1938a: *Scabiosa tenuis* Sprunner. - *Hooker's Icon. Pl.* **34(3)**: tab. 3355 + 3 pp. text.

BURTT, B.L. 1938b: *Crocus laevigatus*. - *Bot. Mag.* **161(1)**: tab. 9515 + 2 pp. text.

BURTT, B.L. 1949: *Crocus sieberi*. - *J. Roy. Hort. Soc.* **74(1)**: 12-17 + 1 plate.

BURTT, B.L. 1950: *Hypericum trichocaulon* Boiss. & Heldr. - *Hooker's Icon. Pl.* **35(4)**: tab. 3499 + 2 pp. text.

BURTT, B.L. 1951a: *Crocus niveus*. - *Bot. Mag.* **168(2)**: tab. 146 + 4 pp. text.

BURTT, B.L. 1951b: The genus *Ricotia*. - *Kew Bull.* **1951(1)**: 123-132 + 2 plates.

BURTT, B.L. 1951c: Two new species of *Colchicum*. - *Kew Bull.* **1950(3)**: 431-434.

BURTT, B.L. 1952: *Delphinium orientale*. - *Bot. Mag.* **169(1)**: tab. 186 + 4 pp. text.

BURTT, B.L. 1954: *Crocus nubigena*. - *Bot. Mag.* **170(2)**: tab. 235 + 5 pp. text.

BURTT, B.L. 1957: The double confusion of *Scabiosa papposa* L. - *Notes Roy. Bot. Gard. Edinburgh* **22(3)**: 279-283.

BURTT, B.L. 1962: *Colchicum parnassicum*. - *Bot. Mag.* **174(1)**: tab. 403 + 3 pp. text.

BURTT, B.L. 1970: The evolution and taxonomic significance of a subterranean ovary in certain monocotyledons. - *Israel J. Bot.* **19(2-3)**: 77-90.

BURTT, B.L. & DAVIS, P.H. 1949: On the flora of the Nearer East: XXIII. Miscellaneous new species and records. - *Kew Bull.* **1949(1)**: 97-115.

BUSCHMANN, A. 1938: Über einige ausdauernde *Cerastium*-Arten aus der Verwandtschaft des *C. tomentosum* Linné. - *Repert. Spec. Nov. Regni Veg.* **43**: 118-143 + 1 plate.

BUSCHMANN, A. 1939: Über einige ausdauernde *Cerastium*-Arten aus der Verwandtschaft des *C. tomentosum* Linné. Ergänzung. - *Repert. Spec. Nov. Regni Veg.* **46**: 33-35.

BUSCHMANN, A. 1942: Zur Klärung des Formenkreises um *Poa badensis* Haenke. - *Oesterr. Bot. Z.* **91 (2-3)**: 81-130.

BUSCHMANN, A. 1952: Ein Beitrag zur systematischen Stellung von *Poa violacea* Bellardi. - *Phyton (Horn)* **4(1-3)**: 132-136.

BUSER, R. 1894: Contribution à la connaissance des Campanulacées. - *Bull. Herb. Boissier* **2(8)**: 501-532 + 5 plates.

BUSER, R. 1940: Kritische Beiträge zur Kenntnis der schweizerischen Weiden. - *Ber. Schweiz. Bot. Ges.* **50**: 567-788.

BUSINSKÝ, R. 1989: Beitrag zur Taxonomie und Nomenklatur von *Pinus heldreichii* Christ und *P. leucodermis* Antoine sowie des Kultivars 'Smidtii'. - *Mitt. Deutsch. Dendrol. Ges.* **79**: 91-106.

BUSINSKÝ, R. 1999: Taxonomic revision of Eurasian pines (genus *Pinus* L.) - Survey of species and infraspecific taxa according to latest knowledge. Towards a monograph of pines in the Old World. - *Acta Průhon.* **68**: 7-86.

BUSSING, A. 2000: Mistletoe: The genus *Viscum*. - CRC Press: 274 pp.

BUSTARD, L. 1987: The tulip twins of Crete. - *Plantsman* **9(3)**: 154-156.

BUTCHER, R.W. 1947: Biological Flora of the British Isles. *Atropa belladonna* L. - *J. Ecol.* **34(2)**: 345-353.

BUTCHER, R.W. 1954: Biological Flora of the British Isles. *Colchicum autumnale* L. - *J. Ecol.* **42(1)**: 249-257.

BUTTLER, K.P. 1967: Zytotaxonomische Untersuchungen an mittel- und südeuropäischen *Draba*- Arten. - *Mitt. Bot. Staatssamml. München* **6(2)**: 275-362.

BUTTLER, K.P. 1976: Die Flora am Stadtberg Pyrgos bei Hagia Marina. - *Documenta, Forsch. Natur- Geisteswiss.* **1**: 38-39.

BUTTLER, K.P. 1977a: Variation in wild populations of annual beet (*Beta*, Chenopodiaceae). - *Pl. Syst. Evol.* **128(1-2)**: 123-136.

BUTTLER, K.P. 1977b: Revison von *Beta* Sektion *Corollinae* (Chenopodiaceae). I. Selbststerile Basisarten. - *Mitt. Bot. Staatssamml. München* **13**: 255-336.

BUTTLER, K.P. 1979: Zur Flora des Ost-Pangaion. In Unger, H.J. & al.: Pangaion - Lagerstättenkundliche, bergbauliche und topographisch-archäologische Untersuchungen. - *Aspekte (Altmühldorf)* **1**: 79-80.

BUTTLER, K.P. 1983: Die *Ophrys ciliata* (*speculum*)-Gruppe, eine Neubewertung (Orchidaceae: Orchideae). - *Jahresber. Naturwiss. Vereins Wuppertal* **36**: 37-57.

BUTTLER, K.P. 1986: Orchideen. Die wildwachsenden Arten und Unterarten Europas, Vorderasiens und Nordafrikas. - München: Mosaik Verlag, Steinbachs Naturführer, 288 pp.

BUTTLER, K.P. 1995: *Sium latifolium* L. In Greuter, W. & Raus, Th. (eds): Med-Checklist notulae, 16. - *Willdenowia* **25(1)**: 175.

BUTTLER, K.P. 2001: Taxonomy of Orchidaceae tribus Orchideae, a traditional approach. - *J. Eur. Orch.* **33(1)**: 7-32.

BUTTLER, K.P. 2006: Zur Benennung der *Carex polyphylla*, zugleich Recherchen zur *Carex guestphalica*. - *Schlechtendalia* **14**: 17-24.

BUTTLER, K.P. & BUSS, H. 1986: *Orchis pinetorum* im Pindusgebirge und weitere Orchideenbeobachtungen auf der Balkanhalbinsel. - *Ber. Arbeitskreis. Heimische Orchid.* **3(1)**: 4-12.

BUTTLER, K.P. & NIESCHALK, CH. 1979: Morphologische und geographische Differenzierung bei *Orchis palustris*. - *Bot. Jahrb. Syst.* **101(1)**: 91-134. [Also distributed as *OPTIMA leaflet* no. 100].

BUTTLER, K.P. & RAUS, TH. 2017: *Platanthera fornicata*. In Raus, Th. & Raab-Straube, E.v. (eds): Euro+Med-Checklist notulae, 7. - *Willdenowia* **47(1)**: 91-92.

BUTTLER, K.P. & THIEME, M. 2013: Florenliste von Deutschland - Gefäßpflanzen. (Version 5 - Juli 2013). - Published at http://www.kp-buttler.de/florenliste/index.htm.

BUYANKIN, V.I. 1974: The distribution of *Amaranthus albus* L. (Amaranthaceae) in the U.S.S.R. [In Russian]. - *Bot. Zhurn. (Moscow & Leningrad)* **59(6)**: 864-866.

BUZUROVIĆ, U. & AL. 2013: On the distribution of *Goniolimon tataricum* (Plumbaginaceae) in Serbia. - *Bot. Serbica* **37(2)**: 167-172.

BUZUROVIĆ, U. & AL. 2015: How many *Goniolimon* species (Plumbaginaceae) do we have in Croatia? In Bogdanović, S. & Jogan, N.: Book of abstracts - 6th Balkan Botanical Congress, p. 53. - Rijeka.

BUZUROVIĆ, U. & AL. 2016: *Goniolimon dalmaticum* Rchb. f. and *G. tataricum* (L.) Boiss. (Plumbaginaceae) in the Croatian flora and their distribution in the Balkan Peninsula. - *Acta Bot. Croat.* **75(2)**: 164-172.

BUZUROVIĆ, U. & AL. 2018: *Goniolimon tataricum* (Plumbaginaceae) and its relatives from SE Europe and the Apennine Peninsula based on morphological and molecular evidence. - *Bot. Serbica* **142 (suppl. 1)**: 58.

BUZUROVIĆ, U. & AL. 2020: Phylogeographic and taxonomic considerations on *Goniolimon tataricum* (Plumbaginaceae) and its relatives from south-eastern Europe and the Apennine Peninsula. - *Pl. Syst. Evol.* **306(2)**: 22 pp.

BYAM-GROUNDS, J.S., HORNÝ, R. & SOJÁK, J. 1988: *Saxifraga grisebachii* versus *S. federici-augusti*. - *Preslia* **60(1)**: 73-74.

BYATT, J.I. 1974: Application of the names *Crataegus calycina* Peterm. and *C. oxyacantha* L. - *Bot. J. Linn. Soc.* **69(1-4)**: 15-21 + 2 plates.

BYATT, J.I. 1976: The genus *Crataegus* (Rosaceae) in Greece. - *Candollea* **31(2)**: 283-301.

BYATT, J.I. 1977: A contribution to the taxonomy and biology of the genus *Crataegus* in Europe. - London: Ph.D. thesis, Westfield College, Univ. London, 255 pp. + plates & maps.

BYATT, J.I., FERGUSON, I.K. & MURRAY, B.G. 1977: Intergeneric hybrids between *Crataegus* L. and *Mespilus* L.: a fresh look at an old problem. - *Bot. J. Linn. Soc.* **74(4)**: 329-343.

BYATT, J.I. & MURRAY, B.G. 1977: Chromosome numbers of some Eurasian species of *Crataegus*. - *Watsonia* **11(4)**: 374-375.

BYFIELD, A. 2000: Important plant areas: Saving Istanbul's exceptional botanical diversity. In Özhatay, N. & al. (eds): Abstracts from the Second Balkan Botanical Congress, Istanbul, Turkey, May 14-18, 2000, p. 8.

BYFIELD, A. & ÖZHATAY, N. 1998: Two new species to Turkey from Turkish Thrace: *Mibora minima* (L.) Desv. (Gramineae) & *Trifolium ornithopodioides* (L.) Sm. (Leguminosae). - *Türk. Bot. Derg.* **22**: 425-429.

CABI, E. & AL. 2009: Morphological and palynological features of the genus *Dasypyrum* (Poaceae) in Turkey. - *Phytol. Balcan.* **15(3)**: 393-400.

CABI, E. & AL. 2016: *Poa densa* (Poaceae), an overlooked Turkish steppe grass, and the evolution of bulbs in *Poa*. - *Willdenowia* **46(2)**: 201-211.

CABI, E., DOĞAN, M. & MAVI, Ö. 2009: Morphological and anatomical properties of the genus *Crithopsis* (Poaceae) in Turkey. - *Biol. Divers. Conservation* **3(2)**: 42-48.

CAFFERTY, S. & GREY-WILSON, CH. 1998: (1354) Proposal to reject the name *Cyclamen europaeum* (Primulaceae). - *Taxon* **47(2)**: 479-480. [approved, cf. Taxon 49(2): 271, 2000].

CAFFERTY, S. & JARVIS, C.E. 2004: Typification of Linnaean plant names in Cyperaceae. - *Taxon* **53 (1)**: 177-181.

CAFFERTY, S., JARVIS, C.E. & TURLAND, N.J. (EDS) 2000: Typification of Linnaean plant names in the Poaceae (Gramineae). - *Taxon* **49(2)**: 239-260.

CAFFERTY, S. & JARVIS, C.E. (EDS) 2002a: Typification of Linnaean plant names in Brassicaceae (Cruciferae). - *Taxon* **51(3)**: 529-537.

CAFFERTY, S. & JARVIS, C.E. (EDS) 2002b: Typification of Linnaean plant names in Rosaceae. - *Taxon* **51(3)**: 539-545.

CAFFERTY, S. & JARVIS, C.E. (EDS) 2002c: Lectotypification of Linnaean plant names in Ericaceae. - *Taxon* **51(4)**: 751-753.

CAFFERTY, S. & JARVIS, C.E. (EDS) 2004a: Typification of Linnaean plant names in Boraginaceae. - *Taxon* **53(3)**: 799-805.

CAFFERTY, S. & JARVIS, C.E. (EDS) 2004b: Typification of Linnaean plant names in Caryophyllaceae. - *Taxon* **53(4)**: 1049-1054.

CAFFERTY, S. & SCHOLZ, H. 1998: (1342) Proposal to reject the name *Secale creticum* (Gramineae), a threat to *Hordeum bulbosum*. - *Taxon* **47(2)**: 457-458. [approved, cf. Taxon 49(2): 268, 2000].

CAFFERTY, S. & SNOGERUP, S. 2000: (1468) Proposal to conserve the name *Rumex alpinus* (Polygonaceae) with a conserved type. - *Taxon* **49(3)**: 571-572. [approved; cf. *Taxon* 51(4): 796, 2002].

CAIN, R.B. 1965: The genus *Primula*. A review of its diversity and origin. - *Quart. Bull. Alpine Gard. Soc. Gr. Brit.* **35(2)**: 128-171.

CAKOVIĆ, D. & AL. 2014: *Colchicum cupanii* Guss. subsp. *glossophyllum* (Heldr.) Rouy, *Datura innoxia* Mill. and *Eclipta prostrata* (L.) L., new floristic records in Montenegro and western Balkan. - *Acta Bot. Croat.* **73(1)**: 255-265.

CAKOVIĆ, D. & AL. 2015: How many taxa? Spatiotemporal evolution and taxonomy of *Amphoricarpos* (Asteraceae, Carduoideae) on the Balkan Peninsula. - *Org. Divers. Evol.* **15(3)**: 429-445.

CAKOVIĆ, D. & AL. 2021: High genetic and morphological diversification of the *Euphorbia verrucosa* alliance (Euphorbiaceae) in the Balkan and Iberian peninsulas. - *Taxon* **70(2)**: 286-307.

CAKOVIĆ, D. & FRAJMAN, B. 2020: Three Tertiary *Euphorbia* species persisted in the forests of the Balkan Peninsula. - *Pl. Syst. Evol.* **306(3)**: 12 pp.

CALESTANI, V. 1905: Conspectus specierum europearum generis Peucedani. - *Boll. Soc. Bot. Ital.* **1905**: 193-201.

CALIGARI, P.D.S. & BRANDHAM, P.E. (EDS) 2001: Wheat taxonomy: the legacy of John Percival. - *The Linnean Special Issue* **3**: 1-190.

CALLAGHAN, D.A. 1998: Biological Flora of the British Isles. *Lythrum hyssopifolium* L. - *J. Ecol.* **86(6)**: 1065-1072.

CALLAGHAN, T.V. 1974: Intraspecific variation in *Phleum alpinum* L. with special reference to polar populations. - *Arctic Alpine Res.* **6(4)**: 361-401.

CALLEN, E.O. 1940: Studies in the genus *Euphrasia* L. - I. - *J. Bot. (London)* **78**: 213-218.

CALLEN, E.O. 1941: Studies in the genus *Euphrasia* L. - II. - *J. Bot. (London)* **79**: 11-13.

CALLEN, E.O. 1959: Studies in the genus *Lotus* (Leguminosae). I. Limits and subdivisions of the genus. - *Canad. J. Bot.* **37(1)**: 157-165.

CALVO, J. & AEDO, C. 2015: A taxonomic revision of the Eurasian/Northwestern African *Senecio doria* group (Compositae). - *Syst. Bot.* **40(3)**: 900-913.

CALVO, J., ÁLVAREZ, I. & AEDO, C. 2014: Three new combinations and a replacement name in Eurasian *Senecio* (Compositae, Senecioneae). - *Novon* **23(2)**: 139-142.

CALVO, J., ÁLVAREZ, I. & AEDO, C. 2015: Systematics of *Senecio* section *Crociseris* (Compositae, Senecioneae). - *Phytotaxa* **211(1)**: 1-105.

CAMA, L. 1980: Flower seeking in Crete. - *J. Scott. Rock Gard. Club* **17(2)**: 84-90.

CAMARDA, I. 1978: Numeri cromosomici per la Flora Italiana: 402-405. - *Inform. Bot. Ital.* **10(1)**: 84-90.

CAMARDA, I. 2015: *Paeonia sandrae* (Paeoniaceae) species nova of Sardinia and relationship with peonies of Corsica and Sicily. - *Fl. Medit.* **25**: 127-136.

CAMBRIA, S. & AL. 2020: *Salix aegaea* (Salicaceae), a new species from island of Ikaria (Greece). - *Phytotaxa* **447(2)**: 127-136.

CAMBRIA, S., BRULLO, C. & BRULLO, S. 2019: *Salix kaptarae* sp. nov. (Salicaceae) from Crete. - *Nordic J. Bot.* **37(7)**: e02335 (9 pp.).

CAMMERLOHER, H. 1935: Ein Beitrag zur Flora von Kaimeni (Santorin). - *Oesterr. Bot. Z.* **84(2)**: 81-90.

CAMPBELL, G.R. 1950: *Mimulus guttatus* and related species. - *Aliso* **2(3)**: 319-337.

CAMPBELL, N.R. 1979: The occurrence of *Orchis robusta* (T. Stephenson) Gölz & Reinhard in Crete. - *Watsonia* **12(3)**: 253-255.

CAMPBELL, N.R. 1982: *Ophrys cretica* and its possible identity with *O. doerfleri* (Orchidaceae). - *Willdenowia* **12(2)**: 231-234.

CAMPBELL, N.R. & CAMPBELL, A.K. 1986: The April-flowering orchids of Crete. 1. The genus *Serapias*. - *Willdenowia* **16(1)**: 47-56.

CAMPION-BOURGET, F. 1967: Observations caryologiques sur quelques espèces du genre *Rhinanthus* L. - *Compt. Rend. Hebd. Séances Acad. Sci., Sér. D* **264(17)**: 2100-2102.

CAMPO, G. & ROMANO, S. 1995: Numeri cromosomici per la Flora Italiana: 1346-1357. - *Inform. Bot. Ital.* **27(1)**: 26-34.

CAMPO, G., ROMANO, S. & MARCENO, C. 1999: Numeri cromosomici per la Flora Italiana: 1401-1408. - *Inform. Bot. Ital.* **30(1-3)**: 47-51.

CAMPOY, J.G. & AL. 2018: Monographs of invasive plants in Europe: *Carpobrotus*. – *Botany Letters* **165(3-4)**: 440-475.

CAMUS, A. 1929: Les châtaigniers. Monographie des genres *Castanea* et *Castaneopsis* (In Encycl. Écon. Sylv., vol. 3). - Paris: P. Lechevalier, 604 pp. + Atlas (109 plates).

CAMUS, A. 1931: Sur quelques Orchidées d'Orient. - *Bull. Soc. Bot. France* **78(1)**: 66.

CAMUS, A. 1936-54: Les chênes. Monographie du genre *Quercus*. - Paris: P. Lechevalier, 3 vols. [cf. TL-2 1: 431-432, 1976].

CAMUS, A. & CAMUS, E.G. 1904-05: Classification des saules d'Europe et monographie des saules de France. - Paris: J. Mersch, 2 vols + atlas [see also *J. Bot. (Morot)* 18-20, 1904-06].

CAMUS, E.G. & CAMUS, A. 1927-29: Iconographie des orchidées d'Europe et du Bassin Méditerranéen. - Paris: P. Lechevalier, 560 pp. + 133 plates.

ČANAK, M. & GAJIĆ, M.R. 1978: Material for the monograph of *Primula vulgaris* Huds. [In Serbo- Croatian with English summary]. - *Glasn. Prir. Mus. Beogradu, Ser. B, Biol. Nauke* **33**: 69-77.

CANDARGY, C.A. 1889: Flore de l'île de Lesbos. Plantes sauvages et cultivées. - Unter-Zürich: A. Diggelmann, 64 pp.

CANDARGY, C.A. 1890: Flore de l'île de Lesbos (Mételin). - *Rev. Méd.-Pharm. Constantinople* **3**: 43-45, 63-66, 97-99, 131-132.

CANDARGY, C.A. 1891: Flore de l'île de Lesbos (Mételin) [cont.]. - *Rev. Méd.-Pharm. Constantinople* **4**: 84-85, 102-104, 127-128, 174-175.

CANDARGY, C.A. 1892a: Flore de l'île de Lesbos (Mételin) [cont.]. - *Rev. Méd.-Pharm. Constantinople* **5**: 58-59.

CANDARGY, P.C. 1892b: Supplément à la flore de l'île de Lesbos (Mételin). - *Rev. Méd.-Pharm. Constantinople* **5**: 106-107, 141, 151-152.

CANDARGY, P.C. 1897: Flore de l'île de Lesbos. - *Bull. Soc. Bot. France* **44**: 140-144, 145-162, 369-373, 449-462.

CANDARGY, P.C. 1898: Flore de l'île de Lesbos [cont.]. - *Bull. Soc. Bot. France* **45**: 108-115, 181-192.

CANDARGY, P.C. 1899: La végétation de l'île de Lesbos (Mytilène). - *Rev. Gén. Bot.* **11**: 268-280, 310-329 + 1 map and 3 plates.

CANDARGY, P.C. 1901: Monografia tis files ton Krithodon. Étude monographique sur la tribu des Hordées ... au Museum d'Histoire Naturelle de Paris en 1897-1898-1899. - *Arch. Biol. Vég. Pure Appl.* **1**: 11-68.

CANDOLLE, A. DE 1830: Monographie des Campanulées. - Paris: viii + 384 pp. + 20 plates.

CANDOLLE, A.P. DE & CANDOLLE, A. DE (EDS) 1824-73: Prodromus systematis naturalis regni vegetabilis. - Paris, etc.: 17 vols. [cf. TL-2 1: 445-449, 1976].

CANEHL, G. 1973: Über Vorkommen von *Ophrys mammosa* var. *helenae* im nördlichen Epirusgebiet. - *Orchidee (Hamburg)* **24(6)**: 258-259.

CANNE, J.M. 1977: A revision of the genus *Galinsoga* (Compositae: Heliantheae). - *Rhodora* **79(819)**: 319-389.

CANNE, J.M. 1978: Circumscription and generic relationships of *Galinsoga* (Compositae, Heliantheae). - *Madroño* **25(2)**: 81-92.

CANNE, J.M. 1983: Cytological and morphological observations in *Galinsoga* and related genera (Asteraceae). - *Rhodora* **85(843)**: 355-366.

CANNON, J.F.M. 1964: Infraspecific variation in *Lathyrus nissolia* L. - *Watsonia* **6**: 28-35.

CANNON, J.F.M. 1967: [Notes on Umbelliferae]. In Heywood, V.H. (ed.): Notulae systematicae ad Floram Europaeam spectantes, no. 6. - *Feddes Repert.* **74(1-2)**: 34-37.

CANNON, M.J. & CANNON, J.F.M. 1984: A revision of the Morinaceae (Magnoliophyta-Dipsacales). - *Bull. Brit. Mus. (Nat. Hist.), Bot.* **12(1)**: 1-35.

CANO, E. & AL. 2017: Morphometric analysis and bioclimatic distribution of *Glebionis coronaria* s.l. (Asteraceae) in the Mediterranean area. - *PhytoKeys* **81**: 103-126.

CANTINO, PH.D., HARLEY, R.M. & WAGSTAFF, S.J. 1992: Genera of Labiatae: status and classification. In Harley, R.M. & Reynolds, T. (eds): Advances in Labiatae science, pp. 511-522.

CAPINERI, R. 1971: Numeri cromosomici per la Flora Italiana: 39-44. - *Inform. Bot. Ital.* **3(1)**: 75-79.

CAPINERI, R. 1987: Numeri cromosomici per la Flora Italiana: 1094-1097. - *Inform. Bot. Ital.* **18(1-3)**: 176-178.

CAPINERI, R., D'AMATO, G. & MARCHI, P. 1976: Numeri cromosomici per la Flora Italiana: 219-231. - *Inform. Bot. Ital.* **8(1)**: 67-74.

CAPINERI, R., D'AMATO, G. & MARCHI, P. 1979: Numeri cromosomici per la Flora Italiana: 534-583. - *Inform. Bot. Ital.* **10(3)**: 421-465.

CAPINERI, R. & GIARDINI, M. 1995: Numeri cromosomici per la Flora Italiana: 1297-1301. - *Inform. Bot. Ital.* **27(2-3)**: 187-190.

CAPINERI, R. & ROSSI, W. 1988: Numeri cromosomici per la Flora Italiana: 1127-1135. - *Inform. Bot. Ital.* **19(3)**: 314-318.

CAPPELLETTI, E.M. & POLDINI, L. 1984: Seed morphology in some European aconites (*Aconitum*, Ranunculaceae). - *Pl. Syst. Evol.* **145(3-4)**: 193-201.

CAPUTO, P., CAMPO, I. & DE LUCA, P. 1996: Morphometric variation in *Lomelosia crenata* (Dipsacaceae). - *Pl. Syst. Evol.* **201(1-4)**: 223-232.

CAPUTO, P. & COZZOLINO, S. 1995: Variabilità nel complesso *Lomelosia crenata* (Cyr.) Greuter & Burdet (Dipsacaceae). - *Inform. Bot. Ital.* **27(2-3)**: 306.

CARDONA, M.A. 1977: Números cromosómicos para la flora española. 40-44. - *Lagascalia* **7(2)**: 212-216.

CARDONA, M.A. 1984: Caryosystématique et différenciation évolutive de quelques "Rubia" méditerranénnes. - *Webbia* **38**: 513-529.

CARDONA, M.A. & CONTANDRIOPOULOS, J. 1978: L'endemisme dans les flores insulaires mediterranéennes. - *Mediterránea (Alicante)* **2**: 49-77.

CARDONA, M.A. & CONTANDRIOPOULOS, J. 1979: Endemism and evolution in the islands of the Western Mediterranean. In: Bramwell, D. (ed.): Plants and islands, pp. 133-169. - London, etc.: Academic Press.

CARDONA, M.A. & CONTANDRIOPOULOS, J. 1980: Números cromosómicos para la flora española. 162-182. - *Lagascalia* **9(2)**: 272-284.

CARDONA, M.A. & CONTANDRIOPOULOS, J. 1983: [Miscellaneous chromosome numbers]. In Löve, Á. (ed.): IOPB chromosome number reports LXXIX. - *Taxon* **32(2)**: 323-324.

CARDONA, M.A., CONTANDRIOPOULOS, J. & SIERRA RÁFOLS, E. 1986: Étude biosystématique d'*Anthyllis hystrix* de Minorque et d'*A. hermanniae* de la Méditerranée orientale et centrale. - *Orsis* **2**: 5-25.

CARDONA, M.A., LLORENS, L. & SIERRA RÁFOLS, E. 1983: Étude biosystématique de *Dorycnium pentaphyllum* Scop. subsp. *fulgurans* (Porta) comb. nova, endémique des Baléares orientales. - *Collect. Bot. (Barcelona)* **14**: 133-150.

CAREY, P.D. & FARRELL, L. 2002: Biological Flora of the British Isles. *Himantoglossum hircinum* (L.) Sprengel. - *J. Ecol.* **90(1)**: 206-218.

CARINE, M.A. & ROBBA, L. 2010: Taxonomy and evolution of the *Convolvulus sabaticus* complex (Convolvulaceae). - *Phytotaxa* **14**: 1-21.

CARLBOM, C. 1969: Evolutionary relationships in the genus *Scrophularia* L. - *Hereditas (Lund)* **61(3)**: 287-301.

CARLISLE, A. & BROWN, A.H.F. 1968: Biological Flora of the British Isles. *Pinus sylvestris* L. - *J. Ecol.* **56(1)**: 269-307.

CARLÓN, L. & AL. 2005: Más, a propósito de algunas *Orobanche* L. y *Phelipanche* Pomel del oeste del Paleártico. - *Documentos Jard. Bot. Atlántico (Gijón)* **3**: 1-71.

CARLÓN, L. & AL. 2008: Más, a propósito de algunas *Phelipanche* Pomel, *Boulardia* F.W. Schultz y *Orobanche* L. (Orobanchaceae) del oeste del Paleártico. - *Documentos Jard. Bot. Atlántico (Gijón)* **6**: 1-128.

CARLSSON, M. & AL. 1974: Chromosome morphology in *Scilla hyacinthoides* L. - *Bot. Not.* **127(1)**: 87-88.

CARLSTRÖM, A. 1984a: New species of *Alyssum*, *Consolida*, *Origanum* and *Umbilicus* from the SE Aegean Sea. - *Willdenowia* **14(1)**: 15-26.

CARLSTRÖM, A. 1984b: A revision of *Cleome* series *Ornithopodioides* Tzvelev (Capparaceae). - *Willdenowia* **14(1)**: 119-130.

CARLSTRÖM, A. 1985: Two new species of *Sedum* (Crassulaceae) from S Greece and SW Turkey. - *Willdenowia* **15(1)**: 107-113.

CARLSTRÖM, A. 1986a: A revision of *Arenaria* sect. *Orientales* ser. *Orientales* incl. ser. *Deflexae* (Caryophyllaceae) in the Aegean and SW Turkey. - *Willdenowia* **15(2)**: 359-374.

CARLSTRÖM, A. 1986b: A revision of the *Campanula drabifolia* complex (Campanulaceae). - *Willdenowia* **15(2)**: 375-387.

CARLSTRÖM, A. 1986c: New taxa and notes from the SE Aegean area and SW Turkey. - *Willdenowia* **16(1)**: 73-78.

CARLSTRÖM, A. 1986d: The phytogeographical position of Rodhos. - *Proc. Roy. Soc. Edinburgh, B, Biol. Sci.* **89**: 79-88.

CARLSTRÖM, A. 1987: A survey of the flora and phytogeography of Rodhos, Simi, Tilos and the Marmaris Peninsula (SE Greece, SW Turkey). - Lund: Ph.D. thesis, Univ. of Lund, 302 + xxii pp.

ČARNI, A. & AL. 2014: Early spring ephemeral therophytic non-nitrophilous grasslands as a habitat of various species of *Romulea* in the southern Balkans. - *Acta Bot. Croat.* **73(1)**: 107-129.

ČARNI, A. & MATEVSKI, V. 2005: Comparison of short-lived ruderal vegetation of the inland and coastal regions in the southern part of the Balkan Peninsula. - *Fitosociologia* **42(1)**: 97-107.

ČARNI, A., MATEVSKI, V. & ŠILIC, U. 2010: Morphological, chorological and ecological plasticity of *Cistus incanus* in the southern Balkans. - *Pl. Biosyst.* **144(3)**: 602-617.

CARNICERO, P. 2017: Evolution, biogeography and systematics of the genus *Cymbalaria* Hill. - Universitat Autònoma de Barcelona: Ph.D. thesis, 230 pp.

CARNICERO, P. & AL. 2017: Different speciation types meet in a Mediterranean genus: The biogeographic history of *Cymbalaria* (Plantaginaceae). - *Taxon* **66(2)**: 393-407.

CARNICERO, P. & AL. 2021: Disentangling relationships among eastern Mediterranean *Cymbalaria* including description of a novel species from the southern Peloponnese (Greece). - *Pl. Syst. Evol.* **307: 13**: doi.org/10.1007/s00606-020-01730-3.

CAROLIN, R.C. 1957: Cytological and hybridization studies in the genus *Dianthus*. - *New Phytol.* **56 (1)**: 81-97.

CARRASCO, M.A. & MARTÍN-BLANCO, C.J. 1995: Consideraciones sobre el género *Kickxia* Dumort. (Scrophulariaceae) en la Península Ibérica. - *Anales Jard. Bot. Madrid* **53(2)**: 213-217.

CARRASCO, M.A. & MONGE, C. 1991: Precisiones sobre el grupo *Aphanes microcarpa* en la Península Ibérica. - *Candollea* **46(1)**: 101-110.

CARRETERO, J.L. 1979: El género *Amaranthus* L. en España. - *Collect. Bot. (Barcelona)* **11**: 105-142.

CARRETERO, J.L. 1981: El género *Echinochloa* Beauv. en el suroeste de Europa. - *Anales Jard. Bot. Madrid* **38(1)**: 91-108.

CARRETERO, J.L. 1984: [Chromosome numbers in *Amaranthus*]. In Löve, Á. (ed.): Chromosome number reports LXXXIV. - *Taxon* **33(3)**: 537.

CARRETERO, J.L. 1990: *Flaveria bidentis* (L.) O. Kuntze [= *F. contrayerba* (Cav.) Pers.] en la Península Ibérica. - *Anales Jard. Bot. Madrid* **47(1)**: 253.

CARROLL, C.P. & JONES, K. 1962: Cytotaxonomic studies in *Holcus*. III. A morphological study of the triploid F1 hybrid between *Holcus lanatus* L. and *H. mollis* L. - *New Phytol.* **61(1)**: 72-84 + 1 plate.

CARTIER, D. 1963: Étude caryologique des différentes sous-espèces du *Plantago atrata* Hoppe. - *Compt. Rend. Hebd. Séances Acad. Sci.* **256(13)**: 2900-2902 + 2 plates.

CARTIER, D. 1971: Étude biosystématique de quelques espèces du genre *Plantago* (Tourn.) L. (sections *Coronopus* DC. et *Oreades* Decne.). I. Historique ... du *Plantago alpina* L. et du *Plantago serpentina* All. - *Rev. Gén. Bot.* **78(929-931)**: 493-556.

CARTIER, D. 1972: Étude biosystématique de quelques espèces du genre *Plantago* (Tourn.) L. (section *Coronopus* DC.). II. La variation ... *Plantago alpina* L. et *Plantago serpentina* All. - *Rev. Gén. Bot.* **79(938-940)**: 201-248.

CARTIER, D. 1973a: Etude biosystématique de quelques espèces du genre *Plantago* (Tourn.) L. (section *Coronopus* DC.). III. Apports de croisement expérimentaux chez les *P. alpina* L. et *P. serpentina* All. ... - *Rev. Gén. Bot.* **80(944-946)**: 5-40.

CARTIER, D. 1973b: Etude biosystématique de quelques espèces du genre *Plantago* (Tourn.) L. IV. La section *Oreades* Decne. - *Rev. Gén. Bot.* **80(947-949)**: 129-149.

CARTIER, D. 1976: [Chromosome numbers in Fabaceae]. In Löve, Á. (ed.): IOPB chromosome number reports LIII. - *Taxon* **25(4)**: 492-494.

CARTIER, D. 1977: [Chromosome numbers from Greece and Turkey]. In Löve, Á. (ed.): IOPB chromosome number reports LVII. - *Taxon* **26(4)**: 451.

CARTIER, D. 1979: Premières prospections caryologiques du genre *Astragalus* L. dans la flore du bassin méditerranéen oriental. - *Rev. Cytol. Biol. Vég. Botaniste* **2(2)**: 169-181.

CARTIER, D. 1981: Contribution à l'étude caryologique du *Lathyrus pratensis* L. - *Bull. Soc. Bot. France, Lett. Bot.* **128(4)**: 287-293.

CARTIER, D. 1989: Contribution à l'étude biosystématique du *Plantago atrata* Hoppe. - *Candollea* **44 (1)**: 249-256.

CARTIER, D. & BLAISE, S. 1981: Répartition des races chromosomiques chez le *Lathyrus pratensis* L. - *Saussurea* **12**: 143-155.

CASPER, S.J. 1962: Revision der Gattung *Pinguicula* in Eurasien. - *Feddes Repert. Spec. Nov. Regni Veg.* **66(1-2)**: 1-148 + 1 fold-out map.

CASPER, S.J. 1963: Gedanken zur Gliederung der Gattung *Pinguicula* L. - *Bot. Jahrb. Syst.* **82(3)**: 321-335.

CASPER, S.J. 1966: Monographie der Gattung *Pinguicula* L. - *Biblioth. Bot.* **127-128**: 209 pp. + 16 plates and 2 maps.

CASPER, S.J. 1970: Die Gattung *Pinguicula* (Lentibulariacae) in Vorderasien. - *Wiss. Z. Friedrich- Schiller-Univ. Jena, Math.-Naturwiss. Reihe* **19(3)**: 275-292.

CASPER, S.J. 1979: Beiträge zur Taxonomie und Chorologie europäischer Wasser- und Sumpfpflanzen. 2. Was ist *Najas marina* L.? - *Feddes Repert.* **90(4)**: 217-238.

CASPER, S.J. 2004: *Pinguicula louisii* Markgraf (Lentibulariaceae) from Albania. Short notice about a nearly unknown and perhaps neglected butterwort. - *Haussknechtia* **10**: 239-245.

CASPER, S.J. 2006: Birth, burial, and rebirth of *Pinguicula hirtiflora* Tenore (Lentibulariaceae) - a reflection of the making of Flora Napolitana (1811-1838). - *Haussknechtia* **11**: 29-40.

CASPER, S.J. & STIMPER, R. 2004: Chromosome numbers in the *Pinguicula crystallina-hirtiflora* aggregate - a preliminary note. - *Haussknechtia* **10**: 247-2249.

CASPERS, N. 1979: Orchideen von Delphi. - *Kosmos (Stuttgart)* **75(4)**: 267-268.

CASTRO, D. DE 1949: Novos números de cromosomas para o género *Cytisus* L. - *Agron. Lusit.* **11(2)**: 85-89 + 1 plate.

CASTRO, D. DE & AL. 2012: Phylogenetic and biogeographical inferences for *Pancratium* (Amaryllidaceae), with an emphasis on the Mediterranean species based on plastid sequence data. - *Bot. J. Linn. Soc.* **170(1)**: 12-28.

CASTROVIEJO, S. 1999a: Las especies mediterráneas del género *Adenocarpus* DC. In Correia, A.I.D. & al. (eds): Livro de resumos. Guia de excursões. V jornadas de taxonomia botânica, 16-19/IX 1999, p. 16.

CASTROVIEJO, S. 1999b: Apuntes sobre algunos *Adenocarpus* (Leguminosae) ibéricos. - *Anales Jard. Bot. Madrid* **57(1)**: 37-46.

CASTROVIEJO, S. & AL. (EDS) 1986: Flora Iberica. Plantas vasculares de la Península Ibérica e Islas Baleares. Vol. I. Lycopodiaceae-Papaveraceae. - Madrid: Real Jardín Botánico, C.S.I.C., pp. i-liv + 1 -575.

CASTROVIEJO, S. & AL. (EDS) 1990: Flora Iberica. Plantas vasculares de la Península Ibérica e Islas Baleares. Vol. II. Platanaceae-Plumbaginaceae (partim). - Madrid: Real Jardín Botánico, C.S.I.C., pp. i-lii + 1-897.

CASTROVIEJO, S. & AL. (EDS) 1993a: Flora Iberica. Plantas vasculares de la Península Ibérica e Islas Baleares. Vol. III. Plumbaginaceae (partim)-Capparaceae. - Madrid: Real Jardín Botanico, C.S.I.C., pp. i-liv + 1-730.

CASTROVIEJO, S. & AL. (EDS) 1993b: Flora Iberica. Plantas vasculares de la Península Ibérica e Islas Baleares. Vol. IV. Cruciferae-Monotropaceae. - Madrid: Real Jardín Botanico, C.S.I.C., pp. i-liv + 1 -730.

CASTROVIEJO, S. & AL. (EDS) 1997a: Flora Iberica. Plantas vasculares de la Península Ibérica e Islas Baleares. Vol. V. Ebenaceae-Saxifragaceae. - Madrid: Real Jardín Botánico, C.S.I.C., pp. i-lv + 1 -320.

CASTROVIEJO, S. & AL. (EDS) 1997b: Flora Iberica. Plantas vasculares de la Península Ibérica e Islas Baleares. Vol. VIII. Haloragaceae-Euphorbiaceae. - Madrid: Real Jardín Botánico, C.S.I.C., pp. i-lv + 1-375.

CASTROVIEJO, S. & AL. (EDS) 1998: Flora Iberica. Plantas vasculares de la Península Ibérica e Islas Baleares. Vol. VI. Rosaceae. - Madrid: Real Jardín Botánico, C.S.I.C., pp. i-xlv + 1-592.

CASTROVIEJO, S. & AL. (EDS) 1999: Flora Iberica. Plantas vasculares de la Península Ibérica e Islas Baleares. Vol. VII (I). Leguminosae (partim). - Madrid: Real Jardín Botánico, C.S.I.C., pp. i-xlv + 1-578.

CASTROVIEJO, S. & AL. (EDS) 2000: Flora Iberica. Plantas vasculares de la Península Ibérica e Islas Baleares. Vol. VII (II). Leguminosae (partim). - Madrid: Real Jardín Botánico, C.S.I.C., pp. i-xxii + 581-1119.

CASTROVIEJO, S. & AL. (EDS) 2001: Flora Iberica. Plantas vasculares de la Península Ibérica e Islas Baleares. Vol. XIV. Myoporaceae-Campanulaceae. - Madrid: Real Jardín Botánico, C.S.I.C., pp. i- xlvii + 1-251.

CASTROVIEJO, S. & AL. (EDS) 2003: Flora Iberica. Plantas vasculares de la Península Ibérica e Islas Baleares. Vol. X. Araliaceae-Umbelliferae. - Madrid: Real Jardín Botánico, C.S.I.C., pp. i-xlv + 1-498.

CASTROVIEJO, S. & AL. (EDS) 2005: Flora Iberica. Plantas vasculares de la Península Ibérica e Islas Baleares. Vol. XXI. Smilacaceae-Orchidaceae. - Madrid: Real Jardín Botánico, C.S.I.C., pp. i-xlv + 1-366.

CASTROVIEJO, S. & AL. (EDS) 2007a: Flora Iberica. Plantas vasculares de la Península Ibérica e Islas Baleares. Vol. XV. Rubiaceae- Dipsacaceae. - Madrid: Real Jardín Botánico, C.S.I.C., pp. i-xlv + 1-449.

CASTROVIEJO, S. & AL. (EDS) 2007b: Flora Iberica. Plantas vasculares de la Península Ibérica e Islas Baleares. Vol. XVIII. Cyperaceae-Pontederiaceae. - Madrid: Real Jardín Botánico, C.S.I.C., pp. i- xlvii + 1-420.

CASTROVIEJO, S. & AL. (EDS) 2009: Flora Iberica. Plantas vasculares de la Península Ibérica e Islas Baleares. Vol. XIII. Plantaginaceae-Scrophulariaceae. - Madrid: Real Jardín Botánico, C.S.I.C., pp. i-xlvii + 1-677.

CASTROVIEJO, S. & AL. (EDS) 2010a: Flora Iberica. Plantas vasculares de la Península Ibérica e Islas Baleares. Vol. XII. Verbenaceae, Labiatae, Callitrichaceae. - Madrid: Real Jardín Botánico, C.S.I.C., pp. i-liv + 1-650.

CASTROVIEJO, S. & AL. (EDS) 2010b: Flora Iberica. Plantas vasculares de la Península Ibérica e Islas Baleares. Vol. XVII. Butomaceae-Juncaceae. - Madrid: Real Jardín Botánico, C.S.I.C., pp. i-xlviii + 1-298.

CASTROVIEJO, S. & AL. (EDS) 2012: Flora Iberica. Plantas vasculares de la Península Ibérica e Islas Baleares. Vol. XI. Gentianaceae-Boraginaceae. - Madrid: Real Jardín Botánico, C.S.I.C., pp. i-xlviii + 1-672.

CASTROVIEJO, S. & AL. (EDS) 2013: Flora Iberica. Plantas vasculares de la Península Ibérica e Islas Baleares. Vol. XX. Liliaceae-Agavaceae. - Madrid: Real Jardín Botánico, C.S.I.C., pp. i-l + 1-651.

CASTROVIEJO, S. & AL. (EDS) 2014: Flora Iberica. Plantas vasculares de la Península Ibérica e Islas Baleares. Vol. XVI (I). Compositae (partim). - Madrid: Real Jardín Botánico, C.S.I.C., pp. i-xlvii + 1-737.

CASTROVIEJO, S. & AL. (EDS) 2015: Flora Iberica. Plantas vasculares de la Península Ibérica e Islas Baleares. Vol. IX. Rhamnaceae-Polygalaceae. - Madrid: Real Jardín Botánico, C.S.I.C., pp. i-xlviii + 1-564.

CASTROVIEJO, S. & AL. (EDS) 2017: Flora Iberica. Plantas vasculares de la Península Ibérica e Islas Baleares. Vol. XVI (II). Compositae (partim). - Madrid: Real Jardín Botánico, C.S.I.C., pp. i-xxiv + 737-1450.

CASTROVIEJO, S. & AL. (EDS) 2020: Flora Iberica. Plantas vasculares de la Península Ibérica e Islas Baleares. Vol. XIX(I). Gramineae (partim). - Madrid: Real Jardín Botánico, C.S.I.C., pp. i-xlvi + 1-738.

CASTROVIEJO, S. & CALVO, R. 1981a: Notas citotaxonómicas sobre *Sedum acre* L. (Crassulaceae). - *Trab. Dept. Bot. Univ. Complut. Madrid* **11**: 49-57.

CASTROVIEJO, S. & CALVO, R. 1981b: Datos citotaxonómicos en *Sedum* serie *Rupestria* Berger. - *Anales Jard. Bot. Madrid* **38(1)**: 37-50.

CASTROVIEJO, S. & COELLO, P. 1980: Datos cariológicos y taxonómicos sobre las Salicorniinae A.J. Scott ibéricas. - *Anales Jard. Bot. Madrid* **37(1)**: 41-73.

CASTROVIEJO, S. & PASCUAL, H. 1995: Notas sobre el género *Lens* Mill. (Leguminosae) en la Península Ibérica e Islas Baleares. - *Anales Jard. Bot. Madrid* **53(2)**: 177-180.

CATALÁN, P. & AL. 2012: Evolution and taxonomic split of the model grass *Brachypodium distachyon*. - *Ann. Bot. (Oxford), n.s.* **109(2)**: 385-405.

CATALÁN, P. & AL. 2016: Updated taxonomic descriptions, iconography, and habitat preferences of *Brachypodium distachyon*, *B. stacei* and *B. hybridum* (Poaceae). - *Anales Jard. Bot. Madrid* **73(1)**: doi: http://dx.doi.org/10.3989/ajbm.2428.

CATTANEO, C. 2015: *Valariana asarifolia*. In Raab-Straube, E. & Raus, Th. (eds): Euro+Med-Checklist notulae, 5. - *Willdenowia* **45(3)**: 459.

CATTANEO, C. 2016a: Report 52. In Vladimirov, V. & al. (eds): New floristic records in the Balkans: 30. - *Phytol. Balcan.* **22(2)**: 265-266.

CATTANEO, C. 2016b: *Campanula laciniata*. - *Biodivers. J.* **7(2)**: I-II.

CATTANEO, C. 2017: Reports 50-73. In Vladimirov, V. & al. (eds): New floristic records in the Balkans: 33. - *Phytol. Balcan.* **23(2)**: 288-291.

CATTANEO, C. 2020: Description of a new *Allium* species from Symi island, (SE Aegean, Greece): *Allium carlstroemi* sp. nov. (Amaryllidaceae). *Parnassiana Arch.* **8**: 103-111.

CATTANEO, C. & GRANO, M 2018b: Reports 23-27. In Vladimirov, V., Aybeke, M. & Tan, Kit (eds): New floristic records in the Balkans: 36. - *Phytol. Balcan.* **24(2)**: 267.

CATTANEO, C. & GRANO, M. 2012a: Indagine preliminare sulla flora dell'isola egea di Alonissos (Sporadi settentrionali, Grecia). - *Boll. Soc. Naturalisti "Silvia Zenari"* **35**: 81-100.

CATTANEO, C. & GRANO, M. 2012b: Osservazioni preliminari sulla flora e sulla vegetazione dell'isola egea di Skyros (Sporadi settentrionali, Grecia). - *Boll. Soc. Naturalisti "Silvia Zenari"* **36**: 47-74.

CATTANEO, C. & GRANO, M. 2014: Note sul paesaggio vegetale e sulla flora vascolare estivia dell'isola egea di Skiathos e delle sue isole circonvicine Tsougriá e Aspróniso (Sporadi settentrionali, Grecia). - *Ann. Mus. Civ. Rovereto* **29**: 243-288.

CATTANEO, C. & GRANO, M. 2015: New contributions on the vascular flora of the Aegean island of Chalki (Archipelago of Rhodes, Aegean Sea). - *Biodivers. J.* **6(4)**: 773-788.

CATTANEO, C. & GRANO, M. 2016a: Contribution to the knowledge of vascular flora on Astypalea Island (Dodecanese, Greece). - *Phytol. Balcan.* **22(3)**: 405-417.

CATTANEO, C. & GRANO, M. 2016b: Reports 73-76. In Vladimirov, V. & Tan, Kit (eds): New floristic records in the Balkans: 31. - *Phytol. Balcan.* **22(3)**: 437.

CATTANEO, C. & GRANO, M. 2017a: Reports 74-76. In Vladimirov, V. & al. (eds): New floristic records in the Balkans: 33. - *Phytol. Balcan.* **23(2)**: 291.

CATTANEO, C. & GRANO, M. 2017b: Reports 67-71. In Vladimirov, V. & al. (eds): New floristic records in the Balkans: 34. - *Phytol. Balcan.* **23(2)**: 419.

CATTANEO, C. & GRANO, M. 2018a: Contribution to the flora of Tilos island (Dodecanese Islands, Greece). - *Parnassiana Arch.* **6**: 41-53.

CATTANEO, C. & GRANO, M. 2018c: Reports 13-14. In Vladimirov, V., Aybeke, M. & Tan, Kit (eds): New floristic records in the Balkans: 35. - *Phytol. Balcan.* **24(1)**: 157.

CATTANEO, C. & GRANO, M. 2018d: Reports 55-60. In Vladimirov, V., Aybeke, M. & Tan, Kit (eds): New floristic records in the Balkans: 37. - *Phytol. Balcan.* **24(3)**: 405-406.

CATTANEO, C. & GRANO, M. 2019a: Checklist updating and analysis of the flora of Symi island and of the nearby island of Seskli (Dodecanese, Greece). - *Bocconea* **28**: 425-463.

CATTANEO, C. & GRANO, M. 2019b: Reports 74-77. In Vladimirov, V., Aybeke, M. & Tan, Kit (eds): New floristic records in the Balkans: 40. - *Phytol. Balcan.* **25(3)**: 304-305.

CATTANEO, C., GRANO, M. & PASTA, S. 2014: Note sul paessagio naturale e sulla flora vascolare di Skopelos (Sporadi settentrionali, Grecia). - *Naturalista Sicil., S. IV* **38(1)**: 3-42.

CATTANEO, C. & GRANO, M. 2020: Reports 27-32. In Vladimirov, V., Aybeke, M. & Tan, Kit: New floristic records in the Balkans: 43. - *Phytol. Balcan.* **26(3)**: 542-544.

CATTANEO, C. & PANITSA, M. 2018: Reports 61-123. In Vladimirov, V., Aybeke, M. & Tan, Kit (eds): New floristic records in the Balkans: 37. - *Phytol. Balcan.* **24(3)**: 406-410.

CATTANEO, C., TAN, KIT & BIEL, B. 2016: Reports 77-78. In Vladimirov, V. & Tan, Kit (eds): New floristic records in the Balkans: 31. - *Phytol. Balcan.* **22(3)**: 438-440.

CAUDERON, Y. 1962: Etude cytogénétique du genre *Agropyrum*. - *Rev. Cytol. Biol. Vég.* **25(3-4)**: 287-301 + 2 plates.

CAUWET-MARC, A.-M. 1976: Biosystématique des espèces vivaces du *Bupleurum* (Umbelliferae) du bassin Méditerranéen occidental. Fasc. 1-3. - Perpignan: Thèse, 426 + 172 + 250 pp.

CAUWET-MARC, A.-M. 1978: [Chromosome numbers in *Bupleurum*]. In Löve, Á. (ed.): IOPB chromosome number reports LXI. - *Taxon* **27(4)**: 385-386.

CAUWET-MARC, A.-M. 1979a: Connaissances caryologiques actuelles sur le genre *Bupleurum* L. (Umbelliferae): nombres chromosomiques et nombres de base. - *Bull. Mus. Natl. Hist. Nat., Sér. 4, B* **1(3)**: 191-211.

CAUWET-MARC, A.-M. 1979b: Contribution de la caryologie et à la connaissance de la systématique et de la phylogénie du genre *Bupleurum* L. - *Candollea* **34(1)**: 49-86.

CAUWET-MARC, A.-M. 1981a: Le genre *Ferula* sur le pourtour du bassin méditerranéen. - *Compt. Rend. Congr. Natl. Soc. Savantes, Sec. Sci.* **106(2)**: 77-87.

CAUWET-MARC, A.-M. 1981b: Le complexe *Ferula communis* L. dans les populations du Sud de France et de la Corse. - *Biol. Écol. Médit.* **8(3-4)**: 101-118.

CAUWET-MARC, A.-M. & AL. 1978: Contribution à l'étude multidisciplinaire de genre *Bupleurum* L. In Cauwet-Marc, A.-M. & Carbonnier, J. (eds): Les Ombellifères, pp. 623-651. - Perpignan.

CAUWET-MARC, A.-M. & BALAYER, M. 1984: Les genres *Orchis* L., *Dactylorhiza* Necker ex Newski, *Neotinea* Reichb. et *Traunsteinera* Reichb.: Caryologie et proposition de phylogénie et d'évolution. - *Bot. Helv.* **94(2)**: 391-406.

CAUWET-MARC, A.-M. & CARBONNIER, J. (EDS) 1978: Actes du deuxième symposium international sur les Ombellifères. Contributions pluridisciplinaires à la systématique (Perpignan, Mai 1977). - Perpignan: xix + 865 pp. (reprinted as vol. 6 of *Monogr. Syst. Bot.*, 1982).

CAUWET-MARC. A.-M. & JURY, S.L. 1978: Données caryologiques sur la tribu des Caucalidées (Umbelliferae). In Cauwet-Marc, A.M. & Carbonnier, J.: Les Ombellifères, pp. 305-323. - Perpignan.

CAUWET, A.-M. 1967a: Contribution à l'étude caryosystématique du genre *Bupleurum* L. - I. - *Bull. Soc. Bot. France* **114(9)**: 371-386.

CAUWET, A.-M. 1967b: Contribution à l'étude caryologique de quelques Ombellifères d'Espagne. - *Naturalia Monspel., Sér. Bot.* **18**: 201-210.

CAUWET, A.-M. 1968: Contribution à l'étude des Ombellifères de la partie orientale des Pyrénées. - *Naturalia Monspel., Sér. Bot.* **19**: 5-27 + 1 plate.

CAUWET, A.-M. 1969: Contribution à l'étude caryosystématique du genre *Bupleurum* L. - II. - *Bull. Soc. Bot. France* **116(1-2)**: 19-28.

CAUWET, A.-M. 1970: Contribution à l'étude caryosystématique du genre *Bupleurum* (Tourn.) L. - Montpellier: Thèse, Fac. Sci., 166 pp.

CAUWET, A.-M. 1971a: Caryosystématique du genre *Bupleurum* L. - *Bot. J. Linn. Soc.* **64, suppl. 1**: 257-265.

CAUWET, A.-M. 1971b: Contribution à l'étude caryosystématique du genre *Bupleurum* (Tourn.) L. - III. - *Bull. Soc. Bot. France* **118(1-2)**: 55-68.

CAUWET, A.-M. 1975: Le genre *Bupleurum* L. dans la partie occidentale du Bassin Méditerranéen. In Guinochet, M. & al. (eds): La flore du Bassin Méditerranéen, pp. 207-222. - Paris: C.N.R.S.

CAVALIERE, D. 1935: Piante delle Isole Egee raccolte dal Console Lorenzo Senni. - *Malpighia* **34**: 181-186.

CAVERS, B.P. & HARPER, J.L. 1964: Biological Flora of the British Isles. *Rumex obtusifolius* L. and *R. crispus* L. - *J. Ecol.* **52(3)**: 737-766.

CAVILLIER, F. 1911: Nouvelles études sur le genre *Doronicum*. - *Annuaire Conserv. Jard. Bot. Genève* **13-14**: 195-368.

CECCHI, L. 2011: A reappraisal of *Phyllolepidum* (Brassicaceae), a neglected genus of the European flora, and its relationships in the tribe Alysseae. - *Pl. Biosyst.* **145(4)**: 818-831.

CECCHI, L. & AL. 2014: Non-monophyly of *Buglossoides* (Boraginaceae: Lithospermeae): Phylogenetic and morphological evidence for the expansion of *Glandora* and reappraisal of *Aegonychon*. - *Taxon* **63(5)**: 1065-1078.

CECCHI, L. & AL. 2018: The genus *Odontarrhena* (Brassicaceae) in Albania: Taxonomy and Nickel accumulation in a critical group of metallophytes from a major serpentine hot-spot. - *Phytotaxa* **351 (1)**: 1-28.

CECCHI, L. & AL. 2020: *Odontarrhena stridii* (Brassicaceae), a new Nickel- hyperaccumulating species from mainland Greece. - *Pl. Syst. Evol.* **306(4)**: https://doi.org/10.1007/s00606-020-01687-3.

CECCHI, L., COPPI, A. & SELVI, F. 2011: Evolutionary dynamics of serpentine adaptation in *Onosma* (Boraginaceae) as revealed by ITS sequence data. - *Pl. Syst. Evol.* **297(3-4)**: 185-199.

CECCHI, L. & HILGER, H.H. 2021: A prickly matter: nomenclatural synopsis of *Onosma* L. and its segregates (Boraginaceae). - *Pl. Biosyst.* **155**: doi: 10.1080/11263504.2021.1998241.

CECCHI, L. & SELVI, F. 2009: Phylogenetic relationships of the monotypic genera *Halacsya* and *Paramoltkia* and the origins of serpentine adaptation in circum-mediterranean Lithospermeae (Boraginaceae): insights from ITS and *matK* DNA sequences. - *Taxon* **58(3)**: 700-714.

CECCHI, L. & SELVI, F. 2014: A synopsis of Boraginaceae subfam. Hydrophylloideae and Heliotropioideae in Italy. - *Pl. Biosyst.* **148(1)**: 2-12.

CECCHI, L. & SELVI, F. 2015: Synopsis of Boraginaceae subfam. Boraginoideae tribe Boragineae in Italy. - *Pl. Biosyst.* **149(4)**: 630-677.

CELA RENZONI, G. 1968: Studio cariologico di *Helictotrichon versicolor* (Villars) Pilger ssp. *praetutianum* (Arc.) nov. comb. - *Giorn. Bot. Ital., n.s.* **102(5)**: 415-425.

CELA RENZONI, G. 1970: *Matthiola tricuspidata* R. Br. (Cruciferae): Analisi cariologica ed embriologica. - *Giorn. Bot. Ital., n.s.* **103(6)**: 531-545.

CELA RENZONI, G. & GARBARI, F. 1970: Il genere *Allium* L. in Italia. I. *Allium pendulinum* Ten. e *A. triquetrum* L. - *Giorn. Bot. Ital., n.s.* **104(1)**: 61-73.

CELA RENZONI, G. & GARBARI, F. 1971: Il genere *Allium* L. in Italia. II. Morfologia cromosomica di alcune specie. - *Atti Soc. Tosc. Sci. Nat. Pisa Mem., Ser. B* **78**: 99-118.

ČELAKOVSKÝ, L. 1867-81: Prodromus der Flora von Böhmen enthaltend die wildwachsenden und allgemein kultivierten Gefässpflanzen des Königreiches. - Praha: 4 vols. [cf. TL-2 1: 476-477, 1976].

ČELAKOVSKÝ, L. 1882: Diagnosen einiger neuer *Thymus*-Arten. - *Flora* **65**: 563-565.

ČELAKOVSKÝ, L. 1883a: Ueber einige Arten der Gattung *Teucrium*. - *Bot. Centralbl.* **14**: 151-155.
ČELAKOVSKÝ, L. 1883b: Ueber einige Arten resp. Rassen der Gattung *Thymus* (Fortsetzung). – *Flora* **66(10)**: 145-160.
ČELAKOVSKÝ, L. 1884: Neue Thymi aus Sintenis Iter Trojanum. - *Flora* **67(29)**: 533-538.
ČELAKOVSKÝ, L. 1885: Ueber einige verkannte orientalische *Carthamus*-Arten. - *Sitzungsber. Königl. Böhm. Ges. Wiss. Prag, Math.-Naturwiss. Cl.* **1885(7)**: 77-96.
ČELAKOVSKÝ, L. 1887a: Beitrag zur Kenntniss der Flora der Athos-Halbinsel. - *Sitzungsber. Königl. Böhm. Ges. Wiss. Prag, Math.-Naturwiss. Cl.* **1887(29)**: 528-547.
ČELAKOVSKÝ, L. 1887b: Ueber einige neue orientalische Pflanzenarten. II. Gattung *Cerastium* L. - *Oesterr. Bot. Z.* **37(10)**: 337-341.
ČELAKOVSKÝ, L. 1893: Ueber das Verhältniss des *Rumex acetoselloides* Balansa zum *R. angiocarpus* Murbeck. - *Sitzungsber. Königl. Böhm. Ges. Wiss. Prag, Math.-Naturwiss. Cl.* **1892(27)**: 391-402.
CELARIER, R.P. 1956: Cytology of *Andropogon distachyus* L. - *Bull. Torrey Bot. Club* **83(3)**: 183-191.
CELARIER, R.P. 1957: The cyto-geography of the *Bothriochloa ischaemum* complex. II. Chromosome behaviour. - *Amer. J. Bot.* **44(9)**: 729-738.
CELARIER, R.P. 1958: Cytotaxonomic notes on the subsection *Halepensia* of the genus *Sorghum*. - *Bull. Torrey Bot. Club* **85(1)**: 49-61.
CELARIER, R.P. 1959: Cytotaxonomy of the Andropogoneae. IV. Subtribe Sorgheae, genus *Sorghum*. - *Cytologia* **23(4)**: 395-418.
CELARIER, R.P. & HARLAN, J.R. 1958: The cytogeography of the *Bothriochloa ischaemum* complex (Gramineae). I. Taxonomy, and geographic distribution. - *J. Linn. Soc., Bot.* **55(363)**: 755-760.
ÇELEBIOĞLU, T. 1977: Etudes morphologiques et cytologiques sur les espèces du *Dorycnium* de la Turquie. - *Biyol. Derg.* **27(2-4)**: 1-77.
ÇELEBIOĞLU, T. & FAVARGER, C. 1982: Contribution à la cytotaxonomie du genre *Minuartia* L. (Caryophyllacées) en Turquie et dans quelques régions voisines. - *Biol. Écol. Médit.* **9(2-3)**: 139-160.
ÇELEBIOĞLU, T. & FAVARGER, C. 1983: [Chromosome numbers in *Minuartia* from Turkey]. In Löve, Á. (ed.): IOPB chromosome number reports LXXXI. - *Taxon* **32(4)**: 666.
ÇELEBIOĞLU, T. & FAVARGER, C. 1986a: Un groupe intéressant de *Cerastium* (Caryophyllaceae) de la flore turque. - *Veröff. Geobot. Inst. E.T.H. Stiftung Rübel Zürich* **87**: 66-78.
ÇELEBIOĞLU, T. & FAVARGER, C. 1986b: A propos du *Minuartia montana* L. et du series *Montanae* Mattf. du genre *Minuartia* L. - *Candollea* **41(2)**: 423-430.
ÇELEBIOĞLU, T. & FAVARGER, C. 1989: Corrélation entre le nombre d'étamines par fleur et le nombre chromosomique chez *Minuartia hybrida* (Vill.) Schischkin (Caryophyllaceae). - *Naturalia Monspel., Sér. Bot.* **53**: 3-15.
ÇELEBIOĞLU, T. & FAVARGER, C. 1991: Nouvelle contribution à la cytotaxonomie de la section *Sabulina* (Reichenb.) Graebner du genre *Minuartia* L. (Caryophyllaceae). - *Naturalia Monspel., Sér. Bot.* **55**: 27-41.
ÇELEBIOĞLU, T. & FAVARGER, C. 1993: Rapports (125-166). In Kamari, G., Felber, F. & Garbari, F. (eds): Mediterranean chromosome number reports - 3. - *Fl. Medit.* **3**: 323-333.
ÇELEBIOĞLU, T., FAVARGER, C. & HUYNH, K.-L. 1983: Contribution à la micromorphologie de la testa des graines du genre *Minuartia* (Caryophyllaceae). I. Sect. *Minuartia*. - *Bull. Mus. Natl. Hist. Nat., B, Adansonia* **5(4)**: 415-435.
CELEP, F., KARAER, F. & DUMAN, H. 2021: Resurrection of *Lamium ponticum* (Lamiaceae) with a new subspecies, *Lamium ponticum* subsp. *anatolicum* from Turkey. - *Phytotaxa* **511(1)**: 71-76.
CELESTI-GRAPOW, L. & AL. 2009: Inventory of the non-native flora of Italy. - *Pl. Biosyst.* **143(2)**: 386-430.
CELLINESE, N. & AL. 2009: Historical biogeography of the endemic Campanulaceae of Crete. - *J. Biogeogr.* **36(7)**: 1253-1269.
CENCI, C.A. 1979: Numero cromosomico e caratteri morfologici di alcuni ecotipi di *Phleum pratense* L. (Gramineae) dell'Italia centrale. - *Giorn. Bot. Ital., n.s.* **113(3)**: 145-155.
CENCI, C.A. 1982: Caratteristiche biometriche e morfologiche di popolazioni naturali tetraploidi di "Dactylis glomerata" L. - *Webbia* **36(1)**: 135-159.
CENCI, C.A. & AL. 1990: *Festuca arundinacea* Schreber (Gramineae) in Italy: morphological, anatomical, karyological and biochemical analyses. - *Webbia* **44(2)**: 255-270.
CENCI, C.A. & FALISTOCCO, E. 1982: Analisi di alcuni caratteri morfologici e citologici in *Dactylis glomerata* L. subsp. *glomerata* e subsp. *hispanica* (Roth) Nyman. - *Giorn. Bot. Ital., n.s.* **116, suppl. 1**: 47-48.
CENCI, C.A., PEGIATI, M.T. & FALISTOCCO, E. 1985: *Phleum pratense* (Gramineae): chromosomal and biometrical analysis of Italian populations. - *Willdenowia* **14(2)**: 343-353.
ČERVENKA, V. 1955: Study of polyploid forms of *Valeriana officinalis* [In Czech with Russian and English summaries]. - *Preslia* **27**: 234-242.
CESCA, G. 1967: Osservazioni citotassonomiche su *Euphorbia apios* L., *E. ceratocarpa* Ten., *E. amygdaloides* L. s.l., *E. characias* L. s.l. - *Arch. Bot. Biogeogr. Ital.* **43(4)**: 280-290 + 11 plates.
CESCA, G. 1969: Ricerche citotassonomiche nelle Euphorbiaceae: *Euphorbia spinosa* L., *E. bivonae* Steud., *E. melitensis* Parl., *E. acanthothamnos* Heldr. & Sart. - *Giorn. Bot. Ital., n.s.* **103(3)**: 205-263.
CESCA, G. 1972: Numeri cromosomici per la Flora Italiana: 98-107. - *Inform. Bot. Ital.* **4(1)**: 45-59.
CESCA, G. & PERUZZI, L. 2001: *Isoëtes* (Lycophytina, Isoetaceae) with terrestrial habitat in Calabria (Italy). New karyological and taxonomical data. - *Fl. Medit.* **11**: 303-309.
ČESCHMEDJIEV, I.V. 1969a: Floristische Materialien [In Bulgarian with Russian and German summaries]. - *Nauchni Trudove Selskost. Inst. "Vasil Kolarov"* **18(1)**: 137-139.
ČESCHMEDJIEV, I.V. 1969b: Karyological studies on several species of onions (genus *Allium* L.) of the Bulgarian flora [In Bulgarian]. In: Second national conference on botany, pp. 377-387. - Sofia: Bulgarian Academy of Sciences.
ČESCHMEDJIEV, I.V. 1969c: A contribution to the systematics of the species of *Sempervivum* L. from Bulgaria [In Russian]. - *Bot. Zhurn. (Moscow & Leningrad)* **54(3)**: 471-475.
ČESCHMEDJIEV, I.V. 1970: Materialien über die Gattung *Merendera* Ram. in Bulgarien [In Bulgarian with Russian and German summaries]. - *Nauchni Trudove Selskost. Inst. "Vasil Kolarov"* **19(1)**: 117-124.
ČESCHMEDJIEV, I.V. 1973a: To the cytotaxonomy of some Bulgarian *Allium* L. species [In Russian]. - *Bot. Zhurn. (Moscow & Leningrad)* **58(6)**: 864-875.
ČESCHMEDJIEV, I.V. 1973b: Zytotaxonomische Untersuchung einiger Wurzelstock-Zwiebelarten von der bulgarischen Flora [In Bulgarian with Russian and German summaries]. - *Nauchni Trudove Selskost. Inst. "Vasil Kolarov"* **22(1)**: 49-55.
ČESCHMEDJIEV, I.V. 1974: Cytotaxonomical study of certain onion species of the section *Allium*. - *Dokl. Bulg. Akad. Nauk* **27(8)**: 1109-1112.
ČESCHMEDJIEV, I.V. 1975a: Cytotaxonomic study of several species of onion from section *Codonoprasum* Reichenb. - *Dokl. Bulg. Akad. Nauk* **28(6)**: 795-798.
ČESCHMEDJIEV, I.V. 1975b: Cytotaxonomic investigation of some onion species from the *Codonoprasum* Reichenb. section [In Bulgarian with Russian and English summaries]. In: Festschrift in honour of Acad. Daki Jordanov, pp. 79-89. - Sofia: Publishing House of the Bulgarian Academy of Sciences.
ČESCHMEDJIEV, I.V. 1976: [Chromosome numbers from Bulgaria]. In Löve, Á. (ed.): IOPB chromosome number reports LIV. - *Taxon* **25(5-6)**: 642-643.
ČESCHMEDJIEV, I.V. 1977a: Floristic materials and critical notes on Bulgarian flora [In Bulgarian with Russian and English summaries]. - *Fitologiya* **7**: 75-83 + 1 plate.
ČESCHMEDJIEV, I.V. 1977b: B-chromosomes in species of the *Allium* L. genus. - *Dokl. Bulg. Akad. Nauk* **30(3)**: 435-437.
ČESCHMEDJIEV, I.V. 1978: New materials from genus *Allium* L. [In Bulgarian with Russian and English summaries]. - *Nauchni Trudove Selskost. Inst. "Vasil Kolarov"* **23(1)**: 65-70.

ČESCHMEDJIEV, I.V. 1979a: Karyosystematic investigation on species of the genus *Allium* L. in Bulgaria [In Bulgarian with Russian and English summaries]. - *Fitologiya* **11**: 40-46.

ČESCHMEDJIEV, I.V. 1979b: Karyologic investigation of *Allium rotundum* L. [In Bulgarian with Russian and English summaries]. - *Nauchni Trudove Selskost. Inst. "Vasil Kolarov"* **24(1)**: 165-168.

ČESCHMEDJIEV, I.V. 1982: Cytotaxonomical study of *Allium flavum* L. and *Allium paczoskianum* Tuzson [In Bulgarian with Russian and English summaries]. - *Nauchni Trudove Selskost. Inst. "Vasil Kolarov"* **27(4)**: 11-23.

ČESCHMEDJIEV, I.V. 1983a: Cytotaxonomic investigation of the genus *Sambucus* L. in Bulgaria [In Bulgarian with Russian and English summaries]. - *Fitologiya* **23**: 30-44.

ČESCHMEDJIEV, I.V. 1983b: [Miscellaneous chromosome numbers from Bulgaria]. In Löve, Á. (ed.): IOPB chromosome number reports LXXX. - *Taxon* **32(3)**: 506.

ČESCHMEDJIEV, I.V. 1984: Cytotaxonomic investigation on *Allium cyrillii* Ten. and *Allium multibulbosum* Jacq. [In Bulgarian with Russian and English summaries]. - *Nauchni Trudove Selskost. Inst. "Vasil Kolarov"* **29(4)**: 115-122.

ČESCHMEDJIEV, I.V. 1988: New materials and chorological data about the Bulgarian flora [In Bulgarian with Russian and English summaries]. - *Fitologiya* **35**: 58-63.

ČESCHMEDJIEV, I.V. 1989: On the species *Allium thracicum* in Bulgaria (Alliaceae). In Phitos, D. (ed.): Abstracts. Sixth OPTIMA meeting, Delphi 10-16 Sept. 1989, p. 84.

ČESCHMEDJIEV, I.V. 1991: *Allium melanantherum* complex in Bulgaria. In Hanelt, P., Hammer, K. & Knüpffer, H. (eds): The genus *Allium* - taxonomic problems and genetic resources, pp. 21-26. - Gatersleben.

ČESCHMEDJIEV, I.V. 1993: *Thymelaea bulgarica* sp. nov. (Thymelaeaceae) and its related species. In Petrova, A.V. (ed.): Abstracts from the VII OPTIMA meeting in Borovetz, Bulgaria, 18-30 July 1993, p. 184.

ČESCHMEDJIEV, I.V. 1994: Reports (313-366). In Kamari, G., Felber, F. & Garbari, F. (eds): Mediterranean chromosome number reports - 4. - *Fl. Medit.* **4**: 269-279.

ČESCHMEDJIEV, I.V. & MARINOV, Y. 2009: *Allium phthioticum*: new species for the Bulgarian flora. - *Phytol. Balcan.* **15(3)**: 385-388.

ČESCHMEDJIEV, I.V. & PETKOVA, L. 1988: Biosystematic investigation on *Allium cepa* L. var. *aggregatum* D. Don [In Bulgarian with Russian and English summaries]. - *Nauchni Trudove Selskost. Inst. "Vasil Kolarov"* **33(4)**: 75-82.

ČESCHMEDJIEV, I.V. & STOJCHEV, G. 2005: Representative of Pontederiaceae Kunth family in Bulgaria. - *Nauchni Trudove Agrar. Univ. Plovdiv* **50(5)**: 167-172.

ČESCHMEDJIEV, I.V. & VASSILEV, R. 2009: Florata na Plovdiv [In Bulgarian]. - Sofia: Bulgarian Biodiversity Foundation, 303 pp.

ČESCHMEDJIEV, I.V. & VODENICHAROV, D. 1998: Materials on the Bulgarian flora. - *Phytol. Balcan.* **4(3)**: 103-113.

CETLOVÁ, V. & AL. 2019: Overlooked morphological variation and a proposal for a new taxonomic circumscription of *Alyssum simplex* (Brassicaceae). - *Phytotaxa* **416(2)**: 149-166.

CETLOVÁ, V. & AL. 2021: Multiple drivers of high species diversity and endemism among *Alyssum* annuals in the Mediterranean: The evolutionary significance of the Aegean hotspot. - *Front. Plant Sci.* **12**: 23 pp. + suppl.

CHABERT, A. 1899: Étude sur le genre *Rhinanthus* L. (suite et fin). - *Bull. Herb. Boissier* **7(7)**: 497-517.

CHADWICK, M.J. 1960: Biological Flora of the British Isles. *Nardus stricta* L. - *J. Ecol.* **48(1)**: 255-267.

CHALLICE, J. & KOVANDA, M. 1985: A chemotaxonomic survey of *Sorbus umbellata* agg. - *Preslia* **57 (2)**: 171-174.

CHAMBEL, M.R. & AL. 2013: Mediterranean pines (*Pinus halepensis* Mill. and *brutia* Ten.). In Pâques, L.E. (ed.): Forest tree breeding in Europe, pp. 229-265. - Dordrecht: Springer Verlag.

CHAMBERLAIN, D.F. & MATTHEWS, V.A. 1969: Materials for a Flora of Turkey: XVIII. Leguminosae: *Astragalus*. - *Notes Roy. Bot. Gard. Edinburgh* **29(3)**: 285-310.

CHAMBERS, H.L. & HUMMER, K.M. 1994: Chromosome counts in the *Mentha* collection at the USDA- ARS national clonal germplasm repository. - *Taxon* **43(3)**: 423-432.

CHANLIDOU, E., KAROUSOU, R. & KOKKINI, S. 1997: The sage plants in Greece: Morphological variation and its taxonomic implications. In: Abstracts from the First Balkan Botanical Congress, Thessaloniki, September 19-22, 1997, p. 20. – Thessaloniki.

CHANLIDOU, E. & KOKKINI, S. 1996: Variation of *Satureja montana* group (Labiatae) in Vikos-Aoos National Park [In Greek with English summary]. In: Proceedings 6th sci. conf. Hellenic Botanical Society, pp. 269-272. - Paralimni, Cyprus.

CHANLIDOU, E. & KOKKINI, S. 1997: On the flora of the Vikos-Aoos National Park (NW Greece). - *Willdenowia* **27(1-2)**: 81-100.

CHAPMAN, V.J. 1947: Biological Flora of the British Isles. *Suaeda maritima* (L.) Dum. and *S. fruticosa* Forsk. - *J. Ecol.* **35(1-2)**: 293-310.

CHAPMAN, V.J. 1950: Biological Flora of the British Isles. *Halimione portulacoides* (L.) Aell. - *J. Ecol.* **38(1)**: 214-222.

CHARALAMBIDIS, S. 2008: Ikones ap' tin alpiki chlorida tou Atho [In Greek]. - *Elliniko Panorama* **65**: 74-83.

CHARALAMPIDOU, CH., ELEFTHERIADOU, E. & THEODOROPOULOS, K. 2021: *Galium uliginosum*. In Raus, Th. & Raab-Straube, E.v. (eds): Euro+Med-Checklist notulae, 14. - *Willdenowia* **51(3)**: 364.

CHARALAMPIDOU, CH. , ELEFTHERIADOU, E. & THEODOROPOULOS, K. 2019: Vascular plant diversity of Western Rhodopes (NE Greece): current state of knowledge. In Daskalakou, E.N. & al. (eds): 16th Conference of the Hellenic Botanical Society – Athens 10-13 October 2019. Program and book of abstracts, p. 110.

CHARILAOU, P. 2018: *Heliotropium curassavicum* L. (Boraginaceae), a new alien species collected in Cyprus. - *Cypricola* **10**: 1-4.

CHARITONIDOU, M., KOUGIOUMOUTZIS, K. & HALLEY, J.M. 2021: An orchid in retrograde: Climate-driven range shift patterns of *Ophrys helenae* in Greece. - *Plants MDPI* **10, 470**: 18 pp.

CHARMET, G, BALFOURIER, F. & CHATARD, V. 1996: Taxonomic relationships and interspecific hybridization in the genus *Lolium* (grasses). - *Ges. Res. & Crop. Evol.* **43**: 319-327.

CHARPIN, A. & MOLERO, J. 1984: *Valerianella orientalis* (Schlecht.) Boiss. et Bal. in Boiss., novadad para la flora Española. - *Collect. Bot. (Barcelona)* **15**: 153-157.

CHARREL, L. 1889: *Colchicum micranthum* Boiss. - *Oesterr. Bot. Z.* **39(12)**: 421-422.

CHARREL, L. 1891: Enumeratio plantarum annis 1888, 1889, 1890 et 1891 in Macedoniâ australi collectarum. - *Oesterr. Bot. Z.* **41(11)**: 374-375.

CHARREL, L. 1892: Enumeratio plantarum annis 1888, 1889, 1890 et 1891 in Macedoniâ australi collectarum [cont.]. - *Oesterr. Bot. Z.* **42(8-12)**: 271-272, 338-341, 380-383, 409-412.

CHASAPIS, M. 2017: Chlorida ke vlastisi tou orous Tzena. [In Greek]. - Thessaloniki: Ph.D. thesis, Aristotelian University, 203 pp.

CHASAPIS, M. & AL. 2016: *Centaurea kotschyana*. In Raab-Straube, E. v. & Raus, Th. (eds): Euro+Med-Checklist notulae, 6. - *Willdenowia* **46(3)**: 426.

CHASAPIS, M. & AL. 2019: Reports 10-24. In Vladimirov, V., Aybeke, M. & Tan, Kit (eds): New floristic records in the Balkans: 39. - *Phytol. Balcan.* **25(2)**: 204-207.

CHASAPIS, M. & AL. 2019a: *Centaurea kotschyana* (Asteraceae), *Iris variegata* (Iridaceae), *Aruncus dioicus* (Rosaceae): three new records for the flora of Greece. In Daskalakou, E.N. & al. (eds): 16th Conference of the Hellenic Botanical Society – Athens 10-13 October 2019. Program and book of abstracts, p. 111.

CHASAPIS, M. & AL. 2019b: New records for the floristic region of North-Central Greece. In Daskalakou, E.N. & al. (eds): 16th Conference of the Hellenic Botanical Society – Athens 10-13 October 2019. Program and book of abstracts, p. 112.

CHASAPIS, M. & AL. 2020a: *Iris variegata*. In Raus, Th. & Raab-Straube, E.v. (eds): Euro+Med-Checklist Notulae, 12. - *Willdenowia* **50(2)**: 318-319.

CHASAPIS, M. & AL. 2020b: *Aruncus dioicus*. In Raus, Th. & Raab-Straube, E.v. (eds): Euro+Med-Checklist Notulae, 12. - *Willdenowia* **50(2)**: 329.

CHASAPIS, M. & AL. 2020c: The vascular flora of Mt Tzena (northern Greece). - *Fl. Medit.* **30**: 55-63 + suppl.

CHASAPIS, M., KARAGIANNAKIDOU, V. & THEODOROPOULOS, K. 2004: Phytosociological research of *Quercus coccifera* L. pseudomaquis on Mount Chortiatis, northern Greece. - *Israel J. Pl. Sci.* **52(4)**: 357-381.

CHASE, A. 1942: The North American species of *Paspalum*. - *Contr. U.S. Natl. Herb.* **28**: 1-310.

CHATER, A.O. 1963: [Notes on *Consolida*]. In Heywood, V.H. (ed.): Notulae systematicae ad Floram Europaeam spectantes, no. 2. - *Feddes Repert. Spec. Nov. Regni Veg.* **68(3)**: 193.

CHATER, A.O. 1964a: *Lychnis viscaria* L. in the Balkan Peninsula. In Heywood, V.H. (ed.): Notulae systematicae ad Floram Europaeam spectantes, no. 3. - *Feddes Repert. Spec. Nov. Regni Veg.* **69 (1)**: 44-46.

CHATER, A.O. 1964b: [Notes on *Saponaria sicula* Rafin.]. In Heywood, V.H. (ed.): Notulae systematicae ad Floram Europaeam spectantes, no. 2. - *Feddes Repert. Spec. Nov. Regni Veg.* **69 (1)**: 50-52.

CHATER, A.O. 1964c: [New combinations in *Consolida*]. In Heywood, V.H. (ed.): Notulae systematicae ad Floram Europaeam spectantes, no. 3. - *Feddes Repert. Spec. Nov. Regni Veg.* **69 (1)**: 55.

CHATER, A.O. 1964d: [New combinations in *Corydalis*]. In Heywood, V.H. (ed.): Notulae systematicae ad Floram Europaeam spectantes, no. 3. - *Feddes Repert. Spec. Nov. Regni Veg.* **69(1)**: 56.

CHATER, A.O. 1968: [Notes on *Astragalus*]. In Heywood, V.H. (ed.): Notulae systematicae ad Floram Europaeam spectantes, no. 7. - *Feddes Repert.* **79(1-2)**: 48-52.

CHATER, A.O. 1974: Taxonomic and nomenclatural notes on *Senecio*. In Heywood, V.H. (ed.): Notulae systematicae ad Floram Europaeam spectantes, no. 15. - *Bot. J. Linn. Soc.* **68(4)**: 272-276.

CHATER, A.O. & MEIKLE, R.D. 1968: [*Astragalus lusitanicus* subsp. *orientalis*, subsp. nov.]. In Heywood, V.H. (ed.): Notulae systematicae ad Floram Europaeam spectantes, no. 7. - *Feddes Repert.* **79(1-2)**: 48.

CHATER, A.O. & WALTERS, S.M. 1964: Notes on European *Silene*. In Heywood, V.H. (ed.): Notulae systematicae ad Floram Europaeam spectantes, no. 3. - *Feddes Repert. Spec. Nov. Regni Veg.* **69 (1)**: 46-49.

CHATER, A.O. & WALTERS, S.M. 1990: *Silene vulgaris* (Moench) Garcke and *S. uniflora* Roth. In Chater, A.O. (ed.): Flora Europaea: Notulae systematicae ad Floram Europaeam spectantes. Series 2. No. 3. - *Bot. J. Linn. Soc.* **103(3)**: 213-216.

CHATZITRIANTAPHYLLOU, M. & AL. 2017: Following the red footprints of *Sedum stefco* Stef. In Dousis, M.A. & Thanos, K.A.: 15th conference of the Hellenic Botanical Society, book of abstracts: p. 131.

CHAUBARD, L.A. & BORY DE SAINT-VINCENT, J.B.G.M. 1838: Nouvelle flore du Péloponnèse et des Cyclades, entièrement revue, corrigée et augmentée par M. Chaubard, pour les Phanérogames et M. Bory de Saint-Vincent, pour les Cryptogames ... - Paris/Strasbourg: F.G. Levrault, 87 pp. + 42 plates.

CHAUDHRI, M.N. 1968: A revision of the Paronychiinae. - *Meded. Bot. Mus. Herb. Rijks Univ. Utrecht* **285**: 1-440.

CHAVDAROV, I.P. & GEORGIEV, G.N. 1973: Karyological study of *Tragopogon porrifolius* L. - *Dokl. Bulg. Akad. Nauk* **26(7)**: 957-960.

CHAWORTH-MUSTERS, J.L. 1932: An expedition to Mount Olympus. - *Proc. Linn. Soc. London* **144**: 65-67.

CHAYTOR, D.A. 1937: A taxonomic study of the genus *Lavandula*. - *J. Linn. Soc., Bot.* **51(338)**: 153-204.

CHAYTOR, D.A. & TURRILL, W.B. 1934: On the flora of the Nearer East: XV. A contribution to the flora of Central Albania. - *Bull. Misc. Inform. Kew* **1934(10)**: 437-443.

CHAYTOR, D.A. & TURRILL, W.B. 1935: The genus *Clypeola* and its intraspecific variation. - *Bull. Misc. Inform. Kew* **1935(1)**: 1-24.

CHEEK, M. 1998: (1371) Proposal to reject the name *Drosera longifolia* (Droseraceae) - *Taxon* **47(3)**: 749-750. [approved, cf. *Taxon* 49(2): 275, 2000].

CHEMINAL, A. & AL. 2020: Medicinal and aromatic Lamiaceae plants in Greece: Linking diversity and distribution patterns with ecosystem services. - *Forests MDPI* **11, 661**: doi:10.3390/f11060661.

CHEN, C.-C. & GIBSON, P.B. 1971: Karyotypes of fifteen *Trifolium* species in section *Amoria*. - *Crop Sci. (Madison)* **11(3)**: 441-445.

CHEN, C.-C. & GIBSON, P.B. 1972: Chromosome relationships of *Trifolium uniflorum* to *T. repens* and *T. occidentale*. - *Canad. J. Genet. Cytol.* **14(3)**: 591-595.

CHENG, R.I. & GRANT, W.F. 1973: Species relationships in the *Lotus corniculatus* group as determined by karyotype and cytophotometric analyses. - *Canad. J. Genet. Cytol.* **15(1)**: 101-115.

CHENNAVEERAIAH, M.S. 1960: Karyomorphologic and cytotaxonomic studies in *Aegilops*. - *Acta Horti Gothob.* **23**: 85-178 + 8 plates.

CHESHMEDZHIEV, I. & MARINOV, Y. 2019: Genus *Hylotelephium* (Crassulaceae) in Bulgaria. - *Phytol. Balcan.* **25(1)**: 47-51.

CHEVALIER, A. 1941: Les *Myosotis* du groupe *Sylvatica* et *Arvensis*. - *Bull. Mus. Natl. Hist. Nat., Sér. 2* **13(3)**: 187-194.

CHIAPPINI, M. & SCRUGLI, A. 1970: Ricerche citotassonomiche in *Leucojum aestivum* L. - *Morisia* **2**: 17-28.

CHIAPPINI, M. & SCRUGLI, A. 1972: Numeri cromosomici per la flora italiana: 126-128. - *Inform. Bot. Ital.* **4(2)**: 131-133 + 1 plate.

CHIARUCCI, A. & AL. 2021: Species-area relationship and small-island effect of vascular plant diversity in a young volcanic archipelago. - *J. Biogeogr.* **48(11)**: 2919-2931.

CHIARUCCI, A. & CONSTANTINIDIS, TH. 2004: A survey of the ultramafic flora and vegetation of Mt. Gerania, Greece. In Boyd, R.S., Baker, A.J.M. & Proctor, J. (eds): Ultramafic Rocks: Their soils, vegetation and fauna, pp. 161-169. - St. Albans, U.K.: Science Reviews.

CHIARUGI, A. 1927: Poliploidia nel genere "Knautia" (Dipsacaceae). - *Nuovo Giorn. Bot. Ital., n.s.* **34 (4)**: 864-871.

CHICHIRICCÒ, G. 1996: Intra- and interspecific reproductive barriers in *Crocus* (Iridaceae). - *Pl. Syst. Evol.* **201(1-4)**: 83-92.

CHICHIRICCÒ, G., FRIZZI, G. & TAMMARO, F. 1979: Numeri cromosomici per la Flora Italiana: 598-601. - *Inform. Bot. Ital.* **11(1)**: 33-36.

CHICHIRICCÒ, G., FRIZZI, G. & TAMMARO, F. 1981: Numeri cromosomici per la Flora Italiana: 652-661. - *Inform. Bot. Ital.* **11(3)**: 307-313.

CHICHIRICCÒ, G. & TAMMARO, F. 1981: Numeri cromosomici per la Flora Italiana: 742-751. - *Inform. Bot. Ital.* **12(2)**: 161-165.

CHICHIRICCÒ, G. & TAMMARO, F. 1982: Numeri cromosomici per la Flora Italiana: 778-786. - *Inform. Bot. Ital.* **12(3)**: 321-325.

CHICHIRICCÒ, G. & TAMMARO, F. 1984: Numeri cromosomici per la Flora Italiana: 910-918. - *Inform. Bot. Ital.* **14(2-3)**: 264-267.

CHILTON, L. 1993a: Plant list for Stoupa, Peloponnisos. - Hunstanton, Marengo Press: 20 pp. [slightly revised edition 2003].

CHILTON, L. 1993b: Plant list for Rhodes. - Hunstanton, Marengo Press: 24 pp. + supplement 1 p. [revised editions 2003 and 2011].

CHILTON, L. 1994: Plant list for Crete. - Hunstanton, Marengo Press: 40 pp.

CHILTON, L. 1995a: Plant list for Georgioupolis, Kavros & Lake Kournas. - Hunstanton, Marengo Press: 16 pp. [slightly revised edition 2003].

CHILTON, L. 1995b: Plant list for Karpathos. - Hunstanton, Marengo Press: 24 pp. [slightly revised edition 2003]

CHILTON, L. 1996a: Provisional plant list for Corfu. - Hunstanton, Marengo Press: 24 pp.

CHILTON, L. 1996b: Plant list for Samos. - Hunstanton, Marengo Press: 24 pp.

CHILTON, L. 1996c: Plant list for Kokkari, Samos. - Hunstanton, Marengo Press: 20 pp. [slightly revised edition 2004].

CHILTON, L. 1996d: Additional records for the Stoupa area 1994-96. - Hunstanton, Marengo Press: 1 p.

CHILTON, L. 1996e: Additional records for the Plakias area 1994-96. - Hunstanton, Marengo Press: 1 p.

CHILTON, L. 1997a: Plant list for Lesvos. - Hunstanton, Marengo Press: 24 pp. [revised edition 2003].

CHILTON, L. 1997b: Plant list for Skala Potamias, Thasos. - Hunstanton, Marengo Press: 12 pp.

CHILTON, L. 1999a: Plant list for Elounda & Aghios Nikolaos, Crete. - Hunstanton, Marengo Press: 12 pp. [slightly revised edition 2003].

CHILTON, L. 1999b: Plant list for Symi. - Hunstanton, Marengo Press: 16 pp. [slightly revised edition 2003; new edition, 20 pp., 2011].

CHILTON, L. 2003a: Plant list for the Pilion (Greece: Thessalia: Magnisia). - Hunstanton, Marengo Press: 24 pp.

CHILTON, L. 2003b: Plant list for Samos [revised edition]. - Hunstanton, Marengo Press: 30 pp.

CHILTON, L. 2003c: Plant list for Lindos & Pefkos, Rhodes [revised edition]. - Hunstanton, Marengo Press: 18 pp. [further revised edition 2011].

CHILTON, L. 2003d: Plant list for Lefkada & Meganisi [revised edition]. - Hunstanton, Marengo Press: 18 pp.

CHILTON, L. 2003e: Plant list for Kos [revised edition]. - Hunstanton, Marengo Press: 26 pp.

CHILTON, L. 2003f: Plant list for Thasos [revised edition]. - Hunstanton, Marengo Press: 32 pp.

CHILTON, L. 2003g: Plant list for Corfu [revised edition]. - Hunstanton, Marengo Press: 28 pp.

CHILTON, L. 2005: Plant list for Parga, including Amoudia and Aherondas. - Hunstanton, Marengo Press: 24 pp.

CHILTON, L. 2009: Reports 37-57. In Vladimirov & al. (eds): New floristic records in the Balkans: 12. - *Phytol. Balcan.* **15(3)**: 437-441.

CHILTON, L. 2010a: Plant list for Thasos (Greece: North Aegean islands). - Hunstanton, Marengo Press: 36 pp.

CHILTON, L. 2010b: Plant list for the Peloponnisos (Greece). - Hunstanton, Marengo Press: 60 pp.

CHILTON, L. & ALLEN, K. 1996: Plant list for Aghios Georgious, Corfu. - Hunstanton, Marengo Press: 24 pp.

CHILTON, L. & TURLAND, N.J. 1993: Plant list for Plakias, Crete. Including list of birds and other animals. - Hunstanton, Marengo Press: 24 pp.

CHILTON, L. & TURLAND, N.J. 1997: Flora of Crete: A supplement. - Retford, England: Marengo Publ., 125 pp.

CHILTON, L. & TURLAND, N.J. 2004: Flora of Crete: Supplement II, Additions 1997-2004 [continuously updated]. - Internet Publ.: 51 pp. + 6 maps (http://www.marengowalks.com/).

CHIOTELIS, C., BAZOS, I. & STRID, A. 2021: Reports 39-41. In Vladimirov, V., Aybeke, M. & Tan, Kit (eds): New floristic records in the Balkans: 45. - *Phytol. Balcan.* **27(2)**: 246.

CHIOVENDA, E. 1939: Intorno all'indigenato della "Periploca graeca" L. in Italia. - *Nuovo Giorn. Bot. Ital., n.s.* **45(1)**: xxvii-xxxv(-xliii).

CHITOS, TH. 1999: Contribution to the study of the flora of Epirus - new species in Greece. In Eleftheriou, E.P. (ed.): Proceedings, 7th Panhellenic scientific conference, Alexandroupolis, pp. 71-74 [In Greek with English abstract]. - Thessaloniki: Univ. Studio Press.

CHITOS, TH. 2005: Fita ton Tzoumerkon [In Greek]. - Ioannina: 33 pp. [privately published].

CHITOS, TH. 2009: Fita tis Ipirou (Flowers of Epirus). [In Greek with English abstract]. - Ioannina: 127 pp. [privately published].

CHITOS, TH. & RAUS, TH. 1998: Notes on *Diplotaxis erucoides*. In Greuter, W. & Raus, Th. (eds): Med-Checklist notulae, 17. - *Willdenowia* **28(1-2)**: 163-174.

CHITOS, TH. & RAUS, TH. 1999: Notes on *Solenanthus scardicus*. In Greuter, W. & Raus, Th. (eds): Med-Checklist notulae, 18. - *Willdenowia* **29(1-2)**: 52.

CHITOS, TH. & THANOPOULOS, R. 1998: Notes on *Cicerbita pancicii, Petteria ramentacea* and *Panicum capillare*. In Greuter, W. & Raus, Th. (eds): Med-Checklist notulae, 17. - *Willdenowia* **28(1-2)**: 163-174.

CHLOROS, N.A. 1884: Waldverhältnisse Griechenlands. - München: Dr. C. Wolf & Sohn, 45 pp.

CHLOUVERAKI, G. 2003: O krinos [In Greek]. - *Fisis [Journal of the Hellenic Society for the Protection of Nature]* **101**: 20-21.

CHMELAŘ, J. & MEUSEL, W. 1986: Die Weiden Europas. 3. Aufl. - Wittenberg: Neue Brehm-Bücherei, 144 pp.

CHOCHLIOUROS, S. 2005: Chloridiki ke fitokinoniologiki erevna tou orous Vermiou - ikologiki prosengisi [In Greek]. - Patras: Ph.D. thesis, Univ. of Patras, 372 pp.

CHOCHLIOUROS, S. & GEORGIADIS, TH. 1997a: Analysis of the flora of Mount Vermio (NC Greece). In: Abstracts from the First Balkan Botanical Congress, Thessaloniki, September 19-22, 1997, p. 9. - Thessaloniki.

CHOCHLIOUROS, S. & GEORGIADIS, TH. 1997b: Additions to the flora of Mount Vermion (N Central Greece). - *Willdenowia* **27(1-2)**: 101-112.

CHOCHLIOUROS, S. & GEORGIADIS, TH. 2000: Floristic and phytogeographical analysis of Mount Vermion (NC Greece). - *Fl. Medit.* **10**: 365-375.

CHODAT, R.H. 1891: Monographia Polygalacearum. Vol. 1. - Genève: Aubert Schuchardt, 143 pp. + 12 plates.

CHODAT, R.H. 1893: Monographia Polygalacearum. Vol. 2. - Genève: Aubert Schuchardt, 500 pp. + 23 plates.

CHODAT, R.H. 1913: Polygalaceae novae vel parum cognitae (VII). - *Bull. Soc. Bot. Genève, Sér. 2* **5**: 108-111.

CHODAT, R.H. 1914: William Barbey-Boissier (1842-1914). Notice biographique. - *Bull. Soc. Bot. Genève, Sér. 2* **6**: 220-240 + 1 planche.

CHOI, B.-H. & OHASHI, H. 1998: (1377) Proposal to conserve the name *Hedysarum* (Leguminosae: Papilionoideae) with a conserved type. - *Taxon* **47(4)**: 577. [approved, cf. Taxon 49(2): 277, 2000].

CHOI, B.-H. & OHASHI, H. 2003: Generic criteria and an infrageneric system for *Hedysarum* and related genera (Papilionoideae-Leguminosae). - *Taxon* **52(3)**: 567-576.

CHOUDHRY, A.S. 1984: Karyomorphological and cytogenetical studies in *Ephedra*. - *J. Sci. Hiroshima Univ., Ser. B, Div. 2, Bot.* **19(1)**: 57-109.

CHOUSOU-POLYDOURI, N. & YANNITSAROS, A. 2008: The flora of Kastros island (Ionian Sea, Greece). In: Praktika 10ou Sinedriou Ellinikis Votanikis Eterias, 5-8 maiou 2005, Ioannina: pp. 737-739 [In Greek with English abstract].

CHOWDHURI, P.K. 1957: Studies in the genus *Silene*. - *Notes Roy. Bot. Gard. Edinburgh* **22(3)**: 221-278.

CHOWDHURI, P.K., DAVIS, P.H. & HOSSAIN, M. 1958: Materials for a Flora of Turkey: III. Ranunculaceae: I. - *Notes Roy. Bot. Gard. Edinburgh* **22(4)**: 403-425.

CHRIST, H. 1863a: Uebersicht der europäischen Abietineen (*Pinus* Linn.). - *Verh. Naturf. Ges. Basel* **3(4)**: 540-557.

CHRIST, H. 1863b: Beiträge zur Kenntnis südeuropäischer *Pinus*-Arten. I-II. - *Flora* **46**: 369-380.

CHRIST, H. 1865: Beiträge zur Kenntnis europäischer *Pinus*-Arten. IV. *Pinus* (*Strobus*) *Peuce* Griseb. - *Flora* **48**: 257-260 + 1 plate.

CHRIST, H. 1867: Beiträge zur Kenntnis der europäischer *Pinus* Arten [cont.]. - *Flora* **50**: 81-84.

CHRIST, H. 1913: Über das Vorkommen des Buchsbaums, *Buxus sempervirens*, in der Schweiz und weiterhin durch Europa und Vorderasien. - *Verh. Naturf. Ges. Basel* **24**: 46-128.

CHRISTENHUSZ, M.J.M. & AL. 2013: Tiptoe through the tulips - cultural history, molecular phylogenetics and classification of *Tulipa* (Liliaceae). - *Bot. J. Linn. Soc.* **172(3)**: 280-328.

CHRISTENHUSZ, M.J.M. & AL. 2019: Phylogenetics, classification and typification of extant horsetails (*Equisetum*, Equisetaceae). - *Bot. J. Linn. Soc.* **189(4)**: 311-352.

CHRISTENHUSZ, M.J.M., FAY, M.F. & GOVAERTS, R. 2013: (2131) Proposal to reject the name *Tulipa praecox* Cav. (Liliaceae). - *Taxon* **62(2)**: 404. [approved, cf. *Taxon* 63(6): 1362, 2014].

CHRISTENHUSZ, M.J.M., ZHANG, XIAN-CHUN & SCHNEIDER, H. 2011: A linear sequence of extant families and genera of lycophytes and ferns. - *Phytotaxa* **19**: 7-54.

CHRISTENSEN, C. 1922: Index to Pehr Forsskål: Flora aegyptiaco-arabica 1775 with a revision of Herbarium Forsskålii. - *Dansk Bot. Ark.* **4(3)**: 1-54.

CHRISTENSEN, K.I. 1984: The morphological variation of some *Crataegus* populations (Rosaceae) in Greece and Yugoslavia. - *Nordic J. Bot.* **4(5)**: 585-595.
CHRISTENSEN, K.I. 1985a: A taxonomic study of *Crataegus* ser. *Kyrtostylae* Pojark. ex Botschantzev in Europe. - *Feddes Repert.* **96(5-6)**: 363-385.
CHRISTENSEN, K.I. 1985b: *Juniperus communis* subsp. *alpina* (Smith) Čelakovský (Cupressaceae). - A nomenclatural comment. - *Taxon* **34(4)**: 686-688.
CHRISTENSEN, K.I. 1991: *Salix xanthicola* (Salicaceae), a new species from northeastern Greece. - *Willdenowia* **21(1-2)**: 105-111.
CHRISTENSEN, K.I. 1992a: Revision of *Crataegus* sect. *Crataegus* and nothosect. *Crataeguineae* (Rosaceae-Maloideae) in the Old World. - *Monogr. Syst. Bot. Missouri Bot. Gard.* **35**: 1-199.
CHRISTENSEN, K.I. 1992b: The structure of some *Crataegus* (Rosaceae) populations in Greece. - *Willdenowia* **22(1-2)**: 65-79.
CHRISTENSEN, K.I. 1995: *Crataegus* (Rosaceae) in the Balkan Peninsula. - *Ann. Mus. Goulandris* **9**: 39-90.
CHRISTENSEN, K.I. 2001: *Crataegus pycnoloba* - a Hawthorn endemic to the Greek mountains [In Danish]. - *Dansk Dendrol. Årsskrift* **19**: 5-10.
CHRISTENSEN, K.I. 2006a: *Salix xanthicola* - distribution, ecology and relationships. In Peev, D.R. (ed.): Book of abstracts, IV Balkan botanical congress, Sofia, Bulgaria, 20-26 June 2006, p. 166.
CHRISTENSEN, K.I. 2006b: *Salix xanthicola* (Salicaceae) - distribution, ecology and relationships. - *Ann. Mus. Goulandris* **11**: 37-79.
CHRISTENSEN, K.I. 2011: *Chamaecytisus* and *Cytisus*. In Greuter, W. & Raus, Th. (eds): Med-Checklist notulae, 30. - *Willdenowia* **41(2)**: 318.
CHRISTENSEN, K.I. & STRID, A. 2010: *Cytisus striatus* (Hill) Rothm. In Greuter, W. & Raus, Th. (eds): Med-Checklist notulae, 29. - *Willdenowia* **40(2)**: 196.
CHRISTENSEN, K.I., ZIELIŃSKI, J. & PETROVA, A.V. 2006: Notes on the geographic distribution and ecology of *Salix xanthicola* (Salicaceae). - *Phytol. Balcan.* **12(2)**: 209-213.
CHRISTENSEN, S.N. 2014: The epiphytic lichen flora of *Platanus orientalis* stands in Greece. - *Willdenowia* **44(2)**: 209-227.
CHRISTIA, C. & AL. 2011: A survey of the benthic aquatic flora in transitional water systems of Greece and Cyprus (Mediterranean Sea). - *Botanica Marina* **54(2)**: 169-178.
CHRISTIA, CH. & PAPASTERGIOU, E.S. 2007: Spatial and temporal variations of aquatic macrophytes and water quality in six coastal lagoons of western Greece. - *Belgian J. Bot.* **140(1)**: 39-50.
CHRISTIAN, P. 1975: The genus *Serapias* in the wild and in cultivation. - *Quart. Bull. Alpine Gard. Soc. Gr. Brit.* **43(3)**: 188-198.
CHRISTODOULAKIS, D. 1984: Neufunde für die Flora der Insel Samos (Griechenland). - *Willdenowia* **14 (1)**: 75-87.
CHRISTODOULAKIS, D. 1986: Die Flora und Vegetation der Insel Samos (Griechenland) [In Greek with German summary]. - Patras: Ph.D. thesis, Univ. of Patras, 382 pp. + 1 map.
CHRISTODOULAKIS, D. 1994: The Greek islets: natural ecosystems of great ecological significance [In Greek with English abstract]. - *Bot. Chron. (Patras)* **11**: 9-14.
CHRISTODOULAKIS, D. 1995: A new subspecies of *Linum gyaricum* (Linaceae) from Greece. - *Nordic J. Bot.* **15(2)**: 145-147.
CHRISTODOULAKIS, D. 1996a: The flora of Ikaria (Greece, E. Aegean Islands). - *Phyton (Horn)* **36(1)**: 63-91.
CHRISTODOULAKIS, D. 1996b: The phytogeographical distribution patterns of the flora of Ikaria (E Aegean, Greece) with the E Mediterranean. - *Flora, Morphol. Geobot. Oekophysiol.* **191(4)**: 393-399.
CHRISTODOULAKIS, D. 1999: Endemic and rare plants of Samos island. Calendar for the year 2000 [In Greek and English]. - Samos: Konstantinos and Maria Zimalis foundation.
CHRISTODOULAKIS, D. 2000: The flora of Samiopoula (E. Aegean Islands, Greece): a biological, chorological and ecological analysis. - *Bot. Chron. (Patras)* **13**: 287-301.
CHRISTODOULAKIS, D. & AL. 1988: Flora, Vegetation und ökologische Bedeutung der Insel Oxeiá (Ionische Inseln, Griechenland). - *Candollea* **43(1)**: 209-222.
CHRISTODOULAKIS, D. & AL. 1990: Flora und Vegetation der Dionysaden-Inseln (Südägäis, Griechenland). - *Willdenowia* **19(2)**: 425-443.
CHRISTODOULAKIS, D. & AL. 2001: New records to the flora of Fournoi (E. Aegean islands, Greece). - *Bocconea* **13**: 491-494.
CHRISTODOULAKIS, D., ARTELARI, R. & GEORGIADIS, TH. 1998: The flora of Fourni (E. Aegean islands, Greece). In Moret, J. (ed.): Abstracts of the IX. OPTIMA meeting, Paris, 11-17 May 1998. - Paris: Mus. Nat. d'Hist. Naturelle.
CHRISTODOULAKIS, D., ECONOMIDOU, E. & GEORGIADIS, TH. 1991: Geobotanische Studie der Grabusen-Inseln (Südägäis, Griechenland). - *Bot. Helv.* **101(1)**: 53-67.
CHRISTODOULAKIS, D. & GEORGIADIS, TH. 1984: Eine neue *Asperula*-Art (Rubiaceae) von der Insel Samos, Griechenland. - *Willdenowia* **13(2)**: 341-344.
CHRISTODOULAKIS, D. & GEORGIADIS, TH. 1990: The vegetation of the island of Samos, Greece. - *Ann. Mus. Goulandris* **8**: 45-80.
CHRISTODOULAKIS, D. & IATROÚ, G. 1994: *Linum phitosianum* spec. nova (Linaceae) aus Griechenland. - *Phyton (Horn)* **33(2)**: 289-294.
CHRISTODOULOU, CH., HAND, R. & IOSIF, K. 2020: *Solenopsis annua* comb. nov., a new taxon for Cyprus. - *Cypricola* **17**: 1-8.
CHRISTODOULOU, CH.S. & HAND, R. 2010: *Silene alexandrina* (Asch.) Danin. In Greuter, W. & Raus, Th. (eds): Med-Checklist notulae, 29. - *Willdenowia* **40(2)**: 190.
CHRISTOFIDES, Y. 2014: An Olympian habitat. - *Alpine Gardener* **82(4)**: 456-467.
CHRISTOFIDES, Y. 2017: Illustrated Flora of Cyprus. - Publ. by the author: 383 pp.
CHRISTOPOULOU, A. & AL. 2021: How effective are the protected areas of the Natura 2000 Network in halting biological invasions? A case study in Greece. - *MDPI Plants* **10, 2113**: https://doi.org/10.3390/plants10102113.
CHRISTOU, E., BAREKA, E.-P. & KAMARI, G. 2008: Karyosystematic study of selected plant taxa from Cyprus. - *Bot. Chron. (Patras)* **19**: 13-20.
CHRISTOV, M. & POPOV, A. 1933: Cytologische Studien über die Gattung *Hieracium*. - *Planta* **20(3)**: 440-447 + 1 plate.
CHRISTOV, M. & TERSIJSKI, D. 1968: A karyological study of *Poa silvicola* Guss. - *Dokl. Akad. Selskost. Nauki Sofia* **24(4)**: 323-325.
CHRONOPOULOS, G. 2002: Flora, vegetation, ecological evaluation and management proposals of the urban and suburban environment of Patras. [In Greek with English summary]. - Patras: Ph.D. thesis, Univ. of Patras, 320 pp.
CHRONOPOULOS, G. & CHRISTODOULAKIS, D. 1996: Contribution to the urban ecology of Greece: The flora of the city of Patras and the surrounding area. - *Bot. Helv.* **106(2)**: 159-176.
CHRONOPOULOS, G. & CHRISTODOULAKIS, D. 1997: An investigation of the flora and vegetation of the city of Patras. In: Abstracts from the First Balkan Botanical Congress, Thessaloniki, September 19-22, 1997, p. 10. - Thessaloniki.
CHRONOPOULOS, G. & CHRISTODOULAKIS, D. 2000: Analysis of the adventive flora of a Greek city: The example of Patras. - *Bot. Helv.* **110(2)**: 171-189.
CHRONOPOULOS, G. & CHRISTODOULAKIS, D. 2004: The flora of the city of Patras (W Greece): An analysis relative to the city's spatial structure and habitat types. - *Bot. Chron. (Patras)* **16**: 115-124.
CHRONOPOULOS, G. & CHRISTODOULAKIS, D. 2006: Contribution to the urban ecology of Greece: The flora of Alexandroupolis (NE Greece) and its vicinity. - *Fresenius Environ. Bull.* **15(11)**: 1460-1472.
CHRONOPOULOS, G., THEOCHAROPOULOS, M. & CHRISTODOULAKIS, D. 2005: Phytosociological study of *Hirschfeldia incana* (L.) Lagraze-Fossat (Cruciferae) communities in mainland Greece. - *Acta Bot. Croat.* **64(1)**: 75-114.
CHRTEK, J. 1960: Neue Arten der Gattung *Polygonum* L. s.str. - *Preslia* **32(3-4)**: 366-368.
CHRTEK, J. 1963: *Phleum subincrassatum* (Griseb.) comb. nova, ein Endemit der Balkanhalbinsel. - *Acta Univ. Carol., Biol.* **1963(3)**: 225-232.

CHRTEK, J. 1965a: Bemerkungen zur Gliederung der Gattung *Trisetum* Pers. - *Bot. Not.* **118(2)**: 210-224.

CHRTEK, J. 1965b: *Parvotrisetum*, eine neue Gattung der Familie Poaceae. - *Preslia* **37**: 200-202.

CHRTEK, J. 1966a: Beitrag zur Kenntnis einiger Arten der Gattung *Trisetum* der Türkei. - *Bot. Not.* **119(4)**: 486-490.

CHRTEK, J. 1966b: *Trisetum rechingeri*, eine neue Art der Flora Griechenlands. - *Acta Univ. Carol., Biol.* **1966(2)**: 91-94.

CHRTEK, J. 1966c: *Poa harveyi*, a new bluegrass species from the Balkan Peninsula. - *Čas. Nár. Mus., Odd. Přír.* **165(1-4)**: 127-129.

CHRTEK, J. 1969: Anmerkungen zur Problematik von *Poa cenisia*-Komplex in Europa. - *Folia Geobot. Phytotax.* **4(2)**: 197-203.

CHRTEK, J. 1970: Anmerkungen zur Art *Bellardiochloa violacea* (Bell.) Chiov. - *Preslia* **42**: 98-99.

CHRTEK, J. 1973a: Bemerkungen zum *Trisetum flavescens* Komplex im Pindus Gebirge. - *Acta Univ. Carol., Biol.* **1971(4)**: 227-230.

CHRTEK, J. 1973b: Taxonomic observations on some species of the genus *Parietaria*. - *Folia Geobot. Phytotax.* **8(4)**: 425-427.

CHRTEK, J. 1979a: Bemerkungen zur Gliederung der Gattung *Urtica* L. - *Folia Geobot. Phytotax.* **14(3)**: 265-266.

CHRTEK, J. 1979b: Bestimmungsschlüssel der in der ČSR vorkommenden Arten der Gattung *Urtica* L. [In Czech with German summary]. - *Zprávy Československ. Bot. Společn. Československ. Akad. Věd* **14 (1)**: 1-7.

CHRTEK, J. 1988: Notices to the dividing of *Acinos suaveolens* (Labiatae) [In Czech with English abstract]. - *Čas. Nár. Mus., Odd. Přír.* **157(1-4)**: 143-144.

CHRTEK, J. 1992a: Taxonomical remarks on the *Stachys recta* group. - *Folia Geobot. Phytotax.* **27(2)**: 177-183.

CHRTEK, J. 1992b: Bemerkungen zur Variabilität der Art *Chondrilla juncea* [In Czech with German summary]. - *Čas. Nár. Mus., Odd. Přír.* **159(1-4)**: 94.

CHRTEK, J. 1995a: *Plantago atrata* subsp. *graeca* in Bulgaria [In Czech with English summary]. - *Čas. Nár. Mus., Odd. Přír.* **164(1-4)**: 16.

CHRTEK, J. 1995b: *Viburnum maculatum*, a neglected plant species from the Balkan peninsula. - *Čas. Nár. Mus., Odd. Přír.* **164(1-4)**: 87-89.

CHRTEK, J. 1996: *Poa harveyi*, a new bluegrass species from the Balkan peninsula. - *Čas. Nár. Mus., Odd. Přír.* **165(1-4)**: 127-129.

CHRTEK, J. & CHRTKOVÁ, A. 1976: Die Gattung *Ceratocephala* in den tschechischen Ländern mit Bemerkungen zu ihrer Gliederung [In Czech with German summary]. - *Preslia* **48(2)**: 113-123.

CHRTEK, J. & CHRTKOVÁ, A. 1979: Taxonomische Bemerkungen zur Art *Trollius europaeus* s.l. in der Tschechoslowakei. - *Preslia* **51(2)**: 97-106.

CHRTEK, J. & CHRTKOVÁ, A. 1983: Bemerkungen zu einigen balkanischen *Oxytropis*-Arten. - *Folia Geobot. Phytotax.* **18(3)**: 309-320.

CHRTEK, J. & HOLUB, J. 1963: Bemerkungen zur Taxonomie und Nomenklatur der Gattungen *Evax* und *Filago* [In Czech with German summary]. - *Preslia* **35**: 13-17.

CHRTEK, J. & JIRÁSEK, V. 1962: Contribution to the systematics of species of the *Poa* L. genus, section *Ochlopoa* (A. et Gr.) V. Jirás. - *Preslia* **34**: 40-68.

CHRTEK, J. & JIRÁSEK, V. 1963: On the taxonomy of the genus "Trisetum" Pers. - *Webbia* **17(2)**: 569-580.

CHRTEK, J. & JIRÁSEK, V. 1966: Zur Systematik von *Poa laxa* Haenke. - *Novit. Bot. Delect. Seminum Horti Bot. Univ. Carol. Prag.* **1966**: 11-17.

CHRTEK, J. & KŘÍSA, B. 1962: A taxonomical study of the species *Luzula spicata* (L.) DC. sensu lato in Europe. - *Bot. Not.* **115(3)**: 293-310.

CHRTEK, J. & KŘÍSA, B. 1964a: On the problems of the species "Luzula spicata" (L.) DC. in Italy. - *Webbia* **19(1)**: 1-10 + 6 plates.

CHRTEK, J. & KŘÍSA, B. 1964b: Notes on the *Luzula spicata* complex in Turkey. - *Notes Roy. Bot. Gard. Edinburgh* **25(2)**: 163-164.

CHRTEK, J. & KŘÍSA, B. 1965: Bemerkungen zu den mediterranen Arten [des] *Luzula spicata*-Komplex. - *Novit. Bot. Delect. Seminum Horti Bot. Univ. Carol. Prag.* **1965**: 27-29.

CHRTEK, J. & KŘÍSA, B. 1967: Beitrag zur Verbreitung der *Luzula alpino-pilosa* (Chaix) Breistr. auf der Balkanhalbinsel. - *Fragm. Balcan.* **6(4)**: 49-51.

CHRTEK, J. & KŘÍSA, B. 1971: On the taxonomy and distribution of the species *Tithymalus seguierianus* (Neck.) Proch. in Slovakia [In Slovakian with English summary]. - *Biológia, A (Bratislava)* **26(10)**: 765-770.

CHRTEK, J. & KŘÍSA, B. 1976: Taxonomic observations on *Polygala rhodopea* and related taxa. - *Preslia* **48(4)**: 299-305.

CHRTEK, J. & KŘÍSA, B. 1990: Anmerkungen zu einigen Arten der *Tithymalus und Chamaesyce* Gattungen. - *Novit. Bot. Univ. Carol.* **6**: 71-76.

CHRTEK, J. & KŘÍSA, B. 1999: A revision of Asian species of the genus *Bufonia* L. - *Acta Univ. Carol., Biol.* **43(2)**: 77-118.

CHRTEK, J. & OSBORNOVÁ-KOSINOVÁ, J. 1986: Notes on the synanthropic plants of Egypt. 2. *Anagallis arvensis* s.l. (Primulaceae). - *Folia Geobot. Phytotax.* **21(4)**: 405-421.

CHRTEK, J. & OSBORNOVÁ, J. 1996a: A proposal for subdivision of the genus *Cenchrus* (Gramineae). - *Acta Univ. Carol., Biol.* **39(2)**: 85-94.

CHRTEK, J. & OSBORNOVÁ, J. 1996b: *Cenchrus ciliaris* and *C. pennisetiformis* in Egypt and adjacent regions. - *Acta Univ. Carol., Biol.* **39(2)**: 95-110.

CHRTEK, J. & POUZAR, Z. 1962: A contribution to the taxonomy of some European species of the genus *Antennaria* Gaertn. - *Acta Univ. Carol., Biol.* **1962(2)**: 105-136.

CHRTEK, J. & SKOČDOPOLOVÁ, B. 1982: Waldstein's collection in Herbarium of the National Museum in Prague. - *Sborn. Nár. Mus. Praze, Řada B, Přír. Vědy* **38(4)**: 201-238.

CHRTEK, J. & SKOČDOPOLOVÁ, B. 1995: Notes on the taxonomy of *Plantago dostalii* Domin [In Czech with English summary]. - *Čas. Nár. Mus., Odd. Přír.* **164(1-4)**: 11-15.

CHRTEK, J. & SLAVÍK, B. 1981: Contribution to the flora of Cyprus. - *Preslia* **53(1)**: 45-65.

CHRTEK, J. & SLAVÍK, B. 1993: Contribution to the flora of Cyprus. 2. - *Fl. Medit.* **3**: 239-259.

CHRTEK, J. & SLAVÍK, B. 1994: Contribution to the flora of Cyprus. 3. - *Fl. Medit.* **4**: 9-20.

CHRTEK, J. & SLAVÍK, B. 2000: Contribution to the flora of Cyprus. 4. - *Fl. Medit.* **10**: 235-259.

CHRTEK, J. & SLAVÍKOVÁ, Z. 1978: Gliederung der Gattung *Adonis* in drei Gattungen. - *Preslia* **50(1)**: 22-25.

CHRTEK, J. & SLAVÍKOVÁ, Z. 1979: Some problems in the genus *Adonis*: *Chrysocyathus* contra *Calathodes*. - *Preslia* **51**: 279-280.

CHRTEK, J. & SOJÁK, J. 1989: Bemerkungen zur Variabilität von *Saxifraga marginata*. - *Folia Geobot. Phytotax.* **24(3)**: 321-324.

CHRTKOVÁ-ŽERTOVÁ, A. 1966a: Bemerkungen zur Taxonomie von *Lotus uliginosus* Schkuhr und *L. pedunculatus* Cav. - *Folia Geobot. Phytotax. Bohemoslov.* **1(1)**: 78-87.

CHRTKOVÁ-ŽERTOVÁ, A. 1966b: Studie über die tschechoslowakischen Arten der Gattung *Lotus*. IV. - *Preslia* **38**: 36-47.

CHRTKOVÁ-ŽERTOVÁ, A. 1969: Bemerkungen zur Taxonomie von *Vicia sepium* L. - *Folia Geobot. Phytotax.* **4(3)**: 303-311.

CHRTKOVÁ-ŽERTOVÁ, A. 1971: The Linnaean material of two species of the genus *Lotus* in the collection Hortus Cliffortianus. - *Ann. Naturhist. Mus. Wien* **75**: 33-37.

CHRTKOVÁ-ŽERTOVÁ, A. 1972: Studien über die tschechoslowakischen Arten der Gattung *Lotus* L. V. *Lotus tenuis* Waldst. & Kit. ex Willd. - *Preslia* **44(2)**: 131-139 + 1 map.

CHRTKOVÁ-ŽERTOVÁ, A. 1973a: Cytotaxonomic study of the *Vicia cracca* complex. 1. Czechoslovak taxa. - *Folia Geobot. Phytotax.* **8(1)**: 67-93.

CHRTKOVÁ-ŽERTOVÁ, A. 1973b: A monographic study of *Lotus corniculatus* L. I. Central and northern Europe. - *Rozpr. Československ. Akad. Věd.* **83(4)**: 1-94 + 1 chart.

CHRTKOVÁ, A. & JAROLÍMOVÁ, V. 1999: Cytotaxonomical study of *Caltha palustris*. - *Preslia* **71(4)**: 349-360.

CHRTKOVÁ, A. & STAVĚLOVÁ, E. 1986: *Coronilla moravica* Chrtková et Stavělová spec. nova from Czechoslovakia. - *Folia Geobot. Phytotax.* **21(3)**: 325-328.

CHURCH, G.L. 1949: A cytotaxonomic study of *Glyceria* and *Puccinellia*. - *Amer. J. Bot.* **36(2)**: 155-165.

CIESLA, W.M., GEILS, B.W. & ADAMS, R.P. 2004: Hosts and geographic distribution of *Arceuthobium oxycedri*. - Internet publ., http://www.fs.fed.us/rm/pubs/rmrs_rn011.pdf.

CIFERRI, R. 1943: La vegetazione delle isole italiane dell'Egeo. - *Arch. Bot. (Forlì)* **19, suppl.**: i-viii.

CIFERRI, R. 1944: Flora e vegetazione delle isole italiane dell'Egeo. - *Atti Ist. Bot. Lab. Crittog. Univ Pavia, Ser. 5, Suppl.* **A**: 1-200.

CIFERRI, R. 1949: La sistematica del Lino. - Bologna: Univ. di Bologna, Fac. di Agraria, 203 pp.

ČINČURA, F. 1970: Bemerkungen zur Zytotaxonomie einiger fremdländischer Arten der Gattung *Vicia* L. (Wicke). - *Acta Fac. Rerum Nat. Univ. Comen., Bot.* **18**: 17-35.

ČINČURA, F. 1973: Angaben über die Zahl und die Morphologie der Chromosomen von *Vicia dumetorum* L. und deren Vergleich mit *Vicia pisiformis* L. - *Acta Fac. Rerum Nat. Univ. Comen., Bot.* **21**: 103-107.

ČINČURA, F. 1979: Einige Bemerkungen zur Zytotaxonomie der Arten *Vicia incana* Gouan und *Vicia tenuifolia* Roth von Lokalitäten in der Slowakei. - *Acta Fac. Rerum Nat. Univ. Comen., Bot.* **27**: 137-149.

ČINČURA, F. 1981: *Vicia cracca* agg. in der Slowakei. - *Acta Fac. Rerum Nat. Univ. Comen., Bot.* **28**: 77-94.

ČINČURA, F. 1985: Beitrag zur Karyologie der Art *Cupressus sempervirens* L. - *Acta Fac. Rerum Nat. Univ. Comen., Bot.* **32**: 39-46.

CINGEL, N.A. VAN DER 1990a: Kreta, orchideeënparadijs [In Dutch]. - *Orchideeën* **52(1)**: 2-10, 15.

CINGEL, N.A. VAN DER 1990b: *Orchis anatolica* × *Orchis prisca*? - *Eurorchis* **2**: 158-160.

CIOCÂRLAN, V. 1975: *Agrostis moldavica* Dobr. et Beldie dans la flore de R.P. Bulgarie ... In: Jordanov, D. & al. (eds): Problems of Balkan flora and vegetation, pp. 244-246. - Sofia: Publishing House of the Bulgarian Academy of Sciences.

CLABBY, G. & OSBORNE, B.A. 1999: Biological Flora of the British Isles. *Mycelis muralis* (L.) Dumort. (*Lactuca muralis* (L.) Gaertner). - *J. Ecol.* **87(1)**: 156-172.

CLAESSENS, J. & KLEYNEN, J. 2006: Anmerkung zur Hybridbildung bei *Platanthera bifolia* und *P. chlorantha*. - *J. Eur. Orch.* **38(1)**: 3-28.

CLAESSENS, M. & CLAESSENS, J. 2000: Een vondst van *Orchis sancta* op Kreta. - *Eurorchis* **12**: 21, 45-46.

CLAPHAM, A.R. 1964: [Note on *Moenchia mantica*]. In Heywood, V.H. (ed.): Notulae systematicae ad Floram Europaeam spectantes, no. 3. - *Feddes Repert. Spec. Nov. Regni Veg.* **69(1)**: 49.

CLAPHAM, A.R., PEARSALL, W.H. & RICHARDS, P.W. 1942: Biological Flora of the British Isles. *Aster tripolium* L. - *J. Ecol.* **30**: 385-395.

CLARK, S.C. 1974: Biological Flora of the British Isles. *Catapodium rigidum* (L.) C.E. Hubbard (*Scleropoa rigida* (L.) Griseb.; *Desmazeria rigida* (L.) Tutin). - *J. Ecol.* **62(3)**: 937-958.

CLARKE, A.E. 1934: The number and morphology of chromosomes in the genus *Melilotus*. - *Univ. Calif. Publ. Bot.* **17**: 435-443.

CLARKE, E.D. 1810-23: Travels in various countries of Europe, Asia and Africa [Appendix II, pp. 716-724 in vol. 2 contains a "List of all plants collected during these travels in Greece, Egypt and the Holy Land"]. - London: Cadell & Davies, 6 vols.

CLARKE, G. 1978: Pollen morphology and generic relationships in Valerianaceae. - *Grana* **17(2)**: 61-75.

CLARKE, G.C.S. & KUPICHA, F.K. 1976: The relationships of the genus *Cicer* L. (Leguminosae): the evidence from pollen morphology. - *Bot. J. Linn. Soc.* **72(1-4)**: 35-44.

CLAUSEN, J. 1931: Cyto-genetic and taxonomic investigations on *Melanium* violets. – *Hereditas (Lund)* **15(3)**: 219-308.

CLAUSEN, R.T. 1938: A monograph of the Ophioglossaceae. - *Mem. Torrey Bot. Club* **19(2)**: 1-177.

CLAUSER, M. & INNOCENTI, A. 2000: The flowers of Crete [English edition]. - Firenze: Case Editrice Bonechi, 96 pp.

CLAYTON, W.D. 1969: A revision of the genus *Hyparrhenia*. - *Kew Bull., Addit. Ser.* **2**: 1-196.

CLAYTON, W.D. 1978: The genus *Phacelurus* (Gramineae). - *Kew Bull.* **33(2)**: 175-179.

CLAYTON, W.D. & RENVOIZE, S.A. 1986: Genera Graminum. Grasses of the World. - *Kew Bull., Addit. Ser.* **13**: i-ix, 1-389.

CLEBSCH, B. 2003: The new book of salvias: sages for every garden, 2nd ed. - Portland, Oregon & Cambridge: Timber Press, 344 pp.

CLELAND, R.E. 1972: *Oenothera*. Cytogenetics and evolution. - London & New York: Academic Press, x + 370 pp.

CLEMENT, W.M. JR. 1962: Chromosome numbers and taxonomic relationships in *Medicago*. - *Crop Sci. (Madison)* **2**: 25-28.

CLEMENTI, J. 1855: Sertulum orientale, seu Recensio plantarum in Olympo Bithynico, in agro Byzantino et Hellenico nonnullisque aliis orientis regionibus, annis 1849-1850. - *Atti Reale Accad. Sci. Torino* **16**: 239-338 + 8 plates [also with separate pagination, 1-103].

CLEMENTI, M. & AL. 2014: Typification of five names listed by Roberto de Visiani in Plantarum Serbicarum Pemptas. - *Phytotaxa* **170(1)**: 57-60.

CLEMENTI, M. & AL. 2015a: Typification of names published by Roberto de Visiani and Josif Pančić in Plantae Serbicae Rariores aut Novae - Decas I. - *Phytotaxa* **202(2)**: 121-134.

CLEMENTI, M. & AL. 2015b: Typification and taxonomical notes on the names published by Roberto de Visiani and Josif Pančić in Plantae Serbicae Rariores aut Novae - Decas II. - *Phytotaxa* **224(1)**: 29-44.

CLEMENTI, M. & VUKOJIČIĆ, S. 2015: Josif Pančić and Roberto de Visiani, joint work and correspondence. In Bogdanović, S. & Jogan, N.: Book of abstracts - 6th Balkan Botanical Congress, p. 20. - Rijeka.

CLENNETT, J.C.B. 2004a: A review of research on *Cyclamen*. - *Herbertia* **57**: 25-35.

CLENNETT, J.C.B. 2004b: An analysis and revision of *Cyclamen* L. with emphasis on subgenus *Gyrophoebe* O. Schwarz. - *Bot. J. Linn. Soc.* **138(4)**: 473-481.

CLERMONT, A., HILGER, H.H. & ZIPPEL, E. 2003: Verbreitung und Differenzierung der mitteleuropäischen Unterarten von *Buglossoides arvensis* (L.) I.M. Johnst. (Boraginaceae). - *Feddes Repert.* **114(1-2)**: 56-68.

CLEVENGER, S. & HEISER, CH.B. 1963: *Helianthus laetiflorus* and *Helianthus rigidus* - hybrids or species? - *Rhodora* **65**: 121-133.

CLIFTON, R.T.F. 1994: Geranium family. Species check list. Ed. IV (issue 2), Part 1: *Erodium*. - Dover, Kent: The Geraniaceae group: 70 pp. (mimeographed).

CLIFTON, R.T.F. 1995: Geranium family. Species check list. Ed. IV (issue 2), Part 2: *Geranium*. - Dover, Kent: The Geraniaceae group: 184 pp. (mimeographed).

CLIFTON, R.T.F. 2004: Geraniales species check list series. Vol. 1, Geraniaceae Knuth, tribe 1 Geranieae, part 2 *Geranium* species checklist. Ed. 5. - Dover, Kent: The Geraniaceae group: 170 pp.

CLIFTON, R.T.F. 2005: The *Erodium* species checklist, ed. 5. - Dover, Kent: The Geraniaceae group: 224 pp.

CLOS, D. 1897: Les *Anagallis* annuels d'Europe au point de vue spécifique. - *Bull. Soc. Bot. France* **44**: 292-307.

COASSINI LOKAR, L. & CORSI, G. 1986: Taxonomical studies on *Seseli elatum* L. and allied species. - I. Systematic implications of some morpho-anatomical characters. - *Stud. Geobot.* **6**: 131-143.

CODIGNOLA, A. & MAFFEI, M. 1984: Numeri cromosomici per la Flora Italiana: 937-940. - *Inform. Bot. Ital.* **15(1)**: 35-38.

COINCY, A. DE 1900: Révision des espèces critiques du genre *Echium*. - *J. Bot. (Morot)* **14(10-11)**: 297-304, 322-330.

COINCY, A. DE 1901: Révision des espèces critiques du genre *Echium* [cont.]. - *J. Bot. (Morot)* **15(9-10)**: 311-329.

COINCY, A. DE 1902: Révision des espèces critiques du genre *Echium* [cont.]. - *J. Bot. (Morot)* **16(2-3)**: 66-68, 107-112.

COINCY, A. DE 1922: L'île de Thasos. - *La Géographie* **38**: 405-426.

COKER, P.D. 1962: Biological Flora of the British Isles. *Corrigiola litoralis* L. - *J. Ecol.* **50(3)**: 833-840.

COKER, P.D. 1966: Biological Flora of the British Isles. *Sibbaldia procumbens* L. [*Potentilla sibbaldii* Hall. fil., *Potentilla procumbens* (L.) Clairv. non Sibth.]. - *J. Ecol.* **54(3)**: 823-831.

COLASANTE, M. 1979: Confronto citotassonomico tra *Dipsacus fullonum* L. e *Dipsacus ferox* Loisel. - *Ann. Bot. (Rome)* **38(1)**: 71-79.

COLASANTE, M. 1983: Genus *Iris*: reconstructed specimen lists of some Italian herbaria. - *Ann. Bot. (Rome)* **41**: 37-65.

COLASANTE, M. 1987: Genus *Iris* subgenus *Iris* section *Iris*: Le Iris barbate e la loro distribuzione. - *Boll. Soc. Ital. Iris* **26**: 46-52.

COLASANTE, M. 1989: Alcune osservazioni sulle Iris barbate spontanee e la loro determinazione in specie. - *Boll. Soc. Ital. Iris* **28**: 39-45.

COLASANTE, M. 1990: *Gynandriris sisyrinchium* (L.) Parl. e polimorfismo di popolazione. - *Boll. Soc. Ital. Iris* **29**: 45-49 + 1 plate.

COLASANTE, M. 1997: Taxonomic note on some Italian bearded Irises: *Iris relicta* Colas. nom. nov. and *Iris bicapitata* Colas. sp. nov. - *Fl. Medit.* **6**: 213-217.

COLASANTE, M., ALTAMURA, L. & CIMA, F. 1989: Osservazioni sul polimorfismo di popolazioni in *Papaver rhoeas* L. - *Giorn. Bot. Ital., n.s.* **123(1-2), suppl. 1**: 113.

COLASANTE, M. & MATHEW, B.F. 1987: Analysis of European Irises based on specimens in the Kew Herbarium (K) including some comments about critical taxonomic groups. - *Ann. Bot. (Rome)* **45(1)**: 163-196.

COLASANTE, M. & RICCI, I. 1972: *Cornucopiae cucullatum* L.: analisi citotassonomica e prima indicazione per la flora italiana. - *Ann. Bot. (Rome)* **31**: 205-211.

COLASANTE, M. & RICCI, I. 1975: *Iris germanica* L. e *Iris marsica* Ricci e Colasante: separazione delle due specie. - *Ann. Bot. (Rome)* **34**: 17-28.

COLASANTE, M. & RUDALL, P.J. 2000: *Veratrum album* and *V. nigrum* (Melanthiaceae) in Italy: micromorphology and systematics. - *Pl. Biosyst.* **134(2)**: 233-240.

COLASANTE, M. & SAUER, W. 1985: Annotated list of *Iris* specimens from the following European herbaria: Geneva (G), Zürich (Z, ZT), Graz (GZU) and Vienna (WU, W) including additional data from the herbaria of IB, KL and B. - *Ann. Bot. (Rome)* **43**: 53-79.

COLASANTE, M. & SAUER, W. 1988: *Iris* specimens investigation and revised lists of some German herbaria: Tübingen (TUB), Munich (M) and Sauer (Sa). - *Ann. Bot. (Rome)* **46**: 35-57.

COLASANTE, M. & VOSA, C.G. 1981: Quinacrine fluorescence analysis of the chromosomes of *Dipsacus fullonum* L. (Dipsacaceae). - *Giorn. Bot. Ital., n.s.* **114(6)**: 267-270.

COLASANTE, M. & VOSA, C.G. 1985: *Iris* specimens investigation and check list from Oxford Fielding-Druce herbarium (OXF). - *Ann. Bot. (Rome)* **43**: 81-100.

COLASANTE, M. & VOSA, C.G. 1987: The genus *Iris* L.: a short synthesis of some taxonomic problems. - *Ann. Bot. (Rome)* **45(1)**: 217-238.

COLASANTE, M.A. & MAURY, A.E. 2018: Iridaceae presenti in Italia. [In Italian and English]. - Roma: Sapienza Università Editrice, 415 pp.

COLES, S., MAXTED, N. & MAESEN, L.J.G. VAN DER 1998: Identification aids for *Cicer* (Leguminosae, Cicereae) taxa. - *Edinburgh J. Bot.* **55(2)**: 243-265.

COLES, S.M. 1971: The *Ranunculus acris* L. complex in Europe. - *Watsonia* **8(3)**: 237-261.

COLES, S.M. 1973: *Ranunculus bulbosus* L. in Europe. - *Watsonia* **9(3)**: 207-228.

COLES, S.M. 1977: *Ranunculus repens* L. in Europe. - *Watsonia* **11(4)**: 353-366.

COLETTIS, O.M. 1963: Die Bergwelt Griechenlands. Ein Führer für Wanderer und Bergsteiger. - Stuttgart: J. Fink Verlag, 133 pp.

COLLINS, R.P. & AL. 1988: Studies on infraspecific variation in *Cyperus longus* L. from Europe. - *New Phytol.* **110(2)**: 279-289.

COLLMAN, J. 2003: Walks with Crete's spring flowers. - Newton Abbot: Beechwood Press, xv + 175 pp.

COLOMBO, P.M. & AL. 1979: Numeri cromosomici per la Flora Italiana: 517-525. - *Inform. Bot. Ital.* **10 (3)**: 406-413.

COLOMBO, P.M., LORENZONI, F.C. & GRIGOLETTO, F. 1983: Pollen grain morphology supports the taxonomical discrimination of Mediterranean oaks (*Quercus*, Fagaceae). - *Pl. Syst. Evol.* **141(3-4)**: 273-284.

COLOMBO, P.M. & MARCENÒ, C. 1984: Étude cytotaxinomique de "Centaurea macroacantha" Guss. et son affinité avec "Centaurea calcitrapa" L. - *Webbia* **38**: 665-669.

COLOMBO, P.M., MARCENÒ, C. & PRINCIOTTA, R. 1979: Numeri cromosomici per la Flora Italiana: 591-597. - *Inform. Bot. Ital.* **11(1)**: 27-31.

COLOMBO, P.M., MARCENÒ, C. & PRINCIOTTA, R. 1981a: Numeri cromosomici per la Flora Italiana: 662-675. - *Inform. Bot. Ital.* **11(3)**: 325-323.

COLOMBO, P.M., MARCENÒ, C. & PRINCIOTTA, R. 1981b: Numeri cromosomici per la Flora Italiana: 760-771. - *Inform. Bot. Ital.* **12(2)**: 173-180.

COLOMBO, P.M., MARCENÒ, C. & PRINCIOTTA, R. 1982: Numeri cromosomici per la Flora Italiana: 794-805. - *Inform. Bot. Ital.* **12(3)**: 333-340.

COMES, H.-P. & KADEREIT, J.W. 1990: Aspects of hybridization between the closely related *Senecio vulgaris* L. and *Senecio vernalis* Waldst. & Kit. - *Flora, Morphol. Geobot. Oekophysiol.* **184(5)**: 381-388.

COMES, H.-P., TRIBSCH, A. & BITTKAU, CH. 2008: Plant speciation in continental island floras as examplified by *Nigella* in the Aegean Archipelago. - *Philos. Trans. Roy. Soc., B, Biol. Sci.* 14 pp. (published online June 2008).

COMPTON, J. 2004: Typification of *Cyclamen creticum* Hildebr. A new combination, *C. creticum* f. *erubescens*. - *Cyclamen* **28(1)**: 32-33.

CONARD, H.S. 1905: The waterlilies. A monograph of the genus *Nymphaea*. - Washington, D.C.: Carnegie Institute, xiii + 279 pp.

CONDOGOURIS, M. 1856: Tableau contenant quelques végétaux indigènes de Chios et un certain nombre d'origine étrangère. - *Compt. Rend. Hebd. Séances Acad. Sci.* **43**: 631-634.

CONEDERA, M. & AL. 2004: Distribution and economic potential of the sweet chestnut (*Castanea sativa* Mill.) in Europe. - *Ecol. Medit.* **30(2)**: 179-193.

CONERT, H.J. 1961: Die Systematik und Anatomie der Arundineae. - Weinheim: Verlag von J. Cramer, 208 pp.

CONERT, H.J. 1969: Dreizahngras (*Danthonia decumbens* De Candolle) und Traubenhafer (*Danthonia alpina* Vest). - *Jahrb. Nassauischen Vereins Naturk.* **100**: 54-72.

CONERT, H.J. 1981: Über das Rohrartige Pfeifengras, *Molinia arundinacea* Schrank. - *Ber. Bayer. Bot. Ges.* **52**: 5-14.

CONERT, H.J. & TÜRPE, A.M. 1974: Revision der Gattung *Schismus* (Poaceae: Arundinoideae: Danthonieae). - *Abh. Senckenb. Naturf. Ges.* **532**: 1-81.

CONRAD, M. 1984: *Viscum album* L. sur dicotylédones en Corse. - *Monde Pl.* **79(415-416)**: 9.

CONSTANTINIDIS, TH. 1997a: Reports (749-753). In Kamari, G., Felber, F. & Garbari, F. (eds): Mediterranean chromosome number reports - 6. - *Fl. Medit.* **6**: 308-312.

CONSTANTINIDIS, TH. 1997b: I chlorida ke i vlastisis ton oreon Gerania, Pateras ke Kitheron [In Greek]. - Athine: Ph.D. thesis, Univ. of Athens, 465 pp.

CONSTANTINIDIS, TH. 1999: *Dianthus haematocalyx* subsp. *phitosianus* (Caryophyllaceae), a new serpentine endemic from Greece. - *Phyton (Horn)* **39(2)**: 277-291.

CONSTANTINIDIS, TH. 2004a: Serpentine areas in Greece: Older and recent investigations. In Vasić, O. (ed.): Abstracts, XI OPTIMA Meeting, Beograd 5-11.IX.2004, p. 49.

CONSTANTINIDIS, TH. 2004b: The floristic diversity of serpentine in Greece 1. An inventory of the Aliki area (Sterea Ellas, Central Greece). - *Phyton (Horn)* **44(1)**: 45-67.

CONSTANTINIDIS, TH. 2004c: *Bufonia paniculata* Delarbre. In Greuter, W. & Raus, Th. (eds): Med- Checklist notulae, 22. - *Willdenowia* **34(1)**: 72.

CONSTANTINIDIS, TH. 2012: *Hedysarum grandiflorum* subsp. *bulgaricum* (Leguminosae) and *Biarum ditschianum* (Araceae), two new records for the Greek flora. - *Phytol. Balcan.* **18(2)**: 155-161.

CONSTANTINIDIS, TH. 2013a: The flora of Kastellorizo Island group (East Aegean Islands, Greece): new records and taxonomic comments. In Gkelis, S. & al. (eds): 13th Panhellenic Scientific Conference, Hellenic Botanical Society, Thessaloniki 3-6 October. Program and book of abstracts, p. 65.

CONSTANTINIDIS, TH. 2013b: The flora of the Kastellorizo island group (East Aegean Islands, Greece): new records and comments. - *Fl. Medit.* **23**: 69-86.

CONSTANTINIDIS, TH. & AL. 2002: The Greek East Aegean islands: Last frontier for the Asiatic flora. In Koyuncu, M. (ed.): VIth Plant Life of Southwest Asia Symposium, Abstracts p. 51. - Van: Yüzüncü Yil University.

CONSTANTINIDIS, TH. & AL. 2013: Floristic data and conservation actions in Karpathos and Saria Islands (Dodekanisa, Greece). In Gkelis, S. & al. (eds): 13th Panhellenic Scientific Conference, Hellenic Botanical Society, Thessaloniki 3-6 October. Program and book of abstracts, p. 98.

CONSTANTINIDIS, TH. & AL. 2018: Where the endemics are found? Distribution patterns of area-restricted vascular plants in the Karpathos island group (Aegean sea, Greece). - *Bot. Serbica* **42 (suppl. 1)**: 86.

CONSTANTINIDIS, TH., BAREKA, E.-P. & KAMARI, G. 2000: Karyotaxonomic approaches to the Greek serpentine flora. In Özhatay, N. & al. (eds): Abstracts from the Second Balkan Botanical Congress, Istanbul, Turkey, May 14-18, 2000, p. 29.

CONSTANTINIDIS, TH., BAREKA, E.-P. & KAMARI, G. 2002a: Karyotaxonomy of Greek serpentine angiosperms. - *Bot. J. Linn. Soc.* **139(1)**: 109-124.

CONSTANTINIDIS, TH., BAREKA, P. & KAMARI, G. 2002b: Karyotypic and morphological differentiation of Greek serpentine angiosperms. In Kamari, G. & al. (eds): Ellinini Votaniki Eteria. 9° Panelliniko Epistimoniko Sinedrio. Praktika, pp. 172-178 [In Greek with English abstract].

CONSTANTINIDIS, TH., CHITOS, TH. & THANOPOULOS, R. 2003: Simvoli sti gnosi tis chloridas tis Ditikis Evritanias me emfasi stis livadikes ektasis [In Greek]. In Platis, P.D. & Papachristou, Th.G. (eds): Proceedings of the 3rd Panhellenic Rangeland Congress in Kapenissi, 4-6 September 2002, pp. 149-157. - Athens.

CONSTANTINIDIS, TH. & GAVALAS, I. 2017: *Trachelium caeruleum* subsp. *caeruleum* (Campanulaceae), a new alien plant in the Greek flora. - *Parnassiana Arch.* **5**: 3-9.

CONSTANTINIDIS, TH. & KALPITZAKIS, E. 2019: *Ferulago trachycarpa* (Apiaceae), a new record for Kriti and Europe. In Daskalakou, E.N. & al. (eds): 16th Conference of the Hellenic Botanical Society – Athens 10-13 October 2019. Program and book of abstracts, p. 79.

CONSTANTINIDIS, TH. & KALPOUTZAKIS, E. 2005: A new species of *Achillea* (Asteraceae: Anthemideae) from south-eastern Peloponnisos, Greece. - *Bot. J. Linn. Soc.* **147(2)**: 249-256.

CONSTANTINIDIS, TH. & KALPOUTZAKIS, E. 2015: Plant guide to Mount Parnon and Moutsos wetland protected area. Endemic, rare and threatened species. - Astros, Arcadia: Management body of Mount Parnon and Moutsos wetland, 309 pp.

CONSTANTINIDIS, TH., KALPOUTZAKIS, E. & KOUGIOUMOUTZIS, K. 2015: The rediscovery of *Stachys virgata* (Lamiaceae), a rare endemic of Peloponnisos, Greece: taxonomy, distribution, karyology and conservation. - *Phytotaxa* **218(3)**: 241-252.

CONSTANTINIDIS, TH., KALPOUTZAKIS, E. & KOUVELIS, V. 2013: A new species of *Euphorbia* L. (Euphorbiaceae) from southern Greece and its phylogenetic placement within *E.* sect. *Patellares*. In Gkelis, S. & al. (eds): 13th Panhellenic Scientific Conference, Hellenic Botanical Society, Thessaloniki 3-6 October. Program and book of abstracts, p. 97.

CONSTANTINIDIS, TH. & KAMARI, G. 1994: Reports (377-386). In Kamari, G., Felber, F. & Garbari, F. (eds): Mediterranean chromosome number reports - 4. - *Fl. Medit.* **4**: 290-295.

CONSTANTINIDIS, TH. & KAMARI, G. 1996: Reports (401-414). In Kamari, G., Felber, F. & Garbari, F. (eds): Mediterranean chromosome number reports - 5. - *Fl. Medit.* **5**: 265-278.

CONSTANTINIDIS, TH. & KAMARI, G. 1998: Notes on the ophiolitic flora of central Greece. In Moret, J. (ed.): Abstracts of the IX. OPTIMA meeting, Paris, 11-17 May 1998. - Paris: Mus. Nat. d'Hist. Naturelle.

CONSTANTINIDIS, TH. & KAMARI, G. 2000: A karyological study of ten taxa of phanerogams (Compositae, Leguminosae, and Umbelliferae) from Greece. - *Bot. Chron. (Patras)* **13**: 117-131.

CONSTANTINIDIS, TH. & KAMARI, G. 2001: Karyotype morphology of *Consolida* species (Ranunculaceae) in Greece: does it concur with existing taxonomic schemes? In Raimondo, F.M. (ed.): Abstracts from the X OPTIMA meeting, Palermo, 13-19 September 2001, p. 141.

CONSTANTINIDIS, TH., KAMARI, G. & PHITOS, D. 1997a: Differentiation of *Centaurea* sect. *Acrolophus* and *Phalolepis* in the mountains of SE Sterea Hellas. In: Abstracts from the First Balkan Botanical Congress, Thessaloniki, September 19-22, 1997, p. 10. – Thessaloniki.

CONSTANTINIDIS, TH., KAMARI, G. & PHITOS, D. 1997b: A cytological study of 28 phanerogams from the mountains of SE Sterea Ellas, Greece. - *Willdenowia* **27(1-2)**: 121-142.

CONSTANTINIDIS, TH., KORAKIS, A. & FAKAS, G. 2014: Not a Cretan area endemic any longer: The discovery of *Crepis tybakiensis* (Asteraceae: Cichorieae) in Samos (East Aegean islands, Greece). - *Parnassiana Arch.* **2**: 65-70.

CONSTANTINIDIS, TH. & PHITOS, D. 2004: *Cephalaria pastricensis* Dörfler & Hayek. In Greuter, W. & Raus, Th. (eds): Med-Checklist notulae, 22. - *Willdenowia* **34(1)**: 74.

CONSTANTINIDIS, TH., PSARAS, G. & KAMARI, G. 2001a: Biosystematic study of the genus *Consolida* (DC.) Gray. In Kamari, Psaras & Constantinidis (eds): Proceedings of the 8th scientific congress of the Greek Botanical Society, October 5-8, 2000, pp. 253-261 [In Greek with English abstract].

CONSTANTINIDIS, TH., PSARAS, G. & KAMARI, G. 2001b: Seed morphology in relation to infrageneric classification of *Consolida* (DC.) Gray (Ranunculaceae). - *Flora, Morphol. Geobot. Oekophysiol.* **196 (2)**: 81-100.

CONSTANTINIDIS, TH. & SCHOLZ, H. 2006: *Anisantha rubens*. In Greuter, W. & Raus, Th. (eds): Med- Checklist notulae, 24. - *Willdenowia* **36(2)**: 727.

CONSTANTINIDIS, TH. & VASSILIADES, D. 1996: *Centaurea musarum* (Compositae): the rediscovery of a rare endemic species. - *Bot. Chron. (Patras)* **12**: 9-14.

CONSTANTINIDIS, TH. & YANNITSAROS, A. 1992: The orchids of the island of Salamis: Distribution, ecology, reproduction, ... [In Greek with English summary]. In: Proc. 12th Annual Meeting Hellen. Soc. Biol. Sci., Mitilini, 27-29.4. 1990, pp. 51-53. – Athine.

CONSTANTINIDIS, TH. & YANNITSAROS, A. 1993: *Centranthus macrosiphon* (Valerianaceae) naturalized in Greece. - *Fl. Medit.* **3**: 279-282.

CONSTANTINIDIS, TH. & YANNITSAROS, A. 1996: A floristic report of mountains Gerania, Pateras, Kitheron and Pastra (Sterea Ellas, Greece). - *Candollea* **51(2)**: 303-322.

CONTANDRIOPOULOS, J. 1964: Contribution à l'étude caryologique des Campanulacées de Grèce. - *Bull. Soc. Bot. France* **111(3-4)**: 222-235.

CONTANDRIOPOULOS, J. 1966: Contribution à l'étude cytotaxinomique des Campanulacées de Grèce. II. - *Bull. Soc. Bot. France* **113(9)**: 453-474 + 1 table.

CONTANDRIOPOULOS, J. 1967: Contribution à l'étude caryologique des Gesnériacées d'Europe et de leur germination. In: Comptes Rendus 91ême Congrès National des Sociétés Savantes Rennes, Section des Sciences, III, pp. 271-280. - Paris: Gauthier-Villars et Bibliothèque National.

CONTANDRIOPOULOS, J. 1968: A propos des nombres chromosomiques des *Salicornia* de la région Méditerranéenne. - *Bull. Mus. Hist. Nat. (Marseille)* **28**: 45-52.

CONTANDRIOPOULOS, J. 1970a: Contribution à l'étude cytotaxinomique des Alysseae Adams de Grèce. *Ber. Schweiz. Bot. Ges.* **79**: 313-334.

CONTANDRIOPOULOS, J. 1970b: Contribution à l'étude cytotaxinomique des Campanulacées du Proche-Orient. - *Bull. Soc. Bot. France* **117(1-2)**: 55-70.

CONTANDRIOPOULOS, J. 1970c: Contribution à l'étude cytotaxinomique des Campanulacées du Proche-Orient, II - Les *Asyneuma* de Turquie. - *Bull. Soc. Bot. France* **117(5-6)**: 209-220.

CONTANDRIOPOULOS, J. 1972: Contribution à l'étude cytotaxinomique des Campanulacées du Proche-Orient, III. - *Bull. Soc. Bot. France* **119(1-2)**: 75-94.

CONTANDRIOPOULOS, J. 1976: Contribution à l'étude cytotaxinomique des Campanulacées du Proche-Orient, IV. - *Bull. Soc. Bot. France* **123(1-2)**: 33-46.

CONTANDRIOPOULOS, J. 1978a: Contribution à l'étude cytotaxinomique des *Sideritis* section *Empedoclea* (Labiatae). - *Pl. Syst. Evol.* **129(4)**: 277-289.

CONTANDRIOPOULOS, J. 1978b: Contribution à l'étude cytobiogéographique du genre *Globularia* L. - *Biol. Écol. Médit.* **5(1)**: 3-13.
CONTANDRIOPOULOS, J. 1980: [Chromosome numbers in Campanulaceae]. In Löve, Á. (ed.): Chromosome number reports LXVI. - *Taxon* **29(1)**: 163-164.
CONTANDRIOPOULOS, J. 1984: Differentiation and evolution of the genus *Campanula* in the Mediterranean region. In Grant, W.F. (ed.): Plant Biosystematics, pp. 141-158. - Toronto, etc.: Academic Press.
CONTANDRIOPOULOS, J. 1990: Specificité de l'endemisme Corse. In: International symposium on biogeographical aspects of insularity, Rome, 18-22 May 1987, pp. 393-416 + 1 table. - Roma: Academia Nazionale dei Lincei.
CONTANDRIOPOULOS, J. 1991: Quelques reflexions a propos de l'endemisme insulaire (La Corse). Methodes de travail. - *Polish Bot. Stud.* **2**: 39-59.
CONTANDRIOPOULOS, J. & AFZAL-RAFII, Z. 1973: Contribution à l'étude cytotaxinomique des *Alyssum* de Turquie. - *Ber. Schweiz. Bot. Ges.* **83(1)**: 14-29.
CONTANDRIOPOULOS, J. & FAVARGER, C. 1975: Problèmes posés par l'endémisme en Méditerranée. In Guinochet, M. & al. (eds): La flore du Bassin Méditerranéen, pp. 175-194. - Paris: C.N.R.S.
CONTANDRIOPOULOS, J. & FAVARGER, C. 1983: Sur quelques espèces de Turquie du genre *Arenaria* L. (étude cytotaxonomique). - *Candollea* **38(2)**: 733-743.
CONTANDRIOPOULOS, J. & GAMISANS, J. 1974: A propos de l'élément arctico-alpin de la flore corse. - *Bull. Soc. Bot. France* **121(5-6)**: 175-204.
CONTANDRIOPOULOS, J. & LANZALAVI, M. 1968: Contribution à l'étude cytotaxinomique des *Hypericum* de Grèce. - *Bull. Soc. Bot. France* **115(1-2)**: 5-13.
CONTANDRIOPOULOS, J. & MARTIN, D. 1967: Contribution à l'étude cytotaxinomique des *Achillea* de Grèce. Irrégularités de la meiose. - *Bull. Soc. Bot. France* **114(7-8)**: 257-275.
CONTANDRIOPOULOS, J., PAMUKÇUOĞLU, A. & QUÉZEL, P. 1972: A propos des *Cicer* vivaces du pourtour Méditerranéen oriental. - *Biol. Gallo-Hellen.* **4(1)**: 3-18.
CONTANDRIOPOULOS, J. & QUÉZEL, P. 1964: A propos de *Veronica contandriopouli* Quézel, espèce nouvelle endémique du mont Kyllini (Grèce méridionale). - *Compt. Rend. Hebd. Séances Acad. Sci.* **259(2)**: 425-427.
CONTANDRIOPOULOS, J. & QUÉZEL, P. 1965: A propos de deux véroniques critiques des montagnes grecques: *Veronica erinoides* Boiss. & Spruner et *V. thessalica* Benth. - *Candollea* **20**: 43-48.
CONTANDRIOPOULOS, J. & QUÉZEL, P. 1967: A propos des phénomènes de reviviscence chez *Ramonda nathalie.* In: Comptes Rendus du 91ême Congrès National des Sociétés Savantes Rennes, Section des Sciences, III, pp. 281-289. - Paris: Gauthier-Villars et Bibliothèque National.
CONTANDRIOPOULOS, J. & QUÉZEL, P. 1973: Deux Dipsacées nouvelles de Grèce. - *Biol. Gallo-Hellen.* **5(1)**: 131-137.
CONTANDRIOPOULOS, J. & QUÉZEL, P. 1974: A propos de l'étude caryologique de quelques *Pinguicula* de Grèce et de Turquie, et en particulier du complexe *Pinguicula hirtiflora* Ten. - *Biol. Écol. Médit.* **1 (1)**: 27-32.
CONTANDRIOPOULOS, J., QUÉZEL, P. & PAMUKÇUOĞLU, A. 1971: Campanulacées nouvelles du pourtour Méditerranéen oriental. - *Ann. Univ. Provence, Sci.* **46**: 53-61.
CONTANDRIOPOULOS, J., QUÉZEL, P. & ZAFFRAN, J. 1965: A propos de deux *Sedum* de la flore grecque. - *Bull. Soc. Bot. France* **112(7-8)**: 462-466.
CONTANDRIOPOULOS, J., QUÉZEL, P. & ZAFFRAN, J. 1973: A propos des Campanules du groupe *aizoon* en Grèce méridionale et en Crète. - *Bull. Soc. Bot. France* **120(7-8)**: 331-340.
CONTANDRIOPOULOS, J. & VERLAQUE, R. 1975: Contribution à l'étude caryologique des *Pterocephalus* méditerranéens. - *Biol. Écol. Médit.* **2(1)**: 9-13.
CONTANDRIOPOULOS, J. & YANNITSAROS, A. 1975: Distribution géographique, écologie et cytotaxinomie du *Linaria hellenica* Turrill (Scrophulariaceae). - *Candollea* **30(2)**: 293-300.
CONTANDRIOPOULOS, J. & ZAFFRAN, J. 1969: A propos de quelques espèces remarquables de la flore grecque. - *Bull. Soc. Bot. France* **116(1-2)**: 103-112.
CONTANDRIOPOULOS, J. & ZEVACO-SCHMITZ, C. 1989: A propos d'un endémique cyrnosarde méconnu: *Scilla corsica* Boullu. In Jeanmonod, D. & Burdet, H.M. (eds): Notes et contributions à la flore de Corse, IV. - *Candollea* **44(1)**: 394-401.
CONTARDO, A. 1969: "Tagetes minuta" L., avventizia nuova per la Toscana. - *Inform. Bot. Ital.* **1**: 114-115.
CONTI, F. 1997: *Minuartia glomerata* subsp. *trichocalycina* comb. & stat. nov. (Caryophyllaceae), a Central Apennine endemic. - *Willdenowia* **27(1-2)**: 73-79.
CONTI, F. 1998: Contributions to the knowledge of *Jurinea mollis* s.l. (Compositae). 1. *Jurinea mollis* subsp. *moschata*, a synonym of *J. mollis* subsp. *mollis*. - *Willdenowia* **28(1-2)**: 47-52.
CONTI, F. 2003: *Minuartia graminifolia* (Caryophyllaceae), a south-east European species. - *Bot. J. Linn. Soc.* **143(4)**: 419-432.
CONTI, F. 2007: Morphometric study and taxonomy of *Genista pulchella* Vis. s.l., a south European species. - *Bot. J. Linn. Soc.* **153(3)**: 245-254.
CONTI, F. 2018: A new combination in *Leontodon* (Asteraceae, Cichorieae). - *Phytotaxa* **360(3)**: 287-291.
CONTI, F. & AL. 2004: An annotated checklist of the Italian vascular flora. In Vasić, O. (ed.): Abstracts, XI OPTIMA Meeting, Beograd 5-11.IX.2004, p. 70.
CONTI, F. & AL. 2011: Contribution to the taxonomy of the Italian and northern Balkanic taxa in the *Centaurea rupestris* group (Asteraceae). - *Ann. Bot. Fenn.* **48(3)**: 193-218.
CONTI, F. & AL. 2012: *Jacobaea vulgaris* subsp. *gotlandica* (Compositae), new for Italy and Montenegro. - *Bot. Serbica* **36(2)**: 145-147.
CONTI, F. & AL. 2015: A new species of *Reichardia* (Asteraceae, Cichorieae) from Albania and re- evaluation of *R. macrophylla*. - *Phytotaxa* **236(2)**: 121-134.
CONTI, F. & AL. 2021: A taxonomic revision of the *Siler montanum* group (Apiaceae) in Italy and the Balkan Peninsula. - *Willdenowia* **51(3)**: 321-347.
CONTI, F. & BARTOLUCCI, F. 2021: *Anthyllis apennina* (Fabaceae), a new species from central Apennine (Italy). - *PhytoKeys* **176**: 111-129.
CONTI, F., REICH, D. & GUTERMANN, W. 2020: Notes on the genus *Echinops* L. (Asteraceae) in SE Europe. - *Adansonia* **42(3)**: 94-104.
CONTI, F. & SANTANGELO, A. 2001: Lectotypification of the names *Arenaria graminifolia*, *Arenaria rosani* Ten. and *Arenaria alpicola* Ten. (Caryophyllaceae). - *Taxon* **50(1)**: 193-199.
CONTI, P. 1897: Classification et distributions des espèces européennes du genre *Matthiola*. - *Bull. Herb. Boissier* **5(1-5)**: 31-59, 315-325 + 1 map.
CONTI, P. 1900: Les espèces du genre *Matthiola*. - *Mém. Herb. Boissier* **18**: 1-86.
CONWAY, V.M. 1942: Biological Flora of the British Isles. *Cladium mariscus* (L.) R. Br. (*C. germanicum* Schrad.; *C. jamaicense* auct. non Crantz; *Mariscus serratus* Gilib.). - *J. Ecol.* **30(1)**: 211-216.
COODE, M.J.E. 1967: Materials towards a Flora of Turkey: XV. Valerianaceae: *Valerianella*. - *Notes Roy. Bot. Gard. Edinburgh* **27(3)**: 219-256.
COODE, M.J.E. & DAVIS, P.H. 1964: A neglected Mediterranean *Fumana*. - *Notes Roy. Bot. Gard. Edinburgh* **26(1)**: 27-34 + 1 plate.
COOK, C.D.K. 1962: Biological Flora of the British Isles. *Sparganium erectum* L. (*S. ramosum* Hudson, nom. illeg.). - *J. Ecol.* **50(1)**: 247-255.
COOK, C.D.K. 1966: A monographic study of *Ranunculus* subgenus *Batrachium* (DC.) A. Gray. - *Mitt. Bot. Staatssamml. München* **6(1)**: 47-237.
COOK, C.D.K. 1967: Emendations to a monographic study of *Ranunculus* subgenus *Batrachium* (DC.) A. Gray. - *Mitt. Bot. Staatssamml. München* **6(2)**: 621-622.
COOK, C.D.K. 1970: Hybridization in the evolution of *Batrachium*. - *Taxon* **19(2)**: 161-166.
COOK, C.D.K. 1983: Aquatic plants endemic to Europe and the Mediterranean. - *Bot. Jahrb. Syst.* **103(4)**: 539-582.
COOK, C.D.K. 1984: One new taxon and two new combinations in *Ranunculus* subgenus *Batrachium* (DC.) A. Gray. - *Anales Jard. Bot. Madrid* **40(2)**: 473.
COOK, C.D.K. 1985: *Sparganium*: some old names and their types. - *Bot. Jahrb. Syst.* **107(1-4)**: 269-276.

COOK, C.D.K. 1990a: Seed dispersal of *Nymphoides peltata* (S.G. Gmelin) O. Kuntze (Menyanthaceae). - *Aquatic Bot.* **37(4)**: 325-340.

COOK, C.D.K. 1990b: Aquatic plant book. - The Hague: SPB Academic Publishing, 228 pp. [2nd revised edition 1996].

COOK, C.D.K. & AL. 1974: Water plants of the world. A manual for the identification of the genera of freshwater macrophytes. - The Hague: Dr. W. Junk Publishers, viii + 561 pp.

COOK, C.D.K. & LÜÖND, R. 1982: A revision of the genus *Hydrocharis* (Hydrocharitaceae). - *Aquatic Bot.* **14(2)**: 177-204.

COOK, C.D.K. & NICHOLLS, M.S. 1986: A monographic study of the genus *Sparganium* (Sparganiaceae). Part 1. Subgenus *Xanthosparganium* Holmberg. - *Bot. Helv.* **96(2)**: 213-267.

COOK, C.D.K. & NICHOLLS, M.S. 1987: A monographic study of the genus *Sparganium* (Sparganiaceae). Part 2. Subgenus *Sparganium*. - *Bot. Helv.* **97(1)**: 1-44.

COOK, C.D.K. & URMI-KÖNIG, K. 1983: A revision of the genus *Stratiotes* (Hydrocharitaceae). - *Aquatic Bot.* **16(3)**: 213-249.

COOK, C.D.K. & URMI-KÖNIG, K. 1985: Revision of the genus *Elodea* (Hydrocharitaceae). - *Aquatic Bot.* **21(2)**: 111-156.

COPE, T.A. & NESBITT, M. 2002: 435. *Lygeum spartum*. Gramineae. - *Curtis's Bot. Mag., ser. 6* **19(1)**: 35-39 + 1 plate.

COPE, T.A. & STACE, C.A. 1978: The *Juncus bufonius* L. aggregate in western Europe. - *Watsonia* **12 (2)**: 113-128.

COPPI, A., SELVI, F. & BIGAZZI, M. 2006: Chromosome studies in Mediterranean species of Boraginaceae. - *Fl. Medit.* **16**: 253-274.

CORLI, A. & AL. 2021: Biological flora of Central Europe: *Marsilea quadrifolia* L. - *Perspect. Pl. Ecol., Evol. Syst.* **53**: doi: https://doi.org/10.1016/j.ppees.2021.125641.

CORREIA, A.I.D. & AL. (EDS) 1999: Livro de resumos. Guia de excursões. V jornadas de taxonomia botânica, 16-19/IX 1999. - Lisboa: 130 pp.

CORRENS, C. 1909: Untersuchungen über die Gattung *Cerastium*. I. Die Verwertung der Haarformen für die Unterscheidung der Arten. - *Oesterr. Bot. Z.* **59(5)**: 169-183.

CORRIAS, B. 1978: Numeri cromosomici per la Flora Italiana: 406-408. - *Inform. Bot. Ital.* **10(1)**: 90-93.

CORRIAS, B. 1981: Numeri cromosomici per la Flora Italiana: 696-700. - *Inform. Bot. Ital.* **12(2)**: 121-123.

CORRIAS, B. 1985: Numeri cromosomici per la Flora Italiana: 977-982. - *Inform. Bot. Ital.* **15(2-3)**: 175-179.

CORSI, G. & AL. 1980: Osservazioni morfologiche, isto-anatomische, cariologiche e fitochimiche su *Melissa officinalis* s.l. (Lamiaceae). - *Atti Soc. Tosc. Sci. Nat. Pisa Mem., Ser. B* **86**: 331-353.

CORSI, G., GARBARI, F. & MAFFEI, F. 1999: Il genere *Urtica* L. (Urticaceae) in Italia. Revisione biosistematica. - *Webbia* **53(2)**: 193-239.

CORSI, G., LOKAR, L. & PAGNI, A.M. 1984: Biological and phytochemical aspects of *Valeriana officinalis*. - *Biochem. Syst. Ecol.* **12(1)**: 57-62.

CORSI, G., MORELLI, I. & PAGNI, A.M. 1981: Studio cariologico anatomico e fitochimico su *Melissa officinalis* L. ssp. *officinalis* e *Melissa officinalis* L. ssp. *altissima* (Sibth. & Sm.) Arcangeli. - *Giorn. Bot. Ital., n.s.* **114(3-4)**: 98.

CORSI, G. & PAGNI, A.M. 1991: Secretory structures and systematic problems on Apiaceae. - *Bot. Chron. (Patras)* **10**: 707-711.

COŞKUNÇELEBI, K., MAKBUL, S. & GÜLTEPE, M. 2015: Presence of *Tragopogon balcanicus* Velen. in Turkey. In Bogdanović, S. & Jogan, N.: Book of abstracts - 6th Balkan Botanical Congress, p. 56. – Rijeka.

ÇOŞKUNÇELEBI, K. & AL. 2020: *Tragopogon abbreviatus* (Asteraceae): a little-known species inferred from morphological and molecular analysis. - *Turk. J. Bot.* **44(3)**: 269-280.

COSTA, R.M.S. & DI GRISTINA, E. 2020: *Xeranthemum cylindraceum* (Asteraceae) in Italy: first record for the Sicilian flora. - *Fl. Medit.* **30**: 235-243.

COSTEA, M., SANDERS, A. & WAINES, G. 2001: Preliminary results towards a revision of the *Amaranthus hybridus* complex (Amaranthaceae). - *Sida* **19**: 931-974.

COSTEA, M. & TARDIF, F.J. 2002: Taxonomy of the most common weedy European *Echinochloa* species (Poaceae: Panicoideae) with special emphasis on characters of the lemma and caryopsis. - *Sida* **20(2)**: 525-548.

COSTICH, D.E. 1989: A note on the infraspecific nomenclature of *Ecballium elaterium* (L.) A. Richard (Cucurbitaceae). - *Anales Jard. Bot. Madrid* **45(2)**: 582.

COTTON, R. & STACE, C.A. 1976: Taxonomy of the genus *Vulpia* (Gramineae). I. Chromosome numbers and geographical distribution of the Old World species. - *Genetica* **46(2)**: 235-255.

COTTON, R. & STACE, C.A. 1977: Morphological and anatomical variation of *Vulpia* (Gramineae). - *Bot. Not.* **130(2)**: 173-187.

COUCOLI, H.D. & SYMEONIDIS, L. 1980: Karyotype analysis on some Greek wild species of *Hordeum* (*marinum* group) and *Taeniatherum*. - *Sci. Ann. Fac. Phys. Math. Aristot. Univ. Thessalonika* **20A**: 77-91.

COUDERC, H. 1975: Contribution à la révision des taxons infraspécifiques de l'*Anthyllis vulneraria* L. en France. - *Rev. Gén. Bot.* **82(968-970)**: 93-118.

COUDERC, H. & AL. 1984a: Variation chromosomique et biosystématique chez plusieurs espèces d'*Ornithogalum* L. - *Webbia* **38**: 671-679.

COUDERC, H. & AL. 1984b: [Chromosome numbers in *Ornithogalum*]. In Löve, Á. (ed.): Chromosome number reports LXXXV. - *Taxon* **33(4)**: 757.

COUDERC, M. 1979: [Chromosome numbers in *Crupina*]. In Löve, Á. (ed.): IOPB chromosome number reports LXV. - *Taxon* **28(5-6)**: 632.

COULOT, P. & AL. 2003: Observations botaniques - Péloponèse - Mai 2001. - *Monde Pl.* **98(481)**: 18-27.

COULTER, TH. 1824: Mémoire sur les Dipsacacées. - *Mém. Soc. Phys. Genève* **2(2)**: 13-60 + 2 plates.

COUR, P. 1972: Cycles végétatifs et phases sporangifères, distribution géographique raisonnée de *Selaginella denticulata* (L.) Spring. - *Bull. Soc. Bot. France* **119(5-6)**: 399-404.

COUSTAUT, D. & DELAY, J. 1968: Etude caryologique du *Glaucium flavum* Crantz. - *Bull. Soc. Bot. N. France* **21(4)**: 163-167 + 2 plates.

COUSTURIER, P. & GANDOGER, M. 1916: Herborisations en Crète (1913-1914). Première partie. - *Bull. Soc. Bot. France* **63**: 1-15.

COUTANCE, A. 1877: L'Olivier. Histoire, botanique, régions, produits, usages, commerce, industrie, etc. - Paris: J. Rotschild, xvi + 436 pp.

COUTINHO, L. DE A. 1940: Raças cariológicas na *V. sativa* L. - *Agron. Lusit.* **2(4)**: 379-403 + 1 plate.

COUTINHO, L. DE A. 1945: Novos subsídios para a cariologia do género *Vicia* L. - *Bol. Soc. Brot., Ser. 2* **19(2)**: 449-455.

COUTINHO, L. DE A. 1956: Subsídios para o estudo cariológico da *Olea europaea* L. - *Genét. Ibér.* **8(1-2)**: 3-26.

COUTINHO, L. DE A. 1957: Note sobre a constitução cromosómica do *Limodorum trabutianum* Batt. e do *Limodorum abortivum* (L.) Sw. - *Agron. Lusit.* **19(3)**: 219-231 + 2 plates.

COUTINHO, L. DE A. & LORENA, M. DA C. 1942: Subsídios para o estudo cariológico do género *Trigonella* L. - *Agron. Lusit.* **4(1)**: 73-86.

COUTINHO, L. DE A. & SANTOS, A. 1943: Novas contribuïções para a cariologia do género *Trigonella* L. - *Agron. Lusit.* **5(4)**: 349-361.

COVILLOT, J. 2011: Quelques espèces orientales récoltées et décrites par E. Boissier: photographiées dans leur milieu naturel. - *Arch. Sci. (Genève)* **64**: 43-56.

COVILLOT, J. & AL. 1997: Excursion de notre société à Santorin (Cyclades). - *Saussurea* **28**: 219-227.

COWAN, R., INGROUILLE, M.J. & LLEDO, M.D. 1998: The taxonomic treatment of agamosperms in the genus *Limonium* Mill. (Plumbaginaceae). - *Folia Geobot.* **33(3)**: 353-366.

COWLEY, J., ÖZHATAY, N. & MATHEW, B.F. 1994: New species of Alliaceae & Hyacinthaceae from Turkey. - *Kew Bull.* **49(3)**: 481-489.

COWLING, W.A. 1986: Collection of wild *Lupinus* in Greece. - *Pl. Genet. Resources Newslett.* **65**: 20-22.

COX, P.A., LAUSHMAN, R.H. & RUCKELSHAUS, M.H. 1992: Surface and submarine pollination in the seagrass *Zostera marina* L. - *Bot. J. Linn. Soc.* **109(2)**: 281-291.

CRAIB, W.G. 1914: *Sempervivum ciliosum* Craib. [In: Decades Kewenses. Plantarum novarum in herbario horti regii conservatum. Decas LXXXIII]. - *Bull. Misc. Inform. Kew* **1914**: 379.

CRANE, M. 1984: Wild tulips of Crete. - *Quart. Bull. Alpine Gard. Soc. Gr. Brit.* **52(2)**: 130-135.

CRÉPIN, F. 1882: Primitiae monographiae rosarum. Matériaux pour servir à l'histoire des roses. - Gent: 856 pp. Originally published in parts 1869-82 [cf. TL-2 1: 566-567, 1976].

CRÉPIN, F. 1888: Observations sur les roses décrites dans le Supplementum Florae Orientalis de Boissier. - *Bull. Soc. Roy. Bot. Belgique* **27(2)**: 97-113.

CRÉPIN, F. 1892a: Les roses de l'île de Thasos et du Mont Athos. - *Bull. Soc. Roy. Bot. Belgique* **31 (2)**: 42-53.

CRÉPIN, F. 1892b: Tableau analytique des roses européennes. - *Bull. Soc. Roy. Bot. Belgique* **31(2)**: 66-92.

CRÉPIN, F. 1897: Les roses recueillies en Thessalie par M. Paul Sintenis en 1896. - *Bull. Herb. Boissier* **5(2)**: 112-116.

CRESPO, M.B. & AL. 2020: Miscellaneous notes on nomenclature and taxonomy of some Old World names in tribe Paniceae (Panicoideae, Poaceae). - *Phytotaxa* **454(4)**: 244-254.

CRESPO, M.B., MATEO, G. & NAVARRO, T. 1994: Una nueva especie del género *Teucrium* L., sección *Polium* (Miller) Schreb. (Lamiaceae), para la flora de la Península Ibérica. - *Acta Bot. Malac.* **19**: 205-208.

CRESPO, M.B. & PENA-MARTÍN, C. 2013: Two new species of *Limonium* (Plumbaginaceae) from Rhodes Island (eastern Aegean area, Greece). - *Phytotaxa* **94(2)**: 30-40.

CRESPO, M.B., SERRA, L. & JUAN, A. 1998: *Solenopsis* (Lobeliaceae): a genus endemic in the Mediterranean region. - *Pl. Syst. Evol.* **210(3-4)**: 211-229.

CRESPO, M.B., SERRA, L. & TURLAND, N.J. 1996: Lectotypification of four names in *Lobelia* (Lobeliaceae). - *Taxon* **45(1)**: 117-120.

CRESTI, L. & AL. 2019: Pleistocene survival in three Mediterranean refugia: origin and diversification of the Italian endemic *Euphorbia gasparrinii* from the *E. verrucosa* alliance (Euphorbiaceae). - *Bot. J. Linn. Soc.* **189(3)**: 262-280.

CRETZOIU, P. 1938: Die europäischen *Syringa*-Arten. - *Pflanzenareale* **4(6)**: 73, Karte 54.

CREUTZBURG, N. 1963: Die paläogeographische Entwicklung der Insel Kreta von Miozän bis zur Gegenwart [In Greek]. - *Kritika Chronika* **15-16**: 336-344.

CREUTZBURG, N. 1966: Die südägäische Inselbrücke. Bau und geologische Vergangenheit. - *Erdkunde* **20(1)**: 20-30 + 1 map.

CRIBB, PH. & CHASE, M.A. 2001: (1481) Proposal to conserve the name *Dactylorhiza* Necker ex Nevski over *Coeloglossum* Hartm. (Orchidaceae). - *Taxon* **50 (2)**: 581-582. [approved, cf. Taxon 53(3): 814, 2004].

CRIBB, PH. & WOOD, J.J. 1981: *Ophrys holoserica*, der korrekte Name für die Hummel-Ragwurz. - *Orchidee (Hamburg)* **32(1)**: 27-28.

CRINS, W.J. & BALL, P.W. 1989: Taxonomy of the *Carex flava* complex (Cyperaceae) in North America and northern Eurasia. II. Taxonomic treatment. - *Canad. J. Bot.* **67(4)**: 1048-1065.

CRISAFULLI, A., PICONE, R.M. & SOLDANO, A. 2013: *Oenothera indecora* (Onagraceae) a neophyte new to Italy. - *Fl. Medit.* **23**: 87-92.

CRISTOFOLINI, G. 1974: Contributo preliminare alla sistematica di *Chamaecytisus hirsutus* e *Chamaecytisus supinus*. - *Giorn. Bot. Ital., n.s.* **108(1-2)**: 55-73.

CRISTOFOLINI, G. 1976: I citisi italiani della sezione "Tubocytisus" DC. - *Webbia* **30(2)**: 257-283.

CRISTOFOLINI, G. 1991: Taxonomic revision of *Cytisus* Desf. sect. *Tubocytisus* DC. (Fabaceae). - *Webbia* **45(2)**: 187-219.

CRISTOFOLINI, G. & FEOLI CHIAPELLA, L. 1977: Serological systematics of the tribe Genisteae (Fabaceae). - *Taxon* **26(1)**: 43-56.

CRISTOFOLINI, G. & JARVIS, C.E. 1991: On the status of *Cytisus hirsutus* and *C. supinus* (Fabaceae). - *Taxon* **40(3)**: 495-498.

CRISTOFOLINI, G. & PIGNATTI, S. 1961: Le Soldanelle italiane (Nota preliminare). - *Nuovo Giorn. Bot. Ital., n.s.* **68(1)**: 157-161.

CRISTOFOLINI, G. & PIGNATTI, S. 1962: Revisione delle forme italiane del genere *Soldanella* L. - *Webbia* **16(2)**: 443-475.

CRISTOFOLINI, G. & TROÌA, A. 2006: A reassessment of the sections of the genus *Cytisus* Desf. (Cytiseae, Leguminosae). - *Taxon* **55(3)**: 733-746.

CRITOPOULOS, P. 1975: The endemic taxa of Crete. In Jordanov, D. & al. (eds): Problems of Balkan flora and vegetation, pp. 169-177. - Sofia: Publishing House of the Bulgarian Academy of Sciences.

CRITOPOULOS, P. 1981: Chorological information in Greece. In Velčev, V.I. & Kožuharov, S.I. (eds): Mapping the flora of the Balkan Peninsula, pp. 57-62. - Sofia: Publishing House of the Bulgarian Academy of Sciences.

CRIVELLI, A.J. & CATSADORAKIS, G. (EDS) 1997: Lake Prespa, northwestern Greece. A unique Balkan wetland. - Dordrecht: Kluwer Academic Publ., xiii + 196 pp. [repr. from *Hydrobiol.* 351].

CRONQUIST, A. 1943: The separation of *Erigeron* from *Conyza*. - *Bull. Torrey Bot. Club* **70(6)**: 629-632.

CROOK, H.C. 1977: The genus *Symphyandra*. - *Quart. Bull. Alpine Gard. Soc. Gr. Brit.* **45(3)**: 246-254.

CROOK, M. & CROOK, H.C. 1981: Plant portraits, 2. *Euphorbia acanthothammnos* (Held. et Sart.). - *Quart. Bull. Alpine Gard. Soc. Gr. Brit.* **49(2)**: 168-171.

CROW, G.E. 1978: A taxonomic revision of *Sagina* (Caryophyllaceae) in North America. - *Rhodora* **80 (821)**: 1-91.

CROWDER, A.A. & AL. 1990: Biological Flora of the British Isles. *Drosera* L. - *J. Ecol.* **78(1)**: 233-267.

CROWL, A.A. & AL. 2015: Evolution and biogeography of the endemic *Roucela* complex (Campanulaceae: *Campanula*) in the Eastern Mediterranean. - *Ecology and Evolution* **5(22)**: 5329-5343.

CSIKI, E., JÁVORKA, A. & KÜMMERLE, J.B. 1926: Adatok Albánia Flórájához. Additamenta ad Floram Albaniae [In Hungarian, German and Latin]. - Budapest: Stephaneum Nyomda és Könyvkiadó R.T., 349 pp. + 21 plates.

CSIKY, J. & PURGER, D. 2013: Herbaceous periwinkle, *Vinca herbacea* Waldst. et Kit. 1799 (Apocynaceae), a new species of the Croatian flora. - *Acta Bot. Croat.* **73(2)**: 399-406.

CSONTOS, P. & AL. 2010: Early distribution and spread of *Ambrosia artemisiifolia* in Central and Eastern Europe. - *Bot. Helv.* **120(1)**: 75-78.

CUBAS, P. 1990: Estudios biosystemáticos en Pteridofitos europeos. In Rita, J. (ed.): Taxonomía, biogeografía y conservación de Pteridofitos, pp. 91-114. - Palma de Mallorca.

CUBAS, P. & AL. 1998: Números cromosómicos para la flora española. 780-792. - *Lagascalia* **20(2)**: 295-302.

CUBAS, P. & AL. 2010: Phylogeny and evolutionary diversification of *Adenoocarpus* DC. (Leguminosae). - *Taxon* **59(3)**: 720-732.

CUBAS, P., ROSSELLÓ, J.A. & PANGUA, E. 1989: A new triploid hybrid in the *Asplenium trichomanes* complex: *A. trichomanes* nothosubsp. *lucanum* (*A. trichomanes* subsp. *inexpectans* × *A. trichomanes* subsp. *quadrivalens*) nothosubsp. nova. - *Candollea* **44(1)**: 181-190.

CUCCUINI, P. 2002: Il genere *Parapholis* C.E. Hubbard (Poaceae) in Italia. Note tassonomiche e palinologiche. - *Webbia* **57(1)**: 7-64.

CUCCUINI, P. & FIORINI, G. 2004: Il genere *Hainardia* Greuter (Poaceae) in Italia. Note tassonomiche, citologiche ed ecologiche. - *Webbia* **59(1)**: 149-175.

CUCCUINI, P. & FIORINI, G. 2020: First contribution to the taxonomic and cytotaxonomic study of *Parapholis* and *Hainardia* (Poaceae) in Albania. - *Fl. Medit.* **30**: 5-18.

CUFODONTIS, G. 1933: Kritische Revision von *Senecio* sectio *Tephroseris*. - *Repert. Spec. Nov. Regni Veg. Beih.* **70**: 1-226 + 3 plates and 2 maps.

CUFODONTIS, G. 1936a: Die von Dr. Th. Just im Jahre 1929 auf den Ionischen Inseln und im nordwestlichen Peloponnes gesammelten Farne und Blütenpflanzen. - *Repert. Spec. Nov. Regni Veg.* **39**: 365-369.

CUFODONTIS, G. 1936b: Revisione monografica delle *Linaria* appartenenti alla sez. *Cymbalaria* Chav. - *Arch. Bot. (Forli)* **12**: 54-81, 135-158, 233-254 + 1 plate.

CUFODONTIS, G. 1947: Die Gattung *Cymbalaria* Hill. Nachträge und Zusammenfassung. - *Bot. Not.* **1947(2)**: 135-156.

CUGNAC, A. DE 1934: Sur quelques *Bromus* et leurs hybrides. III. Données biométriques comparatives sur quelques caractères distinctifs de *Bromus rigidus* Roth, *B. Gussonii* Parlat. et *B. sterilis* L. - *Bull. Soc. Bot. France* **81**: 318-323.

CUGNAC, A. DE 1937a: Sur quelques *Bromus* et leurs hybrides. VI. *B. arduennensis* × *B. arvensis*, hybride expérimental nouveau. Note préliminare. - *Bull. Soc. Bot. France* **84**: 437-440.

CUGNAC, A. DE 1937b: Sur quelques *Bromus* et leur hybrides. VII. *B. sterilis* L. var. *velutinus* Volkart obtenu par synthèse expérimentale à partir du croisement de *B. madritensis* L. par *B. sterilis* L. - *Bull. Soc. Bot. France* **84**: 711-713.

CUGNAC, A. DE 1939: Recherches phylétiques sur le genre *Bromus*. VIII. Une double reconstitution expérimentale, faisant revivre une varieté éteinte, et ses conséquences systématiques. - *Bull. Soc. Bot. France* **86**: 409-419.

CUGNAC, A. DE & CAMUS, A. 1931: Sur quelques *Bromus* et leurs hybrides. I. Révision du *Bromus maximus* Desf. d'après l'étude des peuplements naturels. - *Bull. Soc. Bot. France* **78**: 327-341 + 3 plates.

CUGNAC, A. DE & CAMUS, A. 1936: Sur quelques *Bromus* et leurs hybrides. IV. Deux espèces menacées de disparition: *Bromus (Serrafalcus) grossus* Desf. ex DC.; *Bromus (Michelaria) arduennensis* Dumort. - *Bull. Soc. Bot. France* **83**: 47-68 + 4 plates.

CUGNAC, A. DE & SIMONET, M. 1941: Recherches phylétiques sur le genre *Bromus*. X. Quelques nombres de chromosomes et leur signification phylétique et phylogénétique. - *Bull. Soc. Bot. France* **88**: 513-517.

CULLEN, J. 1963a: The Turkish collections of Paul Sintenis. - *Notes Roy. Bot. Gard. Edinburgh* **25(1)**: 31-39.

CULLEN, J. 1963b: Materials for a Flora of Turkey: VII. - *Notes Roy. Bot. Gard. Edinburgh* **25(1)**: 41-48 + 1 plate.

CULLEN, J. 1976: The *Anthyllis vulneraria* complex: a résumé. - *Notes Roy. Bot. Gard. Edinburgh* **35 (1)**: 1-38.

CULLEN, J. 2010: 662. *Phelypaea boissieri*. Orobanchaceae. - *Curtis's Bot. Mag.* **26(4)**: 379-388.

CULLEN, J. & COODE, M.J.E. 1964: Materials for a Flora of Turkey: IX Berberidaceae. - *Notes Roy. Bot. Gard. Edinburgh* **26(1)**: 35-42.

CULLEN, J. & HEYWOOD, V.H. 1964: Notes on the European species of *Paeonia*. In Heywood, V.H. (ed.): Notulae systematicae ad Floram Europaeam spectantes, no. 3. - *Feddes Repert. Spec. Nov. Regni Veg.* **69(1)**: 32-35.

CULLEN, J. & RATTER, J.A. 1967: Taxonomic and cytological notes on Turkish *Ornithogalum*. - *Notes Roy. Bot. Gard. Edinburgh* **27(3)**: 293-339 + 7 plates.

CULLEN, J. (ED.) 1975: Type specimens of Papaveraceae in European herbaria. - Edinburgh: 66 pp.

CURIEL, D. & AL. 1996: Sexual reproduction of *Zostera noltii* Hornemann in the lagoon of Venice (Italy, north Adriatic). - *Aquatic Bot.* **52(4)**: 313-318.

CURTIS, W. (ED.) 1787a: *Erythronium Dens Canis*. - *Bot. Mag.* **1**: tab. 5 + 1 p. text.

CURTIS, W. (ED.) 1787b: *Iris pumila*. - *Bot. Mag.* **1**: tab. 9 + 1 p. text.

CURTIS, W. (ED.) 1787c: *Nigella damascena*. - *Bot. Mag.* **1**: tab. 22 + 1 p. text.

CURTIS, W. (ED.) 1787d: Lychnis coronaria. - *Bot. Mag.* **1**: tab. 24 + 1 p. text.

CURTIS, W. (ED.) 1787e: *Reseda odorata*. - *Bot. Mag.* **1**: tab. 29 + 1 p. text.

CURTIS, W. (ED.) 1787f: *Lilium chalcedonicum*. - *Bot. Mag.* **1**: tab. 30 + 1 p. text.

CURTIS, W. (ED.) 1788a: *Tolpis barbata*. - *Bot. Mag.* **1**: tab. 35 + 1 p. text.

CURTIS, W. (ED.) 1788b: *Viburnum tinus*. - *Bot. Mag.* **2**: tab. 38 + 1 p. text.

CURTIS, W. (ED.) 1788c: *Cistus incanus*. - *Bot. Mag.* **2**: tab. 43 + 1 p. text.

CURTIS, W. (ED.) 1788d: *Cyclamen persicum*. - *Bot. Mag.* **2**: tab. 44 + 1 p. text.

CURTIS, W. (ED.) 1788e: *Geranium striatum*. - *Bot. Mag.* **2**: tab. 55 + 1 p. text.

CURTIS, W. (ED.) 1789a: *Limonium sinuatum*. - *Bot. Mag.* **2**: tab. 71 + 1 p. text.

CURTIS, W. (ED.) 1789b: *Epilobium angustissimum*. - *Bot. Mag.* **3**: tab. 76 + 1 p. text.

CURTIS, W. (ED.) 1789c: *Colutea arborescens*. - *Bot. Mag.* **3**: tab. 81 + 1 p. text.

CURTIS, W. (ED.) 1789d: *Spartium junceum*. - *Bot. Mag.* **3**: tab. 85 + 1 p. text.

CURTIS, W. (ED.) 1789e: *Campanula speculum*. - *Bot. Mag.* **3**: tab. 102 + 1 p. text.

CURTIS, W. (ED.) 1790a: *Anthyllis tetraphylla*. - *Bot. Mag.* **3**: tab. 108 + 1 p. text.

CURTIS, W. (ED.) 1790b: *Lavatera trimestris*. - *Bot. Mag.* **4**: tab. 109 + 1 p. text.

CURTIS, W. (ED.) 1790c: *Silene pendula*. - *Bot. Mag.* **4**: tab. 114 + 1 p. text.

CURTIS, W. (ED.) 1790d: *Lathyrus sativus*. - *Bot. Mag.* **4**: tab. 115 + 1 p. text.

CURTIS, W. (ED.) 1790e: *Muscari racemosum*. - *Bot. Mag.* **4**: tab. 122 + 1 p. text.

CURTIS, W. (ED.) 1790f: *Anemone hortensis*. - *Bot. Mag.* **4**: tab. 123 + 1 p. text.

CURTIS, W. (ED.) 1790g: *Aubrieta deltoideum*. - *Bot. Mag.* **4**: tab. 126 + 1 p. text

CURTIS, W. (ED.) 1790h: *Alyssum utriculatum*. - *Bot. Mag.* **4**: tab. 130 + 1 p. text.

CURTIS, W. (ED.) 1790i: *Hyacinthus comosus*. - *Bot. Mag.* **4**: tab. 133 + 1 p. text.

CURTIS, W. (ED.) 1790j: *Lupinus luteus*. - *Bot. Mag.* **4**: tab. 140 + 1 p. text.

CURTIS, W. (ED.) 1791a: *Lotus tetragonolobus*. - *Bot. Mag.* **4**: tab. 152 + 1 p. text.

CURTIS, W. (ED.) 1791b: *Hyacinthus botryoiides*. - *Bot. Mag.* **5**: tab. 157 + 1 p. text.

CURTIS, W. (ED.) 1791c: *Alyssum saxatile*. - *Bot. Mag.* **5**: tab. 159 + 1 p. text.

CURTIS, W. (ED.) 1791d: *Cheiranthus maritimus*. - *Bot. Mag.* **5**: tab. 166 + 1 p. text.

CURTIS, W. (ED.) 1791e: *Cytisus laburnum*. - *Bot. Mag.* **5**: tab. 176 + 1 p. text.

CURTIS, W. (ED.) 1792a: *Hypericum coris*. - *Bot. Mag.* **5**: tab. 178 + 1 p. text.

CURTIS, W. (ED.) 1792b: *Syringa vulgaris*. - *Bot. Mag.* **6**: tab. 183 + 1 p. text.

CURTIS, W. (ED.) 1792c: *Coronilla valentina*. - *Bot. Mag.* **6**: tab. 185 + 1 p. text.

CURTIS, W. (ED.) 1792d: *Aster alpinus*. - *Bot. Mag.* **6**: tab. 199 + 1 p. text.

CURTIS, W. (ED.) 1792e: *Antirrhinum alpinum*. - *Bot. Mag.* **6**: tab. 205 + 1 p. text.

CURTIS, W. (ED.) 1792f: *Melissa grandiflora*. - *Bot. Mag.* **6**: tab. 208 + 1 p. text.

CURTIS, W. (ED.) 1792g: *Hibiscus trionum*. - *Bot. Mag.* **6**: tab. 209 + 1 p. text.

CURTIS, W. (ED.) 1793a: *Arabis alpina*. - *Bot. Mag.* **7**: tab. 226 + 1 p. text.

CURTIS, W. (ED.) 1793b: *Fumaria solida*. - *Bot. Mag.* **7**: tab. 231 + 1 p. text.

CURTIS, W. (ED.) 1793c: *Fumaria cava*. - *Bot. Mag.* **7**: tab. 232 + 1 p. text.

CURTIS, W. (ED.) 1793d: *Linum arboreum*. - *Bot. Mag.* **7**: tab. 234 + 1 p. text.

CURTIS, W. (ED.) 1793e: *Oxalis caprina*. - *Bot. Mag.* **7**: tab. 237 + 1 p. text.

CURTIS, W. (ED.) 1793f: *Fagonia cretica*. - *Bot. Mag.* **7**: tab. 241 + 1 p. text.

CURTIS, W. (ED.) 1793g: *Scabiosa atropurpurea*. - *Bot. Mag.* **7**: tab. 247 + 1 p. text.

CURTIS, W. (ED.) 1794a: *Lathyrus articulatus*. - *Bot. Mag.* **8**: tab. 253 + 1 p. text.

CURTIS, W. (ED.) 1794b: *Coronilla varia*. - *Bot. Mag.* **8**: tab. 258 + 1 p. text.

CURTIS, W. (ED.) 1794c: *Ixia bulbocodium*. - *Bot. Mag.* **8**: tab. 265 + 1 p. text.

CURTIS, W. (ED.) 1794d: *Ornithogalum nutans*. - *Bot. Mag.* **8**: tab. 259 + 1 p. text.

CURTIS, W. (ED.) 1794e: *Blitum virgatum*. - *Bot. Mag.* **8**: tab. 276 + 1 p. text.

CURTIS, W. (ED.) 1794f: *Lilium candidum*. - *Bot. Mag.* **8**: tab. 258 + 1 p. text.

CURTIS, W. (ED.) 1795a: *Amaryllis lutea*. - *Bot. Mag.* **9**: tab. 290 + 1 p. text.

CURTIS, W. (ED.) 1795b: *Capparis spinosa*. - *Bot. Mag.* **9**: tab. 291 + 1 p. text.

CURTIS, W. (ED.) 1795c: *Dianthus superbus*. - *Bot. Mag.* **9**: tab. 297 + 1 p. text.

CURTIS, W. (ED.) 1795d: *Origanum dictamnus*. - *Bot. Mag.* **9**: tab. 298 + 1 p. text.

CURTIS, W. (ED.) 1795e: *Linum flavum*. - *Bot. Mag.* **9**: tab. 312 + 1 p. text.

CURTIS, W. (ED.) 1795f: *Daphne cneorum*. - *Bot. Mag.* **9**: tab. 313 + 1 p. text.

CURTIS, W. (ED.) 1796a: *Antirrhinum triphyllum*. - *Bot. Mag.* **9**: tab. 324 + 1 p. text.

CURTIS, W. (ED.) 1796b: *Trifolium incarnatum*. - *Bot. Mag.* **10**: tab. 328 + 1 p. text.

CURTIS, W. (ED.) 1796c: *Ononis natrix*. - *Bot. Mag.* **10**: tab. 329 + 1 p. text.

CURTIS, W. (ED.) 1796d: *Cerinthe major*. - *Bot. Mag.* **10**: tab. 333 + 1 p. text.
CURTIS, W. (ED.) 1796e: *Dorycnium hirsutum*. - *Bot. Mag.* **10**: tab. 336 + 1 p. text.
CURTIS, W. (ED.) 1796f: *Briza maxima*. - *Bot. Mag.* **10**: tab. 357 + 1 p. text.
CURTIS, W. (ED.) 1797a: *Convolvulus althaeoides*. - *Bot. Mag.* **10**: tab. 359 + 1 p. text.
CURTIS, W. (ED.) 1797b: *Astragalus monspessulanus*. - *Bot. Mag.* **11**: tab. 375 + 1 p. text.
CURTIS, W. (ED.) 1797c: *Erodium romanum*. - *Bot. Mag.* **11**: tab. 377 + 1 p. text.
CURTIS, W. (ED.) 1798a: *Alyssum montanum*. - *Bot. Mag.* **12**: tab. 419 + 1 p. text.
CURTIS, W. (ED.) 1798b: *Saxifraga rotundifolia*. - *Bot. Mag.* **12**: tab. 424 + 1 p. text.
CURTIS, W. (ED.) 1799a: *Coronilla emerus*. - *Bot. Mag.* **13**: tab. 445 + 1 p. text.
CURTIS, W. (ED.) 1799b: *Convolvulus cneorum*. - *Bot. Mag.* **13**: tab. 459 + 1 p. text.
CURTIS, W. (ED.) 1799c: *Jasminum fruticans*. - *Bot. Mag.* **13**: tab. 461 + 1 p. text.
CUSMA VELARI, T. & AL. 1999: Reports (1071-1072). In Kamari, G., Felber, F. & Garbari, F. (eds): Mediterranean chromosome number reports - 9. - *Fl. Medit.* **9**: 361-364.
CUSMA VELARI, T. & AL. 2003a: Reports (1312-1314). In Kamari, G., Blanché, C. & Garbari, F. (eds): Mediterranean chromosome number reports - 12. - *Fl. Medit.* **12**: 475-480.
CUSMA VELARI, T. & AL. 2003b: Reports (1369-1371). In: Kamari, G., Blanché, C. & Garbari, F. (eds): Mediterranean chromosome number reports - 13. - *Fl. Medit.* **13**: 397-400.
CUSMA VELARI, T. & AL. 2012: Reports (1761-1762). In Kamari, G., Blanché, C. & Siljak-Yakovlev, S. (eds): Mediterranean chromosome number reports - 22. - *Fl. Medit.* **22**: 221-226.
CUSMA VELARI, T. & FEOLI CHIAPELLA, L. 1984: Numeri cromosomici per la Flora Italiana: 906-909. - *Inform. Bot. Ital.* **14(2-3)**: 259-263.
CUSMA VELARI, T., FEOLI CHIAPELLA, L. & BACCHETTA, G. 2000: Reports (1189-1190). In Kamari, G., Felber, F. & Garbari, F. (eds): Mediterranean chromosome number reports - 10. - *Fl. Medit.* **10**: 402-405.
CUSMA VELARI, T., FEOLI CHIAPELLA, L. & KOSOVEL, V. 2000: Reports (1191-1192). In Kamari, G., Felber, F. & Garbari, F. (eds): Mediterranean chromosome number reports - 10. - *Fl. Medit.* **10**: 405-411.
CUSMA VELARI, T., FEOLI CHIAPELLA, L. & KOSOVEL, V. 2006: Reports (1613-1614). In Kamari, G., Blanché, C. & Garbari, F. (eds): Mediterranean chromosome number reports - 16. - *Fl. Medit.* **16**: 449-452.
CUSMA VELARI, T., FEOLI CHIAPELLA, L. & KOSOVEL, V. 2007: Reports (1620-1622). In Kamari, G., Blanché, C. & Garbari, F. (eds): Mediterranean chromosome number reports - 17. - *Fl. Medit.* **17**: 296-298.
CUSMA VELARI, T., FEOLI CHIAPELLA, L. & MANGIAVACCHI, L. 1997: Reports (840-842). In Kamari, G., Felber, F. & Garbari, F. (eds): Mediterranean chromosome number reports - 7. - *Fl. Medit.* **7**: 236-240.
CUSMA VELARI, T. & LAUSI, D. 1980: Indagini cariologiche in *Stachys recta* L., *S. labiosa* Bertol., *S. subcrenata* Vis., *S. fragilis* Vis. - *Stud. Geobot.* **1(1)**: 149-155.
CYRÉN, O. & HAYEK, A.V. 1928: Der thessalische Olymp. In Karsten, G. & Schenk, H. (eds): Vegetationsbilder, Reihe 18, Heft 6-7. - Jena: Gustav Fischer Verlag, Tafel 31-42 + Text.
CZAPIK, R. 1954: Cytological and embryological studies in *Centaurea scabiosa* L. [In Polish with English summary]. - *Acta Soc. Bot. Poloniae* **23(1)**: 175-194.
CZAPIK, R. 1965: Karyotype analysis of *Ornithogalum umbellatum* L. and *O. Gussonei* Ten. - *Acta Biol. Cracov., Ser. Bot.* **8(1)**: 21-34 + 2 plates.
CZAPIK, R. & NOVOTNÁ, I. 1967: Cyto-taxonomical and genetical problems of the *Arabis hirsuta* (L.) Scop. complex. I. - *Acta Biol. Cracov., Ser. Bot.* **10(2)**: 167-183 + 1 plate.
CZAPIK, R. & NOVOTNÁ, I. 1969: Cyto-taxonomical and genetical problems of the *Arabis hirsuta* (L.) Scop. complex. II. - *Acta Biol. Cracov., Ser. Bot.* **12**: 35-56 + 1 plate.
CZAPIK, R. & NOVOTNÁ, I. 1972: Chromosome numbers of some representatives of the *Arabis hirsuta* complex from Czechoslovakia. - *Preslia* **44(1)**: 1-6 + 2 plates.
CZAPSKA, D. 1959: Cytological and morphological studies in *Plantago media* L. and other species of the genus *Plantago* [In Polish with English summary]. - *Acta Soc. Bot. Poloniae* **28(1)**: 129-143.
CZECZOTT, H. 1932: Distribution of *Fagus orientalis* Lipsky. In Rübel, E. (ed.): Die Buchenwälder Europas. - *Veröff. Geobot. Inst. Rübel Zürich* **8**: 362-387.
CZECZOTT, H. 1938: A contribution to the knowledge of the flora and vegetation of Turkey. - *Repert. Spec. Nov. Regni Veg. Beih.* **107**: 1-281 + plates and maps.
CZECZOTT, H. 1954: The past and present distribution of *Pinus halepensis* Mill. and *Pinus brutia* Ten. In: Huitième congrès international de botanique. Rapports et communications ... sect. 2, 4, 5 et 6, pp. 196-197. - Paris: Pierre André Imp.
CZEFRANOVA, Z.V. 1971a: Conspectus specierum generis *Lens* Mill. [In Russian; English translation of this volume by Amerind Publ. Co., New Delhi, 1978]. - *Novosti Sist. Vysš. Rast.* **8**: 184-191.
CZEFRANOVA, Z.V. 1971b: Conspectus systematicis generis *Lathyrus* L. [In Russian; English translation of this volume by Amerind Publ. Co., New Delhi, 1978]. - *Novosti Sist. Vysš. Rast.* **8**: 191-201.
D'AGATA, C.D.C. & RHIZOPOULOU, S. 2021: Cretan and Greek plants in Italian Renaissance gardens cited in archives. - *Pl. Biosyst.* **155**: doi-org.ludwig.lub.lu.se/10.1080/11263504.2021.2013333.
D'AMATO, F. 1946: Nuovo ricerche embriologiche e cariologiche sul genere *Euphorbia*. - *Nuovo Giorn. Bot. Ital., n.s.* **53(3-4)**: 405-437.
D'AMATO, F. 1955: Revisione citosistematica del genere *Colchicum*. I: *C. autumnale* L., *C. lusitanicum* Brot. e *C. neapolitanum* Ten. - *Caryologia* **7(2)**: 292-349 + 5 plates.
D'AMATO, F. 1957a: Revisione citosistematica del genere *Colchicum*. II. Nuove localitá di *C. autumnale* L., *C. lusitanicum* Brot. e *C. neapolitanum* Ten. e delimitazione dell'areale ... nella peninsola italiana. - *Caryologia* **9(2)**: 315-339.
D'AMATO, F. 1957b: Revisione citosistematica del genere *Colchicum*. III: *C. alpinum* Lam. et DC., *C. Cupanii* Guss., *C. Bivonae* Guss. e chiave analitica per la determ. della specie di *Colchicum* della flora italiana. - *Caryologia* **10(1)**: 111-151 + 5 plates.
D'AMATO, G. 1971: Numeri cromosomici per la Flora Italiana: 67-80. - *Inform. Bot. Ital.* **3(2)**: 138-147.
D'AMATO, G. 1998: Numeri cromosomici per la Flora Italiana: 1396-1400. - *Inform. Bot. Ital.* **29(2-3)**: 280-282.
D'AMATO, G. & PAVESI, A. 1991: Numeri cromosomici per la Flora Italiana: 1247-1250. - *Inform. Bot. Ital.* **22(3)**: 244-246.
D'ARCY, W.G. (ED.) & AL. (EDS) 1986: Solanaceae. Biology and systematics. - New York: Columbia University Press, xii + 603 pp.
D'EMERICO, S. & AL. 1990: Karyological studies of some taxa of the genera *Himantoglossum*, *Orchis*, *Serapias* and *Spiranthes* (Orchidaceae) from Apulia (Italy). - *Caryologia* **43(3-4)**: 267-276.
D'EMERICO, S. & AL. 1995: Preliminary investigation on the biosystematics of *Allium atroviolaceum* Boiss. (Alliaceae). - *Giorn. Bot. Ital., n.s.* **129(2)**: 41.
D'EMERICO, S., BIANCO, P. & MEDAGLI, P. 1993: Chromosome numbers and karyotypes in *Arum* (Araceae). - *Caryologia* **46(2-3)**: 161-170.
D'EMERICO, S., BIANCO, P. & MEDAGLI, P. 1994: Considerazioni citotassonomiche su alcuni generi di Orchidaceae. - *Inform. Bot. Ital.* **25(1)**: 86-89.
D'EMERICO, S., GALASSO, I. & PIGNONE, D. 1998: Karyomorphological considerations on some taxa of the genus *Serapias* L. (Orchidaceae). - *Inform. Bot. Ital.* **29(2-3)**: 311.
D'EMERICO, S., PIGNONE, D. & BIANCO, P. 1995: Indagini citomorfologiche in *Limodorum abortivum* (L.) Swartz (Orchidaceae). - *Inform. Bot. Ital.* **27(2-3)**: 310-311.
D'EMERICO, S., PIGNONE, D. & SCRUGLI, A. 2000: Giemsa C-banded karyotypes in *Serapias* L. (Orchidaceae). - *Bot. J. Linn. Soc.* **133(4)**: 485-492.
D'ERRICO, P. 1956: Studio sistematico delle entità italiane di "Acer Opalus" Mill. - *Webbia* **12(1)**: 41-120 + 41 plates.
D'OVIDIO, R. 1986: Numeri cromosomici per la Flora Italiana: 1011-1021. - *Inform. Bot. Ital.* **16(2-3)**: 251-260.
D'OVIDIO, R. 1987: Numeri cromosomici per la Flora Italiana: 1082-1093. - *Inform. Bot. Ital.* **18(1-3)**: 168-175.

D'OVIDIO, R. & LUCCHESE, F. 1986: Chromosomal races in the *Brachypodium rupestre* complex in Italy. - *Ann. Bot. (Rome)* **44**: 175-180.

D'OVIDIO, R., MARCHI, P. & VISONÀ, L. 1987: Numeri cromosomici per la Flora Italiana: 1053-1063. - *Inform. Bot. Ital.* **18(1-3)**: 145-152.

DABROWSKA, J. 1989: The chromosome numbers of several taxa of the genus *Achillea* L. - *Acta Soc. Bot. Poloniae* **58(2)**: 163-177.

DABROWSKA, J. 1992: Chromosome number and DNA content in taxa of *Achillea* L. inrelation to the distribution of the genus. - *Acta Univ. Wratislav., Prace Bot.* **49**: 1-84.

DABROWSKA, J. 2000: *Achillea crithmifolia* W. et K. in di- and tetraploid forms (2n = 18, 36). In Özhatay, N. & al. (eds): Abstracts from the Second Balkan Botanical Congress, Istanbul, Turkey, May 14-18, 2000, p. 101.

DACO, L. & AL. 2019: Genetic status of the endangered plant species *Gladiolus palustris* in the western part of its distribution area. - *Conserv. Genet.* **20(6)**: 1339-1354.

DADANDI, M.Y. 2004: A new natural hybrid of *Phlomis* L. (Lamiaceae) from South Anatolia. - *Karaca Arbor. Mag.* **7(2)**: 59-66.

DADD, R. 1994: Flowering in *Allium callimischon*. - *New Plantsman* **1(2)**: 106.

DAFIS, S. 1975: Vegetationsgliederung Griechenlands. In Dafis, S. & Landolt, E. (eds): Zur Vegetation und Flora von Griechenland, Band 1. - *Veröff. Geobot. Inst. E.T.H. Stiftung Rübel Zürich* **55**: 23-36.

DAFIS, S. 1985: The palm tree forest of Vaï (Sitia-Crete). - *Sci. Ann. Dept. Forest. Nat. Environm. Aristot. Univ. Thessalonika* **KH(4)**: 141-152.

DAFIS, S. 1989: Vegetation and ecological particularities of Mount Olympus [In Greek]. - *Sci. Ann. Dept. Forest. Nat. Environm. Aristot. Univ. Thessalonika* **LB(1)**: 558-571.

DAFIS, S. 2005: Greek oak woodlands: A species account on their ecological and socio-economic role. - *Bot. Chron. (Patras)* **18(1)**: 75-82.

DAFIS, S. 2010: Ta dasi tis Elladas [In Greek]. - Thessaloniki: Goulandris Natural History Museum, 192 pp.

DAFIS, S. & AL. (EDS) 1996: Directive 92/43/EEC The Greek habitat project NATURA 2000: An overview. - Thessaloniki: xxiv + 893 pp.

DAFIS, S. & JAHN, G. 1975: Zum heutigen Waldbild Griechenlands nach ökologisch-pflanzengeographischen Gesichtspunkten. In Dafis, S. & Landolt, E. (eds): Zur Vegetation und Flora von Griechenland, Band 1. - *Veröff. Geobot. Inst. E.T.H. Stiftung Rübel Zürich* **55**: 99-116.

DAFIS, S. & LANDOLT, E. (EDS) 1975: Zur Vegetation und Flora von Griechenland. Band 1: Einführung; allgemeine Vegetationskunde; Kreta und Ägäische Inseln. - *Veröff. Geobot. Inst. E.T.H. Stiftung Rübel Zürich* **55**: 1-237.

DAFIS, S. & LANDOLT, E. (EDS) 1976: Zur Vegetation und Flora von Griechenland. Band 2: Mazedonien, Thessalien-Epirus, Attika, Peloponnes; einzelne Pflanzengesellschaften; Systematik einzelner Pflanzengruppen; botanische Literatur. - *Veröff. Geobot. Inst. E.T.H. Stiftung Rübel Zürich* **56**: 1-242 + 7 tables.

DAFNI, A. 1979: Orchids of Israel: Notes on distribution, local variation and ecology. - *Mitt. Arbeitskreis Heimische Orchid. Baden-Württemberg* **11(3)**: 206-222.

DAFNI, A. 1987: Pollination in *Orchis* and related genera: evolution from reward to deception. In Arditti, J. (ed.): Orchid biology: reviews and perspectives, IV: 80-104. - Ithaca & London: Comstock Publishing Associates.

DAFNI, A. 1991: Sclerophylly, chemical defense and the evolution of the Mediterranean maqui species. - *Fl. Veg. Mundi* **9**: 167-174.

DAFNI, A. 1996: Autumnal and winter pollination adaptations under Mediterranean conditions. - *Bocconea* **5(1)**: 177-181.

DAFNI, A., COHEN, D. & NOY-MEIR, I. 1981: Life-cycle variation in geophytes. - *Ann. Missouri Bot. Gard.* **68(4)**: 652-660.

DAFNI, A. & DUKAS, R. 1986: Insect and wind pollination in *Urginea maritima* (Liliaceae). - *Pl. Syst. Evol.* **154(1-2)**: 1-10.

DAFNI, A. & HELLER, D. 1982: Adventive flora of Israel - phytogeographical, ecological and agricultural aspects. - *Pl. Syst. Evol.* **140(1)**: 1-18.

DAFNI, A. & IVRI, Y. 1979: Pollination ecology of, and hybridization between, *Orchis coriophora* L. and *O. collina* Sol. ex Russ. (Orchidaceae) in Israel. - *New Phytol.* **83(1)**: 181-187.

DAFNI, A. & IVRI, Y. 1981: The flower biology of *Cephalanthera longifolia* (Orchidaceae) - pollen imitation and facultative floral mimicry. - *Pl. Syst. Evol.* **137(4)**: 229-240.

DAFNI, A., IVRI, Y. & BRANTJES, N.B.M. 1981: Pollination of *Serapias vomeracea* Briq. (Orchidaceae) by imitation of holes for sleeping solitary male bees (Hymenoptera). - *Acta Bot. Neerl.* **30(1-2)**: 69-73.

DAFNI, A., SHMIDA, A. & AVISHAI, M. 1981: Leafless autumnal-flowering geophytes in the Mediterranean region - phytogeographical, ecological and evolutionary aspects. - *Pl. Syst. Evol.* **137 (3)**: 181-193.

DAFNI, A., TALMON, Y. & GERTMANN, Y. 1988: Updated list of the orchids of Israel. - *Israel J. Bot.* **36(3)**: 145-157.

DAHL, Å.E. 1989: Taxonomic and morphological studies in *Hypecoum* sect. *Hypecoum* (Papaveraceae). - *Pl. Syst. Evol.* **163(3-4)**: 227-280.

DAHL, Å.E. 1990a: Biosystematics of *Hypecoum* L. (Papaveraceae). - Göteborg: Ph.D. thesis, Univ. of Göteborg, 12 pp.

DAHL, Å.E. 1990b: Artificial crossing experiments in *Hypecoum* sect. *Hypecoum* (Papaveraceae). - Göteborg: Distributed with Ph.D. thesis, Univ. of Göteborg, 38 pp.

DAHL, Å.E. 1990c: Infrageneric division of *Hypecoum* (Papaveraceae). - *Nordic J. Bot.* **10(2)**: 129-140.

DAHL, Å.E., WASSGREN, A.-B. & BERGSTRÖM, G. 1990: Floral scents in *Hypecoum* sect. *Hypecoum* (Papaveraceae): Chemical composition and relevance to taxonomy and mating system. - *Biochem. Syst. Ecol.* **18(2-3)**: 137-168.

DAHLGREN, G. 1980: Cytological and morphological investigations of the genus *Erodium* L'Hér. in the Aegean. - *Bot. Not.* **133(4)**: 491-514.

DAHLGREN, G. 1991: Karyological investigations in *Ranunculus* subg. *Batrachium* (Ranunculaceae) on the Aegean islands. - *Pl. Syst. Evol.* **177(3-4)**: 193-211.

DAHLGREN, G. 1992: *Ranunculus* subgenus *Batrachium* on the Aegean islands and adjacent areas: nectary types and breeding system. - *Nordic J. Bot.* **12(3)**: 299-310.

DAHLGREN, G. & SVENSSON, L. 1994: Variation in leaves and petals of *Ranunculus* subgenus *Batrachium* on the Aegean Islands, analysed by multivariate analysis. - *Bot. J. Linn. Soc.* **114(3)**: 253-270.

DAHLGREN, K.V.O. 1943: Svedjenävan (*Geranium bohemicum*) och brandnävan (*Geranium lanuginosum*) [In Swedish with German summary]. - *Svensk Bot. Tidskr.* **37(2)**: 127-160.

DAHLGREN, K.V.O. 1945: Neue Mitteilungen über *Geranium bohemicum* und *Geranium lanuginosum* [In Swedish with German summary]. - *Bot. Not.* **1945(4)**: 381-389.

DAHLGREN, K.V.O. 1952: Chromosomes in *Geranium bohemicum* and *G. lanuginosum* with some notes on the hybrid. - *Hereditas (Lund)* **38(3)**: 314-320.

DAHLGREN, K.V.O. 1962: Om *Geranium lanuginosum* Lam. [In Swedish]. - *Svensk Bot. Tidskr.* **56(1)**: 175-182.

DAHLGREN, R., CLIFFORD, H.T. & YEO, P.F. 1985: The families of the monocotyledons. Structure, evolution and taxonomy. - Berlin, etc.: Springer-Verlag, xii + 520 pp.

DAHLGREN, R., KARLSSON, TH. & LASSEN, P. 1971: Studies on the flora of the Balearic Islands. I. Chromosome numbers in Balearic angiosperms. - *Bot. Not.* **124(2)**: 249-269.

DAHLSTEDT, H. 1926: Über einige orientalische *Taraxacum*-Arten. - *Acta Horti Berg.* **9(1)**: 1-36 + 1 plate.

DAISS, H. 2016: Eine spätblühende Sippe von *Cephalanthera* cf. *damasonium*. - *J. Eur. Orch.* **48(2-4)**: 413-414.

DAKSKOBLER, I., VREŠ, B. & PUSCH, J. 2010: *Orobanche pancicii* Beck, a new species to the flora of Slovenia. - *Hacquetia* **9(2)**: 171-176.

DAL CIN D'AGATA, C., SKOULA, M. & BRUNDO, G. 2009: A preliminary inventory of the alien flora of Crete (Greece). - *Bocconea* **23**: 301-315.

DALAMPIRA, E.S., LANARAS, T. & COOK C.-M. 2013: Evidence for the formation of hybrids between *Salvia fruticosa* and *Salvia tomentosa*. In Gkelis, S. & al. (eds): 13th Panhellenic Scientific Conference, Hellenic Botanical Society, Thessaloniki 3-6 October. Program and book of abstracts, p. 65.

DALGAARD, V. 1994: Checklist of chromosome numbers counted in Madeiran flowering plants, with notes on polyploidy, life form, endemism and evolution. - *Nordic J. Bot.* **14(3)**: 241-255.

DALGIÇ, G. 1991: Cytotaxonomic studies on the genus *Muscari* in European Turkey. - *Bot. Chron. (Patras)* **10**: 819-825.

DALGIÇ, G. & BAŠAK, N. 1996: Chromosome numbers of some petaloid monocots from European Turkey. In Öztürk, M.A., Seçmen, Ö. & Görk, G. (eds): Plant life in southwest and central Asia, 1: pp. 54-62. - Izmir: Ege University Press.

DALGIÇ, G. & DANE, F. 1993: A contribution to the flora of Turkey: *Silene fabaroides*. - *Istanbul Üniv. Eczac. Fak. Mecm.* **29(1)**: 27-30.

DALGIÇ, G., DANE, F. & DALGIÇ AKSOY, Ö. 2009: A new record for the flora of Turkey: *Ornithogalum boucheanum* (Hyacinthaceae). In Ivanova, D. (ed.): Proceedings of the IV Balkan botanical congress, pp. 169-173. - Sofia: Institute of Botany, BAS.

DALGIÇ, G. & ÖZHATAY, N. 1993: *Ornithogalum* L. (Hyacinthaceae) in European Turkey. In Petrova, A.V. (ed.): Abstracts from the VII OPTIMA meeting in Borovetz, Bulgaria, 18-30 July 1993, p. 159.

DALLAPORTA, N. 1821: Prospetto delle piante che si trovano nell'isola di Cefalonia e che si possono ad operare a titolo di alimento o di rimedio. - Corfu: Stamperia del Governo, xiii + 148 pp.

DALLIMORE, W. & JACKSON, A.B. 1966: A handbook of Coniferae and Ginkgoaceae. Edition 4. Revised by S.G. Harrison. - London: Edward Arnold Ltd., xix + 729 pp.

DAMANAKIS, M. 1983: Katalogos ton agrostodon tis Elladas [Catalogue of the grasses of Greece; in Greek]. - Athens: Benaki Phytopathological Institute, 58 pp.

DAMANAKIS, M. 1984: Notes on *Coix lacryma-jobi* L. (Poaceae) found as naturalized on Crete [In Greek with English abstract]. - *Bot. Chron. (Patras)* **4(1-2)**: 12-16.

DAMANAKIS, M. 1989: An interactive key for the online identification of the grass weeds of Greece with computer. In Phitos, D. (ed.): Abstracts. Sixth OPTIMA meeting, Delphi 10-16 Sept. 1989, p. 44.

DAMANAKIS, M. 1991: An interactive key for the computer-assisted on line identification of the grass weeds of Greece. - *Bot. Chron. (Patras)* **10**: 347-352.

DAMANAKIS, M. & ECONOMOU, G. 1986: The grasses of Greece. Records of occurrence and distribution. Athens: Benaki Phytopathological Institute, 326 pp. (mimeographed).

DAMANAKIS, M. & MARKAKI, M. 1990: Studies on the biology of *Oxalis pes-caprae* L. under field conditions in Crete [In Greek with English summary]. - *Zizaniology* **2**: 145-154.

DAMANAKIS, M. & MATTHÄS, U. 1987: Endemism and island botany - The flora and vegetation of Crete (Greece) [Guide to post-congress excursion, 16-23 July, 1987]. - Berlin: XIV International Botanical Congress, 21 pp.

DAMANAKIS, M. & SCHOLZ, H. 1990: Phytogeographical notes on the Poaceae of Greece. - *Willdenowia* **19(2)**: 413-423.

DAMANAKIS, M. & YANNITSAROS, A. 1986: The genus *Pennisetum* (Poaceae) in Greece. - *Willdenowia* **15(2)**: 401-406.

DAMBOLDT, J. 1963: *Cystopteris dickieana* Sim und ihr Vorkommen in den Alpen. - *Ber. Bayer. Bot. Ges.* **34**: 64-66.

DAMBOLDT, J. 1965: Zytotaxonomische Revision der isophyllen Campanulae in Europa. - *Bot. Jahrb. Syst.* **84(3)**: 302-358.

DAMBOLDT, J. 1968a: Zur Zytotaxonomie der Gattung *Saxifraga* L. III. - *Ber. Deutsch. Bot. Ges.* **81(1-2)**: 43-52.

DAMBOLDT, J. 1968b: Kritische Bemerkungen zu *Trachelium taygetum* und *Diosphaera hysterantha* (Campanulaceae). - *Oesterr. Bot. Z.* **115(2)**: 144-151.

DAMBOLDT, J. 1968c: Kurzer Nachtrag zur "Zytotaxonomischen Revision der isophyllen Campanulae in Europa". - *Bot. Jahrb. Syst.* **88(2)**: 200-203.

DAMBOLDT, J. 1968d: Vorarbeiten zu einer Revision der Gattung *Asyneuma* (Campanulaceae). I. - *Willdenowia* **5(1)**: 35-54.

DAMBOLDT, J. 1968e: [Chromosome numbers in *Aurinia*, *Matthiola*, *Scorzonera* and *Tragopogon* from Greece]. In Löve, Á. (ed.): IOPB chromosome number reports XVI. - *Taxon* **17(1)**: 96, 103, 104.

DAMBOLDT, J. 1970: Revision der Gattung *Asyneuma*. - *Boissiera* **17**: 1-128 + 2 plates.

DAMBOLDT, J. 1972: Zur Zytotaxonomie der Gattung *Saxifraga* L. IV. Weitere Untersuchungen in der *Saxifraga exarata-moschata*-Gruppe. - *Ber. Deutsch. Bot. Ges.* **84(11)**: 705-712.

DAMBOLDT, J. 1976a: Materials for a Flora of Turkey XXXII: Campanulaceae. - *Notes Roy. Bot. Gard. Edinburgh* **35(1)**: 39-52.

DAMBOLDT, J. 1976b: Beiträge zur Flora Ionica. VI. Karyologisch-systematische Bemerkungen zu einigen Labiaten. - *Candollea* **31(2)**: 273-281.

DAMBOLDT, J. & AL. 1981: Beiträge zur Flora Ionica. VII. Der Formenkreis von *Scabiosa crenata* (Dipsacaceae). - *Phyton (Horn)* **21(1)**: 85-102.

DAMBOLDT, J. & MATTHÄS, U. 1975: Chromosomenzahlen einiger mediterraner und mitteleuropäischer *Centaurea*-Arten (Asteraceae). - *Pl. Syst. Evol.* **123(2)**: 107-115.

DAMBOLDT, J. & MATTHÄS, U. 1979: Karyologische Untersuchungen an *Centaurea* (sect. *Phalolepis*) *pawlowskii* (Compositae). - *Bot. Jahrb. Syst.* **100(3)**: 406-413.

DAMBOLDT, J., MATTHÄS, U. & MELZHEIMER, V. 1974: Bericht über die Griechenland-Exkursion (Kerkyra, Epirus) des Instituts für Systematische Botanik und Pflanzengeographie der Freien Universität Berlin vom 29.6.-6.7. 1974. - 23 pp. (mimeographed).

DAMBOLDT, J. & MELZHEIMER, V. 1974: Über einige *Centaurea*-Arten (Asteraceae) der Athos-Halbinsel. - *Bot. Jahrb. Syst.* **94(3)**: 383-390.

DAMBOLDT, J. & MELZHEIMER, V. 1976: *Brimeura fastigiata* (Viv.) Chouard (Liliaceae) neu für die Flora Griechenlands. - *Bot. Jahrb. Syst.* **95(3)**: 334-338.

DAMBOLDT, J. & PHITOS, D. 1966: Ein Beitrag zur Zytotaxonomie der Gattung *Silene* L. in Griechenland. - *Oesterr. Bot. Z.* **113(2)**: 169-175.

DAMBOLDT, J. & PHITOS, D. 1968: Zur Zytotaxonomie einiger Arten der Gattung *Silene* L. (Caryophyllaceae). - *Verh. Bot. Vereins Prov. Brandenburg* **105**: 44-51.

DAMBOLDT, J. & PHITOS, D. 1970: Beiträge zur Flora Ionica. II. Bemerkungen zu einigen *Silene*-Arten aus Griechenland. - *Oesterr. Bot. Z.* **118(4)**: 341-352.

DAMBOLDT, J. & PHITOS, D. 1971: [Chromosome numbers from Kefallinia, Greece]. In Löve, Á. (ed.): IOPB chromosome number reports XXXIV. - *Taxon* **20(5-6)**: 787.

DAMBOLDT, J. & PHITOS, D. 1972: Beiträge zur Flora Ionica. IV. Studien in der Gattung *Petrorhagia* (Caryophyllaceae). - *Candollea* **27(1)**: 27-40.

DAMBOLDT, J. & PHITOS, D. 1975: Die Karyosystematik der Gattung *Leucojum* L. (Amaryllidaceae) in Griechenland. - *Pl. Syst. Evol.* **123(2)**: 119-131.

DAMBOLDT, J. & PODLECH, D. 1965: Zytotaxonomische Untersuchungen an *Saxifraga*-Sippen der grex *Exarato-moschatae* Engl. et Irmsch. - *Ber. Deutsch. Bot. Ges.* **77(9)**: 332-339.

DAMBOLDT, J. & WULSCHE, A. 1977: Karyologische Untersuchungen an *Urginea maritima* (L.) Baker (Liliaceae) aus Griechenland. - *Mitt. Bot. Staatssamml. München* **13**: 535-544.

DAMIANIDIS, CH. & AL. 2019: *Airopsis tenella*. In Raab-Straube, E. v. & Raus, Th. (eds): Euro+Med-Checklist Notulae, 10. - *Willdenowia* **49(1)**: 104-105.

DAMIANIDIS, CH. & AL. 2019: *Airopsis tenella* (Poaceae), a new record for the flora of Greece. In Daskalakou, E.N. & al. (eds): 16th Conference of the Hellenic Botanical Society – Athens 10-13 October 2019. Program and book of abstracts, p. 64.

DANČÁK, M., DUCHOSLAV, M. & TRÁVNÍČEK, B. 2012: Taxonomy and cytogeography of the *Molinia caerulea* complex in central Europe. - *Preslia* **84(2)**: 351-374.

DANDY, J.E. & CANNON, J.F.M. 1968: [Notes on *Myrrhoides*]. In Heywood, V.H. (ed.): Notulae systematicae ad Floram Europaeam spectantes, no. 7. - *Feddes Repert.* **79(1-2)**: 64-65.

DANE, F., AKSOY, Ö.D. & YILMAZ, G. 2007: Karyological and palynological studies on *Astragalus hamosus* and *A. glycyphyllos* in Turkey. - *Phytol. Balcan.* **13(3)**: 387-391.

DANE, F., DALGIÇ, G. & MERIÇ, Ç. 2005: Karyological studies of five rare and endemic taxa distributed in European Turkey. - *Bot. Chron. (Patras)* **18(2)**: 17-28.

DANE, F., DALGIÇ, G. & DALGIÇ, Ö. 2002: A new genus for the flora of European Turkey and its karyotype: *Biserrula pelecinus* L. In Koyuncu, M. (ed.): VIth Plant Life of Southwest Asia Symposium, Abstracts p. 53. - Van: Yüzüncü Yil University.

DANE, F., DALGIÇ, G. & DALGIÇ, Ö. 2004: *Biserrula* L. (Fabaceae): karyological and palynological survey of the new monotypic genus for the flora of European Turkey. - *Bot. Chron. (Patras)* **16**: 47-52.

DANE, F., DALGIÇ, G. & MERIÇ, Ç. 2002: Karyological studies of some rare and endemic taxa distributed in Turkey. In Koyuncu, M. (ed.): VIth Plant Life of Southwest Asia Symposium, Abstracts p. 54. - Van: Yüzüncü Yil University.

DANE, F. & EKICI, N. 2006: A new record for the flora of Turkey: *Delphinium balcanicum* Pawl. (Ranunculaceae). In Peev, D.R. (ed.): Book of abstracts, IV Balkan botanical congress, Sofia, Bulgaria, 20-26 June 2006, p. 167.

DANE, F. & EKICI, N. 2009: *Delphinium balcanicum* (Ranunculaceae): a contribution to the flora of Turkey. In Ivanova, D. (ed.): Proceedings of the IV Balkan botanical congress, pp. 217-219. - Sofia: Institute of Botany, BAS.

DANE, F. & OLGUN, G. 1996: Caryological study on three *Paeonia* L. species. In Öztürk, M.A., Seçmen, Ö. & Görk, G. (eds): Plant life in southwest and central Asia, 1: pp. 63-69. - Izmir: Ege University Press.

DANE, F. & ÖZHATAY, N. 2000: A new taxon from Turkey: *Trachelium jacquinii* (Sieb.) Boiss. subsp. *dalgiciorum* N. Özhatay & Dane. In Özhatay, N. & al. (eds): Abstracts from the Second Balkan Botanical Congress, Istanbul, Turkey, May 14-18, 2000, p. 31.

DANE, F. & ÖZHATAY, N. 2001: A new taxon from the Turkish flora: *Trachelium jacquinii* (Sieber) Boiss. subsp. *dalgiciorum* N. Özhatay & Dane (Campanulaceae). In Özhatay, N. (ed.): Proceedings of the Second Balkan Botanical Congress, 1: 155-161. - Istanbul.

DANE, F. & YILMAZ, G. 2002: Karyological study of some *Verbascum* L. species. In Koyuncu, M. (ed.): VIth Plant Life of Southwest Asia Symposium, Abstracts p. 55. - Van: Yüzüncü Yil University.

DANE, F. & YILMAZ, G. 2005: New records for the flora of European Turkey: *Verbascum roripifolium* and *V. ovalifolium* subsp. *thracicum* (Scrophulariaceae). - *Bot. Chron. (Patras)* **18(2)**: 11-16.

DANE, F. & YILMAZ, G. 2006: Reports (1462-1465). In Kamari, G., Blanché, C. & Garbari, F. (eds): Mediterranean chromosome number reports - 16. - *Fl. Medit.* **16**: 396-399.

DANE, F. & YILMAZ, G. 2009: A new natural hybrid of *Verbascum* (Scrophulariaceae) from European Turkey. - *Phytol. Balcan.* **15(2)**: 181-184.

DANE, F., YILMAZ, G. & DALGIÇ, G. 2009: Reports 58-60. In Vladimirov & al. (eds): New floristic records in the Balkans: 12. - *Phytol. Balcan.* **15(3)**: 442.

DANERT, S. 1955: Ein Beitrag zur Kenntnis der Gattung *Datura* L. - *Feddes Repert. Spec. Nov. Regni Veg.* **57(3)**: 231-242.

DANESCH, E. & DANESCH, O. 1969: Orchideen Europas. Südeuropa. - Bern & Stuttgart: Verlag Hallwag, 256 pp. incl. 128 colour plates.

DANESCH, E. & DANESCH, O. 1972: Orchideen Europas. *Ophrys*-Hybriden. - Bern & Stuttgart: Verlag Hallwag, 268 pp.

DANESCH, O. & AL. 1975: Hybriden und hybridogene Sippen aus *Ophrys bertolonii* und *O. atrata* (Orchidaceae). - *Pl. Syst. Evol.* **124(2)**: 79-123.

DANESCH, O. & DANESCH, E. 1975: *Ophrys helenae*: Die Geschichte einer Orchidee. - *Kosmos (Stuttgart)* **75(5)**: 185-188.

DANESCH, O. & DANESCH, E. 1976: Hybriden und Hybridschwärme aus *Ophrys argolica* Fleischm. und *Ophrys scolopax* Cav. ssp. *cornuta* (Stev.) E.G. Cam. in Griechenland. In: Proceedings of the 8th world orchid conference, pp. 129-138. - Frankfurt: Deutsche Orchideen Gesellschaft.

DANESCH, O. & DANESCH, E. 1977: *Ophrys helenae* (Orchidaceae), eine lange verkannte Sippe der Balkanhalbinsel. - *Pl. Syst. Evol.* **127(1)**: 11-22.

DANIHELKA, J. & EHRENDORFER, F. 2008: (1853) Proposal to conserve the name *Achillea pannonica* against *A. seidlii* (Asteraceae). - *Taxon* **57(3)**: 1356-1357. [approved, cf. Taxon 60(1): 228, 2011].

DANIHELKA, J., SAUKEL, J. & EHRENDORFER, F. 2006: Typification of the names *Achillea millefolium* var. *collina* and *A. pannonica* (Asteraceae, Anthemideae). - *Taxon* **55(2)**: 507-510.

DANIN, A. 1976a: On three adventive species of *Conyza* (Compositae) in Greece. - *Candollea* **31(1)**: 107-109.

DANIN, A. 1976b: Notes on four adventive composites in Israel. - *Notes Roy. Bot. Gard. Edinburgh* **34 (3)**: 403-410.

DANIN, A. 1983a: *Sorghum halepense* (L.) Pers. and *S. virgatum* (Hack) Stapf in Israel. - *Israel J. Bot.* **32(3)**: 172-173.

DANIN, A. 1984a: *Conyza albida* Sprengel. In Greuter, W. & Raus, Th. (eds): Med-Checklist notulae, 8. - *Willdenowia* **13(2)**: 283.

DANIN, A. 1984b: *Portulaca oleracea* L. subsp. *oleracea*. In Greuter, W. & Raus, Th. (eds): Med- Checklist notulae, 8. - *Willdenowia* **13(2)**: 287.

DANIN, A. 1987: Contributions to the flora of Israel and Sinai. I. Studies in the apopetalous genera *Minuartia, Silene, Polygala* and *Sedum*. - *Israel J. Bot.* **36**: 63-71.

DANIN, A. 1992a: Report on Iter Mediterraneum II. - *Bocconea* **3**: 5-17.

DANIN, A. 1992b: Flora and vegetation of Israel and adjacent areas. - *Bocconea* **3**: 18-42.

DANIN, A. 1992c: Check-list of the vascular plants collected during Iter Mediterraneum II. – *Bocconea* **3**: 43-216.

DANIN, A. 2000: The nomenclature news of Flora Palaestina. - *Fl. Medit.* **10**: 109-172.

DANIN, A. 2004: *Arundo* (Gramineae) in the Mediterranean reconsidered. - *Willdenowia* **34(2)**: 361-369.

DANIN, A. 2005: Distribution atlas of plants in the Flora Palaestina area. - Jerusalem: Israel Academy of Sciences and Humanities, 517 pp.

DANIN, A. 2006: *Arundo mediterranea*. In Greuter, W. & Raus, Th. (eds): Med-Checklist notulae, 24. - *Willdenowia* **36(2)**: 727.

DANIN, A. 2010: *Capparis* in the East Mediterranean countries. - *Fl. Medit.* **20**: 179-185.

DANIN, A. 2011: [Notes on *Portulaca*]. In Greuter, W. & Raab-Straube, E.v.: Euro+Med Notulae, 5. - *Willdenowia* **41(1)**: 131-134.

DANIN, A., BAKER, I. & BAKER, H.G. 1979: Cytogeography and taxonomy of the *Portulaca oleracea* L. polyploid complex. - *Israel J. Bot.* **27(3-4)**: 177-211.

DANIN, A., DOMINA, G. & RAIMONDO, F.M. 2008: Microspecies of the *Portulaca oleracea* aggregate found on major Mediterranean islands (Sicily, Cyprus, Crete, Rhodes). - *Fl. Medit.* **18**: 89-107.

DANIN, A. & NAENNY, W. 2008: Contribution to the recognition of reeds by their rhizome anatomy. - *Fl. Medit.* **18**: 385-392.

DANIN, A., RAUS, TH. & SCHOLZ, H. 2002: Contribution to the flora of Greece: a new species of *Arundo* (Poaceae). - *Willdenowia* **32(2)**: 191-194.

DANIN, A. & SMALL, E. 1989: Contributions to the flora of Israel and Sinai. V. *Trigonella sibthorpii* Boiss., a new record from Israel. - *Israel J. Bot.* **38(2-3)**: 121-124.

DANIN, A. & YOM-TOV, Y. 1990: Ant nests as primary habitats of *Silybum marianum* (Compositae). - *Pl. Syst. Evol.* **169(3-4)**: 209-217.

DANIN, A. & AL. 2001: *Phragmites frutescens* H. Scholz. In Greuter, W. & Raus, Th. (eds): Med- Checklist notulae, 20. - *Willdenowia* **31(2)**: 326.

DANSEREAU, P.M. 1938: Etudes sur les hybrides de Cistes. III. Statistique des caractères du *Cistus monspeliensis* L. - *Bull. Soc. Bot. France* **85**: 668-678.

DANSEREAU, P.M. 1939: Monographie du genre *Cistus* L. - *Boissiera* **4**: 1-90.

DARDIOTI, A. & AL. 1995: *Satureja montana* group: An approach of morphological differentiation ... In: Proc., 5th sci. conf., Hellenic Bot. Soc., Delphi 21-23 October 1994, pp. 289-292 [In Greek with English summary]. - Thessaloniki: Univ. Studio Press.

DARDIOTI, A. & KOKKINI, S. 2005: *Satureja pilosa* subsp. *origanita* (Labiatae): a new taxon from Thrace (NE Greece). - *Bot. Chron. (Patras)* **18(2)**: 109-113.

DARLINGTON, C.D. & WYLIE, A.P. 1955: Chromosome atlas of flowering plants, 2nd ed. - London: George Allen & Unwin Ltd, xix + 520 pp.

DASKALAKOU, E.N. & AL. (EDS) 2019: 16th Conference of the Hellenic Botanical Society – Athens 10-13 October 2019. Program and book of abstracts. - Athens: 118 pp.

DASKALAKOU, E.N. & THANOS, C.A. 2010: Seed and cone morphometric indicators: A new tool for the discrimination between the common Mediterranean pines *Pinus halepensis* Mill. and *P. brutia* Ten. - *Pl. Biosyst.* **144(4)**: 819-825.

DASKALOVA, C. 1977: Fruit and seed development in *Seseli rigidum* W.K. [In Bulgarian with Russian and English summaries]. - *Fitologiya* **7**: 59-74.

DASKALOVA, T. 2005: Polygamy in some species of *Nepeta* (Lamiaceae). - *Phytol. Balcan.* **11(1)**: 79-88.

DAŞKIN, R., YILMAZ, Ö. & KAYNAK, G. 2006: Presence of *Cirsium eriophorum* (L.) Scop. (Asteraceae) in Turkey. - *Turk. J. Bot.* **30(6)**: 461-465.

DATTA, K.B. 1974: Chromosome studies in *Iberis* L. with a view to find out the mechanism of speciation of the genus. - *Cytologia* **39(3)**: 543-551.

DAUBS, E.H. 1965: A monograph of Lemnaceae. - *Illinois Biol. Monogr.* **34**: 1-118 + 21 plates.

DAUMANN, E. 1967: Zur Blütenmorphologie und Bestäubungsökologie von *Veratrum album* subsp. *lobelianum* (Bernh.) Rchb. - *Oesterr. Bot. Z.* **114(2)**: 134-148.

DAUMANN, E. 1971: Zur Bestäubungsökologie von *Aristolochia clematitis* L. - *Preslia* **43(2)**: 105-111.

DAUMANN, E. 1973: Zur Blütenmorphologie und Bestäubungsökologie von *Mercurialis* L. - *Preslia* **44 (4)**: 308-315.

DAVEAU, M.J. 1896: Note sur quelques *Lotus* de la section *Tetragonolobus*. - *Bull. Soc. Bot. France* **43**: 358-369.

DAVID, R.W. 1976: Nomenclature of the British taxa of the *Carex muricata* L. aggregate. - *Watsonia* **11(1)**: 59-65.

DAVID, R.W. & KELCEY, J.G. 1985: Biological Flora of the British Isles. *Carex muricata* L. aggregate. - *J. Ecol.* **73(3)**: 1021-1039.

DAVIDOV, B. 1915: Une récolte botanique dans la Thrace occidentale [In Bulgarian with French summary]. - *Trav. Soc. Bulg. Sci. Nat.* **8**: 43-127.

DAVIES, D. 1992: Alliums. The ornamental onions. - London: B.T. Batsford Ltd, 133 pp.

DAVIES, E.W. 1956a: Some new chromosome numbers in the Cyperaceae. - *Watsonia* **3(5)**: 242-243.

DAVIES, E.W. 1956b: Cytology, evolution and origin of the aneuploid series in the genus *Carex*. - *Hereditas (Lund)* **42(3-4)**: 349-365.

DAVIES, J.A. 1980: Athens apologia. - *Quart. Bull. Alpine Gard. Soc. Gr. Brit.* **48(2)**: 93-96.

DAVIES, P.H. & DAVIES, J.A. 1985a: European orchids. The genus *Ophrys* - Parts 1-3. - *Amer. Orchid Soc. Bull.* **54(2-4)**: 132-141, 282-293, 421-430.

DAVIES, P.H. & DAVIES, J.A. 1985b: European orchids: The genus *Epipactis*. - *Amer. Orchid Soc. Bull.* **54(5)**: 558-571.

DAVIES, P.H. & DAVIES, J.A. 1985c: European orchids: The genus *Cephalanthera*. - *Amer. Orchid Soc. Bull.* **54(6)**: 680-687.

DAVIS, A.P. 1997: (1307) Proposal to conserve the name *Galanthus elwesii* (Amaryllidaceae) with a conserved type. - *Taxon* **46(3)**: 553-554. [approved, cf. Taxon 48(2): 370, 1999].

DAVIS, A.P. 1999: The genus *Galanthus*. - Kew: Royal Botanic Gardens, with Timber Press, 297 pp. + 56 colour plates.

DAVIS, A.P. 2000a: A systematic study of the genus *Galanthus*. In Özhatay, N. & al. (eds): Abstracts from the Second Balkan Botanical Congress, Istanbul, Turkey, May 14-18, 2000, p. 9.

DAVIS, A.P. 2000b: (1471) Proposal to reject the name *Galanthus reflexus* Herb. (Amaryllidaceae) . - *Taxon* **49(4)**: 813-814. [approved; cf. *Taxon* 51(4): 796, 2002].

DAVIS, A.P. 2001: The genus *Galanthus*. In Bishop, M., Davis, A.P. & Grimshaw, J. (eds): Snowdrops - a monograph of cultivated *Galanthus*, pp. 9-63. - Maidenhead, U.K.: The Griffin Press.

DAVIS, A.P. & AL. 1999: CITES bulb checklist for the genera *Cyclamen*, *Galanthus* and *Sternbergia*. - Kew: Royal Botanic Gardens, x + 131 pp.

DAVIS, A.P. & BRICKELL, C.D. 1994: *Galanthus peshmenii*: a new snowdrop from the eastern Aegean. - *New Plantsman* **1(1)**: 14-19.

DAVIS, A.P. & JURY, S.L. 1990: A taxonomic review of *Iris* L. series *Unguiculares* (Diels) Lawrence. - *Bot. J. Linn. Soc.* **103(3)**: 281-300.

DAVIS, A.P. & JURY, S.L. 1991: A new classification of *Iris* L. *Unguiculares* (Diels) Lawrence. - *Iris Year Book* **1991**: 75-84.

DAVIS, A.P. & ÖZHATAY, N. 2001: *Galanthus trojanus*: a new species of *Galanthus* (Amaryllidaceae) from north-western Turkey. - *Bot. J. Linn. Soc.* **137(4)**: 409-412.

DAVIS, P.H. 1937: A collector in Crete. - *Quart. Bull. Alpine Gard. Soc. Gr. Brit.* **5(4)**: 385-405.

DAVIS, P.H. 1939a: Flowers in a Cretan gorge. - *My Garden* **16(62)**: 207-213.

DAVIS, P.H. 1939b: Some plants from the eastern Mediterranean. - *Quart. Bull. Alpine Gard. Soc. Gr. Brit.* **7(1)**: 25-64.

DAVIS, P.H. 1939c: *Omphalodes luciliae*. - *Gard. Ill.* **61(3142)**: 339.

DAVIS, P.H. 1947a: Plants and experiences in the Eastern Mediterranean. - *J. Roy. Hort. Soc.* **72(1)**: 13-17.

DAVIS, P.H. 1947b: On the rocks. - *Quart. Bull. Alpine Gard. Soc. Gr. Brit.* **15(1)**: 37-53.

DAVIS, P.H. 1947c: Fritillaries in the Eastern Mediterranean. - *Lily Year-Book* **11**: 141-144.

DAVIS, P.H. 1951: Cliff vegetation in the Eastern Mediterranean. - *J. Ecol.* **39(1)**: 63-93 + 3 plates.

DAVIS, P.H. 1953: Notes on the summer flora of the Aegean. - *Notes Roy. Bot. Gard. Edinburgh* **21 (3)**: 101-142.

DAVIS, P.H. 1955: Mediterranean itineraries. - *Notes Roy. Bot. Gard. Edinburgh* **21(5)**: 271-278.

DAVIS, P.H. 1957a: Materials for a Flora of Turkey. II: *Linum* Linn. - *Notes Roy. Bot. Gard. Edinburgh* **22(3)**: 135-161 + 5 plates.

DAVIS, P.H. 1957b: New Turkish species of *Tunica*, *Velezia* and *Potentilla*. - *Notes Roy. Bot. Gard. Edinburgh* **22(3)**: 163-171 + 4 plates.

DAVIS, P.H. 1961: Hints for hard-pressed collectors. - *Watsonia* **4(6)**: 283-289.

DAVIS, P.H. 1964: Materials for a Flora of Turkey: VIII Cruciferae I: *Isatis*. - *Notes Roy. Bot. Gard. Edinburgh* **26(1)**: 11-25 + 6 plates.

DAVIS, P.H. 1967: Materials for a Flora of Turkey, XVI: Geraniaceae, Linaceae. - *Notes Roy. Bot. Gard. Edinburgh* **28(1)**: 35-38.

DAVIS, P.H. 1969: Materials for a Flora of Turkey: XIX. Leguminosae: Vicieae. - *Notes Roy. Bot. Gard. Edinburgh* **29(3)**: 311-320.

DAVIS, P.H. 1970: *Geranium* sect. *Tuberosa*, revision and evolutionary interpretation. - *Israel J. Bot.* **19(2-3)**: 91-113.

DAVIS, P.H. 1971: Distribution patterns in Anatolia with particular reference to endemism. In Davis, P. H., Harper, P.C. & Hedge, I.C. (eds): Plant life of South-West Asia, pp. 15-27. - Edinburgh: Botanical Society.

DAVIS, P.H. 1974: Materials for a Flora of Turkey. XXX: Compositae, I [with contributions by several authors]. - *Notes Roy. Bot. Gard. Edinburgh* **33(2)**: 207-264.

DAVIS, P.H. 1975a: Turkey: Present state of floristic knowledge. In Guinochet, M. & al. (eds): La flore du Bassin Mediterranéen, pp. 93-113. - Paris: C.N.R.S.

DAVIS, P.H. 1975b: Materials for a Flora of Turkey XXXI: Compositae, II [with contributions by several authors]. - *Notes Roy. Bot. Gard. Edinburgh* **33(3)**: 409-435.

DAVIS, P.H. 1978: Materials for a Flora of Turkey XXXV: Primulaceae and Scrophulariaceae [with contributions by several authors]. - *Notes Roy. Bot. Gard. Edinburgh* **36(1)**: 1-22.

DAVIS, P.H. 1980: Materials for a Flora of Turkey. XXVII. Labiatae, Plumbaginaceae, Plantaginaceae [with contributions by several authors]. - *Notes Roy. Bot. Gard. Edinburgh* **38(1)**: 23-64.

DAVIS, P.H. 1983: Materials for a Flora of Turkey XXXVIII: Araceae, Dioscoreaceae, Liliaceae [with contributions by several authors]. - *Notes Roy. Bot. Gard. Edinburgh* **41(1)**: 45-54.

DAVIS, P.H. 1985: Why is the flora of Turkey interesting and important? - *Bot. Mag. (Kew Mag.)* **2(4)**: 357-367.

DAVIS, P.H. & AL. 1965: Materials for a Flora of Turkey: X. - *Notes Roy. Bot. Gard. Edinburgh* **26(2)**: 165-201 + 4 plates.

DAVIS, P.H. & EDMONDSON, J.R. 1979: Flora of Turkey: A floristic bibliography. - *Notes Roy. Bot. Gard. Edinburgh* **37(2)**: 273-283.

DAVIS, P.H., HARPER, P.C. & HEDGE, I.C. (EDS) 1971: Plant life of South-West Asia. - Edinburgh: Botanical Society, x + 335 pp.

DAVIS, P.H. & HEDGE, I.C. 1971: Floristic links between N.W. Africa and S.W. Asia. - *Ann. Naturhist. Mus. Wien* **75**: 43-57.

DAVIS, P.H. & HEDGE, I.C. 1975: The Flora of Turkey: Past, present and future. - *Candollea* **30(2)**: 331-351. [Also distributed as *OPTIMA leaflet* no. 12].

DAVIS, P.H. & KHAN, M.S. 1961: *Aristolochia* in the Near East. - *Notes Roy. Bot. Gard. Edinburgh* **23(4)**: 515-546.

DAVIS, P.H. & KHAN, M.S. 1964: Two new Aristolochias from Turkey. - *Notes Roy. Bot. Gard. Edinburgh* **25(2)**: 67-69 + 2 plates.

DAVIS, P.H., MILL, R.R. & TAN, KIT (EDS) 1988: Flora of Turkey and the East Aegean Islands. Vol. 10 (Supplement). - Edinburgh: Edinburgh Univ. Press, xxi + 590 pp.

DAVIS, P.H. & ROBERTS, J. 1955: Materials for a Flora of Turkey. I. - Geraniaceae. - *Notes Roy. Bot. Gard. Edinburgh* **22(1)**: 9-27.

DAVIS, P.H. & STUART, D.C. 1967: Three new species of *Muscari*. - *Lily Year-Book* **30**: 123-125.

DAVIS, P.H. (ED.) 1965: Flora of Turkey and the East Aegean Islands. Vol. 1 [Pteridophyta to Polygalaceae]. - Edinburgh: Edinburgh Univ. Press, xi + 567 pp.

DAVIS, P.H. (ED.) 1967: Flora of Turkey and the East Aegean Islands. Vol. 2 [Portulacaceae to Celastraceae]. - Edinburgh: Edinburgh Univ. Press, xii + 581 pp.

DAVIS, P.H. (ED.) 1970: Flora of Turkey and the East Aegean Islands. Vol. 3 [Leguminosae]. - Edinburgh: Edinburgh Univ. Press, xvii + 628 pp,

DAVIS, P.H. (ED.) 1972: Flora of Turkey and the East Aegean Islands. Vol. 4 [Rosaceae to Dipsacaceae]. - Edinburgh: Edinburgh Univ. Press, xix + 657 pp.

DAVIS, P.H. (ED.) 1975: Flora of Turkey and the East Aegean Islands. Vol. 5 [Compositae]. - Edinburgh: Edinburgh Univ. Press, xix + 890 pp.

DAVIS, P.H. (ED.) 1978: Flora of Turkey and the East Aegean Islands. Vol. 6 [Campanulaceae to Scrophulariaceae]. - Edinburgh: Edinburgh Univ. Press, xix + 825 pp.

DAVIS, P.H. (ED.) 1982: Flora of Turkey and the East Aegean Islands. Vol. 7 [Orobanchaceae to Rubiaceae]. - Edinburgh: Edinburgh Univ. Press, xxi + 947 pp.

DAVIS, P.H. (ED.) 1984: Flora of Turkey and the East Aegean Islands. Vol. 8 [Butomaceae to Typhaceae]. - Edinburgh: Edinburgh Univ. Press, xxi + 632 pp.

DAVIS, P.H. (ED.) 1985: Flora of Turkey and the East Aegean Islands. Vol. 9 [Juncaceae to Gramineae]. - Edinburgh: Edinburgh Univ. Press, xxi + 724 pp.

DAVISON, A.W. 1971: The ecology of *Hordeum murinum* L. II. The ruderal habit. - *J. Ecol.* **59(2)**: 493-506.

DAVY, A.J. 1980: Biological Flora of the British Isles. *Deschampsia caespitosa* (L.) Beauv. (*Aira cespitosa* L., *Deschampsia cespitosa* (L.) Beauv.). - *J. Ecol.* **68(3)**: 1075-1096.

DAVY, A.J., SCOTT, R. & CORDAZZO, C.V. 2006: Biological flora of the British Isles: *Cakile maritima* Scop. - *J. Ecol.* **94(3)**: 695-711.

DE BEER, D. & VERLOOVE, F. 2020: *Crassula alata*. In Raus, Th. & Raab-Straube, E.v. (eds): Euro+Med-Checklist notulae, 12. - *Willdenowia* **50(2)**: 313-314.

DE JONG, T.J., KLINKHAMER, P.G.L. & BOORMAN, L.A. 1990: Biological Flora of the British Isles. *Cynoglossum officinale* L. - *J. Ecol.* **78(4)**: 1123-1144.

DE LEONARDIS, W. & AL. 1983: Numeri cromosomici per la Flora Italiana: 814-830. - *Inform. Bot. Ital.* **13(2-3)**: 158-167.

DE LEONARDIS, W. & AL. 1997: Pollen and seed morphology of three species of *Reseda* L. in Sicily and relationship with their systematic position. - *Pl. Biosyst.* **131(1)**: 4
3-49.

DE MARTIS, B. & LOI, M.C. 1983: Sul genere *Lippia* (Verbenaceae) in Sardegna. - *Atti Soc. Tosc. Sci. Nat. Pisa Mem., Ser. B* **89**: 1-11.

DE MARTIS, B., LOI, M.C. & POLO, M.B. 1984: Il genere "Tamarix" L. (Tamaricaceae) in Sardegna. - *Webbia* **37(2)**: 211-235.

DE MARTIS, B., LOI, M.C. & POLO, M.B. 1986: Contribution to a better understanding of the genus *Tamarix* in Portugal. - *Candollea* **41(2)**: 293-298.

DE NAMUR, C. 1975: A propos des *Gnaphalium* européens: systématique et caryologie. - *Biol. Gallo- Hellen.* **6(1)**: 65-73.

DE NAMUR, C. & VERLAQUE, R. 1976: Contribution à l'étude biogéographique du genre *Helichrysum* Miller. - *Biol. Écol. Médit.* **3(2)**: 17-22.

DE SANTIS, C., PAVONE, P. & ZIZZA, A. 1976: Numeri cromosomici per la Flora Italiana: 232-237. - *Inform. Bot. Ital.* **8(1)**: 74-81.

DE VRIES, H. 2002: *Laurentia gasparrinii* [In Greek]. - *Fisis [Journal of the Hellenic Society for the Protection of Nature]* **97**: 10-11.

DE VRIES, H. 2004: Elate ligo pio konda. Katapliktikes leptomeries gia poli kina fita: i Ligaria (*Vitex agnus-castus* L.) [In Greek]. - *Fisis [Journal of the Hellenic Society for the Protection of Nature]* **105**: 14-15.

DE VRIES, I.M. & JARVIS, C.E. 1987: Typification of seven Linnaean names in the genus *Lactuca* L. (Compositae: Lactuceae). - *Taxon* **36(1)**: 142-154.

DE WET, J.M.J. & HARLAN, J.R. 1970: Biosystematics of *Cynodon* L.C. Rich. (Gramineae). - *Taxon* **19 (4)**: 565-569.

DEBAZAC, E.F. 1969: La végétation de certaines stations de l'Oleo-Ceratonion en Attique. - *Biol. Gallo-Hellen.* **2(1)**: 3-12.

DEBAZAC, E.F. 1970: Contribution à l'étude de la répartition et de l'écologie de *Pinus nigra* en Grèce. - Athènes: Projet UNSF/FAO/GRE - 20/230, 1-19 [n.v.]

DEBAZAC, E.F. 1971: Contribution à la connaissance de la répartition et de l'écologie de *Pinus nigra* Arn. dans la Sud-Est de l'Europe. - *Ann. Sci. Forest.* **28(2)**: 91-139.

DEBAZAC, E.F. & MAVROMMATIS, G. 1971: Les grandes divisions écologiques de la végétation forestière en Grèce continentale. - *Bull. Soc. Bot. France* **118(5-6)**: 429-452.

DEBUSSCHE, M., DEBUSSCHE, G. & GRANDJANNY, M. 2000: Distribution de *Cyclamen repandum* Sibth. & Sm. subsp. *repandum* et écologie en Corse et France continentale. - *Acta Bot. Gallica* **147(2)**: 123-142.

DEBUSSCHE, M., GARNIER, E. & THOMPSON, J.D. 2004: Exploring the causes of variation in phenology and morphology in Mediterranean geophytes: a genus-wide study of *Cyclamen*. - *Bot. J. Linn. Soc.* **145(4)**: 469-484.

DEBUSSCHE, M. & QUÉZEL, P. 1998a: *Cyclamen repandum* is native and alive but endangered in North Africa. - *Cyclamen* **22(2)**: 45-46.

DEBUSSCHE, M. & QUÉZEL, P. 1998b: *Cyclamen repandum* Sibth. & Sm. en Petite Kabylie (Algérie): un témoin biogéographique méconnu au statut taxinomique incertain. - *Acta Bot. Gallica* **144(1)**: 23-33.

DEBUSSCHE, M. & THOMPSON, J.D. 1998: The ecology and evolutionary biology of Western Mediterranean *Cyclamen*. In Moret, J. (ed.): Abstracts of the IX. OPTIMA meeting, Paris, 11-17 May 1998, p. 25. - Paris: Mus. Nat. d'Hist. Naturelle.

DEBUSSCHE, M. & THOMPSON, J.D. 2002: Morphological differentiation among closely related species with disjunct distribution: a case study of Mediterranean *Cyclamen* L. subgen. *Psilanthum* Schwarz (Primulaceae). - *Bot. J. Linn. Soc.* **139(2)**: 133-144.

DECAISNE, J. 1879: Monographie des genres *Ligustrum* et *Syringa*. - *Nouv. Arch. Mus. Hist. Nat., Sér. 2* **2**: 1-45 + 3 plates.

DEFAŸ, J. 1990: *Medicago arborea* L. - *Biocosme Mésogéen* **7(1-2)**: front cover.

DEFAŸ, J. 1991: *Pteris cretica* L. - *Biocosme Mésogéen* **8(3)**: front cover.

DEFAŸ, J. 1992: *Diplotaxis tenuifolia* (L.) DC. - *Biocosme Mésogéen* **9(2-3)**: front cover.

DEFAŸ, J. 1996: Polygala nicaeensis Risso ex Koch. - *Biocosme Mésogéen* **13(1)**: front cover.

DEFILIPPS, R.A. 1973: Index to taxa cited in "A survey of taxonomic and floristic reserach in Europe since 1945", *Webbia* 18, 1963. - *Webbia* **27(2)**: 299-368.

DEGEN, Á.V. 1891a: Bemerkungen über einige orientalische Pflanzenarten. II. *Campanula epigaea* Janka mss. n. sp. - *Oesterr. Bot. Z.* **41(6)**: 194-195.

DEGEN, Á.V. 1891b: Bemerkungen über einige orientalische Pflanzenarten. III. Fünf neue Bürger der europäischen Flora. - *Oesterr. Bot. Z.* **41(7)**: 231-232.

DEGEN, Á.V. 1891c: Ergebnisse einer botanischen Reise nach der Insel Samothrake. - *Oesterr. Bot. Z.* **41(9-10)**: 301-306, 329-338.

DEGEN, Á.V. 1893a: Bemerkungen über einige orientalische Pflanzenarten. VII. *Centaurea affinis* Friv. - *Oesterr. Bot. Z.* **43(2)**: 53-54.

DEGEN, Á.V. 1893b: Bemerkungen über einige orientalische Pflanzenarten. VIII. *Linum Thracicum* (Griseb.). - *Oesterr. Bot. Z.* **43(2)**: 55.

DEGEN, Á.V. 1893c: Bemerkungen über einige orientalische Pflanzenarten. IX. *Cytisus Frivaldszkyanus* n. sp. - *Oesterr. Bot. Z.* **43(12)**: 422-423.

DEGEN, Á.V. 1893d: Bemerkungen über einige orientalische Pflanzenarten. X. Zwei für Europa neue *Cytinus*-Arten in Bulgarien. - *Oesterr. Bot. Z.* **43(12)**: 423-424.

DEGEN, Á.V. 1894: Bemerkungen über einige orientalische Pflanzenarten. XII. Ueber das Vorkommen von *Nigella orientalis* L. und *Nigella sativa* L. in Europa. - *Oesterr. Bot. Z.* **44(2)**: 60-61.

DEGEN, Á.V. 1895a: Bemerkungen über einige orientalische Pflanzenarten. XVII. *Achillea Baldaccii* nov. spec. - *Oesterr. Bot. Z.* **45(1)**: 25.

DEGEN, Á.V. 1895b: Bemerkungen über einige orientalische Pflanzenarten. XVIII. Ueber *Peucedanum macedonicum* Janka und seine Verwandten. - *Oesterr. Bot. Z.* **45(2)**: 63-67.

DEGEN, Á.V. 1895c: Bemerkungen über einige orientalische Pflanzenarten. XX. *Galium Degenii* Baldacci in lit. nov. spec. - *Oesterr. Bot. Z.* **45(4)**: 131.

DEGEN, Á.V. 1895d: Bemerkungen über einige orientalische Pflanzenarten. XXI. Ueber *Saxifraga pseudosancta* Jka und *Saxifraga* "juniperina Ad." Velen. - *Oesterr. Bot. Z.* **45(6)**: 213-215.

DEGEN, Á.V. 1896a: Bemerkungen über einige orientalische Pflanzenarten. XXII-XXVII. - *Oesterr. Bot. Z.* **46(12)**: 413-418.

DEGEN, Á.V. 1896b: Sur une nouvelle espèce du genre *Zygis* Pers. (*Micromeria* Benth.). - *Bull. Herb. Boissier* **4(7)**: 521-524 + 1 plate.

DEGEN, Á.V. 1897: Bemerkungen über einige orientalische Pflanzenarten. XXX. *Echinops orientalis* Trautv. in Europa. - *Oesterr. Bot. Z.* **47(9)**: 313-316.

DEGEN, Á.V. 1898: Bemerkungen über einige orientalische Pflanzenarten. XXXIII. Ueber die systematische Stellung des *Alyssum Dörfleri* m. - *Oesterr. Bot. Z.* **48(3)**: 105-109.

DEGEN, Á.V. 1899: Bemerkungen über einige orientalische Pflanzenarten. XXXVII. *Ranunculus millefoliatus* Vahl und *R. garganicus* Ten. - *Oesterr. Bot. Z.* **49(7)**: 261-262.

DEGEN, Á.V. 1900: Bemerkungen über einige orientalische Pflanzenarten. XL. *Bornmüllera Dieckii* nov. spec. - *Oesterr. Bot. Z.* **50(9)**: 313-314.

DEGEN, Á.V. 1903: Bemerkungen über einige orientalische Pflanzenarten. XLII. Ueber zwei neue *Solenanthus*-Arten in Europa. - *Magyar Bot. Lapok* **2(11-12)**: 311-318.

DEGEN, Á.V. 1905: Bemerkungen über einige orientalische Pflanzenarten. XLIV. *Verbascum Dieckianaum* Borbás et Degen, n. sp. - *Magyar Bot. Lapok* **4(4-5)**: 82-83.

DEGEN, Á.V. 1911: Bemerkungen über einige orientalische Pflanzenarten. LV-LXVII. - *Magyar Bot. Lapok* **10(4-7)**: 108-115.

DEGEN, Á.V. 1912: Bemerkungen über einige orientalische Pflanzenarten. LXVIII. *Silene Regis Ferdinandi* Degen et Urumoff, n. sp. - *Magyar Bot. Lapok* **11(1-4)**: 2 + 1 plate.

DEGEN, Á.V. 1914a: Eugen von Halácsy [obituary]. - *Magyar Bot. Lapok* **13(1-5)**: 3-17.

DEGEN, Á.V. 1914b: Bemerkungen über einige orientalische Pflanzenarten. LXXVI-LXXVII. - *Magyar Bot. Lapok* **13(10-12)**: 309-311.

DEGEN, Á.V. 1917: Bemerkungen über einige orientalische Pflanzenarten. LXXIX. *Centaurea Immanuelis-Löwii* n. sp. - *Magyar Bot. Lapok* **16(1-12)**: 117-120 + 1 plate.

DEGEN, Á.V. 1919: Bemerkungen über einige orientalische Pflanzenarten. LXXX. *Astragalus Jankae* Degen et Bornm. nov. spec. - *Magyar Bot. Lapok* **18(1-12)**: 17-18.

DEGEN, Á.V. 1920a: Bemerkungen über einige orientalische Pflanzenarten. LXXXI. Über einige orientalische *Thymus*-Arten. - *Magyar Bot. Lapok* **19(1-12)**: 15-23.

DEGEN, Á.V. 1920b: Bemerkungen über einige orientalische Pflanzenarten. LXXXII. *Galium Stojanovii* Deg. n. sp. - *Magyar Bot. Lapok* **19(1-12)**: 48.

DEGEN, Á.V. 1922: Bemerkungen über einige orientalische Pflanzenarten. LXXXII-LXXXIII. - *Magyar Bot. Lapok* **21(1-12)**: 64-66.

DEGEN, Á.V. 1925: Bemerkungen über einige orientalische Pflanzenarten. LXXXIII. *Thymus Stojanovii* n. sp. - *Magyar Bot. Lapok* **23(1-12)**: 72-73.

DEGEN, Á.V. 1931: Bemerkungen über einige orientalische Pflanzenarten. XC. Über das Vorkommen der *Celtis Tournefortii* Lam. im kroatischen Küstengebiete. - *Magyar Bot. Lapok* **30(1-12)**: 79-80.

DEGEN, Á.V. 1933a: Bemerkungen über einige orientalische Pflanzenarten. XCI-XCIII. - *Magyar Bot. Lapok* **32(1-6)**: 64-71.

DEGEN, Á.V. 1933b: Bemerkungen über einige orientalische Pflanzenarten. XCIV-CI. - *Magyar Bot. Lapok* **32(7-12)**: 145-152.

DEGEN, Á.V. 1934: Bemerkungen über einige orientalische Pflanzenarten. CII-CX. - *Magyar Bot. Lapok* **33(1-12)**: 69-76.

DEGEN, Á.V. 1936: Flora Velebitica. I. Band. - Budapest: Verlag der Ungar. Akademie der Wissenschaften, xi + 662 pp.

DEGEN, Á.V. 1937: Flora Velebitica. II. Band. - Budapest: Verlag der Ungar. Akademie der Wissenschaften, 667 pp.

DEGEN, Á.V. 1938a: Flora Velebitica. III. Band. - Budapest: Verlag der Ungar. Akademie der Wissenschaften, 598 pp.

DEGEN, Á.V. 1938b: Flora Velebitica. IV. kötet [In Hungarian]. - Budapest: A Magyar Tudománuos Akadémia Kiadása, 99 pp.

DEGEN, Á.V. & DÖRFLER, I. 1897: Beitrag zur Flora Albaniens und Macedoniens. Ergebnisse einer von I. Dörfler im Jahre 1893 unternommenen Reise. - *Denkschr. Kaiserl. Akad. Wiss., Wien. Math.- Naturwiss. Kl.* **64**: 701-748 + 4 plates.

DEGRAEVE, N. 1975a: Contribution à l'étude cytotaxonomique des *Rumex*. - I. Le genre *Rumex* L. sensu stricto. - *Caryologia* **28(2)**: 187-201.

DEGRAEVE, N. 1975b: Contribution à l'étude cytotaxonomique des *Rumex*. - II. Le genre *Bucephalophora* Pau. - *Caryologia* **28(2)**: 203-206.

DEGRAEVE, N. 1976: Contribution à l'étude cytotaxonomique des *Rumex*. IV. Le genre *Acetosa* Mill. - *Cellule* **71(2)**: 231-250 + 3 plates.

DEGRAEVE, N. 1980a: Étude de diverses particularités caryotypiques des genres *Silene*, *Lychnis* et *Melandrium*. - *Bol. Soc. Brot., Ser. 2* **53(1)**: 595-643.

DEGRAEVE, N. 1980b: Contribution à l'étude cytotaxonomique des *Rumex*. III. Le genre *Acetosella* Fourr. - *Genetica* **54(1)**: 29-34.

DEGTJAREVA, G. & AL. 2004: Seed morphology in the genus *Pinguicula* (Lentibulariaceae) and its relation to taxonomy and phylogeny. - *Bot. Jahrb. Syst.* **125(4)**: 431-452.

DEKKER, H. 2003: Opmerkingen over orchideenvondsten op Rhodos in Maart 2002. - *Eurorchis* **15**: 49-54.

DEL CALDO, L. 1971: Numeri cromosomici per la Flora Italiana: 35-38. - *Inform. Bot. Ital.* **3(1)**: 71-73.

DEL CARRATORE, F. & GARBARI, F. 1997: Reports (700-704). In Kamari, G., Felber, F. & Garbari, F. (eds): Mediterranean chromosome number reports - 6. - *Fl. Medit.* **6**: 266-277.

DEL CARRATORE, F. & GARBARI, F. 1998: Indagini biosistematiche sul genere *Salvia* L. sect. *Plethiosphace* Bentham (Labiatae). - *Inform. Bot. Ital.* **29(2-3)**: 297-299.

DEL CARRATORE, GARBARI, F. & JARVIS, C.E. 1998: The application of the Linnaean names *Salvia pratensis, S. agrestis, S. haematodes, S. verbenaca* and *S. clandestina* (Lamiaceae). - *Pl. Biosyst.* **132(2)**: 169-176.

DEL GUACCHIO, E. & AL. 2020: Taxa endemic to Campania (southern Italy): nomenclatural and taxonomic notes. - *Phytotaxa* **449(3)**: 217-231.

DEL GUACCHIO, E. & CAPUTO, P. 2013a: Taxonomic notes on *Asperula* sect. *Cynanchicae* (Rubiaceae). - *Taxon* **62(2)**: 394-400.

DEL GUACCHIO, E. & CAPUTO, P. 2013b: (2136-2137) Proposals to reject the names

Asperula aristata subsp. *scabra* and *A. longiflora* (*A. aristata* subsp. *longiflora*) (Rubiaceae). - *Taxon* **62(2)**: 409-410. [not approved, cf. Taxon 63(6): 1363, 2014].

DEL GUACCHIO, E. & CAPUTO, P. 2020: Splitting *Asperula* (Rubiaceae): a proposal for consistency purposes within sections *Cynanchicae*, *Thliphthisa* and *Hexaphylla*. - *Pl. Biosyst.* **154(5)**: 766-782.

DEL PRETE, C. 1977: Numeri cromosomici per la Flora Italiana: 358-365. - *Inform. Bot. Ital.* **9(2)**: 135-140.

DEL PRETE, C. 1979: Contributi alla conoscenza delle "Orchidaceae" d'Italia. VI. Tavole cromosomiche delle "Orchidaceae" italiane con alcune considerazioni citosistematiche sui generi "Ophrys", "Orchis" e "Serapias". - *Inform. Bot. Ital.* **10(3)**: 379-389.

DEL PRETE, C. 1984a: The genus "Ophrys" L. (Orchidaceae) in Italy: Check-list of the species, subspecies and hybrids. (Contribution to the knowledge of the Italian Orchidaceae XII). - *Webbia* **37 (2)**: 249-257.

DEL PRETE, C. 1984b: The genus "Ophrys" L. (Orchidaceae): a new taxonomic approach. – *Webbia* **38**: 209-220.

DEL PRETE, C. 1999: The OPTIMA project for mapping Mediterranean orchids: the situation in Italy and a provisional checklist. - *Acta Bot. Fenn.* **162**: 145-154.

DEL PRETE, C. & AL. 1999: A database for mapping karyological data of European and Mediterranean orchids. - *Acta Bot. Fenn.* **162**: 155-159.

DEL PRETE, C., GARBARI, F. & GIORDANI, A. 1981: Numeri cromosomici per la Flora Italiana: 690-695. - *Inform. Bot. Ital.* **12(2)**: 117-120.

DEL PRETE, C. & GIORDANI, A. 1978: Numeri cromosomici per la Flora Italiana: 434-446. - *Inform. Bot. Ital.* **10(1)**: 119-129.

DEL PRETE, C. & MAZZOLA, P. 1995: Endemism and speciation in the orchids of Mediterranean islands. - *Ecol. Medit.* **21(1-2)**: 119-134.

DEL PRETE, C., MAZZOLA, P. & MICELI, P. 1991: Karyological differentiation and speciation in C. Mediterranean *Anacamptis* (Orchidaceae). - *Pl. Syst. Evol.* **174(3-4)**: 115-123.

DEL PRETE, C. & MICELI, P. 1994: Reports (387-393). In Kamari, G., Felber, F. & Garbari, F. (eds): Mediterranean chromosome number reports - 4. - *Fl. Medit.* **4**: 296-301.

DEL PRETE, C. & TOSI, G. 1988: Orchidee spontanee d'Italia. Monografia e iconografia. - Milano: U. Mursia editore S.p.A., 172 pp. + 48 plates.

DELAY, J. 1970a: Orophytes, xérothermophiles. - *Inform. Annuelles Caryosyst. Cytogén.* **4**: 1-8.

DELAY, J. 1970b: Particularités du matérial chromatique. - *Inform. Annuelles Caryosyst. Cytogén.* **4**: 25-28.

DELAY, J. 1973: Genre *Plantago* (suite). - *Inform. Annuelles Caryosyst. Cytogén.* **6**: 5-8.

DELAY, J. & PETIT, D. 1973: Matériel chromatique singulier. - *Inform. Annuelles Caryosyst. Cytogén.* **6**: 13-16.

DELCHEVA, M. & PEEV, D.R. 1997: Variability and correlations in populations of *Saxifraga ferdinandi-coburgi* Kell. et Sund. - *Phytol. Balcan.* **3(1)**: 19-25.

DELCHEVA, M., PEEV, D.R. & NIKOLOVA, M. 2010: On the taxonomical identity of *Abies alba* subsp. *borisii-regis* (Mattf.) Koz. et Andr. - morphometry, flavonoids and chorology in Bulgaria. - *Bot. Serbica* **34(2)**: 127-132.

DELCOURT, E. 1972: Contribution à l'étude cytotaxinomique de *Hieracum pilosella* L. - *Bull. Soc. Bot. France* **119(5-6)**: 287-302.

DELESTAING, N. 1954: Contribution à l'étude cytologique du genre *Salvia*. - *Rev. Cytol. Biol. Vég.* **15(3)**: 195-236.

DELEVOY, G. 1948: A propos de *Pinus heldreichii* Christ. - *Trav. Stat. Rech. Eaux Forêts, B* **2**: 1-9 [n. v.].

DELEVOY, G. 1949: A propos de la systématique de *Pinus nigra* Arnold. - *Trav. Stat. Rech. Eaux Forêts, B* **12**: 1-37 + 3 plates.

DELFORGE, P. 1983: Trois hybrides nouveaux d'*Ophrys spruneri* Nyman. - *Orchidophile (Ansières)* **14 (59)**: 496-499.

DELFORGE, P. 1990a: Contribution à la connaissance des orchidées du sud-ouest de Chypre et remarques sur quelques espèces méditerranéennes. - *Naturalistes Belges* **71**: 103-144.

DELFORGE, P. 1990b: Le groupe d'*Ophrys bertolonii* Moretti. - *Mém. Soc. Roy. Bot. Belg.* **11**: 7-29.

DELFORGE, P. 1992a: Contribution à l'étude de trois espèces d'*Ophrys* récemment décrites: *Ophrys cephalonica*, *Ophrys herae* et *Ophrys minoa* (Orchidaceae). - *Naturalistes Belges* **73(3)**: 71-105.

DELFORGE, P. 1992b: Les orchidées de l'île de Leucade (Nomos Lefkada, Nissia Ioniou, Grèce). Observations et additions à la cartographie. - *Naturalistes Belges* **73(3)**: 155-176.

DELFORGE, P. 1993a: Nouvelles observations sur *Ophrys herae* (Orchidaceae). - *Naturalistes Belges* **74(3)**: 107-112.

DELFORGE, P. 1993b: Les orchidées de l'île de Zante (Nomos Zakynthos, Nissia Ioniou, Grèce). Observations et cartographie. - *Naturalistes Belges* **74(3)**: 113-172.

DELFORGE, P. 1994a: Les orchidées des îles d'Andros et de Tinos (Cyclades, Grèce). Observations, cartographie et description d'*Ophrys andria*, une espèce nouvelle du groupe d'*Ophrys bornmuelleri*. - *Naturalistes Belges* **75(4)**: 109-170.

DELFORGE, P. 1994b: Note de synthèse sur la répartition des orchidées des îles ioniennes (Nissia Ioniou, Grèce). - *Naturalistes Belges* **75(4)**: 209-218.

DELFORGE, P. 1994c: Les orchidées des îles de Céphalonie et d'Ithaque (Nomos Kefallinia, Nissia Ioniou, Grèce). Observations et additions à la cartographie. - *Naturalistes Belges* **75(4)**: 219-271.

DELFORGE, P. 1994d: Guide des orchidées d'Europe, d'Afrique du Nord et du Proche-Orient. - Lausanne: Delachaux et Niestlé, 481 pp. [2nd ed. 2001].

DELFORGE, P. 1995a: Quelques observations sur les orchidées de l'île d'Eubée (Nomos Eyboia, Grèce). - *Naturalistes Belges* **76(3)**: 128-143.

DELFORGE, P. 1995b: Les orchidées des îles de Paros et Antiparos (Cyclades, Grèce). Observations, cartographie et description d'*Ophrys parosica*, une nouvelle espèce du sous-groupe d'*Ophrys fusca*. - *Naturalistes Belges* **76(3)**: 144-221.

DELFORGE, P. 1995c: Notes sur les orchidées de l'île d'Ios (Cyclades, Grèce). - *Naturalistes Belges* **76(3)**: 291-304.

DELFORGE, P. 1996: Observations sur les Orchidées du sud-est de la Laconie (Péloponèse, Grèce). - *Naturalistes Belges* **77(4)**: 119-136.

DELFORGE, P. 1997a: Les orchidées de l'île d'Amorgos (Cyclades, Grèce). - *Naturalistes Belges* **78(3)**: 103-152.

DELFORGE, P. 1997b: Description d'*Ophrys aeoli*, d*Ophrys astypalaeica* et d'*Ophrys thesei*, trois nouvelles orchidées des Cyclades (Grèce). - *Naturalistes Belges* **78(3)**: 153-176.

DELFORGE, P. 1997c: Nouveaux hybrides naturels d'orchidées d'Europe. - *Naturalistes Belges* **78(3)**: 177-188.

DELFORGE, P. 1997d: Les orchidées de l'île d'Astypaléa (Dodecanèse, Grèce). - *Naturalistes Belges* **78(3)**: 189-222.

DELFORGE, P. 1998: Note préliminaire sur les orchidées du sud-ouest des Cyclades (Grèce). - *Naturalistes Belges* **79(3)**: 114-116.

DELFORGE, P. 1999a: Contribution à la stabilisation de la nomenclature dans le groupe d'*Ophrys fusca*: désignation d'un néotype pour *Ophrys fusca* Link in Schrader 1800, *Ophrys funerea* Viviani 1824, *Ophrys bilunulata* Risso 1844 et *Ophrys forestieri* ... - *Naturalistes Belges* **80(3)**: 179-229.

DELFORGE, P. 1999b: Contribution taxonomique et nomenclaturale au genre *Himantoglossum* (Orchidaceae). - *Naturalistes Belges* **80(3)**: 387-408.

DELFORGE, P. 1999c: Contribution à la connaissance des *Serapias* des Cyclades (Grèce): *Serapias carica* (H. Baumann & Künkele 1989) P. Delforge 1994 var. *monantha* P. Delforge var. nov. - *Naturalistes Belges* **80(3)**: 409-431.

DELFORGE, P. 2000: *Ophrys leptomera* sp. nova. - *Naturalistes Belges* **81(3)**: 191-192, 217.

DELFORGE, P. 2002: Les orchidées des îles de Milos, Kimolos et Polyaigos (sud-ouest des Cyclades, Grèce). - *Naturalistes Belges* **83**: 67-120.

DELFORGE, P. 2005a: Un *Ophrys* lacedemonien. - *Naturalistes Belges* **85**: 235-244.
DELFORGE, P. 2005b: Contribution à la connaissance du groupe d´*Ophrys tenthredinifera* dans le bassin mediterranean oriental. - *Naturalistes Belges* **86**: 95-140.
DELFORGE, P. 2005c: Contribution à la connaissance du groupe d´*Ophrys blitopertha* en Crète. - *Naturalistes Belges* **86**: 141-146.
DELFORGE, P. 2005d: Note sur *Orchis papilionacea* var. *alibertis*. - *Naturalistes Belges* **86**: 81-90.
DELFORGE, P. 2008a: Note préliminaire sur les orchidées de l'île d'Icaria (Égée orientale, Grèce). - *Naturalistes Belges* **89**: 16-18.
DELFORGE, P. 2008b: Remarques sur *Serapias orientalis* dans le bassin égéen. - *Naturalistes Belges* **89**: 19-38.
DELFORGE, P. 2008c: Note complémentaire sur les orchidées de l'île de Chios (Égée orientale, Grèce). - *Naturalistes Belges* **89**: 62-70.
DELFORGE, P. 2008d: Contribution à la connaissance des orchidées de l'île de Samos (Égée orientale, Grèce). - *Naturalistes Belges* **89**: 71-251.
DELFORGE, P. 2009: Contribution à la connaissance des Orchidées de l'île des Cos (Dodécanèse, Grèce). - *Naturalistes Belges* **90**: 49-232.
DELFORGE, P. 2010a: Un nom pour la variété égéenne de l'*Orchis* papillon. - *Naturalistes Belges* **91**: 15-25.
DELFORGE, P. 2010b: *Orchis ×brigodeana* nothosp. nat. nova. - *Naturalistes Belges* **91**: 26-28.
DELFORGE, P. 2010c: Contribution à la connaissance des Orchidées de l'île de Cythère (Attique, Grèce). - *Naturalistes Belges* **91**: 47-205.
DELFORGE, P. 2011a: *Ophrys ×duchateauana* nothosp. nat. nova. - *Naturalistes Belges* **92**: 30-32.
DELFORGE, P. 2011b: Contribution à la connaissance des Orchidées de l'île de Kéa (Cyclades occidentales, Grèce). - *Naturalistes Belges* **92**: 124-201.
DELFORGE, P. 2012a: *Ophrys ×parvaisiana* nothosp. nat. nov. - *Naturalistes Belges* **93**: 94-96.
DELFORGE, P. 2012b: Contribution à la connaissance des orchidées de l'île d'Icaria (Îles égéennes orientales, Grèce). - *Naturalistes Belges* **93**: 163-241.
DELFORGE, P. 2013a: Contribution à la connaissance des orchidées de l'île de Skyros (Sporades du Nord, Grèce). - *Naturalistes Belges* **94**: 165-244.
DELFORGE, P. 2013b: Nouvelle contribution à la connaissance du groupe d'*Ophrys tenthredinifera* dans le bassin égéen (Grèce): *Ophrys amphidami* et *Ophrys lycomedis* sp. novae. - *Naturalistes Belges* **94**: 281-296.
DELFORGE, P. 2014a: *Ophrys ×lambrechtsiana* nothosp. nat. nova. - *Naturalistes Belges* **95**: 87-92.
DELFORGE, P. 2014b: Les noms d'un morphe égéen hypochrome d'*Anacamptis pyramidalis* et leut traitement dans la 'World Checklist of Selected Plant families'. - *Naturalistes Belges* **95**: 115-124.
DELFORGE, P. 2014c: L'*Ophrys* de Stavros K. - *Naturalistes Belges* **95**: 217-228.
DELFORGE, P. & DELFORGE-ONCKELINX, C. 2014: Contribution à la connaissance des Orchidées de l'île de Kythnos (Cyclades occidentales, Grèce). - *Naturalistes Belges* **95**: 125-216.
DELFORGE, P. & DELFORGE-ONCKELINX, C. 2017: Contribution à la connaissance des Orchidées de la presqu'île de Methana et l l'île de Poros (Golfe Saronique, Attique, Grèce). - *Naturalistes Belges* **98**: 69-172.
DELFORGE, P. & DELFORGE, C. 1984: *Ophrys ×borakisiana* hybr. nat. nov. - *Orchidophile (Ansières)* **15 (63)**: 685-687.
DELFORGE, P., DEVILLERS-TERSCHUREN, J. & DEVILLERS, P. 1991: Contributions taxonomiques et nomenclaturales aux orchidées d'Europe (Orchidaceae). - *Naturalistes Belges* **72(3)**: 99-101.
DELFORGE, P., JOUKOFF, C. & JOUKOFF, A. 1998: *Ophrys icariensis* Hirth et Spaeth dans l'île de Naxos (Cyclades, Grèce). - *Naturalistes Belges* **79(3)**: 107-113.
DELFORGE, P. & KREUTZ, C.A.J. 2003: *Epipactis heraclea*, une nouvelle espèce grecque du group d'*Epipactis helleborine*. - *Naturalistes Belges* **84 (Orchid. 16)**: 19-33.
DELFORGE, P. & SALIARIS, P.A. 2007: Contribution à la connaissance des Orchidées des îles de Chios, Inousses et Psara (Nomos Chiou, Égée orientale, Grèce). - *Naturalistes Belges* **88**: 41-227.
DELIPAVLOV, D.D. 1952: *Lilium rhodopeum* Delip. sp. n. - *Nauchni Trudove Selskost. Inst. "Vasil Kolarov"* **1**: 219-224.
DELIPAVLOV, D.D. 1962a: *Secale rhodopaeum* Delipavlov - A new species of rye from the Rhodope mountains. - *Dokl. Bulg. Akad. Nauk* **15(4)**: 407-410.
DELIPAVLOV, D.D. 1962b: *Sporobolus indicus* R. Br. [In Bulgarian with Russian and German summaries]. - *Nauchni Trudove Selskost. Inst. "Vasil Kolarov"* **11(1)**: 197-198.
DELIPAVLOV, D.D. 1964: *Arenaria rhodopaea* sp.n. - *Dokl. Bulg. Akad. Nauk* **17(7)**: 645-648.
DELIPAVLOV, D.D. 1966: *Amygdalus webbii* Spach [In Bulgarian with Russian and German summaries]. - *Nauchni Trudove Selskost. Inst. "Vasil Kolarov"* **15(2)**: 365-366.
DELIPAVLOV, D.D. 1968: Beitrag zur Erforschung der Gattung *Galanthus* L. (Schneeglöckchen) in Bulgarien [In Bulgarian with Russian and German summaries]. - *Nauchni Trudove Selskost. Inst. "Vasil Kolarov"* **17(1)**: 205-209.
DELIPAVLOV, D.D. 1971a: The genus *Galanthus* (snow-drop) in Bulgaria [In Bulgarian with English summary]. - *Izv. Bot. Inst. (Sofia)* **21**: 161-168.
DELIPAVLOV, D.D. 1971b: Einige für die bulgarische Flora neue und seltene Arten. - *Feddes Repert.* **81(10)**: 713-719.
DELIPAVLOV, D.D. 1976: Neue Arten der Gattung *Tulipa* L. aus Bulgarien. - *Feddes Repert.* **87(1-2)**: 1-4.
DELIPAVLOV, D.D. 1980: Neue Materialien zur Flora Bulgariens. - *Feddes Repert.* **91(1-2)**: 63-67.
DELIPAVLOV, D.D. 1988: Materials and notes on the flora of Bulgaria [In Bulgarian with Russian and English summaries]. - *Fitologiya* **34**: 67-72.
DELIPAVLOV, D.D. 1990: Neue Materialien und chorologische Daten zur Flora von Bulgarien. - *Feddes Repert.* **101(7-8)**: 341-345.
DELIPAVLOV, D.D. 1992: Materials on the flora of Bulgaria. - *Thaiszia* **2**: 59-66.
DELIPAVLOV, D.D. 1998: New taxa and chorological data for the flora of Bulgaria. - *Thaiszia* **8(2)**: 121-128.
DELIPAVLOV, D.D. 1999: Genus *Avena* L. (oats) in the flora of Bulgaria. - *Thaiszia* **9(1)**: 19-26.
DELIPAVLOV, D.D. 2000: *Seseli tortuosum* subsp. *thracicum* (Apiaceae), subsp. nova, and new chorological data from Bulgaria. - *Bot. Chron. (Patras)* **13**: 105-109.
DELIPAVLOV, D.D. & ANGELIEV, V. 1970: Die Arten der Gattung *Galanthus* L. auf der Balkanhalbinsel. - *Arch. Gartenbau* **18(8)**: 427-433.
DELIPAVLOV, D.D. & ČESCHMEDJIEV, I.V. 1983a: The genus *Arum* L. in Bulgaria [In Bulgarian with English abstract]. In Velčev, V.I. & al. (eds): Third national conference of botany, pp. 151-157. - Sofia: Publishing House of the Bulgarian Academy of Sciences.
DELIPAVLOV, D.D. & ČESCHMEDJIEV, I.V. 1983b: New taxa for the flora of Bulgaria [In Bulgarian with English abstract]. In Velčev, V.I. & al. (eds): Third national conference of botany, pp. 158-163. - Sofia: Publishing House of the Bulgarian Academy of Sciences.
DELIPAVLOV, D.D. & ČESCHMEDJIEV, I.V. 1984a: Contribution to the flora of Bulgaria [In Bulgarian with Russian and English summaries]. - *Nauchni Trudove Selskost. Inst. "Vasil Kolarov"* **29(4)**: 101-113.
DELIPAVLOV, D.D. & ČESCHMEDJIEV, I.V. 1984b: New data and some critical reviews for Bulgarian flora [In Bulgarian with Russian and English summaries]. - *Fitologiya* **26**: 60-68.
DELIPAVLOV, D.D. & ČESCHMEDJIEV, I.V. 1989: New materials and critical notes on the flora of Bulgaria [In Bulgarian with Russian and English summaries]. - *Fitologiya* **36**: 58-66.
DELIPAVLOV, D.D., ČESCHMEDJIEV, I.V. & POPOVA, M.T. 1984: New materials and chorological data about the flora of Bulgaria [In Bulgarian with Russian and English summaries]. - *Nauchni Trudove Selskost. Inst. "Vasil Kolarov"* **29(4)**: 95-99.
DELIPAVLOV, D.D. & POPOVA, M.T. 1965: Morphologische und anatomische Untersuchungen der gelbblühende Arten von der Gattung *Lilium* L. in Bulgarien [In Bulgarian with Russian and German summaries]. - *Nauchni Trudove Selskost. Inst. "Vasil Kolarov"* **14(1)**: 69-78.

DELIPAVLOV, D.D., POPOVA, M.T. & ČESCHMEDJIEV, I.V. 1983: A new *Pastinaca* species for the flora of Europe [In Bulgarian with English abstract]. In Velčev V.I. & al. (eds): Third national conference of botany, pp. 164-167. - Sofia: Publishing House of the Bulgarian Academy of Sciences.

DELIPETROU, P. & AL. 2015: Vegetation dynamics of coastal dunes with *Juniperus* spp. in Crete, Gavdos and Chrysi islands (Greece). - *Ecol. Medit.* **41(2)**: 45-60.

DELISLE, D.G. 1962: Taxonomy and distribution of the genus *Cenchrus*. - Iowa State University: Retrospective Theses and Dissertations. Paper 2098, 188 pp.

DELISLE, D.G. 1963: Taxonomy and distribution of the genus *Cenchrus*. - *Iowa State J. Sci.* **37**: 239-351.

DELL, B., HOPKINS, A.J.M. & LAMONT, B.B. 1986: Resilience in mediterranean-type ecosystems. - Dordrecht, etc.: Dr. W. Junk Publishers, viii + 168 pp.

DELLA, A. & IATROÚ, G. 1995: New plant records from Cyprus. - *Kew Bull.* **50(2)**: 387-396.

DEMIRI, M. 1979: Përcaktues Bimësh (Flora e vogël Shqipërisë) [In Albanian]. - Tiranë: Shtëpia Botuese e Librit Shkollor, 4 vols. (898 pp.).

DEMIRI, M. 1983: Flora ekskursioniste e Shqipërisë [In Albanian]. - Tiranë: Shtëpia Botuese e Librit Shkollor, 985 pp.

DEMIRIZ, H. 1963: Über die Verbreitung von *Anemone nemorosa* in der Türkei (Vorläufige Mitteilung) [In Turkish with German summary]. - *Doğa Türk Biol. Derg.* **13(2)**: 62-72.

DEMIRIZ, H. 1964: Georges Vincet Aznavour: 1861-1920. Sein Leben und wissenschaftlicher Nachlass über die Flora von Istanbul [In Turkish with German summary]. - *Doğa Türk Biol. Derg.* **14 (2)**: 49-64.

DEMIRIZ, H. 1969a: Studia ad floram et vegetationem Turciae pertinentia: I. A new *Anemone* for the flora of Turkey: *Anemone ranunculoides* L. - *Istanbul Üniv. Fen Fak. Mecm., B* **34(1-2)**: 119-122.

DEMIRIZ, H. 1969b: Studia ad floram et vegetationem Turciae pertinentia: V. *Telephium* L.: Supplementary material and distribution. - *Istanbul Üniv. Fen Fak. Mecm., B* **34(3-4)**: 183-188.

DEMIRIZ, H. 1969c: Studia ad floram et vegetationem Turciae pertinentia: VI. Taxonomic revision of the genus *Dorycnium* Miller in Turkey: Supplementary material. - *Istanbul Üniv. Fen Fak. Mecm., B* **34(3-4)**: 189-206.

DEMIRIZ, H. 1977: Studia ad floram turcicam: X. A new *Consolida* for the flora of Europe: *Consolida raveyi* (Boiss.) Schröd. - *Istanbul Üniv. Fen Fak. Mecm., B* **42**: 91-93.

DEMIRIZ, H. 1980: [Chromosome numbers in *Consolida* and *Delphinium* from Turkey]. In Löve, Á. (ed.): Chromosome number reports LXVII. - *Taxon* **29(2-3)**: 362.

DEMIRIZ, H. 1981: The chorological information in European Turkey. In Velčev, V.I. & Kožuharov, S.I. (eds): Mapping the flora of the Balkan Peninsula, pp. 69-73. - Sofia: Publishing House of the Bulgarian Academy of Sciences.

DEMIRIZ, H. 1993: An annotated bibliography of Turkish flora and vegetation. - Ankara: Scientific and Technical Research Council of Turkey, xviii + 670 pp.

DEMIRIZ, H. & ÇELEBIOĞLU, T. 1980: [Chromosome numbers in *Dorycnium* from the Istanbul area]. In Löve, Á. (ed.): Chromosome number reports LXVIII. - *Taxon* **29(4)**: 542.

DEMIRIZ, H. & ÇELEBIOĞLU, T. 1981: Chromosome number reports: *Dorycnium*. - *Lotus Newslett.* **12**: 20.

DEMIRIZ, H. & KOKTAY, P. 1980: [Chromosome numbers in *Verbascum* from the Istanbul area]. In Löve, Á. (ed.): Chromosome number reports LXVIII. - *Taxon* **29(4)**: 542.

DEMIRIZ, H. & ÖZHATAY, N. (EDS) 1993: Proceedings of the V OPTIMA meeting, Istanbul, 8-15 September, 1986. - Istanbul Üniversitesi Fen Fakültesi, xxxii + 797 pp. + 3 plates.

DEMIRIZ, H., TUTEL, B. & AYDIN, A. 1969: Studia ad floram et vegetationem Turciae pertinentia: IV. New materials to the Pteridophytes of Turkey: Filicales. - *Istanbul Üniv. Fen Fak. Mecm., B* **34(3-4)**: 137-181.

DEMOLY, J.-P. 1996a: Les hybrides binaires rares du genre *Cistus* L. (Cistaceae). - *Anales Jard. Bot. Madrid* **54(1)**: 241-254.

DEMOLY, J.-P. 1996b: L'identité de *Cistus sintenisii* de Lit. (Cistaceae) et d'un ciste hybride de Corse. - *Biocosme Mésogéen* **13(1)**: 15-21.

DEMUT, E. 1978: Ein neuer *Ophrys*-Bastard von der Insel Kefallinia (Griechenland): *O. bombyliflora* Link × *O. ferrum-equinum* Desf. - *Orchidee (Hamburg)* **29(3)**: 109.

DENALOË, O., MONTMOLLIN, B. DE & OLIVIER, L. 1996: Conservation of Mediterranean island plants. 1. Strategy for action [In French and English]. - Gland and Cambridge: IUCN, 106 pp.

DENČEV, C.M. & KOŽUHAROV, S.I. 1987: New chorological data from the basin of the Vlahinska river (Pirin) [In Bulgarian with Russian and English summaries]. - *Fitologiya* **33**: 57-64.

DENELLE, N. & PUECH, S. 1985: Étude caryosytématique de *Thymelaea hirsuta* (L.) Endl. (Thymelaeaceae) dans deux stations du littoral méditerranéen français (Camargue et Albères). - *Taxon* **34(4)**: 611-616.

DENGLER, H.W. (ED.) 1957: Handbook of hollies. - Washington, D.C.: vi + 193 pp. (special issue of Natl. Hort. Mag.).

DENGLER, J. 1998: Zur Unterscheidung von *Phleum pratense* s.str. und *Phleum bertolonii*. - *Florist. Rundbr.* **32(1)**: 7-12.

DENHAM, A. 2018: Dioscorides' legacy: a classical precursor to travellers in Ottoman lands. In Aščerić-Todd, I. & al. (eds): Travellers in Ottoman lands. The Botanical Legacy, pp. 89-108. - Oxford: ASTENE and Archaeopress Publishing Ltd, 379 pp.

DENK, T. 1999a: The taxonomy of *Fagus* in western Eurasia, 1: *Fagus sylvatica* ssp. *orientalis*. - *Feddes Repert.* **110(3-4)**: 177-200.

DENK, T. 1999b: The taxonomy of *Fagus* in western Eurasia. 1: *Fagus sylvatica* subsp. *sylvatica*. - *Feddes Repert.* **110(5-6)**: 381-412.

DENK, T. & GRIMM, G.W. 2005: Phylogeny and biogeography of *Zelkova* (Ulmaceae sensu stricto) as inferred from leaf morphology, ITS sequence data and the fossil record. - *Bot. J. Linn. Soc.* **147(2)**: 129-157.

DENNEY, M. 1994: *Cyclamen* on Rhodes. - *Cyclamen* **18(2)**: 48-49.

DENNEY, M. 1996a: *C[yclamen] graecum* on Rhodes. - *Cyclamen* **20(2)**: 33.

DENNEY, M. 1996b: The Cyclamen Society expeditions to Crete. - *Cyclamen* **20(2)**: 36-43.

DENNEY, M. 1998: Variations in *Cyclamen graecum* around Athens. - *Cyclamen* **22(1)**: 26-27.

DENNEY, M. 1999: Cyclamen Society expedition to the Dodecanese, 1998. - *Cyclamen* **23(1)**: 15-19 + 4 plates.

DENNEY, M. 2010: The name behind *Cyclamen repandum* Sm. - *Cyclamen* **34**: 54-58.

DERGANC, L. 1902: Geographische Verbreitung der *Daphne Blagayana* Freyer. - *Allg. Bot. Z. Syst.* **8(11-12)**: 176-179, 195-197.

DERGANC, L. 1904: Nachtrag zum Aufsatze über die geographische Verbreitung der *Daphne Blagayana* Freyer. - *Allg. Bot. Z. Syst.* **10(3-4)**: 44-47.

DERMITZAKIS, M.D. 1989: The colonization of Aegean islands in relation with the paleogeographical evolution. Part A [In Greek with French summary]. - *Biol. Gallo-Hellen.* **14(2)**: 99-121.

DERMITZAKIS, M.D. 1990a: The colonization of Aegean islands in relation with the paleogeographical evolution. Part B. - *Biol. Gallo-Hellen.* **17(2)**: 99-130.

DERMITZAKIS, M.D. 1990b: The evolution of the Aegeis during the Late Cenozoic. - *Geolog. Balcan.* **20(3)**: 3-16.

DERRICK, L.N., JERMY, A.C. & PAUL, A.M. 1987: Checklist of European pteridophytes. - *Sommerfeltia* **6**: i-xx, 1-94.

DERRICK, L.N. & SCHEEPEN, J. VAN 1984: The European taxonomic, floristic & biosystematic documentation system. - *Webbia* **38**: 681-685.

DERSCH, G. 1964: Zur Cytologie und Taxonomie der Goldnessel (*Lamium galeobdolon* [L.] L.). - *Ber. Deutsch. Bot. Ges.* **76(9)**: 351-359.

DERSCH, G. 1969: Über das Vorkommen von diploidem Wiesenschaumkraut (*Cardamine pratensis* L.) in Mitteleuropa. - *Ber. Deutsch. Bot. Ges.* **82(3-4)**: 201-207.

DERVIZ-SOKOLOVA, T.G. 1975: *Stachys palustris* L. s.l. (systematics and geography) [In Russian with English summary]. - *Byull. Moskovsk. Obshch. Isp. Prir., Otd. Biol.* **80(3)**: 85-93.

DESCHÂTRES, R. & GREUTER, W. 2001: *Bassia hyssopifolia, Artemisia verlotiorum, Cuscuta campestris, Sedum cepaea, Euphorbia (Chamaesyce) prostrata, Nicandra physalodes, Schoenoplectus lacustris*. In Greuter, W. & Raus, Th. (eds): Med-Checklist notulae, 20. - *Willdenowia* **31(2)**: 320-326.

DESCHÂTRES, R., GUITTONNEAU, G.-G. & ZAFFRAN, J. 1998: Compte rendu du voyage d'études de la Societé botanique de France en Crète (18-25 Mai 1988). - *J. Bot. Soc. Bot. France* **8**: 25-42.

DESCHÂTRES, R., SCHNELLER, J.J. & REICHSTEIN, T. 1978: A tetraploid cytotype of *Asplenium cuneifolium* Viv. in Corsica. - *Brit. Fern Gaz.* **11(6)**: 343-344.

DESFAYES, M. 2004a: Additions to the vascular flora of Albania. - *Ann. Bot., n.s. (Rome)* **4**: 144-158.

DESFAYES, M. 2004b: The specific status of *Cyperus badius* and the subspecies of *Scirpoides holoschoenus* (Cyperaceae), with special reference to Sardinia. - *Fl. Medit.* **14**: 173-188.

DESFONTAINES, R.L. 1798-99: Flora Atlantica sive historia plantarum quae in Atlante, agro Tunetato et Algerien si crescunt. - Paris: L.G. Desgranges, 2 vols. [cf. TL-2 1: 628-629, 1976].

DESFONTAINES, R.L. 1808: Choix de plantes du corollaire des instituts de Tournefort, publiées d'après son herbier, et gravées sur les dessins originaux d'Aubriet. - Paris: F.G. Levrault, 92 pp. + 70 plates.

DESMOTS, D. 2010: *Althenia orientalis* (Tzvelev) Garcia Murillo & Talavera sur l'île de Noirmoutier (Vendée), decouverte ou redecouverte atlantique française? - *E.R.I.C.A.* **23**: 17-22.

DEVESA, J.A. 1979: Números cromosómicos para la flora española. 110-115. - *Lagascalia* **9(1)**: 126-128.

DEVESA, J.A. 1980a: *Carduus pycnocephalus* L. (Compositae) en Portugal. - *Bol. Soc. Brot., Ser. 2* **53(1)**: 671-673.

DEVESA, J.A. 1980b: Números cromosómicos para la flora española. 137-140. - *Lagascalia* **9(2)**: 259-261.

DEVESA, J.A. 1981a: Contribución al estudio cariológico del género *Carduus* en la Península Ibérica. - *Lagascalia* **10(1)**: 65-80.

DEVESA, J.A. 1981b: El género *Evax* Gaertner en Andalucia Occidental. - *Lagascalia* **10(1)**: 113-117.

DEVESA, J.A. 1981c: Números cromosómicos para la flora española. 182-187. - *Lagascalia* **10(2)**: 225-227.

DEVESA, J.A. 1984a: Revisión del género *Scabiosa* en la Península Ibérica e Islas Baleares. - *Lagascalia* **12(2)**: 143-212.

DEVESA, J.A. 1984b: *Pseudoscabiosa*, género nuevo de Dipsacaceae. - *Lagascalia* **12(2)**: 213-221.

DEVESA, J.A. & AL. 1991: Contribución al conocimiento cariológico de las Poaceae en Extremadura (España) - III. - *Bol. Soc. Brot., Ser. 2* **64**: 35-73 + 1 plate.

DEVESA, J.A. & AL. 2003: Nuevas combinaciones en Dipsacaceae. - *Acta Bot. Malac.* **28**: 210-215.

DEVESA, J.A. & LÓPEZ, G. 1997: Nótas taxonómical y nomenclaturales sobre el género *Ononis* (Leguminosae) en la Península Ibérica e Islas Baleares. - *Anales Jard. Bot. Madrid* **55(2)**: 245-260.

DEVESA, J.A., LÓPEZ, J. & GONCALO, L. 2005: Notas taxonómicas sobre el género *Valerianella* Mill. (Valerianaceae) para la Flora Iberica. - *Acta Bot. Malac.* **30**: 41-48.

DEVESA, J.A. & LUQUE, T. 1988: Contribución al conocimiento cariológico de la subfam. Pooideae (Poaceae) en el SW de España. - *Bol. Soc. Brot., Ser. 2* **61**: 281-304.

DEVESA, J.A. & ROMERO ZARCO, C. 1981: Números cromosómicos para la flora española. 188-196. - *Lagascalia* **10(2)**: 227-230.

DEVESA, J.A. & TALAVERA, S. 1981a: Algunas novedades taxonómicas y nomenclaturales para el género *Carduus* en la Península Ibérica. - *Lagascalia* **10(1)**: 59-63.

DEVESA, J.A. & TALAVERA, S. 1981b: Revisión del género *Carduus* (Compositae) en la Península Ibérica e Islas Baleares. - Sevilla: Publicaciones de la Universidad de Sevilla, 118 pp.

DEVESA, J.A. & VALDÉS, B. OTTONELLA, D. 1988: [Chromosome numbers from Sicily]. In Löve, Á. (ed.): Chromosome number reports C. - *Taxon* **37(4)**: 920.

DEVEY, D.S. & AL. 2008: Friends or relatives? Phylogenetics and species delimitation in the controversial European orchid genus *Ophrys*. - *Ann. Bot. (Oxford), n.s.* **101(3)**: 385-402.

DEVILLERS-TERSCHUREN, J. & DEVILLERS, P. 1988: Les *Ophrys* "arachniteformes" du bassin méditerranéen occidental. - *Naturalistes Belges* **69(2)**: 98-112.

DEVILLERS, P. & AL. 2010: Orchids of Lesbos: Distributional and biogeographical notes. - *Naturalistes Belges* **91**: 206-245.

DEVILLERS, P. & DEVILLERS-TERSCHUREN, J. 1994: Essai d'analyse systématique du genre *Ophrys*. - *Naturalistes Belges* **75, hors-série**: 273-400.

DEVILLERS, P. & DEVILLERS-TERSCHUREN, J. 1999a: *Epipactis spiridonovii* et le group d'*Epipactis atrorubens*. - *Naturalistes Belges* **80(3)**: 286-291.

DEVILLERS, P. & DEVILLERS-TERSCHUREN, J. 1999b: Essai de synthese du groupe d'*Epipactis phyllanthes, E. gracilis, E. persica* et de sa représentation dans les hêtraies subméditerranéennes d'Italie, de Grèce, de France, d'Espagne et de Bulgarie. - *Naturalistes Belges* **80(3)**: 283-285, 292-310.

DEVILLERS, P. & DEVILLERS-TERSCHUREN, J. 2004: Petits *Ophrys* du complexe d'*Ophrys fusca* s.l. en Grèce occidentale. - *Naturalistes Belges* **85**: 247-249.

DEVILLERS, P. & DEVILLERS-TERSCHUREN, J. 2009: Rhodian *Ophrys*: Diagnostic characters, relationships and biogeography. - *Naturalistes Belges* **90**: 233-290.

DEVILLERS, P., DEVILLERS-TERSCHUREN, J. & TYTECA, D. 2003: Notes on some of the taxa comprising the group of *Ophrys tenthredinifera* Willdenow. - *J. Eur. Orch.* **35(1)**: 109-161.

DEVILLIERS, P. & AL. 2012: Orchids of Lesbos: Photographic documentation. - *Naturalistes Belges* **93**: 33-62.

DEVOS, N. & AL. 2006: On the monophyly of *Dactylorhiza* Necker ex Nevski (Orchidaceae): is *Coeloglossum viride* (L.) Hartman a *Dactylorhiza*? - *Bot. J. Linn. Soc.* **152(3)**: 261-269.

DEYL, M. 1946: Study of the genus *Sesleria*. - *Opera Bot. Čech.* **3**: 1-257.

DI CASTRI, F. 1981: Mediterranean-type shrublands of the world. In Di Castri, F., Goodall, D.W. & Specht, R.L. (eds): Mediterranean-type shrublands, pp. 1-52. - Amsterdam, etc.: Elsevier.

DI CASTRI, F., GOODALL, D.W. & SPECHT, R.L. (EDS) 1981: Mediterranean-type shrublands [Ecosystems of the World, vol. 11]. - Amsterdam, etc.: Elsevier, xii + 643 pp.

DI CASTRI, F., HANSEN, A.J. & DEBUSSCHE, M. (EDS) 1990: Biological invasions in Europe and the Mediterranean Basin. - Dordrecht, etc.: Kluwer Academic Publishers, ix + 463 pp.

DI CASTRI, F. & MOONEY, H.A. (EDS) 1973: Mediterranean type ecosystems. Origin and structure. - Berlin, etc.: Springer-Verlag, vii + 405 pp.

DI MARTINO, A. 1963: Flora e vegetazione dell'Isola di Pantelleria. - *Lav. Ist. Bot. Giard. Colon. Palermo* **19**: 87-243.

DI MARTINO, A. 1968: Nuove stazioni di *Solanum elaeagnifolium* Cav. var. *leprosum* in Sicilia. - *Lav. Ist. Bot. Giard. Colon. Palermo* **23**: 312-314.

DI MARTINO, A. 1970: Sulla presenza in Sicilia dello *Scorpiurus vermiculata* L. (sua distribuzione in Italia ed areale). - *Lav. Ist. Bot. Giard. Colon. Palermo* **24**: 45-61.

DI MARTINO, A. & RAIMONDO, F.M. 1979: Biological and chorological survey of the Sicilian flora. - *Webbia* **34(1)**: 309-335.

DI MARTINO, A. & TRAPANI, S. 1967: Flora e vegetazione delle Isole di Favignana e Levanzo nell'Arcipelago delle Egadi. I. - Favignana. - *Lav. Ist. Bot. Giard. Colon. Palermo* **22**: 122-228.

DI MARTINO, A. & TRAPANI, S. 1968: Flora e vegetazione delle Isole di Favignana e Levanzo nell'Arcipelago delle Egadi. II. - Levanzo. - *Lav. Ist. Bot. Giard. Colon. Palermo* **23**: 37-152.

DI PASQUALE, G., GARFI, G. & QUÉZEL, P. 1992: Sur la présence d'un *Zelkova* nouveau en Sicile sud- orientale. - *Biocosme Mésogéen* **8(4)-9(1)**: 401-409.

DI PIETRO, R. & AL. 2012: A review of the application of the name *Quercus dalechampii*. - *Taxon* **61 (6)**: 1311-1316.

DI PIETRO, R. & AL. 2015: The *Sesleria* sect. *Argentea* (Poaceae) in the Amphi-Adriatic area. Nomenclatural and coenological remarks. In Bogdanović, S. & Jogan, N.: Book of abstracts - 6th Balkan Botanical Congress, p. 22. - Rijeka.

DI PIETRO, R. & AL. 2017: Nomenclatural and taxonomic notes on *Sesleria* sect. *Argenteae* (Poaceae). - *Phytotaxa* **309(2)**: 101-117.

DI PIETRO, R. & AL. 2021: Nomenclatural and taxonomic notes on some names of *Sesleria* sect. *Argenteae* (Poaceae) from Italy and the Balkans. - *Phytotaxa* **494(1)**: 89-102.

DI PIETRO, R. & SCHOLZ, H. 2006: *Sesleria autumnalis*. In Greuter, W. & Raus, Th. (eds): Med- Checklist notulae, 24. - *Willdenowia* **36(2)**: 729-730.

DIAMANTOPOULOS, J. & MARGARIS, N.S. 1981a: The transition of *Euphorbia acanthothamnos* from the winter to the summer form. - *Flora, Morphol. Geobot. Oekophysiol.* **171(4)**: 315-328.

DIAMANTOPOULOS, J. & MARGARIS, N.S. 1981b: Flowering times and flower colours in the flora of Greece. - *Phyton (Horn)* **21(2)**: 241-244.

DIANA CORRIAS, S. 1977a: Numeri cromosomici per la Flora Italiana: 250. - *Inform. Bot. Ital.* **8(2)**: 203.

DIANA CORRIAS, S. 1977b: Numeri cromosomici per la Flora Italiana: 329-330. - *Inform. Bot. Ital.* **9(1)**: 88-93.

DIANA CORRIAS, S. 1978: Numeri cromosomici per la Flora Italiana: 409-415. - *Inform. Bot. Ital.* **10(1)**: 94-101.

DIANA CORRIAS, S. 1981: Numeri cromosomici per la Flora Italiana: 701-707. - *Inform. Bot. Ital.* **12(2)**: 125-129.

DIANA CORRIAS, S. 1984: Numeri cromosomici per la Flora Italiana: 948-955. - *Inform. Bot. Ital.* **15(1)**: 44-48.

DIANA CORRIAS, S., GARBARI, F. & MARCHI, P. 1986: Numeri cromosomici per la Flora Italiana: 1-1000 [Index]. - *Inform. Bot. Ital.* **16(2-3)**: 219-241. [Also distributed as *OPTIMA leaflet* no. 164].

DIANA CORRIAS, S. & VILLA, R. 1981: Numeri cromosomici per la Flora Italiana: 708-713. - *Inform. Bot. Ital.* **12(2)**: 131-135.

DIANELLIDIS, TH. 1935: To Pilion apo fitologikis apopseos [In Greek]. - *Vouno [Journal of the Hellenic Alpine Club]* **1935**: 33-36.

DIANELLIDIS, TH. 1949: A study of chromosomes of the Orchidaceae [In Greek]. - *Compt. Rend. Acad. Athènes [Praktika Akadimias Athinon]* **23**: 352-359.

DIANELLIDIS, TH. 1951: Cytologische Studien an einigen *Allium*-Arten aus Nord-Griechenland. - *Portug. Acta Biol., Sér. A, Morfol.* **3(2)**: 151-170.

DIANELLIDIS, TH. 1955: Chromosomenzahlen einiger Orchidaceen. - *Sci. Ann. Fac. Phys. Math. Aristot. Univ. Thessalonika* **7**: 3-9.

DIAPOULIS, CH. 1933: Beiträge zur Kenntnis der orientalischen Pomaceen (*Pirus*, *Sorbus*, *Crataegus*). - *Repert. Spec. Nov. Regni Veg.* **34**: 29-72 + 21 plates.

DIAPOULIS, CH. 1934a: Zur Kenntnis der Gattung *Rhamnus*. - *Repert. Spec. Nov. Regni Veg.* **33**: 338-340.

DIAPOULIS, CH. 1934b: Endemika fita tis nisou Kritis. *Zelkova cretica* Spach [In Greek]. - *Dasiki Zoi* **18-19**: 123-127 [n.v.].

DIAPOULIS, CH. 1934c: De la flore grecque. *Staehelina uniflosculosa* Sibth. et Sm. [En grec]. - *Dasiki Zoi* **2(22)**: 207-208 [n.v.].

DIAPOULIS, CH. 1934d: Die Flora von Andros. (Kritische Notiz). - *Repert. Spec. Nov. Regni Veg.* **33**: 340-341.

DIAPOULIS, CH. 1935?: Die Rhamnaceen im Mittelmeergebiet [In Greek with German summary]. - Athine: 56 pp.

DIAPOULIS, CH. 1935a: Apo tin chlorida tou Olimbou [In Greek]. - *Vouno [Journal of the Hellenic Alpine Club]* **1935**: 411-424.

DIAPOULIS, CH. 1935b: I pevki en Elladi. Pevki i Trachia (*Pinus brutia* Tenore) [In Greek]. - *Deltion Fisik. Epist.* **1935**: 51-56.

DIAPOULIS, CH. 1936a: Beitrag zur Kenntnis der Waldvegetation des Olymps und des Pierriagebirges. - *Repert. Spec. Nov. Regni Veg.* **40**: 282-287.

DIAPOULIS, CH. 1936b: *Amaracus dicnamnus* [sic!] Benth. (Amarakos o diktamnos) [In Greek]. - *Dasiki Zoi* **1936**: 7 pp.

DIAPOULIS, CH. 1938: I Dris en Elladi. Peri ton notioteron orion exaploseos Drios tis Makedonikis (*Quercus macedonica* D.C.) [In Greek]. - *Deltion Fisik. Epist.* **45-46**: 13-119.

DIAPOULIS, CH. 1939a: *Quercus aegilops* [In Greek with French summary]. - *Geôrgikon Deltion (Agricultural Bulletin of Greek Agricultural Society)* **1939(5)**: 3-46 + 1 table.

DIAPOULIS, CH. 1939b: Elliniki chloris. Klides prosdiorismou ton ellinikon fiton avtofion ke kalliergimenon. 1 [In Greek]. - Athine: 267 pp.

DIAPOULIS, CH. 1940a: Fitogeografiki episkopisis Vermiou-Pierion-Olimbou. - *Vouno [Journal of the Hellenic Alpine Club]* **1940**: 223-233, 249-254 [n.v.].

DIAPOULIS, CH. 1940b: La flore des praires de la Macédoine occidentale et centrale [In Greek with French summary]. - Athènes: Ministère de l'Agriculture, Publ. Lab. Rech. Forest., 96 pp.

DIAPOULIS, CH. 1948: Elliniki chloris. Klides prosdiorismou ton ellinikon fiton avtofion ke kalliergimenon. 2(1) [In Greek]. - Athine: 528 pp.

DIAPOULIS, CH. 1949: Elliniki chloris. Klides prosdiorismou ton ellinikon fiton avtofion ke kalliergimenon. 2(2) [In Greek]. - Athine: 384 pp.

DIAPOULIS, CH. 1950: I xilokeratea (*Ceratonia siligua* L [sic!]) idia en Elladi. - Athine: 40 pp. + 1 map and 1 chart.

DIAPOULIS, CH. 1954: Le *Styrax* en Grèce. - *Mater. Veg.* **1(1)**: 119-121.

DIAPOULIS, CH. 1958: Apo tin chlorida tis Parnithos [In Greek]. - *Vouno [Journal of the Hellenic Alpine Club]* **1958**: 163-188.

DIAPOULIS, CH. 1961: Endemika fita ton Kikladon nison [In Greek]. - *Epet. Kikladikon Meleton* **1**: 1-16.

DIAPOULIS, CH. 1967: Distribution of *Quercus trojana* in Greece [In Greek with English summary]. - *Biol. Gallo-Hellen.* **1(1)**: 27-31.

DIAPOULIS, CH. 1971: The development of the flora of the volcanic islands Palaia Kammeni and Nea Kammeni. In Kaloyeropoulou, A. (ed.): Acta 1st Internat. Congress Volcano Thera, pp. 238-247. - Athens.

DÍAZ DE LA GUARDIA, C. 1995: Números cromosómicos para la flora española. 769-773. - *Lagascalia* **18(1)**: 119-122.

DÍAZ DE LA GUARDIA, C. & BLANCA LÓPEZ, G. 1985: Sobre *Scorzonera angustifolia* L. (Asteraceae). - *Anales Jard. Bot. Madrid* **42(1)**: 113-116.

DÍAZ DE LA GUARDIA, C. & BLANCA LÓPEZ, G. 1987a: Revisión del género *Scorzonera* L. (Compositae, Lactuceae) en la Península Ibérica. - *Anales Jard. Bot. Madrid* **43(2)**: 271-354.

DÍAZ DE LA GUARDIA, C. & BLANCA LÓPEZ, G. 1987b: Karyology of the *Scorzonera* (Compositae) species from the Iberian Peninsula. - *Pl. Syst. Evol.* **156(1-2)**: 29-42.

DÍAZ DE LA GUARDIA, C. & BLANCA LÓPEZ, G. 1988a: La posición sistemática de *Geropogon* L. (Compositae) en la subtribu Scorzonerinae Dumort. (1). - *Lagascalia* **15(Extra)**: 361-367.

DÍAZ DE LA GUARDIA, C. & BLANCA LÓPEZ, G. 1988b: El género *Geropogon* L. (Compositae, Lactuceae). - *Lazaroa* **9**: 31-44.

DÍAZ DE LA GUARDIA, C. & BLANCA LÓPEZ, G. 1992: Lectotypification of five Linnaean names of *Tragopogon* L. (Compositae). - *Taxon* **41(3)**: 548-551.

DÍAZ LIFANTE, Z. 1990: Números cromosómicos para la flora española. 630-642. - *Lagascalia* **16(2)**: 328-333.

DÍAZ LIFANTE, Z. 1991: *Asphodelus cirerae*, a forgotten species of *Asphodelus* sect. *Verinea* (Liliaceae). Morphological, palynological, karyological and ecogeographical characterization. - *Fl. Medit.* **1**: 87-109.

DÍAZ LIFANTE, Z. 1992: Karyological studies in Asphodelaceae and Anthericaceae. I. Chromosome numbers in *Asphodelus* and *Anthericum* from N Africa. - *Willdenowia* **22(1-2)**: 143-148.

DÍAZ LIFANTE, Z. 1996: A karyological study of *Asphodelus* L. (Asphodelaceae) from the Western Mediterranean. - *Bot. J. Linn. Soc.* **121(4)**: 285-344.

DÍAZ LIFANTE, Z. & ANDRÉS CAMACHO, C. 2007: Morphological variation of *Narcissus serotinus* L. s.l. (Amaryllidaceae) in the Iberian Peninsula. - *Bot. J. Linn. Soc.* **154(2)**: 237-257.

DÍAZ LIFANTE, Z., LUQUE, T. & SANTA BÁRBARA, C. 1992: Chromosome numbers of plants collected during Iter Mediterraneum II in Israel. - *Bocconea* **3**: 229-250.

DÍAZ LIFANTE, Z. & VALDÉS, B. 1993: Characterization and relationships of *Asphodelus aestivus* Brot. In Petrova, A.V. (ed.): Abstracts from the VII OPTIMA meeting in Borovetz, Bulgaria, 18-30 July 1993, p. 87.

DÍAZ LIFANTE, Z. & VALDÉS, B. 1994a: Lectotypification of *Asphodelus ramosus* (Asphodelaceae), a misunderstood Linnaean name. - *Taxon* **43(2)**: 247-251.

DÍAZ LIFANTE, Z. & VALDÉS, B. 1994b: *Asphodelus ayardii*, the correct name for *A. cirerae* (Asphodelaceae). - *Fl. Medit.* **4**: 197-199.

DÍAZ LIFANTE, Z. & VALDÉS, B. 1995: Reproductive biology and hybridization of the species of *Asphodelus* sect. *Verinea* (Pomel) Baker (Asphodelaceae). - *Bot. Helv.* **105(1)**: 97-109.

DÍAZ LIFANTE, Z. & VALDÉS, B. 1996: Revisión del género *Asphodelus* L. (Asphodelaceae) en el Mediterráneo Occidental. - *Boissiera* **52**: 189 pp.

DIAZ, A., AMOIN, M.A. & GIBERNAU, M. 2006: The effectiveness of some mechanisms of reproductive isolation in *Arum maculatum* and *A. italicum* (Araceae). - *Bot. J. Linn. Soc.* **150(3)**: 323-328.

DIAZ, A. & KITE, G.C. 2002: A comparison of the pollination ecology of *Arum maculatum* and *A. italicum* in England. - *Watsonia* **24(2)**: 171-181.

DIEDERICHSEN, A. & RICHARDS, K. 2003: Cultivated flax and the genus *Linum* L.: taxonomy and germplasm conservation. In Muir, A.D. & Westcott, N.D. (eds): Flax, the genus *Linum*, pp. 22-54. - London: Routledge.

DIELS, L. 1917: *Phelipaea Boissieri* Stapf in Macedonien. - *Notizbl. Bot. Gart. Berlin-Dahlem* **6(62)**: 416.

DIEMAR, S. & SEBERG, O. 1989: Biographical and bibliographical notes on C.A. and P.C. Candargy. - *Taxon* **38(4)**: 569-575.

DIETERICH, T. & AL. 2012: Grasslands of the Prespa National Park. - Tirana: 185 pp.

DIETRICH, H. 1971: Blütenmorphologische und palynologische Untersuchungen an *Littorella*. - *Feddes Repert.* **82(2)**: 155-165.

DIETRICH, H. 1975: Cytologische Untersuchungen innerhalb der Familie der Plantaginaceae II. Zusammenstellung bekannter Chromosomenzahlen und der Vergleich zwischen Grundzahl und Lebensdauer. - *Wiss. Z. Friedrich-Schiller-Univ. Jena, Math.-Naturwiss. Reihe* **24(4)**: 437-461.

DIETRICH, H. 1980: Cytologische Untersuchungen innerhalb der Familie der Plantaginaceae III. Cytotaxonomische Ergebnisse. - *Wiss. Z. Friedrich-Schiller-Univ. Jena, Math.-Naturwiss. Reihe* **29 (4)**: 559-587.

DIETRICH, H. 1982: Umkombinationen in der Plantaginaceen-Gattung *Psyllium* Miller. - *Wiss. Z. Friedrich-Schiller-Univ. Jena, Math.-Naturwiss. Reihe* **31(2)**: 195-196.

DIETRICH, J. 1967: Genre *Allium*. Caryotypes de 46 espèces en culture. - *Inform. Annuelles Caryosyst. Cytogén.* **1**: 23-26.

DIETRICH, W. 1967: Die Zytotaxonomie der *Carex*-Sektion *Frigidae* in Europa. - *Feddes Repert.* **75(1-2)**: 1-42.

DIETRICH, W. 1972: [Chromosome numbers in *Carex*]. In Löve, Á. (ed.): IOPB chromosome number reports XXXVI. - *Taxon* **21(2-3)**: 333-335.

DIETRICH, W., WAGNER, W.L. & RAVEN, P.H. 1997: Systematics of *Oenothera* section *Oenothera* subsection *Oenothera*. - *Monogr. Syst. Bot. Missouri Bot. Gard.* **50**: 234 pp.

DÍEZ, M.J. 1983: Pollen morphology of genus *Anchusa* L. (Boraginaceae). Its taxonomic interest. - *Pollen & Spores* **25(3-4)**: 367-381.

DÍEZ, M.J. 1994: A general survey of pollen types in *Anchusa* L. (Boraginaceae) in relation to taxonomy. - *Acta Bot. Gallica* **141(2)**: 233-242.

DÍEZ, M.J. & FERGUSON, I.K. 1996: Studies of the pollen morphology and taxonomy of the tribes Loteae and Coronilleae (Papilionoideae; Leguminosae). 3, *Coronilla* L. and related genera and systematic conclusions. - *Rev. Palaeobot. Palynol.* **94(3-4)**: 239-257.

DIHORU, G. 1970: Morpho-taxonomische Aspekte einiger *Arum*-Arten. - *Rev. Roumaine Biol., Sér. Bot.* **15(2)**: 71-84.

DIHORU, G. 1972: Notes on the taxonomy of *Typha* species in Romania. - *Rev. Roumaine Biol., Sér. Bot.* **17(2)**: 79-86.

DIHORU, G. 1975: Considerations on some species of *Valerianella* Mill. In Jordanov, D. & al. (eds): Problems of Balkan flora and vegetation, pp. 237-243. - Sofia: Publishing House of the Bulgarian Academy of Sciences.

DIHORU, G. & GOGA, I. 2004: *Asperula laevigata* (Rubiaceae) - a new species to the flora of Romania. - *Phytol. Balcan.* **10(2-3)**: 183-185.

DIJK, P. VAN 1995: Biogeography and population biology of hybridization in *Plantago* ... In Valdés, B. & Silvestre, S. (eds): Abstracts. VIII OPTIMA meeting, Sevilla, Spain, 25 September - 1 October, 1995, p. 11.

DIKLIĆ, N. 1971: Systematik und Verbreitung der Art *Anthyllis vulneraria* L. in der SR Serbien [In Serbo-Croatian with German summary]. - *Glasn. Prir. Mus. Beogradu, Ser. B, Biol. Nauke* **26**: 69-77.

DIKLIĆ, N. & MILOJEVIĆ, B. 1976: *Nepeta rtanjensis* Diklić & Milojević spec. nov. - neue Art aus der Gattung *Nepeta* L. [In Serbo-Croatian with German summary]. - *Glasn. Prir. Mus. Beogradu, Ser. B, Biol. Nauke* **31**: 23-35.

DIKLIĆ, N. & NIKOLIĆ, V. 1972: Neue Fundstelle der Art *Viola grisebachiana* Vis. in Serbien [In Serbo-Croatian with German summary]. - *Glasn. Prir. Mus. Beogradu, Ser. B, Biol. Nauke* **27**: 213-217.

DIKLIĆ, N. & NIKOLIĆ, V. 1977: *Nepeta ernesti-mayeri* Diklić & Nikolić sp. nov. - eine neue Art aus der Sektion *Pycnonepeta* Bentham. - *Feddes Repert.* **88(7-8)**: 409-415.

DIKLIĆ, N. & NIKOLIĆ, V. 1978: Neue Angaben über Fundorte von Pflanzenarten in der SR Serbien (VII) [In Serbo-Croatian with German summary]. - *Glasn. Prir. Mus. Beogradu, Ser. B, Biol. Nauke* **33**: 61-67.

DIKLIĆ, N. & NIKOLIĆ, V. 1980: Neue Angaben über Fundorte von Pflanzenarten in der SR Serbien (IX) [In Serbo-Croatian with German summary]. - *Glasn. Prir. Mus. Beogradu, Ser. B, Biol. Nauke* **35**: 17-27.

DIKLIĆ, N. & NIKOLIĆ, V. 1981: *Jurinea mollis* (L.) Reichenb. subsp. *subhastata* (Pančić) Diklić & Nikolić stat. et comb. nov., an interesting plant from northeastern Yugoslavia. - *Anales Jard. Bot. Madrid* **37(2)**: 597-606.

DIKLIĆ, N. & NIKOLIĆ, V. 1982a: The new species of genus *Althaea* L. from the surrounding of Vranje [In Serbo-Croatian with English summary]. - *Glasn. Republ. Zavoda Zaštitu Prir. Prirodnjačke Zbirke Titogradu* **15**: 63-68.

DIKLIĆ, N. & NIKOLIĆ, V. 1982b: On some taxonomic questions in the genus *Poa* L. sect. *Bolbophorum* Ascherson & Graebner (Poaceae Brnh.) [In Serbo-Croatian with English summary]. - *Glasn. Prir. Mus. Beogradu, Ser. B, Biol. Nauke* **37**: 5-12.

DIKLIĆ, N. & STEVANOVIĆ, V. 1993: *Althaea kragujevacensis* Pancic ex Diklic & Stefanovic - Discussion of some taxonomical problems. In Demiriz, H. & Özhatay, N. (eds): Proceedings of the V OPTIMA meeting, pp. 519-525. - Istanbul Üniversitesi Fen Fakültesi.

DIKLIĆ, N., STEVANOVIĆ, V. & NIKETIĆ, M. 1991: The taxonomic status of *Achillea alexandri-regis* Bornm. & Rudski. - *Bot. Chron. (Patras)* **10**: 729-735.

DIKLIĆ, N. & VASIĆ, O. 1998: The investigation of the flora and vegetation of the serpentine area in Serbia. In Moret, J. (ed.): Abstracts of the IX. OPTIMA meeting, Paris, 11-17 May 1998, p. 26. - Paris: Mus. Nat. d'Hist. Naturelle.

DIKLIĆ, N. & VASIĆ, O. 2000: A taxonomic assessment of *Thymus dacicus*, a neglected member of the flora of Yugoslavia. - *Bot. Chron. (Patras)* **13**: 179-186.

DILLENBERGER, M.S. & KADEREIT, J.W. 2012: Two new combinations in *Adenostyles* Cass. (Asteraceae-Senecioneae), conspectus of the genus, and key to its species and subspecies. - *Willdenowia* **42(1)**: 57-61.

DILLENBERGER, M.S. & KADEREIT, J.W. 2013: The phylogeny of the European high mountain genus *Adenostyles* (Asteraceae-Senecioneae) reveals that edaphic shifts coincide with dispersal events. - *Amer. J. Bot.* **100(6)**: 1171-1183.

DILLENBERGER, M.S. & KADEREIT, J.W. 2014: Maximum polyphyly: Multiple origins and delimitation with plesiomorphic characters require a new circumscription of *Minuartia* (Caryophyllaceae). - *Taxon* **63 (1)**: 64-88 + electronic supplement.

DIMADOU, D. 1923: Zizania ton thessalikon agron. - Athine: Tipis Stavrou Christou, 23 pp.

DIMESKA, G. & AL. 1997: *Vicia loiseleurii* (M. B.) D. Litvinov (Fabaceae) a new species for the flora of Republic Macedonia [In Macedonian with English summary]. - *Godišen Zborn. Biol. Fak. Univ. "Kiril i Metodij"* **50**: 47-57.

DIMESKA, G. & AL. 1999: Citotaxonomy of the species *Vicia sylvatica* L. - *Godišen Zborn. Biol. Fak. Univ. "Kiril i Metodij"* **52**: 93-96.

DIMITRAKOPOULOS, P.G. & AL. 2021: Broomrape species parasitizing *Odontarrhena lesbiaca* (Brassicaceae) Individuals act as nickel hyperaccumulators. - *Plants MDPI* **10, 816**: 12 pp.

DIMITRAKOPOULOS, P.G., BAKAS, M. & STRID, A. 2021: Report 24. In Vladimirov, V., Aybeke, M. & Tan, Kit (eds): New floristic records in the Balkans: 44. - *Phytol. Balcan.* **27(1)**: 133.

DIMITRAKOPOULOS, P.G., BAZOS, I. & STRID, A. 2019: *Phelipanche nowackiana*. In Raab-Straube, E. v. & Raus, Th. (eds): Euro+Med-Checklist notulae, 11. - *Willdenowia* **49(3)**: 437-438.

DIMITRELLOS, G. 2005: Geobotaniki erevna tou orous Timfristou (VD Sterea Ellada). Chlorida-vlastisi-axiologisi-diachirisi. [In Greek]. - Patras: Ph.D. thesis, University of Patras, 296 pp.

DIMITRELLOS, G. & CHRISTODOULAKIS, D. 1996: The flora of Mount Timfristos (N.W. Sterea Ellas, Greece). - *Fl. Medit.* **5**: 9-51.

DIMITRELLOS, G. & CHRISTODOULAKIS, D. 1999: The phytogeographical distribution patterns of the flora of Mt Timfristos (N.W. Sterea Ellas, Greece). - *Fl. Medit.* **9**: 215-229.

DIMITROV, A. 1915: Beitrag zur Erforschung der Flora Mazedoniens. - *Trav. Soc. Bulg. Sci. Nat.* **8**: 30-42.

DIMITROV, D.S. 1988: New chorologic data on the flora of southwestern Bulgaria [In Bulgarian with Russian and English summaries]. - *Fitologiya* **34**: 73-76.

DIMITROV, D.S. 1998: A supplement to the flora of the Balkan Peninsula. - *Phytol. Balcan.* **4(3)**: 57-60.

DIMITROV, D.S. 2001: *Alkanna stribrnyi, Trifolium affine, Rhynchocorys elephas, Verbascum pesudonobile, Daucus involucratus, Centranthus longiflorus*. In Greuter, W. & Raus, Th. (eds): Med-Checklist notulae, 20. - *Willdenowia* **31(2)**: 320-325.

DIMITROV, D.S. 2002a: New data on the flora of the floral regions of Mt Slavyanka, Pirin Mts, the valley of river Mesta, and Western and Central Rhodopes. - *Phytol. Balcan.* **8(2)**: 181-184.

DIMITROV, D.S. 2002b: Notes on *Campanula tymphaea* Hausskn. and *Salvia sylvestris* L. In Greuter, W. & Raus, Th. (eds): Med-Checklist notulae, 21. - *Willdenowia* **32(2)**: 196, 200.

DIMITROV, D.S. 2005: *Saponaria calabrica* Guss., *Datura innoxia* Mill. In Greuter, W. & Raus, Th. (eds): Med-Checklist notulae, 23. - *Willdenowia* **35(1)**: 56, 61.

DIMITROV, D.S. 2010a: *Sedum confertifolium* Boiss. In Greuter, W. & Raus, Th. (eds): Med-Checklist notulae, 29. - *Willdenowia* **29(2)**: 195.

DIMITROV, D.S. 2010b: *Juncus hybridus* Brot. In Greuter, W. & Raus, Th. (eds): Med-Checklist notulae, 29. - *Willdenowia* **40(2)**: 202.

DIMITROV, D.S. 2013: *Verbascum chaixii* subsp. *chaixii*. In Raab-Straube, E.v. & Raus, Th.: Notulae ad floram euro-mediterraneam pertinentes No. 30. - *Willdenowia* **43(1)**: 160.

DIMITROV, D.S. 2021: *Tripleurospermum disciforme*. In Raus, Th. & Raab-Straube, E.v. (eds): Euro+Med-Checklist notulae, 14. - *Willdenowia* **51(3)**: 357.

DIMITROV, D.S. & AL. 1997: *Cytinus* (Rafflesiaceae), a new genus for the Bulgarian flora. - *Fl. Medit.* **7**: 49-50.

DIMITROV, D.S. & AL. 2001: *Ophrys argolica* (Orchidaceae), a new species to the Bulgarian flora. - *Phytol. Balcan.* **7(2)**: 199-200.

DIMITROV, D.S. & AL. 2013: New chorological data for the vascular flora of Mesta River valley floristic region (Southwestern Bulgaria). - *Dokl. Bulg. Akad. Nauk* **66(5)**: 701-708.

DIMITROV, D.S. & NIKOLOV, I. 1998: A new taxon and chorological data on the vascular flora of Bulgaria. - *Phytol. Balcan.* **4(3)**: 121-125.

DIMITROV, D.S. & RAUS, TH. 2001: *Aldrovanda vesiculosa* L., *Stratiotes aloides* L. In Greuter, W. & Raus, Th. (eds): Med-Checklist notulae, 20. - *Willdenowia* **31(2)**: 323-326.

DIMITROV, D.S. & TZONEV, R. 2002: On the distribution of *Ambrosia artemisiifolia* L. (Asteraceae) in Bulgaria. - *Phytol. Balcan.* **8(1)**: 31-33.

DIMITROV, D.S. & VUTOV, V. 2010: Reports 77-104. In Vladimirov, V. & al. (eds): New floristic records in the Balkans: 14. - *Phytol. Balcan.* **16(3)**: 425-426.

DIMITROV, D.S. & VUTOV, V.M. 2015: *Poa stiriaca*. In Raab-Straube, E. & Raus, Th. (eds): Euro+Med- Checklist Notulae, 5. - *Willdenowia* **45(3)**: 461.

DIMITROV, S.G. 1962: Die Gattung *Dactylis* in Bulgarien [In Bulgarian with German summary]. - *Nauchni Trudove Selskost. Inst. "Vasil Kolarov"* **11(1)**: 177-190.

DIMITROV, S.G. & AL. 1967: Materialien und Notizen über die Flora Bulgariens [In Bulgarian with Russian and German summaries]. - *Nauchni Trudove Selskost. Inst. "Vasil Kolarov"* **16(1)**: 215-220.

DIMITROV, S.G. & DELIPAVLOV, D.D. 1962: *Sporobolus indicus* R. Br. [In Bulgarian]. - *Nauchni Trudove Selskost. Inst. "Vasil Kolarov"* **11(1)**: 197-198.

DIMITROV, S.G. & POPOVA, M.T. 1983: New species from genus *Malus* Miller for the flora of Bulgaria [In Bulgarian with English abstract]. In Velčev, V.I. & al. (eds): Third national conference of botany, pp. 168-172. - Sofia: Publishing House of the Bulgarian Academy of Sciences.

DIMITROV, TH. 1929: Revue des publications sur le *Pinus peuce* Grsb. depuis 1841 jusqu'a 1928 [In Bulgarian with French summary]. - *Izv. Bulg. Bot. Druzh.* **3**: 293-302.

DIMITROVA, D. 1997: Reports (754-756). In Kamari, G., Felber, F. & Garbari, F. (eds): Mediterranean chromosome number reports - 6. - *Fl. Medit.* **6**: 312-317.

DIMITROVA, D. 1998: Reports (992-993). In Kamari, G., Felber, F. & Garbari, F. (eds): Mediterranean chromosome number reports - 8. - *Fl. Medit.* **8**: 287-290.

DIMITROVA, D. 1999: Reports (1055-1057). In Kamari, G., Felber, F. & Garbari, F. (eds): Mediterranean chromosome number reports - 9. - *Fl. Medit.* **9**: 343-346.

DIMITROVA, D. 2000: Karyotype and DNA-content evolution in ten species of *Crepis* (Asteraceae) distributed in Bulgaria. In Özhatay, N. & al. (eds): Abstracts from the Second Balkan Botanical Congress, Istanbul, Turkey, May 14-18, 2000, p. 101.

DIMITROVA, D. & AL. 1999: Karyotype constancy and genome size variation in Bulgarian *Crepis foetida* s.l. (Asteraceae). - *Pl. Syst. Evol.* **217(3-4)**: 245-257.

DIMOPOULOS, P. 1993: Chloristiki ke fitokinoniologiki erevna tou orous Killini. Ikologiki prosengisi [In Greek with English summary]. - Patras: Ph.D. thesis, Univ. of Patras, 370 pp. + 3 maps.

DIMOPOULOS, P. & AL. 2005a: Thermophilous deciduous forests in Greece - a preliminary survey. - *Bot. Chron. (Patras)* **18(1)**: 83-100.

DIMOPOULOS, P. & AL. 2005b: Vegetation ecology of Kalodiki Fen (NW Greece). – *Biológia (Bratislava)* **60(1)**: 69-82.

DIMOPOULOS, P. & AL. 2010: Vegetation patterns and primary succession on sea-born volcanic islands (Santorini archipelago, Aegean Sea, Greece). - *Phytocoenologia* **40(1)**: 1-14.

DIMOPOULOS, P. & AL. 2013: Vascular plants of Greece: An annotated checklist. - Berlin: Botanischer Garten und Botanisches Museum Berlin-Dahlem; Athens: Hellenic Botanical Society [Englera 31: 1-372].

DIMOPOULOS, P. & AL. 2016: Vascular plants of Greece: An annotated checklist. Supplement. - *Willdenowia* **46(3)**: 301-347.

DIMOPOULOS, P. & AL. 2020: A guide to the alien plants of Greece, with reference to the NATURA 2000 protected areas network. - Univ. of Patras: 112 pp. (separate editions in English and Greek].

DIMOPOULOS, P. & BERGMEIER, E. 1999: Chorology and synchorology of the *Fagus* forests in Greece. In Eleftheriou, E.P. (ed.): Proceedings, 7th Panhellenic scientific conference, Alexandroupolis, pp. 96-101 [In Greek with English abstract]. - Thessaloniki: Univ. Studio Press.

DIMOPOULOS, P. & BERGMEIER, E. 2002: The Aegean small islet specialists: a pluri-methodical approach. In Kamari, G. & al. (eds): Ellinini Votaniki Eteria. 9° Panelliniko Epistimoniko Sinedrio. Praktika, pp. 106-114 [In Greek with English abstract].

DIMOPOULOS, P. & BERGMEIER, E. 2004: Wood pasture in an ancient submediterranean oak forest. - *Ecol. Medit.* **30(2)**: 5-14.

DIMOPOULOS, P., CONSTANTINIDIS, TH. & RAUS, TH. 2019: The Greek Flora and the first volume of the "Flora of Greece". In Daskalakou, E.N. & al. (eds): 16th Conference of the Hellenic Botanical Society - Athens 10-13 October 2019. Program and book of abstracts, p. 18.

DIMOPOULOS, P. & GEORGIADIS, TH. 1990: New floristic reports on Killini mountain. - *Candollea* **45(1)**: 221-233.

DIMOPOULOS, P. & GEORGIADIS, TH. 1992: Floristic and phytogeographical analysis of Mount Killini (NE Peloponnisos, Greece). - *Phyton (Horn)* **32(2)**: 283-305.

DIMOPOULOS, P. & GEORGIADIS, TH. 1995: Present state of the phytosociological research on the Greek mountains, syntaxonomy and future perspectives. - *Ann. Bot. (Rome)* **53**: 119-133.

DINÇ DÜŞEN, O. & SÜMBÜL, H. 2007: A morphological investigation of *Colchicum* L. (Liliaceae) species in the Mediterranean region in Turkey. - *Turk. J. Bot.* **31(5)**: 373-419.

DINÇ, M. & AL. 2001: A new violet record for the flora of Turkey: *Viola jordanii* Hanry. - *OT Sist. Bot. Derg.* **8(1)**: 9-13.

DINÇ, M., YILDIRIMLI, Ş. & ÜNAL, A. 2003: *Viola isaurica* Contandr. & Quézel Violaceae (menekşegiller). - *Karaca Arbor. Mag.* **7(1)**: 29-32.

DINES, T.D. & BONNER, I.R. 2002: A new hybrid horsetail, *Equisetum arvense* × *E. telmateia (E. ×robertsii)* in Britain. - *Watsonia* **24(2)**: 145-157.

DINGLER, H. 1877: *Lathraea rhodopea* n. sp. - *Bot. Zeitung (Berlin)* **35**: 74-76, 93-95.

DINGLER, H. 1881: Beiträge zur orientalischen Flora. - *Flora* **64(24)**: 381-383.

DINGLER, H. 1928: Griechische Rosen von Dr. Joh. Mattfeld gesammelt auf seinen 1924, 1926 und 1927 in Bulgarien, Griechenland und der europäischen Türkei ausgeführten Forschungsreisen. - *Notizbl. Bot. Gart. Berlin-Dahlem* **10(95)**: 479-488.

DINTER, I. 1993: Samos. Pflanzenliste Samos vom 16.-30. April 1993. - Lauffen: 28 pp. (privately printed).

DINTER, I. 1994: Reisetagebuch Frühling auf Samos. Botanische Exkursion vom 02.05.-16.05. 1994 [Natur-Exkursionen, K 9405]. - Lauffen: 29 pp. (privately printed).

DINTER, I. 1995: Botanische Studienwanderreise. Perlen der Ägäis - die Inseln Lesbos und Chios in der Ägäis. - Lauffen: 145 pp. (privately printed).

DINTER, I. 1997: Botanische Studiereise. Insel Korfu (Ionische Inseln, Griechenland) vom 11.-26. April 1997. - Lauffen: 94 pp. (privately printed).

DINTER, I. 1998a: Karpathos. Skriptum zur botanischen Exkursion vom 2.-15. Mai 1998. - Lauffen: 53 pp. (privately printed).

DINTER, I. 1998b: Korfu, Griechenland. - Lauffen: 90 pp. (privately printed).

DINTER, I. 2000: Ein Beitrag zur Verbreitung der *Orchis sancta* auf Karpathos. - *J. Eur. Orch.* **32(3-4)**: 601-608.

DINTER, I. 2001: Rhodos, Griechenland. Botanischer Studienaufenhalt 19. April - 3. Mai 2001. - Privately printed: 90 pp.

DINTER, I. 2003: Griechenland: Peloponnes. Botanische Studienreise 27. März - 11. April 2003. - Privately printed: 44 pp. + appendix and CD disc.

DINTER, I. 2004: Griechenland: Peloponnes. Botanische Studienreise 18. Mai - 1. Juni 2004. - Privately printed: 57 + xx pp. and CD disc with 101 colour photos.

DINTER, I. 2006: Samos, Griechenland. Botanische Studienreise 1.-15. Mai 2006. - Privately printed: 51 pp. + CD with plant lists and photos.

DINTER, I. 2007: Halkidiki. Botanische Studienreise 1.-12. Mai 2007. - Privately printed: 55 pp. + disc with photos etc.

DINTER, I. 2009a: Samos, Griechenland. Private Reise Insel Samos 15.-22. Mai 2009. - Privately printed, 26 + vi pp.

DINTER, I. 2009b: Botanische Studienreise Insel Kos, mit Tagesausflügen nach Kalimnos und Nisiros. - Privately printed, 50 pp. + CD.

DINTER, I. 2009c: *Comperia comperiana* (Steven) Asch. & Graebn. Erstnachweis für Samos, Ostägäische Inseln (Griechenland). - *J. Eur. Orch.* **41(3-4)**: 481-496.

DINTER, I. 2009d: *Paracaryum lithospermifolium* subsp. *cariense* (Boiss.) R.R. Mill, *Galium graecum* subsp. *pseudocanum*. In Greuter, W. & Raus, Th. (eds): Med-Checklist notulae, 28. - *Willdenowia* **39(2)**: 336-340.

DINTER, I. 2011: Nordostgriechenland mit Insel Thasos. Botanische Studienreise vom 3.-17. Juni 2010. - Privately printed: 96 pp. (text, additional photos, etc. also on DVD).

DINTER, I. 2012: Insel Kárpathos, mit Insel Sária. Botanische Studienreise 1.-15. Mai 2012. Exkursionsbericht. - Privately printed: 101 pp.

DINTER, I. 2015: Insel Rhodos, EAe (Gr). Botanische Studienreise vom 21. April - 3. Mai 2015. Exkursionsbericht. - Privately printed: 43 pp. + CD.

DINTER, I. 2017: Insel Korfú, IoI (Gr). Botanische Studienreise vom 9.-20. Mai 2017. Exkursionsbericht. - Ostfildern, privately distributed report: 68 pp. + CD.

DINTER, I. 2018: Das geheime Paradies Griechenlands: Halkidiki. Botanischer Studienurlaub 15-24. Mai 2018. - Ostfildern, privately distributed report: 49 pp. + CD.

DINTER, I. 2019: Insel Zakinthos, IoI (Gr). Botanische Studienreise vom 4.-11. Mai 2019. - Privately printed: 40 pp.

DINTER, I. & ZIMMERMANN, R. 2019: North Cyprus. Botanical excursion 2019. - Privately printed: 140 pp.

DIOSCORIDES, P. 1998: Der Wiener Dioskurides: Codex Medicus Graecus 1 der Österreichischen Nationalbibliothek. Teil 1. Kommentar von Otto Mazal (Glanzlichter der Buchkunst 8/1). - Graz: Akademische Druck- u. Verlagsanstalt.

DIOSCORIDES, P. 1999: Der Wiener Dioskurides: Codex Medicus Graecus 1 der Österreichischen Nationalbibliothek. Teil 2. Kommentar von Otto Mazal (Glanzlichter der Buchkunst 8/2). - Graz: Akademische Druck- u. Verlagsanstalt.

DIOSDADO, J.C. & AL. 1993: Números cromosómicos para la flora española. 691-719. - *Lagascalia* **17(1)**: 173-184.

DIOSDADO, J.C. & PASTOR, J. 1990: Estudio cariosistemático del género *Ranunculus* L. sect. *Ranunculastrum* DC. en la Península Ibérica. - *Lagascalia* **16(2)**: 269-290.

DIOSDADO, J.C. & PASTOR, J. 1991a: Estudio citotaxonómico del género *Ranunculus* sect. *Flammula* (Webb ex Spach) Freyn en la Península Ibérica. - *Candollea* **46(2)**: 303-313.

DIOSDADO, J.C. & PASTOR, J. 1991b: Observaciones cariosistemáticas del género *Ranunculus* L. sect. *Ranunculus* en la Península Ibérica. - *Bol. Soc. Brot., Ser. 2* **64**: 227-239.

DIOSDADO, J.C. & PASTOR, J. 1993a: Karyological studies of some orophyte taxa of the genus *Ranunculus* L. from the Iberian Peninsula. - *Bot. J. Linn. Soc.* **111(1)**: 23-35.

DIOSDADO, J.C. & PASTOR, J. 1993b: Aportación al conocimiento cariológico del género *Ranunculus* L. subgénero *Ficaria* (Schaeffer) L. Benson en la península Ibérica. - *Acta Bot. Malac.* **18**: 77-88.

DIOSDADO, J.C. & PASTOR, J. 1993c: Contribución al conocimiento cariológico del género *Ranunculus* en la Península Ibérica. Secciones *Physophyllum* Freyn y *Hecatonia* (Lour.) DC. - *Bol. Soc. Brot., Ser. 2* **66**: 29-40.

DIOSDADO, J.C. & PASTOR, J. 1996: Consideraciones citotaxonómicas del género *Ranunculus* L. (Ranunculaceae) en la Península Ibérica. - *Anales Jard. Bot. Madrid* **54(1)**: 166-178.

DIOSDADO, J.C., PASTOR, J. & VALDÉS, B. 1993: Contributions to the karyological study of the genus *Ranunculus* L. subgenus *Batrachium* (DC.) A. Gray from the Iberian Peninsula. - *Bot. J. Linn. Soc.* **112(1)**: 75-87.

DIRMENCI, T. 2005: A new subspecies of *Nepeta* (Lamiaceae) from Turkey. - *Bot. J. Linn. Soc.* **147 (2)**: 229-233.

DIRMENCI, T. & ROBSON, N.K.B. 2019: Two new taxa of *Hypericum* (Hypericaceae) from Turkey. - *Kew Bull.* **74(4)**: 9 pp.

DITTRICH, M. 1968a: Karpologische Untersuchungen zur Systematik von *Centaurea* und verwandten Gattungen. - *Bot. Jahrb. Syst.* **88(1-2)**: 70-162 + 4 plates.

DITTRICH, M. 1968b: Morphologische Untersuchungen an den Früchten der Subtribus Cardueae- Centaureinae (Compositae). - *Willdenowia* **5(1)**: 67-107.

DITTRICH, M. 1971: *Lamyropsis* (Charadze) Dittrich - Zur Frucht- und Blütenmorphologie einer kritischen Gruppe aus der *Ptilostemon*-Verwandtschaft. - *Candollea* **26(1)**: 97-102.

DITTRICH, M. 1977: Cynareae - systematic review. In Heywood, V.H., Harborne, J.B. & Turner, B.L. (eds): The biology and chemistry of the Compositae 2: 999-1015. - London, etc.: Academic Press.

DITTRICH, M. 1985: Morphologische und anatomische Untersuchungen an Blüten und Früchten der Gattung *Carlina* (Compositae). I. Teil. - *Bot. Jahrb. Syst.* **107(1-4)**: 591-609.

DIXON, J.M. 1991: Biological Flora of the British Isles. *Avenula pratensis* (L.) Dumort. and *A. pubescens* (Hudson) Dumort. - *J. Ecol.* **79(3)**: 829-845.

DIXON, J.M. 1995: Biological Flora of the British Isles. *Trisetum flavescens* (L.) Beauv. (*T. pratense* Pers., *Avena flavescens* L.). - *J. Ecol.* **83(5)**: 895-909.

DIXON, J.M. 2002: Biological Flora of the British Isles. *Briza media* L. - *J. Ecol.* **90(4)**: 737-752.

DOBOLYI, Z.K. 1983: Die zönologischen Verhältnisse von *Telekia speciosa* (Schreb.) Baumg. in den Karpaten und an der Balkanhalbinsel (Compositae). - *Ann. Hist.-Nat. Mus. Natl. Hung.* **75**: 71-96.

DODDS, J.G. 1953: Biological Flora of the British Isles. *Plantago cornopus* L. - *J. Ecol.* **41(2)**: 467-478.

DOĞAN, B. & DURAN, A. 2009: Distribution and habitat features of the endemic *Jurinea* species for Turkey. In Ivanova, D. (ed.): Proceedings of the IV Balkan botanical congress, pp. 220-223. - Sofia: Institute of Botany, BAS.

DOĞAN, M. 1982: Contribution to the study of Turkish grasses, I. - *Notes Roy. Bot. Gard. Edinburgh* **40(1)**: 75-87.

DOĞAN, M. 1983: Contribution to the study of Turkish grasses, II. - *Notes Roy. Bot. Gard. Edinburgh* **40(3)**: 507-510.

DOĞAN, M. 1984: Chromosome counts of Turkish grasses. - *Willdenowia* **13(2)**: 345-347.

DOĞAN, M. 1988: New taxa of *Alopecurus* (Gramineae) from Turkey. - *Notes Roy. Bot. Gard. Edinburgh* **45(1)**: 111-115.

DOĞAN, M. 1991: A taxonomical revision of the genus *Phleum* L. (Gramineae). - *Karaca Arbor. Mag.* **1(2)**: 53-70.

DOĞAN, M. 1999: A concise taxonomic revision of the genus *Alopecurus* L. (Gramineae). - *Türk. Bot. Derg.* **23(4)**: 245-262.

DOĞAN, M. & AKAYDIN, G. 2007: Synopsis of Turkish *Acantholimon* Boiss. (Plumbaginaceae). - *Bot. J. Linn. Soc.* **154(3)**: 397-419.

DOĞAN, M., AKAYDIN, G. & ERDAL, J. 2020: Phytogeography, ecology and conservation of the genus *Limonium* (Plumbaginaceae) in Turkey and a taxonomic revision - *Pl. Syst. Evol.* **306(6)**: 21 pp.

DOĞU, S., BAĞCI, E. & DINÇ, M. 2009: *Rhaponticoides aytachii* sp. nov. (Asteraceae) from south Anatolia, Turkey. - *Nordic J. Bot.* **27(6)**: 479-482.

DOLCHER, T. 1963: Osservazioni cariologiche su alcune specie del genere *Rhamnus*. - *Giorn. Bot. Ital., n.s.* **70(2)**: 147-150.

DOLCHER, T. & PIGNATTI, S. 1960: Note cariologiche su piante mediterranee (*Biscutella*, *Convolvulus*, *Reichardia*). - *Nuovo Giorn. Bot. Ital., n.s.* **67(1-2)**: 176-184.

DOLCHER, T. & PIGNATTI, S. 1971: Un'ipotesi sull'evoluzione dei *Limonium* del Bacino del Mediterraneo. - *Giorn. Bot. Ital., n.s.* **105(2)**: 95-107.

DOLL, R. 1973a: Revision der Sect. *Erythrosperma* Dahlst. emend. Lindb. f. der Gattung *Taraxacum* Zinn (Teil 1). - *Feddes Repert.* **83(9-10)**: 673-740.

DOLL, R. 1973b: Revision der Sect. *Erythrosperma* Dahlst. emend. Lindb. f. der Gattung *Taraxacum* Zinn. 2. Teil. - *Feddes Repert.* **84(1-2)**: 1-180.

DOLL, R. 1975: Cytotaxonomische Beiträge zur *Taraxacum*-Flora der Türkei. - *Pl. Syst. Evol.* **123(3)**: 199-212.

DOLL, R. 1976a: Apomixis und Speziation in den Gattungen *Taraxacum* und *Chondrilla*. - *Feddes Repert.* **87(7-8)**: 483-492.

DOLL, R. 1976b: Weitere cytotaxonomische Beiträge zur *Taraxacum*-Flora der Türkei. - *Pl. Syst. Evol.* **125(1)**: 21-27.

DOLL, R. 1978: Zur *Taraxacum*-Flora des Rila-Gebirges. - *Feddes Repert.* **88(9-10)**: 629-643.

DOLL, R. 1982: Grundriss der Evolution der Gattung *Taraxacum* Zinn. - *Feddes Repert.* **93(7-8)**: 481-624.

DOMIN, K. 1904: Fragmente zu einer Monographie der Gattung *Koeleria*. - *Magyar Bot. Lapok* **3(6-12)**: 174-187, 254-281, 329-348.

DOMIN, K. 1907: Monographie der Gattung *Koeleria*. - *Biblioth. Bot.* **14/65**: 354 pp. + 22 plates and 3 maps.

DOMIN, K. 1943: Monografická studie o rodu *Dactylis* L. [In Czech]. - *Acta Bot. Bohem.* **14**: 3-147.

DOMINA, G., GREUTER, W. & MAZZOLA, P. 2005: A note on the type of *Orobanche sanguinea* C. Presl (Orobanchaceae), nom. cons. prop. - *Taxon* **54(2)**: 500-502.

DOMINA, G., MARINO, P. & CASTELLANO, G. 2011: The genus *Orobanche* (Orobanchaceae) in Sicily. - *Fl. Medit.* **21**: 205-242.

DOMINA, G., UHLICH, H. & BARONE, G. 2022: *Orobanche australis* Moris ex Bertol. the correct name for *O. thapsoides* Lojac. (Orobanchaceae). - *Phytotaxa* **531(2)**: 91-96.

DOMÍNGUEZ, E. 1976: Revisión de las especies anuales del género *Hippocrepis* L. - *Lagascalia* **5(2)**: 225-261.

DOMÍNGUEZ, E. & CANDAU, P. 1976: Contribución al estudio biosistemático del género *Tetragonolobus* Scop. - *Lagascalia* **6(2)**: 257-263.

DOMÍNGUEZ, E. & GALIANO, E.F. 1974a: Revisión del género *Scorpiurus* L. I. Parte experimental. - *Lagascalia* **4(1)**: 61-84.

DOMÍNGUEZ, E. & GALIANO, E.F. 1974b: Revisión del género *Scorpiurus* L. II. Parte sistemática. - *Lagascalia* **4(2)**: 259-280.

DOMÍNGUEZ, E. & GALIANO, E.F. 1979: Revisión del género *Tetragonolobus* Scop. (Fabaceae). - *Lagascalia* **8(2)**: 189-214.

DOMÍNGUEZ, E. & GIBBS, P.E. 1975: Estudio experimental de la reproducción en las especies annuales europeas del género *Hippocrepis* L. - *Lagascalia* **5(1)**: 35-46.

DOMINO, G., DI GRISTINA, E. & JARVIS, C.E. 2019: Lectotypification of two Linnaean plant names. - *Taxon* **68(6)**: 1327-1328.

DOMKE, W. 1934: Untersuchung über die systematische und geographische Gliederung der Thymelaeaceen nebst einer Neubeschreibung ihrer Gattungen. - *Biblioth. Bot.* **27/111**: 151 pp. + 7 plates, 5 maps and 1 table.

DOMMÉE, B., BOMPAR, J.-L. & DENELLE, N. 1990: Sexual tetramorphism in *Thymelaea hirsuta* (Thymelaeaceae): evidence of the pathway from heterodichogamy to dioecy at the infraspecific level. - *Amer. J. Bot.* **77(11)**: 1449-1462.

DON, D. 1821: A monograph of the genus *Saxifraga*. - *Trans. Linn. Soc. London* **13**: 341-452.

DON, D. 1824: A monograph of the genus *Pyrola*. - *Mem. Wern. Nat. Hist. Soc.* **5**: 220-246.

DON, G. 1832: A monograph of the genus *Allium*. - *Mem. Wern. Nat. Hist. Soc.* **6**: 1-102.

DONADILLE, P. 1964: A propos du *Vincetoxicum nivale* Boiss. & Heldr. de Grèce. - *Bull. Soc. Bot. France* **111(7-8)**: 351-356.

DONCASTER, E.D. 1940: *Dianthus simulans*. - *Quart. Bull. Alpine Gard. Soc. Gr. Brit.* **8(1)**: 66-69.

DÖNMEZ, A.A. 2002: The genus *Crataegus* L. (Rosaceae) with special reference to hybridization and biodiversity in Turkey. In Koyuncu, M. (ed.): VIth Plant Life of Southwest Asia Symposium, Abstracts p. 22. - Van: Yüzüncü Yil University.

DÖNMEZ, A.A. 2014: Nomenclatural, taxonomic and biogeographic novelties in the Turkish *Crataegus* L. (Rosaceae-Maleae) taxa. - *Adansonia, sér. 3* **36(2)**: 245-253.

DÖNMEZ, A.A ., UĞURLU AYDIN, Z. & OYBAK DÖNMEZ, E. 2021: Taxonomic monograph of the tribe Nigelleae (Ranunculaceae): a group including ancient medicinal plants. - *Turk. J. Bot.* **45(5)**: 468-452.

DÖNMEZ, A.A., UĞURLU, Z. & IŞIK, S. 2015: A new species of *Nigella* (Ranunculaceae) from Northeastern Turkey. - *Novon* **23(4)**: 411-415.

DÖNMEZ, A.A. & YILDIRIMLI, Ş. 2000: Taxonomy of the Genus *Prunus* L. (Rosaceae) in Turkey. - *Turk. J. Bot.* **24(3)**: 187-202.

DÖNMEZ, A.A. & YILDIRIMLI, Ş. 2001: Generic delimitation of the Pruneae (Rosaceae) tribe in Turkey. In Özhatay, N. (ed.): Proceedings of the Second Balkan Botanical Congress, 1: 189-194. - Istanbul.

DONNER, J. 1985: Verbreitungskarten zu P.H. Davis "Flora of Turkey, 1-8". - *Linzer Biol. Beitr.* **17(1)**: 1-120.

DOONAN, S. 1983: *Jankaea*. - *Gloxinian* **33(2)**: 30-32.

DOORENBOS, J. 1950: Taxonomy and nomenclature of *Cyclamen*. - *Meded. Landbouwhogeschool* **50 (2)**: 19-29 + 4 plates.

DÖRFELT, H. & KLOTZ, S. 1980: Bestimmungsschlüssel für die mitteleuropäischen *Pseudolysimachium*-Arten. - *Mitt. Florist. Kart.* **6**: 65-69.

DÖRFLER, I. 1901: *Centaurea Halácsyi* n. sp. Eine neue *Centaurea*-Art der griechischen Flora. - *Oesterr. Bot. Z.* **51(6)**: 204.

DÖRFLER, I. 1904: [Bericht über eine botanische Forschungsreise durch Kreta]. - *Oesterr. Bot. Z.* **54(8-12)**: 306-308, 457-462.

DÖRFLER, I. 1905: Mitteilungen aus der Flora Kretas. - *Verh. K.K. Zool.-Bot. Ges. Wien* **55(1)**: 17-20.

DÖRFLER, I. 1916 [Account of an expedition in northern Albania]. - *Oesterr. Bot. Z.* **66(7-9)**: 286-293.

DÖRFLER, I. 1918a: Beschreibung zwei neuer Pflanzenarten aus Albanien. - *Oesterr. Bot. Z.* **67(8-9)**: 310-311.

DÖRFLER, I. 1918b: Beschreibung neuer Pflanzenarten, welche er während der Forschungsreise 1918 in Albanien entdeckt hat. - *Anz. Akad. Wiss. Wien, Math.-Naturwiss. Kl.* **55**: 282-285.

DOROSZENKO, A. 1986: Taxonomic studies on the *Satureja* complex (Labiatae). - Edinburgh: Edinburgh University, 538 pp. + plates [photocopy of unpubl. thesis].

DOROSZEWSKA, A. 1974: The genus *Trollius* L. A taxonomical study. - *Monogr. Bot.* **41**: 1-167 + 16 plates.

DORR, L.J. & BARRIE, F.R. 1993: Typification of the Linnaean names in *Pyrola* (Ericaceae: Pyroloideae). - *Brittonia* **45(2)**: 177-180.

DOSTÁL, J. 1931: Étude préliminaire de l'évolution et de systématique des types du genre *Centaurea* sect. *Cyanus* subsect. *Montanae*. - *Preslia* **10**: 57-69.

DOSTÁL, J. 1973: Preliminary notes on the subtribe Centaureinae. - *Acta Bot. Acad. Sci. Hung.* **19(1-4)**: 73-79.

DOSTÁL, J. 1976: New nomenclatural combinations and taxa of the Compositae subtribe Centaureinae in Europe. In Heywood, V.H. (ed.): Notulae systematicae ad Floram Europaeam spectantes, no. 18. - *Bot. J. Linn. Soc.* **71(3)**: 191-210.

DOSTÁLEK, J. 1979a: Nomenklatur der infraspezifischen Taxa von *Pyrus spinosa* (=*P. amygdaliformis*). - *Preslia* **51(1)**: 31-34.

DOSTÁLEK, J. 1979b: *Pyrus elaeagrifolia* und ihre Hybriden in Bulgarien. - *Folia Geobot. Phytotax.* **14 (2)**: 167-183.

DOSTÁLEK, J. 1980: *Pyrus spinosa* und ihre Hybriden in Südwestbulgarien. - *Folia Geobot. Phytotax.* **15(1)**: 59-73.

DOSTÁLEK, J. 1984: Hybriden von *Pyrus spinosa*. - *Folia Geobot. Phytotax.* **19(1)**: 89-93.

DOSTÁLEK, J. 1997: *Pyrus elaeagrifolia* und ihre Hybriden. - *Feddes Repert.* **108(5-6)**: 345-360.

DOUMAS, P. & CONSTANTINIDIS, TH. 2019: *Aethionema saxatile* subsp. *rhodopaeum* Brassicaceae) and *Saponaria stranjensis* (Caryophyllaceae), two new records for the Greek flora. In Daskalakou, E.N. & al. (eds): 16th Conference of the Hellenic Botanical Society – Athens 10-13 October 2019. Program and book of abstracts, p. 87.

DOUMAS, P., GOULA, K. & CONSTANTINIDIS, TH. 2022: Thirty-two new and noteworthy floristic records from north-eastern Greece. - *Biodiv. Data J.* **10: e81817**: 43 pp. + suppl.

DOUSIS, M.A. & THANOS, K.A. (EDS) 2017: 15th conference of the Hellenic Botanical Society, Chania 14-17 September 2017. Program and book of abstracts. - Athens: 158 pp.

DOWELD, A.B. & SHIPUNOV, A. 2017: (2497) Proposal to reject the name *Plantago indica* (Plantaginaceae). - *Taxon* **66(1)**: 205-206.

DRAPER, D. & AL. 2001: Analysis of infraspecific taxa of *Arum italicum* (Araceae) using morphological and ecological data. - *Bocconea* **13**: 437-443.

DRENKOVSKI, R. 1970: Neue Beiträge zur Kenntnis der Verbreitung der *Pulstatilla halleri* (All.) Willd. subsp. *macedonica* Krause. - *Godišen Zborn. Prir.-Mat. Fak. Univ. Skopje, Biol.* **22**: 223-230.

DRENKOVSKI, R. 1971: Neue Beiträge zur Horologie einiger Sippen der Flora Mazedoniens. - *Fragm. Balcan.* **8(15)**: 129-134.

DRENKOVSKI, R. 1973: Neue Beiträge zur Kenntnis die Arten *Acer campestre* L. und *Acer marsicum* Guss. [In Macedonian with German summary]. - *Godišen Zborn. Prir.-Mat. Fak. Univ. Skopje, Biol.* **25**: 185-203.

DRENKOVSKI, R. 1974a: Beitrag zur Kenntnis der Haselwurz (*Asarum europaeum* L.) in Mazedonien. - *Godišen Zborn. Prir.-Mat. Fak. Univ. Skopje, Biol.* **26**: 215-220.

DRENKOVSKI, R. 1974b: *Acer marsicum* Gussone in der Flora Bulgariens. - *Godišen Zborn. Prir.-Mat. Fak. Univ. Skopje, Biol.* **26**: 221-224.

DRENKOVSKI, R. 1977: *Trollius europaeus* L. (Ranunculaceae) in Mazedonien. - *Glasn. Republ. Zavoda Zaštitu Prir. Prirodnjačke Zbirke Titogradu* **10**: 105-111.

DRENKOVSKI, R. 1984: Basis for horology of species *Crocus scardicus* Košanin (Iridaceae) [In Serbo-Croatian with English summary]. - *Glasn. Republ. Zavoda Zaštitu Prir. Prirodnjačke Zbirke Titogradu* **16**: 27-32.

DRENOVSKI, A.K. 1933a: Ueber die reiche und eigenartige Flora von Alibotuschgebirge (in bulgarisch Mazedonien) [In Bulgarian with German summary]. - Sofia: Eigene Ausgabe, 12 pp.

DRENOVSKI, A.K. 1933b: Zweiter Beitrag zur Flora von Alibotuš [In Bulgarian with German summary]. - Sofia: Eigene Ausgabe, 7 pp.

DRENOVSKI, A.K. 1934a: Die Flora von Alibotuschgebirge in NO. Mazedonien. IV. Pflanzenformationen u. Pflanzenzonen [In Bulgarian with German summary]. - Sofia: Eigene Ausgabe, 32 pp.

DRENOVSKI, A.K. 1934b: Dritter Beitrag zur Flora des Alibotuschgebirges [In Bulgarian]. - *Izv. Bulg. Bot. Druzh.* **6**: 83-86.

DRENOVSKI, A.K. 1936: Sechster Beitrag zur Flora von Alibotuschgebirge (in bulgarisch Macedonien) [In Bulgarian with German summary]. - *Izv. Bulg. Bot. Druzh.* **7**: 74-79.

DRENOVSKI, A.K. 1937: Siebenter Beitrag zur Flora des Alibotusch-Kitkaberges (Neue Fundorte). - Sofia: Eigene Ausgabe, 16 pp.

DROSSOS, E.G. 1976: Beiträge zur Kenntnis der Pflanzenges. Atropetum Belladonnae (Br.-Bl. 1930) Tx. 1931 auf der Halbinsel Chalkidike. In Dafis, S. & Landolt, E. (eds): Zur Vegetation und Flora von Griechenland, Band 2. - *Veröff. Geobot. Inst. E.T.H. Stiftung Rübel Zürich* **56**: 72-86.

DROSSOS, E.G. 1977: Beitrag zur Kenntnis der Pflanzengesellschaft Atropetum belladonnae im griechischen Raum [In Greek with German summary]. - Thessaloniki: Diss., Aristotle Univ., 149 pp. + 22 tables and 1 map.

DROSSOS, E.G. 1992: A floristic study of Mitrikou lake and the lagoons of Nomos Rodhopi in W Thrace (N Greece). - *Willdenowia* **22(1-2)**: 97-117.

DROSSOS, E.G. 1995a: Contribution to the study of the flora and vegetation of the mount range Askio ... In: Proc., 5th sci. conf., Hellenic Bot. Soc., Delphi 21-23 October 1994, pp. 221-224 [In Greek with English summary]. - Thessaloniki: Univ. Studio Press.

DROSSOS, E.G. 1995b: Study of *Lotus aduncus* (Griseb.) Nyman (Leguminosae), a species endemic to the island of Thasos, Greece. - *Ann. Mus. Goulandris* **9**: 125-140.

DROSSOS, E.G. & AL. 1996: Ammophious, halophilous and hydrophilous plant communities of delta of Thessalian Pinios river (Thessalia, Hellas) [In Greek]. - *Sci. Ann. Dept. Forest. Nat. Environm. Aristot. Univ. Thessalonika* **39(1)**: 328-365.

DROSSOS, E.G. & ATHANASIADIS, N. 1989: Contribution to the research of the pastures of Mount Paiko. Flora and vegetation [In Greek with English summary]. - *Sci. Ann. Dept. Forest. Nat. Environm. Aristot. Univ. Thessalonika* **LB/3(13)**: 277-320.

DROSSOS, E.G., KARAGIANNAKIDOU, V. & LAVRENTIADOU, S. 1999: Study of flora and vegetation of the lakes Cheimaditida and Zazari of the province of Florina (W. Macedonia).

In Eleftheriou, E.P. (ed.): Proceedings, 7th Panhellenic scientific conference, Alexandroupolis, pp. 75-79 [In Greek with English abstract]. - Thessaloniki: Univ. Studio Press.

DROSSOS, E.G. & LAVRENTIADOU, S. 2008: Ecological evaluation of the wetland vegetation at the lagoons of Alykes Kitrous Pierias (Macedonia, N. Greece). In: Praktika 10ou Sinedriou Ellinikis Votanikis Eterias, 5-8 maiou 2005, Ioannina: pp. 179-186 [In Greek with English abstract].

DROSSOS, E.G., SYMEONIDIS, L. & KARATAGLIS, S. 1996: A contribution to the study of three taxa of the genus *Lupinus* from north Greece. - *Feddes Repert.* **107(3-4)**: 219-223.

DROUZAS, A.D., CHARITONIDOU, M. & TSIFTSIS, S. 2017: Chloroplast DNA variation in *Epipactis atrorubens* populations from northern Greece. - *Botany Letters* **164(1)**: 55-62.

DRUCE, G.C. 1903: *Valeriana Dioscoridis* Sibth. et Sm. in Corfu. - *J. Bot. (London)* **41**: 350-351.

DRUCE, G.C. 1915: *Potamogeton interruptus* Kit. in Greece. - *J. Bot. (London)* **53**: 373.

DRUCE, G.C. 1916a: [Brief notes on Greek plants]. - *J. Bot. (London)* **54**: 38.

DRUCE, G.C. 1916b: Centaurium maritimum in Corfu. - *J. Bot. (London)* **54**: 66.

DRUCE, G.C. 1917: *Calamintha nebrodensis* Kern. in Greece. - *J. Bot. (London)* **55**: 55.

DRUŠKOVIĆ, B. 1995: [Miscellaneous reports]. In Stace, C.A. (ed.): IOPB chromosome data 9. - *Int. Organ. Pl. Biosyst. Newslett.* **24**: 11-19.

DRUŠKOVIĆ, B. & SUŠNIK, F. 1975: Flora Slovenica: Chromosome number report II [In Slovenian with English summary]. - *Biol. Vestn.* **23(1)**: 9-24.

DRUŠKOVIĆ, B. & SUŠNIK, F. 1976: Chromosome number report III. *Carex*, *Juncus* [In Slovenian with German summary]. - *Biol. Vestn.* **24(2)**: 125-132.

DRUŠKOVIĆ, B. & SUŠNIK, F. 1979: Flora Slovenica: Chromosome number report IV. *Salix* L. [In Slovenian with English summary]. - *Biol. Vestn.* **27(2)**: 115-122.

DU PASQUIER, A.-M., JEANMONOD, D. & YAMAMA, N. 2017: Morphological convergence in the recently diversified *Silene gigantea* complex (Caryophyllaceae) in the Balkan Peninsula and south-western Turkey, with the description of a new subspecies. - *Bot. J. Linn. Soc.* **183(3)**: 474-493.

DU PASQUIER, P.-E., NACIRI, Y. & JEANMONOD, D. 2015: Morphological analysis of the *Silene gigantea* complex (Caryophyllaceae) across the Balkan Peninsula, south-western Turkey and Cyprus. - *Pl. Syst. Evol.* **301(8)**: 2025-2042 + electronic suppl.

DUCKERT-HENRIOD, M.-M. 1991: Rapports (7-25). In Kamari, G., Felber, F. & Garbari, F. (eds): Mediterranean chromosome number reports - 1. - *Fl. Medit.* **1**: 229-237.

DUCKERT-HENRIOD, M.-M. & FAVARGER, C. 1984: *Poa cenisia* All. en Suisse. Étude cytogéographique. - *Diss. Bot.* **72**: 435-451.

DUCKERT-HENRIOD, M.-M. & FAVARGER, C. 1987: Contribution à la cytotaxonomie et à la cytogéographie des *Poa* (Poaceae = Gramineae) de la Suisse (Index des nombres chromosomiques des Spermatophytes de la Suisse: II Poaceae, Genre *Poa*). - *Denkschr. Schweiz. Naturf. Ges.* **100**: 130 pp.

DUDLEY, T.R. 1964a: Studies in *Alyssum*: Near Eastern representatives and their allies, I. - *J. Arnold Arbor.* **45(1)**: 57-100.

DUDLEY, T.R. 1964b: Synopsis of the genus *Alyssum*. - *J. Arnold Arbor.* **45(3)**: 358-373.

DUDLEY, T.R. 1964c: Synopsis of the genus *Aurinia* in Turkey. - *J. Arnold Arbor.* **45(3)**: 390-400.

DUDLEY, T.R. 1965: Studies in *Alyssum*: Near Eastern representatives and their allies, II. Sect. *Meniocus* and sect. *Psilonema*. - *J. Arnold Arbor.* **46(2)**: 181-213 + 4 plates.

DUDLEY, T.R. 1966: A new Greek endemic: *Alyssum vourinonense* and a note regarding *Bornmuellera*. - *Oesterr. Bot. Z.* **113(2)**: 265-270.

DUDLEY, T.R. 1981a: A new *Viburnum* for Bulgaria. - *Fragm. Florist. Geobot.* **27(1-2)**: 163-164.

DUDLEY, T.R. 1981b: The native *Ilex* (Aquifoliaceae) of Turkey, Transcaucasia, Iran, the Georgian Caucasus and Bulgaria. - *Mitt. Deutsch. Dendrol. Ges.* **72**: 97-128.

DUDLEY, T.R. & CHRISTODOULAKIS, D. 1989: *Alyssum samium*: a new endemic from Samos. - *Notes Roy. Bot. Gard. Edinburgh* **45(3)**: 433-438.

DUDLEY, T.R. & CULLEN, J. 1965: Studies in the Old World Alysseae Hayek. - *Feddes Repert.* **71(1-3)**: 218-228.

ĐUG, S. 2000: Ecological and taxonomical distribution of the genus *Pinguicula* L. on Balkan Peninsula. In Özhatay, N. & al. (eds): Abstracts from the Second Balkan Botanical Congress, Istanbul, Turkey, May 14-18, 2000, p. 11.

DUISTERMAAT, H. 1996: Monograph of *Arctium* L. (Asteraceae). Generic delimitation (including *Cousinia* Cass. p.p.), revision of the species, pollen morphology, and hybrids. - Leiden: Rijksherbarium, Ph.D. thesis, xii + 143 pp.

DUISTERMAAT, H. 2003: Typification of *Arctium lappa* L. (Asteraceae). - *Taxon* **52(4)**: 851-852.

DULBERGER, R. 1964: Floral dimorphism and self-incompatibility in *Narcissus tazetta* L. - *Evolution* **18(3)**: 361-363.

DULBERGER, R. 1965: Dimorphism and incompatibility in *Narcissus tazetta* L. - *Israel J. Bot.* **14(2)**: 52-53.

DULBERGER, R. 1969: Chromosome numbers in *Lythrum* in Israel. - *Israel J. Bot.* **17(3)**: 179-183.

DULBERGER, R. 1970a: Floral dimorphism in *Anchusa hybrida* Ten. - *Israel J. Bot.* **19(1)**: 37-41.

DULBERGER, R. 1970b: Tristyly in *Lythrum junceum*. - *New Phytol.* **69(3)**: 751-759.

DULBERGER, R. 1974: Structural dimorphism of stigmatic papillae in distylous *Linum* species. - *Amer. J. Bot.* **61(3)**: 238-243.

DULBERGER, R. & HOROWITZ, A. 1984: Gender polymorphism in flowers of *Silene vulgaris* (Moench) Garcke (Caryophyllaceae). - *Bot. J. Linn. Soc.* **89(2)**: 101-117.

DULIEU, D. & SCHAFFER, P.A. 1973: A propos de la nomenclature des orchidées européennes: *Himantoglossum* versus *Loroglossum*. - *Orchidophile (Ansières)* **4(11)**: 199-201.

DÜLL, R. 1961: Die *Sorbus*-Arten und ihre Bastarde in Bayern und Thüringen. - *Ber. Bayer. Bot. Ges.* **34**: 11-65.

DÜLL, R. 2000: Preliminary list of vascular plants of Samos, Ikaria and Fourni (East Aegean Islands, Greece). - Bad Münstereifel: IDH-Verlag, iv + 136 pp.

DÜLL, R. 2006: The flower beauty of Samos. - Bad Münstereifel: IDH-Verlag, 114 pp.

DÜLL, R. & KALHEBER, H. 2011: Checklist of the cormophytes of Samos, Ikaria and Fourni (East Aegean Islands), 2nd ed. - Privately published: 125 pp.

DUMAN, H. 1998a: A new species of *Teucrium* L. (Labiatae) from south west Anatolia. – *Karaca Arbor. Mag.* **4(3)**: 125-130.

DUMAN, H. 1998b: A new species of *Minuartia* L. (Caryophyllaceae) from south Anatolia. - *Karaca Arbor. Mag.* **4(4)**: 159-162.

DUMAN, H. 1999: Two new species of *Campanula* L. (Campanulaceae) from SW Turkey. - *Edinburgh J. Bot.* **56(3)**: 355-360.

DUMAN, H. 2001: A new species of *Globularia* L. (Globulariaceae) from South Anatolia. - *Bot. J. Linn. Soc.* **137(4)**: 425-428.

DUMAN, H. & DIRMENCI, T. 2017: A new species of *Micromeria* (Lamiaceae) from Köyceğiz (Muğla, southwest of Turkey). - *Turk. J. Bot.* **41(4)**: 383-391.

DUMAN, H., KOYUNCU, M. & ÜNAL, F. 2002: The genus *Sternbergia* Waldst. & Kit. (Amaryllidaceae) in Turkey. - *Karaca Arbor. Mag.* **6(3)**: 115-130.

DUMAN, H. & MILL. R.R. 1999: Two new species of *Potentilla* L. (Rosaceae) from SW Turkey. - *Edinburgh J. Bot.* **56(3)**: 349-354.

DUMONT D'URVILLE, J.S.C. 1822: Enumeratio plantarum quas in insulis archipelagi aut littoribus Ponti- Euxini, annis 1819 et 1820, collegit atque detexit J. Dumont d'Urville. - *Mém. Soc. Linn. Paris* **1**: 255-387.

DUMORTIER, B. 1868: Monographie du genre *Pulmonaria*. - *Bull. Soc. Roy. Bot. Belgique* **7**: 6-36.

DUNBAR, R. 2001: [Notes on *Sternbergia lutea*]. - *New Plantsman* **8(2)**: 72-74.

DUNKEL, F.G. 2015: *Ranunculus pindicola* sp. nov., the only species of the *R. auricomus* complex (Ranunculaceae) in Greece. - *Willdenowia* **45(2)**: 223-230.

DUNKEL, F.G. & GOTTSCHLICH, G. 2018: Notes on some taxa of *Hieracium* and *Pilosella* (Asteraceae), new for the Greek flora. - *Phytol. Balcan.* **24(2)**: 217-224.

DUNKEL, F.G. & GOTTSCHLICH, G. 2021: Notes on some taxa of *Hieracium* and *Pilosella*

(Asteraceae), new for the Greek flora, II and a nomenclatural comment on *Hieracium "chalasinense"*. - *Phytol. Balcan.* **27(1)**: 71-83.
DUNN, A.J. 1997: Biological Flora of the British Isles. *Stachys germanica* L. - *J. Ecol.* **85(4)**: 531-539.
DUNN, P.H. 1976: Distribution of *Carduus nutans*, *C. acanthoides*, *C. pycnocephalus*, and *C. crispus* in the United States. - *Weed Sci.* **24(5)**: 518-524.
DUPASQUIER, P.-E. & AL. 2021: Nomenclatural revision of *Delphinium* subg. *Consolida* (DC.) Huth (Ranunculaceae). - *PhytoKeys* **180**: 81-110.
DURAND, B. 1962: Un complexe polyploïde méconnu: *Mercurialis annua* L. - *Rev. Cytol. Biol. Vég.* **25 (3-4)**: 337-341.
DURAND, B. 1963: Le complexe *Mercurialis annua* L. s.l. Une étude biosystématique. - *Ann. Sci. Nat., Bot. Biol. Vég.* **4(3)**: 579-623.
DURAND, B. 1964: Le complexe *Mercurialis annua* L. s.l. Une étude biosystématique [cont.]. - *Ann. Sci. Nat., Bot. Biol. Vég.* **4(4)**: 625-736.
DURAND, R. & DURAND, B. 1992: Dioécie, monoécie, polyploïdie et spéciation chez les Mercuriales annuelles. - *Bull. Soc. Bot. France, Lett. Bot.* **139(4-5)**: 377-390.
DURKA, W. & AL. 2017: Darwin's legacy in *Platanthera*: are there more than two species in the *Platanthera bifolia/chlorantha* group? - *Pl. Syst. Evol.* **303(3)**: 419-431.
ĐUROVIĆ, S.Z. & AL. 2014: *Silene triflora* (Bornm.) Bornm. (Caryophyllaceae), a neglected species from the Central Balkans. - *Phytotaxa* **172(1)**: 1-12.
ĐUROVIĆ, S.Z. & AL. 2017: Disentangling relationships among the members of the *Silene saxifraga* alliance (Caryophyllaceae): Phylogenetic structure is geographically rather than taxonomically segregated. - *Taxon* **66(2)**: 343-364.
ĐUROVIĆ, S.Z. & AL. 2018: Nomenclatural notes and typification of the names related to *Silene saxifraga* group (Caryophyllaceae). - *Phytotaxa* **334(3)**: 201-214.
ĐUROVIĆ, S.Z., TOMOVIĆ, G. & NIKETIĆ, M. 2018: New taxonomic concept of *Silene saxifraga* group (Caryophyllaceae) - The Balkan Peninsula is the centre of taxonomic diversity of the group. - *Bot. Serbica* **42 (suppl. 1)**: 53.
DUTY, J. 1985: Die *Fagus*-Sippen Europas und ihre geographisch-soziologische Korrelation zur Verbreitung der Assoziationen des Fagion s.l. - *Vegetatio* **59(1-3)**: 177-184.
DUVIGNEAUD, J. 1973: Complément à la distribution et à l'écologie de *Crepis pusilla* (Sommier) Merxm. (Asteraceae). - *Lejeunia, n.s.* **71**: 1-8.
DUWENSEE, H.A. 1998: Ein Beitrag zur Morphologie und Ökologie von *Phleum bertolonii* DC. - *Gleditschia* **26(1-2)**: 11-13.
DUWENSEE, H.A. 1999: Zur Unterscheidung von *Phleum bertolonii* DC. und *Phleum pratense* L. - *Gleditschia* **27(1-2)**: 15-17.
DUWENSEE, H.A. 2000a: Der Wuchsform von *Poa nemoralis* L. - *Gleditschia* **28(1-2)**: 27-30.
DUWENSEE, H.A. 2000b: Eine Bestimmungshilfe zu *Phleum bertolonii* DC. - *Gleditschia* **28(1-2)**: 31-34.
DVOŘÁK, F. 1964a: *Hesperis graeca* sp. nova. - *Oesterr. Bot. Z.* **111(2-3)**: 269-272.
DVOŘÁK, F. 1964b: Taxonomic results of the studies on the chromosome numbers in the genus *Hesperis* L. [In Czech with English summary]. - *Preslia* **36(2)**: 178-184.
DVOŘÁK, F. 1965a: An evolutionally [sic!] younger group of the taxons [sic!] of the *Hesperis* genus on the Balkan Peninsula, in central and south-eastern Europe [In Czech with English summary]. - *Spisy Přír. Fak. Univ. J.E. Purkinje Brně* **23(459)**: 1-20.
DVOŘÁK, F. 1965b: What is *Hesperis macedonica* Adamov.? - *Preslia* **37**: 35-41.
DVOŘÁK, F. 1966a: Ad *Hesperidis* L. generis species in Haemo crescentes adnotationes aliquae. - *Preslia* **38**: 57-64.
DVOŘÁK, F. 1966b: A contribution to the study of the evolution on *Hesperis* series *Matronales* Cvěl. emend. Dvořák. - *Feddes Repert.* **73(2)**: 94-99.
DVOŘÁK, F. 1967a: Further chapter from the study of *Hesperis sylvestris*. - *Phyton (Horn)* **12(1-4)**: 6-30 + 1 plate.
DVOŘÁK, F. 1967b: A note on the genus *Turritis* L. - *Oesterr. Bot. Z.* **114(1)**: 84-87.
DVOŘÁK, F. 1968: Bemerkungen zum einjährigen Silberblatt - *Lunaria annua* L. subsp. *pachyrrhiza* (Borb.) Hayek [In Slovak with German summary]. - *Biológia, A (Bratislava)* **23(7)**: 549-553.
DVOŘÁK, F. 1969: Study of the characters of the genus *Malcolmia* R. Br. Part 2. - *Spisy Přír. Fak. Univ. J.E. Purkinje Brně* **1969(3)**: 75-101.
DVOŘÁK, F. 1970: Some results of the study of the *Lunaria* L. genus [In Slovakian with English summary]. - *Biológia, A (Bratislava)* **25(10)**: 723-730.
DVOŘÁK, F. 1971a: *Hesperis* L. subg. *Mediterranea* Borb. - *Scripta Fac. Sci. Nat. Univ. J.E. Purkynianae Brun., Biol.* **4(1)**: 237-250.
DVOŘÁK, F. 1971b: On the evolutionary relationship in the family Brassicaceae. - *Feddes Repert.* **82 (5)**: 357-372.
DVOŘÁK, F. 1972: Study of the characters of the genus *Malcolmia* R. Br. in Aiton. Part 3. - *Feddes Repert.* **83(4)**: 265-273.
DVOŘÁK, F. 1973a: Infrageneric classification of *Hesperis* L. - *Feddes Repert.* **84(4)**: 259-271.
DVOŘÁK, F. 1973b: Study of the characters of the genus *Malcolmia* R. Br. in Aiton. Part 4. - *Feddes Repert.* **84(4)**: 315-325.
DVOŘÁK, F. 1979: Morphology of chromosomes of the genus *Seseli* L. - *Scripta Fac. Sci. Nat. Univ. J.E. Purkynianae Brun., Biol.* **9(1)**: 19-42.
DVOŘÁK, F. 1980: Karyotype and variability of *Holosteum umbellatum* L. in Bohemia and Moravia. - *Scripta Fac. Sci. Nat. Univ. J.E. Purkynianae Brun., Biol.* **10(6)**: 259-266.
DVOŘÁK, F. 1984: Annotated chromosome counts for *Arenaria leptoclados*, *Arenaria patula* and *Arenaria serpyllifolia*. - *Biológia, A (Bratislava)* **39(9)**: 833-840.
DVOŘÁK, F. 1986a: Some hybrids from the genus *Chenopodium* L. - *Feddes Repert.* **97(11-12)**: 779-804.
DVOŘÁK, F. 1986b: Study of some taxa from the genus *Chenopodium* L. - *Feddes Repert.* **97(11-12)**: 805-830.
DVOŘÁK, F. 1989: Study of some taxa from the range of *Chenopodium ficifolium* Sm. - *Feddes Repert.* **100(9-10)**: 455-479.
DVOŘÁK, F. 1993: Study of some hybrids from *Chenopodium album* agg. - *Feddes Repert.* **104(1-2)**: 1-16.
DVOŘÁK, F. & AL. 1979: [Chromosome numbers from Czechoslovakia]. In Löve, Á. (ed.): IOPB chromosome number reports LXIV. - *Taxon* **28(4)**: 391-392.
DVOŘÁK, F. & AL. 1980: [Chromosome numbers from Czechoslovakia]. In Löve, Á. (ed.): IOPB chromosome number reports LXIX. - *Taxon* **29(4)**: 544.
DVOŘÁK, F. & AL. 1981: [Chromosome numbers from Czechoslovakia]. In Löve, Á. (ed.): IOPB chromosome number reports LXXIII. - *Taxon* **30(4)**: 844.
DVOŘÁK, F. & DADÁKOVÁ, B. 1975: [Chromosome numbers from Czechoslovakia]. In Löve, Á. (ed.): IOPB chromosome number reports XLIX. - *Taxon* **24(4)**: 508-509.
DVOŘÁK, F. & DADÁKOVÁ, B. 1976a: The chromosome morphology of *Hesperis matronalis* subsp. *matronalis* and related diploid taxa. - *Folia Geobot. Phytotax.* **11(3)**: 313-326.
DVOŘÁK, F. & DADÁKOVÁ, B. 1976b: [Chromosome numbers from Czechoslovakia]. In Löve, Á. (ed.): IOPB chromosome number reports LIII. - *Taxon* **25(4)**: 497.
DVOŘÁK, F. & DADÁKOVÁ, B. 1976c: [Chromosome numbers from Czechoslovakia]. In Löve, Á. (ed.): IOPB chromosome number reports LIV. - *Taxon* **25(5-6)**: 645.
DVOŘÁK, F. & DADÁKOVÁ, B. 1977a: [Chromosome numbers from Czechoslovakia]. In Löve, Á. (ed.): IOPB chromosome number reports LVI. - *Taxon* **26(2-3)**: 257.
DVOŘÁK, F. & DADÁKOVÁ, B. 1977b: [Chromosome numbers from Czechoslovakia]. In Löve, Á. (ed.): IOPB chromosome number reports LVIII. - *Taxon* **26(5-6)**: 564-565.
DVOŘÁK, F. & DADÁKOVÁ, B. 1978a: [Chromosome numbers from Czechoslovakia]. In Löve, Á. (ed.): IOPB chromosome number reports LX. - *Taxon* **27(2-3)**: 223.
DVOŘÁK, F. & DADÁKOVÁ, B. 1978b: [Chromosome numbers from Czechoslovakia]. In Löve, Á. (ed.): IOPB chromosome number reports LXI. - *Taxon* **27(4)**: 384.
DVOŘÁK, F. & DADÁKOVÁ, B. 1984: Chromosome counts and chromosome morphology of some selected species. - *Folia Geobot. Phytotax.* **19(1)**: 41-70.

DVOŘÁK, F., DADÁKOVÁ, B. & GRÜLL, F. 1978: Studies of the morphology of chromosomes of some selected species. - *Folia Geobot. Phytotax.* **13(4)**: 343-375.

DVOŘÁK, F., DADÁKOVÁ, B. & GRÜLL, F. 1983: A contribution to the study of *Chenopodium album* agg. - *Folia Geobot. Phytotax.* **18(1)**: 29-43.

DVOŘÁK, F., DADÁKOVÁ, B. & RŮŽIČKA, I. 1979: Chromosome morphology of the Czechoslovak species of the genus *Scorzonera*. - *Folia Geobot. Phytotax.* **14(2)**: 185-199.

DVOŘÁK, F. & KONAŘÍKOVÁ, V. 1970: Study of the characters of the genus *Malcolmia* R. Br. Part 1. - *Feddes Repert.* **81(6-7)**: 387-416.

DVOŘÁK, F., TRNKA, P. & DADÁKOVÁ, B. 1978: Cytotaxonomic study of *Tragopogon* L. in Czechoslovakia. - *Folia Geobot. Phytotax.* **13(3)**: 305-330.

DVOŘÁK, J. & VOREL, J. 1973: Über den Fund einer neuen Hybride *Luzula forsteri* × *luzulina* (*Luzula* ×*romanica* J. Dvořák et Vorel hybr. nova). - *Preslia* **45(1)**: 87-89.

DVOŘÁKOVÁ, M. 1966: Die Gattung *Galeobdolon* in der Tschechoslowakei [In Czech with German summary]. - *Spisy Přír. Fak. Univ. J.E. Purkinje Brně* **30(477)**: 355-371.

DVOŘÁKOVÁ, M. 1974: Bestimmungsschlüssel und Übersicht der tschechoslowakischen Arten und Unterarten der Gattung *Thlaspi* L. [In Czech with German summary]. - *Zprávy Českoslov. Bot. Společn. Českoslov. Akad. Věd* **9(1)**: 1-7.

DVOŘÁKOVÁ, M. 1978: Taxonomische Übersicht der Arten vom *Thlaspi jankae*-Aggregat. - *Preslia* **50 (1)**: 13-21.

DVOŘÁKOVÁ, M. 1983: Zur Kenntnis der karyologischen Verhältnisse von *Thlaspi hungaricum* [In Czech with German summary]. - *Preslia* **55(1)**: 83-85.

DVOŘÁKOVÁ, M. 1985: *Minuartia glaucina*, eine neue Art aus der *Minuartia verna*-Gruppe. - *Preslia* **57 (1)**: 1-8.

DYCE, J.W. 1988: *Polypodium australe* 'Cambricum'. - *Pteridologist* **1(5)**: 217-220.

DYCE, J.W. 1991: Variation in *Blechnum spicant*. - *Pteridologist* **2(2)**: 54-61.

DYER, A.F. & AL. 1976: The karyotype of *Iris pseudacorus* L. - *Trans. Bot. Soc. Edinburgh* **42(4)**: 421-429 + 1 plate.

DYER, A.F., PARKS, J.C. & LINDSAY, S. 2000: Historical review of the uncertain taxonomic status of *Cystopteris dickieana* R. Sim (Dickie's bladder fern). - *Edinburgh J. Bot.* **57(1)**: 71-81.

DYER, W.T.T. 1874: *Crocus cancellatus*. - *Bot. Mag.* **100**: tab. 6103 + 2 pp. text.

DYKES, W.R. 1913: The genus *Iris*. - Cambridge: iii + 125 pp. + 48 plates [reprinted 1974].

DYKES, W.R. 1930: Notes on the tulip species (ed. & ill. E.K. Dykes). - London, 108 pp.

DŽEKOV, S. 1973: *Crataegus orientalis* Pall. in der SR Mazedonien [In Macedonian with German summary]. - *Godišen Zborn. Zemjod.-Šumarsk. Fak. Univ. Skopje, Šumarstvo* **25**: 63-76.

EARNSHAW, F. 1942: Experimental taxonomy. V. Cytological studies in sea plantains allied to *Plantago maritima* L. - *New Phytol.* **41(3)**: 151-164.

EASTWOOD, A. 1934: A revision of *Arctostaphylos* with key and descriptions. - *Leafl. W. Bot.* **1(11)**: 105-127.

EBERLE, G. 1966: Zwei neue Ragwurz- (*Ophrys*-) Mischlinge aus Griechenland. - *Natur & Mus.* **96(9)**: 364-369.

EBERLE, G. 1968: Das Espartogras (*Lygeum spartum* Loefling ex Linné). - *Natur & Mus.* **98(4)**: 149-154.

EBERLE, G. 1969a: Der Löwentrapp, eine krautige Berberidacee der Ostmediterraneis. - *Natur & Mus.* **99(3)**: 113-116.

EBERLE, G. 1969b: Flockenblumen bilden mediterrane Polsterböden. - *Natur & Mus.* **99(10)**: 468-474.

EBERLE, G. 1975a: Die Alpen-Rachenblume (*Tozzia alpina* L.), Voll- und Halbschmarotzer auf Zeit. - *Natur & Mus.* **105(2)**: 46-53.

EBERLE, G. 1975b: Pflanzen am Mittelmeer. Mediterrane Pflanzengemeinschaften Italiens und Griechenlands mit Ausblick auf das ganze Mittelmeergebiet. Zweite, erweiterte Auflage. - Frankfurt am Main: Verlag W. Kramer, 310 pp.

EBERT, I., GREILBUBER, J. & SPETA, F. 1996: Chromosome banding and genome size differentiation in *Prospero* (Hyacinthaceae): diploids. - *Pl. Syst. Evol.* **203(1-2)**: 143-177.

ECCARIUS, W. 2010: Was ist unter *Orchis heroica* E.D. Clarke zu verstehen? Eine Entgegnung. - *Ber. Arbeitskreis. Heimische Orchid.* **27(1)**: 203-221.

ECCARIUS, W. 2016: Die Orchideengattung *Dactylorhiza*. Phylogenie, Taxonomie, Morphologie, Biologie, Verbreitung, Ökologie, Hybridisation. - Eisenach, Selbstverlag: 639 pp.

ECONOMIDOU, E. 1969: Geovotaniki erevna nisou Skiathou. Fitogeografia ton Vorion Sporadon [In Greek with French summary]. - Athènes: Thèse, Univ. Athènes, 154 pp. + 1 table, 8 plates and 2 maps.

ECONOMIDOU, E. 1971: *Verbascum aphentulium* Heldr. (*V. pelium* Hal.). Notes sur sa répartition, son écologie et sa nomenclature. - *Biol. Gallo-Hellen.* **3(2)**: 133-147.

ECONOMIDOU, E. 1973a: Étude et cartographie de la végétation halophile littorale de l'île de Skiathos. - *Biol. Gallo-Hellen.* **4(2)**: 115-137.

ECONOMIDOU, E. 1973b: Contribution à l'étude de la flore et de la phytogéographie de l'île de Skopelos. Phytogéographie des Sporades du Nord, II. - *Biol. Gallo-Hellen.* **5(1)**: 77-121.

ECONOMIDOU, E. 1974: *Euphorbia acanthothamnos* Heldr. & Sart. Recherches sur sa biologique et son écologie. - *Candollea* **29(2)**: 267-279.

ECONOMIDOU, E. 1975a: La végétation des iles de Skiathos et Skopelos (Sporades du Nord). In Dafis, S. & Landolt, E. (eds): Zur Vegetation und Flora von Griechenland, Band 1. - *Veröff. Geobot. Inst. E.T.H. Stiftung Rübel Zürich* **55**: 198-237.

ECONOMIDOU, E. 1975b: La végétation halophytique de l'Attique et sa protection. - *Colloq. Phytosoc.* **4**: 35-41.

ECONOMIDOU, E. 1976a: Bibliographie botanique sur la Grèce (Plantes vasculaires - végétation). In Dafis, S. & Landolt, E. (eds): Vegetation und Flora von Griechenland, Band 2. - *Veröff. Geobot. Inst. E.T.H. Stiftung Rübel Zürich* **56**: 190-242.

ECONOMIDOU, E. 1976b: La repartition des phrygana en Grèce et ses rapports avec le climat et l'influence anthropogene. - *Doc. Phytosoc.* **15-18**: 45-56.

ECONOMIDOU, E. 1981: Le milieu terrestre de l'île de Limnos et ses reliques de forêts. - *Biol. Écol. Médit.* **8(3-4)**: 129-138.

ECONOMIDOU, E. 1995: *Phlomis fruticosa* L. Distribution and biotopes in the continental Greece. In: Proceedings, 5th sci. conf., Hellenic Bot. Soc., Delphi 21-23 October 1994, pp. 201-211 [In Greek with English summary]. - Thessaloniki: Univ. Studio Press.

ECONOMIDOU, E. & YANNITSAROS, A. 1975a: Recherches sur la flore adventice de Grèce. III. Morphologie, développement et phénologie du *Solanum elaeagnifolium* Cav. - *Candollea* **30(1)**: 29-41. [Also distributed as *OPTIMA leaflet* no. 4].

ECONOMIDOU, E. & YANNITSAROS, A. 1975b: Recherches sur la flore adventice de Grèce. IV. Distribution et écologie de deux espèces de *Paspalum* L. en Grèce. In: Proc. Symp. Status, Biology & Control of Grassweeds in Europe, vol. I, pp. 18-23. - Paris: UNESCO.

ECONOMIDOU, E. & YANNITSAROS, A. 1975c: Recherches sur la flore adventice de Grèce. V. Distribution et écologie de *Solanum elaeagnifolium* Cav. - *Biol. Écol. Médit.* **2(4)**: 29-44.

ECONOMOU, G., BILALIS, D. & AVGOULAS, C 2005: Weed flora distribution in Greek cotton fields and its possible influence by herbicides. - *Phytoparasitica* **33(4)**: 406-419.

EDDIE, W.M.M. & INGROUILLE, M.J. 1999: Polymorphism in the Aegean "five-loculed" species of the genus *Campanula* section *Quinqueloculares* (Campanulaceae). - *Nordic J. Bot.* **19(2)**: 153-169.

EDMAN, G. 1929: Zur Entwicklungsgeschichte der Gattung *Oxyria* Hill, nebst zytologischen, embryologischen und systematischen Bemerkungen über einige andere Polygonaceen. - *Acta Horti Berg.* **9(7)**: 165-291.

EDMONDS, J.M. 1977: Taxonomic studies on *Solanum* section *Solanum* (*Maurella*). - *Bot. J. Linn. Soc.* **75(2)**: 141-178.

EDMONDS, J.M. 1978: Numerical taxonomic studies on *Solanum* L. section *Solanum* (*Maurella*). - *Bot. J. Linn. Soc.* **76(1)**: 27-51.

EDMONDS, J.M. 1979: Nomenclatural notes on some species of *Solanum* L. found in Europe. - *Bot. J. Linn. Soc.* **78(3)**: 213-233.

EDMONDSON, J.R. 1973: Notes on *Doronicum* L. in SW Asia. - *Notes Roy. Bot. Gard. Edinburgh* **32(2)**: 255-258.

EDMONDSON, J.R. 1975: The systematic position and distribution of *Poa pirinica*. - *Notes Roy. Bot. Gard. Edinburgh* **34(2)**: 217-220.

EDMONDSON, J.R. 1977: *Omphalodes*. In: Materials for a Flora of Turkey XXXIV. - *Notes Roy. Bot. Gard. Edinburgh* **35(3)**: 300-302.

EDMONDSON, J.R. 1978: Infrageneric taxa in European *Poa* L. In Heywood, V.H. (ed.): Notulae systematicae ad Floram Europaeam spectantes, no. 20. - *Bot. J. Linn. Soc.* **76(4)**: 329-334.

EDMONDSON, J.R. 1982: Additions to the flora of Lesvos and its off-shore islets. - *Ann. Mus. Goulandris* **5**: 33-53.

EDWARDS, A.M. 1998: A *Crocus* foray in the Pindus. - *Quart. Bull. Alpine Gard. Soc. Gr. Brit.* **66(3)**: 361-365.

EDWARDS, A.M. 2000: *Crocus pelistericus* Pulevic, P.C. - *Quart. Bull. Alpine Gard. Soc. Gr. Brit.* **68(4)**: 478-479.

EDWARDS, A.M. 2006: *Sternbergia greuteriana*. - *Alpine Gardener* **74(1)**: 4-5.

EFE, A. 1987: Studies on the morphological and palynological characteristics of *Liquidambar orientalis* Mill. in Turkey. [In Turkish with English summary]. - *Istanbul Üniv. Orman Fak. Derg., A* **37 (2)**: 84-114.

EFE, A. & AL. 2009: *Chamaespartium* Adans. (Leguminosae): a new record for the flora of Turkey. - *Turk. J. Bot.* **33(6)**: 453-456.

EFTHIMIADIS, P. & KARAMANOS, A.J. 2005: Ancestors of some crop plants in the islands of the Eastern Aegean. In Karamanos, A.J. & Thanos, C.A. (eds): Biodiversity and natural heritage in the Aegean. Proceedings of the conference "Theophrastus 2000", pp. 213-233 [In Greek with English summary]. Athina: Afoi Fragoudi O.E.

EFTHIMIADIS, P., ZAMANIS, A. & SKORDA, E. 1985: The two-rowed wild barley from four islands of Greece. - *Sveriges Utsädesfören. Tidskr.* **95**: 63-70.

EFTHIMIOU, G. 2017: *Amorpha fruticosa* L., an alien invasive species in riparian forests. In Dousis, M.A. & Thanos, K.A.: 15th conference of the Hellenic Botanical Society, book of abstracts: p. 78.

EFTHYMIATOU-KATSOUNI, N. & PHITOS, D. 2011: The presence of *Pinus nigra* subsp. *nigra* on Cephalonia (Ionian Islands, Greece). - *Fl. Medit.* **21**: 287-292.

EGGLI, U. 1988: A monographic study of the genus *Rosularia* (Crassulaceae). - *Bradleya* **6, Suppl.**: 1-119.

EGGLI, U. & NYFFELER, R. 1992: *Jovibarba*: A long story with a happy end. - *Bot. Helv.* **102(2)**: 171-173.

EGLI, B.R. 1984: *Bellis longifolia*, ein kretischer Endemit. - *Bot. Helv.* **94(1)**: 61-65.

EGLI, B.R. 1989a: Ecology of dolines in the mountains of Crete (Greece). - *Bielefelder Ökolog. Beitr.* **4**: 59-63.

EGLI, B.R. 1989b: Ecological, mainly edaphic conditions affecting plant speciation; examplified by the mountain flora of Crete. In Phitos, D. (ed.): Abstracts. Sixth OPTIMA meeting, Delphi 10-16 Sept. 1989, p. 42.

EGLI, B.R. 1991: The special flora, ecological and edaphic conditions of dolines in the mountains of Crete. - *Bot. Chron. (Patras)* **10**: 325-335.

EGLI, B.R. 1993: Ökologie der Dolinen im Gebirge Kretas (Griechenland). - Zürich: Ph.D. thesis, 275 pp.

EGLI, B.R. 2000: Forest vegetation of Western Kriti. - *Bot. Chron. (Patras)* **13**: 453-468.

EGLI, B.R. & AL. 1990: *Horstsrissea dolinicola*, a new genus and species of umbels (Umbelliferae, Apiaceae) from Kriti (Greece). - *Willdenowia* **19(2)**: 389-399. [Also distributed as *OPTIMA leaflet* no. 192].

EGOLF, D.R. 1962: A cytological study of the genus *Viburnum*. - *J. Arnold Arbor.* **43(2)**: 132-172.

EGOROVA, T.V. 1981: Generis *Eleocharis* R. Br. (Cyperaceae) flora URSS systema et conspectus [In Russian]. - *Novosti Sist. Vysš. Rast.* **18**: 95-124.

EGOROVA, T.V. & VINOGRADOVA, V.M. 2002: Taxonomic review of the genera *Caucalis* and *Orlaya* (Apiaceae) in the flora of eastern Europe. - *Bot. Zhurn. (Moscow & Leningrad)* **87(11)**: 87-95.

EHRENBERG, L. 1945: Kromosomtalen hos några kärlväxter [In Swedish]. - *Bot. Not.* **1945(4)**: 430-437.

EHRENDORFER, F. 1951: Rassengliederung, Variabilitätszentren und geographische Merkmalsprogressionen als Ausdruck der raum-zeitlichen Entfaltung des Formenkreises *Galium incanum* S.S. - *Oesterr. Bot. Z.* **98(4)**: 427-490.

EHRENDORFER, F. 1952: *Galium rotundifolium* L. und verwandte Sippen in Europa und im Mediterrangebiet. - *Oesterr. Bot. Z.* **99(5)**: 625-631.

EHRENDORFER, F. 1953: Systematische und zytogenetische Untersuchungen an europäischen Rassen des *Achillea millefolium*-Komplexes. - *Oesterr. Bot. Z.* **100(4-5)**: 583-592.

EHRENDORFER, F. 1958a: Critical notes on Turkish Rubiaceae. - *Notes Roy. Bot. Gard. Edinburgh* **22 (4)**: 323-401.

EHRENDORFER, F. 1958b: Ein Variabilitätszentrum als "fossiler" Hybrid-Komplex: Der ostmediterrane *Galium graecum* L. - *G. canum* Req. Formenkreis. Eine Monographie. (Zur Phylogenie der Gattung *Galium*, VI). - *Oesterr. Bot. Z.* **105(1-3)**: 229-279.

EHRENDORFER, F. 1958c: Die geographische und ökologische Entfaltung des europäisch-alpinen Polyploidkomplexes *Galium anisophyllum* Vill. seit Beginn des Quartärs. - *Uppsala Univ. Årsskr.* **6**: 176-181.

EHRENDORFER, F. 1959: Spontane Chromosomenaberrationen und andere Meiosestörungen bei diploiden Sippen des *Achillea millefolium*-Komplexes. (Zur Phylogenie der Gattung *Achillea*, II). - *Chromosoma* **10(4)**: 365-406.

EHRENDORFER, F. 1962a: Notizen zur Systematik und Phylogenie von *Cruciata* Mill. und verwandten Gattungen der Rubiaceae. - *Ann. Naturhist. Mus. Wien* **65**: 11-20 + 1 plate.

EHRENDORFER, F. 1962b: Beiträge zur Phylogenie der Gattung *Knautia* (Dipsacaceae), I. Cytologische Grundlagen und allgemeine Hinweise. - *Oesterr. Bot. Z.* **109(3)**: 276-343.

EHRENDORFER, F. 1962c: Cytotaxonomische Beiträge zur Genese der mitteleuropäischen Flora und Vegetation. - *Ber. Deutsch. Bot. Ges.* **75(4)**: 137-152.

EHRENDORFER, F. 1964: Notizen zur Cytotaxonomie und Evolution der Gattung *Artemisia*. - *Oesterr. Bot. Z.* **111(1)**: 84-142.

EHRENDORFER, F. 1965: Über stammesgeschichtliche Differenzierungsmuster bei den Dipsacaceen. - *Ber. Deutsch. Bot. Ges.* **77, Sondernummer**: 83-94.

EHRENDORFER, F. 1970: Mediterran-mitteleuropäische Florenbeziehungen im Lichte cytotaxonomischer Befunde. - *Feddes Repert.* **81(1-5)**: 3-32.

EHRENDORFER, F. 1975a: Cytosystematik balkanischer Rubiaceae - ein Beitrag zur Geschichte und Differenzierung der Flora und Vegetation des Balkans. In Jordanov, D. & al. (eds): Problems of Balkan flora & veget., pp. 178-186. - Sofia: Publishing House of the Bulgarian Academy of Sciences.

EHRENDORFER, F. 1975b: Infrageneric taxa in European Rubiaceae: *Asperula* and *Galium*. In Heywood, V.H. (ed.): Notulae systematicae ad Floram Europaeam spectantes, no. 16. - *Bot. J. Linn. Soc.* **70(1)**: 2-6.

EHRENDORFER, F. 1975c: Further taxonomic notes on Rubiaceae in Europe. - *Pl. Syst. Evol.* **124(2)**: 173-178.

EHRENDORFER, F. 1975d: Taxonomic notes on *Knautia*. In Heywood, V.H. (ed.): Notulae systematicae ad Floram Europaeam spectantes, no. 17. - *Bot. J. Linn. Soc.* **71(1)**: 39-41.

EHRENDORFER, F. 1980: Hybridisierung, Polyploidie und Evolution bei europäisch-mediterranen Orchideen. - *Jahresber. Naturwiss. Vereins Wuppertal* **33**: 15-34.

EHRENDORFER, F. 1981: Neue Beiträge zur Karyosystematik und Evolution der Gattung *Knautia* (Dipsacaceae) in den Balkanländern. - *Bot. Jahrb. Syst.* **102(1-4)**: 225-238.

EHRENDORFER, F. & AL. 2009: Taxonomic revision, phylogenetics and transcontinental distribution of *Anemone* section *Anemone* (Ranunculaceae). - *Bot. J. Linn. Soc.* **160(3)**: 312-354.

EHRENDORFER, F. & ANČEV, M.E. 1975: *Galium procurrens*, a new diploid relic species of the *G. sylvaticum*-group from the Balkan Peninsula. - *Pl. Syst. Evol.* **124(1)**: 1-6.

EHRENDORFER, F. & GUO, Y.-P. 2005: Changes in the circumscription of the genus *Achillea* (Compositae-Anthemideae) and its subdivision. - *Willdenowia* **35(1)**: 49-54.

EHRENDORFER, F. & GUO, Y.-P. 2006: Multidisciplinary study on *Achillea* sensu lato (Compositae: Anthemideae): new data on systematics and phylogeography. - *Willdenowia* **36(1)**: 69-87.

EHRENDORFER, F. & KRENDL, F. 1974: Notes on Rubiaceae in Europe. In Heywood, V.H. (ed.): Notulae systematicae ad Floram Europaeam spectantes, no. 15. - *Bot. J. Linn. Soc.* **68(4)**: 268-272.

EHRENDORFER, F. & SCHÖNBECK-TEMESY, E. 1975: *Asperula ophiolithica* Ehrend., spec. nov., ein Endemit der Insel Euboea. - *Pl. Syst. Evol.* **123(2)**: 153-156.

EICHBERGER, C. 2001a: Die Baumartige Wolfsmilch, *Euphorbia dendroides* L. Biologie, Ökologie, Pflanzensoziologie und soziokulturelle Stellung einer mediterranen Art. - *Diss. Bot.* **344**: 1-351 + 2 tables and 1 map.

EICHBERGER, C. 2001b: *Euphorbia dendroides* L. (Euphorbiaceae): a monographic view on a Mediterranean species. In Raimondo, F.M. (ed.): Abstracts from the X OPTIMA meeting, Palermo, 13-19 September 2001, p. 197.

EICHBERGER, C. 2003: *Euphorbia dendroides* L. (Euphorbiaceae): a monographic view on a Mediterranean species. - *Bocconea* **16(2)**: 925-933.

EICHBERGER, C. 2009: The vegetation of the Lousi valley (Peloponnesos, Greece): a geobotanical survey. - *Bocconea* **23**: 337-343.

EICHBERGER, C., SIGL, M. & RÜHFEL, H. 2007: Trees and shrubs on classical Greek vases. - *Bocconea* **21**: 117-130.

EIG, A. 1929a: Monographisch-kritische Uebersicht der Gattung *Aegilops*. - *Repert. Spec. Nov. Regni Veg. Beih.* **55**: 1-228 + 15 plates and 3 maps.

EIG, A. 1929b: Zwei neue Gramineengattungen von der *Ventenata*-Gattung gesondert. - *Repert. Spec. Nov. Regni Veg.* **26**: 65-79 + 1 plate.

EIG, A. 1933: On the synonymy of *Cornucopiae alopecuroides* L. - *Proc. Linn. Soc. London* **145(2)**: 69-73.

EIG, A. 1936: *Aegilops* L. - *Pflanzenareale* **4(4)**: 43-50, Karte 38-41.

EIG, A. 1937: Neglected Syrian plants of Banks and Solander. - *J. Bot. (London)* **75**: 185-192.

EIG, A. 1938a: Revision of Oriental species of *Picris* (including *Hagioseris*) of the herbarium of the Hebrew University. - *Palestine J. Bot., Jerusalem Ser.* **1(1)**: 65-79.

EIG, A. 1938b: Taxonomic studies on the Oriental species of the genus *Anthemis*. - *Palestine J. Bot., Jerusalem Ser.* **1(2)**: 161-224.

EIG, A. 1942: Revision of the *Onopordon* species of Palestine, Syria and adjacent countries. - *Palestine J. Bot., Jerusalem Ser.* **2(4)**: 185-199.

EIG, A. 1944: Revision of the oriental *Scrophularia* species of the herbarium of the Hebrew University. - *Palestine J. Bot., Jerusalem Ser.* **3(2)**: 79-93 + 1 plate.

EIG, A. 1945: A revision of the Chenopodiaceae of Palestine and neighbouring countries. – *Palestine J. Bot., Jerusalem Ser.* **3(3)**: 119-137.

EISENMAN, S.H. 2015: Some nomenclatural adjustments and typifications for almond species in the genus *Prunus* sensu lato (Rosaceae). - *Phytotaxa* **222(3)**: 185-198.

EISIKOWITCH, D. 1980: Horned poppy, *Glaucium flavum*, in Israel: Notes on its pollination and distribution. - *Israel J. Bot.* **28(2)**: 87-94.

EITEN, G. 1955: The typification of the names "Oxalis corniculata" L. and "Oxalis stricta" L. - *Taxon* **4 (5)**: 99-105.

EITEN, G. 1963: Taxonomy and regional variation of *Oxalis* section *Corniculatae*. I. Introduction, keys and synopsis of the species. - *Amer. Midl. Naturalist* **69(2)**: 257-309.

EKER, İ., DUMAN, H. & YILDIRIM, H. 2020: *Muscari muglaensis* (Asparagaceae, Scilloideae), a new species from southwestern Anatolia. - *Phytotaxa* **475(4)**: 267-278.

EKER, İ. & ARMAĞAN, M. 2020: Türkiye Florası'nda *Muscari kerkis* (Asparagaceae) 'in varlığı üzerine (On the presence of *Muscari kerkis* (Asparagaceae) in the Flora of Turkey). - *Bağbahçe Bilim Dergisi* **7(3)**: 25-34.

EKIM, T. (ED.) 2010: XIII OPTIMA Meeting, March 22-26 2010, Antalya/Türkiye. Book of Abstracts. - Antalya:184 pp.

EKMAN, S. & KNUTSSON, T. 1994: Nomenclatural notes on *Persicaria*. - *Nordic J. Bot.* **14(1)**: 23-25.

EL BABA, J. 1976: Contribution à l'étude cytotaxinomique et palynologique des *Trifolium* de la Méditerranée orientale. - *Biol. Écol. Médit.* **3(2)**: 23-40.

EL BABA, J. 1977: Contribution à l'étude cytotaxinomique des *Trifolium* du Liban. - *Biol. Écol. Médit.* **4 (1)**: 11-18.

EL BABA, J. 1985: Etude cytotaxonomique de quelques Scrophulariacées méditerranéennes. - *Bull. Soc. Bot. France, Lett. Bot.* **132(3)**: 225-231.

EL HADIDI, M.N. 1978: An introduction to the classification of *Tribulus* L. - *Taeckholmia* **9**: 59-66.

EL HUSSEINI, N. & ZAREH, M. 1994: Systematic studies of Scrophulariaceae in Egypt I. *Linaria* Mill., and *Kickxia* Dumort. - *Feddes Repert.* **105(5-6)**: 317-329.

EL QUALIDI, J., PUECH, S. & NAVARRO, T. 2002: Geographical variation and successive adaptive radiation of yellow-flowered *Teucrium* (Labiatae) in the Mediterranean region. - *Bot. Rev. (Lancaster)* **68(2)**: 209-234.

EL ZEIN, H. & DAGHER-KHARRAT, M.B. 2021: New records of vascular plants for the flora of Lebanon: a rare species rediscovered after seventy years, *Daphne pontica* L. (Thymelaeaceae), and three new occurrences, *Atropa bella-donna* L. (Solanaceae), *Circaea lutetiana* L. (Onagraceae), and *Euonymus latifolius* (L.) Mill. (Celastraceae) - *Check List* **17(2)**: 655-667.

EL-GAZZAR, A. & BADAWI, A.A. 1977: The taxonomic significance of chromosome numbers and geography in *Crataegus* L. - *Phytologia* **35(4)**: 271-275.

EL-GAZZAR, A., EISA, A. M. & KHATTAB, A. A. 2016: Computer-generated key and descriptions of *Phleum* species (Poaceae). - *Webbia* **71(1)**: 25-35.

EL-KAREMY, A.R.Z. 1994: On the taxonomy of the genus *Euphorbia* in Egypt. - *Feddes Repert.* **105(5-6)**: 271-281.

EL-KAREMY, A.R.Z. & HOSNI, H.A. 1997: Systematic revision of Leguminosae in Egypt 2. *Lathyrus* L. - *Fl. Medit.* **6**: 31-42.

EL-KEBLAWY, A. LOVETT-DOUST, J. & LOVETT-DOUST, L. 1996: Gender variation and the evolution of dioecy in *Thymelaea hirsuta* (Thymelaeaceae). - *Canad. J. Bot.* **74(10)**: 1596-1601.

EL-OQLAH, A. 1989: A revision of the genus *Erodium* L'Héritier in the Middle East. - *Feddes Repert.* **100(3-4)**: 97-118.

ELEFTHERIADOU, E. 1992: Die Flora der kaltliebenden Laub- und Koniferenwälder und der hohen Extrawaldland von Elatia Dramas [In Greek with German summary]. - Thessaloniki: Ph.D. thesis, Univ. of Thessaloniki, viii + 166 pp.

ELEFTHERIADOU, E. & AL. 1995: Contribution to the flora of the Pinios delta [In Greek]. - *Sci. Ann. Dept. Forest. Nat. Environm. Aristot. Univ. Thessalonika* **38(1)**: 206-231.

ELEFTHERIADOU, E. & AL. 1998: Geographical distribution of *Arabis procurrens* Waldst. & Kit., *Lilium rhodopaeum* Delip. and *Lathraea rhodopea* Dingler in Greece. In Tsekos, I. & Moustakas, M. (eds): Proceedings of the 1st Balkan Botanical Congress, pp. 57-60. - Dordrecht, etc.: Kluwer Academic Publishers.

ELEFTHERIADOU, E. & AL. 1999: Floristic analysis of forests of *Quercus trojana* Webb in the area of Kozani. In Elefheriou, E.P. (ed.): Proceedings, 7th Panhellenic scientific conference, Alexandroupolis, pp. 316-320 [In Greek with English abstract]. - Thessaloniki: Univ. Studio Press.

ELEFTHERIADOU, E. & RAUS, TH. 1996: The vascular flora of the nature reserve Frakto Virgin Forest of Nomos Dramas (E Makedonia, Greece). - *Willdenowia* **25(2)**: 455-485.

ELEFTHERIADOU, E. & THEODOROPOULOS, K. 2003: Terrestrial flora in the coastal zone of Strymonikos and Ierissos gulfs [In Greek]. - *Sci. Ann. Dept. Forest. Nat. Environm. Aristot. Univ. Thessalonika* **40 (1)**: 285-313.

ELEFTHERIADOU, E. & THEODOROPOULOS, K. 2015: Vascular flora of mires in Rhodopi

Mountain Range. In Bogdanović, S. & Jogan, N.: Book of abstracts - 6th Balkan Botanical Congress, p. 61. - Rijeka.

ELEFTHERIADOU, E., THEODOROPOULOS, K. & ATHANASIADIS, N. 1995: New localities and geographical distribution of some rare species in the Greek flora. In: Proc., 5th sci. conf., Hellenic Bot. Soc., Delphi 21-23 October 1994, pp. 225-230 [In Greek with English summary]. - Thessaloniki: Univ. Studio Press.

ELEFTHERIADOU, E., THEODOROPOULOS, K. & RAUS, TH. 2002: Endemic and rare species from Frakto Virgin Forest of Nomos Dramas (E. Makedonia, Greece). In Kamari, G. & al. (eds): Ellinini Votaniki Eteria. 9° Panelliniko Epistimoniko Sinedrio. Praktika, pp. 115-123 [In Greek with English abstract].

ELEFTHERIADOU, E., THEODOROPOULOS, K. & RAUS, TH. 2004: *Lithospermum glandulosum* Velen. In Greuter, W. & Raus, Th. (eds): Med-Checklist notulae, 22. - *Willdenowia* **34(1)**: 72.

ELEFTHERIADOU, E., THEODOROPOULOS, K. & TSIRIPIDIS, I. 2008: Contribution to the flora of the site "Stena Kalamakion" of the "Natura 2000" network. In: Praktika 10ou Sinedriou Ellinikis Votanikis Eterias, 5-8 maiou 2005, Ioannina: pp. 187-193 [In Greek with English abstract].

ELEFTHERIOU, E. & AL. (EDS) 1995: Proceedings, 5th scientific conference, Hellenic Botanical Society, Delphi 21-23 October 1994. - Thessaloniki: Univ. Studio Press, 380 pp.

ELENA-ROSELLÓ, J.-A. 1981: Cytotaxonomic and evolutionary studies in *Thymus* (Labiatae); relationships of the members of section *Thymus* Jalas (1). - *Anales Jard. Bot. Madrid* **38(1)**: 51-59.

ELKINGTON, T.T. 1963: Biological Flora of the British Isles. *Gentiana verna* L. - *J. Ecol.* **51(3)**: 755-767.

ELKINGTON, T.T. 1964: Biological Flora of the British Isles. *Myosotis alpestris* F.W. Schmidt. - *J. Ecol.* **52(3)**: 709-722.

ELKINGTON, T.T. 1971: Biological Flora of the British Isles. *Dryas octopetala* L. - *J. Ecol.* **59(3)**: 887-905.

ELKINGTON, T.T. 1972: Variation in *Gentiana verna* L. - *New Phytol.* **71(6)**: 1203-1211.

ELLENBAST, F 2008: Erstfund von *Limodorum trabutianum* Battandier auf den südlichen Peloponnes. - *J. Eur. Orch.* **40(1)**: 162-164.

ELLENBAST, F. 2013: Orchideenfunde am Olymp und im südlichen Pindusgebirge - Kurzbericht über eine Reise vom 24.04.-08.05.2013. - *J. Eur. Orch.* **45(2-4)**: 301-316.

ELLIOTT, J. 1992: *Leucojum* - the snowflakes. - *Plantsman* **14(2)**: 70-79.

ELLIOTT, J. 1998a: Fritillarias in the garden. - *Quart. Bull. Alpine Gard. Soc. Gr. Brit.* **66(3)**: 265-275.

ELLIOTT, J. 1998b: Forty years of *Fritillaria*. - *Bull. R.H.S. Lily group* **1997/98**: 23-32.

ELLIS, J.R. 1962: *Fragaria-Potentilla* intergeneric hybridization and evolution in *Fragaria*. - *Proc. Linn. Soc. London* **173(2)**: 99-106.

ELLIS, W.M. 1973: The breeding system and variation in populations of *Poa annua* L. - *Evolution* **27 (4)**: 656-662.

ELLNER, S.P. & SHMIDA, A. 1984: Seed dispersal in relation to habitat in the genus *Picris* (Compositae) in Mediterranean and arid regions. - *Israel J. Bot.* **33(1)**: 25-39.

ELVERS, I. 1932: Chromosomenzahlen in der Gattung *Valerianella* nebst einigen systematischen Bemerkungen. - *Acta Horti Berg.* **11(4)**: 81-87.

ELWES, H.J. 1877-80: A monograph of the genus *Lilium*. Parts 1-7. - London: Taylor & Francis [cf. TL-2 1: 745, 1976].

EM, H. 1953a: *Podocytisus caramanicus* Boiss. et Heldr. in Mazedonien. - *Acta Mus. Maced. Sci. Nat.* **1(4)**: 77-91 + 1 map.

EM, H. 1953b: Neue Mitteilungen über den Griechischen Ahorn (*Acer Heldreichii* Orph. in Boiss.) aus Makedonien [In Serbo-Croatian with German summary]. - *God. Biol. Inst. Sarajevu* **5(1-2)**: 159-168.

EM, H. 1965: Neue Fundorte des Sebenbaums (*Juniperus sabina* L.) in Mazedonien [In Macedonian with German summary]. - *Godišen Zborn. Zemjod.-Šumarsk. Fak. Univ. Skopje, Šumarstvo* **18**: 45-47 + 1 map.

EM, H. 1969: Der Buchsbaum (*Buxus sempervirens* L.) in der Vegetation Mazedoniens. - *Mitt. Ostalpin-Dinarischen Pflanzensoziol. Arbeitsgem.* **9**: 311-322.

EM, H. 1978: Die Gattung *Ephedra* in Makedonien [In Macedonian with German summary]. - *Maked. Akad. Nauk. Umetn., Oddel. Biol. Med. Nauki* **1(2)**: 5-19.

EM, H. 1982a: Die Gattung *Daphne* L. in Mazedonien [In Serbo-Croatian with German summary]. - *Glasn. Republ. Zavoda Zaštitu Prir. Prirodnjačke Zbirke Titogradu* **15**: 49-54.

EM, H. 1982b: *Celtis tournefortii* in Makedonien [In Macedonian with German summary]. - *Šumarski Pregled* **30(3-4)**: 11-21.

EM, H. 1989: Ist der verkahlende Zürgelbaum (*Celtis glabrata* Stev.) autochton auf der Insel Golem Grad im Prespa-See? - *Bios (Thessaloniki)* **1989**: 65-66.

EM, H., DŽEKOV, S. & DIMITROVSKI, T. (EDS) 1974: Wild fruit flora in SR Macedonia [Contributions by several different authors; in Macedonian with English or German summaries]. - *Godišen Zborn. Zemjod.-Šumarsk. Fak. Univ. Skopje, Šumarstvo* **Special Suppl.**: 1-313.

EMIG, W. & KADEREIT, J.W. 1993: The comparative biology of the closely related *Senecio nebrodensis* and *S. viscosus*, a narrow endemic and a widespread ruderal. - *Nordic J. Bot.* **13(4)**: 369-375.

EMMRICH, R. 2002: *Epipactis turcica* auf Samos. - *J. Eur. Orch.* **34(4)**: 797-801.

ENAYET HOSSAIN, A.B.M. 1973: Materials for a Flora of Turkey XXIX: *Tripleurospermum* Schultz Bip. - *Notes Roy. Bot. Gard. Edinburgh* **32(2)**: 251-254.

ENDTMANN, K.J. 1966: Untersuchungen an Sippen der Gelben Taubnessel (*Lamium galeobdolon* (L.) Nathh.). - *Feddes Repert.* **72(2)**: 132-154.

ENDTMANN, K.J. & KLOSS, K.E. 1990: Beiträge zur Verbreitung, Ökologie und Taxonomie von *Lithospermum arvense* subsp. *coerulescens*. - *Verh. Berliner Bot. Vereins* **8**: 149-160.

ENE, L.S.O. 1973: Polyploids in the genus *Iberis*. - *Cytologia* **38(4)**: 699-706.

ENGEL, K. 1972: [Chromosome numbers in *Valeriana*]. In Löve, Á. (ed.): IOPB chromosome number reports XXXVII. - *Taxon* **21(4)**: 495.

ENGEL, K. 1976: Beiträge zur Systematik der Valerianaceae unter besonderer Berücksichtigung cytosystematischer Ergebnisse. - Giessen: Diss., Justus Liebig-Universität, v + 196 pp.

ENGELEN, D. 2020: Inventorying and mapping the orchids of the island of Antikythera (Greece), including a new taxon: *Ophrys aegilica* (Orchidaceae). - *Fl. Medit.* **30**: 273-291.

ENGLER, H.G.A. 1872: Monographie der Gattung *Saxifraga* L. mit besonderer Berücksichtigung der geographischen Verhältnisse ... - Breslau: J.U. Kern's Verlag, iv + 292 pp.

ENGLER, H.G.A. & PRANTL, K.A.E. (EDS) 1887-1915: Die natürlichen Pflanzenfamilien nebst ihren Gattungen und wichtigeren Arten insbesondere den Nutzpflanzen, bearbeitet unter Mitwirkung zahlreicher hervorragender Fachgelehrten. - Leipzig: W. Engelmann, many fascs. [cf. TL-2 1: 764-769, 1976].

ENGLER, H.G.A. & PRANTL, K.A.E. (EDS) 1924-: Die natürlichen Pflanzenfamilien nebst ihren Gattungen und wichtigsten Arten, insbesondere den Nutzpflanzen. Under Mitwirkung zahlreicher hervorragender Fachgelehrten ... Ed. 2. - Leipzig/Berlin: many fascs. [cf. TL-2 1: 769-783, 1976].

ENGLER, H.G.A. (ED.) 1900-: Das Pflanzenreich. Regni vegetabilis conspectus. - Berlin: W. Engelmann, many fascs. [cf. TL-2 1: 785-797, 1976].

ENGSTRAND, L. 1970a: Studies in the Aegean flora. XVIII. Notes and chromosome numbers in Aegean Umbelliferae. - *Bot. Not.* **123(3)**: 384-393.

ENGSTRAND, L. 1970b: The European species of *Scaligeria* (Umbelliferae). - *Bot. Not.* **123(4)**: 505-511.

ENGSTRAND, L. 1973: Generic delimitation of *Bunium*, *Conopodium* and *Geocaryum* (Umbelliferae). - *Bot. Not.* **126(2)**: 146-154.

ENGSTRAND, L. 1977: Biosystematics and taxonomy in *Geocaryum* Cosson (Umbelliferae). - Lund: Ph.D. thesis, Univ. of Lund, 126 pp.

ERBEN, M. 1978: Die Gattung *Limonium* im südwestmediterranen Raum. - *Mitt. Bot. Staatssamml. München* **14(2)**: 361-631.
ERBEN, M. 1979: Karyotype differentiation and its consequences in Mediterranean "Limonium". - *Webbia* **34(1)**: 409-417.
ERBEN, M. 1980: Bemerkungen zur Taxonomie der Gattung *Limonium*. I. - *Mitt. Bot. Staatssamml. München* **16**: 547-563.
ERBEN, M. 1981: Bemerkungen zur Taxonomie der Gattung *Limonium*. II. - *Mitt. Bot. Staatssamml. München* **17**: 485-510.
ERBEN, M. 1984: *Viola merxmuelleri* - eine neue Art der Gattung *Viola* Sektion *Melanium* aus Mittel- Italien (Gargano). - *Mitt. Bot. Staatssamml. München* **20**: 29-38.
ERBEN, M. 1985: Cytotaxonomische Untersuchungen an südosteuropäischen *Viola*-Arten der Sektion *Melanium*. - *Mitt. Bot. Staatssamml. München* **21(2)**: 339-740.
ERBEN, M. 1986a: Bemerkungen zur Taxonomie der Gattung *Limonium* III. - *Mitt. Bot. Staatssamml. München* **22**: 203-220.
ERBEN, M. 1986b: *Viola acrocerauniensis* und *Viola etrusca* - Zwei neue *Viola*-Arten aus der Section Melanium. - *Mitt. Bot. Staatssamml. München* **22**: 493-506.
ERBEN, M. 1988: Bemerkungen zur Taxonomie der Gattung *Limonium*. IV. - *Mitt. Bot. Staatssamml. München* **27**: 381-406.
ERBEN, M. 1989a: Bemerkungen zur Taxonomie der Gattung *Limonium*. V. - *Mitt. Bot. Staatssamml. München* **28**: 313-417.
ERBEN, M. 1989b: *Viola ganiatsasii* - Eine neue Art der Section *Melanium* aus der südlichen Rhodopen. - *Bios (Thessaloniki)* **1989**: 67-73.
ERBEN, M. 1996: The significance of hybridization on the forming of species in the genus *Viola*. - *Bocconea* **5(1)**: 113-118.
ERBEN, M. 2000: *Viola serresiana*, eine neue Art von *Viola* sect. *Melanium* aus Nordost-Griechenland. - *Bot. Chron. (Patras)* **13**: 51-59.
EREN, Ö. 2007: The genus *Rhaponticoides* Vaill. (Asteraceae) in Turkey: a new species and first key. - *Pl. Syst. Evol.* **267(1-4)**: 13-23.
EREN, Ö. & AL. 2008: A new species of *Polygala* L. (Polygalaceae) from south-west Turkey. - *Bot. J. Linn. Soc.* **158(1)**: 82-86.
EREN, Ö., GÖKÇELOĞLU, M. & PAROLLY, G. 2004: The flora and vegetation of Bakırlı Dağı (Western Taurus Mts, Turkey), including annotations on critical taxa of the Taurus range. - *Willdenowia* **34(2)**: 463-503.
ERGUN, F. & DELIORMAN, D. 1995: The life cycle of *Viscum album* L. - *Karaca Arbor. Mag.* **3(2)**: 49-54.
ERIKSEN, B. 1989: Notes on generic and infrageneric delimitation in the Valerianaceae. - *Nordic J. Bot.* **9(2)**: 179-187.
ERKAN UZUNHISARCIKLI, M., DOĞAN GÜNER, E. & EKICI, M. 2018: Synopsis of the genus *Scrophularia* (Scrophulariaceae) in Turkey. - *Phytotaxa* **333(2)**: 151-187.
ERMELINDA, G. & AL. 2015: New data for endemic, rare or protected plants of Shebenik-Jabllanica National Park, Albania. In Bogdanović, S. & Jogan, N.: Book of abstracts - 6th Balkan Botanical Congress, pp. 63-64. – Rijeka.
ERN, H. & BROWICZ, K. 1984: *Acer campestre* L. in the Amanus Mts., Turkey. - *Arbor. Kórnickie* **28**: 3-6.
ERNET, D. 1972: [Chromosome numbers in *Valerianella*]. In Löve, Á. (ed.): IOPB chromosome number reports XXXVII. - *Taxon* **21(4)**: 495-496.
ERNET, D. 1977a: Sprossaufbau und Lebensform von *Valerianella* und *Fedia* (Valerianaceae). - *Pl. Syst. Evol.* **127(4)**: 243-276.
ERNET, D. 1977b: Blütenbau und Fortpflanzungsbiologie von *Valerianella* und *Fedia* (Valerianaceae). - *Pl. Syst. Evol.* **128(1-2)**: 1-22.
ERNET, D. 1978: Fruchtbau und Verbreitungsbiologie von *Valerianella* und *Fedia* (Valerianceae). - *Pl. Syst. Evol.* **130(1-2)**: 85-126.
ERNST, A. 1961: Revision der Gattung *Pinguicula*. - *Bot. Jahrb. Syst.* **80(2)**: 145-194.
ERNST, W.H.O. & MALLOCH, A.J.C. 1994: Biological Flora of the British Isles. *Phleum arenarium* L. (*Phalaris arenaria* Willd.; *Chilochloa arenaria* P. Beauv. - *J. Ecol.* **82(2)**: 403-413.
EROL, O. & KÜÇÜKER, O. 2003: Morpho-anatomical observations on three *Romulea* (Iridaceae) taxa of Turkey. - *Bocconea* **16(2)**: 607-613.
ERTAS, A. 1995: The oaks of Turkey. - *J. Int. Oak Soc.* **6**: 33-42.
ERTEKIN, A.S. & AKBAYIN, H. 2000: A new variety of *Trifolium nigrescens* Viv. (Fabaceae) from Turkey. - *Israel J. Pl. Sci.* **48(1)**: 71-73.
ESCHEMANN-GRUPE, G., HURKA, H. & NEUFFER, B. 2003: Species relationships within *Diplotaxis* (Brassicaceae) and the phylogenetic origin of *D. muralis*. - *Pl. Syst. Evol.* **243(1-2)**: 13-29.
ESCUDERO, M. & LUCEÑO, M. 2009: Systematics and evolution of *Carex* sects. *Spirostachyae* and *Elatae* (Cyperaceae). - *Pl. Syst. Evol.* **279(1-4)**: 163-189.
ESNAULT, M.-A. & HUON, A. 1987: Études morphologiques et caryologiques de *Bromus rigidus* et *Bromus diandrus* Roth: relations taxonomiques. - *Bull. Soc. Bot. France, Lett. Bot.* **134(3)**: 299-305.
ESPEUT, M. 2002: *Conium maculatum* subsp. *viride* (DC.) Espeut, stat. nov. (Apiaceae). - *Candollea* **57(1)**: 131-142.
ESSAD, S. 1954: Contribution à la systematique du genre *Lolium*. - *Ann. Inst. Natl. Rech. Agron., Sér. B, Ann. Amélior. Pl.* **4(3)**: 325-351.
ESSI, L. & AL. 2017: A synopsis of *Briza*, *Brizochloa*, and *Chascolytrum* (Poaceae,Pooideae, Poeae). - *Ann. Missouri Bot. Gard.* **102(3)**: 466-519.
ESSINK, J. & ESSINK, E. 1997: Complex bespreking *Himantoglossum*. - *Eurorchis* **9**: 33-38.
ESSLEMONT, H. 1977: Plant portrait: *Daphne jasminea*. - *J. Scott. Rock Gard. Club* **15(3)**: 224 + 1 plate.
ESTES, J.R. & TYRL, R.J. 1982: The generic concept and generic circumscription in the Triticeae: An end paper. In Esters, J.R., Tyrl, R.J. & Brunken, J.N. (eds): Grasses and grasslands. Systematics and ecology, pp. 145-164. - Norman: Univ. of Oklahoma Press.
ESTES, J.R., TYRL, R.J. & BRUNKEN, J.N. 1982: Grasses and grasslands. Systematics and ecology. - Norman: Univ. of Oklahoma Press, viii + 312 pp.
ESTILAI, A. & KNOWLES, P.F. 1978: Relationship of *Carthamus leucocaulos* to other *Carthamus* species (Compositae). - *Canad. J. Genet. Cytol.* **20(2)**: 221-233.
ETTLINGER, D.M.T. 1988: Notes on orchids seen in Crete - 1978, 1983 & 1988. - Cliftonville: privately published by the author.
ETTLINGER, D.M.T. 1992: Notes sur les orchidées vues à Corfou (Kerkira, Grèce) en 1981 et 1992. - *Naturalistes Belges* **73(3)**: 113-124.
ETTLINGER, D.M.T. 1995: Notes sur les orchidées vues en 1994 dans l'île de Thasos (Nomos Kavála, Grèce). - *Naturalistes Belges* **76(3)**: 222-231.
ETTLINGER, D.M.T. 1996: L'existence d'*Orchis palustris* s.l. en Crète et dans l'île de Kos (Dodécanèse, Grèce). - *Naturalistes Belges* **77(4)**: 111-118.
EVANS, R.L. 1983: A handbook of cultivated Sedums. - Northwood, Middlesex: Science Reviews Ltd, ix + 345 pp. + colour plates.
EVANS, R.L. 1993: *Sedum tristriatum*. - *Sedum Soc. Newsletter* **27**: 17-20.
EVENARI, M., KADOURI, A. & GUTTERMAN, Y. 1977: Eco-physiological investigations on the amphicarpy of *Emex spinosa* (L.) Campd. - *Flora, Morphol. Geobot. Oekophysiol.* **166(3)**: 223-238.
EVERETT, D. 2013: The genus *Tulipa*. Tulips of the world. - Kew Publishing: 380 pp.
EVISON, R. & MATTHEWS, V.A. 1994: *Clematis cirrhosa*. - *New Plantsman* **1(2)**: 95-101.
EVSTATIEVA, L.N. & AL. 1993: A biosystematic study of *Valeriana officinalis* (Valerianaceae) distributed in Bulgaria. - *Pl. Syst. Evol.* **185(3-4)**: 167-179.
FABBRI, F. & FAGIOLI, A. 1971: Numeri cromosomici per la Flora Italiana: 19-22. - *Inform. Bot. Ital.* **3 (1)**: 51-55.
FAGIOLI, A. & FABBRI, F. 1971: Numeri cromosomici per la Flora Italiana: 23-28. - *Inform. Bot. Ital.* **3 (1)**: 55-62.
FAKAS, G, KORAKIS, A. & ALEXIOU, S. 2014: *Colchicum atticum* (Colchicaceae), a new taxon for Samos island, Greece. - *Parnassiana Arch.* **2**: 55-59.

FALCH, M., SCHÖNSWETTER, P. & FRAJMAN, B. 2019: Both vicariance and dispersal have shaped the genetic structure of Eastern Mediterranean *Euphorbia myrsinites* (Euphorbiaceae). - *Perspect. Pl. Ecol., Evol. Syst.* **39**: 9 pp.

FALCIANI, L. 1997: Systematic revision of *Stachys* L. sect. *Eriostomum* (Hoffmans. & Link) Dumort. in Italy. - *Lagascalia* **19(1-2)**: 187-238.

FALCIANI, L., BINI MALECI, L. & MARIOTTI LIPPI, M. 1995: Morphology and distribution of trichomes in Italian species of the *Stachys germanica* group (Labiatae): a taxonomic evaluation. - *Bot. J. Linn. Soc.* **119(3)**: 245-256.

FALCIANI, L. & FIORINI, G. 1997: Reports (679-682). In Kamari, G., Felber, F. & Garbari, F. (eds): Mediterranean chromosome number reports - 6. - *Fl. Medit.* **6**: 243-247.

FALISTOCCO, E. 1987: Cytogenetic investigations and karyological relationships of two *Medicago*: *M. sativa* L. (alfalfa) and *M. arborea* L. - *Caryologia* **40(4)**: 339-346.

FALISTOCCO, E. 1994: Citogenetica ed evoluzione nel genere *Medicago* L. - *Inform. Bot. Ital.* **25(1)**: 81-85.

FALISTOCCO, E. & FALCINELLI, M. 1991: Cytological and morphological studies in *Medicago hispida* Gaertner (= *M. polymorpha* L.). - *Ann. Bot. (Rome)* **49**: 13-25.

FALISTOCCO, E. & MAURELLI, L. 1990: Chromosome study in *Hedysarum coronarium* L. - *Ann. Bot. (Rome)* **48**: 81-86.

FALISTOCCO, E., PICCIRILLI, M. & FALCINELLI, M. 1986: The basic karyotype of *Trifolium subterraneum* subsp. *brachycalycinum*. - *Ann. Bot. (Rome)* **44**: 171-174.

FALISTOCCO, E., PICCIRILLI, M. & FALCINELLI, M. 1987: Cytotaxonomy of *Trifolium subterraneum* L. - *Caryologia* **40(1-2)**: 123-130.

FANLO DOMÍNGUEZ, R. 1975: El género *Valerianella* en la Península Ibérica. II. - *Acta Bot. Malac.* **1**: 47-52.

FANLO DOMÍNGUEZ, R. 1981a: El género *Valerianella* Miller en la Península Ibérica. III. - *Anales Jard. Bot. Madrid* **38(1)**: 61-66.

FANLO DOMÍNGUEZ, R. 1981b: *Valerianella* (Valerianaceae) en la Península Ibérica. - *Lazaroa* **3**: 131-135.

FANLO DOMÍNGUEZ, R. 1985: El género *Fedia* Gaertner (Valerianaceae) en España. - *Lazaroa* **6**: 175-180.

FANLO DOMÍNGUEZ, R. 1986: El género *Centranthus* DC. en España. I. Sección *Calcitrapa* Lange. - *Lagascalia* **14(1)**: 3-8.

FARHAT, P. & AL. 2019: Genome size variation and polyploidy in the geographical range of *Juniperus sabina* L. (Cupressaceae). - *Botany Letters* **166(2)**: 134-143.

FARJON, A. 1992: The taxonomy of multiseeded junipers (*Juniperus* sect. *Sabina*) in southwest Asia and East Africa. - *Edinburgh J. Bot.* **49(3)**: 251-283.

FARJON, A. 2001: World checklist and bibliography of conifers. 2nd ed. - Kew: Royal Botanic Gardens, v + 309 pp.

FARJON, A. 2005a: A bibliography of conifers: selected literature on taxonomy and related disciplines on the Coniferales, 2nd ed. - Kew: Royal Botanic Gardens, 211 pp.

FARJON, A. 2005b: A monograph of Cupressaceae and *Sciadopitys*. - Kew: Royal Botanic Gardens, 643 pp.

FARJON, A. 2005c: Pines: drawings and descriptions of the genus *Pinus*, 2nd ed. - Leiden: Brill, 235 pp.

FARJON, A. 2010: A handbook of the world's conifers. - Leiden-Boston: Brill, 2 vols., 1111 pp.

FARJON, A. & RUSHFORTH, K.D. 1989: A classification of *Abies* Miller (Pinaceae). - *Notes Roy. Bot. Gard. Edinburgh* **46(1)**: 59-79.

FARMAR, L. 1905: *Scilla messeniaca*. - *Bot. Mag.* **131**: tab. 8035 + 2 pp. text.

FARRELL, L. 1985: Biological Flora of the British Isles. *Orchis militaris* L. (*O. galatea* Poir., *O. rivini* Gouan, *O. tephrosanthos* Willd. & Sw.). - *J. Ecol.* **73(3)**: 1041-1053.

FASSETT, N.C. & CALHOUN, B. 1952: Introgression between *Typha latifolia* and *T. angustifolia*. - *Evolution* **6(4)**: 367-378.

FASSOU, D. & AL. 2018: On the *Dianthus fruticosus* and *Dianthus juniperinus* groups in Greece: Distribution patterns, functional diversity and phylogenetic relationships. - *Bot. Serbica* **142 (suppl. 1)**: 58.

FASSOU, D. & AL. 2020: Genetic diversity and range dynamics of *Helleborus odorus* subsp. *cyclophyllus* under different climate change scenarios. - *Forests MDPI* **11, 620**: 18 pp.

FASSOU, G. & AL. 2019: Integrated taxonomy of *Dianthus* (Caryophyllaceae): The Cretan *Dianthus juniperinus-fruticosus* group is monophyletic but the suffruticose life form evolved multiple times. In Daskalakou, E.N. & al. (eds): 16th Conference of the Hellenic Botanical Society – Athens 10-13 October 2019. Program and book of abstracts, p. 46.

FAULKNER, J.S. 1972: Chromosome studies on *Carex* section *Acutae* in north-west Europe. - *Bot. J. Linn. Soc.* **65(3)**: 271-301.

FAURE, J. 1968: Contribution à l'étude cytologique des *Asperula* de Grèce. - Marseille: Thèse, Univ. de Marseille [n.v.].

FAURE, J. & PIETRERA, D. 1969: Contribution à l'étude cytotaxinomique des Rubiacées de Grèce. - *Ann. Fac. Sci. Marseille* **42**: 271-283.

FAURHOLDT, N. 1994: Rhodos - et godt sted at starte [In Danish]. - *Orchideer* **15(1)**: 2-4.

FAVARGER, C. 1953: Notes de caryologie alpine. II. - *Bull. Soc. Neuchâteloise Sci. Nat.* **76**: 133-169.

FAVARGER, C. 1959: Notes de caryologie alpine. III. - *Bull. Soc. Neuchâteloise Sci. Nat.* **82**: 255-285 + 1 plate.

FAVARGER, C. 1962: Contribution à l'étude cytologique des genres *Minuartia* et *Arenaria*. - *Bull. Soc. Neuchâteloise Sci. Nat.* **85**: 53-81 + 2 plates.

FAVARGER, C. 1965: Notes de caryologie alpine. IV. - *Bull. Soc. Neuchâteloise Sci. Nat.* **88**: 5-60 + 1 plate.

FAVARGER, C. 1966a: Contribution à la cytotaxinomie du genre *Petrorhagia* (=*Tunica*). - *Ber. Schweiz. Bot. Ges.* **76**: 270-278.

FAVARGER, C. 1966b: Un nombre chromosomique de base, nouveau pour le genre *Thesium* L. (Santalaceae). - *Bull. Soc. Neuchâteloise Sci. Nat.* **89**: 59-59.

FAVARGER, C. 1967a: Nombres chromosomiques de quelques taxa principalement balkaniques du genre *Minuartia* (L.) Hiern. - *Bot. Jahrb. Syst.* **86(1-4)**: 280-292.

FAVARGER, C. 1967b: Cytologie et distribution des plantes. - *Biol. Rev. Cambridge Philos. Soc.* **42**: 163-206.

FAVARGER, C. 1969a: De caryologia Cerastiorum specierum aliquot imprimis in Peninsula Balcanica crescentium. - *Acta Bot. Croat.* **28**: 63-74.

FAVARGER, C. 1969b: [Chromosome numbers in *Cerastium* and *Herniaria*]. In Löve, Á. (ed.): IOPB chromosome number reports XXII. - *Taxon* **18(4)**: 434.

FAVARGER, C. 1972a: Endemism in the montane floras of Europe. In Valentine, D.H. (ed.): Taxonomy, phytogeography and evolution, pp. 191-204. - London & New York: Academic Press.

FAVARGER, C. 1972b: Sur quelques *Arenaria* d'Europe et d'Asie occidentale. - *Bot. Not.* **125(4)**: 465-476.

FAVARGER, C. 1972c: Nouvelle contribution à l'étude cytologique du genre *Erysimum* L. - *Ann. Sci. Univ. Besançon, Sér. 3, Bot.* **12**: 49-56.

FAVARGER, C. 1973: Cytotaxonomie de quelques oreophytes des Abruzzes. - *Acta Bot. Acad. Sci. Hung.* **19(1-4)**: 81-92.

FAVARGER, C. 1975: Cytotaxonomie et histoire de la flore orophile des Alpes et de quelques autres massifs montagneux d'Europe. - *Lejeunia, n.s.* **77**: 1-45.

FAVARGER, C. 1976a: Un nouveau *Cerastium* de Turquie: *Cerastium dominici* spec. nov. - *Bull. Soc. Neuchâteloise Sci. Nat.* **99**: 81-86 + 1 plate.

FAVARGER, C. 1976b: Nouveau nombre chromosomique "de base" pour le genre *Cerastium* L. et phénomènes endomitotiques chez *Cerastium dominici* Favarger. - *Bull. Soc. Neuchâteloise Sci. Nat.* **99**: 87-92 + 1 plate.

FAVARGER, C. 1977: *Cerastium ligusticum* and *C. dominici* in Turkey. - *Notes Roy. Bot. Gard. Edinburgh* **35(3)**: 315-317.

FAVARGER, C. 1978: Sur quelques *Valeriana* endémiques des Alpes ou des Dinarides. - *Candollea* **33 (1)**: 11-21.

FAVARGER, C. 1987: Commentaires sur la caryologie des espèces de *Minuartia* L. de la Péninsule Ibérique. - *Anales Jard. Bot. Madrid* **44(2)**: 558-564.

FAVARGER, C. 1989: Contribution à la cytotaxonomie des Paronychioideae (Caryophyllaceae). - *Bios (Thessaloniki)* **1989**: 75-85.

FAVARGER, C. 1999: Contribution à la cytogéographie du *Minuartia glomerata* (M. Bieb.) Degen (Caryophyllaceae). - *Bull. Soc. Neuchâteloise Sci. Nat.* **122**: 27-33.

FAVARGER, C. & ÇELEBIOĞLU, T. 1996: Sur la position systématique du *Minuartia juressi* (Willd. ex Schlecht.) Lacaita ssp. *asiatica* McNeill et réflexions sur la section *Plurinerviae* McNeill du genre *Minuartia* (Caryophyllaceae). - *Ann. Naturhist. Mus. Wien, B* **98B Suppl.**: 163-174.

FAVARGER, C. & CORREVON, P. 1967: Mise en évidence de "races chromosomiques" chez l'*Amélanchier*. - *Bull. Soc. Neuchâteloise Sci. Nat.* **90**: 215-218 + 1 plate.

FAVARGER, C., GALLAND, N. & KÜPFER, PH. 1980: Recherches cytotaxonomiques sur la flore orophile du Maroc. - *Naturalia Monspel., Sér. Bot.* **29**: 1-64.

FAVARGER, C. & GARRAUD, L. 1992: Morphologie et écologie de trois cytotypes de *Minuartia hybrida* (Vill.) Schischkin (Etude de populations de la région lyonnaise: France). - *Bauhinia* **10**: 91-98.

FAVARGER, C. & HUYNH, K.-L. 1980: Contribution à la cytotaxonomie des Caryophyllacées méditerranéennes. - *Bol. Soc. Brot., Ser. 2* **53(1)**: 493-514 + 2 plates.

FAVARGER, C. & KRÄHENBÜHL, M. 1996: Nombre chromosomique et affinités probables d'un *Cerastium* (Caryophyllaceae) des Balkans. - *Anales Jard. Bot. Madrid* **54(1)**: 155-165.

FAVARGER, C. & KÜPFER, PH. 1980: Application de la cytotaxonomie a quelques problèmes d'origine ou de mise en place de la flore mediterranéenne. In: Colloque de la Fondation L. Emberger. - *Naturalia Monspel., Sér. Bot.* **No. hors série**: 53-65.

FAVARGER, C. & KÜPFER, PH. 1983: Index des nombres chromosomiques des spermatophytes de la Suisse. Introduction, matériel et méthodes. - *Bot. Helv.* **93(1)**: 3-9.

FAVARGER, C. & STEARN, W.T. 1983: Contribution à la cytotaxonomie de l'*Amelanchier ovalis* Medikus (Rosaceae). - *Bot. J. Linn. Soc.* **87(1)**: 85-103.

FAY, M.F. 2013: 751. *Orchis militaris*. Orchidaceae - *Curtis's Bot. Mag.* **30(1)**: 9-17.

FAY, M.F. & RANKOU, H. 2010: 670. *Anacamptis morio*. Orchidaceae. - *Curtis's Bot. Mag.* **27(2)**: 100-108.

FAY, M.F., SAYERS, B. & TAYLOR, I. 2015: 802. *Dactylorhiza viridis*. Orchidaceae. - *Curtis's Bot. Mag.* **32(1)**: 33-41.

FAY, M.F. & TAYLOR, I. 2015a: 805. *Ophrys fuciflora*. Orchidaceae. - *Curtis's Bot. Mag.* **32(1)**: 42-50.

FAY, M.F. & TAYLOR, I. 2015b: 805. *Orchis anthropophora*. Orchidaceae. - *Curtis's Bot. Mag.* **32(1)**: 63-71.

FAY, M.F. & TAYLOR, I. 2015c: 806. *Orchis purpurea*. Orchidaceae. - *Curtis's Bot. Mag.* **32(1)**: 72-81.

FAY, M.F. & TAYLOR, I. 2015d: 807. *Cephalanthera rubra*. Orchidaceae. - *Curtis's Bot. Mag.* **32(1)**: 82-90.

FAY, M.F., TAYLOR, I. & SAYERS. B. 2015: 804. *Ophrys insectifera*. Orchidaceae. - *Curtis's Bot. Mag.* **32(1)**: 51-62.

FAYED, A.A. & AL. 2019: Taxonomic revision of the genus *Ononis* (Trifolieae, Fabaceae) in Egypt, with the first record of *Ononis viscosa* subsp. *breviflora*. - *Phytotaxa* **408(1)**: 1-29.

FAZAN, L. & AL. 2017: Imprisoned in the Cretan mountains: How relict *Zelkova abelicea* (Ulmaceae) trees cope with Mediterranean climate. - *Sci. Total Environm.* **599-600**: 797-805.

FEARN, G.M. 1972: The distribution of intraspecific chromosome races of *Hippocrepis comosa* L. and their phytogeographical significance. - *New Phytol.* **71(6)**: 1221-1225.

FEARN, G.M. 1973: Biological Flora of the British Isles. *Hippocrepis comosa* L. - *J. Ecol.* **61(3)**: 915-926.

FEDDE, F. 1905: Papaveraceae novae vel notabiles in Herbario Boissier et Barbey-Boissier versantes. - *Bull. Herb. Boissier, Ser. 2* **5(2-4)**: 165-171, 445-448.

FEDDE, F. 1919: Beiträge zur Kenntnis der europäischen Arten der Gattung *Corydalis*. I-II. - *Repert. Spec. Nov. Regni Veg.* **16**: 49-60, 187-192.

FEDDE, F. 1929: Vermischte neue Diagnosen. - *Repert. Spec. Nov. Regni Veg.* **27**: 237-240.

FEDDE, F. 1930: Vermischte neue Diagnosen. - *Repert. Spec. Nov. Regni Veg.* **27**: 395-400.

FEDORONCHUK, N.M. 1979a: The summary of the genus *Trinia* Hoffm. (Apiaceae) [In Russian]. - *Bot. Zhurn. (Moscow & Leningrad)* **64(1)**: 47-50.

FEDORONCHUK, N.M. 1979b: Comparative kariological characteristics of species of the genus *Trinia* Hoffm. and *Rumia crithmifolia* (Willd.) K.-Pol. (Apiaceae) [In Russian]. - *Bot. Zhurn. (Moscow & Leningrad)* **64(3)**: 390-398.

FEDOROV, A.A. 1973: [New combinations in *Campanula*]. In Heywood, V.H. (ed.): Notulae systematicae ad Floram Europaeam spectantes, no. 14. - *Bot. J. Linn. Soc.* **67(3)**: 281.

FEDOROV, A.A. (ED.) 1969: Chromosome numbers of flowering plants [In Russian and English]. - Leningrad: Komarov Bot. Inst., Academy of Sciences of the USSR, 926 pp.

FEDTCHENKO, B. 1914: Vorläufiges Verzeichnis der Arten der Gattung *Tulipa*. - *Bot. Jahrb. Syst.* **50 Suppl.**: 611-617.

FEER, H. 1890: Campanularum novarum decas prima. - *J. Bot. (London)* **28**: 268-274.

FEINBRUN, N. 1933: Über die Variabilität von *Trigonella monspeliaca* L. und die pflanzengeographischen Verhältnisse ihrer Formen. - *Beih. Bot. Centralbl., Abt. 2* **51(2)**: 389-396.

FEINBRUN, N. 1938a: A monographic study on the genus *Bellevalia* Lapeyr. (caryology, taxonomy, geography). - *Palestine J. Bot., Jerusalem Ser.* **1(1)**: 42-54.

FEINBRUN, N. 1938b: A monographic study on the genus *Bellevalia* (caryology, taxonomy, geography) [cont.]. - *Palestine J. Bot., Jerusalem Ser.* **1(2)**: 131-142.

FEINBRUN, N. 1940: A monographic study on the genus *Bellevalia* (caryology, taxonomy, geography) [cont. and end]. - *Palestine J. Bot., Jerusalem Ser.* **1(4)**: 336-409 + 4 plates.

FEINBRUN, N. 1941: The genus *Ornithogalum* in Palestine and neighbouring countries. - *Palestine J. Bot., Jerusalem Ser.* **2(2-3)**: 132-150.

FEINBRUN, N. 1943: *Allium* sectio *Porrum* of Palestine and the neighbouring countries. - *Palestine J. Bot., Jerusalem Ser.* **3(1)**: 1-21.

FEINBRUN, N. 1948: Further studies on *Allium* of Palestine and the neighbouring countries. – *Palestine J. Bot., Jerusalem Ser.* **4(3)**: 144-157.

FEINBRUN, N. 1950: Chromosome counts in Palestinian *Allium* species. - *Palestine J. Bot., Jerusalem Ser.* **5(1)**: 13-16.

FEINBRUN, N. 1953: The genus *Colchicum* of Palestine and neighbouring countries. - *Palestine J. Bot., Jerusalem Ser.* **6(2)**: 71-95.

FEINBRUN, N. 1957a: The genus *Crocus* in Israel and neighbouring countries. - *Kew Bull.* **1957(2)**: 269-285.

FEINBRUN, N. 1957b: Chromosome numbers in the genus *Colchicum*. - *Bull. Res. Council Israel, Sect. B, Biol. Geol.* **6(3-4)**: 283.

FEINBRUN, N. 1958a: Chromosome numbers and evolution in the genus *Colchicum*. - *Evolution* **12(2)**: 173-188.

FEINBRUN, N. 1958b: Chromosome numbers in *Crocus*. - *Genetica* **29**: 172-192.

FEINBRUN, N. 1961: Revision of the genus *Hyacinthella* Schur. - *Bull. Res. Council Israel, Sect. D, Bot.* **10(1-4)**: 324-347.

FEINBRUN, N. 1963a: Chromosomes of some East-Mediterranean *Papaver* species. - *Caryologia* **16 (3)**: 649-652.

FEINBRUN, N. 1963b: Taxonomic studies on *Papaver* sect. *Orthorhoeades* of Palestine and of some other Mediterranean countries. - *Israel J. Bot.* **12(2)**: 74-96.

FEINBRUN, N. 1968: The genus *Lycium* in the Flora Orientalis region. - *Collect. Bot. (Barcelona)* **7(1)**: 359-379.

FEINBRUN, N. 1970: A taxonomic revision of European Cuscutae. - *Israel J. Bot.* **19(1)**: 16-29.

FEINBRUN, N. & SCHMIDA, A. 1978: A new review of the genus *Crocus* in Israel and neighbouring countries. - *Israel J. Bot.* **26(4)**: 172-189.

FEINBRUN, N. & STEARN, W.T. 1958: A revision of *Sternbergia* (Amaryllidaceae) in Palestine. - *Bull. Res. Council Israel, Sect. D, Bot.* **6(3)**: 167-173.

FEINBRUN, N. & STEARN, W.T. 1964: Typification of *Lycium barbarum* L., *L. afrum* L. and *L. europeum* L. - *Israel J. Bot.* **12(3)**: 114-123.
FEINBRUN, N. & TAUB, S. 1964: The *Cuscuta* species of Palestine. - *Israel J. Bot.* **13(1)**: 1-23.
FEKETE, G. 1959: *Stipa bromoides* (L.) Dörfl., eine neuentdeckte Pflanze in Ungarn. - *Acta Bot. Acad. Sci. Hung.* **5(3-4)**: 349-356.
FELBER, F. 1986: Distribution des cytodèmes d'*Anthoxanthum odoratum* L. s. lat. en Suisse. Les relations Alpes - Jura. - *Bot. Helv.* **96(2)**: 145-158.
FELBER, F. 1988a: Distribution des cytodèmes d'*Anthoxanthum odoratum* L. s. lat. en France et dans les régions limitrophes. - *Bull. Soc. Bot. France, Lett. Bot.* **135(3)**: 281-293.
FELBER, F. 1988b: Phénologie de la floraison de populations diploïdes et tétraploïdes d'*Anthoxanthum alpinum* et d'*Anthoxanthum odoratum*. - *Canad. J. Bot.* **66(11)**: 2258-2264.
FELBER, F. 1991: Rapports (26-27). In Kamari, G., Felber, F. & Garbari, F. (eds): Mediterranean chromosome number reports - 1. - *Fl. Medit.* **1**: 236-237.
FELBER, F. 1993: Présence de trois taxons pérennes d'*Anthoxanthum* L. en Corse. In Jeanmonod, D. & Burdet, H.M. (eds): Notes et contributions à la flore de Corse, IX. - *Candollea* **48(2)**: 582-591.
FELDMAN, M. 1991: Utilization of wild gene resources from the Mediterranean in wheat improvement. - *Bot. Chron. (Patras)* **10**: 407-428.
FELDMANN, J. 1951: René Maire (1878-1949). - *Rev. Gén. Bot.* **58**: 65-88.
FENAROLI, L. 1966: Florae Gargaricae Prodromus. - *Webbia* **21(2)**: 839-944 + 22 plates.
FENAROLI. L. 1970: Florae Gargaricae Prodromus [cont.]. - *Webbia* **24(2)**: 435-578.
FENAROLI. L. 1973: Florae Gargaricae Prodromus [cont.]. - *Webbia* **28(2)**: 323-410
FENAROLI. L. 1974: Florae Gargaricae Prodromus [cont.]. - *Webbia* **29(1)**: 123-301.
FENDERSON, G.K. 1986: A synoptic guide to the genus *Primula*. - Lawrence, Kansas: Allen Press, Inc., v + 213 pp.
FENNANE, M. & AL. (EDS) 1999: Flore pratique du Maroc. Manuel de détermination des plantes vasculaires. Volume 1: Pteridophyta, Gymnospermae, Angiospermae (Lauraceae-Neuradaceae). - Rabat: Travaux de l'Institut Scientifique, Sér. Botanique, no. 36, xiv + 558 pp.
FENNANE, M. & IBN TATTOU, M. 1998: Catalogue des plantes vasculaires rares, menacées ou endémiques du Maroc. - *Bocconea* **8**: 1-243.
FENZL, E. 1842: Pugillus plantarum novarum Syriae et Tauri occidentalis primus. - Wien: Friedrich Beck, i + 18 pp. [cf. TL-2 1: 824, 1976].
FENZL, E. 1866: *Sedum magellense* Ten. und *olympicum* Boiss. nebst einer Notiz über *Armeria rumelica* und *canescens* Boiss. - *Verh. K.K. Zool.-Bot. Ges. Wien* **16(4)**: 917-926 + 2 plates.
FEOLI CHIAPELLA, L. 1979: Biosistematica di alcune centauree della sezione *Acrocentron* Cass. - *Giorn. Bot. Ital., n.s.* **113(4)**: 237-252.
FEOLI, E. & POLDINI, L. 1979: Biometria di "Satureja montana" L. s.l. in Italia. - *Webbia* **33(2)**: 205-216.
FERÁKOVÁ, V. 1968: Chromosome study in *Lactuca quercina* L. and *Lactuca chaixii* Vill. - *Folia Geobot. Phytotax.* **3(1)**: 111-116.
FERÁKOVÁ, V. 1970: A biosystematic study of critical species of *Lactuca* sect. *Lactucopsis*. - *Folia Geobot. Phytotax.* **5(4)**: 401-427.
FERÁKOVÁ, V. 1971: Chromosome numbers in some species of the Cichorieae from Slovakia. - *Acta Fac. Rerum Nat. Univ. Comen., Bot.* **19**: 249-255.
FERÁKOVÁ, V. 1976: [Notes on *Lactuca viminea* and *L. alpestris*]. In Heywood, V.H. (ed.): Notulae systematicae ad Floram Europaeam spectantes, no. 19. - *Bot. J. Linn. Soc.* **71(4)**: 267-268.
FERÁKOVÁ, V. 1977: The genus *Lactuca* L. in Europe. - Bratislava: Univerzita Komenského, 122 pp.
FERÁKOVÁ, V. & MURÍN, A. 1976: [Chromosome numbers from Czechoslovakia]. In Löve, Á. (ed.): IOPB chromosome number reports LIII. - *Taxon* **25(4)**: 490-491.
FERÁKOVÁ, V. & VÁCHOVÁ, M. 1976: [Chromosome numbers from Czechoslovakia]. In Löve, Á. (ed.): IOPB chromosome number reports LIII. - *Taxon* **25(4)**: 489-490.
FERGUSON, I.K. 1971: Notes on the genus *Verbascum* (Scrophulariaceae). In Heywood, V.H. (ed.): Notulae systematicae ad Floram Europaeam spectantes, no. 9. - *Bot. J. Linn. Soc.* **64(2)**: 229-233.
FERGUSON, M.E. & AL. 2000: A re-assessment of the taxonomy of *Lens* Mill. (Leguminosae, Papilionoideae, Vicieae). - *Bot. J. Linn. Soc.* **133(1)**: 41-59.
FERLAN, L. 1958: *Limodorum* L.C. Rich. - Saggio critico. - *Agron. Lusit.* **20(3)**: 179-196.
FERNANDES, A. 1931: Estudos nos cromossomas das Liliáceas e Amarilidáceas. - *Bol. Soc. Brot., Ser. 2* **7**: 3-110.
FERNANDES, A. 1933: Note sur les chromosomes de "Pancratium maritimum" L. - *Bol. Soc. Brot., Ser. 2* **8**: 168-175.
FERNANDES, A. 1942: Summary of work on cytology of *Narcissus* L. - *Herbertia* **9**: 126-133.
FERNANDES, A. 1951: Sur la phylogénie des espèces du genre *Narcissus* L. - *Bol. Soc. Brot., Ser. 2* **25**: 113-190 + 2 plates.
FERNANDES, A. 1966: Le problème du *Narcissus tazetta* L. II. Les formes à 20, 21, 30, 31 et 32 chromosomes somatiques. - *Bol. Soc. Brot., Ser. 2* **40**: 277-319 + 3 plates.
FERNANDES, A. 1968a: Sur la caryologie du *Narcissus serotinus* L. - *Collect. Bot. (Barcelona)* **7(1)**: 381-392.
FERNANDES, A. 1968b: Keys to the identification of native and naturalized taxa of the genus *Narcissus* L. - *Daffodil Tulip Year Book* **33**: 37-66.
FERNANDES, A. 1983: Sur l'existence de formes octoploïdes chez l'agrégat du *Rumex acetosella* dans la péninsule Ibérique. - *Revista Biol. (Lisbon)* **12**: 341-362.
FERNANDES, A. 1984: L'agrégat du *Rumex acetosella* au Portugal. - *Mem. Soc. Brot.* **27**: 89-128 + 15 plates.
FERNANDES, A. & FRANÇA, F. 1972: Contribution à la connaissance cytotaxinomique des spermatophyta du Portugal. VI. Plantaginaceae. - *Bol. Soc. Brot., Ser. 2* **46**: 465-501.
FERNANDES, A. & FRANÇA, F. 1977: Le genre *Colchicum* L. au Portugal. - *Bol. Soc. Brot., Ser. 2* **51**: 5-36.
FERNANDES, A. & LEITÃO, M.T. 1971: Contribution à la connaissance cytotaxinomique des Spermatophyta du Portugal. III. Caryophyllaceae. - *Bol. Soc. Brot., Ser. 2* **45**: 143-176.
FERNANDES, A. & LEITÃO, M.T. 1972: Contribution à la connaissance cytotaxinomique des spermatophyta du Portugal. V. Boraginaceae. - *Bol. Soc. Brot., Ser. 2* **46**: 389-405.
FERNANDES, A. & LEITÃO, M.T. 1984a: Contribution à la connaissance cytotaxinomique des spermatophyta du Portugal. XVIII. Lamiaceae. - *Mem. Soc. Brot.* **27**: 27-75.
FERNANDES, A. & LEITÃO, M.T. 1984b: Sur le blocage de la méiose chez des plantes de *Rumex acetosella* L. sensu str. emend. Á. Löve. - *Bol. Soc. Brot., Ser. 2* **57**: 179-200 + 7 plates.
FERNANDES, A. & QUEIRÓS, M. 1969: Contribution à la connaissance cytotaxinomique des Spermatophyta du Portugal. I. Gramineae. - *Bol. Soc. Brot., Ser. 2* **43**: 1-140.
FERNANDES, A. & QUEIRÓS, M. 1971: Contribution à la connaissance cytotaxinomique des Spermatophyta du Portugal. II. Compositae. - *Bol. Soc. Brot., Ser. 2* **45**: 5-121.
FERNANDES, A. & QUEIRÓS, M. 1972: Systèmes génétiques chez *Hedypnois* Scop. I. Microsporogénèse. - *Bol. Soc. Brot., Ser. 2* **46**: 5-62.
FERNANDES, A. & QUEIRÓS, M. 1978: Contribution à la connaissance cytotaxinomique des Spermatophyta du Portugal. IV. Leguminosae (Suppl. 3). - *Bol. Soc. Brot., Ser. 2* **52**: 79-164.
FERNANDES, A., QUEIRÓS, M. & SANTOS, M. DE FÁTIMA 1977: Contribution à la connaissance cytotaxinomique des Spermatophyta du Portugal. XV. Scrophulariaceae. - *Bol. Soc. Brot., Ser. 2* **51**: 37-90.
FERNANDES, A. & SANTOS, M. DE FÁTIMA 1971: Contribution à la connaissance cytotaxinomique des Spermatophyta du Portugal. IV. Leguminosae. - *Bol. Soc. Brot., Ser. 2* **45**: 177-225.
FERNANDES, A. & SANTOS, M. DE FÁTIMA 1975: Contribution à la connaissance cytotaxinomique des Spermatophyta du Portugal. IV. Leguminosae (Suppl. 1). - *Bol. Soc. Brot., Ser. 2* **49**: 173-196.

FERNANDES, A., SANTOS, M. DE FÁTIMA & QUEIRÓS, M. 1977: Contribution à la connaissance cytotaxinomique des Spermatophyta du Portugal. IV. Leguminosae, Suppl. 2. - *Bol. Soc. Brot., Ser. 2* **51**: 137-186.

FERNANDES, R.B. 1959: As plantas portuguesas da secção *Acinos* (Moench) Briquet do género *Satureja* L. - *Bol. Soc. Brot., Ser. 2* **33**: 119-143.

FERNANDES, R.B. 1967a: Notes taxonomiques sur le genre *Lavatera* L. In Heywood, V.H. (ed.): Notulae systematicae ad Floram Europaeam spectantes, no. 6. - *Feddes Repert.* **74(1-2)**: 18-20.

FERNANDES, R.B. 1967b: Sur l'identification et la typification de *Caucalis daucoides* L. et de *Caucalis grandiflora* L. - *Bol. Soc. Brot., Ser. 2* **41**: 395-407.

FERNANDES, R.B. 1968a: O género *Polypodium* L. em Portugal. I. - *Bol. Soc. Brot., Ser. 2* **42**: 35-158 + 13 plates and 27 tables.

FERNANDES, R.B. 1968b: Contribuições para o conhecimento do género *Lavatera* L. - I Notas sobre algumas espécies. - *Collect. Bot. (Barcelona)* **7(1)**: 393-447.

FERNANDES, R.B. 1969a: Contribuições para o conhecimento do género *Lavatera* L. II - Taxonomia. - *Bol. Soc. Port. Ci. Nat., Ser. 2* **12**: 67-103.

FERNANDES, R.B. 1969b: Quelques notes sur le genre *Echium* L. - *Bol. Soc. Brot., Ser. 2* **43**: 145-158 + 4 plates.

FERNANDES, R.B. 1971a: Note sur le *Glechoma hederacea* L. In Heywood, V.H. (ed.): Notulae systematicae ad Floram Europaeam spectantes, no. 8. - *Bot. J. Linn. Soc.* **64(1)**: 72.

FERNANDES, R.B. 1971b: (181) Quelques combinaisons nomenclaturales nouvelles en *Kickxia* Dumort. In Heywood, V.H. (ed.): Notulae systematicae ad Floram Europaeam spectantes, no. 8. - *Bot. J. Linn. Soc.* **64(1)**: 74-75.

FERNANDES, R.B. 1971c: Notes taxonomiques sur le genre *Chaenorhinum* (DC.) Reichenb. In Heywood, V.H. (ed.): Notulae systematicae ad Floram Europaeam spectantes, no. 9. - *Bot. J. Linn. Soc.* **64(2)**: 215-229.

FERNANDES, R.B. 1973a: Notes sur le genre *Buglossoides* Moench. - *Acta Bot. Acad. Sci. Hung.* **19(1-4)**: 93-101.

FERNANDES, R.B. 1973b: Contribution à la connaissance du genre *Chaenorhinum* (DC.) Reichenb. Dessins de fleurs et de graines des taxa européens. - *Bol. Soc. Brot., Ser. 2* **47**: 17-31.

FERNANDES, R.B. 1975a: Identification, typification, affinités et distribution géographique de quelques taxa européens du genre *Anthemis* L. - *Anales Inst. Bot. Cavanilles* **32(2)**: 1409-1488.

FERNANDES, R.B. 1975b: Taxonomic notes on the genus *Anthemis* L. In Heywood, V.H. (ed.): Notulae systematicae ad Floram Europaeam spectantes, no. 16. - *Bot. J. Linn. Soc.* **70(1)**: 6-17.

FERNANDES, R.B. 1983: Sur la typification de l'*Asplenium adiantum-nigrum* L. et de l'*A. onopteris* L. - *Bol. Soc. Brot., Ser. 2* **56**: 59-69.

FERNANDES, R.B. 1984: Identification, typification, affinités et distribution géographique de quelques taxa européens du genre *Anthemis* L.: 2. - *Revista Biol. (Lisbon)* **12(3-4)**: 385-424.

FERNANDES, R.B. 1992: Notas sobre *Lobularia maritima* (L.) Desv. - *Anales Jard. Bot. Madrid* **49(2)**: 311-315.

FERNANDES, R.B. & NOGUEIRA, I. 1971: O género *Filago* L. (*Evax* Gaertn.) em Portugal. - *Bol. Soc. Brot., Ser. 2* **45**: 317-347 + 2 plates.

FERNANDES, R.B. & NOGUEIRA, I. 1978: Nota sobre o género *Evax* Gaertner. - *Bol. Soc. Brot., Ser. 2* **52**: 67-68.

FERNÁNDEZ ALONSO, J.L. 1986: Acerca del hallazgo de *Sternbergia colchiciflora* Waldst. & Kit. en la cuenca del Duero y de su distribución en la Península Ibérica. - *Anales Jard. Bot. Madrid* **42(2)**: 538-539.

FERNÁNDEZ CASAS, J. & FERNÁNDEZ PIQUERAS, J. 1977: Números cromosómicos para la flora española. 24-29. - *Lagascalia* **7(2)**: 203-206.

FERNÁNDEZ CASAS, J. & GARCÍA GUARDIA, G. 1977: Números cromosómicos para la flora española. 35-38. - *Lagascalia* **7(2)**: 209-210.

FERNÁNDEZ CASAS, J. & LAÍNZ, M. 1973: Sobre *Cistanche phelypaea* (L.) P. Cout. - *Anuário Soc. Brot.* **39**: 129-130.

FERNÁNDEZ CASAS, J. & MOLERO, J. 1979: Números cromosómicos para la flora española. 84. - *Lagascalia* **9(1)**: 115.

FERNÁNDEZ CASAS, J., MUÑOZ GARMENDIA, F. & ORTIZ VALBUENA, A. 1979: Números cromosómicos para la flora española. 85-90. - *Lagascalia* **9(1)**: 115-117.

FERNÁNDEZ CASAS, J., PAJARÓN, S. & RODRÍGUEZ PASCUAL, M.L. 1978: Números cromosómicos para la flora española. 60-65. - *Lagascalia* **8(1)**: 109-112 + 2 plates.

FERNÁNDEZ CASAS, J. & RODRÍQUEZ PASCUAL, M.L. 1977: Números cromosómicos para la flora española. 32-34. - *Lagascalia* **7(2)**: 207-208.

FERNÁNDEZ CASAS, J. & RUIZ REJÓN, M. 1974: Estudios cariológicos sobre la flora Española. - *Bol. Soc. Brot., Ser. 2* **48**: 99-105 + 2 plates.

FERNÁNDEZ CASAS, J. & SUSANNA DE LA SERNA, A. 1985: Monografía de la sección *Chamaecyanus* Willk. del género *Centaurea* L. - *Treb. Inst. Bot. Barcelona* **10**: 174 pp.

FERNÁNDEZ PIQUERAS, J. 1976: [Chromosome numbers in *Anthyllis*]. In Löve, Á. (ed.): IOPB chromosome number reports LIV. - *Taxon* **25(5-6)**: 641-642.

FERNÁNDEZ PRIETO, J.A., FERNÁNDEZ-CARVAJAL, M.C. & AEDO, C. 1987: El género *Ammophila* Host en las costas europeas y norteafricanas. - *Candollea* **42(1)**: 399-410.

FERNÁNDEZ-CARVAJAL, M.C. 1981: Revisión del género *Juncus* L. en la Península Ibérica. I. Categorías supraespecíficas y clave para las especies. - *Anales Jard. Bot. Madrid* **38(1)**: 79-89.

FERNÁNDEZ-CARVAJAL, M.C. 1982a: Revisión del género *Juncus* L. en la Península Ibérica. II. Subgéneros *Juncus* y *Genuini* Buchenau. - *Anales Jard. Bot. Madrid* **38(2)**: 417-467.

FERNÁNDEZ-CARVAJAL, M.C. 1982b: Revisión del género *Juncus* L. en la Península Ibérica. III. Subgéneros *Subulati* Buchenau, *Pseudotenageia* Krecz. & Gontsch. y *Poiophylli* Buchenau. - *Anales Jard. Bot. Madrid* **39(1)**: 79-151.

FERNÁNDEZ-CARVAJAL, M.C. 1983: Revisión del género *Juncus* L. en la Península Ibérica. IV. Subgéneros *Juncinella* (Fourr.) Krecz. & Gontsch., *Septati* Buchenau y *Alpini* Buchenau. - *Anales Jard. Bot. Madrid* **39(2)**: 301-379.

FERNÁNDEZ-PERALTA, A.M. & GONZÁLEZ-AGUILERA, J.J. 1982: Cytogenetic and evolutionary studies on the Spanish species of *Reseda* L.: Section *Luteola* Dumort. (Resedaceae). - *Taxon* **31(1)**: 1-8.

FERNÁNDEZ, I., DÍEZ, M.J. & PASTOR, J. 1985: Números cromosómicos para la flora española. 373-381. - *Lagascalia* **13(2)**: 299-302.

FERNÁNDEZ, I.A. & NIETO FELINER, G. 1999: Lectotypification of 16 species names in *Doronicum* (Asteraceae, Senecioneae). - *Taxon* **48(4)**: 801-806.

FERNÁNDEZ, J.-A. & AL. 2011: The world saffron and *Crocus* collection: strategies for establishment, management, characterisation and utilisation. - *Genet. Resources Crop Evol.* **58**: 125-137.

FERNS, F.E.B. 1976: A land called Crete. - *Quart. Bull. Alpine Gard. Soc. Gr. Brit.* **44(3)**: 233-244.

FERNS, F.E.B. 1978: Consider the lilies ... - *Quart. Bull. Alpine Gard. Soc. Gr. Brit.* **46(1)**: 22-45.

FERNS, F.E.B. 1979: Gesneriaceae of alpine affinity. - *Quart. Bull. Alpine Gard. Soc. Gr. Brit.* **47(2)**: 122-150.

FERNS, F.E.B. 1986: Vignettes. Eastern Mediterranean landfalls. - *Quart. Bull. Alpine Gard. Soc. Gr. Brit.* **54**: 43-47, 163-164.

FERNS, F.E.B. 1992: A wild *Crocus* chase to Mugla ... and elsewhere on the way. - *Rock Gard.* **23(1)**: 28-38.

FERRÁNDEZ PALACIO, J.V. 2003: *Astragalus exscapus* L. (Leguminosae), nueva especie para la flora de la Península Ibérica. - *Collect. Bot. (Barcelona)* **26**: 119-124.

FERRARELLA, A.,. RAIMONDO, F.M. & TRAPANI, S. 1978: Numeri cromosomici per la Flora Italiana: 447-456. - *Inform. Bot. Ital.* **10(1)**: 129-140.

FERRARELLA, A. & AL. 1983: Numeri cromosomici per la Flora Italiana: 860-867. - *Inform. Bot. Ital.* **13(2-3)**: 189-194.

FERRARELLA, A., DIA, G. & RAIMONDO, F.M. 1979: Numeri cromosomici per la Flora Italiana: 584-590. - *Inform. Bot. Ital.* **11(1)**: 21-25.

FERRARELLA, A.M. & AL. 1981: Numeri cromosomici per la Flora Italiana: 647-651. - *Inform. Bot. Ital.* **11(3)**: 301-305.

FERRARINI, E. 1987: Note tassonomiche e corologiche su alcune Umbelliferae delle Apli Apuane. - *Webbia* **41(1)**: 45-59.

FERRER-GALLEGO, P.P. 2014: Typification of *Arisarum vulgare* O. Targ. Tozz. (Araceae). - *Candollea* **69(2)**: 115-118.

FERRER-GALLEGO, P.P. 2015: Typification of two names belonging to the Cistaceae family: *Fumana arabica* and *Helianthemum oelandicum* subsp. *italicum*. - *Phytotaxa* **217(3)**: 298-300.

FERRER-GALLEGO, P.P. 2019: Typification of Linnaean specific names in the genus *Campanula* (Campanulaceae). - *Taxon* **68(6)**: 1350-1358.

FERRER-GALLEGO, P.P. 2021a: (2798) Proposal to conserve the name *Cistus laevis* (*Fumana laevis*) against *C. pilosus* (Cistaceae). - *Taxon* **70(1)**: 208-209.

FERRER-GALLEGO, P.P. 2021b: Effective typification of *Ophrys apifera* (Orchidaceae; Orchidoideae). - *Phytotaxa* **521(1)**: 57-62.

FERRER-GALLEGO, P.P. 2021c: (2849) Proposal to conserve the name *Ophrys sphegodes* (Orchidaceae) with a conserved type. - *Taxon* **70(6)**: 1367-1368.

FERRER-GALLEGO, P.P. & AL. 2017: Typification of the Mediterranean endemic conifer *Juniperus turbinata* (Cupressaceae). - *Phytotaxa* **302(2)**: 165-173.

FERRER-GALLEGO, P.P. & AL. 2019: (2704) Proposal to conserve the name *Teucrium polium* (Labiatae) with a conserved type. - *Taxon* **68(4)**: 865-866.

FERRER-GALLEGO, P.P. & BOISSET, F. 2015: Lectotypification of the name *Cistus stipulatus*, basyonim [sic] of *Helianthemum stipulatum* (Cistaceae). - *Phytotaxa* **219(2)**: 195-198.

FERRER-GALLEGO, P.P. & FARJON, A. 2019: (2726) Proposal to conserve the name *Pinus halepensis* (Pinaceae) with a conserved type. - *Taxon* **68(6)**: 1368-1369.

FERRER-GALLEGO, P.P. & LAGUNA LUMBRERAS, E. 2013: Lectotipificación de *Euonymus latifolius* (L.) Mill. (Celastraceae). - *Acta Bot. Malac.* **38**: 168-170.

FERRER-GALLEGO, P.P. & LAGUNA, E. 2013: Tipificación de *Rhamnus lotus* L. (Rhamnaceae). - *Anales Biol. Fac. Biol. Univ. Murcia* **13**: 13-15.

FERRER-GALLEGO, P.P. & LAGUNA, E. 2017: Lectotypification of the Linnaean name *Amaryllis lutea* (Amaryllidaceae). - *Phytotaxa* **309(1)**: 73-78.

FERRER-GALLEGO, P.P. & LAGUNA, E. 2018: Lectotypification of the Linnaean name *Arum dracunculus* (Araceae). - *Phytotaxa* **375(4)**: 290-296.

FERRER-GALLEGO, P.P. & LAGUNA, E. 2021: (2837) Proposal to reject the name *Juniperus lycia* (Cupressaceae). - *Taxon* **70(5)**: 1134-1136.

FERRER-GALLEGO, P.P., LAGUNA, E. & PEDROL, J. 2014: Tipificación de dos nombres linneanos en *Asparagus* L. (Asparagaceae). - *Acta Bot. Malac.* **39**: 219-223.

FERRER-GALLEGO, P.P. & VAN DE BEEK, A. 2021: On *Rubus ulmifolius* (Rosaceae) and related taxa. - *Phytotaxa* **523(2)**: 155-166.

FERRO, G., BENEDETTO, L. DI & CICERO, E. LO 1975: Osservazioni sulla vegetazione rupestre dell'Acaia (Peloponneso). - *Boll. Ist. Bot. Univ. Catania, ser. 4* **12(7-8)**: 85-93 + 1 table and 1 plate.

FERRO, G. & FURNARI, F. 1968: Flora e vegetazione di Stromboli (Isole Eolie). - *Arch. Bot. Biogeogr. Ital.* **44(1-3)**: 21-45, 59-86 + 8 plates.

FERRO, G. & GEORGIADIS, TH. 2005: Primo contributo sulla flora dei corsi d'acqua del Peloponneso (Grecia). - *Quad. Bot. Ambient. Appl.* **16**: 257-274.

FIALA, F.J.L. 1988: Lilacs. The genus *Syringa*. - London: Christopher Helm Ltd, ii + 266 pp.

FICI, S. 1991: Floristic relations between eastern Africa and the Mediterranean region with special reference to northern Somalia. - *Fl. Medit.* **1**: 175-185.

FICI, S. 1992: *Galium rotundifolium* (Rubiaceae) in southern Italy and its relationships with *G. scabrum*. - *Fl. Medit.* **2**: 99-104.

FICI, S. 2001: Intraspecific variation and evolutionary trends in *Capparis spinosa* L. (Capparaceae). - *Pl. Syst. Evol.* **228(3-4)**: 123-141.

FICI, S. 2014: A taxonomic revision of the *Capparis spinosa* group (Capparaceae) from the Mediterranean to Central Asia. - *Phytotaxa* **174(1)**: 1-24.

FICI, S. & AL. 1995: Biosystematic researches on the *Capparis spinosa* L. complex: preliminar anatomical and karyological data. - *Giorn. Bot. Ital., n.s.* **129(2)**: 43.

FICI, S. & GIANGUZZI, L. 1997: Diversity and conservation in wild and cultivated *Capparis* in Sicily. - *Bocconea* **7**: 437-443.

FICI, S. & NATALI, A. 1991: Taxonomic and distributive notes on *Galium rotundifolium* L. (Rubiaceae) in Sicily. - *Bot. Chron. (Patras)* **10**: 723-727.

FICINI, G. & AL. 1981: Numeri cromosomici per la Flora Italiana: 683-689. - *Inform. Bot. Ital.* **12(2)**: 113-116.

FICINI, G., GIORDANI, A. & TOMEI, P.E. 1983: Numeri cromosomici per la Flora Italiana: 836-841. - *Inform. Bot. Ital.* **13(2-3)**: 172-175.

FIEDLER, K.G. 1840: Uebersicht der Gewächse des Königreiches Griechenland. Botanische Abteilung aus dessen Reise durch Griechenland. In: Reise durch alle Theile des Königreiches Griechenland in Auftrag der Königl. Griechischen Regierung in den Jahren 1834-1837. - Dresden: pp. 507-874 + 1 fold-out map.

FIEDLER, S.G. 1941: Albanian circuit. - *Quart. Bull. Alpine Gard. Soc. Gr. Brit.* **9(2-3)**: 152-183, 232-256.

FIELDING, J., TURLAND, N.J. & MATHEW, B.F. 2005: Flowers of Crete. - Kew: Royal Botanic Gardens, xix + 650 pp.

FILATENKO, A. & AL. 2001: Unterscheidungsmerkmale der diploiden Weizenarten. In Hammer, K. & Gladis, Th. (eds): Schriften zu Genetischen Ressourcen 16: 227-233 + table. - Bonn: Zentralstelle für Agrardokumentation und -information.

FILIBECK, G. & AL. 2020: *Festuca valesiaca* Schleich. ex Gaudin newly discovered in the Central Apennines (Italy): a further example of steppe relict in the Abruzzo "dry valleys". - *Pl. Biosyst.* **154 (5)**: 593-600.

FILLIAS, F., GAULLIEZ, A. & GUÉDÈS, M. 1980: *Amaranthus blitum* vs. *A. lividus* (Amaranthaceae). - *Taxon* **29(1)**: 149-150.

FINCH, R.A. 1967: Natural chromosome variation in *Leontodon*. - *Heredity* **22(3)**: 359-386 + 1 plate.

FINKL, A. 1962: Beiträge zur Kenntnis der Flora der Insel Rhodos. - *Acta Albertina Ratisb.* **24**: 101-120 + 3 plates.

FINNIE, TH.J.R. & AL. 2007: Floristic elements in European vascular plants: an analysis based on Atlas Florae Europaeae. - *J. Biogeogr.* **34(11)**: 1848-1872.

FIORI, A. 1896-1908: Flora analitica d'Italia ossia descrizione delle piante vascolari indigene inselvatichite e largamente coltivate in Italia disposte per quadri analitici. - Padova: Tipografia del Seminario, 4 vols. [cf. TL-2 1: 833, 1976].

FIORI, A. 1899-1904: Iconographia florae italicae ossia flora italiana illustrata con 4236 figuri ... - Padova/Udine: 4 fascs. [cf. TL-2 1: 833, 1976].

FIORI, A. 1923-29: Nouva flora analytica d'Italia contenente le descrizione delle piante vascolari indigene inselvatichite e largamente coltivate in Italia. - Padova: 2 vols. [cf. TL-2 1: 834, 1976].

FIORI, A. 1924a: Contributo alla flora di Rodi e catalogo della piante vascolari finora note di quell'isola. - *Annuario Reale Ist. Super. Forest. Naz.* **9**: 141-187.

FIORI, A. 1924b: La flora dell'isola di Rodi. - *Agric. Colon.* **18(5-6)**: 161-168 + 4 plates.

FIORI, A. 1935: Piante raccolte nelle isole dell'Egeo dal Dott. Giuseppe Jannone Ladispoto. - *Nuovo Giorn. Bot. Ital., n.s.* **42(1)**: 242-243.

FIORI, A. 1939: Piante raccolte nelle isole italiane dell'Egeo. Secondo contributo. - *Nuovo Giorn. Bot. Ital., n.s.* **45(1)**: cxxxii-cxxxviii.

FIORINI, G. & CUCCUINI, P. 2002: Note citotassonomiche sul genere *Parapholis* C.E. Hubb. (Poaceae). - *Webbia* **57(1)**: 65-82.

FIORINI, G. & NARDI, E. 1993: Reports (177-178). In Kamari, G., Felber, F. & Garbari, F. (eds): Mediterranean chromosome number reports - 3. - *Fl. Medit.* **3**: 341-344.

FIRBANK, L.G. 1988: Biological Flora of the British Isles. *Agrostemma githago* L. (*Lychnis githago* (L.) Scop.). - *J. Ecol.* **76(4)**: 1232-1246.

FISCHER, F.E.L.V. 1853: Synopsis Astragalorum Tragacantharum. - *Bull. Soc. Imp. Naturalistes Moscou* **26(3)**: 316-486.

FISCHER, G. 1990: Biosystematische Untersuchung von Sippen des *Asplenium trichomanes*-Komplexes. - Bochum: Diplomarbeit [unpubl.].

FISCHER, M.A. 1967: Beiträge zur Cytotaxonomie der *Veronica hederifolia*-Gruppe (Scrophulariaceae). - *Oesterr. Bot. Z.* **114(2)**: 189-233.

FISCHER, M.A. 1969a: Einige Chromosomenzahlen aus den Gattungen *Veronica*, *Pseudolysimachion*, *Paederota*, *Wulfenia* und *Lagotis* (Scrophulariaceae - Veronicinae). - *Oesterr. Bot. Z.* **116(1-5)**: 430-443.

FISCHER, M.A. 1969b: *Veronica caespitosa* Boiss., *V. bombycina* Boiss. & Kotschy und *V. thessalica* Benth. gehören nicht in die Sektion *Veronicastrum* Benth. - *Oesterr. Bot. Z.* **117(1)**: 54-63.

FISCHER, M.A. 1970a: *Veronica quezelii* und *V. saturejoides* Vis. subsp. *munellensis*, zwei neue Sippen der Sektion *Veronicastrum* aus ostmediterranen Gebirgen. - *Oesterr. Bot. Z.* **118(3)**: 201-205.

FISCHER, M.A. 1970b: Zur Cytotaxonomie von *Veronica chamaedrys* L., I.: subsp. *vindobonensis* M. Fischer, eine neue, diploide Sippe. - *Oesterr. Bot. Z.* **118(3)**: 206-215.

FISCHER, M.A. 1972: Neue Taxa, Chromosomenzahlen und Systematik von *Veronica* subsect. *Acinifolia* (Römpp) Stroh. - *Oesterr. Bot. Z.* **120(4-5)**: 413-437.

FISCHER, M.A. 1973a: Zur Cytotaxonomie von *Veronica chamaedrys* L. agg., II.: subsp. *micans* M. Fischer, subsp. nova, eine weitere diploide Sippe. - *Oesterr. Bot. Z.* **121(1-2)**: 73-79.

FISCHER, M.A. 1973b: Observations sur quatre espèces de *Veronica* dans la flora majorquine. - *Bol. Soc. Hist. Nat. Baleares* **18**: 5-20.

FISCHER, M.A. 1973c: Notizen zur Systematik, Chromosomenzahl und Verbreitung einiger *Veronica*-Sippen in Kärnten. - *Carinthia II* **83**: 379-388.

FISCHER, M.A. 1974a: Beitrag zu einer systematischen Neubearbeitung der Gruppe um *Pseudolysimachion spicatum* (L.) Opiz (= *Veronica spicata* L.). - *Phyton (Horn)* **16(1-4)**: 29-47.

FISCHER, M.A. 1974b: *Veronica vindobonensis* M. Fischer (Zur Cytotaxonomie von *Veronica chamaedrys* agg., III.). - *Oesterr. Bot. Z.* **122(4)**: 287-292.

FISCHER, M.A. 1975a: Untersuchungen über den Polyploidkomplex *Veronica cymbalaria* agg. (Scrophulariaceae). - *Pl. Syst. Evol.* **123(2)**: 97-105.

FISCHER, M.A. 1975b: The *Veronica hederifolia* group: Taxonomy, ecology, and phylogeny. In Walters, S.M. & King, C.J. (eds): European floristic and taxonomic studies, pp. 48-60. - Oxford: E.W. Classey Ltd.

FISCHER, M.A. 1985a: *Veronica beccabunga* gliedert sich in drei vikariierende Unterarten. - *Flora, Morphol. Geobot. Oekophysiol.* **176(1-2)**: 117-128.

FISCHER, M.A. 1985b: Zur Chorologie und Systematik der *Veronica hederifolia*-Gruppe in Jugoslawien. - *Rad. Odjel. Prirod. Mat. Nauka (Sarajevo)* **23**: 55-77 + 1 map.

FISCHER, M.A. 1987: On the origin of *Veronica persica* (Scrophulariaceae) - a contribution to the history of a neophytic weed. - *Pl. Syst. Evol.* **155(1-4)**: 105-132.

FISCHER, M.A. 2010: Do plant identification keys enable identification? - *Phytol. Balcan.* **16(2)**: 175-185.

FISCHER, M.A. & BEDALOV, M. 1988: The genera *Paederota* and *Pseudolysimachion* (Scrophulariaceae) in Croatia. - *Acta Bot. Croat.* **47**: 149-156.

FISCHER, M.A., BEDALOV, M. & VALJAK, B. 1989: The genus *Veronica* (Scrophulariaceae) in Croatia. - *Linzer Biol. Beitr.* **21(1)**: 143-172.

FISCHER, M.A. & FISCHER, G. 1981: Distribution patterns of *Veronica* species in the Balkan Peninsula. In Velčev, V.I. & Kožuharov, S.I. (eds): Mapping the flora of the Balkan Peninsula, pp. 175-232. - Sofia: Publishing House of the Bulgarian Academy of Sciences.

FISCHER, M.A. & GREUTER, W. 1976: Über zwei diploide, in Südwest-Anatolien endemische Arten aus dem Komplex der *Veronica cymbalaria* (Scrophulariaceae). - *Pl. Syst. Evol.* **125(4)**: 245-252. [Also distributed as *OPTIMA leaflet* no. 27].

FISCHER, M.A., OSWALD, K. & ADLER, W. 2008: Exkursionsflora für Österreich, Lichtenstein und Südtirol. Bestimmungsbuch für alle in der Republik Österreich, im Fürstentum Lichtenstein und in der autonomen Provinz Bozen/Südtirol (Italien) wildwachsenden sowie die wichtigsten kultivierten Gefässpflanzen (Farnpflanzen und Samenpflanzen) mit Angaben über ihre Ökologie und Verbreitung. - Linz: Land Oberösterreich, Biologiezentrum der Oberösterreichischen Landesmuseum, 3. Auflage, 1391 pp.

FISCHER, M.A., VELJOVIĆ, V. & TATIĆ, B. 1984: *Veronica scardica* - a neglected species of the Serbian flora. - *Bull. Inst. Jard. Bot. Univ. Belgrade, n.s.* **18**: 37-53.

FIŠEROVÁ, D. 1974: The *Poa alpina* L. complex in Bulgaria. - *Preslia* **46(3)**: 213-224.

FIŠEROVÁ, D. 1989: A contribution to the population study of *Lilium albanicum* Grisebach in the mountains of Macedonia, Yugoslavia. - *Fragm. Balcan.* **14(7)**: 59-67.

FITZ, K. 1981: Merkmals-Tabellen für die österreichischen *Geranium*-Arten. - *Verh. Zool.-Bot. Ges. Österreich* **120**: 99-112.

FLEISCHMANN, A. 2015: The intricate *Pinguicula crystallina/hirtiflora*-complex. - *Carniv. Pl. Newslett.* **44 (2)**: 48-61.

FLEISCHMANN, H. 1907: Nachträgliche Bemerkung zu der Abhandlung “Interessante Orchideen aus Korfu”. - *Oesterr. Bot. Z.* **57(2)**: 74.

FLEISCHMANN, H. 1925: Beitrag zur Orchideenflora der Insel Kreta. - *Oesterr. Bot. Z.* **74(7-9)**: 180-194.

FLOHE, J. 2006: Kreta [Crete]. Ein eigenartiges Blumenparadies [A unique paradise of flowers]. - Essen: Buchverlag Professor A.W. Geisler, 182 pp.

FLOHE, J. 2015: Wildblumen des griechischen Festlands. Empfehlenswerte botanische Exkursionen auf dem Festland einschliesslich Peloponnes. - Dormagen: Selbstverlag Johannes Flohe, 374 pp.

FLOTO, E. 1954: Iagttagelser over *Smyrnium perfoliatum* L. [In Danish]. - *Bot. Tidsskr.* **51**: 66-70.

FOGELBERG, S.O. 1938: The cytology of *Cuscuta*. - *Bull. Torrey Bot. Club* **65(9)**: 631-645.

FOGG, G.E. 1950: Biological Flora of the British Isles. *Sinapis arvensis* L. - *J. Ecol.* **38**: 415-429.

FOGGI, B. & AL. 2012: Lectotypification of taxa belonging to the “Festuca circummediterranea” group. - *Candollea* **67(2)**: 221-228.

FOGGI, B., NARDI, E. & ROSSI, G. 2001: Nomenclatural notes and typification in *Sesleria* Scop. (Poaceae). - *Taxon* **50(4)**: 1101-1106.

FOGGI, B. & RICCERI, C. 1991: Segnalazioni floristiche italianae: 615. *Aeluropus littoralis* (Gouan) Parl. (Gramineae). - *Inform. Bot. Ital.* **22(3)**: 247.

FOGGI, B. & RICCERI, C. 1993: On the revaluation of *Silene staminea* Bertol. (Caryophyllaceae). - *Webbia* **47(2)**: 191-201.

FOGGI, B., ROSSI, G. & SIGNORINI, M.A. 1999: The *Festuca violacea* aggregate (Poaceae) in the Alps and Apennines (central southern Europe). - *Canad. J. Bot.* **77(7)**: 989-1013.

FOGGI, B., SCHOLZ, H. & VALDÉS, B. 2005: The Euro+Med treatment of *Festuca* (Gramineae) new names and new combinations in *Festuca* and allied genera. - *Willdenowia* **35(2)**: 241-244.

FOGGI, B. & SIGNORINI, M.A. 2004: Types of *Festuca* (Gramineae) names in the herbarium W: types from Hackel’s herbarium. - *Ann. Naturhist. Mus. Wien* **105B**: 613-644.

FOGGI, B., SIGNORINI, M.A. & ROSSI, G. 2003: The genus *Festuca* in Italy. - *Bocconea* **16(1)**: 55-64.

FOGGI, B., SIGNORINI, M.A. & VITEK, E. 2004: Type names of *Festuca* (Gramineae) in the herbarium W: types from Host’s herbarium. - *Ann. Naturhist. Mus. Wien* **105B**: 597-612.

FOHRINGER, F. 2008: *Ophrys minutula* Gölz & H.R. Reinhard x *Ophrys regis-ferdinandii* (Renz) Buttler, eine neue Hybride aus Chios. - *J. Eur. Orch.* **40(2)**: 397-402.

FOHRINGER, F. & FOHRINGER, M. 2004: *Ophrys* ×*messiniae* F. Fohringer & M. Fohringer - hybr. nat. nov. = *Ophrys argolica* H. Fleischmann × *Ophrys vernixia* subsp. *orientalis* H.F. Paulus - eine neue Hybridenart der Peloponnes. - *Linzer Biol. Beitr.* **36(1)**: 567-571.

FOLEY, M.J.Y. 1999: *Orobanche sanguinea* C. Presl or *O. crinita* Viv.?: the correct name for the widespread Mediterranean coastal plant. - *Candollea* **54(1)**: 89-95.

FOLEY, M.J.Y. 2000a: Orobanchaceae in the "Flora Iberica" area: new taxa, excluded taxa and typification. - *Anales Jard. Bot. Madrid* **58(2)**: 223-233.

FOLEY, M.J.Y. 2000b: The taxonomic position of *Orobanche rechingeri* Gilli (Orobanchaceae) in relation to *Orobanche nowackiana* Markgr. - *Candollea* **55(2)**: 269-276.

FOLEY, M.J.Y. 2000c: A morphological comparison between some British *Orobanche* species (Orobanchaceae) and their closely-related, non-British counterparts fromcontinental Europe: *Orobanche rapum-genistae* Thuill. s.l. - *Watsonia* **23**: 413-419.

FOLEY, M.J.Y. 2001: Typification of *Carex davalliana* Sm. and *Carex trinervis* Degl. (Cyperaceae). - *Candollea* **56(1)**: 131-135.

FOLEY, M.J.Y. 2003: *Salvia veneris* Hedge (Lamiaceae/Labiatae) in Cyprus and its relationship to its taxonomically closest allies. - *Karaca Arbor. Mag.* **7(1)**: 1-12.

FOLEY, M.J.Y. 2008: *Phelipanche schultzioides* (Orobanchaceae), a new species from Peloponnisos, Greece. - *Bot. Chron. (Patras)* **19**: 5-12.

FOLEY, M.J.Y. 2010: *Erigeron uniflorus* subsp. *parnassensis* A.G. Game ex M.J.Y. Foley, subsp. nov. In Greuter, W. & Raus, Th. (eds): Med-Checklist notulae, 29. - *Willdenowia* **40(2)**: 192-193.

FOLEY, M.J.Y. & SOUTHAM, M.J. 2007: *Oenanthe incrassans* Bory & Chaub. (Apiaceae), a distinctive plant of the Aegean region. - *Candollea* **62(2)**: 125-130.

FOLLAK, S., ALDRIAN, U. & SCHWARZ, M. 2014: Spread dynamics of *Abutilon theophrasti* in Central Europe. - *Plant Protect. Sci.* **50(3)**: 157-163.

FOLLIERI, M. 1977: Classification and phylogeny of living and fossil water ferns of the genus "Azolla". - *Webbia* **31(1)**: 97-104.

FORCELLA, F., WOOD, J.T. & DILLON, S.P. 1986: Characteristics distinguishing invasive weeds within *Echium* (Bugloss). - *Weed Res.* **26(5)**: 351-364.

FORD-LLOYD, B.V. & WILLIAMS, J.T. 1976: A revision of *Beta* section *Vulgares* (Chenopodiaceae) with new light on the origin of cultivated beets. - *Bot. J. Linn. Soc.* **71(2)**: 89-102.

FORD, B.A. & BALL, P.W. 1988: A reevaluation of the *Triglochin maritimum* complex (Juncaginaceae) in eastern and central North America and Europe. - *Rhodora* **90(863)**: 313-337.

FORISSIER, R. 1973: Recherches cytotaxonomiques préliminaires sur les genres *Lembotropis*, *Cytisus*, *Chamaecytisus*, *Genista* et *Chamaespartium*. - *Bull. Soc. Neuchâteloise Sci. Nat.* **96**: 51-65.

FORMÁNEK, E. 1890a: Beitrag zur Flora von Serbien, Macedonien und Thessalien. - *Deutsche Bot. Monatsschr.* **8(5-6)**: 65-72.

FORMÁNEK, E. 1890b: Beitrag zur Flora von Serbien, Macedonien und Thessalien [cont.]. - *Deutsche Bot. Monatsschr.* **8(11-12)**: 161-175.

FORMÁNEK, E. 1891a: Beitrag zur Flora von Serbien, Macedonien und Thessalien [cont.]. - *Deutsche Bot. Monatsschr.* **9(2-3)**: 24-32.

FORMÁNEK, E. 1891b: Beitrag zur Flora von Serbien, Macedonien und Thessalien [cont.]. - *Deutsche Bot. Monatsschr.* **9(4-5)**: 63-76.

FORMÁNEK, E. 1891c: Beitrag zur Flora von Serbien, Macedonien und Thessalien [cont.]. - *Deutsche Bot. Monatsschr.* **9(6-7)**: 88-101.

FORMÁNEK, E. 1891d: Beitrag zur Flora des Balkans, Bosporus und Kleinasien. - *Verh. Naturf. Vereins Brünn* **29**: 113-158.

FORMÁNEK, E. 1892: Beitrag zur Flora von Serbien und Macedonien. - *Verh. Naturf. Vereins Brünn* **30**: 50-96.

FORMÁNEK, E. 1893: Beitrag zur Flora von Serbien und Bulgarien. - *Verh. Naturf. Vereins Brünn* **31**: 110-136.

FORMÁNEK, E. 1894: Zweiter Beitrag zur Flora von Serbien und Macedonien. - *Verh. Naturf. Vereins Brünn* **32**: 146-210.

FORMÁNEK, E. 1895: Beitrag zur Flora von Albanien, Korfu und Epirus. - *Verh. Naturf. Vereins Brünn* **33**: 109-159.

FORMÁNEK, E. 1896: Zweiter Beitrag zur Flora von Serbien, Macedonien und Thessalien. - *Verh. Naturf. Vereins Brünn* **34**: 255-365.

FORMÁNEK, E. 1897a: Dritter Beitrag zur Flora von Thessalien. - *Verh. Naturf. Vereins Brünn* **35**: 134-215.

FORMÁNEK, E. 1897b: Einige neue Arten aus Thessalien. - *Deutsche Bot. Monatsschr.* **15**: 73-76, 197-199.

FORMÁNEK, E. 1897c: Berichtigung zum Artikel "Neue Arten aus Thessalien". - *Deutsche Bot. Monatsschr.* **15**: 320-321.

FORMÁNEK, E. 1898a: Dritter Beitrag zur Flora von Serbien und Bulgarien. - *Verh. Naturf. Vereins Brünn* **36**: 6-118.

FORMÁNEK, E. 1898c: Beitrag zur Flora von Griechenland. - *Deutsche Bot. Monatsschr.* **16**: 77-81.

FORMÁNEK, E. 1898d: Zur Flora Thessaliens. - *Deutsche Bot. Monatsschr.* **16**: 172-173.

FORMÁNEK, E. 1899: Fünfter Beitrag zur Flora von Macedonien. - *Verh. Naturf. Vereins Brünn* **37**: 124-220.

FORMÁNEK, E. 1900: Sechster Beitrag zur Flora von Macedonien. - *Verh. Naturf. Vereins Brünn* **38**: 165-240.

FORSSKÅL, P. 1775: Flora Aegyptiaco-Arabica. Sive descriptiones plantarum, quas per Aegyptum inferiorem et Arabium felicem detexit ... Post mortem auctoris edidit Carsten Niebuhr. - Kjøbenhavn: Möller, pp. 1-32, i-cxxvi, 1-219.

FORSTER, P.I. 1990: (982) Proposal to conserve *Marsdenia* R. Br. against *Stephanotis* Thouars (Asclepiadaceae). - *Taxon* **39(2)**: 364-367. [approved, cf. Taxon 42(4): 876, 1993].

FÖRTHER, H. 1998: Die infragenerische Gliederung der Gattung *Heliotropium* L. und ihre Stellung innerhalb der subfam. Heliotropioideae (Schrad.) Arn. (Boraginaceae). - *Sendtnera* **5**: 35-241.

FÖRTHER, H. & PODLECH, D. 1991: Revision der *Ononis natrix*-Gruppe (Leguminosae) von Makaronesien, Nordafrika und dem angrenzenden Westasien. - *Mitt. Bot. Staatssamml. München* **30**: 197-296.

FOSBERG, F.R. 1973: Type specimens of *Buxus sempervires* Linnaeus. - *Boxwood Bull.* **13(2)**: 18-21.

FOSBERG, F.R. 1977: *Paspalum distichum* again. - *Taxon* **26(2-3)**: 201-202.

FOSTER, M. 1893: *Iris Athoa* Foster, n. sp. - *Gard. Chron., Ser. 3* **13**: 711.

FOTIADIS, G. 2004: Kathorismos ton dasikos fitokinoniologikon monadon tou ellinikou timatos tou orous Beles ke tis orosiras ton Krousion. [In Greek]. - Thessaloniki, Ph.D. thesis, Aristotelion University, 281 pp.

FOTIADIS, G. 2006: The vascular flora of valonia oak (*Quercus ithaburensis* subsp. *macrolepis* forest remnants in Thrace (NE Greece). In Peev, D.R. (ed.): Book of abstracts, IV Balkan botanical congress, Sofia, Bulgaria, 20-26 June 2006, p. 219.

FOTIADIS, G. 2013: Description of flora and vegetation of the Graecian juniper wood habitat type in Prespa [In Greek with English summary]. - Society for the Protection of Prespa, 56 pp.

FOTIADIS, G. & AL. 2009: The vascular flora of *Quercus ithaburensis* ssp. *macrolepis* (Fagaceae) forest remnants found in Thrace (NE Greece). In Ivanova, D. (ed.): Proceedings of the IV Balkan botanical congress, pp. 415-421. - Sofia: Institute of Botany, BAS.

FOTIADIS, G. & AL. 2010: Ecology, chorology and commonness of the *Trifolium* species in Greece. - *Feddes Repert.* **121(1-2)**: 65-79.

FOTIADIS, G. & AL. 2012a: Prespa National Park - Varnous mountain. In Vrahnakis, M. & al. (eds): Report of the 9th European Dry Grassland Meeting (EDGM), 19-23 May 2012, pp. 109-120. - Prespa: Information Centre of Prespa National Forest Management Body.

FOTIADIS, G. & AL. 2012b: Prespa National Park - Triklario mountain. In Vrahnakis, M. & al. (eds): Report of the 9th European Dry Grassland Meeting (EDGM), 19-23 May 2012, pp. 122-133. - Prespa: Information Centre of Prespa National Forest Management Body.

FOTIADIS, G. & ATHANASIADIS, N. 2008: The vascular flora of forest and shrubland communities of Krousia and Belles Mts (NE Greece). - *Phytol. Balcan.* **14(3)**: 343-365.

FOTIADIS, G., PANAJIOTIDIS, S. & GERASIMIDIS, A. 2005: Phytosociological approach on the forest vegetation of Mt Verno (W Macedonia, Greece). - *Bot. Chron. (Patras)* **18(1)**: 101-116.

FOURNARAKI, CH. 2010: Conservation of threatened plants of Crete - seed ecology, operation and management of a gene bank. - Athens: National and Kapodistrian Univ. (Ph.D. thesis), 439 pp. [In Greek with English summary].

FOURNARAKI, CH. & AL. 2019: Preliminary results of actions to reinforce natural populations of threatened plant species in Crete. In Daskalakou, E.N. & al. (eds): 16th Conference of the Hellenic Botanical Society – Athens 10-13 October 2019. Program and book of abstracts, p. 47.

FOURNARAKI, CH. & GOTSIOU, P. 2004: SOS gia 8 fita pou iparchoun mono stin Kriti ke kindinevoun amesa me exafanisi [In Greek]. - *Kritiko Panorama* **3**: 66-81.

FOURNARAKI, CH., GOTSIOU, P. & KYPRIOTAKIS, Z. 2002: Apiloumena fita tis Kritis simfona me tin Evropaiki odigia 92/43/EOK gia tin diatirisi ton fisikon ikotopon kathos ke tin agrias panidas ke chloridas [In Greek]. - Chania: M.A.I.Ch. (Enimerotiko filladio), 40 pp.

FOURNARAKI, CH., KYPRIOTAKIS, Z. & GOTSIOU, P. 2003: Seed conservation of the endemic and threatened plants of Crete. In Smith, R.D. & al. (eds): Seed conservation, turning science into practice, pp. 870-878. - Kew: Royal Botanic Gardens.

FOURNARAKI, CH. & THANOS, C.A. 2002: Seeds of *Zelkova abelicea*, an endemic tree of Crete. In: Tree Seeds 2002. Annual meeting of IUFRO "research group for seed physiology and technology", 11-15 September 2002, pp. 83-84. - Chania: M.A.I.Ch.

FOURNIER, E. 1866: Monographie du genre *Hesperis*. - *Bull. Soc. Bot. France* **13**: 326-362.

FOUZDAR, A. & TANDON, S.L. 1975: Cytotaxonomic investigations in the genus *Lathyrus*. - *Nucleus (Calcutta)* **18(1-2)**: 24-33.

FOUZDAR, A. & TANDON, S.L. 1976: Cytogenetical evolution in the genus *Pisum*. - *Cytologia* **41(1)**: 91-104.

FOX, D. 1988: The odd one out [*Lilium rhodopeum*]. - *Bull. R.H.S. Lily Group* **1997/98**: 36-40.

FOX, H.M. 1953: Spring flowers seen in Israel and Greece. - *Garden (New York)* **3**: 151-154.

FOX, M.D. 1990: Mediterranean weeds: exchanges of invasive plants between the five mediterranean regions of the world. In Di Castri, F., Hansen, A.J. & Debussche, M. (eds): Biological invasions ..., pp. 179-200. - Dordrecht, etc.: Kluwer Academic Publishers.

FRAAS, C. 1845: Synopsis plantarum florae classicae oder: Uebersichtliche Darstellung der in den klassischen Schriften der Griechen und Römer vorkommenden Pflanzen ... - München: E.A. Fleischmann, xxxix + 320 pp.

FRAAS, C. 1861: Zur neuen arkadischen Tannenspecies (*Ab. reginae Amaliae*). - *Flora* **44**: 62-63.

FRAGMAN, O., LEVY-YAMAMORI, R. & CHRISTODOULOU, P. 2001: Flowers of the Eastern Mediterranean. - Ruggell: A.R.G. Gantner Verlag, vi + 458 pp.

FRAJMAN, B. 2007a: Taxonomy and reticulate phylogeny of *Heliosperma* and related genera (Sileneae, Caryophyllaceae). [Digital comprehensive summaries of Uppsala dissertations from the Faculty of Science and technology 328]. - Uppsala: Acta Universitatis Upsaliensis, 34 pp.

FRAJMAN, B. 2007b: (1760) Proposal to reject the name *Cucubalus quadrifidus* (*Heliosperma quadrifidum, Silene quadrifida*) (Caryophyllaceae, Sileneae). - *Taxon* **56(1)**: 260-261. [approved, cf. Taxon 58(1): 285, 2009].

FRAJMAN, B. & AL. 2019: Integrating phylogenomics, phylogenetics, morphometrics, relative genome size and ecological niche modelling disentangles the diversification of Eurasian *Euphorbia seguieriana* s. l. (Euphorbiaceae). - *Molec. Phyl. Evol.* **134**: 238-252.

FRAJMAN, B., GRANISZEWSKA, M. & SCHÖNSWETTER, P. 2016: Evolutionary patterns and morphological diversification within the European members of the *Euphorbia illirica* (*E. villosa*) group: one or several species? - *Preslia* **88(3)**: 369-390.

FRAJMAN, B., HEIDARI, N. & OXELMAN, B. 2009: Phylogenetic relationships of *Atocion* and *Viscaria* (Sileneae, Caryophyllaceae) inferred from chloroplast, nuclear ribosomal and low-copy gene DNA sequences. - *Taxon* **58(3)**: 811-824.

FRAJMAN, B., PACHSCHWÖLL, C. & SCHÖNSWETTER, P. 2014: Contributions to the knowledge of the flora of the Dinarides (Balkan Peninsula). - *Phyton (Horn)* **54(1)**: 27-46.

FRAJMAN, B. & RAEBLER, K. 2006: (1737) Proposal to conserve the name *Heliosperma* against *Ixoca* (Caryophyllaceae, Sileneae). - *Taxon* **55(3)**: 801-808. [approved, cf. Taxon 56(4): 1293, 2007].

FRAJMAN, B. & SCHÖNSWETTER, P. 2017: Amphi-Adriatic distributions in plants revisited: Pleistocene trans-Adriatic dispersal in the *Euphorbia barrelieri* group (Euphorbiaceae). - *Bot. J. Linn. Soc.* **185(2)**: 240-252.

FRAJMAN, B. & SCHÖNSWETTER, P. 2018: Disentangling relationships in different groups of *Euphorbia* subgen. *Esula* (Euphorbiaceae) from the Balkan Peninsula. - *Bot. Serbica* **42 (suppl. 1)**: 38.

FRAJMAN, B., TROLLESSON, M. & OXELMAN, B. 2013: Taxonomic revision of *Atocion* and *Viscaria* (Sileneae, Caryophyllaceae). - *Bot. J. Linn. Soc.* **173(2)**: 194-210.

FRANCHET, A. 1890: Monographie du genre *Chrysosplenium* Tourn. - *Nouv. Arch. Mus. Hist. Nat., Sér. 3* **2**: 87-114 + 4 plates.

FRANCHET, A. 1891: Monographie du genre *Chrysosplenium* Tourn. (suite). - *Nouv. Arch. Mus. Hist. Nat., Sér. 3* **3**: 1-32 + 7 plates.

FRANCINI CORTI, E. & PESCI, G. 1971: Numeri cromosomici per la Flora Italiana: 45. - *Inform. Bot. Ital.* **3(1)**: 81-82.

FRANCINI, E. & PARDI-RICCADONNA, P. 1936: Ricerche sulla vegetazione dell'Etruria Marittima. IV. - Ecologia e significato della "Periploca graeca" L. sul Litorale Toscano. - *Nuovo Giorn. Bot. Ital., n.s.* **43(1)**: 167-196 + 2 fold-out tables.

FRANCIS, C.M. & KATZNELSON, J. 1977: Observations on the distribution and ecology of subterranean clovers, some other clovers, and medics in Greece and Crete. - *Austral. Pl. Introd. Rev.* **12(1)**: 17-25.

FRANCO, J. DO AMARAL 1962: Taxonomy of the common juniper. - *Bol. Soc. Brot., Ser. 2* **36**: 101-120.

FRANCO, J. DO AMARAL 1963: Taxonomic notes on *Juniperus oxycedrus* L. and *J. macrocarpa* Sm. In Heywood, V.H. (ed.): Notulae systematicae ad Floram Europaeam spectantes, no. 2. - *Feddes Repert. Spec. Nov. Regni Veg.* **68(3)**: 163-167 + 1 fold-out map.

FRANCO, J. DO AMARAL 1968: [Notes on *Crataegus*]. In Heywood, V.H. (ed.): Notulae systematicae ad Floram Europaeam spectantes, no. 7. - *Feddes Repert.* **79(1-2)**: 37-39.

FRANCO, J. DO AMARAL 1975a: [Notes on *Onopordum*]. In Heywood, V.H. (ed.): Notulae systematicae ad Floram Europaeam spectantes, no. 17. - *Bot. J. Linn. Soc.* **71(1)**: 45-46.

FRANCO, J. DO AMARAL 1975b: [Notes on *Carduus*]. In Heywood, V.H. (ed.): Notulae systematicae ad Floram Europaeam spectantes, no. 17. - *Bot. J. Linn. Soc.* **71(1)**: 48-50.

FRANCO, J. DO AMARAL 1978: Notes on *Dactylorhiza sulphurea*. In Heywood, V.H. (ed.): Notulae systematicae ad Floram Europaeam spectantes, no. 20. - *Bot. J. Linn. Soc.* **76(4)**: 366-367.

FRANCO, J. DO AMARAL & ROCHA AFONSO, M.L. 1971a: [Notes on *Fraxinus angustifolia*]. In Heywood, V.H. (ed.): Notulae systematicae ad Floram Europaeam spectantes, no. 11. - *Bot. J. Linn. Soc.* **64(4)**: 377.

FRANCO, J. DO AMARAL & ROCHA AFONSO, M.L. 1971b: [Notes on *Blackstonia perfoliata*]. In Heywood, V.H. (ed.): Notulae systematicae ad Floram Europaeam spectantes, no. 11. - *Bot. J. Linn. Soc.* **64(4)**: 378.

FRANK, E. 1980: *Cyclamen graecum album*. - *Cyclamen Soc. J.* **4(2)**: 41-42.

FRANK, E. 1992: The Peloponnese in midwinder. - *Cyclamen Soc. J.* **15(2)**: 50-51.

FRANKE, A. & HURKA, H. 2000: Molecular systematics and biogeography of the *Cardamine pratensis* complex (Brassicaceae). - *Pl. Syst. Evol.* **224(3-4)**: 213-234.

FRANZÉN, R. 1979: Helikons alpina flora. [Unpublished flora list; in Swedish]. - Lund: Univ. of Lund, photocopy, 8 pp.

FRANZÉN, R. 1980: Floristic reports from Mount Siniatsikon and Mount Vermion, northern Greece. (Materials for the Mountain Flora of Greece, 6). - *Bot. Not.* **133(4)**: 527-537.

FRANZÉN, R. 1983: *Geranium thessalum* sp. nov. from northern Greece. (Materials for the Mountain Flora of Greece, 13). - *Nordic J. Bot.* **2(6)**: 549-552.

FRANZÉN, R. 1986a: The *Valeriana crinii*-group (Valerianaceae) in Greece. (Materials for the Mountain Flora of Greece, 28). - *Willdenowia* **15(2)**: 351-357.

FRANZÉN, R. 1986b: Taxonomy of the *Achillea clavennae* group and the *A. ageratifolia* group (Asteraceae, Anthemideae) on the Balkan Peninsula. - *Willdenowia* **16(1)**: 13-33.

FRANZÉN, R. 1986c: *Anthemis cretica* (Asteraceae) and related species in Greece. (Materials for the Mountain Flora of Greece, 29). - *Willdenowia* **16(1)**: 35-45.

FRANZÉN, R. 1987: Biosystematics of the *Achillea clavennae* and *A. ageratifolia* groups (Asteraceae). Lund: Ph.D. thesis, Univ. of Lund, 43 pp.

FRANZÉN, R. 1988: Differentiation patterns in *Achillea ageratifolia* (Asteraceae). - *Willdenowia* **18(1)**: 47-62.

FRANZÉN, R. & GUSTAVSSON, L.-Å. 1983: Chromosome numbers in flowering plants from the high mountains of Sterea Ellas, Greece. (Materials for the Mountain Flora of Greece, 19). - *Willdenowia* **13(1)**: 101-106.

FRASER-JENKINS, C.R. 1977: Three species in the *Dryopteris villarii* aggregate (Pteridophyta, Aspidiaceae). - *Candollea* **32(2)**: 305-319.

FRASER-JENKINS, C.R. 1980a: *Dryopteris affinis*: a new treatment for a complex species in the European Pteridophyte flora. - *Willdenowia* **10(1)**: 107-115.

FRASER-JENKINS, C.R. 1980b: Nomenclatural notes on *Dryopteris*: 4. - *Taxon* **29(5-6)**: 607-612.

FRASER-JENKINS, C.R. 1986: A classification of the genus *Dryopteris* (Pteridophyta: Dryopteridaceae). - *Bull. Brit. Mus. (Nat. Hist.), Bot.* **14(3)**: 183-218.

FRASER-JENKINS, C.R. 1989: Some comments on the nomenclature of *Dryopteris*. - *Indian Fern J.* **5(1-2)**: 69-77.

FRASER-JENKINS, C.R. 2007: The species and subspecies in the *Dryopteris affinis* group. - *Brit. Fern Gaz.* **18(1)**: 1-26.

FRASER-JENKINS, C.R. & CORLEY, H.V. 1972: *Dryopteris caucasica* - an ancestral diploid in the male fern aggregate. - *Brit. Fern Gaz.* **10(5)**: 221-231 + 2 plates.

FRASER-JENKINS, C.R. & GIBBY, M. 1980: Two new hybrids in the *Dryopteris villarii* aggregate (Pteridophyta, Dryopteridaceae), and the origin of *D. submontana*. - *Candollea* **35(1)**: 305-310.

FRASER-JENKINS, C.R. & JERMY, A.C. 1976: Nomenclatural notes on *Dryopteris* Adans. - *Taxon* **25(5-6)**: 659-665.

FRASER-JENKINS, C.R. & JERMY, A.C. 1977: Nomenclatural notes on *Dryopteris*: 2. - *Brit. Fern Gaz.* **11 (5)**: 338-340.

FRASER-JENKINS, C.R. & JERMY, A.C. 1978: Nomenclatural notes on *Dryopteris* Adans. - *Taxon* **27(1)**: 129-131.

FREDERIKSEN, S. 1986: Revision of *Taeniatherum* (Poaceae). - *Nordic J. Bot.* **6(4)**: 389-397.

FREDERIKSEN, S. 1991: Taxonomic studies in *Dasypyrum* (Poaceae). - *Nordic J. Bot.* **11(2)**: 135-142.

FREDERIKSEN, S. 1993: Taxonomic studies in some annual genera of the Triticeae (Poaceae). - *Nordic J. Bot.* **13(5)**: 481-493.

FREDERIKSEN, S. & BOTHMER, R.V. 1986: Relationships in *Taeniatherum* (Poaceae). - *Canad. J. Bot.* **64(10)**: 2343-2347.

FREDERIKSEN, S. & BOTHMER, R.V. 1989: Intergeneric hybridization between *Taeniatherum* and different genera of Triticeae, Poaceae. - *Nordic J. Bot.* **9(3)**: 229-240.

FREDERIKSEN, S. & PETERSEN, G. 1997: Morphometrical analyses of *Secale* (Triticeae, Poaceae). - *Nordic J. Bot.* **17(2)**: 185-198.

FREDERIKSEN, S. & PETERSEN, G. 1998: A taxonomic revision of *Secale* (Triticeae, Poaceae). - *Nordic J. Bot.* **18(4)**: 399-420.

FREITAG, H. 1975: The genus *Piptatherum* (Gramineae) in Southwest Asia. - *Notes Roy. Bot. Gard. Edinburgh* **33(3)**: 341-408 + 2 plates.

FREITAG, H. 1985: The genus *Stipa* (Gramineae) in southwest and south Asia. - *Notes Roy. Bot. Gard. Edinburgh* **42(3)**: 355-489.

FREITAG, H. & MAIER-STOLTE, M. 1989a: The *Ephedra fragilis*-group in the Mediterranean. In Phitos, D. (ed.): Abstracts. Sixth OPTIMA meeting, Delphi 10-16 Sept. 1989, p. 94.

FREITAG, H. & MAIER-STOLTE, M. 1989b: The *Ephedra*-species of Forsskål: Identity and typification. - *Taxon* **38(4)**: 545-556.

FRESE, L. & AL. 2009: Survey of *Beta nana* (Boiss. & Heldr.) in Greece. In: Frese, L., Maggioni, L. & Lipman, E. (eds): Report of a working group on *Beta* and the world *Beta* network. 3rd Joint Meeting, 8-11 March 2006, Puerto de la Cruz, Tenerife, Spain, pp. 45-52.

FREUDENSTEIN, J.V. 1996: (1260) Proposal to conserve the name *Corallorhiza* Gagnebin (Orchidaceae) with a conserved spelling. - *Taxon* **45(4)**: 695-696. [approved, cf. Taxon 48(2): 363, 1999].

FREUDENSTEIN, J.V. 1997: A monograph of *Corallorhiza* (Orchidaceae). - *Harvard Pap. Bot.* **10**: 5-51.

FREY, A. 1974: Genus *Amaranthus* L. in Poland [In Polish with English summary]. - *Fragm. Florist. Geobot.* **20(2)**: 143-201 + 20 plates.

FREY, D., BALTISBERGER, M. & EDWARDS, P.J. 2003: Cytology of *Erigeron annuus* s.l. and its consequences in Europe. - *Bot. Helv.* **113(1)**: 1-14.

FREY, L. 1969: Chromosome numbers in the genus *Scabiosa* L. I. - *Fragm. Florist. Geobot.* **15(2)**: 179-185.

FREY, L. 1970: Chromosome numbers in the genus *Scabiosa* L. II. - *Fragm. Florist. Geobot.* **16(3)**: 391-394.

FREY, L. 1973: Karyological differentiation in the genus *Molinia* Schrank in Poland. - *Fragm. Florist. Geobot.* **19(4)**: 389-396.

FREY, L. 1975: Taxonomical studies on the genus *Molinia* Schrank in Poland. - *Fragm. Florist. Geobot.* **21(1)**: 21-50.

FREY, L. 1976: Present distribution of *Molinia arundinacea* Schrank in Europe. - *Fragm. Florist. Geobot.* **22(3)**: 275-279 + 1 map.

FREY, L. 1984: Cytotaxonomical studies on the genus *Deschampsia* P. B. sensu lato in Poland. - *Fragm. Florist. Geobot.* **28(2)**: 117-144 + 2 plates.

FREY, L. 1991a: New chromosome numbers in *Trisetum* (Poaceae). - *Fragm. Florist. Geobot.* **35(1-2)**: 97-99.

FREY, L. 1991b: Taxonomy, karyology and distribution of selected genera of tribe Aveneae (Poaceae) in Poland: I. *Avenula*. - *Fragm. Florist. Geobot.* **35(1-2)**: 101-137.

FREY, L. 1991c: Distribution of *Avena strigosa* (Poaceae) in Europe. - *Fragm. Florist. Geobot.* **36(2)**: 281-288.

FREY, L. 1997a: The eastern limit of European distribution of *Aira caryophyllea* (Poaceae). - *Fragm. Florist. Geobot.* **42(2)**: 255-263.

FREY, L. 1997b: Karyology of the genus *Agrostis* (Poaceae) - a review. - *Fragm. Florist. Geobot.* **42 (2)**: 361-400.

FREY, L. 1999: *Avenella*: a genus of the Aveneae (Poaceae) worthy of recognition. - *Fragm. Florist. Geobot. Suppl.* **7**: 27-32.

FREY, R. 1989: Taxonomische Revision der Gattung *Peucedanum*: Sektion *Peucedanum* und Sektion *Palimbioidea* (Umbelliferae). - *Candollea* **44(1)**: 257-327.

FREYN, J.F. 1878: *Muscari (Bellevalia, Leopoldia) Weissii*. - *Oesterr. Bot. Z.* **28(3)**: 87-88.

FREYN, J.F. 1880: *Trifolium xanthinum* (Sect. *Lagopus* Koch), eine bisher unbeschriebene Art der griechischen Flora. - *Bot. Centralbl.* **1(7-8)**: 308-310.

FREYN, J.F. 1881: Zur Kenntnis einiger Arten der Gattung *Ranunculus*. - *Bot. Centralbl.* **6, Beil. 26**: 1-22 + 2 plates.

FREYN, J.F. 1890: Beiträge zur Kenntnis einiger Arten der Gattung *Ranunculus*. - *Bot. Centralbl.* **41(1-5)**: 1-6, 33-37, 73-78, 129-134.

FREYN, J.F. 1895a: Ueber neue und bemerkenswerthe orientalische Pflanzenarten. - *Bull. Herb. Boissier* **3(1-6)**: 31-40, 75-83, 97-108, 177-193, 302-307.

FREYN, J.F. 1895b: Ueber neue und bemerkenswerthe orientalische Pflanzenarten [cont.]. - *Bull. Herb. Boissier* **3(7-12)**: 345-358, 466-478, 497-511, 643-671.

FREYN, J.F. 1896: Ueber neue und bemerkenswerthe orientalische Pflanzenarten [cont.]. - *Bull. Herb. Boissier* **4(1-3)**: 42-57, 134-144, 178-200.

FREYN, J.F. 1897: Ueber neue und bemerkenswerthe orientalische Pflanzenarten. II. - *Bull. Herb. Boissier* **5(7-10)**: 579-626, 781-803.

FREYN, J.F. 1898: Ueber neue und bemerkenswerthe orientalische Pflanzenarten. III. - *Bull. Herb. Boissier* **6(11-12)**: 881-892, 974-990.

FREZET, C. 1975: Contribution à l'étude botanique et biochimique de *Coris monspeliensis* L. (Primulacées). - Lyon: Univ. Claude-Bernard, Thèse, 117 pp.

FRIDLENDER, A. 2000: Le genre *Arum* en Corse. - *Candollea* **55(2)**: 255-267.

FRIDLENDER, A. & STEPHAN, J. 2020: *Amelanchier ovalis* and *A. parviflora*. In Raus, Th. & Raab-Straube, E.v. (eds): Euro+Med-Checklist notulae, 12. - *Willdenowia* **50(2)**: 326-329.

FRIEBE, B. & AL. 1996: Standard karyotypes of *Aegilops uniaristata, Ae. mutica, Ae. comosa* subspecies *comosa* and *heldreichii* (Poaceae). - *Pl. Syst. Evol.* **202(3-4)**: 199-210.

FRIEDRICH, H.CH. 1973: Zur Cytotaxonomie der Gattung *Crassula*. - *Garcia de Orta, Sér. Bot.* **1(1-2)**: 49-65 + 4 plates.

FRIEDRICHSTHAL, E. 1838: Reise in den südlichen Theilen von Neu-Griechenland. Beiträge zur Charakteristik dieses Landes. Mit einem botanischen Anhange. - Leipzig: Verlag von W. Engelmann, viii + 311 pp.

FRIESEN, N. 1996: A taxonomic and chorological revision of the genus *Allium* L. sect. *Schoenoprasum* Dumort. - *Candollea* **51(2)**: 461-473.

FRIESEN, N. 2002: The *Allium ampeloprasum* polyploid complex and origin of its cultivars. In Koyuncu, M. (ed.): VIth Plant Life of Southwest Asia Symposium, Abstracts p. 67. - Van: Yüzüncü Yil University.

FRITSCH, B. 1973: Karyologische Untersuchungen in der Gattung *Echium* L. - *Bot. Not.* **126(4)**: 450-458.

FRITSCH, C. 1894: Beiträge zur Flora der Balkanhalbinsel mit besonderer Berücksichtigung von Serbien. I. - *Verh. K.K. Zool.-Bot. Ges. Wien* **44(1)**: 93-136.

FRITSCH, C. 1895a: Beiträge zur Flora der Balkanhalbinsel mit besonderer Berücksichtigung von Serbien. II. - *Verh. K.K. Zool.-Bot. Ges. Wien* **44(3-4)**: 301-327.

FRITSCH, C. 1895b: Ueber die Auffindung einer marinen Hydrocharidee im Mittelmeer. - *Verh. K.K. Zool.-Bot. Ges. Wien* **45**: 104-106.

FRITSCH, C. 1895c: Beiträge zur Flora der Balkanhalbinsel mit besonderer Berücksichtigung von Serbien. III. - *Verh. K.K. Zool.-Bot. Ges. Wien* **45**: 367-382.

FRITSCH, C. 1898a: Zur Systematik der Gattung *Sorbus*. I. Die Abgrenzung der Gattung. - *Oesterr. Bot. Z.* **48(1-2)**: 1-4, 47-49.

FRITSCH, C. 1898b: Zur Systematik der Gattung *Sorbus*. II. Die europäischen Arten und Hybriden. (Erste Abtheilung). - *Oesterr. Bot. Z.* **48(5)**: 167-171.

FRITSCH, C. 1899a: Beiträge zur Flora der Balkanhalbinsel mit besonderer Berücksichtigung von Serbien. IV-V. - *Verh. K.K. Zool.-Bot. Ges. Wien* **49**: 221-242, 460-470.

FRITSCH, C. 1899b: Zur Systematik der Gattung *Sorbus*. II. Die europäischen Arten und Hybriden. (Zweite Abtheilung). - *Oesterr. Bot. Z.* **49(11-12)**: 381-385, 426-429.

FRITSCH, C. 1909: Neue Beiträge zur Flora der Balkanhalbinsel, insbesondere Serbiens, Bosniens und der Herzegowina. I. - *Mitt. Naturwiss. Vereines Steiermark* **45(1)**: 131-183.

FRITSCH, C. 1910: Neue Beiträge zur Flora der Balkanhalbinsel, insbesondere Serbiens, Bosniens und der Herzegowina. II. - *Mitt. Naturwiss. Vereines Steiermark* **46(1)**: 294-328.

FRITSCH, C. 1911: Neue Beiträge zur Flora der Balkanhalbinsel, insbesondere Serbiens, Bosniens und der Herzegowina. III. - *Mitt. Naturwiss. Vereines Steiermark* **47**: 145-218.

FRITSCH, C. 1914: Neue Beiträge zur Flora der Balkanhalbinsel, insbesondere Serbiens, Bosniens und der Herzegowina. IV. - *Mitt. Naturwiss. Vereines Steiermark* **50(1)**: 349-385.

FRITSCH, C. 1915: Neue Beiträge zur Flora der Balkanhalbinsel, insbesondere Serbiens, Bosniens und der Herzegowina. V. - *Mitt. Naturwiss. Vereines Steiermark* **51(1)**: 173-215.

FRITSCH, C. 1916: Neue Beiträge zur Flora der Balkanhalbinsel, insbesondere Serbiens, Bosniens und der Herzegowina. VI. - *Mitt. Naturwiss. Vereines Steiermark* **52**: 293-332.

FRITSCH, C. 1917: Neue Beiträge zur Flora der Balkanhalbinsel, insbesondere Serbiens, Bosniens und der Herzegowina. VII. - *Mitt. Naturwiss. Vereines Steiermark* **53**: 211-221.

FRITSCH, C. 1918: Neue Beiträge zur Flora der Balkanhalbinsel, insbesondere Serbiens, Bosniens und der Herzegowina. VIII. - *Mitt. Naturwiss. Vereines Steiermark* **54**: 235-299.

FRITSCH, C. 1929: Zur Kenntnis der *Camelina rumelica* Velenovsky. - *Sitzungsber. Akad. Wiss. Wien, Math.-Naturwiss. Kl., Abt. 1* **138(7)**: 347-370.

FRITSCH, R. & AL. 1977: Vergleichend-anatomische Untersuchungen im Verwandtschaftskreis von *Triticum* L. und *Aegilops* L. (Gramineae). - *Kulturpflanze* **25**: 155-265.

FRITSCH, R.M. 2002: Taxonomy of the genus *Allium* L.: contributions from IPK Gatersleben. - *Herbertia* **56**: 19-50.

FRITSCH, R.M. 2012: Illustrated key to the sections and subsections and brief general circumscription of *Allium* subg. *Melanocrommyum*. - *Phyton (Horn)* **52(1)**: 1-37.

FRIVALDSZKY, E. 1835: Succinctae diagnoses specierum plantarum novarum europaeo-turcicarum in catalogo meo occurentium. - *Flora* **18(1)**: 331-336.

FRIVALDSZKY, E. 1836: Succinctae diagnoses specierum plantarum novarum anno 1835 in Turcia europaea collectarum. - *Flora* **19(2)**: 433-440.

FRIVALDSZKY, E. 1841: Rövid áttekintése egy természetrajzi utazának [In Hungarian]. - *Kir. Magyar Term. Társ. Évk. (Pest)* **1**: 163-184. - [See also Aquila 9: 206-208, 1902].

FRIZZI, G. 1986: Numeri cromosomici per la Flora Italiana: 1003-1006. - *Inform. Bot. Ital.* **16(2-3)**: 243-245.

FRÖDERSTRÖM, H. 1930: The genus *Sedum* L. A systematic essay. Part I. - *Acta Horti Gothob. App.* **5**: 3-75 + 28 plates.

FRÖDERSTRÖM, H. 1931: The genus *Sedum* L. A systematic essay. Part II. - *Acta Horti Gothob. App.* **6**: 3-111 + 65 plates.

FRÖDERSTRÖM, H. 1932: The genus *Sedum* L. A systematic essay. Part III. - *Acta Horti Gothob. App.* **7**: 1-126 + 68 plates.

FRÖDERSTRÖM, H. 1935: The genus *Sedum* L. A systematic essay. Part IV. - *Acta Horti Gothob. App.* **10**: 3-262 + 115 plates.

FRODIN, D.G. & GOVAERTS, R. 2004: World checklist and bibliography of Araliaceae. - Kew: Royal Botanic Gardens, ix + 444 pp.

FRÖHNER, S. 1975: Kritik an der europäischen *Alchemilla*-Taxonomie. - *Feddes Repert.* **86(3)**: 119-169.

FRÖHNER, S. 1986: Zur infragenerischen Gliederung der Gattung *Alchemilla* L. in Eurasien. - *Gleditschia* **14(1)**: 3-49.

FRÖST, S. 1957: The inheritance of the accessory chromosomes in *Centaurea scabiosa*. - *Hereditas (Lund)* **43(2)**: 403-422.

FRY, K. 1992: *Cyclamen creticum* in Crete. - *Cyclamen* **16(2)**: 44.

FRYER, J. & HYLMÖ, B. 2009: Cotoneasters. A comprehensive guide to shrubs for flowers, fruit, and foliage. - Portland & London: Timber Press, 344 pp.

FUCHS-ECKERT, H.P. 1980: Beiträge zur Nomenklatur und Taxonomie der Schweizer Flora. - *Feddes Repert.* **90(9-10)**: 525-689.

FUCHS, H.P. 1961a: Zur Nomenklatur der Gattung *Vincetoxicum* F.C. Medikus, non Th. Walter. - *Verh. Naturf. Ges. Basel* **72(2)**: 343-349.

FUCHS, H.P. 1961b: The genus *Cheilanthes* Swartz and its European species. - *Brit. Fern Gaz.* **9(2)**: 38-48.

FUCHS, H.P. 1962: Nomenklatur, Taxonomie und Systematik der Gattung *Isoëtes* Linnaeus in geschichtlicher Betrachtung. - *Beih. Nova Hedwigia* **3**: 103 pp. + 13 plates.

FUENTE GARCÍA, V. DE LA & SÁNCHEZ-MATA, D. 1986a: Tipificación de algunos táxones ibéricos del género *Festuca* L. (Gramineae) descritos por E. Hackel. - *Candollea* **41(1)**: 163-171.

FUENTE GARCÍA, V. DE LA & SÁNCHEZ-MATA, D. 1986b: Datos taxonómicos sobre el género *Festuca* L. (Gramineae) en la Península Ibérica. - *Candollea* **41(2)**: 441-448.

FUENTE GARCÍA, V. DE LA., SÁNCHEZ-MATA, D. & MORENO SAIZ, J.C. 1988: Sobre el género *Festuca* L. (Gramineae). Tipificaciones en el herbario original de E. Hackel. - *Candollea* **43(2)**: 513-520.

FUENTES-BAZAN, S., UOTILA, P. & BORSCH, TH. 2012: A novel phylogeny-based generic classification for *Chenopodium* sensu lato, and a tribal rearrangement of Chenopodioideae (Chenopodiaceae). - *Willdenowia* **42(1)**: 5-24.

FUKAREK, P. 1941: Vorläufige Mitteilung über die natürlichen Standorte der Panzerkiefer. - *Mitt. Deutsch. Dendrol. Ges.* **54**: 1-12 + 4 plates and 1 map.

FUKAREK, P. 1950a: Materialen über die geographische Verbreitung der Panzerkiefer (*Pinus Heldreichii* Christ) [In Serbo-Croatian with German summary]. - *God. Biol. Inst. Sarajevu* **2(1-2)**: 21-41.

FUKAREK, P. 1950b: Podaci o raširenju molike (*Pinus Peuce* Grisebach) na Balkanskom Poluostrvu [In Serbo-Croatian]. - *God. Biol. Inst. Sarajevu* **2(1-2)**: 43-52.

FUKAREK, P. 1951: Eine neue Varietät der Panzerkiefer aus Serbien und Nord-Montenegro [In Serbo-Croatian with German summary]. - *God. Biol. Inst. Sarajevu* **4(1)**: 41-49.

FUKAREK, P. 1952: Areal der natürlichen Verbreitung der gemeinen Esche (*Fraxinus excelsior* L.) mit besonderer Berücksichtigung ihrer Standorte auf der Balkanhalbinsel [In Serbo-Croatian with German summary]. - *Rad. Poljopr.-Šumarsk. Fak. Univ. Sarajevu, A* **1952**: 41-65.

FUKAREK, P. 1956: Die Verbreitung der Baumhasel (*Corylus colurna* L.) in Bosnien und der Herzegowina [In Serbo-Croatian with German summary]. - *God. Biol. Inst. Sarajevu* **9(1-2)**: 153-176.

FUKAREK, P. 1958a: Einige neue Daten über die Verbreitung des *Daphne Blagajana* [sic!] Kiefersan der Balkanhalbinsel [In Serbo-Croatian with English and German summaries]. - *Godišen Zborn. Zemjod.-Šumarsk. Fak. Univ. Skopje, Šumarstvo* **11**: 635-637.

FUKAREK, P. 1958b: Beitrag zur Kenntniss der systematischen Stellung, Gliederung und der rezenten Verbreitung der Schwarzkiefer [In Serbo-Croatian with German summary]. - *Rad. Poljopr.-Šumarsk. Fak. Univ. Sarajevu, B* **3**: 3-92.

FUKAREK, P. 1965: Expansion et caractéristiques écologiques de l'*Amphoricarpos neumayeri* Vis. Endémique [In Serbo-Croatian with French summary]. - *Glasn. Zemaljsk. Muz. Sarajevu* **3-4**: 159-180.

FUKAREK, P. 1967: Neue Standorte der Panzerkiefer (*Pinus heldreichii* Christ emend. Markgraf). - *Bot. Jahrb. Syst.* **86(1-4)**: 449-462.

FUKAREK, P. 1968: A short description of some trees and shrubs of the Balkan Peninsula. - *Int. Dendrol. Soc. Year Book* **1968**: 25 pp.

FUKAREK, P. 1969a: Betrachtungen über einige dem Gebiete der Balkanländer und Rumänien gemeinsame Baum- und Straucharten. - *Rev. Roumaine Biol., Sér. Bot.* **14(1)**: 33-46.

FUKAREK, P. 1969b: Ueber einige illyrisch-apenninische Baum- und Straucharten. - *Mitt. Ostalpin-Dinarischen Pflanzensoziol. Arbeitsgem.* **9**: 79-90.

FUKAREK, P. 1970a: Bemerkungen zu einigen balkanischen und balkano-karpatischen Baum- und Straucharten. - *Feddes Repert.* **81(1-5)**: 163-170.

FUKAREK, P. 1970b: Die Fichte und die Fichtenwälder an ihren südlichen Arealgrenzen in den Balkanländern. - *Rad. Odjel. Prirod. Mat. Nauka (Sarajevo)* **11**: 147-174.

FUKAREK, P. 1971: Über neuere Ergebnisse der Forschungen des Verbreitungsareals der spitzblättrigen Esche (*Fraxinus angustifolia* Vahl). - *Gozd. Vestn.* **29**: 193-201 [n.v.].

FUKAREK, P. 1973: Die Tannen der Balkanhalbinsel. In: Kazdaği göknari ve Türkiye florasi uluslararasi simpozyomu bildirileri, pp. 83-94. - Istanbul Üniversitesi Orman Fakültesi Yayinlari.

FUKAREK, P. 1975: Unterschiede in der Dendroflora der westlichen und östlichen Gebiete der Balkanhalbinsel. In Jordanov, D. & al. (eds): Problems of Balkan flora and vegetation, pp. 146-161. - Sofia: Publishing House of the Bulgarian Academy of Sciences.

FUKAREK, P. 1978: The distribution and the ecological prediction of the Orchidaceae - *Listera cordata* (L.) Brown in the Balkan Peninsula [In Serbo-Croatian with English summary]. - *Glasn. Zemaljsk. Muz. Sarajevu* **17**: 153-168 + 1 map.

FUKAREK, P. 1979: Contemporary views on the taxonomy and nomenclature of *Pinus leucodermis* Ant. and *Pinus heldreichii* Christ [In Serbo-Croatian with English summary]. - *Glasn. Zemaljsk. Muz. Sarajevu* **18**: 63-87.

FURNARI, F. 1962: Nuova stazione sicula di *Coris monspeliensis* L. - *Boll. Ist. Bot. Univ. Catania, ser. 3* **3**: 48-53.

FÜRNKRANZ, D. 1968: Beiträge zur Systematik und Karyologie der Gattung *Cicer*. I. Die europäischen Arten. - *Oesterr. Bot. Z.* **115(3)**: 400-410.

FÜRNKRANZ, D. 1969a: Eine neue mediterrane Reliktart der Gattung *Taraxacum*. - *Oesterr. Bot. Z.* **117(2)**: 149-156.

FÜRNKRANZ, D. 1969b: Beiträge zur Verbreitungsgeschichte der Gattung *Taraxacum* im zentral- und nordostmediterranen Raum. - *Oesterr. Bot. Z.* **117(2)**: 157-162.

GABRIELIAN, E. 1961: The genus *Sorbus* L. in Turkey. - *Notes Roy. Bot. Gard. Edinburgh* **23(4)**: 483-496.

GADELLA, TH.W.J. 1964: Cytotaxonomic studies in the genus *Campanula*. - *Wentia* **11**: 1-104 + 4 plates and 7 fold-out tables.

GADELLA, TH.W.J. 1966: Some notes on the delimitation of genera in the Campanulaceae. I and II. - *Proc. Kon. Ned. Akad. Wetensch. C* **69(4)**: 502-508, 509-521.

GADELLA, TH.W.J. 1972a: Cytological and hybridization studies in the genus *Symphytum*. - *Symp. Biol. Hung.* **12**: 189-199.

GADELLA, TH.W.J. 1972b: Some notes on *Ornithogalum umbellatum* L. and *Ornithogalum divergens* Bor. - *Acta Bot. Neerl.* **21(3)**: 257-260.

GADELLA, TH.W.J. 1977: Cytotaxonomic studies in *Stellaria graminea* L. in the Netherlands. - *Proc. Kon. Ned. Akad. Wetensch. C* **80(3)**: 161-170.

GADELLA, TH.W.J. 1984a: On the reproduction, hybridization and cytology in *Hieracium caespitosum* Dumort. [In Dutch with English summary]. - *Gorteria* **12(5)**: 108-109.

GADELLA, TH.W.J. 1984b: Cytology and the mode of reproduction of some taxa of *Hieracium* subgenus *Pilosella*. - *Proc. Kon. Ned. Akad. Wetensch. C* **87(4)**: 387-399.

GADELLA, TH.W.J. & KLIPHUIS, E. 1963: Chromosome numbers of flowering plants in the Netherlands. - *Acta Bot. Neerl.* **12(2)**: 195-230.

GADELLA, TH.W.J. & KLIPHUIS, E. 1966: Chromosome numbers of flowering plants in the Netherlands. - *Proc. Kon. Ned. Akad. Wetensch. C* **69(5)**: 541-556.

GADELLA, TH.W.J. & KLIPHUIS, E. 1967: Chromosome numbers of flowering plants in the Netherlands. - *Proc. Kon. Ned. Akad. Wetensch. C* **70(1)**: 7-20.

GADELLA, TH.W.J. & KLIPHUIS, E. 1968a: *Parnassia palustris* in the Netherlands. - *Acta Bot. Neerl.* **17(3)**: 165-172.

GADELLA, TH.W.J. & KLIPHUIS, E. 1968b: [Miscellaneous chromosome numbers]. In Löve, Á. (ed.): ; ; : IOPB chromosome number reports XVI. - *Taxon* **17(2)**: 200-201.

GADELLA, TH.W.J. & KLIPHUIS, E. 1968c: Chromosome numbers of flowering plants in the Netherlands. - *Meded. Bot. Mus. Herb. Rijks Univ. Utrecht* **274**: 168-183.

GADELLA, TH.W.J. & KLIPHUIS, E. 1970: Chromosome studies in some flowering plants collected in the French Alps (Haute Savoie). - *Rev. Gén. Bot.* **77(917-919)**: 487-497.

GADELLA, TH.W.J. & KLIPHUIS, E. 1971: Chromosome numbers of flowering plants in the Netherlands. - *Proc. Kon. Ned. Akad. Wetensch. C* **74**: 335-343.

GADELLA, TH.W.J. & KLIPHUIS, E. 1972: Studies in chromosome numbers of Yugoslavian angiosperms. - *Acta Bot. Croat.* **31**: 91-103.

GADELLA, TH.W.J. & KLIPHUIS, E. 1978: Cytotaxonomic studies in the genus *Symphytum*. VIII. Chromosome numbers and classification of ten European species. - *Proc. Kon. Ned. Akad. Wetensch. C* **81(2)**: 162-172.

GADELLA, TH.W.J. & RAAMSDONK, L.W.D. VAN 1981: Biosystematic studies on *Ornithogalum umbellatum* L. s.l. - *Bol. Soc. Brot., Ser. 2* **53(2)**: 745-791 + 1 plate.

GAFFAROĞLU, N. 1980a: Taxonomische Untersuchung über die *Anthemis cotula* L. und *A. pseudocotula* Boiss., Arten die in Izmir und Umgebung eine grosse Verbreitung zeigt [In Turkish with German abstract]. - *Ege Üniv. Fen Fak. Derg., B* **4(1-4)**: 39-51.

GAFFAROĞLU, N. 1980b: Die Bestimmungsschlüssel und die Verbreitungsplätze der in Westanatolien verbreitenen *Anthemis* L. [In Turkish with German abstract]. - *Ege Üniv. Fen Fak. Derg., B* **4(1-4)**: 155-166.

GAGALIS, G.A. 1974: Elliniki votaniki vivliografia 1822-1972 (Bibliografia botanica graeca 1822-1972). - Athine: 130 pp.

GAGE, E. & AL. 2011: Phylogenetic systematics of *Sternbergia* (Amaryllidaceae) based on plastid and ITS sequence data. - *Bot. J. Linn. Soc.* **166(2)**: 149-162.

GAGE, E. & WILKIN, P. 2008: A morphometric study of species delimitation in *Sternbergia lutea* (Alliaceae: Amaryllidoideae) and its allies *S. sicula* and *S. greuteriana*. - *Bot. J. Linn. Soc.* **158(3)**: 460-469.

GAGNEPAIN, F. 1916: Révision du genre *Circaea*. - *Bull. Soc. Bot. France* **63**: 39-43.

GAJEWSKI, W. 1957: A cytogenetic study of the genus *Geum*. - *Monogr. Bot.* **4**: 1-416 + 4 plates.

GAJEWSKI, W., SWIETLINSKA, Z. & ZUK, J. 1963: Relationship between biosystematics and formal taxonomy of the *Rumex acetosa* group. - *Regnum Veg.* **27**: 16-24.

GAJIĆ, M.R. 1973a: The supplement to the knowledge of the flora of Yugoslavia [In Serbo-Croatian with German summary]. - *Glasn. Prir. Mus. Beogradu, Ser. B, Biol. Nauke* **28**: 5-33.

GAJIĆ, M.R. 1973b: New forms of the species *Phyteuma orbiculare* L. and *Jasione dentata* (D.C.) Hal. [In Serbo-Croatian with English summary]. - *Glasn. Prir. Mus. Beogradu, Ser. B, Biol. Nauke* **28**: 61-67.

GAJIĆ, M.R. 1973c: Variability of shop-sage (*Salvia officinalis* L.). The new subspecies (*S. officinalis* L. subsp. *multiflora* Gajić subsp. nova) [In Serbo-Croatian with English summary]. - *Glasn. Prir. Mus. Beogradu, Ser. C* **7**: 29-51 + 1 table.

GAJIĆ, M.R. 1974: Variability of *Achillea ageratifolia* (Sibth. et Sm.) Boiss. - the new forms *A. ageratifolia* (Sibth. et Sm.) Boiss. [In Serbo-Croatian with English summary]. - *Glasn. Prir. Mus. Beogradu, Ser. B, Biol. Nauke* **29**: 5-16.

GAJIĆ, M.R. 1975a: *Allium ursinum* L. subsp. *ucrainicum* Kleopow et Oxner - the new subspecies for the flora of Yugoslavia [In Serbo-Croatian with English summary]. - *Glasn. Prir. Mus. Beogradu, Ser. C* **8**: 21-23.

GAJIĆ, M.R. 1975b: Material for the flora of Yugoslavia [In Serbo-Croatian with English summary]. - *Glasn. Prir. Mus. Beogradu, Ser. B, Biol. Nauke* **30**: 5-14.

GAJIĆ, M.R. 1976a: Die Variabilität der *Centaurea nervosa* Willd. Ihre neuen infraspezifischen Formen: subsp. *josifovicii* und f. *sessilifolia*. - *Bull. Acad. Serbe Sci., B, Sci. Nat.* **54(14)**: 13-19.

GAJIĆ, M.R. 1976b: Material for the monographs of some plant species [In Serbo-Croatian with English summary]. - *Glasn. Prir. Mus. Beogradu, Ser. B, Biol. Nauke* **31**: 5-17.

GAJIĆ, M.R. 1979: The Balcan element in the flora of Serbia [In Serbo-Croatian with English summary]. - *Glasn. Prir. Mus. Beogradu, Ser. B, Biol. Nauke* **34**: 5-20.

GAJIĆ, M.R. 1980: Relict species in the flora of Serbia [In Serbo-Croatian with English summary]. - *Glasn. Prir. Mus. Beogradu, Ser. B, Biol. Nauke* **35**: 39-42.

GAJIĆ, M.R. & TATIĆ, B. 1981: Chorological information in Yugoslavia. In Velčev, V.I. & Kožuharov, S.I. (eds): Mapping the flora of the Balkan Peninsula, pp. 74-79. - Sofia: Publishing House of the Bulgarian Academy of Sciences

GALANOS, CH. 2015: The alien flora of terrestrial and marine ecosystems of Rodos island (SE Aegean), Greece. - *Willdenowia* **45(2)**: 261-278.

GALANOS, CH. 2016: Reports 110-118. In Vladimirov, V. & Tan, Kit (eds): New floristic records in the Balkans: 31. - *Phytol. Balcan.* **22(3)**: 442-444.

GALANOS, CH. 2018: *Epipactis densifolia* (Orchidaceae), verified for the first time from Rhodes Island, SE Aegean, Greece. - *J. Eur. Orch.* **50(2-4)**: 264-272.

GALANOS, CH. 2020: Reports 54-55. In Vladimirov, V., Aybeke, M. & Tan, Kit: New floristic records in the Balkans: 43. - *Phytol. Balcan.* **26(3)**: 545-547.

GALANOS, CH. & TZANOUDAKIS, D. 2017: *Allium symiacum* (Amaryllidaceae), a new species from Symi Island (SE Aegean, Greece). - *Willdenowia* **47(2)**: 107-113.

GALANOS, CH. & TZANOUDAKIS, D. 2019: *Allium panormitisi* (Amaryllidaceae), a new autumn-flowering species from Symi Island, SE Aegean, Greece. - *Bot. Serbica* **43(2)**: 197-203.

GALASSO, G. & AL. 2018: An updated checklist of the vascular flora alien to Italy. - *Pl. Biosyst.* **152(3)**: 556-592.

GALBANY-CASALS, M. & AL. 2004: Phylogenetic relationships in the Mediterranean *Helichrysum* (Asteraceae, Gnaphalieae) based on nuclear rDNA ITS sequence data. - *Austral. Syst. Bot.* **17(3)**: 241-253.

GALBANY-CASALS, M. & AL. 2006: Typification of names in *Gnaphalium* L. and *Helichrysum* Mill. (Asteraceae), and some taxonomic notes. - *Taxon* **55(2)**: 489-501.

GALBANY-CASALS, M. & AL. 2010: How many Cassini anagrams should there be? Molecular systematics and phylogenetic relationships in the *Filago* group (Asteraceae, Gnaphalieae), with special focus on the genus *Filago*. - *Taxon* **59(6)**: 1671-1689.

GALBANY-CASALS, M., SÁEZ, L. & BENEDÍ, C. 2006: A taxonomic revision of *Helichrysum* sect. *Stoechadina* (Asteraceae, Gnaphalieae). - *Canad. J. Bot.* **84(8)**: 1203-1232.

GALICIA HERBADA, D. 1994: Distribución del género *Thymelaea* Miller (Thymelaeaceae) en la Península Ibérica e Islas Baleares. - *Candollea* **50(1)**: 51-130.

GALICIA-HERBADA, D. 2006: Origin and diversification of *Thymelaea* (Thymelaeaceae): inferences from a phylogenetic study based on ITS (rDNA) sequences. - *Pl. Syst. Evol .* **257(3-4)**: 159-187.

GALIL, J. 1965a: Vegetative dispersal of *Allium ampeloprasum* L. I. Vegetative reproduction. - *Israel J. Bot.* **14(3)**: 135-140.

GALIL, J. 1965b: Vegetative dispersal of *Allium neapolitanum*. - *Amer. J. Bot.* **52(3)**: 282-286.

GALIL, J. 1966: Vegetative dispersal of *Allium ampeloprasum* L. II. Sprouting of bulblets. - *Israel J. Bot.* **14(4)**: 184-191.

GALIL, J. 1968: Vegetative dispersal in *Oxalis cernua*. - *Amer. J. Bot.* **55(1)**: 68-73.

GALIL, J. 1969: Morpho-ecological studies on *Gladiolus segetum* Gawl. - *Israel J. Bot.* **18(2)**: 43-54.

GALIL, J. 1970: Studies on the survival of dicotyledonous cormous weeds in ploughed fields. - *J. Appl. Ecol.* **7(1)**: 61-65 + 2 plates.

GALIL, J. 1979: Morpho-ecological studies on *Arisarum vulgare* Targ.-Tozz. - *Israel J. Bot.* **27(2)**: 77-89.

GALIL, J. 1982: Morpho-ecological studies on geophilic plants. Vegetative dispersal of *Gynandriris sisyrinchium* L. - *Israel J. Bot.* **30(4)**: 165-172.

GALIL, J. & NEEMAN, G. 1977: Pollen transfer and pollination in the common fig (*Ficus carica* L.). - *New Phytol.* **79(1)**: 163-171 + 1 plate.

GALLAND, N. 1988: Recherche sur l'origin de la flore oreophile du Maroc. Etude caryologique et cytogéographique. - *Trav. Inst. Sci. Univ. Mohammed V, Sér. Bot.* **35**: iv + 168 pp.

GALLAND, N. 1989: Phenotypic plasticity within the polyploid complex *Thlaspi perfoliatum* L. agg. (Brassicaceae). In Phitos, D. (ed.): Abstracts. Sixth OPTIMA meeting, Delphi 10-16 Sept. 1989, p. 96.

GALLAND, N. & FAVARGER, C. 1988: Un complexe polyploïde méconnu: *Thlaspi perfoliatum* L. agg. (Brassicaceae). - *Monogr. Inst. Piren. Ecol.* **4**: 205-211.

GALLEGO CIDONCHA, M.J., MUÑOZ GARMENDIA, F. & NAVARRO ARDANA, C. 1993: *Xolantha* Raf., nombre prioritario para *Tuberaria* (Dunal) Spach (Cistaceae). - *Anales Jard. Bot. Madrid* **51(1)**: 169.

GALLEGO MARTÍN, F. 1984: Notas sobre la cariología de *Ecballium elaterium* (L.) Richard. - *Stud. Bot. Univ. Salamanca* **3**: 321-323.

GALLEGO MARTÍN, F., SÁNCHEZ ANTA, M.A. & NAVARRO ANDRÉS, F. 1986: Datos cariológicos de algunas genisteas supramediterráneas. - *Lazaroa* **8**: 97-103.

GALLEGO, M.J. 1981: Números cromosómicos para la flora española. 197-200. - *Lagascalia* **10(2)**: 231-232.

GALLEGO, M.J. 1985: Números cromosómicos para la flora española. 404-408. - *Lagascalia* **13(2)**: 311-313.

GALLEGO, M.J. 1986: Números cromosómicos para la flora española. 464-470. - *Lagascalia* **14(2)**: 286-288.

GALLEGO, M.J. & APARICIO, A. 1990: Números cromosómicos para la flora española. 603-617. - *Lagascalia* **15(2)**: 288-295.

GALLEGO, M.J. & APARICIO, A. 1993: Karyological study in the genus *Tuberaria* sect.

Scorpioides (Cistaceae): taxonomic and evolutionary inferences. - *Pl. Syst. Evol.* **184(1-2)**: 11-25.

GALLEGO, M.J. & TALAVERA, S. 1983: Nota sobre las especies españolas del género *Calendula*. - *Lagascalia* **11(1)**: 100-104.

GALLEGO, M.J., TALAVERA, S. & SILVESTRE, S. 1980: Revisión del género *Reichardia* Roth (Compositae). - *Lagascalia* **9(2)**: 159-217.

GALLO, L. 2002: Remarks on *Sedum litoreum* Guss. - *Sedum Soc. Newsletter* **62**: 79-81.

GALLO, L. 2014: *Sedum ochroleucum* subsp. *mediterraneum* (Crassulaceae), a new Italian endemic. - *Willdenowia* **44(1)**: 27-33.

GALLO, L. 2020: Typification and taxonomic remarks in *Petrosedum* (Crassulaceae). - *Phytotaxa* **440 (1)**: 81-86.

GAMAL-ELDIN, E. 1981: Revision der Gattung *Pulicaria* (Compositae-Inuleae) für Afrika, Makaronesien und Arabien. - *Phanerog. Monogr.* **14**: 1-311 + plates and maps.

GAMAL-ELDIN, E. & WAGENITZ, G. 1984: Eine neue *Centaurea*-Art der Sektion *Phalolepis* (Compositae) aus Nordgriechenland (Chalkidike). - *Willdenowia* **13(2)**: 323-327.

GAMISANS, J. 1971: Contribution à l'étude de la flore de la Corse. II. - *Candollea* **26(2)**: 309-358.

GAMISANS, J. 1988: Compléments au Prodrome de la Flore Corse. Plantaginaceae. - Genève: Conservatoire et Jardin Botanique, 56 pp.

GAMISANS, J. 1990: Compléments au Prodrome de la Flore Corse. Caprifoliaceae. - Genève: Conservatoire et Jardin Botanique, 19 pp.

GAMISANS, J. 1991: Compléments au Prodrome de la Flore Corse. Annexe no. 2. La végétation de la Corse. - Genève: Conservatoire et Jardin Botanique, 391 pp.

GAMISANS, J. & AL. 1994: Les espèces du genre *Romulea* en Corse. - *Candollea* **49(2)**: 509-526.

GAMISANS, J. & HEBRARD, J.-P. 1979: A propos de la végétation des forêts d'Epire et de Macédoine grecque occidentale. - *Doc. Phytosoc. n.s.* **4**: 289-341.

GAMISANS, J. & HEBRARD, J.-P. 1980: A propos de la végétation des forêts en Grèce du nord-est (Macédoine orientale et Thrace occidentale). - *Doc. Phytosoc. n.s.* **5**: 243-289.

GAMISANS, J. & JEANMONOD, D. 1993a: Catalogue des plantes vasculaires de la Corse. Ed. 2. - Genève: Conservatoire et Jardin botanique, 285 pp.

GAMISANS, J. & JEANMONOD, D. 1993b: Compléments au Prodrome de la Flore Corse. Annexe no. 3. Catalogue des plantes vasculaires de la Corse (seconde édition). - Genève: Conservatoire et Jardin Botanique, 258 pp.

GAMISANS, J. & JEANMONOD, D. 1997: Compléments au Prodrome de la Flore Corse. Campanulaceae. Genève: Conservatoire et Jardin Botanique, 43 pp.

GAMISANS, J. & JEANMONOD, D. 1998: Compléments au Prodrome de la Flore Corse. Asteraceae - I. - Genève: Conservatoire et Jardin Botanique, 340 pp.

GAMISANS, J. & MARZOCCHI, J.-F. 1996: La flore endémique de la Corse. - Aix-en-Provence: Edisud, 208 pp.

GAMS, H. 1927: Die Gattung *Trapa* L. - *Pflanzenareale* **1(3)**: 39-41, Karte 25-27.

GAMS, H. 1936: Zur Geschichte, klimatischen Begrenzung und Gliederung der immergrünen Mittelmeerstufe. - *Veröff. Geobot. Inst. Rübel Zürich* **12**: 163-204.

GAMS, H. 1960: Nachträge zur Flora und Vegetation des Olymps. - *Oesterr. Bot. Z.* **107(2)**: 177-193.

GAMS, H. 1975: Vergleichende Betrachtung europäischer Ophiolith-Floren. In Dafis, S. & Landolt, E. (eds): Zur Vegetation und Flora von Griechenland, Band 1. - *Veröff. Geobot. Inst. E.T.H. Stiftung Rübel Zürich* **55**: 117-140.

GANATSAS, P. & AL. 2019: Regeneration ecology of the rare plant species *Verbascum dingleri*: Implications for species conservation. - *Sustainability* **11**: 3305; doi:10.3390/su11123305 www.

GANČEV, I. 1958: Neue Arten und kritische Notizen zur Flora Bulgariens [In Bulgarian with Russian and German summaries]. - *Izv. Bot. Inst. (Sofia)* **6**: 375-383.

GANČEV, I. & BONDEV, I. 1965: Kritische Notizen über zwei Pflanzen [In Bulgarian with Russian and German abstracts]. - *Izv. Bot. Inst. (Sofia)* **15**: 271-272.

GANČEV, S. & KOČEV, H. 1963: Sur la propagation du *Psilurus aristatus* en Bulgarie septentrionale [In Bulgarian with Russian and French summaries]. - *Izv. Bot. Inst. (Sofia)* **11**: 103-110.

GANDHI, K.N. & BROWN, L.E. 1993: *Crepis zacintha* (Asteraceae), adventive in Texas, and the first record in North America. - *Sida* **15(3)**: 547-550.

GANDOGER, M. 1883: Contributiones ad floram terrae slavorum meridionalium, fasc. 1. - Zagreb: Tisak Dioničke Tiskare, 38 pp.

GANDOGER, M. 1883-91: Flora Europae terrarumque adjacentium sive enumeratio plantarum per Europam atque totam regionem mediterraneam cum insulis atlanticis sponte crescentium, novo fundamento instauranda. - Paris, etc.: 27 vols. (lithographed manuscript) [cf. TL-2 1: 910-911, 1976].

GANDOGER, M. 1892-93: Monographia Rosarum Europae et Orientis. - Paris: 4 vols. (lithographed manuscript) [cf. TL-2 1: 911, 1976].

GANDOGER, M. 1915a: A la recherche du genre *Ammanthus* Boiss. Découverte du *Bellium minutum* L. - *Bull. Soc. Bot. France* **62**: 8-12.

GANDOGER, M. 1915b: Plantes nouvelles pour l'île de Crète. - *Bull. Soc. Bot. France* **62**: 151-157.

GANDOGER, M. 1916a: Troisième voyage botanique dans l'île de Crète. - *Bull. Soc. Bot. France* **63**: 219-243.

GANDOGER, M. 1916b: Flora Cretica. - Paris: A. Hermann et fil., 181 pp. (lithographed manuscript) .

GANDOGER, M. 1917a: Les *Campanula* et les Campanulacées de l'île de Crète. - *Bull. Soc. Bot. France* **64**: 1-8.

GANDOGER, M. 1917b: Quatrième voyage botanique en Crète (1917). - *Bull. Soc. Bot. France* **64**: 110-122.

GANDOGER, M. 1920a: Addenda et corrigenda ad floram Cretae. - *Bull. Soc. Bot. France* **67**: 181-185.

GANDOGER, M. 1920b: Florule de Syra (Grèce), d'après les récoltes faites par les frères Octave et Denis. - *Bull. Soc. Bot. France* **67**: 277-284.

GANDOGER, M. 1921a: Plantes endémiques ou rarissimes de l'île de Crète. - *Monde Pl., Sér. 3* **22 (13)**: 4-6.

GANDOGER, M. 1921b: Plantes endémiques ou rarissimes de l'île de Crète (suite). - *Monde Pl., Sér. 3* **22(15)**: 4-5.

GANDOGER, M. 1922: L'ascension du mont Ida (île de Crète). - *Bull. Soc. Bot. France* **69**: 16-20.

GANIATSAS, K.A. 1936: Untersuchungen über die Vegetation auf den Salzböden bei Saloniki. - *Ber. Deutsch. Bot. Ges.* **54(7)**: 430-444.

GANIATSAS, K.A. 1938: Erevne epi tis chloridos tou orous Chortiatou [In Greek]. - *Sci. Ann. Fac. Phys. Math. Aristot. Univ. Thessalonika* **5**: 3-30 + 2 plates and 1 map.

GANIATSAS, K.A. 1939: Votanike erevne epi tou orous Vermiou [In Greek]. - *Sci. Ann. Fac. Phys. Math. Aristot. Univ. Thessalonika* **5**: 225-265 + 2 plates.

GANIATSAS, K.A. 1940: Simvoli is tin gnosin tis chloridos tou orous Timfristou [In Greek]. - *Sci. Ann. Fac. Phys. Math. Aristot. Univ. Thessalonika* **6**: 17-36.

GANIATSAS, K.A. 1955: I chloris ton orinon voskon tou Vermiou (Fitokinoniki meleti) [In Greek]. - *Sci. Ann. Fac. Phys. Math. Aristot. Univ. Thessalonika* **7**: 115-145.

GANIATSAS, K.A. 1963: I vlastisis ke i chloris tis chersonisou tou Agiou Orous [In Greek]. - Thessaloniki: 172 pp. (Reprint from *Athoniki Politia*, pp. 509 ff.).

GANIATSAS, K.A. 1970: I chloris ke i vlastisis tis Limnis ton Ioanninon [In Greek]. - *Ipirotiki Estia* **1970**: 20 pp.

GANIATSAS, K.A. 1971: Votanike erevne epi tis charadras tou Vikou [In Greek]. - *Ipirotiki Estia* **1971**: 29 pp.

GANIATSAS, K.A. 1974: I chloris tis Kritis [In Greek]. - *Amalthia (Agios Nikolaos)* **5(20-21)**: 187-206.

GANIATSAS, K.A. 1975a: Begrüssung und Einführung in die griechische Vegetation und Flora. In Dafis, S. & Landolt, E. (eds): Zur Vegetation und Flora von Griechenland, Band 1. - *Veröff. Geobot. Inst. E.T.H. Stiftung Rübel Zürich* **55**: 18-22.

GANIATSAS, K.A. 1975b: Übersicht über die Flora und die Vegetationsverhältnisse der Athos-Halbinsel. In Jordanov, D. & al. (eds): Problems of Balkan flora and vegetation, pp. 380-390. - Sofia: Publishing House of the Bulgarian Academy of Sciences.

GANIATSAS, K.A. 1976: I vlastisi tis Kritis [In Greek]. - *Kritologia* **2(1)**: 1-38.

GARAJ, V., PAPEŠ, D. & PEVALEK, B. 1981: [Chromosome numbers in *Orchis*]. In Löve, Á. (ed.): Chromosome number reports LXX. - *Taxon* **30(1)**: 69-70.

GARBARI, F. 1967: Contributo allo studio citologico dei *Muscari* italiani. - *Caryologia* **19(4)**: 419-428.

GARBARI, F. 1968: Il genere *Muscari* (Liliaceae): Contributo alla revisione citotassonomica. - *Giorn. Bot. Ital., n.s.* **102(2)**: 87-105.

GARBARI, F. 1969: Nuove osservazioni citologiche sui generi *Muscari* e *Leopoldia*. - *Giorn. Bot. Ital., n.s.* **103(1)**: 1-9.

GARBARI, F. 1970: Il genere *Brimeura* Salisb. (Liliaceae). - *Atti Soc. Tosc. Sci. Nat. Pisa Mem., Ser. B* **72**: 12-36.

GARBARI, F. 1971: B chromosomes in "Listera ovata" (L.) R. Br. ("Orchidaceae"). Preliminary observations in Italian populations. - *Inform. Bot. Ital.* **3(2)**: 216-221.

GARBARI, F. 1973a: Note sul genere *Pseudomuscari* (Liliaceae). - *Webbia* **27(2)**: 369-381.

GARBARI, F. 1973b: Le specie del genere "Leopoldia" Parl. (Liliaceae) in Italia. - *Webbia* **28(1)**: 57-80.

GARBARI, F. 1973c: Il genere *Paspalum* L. (Gramineae) in Italia. - *Atti Soc. Tosc. Sci. Nat. Pisa Mem., Ser. B* **79**: 52-65.

GARBARI, F. 1976: In genere "Allium" L. in Italia. VIII. "Allium atroviolaceum" Boiss., specie nuova per la nostra flora. - *Inform. Bot. Ital.* **7(3)**: 352-354.

GARBARI, F. 1979: Cytotaxonomical and biosystematic aspects of the Mediterranean flora of Italy. - *Webbia* **34(1)**: 337-355.

GARBARI, F. 1984: Some karyological and taxonomic remarks on the Italian "Muscari" (Liliaceae). - *Webbia* **38**: 139-164.

GARBARI, F. 1988a: *Allium aethusanum* (Alliaceae), a new species from the island of Favignana (Aegadean/Egadi Islands, Sicily). - *Atti Soc. Tosc. Sci. Nat. Pisa Mem., Ser. B* **94**: 203-207.

GARBARI, F. 1988b: L'indagine citotassonomica in Italia: cronache, risultati e prospettive. In: 100 anni di ricerche botaniche in Italia 1888-1988, pp. 519-526. - Firenze: Società Botanica Italiana.

GARBARI, F. 1990a: Il genere *Ornithogalum* L. (Hyacinthaceae) in Italia. X. Considerazioni su *O. biflorum* Jord. & Fourr. - *Boll. Soc. Sarda Sci. Nat.* **27**: 253-258. [Also distributed as *OPTIMA leaflet* no. 203].

GARBARI, F. 1990b: L'endemismo vegetale: genesi, tipi e significato biogeografico. - *Studi Trent. Sci. Nat., Acta Biol.* **66**: 113-120.

GARBARI, F. 2004: *Muscari neglectum* Guss. e *M. atlanticum* Boiss. et Reuter (Hyacinthaceae). Tipi, caratteristiche e considerazioni suelle due specie. - *Inform. Bot. Ital.* **35(2)**: 329-336.

GARBARI, F. & AL. 2001: The genus *Ornithogalum* L. (Hyacinthaceae) in Italy: towards a redefinition of infrageneric taxa, with new proposals. In Raimondo, F.M. (ed.): Abstracts from the X OPTIMA meeting, Palermo, 13-19 September 2001, p. 60.

GARBARI, F. & AL. 2003: The genus *Ornithogalum* L. (Hyacinthaceae) in Italy, XIV: towards a refininition of infrageneric taxa, with new proposals. - *Bocconea* **16(1)**: 269-281.

GARBARI, F., BARSOTTI, L. & MICELI, P. 1989: Some karyological, anatomical and phytogeographic aspects of *Allium callimischon* s.l. from Greece. - *Bios (Thessaloniki)* **1989**: 87-97.

GARBARI, F. & CELA RENZONI, G. 1976: Il genere *Allium* in Italia. VII. Il caso di *Allium commutatum* Guss. - *Lav. Soc. Ital. Biogeogr., n.s.* **5**: 67-80.

GARBARI, F. & CORSI, G. 1972: Il genere *Allium* L. in Italia. III. *Allium amethystinum* Tausch in Toscana. - *Inform. Bot. Ital.* **4(2)**: 125-127.

GARBARI, F., CORSI, G. & MASINI, A. 1989: Anatomical investigations on the *Allium cupanii-hirtovaginatum* complex. In Phitos, D. (ed.): Abstracts. Sixth OPTIMA meeting, Delphi 10-16 Sept. 1989, p. 97.

GARBARI, F., CORSI, G. & MASINI, A. 1991: Anatomical investigations in the *Allium cupanii - A. hirtovaginatum* complex. - *Bot. Chron. (Patras)* **10**: 805-808.

GARBARI, F. & DI MARTINO, A. 1972: "Leopoldia gussonei" Parl. (Liliaceae), specie endemica siciliana. - *Webbia* **27(1)**: 289-297.

GARBARI, F. & GIORDANI, A. 1985: *Ornithogalum collinum* Guss. (Liliaceae): typification, caryology and leaf anatomical characters. - *Atti Soc. Tosc. Sci. Nat. Pisa Mem., Ser. B* **91**: 63-70. [Also distributed as *OPTIMA leaflet* no. 151].

GARBARI, F., GIORDANI, A. & ARNOLD, N. 1989: Chromosome numbers for the flora of Cyprus. – *Atti Soc. Tosc. Sci. Nat. Pisa Mem., Ser. B* **95**: 35-40. [Also distributed as *OPTIMA leaflet* no. 180].

GARBARI, F. & GREUTER, W. 1970: On the taxonomy and typification of *Muscari* Miller (Liliaceae) and allied genera, and on the typification of generic names. - *Taxon* **19(3)**: 329-335. [conservation of *Leopoldia*, cf. *Taxon* 21(4): 534, 1972].

GARBARI, F., GREUTER, W. & MICELI, P. 1979: The "Allium cupanii" group: a preliminary taxonomic, caryological and leaf anatomical study. - *Webbia* **34(1)**: 459-480.

GARBARI, F., MICELI, P. & MONTI, G. 1981: Numeri cromosomici per la Flora Italiana: 676-682. - *Inform. Bot. Ital.* **12(2)**: 107-111.

GARBARI, F. & MONTI, G. 1970: Secondo contributo su *Arisarum vulgare* (Araceae). - *Atti Soc. Tosc. Sci. Nat. Pisa Mem., Ser. B* **76**: 113-120.

GARBARI, F. & SENATORI, E. 1975: Il genere *Allium* L. in Italia. VI. Contributo alla citosistematica di alcune specie. - *Atti Soc. Tosc. Sci. Nat. Pisa Mem., Ser. B* **82**: 1-23.

GARBARI, F. & TORNADORE, N. 1970a: The genus *Ornithogalum* L. (Liliaceae). I. *Ornithogalum kochii* Parl.: morphological and caryotypical analysis. - *Atti Soc. Tosc. Sci. Nat. Pisa Mem., Ser. B* **77**: 101-111.

GARBARI, F. & TORNADORE, N. 1970b: Numeri cromosomici per la Flora Italiana: 1-12. - *Inform. Bot. Ital.* **2(2)**: 74-82.

GARBARI, F. & TORNADORE, N. 1970c: On the genus "Leopoldia" Parl. in Italy. - *Giorn. Bot. Ital., n.s.* **103(6)**: 613-614.

GARBARI, F. & TORNADORE, N. 1971: Numeri cromosomici per la Flora Italiana: 83-85. - *Inform. Bot. Ital.* **3(2)**: 153-154.

GARBARI, F. & TORNADORE, N. 1972a: Numeri cromosomici per la Flora Italiana: 108-123. - *Inform. Bot. Ital.* **4(1)**: 60-66.

GARBARI, F. & TORNADORE, N. 1972b: The genus *Ornithogalum* L. (Liliaceae). II. Taxonomy of some Italian entities. - *Giorn. Bot. Ital., n.s.* **106(5)**: 285.

GARBARI, F., TORNADORE, N. & PECORI, E. 1973: Numeri cromosomici per la Flora Italiana: 151-161. - *Inform. Bot. Ital.* **5(2)**: 161-169.

GARBARI, F. (ED.) 2002: Chromosome numbers for the Italian flora. Published on-line: http://www.dsb. unipi.it/chrobase/.

GARBER, E.D. 1950: Cytotaxonomic studies in the genus *Sorghum*. - *Univ. Calif. Publ. Bot.* **23(6)**: 283-361.

GARCÍA HERRÁN, M.C. 1991: Rapports (28-29). In Kamari, G., Felber, F. & Garbari, F. (eds): Mediterranean chromosome number reports - 1. - *Fl. Medit.* **1**: 237-238.

GARCÍA HERRÁN, M.C. 1994: Reports (367-370). In Kamari, G., Felber, F. & Garbari, F. (eds): Mediterranean chromosome number reports - 4. - *Fl. Medit.* **4**: 280-284.

GARCÍA JACAS, N. & SERNA, A.S. DE LA 1992: Karyological notes on *Centaurea* sect. *Acrocentron* (Asteraceae). - *Pl. Syst. Evol.* **179(1-2)**: 1-18.

GARCÍA JACAS, N., SUSANNA, A. & İLARSLAN, R. 1996: Aneuploidy in the Centaureinae (Compositae): is n=7 the end of the series? - *Taxon* **45(1)**: 39-42.

GARCÍA MARTÍN, F. & SILVESTRE, S. 1985: Números cromosómicos para la flora española. 409-421. - *Lagascalia* **13(2)**: 313-317.

GARCÍA MARTÍN, F. & SILVESTRE, S. 1985: Revisión de los géneros *Elaeoselinum* Koch ex

DC., *Margotia* Boiss. y *Distichoselinum* García Martín & Silvestre (Umbelliferae). - *Lagascalia* **13(2)**: 205-237.
GARCIA, C., PASTOR, J. & LUQUE, T. 1989: Contribución al estudio cariológico del género *Rumex* (Polygonaceae). - *Acta Bot. Malac.* **14**: 129-140.
GARCÍA, I. 1977: Números cromosómicos para la flora española. 15-17. - *Lagascalia* **7(2)**: 199-201.
GARCÍA, I. & VALDÉS, B. 1981: Números cromosómicos para la flora española. 225-239. – *Lagascalia* **10(2)**: 241-247.
GARCIA, J.G. 1942: Contribuïção para o estudo cário-sistemático do género *Lavandula* L. - *Bol. Soc. Brot., Ser. 2* **16**: 183-193.
GARCIA, M.A. & CAFFERTY, S. 2005: Revised lectotypification of *Cuscuta epithymum* (L.) L. (Convolvulaceae). - *Taxon* **54(2)**: 477-478.
GARCÍA, M.A. 1998: *Cuscuta rausii* (Convolvulaceae), a new species from Greece. - *Ann. Bot. Fenn.* **35(3)**: 171-174.
GARCÍA, M.A. & CASTROVIEJO, S. 2003: Estudios citotaxonómicos de las especies ibéricas del género *Cuscuta* (Convolvulaceae). - *Anales Jard. Bot. Madrid* **60(1)**: 33-44.
GARDÉ, A. 1949: Estudo citológico de algunas espécies do género *Trigonella* L. - *Genét. Ibér.* **1(1-2)**: 69-83.
GARDÉ, A. & MALHEIROS-GARDÉ, N. 1949: Contribuição para o estudo cariológico da família Umbelliferae. I. - *Agron. Lusit.* **11(2)**: 91-140.
GARDÉ, A. & MALHEIROS-GARDÉ, N. 1952: Chromosome number in *Luzula multiflora* Lej. - *Genét. Ibér.* **4**: 91-94.
GARDÉ, A. & MALHEIROS-GARDÉ, N. 1953: Contribuição para o estudo cariológico da família Umbelliferae. III. - *Brotéria, Ci. Nat.* **23(1)**: 5-35.
GARDOU, C. 1975: Quelques vues synthétiques sur les Centaurées de la section *Acrocentron* (Cass.) O. Hoffm. dans la flore méditerranéenne. In Guinochet, M. & al. (eds): La flore du Bassin Méditerranéen, pp. 537-547. - Paris: C.N.R.S.
GARDOU, C. & TCHEHREHGOCHA, I. 1975: [Chromosome numbers in *Centaurea* from Turkey]. In Löve, Á. (ed.): IOPB chromosome number reports XLIX. - *Taxon* **24(4)**: 502.
GARFI, G. 1997: On the flowering of *Zelkova sicula* (Ulmaceae): additional description and comments. - *Pl. Biosyst.* **131(2)**: 137-142.
GARLAND, M.A. 2021: (2821) Proposal to conserve *Ailanthus*, nom. cons., (Simaroubaceae) as being of feminine gender. - *Taxon* **70(3)**: 681-682.
GARNATJE, T. & AL. 2004: Contribution to the karyological knowledge of *Echinops* (Asteraceae, Cardueae) and related genera. - *Bot. J. Linn. Soc.* **145(3)**: 337-344.
GARNWEIDNER, E. 1986: Florenliste der Exkursion der Bayerischen Botanischen Gesellschaft 1983 nach Griechenland. - *Ber. Bayer. Bot. Ges.* **57**: 121-136.
GARNWEIDNER, E. 1989: Florenliste der Exkursion der Bayerischen Botanischen Gesellschaft 1985 nach Kreta. - *Ber. Bayer. Bot. Ges.* **60**: 157-168.
GARNWEIDNER, E. 1992: Florenliste der Exkursion der Bayerischen Botanischen Gesellschaft 1988 nach Rhodos. - *Ber. Bayer. Bot. Ges.* **63**: 155-174.
GARTMANN, F. 1988: Habitat-related differences between the vicarious fern species *Gymnocarpium dryopteris* and *G. robertianum*. - *Ann. Bot. Fenn.* **25(3)**: 261-274.
GARTNER, H. 1939: Zur systematischen Anordnung einiger Arten der Gattung *Cerastium* L. - *Repert. Spec. Nov. Regni Veg. Beih.* **113**: 1-96 + 19 plates and 7 maps.
GARVE, E. 1984: Botanischer Exkursionsbericht von Rhodos, 13.-27.4. 1984. [Unpublished flora list]. - Photocopy: 14 pp.
GATES, R.R. 1950: Genetics and taxonomy of the cultivated Brassicas and their wild relatives. - *Bull. Torrey Bot. Club* **77(1)**: 19-28.
GATES, R.R. 1958: Taxonomy and genetics of *Oenothera*. Forty years study in the cytology and evolution of the Onagraceae. - Den Haag: Dr. W. Junk, 115 pp.
GAUSSEN, H. 1960: Les gymnospermes actuelles et fossiles. Fasc. 6: Généralités, Genre *Pinus*. - Toulouse: Faculté des Sciences, 272 pp. + 2 tables and 6 plates.
GAUSSEN, H. 1964: Les gymnospermes actuelles et fossiles. Fasc. 7: *Pinus* (suite), *Cedrus* et *Abies*. - Toulouse: Faculté des Sciences, 208 pp. + 8 plates.
GAUSSEN, H. 1966: Les gymnospermes actuelles et fossiles. Fasc. 8: Genres *Pseudolarix*, *Keteleeria*, *Larix*, *Pseudotsuga*, *Pitiytes*, *Picea*, *Cathaya*, *Tsuga*. - Toulouse: Faculté des Sciences, 192 pp. + 6 plates.
GAUSSEN, H. 1968: Les gymnospermes actuelles et fossiles. Fasc. 10: Les Cupressacées. - Toulouse: Faculté des Sciences, 326 pp. + 12 plates.
GAY, J. 1831: Nouvelles espèces de *Crocus*. - *Bull. Sci. Nat. Géol.* **31**: 319-321.
GÁYER, G. 1908: Vorarbeiten zu einer Monographie der europäischen *Aconitum*-Arten. I. - *Magyar Bot. Lapok* **8(5-12)**: 114-206, 310-327.
GÁYER, G. 1930: Nachträge und Berichtigungen zur Bearbeitung der Gattung *Aconitum* in der Ascherson-Graebnerschen Synopsis. - *Magyar Bot. Lapok* **29(1-12)**: 39-48.
GAZER, M. 1993: Revision of *Astragalus* L. sect. *Sesamei* DC. (Leguminosae). - *Sendtnera* **1**: 69-155.
GÉHU, J.-M. 1989: Les salicornes annuelles d'Europe: système taxonomique et essai de clé de détermination. - *Colloq. Phytosoc.* **18**: 227-241.
GÉHU, J.-M. & AL. 1986: Données synsystematiques et synchorologiques sur la végétation du littoral sedimentaire de la Grèce continentale. - *Doc. Phytosoc. n.s.* **10**: 43-92.
GÉHU, J.-M. & AL. 1987: Données sur la végétation littorale de la Crète (Grèce). - *Ecol. Medit.* **13(1-2)**: 93-105.
GÉHU, J.-M. & AL. 1992: Premières données sur la végétation littorale des îles de Rhodos et Karpathos. - *Colloq. Phytosoc.* **19**: 545-582.
GELDEREN, D.M. VAN, JONG, P.C. DE & OTERDOOM, H.J. 1994: Maples of the World. - Portland, Oregon: Timber Press, 458 pp.
GELMI, E. 1889: Contribuzione alla flora dell'isola Corfu. - *Nuovo Giorn. Bot. Ital.* **21(3)**: 446-454.
GELTMAN, D.V. 1988: Genus *Urtica* L. (Urticaceae) in URSS [In Russian]. - *Novosti Sist. Vysš. Rast.* **25**: 68-80.
GEMBARDT, C. & GÖLZ, P. 2002: Nachweis von *Nigritella* in Griechenland. - *J. Eur. Orch.* **34(4)**: 803-806.
GEMEINHOLZER, B. & BACHMANN, K. 2005: Examining morphological and molecular diagnostic character states of *Cichorium intybus* L. (Asteraceae) and *C. spinosum* L. - *Pl. Syst. Evol.* **253(1)**: 105-123.
GENOVA, E.M. & AL. 1992: On the distribution and stocks of *Origanum heracleoticum* L. in Bulgaria. [In Bulgarian with English summary]. - *Fitologiya* **42**: 66-72.
GENTILE, S. & GASTALDO, P. 1976: *Quercus calliprinos* Webb e *Quercus coccifera* L.: Ricerche sull'anatomia fogliare e valutazioni tassonomiche e corologiche. - *Giorn. Bot. Ital., n.s.* **110(1-2)**: 89-115.
GENTILI, R. & AL. 2017: Distribution map of *Ambrosia artemisiifolia* L. (Asteraceae) in Italy. - *Pl. Biosyst.* **151(3)**: 381-386.
GENTSCHEV, G. 1937: Zytologische und embryologische Studien über einige *Hieracium*-Arten. - *Planta* **27(2)**: 165-195.
GEORGAKOPOULOU, A. & AL. 2002: Biosystematic data (cytology - breeding systems) of three *Limonium* species (Plumbaginaceae). In Kamari, G. & al. (eds): Ellinini Votaniki Eteria. 9° Panelliniko Epistimoniko Sinedrio. Praktika, pp. 79-82 [In Greek with English abstract].
GEORGAKOPOULOU, A. & AL. 2006: Breeding systems and cytology in Greek populations of five *Limonium* species (Plumbaginaceae). - *Willdenowia* **36(2)**: 741-750.
GEORGESCU, C.C. 1936: Neue Beiträge zur Systematik der Schwarzföhre. - *Repert. Spec. Nov. Regni Veg.* **41**: 181-187.
GEORGHIOU, K. & AL. 1996: A database for the endemic, rare and threatened plants of Greece [In Greek with English summary]. In: Proceedings 6th sci. conf. Hellenic Botanical Society, pp. 232-235. - Paralimni, Cyprus.
GEORGHIOU, K. & AL. 2008: Conservation biology of the threatened plant *Bupleurum capillare* Boiss. & Heldr. In: Praktika 10ou Sinedriou Ellinikis Votanikis Eterias, 5-8 maiou 2005, Ioannina: pp. 109-116 [In Greek with English abstract].

GEORGHIOU, K. & DELIPETROU, P. 2010: Pattern and traits of the endemic plants of Greece. - *Bot. J. Linn. Soc.* **162(2)**: 130-422.

GEORGIADIS, CH. & YANNITSAROS, A. 2001: New adventive species of the flora of Cyprus. - Second series. In Kamari, Psaras & Constantinidis (eds): Proceedings of the 8th scientific congress of the Greek Botanical Society, October 5-8, 2000, pp. 112-114 [In Greek with English abstract].

GEORGIADIS, TH. 1979: *Centaurea musakii*: a new species from Thessalia (Greece). - *Bot. Not.* **132 (3)**: 311-312.

GEORGIADIS, TH. 1980: Contribution à l'étude phylogénétique du genre *Centaurea* L. (Sectio *Acrolophus* [Cass.] DC.) en Grèce. - Marseille: Thèse, Université de Provence, 286 pp.

GEORGIADIS, TH. 1981a: A propos de *Centaurea ipsaria* Stoj. & Kit. et de *C. thasia* Hayek (sectio *Acrolophus* (Cass.) D.C.) de l'île de Thasos (Grèce). - *Bot. Chron. (Patras)* **1(1)**: 14-17.

GEORGIADIS, TH. 1981b: Problèmes de différenciation et d'introgression dans *Centaurea* subg. *Acrolophus* (Compositae) en Grèce. - *Bot. Jahrb. Syst.* **102(1-4)**: 321-337.

GEORGIADIS, TH. 1983a: Contribution à l'étude cytogéographique du genre *Centaurea* L. (section *Acrolophus* (Cass.) DC.) en Grèce. - *Candollea* **38(1)**: 325-340.

GEORGIADIS, TH. 1983b: Contribution à l'étude de la flore et de la végétation de l'île d'Othoni. - *Candollea* **38(2)**: 503-539.

GEORGIADIS, TH. 1985a: Contribution à l'étude de la flore de l'île Erikoussa (Grèce). - *Willdenowia* **15(1)**: 203-210.

GEORGIADIS, TH. 1985b: Contribution à l'étude palynologique du genre *Centaurea* L. (section *Acrolophus* Cass.) DC. en Grèce. - *Pollen & Spores* **27(1)**: 29-51.

GEORGIADIS, TH. 1999: Notes on *Pycnocomon rutifolium* and *Nothoscordum gracile*. In Greuter, W. & Raus, Th. (eds): Med-Checklist notulae, 18. - *Willdenowia* **29(1-2)**: 54-67.

GEORGIADIS, TH. & AL. 1994: Ikologikes eppitosis ke diachirisi stous igrotopous Kalama - Acheronta - Kalodikiou [In Greek]. - Patra: Tmima Viologias, Panepistimiou Patron, vi + iv + 463 pp. + maps & plates.

GEORGIADIS, TH. & CHATZIKYRIAKOU, G. 1993: *Centaurea akamantis* (Compositae), a new species from Cyprus. - *Willdenowia* **23(1-2)**: 157-162.

GEORGIADIS, TH. & CHRISTODOULAKIS, D. 1984a: Contribution à l'étude cytogéographique des Centaurées de l'île de Samos. - *Candollea* **39(1)**: 307-318.

GEORGIADIS, TH. & CHRISTODOULAKIS, D. 1984b: To dasos Strofilias (N. Achaïas - Ilias) [In Greek]. - *Geôtechnica* **3**: 25-35.

GEORGIADIS, TH., DIMITRELLOS, G. & ROUTSI, E. 1996: *Centaurea messenicolasiana* (Asteraceae), a new species of *C.* sect. *Phalolepis* (Cass.) DC. from Greece. - *Willdenowia* **25(2)**: 561-569.

GEORGIADIS, TH. & DIMOPOULOS, P. 1993: Etude de la végétation supraforestière du Mont Kyllini (Péloponnèse-Grèce). - *Bot. Helv.* **103(2)**: 149-175.

GEORGIADIS, TH., DIMOPOULOS, P. & DIMITRELLOS, G. 1995: Comparative floristic research of the wetlands Acheron-Kalodiki (W Greece). In: Proc., 5th sci. conf., Hellenic Bot. Soc., Delphi 21-23 October 1994, pp. 299-303 [In Greek with English summary]. - Thessaloniki: Univ. Studio Press.

GEORGIADIS, TH., DIMOPOULOS, P. & DIMITRELLOS, G. 1997: The flora and vegetation of the Acheron delta (W Greece) aiming at nature conservation. - *Phyton (Horn)* **37(1)**: 31-60.

GEORGIADIS, TH., ECONOMIDOU, E. & CHRISTODOULAKIS, D. 1990: Flora and vegetation of the Strofilia coastal area (NW Peloponnisos - Greece). - *Phyton (Horn)* **30(1)**: 15-36.

GEORGIADIS, TH., IATROÚ, G. & GEORGIOU, O. 1986: Contribution à l'étude de la flore et de la végétation de l'île de Paxi, Grèce. - *Willdenowia* **15(2)**: 567-602.

GEORGIADIS, TH. & PHITOS, D. 1976: Contribution à l'étude cytotaxonomique du genre *Centaurea* L. (Sectio *Acrolophus* (Cass.) DC.) en Grèce. - *Biol. Écol. Médit.* **3(1)**: 13-16.

GEORGIADIS, TH. & PHITOS, D. 1977: Deux *Centaurea* nouvelles de la section *Phalolepis* (Cass.) D. C. de la Grèce. - *Biol. Écol. Médit.* **4(1)**: 3-9.

GEORGIADIS, TH. & PHITOS, D. 1978: Une *Centaurea* nouvelle du mont Olympe (Grèce) de la section *Phalolepis* (Cass.) DC. *Centaurea litochorea* sp. nov. - *Biol. Écol. Médit.* **5(1)**: 31-34.

GEORGIADIS, TH. & STRID, A. 1999: Notes on *Flaveria bidentis*. In Greuter, W. & Raus, Th. (eds): Med- Checklist notulae, 18. - *Willdenowia* **29(1-2)**: 53.

GEORGIADIS, TH. & YANNITSAROS, A. 1995: New adventive species of the flora of Cyprus. In: Praktika 16ou Sinedriou E.E.V.E., Volos, 5.-7.5. 1994, 2, pp. 1-3 [In Greek with English summary].

GEORGIADOU TVRTKOVIC-SAHIN, E.E. 1982: The systematics of the genus *Helichrysum* Miller in the countries bordering the Mediterranean. - Reading: Ph.D. thesis, Univ. of Reading, 1062 pp.

GEORGIEV, T. 1926: Zwei *Crepis*-Arten aus Bulgarien. - *Izv. Bulg. Bot. Druzh.* **1**: 67-68.

GEORGIEV, T. 1932: Additamenta ad Hieracii bulgarici. - *Izv. Bulg. Bot. Druzh.* **5**: 94-97.

GEORGIEV, T. 1935: Bemerkungen über die Sistematik und Verbreitung Reihen *alpine* und *ervensia* von der *Cerestium* [sic!] [In Bulgarian with German summary]. - *God. Sofiisk. Univ. Agron.-Lesov. Fak., 1* **13**: 402-432.

GEORGIEV, T. 1961: Ergänzung der Flora Bulgariens durch die Gattung *Hieracium* L. [In Bulgarian with Russian and German summaries]. - *Izv. Bot. Inst. (Sofia)* **8**: 169-174.

GEORGIEV, V. 1997: New chorological data about two species of genus *Hypericum* L. from Bulgaria. - *God. Sofiisk. Univ. "Kliment. Ohridski" Biol. Fak., 2 Bot.* **89**: 11-12.

GEORGIEVA, E. & RUSAKOVA, V. 2001: *Gentiana lutea* L. and *Gentiana punctata* L. in the vegetation of Rila and Stara Planina mountains (Bulgaria). In Özhatay, N. (ed.): Proceedings of the Second Balkan Botanical Congress, 1: 239-244. - Istanbul.

GEORGIEVA, E., RUSAKOVA, V. & EVSTATIEVA, L. 2000: *Gentiana lutea* L. and *Gentiana punctata* L. in the vegetation of Rila and Stara Planina mountains (Bulgaria). In Özhatay, N. & al. (eds): Abstracts from the Second Balkan Botanical Congress, Istanbul, Turkey, May 14-18, 2000, p. 38.

GEORGIEVA, U. 2000: *Romulea linaresii* Parl.: a new species for Mt Strandzha. - *Phytol. Balcan.* **6(2-3)**: 177-178.

GEORGIEVSKA, Z.L. 1978: Contribution to the morphological variability of *Quercus coccifera* L. in Macedonia [In Macedonian with English summary]. - *Fragm. Balcan.* **10(16)**: 139-147.

GEORGIEVSKA, Z.L. 1981: Second contribution to get acquinted with, morfological [sic!] variability of *Q. coccifera* L. in Macedonia [In Macedonian with English summary]. - *Fragm. Balcan.* **11(9)**: 75-83.

GEORGIOU, CH. & AL. 2010: Chemical composition of *Echinophora tenuifolia* subsp. *sibthorpiana* essential oil from Greece. - *Rec. Nat. Prodr.* **4(3)**: 167-170.

GEORGIOU, O. 1988: The flora of Kerkira (Ionian Islands, Greece). A contribution. - *Willdenowia* **17(1-2)**: 87-101.

GEORGIOU, O. 1989: *Anthemis werneri* Stoj. & Acht., an endemic species of the Aegean islands (Greece). In Phitos, D. (ed.): Abstracts. Sixth OPTIMA meeting, Delphi 10-16 Sept. 1989, p. 98.

GEORGIOU, O. 1990: Viosistimatiki meleti tis omadas *Anthemis tomentosa* (Asteraceae) stin Ellada [In Greek]. - Patras: Ph.D. thesis, Univ. of Patras, 299 pp.

GEORGIOU, O. 1991: *Anthemis werneri* (Asteraceae), an endemic species of the Aegean islands. - *Bot. Chron. (Patras)* **10**: 741-747.

GEORGIOU, O. 1997: Taxonomic notes on *Anthemis peregrina* (Asteraceae). - *Fl. Medit.* **7**: 101-106.

GEORGIOU, O. 2000: *Malcolmia nana* (Cruciferae) in Greece. - *Bot. Chron. (Patras)* **13**: 369-374.

GEORGIOU, O., DIMITRELLOS, G. & GEORGIADIS, TH. 1997: The species *Moltkia petraea* (Tratt.) Griseb. (Boraginaceae) in Greece. In: Abstracts from the First Balkan Botanical Congress, Thessaloniki, September 19-22, 1997, p. 16. - Thessaloniki.

GEORGIOU, O., DIMITRELLOS, G. & GEORGIADIS, TH. 2000: *Moltkia petraea* (Boraginaceae) in Greece. - *Phyton (Horn)* **40(1)**: 57-69.

GEORGIOU, O., PANITSA, M. & TZANOUDAKIS, D. 2006: *Anthemis scopulorum* (Asteraceae), an "islet specialist" endemic to the Aegean islands (Greece). - *Willdenowia* **36(1)**: 339-349.

GERACI, A. & SCHICCHI, R. 2003: Cytogeographical investigation of *Scilla autumnalis* (Hyacinthaceae) in Sicily. - *Fl. Medit.* **12**: 177-182.
GERAKARIS, V. & AL. 2017: Effectiveness of *Posidonia oceanica* biotic indices for assessing the ecological status of coastal waters in the Saronikos Gulf (Aegean Sea, Eastern Mediterranean). - *Medit. Marine Sci.* **18(1)**: 161-178.
GERAKIS, A., HALDAS, X. & GIANNAKOULIAS, M. 2016: *Anacamptis palustris* subsp. *robusta* (Orchidaceae): a new record to the flora of Cephalonia, Greece. - *J. Eur. Orch.* **48(1)**: 11-18.
GERAKIS, P.A. 1995: Preservation and study of the Greek wetlands. - *Ann. Mus. Goulandris* **9**: 521-529.
GERASIMIDIS, A., FOTIADIS, G. & PANAJIOTIDIS, S. 2009: Contribution to the phytosociological study of low-altitude *Buxus sempervirens* (Buxaceae) formations (Mt Olympos, Greece). In Ivanova, D. (ed.): Proceedings of the IV Balkan botanical congress, pp. 296-302. - Sofia: Institute of Botany, BAS.
GERASIMIDIS, A. & KORAKIS, G. 2009: Contribution to the study of the flora of Mount Mitsikeli, NW Greece. - *Fl. Medit.* **19**: 161-184.
GERASIMIDIS, A., KORAKIS, G. & MINOGLOU, D. 2008: Contribution to the study of the flora of Mt. Mitsikeli. In: Praktika 10ou Sinedriou Ellinikis Votanikis Eterias, 5-8 maiou 2005, Ioannina: pp. 87-96 [In Greek with English abstract].
GERBAUD, O. 1996: *Ophrys* ×*martineae* hyb. nat. nov. (= *O. basilissa* × *O. omegaifera*): Un nouvel hybride du groupe d'*Ophrys omegaifera* decouvert en Crète. - *Rhône-Alpes-Orchid.* **18**: 41-45.
GERBAULT, E.-L. 1920: Sur la constitution du phénotype *Ranunculus sardous*. - *Bull. Soc. Bot. France* **67**: 266-276 + 4 plates.
GERMAN, D.A. & AL-SHEHBAZ, I.A. 2020: (2776) Proposal to reject the name *Lunariaperennis* Mill. (Cruciferae). - *Taxon* **69(5)**: 1120-1121.
GERSTBERGER, P. 1978: Zur Unterscheidung von *Fragaria viridis* Duchesne und *Fragaria vesca* L. im vegetativen Zustand. - *Göttinger Florist. Rundbr.* **12(4)**: 93-97.
GERSTBERGER, P. 1980: Neue Merkmale zur Unterscheidung von *Sorbus domestica* L. und *Sorbus aucuparia* L. - *Göttinger Florist. Rundbr.* **14(1)**: 1-5.
GERVAIS, C. 1965: Nombres chromosomiques chez quelques graminées des Alpes (Note préliminaire). - *Bull. Soc. Neuchâteloise Sci. Nat.* **88**: 61-64.
GERVAIS, C. 1966: Nombres chromosomiques chez quelques graminées alpines. - *Bull. Soc. Neuchâteloise Sci. Nat.* **89**: 87-99 + 1 plate.
GERVAIS, C. 1968: Notes de cytotaxinomie sur quelques *Avena* vivaces. - *Bull. Soc. Neuchâteloise Sci. Nat.* **91**: 105-117.
GERVAIS, C. 1972: Nouvelles déterminations de nombres chromosomiques chez les Avoines vivaces. I. - *Bull. Soc. Neuchâteloise Sci. Nat.* **95**: 57-61.
GERVAIS, C. 1973a: Contribution à l'étude cytologique et taxonomique des avoines vivaces (g. *Helictotrichon* Bess. et *Avenochloa* Holub). - *Denkschr. Schweiz. Naturf. Ges.* **88**: 166 pp.
GERVAIS, C. 1973b: Nouvelles déterminations de nombres chromosomiques chez les Avoines vivaces. II. - *Bull. Soc. Neuchâteloise Sci. Nat.* **96**: 81-87.
GERVAIS, C. 1983: Wide hybridization attempts in the tribe Aveneae Nees. - *Bot. Helv.* **93(2)**: 195-212.
GESLOT, A. 1980: Biométrie des graines et nombres chromosomiques dans la sous-section *Heterophylla* du genre *Campanula* (Campanulaceae). - *Pl. Syst. Evol.* **134(3-4)**: 193-206.
GESLOT, A. 1982: Les campanules de la sous-section *Heterophylla* (Wit.) Fed. dans les Pyrenées: une étude de biosystématique. - Marseille: Thèse, Univ. Aix-Marseille, 278 pp. + 3 plates & 125 tables (2 vols.).
GEWEHR, S., BABALONAS, D. & JÜRGENS, N. 1997: The distribution of aquatic macrophytes in the Aliakmon river (NW-Greece). In: Abstracts from the First Balkan Botanical Congress, Thessaloniki, September 19-22, 1997, p. 12. - Thessaloniki.
GHAHREMANINEJAD, F. 2005: Proposal to transfer *Astracantha dolinicola* (Fabaceae) to *Astragalus*. - *Nordic J. Bot.* **23(6)**: 711-712.
GHAZANFAR, S.A. 1983: Cytological studies in the genus *Silene* L. - *New Phytol.* **93(1)**: 123-127.
GHAZANFAR, S.A. 1990: Hybridization studies in the genus *Silene* sectt. *Siphonomorpha* and *Auriculatae* (Caryophyllaceae). - *Pl. Syst. Evol.* **168(3-4)**: 143-148.
GHEBREHIWET, M. 2001: Taxonomy, phylogeny and biogeography of *Kickxia* and *Nanorrhinum* (Scrophulariaceae). - *Nordic J. Bot.* **20(6)**: 655-689.
GHOSH, A., MALLICK, T. & NASKAR, S. 2018: Lectotypification of *Cyperus pygmaeus* (Cyperaceae) and notes on its distinctness with *C. michelianus*. - *Phytotaxa* **376(2)**: 123-125.
GIACOBBE, A. 1939: Ricerche geografiche et ecologiche sul *Laurus nobilis* L. - *Arch. Bot. (Forlì)* **15**: 33-82 + 2 plates.
GIANNAKIS, TH. & AL. 2016: *Ambrosia psilostachya*. In Raab-Straube, E. v. & Raus, Th. (eds): Euro+Med-Checklist notulae, 6. - *Willdenowia* **46(3)**: 426.
GIANNAKIS, TH. & AL. 2021: *Epilobium alpestre*. In Raus, Th. & Raab-Straube, E.v. (eds): Euro+Med-Checklist Notulae, 14. - *Willdenowia* **51(3)**: 363.
GIANNAKOS, L. 2007: Chloridiki pikilotita ton Oreon Valtou [In Greek]. - Athens: M.Sc. thesis, Department of Systematic Botany, Agricultural University, 203 pp.
GIANNOPOLITIS, C.N. 1998: Notes on *Ipomoea hederacea*. In Greuter, W. & Raus, Th. (eds): Med- Checklist notulae, 17. - *Willdenowia* **28(1-2)**: 163-174.
GIANNOPOLITIS, C.N. 2003: Family Cucurbitaceae. In 'An Identification Guide for the Weeds of Greece, part 1'. - *Georgia – Ktinotrofia* **9**: 158-160.
GIANNOPOULOS, K. & TAN, KIT 2011: Reports 49-51. In Vladimirov, V. & al. (eds): New floristic records in the Balkans: 15. - *Phytol. Balcan.* **17(1)**: 136-138.
GIANNOPOULOS, K. & TAN, KIT 2021a: Report 41. In Vladimirov, V., Aybeke, M. & Tan, Kit (eds): New floristic records in the Balkans: 44. - *Phytol. Balcan.* **27(1)**: 134-135.
GIANNOPOULOS, K. & TAN, KIT 2021b: Contributions to the bulb flora of Ilias (NW Peloponnisos, Greece): Asphodelaceae, Colchicaceae, Fumariaceae and Geraniaceae. - *Phytol. Balcan.* **27(2)**: 231-237.
GIANNOPOULOS, K., TAN, KIT & VOLD, G. 2011a: Reports 20-61. In Vladimirov, V. & Tan, Kit (eds): New floristic records in the Balkans: 16. - *Phytol. Balcan.* **17(2)**: 250-255.
GIANNOPOULOS, K., TAN, KIT & VOLD, G. 2011b: Reports 25-69. In Vladimirov, V., Dane, F. & Tan, Kit (eds): New floristic records in the Balkans: 17. - *Phytol. Balcan.* **17(3)**: 365-370.
GIANNOPOULOS, K., TAN, KIT & VOLD, G. 2012a: Reports 78-98. In Vladimirov, V. & al. (eds): New floristic reports in the Balkans: 18. - *Phytol. Balcan.* **18(1)**: 78-81.
GIANNOPOULOS, K., TAN, KIT & VOLD, G. 2012b: Reports 89-117. In Vladimirov, V., Dane, F. & Tan, Kit (eds): New floristic records in the Balkans: 20. - *Phytol. Balcan.* **18(3)**: 348-351.
GIANNOPOULOS, K., TAN, KIT & VOLD, G. 2013: Report 30. In Vladimirov, V., Dane, F. & Tan, Kit (eds): New floristic records in the Balkans: 23. - *Phytol. Balcan.* **19(3)**: 378-379.
GIANNOPOULOS, K., TAN, KIT & VOLD, G. 2014a: Reports 56-82. In Vladimirov, V. & Tan, Kit (eds): New floristic records in the Balkans: 24. - *Phytol. Balcan.* **20(1)**: 108-111.
GIANNOPOULOS, K., TAN, KIT & VOLD, G. 2014b: Reports 121-142. In Vladimirov, V. & al. (eds): New floristic records in the Balkans: 25. - *Phytol. Balcan.* **20(2-3)**: 197-200.
GIANNOPOULOS, K., TAN, KIT & VOLD, G. 2015a: Reports 191-200. In Vladimirov, V., Dane, F. & Tan, Kit (eds): New floristic records in the Balkans: 26. - *Phytol. Balcan.* **21(1)**: 73-74.
GIANNOPOULOS, K., TAN, KIT & VOLD, G. 2015b: Reports 55-68. In Vladimirov, V. & al. (eds): New floristic records in the Balkans: 27. - *Phytol. Balcan.* **21(2)**: 197-200.
GIANNOPOULOS, K., TAN, KIT & VOLD, G. 2015c: Reports 102-108. In Vladimirov, V., Dane, F. & Tan, Kit (eds): New floristic records in the Balkans: 28. - *Phytol. Balcan.* **21(3)**: 378-379.
GIANNOPOULOS, K., TAN, KIT & VOLD, G. 2016a: Reports 556-60. In Vladimirov, V. & al. (eds): New floristic records in the Balkans: 29. - *Phytol. Balcan.* **22(1)**: 99-101.
GIANNOPOULOS, K., TAN, KIT & VOLD, G. 2016b: Reports 59-75. In Vladimirov, V. & al. (eds): New floristic records in the Balkans: 30. - *Phytol. Balcan.* **22(2)**: 267-269.
GIANNOPOULOS, K., TAN, KIT & VOLD, G. 2017a: Reports 33-40. In Vladimirov, V., Aybeke, M. & Tan, Kit (eds): New floristic records in the Balkans: 32. - *Phytol. Balcan.* **23(1)**: 125-126.
GIANNOPOULOS, K., TAN, KIT & VOLD, G. 2017b: Reports 77-118. In Vladimirov, V. & al. (eds): New floristic records in the Balkans: 33. - *Phytol. Balcan.* **23(2)**: 291-295.

GIANNOPOULOS, K., TAN, KIT & VOLD, G. 2017c: Reports 94-97. In Vladimirov, V. & al. (eds): New floristic records in the Balkans: 34. - *Phytol. Balcan.* **23(3)**: 421-422.

GIANNOPOULOS, K., TAN, KIT & VOLD, G. 2018a: Reports 28-102. In Vladimirov, V., Aybeke, M. & Tan, Kit (eds): New floristic records in the Balkans: 36. - *Phytol. Balcan.* **24(2)**: 267-272.

GIANNOPOULOS, K., TAN, KIT & VOLD, G. 2018b: Reports 124-129. In Vladimirov, V., Aybeke, M. & Tan, Kit (eds): New floristic records in the Balkans: 37. - *Phytol. Balcan.* **24(3)**: 410-413.

GIANNOPOULOS, K., TAN, KIT & VOLD, G. 2018c: Reports 15-28. In Vladimirov, V., Aybeke, M. & Tan, Kit (eds): New floristic records in the Balkans: 35. - *Phytol. Balcan.* **24(1)**: 157-159.

GIANNOPOULOS, K., TAN, KIT & VOLD, G. 2019a: Reports 43-52. In Vladimirov, V. & al. (eds): New floristic records in the Balkans: 38. - *Phytol. Balcan.* **25(1)**: 102-105.

GIANNOPOULOS, K., TAN, KIT & VOLD, G. 2019b: Reports 25-36. In Vladimirov, V., Aybeke, M. & Tan, Kit (eds): New floristic records in the Balkans: 39. - *Phytol. Balcan.* **25(2)**: 207-211.

GIANNOPOULOS, K., TAN, KIT & VOLD, G. 2019c: Reports 78-84. In Vladimirov, V., Aybeke, M. & Tan, Kit (eds): New floristic records in the Balkans: 40. - *Phytol. Balcan.* **25(3)**: 305-306.

GIANNOPOULOS, K., TAN, KIT & VOLD, G. 2020: Reports 24-29. In Vladimirov, V., Aybeke, M. & Tan, Kit (eds): New floristic records in the Balkans: 42. - *Phytol. Balcan.* **26(2)**: 193-194.

GIANNOPOULOS, K., TAN, KIT & VOLD, G. 2021a: Contribution to the bulb flora of Ilias (NW Peloponnese, Greece): Alliaceae. - *Phytol. Balcan.* **27(1)**: 85-95.

GIANNOPOULOS, K., TAN, KIT & VOLD, G. 2021b: Contribution to the bulb flora of Ilias (NW Peloponnese, Greece): Amaryllidaceae, Araceae and Aristolochiaceae. - *Phytol. Balcan.* **27(1)**: 97-106.

GIANNOPOULOS, K., TAN, KIT & VOLD, G. 2021c: Reports 65-67. In Vladimirov, V., Aybeke, M. & Tan, Kit (eds): New floristic records in the Balkans: 45. - *Phytol. Balcan.* **27(2)**: 255-259.

GIARDINA, G., RAIMONDO, F.M. & SPADARO, V. 2007: A catalogue of plants growing in Sicily. - *Bocconea* **20**: 5-582.

GIBBS, P.E. 1966: A revision of the genus *Genista* L. - *Notes Roy. Bot. Gard. Edinburgh* **27(1)**: 11-99 + 2 plates.

GIBBS, P.E. 1967: A revision of the genus *Adenocarpus*. - *Bol. Soc. Brot., Ser. 2* **41**: 67-121.

GIBBS, P.E. 1968a: Taxonomy and distribution of the genus *Calicotome*. - *Notes Roy. Bot. Gard. Edinburgh* **28(3)**: 275-286.

GIBBS, P.E. 1968b: [Notes on *Chamaespartium*]. In Heywood, V.H. (ed.): Notulae systematicae ad Floram Europaeam spectantes, no. 7. - *Feddes Repert.* **79(1-2)**: 54.

GIBBS, P.E. 1971a: Studies of synthetic hybrids of British species of *Senecio*: I. *Senecio viscosus* L. × *S. vulgaris* L. - *Trans. Bot. Soc. Edinburgh* **41(2)**: 213-218 + 1 plate.

GIBBS, P.E. 1971b: Taxonomic studies on the genus *Echium*. I. An outline revision of the Spanish species. - *Lagascalia* **1**: 27-82.

GIBBS, P.E. & DINGWALL, I. 1971: A revision of the genus *Teline*. - *Bol. Soc. Brot., Ser. 2* **45**: 269-316.

GIBBY, M. & AL. 2002: Polyploidy, phylogeography and Pleistocene refugia of the rock fern *Asplenium ceterach*. In Koyuncu, M. (ed.): VIth Plant Life of Southwest Asia Symposium, Abstracts p. 25. - Van: Yüzüncü Yil University.

GIBERNAU, M., MACQUART, D. & PRZETAK, G. 2004: Pollination in *Arum* - a review. - *Aroideana* **27**: 148-166.

GIBEY, M. 1995: Two rare European lilies. - *Quart. Bull. Alpine Gard. Soc. Gr. Brit.* **63(4)**: 348-350.

GIBSON, D.J. & NEWMAN, J.A. 2001: Biological Flora of the British Isles. *Festuca arundinacea* Schreber (*F. elatior* L. subsp. *arundinacea* (Schreber) Hackel). - *J. Ecol.* **89(2)**: 304-324.

GIBSON, P.B. & AL. 1971: Interspecific hybridization of *Trifolium uniflorum* L. - *Crop Sci. (Madison)* **11**: 895-899.

GIBSON, P.B. & CHEN, C.-C. 1971: Reproduction and cytology of *Trifolium uniflorum*. - *Crop Sci. (Madison)* **11(1)**: 69-70.

GIL, J. & CUBERO, J.I. 1993: Multivariate analysis of the *Vicia sativa* L. aggregate. - *Bot. J. Linn. Soc.* **113(4)**: 389-400.

GILBERT, M.G. 1987: The taxonomic position of the genera *Telephium* and *Corrigiola*. - *Taxon* **36(1)**: 47-49.

GILBERT, O.L. 1970: Biological Flora of the British Isles. *Dryopteris villarii* (Bellardi) Woynar (*Asplenium rigidum* Sw.; *Lastrea rigida* (Sw.) C. Presl). - *J. Ecol.* **58(1)**: 301-313.

GILES, B.E. 1984: A comparison between quantitative and biochemical variation in the wild barley *Hordeum murinum*. - *Evolution* **38(1)**: 34-41.

GILES, B.E. & BOTHMER, R.V. 1985: The progenitor of barley (*Hordeum vulgare* ssp. *spontaneum*) – its importance as a gene resource. - *Sveriges Utsädesfören. Tidskr.* **95**: 53-61.

GILES, B.E. & LEFKOVITCH, L.P. 1986: A taxonomic investigation of the *Hordeum murinum* complex (Poaceae). - *Pl. Syst. Evol.* **153(3-4)**: 181-197.

GILL, G.S. & CARSTAIRS, S.A. 1988: Morphological, cytological and ecological discrimination of *Bromus rigidus* from *Bromus diandrus*. - *Weed Res.* **28(5)**: 399-405.

GILL, I.S. 1988: The taxonomy and distribution of *Biarum davisii* Turrill. - Reading: Honours project, 144 pp.

GILL, J.J.B. & AL. 1972: The distribution of chromosome races of *Ranunculus ficaria* L. in the British Isles. - *Ann. Bot. (Oxford), n.s.* **36(1)**: 31-47.

GILL, J.J.B. & CAINE, A. 1988: Cytogenetic studies in *Papaver* L.; the x=6 and x=7 diploids. - *Biol. J. Linn. Soc.* **35(3)**: 261-274.

GILLESPIE, D.J. & MCCOMB, J.A. 1991: Morphology and distribution of species in the *Medicago murex* complex. - *Canad. J. Bot.* **69(12)**: 2655-2662.

GILLESPIE, L.J. & AL. 2018: Phylogeny and taxonomic synopsis of *Poa* subgenus *Pseudopoa* (including *Eremopoa* and *Lindbergella*) (Poaceae, Poeae, Poinae). - *PhytoKeys* **111**: 69-101.

GILLI, A. 1966: Neue Fundorte von Orobanchaceen aus dem Mediterrangebiet. - *Oesterr. Bot. Z.* **113 (2)**: 211-216.

GILLI, A. 1971: Die Orobanchaceen der Türkei. - *Feddes Repert.* **82(6)**: 381-406.

GILLI, A. 1977: Winterblüher auf Rhodos. - *Phyton (Horn)* **18(1-2)**: 95-97.

GILOT-DELHALLE, J., DEGRAEVE, N. & MOUTSCHEN, J. 1976: Cytotaxonomic investigation of the genus *Nigella* (Helleboreae) with C-banding techniques. - *Caryologia* **29(2)**: 139-154.

GILOT, J. 1965: Contribution à l'étude cytotaxonomique des Genisteae et des Loteae. - *Cellule* **65(3)**: 317-347.

GINZBERGER, A. 1896: Über einige *Lathyrus*-Arten aus der Sektion *Eulathyrus* und ihre geographische Verbreitung. - *Sitzungsber. Kaiserl. Akad. Wiss., Math.-Naturwiss. Cl., Abt. 1* **105(1)**: 281-352 + 2 maps and 1 plate.

GINZBERGER, A. 1921: Zur Kenntnis des Formenkreises von *Phagnalon rupestre* (L.) DC. und *Phagnalon graecum* Boiss. et Heldr. - *Oesterr. Bot. Z.* **70(6-8)**: 197-204.

GINZBERGER, A. 1922: Zur Gliederung des Formenkreises von *Reichardia picroides* (L.) Roth. - *Oesterr. Bot. Z.* **71(4-6)**: 73-83.

GINZBERGER, A. 1932: Friedrich Vierhapper. Nachruf. - *Verh. Zool.-Bot. Ges. Wien* **82**: 28 pp. [Sonderabdruck].

GIORDANI, A. & GARBARI, F. 1990: The genus *Ornithogalum* L. (Hyacinthaceae). IX. Characters and typification of *O. kochii* Parl., a misinterpreted species. - *Atti Soc. Tosc. Sci. Nat. Pisa Mem., Ser. B* **96**: 1-8. [Also distributed as *OPTIMA leaflet* no. 198].

GIORDANI, A. & GARBARI, F. 1993: Characters and typification of a misinterpreted entity: *Ornithogalum kochii* Parl. (Liliaceae). In Demiriz, H. & Özhatay, N. (eds): Proceedings of the V OPTIMA meeting, pp. 565-571. - Istanbul Üniversitesi Fen Fakültesi.

GIORDANI, A., MICELI, P. & MONTI, G. 1982: Numeri cromosomici per la Flora Italiana: 787-793. - *Inform. Bot. Ital.* **12(3)**: 327-332.

GIOUTLAKIS, G. & ALEXIOU, S. 2014: Two new localities for *Limoniastrum monopetalum* (Plumbaginaceae) from Greece. - *Parnassiana Arch.* **2**: 13-16.

GIOVANNINI, A., GARBARI, F. & GIORDANI, A. 2001: The chromosome number of *Isoëtes duriei* Bory (Lycopodiophyta, Isoëtaceae). - *Webbia* **56(1)**: 219-221.

GIOVINO, A. & AL. 2015: Taxonomy and conservation of *Pancratium maritimum* (Amaryllidaceae) and relatives in the Central Mediterranean. - *Acta Bot. Gallica* **162(4)**: 289-299.

GIRI, N., TAYLOR, N.L. & COLLINS, G.B. 1981: Chromosome numbers in some *Trifolium* species with a karyotype for *T. physodes*. - *Canad. J. Genet. Cytol.* **23(4)**: 621-626.

GIUGNI, G. 1983: Contributo alla conoscenza dell'*Adiantum capillus-veneris* L. - *Boll. Soc. Ticinese Sci. Nat.* **70**: 81-90.

GIUSEPPI, P.L. 1929: Plant collecting on Olympus. - *Gard. Ill.* **1929**: 836-838.

GIUSEPPI, P.L. 1930: Some Balkan alpines. - *Quart. Bull. Alpine Gard. Soc. Gr. Brit.* **1(3)**: 40-45 + 3 plates.

GIUSEPPI, P.L. 1934: Some Greek alpines. - *Quart. Bull. Alpine Gard. Soc. Gr. Brit.* **2(6)**: 225-227.

GIUSEPPI, P.L. 1935: Collecting plants in the Balkans in 1934. - *Quart. Bull. Alpine Gard. Soc. Gr. Brit.* **3(1)**: 7-16.

GIUSEPPI, P.L. 1938: Shqypnia - the plantsman's heaven. - *Quart. Bull. Alpine Gard. Soc. Gr. Brit.* **6 (1)**: 1-23.

GIUSEPPI, P.L. 1941: Up along! Down along! - *Quart. Bull. Alpine Gard. Soc. Gr. Brit.* **9(2)**: 101-106.

GIUSEPPI, P.L. 1943: Over the hills and far away. - *Quart. Bull. Alpine Gard. Soc. Gr. Brit.* **11(1-2)**: 53-59, 105-115.

GKELIS, S. & AL. (EDS) 2013: 13th Panhellenic Scientific Conference, Hellenic Botanical Society, Thessaloniki 3-6 October. Program and book of abstracts. - Thessaloniki: 157 pp.

GLADKOVA, V.N. 1967: Chromosome numbers of some species of *Cotoneaster* Med. and *Crataegus* L. [In Russian]. - *Bot. Zhurn. (Moscow & Leningrad)* **52(3)**: 354-356.

GLADSTONES, J.S. 1974: Lupins of the Mediterranean region and Africa. - *Techn. Bull. Dept. Agric. Western Australia* **26**: 48 pp.

GLASAU, F. 1939: Monographie der Gattung *Cyclamen* auf morphologisch-cytologischer Grundlage. - *Planta* **30(3)**: 507-550.

GLEN, H.F. & LAVRANOS, J.J. 1991: *Juniperus phoenicea*. - *Fl. Pl. Africa* **51(2)**: tab. 2022 + 2 pp. text.

GLETSOS, M. 2017: Report 41. In Vladimirov, V., Aybeke, M. & Tan, Kit (eds): New floristic records in the Balkans: 32. - *Phytol. Balcan.* **23(1)**: 126-127.

GLÜCK, H. 1934: *Limosella*-Studien. Beiträge zur Systematik, Morphologie und Biologie der Gattung *Limosella*. - *Bot. Jahrb. Syst.* **66(5)**: 488-566 + 3 plates.

GODWIN, H. 1943a: Biological Flora of the British Isles. *Rhamnus cathartica* L. - *J. Ecol.* **31**: 69-76.

GODWIN, H. 1943b: Biological Flora of the British Isles. *Frangula alnus* Miller. - *J. Ecol.* **31**: 77-92.

GOEDECKE, F. & BERGMEIER, E. 2018: Ecology and potential distribution of the Cretan endemic tree species *Zelkova abelicea*. - *J. Medit. Ecol.* **16**: 15-26.

GOEPFERT, D. 1974: Karyotypes and DNA content in species of *Ranunculus* L. and related genera. - *Bot. Not.* **127(4)**: 464-489.

GOETGHEBEUR, P. 1986: Genera Cyperacearum. Een bijdrage tot de kennis van de morfologie, systematiek en fylogenese van de Cyperaceae-genera. - Gent: Ph.D. thesis Rijksuniversiteit Gent.

GOETGHEBEUR, P. 1989: Studies in Cyperaceae. 9. Problems in the lectotypification and infrageneric taxonomy of *Cyperus* L. - *Bull. Soc. Roy. Bot. Belgique* **122(1)**: 103-114.

GÖKTÜRK, R.S. 2009: A new subspecies [of] *Trigonella coerulescens* (Fabaceae) from Turkey. - *Ann. Bot. Fenn.* **46(1)**: 62-64.

GÖKTÜRK, R.S. & SÜMBÜL, H. 2014: A taxonomic revision of the genus *Cephalaria* (Caprifoliaceae) in Turkey. - *Turk. J. Bot.* **38(5)**: 927-968.

GOLDBLATT, P. 1980: Systematics of *Gynandriris* (Iridaceae), a Mediterranean-southern African disjunct. - *Bot. Not.* **133(3)**: 239-260.

GOLDBLATT, P. 1990: Phylogeny and classification of Iridaceae. - *Ann. Missouri Bot. Gard.* **77(4)**: 607-627.

GOLDBLATT, P. 1998: Reduction of *Barnardiella, Galaxia, Gynandriris, Hexaglottis, Homeria,* and *Roggeveldia* in *Moraea* (Iridaceae: Irideae). - *Novon* **8(4)**: 371-377.

GOLDBLATT, P. 2002: Phylogeny and classification of the Iridaceae and the relationships of *Iris*. - *Ann. Bot., n.s. (Rome)* **1(2)**: 13-28.

GOLDBLATT, P. & JOHNSON, D.E. (EDS) 1990: Index to plant chromosome numbers 1986-1987. - *Monogr. Syst. Bot. Missouri Bot. Gard.* **30**: i-ix, 1-243.

GOLDBLATT, P. & JOHNSON, D.E. (EDS) 1991: Index to plant chromosome numbers 1988-1989. - *Monogr. Syst. Bot. Missouri Bot. Gard.* **40**: i-viii, 1-238.

GOLDBLATT, P. & JOHNSON, D.E. (EDS) 1994: Index to plant chromosome numbers 1990-1991. - *Monogr. Syst. Bot. Missouri Bot. Gard.* **51**: i-ix, 1-267.

GOLDBLATT, P. & JOHNSON, D.E. (EDS) 1996: Index to plant chromosome numbers 1992-1993. - *Monogr. Syst. Bot. Missouri Bot. Gard.* **58**: i-x, 1-276.

GOLDBLATT, P. & JOHNSON, D.E. (EDS) 1998: Index to plant chromosome numbers 1994-1995. - *Monogr. Syst. Bot. Missouri Bot. Gard.* **69**: i-ix, 1-208.

GOLDBLATT, P. & JOHNSON, D.E. (EDS) 2000: Index to plant chromosome numbers 1996-1997. - *Monogr. Syst. Bot. Missouri Bot. Gard.* **81**: i-xi, 1-188.

GOLDBLATT, P. & JOHNSON, D.E. (EDS) 2003: Index to plant chromosome numbers 1998-2000. - *Monogr. Syst. Bot. Missouri Bot. Gard.* **94**: i-ix, 1-297.

GOLDBLATT, P. & JOHNSON, D.E. (EDS) 2010: Index to plant chromosome numbers 2004-2006. - *Regnum Veg.* **152**: i-ix + 1-246.

GOLDBLATT, P. & MANNING, J.C. 2008: The *Iris* family. Natural history and classification. - Portland & London: Timber Press, 290 pp.

GOLDBLATT, P. (ED.) 1981: Index to plant chromosome numbers 1975-1978. - *Monogr. Syst. Bot. Missouri Bot. Gard.* **5**: i-vii, 1-553.

GOLDBLATT, P. (ED.) 1984: Index to plant chromosome numbers 1979-1981. - *Monogr. Syst. Bot. Missouri Bot. Gard.* **8**: i-viii, 1-427.

GOLDBLATT, P. (ED.) 1985: Index to plant chromosome numbers 1982-1983. - *Monogr. Syst. Bot. Missouri Bot. Gard.* **13**: i-viii, 1-224.

GOLDBLATT, P. (ED.) 1988: Index to plant chromosome numbers 1984-1985. - *Monogr. Syst. Bot. Missouri Bot. Gard.* **23**: i-ix, 1-264.

GOLICIN, S. 1957: Nota critica de *Vinca herbacea* Waldst. et Kit. - *Bot. Mater. Gerb. Bot. Inst. Komarova Akad. Nauk S.S.S.R.* **18**: 179-182.

GÖLZ, P. & OTT, M. 1995: Die Orchideen der Insel Kithira (Ein Beitrag zum OPTIMA-Projekt "Kartierung der mediterranen Orchideen"). - *J. Eur. Orch.* **27(4)**: 622-658.

GÖLZ, P. & REINHARD, H.R. 1977: Statistische Untersuchungen über einige Arten der Orchideengattung *Serapias* (*S. parviflora* Parl., *S. laxiflora* Chaub. nom. illeg., *S. vomeracea* [Burm. f.] Briq.). - *Orchidee (Hamburg)* **28(3)**: 108-116.

GÖLZ, P. & REINHARD, H.R. 1978: Orchideen auf Kos, Samos und Chios. Ein Beitrag zur Kenntnis der Orchideenflora der Ägäis. - *Orchidee (Hamburg)* **29(3)**: 103-106.

GÖLZ, P. & REINHARD, H.R. 1980: *Serapias* (Orchidaceae). Ergebnisse statistischer und chorologischer Untersuchungen. - *Mitt. Arbeitskreis Heimische Orchid. Baden-Württemberg* **12(3)**: 123-189.

GÖLZ, P. & REINHARD, H.R. 1981: Die Orchideenflora der ostägäischen Inseln Kos, Samos, Chios und Leswos (Griechenland). - *Beih. Veröff. Naturschutz Landschaftspflege Baden-Württemberg* **19**: 5-127.

GÖLZ, P. & REINHARD, H.R. 1983: Orchideen in Nordwestgriechenland: Ein Beitrag zur Kenntnis der Orchideenflora des Balkans (I). - *Mitt. Arbeitskreis Heimische Orchid. Baden-Württemberg* **15(2)**: 161-216.

GÖLZ, P. & REINHARD, H.R. 1984: Die Orchideenflora Albaniens: Ein Beitrag zur Kenntnis der Orchideenflora des Balkans (II). - *Mitt. Arbeitskreis Heimische Orchid. Baden-Württemberg* **16(2)**: 193-394.

GÖLZ, P. & REINHARD, H.R. 1986: Orchideen in Jugoslavien: Ein Beitrag zur Kenntnis der Orchideenflora des Balkans (III). - *Mitt. Arbeitskreis Heimische Orchid. Baden-Württemberg* **18(4)**: 689-827.

GÖLZ, P. & REINHARD, H.R. 1987: Bemerkungen zu Baumann, H. & S. Künkele (1986): Die Gattung *Ophrys* L. - Eine taxonomische Übersicht. - *Mitt. Arbeitskreis Heimische Orchid. Baden-Württemberg* **19(4)**: 866-894.

GÖLZ, P. & REINHARD, H.R. 1989a: Zur Orchideenflora von Lesvos. - *Mitt. Arbeitskreis Heimische Orchid. Baden-Württemberg* **21(1)**: 1-87.

GÖLZ, P. & REINHARD, H.R. 1989b: Über einige Besonderheiten im ostmediterranen *Ophrys scolopax*- Komplex. - *Mitt. Arbeitskreis Heimische Orchid. Baden-Württemberg* **21(4)**: 1040-1067.

GÖLZ, P. & REINHARD, H.R. 1992a: *Dactylorhiza pythagorae* Gölz & Reinhard, spec. nov. von Samos. - *Mitt. Arbeitskreis Heimische Orchid. Baden-Württemberg* **24(1)**: 52-67.

GÖLZ, P. & REINHARD, H.R. 1992b: Untersuchungen über *Dactylorhiza saccifera* (Brongn.) Soó. - *Mitt. Arbeitskreis Heimische Orchid. Baden-Württemberg* **24(1)**: 69-103.

GÖLZ, P. & REINHARD, H.R. 1993: *Serapias*-Probleme under besonderer Berücksichtigung der *Serapias*-Flora der Insel Kerkira (Korfu) (1. Teil). - *Mitt. Arbeitskreis Heimische Orchid. Baden- Württemberg* **25(1)**: 1-58.

GÖLZ, P. & REINHARD, H.R. 1994: *Serapias*-Probleme unter besonderer Berücksichtigung der Serapiasflora von Kerkira (Korfu), ergänzt durch Untersuchungen an der Serapiasflora Zyperns. - *J. Eur. Orch.* **26(3-4)**: 365-425.

GÖLZ, P. & REINHARD, H.R. 1995: Die Orchideenflora der ionischen Inseln Kefallinia und Zakinthos: Neue Beobachtungen und Erkenntnisse. - *J. Eur. Orch.* **27(4)**: 555-621.

GÖLZ, P. & REINHARD, H.R. 1996: Gestaltwandel innerhalb kretischer Orchideen-aggregate im Verlauf der Monate Januar bis Mai. - *J. Eur. Orch.* **28(4)**: 641-701.

GÖLZ, P. & REINHARD, H.R. 1997: Über die Gattung *Dactylorhiza* - Neue Erkenntnisse und neue Fragen. - *J. Eur. Orch.* **29(4)**: 585-640.

GÖLZ, P. & REINHARD, H.R. 2001: Der ostmediterrane und anatolische *Ophrys holoserica*-Komplex - "Splitter" kontra "Lumper". - *J. Eur. Orch.* **33(4)**: 941-1024.

GÓMEZ-CAMPO, C. 1977: Studies on Cruciferae: II. New names for *Rhynchosinapis* species under *Hutera*. - *Anales Inst. Bot. Cavanilles* **34(1)**: 147-149. [Also distributed as *OPTIMA leaflet* no. 59].

GÓMEZ-CAMPO, C. 1978: Studies on Cruciferae: V. Chromosome numbers for twenty-five taxa. - *Anales Inst. Bot. Cavanilles* **35**: 177-182.

GÓMEZ-CAMPO, C. 1980: Morphology and morphotaxonomy in the tribe Brassiceae. In: Tsunoda, S., Hinata, K. & Gómez-Campo, C. (eds): *Brassica* crops and wild allies, pp. 3-31. - Tokyo: Japan Scientific Societies Press.

GÓMEZ-CAMPO, C. 1999: Taxonomy. In Gómez-Campo, C. (ed.): Biology of *Brassica* coenospecies. - Amsterdam, etc.: Elsevier, ix + 489 pp.

GÓMEZ-CAMPO, C. & GUSTAFSSON, M. 1991: Germplasm of wild n=9 Mediterranean species of *Brassica*. - *Bot. Chron. (Patras)* **10**: 429-434.

GÓMEZ-CAMPO, C. & HINATA, K. 1980: A checklist of chromosome numbers in the tribe Brassiceae. In: Tsunoda, S., Hinata, K. & Gómez-Campo, C. (eds): *Brassica* crops and wild allies, pp. 51-63. - Tokyo: Japan Scientific Societies Press.

GÓMEZ-CAMPO, C. & MARTÍNEZ LABORDE, J.B. 1998: Reajustes taxonómicos y nomenclaturales en la tribu Brassiceae (Cruciferae). - *Anales Jard. Bot. Madrid* **56(2)**: 379-381.

GÓMEZ-CAMPO, C. & TORTOSA, M.E. 1974: The taxonomic and evolutionary significance of some juvenile characters in the Brassicaceae. - *Bot. J. Linn. Soc.* **69(1-4)**: 105-124.

GÓMEZ-CAMPO, C. (ED.) 1985: Plant conservation in the Mediterranean area. - Dordrecht, etc.: Dr. W. Junk Publishers, xii + 269 pp. (Geobotany 7).

GÓMEZ-GARCÍA, J. 1970: Notas cariológicas sobre el género *Sideritis* L. en España. - *Anales Inst. Bot. Cavanilles* **27**: 115-124.

GÖMÖRY, D. & AL. 1999: Genetic differentiation and phylogeny of beech on the Balkan peninsula. - *J. Evol. Biol.* **12**: 746-754.

GÖMÖRY, D. & PAULE, L. 2010: Reticulate evolution patterns in western-Eurasian beeches. - *Bot. Helv.* **120**: 63-74.

GÖNÜZ, A. & ÖZÖRGÜCÜ, B. 1999: An investigation on the morphology, anatomy and ecology of *Origanum onites* L. - *Türk. Bot. Derg.* **23(1)**: 19-32.

GONZÁLEZ SIERRA, G. & AL. 1992: Revisión taxonomica de las especies ibericas del género *Onopordum* L. - *Candollea* **47(1)**: 181-213.

GONZÁLEZ ZAPATERO, A., ELENA-ROSELLÓ, J.-A. & NAVARRO ANDRÉS, F. 1988: Números cromosómicos para la flora española. 504-515. - *Lagascalia* **15(1)**: 112-119.

GONZÁLEZ-AGUILERA, J.J. & FERNÁNDEZ-PERALTA, A.M. 1983: The nature of polyploidy in *Reseda* sect. *Leucoreseda* (Resedaceae). - *Pl. Syst. Evol.* **142(3-4)**: 223-237.

GONZÁLEZ-AGUILERA, J.J. & FERNÁNDEZ-PERALTA, A.M. 1984: Phylogenetic relationships in the family Resedaceae L. - *Genetica* **64(3)**: 185-197.

GONZÁLEZ-AGUILERA, J.J., FERNÁNDEZ-PERALTA, A.M. & RUIZ REJÓN, M. 1979: [Chromosome numbers in Resedaceae from Spain]. In Löve, Á. (ed.): IOPB chromosome number reports LXV. - *Taxon* **28(5-6)**: 634-635.

GONZÁLEZ-AGUILERA, J.J., FERNÁNDEZ-PERALTA, A.M. & SAÑUDO, A. 1980: Cytogenetic and evolutive studies on the Spanish species of the family Resedaceae L. - *Bol. Soc. Brot., Ser. 2* **53(1)**: 519-536 + 2 plates.

GONZÁLEZ-AGUILERA, J.J. & RUIZ REJÓN, M. 1976: [Chromosome numbers in *Reseda* from Spain]. In Löve, Á. (ed.): IOPB chromosome number reports LIV. - *Taxon* **25(5-6)**: 631.

GONZÁLEZ-AGUILERA, J.J. & RUIZ REJÓN, M. 1978: [Chromosome numbers in *Reseda* from Spain]. In Löve, Á. (ed.): IOPB chromosome number reports LX. - *Taxon* **27(2-3)**: 225.

GONZÁLEZ-BERNÁLDEZ, F., SÁNCHEZ DE RIVERA, A.M. & ANGULO, M.D. 1973: Estudios cromosómicos en el género *Trifolium*. IV. - *Lagascalia* **3(2)**: 195-203.

GONZÁLEZ, F. & SILVESTRE, S. 1980: Números cromosómicos para la flora española. 141-149. - *Lagascalia* **9(2)**: 261-266.

GONZALO, R., AEDO, C. & GARCÍA, M.A. 2013: Taxonomic revision of the Eurasian *Stipa* subsections *Stipa* and *Tirsae* (Poaceae). - *Syst. Bot.* **38(2)**: 344-378.

GOODE, T. 2003: Crocuses in the garden. - *Alpine Gardener* **71(2)**: 137-143.

GOODSPEED, T.H. 1954: The genus *Nicotiana*. Origin, relationships and evolution of its species in the light of their distribution, morphology and cytogenetics. - *Chron. Bot.* **16**: i-xxi + 1-536.

GOODSPEED, T.H. & THOMPSON, M.C. 1959: Cytotaxonomy of *Nicotiana*. II. - *Bot. Rev. (Lancaster)* **25 (2)**: 385-415.

GORANOVA, V.Y. 2007: Chromosome numbers of some shrubs and trees from the Bulgarian flora. - *Phytol. Balcan.* **13(3)**: 383-386.

GORANOVA, V.Y., STANIMIROVA, P. & ANČEV, M.E. 2006: Reports (1572-1583). In Kamari, G., Blanché, C. & Garbari, F. (eds): Mediterranean chromosome number reports - 16. - *Fl. Medit.* **16**: 425-431.

GORENFLOT, R. 1959: Le polymorphisme de *Plantago coronopus* L., ses manifestations et ses causes. - *Rev. Cytol. Biol. Vég.* **20(4)**: 237-499 + 22 plates.

GORENFLOT, R. 1960: La polyploïdie chez *Plantago coronopus* L. - *Rev. Cytol. Biol. Vég.* **22(2)**: 77-108 + 7 plates.

GORENFLOT, R. 1964: Introgression, polymorphisme et taxonomie chez les Plantaginacées. - *Adansonia* **4(3)**: 393-417.

GORENFLOT, R. 1966: Caryologie d'espèces méditerranéennes et sahariennes du genre *Plantago* L. - *Compt. Rend. Hebd. Séances Acad. Sci., Sér. D* **262**: 878-881 + 1 plate.

GORENFLOT, R. 1976: Le complexe polyploïde du *Phragmites australis* (Cav.) Trin. ex Steud. (= *P. communis* Trin.). - *Bull. Soc. Bot. France* **123(5-6)**: 261-271.

GORENFLOT, R. & AL. 1972: Le complexe polyploïde du *Phragmites communis* Trin. - *Compt. Rend. Hebd. Séances Acad. Sci., Sér. D* **274(10)**: 1501-1504 + 2 plates.

GORENFLOT, R., CARTIER, D. & LENOIR, A. 1975: La polyploïdie du *P[hragmites] australis* (Cav.) Trin. ex Steud. dans le Bassin Méditerranéen. In: Guinochet, M. & al. (eds): La flore du Bassin Méditerranéen, pp. 165-173. - Paris: C.N.R.S.

GORENFLOT, R., RAICU, P. & CARTIER, D. 1972: Caryologie de la Canne de Provence

(*Arundo donax* L.). - *Compt. Rend. Hebd. Séances Acad. Sci., Sér. D* **274(3)**: 391-393 + 1 plate.
GORI, C. 1957: Sull'embriologia e citologia di alcune specie del genere *Reseda*. - *Caryologia* **10(2)**: 391-401.
GORI, C. 1958: Sul cariogramma della *Fritillaria messanensis* Raf. - *Caryologia* **11(1)**: 28-33 + 2 plates.
GÖRK, G. & AL. 1989: The flora of Çeşme (Izmir) peninsula [In Turkish with English summary]. - *Doğa Türk Bot. Derg.* **13(2)**: 249-295.
GORNALL, R.J. 1987: An outline of a revised classification of *Saxifraga* L. - *Bot. J. Linn. Soc.* **95(4)**: 273-292.
GORNALL, R.J., HULTGÅRD, U.-M. & JONSELL, B. 1989: On the lectotypification of *Parnassia palustris* L. - *Watsonia* **17(4)**: 438-439.
GOSLER, A.G., KELLY, C.K. & BLAKEY, J.K. 1994: Phenotypic plasticity in leaf morphology of *Crataegus monogyna* (Rosaceae): an experimental study with taxonomic implications. - *Bot. J. Linn. Soc.* **115(3)**: 211-219.
GOTTLIEB-TANNENHAIN, P.V. 1904: Studien über die Formen der Gattung *Galanthus*. - *Abh. K.K. Zool.-Bot. Ges. Wien* **2(4)**: vii + 95 pp. + 2 plates and 1 map.
GOTTSCHLICH, G. 2015: Karl Hermann Zahn (1865-1940). Leben, Werk und wissenschaftliches Umfeld des Monographen der Gattung *Hieracium* (Habichtkraut, Asteraceae). - *Stapfia* **102**: 1-126.
GOTTSCHLICH, G. 2018a: Et in Arcadia ego ...! - Impressionen einer hieraciologischen Sammelreise nach Griechenland. - *Mitt. Pollichia Pfälz. Vereins Naturk.* **99**: 27-34.
GOTTSCHLICH, G. 2018b: Korrespondenz und andere Archivalien im Nachlass des *Hieracium*-Monographen Karl Hermann Zahn (1865-1940). Verzeichnis und Auswertung. - *Stapfia* **109**: 197-227.
GOTTSCHLICH, G. 2020: Ergebnisse von Herbar-, Feld- und Literaturstudien zur Gattung *Hieracium* s. l. (*Hieracium* s. str. und *Pilosella*) als Vorarbeiten für die "Neue Flora von Bayern". - *Ber. Bayer. Bot. Ges.* **90**: 129-146.
GOTTSCHLICH, G. & AL. 2010: Reports 45-46. In Vladimirov, V., Dane, F. & Tan, Kit (eds): New floristic records in the Balkans: 13. - *Phytol. Balcan.* **16(1)**: 151-152.
GOTTSCHLICH, G. & AL. 2011: *Hieracium spp.* and *Pilosella spp.* In Greuter, W. & Raus, Th. (eds): Med-Checklist notulae, 30. - *Willdenowia* **41(2)**: 314-317.
GOTTSCHLICH, G. & AL. 2013: Three new *Hieracium* taxa (Asteraceae, Cichorieae) from Mt Cholomon, Chalkidiki, Central Macedonia, Greece. - *Willdenowia* **43(1)**: 59-64.
GOTTSCHLICH, G. & BERGMEIER, E. 2010: Report 44. In Vladimirov, V., Dane, F. & Tan, Kit (eds): New floristic records in the Balkans: 13. - *Phytol. Balcan.* **16(1)**: 150-151.
GOTTSCHLICH, G., BIEL, B. & TAN, KIT 2009: Reports 17-22. In Vladimirov, V., Dane, F. & Tan, Kit (eds): New floristic records in the Balkans: 10. - *Phytol. Balcan.* **15(1)**: 120-122.
GOTTSCHLICH, G., DOMINA, G. & DI GRISTINA, E. 2017: *Hieracium umbrosum* subsp. *abietinum* (Asteraceae), a further example of amphi-Adriatic disjunction. - *Pl. Biosyst.* **151(5)**: 792-794.
GOTTSCHLICH, G., DRENCKHAHN, D. & MEIEROTT, L. 2006: *Hieracium greuteri* (Compositae), a local endemic of N Peloponnisos (Greece). - *Willdenowia* **36(1)**: 351-356.
GOTTSCHLICH, G. & DUNKEL, F.G. 2018: New taxa of *Hieracium* and *Pilosella* (Asteraceae) from northern and central Greece. - *Stapfia* **109**: 3-24.
GOTTSCHLICH, G. & DUNKEL, F.G. 2019: New taxa of *Hieracium* and *Pilosella* (Asteraceae) from northern Greece, II. - *Stapfia* **111**: 5-32.
GOTTSCHLICH, G., DUNKEL, F.G. & MEIEROTT, L. 2014: *Hieracium scapigerum* subsp. *falacronese* (Asteraceae, Cichorieae), a new subspecies from Northern Greece. - *Phytol. Balcan.* **20(2-3)**: 171-173.
GÖTZ, E. 1967: Die *Aconitum variegatum*-Gruppe und ihre Bastarde in Europa. - *Feddes Repert.* **76 (1-2)**: 1-62.
GÖTZ, E. 1975: Die Gehölze der Mittelmeerländer. Ein Bestimmungsbuch nach Blattmerkmalen. - Stuttgart: Verlag Eugen Ulmer, 114 pp.
GÖTZ, E. 1994: Zur Biosystematik des *Asplenium trichomanes*-Komplexes in Europa und auf den Makaronesischen Inseln. - Bochum: Diplomarbeit [unpubl.].
GOUKASIAN, A. & NAZAROVA, E. 1998: Reports (1009-1025). In Kamari, G., Felber, F. & Garbari, F. (eds): Mediterranean chromosome number reports - 8. - *Fl. Medit.* **8**: 307-313.
GOULA, K. & AL. 2008: The weed flora in wheat-fields around Thessaloniki: changes since 1936. In: Praktika 10ou Sinedriou Ellinikis Votanikis Eterias, 5-8 maiou 2005, Ioannina: pp. 143-148 [In Greek with English abstract].
GOULA, K. & CONSTANTINIDIS, TH. 2019: Taxonomic diversity of *Tripleurospermum* (Asteraceae) in Greece: morphology and karyology. In Daskalakou, E.N. & al. (eds): 16th Conference of the Hellenic Botanical Society – Athens 10-13 October 2019. Program and book of abstracts, p. 63.
GOULA, K. & CONSTANTINIDIS, TH. 2021: Taxonomic diversity and karyology of *Anthemis rigida* (Anthemideae, Asteraceae) in the Aegean, Greece. - *Phytotaxa* **484(1)**: 129-143.
GOULA, K. & KONSOULAS, G. 2016: *Colchicum chimonanthum* (Colchicaceae): Confirmation of old reports and discovery of new populations. - *Parnassiana Arch.* **4**: 23-26.
GOULANDRIS, A. 1971: A new Flora of Greece. - *Boissiera* **19**: 21-22.
GOULANDRIS, N. 1977: Sophia Topali (1900-1944). - *Ann. Mus. Goulandris* **3**: 35-38.
GOULANDRIS, N. 1978: Leonidas Pinatzis (1891-1964), a notable Greek field-botanist. - *Ann. Mus. Goulandris* **4**: 19-26.
GOULANDRIS, N. 1983: Constantine Goulimis (1886-1963). - *Ann. Mus. Goulandris* **6**: 37-43.
GOULANDRIS, N., GOULIMIS, C.N. & STEARN, W.T. 1968: Wild flowers of Greece. - Kifissia: Goulandris Bot. Mus., xxxii + 212 pp. incl. 103 pp. of colour plates.
GOULANDRIS, N. & TAN, KIT 1989: *Biarum davisii*. In Tan, Kit (ed.): Plant taxonomy, phytogeography and related subjects. The Davis & Hedge festschrift, pp. xviii-xix + 1 plate. - Edinburgh: Edinburgh Univ. Press.
GOULIMIS, C.N. 1948: I chloris tou Chelmou [In Greek]. - *Vouno [Journal of the Hellenic Alpine Club]* **1947-48**: 17-28.
GOULIMIS, C.N. 1953: The richness of the Greek flora. - Athens: 8 pp. [reproduced from the Athenian monthly "Eklogi"].
GOULIMIS, C.N. 1954a: Flora of Mount Gamila [In Greek with English summary]. - *Vouno [Journal of the Hellenic Alpine Club]* **1954**: 126-134.
GOULIMIS, C.N. 1954b: I chloris tis Idras [In Greek]. - *Mellon tis Idras* **14**: 100.
GOULIMIS, C.N. 1955: Some alpines of Greece. - *Quart. Bull. Alpine Gard. Soc. Gr. Brit.* **23(4)**: 328-334.
GOULIMIS, C.N. 1956a: New additions to the Greek flora [In Greek with English summary]. - Athine: 35 pp.
GOULIMIS, C.N. 1956b: Les colchiques de Grèce et la repartition géographique du genre *Colchicum* [In Greek with French summary]. - Athènes: 27 pp. (from *Archia Farmakevtikis*).
GOULIMIS, C.N. 1957a: Some notable Greek plants. - *Quart. Bull. Alpine Gard. Soc. Gr. Brit.* **25(1)**: 105-108.
GOULIMIS, C.N. 1957b: A letter from Greece. - *Quart. Bull. Alpine Gard. Soc. Gr. Brit.* **25(4)**: 286-289.
GOULIMIS, C.N. 1958: Fritillaries of Greece. - *Quart. Bull. Alpine Gard. Soc. Gr. Brit.* **26(2)**: 133-134.
GOULIMIS, C.N. 1959: Report on species of plants requiring protection in Greece and measures for securing their protection. - *Proceedings of the IUCN Technical Meeting, Athens* **5**: 168-188.
GOULIMIS, C.N. 1960a: New additions to the Greek flora. Second series. - Athens: 31 pp.
GOULIMIS, C.N. 1960b: To oros Vourinos ke i chloris tou [In Greek]. - *Vouno [Journal of the Hellenic Alpine Club]* **1960**: 122-131.
GOVAERTS, R. 1995: (1195-1197) Proposals to conserve or reject three species names in *Quercus* L. (Fagaceae). - *Taxon* **44(4)**: 631-633. [all approved; cf. *Taxon* 47(4): 864-65, 1998; cf. also *Taxon* 49(3): 537, 2000].

GOVAERTS, R. 1996: (1261) Proposal to reject the name *Betula alba* (Betulaceae). - *Taxon* **45(4)**: 697-698. [approved, cf. *Taxon* 48(2): 363, 1999].

GOVAERTS, R. 1998: Extinct seed plants of Turkey, with special reference to *Adonis* sect. *Consiligo*. - *Karaca Arbor. Mag.* **4(4)**: 163-166.

GOVAERTS, R. & AVISHAI, M. 2000: The conserved type of the name *Quercus pubescens* Willd. (Fagaceae). - *Taxon* **49(3)**: 537.

GOVAERTS, R. & FRODIN, D.G. 1998: World checklist and bibliography of Fagales (Betulaceae, Corylaceae, Fagaceae and Ticodendraceae). - Kew: Royal Botanic Gardens, vii + 407 pp.

GOVAERTS, R. & FRODIN, D.G. 2002: World checklist and bibliography of Araceae (and Acoraceae). - Kew: Royal Botanic Gardens, xi + 560 pp.

GOVAERTS, R., FRODIN, D.G. & RADCLIFFE-SMITH, A. 2000: World checklist and bibliography of Euphorbiaceae (and Pandaceae). - Kew: Royal Botanic Gardens, 4 vols.

GOVAERTS, R. & NICKRENT, D.L. 2010: 658. *Cytinus ruber*. - *Curtis's Bot. Mag.* **26(4)**: 314-321.

GOYDER, D.J. & NICHOLAS, A. 2001: A revision of *Gomphocarpus* R. Br. (Apocynaceae: Asclepiadeae). - *Kew Bull.* **56(4)**: 769-836.

GRABNER, S., GRIEHSER, B. & HEISELMAYER, P. 1995: Vegetationsmuster in Erosionsrinnen der Phrygana auf den Kykladeninseln Mykonos und Tinos. - *Salzburg. Geogr. Arb.* **29**: 267-306.

GRABNER, S. & HEISELMAYER, P. 2000: Vegetation dynamics of phrygana in erosion rills on Mykonos (Greece). - *Acta Bot. Croat.* **59(1)**: 111-134.

GRADSTEIN, S.R. & KERN, J.H. 1968: A new *Carex* from Crete. - *Acta Bot. Neerl.* **17(4)**: 242-247.

GRADSTEIN, S.R. & SMITTENBERG, J.H. 1977: The hydrophilous vegetation of western Crete. – *Vegetatio* **34(2)**: 65-86.

GRAF, U. 1977: Ein neuer *Ophrys*-Bastard aus Korfu: *O. bombyliflora* Link × *O. attica* (Boiss. & Orph.) Soó. - *Orchidee (Hamburg)* **28(4)**: 149.

GRAHAM, D. 1994: Plant portraits: *Fritillaria tuntasia*. - *Rock Gard.* **24(1)**: 72, 74.

GRAHAM, S.A. & CAVALCANTI, T.B. 2001: New chromosome counts in the Lythraceae and a review of chromosome numbers in the family. - *Syst. Bot.* **26(3)**: 445-458.

GRAMLICH, R. 1991: *Veronica argute-serrata* - neu für Deutschland. - *Florist. Rundbr.* **25(1)**: 3-5.

GRANADO, C., LUQUE, T. & PASTOR, J. 1988: Números cromosómicos para la flora española. 551-555. - *Lagascalia* **15(1)**: 133-136.

GRANEL DE SOLIGNAC, L. 1970: Recherches de caractères nouveaux pour la classification des Orobanchées. - *Naturalia Monspel., Sér. Bot.* **21**: 75-78 + 3 plates.

GRANT, A.L. 1924: A monograph of the genus *Mimulus*. - *Ann. Missouri Bot. Gard.* **11(2-3)**: 99-388 + 1 plate.

GRANT, W.F. 1959: Cytogenetic studies in *Amaranthus*. III. Chromosome numbers and phylogenetic aspects. - *Canad. J. Genet. Cytol.* **1(4)**: 313-328.

GRANT, W.F. 1995: A chromosome atlas and interspecific-intergeneric index for *Lotus* and *Tetragonolobus* (Fabaceae). - *Canad. J. Bot.* **73(11)**: 1787-1809.

GRANT, W.F. & SMALL, E. 1996: The origin of the *Lotus corniculatus* L. (Fabaceae) complex: a synthesis of diverse evidence. - *Canad. J. Bot.* **74(7)**: 975-989.

GRAU, J. 1964: Die Zytotaxonomie der *Myosotis-alpestris*- und der *Myosotis-silvatica*-Gruppe in Europa. - *Oesterr. Bot. Z.* **111(5)**: 561-617.

GRAU, J. 1967a: Primäre und sekundäre Chromosomenbasiszahlen bei *Omphalodes*. - *Oesterr. Bot. Z.* **114(1)**: 66-72.

GRAU, J. 1967b: Cytotaxonomische Bearbeitung der Gattung *Myosotis* L. II. *Myosotis sicula* s.l. - *Mitt. Bot. Staatssamml. München* **6(2)**: 517-530.

GRAU, J. 1968a: Cytotaxonomische Bearbeitung der Gattung *Myosotis* L. III. Die annuellen Sippen. - *Mitt. Bot. Staatssamml. München* **7(1)**: 17-100.

GRAU, J. 1968b: Cytologische Untersuchungen an Boraginaceen. I. - *Mitt. Bot. Staatssamml. München* **7(2)**: 277-294.

GRAU, J. 1970: Cytotaxonomische Bearbeitung der Gattung *Myosotis* L. IV. Ergänzende Studien. - *Mitt. Bot. Staatssamml. München* **8**: 127-136.

GRAU, J. 1971: Cytologische Untersuchungen an Boraginaceae. II. - *Mitt. Bot. Staatssamml. München* **9(1)**: 177-194.

GRAU, J. 1975: Studies in the genus *Myosotis*. In Walters, S.M. & King. C.J. (eds): European floristic and taxonomic studies, pp. 82-89. - Oxford: E.W. Classey Ltd.

GRAU, J. 1976: Die Cytologie südwestmediterraner *Scrophularia*-Arten. - *Mitt. Bot. Staatssamml. München* **12(2)**: 609-654.

GRAU, J. 1977: Astereae - systematic review. In Heywood, V.H., Harborne, J.B. & Turner, B.L. (eds): The biology and chemistry of the Compositae 1: 539-565. - London, etc.: Academic Press.

GRAU, J. 1981: Die Verwandtschaftsbeziehungen der balkanischen Scrophularien. - *Bot. Jahrb. Syst.* **102(1-4)**: 307-313.

GRAU, J. 1984: Nomenklatorische Studien an *Ranunculus* L. sect. *Ranunculastrum* DC. von der Iberischen Halbinsel. - *Mitt. Bot. Staatssamml. München* **20**: 51-60.

GRAU, J. & ERBEN, M. 1988: Chromosomenzahlen griechischer Hieracien. - *Mitt. Bot. Staatssamml. München* **27**: 97-98.

GRAU, J. & LEINS, P. 1968: Pollenkorntypen und Sektionsgliederung der Gattung *Myosotis*. - *Ber. Deutsch. Bot. Ges.* **81(3-4)**: 107-115.

GRAU, J. & LIPPERT, W. 1980: Eine neue *Scrophularia*-Art vom Balkan. - *Mitt. Bot. Staatssamml. München* **16 Beih.**: 19-22.

GRAU, J. & SCHWAB, A. 1982: Mikromerkmale der Blüte zur Gliederung der Gattung *Myosotis*. - *Mitt. Bot. Staatssamml. München* **18**: 9-58.

GRAYUM, M.H. 1990: Evolution and phylogeny of the Araceae. - *Ann. Missouri Bot. Gard.* **77(4)**: 628-697.

GRAZE, H. 1933: Die chromosomalen Verhältnisse in der Sektion *Pseudolysimachia* Koch der Gattung *Veronica*. - *Jahrb. Wiss. Bot.* **77(4)**: 507-559.

GRAZE, H. 1935: Weitere Chromosomenuntersuchungen bei *Veronica*-Arten der Sektion *Pseudolysimachia* Koch. - *Jahrb. Wiss. Bot.* **81(4)**: 609-662.

GRAZI, G. & URECH, K. 1982: Einige morphologische Merkmale der Mistelbeere (*Viscum album* L.) und deren taxonomische Bedeutung. - *Beitr. Biol. Pflanzen* **56(2-3)**: 293-306.

GREBENCHIKOV, O. 1938: On the occurrence of *Fagus orientalis* in Greece. - *Bull. Misc. Inform. Kew* **1938(1)**: 38-45.

GREBENCHIKOV, O. 1954: Über die ursprünglichen (natürlichen) Lokalitäten und Standorte des *Aesculus hippocastanum* L. in Europa [In Slovakian with Russian and German summaries]. - *Biológia (Bratislava)* **9(5)**: 516-530.

GRECESCU, D. 1907: Plante Macedonice din vilaieturile Monastir şi Salonic [In Romanian]. - *Analele Acad. Române, Ser. 2* **29(3)**: 27-136.

GREEN, E.P. & SHORT, F.T. 2003: World atlas of seagrasses. - Berkeley, Calif.: Univ. Calif. Press, xii + 298 pp.

GREEN, M.L. 1918: *Linum elegans*. - *Bot. Mag.* **144**: tab. 8769 + 2 pp. text.

GREEN, P.S. 1961: Studies in the genus *Jasminum* I: Section *Alternifolia*. - *Notes Roy. Bot. Gard. Edinburgh* **23(3)**: 355-384.

GREEN, P.S. 1985: *Fraxinus rotundifolia* Mill., *F. parvifolia* Lam. or *F. angustifolia* Vahl? - *Kew Bull.* **40(1)**: 131-134.

GREEN, P.S. 1988a: (914) Proposal to conserve the name *Fraxinus angustifolia* Vahl against *F. rotundifolia* Mill. (Oleaceae). - *Taxon* **37(2)**: 488-489. [approved, cf. Taxon 39(2): 296, 1990].

GREEN, P.S. 1988b: *Jasminum*. - *Plantsman* **10(3)**: 148-159.

GREEN, P.S. 2002: A revision of *Olea* L. (Oleaceae). - *Kew Bull.* **57(1)**: 91-140.

GREEN, P.S. & KUPICHA, F.K. 1979: Notes on the genus *Olea*. - *Kew Bull.* **34(1)**: 69-75.

GREGER, H. 1975: Laubblatt-Flavonoide und Systematik bei *Matricaria* und *Tripleurospermum* (Asteraceae-Anthemideae). - *Pl. Syst. Evol.* **124(1)**: 35-55.

GREGER, H. & ERNET, D. 1973: Flavonoid-Muster, Systematik und Evolution bei *Valerianella*. - *Phytochemistry* **12(7)**: 1693-1699.

GREGOR, M. & MEIEROTT, L. 2013: Reports 68-76. In Vladimirov, V. & al. (eds): New floristic records in the Balkans: 22. - *Phytol. Balcan.* **19(2)**: 277.
GREGOR, T., MEIEROTT, L. & RAUS, TH. 2016: *Asperula tymphaea* (Rubiaceae) - a new species from Northern Pindus, Greece. - *Phytol. Balcan.* **22(2)**: 255-258.
GREGOR, TH., PONCET, V. & TISON, J.-M. 2009: (1891) Proposal to conserve the name *Potentilla inclinata* against *P. assurgens* (Rosaceae). - *Taxon* **58(2)**: 651-652. [approved, cf. Taxon 60(4): 1203, 2011].
GREGORY, M. & AL. 1998: Nomenclator alliorum: *Allium* names and synonyms: a world guide. - Kew: Royal Botanic Gardens, 83 pp.
GREGORY, W.C. 1941: Phylogenetic and cytological studies in the Ranunculaceae. - *Trans. Amer. Philos. Soc., n.s.* **31(5)**: 443-521 incl. 12 plates.
GREGSON, N.M. 1963: Karyotype of *Theligonum cynocrambe* L. - *Nature* **200(4909)**: 917.
GREIG-SMITH, P. 1948: Biological Flora of the British Isles. *Urtica* L. - *J. Ecol.* **36(2)**: 339-355.
GREILHUBER, J. 1971: Chromosomenzahlen von *Verbascum austriacum*, *V. lanatum* und einigen Rhinanthoideen (Scrophulariaceae). - *Mitt. Bot. Arbeitsgem. Oberösterr. Landesmus. Linz* **3(1)**: 31-35.
GREILHUBER, J. 1982: Trends in der Chromosomenevolution von *Scilla* (Liliaceae). - *Stapfia* **10**: 11-51.
GREILHUBER, J. 1986: On karyotype evolution in *Cyclamen* L. subgen. *Psilanthum* Schwz. (Primulaceae). - *Sauteria* **1**: 211-222.
GREILHUBER, J. 1989a: Karyotype structure and evolution in *Cyclamen* L. subgen. *Psilanthum* Schwz. (Primulaceae). - *Flora, Morphol. Geobot. Oekophysiol.* **183(1-2)**: 103-113.
GREILHUBER, J. 1989b: Taxonomic relevance of karyotype structure in *Cyclamen* L. subgen. *Psilanthum* Schwarz. - *Cyclamen Soc. J.* **13(2)**: 57-62.
GREILHUBER, J. & EHRENDORFER, F. 1975: Chromosome numbers and evolution in *Ophrys* (Orchidaceae). - *Pl. Syst. Evol.* **124(2)**: 125-138.
GREILHUBER, J. & SPETA, F. 1977: Giemsa karyotypes and their evolutionary significance in *Scilla bifolia*, *S. drunensis*, and *S. vindobonensis* (Liliaceae). - *Pl. Syst. Evol.* **127(2-3)**: 171-190.
GREIMLER, J., HERMANOWSKI, B. & JANG, C.-G. 2004: A re-evaluation of morphological characters in European *Gentianella* section *Gentianella* (Gentianaceae). - *Pl. Syst. Evol.* **248(1-4)**: 143-169.
GREMAUD, M. 1978: Contribution à la cytotaxonomie du genre *Carduus*. - *Ber. Schweiz. Bot. Ges.* **87 (3-4)**: 173-181.
GREMAUD, M. 1979: Nouvelle contribution à la cytotaxonomie du genre *Carduus* (Compositae). - *Ber. Schweiz. Bot. Ges.* **89(1-2)**: 133-143.
GREMAUD, M. 1981a: Recherches de taxonomie expérimentale sur le *Carduus defloratus* L. s.l. (Compositae). I. Données taxonomiques. Variation caryologique. - *Rev. Cytol. Biol. Vég. Botaniste* **4 (1)**: 1-75.
GREMAUD, M. 1981b: Recherches de taxonomie expérimentale sur le *Carduus defloratus* L. s.l. (Compositae). II. Variation morphologique. Taxonomie numérique. - *Rev. Cytol. Biol. Vég. Botaniste* **4(2)**: 111-171.
GREMAUD, M. 1981c: Recherches de taxonomie expérimentale sur le *Carduus defloratus* L. s.l. (Compositae). III. Hybridation expérimentale, cytologie des hybrides F1; données carpologiques et palynologiques. - *Rev. Cytol. Biol. Vég. Botaniste* **4(3)**: 207-268.
GRENIER, E. 1994: A propos de *Lamium galeobdolon* (L.) L. - *Monde Pl.* **89(450)**: 11-12.
GRENIER, J.C.M. & GODRON, D.A. 1848-56: Flore de France, ou description des plantes qui croissent naturellement en France et en Corse. - Paris, etc.: 3 vols. [cf. TL-2 1: 999-1000, 1976].
GREUTER, W. 1965: Beiträge zur Flora der Südägäis, 1-7. - *Candollea* **20**: 167-218.
GREUTER, W. 1966: Kreta, Landschaft und Pflanzenwelt. - *Mitth. Naturf. Ges. Bern, N.S.* **23**: xxxi- xxxiii.
GREUTER, W. 1967a: Beiträge zur Flora der Südägäis 8-9. - *Bauhinia* **3(2)**: 243-254.
GREUTER, W. 1967b: Contributiones floristicae austro-aegaeae 10-12. - *Candollea* **22(2)**: 233-253.
GREUTER, W. 1968a: Une espèce de *Crocus* découverte en Cyrénaïque: *Crocus Boulosii*. - *Candollea* **23(1)**: 45-49.
GREUTER, W. 1968b: Notulae nomenclaturales et bibliographicae 1-4. - *Candollea* **23(1)**: 81-99.
GREUTER, W. 1968c: Contributio floristica austro-aegaea 13. - *Candollea* **23(1)**: 143-150.
GREUTER, W. 1968d: Le dattier de Théophraste, spécialité crétoise. - *Mus. Genève* **81**: 14-16.
GREUTER, W. 1968e: Notulae nomenclaturales et bibliographicae 5-6. - *Candollea* **23(2)**: 257-265.
GREUTER, W. 1969: Contributio floristica austro-aegaea 14. - *Candollea* **24(1)**: 45-49.
GREUTER, W. 1970a: The taxonomic position of *Lloydia graeca* (Liliaceae) and related species. - *Israel J. Bot.* **19(2-3)**: 155-160.
GREUTER, W. 1970b: Zur Paläogeographie und Florengeschichte der südlichen Ägäis. - *Feddes Repert.* **81(1-5)**: 233-242.
GREUTER, W. 1970c: Contributio floristica austro-aegaea 15. - *Candollea* **25(2)**: 189-192.
GREUTER, W. 1971a: *Bromus Caroli-Henrici*, eine verkannte ostmediterrane Graminee. - *Ann. Naturhist. Mus. Wien* **75**: 83-89.
GREUTER, W. 1971b: L'apport de l'homme à la flore spontanée de la Crète. - *Boissiera* **19**: 329-337.
GREUTER, W. 1972a: The relict element of the flora of Crete and its evolutionary significance. In Valentine, D.H. (ed.): Taxonomy, phytogeography and evolution, pp. 161-177. - London & New York: Academic Press.
GREUTER, W. 1972b: L'écueil à *Silene Holzmannii*, en Crète, et son peuplement végétal. - *Saussurea* **3**: 157-166.
GREUTER, W. 1972c: Floristic report on the Cretan area. - Distributed at VII Flora Europaea symposium, Coimbra [mimeographed]: 72 pp.
GREUTER, W. 1972d: Betrachtungen zur Pflanzengeographie der Südägäis. - *Opera Bot.* **30**: 49-64.
GREUTER, W. 1972e: Schlussbetrachtungen. - *Opera Bot.* **30**: 80-83.
GREUTER, W. 1973a: Additions to the flora of Crete, 1938-1972. - *Ann. Mus. Goulandris* **1**: 15-83.
GREUTER, W. 1973b: Monographie der Gattung *Ptilostemon* (Compositae). - *Boissiera* **22**: 1-215 + 8 plates.
GREUTER, W. 1974a: Note sur deux variétés grecques d'*Alyssum doerfleri* (Cruciferae) et sur la classification de quelques espèces vivaces de ce genre. - *Candollea* **29(1)**: 135-146.
GREUTER, W. 1974b: Note sur deux espèces nouvelles pour la Corse. - *Bull. Soc. Bot. France* **121(3-4)**: 145-147.
GREUTER, W. 1975a: Some notes on *Bornmuellera* in Greece, and an interspecific hybrid in the Alysseae (Cruciferae). - *Candollea* **30(1)**: 13-20. [Also distributed as *OPTIMA leaflet* no. 2].
GREUTER, W. 1975b: Quisquiliae floristicae graecae, 1-3. - *Candollea* **30(2)**: 323-330. [Also distributed as *OPTIMA leaflet* no. 11].
GREUTER, W. 1975c: First OPTIMA Meeting in Crete - September 1975. Guide to the excursions. - Genève: Secrétariat OPTIMA, 33 pp.
GREUTER, W. 1975d: Floristic studies in Greece. In Walters, S.M. & King, C.J. (eds): European floristic and taxonomic studies, pp. 18-37. - Oxford: E.W. Classey Ltd. [Also distributed as *OPTIMA leaflet* no. 8].
GREUTER, W. 1975e: Die Insel Kreta - eine geobotanische Skizze. In Dafis, S. & Landolt, E. (eds): Zur Vegetation und Flora von Griechenland, Band 1. - *Veröff. Geobot. Inst. E.T.H. Stiftung Rübel Zürich* **55**: 141-197. [Also distributed as *OPTIMA leaflet* no. 9].
GREUTER, W. 1975f: Historical phytogeography of the southern half of the Aegean area. In Jordanov, D. & al. (eds): Problems of Balkan flora and vegetation, pp. 17-21. - Sofia: Publishing House of the Bulgarian Academy of Sciences.
GREUTER, W. 1975g: Floristic report on the Cretan area. - *Mem. Soc. Brot.* **24(1)**: 131-171.
GREUTER, W. 1975h: *Cistus albanicus* Heywood, "neu" für Griechenland. - *Ann. Mus. Goulandris* **2**: 35-47.
GREUTER, W. 1976: The flora of Psara (E. Aegean Islands, Greece) - an annotated catalogue. - *Candollea* **31(2)**: 191-242. [Also distributed as *OPTIMA leaflet* no. 25].
GREUTER, W. 1977a: Chorological additions to the Greek flora. 1. Unpublished records

mapped in "Atlas florae europaeae", volume 3. - *Candollea* **32(1)**: 21-49. [Also distributed as *OPTIMA leaflet* no. 38].
GREUTER, W. 1977b: Catalogue des plantes récoltées lors de l'excursion de la Société botanique [de Genève] en Grèce du 16 au 30 juillet 1977. - Mimeographed: 14 pp.
GREUTER, W. 1979a: Mediterranean conservation as viewed by a plant taxonomist. - *Webbia* **34(1)**: 87-99.
GREUTER, W. 1979b: Quisquiliae floristicae graecae, 4. *Lamyropsis carpini* (Compositae), a new species from NW. Greece. - *Willdenowia* **9(1)**: 57-66.
GREUTER, W. 1979c: The origins and evolution of island floras as exemplified by the Aegean archipelago. In Bramwell, D. (ed.): Plants and islands, pp. 87-106. - London, etc.: Academic Press.
GREUTER, W. 1979d: The flora and phytogeography of Kastellorizo (Dhodhekanisos, Greece). 1. An annotated catalogue of the vascular plant taxa. - *Willdenowia* **8(3)**: 531-611. [Also distributed as *OPTIMA leaflet* no. 87].
GREUTER, W. 1980: The endemic flora of Crete and the significance of its protection. In Antipas, B. (ed.): Praktika sinedrio prostasias panidas-chloridas-viotopon, pp. 91-97. - Athine [n.v.].
GREUTER, W. 1981a: Probleme des Naturschutzes im Mittelmeergebiet. - *Beih. Veröff. Naturschutz Landschaftspflege Baden-Württemberg* **19**: 155-158.
GREUTER, W. 1981b: *Alkanna* and *Cynoglossum* s. lat. In Greuter, W. (ed.): Med-Checklist notulae, 3. - *Willdenowia* **11(1)**: 32-36.
GREUTER, W. 1981c: *Myosotis solange* Greuter & Zaffran, sp. nova. In Greuter, W. (ed.): Med-Checklist notulae, 3. - *Willdenowia* **11(1)**: 38.
GREUTER, W. 1981d: *Noaea* and *Symphytum*. In Greuter, W. (ed.): Med-Checklist notulae, 3. - *Willdenowia* **11(1)**: 38-39.
GREUTER, W. 1981e: *Cornus sanguinea*. In Greuter, W. (ed.): Med-Checklist notulae, 3. - *Willdenowia* **11(1)**: 42.
GREUTER, W. 1981f: *Sedum amplexicaule* subsp. *tenuifolium* (Sm.) Greuter and *Sedum creticum* C. Presl. In Greuter, W. & Raus, Th. (eds): Med-Checklist notulae, 4. - *Willdenowia* **11(2)**: 277.
GREUTER, W. 1984a: Les *Arum* de la Crète. - *Bot. Helv.* **94(1)**: 15-22.
GREUTER, W. 1984b: *Silene euxina* Rupr. In Greuter, W. & Raus, Th. (eds): Med-Checklist notulae, 9. - *Willdenowia* **14(1)**: 45.
GREUTER, W. 1984c: *Silene parnassica* subsp. *vourinensis* Greuter, subsp. nova. In Greuter, W. & Raus, Th. (eds): Med-Checklist notulae, 9. - *Willdenowia* **14(1)**: 47.
GREUTER, W. 1985a: *Satureja* L. [s.l.] and *Scutellaria* L. In Greuter, W. & Raus, Th. (eds): Med- Checklist notulae, 10. - *Willdenowia* **14(2)**: 302-308.
GREUTER, W. 1985b: Dipsacaceae. In Greuter, W. & Raus, Th. (eds): Med-Checklist notulae, 11. - *Willdenowia* **15(1)**: 71-76.
GREUTER, W. 1985c: *Stachys horvaticii* Micevski. In Greuter, W. & Raus, Th. (eds): Med-Checklist notulae, 11. - *Willdenowia* **15(1)**: 78-79.
GREUTER, W. 1985d: *Satureja parnassica* Heldr. & Sart. ex Boiss. In Greuter, W. & Raus, Th. (eds): Med-Checklist notulae, 11. - *Willdenowia* **15(1)**: 77.
GREUTER, W. 1986a: *Draba strasseri* Greuter, sp. nova. In Greuter, W. & Raus, Th. (eds): Med- Checklist notulae, 12. - *Willdenowia* **15(2)**: 418-419.
GREUTER, W. 1986b: *Trifolium clypeatum* L. In Greuter, W. & Raus, Th. (eds): Med-Checklist notulae, 12. - *Willdenowia* **15(2)**: 429.
GREUTER, W. 1987a: *Onobrychis aliacmonia* (Leguminosae) - the unusual story of a rediscovery. - *Pl. Syst. Evol.* **155(1-4)**: 215-217.
GREUTER, W. 1987b: (870) Proposal to reject *Fumaria bulbosa* L. - *Taxon* **36(1)**: 169-173. [approved, cf. Taxon 43(1): 113-114, 1994].
GREUTER, W. 1987c: Some notes on *Lesquereuxia* (Scrophulariaceae). - *Bot. Jahrb. Syst.* **108(2-3)**: 251-257.
GREUTER, W. 1987d: *Orobanche baumanniorum* Greuter, spec. nova. In Greuter, W. & Raus, Th. (eds): Med-Checklist notulae, 14. - *Willdenowia* **16(2)**: 448-449.
GREUTER, W. 1989a: Recent progress of the Med-Checklist scheme. In Phitos, D. (ed.): Abstracts. Sixth OPTIMA meeting, Delphi, 10-16 Sept. 1989, p. 8.
GREUTER, W. 1989b: *Primula acaulis*, *Primula elatior* and the "Flora Anglica" of Linnaeus. - *Candollea* **44(2)**: 562-567.
GREUTER, W. 1989c: *Oenothera indecora* Camb. subsp. *indecora*. In Greuter, W. & Raus, Th. (eds): Med-Checklist notulae, 15. - *Willdenowia* **19(1)**: 37.
GREUTER, W. 1989d: *Adonis microcarpa* subsp. *cretica* (Huth) Vierh. In Greuter, W. & Raus, Th. (eds): Med-Checklist notulae, 15. - *Willdenowia* **19(1)**: 43.
GREUTER, W. 1991: Botanical diversity, endemism, rarity, and extinction in the Mediterranean area: an analysis based on the published volumes of Med-Checklist. - *Bot. Chron. (Patras)* **10**: 63-79.
GREUTER, W. 1994: (1085) Proposal to conserve the name *Silene sieberi* Fenzl with a conserved type. - *Taxon* **43(1)**: 127-128. [approved, cf. Taxon 44(4): 611, 1995].
GREUTER, W. 1995a: (1146) Proposal to conserve the name *Silene apetala* Willd. (Caryophyllaceae) with a conserved type. - *Taxon* **44(1)**: 101-102. [approved, cf. Taxon 46(2): 323-324, 1997].
GREUTER, W. 1995b: (1147) Proposal to conserve the name *Silene gallica* L. (Caryophyllaceae) against several synonyms of equal priority. - *Taxon* **44(1)**: 102-104.
GREUTER, W. 1995c: Studies in Greek Caryophylloideae: *Agrostemma*, *Silene*, and *Vaccaria*. - *Willdenowia* **25(1)**: 105-142.
GREUTER, W. 1995d: *Silene* (Caryophyllaceae) in Greece: a subgeneric and sectional classification. - *Taxon* **44(4)**: 543-581.
GREUTER, W. 1995e: Origin and peculiarities of Mediterranean island floras. - *Ecol. Medit.* **21(1-2)**: 1-10.
GREUTER, W. 1996: (1274) Proposal to conserve the name *Cistus albanicus* (Cistaceae). - *Taxon* **45(4)**: 715-716. [not approved, cf. Taxon 48(2): 363, 1999].
GREUTER, W. 1997a: The early botanical exploration of Greece. In: Abstracts from the First Balkan Botanical Congress, Thessaloniki, September 19-22, 1997, p. 18. - Thessaloniki.
GREUTER, W. 1997b: Save *Asteriscus*, sink *Nauplius* (Compositae). - *Fl. Medit.* **7**: 41-48.
GREUTER, W. 1998: The early botanical exploration of Greece. - Pp. 9-20 in: Tsekos, I. & Moustakas, M. (eds): Progress in botanical research. Proceedings of the 1st Balkan Botanical Congress.
GREUTER, W. 2001a: [Request for decision on homonymy of *Centaurea epirota* Halácsy and *C. epirotica* (Baldacci) Halácsy]. - *Taxon* **50(2)**: 567.
GREUTER, W. 2001b: Diversity of Mediterranean island floras. - *Bocconea* **13**: 55-64.
GREUTER, W. 2001c: Source data for floristics and flora writing in the Balkans: an overview. In Özhatay, N. (ed.): Proceedings of the Second Balkan Botanical Congress, 1: 25-32. - Istanbul.
GREUTER, W. 2003a: The ancient Greek roots of biological sciences. - *Fl. Medit.* **12**: 5-10.
GREUTER, W. 2003b: The Euro+Med treatment of Astereae (Compositae) - generic concepts and required new names. - *Willdenowia* **33(1)**: 45-47.
GREUTER, W. 2003c: The Euro+Med treatment of Cardueae (Compositae) - generic concepts and required new names. - *Willdenowia* **33(1)**: 49-61.
GREUTER, W. 2003d: The Euro+Med treatment of Cichorieae (Compositae) - generic concepts and required new names. - *Willdenowia* **33(2)**: 229-238.
GREUTER, W. 2003e: The Euro+Med treatment of Gnaphalieae and Inuleae (Compositae) - generic concepts and required new names. - *Willdenowia* **33(2)**: 239-244.
GREUTER, W. 2003f: The Euro+Med treatment of Senecioneae and the minor Compositae tribes - generic concepts and required new names, with an addendum to Cardueae. - *Willdenowia* **33(2)**: 245-250.
GREUTER, W. 2004: (1645) Proposal to conserve the name *Ophrys speculum* (Orchidaceae) with a conserved type. - *Taxon* **53(4)**: 1071-1072. [approved, cf. Taxon 55(4): 1101, 2005].
GREUTER, W. 2008a: *Ephedra nebrodensis* subsp. *procera* (Fisch. & C.A. Mey.) K. Richt., *Crepis hellenica* Kamari. In Greuter, W. & Raus, Th. (eds): Med-Checklist notulae, 27. - *Willdenowia* **38(2)**: 465-466.

GREUTER, W. 2008b: On the correct name of the Late Spider Orchid, and its appropriate spelling: *Ophrys holosericea*. - *J. Eur. Orch.* **40(4)**: 657-662.

GREUTER, W. 2009: *Apera intermedia* Hackel, *Allium melanogyne* Greuter, sp. nov., *Allium orientale* Boiss. In Greuter, W. & Raus, Th. (eds): Med-Checklist notulae, 28. - *Willdenowia* **39(2)**: 340-343.

GREUTER, W. 2012: Results of the seventh "Iter Mediterraneum" in the Peloponnese, Greece, May to June 1995. - *Bocconea* **25**: 5-127.

GREUTER, W., AGHABABIAN, M. & WAGENITZ, G. 2005: (1670-1675) Proposals to conserve the names *Bellidiastrum, Berkheya, Euryops, Notobasis, Picnomon* and *Urospermum* (Compositae) against six generic names of Vaillant. - *Taxon* **54(1)**: 196-198.

GREUTER, W. & AL. 1983: *Asplenium bourgaei*. A new addition to the flora of Europe. - *Brit. Fern Gaz.* **12(5)**: 271-274.

GREUTER, W. & AL. 2001: (1509) Proposal to conserve the name *Centaurea* (Compositae) with a conserved type. - *Taxon* **50(4)**: 1201-1205. [approved, cf. Taxon 53(3): 818, 2004].

GREUTER, W. & ATHANASIOU, K. 1990: Excursion to Mount Parnassos (16 October 1990). - Photocopy: 3 pp.

GREUTER, W., BÖHLING, N. & JAHN, R. 2002: The *Cerastium scaposum* group (Caryophyllaceae): three annual taxa endemic to Crete (Greece), two of them new. - *Willdenowia* **32(1)**: 45-54.

GREUTER, W., BURDET, H.M. & LONG, G. 1984: Med-Checklist. A critical inventory of vascular plants of the circum-mediterranean countries. Vol. 1 (Pteridophyta, Gymnospermae, Dicotyledones: Acanthaceae-Cneoraceae), pp. i-xvi, 1-330, xvii-c. - Genève: Conservatoire et Jardin Botanique, Med-Checklist trust of OPTIMA.

GREUTER, W., BURDET, H.M. & LONG, G. 1986: Med-Checklist. A critical inventory of vascular plants of the circum-mediterranean countries. Vol. 3 (Dicotyledones: Convolvulaceae-Labiatae), pp. i-xvi, 1-395, xvii-cxxix. - Genève: Conservatoire et Jardin Botanique, Med-Checklist trust of OPTIMA.

GREUTER, W., BURDET, H.M. & LONG, G. 1989: Med-Checklist. A critical inventory of vascular plants of the circum-mediterranean countries. Vol. 4 (Dicotyledones: Lauraceae-Rhamnaceae), pp. i-xvii, 1-458, xix-cxxix. - Genève: Conservatoire et Jardin Botanique, Med-Checklist trust of OPTIMA.

GREUTER, W. & DITTRICH, M. 1973: Neuer Beitrag zur Kenntnis der Gattung *Lamyropsis* (Compositae): die Identität von *Cirsium microcephalum* Moris. - *Ann. Mus. Goulandris* **1**: 85-98.

GREUTER, W. & GRAU, J. 1970: Zum Vorkommen von drei Unterarten der *Myosotis refracta* Boiss. auf Kreta. - *Candollea* **25(1)**: 7-10.

GREUTER, W., GUTERMANN, W. & TALAVERA, S. 2006: A preliminary conspectus of *Scorzoneroides* (Compositae, Cichorieae) with validation of the required new names. - *Willdenowia* **36(2)**: 689-692.

GREUTER, W. & KARL, R. 2015: *Alkanna macrosiphon, Anchusa pusilla*. In Raab-Straube, E. & Raus, Th. (eds): Euro+Med-Checklist notulae, 5. - *Willdenowia* **45(3)**: 450.

GREUTER, W. & KARL, R. 2019: Ein neues Johanniskraut von den Kykladen (Ägäis, Griechenland): *Hypericum* (sect. *Drosocarpium*) *perfoliatum* subsp. *phitosianum* (Hypericaceae). - *Bot. Chron. (Patras)* **22**: 39-47.

GREUTER, W., KOUMPLI-SOVANTZI, L. & YANNITSAROS, A. 2002: *Eleocharis caduca* (Cyperaceae): a redeemed species of Tropical African origin discovered in Crete (Greece). - *Bot. Chron. (Patras)* **15**: 17-30.

GREUTER, W. & LASSEN, P. 1999: Notes on *Coronilla repanda* subsp. *repanda*. In Greuter, W. & Raus, Th. (eds): Med-Checklist notulae, 18. - *Willdenowia* **29(1-2)**: 58.

GREUTER, W., MATTHÄS, U. & RISSE, H. 1982: Notes on Cardaegean plants. 3. *Medicago strasseri*, a new leguminous shrub from Kriti. - *Willdenowia* **12(2)**: 201-206. [Also distributed as *OPTIMA leaflet* no. 128].

GREUTER, W., MATTHÄS, U. & RISSE, H. 1984: Additions to the flora of Crete, 1973-1983 - I. - *Willdenowia* **14(1)**: 27-36. [Also distributed as *OPTIMA leaflet* no. 139].

GREUTER, W., MATTHÄS, U. & RISSE, H. 1985a: Additions to the flora of Crete, 1973-1983 (1984) - II. - *Willdenowia* **14(2)**: 269-297. [Also distributed as *OPTIMA leaflet* no. 153].

GREUTER, W., MATTHÄS, U. & RISSE, H. 1985b: Additions to the flora of Crete, 1973-1983 (1984) - III. - *Willdenowia* **15(1)**: 23-60. [Also distributed as *OPTIMA leaflet* no. 154].

GREUTER, W., OBERPRIELER, CH. & VOGT, R. 2003: The Euro+Med treatment of Anthemideae (Compositae) - generic concepts and required new names. - *Willdenowia* **33(1)**: 37-43.

GREUTER, W. & PAPANICOLAOU, K. 1979: A new species of *Centaurea* from Mt. Pangaion. (Materials for the Mountain Flora of Greece, 4). - *Bot. Not.* **132(4)**: 471-474. [Also distributed as *OPTIMA leaflet* no. 98].

GREUTER, W., PHITOS, D. & RUNEMARK, H. 1975: Greece and the Greek islands. A report on the available floristic information and on current floristic and phytotax. research. In Guinochet, M. & al. (eds): La flore du Bassin Méditerranéen, pp. 67-89. - Paris: C.N.R.S. [Also distributed as *OPTIMA leaflet* no. 22].

GREUTER, W. & PIEPER, H. 1975: Notiz zur Flora und Biogeographie der landfernen südägäischen Klippe Avgo. - *Candollea* **30(1)**: 7-11. [Also distributed as *OPTIMA leaflet* no. 1].

GREUTER, W., PLEGER, R. & RAUS, TH. 1983: The vascular flora of the Karpathos island group (Dodecanesos, Greece). A preliminary checklist. - *Willdenowia* **13(1)**: 43-78. [Also distributed as *OPTIMA leaflet* no. 131].

GREUTER, W. & RAAB-STRAUBE, E.V. (EDS) 2005: Euro+Med Notulae, 1. - *Willdenowia* **35(2)**: 223-239.

GREUTER, W. & RAAB-STRAUBE, E.V. (EDS) 2006: Euro+Med Notulae, 2 – *Willdenowia* **36(2)**: 707-717.

GREUTER, W. & RAAB-STRAUBE, E.V. (EDS) 2007: Euro+Med Notulae, 3. - *Willdenowia* **37(1)**: 139-189.

GREUTER, W. & RAAB-STRAUBE, E.V. (EDS) 2011: Euro+Med Notulae, 5. - *Willdenowia* **41(1)**: 129-138.

GREUTER, W. & RAUS, TH. 1981: *Crassula alata* (Viv.) A. Berger. In Greuter, W. & Raus, Th. (eds): Med-Checklist notulae, 4. - *Willdenowia* **11(2)**: 276.

GREUTER, W. & RAUS, TH. 1986: Notes on *Sedum, Umbilicus, Alyssum, Carrichtera, Hymenolobus* and *Origanum*. In Greuter, W. & Raus, Th. (eds): Med-Checklist notulae, 13. - *Willdenowia* **16(1)**: 103-106.

GREUTER, W. & RAUS, TH. (EDS) 1981: Med-Checklist notulae, 4. - *Willdenowia* **11(2)**: 271-280. [Also distributed as *OPTIMA leaflet* no. 119].

GREUTER, W. & RAUS, TH. (EDS) 1982a: Med-Checklist notulae, 5. - *Willdenowia* **12(1)**: 33-46. [Also distributed as *OPTIMA leaflet* no. 126].

GREUTER, W. & RAUS, TH. (EDS) 1982b: Med-Checklist notulae, 6. - *Willdenowia* **12(2)**: 183-199. [Also distributed as *OPTIMA leaflet* no. 127].

GREUTER, W. & RAUS, TH. (EDS) 1983: Med-Checklist notulae, 7. - *Willdenowia* **13(1)**: 79-99. [Also distributed as *OPTIMA leaflet* no. 132].

GREUTER, W. & RAUS, TH. (EDS) 1984a: Med-Checklist notulae, 8. - *Willdenowia* **13(2)**: 277-288. [Also distributed as *OPTIMA leaflet* no. 136].

GREUTER, W. & RAUS, TH. (EDS) 1984b: Med-Checklist notulae, 9. - *Willdenowia* **14(1)**: 37-54. [Also distributed as *OPTIMA leaflet* no. 140].

GREUTER, W. & RAUS, TH. (EDS) 1985a: Med-Checklist notulae, 10. - *Willdenowia* **14(2)**: 299-308. [Also distributed as *OPTIMA leaflet* no. 155].

GREUTER, W. & RAUS, TH. (EDS) 1985b: Med-Checklist notulae, 11. - *Willdenowia* **15(1)**: 61-84. [Also distributed as *OPTIMA leaflet* no. 156].

GREUTER, W. & RAUS, TH. (EDS) 1986a: Med-Checklist notulae, 12. - *Willdenowia* **15(2)**: 413-432. [Also distributed as *OPTIMA leaflet* no. 158].

GREUTER, W. & RAUS, TH. (EDS) 1986b: Med-Checklist notulae, 13. - *Willdenowia* **16(1)**: 103-116. [Also distributed as *OPTIMA leaflet* no. 159].

GREUTER, W. & RAUS, TH. (EDS) 1987: Med-Checklist notulae, 14. - *Willdenowia* **16(2)**: 439-452. [Also distributed as *OPTIMA leaflet* no. 167].

GREUTER, W. & RAUS, TH. (EDS) 1989: Med-Checklist notulae, 15. - *Willdenowia* **19(1)**: 27-48. [Also distributed as *OPTIMA leaflet* no. 182].
GREUTER, W. & RAUS, TH. (EDS) 1995: Med-Checklist notulae, 16. - *Willdenowia* **25(1)**: 171-176.
GREUTER, W. & RAUS, TH. (EDS) 1998: Med-Checklist notulae, 17. - *Willdenowia* **28(1-2)**: 163-174.
GREUTER, W. & RAUS, TH. (EDS) 1999: Med-Checklist notulae, 18. - *Willdenowia* **29(1-2)**: 51-67.
GREUTER, W. & RAUS, TH. (EDS) 2000: Med-Checklist notulae, 19. - *Willdenowia* **30(2)**: 229-243.
GREUTER, W. & RAUS, TH. (EDS) 2001: Med-Checklist notulae, 20. - *Willdenowia* **31(2)**: 319-328.
GREUTER, W. & RAUS, TH. (EDS) 2002: Med-Checklist notulae, 21. - *Willdenowia* **32(2)**: 195-208.
GREUTER, W. & RAUS, TH. (EDS) 2004: Med-Checklist notulae, 22. - *Willdenowia* **34(1)**: 71-80.
GREUTER, W. & RAUS, TH. (EDS) 2005: Med-Checklist notulae, 23. - *Willdenowia* **35(1)**: 55-64.
GREUTER, W. & RAUS, TH. (EDS) 2006: Med-Checklist notulae, 24. - *Willdenowia* **36(2)**: 719-730.
GREUTER, W. & RAUS, TH. (EDS) 2007a: Med-Checklist notulae, 25. - *Willdenowia* **37(1)**: 205-213.
GREUTER, W. & RAUS, TH. (EDS) 2007b: Med-Checklist notulae, 26. - *Willdenowia* **37(2)**: 435-444.
GREUTER, W. & RAUS, TH. (EDS) 2008: Med-Checklist notulae, 27. - *Willdenowia* **38(2)**: 465-474.
GREUTER, W. & RAUS, TH. (EDS) 2009: Med-Checklist notulae, 28. - *Willdenowia* **39(2)**: 335-345.
GREUTER, W. & RECHINGER, K.H. 1967: Flora der Insel Kythera gleichzeitig Beginn einer nomenklatorischen Überprüfung der griechischen Gefässpflanzenarten. - *Boissiera* **13**: 1-206.
GREUTER, W. & RECHINGER, K.H. 1972: *Verbascum spathulisepalum* (Scrophulariaceae), eine neue Art aus Nordost-Griechenland. - *Bot. Not.* **125(4)**: 493-496.
GREUTER, W. & RISSE, H. 1981: *Helianthemum lippii* (L.) Dum.-Cours. In Greuter, W. & Raus, Th. (eds): Med-Checklist notulae, 4. - *Willdenowia* **11(2)**: 275.
GREUTER, W. & SCHOLZ, H. 1980: *Agrostis merxmuelleri* (Gramineae), eine neue Art aus dem griechisch-albanischen Grenzgebiet. - *Mitt. Bot. Staatssamml. München* **16 Beih.**: 23-27.
GREUTER, W. & SCHOLZ, H. 1996: *Phragmites* in Crete, *Cenchrus frutescens*, and the nomenclature of the common reed (Gramineae). - *Taxon* **45(3)**: 521-523.
GREUTER, W. & STRID, A. 1981: Notes on Cardaegean plants. 2. A new species of *Ranunculus* sect. *Ranunculus* from the mountains of W Kriti. - *Willdenowia* **11(2)**: 267-269.
GREUTER, W. & ZIMMER, B. 1996: *Valerianella orientalis* (Valerianaceae) in Griechenland. - *Ann. Naturhist. Mus. Wien, B* **98B Suppl**: 291-300.
GREUTER, W. (ED.) 1980a: Med-Checklist notulae, 1. - *Willdenowia* **10(1)**: 13-21. [Also distributed as *OPTIMA leaflet* no. 105].
GREUTER, W. (ED.) 1980b: Med-Checklist notulae, 2. - *Willdenowia* **10(2)**: 227-232. [Also distributed as *OPTIMA leaflet* no. 108].
GREUTER, W. (ED.) 1981: Med-Checklist notulae, 3. - *Willdenowia* **11(1)**: 23-43. [Also distributed as *OPTIMA leaflet* no. 117].
GREY-WILSON, CH. 1969: The heights of Olympus. - *Quart. Bull. Alpine Gard. Soc. Gr. Brit.* **37(3)**: 254-273.
GREY-WILSON, CH. 1971: Plants of Macedonia [In series: Alpines of the world]. - *Quart. Bull. Alpine Gard. Soc. Gr. Brit.* **39(1)**: 15-19.
GREY-WILSON, CH. 1982: *Lilium carniolicum*. - *Bot. Mag.* **184(1)**: tab. 839 + pp. 9-11.
GREY-WILSON, CH. 1988: The genus *Cyclamen*. - London: Royal Botanic Gardens, Kew, etc., 147 pp. + 12 plates
GREY-WILSON, CH. 1989a: *Glaucium flavum*. - *Bot. Mag. (Kew Mag.)* **6(4)**: tab. 133 + pp. 147-151.
GREY-WILSON, CH. 1989b: *Cyclamen* - an identity crisis. - *Cyclamen* Soc. J. **13(2)**: 64-67.
GREY-WILSON, CH. 1989c: Return to Mount Olympus, 1. - *Quart. Bull. Alpine Gard. Soc. Gr. Brit.* **57 (1)**: 24-41.
GREY-WILSON, CH. 1989d: Return to Mount Olympus, 2. - *Quart. Bull. Alpine Gard. Soc. Gr. Brit.* **57 (2)**: 115-129.
GREY-WILSON, CH. 1991: *Cyclamen* - a reappraisal. - *Plantsman* **13(1)**: 1-20 + 1 plate.
GREY-WILSON, CH. 1993a: The autumn glories of Greece. - *Garden (London)* **118(10)**: 440-444.
GREY-WILSON, CH. 1993b: Poppies. A guide to the poppy family in the wild and in cultivation. - Portland, Oregon: Timber Press, 208 pp. + 32 plates.
GREY-WILSON, CH. 1997: *Cyclamen*. A guide for gardeners, horticulturalists and botanists. - London: B.T. Batsford Ltd, 192 pp.
GREY-WILSON, CH. 2000: *Clematis*: the genus. A comprehensive guide for gardeners, horticulturists and botanists. - London: B.T. Batsford Ltd, 224 pp.
GREY-WILSON, CH. 2001: [Notes on *Cyclamen hederifolium*]. - *New Plantsman* **8(4)**: 206.
GREY-WILSON, CH. 2002a: [*Iris cretensis*]. - *Alpine Gardener* **70(4)**: front cover.
GREY-WILSON, CH. 2002b: [*Crocus cancellatus* subsp. *mazziaricus*]. - *Alpine Gardener* **70(4)**: back cover.
GREY-WILSON, CH. 2010: A field guide to the bulbs of Greece. - Pershore, U.K.: Alpine Garden Society, 224 pp.
GREY-WILSON, CH., LEEDS, R. & ROLFE, R. 2020: *Colchicum*. The complete guide. - Peterborough: RHS Media, 576 pp.
GREY-WILSON, CH. & MATHEW, B.F. 1981: Bulbs: the bulbous plants of Europe and their allies. - London: Collins, 285 pp., 48 plates.
GRIEBL, N. 2007: Fuss-Orchideenreise durch den Balkan, 1999. - *Ber. Arbeitskreis. Heimische Orchid.* **23(1-2)**: 67-74.
GRIEHSER, B., GRABNER, S. & HEISELMAYER, P. 1995: Vegetationsentwicklung nach Brandrodung der Phrygana auf den Kykladeninseln Mykonos und Tinos. - *Salzburg. Geogr. Arb.* **29**: 307-336.
GRIERSON, A.J.C. 1967: *Primula vulgaris* ssp. *sibthorpii*. - *Bot. Mag.* **176(2)**: tab. 500 + 5 pp. text.
GRIESINGER, R. 1935: Zytologische und experimentelle Untersuchungen an *Erophila verna*. - *Flora* **129**: 363-379 + 2 plates.
GRIESINGER, R. & KLINKOWSKI, M. 1939: Geographie und Cytologie des europäischen Formenkreises der Gattung *Ornithopus*. - *Züchter* **11(6)**: 147-161.
GRIFFITHS, D.J., ROWLANDS, D.G. & PEREGRINE, W.T.H. 1959: Cytogenetic relationships of certain artificial and natural species of *Avena*. - *J. Agric. Sci. (Cambridge)* **52(2)**: 189-199 + 4 plates.
GRIGORIADIS, N., THEODOROPOULOS, K. & ELEFTHERIADOU, E. 2005: Effects of clear cutting on vegetation of a coppiced *Quercus frainetto* Ten. forest during a four-year period after cutting (Pieria, Greece). - *Bot. Chron. (Patras)* **18(1)**: 135-147.
GRIGORIADOU, K. & AL. 2020: GIS-facilitated effective propagation protocols of the endangered local endemic of Crete *Carlina diae* (Rech. f.) Meusel and A. Kästner (Asteraceae): Serving ex situ conservation needs and its future sustainable utilization as an ornamental. - *Plants MDPI* **9, 1465**: 17 pp.
GRIGSON, G. 1979: Flowers of the Cretan winter. - *Country Life (London)* **166(4295)**: 1514-1516.
GRILLI CAIOLA, M. & BRANDIZZI, F. 1997: Pollen-pistil interaction in *Hermodactylus tuberosus* Mill. (Iridaceae). - *Pl. Biosyst.* **131(3)**: 197-205.
GRIMSHAW, J. 2008: Variation in *Leucojum aestivum*. - *Plantsman* **7(1)**: 50-54.
GRIMSHAW, J.M. & BENNETT, S.T. 1992: Discoveries in *Cyclamen* cytology and their implications. - *Cyclamen Soc. J.* **15(2)**: 41-45.

GRINTZESCO, J. 1910: Monographie du genre *Astrantia*. - *Annuaire Conserv. Jard. Bot. Genève* **13-14**: 66-194.

GRISEBACH, A. 1839: Genera et species gentianearum adjectis observationibus quibusdam phytogeographicis. - Stuttgart, Tübingen: Cotta, viii + 364 pp.

GRISEBACH, A. 1841: Reise durch Rumelien und nach Brussa im Jahre 1839. - Göttingen: Vandenhoeck und Ruprecht. Vol. 1, vi + 362 pp.; vol. 2, 374 pp.

GRISEBACH, A. 1843: Spicilegium Florae Rumelicae et Bithynicae exhibens synopsin plantarum quas in aest. 1839 legit ... Vol. 1. - Brunsvigae: F. Vieweg et filum, xii + 407 pp.

GRISEBACH, A. 1844: Spicilegium Florae Rumelicae et Bithynicae ... Vol. 2 pp. 1-160. - Brunsvigae: F. Vieweg et filum.

GRISEBACH, A. 1846: Spicilegium Florae Rumelicae et Bithynicae ... Vol. 2 pp. 161-548. - Brunsvigae: F. Vieweg et filum.

GROENDIJK-WILDERS, N., BRANDENBURG, W.A. & JARVIS, C.E. 1988: Lectotypification of *Salix alba* L. (Salicaceae). - *Taxon* **37(1)**: 164-167.

GROVE, A. 1933: *Lilium dalmaticum* and *Lilium cattaniae*. - *New Fl. & Silva* **5(3)**: 172-180.

GROVE, A. & COTTON, A.D. 1933-62: A supplement to Elwes' Monograph of the genus *Lilium*. Parts 1-9. - London: Royal Horticultural Society, 38 plates + text.

GROVE, A.T., MOODY, J. & RACKHAM, O. (EDS) 1990: Stability and change in the Cretan landscape. Petromarula 1 [Proceedings of a symposium held at Chania, Crete]. - Cambridge: Corpus Christi College, 88 pp.

GROVES, R.H. 1986: Invasion of mediterranean ecosystems by weeds. In Dell, B., Hopkins, A.J.M. & Lamont, B.B. (eds): Resilience in mediterranean-type ecosystems, pp. 129-145. - Dordrecht, etc.: Dr. W. Junk Publishers.

GROVES, R.H. 1991: The biogeography of mediterranean plant invasions. In Groves, R.H. & Di Castri, F. (eds): Biogeography of mediterranean invasions, pp. 427-438. - Cambridge, etc.: Cambridge Univ. Press.

GROVES, R.H. & DI CASTRI, F. (EDS) 1991: Biogeography of mediterranean invasions. - Cambridge, etc.: Cambridge Univ. Press, xiii + 485 pp.

GROZENA, N.H. & CVETANOVA, Y.G. 2016: Karyological and morpholgical variation within *Blitum virgatum* L. (Chenopodiaceae) in Bulgaria. - *Trakia J. Sci.* **14(1)**: 16-23.

GROZEVA, N. 2007a: Reports (1623-1630). In Kamari, G., Blanché, C. & Garbari, F. (eds): Mediterranean chromosome number reports - 17. - *Fl. Medit.* **17**: 299-307.

GROZEVA, N. 2007b: *Chenopodium pumilio* (Chenopodiaceae): a new species to the Bulgarian flora. - *Phytol. Balcan.* **13(3)**: 331-334.

GROZEVA, N. & STOEVA, M.P. 2006: Reports (1466-1472). In Kamari, G., Blanché, C. & Garbari, F. (eds): Mediterranean chromosome number reports - 16. - *Fl. Medit.* **16**: 400-408.

GROZEVA, N.H. 2018: Karyological study of the genus *Atriplex* L. in Bulgaria. - *Bulg. J. Agric. Sci.* **24 (suppl. I)**: 52-58.

GROZEVA, N.H. & CVETANOVA, Y.G. 2011: *Chenopodium bonus-henricus* L. (Perennial Goosefoot) in Bulgaria: I. Population variability. - *Trakia J. Sci.* **9(3)**: 1-7.

GROZEVA, N.H. & CVETANOVA, Y.G. 2013: Karyological and morphological variations within the genus *Dysphania* (Chenopodiaceae) in Bulgaria. - *Acta Bot. Croat.* **72(1)**: 49-69.

GROZEVA, N.H. & GOSPODINOVA, S. 2016: Karyological study of genus *Bassia* All. in Bulgaria. - *Trakia J. Sci.* **14(2)**: 158-162.

GROZEVA, N.H. & TODOROVA, M. 2014: Karyological and morphological variation within *Bassia hirsuta* (L.) Asch. in Bulgaria. - *Trakia Univ. J. Nat. Sci.* **15(1)**: 17-23.

GRUBE, A. 1982: Liste der von mir in der Zeit vom 3.4. bis 30.4. 1971 auf Kreta gefundenen Orchideenarten. - *Mitt. Arbeitskreis Heimische Orchid. Hessen* **4**: 41-42.

GRUBOV, V.I. 1949: Monographic sketch of the genus *Rhamnus* L. s.l. [In Russian]. - *Trudy Bot. Inst. Akad. Nauk S.S.S.R., Ser. 1, Fl. Sist. Vysš. Rast.* **8**: 241-423.

GRUENBERG-FERTIG, I. & STEARN, W.T. 1972: Nomenclatural remarks on *Trifolium*. - *Israel J. Bot.* **21(1)**: 1-8.

GRUENBERG-FERTIG, I. & ZOHARY, M. 1970: Nomenclatural remarks on some plants of Palestine. - *Israel J. Bot.* **19(2-3)**: 293-304.

GRUENBERG-FERTIG, I., ZOHARY, M. & HEYN, C.C. 1974: (380) Proposal to conserve the generic name 5985 *Hippomarathrum* Link (1821) ... - *Taxon* **23(2-3)**: 438-440. [not approved, cf. Taxon 27(2-3): 286, 1978].

GRULICH, V. 1984a: Generic division of Sedoideae in Europe and the adjacent regions. - *Preslia* **56 (1)**: 29-45.

GRULICH, V. 1984b: Notes on the Czechoslovak species of the genus *Hylotelephium*. - *Preslia* **56(1)**: 47-53.

GRULICH, V. & HODÁLOVÁ, I. 1994: The *Senecio doria* group (Asteraceae: Senecioneae) in central and southeastern Europe. - *Phyton (Horn)* **34(2)**: 247-265.

GUACCHIO, E. DEL, BERNARDO, L. & IAMONICO, D. 2021: The chimeric *Onobrychis tenoreana* (Fabaceae) and lectotypification of the names *O. laconica* and *O. pentelica*. - *Phytotaxa* **505(2)**: 240-244.

GUACCHIO, E. DEL, VALLARIELLO, R. & IAMONICO, D. 2020: Taxonomic investigations on *Malva cretica* s.l. (Malvaceae). - *Anales Jard. Bot. Madrid* **77(1)**: DOI: https://doi.org/10.3989/ajbm.253.

GUACCHIO, E.DEL, CAMBRIA, S. & BRULLO, S. 2021: Typification of the name *Ephedra nebrodensis* (Ephedraceae). - *Phytotaxa* **496(1)**: 90-92.

GUÉDÈS, M. 1976: The case for *Paspalum distichum* and against futile name-changes. - *Taxon* **25(4)**: 512-513.

GUÉDÈS, M. 1981: Against rejecting the name *Paspalum distichum* L.: Comment on proposal 528. - *Taxon* **30(1)**: 301.

GÜEMES, J. & MUÑOS GARMENDIA, F. 2004: Lectotypification of the names of two species of *Fumana* (Cistaceae). - *Taxon* **53(4)**: 1059-1061.

GUERN, M. 1975: Le complexe polyploïde de l'*Hippocrepis comosa* L. - Paris: Univ. Paris-Sud, Diss., 165 pp.

GUERN, M. 1977: Le complexe polyploïde de l'*Hippocrepis comosa* L. I. La mitose. - *Bull. Soc. Bot. France* **124(1-2)**: 69-83.

GUERN, M. & GORENFLOT, R. 1966: Caryologie du genre *Hippocrepis* L. - *Compt. Rend. Hebd. Séances Acad. Sci., Sér. D* **263**: 509-512 + 1 plate.

GUEST, E. (ED.) 1966: Flora of Iraq. Vol. 1, Introduction. - Baghdad: Ministry of Agriculture, xvi + 213 pp.

GUGLER, W. 1903: Ueber *Centaurea Adami* Willd. - *Allg. Bot. Z. Syst.* **9**: 88-91.

GUGLER, W. 1908: Die Centaureen des ungarischen Nationalmuseums. Vorarbeiten zu einer Monographie der Gattung *Centaurea* und der ihr nächst verwandten Genera. - *Ann. Hist.-Nat. Mus. Natl. Hung.* **6**: 15-297 + 1 plate.

GUIGNARD, G. 1987: Caryologie chez trois taxa tétraploïdes du genre *Dactylis* (Poaceae). - *Taxon* **36(1)**: 29-33.

GUILLÉN, A. & RUIZ REJÓN, M. 1984: Structural variability and chromosome numbers variation in natural populations of *Scilla autumnalis* (Liliaceae). - *Pl. Syst. Evol.* **144(3-4)**: 201-207.

GUILLERM, J.L. 1990: The invading weeds within the western Mediterranean basin. In Di Castri, F., Hansen, A.J. & Debussche, M. (eds): Biological invasions in Europe and the Mediterranean Basin, pp. 61-84. - Dordrecht, etc.: Kluwer Academic Publishers.

GUILLERM, J.L. 1991: Weed invasion in agricultural areas. In Groves, R.H. & Di Castri, F. (eds): Biogeography of mediterranean invasions, pp. 379-392. - Cambridge, etc.: Cambridge Univ. Press.

GUINEA, E. 1963: El género *Biscutella* L. - *Anales Inst. Bot. Cavanilles* **21(2)**: 387-405.

GUINEA, E. & CHATER, A.O. 1971: [New combination in *Micromeria*]. In Heywood, V.H. (ed.): Notulae systematicae ad Floram Europaeam spectantes, no. 11. - *Bot. J. Linn. Soc.* **64(4)**: 381.

GUINOCHET, M. 1957: Contribution à l'étude caryologique du genre *Centaurea* L. sens. lat. - *Bull. Soc. Hist. Nat. Afrique N.* **48(3-4)**: 282-300.

GUINOCHET, M. & AL. (EDS) 1975: La flore du Bassin Méditerranéen. Essai de systématique synthétique (C.N.R.S., Colloques internationaux no. 235). - Paris: C.N.R.S., 576 pp.

GUINOCHET, M. & FOISSAC, J. 1962: Sur les caryotypes de quelques espèces du genre *Centaurea* L. et leur signification taxonomique. - *Rev. Cytol. Biol. Vég.* **25(3-4)**: 373-389.

GUINOCHET, M. & LEMÉE, G. 1950: Contribution à la connaissance des races biologiques de *Molinia coerulea* (L.) Moench. - *Rev. Gén. Bot.* **57(680)**: 565-593.

GUIOL, F. 1928: Notes sur les plantes nouvelles pour le mont Parnès en Attique. - *Monde Pl., Sér. 3* **29(57)**: 3-4.

GUITIÁN, J., GUITIÁN, P. & MEDRANO, M. 1998: Floral biology of the distylous Mediterranean shrub *Jasminum fruticans* (Oleaceae). - *Nordic J. Bot.* **18(2)**: 195-201.

GUITTONNEAU, G.-G. 1964: Contribution à l'étude caryologique du genre *Erodium* L'Hér. - I. - *Bull. Soc. Bot. France* **111(1-2)**: 1-4.

GUITTONNEAU, G.-G. 1965a: Contribution à l'étude caryosystématique du genre *Erodium* L'Hér. - II. - *Bull. Soc. Bot. France* **112(1-2)**: 25-32 + 1 plate.

GUITTONNEAU, G.-G. 1965b: Note sur la découverte de la polysomatie dans le genre *Erodium* L'Hér. - *Compt. Rend. Hebd. Séances Acad. Sci.* **260**: 5332-5335 + 2 plates.

GUITTONNEAU, G.-G. 1966: Contribution à l'étude caryosystématique du genre *Erodium* L'Hér. - III. - *Bull. Soc. Bot. France* **113(1-2)**: 3-11.

GUITTONNEAU, G.-G. 1967: Contribution à l'étude caryosystématique du genre *Erodium* L'Hérit. - IV. - *Bull. Soc. Bot. France* **114(1-2)**: 32-42.

GUITTONNEAU, G.-G. 1972: Contribution à l'étude biosystématique du genre *Erodium* L'Hér. dans le bassin méditérranéen occidental. - *Boissiera* **20**: 1-154.

GUITTONNEAU, G.-G. 1975: Contribution à l'étude caryosystématique et phylogénétique des Géraniacées dans le Bassin Méditerrannéen. In Guinochet, M. & al. (eds): La flore du Bassin Méditerranéen, pp. 195-205. - Paris: C.N.R.S.

GUITTONNEAU, G.-G. 1985: Observations critiques sur la caryologie, la morphologie et la chorologie du genre *Erodium* L'Hér. dans le bassin méditerranéen. - *Bull. Soc. Bot. France, Lett. Bot.* **132(4-5)**: 319-325.

GUITTONNEAU, G.-G. 1986a: Hybridations expérimentales et conséquences phylogénétiques dans le genre *Erodium* L'Hér. - *Candollea* **41(2)**: 285-292.

GUITTONNEAU, G.-G. 1986b: [Chromosome numbers in *Erodium*]. In Löve, Á. (ed.): Chromosome number reports XCIII. - *Taxon* **35(4)**: 897.

GUITTONNEAU, G.-G. 1988: Progrès récents sur la systématique expérimentale de quelques espèces du genre *Erodium* L'Hér. (Geraniaceae) de la région méditerranéenne. - *Bull. Soc. Bot. France, Lett. Bot.* **135(4-5)**: 369-375.

GUITTONNEAU, G.-G. 1989: *Erodium brachycarpum* (Godr.) Thell., Mediterranen species sympatric to *E. botrys* (Cav.) Bertol. In Phitos, D. (ed.): Abstracts. Sixth OPTIMA meeting, Delphi 10-16 Sept. 1989, p. 100.

GUITTONNEAU, G.-G. 1991: *Erodium brachycarpum* (Geraniaceae), a Mediterranean species close to *E. botrys*. - *Bot. Chron. (Patras)* **10**: 685-687.

GUITTONNEAU, G.-G. & HOUÉROU, H.-N. DE 1968: Deux nouvelles espèces du genre *Erodium* L'Hérit. découvertes en Libye. - *Bull. Soc. Bot. France* **115(9)**: 591-600.

GUITTONNEAU, G.-G., MONTSERRAT RECODER, P. & VILLAR, L. 1987: L'*Erodium brachycarpum* (Godr.) Thell., espèce méditerranéenne méconnue de la série évolutive de l'*E. botrys* (Cav.) Bertol. (Geraniaceae). - *Candollea* **42(1)**: 417-425.

GUITTONNEAU, G.-G. & ROUX, M. 1977: Sur la taxinomie du genre *Erodium*. - *Cahiers Analyse Données* **2(1)**: 97-113.

GUITTONNEAU, G.-G., ROUX, M. & UZAN, J. 1971: Essai d'application de l'analyse factorielle des correspondances à la subdivision du genre *Erodium* L'Hér. - *Bull. Soc. Bot. France* **118(9)**: 719-732.

GÜLERYÜZ, G. 1998a: *Arabis drabiformis* Boiss. Brassicaceae/Cruciferae (Lahanagiller). - *Karaca Arbor. Mag.* **4(4)**: 185-188.

GÜLERYÜZ, G. 1998b: *Pedicularis olympica* Boiss. Scrophulariaceae (Yüksükotugiller). - *Karaca Arbor. Mag.* **4(4)**: 189-190.

GÜLERYÜZ, G. 1999a: *Achillea multifida* (DC.) Boiss. - *Karaca Arbor. Mag.* **5(1)**: 43-44.

GÜLERYÜZ, G. 1999b: *Aubieta olympica* Boiss. - *Karaca Arbor. Mag.* **5(1)**: 45-48.

GÜLERYÜZ, G. & ARSLAN, H. 1998: *Astragalus sibthorpianus* Boiss. - *Karaca Arbor. Mag.* **4(3)**: 143-144.

GÜLERYÜZ, G. & PIRDAL, M. 1996: Studies in the ecology of some endemics from Uludağ-Türkiye. In Öztürk, M.A., Seçmen, Ö. & Görk, G. (eds): Plant life in southwest and central Asia, 2: pp. 741-749. - Izmir: Ege University Press.

GÜNAK-SÜNTER, G. 1978: [Chromosome numbers in *Dianthus* from the Istanbul area]. In Löve, Á. (ed.): IOPB chromosome number reports LXI. - *Taxon* **27(4)**: 376.

GÜNER, A. 1995: *Iris xanthospuria* B. Mathew & T. Baytop. - *Karaca Arbor. Mag.* **3(2)**: 91-92.

GÜNER, A. & AL. 2000: A continuation of Peter Davis' monumental Flora - Volume 11, second supplement to the Flora of Turkey and the East Aegean Islands. In Özhatay, N. & al. (eds): Abstracts from the Second Balkan Botanical Congress, Istanbul, Turkey, May 14-18, 2000, p. 13.

GÜNER, A. & AL. (EDS) 2001: Flora of Turkey and the East Aegean Islands. Vol. 11 (Supplement 2). - Edinburgh: Edinburgh Univ. Press, xix + 656 pp.

GÜNER, E.D. 2016: *Oenanthe incrassans*: An enigmatic species from Turkey and its comparison with *Oenanthe pimpinelloides* (Apiaceae). - *PhytoKeys* **62**: 101-111.

GÜNER, E.D., TEKŞEN, M. & BANI, B. 2016: *Oenanthe millefolia* (Umbelliferae), a new species record for the Turkish and Greek flora. - *PhytoKeys* **70**: 31-40.

GUNN, C.R., SKRDLA, W.H. & SPENCER, H.C. 1978: Classification of *Medicago sativa* L. using legume characters and flower colors. - *Techn. Bull. U.S.D.A.* **1574**: iv + 84 pp.

GUO, Y.-H. & AL. 1990: The pollination ecology of *Zannichellia palustris* L. (Zannichelliaceae). - *Aquatic Bot.* **38(4)**: 341-356.

GUO, Y.-H. & COOK, C.D.K. 1990: The floral biology of *Groenlandia densa* (L.) Fourreau (Potamogetonaceae). - *Aquatic Bot.* **38(2-3)**: 283-288.

GUO, Y.-P., EHRENDORFER, F. & SAMUEL, R. 2004: Phylogeny and systematics of *Achillea* (Asteraceae-Anthemideae) inferred from nrITS and plastid trnL-F DNA sequences. - *Taxon* **53(3)**: 657-672.

GUPTA, K.M. 1962: *Marsilea* (Botanical Monograph no. 2). - New Delhi: Council of Scientific & Industrial Research, x + 113 pp.

GUPTA, P.K. & BAUM, B.R. 1986: Nomenclature and related taxonomic issues in wheats, triticales and some of their wild relatives. - *Taxon* **35(1)**: 144-149.

GUPTA, P.K. & SRIVASTVA, A.K. 1970: Natural triploidy in *Cynodon dactylon* (L.) Pers. - *Caryologia* **23(1)**: 29-35.

GUPTA, S.K., HAMAL, I.A. & KOUL, A.K. 1991: Chromosome evolution in the genus *Torilis* (Adans) Apiaceae. - *Cytologia* **56(4)**: 597-602.

GUSEV, Y.D. 1972: The survey of genus *Amaranthus* in the U.S.S.R. [In Russian with English summary]. - *Bot. Zhurn. (Moscow & Leningrad)* **57(5)**: 457-464.

GUSSEV, CH. 1993: *Helichrysum plicatum* DC. - a new species for the Bulgarian flora. In Petrova, A.V. (ed.): Abstracts from the VII OPTIMA meeting in Borovetz, Bulgaria, 18-30 July 1993, p. 184.

GUSSEV, CH. & AL. 1998: New chorological data on vascular plants in the Eastern Rhodopes. - *Phytol. Balcan.* **4(1-2)**: 187-195.

GUSSEV, CH. & AL. 2003: Conservation of *Leucojum aestivum* (Amaryllidaceae) in Bulgaria. - *Bocconea* **16(2)**: 815-521.

GUSSEV, CH. & AL. 2020: New floristic and syntaxonomic data from rice fields in Bulgaria. - *Bot. Serbica* **44(1)**: 95-100.

GUSSONE, G. 1826: Plantae rariores quas in itinere per oras Jonii ac Adriatici Maris et per regiones Samnii ac Aprutii. - Neapoli: Ex Regia Typographia.

GUSSONE, G. 1827-28: Florae Siculae Prodromus sive plantarum in Sicilia ulteriori nascentium enumeratio secundum systema Linnaeanum disposita. - Napoli: 2 vols. [cf. TL-2 1: 1025-1026, 1976].

GUSSONE, G. 1842-45: Florae Siculae Synopsis exhibens plantas vasculares in Sicilia insulisque adjacentibus huc usque detectas, secundum systema Linnaeanum dispositas. - Napoli: Tramater, 2 vols. [cf. TL-2 1: 1026, 1976].

GUSTAFSON, A.H. 1933: Cytological studies in the genus *Hieracium*. - *Bot. Gaz.* **94(3)**: 512-533.

GUSTAFSSON, M. 1970: Studies in the Aegean flora. XVII. Variation and distribution of *Atriplex recurva* D'Urv. (Chenopodiaceae). - *Bot. Not.* **123(3)**: 371-383.
GUSTAFSSON, M. & AL. 1976: Meiosis in Greek *Brassica* of the *oleracea* group. - *Bot. Not.* **129(1)**: 73-84.
GUSTAFSSON, M. & LANNÉR-HERRERA, C. 1997a: Overview of the *Brassica oleracea* complex: their distribution and ecological specificities. - *Bocconea* **7**: 27-37.
GUSTAFSSON, M. & LANNÉR-HERRERA, C. 1997b: Diversity in natural populations of wild cabbage (*Brassica oleracea* L.). - *Bocconea* **7**: 95-102.
GUSTAFSSON, M. & SNOGERUP, S. 1972: *Scorzonera scyria*, a new chasmophytic species from Greece. - *Bot. Not.* **125(4)**: 323-328.
GUSTAFSSON, M. & SNOGERUP, S. 1974: Studies in the Aegean flora. XXII. The flora of the island of Skantzoura. - *Bot. Not.* **127(3)**: 364-372.
GUSTAFSSON, M. & SNOGERUP, S. 1983: A new subspecies of *Brassica cretica* from Peloponnisos, Greece. - *Bot. Chron. (Patras)* **3(1-2)**: 7-11.
GUSTAVSSON, L.-Å. 1976: New species of *Anchusa* and *Arenaria* from Sterea Ellas, Greece. - *Bot. Not.* **129(3)**: 273-278.
GUSTAVSSON, L.-Å. 1977: *Drabopsis verna* C. Koch (Brassicaceae) new to Europe. - *Bot. Not.* **130(2)**: 213-214.
GUSTAVSSON, L.-Å. 1978a: Floristic reports from the high mountains of Sterea Ellas, Greece. 1. - *Bot. Not.* **131(1)**: 7-25.
GUSTAVSSON, L.-Å. 1978b: Floristic reports from the high mountains of Sterea Ellas, Greece. 2. - *Bot. Not.* **131(2)**: 201-213.
GUŞULEAC, M. 1927a: Die europäischen Arten der Gattung *Anchusa* Linné. - *Bul. Fac. Şti. Cernăuţi* **1(1)**: 73-123 + 13 plates.
GUŞULEAC, M. 1927b: Die aussereuropäischen Arten der Gattung *Anchusa* Linné. - *Bul. Fac. Şti. Cernăuţi* **1(2)**: 235-325 + 5 plates.
GUŞULEAC, M. 1928: Die monotypischen und artenarmen Gattungen der Anchusae (*Caryolopha*, *Brunnera*, *Hormuzakia*, *Gastrocotyle*, *Phyllocara*, *Trachystemon*, *Procopiana* und *Borago*). - *Bul. Fac. Şti. Cernăuţi* **2(2)**: 394-461 + 6 plates.
GUŞULEAC, M. 1929a: Zur Nomenklatur der *Anchusa officinalis* Linné und *A. italica* Retz. - *Repert. Spec. Nov. Regni Veg.* **26**: 81-84 + 2 plates.
GUŞULEAC, M. 1929b: Species Anchusae generis Linn. hucusque cognitae. - *Repert. Spec. Nov. Regni Veg.* **26**: 286-322 + 18 plates.
GUTERMANN, W. 1995: Catalogus abbreviatus plantarum vascularium ex Ionii Insuli cognitarum. [Editio reformata (Bryophytos includens) ad usum scientiae amabilis studosiorum iternis 1995 mediterranei participium]. - Wien: Inst. Bot. Univ. Wien, 57 pp. (privately distributed).
GUTERMANN, W. 1997a: Catalogus abbreviatus plantarum vascularium ex Ionii Insuli cognitarum. [Editio reformata (Bryophytos includens) ad usum scientiae amabilis studosiorum iternis 1997 mediterranei participium]. - Wien: Inst. Bot. Univ. Wien, 76 pp. (privately distributed).
GUTERMANN, W. 1997b: Ithaka-Herbar. - Vienna: typescript, 11 pp.
GUTERMANN, W. 1999: Catalogus abbreviatus plantarum vascularium ex Ionii Insuli cognitarum. [Editio reformata (Bryophytos includens) ad usum scientiae amabilis studosiorum iternis 1999 mediterranei participium]. - Wien: Inst. Bot. Univ. Wien, 81 pp. (privately distributed).
GUTERMANN, W. 2003: Catalogus abbreviatus plantarum vascularium ex Ionii Insuli cognitarum [updated version]. - Vindobonae [Vienna]: computer printout, 28 pp.
GUTERMANN, W. 2005: *Pardoglossum cheirifolium* (L.) Barbier & Mathez, *Trifolium cinctum* DC. In Greuter, W. & Raus, Th. (eds): Med-Checklist notulae, 23. - *Willdenowia* **35(1)**: 55, 59-60.
GUTERMANN, W. 2006: Notulae Nomenclaturales 19-25 (Neue Namen undKombinationen zur Flora Südtirols). - *Phyton (Horn)* **46(1)**: 71-81.
GUTERMANN, W. 2011: Catalogus abbreviatus plantarum vascularium ex Ionii Insuli cognitarum. [Editio reformata (Bryophytos includens) ad usum scientiae amabilis studosiorum iternis 2011 mediterranei participium]. - Wien: Inst. Bot. Univ. Wien, 84 pp. (privately distributed).
GUTERMANN, W. 2012: A note on *Pteris vittata* L. (Pteridaceae) in Montenegro. - *Acta Bot. Croat.* **71 (2)**: 371-374.
GUTERMANN, W. & AL. 2011: (2005) Proposal to conserve the name *Convallaria latifolia* Jacq. (*Polygonatum latifolium*) (Ruscaceae, Liliaceae s.l.) against *C. latifolia* Mill. - *Taxon* **60(2)**: 595-596. [approved, cf. Taxon 63 (6): 1359, 2014].
GUTERMANN, W. & AL. 2014: The vascular plants of Kálamos (Ionian Islands, Greece): a catalogue. - *Phyton (Horn)* **54(2)**: 161-196.
GUTERMANN, W. & EHRENDORFER, F. 2000: *Asperula naufraga* (Rubiaceae), a new species from Zakinthos (Ionian Islands, Greece), with notes on its ecology, karyology and relationships. (Materials towards a Flora Ionica, 1. - *Bot. Chron. (Patras)* **13**: 61-70.
GUTERMANN, W., EHRENDORFER, F. & FISCHER, M.A. 1974: Neue Namen und kritische Bemerkungen zur Gefässpflanzenflora Mitteleuropas. - *Oesterr. Bot. Z.* **122(4)**: 259-273.
GUTERMANN, W. & MERXMÜLLER, H. 1961: Die europäischen Sippen von *Oxytropis* sectio *Oxytropis*. - *Mitt. Bot. Staatssamml. München* **4(2)**: 199-275.
GUTERMANN, W. & STRID, A. 2005: *Rhamnus orbiculata* Bornm. In Greuter, W. & Raus, Th. (eds): Med-Checklist notulae, 23. - *Willdenowia* **35(1)**: 60-61.
GUTERMANN, W. & TAN, KIT 2006: *Thymelaea gussonei*. In Greuter, W. & Raus, Th. (eds): Med- Checklist notulae, 24. - *Willdenowia* **36(2)**: 726.
GUTIÉRREZ BUSTILLO, A.M. 1981: Revisión del género *Angelica* L. (Umbelliferae) en la Península Ibérica. - *Lazaroa* **3**: 137-161.
GUTIÉRREZ, H.F. 2014: (2259) Proposal to conserve the name *Pennisetum clandestinum* (*Cenchrus clandestinus*) against *P. longistylum* (Poaceae). - *Taxon* **63(1)**: 189-190. [not approved, cf. Taxon 63(6): 1368, 2014].
GUTIÉRREZ-LARRUSCAIN, D. 2018: Phylogeny of the *Inula* group (Asteraceae: Inuleae): Evidence from nuclear and plastid genomes and a recircumscription of *Pentanema*. - Taxon **67**(1): 149-164.
GUTTE, P. & AL. 1987: *Ailanthus altissima* (Mill.) Swingle - eine vergleichend pflanzengeographische Studie. - *Folia Geobot. Phytotax.* **22(3)**: 241-262.
GÜVENÇ, A. & DUMAN, H. 2010: Morphological and anatomical studies of annual taxa of *Sideritis* L. (Lamiaceae), with notes on chorology in Turkey. - *Turk. J. Bot.* **34(2)**: 83-104.
GUYOT, J. 1979: Les hybrides *Viola alba* × *odorata* et *Viola hirta* × *odorata*. - *Monde Pl.* **74(399)**: 1-3.
GUZOWSKA, I. 1966: Microsporogenesis and chromosome number in *Cytinus Hypocistis* L. - *Acta Soc. Bot. Poloniae* **35(3)**: 445-454.
HAAN, I. DE & DOORENBOS, J. 1951: The cytology of the genus *Cyclamen*. - *Meded. Landbouwhogeschool* **51(7)**: 151-166.
HAAS, J., HAAS, J. & HAAS, M. 2000: Orchideen-Exkursion Karpathos. Exkursionsprotokoll - 02.04.1998 bis 14.04.1998. - *J. Eur. Orch.* **32(1)**: 159-168.
HAAS, P. 1950: Systematische Beobachtungen über das Quellgras, *Catabrosa aquatica* P.B. - *Ber. Bayer. Bot. Ges.* **28**: 212-216.
HABECK, F. & REIF, A. 1994: Die Waldgesellschaften der montanen und subalpinen Stufe des Ostabfalls des Olymp, Griechenland. - *Phytocoenologia* **22**: 501-536.
HABELER, E. 1963: Cytotaxonomie von *Cardamine amara* des Alpenostrandes. - *Phyton (Horn)* **10(3-4)**: 161-205.
HACKBARTH, J. 1966. Polyploidie in der Gattung *Lupinus*. - *Acta Agric. Scand., Suppl.* **16**: 253-255.
HACKEL, E. 1878: Zwei kritische Gräser der griechischen Flora. - *Oesterr. Bot. Z.* **28(6)**: 189-192.
HACKEL, E. 1882: Monographia Festucarum Europaearum. - Kassel & Berlin: Verlag von Theodor Fischer, ix + 216 pp. + 4 plates.
HACKEL, E. 1898: *Poa Grimburgii* n. sp. - *Oesterr. Bot. Z.* **48(1)**: 12-14.
HACKEL, E. 1903: Über *Bromus japonicus* Thunb. - *Magyar Bot. Lapok* **2(3)**: 57-63.

HADAČEK, F. 1989: Vergleichende phytochemische Untersuchungen in der Gattung *Peucedanum* (Apiaceae-Apioideae). - *Stapfia* **18**: 1-186.
HADAČEK, F. & SAMUEL, R. 1994: Chromosome counts and chemotaxonomy in *Peucedanum* sect. *Peucedanum* (Apiaceae-Apioideae) from the Balkan Peninsula. - *Willdenowia* **24(1-2)**: 33-48.
HADJIKYRIAKOU, G.N. 2018: *Fuirena pubescens* (Poir.) Kunth (Cyperaceae), a new addition to the native flora of Cyprus. - *Cypricola* **7**: 1-6.
HADJIKYRIAKOU, G.N. & HAND, R. 2008: Notes on *Teucrium* sect. *Polium* (Lamiaceae) in Cyprus. - *Willdenowia* **38(1)**: 111-125.
HADJIKYRIAKOU, G.N. & HAND, R. 2011: *Teucrium salaminium* Hadjik. & Hand (Lamiaceae, *Teucrium* sect. *Polium*), a new species from Cyprus. - *Candollea* **66(2)**: 341-351.
HADJIKYRIAKOU, G.N., HAND, R. & MANISON, G. 2011: *Arum cylindraceum* subsp. *pitsyllianum* (Araceae), a new taxon from Cyprus. - *Novon* **21(4)**: 431-436.
HADŽIABLAHOVIĆ, S. 2012: *Vicia loiseleurii* (M. B.) D. Litvinov (Leguminosae) in the flora of Montenegro. - *Natura Montenegrina* **11(3)**: 425-341.
HAESLER, I. 1982: Lectotypisierung der Arten *Viola hirta* L. und *Viola odorata* L. - *Mitt. Bot. Staatssamml. München* **18**: 289-296.
HAEUPLER, H. (ED.) 1991: Exkursion zum Peloponnes im Rahmen des S-Blocks "Mediterrane Ökosysteme am Beispiel des Peloponnes" (Seminarbeiträge, Exkursionsprotokolle und Artenlisten). - Bochum: Ruhr-Universität, 280 pp.
HAGEMANN, I. 1981: Wuchsformenuntersuchungen an balkanischen *Hypericum*-Sippen. - *Bot. Jahrb. Syst.* **102(1-4)**: 239-254.
HAGEMANN, I. 1983: Wuchsformenuntersuchungen an zentraleuropäischen *Hypericum*-Arten. - *Flora, Morphol. Geobot. Oekophysiol.* **173(1-2)**: 97-142.
HAGEMANN, I. 1987: On subspecific taxa of *Hypericum empetrifolium* (Hypericaceae) from Crete. - *Pl. Syst. Evol.* **155(1-4)**: 165-187.
HAGEMANN, I. 1989: Wuchsformen einiger *Hypericum*-Arten, ein Beitrag zum morphologischen und zum ökologischen Anliegen der Wuchsformen-Forschung. - *Flora, Morphol. Geobot. Oekophysiol.* **183(3-4)**: 225-309.
HAGEMANN, I. 1991: Die ökogeographische Differenzierung innerhalb der Gattung *Hypericum* (Johanniskraut) im östlichen Mittelmeergebiet. - *Sitzungsber. Ges. Naturf. Freunde Berlin, N.S.* **31**: 59-76.
HAGEMANN, I. & MEUSEL, H. 1984: *Hypericum triquetrifolium* Turra, ein Wurzelspross-Geophyt. Wuchsform und Verbreitung. - *Flora, Morphol. Geobot. Oekophysiol.* **175(6)**: 385-405.
HAGEMANN, I. & RAUS, TH. 1984: *Fumana aciphylla* Boiss. In Greuter, W. & Raus, Th. (eds): Med- Checklist notulae, 9. - *Willdenowia* **14(1)**: 52.
HAGER, J. 1985: Pflanzenökologische Untersuchungen in den subalpinen Dornpolsterfluren Kretas. - *Diss. Bot.* **89**: 1-196 + 25 pp. of lists and plates.
HAGERUP, O. 1944: Notes on some boreal polyploids. - *Hereditas (Lund)* **30(1-2)**: 152-160.
HAINES, R.W. & LYE, K.A. 1971: Studies in African Cyperaceae. IV. *Lipocarpha* R. Br., *Hemicarpha* Nees, and *Isolepis* R. Br. - *Bot. Not.* **124**: 473-482.
HÁJEK, M., HÁJKOVÁ, P. & APOSTOLOVA, I. 2005: Notes on the Bulgarian wetland flora, including new national and regional records. - *Phytol. Balcan.* **11(2)**: 173-184.
HÁJEK, M., HÁJKOVÁ, P. & APOSTOLOVA, I. 2006: New wetland vascular plants for Bulgaria. - *Phytol. Balcan.* **12(3)**: 367-370.
HALÁCSY, E.V. 1886: *Goniolimon Heldreichii* n. sp. (*Statice Heldreichii*). Eine neue *Goniolimon*-Art der thessalischen Flora. - *Verh. K.K. Zool.-Bot. Ges. Wien* **36**: 241-242 + 1 plate.
HALÁCSY, E.V. 1888: Beiträge zur Flora der Landschaft Doris, inbesondere des Gebirges Kiona in Griechenland. - *Verh. K.K. Zool.-Bot. Ges. Wien* **38**: 745-764 + 1 plate.
HALÁCSY, E.V. 1890: Beiträge zur Flora der Balkanhalbinsel. I-IV. - *Oesterr. Bot. Z.* **40**: 37-41, 114-116, 164-166, 404-406.
HALÁCSY, E.V. 1891a: Beiträge zur Flora der Balkanhalbinsel. V. - *Oesterr. Bot. Z.* **41(7)**: 221-223.
HALÁCSY, E.V. 1891b: Beiträge zur Flora der Balkanhalbinsel. VI. - *Oesterr. Bot. Z.* **41(11)**: 370-372.
HALÁCSY, E.V. 1891c: Beiträge zur Flora der Balkanhalbinsel. VII. - *Oesterr. Bot. Z.* **41(12)**: 408-409.
HALÁCSY, E.V. 1892a: Beiträge zur Flora der Balkanhalbinsel. VIII. - *Oesterr. Bot. Z.* **42(11-12)**: 368-375, 400-401.
HALÁCSY, E.V. 1892b: Beiträge zur Flora der Balkanhalbinsel. IX. Florula insulae Thasos. - *Oesterr. Bot. Z.* **42(12)**: 412-420.
HALÁCSY, E.V. 1893a: Beiträge zur Flora der Balkanhalbinsel. IX. Florula insulae Thasos (Schluss). - *Oesterr. Bot. Z.* **43(1)**: 22-24.
HALÁCSY, E.V. 1893b: Novitäten aus der Flora Albaniens. - *Verh. K.K. Zool.-Bot. Ges. Wien* **42(4)**: 576-578.
HALÁCSY, E.V. 1893c: Beiträge zur Flora der Balkanhalbinsel. X. *Centaurea Formanekii*. - *Oesterr. Bot. Z.* **43(2)**: 55-56.
HALÁCSY, E.V. 1893d: [Account of a botanical expedition in Greece]. - *Oesterr. Bot. Z.* **43**: 150, 299-300, 328, 436-437.
HALÁCSY, E.V. 1894a: Botanische Ergebnisse einer im Auftrage der hohen Kaiserl. Akademie der Wissenschaften unternommenen Forschungsreise in Griechenland. I. Beitrag zur Flora von Epirus. - *Denkschr. Kaiserl. Akad. Wiss., Wien. Math.-Naturwiss. Kl.* **61**: 217-268 + 3 plates.
HALÁCSY, E.V. 1894b: Botanische Ergebnisse einer im Auftrage der hohen Kaiserl. Akademie der Wissenschaften unternommenen Forschungsreise in Griechenland. II. Beitrag zur Flora von Aetolien und Acarnanien. - *Denkschr. Kaiserl. Akad. Wiss., Wien. Math.-Naturwiss. Kl.* **61**: 309-322 + 2 plates.
HALÁCSY, E.V. 1894c: Botanische Ergebnisse einer im Auftrage der hohen Kaiserl. Akademie der Wissenschaften unternommenen Forschungsreise in Griechenland. III. Beitrag zur Flora von Thessalien. - *Denkschr. Kaiserl. Akad. Wiss., Wien. Math.-Naturwiss. Kl.* **61**: 467-486 + 2 plates.
HALÁCSY, E.V. 1894d: Botanische Ergebnisse einer im Auftrage der hohen Kaiserl. Akademie der Wissenschaften unternommenen Forschungsreise in Griechenland. IV. Beitrag zur Flora von Achaia und Arcadien. - *Denkschr. Kaiserl. Akad. Wiss., Wien. Math.-Naturwiss. Kl.* **61**: 487-535.
HALÁCSY, E.V. 1894e: *Centaurea Charrelii* Halácsy et Dörfler. sp. n. - *Jahreskat. Wiener Bot. Tauschvereins* **1894**: 6.
HALÁCSY, E.V. 1895a: Beitrag zur Flora von Griechenland. - *Oesterr. Bot. Z.* **45**: 121-125, 171-177, 215-221, 257-260.
HALÁCSY, E.V. 1895b: Beitrag zur Flora von Griechenland. (Fortsetzung). - *Oesterr. Bot. Z.* **45**: 337-341, 382-387, 409-412, 460-463.
HALÁCSY, E.V. 1896a: Beitrag zur Flora von Griechenland (Schluss). - *Oesterr. Bot. Z.* **46(1)**: 11-19.
HALÁCSY, E.V. 1896b: Ueber eine neue *Lonicera* aus der Balkanhalbinsel. - *Verh. K.K. Zool.-Bot. Ges. Wien* **46**: 473-474.
HALÁCSY, E.V. 1897a: Florula Sporadum. - *Oesterr. Bot. Z.* **47(2-3)**: 60-62, 92-99.
HALÁCSY, E.V. 1897b: Beitrag zur Flora von Griechenland. - *Oesterr. Bot. Z.* **47(8-9)**: 281-287, 324-326.
HALÁCSY, E.V. 1898a: Die bisher bekannten *Centaurea*-Arten Griechenlands. - *Bull. Herb. Boissier* **6(7-8)**: 565-603, 633-659.
HALÁCSY, E.V. 1898b: Die bisher bekannten *Verbascum*-Arten Griechenlands. - *Verh. K.K. Zool.-Bot. Ges. Wien* **48**: 119-150.
HALÁCSY, E.V. 1898c: Beiträge zur Flora von Griechenland. - *Verh. K.K. Zool.-Bot. Ges. Wien* **48**: 700-714.
HALÁCSY, E.V. 1899a: Florula Strophadum. - *Oesterr. Bot. Z.* **49(1)**: 24-25.
HALÁCSY, E.V. 1899b: Eine neue *Statice*-Art der griechischen Flora. - *Allg. Bot. Z. Syst.* **5(1)**: 1-2.
HALÁCSY, E.V. 1899c: Beiträge zur Flora von Griechenland. Zweiter Theil. - *Verh. K.K. Zool.-Bot. Ges. Wien* **49(2-3)**: 185-195.

HALÁCSY, E.V. 1900: Conspectus Florae Graecae. Vol. 1 pp. 1-576. - Lipsiae [Leipzig]: Guilelmi Engelmann [Reprinted 1969 by Verlag J. Cramer].
HALÁCSY, E.V. 1901: Conspectus Florae Graecae. Vol. 1 pp. 577-825. - Lipsiae [Leipzig]: Guilelmi Engelmann [Reprinted 1969 by Verlag J. Cramer].
HALÁCSY, E.V. 1902a: Conspectus Florae Graecae. Vol. 2 (612 pp.). - Lipsiae [Leipzig]: Guilelmi Engelmann [Reprinted 1969 by Verlag J. Cramer].
HALÁCSY, E.V. 1902b: Theodor von Heldreich [obituary]. - *Magyar Bot. Lapok* **1(11)**: 325-336.
HALÁCSY, E.V. 1904a: Conspectus Florae Graecae. Vol. 3 (519 pp.). - Lipsiae [Leipzig]: Guilelmi Engelmann [Reprinted 1969 by Verlag J. Cramer].
HALÁCSY, E.V. 1904b: Novitäten aus der griechischen Flora. - *Verh. K.K. Zool.-Bot. Ges. Wien* **54**: 483-484.
HALÁCSY, E.V. 1905: Ueber die Entdeckung von *Solenanthus Tournefortii* DC. in Europa. - *Magyar Bot. Lapok* **4(8-11)**: 259-260.
HALÁCSY, E.V. 1906: Aufzählung der von Herrn Prof. Dr. L. Adamović im Jahre 1905 auf der Balkan halbinsel gesammelten Pflanzen. - *Oesterr. Bot. Z.* **56(5-7)**: 205-212, 277-283.
HALÁCSY, E.V. 1908: Conspectus Florae Graecae. Supplementum (132 pp.). - Lipsiae [Leipzig]: Guilelmi Engelmann [Reprinted 1969 by Verlag J. Cramer].
HALÁCSY, E.V. 1910: Aufzählung der von Dr. B. Tuntas auf der Insel Scyros der nördlichen Sporaden im Juni 1908 gesammelten Arten. - *Oesterr. Bot. Z.* **60(3-4)**: 114-118, 141-145.
HALÁCSY, E.V. 1912: Supplementum secundum Conspectus florae graecae. - *Magyar Bot. Lapok* **11(5-8)**: 114-202. [Bound together with vols. 2-3 and suppl. 1 in the reprinted edition].
HALADA, L. 1994: *Ruscus hypoglossum* L. in Slovakia. - *Thaiszia* **4(2)**: 183-195.
HALBMAYR, H. 1988: Dem Enzian auf der Spur. Monographie der Gattung *Gentiana* und nahe verwandter Gattungen. - Wien-Schönbrunn: H. Halbmayr, 481 pp.
HALDA, J.J. 1979: A new intergeneric hybrid in the family Gesneraceae [sic!]. - *Preslia* **51(4)**: 375-376.
HALDA, J.J. 1991: Plants of Mt. Olympus. - *Bull. Amer. Rock. Gard. Soc.* **49(4)**: 261-264 + 4 plates.
HALDA, J.J. 1992: The genus *Primula* in cultivation and in the wild. - Englewood, Col.: Tethys Books, 364 + ix pp.
HALDA, J.J. 1997: Some nomenclatoric changes and new descriptions in the genus *Daphne* L. - *Acta Mus. Richnov., Sect. Nat.* **4(2)**: 67-70.
HALDA, J.J. 1998a: Some taxonomic problems in the genus *Daphne* L. - *Acta Mus. Richnov., Sect. Nat.* **5(4)**: 133-160.
HALDA, J.J. 1998b: Some taxonomic problems in the genus *Daphne* L. 1b. - *Acta Mus. Richnov., Sect. Nat.* **5(5)**: 165-186.
HALDA, J.J. 1999: Some taxonomic problems in the genus *Daphne* L. 2. - *Acta Mus. Richnov., Sect. Nat.* **6(3)**: 195-233.
HALDA, J.J. 2001a: New descriptions and combinations. - *Acta Mus. Richnov., Sect. Nat.* **8(1)**: 31-34.
HALDA, J.J. 2001b: The genus *Daphne*. - Dobre, Czech Republic: SEN, 231 pp.
HALDA, J.J. 2004: The genus *Paeonia*. - Portland: Timber Press, 227 pp.
HALDA, J.J. 2013: Two new intergeneric hybrids in the family Gesneriaceae. - *Int. Rock Gard.* **47**: 2-5.
HALDA, J.J. & HALDOVÁ, J. 1997: The genus *Gentiana*. - Dobre, Czech Republic: SEN, 209 pp.
HALDIMANN, G. 1989: Agamospermie chez la forme submergée de *Callitriche palustris* L. et contribution à l'étude cytologique de cette espèce. - *Bull. Soc. Neuchâteloise Sci. Nat.* **112**: 29-32 + 3 plates.
HALL, A.D. 1937: Polyploidy in *Tulipa*. - *J. Linn. Soc., Bot.* **50(335)**: 481-489.
HALL, A.D. 1938: Three new species of tulips. - *J. Bot. (London)* **76**: 313-319.
HALL, A.D. 1940a: *Tulipa bakeri*. - *Bot. Mag.* **163(1)**: tab. 9607 + 2 pp. text.
HALL, A.D. 1940b: The genus *Tulipa*. With forty illustrations in colour by H.C. Osterstock. - London: Royal Horticultural Society, viii + 171 pp. and 23 + 40 plates.
HALL, T. 2000: *Campanula saxatilis* L. subsp. *cytherea* Rech. f. ex Phitos and *Campanula thessala* Maire. In Rolfe, R. (ed.): Plant awards 1998-1999. - *Quart. Bull. Alpine Gard. Soc. Gr. Brit.* **68(2)**: 263-265.
HALL, T. 2001: *Androcymbium rechingeri* Greuter. In Rolfe, R. (ed.): Plant awards 2000-2001. - *Alpine Gardener* **69(4)**: 505-506.
HALL, T. 2003a: [Photo of *Tulipa cretica*. In: Early spring show, March 2, 2002]. - *Alpine Gardener* **71 (1)**: 42-43.
HALL, T. 2003b: *Scilla talosii* Tzanoud. & Kypr. (syn. *Prospero talosii* (Tzanoud. & Kypr.) F. Speta). - *Alpine Gardener* **71(4)**: 427-428.
HALL, T., TILLIE, N. & CHASE, M.W. 2002: A re-evaluation of the bulbous irises. - *Ann. Bot., n.s. (Rome)* **1(2)**: 123-126.
HALLIDAY, G. 1964: Studies in the *Minuartia verna* complex, I. In Heywood, V.H. (ed.): Notulae systematicae ad Floram Europaeam spectantes, no. 3. - *Feddes Repert. Spec. Nov. Regni Veg.* **69 (1)**: 8-14.
HALLIER, H. 1897a: Bausteine zu einer Monographie der Convolvulaceen. - *Bull. Herb. Boissier* **5(5-11)**: 366-387, 736-754, 804-820, 996-1013.
HALLIER, H. 1897b: Bausteine zu einer Monographie der Convolvulaceen [cont.]. - *Bull. Herb. Boissier* **5(12)**: 1021-1052.
HALLIWELL, B. 1979: History of *Lilium candidum* in cultivation. - *Lilies* **1978/79**: 63-64.
HALX, G. 1972: Über einige Bastarde von *Ophrys cretica* (Vierh.) Nelson. - *Hoppea* **30**: 97-110 + 2 plates. [Also distributed as *OPTIMA leaflet* no. 30].
HAMASHA, H.R., HAGEN, B.V. & RÖSER, M. 2012: *Stipa* (Poaceae) and allies in the Old World: molecular phylogenetics realigns genus circumscription and gives evidence on the origin of American and Australian lineages. - *Pl. Syst. Evol.* **298(2)**: 351-367.
HAMBLER, D.J. 1958: Some taxonomic investigations on the genus *Rhinanthus*. - *Watsonia* **4(3)**: 101-116 + 1 plate.
HAMEL, J.-L. 1953: Contribution à l'étude cyto-taxinomique des Saxifragacées. - *Rev. Cytol. Biol. Vég.* **14(3-4)**: 113-313.
HAMEL, J.-L. 1955: Étude caryologique de quelques *Eryngium*. - *Bull. Soc. Bot. France* **102(9)**: 488-502.
HAMEL, J.-L. 1958: Matériaux pour l'étude caryo-taxinomique des Saxifragacées. IV. Trois saxifrages de la section *Kabschia* Engl. - *Bull. Mus. Natl. Hist. Nat., Sér. 2* **30(2)**: 199-205.
HÄMET-AHTI, L. 1980: *Juncus alpinoarticulatus*: the legitimate name for *Juncus alpinus*. - *Ann. Bot. Fenn.* **17(3)**: 341-342.
HAMILTON, A.P. 1976a: The European gladioli. - *Gladiolus Annual Brit. Gladiolus Soc.* **1976**: 35-39.
HAMILTON, A.P. 1976b: The European gladioli. - *Quart. Bull. Alpine Gard. Soc. Gr. Brit.* **44(2)**: 140-147.
HAMMER, K. 1978: Entwicklungstendenzen blütenökologischer Merkmale bei *Plantago*. - *Flora, Morphol. Geobot. Oekophysiol.* **167(1)**: 41-55.
HAMMER, K. 1980a: Vorarbeiten zur monographischen Darstellung von Wildpflanzensortimenten: *Aegilops* L. - *Kulturpflanze* **28**: 33-180.
HAMMER, K. 1980b: Zur Taxonomie und Nomenklatur der Gattung *Aegilops* L. - *Feddes Repert.* **91(4)**: 225-258.
HAMMER, K. 1984: Bestäubungsökologische Merkmale und Phylogenie von *Hordeum* L. subgen. *Hordeum*. - *Flora, Morphol. Geobot. Oekophysiol.* **175(5)**: 339-344.
HAMMER, K. 1990: Bemerkungen zu *Secale strictum* (K. Presl) K. Presl. - *Feddes Repert.* **101(5-6)**: 285-287.
HAMMER, K., ROMEIKE, A. & TITTEL, C. 1983: Vorarbeiten zur monographischen Darstellung von Wildpflanzensortimenten: *Datura* L., sectiones *Dutra* Bernh., *Ceratocaulis* Bernh. et *Datura*. - *Kulturpflanze* **31**: 13-75.
HAMMES, W. & ARNDT, H.J. 1985: Die Identifikation verschiedener Buchenarten anhand morphologischer Blattmerkmale. - *Allg. Forstz.* **41(45)**: 1207-1212.
HAMMETT, K.R.W. & AL. 1996: New interspecific hybrids in *Lathyrus* (Leguminosae): *Lathyrus annuus* × *L. hierosolymitanus*. - *Bot. J. Linn. Soc.* **122(2)**: 89-101.

HAMMOUD, C. & AL. 2021: Past connections with the mainland structure patterns of insular species richness in a continental-shelf archipelago (Aegean Sea, Greece). - *Ecology and Evolution* **11**: 5451-5458.

HAMPE, E. 1837: Revision der durch Hrn. Dr. Frivaldsky von Frivald in Pest vertheilten, getrockneten Pflanzen aus der Türkei. - *Flora* **20(1)**: 225-234.

HAMPE, E. 1842: II. Correspondenz [account of plants collected by Spruner in Attica]. - *Flora* **25(3)**: 57-64, 74-80.

HAMZAOĞLU, E. & KOÇ, M. 2021: *Dianthus muglensis* nom. nov. and notes on *D. goerkii*, the correct name for *D. leucophaeus* var. *patens* (Caryophyllaceae). - *Phytotaxa* **491(4)**: 291-296.

HAND, R. 2001a: Lectotypification of the Linnaean names *Thalictrum angustifolium* and *T. lucidum* (Ranunculaceae). - *Nordic J. Bot.* **20(5)**: 527-529.

HAND, R. 2001b: Revision der in Europa vorkommenden Arten von *Thalictrum* subsectio *Thalictrum* (Ranunculaceae). - *Bot. Naturschutz Hessen, Beiheft* **9**: 1-358 (Diss., Freie Univ. Berlin).

HAND, R. 2019: *Orobanche hedypnoidis* comb. nov. in Cyprus, new but not unexpected. - *Cypricola* **14**: 1-4.

HAND, R. 2020: Various noteworthy records of flowering plants in Cyprus (1996–2019) and some status clarifications. - *Cypricola* **15**: 1-11.

HAND, R. & AL. 2015: Multiple origins of dendroid shrubs in the eastern Mediterranean *Euphorbia hierosolymitana* group (Euphorbiaceae) with description of a new species, *Euphorbia lemesiana*, from Cyprus. - *Bot. J. Linn. Soc.* **179(2)**: 295-307.

HAND, R. & AL. 2021: *Rostraria hadjikyriakou* (Poaceae), a new gypsophilous and endemic species from Cyprus. - *Fl. Medit.* **31**: 71-82.

HAND, R. & HADJIKYRIAKOU, G.N. 2009: *Cynara makrisii* (Asteraceae, Cardueae), a new artichoke species in Cyprus. - *Willdenowia* **39(1)**: 77-81.

HAND, R., HADJIKYRIAKOU, G.N. & CHRISTODOULOU, CH.S. 2019: Updated numbers of the vascular flora of Cyprus including the endemism rate, - *Cypricola* **13**: 1-6.

HAND, R., HADJIKYRIAKOU, G.N. & ZETZSCHE, H. 2012: *Scaligeria alziarii* (Apiaceae), a new sibling species of *S. napiformis* from Cyprus. - *Willdenowia* **42(2)**: 199-207.

HAND, R. 2018: *Thalictrum* subsectio *Thalictrum*: Nachträge zum Namenskatalog der in Europa vorkommenden Arten. - *Haussknechtia* **14**: 35-42.

HAND, R. (ED.) 2001: Supplementary notes on the flora of Cyprus II. - *Willdenowia* **31(2)**: 383-309.

HAND, R. (ED.) 2003: Supplementary notes on the flora of Cyprus III. - *Willdenowia* **33(2)**: 305-325.

HAND, R. (ED.) 2006: Supplementary notes to the flora of Cyprus V. - *Willdenowia* **36(2)**: 761-809.

HAND, R. (ED.) 2015: Supplementary notes on the flora of Cyprus VIII - *Willdenowia* **45(2)**: 245-259.

HANDEL-MAZZETTI, H.V. 1907: Monographie der Gattung *Taraxacum*. - Leipzig & Wien: Franz Deuticke, xi + 175 pp. + 5 plates and 2 maps.

HANDEL-MAZZETTI, H.V. 1909: Revision der balkanischen und vorderasiatischen *Onobrychis*-Arten aus der Sektion *Eubrychis*. - *Oesterr. Bot. Z.* **59(10-12)**: 369-378, 424-430, 479-488 + 1 map.

HANDEL-MAZZETTI, H.V. 1910: Revision der balkanischen und vorderasiatischen *Onobrychis*-Arten aus der Sektion *Eubrychis* [Fortsetzung]. - *Oesterr. Bot. Z.* **60(1-2)**: 5-12, 64-71 + 1 plate.

HANDEL-MAZZETTI, H.V. 1923: Nachträge zur Monographie der Gattung *Taraxacum*. - *Oesterr. Bot. Z.* **72(6-8)**: 254-275.

HANDEL-MAZZETTI, H.V. 1929: Berichtigung zu Dr. A. Hayeks Arbeit "Ein Beitrag zur Kenntnis der Vegetation und der Flora des thessalischen Olymp". - *Beih. Bot. Centralbl., Abt. 2* **45(3)**: 540.

HANELT, P. 1960: Die Lupinen. Zur Botanik und Geschichte landwirtschaftlich wichtiger Lupinenarten. - Wittenberg: Neue Brehm-Bücherei, 104 pp.

HANELT, P. 1961: Zur Kenntnis von *Carthamus tinctorius* L. - *Kulturpflanze* **9**: 114-145.

HANELT, P. 1963: Monographische Übersicht der Gattung *Carthamus* L. (Compositae). - *Feddes Repert. Spec. Nov. Regni Veg.* **67(1-3)**: 41-180 + 18 plates.

HANELT, P. 1972: Die intraspezifische Variabilität von *Vicia faba* L. und ihre Gliederung. - *Kulturpflanze* **20**: 75-128.

HANELT, P. 1993: Taxonomic problems of the Mediterranean Alliums and their relationships to other groups ... In Petrova, A.V. (ed.): Abstracts from the VII OPTIMA meeting in Borovetz, Bulgaria, 18-30 July 1993, p. 49.

HANELT, P. 1996a: (1258) Proposal to conserve the name *Allium ampeloprasum* against *A. porrum* (Liliaceae). - *Taxon* **45(4)**: 691-692. [approved, cf. Taxon 48(2): 362, 1999].

HANELT, P. 1996b: Taxonomic problems in Mediterranean *Allium*, and relationships with non-Mediterranean *Allium* groups. - *Bocconea* **5(1)**: 259-265.

HANELT, P. 1997a: Gene flow between crops and related taxa - some case studies. - *Bocconea* **7**: 51-61.

HANELT, P. 1997b: European wild relatives of *Prunus* fruit crops. - *Bocconea* **7**: 401-408.

HANELT, P. & AL. 1991: Infrageneric grouping of *Allium* - the Gatersleben approach. In Hanelt, P., Hammer, K. & Knüpffer, H. (eds): The genus *Allium* - taxonomic problems and genetic resources, pp. 21-26. - Gatersleben.

HANELT, P., HAMMER, K. & KNÜPFFER, H. (EDS) 1991: The genus *Allium* - taxonomic problems and genetic resources. Proceedings of an international symposium held at Gatersleben, Germany, June 11-13, 1991. - Gatersleben: Institut für Kulturpflanzenforschung, 359 pp.

HANELT, P. & METTIN, D. 1966: Cytosystematische Untersuchungen in der Artengruppe um *Vicia sativa* L. II. - *Kulturpflanze* **14**: 137-161.

HANELT, P. & METTIN, D. 1970a: Zur systematischen Stellung von *Vicia oroboides* Wulf. - *Kulturpflanze* **18**: 179-188.

HANELT, P. & METTIN, D. 1970b: Über die systematische Stellung temperater und meridionaler Sippen der Gattung *Vicia* L. - *Feddes Repert.* **81(1-5)**: 147-161.

HANELT, P., SCHÄFER, H.I. & SCHULTZE-MOTEL, J. 1972: Die Stellung von *Vicia faba* L. in der Gattung *Vicia* L. und Betrachtungen zur Entstehung dieser Kulturart. - *Kulturpflanze* **20**: 263-275.

HANELT, P. & SCHULTZE-MOTEL, J. 1962: Beobachtungen an einer blaublühenden Sippe von *Lithospermum arvense* L. - *Kulturpflanze* **10**: 122-131.

HANKS, G.R. 2002: Narcissus and daffodil. The genus *Narcissus*. - London and New York: Taylor & Francis, xix + 428 pp.

HANLIDOU, E. & AL. 2018: Changes between current and past composition of the macro- and microflora in the rice fields of the delta of the Axios-Loudias-Aliakmon (Greece). - *Bot. Serbica* **142 (suppl. 1)**: 170.

HANLIDOU, E., KOUTOUVELA, E. & TSOULKANAKI, O. 2019: New alien taxa in the rice-fields of the national park of the delta of Axios-Loudias-Aliakmon. In: Abstracts from the XVI Optima meeting, Athens 2-5 October 2019, p. 133.

HANNA, M.R. 1961: Cytological studies in *Bromus* species, sect. *Bromopsis*. - *Canad. J. Bot.* **39(3)**: 757-773 + 2 plates.

HANS, A.S. 1973: Chromosomal conspectus of the Euphorbiaceae. - *Taxon* **22(5-6)**: 591-636.

HANSEN, A. 1969: A new *Fritillaria* from Rhodes. - *Notes Roy. Bot. Gard. Edinburgh* **29(3)**: 329.

HANSEN, A. 1971: Flora der Inselgruppe Santorin. - *Candollea* **26(1)**: 109-163.

HANSEN, A. 1980a: Eine Liste der Flora der Inseln Kos, Kalymnos, Pserimos, Telendos und Nachbar-Inselchen (Ostägäis, Griechenland). - *Biol. Gallo-Hellen.* **9(1)**: 3-105.

HANSEN, A. 1980b: *Convolvulus sabaticus, Coronilla juncea, Genista sylvestris, Vincetoxicum huteri*. - *Willdenowia* **10(2)**: 231-232.

HANSEN, A. 1982: Additions to and notes on the flora of Corfu and Vidos (Ionian Islands, Greece). - *Bot. Chron. (Patras)* **2(1)**: 18-49.

HANSEN, A. 1986a: [Several new records for Greece]. In Greuter, W. & Raus, Th. (eds): Med-Checklist notulae, 12. - *Willdenowia* **15(2)**: 414-431.

HANSEN, A. 1986b: *Digitaria ciliaris* (Retz.) Koeler. In Greuter, W. & Raus, Th. (eds): Med-Checklist notulae, 13. - *Willdenowia* **16(1)**: 115.

HANSEN, A. & CARRIÓN, J.-S. 1990: *Flaveria bidentis* (L.) Kuntze (Asteraceae), nueva adventicia para España. - *Candollea* **45(1)**: 235-239.

HANSEN, A. & LOHMEYER, W. 1977: New finds of *Cornucopiae cucullatum* (Poaceae) in Europe. - *Bot. Not.* **130(2)**: 203-204.

HANSEN, A. & NIELSEN, H. 1993: Contribution to the flora of Lesvos (Mitilini), Greece. - *Willdenowia* **23 (1-2)**: 143-156.

HANSEN, A., SNOGERUP, S. & RUNGBY, S. 1966: Beiträge zur Kenntnis der Flora der Insel Rhodos. - *Sitzungsberg. Österr. Akad. Wiss., Math.-Naturwiss. Kl., Abt. 1, Biol.* **175(9-10)**: 363-381 + 1 map.

HANSEN, A. & STRID, A. 1989: *Flaveria bidentis* (L.) O. Kuntze. In Greuter, W. & Raus, Th. (eds): Med- Checklist notulae, 15. - *Willdenowia* **19(1)**: 28.

HANSEN, F. 1992: Distribution of *Clematis* species in Europe. - *Clematis Int.* **1992**: 81.

HANSEN, I. 1950: Die europäischen Arten der Gattung *Erica*. - *Bot. Jahrb. Syst.* **75(1)**: 1-81.

HANTZ, J. 1983: *Oxalis corniculata* L. in South-West Asia [In Polish with English summary]. - *Fragm. Florist. Geobot.* **27(4)**: 577-580 + 1 map.

HANTZ, J. 1986: Distribution of *Oxalis pes-caprae* L. in the East Mediterranean region. - *Ann. Mus. Goulandris* **7**: 49-56.

HARA, H. 1957: Synopsis of the genus *Chrysosplenium* L. (Saxifragaceae). - *J. Fac. Sci. Univ. Tokyo, Sect. 3, Bot.* **7(1)**: 1-90.

HARBERD, D.J. 1972: A contribution to the cyto-taxonomy of *Brassica* (Cruciferae) and its allies. - *Bot. J. Linn. Soc.* **65(1)**: 1-23.

HÅRD AF SEGERSTAD, F. 1940: Zur Systematik und Biologie von *Juncus effusus* L. und *Juncus conglomeratus* L. - *Acta Horti Gothob.* **14**: 41-59.

HARDING, W. 1969: The genus *Saxifraga*. - *Quart. Bull. Alpine Gard. Soc. Gr. Brit.* **37(1-4)**: 53-58, 112-138, 210-218, 388-393.

HARDING, W. 1970: The genus *Saxifraga* [cont.]. - *Quart. Bull. Alpine Gard. Soc. Gr. Brit.* **38(1-4)**: 43-65, 194-216, 287-292, 383-389.

HARDING, W. 1983: Plant portraits: *Saxifraga spruneri* Boiss. - *Quart. Bull. Alpine Gard. Soc. Gr. Brit.* **51(2)**: 151-153.

HARDION, L. & AL. 2012a: Revised systematics of Mediterranean *Arundo* (Poaceae) based on AFLP fingerprints and morphology. - *Taxon* **61(6)**: 1217-1226.

HARDION, L. & AL. 2012b: *Arundo micrantha* Lam. (Poaceae), the correct name for *Arundo mauritanica* Desf. and *Arundo mediterranea* Danin. - *Candollea* **67(1)**: 131-135.

HARDION, L. & AL. 2014: Origin of the invasive *Arundo donax* (Poaceae): a trans-Asian expedition in herbaria. - *Ann. Bot. (Oxford), n.s.* **114(3)**: 455-462.

HARDWICK, C. 2002: *Ophrys cretica* (Vierh.) Nels., P.C. - *Alpine Gardener* **70(4)**: 456-457.

HARIDON, L. & AL. 2014: Distinct evolutionary histories of lowland biota on Italian and Balkan peninsulas revealed by the phylogeography of *Arundo plinii* (Poaceae). - *J. Biogeogr.* **41(11)**: 2150-2161.

HARITONIDIS, S. 1989a: Evolution and present situation of the meadows of marine phanerogams in Greek seas. In Phitos, D. (ed.): Abstracts. Sixth OPTIMA meeting, Delphi 10-16 Sept. 1989, p. 41.

HARITONIDIS, S. 1989b: Evolution of marine phanerogam communities in the Greek coasts, during the last twenty years [In Greek with English abstract]. - *Bios (Thessaloniki)* **1989**: 99-104.

HARITONIDIS, S. 1991: Evolution and present situation of the meadows of marine phanerogams in Greek seas. - *Bot. Chron. (Patras)* **10**: 309-313.

HARITONIDIS, S. & DIAPOULIS, A. 1990: Evolution of Greek marine phanerogam meadows over the last 20 years. - *Posidonia Newslett.* **3(2)**: 5-10.

HARITONIDIS, S. & TSEKOS, I. 1976: Marine algae of the Greek west coast. - *Botanica Marina* **19(5)**: 273-286.

HARITONIDOU, P. 1983a: Environmental protection for the Great Marsh at Skinias. Can its vanishing flora and fauna be saved. - *Ann. Mus. Goulandris* **6**: 45-52.

HARITONIDOU, P. 1983b: On the floral morphology of *Posidonia oceanica* (L.) Delile (Potamogetonaceae) in the Aegean. - *Ann. Mus. Goulandris* **6**: 53-62.

HÄRLE, A. 1932: Die Arten und Formen der *Veronica*-Sektion *Pseudolysimachia* Koch auf Grund systematischer und experimenteller Untersuchungen. - *Biblioth. Bot.* **26/104**: 86 pp. + 2 maps and 26 plates.

HARLEY, R.M. 1972: Notes on the genus *Mentha* (Labiatae). In Heywood, V.H. (ed.): Notulae systematicae ad Floram Europaeam spectantes, no. 12. - *Bot. J. Linn. Soc.* **65(2)**: 250-253.

HARLEY, R.M. & BRIGHTON, C.A. 1977: Chromosome numbers in the genus *Mentha* L. - *Bot. J. Linn. Soc.* **74(1)**: 71-96.

HARLEY, R.M. & REYNOLDS, T. (EDS) 1992: Advances in Labiatae science. - Kew: Royal Botanic Gardens, xii + 568 pp.

HARLING, G. 1960: Further embryological and taxonomical studies in *Anthemis* L. and some related genera. - *Svensk Bot. Tidskr.* **54(4)**: 571-590.

HARMATA, K. 1977: Pollen morphology and taxonomy in the genera *Symphytum* L. and *Procopiana* Guşuleac. - *Zesz. Nauk. Uniw. Jagiellon. Prace Bot.* **5**: 6-29 + 1 table and 8 plates.

HARMATA, K. 1981: A supplement to the pollen morphology and taxonomy of the genera *Symphytum* L. and *Procopiana* Guşuleac. - *Zesz. Nauk. Uniw. Jagiellon. Prace Bot.* **8**: 7-10 + 6 plates.

HARPER, J.L. 1957: Biological Flora of the British Isles. *Ranunculus acris* L., *R. repens* L. and *R. bulbosus* L. - *J. Ecol.* **45(1)**: 289-342.

HARPKE, D. & AL. 2014: Phylogeny, geographic distribution, and new taxonomic circumscription of the *Crocus reticulatus* species group (Iridaceae). - *Turk. J. Bot.* **38(6)**: 1182-1198.

HARPKE, D. & AL. 2016: Neotypification of the name *Crocus biflorus* Mill. (Iridaceae) and its consequences in the taxonomy of the genus. - *Phytotaxa* **260(2)**: 131-143.

HARRIMAN, N.A. 1998: (1357) Proposal to conserve the name *Bidens* (Asteraceae) with a conserved gender. - *Taxon* **47(2)**: 485-486. [approved, cf. Taxon 49(2): 272, 2000].

HARRIS, S. 2007: The magnificent Flora Graeca. How the Mediterranean came to the English garden. - Oxford: Bodleian Library, 189 pp.

HARRIS, S.A. 2002: Introduction of Oxford ragwort, *Senecio squalidus* L. (Asteraceae), to the United Kingdom. - *Watsonia* **24(1)**: 31-43.

HARRIS, S.A. 2021: Sibthorp's Flora Graeca expedition and teaching Linnean botany in Oxford Physic Garden. - *Curtis's Bot. Mag.* **38(4)**: 451-471.

HART, H.'T 1971: Cytological and morphological variation in *Sedum acre* L. in western Europe. - *Acta Bot. Neerl.* **20(3)**: 282-290.

HART, H.'T 1972: Chromosome numbers in the series *Rupestria* Berger of the genus *Sedum* L. - *Acta Bot. Neerl.* **21(4)**: 428-435.

HART, H.'T 1974: *Sedum tenuifolium* (Sibth. & Sm.) Strobl subsp. *ibericum* nov. ssp. - *Acta Bot. Neerl.* **23(4)**: 549-554.

HART, H.'T 1975: The taxonomy and cytology of the *Sedum laconicum* aggregate in south-eastern Europe. - *Proc. Kon. Ned. Akad. Wetensch. C* **78(5)**: 435-448.

HART, H.'T 1978: Biosystematic studies in the *acre*-group and the series *Rupestria* Berger of the genus *Sedum* L. (Crassulaceae). - Utrecht: Diss., Univ. Utrecht, 153 pp.

HART, H.'T 1982a: The systematic position of *Sedum tuberosum* Coss. & Let. (Crassulaceae). - *Proc. Kon. Ned. Akad. Wetensch. C* **85(4)**: 497-508.

HART, H.'T 1982b: The white-flowered European *Sedum* species. 1. Principles of a phylogenetic classification of the Sedoideae (Crassulaceae) and the position of the white-flowered *Sedum* species. - *Proc. Kon. Ned. Akad. Wetensch. C* **85(4)**: 663-675.

HART, H.'T 1984a: A reappraisal of *Sedum grisebachii* (Crassulaceae). Typification and infraspecific variation. - *Willdenowia* **13(2)**: 295-307.

HART, H.'T 1984b: *Sedum apoleipon*, a new species of the *Sedum acre* group (Crassulaceae) from Central Greece. (Materials for the Mountain Flora of Greece, 21). - *Willdenowia* **13(2)**: 309-319.

HART, H.'T 1984c: The white-flowered European *Sedum* species. 3. The systematic position of *S. stellatum* L. - *Proc. Kon. Ned. Akad. Wetensch. C* **87(4)**: 401-411.

HART, H.'T 1985a: Chromosome numbers in *Sedum* (Crassulaceae) from Greece. (Materials for the Mountain Flora of Greece, 27). - *Willdenowia* **15(1)**: 115-135.

HART, H.'T 1985b: Sexual reproduction and hybridisation in *Sedum telephium* (Crassulaceae). - *Acta Bot. Neerl.* **34(1)**: 1-4.

HART, H.'T 1985c: The vascular pattern of the flowers of *Sedum anacampseros* (Crassulaceae). - *Acta Bot. Neerl.* **34(1)**: 119-121.

HART, H.'T 1987: Natural hybrids in *Sedum* (Crassulaceae). 1. Two new hybrids of *S.* series *Rupestria* and a new locality of *S.* ×*brevierei*. - *Bot. Jahrb. Syst.* **109(1)**: 1-16.

HART, H.'T 1989: *Sedum creticum* Presl (Crassulaceae): Typification and infraspecific variation. - *Taxon* **38(4)**: 647-651.

HART, H.'T 1990: *Sedum ursi* (Crassulaceae), a new species from Sandras Daği (Turkey). - *Acta Bot. Neerl.* **39(2)**: 203-206.

HART, H.'T 1991: Evolution and classification of the European *Sedum* species (Crassulaceae). - *Fl. Medit.* **1**: 31-61.

HART, H.'T 1993a: The evolution of the *Sedum acre* group (Crassulaceae). In Petrova, A.V. (ed.): Abstracts from the VII OPTIMA meeting in Borovetz, Bulgaria, 18-30 July 1993, p. 17.

HART, H.'T 1993b: About the name *Sedum sexangulare* L. - *Sedum Soc. Newsletter* **27**: 22-24.

HART, H.'T 1995a: Diversity within Mediterranean Crassulaceae. In Valdés, B. & Silvestre, S. (eds): Abstracts. VIII OPTIMA meeting, Sevilla, Spain, 25 September - 1 October, 1995, p. 19.

HART, H.'T 1995b: Variation in *Sedum litoreum* and the *S. rubens* complex (Crassulaceae); new names and combinations. - *OT Sist. Bot. Derg.* **2(2)**: 1-9.

HART, H.'T 1996: The evolution of the *Sedum acre* group (Crassulaceae). - *Bocconea* **5(1)**: 119-128.

HART, H.'T 1997: Greek checklist update. - *Sedum Soc. Newsletter* **41**: 48.

HART, H.'T & AL. 1993: The allopolyploid origin of *Sedum rupestre* subsp. *rupestre* (Crassulaceae). - *Pl. Syst. Evol.* **184(3-4)**: 195-206.

HART, H.'T & ALPINAR, K. 1991a: Biosystematic studies in *Sedum* (Crassulaceae) of Turkey. 1. Notes on four hitherto little known species collected in the western part of Anatolia. - *Willdenowia* **21(1-2)**: 143-156.

HART, H.'T & ALPINAR, K. 1991b: Biosystematic studies in *Sedum* of Turkey (Crassulaceae). III *Sedum euxinum*, a new species from northeastern Anatolia. - *Acta Bot. Neerl.* **40(4)**: 357-361.

HART, H.'T & ALPINAR, K. 1991c: Biosystematic studies in *Sedum* (Crassulaceae) of Turkey: Morphological and cytological variation in *S. pallidum* and *S. rubens*. - *Bot. Chron. (Patras)* **10**: 641-649.

HART, H.'T & ALPINAR, K. 1996: Biosystematic studies in *Sedum* (Crassulaceae) in Turkey. In Öztürk, M.A., Seçmen, Ö. & Görk, G. (eds): Plant life in southwest and central Asia, 1: pp. 71-99. - Izmir: Ege University Press.

HART, H.'T & ARKEL, J. VAN 1985: Quantitative aspects of the influence of day-length and temperature on *Sedum telephium* (Crassulaceae). - *Acta Bot. Neerl.* **34(1)**: 115-118.

HART, H.'T & BERENDSEN, W. 1980: Ornamentation of the testa in *Sedum* (Crassulaceae). - *Pl. Syst. Evol.* **135(1-2)**: 107-117.

HART, H.'T & BERG, A.J. VAN DEN 1982: The white-flowered European *Sedum* species. 2. Cytotaxonomic notes on *S. album* L. and *S. gypsicolum* Boiss. & Reuter. - *Proc. Kon. Ned. Akad. Wetensch. C* **85(4)**: 677-691.

HART, H.'T & ČESCHMEDJIEV, I.V. 1985: A cytotaxonomic note on *Sedum aetnense* (Crassulaceae). - *Acta Bot. Neerl.* **34(3)**: 321-324.

HART, H.'T & EGGLI, U. 1988: Cytotaxonomic studies in *Rosularia* (Crassulaceae). - *Bot. Helv.* **98(2)**: 223-234.

HART, H.'T & EGGLI, U. (EDS) 1995: Evolution and systematics of the Crassulaceae. - Leiden: Backhuys Publishers, 192 pp.

HART, H.'T & JARVIS, C.E. 1993: Typification of Linnaeus's names for European species of *Sedum* subgen. *Sedum* (Crassulaceae). - *Taxon* **42(2)**: 399-410.

HART, H.'T & LOON, J.CHR. VAN 1982: [Chromosome numbers in Crassulaceae]. In Löve, Á. (ed.): Chromosome number reports LXXV. - *Taxon* **31(2)**: 362-363.

HART, H.'T. & EGGLI, U. 2003: Sedums of Europe: stonecrops and wallpeppers. - Lisse: A.A. Balkema, 125 pp.

HARTOG, C. DEN 1970: The sea-grasses of the world. - *Verh. Kon. Ned. Akad. Wetensch., Afd. Natuurk., Sect. 2* **59(1)**: 1-275 + 31 plates.

HARTOG, C. DEN 1972: Range extension of *Halophila stipulacea* (Hydrocharitaceae) in the Mediterranean. - *Blumea* **20(1)**: 154.

HARTOG, C. DEN & AL. 1987: Chromosome numbers of the European seagrasses. - *Pl. Syst. Evol.* **156 (1-2)**: 55-59.

HARTVIG, P. 1978: Floristic notes from the mountains of north-western Greece. - *Ann. Mus. Goulandris* **4**: 27-38.

HARTVIG, P. 1979a: *Cerastium smolikanum* Hartvig, sp. nov. and *C. vourinense* from N Greece. (Materials for the Mountain Flora of Greece, 1). - *Bot. Not.* **132(3)**: 359-361.

HARTVIG, P. 1979b: *Veronica bornmuelleri* (Scrophulariaceae) new to Europe. (Materials for the Mountain Flora of Greece, 3). - *Bot. Not.* **132(3)**: 367-369.

HARTVIG, P. 1982: *Centaurea vlachorum* sp. nov. (subgen. *Jacea*) from NW Greece. (Materials for the Mountain Flora of Greece, 8). - *Nordic J. Bot.* **1(6)**: 705-707.

HARTVIG, P. 1984: Two new species of Apiaceae from Greece. (Materials for the Mountain Flora of Greece, 20). - *Willdenowia* **13(2)**: 289-293.

HARTVIG, P. 1985: *Thamnosciadium* Hartvig, a new genus of Umbelliferae from Greece.- *Willdenowia* **14(2)**: 321-325.

HARTVIG, P. 1987a: Chromosome numbers in Nordic populations of the *Carex muricata* group (Cyperaceae). - *Symb. Bot. Upsal.* **27(2)**: 127-138.

HARTVIG, P. 1987b: A taxonomical revision of *Thymus* sect. *Teucrioides* (Lamiaceae). - *Pl. Syst. Evol.* **155(1-4)**: 197-213.

HARTVIG, P. 1989: *Polygonum papillosum* (Polygonaceae), a new species from central Greece. - *Willdenowia* **19(1)**: 75-78. [Also distributed as *OPTIMA leaflet* no. 186].

HARTVIG, P. 1998: *Campanula pangea*, a new species of *C.* sect. *Involucratae* from Mt Pangeon, NE Greece. - *Willdenowia* **28(1-2)**: 65-68.

HARTWIG, U. 1983: Biosystematische Unteruchungen der Artengruppe von *Betonica officinalis*. - *Ber. Geobot. Inst. E.T.H. Stiftung Rübel* **50**: 31-32. [Summary of a manuscript (71 pp.)].

HASKELL, G. & MARKS, G.E. 1952: Chromosome ecology of British *Galinsoga* species. - *New Phytol.* **51(3)**: 382-387.

HAŠKOVCOVI, C. & HAŠKOVCOVI, Š. 2020: In search of Cretan natives. - *Alpine Gardener* **88(4)**: 414-421.

HAŠKOVCOVI, C. & HAŠKOVCOVI, Š. 2021: A gallery of Crete's finest. - *Alpine Gardener* **89(1)**: 58-65.

HASLAM, S.M. 1972: Biological Flora of the British Isles. *Phragmites communis* Trin. (*Arundo phragmites* L.; ?*Phragmites australis* (Cav.) Trin. ex Steudel). - *J. Ecol.* **60(2)**: 585-610.

HASSEMER, G., MEUDT, H.M. & RØNSTED, N. 2017: Nomenclatural and taxonomic notes on Mediterranean narrow-leaved plantains (*Plantago* section *Maritima*, Plantaginaceae). - *Webbia* **72 (2)**: 197-205.

HATCHER, P.E. 2003: Biological Flora of the British Isles. *Impatiens noli-tangere* L. - *J. Ecol.* **91(1)**: 147-167.

HATTWIG, M. & AL. 1992: Biotoptypenkartierung Keramoti. In Mattes, H. & Lienau, C. (eds): Colloquium Nordost-Griechenland 1, pp. 22-31. - Münster: Inst. Geographie.

HATZISKAKIS, S. , TSIRIPIDIS, I. & PAPAGEORGIOU, A.C. 2011: Leaf morphological variation in beech (*Fagus sylvatica* L.) populations in Greece and its relation to their post-glacial origin. - *Bot. J. Linn. Soc.* **165(4)**: 422-436.

HAUENSCHILD, F. & AL. 2016: Analysis of the cosmopolitan buckthorn genera *Frangula* and *Rhamnus* s.l. supports the description of a new genus, *Ventia*. - *Taxon* **65(1)**: 65-78.

HAUFLER, CH.H. & AL. 1991: Developing a picture of phylogenetic relationships among diploid species in the *Polypodium vulgare* complex. - *Amer. J. Bot.* **78(6) Suppl.**: 152-153.

HAUFLER, CH.H., SOLTIS, D.E. & SOLTIS, P.S. 1995: Phylogeny of the *Polypodium vulgare* complex: insights from chloroplast DNA restriction site data. - *Syst. Bot.* **20(2)**: 110-119.
HAUFLER, CH.H., WINDHAM, M.D. & RABE, E.W. 1995: Reticulate evolution in the *Polypodium vulgare* complex. - *Syst. Bot.* **20(2)**: 89-109.
HAUFLER, CH.H. & ZHONGREN, W. 1991: Chromosomal analyses and the origin of allopolyploid *Polypodium virginianum* (Polypodiaceae). - *Amer. J. Bot.* **78(5)**: 624-629.
HAUG, W. & TRÄNKLE, U. 1986: Ein Beitrag zur Orchideenflora von Griechenland. - *Mitt. Arbeitskreis Heimische Orchid. Baden-Württemberg* **18(4)**: 828-846.
HAUKE, R.L. 1963: A taxonomic monograph of the genus *Equisetum* subgenus *Hippochaete*. - *Beih. Nova Hedwigia* **8**: 1-123 + 12 tables, 3 graphs & 22 plates.
HAUKE, R.L. 1967: A systematic study of *Equisetum arvense*. - *Nova Hedwigia* **13**: 81-109 + 9 plates.
HAUKE, R.L. 1974: The taxonomy of *Equisetum* - an overview. - *New Bot.* **1(1-2)**: 89-95.
HAUKE, R.L. 1978: A taxonomic monograph of *Equisetum* subgenus *Equisetum*. - *Nova Hedwigia* **30**: 385-455.
HAUKE, R.L. 1992: Revisiting *Equisetum ramosissimum*. - *Amer. Fern J.* **82(2)**: 83-84.
HAUSDORF, B. & HENNIG, CH. 2005: The influence of recent geography, palaeogeography and climate on the composition of the fauna of the central Aegean Islands. - *Biol. J. Linn. Soc.* **84(4)**: 785-795.
HAUSER, M.L. 1975: Zytotaxonomische Untersuchungen an *Campanula patula* L. s.l. und *C. Rapunculus* L. in der Schweiz und in Österreich. - *Veröff. Geobot. Inst. E.T.H. Stiftung Rübel Zürich* **53**: 1-73.
HAUSSKNECHT, C. 1873a: Beitrag zur Kenntnis der Arten von *Fumaria* sect. *Sphaerocapnos* DC. - *Flora* **56**: 401-414, 417-425, 441-446, 456-462.
HAUSSKNECHT, C. 1873b: Beitrag zur Kenntnis der Arten von *Fumaria* sect. *Sphaerocapnos* DC. [cont.]. - *Flora* **56**: 485-496, 505-526, 536-544, 546-560.
HAUSSKNECHT, C. 1873c: Beitrag zur Kenntnis der Arten von *Fumaria* sect. *Sphaerocapnos* DC. [Schluss]. - *Flora* **56**: 562-568 + 1 table.
HAUSSKNECHT, C. 1884: Monographie der Gattung *Epilobium*. - Jena: Gustav Fischer Verlag, viii + 318 pp. + 3 charts and 23 plates.
HAUSSKNECHT, C. 1886: *Acanthus Caroli Alexandri*. Ein neuer Bürger der europäischen Flora. - *Gartenflora* **35**: 626-635.
HAUSSKNECHT, C. 1887a: In: Sitzungsbericht, Botanischer Verein für Gesamtthüringen [report on plants collected in Greece]. - *Mitt. Geogr. Ges. (Thüringen) Jena* **5(3)**: 60-63, 70-72.
HAUSSKNECHT, C. 1887b: Kurze Mitteilungen über neue Pflanzen-Arten der griechischen Flora. - *Mitt. Geogr. Ges. (Thüringen) Jena* **5(3)**: 85-88.
HAUSSKNECHT, C. 1888a: *Reseda tymphaea*. In: Kleinere botanische Mitteilungen. - *Mitt. Geogr. Ges. (Thüringen) Jena* **6(2)**: 10.
HAUSSKNECHT, C. 1888b: Notes on *Alkanna pindicola* and *A. methanaea*. In: Kleinere botanische Mitteilungen. - *Mitt. Geogr. Ges. (Thüringen) Jena* **6(2)**: 32.
HAUSSKNECHT, C. 1891: Ueber einige kritische *Rumex*-Arten. - *Mitth. Thüring. Bot. Vereins, N.F.* **1**: 31-35.
HAUSSKNECHT, C. 1893: Symbolae ad floram graecam. Aufzählung der im Sommer 1885 in Griechenland gesammelten Pflanzen. - *Mitth. Thüring. Bot. Vereins, N.F.* **3-4**: 96-116.
HAUSSKNECHT, C. 1894: Symbolae ad floram graecam. Aufzählung der im Sommer 1885 in Griechenland gesammelten Pflanzen [Fortsetzung]. - *Mitth. Thüring. Bot. Vereins, N.F.* **5**: 41-126.
HAUSSKNECHT, C. 1895: Symbolae ad floram graecam. Aufzählung der im Sommer 1885 in Griechenland gesammelten Pflanzen [Fortsetzung]. - *Mitth. Thüring. Bot. Vereins, N.F.* **7**: 25-64.
HAUSSKNECHT, C. 1896: Symbolae ad floram graecam. Aufzählung der im Sommer 1885 in Griechenland gesammelten Pflanzen [Fortsetzung]. - *Mitth. Thüring. Bot. Vereins, N.F.* **8**: 43-54.
HAUSSKNECHT, C. 1897: Symbolae ad floram graecam. Aufzählung der im Sommer 1885 in Griechenland gesammelten Pflanzen [Fortsetzung]. - *Mitth. Thüring. Bot. Vereins, N.F.* **10**: 47-66.
HAUSSKNECHT, C. 1898a: Symbolae ad floram graecam. Aufzählung der im Sommer 1885 in Griechenland gesammelten Pflanzen [Fortsetzung]. - *Mitth. Thüring. Bot. Vereins, N.F.* **11**: 30-65.
HAUSSKNECHT, C. 1898b: Drei neue Cruciferen-Gattungen der orientalischen Flora. - *Mitth. Thüring. Bot. Vereins, N.F.* **11**: 68-76.
HAUSSKNECHT, C. 1900: Symbolae ad floram graecam. Aufzählung der im Sommer 1885 in Griechenland gesammelten Pflanzen [Schluss]. - *Mitth. Thüring. Bot. Vereins, N.F.* **13/14**: 18-77.
HAUTZINGER, L. 1976a: *Orchis prisca* (Orchidaceae), eine neue Art aus Kreta. - *Pl. Syst. Evol.* **124(4)**: 311-313.
HAUTZINGER, L. 1976b: Nomenclatorische und systematische Beiträge zur Familie Orchidaceae. - *Verh. Zool.-Bot. Ges. Wien* **115**: 40-54.
HAUTZINGER, L. 1978: Genus *Orchis* L. (Orchidaceae); Sectio *Robustocalcare* Hautzinger. - *Ann. Naturhist. Mus. Wien* **81**: 31-73.
HAWKES, J.G. 1972: [Notes on *Mandragora officinarum*]. In Heywood, V.H. (ed.): Notulae systematicae ad Floram Europaeam spectantes, no. 13. - *Bot. J. Linn. Soc.* **65(4)**: 356-357.
HAWKES, J.G. 1991: Solanaceae III: taxonomy, chemistry, evolution. - Kew: Royal Botanic Gardens, ix + 483 pp.
HAWKES, J.G., LESTER, R.N. & SKELDING, A.D. (EDS) 1979: The biology and taxonomy of the Solanaceae. - London: Academic Press (for the Linnean Society of London), xvii + 738 pp.
HAWKSWORTH, F.G. & WIENS, D. 1972: Biology and classification of dwarf misteltoes (*Arceuthobium*) [Agriculture Handbook no. 401]. - Washington, D.C.: U.S.D.A., viii + 234 pp.
HAWKSWORTH, F.G. & WIENS, D. 1976: *Arceuthobium oxycedri* and its segregate *A. juniperi-procerae* and *A. azoricum* (Viscaceae). - *Kew Bull.* **31(1)**: 71-80.
HAYEK, A.V. 1905: Monographische Studien über die Gattung *Saxifraga*, I. Die Sektion *Porphyrion* Tausch. - *Denkschr. Kaiserl. Akad. Wiss., Wien. Math.-Naturwiss. Kl.* **77**: 611-709 + 2 plates and 2 maps.
HAYEK, A.V. 1906: Ein Beitrag zur Kenntnis der Flora des Sandschak's Novipazar. - *Magyar Bot. Lapok* **5(8-10)**: 273-281.
HAYEK, A.V. 1911: [Beschreibung von *Tulipa Urumoffii*]. - *Verh. K.K. Zool.-Bot. Ges. Wien* **61(3-4)**: 110.
HAYEK, A.V. 1912: Über die Blütenbiologie von *Cytinus Hypocistis* L. - *Oesterr. Bot. Z.* **62(6)**: 238-240.
HAYEK, A.V. 1913: Centaureae novae et combinationes nominum Centaurearum novae. - *Repert. Spec. Nov. Regni Veg.* **12**: 123-126.
HAYEK, A.V. 1914a: Neue orientalische Pflanzenarten. - *Oesterr. Bot. Z.* **64(8)**: 358-360.
HAYEK, A.V. 1914b: Nachruf an Dr. Eugen v. Halácsy. - *Verh. K.K. Zool.-Bot. Ges. Wien* **64(9-10)**: 333-348.
HAYEK, A.V. 1915: [Über Umbelliferae]. - *Mitt. Naturwiss. Vereines Steiermark* **51(1)**: 201-214.
HAYEK, A.V. 1917a: Beitrag zur Kenntnis der Flora des albanisch-montenegrinischen Grenzgebietes. (Bearbeitung der von I. Dörfler im Jahre 1914 ... gesammelten Farn- und Blütenpflanzen). - *Denkschr. Kaiserl. Akad. Wiss., Wien, Math.-Naturwiss. Kl.* **94**: 127-210 + 7 plates.
HAYEK, A.V. 1917b: Zur Kenntnis der Flora des Berges Žlep bei Ipek. - *Ann. K.K. Naturhist. Hofmus.* **31**: 65-76.
HAYEK, A.V. 1921: Diagnosen neuer von J. Dörfler und H. Zerny in den Jahren 1916 und 1918 in Albanien gesammelter Pflanzenformen. - *Oesterr. Bot. Z.* **70(1-2)**: 12-22.
HAYEK, A.V. 1922: Versuch einer natürlichen Gliederung des Formenkreises der *Minuartia verna* (L.) Hiern. - *Oesterr. Bot. Z.* **71(4-6)**: 89-116.
HAYEK, A.V. 1924a: Plantae novae orientales. - *Repert. Spec. Nov. Regni Veg.* **19**: 333-335.
HAYEK, A.V. 1924b: Prodromus Florae Peninsulae Balcanicae. Band 1. - *Repert. Spec. Nov. Regni Veg. Beih.* **30**: i-ix + 1-352. [Reprinted by Otto Koeltz Verlag 1970-71].

HAYEK, A.V. 1924c: Zweiter Beitrag zur Kenntnis der Flora von Albanien. (Bearbeitung der von J. Dörfler auf zwei in den Jahren 1916 und 1918 ... unternommenen Forschungsreisen ...). - *Denksch. Akad. Wiss. Wien, Math.-Naturwiss. Kl.* **99**: 101-224 + 2 plates.

HAYEK, A.V. 1925a: Prodromus Florae Peninsulae Balcanicae. Band 1 [cont.]. - *Repert. Spec. Nov. Regni Veg. Beih.* **30(1)**: 353-672. [Reprinted by Otto Koeltz Verlag 1970-71].

HAYEK, A.V. 1925b: Plantae novae orientales. II. - *Repert. Spec. Nov. Regni Veg.* **21**: 256-261.

HAYEK, A.V. 1926a: Prodromus Florae Peninsulae Balcanicae. Band 1 [cont.]. - *Repert. Spec. Nov. Regni Veg. Beih.* **30(1)**: 673-960. [Reprinted by Otto Koeltz Verlag 1970-71].

HAYEK, A.V. 1926b: Über einige Orchideen, besonders aus der Mediterranflora. - *Repert. Spec. Nov. Regni Veg.* **22**: 387-390.

HAYEK, A.V. 1927a: Der Blütenstand von *Jankaea Heldreichii* Boiss. - *Magyar Bot. Lapok* **25(1-12)**: 191-195.

HAYEK, A.V. 1927b: Prodromus Florae Peninsulae Balcanicae. Band 1 [cont.]. - *Repert. Spec. Nov. Regni Veg. Beih.* **30(1)**: 961-1193 + 1 map. [Reprinted by Otto Koeltz Verlag 1970-71].

HAYEK, A.V. 1928a: Prodromus Florae Peninsulae Balcanicae. Band 2. - *Repert. Spec. Nov. Regni Veg. Beih.* **30(2)**: 1-96. [Reprinted by Otto Koeltz Verlag 1970-71].

HAYEK, A.V. 1928b: Ein Beitrag zur Kenntnis der Vegetation und der Flora des thessalischen Olymp. - *Beih. Bot. Centralbl., Abt. 2* **45(2)**: 220-328.

HAYEK, A.V. 1929: Prodromus Florae Peninsulae Balcanicae. Band 2 [cont.]. - *Repert. Spec. Nov. Regni Veg. Beih.* **30(2)**: 97-336. [Reprinted by Otto Koeltz Verlag 1970-71].

HAYEK, A.V. 1930: Prodromus Florae Peninsulae Balcanicae. Band 2 [cont.]. - *Repert. Spec. Nov. Regni Veg. Beih.* **30(2)**: 337-576. [Reprinted by Otto Koeltz Verlag 1970-71].

HAYEK, A.V. 1931: Prodromus Florae Peninsulae Balcanicae. Band 2 [cont.]. - *Repert. Spec. Nov. Regni Veg. Beih.* **30(2)**: 577-1152. [Reprinted by Otto Koeltz Verlag 1970-71].

HAYEK, A.V. 1932: Prodromus Florae Peninsulae Balcanicae. Band 3. - *Repert. Spec. Nov. Regni Veg. Beih.* **30(3)**: 1-368. [Reprinted by Otto Koeltz Verlag 1970-71].

HAYEK, A.V. 1933: Prodromus Florae Peninsulae Balcanicae. Band 3 [cont.]. - *Repert. Spec. Nov. Regni Veg. Beih.* **30(3)**: 369-472. [Reprinted by Otto Koeltz Verlag 1970-71].

HAYNES, R.R. 1985: A revision of the clasping-leaved *Potamogeton* (Potamogetonaceae). - *Sida* **11 (2)**: 173-188.

HAYNES, R.R. 1986: Typification of Linnaean species of *Potamogeton* (Potamogetonaceae). - *Taxon* **35(3)**: 563-573.

HEATH, P.V. 1992a: The type of *Rosa pendulina* Linné. - *Calyx* **2(2)**: 81.

HEATH, P.V. 1992b: The type of *Rosa Sedum rupestre* Linné. - *Calyx* **2(2)**: 82.

HEATH, P.V. 1992c: Commentaries on proposals to concerve the Linnaean names *Rubus, Coronilla, Daphne, Erica, Ornithogalum, Biscutella, Adonis, Aegilops, andf Bromus*. - *Calyx* **2(3)**: 93, 95, 97, 99, 116, 118, 120, 122, 124.

HEATH, P.V. 1992d: Commentary on the proposals to conserve *Rosa* Linné and *Anemone* Linné. - *Calyx* **2(4)**: 125-132, 152.

HEATH, P.V. 1993a: Commentary on the proposals to conserve *Sedum* Linné. - *Calyx* **3(1)**: 26.

HEATH, P.V. 1993b: The type of *Rosa sempervirens* Linné. - *Calyx* 3(1): 39-40

HEATH, P.V. 1993c: The type of *Crassula* Linné. - *Calyx* **3(2)**: 80.

HEATH, P.V. 1993d: Commentary on the proposal to conserve *Briza* Linné. - *Calyx* **3(3)**: 100.

HEATH, P.V. 1993e: Commentaries on proposals to conserve the Linnaean generic names *Agrostis*, *Aira*, *Convolvulus*, *Lathyrus*, *Delphinium*, *Melica*, *Salix* and *Viburnum*. - *Calyx* **4(1)**: 15, 16, 17, 23, 24-26, 29, 31, 32.

HEATH, P.V. 1994a: Commentaries on proposals to conserve the Linnaean names *Matricaria*, *Andropogon*, *Sisymbrium*, *Stachys*, *Bupleurum*, *Crepis*, *Cleome* and *Ophrys*. - *Calyx* **4(2)**: 48-50, 52, 55-56, 64-65, 71, 72, 73-76.

HEATH, P.V. 1994b: Commentaries on proposals to conserve the Linnean names *Elymus*, *Thalictrum*, *Apium*, *Angelica*, *Chaerophyllum* and *Psoralea*. - *Calyx* **4(3)**: 81-82, 85, 86, 94-95, 98-99, 107.

HEATH, P.V. 1994c: Commentaries on proposals to conserve the Linnaean names *Erigeron*, *Draba*, *Picris* and *Scutellaria*. - *Calyx* **4(3)**: 108-109, 110-112, 117, 118.

HEATH, P.V. 1994d: Commentaries on proposals to conserve the Linnean names *Carex*, *Cotyledon* and *Scabiosa*. - *Calyx* **5(1)**: 15-16, 20-21, 22-26.

HEATH, P.V. 1994e: The type of *Rosa arvensis* Hudson. - *Calyx* **5(1)**: 38-40.

HEATH, P.V. 1998: Commentary on the proposal to reject the name *Drosera longifolia* Linné. - *Calyx* **6(1)**: 38-39.

HEATH, P.V. 2000: New combinations in Rosa Linné (first instalment). - *Calyx* **6(2)**: 47.

HEATH, R.E. 1977: Plant portraits: *Cytisus procumbens* Sprengel. - *Quart. Bull. Alpine Gard. Soc. Gr. Brit.* **45(1)**: 56, 58, 61.

HÉBERT, L.-P. 1975: Contribution à la cytotaxonomie du genre *Sedum* L. - *Bull. Soc. Neuchâteloise Sci. Nat.* **98**: 59-70.

HÉBERT, L.-P. 1976a: La phytogéographie dynamique et la cytotaxonomie de quelques espèces de *Sedum* L. en Corse. - *Biol. Écol. Médit.* **3(3)**: 65-70.

HÉBERT, L.-P. 1976b: Nouvelle contribution à la cytotaxonomie du genre *Sedum* L. - *Bull. Soc. Neuchâteloise Sci. Nat.* **99**: 97-107.

HÉBERT, L.-P. 1977: Étude cytotaxonomique de quelques espèces du genre *Sedum* L. endémiques dans la région méditerranéenne. - *Bull. Soc. Neuchâteloise Sci. Nat.* **100**: 113-120.

HÉBERT, L.-P. 1983: Analyse d'un complexe chromosomique en Méditerranée: *Sedum* ser. *Rupestria* Berger emend. - *Rev. Cytol. Biol. Vég. Botaniste* **6(3-4)**: 179-224.

HECHT-MARKOU, P. 1999: Das Vorkommen von *Proboscidea louisianica* in Griechenland und zwar auf der Insel Lesbos. - *Ann. Mus. Goulandris* **10**: 73-78.

HEDBERG, I. 1961a: Cytotaxonomic studies in *Anthoxanthum odoratum* L. s. lat. 1. Morphological analysis of herbarium specimens. - *Svensk Bot. Tidskr.* **55(1)**: 118-128 + 1 plate.

HEDBERG, I. 1961b: Chromosome studies in *Helictotrichon* L. - *Bot. Not.* **114(4)**: 389-396.

HEDBERG, I. 1967: Cytotaxonomic studies on *Anthoxanthum odoratum* L. s. lat. II. Investigations of some Swedish and a few Swiss population samples. - *Symb. Bot. Upsal.* **18(5)**: 1-88 + 8 plates.

HEDBERG, I. 1969: Cytotaxonomic studies on *Anthoxanthum odoratum* L. s. lat. III. Investigations of Swiss and Austrian population samples. - *Svensk Bot. Tidskr.* **63(2)**: 233-250 + 2 plates.

HEDBERG, I. 1970: Cytotaxonomic studies on *Anthoxanthum odoratum* L. s. lat. IV. Karyotypes, meiosis and the origin of tetraploid *A. odoratum*. - *Hereditas (Lund)* **64(2)**: 153-176.

HEDBERG, I. 1987: The genesis of tetraploid *Anthoxanthum odoratum*. - *Symb. Bot. Upsal.* **27(2)**: 147-154.

HEDBERG, I. 1990: Morphological, cytotaxonomic and evolutionary studies in *Anthoxanthum odoratum* L. s. lat. - a critical review. - *Sommerfeltia* **11**: 97-107.

HEDBERG, O. 1955: A taxonomic revision of the genus *Sibthorpia* L. - *Bot. Not.* **108(2)**: 161-183.

HEDBERG, O. 1962: Intercontinental crosses in *Arabis alpina* L. - *Caryologia* **15(2)**: 253-260.

HEDBERG, O. 1975: A cytogenetic study of the genus *Sibthorpia* L. (Scrophulariaceae). - *Caryologia* **28(2)**: 251-260.

HEDBERG, O. 1993: *Rumex nepalensis* (Polygonaceae) - a Himalayan element in the afromontane flora. - *Fragm. Florist. Geobot. Suppl.* **2(1)**: 91-96.

HEDGE, I.C. 1957: Studies in East Mediterranean species of *Salvia*. - *Notes Roy. Bot. Gard. Edinburgh* **22(3)**: 173-188.

HEDGE, I.C. 1961a: *Salvia glutinosa* L. and *S. nubicola* Sweet: two previously confused species. - *Notes Roy. Bot. Gard. Edinburgh* **23(3)**: 205-221.

HEDGE, I.C. 1961b: Studies in East Mediterranean species of *Salvia*: IV. - *Notes Roy. Bot. Gard. Edinburgh* **23(4)**: 559-567.

HEDGE, I.C. 1969: Materials for a Flora of Turkey, XVII: *Onobrychis* and *Hedysarum*. - *Notes Roy. Bot. Gard. Edinburgh* **29(1)**: 33-37.

HEDGE, I.C. 1974: A revision of *Salvia* in Africa including Madagascar and the Canary Islands. - *Notes Roy. Bot. Gard. Edinburgh* **33(1)**: 1-121.

HEDGE, I.C. 1986: Labiatae of South-West Asia: diversity, distribution and endemism. - *Proc. Roy. Soc. Edinburgh, B, Biol. Sci.* **89**: 23-35.

HEDGE, I.C. & AL. 1967: Materials for a Flora of Turkey: XII. - *Notes Roy. Bot. Gard. Edinburgh* **27(2)**: 211-217.

HEDGE, I.C. & HENDRYCH, R. 1964: Three new species of *Thesium* from Kurdistan and Samos. - *Preslia* **36**: 240-244.

HEDGE, I.C. & LAMOND, J.M. 1964: A guide to the Turkish genera of Umbelliferae. - *Notes Roy. Bot. Gard. Edinburgh* **25(2)**: 171-177.

HEDGE, I.C. & LAMOND, J.M. 1973: A review of the tribe Echinophoreae (Umbelliferae). - *Notes Roy. Bot. Gard. Edinburgh* **32(2)**: 167-188.

HEDGE, I.C., LAMOND, J.M. & MATTHEWS, V.A. 1971: Materials for a Flora of Turkey XXV: Some new species of Umbelliferae. - *Notes Roy. Bot. Gard. Edinburgh* **31(1)**: 75-79.

HEDGE, I.C. & YALTIRIK, F. 1981: *Quercus ithaburensis* subsp. *macrolepis* (Kotschy) Hedge & Yalt. In Greuter, W. (ed.): Med-Checklist notulae, 3. - *Willdenowia* **11(1)**: 42.

HEDGE, I.C. & YALTIRIK, F. 1994: The oaks of Turkey. - *J. Int. Oak Soc.* **5**: 3-21.

HEDLUND, TH. 1901: Monographie der Gattung *Sorbus*. - *Kongl. Svenska Vetensk. Acad. Handl.* **35 (1)**: 1-147.

HEDRÉN, M. & AL. 2007: Patterns of polyploid evolution in Greek marsh orchids (*Dactylorhiza*; Orchidaceae) as revealed by allozymes, AFLPs, and plastid DNA data. - *Amer. J. Bot.* **94(7)**: 1205-1218.

HEEDE, C.J. VAN DEN & VIANE, R.L.L. 2002: New species and new hybrids in *Asplenium* subgenus *Ceterach* (Aspleniaceae). - *GEP News* **9**: 1-4.

HEGELMAIER, F. 1864: Monographie der Gattung *Callitriche*. - Stuttgart: Verlag von Ebner & Seubert, iv + 64 pp. + 4 plates.

HEGELMAIER, F. 1867: Zur Systematik von *Callitriche*. - *Verh. Bot. Vereins Prov. Brandenburg* **9**: 1-40 + 1 plate.

HEGELMAIER, F. 1868: Die Lemnaceen. Eine monographische Untersuchung. - Leipzig: Verlag von Wilhelm Engelmann, vi + 169 pp. + 16 plates.

HEGELMAIER, F. 1895: Systematische Übersicht der Lemnaceeen. - *Bot. Jahrb. Syst.* **21(3)**: 268-305.

HEGI, G. (ED.) 1906-31: Illustrierte Flora von Mittel-Europa. Mit besonderer Berücksichtigung von Oesterreich, Deutschland und der Schweiz ... - München: J.F. Lehmann, 7 vols. [cf. TL-2 2: 132-136]; 2. ed. 1957-; 3. ed. 1966-

HEHN, V. 1874: Kulturpflanzen und Haustiere in ihrem Übergang aus Asien nach Griechenland und Italien sowie in das übrige Europa. Zweite, umgearbeitete Auflage [also later editions]. - Berlin: Gebrüder Bornträger, xii + 544 pp.

HEILBORN, O. 1922: Die Chromosomenzahlen der Gattung *Carex*. Vorläufige Mitteilung. - *Svensk Bot. Tidskr.* **16(2)**: 271-274.

HEILBORN, O. 1924: Chromosome numbers and dimensions, species-formation and phylogeny in the genus *Carex*. - *Hereditas (Lund)* **5(2)**: 129-216 + 1 plate.

HEILBORN, O. 1927: Chromosome numbers in *Draba*. - *Hereditas (Lund)* **9(1)**: 59-68.

HEILBORN, O. 1928: Chromosome studies in Cyperaceae. - *Hereditas (Lund)* **11(2-3)**: 182-192.

HEILBORN, O. 1932: Aneuploidy and polyploidy in *Carex*. - *Svensk Bot. Tidskr.* **26(1-2)**: 137-146.

HEILBORN, O. 1939: Chromosome studies in Cyperaceae. III-IV. - *Hereditas (Lund)* **25(3)**: 224-240.

HEIMBURGER, M. 1959: Cytotaxonomic studies in the genus *Anemone*. - *Canad. J. Bot.* **37(4)**: 587-612 + 1 plate.

HEIMERL, A. 1884: Monographia sectionis "Ptarmica" Achilleae generis. Die Arten, Unterarten, Varietäten und Hybriden der Section *Ptarmica* des Genus *Achillea*. - *Denkschr. Kaiserl. Akad. Wiss., Wien. Math.-Naturwiss. Kl.* **48**: 113-192 + 3 plates + Index i-iv.

HEIMERL, A. 1926: Achilleen-Studien I-IV. - *Oesterr. Bot. Z.* **75(7-9)**: 132-153.

HEINRICHER, E. 1917a: Berichtigende Mitteilung über die Keimungsbedingungen der Samen von *Arceuthobium Oxycedri* (DC.) M. Bieb. - *Ber. Deutsch. Bot. Ges.* **35(3)**: 204-212.

HEINRICHER, E. 1917b: Die erste Aufzucht einer Rafflesiacee, *Cytinus Hypocistis* L., aus Samen. - *Ber. Deutsch. Bot. Ges.* **35(6)**: 505-512 + 1 plate.

HEINRICHER, E. 1917c: Zur Kenntnis der Blüte von *Cytinus Hypocistis* L. - *Ber. Deutsch. Bot. Ges.* **35(6)**: 513-517 + 1 plate.

HEINRICHER, E. 1931: Monographie der Gattung *Lathraea*. - Jena: Gustav Fischer Verlag, iv + 152 pp. + 5 plates

HEINRICHER, E. 1934: Zur Frage der Artbildung bei *Cytinus Hypocistis* nebst andere Bemerkungen. - *Ber. Deutsch. Bot. Ges.* **52(1)**: 48-53.

HEISELMAYER, P. 1986: Die vegetationskundliche und pflanzengeographische Stellung der Kykladen innerhalb des mediterranen Lebensraumes. In Riedl, H. & Kern, W. (eds): Salzburger Exkursionsberichte 10: 133-147. - Salzburg: Institut für Geographie.

HEISELMAYER, P. 1988: Kreta - Vegetation und Pflanzengeographie einer südägäischen Insel. - *Ber. Naturwiss.-Med. Verein Innsbruck* **75**: 251-260.

HEISELMAYER, P. 1996: Menschlich bedingte Vegetationsveränderungen auf den Kykladen (Ägäis). - *Arch. für Nat. -Lands.* **35**: 161-174.

HEISELMAYER, P., GRABNER, S. & GRIEHSER, B. 1995: Vegetation und Vegetationsentwicklung der Phrygana in Mykonos und Tinos. - *Salzburg. Geogr. Arb.* **29**: 233-266.

HEISELMAYER, P. & PILSL, P. 1983: Vegetationskundliche Studien an Brandflächen unterschiedlichen Alters auf Siphnos. - *Salzburg. Exkursionsber.* **9**: 203-216.

HEISELMAYER, P. & SIEBENBRUNNER, A. 1986: Vegetation und Brachenentwicklung auf Seriphos (Kykladen). In Riedl, H. & Kern, W. (eds): Salzburger Exkursionsberichte 10: 149-157. - Salzburg: Institut für Geographie.

HEITZ, B. 1974: Hétérostylie et spéciation dans le groupe *Linum perenne*. - *Ann. Sci. Nat., Bot. Biol. Vég.* **14(4)**: 385-405.

HEJNOWICZ, D. 1991: Contribution to the flora of Turkey. - *Fragm. Florist. Geobot.* **35(1-2)**: 83-88.

HELBÆK, H. 1971: The origin and migration of rye, *Secale cereale* L.; a paleo-ethnobotanical study. In Davis, P.H., Harper, P.C. & Hedge, I.C. (eds): Plant life of South-West Asia, pp. 265-280. - Edinburgh: Botanical Society.

HELDREICH, TH.V. 1844: Einige Bemerkungen über griechische *Arbutus*-Arten. - *Flora* **27(1)**: 13-15.

HELDREICH, TH.V. 1860a: Ueber die neue arkadische Tanne (*Abies Reginae Amaliae*). - *Gartenflora* **9**: 313-317.

HELDREICH, TH.V. 1860b: Descriptio specierum novarum. In appendix to Index Seminum quae Hortus Botanicus Regiae Universitatis Othonis Athenarum pro mutua commutatione offert, anno 1860 collecta [same species as in Heldreich 1860c]. - Athine.

HELDREICH, TH.V. 1860c: Descriptio specierum novarum [*Campanula leutweinii, Centaurea niederi, Mattia schmidtii*]. - *Ann. Sci. Nat. Bot., Sér. 4* **13**: 379-382.

HELDREICH, TH.V. 1861a: Zur Kenntniss der griechischen Tannen. - *Gartenflora* **10**: 286-288.

HELDREICH, TH.V. 1861b: Ueber Pflanzen der Griechischen, insbesondere der Attischen Flora, die als Zierpflanzen empfehlenswerth sind. - *Gartenflora* **10**: 343-349.

HELDREICH, TH.V. 1862a: *Tulipa Orphanidea* Boiss. und die Tulpen Griechenlands. - *Gartenflora* **11**: 309-311 + 1 plate.

HELDREICH, TH.V. 1862b: Die Nutzpflanzen Griechenlands. Mit besonderer Berücksichtigung der neugriechischen und pelasgischen Vulgarnamen. - Athen: K. Wilberg, viii + 104 pp.

HELDREICH, TH.V. 1862c: Phänologische Notizen aus Griechenland. - *Oesterr. Bot. Z.* **12(12)**: 385-386.

HELDREICH, TH.V. 1863: Phänologische Notizen aus Griechenland. - *Oesterr. Bot. Z.* **13(2-4)**: 48-49, 91-93, 126-127.

HELDREICH, TH.V. 1871: [Account of an excursion to Crete]. - *Oesterr. Bot. Z.* **21(1)**: 29-30.

HELDREICH, TH.V. 1873: *Glaucium serpieri* Heldr. - *Gartenflora* **22**: 323-324 + 1 plate.

HELDREICH, TH.V. 1874: *Tulipa Hageri* Heldr., eine neue Tulpenart der griechischen Flora. - *Gartenflora* **23**: 97-99 + 1 plate.

HELDREICH, TH.V. 1876: Sertulum plantarum novarum vel minus cognitarum Florae Hellenicae. In: Atti del congresso internazionale botanico tenuto in Firenze 1874, pp. 136-137, 227-240. - Firenze.

HELDREICH, TH.V. 1877a: Catalogus systematicus herbarii Theodorii G. Orphanidis ... Fasciculus primus: Leguminosae. - Florentiae: M. Ricci, 79 pp. [not continued].
HELDREICH, TH.V. 1877b: Die Pflanzen der attischen Ebene. In Mommsen, A. (Hrsg.): Griechische Jahreszeiten 5: pp. 471-597. - Schleswig: J. Bergas.
HELDREICH, TH.V. 1877c: Pflanzen-geographische Notizen über drei neue Arten der europäischen Flora. - *Oesterr. Bot. Z.* **27(5)**: 156-157.
HELDREICH, TH.V. 1878a: Über *Silene Ungeri* Fenzl, ihre Synonyma und ihren Verbreitungsbezirk. - *Oesterr. Bot. Z.* **28(1)**: 27-29.
HELDREICH, TH.V. 1878b: Zwei neue Pflanzenarten von den Jonischen Inseln. - *Oesterr. Bot. Z.* **28(2)**: 50-53.
HELDREICH, TH.V. 1878c: Über die Liliaceen-Gattung *Leopoldia* und ihre Arten. - *Bull. Soc. Imp. Naturalistes Moscou* **53(1)**: 56-75.
HELDREICH, TH.V. 1879a: Ein Beitrag zur Flora von Epirus, geliefert von Herrn N.K. Chodzes. - *Verh. Bot. Vereins Prov. Brandenburg* **21**: 61-63.
HELDREICH, TH.V. 1879b: Beiträge zur Kenntniss des Vaterlandes und der geographischen Verbreitung der Rosskastanie, des Nussbaums und der Buche. - *Verh. Bot. Vereins Prov. Brandenburg* **21**: 139-153.
HELDREICH, TH.V. 1879c: *Teucrium Halacsyanum* n. sp. Eine neue *Teucrium*-Art der griechischen Flora. - *Oesterr. Bot. Z.* **29(8)**: 241-242.
HELDREICH, TH.V. 1879d: Eine insectenfressende Pflanze der griechischen Flora. - *Oesterr. Bot. Z.* **29(9)**: 291-292.
HELDREICH, TH.V. 1880a: *Stachys Spreitzenhoferi* n. sp. Eine neue *Stachys*-Art der griechischen Flora. - *Oesterr. Bot. Z.* **30(11)**: 344-346.
HELDREICH, TH.V. 1880b: L'Attique au point de vue des caractères de sa végétation. - Paris: Compte Rendu Congr. Int. Bot. Hortic. 1878, pp. 106-119.
HELDREICH, TH.V. 1880c: Personalnachrichten - Joseph Sartori. - *Bot. Centralbl.* **1(38)**: 1182-1184.
HELDREICH, TH.V. 1880d: Musinitza, eine Idylle vom Korax. Mit topographischen und philologisch- dendrologischen Bemerkungen. - *Arch. Mittel- Neugriech. Philol.* **1**: 89-103.
HELDREICH, TH.V. 1881a: Der Asphodelos, ein griechisches Pflanzenbild. - *Deutsche Gart.* **2**: 379-380.
HELDREICH, TH.V. 1881b: Die Ferulastaude (*Ferula communis* L.). - *Verh. Bot. Vereins Prov. Brandenburg* **23**: xx-xxvii.
HELDREICH, TH.V. 1881c: Ein homerischer Pflanzenname. - *Bot. Centralbl.* **8**: 314-317.
HELDREICH, TH.V. 1881d: Beobachtungen von Dr. J. F. Jul. Schmidt über den Hergang der Keimung bei *Phoenix dactylifera* L. - *Bot. Centralbl.* **8**: 386 + 1 plate.
HELDREICH, TH.V. 1882a: Beispiel von Heterophyllie, beobachtet bei *Ceratonia Siliqua*. - *Sitzungsber. Ges. Naturf. Freunde Berlin* **1882(6)**: 113-115.
HELDREICH, TH.V. 1882b: Nachträgliches über das wilde Vorkommen der Rosskastanie. - *Verh. Bot. Vereins Prov. Brandenburg* **24**: 20-21.
HELDREICH, TH.V. 1882c: Flore de l'île de Céphalonie ou catalogue des plantes qui croissent naturellement et se cultivent le plus fréquemment dans cette île ... - Lausanne: Georges Bridel éditeur, 90 pp.
HELDREICH, TH.V. 1883a: Bericht über die botanischen Ergebnisse einer Bereisung Thessaliens. - *Sitzungsber. Königl. Preuss. Akad. Wiss. Berlin* **6**: 155-164.
HELDREICH, TH.V. 1883b: Peri fitologikis exetaseos tis Thessalias [In Greek]. - *Parnassos* **1883**: 257-266.
HELDREICH, TH.V. 1884a: Peri votanikis ekdromis en Attiki [In Greek]. - *Parnassos* **1884**: 180-182.
HELDREICH, TH.V. 1884b: Peri ioskiamos (*Hyoscyamus*) [In Greek]. - *Periodikon Athines Farmakevt. Eterias* **?**: ? [n.v.].
HELDREICH, TH.V. 1885: Peri likiskou ke tis kalliergias avtou en elladi [In Greek]. - Athine: Tipografiou tou "Mellontos", 7 pp.
HELDREICH, TH.V. 1886: Bemerkungen über die Gattung *Mandragora* und Beschreibung einer neuen Art. - *Mitt. Geogr. Ges. (Thüringen) Jena* **4(3)**: 75-80.
HELDREICH, TH.V. 1887: Th.G. Orphanidis os votanikos [In Greek]. - Athine: 15 pp.
HELDREICH, TH.V. 1889: Die *Malabaila*-Arten der griechischen Flora. - *Oesterr. Bot. Z.* **39(7)**: 241-243.
HELDREICH, TH.V. 1890a: I chloris tou Parnassou [In Greek]. - Athine: Tipografiou Alex. Papageorgiou, 30 pp. (Parnassos 13: 174-192, 254-264).
HELDREICH, TH.V. 1890b: Ueber *Campanula anchusiflora* und *C. tomentosa* der griechischen Flora. - *Bot. Centralbl.* **44(7)**: 209-214.
HELDREICH, TH.V. 1890c: Note sur une nouvelle espèce de *Centaurea* de l'île de Crète. - *Bull. Soc. Bot. France* **37**: 242-244.
HELDREICH, TH.V. 1891: Chloris tou Piliou [In Greek]. - *Palligenesia [a newspaper in Athens]* 31.10. 1891.
HELDREICH, TH.V. 1894: Les onagracées de la flore grecque. - *Monde Pl. Rev. Mens. Bot.* **3**: 141-145, 172-174.
HELDREICH, TH.V. 1897: Publications de Th. de Heldreich. - Le Mans: Typographie E. Monnoyer, 7 pp. [reprinted from *Monde des Plantes*].
HELDREICH, TH.V. 1898a: Flore de l'île d'Égine. - *Bull. Herb. Boissier* **6(3)**: 221-238.
HELDREICH, TH.V. 1898b: Flore de l'île d'Égine (suite). - *Bull. Herb. Boissier* **6(4)**: 289-308.
HELDREICH, TH.V. 1898c: Flore de l'île d'Égine (suite). - *Bull. Herb. Boissier* **6(5)**: 379-400 + 1 map.
HELDREICH, TH.V. 1898d: I chloris tis Eginis [In Greek]. - Athine: Tipografiou tis Estias, 59 pp. + 1 map [also as Year Book of Parnassus Society 2: 265-321].
HELDREICH, TH.V. 1898e: Ergebnisse einer botanischen Excursion auf die Cykladen im Hochsommer 1897. - *Oesterr. Bot. Z.* **48(5)**: 182-185.
HELDREICH, TH.V. 1899a: Une Graminée de l'Atlas retrouvée sur le mont Taygète, en Grèce. - *Bull. Acad. Int. Géogr. Bot.* **8(117-118)**: 230.
HELDREICH, TH.V. 1899b: I chloris tis Thiras [In Greek]. - *Epet. Sillog. Parnassou* **1899**: 35 pp. [also as Year Book of Parnassus Society 3: 259-291].
HELDREICH, TH.V. 1899c: Die Flora der Insel Thera. In Gaertringen, H.v.: Thera, Untersuchungen, Vermessungen und Ausgrabungen in den Jahren 1895-1898, pp. 127-140. - Berlin: Verlag Georg Riemer.
HELDREICH, TH.V. 1900: Peri ton fiton ton parechonton to "ellinikon tsai" [In Greek]. - *Epet. Sillog. Parnassou* **1900**: 6 pp.
HELDREICH, TH.V. 1901a: Simvole pros sintaxin tis chloridos ton Kikladon. Chloris tis Mikonou ke ton parakimenon nison Dilou ke Rinias [In Greek]. - *Epet. Sillog. Parnassou* **5**: 239-259.
HELDREICH, TH.V. 1901b: Prosthike is tin chlorida tis Thiras, kata tin sillogin tou k. Wilski [In Greek]. - *Epet. Sillog. Parnassou* **5**: 4 pp.
HELDREICH, TH.V. 1902a: Die Flora von Thera. In: Hiller v. Gärtringen, F.: Die Insel Thera, Band I, Kap. IV, Nachtrag pp. 119-130. - Berlin.
HELDREICH, TH.V. 1902b: Un noveau *Myosurus* [posthumously published by M. Léveillé]. - *Bull. Acad. Int. Géogr. Bot.* **11(155)**: 296.
HELLER, D. 1989: Conspectus Florae Orientalis. In Phitos, D. (ed.): Abstracts. Sixth OPTIMA meeting, Delphi, 10-16 Sept. 1989, p. 7.
HELLER, D. 1991: Conspectus Florae Orientalis. - *Bot. Chron. (Patras)* **10**: 55-61.
HELLER, D. & HEYN, C.C. 1986: Conspectus Florae Orientalis. An annotated catalogue of the flora of the Middle East. Fasc. 3. - Jerusalem: The Israel Academy of Sciences and Humanities, xii + 160 pp. + 1 map.
HELLER, D. & HEYN, C.C. 1987: Conspectus Florae Orientalis. An annotated catalogue of the flora of the Middle East. Fasc. 4. - Jerusalem: The Israel Academy of Sciences and Humanities, xii + 92 pp. + 1 map.
HELLER, D. & HEYN, C.C. 1990: Conspectus Florae Orientalis. An annotated catalogue of the flora of the Middle East. Fasc. 5. - Jerusalem: The Israel Academy of Sciences and Humanities, xii + 79 pp. + 1 map
HELLER, D. & HEYN, C.C. 1991: Conspectus Florae Orientalis. An annotated catalogue of the

flora of the Middle East. Fasc. 6. - Jerusalem: The Israel Academy of Sciences and Humanities, xii + 191 pp. + 1 map.

HELLER, D. & HEYN, C.C. 1993a: Conspectus Florae Orientalis. An annotated catalogue of the flora of the Middle East. Fasc. 7. - Jerusalem: The Israel Academy of Sciences and Humanities, xii + 53 pp. + 1 map

HELLER, D. & HEYN, C.C. 1993b: Conspectus Florae Orientalis. An annotated catalogue of the flora of the Middle East. Fasc. 8. - Jerusalem: The Israel Academy of Sciences and Humanities, xii + 174 pp. + 1 map.

HELLER, D. & HEYN, C.C. 1994: Conspectus Florae Orientalis. An annotated catalogue of the flora of the Middle East. Fasc. 9. - Jerusalem: The Israel Academy of Sciences and Humanities, xii + 171 pp. + 1 map.

HELLER, D. (ED.) 1987: Bibliography of botanical research of the Middle Eastern region, compiled by Hedda Yudkiss. - *Boissiera* **39**: 1-439.

HELLMANN, V. & RAUS, TH. 1999: Notes on *Carex hordeistichos*. In Greuter, W. & Raus, Th. (eds): Med-Checklist notulae, 18. - *Willdenowia* **29(1-2)**: 64.

HELLMANN, V. & SEYBOLD, S. 1980: *Ajuga piskoi* (Labiatae) - neu für Jugoslawien. - *Pl. Syst. Evol.* **135 (3-4)**: 295-297.

HELLWIG, F.H. 1994: Chromosomenzahlen aus der Tribus Cardueae (Compositae). - *Willdenowia* **24 (1-2)**: 219-248.

HELLWIG, F.H. 2004: Centaureinae (Asteraceae) in the Mediterranean - history of ecogeographical radiation. - *Pl. Syst. Evol.* **246(3-4)**: 137-162.

HELLWIG, F.H. & VICTOR, K. 2018: The botanist Carl Haussknecht (1838-1903) in theOttoman Empire and Persia (1865 and 1866-1869): a biographical sketch and itinerary of his expeditions. In Aščerić-Todd, I. & al. (eds): Travellers in Ottoman lands. The Botanical Legacy, pp. 147-157. - Oxford: ASTENE and Archaeopress Publishing Ltd, 379 pp.

HELM, J. 1972: *Apium graveolens* L. Geschichte der Kultur und Taxonomie. - *Kulturpflanze* **19**: 73-100.

HEMPEL, W. 1970: Taxonomische und chorologische Untersuchungen an Arten von *Melica* L. Subgen. *Melica*. - *Feddes Repert.* **81(1-5)**: 131-145.

HEMPEL, W. 1971a: Die systematische Stellung von *Melica uniflora* Retz. und *Melica rectiflora* Boiss. et Heldr. (*Melica* L. Subgen. *Bulbimelica*, subgen. nov.). (Vorarb. einer Revision der Gattung *Melica* L. - I). - *Feddes Repert.* **81(10)**: 657-686.

HEMPEL, W. 1971b: Die Verwandtschaft der Gattungen *Molinia* Schrank, *Moliniopsis* Hayata und *Schizachne* Hack. mit dem Formenkreis *Melica* L. Subgen. *Bulbimelica* Hempel. (Vorarb. Rev. der Gattung *Melica* L. - II). - *Feddes Repert.* **81(10)**: 687-712.

HEMPEL, W. 1973: Die systematische Stellung von *Melica altissima* L. und des *Melica minuta* L. - *Melica ramosa* Vill. Komplexes (*Melica* L. subgen. *Altimelica* Hempel). - *Feddes Repert.* **84(7-8)**: 533-568.

HEMPEL, W. 2004a: (1637) Proposal to conserve the name *Melica transsilvanica* against *M. caricina* (Poaceae). - *Taxon* **52(3)**: 836. [approved, cf. Taxon 55(4): 1100, 2005].

HEMPEL, W. 2004b: Die infragenerische Gliederung der Gattung *Melica* L. - *Haussknechtia* **10**: 169-198.

HEMPEL, W. 2012: Revision und Phylogenie der Arten der Gattung *Melica* L. (Poaceae) in Eurasien und Nordafrika. - *Feddes Repert.* **122(1-2)**: 1-253.

HENDERSON, N.C. 1962: A taxonomic revision of the genus *Lycopus* (Labiatae). - *Amer. Midl. Naturalist* **68(1)**: 95-138.

HENDRYCH, R. 1964a: A new endemic *Thesium* species from Italy. - *Preslia* **36**: 114-122.

HENDRYCH, R. 1964b: Specierum palaeoarcticarum generis *Thesium* nomina et synonyma. - *Novit. Bot. Delect. Seminum Horti Bot. Univ. Carol. Prag.* **1964**: 7-18.

HENDRYCH, R. 1964c: On the new species *Thesium macedonicum*. - *Biol. Glasn.* **17(1)**: 11-15 + 2 plates.

HENDRYCH, R. 1966a: Systematic study on *Thesium alpinum*. - *Acta Univ. Carol., Biol.* **1966(2)**: 107-138.

HENDRYCH, R. 1966b: Remarks on the species *Trifolium patens*. - *Preslia* **38(2)**: 137-150.

HENDRYCH, R. 1967: *Trifolium retusum* in der Tschechslovakei. - *Preslia* **39**: 30-42.

HENDRYCH, R. 1968: Abriss der chorologischen Entwicklung der Gattung *Thesium* (Santalaceae) in der Paläoarktis. - *Preslia* **40(1)**: 28-51.

HENDRYCH, R. 1969: A treatise of *Thesium arvense*. - *Acta Univ. Carol., Biol.* **1968(3)**: 243-262.

HENDRYCH, R. 1970: The outline of the taxonomy and chorology of *Thesium linophyllon*. - *Acta Univ. Carol., Biol.* **1969(2)**: 119-170.

HENDRYCH, R. 1972: The natural history and systematics of the genus *Thesium* L. - *Acta Univ. Carol., Biol.* **1970(4)**: 293-358.

HENDRYCH, R. 1975a: Was ist *Trifolium grandiflorum* Schreber (1767)? - *Preslia* **47(3)**: 280-281.

HENDRYCH, R. 1975b: *Trifolium spadiceum* und sein Vorkommen in der Tschechoslowakei (*Trifolium*- Studien X). - *Acta Univ. Carol., Biol.* **1974(1)**: 11-37.

HENDRYCH, R. 1976a: *Thesium* in Flora Europaea (1964) and today. - *Preslia* **48(2)**: 107-112.

HENDRYCH, R. 1976b: Vorläufige Mitteilung zur Gattung *Chrysaspis* Desvaux (1818). - *Preslia* **48(3)**: 216-224.

HENDRYCH, R. 1977: Bemerkungen zur Variabilität von *Cruciata glabra* (*Galium vernum*). - *Preslia* **49 (3)**: 193-201.

HENDRYCH, R. 1978a: Remarks on *Thesium parnassi*. - *Preslia* **50(3)**: 199-208.

HENDRYCH, R. 1978b: Erwägungen zur Chorologie und Epiontologie von *Cruciata glabra*. - *Preslia* **50(4)**: 289-304.

HENDRYCH, R. 1979: *Cruciata glabra* auf dem Gebiet der Tschechischen Sozialistischen Republik. - *Preslia* **51(1)**: 1-30.

HENDRYCH, R. 1982: Material and notes about the geography of the highly stenochoric to monotopic endemic species of the European flora. - *Acta Univ. Carol., Biol.* **1980(3-4)**: 335-372.

HENDRYCH, R. 1987: Einige Bemerkungen zu den *Echinops*-Arten in der Tschechoslowakei. - *Preslia* **59(2)**: 135-154.

HENDRYCH, R. 1988a: Die ersten nomenklatorischen Ergänzungen zur *Trifolium*-Monographie von Zohary und Heller (taxa supraspecifica). - *Preslia* **60(3)**: 215-237.

HENDRYCH, R. 1988b: Zweite Reihe nomenklatorischer Ergänzungen zur *Trifolium*-Monographie von Zohary und Heller (nomina excludenda). - *Novit. Bot. Univ. Carol.* **4**: 49-53.

HENDRYCH, R. 1989a: *Physalis alkekengi*, in Europa und in der Tschechoslowakei besonders. - *Acta Univ. Carol., Biol.* **33(1)**: 1-42.

HENDRYCH, R. 1989b: Vierte Reihe der nomenklatorischen Ergänzungen zur *Trifolium*-Monographie von Zohary und Heller (taxa specifica et infraspecifica). - *Acta Univ. Carol., Biol.* **33(4)**: 257-314.

HENDRYCH, R. 1990a: Dritte Reihe der Ergänzungen zur *Trifolium*-Monographie von Zohary und Heller (plantae hybridae). - *Preslia* **62(1)**: 43-60.

HENDRYCH, R. 1990b: Über *Trifolium medium* L. var. *banaticum* Heuffel und über die ihm nahe verwandten Sippen. - *Acta Univ. Carol., Biol.* **34(5-6)**: 321-358.

HENDRYCH, R. 1991: Eine unnötigerweise vergessene Art - *Trifolium bithynicum* Boissier (1849). - *Preslia* **63(1)**: 47-62.

HENDRYCH, R. 1993: Bemerkungen zur balkanischen Art *Trifolium pseudomedium*. - *Preslia* **65(2)**: 131-145.

HENDRYCH, R. 1994a: Bemerkungen zu den *Thesium*-Arten der Türkei. - *Preslia* **66(1)**: 23-40.

HENDRYCH, R. 1994b: *Thesium procumbens*, eine verschollene Art Bulgariens? - *Preslia* **66(3)**: 265-272.

HENDRYCH, R. 1996: *Thesium bergeri*, eine Art des östlichen Mittelmeergebietes. - *Preslia* **68(1)**: 3-40.

HENDRYCH, R. & HENDRYCHOVÁ, H. 1989a: Zur Frage des Vorkommens von *Aposeris foetida* in der Tschechoslowakei. - *Acta Univ. Carol., Biol.* **31(5-6)**: 285-311.

HENDRYCH, R. & HENDRYCHOVÁ, H. 1989b: Die *Pedicularis*-Arten der Tschechoslowakei, früher und jetzt. - *Acta Univ. Carol., Biol.* **32(5)**: 403-456.

HENEEN, W.K. 1977a: Chromosomal polymorphism in isolated populations of *Elymus* (*Agropyron*) in the Aegean. II. *Elymus rechingeri*. - *Hereditas (Lund)* **86(2)**: 211-224.
HENEEN, W.K. 1977b: Chromosomal polymorphism in isolated populations of *Elymus* (*Agropyron*) in the Aegean. III. *Elymus diae*. - *Hereditas (Lund)* **86(2)**: 225-236.
HENEEN, W.K. 1977c: Chromosomal polymorphism in isolated populations of *Elymus* (*Agropyron*) in the Aegean. IV. General discussion. - *Hereditas (Lund)* **86(2)**: 237-244.
HENEEN, W.K. 1977d: Chromosome polymorphism in Aegean *Elymus* species found in small isolated populations. - *Publ. Cairo Univ. Herb.* **7-8**: 21-32.
HENEEN, W.K. & RUNEMARK, H. 1962: Chromosomal polymorphism and morphological diversity in *Elymus rechingeri*. - *Hereditas (Lund)* **48(4)**: 545-564.
HENEEN, W.K. & RUNEMARK, H. 1972a: Chromosomal polymorphism in isolated populations of *Elymus* (*Agropyron*) in the Aegean. I. *Elymus striatulus* sp. nov. - *Bot. Not.* **125(4)**: 419-429.
HENEEN, W.K. & RUNEMARK, H. 1972b: Cytology of the *Elymus* (*Agropyron*) *elongatus* complex. - *Hereditas (Lund)* **70(2)**: 155-164.
HENKE, E. 1983: Ostern in der Sfakia. - *Mitt. Arbeitskreis Heimische Orchid. Hessen* **1(1)**: 10-22.
HENKE, E. 1984: *Orchis ustulata* L. auf der Peloponnes. - *Ber. Arbeitskreis. Heimische Orchid.* **1(1)**: 92.
HENKE, E. 1986: Exkursionen in die Orchideenflora Kretas. - *Ber. Arbeitskreis. Heimische Orchid.* **3(1)**: 13-38.
HENKE, E. 1987a: Botanische Exkursionen zum Olymp auf Lesbos. - *Ber. Arbeitskreis. Heimische Orchid.* **4(1)**: 145-146.
HENKE, E. 1987b: *Orchis sancta* und *Ophrys regis-ferdinandii* auf Chios. - *Ber. Arbeitskreis. Heimische Orchid.* **4(1)**: 147.
HENKE, E. 1988: Orchideenerlebnisse auf Samos. - *Ber. Arbeitskreis. Heimische Orchid.* **5(1-2)**: 47-49.
HENNECKE, M. 2016: Was ist *Ophrys fucifera*? - *Ber. Arbeitskreis. Heimische Orchid.* **33(1)**: 103-109.
HENNECKE, M. & MUNZINGER, S. 2014a: Die neue systematische Gliederung der Gattung *Ophrys*. - *Ber. Arbeitskreis. Heimische Orchid.* **31(1)**: 62-147.
HENNECKE, M. & MUNZINGER, S. 2014b: Die neue systematische Gliederung der Gattung *Ophrys*. - *Ber. Arbeitskreis. Heimische Orchid.* **31(2)**: 200-257.
HENNING, W. 1947: Morphologisch-systematische und genetische Untersuchungen an Arten und Artbastarden der Gattung *Malus*. - *Züchter* **17-18**: 289-349.
HENRARD, J.TH. 1950: Monograph of the genus *Digitaria*. - Leiden: Universitaire Pers, xxi + 999 pp.
HEPER, M. & MALYER, H. 1996: Taxonomical investigations on the genus *Chamaecytisus* Link in Turkey. In Öztürk, M.A., Seçmen, Ö. & Görk, G. (eds): Plant life in southwest and central Asia, 1: pp. 273-294. - Izmir: Ege University Press.
HEPPER, F.N. 1972: *Anemone blanda*. - *Bot. Mag.* **178(3)**: tab. 598 + 3 pp. text.
HEPPER, F.N. 1979: On typifying Linnaean names of Solanaceae. In Hawkes, J.G., Lester, R.N. & Skelding, A.D. (eds): The biology and taxonomy of the Solanaceae, pp. 131-133. - London, etc.: Academic Press.
HEPPER, F.N. & FRIIS, I. 1994: The plants of Pehr Forsskål's Flora Aegyptiaco-Arabica collected on the Royal Danish Expedition to Egypt and the Yemen 1761-63. - Kew: Royal Botanic Gardens, xii + 400 pp.
HEPPER, F.N. & JAEGER, P.-M.L. 1985: The typification of six Linnaean names in *Solanum*. - *Kew Bull.* **40(2)**: 387-391.
HERBERT, W. 1837: Amaryllidaceae; preceded by an attempt to arrange the monocotyledonous orders, and followed by a treatise on cross-bred vegetables, and supplement. - London: J. Ridgway and sons. viii + 428 pp.
HERBERT, W. 1845a: Crocorum synopsi addenda. - *Edwards's Bot. Reg.* **31**: 80-83.
HERBERT, W. 1845b: [Notes on *Cytinus hypocistis*]. - *Edwards's Bot. Reg.* **31**: 85-86.
HERBERT, W. 1845c: [Notes on *Crocus*]. - *Edwards's Bot. Reg.* **31**: 1-8.
HERBERT, W. 1847: A history of the species of *Crocus*. - *J. Hort. Soc. London* **2**: 249-293.
HERBORG, J. 1985: *Senecio ovatus* (P. Gaertn., Mey. & Scherb.) Willd., der korrekte Name für *Senecio fuchsii* C.C. Gmelin. - *Willdenowia* **15(1)**: 183-186.
HERBORG, J. 1987: Die Variabilität und Sippenabgrenzung in der *Senecio nemorensis*-Gruppe (Compositae) im europäischen Teilareal. - *Diss. Bot.* **107**: 1-262 + 1 table.
HERGT, B. 1903: Hofrat Prof. Carl Haussknecht, geb. den 30. Nov. 1838, gest. den 7. Juli 1903 [obituary]. - *Mitt. Thüring. Bot. Vereins, N.F.* **18**: 1-20 + 2 plates.
HERMANN, F. 1929a: Drei neue Pflanzenarten aus Bulgarien. - *Izv. Bulg. Bot. Druzh.* **3**: 43-44.
HERMANN, F. 1929b: Rosae bulgaricae. - *Izv. Bulg. Bot. Druzh.* **3**: 293-294.
HERMANN, F. 1932: *Polygala Rhodopeum*, species propria in Haemi peninsula endemica. - *Repert. Spec. Nov. Regni Veg.* **30**: 249-250.
HERMANN, F. 1935: Ein neuer Weizenbastard aus Thazien. - *Repert. Spec. Nov. Regni Veg.* **39**: 128.
HERMANN, F. 1936: Übersicht über die europäischen Rotten und einige andere Arten der Gattung *Trifolium*. - *Repert. Spec. Nov. Regni Veg.* **39**: 332-351.
HERMANN, F. 1937a: Übersicht über die *Herniaria*-Arten des Berliner Herbars. - *Repert. Spec. Nov. Regni Veg.* **42**: 203-224.
HERMANN, F. 1937b: Einige Nachträge zu *Herniaria*. - *Repert. Spec. Nov. Regni Veg.* **42**: 272.
HERMANN, F. 1938a: *Trifolium*. Conspectus sectionum et subsectionum europaearum generis. - *Repert. Spec. Nov. Regni Veg.* **43**: 316-318.
HERMANN, F. 1938b: *Triticum*. Conspectus sectionum et subsectionum nonnullarum. - *Repert. Spec. Nov. Regni Veg.* **44**: 158-160.
HERMANN, F. 1939a: Über die Verbreitung von *Poa timoleontis* in Bulgarien. - *Izv. Bulg. Bot. Druzh.* **8**: 99.
HERMANN, F. 1939b: Sectiones et subsectiones nonnullae europaeae generis *Allium*. - *Repert. Spec. Nov. Regni Veg.* **46**: 57-58.
HERMANN, F. 1939c: *Festuca*. Conspectus sectionum vel subgenerum europaeorum. - *Repert. Spec. Nov. Regni Veg.* **46**: 166-168.
HERMANN, F. 1939d: *Geranium*. Conspectus sectionum europaearum. - *Repert. Spec. Nov. Regni Veg.* **46**: 191-192.
HERMANN, F. 1941: Zur Unterscheidung unserer heimischen Erdbeerarten. - *Repert. Spec. Nov. Regni Veg.* **50**: 363-365.
HERMANN, F. & STEFANOV, B. 1929: Nachtrag zur Flora des Piringebirges in Bulgarien. - *Izv. Bulg. Bot. Druzh.* **3**: 53-55.
HERMANN, F., STEFANOV, B. & GEORGIEV, T. 1931: Neuer Nachtrag zur Flora Bulgariens. - *Izv. Bulg. Bot. Druzh.* **4**: 25-28.
HERMEY, W. 1976: Orchideen Europas und der Mittelmeerländer. Orchideenbeobachtungen auf Kreta. - *Kakteen Orchid. Rundschau* **1(4-5)**: 53-56, 63-66.
HERMJAKOB, G. 1969a: Orchideen in Attika. - *Festschr. Deutsch. Schule Athen* **1969**: 89-108.
HERMJAKOB, G. 1969b: Braune Ragwurz (*Ophrys fusca* ssp. *fusca*) blüht zur Weinachtszeit in Attika. - *Orchidee (Hamburg)* **20(2)**: 83-84.
HERMJAKOB, G. 1969c: Kleinblütiges Knabenkraut (*Neotinea intacta*) und Kretische Ragwurz (*Ophrys cretica* ssp. *cretica*) auf Ägina. - *Orchidee (Hamburg)* **20(2)**: 84-85.
HERMJAKOB, G. 1973: Eine für Griechenland neue, autogame *Epipactis*-Art: *E. leptochila* (Godf.) Godf. (Orchidaceae). - *Ann. Mus. Goulandris* **1**: 99-103 + 4 plates.
HERMJAKOB, G. 1974: Orchids of Greece and Cyprus: 1. The genus *Ophrys*. - Kifissia: Goulandris Natural History Museum.
HERMJAKOB, G. 1976: Orchideenstandorte in Südeuropa: Aus der Flora Griechenlands. In: Proceedings of the 8th world orchid conference, pp. 95-98. - Frankfurt: Deutsche Orchideen Gesellschaft.
HERMJAKOB, G. 1977: Die aktuelle und potentielle natürliche Hartlaubvegetation Attikas. - Münster: Diss., Westf. Wilhelms-Univ., 114 pp. + plates and tables.

HERNÁNDEZ CARDONA, A.M. 1978: Estudio monográfico de los géneros *Poa* y *Bellardiochloa* en la Península Ibérica e islas Baleares. - *Diss. Bot.* **46**: 1-365.
HERNÁNDEZ, H.M. 1981: Sobre la ecología reproductiva de *Nicotiana glauca* Grah: una maleza de distribución cosmopolita. - *Bol. Soc. Bot. México* **41**: 47-73.
HERRERA, J. 1991: The reproductive biology of a riparian Mediterranean shrub, *Nerium oleander* L. (Apocynaceae). - *Bot. J. Linn. Soc.* **106(2)**: 147-172.
HERRERO, A. & AL. 1997: Números cromosomáticos de plantas occidentales, 742-745. - *Anales Jard. Bot. Madrid* **55(2)**: 429-430.
HERRERO, A., PAJARÓN, S. & PRADA, C. 2001: Isozyme variation and genetic relationships among taxa in the *Asplenium obovatum* group (Aspleniaceae, Pteridophyta). - *Amer. J. Bot.* **88(11)**: 2040-2050.
HERRMANN, N., WEISS, G. & DURKA, W. 2006: Biological flora of Central Europe: *Muscari tenuiflorum* Tausch. - *Flora, Morphol. Geobot. Oekophysiol.* **201(2)**: 81-101.
HERRNSTADT, I. & HEYN, C.C. 1971: Studies in *Heptaptera* (Umbelliferae) II: Taxonomic revision. - *Notes Roy. Bot. Gard. Edinburgh* **31(1)**: 91-107.
HERRNSTADT, I. & HEYN, C.C. 1975: New combinations in *Cachrys* L. as a genus distinct from *Hippomarathrum* Link. - *Notes Roy. Bot. Gard. Edinburgh* **33(3)**: 441-443.
HERRNSTADT, I. & HEYN, C.C. 1977: A monographic study of the genus *Prangos* (Umbelliferae). - *Boissiera* **26**: 1-91.
HERTEL, H. 1986: *Ophrys doerfleri* H. Fleischmann auf dem Peloponnes. - *Ber. Arbeitskreis. Heimische Orchid.* **3(1)**: 57.
HERTEL, S. & HERTEL, K. 2003: Zu Beginn der Orchideenblüte auf Kreta und Karpathos. - *J. Eur. Orch.* **35(3)**: 635-664.
HERTEL, S. & HERTEL, K. 2005: Orchideenreise durch die Inselwelt der Ostägäis. - *J. Eur. Orch.* **37(2)**: 419-466.
HERTEL, S. & KREUTZ, K. 2010: *Serapias orientalis* (Greuter) H. Baumann & Künkele subsp. *orientalis* neu auf Kithira und der Peloponnes. - *J. Eur. Orch.* **42(3-4)**: 519-524.
HERTEL, S. & PAULUS, H.F. 2010: *Ophrys mycenensis* S. Hertel & H.F. Paulus, eine neue Art der *Ophrys oestrifera*-Gruppe in Griechenland. - *J. Eur. Orch.* **42(3-4)**: 453-466.
HERTEL, S. & PRESSER, H. 2010: Neue Erkenntnisse zu den Orchideen in Griechenland. - *Ber. Arbeitskreis. Heimische Orchid.* **27(1)**: 146-202.
HERTEL, S. & PRESSER, H. 2018: Bemerkenswerte Orchideenfunde in Albanien und Süd-Griechenland. - *Ber. Arbeitskreis. Heimische Orchid.* **35(1)**: 122-152.
HERTEL, S., TSIFTSIS, S. & ANTONOPOULOS, Z. 2014: *Epipactis pinovica* (Orchidaceae), a new orchid species of Greece. - *J. Eur. Orch.* **46(3-4)**: 701-714.
HERTEL, S. & WEYLAND, H. 2009: *Ophrys malvasiana* S. Hertel & H. Weyland, eine neue Art der *Ophrys holoserica*-Gruppe in Griechenland. - *J. Eur. Orch.* **41(3-4)**: 681-693.
HERVOUET, J.-M. 1984: *Ophrys bornmuelleri* à Rhodes. - *Orchidophile (Ansières)* **15(64)**: 710.
HERZOG, R. 1935: Ein Beitrag zur Systematik der Gattung *Salvinia*. - *Hedwigia* **74(6)**: 257-284.
HERZOG, TH. 1920: Botanische Studien eines Frontsoldaten in Mazedonien. - *Allg. Bot. Z. Syst.* **24-25**: 8-23.
HESLOP-HARRISON, Y. 1955a: Biological Flora of the British Isles. *Nuphar* Sm. - *J. Ecol.* **43(1)**: 342-364.
HESLOP-HARRISON, Y. 1955b: Biological Flora of the British Isles. *Nymphaea* L. em. Sm. (nom. conserv.). - *J. Ecol.* **43(2)**: 719-734.
HESS, E. 1924: Die Zypresse auf der Insel Kreta. - *Schweiz. Z. Forstwesen* **75(1)**: 14-18.
HESS, H.E. 1955: Systematische und zytogenetische Untersuchungen an einigen *Ranunculus*-Arten aus der *Nemorosus*-Gruppe. - *Ber. Schweiz. Bot. Ges.* **65**: 272-301.
HESS, H.E., LANDOLT, E. & HIRZEL, R. 1967: Flora der Schweiz und angrenzender Gebiete. Band 1: Pteridophyta bis Caryophyllaceae. - Basel & Stuttgart: Birkhäuser Verlag, 858 pp.
HESS, H.E., LANDOLT, E. & HIRZEL, R. 1970: Flora der Schweiz und angrenzender Gebiete. Band 2: Nymphaeaceae bis Primulaceae. - Basel & Stuttgart: Birkhäuser Verlag, 965 pp.
HESS, H.E., LANDOLT, E. & HIRZEL, R. 1972: Flora der Schweiz und angrenzender Gebiete. Band 3: Plumbaginaceae bis Compositae + Nachträge. - Basel & Stuttgart: Birkhäuser Verlag, 876 pp.
HETSCH, W. & VERGOS, S. 1997: Die Buchenwälder des Nord-Pindos. - *Forstarchiv* **68**: 10-18.
HEUBL, G.R. 1984: Systematische Untersuchungen an mitteleuropäischen *Polygala*-Arten. - *Mitt. Bot. Staatssamml. München* **20**: 205-428.
HEUBL, G.R. 1989: Bemerkungen zur Karyologie der Gattung *Lysimachia* L. Chromosomenzahlen und Evolution. - *Mitt. Bot. Staatssamml. München* **28**: 297-311.
HEUFFEL, J. 1858: Enumeratio plantarum in Banatu Temesiensi sponte crescentium et frequentius cultarum ... - *Verh. K.K. Zool.-Bot. Ges. Wien* **8**: 39-240.
HEUSSER, C. 1938: Chromosomenverhältnisse bei schweizerischen basitonen Orchideen. - *Ber. Schweiz. Bot. Ges.* **48**: 562-605.
HEWETT, D.G. 1964: Biological Flora of the British Isles. *Menyanthes trifoliata* L. - *J. Ecol.* **52(3)**: 723-735.
HEYN, C.C. 1956: Some chromosome counts in the genus *Medicago*. - *Caryologia* **9(1)**: 160-165.
HEYN, C.C. 1959a: Evolutionary problems in the genus *Medicago* [abstract]. - *Bull. Res. Council Israel, Sect. D, Bot.* **7(2)**: 112.
HEYN, C.C. 1959b: The genus *Medicago* in Linnaeus's Species Plantarum. - *Bull. Res. Council Israel, Sect. D, Bot.* **7(2)**: 157-174.
HEYN, C.C. 1962a: Studies of bulbous *Poa* in Palestine. 1. The agamic complex of *Poa bulbosa*. - *Bull. Res. Council Israel, Sect. D, Bot.* **11(2)**: 117-126.
HEYN, C.C. 1962b: On the cytotaxonomy of *Onobrychis crista-galli* (L.) Lam. and *O. squarrosa* Viv. - *Bull. Res. Council Israel, Sect. D, Bot.* **11(3)**: 177-182.
HEYN, C.C. 1963: The annual species of *Medicago*. - *Scripta Hierosol.* **12**: v + 154 pp. + 5 plates.
HEYN, C.C. 1964: Cytotaxonomy of two legumes from Israel (*Onobrychis caput-galli* (L.) Lam. and *Hedysarum spinosissimum* L.). - *Israel J. Bot.* **12(4)**: 187-191.
HEYN, C.C. 1966: A study of the *Lotus peregrinus* group. - *Israel J. Bot.* **15(2)**: 37-47.
HEYN, C.C. 1968: An evolutionary study of fruit morphology in the tribe Trigonelleae (Leguminosae). - *Phytomorphology* **18(1)**: 54-59.
HEYN, C.C. 1969: The typification of *Medicago* (Legum.). - *Taxon* **18(6)**: 667-669.
HEYN, C.C. 1970: Studies in *Lotus*. III. The *L. angustissimus* group. - *Israel J. Bot.* **19(2-3)**: 271-292.
HEYN, C.C. 1971: Biosystematic approaches to the solution of taxonomic problems in Israel. In Davis, P.H., Harper, P.C. & Hedge, I.C. (eds): Plant life of South-West Asia, pp. 181-193. - Edinburgh: Botanical Society.
HEYN, C.C. 1981: Tribe 23. Trifolieae (Bronn) Benth. (1865). In Polhill, R.M. & Raven, P.H. (eds): Advances in legume systematics 1: pp. 383-385. - Kew: Royal Botanic Gardens.
HEYN, C.C. & DAFNI, A. 1971: Studies in the genus *Gagea* (Liliaceae) I. The platyspermous species in Israel and neighbouring area. - *Israel J. Bot.* **20(3)**: 214-233.
HEYN, C.C. & DAFNI, A. 1977: Studies in the genus *Gagea* (Liliaceae) II. The non-platyspermous species from the Galilee, the Golan heights and Mt. Hermon. - *Israel J. Bot.* **26(1)**: 11-22.
HEYN, C.C., DAGAN, O. & NACHMAN, B. 1975: The annual *Calendula* species: taxonomy and relationships. - *Israel J. Bot.* **23(4)**: 169-201.
HEYN, C.C. & HERRNSTADT, I. 1967: The *Lotus creticus* group. - *Kew Bull.* **21(2)**: 299-309.
HEYN, C.C. & HERRNSTADT, I. 1977: Seed coat structure of Old World *Lupinus* species. - *Bot. Not.* **130 (4)**: 427-435.
HEYN, C.C. & JOEL, A. 1983: Reproductive relationships between annual species of *Calendula* (Compositae). - *Pl. Syst. Evol.* **143(4)**: 311-329.
HEYN, C.C. & PAZY, B. 1990: The annual species of *Adonis* (Ranunculaceae) - a polyploid complex. - *Pl. Syst. Evol.* **168(3-4)**: 181-193.
HEYN, C.C. & RAVIV, V. 1966: Experimental taxonomic studies in the genus *Scorpiurus* (Papilionaceae). - *Bull. Torrey Bot. Club* **93(4)**: 259-267.

HEYWOOD, C.A. 1983: Meiosis in some species and cultivars of *Crocus* (Iridaceae). - *Pl. Syst. Evol.* **143(3)**: 207-225.

HEYWOOD, V.H. 1951: A new species of *Digitalis* from Greece - *Digitalis macedonica*. - *Kew Bull.* **1951(2)**: 149-151.

HEYWOOD, V.H. 1958: The nomenclature and typification of *Ballota hispanica* (L.) Benth. - *Agron. Lusit.* **20(3)**: 199-204.

HEYWOOD, V.H. 1960: On the identity of *Caucalis platycarpos* L. and *C. daucoides* L. - *Agron. Lusit.* **22(1)**: 11-13.

HEYWOOD, V.H. 1963: [Note on *Rhynchosinapis nivalis*]. In Heywood, V.H. (ed.): Notulae systematicae ad Floram Europaeam spectantes, no. 2. - *Feddes Repert. Spec. Nov. Regni Veg.* **68 (3)**: 196.

HEYWOOD, V.H. 1964a: [Note on *Dryopteris villarii*]. In Heywood, V.H. (ed.): Notulae systematicae ad Floram Europaeam spectantes, no. 3. - *Feddes Repert. Spec. Nov. Regni Veg.* **69(1)**: 44.

HEYWOOD, V.H. 1964b: [*Bornmuellera baldaccii* (Degen) Heywood, comb. nov.]. In Heywood, V.H. (ed.): Notulae systematicae ad Floram Europaeam spectantes, no. 3. - *Feddes Repert. Spec. Nov. Regni Veg.* **69(1)**: 61.

HEYWOOD, V.H. 1968a: A synopsis of the European species of *Cytisus* and allied genera. In Heywood, V.H. (ed.): Notulae systematicae ad Floram Europaeam spectantes, no. 7. - *Feddes Repert.* **79(1-2)**: 20-23.

HEYWOOD, V.H. 1968b: [Notes on *Cistus*]. In Heywood, V.H. (ed.): Notulae systematicae ad Floram Europaeam spectantes, no. 7. - *Feddes Repert.* **79(1-2)**: 59-61.

HEYWOOD, V.H. 1968c: [Notes on *Daucus*]. In Heywood, V.H. (ed.): Notulae systematicae ad Floram Europaeam spectantes, no. 7. - *Feddes Repert.* **79(1-2)**: 66-68.

HEYWOOD, V.H. 1972: [Notes on *Sideritis*]. In Heywood, V.H. (ed.): Notulae systematicae ad Floram Europaeam spectantes, no. 13. - *Bot. J. Linn. Soc.* **65(4)**: 354-355.

HEYWOOD, V.H. 1983: Relationships and evolution of the *Daucus carota* complex. - *Israel J. Bot.* **32 (2)**: 51-65.

HEYWOOD, V.H. 1995: The Mediterranean flora in the context of world biodiversity. - *Ecol. Medit.* **21(1-2)**: 11-18.

HEYWOOD, V.H., HARBORNE, J.B. & TURNER, B.L. (EDS) 1977: The biology and chemistry of the Compositae. - London, etc.: Academic Press, 2 vols., 1189 pp.

HEYWOOD, V.H. & HUMPHRIES, C.J. 1977: Anthemideae - systematic review. In Heywood, V.H., Harborne, J.B. & Turner, B.L. (eds): The biology and chemistry of the Compositae 2: 851-898. - London, etc.: Academic Press.

HEYWOOD, V.H. & ZOHARY, D. 1996: A catalogue of the wild relatives of cultivated plants native to Europe. - *Fl. Medit.* **5**: 375-415.

HEYWOOD, V.H. (ED.) 1971: The biology and chemistry of the Umbelliferae. - *Bot. J. Linn. Soc.* **64, suppl. 1**: 438 pp.

HICKS, G.C. 1929: Cytological studies in *Cyperus*, *Eleocharis*, *Dulichium* and *Eriophorum*. - *Bot. Gaz.* **88(2)**: 132-149 + 2 plates.

HIGTON, R.N. & AKEROYD, J.R. 1991: Variation in *Capparis spinosa* L. in Europe. In Chater, A.O. (ed.): Flora Europaea. Notulae systematicae ad Floram Europaeam spectantes. Series 2. No. 4. - *Bot. J. Linn. Soc.* **106(2)**: 104-112.

HILDEBRAND, F. 1894: Über *Cyclamen pentelici* n. sp.? - *Bot. Jahrb. Syst.* **18 Beibl. 44**: 1-5.

HILDEBRAND, F. 1896: Zur Benennung der *Cyclamen*-Arten. - *Bot. Jahrb. Syst.* **21 Beibl. 54**: 15-19.

HILDEBRAND, F. 1898: Die Gattung *Cyclamen* L., eine systematische und biologische Monographie. - Jena: Gustav Fischer Verlag, 190 pp. + 6 plates.

HILDEBRAND, F. 1904: Über *Cyclamen hiemale* nov. spec. - *Gartenflora* **53**: 70-73.

HILDEBRAND, F. 1906a: Über *Cyclamen Pesudo-graecum*. - *Gartenflora* **55**: 629-634.

HILDEBRAND, F. 1906b: Über einige neue und andere noch nicht lange aufgefundene *Cyclamen*-Arten. - *Beih. Bot. Centralbl., Abt. 2* **19**: 367-384.

HILDEBRAND, F. 1908: Über einige neue *Cyclamen*-Arten. - *Gartenflora* **57**: 291-298.

HILDEBRAND, F. 1999: The genus *Cyclamen* L.: a systematic and biological momograph. [Translation by E. Frank of the German original from 1898]. - London: The Cyclamen Society, 126 pp.

HILGER, H.H. 1981a: Die Perikarpentwicklung geflügelter Klausen aus dem *Paracaryum*-Verwandtschaftskreis (Cynoglosseae, Boraginaceae). - *Pl. Syst. Evol.* **138(1-2)**: 73-88.

HILGER, H.H. 1981b: Entwicklungstendenzen im Fruchtbau der anemochoren Boraginaceengattungen *Mattiastrum* und *Paracaryum*. - *Ber. Deutsch. Bot. Ges.* **94(4)**: 677-687.

HILGER, H.H. 1985: Ontogenie, Morphologie und systematische Bedeutung geflügelter und glochidientragender Cynoglosseae- und Eritricheae-Früchte (Boraginaceae). - *Bot. Jahrb. Syst.* **105 (3)**: 323-378.

HILGER, H.H. 1987: Fruchtbiologische Untersuchungen an Heliotropioideae (Boraginaceae). I. Die Ontogenie der monospermen Früchte von *Heliotropium supinum* L. - *Flora, Morphol. Geobot. Oekophysiol.* **179(4)**: 291-303.

HILGER, H.H. & AL. 2005: The Euro+Med treatment of Boraginaceae in Willdenowia 34 - a response. - *Willdenowia* **35(1)**: 43-48.

HILGER, H.H. & BÖHLE, U.-R. 2000: *Pontechium*: a new genus distinct from *Echium* and *Lobostemon* (Boraginaceae). - *Taxon* **49(4)**: 737-746.

HILGER, H.H. & REESE, H. 1983: Ontogenie der Strahlblüten und der heterokarpen Achänen von *Calendula arvensis* (Asteraceae). - *Beitr. Biol. Pflanzen* **58(1)**: 123-147.

HILL, A.W. 1940: *Verbascum banaticum*. - *Bot. Mag.* **163(1)**: tab. 9608 + 2 pp. text.

HILL, D.A. 1985: Aspects of the Yugoslavian flora. 1. The genus *Crocus* in South Serbia. - *Quart. Bull. Alpine Gard. Soc. Gr. Brit.* **53(2)**: 189-196.

HILL, D.A. 1986: Aspects of the Yugoslavian flora. 3. The genus *Ramonda*. - *Quart. Bull. Alpine Gard. Soc. Gr. Brit.* **54(2)**: 182-187.

HILL, D.J. & PRICE, B. 2000: Biological Flora of the British Isles. *Ornithogalum pyrenaicum* L. - *J. Ecol.* **88(2)**: 354-365.

HILLER, W. & KALTEISEN, M. 1988: Die Orchideen der Insel Karpathos. - *Mitt. Arbeitskreis Heimische Orchid. Baden-Württemberg* **20(3)**: 443-518.

HILLIARD, O.M. 1981a: Some generic concepts in Compositae-Gnaphaliinae. - *Bot. J. Linn. Soc.* **82(3)**: 181-232.

HILLIARD, O.M. 1981b: A revision of *Ifloga* in southern Africa, with comments on the northern hemisphere species. - *Bot. J. Linn. Soc.* **82(3)**: 293-312.

HILLIARD, O.M. & BURTT, B.L. 1981: Names in *Gnaphalium*, *Xeranthemum* and *Helichrysum* published between 1753 and 1800. - *Bot. J. Linn. Soc.* **82(3)**: 233-265.

HILLMAN, W.S. 1961: The Lemnaceae, or duckweeds. A review of the descriptive and experimental literature. - *Bot. Rev. (Lancaster)* **27(2)**: 221-287.

HILU, K.W. 1981: Cytotaxonomical studies in *Tribulus terrestris* and *T. alatus* (Zygophyllaceae). - *Nordic J. Bot.* **1(4)**: 531-534.

HIND, D.J.N. 1988: The biology and systematics of *Moehringia* L. (Caryophyllaceae). - Reading: Ph.D. thesis, Univ. of Reading, 410 pp. + 33 plates.

HIND, D.J.N., STRANGE, K. & RIX, E.M. 2007: 603. *Helichrysum orientale*. - *Curtis's Bot. Mag.* **24(4)**: 251-259.

HIND, M. 2019: 916. *Doronicum caucasicum*. Compositae. - *Curtis's Bot. Mag.* **36(3)**: 202-225.

HINDÁKOVÁ, M. 1986: Karyological study of the Slovak flora X. - *Acta Fac. Rerum Nat. Univ. Comen., Bot.* **33**: 49-50.

HINDÁKOVÁ, M. 1992: Karyological study of the Slovak flora XXVII. - *Acta Fac. Rerum Nat. Univ. Comen., Bot.* **39**: 63-65.

HINDÁKOVÁ, M. & ČINČURA, F. 1968: Angaben über die Zahl und Morphologie der Chromosomen einiger Pflanzenarten aus dem Territorium der Ostslowakei I. - *Acta Fac. Rerum Nat. Univ. Comen., Bot.* **14**: 181-227.

HINDÁKOVÁ, M. & LHOTSKÁ, M. 1980: [Chromosome numbers from Czechoslovakia]. In Löve, Á. (ed.): IOPB chromosome number reports LXIX. - *Taxon* **29(5-6)**: 728.

HINDÁKOVÁ, M. & MÁJOVSKÝ, J. 1976: [Chromosome numbers from Czechoslovakia]. In Löve, Á. (ed.): IOPB chromosome number reports LIII. - *Taxon* **25(4)**: 491.
HINDÁKOVÁ, M. & MÁJOVSKÝ, J. 1977: [Chromosome numbers from Czechoslovakia]. In Löve, Á. (ed.): IOPB chromosome number reports LVI. - *Taxon* **26(2-3)**: 262.
HINDÁKOVÁ, M., MÁJOVSKÝ, J. & SCHWARZOVÁ, T. 1986: Karyological study of the Slovak flora XVII. - *Acta Fac. Rerum Nat. Univ. Comen., Bot.* **33**: 81-84.
HINDÁKOVÁ, M. & SCHWARZOVÁ, T. 1977: [Chromosome numbers from Czechoslovakia]. In Löve, Á. (ed.): IOPB chromosome number reports LVI. - *Taxon* **26(2-3)**: 262.
HINDÁKOVÁ, M. & SCHWARZOVÁ, T. 1978: [Chromosome numbers from Czechoslovakia]. In Löve, Á. (ed.): IOPB chromosome number reports LXI. - *Taxon* **27(4)**: 380.
HINDÁKOVÁ, M. & SCHWARZOVÁ, T. 1980: [Chromosome numbers from Czechoslovakia]. In Löve, Á. (ed.): IOPB chromosome number reports LXIX. - *Taxon* **29(5-6)**: 728.
HINDÁKOVÁ, M. & SCHWARZOVÁ, T. 1987: Karyological study of the Slovak flora XX. - *Acta Fac. Rerum Nat. Univ. Comen., Bot.* **34**: 21-25.
HINDÁKOVÁ, M. & ZÁBORSKÝ, J. 1977: [Chromosome numbers from Czechoslovakia]. In Löve, Á. (ed.): IOPB chromosome number reports LVI. - *Taxon* **26(2-3)**: 262.
HINKE, C. & MANOLESKO, C. 1835: Antrag der im Jahre 1833 und 1834 in der europäischen Türkei, namentlich in Rumelien, auf dem Balkan und Rhodope gesammelten wildwachsenden Pflanzen. - *Flora, Int.-Bl.* **18(1)**: 53-59.
HIROE, M. 1979: Umbelliferae of the World. - Tokyo: Ariake Book Company, 2128 pp.
HIRTH, M. 2002: Zur Systematik einiger *Ophrys*-Arten aus dem *sphegodes-mammosa*-Komplex von Kerkira (Korfu) und NW Griechenland. - *Jahresber. Naturwiss. Vereins Wuppertal* **55**: 163-188.
HIRTH, M. 2005: Neue Untersuchungen zur Orchideenflora von Kerkira (Korfu, Griechenland). - *J. Eur. Orch.* **37(1)**: 147-228.
HIRTH, M. 2011: Über die Variabilität von *Ophrys epirotica* (Renz) J. Devillers-Terschuren & P. Devillers. - *J. Eur. Orch.* **43(1)**: 61-68.
HIRTH, M. 2016: A long-term survey of orchids on the small Greek island of Agathonisi from 1994 to 2013. - *J. Eur. Orch.* **48(1)**: 37-52.
HIRTH, M. & PAULUS, H.F. 2011: *Ophrys samiotissa*, eine neue Art der *O. oestrifera-holosericea*- Gruppe aus Samos (Orchidaceae). - *J. Eur. Orch.* **43(4)**: 863-873.
HIRTH, M. & PAULUS, H.F. 2016: Neue Bestäuber-Beobachtungen in der *Ophrys tenthredinifera*-Artengruppe der Ägäis mit Beschreibung von *Ophrys lychnitis* aus Paros. - *J. Eur. Orch.* **48(2-4)**: 346-388.
HIRTH, M. & PAULUS, H.F. 2020: *Ophrys eretriae*, eine neue Art aus dem *Ophrys mammosa*-Komplex von der Insel Euböa (Griechenland). - *J. Eur. Orch.* **52(1)**: 127-150.
HIRTH, M. & SPÄTH, H. 1989: Die Orchideen der Insel Samos. Ein Beitrag zur Kartierung der Orchideen des Mittelmeerraumes. - *Mitt. Arbeitskreis Heimische Orchid. Baden-Württemberg* **21(4)**: 1068-1135.
HIRTH, M. & SPÄTH, H. 1990: Beitrag zur Orchideenflora der Insel Ikaria. *Ophrys icariensis* - eine neue *Ophrys*-Art. - *Mitt. Arbeitskreis Heimische Orchid. Baden-Württemberg* **22(4)**: 693-729.
HIRTH, M. & SPÄTH, H. 1991: Ein neuer Gattungsbastard zwischen *Orchis morio* und *Serapias orientalis* subsp. *carica* auf Samos. - *Mitt. Arbeitskreis Heimische Orchid. Baden-Württemberg* **23(3)**: 505-509.
HIRTH, M. & SPÄTH, H. 1992: Zur Orchideenflora von Samos. - *Mitt. Arbeitskreis Heimische Orchid. Baden-Württemberg* **24(1)**: 1-51.
HIRTH, M. & SPÄTH, H. 1994: Beitrag zur Orchideenflora der ostaegaeischen Inseln Arkoi, Kalymnos, Leipsoi, Leros, Patmos, Phournoi, Telendos: *Ophrys calypsus* - eine neue Ophrysart. *Serapias patmia* - eine neue Serapiasart. - *J. Eur. Orch.* **26(3-4)**: 426-621.
HIRTH, M. & SPÄTH, H. 1998: Zur Orchideenflora von Chios - *Ophrys homeri*, eine neue Ophrysart. - *J. Eur. Orch.* **30(1)**: 3-80.
HIRTH, M. & SPÄTH, H. 2010: Beitrag zur Orchideenflora der ostägäischen Inseln Chalki, Megisti, Nissyros, Pserimos und Tilos. - *J. Eur. Orch.* **42(3-4)**: 563-608.
HISCOCK, S. 1991: Spili - an orchid cornucopia. - *Quart. Bull. Alpine Gard. Soc. Gr. Brit.* **59(3)**: 291-297.
HITCHCOCK, A.S. & CHASE, A. 1910: The North American species of *Panicum*. - *Contr. U.S. Natl. Herb.* **15**: i-xiv + 1-396.
HJELMQVIST, H. 1950: The flax weeds and the origin of cultivated flax. - *Bot. Not.* **1950(2)**: 257-298.
HJELMQVIST, H. 1959: Studien über Embryologie und Variabilität bei einigen *Aphanes*-Arten. - *Bot. Not.* **112(1)**: 17-64.
HOCHREUTINER, B.P.G. 1900: Révision du genre *Hibiscus*. - *Annuaire Conserv. Jard. Bot. Genève* **4**: 23-191.
HODA, P. 1993a: A geobotanical survey of the black pine (*Pinus nigra* Arn.) forests in Albania. - *Ann. Bot. (Rome)* **51**: 77-92.
HODA, P. 1993b: The vegetation of Albania. - *Quart. Bull. Alpine Gard. Soc. Gr. Brit.* **61(4)**: 421-426.
HODAK, N. 1960: The stands of the tropical species *Eleusine indica* Gaertn. in the flora of Jugoslavia [In Serbo-Croatian with English summary]. - *Acta Bot. Croat.* **18-19**: 65-67.
HODAK, N. 1971: Einige neue Fundorte von *Scaligeria cretica* (Miller) Boiss. im Gebiet des kroatischen Küstenlandes [In Serbo-Croatian with German summary]. - *Acta Bot. Croat.* **30**: 119-121.
HODÁLOVÁ, I. & MARHOLD, K. 1998: Morphometric comparison of *Senecio germanicus* and *S. nemorensis* (Compositae) with a new species from Romania. - *Bot. J. Linn. Soc.* **128(3)**: 277-290.
HODGKIN, E. 1959: *Crocus goulimyi*. - *Quart. Bull. Alpine Gard. Soc. Gr. Brit.* **27(4)**: 360-361.
HODGMAN, M.G. 1983: *Cyclamen repandum*. - *J. Scott. Rock Gard. Club* **18(3)**: 278-279.
HOFFMANN-GROBÉTY, A.E. 1933: Une nouvelle Centaurée du Dodécanèse. - *Bull. Soc. Bot. Genève, Sér. 2* **24**: 220-222.
HOFFMANN, A. 1989: Zyto- und ökotaxonomische Untersuchungen über *Bromus ramosus* Huds. und *Bromus benekenii* (Lange) Trimen. - *Tuexenia* **9**: 3-28.
HOFFMANN, J. 1897: Beitrag zur Kenntnis der Gattung *Odontites*. - *Oesterr. Bot. Z.* **47**: 113-117, 184-187, 233-239, 345-349.
HOFFMANN, M.H. 1999a: The phylogeny of *Actaea* (Ranunculaceae): a biogeographical approach. - *Pl. Syst. Evol.* **216(3-4)**: 251-263.
HOFFMANN, M.H. 1999b: Biogeographical and evolutionary patterns in the genus *Caltha* L. (Ranunculaceae). - *Bot. Jahrb. Syst.* **121(3)**: 403-421.
HÖFKER, H. 1915: Übersicht über die Gattung *Ligustrum*. - *Mitt. Deutsch. Dendrol. Ges.* **24**: 51-66 + 6 plates.
HOFMANN, A. 1918: Die Tätigkeit der österreichischen forstlichen Mission in Griechenland. - *Oesterr. Vierteljahrsschr. Forstwesen, N.F.* **36**: 9-19.
HOFMANN, U. 1968: Untersuchungen an Flora und Vegetation der Ionischen Insel Levkas. - *Vierteljahrsschr. Naturf. Ges. Zürich* **113(3)**: 209-256.
HOFMANN, U. & GÖTTMANN, J. 1990: *Morina* L. und *Triplostegia* Wall. ex DC. im Vergleich mit Valerianaceae und Dipsacaceae. - *Bot. Jahrb. Syst.* **111(4)**: 499-553.
HÖGSTRÖM, S. & FÅHRÆUS, G. 1993: Balkanoxel *Sorbus graeca*, det senaste tillskottet til Gotlands *Sorbus*-flora [In Swedish with English summary]. - *Svensk Bot. Tidskr.* **87(5)**: 265-274.
HØILAND, K. 1984: Diktamnos, *Origanum dictamnus*, en truet medisinplante på Kreta [In Norwegian]. - *Blyttia* **42(4)**: 177-180.
HØILAND, K. 1986: Kreta - hjørnet mellom Europa, Asia og Afrika [In Norwegian]. - *Blyttia* **44(4)**: 165-175.
HOLDEN, J.H.W. 1966: Species relationships in the Avenae. - *Chromosoma* **20(1)**: 75-124.
HOLLINGS, E. & STACE, C.A. 1974: Karyotype variation and evolution in the *Vicia sativa* aggregate. - *New Phytol.* **73(1)**: 195-208.

HOLLINGS, E. & STACE, C.A. 1978: Morphological variation in the *Vicia sativa* L. aggregate. - *Watsonia* **12(1)**: 1-14.

HÖLLWORTH, M. 1979: Botanische Wanderungen auf der Insel Kreta. - *Ber. Naturwiss. Vereins Darmstadt* **2**: 48-56.

HOLMBOE, J. 1914: Studies on the vegetation of Cyprus based on researches during the spring and summer 1905. - *Bergens Mus. Skr., n.s.* **1(2)**: vi + 344 pp. + 6 plates.

HOLMES, M. 2002: I chimoniatiki Iris [In Greek]. - *Fisis [Journal of the Hellenic Society for the Protection of Nature]* **98**: 31-32.

HOLTTUM, R.E. 1968: Typification of the generic names *Gymnocarpium* and *Phegopteris*. - *Taxon* **17(5)**: 529-530.

HOLTTUM, R.E. 1983: The family Thelypteridaceae in Europe. - *Acta Bot. Malac.* **8**: 47-58.

HOLUB, J. 1958: Bemerkungen zur Taxonomie der Gattung *Helictotrichon* Bess. In Klástersky, I. & al. (eds): Philipp Maximilian Opiz und seine Bedeutung für die Pflanzentaxonomie, pp. 101-133. - Praha: Nakladatelství Československé akademie věd.

HOLUB, J. 1959a: Die wichtigsten Ergebnisse des taxonomisch-chorologischen Studiums der tschechoslowakischen *Helictotrichon*-Arten [In Czech with German and Russian summaries]. - *Preslia* **31**: 1-7.

HOLUB, J. 1959b: Taxa et combinationes novae generis *Helictotrichon* Bess. apud Schultes. - *Preslia* **31**: 50-51.

HOLUB, J. 1962: Ein Beitrag zur Abgrenzung der Gattungen in der Tribus Aveneae: die Gattung *Avenochloa* Holub. - *Acta Horti Bot. Prag.* **1962**: 75-86.

HOLUB, J. 1966a: Ergänzungen und Berichtigungen zum ersten Band der "Flora Europaea". – *Preslia* **38**: 78-82.

HOLUB, J. 1966b: Bemerkungen zur Nomenklatur der in der Tschechoslowakei vorkommenden Vertreter des Komplexes *Myosotis palustris*. - *Preslia* **38**: 130-136.

HOLUB, J. 1967: Neue Namen innerhalb der Gattungen *Gentianella* Moench, *Gentianopsis* Ma und *Comastoma* (Wettst.) Tokoyuni. - *Folia Geobot. Phytotax.* **2(1)**: 115-120.

HOLUB, J. 1970a: Brief comments on the second volume of "Flora Europaea". - *Preslia* **42**: 90-95.

HOLUB, J. 1970b: *Lamiastrum* versus *Galeobdolon* and comments on problems of unitary designations in Fabricius's work "Enumeratio methodica plantarum horti medici helmstadiensis". - *Folia Geobot. Phytotax.* **5(1)**: 61-88.

HOLUB, J. 1971: *Fallopia* Adans. 1763 instead of *Bilderdykia* Dum. 1827. - *Folia Geobot. Phytotax.* **6 (2)**: 171-177.

HOLUB, J. 1972a: Taxonomic and nomenclatural remarks on *Chamaenerion* auct. - *Folia Geobot. Phytotax.* **7(1)**: 81-90.

HOLUB, J. 1972b: On the taxonomy and nomenclature of the European marsh fern. – *Taxon* **21(2-3)**: 331-332.

HOLUB, J. 1972c: Bemerkungen zu den tschechoslovakischen Taxa der Familie Equisetaceae [In Czech with German summary]. - *Preslia* **44(2)**: 112-130.

HOLUB, J. 1972d: New nomenclatural combinations in Centaureinae (Asteraceae). - *Folia Geobot. Phytotax.* **7(3)**: 313-316.

HOLUB, J. 1973a: A note on the classification of *Botrychium* Sw. s.l. - *Preslia* **45(3)**: 276-277.

HOLUB, J. 1973b: Brief notes on the third volume of "Flora Europaea". - *Preslia* **45(4)**: 355-360.

HOLUB, J. 1973c: New names in phanerogamae. 2. - *Folia Geobot. Phytotax.* **8(2)**: 155-179.

HOLUB, J. 1975: Taxonomic and floristic progress on the Czechoslovak flora and the contribution of Czechoslovak authors to knowledge of the European flora (1961-1972). - *Mem. Soc. Brot.* **24(1)**: 173-352.

HOLUB, J. 1976a: New names in Phanerogamae. 4. - *Folia Geobot. Phytotax.* **11(1)**: 75-85.

HOLUB, J. 1976b: [New combinations in *Filago* s.l.]. In Heywood, V.H. (ed.): Notulae systematicae ad Floram Europaeam spectantes, no. 19. - *Bot. J. Linn. Soc.* **71(4)**: 271.

HOLUB, J. 1976c: *Juncus minutulus* - a neglected species of the Czechoslovak flora [In Czech with English summary]. - *Preslia* **48(4)**: 329-339.

HOLUB, J. 1976d: A newly adopted restriction of illegitimity in generic names and its consequence for *Avenochloa* Holub 1962. - *Folia Geobot. Phytotax.* **11(3)**: 281-300.

HOLUB, J. 1977a: Notes on some species of *Avenula* and *Helictotrichon*. - *Preslia* **49(3)**: 203-221.

HOLUB, J. 1977b: New names in Phanerogamae. 5. - *Folia Geobot. Phytotax.* **12(3)**: 293-311.

HOLUB, J. 1977c: Brief comments on the fourth volume of "Flora Europaea". - *Preslia* **49(4)**: 311-327.

HOLUB, J. 1977d: New names in Phanerogamae. 6. - Folia Geobot. Phytotax. **12(4)**: 417-432.

HOLUB, J. 1978: Komentarě k vydraným taxonům: *Sanguisorba muricata* Gremli [In Czech]. - *Severočeskou Přír.* **8-9**: 137-139, 141.

HOLUB, J. 1983a: Reclassifications and new names for some European phanerogams. - *Folia Geobot. Phytotax.* **18(2)**: 203-206.

HOLUB, J. 1983b: Some neglected subspecies names from Schübler and Martens, Flora von Würtemberg (1834). - *Folia Geobot. Phytotax.* **18(3)**: 325-328.

HOLUB, J. 1983c: A brief note on Slovak taxa of *Gentianella*. - *Preslia* **55(4)**: 371-373.

HOLUB, J. 1984a: New genera in Phanerogamae (1.-3.). - *Folia Geobot. Phytotax.* **19(1)**: 95-99.

HOLUB, J. 1984b: Some new nomenclatural combinations. I. - *Folia Geobot. Phytotax.* **19(2)**: 213-215.

HOLUB, J. 1984c: Subspecies names in Jávorka, Magyar Flóra (1924-1925). - *Preslia* **56(4)**: 303-318.

HOLUB, J. 1986a: Comments on the "Med-Checklist 1". - *Preslia* **58(4)**: 289-306.

HOLUB, J. 1986b: Notes on "Hieracium hoppeanum" in Czechoslovakia [In Czech with English summary]. - *Zprávy Českoslov. Bot. Společn. Českoslov. Akad. Věd* **21(1)**: 21-28.

HOLUB, J. 1988: Selected subspecies names from Richter, Plantae Europaeae, Vol. 1, 1890. - *Folia Geobot. Phytotax.* **23(4)**: 383-415.

HOLUB, J. 1990: Some taxonomic and nomenclatural changes within *Orobanche* s.l. (Orobanchaceae). - *Preslia* **62(3)**: 193-198.

HOLUB, J. 1998: Reclassifications and new names in vascular plants, 1. - *Preslia* **70(2)**: 97-122.

HOLUB, J. & CHRTEK, J. 1962: Zur Nomenklatur des Gattungsnamens *Filago* L. 1753. - *Taxon* **11(6)**: 195-201.

HOLUB, J., MĚSÍČEK, J. & JAVŮRKOVÁ, V. 1970: Annotated chromosome counts of Czechoslovak plants (1-15). (Materials for "Flóra ČSSR"-1). - *Folia Geobot. Phytotax.* **5(3)**: 339-368.

HOLUB, J., MĚSÍČEK, J. & JAVŮRKOVÁ, V. 1971: Annotated chromosome counts of Czechoslovak plants (16-30). (Materials for "Flóra ČSSR"-2). - *Folia Geobot. Phytotax.* **6(2)**: 179-214 + 2 plates.

HOLUB, J., MĚSÍČEK, J. & JAVŮRKOVÁ, V. 1972: Annotated chromosome counts of Czechoslovak plants (31-60). (Materials for "Flóra ČSSR"-3). - *Folia Geobot. Phytotax.* **7(2)**: 167-202.

HOLUBOVÁ-KLÁSKOVÁ, A. 1964: Bemerkungen zur Gliederung der Gattung *Cytisus* L. s.l. - *Acta Univ. Carol., Biol.* **1964(suppl. 2)**: 1-24.

HOLZAPFEL, S. 1994: A revision of the genus *Picris* (Asteraceae, Lactuceae) s.l. in Australia. - *Willdenowia* **24(1-2)**: 97-218.

HÖLZINGER, C. & HÖLZINGER, J. 1986: Beiträge zur Orchideenflora von Kreta. - *Mitt. Arbeitskreis Heimische Orchid. Baden-Württemberg* **18(1)**: 137-150.

HÖLZINGER, J., KÜNKELE, A. & KÜNKELE, S. 1985: Die Verbreitung der Gattung *Ophrys* L. auf dem griechischen Festland. - *Mitt. Arbeitskreis Heimische Orchid. Baden-Württemberg* **17(1)**: 1-101.

HÖLZINGER, J. & KÜNKELE, S. 1983: Beiträge zur Verbreitung der *Dactylorhiza*-Arten in Griechenland. - *Mitt. Arbeitskreis Heimische Orchid. Baden-Württemberg* **15(4)**: 503-540.

HÖLZINGER, J. & KÜNKELE, S. 1988: *Dactylorhiza macedonica* sp. nova. - *Mitt. Arbeitskreis Heimische Orchid. Baden-Württemberg* **20(1)**: 185-195.

HONDELMANN, W. 2002: Die Kulturpflanzen der griechisch-römischen Welt. Pflanzliche Ressourcen der Antike. - Berlin/Stuttgart: Gebrüder Borntraeger, 134 pp.

HÖNER, D. 1986: *Hippocrepis cyclocarpa* (Leguminosae) new to Europe. - *Willdenowia* **16(1)**: 57-60.

HÖNER, D. 1990: Mehrjährige Beobachtungen kleiner Vegetationsflächen im Raume von Karpathos (Nomos Dhodhekanisou, Griechenland). Ein Beitrag zur Klärung des "Kleininselphänomens". - Berlin: Freie Univ., 185 pp. + Anhang N1-N12 [also as *Diss. Bot.* 173, 1991].

HÖNER, D. & GREUTER, W. 1988: Plant population dynamics and species turnover on small islands near Karpathos (South Aegean, Greece). - *Vegetatio* **77(1-3)**: 129-137.

HONG, DE-YUAN 1982a: A new karyotype for *Narcissus tazetta* L. - *Hereditas (Lund)* **97(1)**: 29-31.

HONG, DE-YUAN 1982b: Cytotype variation and polyploidy in *Scilla autumnalis* L. (Liliaceae). - *Hereditas (Lund)* **97(2)**: 227-235.

HONG, DE-YUAN 2000: A subspecies of *Paeonia mascula* (Paeoniaceae) from W. Asia and SE. Europe. - *Acta Phytotax. Sin.* **38(4)**: 381-385.

HONG, DE-YUAN 2004: *Paeonia* in the Mediterranean-W Asian region: Taxonomy and biogeography. In Vasić, O. (ed.): Abstracts, XI OPTIMA Meeting, Beograd 5-11.IX.2004, p. 56.

HONG, DE-YUAN 2010: Peonies of the world. Taxonomy and phytogeography. - Kew Publishing: x + 302 pp.

HONG, DE-YUAN 2011: Peonies of the world. Polymorphism and diversity. - Kew Publishing: xv + 94 pp.

HONG, DE-YUAN & AL. 2007: *Paeonia daurica* Andrews or *P. mascula* ssp. *triternata* (Pall. ex DC.) Stearn & P.H. Davis (Paeoniaceae)? - *Bot. J. Linn. Soc.* **154(1)**: 1-11.

HONG, DE-YUAN & AL. 2008: Relationships and taxonomy of *Paeonia arietina* G. Anderson complex (Paeoniaceae) and its allies. - *Taxon* **57(3)**: 922-932.

HONG, DE-YUAN, WANG, XIAO-QUAN & ZHANG, DA-MING 2004: *Paeonia saueri* (Paeoniaceae), a new species from the Balkans. - *Taxon* **53(1)**: 83-90.

HONSELL, E. 1959: Ricerche citotassonomiche su *Pinguicula hirtiflora* Ten. - *Ann. Bot. (Rome)* **26(2)**: 177-188.

HOOG, M.H. 1973: On the origin of *Tulipa*. - *Lilies* **1973**: 47-64.

HOOKER, J.D. 1869a: *Iris stylosa*. - *Bot. Mag.* **95**: tab. 5773 + 2 pp. text.

HOOKER, J.D. 1869b: *Crocus Orphanidis*. - *Bot. Mag.* **95**: tab. 5776 + 2 pp. text.

HOOKER, J.D. 1870a: *Ophrys speculum*. - *Bot. Mag.* **96**: tab. 5844 + 2 pp. text.

HOOKER, J.D. 1870b: *Malope malacoides*. - *Bot. Mag.* **96**: tab. 5852 + 2 pp. text.

HOOKER, J.D. 1870c: *Serapias cordigera* & *Serapias lingua*. - *Bot. Mag.* **96**: tab. 5868 A & B + 2 pp. text.

HOOKER, J.D. 1871a: *Prunus cerasifera*. - *Bot. Mag.* **97**: tab. 5934 + 2 pp. text.

HOOKER, J.D. 1871b: *Ophrys lutea*. - *Bot. Mag.* **97**: tab. 5941 + 2 pp. text.

HOOKER, J.D. 1872: *Digitalis laevigata*. - *Bot. Mag.* **98**: tab. 5999 + 2 pp. text.

HOOKER, J.D. 1873a: *Arisarum vulgare*. - *Bot. Mag.* **99**: tab. 6023 + 2 pp. text.

HOOKER, J.D. 1873b: *Sedum dasyphyllum* var. *glanduliferum* - *Bot. Mag.* **99**: tab. 6027 + 2 pp. text.

HOOKER, J.D. 1873c: *Crocus Olivieri*. - *Bot. Mag.* **99**: tab. 6031 + 2 pp. text.

HOOKER, J.D. 1873d: *Crocus Sieberi*. - *Bot. Mag.* **99**: tab. 6036 + 2 pp. text.

HOOKER, J.D. 1873e: *Omphalodes Luciliae*. - *Bot. Mag.* **99**: tab. 6047 + 2 pp. text.

HOOKER, J.D. 1874a: *Xiphon sisyrinchium*. - *Bot. Mag.* **100**: tab. 6096 + 2 pp. text.

HOOKER, J.D. 1874b: *Senecio Doronicum* var. *hosmariensis*. - *Bot. Mag.* **100**: tab. 6101 + 2 pp. text.

HOOKER, J.D. 1874c: *Achillea ageratifolia*. - *Bot. Mag.* **100**: tab. 6117 + 2 pp. text.

HOOKER, J.D. 1874d: *Cerinthe gymnandra*. - *Bot. Mag.* **100**: tab. 6130 + 2 pp. text.

HOOKER, J.D. 1875a: *Crocus chrysanthus*. - *Bot. Mag.* **101**: tab. 6162 + 2 pp. text.

HOOKER, J.D. 1875b: *Galanthus Elwesii*. - *Bot. Mag.* **100**: tab. 6166 + 2 pp. text.

HOOKER, J.D. 1875c: *Crocus Crewii*. - *Bot. Mag.* **101**: tab. 6168 + 2 pp. text.

HOOKER, J.D. 1876: *Bongardia Rauwolfii*. - *Bot. Mag.* **102**: tab. 6244 + 2 pp. text.

HOOKER, J.D. 1877: *Thapsia garganica*. - *Bot. Mag.* **103**: tab. 6293 + 2 pp. text.

HOOKER, J.D. 1879: *Linaria dalmatica*. - *Bot. Mag.* **105**: tab. 6424 + 2 pp. text.

HOOKER, J.D. 1880a: *Hypericum aegyptiacum*. - *Bot. Mag.* **106**: tab. 6481 + 2 pp. text.

HOOKER, J.D. 1880b: *Prunus divaricata*. - *Bot. Mag.* **106**: tab. 6519 + 2 pp. text.

HOOKER, J.D. 1880c: *Scabiosa pterocephala*. - *Bot. Mag.* **106**: tab. 6526 + 2 pp. text.

HOOKER, J.D. 1881: *Statice tatarica*. - *Bot. Mag.* **107**: tab. 6537 + 2 pp. text.

HOOKER, J.D. 1882: *Haberlea rhodopensis*. - *Bot. Mag.* **108**: tab. 6651 + 2 pp. text.

HOOKER, J.D. 1883: *Rosa alpina*. - *Bot. Mag.* **109**: tab. 6724 + 2 pp. text.

HOOKER, J.D. 1884a: *Hypericum empetrifolium*. - *Bot. Mag.* **110**: tab. 6764 + 2 pp. text.

HOOKER, J.D. 1884b: *Pinguicula hirtiflora*. - *Bot. Mag.* **110**: tab. 6785 + 2 pp. text.

HOOKER, J.D. 1885: *Cytisus hirsutus*. - *Bot. Mag.* **111**: tab. 6819 + 2 pp. text.

HOOKER, J.D. 1886: *Cerinthe minor*. - *Bot. Mag.* **112**: tab. 6890 + 2 pp. text.

HOOKER, J.D. 1892a: *Vicia narbonensis*. - *Bot. Mag.* **118**: tab. 7220 + 2 pp. text.

HOOKER, J.D. 1892b: *Ferula tingitana*. - *Bot. Mag.* **118**: tab. 7267 + 3 pp. text.

HOOKER, J.D. 1898a: *Daphne Blagayana*. - *Bot. Mag.* **124**: tab. 7579 + 2 pp. text.

HOOKER, J.D. 1898b: *Symphyandra Wanneri*. - *Bot. Mag.* **124**: tab. 7594 + 2 pp. text.

HOOKER, J.D. 1898c: *Astragalus ponticus*. - *Bot. Mag.* **124**: tab. 7622 + 2 pp. text.

HOOKER, J.D. 1900: *Verbascum longifolium*. - *Bot. Mag.* **126**: tab. 7707 + 2 pp. text.

HOOKER, J.D. (ED.) 1904: *Lonicera etrusca* var. *superba*. - *Bot. Mag.* **130**: tab. 7977 + 2 pp. text.

HOOKER, W.J. 1833: *Vallisneria spiralis*. In Botanical miscellany 3: 87-88 + suppl. tab. 23-24. - London: John Murray.

HOOKER, W.J. (ED.) 1827: *Trifolium alpestre*. - *Bot. Mag.* **54**: tab. 2779 + 2 pp. text.

HOOKER, W.J. (ED.) 1828a: *Orobus sessilifolius*. - *Bot. Mag.* **55**: tab. 2796 + 2 pp. text.

HOOKER, W.J. (ED.) 1828b: *Nicotiana glauca*. - *Bot. Mag.* **55**: tab. 2837 + 2 pp. text.

HOOKER, W.J. (ED.) 1828c: *Saponaria glutinosa*. - *Bot. Mag.* **55**: tab. 2855 + 2 pp. text.

HOOKER, W.J. (ED.) 1828d: *Cynara cardunculus*. - *Bot. Mag.* **55**: tab. 2862 + 2 pp. text.

HOOKER, W.J. (ED.) 1830: *Ranunculus millefoliatus*. - *Bot. Mag.* **57**: tab. 3009 + 2 pp. text.

HOOKER, W.J. (ED.) 1831: *Ornithogalum fimbriatum*. - *Bot. Mag.* **58**: tab. 3077 + 2 pp. text.

HOOKER, W.J. (ED.) 1834a: *Arabis verna*. - *Bot. Mag.* **61**: tab. 3331 + 2 pp. text.

HOOKER, W.J. (ED.) 1834b: *Schinus molle*. - *Bot. Mag.* **61**: tab. 3339 + 2 pp. text.

HOOKER, W.J. (ED.) 1835a: *Ophrys tephrosanthos* var. *densifolius*. - *Bot. Mag.* **62**: tab. 3426 + 2 pp. text.

HOOKER, W.J. (ED.) 1835b: *Paeonia russii*. - *Bot. Mag.* **62**: tab. 3431 + 1 p. text.

HOOKER, W.J. (ED.) 1836: *Telekia speciosa*. - *Bot. Mag.* **63**: tab. 3466 + 2 pp. text.

HOOKER, W.J. (ED.) 1838a: *Centaurea depressa*. - *Bot. Mag.* **65**: tab. 3662 + 2 pp. text.

HOOKER, W.J. (ED.) 1838b: *Merendera caucasica*. - *Bot. Mag.* **65**: tab. 3690 + 2 pp. text.

HOOKER, W.J. (ED.) 1841a: *Crocus speciosus*. - *Bot. Mag.* **67**: tab. 3861 + 3 pp. text.

HOOKER, W.J. (ED.) 1841b: *Crocus annulatus*. - *Bot. Mag.* **67**: tab. 3868 + 1 p. text.

HOOKER, W.J. (ED.) 1841c: *Crocus lagenaeflorus* var. *lacteus*. - *Bot. Mag.* **67**: tab. 3869 + 5 pp. text.

HOOKER, W.J. (ED.) 1858: *Fritillaria graeca*. - *Bot. Mag.* **84**: tab. 5052 + 2 pp. text.

HOOKER, W.J. (ED.) 1861: *Cerinthe retorta*. - *Bot. Mag.* **87**: tab. 5264 + 2 pp. text.

HOOPER, S.S. 1977: Nomenclature of a Mediterranean sedge, *Cyperus capitatus* Vand. (Cyperaceae). - *Israel J. Bot.* **26(2)**: 98-99.

HÖPFLINGER, F. 1964: Beiträge zur Flora von Skutari (Nordalbanien). - *Mitt. Naturwiss. Vereines Steiermark* **94**: 92-107 + 1 map.

HÖPKE, J., MUCINA, L. & ALBACH, D.C. 2019: Phylogenetic and morphometric analysis of *Plantago* section *Coronopus* (Plantaginaceae). - *Taxon* **68(2)**: 315-339.

HOPP, I. & PETRIDIS, J. 1987: Prespa. A continuity in the variety of its flora [In Greek with English summary]. - *Fisis [Journal of the Hellenic Society for the Protection of Nature]* **39**: 51-55, 75-77.

HÖRANDL, E. 1989: *Cystopteris dickieana* R. Sim (Pteridophyta) - neu för Österreich und andere Gebiete. - *Verh. Zool.-Bot. Ges. Österreich* **126**: 179-183.

HÖRANDL, E. 1992: Die Gattung *Salix* in Österreich, mit Berücksichtigung angrenzender Gebiete. - *Abh. Zool.-Bot. Ges. Österreich* **27**: 1-170.

HÖRANDL, E. 1994: Systematik und Verbreitung von *Papaver dubium* L. s.l. in Österreich. – *Linzer Biol. Beitr.* **26(1)**: 407-435.

HORJALES LUACES, M. 1972: Estudio cariológico del género *Adenocarpus* DC. - *Trab. Dept. Bot. Fisiol. Veg. Madrid* **5**: 3-41.

HORN AF RANTZIEN, H. 1946: Taxonomical and phytogeographical studies in *Phleum arenarium* L. - *Bot. Not.* **1946(3)**: 364-386.

HORNÝ, R., WEBR, M. & BYAM-GROUNDS, J.S. 1986: Porophyllum Saxifrages. - Stamford: Byam- Grounds Publications, 372 pp. + many plates.

HOROVITZ, A. & GALIL, J. 1972: Gynodioecism in east Mediterranean *Hirschfeldia incana*. Cruciferae. - *Bot. Gaz.* **133(2)**: 127-131.

HOROVITZ, A., GALIL, J. & ZOHARY, D. 1975: Biological flora of Israel. 6. *Anemone coronaria* L. - *Israel J. Bot.* **24(1)**: 26-41.

HOROVITZ, A. & ZOHARY, D. 1966: Spontaneous variegation for perianth colour in wild *Anemone coronaria*. - *Heredity* **21(3)**: 513-515.

HORRILL, A.D. 1972: Biological Flora of the British Isles. *Melampyrum cristatum* L. - *J. Ecol.* **60(1)**: 235-244.

HORSWOOD, R. 2003a: A scree in a hurry. - *Alpine Gardener* **71(2)**: 147-150.

HORSWOOD, R. 2003b: Impressions of Crete. - *Alpine Gardener* **71(4)**: 322-326.

HORVAT, I. 1947: Two new species of irises from Macedonia [In Serbo-Croatian with English summary]. - *Glasn. Biol. Sekc.* **1**: 17-28 + 2 plates.

HORVAT, I. 1953: Ein Beitrag zur Kenntnis der Verbreitung einiger Hochgebirgspflanzen in Südosteuropa [In Serbo-Croatian with German summary]. - *God. Biol. Inst. Sarajevu* **5(1-2)**: 199-218.

HORVAT, I. 1954: Pflanzengeographische Gliederung Südosteuropas. - *Vegetatio* **5-6**: 434-447.

HORVAT, I. 1960: La végétation alpine de la Macédoine dans l'aspect des recherches contemporaines [In Macedonian with French summary]. - *Acta Mus. Maced. Sci. Nat.* **6(8)**: 163-203.

HORVAT, I. 1962: Die Grenze der mediterranen und mitteleuropäischen Vegetation in Südosteuropa im Lichte neuer pflanzensoziologischer Forschungen. - *Ber. Deutsch. Bot. Ges.* **75(3)**: 91-104.

HORVAT, I., GLAVAČ, V. & ELLENBERG, H. 1974: Vegetation Südosteuropas (Geobotanica Selecta, Band IV). - Stuttgart: Gustav Fischer Verlag, xxxii + 768 pp. + 2 maps.

HORVÁTH, O. & AL. 2020: The phylogenetic position of *Vincetoxicum pannonicum* (Borhidi) Holub supports the species' allopolyploid hybrid origin. - *Acta Soc. Bot. Poloniae* **89(3)**: 14 pp.

HORVATIĆ, S. 1928: Die Formen der Sectio *Leucanthemum* aus der Gattung *Chrysanthemum* in der Flora Jugoslawiens [In Serbo-Croatian with German summary]. - *Acta Bot. Inst. Bot. Univ. Zagreb* **3**: 61-140.

HORVATIĆ, S. 1935: Neuer Beitrag zur Kenntnis der *Leucanthemum*-Formen in der Flora Jugoslaviens. - *Acta Bot. Inst. Bot. Univ. Zagreb* **10**: 61-100.

HORVATIĆ, S. 1963: Genus *Leucanthemum* in Flora Jugoslaviae. - *Acta Bot. Croat.* **22**: 203-218 + 3 plates.

HORVATIĆ, S. 1970: *Scaligeria cretica* (Miller) Boiss. in the plant cover of the Croatian littoral [In Serbo-Croatian with English summary]. - *Acta Bot. Croat.* **29**: 175-182.

HORVATIĆ, S. (ED.) 1967: Flora Analytica Iugoslaviae, I/1 [In Serbo-Croatian]. - Zagreb: Tisak Grafičkog Zavoda, pp. 1-216.

HORVATIĆ, S. (ED.) 1973: Flora Analytica Iugoslaviae, I/2 [In Serbo-Croatian]. - Zagreb: Tisak Grafičkog Zavoda, pp. 217-420.

HOSSAIN, M. 1961: A revision of *Trifolium* in the Nearer East. - *Notes Roy. Bot. Gard. Edinburgh* **23(3)**: 387-481.

HOUTZAGERS, G. 1937: The genus *Populus* and its significance in silviculture [In Dutch with English summary]. - Wageningen: H. Veenman & zonen, ix + 266 pp.

HOWARD, H.W. & LYON, A.G. 1952: Biological Flora of the British Isles. *Nasturtium* R. Br. - *J. Ecol.* **40(1)**: 228-245.

HOWARTH, S.E. & WILLIAMS, J.T. 1972: Biological Flora of the British Isles. *Chrysanthemum segetum* L. - *J. Ecol.* **60(2)**: 573-584.

HRABĚTOVÁ-UHROVÁ, A. 1949: Generis *Hippocrepis* L. revisio. La revision du genre *Hippocrepis*. II [In Latin and French]. - *Práce Morav.-Slez. Akad. Věd Přír.* **21(4)**: 1-54.

HRABĚTOVÁ-UHROVÁ, A. 1950a: Generis *Hippocrepis* L. revisio. La revision du genre *Hippocrepis*. IV [In Latin and French]. - *Práce Morav.-Slez. Akad. Věd Přír.* **22(4)**: 99-158 + 2 plates.

HRABĚTOVÁ-UHROVÁ, A. 1950b: Generis *Hippocrepis* L. revisio. La revision du genre *Hippocrepis*. I [In Latin and French]. - *Práce Morav.-Slez. Akad. Věd Přír.* **22(8)**: 219-250.

HRABĚTOVÁ-UHROVÁ, A. 1950c: Generis *Hippocrepis* L. revisio. (La revision du genre *Hippocrepis*). III [In Latin and French]. - *Práce Morav.-Slez. Akad. Věd Přír.* **22(12)**: 331-356 + 4 plates.

HRABĚTOVÁ, A., DVOŘÁK, F. & DADÁKOVÁ, B. 1983: Chromosome morphology of three species of the genus *Cyclamen* L. - *Biológia, A (Bratislava)* **38(5)**: 401-409.

HRIVNÁK, R. & AL. 2016: Alien wetland annual *Lindernia dubia* (Scrophulariaceae): the first recently mentioned localities in Slovakia and their central European context. -*Biológia (Bratislava)* **71(3)**: 281-286.

HROUDA, L. 1972: *Inula germanica* L. in der Tschechoslowakei. - *Preslia* **44**: 227-243.

HROUDA, L. 1973: Die Arten der Gattung *Inula* L. in der Tschechoslowakei [In Czech with German summary]. - *Zprávy Českoslov. Bot. Společn. Českoslov. Akad. Věd* **8(2)**: 85-95.

HROUDA, L. 1974a: Taxonomie und Verbreitung von *Inula salicina* L. s.l. in der Tschechoslowakei. - *Preslia* **46(1)**: 32-56.

HROUDA, L. 1974b: *Inula oculus-christi* L. in der Tschechoslowakei. - *Preslia* **46(2)**: 172-185.

HROUDA, L. & HROUDOVÁ, Z. 1989: Comments to Czechoslovak water-plantains. 1. Key to determination of Czechoslovak species of the genus *Alisma* [In Czech with English summary]. - *Zprávy Českoslov. Bot. Společn. Českoslov. Akad. Věd* **24(1)**: 1-14.

HROUDOVÁ, Z. & AL. 2007: Taxonomy, distribution and ecology of *Bolboschoenus* in Europe. - *Ann. Bot. Fenn.* **44(2)**: 81-102.

HROUDOVÁ, Z., FRANTIK, T. & ZÁKRAVSKÝ, P. 1998: The differentiation of subspecies in *Bolboschoenus maritimus* based on the inflorescence structure. - *Preslia* **70(2)**: 135-154.

HROUDOVÁ, Z., MORAVCOVÁ, L. & ZÁKRAVSKÝ, P. 1998: Differentiation of the Central European *Bolboschoenus* taxa based on fruit shape and anatomy. - *Thaiszia* **8(2)**: 91-109.

HRUBY, J. 1912: Le genre *Arum*. Aperçu systématique avec considérations spéciales sur les relations phylogénétiques des formes. - *Bull. Soc. Bot. Genève, Sér. 2* **4**: 113-160, 330-371.

HRUBY, J. 1930a: Rubi peninsulae balcanae. - *Repert. Spec. Nov. Regni Veg.* **28**: 140-202.

HRUBY, J. 1930b: *Campanula*-Studien innerhalb der *Vulgares* und ihrer Verwandten. - *Magyar Bot. Lapok* **29(1-12)**: 152-276 + 4 plates.

HRUŠEVAR, D. & AL. 2015: *Echinochloa colona* (L.) Link (Poaceae), a new species in the flora of Croatia. - *Acta Bot. Croat.* **74(1)**: 159-164.

HRUŠKA, K. 1971: *Thymus praecox* Op. in der Flora Kroatiens [In Serbo-Croatian with German summary]. - *Acta Bot. Croat.* **30**: 123-125.

HRUŠKA, K. 1982: Considerazioni ecologiche, fitosociologiche e morfologiche sul genere *Anthriscus* Pers. - *Giorn. Bot. Ital., n.s.* **116(3-4)**: 175-187.

HRUŠKA, K. 1986: Considerazioni sull'evoluzione del genere *Chaerophyllum* L. Umbelliferae. - *Giorn. Bot. Ital., n.s.* **119(1-2)**: 37-47.

HRUŠKA, K. & BELLOMARIA, B. 1984: Numeri cromosomici per la Flora Italiana: 877-881. - *Inform. Bot. Ital.* **14(2-3)**: 238-242.

HUBAC, J.M. 1972: Relations entre les dimensions du pollen et le niveau de polyploïdie dans le complexe du *Campanula rotundifolia* L. s.l. - *Bull. Soc. Bot. France* **119(5-6)**: 279-285.

HUBBARD, F.T. 1915: A taxonomic study of *Setaria italica* and its immediate allies. - *Amer. J. Bot.* **2(4)**: 169-198.

HUBER-MORATH, A. 1955: Verbreitung der Gattungen *Verbascum*, *Celsia* und *Staurophragma* im Orient. - *Bauhinia* **1(1)**: 1-84 + 34 plates.

HUBER-MORATH, A. 1958: Die anatolischen Arten der Gattung *Phlomis* L. - *Bauhinia* **1(2)**: 97-123.

HUBER-MORATH, A. 1963: Beitrag zur Kenntnis der Verbreitung von *Gypsophila* und *Bolanthus* in Anatolien. - *Bauhinia* **2(2)**: 177-191.

HUBER-MORATH, A. 1971a: Die türkischen Verbasceen (*Celsia*, *Staurophragma*, *Verbascum*). - *Denkschr. Schweiz. Naturf. Ges.* **87**: 166 pp. + 18 plates.

HUBER-MORATH, A. 1971b: Novitiae florae anatolicae IX. - *Bauhinia* **4(2)**: 197-222.

HUBER-MORATH, A. 1975a: Bemerkungen zu Jiři Ponerts "Combinationes novae, stati novi et taxa nova non tantum specierum turcicarum". - *Bauhinia* **5(3)**: 153-159.

HUBER-MORATH, A. 1975b: Die türkischen Arten der Gattung *Achillea* L. - *Ber. Schweiz. Bot. Ges.* **84(2)**: 123-172.

HUBER-MORATH, A. 1980: Ergänzungen zu P.H. Davis' "Flora of Turkey and the East Aegean Islands" 1-6 (1965-1978). I. - *Candollea* **35(2)**: 569-608.

HUBER-MORATH, A. 1984: Ergänzungen zu P.H. Davis' "Flora of Turkey and the East Aegean Islands" 7 (1982). - *Candollea* **39(1)**: 323-344.

HUBER-MORATH, A. 1987: Letzte Ergänzungen zu P.H. Davis' "Flora of Turkey and the East Aegean Islands" 1-9 (1965-1985). I. - *Candollea* **42(2)**: 717-769.

HUBER-MORATH, A., COODE, M.J.E. & CULLEN, J. 1967: Materials for a Flora of Turkey: XIII. *Silene* L. - *Notes Roy. Bot. Gard. Edinburgh* **28(1)**: 1-8.

HUBER-MORATH, A., MCNEILL, J. & REEVE, H. 1967: Materials for a Flora of Turkey, XIV: Caryophyllaceae. - *Notes Roy. Bot. Gard. Edinburgh* **28(1)**: 17-23.

HUBER-MORATH, A. & RECHINGER, K.H. 1960: Zur Kenntnis der Gattungen *Verbascum* und *Celsia* in Griechenland. - *Mitt. Thüring. Bot. Ges.* **2(1)**: 42-55.

HUBER-MORATH, A. & ZOHARY, M. 1969: Materials for a Flora of Turkey: XX. Leguminosae: Trifolieae. - *Notes Roy. Bot. Gard. Edinburgh* **29(3)**: 321-323.

HUBER, J.A. 1929: Zur Systematik der Gattung *Sedum* L. - *Ber. Naturwiss. Vereins Landshut* **20**: 9-150 + 1 map.

HUBER, W. 1988: Natürliche Bastardierungen zwischen weissblühenden *Ranunculus*-Arten in den Alpen. - *Veröff. Geobot. Inst. E.T.H. Stiftung Rübel Zürich* **100**: 1-160.

HUBER, W. 1993: Biosystematisch-ökologische Untersuchungen an den *Erigeron*-Arten (Asteraceae) der Alpen. - *Veröff. Geobot. Inst. E.T.H. Stiftung Rübel Zürich* **114**: 143 pp.

HUBER, W. 1994: Typification of *Erigeron glabratus* (Asteraceae). - *Taxon* **44(1)**: 87-89.

HUBER, W. & BALTISBERGER, M. 1989: Chromosome numbers. In: Stace, C.A. (ed.): IOPB Chromosome Data. 1. - *Newslett. Int. Organ. Pl. Biosyst.* **13**: 19-20.

HUBER, W. & ZHANG, H. 1991: Morphologische und chemotaxonomische Untersuchungen an den *Erigeron*-Arten der Alpen. - *Ber. Geobot. Inst. E.T.H. Stiftung Rübel* **57**: 116-164.

HÜBL, E., HOLZNER, W. & GLAUNINGER, H. 1996: Beiträge zur *Centaurea cyanus* L. - *Ann. Naturhist. Mus. Wien, B* **98B Suppl.**: 317-327.

HUDZIOK, G. 1988: Verbesserter Schlüssel zur Unterscheidung von *Epilobium roseum, E. obscurum, E. adenocaulon, E. adnatum* und *E. lamyi*. - *Mitt. Florist. Kart.* **14**: 11-12.

HUGHES, T. & HUGHES, D. 1991: Orchid hunting on Rhodes. - *Quart. Bull. Alpine Gard. Soc. Gr. Brit.* **59(2)**: 176-190.

HÜGIN, G. 1987: Einige Bemerkungen zu wenig bekannten *Amaranthus*-Sippen (Amaranthaceae) Mitteleuropas. - *Willdenowia* **16(2)**: 453-478.

HÜGIN, G. 1995: *Malva neglecta, Galium spurium, Anthriscus nitida, Hordeum murinum* subsp. *glaucum, Molineriella minuta* and *Schismus barbatus*. In Greuter, W. & Raus, Th. (eds): Med- Checklist notulae, 16. - *Willdenowia* **25(1)**: 173-176.

HÜGIN, G. 1998: Die Gattung *Chamaesyce* in Europa. Bestimmungsschlüssel mit taxonomisch-nomenklatorischen Anmerkungen. - *Feddes Repert.* **109(3-4)**: 189-223.

HÜGIN, G. 2012: *Stellaria pallida* - noch immer häufig verkannt. - *Kochia* **6**: 91-117.

HÜGIN, G. & LOHMEYER, W. 1993: Bastardbildung und intraspezifische Sippengliederung bei *Echinops sphaerocephalus* (Asteraceae, Cardueae) in Mitteleuropa. - *Willdenowia* **23(1-2)**: 83-93.

HUISKES, A.H.L. 1979: Biological Flora of the British Isles. *Ammophila arenaria* (L.) Link (*Psamma arenaria* (L.) Roem. et Schult.; *Calamagrostis arenaria* (L.) Roth). - *J. Ecol.* **67(1)**: 363-382.

HÜLBUSCH, K.H. 1978: Pflanzen, von denen in der mitteleuropäischen Literatur selten oder gar keine Abbildungen zu finden sind. Folge VII: *Cerastium dichotomum*. - *Göttinger Florist. Rundbr.* **12(2)**: 66-67.

HULL, V.J. & GROVES, R.H. 1973: Variation in *Chondrilla juncea* L. in south-eastern Australia. - *Austral. J. Bot.* **21(1)**: 113-135.

HULTÉN, E. 1960: Studies in the genus *Dryas*. - *Svensk Bot. Tidskr.* **53(4)**: 507-542.

HULTGÅRD, U.-M. 1987: *Parnassia palustris* L. in Scandinavia. - *Symb. Bot. Upsal.* **28(1)**: v + 128 pp.

HUMBERT-DROZ, C. & FELBER, F. 1992: Étude biométrique des stomates et des grains de pollen comme indicateurs du degré de polyploïdie chez *Anthoxanthum alpinum* Löve et Löve. - *Bull. Soc. Neuchâteloise Sci. Nat.* **115**: 31-45.

HUMPHRIES, C.J. 1978a: Notes on the genus *Phleum* L. In Heywood, V.H. (ed.): Notulae systematicae ad Floram Europaeam spectantes, no. 20. - *Bot. J. Linn. Soc.* **76(4)**: 337-340.

HUMPHRIES, C.J. 1978b: Variation in *Taeniatherum caput-medusae* (L.) Nevski. In Heywood, V.H. (ed.): Notulae systematicae ad Floram Europaeam spectantes, no. 20. - *Bot. J. Linn. Soc.* **76(4)**: 340-344.

HUMPHRIES, C.J. 1978c: [Notes on *Dasypyrum*]. In Heywood, V.H. (ed.): Notulae systematicae ad Floram Europaeam spectantes, no. 20. - *Bot. J. Linn. Soc.* **76(4)**: 361-362.

HUMPHRIES, C.J. 1979a: A revision of the genus *Anacyclus* L. (Compositae: Anthemideae). - *Bull. Brit. Mus. (Nat. Hist.), Bot.* **7(3)**: 83-142.

HUMPHRIES, C.J. 1979b: *Aegilops dichasians* (Zhuk.) Humphries, stat. nov. In Heywood, V.H. (ed.): Notulae systematicae ad Floram Europaeam spectantes, no. 20. Corrigenda et addenda. - *Bot. J. Linn. Soc.* **78(3)**: 236.

HUNKELER, U. & FAVARGER, C. 1967: Contribution à la cytotaxinomie du genre *Pimpinella*. - *Bull. Soc. Neuchâteloise Sci. Nat.* **90**: 219-239.

HUNT, D.R. 1969: *Lithospernum zahnii*. - *Bot. Mag.* **177(1)**: tab. 530 + 3 pp. text.

HUNT, F. 1984: Plant portraits: *Anchusa caespitosa*. - *Rock Gard.* **18(4)**: 399-400.

HUNT, F. 2009: *Fritillaria graeca* Boiss. & Sprun. subsp. *graeca* Boiss. & Sprun. ´Bill Ivey´, A.M. - *Alpine Gardener* **77(1)**: 36-38.

HURKA, H. & NEUFFER, B. 1997: Evolutionary processes in the genus *Capsella* (Brassicaceae). - *Pl. Syst. Evol.* **206(1-4)**: 295-316.

HURKA, H. & WÖHRMANN, K. 1977: Analyse der genetischen Variabilität natürlicher Populationen von *Capsella bursa-pastoris* (Brassicaceae). - *Bot. Jahrb. Syst.* **98(1)**: 120-132.

HUSAIN, S.Z. & AL. 1990: A micromorphological study of some representative genera in the tribe Saturejeae (Lamiaceae). - *Bot. J. Linn. Soc.* **103(1)**: 59-80.

HUTCHINGS, M.J. & PRICE, E.A.C. 1999: Biological Flora of the British Isles. *Glechoma hederacea* L. (*Nepeta glechoma* Benth., *N. hederacea* (L.) Trev.). - *J. Ecol.* **87(2)**: 347-364.

HUTCHINSON, C.S. & SEYMOUR, G.B. 1982: Biological Flora of the British Isles. *Poa annua* L. - *J. Ecol.* **70(3)**: 887-901.

HUTCHINSON, J. 1910: *Saxifraga Grisebachii*. - Bot. Mag. **136**: tab. 8308 + 2 pp. text.

HUTCHINSON, J. 1913: *Saxifraga Stribrnyi*. - Bot. Mag. **139**: tab. 8496 + 2 pp. text.

HUTH, E. 1890: Revision der Arten von *Adonis* und *Knowltonia*. - *Samml. Naturwiss. Vorträge* **3(8)**: 61-73 + 1 plate.

HUTH, E. 1891a: Revision der Arten von *Trollius*. - *Helios* **9(1)**: 1-8.

HUTH, E. 1891b: Monographie der Gattung *Caltha*. - *Abh. Vorträge Gesammtgeb. Naturwiss.* **4(1)**: 1-32 + 1 plate. [Reprinted in *Helios* 9: 55-78, 1892 & 9: 99-103, 1893].

HUTH, E. 1892: Monographie der Gattung *Paeonia*. - *Bot. Jahrb. Syst.* **14**: 258-276.

HUTH, E. 1893: Revision der kleinen Ranunculaceen-Gattungen *Myosurus, Trautvetteria, Hamadryas, Glaucidium, Hydrastis, Eranthis, Coptis, Anemonopsis, Actaea, Cimicifuga* und *Xanthorrhiza*. - *Bot. Jahrb. Syst.* **16**: 278-324 + 2 plates.

HUTH, E. 1895: Monographie der Gattung *Delphinium*. - *Bot. Jahrb. Syst.* **20(3-4)**: 322-499 + 3 plates.

HUTTUNEN, A. & AL. 1992: Late-glacial and Holocene history of flora and vegetation in the Western Rhodopes Mountains, Bulgaria. - *Acta Bot. Fenn.* **144**: 63-80.

HUXLEY, A.J. 1961: An ascent of Parnassus. - *Quart. Bull. Alpine Gard. Soc. Gr. Brit.* **29(2)**: 201-208.

HUXLEY, A.J. 1963a: Flowers in Greece. An outline of the flora. Part I. - *J. Roy. Hort. Soc.* **88(9)**: 387-402 + 4 plates.

HUXLEY, A.J. 1963b: Flowers in Greece. An outline of the flora. Part II. - *J. Roy. Hort. Soc.* **88(10)**: 429-442 + 4 plates.

HUXLEY, A.J. 1963c: Flowers in Greece. An outline of the flora. Part III. - *J. Roy. Hort. Soc.* **88(10)**: 472-488 + 2 plates.

HUXLEY, A.J. 1970: Flowers in Greece. An outline of the flora (revised 1969). - London: Royal Horticultural Society, 44 pp.

HUXLEY, A.J. 1974: The madonna lily. - *J. Roy. Hort. Soc.* **99(1)**: 27-28.

HUXLEY, A.J. 1986: Mountain flowers in colour. Second revised edition. - Poole, etc.: Blandford Press, 433 pp.

HUXLEY, A.J. & TAYLOR, W. 1977: Flowers of Greece and the Aegean. With 483 colour illustrations from photographs and 77 line drawings by Victoria Gordon. - London: Chatto & Windus, 185 pp. + 64 pp. of colour plates. [New edition 1989].

HY, F.C. 1901: Sur le *Peucedanum schottii* Besser. - *Bull. Soc. Bot. France* **48**: 105-107.

HYLMÖ, B. & FRYER, J. 1999: Cotoneasters in Europe. - *Acta Bot. Fenn.* **162**: 179-184.

IAMONICO, D. 2013: Lectotypification of the Linnaean name *Hypochaeris maculata* (Asteraceae). - *Nordic J. Bot.* **31(2)**: 222-224.

IAMONICO, D. 2015: Taxonomic revision of the genus *Amaranthus* (Amaranthaceae) in Italy. - *Phytotaxa* **199(1)**: 1-84.

IAMONICO, D. 2016: A new name in *Sagina*, *Sagina alexandrae* (Caryophyllaceae). - *Phytotaxa* **282 (2)**: 164-165.

IAMONICO, D. 2017: *Agrimonia eupatoria* subsp. *major* stat. nov. (Rosaceae) and notes on the *Agrimonia eupatoria* aggregate. - *Phytotaxa* **298(2)**: 165-172.

IAMONICO, D. 2018: A conspectus of *Silene* sect. *Lasiocalycinae* (Caryophyllaceae). -*Phyton (Horn)* **57**: 113-127.

IAMONICO, D. 2020a: Linnaean names in *Arum* (Araceae): Typification of *A. arisarum* (≡ *Arisarum vulgare) and A. tenuifolium (≡ Biarum tenuifolium). - Taxon* **69(3)**: 578-581.

IAMONICO, D. 2020b: A nomenclatural survey of the genus *Amaranthus* (Amaranthaceae). 11. Dieoecious *Amaranthus* species belonging to the sect. *Saueranthus*. - *Darwiniana* **8**: 567-575.

IAMONICO, D. & AL. 2011: Typification of *Trifolium latinum* Sebast. (Fabaceae) and comparison with similar species. - *Ann. Bot. Fenn.* **48(6)**: 459-464.

IAMONICO, D. & AL. 2017: The intricate nomenclatural questions around *Plantago holosteum* (Plantaginaceae). - *Phytotaxa* **306(1)**: 75-84.

IAMONICO, D. & DOMINA, G. 2015: Nomenclatural notes on the *Polycarpon tetraphyllum* aggregate (Caryophyllaceae). - *Pl. Biosyst.* **149(4)**: 720-727.

IAMONICO, D., FORTINI, P. & DI PIETRO, R. 2019: Typification of the Linnaean names *Plantago afra*, *P. altissima*, and *P. lagopus* (*Plantago* subg. *Psyllium*, Plantaginaceae). - *Phytotaxa* **405(4)**: 210-214.

IAMONICO, D. & JARVIS, C.E. 2012: Lectotypification of two Linnaean names in *Chenopodium* L. (Chenopodiaceae). - *Taxon* **61(4)**: 864-865.

IAMONICO, D. & PANITSA, M. 2013: Lectotypification of the Linnaean name *Bryonia cretica* (Cucurbitaceae). - *Acta Bot. Croat.* **72(1)**: 193-196.

IAMONICO, D. & RAAB-STRAUBE, E.V. 2020: *Silene lydia*. In Raus, Th. & Raab-Straube, E.v. (eds): Euro+Med-Checklist notulae, 12. - *Willdenowia* **50(2)**: 311-312.

IAMONICO, D. & REVEAL, J.L. 2012: Lectotypification of the Linnaean name *Carpinus ostrya* L. (Betulaceae). - *Taxon* **61(4)**: 866.

IATRIDIS, Y. 1986: Blumen von Kreta. - Athen: Herausgabe vom Autor, 142 pp.

IATROÚ, G. 1985a: *Petrorhagia grandiflora* sp. nov. (Caryophyllaceae) from Greece. - *Nordic J. Bot.* **5 (5)**: 441-445.

IATROÚ, G. 1985b: A new species of *Potentilla* (Rosaceae) from Peloponnesus, Greece. – *Candollea* **40(1)**: 121-128.

IATROÚ, G. 1985c: Three new species from Peloponnisos. In: Elliniki Votaniki Eteria, Praktika 4ou epistimonikou simposiou, 23-24 martiou 1985, pp. 85-90 [In Greek with English summary]. - Thessaloniki.

IATROÚ, G. 1986: Simvoli sti meleti tou endimismou tis chloridas tis Peloponnisou [In Greek with English summary]. - Patras: Ph.D. thesis, Univ. of Patras, 310 pp. + 6 plates.

IATROÚ, G. 1989a: *Coris monspeliensis* L. In Greuter, W. & Raus, Th. (eds): Med-Checklist notulae, 15. - *Willdenowia* **19(1)**: 41.

IATROÚ, G. 1989b: *Linum hellenicum* (Linaceae), a new species from Peloponnesos, Greece. - *Willdenowia* **19(1)**: 69-73. [Also distributed as *OPTIMA leaflet* no. 185].

IATROÚ, G. 1995: The endemic flora of the island of Kythera. In: Proceedings, 5th sci. conf., Hellenic Bot. Soc., Delphi 21-23 October 1994, pp. 213-216 [In Greek with English summary]. - Thessaloniki: Univ. Studio Press.

IATROÚ, G. 1998: Phytodiversity, speciation, and microendemism in Mediterranean islands and islets. In Moret, J. (ed.): Abstracts of the IX. OPTIMA meeting, Paris, 11-17 May 1998, p. 33. - Paris: Mus. Nat. d'Hist. Naturelle.

IATROÚ, G. & GEORGIADIS, TH. 1984: A new species of *Asperula* (Rubiaceae) from Peloponnesos, Greece. - *Willdenowia* **14(1)**: 55-59.

IATROÚ, G. & GEORGIADIS, TH. 1985: *Minuartia favargeri*: A new species from Peloponnesus (Greece). - *Candollea* **40(1)**: 129-138.

IATROÚ, G. & GEORGIADIS, TH. 1989: The influence of human activities on the Greek endemic flora [In Greek with English summary]. - *Bios (Thessaloniki)* **1989**: 105-116.

IATROÚ, G. & TAN, KIT 1999: *Onobrychis peloponnesiaca* (Iatrou & Kit Tan) Iatrou & Kit Tan. In Greuter, W. & Raus, Th. (eds): Med-Checklist notulae, 18. - *Willdenowia* **29(1-2)**: 59.

IATROÚ, G., TRIGAS, P. & PETTAS, N. 2007: The vascular flora of Akrokorinthos Castle and its surrounding area (NE Peloponnese, Greece). - *Phytol. Balcan.* **13(1)**: 83-93.

IATROÚ, G. & TZANOUDAKIS, D. 1995: *Allium ritsii* (Alliaceae) a new autumn-flowering species from S. Peloponnisos (Greece). - *Phyton (Horn)* **35(2)**: 247-253.

IBERITE, M. 1996: Contribution to knowledge of the genus *Salicornia* L. (Chenopodiaceae) in Italy. - *Ann. Bot. (Rome)* **54(1)**: 145-154.

IETSWAART, J.H. 1980: A taxonomic revision of the genus *Origanum* (Labiatae). - The Hague, etc.: Leiden University Press, xi + 156 pp. [diss.].

IETSWAART, J.H. 1990: Infraspecific delimitation of the *Cyclamen graecum* complex. - *Cyclamen Soc. J.* **14(2)**: 49-53.

IETSWAART, J.H., FOKKINGA, A. & VROMAN, M. 1972: Delimitation of *Origanum scabrum* Boiss. & Heldr. (Labiatae) by means of morphological criteria. - *Acta Bot. Neerl.* **21(4)**: 439-447.

IGERSHEIM, A. 1993: Floral development and secondary pollen presentation in *Petromarula* Vent. ex Hedwig f. (Campanulaceae). - *Bot. Jahrb. Syst.* **115(3)**: 301-313.

IGERSHEIM, A. 2008: A new locality of *Cardiospermum halicacabum* L. (Sapindaceae) in Crete (Greece). - *Ann. Naturhist. Mus. Wien* **110B**: 263.

IGERSHEIM, A. 2010: Kreta. Auf den Spuren des böhmischen Naturforschers und Reisenden Franz Wilhelm Sieber 1789-1844. - Wien: Verlag des Naturhistorischen Museums, 327 pp.

IGERSHEIM, A. & WECKERLE, C. 1999: A new locality for *Phoenix theophrasti* on Crete. - *Palms* **43(1)**: 16-17.

İKINCI, N., OBERPRIELER, C. & GÜNER, A. 2006: On the origin of European lilies: phylogenetic analysis of *Lilium* section *Liriotypus* (Liliaceae) using sequences of the nuclear ribosomal transcribed spacers. - *Willdenowia* **36(2)**: 647-656.

IKONOMOU-AMILLI, A. (ED.) 2007: O viokosmos tou Imittou [In Greek]. - Athens: Amvrosiou O.E., 279 pp.

ILA, H.B., KAFKAS, S & TOPAKTAS, M. 2003: Chromosome numbers of four *Pistacia* (Anacardiaceae) species. - *J. Hort. Sci. Biotechnol.* **78(1)**: 35-38.

ILARDI, V. & TROÌA, A. 2021: Re-evaluation and typification of *Foeniculum piperitum* (Apiaceae), an underknown medicinal plant and crop wild relative. - *Phytotaxa* **508(2)**: 197-205.

İLARSLAN, R. & YURDAKULOL, E. 1994: A new species of *Erodium* (Geraniaceae) from Turkey. - *Fl. Medit.* **4**: 167-170.

İLÇIM, A. & BEHÇET, L. 2007: *Solanum elaeagnifolium* Cav. in Turkey. - *Turk. J. Bot.* **31(6)**: 577-579.

ILIADOU, E. & AL. 2014a: Flora and factors affecting species diversity in protected "Natura 2000" sites of the Ionian area: the Echinades islet group (Greece). - *Willdenowia* **44(1)**: 121-136.

ILIADOU, E. & AL. 2014b: Flora and factors affecting species diversity in the islet groups of the protected "Natura 2000" sites of the Amvrakikos Gulf and Mesolongi Lagoon (Ionian area, Greece). - *Willdenowia* **44(3)**: 439-450.

ILIADOU, E. & AL. 2014c: Comparing the two Greek archipelagos plant species diversity and endemism patterns highlight the importance of isolation and precipitation as biodiversity drivers. - *J. Biol. Res. Thessalon.* **21(16)**: 10 pp.

ILIADOU, E. & AL. 2018: Taxonomic diversity patterns in the Northern Sporades islet complex (Greece). - *Bot. Serbica* **142 (suppl. 1)**: 75-76.

ILIADOU, E. & AL. 2020: Taxonomic and phylogenetic diversity patterns in the Northern Sporades islets complex (West Aegean, Greece). - *Pl. Syst. Evol.* **306(2)**: doi:10.1007/s00606-020-01660-0.

ILIJANIĆ, L. & ŠEGULJA, N. 1969: *Thelygonum cynocrambe* L. na obroncima limskoga kanala (Istra). - *Acta Bot. Croat.* **28**: 419-421.

ILIJANIĆ, L. & TOPIĆ, J. 1986: *Paspalum dilatatum* Poiret, a new adventitious plant in the flora of Yugoslavia. - *Acta Bot. Croat.* **45**: 141-144 + 1 plate.

ILJIN, M.M. 1930: *Chondrilla*. (Geography, ecology and rubber contents) [In Russian with English summary]. - *Trudy Prikl. Bot.* **24(3)**: 147-185.

ILLUMINATI, O. & AL. 1996: Numeri cromosomici per la Flora Italiana: 1358-1371. - *Inform. Bot. Ital.* **27 (2-3)**: 261-274.

ILNICKI, T. & SZELĄG, Z. 2011: Chromosome numbers in *Hieracium* and *Pilosella* from Central and Southeastern Europe. - *Acta Biol. Cracov., Ser. Bot.* **53(1)**: 102-110.

IMAM, M., CHRTEK, J. & SLAVÍKOVÁ, Z. 1977: Die Gattung *Adonis* in Ägypten. - *Publ. Cairo Univ. Herb.* **7-8**: 261-271.

INCEER, H. 2021a: Lectotypification of the name *Chamaemelum heterolepis*, the basionym of *Tripleurospermum heterolepis* and taxonomic notes on *T. tempskyanum* (Asteraceae). - *Bot. Serbica* **45(2)**: 347-352.

INCEER, H. 2021b: Lectotypification of two Boissier's names in *Chamaemelum* (Anthemideae, Asteraceae). - *Candollea* **76**: 259-260.

INCEER, H. & HAYIRLIOGLU-AYAZ, S. 2010: Chromosome numbers in *Tripleurospermum* Sch. Bip. (Asteraceae) and closely related genera: relationships between ploidy level and stomatal length. - *Pl. Syst. Evol.* **285(3-4)**: 149-157.

INGRAM, R. 1977: Synthesis of the hybrid *Senecio squalidus* L. × *S. vulgaris* L. f. *radiatus* Hegi. - *Heredity* **39(1)**: 171-173.

INGROUILLE, M.J. & PEARSON, J. 1987: The pattern of morphological variation in the *Salicornia europaea* L. aggregate (Chenopodiaceae). - *Watsonia* **16(3)**: 269-281.

INGWERSEN, W.T. 1934: *Saxifraga*. The *Kabschia* section of the genus *Saxifraga*, and subsections as under. - *Quart. Bull. Alpine Gard. Soc. Gr. Brit.* **2(6)**: 245-262 + 8 plates.

INGWERSEN, W.T. 1939: *Lilium Jankae* (Kern.). - *Quart. Bull. Alpine Gard. Soc. Gr. Brit.* **7(2)**: 135-136 + 1 plate.

INGWERSEN, W.T. 1942: Plant hunting in the Pirin Mountains of Bulgaria. - *Quart. Bull. Alpine Gard. Soc. Gr. Brit.* **10(2)**: 112-131.

INGWERSEN, W.T. 1944: *Daphne oleoides*. - *Quart. Bull. Alpine Gard. Soc. Gr. Brit.* **12(4)**: 202-204.

INGWERSEN, W.T. 1979: A summary of alpine species of *Dianthus*. - *Plantsman* **1(3)**: 146-153.

INNANGI, M. & IZZO, A. 2015: *Pinguicula lavalvae* (Lentibulariaceae), a new endemic butterwort from southern Italy diagnosed with the aid of geometric morphometrics. - *Pl. Biosyst.* **149(6)**: 990-999.

INOCENCIO, C. & AL. 2006: A systematic revision of *Capparis* section *Capparis* (Capparaceae). - *Ann. Missouri Bot. Gard.* **93(1)**: 122-149.

IOANNIDIS, V. 2020: I chlorida stin techni ke ton politismo [In Greek]. - Kilkis: Techni Makedoniki Kallitechniki Eteria, 72 pp.

IOANNIDIS, V. & AL. 2019: Reports 37-78. In Vladimirov, V., Aybeke, M. & Tan, Kit (eds): New floristic records in the Balkans: 39. - *Phytol. Balcan.* **25(2)**: 211-215.

IOANNIDIS, V., DOULKERIDOU, D. & KOUTIS, K. 2017: Contribution study to the flora of Kilkis region. In Dousis, M.A. & Thanos, K.A.: 15th conference of the Hellenic Botanical Society, book of abstracts: p. 85.

IOANNIDIS, V., DOULKERIDOU, D. & STRID, A. 2020a: Reports 31-61. In Vladimirov, V., Aybeke, M. & Tan, Kit (eds): New floristic records in the Balkans: 42. - *Phytol. Balcan.* **26(2)**: 195-197.

IOANNIDIS, V., DOULKERIDOU, D. & STRID, A. 2020b: Reports 62-95. In Vladimirov, V., Aybeke, M. & Tan, Kit (eds): New floristic records in the Balkans: 42. - *Phytol. Balcan.* **26(2)**: 197-200.

IOANNIDIS, V., DOULKERIDOU, D. & STRID, A. 2021a: Reports 42-59. In Vladimirov, V., Aybeke, M. & Tan, Kit (eds): New floristic records in the Balkans: 44. - *Phytol. Balcan.* **27(1)**: 135-136.

IOANNIDIS, V., DOULKERIDOU, D. & STRID, A. 2021b: Reports 68-104. In Vladimirov, V., Aybeke, M. & Tan, Kit (eds): New floristic records in the Balkans: 45. - *Phytol. Balcan.* **27(2)**: 259-262.

IOANNIDIS, V. & TZANOUDAKIS, D. 2020: *Allium arampatzisii* (Amaryllidaceae) - a new species from northern continental Greece. - *Phytol. Balcan.* **26(3)**: 509-515.

ISELY, D. 1983: Classification of alfalfa (*Medicago sativa* L.) and relatives. - *Iowa State J. Res.* **57(3)**: 207-220.

ISHERWOOD, R. 1983: Note on the fruiting of *Aristolochia sempervirens* L. - *Garden (London)* **108(4)**: 157-158.

ISHIZAKA, H. 1995: Interspecific hybrids between *Cyclamen persicum* and *C. hederifolium*. - *Cyclamen* **19(2)**: 46-49 + 2 plates.

ISHIZAKA, H. 2003: Cytogenetic studies in *Cyclamen persicum, C. graecum* (Primulaceae) and their hybrids. - *Pl. Syst. Evol.* **239(1-2)**: 1-14.

ITO, Y. & AL. 2017a: Towards a better understanding of the *Ruppia maritima* complex (Ruppiaceae): Notes on the correct application and typification of the names *Ruppia cirrhosa* and *R. spiralis*. - *Taxon* **66(1)**: 167-171.

ITO, Y. & AL. 2017b: Phylogeny of *Najas* (Hydrocharitaceae) revisited: Implications for systematics and evolution. - *Taxon* **66(2)**: 309-323.

IVANINA, L.I. 1955: Rod *Digitalis* L. (naperstianka) i yego prakticheskiye primeneniya [In Russian]. - *Trudy Bot. Inst. Akad. Nauk S.S.S.R., Ser. 1, Fl. Sist. Vysš. Rast.* **11**: 198-308.

IVANOVA, D. 1997: Reports (831-839). In Kamari, G., Felber, F. & Garbari, F. (eds): Mediterranean chromosome number reports - 7. - *Fl. Medit.* **7**: 225-235.

IVANOVA, D. 1998: Reports (961-967). In Kamari, G., Felber, F. & Garbari, F. (eds): Mediterranean chromosome number reports - 8. - *Fl. Medit.* **8**: 262-271.

IVANOVA, D. 1999: Reports (1058-1063). In Kamari, G., Felber, F. & Garbari, F. (eds): Mediterranean chromosome number reports - 9. - *Fl. Medit.* **9**: 346-356.

IVANOVA, D. 2000: *Dryopteris affinis* (Lowe) Fraser-Jenkins (Pteridophyta, Dryopteridaceae) in the Bulgarian flora. In Özhatay, N. & al. (eds): Abstracts from the Second Balkan Botanical Congress, Istanbul, Turkey, May 14-18, 2000, p. 41.

IVANOVA, D. 2003: Reports (1328-1333). In: Kamari, G., Blanché, C. & Garbari, F. (eds): Mediterranean chromosome number reports - 13. - *Fl. Medit.* **13**: 354-361.

IVANOVA, D. 2004a: *Dryopteris affinis* subsp. *borreri* (Pteridophyta: Dryopteridaceae) in the Bulgarian flora. - *Fl. Medit.* **14**: 201-218.

IVANOVA, D. 2004b: Distribution of Dryopteridaceae in Bulgaria. I. *Polystichum, Athyrium, Gymnocarpium*. - *Phytol. Balcan.* **9(3)**: 435-459.

IVANOVA, D. 2006: *Polypodium interjectum* and *P.* ×*mantoniae* (Polypodiaceae: Pteridophyta), new to the Bulgarian flora. - *Phytol. Balcan.* **12(2)**: 191-202.

IVANOVA, D., DIMITROVA, D. & VLADIMIROV, V. 2006: Chromosome numbers of selected woody species from the Bulgarian flora. - *Phytol. Balcan.* **12(1)**: 79-84.

IVANOVA, D. & VLADIMIROV, V. 2007: Chromosome numbers of some woody species from the Bulgarian flora. - *Phytol. Balcan.* **13(2)**: 205-207.

IVANOVA, D., VLADIMIROV, V. & STANIMIROVA, P. 2005: Reports (1445-1456) In: Kamari, G., Blanché, C. & Garbari, F. (eds): Mediterranean chromosome number reports - 15. - *Fl. Medit.* **15**: 719-728.

IVANOVA, D. (ED.) 2009: Plant, fungal and habitat diversity investigation and conservation. Proceedings of IV Balkan botanical congress, Sofia, 20-26 June 2006. - Sofia: Institute of Botany, BAS, 712 pp.

IVEY, W. 1983: Plants of the Mediterranean - Along the donkey paths of Greece. - *J. Scott. Rock Gard. Club* **18(2)**: 135-142 + 1 plate.

IVEY, W. 1984: Further thoughts along the donkey paths of Greece. - *Rock Gard.* **19(1)**: 75-82.

IYER, Y.K. 1991: Rapports (30-37). In Kamari, G., Felber, F. & Garbari, F. (eds): Mediterranean chromosome number reports - 1. - *Fl. Medit.* **1**: 238-240.

IZMAIŁOW, R. 1977: Cytological studies in Polish representatives of the genus *Epilobium* L. - *Acta Biol. Cracov., Ser. Bot.* **20(1-2)**: 101-111.

IZMAIŁOW, R. 1981: Karyological studies in species of *Alchemilla* L. from the series *Calycinae* Bus. (section *Brevicaulon* Rothm.). - *Acta Biol. Cracov., Ser. Bot.* **23**: 117-130.

IZUZQUIZA, A. 1990: Números cromosómicos para la flora española. 618-619. - *Lagascalia* **15(2)**: 295-296.

IZUZQUIZA, A. 1998: Números cromosómicos para la flora española. 806-809. - *Lagascalia* **20(2)**: 308-310.

IZUZQUIZA, A. & NIETO FELINER, G. 1991: Cytotaxonomic notes on the genus *Leontodon* (Asteraceae: Hypochoeridinae). - *Willdenowia* **21(1-2)**: 215-224.

JAARSAMA, T.A. & AL. 1990: Chemo- and karyotaxonomic studies on some rhizomatous species of the genus *Symphytum* (Boraginaceae). - *Pl. Syst. Evol.* **169(1-2)**: 31-39.

JABBOUR, F. & RENNER, S.S. 2011: Resurrection of the genus *Staphisagria* J. Hill, sister to all other Delphinieae (Ranunculaceae). - *PhytoKeys* **7**: 21-26.

JACKSON, A.K. 1936a: *Jurinea taygetea* Halácsy. - *Hooker's Icon. Pl.* **34(1)**: tab. 3301 + 2 pp. text.

JACKSON, A.K. 1936b: *Amracus scaber* (Boiss. et Heldr.) Briq. subsp. *pulcher* (Boiss. et Heldr.) Hayek. - *Hooker's Icon. Pl.* **34(1)**: tab. 3302 + 2 pp. text.

JACKSON, A.K. 1936c: *Teucrium aroanium* Orph. - *Hooker's Icon. Pl.* **34(1)**: tab. 3303 + 2 pp. text.

JACKSON, A.K. 1937: *Arenaria serpentini* A. K. Jackson. - *Hooker's Icon. Pl.* **34(2)**: tab. 3326 + 2 pp. text.

JACKSON, A.K. & TURRILL, W.B. 1937: *Bellevalia Atchleyi* A. K. Jackson et Turrill. - *Hooker's Icon. Pl.* **34(2)**: tab. 3329 + 2 pp. text.

JACKSON, B.P. & BERRY, M.I. 1979: *Mandragora* - taxonomy and chemistry of the European species. In Hawkes, J.G., Lester, R.N. & Skelding, A.D. (eds): The biology and taxonomy of the Solanaceae, pp. 505-512. - London, etc.: Academic Press.

JACKSON, L.E. 1985: Ecological origins of California's Mediterranean grasses. - *J. Biogeogr.* **12(4)**: 349-361.

JACOBSEN, N. & BOTHMER, R.V. 1995: Taxonomy in the *Hordeum murinum* complex (Poaceae). - *Nordic J. Bot.* **15(5)**: 449-458.

JACOBSEN, N., SCHEEPEN, J. VAN & ØRGAARD, M. 1997: The *Crocus chrysanthus-biflorus* cultivars. - *New Plantsman* **4(1)**: 6-38.

JACQUEMART, A.-L. 1996: Biological Flora of the British Isles. *Vaccinium uliginosum* L. - *J. Ecol.* **84(5)**: 771-785.

JACQUIN, N.J.V. 1773-78: Florae Austriacae, sive plantarum selectarum in Austriae archiducatu sponte crescentium, iconis, ad vivum coloratae, et descriptionibus, ac synonymis illustratae. - Viennae: Typis L.J. Kaliwoda, 5 vols. [cf. TL-2 2: 410, 1979].

JACQUIN, N.J.V. 1781-95: Icones plantarum rariorum. - Lugundi Batavorum, etc.: 4 vols. [cf. TL-2 2: 410, 1979].

JAFRI, F. & AL. 2020: A new taxonomic backbone for the infrageneric classification of the species‐rich genus *Silene* (Caryophyllaceae). - *Taxon* **69(2)**: 337-368.

JAGEL, A. 1992: Zur Flora und Vegetation der Insel Elafonisos (Lakonien, Griechenland). - Bochum: Diplomarbeit, Ruhr-Universität, 160 pp. + 3 tables + Ergänzungen (2 pp.).

JÄGER, E.J. 1970: Characteristische Typen mediterran-mitteleuropäischer Pflanzenareale. - *Feddes Repert.* **81(1-5)**: 67-92.

JÄGER, E.J. 1972: Comments on the history and ecology of continental European plants. In Valentine, D.H. (ed.): Taxonomy, phytogeography and evolution, pp. 349-362. - London & New York: Academic Press.

JÄGER, E.J. 1973: Zur Verbreitung und Lebensgeschichte der Wildtulpe (*Tulipa sylvestris* L.) und Bemerkungen zur Chorologie der Gattung *Tulipa* L. - *Hercynia, N.F.* **10(4)**: 429-448.

JÄGER, E.J. 1986: *Epilobium ciliatum* Raf. (*E. adenocaulon* Hausskn.) in Europa. - *Wiss. Z. Martin- Luther-Univ. Halle-Wittenberg, Math.-Naturwiss. Reihe* **35(5)**: 122-134.

JÄGER, E.J. 1988: Möglichkeiten der Prognose synanthroper Pflanzenausbreitungen. - *Flora, Morphol. Geobot. Oekophysiol.* **180(1)**: 101-131.

JÄGER, E.J. 1990: Pflanzenarealgrenzen als Leistungsgrenzen. - *Biol. Rundschau* **28**: 295-306.

JÄGER, E.J. 1991: Verbreitung, Biologie und Wuchsform von *Abutilon theophrasti* Medic. - *Wiss. Z. Martin-Luther-Univ. Halle-Wittenberg, Math.-Naturwiss. Reihe* **40(3)**: 37-48.

JAHN, A. 1974a: Beiträge zur Kenntnis der Sippenstruktur einiger Arten der Gattung *Coronilla* L. unter besonderer Berücksichtigung der Series *Luteae* Uhr. und der Series *Roseae* Uhr. 1. Teil. - *Feddes Repert.* **85(7-8)**: 455-532.

JAHN, A. 1974b: Beiträge zur Kenntnis der Sippenstruktur einiger Arten der Gattung *Coronilla* L. unter besonderer Berücksichtigung der Series *Luteae* Uhr. und der Series *Roseae* Uhr. 2. Teil. - *Feddes Repert.* **85(9-10)**: 663-744.

JAHN, A. 1975: Beiträge zur Kenntnis der Sippenstruktur einiger Arten der Gattung *Coronilla* L. unter besonderer Berücksichtigung der Series *Luteae* Uhr. und *Roseae* Uhr. - *Gleditschia* **2**: 183-184.

JAHN, R. 1998: Notes on *Callitriche truncata* subsp. *occidentalis, Barbarea vulgaris* subsp. *arcuata* and *Ampelodesmos mauritanica*. In Greuter, W. & Raus, Th. (eds): Med-Checklist notulae, 17. - *Willdenowia* **28(1-2)**: 163-174.

JAHN, R. 2000: *Salix pedicellata* Desf. In Greuter, W. & Raus, Th. (eds): Med-Checklist notulae, 19. - *Willdenowia* **30(2)**: 240.

JAHN, R. 2001: The phytodiversity of the flora of Kriti (Greece): an approach to the current state of knowledge. In Raimondo, F.M. (ed.): Abstracts from the X OPTIMA meeting, Palermo, 13-19 September 2001, p. 172.

JAHN, R. 2002: Notes on *Centaurea cyanus* L. In Greuter, W. & Raus, Th. (eds): Med-Checklist notulae, 21. - *Willdenowia* **32(2)**: 196-197.

JAHN, R. 2003: The phytodiversity of the flora of Kriti (Greece) - a survey of the current state of knowledge. - *Bocconea* **16(2)**: 845-851.

JAHN, R. 2009: *Rumex cristatus* DC. In Greuter, W. & Raus, Th. (eds): Med-Checklist notulae, 28. - *Willdenowia* **39(2)**: 339.

JAHN, R. 2013: *Nigella doerfleri* Vierh. In Raab-Straube, E. v. & Raus, Th. (eds): Euro+Med-Checklist notulae, 2. - *Willdenowia* **43(2)**: 245.

JAHN, R. 2014: [Note on *Kickxia lanigera*]. In Raab-Straube, E. v. & Raus, Th. (eds): Euro+Med-Checklist Notulae, 3. - *Willdenowia* **44(2)**: 295.

JAHN, R. 2015: *Oxybasis chenopodioides*, *Linum virgultorum*. In Raab-Straube, E. & Raus, Th. (eds): Euro+Med-Checklist notulae, 5. - *Willdenowia* **45(3)**: 450, 456.

JAHN, R. & HAUSER, K. 2009: *Crocus cartwrightianus* Herb., *Crocus laevigatus* Bory & Chaub. In Greuter, W. & Raus, Th. (eds): Med-Checklist notulae, 28. - *Willdenowia* **39(2)**: 341-342.

JAHN, R. & RAUS, TH. 2014a: [Note on *Epilobium tournefortii*]. In Raab-Straube, E. v. & Raus, Th. (eds): Euro+Med-Checklist notulae, 3. - *Willdenowia* **44(2)**: 293.

JAHN, R. & RAUS, TH. 2014b: [Note on *Scrophularia canina*]. In Raab-Straube, E. v. & Raus, Th. (eds): Euro+Med-Checklist notulae, 3. - *Willdenowia* **44(2)**: 295.

JAHN, R., RAUS, TH. & STRID, A. 2014: [Note on *Tamarix tetrandra*]. In Raab-Straube, E. v. & Raus, Th. (eds): Euro+Med-Checklist notulae, 3. - *Willdenowia* **44(2)**: 295.

JAHN, R. & SCHÖNFELDER, P. 1995: Exkursionsflora für Kreta. - Stuttgart: Verlag Eugen Ulmer, 446 pp.

JAHN, R. (ED.) 1996: Kreta. Botanische Exkursion für Fortgeschrittene, 14.-27. April 1996. - Regensburg: 49 pp. (mimeographed).

JAIN, K.K. 1978a: Cytology of the polymorphic *Plantago lanceolata* L. - *Caryologia* **31(2)**: 179-182.

JAIN, K.K. 1978b: Cytology of the polymorphic *Plantago major* Linn. - *Cytologia* **43(2)**: 345-349.

JAKOVLJEVIĆ, S.J. 1933: Über die Temperatur- und Vegetationsverhältnisse der Insel "Stevi Petar" im Prespasee. - *Bull. Inst. Jard. Bot. Univ. Belgrade* **2(3)**: 183-188.

JAKOVLJEVIĆ, S.J. 1934: Végétation macrophytique du Lac de Prespa. - *Bull. Inst. Jard. Bot. Univ. Belgrade* **3(1-2)**: 35-55.

JAKUBSKA-BUSSE, A. & AL. 2020: A revision of the taxonomy and identification of *Epipactis greuteri* (Orchidaceae, Neottieae). - *Plants MDPI* **9, 783**: 17 pp.

JAKUCS, P. 1959: Über die ostbalkanischen Flieder-Buschwälder. - *Acta Bot. Acad. Sci. Hung.* **5(3-4)**: 357-390 + 1 table.

JALAS, J. 1971a: Notes on *Thymus* L. (Labiatae) in Europe. I. Supraspecific classification and nomenclature. In Heywood, V.H. (ed.): Notulae systematicae ad Floram Europaeam spectantes, no. 9. - *Bot. J. Linn. Soc.* **64(2)**: 199-215.

JALAS, J. 1971b: Notes on *Thymus* L. (Labiatae) in Europe. II. Comments on species and subspecies. In Heywood, V.H. (ed.): Notulae systematicae ad Floram Europaeam spectantes, no. 10. - *Bot. J. Linn. Soc.* **64(3)**: 247-271.

JALAS, J. 1974: Notes on *Thymus* L. (Labiatae) in Europe. III. - *Ann. Bot. Fenn.* **11(4)**: 262-266.

JALAS, J. 1980: Turkish taxa of *Thymus* (Labiatae) described as new or revised. - *Ann. Bot. Fenn.* **17 (3)**: 315-324.

JALAS, J. 1981: State and features of the chorological information used for Atlas Florae Europaeae. In Velčev, V.I. & Kožuharov, S.I. (eds): Mapping the flora of the Balkan Peninsula, pp. 13-20. - Sofia: Publishing House of the Bulgarian Academy of Sciences.

JALAS, J. 1983: Atlas Florae Europaeae notes 1. New nomenclatural combinations in Alsinoideae and Paronychioideae (Caryophyllaceae). - *Ann. Bot. Fenn.* **20(2)**: 109-111.

JALAS, J. 1985: Atlas Florae Europaeae notes 5. New nomenclatural combinations in *Dianthus* and *Aconitum*. - *Ann. Bot. Fenn.* **22(3)**: 219-221.

JALAS, J. 1996: Atlas Florae Europaeae notes. 13. Suggestions on *Alyssum* and *Lepidium* (Cruciferae). - *Ann. Bot. Fenn.* **33(4)**: 283-284.

JALAS, J. & AL. (EDS) 1999: Atlas Florae Europaeae. 12. Resedaceae to Platanaceae. - Helsinki: Akateeminen Kirjakauppa, 250 pp.

JALAS, J. & KALEVA, K. 1966: Chromosome studies in *Thymus* L. (Labiatae). IV. Mitotic numbers of some Balkan, Transylvanian and South Alpine taxa. - *Ann. Bot. Fenn.* **3(1)**: 123-127.

JALAS, J. & KALEVA, K. 1967: Chromosome studies in *Thymus* L. (Labiatae). V. - *Ann. Bot. Fenn.* **4(1)**: 74-80.

JALAS, J. & KALEVA, K. 1970: Supraspezifische Gliederung und Verbreitungstypen in der Gattung *Thymus* L. (Labiatae). - *Feddes Repert.* **81(1-5)**: 93-106.

JALAS, J., KUKKONEN, I. & UOTILA, P. 1976: Notes on the flora of European Turkey. - *Notes Roy. Bot. Gard. Edinburgh* **35(1)**: 69-76.

JALAS, J. & POHJO, T. 1965a: Chromosome studies in *Thymus* L. (Labiatae). II. Some Swiss and Hungarian taxa. - *Ann. Bot. Fenn.* **2(2)**: 165-168.

JALAS, J. & POHJO, T. 1965b: Chromosome studies in *Thymus* L. (Labiatae). III. A few counts on miscellaneous taxa. - *Ann. Bot. Fenn.* **2(2)**: 169-170.

JALAS, J., SUOMINEN, J. & LAMPINEN, R. 1994: Atlas Florae Europaeae. 10. Cruciferae (*Sisymbrium* to *Aubrieta*). - Helsinki: Akateeminen Kirjakauppa, 224 pp.

JALAS, J., SUOMINEN, J. & LAMPINEN, R. (EDS) 1996: Atlas Florae Europaeae. 11. Cruciferae (*Ricotia* to *Raphanus*). - Helsinki: Akateeminen Kirjakauppa, 310 pp.

JALAS, J. & SUOMINEN, J. (EDS) 1972: Atlas Florae Europaeae. 1. Pteridophyta (Psilotaceae to Azollaceae). - Helsinki: Akateeminen Kirjakauppa, 121 pp.

JALAS, J. & SUOMINEN, J. (EDS) 1973: Atlas Florae Europaeae. 2. Gymnospermae (Pinaceae to Ephedraceae). - Helsinki: Akateeminen Kirjakauppa, 40 pp.

JALAS, J. & SUOMINEN, J. (EDS) 1976: Atlas Florae Europaeae. 3. Salicaceae to Balanophoraceae. - Helsinki: Akateeminen Kirjakauppa, 128 pp.

JALAS, J. & SUOMINEN, J. (EDS) 1979: Atlas Florae Europaeae. 4. Polygonaceae. - Helsinki: Akateeminen Kirjakauppa, 71 pp.

JALAS, J. & SUOMINEN, J. (EDS) 1980: Atlas Florae Europaeae. 5. Chenopodiaceae to Basellaceae. - Helsinki: Akateeminen Kirjakauppa, 119 pp.

JALAS, J. & SUOMINEN, J. (EDS) 1983: Atlas Florae Europaeae. 6. Caryophyllaceae (Alsinoideae and Paronychioideae). - Helsinki: Akateeminen Kirjakauppa, 176 pp.

JALAS, J. & SUOMINEN, J. (EDS) 1986: Atlas Florae Europaeae. 7. Caryophyllaceae (Silenoideae). - Helsinki: Akateeminen Kirjakauppa, 229 pp.

JALAS, J. & SUOMINEN, J. (EDS) 1989: Atlas Florae Europaeae. 8. Nymphaeaceae to Ranunculaceae. - Helsinki: Akateeminen Kirjakauppa, 261 pp.

JALAS, J. & SUOMINEN, J. (EDS) 1991: Atlas Florae Europaeae. 9. Paeoniaceae to Capparaceae. - Helsinki: Akateeminen Kirjakauppa, 110 pp.

JALAS, J. & UOTILA, M. 1976: Chromosome studies in *Thymus* L. (Labiatae). VI. Counts on Macedonian and Thracian taxa. - *Ann. Bot. Fenn.* **13(2)**: 61-64.

JANCHEN, E. 1908a: Zur Nomenklatur des gemeinen Sonnenröschens. - *Oesterr. Bot. Z.* **58(10-11)**: 406-413, 426-435.

JANCHEN, E. 1908b: Zwei neue Fumanen. - *Oesterr. Bot. Z.* **58(11)**: 439-440.

JANCHEN, E. 1910: Die *Edraianthus*-Arten der Balkanländer. - *Mitt. Naturwiss. Vereins Univ. Wien* **8 (1)**: 1-40.

JANCHEN, E. 1919: Beitrag zur Floristik von Ost-Montenegro. - *Oesterr. Bot. Z.* **68**: 77-98, 166-179, 254-286, 327-340.

JANCHEN, E. 1920a: Die systematische Gliederung der Gattung *Fumana*. - *Oesterr. Bot. Z.* **69(1-3)**: 1-30.

JANCHEN, E. 1920b: Vorarbeiten zu einer Flora der Umgebung von Škodra in Nord-Albanien. - *Oesterr. Bot. Z.* **69**: 128-146, 167-187, 199-207, 230-261.

JANČIĆ, R. 1989a: Distribution of genus *Mentha* L. (Lamiaceae) in the eastern Mediterranean. In Phitos, D. (ed.): Abstracts. Sixth OPTIMA meeting, Delphi 10-16 Sept. 1989, p. 102.

JANČIĆ, R. 1989b: *Mentha serbica* R. Jančić, sp. nov. - a new species of the genus *Mentha* L. (Lamiaceae). - *Glasn. Prir. Mus. Beogradu, Ser. B, Biol. Nauke* **43-44**: 27-33.

JANCZEWSKI, E. DE 1907: Monographie des Grosseilliers, *Ribes* L. - *Mém. Soc. Phys. Genève* **35(3)**: 199-517.

JANIĆ, M. 1981: *Alchemilla crinita* Buser, new species in the flora from Macedonia [In Macedonian with English summary]. - *Godišen Zborn. Biol. Fak. Univ. "Kiril i Metodij"* **34**: 167-169 + 1 plate.

JANIĆ, M. 1986: *Alchemilla viridiflora* Rothm. (*Elatae* series) a new species of the flora in Macedonia and Yugoslavia. - *Fragm. Balcan.* **13(7)**: 67-71.

JANIĆ, M. 1987: *Alchemilla gorcensis* Pawl. and *Alchemilla gracilis* Opiz, two new species in the flora from Macedonia [In Macedonian with English summary]. - *Godišen Zborn. Biol. Fak. Univ. "Kiril i Metodij"* **36**: 247-250.

JANIĆ, M. 1989: Rare and new *Alchemilla* species for the Macedonian and Yugoslav flora (Rosaceae, Rosidae). - *Fragm. Balcan.* **14(11)**: 101-105.

JANIĆ, M. 1990a: A contribution to the study of the *Alchemilla* genus in the Macedonia flora (I) [In Macedonian with English summary]. - *Godišen Zborn. Biol. Fak. Univ. "Kiril i Metodij"* **41-42**: 197-214.

JANIĆ, M. 1990b: Distribution of some species in SR Macedonia [In Macedonian with English summary]. - *Godišen Zborn. Biol. Fak. Univ. "Kiril i Metodij"* **41-42**: 215-219.

JANIĆ, M. 1992: *Alchemilla pyrenaica* Dufour (series *Calycinae*), a new species of the flora in Macedonia. - *Godišen Zborn. Biol. Fak. Univ. "Kiril i Metodij"* **45**: 175-177.

JANIĆ, M. & AL. 1983: A contribution to the knowledge of horology of species of the genus *Alchemilla* L. in SR Serbia (Stara Planina) [in Serbo-Croatian with English summary]. - *Bull. Inst. Jard. Bot. Univ. Belgrade, n.s.* **17**: 51-59.

JANKA, V. 1870: Bemerkungen zu Boissier's "Flora Orientalis". - *Oesterr. Bot. Z.* **20(4)**: 111-114.

JANKA, V. 1871: Correspondenz. - *Oesterr. Bot. Z.* **21(9)**: 248-252.

JANKA, V. 1872: Plantarum novarum turcicarum breviarum. - *Oesterr. Bot. Z.* **22(6)**: 174-182.

JANKA, V. 1873a: Plantarum novarum turcicarum breviarum. II.- *Oesterr. Bot. Z.* **23(6)**: 194-196.

JANKA, V. 1873b: Plantarum novarum turcicarum breviarium. (Fortsetzung). - *Oesterr. Bot. Z.* **23(7)**: 201-205.

JANKA, V. 1873c: Plantarum novarum turcicarum breviarium. (Schluss). - *Oesterr. Bot. Z.* **23(8)**: 241-244.

JANKA, V. 1877a: Generis *Iris* species nova. - *Természetrajzi Füz.* **1(4)**: 242-245 + 1 plate.

JANKA, V. 1877b: Avenaceae europaeae. - *Természetrajzi Füz.* **2**: 8 pp.

JANKA, V. 1878: Descriptiones plantarum novarum. - *Természetrajzi Füz.* **2(1)**: 4 pp. + 1 plate.

JANKA, V. 1881: Scrophularineae europaeae analyticae elaboratae. - *Természetrajzi Füz.* **4(4)**: 284-320.

JANKA, V. 1882: Violae europaeae s.l. - *Természetrajzi Füz.* **5(2-4)**: 253-293.

JANKA, V. 1883a: Plumbagineae europaeae. - *Természetrajzi Füz.* **6**: 164-175.

JANKA, V. 1883b: Brassiceae europaeae. - *Természetrajzi Füz.* **6**: 176-183.

JANKA, V. 1883c: Cruciferae siliculosae florae europaeae. - *Természetrajzi Füz.* **7**: 106-126.

JANKA, V. 1884a: Genisteae europaeae. - *Természetrajzi Füz.* **8**: 57-73.

JANKA, V. 1884b: Trifolieae & Loteae florae europaeae. - *Természetrajzi Füz.* **8**: 145-170.

JANKA, V. 1884c: Hedysareae europaeae. - *Természetrajzi Füz.* **8**: 290-296.

JANKA, V. 1884d: Astragaleae europaeae. - *Természetrajzi Füz.* **8**: 297-310.

JANKOVIĆ, A. 2006: Rasprostranjenje roda *Pedicularis* L. na balkanskom poluostrvu [In Serbian]. - Beograd: Diploma thesis, Univ. of Beograd, 61 pp.

JANKOVIĆ, I. & AL. 2015: Morphological variability of isophylloid bellflower *Campanula versicolor* Andrews (Campanulaceae) from the Balkan Peninsula. In Bogdanović, S. & Jogan, N.: Book of abstracts - 6th Balkan Botanical Congress, p. 26. - Rijeka.

JANKOVIĆ, I. & AL. 2018: Genetic and morphological differentiation within the broadly defined *Campanula versicolor* (Campanulaceae). - *Bot. Serbica* **142 (suppl. 1)**: 60.

JANKOVIĆ, I. & AL. 2019: Genetic and morphological data reveal new insights into the taxonomy of *Campanula versicolor* s.l. (Campanulaceae). - *Taxon* **68(2)**: 340-369.

JANKOVIĆ, M.M. 1957a: Übersicht der Systematik der Gattung *Trapa* L. in Jugoslavien. - *Glasn. Prir. Muz. Srpske Zemlje, Ser. B, Biol. Nauke* **10**: 83-159.

JANKOVIĆ, M.M. 1957b: Ergebnisse der variations-statistischen Analyse der Gattung *Trapa* L. in Jugoslawien [In Serbo-Croatian with German summary]. - *Glasn. Prir. Muz. Srpske Zemlje, Ser. B, Biol. Nauke* **10**: 161-218.

JANKOVIĆ, M.M. 1982: Contribution to the study of the vegetation of the Šarplanina mountain with particular reference to some conspicuous relict plant species [In Serbo-Croatian with English summary]. - *Bull. Inst. Jard. Bot. Univ. Belgrade, n.s.* **15**: 75-129 + 6 plates.

JANKOVIĆ, M.M. 1991: Contribution to the post-glacial history of the waternut (*Trapa natans* L. sp. sp.) with emphasis on the area of Balkan Peninsula ... [In Serbo-Croatian with English summary]. - *Glasn. Prir. Mus. Beogradu, Ser. B, Biol. Nauke* **46**: 87-101.

JANKOVIĆ, M.M. & MIČIĆ, V. 1953: Polymorphisme foliaire chez une plante aquatique (*Trapa natans* L.) [In Serbo-Croatian with French summary]. - *God. Biol. Inst. Sarajevu* **5(1-2)**: 235-256.

JAROS, U., TRIBSCH, A. & COMES, H.P. 2018: Diversification in continental island archipelagos: new evidence on the roles of fragmentation, colonization and gene flow on the genetic divergence of Aegean *Nigella* (Ranunculaceae). - *Ann. Bot. (Oxford), n.s.* **121**: 241-254.

JARVIS, C.E. 1980: Systematic studies in the genus *Tolpis* (Compositae: Cichorieae). - Reading: Unpubl. Ph.D. thesis [n.v.].

JARVIS, C.E. 1992: The Linnaean plant name typification project. - *Bot. J. Linn. Soc.* **109(4)**: 503-513.

JARVIS, C.E. 2007: Order out of Chaos - Linnaean plant names and their types. - London: Linn. Soc. London, Nat. Hist. Mus, 1017 pp.

JARVIS, C.E. & AL. 1993: A list of Linnaean generic names and their types. - *Regnum Veg.* **127**: 100 pp.

JARVIS, C.E. & AL. 2006: Typification of Linnaean plant names in Apiaceae. - *Taxon* **55(1)**: 207-216.

JARVIS, C.E., CAFFERTY, S. & FORREST, L.L. 2001: Typification of Linnaean plant names in Lamiaceae (Labiatae). - *Taxon* **50(2)**: 507-523.

JARVIS, C.E., IVIMEY-COOK, R.B. & CANNON, P.F. 1983: (696) Proposal to reject *Ononis spinosa* Linnaeus (Leguminosae). - *Taxon* **32(2)**: 314-316. [approved, cf. Taxon 34(4): 662; name later conserved with a new type, cf. *Taxon* 38(2): 299-300, 1989].

JARVIS, C.E. & KNEES, S.G. 1988: Linnaean names in the genus *Athamanta* (Umbelliferae: Apioideae) and their typification. - *Taxon* **37(2)**: 472-477.

JARVIS, C.E. & MCCLINTOCK, D. 1990: Notes on the typification of fourteen Linnaean names for European species of *Erica*, *Calluna* and *Andromeda* (Ericaceae). - *Taxon* **39(3)**: 517-520.

JARVIS, C.E., SPENCER, M.A. & CAFFERTY, S. 2005: Typification of Linnaean plant names in Ranunculaceae. - *Taxon* **54(2)**: 467-471.

JARVIS, C.E., STACE, C.A. & WILKINSON, M.J. 1987: Typification of *Festuca rubra* L., *F. ovina* L. and *F. ovina* var. *vivipara* L. - *Watsonia* **16(3)**: 299-302.

JARVIS, C.E. & TURLAND, N.J. (EDS) 1998: Typification of Linnaean specific and varietal names in the Compositae (Asteraceae). - *Taxon* **47(2)**: 347-370.

JASIEWICZ, A. & MIZIANTY, M. 1975: Chromosome numbers of some Bulgarian plants. - *Fragm. Florist. Geobot.* **21(3)**: 277-287 + 1 plate.

JASIEWICZ, A. & PAWLUS, M. 1991: Some remarks on the genus *Scabiosa* (Dipsacaceae). - *Fragm. Florist. Geobot.* **35(1-2)**: 55-60.

JASIŃSKA, A. & AL. 2017: Relationships among *Abies nebrodensis*, *A. alba* and *A. cephalonica* in the morphological and anatomical needle characteristics. - *Pl. Biosyst.* **151(5)**: 775-782.

JAUBERT, H.-L. & SPACH, E. 1842-57: Illustrationes plantarum orientalium ou Choix de plantes nouvelles ou peu connues de l'Asie occidentale. - Paris: 5 vols. [cf. TL-2 2: 427-428, 1979].

JAUZEIN, PH. 1990: Le genre *Typha* en Corse. In Jeanmonod, D. & Burdet, H.M.: Notes et contributions à la flore de Corse, VI. - *Candollea* **45(1)**: 314-334.

JAUZEIN, PH. & TISON, J.-M. 2005: Le complexe d'*Allium ampeloprasum* L. en France. - *Lejeunia, n.s.* **178**: 1-28.

JÁVORKA, S. 1910: Beiträge zur Kenntnis der ungarischen Flora nebst Revision der europaeischen Vertreter des Formenkreises des *Linum flavum* L. - *Magyar Bot. Lapok* **9(5-9)**: 145-163.

JÁVORKA, S. 1918a: Kritische *Calamintha*-Arten. - *Magyar Bot. Lapok* **17(1-12)**: 45-51.

JÁVORKA, S. 1918b: Additamenta nonnulla ad floram bulgaricam. I. *Silene Urumovi* Jáv. n. sp. - *Magyar Bot. Lapok* **17(1-12)**: 69.

JÁVORKA, S. 1919: *Trisetum albanicum* Jáv. spec. nov. - *Magyar Bot. Lapok* **18(1-12)**: 1-2.

JÁVORKA, S. 1921: Plantae novae albanicae. - *Magyar Bot. Lapok* **20(1-12)**: 60-61.

JÁVORKA, S. 1922a: Plantae novae albanicae II. - *Magyar Bot. Lapok* **21(1-12)**: 17-23.

JÁVORKA, S. 1922b: Plantae in insula Creta a Ludovico Biró lectae. - *Magyar Bot. Lapok* **21(1-12)**: 25-26.

JÁVORKA, S. & CSAPODY, V. 1979: Ikonographie der Flora des südöstlichen Mitteleuropa. - Stuttgart: Gustav Fischer Verlag, 704 + 80 pp.

JAVŮRKOVÁ-KRATOCHVÍLOVÁ, V. & TOMŠOVIC, P. 1972: Chromosome study of the genus *Rorippa* Scop. em. Reichenb. in Czechoslovakia. - *Preslia* **44(2)**: 140-156.

JAVŮRKOVÁ, V. 1979: [Chromosome numbers from Czechoslovakia]. In Löve, Á. (ed.): IOPB chromosome number reports LXIV. - *Taxon* **28(4)**: 400-401.

JAVŮRKOVÁ, V. 1980: [Chromosome numbers from Czechoslovakia]. In Löve, Á. (ed.): IOPB chromosome number reports LXIX. - *Taxon* **29(5-6)**: 713-714.

JAVŮRKOVÁ, V. 1981: [Chromosome numbers from Czechoslovakia]. In Löve, Á. (ed.): IOPB chromosome number reports LXXIII. - *Taxon* **30(4)**: 856-857.

JEANMONOD, D. 1984: Révision de la section *Siphonomorpha* Otth du genre *Silene* L. (Caryophyllaceae) en Méditerranée occidentale. III: aggrégat *italica* et espèces affines. - *Candollea* **39(2)**: 549-639.

JEANMONOD, D. 1985a: Révision de la section *Siphonomorpha* Otth du genre *Silene* L. (Caryophyllaceae) en Méditerranée occidentale. IV: Species caeterae. - *Candollea* **40(1)**: 5-34.

JEANMONOD, D. 1985b: Révision de la section *Siphonomorpha* Otth du genre *Silene* L. (Caryophyllaceae) en Méditerranée occidentale. V: Synthèse. - *Candollea* **40(1)**: 35-56.

JEANMONOD, D. 1988: Dipsacaceae. In Jeanmonod, D. & Burdet, H.M. (eds): Compléments au prodrome de la Flore Corse. - Genève: Conservatoire et Jardin Botanique, 40 pp.

JEANMONOD, D. 1992: Scrophulariaceae. In Jeanmonod, D. & Burdet, H.M. (eds): Compléments au Prodrome de la Flore Corse. - Genève: Conservatoire et Jardin Botanique, 234 pp.

JEANMONOD, D. 2003: Le groupe du *Senecio leucanthemifolius* en Corse, avec description d'une nouvelle espèce: *S. serpentinicola* Jeanm. - *Candollea* **58(2)**: 429-459.

JEANMONOD, D. & AL. 2015: Floristic analyses of the Corsican flora: biogeographical origin and endemism. - *Candollea* **70(1)**: 21-41.

JEANMONOD, D. & BOCQUET, G. 1983: Propositions pour un traitement biosystématique du *Silene nutans* L. (Caryophyllaceae). - *Candollea* **38(1)**: 267-295.

JEANMONOD, D., BOCQUET, G. & BURDET, H.M. (EDS) 1986: Notes et contributions à la flore de Corse. - *Candollea* **41(1)**: 1-61.

JEANMONOD, D. & BURDET, H.M. (EDS) 1987: Notes et contributions à la flore de Corse, II. - *Candollea* **42(1)**: 25-95.

JEANMONOD, D. & BURDET, H.M. (EDS) 1988: Notes et contributions à la flore de Corse, III. - *Candollea* **43(1)**: 335-408.

JEANMONOD, D. & BURDET, H.M. (EDS) 1989a: Notes et contributions à la flore de Corse, IV. - *Candollea* **44(1)**: 337-401.

JEANMONOD, D. & BURDET, H.M. (EDS) 1989b: Notes et contributions à la flore de Corse, V. - *Candollea* **44(2)**: 575-637.

JEANMONOD, D. & BURDET, H.M. (EDS) 1990: Notes et contributions à la flore de Corse, VI. - *Candollea* **45(1)**: 261-340.

JEANMONOD, D. & BURDET, H.M. (EDS) 1991: Notes et contributions à la flore de Corse, VII. - *Candollea* **46(1)**: 175-226.

JEANMONOD, D. & BURDET, H.M. (EDS) 1992: Notes et contributions à la flore de Corse, VIII. - *Candollea* **47(2)**: 267-318.

JEANMONOD, D. & BURDET, H.M. (EDS) 1993: Notes et contributions à la flore de Corse, IX. - *Candollea* **48(2)**: 525-591.

JEANMONOD, D. & BURDET, H.M. (EDS) 1994: Notes et contributions à la flore de Corse, X. - *Candollea* **49(2)**: 571-612.

JEANMONOD, D. & BURDET, H.M. (EDS) 1995: Notes et contributions à la flore de Corse, XI. - *Candollea* **50(2)**: 553-583.

JEANMONOD, D. & BURDET, H.M. (EDS) 1996: Notes et contributions à la flore de Corse, XII. - *Candollea* **51(2)**: 515-557.

JEANMONOD, D. & BURDET, H.M. (EDS) 1997: Notes et contributions à la flore de Corse, XIII. - *Candollea* **52(2)**: 239-279.

JEANMONOD, D. & BURDET, H.M. (EDS) 1998: Notes et contributions à la flore de Corse, XIV. - *Candollea* **53(2)**: 171-210.

JEANMONOD, D. & BURDET, H.M. (EDS) 1999: Notes et contributions à la flore de Corse, XV. - *Candollea* **54(2)**: 385-416.

JEANMONOD, D. & GAMISANS, J. 1987: Compléments au Prodrome de la Flore Corse. Annexe no. 1. Introduction. - Genève: Conservatoire et Jardin Botanique, 28 pp. + maps.

JEANMONOD, D., HABASHI, CH. & MANEN, J.-F. 2005: *Orobanche cyrnea* Jeanm., Habashi & Manen, une nouvelle espèce endémique de Corse. - *Candollea* **60(1)**: 255-270.

JEANMONOD, D. & LAMBINON, J. 1994: La flore de la Corse dans la deuxième édition de "Flora Europaea": commentaires et corrections. - *Candollea* **49(2)**: 527-541.

JEANMONOD, D. & NACIRI, Y. 2021: Révision taxonomique de *Ranunculus* sect. *Batrachium* en Corse: une analyse morphologique et statistique. - *Candollea* **76(2)**: 275-292.

JEANMONOD, D., SCHLÜSSEL, A. & GAMISANS, J. 2004: Compléments au Prodrome de la Flore Corse. Asteraceae II. - Genève: Conservatoire et Jardin Botanique, 256 pp.

JEANMONOD, D., SCHLÜSSEL, A. & GAMISANS, J. 2011a: Status and trends in the alien flora of Corsica. - *EPPO Bulletin* **41**: 85-99.

JEANMONOD, D., SCHLÜSSEL, A. & GAMISANS, J. 2011b: Analyse de la flore Corse: aspects biologiques. - *Candollea* **66(1)**: 5-25.

JEANMONOD, D. & SCHLÜSSEL, A. (EDS) 2001a: Notes et contributions à la flore de Corse, XVII. - *Candollea* **56(1)**: 139-170.

JEANMONOD, D. & SCHLÜSSEL, A. (EDS) 2001b: Notes et contributions à la flore de Corse, XVIII. - *Candollea* **56(2)**: 327-362.

JEANMONOD, D. & SCHLÜSSEL, A. (EDS) 2003: Notes et contributions à la flore de Corse, XIX. - *Candollea* **58(2)**: 273-287.

JEANMONOD, D. & SCHLÜSSEL, A. (EDS) 2004: Notes et contributions à la flore de Corse, XX. - *Candollea* **59(1)**: 65-94.

JEANMONOD, D. & SCHLÜSSEL, A. (EDS) 2006: Notes et contributions à la flore de Corse, XXI. - *Candollea* **61(1)**: 93-134.

JEANMONOD, D. & SCHLÜSSEL, A. (EDS) 2008: Notes et contributions à la flore de Corse, XXII. - *Candollea* **63(1)**: 131-151.

JEANMONOD, D. & SCHLÜSSEL, A. (EDS) 2010: Notes et contributions à la flore de Corse, XXIII. - *Candollea* **65(2)**: 267-290.

JEANMONOD, D. & SCHLÜSSEL, A. (EDS) 2012: Notes et contributions à la flore de Corse, XXIV. - *Candollea* **67(2)**: 239-321.

JEANMONOD, D. (ED.) 2000: Notes et contributions à la flore de Corse, XVI. - *Candollea* **55(1)**: 41-74.

JEANMONOD, D. (ED.) 2015: Notes à la flore de Corse, XXV. - *Candollea* **70(1)**: 109-140.

JEANPERT, E. 1919: Enumération des plantes de Macédoine. - *Bull. Mus. Natl. Hist. Nat.* **25**: 390-397, 517-523, 662-668.

JEFFREY, C. 1969: A review of the genus *Bryonia* L. (Cucurbitaceae). - *Kew Bull.* **23(3)**: 441-461.

JEFFREY, C. 1974: *Echinops ruthenicus*. - *Bot. Mag.* **180(2)**: tab. 677 + pp. 77-81.

JEFFREY, C. 1979: Note on the lectotypification of the names *Cacalia* L., *Matricaria* L. and *Gnaphalium* L. - *Taxon* **28(4)**: 349-351.

JEHLÍK, V. & ROSTAŃSKI, K. 1980: Beitrag zur Taxonomie, Ökologie und Chorologie der *Oenothera*-Arten in der Tschechoslowakei. - *Folia Geobot. Phytotax.* **14(4)**: 377-429.

JELENEVSKY, A.G. 1971: De *Veronica spicata* L. s.l. notulae systematicae - *Novosti Sist. Vysš. Rast.* **8**: 215-227. [In Russian; English translation of this volume by Amerind Publ. Co., New Delhi, 1978].

JELENEVSKY, A.G. & KURANOVA, N.G. 2000: Generis *Cruciata* Mill. (Rubiaceae) revisio taxonomica [In Russian]. - *Novosti Sist. Vysš. Rast.* **32**: 152-162.

JENSEN, K.B. & CHEN, S.-L. 1992: An overview: Systematic relationships of *Elymus* and *Roegneria* (Poaceae). - *Hereditas (Lund)* **116(1-2)**: 127-132.

JERMY, A.C. 1984: Origin and distribution of pteridophytes in the Mediterranean area. - *Webbia* **38**: 397-416.

JERMY, A.C. 1985: Ferns in Europe: some advances since 1965. - *Pteridologist* **1(2)**: 73-77.

JERMY, A.C., CHATER, A.O. & DAVID, R.W. 1982: Sedges of the British Isles. BSBI Handbook no. 1. - Torquay, Devon: Devonshire Press, 268 pp.

JERMY, A.C., CRABBE, J.A. & THOMAS, B.A. (EDS) 1973: The phylogeny and classification of the ferns. - *Bot. J. Linn. Soc.* **67, suppl. 1**: xiv + 284 pp.

JERMY, A.C. & HARPER, L. 1971: Spore morphology of the *Cystopteris fragilis* complex. - *Brit. Fern Gaz.* **10(4)**: 211-213.

JERMY, A.C. & JARVIS, C.E. 1992: The lectotypification of *Asplenium trichomanes-ramosum* L., the correct name for *A. viride* Hudson. In Newton, M.E. (ed.): Flora Europaea: Notulae systematicae ad Floram Europaeam spectantes. Series 2. No. 5. - *Bot. J. Linn. Soc.* **109(3)**: 319-321.

JERMY, A.C. & PAGE, C.N. 1980: Additional field characters separating the subspecies of *Asplenium trichomanes* in Britain. - *Brit. Fern Gaz.* **12(2)**: 112-113.

JERMY, A.C. & TUTIN, T.G. 1968: British sedges. A handbook to the species of *Carex* found growing in the British Isles. - London: Botanical Society of the British Isles, 199 pp.

JERMY, A.C. & VIANE, R. 1992: New subgenera within *Asplenium* L. In Newton, M.E. (ed.): Flora Europaea: Notulae systematicae ad Floram Europaeam spectantes. Series 2. No. 5. - *Bot. J. Linn. Soc.* **109(3)**: 321-322.

JERRENTRUP, H. & RESCH, J. 1989: Der Nestos. Leben zwischen Fluss und Meer. - Radolfzell: Verlag Jürgen Resch, 127 pp.

JESSEN, S. 1995: *Asplenium trichomanes* L. subsp. *hastatum*, stat. nov. - eine neue Unterart des Braunstiel-Streifenfarnes in Europa und vier neue intraspezifische Hybriden (Aspleniaceae, Pteridophyta). - *Ber. Bayer. Bot. Ges.* **65**: 107-132.

JIMÉNEZ-MEJÍAS, P. & AL. 2014: Disentangling the taxonomy of *Carex acuta* s.l. in the Mediterranean basin and the Middle East: Re-evaluation of *C. panormitana* Guss. and *C. kurdica* Kük. ex Hand.- Mazz. - *Pl. Biosyst.* **148(1)**: 64-73.

JIMÉNEZ-MEJÍAS, P. & AL. 2017: Typification of 18th century names in *Carex* sect. *Rhynchocystis* (Cyperaceae): *Carex pendula* and allies. - *Taxon* **66(4)**: 973-975.

JIMÉNEZ-MEJÍAS, P. & LUCEÑO, M. 2009: *Carex castroviejoi* Luceño & Jiménez-Mejías (Cyperaceae), a new species from north Greek mountains. - *Acta Bot. Malac.* **34**: 1-3.

JIRÁSEK, V. 1952: Phytogeographical-systematic study of the genus *Eragrostis* P. Beauv. [In Czech with Russian and English summaries]. - *Preslia* **24**: 281-338.

JIRÁSEK, V. 1954: Příspěvek k systematice k taxonomii československých pýrů *Agropyrum* Gaertn. [In Czech with Russian summary]. - *Preslia* **26**: 159-176.

JIRÁSEK, V. 1964: Einige taxonomische Probleme im Komplex der *Poa pratensis* L. s.l. - *Acta Horti Bot. Prag.* **1963**: 60-68.

JIRÁSEK, V. 1965: Einige taxonomische Probleme der Gattung *Molinia* Schrank (Poaceae). - *Acta Univ. Carol., Biol.* **1965(3)**: 227-243.

JIRÁSEK, V. 1966: Über die systematische Einordnung der Gattung *Molinia* Schrank (Poaceae). - *Preslia* **38**: 23-35.

JIRÁSEK, V. & CHRTEK, J. 1964: Zur Frage des taxonomischen Wertes der Gattung *Cynosurus* L. - *Novit. Bot. Delect. Seminum Horti Bot. Univ. Carol. Prag.* **1964**: 23-27.

JIRÁSEK, V. & CHRTEK, J. 1966: *Brizochloa*, eine neue Grasgattung. - *Novit. Bot. Delect. Seminum Horti Bot. Univ. Carol. Prag.* **1966**: 39-41.

JOACHIMIAK, A. & KULA, A. 1993: Cytotaxonomy and karyotype evolution in *Phleum* sect. *Phleum* (Poaceae) in Poland. - *Pl. Syst. Evol.* **188(1-2)**: 17-30.

JOACHIMIAK, A. & KULA, A. 1996: Karyosystematics of the *Phleum alpinum* polyploid complex (Poaceae). - *Pl. Syst. Evol.* **203(1-2)**: 11-25.

JOACHIMIAK, A. & KULA, A. 1997: Systematics and karyology of the section *Phleum* in the genus *Phleum*. - *J. Appl. Genet.* **38(4)**: 463-470.

JOHNSON, B.L. 1945: Cyto-taxonomic studies in *Oryzopsis*. - *Bot. Gaz.* **107(1)**: 1-32.

JOHNSON, D.M. 1988: (911) Proposal to conserve *Marsilea* L. (Pteridophyta: Marsileaceae) with *Marsilea quadrifolia* as typ. conserv. - *Taxon* **37(2)**: 483-486. [approved, cf. Taxon 42(1): 120, 1993].

JOHNSON, M. 1997: Släktet Klematis. - Södertälje: Magnus Johnsons Plantskola AB, xiv + 881 pp. [In Swedish].

JOHNSON, M. 2001: The genus *Clematis*. - Södertälje: Magnus Johnsons Plantskola AB, xiv + 896 pp.

JOHNSON, M.A.T. 1982: Karyotypes of some Greek species of *Allium*. - *Ann. Mus. Goulandris* **5**: 107-119.

JOHNSON, M.A.T. 1994: Cytology of three geophytes from Turkey. - *Kew Bull.* **49(3)**: 491-498.

JOHNSON, M.A.T. 2003: Polyploidy and karyotype variation in Turkish *Bellevalia* (Hyacinthaceae). - *Bot. J. Linn. Soc.* **143(1)**: 87-98.

JOHNSON, M.A.T. & BRANDHAM, P.E. 1997: New chromosome numbers in petaloid monocotyledons and in other miscellaneous angiosperms. - *Kew Bull.* **52(1)**: 121-138.

JOHNSON, M.A.T., GARBARI, F. & MATHEW, B.F. 1991: A cytotaxonomical approach to species delimitation of a group of early-flowering Turkish *Ornithogalum* species (Hyacinthaceae). - *Bot. Chron. (Patras)* **10**: 827-839.

JOHNSON, M.A.T. & GÜNER, A. 2002: *Iris stenophylla* Hausskn. & Siehe ex Baker from Turkey and its cytology. - *Bot. J. Linn. Soc.* **140(2)**: 115-127.

JOHNSON, M.A.T. & ÖZHATAY, N. 1996: Cytology of *Allium* section *Allium*. In Mathew, B.F.: A review of *Allium* section *Allium*, pp. 17-40. - Kew: Royal Botanic Gardens.

JOHNSON, M.A.T., ÖZHATAY, N. & GARBARI, F. 1996: The genus *Muscari* (Hyacinthaceae) in Turkey: taxonomy, distribution and chromosome analysis. In Öztürk, M.A., Seçmen, Ö. & Görk, G. (eds): Plant life in southwest and central Asia, 1: pp. 34-53. - Izmir: Ege University Press.

JOHNSSON, H. 1941: Cytological studies in the genus *Alopecurus*. - *Acta Univ. Lund., 2* **37(3)**: 1-43 + 2 plates.

JOHNSTON, I.M. 1952: Studies in the Boraginaceae, XXIII. A survey of the genus *Lithospermum*. - *J. Arnold Arbor.* **33(4)**: 299-366.

JOHNSTON, I.M. 1954: Studies in the Boraginaceae, XXVI. Further revaluations of the genera of the Lithospermeae. - *J. Arnold Arbor.* **35(1)**: 1-81.

JOHNSTONE, R. 1982: Rhodes supplement. - *Cyclamen Soc. J.* **6(1)**: 18-19.

JOHNSTONE, R. 1986: Plant portraits: *Fritillaria rhodokanakis*. - *Rock Gard.* **19(4)**: 383-384.

JOHNSTONE, R. 1992: The society's Peloponnese expedition, April 1992. - *Cyclamen* **16(2)**: 45-47.

JOKSIMOVIĆ, Ž. 1975: Neuer Fundort der relikten Art *Eriophorum latifolium* in Serbien [In Serbo-Croatian with English summary]. - *Glasn. Prir. Mus. Beogradu, Ser. B, Biol. Nauke* **30**: 177-181.

JONES, A.W. 1989: *Erica manipuliflora* Salisb. in southern Jugoslavia, October 1988. - *Year Book Heather Soc.* **3(7)**: 36-44.

JONES, B.M.G. 1964: [Notes on *Arabis*]. In Heywood, V.H. (ed.): Notulae systematicae ad Floram Europaeam spectantes, no. 3. - *Feddes Repert. Spec. Nov. Regni Veg.* **69(1)**: 59-61.

JONES, D.A. & TURKINGTON, R. 1986: Biological Flora of the British Isles. *Lotus corniculatus* L. - *J. Ecol.* **74(4)**: 1185-1212.

JONES, E.W. 1944: Biological Flora of the British Isles. *Acer* L. - *J. Ecol.* **32(2)**: 215-252 + 1 plate.

JONES, E.W. 1959: Biological Flora of the British Isles. *Quercus* L. - *J. Ecol.* **47(1)**: 169-222.

JONES, G. & AL. 2000: Distinguishing the effects of agricultural practices relating to fertility and disturbance: a functional ecological approach in archaeobotany. - *J. Archaeol. Sci.* **27**: 1073-1084.

JONES, K. 1958: Cytotaxonomic studies in *Holcus*. I. The chromosome complex in *Holcus mollis* L. - *New Phytol.* **57(2)**: 191-210 + 1 plate.

JONES, K. 1960: The typification of the genus *Agropyron* Gaertn. - *Taxon* **9(2)**: 55-56.

JONES, K. 1962: Chromosomal status, gene exchange and evolution in *Dactylis*. 2. The chromosomal analysis of diploid, tetraploid and hexaploid species and hybrids. - *Genetica* **32**: 272-296.

JONES, K. 1964: Chromosomes and the nature and origin of *Anthoxanthum odoratum* L. - *Chromosoma* **15(3)**: 248-274.

JONES, K. & BORRILL, M. 1962: Chromosomal status, gene exchange and evolution in *Dactylis*. 3. The rôle of the inter-ploid hybrids. - *Genetica* **32**: 296-322.

JONES, K. & BRANDHAM, P.E. (EDS) 1976: Current chromosome research. Proceedings of the Kew Chromosome Conference held at the Jodrell Laboratory, the Royal Botanic Gardens, Kew, England, July 1st-3rd, 1976. - Amsterdam, etc.: North-Holland, xv + 240 pp.

JONES, K. & CARROLL, C.P. 1962: Cytotaxonomic studies in *Holcus* L. II. Morphological relationships in *Holcus mollis* L. - *New Phytol.* **61(1)**: 63-71 + 1 plate.

JONES, K., CARROLL, C.P. & BORRILL, M. 1961: A chromosome atlas of the genus *Dactylis* L. - *Cytologia* **26(3-4)**: 333-343.

JONES, V. & RICHARDS, P.W. 1954: Biological Flora of the British Isles. *Juncus acutus* L. - *J. Ecol.* **42 (2)**: 639-650.

JONES, V. & RICHARDS, P.W. 1956: Biological Flora of the British Isles. *Saxifraga oppositifolia* L. - *J. Ecol.* **44(1)**: 300-316.

JONG, P.C. DE 1976: Flowering and sex expression in *Acer* L. A biosystematic study. - *Meded. Landbouwhogeschool* **76(2)**: 1-201.

JONKER, F.P. 1950: Revisie van de Nederlandse Gentianaceae. I. *Centaurium* Hill [In Dutch]. - *Ned. Kruidk. Arch.* **57**: 169-198.

JONSELL, B. 1968: Studies in the north-west European species of *Rorippa* s. str. - *Symb. Bot. Upsal.* **19(2)**: 1-221 + 11 plates.

JONSELL, B. 1969: *Rorippa islandica* und *R. palustris* - zwei scharf getrennte Arten der Alpenländerflora. - *Ber. Geobot. Inst. E.T.H. Stiftung Rübel* **39**: 52-55.

JONSELL, B. 1975: Hybridization in yellow-flowered European *Rorippa* species. In Walters, S.M. & King, C.J. (eds): European floristic and taxonomic studies, pp. 101-110. - Oxford: E.W. Classey Ltd.

JONSELL, B. 1976: Taxonomy and distribution of *Rorippa* (Cruciferae) in southern U.S.S.R. - *Svensk Bot. Tidskr.* **67**: 281-302.

JONSELL, B. 1978: Nomenclatural and taxonomic notes on *Trisetum* Pers. and *Lophochloa* Reichenb. (Gramineae). In Heywood, V.H. (ed.): Notulae systematicae ad Floram Europaeam spectantes, no. 20. - *Bot. J. Linn. Soc.* **76(4)**: 320-322.

JONSELL, B. & JARVIS, C.E. 1994: Lectotypification of Linnaean names for Flora Nordica Vol. 1 (Lycopodiaceae - Papaveraceae). - *Nordic J. Bot.* **14(2)**: 145-164.

JONSELL, B. & JARVIS, C.E. 2002: Lectotypification of Linnaean names for Flora Nordica (Brassicaceae-Apiaceae). - *Nordic J. Bot.* **22(1)**: 67-86.

JOPE, M. 1998: *Cyclamen hederifolium* var. *confusum* in western Crete - some further observations. - *Cyclamen* **22(2)**: 48 + 4 plates.

JOPE, M. 2003: *Cyclamen hederifolium* var. *confusum* on Crete - an update. - *Cyclamen* **27(2)**: 64-65.

JORDAN, J. 2006: A new site for *Anchusa aegyptiaca* [In Greek and English]. - *Fisis [Journal of the Hellenic Society for the Protection of Nature]* **113**: 29.

JORDAN, J. 2007: Enas agnostos kipos ston Imitto [In Greek]. - *Fisis [Journal of the Hellenic Society for the Protection of Nature]* **116**: 24-27.

JORDAN, J. 2009: [Botanical data] in Standring, K. & al. (eds): Prespa walking guide. - Society for the Protection of Prespa, 203 pp. + maps.

JORDANOV, D. 1924: Über die Phytogeographie des westlichen Balkans [In Bulgarian with German summary]. - *God. Sofiisk. Univ. Fiz.-Mat. Fak.* **20(1)**: 1-104.

JORDANOV, D. 1929: Bemerkungen über die Flora Bulgariens. - *Izv. Bulg. Bot. Druzh.* **3**: 65-75.

JORDANOV, D. & AL. (EDS) 1963: Flora Reipublicae Popularis Bulgaricae. I [Pteridophyta to Gramineae]. - Serdicae [Sofia]: In Aedibus Academiae Scientiarum Bulgaricae, 508 pp. + 1 map.

JORDANOV, D. & AL. (EDS) 1964: Flora Reipublicae Popularis Bulgaricae. II [Cyperaceae to Orchidaceae]. - Serdicae [Sofia]: In Aedibus Academiae Scientiarum Bulgaricae, 425 pp.

JORDANOV, D. & AL. (EDS) 1966: Flora Reipublicae Popularis Bulgaricae. III [Juglandaceae to Amaranthaceae]. - Serdicae [Sofia]: In Aedibus Academiae Scientiarum Bulgaricae, 637 pp. + 1 map.

JORDANOV, D. & AL. (EDS) 1970: Flora Reipublicae Popularis Bulgaricae. IV [Cactaceae to Saxifragaceae]. - Serdicae [Sofia]: In Aedibus Academiae Scientiarum Bulgaricae, 748 pp. + 1 map.

JORDANOV, D. & AL. (EDS) 1973: Flora Reipublicae Popularis Bulgaricae. V [Rosaceae]. - Serdicae [Sofia]: In Aedibus Academiae Scientiarum Bulgaricae, 443 pp. + 1 map.

JORDANOV, D. & AL. (EDS) 1975: Problems of Balkan flora and vegetation. Proceedings of the first international symposium on Balkan flora and vegetation. Varna, June 7-14, 1973. - Sofia: Publishing House of the Bulgarian Academy of Sciences, 441 pp.

JORDANOV, D. & AL. (EDS) 1976: Flora Reipublicae Popularis Bulgaricae. VI [Leguminosae]. - Serdicae [Sofia]: In Aedibus Academiae Scientiarum Bulgaricae, 591 pp. + 1 map.

JORDANOV, D. & AL. (EDS) 1979: Flora Reipublicae Popularis Bulgaricae. VII [Oxalidaceae to Araliaceae]. - Serdicae [Sofia]: In Aedibus Academiae Scientiarum Bulgaricae, 530 pp. + 1 map.

JORDANOV, D. & AL. (EDS) 1982: Flora Reipublicae Popularis Bulgaricae. VIII [Umbelliferae to Hydrophyllaceae]. - Serdicae [Sofia]: In Aedibus Academiae Scientiarum Bulgaricae, 518 pp. + 1 map.

JORDANOV, D. & AL. (EDS) 1989: Flora Reipublicae Popularis Bulgaricae. IX [Rubiaceae to Solanaceae]. - Serdicae [Sofia]: In Aedibus Academiae Scientiarum Bulgaricae, 539 pp. + 1 map.

JORDANOV, D. & MARKOVA, M.L. 1971: New habitat for *Symphyandra wanneri* Heuff. in Bulgaria [In Bulgarian]. - *Izv. Bot. Inst. (Sofia)* **20**: 231.

JORDANOVIĆ, S. & AL. 2009: Genus *Leucojum* L. (Amaryllidaceae) - distribution and threatened status in Serbia. - *Bot. Serbica* **33(1)**: 45-50.

JORDON-THADEN, I. & AL. 2010: Molecular phylogeny and systematics of the genus *Draba* s. l. (Brassicaceae) and identification of its closest related genera. - *Molec. Phyl. Evol.* **55**: 524-540.

JØRGENSEN, R.B. 1982: Biosystematics of *Hordeum bulbosum* L. - *Nordic J. Bot.* **2(5)**: 421-434.

JOSIFOVIĆ, M. (ED.) 1970a: Flora SR Srbije. Flore de la Republique Socialiste de Serbie. Vol. 1 [Lycopodiaceae to Ranunculaceae; in Serbo-Croatian]. - Beograd: Srbska Akademija Nauka, 326 pp.

JOSIFOVIĆ, M. (ED.) 1970b: Flora SR Srbije. Flore de la Republique Socialiste de Serbie. Vol. 2 [Berberidaceae to Caryophyllaceae; in Serbo-Croatian]. - Beograd: Srbska Akademija Nauka, 294 pp.

JOSIFOVIĆ, M. (ED.) 1972a: Flora SR Srbije. Flore de la Republique Socialiste de Serbie. Vol. 3 [Amarantaceae to Thymelaeaceae; in Serbo-Croatian]. - Beograd: Srbska Akademija Nauka, 599 pp.

JOSIFOVIĆ, M. (ED.) 1972b: Flora SR Srbije. Flore de la Republique Socialiste de Serbie. Vol. 4 [Spiraeaceae to Fabaceae; in Serbo-Croatian]. - Beograd: Srbska Akademija Nauka, 584 pp.

JOSIFOVIĆ, M. (ED.) 1973: Flora SR Srbije. Flore de la Republique Socialiste de Serbie. Vol. 5 [Lythraceae to Cuscutaceae; in Serbo-Croatian]. - Beograd: Srbska Akademija Nauka, 640 pp.

JOSIFOVIĆ, M. (ED.) 1974: Flora SR Srbije. Flore de la Republique Socialiste de Serbie. Vol. 6 [Boraginaceae to Campanulaceae]. - Beograd: Srbska Akademija Nauka, 599 pp.

JOSIFOVIĆ, M. (ED.) 1975: Flora SR Srbije. Flore de la Republique Socialiste de Serbie. Vol. 7 [Asteraceae to Asparagaceae]. - Beograd: Srbska Akademija Nauka, 651 pp.

JOSIFOVIĆ, M. (ED.) 1976: Flora SR Srbije. Flore de la Republique Socialiste de Serbie. Vol. 8 [Iridaceae to Typhaceae; in Serbo-Croatian]. - Beograd: Srbska Akademija Nauka, 515 pp.

JOSIFOVIĆ, M. (ED.) 1977: Flora SR Srbije. Flore de la Republique Socialiste de Serbie. Vol. 9 [Dodatak. Supplement; in Serbo-Croatian]. - Beograd: Srbska Akademija Nauka, 257 pp.

JOVANOVIĆ, M. & AL. 2018: Taxonomy of yellow flowered *Sempervivum* species in Balkan Peninsula: Insight based on evidence on leaf epidermal structures. - *Bot. Serbica* **42 (suppl. 1)**: 53.

JOVANOVIĆ, M. & AL. 2019: Morphological variability of endemic *Sempervivum ciliosum* subsp. *ciliosum* (Crassulaceae) from the Balkan Peninsula. - *Biol. Nyssana* **10(2)**: 113-123.

JOVANOVIĆ, M. & RAUS, TH. 2018: Notes on *Opuntia stricta*. In Raab-Straube, E. v. & Raus, Th. (eds): Euro+Med-Checklist notulae, 9. - *Willdenowia* **48(2)**: 199.

JOVET, P. & GUÉDÈS, M. 1972: *Paspalum distichum* L. and *P. paspalodes* (Michx.) Scribn. - *Taxon* **21 (4)**: 546.

JOVSANCEVIĆ, M. 1976: *Quercus Macedonica* A. DC. (syn. *Qu. Trojana* Webb.) dans la flore Yougoslave. In: Kazdaği göknari ve Türkiye florasi uluslararasi simpozyomu bildirileri, pp. 151-163 + 1 foldout map. - Istanbul Üniversitesi Orman Fakültesi Yayinlari.

JOYCE, D. (ED.) 2000: Photographic competition 1999. - *Quart. Bull. Alpine Gard. Soc. Gr. Brit.* **68(2)**: 191-200.

JUAN, A., CRESPO, M.B. & RÍOS, S. 2003: Remarks on *Medicago citrina* (sect. *Dendrotelis*, Leguminosae). - *Fl. Medit.* **13**: 303-316.

JUAN, R. & PASTOR, J. 1993: Contribución al estudio cariológico del género *Euphorbia* L., subgén. *Esula* Pers. - *Bol. Soc. Brot., Scr. 2* **66**: 41-54.

JUAN, R., PASTOR, J. & FERNÁNDEZ, I. 1998a: Morfología y anatomía de frutos y semillas en el género *Kickxia* Dumort. - *Lagascalia* **20(2)**: 211-222.

JUAN, R., PASTOR, J. & FERNÁNDEZ, I. 1998b: Estudio de frutos y semillas en *Parentucellia* Viv. (Scrophulariaceae) y sus implicaciones taxonómicas. - *Acta Bot. Malac.* **23**: 51-57.

JULIÁ BERRUEZO, M.A. & MONTSERRAT MARTÍ, J.P. 1999: Dos táxones nuevos en el género *Puccinellia* (Poaceae). - *Fontqueria* **53**: 3-4.

JUNIKKA, L., UOTILA, P. & LAHTI, T. 2006: A phytogeographical comparison of the major Mediterranean islands on the basis of Atlas Florae Europaeae. - *Willdenowia* **36(1)**: 379-388.

JUNIPER, B.E., ROBBINS, R.J. & JOEL, D.M. 1989: Carnivorous plants. - London, etc.: Academic Press, xii + 353 pp.

JURUKOVA-GRĂNČAROVA, P.D. 1983a: Comparative embryological study of two species of genus *Crepis* L. [In Bulgarian with English abstract]. In Velčev, V.I. & al. (eds): Third national conference of botany, pp. 258-264. - Sofia: Publishing House of the Bulgarian Academy of Sciences.

JURUKOVA-GRĂNČAROVA, P.D. 1983b: Embryological investigation of *Urospermum picroides* (L.) Scop. ex Schm. [In Bulgarian with English abstract]. In Velčev, V.I. & al. (eds): Third national conference of botany, pp. 265-269. - Sofia: Publishing House of the Bulgarian Academy of Sciences.

JURY, S.L. 1987: A new species of the genus *Torilis* Adanson (Apiaceae). - *Bot. J. Linn. Soc.* **95(4)**: 293-299.

JURY, S.L. 1996: Pollination and dispersal in Mediterranean umbellifers. - *Bocconea* **5(1)**: 193-199.

JURY, S.L. 1996a: *Thapsia* in the western Mediterranean. - *Lagascalia* **18(2)**: 280-281.

JURY, S.L. 1996b: A new subspecies of *Torilis arvensis* (Hudson) Link. - *Lagascalia* **18(2)**: 282-285.

KABBADAS, D.S. 1938: I chloris tis Ellados, ikonografimeni ke meta dichotomikon klidon. 1. Pteridofita, Gimnosperma ke Monokotila [In Greek]. - Thessaloniki: Tipis Korneliou Theodoridou, 384 pp.

KADEREIT, G. & AL. 2012: Cryptic taxa should have names: Reflections in the glasswort genus *Salicornia* (Amaranthaceae). - *Taxon* **61(6)**: 1227-1239.

KADEREIT, G. & FREITAG, H. 2011: Molecular phylogeny of Camphorosmeae (Camphorosmoideae, Chenopodiaceae): Implications for biogeography, evolution of C_4-photosynthesis and taxonomy. - *Taxon* **60(1)**: 51-78.

KADEREIT, J.W. 1984a: Studies on the biology of *Senecio vulgaris* L. ssp. *denticulatus* (O.F. Muell.) P. D. Sell. - *New Phytol.* **97(4)**: 681-689.

KADEREIT, J.W. 1984b: The origin of *Senecio vulgaris* (Asteraceae). - *Pl. Syst. Evol.* **145(1-2)**: 135-153.

KADEREIT, J.W. 1986a: A revision of *Papaver* section *Argemonidium*. - *Notes Roy. Bot. Gard. Edinburgh* **44(1)**: 25-43.

KADEREIT, J.W. 1986b: A revision of *Papaver* L. sect. *Papaver* (Papaveraceae). - *Bot. Jahrb. Syst.* **108(1)**: 1-16.

KADEREIT, J.W. 1986c: *Papaver somniferum* L. (Papaveraceae): A triploid hybrid? - *Bot. Jahrb. Syst.* **106(2)**: 221-244.

KADEREIT, J.W. 1987a: The taxonomy, distribution and variability of the genus *Roemeria* Medic. (Papaveraceae). - *Flora, Morphol. Geobot. Oekophysiol.* **179(2)**: 135-153.

KADEREIT, J.W. 1987b: Experimental evidence on the affinities of *Papaver somniferum* (Papaveraceae). - *Pl. Syst. Evol.* **156(3-4)**: 189-195.

KADEREIT, J.W. 1989a: *Papaver setigerum* DC. In Greuter, W. & Raus, Th. (eds): Med-Checklist notulae, 15. - *Willdenowia* **19(1)**: 38.

KADEREIT, J.W. 1989b: A revision of *Papaver* L. section *Rhoeadium* Spach. - *Notes Roy. Bot. Gard. Edinburgh* **45(2)**: 225-286.

KADEREIT, J.W. 1990: Some suggestions on the geographical origin of the central, west and north European synanthropic species of *Papaver* L. - *Bot. J. Linn. Soc.* **103(3)**: 221-231.

KADEREIT, J.W. 1995: The genus *Papaver* in the Mediterranean area. In Valdés, B. & Silvestre, S. (eds): Abstracts. VIII OPTIMA meeting, Sevilla, Spain, 25 September - 1 October, 1995, p. 19.

KADEREIT, J.W. 1997: The genus *Papaver* L. in the Mediterranean area. - *Lagascalia* **19(1-2)**: 83-91.

KADEREIT, J.W., LAUX. P. & DILLENBERGER, M.S. 2021: A conspectus of *Tephroseris* (Asteraceae: Senecioneae) in Europe outside Russia and notes on the decline of the genus. - *Willdenowia* **51(2)**: 271-317.

KAGIAMBAKI, A.N. 2011: Contemporary phytogeographical analysis in the Central and Southern Aegean archipelago. [In Greek with English summary]. - Iraklion: Ph.D. thesis, University of Crete, 298 pp.

KAILIDIS, D.S. & MARKALAS, S.D. 1980: Waldbrände in Griechenland. - *Forst- Holzwirt.* **35(20)**: 411-413.

KAILIS, N & ELEFTHERIADOU, E. 2011: Contribution to the description and distribution of *Salix ×velchevii* (Salicaceae). - *Phytol. Balcan.* **17(3)**: 279-282.

KAJAN, E. 1984: Osterurlaub 1983 auf der Sonneninsel Rhodos. - *Ber. Arbeitskreis. Heimische Orchid.* **1(1)**: 71-75.

KAJAN, E. 2003: Pflanzen Griechenlands - Festland und Peloponnes. - Eching bei München: IHW- Verlag, 270 pp.

KAKIAMPAKI, A. & AL. 2011: Factors affecting plant species richness and endemism in the South Aegean (Greece). - *J. Biol. Res. Thessalon.* **16**: 282-295.

KALHEBER, H. 1991: Unpublished list of collections and field notes, mainly from Evvia (1990). - Photocopy: 74 pp.

KALHEBER, H. 1992: Unpublished list of collections and field notes from Crete (1989). - Photocopy: 66 pp.

KALHEBER, H. 1993: Index ad Iconographiam Florae Europaeae. Heft 1: Pteridophyta, Gymnospermae, Dicotyledones (Acanthaceae-Cneoraceae). - Frankfurt a.M.: Senckenbergische Naturforschende Gesellschaft, 164 + 14 pp.

KALHEBER, H. 1994a: Index ad Iconographiam Florae Europaeae. Heft 3: Dicotyledones (Convolvulaceae-Labiatae). - Frankfurt a.M.: Senckenbergische Naturforschende Gesellschaft, 187 pp.

KALHEBER, H. 1994b: The genus *Alchemilla* L. (Rosaceae) in the Turkish vilayet Rize (northeastern Anatolia) with some remarks on the distribution of the genus in other parts of northern Anatolia. - *Sendtnera* **2**: 389-430.

KALHEBER, H. 1995: Fundliste Amorgos, 4.4.-11.4. 1994. - Mimeographed list, 18 pp.

KALHEBER, H. 2000: Ikaria 5.-11. Juni 1999. - Mimeographed list, 14 pp.

KALHEBER, H. 2001: Index ad Iconographiam Florae Europaeae. Heft 4: Dicotyledones, Lauraceae-Rhamnaceae und Cucurbitaceae (Nachtrag zu Heft 3). - Frankfurt a.M.: Senckenbergische Naturforschende Gesellschaft, 294 pp.

KALHEBER, H. 2006: Bavarian plant collectors in Greece - 1. Franz Xaver Berger, Franz Zuccarini and Carl Nikolaus Fraas. - *Willdenowia* **36(1)**: 565-578.

KALHEBER, H. 2014: [Note on *Legousia*]. In Raab-Straube, E. v. & Raus, Th. (eds): Euro+Med-Checklist Notulae, 3. - *Willdenowia* **44(2)**: 289.

KALIS, N. & ELEFTHERIADOU, E. 2013: On the occurrence of *Salix appendiculata* (Salicaceae) in Greece. - *Willdenowia* **43(2)**: 257-261.

KALLIMANIS, A.S. & AL. 2009: Biogeographical determinants for total and endemic species richness in a continental archipelago. - *Biodivers. Conserv.* **19(5)**: 1225-1235.

KALLIMANIS, A.S. & AL. 2011: Examining the relationship between total species richness and single island palaeo- and neo-endemics. - *Acta Oecol.* **37(2)**: 65-70.

KALNÍKOVÁ, V. & PALPURINA, S. 2015: *Epilobium adenocaulon* and *Epilobium glazioviana* (Onagraceae): new alien species for the Bulgarian flora. - *Phytol. Balcan.* **21(1)**: 21-27.

KALOGEROPOULOS, E & DELIPETROU, P. 2013: Contribution to the flora of Attiki. In Gkelis, S. & al. (eds): 13th Panhellenic Scientific Conference, Hellenic Botanical Society, Thessaloniki 3-6 October. Program and book of abstracts, p. 76.

KALOGEROPOULOS, E. & DELIPETROU, P. 2015: Aegean floristic elements in Attiki: New records for *Silene holzmannii* Heldr. ex Boiss. and *Ophrys basilissa* C. Alibertis, A. Alibertis & H.R. Reinhard. - *Parnassiana Arch.* **3**: 11-17.

KALOGEROPOULOS, E., DELIPETROU, P. & ALIBERTIS, A. 2011: The early *Ophrys fusca* of Lavrion (Attiki, Greece). - *J. Eur. Orch.* **43(2)**: 367-377.

KALOGEROPOULOS, E., DELIPETROU, P. & ALIBERTIS, A. 2012: Contribution to the study of *Ophrys oestrifera* group in Greece: a new subspecies from Attica and Peloponnese. - *J. Eur. Orch.* **44(1)**: 63-82.

KALOPISSIS, J. 1975: *Ophrys sphecodes* Mill. ssp. *hebes* Kalopissis, ssp. nov., eine neue *O. sphecodes*-Sippe aus Griechenland. - *Sci. Ann. Fac. Agric. Forest. Univ. Thessalonika Roussopoulou*: 133-137 + 1 plate.

KALOPISSIS, J. 1976: Some remarks on Greek orchids (Orchidaceae L.) [In Greek with English summary]. - *Fisis [Journal of the Hellenic Society for the Protection of Nature]* **6-7**: 7-9, 21-23.

KALOPISSIS, J. 1977: Some remarks on Greek orchids (Orchidaceae L.) II [In Greek with English summary]. - *Fisis [Journal of the Hellenic Society for the Protection of Nature]* **12**: 12, 35.

KALOPISSIS, J. 1979: *Cephalanthera cucullata* Boiss. et Heldr., the endangered Cretan *Cephalanthera* [In Greek with English summary]. - *Fisis [Journal of the Hellenic Society for the Protection of Nature]* **18**: 26-30, 46.

KALOPISSIS, J. 1981: Eine apochrome Variante von *Serapias parviflora* Parl. subsp. *parviflora* aus Griechenland. - *Bot. Chron. (Patras)* **1(2)**: 56-59.

KALOPISSIS, J. 1984: Episkopisi ton orcheoidon tis Kritis [In Greek]. - Mousio Kritikis Ethnologias: 29 pp.

KALOPISSIS, J. 1988: The orchids of Greece. Inventory and review [In Greek and English]. - Vori: Museum of Cretan Ethnology, 40 pp. + 66 pp. of distribution maps.

KALOPISSIS, J. 1992: Zur Vorkommen von *Orchis militaris* in Griechenland. - *Mitt. Arbeitskreis Heimische Orchid. Baden-Württemberg* **24(2)**: 392-397.

KALOPISSIS, J. & APERGIS, M. 1996: ×*Orchiserapias myrtoa* Kalopissis et Aperghis, nothospec. nat. nov. - *J. Eur. Orch.* **28(4)**: 735-745.

KALOPISSIS, J. & CONSTANTINIDIS, TH. 1993: ×*Barlorchis salaminensis* Kalopissis et Constantinidis nothogen. et nothospec. nat. nov. [=*Barlia robertiana* (Loisel.) W. Greuter × *Orchis italica* Poir.]. - *Mitt. Arbeitskreis Heimische Orchid. Baden-Württemberg* **25(4)**: 488-496.

KALOPISSIS, J. & ROBATSCH, K. 1980: *Epipactis cretica* Kalopissis et Robatsch, spec. nov., ein neuer kretischer Endemit. - *Orchidee (Hamburg)* **31(4)**: 141-143.

KALPOUTZAKIS, E. & CONSTANTINIDIS, TH. 2004: A new species of *Centaurea* (sect. *Phalolepis*, Compositae: Cardueae) from eastern Peloponnisos, Greece. - *Bot. J. Linn. Soc.* **146(3)**: 375-383.

KALPOUTZAKIS, E. & CONSTANTINIDIS, TH. 2005: New data on the distribution of endemic and rare taxa in the flora of east Peloponnisos, Greece. - *Bot. Chron. (Patras)* **18(2)**: 115-136.

KALPOUTZAKIS, E. & CONSTANTINIDIS, TH. 2006: Additions and annotations to the flora of Peloponnisos (S Greece). - *Willdenowia* **36(1)**: 271-284.

KALPOUTZAKIS, E., CONSTANTINIDIS, TH. & TRIGAS, P. 2019: Chorological additions for some noteworthy taxa of the Greek flora. - *Bot. Chron. (Patras)* **22**: 233-249.

KALPOUTZAKIS, E., RAABE, U. & CONSTANTINIDIS, TH. 2017: *Centaurea sicula* (Asteraceae), a new record for the Greek flora. In Dousis, M.A. & Thanos, K.A.: 15th conference of the Hellenic Botanical Society, book of abstracts: p. 85.

KALPOUTZAKIS, E., TRIGAS, P. & CONSTANTINIDIS, TH. 2012: *Allium orestis* sp. nov. (Amaryllidaceae) from Parnon and Taigetos mountains, south Peloponnisos, Greece. - *Nordic J. Bot.* **30(2)**: 195-200.

KALTEISEN, M. & REINHARD, H.R. 1987a: Das Areal von *Ophrys promontorii* O. & E. Danesch. - *Mitt. Arbeitskreis Heimische Orchid. Baden-Württemberg* **19(4)**: 801-821.

KALTEISEN, M. & REINHARD, H.R. 1987b: Zwei neue *Ophrys*-Taxa (Orchidaceae) aus dem Ägäischen Archipel: *Ophrys aegaea* Kalteisen & Reinhard, sp. nov., *Ophrys aegaea* subsp. *lucis* Kalteisen & Reinhard, subsp. nov. - *Mitt. Arbeitskreis Heimische Orchid. Baden-Württemberg* **19(4)**: 895-938.

KALTEISEN, M. & WILLING, E. 1981: Verbreitungskarten der Orchideen von Rhodos. - *Mitt. Arbeitskreis Heimische Orchid. Baden-Württemberg* **13(4)**: 377-446.

KALTSIS, A. & AL. 2013a: *Bupleurum kakiskalae*: a rare, endemic plant of Crete and the discovery of a new (third) subpopulation. In Gkelis, S. & al. (eds): 13th Panhellenic Scientific Conference, Hellenic Botanical Society, Thessaloniki 3-6 October. Program and book of abstracts, p. 77.

KALTSIS, A. & AL. 2013b: Floristic study and conservation actions for plant diversity of Skyros Island. In Gkelis, S. & al. (eds): 13th Panhellenic Scientific Conference, Hellenic Botanical Society, Thessaloniki 3-6 October. Program and book of abstracts, p. 78.

KAMARI, G. 1976: Cytotaxonomical study of the *Crepis neglecta* L. complex in Greece [In Greek with English summary]. - Patras: Ph.D. thesis, Univ. of Patras, 193 pp.

KAMARI, G. 1981: A biosystematic study of the genus *Galanthus* L. in Greece, Part II (Cytology). - *Bot. Chron. (Patras)* **1(2)**: 60-98 + 1 table.

KAMARI, G. 1982: A biosystematic study of the genus *Galanthus* (Amaryllidaceae) in Greece. Part I. Taxonomy. - *Bot. Jahrb. Syst.* **103(1)**: 107-135.

KAMARI, G. 1984a: Notes on some species of *Crepis* subsect. *Subcorymbiformes* in Greece. - *Bot. Jahrb. Syst.* **104(3)**: 401-404.

KAMARI, G. 1984b: Caryosystematic studies on "Fritillaria" L. (Liliaceae) in Greece. 1. - *Webbia* **38**: 723-731.

KAMARI, G. 1985: *Fritillaria sporadum* (Liliaceae), a new species from N Sporades (Greece). - *Willdenowia* **14(2)**: 331-333.

KAMARI, G. 1989: The genus *Fritillaria* in Greece: Taxonomy and karyology. In Phitos, D. (ed.): Abstracts. Sixth OPTIMA meeting, Delphi 10-16 Sept. 1989, p. 34.

KAMARI, G. 1991: The genus *Fritillaria* L. in Greece: taxonomy and karyology. - *Bot. Chron. (Patras)* **10**: 253-270.

KAMARI, G. 1992: Karyosystematic studies on three *Crepis* species (Asteraceae) endemic to Greece. - *Pl. Syst. Evol.* **182(1-2)**: 1-19.

KAMARI, G. 1993: *Fritillaria* species with yellow or yellowish-green flowers in Greece. In Petrova, A.V. (ed.): Abstracts from the VII OPTIMA meeting in Borovetz, Bulgaria, 18-30 July 1993, p. 50.

KAMARI, G. 1995: *Minuartia greuteriana* (Caryophyllaceae), a new species from NE Greece. - *Willdenowia* **25(1)**: 99-104.

KAMARI, G. 1996: *Fritillaria* species (Liliaceae) with yellow or yellowish-green flowers in Greece. - *Bocconea* **5(1)**: 221-238.

KAMARI, G. 2001: Past, present and future studies of the Greek flora. In Özhatay, N. (ed.): Proceedings of the Second Balkan Botanical Congress, 1: 33-36. - Istanbul.

KAMARI, G. 2005: Biodiversity of the genus *Fritillaria* in the Aegean region. In Karamanos, A.J. & Thanos, C.A. (eds): Biodiversity and natural heritage in the Aegean. Proceedings of the conference "Theophrastus 2000", pp. 149-160 [In Greek with English summary]. - Athina: Afoi Fragoudi O.E.

KAMARI, G. & AL. 1988: Flora and vegetation of Yioura, N Sporades, Greece. - *Willdenowia* **17(1-2)**: 59-85.

KAMARI, G. & AL. 2004: Karyosystematic studies of plant taxa from the East Mediterranean region (Greece, Cyprus, Syria). - *Phytol. Balcan.* **9(3)**: 487-502.

KAMARI, G. & ANAGNOSTOPOULOS, A. 1991: Reports (1-6). In Kamari, G., Felber, F. & Garbari, F. (eds): Mediterranean chromosome number reports - 1. - *Fl. Medit.* **1**: 224-229.

KAMARI, G. & ARTELARI, R. 1990: Karyosystematic study of the genus *Sternbergia* (Amaryllidaceae) in Greece. I. South Aegean islands. - *Willdenowia* **19(2)**: 367-388. [Also distributed as *OPTIMA leaflet* no. 191].

KAMARI, G., BLANCHÉ, C. & GARBARI, F. (EDS) 2001: Mediterranean chromosome number reports - 11. - *Fl. Medit.* **11**: 435-483.

KAMARI, G., BLANCHÉ, C. & GARBARI, F. (EDS) 2003a: Mediterranean chromosome number reports - 12. - *Fl. Medit.* **12**: 443-486.

KAMARI, G., BLANCHÉ, C. & GARBARI, F. (EDS) 2003b: Mediterranean chromosome number reports - 13.- *Fl. Medit.* **13**: 347-404.

KAMARI, G., BLANCHÉ, C. & GARBARI, F. (EDS) 2004: Mediterranean chromosome number reports - 14. - *Fl. Medit.* **14**: 423-453.

KAMARI, G., BLANCHÉ, C. & GARBARI, F. (EDS) 2005: Mediterranean chromosome number reports - 15. - *Fl. Medit.* **15**: 693-728.

KAMARI, G., BLANCHÉ, C. & GARBARI, F. (EDS) 2006: Mediterranean chromosome number reports - 16. - *Fl. Medit.* **16**: 385-455.

KAMARI, G., BLANCHÉ, C. & GARBARI, F. (EDS) 2007: Mediterranean chromosome number reports - 17. - *Fl. Medit.* **17**: 291-319.

KAMARI, G., BLANCHÉ, C. & SILYAK-YAKOVLEK, S. (EDS) 2012: Mediterranean chromosome number reports - 22. - *Fl. Medit.* **22**: 211-231.

KAMARI, G. & CONSTANTINIDIS, TH. 1994: The *Minuartia verna* complex in Greece. - *Bot. Chron. (Patras)* **11**: 41-54.

KAMARI, G. & CONSTANTINIDIS, TH. 1999: New data concerning the genus *Aethionema* R.Br. (Cruciferae) in Greece. In Elefheriou, E.P. (ed.): Proceedings, 7th Panhellenic scientific conference, Alexandroupolis, pp. 250-251 [In English and Greek]. - Thessaloniki: Univ. Studio Press.

KAMARI, G., FELBER, F. & GARBARI, F. (EDS) 1991: Mediterranean chromosome number reports - 1. - *Fl. Medit.* **1**: 223-245.

KAMARI, G., FELBER, F. & GARBARI, F. (EDS) 1992: Mediterranean chromosome number reports - 2. - *Fl. Medit.* **2**: 223-278.

KAMARI, G., FELBER, F. & GARBARI, F. (EDS) 1993: Mediterranean chromosome number reports - 3. - *Fl. Medit.* **3**: 323-373.

KAMARI, G., FELBER, F. & GARBARI, F. (EDS) 1994: Mediterranean chromosome number reports - 4. - *Fl. Medit.* **4**: 233-301.

KAMARI, G., FELBER, F. & GARBARI, F. (EDS) 1996: Mediterranean chromosome number reports - 5. - *Fl. Medit.* **5**: 261-373.

KAMARI, G., FELBER, F. & GARBARI, F. (EDS) 1997a: Mediterranean chromosome number reports - 6. - *Fl. Medit.* **6**: 223-337.

KAMARI, G., FELBER, F. & GARBARI, F. (EDS) 1997b: Mediterranean chromosome number reports - 7. - *Fl. Medit.* **7**: 197-275.

KAMARI, G., FELBER, F. & GARBARI, F. (EDS) 1998: Mediterranean chromosome number reports - 8. - *Fl. Medit.* **8**: 213-313.

KAMARI, G., FELBER, F. & GARBARI, F. (EDS) 1999: Mediterranean chromosome number reports - 9. - *Fl. Medit.* **9**: 323-387.

KAMARI, G., FELBER, F. & GARBARI, F. (EDS) 2000: Mediterranean chromosome number reports - 10. - *Fl. Medit.* **10**: 381-430.

KAMARI, G. & GEORGIOU, O. 1981: Cytological notes on two *Ornithogalum* species from Greece. - *Bot. Chron. (Patras)* **1(1)**: 29-34.

KAMARI, G. & GREUTER, W. 2000: *Phitosia*, a new genus for *Crepis crocifolia* (Compositae, Cichorieae), a local endemic of Mount Taigetos (Greece). - *Bot. Chron. (Patras)* **13**: 11-36.

KAMARI, G. & HARTVIG, P. 1988: *Crepis merxmuelleri* (Asteraceae), a new species from NW Greece. - *Willdenowia* **18(1)**: 63-66.

KAMARI, G., KYRIAKOPOULOS, C. & KOFINAS, G. 2010: New finding of *Phitosia crocifolia* (Compositae) in E Peloponnisos. - *Fl. Medit.* **20**: 235-238.

KAMARI, G. & MATTHÄS, U. 1986: Cytotaxonomical contributions on the flora of Crete. III. – *Willdenowia* **15(2)**: 515-520.

KAMARI, G. & PAPATSOU, S. 1973: Chromosome studies in some Mediterranean angiosperms. - *Bot. Not.* **126(2)**: 266-268.

KAMARI, G. & PHITOS, D. 2000: *Fritillaria theophrasti*, a new species from Lesvos, East Aegean Islands, Greece. - *Biol. Gallo-Hellen.* **26**: 69-76.

KAMARI, G. & PHITOS, D. 2006: Karyosystematic study of *Fritillaria messanensis* s.l. (Liliaceae). - *Willdenowia* **36(1)**: 217-233.

KAMARI, G., PSARAS, G. & CONSTANTINIDIS, TH. 2001: Proceedings of the 8th scientific congress of the Greek Botanical Society, October 5-8, 2000. - Patras: 479 pp.

KAMARI, G. & STEVANOVIĆ, V. 1996: *Minuartia juniperina* (Caryophyllaceae) in the Balkan Peninsula. - *Phyton (Horn)* **36(1)**: 93-105.

KAMARI, G. & STRID, A. 1989: *Crepis arcuata* (Asteraceae), a new species from N Greece. - *Willdenowia* **19(1)**: 79-82. [Also distributed as *OPTIMA leaflet* no. 187].

KAMARI, G., ZAHOS, A. & SIAGOU, I. 2017: A new yellow-flowered *Fritillaria* species (Liliaceae) from Mt. Tisseon, continental Greece and its taxonomic relationships. - *Phytotaxa* **328(3)**: 227-242.

KAMARI, G. (ED.) 2002: Ellinini Votaniki Eteria. 9° Panelliniko Epistimoniko Sinedrio. Praktika. - Patras: 366 pp.

KAMELIN, R.V. & TAGAEV, I.U. 1986: A review of species of the genus *Scorzonera* (Asteraceae) [In Russian]. - *Bot. Zhurn. (Moscow & Leningrad)* **71(12)**: 1672-1682.

KAMELIN, R.V. & ZABELINA, N.I. 1987: *Periploca graeca* (Asclepiadaceae) - the new genus and species in the flora of Middle Asia [In Russian]. - *Bot. Zhurn. (Moscow & Leningrad)* **72(4)**: 528-529.

KAMSTRA, K. 2011: Greek mountain flora. Northern Greece, vol. 1. - Privately published, 106 pp.

KANDELER, R. 1975: Species delimitation in the genus *Lemna*. - *Aquatic Bot.* **1(4)**: 365-376.

KANISANLI, M. 1974: Chromosome numbers of some monocotyledons [In Turkish with English summary]. - *Ege Üniv. Fen Fak. Ilmi Raporlar Ser.* **176**: 1-12.

KANTSA, A. & AL. 2013: Urban biodiversity hotspots wait to get discovered: The example of the city of Ioannina, NW Greece. - *Landscape and Urban Planning* **120**: 129-137 + electronic suppl.

KAPASA, M. & AL. 2001: Reports (1236-1243). In Kamari, G., Blanché, G. & Garbari, F. (eds): Mediterranean chromosome number reports. - *Fl. Medit.* **11**: 448-454.

KAPLAN, Z. 2005: *Potamogeton schweinfurthii* A. Benn., a species new for Europe. - *Preslia* **77**: 419-431.

KAPTEYN DEN BOUMEESTER, D. & WILLING, E. 1988: Aktuelle Verbreitung der Orchideen auf Kerkira (Korfu/Griechenland). - *Ber. Arbeitskreis. Heimische Orchid., Beiheft* **2**: 4-128.

KARABLIANIS, TH. 2007: Chloridiki pikilotita ton Akarnanikon Oreon [In Greek].. - Athens: M.Sc. thesis, Department of Systematic Botany, Agricultural University, 228 pp.

KARABLIANIS, TH. & CONSTANTINIDIS, TH. 2019: The taxa of *Crocus* ser. *Flavi* (Iridaceae) in Greece: a taxonomic and karyomorphologic study. - *Bot. Chron. (Patras)* **22**: 171-194.

KARABLIANIS, TH., TSIFTSIS, S. & CONSTANTINIDIS, TH. 2013: The genus *Crocus* (Iridaceae) in Greece: some noteworthy floristic records and karyotypes. - *Phytol. Balcan.* **19(1)**: 53-66.

KARABLIANIS, TH. & CONSTANTINIDIS. TH. 2013: A karyological study of the genus *Crocus* L. (Iridaceae) in Greece. In Gkelis, S. & al. (eds): 13th Panhellenic Scientific Conference, Hellenic Botanical Society, Thessaloniki 3-6 October. Program and book of abstracts, p. 82.

KARAGIANNAKIDOU-IATROPOULOU, V. 1977: Statistische Untersuchung von Samen *Juniperus*-Arten von Griechenland mit ihren Bestimmungsschlüsseln. - *Sci. Ann. Fac. Phys. Math. Aristot. Univ. Thessalonika* **17**: 267-283.

KARAGIANNAKIDOU-IATROPOULOU, V. 1980: Über die vertikale Verbreitung der Orchideen auf dem Berg Chortiatis (Thessaloniki, Griechenland). - *Mitt. Arbeitskreis Heimische Orchid. Baden-Württemberg* **12(2)**: 89-97 [Ergänzung 12(3): 211].

KARAGIANNAKIDOU-IATROPOULOU, V. 1983: Stathmologiki erevna stin Querco-Fageta klasi tou orinou sigrotimatos tou Chortiati [In Greek with German summary]. - *Sci. Ann. Fac. Phys. Math. Aristot. Univ. Thessalonika* **22**: 1-161.

KARAGIANNAKIDOU, V. 1988: Floristische Zusammensetzung und pflanzengeographische Analyse der Vegetationszonen des Massivs Chortiatis in Nordostgriechenland. - *Növényrendsz. Novényföldr. Tansz., Eötvös Loránd Tudományegyet. Budapest* **12**: 163-182.

KARAGIANNAKIDOU, V. 1991: Analysis of the flora of Mount Menikion, NE Greece. - *Saussurea* **22**: 33-42.

KARAGIANNAKIDOU, V. 1993: Site research in beech forests of the Chortiatis mountain range, NE Greece. - *Bot. Helv.* **103(1)**: 23-37.

KARAGIANNAKIDOU, V. 1994: Contribution to the study of mountain[sic!]-subalpine grassland vegetation of Mount Menikion, north eastern Greece. - *Ecol. Medit.* **20(3-4)**: 73-84.

KARAGIANNAKIDOU, V. 1996: Statistische Untersuchungen der Samen von *Juniperus*-Arten in Griechenland. - *Mitt. Deutsch. Dendrol. Ges.* **82**: 129-137.

KARAGIANNAKIDOU, V. & BABALONAS, D. 1981: Verbreitung von Orchideen-Sippen in Nordgriechenland. - *Mitt. Arbeitskreis Heimische Orchid. Baden-Württemberg* **13(2)**: 164-204.

KARAGIANNAKIDOU, V., DROSSOS, E.G. & KONSTANTINOU, M. 1996: Analysis of the family Poaceae (Gramineae) on the subalpine grassland flora of NE. Greece [In Greek with English summary]. - Thessaloniki: Aristotle Univ., 10 pp. (duplicated).

KARAGIANNAKIDOU, V. & KOKKINI, S. 1987: The flora of Mount Menikion in North East Greece. - *Phyton (Horn)* **27(2)**: 267-283.

KARAGIANNAKIDOU, V. & KOKKINI, S. 1988: Vegetation and altitudinal zones of Mount Menikion, NE Greece. - *Bot. Helv.* **98(2)**: 149-160.

KARAGIANNAKIDOU, V., KONSTANTINOU, M. & LAVRENTIADOU, S. 2000: Flora, vegetation and ecological management of the Strymon river delta (NE Greece). In Özhatay, N. & al. (eds): Abstracts from the Second Balkan Botanical Congress, Istanbul, Turkey, May 14-18, 2000, p. 15.

KARAGIANNAKIDOU, V., KONSTANTINOU, M. & LAVRENTIADOU, S. 2001: Flora, vegetation and ecological management of Strymon river delta (NE Greece). In Özhatay, N. (ed.): Proceedings of the Second Balkan Botanical Congress, 1: 37-46. – Istanbul.

KARAGIANNAKIDOU, V., KONSTANTINOU, M. & PAPADEMETRIOU, K. 1995a: Analysis of the flora of subalpine grasslands of E. Macedonia, NE Greece. In: Proc., 5th sci. conf., Hellenic Bot. Soc., Delphi 21-23 October 1994, pp. 293-298 [In Greek with English summary]. - Thessaloniki: Univ. Studio Press.

KARAGIANNAKIDOU, V., KONSTANTINOU, M. & PAPADEMETRIOU, K. 1995b: Floristic and phytogeographical research on the upper montane and the subalpine grassland flora of East Macedonia, Greece. - *Feddes Repert.* **106(3-4)**: 193-213.

KARAGIANNAKIDOU, V., KONSTANTINOU, M. & PAPADEMETRIOU, K. 2000: Analysis of the upper montane and the subalpine grassland flora of East Macedonia, Greece. - *Candollea* **54(2)**: 453-472.

KARAGIANNAKIDOU, V. & RAUS, TH. 1996: Vascular plants from Mount Chortiatis (Makedonia, Greece). - *Willdenowia* **25(2)**: 487-559.

KARAGIANNI, P. 2009: Ikologia ton tipon ikotopon tis apoxiramenis limnis Mourias. [In Greek] - Patras: Ph.D. thesis, Univ. of Patras, 248 pp.

KARAGIANNI, P., TINIAKOU, A. & GEORGIADIS, TH. 2010: Flora of the drained Mouria lake area, SW Greece. - *Bot. Chron. (Patras)* **20**: 63-81.

KARAKOSTA, C., THEODOROPOULOS, K. & PAPANASTASIS, V.P. 2013: Changes in plant species composition in old fields at the University Forest of Taxiarhis in Chalkidiki, northern Greece. In Gkelis, S. & al. (eds): 13th Panhellenic Scientific Conference, Hellenic Botanical Society, Thessaloniki 3-6 October. Program and book of abstracts, p. 80.

KARAMANOĞLU, K. 1964: The species of *Convolvulus* in Turkey. - *Commun. Fac. Sci. Univ. Ankara, Sér. C* **9**: 225-251.

KARAMANOĞLU, K. & KOYUNCU, M. 1974: The systematic researches on *Valeriana* species in Turkey [In Turkish with English summary]. - *Ankara Üniv. Eczac. Fak. Mecm.* **4(1)**: 149-178.

KARAMANOS, A.J. & THANOS, C.A. (EDS) 2005: Biodiversity and natural heritage in the Aegean. Proceedings of the conference "Theophrastus 2000" [In Greek and English]. - Athina: Afoi Fragoudi O.E., 316 pp.

KARATAGLIS, S. 1973: Sigrititi erevna tou kariotipi ke ton ilektroforimaton enzimon esterasis avtofion tinon ellinikon idon tou genous *Aegilops* [In Greek with English summary]. - Thessaloniki: 95 pp.

KARATAGLIS, S. 1975a: Inter- and intraspecific relationship between two native Greek diploid *Aegilops* species by esterase electrophoresis. - *Biochem. Physiol. Pflanzen* **167(5)**: 389-398.

KARATAGLIS, S. 1975b: Karyotype analysis on some diploid native Greek *Aegilops* species. - *Caryologia* **28(1)**: 99-110.

KARATAGLIS, S. 1989: *Aegilops cylindrica* var. *kastorianum* (Poaceae): a new variety from Greece. - *Pl. Syst. Evol.* **163(1-2)**: 13-20.

KARATZAS, I. 2004a: Entopismos enos monadikou ivridiou orchideas [In Greek]. - *Fisis [Journal of the Hellenic Society for the Protection of Nature]* **106**: 8-9.

KARATZAS, I. 2004b: ×*Comptoglossum agiasense*, eine seltene Hybride aus Lesvos. - *J. Eur. Orch.* **36(4)**: 951-956.

KARATZAS, I. 2009: Wiedererscheinen von ×*Comptoglossum agiasense* auf Lesbos. - *J. Eur. Orch.* **41(3-4)**: 497-502.

KARATZAS, I. & KARATZA, A. 2008: Orchidées, agrioloúluda tis Lésvou [In Greek]. - Mitilini: Entelechia, 323 pp.

KARAVOKYROU, E. & AL. 1995: Contribution to the study of the Greek insular flora. In: Proceedings, 5th sci. conf., Hellenic Bot. Soc., Delphi 21-23 October 1994, pp. 195-200 [In Greek with English summary]. - Thessaloniki: Univ. Studio Press.

KARAVOKYROU, E. & TZANOUDAKIS, D. 1989: The genus *Allium* in Greece. II. A cytogeographical study of the eastern Aegean species. In Phitos, D. (ed.): Abstracts. Sixth OPTIMA meeting, Delphi 10-16 Sept. 1989, p. 105.

KARAVOKYROU, E. & TZANOUDAKIS, D. 1991: The genus *Allium* in Greece: II. A cytogeographical study of the E Aegean species. - *Bot. Chron. (Patras)* **10**: 777-784.

KARAVOKYROU, E. & TZANOUDAKIS, D. 1995: Two new species of *Allium* sect. *Codonoprasum* Reichenb. from the East Aegean Islands (Greece). - *Ann. Mus. Goulandris* **9**: 141-146.

KARETSOS, G. 2002: Meleti tis ikologias ke tis vlastisis tou orous Iti [In Greek with English summary]. - Patras: Ph.D. thesis, Univ. of Patras, x + 311 pp. + appendix [also on CD]

KARETSOS, G. & GEORGIADIS, TH. 2008: *Veronica oetaea* L.-Å. Gustavsson: the ecology and the phytosociology of an exclusive endemic species - threats and protection suggestions. In: Praktika 10ou Sinedriou Ellinikis Votanikis Eterias, 5-8 maiou 2005, Ioannina: pp. 271-277 [In Greek with English abstract].

KARETSOS, G., TRIGAS, P. & TSAGARI, K. 2014: Chlorida tis Itis. Endemika ke spania idi [In Greek]. - Ellenikou Georgikou Organismou DIMITRA, 304 pp.

KARIDAS, A., GIANNAKIS, TH. & ANTONOPOULOS, Z. 2017: A new locality for *Colchicum soboliferum* (Colchicaceae) from Macedonia, Greece. - *Parnassiana Arch.* **5**: 57-59.

KARIS, P.O. 2004: Taxonomy, phylogeny and biogeography of *Limonium* sect. *Pteroclados* (Plumbaginaceae), based on morphological data. - *Bot. J. Linn. Soc.* **144(4)**: 461-482.

KARL, R. & AL. 2012: Systematics and evolution of arctic-alpine *Arabis alpina* (Brassicaceae) and its closest relatives in the eastern Mediterranean.- *Amer. J. Bot.* **99(4)**: 778-794.

KARLBERG, S. 1960: Forests and forestry in Turkey. - *Acta Horti Gothob.* **23**: 41-70.

KARLÉN, TH. 1984: *Muscari pulchellum* (Liliaceae) and associated taxa in Greece and W Turkey. - *Willdenowia* **14(1)**: 89-118.

KARLÉN, TH. 1985: Karyotypes and chromosome numbers of five species of *Muscari* (Liliaceae). - *Willdenowia* **14(2)**: 313-320.

KARLÉN, TH. 1987: *Muscari sandrasicum* (Liliaceae), a new species from Turkey. - *Willdenowia* **16(2)**: 375-382.

KAROUSOU, R. & AL. 1996: *Calamintha cretica* (Lamiaceae), a Cretan endemic: Distribution and essential oil composition. - *Nordic J. Bot.* **16(3)**: 247-252.

KAROUSOU, R. & AL. 2008: On the flora of Mount Stratonikon (GR1270005), a NATURA 2000 site of N Greece. - *Fl. Medit.* **18**: 529-550.

KAROUSOU, R., BOSABALIDIS, A.M. & KOKKINI, S. 1992: *Sideritis syriaca* ssp. *syriaca*: Glandular trichome structure and development in relation to systematics. - *Nordic J. Bot.* **12(1)**: 31-37.

KAROUSOU, R., CHANLIDOU, E. & KOKKINI, S. 2000a: The oregano herbs in Greece: *Origanum vulgare* vs. *O. onites*. In Özhatay, N. & al. (eds): Abstracts from the Second Balkan Botanical Congress, Istanbul, Turkey, May 14-18, 2000, p. 124.

KAROUSOU, R., CHANLIDOU, E. & KOKKINI, S. 2000b: The sage plants of Greece: Distribution and infraspecific variation. In Kintzios, S.E. (ed.): Sage. The genus *Salvia*, pp. 27-46. - Amsterdam: Harwood Academic Publishers.

KAROUSOU, R., CHANLIDOU, E. & KOKKINI, S. 2002: Distribution of the hybrid *Origanum ×intercedens* and its parental taxa *O. vulgare* L. and *O. onites* L. In Kamari, G. & al. (eds): Ellinini Votaniki Eteria. 9° Panelliniko Epistimoniko Sinedrio. Praktika, pp. 152-157 [In Greek with English abstract].

KAROUSOU, R., DARDIOTI, A. & KOKKINI, S. 1997: *Origanum vulgare* L. in the Balkan Peninsula and Anatolia: Distribution and morphological variation. In: Abstracts from the First Balkan Botanical Congress, Thessaloniki, September 19-22, 1997, p. 24. – Thessaloniki.

KAROUSOU, R. & KOKKINI, S. 1995: Hybrids between species of the genus *Phlomis* on the island of Crete. In: Proc., 5th sci. conf., Hellenic Bot. Soc., Delphi 21-23 October 1994, pp. 281-284 [In Greek with English summary]. - Thessaloniki: Univ. Studio Press.

KAROUSOU, R. & KOKKINI, S. 1997: Distribution and clinal variation of *Salvia fruticosa* Mill. (Labiatae) on the island of Crete (Greece). - *Willdenowia* **27(1-2)**: 113-120.

KAROUSOU, R., VOKOU, D. & KOKKINI, S. 1998: Distribution and essential oils of *Salvia pomifera* subsp. *pomifera* (Labiatae) on the island of Crete (S Greece). - *Biochem. Syst. Ecol.* **26(8)**: 889-897.

KÁRPÁTI, Z.E. 1958: Die Variabilität der Manna-Esche (*Fraxinus ornus* L.). - *Acta Bot. Acad. Sci. Hung.* **4(1-2)**: 93-112.

KÁRPÁTI, Z.E. 1960: Die *Sorbus*-Arten Ungarns und der angrenzenden Gebiete. - *Feddes Repert. Spec. Nov. Regni Veg.* **62(2-3)**: 71-331.

KÁRPÁTI, Z.E. 1965: Geobotanische Betrachtung zur Taxonomie der europäischen *Sorbus*-Arten. - *Drudea* **5(1)**: 67-74.

KÁRPÁTI, Z.E. 1970: Eine kritisch-taxonomische Übersicht der in Europa wildwachsenden Eschen- Arten und deren Unterarten. - *Feddes Repert.* **81(1-5)**: 171-186.

KARSCHON, R., GRUNEWALD, C. & WEINSTEIN, A. 1979: *Solanum nigrum* L. and *S. luteum* Miller in the flora of Israel. - *Israel J. Bot.* **27(2)**: 94-96.

KARYDAS, A. & KAMARI, G. 2019: Monitoring six local endemic taxa of the Mt. Athos and assessment according to the IUCN Red List categories and criteria. - *Bot. Chron. (Patras)* **22**: 195-208.

KASAPLIGIL, B. 1981: Past and present oaks of Turkey. Part I. - *Phytologia* **49(2)**: 95-146.

KASHA, K.J. & SADASIVAIAH, R.S. 1971: Genome relationships between *Hordeum vulgare* L. and *H. bulbosum* L. - *Chromosoma* **35(3)**: 264-287.

KASSIOUMIS, K. & AL. 2004: Distribution and stand structure of *Taxus baccata* populations in Greece; Results of the first national inventory. - *Ecol. Medit.* **30(2)**: 159-170.

KÄSTNER, A. 1978: Beiträge zur Wuchsformenanalyse und systematischen Gliederung von *Teucrium* L. I. Die Infloreszenzen und Blüten. - *Flora, Morphol. Geobot. Oekophysiol.* **167(6)**: 485-514.

KÄSTNER, A. 1981: Beiträge zur Wuchsformanalyse und systematischen Gliederung von *Teucrium* L. Wuchsformen und Verbreitung von Arten der Sektionen *Teucropis* und *Teucrium*. - *Flora, Morphol. Geobot. Oekophysiol.* **171(5)**: 466-519.

KÄSTNER, A. 1985: Beiträge zur Wuchsformdifferenzierung und systematischen Gliederung von *Teucrium* L. IV. Wuchsformen und Verbreitung von Arten der Sektion *Isotriodon*. - *Flora, Morphol. Geobot. Oekophysiol.* **176(1-2)**: 73-93.

KÄSTNER, A. 1986: Beiträge zur Wuchsformdifferenzierung und systematischen Gliederung von *Teucrium* L. V. Wuchsformen und Verbreitung von Arten der Verwandtschaftskreise von *T. marum* und *T. chamaedrys*. - *Flora, Morphol. Geobot. Oekophysiol.* **178(2)**: 111-138.

KÄSTNER, A. 1989a: Beiträge zur Wuchsformdifferenzierung und systematischen Gliederung von *Teucrium* L. VI. Wuchsformen und Verbreitung von Arten der Verwandtschaftskreise von *T. flavum* und *T. polium*. - *Flora, Morphol. Geobot. Oekophysiol.* **183(3-4)**: 189-224.

KÄSTNER, A. 1989b: Übersicht zur systematischen Gliederung der Gattung *Teucrium* L. — *Biocosme Mésogéen* **6(1-2)**: 63-77.

KÄSTNER, A. 2003: Porträts ausgewählter seltener österreichischer Gefässpflanzenarten: (1) *Bifora radians*, (2) *Euphorbia glareosa*, (3) *Helminthotheca echioides*, (4) *Turgenia latifolia*. - *Neilreichia* **2-3**: 149-163.

KATAYAMA, T. & IKEDA, H. 1975: Cytogenetical studies on *Paspalum distichum* Linn. - *Cytologia* **40(3-4)**: 759-764.

KATI, V. & AL. 2000: Les orchidées de la réserve de Dadia (Grèce), leurs habitats et leur conservation. - *Naturalistes Belges* **81(3)**: 193, 269-282.

KATOPODI, E. & TSIFTSIS, S. 2019: Contribution to the knowledge of the orchid flora of Lefkas island (Ionian Islands, Greece). - *Bot. Chron. (Patras)* **22**: 127-143.

KATSALIROU, E., GRERAKIS, A. & HALDAS, X. 2016: Soil preferences of Mediterranean orchids. - *J. Eur. Orch.* **48(2-4)**: 245-282.

KATSANEVAKIS, S. & TSIAMIS, K. 2009: Records of alien marine species in the shallow coastal waters of Chios Island. - *Medit. Mar. Sci.* **10(2)**: 99-107.

KATSANOS, D. 1931: Die rumelische Kiefer, *Pinus peuce* Griseb., in Griechenland [In Greek with German summary]. - *Sci. Ann. Fac. Phys. Math. Aristot. Univ. Thessalonika* **1**: 377-400.

KATSAROS, N., KOKKINI, S. & CHANLIDOU, E. 2008: Flora of the traditional villages of the National Park of Vikos - Aoos. In: Praktika 10ou Sinedriou Ellinikis Votanikis Eterias, 5-8 Maiou 2005, Ioannina: pp. 279-284 [In Greek with English abstract].

KATSIKOPOULOS, I. 1936: Simvoli is tin meletin tis chloridos tis nisou Samothrakis [In Greek]. - *Geôrgikon Deltion (Agricultural Bulletin of Greek Agricultural Society)* **29**: 225-227, 264-267, 301-304.

KATSIOTIS, D. & TAN, KIT 2015: Report 201. In Vladimirov, V., Dane, F. & Tan, Kit (eds): New floristic records in the Balkans: 26. - *Phytol. Balcan.* **21(1)**: 74-76.

KATSOUNI, N. 2001: Natural History Museum of Cephalonia & Ithaka, Greece. In Raimondo, F.M. (ed.): Abstracts from the X OPTIMA meeting, Palermo, 13-19 September 2001, p. 273.

KATZNELSON, J. 1966: A taxonomic revision of sect. *Calycomorphum* of the genus *Trifolium*. II. The anemochoric species. - *Israel J. Bot.* **14(4)**: 171-183.

KATZNELSON, J. 1970: Edaphic factors in the distribution of subterranean clover in the Mediterranean region. - *Proc. Int. Grassland Congr.* **11**: 192-196.

KATZNELSON, J. 1971: Population studies in annual clovers. - *Israel J. Bot.* **20(4)**: 241-262.

KATZNELSON, J. & MORLEY, F.H.W. 1965a: Speciation processes in *Trifolium subterraneum* L. - *Israel J. Bot.* **14(1)**: 15-35.

KATZNELSON, J. & MORLEY, F.H.W. 1965b: A taxonomic revision of sect. *Calycomorphum* of the genus *Trifolium*. I. The geocarpic species. - *Israel J. Bot.* **14(3)**: 112-134.

KATZNELSON, J. & ZOHARY, D. 1967: Diploid and tetraploid *Hordeum bulbosum*. - *Israel J. Bot.* **16(2)**: 57-62.

KATZNELSON, J. & ZOHARY, D. 1970: Seed dispersal in *Trifolium* sect. *Calycomorphum*. - *Israel J. Bot.* **19(2-3)**: 114-120.

KAUFMANN, W. 2008: Fundliste Ithaka 2008. - http://www.wilkau.li/index_htm_files/GR%20Ithaka% 20Funde.pdf

KAUNDUN, S.S., FADY, B. & LEBRETON, PH. 1997: Genetic differences between *Pinus halepensis, Pinus brutia* and *Pinus eldarica* based on needle flavonoids. - *Biochem. Syst. Ecol.* **25(6)**: 553-562.

KAWANO, S. 1963: Cytogeography and evolution of the *Deschampsia caespitosa* complex. - *Canad. J. Bot.* **41(5)**: 719-742.

KAY, Q.O.N. 1969: The origin and distribution of diploid and tetraploid *Tripleurospermum inodorum* (L.) Schultz Bip. - *Watsonia* **7(3)**: 130-141.

KAY, Q.O.N. 1971a: Biological Flora of the British Isles. *Anthemis cotula* L. - *J. Ecol.* **59(2)**: 623-636.

KAY, Q.O.N. 1971b: Biological Flora of the British Isles. *Anthemis arvensis* L. - *J. Ecol.* **59(2)**: 637-648.

KAY, Q.O.N. & PAGE, J. 1985: Dioecism and pollination in *Ruscus aculeatus*. - *Watsonia* **15(3)**: 261-264.

KAYA, A. 2003: Morphological variations in *Acinos alpinus* and *Acinos rotundifolius* (Labiatae) growing in Turkey. - *Fl. Medit.* **12**: 261-274.

KAYACIK, H. 1954: Pines in Turkey and an investigation about their geographical distribution [In Turkish with English summary]. - *Istanbul Üniv. Orman Fak. Derg., A* **4(1-2)**: 44-64.

KAYACIK, H. 1955: The flora of Belgrad forest [In Turkish with English summary]. - *Istanbul Üniv. Orman Fak. Derg., A* **5(1-2)**: 77-90.

KAYACIK, H. 1963: Pines in Turkey and an investigation about their geographical distribution. II. Aleppo pine (*Pinus halepensis* Mill.) [In Turkish with English and German summaries]. - *Istanbul Üniv. Orman Fak. Derg., A* **13(1)**: 1-10.

KAYACIK, H. 1966: A study on the geographical distribution of *Cupressus sempervirens* L. in Turkey [In Turkish with English and German summaries]. - *Istanbul Üniv. Orman Fak. Derg., A* **16(1)**: 39-65.

KAZAKIS, G. & AL. 2007: Vascular plant diversity and climate change in the alpine zone of the Lefka Ori, Crete. - *Biodivers. Conserv.* **16(6)**: 1603-1615.

KAZAKOU, E. & AL. 2010: Species adaptation in serpentine soils in Lesbos Island (Greece): Metal hyperaccumulation and tolerance. - *Plant Soil* **332**: 369-385.

KAZAN, K. & MUEHLBAUER, F.J. 1991: Allozyme variation and phylogeny in annual species of *Cicer* (Leguminosae). - *Pl. Syst. Evol.* **175(1-2)**: 11-21.

KAZIMIERSKI, T. & KAZIMIERSKA, E.M. 1975: Morphological and cytological differences within the species *Lupinus luteus* L. - *Acta Soc. Bot. Poloniae* **44(2)**: 265-275 + 3 plates.

KAZMI, S.M.A. 1964: Revision der Gattung *Carduus* (Compositae). Teil II. - *Mitt. Bot. Staatssamml. München* **5(2)**: 279-550.

KEATING, R., HOCH, P.C. & RAVEN, P.H. 1982: Perennation in *Epilobium* (Onagraceae) and its relation to classification and ecology. - *Syst. Bot.* **7(4)**: 379-404.

KECHAYKIN, A.A. & AL. 2021: (2820) Proposal to conserve the name *Potentilla* (Rosaceae: Potentilleae) with a conserved type. - *Taxon* **70(3)**: 680-681.

KECHKAYIN, A.A. & SHMAKOV, A.I. 2016: A system of subtribe Potentillinae J. Presl (Rosaceae Juss.). - *Turczaninowia* **19(4)**: 114-128.

KEESING, J.L.S. 1979: Living collections division expedition to the Pindus mountains of northern Greece August-September 1978. - *J. Kew Guild* **9(82)**: 680-686.

KEHM, F.H. 1996: Die Pflanzenformationen im Norden Ithakas (Ionische Inseln). - Mainz: Johannes Gutenberg-Universität, Diplomarbeit, [3]+243 pp.,1 map.

KEIGHERY, G.J. & DODD, J. 1997: Occurrence and spread of sea spurge (*Euphorbia paralias*) along the west coast of Western Australia. - *Nuytsia* **11(2)**: 285-286.

KEISSLER, C.V. 1896: Über eine neue *Daphne*-Art und die geographische Verbreitung derselben, sowie die ihrer nächsten Verwandten. - *Verh. K.K. Zool.-Bot. Ges. Wien* **46**: 214-223.

KEISSLER, C.V. 1898: Die Arten der Gattung *Daphne* aus der Section *Daphnanthes*. - *Bot. Jahrb. Syst.* **25(1-2)**: 29-125 + 4 plates.

KEITEL, CH. & REMM, W. 1991: Die Orchideenflora der Inseln Simi, Tilos und Nisyros. - *Mitt. Arbeitskreis Heimische Orchid. Baden-Württemberg* **23(1)**: 81-106.

KELLENBERGER, W. 1978: *Orchis anatolica* × *Orchis patens* ssp. *nitidifolia*. - *Orchidee (Hamburg)* **29 (3)**: 112.

KELLER, R. 1931: Sysopsis rosarum spontanearum Europae mediae. Übersicht über die mitteleuropäischen Wildrosen mit besonderer Berücksichtigung ihrer schweizerischen Fundorte. - *Denkschr. Schweiz. Naturf. Ges.* **65**: x + 796 pp. + 40 plates.

KELLER, R., SCHLECHTER, R. & SOÓ, R.V. 1925-44: Monographie und Iconographie der Orchideen Europas und des Mittelmeergebietes. - Berlin-Dahlem: Verlag des Repertoriums, 5 vols. [cf. TL-2 2: 514-515, 1979].

KELLERER, J. & SÜNDERMANN, F. 1901: *Saxifraga Ferdinandi Coburgi* nov. spec. - *Allg. Bot. Z. Syst.* **7(7-8)**: 116.

KELLY, L.M. 2001: Taxonomy of *Asarum* section *Asarum* (Aristolochiaceae). - *Syst. Bot.* **26(1)**: 17-53.

KEMMER, U. & KEMMER, A. 1987: Orchideensuche auf Rhodos. - *Mitt. Arbeitskreis Heimische Orchid. Baden-Württemberg* **19(4)**: 853-865.

KENICER, G. & NORTON, S. 2008a: 631. *Lathyrus neurolobus*. - *Curtis's Bot. Mag.* **25(4)**: 310-316.

KENICER, G. & NORTON, S. 2008b: 633. *Lathyrus latifolius*. - *Curtis's Bot. Mag.* **25(4)**: 325-332.

KENT, D.H. 1997: The correct authority for lesser chickweed, *Stellaria pallida* (Caryophyllaceae). - *Watsonia* **21(4)**: 364.

KENTZINGER, M. 1974: Contribution à l'étude cytotaxinomique des Géraniacées du bassin méditerranéen oriental. - *Biol. Gallo-Hellen.* **5(2)**: 191-208.

KENTZINGER, M. 1976: Contribution à l'étude cytotaxinomique des Géraniacées du bassin méditerrannéen oriental. - *Biol. Écol. Médit.* **3(1)**: 9-12.

KEREN, A. & EVENARI, M. 1975: Some ecological aspects of distribution and germination of *Pancratium maritimum* L. - *Israel J. Bot.* **23(4)**: 202-215.

KERESZTY, Z. 1987: Chromosome morphology and DNA content in the systematics of the *Scilla bifolia* aggregate. - *Acta Bot. Hung.* **33(3-4)**: 305-316.

KERESZTY, Z. 1988: Taxonomic evaluation of the *Scilla bifolia* complex in Hungary [In Hungarian with English abstract]. - *Bot. Közlem.* **74-75(1-2)**: 47-72.

KERESZTY, Z. & PODANI, J. 1984: A preliminary numerical taxonomic study of the *Scilla bifolia* aggregate (Liliaceae, Scilloideae) in Hungary. - *Acta Bot. Hung.* **30(3-4)**: 353-362.

KERESZTY, Z. & SZILÁGYI, L. 1984: Cytological investigation of *Scilla bifolia* populations in Hungary. I. - *Acta Bot. Hung.* **30(1-2)**: 53-66.

KERESZTY, Z. & SZILÁGYI, L. 1986: Cytological investigation of *Scilla bifolia* populations in Hungary, II. - *Acta Bot. Hung.* **32(1-4)**: 167-176.

KERFOOT, O. & LAVRANOS, J.J. 1984: Studies in the Flora of Arabia X: *Juniperus phoenicea* L. & *J. excelsa* M. Bieb. - *Notes Roy. Bot. Gard. Edinburgh* **41(3)**: 483-489.

KERGUÉLEN, M. 1975: Les gramineae (Poaceae) de la flora française. Essai de mise au point taxonomique et nomenclaturale. - *Lejeunia, n.s.* **75**: 1-343.

KERGUÉLEN, M. 1983: Les gramineae de France au travers de "Flora Europaea" et de la "Flore" du C.N.R.S. - *Lejeunia, n.s.* **110**: 1-79.

KERGUÉLEN, M. 1987: Données taxonomiques, nomenclaturales et chorologiques pour une révision de la flore de France. - *Lejeunia, n.s.* **120**: 1-263.

KERGUÉLEN, M. & PLONKA, F. 1988: Le genre *Festuca* dans la flora française: taxons nouveaux, observations nomenclaturales et taxinomiques. - *Bull. Soc. Bot. Centre-Ouest, n.s.* **19**: 15-30.

KERGUÉLEN, M. & PLONKA, F. 1989: Les *Festuca* de la flore de France (Corse comprise). - *Bull. Soc. Bot. Centre-Ouest, n.s.* **Num. Spec. 10**: 368 pp.

KERNDORFF, H. 1994: Two new taxa in Turkish *Crocus* (Iridaceae). - *Herbertia* **49(1-2)**: 76-86.

KERNDORFF, H. 1995: Notes on *Crocus* (Iridaceae) in Syria and Jordan. - *Herbertia* **50**: 68-81.

KERNDORFF, H. & AL. 2011: Three new species of *Crocus* (Liliiflorae, Iridaceae) from Turkey. - *Stapfia* **95**: 19-25.
KERNDORFF, H. & AL. 2012: Seven new species of *Crocus* (Liliiflorae, Iridaceae) from Turkey. - *Stapfia* **97**: 3-16.
KERNDORFF, H. & AL. 2013a: A new species of *Crocus* (Liliiflorae, Iridaceae) from Turkey. - *Stapfia* **99**: 141-144.
KERNDORFF, H. & AL. 2013b: *Crocus biflorus* Miller (Liliiflorae, Iridaceae) in Anatolia. Part IV. - *Stapfia* **99**: 159-186.
KERNDORFF, H. & PASCHE, E. 1994: *Crocus mathewii*. A new autumn-flowering *Crocus* from Turkey. - *New Plantsman* **1(2)**: 102-106.
KERNDORFF, H. & PASCHE, E. 1996: Crocuses from Turkey to Jordan. - *Quart. Bull. Alpine Gard. Soc. Gr. Brit.* **64(3)**: 296-312.
KERNDORFF, H. & PASCHE, E. 2003: *Crocus biflorus* in Anatolia. - *Plantsman, n.s.* **2(2)**: 77-89.
KERNDORFF, H. & PASCHE, E. 2004a: Two new taxa of the *Crocus biflorus* aggregate (Liliiflorae, Iridaceae) from Turkey. - *Linzer Biol. Bitr.* **36(1)**: 5-10.
KERNDORFF, H. & PASCHE, E. 2004b: *Crocus biflorus* in Anatolia. Part two. - *Plantsman, n.s.* **3**: 201-215.
KERNDORFF, H. & PASCHE, E. 2006: *Crocus biflorus* (Liliiflorae, Iridaceae) in Anatolia. Part three. - *Linzer Biol. Beitr.* **38**: 165-187.
KERNDORFF, H., PASCHE, E. & HARPKE, D. 2015: The genus *Crocus* (Liliiflorae, Iridaceae): Life-cycle, morphology, phenotypic characteristics, and taxonomical relevant parameters. - *Stapfia* **103**: 27-65.
KERNDORFF, H., PASCHE, E. & HARPKE, D. 2016: The Genus *Crocus* (Liliiflorae, Iridaceae): Taxonomical problems and how to determine a species nowadays? - *Stapfia* **105**: 42-50.
KERNER, A.V. 1878: Monographia Pulmonariarum. - Oeniponte [Innsbruck]: Sumptibus Librariae Acad. Wagnerianae, 53 pp. + 12 plates.
KESERCIOĞLU, T. 1973a: Studies on the cytotaxonomy, anatomy and morphology of Turkish plants. 1. *Liquidambar orientalis* Mill. [In Turkish with English summary]. - *Ege Üniv. Fen Fak. Ilmi Raporlar Ser.* **173**: 1-14.
KESERCIOĞLU, T. 1973b: Studies on the cytotaxonomy, anatomy and morphology of Turkish plants. 2. *Nigella damascena* L. and *Datura stramonium* L. [In Turkish with English summary]. - *Ege Üniv. Fen Fak. Ilmi Raporlar Ser.* **174**: 1-13.
KESERCIOĞLU, T. 1978: Some investigations on *Ononis natrix* L. and *Ononis viscosa* L. [In Turkish with English summary]. - *Bitki Koruma Bült.* **5(2)**: 170-180.
KESKIN, M. 2001: The genus *Trifolium* L. sect. *Chronosemium* Ser. in Turkey. In Özhatay, N. (ed.): Proceedings of the Second Balkan Botanical Congress, 1: 267-278. - Istanbul.
KHALAF, M.K. & STACE, C.A. 2000: Breeding systems and relationships of the *Cerastium tomentosum* group. - *Preslia* **72(2-4)**: 323-344.
KHALFALLAH, N. 1981: Contribution à l'étude biosystématique du genre *Arrhenatherum* Beauv. en France. - Paris: Thèse, Univ. de Paris-Sud, 88 pp.
KHALILOV, I.I. 1993: A synopsis of the genus *Crambe* [In Russian]. - *Bot. Zhurn. (Moscow & St. Petersburg)* **78(1)**: 107-115.
KHAN, A. & KHAN, A.H. 1986: A cytogenetic investigation in *Solanum nigrum* complex. - *Indian J. Bot.* **9(1)**: 14-17.
KHAN, M.A. & STACE, C.A. 1999: Breeding relationships in the genus *Brachypodium* (Poaceae: Pooideae). - *Nordic J. Bot.* **19(3)**: 257-269.
KHAN, M.S. 1964: Taxonomic revision of *Euphorbia* in Turkey. - *Notes Roy. Bot. Gard. Edinburgh* **25 (2)**: 71-161 + 6 plates.
KHANDELWAL, S. 1990: Chromosome evolution in the genus *Ophioglossum* L. - *Bot. J. Linn. Soc.* **102(3)**: 205-217.
KHARE, P.S. & KAUR, S. 1983: Intraspecific polyploidy in *Pteris vittata* Linn. - *Cytologia* **48(1)**: 21-25.
KHOSHOO, T.N. 1958a: Biosystematics of *Sisymbrium irio* complex. I: Modifications in phenotype. - *Caryologia* **11(2)**: 109-132 + 4 plates.
KHOSHOO, T.N. 1958b: Biosystematics of *Sisymbrium irio* complex. II: Morphology of the races. - *Caryologia* **11(2)**: 133-157 + 5 plates.
KHOSHOO, T.N. 1959a: Biosystematics of *Sisymbrium irio* complex. III: Cytology of races. - *Caryologia* **11(3)**: 297-318 + 1 plate.
KHOSHOO, T.N. 1959b: Biosystematics of *Sisymbrium irio* complex. IV: Origin of triploid, hexaploid and octoploid. - *Caryologia* **11(3)**: 319-329 + 3 plates.
KHOSHOO, T.N. 1966: Biosystematics of *Sisymbrium irio* complex. XII: Distributional patterns. - *Caryologia* **19(2)**: 143-150.
KHUSH, G.S. 1963: Cytogenetic and evolutionary studies in *Secale* III. Cytogenetics of weedy ryes and origin of cultivated rye. - *Econ. Bot.* **17(1)**: 60-71.
KIEFT, B. & LOON, J.CHR. VAN 1978: [Miscellaneous chromosome numbers from Italy]. In Löve, Á. (ed.): IOPB chromosome number reports LXII. - *Taxon* **27(5-6)**: 524-525.
KIEM, J. 1985: Beitrag zur Flora Griechenlands. Liste der während der BBG-Exkursion 1983 nach Griechenland gesammelten Gräser. - *Ber. Bayer. Bot. Ges.* **56**: 227-230.
KIEM, J. 1987: Die Gattung *Festuca* in Südtirol und in Nachbargebieten. - *Ber. Bayer. Bot. Ges.* **58**: 53-71.
KIERS, A.M. 2000: Endive, cichory, and their wild relatives. A systematic and phylogenetic study of *Cichorium* (Asteraceae). - *Gorteria, Suppl.* **5**: 1-77.
KIGER, R.W. 1973: Sectional nomenclature in *Papaver* L. - *Taxon* **22(5-6)**: 579-582.
KIGER, R.W. 1985: Revised sectional nomenclature in *Papaver* L. - *Taxon* **34(1)**: 150-152.
KIK, C., LINDERS, TH.E. & BIJLSMA, R. 1992: The distribution of cytotypes in ecologically contrasting populations of the clonal perennial *Agrostis stolonifera*. - *Evol. Trends Pl.* **6(2)**: 93-98.
KIK, C., LINDERS, TH.E. & BIJLSMA, R. 1993: Ploidy level and somatic chromosome number variation in *Agrostis stolonifera*. - *Acta Bot. Neerl.* **42(1)**: 73-80.
KIK, CH. & AL. 2021: Collecting and regenerating populations of the *Allium ampeloprasum* complex from Greece. - *Genetic Resources* **2(3)**: 1-10.
KIMBER, G. & FELDMAN, M. 1987: Wild wheat. An introduction. - *Special Rep. Missouri Agric. Exp. Sta.* **353**: i-ii, 1-142.
KIMNACH, M. 1984: *Sedum creticum*, a little-known succulent from the Greek islands. - *Cact. Succ. J. (Los Angeles)* **56(4)**: 162-164.
KING, R.M. & ROBINSON, H. 1970: *Eupatorium*, a Composite genus of arcto-tertiary distribution. - *Taxon* **19(5)**: 769-774.
KINTZIOS, S.E. (ED.) 2000: Sage. The genus *Salvia* (Medicinal and aromatic plants, 14). - Amsterdam: Harwood Academic Publishers, xix + 296 pp.
KINTZIOS, S.E. (ED.) 2002: Oregano. The genera *Origanum* and *Lippia* (Medicinal and aromatic plants, 25). - New York: Taylor & Francis, xiii + 277 pp.
KINZEL, H. & WEBER, M. 1982: Serpentin-Pflanzen. In Kinzel, H. (ed.): Pflanzenökologie und Mineralstoffwechsel, pp. 381-410. - Stuttgart: Verlag Eugen Ulmer.
KIRÁLY, G. & ALEGRO, A. 2015: Re-evaluation of the *Panicum capillare* complex (Poaceae) in Croatia. - *Acta Bot. Croat.* **74(1)**: 173-179.
KIRÁLY, G., STETÁK, D. & BÁNYÁSZ, Á. 2007: Spread of invasive macrophytes in Hungary. - *Neobiota* **7**: 123-130.
KIRAN, Y. & AL. 2008: Karyological notes on another eight species of *Achillea* (Asteraceae) from Turkey. - *Biologia (Bratislava)* **63(3)**: 343-348.
KIRCHHOFF, A, & PETERMANN, J. 1992a: Vegetationskundliche Untersuchungen im Nestos-Tal (NO-Griechenland). In Mattes, H. & Lienau, C. (eds): Colloquium Nordost-Griechenland 1, pp. 15-21. - Münster: Inst. Geographie.
KIRCHHOFF, A, & PETERMANN, J. 1992b: Die Vegetation im Nestostal in Nordost-Griechenland im Bereich des zukünftigen Stausees. - *Ber. Arbeitsgeb. Entwicklungsforsch. (Münster)* **20**: iii + 77 pp. + 8 fold-out charts.
KIRCHNER, D.E. 2002: Phyto-Geographie von *Arabis alpina* L. (Brassicaceae). - *Bot. Jahrb. Syst.* **124 (2)**: 183-210.

KIREÇ, M. & YARCI, C. 1999: The flora of the agricultural areas in Enez (Edirne) and environs. - *Türk. Bot. Derg.* **23(1)**: 53-62.

KIRKBRIDE, J.H. JR. 1995: (1165) Proposal to reject the name *Lotus glaber* Mill. (Leguminosae) . - *Taxon* **44(3)**: 423-424. [approved, cf. Taxon 55(4): 1094, 2005].

KIRKBRIDE, J.H. JR., WIERSEMA, J.H. & TURLAND, N.J. 2006: (1753) Proposal to conserve the name *Ziziphus jujuba* against *Z. zizyphus* (Rhamnaceae). - *Taxon* **55(4)**: 1049-1050. [approved, cf. Taxon 58(1): 283, 2009].

KIRSCHNER, J. 1980: A new species of the *Luzula campestris-multiflora* complex in Central Europe. - *Folia Geobot. Phytotax.* **14(4)**: 431-435.

KIRSCHNER, J. 1982: Notes on the determination of the members of the *Luzula campestris-multiflora* complex (*L. campestris* agg.) in Czechoslovakia [In Czech with English summary]. - *Zprávy Českoslov. Bot. Společn. Českoslov. Akad. Věd* **17(1)**: 25-37.

KIRSCHNER, J. 1990: *Luzula multiflora* and allied species (Juncaceae): A nomenclatural study. - *Taxon* **39(1)**: 106-114.

KIRSCHNER, J. 1991: An account of natural hybridization within *Luzula* sect. *Luzula* (Juncaceae) in Europe. - *Preslia* **63(2)**: 81-112.

KIRSCHNER, J. 1992: A *Luzula* sect. *Luzula* puzzle near Sofia, Bulgaria. A biodiversity contribution. - *Ann. Bot. Fenn.* **29(3)**: 235-241.

KIRSCHNER, J. 1993a: Taxonomic survey of *Luzula* sect. *Luzula* (Juncaceae) in Europe. – *Folia Geobot. Phytotax.* **28(2)**: 141-182.

KIRSCHNER, J. 1993b: Karyological differentiation of *Luzula* sect. *Luzula* in Europe. - *Thaiszia* **2**: 11-39.

KIRSCHNER, J. 1996: Tetraploid populations of *Luzula multiflora* subsp. *multiflora* (Juncaceae) in Europe. - *Preslia* **67(3-4)**: 219-223.

KIRSCHNER, J. 2001: (1499-1501) Proposals to conserve the names *Juncus arcuatus* and *Luzula multiflora* with conserved types, and *L. nodulosa* against *L. graeca* (Juncaceae). - *Taxon* **50(4)**: 1189-1192. [proposal on *Luzula multiflora* withdrawn, the others approved; cf. Taxon 53(3): 816, 2004].

KIRSCHNER, J. & AL. 2021: *Taraxacum assemanii* represents a new section: A revision of the misinterpreted *Taraxacum primigenium*, and the elucidation of the enigmatic *Taraxacum* section *Primigenia* (Compositae, Crepidinae). - *Phytotaxa* **520(2)**: 117-136.

KIRSCHNER, J. & KAPLAN, Z. 2001: Taxonomic and nomenclatural notes on *Luzula* and *Juncus* (Juncaceae). - *Taxon* **50(4)**: 1107-1113.

KIRSCHNER, J., SONCK, C.E. & ŠTĚPÁNEK, J. 1989: Four new *Taraxacum* species of sect. *Palustria* from Greece. - *Ann. Bot. Fenn.* **26(3)**: 343-348.

KIRSCHNER, J., SONCK, C.E. & ŠTĚPÁNEK, J. 1993: An outline of the *Taraxacum subalpinum* group (*T.* sect. *Palustria*). - *Folia Geobot. Phytotax.* **28(1)**: 51-60.

KIRSCHNER, J. & ŠTĚPÁNEK, J. 1985a: "Taraxacum hoppeanum" and its allies (Studies in *Taraxacum* 4.). - *Preslia* **57(2)**: 111-134.

KIRSCHNER, J. & ŠTĚPÁNEK, J. 1985b: Selected and annotated chromosome counts in *Taraxacum* sect. *Palustria* Dahlst. - *Folia Geobot. Phytotax.* **20(4)**: 407-417.

KIRSCHNER, J. & ŠTĚPÁNEK, J. 1986: Towards a monograph of *Taraxacum* sect. *Palustria* (Studies in *Taraxacum* 5.). - *Preslia* **58(2)**: 97-116.

KIRSCHNER, J. & ŠTĚPÁNEK, J. 1993: *Taraxacum* sect. *Palustria* (Compositae) in Bulgaria. - *Thaiszia* **3 (1)**: 13-29.

KIRSCHNER, J. & ŠTĚPÁNEK, J. 1997: A nomenclatural checklist of supraspecific names in *Taraxacum*. - *Taxon* **46(1)**: 87-98.

KIRSCHNER, J. & ŠTĚPÁNEK, J. 1998a: A monograph of *Taraxacum* sect. *Palustria*. - Průhonice: Inst. Bot., Acad. Sci., Czech Rep., 281 pp.

KIRSCHNER, J. & ŠTĚPÁNEK, J. 1998b: A revision of *Taraxacum* sect. *Piesis* (Compositae). - *Folia Geobot.* **33(4)**: 391-414.

KIRSCHNER, J., ŠTĚPÁNEK, J. & ŠTĚPÁNKOVÁ, J. 1982: [Chromosome numbers from Czechoslovakia]. In Löve, Á. (ed.): IOPB chromosome number reports LXXVI. - *Taxon* **31(3)**: 574-575.

KIRSCHNEROVÁ, L. & KIRSCHNER, J. 1996: A nomenclatural and taxonomic account of *Willemetia* (Compositae, Lactuceae, Crepidinae). - *Taxon* **45(4)**: 627-630.

KIRYAKOV, I. & PETROVA, A.V. 2004: *Buglossoides glandulosa* (Boraginaceae): morphology and distribution in Bulgaria. - *Phytol. Balcan.* **9(3)**: 517-527.

KITANOV, B. 1936: Upon the forms of *Acer Heldreichii* Oprh., which occur in Bulgaria [In Bulgarian with English summary]. - *Trav. Soc. Bulg. Sci. Nat.* **17**: 135-142.

KITANOV, B. 1943: Die Vegetation des Boz-Dagh-Gebirges in Ostmazedonien [In Bulgarian with German summary]. - *God. Sofiisk. Univ. Fiz.-Mat. Fak., 3 (Estestv. Istorija)* **39**: 169-291 + 1 map.

KITANOV, B. 1945: Neues Material zur Kenntnis der Flora des Boz-Dagh Gebirges in Ost-Mazedonien [In Bulgarian with German abstract]. - *God. Sofiisk. Univ. Fiz.-Mat. Fak., 3 (Estestv. Istorija)* **41**: 77-80.

KITANOV, B. 1948: Matériaux concernant la flore de la Macédoine et de la Bulgarie [In Bulgarian with French summary]. - *Godišen Zborn. Filoz. Fak. Univ., Prir.-Mat. Odd.* **1**: 215-222.

KITANOV, B. 1950a: *Crepis macedonica* Kitan. sp. nov. [In Latin and Bulgarian]. - *Izv. Bot. Inst. (Sofia)* **1**: 372-376.

KITANOV, B. 1950b: Kritische Bemerkungen über zwei Pflanzenarten Mazedoniens. - *Izv. Bot. Inst. (Sofia)* **1**: 377-392.

KITANOV, B. 1951a: *Tulipa Pavlovii* Kitan. - a new species from Makedonija [In Latin and Bulgarian]. - *Izv. Bot. Inst. (Sofia)* **2**: 177-180 + 1 plate.

KITANOV, B. 1951b: Kritische Bemerkungen über die Vertreter der Gattung *Arrhenatherum* P.B. auf der Balkanhalbinsel [In Bulgarian with Russian and German summaries]. - *Izv. Bot. Inst. (Sofia)* **2**:195-208.

KITANOV, B. 1960: Literatur über die Flora und Pflanzengeographie Bulgariens 1949-1958 [In Bulgarian]. - Sofia: Publishing House of the Bulgarian Academy of Sciences, 75 pp.

KITANOV, B. 1976: Literatur über die Flora und Pflanzengeographie Bulgariens 1959-1968 [sphalm. "1969"] [In Bulgarian]. - Sofia: Publishing House of the Bulgarian Academy of Sciences, 270 pp.

KITANOV, B. 1984: Literatur über die Flora und Pflanzengeographie Bulgariens 1969-1978 [In Bulgarian]. - Sofia: Publishing House of the Bulgarian Academy of Sciences, 240 pp.

KITANOV, B. & DRENKOVSKI, R. 1975: Über die Taxonomie einiger Sippen der Gattung *Crocus* L. auf der Balkanhalbinsel. In Jordanov, D. & al. (eds): Problems of Balkan flora and vegetation, pp. 212-215 + 1 plate. - Sofia: Publishing House of the Bulgarian Academy of Sciences.

KITANOV, B. & KITANOV, G. 1990: Flora na Pirin [In Bulgarian]. - Sofija: Nauka i Izkustvo, 480 pp.

KJELLQVIST, E. & LÖVE, Á. 1963: Chromosome numbers of some *Carex* species from Spain. - *Bot. Not.* **116(2)**: 241-248.

KLÁŠTERSKÁ, I. 1969a: Cytology and some chromosome numbers of Czechoslovak roses. I. - *Folia Geobot. Phytotax.* **4(2)**: 175-189.

KLÁŠTERSKÁ, I. 1969b: [Chromosome numbers in *Rosa* from Czechoslovakia]. In Löve, Á. (ed.): IOPB chromosome number reports XXI. - *Taxon* **18(3)**: 311-312.

KLÁŠTERSKÁ, I. & KLÁŠTERSKÝ, I. 1974: Cytology and some chromosome numbers of Czechoslovak roses. II. - *Bot. Not.* **127(3)**: 328-337.

KLÁŠTERSKÁ, I. & NATARAJAN, A.T. 1974: Cytological studies of the genus *Rosa* with special reference to the section »Caninae«. - *Hereditas (Lund)* **76(1)**: 97-108.

KLÁŠTERSKÝ, I. 1942: Rosae Asiae anteriosis et peninsulae Balcanicae in itineribus cl. K.H. Rechingeri fil., A. Pilati et aliorum collectae. - *Stud. Bot. Čechoslov.* **5(1-2)**: 59-74 + 3 plates.

KLÁŠTERSKÝ, I. 1967: Les formes de *Rosa pimpinellifolia* L. - *Bull. Jard. Bot. Natl. Belg.* **37(1)**: 37-38.

KLÁŠTERSKÝ, I. 1974a: Die Problematik des *Rosa canina* L.-Komplexes in der Tschechoslowakei. - *Mitt. Bot. Arbeitsgem. Oberösterr. Landesmus. Linz* **6(1)**: 65-74.

KLÁŠTERSKÝ, I. 1974b: Die Problematik des *Rosa agrestis*-Komplexes in der Tschechoslowakei. - *Mitt. Bot. Arbeitsgem. Oberösterr. Landesmus. Linz* **6(1)**: 75-82.

KLÁŠTERSKÝ, I. & AL. (EDS) 1958: Philipp Maximilian Opiz und seine Bedeutung für die Pflanzentaxonomie. - Praha: Nakladatelství Československé akademie věd, 271 pp.

KLÁŠTERSKÝ, I. & NOVOTNÁ, I. 1962: The *Arabis hirsuta* complex [In Czech with English summary]. - *Preslia* **34**: 387-393.

KLATT, F.W. 1866: Die Gattung "Lysimachia" L. - *Abh. Naturwiss. Vereins Hamburg* **4**: 1-45 + 24 plates.

KLAUS, W. 1989: Mediterranean pines and their history. - *Pl. Syst. Evol.* **162(1-4)**: 133-163.

KLEIN, E. 1973: *Orchis sancta* L. × *Orchis coriophora* L. ssp. *fragrans* (Poll.) Camus - die erste intragenerische Hybride der Sektion *Coriophorae* Parlat. - *Orchidee (Hamburg)* **24(5)**: 209-211.

KLEIN, E. 1978a: Eine neue Hybride mit einem Mitglied der *Ophrys*-Subsektion *Aegaea*: *O. cretica* (Vierh.) Nelson × *O. bombyliflora* Link. - *Orchidee (Hamburg)* **29(5)**: 215-217.

KLEIN, E. 1978b: Zwei neue *Ophrys*-Hybriden von der Insel Kreta. - *Hoppea* **36(2)**: 467-471.

KLEIN, E. 1989: Die intragenerischen Hybriden der Gattung *Ophrys* sowie deren intergenerischen Hybriden mit den Gattungen *Anacamptis*, *Aceras* und *Serapias*. - *Ber. Arbeitskreis. Heimische Orchid.* **6(1)**: 12-24.

KLEIN, E. 2005: Versuch einer Gliederung der Gattung *Epipactis* Zinn (Orchidaceae-Neottieae). - *J. Eur. Orch.* **37(1)**: 121-130.

KLEIN, J.C. & AL. 1997: Analyse cytogénétique comparée de trois taxons du genre *Hyoseris* L. - *Lagascalia* **19(1-2)**: 529-536.

KLEINSTEUBER, A., RISTOW, M. & HASSLER, M. (HRSG.) 2016: Flora von Rhodos und Chalki. Band 1: Allgemeiner Teil. Spezieller Teil: Polypodiopsida, Equisetopsida und Lycopodiopsida. Pinopsida und Gnetopsida. Magnoliopsida (Familien A-F). - Karlsruhe: Naturwissenschaftlicher Verlag A. Kleinsteuber, 607 pp.

KLIMENKO, M. & AL. 2007: Morphological variation of *Juniperus oxycedrus* subsp. *oxycedrus* (Cupressaceae) in the Mediterranean region. - *Flora, Morphol. Geobot. Oekophysiol.* **202(2)**: 133-147.

KLINKHAMMER, P.G.L. & DE JONG, T.J. 1993: Biological Flora of the British Isles. *Cirsium vulgare* (Savi) Ten. (*Carduus lanceolatus* L., *Cirsium lanceolatum* (L.) Scop., non Hill). - *J. Ecol.* **81(1)**: 177-191.

KLIPHUIS, E. 1963: Cytological observations in relation to the taxonomy of the orchids of the Netherlands. - *Acta Bot. Neerl.* **12**: 172-194.

KLIPHUIS, E. 1974a: Cytotaxonomic notes on some *Galium* species. A. - *Proc. Kon. Ned. Akad. Wetensch. C* **77(4)**: 345-366.

KLIPHUIS, E. 1974b: Cytotaxonomic studies in *Galium palustre* L. - *Proc. Kon. Ned. Akad. Wetensch. C* **77(4)**: 408-425 + 2 plates.

KLIPHUIS, E. 1977: [Miscellaneous chromosome numbers]. In Löve, Á. (ed.): IOPB chromosome number reports LVI. - *Taxon* **26(2-3)**: 267-268.

KLIPHUIS, E. 1980: Cytotaxonomic studies on *Galium aparine* L. - *Proc. Kon. Ned. Akad. Wetensch. C* **83(1)**: 53-64.

KLIPHUIS, E. 1984: Cytotaxonomic studies on *Galium verum* L. - *Bot. Helv.* **94(2)**: 269-278.

KLIPHUIS, E. 1986: Cytotaxonomic investigations on some species of the genus *Galium* (Rubiaceae) from the Balkans. - *Nordic J. Bot.* **6(1)**: 15-20.

KLIPHUIS, E. & BARKOUDAH, Y.I. 1977: Chromosome numbers in some Syrian angiosperms. - *Acta Bot. Neerl.* **26(3)**: 239-249.

KLIPHUIS, E., HERINGA, J. & HOGEWEG, P. 1986: Cytotaxonomic studies on *Galium palustre* L. Morphological differentiation of diploids, tetraploids and octoploids. - *Acta Bot. Neerl.* **35(4)**: 383-392.

KLIPHUIS, E. & WIEFFERING, J.H. 1972: Chromosome numbers of some angiosperms from the South of France. - *Acta Bot. Neerl.* **21(6)**: 598-604.

KLJUYKOV, E.V. 1987: De generibus *Stefanoffia* H. Wolff et *Carum* L. (Apiaceae) in flora Turcica notulae systematicae. - *Novosti Sist. Vysš. Rast.* **23**: 89-92.

KLJUYKOV, E.V. 1988: A survey of the genus *Bunium* L. Revision of the generic system. - *Byull. Moskovsk. Obshch. Isp. Prir., Otd. Biol.* **93(1)**: 76-88.

KLJUYKOV, E.V. & AL. 2015: *Dichoropetalum schottii*: examination of the type specimen, distribution and comparison with *D. carvifolium-chabraei* (Apiaceae). - *Nat. Hist. Sci.* **2(2)**: 131-136.

KLJUYKOV, E.V. & AL. 2020: A taxonomic survey of monocotylar Apiaceae and the implications of their morphological diversity for their systematics and evolution. - *Bot. J. Linn. Soc.* **192(3)**: 449-473.

KLOET, S.P. VAN DER 1981: On the lectotypification of *Vaccinium* L. - *Taxon* **30(3)**: 646-648.

KLOKOV, M. 1957: Conspectus generis *Melittis* L. [In Russian]. - *Bot. Mater. Gerb. Bot. Inst. Komarova Akad. Nauk S.S.S.R.* **18**: 183-217.

KLOPFER, K. & ROBEL, J. 1989a: Beiträge zur Systematik, Morphologie und Anatomie der Gattung *Amaranthus* L. I. Karpomorphologie und -anatomie ausgewählter Vertreter. - *Gleditschia* **17(1)**: 3-21.

KLOPFER, K. & ROBEL, J. 1989b: Beiträge zur Systematik, Morphologie und Anatomie der Gattung *Amaranthus* L. II. Samenmorphologie und -anatomie ausgewählter Vertreter. - *Gleditschia* **17(2)**: 171-182.

KLOTZ, G. 1962a: Zur Systematik und Nomenklatur einiger *Echium*-Arten. I. - *Wiss. Z. Martin-Luther- Univ. Halle-Wittenberg, Math.-Naturwiss. Reihe* **11(2)**: 293-302.

KLOTZ, G. 1962b: Zur Systematik und Nomenklatur einiger *Echium*-Arten. II. - *Wiss. Z. Martin-Luther-Univ. Halle-Wittenberg, Math.-Naturwiss. Reihe* **11(5)**: 703-711.

KLOTZ, G. 1962c: Zur Systematik und Nomenklatur einiger *Echium*-Arten. III. - *Wiss. Z. Martin-Luther-Univ. Halle-Wittenberg, Math.-Naturwiss. Reihe* **11(9)**: 1087-1104.

KLÖTZLI, F. 1976: Grenzen von Laubwäldern in Europa. - *Ber. Deutsch. Bot. Ges.* **89(2-3)**: 371-380.

KNAPP, H.D. & HACKER, E. 1984: Zur Einbürgerung von *Telekia speciosa* (Schreb.) Baumg. in Mecklenburg. - *Gleditschia* **12**: 85-106.

KNAPP, R. 1964: Vegetations-Untersuchungen im mittleren Griechenland und über Probleme der Entstehung anthropogener Pflanzengesellschaften. - *Ber. Oberhess. Ges. Natur- Heilk., Giessen, N.F., Naturwiss. Abt.* **33(4)**: 373-393.

KNAPP, R. 1965: Die Vegetation von Kefallinia, Griechenland. Geobotanische Untersuchung eines mediterranen Gebietes und einige ihrer Anwendungs-Möglichkeiten in Wirtschaft und Landesplanung. - Koenigstein: Verlag Otto Koeltz, 206 pp.

KNAPP, R. 1967: Neue geobotanische Untersuchungen in Afrika und Griechenland. - *Ber. Deutsch. Bot. Ges.* **79(9)**: 413-422.

KNAPP, S. & AL. 2017: A revision of the *Solanum elaegnifolium* clade (*Elaegnifolium* clade; subgenus *Leptostemonum*, Solanaceae). - *PhytoKeys* **84**: 1-104.

KNAPP, S., BARBOZA, G.E. & SÄRKINEN, T. 2017: (2547) Proposal to conserve *Solanum alatum* (Solanaceae) with a conserved type. - *Taxon* **66(4)**: 988-989. [approved, cf. Taxon 69(2): 392, 2020].

KOACH, J. & GALIL, J. 1987: The breeding system of *Arisarum vulgare* Targ-Tozz. - *Israel J. Bot.* **35 (2)**: 79-90.

KOBYLYANSKI, V.D. 1983a: The system of the genus *Secale* L. [In Russian with English summary]. - *Trudy Prikl. Bot.* **79**: 24-38.

KOBYLYANSKI, V.D. 1983b: Phylogeny of the genus *Secale* L. [In Russian with English summary]. - *Trudy Prikl. Bot.* **79**: 38-45.

KOCA, F. 1985: Karyological studies on *Iris attica* Boiss. & Heldr. and *Iris suaveolens* Boiss. & Reuter. - *Istanbul Üniv. Eczac. Fak. Mecm.* **21**: 69-79.

KOCA, F. 2001: Morphological and kariological studies on *Iris pumila* L. (Iridaceae). In Raimondo, F. M. (ed.): Abstracts from the X OPTIMA meeting, Palermo, 13-19 September 2001, p. 122.

KOCA, F. 2003: Morpho-anatomical and karyological studies on *Iris pumila* L. (Iridaceae). – *Bocconea* **16(2)**: 625-639.

KOÇAK, B., ÖZHATAY, N. & MATHEW, B.F. 2000: The musk hyacinth in Turkey. - *New Plantsman* **7(2)**: 68-73.

KOČEV, H. 1963: On the spread and biology of *Poa silvicola* Guss. in Bulgaria [In Bulgarian with Russian and English summaries]. - *Izv. Bot. Inst. (Sofia)* **11**: 5-47.

KOCH, K. 1879: Bäume und Sträucher des alten Griechenlands. - Stuttgart: Verlag von Ferdinand Enke, xx + 270 pp.

KOCH, M. & HURKA, H. 1999: Isozyme analysis in the polyploid complex *Microthlaspi perfoliatum* (L.) F.K. Meyer: Morphology, biogeography and evolutionary history. - *Flora, Morphol. Geobot. Oekophysiol.* **194(1)**: 33-48.

KOCH, M.A., KARL, R. & GERMAN, D.A. 2017: Underexplored biodiversity of Eastern Mediterranean biota: systematics and evolutionary history of the genus *Aubrieta* (Brassicaceae). - *Ann. Bot. (Oxford), n.s.* **119**: 39-57.

KOCH, W.D.J. 1837-38: Synopsis der deutschen und schweizer Flora, enthaltend die genauer bekannten Pflanzen, welche in Deutschland, der Schweiz, in Preussen und Istrien wild wachsen ... Ed. 1. - Frankfurt am Main: Im Verlage von F. Wilmans, 840 pp. [cf. TL-2 2: 593, 1979].

KOCH, W.D.J. 1846-47: Synopsis der deutschen und schweizer Flora, enthaltend die genauer bekannten Pflanzen, welche in Deutschland, der Schweiz, in Preussen und Istrien wild wachsen ... Ed. 2. - Frankfurt am Main: Im Verlage von F. Wilmans, 2 vols. [cf. TL-2 2: 593, 1979].

KOCH, W.D.J. 1892-1907: Synopsis der deutschen und schweizer Flora ... In Verbindung mit namhaften Botanikern herausgegeben von Professor Dr. E. Hallier, fortgesetzt von R. Wohlfarth ... Ed. 3. - Frankfurt am Main: Verlag von F. Wilmans, 3 vols. [cf. TL-2 2: 593-594, 1979].

KOCHJAROVÁ, J. 1991: Karyological study of the Slovak flora XXIII. - *Acta Fac. Rerum Nat. Univ. Comen., Bot.* **38**: 89-99.

KOCHJAROVÁ, J. 1992: Karyological study of the Slovak flora XXVIII. - *Acta Fac. Rerum Nat. Univ. Comen., Bot.* **39**: 67-74.

KÖCKE, A.V. & AL. 2010: Revision of the Mediterranean and Southern African *Triglochin bulbosa* complex (Juncaginaceae). - *Edinburgh J. Bot.* **67(3)**: 353-398.

KOCYAN, A. & AL. 2021: Addendum to the orchid flora of the island of Kythira (Greece). - *J. Eur. Orch.* **53(1)**: 3-18.

KOCYAN, A. & JOSHI, J. 1992: Die Orchideen von Kea. Ein Beitrag zum OPTIMA-Projekt zur Kartierung der Orchideen des Mittelmeerraumes. - *Mitt. Arbeitskreis Heimische Orchid. Baden- Württemberg* **24(3)**: 457-486.

KOEHNE, AE. 1880: Lythraceae monographice describuntur. - *Bot. Jahrb. Syst.* **1**: 142-178, 240-266.

KOEHNE, AE. 1881a: Lythraceae monographice describuntur [cont.]. - *Bot. Jahrb. Syst.* **1**: 305-335, 436-458.

KOEHNE, AE. 1881b: Lythraceae monographice describuntur [cont.]. - *Bot. Jahrb. Syst.* **2**: 136-176.

KOEHNE, AE. 1882a: Lythraceae monographice describuntur [cont.]. - *Bot. Jahrb. Syst.* **2**: 395-429.

KOEHNE, AE. 1882b: Lythraceae monographice describuntur [cont.]. - *Bot. Jahrb. Syst.* **3**: 129-155, 319-352.

KOEHNE, AE. 1886: Lythraceae monographice describuntur [Schluss]. Die geographische Verbreitung der Lythraceen. - *Bot. Jahrb. Syst.* **7**: 1-61 + 1 map.

KOEHNE, AE. 1907: Lythraceae. Nachträge. - *Bot. Jahrb. Syst.* **41(2)**: 74-110.

KOEHNE, AE. 1912: Die geographische Verbreitung der Kirschen, *Prunus* subgen. *Cerasus*. - *Mitt. Deutsch. Dendrol. Ges.* **21**: 168-183.

KOEVA-TODOROVSKA, J.C. 1975: [Chromosome numbers in *Stachys* from Bulgaria]. In Löve, Á. (ed.): IOPB chromosome number reports L. - *Taxon* **24(5-6)**: 675.

KOEVA-TODOROVSKA, J.C. 1978: *Pontostachys* - a new section of the genus *Stachys* [In Bulgarian with Russian and English summaries]. - *Fitologiya* **10**: 33-40 + 7 plates.

KOEVA-TODOROVSKA, J.C. 1988: Karyological and pollenmorphol. study ... *Stachys* L. in Bulgaria [In Bulgarian with Russian and English summaries]. In Velčev, V.I. & al. (eds): 100th anniversary of Acad. N.A. Stojanov, pp. 138-151. - Sofia: Publishing House of the Bulgarian Academy of Sciences.

KOEVA, J.C. 1977: Cytotaxonomical study of some Balkan endemic taxa of genus *Stachys* L. - *Fitologiya* **6**: 38-46.

KOEVA, J.C. & SHARKOVA, S.T. 1994a: Karyological study of some Bulgarian Ranunculaceae taxa. - *God. Sofiisk. Univ. "Kliment. Ohridski" Biol. Fak., 2 Bot.* **84**: 25.

KOEVA, J.C. & SHARKOVA, S.T. 1994b: Karyological study of some species of *Pulsatilla* Mill. (Ranunculaceae) from Bulgaria. - *God. Sofiisk. Univ. "Kliment. Ohridski" Biol. Fak., 2 Bot.* **86**: 29-34.

KOEVA, J.C., SHARKOVA, S.T. & GEORGIEVA, A. 1993: Karyological study of three *Paeonia* L. species. In Petrova, A.V. (ed.): Abstracts from the VII OPTIMA meeting in Borovetz, Bulgaria, 18-30 July 1993, p. 136.

KOEVA, J.C. & TEGOVA, M.I. 1994: Karyological study of some *Adonis* L. species. - *God. Sofiisk. Univ. "Kliment. Ohridski" Biol. Fak., 2 Bot.* **85**: 205-209.

KOFINAS, G., POLYMENAKOS, K. & TAN, KIT 2019: Reports 53-57. In Vladimirov, V. & al. (eds): New floristic records in the Balkans: 38. - *Phytol. Balcan.* **25(1)**: 106-107.

KOFINAS, G. & TAN, KIT 2019: Reports 88-95. In Vladimirov, V., Aybeke, M. & Tan, Kit (eds): New floristic records in the Balkans: 39. - *Phytol. Balcan.* **25(2)**: 216-219.

KOFINAS, G. & TAN, KIT 2020: Reports 43-48. In Vladimirov, V., Aybeke, M. & Tan, Kit (eds): New floristic records in the Balkans: 41. - *Phytol. Balcan.* **26(1)**: 168-170.

KOFINAS, G. & TAN, KIT 2021: Report 78-85. In Vladimirov, V., Aybeke, M. & Tan, Kit (eds): New floristic records in the Balkans: 44. - *Phytol. Balcan.* **27(1)**: 141-145.

KÖHLEIN, F. 1984: Saxifrages and related genera. Translated by D. Winstanley [German original published in 1980]. - London: B.T. Batsford Ltd, 220 pp. + plates.

KÖHLEIN, F. 1987: *Iris*. Translated by M.C. Peters [German original published in 1981]. - London: Christopher Helm Ltd, 370 pp.

KÖHLEIN, F. 1991: Gentians. Translated by D. Winstanley [German original published in 1986]. - London: Christopher Helm Ltd, 183 pp. (incl. colour plates).

KØHLER, L. & AL. 2018: Lesbos 24. til 31 marts [In Danish]. - Privately distributed report, 92 pp.

KOHLMÜLLER, R. 1991a: *Dactylorhiza cordigera* (Fries) Soo × *Leucorchis frivaldii* (Griseb.) Schlechter in Nordgriechenland. - *Ber. Arbeitskreis. Heimische Orchid.* **8(1)**: 54-55.

KOHLMÜLLER, R. 1991b: Eine Zusammenstellung der Hybriden von *Ophrys speculum* Link und ihrer Verwandten. - *Mitt. Arbeitskreis Heimische Orchid. Baden-Württemberg* **23(2)**: 321-330.

KOHLMÜLLER, R. 1995: *Ophrys × vamvakiae* R. Kohlmüller, nothosp. nat. nov. - Ein neuer Hybrid der Orchideenflora der Insel Kreta. - *Ber. Arbeitskreis. Heimische Orchid.* **12(1)**: 59-62.

KOKKINI-GOUZKOUNI, S. 1983: Taxonomic studies in the genus *Mentha* in Greece [In Greek with English summary]. - Thessaloniki: Ph.D. thesis, Univ. of Thessaloniki, viii + 171 pp.

KOKKINI, S. & AL. 1988: Geographical and altitudinal distribution of the Lamiaceae in Greece. - *Phyton (Horn)* **28(2)**: 215-228.

KOKKINI, S. & BABALONAS, D. 1982: Morphological variation in *Nepeta nuda* L. - *Feddes Repert.* **93(1-2)**: 71-76.

KOKKINI, S., CHANLIDOU, E. & KAROUSOU, R. 2000: Smell and essential oil variation in Labiatae: does it deserve a taxonomist's appreciation? - *Bot. Chron. (Patras)* **13**: 187-199.

KOKKINI, S. & VOKOU, D. 1993: The hybrid *Origanum ×intercedens* from the island of Nisyros (SE Greece) and its parental taxa: comparative study of essential oils and distribution. - *Biochem. Syst. Ecol.* **21(3)**: 397-403.

KOKKINI, S., VOKOU, D. & KAROUSOU, R. 1989: Morphological and chemical variation of *Origanum vulgare* L. in Greece. In Phitos, D. (ed.): Abstracts. Sixth OPTIMA meeting, Delphi 10-16 Sept. 1989, p. 43.

KOKKINI, S., VOKOU, D. & KAROUSOU, R. 1991: Morphological and chemical variation of *Origanum vulgare* L. in Greece. - *Bot. Chron. (Patras)* **10**: 337-346.

KOKKORIS, I. 2014: Study on the flora and vegetation of Mt. Panachaiko. Ecological evaluation, management proposals and establishment of a bio-monitoring program using remote sensing methods and geographical information systems. [In Greek with English abstract]. - Patras: Ph.D. thesis, Univ. of Patras, 495 pp.

KOKKORIS, I. & AL. 2014: The native flora of Mountain Panachaikon (Peloponnese, Greece): new records and diversity. - *J. Biol. Res. Thessalon.* **21(9)**: 8 pp. + electronic supplements.

KOKMOTOS, E. 2008: Floristic & phytosociological research of Boeotia mountains (Elikon-Xerovouni- Neraidolakkoma). Comparative exploration and ecological approach. [In Greek with English summary]. - Patras: Ph.D. thesis, University of Patras, 543 pp.

KOKMOTOS, E. & GEORGIADIS, TH. 2001: Contribution to the flora of mount Elikon (Sterea Ellas, Greece). In Kamari, Psaras & Constantinidis (eds): Proceedings of the 8th scientific congress of the Greek Botanical Society, October 5-8, 2000, pp. 208-214 [In Greek with English abstract].

KOKMOTOS, E. & GEORGIADIS, TH. 2005: The flora of mountains Elikon, Xerovouni and Neraidolakkoma (Boeotia, Sterea Ellas, Greece). - *Fl. Medit.* **15**: 403-451.

KOKTAY, P. 1974: Morphological and cytological studies on the *Verbascum* species of Istanbul area. - *Istanbul Üniv. Fen Fak. Mecm., B* **39(1-2)**: 95-124.

KOKTAY, P. 1978: [Chromosome numbers in *Papaver* from the Istanbul area]. In Löve, Á. (ed.): IOPB chromosome number reports LXII. - *Taxon* **27(5-6)**: 522.

KOLEV, I.D. & SEDMAKOVA-LOUCHANSKA, E. 1973: A karyological study of *Trifolium arvense* L., *T. preslianum* Boiss. and *T. smyrnaeum* Boiss. in Bulgaria. - *Dokl. Selskost. Akad. "George Dimitrov"* **5 (1)**: 39-43.

KOLEV, I.D. & TERSIJSKI, D. 1964: Karyologische und embryologische Charakteristik von *Vicia melanops* S.S. [In Bulgarian with Russian and German summaries]. - *Rasteniev. Nauki* **1(7)**: 43-49.

KOLLMANN, F. 1970: *Allium descendens* L., a nomen ambiguum. - *Taxon* **19(5)**: 789-792.

KOLLMANN, F. 1971a: *Allium ampeloprasum* - a polyploid complex. I. Ploidy levels. - *Israel J. Bot.* **20 (1)**: 13-20.

KOLLMANN, F. 1971b: *Allium* "descendens" in the Mediterranean. - *Notes Roy. Bot. Gard. Edinburgh* **31(1)**: 119-124.

KOLLMANN, F. 1971c: *Allium ampeloprasum* L. in Israel (taxonomy). - *Israel J. Bot.* **20(4)**: 263-272.

KOLLMANN, F. 1972: *Allium ampeloprasum* - a polyploid complex. II. Meiosis and relationships between the ploidy types. - *Caryologia* **25(3)**: 295-312.

KOLLMANN, F. 1973: Karyology of some species of *Allium* section *Molium* in Israel. - *Israel J. Bot.* **22 (2)**: 92-112.

KOLLMANN, F. & FEINBRUN, N. 1968: A cyto-taxonomic study in Palestinian *Anagallis arvensis* L. (With an appendix by B.L. Burtt). - *Notes Roy. Bot. Gard. Edinburgh* **28(2)**: 173-186.

KOLLMANN, F., ÖZHATAY, N. & KOYUNCU, M. 1983: New *Allium* taxa from Turkey. - *Notes Roy. Bot. Gard. Edinburgh* **41(2)**: 245-267.

KOLLMANN, F. & SHMIDA, A. 1977: *Allium* species of Mt. Hermon. I. Taxonomy. - *Israel J. Bot.* **26(3)**: 128-148.

KOLLMANN, F., SHMIDA, A. & COHEN, O. 1991: *Allium tardiflorum* Kollmann & Shmida, a new autumn-flowering species. - *Herbertia* **46(1-2)**: 23-32.

KOLLMANN, F. & STEARN, W.T. 1976: Note [on] *Allium trifoliatum* subsp. *hirsutum*. - *Israel J. Bot.* **24(4)**: 201-204.

KOLLMANN, J. & GRUBB, P.J. 2002: Biological Flora of the British Isles. *Viburnum lantana* L. and *Viburnum opulus* L. (*V. lobatum* Lam., *Opulus vulgaris* Borkh.). - *J. Ecol.* **90(6)**: 1044-1070.

KOLTZENBURG, M. 1999: Bestimmungsschlüssel für in Mitteleuropa heimische und kultivierte Pappelarten und -sorten (*Populus* spec.). - *Florist. Rundbr., Beih.* **6**: 54 pp. + 28 plates.

KOMNINOU, O. & KAROUSOU, R. 2013: Recording and mapping the population of *Artemisia eriantha* Ten. in the NATURA 2000 site Oros Olympos (GR 1250001). In Gkelis, S. & al. (eds): 13th Panhellenic Scientific Conference, Hellenic Botanical Society, Thessaloniki 3-6 October. Program and book of abstracts, p. 87.

KONČALOVÁ, M.N., JEHLÍK, V. & MĚSÍČEK, J. 1986: Chromosome numbers of some *Oenothera* species (subgenus *Oenothera*) from Czechoslovakia. - *Folia Geobot. Phytotax.* **21(2)**: 163-172.

KONČALOVÁ, M.N. & KLÁŠTERSKÝ, I. 1978: Cytology and chromosome numbers of some Czechoslovak roses. III. - *Folia Geobot. Phytotax.* **13(1)**: 67-93.

KONOP, R. 1987: Netřesky. Rody *Sempervivum* a *Jovibarba*. - Praha: Vydal Český zahrádkářský svaz, 327 pp. + plates and maps.

KONOPOVÁ, K. & KONOP, R. 1983: *Sempervivum marmoreum* Griseb. - *Skalničky* **1983(1)**: 6-10.

KONSTANTINIDIS, P. & AL. 2008: Taxonomy and ecology of *Castanea sativa* Mill. forests in Greece. - *Pl. Ecol.* **195(2)**: 235-256.

KONSTANTINIDIS, P., TSIOURLIS, G. & TSIONTIS, A. 2002: Taxonomy and ecology of plant communities of *Quercus frainetto* Ten. (*Q. conferta* Kit.) forests in Greece. - *Israel J. Pl. Sci.* **50(2)**: 145-154.

KONSTANTINOU, M. 1992: Phytosociological studies of the vegetation on metalliferous soils in northern Greece [In Greek with English summary]. - *Sci. Ann. School Biol. Aristot. Univ. Thessalonika* **35**: xv + 242 pp. + tables.

KONSTANTINOU, M. & AL. 1999: Endemic and rare plants of the Agion Oros peninsula. In Elefheriou, E. P. (ed.): Proceedings, 7th Panhellenic scientific conference, Alexandroupolis, pp. 348-353 [In Greek with English abstract]. - Thessaloniki: Univ. Studio Press.

KONSTANTINOU, M. & AL. 2009: A web-enabled interactive application for the development of a knowledge base on Greek flora. - *Bot. Serbica* **33(2)**: 131-136.

KONSTANTINOU, M. & BABALONAS, D. 1992: Geobotanische Untersuchung auf Galmei-Erzhalden auf der Insel Thassos (Griechenland). - *J. Agr. Crop Sci.* **168**: 1-9.

KONTOPANOU, A. & PANITSA, M. 2020: Habitat Islands on the Aegean Islands (Greece): Elevational gradient of chasmophytic diversity, endemism, phytogeographical patterns and need for monitoring and conservation. - *Diversity* **12, 33** : doi:10.3390/d12010033.

KONTOS, P. 1909: Epidrasis epi tin ellinikin dasikin vlastisin tou ellinikou klimatos [On the effect of climate on the Greek forest vegetation; in Greek]. - Athine: 78 pp.

KONTOS, P. 1929a: Beitrag zur Kenntnis der Waldverteilung in Thessalien und Epirus. - *Compt. Rend. Acad. Athènes [Praktika Akadimias Athinon]* **4**: 375-381.

KONTOS, P. 1929b: Ein Beitrag zur Kenntnis der Waldverteilung in Macedonien, Thrazien und Nord- Griechenland. - *Compt. Rend. Acad. Athènes [Praktika Akadimias Athinon]* **4**: 429-442.

KONTOS, P. 1930: Beitrag zur Kenntnis der Waldverteilung auf den griechischen Inseln. - *Compt. Rend. Acad. Athènes [Praktika Akadimias Athinon]* **5**: 300-306.

KOOPMAN, J. 2017: The section *Phaestoglochin* (*Carex*, Cyperaceae) in the Netherlands. - *Gorteria* **39**: 79-87.

KOOPMAN, J. 2022: *Carex* Europaea: The genus *Carex* L. (Cyperaceae) in Europe, 1. Accepted names, hybrids, synonyms, distribution, chromosome numbers. 3rd ed. - Weikersheim: Margraf Publishers, 808 pp. [1st ed. in 2011].

KOOPMAN, J. & AL. 2018: Global distribution of *Carex buekii* (Cyperaceae) reappraised. – *Phytotaxa* **358(2)**: 139-161.

KOOPMAN, J. & AL. 2019: Nomenclatural and taxonomic notes on *Carex lazarei* nom. nov. (Cyperaceae). - *Phytotaxa* **422(3)**: 295-297.

KOOPMAN, J., WIĘCŁAW, H. & WILHELM, M. 2016: Distribution of *Carex pallidula* (Cyperaceae) in Europe. - *Acta Soc. Bot. Poloniae* **85(3)**: 7 pp.

KOOTIN-SANWU, M. & WOODELL, S.R.J. 1970: The cytology of *Caltha palustris* L. (sensu lato): Distribution and behavior of the chromosome races. - *Caryologia* **23(2)**: 225-239.

KORAKIS, A. 2014: *Neotinea lactea* (Orchidaceae), a new taxon for Samos island, Greece. - *Parnassiana Arch.* **2**: 21-23.

KORAKIS, G. 2006: New localities of the vulnerable species *Eriolobus trilobatus* (Poiret) M. Roemer in northeastern Greece. In Peev, D.R. (ed.): Book of abstracts, IV Balkan botanical congress, Sofia, Bulgaria, 20-26 June 2006, p. 224-225.

KORAKIS, G. & AL. 2006: Floristic records from Dadia-Lefkimi-Soufli national park, NE Greece. - *Fl. Medit.* **16**: 11-32.

KORAKIS, G. & AL. 2009: New localities of the vulnerable species *Eriolobus trilobatus* (Rosaceae) in northeastern Greece. In Ivanova, D. (ed.): Proceedings of the IV Balkan botanical congress, pp. 422-426. - Sofia: Institute of Botany, BAS.

KORAKIS, G. & AL. 2017: On the distribution and population status of the threatened species *Malus trilobata* in Greece. In Dousis, M.A. & Thanos, K.A.: 15th conference of the Hellenic Botanical Society, book of abstracts: p. 50.

KORAKIS, G. & ATHANASIADIS, N. 2006: The submediterranean vegetation of Mountain Paiko (north central Greece). In Peev, D.R. (ed.): Book of abstracts, IV Balkan botanical congress, Sofia, Bulgaria, 20-26 June 2006, p. 224.

KORAKIS, G., GERASIMIDIS, A. & KATI, V. 2010: The flora. In Catsadorakis, G. & Källander, H. (eds): The Dadia-Lefkimi-Soufli forest national park, Greece: Biodiversity, management and conservation, pp. 63-83. - Athens: WWF Greece.

KOŠANIN, N. 1911: Eine interessante Pflanze von Jakupica in Makedonien. - *Magyar Bot. Lapok* **10(4-7)**: 115-118.

KOŠANIN, N. 1913: *Narthecium scardicum* spec. nova. - *Oesterr. Bot. Z.* **63(4)**: 141-143.

KOŠANIN, N. 1914: *Dioscorea balcanica* Koš. n. sp. - *Oesterr. Bot. Z.* **64(1-2)**: 37-39 + 1 plate.

KOŠANIN, N. 1921: La couverture végétale des montagnes de la Macédoine occidentale et méridionale [In Serbo-Croatian with French summary]. - *Bull. Inst. Jard. Bot. Univ. Belgrade* **6**: 62-74.

KOŠANIN, N. 1922a: La distribution géographique des deux espèces de *Ramondia* du Balkans. - *Glas Srpske Kral. Akad.* **1922**: 34-49.

KOŠANIN, N. 1922b: *Drosera macedonica* spec. nova [In Serbo-Croatian]. - Reprint from "Spomenik, S.M. Lozanić", pp. 83-85 + 1 plate.

KOŠANIN, N. 1926a: Les espèces nouvelles dans la flore de la Serbie du Sud (Macedoine) [In Serbo-Croatian with Latin diagnoses and French summary]. - *Glas Srpske Kral. Akad.* **54**: 19-29 + 5 plates.

KOŠANIN, N. 1926b: Systématique et géographie de lis *albanicum* et *carniolicum* [In Serbo-Croatian with French summary]. - *Glas Srpske Kral. Akad.* **54**: 35-55.

KOŠANIN, N. 1928: Aus der Flora Südserbiens. I. *Crocus*-Arten. - *Bull. Inst. Jard. Bot. Univ. Belgrade* **1(1)**: 90-95.

KOŠANIN, N. 1930a: Die Koniferen Südserbiens. - *Bull. Inst. Jard. Bot. Univ. Belgrade* **1(3)**: 177-190.

KOŠANIN, N. 1930b: Die Verbreitung der *Castanea sativa* im Königreiche S.H.S. - *Bull. Inst. Jard. Bot. Univ. Belgrade* **1(3)**: 191-198.

KOŠANIN, N. 1930c: Nachträge zur *Dioscorea balcanica* Koš. - *Bull. Inst. Jard. Bot. Univ. Belgrade* **1(3)**: 199-202.

KOŠANIN, N. 1930d: Die Bewegungen der Blüten- und Fruchtstiele bei der Gattung *Cyclamen*. - *Bull. Inst. Jard. Bot. Univ. Belgrade* **1(3)**: 260-286 + 1 plate.

KOŠANIN, N. 1932: Das Vorkommen von *Damasonium Bourgaei* Coss. an den Ufern des Prespasees in Südserbien. - *Bull. Inst. Jard. Bot. Univ. Belgrade* **2(1-2)**: 126-127.

KOSENKO, V.N. 1977: Comparative karyological study of three genera of the family Berberidaceae (*Leontice*, *Bongardia*, *Gymnospermium*) [In Russian]. - *Bot. Zhurn. (Moscow & Leningrad)* **62(2)**: 235-240.

KOSHY, J.K. & MATHEW, P.M. 1985: Cytology of the genus *Cleome* Linn. - *Cytologia* **50(2)**: 283-287.

KOSTADINOVA, S. & DIMITROV, D.S. 2002: New data on the vascular flora of Mt Belasitsa. - *Phytol. Balcan.* **8(3)**: 293-306.

KOSTADINOVSKI, M. 1997: *Festuca valida* (Uechtr.) Penzes and *F. thracica* (Acht.) I. Markgraf-Dbg. (Poaceae), new species for the flora of Republic of Macedonia [In Macedonian with English summary]. - *Godišen Zborn. Biol. Fak. Univ. "Kiril i Metodij"* **50**: 41-46.

KOSTADINOVSKI, M. 1999: Taxonomy and chorology of the genus *Festuca* L. (Poaceae) in the flora of Republic of Macedonia. - Skopje: Résumé of doctoral dissertation, 10 pp.

KOSTRAKIEWICZ-GIERAŁT, K. & AL. 2018: The causes of disappearance of sword lily *Gladiolus imbricatus* L. from natural stands - synthesis of current state of knowledge. - *Int. J. Conserv. Sci.* **9(4)**: 821-834.

KOSTRAKIEWICZ, C. 1958: De notis *Trifolium pannonicum* L. a *T. ochroleucum* L. distinguentibus. - *Fragm. Florist. Geobot.* **4(1-2)**: 199-202.

KOTSCHY, TH. 1858-62: Die Eichen Europas und des Orients. Gesammelt, zum Theil neu entdeckt und mit Hinweisung auf ihre Culturfähigkeit für Mitteleuropa beschrieben. - Wien & Olmütz: 40 folio plates with text in Latin, German and French.

KOTSIRAS, K. & AL. 2020: Integrating plant diversity data into mapping and assessment of ecosystem and their services (MAES). Implementation in Greece: Woodland and forest pilot. - *Forests MDPI* **11, 956**: doi:10.3390/f11090956.

KOUDROS, V. & ALEXIOIU, S. 2014: The first lowland locality for rare *Erythronium dens-canis* L. (Liliaceae) from Greece. - *Parnassiana Arch.* **2**: 17-20.

KOUGIOUMOUTZIS, K. 2014: Meleti protipon viopikilotitas ke fitogeografikon schesion nisiotikon endiatimaton me sichrones methodous. [In Greek] - Patras: Ph.D. thesis, University of Patras, 359 pp.

KOUGIOUMOUTZIS, K. & AL. 2010: Evaluation of the flora and vegetation of Trizonia island - floristic affinities with small Ionian islands. - *Bot. Chron. (Patras)* **20**: 45-61.

KOUGIOUMOUTZIS, K. & AL. 2012a: Contribution to the flora of the South Aegean volcanic arc: The Methana peninsula (Saronic Gulf, Greece). - *Edinburgh J. Bot.* **69(1)**: 53-81.

KOUGIOUMOUTZIS, K. & AL. 2012b: Contribution to the flora of the South Aegean volcanic arc: Anafi island (Kiklades, Greece). - *Willdenowia* **42(1)**: 127-141.

KOUGIOUMOUTZIS, K. & AL. 2014: Contribution to the flora of the South Aegean volcanic arc: Kimolos island (Kiklades, Greece). - *Edinburgh J. Bot.* **71(2)**: 135-160.

KOUGIOUMOUTZIS, K. & AL. 2015: Contribution to the flora and biogeography of the Kiklades: Folegandros island (Kiklades, Greece). - *Edinburgh J. Bot.* **72(3)**: 391-412.

KOUGIOUMOUTZIS, K. & AL. 2017: Network biogeography of a complex island system: the Aegean Archipelago revisited. - *J. Biogeogr.* **44(3)**: 651-660.

KOUGIOUMOUTZIS, K. & AL. 2020a: Plant diversity patterns and conservation implications under climate-change scenarios in the Mediterranean: The case of Crete (Aegean, Greece). - *Diversity* **12, 270**: doi: 10.3390/d12070270.

KOUGIOUMOUTZIS, K. & AL. 2020b: Spatial phylogenetics, biogeographical patterns and conservation implications of the endemic flora of Crete (Aegean, Greece) under climate change scenarios. - *Biology (Basel)* **9, 199**: doi: 10.3390/biology9080199.

KOUGIOUMOUTZIS, K. & AL. 2021a: Plant endemism centres and biodiversity hotspots in Greece. - *Biology MDPI* **10, 72**: 27 pp.

KOUGIOUMOUTZIS, K. & AL. 2021c: Climate-change impacts on the southernmost Mediterranean arctic- alpine plant populations. - *Sustainability MDPI* **13, 2021**: 23 pp.

KOUGIOUMOUTZIS, K., KALPOUTZAKIS, E. & CONSTANTINIDIS, TH. 2017: Multivariate morphometric analysis and taxa delimitation in two narrow Greek endemics: *Astragalus maniaticus* and *Aethionema saxatile* subsp. *corinthiacum*. - *Pl. Biosyst.* **151(1)**: 108-116 + 3 electronic appendages.

KOUGIOUMOUTZIS, K., SIMAIAKIS, M. & TINIAKOU, A. 2014: Network biogeographical analysis of the central Aegean archipelago. - *J. Biogeogr.* **41(10)**: 1848-1858.

KOUGIOUMOUTZIS, K. & TINIAKOU, A. 2015: Ecological factors driving plant species diversity in the South Aegean Volcanic Arc and other central Aegean islands. - *Pl. Ecol. Divers.* **8(2)**: 173-186.

KOUGIOUMOUTZIS, K. & TINIAKOU, A. 2013: Contribution to the flora of the volcanic islands of the Cyclades: Kimolos Island. In Gkelis, S. & al. (eds): 13th Panhellenic Scientific Conference, Hellenic Botanical Society, Thessaloniki 3-6 October. Program and book of abstracts, p. 89.

KOUGIOUMOUTZIS, K. & AL. 2021b: Extinction risk assessment of the Greek endemic flora. - *Biology MDPI* **10,195**: 22 pp.

KOUMPLI-SOVANTZI, L. 1983: Studies on the tracheophytes in the lakes and adjacent hydrobiotopes of Aetoloakarnania, Greece [In Greek with English summary]. - Athens: Univ. of Athens, Ph.D. thesis, 346 pp.

KOUMPLI-SOVANTZI, L. 1986: *Cotula coronopifolia* L., a new adventive species for Greece. - *Biol. Gallo-Hellen.* **11(2)**: 313-318.

KOUMPLI-SOVANTZI, L. 1989: The aquatic flora of Aetoloakarnania (W Greece). - *Willdenowia* **18(2)**: 377-385.

KOUMPLI-SOVANTZI, L. 1991: Coastal lagoon flora of NW Peloponnisos (Greece). - *Willdenowia* **21(1-2)**: 119-129.

KOUMPLI-SOVANTZI, L. 1995: *Althenia filiformis* Petit (Zannichelliaceae) in Greece. - *Phyton (Horn)* **35 (2)**: 243-245.

KOUMPLI-SOVANTZI, L. 2008: Ricefield flora in the area of Acheloos river estuaries (W Greece). In: Praktika 10ou Sinedriou Ellinikis Votanikis Eterias, 5-8 maiou 2005, Ioannina: pp. 319-323 [In Greek with English abstract].

KOUMPLI-SOVANTZI, L., PITSOYANI, O. & TSINOPOULOS, T. 2010: Macrophytes from artificial aquatic ecosystems of Karditsa Prefecture (Thessalia, Greece). - *Fl. Medit.* **20**: 5-18.

KOUMPLI-SOVANTZI, L. & VALLIANATOU, F. 1985: The aquatic vegetation of the lake Lyssimachia (Western Greece). - *Thalassographica* **8**: 33-41.

KOUMPLI-SOVANTZI, L. & VALLIANATOU, I. 1994: Floristic notes from aquatic stands of central Greece (Sterea Ellas). - *Candollea* **49(1)**: 195-207.

KOUMPLI-SOVANTZI, L., VALLIANATOU, I. & YANNITSAROS, A. 1997: A contribution to the hydrophilous flora of Peloponnisos (Greece). - *Feddes Repert.* **108(5-6)**: 453-461.

KOUMPLI-SOVANTZI, L. & YANNITSAROS, A. 1993: A contribution to the coastal flora of the Kikladhes (Greece). - *Willdenowia* **23(1-2)**: 121-135.

KOUMPLI-SOVANTZI, L., YANNITSAROS, A. & DOXA, A. 2001: Floral survey of the Delphi archaeological site. In Kamari, Psaras & Constantinidis (eds): Proceedings of the 8th scientific congress of the Greek Botanical Society, October 5-8, 2000, pp. 230-233 [In Greek with English abstract].

KOUTROUMPA, K., KAROUSOU, R. & PIRINTSOS, S. 2013: Genetic diversity of the hybrid *Origanum* x*intercedens* Rech. in Crete. In Gkelis, S. & al. (eds): 13th Panhellenic Scientific Conference, Hellenic Botanical Society, Thessaloniki 3-6 October. Program and book of abstracts, p. 92.

KOUTSERI, I. & NINOU, CH. 2017: LIFE12 NAT/GR/000539-JunEx. Restoration and conservation of Graecian Juniper woods in Prespa National Park, Greece. Layman's report. [Separate versions in English and Greek]. - Society for the Protection of Prespa, 16 pp.

KOUTSIDOU, E. & MARGARIS, N.S. 1990: The recovery of plant communities in grazing exclosures in the island of Chios. - *Petromarula* **1**: 60-64.

KOUTSOMARKOS, K., TAN, KIT & TSIALTAS, I. 2015: Report 70. In Vladimirov, V. & al. (eds): New floristic records in the Balkans: 27. - *Phytol. Balcan.* **21(2)**: 201.

KOUTSOPOULOS, P. & SARLIS, G. 2003: Contribution to the study of the flora of Vouraikos gorge (Peloponnesos, Greece). - *Fl. Medit.* **12**: 299-314.

KOUTSOTHEODORIS, L. 2008: Registration of the flora within the "Monastery of Elona and Leonidio Gorge" (NATURA 2000 - GR 2520005) site of European ecological network and promotion of the environmental elements with the aim of tourist development in the area. [In Greek]. - Patras: M.Sc. thesis, University of Patras, 103 pp.

KOVAČEV, I. 1962: Beitrag zur Erforschung der Gattung *Sternbergia* in Bulgarien [In Bulgarian with German summary]. - *Nauchni Trudove Selskost. Inst. "Vasil Kolarov"* **11(1)**: 195-198.

KOVAČEV, I. 1964: Systematics and evolution of the genus *Sternbergia* W.K. [In Bulgarian]. - *Dokl. Bulg. Akad. Nauk* **17(3)**: 275-277.

KOVAČEV, I. 1965a: Beitrag zur Erforschung der Systematik von Gattung *Gladiolus* L. in dem Mittelmeer-Gebiet sowie in Mittel- und Südeuropa [In Bulgarian with German summary]. - *Nauchni Trudove Selskost. Inst. "Vasil Kolarov"* **14(1)**: 59-67.

KOVAČEV, I. 1965b: On the system of genus *Amaranthus* L. [In Bulgarian with English summary]. - *Dokl. Bulg. Akad. Nauk* **18(7)**: 647-650.

KOVAČEV, I. 1965c: Generis *Colchicum* L. species nova e Bulgaria [In Latin and Russian]. - *Novosti Sist. Vysš. Rast.* **2**: 87-89.

KOVAČEV, I. 1968: Neuer Beitrag zur Systematik der Gattung *Amaranthus* L. [In Bulgarian with German summary]. - *Nauchni Trudove Selskost. Inst. "Vasil Kolarov"* **17(1)**: 203-204.

KOVAČEVA, D. & DELIPAVLOV, D.D. 1988: On the taxonomic status of *Pyrus bulgarica* Khutath. & Sachok. and *P. elaeagrifolia* Pall. - *Nauchni Trudove Selskost. Inst. "Vasil Kolarov"* **33(4)**: 121-128.

KOVAČIĆ, S. 2004: The genus *Campanula* L. (Campanulaceae) in Croatia, circum-Adriatic and west Balkan region. - *Acta Bot. Croat.* **63(2)**: 171-202.

KOVÁCS-LÁNG, E. 1968: Über die Verbreitung der *Pimpinella saxifraga* Unterarten und Varietäten in der ungarischen und karpatischen Flora. - *Ann. Univ. Sci. Budapest. Rolando Eötvös, Sect. Biol.* **9-10**: 241-245.

KOVÁCS, B. 1910: *Carex echinata* Murr. (1770) ist doch nichts anderes als *C. stellulata* Good. (1794). - *Magyar Bot. Lapok* **9(3-4)**: 126-131.

KOVANDA, M. 1961: Flower and fruit morphology of *Sorbus* in correlation to the taxonomy of the genus. - *Preslia* **33(1)**: 1-16.

KOVANDA, M. 1962: Taxonomical studies in *Sorbus* subg. *Aria*. - *Dendrol. Sborn.* **3**: 23-70 + 2 maps.

KOVANDA, M. 1966a: Some chromosome counts in the *Campanula rotundifolia* complex. - *Preslia* **38**: 48-52.

KOVANDA, M. 1966b: Some chromosome counts in the *Campanula rotundifolia* complex. II. – *Folia Geobot. Phytotax.* **1(2)**: 268-273.

KOVANDA, M. 1968a: Cytotaxonomic studies in the genus *Jasione* L. I. *Jasione montana* L. - *Folia Geobot. Phytotax.* **3(2)**: 193-199.

KOVANDA, M. 1968b: New taxa and combinations in the subsection *Heterophylla* (Witas.) Fed. of the genus *Campanula* L. - *Folia Geobot. Phytotax.* **3(4)**: 407-411.

KOVANDA, M. 1970a: Polyploidy and variation in the *Campanula rotundifolia* complex. Part I. (General). - *Rozpr. Českoslov. Akad. Věd.* **80(2)**: 1-95 + 3 plates.

KOVANDA, M. 1970b: Polyploidy and variation in the *Campanula rotundifolia* complex. Part II (taxonomic). 1. Revision of the groups *Saxicolae*, *Lanceolatae* and *Alpicolae* in Czechoslovakia and adjacent regions. - *Folia Geobot. Phytotax.* **5(2)**: 171-208.

KOVANDA, M. 1977: Polyploidy and variation in the *Campanula rotundifolia* complex. Part II (taxonomic). 2. Revision of the groups *Vulgares* and *Scheuchzerianae* in Czechoslovakia and adjacent regions. - *Folia Geobot. Phytotax.* **12(1)**: 23-89.

KOVANDA, M. 1981: Studies in *Phyteuma*. - *Preslia* **53(3)**: 211-238.

KOVANDA, M. & ANČEV, M.E. 1989: The *Campanula rotundifolia* complex in Bulgaria. - *Preslia* **61(3)**: 193-207.

KOVANDA, M. & POUZAR, Z. 1982: (666) Proposal to conserve *Sorbus* L. 1753 with the species *Sorbus aucuparia* L. as its type. - *Taxon* **31(2)**: 340-341. [not approved, cf. Taxon 34(4): 659, 1985].

KOVÁTS, D. 1975: Boraginaceae type specimens of Herbarium Carpato-Pannonicum. - *Stud. Bot. Hung.* **10**: 123-133 + 6 plates.

KOVÁTS, D. 1976: *Phleum* studies I. Data on the taxonomy and morphology of *Phleum bertolonii* DC. and *Phleum pratense* L. - *Acta Bot. Acad. Sci. Hung.* **22(1-2)**: 107-126.

KOVÁTS, D. 1977: *Phleum* studies II. *Phleum hubbardii*, a new species of Poaceae (Gramineae). - *Acta Bot. Acad. Sci. Hung.* **23(1-2)**: 119-142.

KOVÁTS, D. 1980: Distribution and diversity of *Phleum Hubbardii* and *Phleum pratense* (Poaceae) in the Carpathian Basin. - *Stud. Bot. Hung.* **14**: 107-116.

KOVÁTS, D. 1981: Distribution and diversity of *Phleum alpinum* L. and *Phleum commutatum* Gaud. (Poaceae) in the Carpathians. - *Stud. Bot. Hung.* **15**: 65-76.

KOVÁTS, D. 1982: Poaceae type specimens of Herbarium Carpato-Pannonicum I. (*Erianthus-Agrostis*). - *Stud. Bot. Hung.* **16**: 99-107.

KOVÁTS, D. 1984: Poaceae type specimens of Herbarium Carpato-Pannonicum in Budapest II (*Calamagrostis-Sesleria*). - *Stud. Bot. Hung.* **17**: 61-68.

KOVÁTS, D. 1992: Waldstein and Kitaibel types in the Hungarian Natural History Museum in Budapest. - *Ann. Hist.-Nat. Mus. Natl. Hung.* **84**: 33-53.

KOVÁTS, D. 1997a: Taxa and types of Pál Kitaibel and Sándor Jávorka [In Hungarian with English summary]. - *Bot. Közlem.* **84(1-2)**: 97-102.

KOVÁTS, D. 1997b: Taxa and types of Sándor Jávorka [In Hungarian with English summary]. - *Kitaibelia* **2(2)**: 325-326.

KOVÁTS, D. 1998: Plant types of Sándor Jávorka in the Hungarian Natural History Museum in Budapest. I. - *Ann. Hist.-Nat. Mus. Natl. Hung.* **90**: 115-132.

KOVÁTS, D. 1999: Plant types of Sándor Jávorka in the Hungarian Natural History Museum in Budapest. II. - *Ann. Hist.-Nat. Mus. Natl. Hung.* **91**: 45-60.

KOVÁTS, D. 2000: Plant types of Sándor Jávorka in the Hungarian Natural History Museum in Budapest. III. - *Ann. Hist.-Nat. Mus. Natl. Hung.* **92**: 21-40.

KOWARIK, I. & BÖCKER, R. 1984: Zur Verbreitung, Vergesellschaftung und Einbürgerung des Götterbaumes [*Ailanthus altissima* (Mill.) Swingle]. - *Tuexenia* **4**: 9-29.

KOYAMA, T. 1958: Taxonomic study of the genus *Scirpus* Linné. - *J. Fac. Sci. Univ. Tokyo, Sect. 3, Bot.* **7(6)**: 271-366.

KOYUNCU, M. 1993: *Muscari muscarimi* Medikus. - *Karaca Arbor. Mag.* **2(2)**: 91-92.

KOYUNCU, M. 1995: *Hyacinthus orientalis* L. - *Karaca Arbor. Mag.* **3(1)**: 45-46.

KOYUNCU, M. & ÖZHATAY, N. 1983: Cytotaxonomic studies on some *Allium* species (sect. *Melanocrommyum*) of South Anatolia [In Turkish with English summary]. - *Ankara Üniv. Eczac. Fak. Mecm.* **13(1-2)**: 159-169.

KOYUNCU, M. (ED.) 2002: VIth Plant Life of Southwest Asia Symposium. Program and Abstracts. - Van: Yüzüncü Yil University, 151 pp.

KOZLOWSKI, G., JONES, R.A. & NICHOLLS-VUILLE, F.-L. 2008: Biological Flora of Central Europe: *Baldellia ranunculoides* (Alismataceae). - *Perspect. Pl. Ecol., Evol. Syst.* **10(2)**: 109-142.

KOZOLOWSKI, G. & AL. 2013: The Tertiary relict tree *Zelkova abelicea* (Ulmaceae): distribution, population structure and conservation status on Crete. - *Oryx* **48(1)**: 80-87.

KOŽUHAROV, S.I. 1962: Certain wild species of medick (*Medicago*) widespread in Bulgaria [In Bulgarian with Russian and English summaries]. - *Izv. Bot. Inst. (Sofia)* **10**: 37-46.

KOŽUHAROV, S.I. 1963: Notes on certain species of genus *Luzula* DC. (wood rush) in Bulgaria [In Bulgarian with Russian and English summaries]. - *Izv. Bot. Inst. (Sofia)* **11**: 125-139.

KOŽUHAROV, S.I. 1965: The species of genus *Medicago* L. (lucerne) in Bulgaria [In Bulgarian with Russian and English summaries]. - *Izv. Bot. Inst. (Sofia)* **15**: 119-188.

KOŽUHAROV, S.I. 1968: Revision of the collections of genus *Jurinea* Cass. in the Sofia herbaria [In Bulgarian with Russian and English summaries]. - *Izv. Bot. Inst. (Sofia)* **18**: 67-73.

KOŽUHAROV, S.I. 1975a: On the endemism in the Bulgarian flora. In Jordanov, D. & al. (eds): Problems of Balkan flora and vegetation, pp. 162-168. - Sofia: Publishing House of the Bulgarian Academy of Sciences.

KOŽUHAROV, S.I. 1975b: Notes on some European species of *Echinops* L. and *Jurinea* Cass. In Heywood, V.H. (ed.): Notulae systematicae ad Floram Europaeam spectantes, no. 17. - *Bot. J. Linn. Soc.* **71(1)**: 41-43.

KOŽUHAROV, S.I. 1994: The herbaria of Bulgaria. - *God. Sofiisk. Univ. "Kliment. Ohridski" Biol. Fak., 2 Bot.* **85**: 189-192.

KOŽUHAROV, S.I. 1995a: New grass species from Bulgaria. In Valdés, B. & Silvestre, S. (eds): Abstracts. VIII OPTIMA meeting, Sevilla, Spain, 25 September - 1 October, 1995, p. 128.

KOŽUHAROV, S.I. 1995b: Biosystematic studies of grasses in Bulgaria. - *Giorn. Bot. Ital., n.s.* **129(1)**: 71-77.

KOŽUHAROV, S.I. & AL. (EDS) 1987: Almanac of the Institute of Botany with Botanic Garden of the Bulgarian Academy of Sciences [In Bulgarian with English summary]. - Sofia: Bulgarian Academy of Sciences, 359 pp.

KOŽUHAROV, S.I. & ANČEV, M.E. (EDS) 2012: Flora Reipublicae Bulgaricae. XI [Dipsacaceae to Asteraceae]. - Sofia: Bulgarian Academy of Sciences, 523 pp. + 1 map.

KOŽUHAROV, S.I. & ANDREEV, N. 1981: Proposal for a grid to serve the mapping of the Bulgarian flora. In Velčev, V.I. & Kožuharov, S.I. (eds): Mapping the flora of the Balkan Peninsula, pp. 169-174. - Sofia: Publishing House of the Bulgarian Academy of Sciences.

KOŽUHAROV, S.I., ANDREEV, N. & MARKOVA, M.L. 1981: Stand und Besonderheiten der chorologischen Information in Bulgarien. In Velčev, V.I. & Kožuharov, S.I. (eds): Mapping the flora of the Balkan Peninsula, pp. 47-56. - Sofia: Publishing House of the Bulgarian Academy of Sciences.

KOŽUHAROV, S.I. & KUZMANOV, B.A. 1962: Material and critical notes on Bulgaria's flora [In Bulgarian with Russian and English summaries]. - *Izv. Bot. Inst. (Sofia)* **9**: 181-183.

KOŽUHAROV, S.I. & KUZMANOV, B.A. 1964a: Chromosome numbers of four Bulgarian plants. - *Dokl. Bulg. Akad. Nauk* **17(5)**: 491-494.

KOŽUHAROV, S.I. & KUZMANOV, B.A. 1964b: Chromosome numbers of some Bulgarian plant species. - *God. Sofiisk. Univ. Biol.-Geol.-Geogr. Fak., 1. Biol. Bot.* **57**: 103-109.

KOŽUHAROV, S.I. & KUZMANOV, B.A. 1965a: A contribution to the karyological knowledge of the Bulgarian plants. - *Caryologia* **18(2)**: 349-351.

KOŽUHAROV, S.I. & KUZMANOV, B.A. 1965b: Chromosome numbers of flowering plants in Bulgaria. - *Izv. Bot. Inst. (Sofia)* **15**: 255-258.

KOŽUHAROV, S.I. & KUZMANOV, B.A. 1968a: Cytotaxonomic studies on Bulgarian Gramineae. - *Dokl. Bulg. Akad. Nauk* **21(3)**: 269-272.

KOŽUHAROV, S.I. & KUZMANOV, B.A. 1968b: Cytotaxonomic studies on Bulgarian Gramineae. II. - *Izv. Bot. Inst. (Sofia)* **18**: 61-66.

KOŽUHAROV, S.I. & KUZMANOV, B.A. 1968c: [Chromosome numbers in *Puccinellia* and *Tragopogon* from Bulgaria]. In Löve, Á. (ed.): IOPB chromosome number reports XVI. - *Taxon* **17(2)**: 202.

KOŽUHAROV, S.I. & KUZMANOV, B.A. 1969: Nikolai Andreev Stojanov [obituary]. - *Taxon* **18(3)**: 353.

KOŽUHAROV, S.I. & KUZMANOV, B.A. 1970a: Cytotaxonomic studies on Bulgarian Gramineae. III. - *Izv. Bot. Inst. (Sofia)* **20**: 99-108.

KOŽUHAROV, S.I. & KUZMANOV, B.A. 1970b: On the taxonomic value and systematic position of some endemic Balcan *Arenaria* and *Minuartia* species. - *Fragm. Florist. Geobot.* **16(4)**: 449-452.

KOŽUHAROV, S.I., KUZMANOV, B.A. & MARKOVA, T.A. 1972: [Chromosome numbers in Fabaceae from Bulgaria]. In Löve, Á. (ed.): IOPB chromosome number reports XXXVI. - *Taxon* **21(2-3)**: 336-337.

KOŽUHAROV, S.I. & KUZMANOV, B.A. (EDS) 1978: Evolution of flowering plants and florogenesis. 1. Fabaceae, Scrophulariaceae, Rubiaceae. - Sofia: Publishing House of the Bulgarian Academy of Sciences, 146 pp.

KOŽUHAROV, S.I. & KUZMANOV, B.A. (EDS) 1991: Evolution of flowering plants and florogenesis. 2. Asteraceae, Brassicaceae, Poaceae, Cyperaceae. - Sofia: Bulgarian Academy of Sciences, 160 pp. + 12 plates and 1 fold-out chart.

KOŽUHAROV, S.I. & NICOLOVA, T. 1975: Cytotaxonomic studies on Bulgarian Gramineae. In: Festschrift in honour of Acad. Daki Jordanov, pp. 69-77. - Sofia: Publishing House of the Bulgarian Academy of Sciences.

KOŽUHAROV, S.I. & PETROVA, A.V. 1973a: Cytotaxonomic study of five rare Bulgarian grasses. - *Dokl. Bulg. Akad. Nauk* **26(6)**: 829-832.

KOŽUHAROV, S.I. & PETROVA, A.V. 1973b: [Chromosome numbers in Poaceae from Bulgaria]. In Löve, Á. (ed.): IOPB chromosome number reports XL. - *Taxon* **22(2-3)**: 286-287.

KOŽUHAROV, S.I. & PETROVA, A.V. 1974a: Cytotaxonomic study on Poaceae VI. - *Dokl. Bulg. Akad. Nauk* **27(7)**: 961-964.

KOŽUHAROV, S.I. & PETROVA, A.V. 1974b: [Chromosome numbers in *Silene* and *Plantago*]. In Löve, Á. (ed.): IOPB chromosome number reports XLIV. - *Taxon* **23(2-3)**: 377.

KOŽUHAROV, S.I. & PETROVA, A.V. 1975: [Chromosome numbers in Poaceae from Bulgaria]. In Löve, Á. (ed.): IOPB chromosome number reports XLVIII. - *Taxon* **24(2-3)**: 369.

KOŽUHAROV, S.I. & PETROVA, A.V. 1976: Chromosome studies of the higher plants with Euroasiatic distribution in Bulgaria. - *Fragm. Florist. Geobot.* **22(3)**: 301-322.

KOŽUHAROV, S.I. & PETROVA, A.V. 1981a: Caryological studies on Bulgarian Poaceae. - *Bol. Soc. Brot., Ser. 2* **53(2)**: 1161-1175.

KOŽUHAROV, S.I. & PETROVA, A.V. 1981b: The caryotype of a relic grass species and some notes on its relations. - *Bol. Soc. Brot., Ser. 2* **53(2)**: 1177-1181.

KOŽUHAROV, S.I. & PETROVA, A.V. 1988: Atlas Florae Europaeae notes. 12. News and adjustments concerning Bulgarian Ranunculaceae. - *Ann. Bot. Fenn.* **25(4)**: 389-390.

KOŽUHAROV, S.I. & PETROVA, A.V. 1991: Chromosome numbers of Bulgarian angiosperms. – *Fitologiya* **39**: 72-77.

KOŽUHAROV, S.I., PETROVA, A.V. & EHRENDORFER, F. 1981: Evolutionary patterns in some brome grass species (*Bromus*, Gramineae) of the Balkan Peninsula. - *Bot. Jahrb. Syst.* **102(1-4)**: 381-391.

KOŽUHAROV, S.I., PETROVA, A.V. & JANEVA, E.K. 1983: Cytotaxonomic study of genus *Aegilops* L. in Bulgaria [In Bulgarian with English abstract]. In Velčev, V.I. & al. (eds): Third national conference of botany, pp. 207-215. - Sofia: Publishing House of the Bulgarian Academy of Sciences.

KOŽUHAROV, S.I., PETROVA, A.V. & MARKOVA, M.L. 1974: [Chromosome numbers in *Trifolium*]. In Löve, Á. (ed.): IOPB chromosome number reports XLIV. - *Taxon* **23(2-3)**: 377-378.

KOŽUHAROV, S.I., PETROVA, A.V. & MARKOVA, M.L. 1975: [Chromosome numbers in *Trifolium* from Bulgaria]. In Löve, Á. (ed.): IOPB chromosome number reports XLVII. - *Taxon* **24(1)**: 145-146.

KOŽUHAROV, S.I., PETROVA, A.V. & MARKOVA, T.A. 1973: [Chromosome numbers in Fabaceae from Bulgaria]. In Löve, Á. (ed.): IOPB chromosome number reports XL. - *Taxon* **22(2-3)**: 287-288.

KOŽUHAROV, S.I., PETROVA, A.V. & MARKOVA, T.A. 1975: Cytotaxonomic study of Bulgarian Leguminosae species. I. In: Festschrift in honour of Acad. Daki Jordanov, pp. 31-47. - Sofia: Publishing House of the Bulgarian Academy of Sciences.

KOŽUHAROV, S.I., PETROVA, A.V. & NENOVA, I.T. 1988: Genus *Koeleria* Pers. in the flora of Bulgaria ... [In Bulgarian with English summary]. In Velčev, V.I. & al. (eds): 100th anniversary of Academician Nikolaj A. Stojanov, pp. 64-90. - Sofia: Publishing House of the Bulgarian Academy of Sciences.

KOŽUHAROV, S.I., PETROVA, A.V. & STOEVA, M.P. 1974a: Chromosomal variation in the *Brachypodium* genus in Bulgaria with regard to its evolution and taxonomy. - *Genet. Polon.* **15(1-2)**: 13-23.

KOŽUHAROV, S.I., PETROVA, A.V. & STOEVA, M.P. 1974b: Cytotaxonomic study on Poaceae V. g. *Briza*. - *Dokl. Bulg. Akad. Nauk* **27(3)**: 383-386.

KOŽUHAROV, S.I. & STOEVA, M.P. 1984: Two new species for the genus *Poa* sect. *Bolboforum* Aschers. & Graebn. [In Bulgarian with English abstract]. In: 90th anniversary of Acad. Daki Jordanov, pp. 152-158. - Targovište: "D. Stefanov" Publishing House.

KOŽUHAROV, S.I., STOEVA, M.P. & KUZMANOV, B.A. 1974: Taxonomic study of the annual species of genus *Bromus* L. in Bulgaria [In Bulgarian with Russian and English summaries]. - *Izv. Bot. Inst. (Sofia)* **25**: 27-61.

KOŽUHAROV, S.I. (ED.) 1995: Flora Reipublicae Bulgaricae. X [Scrophulariaceae to Valerianaceae]. - Serdicae [Sofia]: In Aedibus Academiae Scientiarum Bulgaricae, 429 pp. + 1 map.

KOZUHAROVA, E. & AL. 2017: *Onobrychis pindicola* and *O. montana* (Fabaceae) in the Pirin and Slavyanka Mts (SW Bulgaria): can we distinguish between them? - *Phytol. Balcan.* **23(3)**: 371-380.

KOŽUHAROVA, E.K. 1995: Natural hybridization between *Gentiana lutea* L. and *Gentiana punctata* L. (Gentianaceae). - *God. Sofiisk. Univ. "Kliment. Ohridski" Biol. Fak., 2 Bot.* **88**: 35-43.

KOŽUHAROVA, E.K. & AL. 2007: Two rare *Oxytropis* species (Fabaceae) endemic to the Pirin Mts, Bulgaria. - *Phytol. Balcan.* **13(3)**: 335-346.

KOŽUHAROVA, E.K. & AL. 2014: Distribution of *Arum* (Araceae) in Bulgaria. - *Fl. Medit.* **24**: 51-62.

KOŽUHAROVA, K.S. 1993: Taxonomy of *Prunus cerasifera* Ehrh. in Bulgaria. In Petrova, A.V. (ed.): Abstracts from the VII OPTIMA meeting in Borovetz, Bulgaria, 18-30 July 1993, p. 101.

KRACH, B. 1988: *Tephroseris integrifolia* subsp. *vindelicorum*, eine neue Sippe vom Augsburger Lechfeld. - *Mitt. Bot. Staatssamml. München* **27**: 73-86.

KRÄHENBÜHL, M. 1984: Le complexe polyploïde et aneuploïde du *Mercurialis perennis* L. dans le canton de Neuchâtel. - *Bull. Soc. Neuchâteloise Sci. Nat.* **107**: 39-54.

KRÄHENBÜHL, M. & KÜPFER, PH. 1992: Rapports (92-97). In Kamari, G., Felber, F. & Garbari, F. (eds): Mediterranean chromosome number reports - 2. - *Fl. Medit.* **2**: 255-258.

KRÄHENBÜHL, M. & KÜPFER, PH. 1995: Le genre *Mercurialis* (Euphorbiaceae): cytogéographie et évolution du complexe polyploïde des *M. perennis* L., *M. ovata* Sternb. et Hoppe et *M. leiocarpa* Sieb. et Zucc. - *Candollea* **50(2)**: 411-430.

KRAHULCOVÁ, A. 1982: Cytotaxonomic study of *Chelidonium majus* L. s.l. - *Folia Geobot. Phytotax.* **17 (3)**: 237-268.

KRAHULCOVÁ, A. 1988: Selected chromosome counts of the Czechoslovak flora. I. - *Folia Geobot. Phytotax.* **23(4)**: 375-381.

KRAHULCOVÁ, A. 1990: Selected chromosome counts of the Czechoslovak flora. II. - *Folia Geobot. Phytotax.* **25(4)**: 381-388.

KRAHULCOVÁ, A. 1994: Cytogeography of *Geum montanum* (Rosaceae). - *Folia Geobot. Phytotax.* **29 (1)**: 85-90.

KRAHULCOVÁ, A. & AL. 2009: The agamic complex of *Pilosella* (Asteraceae) in Bulgaria and SW Romania: variation in ploidy level and breeding systems. - *Phytol. Balcan.* **15(3)**: 377-384.

KRAJMEROVÁ, D. & AL. 2015: Natural hybridization in eastern-Mediterranean firs: The case of *Abies borisii-regis*. - *Pl. Biosyst.* **150(6)**: 1189-1199.

KRÁL, M. 1966: Die Begrenzung der Gattung *Parageum* Nakai et Hara. - *Preslia* **38**: 151-153.

KRÁL, M. 1968: Systematic position of *Sedum cepaea* L. - *Preslia* **40(1)**: 18-20.

KRÁL, M. 1970: *Rhamnus prunifolius* Sibth. & Sm. found in Dalmatia (Croatia, Yugoslavia). - *Preslia* **44(3)**: 277.

KRÁL, M. 1984: *Pirinia*, a new genus of the Caryophyllaceae. - *Preslia* **56(2)**: 161-163.

KRÁL, M. 1987: *Sedum creticum* - a forgotten species. - *Preslia* **59(4)**: 307-310.

KRÁL, M. 1988: Additional notes on the problem of *Sedum creticum*. - *Preslia* **60(3)**: 245.

KRÁLIK, E. 1986: Karyological study of the Slovak flora XVI. - *Acta Fac. Rerum Nat. Univ. Comen., Bot.* **33**: 77-80.

KRÄMER, E. & KRÄMER, K. 1983: Beiträge zur Orchideenflora der Provinz Attika (Griechenland). - *Mitt. Arbeitskreis Heimische Orchid. Baden-Württemberg* **15(4)**: 541-558.

KRAMINA, T.E. 2006: A contribution to the taxonomic revision of the *Lotus angustissimus*-complex (Leguminosae, Loteae). - *Wulfenia* **13**: 57-92.

KRANJČEV, R. & ŠEŠOK, D. 2016: A revision of the genus *Fritillaria* (Liliaceae) in Croatia. - *Nat. Croat.* **25(2)**: 185-212.

KRASKOVITS, G. & FLEISCHMANN, H. 1907: Interessante Orchideen aus Korfu. - *Oesterr. Bot. Z.* **57(1)**: 4-7 + 1 plate.

KRATTINGER, K. 1978: Biosystematische Untersuchungen innerhalb der Gattung *Typha* L. - *Mitt. Bot. Mus. Univ. Zürich* **298**: 270 pp.

KRAUSE, K. 1926: Über den giftigen Honig des pontischen Kleinasien. - *Naturwissenschaften* **44**: 976-978.

KRAUSE, K. 1958: Zur taxonomischen Gliederung, Verbreitung und Genetik der *Pulsatilla Halleri* (All.) Willd. - *Bot. Jahrb. Syst.* **78(1)**: 1-68.

KRAUSE, P.J. 2000: The alpine zone [of northern Greece]. - *Quart. Bull. Alpine Gard. Soc. Gr. Brit.* **68 (3)**: 380-396.

KRAUSE, P.J. 2002: [Notes from Mt Olympus]. - *Alpine Gardener* **70(2)**: 131-133.

KRAUSE, P.J. 2004: Autumn bulbs of the Peloponnese. - *Alpine Gardener* **72(4)**: 367-384.

KRAUSE, W. & LUDWIG, W. 1956: Zur Kenntnis der Flora und der Vegetation auf Serpentinstandorten des Balkans. 1. *Halacsya sendtneri* (Boiss.) Dörfl. - *Ber. Deutsch. Bot. Ges.* **69(9)**: 417-428 + 1 plate.

KRAUSE, W., LUDWIG, W. & SEIDEL, F. 1963: Zur Kenntnis der Flora und Vegetation auf Serpentinstandorten des Balkans. 6. Vegetationsstudien in der Umgebung von Mantoudi (Euböa). - *Bot. Jahrb. Syst.* **82(4)**: 337-403 + 7 tables (Beilagen).

KREBS, G. 1994a: Taxonomische Untersuchungen in der Subtribus Malvinae. - *Feddes Repert.* **105(1-2)**: 7-18.

KREBS, G. 1994b: Taxonomische Untersuchungen in der Subtribus Malvinae II. *Dinacrusa*. - *Feddes Repert.* **105(5-6)**: 299-315.

KRENDL, F. 1988: Die Arten der *Galium mollugo*-Gruppe in Griechenland. - *Bot. Chron. (Patras)* **6-7**: 5-170.

KRENDL, F. 2014: Die Rubiaceae in Albanien: Zytologie, Ökologie und Verbreitung der bisher nachgewiesenen Taxa. - *Ann. Naturhist. Mus. Wien, B* **116**: 119-151.

KRENDL, F. & POLATSCHEK, A. 1984: Die Gattung *Ononis* L. in Österreich. - *Verh. Zool.-Bot. Ges. Österreich* **122**: 77-91.

KRENDL, F. & VITEK, E. 2007: *Galium ionicum* Krendl, nomen novum pro *Galium mixtum* (Rubiaceae) - *Ann. Naturhist. Mus. Wien* **108B**: 265-266.

KRENN, L. & AL. 2001: Chemotaxonomische Untersuchung der Gattung *Charybdis* Speta (Urgineoideae, Hyacinthaceae). - *Stapfia* **75**: 101-118 + 1 fold-out chart.

KRESS, A. 1969: Zytotaxonomische Untersuchungen an Primulaceen. - *Phyton (Horn)* **13(3-4)**: 211-225.

KRESS, A. 1984a: Primulaceen-Studien 3. Chromosomenzählungen an verschiedenen Primulaceen: Teil A. *Androsace*. - Gröbenzell bei München: 32 pp.

KRESS, A. 1984b: Primulaceen-Studien 7. Chromosomenzählungen an verschiedenen Primulaceen: Teil B. *Soldanella*. - Gröbenzell bei München: ii + 26 pp.

KRESTOVSKAJA, T. 1992: Systematics and phytogeography of *Leonurus* L. In Harley, R.M. & Reynolds, T. (eds): Advances in Labiatae science, pp. 139-148. - Kew: Royal Botanic Gardens.

KRESTOVSKAYA, T.V. 2017: Taxonomic review of species of the *Stachys* section *Candidae* (Lamiaceae). [In Russian with English summary]. - *Novosti Sist. Vysš. Rast.* **48**: 118-122.

KRESTOVSKAYA, T.V. 2019: Taxonomic note on *Stachys mucronata* (Lamiaceae) and its position in the genus system. [In Russian with English summarry]. - *Novosti Sist. Vysš. Rast.* **50**: 154-157.

KRETZSCHMAR, G. & KRETZSCHMAR, H. 1995: *Ophrys grigoriana* - eine neue Art aus Kreta. - *Ber. Arbeitskreis. Heimische Orchid.* **12(1)**: 54-58.

KRETZSCHMAR, G. & KRETZSCHMAR, H. 1996: Orchideen der Insel Naxos. - *Ber. Arbeitskreis. Heimische Orchid.* **13(1)**: 4-30.

KRETZSCHMAR, G. & KRETZSCHMAR, H. 1998: Neue *Ophrys*-Hybriden aus Griechenland und der Ägäis. - *Ber. Arbeitskreis. Heimische Orchid.* **15(1)**: 48-56.

KRETZSCHMAR, G. & KRETZSCHMAR, H. 2001: *Orchis papilionacea* subsp. *alibertis*, eine neue Unterart aus Kreta. - *Ber. Arbeitskreis. Heimische Orchid.* **18(1)**: 128-132.

KRETZSCHMAR, H., ECCARIUS, W. & DIETRICH, H. 2007: The orchid genera *Anacamptis, Orchis* and *Neotinea*. - Bürgel: Echino Media Verlag, 544 pp.

KRETZSCHMAR, H. & JAHN, R. 2002: *Ophrys cretica* subsp. *bicornuta* - eine neue Unterart aus Ostkreta. - *Ber. Arbeitskreis. Heimische Orchid.* **18(2)**: 43-45.

KRETZSCHMAR, H. & KRETZSCHMAR, G. 2003a: *Ophrys Xcapellae-pacis*, eine neue Hybride aus Kreta. - *Ber. Arbeitskreis. Heimische Orchid.* **20(1)**: 69-72.

KRETZSCHMAR, H. & KRETZSCHMAR, G. 2003b: Zur Variationsbreite von *Ophrys ferrum-equinum*. - *Ber. Arbeitskreis. Heimische Orchid.* **20(1)**: 75-79.

KRETZSCHMAR, H., KRETZSCHMAR, G. & ECCARIUS, W. 2001: Orchideen auf Rhodos. - Selbstverlag: 240 pp.

KRETZSCHMAR, H., KRETZSCHMAR, G. & ECCARIUS, W. 2002: Orchideen auf Kreta, Kasos und Karpathos. Ein Feldführer durch die Orchideenflora der zentralen Inseln der Südägäis. - Bad Hersfeld, Hoehl-Druck GmbH, 416 pp.

KRETZSCHMAR, H., KRETZSCHMAR, G. & ECCARIUS, W. 2004: Orchideen Kreta & Dodekanes. Die Orchideenflora der Inseln Kreta, Kasos, Karpathos und Rhodos. - Rethymno: Mediterraneo Editions, 239 pp.

KRETZSCHMAR, H. & KREUTZ, C.A.J. 2001: *Ophrys dodekanensis* - eine neue Art aus der Südostägäis. - *Ber. Arbeitskreis. Heimische Orchid.* **18(1)**: 34-37.

KRETZSCHMAR, H., WENKER, D. & WILLING, E. 1984: Orchideenkartierung der Insel Rhodos - aktuelle Übersicht. - *Ber. Arbeitskreis. Heimische Orchid.* **1(2)**: 130-146.

KREUTZ, C.A.J. 1994: Verslag van een orchideenreis naar Noord-Griekenland. - *Eurorchis* **6**: 81-99.

KREUTZ, C.A.J. 1995: *Orobanche*. Die Sommerwurzarten Europas / The European broomrape species. 1. Mittel- und Nordeuropa / Central and Northern Europe [text in German and English]. - Maastricht: Stichting Natuurpubl. Limburg, 159 pp. incl. many colour plates.

KREUTZ, C.A.J. 1997: *Ophrys rhodia* (H. Baumann & Künkele) P. Delforge in Israel? - *Ber. Arbeitskreis. Heimische Orchid.* **14(1)**: 69-74.

KREUTZ, C.A.J. 1998: Die Orchideen der Türkei. Bechreibung, Ökologie, Verbreitung, Gefährdung, Schutz. - Landgraaf: Selbstverlag, 766 pp.

KREUTZ, C.A.J. 2000: Notizen zu einigen spätblühenden Orchideenarten auf Kreta. - *Ber. Arbeitskreis. Heimische Orchid.* **17(2)**: 106-114.

KREUTZ, C.A.J. 2001: *Ophrys helios*, eine neue Art von Karpathos (Ostägäis). - *J. Eur. Orch.* **33(3)**: 871-880.

KREUTZ, C.A.J. 2002a: Die Orchideen von Rhodos und Karpathos / The orchids of Rhodes and Karpathos [text in German and English]. - Raalte & Landgraaf: Seckel & Kreutz Publishers, 320 pp. incl. many colour plates.

KREUTZ, C.A.J. 2002b: Bestätigung von *Orchis punctulata* Steven ex Lindley für Rhodos. - *Ber. Arbeitskreis. Heimische Orchid.* **19(1)**: 79-81.

KREUTZ, C.A.J. 2004a: Die Orchideen von Zypern. Beschreibung, Lebensweise, Verbreitung, Gefährdung, Schutz und Ikonographie [In German and English]. - Valkenburg aan de Geul: Spektra Druk B.V., 416 pp.

KREUTZ, C.A.J. 2004b: Kompendium der europäischen Orchideen. Catalogue of European orchids. - Landgraaf: Kreutz Publishers, 238 pp.

KREUTZ, C.A.J. 2006: Orchideen Europas - Beiträge, Umkombinationen und kurze Bemerkungen. - *Orchidee (Hamburg)* **57(1)**: 99-104.

KREUTZ, C.A.J. 2007a: Beitrag zur Taxonomie und Nomenklatur europäischer, mediterraner, nordafrikanischer und vorderasiatischer Orchideen. - *Ber. Arbeitskreis. Heimische Orchid.* **24(1)**: 77-141.

KREUTZ, C.A.J. 2007b: Neukombinationen und Ergänzungen zu verschiedenen europäischen Orchideentaxa. - *Ber. Arbeitskreis. Heimische Orchid.* **24(1)**: 142-186.

KREUTZ, C.A.J. 2011: Beitrag zur Kenntnis europäischer, mediterraner und vorderasiatischer Orchideen. - *Ber. Arbeitskreis. Heimische Orchid.* **28(2)**: 263-299.

KREUTZ, C.A.J. 2012: *Ophrys mammosa* Desfontaines subsp. *parviflora* Kreutz & H. Heitz, eine kleinblütige Sippe aus dem *Ophrys mammosa*-Formenkreis. - *Ber. Arbeitskreis. Heimische Orchid.* **29(2)**: 144-151.

KREUTZ, K. 1990a: Beitrag zur Orchideenflora Kretas. - *Mitt. Arbeitskreis Heimische Orchid. Baden- Württemberg* **22(2)**: 358-384.

KREUTZ, K. 1990b: De orchideeënflora op het griekse eiland Kreta. - *Eurorchis* **2**: 107-154.

KREY, W.-D. 1973: Ein bisher nicht bekannter *Ophrys*-Bastard auf Rhodos. - *Orchidee (Hamburg)* **24 (6)**: 257-258.

KREY, W.-D. & AL. 1980: Hellas 1979 - Beitrag zur heutigen Orchideensituation in Griechenland. - *Orchidee (Hamburg)* **31(3)**: 108-112.

KREY, W.-D. & AL. 1989: Tagesbuch einer Mai-Reise zu den Orchideen auf Lesbos. - *Mitt. Arbeitskreis Heimische Orchid. Baden-Württemberg* **21(1)**: 88-102.

KRIEMADI, E., BAREKA, E.-P. & KAMARI, G. 2002: Karyological study of some taxa from Levkas island. In Kamari, G. & al. (eds): Ellinini Votaniki Eteria. 9° Panelliniko Epistimoniko Sinedrio. Praktika, pp. 164-171 [In Greek with English abstract].

KRIEMADI, E., BAREKA, E.-P. & KAMARI, G. 2003: Reports (1278-1283). In Kamari, G.,

Blanché, C. & Garbari, F. (eds): Mediterranean chromosome number reports - 12. - *Fl. Medit.* **12**: 444-450.

KRIGAS, N. 2004: Chlorida ke anthropines drastiriotites stin periochi tis Thessalonikis: Viologiki prosengisi ke istoriki sindhesi. [In Greek]. - Thessaloniki: Ph.D. thesis, Aristotelian University, 382 pp.

KRIGAS, N. & AL. 1999: The vascular flora of the Byzantine walls of Thessaloniki (N Greece). - *Willdenowia* **29(1-2)**: 77-94.

KRIGAS, N. & KOKKINI, S. 2004: A survey of the alien vascular flora of the urban and suburban area of Thessaloniki, N Greece. - *Willdenowia* **34(1)**: 81-99.

KRIGAS, N. & KOKKINI, S. 2005: The indigenous vascular flora of the urban and suburban area of Thessaloniki (N Greece). - *Bot. Chron. (Patras)* **18(2)**: 29-85.

KŘÍSA, B. 1963: Taxonomische Stellung der Art *Juncus alpinus* Vill. s.l. in der europäischen Flora. - *Novit. Bot. Delect. Seminum Horti Bot. Univ. Carol. Prag.* **1963**: 28-35.

KŘÍSA, B. 1965: Beitrag zur Gliederung der Gattung *Pyrola* in holoarktischen Gebieten. - *Novit. Bot. Delect. Seminum Horti Bot. Univ. Carol. Prag.* **1965**: 31-35.

KŘÍSA, B. 1971: Beitrag zur Taxonomie und Chorologie der Gattung *Pyrola* L. - *Bot. Jahrb. Syst.* **90(4)**: 476-508.

KRITIKOS, P. 1955a: Ek ton ellinikon idon kolchikou: Kolchikon to platifillon ke Kolchikon to Koupanion (*Colchicum latifolium* S.S. et *Colchicum Cupani* Guss.) [In Greek with French summary]. - *Compt. Rend. Acad. Athènes [Praktika Akadimias Athinon]* **30**: 223-244.

KRITIKOS, P. 1955b: Ek ton ellinikon idon kolchikou: Kolchikon tou Parnassion (*Colchicum parnassicum* Sart., Orph., Heldr.) ke Kolchikon to glossoides (*Colchicum lingulatum* Boiss. & Sprun.) [In Greek with French summary]. - *Compt. Rend. Acad. Athènes [Praktika Akadimias Athinon]* **30**: 245-255.

KRITIKOS, P. 1961: O mandragoras [In Greek]. - Athine: 56 pp.

KRITSIS, K. & AL. 2021: GRASP-125: A dataset for Greek vascular plant recognition in natural environment. - *Sustainability* **13, 11865**: doi.org/10.3390/su132111865.

KŘIVKA, P. 2010: Josef Velenovský, a great Czech botanist, philosopher and politician. - *Phytol. Balcan.* **16(2)**: 313-318.

KŘIVKA, P. & HOLUBEC, V. 2010: The Balkan collections in the main Czech herbaria. - *Phytol. Balcan.* **16(2)**: 215-220.

KRIVOSEJ, Z. & AL. 1998: Serpentinopbytes [sic!] of Kosovo and Metohia (Serbia). In Moret, J. (ed.): Abstracts of the IX. OPTIMA meeting, Paris, 11-17 May 1998. - Paris: Mus. Nat. d'Hist. Naturelle.

KRIVOŠEJ, Z., AMDŽIĆ, L. & STEVANOVIĆ, V. 1998: *Silene graefferi* Guss. - new species for Yugoslav Flora. - *Univ. Thought, Publ. Nat. Sci.* **5(2)**: 63-65.

KROCHMAL, A. & LAVRENTIADES, G. 1955: Poisonous plants of Greece. - *Econ. Bot.* **9(2)**: 175-189.

KRONFELD, M. 1889: Monographie der Gattung *Typha* Tourn. (Typhinae Agdh., Typhaceae Schur- Engl.). - *Verh. K.K. Zool.-Bot. Ges. Wien* **39**: 89-192 + 2 plates.

KROON, G.H. & ZEILINGA, A.E. 1974: Apomixis and heterogamy in rose rootstocks (*Rosa canina* L.). - *Euphytica* **23(2)**: 345-352.

KROPF, M. 2008: Intraspecific patterns of European mountain plants: a morphometric analysis confirms molecular results in the submediterranean oreophyte *Anthyllis montana* L. - *Taxon* **57(2)**: 511-424.

KRUSE, K., MONSER, U. & PAUSCH, S. 1995: I Flächennutzkartierung, II Vegetation an Ackerrändern, III Bodenkartierung. In Mattes, H. & Lienau, C. (eds): Colloquium Nordost-Griechenland 3, pp. 21-51. - Münster: Inst. Geographie.

KRUSHEVA, R.M. 1975: [Chromosome numbers from Bulgaria]. In Löve, Á. (ed.): IOPB chromosome number reports L. - *Taxon* **24(5-6)**: 676-677.

KRUSHEVA, R.M. 1986: [Chromosome numbers from Bulgaria]. In Löve, Á. (ed.): Chromosome number reports XCII. - *Taxon* **35(3)**: 613.

KRZAK, J. 1982: Chromosome numbers within *Polygonum aviculare* s.l. from Poland. - *Acta Biol. Cracov., Ser. Bot.* **24**: 1-10 + 1 plate.

KUBÁT, K. 1980: Bemerkungen zu einigen tschechoslowakischen Arten der Gattung *Papaver*. - *Preslia* **52(2)**: 103-115.

KUBÁT, K. 1985: Bemerkungen zu einigen tschechoslowakischen Arten der Gattung *Rumex* s.str. - *Preslia* **57(3)**: 205-217.

KUČERA, J., LIHOVÁ, J. & MARHOLD, K. 2006: Taxonomy and phylogeography of *Cardamine impatiens* and *C. pectinata* (Brassicaceae). - *Bot. J. Linn. Soc.* **152(2)**: 169-195.

KUČERA, J. & SLOVÁK, M. 2012: [Chromosome numbers in *Picris*]. - *Taxon* **61(4)**: 898 + electronic suppl.

KUČERA, J., VALKO, I. & MARHOLD, K. 2005: On-line database of the chromosome numbers of the genus *Cardamine* (Brassicaceae). - *Biológia (Bratislava)* **60(4)**: 473-476.

KÜÇÜKER, O. 1983: [Chromosome numbers in *Colchicum* from the Istanbul area]. In Löve, Á. (ed.): IOPB chromosome number reports LXXXIV. - *Taxon* **33(3)**: 536.

KÜÇÜKER, O. 1985: The morphological, anatomical and cytological studies of some *Colchicum* species of Istanbul area. - *Istanbul Üniv. Fen Fak. Mecm., B* **50**: 87-111.

KÜÇÜKER, O. 1996: Contributions to the knowledge of some endangered *Colchicum* species of Turkey. - *Fl. Medit.* **5**: 211-219.

KUGLER, H. 1930: Blütenökologische Untersuchungen an *Bryonia dioica* Jacq. - *Flora* **124**: 94-118.

KUH, M., YILDIZ, K. & MINARECI, E. 2017: A taxonomic study of the *Silene* sections *Behenantha* and *Dichotomae* (Caryophyllaceae) in Turkey based on the micromorphology of their seed and pollen. - *Turk. J. Bot.* **41(5)**: 493-504.

KUHN, M. 1881: Uebersicht über die Arten der Gattung *Adiantum*. - *Jahrb. Königl. Bot. Gart. Berlin* **1**: 337-351.

KÜHN, R., PEDERSEN, H.Æ. & CRIBB, PH. 2019: Field guide to the orchids of Europe and the Mediterranean. - Kew Publishing: x + 430 pp.

KUIPER, P.J.C. & BOS, M. (EDS) 1992: *Plantago*: A multidisciplinary study. - Berlin, etc.: Springer- Verlag, xiii + 368 pp.

KÜKENTHAL, G. 1938: Vorarbeiten zu einer Monographie der Rhynchosporideae I-III. - *Repert. Spec. Nov. Regni Veg.* **44**: 1-32, 65-101, 161-195.

KÜKENTHAL, G. 1942: Vorarbeiten zu einer Monographie der Rhynchosporoideae XII & XIII. - *Repert. Spec. Nov. Regni Veg.* **51**: 1-17, 139-193.

KÜKENTHAL, G. 1949: Vorarbeiten zu einer Monographie der Rhynchosporoideae: *Rhynchospora*. - *Bot. Jahrb. Syst.* **74(3)**: 375-509.

KÜKENTHAL, G. 1950: Vorarbeiten zu einer Monographie der Rhynchosporoideae: *Rhynchospora* [cont.]. - *Bot. Jahrb. Syst.* **75(1-2)**: 90-195.

KÜKENTHAL, G. 1951: Vorarbeiten zu einer Monographie der Rhynchosporoideae: *Rhynchospora* [cont.]. - *Bot. Jahrb. Syst.* **75(3)**: 273-314.

KUKKONEN, I. 1984: On the inflorescence structure in the family Cyperaceae. - *Ann. Bot. Fenn.* **21(3)**: 257-264.

KUKKONEN, I. & UOTILA, P. 1977: On the taxonomy, morphology and distribution of *Asarum europaeum* (Aristolochiaceae). - *Ann. Bot. Fenn.* **14(3)**: 131-142.

KUKKONEN, I. & UOTILA, P. 1979: "Asarum europaeum" L. in southern Europe. - *Webbia* **34(1)**: 507-511.

KULA, A. 1989: The investigations of the karyotype of *Phleum pratense* L. with the C-banding method and the problem of the origin of this species [In Polish with English summary]. - *Fragm. Florist. Geobot.* **33(3-4)**: 257-265.

KULL, U. 1974: Streifzüge durch die Vegetation Kretas. - *Natur & Mus.* **104(4)**: 120-127.

KULL, U. 1982: Grosse Botanische Exkursion der Universität Stuttgart. Athen - Kreta. 22.3. 1981 bis 12.4. 1981. Pflanzenliste. - Mimeographed: 23 pp.

KULL, U. & DIAMANTOGLOU, S. 1998: Kreta. Allgemeiner Exkursionsbericht. Kumulative Pflanzenliste der Exkursionen 1974-1995 zugleich Führer zur botanisch-geologischen Exkursion der Gesellschaft für Naturkunde in Württemberg April 1998. - *Arb. Mitteil. Biol. Inst. Univ. Stuttgart* **28**: viii + 365 pp.

KÜMMERLE, J.B. 1921: *Asplenium Bornmülleri* Kümm. spec. nova. - *Bot. Közlem.* **19**: 81-83.

KÜMMERLE, J.B. 1922: Pteridologische Mitteilungen. Zwei neue Farne aus Albanien. - *Magyar Bot. Lapok* **21(1-12)**: 1-5.
KÜMMERLE, J.B. 1926: Beiträge zur Kenntnis der Pteridophyten der Balkanhalbinsel. - *Bot. Közlem.* **25**: 143-148.
KUNEV, G. 2020: *Fumana bonapartei* and *F. aciphylla* (Cistaceae), new additions to the Bulgarian flora. - *Fl. Medit.* **30**: 339-345.
KUNEV, G. 2021: *Allium melanogyne*. In Raus, Th. & Raab-Straube, E.v. (eds): Euro+Med-Checklist notulae, 13. - *Willdenowia* **51(1)**: 142-143.
KUNEV, G. & AL. 2020: Phytosociological study of submontane genistoid scrub communities from the Southeastern Balkans. - *Acta. Bot. Croat.* **79(2)**: 170-184.
KUNEV, G. & KOSTADINOV, I. 2021: *Silene tenuiflora*. In Raus, Th. & Raab-Straube, E.v. (eds): Euro+Med-Checklist notulae, 13. - *Willdenowia* **51(1)**: 144-145.
KUNKEL, G. 1996: Ökologische Untersuchungen im Aladjagiola. - *Ber. Arbeitsgeb. Entwicklungsforsch. (Münster)* **25**: 6-25, 121-129.
KÜNKELE, S. 1979a: Revision der von M. Gandoger und P. Cousturier auf Kreta gesammelten Orchideen. - *Mitt. Arbeitskreis Heimische Orchid. Baden-Württemberg* **11(3)**: 173-205.
KÜNKELE, S. 1979b: Historischer Überblick zur Erforschung der Orchideen von Kreta. Vorarbeiten zur Orchideenflora von Kreta (II). - *Mitt. Arbeitskreis Heimische Orchid. Baden-Württemberg* **11(4)**: 283-309.
KÜNKELE, S. 1983: Zum Stand der Orchideenkartierung in Griechenland. - *Mitt. Arbeitskreis Heimische Orchid. Baden-Württemberg* **15(1)**: 11-42.
KÜNKELE, S. 1985: Le genre *Ophrys* en Grèce. In: 8e Coll. Orch., Soc. Franc. Orphidophilie, Paris 1984, pp. 81-100. - Paris.
KÜNKELE, S. & PAYSAN, K. 1981: Die Orchideenflora von Euböa (Griechenland). - *Beih. Veröff. Naturschutz Landschaftspflege Baden-Württemberg* **23**: 1-138.
KUNTZE, O. 1885: Monographie der Gattung *Clematis*. - *Verh. Bot. Vereins Prov. Brandenburg* **26(2)**: 85-202.
KUNTZE, O. 1887: Nachtrag zur *Clematis*-Monographie. - *Verh. K.K. Zool.-Bot. Ges. Wien* **37**: 47-50.
KUNZ, H. 1959: Bemerkungen zu einigen *Rhinanthus*-Sippen. - *Phyton (Horn)* **8(3-4)**: 243-258.
KÜPFER, PH. 1969: Recherches cytotaxinomiques sur la flore des montagnes de la péninsule ibérique. - *Bull. Soc. Neuchâteloise Sci. Nat.* **92**: 31-48.
KÜPFER, PH. 1972: Cytotaxonomie et cytogéographie de quelques groupes d'orophytes du bassin occidental de la Méditerranée et des Alpes. - *Compt. Rend. Hebd. Séances Acad. Sci., Sér. D* **275 (16)**: 1753-1756.
KÜPFER, PH. 1974: Recherches sur les liens de parenté entre la flore orophile des Alpes et celle des Pyrénées. - *Boissiera* **23**: 1-322 + 10 plates.
KÜPFER, PH. 1993: *Alyssum bilimekii* Willk. = *Aurinia saxatilis* (L.) Desv. (Cruciferae). - *Anales Jard. Bot. Madrid* **51(1)**: 169-170.
KÜPFER, PH. & RAIS, J.-R. 1983: Index des nombres chromosomiques des spermatophytes de la Suisse. I. Saxifragaceae. - *Bot. Helv.* **93(1)**: 11-25.
KUPICHA, F.K. 1976: The infrageneric structure of *Vicia*. - *Notes Roy. Bot. Gard. Edinburgh* **34(3)**: 287-326.
KUPICHA, F.K. 1977: The delimitation of the tribe Vicieae (Leguminosae) and the relationships of *Cicer* L. - *Bot. J. Linn. Soc.* **74(2)**: 131-162.
KUPICHA, F.K. 1981a: Tribe 21. Vicieae (Adans.) DC. (1825), nom. conserv. prop. In Polhill, R.M. & Raven, P.H. (eds): Advances in legume systematics 1: pp. 377-381. - Kew: Royal Botanic Gardens.
KUPICHA, F.K. 1981b: Tribe 22. Circeae Alefeld (1859). In Polhill, R.M. & Raven, P.H. (eds): Advances in legume systematics 1: pp. 382-xxx. - Kew: Royal Botanic Gardens.
KUPICHA, F.K. 1983: The infrageneric structure of *Lathyrus*. - *Notes Roy. Bot. Gard. Edinburgh* **41(2)**: 209-244.
KUPRIANOVA, L. 1948: The genus *Glechoma* L. and its species [In Russian]. - *Bot. Zhurn. (Moscow & Leningrad)* **33(2)**: 230-238.
KURLOVICH, B.S. (ED.) 2002: Lupins: geography, classification, genetic resources and breeding. - St. Petersburg: Intan, xi + 468 pp.
KURTTO, A. 1983: Atlas Florae Europaeae notes. 3. Chorology and taxonomy of some SE Europaean taxa of the Caryophyllaceae, subfamilies Alsinoideae and Paronychioideae. - *Ann. Bot. Fenn.* **20(4)**: 357-360.
KURTTO, A. 1985a: Atlas Florae Europaeae notes. 4. Chorology and taxonomy of some S European taxa of the Caryophyllaceae, subfamily Silenoideae. - *Ann. Bot. Fenn.* **22(1)**: 49-51.
KURTTO, A. 1985b: Taxonomy of *Symphytum ottomanum*, *S. pseudobulbosum* and *S. orientale* (Boraginaceae). - *Ann. Bot. Fenn.* **22(4)**: 319-331.
KURTTO, A. & AL. (EDS) 2010: Atlas Florae Europaeae. 15. Rosaceae (*Rubus*). - Helsinki: Akateeminen kirjakauppa, 362 pp.
KURTTO, A. & ERIKSSON, T. 2003: Atlas Florae Europaeae notes. 15. Generic delimitation and nomenclatural adjustments in Potentilleae (Rosaceae). - *Ann. Bot. Fenn.* **40(2)**: 135-141.
KURTTO, A., FRÖHNER, S. & LAMPINEN, R. (EDS) 2007: Atlas Florae Europaeae. 14. Rosaceae (*Alchemilla* and *Aphanes*). - Helsinki: Akateeminen kirjakauppa, 200 pp.
KURTTO, A., LAMPINEN, R. & JUNIKKA, L. (EDS) 2004: Atlas Florae Europaeae. 13. Rosaceae (*Spiraea* to *Fragaria*, excl. *Rubus*. - Helsinki: 320 pp.
KURTTO, A., SENNIKOV, A.N. & LAMPINEN, R. Atlas Florae Europaeae. 16. Rosaceae (*Cydonia* to *Prunus prostrata*, excl. *Sorbus*). - Helsiniki: Societas Biologica Fennica Vanamo, 168 pp.
KURTTO, A., SENNIKOV, A.N. & LAMPINEN, R. 2018: Atlas Florae Europaeae. 17. Rosaceae (*Sorbus* s. lato). - Helsiniki: Societas Biologica Fennica Vanamo, 132 pp.
KURZE, H. & KURZE, O. 1990: Ein neuer Gattungsbastard zwischen *Anacamptis pyramidalis* (L.) Rich. und *Orchis laxiflora* Lam. - *Mitt. Arbeitskreis Heimische Orchid. Baden-Württemberg* **22(4)**: 740-746.
KURZE, O. & KURZE, H. 1993: Neuer Fundort von *Ophrys doerfleri* H. Fleischmann auf dem südöstlichen Peloponnes. - *Mitt. Arbeitskreis Heimische Orchid. Baden-Württemberg* **25(4)**: 502-503.
KUŠAN, F. 1953a: Über die Verbreitung und Verwandschaftszuhörigkeit des Zwergwacholders in Jugoslawien [In Serbo-Croatian with German and French summaries]. - *God. Biol. Inst. Sarajevu* **5(1-2)**: 287-302.
KUŠAN, F. 1953b: Ein Beitrag zur Flora von montenegrinisch-albanischen und makedonischen Gebirge [In Serbo-Croatian with German summary]. - *Glasn. Hrvatsk. Prir. Društva* **11/B**: 178-190.
KUŠAN, F. 1970: *Alyssum samoborense* Horv., eine isolierte Sippe von *Alyssum montanum* s. l. und ihre Stellung innerhalb verwandter Sippen im Südosten Europas [In Serbo-Croatian with German summary]. - *Acta Bot. Croat.* **29**: 183-196 + 1 map.
KUSHNIR, U., GALIL, J. & FELDMAN, M. 1977a: Cytology and distribution of *Ornithogalum* in Israel. I. Section *Heliocharmos* Bak. - *Israel J. Bot.* **26(2)**: 63-82.
KUSHNIR, U., GALIL, J. & FELDMAN, M. 1977b: Cytology and distribution of *Ornithogalum* in Israel. II. Section *Beryllis* (Salisb.) Bak. - *Israel J. Bot.* **26(2)**: 83-92.
KUTA, E. 1980: Karyological studies on the genus *Vicia* L. I. - *Acta Biol. Cracov., Ser. Bot.* **22(1)**: 81-99 + 3 plates.
KUTLUK, H. & AYTUĞ, B. 2001: Davis' Flora of Turkey. In Özhatay, N. (ed.): Proceedings of the Second Balkan Botanical Congress, 1: 289-294. - Istanbul.
KUZMANOV, B.A. 1961: Materials and critical comments on the genus *Euphorbia* L. in Bulgaria [In Bulgarian with Russian and English summaries]. - *Izv. Bot. Inst. (Sofia)* **8**: 145-158.
KUZMANOV, B.A. 1963: A taxonomic study of the species of genus *Euphorbia* L. widespread in Bulgaria [In Bulgarian with Russian and English summaries]. - *Izv. Bot. Inst. (Sofia)* **12**: 101-186.
KUZMANOV, B.A. 1964: On the origin of *Euphorbia* subg. *Esula* in Europe (Euphorbiaceae). – *Blumea* **12(2)**: 369-379.
KUZMANOV, B.A. 1971: A catalogue of the taxa (Pteridophyta-Angiospermae) described by

Bulgarian botanists from neighbouring countries (1888-1959). - *Fragm. Florist. Geobot.* **17(2)**: 177-210.
KUZMANOV, B.A. 1973: [Chromosome numbers from Bulgaria]. In Löve, Á. (ed.): IOPB chromosome number reports XLII. - *Taxon* **22(5-6)**: 651.
KUZMANOV, B.A. 1974: [Chromosome numbers from Bulgaria]. In Löve, Á. (ed.): IOPB chromosome number reports XLVI. - *Taxon* **23(5-6)**: 808.
KUZMANOV, B.A. 1975a: Taxonomic studies on vascular plants in Bulgaria 1961-1971. - *Mem. Soc. Brot.* **24(1)**: 25-79.
KUZMANOV, B.A. 1975b: A catalogue of the taxa (Pteridophyta-Angiospermae) described by Bulgarian botanists from neighbouring countries. Short supplement (-1969). - *Fragm. Florist. Geobot.* **21(1)**: 51-52.
KUZMANOV, B.A. 1975c: [Chromosome numbers from Bulgaria]. In Löve, Á. (ed.): IOPB chromosome number reports XLIX. - *Taxon* **24(4)**: 504.
KUZMANOV, B.A. 1978a: Cytotaxonomic investigation of Bulgarian leguminous plants [In Bulgarian with Russian and English summaries]. In Kožuharov, S.I. & Kuzmanov, B.A. (eds): Evolution of flowering plants ..., pp. 11-71. - Sofia: Publishing House of the Bulgarian Academy of Sciences.
KUZMANOV, B.A. 1978b: About the "Red book of rare Bulgarian plants" [In Bulgarian with Russian and English summaries]. - *Fitologiya* **9**: 17-33.
KUZMANOV, B.A. 1979: "Flora Europaea" and the taxonomic studies on vascular plants in Bulgaria. - *Candollea* **34(1)**: 11-19.
KUZMANOV, B.A. 1981a: *Erigeron atticus* Vill. - New species for the flora of Turkey. - *Fitologiya* **18**: 60.
KUZMANOV, B.A. 1981b: Mapping and protection of the threatened plants in Bulgarian flora. In Velčev, V.I. & Kožuharov, S.I. (eds): Mapping the flora of the Balkan Peninsula, pp. 95-156. - Sofia: Publishing House of the Bulgarian Academy of Sciences.
KUZMANOV, B.A. 1981c: On the project of mapping the medicinal plants in Bulgaria. In Velčev, V.I. & Kožuharov, S.I. (eds): Mapping the flora of the Balkan Peninsula, pp. 157-159. - Sofia: Publishing House of the Bulgarian Academy of Sciences.
KUZMANOV, B.A. 1981d: Balkan endemism and the problem of species conservation, with particular reference of the Bulgarian flora. - *Bot. Jahrb. Syst.* **102(1-4)**: 255-270.
KUZMANOV, B.A. 1985: [Chromosome numbers from Bulgaria]. In Löve, Á. (ed.): Chromosome number reports LXXXVII. - *Taxon* **34(2)**: 346.
KUZMANOV, B.A. 1986: Polyploidy and evolutionary pattern in genus *Thalictrum* L. - *Fitologiya* **31**: 14-20.
KUZMANOV, B.A. 1988: *Cirsium stojanovii* Kuzm., species nova. In Velčev, V.I. & al. (eds): 100th anniversary of Academician Nikolaj A. Stojanov, pp. 60-63. - Sofia: Publishing House of the Bulgarian Academy of Sciences.
KUZMANOV, B.A. 1991: Evolutionary trends of Bulgarian Compositae [In Bulgarian with English summary]. In Kožuharov, S.I. & Kuzmanov, B.A. (eds): Evolution of flowering plants ... 2, pp. 7-84 + 12 plates and 1 chart. - Sofia: Publishing House of the Bulgarian Academy of Sciences.
KUZMANOV, B.A. 1993: Chromosome numbers of Bulgarian angiosperms: An introduction to a chromosome atlas of the Bulgarian flora. - *Fl. Medit.* **3**: 19-163.
KUZMANOV, B.A. & AL. 1981a: [Chromosome numbers in Asteraceae]. In Löve, Á. (ed.): Chromosome number reports LXXII. - *Taxon* **30(3)**: 701-702.
KUZMANOV, B.A. & AL. 1981b: Variation in *Leucanthemum vulgare* complex in Bulgaria. 2. A comparative morphologic and anatomic investigation [In Bulgarian with Russian and English summaries]. - *Fitologiya* **16**: 41-69.
KUZMANOV, B.A. & AL. 1983: [Chromosome numbers in Asteraceae from Bulgaria]. In Löve, Á. (ed.): IOPB chromosome number reports LXXX. - *Taxon* **32(3)**: 505.
KUZMANOV, B.A. & AL. 1984a: Chemotaxonomic study on genus *Plantago* L. - *Fitologiya* **24**: 29-34.
KUZMANOV, B.A. & AL. 1984b: Chemotaxonomic studies on the Bulgarian species of the genus *Corydalis* Vent. 3. Cluster analysis [In Bulgarian with Russian and English summaries]. - *Fitologiya* **25**: 3-12.
KUZMANOV, B.A. & AL. 1986: Variation of the Bulgarian species of the genus *Corydalis* Vent. 4. Morphological study [In Bulgarian with Russian and English summaries]. - *Fitologiya* **32**: 3-18.
KUZMANOV, B.A. & AL. 1987: Biosystematic studies on *Peucedanum* L. in Bulgaria. 1. Karyological study. - *Fitologiya* **33**: 3-13.
KUZMANOV, B.A. & AL. 1990: Chromosome numbers of Bulgarian angiosperms. - *Fitologiya* **38**: 92.
KUZMANOV, B.A. & ANČEV, M.E. 1973: [Chromosome numbers in Asteraceae from Bulgaria]. In Löve, Á. (ed.): IOPB chromosome number reports XLI. - *Taxon* **22(4)**: 461.
KUZMANOV, B.A., ANČEV, M.E. & MARKOVA, T.A. 1973: [Chromosome numbers in Fabaceae from Bulgaria]. In Löve, Á. (ed.): IOPB chromosome number reports XLI. - *Taxon* **22(4)**: 461.
KUZMANOV, B.A., ANDREEV, N. & GEORGIEVA, S.B. 1977: [Chromosome numbers in *Peucedanum* from Bulgaria]. In Löve, Á. (ed.): IOPB chromosome number reports LVII. - *Taxon* **26(4)**: 448.
KUZMANOV, B.A., ANDREEV, N. & KOZOVSKA, V. 1981: Chemotaxonomic study on Bulgarian species of *Peucedanum* L. I. - *Anales Jard. Bot. Madrid* **37(2)**: 779-788.
KUZMANOV, B.A., ESTATIEVA, L. & GEORGIEVA, S.B. 1981: [Chromosome numbers in *Thalictrum*]. In Löve, Á. (ed.): Chromosome number reports LXXII. - *Taxon* **30(3)**: 702.
KUZMANOV, B.A. & GEORGIEVA, S.B. 1976: [Chromosome numbers from Bulgaria]. In Löve, Á. (ed.): : : IOPB chromosome number reports LIII. - *Taxon* **25(4)**: 500.
KUZMANOV, B.A. & GEORGIEVA, S.B. 1977a: [Chromosome numbers from Bulgaria]. In Löve, Á. (ed.): IOPB chromosome number reports LVII. - *Taxon* **26(4)**: 448.
KUZMANOV, B.A. & GEORGIEVA, S.B. 1977b: [Chromosome numbers from Bulgaria]. In Löve, Á. (ed.): IOPB chromosome number reports LVIII. - *Taxon* **26(5-6)**: 558.
KUZMANOV, B.A. & GEORGIEVA, S.B. 1980: [Chromosome numbers in Asteraceae from Bulgaria]. In Löve, Á. (ed.): Chromosome number reports LXIX. - *Taxon* **29(5-6)**: 715.
KUZMANOV, B.A. & GEORGIEVA, S.B. 1983: [Chromosome numbers in Asteraceae from Bulgaria]. In Löve, Á. (ed.): IOPB chromosome number reports LXXXI. - *Taxon* **32(4)**: 665.
KUZMANOV, B.A. & GEORGIEVA, S.B. 1987: [Chromosome numbers in *Centaurea* from Bulgaria]. In Löve, Á. (ed.): Chromosome number reports XCIV. - *Taxon* **36(1)**: 284.
KUZMANOV, B.A. & GEORGIEVA, S.B. 1990: Cytotaxonomy of Bulgarian *Centaurea* species. - *Razprave Razreda Sazu* **31(7)**: 105-125.
KUZMANOV, B.A., GEORGIEVA, S.B. & KESTENOVA, R.H. 1988: Variation in *Leucanthemum* vulgare group in Bulgaria. Karyological study. In Velčev, V.I. & al. (eds): 100th anniversary of Academician Nikolaj A. Stojanov, pp. 115-124. - Sofia: Publishing House of the Bulgarian Academy of Sciences.
KUZMANOV, B.A., GEORGIEVA, S.B. & NIKOLOVA, V.T. 1979: [Chromosome numbers in Asteraceae from Bulgaria]. In Löve, Á. (ed.): IOPB chromosome number reports LXIV. - *Taxon* **28(4)**: 408.
KUZMANOV, B.A., GEORGIEVA, S.B. & NIKOLOVA, V.T. 1986: [Chromosome numbers in Bulgarian Asteraceae; in Bulgarian]. - *Fitologiya* **31**: 71-74.
KUZMANOV, B.A., GEORGIEVA, S.B. & NIKOLOVA, V.T. 1991: Karyological study of Bulgarian plants from the family Compositae. IV. Tribus Cardueae Cass. [In Bulgarian with Russian and English summaries]. - *Fitologiya* **39**: 3-33.
KUZMANOV, B.A., GEORGIEVA, S.B. & NIKOLOVA, V.T. 1993: Karyological study of Bulgarian Asteraceae. VII. - *Fitologiya* **44**: 16-31.
KUZMANOV, B.A. & JURUKOVA-GRĂNČAROVA, P.D. 1983: Evolutionary trends in the Bulgarian Compositae plants of Cichorioideae, Asteraceae [In Bulgarian with English and Russian summaries]. - *Fitologiya* **23**: 3-29.
KUZMANOV, B.A., JURUKOVA-GRĂNČAROVA, P.D. & GEORGIEVA, S.B. 1992: Karyological study of Bulgarian angiosperms. - *Fitologiya* **42**: 73-77.

KUZMANOV, B.A., JURUKOVA-GRĂNČAROVA, P.D. & GEORGIEVA, S.B. 1993: Karyological study of Bulgarian Asteraceae. VI. - *Fitologiya* **44**: 3-15.

KUZMANOV, B.A. & JURUKOVA, P.D. 1977: [Chromosome numbers from Bulgaria]. In Löve, Á. (ed.): IOPB chromosome number reports LVIII. - *Taxon* **26(5-6)**: 558.

KUZMANOV, B.A. & KOŽUHAROV, S.I. 1973: Karyological study of Bulgarian Compositae. 2. - *Izv. Bot. Inst. (Sofia)* **24**: 125-137.

KUZMANOV, B.A., KOŽUHAROV, S.I. & IVANOVA, H. 1966: Biometric studies on the genus *Betula* in Bulgaria. - *Fragm. Florist. Geobot.* **12(4)**: 329-336.

KUZMANOV, B.A., KOŽUHAROV, S.I. & SIMEONOV, S.Y. 1972: Taxonomic studies of the Bulgarian species of the genus *Tragopogon* L. [In Bulgarian with Russian and English summaries]. - *Izv. Bot. Inst. (Sofia)* **22**: 5-32.

KUZMANOV, B.A. & MARKOVA, T.A. 1973: [Chromosome numbers in Fabaceae from Bulgaria]. In Löve, Á. (ed.): IOPB chromosome number reports XL. - *Taxon* **22(2-3)**: 288-289.

KUZMANOV, B.A. & NIKOLOVA, V.T. 1977: [Chromosome numbers from Bulgaria]. In Löve, Á. (ed.): IOPB chromosome number reports LVIII. - *Taxon* **26(5-6)**: 559.

KUZMANOV, B.A. & NIKOLOVA, V.T. 1980: [Chromosome numbers in Asteraceae from Bulgaria]. In Löve, Á. (ed.): Chromosome number reports LXIX. - *Taxon* **29(5-6)**: 715.

KUZMANOV, B.A., NINOVA, D.J. & GEORGIEVA, S.B. 1979: [Chromosome numbers in *Centaurea* from Bulgaria]. In Löve, Á. (ed.): IOPB chromosome number reports LXIV. - *Taxon* **28(4)**: 408.

KUZMANOV, B.A., NINOVA, D.J. & GEORGIEVA, S.B. 1981: [Chromosome numbers in *Centaurea*]. In Löve, Á. (ed.): Chromosome number reports LXXII. - *Taxon* **30(3)**: 702.

KUZMANOV, B.A., NINOVA, D.J. & KESTENOVA, R.H. 1980: Variability of *Leucanthemum vulgare* complex in Bulgaria. 2. Comparative morphological and anatomic investigation [In Bulgarian with English abstract]. - *Fitologiya* **15**: 3-21.

KUZMANOV, B.A., ROBEVA, P.N. & GEORGIEVA, S.B. 1987: Cytotaxonomic study of Bulgarian species of *Thalictrum* L. - *Fitologiya* **33**: 14-22.

KUZMANOV, B.A. & STANČEV, G. 1972: [Chromosome numbers in *Trifolium* from Bulgaria]. In Löve, Á. (ed.): IOPB chromosome number reports XXXVIII. - *Taxon* **21(5-6)**: 681. [cf. also correction in *Taxon* 22(2-3): 291, 1973].

KUZMANOV, B.A., THIN, N.N. & GEORGIEVA, S.B. 1980: [Chromosome numbers in *Anthemis* from Bulgaria]. In Löve, Á. (ed.): Chromosome number reports LXIX. - *Taxon* **29(5-6)**: 714-715.

KUZMANOV, B.A., THIN, N.N. & GEORGIEVA, S.B. 1981: A cytotaxonomic study on Bulgarian *Anthemis* species. - *Candollea* **36(1)**: 19-76.

KUZMANOV, B.A., ZDRAVKOVA, M.A. & GEORGIEVA, S.B. 1989: Variability of the Bulgarian yarrow species from the group *Achillea millefolium* and *A. crithmifolia*. 1. Morphological and karyological study [In Bulgarian with Russian and English summaries]. - *Fitologiya* **36**: 26-43.

KUZMANOVIĆ, N. & AL. 2013: Retypification of the name *Eryngium palmatum* (Apiaceae). – *Phytotaxa* **105(2)**: 58-60.

KUZMANOVIĆ, N. & AL. 2013a: Genetic, cytological and morphological differentiation within the Balkan- Carpathian *Sesleria rigida* sensu Fl. Eur. (Poaceae): A taxonomically intricate tetraploid-octoploid complex. - *Taxon* **62(3)**: 458-472.

KUZMANOVIĆ, N. & AL. 2013b: *Sesleria serbica* (Poaceae), a neglected species of the Balkan Peninsula. - *Bot. Serbica* **37(2)**: 113-120.

KUZMANOVIĆ, N. & AL. 2014: Review of nomenclatural notes in the European Gesneriaceae. - *Bot. Serbica* **38(2)**: 247-250.

KUZMANOVIĆ, N. & LAKUŠIĆ, D. 2015: Insight into morpho-anatomical variability of *Festuca violacea* group (Poaceae) on Balkan Peninsula. In Bogdanović, S. & Jogan, N.: Book of abstracts - 6th Balkan Botanical Congress, p. 82. - Rijeka.

KUZMANOVIĆ, N. & AL. 2021: Long neglected diversity in the Accursed Mountains (western Balkan Peninsula): *Ranunculus bertisceus* is a genetically and morphologically divergent new species. - *Bot. J. Linn. Soc.* **196(3)**: 384-406.

KYNČLOVÁ, M. 1970: Comparative morphology of achenes of the tribe Anthemideae Cass. (family Asteraceae) and its taxonomic significance. - *Preslia* **42(1)**: 33-53.

KYPRIOTAKIS, Z. 2001: The flora of Milos (Cyclades islands, Greece). In Raimondo, F.M. (ed.): Abstracts from the X OPTIMA meeting, Palermo, 13-19 September 2001, p. 258.

KYPRIOTAKIS, Z., ANTALOUDAKI, E. & TZANOUDAKIS, D. 2017: *Ornithogalum insulare* (Hyacinthaceae): A new species from the Cretan area (S. Aegean, Greece). In Dousis, M.A. & Thanos, K.A.: 15th conference of the Hellenic Botanical Society, book of abstracts: p. 96.

KYPRIOTAKIS, Z., ANTALOUDAKI, E. & TZANOUDAKIS, D. 2018: *Ornithogalum insulare* (Hyacinthaceae): A new species from the Cretan area (S. Aegean, Greece). - *Bot. Serbica* **42(1)**: 117-122.

KYPRIOTAKIS, Z. & ANTALOUDAKI. E. 2013: Contribution to the study of chasmophytic flora of the Venetian walls and fortresses of Crete. In Gkelis, S. & al. (eds): 13th Panhellenic Scientific Conference, Hellenic Botanical Society, Thessaloniki 3-6 October. Program and book of abstracts, p. 95.

KYPRIOTAKIS, Z. & ARTELARI, R. 1998: *Limonium cornarianum* (Plumbaginaceae), a new species from Crete (Aegean area, Greece). - *Phyton (Horn)* **38(1)**: 143-147.

KYPRIOTAKIS, Z., DIMOPOULOS, P. & TZANOUDAKIS, D. 1999: Contribution to the survey of the chasmophytic vegetation of Crete: rock-cliffs and gorges. In Elefheriou, E.P. (ed.): Proceedings, 7th Panhellenic scientific conference, Alexandroupolis, pp. 102-107 [In Greek with English abstract]. - Thessaloniki: Univ. Studio Press.

KYPRIOTAKIS, Z., FOURNARAKI, CH. & GOTSIOU, P. 2001: Protection and conservation of endemic and threatened taxa of the native flora of Crete and of the local varieties of cultivated fruits and vegetables. In Kamari, Psaras & Constantinidis (eds): Proc. 8th scientific congress of the Greek Bot. Society, Oct. 5-8, 2000, pp. 240-252. [In Greek with English abstract].

KYPRIOTAKIS, Z. & KALOPISSIS, J. 1988: Neue Standorte der *Cephalanthera cucullata* Boiss. & Heldr. - *Mitt. Arbeitskreis Heimische Orchid. Baden-Württemberg* **20(1)**: 69-73.

KYPRIOTAKIS, Z. & TAN, KIT 2010: Report 47. In Vladimirov, V., Dane, F. & Tan, Kit (eds): New floristic records in the Balkans: 13. - *Phytol. Balcan.* **16(1)**: 152.

KYPRIOTAKIS, Z. & TZANOUDAKIS, D. 1998: Contribution to the study of the Greek insular flora: The chasmophytes of the Cretan area. In Moret, J. (ed.): Abstracts of the IX. OPTIMA meeting, Paris, 11-17 May 1998. - Paris: Mus. Nat. d'Hist. Naturelle.

KYPRIOTAKIS, Z. & TZANOUDAKIS, D. 1999: A new species of *Bellevalia* from eastern Crete and its confusion with *Muscari macrocarpum* Sweet. - *Bot. Helv.* **109(1)**: 85-90.

KYPRIOTAKIS, Z. & TZANOUDAKIS, D. 2000: Contribution to the knowledge and the conservation of the Greek insular flora: The case of the islet of Dia (Cretan area, Greece). In Özhatay, N. & al. (eds): Abstracts from the Second Balkan Botanical Congress, Istanbul, Turkey, May 14-18, 2000, p. 45.

KYPRIOTAKIS, Z. & TZANOUDAKIS, D. 2001: Contribution to the study of the Greek insular flora: The chasmophytic flora of Crete. - *Bocconea* **13**: 495-503.

KYRIAKOPOULOS, CH. & AL. 2018: *Potentilla greuteriana* (Rosaceae), a new species from Mt. Taigetos (S Peloponnisos, Greece). - *Fl. Medit.* **28**: 351-364.

KYRIAKOPOULOS, CH., BAREKA, E.-P. & KAMARI, G. 2014: Reports (1840-1844). In Kamari, G., Blanché, C. & Siljak-Yakovlev, S. (eds): Mediterranean chromosome number reports - 24. - *Fl. Medit.* **24**: 287-291.

KYRIAKOPOULOS, CH., BAREKA, P. & KAMARI, G. 2016: Karyological data of some endemic taxa from Mt Taigetos, Greece. - *Fl. Medit.* **26**: 224-228.

KYRIAKOPOULOS, CH., LIVERI, E. & PHITOS, D. 2017: *Campanula kamariana* (section *Quinquelocularis*), a new species from S Peloponnisos, Greece. - *Fl. Medit.* **27**: 131-135.

KYRIAKOPOULOS, CH.K. & KAMARI, G. 2016: The rediscovery of *Aquilegia ottonis* subsp. *taygetea* (Ranunculaceae), an endemic taxon of S. Peloponnisos, Greece. - *Bot. Chron. (Patras)* **21**: 75-82.

LA VALVA, V. 1977: Numeri cromosomici per la Flora Italiana: 251-254. - *Inform. Bot. Ital.* **8(2)**: 204-207.

LA VALVA, V. 1993: La collezione Gussone Sicilia. - *Webbia* **48**: 515-537.

LA VALVA, V. & AL. 1985: Morphology and alkaloid chemistry of *Papaver setigerum* DC. (Papaveraceae). - *Taxon* **34(2)**: 191-196.
LA VALVA, V., MORALDO, B. & CAPUTO, G. 1981: *Vincetoxicum hirundinaria* Medicus subsp. *luteolum* (Jord. et Fourr.) La Valva, Moraldo et Caputo stat. et comb. nov. (Asclepiadaceae), entità nuova per la flora italiana. - *Delpinoa, n.s.* **21-22**: 91-96.
LACAITA, C.C. 1912a: Plante Italiane critiche o rare VI. *Astragalus hypoglottis*. - *Nuovo Giorn. Bot. Ital., n.s.* **19(3)**: 309-326.
LACAITA, C.C. 1912b: What is *Astragalus hypoglottis* L.? - *J. Bot. (London)* **50**: 217-229.
LACAITA, C.C. 1917: *Calamintha nebrodensis* Kern. & Strobl. - *J. Bot. (London)* **55**: 112.
LACAITA, C.C. 1918: Two critical plants of the Greek flora. - *J. Linn. Soc., Bot.* **44(296)**: 125-129.
LACAITA, C.C. 1919: A revision of some critical species of *Echium*, as examplified in the Linnean and other herbaria; with a description of *Echium judaeum*, a new species from Palestine. - *J. Linn. Soc.,Bot.* **44(299)**: 363-438.
LACAITA, C.C. 1920: *Quercus aegilops*. - *Bull. Misc. Inform. Kew* **1920(3)**: 100-105.
LACAITA, C.C. 1924a: Plante italiane critiche o rare (XCI-XCVII). - *Nuovo Giorn. Bot. Ital., n.s.* **31(1)**: 18-35 + 3 fold-out plates.
LACAITA, C.C. 1924b: The Onosmas of Linnaeus and Sibthorp, with a note on those of Tournefort's herbarium. - *J. Linn. Soc., Bot.* **46(310)**: 387-400 + 1 plate.
LACK, H.W. 1974: Die Gattung *Picris* L. sensu lato im ostmediterran-westasiatischen Raum. - Wien: VWGÖ (Diss. Univ. Wien, 116), 184 pp. + plates and maps.
LACK, H.W. 1975a: A note on *Helminthotheca* Zinn (Compositae). - *Taxon* **24(1)**: 111-112.
LACK, H.W. 1975b: Type specimens of the Linnaean species of *Picris* L. (Compositae). - *Taxon* **24 (1)**: 113-116.
LACK, H.W. 1979: Current projects on the Mediterranean flora. - Berlin: OPTIMA, 118 pp.
LACK, H.W. 1980: Das Herbar der zoologisch-botanischen Gesellschaft in Wien. - *Willdenowia* **10(1)**: 67-81.
LACK, H.W. 1984: Current projects on the Mediterranean flora. 2nd ed. - Berlin: OPTIMA, 152 pp.
LACK, H.W. 1986: Chronological sequence of publications by K.H. Rechinger since 1971. - *Proc. Roy. Soc. Edinburgh, B, Biol. Sci.* **89**: 3-5.
LACK, H.W. 1987: Karl Heinz Rechinger - a Grand Old Man in botany. - *Pl. Syst. Evol.* **155(1-4)**: 7-14.
LACK, H.W. 1991: In den Schluchten des Balkans. Die Zar Boris III.-Tanne im Botanischen Garten. - *Museums Journal, Bot. Gart. & Bot. Mus. Berlin-Dahlem* **5(1)**: 46-49.
LACK, H.W. 1996: Die frühe botanische Erforschung der Insel Kreta. - *Ann. Naturhist. Mus. Wien, B* **98B Suppl.**: 183-236.
LACK, H.W. 1997a: The Sibthorpian herbarium at Oxford - guidelines for its use. - *Taxon* **46(2)**: 253-263.
LACK, H.W. 1997b: Die Frontispize von John Sibthorps "Flora Graeca". - *Ann. Naturhist. Mus. Wien, B* **99B**: 615-654.
LACK, H.W. 1998: Lilac and horse-chestnut: Discovery and rediscovery. In Moret, J. (ed.): Abstracts of the IX. OPTIMA meeting, Paris, 11-17 May 1998. - Paris: Mus. Nat. d'Hist. Naturelle.
LACK, H.W. 1999a: The Flora Graeca story - Oxford's finest botanical treasure. Sibthorp, Bauer, and Hawkins in the Levant - an exhibition at the Bodleian Library 19 July 1999 to 25 September 1999. - Oxford: Bodleian Library, vi + 50 pp.
LACK, H.W. 1999b: Obituary: Karl Heinz Rechinger (1906-1998). - *Taxon* **48(2)**: 419-426.
LACK, H.W. 1999c: Vom Castanienbaum. Ein Souvenir der Botschafter in Istanbul. – *Museums Journal, Bot. Gart. & Bot. Mus. Berlin-Dahlem* **13(4)**: 88-89.
LACK, H.W. 1999d: John Sibthorp und John Hawkins auf den ionischen Inseln. -*Haussknechtia* **7**: 25-28.
LACK, H.W. 2000a: An early woodcut of a plant collected in Greece. - *Bot. Chron. (Patras)* **13**: 249-254.
LACK, H.W. 2000b: Lilac and horse-chestnut: Discovery and rediscovery. - *Curtis's Bot. Mag., ser. 6* **17(2)**: 109-141.
LACK, H.W. 2000c: Karl Heinz Rechinger - a life for botany. - *Fl. Medit.* **10**: 11-64.
LACK, H.W. 2001: Lilac and horse-chestnut: discovery and rediscovery. - *Bocconea* **13**: 613-616.
LACK, H.W. 2015: The Bauers: Joseph, Franz & Ferdinand. Masters of botanical illustration. - Munich: Prestel Verlag, 496 pp.
LACK, H.W. 2019: Michiel's Codice erbario (c. 1550) and the flora of Greece. - *Bot. Chron. (Patras)* **22**: 63-72.
LACK, H.W. & BARINA, Z. 2020: The early botanical exploration of Albania (1839-1945). - *Willdenowia* **50(3)**: 519-558 + suppl.
LACK, H.W. & MABBERLEY, D.J. 1999: The Flora Graeca story. Sibthorp, Bauer, and Hawkins in the Levant. - Oxford, etc.: Oxford Univ. Press, xxxi + 327 pp.
LACK, H.W. & SYDOW, C.-O.V. 1984a: Dörflers Sammlung von Botanikerbriefen in der Universitätsbibliothek Uppsala. I. Einführung. Verzeichnis der Briefschreiber A-F. - *Willdenowia* **13(2)**: 397-428.
LACK, H.W. & SYDOW, C.-O.V. 1984b: Dörflers Sammlung von Botanikerbriefen in der Universitätsbibliothek Uppsala. II. Verzeichnis der Briefschreiber G-R. - *Willdenowia* **14(1)**: 203-225.
LACK, H.W. & SYDOW, C.-O.V. 1985: Dörflers Sammlung von Botanikerbriefen in der Universitätsbibliothek Uppsala. III. Verzeichnis der Briefschreiber S-Z. Verzeichnis der Briefempfänger. Miscellanea. - *Willdenowia* **14(2)**: 435-456.
LADIZINSKY, G. 1971: Biological flora of Israel. 2. *Avena* L. - *Israel J. Bot.* **20(2)**: 133-151.
LADIZINSKY, G. 1974: Genome relationships in the diploid oats. - *Chromosoma* **47(1)**: 109-117.
LADIZINSKY, G. 1975: On the origin of the broad bean, *Vivia faba* L. - *Israel J. Bot.* **24(2-3)**: 80-88.
LADIZINSKY, G. 1978: Chromosomal polymorphism in wild populations of *Vicia sativa* L. - *Caryologia* **31(2)**: 233-241.
LADIZINSKY, G. 1986: A new *Lens* from the Middle-East. - *Notes Roy. Bot. Gard. Edinburgh* **43(3)**: 489-492.
LADIZINSKY, G. & ABBO, SH. 1996: Genetic diversity in the genus *Lens*. In Pickersgill, B. & Lock, J.M. (eds): Advances in legume systematics. Part 8: Legumes of economic importance, pp. 53-64. - Kew: Royal Botanic Gardens.
LADIZINSKY, G. & ADLER, A. 1976a: Genetic relationships among the annual species of *Cicer* L. - *Theor. Appl. Genet.* **48(4)**: 197-203.
LADIZINSKY, G. & ADLER, A. 1976b: The origin of chickpea *Cicer arietinum* L. - *Euphytica* **25(1)**: 211-217.
LADIZINSKY, G. & AL. 1984: The biological species of the genus *Lens* L. - *Bot. Gaz.* **145(2)**: 253-261.
LADIZINSKY, G., BRAUN, D. & MUEHLBAUER, F.J. 1983: Evidence for domestication of *Lens nigricans* (M. Bieb.) Godron in S. Europe. - *Bot. J. Linn. Soc.* **87(2)**: 169-176.
LADIZINSKY, G. & VOSA, C.G. 1986: Karyotype and C-banding in *Trigonella* section *Foenum-graecum* (Fabaceae). - *Pl. Syst. Evol.* **153(1-2)**: 1-5.
LADIZINSKY, G. & ZOHARY, D. 1968: Genetic relationships between diploids and tetraploids in series *Eubarbatae* of *Avena*. - *Canad. J. Genet. Cytol.* **10(1)**: 68-81.
LADIZINSKY, G. & ZOHARY, D. 1971: Notes on species delimitation, species relationships and polyploidy in *Avena* L. - *Euphytica* **20(3)**: 380-395.
LAFRANCHIS, T. 2007: Northern Peloponnese delights. - *Alpine Gardener* **75(2)**: 200-221.
LAFRANCHIS, T. & SFIKAS, G. 2009: Flowers of Greece. - Paris: Diatheo, vol. 1 (431 pp.) & vol. 2 (446 pp.) + DVD [Reprint 2021].
LAGARDE, F. 1997: *Coridothymus capitatus* (L.) Reichemb. fil in Österr. Bol. [sic!] (1857) [Syn. *Satureja capitata* L. (1753), *Thymbra capitata* (L.) Cav. (1803), *Thymus capitatus* (L.) Hoffmans et Link (1809)]. - *Pl. Mont. Rocaille* **12(181)**: 156-157.
LAGIOU, E. & AL. 1997: The vascular flora of the walls of Thessaloniki. In: Abstracts from the First Balkan Botanical Congress, Thessaloniki, September 19-22, 1997, p. 28. - Thessaloniki.

LAGOUDAKIS, N. & AL. 2002: Plant species with alkaloids in the city of Thessaloniki (N Greece): distribution and alkaloid content. - *Bot. Chron. (Patras)* **15**: 35-44.

LAHONDÈRE, C. 1987: Contribution à l'étude du genre *Salicornia* L. en Corse. - *Bull. Soc. Bot. Centre-Ouest, n.s.* **18**: 73-78.

LAHONDÈRE, C. 1988: Seconde contribution à l'ètude du genre *Salicornia* L. en Corse. - *Bull. Soc. Bot. Centre-Ouest, n.s.* **19**: 105-108.

LAÍNZ, M. 1984: *Isoetes* "histrix" Bory. - *Anales Jard. Bot. Madrid* **40(2)**: 478-479.

LAÍNZ, M. 1992: Algo sobre las indicaciones lusitanas de *Viola hymettia* Boiss. & Heldr. - *Anales Jard. Bot. Madrid* **49(2)**: 303.

LAÍNZ, M. 1994: Sobre las desventuras del caracterizadísimo *Geum heterocarpum* Boiss. (1838). - *Anales Jard. Bot. Madrid* **51(2)**: 320-321.

LAÍNZ, M. 1995: *Thesium humifusum* DC. in Lam. & DC., Fl. Franc. ed. 3, 5: 366 (1815) versus *Th. divaricatum* Jan ex Mert. & W.D.J. Koch in Röhling, Deutschl. Fl. ed. 3, 2: 285 (1826). - *Anales Jard. Bot. Madrid* **52(2)**: 247-248.

LAKEBERG, D. 1987: Über *Helictotrichon agropyroides* (Boissier) Henrard 1940 aus S-Griechenland (Poaceae: Pooideae: Aveneae). - *Senckenberg. Biol.* **67(4-6)**: 425-429.

LAKON, M. 1907: Drei Diagnosen aus Heldreichs "Simvole pros sintaxin chloridos ton Kikladon: I chloris tis Mikonou ke ton parakimenon nison Dilou ke Rinias" [Extracted from *Parnassos* 1901: 239-255]. - *Repert. Spec. Nov. Regni Veg.* **4**: 302.

LAKOWITZ, K. 1929: Verzeichnis der in Griechenland und auf der Insel Kreta während der Vereinsstudienfahrt im April 1928 gesammelten Pflanzen. - *Ber. Versamml. Westpreuss. Bot.-Zool. Vereins Danzig* **51**: 61-68.

LAKUŠIĆ, D. & AL. 2006: Morpho-anatomical differentiation of the Balkan populations of the species *Teucrium flavum* L. (Lamiaceae). - *Flora* **201(2)**: 108-119.

LAKUŠIĆ, D. & AL. 2012: Distribution of *Lunaria telekiana* (Brassicaceae), a poorly known species of European concern. - *Bot. Serbica* **36(2)**: 139-144.

LAKUŠIĆ, D. & AL. 2013a: Molecular phylogeny of the *Campanula pyramidalis* complex (Campanulaceae) inferred from chloroplast and nuclear non-coding sequences and its taxonomic implications. - *Taxon* **62(3)**: 505-524.

LAKUŠIĆ, D. & AL. 2013b: *Edraianthus canescens* (Campanulaceae), a new species from the Central Balkan Peninsula. - *Phytotaxa* **118(1)**: 22-28.

LAKUŠIĆ, D. & AL. 2006: *Cardamine panicicii* (Brassicaceae), a neglected species of the Balkan Peninsula. - *Willdenowia* **36(1)**: 177-191.

LAKUŠIĆ, D. & CONTI, F. 2004: *Asyneuma pichleri* (Campanulaceae), a neglected species of the Balkan Peninsula. - *Pl. Syst. Evol.* **247(1-2)**: 23-36.

LAKUŠIĆ, D. & LAKUŠIĆ, B. 2014: Morpho-anatomical differentiation of the species *Teucrium montanum* (Lamiaceae) in the Central Balkan Peninsula. - *Bot. Serbica* **38(1)**: 109-120.

LAKUŠIĆ, R. 1960: Areali evropskih vrsta roda *Wulfenia* [In Serbo-Croatian]. - *God. Biol. Inst. Univ. Sarajevu* **13(1-2)**: 21-40.

LAKUŠIĆ, R. 1964: The mountain maple (*Acer heldreichii* Orph.) [In Serbo-Croatian with English summary]. - *God. Biol. Inst. Univ. Sarajevu* **17**: 117-143 + 1 map and 1 table.

LAKUŠIĆ, R. 1969: *Utricularia vulgaris* L. - eine neue carnivore Pflanze in der Flora der Crna Gora [In Serbo-Croatian with German summary]. - *Glasn. Republ. Zavoda Zaštitu Prir. Prirodnjačke Zbirke Titogradu* **2**: 85-86.

LAKUŠIĆ, R. 1973: *Rumex balcanicus* Rech. fil., neues tertiers Relikt in der Flora von Montenegro [In Serbo-Croatian with German summary]. - *Glasn. Republ. Zavoda Zaštitu Prir. Prirodnjačke Zbirke Titogradu* **5**: 29-32.

LAKUŠIĆ, R. 1974: Das natürliche System der Populationen und der Arten in der Gattung *Edraianthus* DC. [In Serbo-Croatian with German summary]. - *God. Biol. Inst. Univ. Sarajevu* **27**: 1-130.

LAKUŠIĆ, R. 1975: *Valeriana braunii-blanquetii* Lakušić species nova [In Serbo-Croatian and Latin with German summary]. - *Glasn. Republ. Zavoda Zaštitu Prir. Prirodnjačke Zbirke Titogradu* **8**: 101-107.

LAKUŠIĆ, R. 1978: Die chorologisch-ökologische und morphologisch-zytologische Differenzierung der europäischen Arten der Gattung *Wulfenia* Jacq. - *Bot. Jahrb. Syst.* **99(4)**: 443-461.

LAKUŠIĆ, R. & GRGIĆ, P. 1969: Distribution and ecology of the species *Valeriana pančićii* Hal. et Bald., *Valeriana bertiscea* Pančić, *Asperula dörfleri* Wettstein and *Gentiana levicalyx* Rohlena [In Serbo-Croatian with English summary]. - *Ekologija (Belgrade)* **4(2)**: 237-245.

LAKUŠIĆ, R. & GRGIĆ, P. 1971: Ecology and distribution of the endemic species *Narthecium scardicum* Koš., *Pinguicula balcanica* Cas., *Gymnadenia friwaldii* Hampe and *Silene asterias* Gris. [In Serbo-Croatian with English summary]. - *Ekologija (Belgrade)* **6(2)**: 337-350.

LAKUŠIĆ, R. & MARKIŠIĆ, H. 1982: A new species of the genus *Hedysarum* on Prokletije Mountains (Jugoslavia) [In Serbo-Croatian with English and German abstracts]. - *Glasn. Zemaljsk. Muz. Sarajevu* **21**: 65-72.

LAKUŠIĆ, R. & MEĐEDOVIĆ, S. 1976: Ecological and cytogenetic characteristcs of *Achillea clavenae* L. and *Achillea ageratifolia* (S.S.) Boiss. subsp. *aizoon* (Gris.) Heim. [In Serbo-Croatian with English summary]. - *God. Biol. Inst. Univ. Sarajevu* **29**: 69-77.

LAKUŠIĆ, R. & PAVLOVIĆ, D. 1974: Five new species in flora of Montenegro [In Serbo-Croatian with English summary]. - *Glasn. Republ. Zavoda Zaštitu Prir. Prirodnjačke Zbirke Titogradu* **6**: 59-61.

LAKUŠIĆ, R. & PULEVIĆ, V. 1980: The distribution and ecology of species *Daphne malyana* Blečić [In Serbo-Croatian with English summary]. - *Glasn. Republ. Zavoda Zaštitu Prir. Prirodnjačke Zbirke Titogradu* **13**: 23-27.

LAKUŠIĆ, R. & STEVANOVIĆ, V. 1995: *Draba bertiscea* (*D.* sect. *Aizopsis*, Brassicaceae), a new species from Montenegro (Yugoslavia). - *Willdenowia* **25(1)**: 75-80.

LALL, S.S. 1970: Materials for a Flora of Turkey: XXI. *Scrophularia*. - *Notes Roy. Bot. Gard. Edinburgh* **30(1)**: 129-140 + 1 plate.

LAMARCK, J.P. DE & POIRET, J.L.M. 1783-1817: Encyclopédie méthodique. Botanique. - Paris/Liège: 13 vols. [cf. TL-2 2: 732-733, 1979].

LAMB, R.P.H. 1982: A brief encounter with the ferns of Corfu. - *Bull. Brit. Pteridol. Soc.* **2(4)**: 203-204.

LAMBDON, PH.D. & AL. 2008: Alien flora of Europe: species diversity, temporal trends, geographical patterns and research needs. - *Preslia* **80(2)**: 101-149.

LAMBERTINI, C. & AL. 2006: A phylogeographic study of the cosmopolitan genus *Phragmites* (Poaceae) based on AFLPs. - *Pl. Syst. Evol.* **258(3-4)**: 161-182.

LAMBINON, J. 1981a: (588) Proposition de rejet *Aegilops ovata* L., Sp. Pl., ed. 1: 1050. 1753 (Poaceae). - *Taxon* **30(1)**: 361-362. [not approved, cf. Taxon 35(3): 557-558, 1986].

LAMBINON, J. 1981b: (589) Proposition de rejet *Bromus hordeaceus* L., Sp. Pl., ed. 1: 77. 1753 (Poaceae). - *Taxon* **30(1)**: 362. [not approved, cf. Taxon 35(3): 558, 1986].

LAMBINON, J. 1981c: (590) Proposition de rejet *Carex muricata* L., Sp. Pl., ed. 1: 974. 1753 (Cyperaceae). - *Taxon* **30(1)**: 362. [not approved, cf. Taxon 35(3): 558, 1986].

LAMBINON, J. 1981d: (593) Proposition de rejet *Dipsacus fullonum* L., Sp. Pl., ed. 1: 97. 1753 (Dipsacaceae). - *Taxon* **30(1)**: 362-363. [not approved, cf. Taxon 35(3): 559, 1986].

LAMBINON, J. 1981e: (594) Proposition de rejet *Euphorbia verrucosa* L., Sp. Pl., ed. 1: 459. 1753 (Euphorbiaceae). - *Taxon* **30(1)**: 363. [not approved, cf. Taxon 35(3): 559, 1986].

LAMBINON, J. 1981f: (595) Proposition de rejet *Medicago polymorpha* L., Sp. Pl., ed. 1: 779. 1753 (Leguminosae-Papilionoideae). - *Taxon* **30(1)**: 363. [not approved, cf. Taxon 35(3): 560-561, 1986].

LAMBINON, J. 1981g: (596) Proposition de rejet *Phleum exaratum* Hochst. ex Griseb., Spicil. Fl. Rumel. 2: 463. 1845 (ou 1846?) (Poaceae). - *Taxon* **30(1)**: 363. [not approved, cf. Taxon 35(3): 561, 1986].

LAMBINON, J. 1981h: (597) Proposition de rejet *Potamogeton pusillus* L., Sp. Pl., ed. 1: 127. 1753 (Potamogetonaceae). - *Taxon* **30(1)**: 363. [not approved, cf. Taxon 35(3): 561-562, 1986].

LAMBINON, J. 1981i: (599) Proposition de rejet *Spergularia media* (L.) C. Presl, Fl. Sic. 1: 161.

1826 (Caryophyllaceae) - *Arenaria media* L., Spec. Pl., ed. 2: 606. 1762. - *Taxon* **30(1)**: 364. [not approved, cf. Taxon 35(3): 562, 1986].
LAMBINON, J. 1981j: (600) Proposition de rejet *Vulpia membranacea* (L.) Dum., Obs. Gram. Belg., 100. 1824 (Poaceae) - *Stipa membranacea* L., Spec. Pl., ed. 1: 560. 1753. - *Taxon* **30(1)**: 364. [not approved, cf. Taxon 35(3): 562-563, 1986].
LAMBINON, J. 1991: Le problème de la variabilité de *Plantago major* L. en Corse. In Jeanmonod, D. & Burdet, H.M. (eds): Notes et contributions à la flore de Corse, VII. - *Candollea* **46(1)**: 210-217.
LAMBINON, J. & AL. 1992: Le groupe de *Carex flava* L. en Corse. - *Candollea* **47(2)**: 306-311.
LAMBINON, J. & DESCHÂTRES, R. 1991: Les *Anthoxanthum* annuels de Corse. In Jeanmonod, D. & Burdet, H.M. (eds): Notes et contributions à la flore de Corse, VII. - *Candollea* **46(1)**: 217-221.
LAMBROPOULOU, E., KOUGIOUMOUTZIS, K. & SPANOU, S. 2017: Floristic diversity and composition along altitudinal gradient in Mount Panachaiko, Achaia. In Dousis, M.A. & Thanos, K.A.: 15th conference of the Hellenic Botanical Society, book of abstracts: p. 98.
LÄMMERMAYR, L. 1930: Die Gattung *Hedera*. - *Pflanzenareale* **2(7)**: 64-69, Karte 65-68.
LAMMERS, TH.G. 1996: (1252) Proposal to conserve the name *Edraianthus* against *Pilorea* (Campanulaceae). - *Taxon* **45(3)**: 563-564. [approved, cf. Taxon 48(2): 362, 1999].
LAMMERS, TH.G. 1997: (1321) Proposal to conserve the name *Laurentia* (Campanulaceae) with a conserved type. - *Taxon* **46(4)**: 795-799. [not approved; cf. *Taxon* 49(2): 264-265, 2000].
LAMPRECHT, H.V. 1962: Cyto-systematische Untersuchungen an *Anemone nemorosa* L. und *Anemone ranunculoides* L. - *Beitr. Biol. Pflanzen* **37(1)**: 107-146.
LAMPRECHT, H.V. 1974: Monographie der Gattung *Pisum*. Die genetischen Grundlagen für die Züchtung der Erbse und der verwandten Leguminosen [edited by K. Mecenovic after the death of the author]. - Graz: Steiermärkischen Landesdruckerei, xii + 655 pp.
LANCASTER, R. 1992: *Pinus leucodermis*, well worth spending a night in jail for. - *Garden (London)* **117(1)**: 20-21.
LANDERER, X. 1853: Botanische Notizen aus Griechenland. - *Oesterr. Bot. Wochenbl.* **3**: 278-279, 317-319.
LANDERER, X. 1856a: Botanische Notizen aus Griechenland. - *Flora* **39**: 449-458, 647-652, 753-756.
LANDERER, X. 1856b: *Quercus coccifera* und dessen Schildläuse in Griechenland. - *Oesterr. Bot. Wochenbl.* **6**: 411-412.
LANDERER, X. 1856c: Botanische Notizen aus Griechenland. - *Oesterr. Bot. Wochenbl.* **6**: 58-60.
LANDERER, X. 1856d: Botanische Notizen aus Griechenland. - *Oesterr. Bot. Wochenbl.* **6(16)**: 123-125.
LANDERER, X. 1856e: Botanische Notizen aus Griechenland. - *Oesterr. Bot. Wochenbl.* **6(24)**: 187-190.
LANDERER, X. 1857: Über die Forstgewächse in Griechenland. - *Flora* **40**: 385-396.
LANDERER, X. 1859: Botanische Mittheilungen aus Griechenland. - *Flora* **42**: 193-205, 238-239, 516-523.
LANDI, M. 1933: Osservazioni e ricerche sulla "Lunaria pachyrrhiza" Borbás. - *Arch. Bot. (Forlì)* **9**: 104-117.
LANDOLT, E. 1954: Die Artengruppe des *Ranunculus montanus* Willd. in den Alpen und im Jura (Zytologisch-systematische Untersuchungen). - *Ber. Schweiz. Bot. Ges.* **64**: 9-83 + 2 plates.
LANDOLT, E. 1956: Die Artengruppe des *Ranunculus montanus* in den Pyrenäen und anderen europäischen Gebirgen westlich der Alpen. - *Ber. Schweiz. Bot. Ges.* **66**: 92-117.
LANDOLT, E. 1975: Morphological differentiation and geographical distribution of the *Lemna gibba - Lemna minor* group. - *Aquatic Bot.* **1(4)**: 345-363.
LANDOLT, E. 1980a: Key to the determination of taxa within the family of Lemnaceae. In Landolt, E. (ed.): Biosystematische Untersuchungen in der Familie der Wasserlinsen (Lemnaceae). Vol. 1. - *Veröff. Geobot. Inst. E.T.H. Stiftung Rübel Zürich* **70**: 13-21.
LANDOLT, E. 1980b: Bibliographie der Familie der Lemnaceae. In Landolt, E. (ed.): Biosystematische Untersuchungen in der Familie der Wassenlinsen (Lemnaceae). Vol. 1. - *Veröff. Geobot. Inst. E.T.H. Stiftung Rübel Zürich* **70**: 142-204.
LANDOLT, E. 1982: Distribution pattern and ecophysiological characteristics of the European species of the Lemnaceae. - *Ber. Geobot. Inst. E.T.H. Stiftung Rübel* **49**: 127-145.
LANDOLT, E. 1984a: Über die Artengruppe der *Cardamine pratensis* L. in der Schweiz. - *Diss. Bot.* **72**: 481-497.
LANDOLT, E. 1984b: Verbreitungsmuster in der Familie der Lemnaceae und ihre ökologische Deutung. - *Verh. Ges. Ökol.* **12**: 241-253.
LANDOLT, E. 1986: The family of Lemnaceae - a monographic study. Vol. 1. - *Veröff. Geobot. Inst. E.T.H. Stiftung Rübel Zürich* **71**: 1-566.
LANDOLT, E. 1989: Geographisch-ökologisches Vorkommen der Lemnaceae-Arten in der Schweiz im Rahmen ihrer Gesamtverbreitung. - *Flora, Morphol. Geobot. Oekophysiol.* **182(1-2)**: 87-98.
LANDOLT, E. 1990: Ueber zwei seit kurzer Zeit in Europa neu beobachtete *Lemna*-Arten. – *Razprave IV. Razreda Sazu* **31(8)**: 127-135.
LANDOLT, E. 1994: Taxonomy and ecology of the section *Wolffia* of the genus *Wolffia* (Lemnaceae). - *Ber. Geobot. Inst. E.T.H. Stiftung Rübel* **60**: 137-151.
LANDOLT, E. & AL. 1975: Konkurrenzunterschungen zwischen nah verwandten Arten von *Scabiosa columbaria* L. s.l. - *Ber. Geobot. Inst. E.T.H. Stiftung Rübel* **43**: 83-142.
LANDOLT, E. & URBANSKA-WORYTKIEWICZ, K. 1972: Zytotaxonomische Untersuchungen an *Cardamine pratensis* L. s.l. im Bereich der Schweizer Alpen und des Jura. - *Ber. Deutsch. Bot. Ges.* **84(11)**: 683-690.
LANDOLT, E. & URBANSKA-WORYTKIEWICZ, K. 1980: List of studied Lemnaceae samples: origin and chromosome numbers. In Landolt, E. (ed.): Biosystematische Untersuchungen in der Familie der Wasserlinsen (Lemnaceae). Vol. 1. - *Veröff. Geobot. Inst. E.T.H. Stiftung Rübel Zürich* **70**: 205-247.
LANDOLT, E. (ED.) 1980: Biosystematische Untersuchungen in der Familie der Wasserlinsen (Lemnaceae). Vol. 1. - *Veröff. Geobot. Inst. E.T.H. Stiftung Rübel Zürich* **70**: 1-247.
LANDSTRÖM, TH. 1989: The species of *Ornithogalum* L. subg. *Ornithogalum* (Hyacinthaceae) in Greece. - Lund: Ph.D. thesis, Univ. of Lund, 92 pp.
LANE, D. 1984: Factors affecting the development of populations of *Oxalis pes-caprae* L. - *Weed Res.* **24(3)**: 219-225.
LANG, A.V. 1940: Untersuchungen über einige Verwandtschafts- und Abstammungsfragen in der Gattung *Stachys* L. auf cytogenetischer Grundlage. - *Biblioth. Bot.* **29/118**: 94 pp. + 8 plates and 1 map.
LANG, G. 1970: Florengeschichte und mediterran-mitteleuropäische Florenbeziehungen. - *Feddes Repert.* **81(1-5)**: 315-335.
LANG, W. 1971: Die Edelkastanie, ihre Verbreitung und ihre Beziehung zu den naturgegebenen Grundlagen. - *Mitt. Pollichia Pfälz. Vereins Naturk.* **132**: 86-160.
LANGANGEN, A. 2010: *Chara hellenica* spec. nov. and other charophytes collected in Corfu and Amorgos islands (Greece) in 2008. - *Studia Bot. Hung.* **41**: 21-30.
LANGE, D. 1989: The genus *Helictotrichon* (Poaceae) in Greece. In Phitos, D. (ed.): Abstracts. Sixth OPTIMA meeting, Delphi 10-16 Sept. 1989, p. 107.
LANGE, D. 1991: *Helictotrichon agropyroides* (Boiss.) Henrard (Poaceae) - an endemic in S and SE Greece. - *Bot. Chron. (Patras)* **10**: 853-859.
LANGE, D. 1995a: Untersuchungen zur Systematik und Taxonomie der Gattung *Helictotrichon* Besser ex J.A. Schultes & J.H. Schultes (Poaceae) in Südosteuropa und Vorderasien. - *Biblioth. Bot.* **144**: i- iv + 1-238.
LANGE, D. 1995b: *Gaudiniopsis macra* (M. Bieb.) Eig. In Greuter, W. & Raus, Th. (eds): Med-Checklist notulae, 16. - *Willdenowia* **25(1)**: 176.
LANGENFELD, W. 1971: Die Evolution der Gattung *Malus* Mill. - *Wiss. Z. Univ. Rostock, Reihe Math.* **20 (1)**: 49-51.
LÄNGER, R. & SAUKEL, J. 1993: Systematics of *Primula veris* (Primulaceae). - *Pl. Syst. Evol.* **188(1-2)**: 31-55.

LANGERON, M. 1927: Annotations et additions à la flore crétoise. - *Bull. Soc. Bot. France* **74**: 130-139.

LANGHE, J.-E. DE & D'HOSE, R. 1980: Les orchidées de Péloponnèse (Grèce): prospections faites en 1978 et 1979. - *Bull. Soc. Roy. Bot. Belgique* **113(1)**: 105-118.

LANSDOWN, R. 2006: Notes on the water-starworts (*Callitriche*) recorded in Europe. - *Watsonia* **26**: 105-120.

LANSDOWN, R. & AL. 2017: New distribution and taxonomic information on *Callitriche* (Plantaginaceae) in the Mediterranean region. - *Phytotaxa* **313(1)**: 91-104.

LANSDOWN, R. & JARVIS, C.E. 2004: Linnaean names in *Callitriche* L. (Callitrichaceae) and their typification. - *Taxon* **53(1)**: 169-172.

LANSDOWN, R., KEFALAS, K. & BAZOS, I. 2016: New information on the status and distribution of *Callitriche pulchra* (Plantaginaceae), including a first record from Cyprus. - *Willdenowia* **46(3)**: 379-385.

LANSDOWN, R. & STRID, A. 2011: *Callitriche platycarpa* Kuetz. In Greuter, W. & Raus, Th. (eds): Med- Checklist notulae, 30. - *Willdenowia* **41(2)**: 312.

LANSDOWN, R. & TURLAND, N.J. 2010: *Callitriche lenisulca* Clavaud. In Greuter, W. & Raus, Th. (eds): Med-Checklist notulae, 29. - *Willdenowia* **40(2)**: 190.

LANSDOWN, R.V. 2008: Water-starworts (*Callitriche*) of Europe. BSBI Handbook No. 11. - London: Botanical Society of the British Isles, 184 pp.

LANZA, D. 1923: Monografia del genere *Calendula* L. - *Atti Reale Accad. (Palermo), Ser. 3* **12**: 1-166 + 10 plates.

LANZA, D. 1930: *Centaurea calcitrapa* L. e specie affini. - *Atti Reale Accad. (Palermo), Ser. 3* **15**: 90 pp. + 4 plates.

LÄPPLE, H. 1992: *Ophrys × marmarensis* subsp. *saskiana*. Eine neue *Ophrys*-Hybride aus Rhodos. - *Mitt. Arbeitskreis Heimische Orchid. Baden-Württemberg* **24(4)**: 626-632.

LARSEN, K. 1955: Cytotaxonomical studies on the Mediterranean flora. - *Bot. Not.* **108(2)**: 263-275.

LARSEN, K. 1956: Chromosome studies in some Mediterranean and south European flowering plants. - *Bot. Not.* **109(3)**: 293-307.

LARSEN, K. 1957a: Cytological observations on some species of *Globularia*. - *Bot. Not.* **110(2)**: 265-270.

LARSEN, K. 1957b: Cryptospecies in *Lathyrus pratensis*. - *Bot. Tidsskr.* **53**: 291-294.

LARSEN, K. 1958a: Cytological and experimental studies on the genus *Erodium* with special reference to the collective species *E. cicutarium* (L.) L'Hér. - *Biol. Meddel. Kongel. Danske Vidensk. Selsk.* **23 (6)**: 1-25.

LARSEN, K. 1958b: Cytotaxonomical studies in *Lotus*. IV. Some cases of polyploidy. - *Bot. Tidsskr.* **54**: 44-57.

LARSEN, K. & LÆGAARD, S. 1971: Chromosome studies of the Sicilian flora. - *Bot. Tidsskr.* **66**: 249-268.

LARSEN, K. & ŽERTOVÁ, A. 1963: On the variation pattern of *Lotus corniculatus* in eastern Europe. - *Bot. Tidsskr.* **59**: 177-194.

LARSEN, K. & ŽERTOVÁ, A. 1969: A taxonomic study on *Lotus castellanus* Boiss. et Reuter. - *Feddes Repert.* **80(4-6)**: 305-314.

LASSEN, P. 1983: *Medicago heyniana* Greuter. In Greuter, W. & Raus, Th. (eds): Med-Checklist notulae, 7. - *Willdenowia* **13(1)**: 98.

LASSEN, P. 1987: Notes on *Hippocrepis*, *Lotus* and *Melilotus*. In Greuter, W. & Raus, Th. (eds): Med-Checklist notulae, 14. - *Willdenowia* **16(2)**: 443-446.

LASSEN, P. 1989: A new delimitation of the genera *Coronilla*, *Hippocrepis*, and *Securigera* (Fabaceae). - *Willdenowia* **19(1)**: 49-62. [Also distributed as *OPTIMA leaflet* no. 183].

LASSEN, P. 1996: *Trifolium andricum* (Fabaceae), a new species from Greece. - *Ann. Naturhist. Mus. Wien, B* **98B Suppl.**: 301-303.

LASSEN, P. 1999: Notes on *Euphorbia pterococca, Coronilla valentina, Lathyrus pannonicus, Lotus corniculatus, Lotus pedunculatus, Medicago blancheana, Trifolium* spp., *Trigonella cephalotes* and *Vicia* spp. In Greuter, W. & Raus, Th. (eds): Med-Checklist notulae, 18. - *Willdenowia* **29(1-2)**: 51-67.

LASSEN, P. 2011: *Trigonella*. In Greuter, W. & Raus, Th. (eds): Med-Checklist notulae, 30. - *Willdenowia* **41(2)**: 318.

LASSEN, P. & STRID, A. 2014: [Note on *Lathyrus gorgoni*]. In Raab-Straube, E. v. & Raus, Th. (eds): Euro+Med-Checklist notulae, 3. - *Willdenowia* **44(2)**: 293.

LASSEN, P., TSIALTAS, I. & TAN, KIT 2014: Reports 146-148. In Vladimirov, V. & al. (eds): New floristic records in the Balkans: 25. - *Phytol. Balcan.* **20(2-3)**: 201-202.

LASTRUCCI, L., FRIGNANI, F. & KAPLAN, Z. 2010: *Potamogeton schweinfurthii* and similar broad-leaved species in Italy. - *Webbia* **65(1)**: 147-160.

LATOWSKI, K. & CZARNA, A. 1996: Morphological, anatomical and geographical characteristics of *Eragrostis cilianensis* (Poaceae). - *Fragm. Florist. Geobot.* **41(2)**: 579-580.

LATSARI, M. & KOKKINI, S. 1995: *Calamintha nepeta-officinalis* complex: Geographic-climatic pattern ... In: Proc., 5th sci. conf., Hellenic Bot. Soc., Delphi 21-23 October 1994, pp. 285-288 [In Greek with English summary]. - Thessaloniki: Univ. Studio Press.

LATTANZI, E. 2008: The distribution of three species of the genus *Calicotome* in Italy. - *Fl. Medit.* **18**: 123-125.

LAULUND, P. 1989: En sammenligning af flora og vegetation på kalk og serpentin på bjerget Kallidromon i Grækenland. [Comparison of flora and vegetation on limestone and serpentine on Mt. Kallidromon; in Danish]. - Copenhagen: Unpubl. M.Sc. thesis, Bot. Lab., Univ. of Copenhagen, 196 pp.

LAUNERT, E. 1968: A monographic survey of the genus *Marsilea* Linnaeus. - *Senckenberg. Biol.* **49(3-4)**: 273-315.

LAUTENSCHLAGER, E. 1979: Bestimmungsschlüssel für die wildwachsenden Weiden der Schweiz. - *Bauhinia* **6(3)**: 331-352.

LAUTENSCHLAGER, E. 1983: Atlas der schweizer weiden (Gattung *Salix* L.). - *Bauhinia, Beiheft* **1983**: 1-103.

LAUTENSCHLAGER, E. 1989: Die Weiden der Schweiz und angrenzender Gebiete. Bestimmungsschlüssel und Artbeschreibungen für die Gattung *Salix*. - Basel, etc.: Birkhäuser Verlag, 136 pp.

LAVRENTIADES, G. 1956: Erevne epi tis idroviou ke eloviou chloridos ton emvriofiton tis ellinikis Makedonias [In Greek]. - Thessaloniki: 88 pp. + 1 map.

LAVRENTIADES, G. 1961a: Chloristiki, fitogeografiki ke fitokinoniologiki erevna tis chersonisou tis Kassandras [In Greek with English summary]. - Thessaloniki: 157 pp. + 1 map.

LAVRENTIADES, G. 1961b: Contribution to the knowledge of the medicinal plants of Greece. - *Pl. Med. (Stuttgart)* **9(2)**: 164-169.

LAVRENTIADES, G. 1963: On the vegetation of the Keramoti coasts. - *Boll. Ist. Bot. Univ. Catania, ser. 3* **4**: 83-103 + 4 tables.

LAVRENTIADES, G. 1964: The ammophilous vegetation of the western Peloponnesos coasts. - *Vegetatio* **12(3-4)**: 223-287 + 2 fold-out tables.

LAVRENTIADES, G. 1969: Studies on the flora and vegetation of the Ormos Archangelou in Rhodos Island. - *Vegetatio* **19**: 308-329.

LAVRENTIADES, G. 1971: On the vegetation of sand dunes of Greek coasts. - *Colloq. Phytosoc.* **1**: 92-98.

LAVRENTIADES, G. 1973: A survey of flora [and] vegetation on rice-plantation in Gallikos river mouth (Kalochori, by Thessaloniki) [In Greek with English summary]. - *Hellen. Okeanol. Limnol. Inst. Okeanogr. Halieut. Ereunon* **11**: 145-165.

LAVRENTIADES, G. 1975a: On the vegetation of the Porto-Lagos coasts. In Jordanov, D. & al. (eds): Problems of Balkan flora and vegetation, pp. 365-379. - Sofia: Publishing House of the Bulgarian Academy of Sciences.

LAVRENTIADES, G. 1975b: Substratum of some vegetation types in Greece. In Dierschke, H. (ed.): Berichte der internationalen Symposien der internationalen Vereinigung für Vegetationskunde, pp. 423-431 + 1 table. - Vaduz: Verlag J. Cramer.

LAVRENTIADES, G. 1975c: Succession in the vegetation of sand dunes of Greek coasts and the role of most important dynamogenetic value species. In Tüxen, R. (ed.): Sukzessionsforschung, pp. 293-311. - Vaduz: Verlag J. Cramer.

LAVRENTIADES, G. 1976: On the vegetation of Patras area. In Dafis, S. & Landolt, E. (eds): Zur Vegetation und Flora von Griechenland, Band 2. - *Veröff. Geobot. Inst. E.T.H. Stiftung Rübel Zürich* **56**: 59-71.

LAVRENTIADES, G. 1979: Über die Unkrautgesellschaften in Feldern von Oräokastron, Reg. Bez. Saloniki. - *Doc. Phytosoc. n.s.* **4**: 571-584.

LAVRENTIADES, G. 1980: On the grain-field weeds of the American Farm School of Thessaloniki. - *Phytocoenologia* **7**: 318-335.

LAVRENTIADES, G. 1993: Dry costal ecosystems of Greece. In van der Maarel, E. (ed.): Ecosystems of the world. 2A. Dry coastal ecosystems. Polar regions and Europe, pp. 429-441. - Elsevier.

LAVRENTIADES, G. 1999: Plant associations of the suburbus of Thessaloniki. I. Ass. *Solanum elaeagnifolium-Hordeum murinum*, ass. nov. In Eleftheriou, E.P. (ed.): Proceedings, 7th Panhellenic scientific conference, Alexandroupolis, pp. 58-65 [In Greek with English abstract]. - Thessaloniki: Univ. Studio Press.

LAVRENTIADES, G. & BABALONAS, D. 1976: Über die Vegetation des östlichen Kalava-Küsten (Nordgriechenland). II. Sandige Küstenstreifenvegetation. - *Sci. Ann. Fac. Phys. Math. Aristot. Univ. Thessalonika* **16**: 309-324.

LAVRENTIADES, G. & PAPANICOLAOU, K. 1978: *Euphorbia anthula*, a new locally endemic species from Mt. Athos. - *Sci. Ann. Fac. Phys. Math. Aristot. Univ. Thessalonika* **18**: 119-128.

LAVRENTIADES, G. & PAVLIDES, G. 1985: Contribution to the knowledge of aquatic and helophytic plant communities of the lake Mikri Prespa [In Greek with English abstract]. In: Proceedings of 4th symposium, Greek Soc. Biol. Sci., pp. 145-155. – Thessaloniki.

LAW, R., BRADSHAW, A.D. & PUTWAIN, P.D. 1977: Life-history variation in *Poa annua*. - *Evolution* **31(2)**: 233-246.

LAWALRÉE, A. 1947: Les *Ambrosia* adventices en Europe Occidentale. - *Bull. Jard. Bot. État. Bruxelles* **18(3-4)**: 305-315.

LAWALRÉE, A. 1955: Note complémentaire sur les *Ambrosia* adventices en Europe occidentale. - *Bull. Soc. Roy. Bot. Belgique* **87(2)**: 207-208.

LAWRENCE, G.H.M. & AL. (EDS) 1968: B-P-H. Botanico-Periodicum-Huntianum. - Pittsburgh, Pa.: Hunt Botanical Library, 1063 pp.

LAZARE, J.-J. 1984: Contribution à l'étude biosystématique et écologique du complexe orophile *Carex sempervirens* Vill. (Cyperaceae). - Paris: Thèse, Université de Paris VII, xviii + 441 pp.

LAZARE, J.-J. 1986: Contribution à l'etude biosystematique et écologique du complexe orophile *Carex sempervirens* Vill. (Cyperaceae). Essai de synthèse dans la ride montagneuse nord-méditérraneene. - *Prineos* **128**: 23-64.

LAZARE, J.-J. 1988: Contribution à l'étude biosystématique et écologique du complexe orophile *Carex sempervirens* Vill. (Cyperaceae). Historique taxonomique, caryologie et biologie de la reproduction. - *Lejeunia, n.s.* **126**: 1-46.

LAZARE, J.-J. 1993: Les complexes *Carex ferruginea* Scop. et *Carex sempervirens* Vill. dans la ride montagneuse nord-méditerranéenne. In Demiriz, H. & Özhatay, N. (eds): Proceedings of the V OPTIMA meeting, pp. 619-626. - Istanbul Üniversitesi Fen Fakültesi.

LAZARE, J.-J. & CHARPIN, A. 1996: Lectotypification of names of taxa described by J. Thore. - *Taxon* **45(1)**: 99-103.

LAZAREVIĆ, M. & AL. 2015: Patterns of cytotype distribution and genome size variation in the genus *Sesleria* Scop. (Poaceae). - *Bot. J. Linn. Soc.* **179(1)**: 126-143.

LAZAREVIĆ, M., RAKIĆ, T. & ŠINŽAR-SEKULIĆ, J. 2014: Morphological differentiation between the flowers of *Ramonda serbica*, *R. nathaliae* and their hybrid. - *Bot. Serbica* **38(1)**: 91-98.

LAZARIDOU, M. 1999: Mapping the coastal zone of E.O.T. camping in Epanomi, Thermaikos Gulf. In: Proceedings of 21st Panhellenic Conference of EEBE, Syros, May 28-31, 1999, pp. 151-152 [In Greek and English].

LAZARINA, M. & AL. 2019: Patterns and drivers of species richness and turnover of neo-endemic and palaeo-endemic vascular plants in a Mediterranean hotspot: The case of Crete, Greece. - *J. Biol. Res. Thessalon.* **26**: 1-13.

LAZZARO, L. & AL. 2010: Lectotypification of the names *Plantago weldenii* and *P. commutata* (Plantaginaceae). - *Phytotaxa* **429(1)**: 80-84.

LE FLOC'H, E. 1991: Invasive plants of the Mediterranean basin. In Groves, R.H. & Di Castri, F. (eds): Biogeography of mediterranean invasions, pp. 67-80. - Cambridge, etc.: Cambridge Univ. Press.

LE HOUÉROU, H.N. 1981: Impacts of man and his animals on Mediterranean vegetation. In Di Castri, F., Goodall, D.W. & Specht, R.L. (eds): Mediterranean-type shrublands, pp. 479-521. - Amsterdam, etc.: Elsevier.

LE HOUÉROU, H.N. 1991: Plant invasions in the rangelands of the isoclimatic mediterranean zone. In Groves, R.H. & Di Castri, F. (eds): Biogeography of mediterranean invasions, pp. 393-404. - Cambridge, etc.: Cambridge Univ. Press.

LEADLAY, E.H. & HEYWOOD, V.H. 1990: The biology and systematics of the genus *Coincya* Porta et Rigo ex Rouy (Cruciferae). - *Bot. J. Linn. Soc.* **102(4)**: 313-398.

LEAL PÉREZ-CHAO, J. & AL. 1980: Números cromosómicos para la flora española. 155-161. - *Lagascalia* **9(2)**: 269-272.

LEBEDA, A. & AL. 2004: Geographical distribution of wild *Lactuca* species (Asteraceae, Lactuceae). - *Bot. Rev. (Lancaster)* **70(3)**: 328-356.

LEBEL, E. 1868: Révision du genre *Spergularia*. Les spergularies françaises et 2 espèces des Canaries. - *Mém. Soc. Sci. Nat. Cherbourg* **14**: 43-75.

LEBLEBICI, E. 1974: The calaminthoid genera in Turkey: New names in *Acinos* and *Cyclotrichium*, including a new species from N. Iraq. - *Bitki Koruma Bült.* **1(3)**: 403-408.

LEBLEBICI, E., DUMAN, H. & AYTAÇ, Z. 1993: *Polygonum ekimianum* (Polygonaceae), a new species from Anatolia. - *Willdenowia* **23(1-2)**: 163-166.

LEBLEBICI, E. & SEÇMEN, Ö. 1996: Aquatic flora of the Marmara region, NW Turkey. - *Willdenowia* **25 (2)**: 613-626.

LEBRETON, A. 2013: *Myriophyllum heterophyllum* Michaux [Haloragaceae] en Haute-Vienne (Limousin, France), et situation de cette plante invasive en France et en Europe. - *Bull. OEPP* **43(1)**: 180-192.

LEBRETON, PH. & BARBERO, M. 1998: Systematic study of the subgenus *Oxycedrus* (*Oxycedroides* section) of the genus *Juniperus* (Cupressaceae) - *Ecol. Medit.* **24(1)**: 53-61.

LEBRETON, PH. & RIVERA, D. 1989: Analyse du taxon *Juniperus phoenicea* L. sur des bases biochimiques et biométriques. - *Naturalia Monspel., Sér. Bot.* **53**: 17-41.

LEBRETON, PH. & THIVEND, S. 1981: Sur une sous-espèce du Genévrier de Phénicie, *Juniperus phoenicea* L., définie à partir de critères biochemiques. - *Naturalia Monspel., Sér. Bot.* **47**: 1-12.

LECOYER, J.-C. 1885: Monographie du genre *Thalictrum*. - *Bull. Soc. Roy. Bot. Belgique* **24(1)**: 78-324 + 3 plates.

LEE, A.T. & GLADSTONES, J.S. 1979: (474-475) Proposals to reject *Lupinus varius* and *Lupinus hirsutus* of Europe (Fabaceae). - *Taxon* **28(5-6)**: 616-621. [approved, cf. Taxon 32(4): 624, 1983].

LEE, B.-Y., LEVIN, G.A. & DOWNIE, S.R. 2001: Relationships within the spiny-fruited Umbellifers (Scandiceae subtribes Daucinae and Torilidinae) as assessed by phylogenetic analysis of morphological characters. - *Syst. Bot.* **26(3)**: 622-642.

LEES, D.C. & AL. 2011: Tracking origins of invasive herbivores through herbaria and archival DNA: the case of the horse-chestnut leaf miner. - *Front. Ecol. Envir.* **9(6)**: 322-328 + appendix.

LEEUWENBERG, A.J.M. 1983: Series of revisions of Apocynaceae, XIII. Notes of *Nerium* L. and *Tabernaemontana* L. - *Meded. Landbouwhogeschool* **83(7)**: 55-60.

LEGAKIS, A. & KYPRIOTAKIS, Z. 1994: A biogeographical analysis of the island of Crete, Greece. - *J. Biogeogr.* **21(4)**: 441-445.

LEGENDRE, L. 2000: The genus *Pinguicula* L. (Lentibulariaceae): an overview. - *Acta Bot. Gallica* **147 (1)**: 77-95.

LEHMANN, E. 1908: Geschichte und Geographie der *Veronica*-Gruppe *Agrestis*. - *Bull. Herb. Boissier, Ser. 2* **8(4-9)**: 229-244, 337-352, 410-425, 644-660.

LEHMANN, E. 1909: Über Zwischenrassen in der *Veronica*-Gruppe *agrestis*. - *Z. Indukt. Abstammungs- Vererbungsl.* **2(3)**: 145-208 + 1 plate.

LEHMANN, E. 1930a: Geschichte und Geographie der *Veronica*-Gruppe *Megasperma*. - *Biblioth. Bot.* **25/99**: 55 pp. + 1 plate.

LEHMANN, J.G.C. 1856: Revisio potentillarum iconibus illustrata. - *Nova Actorum Acad. Caes. Leop.-Carol. Nat. Cur.* **32, suppl.**: xiv + 230 pp. + 64 plates.

LEHMANN, L. & JESSEN, S. 2006: *Edraianthus*: Becher- oder Büschelglocke (Familie: Campanulaceae - Glockenblumengewächse). - *Sammelblätter Gebirgspflanzen* **3.28.5**: 1-10.

LEIBFRITZ, S. & SCHULER, A. 1997: Pflanzenbestands- und Standortsaufnahmen im Flussgebiet des Nestos / NO-Griechenland. - Ulm: Diplomarbeit, 193 pp.

LEIN, A. 1948: Natürliche Standorte von diploidem und tetraploidem *Hordeum bulbosum* L. - *Züchter* **19(1-3)**: 6-9.

LEINS, P. & BONNERY-BRACHTENDORF, R. 1977: Entwicklungsgeschichtliche Untersuchungen an Blüten von *Datisca cannabina* (Datiscaceae). - *Beitr. Biol. Pflanzen* **53(1)**: 143-155.

LEINS, P. & MERXMÜLLER, H. 1966: Zur Gliederung der *Oxytropis campestris*-Gruppe. - *Mitt. Bot. Staatssamml. München* **6(1)**: 19-31.

LEITÃO, M.T. & ALVES, M.C. 1976: Contribuição para o conhecimento citotaxonómico das Spermatophyta de Portugal. XIV. Cistaceae. - *Bol. Soc. Brot., Ser. 2* **50**: 247-263.

LEMPERG, F. 1933: Observations on the genus *Edraianthus*. - *New Fl. & Silva* **5(3)**: 136-144 + 2 plates.

LEMPERG, F. 1936: Studies in the perennial species of the genus *Dianthus* L. - I. - *Acta Horti Gothob.* **11**: 71-134.

LEMPIÄINEN, T. 1982: Morphological and chemical variation among Eurasian populations of *Filipendula vulgaris* (Rosaceae). - *Ann. Bot. Fenn.* **19(3)**: 127-146.

LENHERR, A. 1983: Biosystematische und chemotaxonomische Untersuchungen in der Artengruppe *Stachys recta* L. - Zürich: Ph.D. thesis, ETH Zürich (Diss. ETH Nr. 7453), 131 pp.

LENHERR, A. & BALTISBERGER, M. 1984: *Stachys beckeana* (Labiatae) in Albanien und Jugoslawien. - *Pl. Syst. Evol.* **145(1-2)**: 97-104.

LENSKI, I. 1966: Chromosomenzahlen in der Gattung *Sibthorpia* (Scrophulariaceae). - *Naturwissenschaften* **53(24)**: 710-711.

LENTINI, F. & RAIMONDO, F.M. 1984: Contribution à la connaissance du genre "Colchicum" L. en Sicile. - *Webbia* **38**: 747-755.

LENTINI, F., ROMANO, S. & RAIMONDO, F.M. 1990: Numeri cromosomici per la Flora Italiana: 1185-1196. - *Inform. Bot. Ital.* **20(2-3)**: 637-646.

LENTON, N. & TURLAND, N.J. 2010: *Cyanus depressus* (M. Bieb.) Soják. In Greuter, W. & Raus, Th. (eds): Med-Checklist notulae, 29. - *Willdenowia* **40(2)**: 191-192.

LEÓN, E. & AL. 2014: El agregado de *Hordeum murinum* (Poaceaee) en "Flora Iberica". - *Acta Bot. Malac.* **39**: 311-319.

LEONARD, J. 1977: Le genre irano-touranien *Drabopsis* et son unique espèce *D. nuda* (Cruciferae). - *Publ. Cairo Univ. Herb.* **7-8**: 295-297.

LEONCINI, M.L. 1951: Biotipi cariologici e sistematici di *Caltha* in Italia. - *Caryologia* **3(3)**: 336-350.

LEONCINI, M.L. 1952: Nuove osservazioni cariologiche sul gen. *Caltha*. - *Caryologia* **4(3)**: 367-371.

LEONOVA, T.G. 1976: Conspectus specierum generis *Typha* L. florae partis europaeae URSS [In Russian]. - *Novosti Sist. Vysš. Rast.* **13**: 8-15.

LEONTARITOU, P. & AL. 2021: Exploration of genetic, morphological and essential oil variation reveals tools for the authentication and breeding of *Salvia pomifera* subsp. *calycina* (Sm.) Hayek. - *Phytochemistry* **191**: 9 pp.

LEPPER, J.D. 1981: Orchids of Greece. - Devon: Arthur H. Stockwell Ltd, 59 pp. + 8 plates .

LEPPER, J.D. 1990: Mediterranean orchids. - Ilfracombe: A.H. Stockwell Ltd, 86 pp.

LEPPER, L. 1975: Nachtrag zur Zytologie von *Cyclamen* L. - *Wiss. Z. Friedrich-Schiller-Univ. Jena, Math.-Naturwiss. Reihe* **24(4)**: 429-436.

LEPŠÍ, M. & AL. 2019: *Stellaria ruderalis*, a new species in the *Stellaria media* group from central Europe. - *Preslia* **91(3)**: 391-420.

LEROY, J.-F. 1982: Origine et évolution du genre *Platanus* (Platanaceae). - *Compt. Rend. Hebd. Séances Acad. Sci., Sér. 3, Sci. Vie* **295(3)**: 251-254.

LES, D.H. 1986: The evolution of achene morphology in *Ceratophyllum* (Ceratophyllaceae), I. Fruit-spine variation and relationships of *C. demersum*, *C. submersum* and *C. apiculatum*. - *Syst. Bot.* **11 (4)**: 549-558.

LESINS, K. & LESINS, I. 1963: Some little-known Medicagos and their chromosome complements. II. Species from Turkey. - *Canad. J. Genet. Cytol.* **5(2)**: 133-137.

LESINS, K. & LESINS, I. 1964: Diploid *Medicago falcata* L. - *Canad. J. Genet. Cytol.* **6(2)**: 152-163.

LESINS, K. & LESINS, I. 1965: Some little-known Medicagos and their chromosome complements. III. Some Mediterranean species. - *Canad. J. Genet. Cytol.* **7(1)**: 97-102.

LESINS, K. & LESINS, I. 1979: Genus *Medicago* (Leguminosae). A taxogenetic study. - The Hague, etc.: Dr. W. Junk Publishers, xi + 228 pp.

LESINS, K., LESINS, I. & GILLIES, C.B. 1970: *Medicago murex* with 2n=16 and 2n=14 chromosome complements. - *Chromosoma* **30(1)**: 109-122.

LESINS, K., SADASIVAIAH, R.S. & SINGH, S.M. 1976: Relationship of taxa in genus *Medicago* as revealed by hybridization: VIII. Section *Rotatae*. - *Canad. J. Genet. Cytol.* **18(2)**: 345-355.

LESINS, K., SINGH, S.M. & ERAC, A. 1971: Relationship of taxa in the genus *Medicago* as revealed by hybridization. V. Section *Intertextae*. - *Canad. J. Genet. Cytol.* **13(2)**: 335-346.

LETZ, D.R. 2009: A new species of the *Sempervivum marmoreum* group in Central Europe. – *Preslia* **81(3)**: 293-308.

LETZ, D.R. & AL. 2012: Taxonomy and distribution of *Cerastium pumilum* and *C. glutinosum* in Central Europe. - *Preslia* **84(1)**: 33-69.

LETZ, D.R. & MARHOLD, K. 1996: Lectotypification of some names in *Jovibarba* and *Sempervivum* (Crassulaceae). - *Taxon* **45(1)**: 111-116.

LEUENBERGER, B.E. 1991: Interpretation and typification of *Cactus ficus-indica* L. and *Opuntia ficus-indica* (L.) Miller (Cactaceae). - *Taxon* **40(4)**: 621-627.

LEUTE, G.-H. 1969: Untersuchungen über den Verwandtschaftskreis der Gattung *Ligusticum* L. (Umbelliferae). I. Teil. - *Ann. Naturhist. Mus. Wien* **73**: 55-98.

LEUTE, G.-H. 1970: Untersuchungen über den Verwandtschaftskreis der Gattung *Ligusticum* L. (Umbelliferae). II. Teil. - *Ann. Naturhist. Mus. Wien* **74**: 457-519.

LEUTE, G.-H. 1971: Die Arten der Gattung *Conium* L. (Umbelliferae). - *Ann. Naturhist. Mus. Wien* **75**: 91-98.

LEUTE, G.-H. 1972: Bemerkungen zur Gattung *Peucedanum* L. in Griechenland. - *Oesterr. Bot. Z.* **120(1-2)**: 29-32.

LEUTE, G.-H. 1973: Chromosomenzahlen dreier Apiaceen aus Österreich. - *Oesterr. Bot. Z.* **121(1-2)**: 143-144.

LEUTE, G.-H. 1978: Vorläufige Mitteilung über zwei Sippen aus dem *Campanula patula*-Aggregat und ihre Verbreitung in Kärnten. - *Carinthia II* **88**: 243-255.

LEUZINGER, M. & AL. 2015: Molecular diversity, phylogeography and genetic relationships of the *Silene paradoxa* group of section *Siphonomorpha* (Caryophyllaceae). - *Pl. Syst. Evol.* **301(1)**: 265-278 + electronic suppl.

LEVCHENKO, M.F. 1974: The life cycle of *Chelidonium majus* [In Russian]. - *Bot. Zhurn. (Moscow & Leningrad)* **59(1)**: 82-96.

LÉVEILLÉ, H. 1910-11: Iconographie du genre *Epilobium*. - Le Mans: Imprimerie Monnoyer, 328 pp.

LÉVEILLÉ, H. 1917: Souvenir de guerre [enumeration of plants collected in the vicinity of Thessaloniki]. - *Bull. Géogr. Bot.* **27**: 97-99.

LEVEN, S. 2002: The autumn snowdrop. - *Rock Gard.* **27(3)**: 238-239.

LEVEN, S. 2011: Enthusiasm and experience ... *Cyclamen graecum* in its own home and mine. - *Rock Gard.* **127**: 54-63.

LÉVÊQUE, M. & GORENFLOT, R. 1969: Prospections caryologiques dans la flore littorale du Boulonnais. - *Bull. Soc. Bot. N. France* **22(1)**: 27-58 + 1 plate.

LEVICHEV, I.G. 1999: Phytogeographical analysis of the genus *Gagea* Salisb. (Liliaceae). – *Komarovia* **1**: 45-57.

LEVICHEV, I.G. & TISON, J.-M. 2004a: Typification of the Caucasian *Gagea* (Liliaceae) taxa described by Karl Koch. - *Candollea* **59(1)**: 119-133.

LEVICHEV, I.G. & TISON, J.-M. 2004b: Etude nomenclaturale de *Gagea reticulata* (Pall.) Schult. & Schult. f. (Liliaceae), de ses variétés β *tenuifolia* Boiss. et γ *fibrosa* Boiss., et de *Gagea rigida* Boiss. & Spruner. - *Candollea* **59(2)**: 309-324.

LEVIER, E. 1884: Les tulipes de l'Europe. - *Bull. Soc. Neuchâteloise Sci. Nat.* **14**: 201-312 + 9 plates.

LEWIN, R.A. 1948: Biological Flora of the British Isles. *Sonchus* L. - *J. Ecol.* **36(1)**: 203-223.

LEWIS, L. 2015: Some observations on orchids flowering on the Greek island of Thassos in late May. - *J. Eur. Orch.* **47(2-4)**: 421-432.

LEWIS, L. & KREUTZ, C.A.J. 2013: On the correct name of the early-flowering form of the Pink Butterfly Orchid *Anacamptis (Orchis) papilionacea* in the Eastern Aegean. - *J. Eur. Orch.* **45(1)**: 59-76.

LEWIS, W.H. & OLIVER, R.L. 1965: Realignment of *Calystegia* and *Convolvulus* (Convolvulaceae). - *Ann. Missouri Bot. Gard.* **52(2)**: 217-222.

LIDÉN, M. 1981: (547) Proposal to change the typification of *Corydalis*, nomen conservandum . - *Taxon* **30(1)**: 322-325. [approved, cf. Taxon 33(4): 705, 1984].

LIDÉN, M. 1986: Synopsis of Fumarioideae (Papaveraceae) with a monograph of the tribe Fumarieae. - *Opera Bot.* **88**: 1-133.

LIDÉN, M. 1989: Tuberous *Corydalis* in the Mediterranean checklist area. - *Notes Roy. Bot. Gard. Edinburgh* **45(2)**: 349-363.

LIDÉN, M. 1991: New tuberous species of *Corydalis* (Papaveraceae). - *Willdenowia* **21(1-2)**: 175-179.

LIDÉN, M. 1996: New taxa of tuberous *Corydalis* (Fumariaceae). - *Willdenowia* **26(1-2)**: 23-35.

LIDÉN, M. & ZETTERLUND, H. 1988: Notes on the genus *Corydalis*. - *Quart. Bull. Alpine Gard. Soc. Gr. Brit.* **56(2)**: 146-169.

LIDÉN, M. & ZETTERLUND, H. 1997: *Corydalis*: A gardener's guide and a monograph of the tuberous species. - Avon Banks, Pershore: AGS Publications, 140 pp. + plates.

LIEBEL, H.T. & SEGRETO, R. 2009: Per Charterflug ins Orchideenparadies: Ein ungewöhnlicher Standort am Flughafen von Chania (Westkreta). - *Ber. Arbeitskreis. Heimische Orchid.* **26(1)**: 138-142.

LIEBERTZ, J. 1981: Die Orchideen der Insel Skiáthos. - *Orchidee (Hamburg)* **32(2)**: 59.

LIEDE, S. 1996: *Cynanchum - Rhodostegiella - Vincetoxicum - Tylophora* (Asclepiadaceae): new considerations on an old problem. - *Taxon* **45(2)**: 193-211.

LIENAU, C. (ED.) 1989: "Europapark" in Nordost-Griechenland? Beiträge eines Symposiums am 26. und 27.6. 1988 in Münster zur Bewahrung des europäischen Naturerbes in Nordost-Griechenland. - *Ber. Arbeitsgeb. Entwicklungsforsch. (Münster)* **15**: 95 pp.

LIEUTIER, H. 1993: Tourisme Sedumaniaque en Crete. - *Sedum Soc. Newsletter* **27**: 11-17.

LIÈVRE, A. LE 1992: A tale of two botanists [Heldreich, Orphanides]. - *Garden (London)* **117(3)**: 138-140.

LIÈVRE, A. LE 1994: A view of Edmond Boissier. - *Bot. Mag. (Kew Mag.)* **11(3)**: 131-143.

LIMPRICHT, W. 1924: Studien über die Gattung *Pedicularis*. - *Repert. Spec. Nov. Regni Veg.* **20**: 161-265.

LINCZEVSKI, J.A. 1970: (321) Amendment to proposal no. 258 to conserve the generic name 6348. *Acantholimon* Boissier ... - *Taxon* **19(6)**: 949. [approved, cf. Taxon 21(4): 535, 1972].

LINDBERG, H. 1946: Iter Cyprium. Contributio ad cognitionem florae insulae Cypri. - *Acta Soc. Sci. Fenn., Ser. B., Opera Biol.* **2(7)**: 1-37 + 12 plates.

LINDE-LAURSEN, I., BOTHMER, R.V. & JACOBSEN, N. 1989: Giemsa C-banded karyotypes of *Hordeum marinum* and *H. murinum*. - *Genome* **32**: 629-639.

LINDE-LAURSEN, I., BOTHMER, R.V. & JACOBSEN, N. 1990: Giemsa C-banded karyotypes of diploid and tetraploid *Hordeum bulbosum* (Poaceae). - *Pl. Syst. Evol.* **172(1-4)**: 141-150.

LINDE-LAURSEN, I. & FREDERIKSEN, S. 1991: Comparison of the Giemsa C-banded karyotypes of *Dasypyrum villosum* (2x) and *D. breviaristatum* (4x) from Greece. - *Hereditas (Lund)* **114(3)**: 237-244.

LINDERMAYER, A. 1855: Euboea. Eine naturhistorische Skizze. - *Bull. Soc. Imp. Naturalistes Moscou* **28(2)**: 401-451.

LINDHOF, A. 1996: Die Pflanzenformationen des südlichen Ithakas (Ionische Inseln). - Diplomarbeit. Johannes Gutenberg-Universität Mainz, Mainz. [3] +143 pp.

LINDIG, C. & LINDIG, D. 1991: *Dactylorhiza kalopissii*: Erstnachweis für Bulgarien. – *Orchidee (Hamburg)* **42(1)**: 34-36.

LINDLEY, J. 1827a: *Ophrys fusca*. - *Edwards's Bot. Reg.* **13**: Plate 1071 + 2 pp. text.

LINDLEY, J. 1827b: *Geum coccineum*. - *Edwards's Bot. Reg.* **13**: Plate 1088 + 2 pp. text.

LINDLEY, J. 1827c: *Ophrys tenthredinifera* var. *minor*. - *Edwards's Bot. Reg.* **13**: Plate 1093 + 2 pp. text.

LINDLEY, J. 1837a: *Silene chloraefolia*. - *Edwards's Bot. Reg.* **23**: Plate 1989 + 1 p. text.

LINDLEY, J. 1837b: *Sternbergia colchiciflora*. - *Edwards's Bot. Reg.* **23**: Plate 2008 + 2 pp. text.

LINDLEY, J. 1845: *Ornithogalum nanum*. - *Edwards's Bot. Reg.* **31**: Plate 39 + 1 p. text.

LINDLEY, J. 1846: *Ophrys cornuta*. - *Edwards's Bot. Reg.* **32**: Plate 52 + 1 p. text.

LINDLEY, J. 1847a: *Ophrys fuciflora*. - *Edwards's Bot. Reg.* **33**: Plate 25 + 2 pp. text.

LINDLEY, J. 1847b: *Trichonemata graeca*. - *Edwards's Bot. Reg.* **33**: Plate 40 + 2 pp. text.

LINDLEY, J. 1847c: Ophrydes graecae. - *Edwards's Bot. Reg.* **33**: Plate 46 + 2 pp. text.

LINDMAN, C.A.M. 1912: Wie ist die Kollektivart *Polygonum aviculare* zu spalten? - *Svensk Bot. Tidskr.* **6(3)**: 673-696 + 4 plates.

LINDQUIST, B. 1932: Taxonomical remarks on *Juncus alpinus* Villars and some related species. - *Bot. Not.* **1932(5)**: 313-372.

LINDQVIST, K. 1960: Cytogenetic studies in the *Serriola* group of *Lactuca*. - *Hereditas (Lund)* **46(1-2)**: 75-151.

LINK, H.F. 1834: Symbolae ad floram graecam. - *Linnaea* **9**: 129-141, 567-588 + 1 plate.

LINK, H.F. 1841: Abietinae horti regii botanici berolinensis cultae. - *Linnaea* **15**: 481-545.

LINK, H.F. 1843: De gramineis Europae australis observationes botanicae.- *Linnaea* **17**: 385-407.

LINNAEUS, C. 1753: Species Plantarum, exhibentes plantas rite cognitas, ad genera relatas, cum differentiis specificis, nominibus trivialibus, synonymis selectis, locis natalibus, secundum systema sexuale digestas. - Holmiae: 2 vols. [several fascimile editions, cf. TL-2 3: 93-95, 1981].

LINNAEUS, C. 1762-63: Species Plantarum, exhibentes plantas rite cognitas, ad genera relatas, cum differentiis specificis, nominibus trivialibus, synonymis selectis, locis natalibus, secundum systema sexuale digestas. Ed. 2. - Holmiae: 2 vols. [cf. TL-2 3: 104, 1981].

LINNAEUS, C. 1767[-71]: Mantissa plantarum [et Mantissa altera]. - Holmiae: Laurent. Salvii, 142 pp. [cf. TL-2 3: 107, 1981].

LIPKIN, Y. 1975a: *Halophila stipulacea*, a review of a successful immigration. - *Aquatic Bot.* **1(3)**: 203-215.

LIPKIN, Y. 1975b: *Halophila stipulacea* in Cyprus and Rhodes, 1967-1970. - *Aquatic Bot.* **1(4)**: 309-320.

LIPKIN, Y. 1976: Note on the male flower of *Halophila stipulacea*. - *Israel J. Bot.* **24(4)**: 198-200.

LIPPERT, W. 1984: Zur Kenntnis des *Aphanes microcarpa*-Komplexes. - *Mitt. Bot. Staatssamml. München* **20**: 451-463.

LIPPERT, W. 1985: Pflanzenliste der Exkursion des Institutes für Systematische Botanik der Universität München nach Nordgriechenland. [Unpublished flora list]. - Photocopy: 15 pp.

LIPPOLD, H. 1968: Die europäischen Arten der Gattung *Acer* aus der Sektion *Platanoidea* Pax emend. Momotani. - *Wiss. Z. Friedrich-Schiller-Univ. Jena, Math.-Naturwiss. Reihe* **17**: 341-347.

LIPSCHITZ, S.J. 1935: Fragmenta monographiae generis *Scorzonera* [In Russian]. - Moscow: The State United Technical-Scientific Bookstore Ed. "Onti", 164 pp.

LIPSCHITZ, S.J. 1939: Fragmenta monographiae generis *Scorzonera*. II [In Russian]. - Moscow: Societas Naturae Curiosorum Mosquensis, 158 pp.

LIPSKY, W. 1898: Florae Caucasicae imprimis Colchicae. - *Trudy Imp. S.-Peterburgsk. Bot. Sada* **14(10)**: 245-316.

LIS BALCHIN, M. 2002: The taxonomy of the genus *Lavandula* L. In Lis Balchin, M. (ed.): Lavender: the genus *Lavandula*, pp. 2-34. - London: Taylor & Francis.

LISTON, A. 1986: A new species of *Petrorhagia* (Caryophyllaceae). - *Candollea* **41(1)**: 179-181.

LISTON, A., RIESEBERG, L.H. & ELIAS, TH.S. 1989: Morphological stasis and molecular divergence in the intercontinental disjunct genus *Datisca* (Datiscaceae). - *Aliso* **12(3)**: 525-542.

LITARDIÈRE, R. DE 1935: Observations sur un *Festuca* du Mont Olympe de Thessalie: *F. ovina* L. subsp. *frigida* Hack. var. *glacialis* (Miég.) Hack. subvar. *olympica* (Vetter) R. Lit. - *Bull. Soc. Hist. Nat. Afrique N.* **26 bis**: 40-44.

LITARDIÈRE, R. DE 1939: Sur les caractères chromosomiques et la systématique des *Poa* du groupe du *Poa annua* L. - *Rev. Cytol. Cytophysiol. Vég.* **4(1)**: 81-85.

LITARDIÈRE, R. DE 1942: Festucae graeciae. Énumération des *Festuca* récoltés par C. de Regel en Grèce. - *Candollea* **9**: 138-142.

LITARDIÈRE, R. DE 1943: Recherches caryologiques et caryo-taxonomiques sur les Boraginacées. II. Nombres chromosomiques dans le genre *Echium*. - *Boissiera* **7**: 155-165.

LITARDIÈRE, R. DE 1945a: Recherches taxinomiques et caryologiques sur le *Melissa officinalis* L. - *Rev. Bot. Appl. Agric. Trop.* **25(275-276)**: 16-18.

LITARDIÈRE, R. DE 1945b: Contribution à l'étude du genre *Festuca*. - *Candollea* **10**: 103-146.

LITAV, M. & ORSHAN, G. 1971: Biological flora of Israel. 1. *Sarcopoterium spinosum* (L.) Sp. - *Israel J. Bot.* **20(1)**: 48-64.

LITTLE, D.P. & BARRINGTON, D.S. 2003: Major evolutionary events in the origin and diversification of the fern genus *Polystichum* (Dryopteridaceae). - *Amer. J. Bot.* **90(3)**: 508-514.

LIU, T.-S. 1971: A monograph of the genus *Abies*. - Taipeh: Taiwan Nat. Univ., Dept. of Forestry, 608 pp.

LIVANIOU-TINIAKOU, A. 1983: Cytotaxonomical contribution to the study of the genus *Viola* (Violaceae) in Greece [In Greek with English abstract]. - *Bot. Chron. (Patras)* **3(1-2)**: 22-29.

LIVANIOU-TINIAKOU, A. 1989: On some *Viola* species (sectio *Viola*) from Greece. In Phitos, D. (ed.): Abstracts. Sixth OPTIMA meeting, Delphi 10-16 Sept. 1989, p. 108.

LIVANIOU-TINIAKOU, A. 1991: Viosistimatiki meleti tou genous *Viola* sectio *Viola* (Violaceae) stin Ellada [In Greek with English summary]. - Patras: Ph.D. thesis, Univ. of Patras, 336 pp.

LIVANIOU-TINIAKOU, A. & AL. 1998: The flora of Serifos (Kiklades islands, Greece). In Moret, J. (ed.): Abstracts of the IX. OPTIMA meeting, Paris, 11-17 May 1998. - Paris: Mus. Nat. d'Hist. Naturelle.

LIVANIOU-TINIAKOU, A. & AL. 2003: Floristic dynamics in correlation with the type of substrate and human activities: The example of Serifos (Kiklades islands, Greece). - *Fresenius Environ. Bull.* **12(12)**: 1524-1533.

LIVERI, C. & AL. 2020: Another piece of the puzzle, another brick in the wall: The inevitable fate of *Campanula* section *Quinqueloculares* (Campanulaceae: Campanuloideae). - *Taxon* **69(6)**: 1239-1258.

LIVERI, E. 2021: Phylogeny and biogeography of *Campanula* L. section *Quinqueloculares* (Boiss.) Phitos (Campanuloideae, Campanulaceae). - Patras: Ph.D. thesis, Univ. of Patras, 246 pp.

LIVERI, E., BAREKA, E.-P. & KAMARI, G. 2012: Reports (1753-1758). In Kamari, G., Blanché, C. & Siljak-Yakovlev, S. (eds): Mediterranean chromosome number reports - 22. - *Fl. Medit.* **22**: 212-220.

LIVERI, E., BAREKA, E.-P. & KAMARI, G. 2020: Karyosystematic study of some taxa from *Campanula* section *Quiqueloculares* (Campanulaceae). I. - *Fl. Medit.* **30**: 440-455.

LIVERI, E., BAREKA, P. & KAMARI, G. 2014: Reports (1824-1825). In Kamari, G., Blanché, C. & Siljak- Yakovlev, S. (eds): Mediterranean chromosome number reports - 24. - *Fl. Medit.* **24**: 274-276.

LIVERI, E., BAREKA, P. & KAMARI, G. 2015: Reports (1859-1862). In Kamari, G., Blanché, C. & Siljak- Yakovlev, S. (eds): Mediterranean chromosome number reports - 25. - *Fl. Medit.* **25**: 156-158.

LIVERI, E., BAREKA, P. & KAMARI, G. 2018: Taxonomic study on the Greek endemic genus *Hymenonema* (Asteraceae: Cichorieae), using morphological and karyological traits. - *Willdenowia* **48(1)**: 5-21.

LIVERI, E., KATAPODI, E. & KAMARI, G. 2019: Karyosystematic study of some taxa from the Ionian floristic region (Greece). II. - *Fl. Medit.* **29**: 308-320.

LIVERI, E., PHITOS, D. & KAMARI, G. 2021: Karyological study of some plant taxa from Greece. - *Fl. Medit.* **31**: 346-354.

LJUNGSTRAND, E. & STRID, A. 2008: *Matricaria discoidea* DC., *Leymus arenarius* (L.) Hochst., *Puccinellia fasciculata* (Torr.) E.P. Bicknell subsp *fasciculata*. In Greuter, W. & Raus, Th. (eds): Med-Checklist notulae, 27. - *Willdenowia* **38(2)**: 466-471.

LJUNGSTRAND, E. & STRID, A. 2010: *Deschampsia media* (Gouan) Roem. & Schult.. In Greuter, W. & Raus, Th. (eds): Med-Checklist notulae, 29. - *Willdenowia* **40(2)**: 200.

LLEDÓ, M.D. & AL. 2001: *Leucojum*: one small genus, three data sets, how many answers? In Raimondo, F.M. (ed.): Abstracts from the X OPTIMA meeting, Palermo, 13-19 September 2001, p. 37.

LLEDÓ, M.D. & AL. 2004: Phylogenetic analysis of *Leucojum* and *Galanthus* (Amaryllidaceae) based on plastid *mat*K and nuclear ribosomal spacer (ITS) DNA sequences and morphology. - *Pl. Syst. Evol.* **246(3-4)**: 223-243.

LLEDÓ, M.D. & CRESPO, M.B. 1996: Más datos sobre *Leucojum valentinum* Pau (Amaryllidaceae). - *Acta Bot. Malac.* **21**: 289-290.

LLORET, F.J. 1993: Sobre nomenclatura del género *Bromus* L. (Poaceae). - *Collect. Bot. (Barcelona)* **22**: 151.

LO PRESTI, R.M. & OBERPRIELER, C. 2009: Evolutionary history, biogeography and eco-climatological differentiation of the genus *Anthemis* L. (Compositae, Anthemideae) in the circum-Mediterranean area. - *J. Biogeogr.* **36(7)**: 1313-1332.

LOCK, J.M. 1989: Legumes of Africa. A check-list. - Kew: Royal Botanic Gardens, vii + 619 pp.

LOCONTE, H. & ESTES, J.R. 1989: Generic relationships within *Leonticeae* (Berberidaceae). - *Canad. J. Bot.* **67(8)**: 2310-2316.

LODGE, R.W. 1959: Biological Flora of the British Isles. *Cynosurus cristatus* L. - *J. Ecol.* **47(2)**: 511-518.

LOENHOUD, P.J. VAN & STERK, A.A. 1976: A study of the *Juncus bufonius* complex in the Netherlands. - *Acta Bot. Neerl.* **25(3)**: 193-204.

LOESENER, O. 1927: Übersicht über die Arten der Gattung *Veratrum*. Teil I. - *Repert. Spec. Nov. Regni Veg.* **24**: 61-72.

LOESENER, T. 1919: Über die Aquifoliaceen, besonders über *Ilex*. - *Mitt. Deutsch. Dendrol. Ges.* **28**: 1-66 + 4 plates.

LOIDL, J. 1982: B-chromosomes in *Allium flavum* (Liliaceae) and some related species. - *Pl. Syst. Evol.* **139(3-4)**: 197-207.

LÖKI, V., FEKETE, R. & MOLNAR, A.V. 2020: *Ophrys helenae*. In Raus, Th. & Raab-Straube, E.v. (eds): Euro+Med-Checklist notulae, 12. - *Willdenowia* **50(2)**: 320-321.

LOMBARDO, E. & AL. 2020: Distribution, ecological role and symbioses of selected shrubby species in the Mediterranean Basin: a review. - *Pl. Biosyst.* **154(4)**: 438-454.

LONSING, A. 1939: Über einjährige europäische *Cerastium*-Arten aus der Verwandtschaft der Gruppen "Ciliatopetala" Fenzl und "Cryptodon" Pax. - *Repert. Spec. Nov. Regni Veg.* **46**: 139-165 + 5 plates.

LOON, J.CHR. VAN 1980: [Miscellaneous chromosome numbers from Greece, Italy and Jugoslavia]. In Löve, Á. (ed.): Chromosome number reports LXIX. - *Taxon* **29(5-6)**: 718-720.

LOON, J.CHR. VAN 1982: [Miscellaneous chromosome numbers, especially from Greece]. In Löve, Á. (ed.): Chromosome number reports LXXVII. - *Taxon* **31(4)**: 763-764.

LOON, J.CHR. VAN 1984a: Chromosome numbers in *Geranium* from Europe. I. The perennial species. - *Proc. Kon. Ned. Akad. Wetensch. C* **87(3)**: 263-277.

LOON, J.CHR. VAN 1984b: Chromosome numbers in *Geranium* from Europe. II. The annual species. - *Proc. Kon. Ned. Akad. Wetensch. C* **87(3)**: 279-296.

LOON, J.CHR. VAN 1984c: Hybridization experiments in *Geranium*. - *Genetica* **65(2)**: 167-171.

LOON, J.CHR. VAN 1987: A cytotaxonomical atlas of the Balkan flora. - Berlin & Stuttgart: Gebrüder Bornträger Verlagsbuchhandlung, xxvii + 416 pp.

LOON, J.CHR. VAN & JONG, H. DE 1978: [Miscellaneous chromosome numbers]. In Löve, Á. (ed.): IOPB chromosome number reports LIX. - *Taxon* **27(1)**: 56-60.

LOON, J.CHR. VAN & KIEFT, B. 1980: [Miscellaneous chromosome numbers from Jugoslavia]. In Löve, Á. (ed.): Chromosome number reports LXVIII. - *Taxon* **29(4)**: 538-542.

LOON, J.CHR. VAN & OUDEMANS, J.J.M.H. 1982: [Miscellaneous chromosome numbers from Greece]. In Löve, Á. (ed.): Chromosome number reports LXXV. - *Taxon* **31(2)**: 343-344.

LOON, J.CHR. VAN & SETTEN, A.K. VAN 1982: [Miscellaneous chromosome numbers from Bulgaria]. In Löve, Á. (ed.): Chromosome number reports LXXVI. - *Taxon* **31(3)**: 589-592.

LOON, J.CHR. VAN & SNELDERS, H.C.M. 1979: [Miscellaneous chromosome numbers from Greece]. In Löve, Á. (ed.): IOPB chromosome number reports LXV. - *Taxon* **28(5-6)**: 632-634.

LOON, J.CHR. VAN., GADELLA, TH.W.J. & KLIPHUIS, E. 1971: Cytological studies in some flowering plants from southern France. - *Acta Bot. Neerl.* **20(1)**: 157-166.

LOOS, B.P. 1993: Morphological variation in *Lolium* (Poaceae) as a measure of species relationships. - *Pl. Syst. Evol.* **188(1-2)**: 87-99.

LOOS, B.P. & JARVIS, C.E. 1992: The typification of *Lolium perenne* L. and *Lolium temulentum* L. (Poaceae). - *Bot. J. Linn. Soc.* **108(4)**: 399-408.

LOOS, G.H. 1995: Bestimmungsschlüssel für die *Leonurus cardiaca*-Gruppe in Deutschland. - *Florist. Rundbr.* **29(1)**: 30-33.

LOOS, G.H. 1996: Zur Identität von *Carex leersiana* Rauschert, *C. chabertii* F.W. Schultz, *C. polyphylla* Kar. & Kir. und *C. guestphalica* (Boenn. ex Rchb.) Boenn. ex O.F. Lang. - *Feddes Repert.* **107(1-2)**: 61-74.

LOOS, G.H. 1997: Zur Taxonomie der Goldnesseln (*Lamium* L. subgenus *Galeobdolon* (Adans.) Aschers.). - *Florist. Rundbr.* **31(1)**: 39-50.

LOPANE, F. 1951: Sul numero cromosomico di *Periploca graeca* L. (Asclepiadaceae). - *Caryologia* **4 (1)**: 44-46.

LÓPEZ ALONSO, D. & PASCUAL REGUERA, L. 1989: Population structure and pattern of geographic variation in *Muscari comosum* along its range of distribution. - *Genetica* **78(1)**: 39-49.

LÓPEZ GONZÁLEZ, G. 1979: Sobre la nomenclatura de las especies españolas del género *Cheilanthes* Swartz. - *Anales Jard. Bot. Madrid* **36**: 69-75.

LÓPEZ GONZÁLEZ, G. 1990: Acerca de la clasificación natural del género *Carthamus* L., s.l. - *Anales Jard. Bot. Madrid* **47(1)**: 11-34.

LÓPEZ GONZÁLEZ, G. 1993: Notas al género *Umbilicus* DC. (Crassulaceae). - *Anales Jard. Bot. Madrid* **51(1)**: 170-171.

LÓPEZ GONZÁLEZ, G. 1994a: Nota sobre el género *Anthoxanthum* L. (Gramineae). - *Anales Jard. Bot. Madrid* **51(2)**: 309-312.

LÓPEZ GONZÁLEZ, G. 1994b: Nota sobre el género *Hyparrhenia* Andersson ex E. Fourn. (Gramineae). - *Anales Jard. Bot. Madrid* **51(2)**: 312-313.

LÓPEZ GONZÁLEZ, G. 1994c: Notas sobre el género *Onosma* L. (Boraginaceae) en el Mediterráneo occidental. - *Anales Jard. Bot. Madrid* **52(1)**: 43-52.

LÓPEZ GONZÁLEZ, G. 1994d: *Rorippa pyrenaica* (All.) Rchb. o *R. stylosa* (Pers.) Mansf. & Rothm.? (Cruciferae). - *Anales Jard. Bot. Madrid* **52(1)**: 98-102.

LÓPEZ GONZÁLEZ, G. 1995a: Un nombre nuevo para *Sedum candollei* Raym.-Hamet, nom. illeg. [*Mucizonia sedoides* (DC.) D.A. Webb]. - *Anales Jard. Bot. Madrid* **52(2)**: 221-223.

LÓPEZ GONZÁLEZ, G. 1995b: *Hymenolobus procumbens* (L.) Nutt. ex Torr. & A. Gray, nom. inval. (Cruciferae). - *Anales Jard. Bot. Madrid* **53(1)**: 127-128.

LÓPEZ GONZÁLEZ, G. 1997: Nota sobre *Anchusa undulata* L., s.l. (Boraginaceae) y su nomenclatura. - *Anales Jard. Bot. Madrid* **55(1)**: 166-167.

LÓPEZ GONZÁLEZ, G. & BAYER, E. 1988a: El género *Ziziphora* L. (Labiatae) en la Mediterráneo occidental y sus relaciones con *Acinos* Miller, parentesco o convergencia? - *Lagascalia* **15(Extra)**: 49-64.

LÓPEZ GONZÁLEZ, G. & BAYER, E. 1988b: El género *Ziziphora* L. (Labiatae) en el Mediterráneo occidental. - *Acta Bot. Malac.* **13**: 151-162.

LÓPEZ GONZÁLEZ, G. & NIETO FELINER, G. 1986: Apuntes para un tratamiento taxonómico del género *Arenaria* L. en la Península Ibérica y Baleares. - *Anales Jard. Bot. Madrid* **42(2)**: 343-361.

LÓPEZ GONZÁLEZ, G. & ROMO, A.M. 1988: *Dianthus dubius* Raf., un nombre prioritario para *Petrorhagia velutina* (Guss.) P.W. Ball & Heywood (Caryophyllaceae). - *Anales Jard. Bot. Madrid* **45 (1)**: 363.

LÓPEZ PACHERO, M.J. & AL. 2002: Números cromosómicos para la flora española. 810-814. - *Lagascalia* **22**: 153-155.

LÓPEZ PANCHERO, M.J., PUENTE GARCÍA, E. & PENAS MERINO, Á. 1999: Números cromosomáticos de plantas occidentales, 813-817. - *Anales Jard. Bot. Madrid* **57(1)**: 139.

LÓPEZ-VINYALLONGA, S. & AL. 2015: Mountains and refuges: Genetic structure and evolutionary history in closely related, endemic *Centaurea* in continental Greece. - *Molec. Phyl. Evol.* **92**: 243-154.

LÓPEZ, E. & DEVESA, J.A. 2011: Revisión taxonómica del complejo *Centaurea alba* L. (Asteraceae) en la Península Ibérica - *Collect. Bot. (Barcelona)* **30**: 37-52.

LÓPEZ, G. 1979: Algunas consideraciones sobre los linos del grupo *Linum tenuifolium* L. en España. - *Mém. Soc. Bot. Genève* **1**: 99-109.

LORCH, J. 1962: A revision of *Crypsis* Ait. s.l. (Gramineae). - *Bull. Res. Council Israel, Sect. D, Bot.* **11(2)**: 91-110.

LORD, E.M. 1982: Floral morphogenesis in *Lamium amplexicaule* L. (Labiatae) with a model for the evolution of the cleistogamous flower. - *Bot. Gaz.* **143(1)**: 63-72.

LORENZ, K. & LORENZ, R. 2007: *Orchis morio* subsp. *caucasica* × *Serapias bergonii*, ein neuer Bastard der Nothogattung *Orchiserapias*. - *J. Eur. Orch.* **39(3-4)**: 527-542.

LORENZ, R. 2019: Zum Korrekten Namen von *Himantoglossum calcaratum* subsp. *rumelicum* in Griechenland. - *J. Eur. Orch.* **51(3-4)**: 508-511.

LORENZ, R. & GEMBARDT, C. 1987: Die Orchideenflora des Gargano (Italien). Ein Beitrag zum OPTIMA-Projekt "Kartierung der mediterranen Orchideen". - *Mitt. Arbeitskreis Heimische Orchid. Baden-Württemberg* **19(3)**: 385-756.

LORENZO, P. & AL. 2008: The genus *Sternbergia* Waldst. et Kit. (Amaryllidaceae) in Italy. Contribution to the cytotaxonomical and morpho-anatomical knowledge. - *Caryologia* **61(1)**: 107-113.

LORENZONI, F.C. & AL. 1989: Morphological characteristics of the "spiny oaks" in three Hellenic stations. In Phitos, D. (ed.): Abstracts. Sixth OPTIMA meeting, Delphi 10-16 Sept. 1989, p. 85.

LORENZONI, F.C. & AL. 1991: Morphological characteristics of *Quercus coccifera* s.l. in three Hellenic populations. - *Bot. Chron. (Patras)* **10**: 617-634.

LORENZONI, F.C. & DEFRANCESCO, E. 1984: Taxonomic importance of the pollen grain morphology in "Quercus coccifera" s.l. - *Webbia* **38**: 659-664.

LORENZONI, G.G. & AL. 1977: Distribuzione e importanza fitogeografica di un relitto floristico del litorale Salentino: *Sarcopoterium spinosum* (L.) Spach. - *Thalassia Salentina* **7**: 63-70.

LOSA ESPAÑA, T.M. 1963: Especies españoles del género *Chaenorrhinum* Lge. - *Anales Inst. Bot. Cavanilles* **21(2)**: 543-566 + 6 plates.

LOSA, T.M. 1962: Los "Plantagos" españolas. - *Anales Inst. Bot. Cavanilles* **20**: 5-50.

LOUDON, J.C. 1838: Descriptive and historical notice of the *Abies cephalonica*. - *Gard. Mag. & Reg. Rural Domest. Improv.* **14**: 81-85 tt. 6-7.

LOUSLEY, J.E. 1939: Notes on British Rumices: I. - *Rep. Bot. Soc. Exch. Club Brit. Isles* **12(1)**: 118-157.

LOUSLEY, J.E. 1944: Notes on British Rumices: II. - *Rep. Bot. Soc. Exch. Club Brit. Isles* **12(5)**: 547-585.

LOVAŠEN-EBERHARDT, Ž. 1970: Cytotaxonomische Untersuchungen der Art *Plantago*

holosteum Scop. s.l. [In Serbo-Croatian with German summary]. - *Acta Bot. Croat.* **29**: 197-201.
LÖVE, Á. 1961: Some notes on *Myriophyllum spicatum*. - *Rhodora* **63(749)**: 139-145.
LÖVE, Á. 1962: Biosystematische Analyse der *Elytrigia Junceae* Gruppe. - *Kulturpflanze, Beih.* **3**: 74-85.
LÖVE, Á. 1967: [Chromosome numbers in *Rumex*. s.l.]. In Löve, Á. (ed.): IOPB chromosome number reports XIII. - *Taxon* **16(5)**: 448-454.
LÖVE, Á. 1983: The taxonomy of *Acetosella*. - *Bot. Helv.* **93(2)**: 145-168.
LÖVE, Á. 1984: Conspectus of the Triticeae. - *Feddes Repert.* **95(7-8)**: 425-521.
LÖVE, Á. 1986: Some taxonomical adjustments in eurasiatic wheatgrasses. - *Veröff. Geobot. Inst. E.T.H. Stiftung Rübel Zürich* **87**: 43-52.
LÖVE, Á. & KAPOOR, B.M. 1967a: New combinations in *Acetosa* and *Bucephalophora*. - *Taxon* **16(6)**: 519-522.
LÖVE, Á. & KAPOOR, B.M. 1967b: A chromosome atlas of the collective genus *Rumex*. - *Cytologia* **32 (3-4)**: 328-342.
LÖVE, Á. & KJELLQVIST, E. 1964: Chromosome numbers of some Iberian Cistaceae. - *Portug. Acta Biol., Sér. A, Morfol.* **8(1-2)**: 69-80.
LÖVE, Á. & KJELLQVIST, E. 1972: Cytotaxonomy of Spanish plants. I. Introduction. Pteridophyta and Gymnospermae. - *Lagascalia* **2(1)**: 23-35.
LÖVE, Á. & KJELLQVIST, E. 1973: Cytotaxonomy of Spanish plants. II. Monocotyledons. - *Lagascalia* **3 (2)**: 147-182.
LÖVE, Á. & KJELLQVIST, E. 1974a: Cytotaxonomy of Spanish plants. III. Dicotyledons: Salicaceae - Rosaceae. - *Lagascalia* **4(1)**: 3-32.
LÖVE, Á. & KJELLQVIST, E. 1974b: Cytotaxonomy of Spanish plants. IV. Dicotyledons: Caesalpiniaceae - Asteraceae. - *Lagascalia* **4(2)**: 153-211.
LÖVE, Á. & LÖVE, D. 1961: Chromosome numbers of central and northwest European plant species. - *Opera Bot.* **5**: 1-581.
LÖVE, Á. & LÖVE, D. 1966: The variation of *Blechnum spicant*. - *Bot. Tidsskr.* **62**: 186-196.
LÖVE, Á. & LÖVE, D. 1968a: The diploid perennial *Anthoxanthum*. - *Sci. Iceland* **1968**: 26-30.
LÖVE, Á. & LÖVE, D. 1968b: Cytotaxonomy of *Blechnum spicant*. - *Collect. Bot. (Barcelona)* **7(2)**: 665-676.
LÖVE, Á. & LÖVE, D. 1969: Remarks on the cytotaxonomy of Mediterranean plants. In: V Simposio de Flora Europaea. - Sevilla: Publicaciones de la Universidad de Sevilla.
LÖVE, Á. & LÖVE, D. 1973: Cytotaxonomy of the boreal taxa of *Phyllitis*. - *Acta Bot. Acad. Sci. Hung.* **19(1-4)**: 201-206.
LÖVE, Á. & LÖVE, D. 1974a: Nomenclatural adjustments in the Yugoslavian flora. II. Pteridophytes and dicotyledons. - *Preslia* **46(2)**: 123-138.
LÖVE, Á. & LÖVE, D. 1974b: Cytotaxonomical atlas of the Slovenian flora. - Lehre: Verlag von J. Cramer, xx + 1241 pp.
LÖVE, Á. & LÖVE, D. 1975a: Nomenclatural adjustments in some European monocotyledons. - *Folia Geobot. Phytotax.* **10(3)**: 271-276.
LÖVE, Á. & LÖVE, D. 1975b: The Spanish Gentians. - *Anales Inst. Bot. Cavanilles* **32(2)**: 221-232.
LÖVE, Á. & LÖVE, D. 1982: [Miscellaneous chromosome numbers from Italy]. In Löve, Á. (ed.): Chromosome number reports LXXVI. - *Taxon* **31(3)**: 583-587.
LÖVE, Á., LÖVE, D. & PICHI SERMOLLI, R.E.G. 1977: Cytotaxonomical atlas of the Pteridophyta. - Vaduz: J. Cramer, xviii + 398 pp.
LÖVE, Á. (ED.) 1964-88: IOPB chromosome number reports I-C. - *Taxon* **13-33**
LÖVE, D. 1974: Note on *Xanthium strumarium* L. In Heywood, V.H. (ed.): Notulae systematicae ad Floram Europaeam spectantes, no. 15. - *Bot. J. Linn. Soc.* **68(4)**: 277-278.
LÖVE, D. & BOŞCAIU, N. 1966: *Vaccinium gaultherioides* Bigel., an arctic-alpine species. - *Rev. Roumaine Biol., Sér. Bot.* **11(4)**: 295-305.
LÖVE, D. & DANSEREAU, P. 1959: Biosystematic studies on *Xanthium*: taxonomic appraisal and ecological status. - *Canad. J. Bot.* **37(2)**: 173-208.
LOVELIUS, O.L. 1979: Synopsis generis *Atraphaxis* L. (Polygonaceae) [In Russian]. - *Novosti Sist. Vysš. Rast.* **15**: 114-128.
LOVIS, J.D. 1955: *Asplenium adulterinum* and its probable parents. - *Proc. Bot. Soc. Brit. Isles* **1(3)**: 389-390.
LOVIS, J.D. 1964: The taxonomy of *Asplenium trichomanes* in Europe. - *Brit. Fern Gaz.* **9(5)**: 147-160.
LOVIS, J.D. 1968: Artificial reconstruction of a species of fern, *Asplenium adulterinum*. - *Nature* **217 (5134)**: 1163-1165.
LOVIS, J.D. 1970: The synthesis of a new *Asplenium*. - *Brit. Fern Gaz.* **10(3)**: 153-157 + 1 plate.
LOVIS, J.D. 1973: A biosystematic approach to phylogenetic problems and its application to the Aspleniaceae. In Jermy, A.C., Crabbe, J.A. & Thomas, B.A. (eds): The phylogeny and classification of the ferns. - *Bot. J. Linn. Soc.* **67, suppl. 1**: 211-228 + 2 plates.
LOVIS, J.D., MELZER, H. & REICHSTEIN, T. 1965: *Asplenium* ×*adulteriniforme* hybr. nov. = diploides *Asplenium trichomanes* L. × *A. viride* Hudson. - *Bauhinia* **2(3)**: 231-237.
LOVIS, J.D., RASBACH, H. & REICHSTEIN, T. 1989: *Asplenium trichomanes* L. nothosubsp. *melzeri* nothosubsp. nov. The triploid hybrid between *A. trichomanes* subsp. *inexpectans* and subsp. *quadrivalens*. - *Candollea* **44(2)**: 543-553.
LOVIS, J.D. & REICHSTEIN, T. 1964: A diploid form of *Asplenium ruta-muraria*. - *Brit. Fern Gaz.* **9(5)**: 141-146 + 1 plate.
LOVIS, J.D. & REICHSTEIN, T. 1968a: Über das spontane Entstehen von *Asplenium adulterinum* aus einem natürlichen Bastard. - *Naturwissenschaften* **55(3)**: 117-120.
LOVIS, J.D. & REICHSTEIN, T. 1968b: Die zwei diploiden *Asplenium trichomanes* × *viride*-Bastarde und ihre Fähigkeit zur spontanen Chromosomenverdoppelung. - *Bauhinia* **4(1)**: 53-63.
LOVIS, J.D. & REICHSTEIN, T. 1985: *Asplenium trichomanes* subsp. *pachyrachis* (Aspleniaceae, Pteridophyta), and a note on the typification of *A. trichomanes*. - *Willdenowia* **15(1)**: 187-201.
LOVKA, M. 1975: A contribution to the cytotaxonomy of Yugoslav Spermatophyta, Part I: Liliaceae s. lat. [In Slovenian with English summary]. - *Biol. Vestn.* **23(1)**: 25-40.
LOVKA, M. 1995: [Miscellaneous reports]. In Stace, C.A. (ed.): IOPB chromosome data 9. - *Int. Organ. Pl. Biosyst. Newslett.* **24**: 21-23.
LOVKA, M. & AL. 1971: [Miscellaneous chromosome numbers from Jugoslavia]. In Löve, Á. (ed.): IOPB chromosome number reports XXXIV. - *Taxon* **20(5-6)**: 788-791.
LOVKA, M. & AL. 1972: [Miscellaneous chromosome numbers from Jugoslavia]. In Löve, Á. (ed.): IOPB chromosome number reports XXXVI. - *Taxon* **21(2-3)**: 337-339.
LOVKA, M. & SUŠNIK, F. 1973: [Chromosome numbers from Jugoslavia]. In Löve, Á. (ed.): IOPB chromosome number reports XL. - *Taxon* **22(2-3)**: 289.
LÖVKVIST, B. 1956: The *Cardamine pratensis* complex. Outlines of its cytogenetics and taxonomy. - *Symb. Bot. Upsal.* **14(2)**: 1-131 + 16 maps.
LOVRIĆ, A.-Ž. 1968: Contribution à l'étude des Centaurées illyriques avec une aperçu spécial de la section *Pterolophus* (Cass.) DC. [In Serbo-Croatian with French summary]. - *Acta Bot. Croat.* **26-27**: 263-278.
LOVRIĆ, A.-Ž. 1982a: [Chromosome numbers in *Centaurea* from Jugoslavia]. In Löve, Á. (ed.): Chromosome number reports LXXVII. - *Taxon* **31(4)**: 762-763.
LOVRIĆ, A.-Ž. 1982b: [Chromosome numbers in *Centaurea* from Jugoslavia]. In Löve, Á. (ed.): Chromosome number reports LXXVII. - *Taxon* **31(4)**: 774-775.
LOVRIĆ, A.-Ž. 1985: Biology and distribution of *Asplenium* s.l. in the West Balkans. - *Proc. Roy. Soc. Edinburgh, B, Biol. Sci.* **86**: 472-473.
LOVRIĆ, A.-Ž. & RAC, M. 1991: A comparison of the coastal vegetation of Greece, Yugoslavia, Bulgaria and Turkey. - *Bot. Chron. (Patras)* **10**: 315-324.
LOWDEN, R.M. 1982: An approach to the taxonomy of *Vallisneria* L. (Hydrocharitaceae). - *Aquatic Bot.* **13(3)**: 269-298.
LOWE, A.J. & ABBOTT, R.J. 2003: A new British species, *Senecio eboracensis* (Asteraceae), another hybrid derivative of *S. vulgaris* L. and *S. squalidus* L. - *Watsonia* **24(3)**: 375-388.
LOWE, D.B. 1941: *Campanula oreadum*. - *Quart. Bull. Alpine Gard. Soc. Gr. Brit.* **9(2)**: 106.

LOWE, D.B. 1953: The violet crown. - *Quart. Bull. Alpine Gard. Soc. Gr. Brit.* **21(4)**: 268-269.

LOWE, D.B. 1988: Soldanellas: a review. - *Quart. Bull. Alpine Gard. Soc. Gr. Brit.* **56(3)**: 205-220.

LOWE, M.R. 1999: Notes sur les orchidées des Sporades du Nord (Grèce). - *Naturalistes Belges* **80(3)**: 155-172.

LOWE, M.R. 2011: Studies in *Ophrys* L. sectio *Pseudophrys* Godfery - II. *Andrena flavipes* Pz. pollinated taxa. - *J. Eur. Orch.* **43(3)**: 455-497.

LÜBBERT, G. 1951: Vergleichende cytologische, morphologische und physiologische Untersuchungen innerhalb der Gattung *Diplotaxis*. - *Beitr. Biol. Pflanzen* **28(3)**: 254-295.

LUCAS, G.LL. & WALTERS, S.M. (EDS) 1977: Preprint of 40 red data sheets on rare & threatened species in Europe. - Kew: Royal Botanic Gardens.

LUCCHESE, F. 1990: Revision and distribution of *Brachypodium phoenicoides* (L.) Roemer et Schultes in Italy. - *Ann. Bot. (Rome)* **48**: 163-177.

LUCCHESE, F. & LATTANZI, E. 1989: *Euphorbia phymatosperma* Boiss. et Gaill. subsp. *cernua* (Boiss.) Vindt nel Lazio: suo significato ecologico e fitogeografico. - *Arch. Bot. Ital.* **64**: 84-92.

LUCEÑO, M. 1992a: Cytotaxonomic studies in Iberian and Macaronesian species of *Carex* (Cyperaceae). - *Willdenowia* **22(1-2)**: 149-165.

LUCEÑO, M. 1992b: Estudios en la sección *Spirostachyae* (Drejer) Bailey del género *Carex*. II. *C. lainzii* Luceño & al.; *C. troodi* Turrill y *C. idaea* Greuter et al. (Cyperaceae). - *Anales Jard. Bot. Madrid* **50(2)**: 221-227.

LUCEÑO, M. 1993: Chromosome studies on *Carex* L. section *Mitratae* Kükenth. (Cyperaceae) in the Iberian Peninsula. - *Cytologia* **58(3)**: 321-330.

LUCEÑO, M. 1994: Monografía del género *Carex* en la Península Ibérica e Islas Baleares. - *Ruizia* **14**: 1-139.

LUCEÑO, M. & AEDO, C. 1994: Taxonomic revision of the Iberian species of *Carex* L. section *Phacocystis* Dumort. (Cyperaceae). - *Bot. J. Linn. Soc.* **114(2)**: 183-214.

LUCEÑO, M. & SEGURA ZUBIZARRETA, A. 1986: *Carex curta* Good. (= *C. canescens* auct.) en el sistema Ibérico. - *Anales Jard. Bot. Madrid* **43(1)**: 189-190.

LÜCKEL, E. 1971: Bemerkungen zur Nestwurz, zur Systematik und zur Verhaltensforschung. - *Orchidee (Hamburg)* **22(5)**: 210-214.

LÜCKEL, E. 1983: Bemerkungen zum Namen *Ophrys holoserica*. - *Orchidee (Hamburg)* **34(5)**: 179-180.

LUMARET, R. 1981: Structure génétique d'un complexe polyploïd: *Dactylis glomerata* L. (fam. Graminacées). Relations entre le polymorphisme enzymatique et certains aspects de la biologie, de l'écologie de d'évolution. - Montpellier: Thèse, vi + 168 pp.

LUMARET, R. 1988: Cytology, genetics and evolution in the genus *Dactylis*. - *Crit. Rev. Pl. Sci.* **7(1)**: 55-91.

LUMARET, R. 1997: Polyploidy and the critical size of natural populations: the case of cocksfoot (*Dactylis glomerata* L.), a grass used as a fodder plant. - *Bocconea* **7**: 133-139.

LUMARET, R. & AL. 1997: Cultivated olive and oleaster: two very closely connected partners of the same species (*Olea europaea*). Evidence from enzyme polymorphism. - *Bocconea* **7**: 39-42.

LUNGEANU, I. 1971: [Chromosome numbers in *Ornithogalum* from Romania]. In Löve, Á. (ed.): IOPB chromosome number reports XXXIII. - *Taxon* **20(4)**: 610.

LUNGEANU, I. 1975: [Chromosome numbers from Romania]. In Löve, Á. (ed.): IOPB chromosome number reports XLIX. - *Taxon* **24(4)**: 503-504.

LÜÖND, B.W. & LÜÖND-SCHLUMPF, R. 1979: Biosystematische Untersuchungen an *Ajuga* L. - *Mitt. Bot. Mus. Univ. Zürich* **310**: 234 pp.

LÜÖND, B.W. & LÜÖND, R. 1980: Hybridization in *Ajuga* populations. - *Candollea* **35(1)**: 87-109.

LUQUE, T. 1980: Números cromossómicos de algunas Boraginaceas de Portugal. - *Bol. Soc. Brot., Ser. 2* **53(1)**: 663-670.

LUQUE, T. 1981: Números cromosómicos para la flora española. 210-213. - *Lagascalia* **10(2)**: 236-237.

LUQUE, T. 1983a: Estudio cariológico de Boragináceas españolas, I. *Anchusa*. - *Lagascalia* **12(1)**: 81-97.

LUQUE, T. 1983b: Números cromosómicos para la flora española. 284-289. - *Lagascalia* **12(1)**: 128-130.

LUQUE, T. 1984a: Números cromosómicos para la flora española. 314-320. - *Lagascalia* **12(2)**: 284-286.

LUQUE, T. 1984b: Estudio cariológico de Boragináceas españolas. II. *Echium* L. de España peninsular e Islas Baleares. - *Lagascalia* **13(1)**: 17-38.

LUQUE, T. 1985: Números cromosómicos para la flora española. 363-367. - *Lagascalia* **13(2)**: 293-295.

LUQUE, T. 1989: Estudio cariológico de Boraginaceae españolas. IV - *Pulmonaria* L., *Symphytum* L., *Pentaglottis* Tausch y *Borago* L. (Boragineae). - *Bol. Soc. Brot., Ser. 2* **62**: 211-220.

LUQUE, T. 1990: Karyological studies on Spanish Boraginaceae. V: *Onosma* L., *Cerinthe* L. and *Alkanna* Tausch (Lithospermeae). - *Bot. J. Linn. Soc.* **102(3)**: 261-272.

LUQUE, T. 1992: Karylogical studies on Spanish Boraginaceae. VI. Contribution to the tribe Eritrichieae. - *Bot. J. Linn. Soc.* **110(1)**: 77-94.

LUQUE, T. 1995: Karyology of *Nonea* Medicus (Boraginaceae) in Spain; relationships between genera of Boragineae Barbier & Mathez (Anchuseae DC.). - *Bot. J. Linn. Soc.* **117(4)**: 321-331.

LUQUE, T. 1996: Karyological studies on Spanish Boraginaceae. VII. *Heliotropium* L. - *Willdenowia* **25 (2)**: 659-668.

LUQUE, T. & DEVESA, J.A. 1985: Números cromosómicos para la flora española. 382-393. - *Lagascalia* **13(2)**: 303-306.

LUQUE, T. & DEVESA, J.A. 1986: Contribución al estudio citotaxonómico del género *Lavatera* (Malvaceae) en España. - *Lagascalia* **14(2)**: 227-239.

LUQUE, T. & DÍAZ LIFANTE, Z. 1991: Chromosome numbers of plants collected during Iter Mediterraneum I in the SE of Spain. - *Bocconea* **1**: 303-364.

LUQUE, T. & DÍAZ LIFANTE, Z. 1994: Caryological studies of the genus *Convolvulus* L. - *Candollea* **49 (1)**: 233-243.

LUQUE, T. & MEJÍAS, J.A. 1986: Números cromosómicos para la flora española. 491-496. - *Lagascalia* **14(2)**: 301-304.

LUQUE, T., MEJÍAS, J.A. & DÍAZ LIFANTE, Z. 1988: Números cromosómicos para la flora española. 544-550. - *Lagascalia* **15(1)**: 130-133.

LUQUE, T., ROMERO ZARCO, C. & DEVESA, J.A. 1984: Números cromosómicos para la flora española. 321-330. - *Lagascalia* **12(2)**: 286-290.

LUQUE, T. & VALDÉS, B. 1984: Karyological studies on Spanish Boraginaceae: *Lithospermum* L. sensu lato. - *Bot. J. Linn. Soc.* **88(4)**: 335-350.

LUQUE, T. & VALDÉS, B. 1986: Caryological study of Spanish Boraginaceae III. *Cynoglossum* L. s.str. - *Willdenowia* **15(2)**: 485-496.

LUSCOMBE, E.G.B. 1972: *Iris pumila* Linnaeus. Descriptive notes of its botanical characteristics including morphology, variability, the geographical range of its natural distribution ...: a résumé of available data. Part 1. - *Iris Year Book* **1972**: 149-157.

LUSCOMBE, E.G.B. 1973: *Iris pumila* Linnaeus. Descriptive notes of its botanical characteristics including morphology, variability, the geographical range of its natural distribution ...: a résumé of available data. Part 2. - *Iris Year Book* **1973**: 71-79.

LUSCOMBE, E.G.B. 1974: *Iris pumila* Linnaeus. Descriptive notes of its botanical characteristics including morphology, variability, the geographical range of its natural distribution ...: a résumé of available data. Part 3. - *Iris Year Book* **1974**: 66-78.

LUSINA, G. 1922: Contributo alla distribuzione geografica e all'ecologia dello "Scolopendrium hybridum" Milde. - *Ann. Bot. (Rome)* **15**: 87-95.

LUZZATTO, G. 1932: *Anemone hortensis* L., *A. pavonina* Lam., *A. fulgens* Gay e *A. regina* Risso. Studio sistematico e fitogeografico. - *Arch. Bot. (Forlì)* **8**: 256-274.

LUZZATTO, G. 1933: *Anemone hortensis* L., *A. pavonina* Lam., *A. fulgens* Gay e *A. regina* Risso. Studio sistematico e fitogeografico. Seconda parte. - *Arch. Bot. (Forlì)* **9**: 205-236.

LUZZATTO, G. 1934: *Anemone hortensis* L., *A. pavonina* Lam., *A. fulgens* Gay e *A. regina* Risso. Studio sistematico e fitogeografico (cont.). - *Arch. Bot. (Forlì)* **10**: 101-118.

LYNCH, R.I. 1890: A new classification of the genus *Paeonia*. - *J. Roy. Hort. Soc.* **12(3)**: 428-445.

LYSKOV, D. 2020: Typification of the name *Ferulago galbanifera* (Apiaceae) and its priority over the name *F. campestris*. - *Phytotaxa* **442(1)**: 39-40.

LYSKOV, D. & AL. 2018: Molecular data to elucidate the taxonomy of *Seseli* sect. *Seseli* in the east Mediterranean and southern Europe. - *Nordic J. Bot.* **36(9)**: 9 pp.

MA, J.S. 2001: A revision of *Euonymus* (Celastraceae). - *Thaiszia* **11(1-2)**: 1-264.

MABBERLEY, D.J. 2002: *Potentilla* and *Fragaria* (Rosaceae) reunited. - *Telopea* **9(4)**: 793-801.

MACHULE, M. 1957: Die mitteleuropäischen *Thymus*-Arten, Formen und Bastarde. - *Mitt. Thüring. Bot. Ges.* **1(4)**: 13-89.

MACHULE, M. 1960: Die mitteleuropäischen *Thymus*-Arten, Formen und Bastarde. Nachtrag. - *Mitt. Thüring. Bot. Ges.* **2(1)**: 176-207.

MACWATT, J. 1923: The Primulas of Europe. - London: Country Life Ltd, xvi + 208 pp. + plates.

MADHANI, H. & AL. 2018: Untangling phylogenetic patterns and taxonomic confusion in tribe Caryophylleae (Caryophyllaceae) with special focus on generic boundaries. - *Taxon* **67(1)**: 83-112.

MAESEN, L.J.G. VAN DER 1972: *Cicer* L., a monograph of the genus, with special reference to the chickpea (*Cicer arietinum* L.), its ecology and cultivation. - *Meded. Landbouwhogeschool* **72(10)**: 1-342.

MAFFEI, M., DEIRINO, L. & CODIGNOLA, A. 1987: Numeri cromosomici per la Flora Italiana: 1064-1069. - *Inform. Bot. Ital.* **18(1-3)**: 153-158.

MAGGINI, F. 1972: The chromosome complement of *Bellevalia dubia* (Guss.) R. et S. and the problem of *Bellevalia webbiana* Parl. - *Ann. Bot. (Rome)* **31**: 115-123.

MAGRINI, S. & SCOPPOLA, A. 2015: Cytological status of *Viola kitaibeliana* (Section Melanium, Violaceae) in Europe. - *Phytotaxa* **238(3)**: 288-292.

MAHAL, C. & PAL, M. 1984: *Antirrhinum orontium* complex: biosystematic studies. - *Proc. Indian Acad. Sci., Pl. Sci.* **93(4)**: 455-458.

MAHMOUD, A., GRIME, J.P. & FURNESS, S.B. 1975: Polymorphism in *Arrhenatherum elatius* (L.) Beauv. ex J. & C. Presl. - *New Phytol.* **75(2)**: 269-276.

MAÏA, N. & VENARD, P. 1976: Contribution à l'étude cytotaxonomique d'espèces méditerranéennes d'*Anemone* et de leurs hybrides. - *Canad. J. Genet. Cytol.* **18(1)**: 151-168.

MAIKE, W. 1995: Vegetationsuntersuchungen in den Lagunen bei Agiasma, Nordost-Griechenland. In Mattes, H. & Lienau, C. (eds): Colloquium Nordost-Griechenland 3: pp. 2-20. - Münster: Inst. Geographie.

MAILLEFER, A. 1940: Herborisations pendant une croisière dans l'Adriatique et autour de la Grèce, en 1938. - *Bull. Soc. Vaud. Sci. Nat.* **61(251)**: 1-12.

MAIRE, R. 1921a: Contribution à l'étude de la végétation et de la flora de l'île de Skyros. - *Bull. Soc. Bot. France* **68**: 68-74.

MAIRE, R. 1921b: Contribution à l'étude de la flore grecque. - *Bull. Soc. Bot. France* **68**: 370-380.

MAIRE, R. 1939: Sertulum orientale. - *Notul. Syst. (Paris)* **8(1)**: 24-26.

MAIRE, R. & PETITMENGIN, M. 1907: Étude des plantes vasculaires récoltées en Grèce (1904). In Maire, R.: Matériaux pour servir à l'étude de la flore et de la géographie botanique de l'orient. Fasc. 2, 70 pp. + 4 plates. - Nancy: Imprimerie Berger-Levrault et Co.

MAIRE, R. & PETITMENGIN, M. 1908: Étude des plantes vasculaires récoltées en Grèce (1906). Materiaux pour servir à l'étude de la flore et de la géographie botanique de l'orient (Missions du ministère de l'instruction publique en 1904 et en 1906). Quatrième fascicule. - *Bull. Soc. Sci. Nancy Ser. 3* **9**: 151-266 & 360-481. [separate reprint, pp. 3-239].

MAJOR, C.-J. FORSYTH & BARBEY, W. 1894a: Halki. Étude botanique. - Lausanne: Georges Bridel, 7 pp. + 1 plate.

MAJOR, C.-J. FORSYTH & BARBEY, W. 1894b: Saria. Étude botanique. - *Bull. Herb. Boissier* **2(4)**: 241-246 + 1 plate.

MAJOR, C.-J. FORSYTH & BARBEY, W. 1894c: Kasos. Étude botanique. - *Bull. Herb. Boissier* **2(5)**: 329-341.

MAJOR, C.-J. FORSYTH & BARBEY, W. 1894d: Kos. Étude botanique. - *Bull. Herb. Boissier* **2(6)**: 404-416.

MAJOR, C.-J. FORSYTH & BARBEY, W. 1895a: Amoi. Étude botanique. - *Bull. Herb. Boissier* **3(1)**: 30.

MAJOR, C.-J. FORSYTH & BARBEY, W. 1895b: Syra. Matériaux pour la flore de Syra. - *Bull. Herb. Boissier* **3(2)**: 87-88.

MAJOR, C.-J. FORSYTH & BARBEY, W. 1895c: Telandos. Étude botanique. - *Bull. Herb. Boissier* **3(4)**: 174-176.

MAJOR, C.-J. FORSYTH & BARBEY, W. 1896: Kalymnos. Étude botanique. - *Bull. Herb. Boissier* **4(1)**: 20-39.

MAJOR, C.-J. FORSYTH & BARBEY, W. 1897a: Ikaria. Étude botanique. - *Bull. Herb. Boissier* **5(4)**: 279-284.

MAJOR, C.-J. FORSYTH & BARBEY, W. 1897b: Sertum cerigense. Étude botanique. - *Bull. Herb. Boissier* **5(5)**: 398-400.

MÁJOVSKÝ, J. & AL. 1970a: Index of chromosome numbers of Slovakian flora (Part 1). - *Acta Fac. Rerum Nat. Univ. Comen., Bot.* **16**: 1-26.

MÁJOVSKÝ, J. & AL. 1970b: Index of chromosome numbers of Slovakian flora (Part 2). - *Acta Fac. Rerum Nat. Univ. Comen., Bot.* **18**: 45-60.

MÁJOVSKÝ, J. & AL. 1974a: Index of chromosome numbers of Slovakian flora (Part 3). - *Acta Fac. Rerum Nat. Univ. Comen., Bot.* **22**: 1-20.

MÁJOVSKÝ, J. & AL. 1974b: Index of chromosome numbers of Slovakian flora (Part 4). - *Acta Fac. Rerum Nat. Univ. Comen., Bot.* **23**: 1-23.

MÁJOVSKÝ, J. & AL. 1976: Index of chromosome numbers of Slovakian Flora (Part 5). - *Acta Fac. Rerum Nat. Univ. Comen., Bot.* **25**: 1-18.

MÁJOVSKÝ, J. & AL. 1978: Index of Chromosome Numbers of Slovakian Flora (Part 6). - *Acta Fac. Rerum Nat. Univ. Comen., Bot.* **26**: 1-42.

MÁJOVSKÝ, J. & MURÍN, A. 1980: Cytotaxonomy of the genus *Pulmonaria* L. in Slovakia. - *Bol. Soc. Brot., Ser. 2* **53(1)**: 725-739.

MÁJOVSKÝ, J., MURÍN, A. & UHRÍKOVÁ, A. 1984: Gattung *Muscari* Miller in der Slowakei. - *Acta Fac. Rerum Nat. Univ. Comen., Bot.* **31**: 1-17.

MÁJOVSKÝ, J. & UHRÍKOVÁ, A. 1985: *Sedum telephium* aggr. in der Slowakei. - *Acta Fac. Rerum Nat. Univ. Comen., Bot.* **32**: 25-38.

MÁJOVSKÝ, J. & UHRÍKOVÁ, A. 1990a: Karyosystematische Studium der Gattung *Sorbus* L. emend. Cr. in der Slowakei I. - *Acta Fac. Rerum Nat. Univ. Comen., Bot.* **37**: 5-15.

MÁJOVSKÝ, J. & UHRÍKOVÁ, A. 1990b: *Mercurialis perennis* L. in der Slowakei. - *Acta Fac. Rerum Nat. Univ. Comen., Bot.* **37**: 41-49.

MÁJOVSKÝ, J. & VÁCHOVÁ, M. 1986: Karyological study of the Slovak flora XIII. - *Acta Fac. Rerum Nat. Univ. Comen., Bot.* **33**: 63-67.

MALAKATES, S. 1927: Die Flora von Kea. - *Repert. Spec. Nov. Regni Veg.* **24**: 193-199 + 1 map.

MALAKATES, S. 1928a: Chloris Kimolou ke ton peri avtin nisidon [In Greek]. - Athine: Ethnikou Tipografiou, 26 pp.

MALAKATES, S. 1928b: Simvole pros sintaxin tis chloridos ton Kikladon. Chloris Serifou [In Greek]. - Athine: Ethnikou Tipografiou, 16 pp.

MALAKATES, S. 1933a: Simvole pros sintaxin tis chloridos ton Kikladon. Chloris Sifnou [In Greek]. - Athine (reprinted from *Ekpedevtika Chronika*): 40 pp.

MALAKATES, S. 1933b: Die Flora von Andros. - *Repert. Spec. Nov. Regni Veg.* **33**: 81-101.

MALAKATES, S. 1935: Ta Gramineae tis Ellados [In Greek]. - Athine: 64 pp.

MALECI, L.B. & MAUGINI, E. 1981: *Iris pseudopumila* Tin. and *Iris attica* Boiss. & Heldr.: two very similar karyotypes. - *Bol. Soc. Brot., Ser. 2* **53(2)**: 921-931 + 2 plates.

MAŁECKA, J. 1972: Further cyto-taxonomic studies in the genus *Taraxacum* section *Palustria* Dahlstedt. - *Acta Biol. Cracov., Ser. Bot.* **15(2)**: 113-126 + 6 plates.

MAŁECKA, J. 1981: Further cyto-taxonomic studies in the genus *Taraxacum* section *Palustria* Dahlstedt. Part IV. The species from Macedonia. - *Acta Biol. Cracov., Ser. Bot.* **23**: 107-115 + 4 plates.

MAŁECKA, J. & POPEK, R. 1982: Karyological studies in the Polish representatives of the genus *Rosa* L. I. - *Acta Biol. Cracov., Ser. Bot.* **24**: 79-90.

MAŁECKA, J. & POPEK, R. 1984: Karyological studies in the Polish representatives of the genus *Rosa* L. II. - *Acta Biol. Cracov., Ser. Bot.* **26**: 43-54.

MAŁECKA, J., POPEK, R. & FACSAR, G. 1991: Cyto-taxonomical studies in the genus *Rosa* L.: representatives from Hungary. - *Acta Biol. Cracov., Ser. Bot.* **32**: 189-196.

MALÉCOT, V. 2002: Lectotypification of Linnaean names in *Viburnum* L. (Viburnaceae). - *Taxon* **51(4)**: 747-750.

MALHEIROS-GARDÉ, N. & GARDÉ, A. 1951: Agmatoploidia no genero "Luzula" DC. - *Genét. Ibér.* **3(1-2)**: 155-176.

MALHEIROS, N. & GARDÉ, A. 1947: Contribuições para o estudo citológico do género *Luzula* Link. - *Agron. Lusit.* **9(1)**: 75-79 + 1 plate.

MALIK, C.P. & GROVER, I.S. 1973a: Cytogenetic studies in *Papaver*. II. Hybrids among species with 7 as the haploid chromosome number. - *Caryologia* **26(1)**: 13-25.

MALIK, C.P. & GROVER, I.S. 1973b: Cytogenetic studies in *Papaver*. 3. Induced polyploids in some *Papaver* species. - *Chromosome Inform. Serv.* **15**: 27-29.

MALIK, C.P. & THOMAS, P.T. 1966: Karyotypic studies in some *Lolium* and *Festuca* species. - *Caryologia* **19(2)**: 167-196.

MALINOWSKI, E. 1910a: Monographie du genre *Biscutella* L. 1. Classification et distribution géographique. - *Bull. Int. Acad. Sci. Cracovie, Cl. Sci. Math.* **1910**: 111-139.

MALINOWSKI, E. 1910b: Les espèces du genre *Crucianella* L. - *Bull. Soc. Bot. Genève, Sér. 2* **2**: 9-16.

MALO, S. & SHUKA, L. 2008: New records on the flora of the Gjirokastra region (south Albania). - *Natura Montenegrina* **7(3)**: 369-373.

MALO, S. & SHUKA, L. 2013: Distribution of *Colchicum doerfleri* Halácsy, *Colchicum triphyllum* Kunze and *Colchicum bivonae* Guss. in Albania. - *Int. J. Ecosyst. Ecol. Sci.* **3(2)**: 273-278.

MALÝ, K. 1907: Bemerkungen über die Arten der Gattung *Heliosperma* aus der Verwandtschaft des *H. Retzdorffianum*. - *Wiss. Mitt. Bosnien & Herzegovina* **10**: 628-634.

MALÝ, K. 1910: Prilozi za floru Bosne i Hercegovine. II. - *Glasn. Zemaljsk. Muz. Bosni Hercegovini* **22(4)**: 685-694.

MALÝ, K. 1928: Ein Beitrag zur Kenntnis der geographischen Verbreitung der Gattung *Koeleria* in den Balkan-Ländern. - *Bull. Inst. Jard. Bot. Univ. Belgrade* **1(1)**: 43-59.

MALÝ, K. 1930: *Satureja*-Arten Südserbiens. - *Bull. Inst. Jard. Bot. Univ. Belgrade* **1(3)**: 229-232.

MALÝ, K. 1932: Carices der Balkanhalbinsel. - *Bull. Inst. Jard. Bot. Univ. Belgrade* **2**: 59-66.

MALÝ, K. 1939: *Alchemilla acutiloba*, ein orientalisches Florenelement in Bosnien. - *Repert. Spec. Nov. Regni Veg.* **46**: 49-52.

MALÝ, K. 1950: Flora Bosnae et Hercegovinae. IV. Sympetalae (Gamopetalae). Pars 1 [continuation of the work of G. Beck v. Mannagetta]. - Sarajevo: Dražavna Štamparija, 73 pp.

MALÝ, K. 1951: Vorläufige Mitteilung über eine für die Balkanhalbinsel kritische *Achillea* [In Serbo- Croatian with German summary]. - *God. Biol. Inst. Sarajevu* **3(1-2)**: 5-8.

MALYER, H., HEPER, M. & BIÇAKÇI, A. 1995: Distributions of *Chamaecytisus* (Fabaceae/Genisteae) species in Turkey [In Turkish with English abstract]. - *OT Sist. Bot. Derg.* **2(1)**: 133-146.

MALYUTIN, N.I. 1973: The phylogeny and the taxonomy of the *Delphinium* L. genus [In Russian with English summary]. - *Bot. Zhurn. (Moscow & Leningrad)* **58(12)**: 1710-1722.

MALZEV, A.I. 1930: Wild and cultivated oats: Sectio *Euavena* Griseb. [In Russian with English summary]. - *Trudy Prikl. Bot. Selekt. Suppl.* **38**: 1-522 + 100 plates.

MANAFZADEH, S., SALVO, G. & CONTI, E. 2014: A tale of migrations from east to west: the Irano-Turanian floristic region as a source of Mediterranean xerophytes. - *J. Biogeogr.* **41(2)**: 366-379.

MANALIS, P. 1995: Perivallontiki ekpedevsi ke i chlorida tou orous Egaleo (Aformi gia mia erevna) [In Greek]. - *Fisis [Journal of the Hellenic Society for the Protection of Nature]* **71**: 22-25.

MANDENOVA, I.P. 1962: Taxonomic review of Turkish species of *Heracleum*. - *Notes Roy. Bot. Gard. Edinburgh* **24(2)**: 173-181.

MANDENOVA, I.P. 1970: A revision of *Rosa* in Turkey. - *Notes Roy. Bot. Gard. Edinburgh* **30(2)**: 327-340.

MANIOS, T. & AL. 2002: Plant species in a two-year-old free water surface constructed wetland treating domestic wastewater in the island of Crete. - *J. Environm. Sci. Health, A* **37(7)**: 1327-1335.

MANISON, G., ZELTNER, L. & BRETAGNOLLE, F. 2005: Phylogenetic patterns and polyploid evolution within the Mediterranean genus *Centaurium* (Gentianaceae - Chironieae). - *Taxon* **54(4)**: 931-950.

MANITZ, H. 1984: Die Cytologie der Convolvulaceae und Cuscutaceae. I. Zusammenstellung der bekannten Chromosomenzahlen. - *Wiss. Z. Friedrich-Schiller-Univ. Jena, Math.-Naturwiss. Reihe* **32 (6)**: 915-944.

MANN TAYLOR, J. 1998: *Phlomis*, the neglected genus. A guide for gardeners and horticulturists. - Gloucester: NCCPG, 88 pp.

MANNING, J., FOREST, F. & VINNERSTEN, A. 2007: The genus *Colchicum* L. redefined to include *Androcymbium* Willd. based on molecular evidence. - *Taxon* **56(3)**: 872-882.

MANNS, U. & ANDERBERG, A.A. 2009: New combinations and names in *Lysimachia* (Myrsinaceae) for species of *Anagallis*, *Pelletiera* and *Trientalis*. - *Willdenowia* **39(1)**: 49-54.

MANOLAKI, P., TSAKIRI, E. & PAPASTERGIADOU, E. 2011: Inventory of aquatic and riparian flora of Acheron and Louros rivers, and Zirou lake in western Greece. - *Fresenius Environ. Bull.* **20(4)**: 861-874.

MANSFELD, R. 1925: Vorarbeiten zu einer Monographie der Gattung *Ligustrum*. - *Bot. Jahrb. Syst.* **59 Beibl. 132**: 19-75.

MANSFELD, R. 1938: Zur Nomenklatur der europäischen Orchideengattungen. - *Repert. Spec. Nov. Regni Veg.* **44**: 145-154.

MANSION, G. 2004: A new classification of the polyphyletic genus *Centaurium* Hill (Chironiinae, Gentianaceae): description of the New World endemic *Zeltnera*, and reinstatement of *Gyrandra* Griseb. and *Schenkia* Griseb. - *Taxon* **53(3)**: 719-740.

MANSION, G. & BLAISE, S. 1995: [Chromosome numbers of *Lotus*]. In Stace, C.A. (ed.): IOPB chromosome data 9. - *Int. Organ. Pl. Biosyst. Newslett.* **24**: 23.

MANSION, G., ZELTNER, L. & BRETAGNOLLE, F. 2005: Phylogenetic patterns and polyploid evolution within the Mediterranean genus *Centaurium* (Gentianaceae-Chironieae). - *Taxon* **54(4)**: 931-950.

MANTON, I. & REICHSTEIN, T. 1962: Diploides *Asplenium obovatum* Viv. - *Bauhinia* **2(1)**: 79-91.

MANUEL, R.L. 1989a: Wild orchids of Rhodes. Part 1. - *Orchid Rev.* **97(1143)**: 4-10.

MANUEL, R.L. 1989b: Wild orchids of Rhodes. Part 2. - *Orchid Rev.* **97(1144)**: 56-62.

MANUEL, R.L. 1996: Orchidées de Crète. Une compilation de mentions récentes. – *Naturalistes Belges* **77(4)**: 137-170.

MARAIS, W. 1976: Taxonomic notes on *Romulea* (Iridaceae) from the Mediterranean region. - *Kew Bull.* **30(4)**: 707-708.

MARAIS, W. 1980: Notes on *Tulipa* (Liliaceae). - *Kew Bull.* **35(2)**: 257-259.

MARANNE, I. 1913: Les *Erophila* DC. - *Bull. Soc. Bot. France* **60**: 276-281, 345-353, 379-389, 422-425.

MARCENÒ, C. & AL. 1989: The approach to *Crepis* studies. In Phitos, D. (ed.): Abstracts. Sixth OPTIMA meeting, Delphi 10-16 Sept. 1989, p. 111.

MARCENÒ, C. & AL. 1991: A contribution to the *Crepis vesicaria* studies. - *Bot. Chron. (Patras)* **10**: 749-760.

MARCENÒ, C., OTTONELLO, D. & ROMANO, S. 1995: *Prunus webbii* (Spach) Vierh. (Rosaceae), specie nuova per la flora di Sicilia. - *Webbia* **50(1)**: 37-43.

MARCHANT, R. 2016: Wild flowers of the Mani. A comprehensive photographic identification guide for the flowers of the Mani Peninsula in Southern Greece. - NatureFocus Publishing: 262 pp.

MARCHANT, R. 2019: Wild flowers of the Mani. Second edition. A comprehensive photographic identification guide for the flowers of the Mani Peninsula in southern Greece. - NatureFocus Publishing: 280 pp.

MARCHETTI, D. 2005: Le pteridofite d'Italia. - *Ann. Mus. Civico Rovereto* **19**: 71-231.

MARCHI, P. 1971a: Numeri cromosomici per la Flora Italiana: 46-56. - *Inform. Bot. Ital.* **3(1)**: 82-94.

MARCHI, P. 1971b: Numeri cromosomici per la Flora Italiana: 57-66. - *Inform. Bot. Ital.* **3(2)**: 124-138.

MARCHI, P. 1972: Per una revisione citotassonomica della Flora Italiana. Genere *Leucanthemum* Adans. em. Brig. et Cav. (Compositae): Osservazioni su specie appartenenti al ciclo del *L. vulgare* Lamk. - *Ann. Bot. (Rome)* **29**: 259-300.

MARCHI, P. 1984: Karyotype structure of "Leucanthemum vulgare" Lam. aggr. (Compositae) polyploid species and the problem of their origin. - *Webbia* **38**: 541-544.

MARCHI, P., CAPINERI, R. & D'AMATO, G. 1975: Numeri cromosomici per la Flora Italiana: 182-189. - *Inform. Bot. Ital.* **6(3)**: 303-312.

MARCHI, P., CAPINERI, R. & D'AMATO, G. 1976: Numeri cromosomici per la Flora Italiana: 208-218. - *Inform. Bot. Ital.* **7(3)**: 377-389.

MARCHI, P. & D'AMATO, G. 1973: Numeri cromosomici per la Flora Italiana: 145-150. - *Inform. Bot. Ital.* **5(1)**: 93-100.

MARCHI, P. & VISONÀ, L. 1984: Numeri cromosomici per la Flora Italiana: 889-905. - *Inform. Bot. Ital.* **14(2-3)**: 248-258.

MARCHIONI ORTU, A. 1984: Numeri cromosomici per la Flora Italiana: 873-876. - *Inform. Bot. Ital.* **14(2-3)**: 234-237.

MARCHIONI, A. 1972: Sulla presenza in Sardegna di *Cotula coronopifolia* L. - *Ann. Bot. (Rome)* **29**: 179-182.

MARCHIORI, S. 1996: *Centaurea pumilio* L. (Asteraceae), una nuova specie per la floraitaliana. - *Thalassia Salentina* **22**: 41-45.

MARCUCCI, R. 2005: Reports (1415-1427) In: Kamari, G., Blanché, C. & Garbari, F. (eds): Mediterranean chromosome number reports - 15. - *Fl. Medit.* **15**: 694-702.

MARCUCCI, R. & TORNADORE, N. 1994: Variabilità cariologica di *Allium roseum* L. (Alliaceae). - *Inform. Bot. Ital.* **25(1)**: 90-95.

MARCUCCI, R. & TORNADORE, N. 1997a: Cytological and taxonomical notes on *Allium arvense* Guss. (Alliaceae) in Italy. - *Webbia* **51(2)**: 189-199.

MARCUCCI, R. & TORNADORE, N. 1997b: Intraspecific variation of *Allium roseum* L. (Alliaceae). - *Webbia* **52(1)**: 137-154.

MARCUCCI, R. & TORNADORE, N. 1997c: Numeri cromosomici per la Flora Italiana: 1378-1390. - *Inform. Bot. Ital.* **28(2)**: 256-266.

MARCUCCI, R. & TORNADORE, N. 1999: Reports (1089-1098). In Kamari, G., Felber, F. & Garbari, F. (eds): Mediterranean chromosome number reports - 9. - *Fl. Medit.* **9**: 372-378.

MARCUSSEN, T. 2003: Evolution, phylogeography, and taxonomy within the *Viola alba* complex (Violaceae). - *Pl. Syst. Evol.* **237(1-2)**: 51-74.

MARCUSSEN, T. & BORGEN, L. 2000: Allozymic variation and relationships within *Viola* subsection *Viola* (Violaceae). - *Pl. Syst. Evol.* **223(1-2)**: 29-57.

MARCUSSEN, T. & BORGEN, L. 2003: *Viola sieheana* is two species: the recognition of *V. caspia*. - Manuscript: 18 pp.

MARCUSSEN, T. & BORGEN, L. 2011: Species delimitation in the Ponto-Caucasian *Viola sieheana* complex, based on evidence from allozymes, morphology, ploidy levels, and crossing experiments. - *Pl. Syst. Evol.* **291(3)**: 183-196.

MARCUSSEN, T., BORGEN, L. & NORDAL, I. 2003: Population differentiation and hybridisation in the Greek endemism *Viola chelmea* ssp. *chelmea*. - Manuscript: 8 pp.

MARCYSIAK, K. 2014: Geographical differentiation of *Dryas octopetala* in Europe based on morphological features. - *Dendrobiology* **72**: 113-123.

MARCYSIAK, K. & AL. 2007: Numerical taxonomy of *Juniperus thurifera, J. excelsa* and *J. foetidissima* (Cupressaceae) based on morphological characters. - *Bot. J. Linn. Soc.* **155(4)**: 483-495.

MARÉCHAL, R., MASCHERPA, J.-M. & STAINER, F. 1978: Étude taxonomique d'un groupe complexe d'espèces des genres *Phaceolus* et *Vigna* (Papilionaceae) sur la base données morphologiques et polliniques, traitées par l'analyse informatique. - *Boissiera* **28**: 1-273.

MARGARIS, N.S., KOEDAM, A. & VOKOU, D. (EDS) 1982: Aromatic plants: Basic and applied aspects. Proceedings of an international symposium ... held at Kallithea (Chalkidiki) Greece, 14-19 September 1981. - The Hague, etc.: Martinus Nijhoff Publishers, xi + 283 pp.

MARGELÍ, M. & AL. 1996: Reports (518-523). In Kamari, G., Felber, F. & Garbari, F. (eds): Mediterranean chromosome number reports - 5. - *Fl. Medit.* **5**: 334-337.

MARGELÍ, M., PEDROLA, J. & VALLÉS, J. 1995: Cytogenetical studies in the genus *Androcymbium* Willd. (Colchicaceae). In Valdés, B. & Silvestre, S. (eds): Abstracts. VIII OPTIMA meeting, Sevilla, Spain, 25 September - 1 October, 1995, p. 76.

MARGOT, H. & REUTER, F.-G. 1839: Essai d'une flore de l'île de Zante. - *Mém. Soc. Phys. Genève* **8(2)**: 250-314 + 5 plates. [cf. TL-2 3: 293, 1981].

MARGOT, H. & REUTER, F.-G. 1841: Essai d'une flore de l'île de Zante [cont.]. - *Mém. Soc. Phys. Genève* **9(1)**: 1-56 + 1 plate. [cf. TL-2 3: 293, 1981].

MARGOT, J. 1994: La position géobotanique de Rhodes dans l'arc insulaire sud-égéen. - *Naturalistes Belges* **75(1)**: 1-25.

MARHOLD, K. 1989: Critical notes to *Cardamine pratensis* agg. in NE Hungary. - *Acta Bot. Hung.* **35(1-4)**: 279-285.

MARHOLD, K. 1994a: Chromosome numbers of the genus *Cardamine* L. (Cruciferae) in the Carpathians and in Pannonia. - *Phyton (Horn)* **34(1)**: 19-34.

MARHOLD, K. 1994b: Taxonomy of the genus *Cardamine* L. (Cruciferae) in the Carpathians and Pannonia. I. *Cardamine pratensis* group. - *Folia Geobot. Phytotax.* **29(3)**: 335-374.

MARHOLD, K. 1995a: Taxonomy of the genus *Cardamine* L. (Cruciferae) in the Carpathians and Pannonia. II. *Cardamine amara* L. - *Folia Geobot. Phytotax.* **30(1)**: 63-80.

MARHOLD, K. 1995b: Taxonomy of the genus *Cardamine* L. (Cruciferae) in the Carpathians and Pannonia. III. - *Folia Geobot. Phytotax.* **30(4)**: 397-434.

MARHOLD, K. 1996: Typification of the Linnaean names of the genus *Cardamine* (Cruciferae). - *Bot. J. Linn. Soc.* **121(2)**: 111-131.

MARHOLD, K. 1998: Morphometric comparison of diploid populations of *Cardamine amara* (Brassicaceae) from central Europe and the Balkan Peninsula. - *Thaiszia* **8(1)**: 19-32.

MARHOLD, K. 2001: Lectotypification of names of the European representatives of *Cardamine* subg. *Dentaria* (Cruciferae). - *Willdenowia* **31(1)**: 43-49.

MARHOLD, K. & AL. 2004: The *Bolboschoenus maritimus* group (Cyperaceae) in central Europe, including *B. laticarpus*, spec. nova. - *Phyton (Horn)* **44(1)**: 1-21.

MARHOLD, K. & AL. 2005: Morphological and karyological differentiation of *Gymnadenia densiflora* and *G. conopsea* in the Czech Republic and Slovakia. - *Preslia* **77(2)**: 159-176.

MARHOLD, K. & AL. 2016: *Cardamine occulta*, the correct species name for invasive Asian plants previously classified as *C. flexuosa*, and its occurrence in Europe. - *PhytoKeys* **62**: 57-72.

MARHOLD, K. & ANČEV, M.E. 1995: Morphological and karyological variation of *Cardamine amara* (Cruciferae). In Valdés, B. & Silvestre, S. (eds): Abstracts. VIII OPTIMA meeting, Sevilla, Spain, 25 September - 1 October, 1995, p. 110.

MARHOLD, K. & ANČEV, M.E. 1999: *Cardamine penzesii*, a rediscovered taxon of the *C. pratensis* group (Cruciferae). - *Ann. Bot. Fenn.* **36(3)**: 171-180.

MARHOLD, K., ANČEV, M.E. & TAN, KIT 1996: A new subspecies of *Cardamine amara* (Brassicaceae) from Bulgaria and Greece. - *Ann. Bot. Fenn.* **33(3)**: 199-204.

MARHOLD, K., DUCHÁČEK, M. & HROUDOVÁ, Z. 2006: Typification of three names in the *Bolboschoenus maritimus* group. - *Willdenowia* **36(1)**: 103-113.

MARHOLD, K., HUTHMANN, M. & HURKA, H. 2002: Evolutionary history of the polyploid complex of *Cardamine amara*: isozyme evidence. - *Pl. Syst. Evol.* **233(1-2)**: 15-28.

MARHOLD, K. & MÁRTONFI, P. 2001: Typification of two Linnaean species names of the genus *Thlaspi* (Brassicaceae). - *Novon* **11(2)**: 189-192.

MARHOLD, K., PERNÝ, & LIHOVÁ, J. 2004: A new taxonomic treatment of some *Cardamine* taxa in the Balkan Peninsula. In Vasić, O. (ed.): Abstracts, XI OPTIMA Meeting, Beograd 5-11. IX.2004, p. 22.

MARHOLD, K. & RAYNER, T.G.J. 1994: Typification of the names of two species of the *Cardamine pratensis* group (Brassicaceae). - *Taxon* **43(1)**: 77-83.

MARHOLD, K. & TAN, KIT 2000: The distribution of *Cardamine matthioli* (Brassicaceae) in Greece. - *Thaiszia* **9(2)**: 109-112.

MARHOLD, K. & VALACHOVIČ, M. 1998: Coenotic differentiation of the infraspecific taxa of *Cardamine amara* (Brassicaceae) in Central Europe and the Balkan Peninsula. - *Thaiszia* **8(2)**: 147-161.

MARHOLD, K. & ZÁBORSKÝ, J. 1986: A new species of *Cardamine pratensis* agg. from eastern Slovakia. - *Preslia* **58(3)**: 193-198.

MARHOLD, K., ZOZOMOVÁ-LIHOVÁ, J. & ŠPANIEL, S. 2011: (1993) Proposal to conserve the name *Alyssum montanum* (Cruciferae) with a conserved type. - *Taxon* **60(1)**: 237-238. [approved, cf. Taxon 61(5): 1110, 2012].

MARHOLD, K. & AL. (EDS) 2005: Karyological database of ferns and flowering plants of Slovakia. Available on-line: http://147.213.100.121/webapp/.

MARHOLD, K. (ED.) 1990: List of taxonomical projects concerning the Carpathian and Pannonian flora. - Bratislava: 79 pp.

MARIANI, A. 1975: Cytogenetic research on hexaploid alfalfa, *Medicago sativa* L. - *Caryologia* **28(3)**: 359-373.

MARIN, P.D., PETKOVIĆ, B. & DULETIĆ, S. 1994: Nutlet sculpturing of selected *Teucrium* species (Lamiaceae): a character of taxonomic significance. - *Pl. Syst. Evol.* **192(3-4)**: 199-214.

MARIN, P.D. & TATIĆ, B. 2001: Serpentine soil and plant diversity, with emphasis of Balkan Peninsula. - *Bocconea* **13**: 145-150.

MARINOV, Y. & STOYANOV, S. 2019: First record of *Dorycnium haussknechtii* (Fabaceae) for the Balkans. - *Phytol. Balcan.* **25(1)**: 63-68.

MARKANTONATOU, A. 1991: Simvoli sti chlorida ke vlastisi tis nisou Ithakis. [In Greek]. - Georgiko Panepistimio Athinon: Diploma thesis, 58 pp.

MARKANTONATOU, A., SARLIS, G. & CONSTANTINIDIS, TH. 2002: The flora of Ithaki island (Ionian Sea, Greece): a work in progress. In Kamari, G. & al. (eds): Ellinini Votaniki Eteria. 9° Panelliniko Epistimoniko Sinedrio. Praktika, pp. 182-188 [In Greek with English abstract].

MARKGRAF-DANNENBERG, I. 1976: Die Gattung *Festuca* in Griechenland. In Dafis, S. & Landolt, E. (eds): Zur Vegetation und Flora von Griechenland, Band 2. - *Veröff. Geobot. Inst. E.T.H. Stiftung Rübel Zürich* **56**: 92-182. [Also distributed as *OPTIMA leaflet* no. 24].

MARKGRAF-DANNENBERG, I. 1978: New taxa and names in European *Festuca* (Gramineae). In Heywood, V.H. (ed.): Notulae systematicae ad Floram Europaeam spectantes, no. 20. - *Bot. J. Linn. Soc.* **76(4)**: 322-328.

MARKGRAF-DANNENBERG, I. 1981: The genus *Festuca* (Gramineae) in Turkey: new taxa and new names. - *Willdenowia* **11(2)**: 201-210.

MARKGRAF, F. 1920: Botanische Kriegsbeobachtungen in Thrazien. - *Verh. Bot. Vereins Prov. Brandenburg* **62**: 8-15.

MARKGRAF, F. 1926: Bemerkenswerte neue Pflanzenarten aus Albanien. - *Ber. Deutsch. Bot. Ges.* **44 (7)**: 420-432 + 2 plates.

MARKGRAF, F. 1927: An den Grenzen des Mittelmeergebiets. Pflanzengeographie von Mittelalbanien. - *Repert. Spec. Nov. Regni Veg. Beih.* **45**: i-viii, 1-217 + 7 plates and 1 map.

MARKGRAF, F. 1930: In Albaniens Bergen. - Stuttgart: Strecker und Schröder, xiv + 244 pp. + 16 plates.

MARKGRAF, F. 1931a: *Soldanella pindicola* Hausskn. - *Notizbl. Bot. Gart. Berlin-Dahlem* **11(103)**: 219-223.

MARKGRAF, F. 1931b: Die Panzerkiefer. - *Mitt. Deutsch. Dendrol. Ges.* **43**: 250-255 + 2 plates.

MARKGRAF, F. 1931c: Pflanzen aus Albanien 1928. - *Denksch. Akad. Wiss. Wien, Math.-Naturwiss. Kl.* **102**: 317-360 + 1 plate and 1 map.

MARKGRAF, F. 1932: Pflanzengeographie von Albanien. Ihre Bedeutung für Vegetation und Flora der Mittelmeerländer. - *Biblioth. Bot.* **26/105**: 132 pp. + 7 plates and 1 map.

MARKGRAF, F. 1942: Kurzer Reisebericht aus Nordgriechenland und Albanien. - *Notizbl. Bot. Gart. Berlin-Dahlem* **15(5)**: 662-666.

MARKGRAF, F. 1943: Die Südgrenze mitteleuropäischer Vegetation auf der Balkanhalbinsel. - *Ber. Deutsch. Bot. Ges.* **60, Generalvers.-Heft**: 118-127 + 2 plates.

MARKGRAF, F. 1949a: Eine neue Höhenstufenkarte der Vegetation Albaniens. - *Ber. Geobot. Forschungsinst. Rübel Zürich* **1948**: 109-119 + 1 map.

MARKGRAF, F. 1949b: Die Verbreitung von *Brachypodium serpentini* Hubb. - *Bot. Jahrb. Syst.* **74(2)**: 268-270.

MARKGRAF, F. 1952a: Zur Abgrenzung der Mittelmeervegetation. - *Vegetatio* **3(4-5)**: 324-325.

MARKGRAF, F. 1952b: Johannes Mattfeld [obituary]. - *Taxon* **1(3)**: 33-34.

MARKGRAF, F. 1953: Differenzierung in der Mediterranflora. - *God. Biol. Inst. Sarajevu* **5(1-2)**: 303-310.

MARKGRAF, F. 1959: Höhenstufen der Waldvegetation im ostmediterranen Raum. - *Ber. Geobot. Forschungsinst. Rübel Zürich* **1959**: 75.

MARKGRAF, F. 1970: Die floristische Stellung und Gliederung Albaniens. - *Feddes Repert.* **81(1-5)**: 215-222.

MARKGRAF, F. 1971: Beobachtungen an den europäischen Arten der Gattung *Vincetoxicum*. In Heywood, V.H. (ed.): Notulae systematicae ad Floram Europaeam spectantes, no. 11. - *Bot. J. Linn. Soc.* **64(4)**: 370-376.

MARKGRAF, F. 1975: Floristic report for Albania. - *Mem. Soc. Brot.* **24(1)**: 5-7.

MARKIŠIĆ, H. 1988: Contribution to the knowledge of flora of Montenegro [In Serbo-Croatian with English summary]. - *Glasn. Republ. Zavoda Zaštitu Prir. Prirodnjačke Zbirke Titogradu* **19**: 11-14.

MARKOVA, M.L. 1961a: Neue Angaben über *Rubus tomentosus* Borkh. in Bulgarien. - *Dokl. Bulg. Akad. Nauk* **14(6)**: 619-621.

MARKOVA, M.L. 1961b: Kritische Notizen über die Gattung *Rubus* L. [In Bulgarian with Russian and German summaries]. - *Izv. Bot. Inst. (Sofia)* **8**: 159-168.

MARKOVA, M.L. 1962a: Bemerkungen zu der Abhandlung "Rubi Peninsulae Balcanae" von Dr. J. Hruby und den von ihm im Rila-Gebirge gesammelten Herbariummaterialien [In Bulgarian with Russian and German summaries]. - *Izv. Bot. Inst. (Sofia)* **9**: 79-90.

MARKOVA, M.L. 1962b: Neue Arten, Varietäten und Fundorte der Gattung *Rubus* L. in Bulgarien [In Bulgarian with Russian and German summaries]. - *Izv. Bot. Inst. (Sofia)* **10**: 199-200.

MARKOVA, M.L. 1968a: Karyological study of *Epimedium pubigerum* (DC.) Morren & Decne and *Chenopodium multifidum* L. - *Dokl. Bulg. Akad. Nauk* **21(1)**: 51-53.

MARKOVA, M.L. 1968b: Chromosome number of four Bulgarian taxa of the genus *Rubus* L. - *Izv. Bot. Inst. (Sofia)* **18**: 75-77.

MARKOVA, M.L. 1969: Comparative karyological study on *Leontice leontopetalum* L. [In Bulgarian]. In: Second national conference on botany, pp. 389-392. - Sofia: Publishing House of the Bulgarian Academy of Sciences.

MARKOVA, M.L. 1970a: Karyological study of *Verbena officinalis* L. var. *rumelica* (Vel.) Stoj. et Stef. and of *Haberlea rhodopensis* Friv. - *Dokl. Bulg. Akad. Nauk* **23(10)**: 1287-1289.

MARKOVA, M.L. 1970b: Chromosome numbers of fifteen Bulgarian taxa. - *Izv. Bot. Inst. (Sofia)* **20**: 81-92.

MARKOVA, M.L. 1971: [Chromosome numbers in *Potentilla* from Bulgaria]. In Löve, Á. (ed.): IOPB chromosome number reports XXXIII. - *Taxon* **20(4)**: 610-611.

MARKOVA, M.L. 1972a: Chromozomenzahl der Bulgarischen Vertreter der Gattung *Potentilla* L. I. - *Izv. Bot. Inst. (Sofia)* **22**: 143-148.

MARKOVA, M.L. 1972b: [Chromosome numbers from Bulgaria]. In Löve, Á. (ed.): IOPB chromosome number reports XXXVI. - *Taxon* **21(2-3)**: 339.

MARKOVA, M.L. 1975a: Karyosystematische Untersuchungen an den Cistaceae Bulgariens. - *Pl. Syst. Evol.* **123(4)**: 283-315.

MARKOVA, M.L. 1975b: Über die Chromosomenevolution der Vertreter der Cistaceae in der VR Bulgarien. In Jordanov, D. & al. (eds): Problems of Balkan flora and vegetation, pp. 194-201. - Sofia: Publishing House of the Bulgarian Academy of Sciences.

MARKOVA, M.L. 1982a: [Chromosome numbers from Bulgaria]. In Löve, Á. (ed.): Chromosome number reports LXXVII. - *Taxon* **31(4)**: 761.

MARKOVA, M.L. 1982b: [Chromosome numbers in Lamiaceae from Bulgaria]. In Löve, Á. (ed.): Chromosome number reports LXXVII. - *Taxon* **31(4)**: 776-777.

MARKOVA, M.L. 1983a: [Chromosome numbers from Bulgaria]. In Löve, Á. (ed.): IOPB chromosome number reports LXXVIII. - *Taxon* **32(1)**: 140.

MARKOVA, M.L. 1983b: [Chromosome numbers in Boraginaceae and Lamiaceae from Bulgaria]. In Löve, Á. (ed.): IOPB chromosome number reports LXXX. - *Taxon* **32(3)**: 509-510.

MARKOVA, M.L. 1987: Karyological study of Bulgarian taxa from the genus *Thymus* L. I [In Bulgarian with English summary]. In Kuzmanov, B.A. & al. (eds): Proceedings, fourth national conference on botany, pp. 38-45. - Sofia: Publishing House of the Bulgarian Academy of Sciences.

MARKOVA, M.L. 1989: Chromosome numbers of Bulgarian angiosperms. - *Fitologiya* **36**: 67-68.

MARKOVA, M.L. 1992: Bogdan Kuzmanov (1934-1991). - *Fl. Medit.* **2**: 5-7.

MARKOVA, M.L. 1993: Karyology of some Bulgarian *Alkanna* species (Boraginaceae). In Petrova, A.V. (ed.): Abstracts from the VII OPTIMA meeting in Borovetz, Bulgaria, 18-30 July 1993, p. 137.

MARKOVA, M.L. & AL. 1974: Karyologische Untersuchungen der in Bulgarien wildwachsenden Vertreter der Gattung *Ornithogalum* L. I. - *Izv. Bot. Inst. (Sofia)* **25**: 63-92.

MARKOVA, M.L. & ANČEV, M.E. 1973: Karyological study of the Balkan endemics *Viola delphinantha* Boiss. and *Galium kerneri* Deg. - *Dokl. Bulg. Akad. Nauk* **26(6)**: 827-828.

MARKOVA, M.L. & ČERNEVA, Ž.M. 1982: [Chromosome numbers from Bulgaria]. In Löve, Á. (ed.): Chromosome number reports LXXVII. - *Taxon* **31(4)**: 777.

MARKOVA, M.L. & ČERNEVA, Ž.M. 1984: Karyological studies on two new species to the Bulgarian flora [In Bulgarian with English abstract]. - *Fitologiya* **27**: 35-39.

MARKOVA, M.L. & GORANOVA, V.Y. 1993: Karyological study of representatives of family Malvaceae in Bulgaria. - *Fitologiya* **46**: 12-32.

MARKOVA, M.L. & GORANOVA, V.Y. 1994a: Karyological study of the representatives of genus *Potentilla* L. in Bulgaria. - *God. Sofiisk. Univ. "Kliment. Ohridski" Biol. Fak., 2 Bot.* **85**: 199-204.

MARKOVA, M.L. & GORANOVA, V.Y. 1994b: Reports (240-266). In Kamari, G., Felber, F. & Garbari, F. (eds): Mediterranean chromosome number reports - 4. - *Fl. Medit.* **4**: 233-254.

MARKOVA, M.L. & GORANOVA, V.Y. 1996: Reports (435-473). In Kamari, G., Felber, F. & Garbari, F. (eds): Mediterranean chromosome number reports - 5. - *Fl. Medit.* **5**: 289-317.

MARKOVA, M.L. & GORANOVA, V.Y. 1997: Reports (757-761). In Kamari, G., Felber, F. & Garbari, F. (eds): Mediterranean chromosome number reports - 6. - *Fl. Medit.* **6**: 317-323.

MARKOVA, M.L. & IVANOVA, P.S. 1970: Karyologische Untersuchung der Vertreter der Fam. Borraginaceae, Labiatae und Scrophulariaceae in Bulgarien. - *Izv. Bot. Inst. (Sofia)* **20**: 93-98.

MARKOVA, M.L. & IVANOVA, P.S. 1971a: Karyologische Untersuchungen der Vertreter der Fam. Borraginaceae, Labiatae und Scrophulariaceae in Bulgarien. II. - *Izv. Bot. Inst. (Sofia)* **21**: 123-131.

MARKOVA, M.L. & IVANOVA, P.S. 1971b: [Chromosome numbers from Bulgaria]. In Löve, Á. (ed.): IOPB chromosome number reports XXXIII. - *Taxon* **20(4)**: 611.

MARKOVA, M.L. & IVANOVA, P.S. 1972: Karyologische Untersuchung der Vertreter der Fam. Boraginaceae, Labiatae und Scrophulariaceae in Bulgarien. III. - *Izv. Bot. Inst. (Sofia)* **22**: 149-157.

MARKOVA, M.L. & IVANOVA, P.S. 1973: [Chromosome numbers from Bulgaria]. In Löve, Á. (ed.): IOPB chromosome number reports XL. - *Taxon* **22(2-3)**: 289-290.

MARKOVA, M.L. & IVANOVA, P.S. 1974: [Chromosome numbers in Boraginaceae and Lamiaceae]. In Löve, Á. (ed.): IOPB chromosome number reports XLIV. - *Taxon* **23(2-3)**: 378.

MARKOVA, M.L. & IVANOVA, P.S. 1975: Karyologische Untersuchung der Vertreter der Familie Boraginaceae, Labiatae und Scrophulariaceae. IV. In: Festschrift in honour of Acad. Daki Jordanov, pp. 61-67. - Sofia: Publishing House of the Bulgarian Academy of Sciences.

MARKOVA, M.L. & IVANOVA, P.S. 1982a: Karyological study in the genus *Salvia* L. in Bulgaria. I [In Bulgarian with English abstract]. - *Fitologiya* **19**: 24-42.

MARKOVA, M.L. & IVANOVA, P.S. 1982b: Karyological study of the genus *Salvia* L. in Bulgaria. II [In Bulgarian with English summary]. - *Fitologiya* **20**: 3-19.

MARKOVA, M.L., RADENKOVA, J. & IVANOVA, P.S. 1972: [Chromosome numbers in *Ornithogalum* from Bulgaria]. In Löve, Á. (ed.): IOPB chromosome number reports XXXVI. - *Taxon* **21(2-3)**: 339-340.

MARKOVA, M.L. & ROBEVA, P.N. 1972: [Chromosome numbers in *Seseli* from Bulgaria]. In Löve, Á. (ed.): IOPB chromosome number reports XXXVI. - *Taxon* **21(2-3)**: 340.

MARKOVA, M.L. & ROBEVA, P.N. 1973: Karyological study of three Bulgarian taxa of the genus *Seseli* L. - *Izv. Bot. Inst. (Sofia)* **24**: 199-213.

MARKOVA, M.L. & THI THU, N. 1974: [Chromosome numbers in Lamiaceae from Bulgaria]. In Löve, Á. (ed.): IOPB chromosome number reports XLIII. - *Taxon* **23(1)**: 193-194.

MAROULIS, G. 2003: Flora and vegetation of Mount Erimanthos ecosystems. [In Greek with English summary]. - Patras: Ph.D. thesis, Univ. of Patras, iii + 405 + xiv pp.

MAROULIS, G. & ARTELARI, R. 2001: New records for the flora of Mount Erimanthos (NW Peloponnisos, Greece). - *Fl. Medit.* **11**: 311-331.

MAROULIS, G. & ARTELARI, R. 2005: A floristic report from Mount Erimanthos (NW-Peloponnisos, Greece). - *Fl. Medit.* **15**: 109-120.

MAROULIS, G., ARTELARI, R. & GEORGIADIS, TH. 2002: Ecological - synecological study of five important taxa from the flora of Mt Erimanthos (NW Peloponnisos). In Kamari, G. & al. (eds): Ellinini Votaniki Eteria. 9° Panelliniko Epistimoniko Sinedrio. Praktika, pp. 189-197 [In Greek with English abstract].

MARSCHALL V. BIEBERSTEIN, F.A. 1808-19: Flora Taurico-Caucasica exhibens stirpes phaenerogamas, in chersoneso taurica et regionibus caucasicis sponte crescentes ... - Charkouiae (Charkow): Typis academicis, 3 vols. [cf. TL-2 3: 306, 1981].

MARSDEN-JONES, E.M. & TURRILL, W.B. 1933: Studies in variation of *Anthyllis vulneraria*. - *J. Genet.* **27 (2)**: 261-285 + 4 plates.

MARSDEN-JONES, E.M. & WEISS, F.E. 1938: The essential differences between *Anagallis arvensis* Linn. and *Anagallis femina* Mill. - *Proc. Linn. Soc. London* **150(3)**: 146-155.

MARSDEN-JONES, E.M. & WEISS, F.E. 1960: The genetics and pollination of *Anagallis arvensis* subsp. *arvensis* and *Anagallis arvensis* subsp. *foemina*. - *Proc. Linn. Soc. London* **171(1-2)**: 27-29.

MARSH, A.S. 1914: *Azolla* in Britain and in Europe. - *J. Bot. (London)* **52**: 209-213.

MARTELLI, U. 1888: Sulla *Quercus macedonica*. - *Nuovo Giorn. Bot. Ital.* **20(3)**: 427-431.

MARTELLOS, S. & AL. 2020: FlorItaly - the portal to the Flora of Italy. - *PhytoKeys* **156**: 55-71.

MARTÍN CIUDAD, A. & AL. 1991: Números cromosomáticos de plantas vasculares ibéricas. - Madrid: Real Jardín Botánico (Archivos de Flora Iberica, 1), v + 202 pp.

MARTÍN-BLANCO, C.J. 1993: Sinopsis del género *Valerianella* Miller (Valerianaceae) en la Península Ibérica. - *Bot. Complut.* **18**: 151-156.

MARTÍN-BRAVO, S. 2011: *Reseda inodora*. In Greuter, W. & Raus, Th. (eds): Med-Checklist notulae, 30. - *Willdenowia* **41(2)**: 322.

MARTÍN-BRAVO, S. & JIMÉNEZ-MEJÍAS, P. 2013: *Reseda minoica* (Resedaceae), a new species from the eastern Mediterranean region. - *Ann. Bot. Fenn.* **50(1-2)**: 55-60.

MARTIN-NOUGET, D. 1969: Contribution à l'étude cytotaxinomique des *Achillea* et des *Pyrethrum* de Turquie. - *Bull. Soc. Bot. France* **116**: 197-206.

MARTIN, A. & MATHEZ, J. 1991: Polymorphisme et taxinomie chez les Valerianaceae: quelques indications sur les valerianelles proches de *Valerianella coronata*. - *Naturalia Monspel., Sér. Bot.* **55**: 61-75.

MARTIN, B. 2003: *Ostrya carpinifolia* Scop. - *Biocosme Mésogéen* **19(4)**: front cover.

MARTIN, E. & AL. 2010: Karyology of ten Turkish *Trigonella* L. (Leguminosae) species from section *Cylindricae* Boiss. - *Turk. J. Bot.* **34(6)**: 485-494.

MARTIN, E. & AL. 2011: Karyological studies of *Clinopodium* L. (sect. *Pesudomelissa*) and *Micromeria* Benth. s.str. (Lamiaceae) from Turkey. - *Caryologia* **64(4)**: 398-404.

MARTINČIĆ, A. 1973: Die Gattung *Pilularia* L. in Flora Jugoslawiens [In Slovenian with German summary]. - *Biol. Vestn.* **21(1)**: 63-66.

MARTINČIĆ, A. 1990: A contribution to the knowledge of *Alchemilla* genus in Yugoslavia [In Slovenian with English summary]. - *Biol. Vestn.* **38(3)**: 23-28.

MARTÍNEZ-AZORÍN, M. & CRESPO, M.B. 2016: (2483) Proposal to conserve the name *Scilla* (Hyacinthaceae) with a conserved type. - *Taxon* **65(6)**: 1427-1428. [approved, cf. Taxon 68(4): 848, 2019].

MARTÍNEZ-AZORÍN, M., CRESPO, M.B. & JUAN, A. 2009: Nomenclature and taxonomy of *Ornithogalum divergens* Boreau (Hyacinthaceae) and related taxa of the polyploid complex of *Ornithogalum umbellatum*. - *Candollea* **64(2)**: 163-169.

MARTÍNEZ-AZORÍN, M., CRESPO, M.B. & JUAN, A. 2010: Taxonomic revision of *Ornithogalum* subg. *Ornithogalum* (Hyacinthaceae) in the Iberian Peninsula and Balearic Islands. - *Pl. Syst. Evol.* **289(2)**: 181-211.

MARTÍNEZ-AZORÍN, M., WITTMANN, H. & CRESPO, M.B. 2012: (2080) Proposal to conserve the name *Ornithogalum narbonense* (Hyacinthaceae, Ornithogaloideae) with a conserved type. - *Taxon* **61(4)**: 881. [approved, cf. Taxon 62(6): 1320, 2013].

MARTÍNEZ-FLORES, F. & AL. 2016: Lectotype designation for seven species names in the *Daucus guttatus* complex (Apiaceae) from the central and eastern Mediterranean basin. - *Syst. Bot.* **41(2)**: 465-478.

MARTÍNEZ-LABORDE, J.B. 1988: El género *Diplotaxis* (Cruciferae) en España. - *Lagascalia* **15(Extra)**: 243-248.

MARTÍNEZ-LABORDE, J.B. 1991: Notes on the taxonomy of *Diplotaxis* DC. (Brassicaceae). - *Bot. J. Linn. Soc.* **106(1)**: 67-71.

MARTÍNEZ-ORTEGA, M.M. & AL. 2002: Revised typification of *Veronica agrestis* L. (Scrophulariaceae). - *Taxon* **51(4)**: 763-764.

MARTÍNEZ-ORTEGA, M.M., CAFFERTY, S. & RICO, E. 2001: Typification of some specific names in *Veronica* (Scrophulariaceae) described by C. Linnaeus and N.J. Jacquin. - *Taxon* **50(2)**: 551-556.

MARTÍNEZ-ORTEGA, M.M., HERRERO, A. & MUÑOZ-CENTENO, L.M. 2006: (1727) Proposal to reject the name *Veronica latifolia* (Scrophulariaceae). - *Taxon* **55(2)**: 538-539. [approved, cf. Taxon 56(4): 1293, 2007].

MARTÍNEZ-ORTEGA, M.M. & RICO, E. 2001a: Taxonomy of *Veronica* subsect. *Serpyllifoliae* (Scrophulariaceae). - *Bot. J. Linn. Soc.* **135(2)**: 179-194.

MARTÍNEZ-ORTEGA, M.M. & RICO, E. 2001b: Typification of 10 specific names in *Veronica* L. (Scrophulariaceae, Veroniceae). - *Taxon* **50(2)**: 543-549.

MARTÍNEZ-ORTEGA, M.M., RICO, E. & FISCHER, M.A. 2001: Typification of the names *Veronica jacquinii* Baumg. and *Veronica dentata* F.W. Schmidt (Scrophulariaceae). - *Taxon* **50(1)**: 187-189.

MARTÍNEZ, M. & AL. 1998: Lectotypification and epitypification of the name *Veronica multifida* (Scrophulariaceae). - *Taxon* **47(2)**: 437-439.

MARTINI, F. 1990: New localities of *Ballota acetabulosa* (L.) Bentham in Yugoslavia. - *Acta Bot. Croat.* **49**: 101-105.

MARTINI, F. & PAIERO, P. 1988: I salici d'Italia. Guida al riconoscimento e all'utilizzazione pratica [2nd ed.]. - Trieste: Edizioni Lint, 161 pp.

MARTINI, F. & POLDINI, L. 1990: *Daphne blagayana* Freyer (Thymelaeaceae), nuova per la flora d'Italia. - *Webbia* **44(2)**: 295-306.

MARTINIS, A. & AL. 2018a: Spatial analysis of orchids diversity unveils hot-spots: The case of Zante Island. - *J. Agric. Informatics* **1**: 26-40.

MARTINIS, A. & AL. 2018b: Ecological assessment of *Juniperus turbinata* Guss. forest on the Strofades Islands, Ionian Sea, Greece. - *J. Forest Sci.* **64**: 345-352.

MARTINIS, Z. 1973: Taxa nova et combinationes novae generis *Helleborus* L. in Flora Jugoslaviae. In Trinajstić, I. (ed.): Suppl. Fl. Anal. Ital. 1: 13-16. - Zagreb.

MARTINOLI, G. 1949a: Cariologia comparata del *Pancratium maritimum* L. e del *Pancratium illyricum* L., in rapporto alla validità del genere *Halmyra*. - *Caryologia* **1(2)**: 122-130.

MARTINOLI, G. 1949b: Ricerche citotassonomiche sui generi *Urginea* e *Scilla* della flora sarda. - *Caryologia* **1(3)**: 329-357 + 1 plate.

MARTINOLI, G. 1950: *Satureja thymbra* L., elemento mediterraneo-orientale della Sardegna. - *Nuovo Giorn. Bot. Ital., n.s.* **56(4)**: 576-592.

MARTINOLI, G. 1951: Studio cariologico sul genere *Ruscus* (Asparagaceae). - *Caryologia* **4(1)**: 86-97.

MARTINOLI, G. 1953: Studio citotassonomico dei generi *Hyoseris* e *Robertia* con particolare riferimento all'*Hyoseris taurina* G. Martinoli sp. nov. (Asteraceae). - *Caryologia* **5(3)**: 253-281 + 3 plates.

MARTINOLI, G. 1955a: Cariologi di alcune specie del genere *Allium* (Liliaceae) delle Sardegna. - *Caryologia* **7(1)**: 145-156 + 1 plate.

MARTINOLI, G. 1955b: Citotassonomia di alcune specie del genere *Avena* della Sardegna. - *Caryologia* **7(1)**: 191-204 + 1 plate.

MARTINOLI, G. 1959: Tassonomia ed ecologia delle specie del genere *Anagallis* della Sardegna. - *Webbia* **15(1)**: 1-45 + 6 plates.

MARTINOLI, G. 1969: *Poterium spinosum* L. Elemento del mediterraneo orientale a Calamosca (Capo S. Elia, Sardegna). - *Giorn. Bot. Ital., n.s.* **103(4)**: 325-340.

MARTINOVSKÝ, J.O. 1967a: Neue submediterrane *Stipa*-Arten und die taxonomische Einteilung der Federgrassippen der Serie *Pulcherrimae* Martinovský. XI. Beitrag zur Kenntnis der Federgräser der Sektion *Pennatae*. - *Preslia* **39**: 260-275.

MARTINOVSKÝ, J.O. 1967b: Zur Taxonomie und Verbreitung von *Stipa paradoxa* (Junge) P. Smirnov und *Stipa crassiculmis* P. Smirnov. - *Bot. Jahrb. Syst.* **87(4)**: 383-398.

MARTINOVSKÝ, J.O. 1970: Über drei neue *Stipa*-Sippen aus dem Verwandtschaftskreis *Stipa joannis* s.l. XXII. Beitrag zur Kenntnis der *Stipa*-Sippen. - *Oesterr. Bot. Z.* **118(1-2)**: 171-181.

MARTINOVSKÝ, J.O. 1971: *Stipa mayeri* Martinovský, eine neue balkanische Federgrassippe. XXIII. Beitrag zur Kenntnis der *Stipa*-Gattung. - *Acta Bot. Croat.* **30**: 141-146.

MARTINOVSKÝ, J.O. 1972: Studien über einige submediterrane Federgrassippen. XXIV. Beitrag zur Kenntnis der Gattung *Stipa*. - *Preslia* **44(1)**: 7-23.

MARTINOVSKÝ, J.O. 1975: Taxonomische Studie über die *Stipa*-Serie *Dasyphyllae*. XXV. Beitrag zur Kenntnis der Gattung *Stipa*. - *Preslia* **47(3)**: 249-261.

MARTINOVSKÝ, J.O. & MORALDO, B. 1980: Studia taxonomica ad taxa sectionis *Stipa* in regione mediterranea atque submediterranea occurentia. - *Preslia* **52(1)**: 13-34.

MARTINOVSKÝ, J.O. & SKALICKÝ, V. 1969: Zur Nomenklatur einiger *Stipa*-Sippen der *Pennatae*-Gruppe. XVI. Beitrag zur Kenntnis der europäischen Federgrassippen. - *Preslia* **41**: 327-341.

MARTINS, L. 2006: Systematics and biogeography of *Klasea* (Asteraceae-Cardueae) and a synopsis of the genus. - *Bot. J. Linn. Soc.* **152(4)**: 435-464.

MÁRTONFI, P. 1997: Nomenclatural survey of the genus *Thymus* sect. *Serpyllum* in the Carpathians and Pannonia. - *Thaiszia* **7(2)**: 111-181.

MASCI, S.M., MARCHI, P. & VISONÀ, L. 1987: Numeri cromosomici per la Flora Italiana: 1098-1105. - *Inform. Bot. Ital.* **19(2)**: 167-172.

MASCIA, F. & AL. 2013: *Arundo micrantha*, a new reed species for Italy, threatened in the freshwater habitat by the congeneric invasive *A. donax*. - *Pl. Biosyst.* **147(3)**: 717-729.

MASCLANS, F. 1968: El género *Coris* en la région mediterránea occidental. - *Collect. Bot. (Barcelona)* **7(2)**: 749-758.

MASSIAS, J. 1919: Les forêts de Grèce. - *Rev. Eaux Forêts* **57**: 237-247.

MAST DE MAEGHT, J. & AL. 2005: *Ophrys polyxo*, a scolopaxoid *Ophrys* from Rhodes. - *Naturalistes Belges* **86**: 147-156.

MASTENBROEK, O. & AL. 1984: Corresponding patterns of geographic variation among populations of *Silene latifolia* (= *S. alba* = *S. pratensis*) (Caryophyllaceae). - *Pl. Syst. Evol.* **145(3-4)**: 227-242.

MASTERS, M.T. 1904: A general view of the genus *Pinus*. - *J. Linn. Soc., Bot.* **35(248)**: 560-659 + 4 plates.

MASTROGIANNI, A. & AL. 2019: *Staphylea pinnata*. In Raab-Straube, E. v. & Raus, Th. (eds): Euro+Med-Checklist notulae, 11. - *Willdenowia* **49(3)**: 440.

MATEU, I. 1986: Revisión del género *Phlomis* L. (Labiatae) en la Península Ibérica e Islas Baleares. - *Acta Bot. Malac.* **11**: 177-204.

MATEVSKI, V. 1983: Beitrag zur Kenntnis einiger Taxa der Gattung *Thymus* L. (Labiatae) im Becken von Skopje [In Macedonian with German summary]. - *Godišen Zborn. Biol. Fak. Univ. "Kiril i Metodij"* **36**: 135-147.

MATEVSKI, V. 1985: *Thymus balcanicus* Borbas in der Flora der SR Makedonien [In Macedonian with German summary]. - *Maked. Akad. Nauk. Umetn., Oddel. Biol. Med. Nauki* **6(1-2)**: 41-49.

MATEVSKI, V. 1987: *Thymus tosevii* Vel. - Komplex in der Flora der SR Makedonien [In Macedonian with German summary]. - *Godišen Zborn. Biol. Fak. Univ. "Kiril i Metodij"* **39-40**: 215-240.

MATEVSKI, V. 1989: Taxonomie und Chorologie von *Thymus albanus* H. Br. in der Flora Jugoslawiens [In Serbo-Croatian with German abstract]. In: Minerali, stijene, izumrli i živi svijet BiH, pp. 291-299. - Sarajevo: Zemaljski Muzej Bosne i Hercegovine.

MATEVSKI, V. 1990: *Thymus rohlenae* Vel. und *Thymus grisebachii* Ronn. in der Flora der SR Makedonien [In Macedonian with German abstract]. - *Maked. Akad. Nauk. Umetn., Oddel. Biol. Med. Nauki* **8(1-2)**: 57-64.

MATEVSKI, V. 1991: Beitrag zur Taxonomie und Chorologie einiger Arten der Gattung *Thymus* L. Sect. *Marginati* (A. Kerner) A. Kerner Subsect. *Marginati* in der Flora Makedoniens [In Macedonian with German abstract]. - *Maked. Akad. Nauk. Umetn., Oddel. Biol. Med. Nauki* **9(1-2)**: 51-63.

MATEVSKI, V. 1995: The flora of Dobra Voda mountain [In Macedonian with English summary]. - Skopje: Univ. "Kiril i Metodij", 46 pp.

MATEVSKI, V. 2009: Floristički i vegetaciski istražuvanja na teritorijata na Makedonija (1839-1945) [Floristic and vegetational investigations on the territory of Macedonia (1839–1945)]. - Pp. 315-342 in: Grozdanov, C., Tomovski, K. & Topolinjska, Z. (eds), Otkrivanjeto i proučuvanjeto na Makedonija vo Evropskata nauka do formiranjeto na Makedonskite državni instituciji [Discovering and exploring Macedonia in European science until the establishment of Macedonian state institutions]. - Skopje: Makedonska Akademija na naukite i umetnostite [Macedonian Academy of Sciences and Arts].

MATEVSKI, V. 2018: Hochgebirgs-(Sub-)endemiten in der Flora der Republik Makedonien (unter besonderer Berücksichtigung des Galičica-Gebirges). In: Wissenschaftliche Konferenz Ohrid 20-21 September 2015, Auseinandersetzungen Band 4 pp. 249-303. - Skopje.

MATEVSKI, V. & AL. 2008: Flora and vegetation of the Macedonian steppe. - Ljubljana: ZRC SAZU, Biološki inštitut Jovana Hadžija, 99 pp.

MATEVSKI, V. & (ED.) 2007: Collection of papers devoted to Academician Kiril Micevski on the occassion [sic!] of the 80 years of his bearth [sic!]. - Skopje: Makedoniska Akademija Naukite Umetnostite, 346 pp.

MATEVSKI, V. & KOSTADINOVSKI, M. 1990: *Galinsoga ciliata* (Rafin.) Blake. Neue Art in der Adventivflora Makedoniens [In Macedonian with German summary]. - *Godišen Zborn. Biol. Fak. Univ. "Kiril i Metodij"* **41-42**: 191-195.

MATEVSKI, V. & KOSTADINOVSKI, M. 1996: Review of plant species with locus classicus within the area of three national parks in the Republic of Macedonia. In: National parks and their role in biodiversity protection on Balkan Peninsula, pp. 89-98 [In Macedonian with English abstract].

MATEVSKI, V. & KOSTADINOVSKI, M. 1997a: Beitrag zur Flora von Republik Makedonien II [In Macedonian with German summary]. - *Godišen Zborn. Biol. Fak. Univ. "Kiril i Metodij"* **50**: 25-39.

MATEVSKI, V. & KOSTADINOVSKI, M. 1997b: *Hymenocarpos circinnatus* (L.) Savi, a new species of the flora of Republic of Macedonia. In: Zbornik radova sa V Simpozijuma o flori jugoistočne Srbije, Zaječar, pp. 41-44.

MATEVSKI, V. & KOSTADINOVSKI, M. 1999: Contribution of the flora of the Republic of Macedonia. III [In Macedonian with English abstract]. - *Godišen Zborn. Biol. Fak. Univ. "Kiril i Metodij"* **52**: 102-108.

MATEVSKI, V., MICEVSKI, K. & KOSTADINOVSKI, M. 1992: Beitrag zur Flora von Republik Makedonien. - *Godišen Zborn. Biol. Fak. Univ. "Kiril i Metodij"* **45**: 167-174.

MATEVSKI, V., MICEVSKI, K. & SEKOVSKI, Ž. 1990: *Thymus karadzicensis* Matevski & Micevski spec. nov. in der Flora von Makedonien. - *Razprave IV. Razreda Sazu* **31(11)**: 169-178.

MATEVSKI, V., SEKOVSKI, Ž. & KOSTADINOVSKI, M. 1993: Seltene und weniger bekannte Pflanzen in der Flora der Republik Makedonien [In Macedonian with German summary]. - *Godišen Zborn. Biol. Fak. Univ. "Kiril i Metodij"* **46**: 115-118.

MATEVSKI, V., SEKOVSKI, Ž. & KOSTADINOVSKI, M. 1994a: Beitrag zur Flora von Republik Makedonien [In Macedonian with German summary]. - *Godišen Zborn. Biol. Fak. Univ. "Kiril i Metodij"* **47**: 81-89.

MATEVSKI, V., SEKOVSKI, Ž. & KOSTADINOVSKI, M. 1994b: A new horology data for *Silene paeoniensis* Bornm. in the flora of the Republic of Macedonia [In Macedonian with English summary]. - *Ekol. Zašt. Život. Sred. (Skopje)* **2(1)**: 17-19.

MATEVSKI, V., SEKOVSKI, Ž. & KOSTADINOVSKI, M. 1995: Beitrag zur Chorologie von *Sideritis lanata* (L.) Boiss. und *Sideritis raeseri* Boiss. & Heldr. in der Flora von Republik Makedonien [In Macedonian with German abstract]. - *Godišen Zborn. Biol. Fak. Univ. "Kiril i Metodij"* **48**: 43-48.

MATEVSKI, V., SEKOVSKI, Ž. & KOSTADINOVSKI, M. 1996: Contribution to the chorology and cytotaxonomy of *Ranunculus cacuminis* Strid et Papanicolaou of the flora of Republic of Macedonia. In: Abstracts, Fifth international congress of systematic and evolutionary biology, p. 269.

MÁTHÉ, I. & MÁTHÉ, I. JR. 1972: Contribution to variability of *Solanum dulcamara* L. [In Hungarian with English summary]. - *Bot. Közlem.* **59(2)**: 129-134.

MATHEW, B.F. 1967: A gardener's guide to Hellebores. - *Quart. Bull. Alpine Gard. Soc. Gr. Brit.* **35(1)**: 1-32.

MATHEW, B.F. 1973: *Crocus olivieri*. - *Bot. Mag.* **179(3)**: tab. 639 + 5 pp. text.

MATHEW, B.F. 1974: Ornithogalums for the garden. - *Lilies* **1973/74**: 85-87.

MATHEW, B.F. 1975a: Greek crocuses. - *Quart. Bull. Alpine Gard. Soc. Gr. Brit.* **43(1)**: 47-55.

MATHEW, B.F. 1975b: Plant awards 1974-75: *Fritillaria davisii* Turrill P.C. - *Quart. Bull. Alpine Gard. Soc. Gr. Brit.* **43(4)**: 314.

MATHEW, B.F. 1976a: *Crocus cvijicii*. - *Bot. Mag.* **181(2)**: tab. 721 + pp. 73-75.

MATHEW, B.F. 1976b: *Crocus olivieri* and its allies (Iridaceae). - *Kew Bull.* **31(2)**: 201-208 + 3 plates.

MATHEW, B.F. 1977a: *Crocus sativus* and its allies (Iridaceae). - *Pl. Syst. Evol.* **128(1-2)**: 89-103.

MATHEW, B.F. 1977b: *Crocus goulimyi* Turrill. - *Quart. Bull. Alpine Gard. Soc. Gr. Brit.* **45(4)**: 321-322.

MATHEW, B.F. 1978: The larger bulbs. - London: B.T. Batsford & The Royal Horticultural Society, 156 pp. + plates.

MATHEW, B.F. 1980: Plant awards 1979-80: *Crocus scardicus* Košanin. - *Quart. Bull. Alpine Gard. Soc. Gr. Brit.* **48(4)**: 299.

MATHEW, B.F. 1981: The Iris. - London: B.T. Batsford Ltd, 202 pp. + 32 plates.

MATHEW, B.F. 1982a: The Crocus. A revision of the genus *Crocus* (Iridacae). - London: B.T. Batsford Ltd, 127 pp. + 96 plates.

MATHEW, B.F. 1982b: Plant awards, 1981-1982. *Biarum davisii* Turrill. - *Quart. Bull. Alpine Gard. Soc. Gr. Brit.* **50(4)**: 288-289, 324.

MATHEW, B.F. 1982c: Plant awards, 1981-1982. *Crocus biflorus* subsp. *melantherus* (Boiss. & Orph. ex) Mathew. - *Quart. Bull. Alpine Gard. Soc. Gr. Brit.* **50(4)**: 289-290.

MATHEW, B.F. 1982d: Plant awards, 1981-1982. *Crocus oreocreticus* B.L. Burtt. - *Quart. Bull. Alpine Gard. Soc. Gr. Brit.* **50(4)**: 314.

MATHEW, B.F. 1982e: Plant awards, 1981-1982. *Cyclamen graecum* Link forma *album* R. & E. Frank P.C. - *Quart. Bull. Alpine Gard. Soc. Gr. Brit.* **50(4)**: 327.

MATHEW, B.F. 1983a: The Greek species of *Crocus* (Iridaceae), a taxonomic survey. - *Ann. Mus. Goulandris* **6**: 63-86.

MATHEW, B.F. 1983b: A review of the genus *Sternbergia*. - *Plantsman* **5(1)**: 1-16 + 1 plate.

MATHEW, B.F. 1984: Rare and little-known crocuses in cultivation. - *Bot. Mag. (Kew Mag.)* **1(2)**: 68-74.

MATHEW, B.F. 1985: *Crocus hadriaticus* and its variants. - *Bot. Mag. (Kew Mag.)* **2(3)**: 309-311.

MATHEW, B.F. 1987: The smaller bulbs. - London: B.T. Batsford Ltd., 190 pp. + 32 plates.

MATHEW, B.F. 1989a: A review of *Veratrum*. - *Plantsman* **11(1)**: 35-61.

MATHEW, B.F. 1989b: Hellebores. - St. John's Woking, Surrey: Alpine Garden Society, ix + 180 pp.

MATHEW, B.F. 1992: A taxonomic and horticultural review of *Erythronium* L. (Liliaceae). - *Bot. J. Linn. Soc.* **109(4)**: 453-471.

MATHEW, B.F. 1993: A new *Crocus* variant. - *Quart. Bull. Alpine Gard. Soc. Gr. Brit.* **61(3)**: 314-318.

MATHEW, B.F. 1994: *Bellevalia hyacinthoides*. - *Bot. Mag. (Kew Mag.)* **11(3)**: tab. 252 + pp. 128-131.

MATHEW, B.F. 1995a: A new variant of *Crocus goulimyi* (Iridaceae). - *Ann. Mus. Goulandris* **9**: 165-168.

MATHEW, B.F. 1995b: An interesting new autumn-flowering *Crocus* from Turkey. - *New Plantsman* **2 (3)**: 182-184.

MATHEW, B.F. 1996: A review of *Allium* section *Allium*. - Kew: Royal Botanic Gardens, ix + 176 pp. + 16 plates.

MATHEW, B.F. 1998: The Lily group *Fritillaria* exhibit. - *Bull. R.H.S. Lily Group* **1997/98**: 41-44.

MATHEW, B.F. 1999: Botany, taxonomy and cytology of *C. sativus* L. and its allies. In Negbi, M. (ed.): Saffron, *Crocus sativus* L., pp. 19-34. - Amsterdam: Harwood Academic Publishers.

MATHEW, B.F. 2000a: Peloponnisos (Greece) - A *Crocus* paradise. - *Bot. Chron. (Patras)* **13**: 255-265.

MATHEW, B.F. 2000b: BN editor makes nomenclatural slip in *Crocus*. - *Bulbs* **32**: 7.

MATHEW, B.F. 2002a: Classification of the genus *Narcissus*, In Hanks, G.H. (ed.): Narcissus and daffodil. The genus *Narcissus*, pp. 30-52. - London and New York: Taylor & Francis, xix + 428 pp.

MATHEW, B.F. 2002b: *Crocus* up-date. - *Plantsman, n.s.* **1(1)**: 44-56.

MATHEW, B.F. 2002c: *Romulea*. - *Alpine Gardener* **70(1)**: 86-94.

MATHEW, B.F. 2002d: Some aspects of the 'Juno group' of irises. - *Ann. Bot., n.s. (Rome)* **1(2)**: 113-122.

MATHEW, B.F. 2005: Hardy Hyacinthaceae. Part 1: *Muscari*. - *Plantsman, n.s.* **4(1)**: 40-53.

MATHEW, B.F. & BRIGHTON, C.A. 1975a: Studies in the genus *Crocus* L. In Jordanov, D. & al. (eds): Problems of Balkan flora and vegetation, pp. 209-211. - Sofia: Publishing House of the Bulgarian Academy of Sciences.

MATHEW, B.F. & BRIGHTON, C.A. 1975b: Studies in the genus *Crocus*. In Walters, S.M. & King, C.J. (eds): European floristic and taxonomic studies, pp. 76-81. - Oxford: E.W. Classey Ltd.

MATHEW, B.F. & BRIGHTON, C.A. 1977: *Crocus tournefortei* and its allies (Iridaceae). - *Kew Bull.* **31(4)**: 775-784 + 2 plates.

MATHEW, B.F., BRIGHTON, C.A. & BAYTOP, T. 1979: Taxonomic and cytological notes on Asiatic *Crocus*. - *Notes Roy. Bot. Gard. Edinburgh* **37(3)**: 469-473.

MATHEW, B.F. & ÖZHATAY, N. 2001: The *Cyclamen* of Turkey. A guide to the species of *Cyclamen* growing in Turkey. - London: The Cyclamen Society, 33 pp.

MATHEW, B.F. (ED.) 2013: Genus *Cyclamen* in science, cultivation, art and culture. - Kew: Royal Botanic Gardens, 574 pp.

MATHEZ, J. 1975: Contribution de la palynomorphologie à la connaissance des Cynoglossées du Bassin Méditerranéen occidental. In Guinochet, M. & al. (eds): La flore du Bassin Méditerranéen, pp. 279-287. - Paris: C.N.R.S.

MATHEZ, J. 1984: Introduction à une révision du genre *Fedia* Gaertn. emend. Moench. - *Mem. Soc. Brot.* **27**: 129-175 + 4 plates.

MATHEZ, J. & XENA DE ENRECH, N. 1985a: Le polymorphisme génétique de la morphologie des fruits du genre *Fedia* Gaertn. (Valerianaceae). I. Détermination du mécanisme de contrôle génétique chez les espèces *F. cornucopiae* & *F. graciliflora*. - *Candollea* **40(2)**: 425-434.

MATHEZ, J. & XENA DE ENRECH, N. 1985b: Heterocarpy, fruit polymorphism and discriminating dissemination in the genus *Fedia* (Valerianaceae). In Jacquard, P. & al. (eds): Genetic differentiation and dispersal in plants, pp. 431-441. - Berlin & Heidelberg: Springer-Verlag.

MATSOUKAS, P. 1990: Plant list of Mount Vounasa [unpubl.]. - Photocopy: 3 pp.

MATTATIA, J. & FEINBRUN-DOTHAN, N. 1987: The nomenclature of *Aegilops biuncialis* Vis. - *Israel J. Bot.* **35(1)**: 53-54.

MATTEI, G.E. 1911: Un Trifoglio nuovo per l'Europa. - *Malpighia* **24(3)**: 241-257.

MATTEI, G.E. & LOJACONO POJERO, M. 1908: Contribuzione alla flora dell'isola di Tenos (Cicladi). - *Boll. Reale Orto Bot. Palermo* **7(1-3)**: 70-72.

MATTES, H. & LIENAU, C. (EDS) 1992: Colloquium Nordost-Griechenland. 1. Beiträge eines Colloquiums am Institut für Geographie (1.2. 1992). - Münster: Inst. Geographie, iv + 72 pp.

MATTES, H. & LIENAU, C. (EDS) 1993: Colloquium Nordost-Griechenland. 2. Beiträge eines Colloquiums im Haus Rothenberge (26.-28.11. 1992). - Münster: Inst. Geographie, iv + 86 pp.

MATTES, H. & LIENAU, C. (EDS) 1995: Colloquium Nordost-Griechenland. 3. Beiträge eines Colloquiums im Haus Rothenberge (25. u. 26.2. 1994). - Münster: Inst. Geographie, 121 pp.

MATTES, H. & LIENAU, C. (EDS) 1996: Das Aladjagiola im Nestosdelta in Nordost-Griechenland. Beiträge zur Fauna, Landnutzung und Naturschutz. - *Ber. Arbeitsgeb. Entwicklungsforsch. (Münster)* **25**: ii + 139 pp. + Beilage.

MATTFELD, J. 1921: Enumeratio specierum generis *Minuartia* (L.) emend. Hiern. - *Bot. Jahrb. Syst.* **57 Beibl. 126**: 27-33.

MATTFELD, J. 1922a: Beitrag zur Kenntnis der systematischen Gliederung und geographischen Verbreitung der Gattung *Minuartia*. - *Bot. Jahrb. Syst.* **57 Beibl. 127**: 13-63.

MATTFELD, J. 1922b: Geographisch-genetische Untersuchungen über die Gattung *Minuartia* (L.) Hiern. - *Repert. Spec. Nov. Regni Veg. Beih.* **15**: 1-228 + 5 maps.

MATTFELD, J. 1925a: Die in Europa und dem Mittelmeergebiet wildwachsenden Tannen. - *Mitt. Deutsch. Dendrol. Ges.* **35**: 1-37 + 10 plates.

MATTFELD, J. 1925b: Zur Kenntnis der Formenkreise der europäischen und kleinasiatischen Tannen. - *Notizbl. Bot. Gart. Berlin-Dahlem* **9(84)**: 229-246.

MATTFELD, J. 1925c: [Bulgarian itinerary, 1924]. In Markgraf, F. & Arndt, A.: Tagesordnung der Sitzungen im Vereinsjahr 1924/25. Sitzung am 15. Mai 1925. - *Verh. Bot. Vereins Prov. Brandenburg* **67**: 182-186.

MATTFELD, J. 1926a: Die europäischen und mediterranen *Abies*-Arten. - *Pflanzenareale* **1(2)**: 22-29, Karte 14.

MATTFELD, J. 1926b: Eine neue Königskerze aus Thrazien, *Verbascum Dingleri*. - *Izv. Bulg. Bot. Druzh.* **1**: 101-102 + 1 plate.

MATTFELD, J. 1927a: Aus Wald und Macchie in Griechenland. - *Mitt. Deutsch. Dendrol. Ges.* **38**: 106-151 + 4 plates.

MATTFELD, J. 1927b: A botanical journey in Greece in the summer of 1926. - *J. Arnold Arbor.* **8(3-4)**: 133-149, 204-233 + 4 plates.

MATTFELD, J. 1929a: Die pflanzengeographische Stellung Ost-Thrakiens. - *Verh. Bot. Vereins Prov. Brandenburg* **71**: 1-37 + 11 plates and 1 map.

MATTFELD, J. 1929b: *Minuartia* (L.) Hiern. - *Pflanzenareale* **2(6)**: 43-57, Karte 54-61.

MATTFELD, J. 1930: Über hybridogene Sippen der Tannen nachgewiesen an den Formen der Balkanhalbinsel, zugleich ein Beitrag zur Waldgeschichte der Balkanhalbinsel. - *Biblioth. Bot.* **25/100**: 84 pp. + 2 plates.

MATTFELD, J. 1936: Die Buchen der Chalkidike. - *Izv. Bulg. Bot. Druzh.* **7**: 63-73.

MATTFELD, J. 1937: Die Arten der Gattung *Aubrieta* Adanson. - *Blätt. Staudenk.* **1**: 3 + 10 pp.

MATTFELD, J. 1939: The species of the genus *Aubrieta* Adanson [translated from *Blätt. Staudenk.* 1, 1937]. - *Quart. Bull. Alpine Gard. Soc. Gr. Brit.* **7(2-3)**: 157-181, 217-227.

MATTHÄS, U. 1976: Zur Cytotaxonomie von *Centaurea subciliaris* Boiss. & Heldr. (Sektion *Phalolepis* [Cass.] DC.) und verwandter Sippen im europäischen Mediterrangebiet. I. - *Bot. Jahrb. Syst.* **95(4)**: 418-434.

MATTHÄS, U. 1981: Differenzierungsmuster bei *Centaurea* sect. *Phalolepis* (Compositae). - *Bot. Jahrb. Syst.* **102(1-4)**: 315-319.

MATTHÄS, U. 1988: Die laubwerfenden Eichenwälder Kretas. - *Diss. Bot.* **119**: 1-172 + 1 map.

MATTHÄS, U. 1989: Problems in phytogeography and taxonomy of the deciduous oaks on Crete. In Phitos, D. (ed.): Abstracts. Sixth OPTIMA meeting, Delphi 10-16 Sept. 1989, p. 113.

MATTHÄS, U. 1991: Phytogeographical and taxonomic problems of the deciduous oaks of Crete. - *Bot. Chron. (Patras)* **10**: 635-640.

MATTHEWS, J.F., KETRON, D.W. & ZANE, S.F. 1993: The biology and taxonomy of the *Portulaca oleracea* L. (Portulacaceae) complex in North America. - *Rhodora* **95(882)**: 166-183.

MATTHEWS, J.F., KETRON, D.W. & ZANE, S.F. 1994: The seed surface morphology and cytology of six species of *Portulaca* (Portulacaceae). - *Castanea* **59(4)**: 331-337.

MATTHEWS, V.A. 1984: *Lilium pyrenaicum* - a complex species. - *Bot. Mag. (Kew Mag.)* **1(1)**: 36-43.

MATTHEWS, V.A. 1998: The turban buttercup - *Ranunculus asiaticus*. - *New Plantsman* **5(2)**: 68-73.

MATUSZKIEWICZ, A. & MATUSZKIEWICZ, W. 1948: A contribution to the taxonomy of the genus *Molinia* Schrk. [In Polish with English summary]. - *Ann. Univ. Mariae Curie-Skłodowska, Sect. E* **3**: 347-367.

MATUSZKIEWICZ, W. 1948: Materials to the monography of the genus *Calamagrostis* Adans. [In Polish with English summary]. - *Ann. Univ. Mariae-Curie-Skłodowska, Sect. C, Biol.* **3(7)**: 223-258.

MATVEJEVA, J. 1969: *Sternbergia colchiciflora* W. K. in the Skopje valley [In Macedonian with English summary]. - *Godišen Zborn. Prir.-Mat. Fak. Univ. Skopje, Biol.* **21**: 185-187.

MATVEJEVA, J. 1970: Contribution à la connaissance de la flore de Macédoine [In Macedonian with French summary]. - *Fragm. Balcan.* **7(19)**: 177-180.

MATVEJEVA, J. 1971a: *Lotus angustissimus* L. var. *uniflorus* Matvejeva var. nouvelle. - *Fragm. Balcan.* **8(8)**: 57-60.

MATVEJEVA, J. 1971b: *Linaria cymbalaria* (L.) Mill. A species new for the flora of Macedonia. - *Fragm. Balcan.* **8(17)**: 145-146.

MATVEJEVA, J. 1971c: The genus *Satureia* s.l. in the flora of Skopje's valley [In Macedonian with English summary]. - *Godišen Zborn. Prir.-Mat. Fak. Univ. Skopje, Biol.* **23**: 201-216.

MAUGE, CH. 1996: En Ariège: De la *Bellevalia romana*, et de quelques autres plantes d'interêt. - *Monde Pl.* **91(456)**: 26.

MAUGINI, E. 1953a: Nuovi reperti di biotipi diploidi di *Urginea maritima* Bak. (Liliaceae). - *Caryologia* **5 (2)**: 249-252.

MAUGINI, E. 1953b: Contributo alla conoscenza cariologica e sistematica del genere *Caltha* in Italia. - *Caryologia* **5(3)**: 297-305.

MAUGINI, E. 1953c: Citosistematica di alcuni *Narcissus* della flora italiana appartenenti alla sect. *Hermione*. - *Caryologia* **5(3)**: 313-341.

MAUGINI, E. 1957: Ricerche cito-sistematiche sul genere *Caltha* in Italia. - *Caryologia* **9(3)**: 408-435 + 9 plates.

MAUGINI, E. 1960: Ricerche sulla citogeografia e sulla tassonomia dell'*Urginea maritima* Bak. - *Caryologia* **13(1)**: 151-163.

MAUGINI, E. 1970: Ricerche cariologiche su *Lagurus ovatus* L. - *Caryologia* **23(2)**: 259-271.

MAUGINI, E. & BINI MALECI, L. 1974: Alcune notazioni sulla citogeografia di "Urginea maritima" (L.) Bak.: segnalazione di esemplari pentaploidi. - *Webbia* **29(1)**: 309-315.

MAUGINI, E. & MOSKOVA, R. 1971: Numeri cromosomici per la Flora Italiana: 29-31. - *Inform. Bot. Ital.* **3(1)**: 62-66.

MAURIÈRES, A. 1983: Hypothèse sur les origines de l'*Ophrys attica*. - *Orchidophile (Ansières)* **14(57)**: 410-411.

MAVRODIEV, E., LAKTIONOV, A. & ALEXEEV, Y. 2015: On the new dogbanes of south-eastern European Russia due to the acceptance of the genus *Poacynum* Baill. (Apocynaceae). [In Russian with English summary]. - *Novosti Sist. Vysš. Rast.* **46**: 157-163.

MAVROMMATIS, G. 1969: Note sur la répartition de *Ruscus hypoglossum* en Grèce. - *Biol. Gallo- Hellen.* **2(1)**: 73-77.

MAVROMMATIS, G. 1971: Recherches phytosociologiques et écologiques dans le massif de l'Ossa (Grèce) en vue de sa gestion forestière. - Montpellier: Thèse, Univ. Sci. Techn. Languedoc, 145 pp. + maps and tables

MAVROMMATIS, G. 1973: The ecology of the "Vai" palm forest of Crete (*Phoenix theophrastii* Gr.) [In Greek]. - *Dasos* **25(59/60)**: 21-24.

MAVROMMATIS, G. 1974: Some observations on the ecology of the forest vegetation of Lesvos island [In Greek with English summary]. - *Dasos* **63/64**: 42-47.

MAVROMMATIS, G. 1976: Dierevnisis tou ikosistimatos ethnikou drimou Samarias Kritis [In Greek]. - Athine: 28 pp.

MAW, G. 1886: A monograph of the genus *Crocus*. - London: Dulau & Co., 326 pp. + xx + 10 pp. + 71 plates, 1 map & 2 charts.

MAXIMOVA, S.A. 2001: Carpology and taxonomy of the endemic genera *Hymenonema* Cass. and *Rothmaleria* Font Quer (Compositae) of the Mediterranean basin. In Raimondo, F.M. (ed.): Abstracts from the X OPTIMA meeting, Palermo, 13-19 September 2001, p. 124.

MAXTED, N. 1989: A new *Vicia* from southwest Turkey. - *Notes Roy. Bot. Gard. Edinburgh* **45(3)**: 453-456.

MAXTED, N. 1990a: A revision of *Vicia* subgenus *Vicia* using database techniques. - Southampton: Ph.D. thesis, viii + 708 pp.

MAXTED, N. 1990b: An herbarium based eco-geographical study of *Vicia* subgenus *Vicia*. - Southampton: Dept. of Biology, 197 pp.

MAXTED, N. 1993: A phenetic investigation of *Vicia* L. subgenus *Vicia* (Leguminosae, Vicieae). - *Bot. J. Linn. Soc.* **111(2)**: 155-182.

MAXTED, N. 1994: A phenetic investigation of *Vicia* section *Peregrinae* (Leguminosae: Papilionoideae: Vicieae). - *Edinburgh J. Bot.* **51(1)**: 75-97.

MAXTED, N. 1995: An ecogeographical study of *Vicia* subgenus *Vicia* [Systematic and ecogeographic studies on crop genepools, 8]. - Rome: International Plant Genetic Resources Institute, v + 184 pp.

MAXTED, N., CALLIMASSIA, M.A. & BENNETT, M.D. 1991: Cytotaxonomic studies of eastern Mediterranean *Vicia* species (Leguminosae). - *Pl. Syst. Evol.* **177(3-4)**: 221-234.

MAXTED, N. & DOUGLAS, C. 1997: A phenetic investigation of *Vicia* section *Hypechusa* (Alef.) Achers. & Graebner (Leguminosae, Papilionoideae, Vicieae). - *Lagascalia* **19(1-2)**: 345-369.

MAXTED, N., KHATTAB, A.M.A. & BISBY, F.A. 1991: The newly discovered relatives of *Vicia faba* L. do little to resolve the enigma of its origin. - *Bot. Chron. (Patras)* **10**: 435-465.

MAXTED, N., OBARI, H. & TAN, A. 1990: Vicieae (Leguminosae): new and interesting endemic species from eastern Mediterranean. - *Pl. Genet. Resources Newslett.* **78-79**: 21-25.

MAXWELL, B. 2001: *Fritillaria pontica* Wahlenb. In Rolfe, R. (ed.): Plant awards 2000-2001. - *Alpine Gardener* **69(4)**: 483.

MAYER, A. 1995: Comparative study of the coastal vegetation of Sardinia (Italy) and Crete (Greece) with respect to the effects of human influence. - Eching bei München: IHW Verlag, 264 pp. + 20 tables (Libri Botanici 15).

MAYER, E. 1958: *Rhododendron luteum* Sweet am südöstlichen Alpenrand [In Slovenian with German summary]. - *Razpr. Slov. Akad. Znan. Umetn., Razr. Prir. Med. Vede, Oddel. Prir. Vede* **4**: 39-83.

MAYER, E. 1963: Übersicht der Pteridofyten Jugoslawiens [In Slovenian with German summary]. - *Razpr. Slov. Akad. Znan. Umetn., Razr. Prir. Med. Vede, Oddel. Prir. Vede* **6**: 45-73.

MAYER, E. 1964: Catalogus Flora Jugoslaviae. I/1. Pteridophyta. - Ljubljana: Acad. Sci. Art. Slovenica, 34 pp.

MAYER, E. 1968a: Notulae ad floram Jugoslaviae II. Conspectus generis *Melampyrum* L. - *Bull. Inst. Jard. Bot. Univ. Belgrade, n.s.* **3**: 239-254.

MAYER, E. 1968b: Zur Kenntnis der Gattung *Gentianella* Moench in Jugoslawien. II. Der *G. aspera-*, *G. germanica-* und *G. austriaca*-Komplex. - *Biol. Vestn.* **16**: 23-28.

MAYER, E. 1969: Notulae ad floram Jugoslaviae III. Conspectus generis *Pedicularis* L. - *Acta Bot. Croat.* **28**: 437-448.

MAYER, E. 1970: Zur Bewertung der einblütigen Sippen im *Edraianthus graminifolius*-Komplex. - *Fragm. Florist. Geobot.* **16(1)**: 109-113.

MAYER, E. 1971a: Zur Bewertung und Benennung einiger europäischen *Pedicularis*-Sippen. - *Oesterr. Bot. Z.* **119(1-3)**: 323-327.

MAYER, E. 1971b: Notulae ad floram Jugoslaviae. IV. Conspectus generis *Rhinanthus* L. - *Glasn. Prir. Mus. Beogradu, Ser. B, Biol. Nauke* **25**: 225-238.

MAYER, E. 1972a: Notulae ad floram Jugoslaviae. V. Conspectus generis *Odontites* Ludw. - *Glasn. Republ. Zavoda Zaštitu Prir. Prirodnjačke Zbirke Titogradu* **4**: 5-14.

MAYER, E. 1972b: Zur Entwicklung der Gattung *Pedicularis* L. im europäischen Raume. - *Symp. Biol. Hung.* **12**: 135-139.

MAYER, E. 1973: Notizen zur Flora von Serbien und Makedonien. - *Glasn. Prir. Mus. Beogradu, Ser. B, Biol. Nauke* **28**: 69-79.

MAYER, E. 1976: Zur Kenntnis des *Saponaria sicula*-Komplexes auf der Balkanhalbinsel. - *Acta Bot. Croat.* **35**: 239-244.

MAYER, E. 1981: Floristische Notizen aus Serbien. II. *Podocytisus caramanicus* Boiss. & Heldr. - *Acta Bot. Croat.* **40**: 233-238.

MAYER, E. & BLEČIĆ, V. 1969: Zur Taxonomie und Chorologie von *Edraianthus* sectio *Uniflori*. - *Phyton (Horn)* **13(3-4)**: 241-247.

MAYER, E. & DIKLIĆ, N. 1971: Zur Verbreitung von *Asplenium lepidum* Presl in Serbien [In Serbo- Croatian with English summary]. - *Glasn. Prir. Mus. Beogradu, Ser. B, Biol. Nauke* **26**: 181-185.

MAYER, E. & MICEVSKI, K. 1964: Zur Bewertung der Sippen *Hordeum marinum* Huds. und *Hordeum hystrix* Roth [In Slovenian with German summary]. - *Biol. Vestn.* **12**: 51-58.

MAYER, E. & MICEVSKI, K. 1970: Zur Taxonomie und Chorologie von *Tulipa scardica* Bornm. - *Feddes Repert.* **80(7-8)**: 591-598.

MAYER, E. & MICEVSKI, K. 1982: Zur taxonomischen Bewertung von *Eryngium wiegandii* Adamović. - *Acta Bot. Croat.* **41**: 181-186 + 2 plates.

MAYER, E. & TRPIN, D. 1965: Der *Dianthus sylvestris*-Komplex in Jugoslawien [In Slovenian with German summary]. - *Biol. Vestn.* **13**: 53-59.

MAYER, E. & WRABER, T. 1974: Die Gattung *Haplophyllum* Juss. in Jugoslawien. - *Phyton (Horn)* **16(1-4)**: 117-125.

MAYER, V. 1995: The epicalyx in fruits of *Scabiosa* and *Tremastelma* (Dipsacaceae): Anatomy and ecological significance. - *Bot. Jahrb. Syst.* **117(3)**: 333-357.

MAYER, V. & EHRENDORFER, F. 2000: Fruit differentiation, palynology, and systematics in *Pterocephalus* Adanson and *Pterocephalodes*, gen. nov. (Dipsacaceae). - *Bot. J. Linn. Soc.* **132(1)**: 47-78.

MAZUR, M. & AL. 2016: Morphological and ecological comparison of populations of *Juniperus turbinata* Guss. and *J. phoenicea* L. from the Mediterranean region. - *Pl. Biosyst.* **150(2)**: 313-322.

MAZZIARI, D. 1834-35: Flora Corcirense [In Italian and Greek]. In: Ionian Anthology 2: 424-469, 3: 669-703, 4: 940-961, 5: 180-227. - Kerkira.

MAZZOLA, P., GRISAFI, F. & ROMANO, S. 1983: Numeri cromosomici per la Flora Italiana: 850-859. - *Inform. Bot. Ital.* **13(2-3)**: 182-188.

MAZZOLA, P., GRISAFI, F. & ROMANO, S. 1984: Numeri cromosomici per la Flora Italiana: 919-928. - *Inform. Bot. Ital.* **14(2-3)**: 268-274.

MAZZOLA, P., MICELI, G. & NOT, R. 1984: Numeri cromosomici per la Flora Italiana: 929-936. - *Inform. Bot. Ital.* **14(2-3)**: 275-279.

MAZZOLENI, S. 1989: Fire and Mediterranean plants: Germination responses to heat exposure. - *Ann. Bot. (Rome)* **47**: 227-233.

MCALLISTER, H. 2010: The genus *Sorbus*. Mountain ash and other rowans. - Kew Publishing: 256 pp.

MCCLINTOCK, D. 1971: Recent development in the knowledge of European Ericas. - *Bot. Jahrb. Syst.* **90(4)**: 509-523.

MCCLINTOCK, D. 1989: The heathers of Europe and adjacent areas. - *Bot. J. Linn. Soc.* **101(3)**: 279-289.

MCCLINTOCK, D. & WIJNANDS, O. 1982: *Erica carnea* or *E. herbacea*? - *Taxon* **31(2)**: 319-320.

MCGREGOR, M. 2008: Saxifrages: A definitive guide to the 2000 species, hybrids & cultivars. - Timber Press: 384 pp.

MCGREGOR, M. & HARDING, W. 1998: Saxifrages. The complete list of species. - Hutton, Driffield, U.K.: The Saxifrage Society, ii + 176 pp.

MCLEWIN, W. & MATHEW, B.F. 1995: Hellebores: the first of a series of articles discussing the genus *Helleborus*. - *New Plantsman* **2(2)**: 112-122.

MCMILLAN, C. 1975: The *Xanthium strumarium* complexes in Australia. - *Austral. J. Bot.* **23(1)**: 173-192.

MCNAUGHTON, I.H. & HARPER, J.L. 1964: Biological Flora of the British Isles. *Papaver* L. - *J. Ecol.* **52 (3)**: 767-793.

MCNEAL, J.R. & AL. 2013: Phylogeny and origins of holoparasitism in Orobanchaceae. - *Amer. J. Bot.* **100(5)**: 971-983.

MCNEILL, J. 1962: Taxonomic studies in the Alsinoideae: I. Generic and infra-generic groups. - *Notes Roy. Bot. Gard. Edinburgh* **24(2)**: 79-155.

MCNEILL, J. 1963: Taxonomic studies in the Alsinoideae: II. A revision of the species in the Orient. - *Notes Roy. Bot. Gard. Edinburgh* **24(3)**: 241-404.

MCNEILL, J. 1968: Taxonomic and nomenclatural notes on *Polygala* in Europe. In Heywood, V.H. (ed.): Notulae systematicae ad Floram Europaeam spectantes, no. 7. - *Feddes Repert.* **79(1-2)**: 23-34.

MCNEILL, J. 1969: New species of *Minuartia* (Caryophyllaceae). - *Notes Roy. Bot. Gard. Edinburgh* **29(3)**: 325-327.

MCNEILL, J. 1973a: *Gypsophila* and *Stellaria*: An unexpected problem in generic delimination. - *Notes Roy. Bot. Gard. Edinburgh* **32(3)**: 389-395.

MCNEILL, J. 1973b: Note on *Solidago* L. In Heywood, V.H. (ed.): Notulae systematicae ad Floram Europaeam spectantes, no. 14. - *Bot. J. Linn. Soc.* **67(3)**: 279-280.

MCNEILL, J. 1978: *Silene alba* and *S. dioica* in North America and the generic delimitation of *Lychnis*, *Melandrium* and *Silene* (Caryophyllaceae). - *Canad. J. Bot.* **56(3)**: 297-308.

MCNEILL, J. 1981a: The taxonomy and distribution in eastern Canada of *Polygonum arenastrum* (4x=40) and *P. monspeliense* (6x=60), introduced members of the *P. aviculare* complex. - *Canad. J. Bot.* **59(12)**: 2744-2751.

MCNEILL, J. 1981b: Nomenclatural problems in *Polygonum*. - *Taxon* **30(3)**: 630-641.

MCNEILL, J. & AL. 1983: Taxonomic and nomenclatural notes on *Atriplex* L. (Chenopodiaceae). - *Taxon* **32(4)**: 549-556.

MCNEILL, J. & BASSETT, I.J. 1974: Pollen morphology and the infrageneric classification of *Minuartia* (Caryophyllaceae). - *Canad. J. Bot.* **52(6)**: 1225-1231.

MCNEILL, J. & CROMPTON, C.W. 1978: Pollen dimorphism in *Silene alba* (Caryophyllaceae). - *Canad. J. Bot.* **56(10)**: 1280-1286.

MCNEILL, J. & PRENTICE, H.C. 1981: *Silene pratensis* (Rafn) Godron & Gren., the correct name for the white campion or white cockle (*Silene alba* (Miller) E.H.L. Krause, nom. illeg.). - *Taxon* **30(1)**: 27-32.

MCVEAN, D.N. 1953: Biological Flora of the British Isles. *Alnus glutinosa* (L.) Gaertn. (*A. rotundifolia* Stokes). - *J. Ecol.* **41(2)**: 447-466.

MEACHAM, CH.A. 1980: Phylogeny of the Berberidaceae with an evaluation of classifications. - *Syst. Bot.* **5(2)**: 149-172.

MEADOWS, G.E.M. 1975: Plant portraits: *Ophrys sphegodes* Mill. - *Quart. Bull. Alpine Gard. Soc. Gr. Brit.* **43(2)**: 102, 104-106.

MÉDAIL, F. 2021: Plant biogeography and vegetation patterns of the Mediterranean Islands. - *Bot. Rev. (Lancaster)* **87**: doi.org/10.1007/s12229-021-09245-3.

MÉDAIL, F. & ABOUCAYA, A. 1993: *Thymelaea* ×*conradiae* Aboucaya & Médail (= *T. hirsuta* (L.) Endl. × *T. tartonraira* (L.) All. subsp. *tartonraira*, hybride nouveau découvert en Corse. - *Candollea* **48(2)**: 576-582.

MÉDAIL, F. & AL. 2001: Systematics, ecology and phylogeographic significance of *Olea europaea* L. ssp. *maroccana* (Greuter & Burdet) P. Vargas et al., a relictual olive tree in south-west Morocco. - *Bot. J. Linn. Soc.* **137(3)**: 249-266.

MÉDAIL, F. & DIADEMA, K. 2009: Glacial refugia influence plant diversity patterns in the Mediterranean Basin. - *J. Biogeogr.* **36(7)**: 1333-1345.

MÉDAIL, F. & QUÉZEL, P. 1997: Hot spot analysis for conservation of plant biodiversity in the Mediterranean basin. - *Ann. Missouri Bot. Gard.* **84(1)**: 112-127.

MEĐEDOVIĆ, S. 1975: Chromosome complement of the species *Leucojum aestivum* L. from the population of carst fields in Bosnia and Herzegovina [In Serbo-Croatian with English summary]. - *God. Biol. Inst. Univ. Sarajevu* **28**: 199-206.

MEĐEDOVIĆ, S. 1976: Comparative analysis of the chromosome complements of the species *Cephalaria pastricensis* (Dörfler & Hay.) and *Cephalaria leucantha* (L.) Schrad. [In Serbo-Croatian with English summary]. - *God. Biol. Inst. Univ. Sarajevu* **29**: 79-86.

MEĐEDOVIĆ, S. 1984: Cytogenetic variability and possible direction of karyotype evolution in the genus *Achillea* of the Dinarids [In Serbo-Croatian with English summary]. - *God. Biol. Inst. Univ. Sarajevu* **37**: 61-77.

MEĐEDOVIĆ, S. & ŠILJAK, S. 1978: Interpopulational variability of some chromosome complement characteristics in *Achillea lingulata* from the mountains of the Central and Western Bosnia [In Serbo- Croatian with English summary]. - *Glasn. Zemaljsk. Muz. Sarajevu* **17**: 169-177.

MEDWECKA-KORNAŚ, A. & AL. 1986: The oak forest Quercetum frainetto-cerris on the foothills of the mountain range Galičica (Macedonia). - *Acta Soc. Bot. Poloniae* **55(3)**: 325-342.

MEHRA, P.N. & SANDHU, R.S. 1976: Morphology of the fern *Anogramma leptophylla*. - *Phytomorphology* **26(1)**: 60-76.

MEI, G. & AL. 2021: *Cystopteris dickieana* R. Sim (Cystopteridaceae), a new fern in the continental Balkans flora. - *Pl. Biosyst.* **155(1)**: 1-4.

MEIKLE, R.D. 1954: A survey of the flora of Chios. - *Kew Bull.* **1954(1)**: 85-199.

MEIKLE, R.D. 1956: Notes on the flora of Cyprus: I. - *Kew Bull.* **1956(3)**: 545-549.

MEIKLE, R.D. 1959: The batrachian Ranunculi of the Orient. - *Notes Roy. Bot. Gard. Edinburgh* **23(1)**: 13-21.

MEIKLE, R.D. 1967: *Campanula hawkinsiana*. - *Bot. Mag.* **176(2)**: tab. 505 + 4 pp. text.

MEIKLE, R.D. 1970: The nomenclature and taxonomy of *Helianthemum lavandulifolium* auctt. - *Israel J. Bot.* **19(2-3)**: 249-253.

MEIKLE, R.D. 1972: *Cyclamen graecum*. - *Bot. Mag.* **178(3)**: tab. 603 + 4 pp. text.

MEIKLE, R.D. 1977a: Flora of Cyprus, vol. 1. - London: Bentham-Moxon Trust, Royal Botanic Gardens, Kew, pp. i-xii + 1-832.

MEIKLE, R.D. 1977b: *Colchicum burttii*. - *Bot. Mag.* **181(3)**: tab. 735 + pp. 131-135.

MEIKLE, R.D. 1977c: *Ranunculus asiaticus*. - *Bot. Mag.* **181(4)**: tab. 741 + pp. 159-162.

MEIKLE, R.D. 1978: (461) A proposal for the rejection of the name *Scabiosa papposa* L. - *Taxon* **27(5-6)**: 560-561. [approved, cf. Taxon 31(3): 542, 1982].

MEIKLE, R.D. 1979a: *Rhagadiolus*: A misunderstood genus. - *Taxon* **28(1-3)**: 133-141.

MEIKLE, R.D. 1979b: Some notes on *Cyclamen*. In Bailey, R.H. (ed.): Proceedings of the conference of the Cyclamen Society, pp. 3-12. - London: Barbican Associates.

MEIKLE, R.D. 1980a: *Cyclamen rhodium*. - *Cyclamen Soc. J.* **4(2)**: 45.

MEIKLE, R.D. 1980b: (516) Proposal to conserve 9566 *Rhagadiolus* Juss. over *Rhagadiolus* Zinn (Compositae). - *Taxon* **29(1)**: 159-160. [not approved, cf. Taxon 31(3): 544, 1982].

MEIKLE, R.D. 1980c: Some notes on *Laurentia* Adanson (Campanulaceae). - *Kew Bull.* **34(2)**: 373-375.

MEIKLE, R.D. 1980d: *Alkanna tinctoria*: a nomenclatural tangle. - *Kew Bull.* **34(4)**: 821-824.

MEIKLE, R.D. 1980e: *Cyclamen hederifolium*. - *Bot. Mag.* **182(4)**: tab. 783 + pp. 137-141.

MEIKLE, R.D. 1982: (671) Proposal to conserve 1093 *Bellevalia* Lapeyr. (1808) [Liliaceae] against *Bellevalia* Scopoli (1777) [Verbenaceae]. - *Taxon* **31(3)**: 568-569. [approved, cf. Taxon 34(4): 660, 1985].

MEIKLE, R.D. 1983: Additions to the flora of Cyprus. - *Ann. Mus. Goulandris* **6**: 87-94.

MEIKLE, R.D. 1985: Flora of Cyprus, vol. 2. - London: Bentham-Moxon Trust, Royal Botanic Gardens, Kew, pp. i-xiii + 833-1969.

MEIKLE, R.D. 1994: Boissier, Floras and types: Some observations. - *Bot. Mag. (Kew Mag.)* **11(3)**: 143-146.

MEIKLE, R.D. & AL. 1974: *Colchicum*: A lily group discussion. - *Lilies* **1973/74**: 88-96.

MEILI-FREI, E. 1965: Cytogenetik und Cytotaxonomie einheimischer Arten von *Epipactis, Listera, Neottia* (Orchidaceae). - *Ber. Schweiz. Bot. Ges.* **75**: 219-292 + 16 plates.

MEJÍAS, J.A. 1986: Números cromosómicos para la flora española. 471-478. - *Lagascalia* **14(2)**: 289-292.

MEJÍAS, J.A. 1993a: Cytotaxonomic studies in the Iberian taxa of the genus *Lactuca* (Compositae). - *Bot. Helv.* **103(1)**: 113-130.

MEJÍAS, J.A. 1993b: Reports (167-170). In Kamari, G., Felber, F. & Garbari, F. (eds): Mediterranean chromosome number reports - 3. - *Fl. Medit.* **3**: 333-338.

MEJÍAS, J.A. 1998: Reports (936-940). In Kamari, G., Felber, F. & Garbari, F. (eds): Mediterranean chromosome number reports - 8. - *Fl. Medit.* **8**: 245-251.

MELDERIS, A. 1932: Genetical and taxonomical studies in the genus *Erythraea* Rich. I. - *Acta Hort. Bot. Univ. Latv.* **6**: 123-158.

MELDERIS, A. 1972: Taxonomic studies on the European species of the genus *Centaurium* Hill. In Heywood, V.H. (ed.): Notulae systematicae ad Floram Europaeam spectantes, no. 12. - *Bot. J. Linn. Soc.* **65(2)**: 224-250.

MELDERIS, A. 1978a: Taxonomic notes on the genus *Festucopsis* (C.E. Hubbard) Melderis. In Heywood, V.H. (ed.): Notulae systematicae ad Floram Europaeam spectantes, no. 20. - *Bot. J. Linn. Soc.* **76(4)**: 316-320.

MELDERIS, A. 1978b: Taxonomic notes on the tribe Triticeae (Gramineae), with special reference to the genera *Elymus* L. sensu lato, and *Agropyron* Gaertner sensu lato. In Heywood, V.H. (ed.): Notulae systematicae ad Floram Europaeam spectantes, no. 20. - *Bot. J. Linn. Soc.* **76(4)**: 369-384.

MELDERIS, A. 1984: New taxa and combinations in Gramineae tribe Triticeae from Turkey. – *Notes Roy. Bot. Gard. Edinburgh* **42(1)**: 77-82.

MELIKOKI, K. & AL. 2013: The genus Hieracium s.l. in Prefecture of Chalkidiki. In Gkelis, S. & al. (eds): 13th Panhellenic Scientific Conference, Hellenic Botanical Society, Thessaloniki 3-6 October. Program and book of abstracts, p. 110.

MELKÓ, E. 1976: Numerical taxonomic studies on *Iris pumila* L. by cluster analysis. - *Acta Bot. Acad. Sci. Hung.* **22(3-4)**: 403-414.

MELKÓ, E. 1977: Morphologische Untersuchungen an der Art *Iris pumila* L. [In Hungarian with German summary]. - *Bot. Közlem.* **64(3)**: 155-164.

MELKÓ, E. 1978: Angaben zur Systematik und Verbreitung von *Iris pumila* L. [In Hungarian with German summary]. - *Bot. Közlem.* **65(1)**: 15-26.

MELNIKOV, D. 2016: On the taxonomic status of the genus *Acinos* (Lamiaceae). - *Bot. Zhurn. (Moscow & Leningrad)* **101(1)**: 80-94.

MELVILLE, R. 1937: *Corylus colurna*. - *Bot. Mag.* **160(1)**: tab. 9469 + 4 pp. text.

MELVILLE, R. 1952: *Pinus nigra* var. *leucodermis*. - *Bot. Mag.* **169(2)**: tab. 190 + 4 pp. text.

MELVILLE, R. 1958: *Ulmus canescens*: An Eastern Mediterranean elm. - *Kew Bull.* **1957(3)**: 499-502.

MELVILLE, R. 1983: The affinity of *Paeonia* and a second genus of Paeoniaceae. - *Kew Bull.* **38(1)**: 87-105.

MELZER, H. 1987a: *Chenopodium bonus-henricus* × *foliosum* = *Ch.* ×*tkalcsicsii* Melzer hybr. nov. - eine neue Hybride zweier sehr unterschiedlicher Arten. - *Ber. Bayer. Bot. Ges.* **58**: 73-77.

MELZER, H. 1987b: *Asplenium ruta-muraria* L. subsp. *dolomiticum* Lovis et Reichstein, die Dolomit-Mauerraute - neu für Österreich. - *Linzer Biol. Beitr.* **19(2)**: 295-297.

MELZHEIMER, V. 1974: Bemerkungen zur Cytologie einiger Arten der Gattung *Silene* L. von der Balkan-Halbinsel. - *Candollea* **29(2)**: 337-343.

MELZHEIMER, V. 1977: Biosystematische Revision einiger *Silene*-Arten (Caryophyllaceae) der Balkan- Halbinsel (Griechenland). - *Bot. Jahrb. Syst.* **98(1)**: 1-92.

MELZHEIMER, V. 1978: Bemerkungen zur Cytologie einiger Arten der Gattung *Silene* (Caryophyllaceae) aus Zentralgriechenland und Kreta. - *Pl. Syst. Evol.* **130(3-4)**: 203-207.

MELZHEIMER, V. 1979: Biosystematic researches in some groups of the genus "Silene" from the Balkan Peninsula (Greece). - *Webbia* **34(1)**: 517-522.

MELZHEIMER, V. 1980: Revision einiger balkanischer Arten von *Silene* sect. *Inflatae* (Caryophyllaceae). - *Bot. Jahrb. Syst.* **101(2)**: 153-190.

MELZHEIMER, V. 1981a: Zum Polymorphismus von *Silene italica* (Caryophyllaceae) im Süd-Balkan. - *Bot. Jahrb. Syst.* **102(1-4)**: 285-295.

MELZHEIMER, V. 1981b: *Silene multicaulis* Guss. subsp. *cretica* Melzh., subspec. nov., von der Insel Kreta (Griechenland). - *Phyton (Horn)* **21(1)**: 131-136 + 1 plate.

MELZHEIMER, V. 1983: *Silene conglomeratica* (Caryophyllaceae), eine neue Art aus S-Griechenland (Peloponnes). - *Willdenowia* **13(1)**: 123-127.

MELZHEIMER, V. 1987: On the taxonomic position of *Silene thebana* (Caryophyllaceae). - *Pl. Syst. Evol.* **155(1-4)**: 251-256.

MELZHEIMER, V. & BAYTOP, A. 1980: A new *Silene* from Turkey. - *Notes Roy. Bot. Gard. Edinburgh* **38(3)**: 449-451.

MELZHEIMER, V. & CHRISTODOULAKIS, D. 1982: Eine neue *Silene*-Art aus der Insel Samos (Griechenland). - *Bot. Chron. (Patras)* **2(1)**: 50-52.

MELZHEIMER, V. & DAMBOLDT, J. 1973: Zur Morphologie und Cytologie tetraploider Sippen von *Silene vulgaris* (Caryophyllaceae). - *Willdenowia* **7(1)**: 83-100.

MELZHEIMER, V. & GREUTER, W. 1979: Über zwei bemerkenswerte Arten der Gattung *Silene* (Caryophyllaceae) aus Nord-Griechenland. - *Willdenowia* **8(3)**: 613-623.

MELZHEIMER, V. & GREUTER, W. 1982: Quisquiliae floristicae graecae, 5. *Silene intonsa* (Caryophyllaceae), a new species from NW Greece. - *Willdenowia* **12(1)**: 29-31. [Also distributed as *OPTIMA leaflet* no. 125].

MELZHEIMER, V. & POLATSCHEK, A. 1992: Révision des *Silene (Heliosperma) pusilla* agg. (Caryophyllaceae) für die Flora Österreichs. - *Phyton (Horn)* **31(2)**: 281-306.

MELZHEIMER, V. & ULRICH, R. 1994: *Silene fabaria* (L.) Sm. (Caryophyllaceae), neu für die Flora von Kreta (Griechenland). - *Phyton (Horn)* **33(2)**: 231-236.

MENEMEN, Y. & JURY, S.L. 2001a: A systematic study of the genus *Malabaila* Hoffm. (Umbelliferae) comparing with the closely related genera. In Özhatay, N. (ed.): Proceedings of the Second Balkan Botanical Congress, 1: 299-312. - Istanbul.

MENEMEN, Y. & JURY, S.L. 2001b: A taxonomic revision of the genus *Pastinaca* L. (Umbelliferae). - *Israel J. Pl. Sci.* **49(1)**: 67-77.

MENEMEN, Y., JURY, S.L. & KEITH-LUCAS, D.M. 2000: A systematic study of the genus *Malabaila* Hoffm. and closely related genera of the family Umbelliferae. In Özhatay, N. & al. (eds): Abstracts from the Second Balkan Botanical Congress, Istanbul, Turkey, May 14-18, 2000, p. 17.

MENNEMA, J. 1971: *Lamium galeobdolon* (L.) L. in Nederland [In Dutch with English summary]. - *Gorteria* **5(7-10)**: 193-198.

MENNEMA, J. 1989: A taxonomic revision of *Lamium* (Lamiaceae). - *Leiden Bot. Ser.* **11**: i-vii, 1-198.

MENZEL, M.Y. 1951: The cytotaxonomy and genetics of *Physalis*. - *Proc. Amer. Philos. Soc.* **95(2)**: 132-183.

MERIÇ, Ç & GÜLER, N. 2008: Nuclear DNA content of *Dianthus ingoldbyi* (Caryophyllaceae). - *Phytol. Balcan.* **14(3)**: 413-416.

MERIÇ, Ç. & DANE, F. 2001: A new record for the flora of European Turkey and a new karyotype: *Vicia articulata* Hornem. (Fabaceae). In Özhatay, N. (ed.): Proceedings of the Second Balkan Botanical Congress, 1: 313-318. - Istanbul.

MERIÇ, Ç. & DANE, F. 2005: A new alien species for the flora of European Turkey: *Galinsoga parviflora* (Asteraceae). - *Phytol. Balcan.* **11(1)**: 63-66.

MERIÇLI, A.H. 1983: *Silybum marianum* in der Türkei. - *Istanbul Üniv. Eczac. Fak. Mecm.* **19**: 57-63.

MERMOUD, M. 1980: Voyages d'Edmond Boissier en Orient en 1842 et 1846. - *Candollea* **35(1)**: 71-85.

MERMYGKAS, D. 2019: The Herbarium of the Goulandris Natural History Museum (ATH) - History and content. In Daskalakou, E.N. & al. (eds): 16th Conference of the Hellenic Botanical Society – Athens 10-13 October 2019. Program and book of abstracts, p. 83.

MERMYGKAS, D. & SKOUROS, TH. 2017: Report 44. In Vladimirov, V., Aybeke, M. & Tan, Kit (eds): New floristic records in the Balkans: 32. - *Phytol. Balcan.* **23(1)**: 127-128.

MERMYGKAS, D., TAN, KIT & YANNITSAROS, A. 2010: A new species of *Iris* (Iridaceae) from the northern Peloponnese (Greece). - *Phytol. Balcan.* **16(2)**: 263-266.

MERMYGKAS, D. & YANNITSAROS, A. 2015: A floristic report from Mt Saitas, North Peloponnese, Greece. - *Phytol. Balcan.* **21(3)**: 315-349.

MERMYGKAS, D., ZIKOS, A. & CONSTANTINIDIS, TH. 2021: Biological traits, habitat preferences and endemism in the flora of Peloponnisos, Greece. - *Fl. Medit.* **31**: 37-52 + suppl.

MERRYWEATHER, J. 1995a: The errant *votanologos* in north-east Corfu. - *Pteridologist* **2(6)**: 242-243.

MERRYWEATHER, J. 1995b: Fern hunting in Eastern Crete. - *Pteridologist* **2(6)**: 262-264.

MERT, H.H., DOĞAN,Y. & BAŠLAR, S. 1996: Studies on the ecology of *Spartium junceum* L. in West Anatolia. In Öztürk, M.A., Seçmen, Ö. & Görk, G. (eds): Plant life in southwest and central Asia, 2: pp. 713-724. - Izmir: Ege University Press.

MERXMÜLLER, H. 1968: *Melitella* (Cichoriaceae) - über ein Vorkommen in Australien und die taxonomische Einreihung. - *Mitt. Bot. Staatssamml. München* **7**: 271-275.

MERXMÜLLER, H. 1974: Veilchenstudien I-IV. - *Phyton (Horn)* **16(1-4)**: 137-158.

MERXMÜLLER, H. & DAMBOLDT, J. 1962: Die Chromosomenzahlen einiger rupicoler mediterraner Campanulae. - *Ber. Deutsch. Bot. Ges.* **75(7)**: 233-236.

MERXMÜLLER, H., FRIEDRICH, H.CH. & GRAU, J. 1971: Cytotaxonomische Untersuchungen zur Gattungstruktur von *Crassula*. - *Ann. Naturhist. Mus. Wien* **75**: 111-119.

MERXMÜLLER, H. & GRAU, J. 1963: Chromosomenzahlen aus der Gattung *Myosotis* L. - *Ber. Deutsch. Bot. Ges.* **76(1)**: 23-29.

MERXMÜLLER, H. & GRAU, J. 1967: *Moehringia*-Studien. - *Mitt. Bot. Staatssamml. München* **6(2)**: 257-273.

MERXMÜLLER, H. & GRAU, J. 1969: Dysploidie bei *Pulmonaria*. - *Rev. Roumaine Biol., Sér. Bot.* **14(1)**: 57-63.

MERXMÜLLER, H. & GRAU, J. 1977: Fruchtanatomische Untersuchungen in der *Inula*-Gruppe (Asteraceae). - *Publ. Cairo Univ. Herb.* **7-8**: 9-20.

MERXMÜLLER, H., LEINS, P. & ROESSLER, H. 1977: Inuleae - systematic review. In Heywood, V.H., Harborne, J.B. & Turner, B.L. (eds): The biology and chemistry of the Compositae 1: 577-602. - London, etc.: Academic Press.

MERXMÜLLER, H. & LIPPERT, W. 1977: Veilchenstudien V-VII. - *Mitt. Bot. Staatssamml. München* **13**: 503-533. [Also distributed as *OPTIMA leaflet* no. 57].

MERXMÜLLER, H. & PODLECH, D. 1961: Über die europäischen Vertreter von *Helleborus* sect. *Helleborus*. In Heywood, V.H. (ed.): Notulae systematicae ad Floram Europaeam spectantes, no. 1. - *Feddes Repert. Spec. Nov. Regni Veg.* **64(1)**: 2-8.

MERXMÜLLER, H. & PODLECH, D. 1976: *Fimbristylis ferruginea* in Kreta. - *Mitt. Bot. Staatssamml. München* **12(2)**: 683-684.

MERXMÜLLER, H. & STRID, A. 1977: A new species in the *Cerastium alpinum* group from Mt Olympus, Greece. - *Bot. Not.* **130(4)**: 469-472.
MESHINEV, T. & APOSTOLOVA, I. 1998: Distribution of higher plants biodiversity and the problem of habitat diversity in Bulgaria. - *Phytol. Balcan.* **4(1-2)**: 151-159.
MĚSÍČEK, J. 1967: The chromosome morphology of *Arabidopsis thaliana* (L.) Heynh. and some remarks on the problem of *Hylandra suecica* (Fr.) Löve. - *Folia Geobot. Phytotax.* **2(2)**: 433-436.
MĚSÍČEK, J. & HROUDA, L. 1975: Chromosome numbers in Czechoslovak species of *Gagea* (Liliaceae). - *Folia Geobot. Phytotax.* **9(4)**: 359-368.
MĚSÍČEK, J. & JAROLÍMOVÁ, V. 1992: List of chromosome numbers of the Czech vascular plants. - Praha: Czechoslovak Academy of Sciences, 144 pp.
MESSERI, A. 1942: Studio sistematico e fitogeografico di "Lagurus ovatus" L. - *Nuovo Giorn. Bot. Ital., n.s.* **49(2)**: 133-204.
METCALFE. D.J. 2005: *Hedera helix* L. - *J. Ecol.* **93(3)**: 632-648.
METHERELL, C. & RUMSEY, F.J. 2018: Eyebrights ('Euphrasia') of the UK and Ireland. - Bristol: BSBI handbook No. 18, 220 pp.
METTIN, D. & HANELT. P. 1964: Cytosystematische Untersuchungen in der Artengruppe um *Vicia sativa* L. - *Kulturpflanze* **12**: 163-225.
METTIN, D. & HANELT, P. 1968: Bemerkungen zur Karyologie und Systematik einiger Sippen der Gattung *Vicia* L. - *Feddes Repert.* **77(1)**: 11-30.
METTIN, D. & HANELT, P. 1973: Über Speziationsvorgänge in der Gattung *Vicia* L. - *Kulturpflanze* **21**: 25-54.
MEUSEL, H. 1955: Über Wuchsform und Wuchsdauer mediterraner Einjahrspflanzen. - *Wiss. Z. Martin-Luther-Univ. Halle-Wittenberg, Math.-Naturwiss. Reihe* **4(3)**: 643-650.
MEUSEL, H. 1957: Wuchsformen, Vorkommen und Verbreitung der annuellen *Asteriscus*-Arten. - *Beitr. Biol. Pflanzen* **34(1)**: 89-111.
MEUSEL, H. 1962: Die mediterran-mitteleuropäischen Florenbeziehungen auf Grund vergleichender chorologischer Untersuchungen. - *Ber. Deutsch. Bot. Ges.* **75(4)**: 107-118.
MEUSEL, H. 1969a: Eine neue Unterart von *Euphorbia amygdaloides* L. mit stark abweichender Wuchsform. - *Wiss. Z. Martin-Luther-Univ. Halle-Wittenberg, Math.-Naturwiss. Reihe* **18(1)**: 131-135.
MEUSEL, H. 1969b: Zur ökogeographischen Stellung von *Daphne blagayana*. - *Rev. Roumaine Biol., Sér. Bot.* **14(1)**: 51-56.
MEUSEL, H. 1970: Wuchsformenreihen mediterran-mitteleuropäischer Angiospermen-Taxa. - *Feddes Repert.* **81(1-5)**: 41-59.
MEUSEL, H. & AL. 1971: Die Schachtelhalme Europas. - Wittenberg: A. Ziemsen Verlag, 84 pp.
MEUSEL, H. & AL. 1978: Vergleichende Chorologie der zentraleuropäischen Flora. Band II: Text (pp. i- xi & 1-418) und Karten (pp. 259-421). - Jena: Gustav Fischer Verlag.
MEUSEL, H. & EHRENDORFER, F. 1984: Pflanzenverbreitung in Raum und Zeit (am Beispiel der Mittelmeerländer). - *Nova Acta Leop., N.F.* **54(244)**: 185-210.
MEUSEL, H. & JÄGER, E.J. 1962: Über die Verbreitung einiger Papilionaceen-Gattungen. - *Kulturpflanze, Beih.* **3**: 249-262.
MEUSEL, H. & JÄGER, E.J. 1989: Ecogeographical differentiation of the submediterranean deciduous forest flora. - *Pl. Syst. Evol.* **162(1-4)**: 315-329.
MEUSEL, H. & JÄGER, E.J. 1992: Vergleichende Chorologie der zentraleuropäischen Flora. Band III: Text (pp. i-ix & 1-333), Karten, Literatur und Register (pp. i-ix & 422-688). - Jena: Gustav Fischer Verlag.
MEUSEL, H., JÄGER, E.J. & WEINERT, E. 1965: Vergleichende Chorologie der zentraleuropäischen Flora. Text (583 pp.) und Karten (258 pp.). - Jena: Gustav Fischer Verlag.
MEUSEL, H. & KÄSTNER, A. 1972: Übersicht zur systematischen Gliederung der Gattung *Carlina*. - *Feddes Repert.* **83(4)**: 213-232 + 3 plates.
MEUSEL, H. & KÄSTNER, A. 1977: Über einige neue oder neu einzustufende Taxa von *Carlina* sect. *Corymbosae*. - *Feddes Repert.* **88(7-8)**: 401-408.
MEUSEL, H. & KÄSTNER, A. 1990: Lebensgeschichte der Gold- und Silberdisteln. Monographie der mediterran-mitteleuropäischen Compositen-Gattung *Carlina*. Band 1. Merkmalsspektren und Lebensräume der Gattung. - *Denkschr. Österr. Akad. Wiss., Math.-Naturwiss. Kl.* **127**: 294 pp. + 32 plates.
MEUSEL, H. & KÄSTNER, A. 1994: Lebensgeschichte der Gold- und Silberdisteln. Monographie der mediterran-mitteleuropäischen Compositen-Gattung *Carlina*. Band 2. Artenvielfalt und Stammegeschichte der Gattung. - *Denkschr. Österr. Akad. Wiss., Math.-Naturwiss. Kl.* **128**: 657 pp. + 32 plates.
MEUSEL, H., KÄSTNER, A. & RAUS, TH. 1984: Zur systematischen und ökogeographischen Stellung von *Carlina tragacanthifolia* Klatt. - *Flora, Morphol. Geobot. Oekophysiol.* **175(3)**: 145-182.
MEUSEL, H. & KNAPP, H.D. 1983: Ökogeographische Analyse der Areale einiger mediterraner und mediterran-mitteleuropäischer Orchideen. - *Jahresber. Naturwiss. Vereins Wuppertal* **36**: 80-94.
MEUSEL, H. & MÖRCHEN, G. 1977: Zur ökogeographischen und morphologischen Differenzierung einiger *Scrophularia*-Arten. - *Flora, Morphol. Geobot. Oekophysiol.* **166(1)**: 1-20.
MEUSEL, H. & OHLE, H. 1966: Zur Taxonomie und Cytologie der Gattung *Calendula*. - *Oesterr. Bot. Z.* **113(2)**: 191-210.
MEUSEL, H. & WERNER, K. 1962: Über die Gliederung von *Carlina acaulis* L. und *Carlina vulgaris* L. - *Wiss. Z. Martin-Luther-Univ. Halle-Wittenberg, Math.-Naturwiss. Reihe* **11(2)**: 279-292.
MEYER, C.A. 1842: Revision der Arten der Gattung *Agrimonia*. - *Bull. Sci. Acad. Imp. Sci. Saint-Pétersbourg* **10**: 336-349.
MEYER, D. 1980: Numerisch-taxonomische Untersuchungen an *Pisum sativum* L. - *Kulturpflanze* **28**: 285-340.
MEYER, D.E. 1952: Untersuchungen über Bastardierung in der Gattung *Asplenium*. - *Biblioth. Bot.* **30/123**: 34 pp. + 1 plate.
MEYER, D.E. 1957: Zur Zytologie der Asplenien Mitteleuropas (I-XV). - *Ber. Deutsch. Bot. Ges.* **70(2)**: 57-66.
MEYER, D.E. 1958a: Die Chromosomenzahlen der Asplenien Mitteleuropas. - *Willdenowia* **2(1)**: 41-52.
MEYER, D.E. 1958b: Zur Zytologie der Asplenien Mitteleuropas (XVI-XX). - *Ber. Deutsch. Bot. Ges.* **71(1)**: 11-20.
MEYER, D.E. 1959: Zur Zytologie der Asplenien Mitteleuropas (XXI-XXIII). - *Ber. Deutsch. Bot. Ges.* **72(1)**: 37-48.
MEYER, D.E. 1960a: Zur Gattung *Polystichum* in Mitteleuropa. - *Willdenowia* **2(3)**: 336-342.
MEYER, D.E. 1960b: Zur Zytologie der Asplenien Mitteleuropas (XXIV-XXVIII). - *Ber. Deutsch. Bot. Ges.* **73(9)**: 386-394.
MEYER, D.E. 1962a: Zur Zytologie der Asplenien Mitteleuropas (XXIX. Abschluss). - *Ber. Deutsch. Bot. Ges.* **74(10)**: 448-461.
MEYER, D.E. 1962b: Neue und seltene Asplenien Europas. - *Ber. Deutsch. Bot. Ges.* **75(1)**: 24-34.
MEYER, D.E. 1963: Neue und seltene Asplenien Europas. 2. Mitteilung. - *Ber. Deutsch. Bot. Ges.* **76 (1)**: 13-22.
MEYER, D.E. 1964: Über neue und seltene Asplenien Europas. 3. Mitteilung. - *Ber. Deutsch. Bot. Ges.* **77(1)**: 3-13.
MEYER, D.E. 1967: Über neue und seltene Asplenien Europas. 4. Mitteilung. - *Ber. Deutsch. Bot. Ges.* **80(1)**: 28-39.
MEYER, F.H. 1969: *Quercus ilex* in Kleinasien. - *Ber. Deutsch. Bot. Ges.* **82(7-8)**: 505-519.
MEYER, F.H. 1991: Turkish *Acer* species. - *Karaca Arbor. Mag.* **1(3)**: 91-95 + 3 plates.
MEYER, F.K. 1973: Conspectus der "Thlaspi"-Arten Europas, Afrikas und Vorderasiens. - *Feddes Repert.* **84(5-6)**: 449-469.

MEYER, F.K. 1978: *Hypericum haplophylloides* Hal. et Bald. - eine seltene Pflanze der westlichen Balkan-Halbinsel. - *Wiss. Z. Friedrich-Schiller-Univ. Jena, Math.-Naturwiss. Reihe* **27(1)**: 67-78.

MEYER, F.K. 1979: Kritische Revision der "Thlaspi"-Arten Europas, Afrikas und Vorderasiens. I. Geschichte, Morphologie und Chorologie. - *Feddes Repert.* **90(3)**: 129-154.

MEYER, F.K. 1982: Was ist *Hutchinsia* R. Br. in Ait.? - *Wiss. Z. Friedrich-Schiller-Univ. Jena, Math.- Naturwiss. Reihe* **31(2)**: 267-276.

MEYER, F.K. 1985: Beitrag zur Kenntnis ost- und südosteuropäischer *Soldanella*-Arten. - *Haussknechtia* **2**: 7-41.

MEYER, F.K. 1986: Eine neue *Noccaea*-Art (Cruciferae) von Kreta. - *Willdenowia* **15(2)**: 389-391.

MEYER, F.K. 1987: Die europäischen *Acantholimon*-Sippen, ihre Nachbarn und nächsten Verwandten. - *Haussknechtia* **3**: 3-48.

MEYER, F.K. 1991: Seed-coat anatomy as a character for a new classification of *Thlaspi*. - *Fl. Veg. Mundi* **9**: 9-15.

MEYER, F.K. 1999: Das Herbarium Haussknecht - Geschichte und Gegenwart. -*Haussknechtia* **7**: 3-14.

MEYER, F.K. 2001a: Kritische Revision der "*Thlaspi*"-Arten Europas, Afrikas und Vorderasiens. Spezieller Teil. I. *Thlaspi* L. - *Haussknechtia* **8**: 3-42.

MEYER, F.K. 2001b: Kritische Revision der "*Thlaspi*"-Arten Europas, Afrikas und Vorderasiens. Spezieller Teil. II. *Neurotropis* (DC.) F.K. Mey. - *Haussknechtia* **8**: 43-58.

MEYER, F.K. 2003a: Kritische Revision der "*Thlaspi*"-Arten Europas, Afrikas und Vorderasiens. Spezieller Teil. III. *Microthlaspi* F.K. Mey. - *Haussknechtia* **9**: 3-59.

MEYER, F.K. 2003b: Kritische Revision der "*Thlaspi*"-Arten Europas, Afrikas und Vorderasiens. Spezieller Teil. IV. *Thalpiceras* F.K. Mey. - *Haussknechtia* **9**: 61-113.

MEYER, F.K. 2003c: Kritische Revision der "*Thlaspi*"-Arten Europas, Afrikas und Vorderasiens. Spezieller Teil. V. *Noccidium* F.K. Mey. - *Haussknechtia* **9**: 115-124.

MEYER, F.K. 2003d: Kritische Revision der "*Thlaspi*"-Arten Europas, Afrikas und Vorderasiens. Spezieller Teil. VI. *Kotschyella* F.K. Mey. - *Haussknechtia* **9**: 125-134.

MEYER, F.K. 2011: Beiträge zur Flora von Albanien. - *Haussknechtia, Beiheft* **15**: 1-220.

MEYER, K. 1923: Kulturgeschichtliche und systematische Beiträge zur Gattung *Prunus*. - *Repert. Spec. Nov. Regni Veg. Beih.* **22**: 1-64 + 2 plates.

MEYER, S. & BAZOS, I. 2020: *Bellevalia speciosa*. In Raus, Th. & Raab-Straube, E.v. (eds): Euro+Med-Checklist notulae, 12. - *Willdenowia* **50(2)**: 307-308.

MEYER, S. & WILLING, B. 2010: *Adonis aestivalis* L. subsp. *aestivalis* In Greuter, W. & Raus, Th. (eds): Med-Checklist notulae, 29. - *Willdenowia* **40(2)**: 198.

MICELI, P., FICINI, G. & GARBARI, F. 1984: The genus "Allium" L. in Italy. XIII. Morphological, caryological and leaf anatomical study of some C-W Mediterranean triploid populations of *Allium trifoliatum* Cyr. - *Webbia* **38**: 793-803.

MICELI, P. & GARBARI, F. 1977: Numeri cromosomici per la Flora Italiana: 255-262. - *Inform. Bot. Ital.* **8 (2)**: 207-216.

MICELI, P. & GARBARI, F. 1979: Cromosomi ed anatomia fogliare di quattro *Allium* diploidi di Grecia. - *Atti Soc. Tosc. Sci. Nat. Pisa Mem., Ser. B* **86**: 37-51.

MICELI, P. & GARBARI, F. 1987: *Allium eivissanum* (Alliaceae), a new species from Eivissa (Balearic Isles, Spain). - *Willdenowia* **16(2)**: 383-390.

MICELI, P., SPAZZAFUMO, A. & GARBARI, F. 1978: Numeri cromosomici per la Flora Italiana: 423-433. - *Inform. Bot. Ital.* **10(1)**: 110-119.

MICEV, K. 1956: Bibliographie der Flora und der Vegetation Macedoniens. - *Godišen Zborn. Filoz. Fak. Univ., Prir.-Mat. Odd.* **7**: 1-32.

MICEVSKI, K. 1956: Eine Überprüfung der Verbreitungsgebiete von *Ramonda nathaliae* Panč. et Petrov. und *Ramonda serbica* Panč. in Mazedonien ... Merkmalen der beiden Arten [In Macedonian with German summary]. - *Godišen Zborn. Filoz. Fak. Univ., Prir.-Mat. Odd.* **9(10)**: 121-142.

MICEVSKI, K. 1962: Beitrag zur Kenntnis der Flora Mazedoniens. II. - *Godišen Zborn. Prir.-Mat. Fak. Univ. Skopje, Biol.* **13**: 163-174.

MICEVSKI, K. 1963a: III Beitrag zur Kenntnis der Flora Mazedoniens. *Chenopodium ambrosioides* L. var. *anthelminticum* (L.) Aellen in der Flora Mazedoniens [In Macedonian with English summary]. - *Godišen Zborn. Prir.-Mat. Fak. Univ. Skopje, Biol.* **14**: 183-184.

MICEVSKI, K. 1963b: Die Wasser- und Sumpfvegetation des Dorjan-Sees [In Macedonian with German summary]. - *Acta Mus. Maced. Sci. Nat.* **8**: 175-195.

MICEVSKI, K. 1965: Die Halophytenvegetation von Ovče Polje [In Macedonian with German summary]. - *Acta Mus. Maced. Sci. Nat.* **10(3)**: 67-90 + fold-out tables.

MICEVSKI, K. 1968: Einige interessante Pflanzensippen für die Flora von Mazedonien. - *Fragm. Balcan.* **6(12)**: 121-128.

MICEVSKI, K. 1969a: Die Wasservegetation der Seen von Ohrid und Prespa. - *Acta Mus. Maced. Sci. Nat.* **11(4)**: 61-80 + 1 fold-out table.

MICEVSKI, K. 1969b: *Stachys iva* Griseb. und *Stachys horvaticii* Micevski spec. nov. in der Flora von Mazedonien [In Serbo-Croatian with Latin diagnosis and German summary]. - *Acta Bot. Croat.* **28**: 449-454.

MICEVSKI, K. 1969c: Beitrag zur Kenntnis der Flora von Makedonien. IV. - *Godišen Zborn. Prir.-Mat. Fak. Univ. Skopje, Biol.* **21**: 109-117.

MICEVSKI, K. 1970a: *Astragalus sericophyllus* Griseb. emend. K. Micevski und *Astragalus mayeri* K. Micevski spec. nov. in der Flora von Mazedonien. - *Fragm. Balcan.* **7(17)**: 157-166.

MICEVSKI, K. 1970b: Beitrag zur Kenntnis der Flora von Mazedonien. V [In Macedonian with German summary]. - *Godišen Zborn. Prir.-Mat. Fak. Univ. Skopje, Biol.* **22**: 167-178.

MICEVSKI, K. 1971: *Astragalus gracaninii* Micevski spec. nov. in der Flora von Mazedonien [In Macedonian with Latin diagnosis and German summary]. - *Godišen Zborn. Prir.-Mat. Fak. Univ. Skopje, Biol.* **23**: 121-130 + 1 plate.

MICEVSKI, K. 1972: *Astragalus physocalyx* Fisch. und *Astragalus ponticus* Pall., zwei neue Arten für die Flora von Mazedonien und Jugoslawien [In Macedonian with German summary]. - *Godišen Zborn. Prir.-Mat. Fak. Univ. Skopje, Biol.* **24**: 67-72.

MICEVSKI, K. 1973a: Beitrag zur Taxonomie und Chorologie des *Medicago prostrata*-agg. in Mazedonien [In Macedonian with German summary]. - *Godišen Zborn. Prir.-Mat. Fak. Univ. Skopje, Biol.* **25**: 137-147 + 4 plates.

MICEVSKI, K. 1973b: Einige unbekannte und seltene *Astragalus*- und *Oxytropis*-Arten der Flora von Mazedonien [In Macedonian with German summary]. - *Godišen Zborn. Prir.-Mat. Fak. Univ. Skopje, Biol.* **25**: 157-160.

MICEVSKI, K. 1973c: *Erodium guicciardii* Heldr. - Eine neue Art für die Flora von Mazedonien und Jugoslawien [In Macedonian with German summary]. - *Godišen Zborn. Prir.-Mat. Fak. Univ. Skopje, Biol.* **25**: 161-163.

MICEVSKI, K. 1974a: Beitrag zur Kenntnis der Flora von Mazedonien. VI [In Macedonian with German summary]. - *Godišen Zborn. Prir.-Mat. Fak. Univ. Skopje, Biol.* **26**: 125-129.

MICEVSKI, K. 1974b: Einige unbekannte Arten der Flora von Mazedonien und Jugoslawien [In Macedonian with German summary]. - *Godišen Zborn. Prir.-Mat. Fak. Univ. Skopje, Biol.* **26**: 131-135.

MICEVSKI, K. 1974c: Beitrag zur Kenntnis einiger Farne aus der Flora von Mazedonien [In Macedonian with German summary]. - *Godišen Zborn. Prir.-Mat. Fak. Univ. Skopje, Biol.* **26**: 137-142.

MICEVSKI, K. 1975a: Die Fam. Lycopodiaceae in der Flora von Makedonien [In Macedonian with German summary]. - *Godišen Zborn. Prir.-Mat. Fak. Univ. Skopje, Biol.* **27-28**: 169-173.

MICEVSKI, K. 1975b: Beitrag zur Kenntnis einiger Sippen von *Centaurea* L. sect. *Acrocentron* Cass. in der Flora von Makedonien [In Macedonian with German summary]. - *Godišen Zborn. Prir.-Mat. Fak. Univ. Skopje, Biol.* **27-28**: 175-185.

MICEVSKI, K. 1976: Beitrag zur Taxonomie und Chorologie von *Centaurea marmorea* Bornm. et Soška in der Flora Mazedoniens [In Macedonian with German summary]. - *Posebno Izd., Mus. Maced. Sci. Nat.* **7**: 81-87.

MICEVSKI, K. 1977a: Zwei neue Caryophyllaceen aus Mazedonien [In Macedonian with Latin diagnoses and German summary]. - *Fragm. Balcan.* **10(4)**: 29-39.

MICEVSKI, K. 1977b: Polymorphismus und Verbreitung von *Echinops bannaticus* Rochel ex Schrad. in der Flora von Makedonien [In Macedonian with German summary]. - *Maked. Akad. Nauk. Umetn., Pril. Oddel. Prirod.-Mat.* **9(2)**: 59-68 + 4 plates.

MICEVSKI, K. 1978a: Seltene und unbekannte Arten in der Flora Makedoniens [In Macedonian with German summary]. - *Godišen Zborn. Biol. Fak. Univ. "Kiril i Metodij"* **31**: 151-165.

MICEVSKI, K. 1978b: *Centaurea kozjakensis*, eine neue Art der Sektion *Phalolepis* (Cass.) DC. aus Makedonien [In Macedonian with Latin diagnosis and German summary]. - *Maked. Akad. Nauk. Umetn., Pril. Oddel. Prirod.-Mat.* **10(1)**: 53-58 + 1 plate.

MICEVSKI, K. 1978c: Polymorphismus der Arten der Sect. *Phalolepis* (Cass.) DC. der Gattung *Centaurea* (Asteraceae) in Makedonien [In Macedonian with German summary]. - *Maked. Akad. Nauk. Umetn., Pril. Oddel. Prirod.-Mat.* **10(2)**: 55-65.

MICEVSKI, K. 1978d: Endemismus und Reliktenfrage in der Flora und Vegetation von Makedonien [In Macedonian with German summary]. - *Maked. Akad. Nauk. Umetn., Pril. Oddel. Prirod.-Mat.* **10(2)**: 79-93.

MICEVSKI, K. 1978e: *Fritillaria ionica* Hal. var. *ochridana* Micevski var. nov. in der Flora von Makedonien [In Serbo-Croatian with German summary]. - *Acta Bot. Croat.* **37**: 211-214.

MICEVSKI, K. 1979a: Beitrag zur Taxonomie und Chorologie der Arten aus der sect. *Corothamnus* (Koch) Nyman der Gattung *Cytisus* L. (Fabaceae) in der Flora von Makedonien [In Macedonian with German summary]. - *Maked. Akad. Nauk. Umetn., Pril. Oddel. Prirod.-Mat.* **11(1)**: 65-77.

MICEVSKI, K. 1979b: Was ist *Euphorbia pindicola* Hausskn.? [In Macedonian with German summary]. - *Maked. Akad. Nauk. Umetn., Pril. Oddel. Prirod.-Mat.* **11(2)**: 59-70.

MICEVSKI, K. 1980a: Beitrag zur Taxonomie und Chorologie von *Ranunculus demissus* DC. in der Flora von Makedonien [In Macedonian with German summary]. - *Godišen Zborn. Biol. Fak. Univ. "Kiril i Metodij"* **33**: 111-117.

MICEVSKI, K. 1980b: Beitrag zur Taxonomie und Chorologie von *Erodium absinthoides* Willd. in der Flora von Makedonien [In Macedonian with German summary]. - *Godišen Zborn. Biol. Fak. Univ. "Kiril i Metodij"* **33**: 119-131.

MICEVSKI, K. 1980c: *Erodium guicciardii* Heldr. und dessen polymorphismus im Bereiche von Makedonien [In Serbo-Croatian with German summary]. - *Acta Bot. Croat.* **39**: 165-170.

MICEVSKI, K. 1980d: *Knautia degenii* Borb. und *Knautia magnifica* Boiss. et Orph. in der Flora von Makedonien. - *Fragm. Balcan.* **10(18)**: 155-165.

MICEVSKI, K. 1980e: *Geranium cinereum* Cav. in der Flora von Makedonien [In Macedonian with German summary]. - *Maked. Akad. Nauk. Umetn., Oddel. Biol. Med. Nauki* **1(2)**: 21-33.

MICEVSKI, K. 1981a: Kritische Übersicht der Gattung *Laserpitium* L. (Apiaceae) in der Flora von Makedonien. - *Godišen Zborn. Biol. Fak. Univ. "Kiril i Metodij"* **34**: 23-32 + 1 plate.

MICEVSKI, K. 1981b: *Knautia caroli-rechingeri* Micevski spec. nov. in der Flora von Makedonien [In Serbo-Croatian with Latin diagnosis and German summary]. - *Acta Bot. Croat.* **40**: 239-243.

MICEVSKI, K. 1981c: Beitrag zur Taxonomie und Chorologie der Sekt. *Pseudothymbra* Benth. der Gattung *Thymus* L. (Labiatae) in der Flora von Makedonien [In Macedonian with German summary]. - *Maked. Akad. Nauk. Umetn., Oddel. Biol. Med. Nauki* **2(1-2)**: 17-42.

MICEVSKI, K. 1982: Flora und Vegetation von Makedonien und das Problem ihres Schutzes [In Macedonian with German summary]. - *Maked. Akad. Nauk. Umetn., Oddel. Biol. Med. Nauki* **3(1)**: 77-92.

MICEVSKI, K. 1983a: Die Gattung *Pulsatilla* Mill. in der Flora von Makedonien [In Macedonian with German summary]. - *Godišen Zborn. Biol. Fak. Univ. "Kiril i Metodij"* **36**: 111-117.

MICEVSKI, K. 1983b: Die Gattung *Aconitum* L. in der Flora von Makedonien [In Macedonian with German summary]. - *Godišen Zborn. Biol. Fak. Univ. "Kiril i Metodij"* **36**: 119-126.

MICEVSKI, K. 1983c: Beitrag zur Kenntnis der Flora von Makedonien. VII [In Macedonian with German summary]. - *Godišen Zborn. Biol. Fak. Univ. "Kiril i Metodij"* **36**: 127-134.

MICEVSKI, K. 1983d: Beitrag zur Taxonomie und Chorologie der *Ranunculus montanus*-Gruppe in der Flora von Makedonien [In Macedonian with German summary]. - *Maked. Akad. Nauk. Umetn., Oddel. Biol. Med. Nauki* **4(1-2)**: 125-141.

MICEVSKI, K. 1984: *Achillea corabensis* (Heimerl) Micevski stat. & comb. nov. in der Flora der SR Makedonien [In Macedonian with German summary]. - *Maked. Akad. Nauk. Umetn., Oddel. Biol. Med. Nauki* **5(2)**: 5-10.

MICEVSKI, K. 1985a: Zwei neue Arten der Gattung *Centaurea* L. (Asteraceae), subgen. *Acrolophus* (Cass.) Dobrocz. sect. *Arenariae* (Hayek) Dostal. - *Acta Bot. Croat.* **44**: 83-89 + 2 plates.

MICEVSKI, K. 1985b: *Alyssum bargalense* Micevski spec. nova in der Flora von Makedonien [In Macedonian with Latin diagnosis and German summary]. - *Fragm. Balcan.* **12(10)**: 105-112.

MICEVSKI, K. 1985c: Flora na SR Makedonija, 1(1) [In Macedonian]. - Skopje: Makedoniska Akademija Naukite Umetnostite, pp. [4+] 5-152.

MICEVSKI, K. 1985d: Die Gattung *Berberis* in der Flora von Makedonien [In Macedonian with German summary]. - *Godišen Zborn. Biol. Fak. Univ. "Kiril i Metodij"* **37-38**: 11-17.

MICEVSKI, K. 1986: Beitrag zur Taxonomie und Chorologie von *Ranunculus degenii* Kümm. et Jáv. in der Flora der SR Makedonien [In Macedonian with German summary]. - *Maked. Akad. Nauk. Umetn., Oddel. Biol. Med. Nauki* **5(1)**: 5-9.

MICEVSKI, K. 1987: Beitrag zur Kenntnis der Flora von Makedonien. VIII [In Macedonian with German abstract]. - *Godišen Zborn. Biol. Fak. Univ. "Kiril i Metodij"* **39-40**: 193-202.

MICEVSKI, K. 1989: *Osmunda regalis* L. und ihre Zuhörigkeit der Sumpfvegetation von Jugoslawien [In Serbo-Croatian with German summary]. In: Minerali, stijene, izumrli i živi svijet BiH, pp. 403-409. - Sarajevo: Zemaljski Muzej Bosne i Hercegovine.

MICEVSKI, K. 1990a: Novitäten in der Gattung *Dianthus* L. (Caryophyllaceae) in der Flora der SR Makedonien [In Macedonian with Latin diagnoses and German summary]. - *Maked. Akad. Nauk. Umetn., Oddel. Biol. Med. Nauki* **7(1-2)**: 31-46.

MICEVSKI, K. 1990b: Beitrag zur Kenntnis der Gattung *Centaurea* L. (Asteraceae) in SR Makedonien [In Macedonian]. - *Maked. Akad. Nauk. Umetn., Oddel. Biol. Med. Nauki* **7(1-2)**: 47-56.

MICEVSKI, K. 1990c: Beitrag zur Kenntnis der Flora der SR Makedonien. IX [In Macedonian with German summary]. - *Godišen Zborn. Biol. Fak. Univ. "Kiril i Metodij"* **41-42**: 177-186.

MICEVSKI, K. 1990d: Ein neues *Bupleurum* (Umbelliferae) aus Makedonien - *Bupleurum mayeri* Micevski spec. nov. - *Razprave IV. Razreda Sazu* **31(12)**: 179-183.

MICEVSKI, K. 1991a: Neuheiten aus der Familie Labiatae der Flora von Makedonien [In Macedonian with Latin diagnoses and German summary]. - *Maked. Akad. Nauk. Umetn., Oddel. Biol. Med. Nauki* **9(1-2)**: 17-37.

MICEVSKI, K. 1991b: Neuheiten aus der Familie Caryophyllaceae in der Flora von Makedonien [In Macedonian with Latin diagnoses and German summary]. - *Maked. Akad. Nauk. Umetn., Oddel. Biol. Med. Nauki* **9(1-2)**: 39-50.

MICEVSKI, K. 1991c: *Sporobolus indicus* (L.) R. Br., eine neue Art in der Flora der Republik Makedonien [In Macedonian with German summary]. - *Godišen Zborn. Biol. Fak. Univ. "Kiril i Metodij"* **43-44**: 179-183.

MICEVSKI, K. 1993: Flora na Republika Makedonija, 1(2) [In Macedonian]. - Skopje: Makedoniska Akademija Naukite Umetnostite, pp. [4+] 153-394.

MICEVSKI, K. 1994a: Die Gattung *Tamarix* L. (Tamaricaceae) in der Flora der Republik Makedonien [In Macedonian with German summary]. - *Maked. Akad. Nauk. Umetn., Oddel. Biol. Med. Nauki* **15 (1-2)**: 37-40.

MICEVSKI, K. 1994b: Die Sect. *Odontarrhena* (C.A. Meyer) Koch in der Gattung *Alyssum* L. (Cruciferae) in der Flora der Republik Makedonien [In Macedonian with German summary]. - *Maked. Akad. Nauk. Umetn., Oddel. Biol. Med. Nauki* **15(1-2)**: 41-58.

MICEVSKI, K. 1995: Flora na Republika Makedonija, 1(3) [In Macedonian]. - Skopje: Makedoniska Akademija Naukite Umetnostite, pp. [4+] 401-775.

MICEVSKI, K. 1998: Flora na Republika Makedonija, 1(4) [In Macedonian]. - Skopje: Makedoniska Akademija Naukite Umetnostite, pp. 781-1113.

MICEVSKI, K. 2001: Flora na Republika Makedonija, 1(5) [In Macedonian]. - Skopje: Makedoniska Akademija Naukite Umetnostite, pp. [4+] 1121-1430.

MICEVSKI, K. 2005: Flora na Republika Makedonija, 1(6) [In Macedonian]. - Skopje: Makedoniska Akademija Naukite Umetnostite, pp. [4+] 1437-1715.

MICEVSKI, K. & KOSTADINOVSKI, M. 1991: Beitrag zur Flora von Makedonien. I. - *Godišen Zborn. Biol. Fak. Univ. "Kiril i Metodij"* **43-44**: 191-198.

MICEVSKI, K. & MATEVSKI, V. 1980: *Thymus skopjensis* Micevski & Matevski, spec. nov. in der Flora von Makedonien [In Macedonian with Latin diagnosis and German summary]. - *Maked. Akad. Nauk. Umetn., Oddel. Biol. Med. Nauki* **1(2)**: 35-38 + 1 plate.

MICEVSKI, K. & MATEVSKI, V. 1983: Seltene und veniger bekannte Pflanzen in der Flora von Makedonien. I [In Macedonian with German summary]. - *Godišen Zborn. Biol. Fak. Univ. "Kiril i Metodij"* **36**: 149-153.

MICEVSKI, K. & MATEVSKI, V. 1985: *Thymus ciliatopubescens* (Hal.) Hal. var. *bistrae* Micevski et Matevski var. nov. in der Flora von Makedonien [In Macedonian with German summary]. - *Maked. Akad. Nauk. Umetn., Oddel. Biol. Med. Nauki* **6(1-2)**: 21-25.

MICEVSKI, K. & MATEVSKI, V. 1986: Taxonomische und chorologische Besonderheit der Gattung *Thymus* L. des Gebirges Bistra in Makedonien [In Serbo-Croatian with German summary]. - *Biosistematika* **12(1)**: 23-29.

MICEVSKI, K. & MATEVSKI, V. 1987a: Seltene und wenig bekannte Pflanzen in der Flora von Makedonien. II [In Macedonian with German summary]. - *Godišen Zborn. Biol. Fak. Univ. "Kiril i Metodij"* **39-40**: 203-208.

MICEVSKI, K. & MATEVSKI, V. 1987b: *Kitaibelia vitifolia* Willd. in der Flora der SR Makedonien [In Macedonian with German summary]. - *Godišen Zborn. Biol. Fak. Univ. "Kiril i Metodij"* **39-40**: 209-213.

MICEVSKI, K. & MATEVSKI, V. 1990: *Astragalus austriacus* Jacq. in der Flora der SR Makedonien [In Macedonian with German summary]. - *Godišen Zborn. Biol. Fak. Univ. "Kiril i Metodij"* **41-42**: 187-189.

MICEVSKI, K. & MATEVSKI, V. 1991: *Cirsium erisethales* (Jacq.) Scop. (Compositae) in der Flora der Republik Makedonien [In Macedonian with German summary]. - *Godišen Zborn. Biol. Fak. Univ. "Kiril i Metodij"* **43-44**: 185-190.

MICEVSKI, K. & MATEVSKI, V. 1992: *Conringia austriaca* Adanson, eine neue Art in der Flora der Republik Makedonien. - *Godišen Zborn. Biol. Fak. Univ. "Kiril i Metodij"* **45**: 163-165.

MICEVSKI, K. & MATEVSKI, V. 1993: *Delphinium albiflorum* DC. (Ranunculaceae) in der Flora der Republik Makedonien [In Macedonian with German summary]. - *Godišen Zborn. Biol. Fak. Univ. "Kiril i Metodij"* **45**: 111-114.

MICEVSKI, K. & MAYER, E. 1970: Zur Kenntnis der *Saxifraga grisebachii* Degen et Dörfler. - *Feddes Repert.* **80(7-8)**: 599-605.

MICEVSKI, K. & MAYER, E. 1980a: *Lilium heldreichii* Freyn in der Flora von Makedonien [In Macedonian with German summary]. - *Godišen Zborn. Biol. Fak. Univ. "Kiril i Metodij"* **33**: 97-102.

MICEVSKI, K. & MAYER, E. 1980b: *Gnaphalium hoppeanum* Koch im Gebirge Galićica in Makedonien [In Macedonian with German summary]. - *Godišen Zborn. Biol. Fak. Univ. "Kiril i Metodij"* **33**: 103-109.

MICEVSKI, K. & MAYER, E. 1982: *Ephedra major* Host in der Flora von Makedonien [In Serbo-Croatian with German summary]. - *Glasn. Republ. Zavoda Zaštitu Prir. Prirodnjačke Zbirke Titogradu* **15**: 55-62.

MICEVSKI, K. & MAYER, E. 1983: *Genista fukarekiana* K. Micevski & E. Mayer spec. nova in der Flora von Makedonien. - *Rad. Odjel. Prirod. Mat. Nauka (Sarajevo)* **21**: 193-197.

MICEVSKI, K. & MAYER, E. 1984: *Astragalus* L. subg. *Tragacantha* Bunge in der Flora von Jugoslawien. - *Acta Bot. Croat.* **43**: 307-314 + 1 plate.

MICEVSKI, K. & MAYER, E. 1985: *Ferulago macedonica* Micevski & E. Mayer spec. nova - eine neue endemische Art der Balkanhalbinsel. - *Acta Bot. Croat.* **44**: 77-81 + 2 plates.

MICHAEL, P.W. 1964: The identity and origin of varieties of *Oxalis pes-caprae* L. naturalized in Australia. - *Trans. Roy. Soc. South Australia* **88**: 167-173.

MIČIETA, K. 1981: Zytotaxonomische Probleme einiger Pflanzensippen des Javorníky-Gebirges. - *Acta Fac. Rerum Nat. Univ. Comen., Bot.* **28**: 95-104.

MIČIETA, K. 1982: Zytotaxonomischer Beitrag zur Flora des Javorníky-Gebirges II. - *Acta Fac. Rerum Nat. Univ. Comen., Bot.* **29**: 55-61.

MIČIETA, K. 1983a: Contribution to the chromosome numbers of some species of the genus *Juncus* in Slovakia. - *Folia Geobot. Phytotax.* **18(2)**: 195-198.

MIČIETA, K. 1983b: Karyotaxonomical studies in the *Juncus bufonius* aggregate in Slovakia. – *Folia Geobot. Phytotax.* **18(3)**: 329-331.

MIČIETA, K. 1985: Clavis analytica specierum Slovacarum generis *Juncus* L. - *Acta Fac. Rerum Nat. Univ. Comen., Bot.* **32**: 21-25.

MIČIETA, K. 1986: Karyological study of the Slovak flora XI. - *Acta Fac. Rerum Nat. Univ. Comen., Bot.* **33**: 51-55.

MIÈGE, J. & GREUTER, W. 1973: Nombres chromosomiques de quelques plantes récoltées en Crète. - *Ann. Mus. Goulandris* **1**: 105-111 + 1 plate.

MÍGUEZ, M. , MARTÍN-BRAVO, S. & JIMÉNEZ-MEJÍAS, P. 2018: Reconciling morphology and phylogeny allows an integrative taxonomic revision of the giant sedges of *Carex* section *Rhynchocystis* (Cyperaceae). - *Bot. J. Linn. Soc.* **188(1)**: 34-58.

MIJUŠKOVIĆ, M. 1973: *Arceuthobium oxycedri* (DC.) M.B. (Loranthaceae), plante semiparasite non signalée auparavant au Montenegro [In Serbo-Croatian with French summary]. - *Glasn. Republ. Zavoda Zaštitu Prir. Prirodnjačke Zbirke Titogradu* **5**: 25-28.

MILIARAKIS, S. 1910: Ta dimodi onomata ton fiton prosdiorizomena epistimonikos [On the vernacular names of Greek plants; in Greek, based on material left by Heldreich]. - Athine: xvii + 236 pp.

MILL, R.R. 1979: Taxonomic studies on generic limits in the family Boraginaceae (tribe Cynoglosseae). - St. Andrews: Ph.D. thesis, Univ. of St. Andrews, 787 pp.

MILL, R.R. 1996: The correct authority for the lesser chickweed, *Stellaria pallida* (Caryophyllaceae). - *Watsonia* **21(2)**: 204-205.

MILL, R.R. 2011: A new subspecies of *Omphalodes luciliae* (Boraginaceae) from Turkey. - *Edinburgh J. Bot.* **68(2)**: 159-163.

MILL, R.R. & TAN, KIT 1989: Ian Charleson Hedge: Biographical sketch; list of publications. In Tan, Kit (ed.): Plant taxonomy, phytogeography and related subjects. The Davis & Hedge festschrift, pp. xx-xxvi. - Edinburgh: Edinburgh Univ. Press.

MILLAKU, F. & ELEZAJ, I. 2015: *Tulipa luanica* (Liliaceae), a new species from southern Kosovo. - *Ann. Bot. Fenn.* **52(3-4)**: 315-320.

MILLER, H.F.R. 1935: An introduction to the genus *Sempervivum*. - *Quart. Bull. Alpine Gard. Soc. Gr. Brit.* **3(4)**: 254-290 + 2 plates.

MILLER, PH. 1768: The gardeners dictionary: containing the best and newest methods of cultivating and improving the kitchen, fruit, flower garden and nursery ... Ed. 8. - London [cf. TL-2 3: 495, 1981].

MILLETTI, N. 1987: Revisione sistematica del genere *Globularia* L. (Globulariaceae) in Italia. - Firenze: Ph.D. diss., Univ. degli Studi di Firenze.

MILLETTI, N. & JARVIS, C.E. 1987: Linnaean names in the genus *Globularia* L. (Globulariaceae) and their typification. - *Taxon* **36(3)**: 635-639.

MILLWARD, D. 2000: Mountainous northern Greece and its plant communities. - *Quart. Bull. Alpine Gard. Soc. Gr. Brit.* **68(3)**: 359-373.

MILNE-REDHEAD, E. 1936: *Campanula formanekiana*. - *Bot. Mag.* **159(2)**: tab. 9436 + 4 pp. text.

MILNE-REDHEAD, E. 1939: *Campanula incurva*. - *Bot. Mag.* **161(4)**: tab. 9556 + 2 pp. text.

MILNE-REDHEAD, E. & TURRILL, W.B. 1929: *Campanula orphanidea* Boiss. - *Bull. Misc. Inform. Kew* **1929(4)**: 142.

MINARECI, E., ALTAN, Y. & AKTAN, T. 2011: A new record from Turkey: *Consolida samia* P.H. Davis (Ranunculaceae). - *J. Anim. Pl. Sci.* **21(3)**: 553-555.

MINISSALE, P., MOLINA, J.A. & SCIANDRELLO, S. 2017: *Pilularia minuta* Durieu (Marsileaceae) discovered in south-eastern Sicily: new insights into its ecology, distribution and conservation status. - *Acta Bot. Gallica, Botany Letters* **164(3)**: 197-208.

MINOGLOU, D. 2000: I chlorida tou panepistimiakou dasous Pertouliou Trikalon. [In Greek]. - Thessaloniki: Ph.D. thesis, Aristotelian University, 190 pp.

MINOGLOU, D. & ATHANASIADIS, N. 1999: Meadows of Pertouli - Analysis of their flora and its forage value. In Elefheriou, E.P. (ed.): Proceedings, 7th Panhellenic scientific conference, Alexandroupolis, pp. 336-341 [In Greek with English abstract]. - Thessaloniki: Univ. Studio Press.

MINOGLOU, D. & ATHANASIADIS, N. 2001: Endemic and protected taxa of the University forest of Petrouli-Trikala. In Kamari, Psaras & Constantinidis (eds): Proceedings of the 8th scientific congress of the Greek Botanical Society, October 5-8, 2000, pp. 292-301 [In Greek with English abstract].

MIREK, Z. 1981: Genus *Camelina* in Poland - Taxonomy, distribution and habitats. - *Fragm. Florist. Geobot.* **27(3)**: 445-507 + 1 table.

MIREK, Z. & FISCHER, M.A. 1986: Additions to the ecogeography of *Veronica vindobonensis* with special reference to Poland. (Cytotaxonomy of *Veronica chamaedrys* agg., IV). - *Phyton (Horn)* **26 (1)**: 107-129.

MIRKOVIĆ, D. 1966: Cytotaxonomical investigations of species of the genus *Leucanthemum* Adans. em. Briqu. et Cav. from Yugoslavia [In Serbo-Croatian with English summary]. - *Acta Bot. Croat.* **25**: 137-152 + 1 plate.

MIRKOVIĆ, D. 1969: Cytotaxonomical and cytogenetical investigation of species of the genus *Leucanthemum* from Yugoslavia - II. Comparative studies on diploids: *L. rotundifolium* (W. K.) DC. and *L. praecox* H-ić. - *Acta Bot. Croat.* **28**: 245-252 + 4 plates.

MIROV, N.T. 1955: Relationships between *Pinus halepensis* and other *Insignes* pines of the Mediterranean region. - *Bull. Res. Council Israel, Sect. D, Bot.* **5(1)**: 65-72.

MIROV, N.T. 1967: The genus *Pinus*. - New York: The Ronald Press Co., viii + 602 pp.

MISSETZIS, G.N. & SALIARIS, P.A. 2009: Erstnachweis von *Comperia comperiana* auf Chios (Ägäis, Griechenland). - *J. Eur. Orch.* **41(3-4)**: 695-696.

MITCHELL, B. 1987: *Paris* - part I. - *Plantsman* **9(2)**: 81-89.

MITCHELL, P.J. 1973: The *Sempervivum* and *Jovibarba* handbook. - Burgess Hill: Sempervivum Society, 44 pp.

MITCHELL, P.J. 1976: Balkan Sempervivums. - *J. Sempervivum Soc.* **7(1-3)**: 13-20, 41-44, 64-68.

MITCHELL, P.J. 1978a: *Sempervivum ballsii* Wale. - *J. Sempervivum Soc.* **9(1)**: 22-26.

MITCHELL, P.J. 1978b: *Sempervivum cilosum* Wale. - *J. Sempervivum Soc.* **9(3)**: 76-79.

MITCHELL, P.J. 1978b: *Sempervivum cilosum* Wale. - *J. Sempervivum Soc.* **10(1)**: 19-20.

MITCHELL, P.J. 1983: The graecian houseleek. - *Houselekes* **14(2)**: 52-58.

MITRA, J. 1956: Karyotype analysis of bearded *Iris*. - *Bot. Gaz.* **117(4)**: 265-293.

MITRAKOS, K. 1989: The botany of *Ceratonia*. - *Bios (Thessaloniki)* **1989**: 149-157.

MITRAKOS, K., DIAMANTOGLOU, S. & CHRISTODOULAKIS, N.S. 1991: Distribution patterns of four evergreen sclerophyllous species in Greece. - *Bot. Jahrb. Syst.* **113(1)**: 1-6.

MITRUSHI, I. 1966: Dendroflora e Shqipërisë [In Albanian]. - Tiranë: Universiteti Shtetëror, 519 pp.

MITSOPOULOS, D.J. & PANETSOS, C.P. 1987: Origin of variation in fir forests of Greece. - *Silvae Genet.* **36(1)**: 1-15.

MITSUOKA, S. & EHRENDORFER, F. 1972: Cytogenetics and evolution of *Matricaria* and related genera (Asteraceae-Anthemideae). - *Oesterr. Bot. Z.* **120(1-2)**: 155-200.

MIZIANTY, M. 1990: Biosystematic studies on *Dactylis* L. 1. Review of the previous studies. 1.2. Cytology, genetics, experimental studies, and evolution. - *Acta Soc. Bot. Poloniae* **59(1-4)**: 105-118.

MLAKAR, J. 1987: The genus *Polypodium* in Slovenia [In Slovenian with English summary]. - *Biol. Vestn.* **35(1)**: 45-58.

MNAJED, H., COMBIER, H. & LEBRETON, PH. 1972: Différenciation des espèces *Sedum acre* L. et *Sedum sexangulare* (auct. non L.). - *Bull. Mens. Soc. Linn. Soc. Bot. Lyon* **41(1)**: 6-9.

MÖBIUM, M. 1933: Pflanzen der altkretischen Kunst. - *Bot. Jahrb. Syst.* **66(1)**: 105-112.

MOGGI, G., MILLETTI, N. & PAOLI, P. 1987: Bibliografia geobotanica toscana. - Firenze: Museo Botanica dell'Università, 129 pp.

MOGIE, M. & RICHARDS, A.J. 1983: Satellited chromosomes, systematics and phylogeny in *Taraxacum* (Asteraceae). - *Pl. Syst. Evol.* **141(3-4)**: 219-229.

MOHR, O.W.P. & MEEUSE, A.D.J. 1991: Note on some characteristics of *Ephedra campylopoda* C.A. Mey. - *Israel J. Bot.* **40(4)**: 315-322.

MOLERO BRIONES, J. 1993: (1073) Proposal to conserve *Euphorbia falcata* L. (Euphorbiaceae) with a conserved type - *Taxon* **42(3)**: 715-717. [approved, cf. Taxon 44(4): 610, 1995].

MOLERO, J. & ROVIRA, A.M. 1987: Taxonomía del grupo "Fumana thymifolia" (Cistaceae). - *Candollea* **42(2)**: 501-531.

MOLERO, J. & ROVIRA, A.M. 1992: *Euphorbia* L. subsect. *Esula* (Boiss. in DC.) Pax in the Iberian Peninsula. Leaf surface, chromosome numbers and taxonomic treatment. - *Collect. Bot. (Barcelona)* **21**: 121-181.

MOLERO, J. & ROVIRA, A.M. 1998: (1349) Proposal to conserve the name *Euphorbia obtusifolia* Poir. (Euphorbiaceae) against an earlier homonym. - *Taxon* **47(2)**: 469-470.

MOLERO, J. & ROVIRA, A.M. 2005: Typification of some Macaronesian and Mediterranean dendroid spurges. - *Taxon* **54(2)**: 472-474.

MOLERO, J., ROVIRA, A.M. & VICENS, J. 1996: *Euphorbia* L. sect. *Cymatospermum* (Porkh.) Porkh. (Euphorbiaceae) en la Península Ibérica. Morfología de las semillas. Precisiones taxonómicas y corológicas sobre algunos táxones críticos. - *Anales Jard. Bot. Madrid* **54(1)**: 207-229.

MOLINA, A., ACEDO, C. & LLAMAS, F. 2006: (1722) Proposal to conserve the name *Carex rostrata* against *C. inflata* (Cyperaceae) with a conserved type. - *Taxon* **55(2)**: 532-533. [approved, cf. Taxon 56(4): 1292, 2007].

MOLINA, A., ACEDO, C. & LLAMAS, F. 2008a: Taxonomy and new taxa of the *Carex divulsa* aggregate in Eurasia (section *Phaestoglochin, Cyperaceae*. - *Bot. J. Linn. Soc.* **156(3)**: 385-409.

MOLINA, A., ACEDO, C. & LLAMAS, F. 2008b: (1817) Proposal to conserve the name *Carex leersii* F.W. Schultz against *C. leersii* Willd. and *C. chabertii* (Cyperaceae). - *Taxon* **57(2)**: 648-649. [approved, cf. Taxon 59(4): 1274, 2010].

MOLINA, A. & AL. 2006: Typification of some of Hudson's plant names in *Carex* L. - *Taxon* **55(4)**: 1009-1013.

MØLLER, K.B. 1994: A floristic inventory of the East Aegean island Patmos. - Copenhagen: Unpubl. M.Sc. thesis, 122 pp. + appendix 1-4.

MOLNAR, A.V. 1998: Über die Bestäubung von *Ophrys helenae* Renz. - *J. Eur. Orch.* **30(3)**: 588-590.

MOLNÁR, A.V. & AL. 2012: *Himantoglossum jankae* (Orchidaceae: Orchideae), a new name for a long- misnamed lizard orchid. - *Phytotaxa* **73**: 8-12.

MONASTERIO-HUELIN, E. & WEBER, H.E. 1996: Taxonomy and nomenclature of *Rubus ulmifolius* and *Rubus sanctus* (Rosaceae). - *Edinburgh J. Bot.* **53(3)**: 311-322.

MONNET, A.-C. & AL. 2021: WOODIV, a database of occurrences, functional traits, and phylogenetic data for all Euro-Mediterranean trees. - *Scientific Data* **8, 89**: doi.org/10.1038/s41597-021-00873-3.

MONNIER, P. 1976: Systématique et biosystématique du genre *Spergularia* dans le bassin méditerranéen occidental. - *Naturalia Monspel., Sér. Bot.* **25**: 65-87.

MONRO, A. & SPENCER, M.A. 2005: Typification of Linnaean plant names in Urticaceae. - *Taxon* **54(3)**: 796-798.

MONTI, G. & GARBARI, F. 1974: Appunti citotassonomici sul genere *Biarum* Schott (Araceae) in Italia. - *Giorn. Bot. Ital., n.s.* **108(1-2)**: 19-26.

MONTI, G., PAGNI, A.M. & VIEGI, L. 1978: Numeri cromosomici per la Flora Italiana: 416-422. - *Inform. Bot. Ital.* **10(1)**: 101-110.

MONTMOLLIN, B. DE 1982: Étude cytotaxonomique de la flore endémique de la Crète. I. Note préliminaire. - *Bull. Soc. Neuchâteloise Sci. Nat.* **105**: 65-77.

MONTMOLLIN, B. DE 1984: Étude cytotaxonomique de la flore de la Crète. II. Nombres chromosomiques. - *Bot. Helv.* **94(2)**: 261-267.

MONTMOLLIN, B. DE 1986: Étude cytotaxonomique de la flore de la Crète. III. Nombres chromosomiques. - *Candollea* **41(2)**: 431-439.

MONTMOLLIN, B. DE 1989: Endémisme intra-crétois et trans-égéo-crétois, origine et différenciation. In Phitos, D. (ed.): Abstracts. Sixth OPTIMA meeting, Delphi, 10-16 Sept. 1989, p. 17.

MONTMOLLIN, B. DE 1991: Endémisme intra-crétois et trans-égéo-crétois - origine et différenciation. - *Bot. Chron. (Patras)* **10**: 125-135.

MONTMOLLIN, B. DE & IATROÚ, G. 1995: Connaissance et conservation de la flore de l'île de Crète. - *Ecol. Medit.* **21(1-2)**: 173-184.

MONTSERRAT MARTÍNEZ, M. & AL. 2000: (1442) Proposal to conserve the name *Veronica agrestis* L. (Scrophulariaceae) with a conserved type. - *Taxon* **49(1)**: 99-100. [withdrawn, cf. Taxon 51(4): 795, 2002].

MONTSERRAT RECODER, P. 1963: El género *Luzula* en España. - *Anales Inst. Bot. Cavanilles* **21(2)**: 407-523 + 17 plates.

MONTSERRAT RECODER, P. 1988: La corología y especialización en algunas *Minuartia*. - *Lazaroa* **9**: 189-200.

MOODY, M.L. & LES, D.H. 2010: Systematics of the aquatic angiosperm genus *Myriophyllum* (Haloragaceae). - *Syst. Bot.* **35(1)**: 121-139.

MOONEY, H.A. 1988: Lessons from mediterranean-climate regions. In Wilson, E.O. (ed.): Biodiversity. National Academy Press.

MOONEY, H.A. & DUNN, E.L. 1970: Convergent evolution of Mediterranean-climate evergreen sclerophyll shrubs. - *Evolution* **24(2)**: 292-303.

MOORE, A.J. & DILLENBERGER, M.S. 2017: A conspectus of the genus *Cherleria* (*Minuartia* s.l., Caryophyllaceae). - *Willdenowia* **47(1)**: 5-15.

MOORE, D.M. 1963: The subspecies of *Montia fontana* L. - *Bot. Not.* **116(1)**: 16-30.

MOORE, D.M. 1971: Chromosome studies in the Umbelliferae. - *Bot. J. Linn. Soc.* **64, suppl. 1**: 233-255.

MOORE, D.M. 1982: Flora Europaea check-list and chromosome index. - Cambridge, etc.: Cambridge Univ. Press, x + 423 pp.

MOORE, P. 1993: Peloponnese expedition, April 1993. - *Cyclamen* **17(2)**: 50-53.

MOORE, P. 1996: *Cyclamen repandum* subsp. *peloponnesiacum* forma *vividum*. - *Cyclamen* **20(1)**: 13-14.

MOORE, P. & FRANK, E. 1996: *Cyclamen graecum*. - *Cyclamen* **20(2)**: 44-46 + 2 plates.

MOORE, P. & JOPE, M. 2011: The *Cyclamen* of Greece. A guide to the species of *Cyclamen* growing in Greece. - Cyclamen Society: 40 pp.

MOORE, P., JOPE, M. & MOORE, N. 2001: *Cyclamen hederifolium* var. *confusum* in Crete. – *Cyclamen* **25(1)**: 25-26.

MOORE, R.J. (ED.) 1970: Index to plant chromosome numbers for 1968. - *Regnum Veg.* **68**: 1-119.

MOORE, R.J. (ED.) 1971: Index to plant chromosome numbers for 1969. - *Regnum Veg.* **77**: 1-116.

MOORE, R.J. (ED.) 1972: Index to plant chromosome numbers for 1970. - *Regnum Veg.* **84**: 1-138.

MOORE, R.J. (ED.) 1973: Index to plant chromosome numbers for 1971. - *Regnum Veg.* **90**: 1-539.

MOORE, R.J. (ED.) 1974: Index to plant chromosome numbers for 1972. - *Regnum Veg.* **91**: 1-108.

MOORE, R.J. (ED.) 1977: Index to plant chromosome numbers 1973/74. - *Regnum Veg.* **96**: 1-257.

MOORE, S. 1878: [Description of *Centaurea ebenoides* Heldr. ex S. Moore]. - *J. Bot. (London)* **16**: 133.

MORALDO, B. 1986: Il genere *Stipa* L. (Gramineae) in Italia. - *Webbia* **40(2)**: 203-278.

MORALDO, B. & RICCERI, C. 2003: Alcune novità tassonomico-nomenclaturali sul genere *Stipa* L. (Poaceae) in Italia. - *Webbia* **58(1)**: 103-111.

MORALES VALVERDE, R. 1986: Taxonomía de los géneros *Thymus* (excluida la sección *Serpyllum*) y *Thymbra* en la Península Ibérica. - *Ruizia* **3**: 1-324.

MORALES VALVERDE, R. 1987: El género *Thymbra* L. (Labiatae). - *Anales Jard. Bot. Madrid* **44(2)**: 349-380.

MORALES VALVERDE, R. 1993: Sinopsis y distribución del género *Micromeria* Bentham. - *Bot. Complut.* **18**: 157-168.

MORALES VALVERDE, R. 1994a: Nota sobre el género *Clinopodium* L. (Labiatae). - *Anales Jard. Bot. Madrid* **51(2)**: 307-308.

MORALES VALVERDE, R. 1994b: Números cromosómicos para la flora española. 764-768. – *Lagascalia* **17(2)**: 388-391.

MORALES VALVERDE, R. 1995: Synopsis of the genus *Thymus* L. in the Mediterranean area. In Valdés, B. & Silvestre, S. (eds): Abstracts. VIII OPTIMA meeting, Sevilla, Spain, 25 September - 1 October, 1995, p. 30.

MORALES VALVERDE, R. 1997: Synopsis of the genus *Thymus* L. in the Mediterranean area. - *Lagascalia* **19(1-2)**: 249-261.

MORALES VALVERDE, R. 2002: The history, botany and taxonomy of the genus *Thymus*. In Stahl- Biskup, E. & Saez, F. (eds): Thyme: the genus *Thymus*, pp. 1-43. - London: Taylor & Francis.

MORALES VALVERDE, R. & CASTILLO, J. 2004: El género *Sternbergia* (Amaryllidaceae) en la Península Ibérica. - *Anales Jard. Bot. Madrid* **61(2)**: 119-128.

MORALES VALVERDE, R. & LUQUE, M.N. 1997: El género *Calamintha* (Labiatae) en la Península Ibérica e Islas Baleares. - *Anales Jard. Bot. Madrid* **55(2)**: 261-276.

MORALES, A.Á., NAVARRO ANDRÉS, F. & ANGELOS SÁNCHEZ, A. 2012: Datos corológicos y morfológicos de las especies del género *Ambrosia* L. (Compositae) presentes en la Península Ibérica. - *Bot. Complut.* **36**: 85-96.

MORARIU, I. 1981: Genul *Gentianella* Moench în flora României [In Rumanian]. - *Stud. Cercet. Biol. (Bucharest), Ser. Biol. Veg.* **33(2)**: 109-114.

MORAZZONI, P. & BOMBARDELLI, E. 1995: *Silybum marianum (Carduus marianus)*. - *Fitoterapia* **66(1)**: 3-42.

MORENO MORAL, G., SÁNCHEZ PEDRAJA, Ó. & PIWOWARCZYK, R 2018: Contributions to the knowledge of *Cistanche* (Orobanchaceae) in the Western Palearctic. - *Phyton (Horn)* **57(1-2)**: 19-36.

MORENO SAIZ, J.C., PATARO, L. & SOTOMAYOR, P. 2009: Atlas de los pteridófitos de la Península Ibérica e e Islas Baleares. - *Acta Bot. Malac.* **50**: 5-55.

MORENO, M. 1984a: Notas de taxonomia y nomenclatura: *Iberis candolleana* Jordan, *I. epirota* Halácsy e *I. carnosa* Waldst. & Kit., non Willd. (Cruciferae). - *Taxon* **33(1)**: 99-100.

MORENO, M. 1984b: Aproximación taxonómica a las poblaciones españolas de *Iberis carnosa* Willd. (= *Iberis pruitii* Tineo). - *Anales Jard. Bot. Madrid* **41(1)**: 43-57.

MORENO, M. 1985: Números cromosómicos para la flora española. 394-403. - *Lagascalia* **13(2)**: 307-311.

MORET, J. 1991: Les stratégies de reproduction du complexe polyploïde *Ornithogalum umbellatum* (Hyacinthaceae) en France. - *Bull. Mus. Natl. Hist. Nat., B, Adansonia* **13(1-2)**: 25-46.

MORET, J. & AL. 1992: Gynodioecy, herkogamy, and sex-ratio in *Romulea bulbocodium* var. *dioica* (Iridaceae). - *Evol. Trends Pl.* **6(2)**: 99-109.

MORET, J. & COUDERC, H. 1986: Contribution of caryology to the systematic knowledge of the *Ornithogalum* L. genus in North Africa: The *Heliocharmos* Baker sub-genus. - *Caryologia* **39(3-4)**: 259-272.

MORET, J. & FAVEREAU, Y. 1991: Balance reproduction sexuée / multiplication végétative dans une complexe polyploïde du genre *Ornithogalum* (Liliaceae) en Méditerranée occidentale. - *Bull. Soc. Bot. France, Lett. Bot.* **138(3)**: 201-214.

MORET, J. (ED.) 1998: Abstracts of the IX. OPTIMA meeting, Paris, 11-17 May 1998. - Paris: Mus. Nat. d'Hist. Naturelle, 158 pp.

MORGAN-RICHARDS, M. & WOLFF, K. 1999: Genetic structure and differentiation of *Plantago major* reveals a pair of sympatric sister species. - *Molec. Ecol.* **8(6)**: 1027-1036.

MÖRIKE, T. 2018: Botanical explorations by Frederik Hasselquist (1749-1752) and Pehr Forsskål (1761-1763): Linné's apostles in the Holy Land. In Aščerić-Todd, I. & al. (eds): Travellers in Ottoman lands. The Botanical Legacy, pp. 109-121. - Oxford: ASTENE and Archaeopress Publishing Ltd, 379 pp.

MORIS, G.G. 1837-59: Flora sardoa seu historia plantarum in Sardinia et adjacentibus insulis vel sponte nascentium vel ad utilitatem excultarum. - Torino: 3 vols [cf. TL-2 3: 586, 1981].

MORISSET, P. 1978: Chromosome numbers in *Ononis* L. series *Vulgares* Širj. - *Watsonia* **12(2)**: 145-153.

MORLEY, F.H.W. 1983: Validation of combinations in the *Trifolium subterraneum* complex (Fabaceae). - *Taxon* **32(3)**: 466.

MORRONE, O. & AL. 2014: Revision of the Old World Species of *Setaria* (Poaceae: Panicoideae: Paniceae). - *Syst. Bot. Monogr.* **96**: 1-161.

MORSCHEK, G. & MORSCHEK, K. 1996: Orchids of Cyprus. A guide to the orchid flora of Cyprus [text in two columns, in English and German]. - Published by the authors: 185 pp.

MORSCHHAUSER, T. & PAPP, B. 1996: Botanical collection trip in Greece [In Hungarian with English abstract]. - *Bot. Közlem.* **81(1)**: 13-20.

MORTON, C.V. 1969: The fern genus *Gymnocarpium* Newm. and its typification. - *Taxon* **18(6)**: 661-662.

MORTON, F. 1914: Beiträge zur Kenntnis der Pteridophytengattung *Phyllitis*. - *Oesterr. Bot. Z.* **64(1-2)**: 19-36.

MORY, B. 1979a: Beiträge zur Kenntnis der Sippenstruktur der Gattung *Glaucium* Miller (Papaveraceae). - *Feddes Repert.* **89(9-10)**: 499-594 + 4 plates.

MORY, B. 1979b: Ergebnisse einer Revision und Bestimmungsschlüssel der Gattung *Glaucium* Miller (Papaveraceae). - *Gleditschia* **7**: 117-126.

MÖSCHL, W. 1933: Zwei neue *Cerastium*-Arten der Balkanhalbinsel. - *Oesterr. Bot. Z.* **82(3)**: 226-234.

MÖSCHL, W. 1936a: Eine neue *Cerastium*-Art der Balkanhalbinsel. - *Repert. Spec. Nov. Regni Veg.* **39**: 374-376 + 1 plate.

MÖSCHL, W. 1936b: Über einjährige europäische Arten der Gattung *Cerastium* (*Orthodon-Fugacia-Leiopetala*). - *Repert. Spec. Nov. Regni Veg.* **41**: 153-163.

MÖSCHL, W. 1938: Morphologie einjähriger europäischer Arten der Gattung *Cerastium* (*Orthodon-Fugacia-Leiopetala*). - *Oesterr. Bot. Z.* **87(4)**: 249-272.

MÖSCHL, W. 1943: Die Sippe des *Cerastium ramosissimum* Boissier. - *Oesterr. Bot. Z.* **92(4)**: 161-182.

MÖSCHL, W. 1949a: *Cerastium semidecandrum* Linné, sensu latiore. - *Mem. Soc. Brot.* **5**: 5-123.

MÖSCHL, W. 1949b: *Cerastium gracile* Dufour. - *Collect. Bot. (Barcelona)* **2(2)**: 165-198.

MÖSCHL, W. 1951: Cerastia lusitaniae archipelagorumque "Açores" et "Madeira". - *Agron. Lusit.* **13 (1)**: 23-66.

MÖSCHL, W. 1961: *Cerastium vourinense* Möschl & Rechinger species nova. - *Bol. Soc. Brot., Ser. 2* **35**: 129-134 + 2 plates.

MÖSCHL, W. 1962a: *Cerastium epiroticum* Möschl & Rechinger, species nova. - *Bol. Soc. Brot., Ser. 2* **36**: 41-46 + 2 plates.

MÖSCHL, W. 1962b: *Cerastium runemarkii* Möschl & Rechinger, nom. novum. - *Anz. Österr. Akad. Wiss., Math.-Naturwiss. Kl.* **99(14)**: 231-232.

MÖSCHL, W. 1973: Über die Cerastien Österreichs. - *Mitt. Naturwiss. Vereines Steiermark* **103**: 141-169.

MOSKOVA, R.D. & GEORGIEV, G.N. 1972: Karyological data on *Scilla bifolia* L. - *Dokl. Bulg. Akad. Nauk* **25(1)**: 111-112.

MOSQUIN, T. 1966: A new taxonomy for *Epilobium angustifolium* L. (Onagraceae). - *Brittonia* **18(2)**: 167-188.

MOSSA, L. & SCRUGLI, A. 1970: Osservazioni cariologiche in *Allium chamaemoly* L. - *Morisia* **2**: 53-62.

MOSYAKIN, S.L. 2017a: Taxonomic and nomenclatural notes on Pontic-Mediterranean coastal and some Australasian taxa of *Salsola* (Chenopodiaceae). - *Ukrainian Bot. J.* **74(6)**: 521-531.

MOSYAKIN, S.L. 2017b: Notes on taxonomy and nomenclature of *Chenopodium acerifolium* and *C. betaceum* (*C. strictum* auct.) (Chenopodiaceae). - *Phytotaxa* **324(2)**: 139-154.

MOSYAKIN, S.L. & BROCK, J.R. 2021: On the proper type designation for *Camelina microcarpa*, a wild relative and possible progenitor of the crop species *C. sativa* (Brassicaceae). - *Candollea* **76(1)**: 55-63.

MOSYAKIN, S.L., FREITAG, H. & RILKE, S. 2017: *Kali* versus *Salsola*: the instructive story of a questionable nomenclatural resurrection. - *Israel J. Pl. Sci.* **64(1-2)**: 18-30.

MOTZKUS, B. 1974: Beiträge zur Kenntnis einiger Chromosomenzahlen der Gattung *Coronilla* L. - *Feddes Repert.* **84(9-10)**: 741-746.

MOTZKUS, B. 1976: Beitrag zur Kenntnis der Chromosomenzahlen und der Meioseablaufs in der Gattung *Coronilla*. - *Gleditschia* **4**: 23-30.

MOTZKUS, B. 1980: Die taxonomische Gliederung der Gattung *Coronilla*. In Vent, W. (ed.): Beiträge zu Prinzipien und Problemen det Systematik und Evolutionsforschung, pp. 45-61. - Berlin: Humboldt-Universität.

MOULALIS, D. & ILLIES, Z.M. 1975: Vergleichende zytologische Untersuchungen der Chromosomenstruktur von *Abies borisii regis* Mattf., *A. cephalonica* Loud. und *A. alba* Mill. - *Silvae Genet.* **24(4)**: 115-118.

MOULOPOULOS, CH. 1965: The beech woods of Greece. Part I. The beech species and their distribution in Greece [In Greek with English summary]. - Thessaloniki: 88 pp. + 19 plates and 2 maps.

MOUNTFORT, C.C. 1948: In search of *Crocus*. - *Quart. Bull. Alpine Gard. Soc. Gr. Brit.* **16(2)**: 107-117.

MOUNTFORT, C.C. 1959: *Crocus sieberi*. - *Quart. Bull. Alpine Gard. Soc. Gr. Brit.* **27(4)**: 362-363.

MOUNTOUSIS, I.A. 2008: Flora and vegetation of Mount Varnouda pasturelands and seasonal variations in nutritive value of grazable material. [In Greek with English summary]. - Thessaloniki: Aristotelian University, 217 pp.

MRKVICKA, A.CH. 1991: *Spiranthes aestivalis* (Poir.) Rich. - Beobachtungen zu Keimung, Entwicklung und Ökologie. - *Mitt. Arbeitskreis Heimische Orchid. Baden-Württemberg* **23(3)**: 473-486.

MRKVICKA, A.CH. 1992a: Liste der Chromosomenzahlen europäischer Orchideen. - *Mitt. Arbeitskreis Heimische Orchid. Baden-Württemberg* **24(1)**: 125-140.

MRKVICKA, A.CH. 1992b: Erstnachweis von *Epipactis greuteri* H. Baumann & Künkele in Österreich. - *Mitt. Arbeitskreis Heimische Orchid. Baden-Württemberg* **24(3)**: 450-456.

MUASYA, A.M. & SIMPSON, D.A. 2002: A monograph of the genus *Isolepis* R. Br. (Cyperaceae). - *Kew Bull.* **57(2)**: 257-362.

MUCINA, L. & DIMOPOULOS, P. 2000: New locality of *Quercus trojana* subsp. *euboica* (Fagaceae). - *Fl. Medit.* **10**: 261-264.

MUELLNER, A.N., VASSILIADES, D. & RENNER, S.S. 2007: Placing Biebersteiniaceae, a herbaceous clade of Sapindales, in a temporal and geographic context. - *Pl. Syst. Evol.* **266(3-4)**: 233-252.

MUGNAI, M. & AL. 2021: Synopsis of *Euphorbia* section *Anisophyllum* (Euphorbiaceae) in Italy, with an insight on variation of distribution over time in Tuscany - *Phytotaxa* **485(1)**: 1-65.

MUGNIER, C. 1981: [Chromosome numbers in *Linum*]. In Löve, Á. (ed.): Chromosome number reports LXXIII. - *Taxon* **30(4)**: 851.

MUGNIER, C. 1983: [Chromosome numbers in *Linum*]. In Löve, Á. (ed.): IOPB chromosome number reports LXXIX. - *Taxon* **32(2)**: 323.

MUGNIER, C. & ŠILJAK-YAKOVLEV, S. 1987: Karyological study in some Yugoslavian populations of *Hypochoeris* (Compositae). - *Caryologia* **40(4)**: 319-325.

MUIR, A.D. & WESTCOTT, N.D. (EDS) 2003: Flax: the genus *Linum*. - London: Routledge, xii + 307 pp.

MUIRHEAD, C.W. 1969: Turkish species of *Sempervivum*. - *Notes Roy. Bot. Gard. Edinburgh* **29(1)**: 15-28 + 3 plates.

MUKHERJEE, S.K. 1957: Origin and distribution of *Saccharum*. - *Bot. Gaz.* **119(1)**: 55-61.

MULLAJ, A. 1987: [Some new species of the flora of Albania]. - *Sul. Shken. Natyres (Tiranë)?* **41(2)**: 105-107 [n.v.].

MULLAJ, A. & AL. 2010: New records for the Albanian flora. - *Bot. Serbica* **34(2)**: 163-167.

MULLAJ, A. & AL. 2017: Report 125. In Vladimirov, V. & al. (eds): New floristic records in the Balkans: 33. - *Phytol. Balcan.* **23(2)**: 297-298.

MULLAJ, A. & TAN, KIT 2000: The flora of Albania. In Özhatay, N. & al. (eds): Abstracts from the Second Balkan Botanical Congress, Istanbul, Turkey, May 14-18, 2000, p. 47.

MULLAJ, A. & TAN, KIT 2010: *Erica multiflora* (Ericaceae), *Onosma pygmaea* (Boraginaceae) and *Typha minima* (Typhaceae) in Albania. - *Phytol. Balcan.* **16(2)**: 267-269.

MÜLLER, E. 1970: Rhodos: statt Rosen Orchideen. - *Kosmos (Stuttgart)* **66(11)**: 470-472.

MÜLLER, G. 1982: Contribution à la cytotaxonomie de la section *Cyclostigma* Griseb. du genre *Gentiana* L. - *Feddes Repert.* **93(9-10)**: 625-722.

MULLER, J. 1858: Monographie de la famille des Résédacées. - *Neue Denkschr. Allg. Schweiz. Ges. Gesammten Naturwiss.* **16**: 1-239 + 10 plates.

MÜLLER, M. & BALTISBERGER, M. 1984: Cytotaxonomische Untersuchungen in der Artengruppe des *Ranunculus alpestris* (Ranunculaceae). - *Pl. Syst. Evol.* **145(3-4)**: 269-289.

MÜLLEROTT, M. 1976: Zum Vorkommen und zur Morphologie von *Gagea*-Arten insbesondere von *Gagea villosa* (MB.) Duby. - *Hoppea* **35**: 279-286.

MULLIGAN, G.A. 1961: Chromosome numbers of Canadian weeds. III. - *Canad. J. Bot.* **39(5)**: 1057-1066.

MULLIGAN, G.A. 1984: Chromosome numbers of some plants native and naturalized in Canada. - *Naturaliste Canad.* **111(4)**: 447-449.

MULLIGAN, G.A. & FRANKTON, C. 1962: Taxonomy of the genus *Cardaria* with particular reference to the species introduced into North America. - *Canad. J. Bot.* **40(11)**: 1411-1425 + 1 plate.

MULLIGAN, G.A., MUNRO, D.B. & MCNEILL, J. 1983: The status of *Stachys palustris* (Labiatae) in North America. - *Canad. J. Bot.* **61(3)**: 679-682.

MUMINOVIĆ, S., KAVGIĆ, I. & ŠAJNA, N. 2017: Persistence of *Centaurea pumilio* L., a rare sand dune species. - *Ecol. Medit.* **40(1)**: 19-25.

MUMMENHOFF, K. 1995: Should *Cardaria draba* (L.) Desv. be classified within the genus *Lepidium* L. (Brassicaceae)? Evidence from subunit polypeptide composition of RUBISCO. - *Feddes Repert.* **106 (1-2)**: 25-28.

MUMMENHOFF, K. & HURKA, H. 1990: Evolution of the tetraploid *Capsella bursa-pastoris* (Brassicaceae): isoelectric focusing analysis of Rubisco. - *Pl. Syst. Evol.* **172(1-4)**: 205-213.

MUMMENHOFF, K. & ZUNK, K. 1991: Should *Thlaspi* (Brassicaceae) be split? Preliminary evidence from isoelectric focusing analysis of Rubisco. - *Taxon* **40(3)**: 427-434.

MUÑOS GARMENDIA, F. & AEDO, C. 1998: (1335) Proposal to conserve the name *Pyracantha coccinea* against *Mespilus pauciflora* (Rosaceae). - *Taxon* **47(1)**: 171-172. [approved, cf. Taxon 49(2): 268, 2000].

MUÑOZ RODRÍGUEZ, A.F. 1992: Revisión del género *Trifolium* sect. *Trifolium* en la Península Ibérica e Islas Baleares. - *Acta Bot. Malac.* **17**: 79-118.

MUÑOZ RODRÍGUEZ, A.F. 1993: *Trifolium* sect. *Trifolium*: II. Estudio cariológico. - *Acta Bot. Malac.* **18**: 89-118.

MUÑOZ RODRÍGUEZ, A.F. 1995a: *Trifolium* sect. *Paramesus* y sect. *Trifoliastrum* en la Península Ibérica. - *Stud. Bot. Univ. Salamanca* **14**: 47-102.

MUÑOZ RODRÍGUEZ, A.F. 1995b: *Trifolium* sect. *Paramesus* y sect. *Trifoliastrum* en la Península Ibérica. II. Estudio cariológico. - *Stud. Bot. Univ. Salamanca* **14**: 103-128.

MÜNTZING, A. 1930a: Outlines to a genetic monograph of the genus *Galeopsis*, with special reference to the nature and inheritance of partial sterility. - *Hereditas (Lund)* **13(2)**: 185-341.

MÜNTZING, A. 1930b: Über Chromosomenvermehrung in *Galeopsis*-Kreutzungen und ihre phylogenetische Bedeutung. - *Hereditas (Lund)* **14(2)**: 153-172.

MUNZ, PH.A. 1946: *Aquilegia*. The cultivated and wild columbines. - *Gentes Herb.* **7(1)**: 1-150.

MUNZ, PH.A. 1967a: A synopsis of the Asian species of *Consolida* (Ranunculaceae). - *J. Arnold Arbor.* **48(2)**: 159-202.

MUNZ, PH.A. 1967b: A synopsis of the Asian species of *Delphinium*, sensu stricto. - *J. Arnold Arbor.* **48(3-4)**: 249-302, 476-545.

MUNZ, PH.A. 1968: A synopsis of the Asian species of *Delphinium*, sensu stricto [cont.]. - *J. Arnold Arbor.* **49(1-2)**: 73-166, 233-259.

MUNZINGER, S. 1994: *Ophrys gottfriediana* Renz: Neufund für die Insel Rhodos? - *J. Eur. Orch.* **26(3-4)**: 622-627.

MUNZINGER, S. 1995: Macchie, Garrigue, Phrygana - Beschreibung und kritische Anmerkungen zur gegenseitigen Abgrenzung. - *J. Eur. Orch.* **27(2)**: 319-336.

MUNZINGER, S. & HENNECKE, M. 2014a: Differentialdiagnostische Erwägungen zur Anwendung von Artkonzepten für die Gattung *Ophrys*. - *Ber. Arbeitskreis. Heimische Orchid.* **31(1)**: 6-35.

MUNZINGER, S. & HENNECKE, M. 2014b: Fazit aus Beobachtungen von Chios. *Ophrys homeri* - eine Art hybridogenen Ursprungs. - *Ber. Arbeitskreis. Heimische Orchid.* **31(1)**: 53-61.

MURATOVIČ, E. & AL. 2005: Does *Lilium bosniacum* merit species rank? A classical and molecular- cytogenetic analysis. - *Pl. Syst. Evol.* **252(1-2)**: 97-109.

MURAVJOVA, O.A. 1936: The genus *Sibbaldia* L. and its species [In Russian with English summary]. - *Trudy Bot. Inst. Akad. Nauk S.S.S.R., Ser. 1, Fl. Sist. Vysš. Rast.* **2**: 217-241 + 1 map.

MURBECK, S. 1891: Beiträge zur Kenntniss der Flora von Südbosnien und der Hercegovina. – *Acta Univ. Lund.* **27**: 1-182.

MURBECK, S. 1892: Studien über Gentianen aus der Gruppe *Endotricha* Froel. - *Acta Horti Berg.* **2(3)**: 3-28 + 1 map.

MURBECK, S. 1898a: Ueber eine neue *Alectorolophus*-Art und das Vorkommen saison-trimorpher Arten-Gruppen innerhalb der Gattung. - *Oesterr. Bot. Z.* **48(2)**: 41-46, 90-93 + 1 plate.

MURBECK, S. 1898b: Contributions a la connaissance de la flore du nord-ouest de l'Afrique et plus spécialment de la Tunisie. II. Primulaceae - Labiatae. - *Acta Univ. Lund.* **34**: 1-41.

MURBECK, S. 1907: Die *Vesicarius*-Gruppe der Gattung *Rumex*. - *Acta Univ. Lund., 2* **2(14)**: 1-31 + 2 Tafeln.

MURBECK, S. 1925: Monographie der Gattung *Celsia*. - *Acta Univ. Lund., 2* **22(1)**: 1-239 + 16 plates.

MURBECK, S. 1930a: Neue Bastarde zwischen *Celsia* und *Verbascum*. - *Bot. Not.* **1930(1)**: 1-12.

MURBECK, S. 1930b: Die in den Sammlungen der Universität zu Beograd enthaltenen jugoslavischen *Verbascum*-Formen. - *Bull. Inst. Jard. Bot. Univ. Belgrade* **1(3)**: 215-228.

MURBECK, S. 1933: Monographie der Gattung *Verbascum*. - *Acta Univ. Lund., 2* **29(2)**: 1-630 + 31 plates.

MURBECK, S. 1936: Nachträge zur Monographie der Gattung *Verbascum*. - *Acta Univ. Lund., 2* **32(1)**: 1-46 + 8 plates.

MURBECK, S. 1939: Weitere Studien über die Gattungen *Verbascum* und *Celsia*. - *Acta Univ. Lund., 2* **35(1)**: 1-71 + 4 plates.

MURÍN, A. 1982: Cytotaxonomical study on the genus *Polypodium* in Slovakia [In Slovakian with English and Russian summaries]. - *Biológia, A (Bratislava)* **37(1)**: 83-87.

MURÍN, A. 1986: Karyological study of the Slovak flora XVIII. - *Acta Fac. Rerum Nat. Univ. Comen., Bot.* **33**: 85-92.

MURÍN, A. 1992: Karyological study of the Slovak flora XXIV. - *Acta Fac. Rerum Nat. Univ. Comen., Bot.* **39**: 45-51.

MURÍN, A. & FERÁKOVÁ, V. 1981: Caryological study of Slovakian flora III. - *Acta Fac. Rerum Nat. Univ. Comen., Bot.* **28**: 59-61.

MURÍN, A. & FERÁKOVÁ, V. 1986: Karyological study of the Slovak flora VIII. - *Acta Fac. Rerum Nat. Univ. Comen., Bot.* **33**: 41-43.

MURÍN, A. & FERÁKOVÁ, V. 1988: Karyological variability of the species *Allium scorodoprasum* L. - *Acta Fac. Rerum Nat. Univ. Comen., Bot.* **36**: 65-77.

MURÍN, A. & MÁJOVSKÝ, J. 1976: [Chromosome numbers from Czechoslovakia]. In Löve, Á. (ed.): IOPB chromosome number reports LIII. - *Taxon* **25(4)**: 487-488.

MURÍN, A. & MÁJOVSKÝ, J. 1978: [Chromosome numbers from Czechoslovakia]. In Löve, Á. (ed.): IOPB chromosome number reports LXI. - *Taxon* **27(4)**: 376-378.

MURÍN, A. & MÁJOVSKÝ, J. 1979: Karyological study of Slovakian flora I. - *Acta Fac. Rerum Nat. Univ. Comen., Bot.* **27**: 127-133.
MURÍN, A. & MÁJOVSKÝ, J. 1982: Die Bedeutung der Polyploidie in der Entwicklung der in der Slowakei wachsenden Arten der Gattung *Symphytum* L. - *Acta Fac. Rerum Nat. Univ. Comen., Bot.* **29**: 1-25.
MURÍN, A. & MÁJOVSKÝ, J. 1983: Karyological study of Slovakian flora IV. - *Acta Fac. Rerum Nat. Univ. Comen., Bot.* **30**: 1-16.
MURÍN, A. & MÁJOVSKÝ, J. 1987: Karyological study of the Slovak flora XIX. - *Acta Fac. Rerum Nat. Univ. Comen., Bot.* **34**: 3-20.
MURÍN, A. & MÁJOVSKÝ, J. 1992: Karyological study of the Slovak flora XXV. - *Acta Fac. Rerum Nat. Univ. Comen., Bot.* **39**: 53-57.
MURÍN, A. & NEISCHLOVÁ, E. 1973: Chromosomenzahlen einiger Taxa aus dem Verwandtschaftskreis von *Genista tinctoria* L. - *Biológia, A (Bratislava)* **28(7)**: 493-497.
MURÍN, A. & PACLOVÁ, L. 1979: [Chromosome numbers from Czechoslovakia]. In Löve, Á. (ed.): IOPB chromosome number reports LXIV. - *Taxon* **28(4)**: 403-405.
MURÍN, A. & PACLOVÁ, L. 1986: Karyological study of the Slovak flora IX. - *Acta Fac. Rerum Nat. Univ. Comen., Bot.* **33**: 45-48.
MURÍN, A. & SCHWARZOVÁ, T. 1979: Karyological study of Slovakian flora II. - *Acta Fac. Rerum Nat. Univ. Comen., Bot.* **27**: 135-136.
MURÍN, A. & SVOBODOVÁ, Z. 1992: Karyological study of the Slovak flora XXVI. - *Acta Fac. Rerum Nat. Univ. Comen., Bot.* **39**: 59-62.
MURÍN, A. & ZÁBORSKÝ, J. 1976: [Chromosome numbers from Czechoslovakia]. In Löve, Á. (ed.): IOPB chromosome number reports LIII. - *Taxon* **25(4)**: 489.
MURR, J. 1890: Die Pflanzenwelt in der griechischen Mythologie. - Innsbruck: Verlag der Wagner'schen Universitäts-Buchhandlung, viii + 324 pp.
MURR, J. 1904: Additamenta ad genus *Chenopodium*. - *Magyar Bot. Lapok* **3(1-2)**: 37-39 + 1 plate.
MURR, J. 1905: Indirekte Beiträge zur Flora Graeca. - *Magyar Bot. Lapok* **4(1-3)**: 4-8.
MURR, J. 1913: Zur Flora Graeca. - *Magyar Bot. Lapok* **12(1-5)**: 107-108.
MURRAY, A.E. 1970: A monograph of the Aceraceae. A thesis in horticulture. - Pennsylvania State Univ.: Ph.D. thesis, v + 337 pp.
MURRAY, A.E. 1978: *Acer* notes No. 8. - *Kalmia* **8(2-3)**: 17-20.
MURRAY, A.E. 1979: Afrasian and European maples. - *Kalmia* **9**: 2-40.
MURRAY, A.E. 1982: Infrageneric taxa in *Acer* (Aceraceae). - *Kalmia* **12**: 2-28.
MURRAY, B.G. 1974: Breeding systems and floral biology in the genus *Briza*. - *Heredity* **33(2)**: 285-292.
MURRAY, B.G. 1975a: The cytology of the genus *Briza* L. (Gramineae). I. Chromosome numbers, karyotypes and nuclear DNA variation. - *Chromosoma* **49(3)**: 299-308.
MURRAY, B.G. 1975b: The cytology of the genus *Briza* L. (Gramineae). II. Chiasma frequency, polyploidy and interchange heterozygosity. - *Chromosoma* **57(1)**: 81-93.
MURRAY, B.G. 1976: The cytology of the genus *Briza* L. (Gramineae). III. B chromosomes. - *Chromosoma* **59(1)**: 73-81.
MURRAY, B.G. 1979: Unusual chromosome pairing and B chromosomes in *Briza spicata* (Poaceae). - *Pl. Syst. Evol.* **132(4)**: 245-253.
MURRAY, J. 1983: *Cyclamen* in South West Turkey. - *Cyclamen Soc. J.* **7(1)**: 15.
MURRAY, J.A. 1784: Caroli à Linné equitis Systema vegetabilium secundum classes ordines genera species cum characteribus et differentiis ... Ed. 14. - Gottingae: Typis et impensis Jo. Christ. Dietrich [cf. TL-2 3: 670, 1981].
MURTHY, H.N. 1988: Karyomorphological studies in *Bellis perennis* L. - *Chromosome Inform. Serv.* **45**: 3-4.
MUSANO, L. & MAGGINI, F. 1977: Triploidia in "Bellevalia dubia" (Guss.) Roemer et Schultes. - *Inform. Bot. Ital.* **8(2)**: 186-187.
MUSARELLA, C.M., BRULLO, S. & GIUSSO DEL GALDO, G. 2020: Contribution to the orophilous cushion-like vegetation of central-southern and Insular Greece. - *Plants MDPI* **9,1678**: 1-141.
MUSCHLER, R. 1908: Die Gattung *Coronopus* (L.) Gaertn. - *Bot. Jahrb. Syst.* **41(3)**: 111-147.
MUSIAŁ, K., JANAS, A. & SZELĄG, Z. 2016: Chromosome numbers in *Hieracium* (Asteraceae) from Central and Southeastern Europe II. - *Acta Biol. Cracov., Ser. Bot.* **58(1)**: 119-123.
MUSIAŁ, K. , JANAS, A. & SZELĄG, Z. 2017: Chromosome numbers in *Hieracium* (Asteraceae) from Central and Southeastern Europe III. - *Acta Biol. Cracov., Ser. Bot.* **59(2)**: 83-87.
MUSIAŁ, K., JANAS, A. & SZELĄG, Z. 2018: Chromosome numbers in *Hieracium* (Asteraceae) from Central and Southeastern Europe IV. - *Acta Biol. Cracov., Ser. Bot.* **60(2)**: 75-79.
MUSIAŁ, K. & SZELĄG, Z. 2015: Chromosome numbers in *Hieracium* (Asteraceae) from Central and Southeastern Europe I. - *Acta Biol. Cracov., Ser. Bot.* **57(2)**: 115-120.
MUSIAŁ, K. & SZELĄG, Z. 2019: Chromosome numbers in *Hieracium* (Asteraceae) from Central and Southeastern Europe V. - *Acta Biol. Cracov., Ser. Bot.* **61(2)**: 63-68.
MUSSACK, A. 1933: Untersuchungen über *Cystopteris fragilis*. - *Beih. Bot. Centralbl., Abt. 2* **51(1)**: 204-254.
MUSSELMAN, L.J. 1986: Taxonomy of *Orobanche*. In Borg, S.J. ter (ed.): Proceedings of a workshop on biology and control of Orobanchaceae, pp. 2-10. - Wageningen: LH/VPO.
MUSSELMAN, L.J. & SAOUD, N.S. 2004: The type specimens of George Edward Post in Beirut and Geneva. - *Türk. Bot. Derg.* **28(1-2)**: 155-160.
MYERSCOUGH, P.J. 1980: Biological Flora of the British Isles. *Epilobium angustifolium* L. (*Chamaenerion angustifolium* (L.) Scop.). - *J. Ecol.* **68(3)**: 1047-1074.
MYKLESTAD, Å. & BIRKS, H.J.B. 1993: A numerical analysis of the distribution patterns of *Salix* L. species in Europe. - *J. Biogeogr.* **20(1)**: 1-32.
MYLONAKI, I. 2004: Die Reisen der Universität Wien nach Griechenland. - Pp. 315-323 in: Hörander W. (ed.): Wiener Byzantinistik und Neogräzistik. Beiträge zum Symposium Vierzig Jahre Institut für Byzantinistik und Neogräzistik der Universität Wien in Gedenken an Herbert Hunger (Wien, 4.-7. Dezember 2002). - Wien: Verlag der Österreichischen Akademie der Wissenschaften (= Byzantina et Neograeca Vindobonensia 24).
NACHYCHKO, V.O., SOSNOVSKY, Y.V. & HONCHARENKO, V.I. 2019: First record of Balkan *Thymus jankae* (Lamiaceae) from Ukraine, with taxonomic remarks on the species. - *Botany Letters* **166(1)**: 41-50.
NADEZHINA, T. 1979: De speciebus nonnullis e sectione *Pseudoglycyrrhiza* Krug. generis *Glycyrrhiza* L. [In Russian]. - *Novosti Sist. Vysš. Rast.* **15**: 174-181.
NADJI, A.U.R. 1892: Empire Ottoman géographie botanique. Faits nouveaux relatifs à la province de Salonique. - Salonique: Imp. S. Muratori, 48 pp.
NÄGELI, C.V. & PETER, A. 1885-89: Die Hieracien Mittel-Europas. Vol. 1 (xi + 931 pp.) 1885; Vol. 2 pp. 1-240, 1886; Vol. 2 pp. 241-340, 1889. - München: Verlag von R. Oldenbourg.
NAKIPOĞLU, M. 1993a: Karyological studies on *Salvia* L. species of Turkey. I. *S. fruticosa* Mill., *S. tomentosa* Mill., *S. officinalis* L., *S. smyrnaea* Boiss. (Lamiaceae) [In Turkish with English abstract]. - *Doğa Türk. Bot. Derg.* **17(1)**: 21-25.
NAKIPOĞLU, M. 1993b: Karyological studies on some *Salvia* L. species of Turkey. II. *S. viridis* L., *S. glutinosa* L., *S. virgata* Jacq., *S. verbenaca* L., *S. argentea* L. [In Turkish with English abstract]. - *Doğa Türk. Bot. Derg.* **17(3)**: 157-161.
NANDI, O.I. 1998: Floral development and systematics of Cistaceae. - *Pl. Syst. Evol.* **212(1-2)**: 107-134.
NARDI, E. 1972: "Asplenium adulterinum" Milde in Italia. - *Webbia* **26(2)**: 463-490 + 2 plates.
NARDI, E. 1973: "Asplenium adulterinum" Milde nei pressi di Stresa (Lago Maggiore). - *Webbia* **28(1)**: 37-47.
NARDI, E. 1974: Problemi sistematici e distributivi di "Cystopteris dickieana" s.l. in Italia. - *Webbia* **29(1)**: 329-360.
NARDI, E. 1976a: La distribuzione italiana di "Dryopteris pallida" (Bory) Fomin. - *Webbia* **30(1)**: 3-32. [Also distributed as *OPTIMA leaflet* no. 28].
NARDI, E. 1976b: "Dryopteris assimilis" S. Walker in Italia. - *Webbia* **30(2)**: 457-478. [Also distributed as *OPTIMA leaflet* no. 36].
NARDI, E. 1977a: Note sistematiche sul "Polypodium australe" s.l. delle Isole Atlantiche (Azzorre, Madera, Canarie). - *Webbia* **31(1)**: 79-96.

NARDI, E. 1977b: Commentaria pteridologica. I. De nonnullis filicibus paleomediterraneae regionis. - *Webbia* **32(1)**: 95-100.

NARDI, E. 1979a: Commentaria pteridologica. II. De legitimo usu nominis Polypodii cambrici L. - *Webbia* **33(2)**: 425-433.

NARDI, E. 1979b: Commentaria pteridologica. III. Notulae chorologicae atque cytologicae Italicae. - *Webbia* **33(2)**: 435-447.

NARDI, E. 1983a: Proposal to reject the name *Aristolochia longa* L. (Aristolochiaceae). - *Taxon* **32(4)**: 654-656. [cf. also *Taxon* 36(2): 432, 1987 and 37(4): 978, 1988].

NARDI, E. 1983b: Commentaria pteridologica. IV. De "Asplenio balearico" Shivas in Italia reperto. - *Webbia* **36(2)**: 217-223.

NARDI, E. 1984: The genus "Aristolochia" L. (Aristolochiaceae) in Italy. - *Webbia* **38**: 221-300. [Also distributed as *OPTIMA leaflet* no. 152].

NARDI, E. 1985: Sulla tazzonomia di *Aristolochia insularis* Nardi et Arr. - *Webbia* **39(1)**: 119-127.

NARDI, E. 1987: Il genere *Aristolochia* L. in Corsica. - *Webbia* **41(2)**: 225-239.

NARDI, E. 1988: (920) Renewed proposal to reject *Aristolochia longa* L. (Aristolochiaceae). - *Taxon* **37(4)**: 978-980. [approved, cf. Taxon 54(4): 1093, 2005].

NARDI, E. 1989: De speciebus *Aristolochiae pallidae* gregis (Aristolochiaceae) in Graecia crescentibus. - *Willdenowia* **18(2)**: 367-375.

NARDI, E. 1991: The genus *Aristolochia* L. (Aristolochiaceae) in Greece. - *Webbia* **45(1)**: 31-69. [Also distributed as *OPTIMA leaflet* no. 215].

NARDI, E. 1993: Systematic revision of the *Aristolochia auricularia* group (Aristolochiaceae). - *Fl. Medit.* **3**: 223-232.

NARDI, E. 2014: Nomenclatural notes on *Aquilegia* L. (Ranunculaceae) from Europe. - *Webbia* **69(1)**: 105.

NARDI, E. 2015: The genus *Aquilegia* L. (Ranunculaceae) in Italy. - Firenze: Edizioni Polistampa, 685 pp.

NARDI, E. & ARRIGONI, P.V. 1983: Le piante endemiche della Sardegna: 134-135. - *Boll. Soc. Sarda Sci. Nat.* **22**: 347-356.

NARDI, E. & GARBARI, F. 1989: *Asplenium sagittatum* (DC.) Bange (Aspleniaceae) new to the Jordanian flora. - *Atti Soc. Tosc. Sci. Nat. Pisa Mem., Ser. B* **95**: 29-33. [Also distributed as *OPTIMA leaflet* no. 179].

NARDI, E. & MORI, B. 1977: "Dryopteris pseudomas" (Wollaston) Holub & Pouzar triploide nell'Appennino Toscano. - *Webbia* **31(1)**: 35-42.

NARDI, E. & NARDI, C.N. 1987: Taxonomic and chorological notes on the genus *Aristolochia* L. (Aristolochiaceae) from the central and eastern Mediterranean area. - *Bot. Helv.* **97(2)**: 155-165.

NARDI, E. & RAFFAELLI, M. 1977: Un endemismo italiano da rivalutare: "Polygonum tenoreanum" nom. nov. (= *P. elegans* Ten.). - *Webbia* **31(2)**: 513-527.

NARDI, E., RASBACH, H. & REICHSTEIN, T. 1978: Identification of "Cheilanthes fragrans" var. "gennarii" Fiori with "C. guanchica" Bolle and remarks on related taxa. - *Webbia* **33(1)**: 1-18.

NARDI, E., RASBACH, H. & REICHSTEIN, T. 1979: "Cheilanthes tinaei" Tod., an earlier name for "C. corsica" Reichstein & Vida, and related species in Sicily. - *Webbia* **33(2)**: 449-456.

NARDI, E. & REICHSTEIN, T. 1985: Nomenclatural notes on *Cheilanthes pteridioides* (Reichard) C. Chr. (Sinopteridacae, Pteridophyta). - *Webbia* **39(1)**: 135-139.

NARDI, E. & REICHSTEIN, T. 1986: (807) Proposal to reject *Polypodium pteridioides* Reichard and all combinations based on it (Sinopteridaceae: Pteridophyta). - *Taxon* **35(1)**: 172-174. [not approved, cf. Taxon 36(1): 166-167, 1987 and 36(4): 740-741, 1987].

NARDI, E. & TOMMEI, A. 1976: Osservazioni biosistematiche sul genere "Polypodium" L. in Italia. - *Webbia* **30(2)**: 219-256. [Also distributed as *OPTIMA leaflet* no. 32].

NATALI, A. 1990: *Sherardia arvensis* L. var. *maritima* Griseb. dans la domaine cyrno-sarde. In Jeanmonod, D. & Burdet, H.M.: Notes et contributions à la flore de Corse, VI. - *Candollea* **45(1)**: 302-308.

NATALI, A. & JEANMONOD, D. 1996: Compléments au Prodrome de la Flore Corse. Annexe no. 4. Flore analytique des plantes introduites en Corse. - Genève: Conservatoire et Jardin Botanique, 211 pp.

NATARAJAN, G. 1978: [Miscellaneous chromosome numbers from the Montpellier area]. In Löve, Á. (ed.): IOPB chromosome number reports LXII. - *Taxon* **27(5-6)**: 526-531.

NATARAJAN, G. 1979: [Miscellaneous chromosome numbers from the Montpellier area]. In Löve, Á. (ed.): IOPB chromosome number reports LXV. - *Taxon* **28(5-6)**: 629.

NATARAJAN, G. 1981: [Chromosome numbers from the Montpellier area]. In Löve, Á. (ed.): Chromosome number reports LXXII. - *Taxon* **30(3)**: 698-699.

NATHO, G. 1959: Variationsbreite und Bastardbildung bei mitteleuropäischen Birkensippen. - *Feddes Repert. Spec. Nov. Regni Veg.* **61(3)**: 211-273.

NAUENBURG, J.D. 1986: Untersuchungen zur Variabilität, Ökologie und Systematik der *Viola tricolor*-Gruppe in Mitteleuropa. - Göttingen: Diss., 124 pp.

NAUENBURG, J.D. 1991: Taxonomie und Korrekturen zur Nomenklatur von *Viola tricolor* s.l. (Violaceae) in Mitteleuropa. - *Willdenowia* **21(1-2)**: 51-56.

NAVARRO, T. 1995: Revisión del género *Teucrium* L. sección *Polium* (Mill.) Schreb., (Lamiaceae) en la Península Ibérica y Baleares. - *Acta Bot. Malac.* **20**: 173-265.

NAVARRO, T. & EL QUALIDI, J. 1999: The genus *Teucrium* L.: A model of biodiversity in the Mediterranean area. In Correia, A.I.D. & al. (eds): Livro de resumos. Guia de excursões. V jornadas de taxonomia botânica, 16-19/IX 1999, p. 59.

NAVARRO, T. & EL QUALIDI, J. 2000: Synopsis of *Teucrium* L. (Labiatae) in the Mediterranean region and surrounding areas. - *Fl. Medit.* **10**: 349-363.

NAVARRO, T. & ROSÚA, J.L. 1990a: Nomenclatural and taxonomic notes on the *Teucrium* section *Polium* (Miller) Schreber (Lamiaceae) in the Iberian peninsula. - *Candollea* **45(2)**: 581-589.

NAVARRO, T. & ROSÚA, J.L. 1990b: Tipificación de *Teucrium polium* (Lamiaceae). - *Anales Jard. Bot. Madrid* **47(1)**: 35-41.

NAVASHIN, M.S. 1985: Problemy kariologii i tsitogenetiki v isseldovaniyakh na vidakh roda *Crepis* [In Russian]. - Moskva: Nauka, 347 pp.

NAVEH, Z. 1967: Mediterranean ecosystems and vegetation types in California and Israel. - *Ecology* **48(3)**: 445-459.

NAVEH, Z. 1975: The evolutionary significance of fire in the Mediterranean region. - *Vegetatio* **29(3)**: 199-208.

NAVEH, Z. 1991: The role of fire in Mediterranean vegetation. - *Bot. Chron. (Patras)* **10**: 385-405.

NAVEH, Z. & VERNET, J.-L. 1991: The palaeohistory of the Mediterranean biota. In Groves, R.H. & Di Castri, F. (eds): Biogeography of mediterranean invasions, pp. 19-32. - Cambridge, etc.: Cambridge Univ. Press.

NAVEH, Z. & WHITTAKER, R.H. 1979: Structural and floristic diversity of shrublands and woodlands in northern Israel and other Mediterranean areas. - *Vegetatio* **41(3)**: 171-180.

NAYLOR, R.E.L. 1972: Biological Flora of the British Isles. *Alopecurus myosuroides* Huds. (*A. agrestis* L.). - *J. Ecol.* **60(2)**: 611-622.

NAZAROVA, E.A. 1994: Karyological study of the genus *Steptorhamphus* Bunge (Lactuceae, Asteraceae). - *Caryologia* **47(2)**: 151-158.

NAZAROVA, E.A. 1998: Karyosystematic investigation of the genus *Scorzonera* L. s.l. (Lactuceae, Asteraceae). - *Caryologia* **50(3-4)**: 239-261.

NAZAROVA, E.A. & GOUKASIAN, A. 1996: Reports (528-542). In Kamari, G., Felber, F. & Garbari, F. (eds): Mediterranean chromosome number reports - 5. - *Fl. Medit.* **5**: 340-345.

NEDELCHEVA, A.M. 1995: Karyological study of some species of genus *Achillea* L. sect. *Filipendulinae* (DC.) Afan. (Asteraceae). - *God. Sofiisk. Univ. "Kliment. Ohridski" Biol. Fak., 2 Bot.* **88**: 55-61.

NEDELCHEVA, A.M. 1996: Reports (479-480). In Kamari, G., Felber, F. & Garbari, F. (eds): Mediterranean chromosome number reports - 5. - *Fl. Medit.* **5**: 320-323.

NEGARESH, K. 2018: Typification of *Onopordum heteracanthum* var. *carmanicum* and *O. myriacanthum* (Asteraceae, Cardueae-Carduinae). - *Ann. Naturhist. Mus. Wien, B* **120**: 141-144.

NEGARESH, K. & RAHIMINEJAD, M.R. 2018: A revision of *Centaurea* sect. *Cynaroides* (Asteraceae, Cardueae-Centaureinae). - *Phytotaxa* **363(1)**: 1-131.

NEGBI, M. (ED.) 1999: Saffron, *Crocus sativus* L. - Amsterdam: Harwood Academic Publishers, 172 pp.

NEGODI, G. 1936: Contributo alla cariologia di “Stellaria media” (L.) Cir. - *Nuovo Giorn. Bot. Ital., n.s.* **43(2)**: 416-424.

NEGODI, G. 1938a: Cariologia dei generi “Aposeris” e “Hyoseris” (Compositae-Cichorieae, tribus Cichorinae). - *Atti Reale Accad. Naz. Lincei, Rendiconti Cl. Sci. Fis., Ser. 6* **27**: 117-120.

NEGODI, G. 1938b: Cariologia del gen. *Urospermum* (Compositae - Liguliflorae - Leontodontinae). - *Arch. Bot. (Forlì)* **14**: 1-19.

NEGODI, G. 1938c: Caratteri nucleari e morfologici in alcune Calendulae (Compositae-Tubuliflorae). - *Ann. Bot. (Rome)* **21(3)**: 510-518.

NEGRI, G. 1953: Adriano Fiori. - *Nuovo Giorn. Bot. Ital., n.s.* **60**: 697-705.

NEGRI, G. 1958: Renato Pampanini (1875-1949). - *Nuovo Giorn. Bot. Ital., n.s.* **65**: 771-805.

NELMES, E. 1939a: Notes on *Carex*: V. Two near relatives of *Carex sylvatica* Huds. - *Bull. Misc. Inform. Kew* **1939**: 98-100.

NELMES, E. 1939b: Notes on British Carices. - *J. Bot. (London)* **77**: 259-266, 301-304.

NELMES, E. 1947: Two critical groups of British sedges. - *Rep. Bot. Soc. Exch. Club Brit. Isles* **13(1)**: 95-105.

NELMES, E. 1948: Notes on British Carices - VII. - *Rep. Bot. Soc. Exch. Club Brit. Isles* **13(3)**: 334-337.

NELMES, E. 1959: De Caricibus Britannicis notae, IX [In Russian and English]. - *Bot. Mater. Gerb. Bot. Inst. Komarova Akad. Nauk S.S.S.R.* **19**: 75-78.

NELSON, E. 1962: Gestaltwandlung und Artbildung erörtert am Beispiel der Orchidaceen Europas und der Mittelmeerländer insbesondere der Gattung *Ophrys*, mit einer Monographie und Ikonographie der Gattung *Ophrys*. - Chernex-Montreux: 250 pp. + 58 plates and 8 maps (2 vols.).

NELSON, E. 1968: Monographie und Ikonographie der Orchideen-Gattungen *Serapias*, *Aceras*, *Loroglossum*, *Barlia*. - Chernex-Montreux: 79 pp. (Textband) + 36 unmounted plates and 6 maps on 2 pp.

NELSON, E. 1976: Monographie und Ikonographie der Orchidaceen-Gattung *Dactylorhiza*. Folge zu ”Gestaltwandlung und Artbildung erörtert am Beispiel der Orchidaceen Europas und der Mittelmeerländer”. - Zürich: Verlag Speich AG, 127 pp. (Textband) + 87 unmounted plates.

NELSON, E. 1979: Nachtrag zu E. Nelson, “Monographie und Ikonographie der Orchideengattung *Dactylorhiza*”, Dezember 1976. - *Taxon* **28(5-6)**: 592-593.

NELSON, E.C. 2007: The original material of two Turkish species of *Erica* (Ericaceae) described and named by Richard Anthony Salisbury (1761-1829). - *Turk. J. Bot.* **31(5)**: 463-466.

NELSON, E.C. & WULFF, E.M.T. 2007: (1786) Proposal to conserve the name *Erica manipuliflora* against *E. forskalii* (Ericaceae). - *Taxon* **56(3)**: 959-961. [approved, cf. Taxon 58(1): 287, 2009].

NEMATI, Z. & AL. 2018: On the origin of saffron: Phylogeny of *Crocus* series *Crocus*. - *Bot. Serbica* **142 (suppl. 1)**: 63.

NEMLI, Y. 1977: Recherche cytotaxonomique sur le genre *Chrysanthemum* dans la région d’Egée [In Turkish with French summary]. - *Bitki Koruma Bült.* **4(3)**: 376-380.

NERANTZIS, X., KOUDROS, V. & ALEXIOU, S. 2014: New localities for *Colchicum atticum* (Colchicaceae) from NE Greece. - *Parnassiana Arch.* **2**: 53-54.

NESOM, G.L. 1989: The separation of *Trimorpha* (Compositae: Astereae) from *Erigeron*. - *Phytologia* **67(1)**: 61-66.

NESOM, G.L. 1990: Taxonomic summary of *Omalotheca* (Asteraceae: Inuleae). - *Phytologia* **68(4)**: 241-246.

NESTLER, H. 1933: Beiträge zur systematischen Kenntnis der Gattung *Linum*. - *Beih. Bot. Centralbl., Abt. 2* **50(2)**: 497-551 + 4 plates.

NEUFFER, B. & ESCHNER, S. 1995: Life-history traits and ploidy levels in the genus *Capsella* (Brassicaceae). - *Canad. J. Bot.* **73(9)**: 1354-1365.

NEUMANN, A. & POLATSCHEK, A. 1972: Cytotaxonomischer Beitrag zur Gattung *Salix*. - *Ann. Naturhist. Mus. Wien* **76**: 619-633.

NEUMANN, H. 1995: Die Orchidee des Jahres 1995 - *Ophrys apifera* Hudson, Bienen-Ragwurz, Bienentragende Ragwurz. - *Ber. Arbeitskreis. Heimische Orchid.* **12(1)**: 81-90.

NEUMAYER, H. 1923a: Die Frage der Gattungsabgrenzung innerhalb der Silenoideen. - *Verh. Zool.-Bot. Ges. Wien* **72**: 53-59.

NEUMAYER, H. 1923b: Einige Fragen der speziellen Systematik, erläutert an einer Gruppe der Gattung *Silene*. - *Oesterr. Bot. Z.* **72(6-8)**: 276-287.

NEUROTH, R. 1990: Cytotaxonomische Studien an europäischen Sippen des *Polypodium vulgare*- Komplexes. - Kaiserslautern: Diplomarbeit [unpubl.].

NEUROTH, R. 1996: Biosystematik und Evolution des *Polypodium vulgare*-Komplexes (Polypodiaceae: Pteridophyta). - *Diss. Bot.* **256**: v + 211 pp.

NEVES, J. DE BARROS 1939: Contribution à l’étude caryologique du genre *Leucojum* L. - *Bol. Soc. Brot., Ser. 2* **13**: 545-572.

NEVES, J. DE BARROS 1942: Sobre a cariologia de *Ranunculus ficaria* L. - *Bol. Soc. Brot., Ser. 2* **16**: 169-181.

NEVES, J. DE BARROS 1944: Contribuição para o estudo cariosystemático das espécies portuguesas do género *Ranunculus* L. - Coimbra: Diss., Univ. Coimbra, 200 pp.

NEVES, J. DE BARROS 1952: Estudos cariológicos no género *Ornithogalum* L. - *Bol. Soc. Brot., Ser. 2* **26**: 5-192.

NEVES, J. DE BARROS 1973: Contribution à la connaissance cytotaxinomique des Spermatophyta du Portugal: VIII. Liliaceae. - *Bol. Soc. Brot., Ser. 2* **47**: 157-212.

NEVO, E. & AL. 1979: Genetic diversity and environmental associations of wild barley, *Hordeum spontaneum*, in Israel. - *Evolution* **33(3)**: 815-833.

NEVSKI, A. 1941: Monographiae genus *Hordeum* [In Russian]. - *Trudy Bot. Inst. Akad. Nauk S.S.S. R., Ser. 1, Fl. Sist. Vysš. Rast.* **5**: 64-255.

NEW, J.K. 1961: Biological Flora of the British Isles. *Spergula arvensis* L. - *J. Ecol.* **49(1)**: 205-215.

NEWGER, K. 1994a: “Rhodos” - Insel der Orchideen. - *J. Orchideenfr.* **1**: 28-35.

NEWGER, K. 1994b: “Rhodos” - Insel der Orchideen [cont.]. - *J. Orchideenfr.* **2**: 61-69.

NEWSHOLME, C. 1992: Willows: the genus *Salix*. - London: B.T. Batsford Ltd, 224 pp. + 32 pp. of colour plates.

NEWTON, L.E. 1984: Terminology of structures associated with pollinia of the Asclepiadaceae. - *Taxon* **33(4)**: 619-621.

NICHOLLS, M.S. 1986a: A systematic study of the *Linum tenuifolium* group (Linaceae). - *Bot. J. Linn. Soc.* **91(4)**: 473-490.

NICHOLLS, M.S. 1986b: Variation and evolution in *Linum tenuifolium* (Linaceae). - *Pl. Syst. Evol.* **153 (3-4)**: 243-258.

NICHOLLS, P. 1994: *C[yclamen] hederifolium* on Crete. - *Cyclamen* **18(1)**: 8.

NICHOLLS, P. 1999: *Cyclamen hederifolium* var. *confusum* on Zakinthos. - *Cyclamen* **23(2)**: 45.

NICOLSON, D.H. 1975: Lectotypification of genera of Araceae. - *Taxon* **24(4)**: 467-468.

NICOLSON, D.H. & STEYSKAL, G.C. 1976: The masculine gender of the generic name *Styrax* Linnaeus (Styracaceae). - *Taxon* **25(5-6)**: 581-587.

NIEBUHR, A.D. 1970: Herbs of Greece. - Athens: J. Makris, 140 pp.

NIELSEN, H. 1989: Contribution to the flora of Lesbos (Mitilini). - Copenhagen: M.Sc. thesis, Bot. Lab., Univ. Copenhagen, 62 pp. + appendix 229 pp.

NIELSEN, I. 1975: Chromosome counts in the genus *Trifolium*. - *Bot. Tidsskr.* **70(2-3)**: 180-183.

NIETO FELINER, G. 1995: Hybridization in the genus *Epilobium* (Onagraceae) in the Iberian Peninsula. - *Anales Jard. Bot. Madrid* **52(2)**: 241-247.

NIETO FELINER, G. 1996: Notes on *Epilobium* (Onagraceae) from the Western Mediterranean. - *Anales Jard. Bot. Madrid* **54(1)**: 255-264.

NIETO FELINER, G. & AEDO, C. 1995: A cladistic analysis of *Geranium* subgen. *Erodioidea* (Picard) Yeo (Geraniaceae). - *Bot. J. Linn. Soc.* **119(3)**: 195-212.

NIJS, J.C.M. DEN 1974: Biosystematic studies of the *Rumex acetosella complex*. I. Angiocarpy and chromosome numbers in France. - *Acta Bot. Neerl.* **23(5-6)**: 655-675.

NIJS, J.C.M. DEN 1976: Biosystematic studies of the *Rumex acetosella*-complex II. The Alpine region. - *Acta Bot. Neerl.* **25(6)**: 417-447.

NIJS, J.C.M. DEN 1983: Biosystematic studies of the *Rumex acetosella* complex (Polygonaceae). VI. South-eastern Europe, including a phylogenetic survey. - *Bot. Jahrb. Syst.* **104(1)**: 33-90.

NIJS, J.C.M. DEN 1984: Biosystematic studies of the *Rumex acetosella* complex (Polygonaceae). VIII. A taxonomic revision. - *Feddes Repert.* **95(1-2)**: 43-66.

NIJS, J.C.M. DEN 1995: *Taraxacum*: Ploidy levels, hybridization and speciation ... In Valdés, B. & Silvestre, S. (eds): Abstracts. VIII OPTIMA meeting, Sevilla, Spain, 25 September - 1 October, 1995, p. 11.

NIJS, J.C.M. DEN & HULST, A.W. VAN DER 1982: Biosystematic studies of the *Rumex acetosella* complex. V. Cytogeography and morphology in the Czech Socialistic Republic and a part of Lower Austria (Austria). - *Folia Geobot. Phytotax.* **17(1)**: 49-62.

NIJS, J.C.M. DEN & HULST, A.W. VAN DER 1988: Cytogeography of *Taraxacum* section *Erythrosperma*: Diploid sexuals in SE and SW Europe. - *Bot. Jahrb. Syst.* **110(1)**: 83-93.

NIJS, J.C.M. DEN & PANHORST-SANGSTER, T. 1980: Biosystematic studies of the *Rumex acetosella*-complex III. A note on the synsystematic position of "R. tenuifolius (Wallr.) Löve". - *Acta Bot. Neerl.* **29(2-3)**: 179-192.

NIJS, J.C.M. DEN., HOOGHIEMSTRA, H. & SCHALK, P.H. 1980: Biosystematic studies of the *Rumex acetosella* complex (Polygonaceae). IV. Pollen morphology and the possibilities of identification of cytotypes in pollen analysis. - *Phyton (Horn)* **20(3-4)**: 307-323.

NIJS, J.C.M. DEN., SORGDRAGER, K. & STOOP, J. 1985: Biosystematic studies of the *Rumex acetosella* complex. IX. Cytogeography of the complex in the Iberian Peninsula and taxonomic discussion. - *Bot. Helv.* **95(2)**: 141-156.

NIKETIĆ, M. 1992: A taxonomical-chorological differentiation of species *Aquilegia grata* F. Maly ex Zimmeter and *Aquilegia ottonis* Orph. ex Boiss. (Ranunculaceae). - *Glasn. Prir. Mus. Beogradu, Ser. B, Biol. Nauke* **47**: 53-66.

NIKETIĆ, M. 1994: *Cerastium neoscardicum*, a new species of ser. *Alpina* from Mt Shar-Planina, Serbia. - *Bull. Inst. Jard. Bot. Univ. Belgrade* **26/27**: 63-70.

NIKETIĆ, M. 1999: *Cerastium* subsection of the genus *Cerastium* L. (Caryophyllaceae) on Balkan Peninsula. Annotated check-list. - *Glasn. Prir. Mus. Beogradu, Ser. B, Biol. Nauke* **49-50**: 39-61.

NIKETIĆ, M. 2007a: Endemic representatives of genus *Cerastium* L. in Southeastern Europe - Taxonomy, chorology and ecology [In Serbian with English abstract]. - Beograd: Ph.D. thesis, Univ. of Beograd, xi + 249 pp.

NIKETIĆ, M. 2007b: *Cerastium banaticum* (Caryophyllaceae) in Macedonia - distribution and infraspecific differentiation. In Matevski, V. (ed.): Collection of papers devoted to Academician Kiril Micevski, pp.171-190. - Skopje: Macedonian Academy of Sciences and Arts.

NIKETIĆ, M. 2014: Nomenclature review of the plants published by Josif Pančić (*Nomenclator Pancicianus novus*). - *Bot. Serbica* **38(2)**: 209-236.

NIKETIĆ, M. & AL. 2014: New species for the vascular flora of Republic of Macedonia and their distribution in the Balkan Peninsula. - *Bot. Serbica* **38(1)**: 57-67.

NIKETIĆ, M. & AL. 2018: Material on the annotated checklist of vascular flora of Serbia. Nomenclatural, taxonomic and floristic notes. I. - *Bull. Nat. Hist. Mus. Beograd* **11**: 101-180.

NIKETIĆ, M. & AL. 2021: Diversification within ploidy-variable Balkan endemic *Cerastium decalvans* (Caryophyllaceae) reconstructed based on genetic, morphological and ecological evidence. - *Bot. J. Linn. Soc.* **XX**: 1-31.

NIKETIĆ, M. & DIKLIĆ, N. 1990: *Potentilla mollis* Pančić - some morphological-chorological features and its systematic value. - *Razprave IV. Razreda Sazu* **31(13)**: 185-199.

NIKETIĆ, M. & STEVANOVIĆ, V. 2007: A new species of *Heliosperma* (Caryophyllaceae) from Serbia and Montenegro. - *Bot. J. Linn. Soc.* **154(1)**: 55-63.

NIKETIĆ, M., STEVANOVIĆ, V. & TOMOVIĆ, G. 2007: Nomenclatural and taxonomic notes on the flora of Serbia and the Balkan Peninsula. I. Caryophyllaceae. - *Arch. Biol. Sci., Belgrade* **59(4)**: 387-396.

NIKETIĆ, M. & TOMEVIĆ, G. 2018: An annotated checklist of vascular flora of Serbia. 1. Lycopodiopsida, Polypodiopsida, Gnetopsida, Pinopsida and Liliopsida. [In Serbian and English]. - Belgrade: Serbian Academy of Sciences and Arts, Monographs vol. DCXC, Dept. of Chemical and Biological Sciences, Book 10, 294 pp.

NIKETIĆ, M. & TOMOVIĆ, G. 2008: Taxonomy and nomenclature of the *Linaria genistifolia* complex (Plantaginaceae-Antirrhineae) in S.E. Europe and Anatolia. - *Taxon* **57(2)**: 619-629.

NIKETIĆ, M., TOMOVIĆ, G. & SILJAK-YAKOVLEV, S. 2018: A new spontaneous hybrid between the cultivated and wild *Iris* species from Serbia. - *Bull. Nat. Hist. Mus. Beograd* **11**: 189-210.

NIKITIDIS, N. 2019: Flora Amorgina. Spania ke endemika fita tis Amorgou ke ton Mikroni Kikladon. [In Greek]. - Athens: https://openbook.gr/flora-amorgina/, 184 pp.

NIKLFELD, H. 1981: Aspects of mapping regional floras (A review based on mapping the flora of Central Europe). In Velčev, V.I. & Kožuharov, S.I. (eds): Mapping the flora of the Balkan Peninsula, pp. 21-43. - Sofia: Publishing House of the Bulgarian Academy of Sciences,

NIKOLAIDOU, S. & YANNITSAROS, A. 1992: Floristic investigation of the island Spetses [In Greek with English summary]. In: Proceedings, 12th Annual Meeting of the Hellenic Society of Biological Sciences, pp. 174-176. - Athine.

NIKOLIĆ, D. & AL. 2014: Lectotypification of some names in the *Jovibarba heuffelii* group (Crassulaceae) - *Phytotaxa* **174(4)**: 206-222.

NIKOLIĆ, T. 1989: Phenotypic variability of leaflets in *Thalictrum minus* complex (Ranunculaceae). - *Acta Bot. Croat.* **48**: 177-187.

NIKOLIĆ, T. 1991a: Phenotypic variability of fruit and its taxonomic significance in *Thalictrum minus* complex and species *Thalictrum velebiticum* Deg. - *Feddes Repert.* **102(1-2)**: 69-74.

NIKOLIĆ, T. 1991b: Phenotypic plasticity and taxonomic significance in *Thalictrum minus*-complex. - *Feddes Repert.* **102(5-6)**: 345-349.

NIKOLIĆ, T. 2020: Flora Croatica. Vaskularna flora Republike Hrvatske [In Croatian].

NIKOLIĆ, T. & AL. 2015: Endemi u hrvatskoj flori [In Croatian]. - Zagreb: 492 pp.

NIKOLIĆ, T., FOIS, M. & MILAŠINOVIĆ, B. 2020: The endemic and range restricted vascular plants of Croatia: diversity, distribution patterns and their conservation status. - *Phytotaxa* **436(2)**: 125-140.

NIKOLIĆ, V. & DIKLIĆ, N. 1973: Neue Fundangaben von Pflanzenarten aus Serbien (III) [In Serbo- Croatian with German summary]. - *Glasn. Prir. Mus. Beogradu, Ser. B, Biol. Nauke* **28**: 35-40.

NIKOLIĆ, V. & DIKLIĆ, N. 1975: Neue Fundangaben von Pflanzenarten aus Serbien (V) [In Serbo- Croatian with English summary]. - *Glasn. Prir. Mus. Beogradu, Ser. B, Biol. Nauke* **30**: 15-22.

NIKOLIĆ, V. & DIKLIĆ, N. 1977: Neue Angaben über Fundstellen von Pflanzenarten in Serbien (VI) [In Serbo-Croatian with German summary]. - *Glasn. Prir. Mus. Beogradu, Ser. B, Biol. Nauke* **32**: 17-26.

NIKOLIĆ, V. & DIKLIĆ, N. 1978: Einige neue Pflanzenarten in der Flora der SR Serbien [In Serbo- Croatian with German ummary]. - *Glasn. Prir. Mus. Beogradu, Ser. B, Biol. Nauke* **33**: 95-102.

NIKOLIĆ, V. & DIKLIĆ, N. 1979: Neue Angaben über Fundstellen von Pflanzenarten in Serbien (VIII) [In Serbo-Croatian with German summary]. - *Glasn. Prir. Mus. Beogradu, Ser. B, Biol. Nauke* **34**: 31-44.

NIKOLIĆ, V. & MAYER, E. 1987: Zur taxonomischen Bewertung von *Hypecoum pseudograndiflorum* Petrović (Papaveraceae) [In Serbo-Croatian with German summary]. - *Glasn. Prir. Mus. Beogradu, Ser. B, Biol. Nauke* **42**: 9-14.

NIKOLOV, N.A. 1991: Chromosome numbers of Bulgarian angiosperms from North Pirin Mountain: Reserve "Bajuvi Dupki - Džindžirica". - *Fitologiya* **41**: 70-72.

NIKOLOVSKI, T. & MATVEJEVA, J. 1971: *Cotyledon chloranthus* (Heldr. Sart.) Hal. as a new

species in the flora from SRM [In Macedonian with English summary]. - *Fragm. Balcan.* **8(20)**: 173-176.
NILSSON, H. 1938: *Anagallis arvensis* L. s.l. und die Natur ihrer Farbenvarianten. - *Hereditas (Lund)* **24(1-2)**: 97-109.
NILSSON, L.A. 1978a: Pollination ecology and adaptation in *Platanthera chlorantha* (Orchidaceae). - *Bot. Not.* **131(1)**: 35-51.
NILSSON, L.A. 1978b: Pollination ecology of *Epipactis palustris* (Orchidaceae). - *Bot. Not.* **131(3)**: 355-368.
NILSSON, L.A. 1980: The pollination ecology of *Dactylorhiza sambucina* (Orchidaceae). - *Bot. Not.* **133(3)**: 367-385.
NILSSON, Ö. 1966: Studies in *Montia* L., *Claytonia* L. and allied genera. II. Some chromosome numbers. - *Bot. Not.* **119(3)**: 464-468.
NILSSON, Ö. 1971: Studies in *Montia* L., *Claytonia* L. and allied genera. Taxonomy, morphology, pollen morphology, cytology, reproduction and distribution. - Uppsala: Thesis (Acta Univ. Upsal. 185: 6 pp.).
NILSSON, Ö. & LASSEN, P. 1971: Chromosome numbers of vascular plants from Austria, Mallorca and Jugoslavia. - *Bot. Not.* **124(2)**: 270-276.
NINOVA, D.J. & AL. 1984a: Variability of *Glechoma hirsuta* Waldst. & Kit. and *Glechoma hederacea* L. in Bulgaria. Anatomical study. I [In Bulgarian with English abstract]. - *Fitologiya* **24**: 3-28.
NINOVA, D.J. & AL. 1984b: Variability of *Glechoma hirsuta* Waldst. & Kit. and *Glechoma hederacea* L. in Bulgaria. An anatomical investigation. II [In Bulgarian with Russian and English summaries]. - *Fitologiya* **25**: 13-40.
NINOVA, D.J. & AL. 1984c: Morphological variability of *Glechoma hirsuta* Wald. & Kit. and *Glechoma hederacea* L. [In Bulgarian with Russian and English summaries]. - *Fitologiya* **26**: 3-59.
NITZELIUS, T.G. 1968: *Abies*. A review of the firs in the Mediterranean. - *Lustgården* **49**: 146-189.
NOACK, H. 1984: Das Portrait: *Quercus frainetto* Tenore, die ungarische Eiche. - *Mitt. Deutsch. Dendrol. Ges.* **75**: 167-171.
NOBIS, M., KLICHOWSKA, E. & NOWAK, A. 2017: (2510) Proposal to conserve the name *Stipa pulcherrima* (Poaceae) with a conserved type. - *Taxon* **66(2)**: 518-519. [approved, cf. Taxon 69(2): 392, 2020].
NOLD, R. 2003: Columbines: *Aquilegia*, *Paraquilegia* and *Semiaquilegia*. - Timber Press, 192 pp.
NORDBORG, G. 1963: Studies in *Sanguisorba officinalis* L. - *Bot. Not.* **116(2)**: 267-288.
NORDBORG, G. 1966: *Sanguisorba* L., *Sarcopoterium* Spach, and *Bencomia* Webb et Berth. Delimination and subdivision of the genera. - *Opera Bot.* **11(2)**: 1-103 + 6 plates.
NORDBORG, G. 1967: The genus *Sanguisorba* section *Poterium*. Experimental studies and taxonomy. - *Opera Bot.* **16**: 1-166.
NORDENSKIÖLD, H. 1937: Intra- and interspecific hybrids of *Phleum pratense* and *P. alpinum*. - *Hereditas (Lund)* **23(3)**: 304-316.
NORDENSKIÖLD, H. 1945: Cyto-genetic studies in the genus *Phleum*. - *Acta Agric. Suec.* **1(1)**: 1-138.
NORDENSKIÖLD, H. 1949a: Synthesis of *Phleum pratense* L. from *P. nodosum* L. - *Hereditas (Lund)* **35 (2)**: 190-202.
NORDENSKIÖLD, H. 1949b: The somatic chromosome numbers of some *Luzula* species. - *Bot. Not.* **1949(1)**: 81-92.
NORDENSKIÖLD, H. 1951: Cyto-taxonomical studies in the genus *Luzula*. I. Somatic chromosomes and chromosome numbers. - *Hereditas (Lund)* **37(3)**: 325-355.
NORDENSKIÖLD, H. 1956: Cyto-taxonomic studies in the genus *Luzula*. II. Hybridization experiments in the *campestris-multiflora* complex. - *Hereditas (Lund)* **42(1-2)**: 7-73.
NORDENSKIÖLD, H. 1971: Hybridization experiments in the genus *Luzula*. IV. Studies with taxa of the *campestris-multiflora* complex from the northern and southern hemispheres. - *Hereditas (Lund)* **68 (1)**: 47-59.
NORDENSTAM, B. 1960: Studies in the Aegean flora. II. The genus *Lyrolepis*. - *Bot. Not.* **113(4)**: 451-457.
NORDENSTAM, B. 1971: Cytogeography of the genus *Hedypnois* (Compositae). - *Bot. Not.* **124(4)**: 483-489.
NORDENSTAM, B. 1977: Senecioneae and Liabeae - systematic review. In Heywood, V.H., Harborne, J.B. & Turner, B.L. (eds): The biology and chemistry of the Compositae 2: 799-830. - London, etc.: Academic Press.
NORDENSTAM, B. 1978: Taxonomic studies in the tribe Senecioneae (Compositae). - *Opera Bot.* **44**: 1-83.
NORDENSTAM, B. 2005: (1696) Proposal to conserve the name *Senecio leucanthemifolius* against *S. varicosus* (Compositae). - *Taxon* **54(2)**: 551-552. [approved, cf. Taxon 56(4): 1289, 2007].
NORDHAGEN, R. 1932: Über die Einrollung der Fruchtstiele bei der Gattung *Cyclamen* und ihre biologische Bedeutung. - *Beih. Bot. Centralbl., Abt. 2* **49, Erg.-Bd**: 359-395.
NORLINDH, T. 1972: Notes on the variation and taxonomy in the *Scirpus maritimus* complex. - *Bot. Not.* **125(4)**: 397-405.
NORLINDH, T. 1977: Calenduleae - systematic review. In Heywood, V.H., Harborne, J.B. & Turner, B. L. (eds): The biology and chemistry of the Compositae 2: 961-987. - London, etc.: Academic Press.
NORMAN, T. 1976: *Ophrys*: a progress report. - *Quart. Bull. Alpine Gard. Soc. Gr. Brit.* **44(1)**: 42-60.
NORTH, C. 1990: Into Albania. - *Rock Gard.* **22(1)**: 63-71.
NORTH, C. 2003: Wild Flowers around the Mediterranean. - CD-Rom: c. 264 pp.
NORTH, C. & NORTH, M. 1983: Spring flowers on the island of Corfu. - *J. Scott. Rock Gard. Club* **18(2)**: 143-149 + 1 plate.
NORTH, C. & NORTH, M. 1986: Spring flowers of Rhodes. - *Rock Gard.* **19(4)**: 406-408.
NORTH, C. & NORTH, M. 1990: The Gargano. - *Rock Gard.* **21(4)**: 338-345.
NORTHWAY, M. 1971: Alpines in colour: *Asperula suberosa* Linn. - *Quart. Bull. Alpine Gard. Soc. Gr. Brit.* **39(2)**: 104-105.
NORVERTO, C.A., GONZÁLEZ ANDRÉS, F. & ORTIZ, J.M. 1994: Leaf and stem anatomy of species of *Cytisophyllum*, *Cytisus*, *Chamaecytisus*, *Genista*, and *Genista* sect. *Teline* (Fabaceae: Genisteae) as an aid for taxonomy. - *Israel J. Pl. Sci.* **42(3)**: 213-225.
NOUVIANT, J. 1997: Recherches sur *Ephedra* en Europe I: Typification de *Ephedra distachya*. - *Bull. Murith. Soc. Valais, Sci. Nat.* **114**: 127-134.
NOUVIANT, J. 1999a: Recherches sur *Ephedra* en Europe IV. Critique de *Ephedra nebrodensis*. - *Bull. Murith. Soc. Valais, Sci. Nat.* **116**: 11 pp.
NOUVIANT, J. 1999b: Recherches sur *Ephedra* en Europe V: Typification de *Ephedra monostachya* et clé de détermination des *Ephedra* européens. - *Bull. Murith. Soc. Valais, Sci. Nat.* **116**: 10 pp.
NOVÁK, F.A. 1924: Generi Euphorbiae species nova balcanica. - *Acta Bot. Bohem.* **3**: 35-40 + 1 plate.
NOVÁK, F.A. 1926: Monografická studie o *Dianthus Kitaibelii* (Janka) s.l. [In Czech with Latin descriptions]. - *Spisy Přír. Fak. Karlovy Univ.* **71**: 1-41.
NOVÁK, F.A. 1928: Dianthi fimbriati europaei. - *Repert. Spec. Nov. Regni Veg.* **25**: 38-47, 204-208.
NOVÁK, F.A. 1929: *Ligusticum olympicum*. Generis Ligustici species nova Balcanica. - *Acta Bot. Bohem.* **8**: 5-6 + 1 fig.
NOVÁK, F.A. 1934: *Pinus pindica* et *Pinus magellensis*. - *Bull. Int. Acad. Tchéque Sci., Cl. Sci. Math. Nat. Méd.* **1934**: 4 pp.
NOVÁK, F.A. 1936: A Goniolimonis varietates et formae bulgaricae ad huc cognitae. - *Izv. Bulg. Bot. Druzh.* **7**: 56-62.
NOVÁK, F.A. 1938: De Armeriis balcanicis nonnullis. - *Věstn. Král. České Společn. Nauk, Tř. Mat.-Přír.* **1938**: 1-25 + 1 plate and 1 map.
NOVÁK, F.A. 1939: De Armeriis balcanicis nonnullis, II. - *Věstn. Král. České Společn. Nauk. Tř. Mat.-Přír.* **1939**: 1-24.

NOVOTNÁ, I. 1962: Untersuchungen über die Chromosomenzahl innerhalb *Arabis hirsuta* (Komplex). - *Preslia* **34(3)**: 249-254.

NOVOTNÁ, I. 1965: Experimentelle Bearbeitung der Schotenlänge bei *Arabis hirsuta*-Komplex. - *Preslia* **37**: 144-155.

NOVOTNÁ, I. 1968: Experimentelle Bearbeitung der Schotenlänge bei *Arabis hirsuta*-Komplex. II. - *Folia Geobot. Phytotax.* **3(2)**: 201-216.

NOVOTNÁ, I. 1971: Experimentelle Analyse von fünf Habitus-Merkmalen an Repräsentanten des *Arabis hirsuta*-Komplexes. - *Stud. Českoslov. Akad. Věd.* **6**: 1-88.

NOVOTNÁ, I. & CZAPIK, R. 1971: Investigations on some hybrids from the *Arabis hirsuta* complex. - *Folia Geobot. Phytotax.* **6(4)**: 419-440.

NOVOTNÁ, I. & CZAPIK, R. 1975: Studies on the progenies of hybrids from the *Arabis hirsuta* complex. - *Folia Geobot. Phytotax.* **9(4)**: 341-357.

NOWAKOWSKA, T. & ŻEGLICKA, B. 1972: Karyological studies in *Polygonatum* Mill. I. Karyotype analysis of *P. multiflorum* (L.) All. and *P. odoratum* (Mill.) Druce. - *Acta Biol. Cracov., Ser. Bot.* **15(1)**: 77-86 + 2 plates.

NUR, U. & ZOHARY, D. 1959: Distribution patterns of diploid and tetraploid forms of *Dactylis glomerata* L. in Israel. - *Bull. Res. Council Israel, Sect. D, Bot.* **7(1)**: 13-22.

NYÁRADY, A. & SZABO, A.T. 1981: The chorological information in Romania. In Velčev, V.I. & Kožuharov, S.I. (eds): Mapping the Flora of the Balkan Peninsula, pp. 63-68. - Sofia: Publishing House of the Bulgarian Academy of Sciences.

NYÁRÁDY, E.J. 1928a: Vorstudium über einige Arten der Section *Odontarrhena* der Gattung *Alyssum*. - *Bul. Grăd. Bot. Univ. Cluj* **7(1-2)**: 3-51 + 10 plates.

NYÁRÁDY, E.J. 1928b: Vorstudium über einige Arten der Section *Odontarrhena* der Gattung *Alyssum* (Fortsetzung). - *Bul. Grăd. Bot. Univ. Cluj* **7(3-4)**: 65-160.

NYÁRÁDY, E.J. 1928c: Vorstudium über einige Arten der Section *Odontarrhena* der Gattung *Alyssum* (Fortsetzung). - *Bul. Grăd. Bot. Univ. Cluj* **8(2-4)**: 152-156.

NYÁRÁDY, E.J. 1929a: Vorstudium über einige Arten der Section *Odontarrhena* der Gattung *Alyssum* (Fortsetzung). - *Bul. Grăd. Bot. Univ. Cluj* **9(1-2)**: 1-68.

NYÁRÁDY, E.J. 1929b: Neue Beiträge zur Kenntnis der balkanischen *Alyssum*-Arten. - *Repert. Spec. Nov. Regni Veg.* **27**: 392-395 + 1 plate.

NYÁRÁDY, E.J. 1949: Synopsis specierum, variatorum et formarum sectionis *Odontarrhenae* generis *Alyssum*. - *Analele Acad. Republ. Populare Române, ser. A* **1(3)**: 67-194 + 6 plates.

NYDEGGER-HÜGLI, M. 1989: Erste Ergänzungen zu P. H. Davis' "Flora of Turkey and the East Aegean Islands" 1-10 (1965-1988). - *Bauhinia* **9(2)**: 141-152.

NYDEGGER-HÜGLI, M. 1990: Zweite Ergänzungen zu P. H. Davis' "Flora of Turkey and the East Aegean Islands" 1-10 (1965-1988). - *Bauhinia* **9(3)**: 205-219.

NYDEGGER-HÜGLI, M. 1991: Dritte Ergänzungen zu P. H. Davis' "Flora of Turkey and the East Aegean Islands" 1-10 (1965-1988). - *Bauhinia* **9(4)**: 273-290.

NYDEGGER-HÜGLI, M. 1992: Vierte Ergänzungen zu P. H. Davis' "Flora of Turkey and the East Aegean Islands" 1-10 (1965-1988). - *Bauhinia* **10**: 59-74.

NYDEGGER-HÜGLI, M. 1993: Fünfte Ergänzungen zu P. H. Davis' "Flora of Turkey and the East Aegean Islands" 1-10 (1965-1988). - *Bauhinia* **11(1)**: 19-34.

NYDEGGER-HÜGLI, M. 1994: Sechste Ergänzungen zu P. H. Davis' "Flora of Turkey and the East Aegean Islands" 1-10 (1965-1988). - *Bauhinia* **11(2)**: 103-119.

NYDEGGER-HÜGLI, M. 1995: Siebte Ergänzungen zu P. H. Davis' "Flora of Turkey and the East Aegean Islands" 1-10 (1965-1988). - *Bauhinia* **11(3)**: 177-190.

NYDEGGER-HÜGLI, M. 1996: Achte Ergänzungen zu P. H. Davis' "Flora of Turkey and the East Aegean Islands" 1-10 (1965-1988). - *Bauhinia* **11(4)**: 221-238.

NYDEGGER-HÜGLI, M. 1998: Neunte Ergänzungen zu P. H. Davis' "Flora of Turkey and the East Aegean Islands" 1-10 (1965-1988). - *Bauhinia* **12**: 81-107.

NYDEGGER-HÜGLI, M. 1999: Zehnte Ergänzungen zu P. H. Davis' "Flora of Turkey and the East Aegean Islands" 1-10 (1965-1988). - *Bauhinia* **13**: 41-60.

NYDEGGER-HÜGLI, M. 2000: Elfte Ergänzungen zu P. H. Davis' "Flora of Turkey and the East Aegean Islands" 1-10 (1965-1988). - *Bauhinia* **14**: 93-122.

NYDEGGER-HÜGLI, M. 2001: Zwölfte Ergänzungen zu P. H. Davis' "Flora of Turkey and the East Aegean Islands" 1-10 (1965-1988). - *Bauhinia* **15**: 97-114.

NYDEGGER-HÜGLI, M. 2002: Dreizehnte und letzte Ergänzungen zu P. H. Davis' "Flora of Turkey and the East Aegean Islands" 1-10 (1965-1988). - *Bauhinia* **16**: 33-55.

NYGREN, A. 1946: The genesis of some Scandinavian species of *Calamagrostis*. - *Hereditas (Lund)* **32(2)**: 131-262.

NYGREN, A. 1957: *Poa timoleontis* Heldr., a new diploid species of the section *Bolbophorum* A. & Gr. with accessory chromosomes only in the meiosis. - *Kungl. Lantbrukshögskolans Ann.* **23**: 489-495.

NYGREN, A. 1962a: Artificial and natural hyrbidization in European *Calamagrostis*. - *Symb. Bot. Upsal.* **17(3)**: 1-105.

NYGREN, A. 1962b: On *Poa badensis, Poa concinna, Poa pumila, Poa xerophila* and the possible origin of *Poa alpina*. - *Kungl. Vetenskapssamh. Uppsala Årsbok* **6**: 1-29.

NYMAN, C.F. 1855: Sylloge Florae Europaeae seu plantarum vascularium Europae indigenarum enumeratio adjectis synonymis gravioribus et indicata singularum distributione geographica. - Oerebroae (Örebro): N.M. Lindh, xxiv + 442 pp. [cf. TL-2 3: 798, 1981].

NYMAN, C.F. 1878-90: Conspectus Florae Europaeae, seu enumeratio methodica plantarum phanerogamarum Europae indigenarum, indicatio distributionis geographicae singularum ... - Örebro: Officinae Bohlinianae, 4 parts and 2 suppl. [cf. TL-2 3: 798, 1981].

O'KANE, S.L. JR. & AL-SHEHBAZ, I.A. 1997: A synopsis of *Arabidopsis* (Brassicaceae). - *Novon* **7(3)**: 323-327.

O'LEARY, N. & MÚLGURA, M.E. 2012: A taxonomic revision of the genus *Phyla* (Verbenaceae). - *Ann. Missouri Bot. Gard.* **98(4)**: 578-596.

OBERDORFER, E. 1948: Gliederung und Umgrenzung der Mittelmeervegetation auf der Balkanhalbinsel. - *Ber. Geobot. Forschungsinst. Rübel Zürich* **1947**: 84-111.

OBERDORFER, E. 1952: Beitrag zur Kenntnis der nordägäischen Küstenvegetation. - *Vegetatio* **3(6)**: 329-349 + 3 plates.

OBERDORFER, E. 1954a: Über Unkrautgesellschaften der Balkanhalbinsel. - *Vegetatio* **4(6)**: 379-411 + 3 plates & 9 fold-out tables.

OBERDORFER, E. 1954b: Nordägäische Kraut- und Zwergstrauchfluren im Vergleich mit den entsprechenden Vegetationseinheiten des westlichen Mittelmeergebietes. - *Vegetatio* **5-6**: 88-96.

OBERDORFER, E. 1992: Einige Bemerkungen zur Taxonomie und Ökologie der *Luzula luzuloides* (Lam.) Dandy et Wilmott ssp. *cuprina* (Roch. ex Aschers. & Graebner) Chrtek et Křísa. - *Bauhinia* **10**: 17-22.

OBERPRIELER, CH. 1998: The systematics of *Anthemis* L. (Compositae, Anthemideae) in W and C North Africa. - *Bocconea* **9**: 1-328.

OBERPRIELER, CH. & VOGT, R. 1994: Reports (294-312). In Kamari, G., Felber, F. & Garbari, F. (eds): Mediterranean chromosome number reports - 4. - *Fl. Medit.* **4**: 262-269.

OBERPRIELER, CH. & VOGT, R. 1996: Chromosome numbers of North African phanerogams. VI. Some counts in Leguminosae. - *Willdenowia* **25(2)**: 669-680.

OBERPRIELER, CH. & VOGT, R. 1997: Reports (692-699). In Kamari, G., Felber, F. & Garbari, F. (eds): Mediterranean chromosome number reports - 6. - *Fl. Medit.* **6**: 262-266.

OBERPRIELER, CH. & VOGT, R. 2006: The taxonomic position of *Matricaria macrotis* (Compositae- Anthemideae). - *Willdenowia* **36(1)**: 329-338.

OBESO, J.R. 1992: Pollination ecology and seed set in *Asphodelus albus* (Liliaceae) in northern Spain. - *Flora, Morphol. Geobot. Oekophysiol.* **187(3-4)**: 219-226.

OBESO, J.R. 1996: Producción de frutos y semillas en *Ilex aquifolium* L. (Aquifoliaceae). - *Anales Jard. Bot. Madrid* **54(1)**: 533-539.

OBÓN DE CASTRO, C. & RIVERA NUÑES, D. 1994: A taxonomic revision of the section *Sideritis* (genus *Sideritis*) (Labiatae). - Berlin: J. Cramer, x + 640 pp.

OBRADOVIĆ, M. 1968: O varijabilnosti lista kod vrste *Chenopodium opulifolium* Schrad. [In Serbo- Croatian]. - *Glasn. Zemaljsk. Muz. Sarajevu* **7**: 215-218.

OBRADOVIĆ, M. 1971: Über die Variabilität des Blattes bei der Art *Chenopodium opulifolium* Schrad. - *Wiss. Mitt. Bosnisch-Herzegowinischen Landesmus., C* **1**: 217-220.

OCKENDON, D.J. 1967: Notes on the *Linum perenne* group. In Heywood, V.H. (ed.): Notulae systematicae ad Floram Europaeam spectantes, no. 6. - *Feddes Repert.* **74(1-2)**: 20-22.
OCKENDON, D.J. 1968: Biosystematic studies in the *Linum perenne* group. - *New Phytol.* **67(4)**: 787-813 + 2 plates.
OCKENDON, D.J. 1971: Taxonomy of the *Linum perenne* group in Europe. - *Watsonia* **8(3)**: 205-235.
OCKENDON, D.J. 1976: [*Valeriana saxatilis* subsp. *pancicii*, comb. nov.]. In Heywood, V.H. (ed.): Notulae systematicae ad Floram Europaeam spectantes, no. 19. - *Bot. J. Linn. Soc.* **71(4)**: 274.
OFLAS, S. 1974: An investigation related to the distribution of *Styrax officinalis* in West Anatolia and to its phytosociologic, morphologic, anatomic features and economic potentials [In Turkish with English summary]. - *Ege Üniv. Fen Fak. Ilmi Raporlar Ser.* **128**: 1-68.
OGANEZOVA, G.H. 2014: On the treatment of *Merendera* and *Bulbocodium* (Colchicaceae) as separate genera. - *Fl. Medit.* **24**: 79-92.
OGILVIE-GRANT, M. 1939a: Plants and western Crete. - *New Fl. & Silva* **12(1)**: 26-29.
OGILVIE-GRANT, M. 1939b: Some Greek fritillaries. - *Lily Year-Book* **1939**: 144-146.
OGILVIE-GRANT, M. 1945: On the flora of Greece. - *J. Roy. Hort. Soc.* **70(10)**: 298-305.
OGILVIE-GRANT, M. 1947: A note on the tulips of Greece. - *Daffodil Tulip Year Book* **13**: 116-118.
OGILVIE-GRANT, M. 1961: Interesting plants of the Eastern Mediterranean. - *Quart. Bull. Alpine Gard. Soc. Gr. Brit.* **29(3-4)**: 51-64.
OĞUZ, G. 1972: Morphological and taxonomical studies on *Glycyrrhiza* species of Turkey [In Turkish with English summary]. - *Ege Üniv. Fen Fak. Ilmi Raporlar Ser.* **114**: i-iv + 1-90.
OĞUZ, G. & AL. 1973: *Pinguicula crystallina* Bauer et Sibth.; a new record for Turkey. - *Ege Üniv. Fen Fak. Ilmi Raporlar Ser.* **170**: 1-5.
OHBA, H. 1977: The taxonomic status of *Sedum telephium* and its allied species (Crassulaceae). - *Bot. Mag. (Tokyo)* **90**: 41-56.
OHBA, H. 1978: Generic and infrageneric classification of the Old World Sedoideae (Crassulaceae). - *J. Fac. Sci. Univ. Tokyo, Sect. 3, Bot.* **12(4)**: 139-198.
OHLE, H. 1974: Beiträge zur Revision der Gattung *Calendula*. II. Taxonomische Revision der südeuropäischen perennierenden *Calendula*-Sippen. - *Feddes Repert.* **85(4)**: 245-283 + 8 plates.
OHTA, S. 1991: Phylogenetic relationship of *Aegilops mutica* Boiss. with the diploid species of congeneric *Aegilops-Triticum* complex, based on the new method of genome analysis using its B- chromosomes. - *Mem. Coll. Agric. Kyoto Univ.* **137**: 1-116.
OHTA, S. 1995: Morphological differentiation and geographical distribution of *Aegilops biuncialis* Vis. in the Aegean region. In: Proc. 8th Intern. Wheat Genet. Symp., Beijing, China, 99-104.
OJA, T., JAASKA, V. & VISLAP, V. 2003: Breeding system, evolution and taxonomy of *Bromus arvensis*, *B. japonicus* and *B. squarrosus* (Poaceae). - *Pl. Syst. Evol.* **242(1-4)**: 101-117.
OJA, T. & TALVE, T. 2012: Genetic diversity and differentiation in six species of the genus *Rhinanthus* (Orobanchaceae). - *Pl. Syst. Evol .* **298(5)**: 901-911.
OKEKE, S. 1978: Morphological variation of bracts, bracteoles and fruits in *Daucus*. In Cauwet-Marc, A.M. & Carbonnier, J.: Les Ombellifères, pp. 161-174. - Perpignan.
OLIVA, R. 1978: Números cromosómicos para la flora española. 45-59. - *Lagascalia* **8(1)**: 105-109.
OLIVIERI, I., GOUYON, P.-H. & PROSPERI, J.-M. 1991: Life cycles of some Mediterranean invasive plants. In Groves, R.H. & Di Castri, F. (eds): Biogeography of mediterranean invasions, pp. 145-157. - Cambridge, etc.: Cambridge Univ. Press.
OLOWOKUDEJO, J.D. & HEYWOOD, V.H. 1984: Cytotaxonomy and breeding system of the genus *Biscutella* (Cruciferae). - *Pl. Syst. Evol.* **145(3-4)**: 291-309.
OLŠAVSKÁ, K. & AL. 2011: Biosystematic study of the *Cyanus triumfetti* group in Central Europe. - *Preslia* **83(1)**: 59-88.
OLŠAVSKÁ, K. & AL. 2013: Cytogeography of European perennial species of *Cyanus* (Asteraceae). - *Bot. J. Linn. Soc.* **173(2)**: 230-257.
OLŠAVSKÁ, K. & AL. 2016: On the origins of Balkan endemics: the complex evolutionary history of the *Cyanus napulifer* group (Asteraceae). - *Ann. Bot. (Oxford), n.s.* **118(6)**: 1071-1088.
ONNO, M. 1932: Geographisch-morphologische Studien über *Aster alpinus* L. und verwandte Arten. - *Biblioth. Bot.* **26/106**: 83 pp. + 6 plates and 6 maps.
ONNO, M. 1933: Die Wildformen aus dem Verwandtschaftskreis "Brassica oleracea L.". - *Oesterr. Bot. Z.* **82(4)**: 309-334.
ONNO, M. 1937: Die Wildformen von *Daucus* Sect. *Carota*. - *Beih. Bot. Centralbl., Abt. 2* **56(1-2)**: 83-136.
OOST, E.H. & AL. 1987: Lectotypification of *Brassica rapa* L., *B. campestris* L. and neotypification of *B. chinensis* L. (Cruciferae). - *Taxon* **36(3)**: 625-634.
OOST, E.H., BRANDENBURG, W.A. & JARVIS, C.E. 1989: Typification of *Brassica oleracea* L. (Cruciferae) and its Linnaean varieties. - *Bot. J. Linn. Soc.* **101(4)**: 329-345.
OPRAVIL, E. 1983: *Xanthium strumarium* L. - ein europäischer Archäophyt? - *Flora, Morphol. Geobot. Oekophysiol.* **173(1-2)**: 71-79.
ORCAN, N. & BINZET, R. 2006: A new record for the flora of Turkey: *Alyssum idaeum* Boiss. & Heldr. - *Pakistan J. Bot.* **38(4)**: 931-933.
ORDEN, C.V. 1974: April in Griekenland. - *Orchideeën* **36(5)**: 142-144.
ØRGAARD, M. & JACOBSEN, N. 2004: Iagttagelser af *Crocus cartwrightianus* - en nær slægtning til safrankrokus [In Danish]. - *Alpine Have* **82**: 10-14.
ORNDUFF, R. 1970: Cytogeography of *Nymphoides* (Menyanthaceae). - *Taxon* **19(5)**: 715-719.
ORNDUFF, R. 1975a: Heterostyly and pollen flow in *Hypericum aegypticum* (Guttiferae). - *Bot. J. Linn. Soc.* **71(1)**: 51-57.
ORNDUFF, R. 1975b: Pollen flow in *Lythrum junceum*, a tristylous species. - *New Phytol.* **75(1)**: 161-166.
ORNDUFF, R. 1987: Reproductive systems and chromosome races of *Oxalis pes-caprae* L. and their bearing on the genesis of a noxious weed. - *Ann. Missouri Bot. Gard.* **74(1)**: 79-84.
ORNDUFF, R. (ED.) 1967: Index to plant chromosome numbers for 1965. - *Regnum Veg.* **50**: 1-128.
ORNDUFF, R. (ED.) 1968: Index to plant chromosome numbers for 1966. - *Regnum Veg.* **55**: 1-126.
ORNDUFF, R. (ED.) 1969: Index to plant chromosome numbers for 1967. - *Regnum Veg.* **59**: 1-129.
ORPHANIDES, TH. 1850: Prospectus flora graeca exsiccata. Collection de plantes rares et intéressantes de la Grèce. - Athènes: C. Nicolaides Philadelphien, 16 pp.
ORPHANIDES, TH. 1866: Enumeratio chloridis hellenicae. - Athine [n.v.]
ORPHANIDES, TH. 1867: Sur l'état actuel de la flore grecque. - Paris: Act. Congr. Int. Bot. Paris, pp. 70-78.
ORPHANIDES, TH. 1868: Peri tis avtofious ellinikis vlastiseos (On the indigenous Greek flora) [In Greek]. - Athine: Tipis A. Ktena & P. Soutsa, 63 pp.
ORPHANIDES, TH. 1870a: Sur le profit que l'horticulture européenne peut retirer de la flore grecque. In: Bulletin du Congrès international de botanique et d'horticulture de St. Pétersbourg le 6/18 le 8/20 et le 10/22 mai 1869, pp. 112-129. - St. Pétersbourg.
ORPHANIDES, TH. 1870b: Sur quelques plantes rares ou nouvelles de la Grèce. In: Bulletin du Congrès international de botanique et d'horticulture de St. Pétersbourg le 6/18 le 8/20 et le 10/22 mai 1869, pp. 130-142. - St. Pétersbourg.
ORPHANIDES, TH. 1876a: Dissertation sur les caractères spécifiques du genre *Colchicum*, et sur quelques espèces nouvellement découvertes en Grèce. In: Atti del congresso internaz. botanico tenuto in Firenze 1874, pp. 27-36. - Firenze.
ORPHANIDES, TH. 1876b: [Presentation of new and rare plant species from Greece]. In: Atti del congresso internazionale botanico tenuto in Firenze 1874, pp. 214-216. - Firenze.
ORSHANSKY, G. 1938: Seasonal leaf dimorphism in *Ononis natrix* L. - *Palestine J. Bot., Jerusalem Ser.* **1(2)**: 233-234.
ORTEGA OLIVENCIA, A. & CATALÁN, P. 2009: Systematics and evolutionary history of the circum-Mediterranean genus *Anagyris* L. (Fabaceae) based on morphological and molecular data. - *Taxon* **58(4)**: 1290-1306.
ORTEGA OLIVENCIA, A. & DEVESA, J.A. 1992: Revisión del género *Scrophularia* L. (Scrophulariaceae) en la Península Ibérica e Islas Baleares. - *Ruizia* **11**: 1-157.
ORTEGA OLIVENCIA, A. & DEVESA, J.A. 2002: (1527) Proposal to conserve the name

Scrophularia auriculata (Scrophulariaceae) with a conserved type. - *Taxon* **51(1)**: 201-202. [approved, cf. Taxon 53: 821, 2004].

ORTIZ, L., GONZÁLEZ, A. & CHUECA, M.C. 1985: On the presence of diploid and tetraploid forms of *Hordeum bulbosum* L. in Spain. - *Anales Jard. Bot. Madrid* **41(2)**: 361-365.

ORTÚÑEZ RUBIO, E. 1993: Biosistemática de la seccion *Ovinae* Fr. (intravaginales) del género *Festuca* en la Península Ibérica. - Madrid: Universidad Autonoma de Madrid, 261 pp.

ORTÚÑEZ RUBIO, E. & FUENTE GARCÍA, V. DE LA 1996: Reports (394-400). In Kamari, G., Felber, F. & Garbari, F. (eds): Mediterranean chromosome number reports - 5. - *Fl. Medit.* **5**: 261-265.

OSBORNOVÁ, J. 1984: Habitat differentiation of neophytic populations of *Solanum elaeagnifolium* Cav. in Greek Macedonia. - *Acta Bot. Slov., A, Suppl.* **1**: 253-255.

OSCHMANN, J. 2000: Morphologische und molekularsystematische Untersuchungen an der *Centaurea stoebe* L.-Gruppe (Asteraceae-Cardueae) in Europa. - *Diss. Bot.* **324**: 242 pp.

OSTENFELD, C.H. 1927: Meeresgräser II. Marine Potamogetonaceae. - *Pflanzenareale* **1(4)**: 46-50; Karte 34, 37-39.

ÖSTERGREN, G. 1940a: On the morphology of *Agropyron junceum* (L.) PB., *A. repens* (L.) PB. and their spontaneous hybrids. - *Bot. Not.* **1940(2)**: 133-143.

ÖSTERGREN, G. 1940b: Cytology of *Agropyron junceum*, *A. repens* and their spontaneous hybrids. - *Hereditas (Lund)* **26(3-4)**: 305-316.

ÖSTERGREN, G. 1942: Chromosome numbers in *Anthoxanthum*. - *Hereditas (Lund)* **28(1-2)**: 242-243.

OSTERMEYER, F. 1887: Beitrag zur Flora der jonischen Inseln Corfu, Sta. Maura, Zante und Cerigo. - *Verh. K.K. Zool.-Bot. Ges. Wien* **37**: 651-672.

OSTERMEYER, F. 1890: Beitrag zur Flora von Kreta. - *Verh. K.K. Zool.-Bot. Ges. Wien* **40**: 291-300.

OSVAČILOVÁ, V. 1983: Some nomenclatoric and taxonomic notes to the genus *Thalictrum* L. - *Biológia, A (Bratislava)* **38(5)**: 455-459.

OTT, G. 1990: *Biarum tenuifolium* (L.) Schott aus dem Mittelmeerraum. - *Palmengarten* **54(3)**: 175, 178.

OTTONELLO, D., ROMANO, S. & ALLIATA, N. 1986: Numeri cromosomici per la Flora Italiana: 1037-1048. - *Inform. Bot. Ital.* **17(1-3)**: 91-98.

OUDEJANS, R.C.H.M. 1990: World catalogue of species names published in the tribe Euphorbieae (Euphorbiaceae) with their geographical distribution. - Utrecht: Computer printout, viii + 444 pp.

OUDEJANS, R.C.H.M. 1992: New combinations for infraspecific taxa in the genus *Euphorbia* L. (Euphorbiaceae). - *Collect. Bot. (Barcelona)* **21**: 183-189.

OUDEJANS, R.C.H.M. 1993: World catalogue of species names published in the tribe Euphorbieae (Euphorbiaceae) with their geographical distribution. Cumulative supplement, I. - Scherpenzeel: Computer printout, 78 pp.

OUDEJANS, R.C.H.M. & MOLERO, J. (EDS) 1992: Current research in the taxonomy of genus *Euphorbia* L. s.l. (Euphorbiaceae). - *Collect. Bot. (Barcelona)* **21**: 1-253.

OVADIAHU-YAVIN, Z. 1969: Cytotaxonomy of the genus *Bromus* of Palestine. - *Israel J. Bot.* **18(4)**: 195-216.

OVERBECK, J., ANAGNOSTIDIS, K. & ECONOMOU-AMILLI, A. 1982: A limnological survey of three Greek lakes: Trichonis, Lyssimachia and Amvrakia. - *Arch. Hydrobiol.* **95(1-4)**: 365-394.

OVINGTON, J.D. & SCURFIELD, G. 1956: Biological Flora of the British Isles. *Holcus mollis* L. - *J. Ecol.* **44(1)**: 272-280.

OWNBEY, M. 1950: Natural hybridization and amphiploidy in the genus *Tragopogon*. - *Amer. J. Bot.* **37(7)**: 487-499.

OWNBEY, M. & MCCOLLUM, G.D. 1954: The chromosomes of *Tragopogon*. - *Rhodora* **56(661)**: 7-21.

OXELMAN, B. 1984: Studies in *Silene* L. sect. *Atocion* Otth (Caryophyllaceae), with special reference to the *sedoides*-group. - Göteborg: Univ. of Göteborg, mimeographed report, 48 pp.

OXELMAN, B. 1987: List of collections from Greece. [Unpublished flora list]. - Photocopy: 37 pp.

OXELMAN, B. 1990a: Pollinator sharing and its consequences. A case study on *Silene integripetala*, an endemic of southern Greece. - Göteborg: Univ. of Göteborg, 11 pp. (manuscript).

OXELMAN, B. 1990b: Phenotypic and genotypic effects of inbreeding in *Silene integripetala*, a gynomonoecious annual. - Göteborg: Univ. of Göteborg, 6 pp. (manuscript).

OXELMAN, B. 1995a: A revision of the *Silene sedoides*-group (Caryophyllaceae). - *Willdenowia* **25(1)**: 143-169.

OXELMAN, B. 1995b: Systematic studies in the genus *Silene* L. (Caryophyllaceae). Morphological and molecular approaches. - Göteborg: Ph.D. thesis [summary + 6 published and unpublished papers].

OXELMAN, B. 1995c: RAPD patterns, nrDNA ITS sequences, and morphological patterns in the *Silene sedoides*-group (Caryophyllaceae). - Göteborg: 25 pp. [part of Oxelman 1995b].

OXELMAN, B. & AL. 2001: A revised generic classification of the tribe Sileneae (Caryophyllaceae). - *Nordic J. Bot.* **20(6)**: 743-748.

OXELMAN, B. & LIDÉN, M. 1987: (874) Proposal to reject the name *Silene rubella* L. (Caryophyllaceae) . - *Taxon* **36(2)**: 477-478. [approved, cf. Taxon 42(3): 687-688, 1993].

OXELMAN, B. & LIDÉN, M. 1995: Generic boundaries in the tribe Sileneae (Caryophyllaceae) as inferred from nuclear rDNA sequences. - *Taxon* **44(4)**: 525-542.

OXELMAN, B., LIDÉN, M. & JONSELL, B. 2001: (1476) Proposal to conserve the name *Viscaria* against *Steris* (Caryophyllaceae, Sileneae). - *Taxon* **50(1)**: 281-282. [approved, cf. Taxon 51(4): 797, 2002].

ÖZ, S., TÜMEN, G. & MALYER, H. 1995: Karyological studies on *Sideritis* species grown up in Balikesir, Kazdağ region [In Turkish with English abstract]. - *OT Sist. Bot. Derg.* **2(2)**: 73-82.

ÖZCAN, T. & KUŞ, S. 2001: Taxonomical investigation of some endemic *Bupleurum* L. species for Turkey. In Özhatay, N. (ed.): Proceedings of the Second Balkan Botanical Congress, 1: 319-330. - Istanbul.

ÖZÇANDIR, A., AYKURT, C. & ÖZÜDOĞRU, B. 2019: *Ricotia candiriana* (Brassicaceae), a new species from southern Anatolia, Turkey. - *Phytotaxa* **388(4)**: 287-194.

ÖZDEMIR, B.B.C. & SEPET, H. 2011: Karyotype analysis on the species of *Bellis* L. (Asteraceae) in Turkey. - *Caryologia* **64(3)**: 251-255.

ÖZDEMIR, C. & ŞENEL, G. 1999: The morphological, anatomical and karyological properties of *Salvia sclarea* L. - *Türk. Bot. Derg.* **23(1)**: 7-18.

ÖZDEMIR, F., ZEYBEK, N. & ÖZTÜRK, M.A. 1984: Morphological, anatomical and ecological observations on three *Briza* L. species. - *Ege Üniv. Fen Fak. Derg., B* **7(1)**: 111-121.

OZENDA, P. 1950: L'aire de répartition de l'*Euphorbia dendroides* et sa valeur biogéographique. - *Bull. Soc. Bot. France* **97, 77. Sess. extr.**: 172-181.

OZENDA, P. 1975a: Sur la définition d'un etage de végétation supraméditerranéen en Grèce. In Dafis, S. & Landolt, E. (eds): Zur Vegetation und Flora von Griechenland, Band 1. - *Veröff. Geobot. Inst. E.T.H. Stiftung Rübel Zürich* **55**: 84-98.

OZENDA, P. 1975b: Les limites de la végétation méditerranéenne en montagne, en relation avec le projet de Flora Mediterranea. In Guinochet, M. & al. (eds): La flore du Bassin Méditerranéen, pp. 335-343. - Paris: C.N.R.S.

OZENDA, P. 1975c: Sur les étages de végétation dans les montagnes du Bassin Méditerranéen. - *Doc. Cartogr. Ecol.* **16**: 1-32.

ÖZGIŞI, K., OCAK, A. & ÖZÜGOĞRU, B. 2018: *Noccaea birolmutlui*, a new crucifer species from South West Anatolia, Turkey. - *Phytotaxa* **345(1)**: 359-367.

ÖZHATAY, N. 1983: Cytotaxonomic studies on the genus *Allium* in European Turkey and around Istanbul. I. Sect. *Molium* and *Scorodon*. - *Istanbul Üniv. Eczac. Fak. Mecm.* **19**: 25-36.

ÖZHATAY, N. 1984a: Cytotaxonomic studies on the genus *Allium* in European Turkey and around Istanbul. II. Sect. *Conodoprasum*. - *Istanbul Üniv. Eczac. Fak. Mecm.* **20**: 29-42.

ÖZHATAY, N. 1984b: Cytotaxonomic studies on the genus *Allium* in European Turkey and around Istanbul. III. Sect. *Allium* and sect. *Melanocrommyum*. - *Istanbul Üniv. Eczac. Fak. Mecm.* **20**: 43-65.

ÖZHATAY, N. 1986a: *Allium* species in North Anatolia and their chromosome numbers [In Turkish with English summary]. - *Doğa Bilim. Derg., A* **10(3)**: 452-458.

ÖZHATAY, N. 1986b: Two new *Allium* species from Turkey. - *Notes Roy. Bot. Gard. Edinburgh* **44(1)**: 147-150.

ÖZHATAY, N. 1990: The genus *Allium* in European Turkey and around Istanbul. - *Ann. Mus. Goulandris* **8**: 115-128.

ÖZHATAY, N. 1991: Cytotaxonomy of Turkish *Allium species of section Allium.* In Hanelt, P., Hammer, K. & Knüpffer, H. (eds): The genus *Allium* - taxonomic problems and genetic resources, pp. 21-26. - Gatersleben.

ÖZHATAY, N. 1993: *Allium* in Turkey: distribution, diversity, endemism and chromosome number. In: Demiriz, H. & Özhatay, N. (eds): Proceedings of the V OPTIMA meeting, Istanbul, 8-15 September, 1986, pp. 247-271. - Istanbul Üniversitesi Fen Fakültesi.

ÖZHATAY, N. 2000: Abstracts from the Second Balkan Botanical Congress, Istanbul, Turkey, May 14-18, 2000. - Istanbul: 249 pp.

ÖZHATAY, N. 2001a: Petaloid monocots recently described from the Mediterranean region in Turkey. In Raimondo, F.M. (ed.): Abstracts from the X OPTIMA meeting, Palermo, 13-19 September 2001, p. 260.

ÖZHATAY, N. 2001b: The flora of European Turkey. In Özhatay, N. (ed.): Proceedings of the Second Balkan Botanical Congress, 1: 37-46. - Istanbul.

ÖZHATAY, N. 2002: Diversity of bulbous monocots in Turkey with special reference. Chromosome numbers. - *Pure Appl. Chem.* **74(4)**: 547-555.

ÖZHATAY, N. & AKALIN, E. 2000: A new species of *Ferulago* W. Koch (Umbelliferae) from north-west Turkey. - *Bot. J. Linn. Soc.* **133(4)**: 535-542.

ÖZHATAY, N. & AL. 1991: A new hexaploid *Bellevalia* (Hyacinthaceae) from European Turkey. - *Bot. J. Linn. Soc.* **107(1)**: 89-99.

ÖZHATAY, N. & AL. 1997: Contributions between 1985-1996 to the flora of European Turkey. In: Abstracts from the First Balkan Botanical Congress, Thessaloniki, September 19-22, 1997, p. 33. - Thessaloniki.

ÖZHATAY, N. & AL. 2000: Contributions to the flora of European Turkey between 1985 and 1988: a synopsis. - *Bot. Chron. (Patras)* **13**: 383-388.

ÖZHATAY, N. & AL. 2015: A new *Fritillaria* species from Mediterranean region of Turkey; *Fritillaria asumaniae*. - *Fl. Medit.* **25 (Special Issue)**: 199-208.

ÖZHATAY, N. & AL. 2018: *Allium istanbulense*, a new autumnal species of *A.* sect. *Codonoprasum* (Amaryllidaceae) from Turkey and its taxonomic position among allied species. - *Phytotaxa* **334(2)**: 152-166.

ÖZHATAY, N. & ALPINAR, K. 1980: A new record for the Turkish flora: *Gypsophila glomerata* Pallas ex Bieb. - *Istanbul Üniv. Eczac. Fak. Mecm.* **16**: 10-13.

ÖZHATAY, N. & BYFIELD, A. 1995: A new *Fritillaria* (Liliaceae) from south-west Turkey. - *Karaca Arbor. Mag.* **3(1)**: 7-16.

ÖZHATAY, N. & DALGIÇ, G. 1989: *Muscari tenuiflorum* in European Turkey: Description, distribution, karyotype. - *Istanbul Üniv. Eczac. Fak. Mecm.* **25**: 17-21.

ÖZHATAY, N., DANE, F. & DALGIÇ, G. 1993: A new genus for the flora of Turkey: *Trachelium jacquinii*. - *Istanbul Üniv. Eczac. Fak. Mecm.* **29(1)**: 9-12.

ÖZHATAY, N. & GENÇ, I. 2013: *Allium cyrilli* complex (sect. *Melanocrommyum*) in Turkey. - *Turk. J. Bot.* **37(1)**: 39-45.

ÖZHATAY, N. & JOHNSON, M.A.T. 1993: Some karyological and taxonomic remarks on Turkish *Allium*, *Ornithogalum*, *Bellevalia*, *Muscari*. In Petrova, A.V. (ed.): Abstracts, VII OPTIMA meeting in Borovetz, Bulgaria, 18-30 July 1993, p. 51.

ÖZHATAY, N. & JOHNSON, M.A.T. 1996: Some karyological remarks on Turkish *Allium* sect. *Allium*, *Bellevalia*, *Muscari*, and *Ornithogalum* subg. *Ornithogalum*. - *Bocconea* **5(1)**: 239-249.

ÖZHATAY, N., JOHNSON, M.A.T. & MATHEW, B.F. 1989: A new hexaploid *Bellevalia* species from European Turkey. In Phitos, D. (ed.): Abstracts. Sixth OPTIMA meeting, Delphi 10-16 Sept. 1989, p. 121.

ÖZHATAY, N., JOHNSON, M.A.T. & MATHEW, B.F. 1991: Chromosome numbers of Turkish *Bellevalia* species, including a new hexaploid from European Turkey. - *Bot. Chron. (Patras)* **10**: 813-818.

ÖZHATAY, N., KOÇYIĞIT, M. & AKALIN, E. 2010: *Allium rumelicum*, sect. Codonoprasum, a new species from European Turkey. - *Phytol. Balcan.* **16(3)**: 355-359.

ÖZHATAY, N. & KÜLTÜR, Ş. 2006: Check-list of additional taxa to the Supplement Flora of Turkey III. - *Turk. J. Bot.* **30(4)**: 281-316.

ÖZHATAY, N., KÜLTÜR, Ş. & AKSOY, A. 1994: Check-list of additional taxa to the Supplement Flora of Turkey. - *Türk. Bot. Derg.* **18**: 497-514.

ÖZHATAY, N., KÜLTÜR, Ş. & AKSOY, A. 1999: Check-list of additional taxa to the Supplement Flora of Turkey, II. - *Türk. Bot. Derg.* **23**: 151-169.

ÖZHATAY, N. & MERIÇLI, A.H. 1982: The chromosomes of *Silybum marianum* in Turkey. - *Istanbul Üniv. Eczac. Fak. Mecm.* **18**: 123-126.

ÖZHATAY, N. & ÖZHATAY, E. 1980: A new record for the Turkish flora: *Saxifraga graeca* Boiss. & Heldr. - *Istanbul Üniv. Eczac. Fak. Mecm.* **16**: 14-16.

ÖZHATAY, N. & ÖZHATAY, E. 1995: A new white *Paeonia* L. from NW Turkey: *P. mascula* Miller subsp. *bodurii* N. Özhatay. - *Karaca Arbor. Mag.* **3(1)**: 17-26.

ÖZHATAY, N. & ÖZHATAY, E. 1981: A new record for the Turkish flora: *Allium callimischon* Link ssp. *haemostictum* Stearn. - *J. Fac. Pharm. Istanbul* **17**: 103-106.

ÖZHATAY, N., PAGE, M. & SINNOTT, M. 2000: *Paeonia turcica* (Paeoniaceae). - *Curtis's Bot. Mag., ser. 6* **17(2)**: 92-98.

ÖZTÜRK, M.A. 1979: Preliminary observations on the edaphic and biotic relations of *Myrtus communis* L. - *Ege Üniv. Fen Fak. Derg., B* **3(1-4)**: 137-142.

ÖZTÜRK, M.A. & AL. 1983a: A general account of the distribution of the cruciferous plants in Turkey. - *Ege Üniv. Fen Fak. Derg., B* **6(1)**: 87-98.

ÖZTÜRK, M.A. & AL. 1983b: Ecological studies of macchia elements in Aegean region of Turkey. - *Mem. Fac. Integr. Arts Sci. Hiroshima Univ., Stud. Environm. Sci.* **8**: 51-86.

ÖZTÜRK, M.A. & AL. 1990: Aegean region, Turkey. Plants and landscape. - Izmir: 176 pp.

ÖZTÜRK, M.A. & AL. 1991: A general account of high mountain flora and vegetation of Mediterranean part of Turkey. - *Ege Üniv. Fen Fak. Derg., B* **13(1)**: 1-8.

ÖZTÜRK, M.A. & AL. 2008: Ecology of tertiary relict endemic *Liquidambar orientalis* Mill. forests. - *Forest Ecol. Managem.* **256(4)**: 510-518.

ÖZTÜRK, M.A. & FISCHER, M.A. 1982: Karyosystematics of *Veronica* sect. *Beccabunga* (Scrophulariaceae) with special reference to the taxa in Turkey. - *Pl. Syst. Evol.* **140(4)**: 307-319.

ÖZTÜRK, M.A. & GÖRK, G. 1979: Studies on the taxonomy and morphology of West Anatolian species of *Mentha* [In Turkish with English summary]. - *Ege Üniv. Fen Fak. Derg., B* **3(1-4)**: 43-56.

ÖZTÜRK, M.A. & PIRDAL, M. 1986: Studies on the germination of *Asphodelus aestivus* Brot. - *Biotronics* **15**: 55-60.

ÖZTÜRK, M.A., SEÇMEN, Ö. & GÖRK, G. (EDS) 1996: Plant life in southwest and central Asia. - Izmir: Ege Univ. Press. Vol. 1 pp. i-x, 1-500; vol. 2 pp. i-viii, 501-1093.

ÖZTÜRK, M.A. & VARDAR, Y. 1974: Distribution and economical prospects of *Myrtus communis* L. - *Bitki Koruma Bült.* **1(1)**: 100-107.

ÖZÜDOĞRU, B. & AL. 2015: Phylogeny, diversification and biogeographic implications of the eastern Mediterranean endemic genus *Ricotia* (Brassicaceae). - *Taxon* **64(4)**: 727-740.

ÖZÜDOĞRU, B. & AL. 2019: Phylogeny of the genus *Noccaea* (Brassicaceae) and a critical review of Its generic circumscription. - *Ann. Missouri Bot. Gard.* **104(3)**: 339-354.

ÖZÜDOĞRU, B. & MUMMENHOFF, K. 2020: Phylogenetic and biogeographical history confirm the Anatolian origin of *Bornmuellera* (Brassicaceae) and clade divergence between Anatolia and the Balkans in the Plio- Pleistocene transition. - *Turk. J. Bot.* **44(6)**: 593-603.

PACHSCHWÖLL, C., REICH, D. & TAN, KIT 2019: The botanical legacy of Mihael Dimonie (1870-1935), an almost forgotten plant collector in the southern Balkan Peninsula before the First World War. - *Willdenowia* **49(2)**: 257-279.

PACKER, J.G. & DENFORD, K.E. 1974: A contribution to the taxonomy of *Arctostaphylos uva-ursi*. - *Canad. J. Bot.* **52(4)**: 743-753.

PACKHAM, J.R. 1978: Biological Flora of the British Isles. *Oxalis acetosella* L. - *J. Ecol.* **66(2)**: 669-693.

PACKHAM, J.R. 1983: Biological Flora of the British Isles. *Lamiastrum galeobdolon* (L.) Ehrend. & Polatschek (*Galeobdolon luteum* Hudson; *Lamium galeobdolon* (L.) Nath.). - *J. Ecol.* **71(3)**: 975-997.

PACLT, J. 1952: Über die Identität von *Parietaria ramiflora* Mönch mit *Parietaria erecta* Mertens et Koch (= *P. officinalis* L.). - *Phyton (Horn)* **4(1-3)**: 46-50.

PACLT, J. 1954: Nachtrag zu meiner *Parietaria*-Studie. - *Phyton (Horn)* **5(3)**: 242-246.

PACLT, J. 1959: Über die Variabilität von *Parietaria officinalis*. - *Phyton (Horn)* **8(1-2)**: 171-174.

PACLT, J. 1998a: (1334) Proposal to amend the gender of *Nuphar*, nom. cons. (Nymphaeaceae), to neuter. - *Taxon* **47(1)**: 167-169. [not approved, cf. Taxon 49(2): 267, 2000].

PACLT, J. 1998b: (1351) Proposal to amend the gender of *Euonymus*, nom. cons. (Celastraceae), to feminine. - *Taxon* **47(2)**: 473-474. [not approved, cf. Taxon 49(2): 270, 2000].

PAGE, C.N. 1972: An assessment of inter-specific relationships in *Equisetum* subgenus *Equisetum*. - *New Phytol.* **71(2)**: 355-369 + 4 plates.

PAGE, C.N. 1976: The taxonomy and phytogeography of bracken - a review. - *Bot. J. Linn. Soc.* **73(1-3)**: 1-34.

PAGE, C.N. 1989: Three subspecies of bracken, *Pteridium aquilinum* (L.) Kuhn, in Britain. – *Watsonia* **17(4)**: 429-434.

PAGE, C.N. 1990: Hybrids in the genus *Equisetum* in Europe: An updated annotation. In Rita, J. (ed.): Taxonomía, biogeografía y conservación de Pteridofitos, pp. 151-156. - Palma de Mallorca.

PAGE, C.N. & MILL, R.R. 1994: Scottish bracken (*Pteridium*): new taxa and a new combination. - *Bot. J. Scotland* **47(1)**: 139-140.

PAGE, C.N. & MILL, R.R. 1995: The taxa of Scottish bracken in a European perspective. - *Bot. J. Scotland* **47(2)**: 229-247.

PAGLIA, E. 1909: Osservazioni sul *Biarum tenuifolium* (L.) Schott. - *Rivista Ital. Sci. Nat.* **29**: 22-25.

PAGNI, A.M. & AL. 1990: Études morpho-anatomiques et phytochimiques de *Calamintha nepeta* (L.) Savi (Labiées). - *Pl. Méd. Phytothérap.* **24(3)**: 203-213.

PAGNI, A.M., CORSI, G. & CAPPELLETTI, E.M. 1986: Fruit morpho-anatomical aspects and secretory structures in three related *Athamanta* species (Umbelliferae). - *Bot. Jahrb. Syst.* **106(2)**: 211-220.

PAHLEVANI, A.H. & RIINA, R. 2011: A synopsis of *Euphorbia* subgen. *Chamaesyce* in Iran. - *Ann. Bot. Fenn.* **48(4)**: 304-316.

PAIVA, J. 1989: El género *Polygala* en el Mediterráneo Occidental. - *Collect. Bot. (Barcelona)* **17(2)**: 191-203.

PAIVA, J., CIRUJANO, S. & VILLANEUVA, E. 1986: *Montia fontana* L. (Portulacaceae) en la Península Ibérica. - *Bol. Soc. Brot., Ser. 2* **59**: 321-332.

PAJARON, S. (ED.) 1993: Bibliografía botánica Ibérica, 1991. - *Bot. Complut.* **18**: 363-420.

PAJARÓN, S. (ED.) 1994: Bibliografía botánica Ibérica, 1992-1993. - *Bot. Complut.* **19**: 201-261.

PALASHEV, I. & NIKOLOV, V. 1979: On the distribution, ecology and biological peculiarities of the dendriform hazel (*Corylus colurna* L.) in Bulgaria [In Bulgarian with English and Russian summaries]. - *Gorskost. Nauka* **16(5)**: 26-42.

PALAVEEVA-KOVATSCHEVSKA, M. 1967: Einige biologische Besonderheiten der gewöhnlichen Osterluzei *Aristolochia clematitis* L. [In Bulgarian with Russian and German summaries]. - *Izv. Bot. Inst. (Sofia)* **17**: 199-203.

PALHINHA, R. DE TELLES 1942: Algumas considerações sôbre a distribuïção geográfica e a ecologia do *Arceuthobium oxycedri* (DC.) Marsch.-Bieb. - *Bol. Soc. Brot., Ser. 2* **16**: 137-143.

PALMER, J.H. & SAGAR, G.R. 1963: Biological Flora of the British Isles. *Agropyron repens* (L.) Beauv. (*Triticum repens* L., *Elytrigia repens* (L.) Nevski). - *J. Ecol.* **51(3)**: 783-794.

PALMER, L. & MEIKLE, R.D. 1965: A new *Cyclamen* from Rhodes. *Cyclamen repandum* var. *rhodense*. - *J. Roy. Hort. Soc.* **90(7)**: 291-293 + 1 plate.

PALOMEQUE, T. & SAÑUDO PALAZUELES, A. 1981: [Chromosome numbers in *Cistanche* and *Orobanche*]. In Löve, Á. (ed.): Chromosome number reports LXXI. - *Taxon* **30(2)**: 507-508.

PAMPANINI, R. 1912: La *Genista sericea* Wulf. e la sua distribuzione in Italia. - *Nuovo Giorn. Bot. Ital., n.s.* **19(3)**: 327-348.

PAMPANINI, R. 1923: L'esplorazione botanica dell'isola di Rodi dal 1761 al 1922. - *L'Universo (Firenze)* **4(11-12)**: 859-871, 955-971 + 3 plates and 1 map.

PAMPANINI, R. 1925a: Un manipolo di pianta della Tripolitana e dell'isola di Rodi. - *Boll. Soc. Bot. Ital.* **1925**: 22-24.

PAMPANINI, R. 1925b: Contributo alla conoscenza della flora dell'isola di Rodi. - *Boll. Soc. Bot. Ital.* **1925**: 140-146.

PAMPANINI, R. 1926: L'esplorazione botanica del Dodecaneso dal 1787 al 1924. - *Nuovo Giorn. Bot. Ital., n.s.* **33(1)**: 20-38.

PAMPANINI, R. 1927: Secondo contributo alla conoscenza della flora dell'isola di Rodi. - *Nuovo Giorn. Bot. Ital., n.s.* **34(1)**: 369-380.

PAMPANINI, R. 1930: Prodromo della Flora Cirenaica. - Forlì: Tipografia Valbonesi, xxxviii + 575 pp. + 6 plates.

PAMPANINI, R. 1936: L'*Arbutus* della Cirenaica. - *Arch. Bot. (Forlì)* **12**: 109-134 + 1 plate.

PAMUKÇUOĞLU, B. 1964: Forêts de *Liquidambar* orientalis en Turquie [In Turkish with French summary]. - *Doğa Türk Biol. Derg.* **14(2)**: 71-76.

PANAYOTIDIS, P. & MONTESANTO, B. 1995: Recent expansion of tropical marine macrophytes in the Mediterranean. In: Proceedings, 5th sci. conf., Hellenic Bot. Soc., Delphi 21-23 October 1994, pp. 121-124 [In Greek with English summary]. - Thessaloniki: Univ. Studio Press.

PANČIĆ, J. 1856 : Verzeichniss der in Serbien wildwachsenden Phanerogamen, nebst den Diagnosen einiger neuer Arten. - *Verh. Zool.-Bot. Vereins Wien* **6**: 475-598.

PANČIĆ, J. 1874: Flora Kneževine Srbije ... Flora principatus Serbiae. - Beograd: Državio Štamiariji, xxxiv + 798 pp.

PANČIĆ, J. 1875: Elenchus plantarum vascularium quas aestate a. 1873 in Crna Gora [Montenegro] legit. Dr. Jos. Pančić. - Belgradi: vii + 106 pp.

PANČIĆ, J. 1883: Elementa ad floram principatus Bulgariae. - Beograd: i + 71 pp. [cf. TL-2 4: 43, 1983].

PANETSOS, C.P. 1975: Natural hybridization between *Pinus halepensis* and *Pinus brutia* in Greece. - *Silvae Genet.* **24(5-6)**: 163-168.

PANETSOS, C.P. 1987: Monograph of *Abies cephalonica* Loudon. - *Anali Šumarstvo* **7(1)**: 1-22.

PANGUA, E., PÉREZ-RUZAFA, I. & PAJARON, S. 2011: Gametophyte features in a peculiar annual fern, *Anogramma leptophylla*. - *Ann. Bot. Fenn.* **48(6)**: 465-472.

PANIGRAHI, G. 1965: Preliminary studies in the cytotaxonomy of the *Dryopteris villarsii* (Bell.) Woynar complex in Europe. - *Amer. Fern J.* **55(1)**: 1-8.

PANIGRAHI, G. 1987a: Comments on Prop. 807 to reject *Polypodium pteridioides* Reichard (Sinopteridaceae: Pteridophyta). - *Taxon* **36(1)**: 166-167.

PANIGRAHI, G. 1987b: (867) Proposal to delete *Allosorus Bernhardi*, nom. rej., and to emend typification of *Cheilanthes* Swartz, nom. cons. (Pteridophyta). - *Taxon* **36(1)**: 167-168. [not approved, cf. Taxon 42(1): 120, 1993].

PANIGRAHI, G. & BASU, S.K. 1982: Nomenclatural notes on two species of *Cheilanthes* Sw. (Sinopteridaceae). - *Taxon* **31(1)**: 102-104.

PANITSA, M. 1997: Simvoli sti gnosi tis chloridas ke tis vlastisis ton nisidon tou anatolikou Egeou [In Greek with English summary]. - Patras: Ph.D. thesis, Univ. of Patras, xiv + 345 pp.

PANITSA, M. & AL. 1994: Contribution to the study of the Greek flora: Flora and vegetation of the Enousses (Oinousses) islands (E. Aegean area). - *Flora, Morphol. Geobot. Oekophysiol.* **189(1)**: 69-78.

PANITSA, M. & AL. 2002: Contribution to the study of the flora and vegetation of the Kithira island group: Offshore islets of Kithira. In Kamari, G. & al. (eds): Ellinini Votaniki Eteria. 9° Panelliniko Epistimoniko Sinedrio. Praktika, pp. 228-232 [In Greek w. English abstract].

PANITSA, M. & AL. 2003: Floristic investigation of Lemnos island (NE Aegean area, Greece). - *Willdenowia* **33(1)**: 79-105.

PANITSA, M. & AL. 2004: Contribution to the study of the flora and vegetation of the Kithira island group: Offshore islets of Kithira (S Aegean, Greece). - *Willdenowia* **34(2)**: 101-115.

PANITSA, M. & AL. 2006: Patterns of species richness on very small islands: the plants of the Aegean archipelago. - *J. Biogeogr.* **33(7)**: 1223-1234.

PANITSA, M. & AL. 2010: Factors affecting plant species richness and endemism on land-bridge islands - an example from the East Aegean archipelago. - *Acta Oecol.* **36**: 431-437.

PANITSA, M. & AL. 2018: Herbarium of vascular plants collection of the University of Patras (Greece). - *Bot. Serbica* **142 (suppl. 1)**: 176.

PANITSA, M. & AL. 2019: Plant and habitat diversity – preliminary results from the Terra Lemnia project on Lemnos island (Greece). In Daskalakou, E.N. & al. (eds): 16th Conference of the Hellenic Botanical Society – Athens 10-13 October 2019. Program and book of abstracts, p. 36.

PANITSA, M. & AL. 2020: Distribution patterns of ruderal plant diversity in Greece. - *Biodivers. Conserv.* **29**: 869-891.

PANITSA, M. & AL. 2021: Natural and cultural heritage interaction: aspects of plant diversity in three East Peloponnesian castles (Greece) and conservation evaluation. - *Pl. Biosyst.* **155**: https://doi-org. ludwig.lub.lu.se/10.1080/11263504.2021.1889701.

PANITSA, M., DIMOPOULOS, P. & TZANOUDAKIS, D. 1992: Floristic and phytogeographical study of Inouses (Nomos Chiou, E Aegean Sea) [In Greek with English summary]. In: Proc. 12th Annual Meeting Hellen. Soc. Biol. Sci., Mitilini 27.-29.4. 1990, pp. 177-178. - Athine.

PANITSA, M. & ILIADOU, E. 2013: Flora and phytogeography of the Ionian islands (Greece). Pp. 243-260 in: Cardona Pons E. & al. (eds): 2nd Botanical Conference in Menorca. Proceedings and abstracts. Islands and plants: preservation and understanding of flora on Mediterranean Islands. - Menorca: Institut Menorquí d'Estudis, Consell Insular de Menorca. 412 pp.

PANITSA, M., KAGIAMPAKI, A. & KOUGIOUMOUTZIS, K. 2018: Plant diversity and biogeography of the Aegean archipelago: A new synthesis. In: Sfenthourakis, S. & al. (eds): Biogeography and biodiversity of the Aegean. - Nicosia, Cyprus: Broken Hill Publishers, pp. 223-244.

PANITSA, M. & KONTOPANOU, A. 2017: Diversity of chasmophytes in the vascular flora of Greece: floristic analysis and phytogeographical patterns. - *Bot. Serbica* **41(2)**: 199-211.

PANITSA, M. & TZANOUDAKIS, D. 1992: Floristic and phytogeographical studies of North Dodecanesos. In Thanos, C.A. (ed.): MEDECOS VI. Plant-animal interactions in Mediterranean-type ecosystems, pp. 367-374. - Athens.

PANITSA, M. & TZANOUDAKIS, D. 1996: Contribution to the knowledge and the management of the biodiversity in the East Aegean area: The small islands group Kalolimni-Imia [In Greek with English summary]. In: Proceedings 6th sci. conf. Hellenic Botanical Society, pp. 105-111. - Paralimni, Cyprus.

PANITSA, M. & TZANOUDAKIS, D. 1997: The floristic diversity of the Greek flora as exemplified by the exploration of some East Aegean islets. In: Abstracts from the First Balkan Botanical Congress, Thessaloniki, September 19-22, 1997, p. 36. - Thessaloniki.

PANITSA, M. & TZANOUDAKIS, D. 1998: Contribution to the study of the Greek flora: Flora and vegetation of the E Aegean islands Agathonisi and Pharmakonisi. - *Willdenowia* **28(1-2)**: 95-116.

PANITSA, M. & TZANOUDAKIS, D. 2001a: Flora, vegetation and important habitat types of Tilos island group (East Aegean). In Kamari, Psaras & Constantinidis (eds): Proceedings of the 8th scientific congress of the Greek Botanical Society, October 5-8, 2000, pp. 342-351 [In Greek with English abstract].

PANITSA, M. & TZANOUDAKIS, D. 2001b: A floristic investigation of the islet groups Arki and Lipsi (East Aegean area, Greece). - *Folia Geobot.* **36(3)**: 265-279.

PANITSA, M. & TZANOUDAKIS, D. 2010: Floristic diversity on small islands and islets: Leros islets' group (East Aegean area, Greece). - *Phytol. Balcan.* **16(2)**: 271-284.

PANITSA, M., TZANOUDAKIS, D. & SFENTHOURAKIS, S. 2008: Turnover of plants on small islets of the eastern Aegean Sea within two decades. - *J. Biogeogr.* **35(6)**: 1049-1061.

PANOV, P.P. 1973a: Diagnoses plantarum novarum ser. I. - *Dokl. Bulg. Akad. Nauk* **26(8)**: 1057-1059.

PANOV, P.P. 1973b: Diagnoses plantarum novarum ser. II. - *Dokl. Bulg. Akad. Nauk* **26(9)**: 1227-1230.

PANOV, P.P. 1974: *Minuartia jordanovii* and *M. kitanovii*: two new species of sectio *Minuartia*. - *Dokl. Bulg. Akad. Nauk* **27(10)**: 1423-1426.

PANOV, P.P. 1996: New taxa for the Bulgarian flora. - *Fitologiya* **48**: 3-10.

PANOV, P.P. & JORDANOV, D. 1966: Bemerkungen zu *Allium margaritaceum* Sibth. et Sm. und *Allium guttatum* Stev. in Bulgarien [In Bulgarian with Russian and German summaries]. - *Izv. Bot. Inst. (Sofia)* **16**: 239-245.

PANTELAS, V., PAPACHRISTOPHOROU, T. & CHRISTODOULOU, P. 1993: Cyprus flora in colour. The endemics. - Athens: 104 pp.

PANTERA, A. & AL. 2008: Distribution and phytogeographical analysis of *Quercus ithaburensis* ssp. *macrolepis* in Greece. - *Ecol. Medit.* **34**: 73-82.

PANTERA, A. & PAPANASTASIS, V. 2003: Inventory of Valonia oak (*Quercus ithaburensis* Decaisne subsp. *macrolepis* (Kotschy) Hedge & Yalt.) in Greece. [In Greek with English abstract]. - *Geôtechn. Epistimon. Themata* **1/2003**: 34-43.

PAPADEMETRIOU, K., KARAGIANNAKIDOU, V. & PAPANASTASIS, V. 1996: Vegetation and production of the subalpine rangelands of the mountain Pangeon [In Greek with English summary]. - Thessaloniki: 10 pp.

PAPADROSOU, A. & CONSTANTINIDIS, TH. 2017: Spontaneous plant species in agricultural land: an investigation in Magnisia prefecture, east-central Greece. In Dousis, M.A. & Thanos, K.A.: 15th conference of the Hellenic Botanical Society, book of abstracts: p. 109.

PAPAGEORGIOU, A.C. & AL. 2014: Complex fine-scale phylogeographical patterns in a putative refugial region for *Fagus sylvatica* (Fagaceae). - *Bot. J. Linn. Soc.* **174(4)**: 516-528.

PAPAIOANNOU, I. 1935a: Ta oria tis geografikis exaploseos tis Chalepiou Pevkis ke tis Trachias Pevkis is tin V.A. Chalkidikin ke e dasike fitokinonike enosis avton [In Greek]. - *Deltion Fisik. Epist.* **1935 (Oct.-Nov.)**: 13 pp.

PAPAIOANNOU, I. 1935b: I *Pinus brutia* Ten. ke o k. Diapoulis (kritiki meleti) [In Greek]. - *Deltion Fisik. Epist.* **1935 (Dec.)**: 11 pp.

PAPAIOANNOU, I. 1935c: I Chalepios Pevki (*Pinus halepensis* Mill.), i Pevki tis Pitiountos (*Pinus pithyusa* Stev.) ke i Trachia Pevki (*Pinus brutia* Ten.) en Elladi [In Greek]. - *Dasiki Zoi* **1935**: 48 pp.

PAPAIOANNOU, I. 1936a: Über Artbastarde zwischen *Pinus brutia* Ten. und *Pinus halepensis* Mill. in Nordostchalkidiki (Griechenland). - *Forstwiss. Centralbl.* **58(2)**: 194-205.

PAPAIOANNOU, I. 1936b: Eine neue Varietät von *Pinus Brutia* Ten., *Pinus brutia* Ten. var. *Agraphiotii*. - *Compt. Rend. Acad. Athènes [Praktika Akadimias Athinon]* **11**: 14-24.

PAPAIOANNOU, I. 1939: *Quercus agriobalanidea*, spec. nov. Eine neue Eichenart in Griechenland, ihre pflanzengeographische Stellung und ihre forstwirtschaftliche Bedeutung. - *Compt. Rend. Acad. Athènes [Praktika Akadimias Athinon]* **14**: 466-474.

PAPAIOANNOU, I. 1940: Contribution sur la répartition géographique du *Quercus trojana* Webb, en Grèce [In Greek with French summary]. - *Sci. Ann. Fac. Agric. Forest. Univ. Thessalonika* **1939-40**: 1-17 + 1 plate.

PAPAIOANNOU, I. 1949a: Encore une nouvelle espèce de chène en Grèce, *Quercus euboïca*. - *Compt. Rend. Acad. Athènes [Praktika Akadimias Athinon]* **23**: 336-352.

PAPAIOANNOU, I. 1949b: Petits peuplements de *Rhododendron luteum* Sweet dans l'île de Lesbos [In Greek with French summary]. - *Dasos* **8-9**: 16 pp.

PAPAIOANNOU, I. 1950: De petits groupes de rhododendron (*R. luteum* Sweet) sur Lesbos [In Greek with French summary]. - *Compt. Rend. Acad. Athènes [Praktika Akadimias Athinon]* **24**: 46-57.

PAPAIOANNOU, I. 1957: Die Panzerkiefer (*Pinus Heldreichii* Christ), ihre geographische Verbreitung und die Waldformationen im griechischen Teil des südlichen Orvilosgebirge [In

Greek with German summary]. - *Vouno [Journal of the Hellenic Alpine Club]* **1957**: 97-110.

PAPAIOANNOU, I. 1973: Die zirkummediterrane Verbreitung des *Cistus laurifolius* L. besonders in Thrazien. In: Kazdaği göknari ve Türkiye florasi uluslararasi simpozyomu bildirileri, pp. 103-134. - Istanbul Üniversitesi Orman Fakültesi Yayinlari.

PAPANASTASIS, V. & AL. 2019: Agrioloulouda tou Siniatsikou. [in Greek].

PAPANASTASIS, V.P. 1997: Distribution pattern of C4 grasses *Dichanthium ischaemum* and *Chrysopogon gryllus* in N. Greece. In: Abstracts from the First Balkan Botanical Congress, Thessaloniki, September 19-22, 1997, p. 37. - Thessaloniki.

PAPANASTASIS, V.P. & PAPADEMETRIOU, K. 1983: The most important grasses of rangelands in Greece [In Greek with English abstract]. - Athine: 185 pp.

PAPANDROPOULOS, D. 2010: Nea thesi tis *Pinguicula crystallina* subsp. *hirtiflora*. [New locality for *Pinguicula crystallina* subsp. *hirtiflora*]. - *Fisis [Journal of the Hellenic Society for the Protection of Nature]* **129**: 47.

PAPANICOLAOU, K. 1979: A new species of *Chondrilla* from Mt Pangaion. - *Bot. Not.* **132(3)**: 363-365.

PAPANICOLAOU, K. 1980: Inheritance of dissected leaflets in *Vicia grandiflora* Scop. - *Bot. Not.* **133(2)**: 165-167.

PAPANICOLAOU, K. 1983: *Isatis vermia* sp. nov. from north central Greece. (Materials for the Mountain Flora of Greece, 14). - *Nordic J. Bot.* **2(6)**: 553-556.

PAPANICOLAOU, K. 1984: [Chromosome numbers from Mt Athos and Mt Pangeo]. In Löve, Á. (ed.): Chromosome number reports LXXXII. - *Taxon* **33(1)**: 130-131.

PAPANICOLAOU, K. 1986: Contribution to the flora of Mount Pangaion (Pangeon), North East Greece. - *Ann. Mus. Goulandris* **7**: 67-156.

PAPANICOLAOU, K. & AL. 2009: Island of Paxi. Phytoprosopography [In Greek and English]. – Paxi Public Benefit Company, Gaios, 390 pp.

PAPANICOLAOU, K., BABALONAS, D. & KOKKINI, S. 1983: Distribution patterns of some Greek mountain endemic plants in relation to geological substrate. - *Flora, Morphol. Geobot. Oekophysiol.* **174(5-6)**: 405-437.

PAPANICOLAOU, K. & KOKKINI, S. 1982a: *Armeria* (Plumbaginaceae) in the Greek mountains. - *Willdenowia* **12(2)**: 213-219.

PAPANICOLAOU, K. & KOKKINI, S. 1982b: A new species of *Armeria* (Plumbaginaceae) from Euboea, Greece. - *Willdenowia* **12(2)**: 221-225.

PAPANICOLAOU, K. & KOKKINI, S. 1982c: A taxonomic revision of *Sideritis* L. sect. *Empedoclia* (Rafin.) Bentham (Labiatae) in Greece. In Margaris, N.S., Koedam, A. & Vokou, D. (eds): Aromatic plants: Basic and applied aspects, pp. 101-128. - The Hague, etc.: Martinus Nijhoff Publishers.

PAPANICOLAOU, K. & KOKKINI, S. 1983a: A new species of *Ranunculus* from Greece. - *Feddes Repert.* **94(9-10)**: 591-593.

PAPANICOLAOU, K. & KOKKINI, S. 1983b: A homoeotic mutation as well as inheritance of hairy sepals and anthers in *Ranunculus millefoliatus* Vahl. - *Feddes Repert.* **94(9-10)**: 631-634.

PAPANICOLAOU, K. & KOKKINI, S. 1983c: Inheritance of dissected outer perianth-segments in *Tulipa sylvestris* L. subsp. *australis* (Link) Pamp. (Liliaceae). - *Feddes Repert.* **94(9-10)**: 635-637.

PAPANICOLAOU, K. & ZAHAROF, E. 1980: *Crocus* in Greece: New taxa and chromosome numbers. - *Bot. Not.* **133(2)**: 155-163.

PAPANICOLAOU, K. & ZAHAROF, E. 1983: Cytological notes and taxonomic comments on four *Galanthus* L. taxa from Greece. - *Israel J. Bot.* **32(1)**: 22-32.

PAPANIKOLAOU, A. & PANITSA, M. 2020: Plant species richness and composition of a habitat island within Lake Kastoria and comparison with those of a true island within the protected Pamvotis lake (NW Greece). - *Biodiv. Data J.* **8**: e48704 (23 pp.).

PAPANIKOLAOU, G.D. & SARLIS, G.P. 1991: Phytosociological studies in the Tatoi district (Attica, Greece). - *Vegetatio* **93(2)**: 81-90.

PAPARISTO, K. & AL. (EDS) 1988: Flora e Shqipërisë. Vol. 1. Lycopodiaceae-Platanaceae [In Albanian]. - Tiranë: Akad. Shkencave e RPS të Shqipërisë, 457 pp.

PAPARISTO, K. & QOSJA, XH. 1983: Një nenspecie e re e gjinisë *Leucojum* në Shqiperi *L. valentinum* Pau. subsp. *vlorense* (subsp. nov.) [In Albanian]. - *Bul. Shken. Natyres (Tirane)* **1**: 93-100.

PAPARISTO, K., QOSJA, XH. & DEMIRI, M. 1962: Flora e Tiranës [In Albanian]. - Tiranë: Universiteti Shtetëror, 520 pp.

PAPARISTO, K., QOSJA, XH. & DEMIRI, M. 1965: Flora e Tiranës (Ikonografia). - Tiranë: Universiteti Shtetëror, 515 pp.

PAPASTERGIADOU, E. 1990: Phytosociological and ecological studies of aquatic macrophytes (hydrophytes) in northern Greece. - *Sci. Ann. School Biol. Aristot. Univ. Thessalonika* **24**: xi + 266 pp. + 68 pp. of figs & tables.

PAPASTERGIADOU, E. 1995: Aquatic vegetation of the Greek wetlands. - *Ann. Bot. (Rome)* **53**: 81-85.

PAPASTERGIADOU, E. & BABALONAS, D. 1993a: Aquatic flora of N Greece. I. Hydrophytes. - *Willdenowia* **23(1-2)**: 137-142.

PAPASTERGIADOU, E. & BABALONAS, D. 1993b: Ecological studies on aquatic macrophytes of a dam lake Lake Kerkini, Greece. - *Arch. Hydrobiol., Suppl.* **90(2)**: 187-206.

PAPASTERGIOU, E.S. & AL. 2007: Land use changes and associated environmental impacts on the Mediterranean shallow Lake Stymfalia, Greece. Pp. 361-372 in Gulati, R.D. & al. (eds): Shallow lakes in a changing world. Proceedings of the 5th international symposium on shallow lakes, held at Dalfsen, The Netherlands, 5-9 June 2005. - Springer Verlag.

PAPATSOU, S. 1975: Flora and vegetation of the island Nisyros and the surrounding islets [In Greek with English summary]. - Patras: Ph.D. thesis, Univ. of Patras, 138 pp. + 1 map.

PAPATSOU, S. & PHITOS, D. 1975: Two new taxa from the eastern Aegean. - *Notes Roy. Bot. Gard. Edinburgh* **34(2)**: 203-204.

PAPEŠ, D. 1971: B chromosomes of the genus *Leucanthemum* in Yugoslavia. - *Genetika (Belgrade)* **3(2)**: 261-270.

PAPEŠ, D. 1972a: Pentaploids in populations of various *Leucanthemum* species [In Serbo-Croatian with English summary]. - *Acta Bot. Croat.* **31**: 71-79 + 2 plates.

PAPEŠ, D. 1972b: Variation of chromosome numbers in *Leucanthemum* heptaploids [In Serbo-Croatian with English summary]. - *Acta Bot. Croat.* **31**: 81-86 + 2 plates.

PAPEŠ, D. 1975a: Cytotaxonomy of the genus *Leucanthemum* in Yugoslavia. In Walters, S.M. & King, C.J. (eds): European floristic and taxonomic studies, pp. 90-100. - Oxford: E.W. Classey Ltd.

PAPEŠ, D. 1975b: Cytotaxonomy and some evolutionary aspects of the genus *Leucanthemum* in Yugoslavia. In: Jordanov, D. & al. (eds): Problems of Balkan flora and vegetation, pp. 187-193. - Sofia: Publishing House of the Bulgarian Academy of Sciences.

PAPEŠ, D. & BOSILJEVAC, V. 1984: Tetraploid populations of bulbous barley (*Hordeum bulbosum* L.). - *Acta Bot. Croat.* **43**: 335-340 + 2 plates.

PAPEŠ, D. & RADIĆ, J. 1982: [Chromosome numbers in *Centaurea* from Jugoslavia]. In Löve, Á. (ed.): Chromosome number reports LXXVII. - *Taxon* **31(4)**: 769-770.

PAPEŠ, D. & ŠILIĆ, Č. 1981: [Chromosome numbers in Lamiaceae]. In Löve, Á. (ed.): Chromosome number reports LXX. - *Taxon* **30(1)**: 70.

PAPEŠ, D. & TRINAJSTIĆ, I. 1981: [Chromosome numbers in *Ficaria* (*Ranunculus*)]. In Löve, Á. (ed.): Chromosome number reports LXX. - *Taxon* **30(1)**: 70.

PAPI, R.M., SPANOS, K.A. & KYRIAKIDIS, D.A. 2012: Genetic variation of *Fraxinusangustifolia* natural populations in Greece based on nuclear and chloroplast microsatellite markers. - *Eur. J. Forest Res.* **131**: 1151-1161.

PAPIOMITOGLOU, V. 2006: Wild flowers of Greece. - Mediterraneo editions, 260 pp.

PAPP, C. 1932: Monographie der europäischen Arten der Gattung *Melica* L. - *Bot. Jahrb. Syst.* **65(2-3)**: 275-348.

PARADIS, G. 1994: *Glinus lotoides* L. (Molluginaceae), espèce nouvelle pour la flore de Corse et de France. Localisation. Synécologie. - *Acta Bot. Gallica* **140(7)**: 819-826.

PARDI, P. 1933a: Contributo alla cariologia delle Asclepiadaceae. - *Atti Reale Accad. Naz. Lincei, Rendiconti Cl. Sci. Fis., Ser. 6* **17**: 1101-1104.

PARDI, P. 1933b: Studi sulla cariologia delle Asclepiadaceae. - *Nuovo Giorn. Bot. Ital., n.s.* **40(4)**: 576-589.

PARDO, C. 1981: Estudio sistemático del género *Seseli* L. (Umbelliferae) en la Península Ibérica. - *Lazaroa* **3**: 163-188.

PARENT, G.H. 1980: Le genre *Sisyrinchium* L. (Iridaceae) en Europe. Un bilan provisoire. - *Lejeunia, n.s.* **99**: 1-40.

PARENT, G.H. 1987: Phytogéographie des Monts Aroania (= Chelmos) (Grèce, Péloponnèse). Aperçu général des centres d'intérêt scientifique de ce massif. - *Linneana Belg.* **11(3)**: 103-148.

PARENT, G.H. 2005: Données floristiques inédites sur les montagnes grecques. - *Syst. Geogr. Pl.* **75**: 195-238.

PARENT, G.H. 2006: *Adoxa moschatellina* aussi présent en Grèce. - *Adoxa* **50**: 1-6.

PARFENOV, V.I. 1971: De systematica intraspecifica *Picea abies* (L.) Karst. [In Russian; English translation of this volume by Amerind Publ. Co., New Delhi, 1978]. - *Novosti Sist. Vysš. Rast.* **8**: 4-9.

PARK, J.-M. & AL. 2006: Phylogeny and biogeography of isophyllous species of *Campanula* (Campanulaceae) in the Mediterranean area. - *Syst. Bot.* **31(4)**: 862-880.

PARKER, J.S. 1975: Aneuploidy and isolation in two *Hypochoeris* species. - *Chromosoma* **52(1)**: 89-101.

PARKER, J.S., AINSWORTH, C.C. & TAYLOR, S. 1978: The B-chromosome system of *Hypochoeris maculata*. - *Chromosoma* **67(2)**: 123-143.

PARKER, J.S. & AL. 1991: Chromosomal structure of populations of *Scilla autumnalis* in the Iberian Peninsula. - *Heredity* **67(3)**: 287-297.

PARKS, J. 1995: What is *Cystopteris dickieana*? - *Pteridologist* **2(6)**: 290-292.

PARLATORE, F. 1839: Flora Panormitana sive plantarum prope Panormum sponte nascentium enumertatio auctore Philippo Parlatore ... - Panormi [Palermo]: Typographeo Petri Pensante, xxxi + 128 pp.

PARLATORE, F. 1845: Flora Palermitana ossia descrizione della piante che crescono spontanee nella valle di Palermo. - Firenze: Società tipografica, xxii + 442 pp. [reprinted 1992 in Palermo].

PARLATORE, F. 1846: Catalogo delle piante dal Sig. Heldreich nel suo viaggio nel Peloponneso nell'anno 1844 secondo le determinazioni di Boissier e Heldreich. - *Giorn. Bot. Ital.* **2(1, 3)**: [32-]33-50.

PARLATORE, F. 1848-96: Flora Italiana, ossia descrizione della piante che crescono spontanee o vegetano come tali in Italia e nelle isole ad essa aggiacenti ... - Firenze: Tipografia Le Monnier, 10 vols. [cf. TL-2 4: 69-70, 1983].

PARNELL, J. 1987: Variation in *Jasione montana* L. (Campanulaceae) and related species in Europe and North Africa. - *Watsonia* **16(3)**: 249-267.

PARNELL, J. 1988: Revision of the genus *Jovibarba* and consideration of the *Sempervivum* [sphalm. *Sedum*] *tectorum/marmoreum* complex and *S. montanum* subsp. *carpaticum*. - *Acta Bot. Hung.* **34(1-2)**: 209-224.

PARNELL, J. 1991: Pollen morphology of *Jovibarba* Opiz and *Sempervivum* L. (Crassulaceae). - *Kew Bull.* **46(4)**: 733-738.

PARNELL, J. & FAVARGER, C. 1990: Notes on *Sempervivum* L. and *Jovibarba* Opiz. In Chater, A.O. (ed.): Flora Europaea: Notulae systematicae ad Floram Europaeam spectantes. Series 2. No. 3. - *Bot. J. Linn. Soc.* **103(3)**: 216-220.

PAROLLY, G. 1992: Die Orchideenflora Montenegros. - *Mitt. Arbeitskreis Heimische Orchid. Baden- Württemberg* **24(2)**: 141-391.

PAROLLY, G. 1995: Ergänzungen zur Flora von Montenegro. - *Willdenowia* **25(1)**: 57-74.

PAROLLY, G., EREN, Ö. & NORDT, B. 2005: *Lomelosia solymica* (Dipsacaceae), a new chasmophyte from the Western Taurus Mts, Turkey. - *Willdenowia* **35(1)**: 107-115.

PAROLLY, G. & EREN, Ö. (EDS) 2006: Contributions to the flora of Turkey, 1. - *Willdenowia* **36(2)**: 823-844.

PAROLLY, G. & EREN, Ö. (EDS) 2007: Contributions to the flora of Turkey, 2. - *Willdenowia* **37(1)**: 243-271.

PAROLLY, G. & HEIN, P. 2000: *Arabis lycia* (Cruciferae), a new chasmophyte from the Taurus Mts, Turkey, and notes on related species. - *Willdenowia* **30(2)**: 293-304.

PAROLLY, G. & KILIAN, N. 2003: *Scorzonera karabelensis* (Compositae), a new species from SW Anatolia, with a key to the subscapigerous *Scorzonera* in Turkey. - *Willdenowia* **33(2)**: 327-335.

PAROLLY, G. & NORDT, B. 2001: *Seseli hartvigii* (Apiaceae), a new name for *S. ramosissimum* Hartvig & Strid, with carpological and ecological notes on this species. - *Willdenowia* **31(1)**: 87-92.

PAROLLY, G. & NORDT, B. 2002: A new chasmophytic species of *Potentilla* (Rosaceae) from S Anatolia, including some taxonomic remarks on *P.* subg. *Fragariastrum* in the E Mediterranean. - *Willdenowia* **32(1)**: 73-84.

PAROLLY, G. & NORDT, B. 2005: A further new *Peucedanum* species (Apiaceae) from the Taurus Mts, Turkey. - *Willdenowia* **35(1)**: 97-105.

PAROLLY, G. & TAN, KIT 2006: A new species of *Hesperis* (Brassicaceae) from SW Anatolia, Turkey. - *Willdenowia* **36(2)**: 851-856.

PARRA, R. & AL. 1998: Números cromosómicos para la flora española. 793-805. - *Lagascalia* **20(2)**: 302-308.

PARRIS, B.S. & FRASER-JENKINS, C.R. 1980: A provisional checklist of Turkish Pteridophyta. - *Notes Roy. Bot. Gard. Edinburgh* **38(2)**: 273-281. [Also distributed as *OPTIMA leaflet* no. 115].

PARRNELL, J. 1985: Biological Flora of the British Isles. *Jasione montana* L. - *J. Ecol.* **73(1)**: 341-358.

PARTRIDGE, J.W. 2001: Biological Flora of the British Isles. *Persicaria amphibia* (L.) Gray (*Polygonum amphibium* L.). - *J. Ecol.* **89(3)**: 487-501.

PARTSCH, J. 1886: Bericht über die wissenschaftlichen Ergebnisse seiner Reisen auf den Inseln des lonischen Meeres. - *Sitzungsber. Königl. Preuss. Akad. Wiss. Berlin* **36**: 615-628.

PARVAIS, C. & DELFORGE, P. 2014: Addition à la flore orchdéenne de l'île de Cythère (Attique, Grèce). - *Naturalistes Belges* **95**: 79-86.

PASCHALIDOU, D. & HANLIDOU, E. 2019: Phytodiversity in two former army camps of Thessaloniki, Kodra and Pavlos Melas. In Daskalakou, E.N. & al. (eds): 16th Conference of the Hellenic Botanical Society – Athens 10-13 October 2019. Program and book of abstracts, p. 97.

PASCHE, E. 1994: A new *Crocus* (Iridaceae) from Turkey. - *Herbertia* **49(1-2)**: 67-75.

PASCHE, E. & KERNDORFF, H. 2002: Die Gattung *Sternbergia* Waldst. & Kit. (Asparagales, Amaryllidaceae) im Vergleich, unter besonderer Berücksichtigung der wiederentdeckten *Sternbergia schubertii* Schenk. - *Stapfia* **80**: 395-416.

PASCHER, A. 1959: Über *Atropa*. - *Flora* **148(1)**: 84-109.

PASCHER, A.A. 1904: Übersicht über die Arten der Gattung *Gagea*. - *Sitzungsber. Deutsch. Naturwiss.-Med. Vereins Böhmen "Lotos" Prag, ser. 2* **24**: 109-133.

PASCHER, A.A. 1906: Zur Kenntnis zweier mediterraner Arten der Gattung *Gagea* (*Gagea foliosa* R. Sch., *Gagea peduncularis* Pasch.). - *Beih. Bot. Centralbl., Abt. 2* **20**: 76-107.

PASHALIEV, I. 1995: Contribution to the flora of Southwestern Bulgaria. - *Phytol. Balcan.* **1(2)**: 103-104.

PASSALACQUA, N.G. & BERNARDO, L. 2004: The genus *Paeonia* L. in Italy: taxonomic survey and revision. - *Webbia* **59(2)**: 215-268.

PASTOR DÍAZ, J.E. (ED.) 1992: Atlas cromosómico de la flora vascular de Andalucia occidental. - Sevilla: Publicaciones de la Universidad de Sevilla, 542 pp.

PASTOR, J. 1979: Números cromosómicos para la flora española. 116-120. - *Lagascalia* **9(1)**: 128-130.

PASTOR, J. 1981: Números cromosómicos para la flora española. 220-224. - *Lagascalia* **10(2)**: 239-241.

PASTOR, J. 1984: Números cromosómicos para la flora española. 300-303. - *Lagascalia* **12(2)**: 279-280.

PASTOR, J. 1985: Números cromosómicos para la flora española. 368-372. - *Lagascalia* **13(2)**: 296-299.

PASTOR, J. & AL. 1990: Números cromosómicos para la flora española. 556-591. - *Lagascalia* **15(2)**: 269-282.

PASTOR, J., FERNÁNDEZ, I. & DÍEZ, M.J. 1984: Números cromosómicos para la flora española. 304-313. - *Lagascalia* **12(2)**: 281-284.

PASTOR, J., FERNÁNDEZ, I. & DÍEZ, M.J. 1988: Números cromosómicos para la flora española. 528-543. - *Lagascalia* **15(1)**: 124-129.

PASTOR, J. & VALDÉS, B. 1983: Revision del genero *Allium* (Liliaceae) en la Peninsula Iberica & Islas Balearas. - Sevilla: Publicaciones de la Universidad de Sevilla, 182 pp.

PASTOR, J. & VALDÉS, B. 1986a: Bulb structure in some species of *Allium* (Liliaceae) of the Iberian Peninsula. - *Ann. Mus. Goulandris* **7**: 249-261.

PASTOR, J. & VALDÉS, B. 1986b: Números cromosómicos para la flora española. 485-490. - *Lagascalia* **14(2)**: 297-301.

PASTOR, J. & VALDÉS, B. 1988: Citotaxonomia de *Allium chamaemoly* L. - *Lagascalia* **15(Extra)**: 423-431.

PASTRIKOS, G. 2018: *Carthamus caeruleus* L. (Cardueae, Asteraceae). a new record for Rodos island, Greece. - *Parnassiana Arch.* **6**: 11-13.

PASTRIKOS, G. & GALANOS, CH. 2014: *Arum creticum* (Araceae), first record from the island of Rodos, Greece. - *Parnassiana Arch.* **2**: 61-64.

PATELI, M. & AL. 2003: Vascular plants in the suburban area of Thessaloniki (N Greece). I. The industrial park of Sindos. - *Fl. Medit.* **12**: 323-339.

PATELI, M., KRIGAS, N. & KOKKINI, S. 1999: Flora of the industrial area of the city of Thessaloniki (Sindos area). In Elefheriou, E.P. (ed.): Proceedings, 7th Panhellenic scientific conference, Alexandroupolis, pp. 307-310 [In Greek with English abstract]. - Thessaloniki: Univ. Studio Press.

PATELI, M., KRIGAS, N. & KOKKINI, S. 2002: "Gains" and "losses" of the flora of Aristotle University of Thessaloniki campus - from 1992 to 2001. In Kamari, G. & al. (eds): Ellinini Votaniki Eteria. 9° Panelliniko Epistimoniko Sinedrio. Praktika, pp. 273-278 [In Greek with English abstract].

PATERSON, A. 1979: Plants about the Parthenon. - *Garden (London)* **104(2)**: 63-65.

PATON, A. 1990a: A global taxonomic investigation of *Scutellaria* (Labiatae). - *Kew Bull.* **45(3)**: 399-450.

PATON, A. 1990b: The phytogeography of *Scutellaria* L. - *Notes Roy. Bot. Gard. Edinburgh* **46(3)**: 345-359.

PATON, A. 1994: Three membranous-bracted species of *Origanum*. - *Bot. Mag. (Kew Mag.)* **11(3)**: tab. 248-250 + pp. 109-117.

PATON, D.N. 1966: Crete's natural wealth of flowers. - *Gard. Chron., ser. 3* **160(19-20)**: 14-15, 20-21.

PATZAK, A.W. 1958: Revision der Gattung *Ballota* Section *Ballota*. - *Ann. Naturhist. Mus. Wien* **62**: 57-86.

PATZAK, A.W. 1959: Revision der Gattung *Ballota* Section *Acanthoprasium* und Section *Beringeria*. - *Ann. Naturhist. Mus. Wien* **63**: 33-81.

PATZAK, A.W. 1961: Zwei neue *Ballota*-Arten aus der Türkei nebst einem Nachtrag zur Gattung *Ballota*. - *Ann. Naturhist. Mus. Wien* **64**: 42-56.

PATZAK, A.W. 1962: Der Formenkreis der *Scabiosa palaestina* L. und *Scabiosa rotata* M.B. - *Ann. Naturhist. Mus. Wien* **65**: 21-28.

PAUL, P. 1975: Biologie et écologie du *Tuberaria guttata* (L.) Fourreau (= *Helianthemum guttatum* (L.) Miller). - *Bull. Soc. Bot. France* **122(9)**: 367-370.

PAULE, L. 1997: Gene conservation in European beech (*Fagus sylvatica*). - *Bocconea* **7**: 367-381.

PAULUS, H.F. 1988: Beobachtungen und Experimente zur Pseudo-Kopulation auf *Ophrys* Arten (Orchidaceae) Kretas (II) mit einer Beschreibung von *Ophrys sitiaca* nov. spec. aus der *Ophrys fusca-omegaifera*-Formenkreis. - *Mitt. Arbeitskreis Heimische Orchid. Baden-Württemberg* **20(4)**: 817-882.

PAULUS, H.F. 1994: Untersuchungen am *Ophrys cretica*-Komplex mit Beschreibung von *Ophrys ariadnae* H.F. Paulus spec. nov. (Orchidaceae). - *J. Eur. Orch.* **26(3-4)**: 628-643.

PAULUS, H.F. 1998: Der *Ophrys fusca* s.str.-Komplex auf Kreta und anderer Ägäisinseln mit Beschreibung von *O. blitopertha, O. creberrima, O. cinerophila, O. cressa, O. thriptiensis* und *O. creticola* spp. nov. - *J. Eur. Orch.* **30(1)**: 157-201.

PAULUS, H.F. 2001: Daten zur Bestäubungsbiologie und Systematik der Gattung *Ophrys* in Rhodos (Griechenland) mit Beschreibung von *O. parvula, O. persephonae, O. lindia, O. eptapigiensis* aus der *O. fusca*-Gruppe und *O. cornutula* aus der *O. oestrifera*-Gruppe. - *Ber. Arbeitskreis. Heimische Orchid.* **18(1)**: 38-86 [+ 18(2): 239].

PAULUS, H.F. 2002: Daten zur Bestäubungsbiologie und Systematik der Gattung *Ophrys* in Rhodos (Griechenland) II. Über *Ophrys holoserica* s.lat.: *Ophrys episcopalis, Op. maxima* und *Ophrys halia* spec. nov. (Orchidaceae und Insecta, Apoidea). - *Ber. Arbeitskreis. Heimische Orchid.* **18(2)**: 46-63.

PAULUS, H.F. 2008: In memoriam Hans R. Reinhard: *Ophrys reinhardiorum* H.F. Paulus spec. nov. und ihr Bestäuber *Eupavlovskia funeraria*. Zur Bestäubungsbiologie und Taxonomie einer neuen Art der *Ophrys reinholdii*-Gruppe aus Nordostgriechenland (Orchidaceae; Insecta, Apoidea). - *Ber. Arbeitskreis. Heimische Orchid.* **25(1)**: 6-30.

PAULUS, H.F. 2011a: Attackieren oder Kopulieren? Das ambivalente Verhalten von *Xylocopa* gegenüber der Sexualtäuschung von *Ophrys grigoriana* in Kreta (Orchidaceae und Insecta, Apoidea). - *J. Eur. Orch.* **43(1)**: 35-60.

PAULUS, H.F. 2011b: Zur Bestäubungsbiologie einiger *Ophrys*-Arten in Nordthessalien mit Beschreibung von *Ophrys olympiotissa* aus der *Ophrys argolica - ferrum-equinum*-Gruppe (Orchidaceae und Insecta, Apoidea). - *J. Eur. Orch.* **43(3)**: 498-526.

PAULUS, H.F., ALIBERTIS, C. & ALIBERTIS, A. 1990: *Ophrys mesaritica* H.F. Paulus und C. & A. Alibertis spec. nov. aus Kreta, eine neue Art aus dem *Ophrys fusca-iricolor*-Artenkreis. - *Mitt. Arbeitskreis Heimische Orchid. Baden-Württemberg* **22(4)**: 772-787.

PAULUS, H.F. & GACK, C. 1992: Die Gattung *Ophrys* (Orchidaceae) auf der Kykladeninsel Naxos: Daten zur Bestäubungsbiologie und zur Floristik. - *Mitt. Arbeitskreis Heimische Orchid. Baden-Württemberg* **24(3)**: 403-449.

PAULUS, H.F. & HIRTH, M. 2009a: *Ophrys apollonae* spec. nov., eine neue Art der *Ophrys omegaifera*-Gruppe aus Rhodos, Samos und Chios. - *J. Eur. Orch.* **41(3-4)**: 643-662.

PAULUS, H.F. & HIRTH, M. 2009b: Bestäubung und Systematik der *Ophrys holoserica-episcopalis*-Gruppe: *Ophrys saliarisii* spec. nov. aus Chios und Rhodos. - *J. Eur. Orch.* **41(3-4)**: 663-680.

PAULUS, H.F. & HIRTH, M. 2011: *Ophrys heterochila*, *Ophrys dodekanensis*, *Ophrys minutula*, *Ophrys oreas* - Versuch einer Neubewertung. - *J. Eur. Orch.* **43(3)**: 651-679.

PAULUS, H.F. & HIRTH, M. 2012: Bestäubungsbiologie und Systematik der *Ophrys tenthredinifera*- Artengruppe in der Ostägäis (Orchidaceae, Insecta). - *J. Eur. Orch.* **44(3)**: 625-686.

PAULUS, H.F. & HIRTH, M. 2014: Zur Bestäubungsbiologie der Gattung *Ophrys* auf den Ionischen Inseln mit speziellen Bemerkungen zum Status von *Ophrys punctulata-O. leucadica* sowie Beschreibungen von zwei neuen Arten aus der *O. lutea*- und *O. oestrifera*-Gruppe. - *J. Eur. Orch.* **46 (2)**: 233-304.

PAULUS, H.F. & HIRTH, M. 2015: *Ophrys parosica-phaseliana-pelinaea-meropes* on the Eastern Aegean Islands - pollination biology of a confusing *O. fusca* species group (Orchidaceae, Insecta- Apoidea). - *J. Eur. Orch.* **47(2-4)**: 365-404.

PAULUS, H.F. & HIRTH, M. 2016: Eine neue *Ophrys fusca*-Sippe aus Nordgriechenland: *Ophrys bilunulata* subsp. *kalirachiensis* (Orchidaceae und Insecta, Apoidea). - *J. Eur. Orch.* **48(1)**: 53-70.

PAULUS, H.F. & HIRTH, M. 2017: Bestäubungsbiologie und Systematik des *Ophrys mammosa*-Komplex im östlichen Mittelmeerraum mit Neubeschreibungen von *Ophrys prespaensis* und *O. willingii* vom griechischen Festland. - *J. Eur. Orch.* **49(2)**: 221-314.

PAULUS, H.F. & SALKOWSKI, H.-E. 2007: Bestäubungsbiologische Untersuchungen an Winterorchideen auf der Ägäis-Insel Kos. - *Ber. Arbeitskreis. Heimische Orchid.* **24(2)**: 4-29.

PAULY, D. 1962: Aperçu sur l'écologie du Sapin de Cephalonie et de ses hybrides. - *Rev. Forest. Franç.* **14**: 755-769 [n.v.].

PAUN, O. & AL. 2005: Phylogenetic relationships and biogeography of *Ranunculus* and allied genera (Ranunculaceae) in the Mediterranean region and in the European Alpine System. - *Taxon* **54(4)**: 911-930.

PAUNERO RUIZ, E. 1946: Las especies españolas del género *Agrostis*. - *Anales Jard. Bot. Madrid* **7**: 561-644.

PAUNERO RUIZ, E. 1948: Revisión de las especies españolas del género *Phalaris*. - *Anales Jard. Bot. Madrid* **8**: 475-522.

PAUNERO RUIZ, E. 1950: Las especies españolas del género *Trisetaria* Forsk. - *Anales Jard. Bot. Madrid* **9**: 503-582.

PAUNERO RUIZ, E. 1952: Las especies españolas del género *Alopecurus*. - *Anales Inst. Bot. Cavanilles* **10(2)**: 301-345.

PAUNERO RUIZ, E. 1954: Las especies españolas del género *Anthoxanthum*. - *Anales Inst. Bot. Cavanilles* **12(1)**: 401-442.

PAUNERO RUIZ, E. 1956: Las Aveneas españolas. II. - *Anales Inst. Bot. Cavanilles* **14(1)**: 187-251.

PAUNERO RUIZ, E. 1957a: Las Aveneas españolas. III. - *Anales Inst. Bot. Cavanilles* **15**: 377-415.

PAUNERO RUIZ, E. 1957b: Las Andropogoneas españolas. - *Anales Inst. Bot. Cavanilles* **15**: 417-459.

PAUNERO RUIZ, E. 1959: Las Aveneas españolas. IV. - *Anales Inst. Bot. Cavanilles* **17(1)**: 257-376.

PAUNERO RUIZ, E. 1962: Las Paniceas españolas. - *Anales Inst. Bot. Cavanilles* **20**: 51-90 + 22 plates.

PAUNERO RUIZ, E. 1964a: Notas sobre gramíneas. II. Consideraciones acerca de las especies españolas del género *Vulpia* Gmel. - *Anales Inst. Bot. Cavanilles* **22**: 81-114 + 37 plates.

PAUNERO RUIZ, E. 1964b: Notas sobre gramíneas. III. Consideraciones acerca de las especies españolas del género *Parapholis*. - *Anales Inst. Bot. Cavanilles* **22**: 187-201 + 16 plates.

PAUNERO RUIZ, E. 1967: Notas sobre gramínes. V. Datos acerca del género *Catapodium*. - *Anales Inst. Bot. Cavanilles* **25**: 207-228 + 12 plates.

PAVLETIĆ, Z. 1976: *Valantia hispida* L. in der Flora Jugoslawiens [In Serbo-Croatian with German summary]. - *Acta Bot. Croat.* **35**: 245-247.

PAVLETIĆ, Z. 1985a: Current nomenclature state of the Adamović's taxa - the base for their revalorization [In Serbo-Croatian with English summary]. - *Glasn. Prir. Mus. Beogradu, Ser. B, Biol. Nauke* **40**: 17-25.

PAVLETIĆ, Z. 1985b: Recent views on the kinship circle of *Silene multicaulis* Guss. s.l. [In Serbo- Croatian with English summary]. - *Biosistematika* **10(2)**: 65-68.

PAVLETIĆ, Z. 1989: Contribution to the taxonomy of the species *Daucus carota* L. (Apiaceae) [In Serbo-Croatian with English summary]. - *Biosistematika* **14(1)**: 23-29.

PAVLIDES, G. 1976a: On the flora and vegetation of the Sithonia peninsula (Chalkidiki, Greece) [In Greek with English summary]. - Thessaloniki: 300 pp. + 16 plates and 1 map.

PAVLIDES, G. 1976b: Untersuchungen über die Vegetationsverhältnisse der Küste der Sithonia-Halbinsel. In Dafis, S. & Landolt, E. (eds): Zur Vegetation und Flora von Griechenland, Band 2. - *Veröff. Geobot. Inst. E.T.H. Stiftung Rübel Zürich* **56**: 5-20.

PAVLIDES, G. 1982: Geobotanical study of the mountain range of Vertiscos. I. Flora and vegetation [In Greek with English summary]. - Thessaloniki: 173 pp. + 1 map.

PAVLIDES, G. 1985: Geobotanical study of the national park of lakes Prespa (NW Greece). Part A'. Ecology, flora, phytogeography, vegetation [In Greek with English summary]. - Thessaloniki: xi + 308 pp. + 1 map.

PAVLIDES, G. 1989: The vegetation of the aquatic plants of artificial Agras lake [In Greek with English summary]. - *Bios (Thessaloniki)* **1989**: 159-170.

PAVLIDES, G. 1997a: The flora of Prespa National Park with emphasis on species of conservation interest. In Crivelli, A.J. & Catsadorakis, G. (eds): Lake Prespa, northwestern Greece: A unique Balkan wetland, pp. 35-40. - Dordrecht, etc.: Kluwer Academic Publishers [repr. from *Hydrobiologia* 351].

PAVLIDES, G. 1997b: Aquatic and terrestrial vegetation of the Prespa area. In Crivelli, A.J. & Catsadorakis, G. (eds): Lake Prespa, northwestern Greece: A unique Balkan wetland, pp. 41-60. - Dordrecht, etc.: Kluwer Academic Publishers [repr. from *Hydrobiologia* 351].

PAVLOPOULOU, M.G. & PANITSA, M. 2019: Archaeological sites as habitat islands: Plant species diversity of Ancient Messini (SW Peloponnisos, Greece). In Daskalakou, E.N. & al. (eds): 16th Conference of the Hellenic Botanical Society – Athens 10-13 October 2019. Program and book of abstracts, p. 98.

PAVLOVA, D.K. 1996: Reports (474-478). In Kamari, G., Felber, F. & Garbari, F. (eds): Mediterranean chromosome number reports - 5. - *Fl. Medit.* **5**: 317-320.

PAVLOVA, D.K. 1997: Reports (762-766). In Kamari, G., Felber, F. & Garbari, F. (eds): Mediterranean chromosome number reports - 6. - *Fl. Medit.* **6**: 323-328.

PAVLOVA, D.K. 1999: Reports (1064-1065). In Kamari, G., Felber, F. & Garbari, F. (eds): Mediterranean chromosome number reports - 9. - *Fl. Medit.* **9**: 356-359.

PAVLOVA, D.K. 2004: On the evolutionary mechanisms and florogenetic relationships of the species of subgenus *Cercidothrix* (Astragalus, Fabaceae) in the Bulgarian flora. - *Phytol. Balcan.* **9(3)**: 461-470.

PAVLOVA, D.K. 2007a: A new species of *Aethionema* (Brassicaceae) from the Bulgarian flora. - *Bot. J. Linn. Soc.* **155(4)**: 533-540.

PAVLOVA, D.K. 2007b: Endemic and rare plants growing on serpentines in the Rhodopes mountains (Bulgaria). In Matevski, V. (ed.): Collection of papers devoted to Academician Kiril Micevski, pp.157-170. - Skopje: Macedonian Academy of Sciences and Arts.

PAVLOVA, D.K. 2009: *Onosma bulgarica* sp. nov. (Boraginaceae-Lithospermeae) found on serpentine in Bulgaria. - *Nordic J. Bot.* **27(3)**: 216-221.

PAVLOVA, D.K. 2010: A survey of the serpentine flora in the West Bulgarian Frontier Mts (Mt Vlahina and Mt Ograzhden). - *Phytol. Balcan.* **16(1)**: 97-107.

PAVLOVA, D.K. & AL. 1997: New chorological data and critical notes for the flora of the Eastern Rhodopes Mts. (Southern Bulgaria). - *OT Sist. Bot. Derg.* **4(1)**: 3-8.

PAVLOVA, D.K. & ANČEV, M.E. 2006: *Aethionema rhodopaeum* - a new species from the Bulgarian flora. In Peev, D.R. (ed.): Book of abstracts, IV Balkan botanical congress, Sofia, Bulgaria, 20-26 June 2006, p. 271.

PAVLOVA, D.K. & DIMITROV, D.S. 2001: A contribution to the flora of the serpentine areas in the Rodopes Mts. (Southern Bulgaria). - *OT Sist. Bot. Derg.* **8(2)**: 13-18.

PAVLOVA, D.K., DIMITROV, D.S. & NIKOLOVA, M. 1999: *Oxytropis kozuharovii* (Fabaceae), a new species from Bulgaria. - *Willdenowia* **29(1-2)**: 69-75.

PAVLOVA, D.K. & KOŽUHAROV, S.I. 1993: Chromosome numbers of Bulgarian angiosperms. – *Fitologiya* **44**: 75.

PAVLOVA, D.K. & KOŽUHAROV, S.I. 1995a: New taxa and new taxonomic combinations in the genus *Astragalus* L. in flora of Bulgaria. - *OT Sist. Bot. Derg.* **1(2)**: 17-25.

PAVLOVA, D.K. & KOŽUHAROV, S.I. 1995b: New data on the genus *Astracantha* Podl. in the Balkan Peninsula. - *God. Sofiisk. Univ. "Kliment. Ohridski" Biol. Fak., 2 Bot.* **88**: 51-53.

PAVLOVA, D.K., KOŽUHAROVA, E.K. & DIMITROV, D.S. 2003: A floristic catalogue of serpentine areas in the eastern Rhodope Mountains (Bulgaria). - *Polish Bot. J.* **48(1)**: 21-41.

PAVLOVA, D.K. & TOSHEVA, A. 2000: Reports (1208-1212) In Kamari, G., Felber, F. & Garbari, F. (eds): Mediterranean chromosome number reports - 10. - *Fl. Medit.* **10**: 419-423.

PAVLOVA, D.K. & TOSHEVA, A. 2001: Reports (1244-1247) In Kamari, G., Blanché, C. & Garbari, F. (eds): Mediterranean chromosome number reports. - *Fl. Medit.* **11**: 455-459.

PAVLOVA, D.K. & TOSHEVA, A. 2002: Karyological study of *Melilotus alba* Med. (Fabaceae) populations in Bulgaria. - *Caryologia* **55(2)**: 105-110.

PAVLOVA, D.K. & TOSHEVA, A. 2003a: Reports (1284-1287). In Kamari, G., Blanché, C. & Garbari, F. (eds): Mediterranean chromosome number reports - 12. - *Fl. Medit.* **12**: 450-454.

PAVLOVA, D.K. & TOSHEVA, A. 2003b: Reports (1334-1338). In: Kamari, G., Blanché, C. & Garbari, F. (eds): Mediterranean chromosome number reports - 13. - *Fl. Medit.* **13**: 361-365.

PAVLOVA, D.K. & TOSHEVA, A. 2004: Reports (1382-1386). In Kamari, G., Blanché, C. & Garbari, F. (eds): Mediterranean chromosome number reports - 14. - *Fl. Medit.* **14**: 429-433.

PAVLOVIĆ, Z. 1953: Contribution à la connaissance de la flore serpentinicole de la montagne Ozren, près de Sjenica (Serbie) [In Serbo-Croatian with French summary]. - *Glasn. Prir. Muz. Srpske Zemlje, Ser. B, Biol. Nauke* **5-6**: 3-19.

PAVLOVIĆ, Z. 1962: Characteristische Elemente der serpentinischen Flora und Vegetation Serbiens [In Serbo-Croatian with German summary]. - *Glasn. Prir. Mus. Beogradu, Ser. B, Biol. Nauke* **18**: 3-20.

PAVON, D., VÉLA, E. & MÉDAIL, F. 2020: Are Mediterranean trees well known? "Juniperus turbinata" (Cupressaceae), a common but misunderstood taxon. - *Ecol. Medit.* **46(2)**: 77-104.

PAVONE, P., TERRASI, C.M. & ZIZZA, A. 1981: [Chromosome numbers from Sicily]. In Löve, Á. (ed.): Chromosome number reports LXXII. - *Taxon* **30(3)**: 695-696.

PAWŁOWSKA, S. 1963: De Soldanellis, quae in parte septentrionali Carpatorum crescunt. - *Fragm. Florist. Geobot.* **9(1)**: 3-30 + 2 plates.

PAWŁOWSKI, B. 1953: Alchemillae in tribus itineribus balcanicis lectae. - *Acta Soc. Bot. Poloniae* **22(1)**: 245-258.

PAWŁOWSKI, B. 1958: De distributione geographica *Alchemillae gorcensis* Pawł. - *Fragm. Florist. Geobot.* **4(1-2)**: 133-138 + 1 table and 2 plates.

PAWŁOWSKI, B. 1959: De positione systematica *Potentillae Dörfleri* Wettst. - *Fragm. Florist. Geobot.* **5 (3)**: 399-407.

PAWŁOWSKI, B. 1961: Observationes ad genus *Symphytum* L. pertinentes. I. De systemate generis. - *Fragm. Florist. Geobot.* **7(2)**: 327-356 + 1 table.

PAWŁOWSKI, B. 1963a: Dispositio systematica specierum europaearum generis *Delphinium* L. - *Fragm. Florist. Geobot.* **9(4)**: 429-446 + 4 plates.

PAWŁOWSKI, B. 1963b: Hieracia balcanica nova vel minus cognita. - *Acta Soc. Bot. Poloniae* **32(3)**: 473-491 + 2 plates.

PAWŁOWSKI, B. 1965: De generis *Potentilla* L. serie *Crassinerviae* (Th. Wolf) B. Pawł. nec non de taxis affinibus. - *Fragm. Florist. Geobot.* **11(1)**: 53-91.

PAWŁOWSKI, B. 1969: De speciebus aliquot balcanicis generis *Erigeron* L. - *Acta Bot. Croat.* **28**: 285-291.

PAWŁOWSKI, B. 1971a: *Symphyta* mediterrana nova vel minus cognita. - *Fragm. Florist. Geobot.* **17 (1)**: 17-37.

PAWŁOWSKI, B. 1971b: De genere *Procopiana* Guşuleac. - *Fragm. Florist. Geobot.* **17(1)**: 39-58.

PAWŁOWSKI, B. 1972: De *Alchemillis* turcicis e subsectionibus *Chirophyllum* Rothm. et *Calycanthum* Rothm. sectionis *Alchemilla*. - *Fragm. Florist. Geobot.* **18(1)**: 3-44.

PAX, F. 1885: Monographie der Gattung *Acer*. - *Bot. Jahrb. Syst.* **6**: 287-374 + 1 plate.

PAX, F. 1886: Monographie der Gattung *Acer*. Spezieller Teil (Fortsetzung und Schluss). - *Bot. Jahrb. Syst.* **7**: 177-263.

PAX, F. 1889a: Monographische Übersicht über die Arten der Gattung *Primula*. - *Bot. Jahrb. Syst.* **10**: 75-241.

PAX, F. 1889b: Nachträge und Ergänzungen zu der Monographie der Gattung *Acer*. - *Bot. Jahrb. Syst.* **11**: 72-83.

PAX, F. 1893: Weitere Nachträge zur Monographie der Gattung *Acer*. - *Bot. Jahrb. Syst.* **16**: 393-404.

PAX, F. 1928: Hippocastanaceae DC. - *Pflanzenareale* **2(1)**: 8, Karte 8.

PAX, F. 1930: Nachträge und Verbesserungen zu den Europa-Karten: *Buxus*. - *Pflanzenareale* **2(8)**: 88, Karte 80 (Nachtragskarte 1b).

PAZY, B. & AL. 1977: Studies in populations of the Old World *Lupinus* species. I. Chromosomes of the East-Mediterranean lupines. - *Israel J. Bot.* **26(3)**: 115-127.

PAZY, B. & PLITMANN, U. 1994: Holocentric chromosome behavior in *Cuscuta* (Cuscutaceae). - *Pl. Syst. Evol.* **191(1-2)**: 105-109.

PAZY, B. & PLITMANN, U. 1995: Chromosome divergence in the genus *Cuscuta* and its systematic implications. - *Caryologia* **48(2)**: 173-180.

PAZY, B. & ZOHARY, D. 1965: The process of introgression between *Aegilops* polyploids: natural hybridization between *A. variabilis*, *A. ovata* and *A. biuncialis*. - *Evolution* **19(3)**: 385-394.

PEARCE, N.R. 2006: John Stuart Mill's botanical collections from Greece (a private passion). - *Phytol. Balcan.* **12(2)**: 149-164.

PEARL, R. 1907: Variation and differentiation in *Ceratophyllum*. - *Publ. Carnegie Inst. Washington* **58**: 136 pp. + 2 fold-out charts.

PECK, P. 1998a: *Cyclamen hederifolium* on Thassos. - *Cyclamen* **22(1)**: 29-31.

PECK, P. 1998b: Spring in the Mani. - *Cyclamen* **22(2)**: 49-50.

PEDASHENKO, H., APOSTOLOVA, I. & VASSILEV, K. 2012: *Amorpha fruticosa* invasibility in different habitats in lower Danube. - *Phytol. Balcan.* **18(3)**: 285-291.

PEDERSEN, H.Æ. 2006: Systematics and evolution of the *Dactylorhiza romana/sambucina* polyploid complex (Orchidaceae). - *Bot. J. Linn. Soc.* **152(4)**: 405-434.

PEDERSEN, H.Æ. 2011: Lectotypification of the name *Epipactis leptochila* (Orchidaceae). - *Nordic J. Bot.* **29(2)**: 182-183.

PEDERSEN, H.Æ. & FAURHOLDT, N. 1997a: A critical approach to *Ophrys calypsus* (Orchidaceae) and to the records of *O. holoserica* subsp. *apulica* from Greece. - *Fl. Medit.* **7**: 153-162.

PEDERSEN, H.Æ. & FAURHOLDT, N. 1997b: Beiträge zur Orchideenflora der ostmediterranen Inseln Rhodos und Zypern. - *Orchidee (Hamburg)* **48(6)**: 232-236.

PEDERSEN, H.Æ. & FAURHOLDT, N. 2001: *Ophrys* - Versuchsweise Definitionen der Kategorien Art, Unterart und Varietät in der Gattung und einige daraus resultierende taxonomische Änderungen. - *Orchidee (Hamburg)* **53(3)**: 341-346.

PEDERSEN, H.Æ. & FAURHOLDT, N. 2005: Two new combinations in *Ophrys* (Orchidaceae). - *J. Eur. Orch.* **37(2)**: 287-290.

PEDERSEN, H.Æ. & FAURHOLDT, N. 2007: *Ophrys*. The bee orchids of Europe. - Kew Publishing: 297 pp.

PEDROL, J. 1992: (1050) Proposal to amend 2261 *Suaeda*, nom. cons. (Chenopodiaceae). - *Taxon* **41(2)**: 337-338. [approved, cf. Taxon 43(2): 274, 1994].

PEDROL, J. & CASTROVIEJO, S. 1988: A propósito del tratamiento taxonómico y nomenclatural del género *Suaeda* Forsskål ex Scop. (Chenopodiaceae) en "Flora Iberica". - *Anales Jard. Bot. Madrid* **45(1)**: 93-102.

PEDROLA-MONFORT, J. & CAUJAPÉ-CASTELLS, J. 1996: Genetic and morphological divergence in the *Androcymbium gramineum* complex (Colchicaceae). - *Pl. Syst. Evol.* **201(1-4)**: 149-162.

PEDROTTI, F. & CORTINI PEDROTTI, C. 1971: Numeri cromosomici per la Flora Italiana: 13-18. - *Inform. Bot. Ital.* **3(1)**: 47-51.

PEER, T. 1983: Zum Vorkommen von *Genista radiata* (L.) Scop. in Südtirol. - *Ber. Bayer. Bot. Ges.* **54**: 127-134.

PEEV, D.R. 1972a: New taxa and ploidy levels of some Bulgarian *Veronica* species. - *Dokl. Bulg. Akad. Nauk* **25(6)**: 811-814.

PEEV, D.R. 1972b: [Chromosome numbers in *Veronica* from Bulgaria]. In Löve, Á. (ed.): IOPB chromosome number reports XXXVII. - *Taxon* **21(4)**: 499-500.

PEEV, D.R. 1975: Genus *Veronica* in Bulgaria - some new taxa and chromosome numbers. - *Fitologiya* **2**: 42-46.

PEEV, D.R. 1976a: Chromosome numbers and critical notes on the taxonomy of some Bulgarian Primulaceae species. - *Fitologiya* **4**: 16-24.

PEEV, D.R. 1976b: [Chromosome numbers from Bulgaria]. In Löve, Á. (ed.): IOPB chromosome number reports LIV. - *Taxon* **25(5-6)**: 643-644.

PEEV, D.R. 1977a: [Chromosome numbers from Bulgaria]. In Löve, Á. (ed.): IOPB chromosome number reports LV. - *Taxon* **26(1)**: 107.

PEEV, D.R. 1977b: Chromosome numbers and critical notes on the taxonomy of some

Primulaceae species in Bulgaria [In Bulgarian with Russian and English summaries]. - *Fitologiya* **7**: 3-9.

PEEV, D.R. 1978: Taxonomy and microevolution of ... *Veronica* L. in Bulgaria [In Bulgarian with English and Russian summaries]. In Kožuharov, S.I. & Kuzmanov, B.A. (eds): Evolution of flowering plants ..., pp. 72-106. - Sofia: Publishing House of the Bulgarian Academy of Sciences.

PEEV, D.R. 1982: Different substitution tendencies of leaf flavones in the *Veronica hederifolia* group (Scrophulariaceae). - *Pl. Syst. Evol.* **140(2-3)**: 235-242.

PEEV, D.R. & AL. 2009a: The pink flowering *Crepis rubra* (Asteraceae) - new for the Bulgarian flora. - *Phytol. Balcan.* **15(1)**: 59-62.

PEEV, D.R. & AL. 2009b: Balkan vascular flora in numbers. In Ivanova, D. (ed.): Proceedings of the IV Balkan botanical congress, pp. 140-143. - Sofia: Institute of Botany, BAS.

PEEV, D.R. & AL. (EDS) 2015: Red Data Book of the Republic of Bulgaria. Volume 1. Plants and Fungi. - Sofia: BAS & MoEW, 881 pp.

PEEV, D.R. & ANDREEV, N. 1978: [Chromosome numbers in Apiaceae from Bulgaria]. In Löve, Á. (ed.): IOPB chromosome number reports LXII. - *Taxon* **27(5-6)**: 534-535.

PEEV, D.R. & KOŽUHAROV, S.I. 1991: Variability ... of three annual species of genus *Bromus* L. in Bulgaria [In Bulgarian with English summary]. In Kožuharov, S.I. & Kuzmanov, B.A. (eds): Evolution of flowering plants ... 2, pp. 119-139. - Sofia: Publishing House of the Bulgarian Academy of Sciences.

PEEV, D.R. (ED.) 2006: Book of abstracts, IV Balkan botanical congress, Sofia, Bulgaria, 20-26 June 2006. - Sofia: Bulgarian Academy of Sciences, 314 pp.

PEITZ, E. 1980: Die Orchideen der Insel Korfu. - *Schriftenreihe Heimvolkshochschule Schloss Dhaun* **5**: 1-26.

PEITZ, E. 1983: Die Orchideen der Insel Korfu (Ergänzungen zum gleichen Beitrag in der "Schriftenreihe der Heimvolkshochschule Schloss Dhaun", Heft 5/1980). - *Dhauner Echo* **42**: 29-30.

PEJČINOVIĆ, M. 1978: Karyologic investigation of some species of the genus of *Aquilegia* [In Serbo-Croatian with English summary]. - *God. Biol. Inst. Univ. Sarajevu* **31**: 153-157.

PEJČINOVIĆ, M. 1990: Cytotaxonomic analysis of the species of the genera *Orlaya* Hoffm. and *Caucalis* L. [In Serbo-Croatian with English summary]. - *Biosistematika* **15(2)**: 127-131.

PENEV, I. & KOŽUHAROV, S.I. 1968: A new foxtail grass from Bulgaria. - *Notes Roy. Bot. Gard. Edinburgh* **28(2)**: 187-188.

PENEV, N. & GEORGIEV, A. 1958: Die Peuce-(Rumelische) Kiefer im Zentralbalkangebirge [In Bulgarian with Russian and German summaries]. - *Izv. Bot. Inst. (Sofia)* **6**: 215-282.

PENEVA, P.C. & AL. 1988: Morphological and karyological analysis of native populations from chamomile (*Matricaria recutita* L.) [In Bulgarian with Russian and English summaries]. - *Fitologiya* **34**: 28-42.

PENEVA, P.C., POPOVA, E.D. & KUZMANOV, B.A. 1989: Morphological variability of the chamomile. I. Cluster analysis [In Bulgarian with Russian and English summaries]. - *Fitologiya* **36**: 15-25.

PĚNKOVÁ, I. 1986: Contribution to the taxonomy of *Plantago major* L. s.l. [In Czech with English summary]. - *Preslia* **58(2)**: 117-139.

PENNACCHINI, V. & BONIN, G. 1975: *Pinus leucodermis* Ant. et *Pinus nigra* Arn. en Calabre septentrionale. - *Ecol. Medit.* **1**: 35-61.

PÉNZES, A. 1939: Data on the spread and physiology of the plant species of the Balay Island. [Additamenta ad distributionem et oecologiam florae Balcanicae]. (In Hungarian). - *Borbásia* **1**: 161-171.

PÉNZES, A. 1944: Über die Ökologie und Systematik der Gruppe von *Salvia verticillata* L. mit besonderer Berücksichtigung der Wasserverdampfung durch die Pflanzenhaare. - *Borbásia* **5-6(1-3)**: 1-31.

PÉNZES, A. 1965: Plantae novae ex Albania, Bulgaria et Hungaria. - *Ann. Hist.-Nat. Mus. Natl. Hung.* **57**: 169-177.

PÉNZES, A. 1968: Beiträge zur Flora der Balkanhalbinsel. - *Feddes Repert.* **77(1)**: 1-10.

PÉNZES, A. 1970: *Picea* und *Juniperus* Studien [In Hungarian with German summary]. - *Bot. Közlem.* **57(1)**: 45-50.

PERAKI, E., KOPAKA, K. & PIRINTSOS, S. (EDS) 2011: Program and Book of Abstracts, 12th congress of Hellenic Botanical Society. - Rethimno: 198 pp.

PERDETZOGLOU, D. 1994: Farmakognostiki, chimiotaxinomiki ke morfologiki meleti ton taxinomikon monadon tou genous *Scabiosa* L. s.l. stin Kriti [In Greek]. - Athine: 2 vols., 420 + 342 pp.

PERDETZOGLOU, D. & AL. 1994: Comparative phytochemical and morphological study of two species of the *Scabiosa* L. genus. - *Feddes Repert.* **105(3-4)**: 157-165.

PERDETZOGLOU, D. & AL. 1996: A comparative study of eight taxa of *Lomelosia* Raf. (Dipsacaceae) from Greece, according to their fatty acid and sterol composition and antibacterial activity. - *Feddes Repert.* **107(1-2)**: 37-42.

PERDETZOGLOU, D. & TAN, KIT 1995: A new species of *Cephalaria* (Dipsacaceae) from northern Greece. - *Ann. Bot. Fenn.* **32(4)**: 227-228.

PERDIGÓ I ARISÓ, M.-T. 1981: Contribution à l'étude cytotaxinomique du genre *Eryngium* L. - *Biol. Écol. Médit.* **8(1)**: 3-12.

PERDIGÓ I ARISÓ, M.-T. & LLAURADÓ MIRAVALL, M. 1985: Nueva aportación al estudio biosistemático del género *Eryngium* L. - *Lazaroa* **6**: 189-198.

PEREIRA, A. DE LEMOS 1942: Contribuïção ao conhecimento cariológio do género *Nigella* L. II. - *Bol. Soc. Brot., Ser. 2* **16**: 5-40.

PEREIRA, A.L. 1999: Taxonomy of the genus *Azolla* in Portugal. In Correia, A.I.D. & al. (eds): Livro de resumos. Guia de excursões. V jornadas de taxonomia botânica, 16-19/IX 1999, p. 52.

PEREIRA, A.L. & AL. 2001a: Taxonomy of the genus *Azolla* Lam. in Portugal. - *Portug. Acta Biol.* **19(1-4)**: 277-282.

PEREIRA, A.L. & AL. 2001b: Taxonomic re-evaluation of the *Azolla* genus in Portugal. - *Pl. Biosyst.* **135 (3)**: 285-294.

PÉREZ CARRO, F.J., DÍEZ GONZÁLEZ, T.E. & FERNÁNDEZ ARECES, M.P. 1990: Datos geobotánicos, taxonómicos y corológicos sobre *Asplenium trichomanes* L. subsp. *pachyrachis* (Christ) Lovis & Reichst. - *Monogr. Inst. Piren. Ecol.* **5**: 565-576.

PÉREZ CARRO, F.J. & FERNÁNDEZ ARECES, M.P. 1991: *Anogramma leptophylla* (Sinopteridaceae: Pteridophyta), an aneuploid taxon? - *Brit. Fern Gaz.* **14(2)**: 55-56.

PÉREZ CHISCANO, J.L. 1996: Aportaciones al conocimiento de *Urginea maritima* (L.) Baker (Liliaceae). - *Anales Jard. Bot. Madrid* **54(1)**: 392-398.

PÉREZ DE PAZ, P.L. 1994: Re the genus *Micromeria* Benth. (Labiatae) and nomenclatural inconsistency. - *Taxon* **43(2)**: 241-242.

PÉREZ DE PAZ, P.L. & NEGRÍN SOSA, L. 1992: Revisión taxonómica de *Sideritis* L. subgénero *Marrubiastrum* (Moench) Mend.-Heuer (Endemismo Macaronésico). - *Phanerog. Monogr.* **20**: 1-327.

PÉREZ MORALES, C. & AL. 1997: Typification and definition of *Doronicum austriacum* Jacq. (Asteraceae). - *Compositae Newslett.* **30**: 1-4.

PERI, O. 2015: Bulbs of the Eastern Mediterranean. - Pershore, U.K.: Alpine Garden Society, 256 pp.

PERILLE, M., CORTIZO, C. & SAHUQUILLO, E. 2003: About the taxonomic problems in *Orchis coriophora* (Orchidaceae). - *Bocconea* **16(2)**: 669-674.

PERILLE, M. & SAHUQUILLO, E. 2001: About the taxonomic problems in *Orchis coriophora* L. (Orchidaceae). In Raimondo, F.M. (ed.): Abstracts from the X OPTIMA meeting, Palermo, 13-19 September 2001, p. 130.

PERKO, M.L. 1998: A short note on the orchids of Cephalonia. - *Fisis [Journal of the Hellenic Society for the Protection of Nature]* **82**: 17, 43-46.

PERKO, M.L. 2000: Ergänzungen zur Orchideenflora der griechischen Inseln Samos, Karpathos und Rhodos. - *Ber. Arbeitskreis. Heimische Orchid.* **17(2)**: 79-83.

PERLA, F. & AL. 2018: Geographical distribution of diploid and tetraploid populations of *Juniperus sabina* L. - *Bot. Serbica* **142 (suppl. 1)**: 76.

PERNÝ, M., TRIBSCH, A. & ANČEV, M.E. 2004: Infraspecific differentiation in the Balkan diploid *Cardamine acris* (Brassicaceae): molecular and morphological evidence. - *Folia Geobot. Phytotax.* **39(4)**: 405-429.

PERRENOUD, R. & FAVARGER, C. 1971: Sur l'existence d'hybrides entre le colchique des Alpes (*Colchicum alpinum* D.C.) et le colchique d'automne (*Colchicum autumnale* L.) dans les Alpes françaises. - *Bull. Soc. Neuchâteloise Sci. Nat.* **94**: 21-27.

PERRING, F.H. & GARDINER, B.G. 1976: The biology of bracken. - *Bot. J. Linn. Soc.* **73(1-3)**: vii + 302 pp.

PERRY, B.A. 1943: Chromosome number and phylogenetic relationships in the Euphorbiaceae. - *Amer. J. Bot.* **30(7)**: 527-543.

PERRY, G. & MCNEILL, J. 1986: The nomenclature of *Eragrostis cilianensis* (Poaceae) and the contribution of Bellardi to Allioni's Flora Pedemontana. - *Taxon* **35(4)**: 696-701.

PERRY, G. & MCNEILL, J. 1987: (878) Proposal to reject the name *Echium lycopsis* L. (Boraginaceae) with notes on the correct name for the weed Paterson's Curse (Purple Viper's Bugloss). - *Taxon* **36(2)**: 483-492. [approved, cf. Taxon 39: 294-295, 1990].

PERRY, J.M. & PERRY, A.M. 2014: A new locality for *Cypripedium calceolus* from the Rodopi Mountains, Bulgaria. - *J. Eur. Orch.* **46(1)**: 31-40.

PERSSON, D. 1981: Biosystematics of *Stachys swainsonii* Benth. (Lamiaceae) and its relations to some other chasmophytic *Stachys* species. - Lund: Ph.D. thesis, Univ. of Lund, 174 pp.

PERSSON, J. 1971: Studies in the Aegean flora. XIX. Notes on *Alyssum* and some other genera of Cruciferae. - *Bot. Not.* **124(3)**: 399-418.

PERSSON, J. & PERSSON, K. 1998: *Fritillaria sororum* (Liliaceae): a new species from Turkey. - *New Plantsman* **5(2)**: 114-116.

PERSSON, J. & STRID, A. 1982: A new species of *Astragalus* (Fabaceae) from Mt. Taygetos. (Materials for the Mountain Flora of Greece, 15). - *Willdenowia* **12(2)**: 207-209.

PERSSON, K. 1974: Biosystematic studies in the *Artemisia maritima* complex in Europe. - *Opera Bot.* **35**: 1-188.

PERSSON, K. 1988: New species of *Colchicum* (Colchicaceae) from the Greek mountains. (Materials for the Mountain Flora of Greece, 30). - *Willdenowia* **18(1)**: 29-46.

PERSSON, K. 1989: The genus *Tulipa* in Greece: Taxonomy, karyology and hybridization. In Phitos, D. (ed.): Abstracts. Sixth OPTIMA meeting, Delphi 10-16 Sept. 1989, p. 32.

PERSSON, K. 1993: Reproductive strategies and evolution in *Colchicum*. In: Demiriz, H. & Özhatay, N. (eds): Proceedings of the V OPTIMA meeting, Istanbul, 8-15.9. 1986, pp. 397-414. - Istanbul Üniversitesi Fen Fakültesi.

PERSSON, K. 1998: Comments on some tessellated *Colchicum* species in the East Mediterranean area. - *Candollea* **53(2)**: 399-418.

PERSSON, K. 1999a: The genus *Colchicum* in Turkey. I. New species. - *Edinburgh J. Bot.* **56(1)**: 85-102.

PERSSON, K. 1999b: The genus *Colchicum* in Turkey. II. Revision of the large-leaved autumnal species. - *Edinburgh J. Bot.* **56(1)**: 103-142.

PERSSON, K. 1999c: New and revised species of *Colchicum* (Colchicaceae) from the Balkan Peninsula. - *Pl. Syst. Evol.* **217(1-2)**: 55-80.

PERSSON, K. 2000: Two new bulbous species from the central Taurus Mountains in Turkey. - *New Plantsman* **7(4)**: 200-208.

PERSSON, K. 2001: A new soboliferous species of *Colchicum* in Turkey. - *Bot. J. Linn. Soc.* **135(1)**: 85-88.

PERSSON, K. 2005: A new Turkish species of *Colchicum* (Colchicaceae) related to *C. boissieri*. - *Edinburgh J. Bot.* **62(3)**: 181-192.

PERSSON, K. 2007a: Nomenclatural synopsis of the genus *Colchicum* (Colchicaceae), with some new species and combinations. - *Bot. Jahrb. Syst.* **127(2)**: 165-242.

PERSSON, K. 2007b: A note on *Tulipa cinnabarina* - *Plantsman* **6(1)**: 58.

PERSSON, K. 2009: Colchicaceae. In Marhold, K. (ed.): IAPT/IOPB chromosome data 7. - *Taxon* **58 (1)**: 181-183 + electronic supplement.

PERSSON, K. & PERSSON, J. 2000: *Hyacinthella* Schur (Hyacinthaceae) in the Balkan countries. - *Candollea* **55(2)**: 213-225.

PERSSON, K. & WENDELBO, P. 1979: *Bellevalia hyacinthoides*, a new name for *Strangweja spicata* (Liliaceae). - *Bot. Not.* **132(1)**: 65-70.

PERSSON, K. & WENDELBO, P. 1981: Taxonomy and cytology of the genus *Hyacinthella* (Liliaceae-Scilloideae) with special reference to the species in S.W. Asia. Part I. - *Candollea* **36(2)**: 513-541.

PERSSON, K. & WENDELBO, P. 1982: Taxonomy and cytology of the genus *Hyacinthella* (Liliaceae-Scilloideae) with special reference to the species in S.W. Asia. Part II. - *Candollea* **37(1)**: 157-175.

PERUZZI, L. 2003a: Contribution to the cytotaxonomical knowledge of *Gagea* Salisb. (Liliaceae) sect. *Foliatae* A. Terracc. and synthesis of karyological data. - *Caryologia* **56(1)**: 115-128.

PERUZZI, L. 2003b: Reports (1344-1347). In: Kamari, G., Blanché, C. & Garbari, F. (eds): Mediterranean chromosome number reports - 13. - *Fl. Medit.* **13**: 370-373.

PERUZZI, L. 2004: Contribution to the cytotaxonomical knowledge of the genus *Pinguicula* L. (Lentibulariaceae): a synthesis of karyological data. - *Carniv. Pl. Newslett.* **33(4)**: 103-110.

PERUZZI, L. 2007: Some observations on *Pinguicula balcanica* Casper and *P. hirtiflora* Ten. (Lentibulariaceae) from Balkans. - *Carniv. Pl. Newslett.* **36(4)**: 100-103.

PERUZZI, L. & AL. 2008: On the phylogenetic position and taxonomic value of *Gagea trinervia* (Viv.) Greuter and *Gagea* sect. *Anthericoides* A. Terracc. (Liliaceae). - *Taxon* **57(3)**: 1201-1214.

PERUZZI, L. & AL. 2015: An inventory of the names of vascular plants endemic to Italy, their loci classici and types. - *Phytotaxa* **196(1)**: 1-217.

PERUZZI, L. & AL. 2017: *Fritillaria messanensis* subsp. *gracilis* (Liliaceae), a new record for the Italian flora (S Italy). - *Phytotaxa* **307(2)**: 167-170.

PERUZZI, L. & AL. 2019: An inventory of the names of native, non-endemic vascular plants described from Italy, their loci classici and types. - *Phytotaxa* **410**: 1-215.

PERUZZI, L. & BARTOLUCCI, F. 2009: Typification of the names within *Fritillaria montana* complex (Liliaceae) from the central Mediterranean area. - *Candollea* **64(1)**: 133-142.

PERUZZI, L., CAPARELLI, K.F. & BARTOLUCCI, F. 2009: *Fritillaria messanensis* subsp. *neglecta* (Parl.) Nyman, a fourth subspecies within *Fritillaria messanensis* Raf. (Liliaceae) from NW Balkans. - *Candollea* **64(2)**: 237-244.

PERUZZI, L. & CARTA, A. 2013: A taxonomic revision of *Silene nocturna* species complex (Caryophyllaceae) in Italy. - *Phytotaxa* **88(3)**: 38-48.

PERUZZI, L. & CESCA, G. 2003a: Reports (1295-1304). In Kamari, G., Blanché, C. & Garbari, F. (eds): Mediterranean chromosome number reports - 12. - *Fl. Medit.* **12**: 461-470.

PERUZZI, L. & CESCA, G. 2003b: Reports (1348-1356). In: Kamari, G., Blanché, C. & Garbari, F. (eds): Mediterranean chromosome number reports - 13. - *Fl. Medit.* **13**: 373-382.

PERUZZI, L. & CESCA, G. 2004: Chromosome numbers of flowering plants from Calabria, S Italy, II. - *Willdenowia* **34(2)**: 353-360.

PERUZZI, L., CESCA, G. & PUNTILLO, D. 2003: *Isoetes* (Isoetaceae), *Ophioglossum* and *Botrychium* (Ophioglossaceae) in Calabria (S Italy): more karyological and taxonomical data. - *Caryologia* **56(3)**: 359-363.

PERUZZI, L. & GARGANO, D. 2003: Reports (1357-1361). In: Kamari, G., Blanché, C. & Garbari, F. (eds): Mediterranean chromosome number reports - 13. - *Fl. Medit.* **13**: 382-387.

PERUZZI, L., IIRITI, G. & FRIGNANI, F. 2011: Contribution to the karyological knowledge of Mediterranean *Romulea* species (Iridaceae). - *Folia Geobot.* **46(1)**: 87-94.

PERUZZI, L. & PASSALACQUA, N.G. 2002: Biosystematic and taxonomic considerations about Italian units of the genus *Ornithogalum* (Hyacinthaceae) showing reflexed pedicels. - *Webbia* **57(2)**: 193-216.

PERUZZI, L. & PASSALACQUA, N.G. 2003: Reports (1362-1365). In: Kamari, G., Blanché, C. & Garbari, F. (eds): Mediterranean chromosome number reports - 13. - *Fl. Medit.* **13**: 387-392.

PERUZZI, L., PASSALACQUA, N.G. & CESCA, G. 2004: *Pinguicula crystallina* Sibth. et Smith subsp. *hirtiflora* (Ten.) Strid (Lentibulariaceae) in Calabria (southern Italy). Cytotaxonomical study and ex situ conservation in the Botanic Garden of Calabria University. - *Carniv. Pl. Newslett.* **33(3)**: 68-74.

PERUZZI, L., PASSALACQUA, N.G. & JARVIS, C.E. 2004: Lectotypification of *Aizoon hispanicum, Plantago albicans*, and *Staphylea pinnata*, names of three Linnaean species occurring in Calabria (S. Italy). - *Taxon* **53(2)**: 540-542.

PERUZZI, L., PASSALACQUA, N.G. & JARVIS, C.E. 2006: Typification of the accepted names in the *Jacobaea maritima* group (Asteraceae). - *Taxon* **55(4)**: 1001-1004.

PERUZZI, L., SCUDERI, L. & RAIMONDO, F.M. 2009: Distribution of the genus *Gagea* (Liliaceae) in Sicily. - *Fl. Medit.* **19**: 25-47.

PERUZZI, L. & TISON, J.-M. 2004: Typification and taxonomic status of eleven taxa of *Gagea* Salisb. (Liliaceae) described by Achille and Nicola Terraciano and conserved at Napoli (NAP). - *Candollea* **59(2)**: 325-346.

PERUZZI, L. & TISON, J.-M. 2007: Typification of seven critical Mediterranean *Gagea* Salisb. (Liliaceae) taxa. - *Candollea* **62(2)**: 173-187.

PESARESI, S. & AL. 2017: The *Pinus halepensis* Mill. forests in the central-eastern European Mediterranean basin. - *Pl. Biosyst.* **151(3)**: 512-529.

PESCI, G. 1971: Numeri cromosomici per la Flora Italiana: 81. - *Inform. Bot. Ital.* **3(2)**: 148-149.

PEŞMEN, H. 1968: Apportationes ad floram turcicam, 1-2. - *Candollea* **23(2)**: 267-273.

PEŞMEN, H. 1971: Materials for a flora of Turkey XXIV: *Ferula* and *Ferulago*. - *Notes Roy. Bot. Gard. Edinburgh* **31(1)**: 69-74 + 2 plates.

PETANIDOU, TH. & AL. 1995: Constraints on flowering phenology in a phryganic (East Mediterranean shrub) community. - *Amer. J. Bot.* **82(5)**: 607-620.

PETER, A. 1884: Über spontane und künstliche Gartenbastarde der Gattung *Hieracium* sect. *Piloselloidea*. - *Bot. Jahrb. Syst.* **5**: 203-238, 239-286, 448-496.

PETER, A. 1885: Über spontane und künstliche Gartenbastarde der Gattung *Hieracium* sect. *Piloselloidea* (Schluss). - *Bot. Jahrb. Syst.* **6**: 111-136.

PETER, R. 1989: Ergänzungen zur Orchideenflora von Rhodos. - *Mitt. Arbeitskreis Heimische Orchid. Baden-Württemberg* **21(2)**: 279-350.

PETEREK, M. 2004: Beiträge zur Orchideenflora von Korfu (Kerkira), Nordwestgriechenland. - *J. Eur. Orch.* **36(4)**: 993-1008.

PETERKEN, G.F. & LLOYD, P.S. 1967: Biological Flora of the British Isles. *Ilex aquifolium* L. - *J. Ecol.* **55(3)**: 841-858.

PETERMANN, J. 1993: Zur Struktur und Dynamik ostmakedonischer Eichenwälder (*Quercion frainetto*). In Mattes, H. & Lienau, C. (eds): Colloquium Nordost-Griechenland 2: pp. 32-36. - Münster: Inst. Geography.

PETERMANN, J. 1997: Vegetation, Struktur und Dynamik ostmakedonischer Eichenwälder. - Münster: Diss., Westf. Wilhelms-Univ., 186 pp. + plates and tables.

PETERMANN, J. 1999: Winterkahle Eichenwälder im Westen der griechischen Rhodopen. Vegetation, Struktur und Dynamik. - *Arb. Inst. Landschaftsökol. Westf. Wilhelms-Univ. Münster* **5**: iv + 151 pp.

PETERSEN, G. 1989: Cytology and systematics of Araceae. - *Nordic J. Bot.* **9(2)**: 119-166.

PETERSEN, G. 1991: Intergeneric hybridization between *Hordeum* and *Secale* (Poaceae). I. Crosses and development of hybrids. - *Nordic J. Bot.* **11(3)**: 253-270.

PETERSEN, G. 1994: Chromosome numbers of the genera of Araceae. - *Aroideana* **16**: 37-46.

PETERSON, A. & AL. 2010: *Gagea bohemica* (Liliaceae), a highly variable monotypic species within *Gagea* sect. *Didymobulbos*. - *Pl. Biosyst.* **144(2)**: 308-322.

PETERSON, D. 1936: *Stellaria*-Studien. Zur Zytologie, Genetik, Ökologie und Systematik der Gattung *Stellaria*, insbesondere der *media*-Gruppe. - *Bot. Not.* **1936(3-4)**: 281-419.

PETERSON, P.M. & AL. 2014: A molecular phylogeny and new subgeneric classification of *Sporobolus* (Poaceae: Chloridoideae: Sporobolinae). - *Taxon* **63(6)**: 1212-1243.

PETERSON, P.M. & AL. 2020: A phylogeny of species near *Agrotis* supporting the recognition of two new genera, *Agrotula* and *Alpagrotis* (Poaceae, Pooideae, Agrostidinae) from Europe. - *PhytoKeys* **167**: 57-82.

PETKOFF, S. 1910: La flore aquatique et algologique de la Macédoine du S.-O. [In Bulgarian with French summary]. - Philippopoli: Imprimerie Chr. G. Danoff, 189 pp. + 4 plates,

PETKOVIĆ, B. & AL. 1985: A new locality of Serbian ramonda (*Ramonda serbica* Panč.) in the canyon of the river Godulja Ibar's left tributary [In Serbo-Croatian with English summary]. - *Bull. Inst. Jard. Bot. Univ. Belgrade, n.s.* **19**: 169-174.

PETKOVIĆ, B. & AL. 1986: Eine neue Fundort serbischen Ramonda (*Ramonda serbica* Panč.) in Schlucht Crna Reka rechts Nebenfluss des Ibar [In Serbo-Croatian with German summary]. - *Bull. Inst. Jard. Bot. Univ. Belgrade, n.s.* **20**: 65-69.

PETKOVIĆ, B. & AL. 1988: A new locality of Serbian Ramonda (*Ramonda serbica* Panč.) on the northeast of Montenegro [In Serbo-Croatian with English summary]. - *Glasn. Republ. Zavoda Zaštitu Prir. Prirodnjačke Zbirke Titogradu* **19**: 5-10.

PETKOVIĆ, B., DELIĆ, G. & TATIĆ, B. 1997: Variation in *Verbascum phoeniceum* (Scrophulariaceae) in Serbia as affected by geological substratum. - *Bocconea* **5(2)**: 647-654.

PETRAK, F. 1910a: Über neue oder wenig bekannte Cirsien aus dem Oriente. - *Oesterr. Bot. Z.* **60**: 351-356, 393-396, 436-441, 459-463.

PETRAK, F. 1910b: Über den Formenkreis des *Cirsium Sintenisii* Freyn. - *Oesterr. Bot. Z.* **60(12)**: 463-469.

PETRAK, F. 1912: Der Formenkreis des *Cirsium eriophorum* (L.) Scop. in Europa. - *Biblioth. Bot.* **18/78**: 92 pp. + 6 plates and 1 map.

PETRAK, F. 1960: Über einige Arten und Bastarde der Gattung *Cirsium*. - *Mitt. Thüring. Bot. Ges.* **2 (1)**: 13-41.

PETRAKIS, P.V., ROUSSIS, V. & ORTIZ, A. 2000: Monoterpenoid diversity in relation to morphology of *Pinus brutia* and *Pinus halepensis* in an East Mediterranean area (Attiki, Greece): implications for pine evolution. - *Edinburgh J. Bot.* **57(3)**: 349-375.

PETROS, N. 2006: Ditiki Rodopi [In Greek]. - *Fisis [Journal of the Hellenic Society for the Protection of Nature]* **114**: 22-27.

PETROU, M. 2015: Sorrow children - the strange *Ophrys* of Tzia (Cyclades, Greece). - *J. Eur. Orch.* **47(2-4)**: 323-337.

PETROU, N., PETROU, M. & GIANNAKOULIAS, M. 2011: Orchids of Greece. - KOAN Publishing House: 320 pp.

PETROV, S. & AL. 1988: Palynomorphological characteristics of representatives of fam. Iridaceae Lindl. found in Bulgaria. In Velčev, V.I. & al. (eds): 100th anniversary of Acad. Nikolaj A. Stojanov, pp. 152-169 + 22 plates. - Sofia: Publishing House of the Bulgarian Academy of Sciences.

PETROVA, A. 2010: First records of *Vincetoxicum nigrum* and *Avena byzantina* in Bulgaria. - *Phytol. Balcan.* **16(1)**: 75-78.

PETROVA, A. & AL. 2005: *Vulpia fasciculata*: a new species for the Bulgarian flora. - *Phytol. Balcan.* **11 (2)**: 133-136.

PETROVA, A. & VENKOVA, D.Y. 2008: *Epipactis exilis* and *E. greuteri* (Orchidaceae) in the Bulgarian flora. - *Phytol. Balcan.* **19(1)**: 69-93.

PETROVA, A. & VLADIMIROV, V. 2007: Recent (1994-2004) taxonomic studies on the Bulgarian flora. - *Bocconea* **21**: 7-25.

PETROVA, A. & VLADIMIROV, V. 2010: Balkan endemics in the Bulgarian flora. - *Phytol. Balcan.* **16(2)**: 293-311.

PETROVA, A. & VLADIMIROV, V. 2018: Recent progress in floristic and taxonomic studies in Bulgaria. - *Bot. Serbica* **42(1)**: 35-69.

PETROVA, A. & VLADIMIROV, V. 2020: Chromosome atlas of the Bulgarian vascular plants. - *Phytol. Balcan.* **26(2)**: 217-427.

PETROVA, A., VLADIMIROV, V. & GEORGIEV, V. 2013: Invasive alien species of vascular plants in Bulgaria. - Sofia: Bulgarian Academy of Sciences, 319 pp.

PETROVA, A.S. 2007: *Centaurea jankae* and *C. trinervia* (Asteraceae): new taxa for the Bulgarian flora. - *Phytol. Balcan.* **13(3)**: 353-358.

PETROVA, A.S. & AL. 2006: *Vulpia fasciculata*: a new species for the Bulgarian flora. - *Phytol. Balcan.* **11(2)**: 133-136.

PETROVA, A.S., GERASIMOVA, I. & VASSILEV, R. 1998: Contribution to the flora of the Eastern Rhodope Mountains, Bulgaria [In Bulgarian with English summary]. - *Hist. Nat. Bulg.* **9**: 115-127.

PETROVA, A.S., GERASIMOVA, I. & VENKOVA, D.Y. 1999: New data for the flora of Eastern Rhodope Mountains, Bulgaria [In Bulgarian with English summary]. - *Hist. Nat. Bulg.* **10**: 117-123.

PETROVA, A.S. & VENKOVA, D.Y. 2006a: *Epipactis leptochila* (Orchidaceae): a new species for the Bulgarian flora. - *Phytol. Balcan.* **12(1)**: 75-78.

PETROVA, A.S. & VENKOVA, D.Y. 2006b: *Epipactis pontica* (Orchidaceae): a new species for the Bulgarian flora. - *Phytol. Balcan.* **12(2)**: 249-253.

PETROVA, A.S. & VLADIMIROV, V. 2009: Two alien species of *Bidens* (Asteraceae) new to the Bulgarian flora. - *Phytol. Balcan.* **15(3)**: 367-371.

PETROVA, A.V. 1969: On the pollen of three species from the *Linum flavum* group in Bulgaria. - *Dokl. Bulg. Akad. Nauk* **22(10)**: 1173-1176.

PETROVA, A.V. 1972: [Chromosome numbers in *Linum* from Bulgaria]. In Löve, Á. (ed.): IOPB chromosome number reports XXXV. - *Taxon* **21(1)**: 164.

PETROVA, A.V. 1973a: Taxonomic study of the wild species of the genus *Linum* L. (flax) in Bulgaria [In Bulgarian with Russian and English summaries]. - *Izv. Bot. Inst. (Sofia)* **23**: 15-76.

PETROVA, A.V. 1973b: Taxonomic study of the wild species from genus *Linum* L. (flax) in Bulgaria. II. (Palynological and karyological characteristics) [In Bulgarian with Russian and English summaries]. - *Izv. Bot. Inst. (Sofia)* **24**: 5-29.

PETROVA, A.V. 1975: [Chromosome numbers from Bulgaria]. In Löve, Á. (ed.): IOPB chromosome number reports XLIX. - *Taxon* **24(4)**: 510-511.

PETROVA, A.V. 1996a: Reports (415-434). In Kamari, G., Felber, F. & Garbari, F. (eds): Mediterranean chromosome number reports - 5. - *Fl. Medit.* **5**: 279-288.

PETROVA, A.V. 1996b: Karyological investigations in Bulgaria during the last decade (1983-1993). - *Bocconea* **5(1)**: 21-31.

PETROVA, A.V. 1996c: Josef Velenovský - his biography, scientific activity and contribution to the Bulgarian botany. - *Phytol. Balcan.* **2(2)**: 3-24.

PETROVA, A.V. 2000: Karyological study of some species of *Sesleria* (Poaceae) growing in Bulgaria. - *Bot. Chron. (Patras)* **13**: 133-140.

PETROVA, A.V. 2001: The flora of Bulgaria, past present and future. In Özhatay, N. (ed.): Proceedings of the Second Balkan Botanical Congress, 1: 47-52. - Istanbul.

PETROVA, A.V. 2002: Biosystematic and floristic studies in Bulgaria for the period 1993-2001. In Temniskova, D. (ed.): Proceedings of the sixth national conference of botany, Sofia, June 18-20, 2001, pp. 27-46. - Sofia: Sofia Univ. "St. Kliment. Ohridski" Press.

PETROVA, A.V. 2006: *Geranium aristatum* (Geraniaceae): a new species for the Bulgarian flora. - *Phytol. Balcan.* **12(2)**: 215-220.

PETROVA, A.V. & KOŽUHAROV, S.I. 1982a: Cytotaxonomic study of genus *Trifolium* L. in Bulgaria. I. - *Fitologiya* **19**: 3-23.

PETROVA, A.V. & KOŽUHAROV, S.I. 1982b: Cytotaxonomic study of genus *Trifolium* L. in Bulgaria. II. - *Fitologiya* **20**: 20-41.

PETROVA, A.V. & KOŽUHAROV, S.I. 1983: Cytotaxonomic study of genus *Agropyron* Gaertn. in Bulgaria [In Bulgarian with English abstract]. In Velčev, V.I. & al. (eds): Third national conference of botany, pp. 215-222. - Sofia: Publishing House of the Bulgarian Academy of Sciences.

PETROVA, A.V. & KOŽUHAROV, S.I. 1989: Karyological characteristic of the Bulgarian rye species (*Secale* L.). In: Fourth national conference on cytogenetics, pp. 189-191. - Vratza, Bulgaria.

PETROVA, A.V., KOŽUHAROV, S.I. & EHRENDORFER, F. 1993: A new species in the *Bromus riparius* group. In Petrova, A.V. (ed.): Abstracts from the VII OPTIMA meeting in Borovetz, Bulgaria, 18-30 July 1993, p. 186.

PETROVA, A.V., KOŽUHAROV, S.I. & EHRENDORFER, F. 1997: Karyosystematic notes on *Bromus* (Gramineae), and a new species of the *B. riparius* polyploid complex from Bulgaria. - *Bocconea* **5(2)**: 775-780.

PETROVA, A.V. & STANIMIROVA, P. 2001a: Reports (1248-1253). In Kamari, G., Blanché, C. & Garbari, F. (eds): Mediterranean chromosome number reports. - *Fl. Medit.* **11**: 460-466.

PETROVA, A.V. & STANIMIROVA, P. 2001b: Karyological study of some *Geranium* species growing in Bulgaria. In Raimondo, F.M. (ed.): Abstracts from the X OPTIMA meeting, Palermo, 13-19 September 2001, p. 131.

PETROVA, A.V. & STANIMIROVA, P. 2003a: Reports (1288-1294). In Kamari, G., Blanché, C. & Garbari, F. (eds): Mediterranean chromosome number reports - 12. - *Fl. Medit.* **12**: 454-460.

PETROVA, A.V. & STANIMIROVA, P. 2003b: Karyological study of some *Geranium* (Geraniaceae) species growing in Bulgaria. - *Bocconea* **16(2)**: 675-682.

PETROVA, A.V. & STOYANOVA, K. 1997: Reports (783-802). In Kamari, G., Felber, F. & Garbari, F. (eds): Mediterranean chromosome number reports - 7. - *Fl. Medit.* **7**: 204-213.

PETROVA, A.V. & STOYANOVA, K. 1998: Reports (941-960). In Kamari, G., Felber, F. & Garbari, F. (eds): Mediterranean chromosome number reports - 8. - *Fl. Medit.* **8**: 251-262.

PETROVA, A.V. & VLADIMIROV, V. (EDS) 2009: Red List of Bulgarian vascular plants. - *Phytol. Balcan.* **15(1)**: 63-94.

PETROVA, A.V., ZIELIŃSKI, J. & NATCHEVA, R. 2006: Reports (1584-1603). In Kamari, G., Blanché, C. & Garbari, F. (eds): Mediterranean chromosome number reports - 16. - *Fl. Medit.* **16**: 431-442.

PETROVA, A.V., ZIELIŃSKI, J. & NATCHEVA, R. 2007: Chromosome numbers of woody plants from Bulgaria. - *Phytol. Balcan.* **13(3)**: 371-378.

PETROVA, A.V. (ED.) 1993: Abstracts from the VII OPTIMA meeting in Borovetz, Bulgaria, 18-30 July 1993. - Sofia: Bulgarian Academy of Sciences, Institute of Botany, 186 pp.

PETROVA, A.V. (ED.) 2006: Atlas of Bulgarian endemic plants. - Sofia: Gea-Libris Publishing House, 399 pp.

PETROVIĆ, D. & STEŠEVIĆ, D. 2011: Shift of the western boundary of the distribution area of *Micromeria cristata* (Hampe) Griseb. and *Steptorhamphus tuberosus* (Jacq.) Grossh. - *Acta Bot. Croat.* **70(2)**: 259-267.

PETTERSSON, B. 1962: Om *Ceterach officinarum* och dess ekologi [In Swedish]. - *Bot. Not.* **115(2)**: 237-240.

PFEIFER, M. & AL. 2009: Phylogeography and genetic structure of the orchid *Himantoglossum hircinum* (L.) Spreng. across its European central-marginal gradient. - *J. Biogeogr.* **36(12)**: 2353-2365.

PFEIFFER, N.E. 1922: Monograph of the Isoetaceae. - *Ann. Missouri Bot. Gard.* **9(2)**: 79-232 + 19 plates.

PFITZENMEYER, C.D.C. 1962: Biological Flora of the British Isles. *Arrhenatherum elatius* (L.) J. & C. Presl (*A. avenaceum* Beauv.). - *J. Ecol.* **50(1)**: 235-245.

PFOSSER, M. & SPETA, F. 2001a: From *Scilla* to *Carybdis* - is our voyage safer now? In Raimondo, F. M. (ed.): Abstracts from the X OPTIMA meeting, Palermo, 13-19 September 2001, p. 36.

PFOSSER, M. & SPETA, F. 2001b: Bufadienolide und DNA-Sequenzen: Über Zusammenhalt und Aufteilung der Urgineoideae (Hyacinthaceae). - *Stapfia* **75**: 177-250.

PFOSSER, M. & SPETA, F. 2004: From *Scilla* to *Carybdis* - is our voyage safer now? - *Pl. Syst. Evol.* **246(3-4)**: 245-263.

PHEREOS, C. & YANNITSAROS, A. 2001: The distribution and ecology of *Lilium candidum* L. in Monemvasia area. In Kamari, Psaras & Constantinidis (eds): Proceedings of the 8th scientific congress of the Greek Botanical Society, October 5-8, 2000, pp. 448-453 [In Greek with English abstract].

PHEREOS, C., YANNITSAROS, A. & SCHOLZ, H. 2004: *Cenchrus ciliaris* L. (Gramineae) in Greece. - *Bot. Chron. (Patras)* **16**: 53-60.

PHILIPPSON, A. 1890: Der Wald in Griechenland. - *Naturwiss. Wochenschr.* **5**: 334-336.

PHILIPPSON, A. 1894: Über das Vorkommen der Rosskastanie und der Buche in Nordgriechenland. - *Naturwiss. Wochenschr.* **9(35)**: 421-423.

PHILIPPSON, A. 1895a: Reisen und Forschungen in Nordgriechenland. - *Z. Ges. Erdk. Berlin* **30**: 135-225, 417-498 + 5 plates.

PHILIPPSON, A. 1895b: Zur Vegetationskarte des Peloponnes. - *Petermann's Mitth. Justus Pethers' Geogr. Anst.* **1895**: 273-279 + 1 plate.

PHILIPPSON, A. 1950-59: Die griechischen Landschaften (unter Mitwirkung von Lehmann, H. & Kirsten, E.). - Frankfurt am Main: 4 Bände.
PHILIPSON, W.R. 1937: A revision of the British species of the genus *Agrostis* Linn. - *J. Linn. Soc., Bot.* **51(338)**: 73-151 + 21 plates.
PHILLIPS, M.E. 1954: Biological Flora of the British Isles. *Eriophorum angustifolium* Roth (*E. Polystachion* L.). - *J. Ecol.* **42(2)**: 612-622.
PHILLIPS, R. & RIX, E.M. 1989: Bulbs. - London: Pan Books Ltd, 255 pp. incl. many colour photographs.
PHITOS, D. 1960a: Phytogeographical observations on the massif Tymphrestos-Oxia [In Greek]. - *Ann. Forest.?* **25**: 1099-1106 [n.v.].
PHITOS, D. 1960b: Pflanzengeographische Untersuchung der zentrale Euboea [In Greek with German summary]. - Athen: Thesis, 106 pp.
PHITOS, D. 1962: Beitrag zur Kenntnis der Flora von Nord-Pindos. - *Mitt. Bot. Staatssamml. München* **4(3)**: 285-293.
PHITOS, D. 1963a: Beiträge zur Kenntnis der *Campanula rupestris*-Gruppe. - *Phyton (Horn)* **10(1-2)**: 124-127.
PHITOS, D. 1963b: Eine neue Art der Gattung *Campanula* aus der Ägäis. - *Mitt. Bot. Staatssamml. München* **5(1)**: 121-124.
PHITOS, D. 1964a: Eine neue Art der Gattung *Centaurea* aus der Ägäis. - *Ann. Naturhist. Mus. Wien* **67**: 165-167.
PHITOS, D. 1964b: Trilokuläre *Campanula*-Arten der Ägäis. - *Oesterr. Bot. Z.* **111(2-3)**: 208-230.
PHITOS, D. 1964c: Beiträge zur Flora Hellenica. - *Verh. Zool.-Bot. Ges. Wien* **103-104**: 223-230.
PHITOS, D. 1964d: Beiträge zur Kenntnis der südägäischen *Campanula*-Arten. - *Ber. Deutsch. Bot. Ges.* **77(2)**: 49-54.
PHITOS, D. 1965a: Die quinqueokulären *Campanula*-Arten. - *Oesterr. Bot. Z.* **112(4)**: 449-498.
PHITOS, D. 1965b: Wild flowers of Greece. - Athens: Filodasiki Enosis Athinon, 87 pp. [separate edition in Greek].
PHITOS, D. 1966a: Drei neue Arten aus Griechenland. - *Oesterr. Bot. Z.* **113(2)**: 271-272.
PHITOS, D. 1966b: Die Gattung *Symphyandra* in der Ägäis. - *Ber. Deutsch. Bot. Ges.* **79(5)**: 246-249.
PHITOS, D. 1967: Florula Sporadum. - *Phyton (Horn)* **12(1-4)**: 102-149.
PHITOS, D. 1969: *Campanula pinatzi*, ena neon idos tis ellinikis chloridos [In Greek]. - *Vouno [Journal of the Hellenic Alpine Club]* **1969**: 2-3.
PHITOS, D. 1970a: Die Gattung *Aubrieta* in Griechenland. - *Candollea* **25(1)**: 69-87.
PHITOS, D. 1970b: Zur Polyploidie in der Gattung *Centaurea* L. Sektion *Acrocentron* (Compositae). - *Ber. Deutsch. Bot. Ges.* **83(2)**: 69-73.
PHITOS, D. 1975: Taxonomic and floristic research in Greece during the last decade 1961-1971. - *Mem. Soc. Brot.* **24(2)**: 579-597.
PHITOS, D. 1980: Contribution to the cytotaxonomic study of the genus *Ornithogalum* L. in Greece [In Greek with French and English summaries]. - *Biol. Gallo-Hellen.* **9(1)**: 147-156.
PHITOS, D. 1981: The genus *Bolanthus* (Caryophyllaceae) in Greece. - *Bot. Chron. (Patras)* **1(1)**: 35-45.
PHITOS, D. 1982: *Silene greuteri* (Caryophyllaceae), species nova. - *Bot. Chron. (Patras)* **2(1)**: 53-54.
PHITOS, D. 1988a: Chromosome numbers in some species of the Greek flora [In Greek with English abstract]. - *Bot. Chron. (Patras)* **8(1-2)**: 45-50.
PHITOS, D. 1988b: The Schinias wood - a lost vision [In Greek with English summary]. - *Fisis [Journal of the Hellenic Society for the Protection of Nature]* **43**: 7-12, 36.
PHITOS, D. 1990: Ioannis Politis (1886-1968) [In Greek with English abstract]. - *Bot. Chron. (Patras)* **9 (1-2)**: 5-12.
PHITOS, D. 1992: *Centaurea sibthorpii* Halácsy (Compositae-Cardueae) - a disputable taxon. - *Phyton (Horn)* **32(2)**: 277-281 + 1 plate.
PHITOS, D. 1995: Vivlio Apiloumenon Idon (Red Data Book) tis ellinikis chloridas. In: Proceedings, 5th scientific conference, Hellenic Botanical Society, Delphi 21-23 October 1994, pp. 55-59 [In Greek]. - Thessaloniki: Univ. Studio Press.
PHITOS, D. 2004: The *Pinus pinea* forest at Shoinias, Marathonas and the Megalos Elos [In Greek with English summary]. - *Bot. Chron. (Patras)* **16**: 19-22.
PHITOS, D. 2010: *Lilium candidum* L. in Greece from a phytogeographical, historical and religious aspect [In Greek with English abstract]. - *Bot. Chron. (Patras)* **20**: 99-128.
PHITOS, D. 2016a: The medicinal plant *Helleborus* (Ranunculaceae) and Phokian Antikyra. [In Greek with English Abstract]. - *Bot. Chron. (Patras)* **21**: 21-42.
PHITOS, D. 2016b: The sectio *Quinqueloculares* of the genus *Campanula* (Campanulaceae) and its polymorphism - Taxonomic review of the last decades. - *Bot. Chron. (Patras)* **21**: 91-102.
PHITOS, D. 2021: Reinstating *Campanula nisyria* as a distinct species of sect. *Quinqueloculares* (Campanulaceae). - *Fl. Medit.* **31**: 101-107.
PHITOS, D. & AL. 2015: The mountain Aenos on Cephalonia island. History - physiography - biodiversity. [English edition]. - Cephalonia: 318 pp.
PHITOS, D. & AL. (EDS) 1996: The red data book of rare and threatened plants of Greece. - Athens: WWF, xlvii + 527 pp.
PHITOS, D. & ARTELARI, R. 1981: A new species of *Limonium* from Kefallinia (Greece). - *Bot. Chron. (Patras)* **1(1)**: 18-21.
PHITOS, D. & CONSTANTINIDIS, TH. 1993: A new species of *Centaurea* sect. *Phalolepis* from Greece. - *Fl. Medit.* **3**: 273-275.
PHITOS, D., CONSTANTINIDIS, TH. & KAMARI, G. (EDS) 2009a: The Red Data Book of rare and threatened plants of Greece, vol. 1 A-D [In Greek]. - Patras: Elliniki Votaniki Eteria, xxvi + 405 pp.
PHITOS, D., CONSTANTINIDIS, TH. & KAMARI, G. (EDS) 2009b: The Red Data Book of rare and threatened plants of Greece, vol. 2 E-Z [In Greek]. - Patras: Elliniki Votaniki Eteria, 413 pp.
PHITOS, D. & DAMBOLDT, J. 1970: Beiträge zur Flora Ionica. I. Die *Stachys swainsonii*-Gruppe (Labiatae). - *Ber. Deutsch. Bot. Ges.* **82(9)**: 595-601.
PHITOS, D. & DAMBOLDT, J. 1971: Beiträge zur Flora Ionica. III. Cytotaxonomische Bemerkungen zu einigen griechischen Compositen. - *Ann. Naturhist. Mus. Wien* **75**: 157-162.
PHITOS, D. & DAMBOLDT, J. 1976: Ein Beitrag zur Kenntnis der Gattung *Centaurea* L. in Griechenland. In Dafis, S. & Landolt, E. (eds): Zur Vegetation und Flora von Griechenland, Band 2. - *Veröff. Geobot. Inst. E.T.H. Stiftung Rübel Zürich* **56**: 183-189.
PHITOS, D. & DAMBOLDT, J. 1985: Die Flora der Insel Kefallinia (Griechenland) [In Greek with German introduction]. - *Bot. Chron. (Patras)* **5(1-2)**: 1-204 + 2 maps.
PHITOS, D. & GEORGIADIS, TH. 1981: A propos du groupe de *Centaurea achaia* Boiss. et Heldr. de la section *Acrocentron* (Cass.) DC. - *Bot. Chron. (Patras)* **1(2)**: 99-114.
PHITOS, D. & GREUTER, W. 1993: *Saponaria jagelii*, a new species from the island of Elafonisos (Peloponnisos, Greece). - *Fl. Medit.* **3**: 277-278.
PHITOS, D. & GREUTER, W. (EDS) 1991: Proceedings of the VI OPTIMA meeting, Delphi, 10-16. Sept. 1989. - *Bot. Chron. (Patras)* **10**: 987 pp.
PHITOS, D. & KAMARI, G. 1973: Cytologische Untersuchungen an einigen griechischen Blütenpflanzen (*Centaurea*, *Aethionema*). - *Biol. Gallo-Hellen.* **4(2)**: 147-151.
PHITOS, D. & KAMARI, G. 1974: Zytotaxonomische Beiträge zur Flora von Kreta. I. - *Bot. Not.* **127(3)**: 302-308.
PHITOS, D. & KAMARI, G. 1983: Chromosomes numbers in *Crocus* species (Iridaceae) from Greece [In Greek with English abstract]. - *Bot. Chron. (Patras)* **3(1-2)**: 30-32.
PHITOS, D. & KAMARI, G. 1984: Cytotaxonomical contributions on the flora of Crete. II. - *Candollea* **39(2)**: 447-451.
PHITOS, D. & KAMARI, G. 1988: Contribution to the cytotaxonomy of the genus *Campanula* (Campanulaceae) in Greece. - *Willdenowia* **17(1-2)**: 103-105.
PHITOS, D. & KAMARI, G. 1990: *Gymnospermium altaicum* subsp. *odessanum*

(Berberidaceae) – an unusual geographical distribution. - *Razprave IV. Razreda Sazu* **31(14)**: 201-207.
PHITOS, D. & KAMARI, G. 1992: *Bufonia euboica* (Caryophyllaceae), a new species from Greece. - *Willdenowia* **22(1-2)**: 81-83.
PHITOS, D., KAMARI, G. & ATHANASIOU, K. 1990: Chromosome numbers in some species of the Greek flora. - *Bot. Chron. (Patras)* **9(1-2)**: 41-47.
PHITOS, D., KAMARI, G. & KATSOUNI, N. 2016: Two rare taxa of the genus *Fumaria* (Fumariaceae) for the Greek flora. - *Bot. Chron. (Patras)* **21**: 83-90.
PHITOS, D., KATSOUNI, N. & KAMARI, G. 2004: Some interesting species from the flora of Kefallinia and Ithaki (Ionian islands, Greece) [In Greek with English summary]. - *Bot. Chron. (Patras)* **16**: 97-101.
PHITOS, D. & SNOGERUP, S. 1973: A new species of *Aethionema* from Skiros, Greece. - *Bot. Not.* **126 (2)**: 142-145.
PHITOS, D. & STRID, A. 1994: *Arenaria leucadia*, a new species from the island of Levkas (W Greece). - *Bot. Chron. (Patras)* **11**: 55-57.
PHITOS, D., TURLAND, N.J. & BERGMEIER, E. 2011: A new subspecies of *Bolanthus creutzburgii* (Caryophyllaceae) from coastal SW Kriti (Greece). - *Fl. Medit.* **21**: 317-323.
PHITOS, D. & TZANOUDAKIS, D. 1981: A new species of *Allium* from Euboea (Greece). - *Bot. Chron. (Patras)* **1(1)**: 11-13.
PHITOS, D. (ED.) 1989: Abstracts. Sixth OPTIMA meeting, Delphi, 10-16 Sept. 1989. - Patras: 157 pp. (mimeographed).
PHOUPHAS, CH. 1940a: La botanique en Grèce. - *Actes Inst. Bot. Univ. Athènes* **1**: 15-25.
PHOUPHAS, CH. 1940b: Dr. D. Zaganiaris [obituary]. - *Actes Inst. Bot. Univ. Athènes* **1**: 235-236.
PHOUPHAS, CH. 1968: Contribution à l'étude de la flore de l'île "Spetsai". - *Bull. Soc. Bot. France* **115(9)**: 583-590.
PICCI, V. 1970: Ricerche embriologiche sul genere *Thalictrum*. II. Embriologia e cariologia di *Thalictrum lucidum* L. e di *Thalictrum minus* L. subsp. *minus*. - *Giorn. Bot. Ital., n.s.* **103(6)**: 475-483.
PICCI, V. 1972: *Catananche lutea* L., specie nuova per la Sardegna. Brevi note sulla sua ecologia. - *Ann. Bot. (Rome)* **30**: 171-176.
PICHI SERMOLLI, R.E.G. 1954: Report of the special committee for Pteridophyta and Phanerogamae [conservation of generic names]. - *Taxon* **3(8)**: 240-243.
PICHI SERMOLLI, R.E.G. 1976: Report of the committee for Pteridophyta [on the citation of *Ceterach*]. - *Taxon* **25(1)**: 176.
PICHI SERMOLLI, R.E.G. 1979: A survey of the pteridological flora of the Mediterranean region. - *Webbia* **34(1)**: 175-242.
PICHI SERMOLLI, R.E.G. 1983a: Fragmenta Pteridologiae - VIII. - *Webbia* **37(1)**: 111-140.
PICHI SERMOLLI, R.E.G. 1983b: Notes on Adanson's fern genera "Ceterac" and "Scolopendrium". - *Webbia* **37(1)**: 159-169.
PICHI SERMOLLI, R.E.G. 1985: The fern genus *Cosentinia*. - *Webbia* **39(1)**: 179-189.
PICHI SERMOLLI, R.E.G. 1989: Again on the typification of the generic name *Notholaena* R. Brown. - *Webbia* **43(2)**: 301-310.
PICHI SERMOLLI, R.E.G. 1990a: Speciazione e distribuzione geografica nelle Pteridophyta. - *Anales Jard. Bot. Madrid* **46(2)**: 489-518.
PICHI SERMOLLI, R.E.G. 1990b: Le pteridofite europee: La loro tassonomia e nomenclatura oggi. In Rita, J. (ed.): Taxonomía, biogeografía y conservación de Pteridofítos, pp. 11-27. - Palma de Mallorca.
PICHI SERMOLLI, R.E.G. & AL. 1991: Considerazioni sull'affinitá ed origine della flora pteridologica della Regione Mediterranea. - *Acta Bot. Malac.* **16(1)**: 235-280.
PICHI SERMOLLI, R.E.G. & CHIARINO-MASPES, V. 1963: Ricerche geobotaniche su "Notholaena Marantae" in Italia. - *Webbia* **17(2)**: 407-451.
PICHI SERMOLLI, R.E.G. & MOGGI, G. 1975: Report on the progress of floristic research in Italy since 1961. - *Mem. Soc. Brot.* **24(2)**: 623-746.
PICHON, M. 1951: Classification des Apocynacées: XXII, Les espèces du genre *Vinca*. - *Bull. Mus. Natl. Hist. Nat., Sér. 2* **23(4)**: 439-444.
PICKERSGILL, B. & LOCK, J.M. (EDS) 1996: Advances in legume systematics. Part 8: Legumes of economic importance. - Kew: Royal Botanic Gardens, viii + 143 pp.
PIEHL, M.A. 1965: The natural history and taxonomy of *Comandra* (Santalaceae). - *Mem. Torrey Bot. Club* **22(1)**: 1-97.
PIERI, M.T. 1808: Della Corciresi Flora. Centuria prima. - Corcirae: Nella Tipografia Nazionale, appresso Dionisi Sarandopulo, 175 pp.
PIERI, M.T. 1814: Della Corciresi Flora. Centurie prima, seconda, e terza, ossia steria di Piante trecento, appartenenti al suolo dell'Isola di Corfu. - Corcirae: Nella Stamperia del Governo, 141 pp.
PIERI, M.T. 1824: Flora Corcirensis centuriae prima et secunda, sive enumeratio ducentarum plantarum quas in insula Corcirae invenit. - Corcirae: Typis Publicis, 78 pp. [cf. *J. Bot. (London)* 31: 355-357, 1893].
PIETRERA, D. 1968: Contribution à l'étude cytologique des *Galium* de Grèce. - Marseille: Thèse, Univ. de Marseille [n.v.].
PIETRO, R. DI, BERGMEIER, E. & DIMOPOULOS, P. 2005: *Sesleria autumnalis* (Gramineae) a new addition to the flora of Greece. - *Fl. Medit.* **15**: 343-349.
PIFKÓ, D. 2004: Plant types of Sándor Jávorka in the Hungarian Natural History Museum in Budapest. IV. - *Ann. Hist.-Nat. Mus. Natl. Hung.* **96**: 63-71.
PIFKÓ, D. 2015: Index of scientific names of *Chamaecytisus* (Leguminosae) taxa. - *Studia Bot. Hung.* **46(2)**: 175-203.
PIFKÓ, D. 2018: Morphological studies on *Chamaecytisus lasiosemium* (Boiss.) Pifkó and other Balkan *Chamaecytisus* species. - *Bot. Serbica* **42 (suppl. 1)**: 49.
PIFKÓ, D. & BARINA, Z. 2016: The new *Chamaecytisus* species (Leguminosae-Papilionoideae) from Albania, with an overview on the *Ch. ratisbonensis* and *Ch. eriocarpus* species groups. - *Studia Bot. Hung.* **47(1)**: 163-178.
PIGNATTI, E. & PIGNATTI, S. 1985: Mediterranean type vegetation of SW. Australia, Chile and the Mediterranean basin, a comparison. - *Ann. Bot. (Rome)* **43**: 227-243.
PIGNATTI, S. 1971: Studi sui *Limonium*. VIII. In Heywood, V.H. (ed.): Notulae systematicae ad Floram Europaeam spectantes, no. 11. - *Bot. J. Linn. Soc.* **64(4)**: 361-370.
PIGNATTI, S. 1973: Note critiche sulla Flora d'Italia. I. Appunti miscellanei. - *Giorn. Bot. Ital., n.s.* **107 (5)**: 207-221.
PIGNATTI, S. 1974: Note critiche sulla flora d'Italia. II. Il gruppo di *Minuartia verna*. - *Giorn. Bot. Ital., n. s.* **108(3-4)**: 95-104.
PIGNATTI, S. 1977: Note critiche sulla Flora d'Italia. V. Nuovi appunti miscellanei. - *Giorn. Bot. Ital., n. s.* **111(1-2)**: 45-61.
PIGNATTI, S. 1978: Evolutionary trends in Mediterranean flora and vegetation. - *Vegetatio* **37(3)**: 175-185.
PIGNATTI, S. 1979a: Plant geographical and morphological evidences in the evolution of the Mediterranean flora (with particular reference to the Italian representatives). - *Webbia* **34(1)**: 243-255.
PIGNATTI, S. 1979b: Note critiche sulla Flora d'Italia. VI. Ultimi appunti miscellanei. - *Giorn. Bot. Ital., n.s.* **113(5-6)**: 359-368.
PIGNATTI, S. 1980: Il gruppo di *Acer opalus* auct. - *Nat. Montagna* **27(3)**: 189-192.
PIGNATTI, S. 1982a: Flora d'Italia. Vol. 1 (Introduzione, Pteridophyta - Leguminosae), 790 pp. - Bologna: Edagricole.
PIGNATTI, S. 1982b: Flora d'Italia. Vol. 2 (Oxalidaceae - Campanulaceae), 732 pp. - Bologna: Edagricole.
PIGNATTI, S. 1982c: Flora d'Italia. Vol. 3 (Compositae e Monocotyledoneae), 780 pp. - Bologna: Edagricole.
PIGNATTI, S. 1982d: New species of *Limonium* from Italy and Tunesia. - *Webbia* **36(1)**: 47-56.
PIGNATTI, S. 2017a: Flora d'Italia, seconda edizione. Vol. 1. - Bologna: Edagricole, xlvii + 1064 pp.
PIGNATTI, S. 2017b: Flora d'Italia, seconda edizione. Vol. 2. - Bologna: Edagricole, xviii + 1178 pp.

PIGNATTI, S. 2018: Flora d'Italia, seconda edizione. Vol. 3. - Bologna: Edagricole, xix + 1288 pp.

PIGNATTI, S. 2019: Flora d'Italia, seconda edizione. Vol. 4. - Bologna: Edagricole, cxlv + 1054 pp. + Flora d'Italia digitale.

PIGNOTTI, L. 2003: Reports (1366-1367). In: Kamari, G., Blanché, C. & Garbari, F. (eds): Mediterranean chromosome number reports - 13. - *Fl. Medit.* **13**: 392-394.

PIGNOTTI, L. & FIORINI, G. 1998: Reports (899-904). In Kamari, G., Felber, F. & Garbari, F. (eds): Mediterranean chromosome number reports - 8. - *Fl. Medit.* **8**: 214-221.

PIGOTT, A. 1986: Genome analysis and possible combinations in the *Dryopteris filix-mas* group. - *Pteridologist* **1(3)**: 99-104.

PIGOTT, C.D. 1955: Biological Flora of the British Isles. *Thymus* L. - *J. Ecol.* **43(1)**: 365-387.

PIGOTT, C.D. 1991: Biological Flora of the British Isles. *Tilia cordata* Miller (*T. europaea* L. pro parte, *T. parvifolia* Ehrh. ex Hoffm., *T. sylvestris* Desf., *T. foeminea folio minore* Bauhin). - *J. Ecol.* **79(4)**: 1147-1207.

PIGOTT, C.D. 1997: (1295-1296) Two proposals to maintain the names *Tilia cordata* and *T. platyphyllos* (Tiliaceae) in their current use. - *Taxon* **46 (2)**: 351-353. [both approved, cf. Taxon 48(2): 368, 1999].

PIGOTT, C.D. 2002: A review of chromosome numbers in the genus *Tilia* (Tiliaceae). - *Edinburgh J. Bot.* **59(2)**: 239-246.

PIGOTT, C.D. 2012: Lime-trees and basswoods. A biological monograph of the genus *Tilia*. - Cambridge: Cambridge Univ. Press, xiii + 395 pp.

PIGOTT, C.D. & SELL, P.D. 1995: Nomenclature of the European species of *Tilia*. I. *Tilia europaea*. - *Kew Bull.* **50(1)**: 135-139.

PIIRAINEN, M. 1991: Flora Nordica notes. 1. *Salicornia* (Chenopodiaceae) in northern Europe: Typification and taxonomic notes. - *Ann. Bot. Fenn.* **28(1)**: 81-85.

PIIRAINEN, M., LIEBISCH, O. & KADEREIT, G. 2017: Phylogeny, biogeography, systematics and taxonomy of Salicornioideae (Amaranthaceae/Chenopodiaceae) – A cosmopolitan, highly specialized hygrohalophyte lineage dating back to the Oligocene. - *Taxon* **66(1)**: 109-132.

PILGER, R. 1916: Die Taxales. - *Mitt. Deutsch. Dendrol. Ges.* **25**: 1-28 + 8 plates.

PILGER, R. 1922a: Über die Formen von *Plantago major* L. - *Repert. Spec. Nov. Regni Veg.* **18**: 257-283.

PILGER, R. 1922b: Beiträge zur Kenntnis der Gattung *Plantago*. I. Die altweltlichen Arten der Sektion *Leucopsyllium* Decne. - *Repert. Spec. Nov. Regni Veg.* **18**: 449-475.

PILGER, R. 1923: Beiträge zur Kenntnis der Gattung *Plantago*. - *Repert. Spec. Nov. Regni Veg.* **19**: 105-112, 114-119.

PILGER, R. 1925: Beiträge zur Kenntnis der Gattung *Plantago*. VI. *Plantago gentianoides* Sibth. & Sm. - *Repert. Spec. Nov. Regni Veg.* **21**: 97-102.

PILGER, R. 1926: Beiträge zur Kenntnis der Gattung *Plantago*. VII. Sektion *Oreades* Decne. - *Repert. Spec. Nov. Regni Veg.* **23**: 241-270.

PILGER, R. 1928: Beiträge zur Kenntnis der Gattung *Plantago*. VIII. Weitere Bemerkungen über die altweltlichen Arten der Sektion *Leucopsyllium* Decne. - *Repert. Spec. Nov. Regni Veg.* **24**: 317-322.

PILGER, R. 1930: *Plantago coronopus* und verwandte Arten. - *Repert. Spec. Nov. Regni Veg.* **28**: 262-322.

PILGER, R. 1931: Die Gattung *Juniperus* L. - *Mitt. Deutsch. Dendrol. Ges.* **43**: 255-269 + 4 plates.

PILGER, R. 1933: Beiträge zur Kenntnis der Gattung *Plantago*. IX. Bemerkungen zu *Plantago maritima* L. und Verwandten. - *Repert. Spec. Nov. Regni Veg.* **34**: 147-166.

PILOUS, V. 1982: Za flórou thesalského Olympu. - *Skalničky* **1982(4)**: 111-118.

PILS, G. 1982: Über die Aussagemöglichkeiten verschiedener Methoden zum Studium von Systematik und Phylogenie der Gattung *Festuca* L. (Poaceae). - *Stapfia* **10**: 71-80.

PILS, G. 2016: Illustrated Flora of Albania. - Eigenverlag: 576 pp.

PIMENOV, M.G. 1979: De generis *Seseli* L. notulae systmaticae. II. Adumbratio specierum florae URSS [In Russian]. - *Novosti Sist. Vysš. Rast.* **15**: 188-200.

PIMENOV, M.G. 1992: *Ormosolenia* restored. - *Edinburgh J. Bot.* **49(2)**: 219-223.

PIMENOV, M.G. & AL. 1993: Computer database on the karyotype of the world Umbelliferae. - *Bot. Zhurn. (Moscow & St. Petersburg)* **78(10)**: 65-72.

PIMENOV, M.G., ALEXEEVA, T.V. & KLJUYKOV, E.V. 1998: Reports (905-935). In Kamari, G., Felber, F. & Garbari, F. (eds): Mediterranean chromosome number reports - 8. - *Fl. Medit.* **8**: 221-245.

PIMENOV, M.G. & JACQUEMOUD, F. 2020: Nomenclatural types of the Umbelliferae in P.E. Boissier's oriental herbarium (G-BOIS). - *Phytotaxa* **466(1)**: 1-91.

PIMENOV, M.G. & KLJUYKOV, E.V. 1996: *Tamamschjanella*, a new genus of the Umbelliferae. [In Russian]. - *Bot. Zhurn. (Moscow & Leningrad)* **81(8)**: 74-79.

PIMENOV, M.G., KLJUYKOV, E.V. & OSTROUMOVA, T.A. 2007: Critical taxonomic analysis of *Dichoropetalum*, *Johrenia*, *Zeravschania* and related genera of Umbelliferae-Apioideae-Peucedaneae. - *Willdenowia* **37(2)**: 465-502.

PIMENOV, M.G. & KLUYKOV, E.V. 1980: Notes on *Muretia* and *Stefanoffia*. - *Notes Roy. Bot. Gard. Edinburgh* **38(2)**: 267-272.

PIMENOV, M.G. & KLUYKOV, E.V. 2010: Two new species of *Seseli* (Umbelliferae) from Turkey. - *Fl. Medit.* **20**: 19-27.

PIMENOV, M.G. & LAVROVA, T.V. 1994: Second species of the genus *Tamamschjania* M. Pimen. et Kljuykov (Umbelliferae). - *Feddes Repert.* **105(7-8)**: 433-436.

PIMENOV, M.G. & LEONOV, M.V. 1993: The genera of the Umbelliferae: a nomenclator. - Kew: Royal Botanic Gardens and Moscow: Bot. Garden of Moscow Univ., 156 pp.

PIMENOV, M.G. & LEONOV, M.V. 2004: The Asian Umbelliferae Biodiversity Database (ASIUM) with particular reference to the south-west Asian taxa. - *Türk. Bot. Derg.* **28**: 139-145.

PIMENOV, M.G. & OSTROUMOVA, T.A. 1994: The genus *Malabaila* Hoffm. (Umbelliferae: Tordylieae): A carpological investigation and taxonomic implications. - *Feddes Repert.* **105(3-4)**: 141-155.

PIMENOV, M.G. & POGORELOVA, O.V. 1974: The taxonomic position of the genera *Ammi* L. and *Visnaga* Mill. - *Rastitel'n. Resursy* **10(2)**: 216-219.

PIMENOV, M.G. & SDOBNINA, L.I. 1975: On the taxonomy of the genus *Seseli* L. I. Revision of the genus *Libanotis* Hill (Umbelliferae). [In Russian with English summary]. - *Bot. Zhurn. (Moscow & Leningrad)* **60(8)**: 1108-1122.

PIMENOV, M.G. & SHNER, J.V. 2008: Reports (1650-1663). In Kamari, G., Blanché, C. & Šiljak-Yakovlev, S. (eds): Mediterranean chromosome number reports - 18. - *Fl. Medit.* **18**: 573-582.

PIMENOV, M.G. & SUTORÝ, K. 2014: Name typification of Umbelliferae taxa described by Joseph Franz Freyn. - *Pl. Biosyst.* **148(6)**: 1102-111.

PIMENOV, M.G. & TIKHOMIROV, V.N. 1983: The taxonomic problems in the genera *Prangos* Lindl., *Cachrys* L., *Cryptodiscus* Schrenk and *Hippomarathrum* Hoffmgg. et Link (Umbelliferae-Apioideae). - *Feddes Repert.* **94(3-4)**: 145-164.

PIMENOV, M.G. & TOMKOVICH, L.P. 1979: Lectotypification of the genus *Ferulago* Koch (Umbelliferae). - *Taxon* **28(4)**: 409-411.

PIMENTAL PEREIRA, M. 2006: Estudio taxonomico y biogeografico de las especies europeas del genero *Anthoxanthum* L. - Coruna: Universidade de Coruna, Ph.D. thesis, 266 pp.

PIMENTEL PEREIRA, M. & AL. 2007: European sweet vernal grasses (*Anthoxanthum*: Poaceae, Pooideae, Aveneae): A morphometric taxonomical approach. - *Syst. Bot.* **32(1)**: 43-59.

PINTER, I. & AL. 2002: Phylogenetic and biosystematic relationships in four highly disjunct polyploid complexes in the subgenera *Ceterach* and *Phyllitis* in *Asplenium* (Aspleniaceae). - *Org. Divers. Evol.* **2(4)**: 299-311.

PIRINI, CH. 2011: The ecosystems of lakes Vegotida and Petron: Flora, vegatation and plant geography. [In Greek with English summary]. - Thessaloniki: Ph.D. thesis, Aristotelian University, 363 pp.

PIRINI, CH. & AL. 2009: *Artemisia campestris* inland vegetation type in the "NATURA 2000" network site "Limnes Vegoritida - Petron". In Ivanova, D. (ed.): Proceedings of the IV Balkan botanical congress, pp. 314-320. - Sofia: Institute of Botany, BAS.

PIRINI, CH. & AL. 2010: Pseudo-steppic and aquatic flora of the Natura 2000 network site Limnes Vegoritida-Petron (North-Central Greece). - *Phytol. Balcan.* **16(1)**: 109-129.

PIRINI, CH. & BABALONAS, D. 2002: Flora and vegetation of steppic grasslands around Petron lake, Macedonia, Greece. In Kamari, G. & al. (eds): Ellinini Votaniki Eteria. 9° Panelliniko Epistimoniko Sinedrio. Praktika, pp. 280-285 [In Greek with English abstract].

PIRINI, CH., TSIRIPIDIS, I. & BERGMEIER, E. 2014: Steppe-like grassland vegetation in the hills around the lakes of Vegoritida and Petron, North-Central Greece. - *Hacquetia* **13(1)**: 121-169.

PIRKER, B. & GREUTER, W. 1993: *Silene* sect. *Conoimorpha* in Greece. In Petrova, A.V. (ed.): Abstracts from the VII OPTIMA meeting in Borovetz, Bulgaria, 18-30 July 1993, p. 105.

PISTRICK, K. 1986: Systematische Untersuchungen in der Gattung *Raphanus* L. - Gatersleben: Diss., Institut für Kulturpflanzenforschung [n.v.].

PISTRICK, K. 1987: Untersuchungen zur Systematik der Gattung *Raphanus* L. - *Kulturpflanze* **35**: 225-321.

PISTRICK, K. & JARVIS, C.E. 1987: Zur Typisierung von *Raphanus sativus* L. (Cruciferae). - *Feddes Repert.* **98(9-10)**: 477-483.

PITTONI, H. 1974: Behaarung und Chromosomenzahlen sternhaariger *Leontodon*-Sippen. - *Phyton (Horn)* **16(1-4)**: 165-188.

PIWOWARSZYK, R. & AL. 2019: Holoparasitic Orobanchaceae (*Cistanche*, *Diphelypaea*, *Orobanche*, *Phelipanche*) in Armenia: distribution, habitats, host range and taxonomic problems - *Phytotaxa* **386(1)**: 1-106.

PIZARRO, J. 1995: Contribución al estudio taxonómico de *Ranunculus* L. subgen. *Batrachium* (D.C.) A. Gray (Ranunculaceae). - *Lazaroa* **15**: 21-113.

PIZZOLONGO, P. 1959: Ricerche sulla cariologia del genere *Centranthus* e loro importanza tassonomica. - *Delpinoa, n.s.* **1**: 149-164 + 2 plates.

PIZZOLONGO, P. 1962: Osservazioni cariologiche su *Arundo donax* L. e *Arundo plinii* Turra. - *Ann. Bot. (Rome)* **27(2)**: 173-188.

PIZZOLONGO, P. 1964: Considerazioni sul biotipo a 22 cromosomi di *Sternbergia lutea* Ker-Gawl. rinvenuti in una nuova stazione. - *Delpinoa, n.s.* **5**: 105-114 + 1 plate.

PLANCHUELO, A.M. 1996: Relationship between South American and European species of *Lupinus*. In Pickersgill, B. & Lock, J.M. (eds): Advances in legume systematics. Part 8: Legumes of economic importance, pp. 109-116. - Kew: Royal Botanic Gardens.

PLATAKIS, E.K. 1951: O diktamnos tis Kritis (*Origanum dictamnus* L.) [In Greek]. - Athine: 39 pp.

PLATAKIS, E.K. 1966: Aithalis platanos en Kriti [In Greek]. - *Kritiki Estia* **160/161**: 169-172, 210-213.

PLATAKIS, E.K. 1975: O diktamnos tis Kritis [In Greek]. - Iraklion: 62 pp.

PLATZER, H. 1962: Untersuchungen über die phänotypische und karyotypische Variabilität der europäischen Unkrauthirsen aus den Gattungen *Setaria*, *Digitaria* und *Echinochloa*. - *Z. Pflanzenzücht.* **47**: 330-368.

PLITMANN, U. 1973a: Taxonomic studies in *Centaurea* sect. *Calcitrapa*. I. New taxa in Palestine. - *Israel J. Bot.* **22(1)**: 47-66.

PLITMANN, U. 1973b: Biological flora of Israel. 4. *Vicia sativa* subsp. *amphicarpa* (Dorth.) Aschers. & Graebn. - *Israel J. Bot.* **22(3)**: 178-194.

PLITMANN, U. 1975: Taxonomic studies in *Centaurea* sect. *Calcitrapa*. II. The section in the Middle East. - *Israel J. Bot.* **24(1)**: 10-25.

PLITMANN, U. 1976: Taxonomic studies in *Centaurea* sect. *Calcitrapa*. III. Cytotaxonomic notes. - *Israel J. Bot.* **25(1-2)**: 84-89.

PLITMANN, U. 1981: Evolutionary history of the Old World lupines. - *Taxon* **30(2)**: 430-437.

PLITMANN, U. & HEYN, C.C. 1978: The genus *Lupinus* in the Old World - problems and prospects. - *Israel J. Bot.* **27(1)**: 30-31.

PLITMANN, U. & PAZY, B. 1984: Cytogeographical distribution of the Old World "Lupinus". - *Webbia* **38**: 531-539.

PLOCEK, A. 1978: New species of *Alchemilla* ser. *Hirsutae*. - *Folia Geobot. Phytotax.* **13(1)**: 19-31.

PLOCEK, A. 1983: Fifteen new species and varieties of *Alchemilla* (Rosaceae). - *Folia Geobot. Phytotax.* **18(4)**: 415-432.

PLOCEK, A. 1990: New taxa and combinations in *Alchemilla* (Rosaceae) (2). - *Folia Geobot. Phytotax.* **25(4)**: 389-406.

PLOCEK, A. 1998: Three new species of *Alchemilla* in the Balkans. - *Candollea* **53(2)**: 309-320.

POBEDIMOVA, E.G. 1964: Genus *Cakile* Mill. (pars specialis) [In Russian]. - *Novosti Sist. Vysš. Rast.* **1**: 90-128.

PODDUBNAJA-ARNOLDI, W.A. 1933: Geschlechtliche und ungeschlechtliche Fortpflanzung bei einigen *Chondrilla*-Arten. - *Planta* **19(1)**: 46-86.

PODLECH, D. 1962: Beitrag zur Kenntnis der Subsektion *Heterophylla* (Witas.) Fed. der Gattung *Campanula* L. - *Ber. Deutsch. Bot. Ges.* **75(7)**: 237-244.

PODLECH, D. 1965: Revision der europäischen und nordafrikanischen Vertreter der Subsect. *Heterophylla* (Wit.) Fed. der Gattung *Campanula*. - *Feddes Repert.* **71(1-3)**: 50-187.

PODLECH, D. 1967: Pflanzenliste der Pfingstexkursion 1967 des Instituts für Systematische Botanik der Universität München nach Griechenland. [Unpublished flora list]. - Photocopy: 14 pp.

PODLECH, D. 1974: Pflanzen aus Griechenland. (Gebiet des Olymp - Mazedonien und Thessalien). [Unpublished flora list]. - Photocopy: 7 pp.

PODLECH, D. 1982: Neue Aspekte zur Evolution und Gliederung der Gattung *Astragalus* L. - *Mitt. Bot. Staatssamml. München* **18**: 359-378.

PODLECH, D. 1983: Zur Taxonomie und Nomenklatur der tragacanthoiden Astragali. - *Mitt. Bot. Staatssamml. München* **19**: 1-23. [Also distributed as *OPTIMA leaflet* no. 135].

PODLECH, D. 1986: Taxonomic and phytogeographical problems in *Astragalus* of the Old World and South-West Asia. - *Proc. Roy. Soc. Edinburgh, B, Biol. Sci.* **89**: 37-43.

PODLECH, D. 1988: Revision von *Astragalus* sect. *Caprini* DC. (Leguminosae). - *Mitt. Bot. Staatssamml. München* **25**: 1-924.

PODLECH, D. 1990a: Die Typifizierung der altweltlichen Sektionen der Gattung *Astragalus* L. (Leguminosae). - *Mitt. Bot. Staatssamml. München* **29**: 461-494.

PODLECH, D. 1990b: Revision von *Astragalus* L. sect. *Platyglottis* Bunge (Leguminosae). - *Mitt. Bot. Staatssamml. München* **29**: 541-571.

PODLECH, D. 1991: The systematics of the annual species of the genus *Astragalus* L. (Leguminosae). - *Fl. Veg. Mundi* **9**: 1-8.

PODLECH, D. 1993: Beiträge zur Kenntnis der Gattung *Astragalus* L. (Leguminosae) III-VI. - *Sendtnera* **1**: 267-272.

PODLECH, D. 1994: Revision der altweltlichen annuellen Arten der Gattung *Astragalus* L. (Leguminosae). - *Sendtnera* **2**: 39-170.

PODLECH, D. 1998: Typification of Astragalus species II. Species mainly of the herbaria of Paris (P) and Geneva (G). - *Sendtnera* **5**: 247-263.

PODLECH, D. 1999: Typification of Astragalus species III. (Leguminosae). - *Sendtnera* **6**: 175-191.

PODLECH, D. 2003: Thesaurus Astragalorum. - Internet publ.: www.botanik.biologie.uni-muenchen. de/botsyst/thesaa-d.html.

PODLECH, D. 2008: The genus *Astragalus* L. (Fabaceae) in Europe with exclusion of the former Soviet Union. - *Feddes Repert.* **119(5-6)**: 310-387.

PODLECH, D. & DAMBOLDT, J. 1964: Zytotaxonomische Beiträge zur Kenntnis der Campanulaceen in Europa. - *Ber. Deutsch. Bot. Ges.* **76(9)**: 360-369.

PODLECH, D. & LI, ? 1973: Exkursion nach Nordgriechenland (Olymp) und Südjugoslawien. Unpublished flora list. - Photocopy: 8 pp.

PODLECH, D. & SYTIN, A. 1996: Typification of Russian and some other species of *Astragalus* l. - *Sendtnera* **3**: 149-176.

PODLECH, D. & ZARRE, SH. 2013: A taxonomic revision of the genus *Astragalus* L. (Leguminosae) in the Old World. - Wien: Naturhistorisches Museum, 3 vols, 2439 pp.

PODOBNIK, A. 1978: Beitrag zur Kenntnis der morphologischen Variabilität und der Verbreitung des *Aquilegia vulgaris*-Aggregates in Slowenien [In Slovenian with German summary]. - *Biol. Vestn.* **26 (1)**: 27-40.

PODOBNIK, A. 1986: A new species of genus *Aquilegia* L. from the Balkan Peninsula. – *Biosistematika* **12(1)**: 15-21.

PODPĚRA, J. 1902: Ein Beitrag zu den Vegetationsverhältnissen von Südbulgarien (Ostrumelien). - *Verh. K.K. Zool.-Bot. Ges. Wien* **52**: 608-694.

POELLNITZ, K.V. 1934: Versuch einer Monographie der Gattung *Portulaca* L. - *Repert. Spec. Nov. Regni Veg.* **37**: 240-320.

POELLNITZ, K.V. 1938a: Zur Kenntnis von *Corydalis* Sekt. *Pes-gallinaceus* Irmisch. - *Repert. Spec. Nov. Regni Veg.* **44**: 154-157.

POELLNITZ, K.V. 1938b: Zur Kenntnis von *Corydalis* Sekt. *Pes-gallinaceus* Irmisch [cont.]. - *Repert. Spec. Nov. Regni Veg.* **45**: 96-112.

POGAN, E. 1963: Karyological studies in *Alisma* L. The origin and the karyotypic evolution of *A. lanceolatum* With. - *Acta Biol. Cracov., Ser. Bot.* **5(1)**: 1-28 + 5 plates.

POGAN, E. 1972: Studies in *Eleocharis* R. Br. I. Chromosome numbers of *E. palustris* (L.) R. et S. and *E. uniglumis* (Link) Schult. - *Acta Biol. Cracov., Ser. Bot.* **15(1)**: 69-75 + 1 plate.

POGAN, E. & AL. 1982a: Further studies in chromosome numbers of Polish angiosperms. Part XV. - *Acta Biol. Cracov., Ser. Bot.* **24**: 91-126 + 1 plate.

POGAN, E. & AL. 1982b: Further studies in chromosome numbers of Polish angiosperms. Part XVI. - *Acta Biol. Cracov., Ser. Bot.* **24**: 159-189.

POGAN, E. & AL. 1985: Further studies in chromosome numbers of Polish angiosperms. Part XVIII. - *Acta Biol. Cracov., Ser. Bot.* **27**: 57-74 + 1 plate.

POGAN, E. & AL. 1987: Further studies in chromosome numbers of Polish angiosperms. Part XX. - *Acta Biol. Cracov., Ser. Bot.* **29**: 1-17 + 1 plate.

POGAN, E. & AL. 1989: Further studies in chromosome numbers of Polish angiosperms. Part XXI. - *Acta Biol. Cracov., Ser. Bot.* **30**: 119-136.

POGAN, E. & AL. 1990: Further studies in chromosome numbers of Polish angiosperms. Part XXII. - *Acta Biol. Cracov., Ser. Bot.* **31**: 1-17.

POGAN, E. & WCISŁO, H. 1973: Studies in *Ranunculus ficaria* L. I. Karyological analysis of *R. ficaria* L. ssp. *bulbifer* (Marsden-Jones) Lawalrée and *R. ficaria* L. ssp. *calthifolius* (Rchb.) Arc. - *Acta Biol. Cracov., Ser. Bot.* **16(1)**: 135-143 + 1 plate.

POGAN, E. & WCISŁO, H. 1974: Studies in *Ranunculus ficaria* L. II. Further karyological studies. - *Acta Biol. Cracov., Ser. Bot.* **17(2)**: 165-174.

POGAN, E. & WCISŁO, H. 1975: Studies in *Ranunculus ficaria* L. III. Karyotype analysis. - *Acta Biol. Cracov., Ser. Bot.* **18(2)**: 79-99 + 2 plates.

POGAN, E. & WCISŁO, H. 1981a: Studies in *Ranunculus ficaria* L. IV. Cytoembryological studies. - *Acta Biol. Cracov., Ser. Bot.* **23**: 37-54 + 3 plates.

POGAN, E. & WCISŁO, H. 1981b: Studies in *Ranunculus ficaria* L. V. Cytoembryological analysis of *Ranunculus ficaria* L. subsp. *calthifolius* (Rchb.) Arc. with B-chromosomes. - *Acta Biol. Cracov., Ser. Bot.* **23**: 83-96 + 1 plate.

POGAN, E. & WCISŁO, H. 1986: Studies in *Ranunculus ficaria* L. VII. Additions to chromosome numbers. - *Acta Biol. Cracov., Ser. Bot.* **28**: 87-92.

POGAN, E. & WCISŁO, H. 1991: Chromosome numbers of Polish angiosperms. - *Acta Biol. Cracov., Ser. Bot.* **32**: 1-169.

POGGIO, L. & AL. 1994: The chromosomes of *Coriandrum sativum* L. - *Cytologia* **59(1)**: 17-23.

POGLIANI, M. 1971: Numeri cromosomici per la Flora Italiana: 86-91. - *Inform. Bot. Ital.* **3(2)**: 155-157.

POGLIANI, M. 1972: Osservazioni cariologiche in *Anthyllis barba-jovis* L. - *Ann. Bot. (Rome)* **30**: 207-213.

POGLIANI, M. & DEL GROSSO, F. 1972: Numeri cromosomici per la flora italiana: 129-130. - *Inform. Bot. Ital.* **4(3)**: 224.

POHL, G. & LENSKI, I. 2004: Zur Verbreitung und Vergesellschaftung von *Pennisetum orientale* Rich. in Nordeuböa (Griechenland). - *Senckenberg. Biol.* **83(2)**: 1-15.

POHLMANN, J. 1977: Karyologie der Gattung *Silene*. I. Neue Chromosomenzählungen. - *Mitt. Inst. Allg. Bot. Hamburg* **15**: 69-73 + 1 plate.

POJARKOVA, A.I. 1933: Botanico-geographical survey of the maples in USSR, in connection with the history of the whole genus *Acer* L. [In Russian with English summary]. - *Trudy Bot. Inst. Akad. Nauk S.S.S.R., Ser. 1, Fl. Sist. Vysš. Rast.* **1**: 225-374 + 12 maps.

POJARKOVA, A.I. 1954: Addenda ad elaborationem generis *Cotoneaster* Med. in "Flora URSS". I [In Russian]. - *Bot. Mater. Gerb. Bot. Inst. Komarova Akad. Nauk S.S.S.R.* **16**: 109-132.

POJARKOVA, A.I. 1960: Generis Crataegi L. species novae orientali-mediterraneae [In Russian]. - *Bot. Mater. Gerb. Bot. Inst. Komarova Akad. Nauk S.S.S.R.* **20**: 186-193.

POLAT, N. & AL. 2018: Chromosome counts and karyotype analysis of some representatives of genus *Cirsium* Mill. (Asteraceae) in Turkey. - *Caryologia* **71(2)**: 133-138.

POLATSCHEK, A. 1972: Beitrag zur Cytotaxonomie der Gattung *Thlaspi*. - *Oesterr. Bot. Z.* **120(1-2)**: 201-206.

POLATSCHEK, A. 1973: Die Vertreter der Gattung *Erysimum* auf Kreta. - *Ann. Mus. Goulandris* **1**: 113-126.

POLATSCHEK, A. 1974: Systematisch-nomenklatorische Vorarbeit zur Gattung *Erysimum* in Italien. - *Ann. Naturhist. Mus. Wien* **78**: 171-182.

POLATSCHEK, A. 1978: *Erysimum cephalonicum* (Brassicaceae), ein neuer Endemit der Insel Kefallinia (Griechenland). - *Pl. Syst. Evol.* **130(3-4)**: 219-221.

POLATSCHEK, A. 1982: *Erysimum canum* und *E. hayekii* (Brassicaceae). - *Pl. Syst. Evol.* **140(4)**: 321-323.

POLATSCHEK, A. 1983a: Chromosomenzahlen und Hinweise auf Systematik und Verbreitung von Brassicaceae-Arten aus Europa, Nordafrika, Asien und Australien. - *Phyton (Horn)* **23(1)**: 127-139.

POLATSCHEK, A. 1983b: *Erysimum*. In Greuter, W. & Raus, Th. (eds): Med-Checklist notulae, 7. - *Willdenowia* **13(1)**: 88-93.

POLATSCHEK, A. 1994: Nomenklatorischer Beitrag zur Gattung *Erysimum* (Brassicaceae). - *Phyton (Horn)* **34(2)**: 189-202.

POLATSCHEK, A. 1997: *Erysimum* (Brassicaceae): Chromosomenzählungen griechischer Arten. - *Linzer Biol. Beitr.* **29(1)**: 545-553.

POLATSCHEK, A. 2013: Revision der Gattung *Erysimum* (Cruciferae): Teil 5. Nord-, West-, Zentraleuropa, Rumänien und westliche Balkan-Halbinsel bis Albanien. - *Ann. Naturhist. Mus. Wien, B* **115**: 75-218.

POLDINI, L. 1970: "Festuca calva" (Hackel) Richter e "Gentiana lutea" L. subsp. "symphyandra" Murb. entità nuove per la flora italiana. - *Webbia* **25(1)**: 191-198.

POLDINI, L. 1973: Ad floram italicam notulae taxonomicae et geobotanicae. 8. "Crambe tatarica" Sebeók. - *Webbia* **28(1)**: 31-36.

POLHILL, R.M. 1972: *Amorpha fruticosa*. - *Bot. Mag.* **178(3)**: tab. 604 + 4 pp. text.

POLHILL, R.M. 1976: Genisteae (Adans.) Benth. and related tribes (Leguminosae). - *Bot. Syst.* **1**: 143-368.

POLHILL, R.M. 1981a: Tribe 16. Galegeae (Bronn) Torrey & Gray (1838). In Polhill, R.M. & Raven, P. H. (eds): Advances in legume systematics 1: pp. 357-363. - Kew: Royal Botanic Gardens.

POLHILL, R.M. 1981b: Tribe 18. Hedysareae DC. (1825). In Polhill, R.M. & Raven, P.H. (eds): Advances in legume systematics 1: pp. 367-370. - Kew: Royal Botanic Gardens.

POLHILL, R.M. 1981c: Tribe 19. Loteae DC. (1825). In Polhill, R.M. & Raven, P.H. (eds): Advances in legume systematics 1: pp. 371-375. - Kew: Royal Botanic Gardens.

POLHILL, R.M. & AL. 1979: (459) Proposal to conserve and retypify the name *Cytisus* (Leguminosae) .. - *Taxon* **27(5-6)**: 556-559. [approved, cf. Taxon 31(3): 541, 1982].

POLHILL, R.M. & RAVEN, P.H. 1981a: Advances in legume systematics. Part 1. - London: Royal Botanic Gardens, Kew, pp. xvi + 1-425 + xxi.

POLHILL, R.M. & RAVEN, P.H. 1981b: Advances in legume systematics. Part 2. - London: Royal Botanic Gardens, Kew, pp. i-v + 427-1049.

POLI, E., MAUGERI, G. & D'URSO, A. 1973: La "Celtis tournefortii" Lam. sull'Etna. - *Inform. Bot. Ital.* **5 (2)**: 157-158.

POLITIS, J. 1926: Peri tis parousias Alifilou tis parafilloidous (*Halophila stipulacea* (Forsk)

Aschers) en tes ellinikes thalasses [In Greek with French summary]. - *Compt. Rend. Acad. Athènes [Praktika Akadimias Athinon]* **1**: 111-113.
POLITIS, J. 1953: Contribution à l'étude de la flore de la Chalcidique. - *Compt. Rend. Acad. Athènes [Praktika Akadimias Athinon]* **19**: 1-97.
POLUNIN, O. 1976: Trees and bushes of Europe. With drawings by B. Everard. - London, etc.: Oxford University Press, xvi + 208 pp.
POLUNIN, O. 1980: Flowers of Greece and the Balkans. A field guide. - Oxford Univ. Press: xv + 592 pp. + 80 pp. of plates.
POLUNIN, O. & HUXLEY, A.J. 1974: Flowers of the Mediterranean. - London: Chatto & Windus, xii + 260 pp. + 64 pp. of plates.
POLYMENAKOS, K. & AL. 2010: Reports 154-157. In Vladimirov, V. & al. (eds): New floristic records in the Balkans: 14. - *Phytol. Balcan.* **16(3)**: 433-434.
POLYMENAKOS, K. & AL. 2011: Reports 73-83. In Vladimirov, V. & al. (eds): New floristic records in the Balkans: 15. - *Phytol. Balcan.* **17(1)**: 145-148.
POLYMENAKOS, K. & CONSTANTINIDIS, TH. 2017: *Hedysarum pestalozzae* (Fabaceae), a floristic record new for Greece and Europe. In Dousis, M.A. & Thanos, K.A.: 15th conference of the Hellenic Botanical Society, book of abstracts: p. 55.
POLYMENAKOS, K. & TAN, KIT 2011: Reports 94-107. In Vladimirov, V, Dane, F. & Tan, Kit (eds): New floristic reports in the Balkans: 17. - *Phytol. Balcan.* **17(3)**: 374-376.
POLYMENAKOS, K. & TAN, KIT 2012a: Reports 106-113. In Vladimirov, V. & al. (eds): New floristic records in the Balkans: 18. - *Phytol. Balcan.* **18(1)**: 82-84.
POLYMENAKOS, K. & TAN, KIT 2012b: Reports 85-104. In Vladimirov, V., Dane, F. & Tan, Kit (eds): New floristic records in the Balkans: 19. - *Phytol. Balcan.* **18(2)**: 218-221.
POLYMENAKOS, K. & TAN, KIT 2012c: Reports 123-141. In Vladimirov, V., Dane, F. & Tan, Kit (eds): New floristic records in the Balkans: 20. - *Phytol. Balcan.* **18(3)**: 353-356.
POLYMENAKOS, K. & TAN, KIT 2013a: Reports 103-114. In Vladimirov, V., Dane, F. & Tan, Kit (eds): New floristic records in the Balkans: 21. - *Phytol. Balcan.* **19(1)**: 143-146.
POLYMENAKOS, K. & TAN, KIT 2013b: Reports 100-108. In Vladimirov, V. & al. (eds): New floristic records in the Balkans: 22. - *Phytol. Balcan.* **19(2)**: 286-288.
POLYMENAKOS, K. & TAN, KIT 2013c: Reports 65-80. In Vladimirov, V., Dane, F. & Tan, Kit (eds): New floristic records in the Balkans: 23. - *Phytol. Balcan.* **19(3)**: 385-388.
POLYMENAKOS, K. & TAN, KIT 2014a: Reports 83-103. In Vladimirov, V. & Tan, Kit (eds): New floristic records in the Balkans: 24. - *Phytol. Balcan.* **20(1)**: 112-115.
POLYMENAKOS, K. & TAN, KIT 2014b: Reports 149-204. In Vladimirov, V. & al. (eds): New floristic records in the Balkans: 25. - *Phytol. Balcan.* **20(2-3)**: 286-292.
POLYMENAKOS, K. & TAN, KIT 2015a: Reports 217-234. In Vladimirov., V, Dane, F. & Tan, Kit (eds): New floristic records in the Balkans: 26. - *Phytol. Balcan.* **21(1)**: 79-81.
POLYMENAKOS, K. & TAN, KIT 2015b: Reports 71-76. In Vladimirov, V. & al. (eds): New floristic records in the Balkans: 27. - *Phytol. Balcan.* **21(2)**: 201-203.
POLYMENAKOS, K. & TAN, KIT 2015c: Reports 109-112. In Vladimirov, V., Dane, F. & Tan, Kit (eds): New floristic records in the Balkans: 28. - *Phytol. Balcan.* **21(3)**: 379-380.
POLYMENAKOS, K. & TAN, KIT 2016a: Reports 61-65. In Vladimirov, V. & al. (eds): New floristic records in the Balkans: 29. - *Phytol. Balcan.* **22(1)**: 101-104.
POLYMENAKOS, K. & TAN, KIT 2016b: Reports 140-144. In Vladimirov, V. & Tan, Kit (eds): New floristic records in the Balkans: 31. - *Phytol. Balcan.* **22(3)**: 449-451.
POLYMENAKOS, K. & TAN, KIT 2017a: Reports 50-55. In Vladimirov, V., Aybeke, M. & Tan, Kit (eds): New floristic records in the Balkans: 32. - *Phytol. Balcan.* **23(1)**: 130-131.
POLYMENAKOS, K. & TAN, KIT 2017b: Reports 144-149. In Vladimirov, V. & al. (eds): New floristic records in the Balkans: 33. - *Phytol. Balcan.* **23(2)**: 301-302.
POLYMENAKOS, K. & TAN, KIT 2017c: Reports 125-130. In Vladimirov, V. & al. (eds): New floristic records in the Balkans: 34. - *Phytol. Balcan.* **23(3)**: 428-431.
POLYMENAKOS, K. & TAN, KIT 2018a: Reports 42-45. In Vladimirov, V., Aybeke, M. & Tan, Kit (eds): New floristic records in the Balkans: 35. - *Phytol. Balcan.* **24(1)**: 164-165.
POLYMENAKOS, K. & TAN, KIT 2018b: Reports 120-123. In Vladimirov, V., Aybeke, M. & Tan, Kit (eds): New floristic records in the Balkans: 36. - *Phytol. Balcan.* **24(2)**: 276-277.
POLYMENAKOS, K. & TAN, KIT 2019: Reports 58-78. In Vladimirov, V. & al. (eds): New floristic records in the Balkans: 38. - *Phytol. Balcan.* **25(1)**: 107-111.
POLYMENAKOS, K. & TAN, KIT 2020a: Floristic notes for the plateau of Dervenochoria and Mts Parnitha, Pastra and Kitheronas (nomos Viotias, Sterea Ellas, Central Greece). - *Phytol. Balcan.* **26(1)**: 131-149.
POLYMENAKOS, K. & TAN, KIT 2020b: Reports 96-102. In Vladimirov, V., Aybeke, M. & Tan, Kit (eds): New floristic records in the Balkans: 42. - *Phytol. Balcan.* **26(2)**: 200-202.
POLYMENAKOS, K. & TAN, KIT 2021: Reports 60-74. In Vladimirov, V., Aybeke, M. & Tan, Kit (eds): New floristic records in the Balkans: 44. - *Phytol. Balcan.* **27(1)**: 136-138.
POLYMENAKOS, K., TAN, KIT & PANTAVOS, V. 2021: Notes on the flora of Attiki (Sterea Ellas, Greece). - *Nordic J. Bot.* doi-org.ludwig.lub.lu.se/10.1111/njb.03444.
PONCET, V. 1999: L'herbier Dominique Villars. Témoin de la flore du Dauphiné. - Grenoble: Muséum d'histoire naturelle, 200 pp.
PONERT, J. 1973: Neue taxonomische Kombinationen, Kategorien und Taxa, vor allem der türkischen Arten. - *Feddes Repert.* **83(9-10)**: 617-644. [cf. Huber-Morath in *Bauhinia* 5(3): 153-159, 1975].
PONERT, J. 1975: Convallariae generis conspectus. Übersicht der Gattung *Convallaria*. - *Feddes Repert.* **86(9-10)**: 543-558.
PONS, A. 1981: The history of the Mediterranean shrublands. In Di Castri, F., Goodall, D.W. & Specht, R.L. (eds): Mediterranean-type shrublands, pp. 131-138. - Amsterdam, etc.: Elsevier.
PONS, A. & QUÉZEL, P. 1985: The history of the flora and vegetation and past and present human disturbance in the Mediterranean region. In Gómez-Campo, C. (ed.): Plant conservation in the Mediterranean area, pp. 25-43. - Dordrecht, etc.: Dr. W. Junk Publishers.
PONS, G. 1898: Saggio di una revista critica delle specie Italiane del genere *Ranunculus* L. - *Nuovo Giorn. Bot. Ital., n.s.* **5(2-4)**: 210-254, 353-392.
PONS, G. 1901: Saggio di una revista critica delle specie Italiane del genere *Ranunculus* L. [cont.]. - *Nuovo Giorn. Bot. Ital., n.s.* **8(1)**: 5-27.
POPATANASOV, A. 2015: Reports 77-79 In Vladimirov, V. & al. (eds): New floristic records in the Balkans: 27. - *Phytol. Balcan.* **21(2)**: 203-204.
POPEK, R., FACSAR, G. & MAŁECKA, J. 1991: Cyto-taxonomische Untersuchungen an der Gattung *Rosa* (Rosaceae) - die Arten aus Ungarn und anderen Gebieten. - *Fragm. Florist. Geobot.* **36(1)**: 81-87.
POPIELA, A., ŁYSKO, A. & MOLNÁR, A. 2013: Recent distribution of the Euro-Siberian-sub-Mediterranean *Elatine alsinastrum* L. (Elatinaceae). - *Acta Bot. Croat.* **72(2)**: 375-386.
POPIELA, A. & ŁYSKO, A. 2010: The distribution of *Elatine macropoda* Guss. (Elatinaceae) - *Acta Soc. Bot. Poloniae* **79(1)**: 81-86.
POPOV, M.G. 1929: The genus *Cicer* and its species [In Russian with English summary]. - *Trudy Prikl. Bot.* **21(1)**: 1-239.
POPOVA, M.T. 1963: Anatomische und kariologische Untersuchungen der Arten der Gattung *Dactylis* L. in Bulgarien [In Bulgarian with German summary]. - *Nauchni Trudove Selskost. Inst. "Vasil Kolarov"* **12(1)**: 113-122.
POPOVA, M.T. 1967: Caryological investigations of *Lilium rhodopaeum* Delip. - *Dokl. Bulg. Akad. Nauk* **20(8)**: 849-851.
POPOVA, M.T. 1968: Chromosomenzahl einiger Pflanzen von der bulgarischen Flora [In Bulgarian with German summary]. - *Nauchni Trudove Selskost. Inst. "Vasil Kolarov"* **17(1)**: 189-194.
POPOVA, M.T. 1970: Some remarks on the lilies of the Balkans. - *Lily Year-Book* **33**: 66-70.
POPOVA, M.T. 1972a: [Chromosome numbers from Bulgaria]. In Löve, Á. (ed.): IOPB chromosome number reports XXXV. - *Taxon* **21(1)**: 164-165.
POPOVA, M.T. 1972b: Cytotaxonomical study on Bulgarian *Galanthus* species. - *Dokl. Bulg. Akad. Nauk* **25(3)**: 377-380.

POPOVA, M.T. 1972c: Cytotaxonomic study of Bulgarian plants of the Liliaceae family. - *Dokl. Bulg. Akad. Nauk* **25(5)**: 669-672.

POPOVA, M.T. 1973: Cytotaxonomic investigation in genus *Ranunculus* L. [In Bulgarian with Russian and English summaries]. - *Izv. Bot. Inst. (Sofia)* **24**: 233-239.

POPOVA, M.T. 1981: Distribution of g. *Lilium* on the Balkan Peninsula. In Velčev, V.I. & Kožuharov, S.I. (eds): Mapping the flora of the Balkan Peninsula, pp. 233-247. - Sofia: Publishing House of the Bulgarian Academy of Sciences.

POPOVA, M.T. 1982: Karyological studies in *Crocus* L. - *Dokl. Bulg. Akad. Nauk* **35(5)**: 681-683.

POPOVA, M.T. 1983: *Rosa smoljanensis* Popova sp. nova (sect. *Caninae* Crep.) [In Bulgarian with English abstract]. In Velčev, V.I. & al. (eds): Third national conference of botany, pp. 173-174. - Sofia: Publishing House of the Bulgarian Academy of Sciences.

POPOVA, M.T. 1988: De specie *Onosma visianii* C.C. Clementi (Boraginaceae) nota [In Russian]. - *Novosti Sist. Vysš. Rast.* **25**: 126-130.

POPOVA, M.T. 1989: *Muscari* Miller in Bulgaria - taxa and chromosome numbers. In Phitos, D. (ed.): Abstracts. Sixth OPTIMA meeting, Delphi 10-16 Sept. 1989, p. 125.

POPOVA, M.T. 1997: Biodiversity in genus *Lilium* L. on the Balkan Peninsula. In: Abstracts from the First Balkan Botanical Congress, Thessaloniki, September 19-22, 1997, p. 40. - Thessaloniki.

POPOVA, M.T. & ANASTASSOV, H. 1993: A biosystematic study of *Ornithogalum* L. species. In Petrova, A.V. (ed.): Abstracts from the VII OPTIMA meeting in Borovetz, Bulgaria, 18-30 July 1993, p. 158.

POPOVA, M.T. & ČESCHMEDJIEV, I.V. 1966: Cytotaxonomical studies of the genus *Colchicum* L. distributed in Bulgaria. I. *Colchicum diampolis* Delip. et Ceschm. [In Bulgarian with English summary]. - *Nauchni Trudove Selskost. Inst. "Vasil Kolarov"* **15(1)**: 43-48.

POPOVA, M.T. & ČESCHMEDJIEV, I.V. 1975: [Chromosome numbers in *Iris* from Bulgaria]. In Löve, Á. (ed.): IOPB chromosome number reports XLVII. - *Taxon* **24(1)**: 143.

POPOVA, M.T. & ČESCHMEDJIEV, I.V. 1976: Karyologic investigation of the genus *Iris* L. in Bulgaria [In Bulgarian with English summaries]. - *Fitologiya* **5**: 57-65.

POPOVA, M.T. & ČESCHMEDJIEV, I.V. 1978: [Miscellaneous chromosome numbers from Bulgaria]. In Löve, Á. (ed.): IOPB chromosome number reports LXI. - *Taxon* **27(4)**: 384-385.

POPOVA, M.T. & ČESCHMEDJIEV, I.V. 1980: Caryological characteristics of two species of genus *Iris* L. [In Bulgarian with Russian and English summaries]. - *Nauchni Trudove Selskost. Inst. "Vasil Kolarov"* **25(4)**: 41-47.

POPOVA, M.T. & KOŽUHAROVA, K.S. 1983: Studies on the hybridization of *Rosa canina* L. and *Rosa pendulina* L. [In Bulgarian with English abstract]. In Velčev, V.I. & al. (eds): Third national conference of botany, pp. 183-189. - Sofia: Publishing House of the Bulgarian Academy of Sciences.

POPOVA, M.T. & KUZMANOV, B.A. 1986: [Chromosome numbers in Bulgarian *Valeriana*. [In Bulgarian]. - *Fitologiya* **31**: 75.

POPOVIĆ, I.R. & OBRATOV-PETKOVIĆ, D.D. 2009: Serpentine flora of some regions in Serbia. In Ivanova, D. (ed.): Proceedings of the IV Balkan botanical congress, pp. 403-407. - Sofia: Institute of Botany, BAS.

POPP, M. & OXELMAN, B. 2001: Inferring the history of the polyploid *Silene aegaea* (Caryophyllaceae) using plastid and homoeologous nuclear DNA sequences. - *Molec. Phyl. Evol.* **20**: 474-481.

PORCEDDU, E. & LAFIANDRA, D. 1986: Origin and evolution of wheats. In Barigozzi, C. (ed.): The origin and domestication of cultivated plants, pp. 143-178. - Amsterdam, etc.: Elsevier.

PORTAL, R. 1999: *Festuca* de France. - Le Puy-en-Velay: 371 pp.

PORTELA, I. & MORENO, M. 1985: Contribución al estudio de la semilla en el género *Iberis* L. II. Anatomía de la cubierta. - *Bol. Soc. Brot., Ser. 2* **58**: 63-83.

POSZ, E. 2014: Rodzaj *Euphrasia* L. w Polsce. Taksonomia i rozmieszczenie. – Genus of *Euphrasia* in Poland. Taxonomy and distribution. - Kraków: Instytut Botaniki Uniwersytetu Jagiellońskiego, 195 pp.

POTZTAL, E. 1951: Anatomisch-systematische Untersuchungen an den Gattungen *Arrhenatherum* und *Helictotrichon*. - *Bot. Jahrb. Syst.* **75(3)**: 321-332 + 3 plates.

POULIS, G. & ZERVAS, D. 2017: First confirmed record of *Elodea canadensis* (Hydrocharitaceae) in Greece. - *Hacquetia* **16(2)**: 175-179.

POWELL, A.M. 1979: Systematics of *Flaveria* (Flaveriinae-Asteraceae). - *Ann. Missouri Bot. Gard.* **65 (2)**: 590-636.

POŽGAJ, J. 1986: *Quercus frainetto* Ten., its natural area and the occurrence recorded in Slovakia [In Czech with English summary]. - *Preslia* **58(3)**: 211-221.

PRADA, C. 1983: El género *Isoetes* L. en la Península Ibérica. - *Acta Bot. Malac.* **8**: 73-99.

PRADA, C. & ROLLERI, C.H. 2005: A new species of *Isoetes* (Isoetaceae) from Turkey, with a study of microphyll intercellular pectic protuberances and their potential taxonomic value. - *Bot. J. Linn. Soc.* **147(2)**: 213-228.

PRAEGER, R.LL. 1921: An account of the genus *Sedum* as found in cultivation. - *J. Roy. Hort. Soc.* **46**: 1-314.

PRAEGER, R.LL. 1930: Two new Semperviva from Macedonia. - *Bull. Inst. Jard. Bot. Univ. Belgrade* **1(3)**: 209-214.

PRAEGER, R.LL. 1932: An account of the *Sempervivum* group. - London: Royal Horticultural Society, 265 pp.

PRAIN, D. 1895: A revision of the genus *Chelidonium*. - *Bull. Herb. Boissier* **3(11)**: 570-587.

PRAIN, D. 1918: The genus *Chrozophora*. - *Bull. Misc. Inform. Kew* **1918(2-3**: 29-120.

PRAKASH, S. 1974: Haploid meiosis and origin of *Brassica tournefortii* Gouan. - *Euphytica* **23(3)**: 591-595.

PRAKASH, S. & HINATA, K. 1980: Taxonomy, cytogenetics and origin of crop Brassicas, a review. - *Opera Bot.* **55**: 1-57.

PRATT, K. & JEFFERSON-BROWN, M. 1997: The gardener's guide to growing fritillaries. - Portland, Oregon: Timber Press, 160 pp.

PRAVDIN, L.F. 1969: Scots pine. Variation, intraspecific taxonomy and selection [Translated; Russian original published in 1964]. - Jerusalem: Keter Press, iv + 208 pp.

PRAŻMO, W. 1960: Genetic studies on the genus *Aquilegia* L. I. Crosses between *Aquilegia vulgaris* L. and *Aquilegia ecalcarata* Maxim. - *Acta Soc. Bot. Poloniae* **29(1)**: 57-77.

PRAŻMO, W. 1965a: Cytogenetic studies on the genus *Aquilegia*. III. Inheritance of the traits distinguishing different complexes in the genus *Aquilegia*. - *Acta Soc. Bot. Poloniae* **34(3)**: 403-437.

PRAŻMO, W. 1965b: Cytogenetic studies on the genus *Aquilegia*. IV. Fertility relationships among the *Aquilegia* species. - *Acta Soc. Bot. Poloniae* **34(4)**: 667-685.

PREKORŠEK, B. 1971: Beitrag zur Taxonomie der Arten des *Gentianella austriaca*-Komplexes [In Slovenian with German summary]. - *Biol. Vestn.* **19**: 73-81.

PRENTICE, H.C. 1979a: Experimental taxonomy of *Silene* section *Elisanthe* (Caryophyllaceae): crossing experiments. - *Bot. J. Linn. Soc.* **77(3)**: 203-216.

PRENTICE, H.C. 1979b: Numerical analysis of infraspecific variation in European *Silene alba* and *S. dioica* (Caryophyllaceae). - *Bot. J. Linn. Soc.* **78(3)**: 181-212.

PRESL, J. & PRESL, C. 1822: Deliciae pragensis, historiam naturalis spectantes ... - Pragae: Sumptibus Calvae, viii + 244 pp. [cf. TL-2 4: 390, 1983].

PRESL, K.B. 1826: Flora Sicula, exhibens plantas vasculosas in Sicilia aut sponte crescentes aut frequentissime cultas, secundum systema naturale digestas. - Pragae [Phaha]: Sumptibus A. Borrosch, xlvi + 216 pp.

PRESS, J.R. 1988: Intraspecific variation in *Rumex bucephalophorus* L. In Chater, A.O. (ed.): Flora Europaea. Notulae systematicae ad Floram Europaeam spectantes. Series 2. No. 2. - *Bot. J. Linn. Soc.* **97(4)**: 344-355.

PRESSER, H. 2002: Bemerkungen zu einigen italienischen und griechischen Orchideen. - *Ber. Arbeitskreis. Heimische Orchid.* **18(2)**: 132-139.

PRESSER, H. 2004: Bemerkungen zu einigen Orchideen in Griechenland. - *J. Eur. Orch.* **36(4)**: 969-991.

PRESSER, H. 2006: Bemerkungen zur Orchideenflora von Kreta. - *J. Eur. Orch.* **38(1)**: 39-56.

PRESSER, H. 2015: Eine Bestäuberbeobachtung auf *Gymnadenia frivaldii*. - *J. Eur. Orch.* **47(2-4)**: 303-308.

PRESSER, H. 2018: Eine neue Art aus dem *Ophrys oestrifera*-Formenkreis in Nordgriechenland. - *Ber. Arbeitskreis. Heimische Orchid.* **35(1)**: 171-184.

PRESSER, H. & HERTEL, S. 2012: Zur Kenntnis der Orchideen in Griechenland. - *Ber. Arbeitskreis. Heimische Orchid.* **29(1)**: 223-253.

PRETEL MARTÍNEZ, A. 1974: [Chromosome numbers in *Astragalus* from Spain]. In Löve, Á. (ed.): IOPB chromosome number reports XLVI. - *Taxon* **23(5-6)**: 804-805.

PRIME, C.T. 1954: Biological Flora of the British Isles. *Arum neglectum* (Towns.) Ridley (*Arum italicum* var. *neglectum* Townsend; *Arum italicum* auct. angl. non Mill.). - *J. Ecol.* **42(1)**: 241-248.

PRIME, C.T. 1960: Lords and Ladies. - London: Collins, xiv + 241 pp. + 7 plates [reprinted 1981].

PRIME, C.T. 1961: Taxonomy and nomenclature in some species of the genus *Arum* L. - *Watsonia* **5 (2)**: 106-109.

PRINA, A. 2000: A taxonomic revision of *Crambe*, sect. *Leptocrambe* (Brassicaceae). - *Bot. J. Linn. Soc.* **133(4)**: 509-524.

PRINGLE, J.S. 1987: (877) Proposal to reject the name *Centaurium minus* Moench. - *Taxon* **36(2)**: 482-483. [approved, cf. T*axon* 39(2): 294, 1990].

PRINGLE, M. 1981: Orchid hunting in Corfu and Western Greece. - *Bull. Canterbury Alpine Gard. Soc.* **1(1)**: 27-29.

PRISZTER, SZ. 1958: Über die bisher bekannten Bastarde der Gattung *Amaranthus*. - *Bauhinia* **1(2)**: 126-135.

PRISZTER, SZ. 1991: Die Vorkommnisse von *Ceratocephala falcata* (L.) Pers. in Ungarn [In Hungarian with German abstract]. - *Bot. Közlem.* **77(1-2)**: 31-35.

PRISZTER, SZ. & BORHIDI, A. 1967: Revision der Formen von *Lilium Martagon* L. - *Acta Bot. Acad. Sci. Hung.* **13(1-2)**: 175-200.

PRISZTER, SZ. & ISÉPY, I. 1974: Chorologische und phänologische Untersuchungen an mediterranen Geophyten. II. *Galanthus*. - *Ann. Univ. Sci. Budapest. Rolando Eötvös, Sect. Biol.* **16**: 87-101.

PRITCHARD, A.J. 1968: The somatic chromosomes of *Trifolium cherleri* L., *T. hirtum* All., *T. ligusticum* Balb., and *T. scabrum* L. - *Caryologia* **20(4)**: 323-331.

PRITCHARD, A.J. 1977: *Gentianella* and *Gentiana*. In: Materials for a Flora of Turkey XXIV. - *Notes Roy. Bot. Gard. Edinburgh* **35(3)**: 312-313.

PRITZEL, E. 1908: Vegetationsbilder aus dem mittleren und südlichen Griechenland. - *Bot. Jahrb. Syst.* **41(3)**: 180-214 + 9 plates.

PRITZEL, E. 1914: Süd- und Mittelgriechenland. Lichtbilder zur Pflanzengeographie und Biologie. 31.-32. Reihe. - *Repert. Spec. Nov. Regni Veg.* **14**: 46-48.

PRIVAT, G., PELLECUER, J. & MOTTE, M.-E. 1977: Etude écologique de *Satureia montana* L. (Labiées). - *Naturalia Monspel., Sér. Bot.* **26**: 55-72.

PROCTOR, M.C.F. 1954: The cytology of the British species of *Helianthemum* and their allies. - *Proc. Bot. Soc. Brit. Isles* **1(1)**: 87-88.

PROCTOR, M.C.F. 1955: Some chromosome counts in the European Cistaceae. - *Watsonia* **3(3)**: 154-159 + 1 plate.

PROCTOR, M.C.F. 1956: Biological Flora of the British Isles. *Helianthemum* Mill. - *J. Ecol.* **44(2)**: 675-692.

PROCTOR, M.C.F. 1960: Biological Flora of the British Isles. *Tuberaria guttata* (L.) Fourreau. - *J. Ecol.* **48(1)**: 243-253.

PROCTOR, M.C.F. 1968: [Notes on *Helianthemum*]. In Heywood, V.H. (ed.): Notulae systematicae ad Floram Europaeam spectantes, no. 7. - *Feddes Repert.* **79(1-2)**: 57-59.

PRODANOVIĆ, D. & AL. 2010: Contribution to the knowledge of serpentine flora in Kosovo's part of the Ibar river valley. - *Bot. Serbica* **34(2)**: 81-86.

PROFAS, G., TRIGAS, P. & SCHOLZ, H. 2010: *Puccinellia bilykiana* Klok. In Greuter, W. & Raus, Th. (eds): Med-Checklist notulae, 29. - *Willdenowia* **40(2)**: 201.

PROFUMO, P. 1965: "Cystopteris dickieana" Sim in Italia. - *Webbia* **20(2)**: 735-744.

PROSKURYAKOVA, G.M. & BELYANINA, N.B. 1985: K sistematike roda *Mandragora* i ego istorii [In Russian]. - *Biol. Nauki (Moscow)* **2(254)**: 76-86.

PRYER, K.M. & HAUFLER, CH.H. 1993: Isozymic and chromosomal evidence for the allotetraploid origin of *Gymnocarpium dryopteris* (Dryopteridaceae). - *Syst. Bot.* **18(1)**: 150-172.

PRZYWARA, L. 1979: Karyological studies in the genus *Myosotis* L. in Poland. - *Acta Biol. Cracov., Ser. Bot.* **21(2)**: 125-141.

PRZYWARA, L. 1983: Further karyological studies on the series *Palustres* M. Pop. of the genus *Myosotis* L. from Poland. - *Acta Biol. Cracov., Ser. Bot.* **25**: 85-101 + 3 plates.

PRZYWARA, L. 1988: Biosystematic studies on the series *Palustres* M. Pop. of the genus *Myosotis* L. from Poland. - *Fragm. Florist. Geobot.* **31-32(1-2)**: 77-116 + 4 plates.

PRZYWARA, L. & ROSTAŃSKI, K. 1981: Chromosome numbers in the genus *Oenothera* L. from Poland. - *Acta Biol. Cracov., Ser. Bot.* **22(2)**: 155-162.

PRZYWARA, L. & SCHMAGER, J. 1967: Cytotaxonomical studies in the species of the genus *Lotus* L. - *Acta Biol. Cracov., Ser. Bot.* **10(1)**: 113-126 + 2 plates.

PRZYWARA, L. & SCHMAGER, J. 1968: Cytotaxonomical studies on *Chrysanthemum leucanthemum* L. sensu lato. - *Acta Biol. Cracov., Ser. Bot.* **11(2)**: 105-116 + 2 plates.

PUECH, S. 1968: Hétérocarpie rythmique dans une population cévenole d'*Aethionema saxatile* (L.) R. Br.: Premiers résultats d'observations effectuées en milieu naturel et en culture. - *Bull. Soc. Bot. France* **115(9)**: 553-564.

PUECH, S. 1975: Contribution caryologique à l'étude des *Teucrium* de la section *Polium* du bassin méditeranéen occidental. In Guinochet, M. & al. (eds): La flore du Bassin Méditerranéen, pp. 223-244. - Paris: C.N.R.S.

PUECH, S. 1976a: Recherches de biosystématique sur les *Teucrium* (Labiées) de la section *Polium* du bassin méditerranéen occidental (Espagne et France). - Languedoc: Thèse; tome 1 (texte et figures) 138 pp.; tome 2, 36 planches.

PUECH, S. 1976b: [Chromosome numbers in *Teucrium*]. In Löve, Á. (ed.): IOPB chromosome number reports LIV. - *Taxon* **25(5-6)**: 642.

PUECH, S. 1978: Les *Teucrium* de la section *Polium*, au Portugal. - *Bol. Soc. Brot., Ser. 2* **52**: 37-50.

PUECH, S. 1982: Contribution à l'étude des *Teucrium* de la section *Polium* en Tunisie. - *Bull. Soc. Bot. France, Lett. Bot.* **129(1)**: 41-52.

PUECH, S. 1984: Les *Teucrium* (Labiées) de la section *Polium* du bassin méditerranéen occidental (France et péninsule ibérique). - *Naturalia Monspel., Sér. Bot.* **No. hors série**: 1-71.

PUENTE GARCÍA, E. & AL. 1995: Aportaciones al conocimiento del genero *Spergula* L. - *Lagascalia* **18 (1)**: 15-24.

PUFF, CH. 1976: Über die *Galium palustre*-Gruppe (*Galium* Sect. *Aparinoides*, Rubiaceae) in Österreich. I. Einige Chromosomenzahlen von *G. palustre* und *G. elongatum*. - *Linzer Biol. Beitr.* **8 (1)**: 3-11.

PUGLISI, S. 1997: Gene conservation in pine species. - *Bocconea* **7**: 345-356.

PUGSLEY, H.W. 1915: *Narcissus poeticus* and its allies. - *J. Bot. (London)* **53 suppl. 2**: 1-44 + 2 plates.

PUGSLEY, H.W. 1919: A revision of the genera *Fumaria* and *Rupicapnos*. - *J. Linn. Soc., Bot.* **44(298)**: 233-355 + 8 plates.

PUGSLEY, H.W. 1927: Further notes on the genera *Fumaria* and *Rupicapnos*. - *J. Linn. Soc., Bot.* **47 (317)**: 427-469 + 3 plates.

PUGSLEY, H.W. 1931a: *Narcissus hellenicus* Pugsl. - *J. Bot. (London)* **69**: 195-196.

PUGSLEY, H.W. 1931b: The forms of *Symphytum tuberosum* L. - *J. Bot. (London)* **69**: 89-97.

PUGSLEY, H.W. 1932: Further notes on the genera *Fumaria* and *Rupicapnos*. - II. - *J. Linn. Soc., Bot.* **49(327)**: 93-113 + 1 plate.

PUGSLEY, H.W. 1934: Further notes on the genera *Fumaria* and *Rupicapnos*. - III. - *J. Linn. Soc., Bot.* **49(331)**: 517-529 + 1 plate.

PUGSLEY, H.W. 1937a: Further notes on the genera *Fumaria* and *Rupicapnos*. - IV. - *J. Linn. Soc., Bot.* **50(335)**: 541-559 + 2 plates.

PUGSLEY, H.W. 1937b: Notes on Poet's Narcissi. - *J. Bot. (London)* **75**: 53-58.

PUGSLEY, H.W. 1940: Notes on *Orobanche* L. - *J. Bot. (London)* **78**: 105-116.

PUIZINA, J., SOLIČ, M.E. & PAPEŠ, D. 1996: Reports (524-527). In Kamari, G., Felber, F. & Garbari, F. (eds): Mediterranean chromosome number reports - 5. - *Fl. Medit.* **5**: 337-340.

PUJADAS SALVÀ, A. 1992: Algunos *Sisymbrium* (Cruciferae) a excluir de la flora ibérica, y otras precisiones corológicas. - *Anales Jard. Bot. Madrid* **50(1)**: 139-140.

PUJADAS SALVÀ, A. 2002: El complejo de *Daucus carota* L. (Apiaceae) en la Flora Ibérica. - *Anales Jard. Bot. Madrid* **59(2)**: 368-375.

PUJADAS SALVÀ, A. 2013: El género *Erigeron* L. (Asteraceae) en la Flora Iberica. - *Acta Bot. Malac.* **38**: 241-253.

PULEVIĆ, V. 1970: *Euphorbia dendroides* L. und *Euphorbia wulfenii* Hoppe in der Flora von Montenegro [In Serbo-Croatian with German summary]. - *Glasn. Republ. Zavoda Zaštitu Prir. Prirodnjačke Zbirke Titogradu* **3**: 67-69.

PULEVIĆ, V. 1972: *Hyacinthella dalmatica* (Baker) Hayek in der Flora von Montenegro [In Serbo- Croatian with German summary]. - *Glasn. Republ. Zavoda Zaštitu Prir. Prirodnjačke Zbirke Titogradu* **4**: 73-76.

PULEVIĆ, V. 1974a: Beitrag zur Flora von Montenegro [In Serbo-Croatian with German summary]. - *Glasn. Republ. Zavoda Zaštitu Prir. Prirodnjačke Zbirke Titogradu* **6**: 77-83.

PULEVIĆ, V. 1974b: *Colchicum hungaricum* Janka in der Flora von Montenegro [In Serbo-Croatian with German summary]. - *Glasn. Republ. Zavoda Zaštitu Prir. Prirodnjačke Zbirke Titogradu* **7**: 59-61.

PULEVIĆ, V. 1976a: *Daphne laureola* L. Rare species in dendroflora of Montenegro [In Serbo-Croatian with English summary]. - *Poljopr. Šumarstvo* **22(4)**: 45-48.

PULEVIĆ, V. 1976b: Two new species of the genus *Crocus* from Yugoslavia. - *Glasn. Republ. Zavoda Zaštitu Prir. Prirodnjačke Zbirke Titogradu* **9**: 39-43.

PULEVIĆ, V. 1976c: Some new and rare plants in the flora of Montenegro [In Serbo-Croatian with English summary]. - *Glasn. Republ. Zavoda Zaštitu Prir. Prirodnjačke Zbirke Titogradu* **9**: 99-102.

PULEVIĆ, V. 1977a: A contribution to the knowledge of taxonomy and horology of some species of *Crocus* L. from Yugoslavia (*C. weldenii*, *C. alexandri* and *C. adamii* sens. lat.) [In Serbo-Croatian with English summary]. - *Glasn. Republ. Zavoda Zaštitu Prir. Prirodnjačke Zbirke Titogradu* **10**: 81-99.

PULEVIĆ, V. 1977b: The contribution to the knowledge of taxonomy and horology of *Crocus tommasinianus* Herbert (Iridaceae) [In Serbo-Croatian with English summary]. - *Poljopr. Šumarstvo* **23(1)**: 53-60.

PULEVIĆ, V. 1978: *Crocus thomasii* Ten. and *Crocus pallasii* Goldb. in the flora of Yugoslavia [In Serbo-Croatian with English summary]. - *Glasn. Republ. Zavoda Zaštitu Prir. Prirodnjačke Zbirke Titogradu* **11**: 133-138.

PULEVIĆ, V. 1979: The contribution to the knowledge of genus *Crocus* L. in Yugoslavia (division *Nudiflori* section *Reticulati*) [In Serbo-Croatian with English summary]. - *Glasn. Republ. Zavoda Zaštitu Prir. Prirodnjačke Zbirke Titogradu* **12**: 195-212.

PULEVIĆ, V. 1980: The bibliography of flora and vegetation of Montenegro [In Serbo-Croatian]. - Titograd: The Montenegrin Academy of Sciences and Arts, 235 pp.

PULEVIĆ, V. 1981: On areals of species *Herniaria nigrimontium* Hermann and *Allium phthioticum* Boiss. & Heldr. [In Serbo-Croatian with English summary]. - *Glasn. Republ. Zavoda Zaštitu Prir. Prirodnjačke Zbirke Titogradu* **14**: 23-29.

PULEVIĆ, V. 1982: Genus *Romulea* Maratti in flora of Yugoslavia [In Serbo-Croatian with English summary]. - *Glasn. Republ. Zavoda Zaštitu Prir. Prirodnjačke Zbirke Titogradu* **15**: 85-90.

PULEVIĆ, V. 1983a: The contribution to the knowledge of morphology and distribution of the species *Crocus malyi* Visiani [In Serbo-Croatian with English and Italian summaries]. - *Povrem. Izd. Muz. Grada Šibenika* **10**: 309-313.

PULEVIĆ, V. 1983b: On distribution of species *Crocus weldenii* Hoppe & Fürnrohr and *Crocus chrysanthus* Herbert in Yugoslavia [In Macedonian with English summary]. - *Maked. Akad. Nauk. Umetn., Oddel. Biol. Med. Nauki* **4(1-2)**: 159-166.

PULEVIĆ, V. 1984: *Euphorbia prostrata* Aiton. A new weed species in the flora of Yugoslavia. In: Second congress on weeds, pp. 113-117 [In Serbo-Croatian with English summary]. - Osijek.

PULEVIĆ, V. 1985: Some remarks on species *Crocus banaticus* Gay ... in the flora of Serbia In: Zbornik radova Simpoz. "Stogodišnjica Flore okoline Niša", pp. 11-14 [In Serbo-Croatian with English summary]. - Niš.

PULEVIĆ, V. 1987: Supplement to bibliography of flora and vegetation of Montenegro [In Serbo-Croatian with English abstract]. - *Glasn. Republ. Zavoda Zaštitu Prir. Prirodnjačke Zbirke Titogradu* **18**: 1-106.

PULEVIĆ, V. 2005: Material for vascular flora of Montenegro. A supplement to "Conspectus Florae Montenegrinae" (J. Rohlena) [In Serbo-Croatian with English summary]. - Podgorica: Republic Institute for the Protection of Nature, 218 pp.

PULEVIĆ, V. & KARAMAN, V. 1996: Additional literature data on the flora of Crna Gora (Montenegro). - *Bocconea* **5(1)**: 53-54.

PULEVIĆ, V. & MALAJ, V. 1985: Antonio Baldacci (1867-1950) [with list of publications; in Serbo- Croatian with Albanian and Italian summaries]. - *Bibliogr. Bjesn.* **14(1)**: 43-71.

PUNDEVA, R.S. 1974: Karyological study on two varieties of *Anthoxanthum odoratum* L. - var. *odoratum* and var. *montanum* Aschers. et Graebn. - *Dokl. Bulg. Akad. Nauk* **27(3)**: 395-398.

PUNTILLO, D. 1997: A new species of *Centaurea* sect. *Phalolepis* from Calabria (S. Italy). - *Fl. Medit.* **6**: 219-222.

PUTIYEVSKY, E. & KATZNELSON, J. 1972: Cytology and crossability of several Mediterranean *Trifolium* species. - *Israel J. Bot.* **21(3)**: 179-181.

PÜTZ, N. 1994: Vegetative spreading of *Oxalis pes-caprae* (Oxalidaceae). - *Pl. Syst. Evol.* **191(1-2)**: 57-67.

PYŠEK, P., SÁDLO, J. & MANDÁK, B. 2002: Catalogue of alien plants of the Czech Republic. - *Preslia* **74**: 97-186.

QIAN, G.-Z. & AL. 2008: Taxonomic study of *Malus* section *Florentinae* (Rosaceae). - *Bot. J. Linn. Soc.* **158(2)**: 223-227.

QOSJA, X. & AL. (EDS) 1992: Flora e Shqipërisë. Vol. 2. Rosaceae-Umbelliferae [In Albanian]. - Tinanë: Akad. Shkencave Rep. Shqipërisë, 446 pp.

QOSJA, X. & AL. (EDS) 1996: Flora e Shqipërisë. Vol. 3. Pirolaceae-Campanulaceae [In Albanian]. - Tinanë: Akad. Shkencave Rep. Shqipërisë, 331 pp.

QUEIRÓS, M. 1972: Contribuição ao conhecimento citotaxonómico das Spermatophyta de Portugal. VII. Umbelliferae. - *Anuário Soc. Brot.* **38**: 293-314.

QUEIRÓS, M. 1973a: Contribuição para o conhecimento citotaxonómico das Spermatophyta de Portugal. I. Gramineae, supl. 1. - *Bol. Soc. Brot., Ser. 2* **47**: 77-103.

QUEIRÓS, M. 1973b: Contribuição para o conhecimento citotaxonómico das Spermatophyta de Portugal. II. Compositae, supl. 1. - *Bol. Soc. Brot., Ser. 2* **47**: 299-314.

QUEIRÓS, M. 1973c: Contribuição para o conhecimento citotaxonómico das Spermatophyta de Portugal. IX. Cruciferae. - *Bol. Soc. Brot., Ser. 2* **47**: 315-335.

QUEIRÓS, M. 1974a: Contribuição para o conhecimento citotaxonómico das Spermatophyta de Portugal. I. Gramineae, supl. 2. - *Bol. Soc. Brot., Ser. 2* **48**: 81-98.

QUEIRÓS, M. 1974b: Contribuição para o conhecimento citotaxonómico das Spermatophyta de Portugal. VII. Umbelliferae, supl. 1. - *Bol. Soc. Brot., Ser. 2* **48**: 171-186.

QUEIRÓS, M. 1975a: Contribuição para o conhecimento citotaxonómico das Spermatophyta de Portugal. X. Chenopodiaceae. - *Bol. Soc. Brot., Ser. 2* **49**: 121-142.

QUEIRÓS, M. 1975b: Contribuição para o conhecimento citotaxonómico das Spermatophyta de Portugal. XI. Euphorbiaceae. - *Bol. Soc. Brot., Ser. 2* **49**: 143-161.

QUEIRÓS, M. 1976: Contribuição para o conhecimento citotaxonómico das Spermatophyta de Portugal. XII. Onagraceae. - *Bol. Soc. Brot., Ser. 2* **50**: 107-116.

QUEIRÓS, M. 1977: Contribuição para o conhecimento citotaxonómico das Spermatophyta de Portugal. XVI. Malvaceae. - *Bol. Soc. Brot., Ser. 2* **51**: 187-199.

QUEIRÓS, M. 1978a: Sur la caryologie des Ombellifères du Portugal. In Cauwet-Marc, A.M. & Carbonnier, J.: Les Ombellifères, pp. 325-334. - Perpignan.

QUEIRÓS, M. 1978b: Números cromossómicos para a flora portuguesa. 1-15. - *Bol. Soc. Brot., Ser. 2* **52**: 69-77.

QUEIRÓS, M. 1980a: Números cromossómicos para a flora portuguesa. 16-37. - *Bol. Soc. Brot., Ser. 2* **53(1)**: 15-28.

QUEIRÓS, M. 1980b: Números cromossómicos para a flora portuguesa. 38-63. - *Bol. Soc. Brot., Ser. 2* **54**: 47-64.

QUEIRÓS, M. 1980c: Contribuição para o conhecimento citotaxonómico das Spermatophyta de Portugal. XVII. Papaveraceae. - *Bol. Soc. Brot., Ser. 2* **54**: 153-171.

QUEIRÓS, M. 1986a: Notas cariológicas em Rubiaceae portuguesas. - *Bol. Soc. Brot., Ser. 2* **59**: 233-243.

QUEIRÓS, M. 1986b: Indice alfabético dos taxa referidos na série "Números cromossómicos para flora portuguesa". 1-103. - *Bol. Soc. Brot., Ser. 2* **59**: 273-276.

QUEIRÓS, M. 1989: Estudos citotaxonómicos em *Amaranthus* de Portugal. - *Lazaroa* **11**: 9-17.

QUEIRÓS, M. 1990: Notas cariológicas em algumas Ranunculaceae portuguesas. - *Collect. Bot. (Barcelona)* **18**: 45-57.

QUEIRÓS, M. 1993: Números cromossómicos para a flora portuguesa. 116-127. - *Bol. Soc. Brot., Ser. 2* **66**: 117-124 + 1 plate.

QUEIRÓS, M. 1994: Números cromossómicos para a flora portuguesa. 128-138. - *Bol. Soc. Brot., Ser. 2* **66**: 317-321.

QUEIRÓS, M., ORMONDE, J. & NOGUIERA, I. 1990: Contribuição para o estudo citotaxonómico de *Anogramma leptophylla* (L.) Link. In Rita, J. (ed.): Taxonomía, biogeografía y conservación de Pteridofítos, pp. 235-242. - Palma de Mallorca.

QUEIRÓS, M. & VIERA, C. 1993: Catálogo dos taxa referidos na série "Contribuição para o conhecimento citotaxonómico das Spermatophyta de Portugal" - I. - *Bol. Soc. Brot., Ser. 2* **65**: 173-194.

QUESENBERRY, K.H. & TAYLOR, N.L. 1976: Interspecific hybridization in *Trifolium* L., sect. *Trifolium* Zoh. I. Diploid hybrids among *T. alpestre* L., *T. rubens* L., *T. heldreichianum* Hausskn., and *T. noricum* Wulf. - *Crop Sci. (Madison)* **16(3)**: 382-386.

QUÉZEL, P. 1964: Végétation des hautes montagnes de la Grèce méridionale. - *Vegetatio* **12(5-6)**: 289-385 + 34 tables.

QUÉZEL, P. 1967a: La végétation des hauts sommets du Pinde et de l'Olympe de Thessalie. - *Vegetatio* **14(1-4)**: 127-228 + 19 tables.

QUÉZEL, P. 1967b: A propos de quelques hêtraies de Macédoine grecque. - *Bull. Soc. Bot. France* **114(5-6)**: 200-210 + 1 table.

QUÉZEL, P. 1968: Signification phytosociologique des Gesnériacées grecques. - *Collect. Bot. (Barcelona)* **7(2)**: 947-973.

QUÉZEL, P. 1969: La végétation du massif du Bela Voda (Macédoine nord-occidentale). - *Biol. Gallo- Hellen.* **2(2)**: 93-112.

QUÉZEL, P. 1973: Contribution à l'étude de la végétation du Vardoussia. - *Biol. Gallo-Hellen.* **5(1)**: 139-166.

QUÉZEL, P. 1985: Definition of the Mediterranean region and the origin of its flora. In Gómez-Campo, C. (ed.): Plant conservation in the Mediterranean area, pp. 9-24. - Dordrecht, etc.: Dr. W. Junk Publishers.

QUÉZEL, P. 1989: Contribution à l'étude phytosociologique des pelouses ecorchées culminales du massif du Falakron. - *Bios (Thessaloniki)* **1989**: 187-193.

QUÉZEL, P. 1995: La flore du bassin méditerranéen: origine, mise en place, endémisme. - *Ecol. Medit.* **21(1-2)**: 19-39.

QUÉZEL, P. & AL. 1990: Recent plant invasions in the circum-mediterranean region. In Di Castri, F., Hansen, A.J. & Debussche, M. (eds): Biological invasions in Europe and the Mediterranean Basin, pp. 51-60. - Dordrecht, etc.: Kluwer Academic Publishers.

QUÉZEL, P. & BARBERO, M. 1982: Definition and characterisation of mediterranean-type ecosystems. - *Ecol. Medit.* **8**: 15-29.

QUÉZEL, P. & BARBERO, M. 1985: Carte de la végétation potentielle de la Région Méditerranéenne. Feuille No. 1: Méditerranée orientale. - Paris: Éditions du CNRS: 69 pp. + 1 map (1:2500000).

QUÉZEL, P. & CONTANDRIOPOULOS, J. 1965a: Contribution à l'étude de la flore des hautes montagnes de Grèce. - *Naturalia Monspel., Sér. Bot.* **16**: 89-149.

QUÉZEL, P. & CONTANDRIOPOULOS, J. 1965b: Contribution à l'étude de la flore du Pinde central et septentrional et de l'Olympe de Thessalie. - *Candollea* **20**: 51-90.

QUÉZEL, P. & CONTANDRIOPOULOS, J. 1965c: A propos de la végétation des forêts des hêtres dans le massif du Pinde. - *Bull. Soc. Bot. France* **112(5-6)**: 312-319 + 1 table.

QUÉZEL, P. & CONTANDRIOPOULOS, J. 1966: *Teucrium corcyranum*, espèce nouvelle de l'île de Corfou. - *Bull. Soc. Bot. France* **113(5-6)**: 351-353.

QUÉZEL, P. & CONTANDRIOPOULOS, J. 1967: A propos de quelques espèces nouvelles des montagnes grecques. - *Taxon* **16(3)**: 239-240.

QUÉZEL, P. & CONTANDRIOPOULOS, J. 1968: Contribution à l'étude de la flore de la Macédoine grecque. - *Candollea* **23(1)**: 17-37.

QUÉZEL, P., CONTANDRIOPOULOS, J. & PAMUKÇUOĞLU, A. 1970: Contribution à l'étude de la flore des hautes montagnes de l'Anatolie occidentale. - *Candollea* **25(2)**: 341-387.

QUÉZEL, P. & KATRABASSA, M. 1974: Premier aperçu sur la végétation du Chelmos (Peloponèse). - *Biol. Écol. Médit.* **1(1)**: 11-25.

QUINTANAR, A. & AL. 2020: (2729) Proposal to conserve the name *Andropogon ischaemum* (*Bothriochloa ischaemum*) (Gramineae) with a conserved type. - *Taxon* **69(1)**: 198-199.

QUINTANAR, A. & CASTROVIEJO, S. 2008: (1836) Proposal to reject the name *Koeleria nitida* (Gramineae). - *Taxon* **57(3)**: 996. [approved, cf. Taxon 60(1): 226-227, 2011].

QUINTANAR, A. & CASTROVIEJO, S. 2010: (1947) Proposal to conserve *Trisetum* against *Trisetaria* (Gramineae). - *Taxon* **59(5)**: 1602-1603.

QUINTANAR, A., GLAZKOVA, E. & CASTROVIEJO, S. 2009a: On the identity and typification of *Koeleria lobata* (M. Bieb.) Roem. & Schult. (Pooideae, Gramineae). - *Taxon* **58(2)**: 617-620.

QUINTANAR, A., GLAZKOVA, E. & CASTROVIEJO, S. 2009b: (1895) Proposal to reject the name *Dactylis lobata* (*Koeleria lobata*) (Gramineae). - *Taxon* **58(2)**: 656. [approved, cf. Taxon 60(4): 1203, 2011].

RAAB-STRAUBE, E.V. & RAUS, TH. (EDS) 2015a: Euro+Med-Checklist notulae, 4. - *Willdenowia* **45(1)**: 119-129.

RAAB-STRAUBE, E.V. & RAUS, TH. (EDS) 2015b: Euro+Med-Checklist notulae, 5. - *Willdenowia* **45(3)**: 449-464.

RAAB-STRAUBE, E.V. & RAUS, TH. (EDS) 2016: Euro+Med-Checklist notulae, 6. - *Willdenowia* **46(3)**: 423-442.

RAAB-STRAUBE, E.V. & RAUS, TH. (EDS) 2017: Euro+Med-Checklist notulae, 7. - *Willdenowia* **47(1)**: 89-96.

RAAB-STRAUBE, E.V. & RAUS, TH. (EDS) 2020: Euro+Med-Checklist notulae, 12. - *Willdenowia* **50(2)**: 305-341.

RAAB-STRAUBE, E.V. & SCHOLZ, H. 2013: *Bromus arvensis* subsp. *parviflorus*. In Raab-Straube, E.v. & Raus, Th.: Notulae ad floram euro-mediterraneam pertinentes No. 30. - *Willdenowia* **43(1)**: 160-161.

RAABE, U. 2001: *Bergia capensis* L. In Greuter, W. & Raus, Th. (eds): Med-Checklist notulae, 20. - *Willdenowia* **31(2)**: 323.

RAABE, U. 2004: *Arctotheca calendula* (L.) Levyns. In Greuter, W. & Raus, Th. (eds): Med-Checklist notulae, 22. - *Willdenowia* **34(1)**: 73.

RAABE, U. 2019a: *Silene muscipula* in Griechenland wieder aufgefunden. - *Natur u. Heimat (Münster)* **79(4)**: 139-146.

RAABE, U. 2019b: Die Gattung *Lindernia* (Linderniaceae) im Burgenland, Österreich. - *Neilreichia* **10**: 171-183.

RAABE, U. & AL. 2009: *Polygala rausiana* (Polygalaceae), a new species from the northern Peloponnese, Greece. - *Willdenowia* **39(1)**: 69-75.

RAABE, U. & KOUMPLI-SOVANTZI, L. 2001: Contribution to the flora of the Strophylia area (NW. Peloponnisos). In Kamari, Psaras & Constantinidis (eds): Proceedings of the 8th scientific congress of the Greek Botanical Society, October 5-8, 2000, pp. 368-371 [In Greek with English abstract].

RAABE, U. & KOUMPLI-SOVANTZI, L. 2002: Studies on the flora of some endangered wetlands of Argolis (Greece). In Kamari, G. & al. (eds): Ellinini Votaniki Eteria. 9° Panelliniko Epistimoniko Sinedrio. Praktika, pp. 286-289 [In Greek with English abstract].

RAABE, U., RÄTZEL, S. & RAUS, TH. 2010: *Hornungia pauciflora* (W.D.J. Koch) Banfi & al. In Greuter, W. & Raus, Th. (eds): Med-Checklist notulae, 29. - *Willdenowia* **40(2)**: 196.

RAABE, U. & RAUS, TH. 2001: *Heteranthera limosa* (Sw.) Willd. In Greuter, W. & Raus, Th. (eds): Med- Checklist notulae, 20. - *Willdenowia* **31(2)**: 327.

RAABE, U. & RAUS, TH. 2002: Notes on *Rotala ramosior* (L.) Koehne. In Greuter, W. & Raus, Th. (eds): Med-Checklist notulae, 21. - *Willdenowia* **32(2)**: 202-203.

RAABE, U. & RAUS, TH. 2016: *Oenothera lindheimeri*. In Raab-Straube, E. v. & Raus, Th. (eds): Euro+Med-Checklist notulae, 6. - *Willdenowia* **46(3)**: 430.

RAAMSDONK, L.W.D. VAN 1982: Biosystematic studies on the *umbellatum-angustifolium* complex in the genus *Ornithogalum* L. I. Typification and taxonomy. - *Proc. Kon. Ned. Akad. Wetensch. C* **85(4)**: 563-574.

RAAMSDONK, L.W.D. VAN 1984: Biosystematic studies on the *umbellatum-angustifolium* complex in the genus *Ornithogalum* L. - Utrecht: Ph.D. thesis, Univ. of Utrecht, 143 pp.

RAAMSDONK, L.W.D. VAN 1985: Pollenfertility and seed formation in the *Ornithogalum umbellatum/angustifolium* complex (Liliaceae/Scilloideae). - *Pl. Syst. Evol.* **149(3-4)**: 287-297.

RAAMSDONK, L.W.D. VAN 1986: Biosystematic studies on the *umbellatum-angustifolium* complex of the genus *Ornithogalum* (Liliaceae). II. Genome characterization and evolution. - *Nordic J. Bot.* **6(5)**: 525-544.

RAAMSDONK, L.W.D. VAN & DE VRIES, T. 1989: Biosystematic studies in European species of *Gladiolus* (Iridaceae). - *Pl. Syst. Evol.* **165(3-4)**: 189-198.

RAAMSDONK, L.W.D. VAN & DE VRIES, T. 1992: Biosystematic studies in *Tulipa* sect. *Eriostemones* (Liliaceae). - *Pl. Syst. Evol.* **179(1-2)**: 27-41.

RAAMSDONK, L.W.D. VAN & DE VRIES, T. 1995: Species relationships and taxonomy in *Tulipa* subg. *Tulipa* (Liliaceae). - *Pl. Syst. Evol.* **195(1-2)**: 13-44.

RAAMSDONK, L.W.D. VAN., EIJK, J.P. VAN & EIKELBOOM, W. 1995: Crossability analysis in subgenus *Tulipa* of the genus *Tulipa* L. - *Bot. J. Linn. Soc.* **117(2)**: 147-158.

RABEI, S. & ABDEL KHALIK, K. 2012: Conventional keys for Convolvulaceae in the flora of Egypt. - *Fl. Medit.* **22**: 45-61.

RABELER, R.K. 1984a: An unnecessary infraspecific combination in *Medicago* (Fabaceae). - *Taxon* **33 (2)**: 321-322.

RABELER, R.K. 1984b: Notes on nomenclature within *Petrorhagia* (Caryophyllaceae). - *Taxon* **33(4)**: 714-716.

RABELER, R.K. 1985: Isonyms of *Petrorhagia obcordata* (Caryophyllaceae). - *Taxon* **34(2)**: 293.

RABELER, R.K. 1992: (1034) Proposal to conserve 2490 *Silene* L. against *Lychnis* L. (Caryophyllaceae). - *Taxon* **41(1)**: 126-128. [approved, cf. Taxon 43(2): 272, 1994].

RABELER, R.K. 2005: Publication of the combination *Spergularia diandra* (Caryophyllaceae): discovery of an earlier indelible autographic label. - *Taxon* **54(1)**: 147-148.

RACKHAM, O. 1978: The flora and vegetation of Thera and Crete before and after the great eruption. In Doumas, C. (ed.): Thera and the Aegean World I, pp. 755-764. - London.

RACKHAM, O. 1990a: The greening of Myrtos. In Bottema, S., Entjes-Nieborg, G. & van Zeist, W. (eds): Man's role in the shaping of the eastern Mediterranean landscape, pp. 341-348. - Rotterdam: Balkema publishers.

RACKHAM, O. 1990b: Vegetation history of Crete. In Grove, A.T., Moody, J. & Rackham, O. (eds): Stability and change in the Cretan landscape. Petromarula 1, pp. 29-39. - Cambridge: Corpus Christi College.

RADDI, S. & SÜMER, S. 1999: Genetic diversity in natural *Cupressus sempervirens* L. populations in Turkey. - *Biochem. Syst. Ecol.* **27(8)**: 799-814.

RADZHI, A.D. 1972: De genere *Vicia* L. notula systematica [In Russian]. - *Novosti Sist. Vysš. Rast.* **9**: 217-223.

RAFFAELLI, M. 1977a: Consistenza numerica e distribuzione in Italia delle specie del genere *Parietaria* L. - *Giorn. Bot. Ital., n.s.* **111(6)**: 364.

RAFFAELLI, M. 1977b: Ad floram italicam notulae taxonomicae et geobotanicae. 22. Note corologiche sulle specie italiane del genere *Parietaria* L. - *Webbia* **31(1)**: 49-68.

RAFFAELLI, M. 1979a: Contributi alla conoscenza del genere *Polygonum* L. 1. *Polygonum romanum* Jacq. e *Polygonum gallicum* sp. nov. - *Webbia* **33(2)**: 307-325.

RAFFAELLI, M. 1979b: Contributi alla conoscenza del genere *Polygonum* L. 2. *Polygonum bellardii* All. - *Webbia* **33(2)**: 327-342.

RAFFAELLI, M. 1982: Contributi alla conoscenza del genere *Polygonum* L. 4. Le specie italiane della sect. *Polygonum*. - *Webbia* **35(2)**: 361-406.

RAFFAELLI, M. 1983: Lectotypification of *Polygonum maritimum* L. (Polygonaceae). - *Taxon* **32(1)**: 115-117.

RAFFAELLI, M. 1991: *Biscutella* L. ser. *Lyratae* Malin. (Cruciferae) in Italia. - Discussione sui caratteri morfologici e tassonomia. - *Webbia* **45(1)**: 1-30.

RAFFAELLI, M. 1992: *Biscutella* sect. *Iondraba* (Cruciferae) in the Mediterranean area. – *Willdenowia* **22(1-2)**: 19-36.

RAFFAELLI, M. 1993: Acquisizioni recenti e prospettive di ricerca sul genere *Biscutella* L. (Cruciferae) in Italia (con riferimenti all'area europea e mediterranea). - *Webbia* **48**: 761-771.

RAFFAELLI, M. & FIORINI, G. 1986: Numeri cromosomici per la Flora Italiana: 1049-1051. - *Inform. Bot. Ital.* **17(1-3)**: 99-101.

RAFFAELLI, M. & RICCERI, C. 1988: Su alcune specie mediterranee del genere *Euphorbia* L.: *E. bivonae* Steud., *E. papillaris* (Boiss.) Raffaelli et Ricceri, stat. nov., *E. melitensis* Parl. - *Webbia* **42 (1)**: 1-13.

RAHN, K. 1954: Experimental and cytological studies in *Plantago media*. - *Bot. Tidsskr.* **51**: 300-307.

'RAHN, K. 1957: Chromosome numbers in *Plantago*. - *Bot. Tidsskr.* **53**: 369-378.

RAHN, K. 1966: [Chromosome numbers in *Plantago*]. In Löve, Á. (ed.): IOPB chromosome number reports VI. - *Taxon* **15(3)**: 122-127.

RAHN, K. 1976: [Chromosome numbers in *Plantago*]. In Löve, Á. (ed.): IOPB chromosome number reports LIII. - *Taxon* **25(4)**: 499-500.

RAHN, K. 1978: Nomenclatorial changes within the genus *Plantago* L. Infraspecific taxa and subdivisions of the genus. - *Bot. Tidsskr.* **73(2)**: 106-111.

RAHN, K. 1996: A phylogenetic study of the Plantaginaceae. - *Bot. J. Linn. Soc.* **120(2)**: 145-198.

RAIMONDO, F.M. 1993: *Limonium optimae*, a new species from central Sicily. - *Fl. Medit.* **3**: 13-18.

RAIMONDO, F.M. 1997: Les membres italiens du complexe de *Brassica oleracea*: leur distribution et spécificités écologiques. - *Bocconea* **7**: 103-106.

RAIMONDO, F.M. & AL. 1981: Numeri cromosomici per la Flora Italiana: 772-777. - *Inform. Bot. Ital.* **12 (2)**: 313-319.

RAIMONDO, F.M. & GARBARI, F. 1976: Numeri cromosomici per la Flora Italiana: 199-207. - *Inform. Bot. Ital.* **7(3)**: 366-377.

RAIMONDO, F.M., MAZZOLA, P. & OTTONELLO, D. 1991: On the taxonomy and distribution of *Brassica* sect. *Brassica* (Cruciferae) in Sicily. - *Fl. Medit.* **1**: 63-86.

RAIMONDO, F.M., MAZZOLA, P. & SPADARO, V. 2015: A new species of *Smyrnium* (Apiaceae) related to *S. perfoliatum*. - *Fl. Medit.* **25**: 137-142.

RAIMONDO, F.M. & PIGNATTI, S. 1986: Una nuova species di *Limonium* (Plumbaginaceae) della Sicilia. - *Webbia* **39(2)**: 417-421.

RAIMONDO, F.M., ROSSITTO, M. & OTTONELLO, D. 1984: Numeri cromosomici per la Flora Italiana: 967-976. - *Inform. Bot. Ital.* **15(1)**: 58-65.

RAIMONDO, F.M., ROSSITTO, M. & OTTONELLO, D. 1985: Numeri cromosomici per la Flora Italiana: 983-992. - *Inform. Bot. Ital.* **15(2-3)**: 180-187.

RAIMONDO, F.M., ROSSITTO, M. & VILLARI, R. 1982: Bibliografia geobotanica siciliana. - Palermo: Consiglio Nazionale delle Ricerche, 159 pp.

RAINA, S.N., YAMAMOTO, K. & MURAKAMI, M. 1989: Intraspecific hybridization and its bearing on chromosome evolution in *Vicia narbonensis* (Fabaceae). - *Pl. Syst. Evol.* **167(3-4)**: 201-217.

RAJHATHY, T. & MORRISON, J.W. 1962: Cytogenetic studies in the genus *Hordeum*. VI. The *Murinum*- complex. - *Canad. J. Genet. Cytol.* **4(2)**: 240-247.

RAJHATHY, T. & THOMAS, H. 1974: Cytogenetics of oats (*Avena* L.). - Ottawa: Genetics Society of Canada, 90 pp.

RAJPUT, M.T.M., TAHIR, S.S. & HUSAIN, S.Z. 1994: The distribution and synonymy of *Sibbaldia procumbens* L. (a type species of genus *Sibbaldia*) Rosaceae. III. - *Candollea* **49(1)**: 133-139.

RAKAJ, M. & ROSTAŃSKI, K. 2009: New species of the genus *Oenothera* L. from coasts of the Albania and Montenegro. - *Natura Montenegrina* **8(3)**: 163-171.

RAKOSY, D., PAULUS, H.F. & HIRTH, M. 2020: *Ophrys dimidiata*, eine neue, bisher unbeschriebene Art aus dem *Ophrys tenthredinifera*-Komplex auf Kreta (Orchidaceeae). - *J. Eur. Orch.* **52(1)**: 207-226.

RAMÓN-LACA, L. & MABBERLEY, D.J. 2004: The ecological status of the carob-tree (*Ceratonia siliqua*, Leguminosae) in the Mediterranean. - *Bot. J. Linn. Soc.* **144(4)**: 431-436.

RAMOS NUÑES, A. 1983: Estudio biosistemático del género *Hypericum* L. (Guttiferae) en la Península Ibérica e Islas Baleares. I. Caracteres seminales. - *Trab. Dept. Bot. Univ. Complut. Madrid* **12**: 45-62.

RAMOS NUÑES, A. 1986: Estudio biosistemático del género *Hypericum* L. (Guttiferae) en la Península Ibérica e Islas Baleares: 2. Biotipo, ramificación e inflorescencias. - *Acta Bot. Malac.* **11**: 157-176.

RANDALL, R.E. 2003: Biological Flora of the British Isles. *Smyrnium olusatrum* L. - *J. Ecol.* **91(2)**: 325-340.

RANDALL, R.E. 2004: Biological Flora of the British Isles. *Viola kitaibeliana* Schult(es). - *J. Ecol.* **92(2)**: 361-369.

RANDALL, R.E. & TORNTON, G. 1996: Biological Flora of the British Isles. *Peucedanum officinale* L. - *J. Ecol.* **84(3)**: 475-485.

RANĐELOVIĆ, N. 1977: Verbreitung von Pflanzenarten in Südostserbien (II) [In Serbo-Croatian with German summary]. - *Glasn. Prir. Mus. Beogradu, Ser. B, Biol. Nauke* **32**: 81-86.

RANĐELOVIĆ, N. 1979: Verbreitung von Pflanzenarten in Südostserbien (III) [In Serbo-Croatian with German summary]. - *Glasn. Prir. Mus. Beogradu, Ser. B, Biol. Nauke* **34**: 127-130.

RANĐELOVIĆ, N. 1980: *Campanula versicolor* Andrews. Eine neue Art in der Flora Serbiens [In Serbo- Croatian with German summary]. - *Glasn. Prir. Mus. Beogradu, Ser. B, Biol. Nauke* **35**: 53-55.

RANĐELOVIĆ, N. & AL. 1990: A new species of *Crocus* from Yugoslavia. - *Bot. Mag. (Kew Mag.)* **7(4)**: 182-186.

RANĐELOVIĆ, N., BUDAK, V. & STAMENKOVIĆ, V. 1980: Horologische und systematische Betrachtungen der Arten von *Teesdalia* und *Clypeola* [In Serbo-Croatian with German summary]. - *Glasn. Prir. Mus. Beogradu, Ser. B, Biol. Nauke* **35**: 47-52.

RANĐELOVIĆ, N., HILL, D.A. & RANĐELOVIĆ, V. 1990: The genus *Crocus* L. in Serbia. - Beograd: Serbian Academy of Sciences and Arts, 52 pp. + 16 plates.

RANĐELOVIĆ, N. & JOVANOVIĆ, V. 1978: *Trifolium heldreichianum* Hausskn. Eine neue Art in der Flora Serbiens [In Serbo-Croatian with German summary]. - *Glasn. Prir. Mus. Beogradu, Ser. B, Biol. Nauke* **33**: 163-165.

RANĐELOVIĆ, N., RANĐELOVIĆ, V. & HRISTOVSKI, N. 2012: *Crocus jablanicensis* (Iridaceae), a new species from the Republic of Macedonia, Balkan Peninsula. - *Ann. Bot. Fenn.* **49(1-2)**: 99-102.

RANĐELOVIĆ, N., REXHEPI, F. & RUŽIĆ, M. 1977: *Trifolium phleoides* Pourr. ex Willd. Neue Pflanze in Flora von Serbien [In Serbo-Croatian with German summary]. - *Glasn. Prir. Mus. Beogradu, Ser. B, Biol. Nauke* **32**: 41-43.

RANĐELOVIĆ, N. & SEKOVSKI, Ž. 2004: The genus *Crocus* L. in Macedonia. In Vasić, O. (ed.): Abstracts, XI OPTIMA Meeting, Beograd 5-11.IX.2004, p. 108.

RANĐELOVIĆ, N. & STAMENKOVIĆ, V. 1978: *Trifolium cherleri* L. nova vrsta u flori Srbije [In Serbo- Croatian]. - *Glasn. Prir. Mus. Beogradu, Ser. B, Biol. Nauke* **33**: 175-177.

RANĐELOVIĆ, V. & AL. 2010: Phytogeographical and phytocoenological analysis of the threatened plant taxa in the flora of the Vlasina plateau (SE Serbia). - *Biol. Nyssana* **1(1-2)**: 1-7.

RANĐELOVIĆ, V., MILJKOVIĆ, M. & HARPE, D. 2015: *Crocus novicae* sp. nov. from southern Albania (SW Balkan Peninsula). In Bogdanović, S. & Jogan, N.: Book of abstracts - 6th Balkan Botanical Congress, p. 33. - Rijeka.

RANĐELOVIĆ, V., RANĐELOVIĆ, N. & ZLATKOVIĆ, B. 1997: *Geum rhodopaeum* Stoj. & Stef. in flora of Serbia. In: Abstracts from the First Balkan Botanical Congress, Thessaloniki, September 19-22, 1997, p. 41. - Thessaloniki.

RANDJELOVIĆ, N., SEKOVSKI, Ž. & GORDANA, D. 2007: The systematic horological and genetic research into the genus *Crocus* L. in Macedonia. In Matevski, V. (ed.): Collection of papers devoted to Academician Kiril Micevski, pp. 98-131. - Skopje: Macedonian Academy of Sciences and Arts.

RANDOLPH, L.F. 1955: The geographic distribution of European and Eastern Mediterranean species of bearded *Iris*. - *Iris Year Book* **1955**: 35-46.

RANDOLPH, L.F. & MITRA, J. 1956: Chromosome numbers of *Iris* species. - *Bull. Amer. Iris Soc.* **140**: 50-60.

RANDOLPH, L.F. & MITRA, J. 1959: Karyotypes of *Iris pumila* and related species. - *Amer. J. Bot.* **46(2)**: 93-102.

RANDOLPH, L.F. & RECHINGER, K.H. 1954: Die geographische Verbreitung einiger europäischer und mediterraner *Iris*-Arten. - *Verh. Zool.-Bot. Ges. Wien* **94**: 82-96.

RANJBAR, R & KHALVATI, S. 2019: World checklist of *Moltkia* (Boraginaceae) with notes on types. - *Phytotaxa* **408(3)**: 143-160.

RAO, C.K. & JARVIS, C.E. 1986: Lectotypification, taxonomy and nomenclature of *Morus alba*, *M. tatarica* and *M. indica* (Moraceae). - *Taxon* **35(4)**: 705-708.

RAPTIS, D. & AL. 2018: TAU Herbarium: Digitisation of specimens from the early era of the botanical exploration of Greece. - *Bot. Serbica* **142 (suppl. 1)**: 176-177.

RAPTIS, D., KAROUSOU, R. & HANLIDOU, E. 2019: Type specimens in the herbarium of Aristotle University of Thessaloniki (TAU). - *Bot. Chron. (Patras)* **22**: 307-318.

RASBACH, H. & AL. 1986: *Asplenium cuneifolium* Viv. (diploid) from the type locality (Aspleniaceae, Pteridophyta) with an appendix on related plants from other places in south-western and central Europe. - *Candollea* **41(1)**: 219-244.

RASBACH, H., RASBACH, K. & REICHSTEIN, T. 1977: *Cheilanthes guanchica* Bolle in Europa. - *Ber. Deutsch. Bot. Ges.* **90(3)**: 527-530.

RASBACH, H. & REICHSTEIN, T. 1982: Four natural hybrids in the genus *Cheilanthes* (Sinopteridaceae, Pteridophyta). - *Webbia* **35(2)**: 261-273.

RASBACH, H. & REICHSTEIN, T. 1990: The chromosome number of *Anogramma leptophylla* (Adianthaceae: Pteridophyta) from Europe. - *Brit. Fern Gaz.* **13(6)**: 341-348.

RASBACH, H., REICHSTEIN, T. & SCHNELLER, J.J. 1982: Cytological examination of *Dryopteris villarii* (Bell.) Woynar ex Schinz & Thellung from the locus classicus (type locality). - *Bot. Helv.* **92(1)**: 33-40.

RASBACH, H., REICHSTEIN, T. & SCHNELLER, J.J. 1983: Five further natural hybrids in the genus "Cheilanthes" Sw. (Sinopteridaceae, Pteridophyta). - *Webbia* **37(1)**: 43-62.

RASBACH, H., REICHSTEIN, T. & SCHNELLER, J.J. 1991: Hybrids and polyploidy in the genus *Athyrium* (Pteridophyta) in Europe. 2. Origin and description of two triploid hybrids and synthesis of allotetraploids. - *Bot. Helv.* **101(2)**: 209-225.

RASPÉ, O., FINDLAY, C. & JACQUEMART, A.-L. 2000: Biological Flora of the British Isles. *Sorbus aucuparia* L. - *J. Ecol.* **88(5)**: 910-930.

RAT, M. 2015: Herbarium revision reveals neglected and misinterpreted *Ornithogalum* L. species in the Balkan Peninsula. In Bogdanović, S. & Jogan, N.: Book of abstracts - 6th Balkan Botanical Congress, p. 34. - Rijeka.

RATCLIFFE, D. 1959: Biological Flora of the British Isles. *Hornungia petraea* (L.) Rchb.

(*Lepidium petraeum* L.; *Hutchinsia petraea* (L.) R. Br.). - *J. Ecol.* **47(1)**: 241-247.

RATCLIFFE, D. 1960: Biological Flora of the British Isles. *Draba muralis* L. - *J. Ecol.* **48**: 737-744.

RATTER, J.A. 1964: Cytogenetic studies in *Spergularia*. I. Cytology of some Old World species. - *Notes Roy. Bot. Gard. Edinburgh* **25(3)**: 293-302 + 1 plate.

RATTER, J.A. 1965a: Cytogenetic studies in *Spergularia*. II: An attempt to discover cytogenetic relationships of some species. - *Notes Roy. Bot. Gard. Edinburgh* **26(2)**: 203-223 + 4 plates.

RATTER, J.A. 1965b: Cytogenetic studies in *Spergularia*: III. Some interspecific hybrids involving *S. marina* (L.) Griseb. - *Notes Roy. Bot. Gard. Edinburgh* **26(2)**: 224-236 + 1 plate.

RATTER, J.A. 1969a: Cytogenetic studies in *Spergularia*: IV. Some further interspecific hybrids. - *Notes Roy. Bot. Gard. Edinburgh* **29(2)**: 213-223 + 1 plate.

RATTER, J.A. 1969b: Cytogenetic studies in *Spergularia*: V. Some interspecific hybrids involving *S. media* (L.) C. Presl. - *Notes Roy. Bot. Gard. Edinburgh* **29(2)**: 225-232 + 1 plate.

RATTER, J.A. 1973a: Cytogenetic studies in *Spergularia*: VII. Cryptic speciation in *S. media* (L.) C. Presl and *S. marina* (L.) Griseb. - *Notes Roy. Bot. Gard. Edinburgh* **32(2)**: 291-296.

RATTER, J.A. 1973b: Cytogenetic studies in *Spergularia*: VIII. Barriers to the production of viable interspecific hybrids. - *Notes Roy. Bot. Gard. Edinburgh* **32(2)**: 297-301.

RATTER, J.A. 1976: Cytogenetic studies in *Spergularia*: IX. Summary and conclusions. - *Notes Roy. Bot. Gard. Edinburgh* **34(3)**: 411-428.

RATTER, J.A. 1992: A new combination in *Spergularia* (Pers.) J. & C. Presl. In Newton, M.E. (ed.): Flora Europaea: Notulae systematicae ad Floram Europaeam spectantes. Series 2. No. 5. - *Bot. J. Linn. Soc.* **109(3)**: 322-323.

RÄTZEL, S. & AL. 2016: *Phelipanche olbiensis*. In Raab-Straube, E. v. & Raus, Th. (eds): Euro+Med-Checklist Notulae, 6. - *Willdenowia* **46(3)**: 432-433.

RÄTZEL, S. & AL. 2018: *Phelipanche gussoneana* (Lojac.) Domina, Raab-Straube, Rätzel & Uhlich, comb. nov. In Raab-Straube, E. v. & Raus, Th. (eds): Euro+Med-Checklist notulae, 9. - *Willdenowia* **48(2)**: 209-210.

RÄTZEL, S. & AL. 2020: *Phelipanche gussoneana*. In Raus, Th. & Raab-Straube, E.v. (eds): Euro+Med-Checklist notulae, 12. - *Willdenowia* **50(2)**: 323-326.

RÄTZEL, S., KALAENTZIS, K. & RISTOW, M. 2021: *Chasmanthe floribunda*. In Raus, Th. & Raab-Straube, E.v. (eds): Euro+Med-Checklist notulae, 13. - *Willdenowia* **51(1)**: 151.

RÄTZEL, S., RISTOW, M. & UHLICH, H. 2017: Bemerkungen zu ausgewählten Vertretern der Gattung *Phelipanche* Pomel im östlichen Mittelmeergebiet mit der Beschreibung von *Phelipanche hedypnoidis* Rätzel, Ristow & Uhlich, sp. nov. - *Carinthia II* **207**: 643-684.

RÄTZEL, S., RISTOW, M. & UHLICH, H. 2018: *Phelipanche orientalis* (Beck) Soják. In Raab-Straube, E. v. & Raus, Th. (eds): Euro+Med-Checklist notulae, 9. - *Willdenowia* **48(2)**: 211-213.

RÄTZEL, S., RISTOW, M. & UHLICH, H. 2019: *Orobanche ballotae*. In Raab-Straube, E. v. & Raus, Th. (eds): Euro+Med-Checklist notulae, 11. - *Willdenowia* **49(3)**: 435-436.

RÄTZEL, S. & UHLICH, H. 2018: *Orobanche litorea* Guss. (= *O. fuliginosa* Jord.; = *O. sabulicola* Lojac.). In Raab-Straube, E. v. & Raus, Th. (eds): Euro+Med-Checklist notulae, 9. - *Willdenowia* **48(2)**: 207-209.

RAUH, W. 1949: Klimatologie und Vegetationsverhältnisse der Athos-Halbinsel und der ostägäischen Inseln Lemnos, Evstratios, Mytiline und Chios. - *Sitzungsber. Heidelberger Akad. Wiss., Math.- Naturwiss. Kl.* **1949(12)**: 511-615.

RAULIN, V. 1869: Description physique de l'île de Crète. Livre IV: Botanique. - Paris: F. Savy, Libraire-Éditeur, pp. 691-974 + 18 plates and 2 maps.

RAUS, TH. 1979a: Die Vegetation Ostthessaliens (Griechenland). I. Vegetationzonen und Höhenstufen. - *Bot. Jahrb. Syst.* **100(4)**: 564-601.

RAUS, TH. 1979b: Die Vegetation Ostthessaliens (Griechenland). II. Quercetea ilicis und Cisto-Micromerietea. - *Bot. Jahrb. Syst.* **101(1)**: 17-82.

RAUS, TH. 1980: Die Vegetation Ostthessaliens (Griechenland). III. Querco-Fagetea und azonale Gehölzgesellschaften. - *Bot. Jahrb. Syst.* **101(3)**: 313-361.

RAUS, TH. 1981: Human interference with zonal vegetation in the Thessalian coastal section of the Aegean. In Frey, W. & Uerpmann, H.-P. (Hrsg.): Beiträge zur Umweltgeschichte des Vorderen Orients. - *Beih. Tübinger Atlas Vorderen Orients, A* **8**: 40-50.

RAUS, TH. 1982a: Phytogeographical circumscription of the Mediterranean area on the Balkan peninsula and the problem of Thessalian pseudomaquis. - *Ecol. Medit.* **8(1-2)**: 197-201.

RAUS, TH. 1982b: *Azolla caroliniana* Willd. In Greuter, W. & Raus, Th. (eds): Med-Checklist notulae, 6. - *Willdenowia* **12(2)**: 183.

RAUS, TH. 1982c: *Amsinckia lycopsoides* (Lehm.) Lehm. In Greuter, W. & Raus, Th. (eds): Med- Checklist notulae, 6. - *Willdenowia* **12(2)**: 184.

RAUS, TH. 1983a: Ein für Europa neuer Krokus. - *Bot. Chron. (Patras)* **3(1-2)**: 12-14.

RAUS, TH. 1983b: Wechselnasse Wiesen in Griechenland. - *Tuexenia* **3**: 259-270.

RAUS, TH. 1984a: *Asperugo procumbens* L. In Greuter, W. & Raus, Th. (eds): Med-Checklist notulae, 8. - *Willdenowia* **13(2)**: 278.

RAUS, TH. 1984b: *Bassia hirsuta* (L.) Ascherson. In Greuter, W. & Raus, Th. (eds): Med-Checklist notulae, 8. - *Willdenowia* **13(2)**: 282.

RAUS, TH. 1984c: *Silene fabaria* (L.) Sm. In Greuter, W. & Raus, Th. (eds): Med-Checklist notulae, 9. - *Willdenowia* **14(1)**: 46.

RAUS, TH. 1986a: Floren- und Vegetationsdynamik auf der Vulkaninsel Nea Kaimeni (Santorin-Archipel, Kykladen, Griechenland). - *Abh. Westfäl. Mus. Naturk.* **48**: 373-394.

RAUS, TH. 1986b: *Sideritis athoa* Papanicolaou & Kokkini. In Greuter, W. & Raus, Th. (eds): Med- Checklist notulae, 13. - *Willdenowia* **16(1)**: 107.

RAUS, TH. 1987a: *Soldanella* (Primulaceae) in Griechenland. - *Willdenowia* **16(2)**: 335-342.

RAUS, TH. 1987b: Gipfelvegetation des Ossa-Gebirges im Vergleich zum Olymp (Thessalien, Griechenland). - *Münstersche Geogr. Arbeiten* **27**: 199-206.

RAUS, TH. 1988: Vascular plant colonization and vegetation development on sea-born volcanic islands in the Aegean (Greece). - *Vegetatio* **77(1-3)**: 139-147.

RAUS, TH. 1989a: Notes on rare vascular wetland plants of Greece. In Phitos, D. (ed.): Abstracts. Sixth OPTIMA meeting, Delphi 10-16 Sept. 1989, p. 130.

RAUS, TH. 1989b: *Astragalus peregrinus* Vahl subsp. *peregrinus*. In Greuter, W. & Raus, Th. (eds): Med-Checklist notulae, 15. - *Willdenowia* **19(1)**: 30.

RAUS, TH. 1989c: *Mirabilis jalapa* L. In Greuter, W. & Raus, Th. (eds): Med-Checklist notulae, 15. - *Willdenowia* **19(1)**: 37.

RAUS, TH. 1989d: *Adonis micrccarpa* DC. subsp. *microcarpa*. In Greuter, W. & Raus, Th. (eds): Med- Checklist notulae, 15. - *Willdenowia* **19(1)**: 43.

RAUS, TH. 1990: Die Flora von Armathia und der Kleininseln um Kasos (Dodekanes, Griechenland). - *Bot. Chron. (Patras)* **9(1-2)**: 19-39.

RAUS, TH. 1991a: Notes on rare vascular wetland plants of Greece. - *Bot. Chron. (Patras)* **10**: 567-578.

RAUS, TH. 1991b: Asia or Europe? - The phytogeographical position of the Karpathos archipelago (SE Aegean, Greece). - *Fl. Veg. Mundi* **9**: 301-310.

RAUS, TH. 1991c: Die Flora (Farne und Blütenpflanzen) des Santorin-Archipels. In Schmalfuss, H. (ed.): Santorin - Leben auf Schutt und Asche. Ein naturkundlicher Reiseführer, pp. 109-124. - Weikersheim: Verlag Josef Markgraf.

RAUS, TH. 1994: Zur Winterphänologie zentralägäischer Farn- und Blütenpflanzen. - *Phytocoenologia* **24**: 559-572.

RAUS, TH. 1995a: *Ammannia* (Lythraceae) in Greece and the Balkans. In Valdés, B. & Silvestre, S. (eds): Abstracts. VIII OPTIMA meeting, Sevilla, Spain, 25 September - 1 October, 1995, p. 117.

RAUS, TH. 1995b: *Anthemis ruthenica* M. Bieb., *Amorpha fruticosa* L. and *Verbascum adrianopolitanum* Podp. In Greuter, W. & Raus, Th. (eds): Med-Checklist notulae, 16. - *Willdenowia* **25(1)**: 172-174.

RAUS, TH. 1996a: The boreal and central European element in the forest flora of Greece. - *Bocconea* **5(1)**: 63-76.

RAUS, TH. 1996b: Additions and amendments to the flora of the Karpathos island group (Dodekanesos, Greece). - *Bot. Chron. (Patras)* **12**: 21-53.

RAUS, TH. 1996c: Flora von Paros und Antiparos (Kykladen, Griechenland). - *Ann. Naturhist. Mus. Wien, B* **98B Suppl.**: 237-278.

RAUS, TH. 1997a: *Amaranthus* (Amaranthaceae) in Greece and the Aegean. In: Abstracts from the First Balkan Botanical Congress, Thessaloniki, September 19-22, 1997, p. 42. - Thessaloniki.

RAUS, TH. 1997b: *Ammannia* (Lythraceae) in Greece and the Balkans. - *Lagascalia* **19(1-2)**: 851-856.

RAUS, TH. 1997c: Zur Flora und Pflanzengeographie Nordost-Griechenlands - Befunde und Bewertungen. - *Ber. Arbeitsgeb. Entwicklungsforsch. (Münster)* **27**: 26-40.

RAUS, TH. 1997d: Zwergbinsen-Gesellschaften des Fimbristylidenion bisumbellatae in Griechenland - Kenntnisstand und Forschungsbedarf. - *Acta Bot. Hung.* **40(1-4)**: 203-214.

RAUS, TH. 1998: Notes on *Lappula barbata, Euphorbia maculata, Euphorbia prostrata* and *Peucedanum arenarium* subsp. *neumayeri*. In Greuter, W. & Raus, Th. (eds): Med-Checklist notulae, 17. - *Willdenowia* **28(1-2)**: 163-174.

RAUS, TH. 1999: Heinrich Kuhbier und die geobotanische Erforschung der Insel Korfu (Jonische Inseln, Griechenland). - *Abh. Naturwiss. Verein Bremen* **44(2-3)**: 397-416.

RAUS, TH. 2001: *Bacopa* (Scrophulariaceae) in Greece and adjacent countries. In Raimondo, F.M. (ed.): Abstracts from the X OPTIMA meeting, Palermo, 13-19 September 2001, p. 150.

RAUS, TH. 2002: Notes on *Amaranthus quitensis* Kunth and *Anredera cordifolia* (Ten.) Steenis. In Greuter, W. & Raus, Th. (eds): Med-Checklist notulae, 21. - *Willdenowia* **32(2)**: 195-196.

RAUS, TH. 2003: *Bacopa* (Scrophulariaceae) in Greece and adjacent countries. - *Bocconea* **16(2)**: 745-749.

RAUS, TH. 2004: *Cyperus pannonicus* Jacq. In Greuter, W. & Raus, Th. (eds): Med-Checklist notulae, 22. - *Willdenowia* **34(1)**: 77.

RAUS, TH. 2005: *Viola samothracica* (Degen) Raus, comb. & stat. nov. In Greuter, W. & Raus, Th. (eds): Med-Checklist notulae, 23. - *Willdenowia* **35(1)**: 61.

RAUS, TH. 2006a: Found and lost - Staphyleaceae in Greece. - *Willdenowia* **36(1)**: 311-315.

RAUS, TH. 2006b: *Amaranthus watsonii, Oenothera indecora, Oenothera laciniata, Soldanella chrysosticta, Lycium chinense*. In Greuter, W. & Raus, Th. (eds): Med-Checklist notulae, 24. - *Willdenowia* **36(2)**: 720, 724-726.

RAUS, TH. 2007: *Ribes odoratum, Aloysia citriodora*. In Greuter, W. & Raus, Th. (eds): Med-Checklist notulae, 25. - *Willdenowia* **37(1)**: 210-211.

RAUS, TH. 2011: [Notes on *Alnus alnobetula*]. In Greuter, W. & Raab-Straube, E.v.: Euro+Med Notulae, 5. - *Willdenowia* **41(1)**: 129-130.

RAUS, TH. 2012a: Gefässpflanzen von Milos (Kykladen, Griechenland) - eine floristische Handreichung. - *Verh. Zool.-Bot. Ges. Österreich* **148/149**: 197-235.

RAUS, TH. 2012b: *Ornithogalum ulixis* (Speta) Raus, comb. nov. In Greuter, W. & Raus, Th. (eds): Med-Checklist notulae, 31. - *Willdenowia* **42(2)**: 294.

RAUS, TH. 2013a: *Quercus petraea* subsp. *polycarpa*. In Raab-Straube, E.v & Raus, Th.: Notulae ad floram euro-mediterraneam pertinentes No. 30. - *Willdenowia* **43(1)**: 159.

RAUS, TH. 2013b: *Sternbergia colchiciflora* Waldst. & Kit. In Raab-Straube, E. v. & Raus, Th. (eds): Euro+Med-Checklist notulae, 2. - *Willdenowia* **43(2)**: 245.

RAUS, TH. 2013c: *Festuca jeanpertii* subsp. *gracilis* (Hack.) Raus, comb. nov. In Raab-Straube, E. v. & Raus, Th. (eds): Euro+Med-Checklist notulae, 2. - *Willdenowia* **43(2)**: 246.

RAUS, TH. 2014: [Notes on *Campanula*, *Symphyandra*, *Draba*, *Noccaea*, *Genista* and *Pyrus*]. In Raab-Straube, E. v. & Raus, Th. (eds): Euro+Med-Checklist notulae, 3. - *Willdenowia* **44(2)**: 288-294.

RAUS, TH. 2018a: *Sempervivum vereregínae-amaliae* Raus, nom. nov. In Raab-Straube, E. v. & Raus, Th. (eds): Euro+Med-Checklist notulae, 9. - *Willdenowia* **48(2)**: 204-205.

RAUS, TH. 2018b: Santorin: Vegetation und Vulkanismus. - *Hellenika Jahrb. Griech. Kultur Deutsch- Griech. Beziehungen Ser. 2* **13**: 26-45.

RAUS, TH. 2019: Araliaceae in Greece revisited for the first volume of the "Flora of Greece". In Daskalakou, E.N. & al. (eds): 16th Conference of the Hellenic Botanical Society – Athens 10-13 October 2019. Program and book of abstracts, p. 8.

RAUS, TH. 2021: *Tulipa clusiana*. In Raus, Th. & Raab-Straube, E.v. (eds): Euro+Med-Checklist notulae, 13. - *Willdenowia* **51(1)**: 152.

RAUS, TH. 2021: *Rhinanthus rumelicus*. In Raus, Th. & Raab-Straube, E.v. (eds): Euro+Med-Checklist notulae, 14. - *Willdenowia* **51(3)**: 364.

RAUS, TH., DIMOPOULOS, P. & KARADIMOU, E. 2017: Vascular plant diversity on the Christiana island group (Cyclades, Greece): current state of floristic knowledge. In Dousis, M.A. & Thanos, K.A.: 15th conference of the Hellenic Botanical Society, book of abstracts: p. 67.

RAUS, TH., DIMOPOULOS, P. & KARADIMOU, E. 2018: Vascular plant diversity on the Christiana island group and Aspronisi island (Cyclades, Greece). - *Bot. Serbica* **142 (suppl. 1)**: 74-75.

RAUS, TH., ELEFTHERIADOU, E. & THEODOROPOULOS, K. 2002: Notes on *Heteranthera reniformis* Ruiz & Pav. and *H. rotundifolia* (Kunth) Griseb. In Greuter, W. & Raus, Th. (eds): Med-Checklist notulae, 21. - *Willdenowia* **32(2)**: 207-208.

RAUS, TH., KARADIMOU, E. & DIMOPOULOS, P. 2019: Taxonomic and functional plant diversity of the Santorini-Christiana island group (Aegean Sea, Greece). - *Willdenowia* **49(3)**: 363-381 + suppl..

RAUS, TH. & RAAB-STRAUBE, E.V. 2021: *Sempervivum vereregínae-amaliae*. In Raus, Th. & Raab- Straube, E.v. (eds): Euro+Med-Checklist notulae, 13. - *Willdenowia* **51(1)**: 148.

RAUS, TH. & RAAB-STRAUBE, E.V. (EDS) 2021: Euro+Med-Checklist notulae, 13. - *Willdenowia* **51(1)**: 141-168.

RAUS, TH. & RAABE, U. 2002a: Paddy weeds - a neglected subject of Greek floristics. In Kamari, G. & al. (eds): Ellinini Votaniki Eteria. 9° Panelliniko Epistimoniko Sinedrio. Praktika, pp. 290-300 [In Greek with English abstract].

RAUS, TH. & RAABE, U. 2002b: Notes on *Ammannia baccifera* L., *Ammannia senegalensis* L. and *Eleocharis parvula*. In Greuter, W. & Raus, Th. (eds): Med-Checklist notulae, 21. - *Willdenowia* **32 (2)**: 202, 204.

RAUS, TH. & RAABE, U. 2006: *Amaranthus palmeri*. In Greuter, W. & Raus, Th. (eds): Med-Checklist notulae, 24. - *Willdenowia* **36(2)**: 719-720.

RAUS, TH. & RAABE, U. 2016: *Proboscidea lousianica*. In Raab-Straube, E. v. & Raus, Th. (eds): Euro+Med-Checklist notulae, 6. - *Willdenowia* **46(3)**: 429-430.

RAUS, TH., RISTOW, M. & SCHOLZ, H. 2008: *Catapodium balearicum* (Willk.) H. Scholz. In Greuter, W. & Raus, Th. (eds): Med-Checklist notulae, 27. - *Willdenowia* **38(2)**: 470.

RAUS, TH. & SCHOLZ, H. 2002: Notes on *Diplachne fusca* (L.) P. Beauv. and *Echinochloa crus-galli* subsp. *hispidula* (Retz.) Honda. In Greuter, W. & Raus, Th. (eds): Med-Checklist notulae, 21. - *Willdenowia* **32(2)**: 205-206.

RAUS, TH. & SCHULER, A. 2005: *Cenchrus incertus* M.A. Curtis. In Greuter, W. & Raus, Th. (eds): Med-Checklist notulae, 23. - *Willdenowia* **35(1)**: 62-63.

RAUS, TH. & SCHULER, A. 2021: *Spartina versicolor*. In Raus, Th. & Raab-Straube, E.v. (eds): Euro+Med-Checklist notulae, 13. - *Willdenowia* **51(1)**: 150.

RAUS, TH. & STRID, A. 2006: *Alopecurus arundinaceus*. In Greuter, W. & Raus, Th. (eds): Med- Checklist notulae, 24. - *Willdenowia* **36(2)**: 727.

RAUS, TH. & STRID, A. 2019: *Clinopodium dalmaticum* subsp. *bulgaricum*. In Raab-Straube, E. v. & Raus, Th. (eds): Euro+Med-Checklist notulae, 10. - *Willdenowia* **49(1)**: 105-106.

RAUS, TH. & STRID, A. 2020: New combinations in some Balkan *Centaurea* taxa. - *Phytol. Balcan.* **26 (3)**: 505-507.

RAUS, TH. & STRID, A. 2021: *Atriplex tatarica*. In Raus, Th. & Raab-Straube, E.v. (eds): Euro+Med-Checklist notulae, 13. - *Willdenowia* **51(1)**: 145.

RAUS, TH., STRID, A. & TAN, KIT 1999: Notes on *Leersia oryzoides*. In Greuter, W. & Raus, Th. (eds): Med-Checklist notulae, 18. - *Willdenowia* **29(1-2)**: 65-66.

RAUS, TH., STRID, A. & TAN, KIT 2002: Notes on *Stachys obliqua* L., *Trifolium retusum* L. and *Lindernia procumbens* (Koch) Borbás. In Greuter, W. & Raus, Th. (eds): Med-Checklist notulae, 21. - *Willdenowia* **32(2)**: 200-204.

RAUS, TH., STRID, A. & YANNITSAROS, A. 2004: *Acer negundo* L. In Greuter, W. & Raus, Th. (eds): Med- Checklist notulae, 22. - *Willdenowia* **34(1)**: 71-72.

RAUS, TH., TAN, KIT & STRID, A. 1998: Notes on *Bombycilaena erecta*. In Greuter, W. & Raus, Th. (eds): Med-Checklist notulae, 17. - *Willdenowia* **28(1-2)**: 163-174.

RAUS, TH. & VOLPERS, TH. 1986a: *Knautia midzorensis* Form. In Greuter, W. & Raus, Th. (eds): Med- Checklist notulae, 12. - *Willdenowia* **15(2)**: 420-421.

RAUS, TH. & VOLPERS, TH. 1986b: *Carex digitata* L. In Greuter, W. & Raus, Th. (eds): Med-Checklist notulae, 12. - *Willdenowia* **15(2)**: 432.

RAUS, TH. & WILLING, E. 2010: *Nigella orientalis* (Forssk.) Webb. In Greuter, W. & Raus, Th. (eds): Med-Checklist notulae, 29. - *Willdenowia* **40(2)**: 198.

RAUS, TH. & WILLING, E. 2015: *Bidens frondosus*. In Raab-Straube,E. & Raus, Th. (eds): Euro+Med-Checklist notulae, 5. - *Willdenowia* **45(3)**: 451.

RAUS, TH., WILLING, E. & YENA, A. 2010: *Melilotus wolgicus* Poir. In Greuter, W. & Raus, Th. (eds): Med-Checklist notulae, 29. - *Willdenowia* **40(2)**: 197.

RAUSCHERT, S. 1966: Zur Nomenklatur der Farn- und Blütenpflanzen Deutschlands. - *Feddes Repert.* **73**: 47-58, 224-225.

RAUSCHERT, S. 1967a: (203) Vorschlag zur Konservierung des Gattungsnamens *Polystichum* Roth 1799 ... - *Taxon* **16(3)**: 243-245. [approved, cf. Taxon 17(4): 450, 1968].

RAUSCHERT, S. 1967b: (204) Vorschlag zur Konservierung des Gattungsnamens 1399. *Himantoglossum* Spreng. emend. Koch (Orchidaceae). - *Taxon* **16(3)**: 245-246. [approved, cf. Taxon 17(4): 464, 1968].

RAUSCHERT, S. 1968a: (222) Vorschlag zur Konservierung von 2539. *Delphinium* L. mit dem Typus *D. peregrinum* L. (Ranunculaceae). - *Taxon* **17(2)**: 231. [approved, cf. Taxon 18(4): 477, 1969].

RAUSCHERT, S. 1968b: (226) Vorschlag zur Konservierung von 8485. *Asperula* L. mit dem Typus *A. arvensis* L. (Rubiaceae). - *Taxon* **17(2)**: 233-234. [not approved, cf. Taxon 18(4): 478, 1969].

RAUSCHERT, S. 1973: (351) Proposal to conserve the generic name *Digitaria* Rich. in Pers. (1805) vs. *Digitaria* Adans. (1763) and *Syntherisma* Walter (1788) (Gramineae). - *Taxon* **22(1)**: 159-161. [not approved, cf. Taxon 24(1): 245, 1975].

RAUSCHERT, S. 1974: Nomenklatorische Probleme in der Gattung *Matricaria* L. - *Folia Geobot. Phytotax.* **9(3)**: 249-260.

RAUSCHERT, S. 1976a: Zur Nomenklatur und Chorologie des Formenkreises von *Onosma pseudoarenarium* Schur s.lat. - *Folia Geobot. Phytotax.* **11(3)**: 269-279.

RAUSCHERT, S. 1976b: (387) Vorschlag zur Konservierung des Gattungsnamens 6509a *Gentianella* Moench (1794) vs. *Amarella* Gilib. (1781) [Gentianaceae]. -*Taxon* **25(1)**: 192-193. [approved, cf. Taxon 27(2-3): 286, 1978].

RAUSCHERT, S. 1983: Zur Nomenklatur der Farn- und Blütenpflanzen aus dem Gebiet der DDR und BRD (VIII). - *Feddes Repert.* **94(5)**: 289-302.

RAUSCHERT, S. 1986: (857) Proposal to conserve 8978 *Antennaria* (Compositae) vs. sanctioned *Antennaria* (Fungi). - *Taxon* **35(4)**: 749-750. [approved, cf. Taxon 38(2): 301-302, 1989; cf., however, *Taxon* 41(1): 102-103, 1992].

RAVANO, C. 1937: Ricerche sul polimorfismo, sulla biologia e distribuzione di *Silene sericea* All., *S. colorata* Poir. e specie affini. - *Arch. Bot. (Forli)* **13**: 263-289.

RAVANO, C. 1939: Ricerche sul polimorfismo, sulla biologia e distribuzione di *Silene sericea* All., *S. colorata* Poir. e specie affini (cont.). - *Arch. Bot. (Forli)* **15**: 187-218.

RAVEN, J. 1972a: Crete in March, part 1. - *Quart. Bull. Alpine Gard. Soc. Gr. Brit.* **40(3)**: 188-196.

RAVEN, J. 1972b: Crete in March, part 2. - *Quart. Bull. Alpine Gard. Soc. Gr. Brit.* **40(4)**: 301-309.

RAVEN, J. & STEARN, W.T. 1990: Plants and plant lore in ancient Greece. - *Ann. Mus. Goulandris* **8**: 129-180.

RAVEN, P.H. 1962: The genus *Epilobium* in Turkey. - *Notes Roy. Bot. Gard. Edinburgh* **24(2)**: 183-203.

RAVEN, P.H. 1964: The Old World species of *Ludwigia* (including *Jussiaea*), with a synopsis of the genus (Onagraceae). - *Reinwardtia* **6**: 327-427.

RAVEN, P.H. 1971: The relationships between 'mediterranean' floras. In Davis, P.H., Harper, P.C. & Hedge, I.C. (eds): Plant life of South-West Asia, pp. 119-134. - Edinburgh: Botanical Society.

RAVEN, P.H. 1973: The evolution of mediterranean floras. In Di Castri, F. & Mooney, H.A. (eds): Mediterranean type ecosystems, pp. 213-224. - Berlin, etc.: Springer-Verlag.

RAVEN, P.H. 1977: Generic and sectional delimitation in Onagraceae, tribe Epilobieae. - *Ann. Missouri Bot. Gard.* **63(2)**: 326-340.

RAVEN, P.H., DIETRICH, W. & STUBBE, W. 1980: An outline of the systematics of *Oenothera* subsect. *Euoenothera* (Onagraceae). - *Syst. Bot.* **4(3)**: 242-252.

RAVENNA, P. 2001a: *Sternbergia minoica*, a new species of Amaryllidaceae from Greece. - *Onira* **5 (10)**: 39-41.

RAVENNA, P. 2001b: New or critical *Muscari* species from Israel (Hyacinthaceae). - *Onira* **6(2)**: 19-23.

RAVENNA, P. 2002a: On *Muscari nivale*, a valid species mainly from Iran, and the absence of *M. neglectum* in this country (Hyacinthaceae). - *Onira* **6(6)**: 44-47 + 1 plate.

RAVENNA, P. 2002b: Revisional studies in the genus *Muscari* (Hyacinthaceae). - *Onira* **7(5)**: 13-19.

RAVENNA, P. 2002c: Revisional studies in the genus *Muscari* - II. *Muscari juncifolium* antedating *M. neglectum* (Hyacinthaceae). - *Onira* **7(10)**: 54-56.

RAVNIK, V. 1961: Die Resupination, ihre Ausführung und Bedeutung für die systematische Gliederung der Gattungen *Nigritella* and *Gymnadenia* [In Slovenian with German summary]. - *Biol. Vestn.* **8**: 9-15.

RAVNIK, V. 1965: Die morphologisch-systematische und chorologische Problematik der Art *Globularia cordifolia* L., s. lat. (fam. Globulariaceae) [In Slovenian with German summary]. - *Razpr. Slov. Akad. Znan. Umetn., Razr. Prir. Med. Vede, Oddel. Prir. Vede* **8**: 5-41.

RAVNIK, V. 1988: *Plantago argentea* Chaix in Vill. subsp. *liburnica* V. Ravnik subsp. nova [In Slovenian with English summary]. - *Biol. Vestn.* **36(3)**: 53-62.

RAWDON, P. 1954: The interpretation of *Dianthus viscidus* Bory et Chaubard. - *Kew Bull.* **1953(4)**: 543-544.

RAY, M.F. 1995: Systematics of *Lavatera* and *Malva* (Malvaceae: Malveae): a new perspective. - *Pl. Syst. Evol.* **198(1-2)**: 29-53.

RAYCHEVA, T. 2009a: Natural hybrids of *Rumex* subgenus *Rumex* (Polygonaceae) in Bulgaria. In Ivanova, D. (ed.): Proceedings of the IV Balkan botanical congress, pp. 239-244. - Sofia: Institute of Botany, BAS.

RAYCHEVA, T. 2009b: Critical reassessment of the distribution of some taxa of *Rumex* subgenus *Rumex* (Polygonaceae) in Bulgaria - 2. - *Phytol. Balcan.* **15(2)**: 155-169.

RAYCHEVA, T. & DIMITROVA, D. 2007: Critical reassessment of the distribution of some taxa of *Rumex* subg. *Rumex* (Polygonaceae) in Bulgaria. - *Phytol. Balcan.* **13(2)**: 141-151.

RAYCHEVA, T., TEMSCH, E.M. & DIMITROVA, D. 2007: *Rumex pulcher* (Polygonaceae) in the Bulgarian flora: distribution, morphology and karyology. - *Phytol. Balcan.* **13(3)**: 321-330.

RAYCHEVA, T.G., STOYANOV, K.H. & STOYANOV, P.S. 2021: *Rumex kerneri* Borbás (Polygonaceae) in the Bulgarian flora - Morphology, leaf epidermis, pollen morphology, and karyology. - *Ecol. Balkan.* **13(1)**: 119-130.

RAYMOND-HAMET, M. 1929: Contribution à l'étude phytogeographique du genre *Sedum*. - *Candollea* **4**: 1-52.

RAYNAUD, CH. 1977: Recherche des caractères permettant la révision du genre *Helianthemum* Mill. dans la section *Brachypetalum* Dun. - *Naturalia Monspel., Sér. Bot.* **26**: 181-210.

RAYNAUD, CH. 1985: Contribution à l'étude de certaines espèces du genre *Helianthemum* sect. *Helianthemum*. Note préliminaire. - *Anales Jard. Bot. Madrid* **41(2)**: 303-311.

RAYNAUD, CH. 1987: *Atlanthemum* Raynaud, un nouveau genre pour la famille des Cistaceae. - *Anales Jard. Bot. Madrid* **44(2)**: 309-317.

REALES, A. & AL. 2004: Numerical taxonomy study of *Salvia* sect. *Salvia* (Labiatae). - *Bot. J. Linn. Soc.* **145(3)**: 353-371.

REBELE, F. & LEHMANN, C. 2001: Biological flora of Central Europe: *Calamagrostis epigejos* (L.) Roth. - *Flora, Morphol. Geobot. Oekophysiol.* **196(5)**: 325-344.

RECHINGER, K. 1899: Ueber einen neuen hybriden *Rumex* aus Griechenland. - *Verh. K.K. Zool.-Bot. Ges. Wien* **49(2-3)**: 105-107.

RECHINGER, K. 1914: Korfu. In Karsten, G. & Schenk, H. (eds): Vegetationsbilder, 12. Reihe, Heft 4. - Jena: Gustav Fischer Verlag, Tafel 19-24 + Text.

RECHINGER, K.H. 1929: Beitrag zur Kenntnis der Flora der ägäischen Inseln und Ost-Griechenlands. - *Ann. Naturhist. Mus. Wien* **43**: 269-340 + 2 plates.

RECHINGER, K.H. 1932a: Vorarbeiten zu einer Monographie der Gattung *Rumex*. I. - *Beih. Bot. Centralbl., Abt. 2* **49(1)**: 1-132.

RECHINGER, K.H. 1932b: Einige neue balkanische Pflanzen. - *Repert. Spec. Nov. Regni Veg.* **31**: 158-160.

RECHINGER, K.H. 1933a: Vorarbeiten zu einer Monographie der Gattung *Rumex*. II. Die Arten der Subsektion *Patientiae*. - *Repert. Spec. Nov. Regni Veg.* **31**: 225-283 + 2 plates and 1 map.

RECHINGER, K.H. 1933b: Neue Pflanzen aus dem Alibotusch-Gebirge (Bulg. NO-Mazedonien). - *Magyar Bot. Lapok* **32(7-12)**: 152-153.

RECHINGER, K.H. 1933c: Ergebnisse einer botanischen Reise nach Bulgarien. - *Magyar Bot. Lapok* **32**: 5-58 + 1 plate.

RECHINGER, K.H. 1934a: Ein neuer *Rumex* aus der Nordalbanischen Alpen. - *Magyar Bot. Lapok* **33(-12)**: 5-7 + 1 plate.

RECHINGER, K.H. 1934b: Dreizehn neue Pflanzenarten aus Griechenland. - *Magyar Bot. Lapok* **33(1-12)**: 8-22 + 6 plates.

RECHINGER, K.H. 1935: Ergebnisse einer botanischen Reise in den Bertiscus (Nordalbanische Alpen). - *Repert. Spec. Nov. Regni Veg.* **38**: 137-152, 319-389.

RECHINGER, K.H. 1936a: Einige neue Pflanzen aus der südöstlichen Aegaeis. - *Ann. Naturhist. Mus. Wien* **47**: 147-152.

RECHINGER, K.H. 1936b: Ergebnisse einer botanischen Sommerreise nach dem Ägäischen Archipel und Ostgriechenland. - *Beih. Bot. Centralbl., Abt. 2* **54(3)**: 577-688 + 2 plates.

RECHINGER, K.H. 1936c: Neue Pflanzen aus der Aegaeis. - *Oesterr. Bot. Z.* **85(1)**: 56-64.

RECHINGER, K.H. 1937: Revision des Formenkreises der *Stachys cretica*. - *Ann. Naturhist. Mus. Wien* **48**: 167-178.

RECHINGER, K.H. 1938a: Plantae novae Aegaeae. - *Repert. Spec. Nov. Regni Veg.* **43**: 144-151.

RECHINGER, K.H. 1938b: Der Formenkreis der *Inula candida* und seine Verbreitung. - *Oesterr. Bot. Z.* **87(2)**: 81-100.

RECHINGER, K.H. 1938c: Enumeratio Florae Constantinopolitanae. Aufzählung der nach dem Erscheinen von Boissiers Flora Orientalis aus der Umgebung von Konstantinopel bekannt gewordenen Farn- und Blütenpflanzen. - *Repert. Spec. Nov. Regni Veg. Beih.* **98**: 1-73 + 1 map.

RECHINGER, K.H. 1939a: Zur Flora von Ostmazedonien und Westthrazien. - *Bot. Jahrb. Syst.* **69(4)**: 419-552 + 3 plates.

RECHINGER, K.H. 1939b: Plantae novae Aegaeae. II. - *Repert. Spec. Nov. Regni Veg.* **47**: 49-51.

RECHINGER, K.H. 1939c: Zur Flora von Albanien und Mazedonien. - *Repert. Spec. Nov. Regni Veg.* **47**: 165-179.

RECHINGER, K.H. 1939d: Versuch einer natürlichen Gliederung des Formenkreises von *Rumex bucephalophorus* L. (Vorarbeiten zu einer Monographie der Gattung *Rumex*, VI.). - *Bot. Not.* **1939 (4)**: 485-504.

RECHINGER, K.H. 1939e: Der Formenkreis der *Inula candida* und seine Verbreitung. Nachtrag. - *Oesterr. Bot. Z.* **88(3)**: 228-229.

RECHINGER, K.H. 1941a: Monographische Studie über *Teucrium* sect. *Chamaedrys*. - *Bot. Arch.* **42**: 335-420.

RECHINGER, K.H. 1941b: Neue und kritische Labiaten aus dem Orient und Mittelmeergebiet. - *Bot. Jahrb. Syst.* **71(4)**: 526-546.

RECHINGER, K.H. 1942a: *Scutellaria* sect. *Vulgares* subsect. *Peregrinae* im Mittelmeergebiet und Orient. - *Bot. Arch.* **43**: 1-70.

RECHINGER, K.H. 1942b: Östliche Ägäische Inseln. In Walter, H. (ed.). - *Vegetationsbilder* **26(3)**: Tafel 13-18 + text.

RECHINGER, K.H. 1943a: Die *Rumex*-Arten der Balkanhalbinsel. - *Mitt. Thüring. Bot. Vereins, N.F.* **50**: 193-217.

RECHINGER, K.H. 1943b: Neue Beiträge zur Flora von Kreta. (Ergebnisse einer biologischen Forschungsreise nach dem Peloponnes und nach Kreta 1942, im Auftrage des Oberkommandos der Wehrmacht und des Reichsforschungsrates, 6). - *Denksch. Akad. Wiss. Wien, Math.-Naturwiss. Kl.* **105(2)**: 184 pp. + 1 map.

RECHINGER, K.H. 1944a: Flora Aegaea. Flora der Inseln und Halbinseln des ägäischen Meeres. - *Denksch. Akad. Wiss. Wien, Math.-Naturwiss. Kl.* **105 (1)**: xx + 924 pp. + 25 plates and 3 maps. [Reprinted by Otto Koeltz Antiquariat 1973].

RECHINGER, K.H. 1944b: Bericht über eine botanische Forschungsreise nach Kreta. - *Anz. Akad. Wiss. Wien, Math.-Naturwiss. Kl.* **25(12)**: 9 pp.

RECHINGER, K.H. 1947: Der Polymorphismus in der ägäischen Flora. (Kritische Studien zur Systematik, Genetik und Ökologie mediterraner Formenkreise). - *Oesterr. Bot. Z.* **94(1-2)**: 152-234.

RECHINGER, K.H. 1949a: Florae Aegaeae supplementum (mit Beiträgen von B.L. Burtt und P.H. Davis) - *Phyton (Horn)* **1(2-4)**: 194-228. [attached to the reprinted edition of Flora Aegaea].

RECHINGER, K.H. 1949b: Plantarum novarum orientalium descriptiones. - *Oesterr. Bot. Z.* **95(4)**: 422-427.

RECHINGER, K.H. 1949c: Reliquiae Samuelssonianae. - *Ark. Bot., Ser. 2* **1(5)**: 301-327 + 20 plates.

RECHINGER, K.H. 1950: Grundzüge der Pflanzenverbreitung in der Ägäis I-III. - *Vegetatio* **2**: 55-119, 239-308, 365-386.

RECHINGER, K.H. 1955: Zur Flora der Kykladen. - *Anz. Österr. Akad. Wiss., Math.-Naturwiss. Kl.* **92 (2)**: 15-21.

RECHINGER, K.H. 1956: Plantae novae ex insula Euboea. - *Anz. Österr. Akad. Wiss., Math.-Naturwiss. Kl.* **93(9)**: 95-103.

RECHINGER, K.H. 1957a: Plantae novae graeco-macedonicae, imprimis serpentinicolae. - *Anz. Österr. Akad. Wiss., Math.-Naturwiss. Kl.* **94(2)**: 21-27.

RECHINGER, K.H. 1957b: Plantarum species novae euboicae. - *Oesterr. Bot. Z.* **104(1-2)**: 175-176.

RECHINGER, K.H. 1960a: Zwei neue Arten der Gattung *Anchusa* aus Griechenland. - *Oesterr. Bot. Z.* **107(5)**: 471-473.

RECHINGER, K.H. 1960b: Theodor Kotschy, ein Pionier der botanischen Orientforschung. - *Taxon* **9 (2)**: 33-35.

RECHINGER, K.H. 1961a: Die Flora von Euboea. I. Teil. - *Bot. Jahrb. Syst.* **80(3)**: 294-382 + 1 table.

RECHINGER, K.H. 1961b: Die Flora von Euboea (Schluss). - *Bot. Jahrb. Syst.* **80(4)**: 383-465 + 12 plates.

RECHINGER, K.H. 1961c: Die systematische Behandlung polymorpher Formenkreise am Beispiel der ostmediterranen Flora. In Heywood, V.H. (ed.): Problems of taxonomy and distribution in the European flora. - *Feddes Repert. Spec. Nov. Regni Veg.* **63(2)**: 168-176.

RECHINGER, K.H. 1962: Zur Kenntnis orientalischer Labiaten. - *Kulturpflanze, Beih.* **3**: 47-73.

RECHINGER, K.H. 1963: Bericht über die botanische Erforschung von Griechenland. - *Webbia* **18**: 237-259.

RECHINGER, K.H. 1964: Flora of lowland Iraq. - Weinheim: Verlag von J. Cramer, 746 pp.

RECHINGER, K.H. 1965a: Der Endemismus in der griechischen Flora. - *Rev. Roumaine Biol., Sér. Bot.* **10(1-2)**: 135-138.

RECHINGER, K.H. 1965b: Zwei neue Arten aus der griechischen Flora. - *Oesterr. Bot. Z.* **112(1-2)**: 186-187.

RECHINGER, K.H. 1965c: Zur Kenntnis der europäischen Arten der Gattung *Alkanna*. - *Ann. Naturhist. Mus. Wien* **68**: 191-220.

RECHINGER, K.H. 1971a: *Rumex nebroides* Campd.: eine verkannte Art aus dem Subgenus *Acetosa* und ihre Verbreitung. - *Candollea* **26(1)**: 173-181 + 2 plates.

RECHINGER, K.H. 1971b: *Potentilla Goulandrii*, *Dianthus Stamatiadae*, *Onosma psammophilum*, drei neue Arten der griechischen Flora. - *Bot. Not.* **124(1)**: 75-81.

RECHINGER, K.H. 1971c: *Lithospermum Goulandriorum* Rech. f., eine neue, bemerkenswerte Art der griechischen Flora. - *Bot. Not.* **124(3)**: 355-358.

RECHINGER, K.H. 1973: *Onobrychis aliacmonia*, eine neue griechisch-mazedonische Art aus der Sektion *Hymenobrychis*. - *Ann. Mus. Goulandris* **1**: 127-131.

RECHINGER, K.H. 1974: *Aetheorrhiza bulbosa* (L.) Cass. und ihre geographischen Rassen. - *Phyton (Horn)* **16(1-4)**: 211-220.

RECHINGER, K.H. 1975: *Centaurea prespana*, eine neue gelbblühende Art aus der Sektion *Acrolophus*. - *Ann. Mus. Goulandris* **2**: 55-58.

RECHINGER, K.H. 1978: Meine botanischen Forschungen in Griechenland 1927-1976. - *Ann. Mus. Goulandris* **4**: 39-82.

RECHINGER, K.H. & RANDOLPH, L.F. 1954: Ergebnisse einer botanischen Reise in den Bertiscus (Nordalbanische Alpen). - *Verh. Zool.-Bot. Ges. Wien* **94**: 82-96.

RECHINGER, K.H. & RECHINGER-MOSER, F. 1951: Phytogeographia Aegaea. - *Denkschr. Akad. Wiss. Wien, Math.-Naturwiss. Kl.* **105(2)**: 1-208 + tables.

REDONDO, N. & HORJALES, M. 1989: Datos cariológicos del gen. *Saxifraga* L. en el NW Ibérico. - *Bol. Soc. Brot., Ser. 2* **62**: 195-203.

REDOUTÉ, P.J. 1982: Lilies and related flowers [A selection of 108 plates from Redouté's "Les Liliacees" published 1802-16, with new English text by B. Mathew]. - London: Michael Joseph, 240 pp.

REDURON, J.-P. 2007a: Ombellifères de France. 1. Monographie des Ombellifères (Apiaceae) et plantes alliées, indigènes, naturalisées, subspontanées, adventices ou cultivées de la flore française. - *Bull. Soc. Bot. Centre-Ouest, n.s., Numéro special* **26**: 1-564.

REDURON, J.-P. 2007b: Ombellifères de France. 2. Monographie des Ombellifères (Apiaceae) et plantes alliées, indigènes, naturalisées, subspontanées, adventices ou cultivées de la flore française. - *Bull. Soc. Bot. Centre-Ouest, n.s., Numéro special* **27**: 565-1142.

REDURON, J.-P. 2007c: Ombellifères de France. 3. Monographie des Ombellifères (Apiaceae) et plantes alliées, indigènes, naturalisées, subspontanées, adventices ou cultivées de la flore française. - *Bull. Soc. Bot. Centre-Ouest, n.s., Numéro special* **28**: 1143-1726.

REDURON, J.-P. 2008a: Ombellifères de France. 4. Monographie des Ombellifères (Apiaceae) et plantes alliées, indigènes, naturalisées, subspontanées, adventices ou cultivées de la flore française. - *Bull. Soc. Bot. Centre-Ouest, n.s., Numéro special* **29**: 1727-2348.

REDURON, J.-P. 2008b: Ombellifères de France. 5. Monographie des Ombellifères (Apiaceae) et plantes alliées, indigènes, naturalisées, subspontanées, adventices ou cultivées de la flore française. - *Bull. Soc. Bot. Centre-Ouest, n.s., Numéro special* **30**: 2349-3004.

REDURON, J.-P., CHARPIN, A. & PIMENOV, M.G. 1997: Contribution à la nomenclature générique des Apiaceae (Ombellifères). - *J. Bot. Soc. Bot. France* **1**: 91-104.

REED, C.F. 1953: Index Isoëtales. - *Bol. Soc. Brot., Ser. 2* **27**: 5-72.

REED, C.F. 1954: Index Marsileata et Salviniata. - *Bol. Soc. Brot., Ser. 2* **28**: 5-61.

REED, C.F. 1962: Marsileaceae, Azollaceae e Isoetaceae de Portugal. - *Bol. Soc. Brot., Ser. 2* **36**: 73-94.

RÉER, U. & PODLECH, D. 1986: Die europäischen Vertreter der Gattung *Astracantha* Podl. (Leguminosae). - *Mitt. Bot. Staatssamml. München* **22**: 513-569.

REESE, G. 1962: Systematik und Cytologie der *Ruppia maritima*. - *Ber. Deutsch. Bot. Ges.* **75(9)**: 365.

REESE, G. 1967: Cytologische und taxonomische Untersuchungen an *Zannichellia palustris* L. - *Biol. Zentralbl.* **86(Suppl.)**: 277-306.

REESE, G. 1969: Cytotaxonomische Untersuchungen an di- und tetraploiden Sippen von *Eryngium campestre* L. und einem Bastard mit 2n=15. - *Oesterr. Bot. Z.* **117(3)**: 223-247.

REEVE, J. 1996: *Cyclamen hederifolium* on Samos. - *Cyclamen* **20(1)**: 15-16.

REEVE, J. 2000: The Greek island of Angistri. - *Cyclamen* **24(1)**: 10.

REEVES, R.D., BROOKS, R.R. & DUDLEY, T.R. 1983: Uptake of nickel by species of *Alyssum*, *Bornmuellera*, and other genera of Old World tribus Alysseae. - *Taxon* **32(2)**: 184-192.

REEVES, R.D., BROOKS, R.R. & PRESS, J.R. 1980: Nickel accumulation by species of *Peltaria* Jacq. (Cruciferae). - *Taxon* **29(5-6)**: 629-633.

REGEL, C.V. 1933: Die Vegetationsverhältnisse einiger Gebirge im östlichen Teile des Mittelmeergebietes. - *Repert. Spec. Nov. Regni Veg. Beih.* **71**: 1-2.

REGEL, C.V. 1935: La végétation du Pinde et du Taygète. - *Bull. Soc. Bot. Genève, Sér. 2* **27**: 97-100.

REGEL, C.V. 1937a: Die Wälder Griechenlands. - *Verh. Schweiz. Naturf. Ges.* **118**: 154-156.

REGEL, C.V. 1937b: Über die Grenze zwischen Mittelmeergebiet und Mitteleuropa in Griechenland (Vorläufige Mitteilung). - *Ber. Deutsch. Bot. Ges.* **55(1)**: 82-91.

REGEL, C.V. 1938: Über die Depression der Waldgrenze in Griechenland. - *Repert. Spec. Nov. Regni Veg. Beih.* **100**: 28-37.

REGEL, C.V. 1939a: Geobotanische Beobachtungen auf einer Reise in Marokko und in der Sahara. - *Veröff. Geobot. Inst. E.T.H. Stiftung Rübel Zürich* **14**: 192-216.

REGEL, C.V. 1939b: Pflanzengeographisches von der Balkanhalbinsel. - *Repert. Spec. Nov. Regni Veg. Beih.* **111**: 74-84.

REGEL, C.V. 1940: Über die Grenze zwischen Mittelmeergebiet und Mitteleuropa in Griechenland (Zweiter Beitrag). - *Ber. Deutsch. Bot. Ges.* **58(4)**: 155-165.

REGEL, C.V. 1941: Florae Graecae notulae, I. - *Candollea* **8**: 222-242.

REGEL, C.V. 1942a: Florae Graecae notulae, II. - *Candollea* **9**: 104-137.

REGEL, C.V. 1942b: Über die Umkehrung der Vegetationsstufen in Griechenland. - *Ber. Schweiz. Bot. Ges.* **52**: 621-634.

REGEL, C.V. 1942c: Kreuz und quer durch den Pindus. I. - *Mitt. Deutsch. Dendrol. Ges.* **55**: 259-283 + 7 plates.

REGEL, C.V. 1943a: Pflanzengeografische Studien aus Griechenland und Westanatolien. - *Bot. Jahrb. Syst.* **73(1)**: 2-98 + 7 plates.

REGEL, C.V. 1943b: La végétation du mont Oeta en Grèce. - *Boissiera* **7**: 402-413.

REGEL, C.V. 1943c: Die pflanzengeographische Stellung der Krim. - *Wiener Bot. Z.* **92(1-2)**: 25-49.

REGEL, C.V. 1944: Florula montis Oeta in Graecia. - *Feddes Repert. Spec. Nov. Regni Veg.* **54(1)**: 34-55.

REGEL, C.V. 1947a: Vegetationsprobleme aus der Ostmediterraneis. - *Ber. Schweiz. Bot. Ges.* **58**: 45-60.

REGEL, C.V. 1947b: Zur Frage der Grenze zwischen dem Mittelmeergebiet und Mitteleuropa auf der Balkanhalbinsel (Dritte Folge). - *Ber. Geobot. Forschungsinst. Rübel Zürich* **1946**: 15-22.

REGEL, C.V. 1949a: Etudes biométriques sur le genre *Phillyrea*. - *Mém. Soc. Bot. France* **31**: 20-38.

REGEL, C.V. 1949b: Floristische Notizen aus Griechenland (Florae Graecae notulae III). - *Oesterr. Bot. Z.* **95(3)**: 243-250.

REGEL, C.V. 1952: Flore du mont Oeta en Grèce. - *Mém. Soc. Bot. France* **33**: 122-143.

REGEL, C.V. 1953: Studien über die Florenelemente in Griechenland. Die Florenelemente des Oeta. - *God. Biol. Inst. Sarajevu* **5(1-2)**: 339-348.

REGEL, C.V. 1955: Die Verbreitung einiger europäisch-asiatischer *Tragopogon*-Arten. - *Ber. Schweiz. Bot. Ges.* **65**: 251-262.

REGEL, C.V. 1959: Vegetationszonen und Vegetationsstufen in der Türkei. - *Feddes Repert. Spec. Nov. Regni Veg. Beih.* **138**: 230-282.

REGEL, C.V. 1963: Zur Frage der Grenze zwischen dem Mittelmeergebiet und Mitteleuropa. - *Oesterr. Bot. Z.* **110(5)**: 498-510.

REGEL, C.V. 1964: Über *Centaurea cyanus* in der Türkei. - *Ege Üniv. Fen Fak. Ilmi Raporlar Ser.* **17**: 1-5.

REGEL, E. 1875: Alliorum adhuc cognitorum monographia. - *Trudy Imp. S.-Peterburgsk. Bot. Sada* **3 (2)**: 1-266.

REGEL, E. 1877: Tentamen rosarum monographiae. - St. Petersburg: 114 pp.

REGULA-BEVILACQUA, L. 1969: *Salvia peloponnesiaca* Boiss. et Heldr. - new species in the Croatian flora [In Serbo-Croatian with English summary]. - *Acta Bot. Croat.* **28**: 467-468 + 2 plates.

REHDER, A. 1903: Synopsis of the genus *Lonicera*. - *Rep. (Annual) Missouri Bot. Gard.* **14**: 27-232 + 20 plates.

REICH, D. & AL. 2021a: The type specimens in Eugen von Halácsy´s Herbarium Graecum. - *Phytotaxa* **493(1)**: 1-156.

REICH, D. & AL. 2021b: Biogeography of amphi-adriatic *Gentianella crispata* (Gentianaceae): a northern refugium and recent trans-adriatic migration. - *Pl. Biosyst.* **155**: 15 pp.

REICH, D. & TAN, KIT 2020: Report 65. In Vladimirov, V., Aybeke, M. & Tan, Kit: New floristic records in the Balkans: 43. - *Phytol. Balcan.* **26(3)**: 550-551.

REICHARDT, H. W. 1861: Cryptogamas vasculares. In: Heufler [L. v.] (Ed.): Specimen florae cryptogamae septem insularum editum juxta plantas Mazziaranas herbarii Heufleriani et speciatim quoad filices herbarii Tommasiniani. - *Verh. K. K. Zool.-Bot. Ges. Wien* **11**: 412.

REICHENBACH, H.G. FIL. 1850-51: Orchideae in flora germanica recensitae additis Orchideis Europae reliquae, reliqui Rossicii imperi, Algerii ergo Tentamen Orchidiographiae Europaeae iconibus illustratum. - Lipsiae [Leipzig]: F. Hofmeister, x + 194 pp. + 170 plates.

REICHERT, H. 1986: Kritische Anmerkungen zur Beschreibung und Verschlüsselung der engeren *Rosa canina*-Gruppe in der Flora Europaea. - *Göttinger Florist. Rundbr.* **19(2)**: 66-70.

REICHERT, H., GREGOR, TH. & MEIEROTT, L. 2018: *Euphorbia saratoi* (= *E. podperae*, *E. pseudovirgata* auct., *E. virgata* var. *orientalis*, *E. virgultosa*) - in Mitteleuropa und Nordamerika ein Neophyt unklarer Herkunft. - *Kochia* **11**: 1-36.

REICHSTEIN, T. 1981: Hybrids in European Aspleniaceae (Pteridophyta). - *Bot. Helv.* **91**: 89-139.

REICHSTEIN, T. 1982: Hybrids in European Aspleniaceae (Pteridophyta). Addenda and corrigenda. - *Bot. Helv.* **92(1)**: 41-42.

REICHSTEIN, T. & AL. 1973: Die Asplenien der Insel Kreta. - *Ann. Mus. Goulandris* **1**: 133-163 + 22 plates.

REIF, A. & LÖBLICH-ILLE, K. 1999: Sind die Rotbuchenwälder im Pieria-Gebirge (Nordgriechenland) höhenzonal oder extrazonal? Eine Studie zum Übergang zwischen temperaten und submediterranen Wäldern in Nordgriechenland. - *Phytocoenologia* **29(1)**: 87-146.

REIMANN, H. 1913: Der Tannenwald auf Kephallonia. - *Adria (Triest)* **5**: 743-750.

REINEKE, D. 1987: *Epipactis helleborine* - ein Beispiel für die Unzulänglichkeit der heutigen Evolutionstheorien? - *Mitt. Arbeitskreis Heimische Orchid. Baden-Württemberg* **19(4)**: 822-834.

REINEKE, D. & RIETDORF, K. 1987: Zur Phänologie von *Ophrys* spec. und *Orchis ustulata*. - *Mitt. Arbeitskreis Heimische Orchid. Baden-Württemberg* **19(4)**: 835-840.

REINEKE, D. & RIETDORF, K. 1991: Zur Phänologie von *Anacamptis pyramidalis* (L.) Rich. und *Orchis ustulata* L. - *Mitt. Arbeitskreis Heimische Orchid. Baden-Württemberg* **23(4)**: 521-556.

REINHARD, H.R. 1987: Untersuchungen an *Ophrys holoserica* (Burm. fil.) W. Greuter subsp. *elatior* (Gumprecht) Gumprecht (Orchidaceae). - *Mitt. Arbeitskreis Heimische Orchid. Baden-Württemberg* **19(4)**: 769-800.

REINKE, J. 1879: A. Grisebach. Ein Nachruf. - *Bot. Zeitung (Berlin)* **37(33)**: 521-534.

REMMY, K. & GRUBER, F. 1993: Untersuchungen zur Verbreitung und Morphologie des Wild-Apfels (*Malus sylvestris* (L.) Mill.). - *Mitt. Deutsch. Dendrol. Ges.* **81**: 71-94.

RENDLE, A.B. 1899: A systematic revision of the genus *Najas*. - *Trans. Linn. Soc. London, Bot.* **5**: 379-436.

RENNER, O. 1942: Europäische Wildarten von *Oenothera*. - *Ber. Deutsch. Bot. Ges.* **60(9)**: 448-466.

RENNER, O. 1950: Europäische Wildarten von *Oenothera*. II. - *Ber. Deutsch. Bot. Ges.* **63(5)**: 129-138.

RENNER, O. 1956: Europäische Wildarten von *Oenothera*. III. - *Planta* **47(3)**: 219-254.

RENNER, O. & HIRMER, U. 1956: Zur Kenntnis von *Oenothera*. I. Über *Oe. conferta* n. sp. II. Über künstliche Polyploide. - *Biol. Zentralbl.* **75(9-10)**: 513-531.

RENVOIZE, S.A. & CLAYTON, W.D. 1980: (528) Proposal to reject the name *Paspalum distichum* L., Syst. Nat. ed. 10, 2: 855 (1759). - *Taxon* **29(2-3)**: 339-340. [not approved, cf. Taxon 30(1): 301, 1981, and 32(2): 281, 1983].

RENZ, J. 1928: Zur Kenntnis der griechischen Orchideen. - *Repert. Spec. Nov. Regni Veg.* **25**: 225-270 + 11 plates.

RENZ, J. 1929: Über neue Orchideen von Rhodos, Cypern und Syrien. - *Repert. Spec. Nov. Regni Veg.* **27**: 193-219 + 3 plates.

RENZ, J. 1930: Beiträge zur Orchideenflora der Insel Kreta. - *Repert. Spec. Nov. Regni Veg.* **28**: 241-262 + 1 plate.

RENZ, J. 1932a: Die Orchideenflora von Ostkreta (Sitia). - *Repert. Spec. Nov. Regni Veg.* **30**: 97-118 + 1 plate.

RENZ, J. 1932b: Über einige griechische Orchideen. - *Repert. Spec. Nov. Regni Veg.* **30**: 118-121.

RENZ, J. & TAUBENHEIM, G. 1983: Materials for a Flora of Turkey XXXIX: Orchidaceae. - *Notes Roy. Bot. Gard. Edinburgh* **41(2)**: 269-277.

REŠETNIK, I. & AL. 2007: Molecular phylogeny and systematics of the *Lilium carniolicum* group (Liliaceae) based on nuclear ITS sequences. - *Pl. Syst. Evol.* **265(1-2)**: 45-58.

REŠETNIK, I. & AL. 2015: Polyploid evolution and diversification in *Knautia* (Dipsacaceae). In Bogdanović, S. & Jogan, N.: Book of abstracts - 6th Balkan Botanical Congress, pp. 34-35. - Rijeka.

REŠETNIK, I. & BOGDANOVIĆ, S. 2013: Typification of *Lilium jankae* A. Kern. and *Lilium martagon* var. *cattaniae* Vis. - *Candollea* **68(1)**: 151-154.

REVEAL, J.L. 1970: *Sparganium simplex* Huds., a superfluous name. - *Taxon* **19(5)**: 796-797.

REVEAL, J.L. & AL. 1991: (996) Proposal to reject *Festuca elatior* L. with comments on the typification of *F. pratensis* and *F. arundinacea* (Poaceae). - *Taxon* **40(1)**: 135-137. [approved, cf. Taxon 43(1): 115, 1994].

REVEAL, J.L. & AL. 1992: On the typification of seven names in *Anemone* (Ranunculaceae) proposed by Linnaeus. - *Bartonia* **57**: 28-31.

REXHEPI, F. 1977: Neue Arten in der Flora von Serbien und Kosovo [In Serbo-Croatian with German summary]. - *Glasn. Prir. Mus. Beogradu, Ser. B, Biol. Nauke* **32**: 75-78.

REXHEPI, F. 1986: *Centaurea albertii* Rexhepi sp. nov. New kind of gender *Centaurea* [In Serbo-Croatian with English summary]. - *Glasn. Prir. Mus. Beogradu, Ser. B, Biol. Nauke* **41**: 27-36.

REXHEPI, F. 1995: Genus *Potentilla* L. in the Balkans. In Valdés, B. & Silvestre, S. (eds): Abstracts. VIII OPTIMA meeting, Sevilla, Spain, 25 September - 1 October, 1995, p. 114.

REYNAUD, C. 1973: Contribution à l'étude cytotaxinomique du genre *Hypericum* L. en Turquie. I. - *Bull. Soc. Bot. France* **120(5-6)**: 201-216.

REYNAUD, C. 1975: Contribution à l'étude cytotaxinomique de quelques *Hypericum* méditerranéens. - *Biol. Écol. Médit.* **2(2)**: 3-8.

REYNAUD, C. 1980a: Contribution à l'étude cytotaxinomique du genre *Hypericum* L. en Grèce. - *Bull. Soc. Bot. France, Lett. Bot.* **127(4)**: 345-353.

REYNAUD, C. 1980b: [Chromosome numbers in *Hypericum*]. In Löve, Á. (ed.): Chromosome number reports LXVI. - *Taxon* **29(1)**: 164-165.

REYNAUD, C. 1981: Contribution à l'étude cytotaxonomique du genre *Hypericum* L. en Turquie. II. - *Biol. Écol. Médit.* **8(3-4)**: 181-191.

REYNAUD, C. 1986: Étude cytotaxonomique des Millepertuis du bassin méditerranéen et des Iles Canaries. - *Bull. Soc. Bot. France, Lett. Bot.* **133(2)**: 167-177.

REYNAUD, C., FILOSA, D. & VERLAQUE, R. 1992: Rapports (98-106). In Kamari, G., Felber, F. & Garbari, F. (eds): Mediterranean chromosome number reports - 2. - *Fl. Medit.* **2**: 258-264.

REYNAUD, C., VERLAQUE, R. & FILOSA, D. 1993: Rapports (234-239). In Kamari, G., Felber, F. & Garbari, F. (eds): Mediterranean chromosome number reports - 3. - *Fl. Medit.* **3**: 367-373.

REŽETNIK, I., SCHNEEWEISS, G.M. & LIBER, Z. 2014: Two new combinations in *Bornmuellera* (Brassicaceae). - *Phytotaxa* **159(4)**: 298-300.

RICCERI, C. 1982: Note tassonomiche e corologiche sul genere *Eragrostis* Wolf in Italia. - *Webbia* **35 (2)**: 323-354.

RICCERI, C. & ARRIGONI, P.V. 2000: L'aggregato di *Portulaca oleracea* L. (Portulacaceae) in Italia. - *Parlatorea* **4**: 91-97.

RICCI, I. 1958: Osservazioni sullo *Juncus acutus* L. - *Ann. Bot. (Rome)* **26(1)**: 71-78 + 1 plate.

RICCI, I. 1964: *Cichorium intybus* L. subsp. *spicatum* subsp. nova. Studio cittotassinomico. - *Ann. Bot. (Rome)* **28(1)**: 219-223 + 6 plates.

RICCI, I. 1972: *Iris sisyrinchium* L.: analisi citotassonomiche. - *Ann. Bot. (Rome)* **30**: 187-206.

RICCI, I. & COLASANTE, M. 1974: *Iris graminea* L.: analisi cariologica. - *Giorn. Bot. Ital., n.s.* **108(1-2)**: 75-80.

RICH, T.C.G. & NICHOLLS VUILLE, F.L. 2001: Taxonomy and distribution of European *Damasonium* (Alismataceae). - *Edinburgh J. Bot.* **58(1)**: 44-55.

RICHARDS, A.J. 1973: The origin of *Taraxacum* agamospecies. - *Bot. J. Linn. Soc.* **66(3)**: 189-211.

RICHARDS, A.J. 1982: Parnassos, mountain of flowers. - *Quart. Bull. Alpine Gard. Soc. Gr. Brit.* **50(4)**: 333-338.

RICHARDS, A.J. 1985: Sectional nomenclature in *Taraxacum* (Asteraceae). - *Taxon* **34(4)**: 633-644.

RICHARDS, A.J. 1986: A tale of two lilies. - *Quart. Bull. Alpine Gard. Soc. Gr. Brit.* **54(3)**: 210-220.

RICHARDS, A.J. 1993: *Primula* (illustrated by Brigid Edwards). - London: B.T. Batsford Ltd, 299 pp.

RICHARDS, A.J. 2000: The AGS expedition to Greece 1999. - *Quart. Bull. Alpine Gard. Soc. Gr. Brit.* **68(3)**: 314-358.

RICHARDS, A.J. 2001a: [*Fritillaria ehrhartii*]. - *Alpine Gardener* **69(1)**: 65-66.

RICHARDS, A.J. 2001b: *Crocus cancellatus* Herb. subsp. *mazziaricus* (Herb.) B. Mathew. In Rolfe, R. (ed.): Plant awards 2000-2001. - *Alpine Gardener* **69(4)**: 503-504.

RICHARDS, A.J. 2001c: *Aubrieta glabrescens* Turrill. In Rolfe, R. (ed.): Plant awards 2000-2001. - *Alpine Gardener* **69(4)**: 529-530.

RICHARDS, A.J. 2003a: *Stachys spreitzenhoferi* Heldreich subsp. *virella* D. Persson. – *Alpine Gardener* **71(4)**: 426-427.

RICHARDS, A.J. 2003b: [Photo of *Fritillaria euboeica*. In: Blackpool show, March 16, 2002]. - *Alpine Gardener* **71(1)**: 45-47.

RICHARDS, A.J. 2008: Mountain flower walks. The Greek mainland. - Alpine Garden Society: 216 pp.

RICHARDS, A.J. 2015: *Taraxacum limnoticum* (Asteraceae) from the Aegean, Greece. - *Phytol. Balcan.* **21(2)**: 149-154.

RICHARDS, A.J. & BOOTH, T.A. 1976: Karyological indications of evolution in *Hordeum murinum* L. sensu lato. In Jones, K. & Brandham, P.E. (eds): Current chromosome research, pp. 167-174. - Amsterdam, etc.: North-Holland.

RICHARDS, P.W. & CLAPHAM, A.R. 1941: Biological Flora of the British Isles. *Juncus* L. - *J. Ecol.* **29(2)**: 362-391.

RICHARDSON, I.B.K. 1976a: A revision of the genus *Centranthus* (Valerianaceae). - *Bot. J. Linn. Soc.* **71(3)**: 211-234.

RICHARDSON, I.B.K. 1976b: [New combinations in *Tragopogon*]. In Heywood, V.H. (ed.): Notulae systematicae ad Floram Europaeam spectantes, no. 19. - *Bot. J. Linn. Soc.* **71(4)**: 270.

RICHARDSON, I.B.K. 1976c: [New combinations in *Achillea*]. In Heywood, V.H. (ed.): Notulae systematicae ad Floram Europaeam spectantes, no. 19. - *Bot. J. Linn. Soc.* **71(4)**: 271-272.

RICHARDSON, P.M. 1983: Phenolic chemistry distinguishes *Asplenium adiantum-nigrum* L. from *A. cuneifolium* Viv. - *Watsonia* **14(4)**: 414-415.

RICHENS, R.H. 1968: The correct designation of the European field elm. In Heywood, V.H. (ed.): Notulae systematicae ad Floram Europaeam spectantes, no. 7. - *Feddes Repert.* **79(1-2)**: 1-2.

RICHENS, R.H. 1977: New designations in *Ulmus minor* Mill. - *Taxon* **26(5-6)**: 583-584.

RICHENS, R.H. 1980: On fine distinction in *Ulmus* L. - *Taxon* **29(2-3)**: 305-312.

RICHENS, R.H. 1983: Elm. - Cambridge, etc.: Cambridge Univ. Press, xii + 347 pp.

RICO HERNÁNDEZ, E. 1987a: Sobre *Cerastium gracile* Léon Dufour y *C. ramosissimum* Boiss. - *Anales Jard. Bot. Madrid* **43(2)**: 469-470.

RICO HERNÁNDEZ, E. 1987b: Notas sobre *Cerastium* L. - *Anales Jard. Bot. Madrid* **44(2)**: 567-568.

RICO, E. & AL. 2006: (1717) Proposal to reject the name *Lindernia palustris* (Scrophulariaceae) . - *Taxon* **55(1)**: 237-238. [approved, cf. Taxon 56(4): 1291, 2007].

RIECHELMANN, A. 1990: *Ophrys* ×*kattaviensis*. Eine neue *Ophrys*-Hybride von Rhodos. - *Ber. Arbeitskreis. Heimische Orchid.* **7(1)**: 9-11.

RIECHELMANN, A. 1997: Einige Anmerkungen zur Orchideenflora von Rhodos. - *Ber. Arbeitskreis. Heimische Orchid.* **14(1)**: 62-69.

RIECHELMANN, A. 1998a: Orchideen-Exkursion Rhodos. - *J. Eur. Orch.* **30(3)**: 663-671.

RIECHELMANN, A. 1998b: *Ophrys umbilicata* Desf. auf Rhodos. - *Orchidee (Hamburg)* **49(4)**: 184-185.

RIECHELMANN, A. 1999a: *Ophrys calypsus* Hirth & Spaeth und *Ophrys phaseliana* D. & U. Rückbrodt - zwei neue Orchideenarten auf Kreta. - *Ber. Arbeitskreis. Heimische Orchid.* **16(1)**: 66-71.

RIECHELMANN, A. 1999b: Orchideen-Exkursion Ostkreta vom 04. April bis 18. April 1998. - *J. Eur. Orch.* **31(4)**: 977-985.

RIECHELMANN, A. 2000: Einige Anmerkungen zur Orchideenflora von Euböa (Griechenland). - *Ber. Arbeitskreis. Heimische Orchid.* **17(2)**: 34-50.

RIECHELMANN, A. 2001: Orchideen-Exkursion Kreta (West) vom 17. April 2000 bis 01. Mai 2000. - *J. Eur. Orch.* **33(4)**: 1100-1110.

RIECHELMANN, A. 2004: Einige Anmerkungen zur Orchideenflora von Korfu (Kerkira). - *J. Eur. Orch.* **36(2)**: 527-544.

RIECHELMANN, A. 2008: Orchideen-Exkursion Kreta (Ost) vom 08. April 2006 bis 22. April 2006 - Exkursionsprotokoll. - *J. Eur. Orch.* **40(4)**: 751-763.

RIECK, G. 1998: *Paeonia clusii* subsp. *clusii*. - *Pl. Mont. Rocaille* **12(187)**: front cover.

RIEDL, H. 1963a: *Anchusa* subgen. *Chamanchusa* subgen. nov. und das System der Borraginoideae-Anchuseae. - *Oesterr. Bot. Z.* **110(5)**: 543-546.

RIEDL, H. 1963b: Revision der einjährigen Arten von *Adonis* L. - *Ann. Naturhist. Mus. Wien* **66**: 51-90.

RIEDL, H. 1966: Die Gattung *Heliotropium* in Europa. - *Ann. Naturhist. Mus. Wien* **69**: 81-93.

RIEDL, H. 1967a: Die infraspezifischen Einheiten von *Arum maculatum* in Mitteleuropa. - *Phyton (Horn)* **12(1-4)**: 159-168.

RIEDL, H. 1967b: Die Gattung *Ephedra* in Europa. - *Sci. Pharm.?* **35(3)**: 225-228.

RIEDL, H. 1970: *Euphorbia verrucosa* and *E. stricta* to be rejected? - *Taxon* **19(5)**: 798-800.

RIEDL, H. 1971: Übersicht über die ausdauernden und zweijährigen Arten der Gattung *Arnebia* Forssk. - *Ann. Naturhist. Mus. Wien* **75**: 209-222.

RIEDL, H. 1974: Two new species of *Onosma*. - *Notes Roy. Bot. Gard. Edinburgh* **33(2)**: 301-303.

RIEDL, H. 1980a: Notes on *Ephedra*. - *Notes Roy. Bot. Gard. Edinburgh* **38(2)**: 291-295.

RIEDL, H. 1980b: Tentative keys for the identification of species in *Biarum* and *Eminium*, with notes on some taxa included in *Biarum*. - *Aroideana* **3(1)**: 24-31.

RIEDL, H. 2000: Studies on the genus *Lappula* (Boraginaceae) II. - Notes on some predominantly European species. - *Bot. Chron. (Patras)* **13**: 375-381.

RIEDL, H., BINZET, R. & ORCAN, N. 2005: A new species of *Onosma* (Boraginaceae-Lithospermeae) from southern Turkey. - *Edinburgh J. Bot.* **61(2-3)**: 127-130.

RIEGEL, G. & RIEGEL, I. 2004: Ergänzung bisheriger Kenntnisse über die Verbreitung von *Orchis papilionacea* subsp. *schirwanica* auf Rhodos. - *Ber. Arbeitskreis. Heimische Orchid.* **20(2)**: 95-96.

RIEK, R. 1935: Systematische und pflanzengeographische Untersuchungen in der *Veronica*-Sektion *Chamaedrys* Griseb. - *Repert. Spec. Nov. Regni Veg. Beih.* **79**: i-viii, 1-68 + 16 plates.

RIKLI, M. 1901: Die Gattung *Dorycnium* Vill. - *Bot. Jahrb. Syst.* **31(3)**: 314-404 + 4 plates.

RIKLI, M. 1915a: Zur Kenntnis der Flora der Insel Kreta. - *Verh. Schweiz. Naturf. Ges.* **97(2)**: 213.

RIKLI, M. 1915b: Kreta und Sizilien. In Karsten, G. & Schenk, H. (eds): Vegetationsbilder, 13. Reihe, Heft 1-2. - Jena: Gustav Fischer Verlag, 12 Tafeln mit Text.

RIKLI, M. 1917: Eine Fruehlingsfahrt nach Kreta. - *Neujahrsbl. Naturf. Ges. Zürich* **119**: 52 pp. + 6 plates.

RIKLI, M. 1926: *Pinus pinea* L. - *Pflanzenareale* **1(1)**: 16-17, Karte 9.

RIKLI, M. 1930: Nachträge und Verbesserungen zu den Europa-Karten: *Pinus*. - *Pflanzenareale* **2(8)**: 88.

RIKLI, M. 1933: *Erica arborea* L. - *Pflanzenareale* **3(8)**: 95-98, Karte 79-80.

RIKLI, M. 1942-43: Das Pflanzenkleid der Mittelmeerländer. Erster Band. - Bern: Verlag Hans Huber, pp. i-xvi + 1-436 pp. + plates.

RIKLI, M. 1944-47: Das Pflanzenkleid der Mittelmeerländer. Zweiter Band. - Bern: Verlag Hans Huber, pp. i-xv + 437-1093 pp. + plates.

RIKLI, M. 1948: Das Pflanzenkleid der Mittelmeerländer. Dritter Band. - Bern: Verlag Hans Huber, pp. i-viii + 1095-1418 pp. + plates.

RIKLI, M. & RÜBEL, E. 1923: Über Flora und Vegetation von Kreta und Griechenland. - *Vierteljahrsschr. Naturf. Ges. Zürich* **68(1-2)**: 103-227.

RILKE, S. 1999a: Species diversity and polymorphism in *Salsola* sect. *Salsola* sensu lato (Chenopodiaceae). - *Syst. & Geogr. Pl.* **68(1-2)**: 305-314.

RILKE, S. 1999b: Revision der Sektion *Salsola* s.l. der Gattung *Salsola* (Chenopodiaceae). - *Biblioth. Bot.* **149**: 189 pp.

RISTOW, M. 1999: Notes on *Cyperus laevigatus*. In Greuter, W. & Raus, Th. (eds): Med-Checklist notulae, 18. - *Willdenowia* **29(1-2)**: 65.

RISTOW, M. 2013: *Botriochloa pertusa* (L.) A. Camus. In Raab-Straube, E. v. & Raus, Th. (eds): Euro+Med-Checklist notulae, 2. - *Willdenowia* **43(2)**: 246.

RISTOW, M., KRAUSE, J. & BERGMEIER, E. 2021: *Saccharum spontaneum*. In Raus, Th. & Raab-Straube, E.v. (eds): Euro+Med-Checklist notulae, 14. - *Willdenowia* **51(3)**: 361.

RISTOW, M. & SCHOLZ, H. 2013: *Secale strictum* subsp. *anatolicum*. In Raab-Straube, E.v. & Raus, Th.: Notulae ad floram euro-mediterraneam pertinentes No. 30. - *Willdenowia* **43(1)**: 162.

RITA, J. (ED.) 1990: Taxonomía, biogeografía y conservación de Pteridófitos. - Palma de Mallorca: Soc. Hist. Nat. Balears, 253 pp.

RITCHIE, J.C. 1955: Biological Flora of the British Isles. *Vaccinium vitis-idaea* L. - *J. Ecol.* **43(2)**: 701-708.

RITCHIE, J.C. 1956: Biological Flora of the British Isles. *Vaccinium myrtillus* L. - *J. Ecol.* **44(1)**: 291-299.

RITTER-STUDNIČKA, H. 1956: Beitrag zur Ökologie der Serpentinflora in Bosnien. - *Vegetatio* **7(2)**: 89-98.

RITTER-STUDNIČKA, H. 1965: Standortsuntersuchungen und Neufunde von *Halacsya sendtneri* (Boiss.) Dörfl. in Bosnien. - *Oesterr. Bot. Z.* **112(3)**: 371-391.

RITTER-STUDNIČKA, H. 1970a: Die Flora der Serpentinvorkommen in Bosnien. - *Biblioth. Bot.* **130**: 100 pp.

RITTER-STUDNIČKA, H. 1970b: Die Vegetation der Serpentinvorkommen in Bosnien. - *Vegetatio* **21(1-3)**: 75-156.

RIVAS PONCE, M.A. 1988: Nuevos datos para la diagnosis de *Bromus rubens* L. y *B. madritensis* L. (Poaceae). - *Lagascalia* **15(1)**: 89-93.

RIVAS-MARTÍNEZ, S. & BELMONTE, D. 1989: Sinopsis de *Adenocarpus* DC. (Leguminosae). - *Opusc. Bot. Pharm. Complut.* **5**: 69-78.

RIVAS-MARTÍNEZ, S. & HERRERA, M. 1996: Datos sobre *Salicornia* L. (Chenopodiaceae) en España. - *Anales Jard. Bot. Madrid* **54(1)**: 149-154.

RIVERA, D. & AL. 2014: Carpological analysis of *Phoenix* (Arecaceae): contributions to the taxonomy and evolutionary history of the genus. - *Bot. J. Linn. Soc.* **175(1)**: 74-122.

RIX, A. 2018: George Maw (1832-1912): his *Crocus* monograph, travels, and correspondents from the Ottoman Empire. In Aščerić-Todd, I. & al. (eds): Travellers in Ottoman lands. The Botanical Legacy, pp. 205-226. - Oxford: ASTENE and Archaeopress Publishing Ltd, 379 pp.

RIX, E.M. 1971: The taxonomy of the genus *Fritillaria* in the eastern Mediterranean area. - Cambridge: Unpubl. Ph.D. thesis, Univ. of Cambridge, U.K. [n.v.].

RIX, E.M. 1974: *Fritillaria graeca*, *F. messanensis* and related species in Greece and Bulgaria. - *Lilies* **1973-74**: 75-79.

RIX, E.M. 1975a: Notes on *Fritillaria* (Liliaceae) in the eastern Mediterranean region, I & II. – *Kew Bull.* **29(4)**: 633-654.

RIX, E.M. 1975b: Notes on *Fritillaria* (Liliaceae) in the eastern Mediterranean region, III. - *Kew Bull.* **30(1)**: 153-162.

RIX, E.M. 1975c: Fritillary co-operative exhibition. - *Lilies* **1975**: 45-53.

RIX, E.M. 1978: [New combinations in *Fritillaria*]. In Heywood, V.H. (ed.): Notulae systematicae ad Floram Europaeam spectantes, no. 20. - *Bot. J. Linn. Soc.* **76(4)**: 356.

RIX, E.M. 1979a: Notes on *Fritillaria* (Liliaceae) in the eastern Mediterranean region: IV. - *Kew Bull.* **33(4)**: 585-600.

RIX, E.M. 1979b: Some fritillaries and their habit. - *Lilies* **1978/79**: 89-94.

RIX, E.M. 1980: The genus *Acanthus* L. An introduction to the hardy species. - *Plantsman* **2(3)**: 132-140.

RIX, E.M. 1982: The herbaceous Berberidaceae. - *Plantsman* **4(1)**: 1-15 + 1 plate.

RIX, E.M. 1988: *Fritillaria whittallii*. - *New Plantsman* **5(2)**: 110-113.

RIX, E.M. 2000: *Fritillaria tuntasiana*. - *Curtis's Bot. Mag., ser. 6* **17(3)**: 176-170 + 1 plate.

RIX, E.M. & FAY, M.F. 2017: 857. *Platanus orientalis*. Platanaceae. - *Curtis's Bot. Mag.* **34(1)**: 29-40.

RIX, E.M. & STRANGE, K. 2007: *Fritillaria gussichae* [sic!]. - *Curtis's Bot. Mag.* **24(1)**:

RIX, H. 2021: In the lap of the gods. - *Alpine Gardener* **89(4)**: 398-415.

RIZOVSKI, R. 1977: Beitrag zur Kenntnis der Verbreitung von *Dioscorea balcanica* Koš. - *Glasn. Republ. Zavoda Zaštitu Prir. Prirodnjačke Zbirke Titogradu* **10**: 101-104.

RIZZI LONGO, L. & FEOLI CHIAPELLA, L. 1993: Palynotaxonomy of *Genista sericea* group. In Demiriz, H. & Özhatay, N. (eds): Proceedings of the V OPTIMA meeting, Istanbul, 8-15 September 1986, pp. 745-760. - Istanbul Üniversitesi Fen Fakültesi.

RIZZI LONGO, L. & FEOLI CHIAPELLA, L. 2007: Pollen morphology and taxonomy of *Genista* sects. *Acanthospartum* Spach and *Fasselospartum* P. Gibbs (Genisteae, Fabaceae). - *Fl. Medit.* **17**: 165-183.

RIZZI LONGO, L. & FEOLI CHIAPELLA, L. 2009: Pollen morphology and taxonomy of *Genista* sect. *Spartioides* (Genisteae, Fabaceae). - *Fl. Medit.* **19**: 141-160.

RIZZOTTO, M. 1979: Ricerche tassonomiche e corologiche sulle Cistaceae. 1. Il genere *Cistus* L. in Italia. - *Webbia* **33(2)**: 343-378.

RIZZOTTO, M. 1986: Numeri cromosomici per la Flora Italiana: 1008-1010. - *Inform. Bot. Ital.* **16(2-3)**: 248-250.

RIZZOTTO, M. 1992: Numeri cromosomici per la Flora Italiana: 1270-1272. - *Inform. Bot. Ital.* **23(2-3)**: 119-122.

RIZZOTTO, M. & SIGNORINI, M.A. 1986: Numeri cromosomici per la Flora Italiana: 1007. - *Inform. Bot. Ital.* **16(2-3)**: 246-247.

ROALSON, E.H. 2021: A revised synonymy, typification, and key to species of *Cleome* sensu stricto (Cleomaceae). - *Phytotaxa* **496(1)**: 54-68.

ROBATSCH, K. 1978: *Cephalanthera damasonium* - Neufund für Kreta. - *Orchidee (Hamburg)* **29(3)**: 110-111.

ROBATSCH, K. 1988: Beiträge zur Kenntnis der europäischen *Epipactis*-Arten (Orchidaceae). - *Linzer Biol. Beitr.* **20(1)**: 161-172. [Also distributed as *OPTIMA leaflet* no. 172].

ROBATSCH, K. 1989: Beiträge zur Kenntnis der europäischen *Epipactis*-Arten (Orchidaceae). - *Linzer Biol. Beitr.* **21(1)**: 295-302. [Also distributed as *OPTIMA leaflet* no. 178].

ROBATSCH, K. 1990: Beitrag zur Kenntnis der europäischen *Epipactis*-Arten (Orchidaceae). - *Linzer Biol. Beitr.* **22(1)**: 143-149.

ROBATSCH, K. 1995: Beiträge zur Kenntnis der europäischen *Epipactis*-Arten (Orchidaceae) und zur Evolution der Autogamie bei europäischen und asiatischen Gattungen der Neottioideae. - *J. Eur. Orch.* **27(1)**: 125-177.

ROBERTS, A.V. 1977: Relationship between species in the genus *Rosa*, section *Pimpinellifoliae*. - *Bot. J. Linn. Soc.* **74(4)**: 309-328.
ROBERTS, R.H. 1970: A revision of some of the taxonomic characters of *Polypodium australe* Fée. - *Watsonia* **8(2)**: 121-134.
ROBERTS, R.H. 1979: Spore size in *Asplenium adiantum-nigrum* L. and *A. onopteris* L. - *Watsonia* **12(3)**: 233-238.
ROBERTSON, I.H. 1981: Chromosome numbers in *Brachypodium* Beauv. (Gramineae). - *Genetica* **56 (1)**: 55-60.
ROBERTY, G. & VAUTIER, S. 1964: Les genres de Polygonacées. - *Boissiera* **10**: 7-128.
ROBERTZ, A. & ROBERTZ, E. 1990: Ergänzende Anmerkung zu *Ophrys lapethica* Gölz & Reinhard. - *Mitt. Arbeitskreis Heimische Orchid. Baden-Württemberg* **22(2)**: 385-386.
ROBINSON, A. 1998: A holiday in Greece. - *Geraniaceae Group News* **70**: 15-16.
ROBINSON, A. & TAN, KIT 2018: Report 46. In Vladimirov, V., Aybeke, M. & Tan, Kit (eds): New floristic records in the Balkans: 35. - *Phytol. Balcan.* **24(1)**: 165-167.
ROBINSON, H. & KING, R.M. 1977: Eupatorieae - systematic review. In Heywood, V.H., Harborne, J.B. & Turner, B.L. (eds): The biology and chemistry of the Compositae 1: 437-485. - London, etc.: Academic Press.
ROBINSON, J.P., HARRIS, S.A & JUNIPER, B.E. 2001: Taxonomy of the genus *Malus* Mill. (Rosaceae) with emphasis on the cultivated apple, *Malus domestica* Borkh. - *Pl. Syst. Evol.* **226(1-2)**: 35-58.
ROBINSON, M.A. 1990: Primulas. The complete guide. - Swindon: Crowood Press, 271 pp.
ROBSON, N.K.B. 1967a: [Notes on *Hypericum*]. In Heywood, V.H. (ed.): Notulae systematicae ad Floram Europaeam spectantes, no. 6. - *Feddes Repert.* **74(1-2)**: 22-24.
ROBSON, N.K.B. 1967b: Materials for a Flora of Turkey: XI. Notes on Turkish species of *Hypericum*. - *Notes Roy. Bot. Gard. Edinburgh* **27(2)**: 185-204.
ROBSON, N.K.B. 1977: Studies in the genus *Hypericum* L. (Guttiferae). 1. Infrageneric classification. - *Bull. Brit. Mus. (Nat. Hist.), Bot.* **5(6)**: 291-355.
ROBSON, N.K.B. 1980: *Hypericum olympicum* - wild and cultivated. - *Plantsman* **1(4)**: 193-200.
ROBSON, N.K.B. 1981: Studies in the genus *Hypericum* L. (Guttiferae). 2. Characters of the genus. - *Bull. Brit. Mus. (Nat. Hist.), Bot.* **8(2)**: 55-226.
ROBSON, N.K.B. 1985: Studies in the genus *Hypericum* L. (Guttiferae). 3. Sections 1. *Campylosporus* to 6a. *Umbraculoides*. - *Bull. Brit. Mus. (Nat. Hist.), Bot.* **12(4)**: 163-325.
ROBSON, N.K.B. 1986: A review of *Hypericum* sect. *Hirtella*. - *Notes Roy. Bot. Gard. Edinburgh* **43(2)**: 255-265.
ROBSON, N.K.B. 1990: (972) Proposal to reject the name *Hypericum quadrangulum* L. (Guttiferae) . - *Taxon* **39(1)**: 135-137. [approved, cf. Taxon 43(1): 114, 1994].
ROBSON, N.K.B. 1993: Studies in *Hypericum*: validation of new names. - *Bull. Nat. Hist. Mus. London, Bot.* **23(2)**: 67-70.
ROBSON, N.K.B. 1996: Studies in the genus *Hypericum* L. (Guttiferae). 6. Sections 20. *Myriandra* to 28. *Elodes*. - *Bull. Nat. Hist. Mus. London, Bot.* **26(2)**: 75-217.
ROBSON, N.K.B. 2001: Studies in the genus *Hypericum* L. (Guttiferae) 4(1). Section 7. *Roscyna* to 9. *Hypericum* sensu lato (part 1). - *Bull. Nat. Hist. Mus. London, Bot.* **31(2)**: 37-88.
ROBSON, N.K.B. 2002: Studies in *Hypericum* L. (Guttiferae) 4(2). Section 9. *Hypericum* sensu lato (part 2): subsection 1. *Hypericum* series 1. *Hypericum*. - *Bull. Nat. Hist. Mus. London, Bot.* **32(2)**: 61-123.
ROBSON, N.K.B. 2010a: Studies in the genus *Hypericum* L. (Hypericaceae) 5(1). Sections 10. *Olympia* to 15/16. *Crossophyllum*. - *Phytotaxa* **4**: 5-126.
ROBSON, N.K.B. 2010b: Studies in the genus *Hypericum* L. (Hypericaceae) 5(2). Sections 17. *Hirtella* to 19. *Coridium*. - *Phytotaxon* **4**: 127-258.
ROCHEL, A. 1833: Correspondenz ueber eine naturhistorische Reise in die europäische Türkei. - *Flora* **16(2)**: 493-496.
RODGERS, D. & RODGERS, C. 1986: The orchids of Corfu. - *Quart. Bull. Alpine Gard. Soc. Gr. Brit.* **54(4)**: 377-383.
RODIONENKO, G.I. 1987: The genus *Iris* L. Questions of morphology, biology, evolution and systematics [translated; Russian original publ. in 1961]. - Bridgewater, Somerset: Bigwood & Staple Ltd, iv + 222 pp.
RODMAN, J.E. 1974: Systematics and evolution of the genus *Cakile* (Cruciferae). - *Contr. Gray Herb.* **205**: 3-146.
RODRIGUES, J.E. DE MESQUITA 1953: Sobre a cariologia de *Scrophularia canina* L. - *Ciencias (Madrid)* **21(1)**: 259-272.
RODRIGUEZ, F.B. 1946: Revisión crítica le las especies del género "Hippocrepis" de la Península e Islas Baleares. - *Anales Jard. Bot. Madrid* **7**: 197-320.
ROEMER, J.J. & SCHULTES, J.A. 1817-30: Caroli a Linné equitis Systema Vegetabilium secundum classes ordines genera species ... - Stuttgardtiae: Sumtibus J.G. Cottae, 7 vols. [cf. TL-2 4: 847-848, 1983].
ROESSLER, H. 1968: Pflanzen aus Nordgriechenland. [Unpublished flora list]. - Photocopy: 15 pp.
ROGER-SMITH, H. 1941: Botanising on an Hellenic Traveller's tour. - *Quart. Bull. Alpine Gard. Soc. Gr. Brit.* **9(2)**: 120-123 + 1 plate.
ROGERS, C.M. 1982: The systematics of *Linum* sect. *Linopsis* (Linaceae). - *Pl. Syst. Evol.* **140(2-3)**: 225-234.
ROGERS, C.M. 1984: A further note on the relationships of the European *Linum hologynum* and the Australian species of *Linum* (Linaceae). - *Pl. Syst. Evol.* **147(3-4)**: 327-328.
ROGERS, C.M., MILDNER, R. & HARRIS, B.D. 1972: Some additional chromosome numbers in the Linaceae. - *Brittonia* **24(3)**: 313-316.
ROGERS, Z.S. & SPENCER, M.A. 2006: Typification of plant names in Thymelaeaceae published by Linnaeus and Linnaeus filius. - *Taxon* **55(2)**: 483-488.
ROHLENA, J. 1942: Conspectus florae montenegrinae. - *Preslia* **20-21**: 1-506.
ROHMER, M. 2018a: *Ophrys argolica* im Nordwest-Epirus (Griechenland), weit ausserhalb ihres bekannten Verbreitungsgebiet. - *J. Eur. Orch.* **50(2-4)**: 197-201.
ROHMER, M. 2018b: *Limodorum trabutianum*, first report from Western Macedonia (Greece). - *J. Eur. Orch.* **50(2-4)**: 339-344.
ROHRBACH, P. 1868: Monographie der Gattung *Silene*. - Leipzig: Verlag von Wilhelm Engelmann, viii + 249 pp. + 2 plates.
ROHRBACH, P. 1870: Ueber die europäischen Arten der Gattung *Typha*. - *Verh. Bot. Vereins Prov. Brandenburg* **11(2)**: 67-104 + 1 plate.
RÖHRIG, E. 1957: Über die Schwarzkiefer (*Pinus nigra* Arnold) und ihre Formen. 1. Teil: Die natürlichen Standorte. - *Silvae Genet.* **6**: 39-53.
ROHWEDER, H. 1934: Beiträge zur Systematik und Phylogenie des Genus *Dianthus* unter Berücksichtigung der karyologischen Verhältnisse. - *Bot. Jahrb. Syst.* **66(3)**: 249-368.
ROHWEDER, H. 1939: Weitere Beiträge zur Systematik und Phylogenie der Caryophyllaceen unter besonderer Berücksichtigung der karyologischen Verhältnisse. - *Beih. Bot. Centralbl., Abt. 2* **59(1)**: 1-58.
ROHWEDER, O. & URMI, E. 1978: Centrospermen-Studien: 10. Untersuchungen über den Bau der Blüten und Früchte von *Cucubalus baccifer* L. und *Drypis spinosa* L. (Caryophyllaceae-Silenoideae). - *Bot. Jahrb. Syst.* **100(1)**: 1-25.
ROJAS-ANDRÉS, B.M. & MARTÍNEZ-ORTEGA, M.M. 2016: Taxonomic revision of *Veronica* subsection *Pentasepalae* (*Veronica*, Plantaginaceae sensu APG III). - *Phytotaxa* **285**: 1-100.
ROLFE, R. 1998: A summit circuit. - *Quart. Bull. Alpine Gard. Soc. Gr. Brit.* **66(2)**: 195-206.
ROLFE, R. 2000: *Crocus ×gotoburgensis*. In Rolfe R. (ed.): Plant awards 1998-1999. - *Quart. Bull. Alpine Gard. Soc. Gr. Brit.* **68(2)**: 228-230.
ROLFE, R. 2001: [*Anchusa cespitosa*]. - *Alpine Gardener* **69(1)**: 71-72 + front cover.
ROLFE, R. 2002a: Selections from the MESE menu [*Campanula formanekiana, Silene atropurpurea*]. - *Alpine Gardener* **70(4)**: 401-403.
ROLFE, R. 2002b: *Aquilegia ottonis* Orph. ex Boiss. subsp. *amaliae* (Orph. ex Boiss.) Strid, P.C. - *Alpine Gardener* **70(4)**: 479-481.
ROLFE, R. 2004: *Viola* section *Melanium* - a scattershot survey. - *Plantsman, n.s.* **3(3)**: 154-161.
ROLFE, R. 2011: *Fritillaria epirotica* Turrill ex Rix, P.C. - *Alpine Gardener* **80(1)**: 49-50.

ROLFE, R. (ED.) 2004a: Plant Awards 2003-2004. *Ornithogalum sibthorpii* Greuter, P.C. - *Alpine Gardener* **72(4)**: 408-409.

ROLFE, R. (ED.) 2004b: Plant Awards 2003-2004. *Orchis provincialis* Ten., P.C. - *Alpine Gardener* **72(4)**: 421-422.

ROLFE, R.A. 1918: *Rosa glutinosa*. - *Bull. Misc. Inform. Kew* **1918(4)**: 153-156.

ROMA-MARZIO, F. & AL. 2017: Taxonomy of prickly juniper (*Juniperus oxycedrus* group): A phytochemical-morphometric combined approach at the contact zone of two cryptospecies. - *Phytochemistry* **141**: 48-60.

ROMANO, S. & CAMPO, G. 1995: Numeri cromosomici per la Flora Italiana: 1336-1340. - *Inform. Bot. Ital.* **27(1)**: 17-20.

ROMANO, S., CAMPO, G. & COLOMBO, P. 1997: Numeri cromosomici per la Flora Italiana: 1372-1377. - *Inform. Bot. Ital.* **28(2)**: 251-255.

ROMANO, S., COLOMBO, P. & MARCENÒ, C. 1992: Numeri cromosomici per la Flora Italiana: 1273-1283. - *Inform. Bot. Ital.* **23(2-3)**: 123-132.

ROMANO, S., MAZZOLA, P. & RAIMONDO, F.M. 1987a: Numeri cromosomici per la Flora Italiana: 1070-1081. - *Inform. Bot. Ital.* **18(1-3)**: 159-167.

ROMANO, S., MAZZOLA, P. & RAIMONDO, F.M. 1987b: Numeri cromosomici per la Flora Italiana: 1106-1117. - *Inform. Bot. Ital.* **19(2)**: 173-180.

ROMANO, S., OTTONELLO, D. & MARCENÒ, C. 1995: Numeri cromosomici per la Flora Italiana: 1302-1313. - *Inform. Bot. Ital.* **26(2-3)**: 191-199.

ROMASCHENKO, K. & AL. 2004: New chromosome counts in the *Centaurea Jacea* group (Asteraceae, Cardueae) and some related taxa. - *Bot. J. Linn. Soc.* **145(3)**: 345-352.

ROMERO GARCÍA, A.T. & BLANCA LÓPEZ, G. 1988: Contribución al estudio cariosistemático del genero *Agrostis* L. (Poaceae) en la Península Ibérica. - *Bol. Soc. Brot., Ser. 2* **61**: 81-104.

ROMERO GARCÍA, A.T., BLANCA LÓPEZ, G. & MORALES TORRES, C. 1986: El complejo de *Agrostis canina* L. (Poaceae) en la Península Ibérica. - *Anales Jard. Bot. Madrid* **43(1)**: 47-55.

ROMERO GARCÍA, A.T., BLANCA LÓPEZ, G. & MORALES TORRES, C. 1988: Revisión del género *Agrostis* L. (Poaceae) en la Península Ibérica. - *Ruizia* **7**: 1-160.

ROMERO GARCÍA, A.T. & ORTEGA OLIVENCIA, A. 1988: Números cromosómicos para la flora española. 497-503. - *Lagascalia* **15(1)**: 109-112.

ROMERO ZARCO, C. 1984a: Números cromosómicos para la flora española. 337-341. - *Lagascalia* **12 (2)**: 292-294.

ROMERO ZARCO, C. 1984b: Revisón del género *Helictotrichon* Besser ex Schultes & Schultes fil. (Gramineae) en la Península Ibérica. I. Estudio taxonómico. - *Anales Jard. Bot. Madrid* **41(1)**: 97-124.

ROMERO ZARCO, C. 1984c: Revisón taxonómica del género *Avenula* (Dumort.) Dumort. (Gramineae) en la Península Ibérica e Islas Baleares. - *Lagascalia* **13(1)**: 39-146.

ROMERO ZARCO, C. 1985a: Revisión del género *Arrhenatherum* Beauv. (Gramineae) en la Península Ibérica. - *Acta Bot. Malac.* **10**: 123-154.

ROMERO ZARCO, C. 1985b: Revisión del género *Helictotrichon* Bess. ex Schultes & Schultes fil. (Gramineae) en la Península Ibérica. II. Estudios experimentales. - *Anales Jard. Bot. Madrid* **42(1)**: 133-154.

ROMERO ZARCO, C. 1988: Números cromosómicos para la flora española. 516-527. - *Lagascalia* **15 (1)**: 119-124.

ROMERO ZARCO, C. 1990: Las avenas del grupo *barbata* en la Península Ibérica y Baleares. - *Lagascalia* **16(2)**: 243-268.

ROMERO ZARCO, C. 1996: Sinopsis del genero *Avena* L. (Poaceae, Aveneae) en España peninsular y Baleares. - *Lagascalia* **18(2)**: 171-197.

ROMERO ZARCO, C. 1998: Lectotipificación de *Vicia amphicarpa* L., *V. benghalensis* L. y *V. sativa* var. *nigra* L. (Leguminosae). - *Anales Jard. Bot. Madrid* **56(1)**: 178-179.

ROMERO ZARCO, C. 2011: *Helictochloa* Romero Zarco (Poaceae), a new genus of oat grass. - *Candollea* **66(1)**: 87-103.

ROMINGER, J.M. 1962: Taxonomy of *Setaria* (Gramineae) in North America. - *Illinois Biol. Monogr.* **29**: viii + 132 pp.

ROMO, A.M. 1987: *Stellaria nemorum* L. en la Península Ibérica. - *Anales Jard. Bot. Madrid* **44(2)**: 564-567.

ROMO, A.M., DIDUKH, Y. & BORATYŃSKI, A. 2004: *Thesium* (Santalaceae) in Crimea, Ukraine. - *Ann. Bot. Fenn.* **41(4)**: 273-281.

RÖMPP, H. 1928: Die Verwandtschaftsverhältnisse in der Gattung *Veronica*. - *Repert. Spec. Nov. Regni Veg. Beih.* **50**: 1-172.

RONNIGER, K. 1924: Beiträge zur Kenntnis der *Thymus*-Flora der Balkanhalbinsel. I-II. - *Repert. Spec. Nov. Regni Veg.* **20**: 334-336, 385-390.

RONNIGER, K. 1928: *Sorbus florentina* (Zuccagni) Nyman in Albanien. - *Verh. Zool.-Bot. Ges. Wien* **77**: 99-102.

RONNIGER, K. 1932: Zur Flora Südserbiens. *Thymus*-Belege aus dem Herbar der Universität Beograd, welche für Hayek's "Prodromus" Verwendung fanden. - *Bull. Inst. Jard. Bot. Univ. Belgrade* **2**: 67-71.

RONNIGER, K. 1938: Eine neue *Thymus*-Art aus Mazedonien. - *Repert. Spec. Nov. Regni Veg. Beih.* **100**: 171-172.

RONNIGER, K. 1941: Flora der Insel Zante. - *Verh. Zool.-Bot. Ges. Wien* **88/89**: 13-108.

RONSE DECRAENE, L.-P. & AKEROYD, J.R. 1988: Generic limits in *Polygonum* and related genera (Polygonaceae) on the basis of floral characters. - *Bot. J. Linn. Soc.* **98(4)**: 321-371.

RONSE, A.C. & AL. 2010: Taxonomic revision of European *Apium* L. *s.l.*: *Helosciadium* W.D.J. Koch restored. - *Pl. Syst. Evol.* **287(1)**: 1-17.

ROOKS, F. & AL. 2011: The elusive *Juncus minutulus*: a failure to separate tetra- andhexaploid individuals of the *Juncus bufonius* complex in a morphometric comparison of cytometrically defined groups. - *Preslia* **83**: 565-589.

ROOSE, M.L. & GOTTLIEB, L.D. 1976: Genetic and biochemical consequences of polyploidy in *Tragopogon*. - *Evolution* **30(4)**: 818-830.

ROSE, F. 1948: Biological Flora of the British Isles. *Orchis purpurea* Huds. - *J. Ecol.* **36(2)**: 366-377.

ROSE, F. 1989a: Key to annual *Salicornia* species of south England and north France. - *B.S.B.I. News* **53**: 12-16.

ROSE, F. 1989b: Colour identification guide to the grasses, sedges, rushes and ferns of the British Isles and north-western Europe. - London: Viking, 240 pp.

ROSENBAUMOVÁ, R., PLAČKOVÁ, I. & SUDA, J. 2004: Variation in *Lamium* subg. *Galeobdolon* (Lamiaceae) - insights from ploidy levels, morphology and isozymes. - *Pl. Syst. Evol.* **244(3-4)**: 219-244.

ROSENHEIM, P. 1939: European species of the genus *Fritillaria*. - *Quart. Bull. Alpine Gard. Soc. Gr. Brit.* **7(1-2)**: 1-15, 110-124.

RÖSER, M. 1989: Karyologische, systematische und chorologische Untersuchungen an der Gattung *Helictotrichon* Besser ex Schultes & Schultes (Poaceae) im westlichen Mittelmeergebiet. - *Diss. Bot.* **145**: 1-250.

RÖSER, M. 1995: Typification of the names of two widespread Eurasian grass species: *Avena pratensis* L. and *A. pubescens* Huds. (Poaceae: *Helictotrichon*). - *Taxon* **44(3)**: 395-399.

RÖSER, M. 1996: Ecogeography of the grass genus *Helictotrichon* (Poaceae: Aveneae) in the Mediterranean and adjacent regions. - *Pl. Syst. Evol.* **203(3-4)**: 181-280.

RÖSER, M. 1997: Patterns of diversification in Mediterranean oat grasses (Poaceae: Aveneae). - *Lagascalia* **19(1-2)**: 101-120.

RÖSER, M. 2012: *Stipellula*, a new genus, and new combinations in feather grasses (Poaceaae tribe Stipeae). - *Schlechtendalia* **24**: 91-94.

RÖSER, M. & AL. 2009: Generic realignments in the grass tribe Aveneae (Poaceaee). - *Schlechtendalia* **19**: 27-38.

ROSHEVITZ, R.I. 1947: Monographie of the genus *Secale* L. [In Russian with English summary]. - *Trudy Bot. Inst. Akad. Nauk S.S.S.R., Ser. 1, Fl. Sist. Vysš. Rast.* **6**: 105-163.

ROSS, R. 1966: The generic names published by N. M. von Wolf. - *Acta Bot. Neerl.* **15(1)**: 147-161.

ROSS, R. 1967: On some Linnaean species of *Erica*. - *J. Linn. Soc., Bot.* **60(381)**: 61-73 + 1 plate.

ROSSI, W. & AL. 1995: Genetic divergence between *Dactylorhiza romana* and *D. saccifera* (Orchidaceae) with description of their natural hybrid. - *Webbia* **50(1)**: 1-15.

ROSSITTO, M., OTTONELLO, D. & FICI, S. 1985: Numeri cromosomici per la Flora Italiana: 993-1002. - *Inform. Bot. Ital.* **15(2-3)**: 188-194.

RÖSSLER-HAUBER, L. 1946: Zur Kenntnis von *Euphorbia tauriensis* Allioni sensu ampl. - *Ber. Schweiz. Bot. Ges.* **56**: 271-301.

RÖSSLER, H. 1983: Das Alpenlieschgras (*Phleum alpinum* L.) in Bayern. - *Ber. Bayer. Bot. Ges.* **54**: 53-56.

RÖSSLER, L. 1943: Vergleichende Morphologie der Samen europäischer *Euphorbia*-Arten. - *Beih. Bot. Centralbl., Abt. 2* **62(1)**: 97-174 + 4 plates.

RÖSSLER, W. 1953: Scleranthi lusitaniae. - *Agron. Lusit.* **15(2)**: 97-138.

RÖSSLER, W. 1955: Die *Scleranthus*-Arten Österreichs und seiner Nachbarländer. - *Oesterr. Bot. Z.* **102(1)**: 30-72.

ROSTAŃSKI, K. 1965: Some new taxa in the genus *Oenothera* L. subgenus *Oenothera*. - *Fragm. Florist. Geobot.* **11(4)**: 499-523.

ROSTAŃSKI, K. 1966: Die Arten der Gattung *Oenothera* L. in Ungarn. - *Acta Bot. Acad. Sci. Hung.* **12(3-4)**: 337-249.

ROSTAŃSKI, K. 1985: Zur Gliederung der Subsektion *Oenothera* (Sektion *Oenothera*, *Oenothera* L., Onagraceae). - *Feddes Repert.* **96(1-2)**: 3-14.

ROSTAŃSKI, K. 1991: The representatives of the genus *Oenothera* L. in Portugal. - *Bol. Soc. Brot., Ser. 2* **64**: 5-33.

ROSTAŃSKI, K. & AL. 2010: Evening-Primroses (*Oenothera*) occurring in Europe [In English and Polish]. - Kraków: W. Szafer Institute of Botany, 157 pp.

ROSTAŃSKI, K. & FORSTNER, W. 1982: Die Gattung *Oenothera* (Onagraceae) in Österreich. - *Phyton (Horn)* **22(1)**: 87-113.

ROSÚA, J.L. 1985: Notas cariosistemáticas del género *Rosmarinus* L. (Lamiaceae) en la Península Ibérica. - *Anales Jard. Bot. Madrid* **42(1)**: 93-99.

ROSÚA, J.L. & BLANCA LÓPEZ, G. 1985: Notas cariosistemáticas de la sección *Salvia* del género *Salvia* L. (Lamiaceae). - *Anales Jard. Bot. Madrid* **42(1)**: 101-112.

ROSÚA, J.L. & BLANCA LÓPEZ, G. 1986a: Revisión del género *Salvia* L. (Lamiaceae) en el Mediterráneo Occidental: la sección *Salvia*. - *Acta Bot. Malac.* **11**: 227-272.

ROSÚA, J.L. & BLANCA LÓPEZ, G. 1986b: On the typification of *Salvia sclarea* and *S. argentea* (Lamiaceae). - *Taxon* **35(4)**: 718-719.

ROSÚA, J.L. & BLANCA LÓPEZ, G. 1989: Revisón del género *Salvia* L. sect. *Aethiopis* Benth. (Lamiaceae) en el Mediterráneo Occidental. - *Collect. Bot. (Barcelona)* **17(2)**: 205-236.

ROSÚA, J.L. & BLANCA LÓPEZ, G. 1990: Acerca de la distribución de la sección *Salvia* (género *Salvia* L., Lamiaceae) en la Región Mediterránea Occidental y sus relaciones de vicarianza con el este del Mediterráneo. - *Lagascalia* **15(2)**: 137-143.

ROSÚA, J.L. & JARVIS, C.E. 1987: Sobre la lectotipificación de *Salvia officinalis* L. (Lamiaceae). - *Taxon* **36(3)**: 634-635.

ROSÚA, J.L. & NAVARRO, T. 1986: Números cromosómicos para la flora española. 456-463. - *Lagascalia* **14(2)**: 282-286.

ROTHERA, S.L. & DAVY, A.J. 1986: Polyploidy and habitat differentiation in *Deschampsia cespitosa*. - *New Phytol.* **102(3)**: 449-467.

RÖTHLISBERGER, J. 2010: Le genre *Chamaesyce* (Euphorbes stipulées) à Rhodes (Grèce). - *Saussurea* **40**: 101-105.

ROTHMALER, W. 1934: Systematische Vorarbeiten zu einer Monographie der Gattung *Alchemilla* L. I. - *Repert. Spec. Nov. Regni Veg.* **33**: 342-350.

ROTHMALER, W. 1935: Systematische Vorarbeiten zu einer Monographie der Gattung *Alchemilla* (L.) Scop. emend. II. Die systematische Gliederung der Gattung. - *Repert. Spec. Nov. Regni Veg.* **38**: 33-43.

ROTHMALER, W. 1936: Systematische Vorarbeiten zu einer Monographie der Gattung *Alchemilla* (L.) Scop. IV. Die Gruppen der Untergattung *Eualchemilla* (Focke) Buser. - *Repert. Spec. Nov. Regni Veg.* **40**: 208-212.

ROTHMALER, W. 1937: Systematische Vorarbeiten zu einer Monographie der Gattung *Alchemilla* (L.) Scop. VII. Aufteilung der Gattung und Nomenklatur. - *Repert. Spec. Nov. Regni Veg.* **42**: 164-173.

ROTHMALER, W. 1939a: Systematische Vorarbeiten zu einer Monographie der Gattung *Alchemilla* (L.) Scop. VIII. Neue Arten aus der Subsektion *Heliodrosium* Rothm. - *Repert. Spec. Nov. Regni Veg.* **46**: 122-132.

ROTHMALER, W. 1939b: Systematik und Geographie der Subsektion *Calycanthum* der Gattung *Alchemilla* L. - *Repert. Spec. Nov. Regni Veg. Beih.* **100**: 59-93.

ROTHMALER, W. 1941: Systematische Vorarbeiten zu einer Monographie der Gattung *Alchemilla*. IX. Über *Alchemilla*-Arten aus Osteuropa und Asien. - *Repert. Spec. Nov. Regni Veg.* **50**: 78-80, 245-255.

ROTHMALER, W. 1943a: Die Waldverhältnisse im Peloponnes. - *Intersylva* **3**: 329-342.

ROTHMALER, W. 1943b: Zur Gliederung der Antirrhineae. - *Feddes Repert. Spec. Nov. Regni Veg.* **52**: 16-39.

ROTHMALER, W. 1944a: Floristische Ergebnisse einer Reise nach dem Peloponnes. - *Bot. Jahrb. Syst.* **73(4)**: 418-452 + 5 plates.

ROTHMALER, W. 1944b: Die Gliederung der Gattung *Cytisus* L. - *Feddes Repert. Spec. Nov. Regni Veg.* **53(6-11)**: 137-150.

ROTHMALER, W. 1944c: Zur Nomenklatur der europäischen *Alchemilla*-Arten. - *Svensk Bot. Tidskr.* **38 (1)**: 102-112.

ROTHMALER, W. 1949: Revision der Genisteen. II. Die Gattungen *Erinacea*, *Spartium* und *Calicotome*. - *Bot. Jahrb. Syst.* **74(3)**: 271-287.

ROTHMALER, W. 1956: Taxonomische Monographie der Gattung *Antirrhinum*. - *Feddes Repert. Spec. Nov. Regni Veg. Beih.* **136**: 1-124 + 10 plates.

ROTHMALER, W. 1962: Systematische Vorarbeiten zu einer Monographie der Gattung *Alchemilla* L. X. Die mitteleuropäischen Arten. - *Feddes Repert. Spec. Nov. Regni Veg.* **66(3)**: 194-234.

ROTHMALER, W. & SCHNEIDER, U. 1962: Die Gattung *Polypodium* in Europa. - *Kulturpflanze, Beih.* **3**: 234-248.

ROTI-MICHELOZZI, G. 1984: Biosystematic investigations on north western Italian populations of the "Vicia cracca" aggregate. Preliminary observations. - *Webbia* **38**: 815-827.

ROTI-MICHELOZZI, G. 1986: Biosystematic studies on the *Vicia villosa* complex in Europe. – *Candollea* **41(2)**: 399-411.

ROTI-MICHELOZZI, G. & ALLIONE, S. 1988: Numeri cromosomici per la Flora Italiana: 1118-1126. - *Inform. Bot. Ital.* **19(3)**: 305-313.

ROTI-MICHELOZZI, G. & BARBERIS, G. 1990: Biosystematic notes on *Vicia incisa* M. Bieb. (Leguminosae), a rare species for Italy. - *Atti Soc. Tosc. Sci. Nat. Pisa Mem., Ser. B* **96**: 219-225.

ROTI-MICHELOZZI, G. & CAFFARO, L. 1984: Cytotaxonomic studies in wild populations of *Vicia hybrida* L. and *Vicia lutea* L. - *Folia Geobot. Phytotax.* **19(2)**: 169-176.

ROTI-MICHELOZZI, G. & CAFFARO, L. 1989: Karyometric analysis on wild Italian populations of genus *Vicia* section *Ervum* (Leguminosae). - *Atti Soc. Tosc. Sci. Nat. Pisa Mem., Ser. B* **95**: 107-124.

ROTI-MICHELOZZI, G., CAFFARO, L. & BEVILACQUA, L. 1989: New data about *Vicia loiseleurii* (M. Bieb.) Litw., correct binomial for *Vicia meyeri* Boiss. - *Candollea* **44(1)**: 103-117.

ROTI-MICHELOZZI, G. & FERRO, M. 1988: Numeri cromosomici per la Flora Italiana: 1156-1163. - *Inform. Bot. Ital.* **19(3)**: 333-339.

ROTI-MICHELOZZI, G. & RIGGIO BEVILACQUA, L. 1990: Biosystematic notes on the "Lathyrus sylvestris" complex in Italy. - *Candollea* **45(2)**: 599-608.

ROTI-MICHELOZZI, G. & SERRATO-VALENI, G. 1986: Seed characteristics in Italian species of genus *Vicia* section *Ervum* and their diagnostic value. - *Seed Sci. Technol.* **14(2)**: 391-402.

ROTI-MICHELOZZI, G. & SERRATO, G. 1981: Numeri cromosomici per la Flora Italiana: 752-759. - *Inform. Bot. Ital.* **12(2)**: 167-172.

ROTTENBERG, A. 2020: The origin of *Vicia faba* (Fabaceae): a quest of five decades. - *Fl. Medit.* **30**: 365-38.

ROTTENBERG, A. & ZOHARY, D. 1996: The wild ancestry of the cultivated artichoke. - *Genet. Resources Crop Evol.* **43**: 53-58.

ROTTENSTEINER, W.K. 2018: *Aconitum* subsectio *Burnatii* subsectio nova. - *Bot. Serbica* **42 (suppl. 1)**: 40.

RÖTTGER, B. 1990: Beiträge zur Kartierung der Orchideenflora von Rhodos. - *Mitt. Arbeitskreis Heimische Orchid. Baden-Württemberg* **22(2)**: 387-404.

ROUSAKOVA, V. & VITKOVA, A. 1996: Habitat and population characteristics of *Ruta graveolens* L. in Bulgaria. - *Fitologiya* **48**: 24-38.

ROUSI, A. 1961: Cytotaxonomical studies on *Vicia cracca* L. and *V. tenuifolia* Roth. I. Chromosome numbers and karyotype evolution. - *Hereditas (Lund)* **47(1)**: 81-110.

ROUSI, A. 1965: Biosystematic studies on the species aggregate *Potentilla anserina* L. - *Ann. Bot. Fenn.* **2(1)**: 47-112.

ROUSI, A. 1973a: Cytotaxonomical studies on *Vicia cracca* L. and *V. tenuifolia* Roth. III. The relation between karyotype and morphology. - *Ann. Bot. Fenn.* **10(2)**: 89-96.

ROUSI, A. 1973b: Studies on the cytotaxonomy and mode of reproduction of *Leontodon* (Compositae). - *Ann. Bot. Fenn.* **10(3)**: 201-215.

ROUTSI, E. 1993: Viosistematiki meleti tis section *Acrocentron* (Cass.) DC. tou genous *Centaurea* L. stin Ellada [In Greek]. - Patras: Ph.D. thesis, Univ. of Patras, 337 pp. + tables and charts.

ROUTSI, E. & GEORGIADIS, TH. 1988: [Chromosome numbers in *Centaurea* from Greece]. In Löve, Á. (ed.): Chromosome number reports XCIX. - *Taxon* **37(2)**: 399.

ROUTSI, E. & GEORGIADIS, TH. 1994a: Systematic review of *Centaurea rupestris* L., section *Acrocentron* (Cass.) DC., in Greece. - *Candollea* **49(2)**: 359-368.

ROUTSI, E. & GEORGIADIS, TH. 1994b: Contribution to the systematics of the genus *Centaurea* section *Acrocentron* (Asteraceae) in Greece. - *Nordic J. Bot.* **14(4)**: 369-378.

ROUTSI, E. & GEORGIADIS, TH. 1999: Cytogeographical study of *Centaurea* L. sect. *Acrocentron* (Cass.) DC. (Asteraceae) in Greece. - *Bot. Helv.* **109(2)**: 139-151.

ROUVILLE, P. DE 1853: Monographie du genre *Lolium*. - Montpellier: viii + 57 pp.

ROUX, J. & BOULOS, L. 1972: Révision systématique du genre *Sonchus* L. s.l. II. Étude caryologique. - *Bot. Not.* **125(4)**: 306-309.

ROUY, G. 1894: Sur quatre plantes rarissimes de la flore européenne. - *Bull. Soc. Bot. France* **41**: 401-402.

ROUY, G. 1896: Révision du genre *Onopordon*. - *Bull. Soc. Bot. France* **43**: 577-599.

ROWLAND, N. 1991: Orchids in southern Crete. - *Quart. Bull. Alpine Gard. Soc. Gr. Brit.* **59(3)**: 297-304.

ROWLEY, G.D. 1967: Chromosome studies and evolution in *Rosa*. - *Bull. Jard. Bot. Natl. Belg.* **37(1)**: 45-52.

ROWLEY, G.D. 1976: Typification of the genus *Rosa* L. - *Taxon* **25(1)**: 181.

ROWLEY, G.D. 1988: *Sedum*, one genus or forty-one? - *Sedum Soc. Newsletter* **5**: 6-11.

ROY, C. 2002: Le genre *Edraianthus*. - *Pl. Mont. Rocaille* **13(202)**: 401-403.

ROY, J., NAVAS, M,L. & SONIÈ, L. 1991: Invasion by annual brome grasses: a case study challanging the homoclime approach to invasions. In Groves, R.H. & Di Castri, F (eds): Biogeography of mediterranean invasions, pp. 207-224. – Cambridge, etc.: Cambridge Univ. Press.

ROZMUS, M. 1958: Cytological investigations on *Anthoxanthum alpinum* L. et L. a new species of the flora of Poland. - *Acta Biol. Cracov., Ser. Bot.* **1(3)**: 171-184 + 1 plate.

ROZMUS, M. 1960: The taxonomical rank of *Anthoxanthum alpinum* L. et L. in the light of anatomical studies. - *Acta Biol. Cracov., Ser. Bot.* **3(2)**: 81-90.

RUAS, C.F. & AL. 2001: Cytogenetic studies in the genus *Atriplex* (Chenopodiaceae). - *Caryologia* **54 (2)**: 129-145.

RÜBEL, E. (ED.) 1932: Die Buchenwälder Europas. - *Veröff. Geobot. Inst. Rübel Zürich* **8**: 509 pp.

RÜCKBRODT, D. & RÜCKBRODT, U. 1996: *Ophrys phaseliana*, eine neue *Ophrys* aus dem Komplex *Ophrys fusca* s. str. aus der südwestlichen Türkei. - *J. Eur. Orch.* **28(2)**: 391-403.

RÜCKBRODT, D., RÜCKBRODT, U. & GÜGEL, E. 1998: Orchideen-Neufunde für die Kykladeninseln Naxos und Paros. - *Ber. Arbeitskreis. Heimische Orchid.* **14(2)**: 29-31.

RÜCKBRODT, U. 1971: Botanisieren auf Kreta. - *Kosmos (Stuttgart)* **67**: 484-487.

RÜCKBRODT, U. & AL. 1990: Versuch einer Gliederung des *Ophrys-fusca*-Komplexes auf Rhodos und Beschreibung von *Ophrys attaviria* Rückbrodt & Wenker spec. nov. - *Ber. Arbeitskreis. Heimische Orchid.* **7(2)**: 7-17.

RÜCKBRODT, U. & AL. 1992: Bemerkungen zu den in der Türkei vorkommenden Orchideenarten und ihrer Verbreitung. - *Ber. Arbeitskreis. Heimische Orchid.* **9(1)**: 4-103, 161-176.

RÜCKBRODT, U. & RÜCKBRODT, D. 1987: *Himantoglossum affine* (Boiss.) Schltr. auf Kreta? - *Ber. Arbeitskreis. Heimische Orchid.* **4(2)**: 314-321.

RÜCKBRODT, U. & RÜCKBRODT, D. 1990: *Ophrys transhyrcana* Czerniak., Neufund für Rhodos. - *Mitt. Arbeitskreis Heimische Orchid. Baden-Württemberg* **22(4)**: 736-739.

RUDALL, P. & MATHEW, B.F. 1990: Leaf anatomy in *Crocus* (Iridaceae). - *Kew Bull.* **45(3)**: 535-544.

RUDALL, P., OWENS, S.J. & KENTON, A.Y. 1984: Embryology and breeding systems of *Crocus* (Iridaceae): a study in causes of chromosome variation. - *Pl. Syst. Evol.* **148(1-2)**: 119-134.

RÜDENBERG, L. & GREEN, P.S. 1966: A karyological survey of *Lonicera*, I. - *J. Arnold Arbor.* **47(3)**: 222-247.

RÜDENBERG, L. & GREEN, P.S. 1969: A karyological survey of *Lonicera*, II. - *J. Arnold Arbor.* **50(3)**: 449-461.

RUHE, W. 1936: Die Areale der mitteleuropäischen *Acer*-Arten. - *Repert. Spec. Nov. Regni Veg. Beih.* **86**: 95-106 + 8 maps.

RUÍZ DE CLAVIJO, E. 1993: Números cromosómicos para la flora española. 664-690. - *Lagascalia* **17 (1)**: 161-172.

RUÍZ DE CLAVIJO, E. & AL. 1994: Números cromosómicos para la flora española. 748-763. – *Lagascalia* **17(2)**: 379-388.

RUÍZ DE CLAVIJO, E., MUÑOZ, J. & SALVO, A.E. 1984: Sobre la presencia de *Azolla filiculoides* Lam. en España. - *Acta Bot. Malac.* **9**: 129-132.

RUIZ REJÓN, C. & AL. 1990: *Asphodelus tenuifolius* and *A. fistulosus* (Liliaceae) are morphologically, genetically and biologically different species. - *Pl. Syst. Evol.* **169(1-2)**: 1-12.

RUIZ REJÓN, C., LOZANO, R. & RUIZ REJÓN, M. 1986: Números cromosómicos para la flora española. 479-484. - *Lagascalia* **14(2)**: 292-297.

RUIZ REJÓN, M. & AL. 1980: Números cromosómicos para la flora española. 121-126. - *Lagascalia* **9 (2)**: 249-254 + 2 plates.

RUIZ REJÓN, M. & AL. 1981: Números cromosómicos para la flora española. 240-244. - *Lagascalia* **10 (2)**: 247-252.

RUIZ REJÓN, M., FERNÁNDEZ PIQUERAS, J. & OLIVER JIMÉNEZ, J.L. 1978: Números cromosómicos para la flora española. 71. - *Lagascalia* **8(1)**: 118-120.

RUIZ REJÓN, M. & OLIVER JIMÉNEZ, J.L. 1978: Números cromosómicos para la flora española. 68-69. - *Lagascalia* **8(1)**: 113-117 + 1 plate.

RUIZ REJÓN, M., OLIVER JIMÉNEZ, J.L. & POSSE HERRANZ, M.F. 1978: Números cromosómicos para la flora española. 70. - *Lagascalia* **8(1)**: 117-118 + 1 plate.

RUIZ REJÓN, M. & OLIVER, J.L. 1980: Cytogenetic and electrophoretic evidence for a polyploid origin of the basic chromosome number x=14 in the genus *Asphodelus* (Liliaceae). - *Pl. Syst. Evol.* **136(3-4)**: 259-265.

RUIZ REJÓN, M., OLIVER, J.L. & RUIZ REJÓN, C. 1980: Variabilidad cromosómica en *Scilla autumnalis* L. (Liliaceae) de la Península Ibérica. - *Bol. Soc. Brot., Ser. 2* **53(1)**: 555-562.

RUIZ, J.L. 1993: Reports (204-216). In Kamari, G., Felber, F. & Garbari, F. (eds): Mediterranean chromosome number reports - 3. - *Fl. Medit.* **3**: 354-358.

RUIZ, J.L. 1996: Reports (559-589). In Kamari, G., Felber, F. & Garbari, F. (eds): Mediterranean chromosome number reports - 5. - *Fl. Medit.* **5**: 363-373.

RUKŠĀNS, J. 2010: Crocuses. A complete guide to the genus. - Portland & London: Timber Press, 216 pp.

RUKŠĀNS, J. 2013: Seven new crocuses described. - *Alpine Gardener* **81(2)**: 188-193. [Longer version available at AGS website].

RUKŠĀNS, J. 2014a: *Crocus danfordiae* Maw and *C. chrysanthus* (Herbert) Herbert (Iridaceae) and their allies in Turkey and Iran. - *Int. Rock Gard.* **52**: 2-31.

RUKŠĀNS, J. 2014b: Crocuses from "The Holy Mountain" of Athos, Greece. - *Int. Rock Gard.* **59**: 4-15.

RUKŠĀNS, J. 2015: Some new *Crocus* taxa (Iridaceae) from Western Turkey and East Aegean Islands. - *Int. Rock Gard.* **64**: 1-36.

RUKŠĀNS, J. 2016: Some taxonomic problems in the genus *Crocus*. - *Int. Rock Gard.* **74**: 17 pp.

RUKŠĀNS, J. 2017a: Two new *Crocus* species allied to *Crocus nubigena* from the Eastern Aegean Islands. - *Int. Rock Gard.* **90**: 22 pp.

RUKŠĀNS, J. 2017b: The world of *Crocus*. - Riga: Latvian Academy of Sciences, 568 pp.

RUKŠĀNS, J. 2017c: What is *Crocus nubigena* Herb.? - *Int. Rock Gard.* **95**: 18-28.

RUKŠĀNS, J. 2021: Three new *Crocus* (Iridaceae) species from Turkey and the East Aegean Islands. - *Int. Rock Gard.* **12**: 38 pp.

RUMMELSPACHER, J. 1965: Beitrag zur Kenntnis der *Pulsatilla montana* (Hoppe) Reichenbach. - *Feddes Repert.* **71(1-3)**: 1-49.

RUMSEY, F. 2010: 661. *Orobanche caryophyllacea*. - *Curtis's Bot. Mag.* **26(4)**: 369-378.

RUMSEY, F., TURLAND, N.J. & KAMARI, G. 2010: *Orobanche amethystea* Thuill. subsp. *amethystea*. In Greuter, W. & Raus, Th. (eds): Med-Checklist notulae, 29. - *Willdenowia* **40(2)**: 197-198.

RUMSEY, F.J. & JURY, S.L. 1991: An account of *Orobanche* L. in Britain and Ireland. - *Watsonia* **18(3)**: 257-295.

RUMSEY, F.J., VOGEL, J.C. & STAFFORD, P.J. 2001: The genus *Isoetes* (Isoetaceae: Pteridophyta) in Mallorca, Balearic Islands. - *Brit. Fern Gaz.* **16(5)**: 205-208.

RUNEMARK, H. 1961: Studies in the Aegean flora. III. *Cerastium coronense* sp. nov. - *Bot. Not.* **114(4)**: 453-456.

RUNEMARK, H. 1962a: A revision of *Parapholis* and *Monerma* in the Mediterranean. - *Bot. Not.* **115(1)**: 1-17.

RUNEMARK, H. 1962b: A revision of *Pteris dentata* and related species. - *Bot. Not.* **115(2)**: 177-195.

RUNEMARK, H. 1963: Studies in the Aegean flora. V. *Iberis arbuscula* sp. nov. - *Bot. Not.* **116(2)**: 323-325.

RUNEMARK, H. 1967a: Studies in the Aegean flora. X. Cytologic and morphologic notes on *Plantago*. - *Bot. Not.* **120(1)**: 9-16.

RUNEMARK, H. 1967b: Studies in the Aegean flora. XI. *Procopiana* (Boraginaceae) included into *Symphytum*. - *Bot. Not.* **120(1)**: 84-94.

RUNEMARK, H. 1967c: Studies in the Aegean flora. XII. Cytologic and morphologic investigations in *Centaurea*. - *Bot. Not.* **120(2)**: 161-176.

RUNEMARK, H. 1967d: Addendum to "Studies in the Aegean flora XII". - *Bot. Not.* **120(4)**: 486.

RUNEMARK, H. 1968: Studies in the Aegean flora. XIII. *Tordylium* L. (Umbelliferae). - *Bot. Not.* **121(2)**: 233-258.

RUNEMARK, H. 1969: Reproductive drift, a neglected principle in reproductive biology. - *Bot. Not.* **122(1)**: 90-129.

RUNEMARK, H. 1970a: The plant geography of the central Aegean. - *Feddes Repert.* **81(1-5)**: 229-231.

RUNEMARK, H. 1970b: The role of small populations for the differentiation in plants. - *Taxon* **19(2)**: 196-201.

RUNEMARK, H. 1971a: Distribution patterns in the Aegean. In Davis, P.H., Harper, P.C. & Hedge, I.C. (eds): Plant life of South-West Asia, pp. 3-12. - Edinburgh: Botanical Society.

RUNEMARK, H. 1971b: Investigations of the flora in the central Aegean. - *Boissiera* **19**: 169-179.

RUNEMARK, H. 1972: The phytogeography of the central Aegean. In Strid, A. (ed.): Evolution in the Aegean. - *Opera Bot.* **30**: 20-28.

RUNEMARK, H. 1974: Studies in the Aegean flora. XXIII. *Goniolimon* (Plumbaginaceae). - *Bot. Not.* **127(4)**: 540-545.

RUNEMARK, H. 1975: Karyotype differentiation in Mediterranean species complexes. In Guinochet, M. & al. (eds): La flore du Bassin Méditerranéen, pp. 159-163. - Paris: C.N.R.S.

RUNEMARK, H. 1980a: Studies in the Aegean flora. XXIII. The *Dianthus fruticosus* complex (Caryophyllaceae). - *Bot. Not.* **133(4)**: 475-490.

RUNEMARK, H. 1980b: Principles for a "Flora of the Central Aegean". [Unpublished manuscript]. - Photocopy: 3 pp.

RUNEMARK, H. 1980c: Sample manuscript for the Aegean flora [unpubl.]. - Photocopy: 41 pp.

RUNEMARK, H. 1984a: *Cerastium diffusum* Pers. In Greuter, W. & Raus, Th. (eds): Med-Checklist notulae, 9. - *Willdenowia* **14(1)**: 42.

RUNEMARK, H. 1984b: *Silene adelphiae* Runemark, sp. nova. In Greuter, W. & Raus, Th. (eds): Med-Checklist notulae, 9. - *Willdenowia* **14(1)**: 45.

RUNEMARK, H. 1984c: *Atriplex tornabeni* Tin. In Greuter, W. & Raus, Th. (eds): Med-Checklist notulae, 9. - *Willdenowia* **14(1)**: 50-51.

RUNEMARK, H. 1997: Reports (590-678). In Kamari, G., Felber, F. & Garbari, F. (eds): Mediterranean chromosome number reports - 6. - *Fl. Medit.* **6**: 223-243.

RUNEMARK, H. 2000: Reports (1110-1188) In Kamari, G., Felber, F. & Garbari, F. (eds): Mediterranean chromosome number reports - 10. - *Fl. Medit.* **10**: 386-402.

RUNEMARK, H. 2006: Reports (1473-1571). In Kamari, G., Blanché, C. & Garbari, F. (eds): Mediterranean chromosome number reports - 16. - *Fl. Medit.* **16**: 408-425.

RUNEMARK, H. & GREUTER, W. 1981: Notes on Cardaegean plants. 1. The *Sedum litoreum* group. - *Willdenowia* **11(1)**: 13-21. [Also distributed as *OPTIMA leaflet* no. 116].

RUNEMARK, H. & HENEEN, W.K. 1968: *Elymus* and *Agropyron*, a problem of generic delimitation. - *Bot. Not.* **121(1)**: 51-79.

RUNEMARK, H. & PHITOS, D. 1996: A contribution to the distribution of the genus *Campanula* (sectio *Quinqueloculares*) in Greece. - *Bot. Chron. (Patras)* **12**: 5-8.

RUNEMARK, H., SNOGERUP, S. & NORDENSTAM, B. 1960: Studies in the Aegean flora. I. Floristic notes. - *Bot. Not.* **113(4)**: 421-450.

RUŠČIĆ, M. & NIKOLIĆ, T. 2011: *Ammi visnaga* (L.) Lam. (Apiaceae), a new taxon in Croatian flora. - *Acta Bot. Croat.* **70(2)**: 301-306.

RUSH, R. 1984: Variation in *Blechnum spicant*. - *Pteridologist* **1(1)**: 42-43.

RUSHTON, B.S. 1977a: Cytological studies in *Plantago coronopus* L. - *Watsonia* **11(4)**: 387-388.

RUSHTON, B.S. 1977b: Morphological variation in some populations of *Plantago coronopus* L. - *Irish Naturalists' J.* **19(3)**: 61-66.

RUSS, B. & ROLFE, R. 2002: ×*Jancaemonda* 'Balkan Beauty', P.C. - *Alpine Gardener* **70(4)**: 471-472.

RUTISHAUSER, R. & AL. 1998: *Theligonum cynocrambe*: developmental morphology of a peculiar rubiaceous herb. - *Pl. Syst. Evol.* **210(1-2)**: 1-24.

RUŽIĆ, M. & KRIVOŠEJ, Z. 1989: *Alyssum cuneifolium* Ten. Nova vrsta u flori SR Srbije [In Serbo- Croatian]. - *Glasn. Prir. Mus. Beogradu, Ser. B, Biol. Nauke* **43-44**: 185-186.

RUŽIĆ, M., REXHEPI, F. & KRIVOŠEJ, Z. 1989: *Cerastium petricola* Panč. Nova vrsta u flori SR Srbije [In Serbo-Croatian]. - *Glasn. Prir. Mus. Beogradu, Ser. B, Biol. Nauke* **43-44**: 187-188.

RYBERG, M. 1950: Studies in the taxonomy and fertility of some Scandinavian *Corydalis* species of the sect. *Pes-gallinaceus* Irmisch. - *Acta Horti Berg.* **15(10)**: 207-283 + 1 plate.

RYBERG, M. 1954: *Corydalis solida* (L.) Sw. subsp. *laxa* Fr., and *Corydalis pumila* (Host) Rchb. Supplementary notes. - *Acta Horti Berg.* **16(7)**: 233-240.

RYBERG, M. 1955: A taxonomical survey of the genus *Corydalis* Ventenat with reference to cultivated species. - *Acta Horti Berg.* **17(5)**: 115-175 + 8 plates.

RYBERG, M. 1959: A morphological study of *Corydalis nobilis*, *C. cava*, *C. solida*, and some allied species, with special reference to their underground organs. - *Acta Horti Berg.* **19(3)**: 15-119 + 16 plates.

RYBERG, M. 1960: A morphological study of the Fumariaceae and the taxonomic significance of the characters examined. - *Acta Horti Berg.* **19(4)**: 121-248 + 12 plates.

RYCHLEWSKI, J. 1967: Karyological studies on *Nardus stricta* L. - *Acta Biol. Cracov., Ser. Bot.* **10(1)**: 55-72 + 2 plates.

RYCHLEWSKI, J. 1970: Karyology of three species of the genus *Bromus* L. - *Acta Biol. Cracov., Ser. Bot.* **13(1)**: 23-35 + 2 plates.

RYDER, M.L. 1996: Is Fuller's Teasel (*Dipsacus sativus*) really a distinct species? - *Linnean* **11(4)**: 21-27.

RYFF, L.E. & SVIRIN, S.A. 2020: *Aegilops peregrina*. In Raus, Th. & Raab-Straube, E.v. (eds): Euro+Med-Checklist notulae, 12. - *Willdenowia* **50(2)**: 314-315.

RYKA, C. 1954: Cytological studies in *Vicia cracca* L. and *Vicia tenuifolia* Roth [In Polish with English summary]. - *Acta Soc. Bot. Poloniae* **23(2)**: 321-333.

SA'AD, F. EL Z.M.A. 1967: The *Convolvulus* species of the Canary Isles, the Mediterranean region and the Near and Middle East. - *Meded. Bot. Mus. Herb. Rijks Univ. Utrecht* **281**: 1-288.

SAAED, M.W.B, EL-BARASI, Y.M & RAHIL, R.O. 2021: An updated checklist and quantitative analysis of the Marmarica Plateau flora, in the north-eastern part of Libya. - *Phytotaxa* **509(1)**: 1-55.

SABATO, S. 1990: Remarks on the publication dates of Tenore's Flora Napolitana. - *Taxon* **39(3)**: 409-416.

SADIKOĞLU, N. & ÖZHATAY, N. 2001: Evaluation of chromosome counts of flora of Turkey. In Özhatay, N. (ed.): Proceedings of the Second Balkan Botanical Congress, 1: 331-336. – Istanbul.

SÁENZ DE RIVAS, C. 1974: Datos sobre el género *Pseudorlaya* (Murb.) Murb. (Umbelliferae). - *Anales Inst. Bot. Cavanilles* **31(2)**: 191-204.

SÁENZ LAÍN, C. 1981: Research on *Daucus* L. (Umbelliferae). - *Anales Jard. Bot. Madrid* **37(2)**: 481-534.

SÁEZ, L., AEDO, C. & ROMERO-ZARCO, C. 2017: (2495) Proposal to conserve the name *Avena sterilis* (Poaceae) with a conserved type. - *Taxon* **66(1)**: 203-204. [approved, cf. Taxon 68(4): 850, 2019].

SÁEZ, LL. & ALDASORO, J.J. 2003: A taxonomic revision of *Campanula* L. subgenus *Sicyocodon* (Feer) Damboldt and subgenus *Megalocalyx* Damboldt (Campanulaceae). - *Bot. J. Linn. Soc.* **141(2)**: 215-241.

SAFFORD, W.E. 1921: Synopsis of the genus *Datura*. - *J. Washington Acad. Sci.* **11(8)**: 173-189.

SAFINA, L.K. & PIMENOV, M.G. 1990: Carpology of the species of type subgenus of the genus *Ferula* and some problems of their systematics. - *Feddes Repert.* **101(3-4)**: 135-151.

SAGAR, G.R. & HARPER, J.L. 1964: Biological Flora of the British Isles. *Plantago major* L., *P. media* L. and *P. lanceolata* L. - *J. Ecol.* **52(1)**: 189-221.

SAGORSKI, E. 1908a: Über den Formenkreis der *Anthyllis vulneraria* L. - *Allg. Bot. Z. Syst.* **14**: 40-43, 55-58, 89-93, 124-134, 154-157.

SAGORSKI, E. 1908b: Über den Formenkreis der *Anthyllis vulneraria* L. [cont.]. - *Allg. Bot. Z. Syst.* **14**: 172-175, 184-189, 204-205.

SAGORSKI, E. 1909: Über den Formenkreis der *Anthyllis vulneraria* L. [cont.]. - *Allg. Bot. Z. Syst.* **15(1-2)**: 7-11, 19-23.

ŞAHIN, A. & BABAÇ, M.T. 1990: Cytotaxonomic investigations on some *Vicia* L. species in East and Southeast Anatolia - I [In Turkish with English abstract]. - *Doğa Türk. Bot. Derg.* **14(2)**: 124-138.

ŞAHIN, A., GENÇ, H. & BAĞCI, E. 1998: Cytotaxonomic investigations on some *Lathyrus* species growing in the western Mediterranean and southern Aegean regions in Turkey. - *Acta Bot. Hung.* **41 (1-4)**: 229-241.

ŞAHIN, A., GENÇ, H. & BAĞCI, E. 2000: Cytotaxonomic investigations on some *Lathyrus* L. species growing in the eastern Mediterranean and southern Aegean regions - II. - *Acta Bot. Gallica* **147(3)**: 243-256.

SAIDABADI, H. & PANAHI, M.S.-C. 1975: Caryologie du genre *Ceratocephalus* Moench. - *Compt. Rend. Hebd. Séances Acad. Sci., Sér. D* **280(5)**: 849-852.

SAINT-YVES, A. 1921: Sur quelques *Festuca* de Grèce. - *Bull. Soc. Bot. France* **68**: 380-384.

SAINT-YVES, A. 1924: Festucarum varietates novae (Subg. *Eu-Festuca*). - *Bull. Soc. Bot. France* **71 (1)**: 28-43.

SAINT-YVES, A. 1928: Contribution à l'étude des *Festuca* (Subgen. *Eu-Festuca*) de l'Orient. Asie et région méditerranéenne voisine. - *Candollea* **3**: 321-466.

SAINT-YVES, A. 1931: Contribution à l'étude des *Avena* sect. *Avenastrum* (Eurasie et région méditerranéenne). - *Candollea* **4**: 353-504 + 1 plate.

SAINT-YVES, A. 1932: Festucae novae et loci novi Festucarum jam cognitarum (subgen. *Eu-Festuca*). - *Candollea* **5**: 101-141.

SAINT-YVES, A. 1934: Contribution à l'étude des *Brachypodium* (Europe et région méditerranéenne). - *Candollea* **5**: 427-493 + 5 plates.

SAKAMOTO, S. & KOBAYASHI, H. 1982: Variation and distribution of cultivated plants, their wild relatives and weeds native to Turkey, Greece and Romania. In Tani, Y. (ed.): Agrico-pastoral peoples in southwestern Eurasia, II, pp. 41-104. - Kyoto.

SAKELLARAKIS, F.N. & AL. 2019: Flora of Prespa area database, NW Greece. In Daskalakou, E.N. & al. (eds): 16th Conference of the Hellenic Botanical Society – Athens 10-13 October 2019. Program and book of abstracts, p. 102.

SALAS PASCUAL, M. & VERLOOVE, F. 2019: *Opuntia*. In Raab-Straube, E. v. & Raus, Th. (eds): Euro+Med-Checklist notulae, 11. - *Willdenowia* **49(3)**: 424-426.

SALES, F. 1992: (1053) Proposal to conserve *Bromus sterilis* L. (Poaceae) with a conserved type . . - *Taxon* **41(3)**: 584-585. [approved, cf. Taxon 43(2): 275, 1994].

SALES, F. 1993a: Taxonomy and nomenclature of *Bromus* sect. *Genea*. - *Edinburgh J. Bot.* **50(1)**: 1-31.

SALES, F. 1993b: Notes on some annual brome-grasses in Turkey. - *Karaca Arbor. Mag.* **2(1)**: 21-28.

SALES, F. 1994a: Evolutionary tendencies in some annual species of *Bromus (Bromus* L. sect. *Genea* Dum. (Poaceae)). - *Bot. J. Linn. Soc.* **115(3)**: 197-210.

SALES, F. 1994b: A reassessment of the *Bromus madritensis* complex (Poaceae): a multivariate approach. - *Israel J. Pl. Sci.* **42(3)**: 245-255.

SALES, F. & HEDGE, I.C. 1993: *Melilotus* Miller (Leguminosae): Typification and nomenclature. - *Anales Jard. Bot. Madrid* **51(1)**: 171-175.

SALES, F. & HEDGE, I.C. 1994: *Medicago* L. (Leguminosae): Typification and nomenclature. - *Anales Jard. Bot. Madrid* **51(2)**: 321-323.

SALES, F. & HEDGE, I.C. 1995: (1171) Proposal to conserve *Tuberaria* (Dunal) Spach against *Xolantha* Raf. (Cistaceae). - *Taxon* **44(3)**: 437-438. [approved, cf. Taxon 47(2): 443, 1998].

SALES, F. & HEDGE, I.C. 1996: Biogeographical aspects of selected SW Asiatic woody taxa. - *Ann. Naturhist. Mus. Wien, B* **98B Suppl.**: 149-161.

SALES, F. & HEDGE, I.C. 2001: Nomenclature and typification of Western European *Jasione* (Campanulaceae). - *Anales Jard. Bot. Madrid* **59(1)**: 163-172.

SALES, F. & SMITH, P.M. 1990: A new species in the genus *Bromus*. - *Edinburgh J. Bot.* **47(3)**: 361-366.

SALIARIS, A., SALIARIS, P. & ALIBERTIS, A. 2010: *Ophrys fusca* subsp. *sancti-isidorii*, a new subspecies from Chios (Greece). - *J. Eur. Orch.* **42(2)**: 327-332.

SALIARIS, P., SALIARIS, A. & ALIBERTIS, A. 2011: *Ophrys tenthredinifera* subsp. *sanctae-marcellae*, a new subspecies from Chios (Greece). - *J. Eur. Orch.* **43(3)**: 603-608.

SALIARIS, P.A. 1995: Ta agrioloulouda tis Chiou, tomos A' [In Greek with English summary]. - Chios: Ekdosi Dimou Omiroupolis, 127 pp.

SALIARIS, P.A. 2001: The orchids of Chios, Greece: status and conservation. - *J. Eur. Orch.* **33(1)**: 443-446.

SALIARIS, P.A. 2002: Orchidees tis Chiou [In Greek]. - Ekdosi Dimou Kardamilon: 212 pp.

SALIARIS, P.A. 2006: Nea tis ellinikis chloridas [In Greek]. - *Fisis [Journal of the Hellenic Society for the Protection of Nature]* **115**: 21.

SALIARIS, P.A. 2007: Natura Chia. Odiporiko sti chlorida ke panida tis Chiou [In Greek]. - Chios: 308 pp. incl. many colour photos.

SALIARIS, P.A. 2009: *Himantoglossum comperianum* (Steven) P. Delforge on Chios (Eastern Aegean Islands, Greece). - *Naturalistes Belges* **90**: 46-48.

SALISBURY, E.J. 1974: The reproduction of *Juncus tenuis* (*Juncus macer*) and its dispersal. - *Trans. Bot. Soc. Edinburgh* **42(2)**: 187-190.

SALMAKI, Y. & AL. 2016: Phylogeny of non-monophyletic *Teucrium* (Lamiaceae: Ajugoideae): Implications for character evolution and taxonomy. - *Taxon* **65(4)**: 805-822.

SALMERI, C. 1998: *Allium brulloi* (Alliaceae), a new species from Astypalea (Aegean Islands, Greece). - *Willdenowia* **28(1-2)**: 69-75.

SALMERI, C. & AL. 2018: Taxonomic and phylogenetic relations of *Allium chamaespathum* Boiss., a remarkable species of Balkan flora. - *Bot. Serbica* **42 (suppl. 1)**: 43.

SALOMON, B. & LU, B.-R. 1992: Genomic groups, morphology and sectional delimitation in Eurasian *Elymus* (Poaceae, Triticeae). - *Pl. Syst. Evol.* **180(1-2)**: 1-13.

SALTER, T.M. 1939: Some notes on the correct identity of *Oxalis pes-caprae*, Linn. and *Oxalis purpurea*, Linn. with some remarks on the limitations of a species in the genus *Oxalis*. - *J. S. African Bot.* **1939**: 47-52.

SALVO, A.E. & AL. 1986: Estudio biosistemático del género *Polystichum* Roth (Aspidiaceae, Pteridophyta) en la Península Ibérica. - *Bol. Soc. Brot., Ser. 2* **59**: 113-165.

SAMARAS, D.A. & AL. 2017: Contribution to the flora of Mt Goulinas (Sterea Ellas, Greece). - *Fl. Medit.* **27**: 241-266.

SAMARAS, D.A. & AL. 2017: The woody flora of Mt Goulinas. In Dousis, M.A. & Thanos, K.A.: 15th conference of the Hellenic Botanical Society, book of abstracts: p. 120.

SAMARAS, T., HATZAKOS, G. & TAN, KIT 2020: Reports 103-105. In Vladimirov, V., Aybeke, M. & Tan, Kit (eds): New floristic records in the Balkans: 42. - *Phytol. Balcan.* **26(2)**: 202-204.

SAMARAS, TH. & AL. 2019: Report 96. In Vladimirov, V., Aybeke, M. & Tan, Kit (eds): New floristic records in the Balkans: 39. - *Phytol. Balcan.* **25(2)**: 219-220.

SAMAROPOULOS, S., BALIOUSIS, E. & BAREKA, P. 2019: Karyological study of eight *Trifolium* L. taxa from Greece. - *Bot. Chron. (Patras)* **22**: 159-170.

SAMAROPOULOU, S. & AL. 2013: Karyological studies on some endemic and rare species of Kephalonia, Ionian Islands, Greece. - *Fl. Medit.* **23**: 215-221.

SAMAROPOULOU, S. & AL. 2019: Seed morphology in the genus *Fritillaria* (Liliaceae) from Greece and its taxonomic significance. - *Phytotaxa* **416(4)**: 223-237.

SAMAROPOULOU, S. & AL. 2021: Studies on the pollen morphology of *Fritillaria* (Liliaceae) taxa from Greece. - *Pl. Biosyst.* **155**: 12 pp.

SAMAROPOULOU, S., BAREKA, E.-P. & KAMARI, G. 2013: Reports (1817-1823). In Kamari, G, Blanché, C. & Siljak-Yakovlev, S. (eds): Mediterranean chromosome number reports – 23. - *Fl. Medit.* **23**: 285-291.

SAMAROPOULOU, S., BAREKA, P. & KAMARI, G. 2010: Reports (1728-1735). In Kamari, G., Blanché, C. & Siljak-Yakovlev, S. (eds): Mediterranean chromosome number reports - 20. - *Fl. Medit.* **20**: 278-286.

SAMAROPOULOU, S., BAREKA, P. & KAMARI, G. 2015: Reports (1864-1867). In Kamari, G., Blanché, C. & Siljak-Yakovlev, S. (eds): Mediterranean chromosome number reports - 25. - *Fl. Medit.* **25**: 164-167.

SAMAROPOULOU, S., BAREKA, P. & KAMARI, G. 2016: Karyomorphometric analysis of *Fritillaria montana* group in Greece. - *Comp. Cytogen.* **10(4)**: 679-695.

SAMAROPOULOU, S., BAREKA, P. & KAMARI, G. 2019: Hybridization and karyotype variability of three endemic *Fritillaria* L. (Liliaceae) in Argolis Peninsula (Greece). - *Pl. Biosyst.* **154(3)**: 348-360.

SAMIOS, K.M. 1908: Ta dasi tis Kefallinias [In Greek]. - Athine: Ethnikou Tipografiou, 316 pp. + 1 fold- out map.

SAMUELSSON, G. 1923: Eine Exkursion nach Delphi und auf den Parnassos. - *Vierteljahrsschr. Naturf. Ges. Zürich* **68**: 176-179.

SAMUELSSON, G. 1925: Die *Callitriche*-Arten der Schweiz. - *Veröff. Geobot. Inst. Rübel Zürich* **3**: 603-628.

SAMUELSSON, G. 1932: Die Arten der Gattung *Alisma*. - *Ark. Bot.* **24A(7)**: 1-46 + 6 plates.

SAMUELSSON, G. 1933: *Alisma* L. - *Pflanzenareale* **3(8)**: 88-90, Karte 73-75.

SAMUELSSON, G. 1934: Symbolae ad floram graecam. - *Ark. Bot.* **26A(5)**: 1-29.

SAMUELSSON, G. 1939: Bemerkungen über einige *Rumex*-Sippen aus der *Vesicarius*-Gruppe. - *Bot. Not.* **1939(4)**: 505-527.

SÁNCHEZ AGUDO, J.A. & AL. 2012: A contribution towards clarifying the nomenclature of *Veronica* L. (Plantaginaceae). - *Taxon* **61(4)**: 867-870.

SÁNCHEZ AGUDO, J.A., SÁNCHEZ SÁNCHEZ, J. & HERNÁNDEZ, E.R. 1999: Estudio comparativo de *Veronica chamaepithyoides* Lam. y *V. grisebachii* Walters. In Correia, A.I.D. & al. (eds): Livro de resumos. Guia de excursões. V jornadas de taxonomia botânica, 16-19/IX 1999, p. 60.

SÁNCHEZ ANTA, M.A. & GALLEGO MARTÍN, F. 1985: Contribución al estudio cariológico del género *Cistus* L. (1). - *Stud. Bot. Univ. Salamanca* **4**: 103-107.

SÁNCHEZ PEDRAJA, O. & AL. 2020: Index of Orobanchaceae. - http://www.farmalierganes.com/Otrospdf/publica/Orobanchaceae%20Index.htm

SÁNCHEZ-CUXART, A. & BERNAL, M. 1999: El género *Ferula* (Umbelliferae) en la Península Ibérica y las Islas Baleares. In Correia, A.I.D. & al. (eds): Livro de resumos. Guia de excursões. V jornadas de taxonomia botânica, 16-19/IX 1999, p. 56.

SÁNCHEZ-GÓMEZ, P. & AL. 2008: *Anchusa aegyptiaca* (Boraginaceae), a new species for the Iberian flora. - *Fl. Medit.* **18**: 373-378.

SÁNCHEZ-JIMÉNEZ, I. & AL. 2012: Genome size and chromosome number in *Echinops* (Asteraceae, Cardueae) in the Aegean and Balkan regions: technical aspects of nuclear DNA amount assessment and genome evolution in a phylogenetic frame. - *Pl. Syst. Evol.* **298(6)**: 1085-1099.

SÁNCHEZ-JIMÉNEZ, I., HIDALGO, O. & GARNATJE, T. 2012: *Echinops spinosissimus* subsp. *neumayeri* (Vis.) Kožuharov (Asteraceae, Cardueae): a new record for the flora of Greece. - *Adansonia, sér. 3* **34(1)**: 129-132.

SÁNCHEZ-YÉLAMO, M.D. 1994: A chemosystematic survey of flavonoids in the Brassicinae: *Diplotaxis*. - *Bot. J. Linn. Soc.* **115(1)**: 9-18.

SANDA, V. & FIŞTEAG, G. 1992: Arealul unor specii ale familiei Caryophyllaceae în Carpaţii României [In Romanian]. - *Stud. Cercet. Biol. (Bucharest), Ser. Biol. Veg.* **44(2)**: 141-152.

SANDA, V., NEGREAN, G. & ŞTEFAN, N. 1992: Catatera specilior genului *Valerianella* Miller din Flora României [In Romanian]. - *Stud. Cercet. Biol. (Bucharest), Ser. Biol. Veg.* **44(2)**: 115-139.

SANDBRINK, J.M., BREDERODE, J. VAN & GADELLA, TH.W.J. 1990: Phylogenetic relationships in the genus *Symphytum* L. (Boraginaceae). - *Proc. Kon. Ned. Akad. Wetensch. C* **93(3)**: 295-334.

SANDERS, R.W. & CANTINO, PH.D. 1984: Nomenclature of the subdivisions of the Lamiaceae. - *Taxon* **33(1)**: 64-72.

SANDS, M.J.S. 1980: Paxos. Walking and wild flowers. - Walton-on-Thames: The Greek Islands Club, 33 pp.

SANDS, M.J.S. 1991: Paxos. Walking and wild flowers, 2nd ed. - Walton-on-Thames: The Greek Islands Club, 41 pp. + 2 maps.

SANDWITH, N.Y. 1942: *Prangos ferulacea*. - *Bot. Mag.* **163(4)**: tab. 9643 + 4 pp. text.

SANDWITH, N.Y. & SIMPSON, N.D. 1941: Additions to the flora of Cyrenaica. - *J. Bot. (London)* **79**: 33-44.

SANTA BÁRBARA, C. 1994: Números cromosómicos para la flora española. 720-747. - *Lagascalia* **17 (2)**: 367-379.

SANTORINEOS, I. 1935: To ethniko parko ston Olimbo [In Greek]. - *Vouno [Journal of the Hellenic Alpine Club]* **1935**: 193-200.

SANTORINEOS, I. 1938: I idrisi tou ethnikou drimou tou Olimbou [In Greek]. - *Vouno [Journal of the Hellenic Alpine Club]* **1938**: 189-191.

SANTOS GUERRA, A. 1992: Acerca de la identidad de *Frankenia hirsuta* L., Sp. Pl.: 331 (1753) (Frankeniaceae). - *Anales Jard. Bot. Madrid* **50(1)**: 136.

SAÑUDO PALAZUELES, A. 1973: Variabilidad cromosómica de las genisteas de la flora española en relación con su ecología. 1. Número y comportamiento de los cromosomas durante la meiosis. E. Género *Cytisus* L. - *Bol. Real Soc. Esp. Hist. Nat., Secc. Biol.* **71(3-4)**: 341-355.

SAÑUDO PALAZUELES, A. 1974: Variabilidad cromosómica de las genisteas de la flora española en relación con su ecología ... F. Géneros: *Chamaespartium* Adanson y *Echinospartium* (Spach) Rothm. - *Anales Inst. Bot. Cavanilles* **31(1)**: 165-174.

SAÑUDO PALAZUELES, A. 1979: Chromosome variability in the Genisteae (Adams.) Benth. (Leguminosae). - *Webbia* **34(1)**: 363-408.

SAÑUDO PALAZUELES, A., PRETEL MARTÍNEZ, A. & RUIZ REJÓN, M. 1976: [Chromosome numbers in *Ononis* from Spain]. In Löve, Á. (ed.): IOPB chromosome number reports LI. - *Taxon* **25(1)**: 163-164.

SAÑUDO PALAZUELES, A., RUIZ REJÓN, M. & PRETEL MARTÍNEZ, A. 1979: Variabilité chromosomique chez les espèces d'*Ononis* de la flore espagnole. Note préliminaire. - *Webbia* **34(1)**: 535-542.

SAOUD, N.S., KNIO, K. & JURY, S.L. 2007: Phenetic analysis of *Anemone coronaria* (Ranunculaceae) and related species. - *Bot. J. Linn. Soc.* **153(4)**: 417-438.

SARDARA, M. 1984: Ricerche su *Trifolium subterraneum* L., *T. brachycalycinum* Katzn. & Morley e *T. yanninicum* Katz. & Morley in Sardegna. - *Inform. Bot. Ital.* **14(2-3)**: 225-228.

SARFATTI, G. 1959: Prodromo della flora della Sila (Calabria). - *Webbia* **15(1)**: 169-248.

SARFATTI, G. 1965: Prodromo della flora della Sila (Calabria) (continuazione e fine). - *Webbia* **20(2)**: 355-425.

SARIĆ, M.R. & DIKLIĆ, N. 1986: Flore SR Srbije. Flore de la Republique Socialiste de Serbie. Vol. 10. Dodatak [supplement] 2. - Beograd: Srbska Akad. Nauka Umetn., xxxii + 400 pp.

SARIĆ, M.R. (ED.) 1992: The Flora of Serbia, vol. 1 [Vascular cryptogams to Ranunculaceae; in Serbo-Croatian]. - Beograd: Serbian Academy of the Sciences and the Arts, xv + 429 pp.

SARIKA-HATZINIKOLAOU, M. 1999: Floristic and phytosociological study of aquatic ecosystems of Epirus (NW Greece) [In Greek with English summary]. - Athens: National and Capodistrian Univ. [Ph.D. thesis], 495 pp. + 1 folded chart.

SARIKA-HATZINIKOLAOU, M., BABALONAS, D. & YANNITSAROS, A. 1999: Phytosociological study of the helophytic vegetation in aquatic ecosystems of Ipeiros (NW Greece). In Elefheriou, E.P. (ed.): Proceedings, 7th Panhellenic scientific conference, Alexandroupolis, pp. 134-141 [In Greek with English abstract]. - Thessaloniki: Univ. Studio Press.

SARIKA-HATZINIKOLAOU, M., KOUMPLI-SOVANTZI, L. & YANNITSAROS, A. 1992: A preliminary study on the aquatic and hydrophilous flora of Epirus [In Greek with English summary]. In: Proc. 12th Annual Meeting Hellen. Soc. Biol. Sci., Mitilini, 27-29.4. 1990, pp. 196-198. - Athine.

SARIKA-HATZINIKOLAOU, M., KOUMPLI-SOVANTZI, L. & YANNITSAROS, A. 1993: New information on the distribution of several rare hydrophytes in Greece [In Greek with English summary]. In: Proc. 15th Conference Hellen. Soc. Biol. Sci., Florina-Kastoria, 21-24.4. 1993, pp. 135-137. - Florina-Kastoria.

SARIKA-HATZINIKOLAOU, M., KOUMPLI-SOVANTZI, L. & YANNITSAROS, A. 1994: *Myriophyllum alterniflorum* DC. (Haloragaceae), a new record for the Greek flora. - *Phyton (Horn)* **34(2)**: 243-246.

SARIKA-HATZINIKOLAOU, M., KOUMPLI-SOVANTZI, L. & YANNITSAROS, A. 1996: The vascular flora of Lake Kalodhiki (Ipiros, NW Greece). - *Webbia* **50(2)**: 223-236.

SARIKA-HATZINIKOLAOU, M., KOUMPLI-SOVANTZI, L. & YANNITSAROS, A. 1997: Macrophytes in four alpine aquatic ecosystems of N. Pindos (Ipiros, Greece). - *Phyton (Horn)* **37(1)**: 19-30.

SARIKA-HATZINIKOLAOU, M., YANNITSAROS, A. & BABALONAS, D. 2003: The macrophytic vegetation of seven aquatic ecosystems of Epirus (NW Greece). - *Phytocoenologia* **33(1)**: 93-151.

SARIKA-HATZINIKOLAOU, M., YANNITSAROS, A. & KOUMPLI-SOVANTZI, L. 1995: Floristic investigation of the lake Pamvotis (Ipiros). In: Praktika 16ou Sinedriou E.E.V.E., Volos, 5.-7.5. 1994, 2, pp. 10-14 [In Greek with English summary].

SARIKA, M. 2008: Aquatic and hygrophilous flora of Greece. Bibliographical review and addition of new records. In: Praktika 10ou Sinedriou Ellinikis Votanikis Eterias, 5-8 maiou 2005, Ioannina: pp. 549-559 [In Greek with English abstract].

SARIKA, M. 2012: Flora and vegetation of some coastal ecosystems of Sterea Ellas and eastern continental Greece. - *Lazaroa* **33**: 65-69.

SARIKA, M. & AL. 2015: Flora and vegetation of the European-network "Natura 2000" habitats of Naxos island (GR4220014) and of nearby islets Mikres Kyklades (GR4220013), Central Aegean (Greece). - *Plant Sociology* **52(2)**: 3-56.

SARIKA, M. & AL. 2019: Temporal turnover of the flora of lake islands: The island of Lake Pamvotis (Epirus, Greece). - *Biodiv. Data J.* **7**: e37023, 24 pp.

SARIKA, M. & AL. 2020: Plant communities and habitat types in the protected area of Lake Pamvotis (Epirus, Northwestern Greece). - *Medit. Bot.* **41(2)**: 129-148 + suppl..

SARIKA, M., DIMOPOULOS, P. & YANNITSAROS, A. 2005: Contribution to the knowledge of the wetland flora and vegetation of Amvrakikos Gulf, W Greece. - *Willdenowia* **35(1)**: 69-85.

SARIKA, M.A. & AL. 2018: Vegetation units of wetland and terrestrial habitats: the case study of Spercheios River and Maliakos Gulf (Sterea Ellas, Greece), a Natura 2000 Site. - *Hacquetia* **17(2)**: 189-220.

SÄRKINEN, T. & AL. 2018: A revision of the Old World Black Nightshades (Morelloid clade of *Solanum* L., Solanaceae). - *PhytoKeys* **106**: 1-223.

SARLIS, G.P. 1977: Succession of vegetation after forest fires in Attica (Greece). - *Sci. Works Agric. College Athens* **1977**: 98-107.

SARLIS, G.P. 1980: The flora of Mount Egaleo (Attica, Greece). - *Phyton (Horn)* **20(3-4)**: 261-278.

SARLIS, G.P. 1981: The flora and vegetation of the islands and islets of the southern Euboean gulf (Greece) [In Greek, English abstract separate]. - Athine: 208 pp. + 3 maps.

SARLIS, G.P. 1984: The flora and vegetation of the islands and islets of the southern Euboean gulf. - *Rad. Odjel. Prirod. Mat. Nauka (Sarajevo)* **23**: 175-183 + 4 maps.

SARLIS, G.P. 1987: *Zelkova abelicea* (Lam.) Boiss., an endemic species of Crete (Greece). – *Webbia* **41(2)**: 247-255.

SARLIS, G.P. 1988a: Simvoli sti gnosi ton fitokinonion tou Ethnikou Drimou Prespon [In Greek]. - *Geôtechnica* **1**: 39-45.

SARLIS, G.P. 1988b: Meleti ton vlastiseos ton Akarnanikon oreon (Pergandi, Boumistos) [In Greek]. - *Geôtechnica* **2**: 7-22.

SARLIS, G.P. 1990: Parochthia (livadiki) elovia ke idrovia chlorida ke vlastisi tis mikrolimnis Avilarias Etoloakarnanias [In Greek]. - *Geôrgiki Erevna* **14**: 93-111.

SARLIS, G.P. 1993: The flora and vegetation of the islands and islets of the southern Euboean gulf (Greece). In: Demiriz, H. & Özhatay, N. (eds): Proceedings of the V OPTIMA meeting, Istanbul, 8-15.9. 1986, pp. 765-777. - Istanbul Üniversitesi Fen Fakültesi.

SARLIS, G.P. 1994a: Contribution to the study of the flora of Attica (Greece). - *Lagascalia* **17(2)**: 229-256.

SARLIS, G.P. 1994b: Ta fita tis Siras. - *Siriana Grammata* **26-27**: 98-212.

SARVELA, J. 1978: A synopsis of the fern genus *Gymnocarpium*. - *Ann. Bot. Fenn.* **15(2)**: 101-106.

SARVELA, J., BRITTON, D.M. & PRYER, K. 1981: Studies on the *Gymnocarpium robertianum* complex in North America. - *Rhodora* **83(835)**: 421-431.

SATIL, F., DIRMENCI, T. & TÜMEN, G. 2003: The rediscovery of an extinct species: *Satureja coerulea* Janka (Lamiaceae/Labiatae). - *Karaca Arbor. Mag.* **7(1)**: 19-22.

SATTA, V. 2001: Distribution of *Bellevalia romana* (L.) Rchb. (Liliaceae) in Sardinia. In Raimondo, F. M. (ed.): Abstracts from the X OPTIMA meeting, Palermo, 13-19 September 2001, p. 151.

SAUER, E. 1969: Exkursion des Botanischen Institutes der Universität des Saarlandes zum Thessalischen Olymp, Griechenland. [Unpublished flora list]. - Photocopy: 9 pp.

SAUER, E. 1970: Feldnotitzen: Olympexkursion Pfingsten 1970. [Unpublished flora list]. - Photocopy: 2 pp.

SAUER, E. 1975: Fundortnummern der Exkursion zum Thessalischen Olymp 1969. [Unpublished flora list]. - Photocopy: 26 pp.

SAUER, W. 1965: Die *Moehringia bavarica*-Gruppe. - *Bot. Jahrb. Syst.* **84(3)**: 254-301 + 2 plates and 1 map.

SAUER, W. 1975: Karyo-systematische Untersuchungen an der Gattung *Pulmonaria* (Boraginaceae): Chromosomen-Zahlen, Karyotyp-Analysen und allgemeine Hinweise auf die Entwicklungsgeschichte. - *Biblioth. Bot.* **131**: 85 pp.

SAUER, W. 1984: Die Gattung *Avenula* (Dumort.) Dumort., Poaceae, auf der Balkanhalbinsel. - *Acta Bot. Croat.* **43**: 315-328.

SAUER, W. 1987: The *Pulmonaria dacica* group: its affinities with central and south-east European allies and with the genus *Paraskevia* (Boraginaceae). - *Pl. Syst. Evol.* **155(1-4)**: 257-276.

SAUER, W. & CHMELITSCHEK, H. 1976: Beiträge zur Kenntnis ausdauernder Wildhafer: Die Gattung *Avenula* (Dumort.) Dumort. in den Ostalpen. - *Mitt. Bot. Staatssamml. München* **12(2)**: 513-608.

SAUER, W. & HEUBL, G.R. 1984: Beiträge zur Kenntnis ausdauernder Wildhafer: 2. Karyotypanalysen an west- und osteuropäischen sowie an alpinen Wildhaferarten der Gattung *Avenula* (Dumort.) Dumort. (Poaceae). - *Phyton (Horn)* **24(2)**: 193-223.

SAUER, W. & LEEP, H.J. 1979: Karyologische Untersuchungen an anatolischen und südost-europäischen Zwergiris-Sippen: *Iris attica*, *I. mellita* und *I. reichenbachii* (Iridaceae). - *Pl. Syst. Evol.* **131(1-2)**: 81-106.

SAUER, W. & SAUER, G. 1980: *Paraskevia* gen. nov. mit *P. cesatiana* comb. nov. (Boraginaceae), eine endemische Gattung Griechenlands. - *Phyton (Horn)* **20(3-4)**: 285-306 + 4 plates.

SAUER, W. & STEGMEIER, R. 1979: Beiträge zur Kenntnis südost-europäischer und anatolischer Bart- Iris-Arten. Karyotypanalysen von *I. attica*, *I. mellita*, *I. glockiana*, *I. reichenbachii*, *I. illyrica* ... - *Ber. Deutsch. Bot. Ges.* **92(2-3)**: 663-687.

SAUKEL, J. & AL. 2004: Comments on the biosystematics of *Achillea* (Asteraceae-Anthemideae) in Bulgaria. - *Phytol. Balcan.* **9(3)**: 361-400.

SAUNDERS, D.E. 1959: *Cyclamen*. - *Quart. Bull. Alpine Gard. Soc. Gr. Brit.* **27(1)**: 18-76.

SAUNDERS, D.E., MEIKLE, R.D. & GREY-WILSON, CH. 1975: *Cyclamen*. The genus in the wild & in cultivation. A gardener's guide to the genus. 2nd revised ed. - St. John's Woking, Surrey: Alpine Garden Society, 49 pp.

SAUVAGEAU, C. 1908: Bory de Saint-Vincent d'après sa correspondance publiée par M. Lauzun. - Extrait du Journal de Botanique, 2. série, tome I: 24 pp.

SAVELLI, P.R., ALESSANDRINI, A. & LIVERANI, P. 1989: *Epipactis purpurata* Smith ed *E. greuteri* H. Baumann & Kuenkele in Italia, con nuove località di *E. persica* (Soó) Nannfeldt (Orchidaceae). - *Arch. Bot. Biogeogr. Ital.* **64(3-4)**: 192-198.

SAVVIDES, L. 2000: Edible wild plants of the Cyprus flora. - Nicosia: Private publ., 116 pp.

SAXENA, R.B. 2010: Botany, taxonomy and cytology of *Crocus sativus* series. - *Ayu* **31(3)**: 371-381.

SAYA, I. 1971: Numeri cromosomici per la Flora Italiana: 32. - *Inform. Bot. Ital.* **3(1)**: 66-67.

SCALTSOYIANNES, A., TSAKTSIRA, M. & DROUZAS, A.D. 1999: Allozyme differentiation in the Mediterranean firs (*Abies*, Pinaceae). A first comparative study with phylogenetic implications. - *Pl. Syst. Evol.* **216(3-4)**: 289-307.

SCANNERINI, S. 1971: Numeri cromosomici per la Flora Italiana: 33-34. - *Inform. Bot. Ital.* **3(1)**: 67-71.

SCHACHT, W. 1933: An alpine plant-paradise in Macedonia. - *New Fl. & Silva* **5(3)**: 164-171 + 6 plates.

SCHACHT, W. 1937: A botanical excursion to the Athos peninsula. - *New Fl. & Silva* **9(3)**: 149-155 + 5 plates.

SCHACHT, W. 1938a: Mount Olympus. - *Fl. & Sylva* **10**: 220-231 + 4 plates.

SCHACHT, W. 1938b: Die europäischen Gesneraceen. - *Blätt. Staudenk.* **1**: 8 pp.

SCHACHT, W. 1948: Balkan memories. - *Quart. Bull. Alpine Garden Soc. Gr. Brit.* **16(3)**: 178-189.

SCHACHT, W. 1951: European Gesneraceae. - *Quart. Bull. Alpine Garden Soc. Gr. Brit.* **19(2)**: 159-166.

SCHACHT, W. 1969: *Lilium rhodopeum*. - *Lily Year-Book* **32**: 125-126 + 2 plates.

SCHÄFER, H.I. 1973: Zur Taxonomie der *Vicia narbonensis*-Gruppe. - *Kulturpflanze* **21**: 212-273.

SCHÄFER, P.A. 1971: Vorläufiger Beitrag zur *Orchis laxiflora*-Gruppe. - *Orchidee (Hamburg)* **22(5)**: 200-205.

SCHAFFNER, W. 1968: Untersuchungen zur Wuchsform und Sprossgestalt des Flügelginsters, *Cytisus sagittalis* (L.) Koch. - *Bot. Jahrb. Syst.* **88(4)**: 469-514.

SCHAICH, H. & AL. 2015: Land change in Eastern Mediterranean wood-pasture landscapes: The case of deciduous oak woodlands in Lesvos (Greece). - *Environ. Manegem.* **56(1)**: 110-126.

SCHANZER, I.A. 1989: On the geographical variability and evolution of *Filipendula ulmaria* (L.) Maxim. and allied species [In Russian with English summary]. - *Byull. Moskovsk. Obshch. Isp. Prir., Otd. Biol.* **94(6)**: 59-69.

SCHANZER, I.A. 1994: Taxonomic revision of the genus *Filipendula* Mill. (Rosaceae). - *J. Jap. Bot.* **69 (5)**: 290-319.

SCHARFETTER, R. 1912: Die Gattung *Saponaria* Subgenus *Saponariella* Simmler. Eine pflanzengeographisch-genetische Untersuchung. - *Oesterr. Bot. Z.* **62**: 1-8, 74-88, 109-114.

SCHATTANEK, A. 1982: Further observations of *Cyclamen* on Rhodes. - *Cyclamen Soc. J.* **6(1)**: 11.

SCHENK, J.A. 1841: Genera et species cyperacearum, quae in regno Graeco, archipelago et in insulis ioniis, nec non insula Creta crescentes ex operibus Sibthorpii, d'Urvillei ... - München: Typis Doctoris Caroli Wolf, v + 24 pp. [cf. TL-2 5: 135, 1985].

SCHENKLER, G. 1936: Systematische Untersuchungen über die Sektion *Beccabunga* der Gattung *Veronica*. - *Repert. Spec. Nov. Regni Veg. Beih.* **90**: i-v, 1-40 + 1 plate.

SCHIEMANN, E. 1928a: Zytologische und pflanzengeographische Beiträge zur Gattung *Aegilops* (II. Mitteilung). - *Ber. Deutsch. Bot. Ges.* **46, Generalvers.-Heft**: 107-128 + 6 plates.

SCHIEMANN, E. 1928b: Chromosomenzahlen in der Gattung *Aegilops*. - *Ber. Deutsch. Bot. Ges.* **46(5)**: 324-328 + 2 plates.

SCHIEMANN, E. 1929: Zytologische Beiträge zur Gattung *Aegilops*. Chromosomenzahlen und Morphologie. - *Ber. Deutsch. Bot. Ges.* **47(3)**: 164-181 + 1 plate.

SCHIFFNER, V. 1889: Die Gattung *Helleborus*. Eine monographische Skizze. - *Bot. Jahrb. Syst.* **11**: 97-122.

SCHIFFNER, V. 1890: Monographia Hellebororum. Kritische Beschreibung aller bisher bekannt gewordenen Formen der Gattung *Helleborus*. - *Nova Acta Acad. Caes. Leop.-Carol. German. Nat. Cur.* **56(1)**: 1-198 + 8 plates.

SCHIFFNER, V. 1904: Josef Freyn [obituary]. - *Ber. Deutsch. Bot. Ges.* **22**: 15-21.

SCHINDLER, J. 1907: Studien über einige mittel- und südeuropäische Arten der Gattung *Pinguicula*. - *Oesterr. Bot. Z.* **57(11-12)**: 409-421, 458-469 + 4 plates.

SCHINDLER, J. 1908: Studien über einige mittel- und südeuropäische Arten der Gattung *Pinguicula* (Fortsetzung). - *Oesterr. Bot. Z.* **58(1-2)**: 13-18, 61-69 + 4 plates.

SCHINZ, H. & KELLER, R. 1899-1900: Flora der Schweiz. Zum Gebrauch auf Exkursionen, in Schulen und beim Selbstunterricht ... Ed. 1 [several later editions]. - Zürich: Verlag von Albert Raustein, vi + 628 pp. [cf. TL-2 5: 179-180, 1985].

SCHIPPMANN, U. 1986: Die Gattung *Brachypodium* Palisot de Beauvois (Gramineae, Pooideae) in Europa. - *Courier Forschungsinst. Senckenberg* **85**: 175-177.

SCHIPPMANN, U. 1989: Foliar anatomy in the taxonomy of *Brachypodium* (Poaceae). In Phitos, D. (ed.): Abstracts. Sixth OPTIMA meeting, Delphi 10-16 Sept. 1989, p. 134.

SCHIPPMANN, U. 1991: Revision der europäischen Arten der Gattung *Brachypodium* Palisot de Beauvois (Poaceae). - *Boissiera* **45**: 1-250.

SCHIPPMANN, U. & GUTH, S. 1989: Über *Festucopsis sancta* (Poaceae) und Jankas Balkanreise im Jahr 1871. - *Willdenowia* **19(1)**: 99-103.

SCHIPPMANN, U. & JARVIS, C.E. 1988: Typification of three Linnaean names associated with the genus *Brachypodium* (Poaceae). - *Taxon* **37(1)**: 158-164.

SCHIRAREND, C. & OLABI, M.N. 1994: Revision of the genus *Paliurus* Tourn. ex Mill. (Rhamnaceae). - *Bot. Jahrb. Syst.* **116(3)**: 333-359.

SCHLARBAUM, S.E., SMALL, E. & JOHNSON, L.B. 1984: Karyotypic evolution, morphological variability and phylogeny in *Medicago* sect. *Intertextae*. - *Pl. Syst. Evol.* **145(3-4)**: 203-222.

SCHLECHTER, R. 1918: Mitteilungen über einige europäische und mediterrane Orchideen. I. - *Repert. Spec. Nov. Regni Veg.* **15**: 273-302.

SCHLECHTER, R. 1919: Mitteilungen über europäische und mediterrane Orchideen. II. Die Gattung *Gymnadenia* R. Br. und ihre nähreren Verwandten. - *Repert. Spec. Nov. Regni Veg.* **16**: 257-292.

SCHLECHTER, R. 1920: Mitteilungen über europäische und mediterrane Orchideen. III. - *Repert. Spec. Nov. Regni Veg.* **16**: 369-379.

SCHLECHTER, R. 1923: Mitteilungen über europäische und mediterrane Orchideen. IV. - *Repert. Spec. Nov. Regni Veg.* **19**: 33-48.

SCHLENKER, G. 1936a: Experimentelle Untersuchungen in der Sektion *Beccabunga* Griseb. der Gattung *Veronica*. - *Flora* **130(3)**: 305-350 + 1 plate.

SCHLENKER, G. 1936b: Systematische Untersuchungen über die Sektion *Beccabunga* der Gattung *Veronica*. - *Repert. Spec. Nov. Regni Veg. Beih.* **90**: i-v, 1-40 + 2 plates.

SCHLOSSER, J.C. & VUKTINOVIĆ, L. 1869: Flora Croatica exhibens stirpes phanerogamas et vasculares cryptogamas quae in Croatia Slavonia et Dalmatia sponte crescunt ... - Zagreb: Albrect & Fiedler, cxli + 1362 pp. [cf. TL-2 5: 226, 1985].

SCHLÜSSEL, A., JEANMONOD, D. & GAMISANS, J. 2014: Floristic analyses of the Corsican flora: distribution in ecological compartments (vegetation belts, habitats and substrates). - *Candollea* **69(1)**: 25-43.

SCHMALFUSS, H., STIEDEL, C. & SCHLEGEL, M. 1981: The fauna of the Aegean island of Thira, Greece: 1. - *Stuttgarter Beitr. Naturk., A* **347**: 1-14.

SCHMEGEL, C. 2012: Systematische Aufstellung von Blühkalendern am Beispiel der Orchideen von Lesbos (Griechenland). - *Ber. Arbeitskreis. Heimische Orchid.* **19(1)**: 262-277.

SCHMEJA, O. 1931: Beitrag zur Kenntnis der Gattung *Jasione*. - *Beih. Bot. Centralbl., Abt. 2* **48(1)**: 1-51 + 2 plates.

SCHMID, B.W. 1980: *Carex flava* L. s.l. im Lichte der r-Selektion. - Zürich: Ph.D. thesis, Juris Druck & Verlag, 361 pp. + 1 table.

SCHMID, B.W. 1982: Karyology and hybridization in the *Carex flava* complex in Switzerland. - *Feddes Repert.* **93(1-2)**: 23-59.

SCHMID, B.W. 1983: Notes on the nomenclature and taxonomy of the *Carex flava* group in Europe. - *Watsonia* **14(4)**: 309-319.

SCHMID, E. 1949: Prinzipien der natürlichen Gliederung der Vegetation des Mediterrangebietes. - *Ber. Schweiz. Bot. Ges.* **59**: 169-200 + 1 fold-out map.

SCHMID, E. 1970: Die Abgrenzung der Vegetationsgürtel im Mittelmeergebiet. - *Feddes Repert.* **81(1-5)**: 203-213.

SCHMID, E. 1975: Die Vegetationsgürtel Griechenlands. In Dafis, S. & Landolt, E. (eds): Zur Vegetation und Flora von Griechenland, Band 1. - *Veröff. Geobot. Inst. E.T.H. Stiftung Rübel Zürich* **55**: 37-71.

SCHMID, K. 1983: Untersuchungen an *Polygonum aviculare* s.l. in Bayern. - *Mitt. Bot. Staatssamml. München* **19**: 29-149.

SCHMIDT-VOGT, H. 1977: Die Fichte. Band I: Taxonomie, Verbreitung, Morphologie, Ökologie, Waldgesellschaften. - Hamburg & Berlin: Verlag Paul Parey, xviii + 647 pp.

SCHMIDT, A. 1961a: Zytotaxonomische Untersuchungen an europäischen *Viola*-Arten der Sektion *Nomimium*. - *Oesterr. Bot. Z.* **108(1)**: 20-88.

SCHMIDT, A. 1961b: Zytotaxonomische Untersuchungen an *Viola*-Arten der Sekt. *Melanium* (Vorläufige Mitteilung). - *Ber. Bayer. Bot. Ges.* **34**: 93-95.

SCHMIDT, A. 1962: Eine neue Grundzahl in der Gattung *Viola* (Zytotaxonomische Untersuchungen an *Viola parvula* Tin. und *Viola occulta* Lehm.). - *Ber. Deutsch. Bot. Ges.* **75(3)**: 78-84.

SCHMIDT, A. 1963: Zytotaxonomische Untersuchungen an griechischen *Viola*-Arten der Sektion *Melanium*. - *Oesterr. Bot. Z.* **110(2-3)**: 285-293.

SCHMIDT, A. 1964a: Chromosomenzahlen südeuropäischer *Viola*-Arten der Sektion *Melanium*. - *Flora* **154(1)**: 158-162.

SCHMIDT, A. 1964b: Zur systematischen Stellung von *Viola chelmea* Boiss. & Heldr. ssp. *chelmea* und *V. delphinantha* Boiss. - *Ber. Deutsch. Bot. Ges.* **77(7)**: 256-261.

SCHMIDT, A. 1965: Zytotaxonomische Beiträge zu einer Neugliederung der Sektion *Melanium* der Gattung *Viola*. - *Ber. Deutsch. Bot. Ges.* **77, Sondernummer**: 94-99.

SCHMIDT, B. 1979a: Beiträge zur Kenntnis der Sippenstruktur der Gattung *Coronilla* L. – *Feddes Repert.* **90(5-6)**: 257-361 + 1 fold-out table.

SCHMIDT, B. 1979b: Zur Determination von *Coronilla*-Sippen. - *Gleditschia* **7**: 33-41.

SCHMIDT, B. 1979c: *Securigera securidaca* (L.) Degen et Dörfler - eine Spezies der Gattung *Coronilla* L. - *Feddes Repert.* **89(9-10)**: 597-600.

SCHMIDT, D. & MALSO, S. 2020: A continuing trend: *Plantago coronopus* spreads also along the roads in Bosnia and Herzegovina. - *Phytol. Balcan.* **26(3)**: 479-483.

SCHMIDT, P. 1973: Übersicht über die mitteleuropäischen Arten der Gattung *Thymus* L. – *Feddes Repert.* **83(9-10)**: 663-671.

SCHMIDT, P. 1977: Revision der *Thymus-Arten* von Ph. M. Opiz. - *Folia Geobot. Phytotax.* **12(4)**: 377-416.

SCHMIDT, P. 1989: Beitrag zur Systematik und Evolution der Gattung *Picea*. - *Flora, Morphol. Geobot. Oekophysiol.* **182(5-6)**: 435-461.

SCHMIDT, W. 1975: *Nigritella* Rich. - *Mitt. Arbeitskreis Heimische Orchid. Baden-Württemberg* **7**: 94-100.

SCHMITT, E. 1999: Les pivoines. - *Pl. Mont. Rocaille* **12(192)**: 627-635.

SCHMITT, E. 2000a: Étude systematique du genre *Paeonia* L. [suite]. - *Pl. Mont. Rocaille* **13(193)**: 5-6.

SCHMITT, E. 2000b: Les pivoines. Étude systematique du genre *Paeonia* L. [fin]. - *Pl. Mont. Rocaille* **13(194)**: 70-80.

SCHMITT, E. 2003: Typification of the Linnean names of the genus *Paeonia* L. - *Candollea* **58(1)**: 183-188.

SCHMITZ, J. & FROEBE, H.A. 1988: Ein Schlüssel für die Umbelliferengattungen Mitteleuropas. - *Bot. Jahrb. Syst.* **109(4)**: 451-467.

SCHMUCKER, TH. 1930: Endemismen und Charakterpflanzen von Kreta. In Karsten, G. (ed.): Vegetationsbilder 21. Reihe, Heft 5. - Jena: Gustav Fischer Verlag, Tafel 25-30 + Text.

SCHMUCKER, TH. 1935: Über eine *Ophrys*-Population aus Kreta. - *Beih. Bot. Centralbl., Abt. 2* **53(2)**: 427-438 + 3 plates.

SCHNEDLER, W. 1977: Behaarung bei *Herniaria glabra* L. - *Göttinger Florist. Rundbr.* **11(2)**: 27-29.

SCHNEEWEISS, G.M. 2000: Die kurzlebigen Arten der Gattung *Alyssum* (Brassicaceae) in Österreich. - *Ann. Naturhist. Mus. Wien* **102B**: 389-407.

SCHNEIDER, C. 1916a: Beiträge zur Kenntnis der Gattung *Ulmus*. I. Gliederung der Gattung und Übersichr der Arten. - *Oesterr. Bot. Z.* **66(1-2)**: 21-34.

SCHNEIDER, C. 1916b: Beiträge zur Kenntnis der Gattung *Ulmus*. I. Über die richtige Benennung der europäischen Ulmen-Arten. - *Oesterr. Bot. Z.* **66(3-4)**: 65-82.

SCHNEIDER, G. 1987: Neufund von *Orchis sancta* L. auf Kreta. - *Mitt. Arbeitskreis Heimische Orchid. Baden-Württemberg* **19(1)**: 95-96.

SCHNEIDER, G. 1989: Ein Fund von *Orchis sancta* L. × *Orchis coriophora* L. auf Kreta. - *Mitt. Arbeitskreis Heimische Orchid. Baden-Württemberg* **21(1)**: 126-132.

SCHNEIDER, H. 1913: Morphologische und entwicklungsgeschichtliche Untersuchungen an *Thelygonum Cynocrambe* L. - *Flora* **106(1)**: 1-41.

SCHNELLER, J.J. 1975a: Untersuchungen an einheimischen Farnen insbesondere der *Dryopteris filix mas*-Gruppe. 1. Teil. - *Ber. Schweiz. Bot. Ges.* **84(3)**: 195-217.

SCHNELLER, J.J. 1975b: Untersuchungen an einheimischen Farnen insbesondere der *Dryopteris filix mas*-Gruppe. 2. Teil. Cytologische Untersuchungen. - *Ber. Schweiz. Bot. Ges.* **85(1)**: 1-17.

SCHNELLER, J.J. 1979: Biosystematic investigations on the lady fern (*Athyrium filix-femina*). - *Pl. Syst. Evol.* **132(4)**: 255-277.

SCHNELLER, J.J. 1983: Haploid sporophytes in *Asplenium filix-femina* (L.) Roth and evidence for somatic diploidization in one of them. - *Bot. Helv.* **93(1)**: 85-90.

SCHNELLER, J.J. & RASBACH, H. 1984: Hybrids and polyploidy in the genus *Athyrium* (Pteridophyta) in Europe. - *Bot. Helv.* **94(1)**: 81-99.

SCHNELLER, J.J. & SCHMID, B.W. 1982: Investigations on the intraspecific variability in *Athyrium filix- femina* (L.) Roth. - *Bull. Mus. Natl. Hist. Nat., B, Adansonia* **4(3-4)**: 215-228.

SCHOCH-BODMER, H. 1935: *Lythrum salicaria* L. - *Pflanzenareale* **4(3)**: 29-33, Karte 29-30.

SCHOLTE, G.A.M. 1978a: Biosystematic studies in the collective species *Stellaria media* (L.) Vill. (I). - *Proc. Kon. Ned. Akad. Wetensch. C* **81(4)**: 442-456.

SCHOLTE, G.A.M. 1978b: Biosystematic studies in the collective species *Stellaria media* (L.) Vill. (II). - *Proc. Kon. Ned. Akad. Wetensch. C* **81(4)**: 457-465.

SCHOLZ, H. 1955: *Parietaria erecta* Mertens & Koch und *Parietaria ramiflora* Moench. – *Phyton (Horn)* **6(1-2)**: 31-32.

SCHOLZ, H. 1959a: Die Systematik des europäischen *Polygonum aviculare* L. I. Die Zweiteilung des *P. aviculare* nach Lindman und der Formenkreis des *P. aequale* Lindman. - *Ber. Deutsch. Bot. Ges.* **71(10)**: 427-434.

SCHOLZ, H. 1959b: Die Systematik des europäischen *Polygonum aviculare* L. II. Die Arten und Sippen aus der Verwandtschaft des *Polygonum heterophyllum* Lindman. - *Ber. Deutsch. Bot. Ges.* **72(1)**: 63-72.

SCHOLZ, H. 1960: Bestimmungsschlüssel für die Sammelart *Polygonum aviculare* L. - *Verh. Bot. Vereins Prov. Brandenburg* **98-100**: 180-182.

SCHOLZ, H. 1961: Der Formenkreis der *Setaria italica* (L.) P.B. Ein Beitrag zur Kulturpflanzenkunde. - *Ber. Deutsch. Bot. Ges.* **73(10)**: 471-476 + 1 plate.

SCHOLZ, H. 1968a: Die Artbestimmung im *Brachypodium pinnatum*-Komplex. - *Willdenowia* **5(1)**: 113-118.

SCHOLZ, H. 1968b: Systematik und Verbreitung einiger Taxa der Gattung *Poa*, Sektion *Ochlopoa*, im Mittelmeergebiet. - *Ber. Deutsch. Bot. Ges.* **81(1-2)**: 17-21.

SCHOLZ, H. 1970: Zur Systematik der Gattung *Bromus* L. Subgenus *Bromus* (Gramineae). - *Willdenowia* **6(1)**: 139-159.

SCHOLZ, H. 1974a: Über die Umgrenzung der Gattung *Desmazeria* (Gramineae). - *Bot. Jahrb. Syst.* **94(4)**: 556-561.

SCHOLZ, H. 1974b: *Bromus chrysopogon* Viviani - eine wieder als Art bewertete Graminee aus dem Mittelmeergebiet. - *Willdenowia* **7(2)**: 409-414.

SCHOLZ, H. 1974c: Liste der Gräser Libyens. - *Willdenowia* **7(2)**: 419-458.

SCHOLZ, H. 1977: Bemerkungen zur Merkmalsgeographie des *Polygonum aviculare*, insbesondere des *P. arenastrum*. - *Verh. Bot. Vereins Prov. Brandenburg* **113**: 13-22.

SCHOLZ, H. 1981: Bemerkungen über *Bromus madritensis* und *B. rubens* (Gramineae). – *Willdenowia* **11(2)**: 249-258.

SCHOLZ, H. 1982: Eine neue *Danthoniastrum*-Art (Gramineae) aus Albanien. - *Willdenowia* **12(1)**: 47-49.

SCHOLZ, H. 1983a: *Poa*-Studien 1. - Neue Taxa des *Poa bulbosa*-Komplexes aus dem Mittelmeergebiet. - *Bot. Chron. (Patras)* **3(1-2)**: 15-21.

SCHOLZ, H. 1983b: Die Unkraut-Hirse (*Panicum miliaceum* subsp. *ruderale*) - neue Tatsachen und Befunde. - *Pl. Syst. Evol.* **143(3)**: 233-244.

SCHOLZ, H. 1983c: Eine neue *Gaudinopsis*-Art (Gramineae) aus der Türkei. - *Willdenowia* **13(1)**: 111-113.

SCHOLZ, H. 1984a: *Stipa thessala* Hausskn. emend. (Gramineae). - *Willdenowia* **14(1)**: 131-134.

SCHOLZ, H. 1984b: *Poa*-Studien 3. - *Poa sterilis* in Griechenland. - *Bot. Chron. (Patras)* **4(1-2)**: 7-11.

SCHOLZ, H. 1985a: *Poa*-Studien 2. - Systematische Stellung und Synonymie der *P. dolosa* Boiss. & Heldr. - *Willdenowia* **14(2)**: 327-330.

SCHOLZ, H. 1985b: *Poa*-Studien 4. - Über *Poa hackelii* Post und *P. pelasgis* sp. nova (Gramineae). - *Willdenowia* **15(1)**: 91-97.

SCHOLZ, H. 1986a: *Poa* studies 5. - The genus *Poa* (Gramineae) in Greece: Annotated check-list and key to the species. - *Willdenowia* **15(2)**: 393-400.

SCHOLZ, H. 1986b: Bemerkungen zur Flora Griechenlands: *Gastridium phleoides* und *G. ventricosum* (Poaceae). - *Willdenowia* **16(1)**: 65-68.

SCHOLZ, H. 1986c: Die Entstehung der Unkraut-Gerste, *Hordeum vulgare* subsp. *agriocrithon* emend. - *Bot. Jahrb. Syst.* **106(3)**: 419-426.

SCHOLZ, H. 1987a: *Poa*-Studien 6. - Die ökologische und taxonomische Wertigkeit der Zwiebelform bei den altweltlichen *Poa*-Taxa der subsect. *Bulbosae* (Gramineae). - *Willdenowia* **16(2)**: 403-407.

SCHOLZ, H. 1987b: Delimitation and classification of *Bromus fasciculatus* (Poaceae). - *Pl. Syst. Evol.* **155(1-4)**: 277-282.

SCHOLZ, H. 1987c: Neue Synonyme der *Festuca arundinacea* (Poaceae). - *Bot. Jahrb. Syst.* **108(2-3)**: 293-299.

SCHOLZ, H. 1989a: Floristisches und Taxonomisches über einige Gramineae von Rhodos und Symi (Ost-Ägäis, Griechenland). - *Willdenowia* **19(1)**: 105-110.

SCHOLZ, H. 1989b: Neue Taxa der Gattung *Stipa* sect. *Stipa* (Gramineae) aus dem Mittelmeergebiet. - *Willdenowia* **19(1)**: 127-132.

SCHOLZ, H. 1990: Neue und wenig bekannte mediterrane Gramineen-Taxa. - *Willdenowia* **19(2)**: 405-412. [Also distributed as *OPTIMA leaflet* no. 194].

SCHOLZ, H. 1991a: Einheimische Unkräuter ohne Naturstandorte ("Heimatlose" oder obligatorische Unkräuter). - *Fl. Veg. Mundi* **9**: 105-112.

SCHOLZ, H. 1991b: Die Systematik der *Avena sterilis* und *A. fatua* (Gramineae). Eine kritische Studie. - *Willdenowia* **20(1-2)**: 103-112.

SCHOLZ, H. 1991c: Notizen zur Gramineenflora von Kos und Nisyros (Ost-Ägäis, Griechenland). - *Willdenowia* **21(1-2)**: 131-141.

SCHOLZ, H. 1993: *Festuca rivularis*, *Poa maroccana* und *Stipa monticola* (Gramineae) in Griechenland. - *Willdenowia* **23(1-2)**: 113-119.

SCHOLZ, H. 1995a: *Bromus regnii* (Gramineae), a new endemic serpentine annual brome-grass from Cyprus. - *Willdenowia* **25(1)**: 235-238.

SCHOLZ, H. 1995b: *Monerma* P. Beauv. (Poaceae) - not an illegitimate name. - *Feddes Repert.* **106(3-4)**: 169-171.

SCHOLZ, H. 1996a: On annual *Poa* weeds, especially *P. maroccana* (Gramineae). - *Bot. Chron. (Patras)* **12**: 15-19.

SCHOLZ, H. 1996b: Die Taxonomie der *Stipa fontanesii* Parl., und *S. maroccana*, eine neue Art aus Marokko (Gramineae). - *Willdenowia* **26(1-2)**: 225-228.

SCHOLZ, H. 1997: A new annual brome-grass (*Bromus*) from Greece. - *Feddes Repert.* **108(1-2)**: 97-100.

SCHOLZ, H. 1998: Notes on *Elytrigia obtusiflora* subsp. *graeca*. In Greuter, W. & Raus, Th. (eds): Med-Checklist notulae, 17. - *Willdenowia* **28(1-2)**: 163-174.

SCHOLZ, H. 1999a: Short notes on *Phleum* sect. *Achnodon* (Gramineae). - *Willdenowia* **29(1-2)**: 45-49.

SCHOLZ, H. 1999b: Notes on *Vulpia unilateralis*. In Greuter, W. & Raus, Th. (eds): Med-Checklist notulae, 18. - *Willdenowia* **29(1-2)**: 66.

SCHOLZ, H. 2000a: *Eragrostis leptocarpa* Benth., *Panicum hillmanii* Chase. In Greuter, W. & Raus, Th. (eds): Med-Checklist notulae, 19. - *Willdenowia* **30(2)**: 242.

SCHOLZ, H. 2000b: Alte und neue Taxa der Gattung *Catapodium* (Gramineae). - *Bot. Chron. (Patras)* **13**: 95-104.

SCHOLZ, H. 2002: Notes on *Bromus optimae* H. Scholz and *Paspalum notatum* Flüggé. In Greuter, W. & Raus, Th. (eds): Med-Checklist notulae, 21. - *Willdenowia* **32(2)**: 205-206.

SCHOLZ, H. 2004: *Eragrostis pectinacea* (Michx.) Nees, *Vulpia unilateralis* (L.) Stace. In Greuter, W. & Raus, Th. (eds): Med-Checklist notulae, 22. - *Willdenowia* **34(1)**: 78-80.

SCHOLZ, H. 2005: *Bromus japonicus* subsp. *anatolicus* (Boiss. & Heldr.) Pénez, *Catapodium hemipoa* subsp. *occidentale* (Paunero) H. Scholz & S. Scholz. In Greuter, W. & Raus, Th. (eds): Med-Checklist notulae, 23. - *Willdenowia* **35(1)**: 61-62.

SCHOLZ, H. 2007a: *Arrhenatherum palaestinum* subsp. *macedonicum* H. Scholz, subsp. nov. In Greuter, W. & Raus, Th. (eds): Med-Checklist notulae, 25. - *Willdenowia* **37(1)**: 213.

SCHOLZ, H. 2007b: Notes on *Aegilops* (Poaceae). - *Willdenowia* **37(2)**: 431-434.

SCHOLZ, H. 2008a: Some comments on the genus *Bromus* (Poaceae) and three new species. - *Willdenowia* **38(2)**: 411-422.

SCHOLZ, H. 2008b: *Bromus regnii* H. Scholz. In Greuter, W. & Raus, Th. (eds): Med-Checklist notulae, 27. - *Willdenowia* **38(2)**: 470.

SCHOLZ, H. 2010: *Bromus squarrosus* subsp. *consimilis* H. Scholz. In Greuter, W. & Raus, Th. (eds): Med-Checklist notulae, 29. - *Willdenowia* **40(2)**: 199-200.

SCHOLZ, H. & BAZOS, I. 2004: *Bromus regnii* H. Scholz. In Greuter, W. & Raus, Th. (eds): Med- Checklist notulae, 22. - *Willdenowia* **34(1)**: 78.

SCHOLZ, H. & BIEL, B. 2011: *Eragrostis multiglandulosa* H. Scholz, sp. nov. In Greuter, W. & Raus, Th. (eds): Med-Checklist notulae, 30. - *Willdenowia* **41(2)**: 324.

SCHOLZ, H. & BÖHLING, N. 2000: *Phragmites frutescens* (Gramineae) re-visited. The discovery of an overlooked, woody grass in Greece, especially Crete. - *Willdenowia* **30(2)**: 251-261 + electronic suppl.

SCHOLZ, H. & JOGAN, N. 2006: *Bromus hordeaceus*. In Greuter, W. & Raus, Th. (eds): Med-Checklist notulae, 24. - *Willdenowia* **36(2)**: 728.

SCHOLZ, H. & KRIGAS, N. 2004: *Nasella neesiana* (Trin. & Rupr.) Barkworth. In Greuter, W. & Raus, Th. (eds): Med-Checklist notulae, 22. - *Willdenowia* **34(1)**: 78.

SCHOLZ, H. & MIKOLÁŠ, V. 1991: The weedy representatives of Proso Millet (*Panicum miliaceum*, Poaceae) in Central Europe. - *Thaiszia* **1**: 31-41.

SCHOLZ, H. & OBERPRIELER, C. VOGT, R. 1998: Chromosome numbers of North African phanerogams. Some notes on North African Gramineae. - *Lagascalia* **20(2)**: 265-275.

SCHOLZ, H. & RAUS, TH. 1997: Zwei neue Unterarten des *Hordeum murinum* (Gramineae) aus Griechenland und Spanien. - *Feddes Repert.* **108(7-8)**: 527-531.

SCHOLZ, H. & RAUS, TH. 2001: New distribution and morphological data of *Vulpia membranacea*. - *Willdenowia* **31(2)**: 309-313.

SCHOLZ, H. & RAUS, TH. 2006a: Contribution to the flora of Greece: a new species of *Achnatherum* (Poaceae). - *Willdenowia* **36(1)**: 373-378.

SCHOLZ, H. & RAUS, TH. 2006b: *Eragrostis minor*. In Greuter, W. & Raus, Th. (eds): Med-Checklist notulae, 24. - *Willdenowia* **36(2)**: 728.

SCHOLZ, H. & SLAGEREN, M.W. VAN 1994: (1089) Proposal to conserve *Aegilops caudata* (Gramineae) with a conserved type. - *Taxon* **43(2)**: 293-296. [not approved, cf. Taxon 45(4): 675, 1996].

SCHOLZ, H. & STRID, A. 1992: *Festuca stygia* (Poaceae), a new species from Peloponnisos, Greece. - *Willdenowia* **22(1-2)**: 85-88.

SCHOLZ, H. & WILLING, E. 2011: *Vulpia ligustica*. In Greuter, W. & Raus, Th. (eds): Med-Checklist notulae, 30. - *Willdenowia* **41(2)**: 325.

SCHÖNBECK-TEMESY, E. & EHRENDORFER, F. 1979: Materials for a Flora of Turkey. XXXVI. Rubiaceae II. - *Notes Roy. Bot. Gard. Edinburgh* **37(2)**: 243-264.

SCHÖNBECK-TEMESY, E. & EHRENDORFER, F. 1980: New endemic taxa of *Asperula* sect. *Cynanchicae* (Rubiaceae) from the East Aegean Islands. - *Pl. Syst. Evol.* **134(1-2)**: 133-135.

SCHÖNBECK-TEMESY, E. & EHRENDORFER, F. 1985: *Asperula gorganica* und *A. semanensis*, zwei neue Arten aus dem Orient, und die palaeo-mediterrane Sektion *Thliphthisa* (Griseb.) Ehrend. (Rubiaceae). - *Bot. Jahrb. Syst.* **107(1-4)**: 75-93.

SCHÖNBECK-TEMESY, E. & EHRENDORFER, F. 1989: The perennial taxa of *Crucianella* (Rubiaceae) in SW Asia and their eco-geographical differentiation. - *Pl. Syst. Evol.* **165(1-2)**: 101-136.

SCHÖNFELDER, I. & SCHÖNFELDER, P. 1990: Die Kosmos-Mittelmeerflora. - Stuttgart: Kosmos, Franckh'sche Verlagshandlung, 318 pp. incl. many plates.

SCHÖNFELDER, P. 1968: Chromosomenzahlen einiger Arten der Gattung *Biscutella* L. - *Oesterr. Bot. Z.* **115(3)**: 363-371.

SCHÖNFELDER, P. & AL. 1987: Kreta. Arbeitsunterlagen für den APV-Fortbildungslehrgang. - Regensburg: Institut für Botanik der Univ., 139 pp. [privately distributed].

SCHÖNFELDER, P. & AL. 1988: Kreta. Exkursion für Fortgeschrittene, 16.-30. April 1988. - Regensburg: Institut für Botanik der Univ., 86 pp.

SCHÖNFELDER, P. & SCHÖNFELDER, I. 2008: Die neue Kosmos-Mittelmeerflora. - Stuttgart: Kosmos, Franckh'sche Verlagshandlung, 477 pp. incl. many plates.

SCHÖNFELDER, P. & WEGMÜLLER, S. 1974: Zur Unterscheidung und Verbreitung der Sippen von *Lamiastrum galeobdolon* agg. in Süddeutschland. - *Göttinger Florist. Rundbr.* **8(1)**: 24-34.

SCHÖNFELDER, P. (ED.) 1988: Prodromus florae creticae. - Regensburg: Institut für Botanik der Universität, iv + 316 pp.

SCHOT, M.H. 1997: Orchideeënvariaties op Rhodos. - *Eurorchis* **9**: 21-32.

SCHOTSMAN, H.D. 1961: Notes on some Portugeses species of *Callitriche*. - *Bol. Soc. Brot., Ser. 2* **35**: 95-127 + 2 maps.

SCHOTSMAN, H.D. 1967: Les callitriches. Espèces de France et taxa nouveaux d'Europe. - Paris: P. Lechevalier, 152 pp. + 20 plates.

SCHOTSMAN, H.D. 1977: Callitriches de la région méditerranéenne. Nouvelles observations. - *Bull. Centre Études Rech. Sci. Biarritz* **11(3)**: 241-312.

SCHOTT, H.W. 1854: *Anemone blanda* Schtt. et Kotschy. - *Oesterr. Bot. Wochenbl.* **4(16)**: 129-130.

SCHOTT, H.W. 1857: Ueber *Arum Dioscoridis*. - *Oesterr. Bot. Wochenbl.* **7(22)**: 173-175.

SCHOTT, H.W. 1859: Aroideen-Skizzen. - *Oesterr. Bot. Z.* **9(3)**: 98-101.

SCHOUTEN, A.R. 1976: Lijst van planten gevonden tijdens K.N.N.V. reis naar Rodos, 15 t/m 29 april 1976. [Unpublished flora list]. - Photocopy: 18 pp.

SCHOUTEN, A.R. 1977: Lijst van planten gevonden tijdens K.N.N.V.-reis naar Lakonia, ZO-Peloponnesos, stpl. Sparta, 20 april r/m 4 mei 1977. [Unpublished flora list]. - Photocopy: 20 pp.

SCHOUTEN, A.R. 1978: Lijst van planten gevonden tijdens K.N.N.V.-reis naar Kreta 18 april t/m 8 mei 1978. [Unpublished flora list]. - Photocopy: 23 pp.

SCHOUTEN, A.R. 1979: Lijst van planten gevonden tijdens K.N.N.V.-reis naar Ipiros 24 april t/m 11 mai 1979. [Unpublished flora list]. - Photocopy: 19 pp.

SCHOUTEN, A.R. 1980: Lijst van planten gevonden tijdens K.N.N.V.-reis naar Ilia en Achaia, 21 april t/m 9 mei 1980. [Unpublished flora list]. - Photocopy: 19 pp.

SCHREIBER, H.J. 1998: Waldgrenznahe Buchenwälder und Grasländer des Falakron und Pangäon in Nordostgriechenland. Syntaxonomie, Struktur und Dynamik. - *Arb. Inst. Landschaftsökol. Westf. Wilhelms-Univ. Münster* **4**: viii + 171 pp.

SCHRÖDER, F.-G. 1969: Zur Klassifizierung der Anthropochoren. - *Vegetatio* **16(5-6)**: 225-238.

SCHRÖDER, F.-G. 2003: Wer importierte die Robinie nach Europa? - *Mitt. Deutsch. Dendrol. Ges.* **88**: 121-136.

SCHROPP, K. 2008: Mount Timfi surprise. - *Sedum Soc. Newsletter* **84**: 42.

SCHUBERT, V. & SCHUBERT, A. 2001: Die Verbreitung von *Aegilops*-Arten auf 37 Inseln der Ägäis. In Hammer, K. & Gladis, Th. (eds): Schriften zu Genetischen Ressourcen 16: 234-241. - Bonn: Zentralstelle für Agrardokumentation und -information.

SCHULER, A. 2000: Beobachtungen zur Flora und Vegetation des Nestos-Aue in Nordost-Griechenland. - *Tuexenia* **20**: 419-427.

SCHULER, A. 2004: Characterarten mitteleuropäischer Pflanzengesellschaften an ihrer Verbreitungsgrenze in Griechenland. Eine Studie zum ökologischen und syntaxonomischen Verhalten am Arealrand. - *Bot. Chron. (Patras)* **17**: 1-168.

SCHULER, A. 2005: *Hieracium camkorijense* Zahn, *Trifolium simplex* L. subsp. *simplex*, *Triglochin palustris* L. In Greuter, W. & Raus, Th. (eds): Med-Checklist notulae, 23. - *Willdenowia* **35(1)**: 58, 60, 64.

SCHULER, A. 2007: Contribution to the flora of northern and central Greece. - *Willdenowia* **37(1)**: 229-241.

SCHULER, A. & SNOGERUP, S. 2006: *Epilobium adenocaulon*. In Greuter, W. & Raus, Th. (eds): Med- Checklist notulae, 24. - *Willdenowia* **36(2)**: 724.

SCHULER, B. 1986: Taxonomie, Ökologie und Verbreitung von *Avena sterilis* und anderen *Avena*-Arten unter besonderer Berücksichtigung des Getreidebaus im westlichen Mittelmeergebiet. - *Plitis* **4 (5)**: 1-266.

SCHULTZ-BIPONTINIUS, C.H. 1842: [Descriptions of specimens of Compositae sent by Dr. Fraas from Greece]. - *Flora* **25**: 172-174.

SCHULTZ, F.W. 1843: Andeutungen zur Kenntnis einiger Orobanchen Griechenlands. Fragment aus den Vorarbeiten zur Monographie der Gattung *Orobanche*. - *Flora* **26(1)**: 125-132.

SCHULTZ, F.W. & SCHULTZ-BIPONTINIUS, C.H. 1862: *Pilosella* als eigene Gattung aufgestellt. - *Flora* **45**: 417-441.

SCHULTZE-MOTEL, J. 1987: Numerisch-taxonomische Studien an *Triticum* L. und *Aegilops* L. zur Theorie der Klassifizierung von Kulturplanzen. - *Kulturpflanze* **35**: 53-97.

SCHULZ-SCHAEFFNER, J. & JURASITS, P. 1962: Biosystematic investigations in the genus *Agropyron*. I. Cytological studies of species karyotypes. - *Amer. J. Bot.* **49(9)**: 940-953.

SCHULZ-SCHAEFFNER, J. & JURASTIS, P. 1967: Biosystematic investigations in the genus *Agropyron*. IV. Species karyotype analysis, phytogeographic and other biosystematic studies. - *Z. Pflanzenzücht.* **57**: 146-166.

SCHULZ, A. 1918: Abstammung und Heimat des Roggens. - *Ber. Deutsch. Bot. Ges.* **36(1)**: 39-47.

SCHULZ, D.L. 1984: Zur Ausbreitungsgeschichte der *Galinsoga*-Arten in Europa. - *Acta Bot. Slov., A, Suppl.* **1**: 285-296.

SCHULZ, O.E. 1901: Monographie der Gattung *Melilotus*. - *Bot. Jahrb. Syst.* **29(5)**: 660-735 + 3 plates.

SCHULZ, O.E. 1903: Monographie der Gattung *Cardamine*. - *Bot. Jahrb. Syst.* **32(2-3)**: 280-623 + 4 plates.

SCHULZ, O.E. 1933: Kurze Notizen über neue Gattungen, Sektionen und Arten der Cruciferen. - *Bot. Jahrb. Syst.* **66(1)**: 91-102.

SCHULZ, R. 1904: Monographie der Gattung *Phyteuma*. - Geisenheim-am-Rhein: Verlag von J. Schneck, 204 pp. + 3 maps.

SCHULZE, W. 1965: *Siphonostylis*, eine neue Gattung der Iridaceae. - *Oesterr. Bot. Z.* **112(3)**: 331-343.

SCHUMACHER, A. 1945: Die Moorlilien (*Narthecium*)-Arten Europas. - *Arch. Hydrobiol.* **41**: 112-195 + 3 maps.

SCHUSTER, R. 1967: Taxonomische Untersuchungen über die Serie *Palustres* M. Pop. der Gattung *Myosotis* L. - *Feddes Repert.* **74(1-2)**: 39-98 + 6 plates.

SCHÜTT, P. 1991: Tannenarten Europas und Kleinasiens. - Basel, Boston, Berlin: Birkhäuser Verlag, 132 pp.

SCHWARZ, O. 1934: Additamentum ad florulam Lydiae. I-II. - *Repert. Spec. Nov. Regni Veg.* **36**: 65-96, 129-150.

SCHWARZ, O. 1935: Über die ostmediterranen *Cephalanthera*-Arten mit gespornten Blüten. - *Repert. Spec. Nov. Regni Veg.* **38**: 313-316 + 1 plate.

SCHWARZ, O. 1936: Über die Systematik und Nomenklatur der europäischen Schwarzkiefern. - *Notizbl. Bot. Gart. Berlin-Dahlem* **13(117)**: 226-243.

SCHWARZ, O. 1936-39: Monographie der Eichen Europas und des Mittelmeergebietes. - *Repert. Spec. Nov. Regni Veg.* **Sonderbeih. D**: 1-200 + 64 plates.

SCHWARZ, O. 1938a: Die Gattung *Globularia*. - *Bot. Jahrb. Syst.* **69(3)**: 318-373 + 8 plates.

SCHWARZ, O. 1938a: Nachträgliche Notiz zur Nomenklatur der europäischen Schwarzkiefer- - *Notizbl. Bot. Gart. Berlin-Dahlem* **14(121)**: 135-136.

SCHWARZ, O. 1938c: *Cyclamen*-Studien. - *Gartenflora, N.F.* **1938**: 11-38 + 2 plates.

SCHWARZ, O. 1939a: Zweiter Nachtrag zur Systematik und Nomenklatur der Schwarzkiefern. - *Notizbl. Bot. Gart. Berlin-Dahlem* **14(124)**: 381-384.

SCHWARZ, O. 1939b: Zur Nomenklatur von *Ramondia*. - *Repert. Spec. Nov. Regni Veg.* **46**: 187.

SCHWARZ, O. 1944: Zwei neue südwestanatolische *Papaver* der Sect. *Pilosa*. - *Feddes Repert. Spec. Nov. Regni Veg.* **53(1-5)**: 76-80.

SCHWARZ, O. 1951: Taxonomy and nomenclature of *Cyclamen*. - *Quart. Bull. Alpine Gard. Soc. Gr. Brit.* **19(4)**: 396-404.

SCHWARZ, O. 1955: Systematische Monographie der Gattung *Cyclamen* L. - *Feddes Repert. Spec. Nov. Regni Veg.* **58**: 234-283.

SCHWARZ, O. 1963a: Chromosomenzahlen, Lebensformen und Evolution der Gattung *Globularia*. - *Drudea* **3(1)**: 5-16.

SCHWARZ, O. 1963b: Tentative key to the wild species of *Galanthus* L. [with appendices by F. Stern and G. Parker]. - *Quart. Bull. Alpine Gard. Soc. Gr. Brit.* **31(2)**: 131-141.

SCHWARZ, O. 1964: Systematische Monographie der Gattung *Cyclamen* L. Teil II. - *Feddes Repert. Spec. Nov. Regni Veg.* **69(2)**: 73-103.

SCHWARZ, O. 1965: Die kritischen *Hypericum*-Arten der mitteleuropäischen Flora. - *Drudea* **5(1)**: 59-66.

SCHWARZ, O. 1967: The genus *Globularia*. - *Quart. Bull. Alpine Gard. Soc. Gr. Brit.* **35(4)**: 303-316.

SCHWARZ, O. 1968: Beiträge zur Kenntnis der Gattung *Primula*. - *Wiss. Z. Friedrich-Schiller-Univ. Jena, Math.-Naturwiss. Reihe* **17**: 307-332.

SCHWARZ, O. & BÄSSLER, M. 1964: *Danthonia (Sieglingia) decumbens*, ein bemerkenswerter Fall chromosomaler und ökologischer Divergenz. - *Oesterr. Bot. Z.* **111(2-3)**: 193-207.

SCHWARZ, O. & LEPPER, L. 1975: Zwei neue *Cyclamen* aus dem östlichen Mittelmeergebiet. - *Feddes Repert.* **86(9-10)**: 491-497.

SCHWARZ, U. 1981: Vegetationskundliche Beobachtungen der Phrygana auf Syros. – *Salzburger Exkursionsber.* **7**: 153-162.

SCHWARZACHER, T. & SCHWEIZER, D. 1982: Karyotype analysis and heterochromatin differentiation with Giemsa C-banding and fluorescent counterstaining in *Cephalanthera* (Orchidaceae). - *Pl. Syst. Evol.* **141(2)**: 91-113.

SCHWARZENBACH, F. 1922: Untersuchungen über die Sterilität von *Cardamine bulbifera* (L.) Crantz unter der Annahme eines hybriden Ursprungs dieser Art. - *Flora* **115**: 393-514 + 3 plates.

SCHWARZOVÁ, T. 1980: [Chromosome numbers in *Chenopodium* from Czechoslovakia]. In Löve, Á. (ed.): IOPB chromosome number reports LXIX. - *Taxon* **29(5-6)**: 728.

SCHWERIN, F.V. 1909: Monographie der Gattung *Sambucus*. - *Mitt. Deutsch. Dendrol. Ges.* **18**: 1-56 + 1 plate and 4 maps.

SCIANDRELLO, S. & TOMASELLI, V. 2011: Distribution and ecology of *Cornucopiae cucullatum* L. (Poaceae) in the Mediterranean area and evaluation of its conservation status in Italy. - *Acta Bot. Gallica* **158(3)**: 401-407.

SCOPECE, G., COZZOLINO, S. & BATEMAN, R.M. 2010: Just what is a genus? Comparing levels of postzygotic isolation to test alternative hypotheses in Orchidaceae subtribe Orchidinae. - *Taxon* **59(6)**: 1754-1764.

SCOPOLI, J.A. 1771-72: Flora Carniolica, editio secunda aucta et reformata. - Wien: Impensis Joannis Pauli Krauss, 2 vols. [cf. TL-2 5: 455-456, 1985].

SCOPPOLA, A. 2019: An annotated key to the species of *Gastridium* (Poaceae) with distributional novelties to the Italian territory. - *Nat. Hist. Sci.* **6(2)**: 29-36.

SCOPPOLA, A. & AL. 2020: Phylogenetic relationships and taxonomic issues in *Gastridium* (Poaceae) inferred from plastid and nuclear DNA sequence analysis. - *Pl. Biosyst.* **154(6)**: 947-960.

SCOPPOLA, A. & BANFI, E. 2021: Lectotypification of the name *Agrotis ventricosa* Gouan (Poaceae). - *Phytotaxa* **490(1)**: 91-98.

SCOPPOLA, A. & LATTANZI, E. 1993: Distribution of genus *Pseudolysimachion* Opiz in Italy. In Petrova, A.V. (ed.): Abstracts from the VII OPTIMA meeting in Borovetz, Bulgaria, 18-30 July 1993, p. 107.

SCOPPOLA, A., MAGRINI, S. & CELESTINI, M. 2011: *Lupinus* sect. *Albus* (Fabaceae): Taxonomic criticism and conservation value. The case of wild populations in central Italy. - *Pl. Biosyst.* **145(3)**: 514-526.

SCOTT, A.J. 1978a: A review of the classification of *Chenopodium* L. and related genera (Chenopodiaceae). - *Bot. Jahrb. Syst.* **100(2)**: 205-220.

SCOTT, A.J. 1978b: Reinstatement and revision of Salicorniaceae J. Agardh (Caryophyllales). - *Bot. J. Linn. Soc.* **75(4)**: 357-374.

SCOTT, G.A.M. 1963: Biological Flora of the British Isles. *Glaucium flavum* Crantz. - *J. Ecol.* **51(3)**: 743-754.

SCOTT, M. 1984: Plant portraits: *Aster alpinus*. - *Rock Gard.* **19(1)**: 37.

SCRUGLI, A. 1972: Numeri cromosomici per la Flora Italiana: 124-125. - *Inform. Bot. Ital.* **4(2)**: 128-129.

SCRUGLI, A. 1974a: Numeri cromosomici per la Flora Italiana: 162-166. - *Inform. Bot. Ital.* **5(3)**: 264-270.

SCRUGLI, A. 1974b: Numeri cromosomici per la Flora Italiana: 167-171. - *Inform. Bot. Ital.* **6(1)**: 37-43.

SCRUGLI, A. 1977: Numeri cromosomici per la Flora Italiana: 331-347. - *Inform. Bot. Ital.* **9(2)**: 116-125.

SCRUGLI, A. 1979: Numeri cromosomici per la Flora Italiana: 526-533. - *Inform. Bot. Ital.* **10(3)**: 414-421.

SCRUGLI, A. 1981: Numeri cromosomici per la Flora Italiana: 728-732. - *Inform. Bot. Ital.* **12(2)**: 149-153.

SCRUGLI, A. 1983: Sul numero cromosomico di *Orchis laxiflora* Lam. - *Inform. Bot. Ital.* **13(2-3)**: 119-121.

SCRUGLI, A. 1984: Numeri cromosomici per la Flora Italiana: 868-872. - *Inform. Bot. Ital.* **14(2-3)**: 229-233.

SCRUGLI, A. & BOCCHIERI, E. 1977a: Numeri cromosomici per la Flora Italiana: 263-269. - *Inform. Bot. Ital.* **8(2)**: 216-223.

SCRUGLI, A. & BOCCHIERI, E. 1977b: Numeri cromosomici per la Flora Italiana: 348-357. - *Inform. Bot. Ital.* **9(2)**: 127-133.

SCRUGLI, A., DE MARTIS, B. & MULAS, B. 1976: Numeri cromosomici per la Flora Italiana: 238-249. - *Inform. Bot. Ital.* **8(1)**: 82-91.

SCRUGLI, A. & MOSSA, L. 1972: Sulla esistenza di due varietà di *Ranunculus bullatus* L. Osservazioni cariologiche ed ecologiche su materiale della Sardegna. - *Giorn. Bot. Ital., n.s.* **106(1)**: 11-19.

SCRUGLI, A., MOSSA, L. & GRASSO, M.P. 1983: *Ajuga orientalis* L. (Labiatae) nuova per la flora di Sardegna. Considerazioni fitogeografiche. - *Candollea* **38(1)**: 125-129.

SCRUGLI, A., MULAS, B. & DE MARTIS, B. 1975: Numeri cromosomici per la Flora Italiana: 190-195. - *Inform. Bot. Ital.* **6(3)**: 312-318.

SCURFIELD, G. 1954: Biological Flora of the British Isles. *Deschampsia flexuosa* (L.) Trin. - *J. Ecol.* **42 (1)**: 225-233.

SCURFIELD, G. 1962: Biological Flora of the British Isles. *Cardaria draba* (L.) Desv. (*Lepidium draba* L.). - *J. Ecol.* **50(2)**: 489-499.

SEALY, J.R. 1939: *Merendera sobolifera*. - *Bot. Mag.* **162(2)**: tab. 9576 + 3 pp. text.

SEALY, J.R. & WEBB, D.A. 1950: Biological Flora of the British Isles. *Arbutus* L. - *J. Ecol.* **38(1)**: 223-236.

SEAVEY, S.R. & BOUFFORD, D.E. 1983: Observations of chromosomes in *Circaea* (Onagraceae). - *Amer. J. Bot.* **70(10)**: 1476-1481.

SEAVEY, S.R. & RAVEN, P.H. 1977a: Chromosomal evolution in *Epilobium* sect. *Epilobium* (Onagraceae). - *Pl. Syst. Evol.* **127(2-3)**: 107-119.

SEAVEY, S.R. & RAVEN, P.H. 1977b: Chromosomal evolution in *Epilobium* sect. *Epilobium* (Onagraceae), II. - *Pl. Syst. Evol.* **128(3-4)**: 195-200.

SEAVEY, S.R. & RAVEN, P.H. 1978: Chromosomal evolution in *Epilobium* sect. *Epilobium* (Onagraceae), III. - *Pl. Syst. Evol.* **130(1-2)**: 79-83.

SEBALD, O. 1979: Liste der auf einer Reise vom 30.4 bis 28.5. 1979 in Griechenland gesammelten bzw. beobachteten Pflanzenarten. [Unpublished flora list]. - Photocopy: 16 pp.

SEBERG, O. & AL. 1991: *Peridictyon*, a new genus from the Balkan peninsula, and its relationship with *Festucopsis* (Poaceae). - *Willdenowia* **21(1-2)**: 87-104.

SECKEL, B.J. 1985: Laat bloeiende orchideen op Rodos [In Dutch]. - *Orchideeën* **47(6)**: 200-204.

SECKEL, B.J. 1987: Laat bloeiende orchideen op Lesbos [In Dutch]. - *Orchideeën* **49(5)**: 168-170.

SEÇMEN, Ö. & LEBLEBICI, E. 1984: Aquatic flora of Western Anatolia. - *Willdenowia* **14(1)**: 165-178. [Supplement in Willdenowia 14(2): 570, 1985].

SEÇMEN, Ö. & LEBLEBICI, E. 1991: Aquatic flora of Thrace (Turkey). - *Willdenowia* **20(1-2)**: 53-66. [Also distributed as *OPTIMA leaflet* no. 212].

SEÇMEN, Ö. 1977: The flora and vegetation analysis of Gökçeada and Bozcaara islands. [In Turkish with English summary]. - Izmir: 167 pp.

SEEHOLZER, CH. 1993: Biosystematische Untersuchungen an schweizerischen *Drosera*-Arten. - *Bot. Helv.* **103(1)**: 39-53.

SEETHARAM, A. 1972: Interspecific hybridization in *Linum*. - *Euphytica* **21**: 489-495.

SEETHARAM, A. & SRINIVASACHAR, D. 1972: Cytomorphological studies in the genus *Linum*. - *Cytologia* **37(4)**: 661-671.

SEFEROVA, I.V. 1995: The synopsis of the system of genus *Cicer* (Fabaceae) [In Russian with English summary]. - *Bot. Zhurn. (Moscow & St. Petersburg)* **80(8)**: 96-104.

SEGELBERG, I. 1966: *Erodium hirtum* (Forsk.) Willd. in Crete. - *Bot. Not.* **119(2)**: 373-375.

ŠEGOTA, V. & AL. 2019: Status assessment of the rare aquatic plant *Groenlandia densa* (L.) Fourr. (Potamogetonaceae) in the Western Balkans. - *Botany Letters* **166(2)**: 125-133.

SEISER, K. 2002: *Ophrys*-Funde auf Naxos und der Peloponnes. - *Ber. Arbeitskreis. Heimische Orchid.* **18(2)**: 124-125.

SEISUMS, A. 1998a: Identity and typification of *Allium magicum, A. nigrum*, and *A. roseum* (Alliaceae). - *Taxon* **47(3)**: 711-716.

SEISUMS, A. 1998b: (1369) Proposal to conserve the name *Allium nigrum*, with a conserved type, against *A. magicum* (Liliaceae). - *Taxon* **47(3)**: 745-746. [approved, cf. Taxon 49(2): 274, 2000].

SEITZ, W. 1969: Die Taxonomie der *Aconitum napellus*-Gruppe in Europa. - *Feddes Repert.* **80(1)**: 1-76.

SEITZ, W. 1970: Zytotaxonomische Studien an Valerianaceen I. - *Ber. Deutsch. Bot. Ges.* **82(10-11)**: 651-655.

SEITZ, W. 1972: Beitrag zur Zytotaxonomie von *Ranunculus platanifolius* L. und *R. aconitifolius* L. - *Beitr. Biol. Pflanzen* **48(2)**: 255-264.

SEITZ, W. 1978: Botanische Besonderheiten von der Insel Kreta. - *Jahrb. Vereins Schutz Bergwelt* **43**: 197-217.

SEITZ, W., ZINSMEISTER, H.D. & ABICHT, M. 1972: Beitrag zur Systematik der Gattung *Aconitum* in Europa. - *Bot. Jahrb. Syst.* **92(4)**: 490-507.

SEKOVSKI, Ž. 1993: Chromosome atlas of some Macedonian angiosperms. IX. - *Godišen Zborn. Biol. Fak. Univ. "Kiril i Metodij"* **45**: 207-222.

SEKOVSKI, Ž. & AL. 2007: Cytotaxonomical analysis of *Scorzonera mariovoensis* Micevski. In Matevski, V. (ed.): Collection of papers devoted to Academician Kiril Micevski, pp.133-140. - Skopje: Macedonian Academy of Sciences and Arts.

SELIK, M. 1959: *Pinus brutia* in der Türkei. - *Forstwiss. Centralbl.* **78(1-2)**: 43-58.

SELISKY, H. 2002: *Ophrys aegaea* auf den Kykladeninseln Amorgos und Iraklia. - *J. Eur. Orch.* **34(4)**: 739-742.

SELL, P.D. 1963a: Notes on the European species of *Scleranthus*. In Heywood, V.H. (ed.): Notulae systematicae ad Floram Europaeam spectantes, no. 2. - *Feddes Repert. Spec. Nov. Regni Veg.* **68 (3)**: 167-169.

SELL, P.D. 1963b: Taxonomic and nomenclatural notes on European *Fumaria* species. In Heywood, V.H. (ed.): Notulae systematicae ad Floram Europaeam spectantes, no. 2. - *Feddes Repert. Spec. Nov. Regni Veg.* **68(3)**: 174-178.

SELL, P.D. 1976: Taxonomic and nomenclatural notes on the Compositae subfam. Cichorioideae. In Heywood, V.H. (ed.): Notulae systematicae ad Floram Europaeam spectantes, no. 19. - *Bot. J. Linn. Soc.* **71(4)**: 236-267.

SELL, P.D. 1981: *Lapsana intermedia* Bieb. or *Lapsana communis* L. subsp. *intermedia* (Bieb.) Hayek? - *Watsonia* **13(4)**: 299-302.

SELL, P.D. 1991: A new subspecies of *Ranunculus ficaria* L. In Chater, A.O. (ed.): Flora Europaea. Notulae systematicae ad Floram Europaeam spectantes. Series 2. No. 4. - *Bot. J. Linn. Soc.* **106(2)**: 117-118.

SELL, P.D. 1994: *Ranunculus ficaria* L. sensu lato. - *Watsonia* **20(1)**: 41-50.

SELL, P.D. & WHITEHEAD, F.H. 1964: Notes on the annual species of *Cerastium* in Europe. In Heywood, V.H. (ed.): Notulae systematicae ad Floram Europaeam spectantes, no. 3. – *Feddes Repert. Spec. Nov. Regni Veg.* **69(1)**: 14-24.

SELVI, F. 1998: Floral biometrics of the *Anchusa undulata* L. group (Boraginaceae) from the central- eastern Mediterranean. - *Bot. J. Linn. Soc.* **128(3)**: 251-270.

SELVI, F. 2008: Epitypification and identity of *Cynoglossum montanum* L. (Boraginaceae). - *Taxon* **57 (1)**: 295-299.

SELVI, F. 2009: New findings of *Anthemis cretica* (Asteraceae) on serpentine outcrops of Tuscany (C Italy). - *Fl. Medit.* **19**: 119-128.

SELVI, F. 2011: *Paramoltkia doerfleri* (Wettst.) Greuter & Burdet. In Greuter, W. & Raus, Th. (eds): Med-Checklist notulae, 30. - *Willdenowia* **41(2)**: 312.

SELVI, F. & AL. 2004: The phylogenetic relationships of *Cynoglottis* (Boraginaceae-Boragineae) inferred from ITS, 5.8S and trnL sequences. - *Pl. Syst. Evol.* **246(3-4)**: 195-209.

SELVI, F. & AL. 2006: Molecular phylogeny, morphology and taxonomic re-circumscription of the generic complex *Nonea/Elizaldia/Pulmonaria/Paraskevia* (Boraginaceae-Boragineae). - *Taxon* **55 (4)**: 907-918.

SELVI, F. & BIGAZZI, M. 1998: *Anchusa* L. and allied genera (Boraginaceae) in Italy. - *Pl. Biosyst.* **132 (2)**: 113-142.

SELVI, F. & BIGAZZI, M. 1999: *Nonea pallens* (Boraginaceae), a new addition to the flora of Turkey. - *Edinburgh J. Bot.* **56(3)**: 361-369.

SELVI, F. & BIGAZZI, M. 2000: Removal of *Anchusa macedonica* from *Anchusa* (Boraginaceae): evidence from phenetics and karyotypic analysis. - *Taxon* **49(4)**: 765-778.

SELVI, F. & BIGAZZI, M. 2001: The *Nonea pulla* group (Boraginaceae) in Turkey. - *Pl. Syst. Evol.* **227(1-2)**: 1-26.

SELVI, F. & BIGAZZI, M. 2002: Chromosome studies in Turkish species of *Nonea* (Boraginaceae): the role of polyploidy and descending dysploidy in the evolution of the genus. - *Edinburgh J. Bot.* **59(3)**: 405-420.

SELVI, F. & BIGAZZI, M. 2003: Revision of the genus *Anchusa* (Boraginaceae-Boragineae) in Greece. - *Bot. J. Linn. Soc.* **142(2)**: 431-454.

SELVI, F., BIGAZZI, M. & BRUNO, F. 1997: *Anchusa formosa* (Boraginaceae), a new species from Southern Sardinia (Italy). - *Pl. Biosyst.* **131(2)**: 103-111.

SELVI, F. & CECCHI, L. 2009: Typification of names of Euro-Mediterranean taxa of Boraginaceae described by Italian botanists. - *Taxon* **58(2)**: 621-626.

SELVI, F., CECCHI, L. & COPPI, A. 2009: Phylogeny, karyotype evolution and taxonomy of *Cerinthe* L. (Boraginaceae). - *Taxon* **58(4)**: 1307-1325.

SELVI, F. & GREUTER, W. 2012: (2056) Proposal to conserve the name *Myosotis sicula* against *M. gussonei* (Boraginaceae). - *Taxon* **61(2)**: 467-468. [approved, cf. Taxon 62(6): 1318, 2013].

SELVI, F., HILGER, H.H. & BIGAZZI, M. 2001: Recent advances in the systematics of the tribe Boragineae (Boraginaceae). In Raimondo, F.M. (ed.): Abstracts from the X OPTIMA meeting, Palermo, 13-19 September 2001, p. 11.

SELVI, F. & JARVIS, C.E. 2011: Typification of the name *Cynoglossum creticum* Mill. (Boraginaceae). - *Taxon* **60(5)**: 1477.

SELVI, F., NARDI, E. & BIGAZZI, M. 1996: The ultimate types of *Anchusa* L. and *Lycopsis* L. (Boraginaceae). - *Taxon* **45(2)**: 305-307.

SELVI, F., PAPINI, A. & BIGAZZI, M. 2002: Systematics of *Nonea* (Boraginaceae-Boragineae): new insights from phenetic and cladistic analyses. - *Taxon* **51(4)**: 719-730.

SELVI, F. & SUTORÝ, K. 2012: A synopsis of the genus *Cynoglossum* (Boraginaceae-Cynoglosseae) in Italy. - *Pl. Biosyst.* **146(2)**: 461-479.

SENAY, P. 1936: Contribution à l'étude du genre *Arctium*. - *Bull. Soc. Bot. France* **83**: 330-343.

SENAY, P. 1950: Le groupe des *Carex flava* et *C. Oederi*. - *Bull. Mus. Natl. Hist. Nat., Sér. 2* **22(5-6)**: 618-624, 790-796.

SENAY, P. 1951: Le groupe des *Carex flava* et *C. oederi* (suite et fin). - *Bull. Mus. Natl. Hist. Nat., Sér. 2* **23(1)**: 146-152.

SENCER, H.A. & HAWKES, J.G. 1980: On the origin of cultivated rye. - *Biol. J. Linn. Soc.* **13(4)**: 299-313.

SENESSE, G. 1973: Recherche de caractères distinctifs entre deux taxums de l'espèce collective *Asplenium adiantum-nigrum*: *A. adiantum-nigrum* L. et *A. onopteris* L. - Leur intérêt pour l'étude de ce complexe. - *Naturalia Monspel., Sér. Bot.* **23-24**: 231-266 + 3 plates.

SENGHAS, K. 1968: Taxonomische Übersicht der Gattung *Dactylorhiza* Necker ex Nevski. In Senghas, K. & Sundermann, H. (eds): Probleme der Orchideengattung *Dactylorhiza*. - *Jahresber. Naturwiss. Vereins Wuppertal* **21-22**: 32-67.

SENGHAS, K. 1970: Übersicht zur Systematik und Taxonomie der Gattung *Epipactis*. In Senghas, K. & Sundermann, H. (eds): Probleme der Orchideengattung *Epipactis*. - *Jahresber. Naturwiss. Vereins Wuppertal* **23**: 26-37.

SENGHAS, K. 1972a: *Leucorchis* versus *Pseudorchis*. - *Orchidee (Hamburg)* **23(5)**: 203-205.

SENGHAS, K. 1972b: Die europäischen und mediterranen Sippen der Gattung *Orchis*. - *Jahresber. Naturwiss. Vereins Wuppertal* **25**: 18-21.

SENGHAS, K. & SUNDERMANN, H. (EDS) 1968: Probleme der Orchideengattung *Dactylorhiza*. - *Jahresber. Naturwiss. Vereins Wuppertal* **21-22**: 138 pp. + 4 plates.

SENGHAS, K. & SUNDERMANN, H. (EDS) 1970: Probleme der Orchideengattung *Epipactis*. - *Jahresber. Naturwiss. Vereins Wuppertal* **23**: 132 pp. + 1 plate.

SENN, H.A. 1938: Experimental data for a revision of the genus *Lathyrus*. - *Amer. J. Bot.* **25(2)**: 67-78.

SENNIKOV, A. 2011: (2014) Proposal to conserve the name *Mespilus tomentosa* (*Cotoneaster tomentosus*) against *Mespilus orientalis* (Rosaceae). - *Taxon* **60(2)**: 606. [approved, cf. Taxon 61(5): 1112, 2012].

SENNIKOV, A.N. 2016: (2478) Proposal to conserve the name *Myriophyllum spicatum* (Haloragaceae) with a conserved type. - *Taxon* **65(5)**: 1178-1179.

SENNIKOV, A.N. & SOMLYAY, L. 2011: Atlas Florae Europaeae notes 17: Typification of *Cotoneaster tomentosus* (Rosaceae) and its synonyms. - *Taxon* **60(2)**: 579-584.

SERAFIMOV, S.S. 1983: *Amygdalus ×delipavlovii* sp. hybrida [In Bulgarian with English abstract]. In Velčev, V.I. & al. (eds): Third national conference of botany, pp. 175-182. - Sofia: Publishing House of the Bulgarian Academy of Sciences.

SEREGIN, A.P. & TAN, KIT 2011: Reports 86-102. In Vladimirov, V. & al. (eds): New floristic records in the Balkans: 15. - *Phytol. Balcan.* **17(1)**: 149-150.

SERGUEEFF, M. 1908: Répartition géographique du genre *Iberis* L. - *Bull. Herb. Boissier, Ser. 2* **8(9)**: 609-622.

SERRA, CH. & AL. 2001: Reports (1254-1263). In Kamari, G., Blanché, C. & Garbari, F. (eds): Mediterranean chromosome number reports. - *Fl. Medit.* **11**: 466-473.

SERVAIS, P. & SEBA, P. 2017: Tilo Botanica. Flora of Tilos and of the Dodecanese. - Attert, Belgium: 376 pp.

SERVICE, N. 1984: *Iris unguicularis*. Observations on some Greek forms of the species. - *Iris Year Book* **1984**: 58-62.

SERVICE, N. 1988: Bearded Irises: comments on the species in section *Iris*. - *Plantsman* **10(1)**: 6-26.

SERVICE, N. 1990: Winter-flowering irises: series *Unguiculares*. - *Plantsman* **12(1)**: 1-9 + 1 plate.

SERVICE, N. 1996: Natural variation in *Hermodactylus tuberosus*, the widow iris. - *New Plantsman* **3 (2)**: 75-79.

SERVICE, N. 1999: A new subspecies of *Iris pseudopumila*. - *New Plantsman* **6(3)**: 186-188.

SERVICE, N. 2002: On the distribution of *Iris germanica*. - *Ann. Bot., n.s. (Rome)* **1(2)**: 177.
SESSA, E.B. & AL. 2018: Evolution and reciprocal origins in Mediterranean ferns: The *Asplenium obovatum* and *A. adiantum-nigrum* complexes. - *Ann. Missouri Bot. Gard.* **103(2)**: 175-187.
SETCHELL, W.A. 1946: The genus *Ruppia*. - *Proc. Calif. Acad. Sci., Ser. 4* **25(18)**: 469-477.
SEYBOLD, S. 1981: *Leucorchis* oder *Pseudorchis* - ein ungelöstes Problem? - *Mitt. Arbeitskreis Heimische Orchid. Baden-Württemberg* **13(2)**: 157-163.
SFIKAS, G. 1974: Flowers of Mount Olympos. - Athine: Ekdosis Adelf. Tolidi, 63 pp .
SFIKAS, G. 1976: Wild flowers of Greece. - Athens: Efstathiadis group, 125 pp.
SFIKAS, G. 1978a: Simvoli stin erevna tou genous *Viola* ston elliniko choro. Meros A [In Greek]. - Athine: 8 pp.
SFIKAS, G. 1978b: *Geum reptans*, a new species for the Greek flora [In Greek with English summary]. - *Fisis [Journal of the Hellenic Society for the Protection of Nature]* **14**: 15, 41.
SFIKAS, G. 1978c: Simvoli stin erevna tou genous *Viola* ston elliniko choro. Meros C [In Greek]. - Athine: 10 pp.
SFIKAS, G. 1978d: The existence of *Viola pseudomirabilis* in Greece [In Greek with English summary]. - *Fisis [Journal of the Hellenic Society for the Protection of Nature]* **15**: 13-15, 37-38.
SFIKAS, G. 1978e: Self-propagating trees and shrubs of Greece. With 400 illustrations in colour by the author. English version by E. Sutton [New edition 1991; also in German]. - Athens: Efstathiadis group, 213 pp.
SFIKAS, G. 1978f: *Lilium chalcedonicum* and *Lilium heldreichii* [In Greek with English summary]. - *Fisis [Journal of the Hellenic Society for the Protection of Nature]* **16**: 9-14, 27-28.
SFIKAS, G. 1979a: The mountains of Greece. - Athens: Efstathiadis group, 204 pp.
SFIKAS, G. 1979b: Medicinal plants of Greece. - Athens: Efstathiadis group, 142 pp.
SFIKAS, G. 1979c: Threatened plants of our mountains [In Greek with English summary]. – *Fisis [Journal of the Hellenic Society for the Protection of Nature]* **18**: 11-13, 38, 42-44.
SFIKAS, G. 1980: I elliniki galanthi [In Greek]. - *Fisi ke Zoi* **Summer 1980**: 14-15.
SFIKAS, G. 1982: T'agriolouloudia ton vounon mas. Agriomenexesed ki agriopansedes (genos *Viola*) [In Greek]. - Athine: 21 pp.
SFIKAS, G. 1983a: Louloudia ton ellinikon vounon. *Viola rupestris* [In Greek]. - *Korphes* **1983**: 27.
SFIKAS, G. 1983b: Louloudia ton ellinikon vounon. Nees thesis gnoston tis ellinikis chloridas [In Greek]. - *Korphes* **Nov.-Dec. 1983**: 1 p.
SFIKAS, G. 1983c: Merika spania fita ton ellinikon vounon [In Greek]. - *Ikologia ke Perivallon [Ecology and Environment]* **May/June 1983**: 56-57.
SFIKAS, G. 1983d: T'agriolouloudia ton vounon mas. Nees thesis tis *Achillea abrotanoides* sta ellinika vouna [In Greek]. - *Vouna* **January 1983**: 4-5.
SFIKAS, G. 1983e: Louloudia ton ellinikon vounon. *Crocus veluchensis* [In Greek]. - *Korphes* **July- August**: 45.
SFIKAS, G. 1983f: Louloudia ton ellinikon vounon. *Campanula bononiensis* [In Greek]. – *Korphes* **1983**: 45.
SFIKAS, G. 1984a: Some plants of Mt. Orvilos [In Greek with English summary]. - *Fisis [Journal of the Hellenic Society for the Protection of Nature]* **27**: 12-14, 33.
SFIKAS, G. 1984b: Louloudia ton ellinikon vounon. *Colchicum bivonae*, *Colchicum variegatum*, *C. macrophyllum* [In Greek]. - *Korphes* **Jan. 1984**: 45.
SFIKAS, G. 1984c: Ta fita tou Falakrou. Simvoli sti meleti tis chloridas tou [In Greek]. - Drama: Ekdosi E.O.S. Dramas, 20 pp.
SFIKAS, G. 1984d: Louloudia ton ellinikon vounon. Apo ti chlorida tis Zagradenias [In Greek]. - *Korphes* **Sept.-Oct. 1984**: 41.
SFIKAS, G. 1984e: Ta fita tis ditikis Rodopis [In Greek]. - *Anavassi* **11**: 32-33.
SFIKAS, G. 1984f: Apo ti chlorida tis Tzenas [In Greek]. - Korphes **Jul.-Aug.** 1984: 44.
SFIKAS, G. 1984g: Zagori. Piso ap'ta vouna [In Greek]. - Athine. 80 pp.
SFIKAS, G. 1984h: Flowers of Greece. With 112 Greek wild flowers, illustrated by the author. Translated by N. Fenn-Casitoris. - Athens: E. Mitsialis Ltd., 158 pp.
SFIKAS, G. 1985a: Louloudia ton ellinikon vounon. *Colchicum boissieri*, *Colchicum psaridis* [In Greek]. - *Korphes* **May-June 1985**: 45.
SFIKAS, G. 1985b: The flora of the National Park of Pindos Range. - Photocopy: 19 pp.
SFIKAS, G. 1985c: I elliniki fisi mesa stous eones [In Greek]. - Athine: 112 pp.
SFIKAS, G. 1985d: Violas in Greece [In Greek with English summary]. - *Fisis [Journal of the Hellenic Society for the Protection of Nature]* **29**: 15-17, 24.
SFIKAS, G. 1985e: Parnon - A mountain of the Peloponnesus with rare plants [In Greek with English summary]. - *Fisis [Journal of the Hellenic Society for the Protection of Nature]* **31**: 17-18, 40.
SFIKAS, G. 1985f: I chlorida tis ditikis Rodopis. Simvoli sti meleti tis [In Greek]. - Drama: Ekdosi E.O. S, Dramas: 16 pp.
SFIKAS, G. 1986a: Greek flora [In Greek with English summary]. - *Fisis [Journal of the Hellenic Society for the Protection of Nature]* **32**: 8-11, 27-28.
SFIKAS, G. 1986b: About the flora of Oitis [In Greek with English summary]. - *Fisis [Journal of the Hellenic Society for the Protection of Nature]* **33**: 9-19, 47.
SFIKAS, G. 1986c: Nees prosfiges sti chlorida tis Kritis [In Greek]. - *Fisis [Journal of the Hellenic Society for the Protection of Nature]* **33**: 20-21.
SFIKAS, G. 1986d: The species of *Viola* in Greece and the studies of M. Erben [In Greek with English summary]. - *Fisis [Journal of the Hellenic Society for the Protection of Nature]* **34**: 19-23, 60.
SFIKAS, G. 1986e: I exaplosi trion endimikon ellinikon Krokon [In Greek]. - *Fisis [Journal of the Hellenic Society for the Protection of Nature]* **35**: 21-24.
SFIKAS, G. 1987a: Nees thesis gnoston fiton ton ellinikon vounon [In Greek]. - *Fisis [Journal of the Hellenic Society for the Protection of Nature]* **36**: 13-19.
SFIKAS, G. 1987b: Nees thesis gnoston fiton ton ellinikon vounon. Geni *Daphne*, *Linum*, *Geum*, *Aubrieta*, *Arabis*, *Aurinia* ke *Malkolmia* [sic!] [In Greek]. - *Fisis [Journal of the Hellenic Society for the Protection of Nature]* **39**: 31-32.
SFIKAS, G. 1988a: Vulnerable plants of Greece. Part I [In Greek with English summary]. – *Fisis [Journal of the Hellenic Society for the Protection of Nature]* **40**: 15-20, 58-60.
SFIKAS, G. 1988b: Vulnerable plants of Greece. Part II [In Greek with English summary]. – *Fisis [Journal of the Hellenic Society for the Protection of Nature]* **41**: 13-15, 35-37.
SFIKAS, G. 1988c: The three forms of *Lilium candidum* in Greece [In Greek with English summary]. - *Fisis [Journal of the Hellenic Society for the Protection of Nature]* **42**: 10-11, 31.
SFIKAS, G. 1988d: Colchicums of Greece [In Greek with English summary]. - *Fisis [Journal of the Hellenic Society for the Protection of Nature]* **43**: 16-23, 36.
SFIKAS, G. 1989a: Botanical research in Cephallonia and Ithaca [In Greek with English summary]. - *Fisis [Journal of the Hellenic Society for the Protection of Nature]* **46/47**: 10-11, 36.
SFIKAS, G. 1989b: Die wilden Blumen Kretas. - Athens: Efstathiadis group, 320 pp.
SFIKAS, G. 1991a: Great naturalists who have worked in Greece. Theodore Heldreich [In Greek with English summary]. - *Fisis [Journal of the Hellenic Society for the Protection of Nature]* **51**: 26-27, 36-37.
SFIKAS, G. 1991b: Great naturalists who have worked in Greece. Theodore Orfanides [In Greek with English summary]. - *Fisis [Journal of the Hellenic Society for the Protection of Nature]* **52**: 31-32, 42.
SFIKAS, G. 1991c: *Pinus montana* on Orvilos [In Greek with English summary]. - *Fisis [Journal of the Hellenic Society for the Protection of Nature]* **52**: 40, 46.
SFIKAS, G. 1992a: The subspecies of *Cyclamen repandum* in Greece. A proposal for taxonomic revision. - *Fisis [Journal of the Hellenic Society for the Protection of Nature]* **56**: 14-15, 24-27.
SFIKAS, G. 1992b: Great naturalists who have worked in Greece. Constantinos Goulimis [In Greek with English summary]. - *Fisis [Journal of the Hellenic Society for the Protection of Nature]* **56**: 20-21, 28.

SFIKAS, G. 1994a: New sites for some species of the Greek flora. - *Anthophoros [privately distributed newsletter]* **1-1994**: 3-4.

SFIKAS, G. 1994b: *Leucojum valentinum* on the island of Ithaki. - *Anthophoros [privately distributed newsletter]* **2-1994**: 1-2.

SFIKAS, G. 1994c: New species for the flora of Kefalonia. - *Anthophoros [privately distributed newsletter]* **2-1994**: 3.

SFIKAS, G. 1994d: [Various records for *Tulipa*, *Narcissus* and *Lilium*]. - *Anthophoros [privately distributed newsletter]* **3-1994**: 1-2.

SFIKAS, G. 1994e: *Phacelia tanacetifolia* Benth. A new species for the Greek flora on Hios. - *Anthophoros [privately distributed newsletter]* **3-1994**: 3.

SFIKAS, G. 1994f: An unknown type of *Campanula* in the Moni Kapsa gorge (southeastern Crete). - *Anthophoros [privately distributed newsletter]* **3-1994**: 3-4.

SFIKAS, G. 1994g: The tulips of Hios. - *Anthophoros [privately distributed newsletter]* **4-1994**: 1-3.

SFIKAS, G. 1994h: Wild flowers of Cyprus. - Athens: Efstathiadis group S.A., 320 pp.

SFIKAS, G. 1995a: The Mt. Parnon *Viola*. - *Anthophoros [privately distributed newsletter]* **2-1995**: 2-4.

SFIKAS, G. 1995b: The ranges of some species of the *Colchicum* genus. - *Anthophoros [privately distributed newsletter]* **2-1995**: 4-5.

SFIKAS, G. 1995c: *Lilium chalcedonicum* and *Paeonia mascula* subsp. *hellenica* on Mt. Parnitha. - *Anthophoros [privately distributed newsletter]* **2-1995**: 6.

SFIKAS, G. 1995d: The *Hypericum* genus and the herbarium of Anthoforos. - *Anthophoros [privately distributed newsletter]* **2-1995**: 6-8.

SFIKAS, G. 1995e: New sites for *Campanula rupestris* Sibth. & Sm. and *Campanula thessala* Maire. - *Anthophoros [privately distributed newsletter]* **3-1995**: 2-4.

SFIKAS, G. 1995f: *Colchicum sfikasianum*. - *Anthophoros [privately distributed newsletter]* **3-1995**: 5.

SFIKAS, G. 1995g: *Antirrhinum siculum* Miller in Greece. - *Anthophoros [privately distributed newsletter]* **3-1995**: 6-7.

SFIKAS, G. 1995h: *Impatiens balfourii* growing wild on Mt. Pelion. - *Anthophoros [privately distributed newsletter]* **3-1995**: 6-7.

SFIKAS, G. 1995i: Three species of *Crocus*, new to the island of Hios. - *Anthophoros [privately distributed newsletter]* **3-1995**: 8.

SFIKAS, G. 1996a: Information on rare species of the Greek flora. - *Anthophoros [privately distributed newsletter]* **1-1996**: 1-8.

SFIKAS, G. 1996b: I avtofiis toulipes tis patridas mas [In Greek]. - *Fisis [Journal of the Hellenic Society for the Protection of Nature]* **73**: 3-7.

SFIKAS, G. 1996c: I agrii krini (The wild lilies of Greece). - *Fisis [Journal of the Hellenic Society for the Protection of Nature]* **74**: 3-6, 40-41.

SFIKAS, G. 1996d: New sites for species included in the Red Data Book on the Greek flora. - *Anthophoros [privately distributed newsletter]* **2-1996**: 2.

SFIKAS, G. 1996e: New sites for known species of the Greek flora. - *Anthophoros [privately distributed newsletter]* **2-1996**: 3-4.

SFIKAS, G. 1997a: Popular names of the plants of Mani [based on Zaganiaris 1933]. - *Anthophoros [privately distributed newsletter]* **2/3-1997**: 1-14.

SFIKAS, G. 1997b: Ekthesi gia ti chlorida tis periochis giro apo tin techniti limni Plastira (Megdova) tou nomou Karditsas 1996 [In Greek]. - Privately distributed report, 30 pp.

SFIKAS, G. 1997c: Phytogeographical information on some species of the Greek flora. – *Anthophoros [privately distributed newsletter]* **4-1997**: 2-4.

SFIKAS, G. 1997d: The Cyclamens [In Greek with English summary]. - *Fisis [Journal of the Hellenic Society for the Protection of Nature]* **77**: 3-6, 47.

SFIKAS, G. 1997e: The Greek Crocuses [In Greek with English summary]. - *Fisis [Journal of the Hellenic Society for the Protection of Nature]* **79**: 23-30, 45-47.

SFIKAS, G. 1997f: Ta nea tis ellinikis chloridas [In Greek]. - *Fisis [Journal of the Hellenic Society for the Protection of Nature]* **79**: 37-38.

SFIKAS, G. 1998a: The Greek oaks [In Greek with English summary]. - *Fisis [Journal of the Hellenic Society for the Protection of Nature]* **82**: 29-39, 46.

SFIKAS, G. 1998b: Preliminary list of the flora of Mt. Haindou, Rodopi. - *Anthophoros [privately distributed newsletter]* **2-1998**: 1-8.

SFIKAS, G. 1998c: The endemic plants of Crete. - *Anthophoros [privately distributed newsletter]* **3-1998**: 1-7.

SFIKAS, G. 1999a: Orvilos - Falakro. Dio aprostatevti orini viotopi tou Nomou Dramas [In Greek]. - *Fisis [Journal of the Hellenic Society for the Protection of Nature]* **84**: 19-22.

SFIKAS, G. 1999b: Ta fita tou Imittou [In Greek]. - Privately distributed report, 14 pp.

SFIKAS, G. 1999c: The Greek Narcissi [In Greek with English summary]. - *Fisis [Journal of the Hellenic Society for the Protection of Nature]* **85**: 4-6, 47.

SFIKAS, G. 2000: Ta nea tis ellinikis chloridas [In Greek]. - *Fisis [Journal of the Hellenic Society for the Protection of Nature]* **90**: 19-22.

SFIKAS, G. 2001a: I ellinikes anemones [In Greek]. - *Fisis [Journal of the Hellenic Society for the Protection of Nature]* **92**: 23-27.

SFIKAS, G. 2001b: *Viola vourinensis*. - *Fisis [Journal of the Hellenic Society for the Protection of Nature]* **93**: front cover.

SFIKAS, G. 2001c: Nea tis Ellinikis chloridas [In Greek]. - *Fisis [Journal of the Hellenic Society for the Protection of Nature]* **93**: 46.

SFIKAS, G. 2001d: I agriopansedes stin Ellada. Meros A' [In Greek]. - *Fisis [Journal of the Hellenic Society for the Protection of Nature]* **95**: 23-30.

SFIKAS, G. 2001e: I votaniki paradisi tis Elladas [In Greek]. - Toubis A.E: 256 pp.

SFIKAS, G. 2002a: I agriopansedes stin Ellada. Meros B' [In Greek]. - *Fisis [Journal of the Hellenic Society for the Protection of Nature]* **96**: 23-30.

SFIKAS, G. 2002b: I spanii paraliaki viotopi tis Rodou [In Greek]. - *Fisis [Journal of the Hellenic Society for the Protection of Nature]* **97**: 19-22.

SFIKAS, G. 2003a: I paparounes [In Greek]. - *Fisis [Journal of the Hellenic Society for the Protection of Nature]* **100**: 15-18.

SFIKAS, G. 2003b: Neo idos kolchikou sto oros Lirkio [In Greek]. - *Fisis [Journal of the Hellenic Society for the Protection of Nature]* **100**: 39.

SFIKAS, G. 2003c: Nea tis ellinikis chloridas [In Greek]. - *Fisis [Journal of the Hellenic Society for the Protection of Nature]* **103**: 29.

SFIKAS, G. 2005a: Ena ethniko parko sti Rodo [In Greek]. - *Fisis [Journal of the Hellenic Society for the Protection of Nature]* **108**: 4-6.

SFIKAS, G. 2005b: News about Greek flora [In Greek with English summary]. - *Fisis [Journal of the Hellenic Society for the Protection of Nature]* **110**: 17-18.

SFIKAS, G. 2005c: Ta fita tis ikogenias Kouvolboulides stin Ellada. - *Fisis [Journal of the Hellenic Society for the Protection of Nature]* **111**: 23-28.

SFIKAS, G. 2006a: Jean Baptiste Bory de Saint Vincent, enas spoudeos erevnitis tis ellinikis chloridas [In Greek]. - *Fisis [Journal of the Hellenic Society for the Protection of Nature]* **112**: 4-6.

SFIKAS, G. 2006b: Nea topothesia tou *Lilium rhodopaeum* sto oros Papikio [In Greek]. - *Fisis [Journal of the Hellenic Society for the Protection of Nature]* **113**: 29.

SFIKAS, G. 2007: Votanikes exerevnisis sto Chelmo. Akolouthontas ta vimata tou Orphanidi ke tou Goulimi [In Greek]. - *Fisis [Journal of the Hellenic Society for the Protection of Nature]* **117**: 4-7.

SFIKAS, G. 2008a: I monokarpikes kampanoules ton vrachon stin Ellada [In Greek]. - Athens: Privately printed, 34 pp.

SFIKAS, G. 2008b: Katalogos fiton tou orous Penteli Attikis [In Greek]. - Athens: Privately printed, 13 pp.

SFIKAS, G. 2008c: Katalogos fiton tis Zakinthou [In Greek]. - Athens: Privately printed, 18 pp.

SFIKAS, G. 2010: Ta kina onomata ton fiton tis Elladas. The common names of the plants of Greece. - Mimeographed list, 75 pp.

SFIKAS, G. 2015: I *Abies cephalonica* ston Imittó [In Greek]. - *Fisis [Journal of the Hellenic Society for the Protection of Nature]* **150**: 48.

SFIKAS, G. & APERGIS, M. 2004: News about Greek flora [In Greek with English summary]. - *Fisis [Journal of the Hellenic Society for the Protection of Nature]* **106**: 20-21.

SFIKAS, G. & APERGIS, M. 2005: Nea tis ellinikis chloridas [In Greek]. - *Fisis [Journal of the Hellenic Society for the Protection of Nature]* **111**: 17.

SFIKAS, G. & SFIKAS, CH. 1995: *Laburnum anagyroides* in the Southern Pindos range. - *Anthophoros [privately distributed newsletter]* **3-1995**: 2.

SFIKAS, G. & TAN, KIT 2007: Nea idi prostithente stin Elliniki chlorida [In Greek]. - *Fisis [Journal of the Hellenic Society for the Protection of Nature]* **117**: 23.

SFIKAS, G. & TAN, KIT 2013: Reports 115-121. In Vladimirov, V., Dane, F. & Tan, Kit (eds): New floristic records in the Balkans: 21. - *Phytol. Balcan.* **19(1)**: 146-147.

SFIKAS, G. (ED.) 2005: Novelties for the Greek flora [In Greek with English summary]. - *Fisis [Journal of the Hellenic Society for the Protection of Nature]* **109**: 20-21.

SHAGARODSKY, T., AGUINAGALDE, I. & SÁNCHEZ-YÉLAMO, M.D. 1989: Infra-specific variability of *Brassica cretica* Lam. In Phitos, D. (ed.): Abstracts. Sixth OPTIMA meeting, Delphi 10-16 Sept. 1989, p. 136.

SHAN, R.H. & CONSTANCE, L. 1951: The genus *Sanicula* (Umbelliferae) in the old world and the new. - *Univ. Calif. Publ. Bot.* **25(1)**: 1-78.

SHARKOVA, S.T. 1997: Reports (767-772). In Kamari, G., Felber, F. & Garbari, F. (eds): Mediterranean chromosome number reports - 6. - *Fl. Medit.* **6**: 328-333.

SHARKOVA, S.T. & PEEV, D.R. 1997: Karyological study of some Bulgarian *Centaurea* (Asteraceae). - *Phytol. Balcan.* **3(2-3)**: 83-87.

SHARMA, A.K. & SHARMA, A. 1961: Chromosome studies of some varieties of *Narcissus tazetta* L. - *Caryologia* **14(1)**: 97-106.

SHARP, D. 2012: In quest of autumn bulbs in the Peloponnese. - *Rock Gard.* **129**: 20-33.

SHAW, D. & TAN, KIT 2015: Reports 80-81. In Vladimirov, V. & al. (eds): New floristic records in the Balkans: 27. - *Phytol. Balcan.* **21(2)**: 205.

SHAW, G.R. 1914: The genus *Pinus*. - *Publ. Arnold Arbor.* **5**: 1-96.

SHAW, H.K.A. 1982: The prefaces of Halácsy's "Conspectus Florae Graecae" and Hayek's "Prodromus Florae Peninsulae Balcanicae". - *Ann. Mus. Goulandris* **5**: 11-21.

SHAW, H.K.A. & TURRILL, W.B. 1926: Revision of Sibthorp's plants at Kew. - *Bull. Misc. Inform. Kew* **1926(3)**: 120-128.

SHEADER, M. & SHEADER, A.-L. 2003: *Onosma erecta* Sibth. & Sm. - *Alpine Gardener* **71(4)**: 413-414.

SHEASBY, P. 2000: The herbaceous flora [of the mountains of northern Greece]. - *Quart. Bull. Alpine Gard. Soc. Gr. Brit.* **68(3)**: 374-379.

SHEPPARD, A.W. 1991: Biological Flora of the British Isles. *Heracleum sphondylium* L. - *J. Ecol.* **79(1)**: 235-258.

SHETLER, S.G. 1982: Variation and evolution of the nearctic harebells (*Campanula* subsect. *Heterophylla*). - *Phanerog. Monogr.* **11**: 1-516.

SHIFMAN, U. 2011: The *Ophrys umbilicata* group in Israel. - *J. Eur. Orch.* **43(2)**: 401-430.

SHIMWELL, D.W. 1980: Studies in the plant sociology of fern-dominated wall communities on the island of Corfu. - *Brit. Fern Gaz.* **12(2)**: 75-82.

SHIRREFFS, D.A. 1985: Biological Flora of the British Isles. *Anemone nemorosa* L. - *J. Ecol.* **73(3)**: 1005-1020.

SHIRREFFS, D.A. 1986: Cytological studies of *Anemone nemorosa* L. (Ranunculaceae). - *Bot. J. Linn. Soc.* **92(3)**: 255-262.

SHIVAS, M.G. 1961a: Contributions to the cytology and taxonomy of species of *Polypodium* in Europe and America. I. Cytology. - *J. Linn. Soc., Bot.* **58(370)**: 13-25 + 3 plates.

SHIVAS, M.G. 1961b: Contributions to the cytology and taxonomy of species of *Polypodium* in Europe and America. II. Taxonomy. - *J. Linn. Soc., Bot.* **58(370)**: 27-38 + 1 plate.

SHIVAS, M.G. 1962: The *Polypodium vulgare* complex. - *Brit. Fern Gaz.* **9(3)**: 65-70.

SHIVAS, M.G. 1969: A cytotaxonomic study of the *Asplenium adiantum-nigrum* complex. - *Brit. Fern Gaz.* **10(2)**: 68-80 + 2 plates.

SHMIDA, A. 1980: Kermes oaks in the land of Israel. - *Israel Land Nat.* **6(1)**: 9-16.

SHMIDA, A. 1981: Red display flowers of the Mediterranean. - *Israel Land Nat.* **6(2)**: 106-115.

SHMIDA, A. 1982: Mediterranean vegetation in California and Israel: Similarities and differences. - *Israel J. Bot.* **30(3)**: 105-123.

SHMIDA, A. & HELLER, D. 1983: Michael Zohary - Giant of Israel Botany. - *Israel Land Nat.* **Fall 1983**: 33-35.

SHMIDA, A. & KOLLMANN, F. 1977: *Allium* species of Mt. Hermon. II. Distribution, variation and polyploidy correlated with vertical zonation. - *Israel J. Bot.* **26(3)**: 149-159.

SHNER, J.V. & AL. 2020: IAPT chromosome data 32/11. - *Taxon* **69(5)**: E19-E21.

SHNER, J.V. & OSTROUMOVA, T.A. 2004: Reports (1387-1390). In Kamari, G., Blanché, C. & Garbari, F. (eds): Mediterranean chromosome number reports - 14. - *Fl. Medit.* **14**: 433-435.

SHNER, J.V., PIMENOV, M.G. & KLJUYKOV, E.V. 2004: Reports (1391-1409). In Kamari, G., Blanché, C. & Garbari, F. (eds): Mediterranean chromosome number reports - 14. - *Fl. Medit.* **14**: 435-447.

SHUKA, L. 2009: New taxonomic data for the flora of Albania recorded on the serpentine substrate (southeast Albania). - *Natura Montenegrina* **8(1)**: 5-10.

SHUKA, L. 2010: *Bellevalia hyacinthoides* (Bertol.) K. Persson, new species for the Albanian flora. - *Natura Montenegrina* **9(3)**: 417-420.

SHUKA, L. & AL. 2007: The genus *Pinguicula* (Lentibulariaceae) in Albania - a critical review. - *Wulfenia* **14**: 15-65.

SHUKA, L., MALO, S. & TAN, KIT 2011: New chorological data and floristic notes for Albania. - *Bot. Serbica* **35(2)**: 157-162.

SHUKA, L. & TAN, KIT 2009: On the distribution of *Cerastium smolikanum* (Caryophyllaceae) and *Centaurea vlachorum* (Asteraceae) in the Balkan Peninsula. - *Phytol. Balcan.* **15(3)**: 347-350.

SHUKA, L. & TAN, KIT 2013: New records for Albania based on taxa from the Prespa National Park. - *Biodiv. Data J.* **1**: e1014, 24 pp.

SHUKA, L., TAN, KIT & KRASNIQI, E. 2012: *Tulipa kosovarica* (Liliaceae), a new species of tulip from Kosovo. - *Phytotaxa* **62**: 1-9.

SHUKA, L., TAN, KIT & ŠILJAK-YAKOVLEV, S. 2010: *Tulipa albanica* (Liliaceae), a new species from northeastern Albania. - *Phytotaxa* **10**: 17-25.

SIARKA, K. & AL. 2018: Digitisation of Dimitrios Zaganiaris' historical botanical collections deposited in TAU (Greece). - *Bot. Serbica* **142 (suppl. 1)**: 179.

SIBILIO, E. 1962: Osservazioni sistematiche e cariologiche sul *Thymus capitatus* Hoffmgg. et Lk. - *Delpinoa, n.s.* **3**: 225-237 + 1 plate.

SIBTHORP, J. & SMITH, J.E. 1806a: Florae Graecae Prodromus. Vol. 1(1) pp. i-xvi, 1-218. - Londini: Typis Richardi Taylor et socii.

SIBTHORP, J. & SMITH, J.E. 1806b: Flora Graeca. Vol. 1(1) pp. i-viii, 1-36 & plates 1-50. - Londini: Typis Richardi Taylor et socii.

SIBTHORP, J. & SMITH, J.E. 1808: Flora Graeca. Vol. 1(2) pp. 37-82 & plates 51-100. - Londini: Typis Richardi Taylor et socii.

SIBTHORP, J. & SMITH, J.E. 1809: Florae Graecae Prodromus. Vol 1(2) pp. 219-442. - Londini: Typis Richardi Taylor et socii.

SIBTHORP, J. & SMITH, J.E. 1813a: Florae Graecae Prodromus. Vol. 2(1) pp. 1-210. - Londini: Typis Richardi Taylor et socii.

SIBTHORP, J. & SMITH, J.E. 1813b: Flora Graeca. Vol. 2(1) pp. 1-40 & plates 101-150. - Londini: Typis Richardi Taylor et socii.

SIBTHORP, J. & SMITH, J.E. 1816a: Florae Graecae Prodromus. Vol 2(2) pp. 211-422. - Londini: Typis Richardi Taylor et socii.

SIBTHORP, J. & SMITH, J.E. 1816b: Flora Graeca. Vol. 2(2) pp. 41-83 & plates 151-200. - Londini: Typis Richardi Taylor et socii.

SIBTHORP, J. & SMITH, J.E. 1819: Flora Graeca. Vol. 3(1) pp. 1-46 & plates 201-250. - Londini: Typis Richardi Taylor et socii.

SIBTHORP, J. & SMITH, J.E. 1821: Flora Graeca. Vol. 3(2) pp. 47-93 & plates 251-300. - Londini: Typis Richardi Taylor et socii.

SIBTHORP, J. & SMITH, J.E. 1823: Flora Graeca. Vol. 4(1) pp. 1-44 & plates 301-350. - Londini: Typis Richardi Taylor et socii.

SIBTHORP, J. & SMITH, J.E. 1824: Flora Graeca. Vol. 4(2) pp. 45-88 & plates 351-400. - Londini: Typis Richardi Taylor et socii.

SIBTHORP, J. & SMITH, J.E. 1825a: Flora Graeca. Vol. 5(1) pp. 1-38 & plates 401-450. - Londini: Typis Richardi Taylor et socii.

SIBTHORP, J. & SMITH, J.E. 1825b: Flora Graeca. Vol. 5(2) pp. 39-81 & plates 451-500. - Londini: Typis Richardi Taylor et socii.

SIBTHORP, J. & SMITH, J.E. 1825c: Flora Graeca. Vol. 6(1) pp. 1-42 & plates 501-550. - Londini: Typis Richardi Taylor et socii.

SIBTHORP, J. & SMITH, J.E. 1827: Flora Graeca. Vol. 6(2) pp. 43-80 & plates 551-600. - Londini: Typis Richardi Taylor et socii.

SIBTHORP, J. & SMITH, J.E. 1831: Flora Graeca. Vol. 7(1) pp. 1-46 & plates 601-650. - Londini: Typis Richardi Taylor et socii.

SIBTHORP, J. & SMITH, J.E. 1832a: Flora Graeca. Vol. 7(2) pp. 47-88 & plates 651-700. - Londini: Typis Richardi Taylor et socii.

SIBTHORP, J. & SMITH, J.E. 1832b: Flora Graeca. Vol. 8(1) pp. 1-36 & plates 701-750 [co-author J. Lindley]. - Londini: Typis Richardi Taylor et socii.

SIBTHORP, J. & SMITH, J.E. 1835: Flora Graeca. Vol. 8(2) pp. 37-75 & plates 751-800 [co-author J. Lindley]. - Londini: Typis Richardi Taylor et socii.

SIBTHORP, J. & SMITH, J.E. 1837: Flora Graeca. Vol. 9(1) pp. 1-38 & plates 801-850 [co-author J. Lindley]. - Londini: Typis Richardi Taylor et socii.

SIBTHORP, J. & SMITH, J.E. 1839: Flora Graeca. Vol. 9(2) pp. 39-77 & plates 851-900 [co-author J. Lindley]. - Londini: Typis Richardi Taylor et socii.

SIBTHORP, J. & SMITH, J.E. 1840a: Flora Graeca. Vol. 10(1) pp. 1-40 & plates 901-950 [co-author J. Lindley]. - Londini: Typis Richardi Taylor et socii.

SIBTHORP, J. & SMITH, J.E. 1840b: Flora Graeca. Vol. 10(2) pp. 41-106 & plates 951-966 [co-author J. Lindley]. - Londini: Typis Richardi Taylor et socii.

ŠÍDA, O. 1998: Taxonomic problems in *Erigeron* sect. *Trimorphia* (Compositae) in Eurasia. – *Preslia* **70(3)**: 259-269.

ŠÍDA, O. 2000: Type material referring to Flora Hellenica deposited in PRC. 1. Caryophyllaceae. - *Novit. Bot. Univ. Carol.* **14**: 25-32.

SIDJIMOVA, B. 2009: Morphometrical variability in Bulgarian *Galanthus elwesii* (Amaryllidaceae). In Ivanova, D. (ed.): Proceedings of the IV Balkan botanical congress, pp. 205-210. - Sofia: Institute of Botany, BAS.

SIDJIMOVA, B.Z. 2021: Genus *Galanthus* (Amaryllidaceae) in Bulgaria: Notes about taxonomy, chorology and ecology. - *Ecol. Balkan.* **13(2)**: 75-93.

SIEBER, F.W. 1818: [Correspondenz aus Creta]. - *Flora* **1**: 269-277.

SIEBER, F.W. 1821: Avis de plantes. Herbarium creticum, aegyptiacum, palaestinense. - Pragae: 13 pp.

SIEBER, F.W. 1822a: Bemerkungen über *Cenchrus frutescens* L., *Eryngium trifolium* Alpin. und *Campanula pelviformis* Lamarck. - *Flora* **5(1)**: 14-16.

SIEBER, F.W. 1822b: Neue und seltene Gewächse. - Flora **5(1)**: 241-248.

SIEBER, F.W. 1823a: Reise nach der Insel Kreta [review]. - *Flora* **6(2)**: 593-606.

SIEBER, F.W. 1823b: Reise nach der Insel Kreta im griechischen Archipelagus im Jahre 1817. Vol. 1. - Leipzig & Sorau: bey Friedrich Fleischer, xxii + 548 pp.

SIEBER, F.W. 1823c: Reise nach der Insel Kreta im griechischen Archipelagus im Jahre 1817. Vol. 2. - Leipzig & Sorau: bey Friedrich Fleischer, iv + 328 pp. + 14 plates.

SIEBER, V.K. & MURRAY, B.G. 1980: The cytology of the genus *Alopecurus* (Gramineae). - *Bot. J. Linn. Soc.* **79(4)**: 343-355.

SIEBER, V.K. & MURRAY, B.G. 1981a: Structural and numerical chromosomal polymorphism in natural populations of *Alopecurus* (Poaceae). - *Pl. Syst. Evol.* **139(1-2)**: 121-136.

SIEBER, V.K. & MURRAY, B.G. 1981b: Hybridization between tetraploid species of *Alopecurus* L. (Poaceae): morphological studies of natural and artificial hybrids. - *Bot. J. Linn. Soc.* **83(4)**: 285-292.

SIEBER, V.K. & MURRAY, B.G. 1981c: Hybridization between tetraploid species of *Alopecurus* L. (Poaceae): chromosome behavior of natural and artificial hybrids. - *Bot. J. Linn. Soc.* **83(4)**: 293-310.

SIERING, G. & HENNIG, B. 1990: Mitteilung zu einem Vorkommen von *Epipactis persica* (Soó) Nannf. im Rila-Gebirge Bulgariens. - *Ber. Arbeitskreis. Heimische Orchid.* **7(2)**: 4-6.

SIGNORINI, M.A. & FOGGI, B. 1997: A survey of the genus *Festuca* L. (Poaceae) in Italy. III. Nomenclatural notes on some *Festuca* belonging to *Festuca violacea*-group. - *Candollea* **52(2)**: 409-427.

SIGNORINI, M.A., FOGGI, B. & MORI, B. 2001: Reports (1264-1270). In Kamari, G., Blanché, C. & Garbari, F. (eds): Mediterranean chromosome number reports. - *Fl. Medit.* **11**: 473-478.

SIGNORINI, M.A. & MORI, B. 1994: Caryological observations on *Anemone pavonina* Lam. (Ranunculaceae) from Tuscany (Italy). - *Caryologia* **47(2)**: 179-182.

ŞIK, L. & KÜÇÜKER, O. 1998: Reports (997-1000). In Kamari, G., Felber, F. & Garbari, F. (eds): Mediterranean chromosome number reports - 8. - *Fl. Medit.* **8**: 296-298.

SILBA, J. 1983: Addendum to a revision of *Cupressus* L. (Cupressaceae). - *Phytologia* **52(5)**: 349-361.

ŠILIĆ, Č. 1967: *Sternbergia colchiciflora* W.K. var. *dalmatica* Rchb. in der Flora Jugoslawiens [In Serbo-Croatian with German summary]. - *Glasn. Zemaljsk. Muz. Sarajevu* **5**: 115-121.

ŠILIĆ, Č. 1968: *Oreoherzogia pumila* (Turra) Vent subsp. *illyrica* Šilić - ein neues endemisches Taxon des westlichen Teiles der Balkan-Halbinsel [In Serbo-Croatian with German summary]. - *Glasn. Zemaljsk. Muz. Sarajevu* **6**: 61-79.

ŠILIĆ, Č. 1972: Beitrag zur Ermittlung des Taxon *Phyllitis scolopendrium* (L.) Newmann f. *cavernarium* Schiffner et Morton in der Höhlenflora Jugoslawiens. - *Wiss. Mitt. Bosnisch-Herzegowinischen Landesmus., C* **2**: 73-81.

ŠILIĆ, Č. 1974: Conspectus generum *Satureja* L., *Calamintha* Moench, *Micromeria* Bentham, *Acinos* Moench et *Clinopodium* L. florae Jugoslaviae. - *Glasn. Zemaljsk. Muz. Sarajevu* **13**: 105-128.

ŠILIĆ, Č. 1978: The spreading and ecology of the species of the genus *Micromeria* Bentham in the flora of Yugoslavia [In Serbo-Croatian with English summary]. - *God. Biol. Inst. Univ. Sarajevu* **31**: 169-182.

ŠILIĆ, Č. 1979: Monographie der Gattungen *Satureja* L., *Calamintha* Miller, *Micromeria* Bentham, *Acinos* Miller und *Clinopodium* L. in der Flora Jugoslawiens [In Serbo-Croatian with German summary]. - Sarajevo: Zemaljski Muzej BiH, 440 pp.

ŠILIĆ, Č. 1990: The taxonomy and horology of the *Rhamnus illyrica* Griseb. ap. Pant. (= *Rh. orbiculata* Bornmüller) and the *Rh. intermedia* Steud. & Hochst. (Rhamnaceae) ... [In Serbo-Croatian with English summary]. - *Glasn. Zemaljsk. Muz. Sarajevu* **29**: 11-38.

ŠILIĆ, Č. & ŠOLIĆ, M.E. 2002: The taxonomy, chorology and ecology of *Stachys menthifolia* Vis. (Lamiaceae) in the north-west part of its distribution area. - *Acta Bot. Croat.* **61(1)**: 51-56.

ŠILJAK-YAKOVLEV, S. 1980: [Chromosome numbers in Jugoslavian plants]. In Löve, Á. (ed.): Chromosome number reports LXVII. - *Taxon* **29(2-3)**: 347.

ŠILJAK-YAKOVLEV, S. 1981: [Chromosome numbers in Jugoslavian plants]. In Löve, Á. (ed.): Chromosome number reports LXXIII. - *Taxon* **30(4)**: 843-844.

ŠILJAK-YAKOVLEV, S. 1982a: Analyse comparative des caryotypes de deux espèces du genre *Reichardia* Roth (*R. macrophylla* Vis. & Pančić et *R. picroides* (L.) Roth) et leur relation taxonomique. - *Caryologia* **34(3)**: 267-274.

ŠILJAK-YAKOVLEV, S. 1982b: [Miscellaneous chromosome numbers, especially from Jugoslavia]. In Löve, Á. (ed.): Chromosome number reports LXXVII. - *Taxon* **31(4)**: 768.

ŠILJAK-YAKOVLEV, S. 1983: Comparative cytogenetic study of diploid and tetraploid populations of the species *Crepis aurea* (L.) Cass. In: Symposium de genetique de populations, pp. 46-49. - Sarajevo.

ŠILJAK-YAKOVLEV, S. 1986: Etude cytologique et palynologique de Compositae endemiques ou reliques de la flore Yougoslave. - Orsay: Univ. de Paris-Sud, thèse, 217 pp. + 52 plates.

ŠILJAK-YAKOVLEV, S. 1993: La dysploidie, un des mecanismes de l'evolution du caryotype. In Petrova, A.V. (ed.): Abstracts from the VII OPTIMA meeting in Borovetz, Bulgaria, 18-30 July 1993, p. 27.

ŠILJAK-YAKOVLEV, S. 1996: La dysploïdie et l'évolution du caryotype. - *Bocconea* **5(1)**: 211-220.

ŠILJAK-YAKOVLEV, S. 2020: Karyological study of diploid and tetraploid *Crepis aurea* (Asteraceae). - *Fl. Medit.* **30**: 456-460.

ŠILJAK-YAKOVLEV, S. & AL. 2008: Genome size variation and polyploidy in the resurrection plant genus *Ramonda*: Cytogeography of living fossils. - *Envir. Exper. Bot.* **62**: 101-112.

ŠILJAK-YAKOVLEV, S. & AL. 2010: Towards a genome size and chromosome number database of Balkan flora: C-values in 343 taxa of with novel values for 242. - Adv. Sci. Lett. (U.S.A.) **3(2)**: 190-213.

ŠILJAK-YAKOVLEV, S. & AL. 2011: Genome size and chromosome number of *Micromeria acropolitana* (Lamiaceae), a steno-endemic from Greece. - *Phytol. Balcan.* **17(2)**: 201-204.

ŠILJAK, S. 1977: [Chromosome numbers in Asteraceae from Jugoslavia]. In Löve, Á. (ed.): IOPB chromosome number reports LVII. - *Taxon* **26(4)**: 447-448.

SILVESTRE, S. 1972: Estudio taxonómico de los géneros *Conopodium* Koch y *Bunium* L. en la Península Ibérica. I. Parte experimental. - *Lagascalia* **2(2)**: 143-173.

SILVESTRE, S. 1973: Estudio taxonómico de los géneros *Conopodium* Koch y *Bunium* L. en la Península Ibérica. II. Parte sistemática. - *Lagascalia* **3(1)**: 3-48.

SILVESTRE, S. 1980: Números cromosómicos para la flora española. 150-154. - *Lagascalia* **9(2)**: 266-269.

SILVESTRE, S. 1984: Números cromosómicos para la flora española. 353-362. - *Lagascalia* **12(2)**: 298-303.

SILVESTRE, S. 1986: Números cromosómicos para la flora española. 435-455. - *Lagascalia* **14(2)**: 272-281.

SILVESTRE, S. 1990: Números cromosómicos para la flora española. 592-602. - *Lagascalia* **15(2)**: 283-288.

SILVESTRE, S. 1993: Números cromosómicos para la flora española. 643-663. - *Lagascalia* **17(1)**: 151-160.

SIMMLER, G. 1910: Monographie der Gattung *Saponaria*. - *Denkschr. Kaiserl. Akad. Wiss., Wien. Math.-Naturwiss. Kl.* **85**: 433-509 + 2 plates.

SIMMONDS, N.W. 1945: Biological Flora of the British Isles. *Polygonum* L. em. Gaertn. - *J. Ecol.* **33(1)**: 117-143.

SIMMONS, J.B. 1976: The botanic garden as a documented living collection and its role in plant resource maintenance. - *OPTIMA Leafl.* **18**: 1-12.

SIMMONS, J.B. 1979: The Cretan zelkova. - *Garden (London)* **104(6)**: 249-251.

SIMOGLOU, K., DORDAS, CH.A. & KOUTSOS, TH.V. 2015: Report 239. In Vladimirov, V., Dane, F. & Tan, Kit (eds): New floristic records in the Balkans: 28. - *Phytol. Balcan.* **21(3)**: 397-398.

SIMOGLOU, K.B., KOUTSOS, T.V. & TAN, KIT 2019: Report 82. In Vladimirov, V. & al. (eds): New floristic records in the Balkans: 38. - *Phytol. Balcan.* **25(1)**: 113.

SIMOLA, L.K. 1964: On the cytology of *Lathyrus pratensis*. - *Ann. Acad. Sci. Fenn., Ser. A, IV, Biol.* **78**: 1-19.

SIMON PALLISÉ, J. 1997: Una nueva subespecie para *Euphorbia flavicoma* DC. (Euphorbiaceae). - *Anales Jard. Bot. Madrid* **55(1)**: 199-200.

SIMON, C. 1978: *Bellium corsicum* n. sp. sowie Notizen über die Gattung *Bellium* (Asteraceae). - *Bauhinia* **6(2)**: 279-284.

SIMÓN, J. & AL. 1996: Reports (481-490). In Kamari, G., Felber, F. & Garbari, F. (eds): Mediterranean chromosome number reports - 5. - *Fl. Medit.* **5**: 323-331.

SIMÓN, J., MOLERO, J. & BLANCHÉ, C. 1992: Fruit and seed morphology of *Euphorbia* aggr. *flavicoma*. Taxonomic implications. - *Collect. Bot. (Barcelona)* **21**: 211-242.

SIMÓN, L.E. 1996: Notas sobre *Chenopodium* L. subgen. *Ambrosia* A.J. Scott (Chenopodiaceae). 1. Taxonomía. 2. Fitogeografía: áreas disyuntas. - *Anales Jard. Bot. Madrid* **54(1)**: 137-148.

SIMON, T. 1958: Über die alpinen Pflanzengesellschaften des Pirin-Gebirges. - *Acta Bot. Acad. Sci. Hung.* **4(1-2)**: 159-189.

SIMON, T. & VIDA, G. 1966: Neue Angaben zur Verbreitung der *Dryopteris assimilis* S. Walker in Europa. - *Ann. Univ. Sci. Budapest. Rolando Eötvös, Sect. Biol.* **8**: 275-284.

SIMONET, M. 1934: Nouvelles recherches cytologiques et génétiques chez les *Iris*. - *Ann. Sci. Nat. Bot., Sér. 10* **16**: 229-383.

SIMONET, M. 1935: Observations sur quelques espèces et hybrides de l'*Agropyrum*. I. Revision de l'*Agropyrum junceum* (L.) P.B. et de l'*Agropyrum elongatum* (Host) P.B. d'après l'étude cytologique. - *Bull. Soc. Bot. France* **82**: 624-632.

SIMPSON, D.A., MUTHAMA MUASYA, A. & CAFFERTY, S. 2001: Typification of *Scirpus fluitans* and *Scirpus setaceus* (Cyperaceae). - *Kew Bull.* **56(4)**: 1011-1012.

SIMPSON, G. 1962: The A.G.S. tour 1962. Rhodes. - *Quart. Bull. Alpine Gard. Soc. Gr. Brit.* **30(4)**: 330-335.

SIMPSON, G. 1978: Some *Cyclamen* in Jugoslavia. - *Cyclamen Soc. J.* **2(2)**: 21-24.

SIMPSON, G. 1981: *Cyclamen* on Rhodes. - *Cyclamen Soc. J.* **5(2)**: 44-47.

SIMPSON, G. 1978: Further notes on *Cyclamen* in Jugoslavia. - *Cyclamen Soc. J.* **7(1)**: 13-15.

SIMPSON, G. 1984: *Cyclamen* in Jugoslavia. A postscript. - *Cyclamen Soc. J.* **8(1)**: 19-20.

SIMS, J. (ED.) 1800a: *Erica mediterranea*. - *Bot. Mag.* **14**: tab. 471 + 1 p. text.

SIMS, J. (ED.) 1800b: *Geropogon glabrum*. - *Bot. Mag.* **14**: tab. 479 + 1 p. text.

SIMS, J. (ED.) 1800c: *Gentiana verna*. - *Bot. Mag.* **14**: tab. 491 + 1 p. text.

SIMS, J. (ED.) 1801a: *Lithospermum orientale*. - *Bot. Mag.* **15**: tab. 515 + 1 p. text.

SIMS, J. (ED.) 1801b: *Lavatera thuringiaca*. - *Bot. Mag.* **15**: tab. 517 + 1 p. text.

SIMS, J. (ED.) 1801c: *Iris tuberosa*. - *Bot. Mag.* **15**: tab. 531 + 1 p. text.

SIMS, J. (ED.) 1802a: *Trifolium spadiceum*. - *Bot. Mag.* **16**: tab. 557 + 1 p. text.

SIMS, J. (ED.) 1802b: *Clematis viticella*. - *Bot. Mag.* **16**: tab. 565 + 1 p. text.

SIMS, J. (ED.) 1803a: *Gentiana ciliata*. - *Bot. Mag.* **17**: tab. 639 + 1 p. text.

SIMS, J. (ED.) 1803b: *Lonicera implexa*. - *Bot. Mag.* **17**: tab. 640 + 1 p. text.

SIMS, J. (ED.) 1804a: *Gladiolus segetum*. - *Bot. Mag.* **19**: tab. 719 + 1 p. text.

SIMS, J. (ED.) 1804b: *Cheiranthus tristis*. - *Bot. Mag.* **19**: tab. 729 + 1 p. text.

SIMS, J. (ED.) 1804c: *Hesperis tristis*. - *Bot. Mag.* **19**: tab. 730 + 2 pp. text.

SIMS, J. (ED.) 1804d: *Muscari moschatum*. - *Bot. Mag.* **19**: tab. 734 + 2 pp. text.

SIMS, J. (ED.) 1804e: *Androsace villosa*. - *Bot. Mag.* **19**: tab. 743 + 2 pp. text.

SIMS, J. (ED.) 1804f: *Scilla bifolia*. - *Bot. Mag.* **20**: tab. 746 + 2 pp. text.

SIMS, J. (ED.) 1804g: *Asphodelus luteus*. - *Bot. Mag.* **20**: tab. 773 + 1 p. text.

SIMS, J. (ED.) 1804h: *Asphodelus ramosus*. - *Bot. Mag.* **21**: tab. 799 + 2 pp. text.

SIMS, J. (ED.) 1805a: *Silene chloraefolia*. - *Bot. Mag.* **21**: tab. 807 + 1 p. text.

SIMS, J. (ED.) 1805b: *Campanula capitata*. - *Bot. Mag.* **21**: tab. 811 + 1 p. text.

SIMS, J. (ED.) 1805c: *Anemone coronaria*. - *Bot. Mag.* **22**: tab. 841 + 2 pp. text.

SIMS, J. (ED.) 1805d: *Crocus biflorus*. - *Bot. Mag.* **22**: tab. 845 + 1 p. text.

SIMS, J. (ED.) 1805e: *Fritillaria obliqua*. - *Bot. Mag.* **22**: tab. 857 + 1 p. text.

SIMS, J. (ED.) 1805f: *Allium triquetrum*. - *Bot. Mag.* **22**: tab. 869 + 1 p. text.

SIMS, J. (ED.) 1805g: *Verbascum phoeniceum*. - *Bot. Mag.* **23**: tab. 885 + 1 p. text.

SIMS, J. (ED.) 1805h: *Lilium martagon*. - *Bot. Mag.* **23**: tab. 893 + 1 p. text.

SIMS, J. (ED.) 1806a: *Phalangium liliago*. - *Bot. Mag.* **23**: tab. 914 + 2 pp. text.

SIMS, J. (ED.) 1806b: *Ornithogalum squilla*. - *Bot. Mag.* **23**: tab. 918 + 2 pp. text.

SIMS, J. (ED.) 1806c: *Scilla autumnalis*. - *Bot. Mag.* **23**: tab. 919 + 2 pp. text.

SIMS, J. (ED.) 1806d: *Narcissus tazetta*. - *Bot. Mag.* **24**: tab. 925 + 2 pp. text.

SIMS, J. (ED.) 1806e: *Phytolacca americana*. - *Bot. Mag.* **24**: tab. 931 + 2 pp. text.

SIMS, J. (ED.) 1806f: *Echinops ritro*. - *Bot. Mag.* **24**: tab. 932 + 2 pp. text.

SIMS, J. (ED.) 1806g: *Scilla romana*. - *Bot. Mag.* **24**: tab. 939 + 2 pp. text.
SIMS, J. (ED.) 1806h: *Veratrum nigrum*. - *Bot. Mag.* **24**: tab. 963 + 1 p. text.
SIMS, J. (ED.) 1806i: *Celsia cretica*. - *Bot. Mag.* **24**: tab. 964 + 2 pp. text.
SIMS, J. (ED.) 1806j: *Allium paniculatum*. - *Bot. Mag.* **25**: tab. 973 + 1 p. text.
SIMS, J. (ED.) 1806k: *Allium roseum*. - *Bot. Mag.* **25**: tab. 978 + 2 pp. text.
SIMS, J. (ED.) 1807a: *Asphodelus fistulosus*. - *Bot. Mag.* **25**: tab. 984 + 2 pp. text.
SIMS, J. (ED.) 1807b: *Salvia forskoehlei*. - *Bot. Mag.* **25**: tab. 988 + 2 pp. text.
SIMS, J. (ED.) 1807c: *Cyclamen hederifolium*. - *Bot. Mag.* **25**: tab. 1001 + 1 p. text.
SIMS, J. (ED.) 1807d: *Colchicum variegatum*. - *Bot. Mag.* **26**: tab. 1028 + 1 p. text.
SIMS, J. (ED.) 1807e: *Paeonia peregrina*. - *Bot. Mag.* **26**: tab. 1050 + 2 pp. text.
SIMS, J. (ED.) 1807f: *Clematis cirrhosa*. - *Bot. Mag.* **26**: tab. 1070 + 2 pp. text.
SIMS, J. (ED.) 1808a: *Gentiana asclepiadea*. - *Bot. Mag.* **27**: tab. 1078 + 2 pp. text.
SIMS, J. (ED.) 1808b: *Linum austriacum*. - *Bot. Mag.* **27**: tab. 1086 + 1 p. text.
SIMS, J. (ED.) 1808c: *Anthyllis cretica*. - *Bot. Mag.* **27**: tab. 1092 + 2 pp. text.
SIMS, J. (ED.) 1808d: *Aristolochia sempervirens*. - *Bot. Mag.* **28**: tab. 1116 + 1 p. text.
SIMS, J. (ED.) 1808e: *Cercis siliqiuastrum*. - *Bot. Mag.* **28**: tab. 1138 + 2 pp. text.
SIMS, J. (ED.) 1808f: *Allium schoenoprasum*. - *Bot. Mag.* **28**: tab. 1141 + 2 pp. text.
SIMS, J. (ED.) 1808g: *Allium magicum*. - *Bot. Mag.* **28**: tab. 1148 + 2 pp. text.
SIMS, J. (ED.) 1808h: *Digitalis lanata*. - *Bot. Mag.* **29**: tab. 1159 + 1 p. text.
SIMS, J. (ED.) 1809a: *Tulipa sylvestris*. - *Bot. Mag.* **30**: tab. 1202 + 2 pp. text.
SIMS, J. (ED.) 1809b: *Allium Chamae-Moly*. - *Bot. Mag.* **30**: tab. 1203 + 2 pp. text.
SIMS, J. (ED.) 1809c: *Iris pumila*. - *Bot. Mag.* **30**: tab. 1209 + 2 pp. text.
SIMS, J. (ED.) 1809d: *Leucojum aestivum*. - *Bot. Mag.* **30**: tab. 1210 + 2 pp. text.
SIMS, J. (ED.) 1810a: *Iris pumila*. - *Bot. Mag.* **31**: tab. 1261 + 2 pp. text.
SIMS, J. (ED.) 1810b: *Garidella nigellastrum*. - *Bot. Mag.* **31**: tab. 1266 + 2 pp. text.
SIMS, J. (ED.) 1810c: *Achillea Clavenae*. - *Bot. Mag.* **32**: tab. 1287 + 2 pp. text.
SIMS, J. (ED.) 1810d: *Allium flavum*. - *Bot. Mag.* **33**: tab. 1330 + 2 pp. text.
SIMS, J. (ED.) 1811a: *Allium ampeloprasum*. - *Bot. Mag.* **34**: tab. 1385 + 2 pp. text.
SIMS, J. (ED.) 1811b: *Moraea sisyrinchium*. - *Bot. Mag.* **34**: tab. 1407 + 2 pp. text.
SIMS, J. (ED.) 1811c: *Allium paniculatum*. - *Bot. Mag.* **35**: tab. 1432 + 2 pp. text.
SIMS, J. (ED.) 1812a: *Cytisus leucanthus*. - *Bot. Mag.* **35**: tab. 1438 + 1 p. text.
SIMS, J. (ED.) 1812b: *Triglochin bulbosum*. - *Bot. Mag.* **35**: tab. 1445 + 2 pp. text.
SIMS, J. (ED.) 1813a: *Phlomis tuberosa*. - *Bot. Mag.* **35**: tab. 1555 + 2 pp. text.
SIMS, J. (ED.) 1813b: *Muscari moschatum* var. *flavum*. - *Bot. Mag.* **38**: tab. 1565 + 2 pp. text.
SIMS, J. (ED.) 1814a: *Lilium martagon*. - *Bot. Mag.* **39**: tab. 1634 + 2 pp. text.
SIMS, J. (ED.) 1814b: *Iberis saxatilis* var. *corifolia*. - *Bot. Mag.* **40**: tab. 1642 + 2 pp. text.
SIMS, J. (ED.) 1814c: *Veronica multifida*. - *Bot. Mag.* **40**: tab. 1679 + 2 pp. text.
SIMS, J. (ED.) 1815a: *Dianthus virgineus*. - *Bot. Mag.* **42**: tab. 1740 + 2 pp. text.
SIMS, J. (ED.) 1815b: *Allium sphaerocephalon*. - *Bot. Mag.* **42**: tab. 1764 + 2 pp. text.
SIMS, J. (ED.) 1815c: *Verbascum bipinnatifidum*. - *Bot. Mag.* **43**: tab. 1777 + 2 pp. text.
SIMS, J. (ED.) 1816a: *Acanthus spinosus*. - *Bot. Mag.* **43**: tab. 1808 + 2 pp. text.
SIMS, J. (ED.) 1816b: *Thalictrum aquilegifolium*. - *Bot. Mag.* **43**: tab. 1818 + 2 pp. text.
SIMS, J. (ED.) 1816c: *Digitalis ferruginea*. - *Bot. Mag.* **43**: tab. 1828 + 2 pp. text.
SIMS, J. (ED.) 1816d: *Punica granatum*. - *Bot. Mag.* **43**: tab. 1832A+B + 2 pp. text.
SIMS, J. (ED.) 1816e: *Phlomis fruticosa*. - *Bot. Mag.* **43**: tab. 1843 + 2 pp. text.
SIMS, J. (ED.) 1816f: *Hypericum olympicum* var. *latifolium*. - *Bot. Mag.* **43**: tab. 1843 + 2 pp. text.
SIMS, J. (ED.) 1817a: *Phlomis samia*. - *Bot. Mag.* **44**: tab. 1891 + 2 pp. text.
SIMS, J. (ED.) 1817b: *Zizyphus paliurus*. - *Bot. Mag.* **44**: tab. 1893 + 2 pp. text.
SIMS, J. (ED.) 1817c: *Anchusa angustifolia*. - *Bot. Mag.* **44**: tab. 1897 + 2 pp. text.
SIMS, J. (ED.) 1817d: *Momordica elaterium*. - *Bot. Mag.* **44**: tab. 1914 + 2 pp. text.
SIMS, J. (ED.) 1817e: *Daphne oleoides*. - *Bot. Mag.* **44**: tab. 1917 + 2 pp. text.
SIMS, J. (ED.) 1817f: *Anthyllis barba jovis*. - *Bot. Mag.* **44**: tab. 1927 + 2 pp. text.
SIMS, J. (ED.) 1817g: *Ophrys tenthredinifera*. - *Bot. Mag.* **44**: tab. 1930 + 2 pp. text.
SIMS, J. (ED.) 1817h: *Orchis acuminata*. - *Bot. Mag.* **44**: tab. 1932 + 2 pp. text.
SIMS, J. (ED.) 1817i: *Echium creticum*. - *Bot. Mag.* **44**: tab. 1934 + 2 pp. text.
SIMS, J. (ED.) 1817j: *Lathyrus grandiflorus*. - *Bot. Mag.* **44**: tab. 1938 + 2 pp. text.
SIMS, J. (ED.) 1817k: *Passerina hirsuta*. - *Bot. Mag.* **45**: tab. 1949 + 2 pp. text.
SIMS, J. (ED.) 1818a: *Celsia arcturus*. - *Bot. Mag.* **45**: tab. 1962 + 2 pp. text.
SIMS, J. (ED.) 1818b: *Pistacia lentiscus*. - *Bot. Mag.* **45**: tab. 1967 + 2 pp. text.
SIMS, J. (ED.) 1818c: *Silene supina*. - *Bot. Mag.* **45**: tab. 1997 + 2 pp. text.
SIMS, J. (ED.) 1818d: *Vinca herbacea*. - *Bot. Mag.* **45**: tab. 2002 + 2 pp. text.
SIMS, J. (ED.) 1818e: *Arbutus andrachne*. - *Bot. Mag.* **46**: tab. 2024 + 2 pp. text.
SIMS, J. (ED.) 1818f: *Thalictrum aquilegifolium formosum*. - *Bot. Mag.* **46**: tab. 2025 + 2 pp. text.
SIMS, J. (ED.) 1819a: *Arabis caucasica*. - *Bot. Mag.* **46**: tab. 2046 + 1 p. text.
SIMS, J. (ED.) 1819b: *Rosa arvensis*. - *Bot. Mag.* **46**: tab. 2054 + 2 pp. text.
SIMS, J. (ED.) 1820a: *Anchusa undulata*. - *Bot. Mag.* **47**: tab. 2119 + 1 p. text.
SIMS, J. (ED.) 1820b: *Scutellaria orientalis*. - *Bot. Mag.* **47**: tab. 2120 + 2 pp. text.
SIMS, J. (ED.) 1820c: *Coris monspeliensis*. - *Bot. Mag.* **47**: tab. 2131 + 2 pp. text.
SIMS, J. (ED.) 1820d: *Cynoglossum pictum*. - *Bot. Mag.* **47**: tab. 2134 + 2 pp. text.
SIMS, J. (ED.) 1820e: *Plumbago europaea*. - *Bot. Mag.* **47**: tab. 2139 + 2 pp. text.
SIMS, J. (ED.) 1820f: *Vicia tenuifolia*. - *Bot. Mag.* **47**: tab. 2141 + 2 pp. text.
SIMS, J. (ED.) 1820g: *Carthamus lanatus*. - *Bot. Mag.* **47**: tab. 2142 + 2 pp. text.
SIMS, J. (ED.) 1820h: *Asperula arcadiensis*. - *Bot. Mag.* **47**: tab. 2146 + 1 p. text.
SIMS, J. (ED.) 1820i: *Glycyrrhiza echinata*. - *Bot. Mag.* **47**: tab. 2154 + 1 p. text.
SIMS, J. (ED.) 1820j: *Bellis annua*. - *Bot. Mag.* **47**: tab. 2174 + 2 pp. text.
SIMS, J. (ED.) 1820k: *Linaria genistifolia* var. *procera*. - *Bot. Mag.* **47**: tab. 2183 + 2 pp. text.
SIMS, J. (ED.) 1820l: *Anchusa italica*. - *Bot. Mag.* **48**: tab. 2197 + 2 pp. text.
SIMS, J. (ED.) 1821a: *Veronica orchidea*. - *Bot. Mag.* **48**: tab. 2210 + 2 pp. text.
SIMS, J. (ED.) 1821b: *Drypis spinosa*. - *Bot. Mag.* **48**: tab. 2216 + 2 pp. text.
SIMS, J. (ED.) 1821c: *Euphorbia epithymoides*. - *Bot. Mag.* **48**: tab. 2258 + 2 pp. text.
SIMS, J. (ED.) 1821d: *Spartium radiatum*. - *Bot. Mag.* **48**: tab. 2260 + 2 pp. text.
SIMS, J. (ED.) 1821e: *Orobus niger*. - *Bot. Mag.* **48**: tab. 2261 + 1 p. text.
SIMS, J. (ED.) 1821f: *Cicer arietinum*. - *Bot. Mag.* **49**: tab. 2274 + 2 pp. text.
SIMS, J. (ED.) 1821g: *Arum tenuifolium*. - *Bot. Mag.* **49**: tab. 2282 + 2 pp. text.
SIMS, J. (ED.) 1822a: *Cnicus afer*. - *Bot. Mag.* **49**: tab. 2287 + 2 pp. text.
SIMS, J. (ED.) 1822b: *Periploca graeca*. - *Bot. Mag.* **49**: tab. 2289 + 2 pp. text.
SIMS, J. (ED.) 1822c: *Saponaria vaccaria*. - *Bot. Mag.* **49**: tab. 2290 + 2 pp. text.
SIMS, J. (ED.) 1822d: *Malva alcea*. - *Bot. Mag.* **49**: tab. 2297 + 2 pp. text.
SIMS, J. (ED.) 1822e: *Malva moschata* var. *undulata*. - *Bot. Mag.* **49**: tab. 2298 + 2 pp. text.
SIMS, J. (ED.) 1822f: *Ruta angustifolia*. - *Bot. Mag.* **49**: tab. 2311 + 2 pp. text.
SIMS, J. (ED.) 1822g: *Sisyrinchium laxum*. - *Bot. Mag.* **49**: tab. 2312 + 1 p. text.
SIMS, J. (ED.) 1822h: *Mespilus odoratissima*. - *Bot. Mag.* **49**: tab. 2314 + 2 pp. text.
SIMS, J. (ED.) 1822i: *Arbutus unedo* var. *integerrima*. - *Bot. Mag.* **49**: tab. 2319 + 1 p. text.
SIMS, J. (ED.) 1822j: *Salvia bracteata*. - *Bot. Mag.* **49**: tab. 2320 + 2 pp. text.
SIMS, J. (ED.) 1822k: *Anchusa barrelieri*. - *Bot. Mag.* **49**: tab. 2349 + 2 pp. text.
SIMS, J. (ED.) 1822l: *Ribes multiflorum*. - *Bot. Mag.* **50**: tab. 2368 + 2 pp. text.
SIMS, J. (ED.) 1823a: *Euonymus latifolius*. - *Bot. Mag.* **50**: tab. 2384 + 2 pp. text.
SIMS, J. (ED.) 1823b: *Cynanchum nigrum*. - *Bot. Mag.* **50**: tab. 2390 + 2 pp. text.
SIMS, J. (ED.) 1823c: *Geranium macrorhizon*. - *Bot. Mag.* **50**: tab. 2420 + 2 pp. text.
SIMS, J. (ED.) 1823d: *Pyrus amelanchier*. - *Bot. Mag.* **50**: tab. 2430 + 2 pp. text.
SIMS, J. (ED.) 1823e: *Arum italicum*. - *Bot. Mag.* **50**: tab. 2432 + 2 pp. text.
SIMS, J. (ED.) 1823f: *Phlomis herba venti*. - *Bot. Mag.* **51**: tab. 2449 + 2 pp. text.
SIMS, J. (ED.) 1824a: *Coix Lacryma*. - *Bot. Mag.* **51**: tab. 2479 + 2 pp. text.
SIMS, J. (ED.) 1824b: *Serratula simplex*. - *Bot. Mag.* **51**: tab. 2482 + 2 pp. text.
SIMS, J. (ED.) 1824c: *Centaurea spinosa*. - *Bot. Mag.* **51**: tab. 2493 + 2 pp. text.
SIMS, J. (ED.) 1824d: *Ornithogalum narbonense*. - *Bot. Mag.* **51**: tab. 2510 + 2 pp. text.

SIMS, J. (ED.) 1824e: *Bellis sylvestris*. - *Bot. Mag.* **51**: tab. 2511 + 2 pp. text.

SIMS, J. (ED.) 1824f: *Paliurus virgatus*. - *Bot. Mag.* **52**: tab. 2535 + 2 pp. text.

SIMS, J. (ED.) 1825a: *Orchis sulphurea*. - *Bot. Mag.* **52**: tab. 2569 + 1 p. text.

SIMS, J. (ED.) 1825b: *Anthyllis hermanniae*. - *Bot. Mag.* **52**: tab. 2576 + 2 pp. text.

SIMS, J. (ED.) 1825c: *Campanula Lorei*. - *Bot. Mag.* **52**: tab. 2581 + 2 pp. text.

SIMS, J. (ED.) 1826a: *Asphodelus tenuior*. - *Bot. Mag.* **53**: tab. 2626 + 1 p. text.

SIMS, J. (ED.) 1826b: *Campanula ruthenica*. - *Bot. Mag.* **53**: tab. 2653 + 2 pp. text.

SIMS, J. (ED.) 1826c: *Astragalus onobrychis* var. *tenuifolius*. - *Bot. Mag.* **53**: tab. 2665 + 1 p. text.

SIMS, J. (ED.) 1826d: *Cornus mascula*. - *Bot. Mag.* **53**: tab. 2675 + 2 pp. text.

SING, B.P. & KAUR, I. 1999: Systematic position of the genus *Peganum*. - *J. Econ. Taxon. Bot.* **22(3)**: 705-708.

SINGH, A. 1973: Studies on the interspecific hybrids of *Trigonella corniculata* L., *T. hamosa* L. and *T. cretica* L. - *Genetica* **44(2)**: 264-269.

SINGH, A. & SINGH, D. 1976: Karyotype studies in *Trigonella*. - *Nucleus (Calcutta)* **19(1)**: 13-16.

SINGH, T.P. 1995: Alterations in the basic chromosome numbers as a means of speciation in Labiatae. - *Feddes Repert.* **106(1-2)**: 39-47.

ŠINGLIAROVÁ, B., ZOZOMOVÁ-LIHOVA, J. & MRÁZ, P. 2019: Polytopic origin and scale-dependent spatial segregation of cytotypes in primary diploid-autoploid contact zones of *Pilosella rhodopea* (Asteraceae). - *Bot. J. Linn. Soc.* **126(2)**: 360-379.

SINTENIS, P. 1896: [Account of a botanical expedition in the Pindos]. - *Oesterr. Bot. Z.* **46(11)**: 409-410.

SINTENIS, P. & BORNMÜLLER, J. 1891: [Account of an expedition to Athos, Thasos and Olimbos]. - *Oesterr. Bot. Z.* **41**: 218, 431.

ŠINŽAR-SEKULIĆ, J. & AL. 2015: Bioclimatic profiles and distribution models of three paleoendemic *Ramonda* species (Gesneriaceae). In Bogdanović, S. & Jogan, N.: Book of abstracts - 6th Balkan Botanical Congress, p. 38. - Rijeka.

ŞIRANECI, Ş. & ÖZHATAY, N. 1997: The genus *Rosa* in European Turkey. In: Abstracts from the First Balkan Botanical Congress, Thessaloniki, September 19-22, 1997, p. 46. - Thessaloniki.

ŞIRIN, E. & AL. 2021: A new species of *Centaurea* L. subgen. *Cyanus* Mill. (Asteraceae) from Turkey. - *Bot. Serbica* **45(1)**: 13-22.

ŠIRJAEV, G. 1925: *Onobrychis* generis revisio critica. Pars prima. - *Spisy Přír. Fak. Masarykovy Univ.* **56**: 1-197 + 17 plates.

ŠIRJAEV, G. 1926: *Onobrychis* generis revisio critica. Partes secunda et tertia. - *Spisy Přír. Fak. Masarykovy Univ.* **76**: 1-165.

ŠIRJAEV, G. 1927: Die Verbreitung der Untergattung *Euonobrychis* der Gattung *Onobrychis* Tournef. - *Pflanzenareale* **1(5)**: 51-54, Karte 41-44.

ŠIRJAEV, G. 1928: Generis *Trigonella* L. revisio critica. I. - *Spisy Přír. Fak. Masarykovy Univ.* **102**: 1-57 + 2 plates.

ŠIRJAEV, G. 1929: Generis *Trigonella* L. revisio critica. II. - *Spisy Přír. Fak. Masarykovy Univ.* **110**: 1-37 + 3 plates.

ŠIRJAEV, G. 1930a: Generis *Trigonella* L. revisio critica. III. - *Spisy Přír. Fak. Masarykovy Univ.* **128**: 1-31 + 1 plate.

ŠIRJAEV, G. 1930b: *Trigonella*. - *Pflanzenareale* **2(8)**: 77-86, Karte 71, 72, 74-79.

ŠIRJAEV, G. 1931a: Supplementum ad monographiam "*Onobrychis* generis revisio critica". - *Izv. Bulg. Bot. Druzh.* **4**: 7-24.

ŠIRJAEV, G. 1931b: Generis *Trigonella* L. revisio critica. IV. - *Spisy Přír. Fak. Masarykovy Univ.* **136**: 1-33 + 1 plate.

ŠIRJAEV, G. 1932a: Generis *Ononis* L. revisio critica. - *Beih. Bot. Centralbl., Abt. 2* **49(2-3)**: 381-665 + 4 plates.

ŠIRJAEV, G. 1932b: Generis *Trigonella* L. revisio critica. V. - *Spisy Přír. Fak. Masarykovy Univ.* **148**: 1-43 + 1 plate.

ŠIRJAEV, G. 1933: Generis *Trigonella* L. revisio critica. VI. - *Spisy Přír. Fak. Masarykovy Univ.* **170**: 1-37.

ŠIRJAEV, G. 1934a: *Ononis* L. I. Sectio *Natrix* Gris. em. - *Pflanzenareale* **4(2)**: 11-15, Karte 11-20.

ŠIRJAEV, G. 1934b: Generis *Trigonella* L. revisio critica. Pars II, 1. - *Spisy Přír. Fak. Masarykovy Univ.* **192**: 1-15.

ŠIRJAEV, G. 1935: Die Entwicklungsgeschichte der Gattung *Trigonella*. - *Bull. Ass. Russe Rech. Sci. Prague* **2(9)**: 135-162.

ŠIRJAEV, G. 1936: Additamentum ad floram peninsulae Gallipoliensis. - *Beih. Bot. Centralbl., Abt. 2* **54(3)**: 565-576.

ŠIRJAEV, G. 1937: Supplementum II ad *Onobrychis* generis revisio critica. - *Spisy Přír. Fak. Masarykovy Univ.* **242**: 1-14 + 1 plate.

ŠIRJAEV, G. 1939: Conspectus Tragacantharum (*Astragalus* L. subgenus *Tragacantha* Bge.). - *Repert. Spec. Nov. Regni Veg.* **47**: 194-208, 225-261 + 2 plates.

SJÖRS, H. 1988: Vattenklövern, *Menyanthes trifoliata* - en minimonografi [In Swedish with English summary]. - *Svensk Bot. Tidskr.* **82(1)**: 51-64.

SJÖRS, H. 1989: Blåbär, *Vaccinium myrtillus* - ett växtporträtt [In Swedish with English summary]. - *Svensk Bot. Tidskr.* **83(6)**: 411-428.

SKALICKÁ, A. 1967: Taxonomische Studie über die Arten der Gattung *Corothamnus* (W.D.J. Koch) C.B. Presl. - *Preslia* **39**: 10-29.

SKALICKÁ, A. 1969: Taxonomische Revision der Gattung *Lembotropis* Griseb. - *Acta Univ. Carol., Biol.* **1968(3)**: 263-277.

SKALICKÝ, V. 1959: Einige taxonomische und phytogeographische Bemerkungen zu den tschechoslowakischen Arten der Gattung *Anthericum* L. - *Acta Univ. Carol., Biol.* **1959(2)**: 117-157.

SKALICKÝ, V. 1962: Ein Beitrag zur Erkenntnis der europäischen Arten der Gattung *Agrimonia* L. - *Acta Horti Bot. Prag.* **1962**: 87-108.

SKALICKÝ, V. 1968a: Studie über die europäischen Arten der Gattung *Lycopus* L. - *Sborn. Nár. Mus. Praze, Řada B, Přír. Vědy* **24(5)**: 185-216 + 6 plates and 6 maps.

SKALICKÝ, V. 1968b: [Notes on *Agrimonia* and *Aremonia*]. In Heywood, V.H. (ed.): Notulae systematicae ad Floram Europaeam spectantes, no. 7. - *Feddes Repert.* **79(1-2)**: 35-37.

SKALICKÝ, V. 1969: Was ist *Agrimonia odorata* Mill.? - *Feddes Repert.* **79(6)**: 355-361 + 3 plates.

SKALICKÝ, V. 1971: Diagnoses, conspectus et claves analyticae taxorum infragenericorum novorum generis *Agrimonia* L. adhuc rite non publicatorum. - *Novit. Bot. Inst. Horto Bot. Univ. Carol. Prag.* **1971**: 13-22.

SKALICKÝ, V. 1974: Bemerkungen zur Verbreitung der Mistel in der Tschechoslowakei [In Czech with German summary]. - *Zprávy Českoslov. Bot. Společn. Českoslov. Akad. Věd* **9(1)**: 32-34.

SKALICKÝ, V. 1982: Notulae systematicae, diagnosticae et nomenclatoricae ad *Aconitum* generis investigationem pertinentes. - *Preslia* **54(2)**: 115-122.

SKALICKÝ, V. 1985: Taxonomic and nomenclatoric comments on the *Anemone* L. (s.l.) genus. - *Novit. Bot. Univ. Carol.* **2**: 101-110.

SKALIŃSKA, M. 1947: Polyploidy in *Valeriana officinalis* Linn. in relation to its ecology and distribution. - *J. Linn. Soc., Bot.* **53(350)**: 159-186 + 4 plates.

SKALIŃSKA, M. 1964: Studies in the origin of some European species of *Aquilegia*. - *Acta Biol. Cracov., Ser. Bot.* **7(1)**: 1-23.

SKALIŃSKA, M. 1966: Cyto-taxonomical studies in *Crocus*. - *Acta Biol. Cracov., Ser. Bot.* **9(2)**: 137-152 + 2 plates.

SKALIŃSKA, M. & AL. 1957: Further studies in chromosome numbers of Polish angiosperms. - *Acta Soc. Bot. Poloniae* **26(1)**: 215-246 + 2 plates.

SKALIŃSKA, M. & AL. 1959: Further studies in chromosome numbers of Polish angiosperms (dicotyledons). - *Acta Soc. Bot. Poloniae* **28(3)**: 487-529.

SKALIŃSKA, M. & AL. 1964: Additions to chromosome numbers of Polish angiosperms (Fifth contribution). - *Acta Soc. Bot. Poloniae* **33(1)**: 45-76.

SKALIŃSKA, M. & AL. 1968: Further studies in chromosome numbers of Polish Angiosperms. Seventh contribution. - *Acta Biol. Cracov., Ser. Bot.* **11(2)**: 199-224 + 3 plates.

SKALIŃSKA, M. & AL. 1971: Further studies in the chromosome numbers of Polish angiosperms. Ninth contribution. - *Acta Biol. Cracov., Ser. Bot.* **14(1)**: 199-213 + 1 plate.

SKALIŃSKA, M. & AL. 1974: Further studies in chromosome numbers of Polish angiosperms. Tenth contribution. - *Acta Biol. Cracov., Ser. Bot.* **17(2)**: 133-163.

SKALIŃSKA, M. & AL. 1976: Further studies in chromosome numbers of Polish angiosperms. Eleventh contribution. - *Acta Biol. Cracov., Ser. Bot.* **19(2)**: 107-146 + 2 plates.

SKALIŃSKA, M. & AL. 1978: Further studies in chromosome numbers of Polish angiosperms. Twelfth contribution. - *Acta Biol. Cracov., Ser. Bot.* **21(1)**: 31-63.

SKALIŃSKA, M. & KUBIEŃ, E. 1972: Cytological and embryological studies in *Hieracium pratense* Tausch. - *Acta Biol. Cracov., Ser. Bot.* **15(1)**: 39-50 + 5 plates.

SKALIŃSKA, M. & POGAN, E. 1966: Further studies in chromosome numbers of Polish angiosperms. Sixth contribution. - *Acta Biol. Cracov., Ser. Bot.* **9(1)**: 31-58 + 3 plates.

SKALIŃSKA, M. & POGAN, E. 1973: A list of chromosome numbers of Polish angiosperms. - *Acta Biol. Cracov., Ser. Bot.* **16(2)**: 145-201.

SKAN, S.A. 1918: *Ramonda serbica*. - *Bot. Mag.* **144**: tab. 8765 + 2 pp. text.

SKLAVUNOS, C.G. 1919: Die Forstverhältnisse im heutigen Griechenland. - *Forstwiss. Centralbl.* **41**: 81-90, 173-184, 249-268.

SKOG, L.E. 1983: (714) Proposal to conserve the spelling *Jankaea* (Gesneriaceae). - *Taxon* **32(3)**: 491-492. [approved, but one vote short of the necessary qualified majority, cf. *Taxon* 36(1): 76, 1987].

SKOKANOVÁ, K. & AL. 2009: *Centaurea triumfettii* subsp. *triumfettii* (Compositae) does not occur in the western Carpathians and adjacent parts of Pannonia: karyological and morphological evidence. - *Nordic J. Bot.* **27(1)**: 21-36.

SKOKANOVÁ, K. & AL. 2019: The *Cyanus tuberosus* group (Asteraceae) in the Balkans: biological entities require correct names. - *Pl. Syst. Evol.* **305(8)**: 569-596.

SKOKANOVÁ, K. & KOUTECKÝ, P. 2018: (2661) Proposal to conserve the name *Centaurea axillaris* (Compositae) with a conserved type. - *Taxon* **67(6)**: 1224-1225.

SKOKANOVÁ, K., SLOVÁK, M. & KUČERA, J. 2018: Two different examples of hybridization in the genus *Cyanus* from Greece. - *Bot. Serbica* **142 (suppl. 1)**: 55.

ŠKONDRIĆ, S., ALEKSIĆ, J.M. & LAKUŠIĆ, D. 2014: *Campanula cichoracea* (Campanulaceae), a neglected species from the Balkan-Carpathian *C. lingulata* complex as inferred from molecular and morphological characters. - *Willdenowia* **44(1)**: 77-96.

SKOULA, M., ABIDI, C. & KOKKALOU, E. 1996: Essential oil variation of *Lavandula stoechas* L. ssp. *stoechas* growing wild in Crete (Greece). - *Biochem. Syst. Ecol.* **24(3)**: 255-260.

SKOULA, M., DAL CIN D'AGATA, C. & SARPAKI, A. 2009: Contribution to the ethnobotany of Crete, Greece. - *Bocconea* **23**: 479-487.

SKOULA, M. & HARBORNE, J.B. 2002: The taxonomy and chemistry of *Origanum*. In Kintzios, S.E. (ed.): Oregano: the genera *Origanum* and *Lippia*, pp. 67-108. - London: Taylor & Francis.

SKOVSTED, A. 1934: Cytological studies in the tribe Saxifrageae. - *Dansk Bot. Ark.* **8(5)**: 1-52.

SKREDE, S. 1998: New records of *Ophrys speculum* Link, *Orchis provincialis* Lam. & DC. and *Cephalanthera cucullata* Boiss. & Heldr. in Crete. - *J. Eur. Orch.* **30(3)**: 619-623.

SKVORTSOV, A.K. 1971: Euro-amerikanische verwandtschaftliche Beziehungen zweier mediterraner *Salix*-Arten. - *Feddes Repert.* **82(6)**: 407-420.

SKVORTSOV, A.K. 1979: Taxonomy and distribution of *Circaea* (Onagraceae) in the USSR. - *Ann. Missouri Bot. Gard.* **66(4)**: 880-892.

SLAGEREN, M.W. VAN 1994: Wild wheats: A monograph of *Aegilops* L. and *Amblyopyrum* (Jaub. & Spach) Eig (Poaceae). - *Wageningen Agric. Univ. Pap.* **94-7**: xiv + 512 pp.

SLATER, F.M. 1990: Biological Flora of the British Isles. *Gagea bohemica* (Zauschner) J.A. & J.H. Schultes (*G. saxatilis* Koch). - *J. Ecol.* **78(2)**: 535-546.

SLAVÍK, B. 1986: *Epilobium dodonaei* in der Tschechoslowakei. - *Preslia* **58(4)**: 307-338.

SLAVÍK, B. & HEJNÝ, S. 1971: [Notes on the distribution and ecology of *Leucojum aestivum* L. in Europe]. - *Živa* **19**: 83-86 [n.v.].

SLAVÍKOVÁ, Z. 1963a: The systematic position of the species *Galeopsis bifida* Boenn. - *Acta Univ. Carol., Biol.* **1963(3)**: 255-263.

SLAVÍKOVÁ, Z. 1963b: Bemerkungen zur Taxonomie der Gattung *Galeopsis* L. - *Novit. Bot. Delect. Seminum Horti Bot. Univ. Carol. Prag.* **1963**: 39-43.

SLAVÍKOVÁ, Z. 1972: Zur Blütenmorphologie von *Adonis vernalis* L. - *Oesterr. Bot. Z.* **119(4-5)**: 447-453.

SLAVÍKOVÁ, Z. 1982a: Zur Blütenmorphologie einiger *Ranunculus*-Arten von Zypern. - *Preslia* **54(2)**: 105-113.

SLAVÍKOVÁ, Z. 1982b: Zur Kronblattmorphologie einiger *Adonis*-Arten. - *Acta Univ. Carol., Biol.* **1980(3-4)**: 373-380.

SLAVNIĆ, Ž. 1968: Über die Morphologie der Sommer- und Herbstsprosse in der Art *Dorycnium herbaceum* Vill. [In Serbo-Croatian with German summary]. - *Glasn. Zemaljsk. Muz. Sarajevu* **6**: 53-60.

SLAVNIĆ, Ž. 1969: Sur la variabilité morphologique de *Silene sendtneri* Boiss. [In Serbo-Croatian with French summary]. - *Acta Bot. Croat.* **28**: 337-347.

SLAVNIĆ, Ž. 1975: On speciation in *Acer tataricum* Pax in Jugoslavia. - *Wiss. Mitt. Bosnisch-Herzegowinischen Landesmus., C* **4-5**: 125-131.

SLEEP, A. 1983: On the genus *Asplenium* in the Iberian Peninsula. - *Acta Bot. Malac.* **8**: 11-46.

SLEEP, D. 2010: Oh what a lovely weed ... [*Ranunculus asiaticus*]. - *Rock Gard.* **125**: 96-103.

ŠLENKER, M. & AL. 2018a: Introduction of *Cardamine occulta* into Europe and origin of this species. - *Bot. Serbica* **42 (suppl. 1)**: 41.

ŠLENKER, M. & AL. 2018b: Morphology and genome size of the widespread weed *Cardamine occulta*: how it differs from cleistogamic *C. kokaiensis* and other closely related taxa in Europe and Asia. - *Bot. J. Linn. Soc.* **187(3)**: 456-482.

ŠLENKER, M. & AL. 2021: Taxonomic position and circumscription of *Cardamine barbaraeoides* (Brassicaceae), a systematically challenging taxon from the Balkan Peninsula. - *Phytotaxa* **502(2)**: 111-132.

SLEUMER, H. 1941: Vaccinioideen-Studien. - *Bot. Jahrb. Syst.* **71(4)**: 375-510.

SLIM, G.K. & KOOPMANS, A. 1968: Some data on the karyotypes of *Stellaria nemorum* L. ssp. *nemorum* and *Stellaria nemorum* L. ssp. *glochidisperma* Murb. in the Netherlands. - *Acta Bot. Neerl.* **17(2)**: 90-92.

SLOVÁK, M. & AL. 2012: The morphological and genetic variation in the polymorphic species *Picris hieracioides* (Compositae, Lactuceae) in Europe strongly contrasts with traditional taxonomic concepts. - *Syst. Bot.* **37(1)**: 258-268.

SLOVÁK, M. & MARHOLD, K. 2004: Taxonomy of *Picris hieracioides* (Compositae). In Vasić, O. (ed.): Abstracts, XI OPTIMA Meeting, Beograd 5-11.IX.2004, p. 88.

SLUIS, W.G. VAN DER 1985: Chemotaxonomical investigations of the genera *Blackstonia* and *Centaurium* (Gentianaceae). - *Pl. Syst. Evol.* **149(3-4)**: 253-286.

SMALL, E. 1978a: A numerical taxonomic analysis of the *Daucus carota* complex. - *Canad. J. Bot.* **56 (3)**: 248-276.

SMALL, E. 1978b: A numerical and nomenclatural analysis of morpho-geographic taxa of *Humulus*. - *Syst. Bot.* **3(1)**: 37-76.

SMALL, E. 1980: The relationships of hop cultivars and wild variants of *Humulus lupulus*. - *Canad. J. Bot.* **58(6)**: 676-686.

SMALL, E. 1981: A numerical analysis of major groupings in *Medicago* employing traditionally used characters. - *Canad. J. Bot.* **59(9)**: 1553-1557.

SMALL, E. 1983: Pollen ploidy-prediction in the *Medicago sativa* complex. - *Pollen & Spores* **25(3-4)**: 305-320.

SMALL, E. 1984: A clarification of the *Medicago blancheana-bonarotiana-rotata* complex. - *Canad. J. Bot.* **62(8)**: 1693-1696.

SMALL, E. 1985: Morphological differentiation in *Medicago sativa* s.l. in relation to ploidy. - *Canad. J. Bot.* **63(10)**: 1747-1752.

SMALL, E. 1986: Taxonomy of glandular wild alfalfa (*Medicago sativa*). - *Canad. J. Bot.* **64(9)**: 2125-2129.

SMALL, E. 1987a: A taxonomic study of the 'medicagoid' *Trigonella* (Leguminosae). - *Canad. J. Bot.* **65(6)**: 1199-1211.

SMALL, E. 1987b: Generic changes in Trifolieae subtribe Trigonellinae. In Stirton, C.H. (ed.): Advances in legume systematics, pp. 169-181. - Kew: Royal Botanic Gardens.

SMALL, E. 1988: Pollen-ovule patterns in tribe Trifolieae (Leguminosae). - *Pl. Syst. Evol.* **160(3-4)**: 195-205.

SMALL, E. 1989: Polythetic generic separation in tribe Trifolieae subtribe Trigonellinae (Leguminosae). - *Canad. J. Bot.* **67(5)**: 1480-1492.

SMALL, E. 1990: *Medicago rigiduloides*, a new species segregated from *M. rigidula*. - *Canad. J. Bot.* **68(12)**: 2614-2617.

SMALL, E. 1994: Additional clarification of the *Medicago blancheana-bonarotiana-rotata* complex. - *Canad. J. Bot.* **72(6)**: 829-831.

SMALL, E. 2011: Alfalfa and relatives. Evolution and classification of *Medicago*. - Ottawa: National Research Council of Canada, 727 pp.

SMALL, E., BASSETT, I.J. & CROMPTON, C.W. 1981: Pollen variation in tribe Trigonelleae (Leguminosae) with special reference to *Medicago*. - *Pollen & Spores* **23(3-4)**: 295-320.

SMALL, E. & BAUCHAN, G.R. 1984: Chromosome numbers of the *Medicago sativa* complex in Turkey. - *Canad. J. Bot.* **62(4)**: 749-752.

SMALL, E. & BROOKES, B.S. 1983: The systematic value of stigma morphology in the legume tribe Trifolieae with particular reference to *Medicago*. - *Canad. J. Bot.* **61(9)**: 2388-2404.

SMALL, E. & BROOKES, B.S. 1985: *Medicago lesinii*, a new Mediterranean species. - *Canad. J. Bot.* **63 (4)**: 728-734.

SMALL, E. & BROOKES, B.S. 1990a: A numerical taxonomic analysis of the *Medicago littoralis* - *M. truncatula* complex. - *Canad. J. Bot.* **68(8)**: 1667-1674.

SMALL, E. & BROOKES, B.S. 1990b: A taxonomic simplification of *Medicago italica*. - *Canad. J. Bot.* **68 (10)**: 2103-2111.

SMALL, E., BROOKES, B.S. & CRAWFORD, E.J. 1990: Intercontinental differentiation in *Medicago rigidula*. - *Canad. J. Bot.* **68(12)**: 2607-2613.

SMALL, E., BROOKES, B.S. & LASSEN, P. 1990: Circumscription of the genus *Medicago* (Leguminosae) by seed characters. - *Canad. J. Bot.* **68(3)**: 613-629.

SMALL, E., CROMPTON, C.W. & BROOKES, B.S. 1981: The taxonomic value of floral characters in tribe Trigonelleae (Leguminosae), with special reference to *Medicago*. - *Canad. J. Bot.* **59(9)**: 1578-1598.

SMALL, E. & CRONQUIST, A. 1976: A practical and natural taxonomy for *Cannabis*. - *Taxon* **25(4)**: 405-435.

SMALL, E., GRANT, W.F. & CROMPTON, C.W. 1984: A taxonomic study of the *Lotus corniculatus* complex in Turkey. - *Canad. J. Bot.* **62(5)**: 1044-1053.

SMALL, E. & JOMPHE, M. 1989a: A note on *Trigonella* (*Melilotus*) *bicolor*. - *Canad. J. Bot.* **67(5)**: 1604-1606.

SMALL, E. & JOMPHE, M. 1989b: A synopsis of the genus *Medicago* (Leguminosae). - *Canad. J. Bot.* **67(11)**: 3260-3294 + 1 fold-out illustr. key.

SMALL, E., JOMPHE, M. & BROOKES, B.S. 1991: *Medicago truncatula* f. *laxicycla* (Leguminosae), a new taxon with loose fruit coiling promoting rapid germination of the fruit-retained seeds. - *Pl. Syst. Evol.* **174(1-2)**: 37-46.

SMALL, E., LASSEN, P. & BROOKES, B.S. 1987: An expanded circumscription of *Medicago* (Leguminosae, Trifolieae) based on explosive flower tripping. - *Willdenowia* **16(2)**: 415-437.

SMALL, E. & LEFKOVITCH, L.P. 1986: Relationships among morphology, geography and interfertility in *Medicago*. - *Canad. J. Bot.* **64(1)**: 45-52.

SMALL, E., LEFKOVITCH, L.P. & BROOKES, B.S. 1989: Preferential infestation of fruits within the *Medicago truncatula* - *M. littoralis* (Leguminosae) complex by the beal weevil *Bruchidius bimaculatus*. - *Pl. Syst. Evol.* **163(1-2)**: 35-41.

SMEJKAL, M. 1961: Taxonomisches Studium der Art *Ajuga chamaepitys* (L.) Schreb. ampl. Briq. in der Tschechoslowakei [In Czech with German summary]. - *Preslia* **33(4)**: 386-398.

SMEJKAL, M. 1965: Taxonomische Studie über die tschechoslowakischen Arten der Gattung *Scleranthus* L. [In Czech with German summary]. - *Folia Fac. Sci. Nat. Univ. Purkynianae Brun., Biol. 1* **6(4)**: 1-74 + 7 plates.

SMEJKAL, M. 1967: Bestimmungsschlüssel und Übersicht der tschechoslowakischen Arten und Unterarten der Gattung *Cerastium* L. [In Czech]. - *Zprávy Českoslov. Bot. Společn. Českoslov. Akad. Věd* **2(3)**: 137-146.

SMEJKAL, M. 1971: Revison der tschechoslowakischen Arten der Gattung *Camelina* Crantz (Cruciferae). - *Preslia* **43(4)**: 318-337.

SMEJKAL, M. 1975: *Galeobdolon argentatum* sp. nova, ein neuer Vertreter der Kollektivart *Galeobdolon luteum* (Lamiaceae). - *Preslia* **47(3)**: 241-248.

SMEJKAL, M. 1978: Die Gattung *Symphytum* in der Tschechoslowakei [In Czech with German summary]. - *Zprávy Českoslov. Bot. Společn. Českoslov. Akad. Věd* **13(3)**: 145-161.

SMEJKAL, M. & HELANOVÁ-ZICHOVÁ, E. 1974: Tschechoslowakische Arten der Sektion *Beccabunga* Griseb. der Gattung *Veronica* L. [In Czech with German summary]. - *Preslia* **46(2)**: 157-166.

SMIRIS, P. 1980: Standortskundliche und waldbauliche Untersuchungen von naturnahen Buchenwäldern im Voras-Gebirge (Nordgriechenland). - Göttingen: Diss., 106 pp.

SMIRIS, P. 1985: I domi tou Parthenou Dhasous tou Paranestiou. - *Sci. Ann. Dept. Forest. Nat. Environm. Aristot. Univ. Thessalonika* **28**: 595-670.

SMIRIS, P. & DAFIS, S. 1993: Ökologische Wald-Ökosysteme im nördlichen Pindosgebirge. In Mattes, H. & Lienau, C. (eds): Colloquium Nordost-Griechenland 2, pp. 44-49. - Münster: Inst. Geographie.

SMISSEN, R.D. & GARNOCK-JONES, P.J. 2002: Relationships, classification and evolution of *Scleranthus* (Caryophyllaceae) as inferred from analysis of morphological characters. - *Bot. J. Linn. Soc.* **140(1)**: 15-29.

SMIT, P.G. 1973: A revision of *Caltha* (Ranunculaceae). - *Blumea* **21(1)**: 119-150 + 2 plates.

SMITH, A.C. 1979: Comments on *Sempervivum ciliosum* Craib. - *J. Sempervivum Soc.* **10(1)**: 22-23.

SMITH, A.R. 1968: [Notes on *Euphorbia*]. In Heywood, V.H. (ed.): Notulae systematicae ad Floram Europaeam spectantes, no. 7. - *Feddes Repert.* **79(1-2)**: 55.

SMITH, A.R. & AL. 2006: A classification for extant ferns. - *Taxon* **55(3)**: 705-731.

SMITH, C.P. 1938-53: Species Lupinorum. - Saratoga, Calif.: 768 pp.

SMITH, G.F., BURROW, B. & LOWE, D.B. 1984: Primulas of Europe and North America. - St. John's Woking, Surrey: Alpine Garden Society, 251 pp. + 32 plates.

SMITH, G.F. & LOWE, D.B. 1997: The genus *Androsace*. A monograph for gardeners and botanists. - Pershore, U.K.: Alpine Garden Society, 207 pp.

SMITH, J.E. 1804: An illustration of the grass called by Linnaeus *Cornucopiae alopecuroides*. - *Trans. Linn. Soc. London* **7**: 245-248 + 1 plate.

SMITH, J.E. 1808: An inquiry into the genus of the tree called by Pona *Abelicea cretica*. - *Trans. Linn. Soc. London* **9**: 126-130.

SMITH, P.M. 1970: Taxonomy and nomenclature of the brome-grasses (*Bromus* L. s.l.). - *Notes Roy. Bot. Gard. Edinburgh* **30(2)**: 361-375.

SMITH, P.M. 1978: [Notes on *Bromus*]. In Heywood, V.H. (ed.): Notulae systematicae ad Floram Europaeam spectantes, no. 20. - *Bot. J. Linn. Soc.* **76(4)**: 360-361.

SMITH, P.M. 1981: Ecotypes and subspecies in annual brome-grasses (*Bromus*, Gramineae). - *Bot. Jahrb. Syst.* **102(1-4)**: 497-509.

SMITH, P.M. 1985: Observations on Turkish brome-grasses. I. Some new taxa, new combinations and notes on typification. - *Notes Roy. Bot. Gard. Edinburgh* **42(3)**: 491-501.

SMITH, P.M. 1986: Native or introduced? Problems in the taxonomy and plant geography of some widely introduced annual brome-grasses. - *Proc. Roy. Soc. Edinburgh, B, Biol. Sci.* **89**: 273-281.

SMITH, P.M. & SALES, F. 1993: *Bromus* L. sect. *Bromus*: Taxonomy and relationship of some species with small spikelets. - *Edinburgh J. Bot.* **50(2)**: 149-171.

SMITH, S.G. 1932: Cytology of *Anchusa* and its relation to the taxonomy of the genus. - *Bot. Gaz.* **94 (2)**: 394-403.

SMITH, S.G. & KUKKONEN, I. 1999: A new lectotype for *Scirpus maritimus* (Cyperaceae). - *Taxon* **48(2)**: 355-357.

SMITH, U.K. 1979: Biological Flora of the British Isles. *Senecio integrifolius* (L.) Clairv. (*Senecio campestris* (Retz.) DC.). - *J. Ecol.* **67(3)**: 1109-1124.

SMITH, W.W., FORREST, G. & FLETCHER, H.R. 1977: The genus *Primula* [a facsimile reprint of 22 papers published in *Trans. & Proc. Bot. Soc. Edinburgh* and other journals 1929-1950]. - Vaduz: J. Cramer, iii + 835 pp. [cf. TL-2 5: 705-706, 1985].

SMYTHIES, B.E. & MATHEW, B.F. 1976: *Merendera attica* (Liliaceae) in Spain. - *Candollea* **31(2)**: 243-245. [Also distributed as *OPTIMA leaflet* no. 26].

SNOEIJER, W. 1991: *Clematis* index. - Boskoop, Netherlands: 144 pp.

SNOEIJER, W. 1992: A suggested classification for the genus *Clematis*. - *Clematis Int.* **1992**: 7-20.

SNOGERUP, B. 1980: The genus *Reichardia* (Asteraceae) in the Aegean area. - *Bot. Not.* **133(4)**: 515-520.

SNOGERUP, B. 1983: The genus *Odontites* (Scrophulariaceae) in northwest Europe. - Lund: Ph.D. thesis, Univ. of Lund [10 pp. + reprints of previously publ. papers].

SNOGERUP, B. & SNOGERUP, S. 2006: *Paramoltkia doerfleri*. In Greuter, W. & Raus, Th. (eds): Med- Checklist notulae, 24. - *Willdenowia* **36(2)**: 721.

SNOGERUP, S. 1962: Studies in the Aegean flora. IV. *Bupleurum flavum* Forsk. and related species. - *Bot. Not.* **115(4)**: 357-375.

SNOGERUP, S. 1963: Studies in the genus *Juncus*. III. Observations on the diversity of chromosome numbers. - *Bot. Not.* **116(2)**: 142-156.

SNOGERUP, S. 1967a: Studies in the Aegean flora. VIII. *Erysimum* sect. *Cheiranthus*. A. Taxonomy. - *Opera Bot.* **13**: 1-70.

SNOGERUP, S. 1967b: Studies in the Aegean flora. IX. *Erysimum* sect. *Cheiranthus*. B. Variation and evolution in the small-population system. - *Opera Bot.* **14**: 1-86.

SNOGERUP, S. 1970: Studies in the genus *Juncus* L. IV. The typification of *Juncus conglomeratus* L. - *Bot. Not.* **123(3)**: 424-429.

SNOGERUP, S. 1971a: New species of *Bupleurum* from Turkey. - *Bot. Not.* **124(3)**: 359-375.

SNOGERUP, S. 1971b: Evolutionary and plant geographical aspects of chasmophytic communities. In Davis, P.H., Harper, P.C. & Hedge, I.C. (eds): Plant life of South-West Asia, pp. 157-170. - Edinburgh: Botanical Society.

SNOGERUP, S. 1978a: Notes on *Juncus* for Flora Europaea. - *Bot. Not.* **131(2)**: 185-187.

SNOGERUP, S. 1978b: A revision of the *Juncus atratus* group. - *Bot. Not.* **131(2)**: 189-196.

SNOGERUP, S. 1979: Experimental and cytological studies of the "Brassica oleracea" group. - *Webbia* **34(1)**: 357-362.

SNOGERUP, S. 1980a: The wild forms of the *Brassica oleracea* group (2n=18) and ... relations to the cultivated ones. In Tsunoda, S., Hinata, K. & Gómez-Campo, C. (eds): *Brassica* crops and wild allies, pp. 121-132. - Tokyo: Japan Scientific Societies Press.

SNOGERUP, S. 1980b: The Aegean endemics, distribution and present situation. [Unpublished report]. - Photocopy: 22 pp.

SNOGERUP, S. 1982a: A new species of *Epilobium* (Onagraceae) from Northern Greece. (Materials for the Mountain Flora of Greece, 18). - *Willdenowia* **12(2)**: 227-229.

SNOGERUP, S. 1982b: A *Juncus* hybrid new to Greece: *J. fontanesii* Gay × *J. striatus* Schousboe. - *Ann. Mus. Goulandris* **5**: 121-122.

SNOGERUP, S. 1985a: A new annual *Bupleurum* from Kriti. - *Willdenowia* **14(2)**: 309-311.

SNOGERUP, S. 1985b: The Mediterranean islands. In Gómez-Campo, C. (ed.): Plant conservation in the Mediterranean area, pp. 159-173. - Dordrecht, etc.: Dr. W. Junk Publishers.

SNOGERUP, S. 1989a: A new annual *Bupleurum* from southern Peloponnisos. In Tan, Kit (ed.): Plant taxonomy, phytogeography and related subjects. The Davis & Hedge festschrift, pp. 23-26. - Edinburgh: Edinburgh Univ. Press.

SNOGERUP, S. 1989b: Endemism and disjunct populations on Yioura and Kira Panagia - an urgent case for conservation. In Phitos, D. (ed.): Abstracts. Sixth OPTIMA meeting, Delphi 10-16 Sept. 1989, p. 39.

SNOGERUP, S. 1991: Endemism and disjunct populations on Yioura and Kira Panagia - an urgent case for conservation. - *Bot. Chron. (Patras)* **10**: 293-297.

SNOGERUP, S. 1993a: A revision of *Juncus* subgen. *Juncus*. - *Willdenowia* **23(1-2)**: 23-73.

SNOGERUP, S. 1993b: Cytogenetic differentiation and reproduction in *Brassica*, sect. *Brassica*. In Demiriz, H. & Özhatay, N. (eds): Proceedings of the V OPTIMA meeting, Istanbul, 8-15 September, 1986, pp. 377-393. - Istanbul Üniversitesi Fen Fakültesi.

SNOGERUP, S. 1994: Reports (267-284). In Kamari, G., Felber, F. & Garbari, F. (eds): Mediterranean chromosome number reports - 4. - *Fl. Medit.* **4**: 254-258.

SNOGERUP, S. 1996: Reports (491-517). In Kamari, G., Felber, F. & Garbari, F. (eds): Mediterranean chromosome number reports - 5. - *Fl. Medit.* **5**: 331-334.

SNOGERUP, S. 2008: *Rumex vesicarius* L. In Greuter, W. & Raus, Th. (eds): Med-Checklist notulae, 27. - *Willdenowia* **38(2)**: 468.

SNOGERUP, S. & AL. 1989: Flora and vegetation of Kira Panagia, N Sporades, Greece. In Phitos, D. (ed.): Abstracts. Sixth OPTIMA meeting, Delphi 10-16 Sept. 1989, p. 139.

SNOGERUP, S. & AL. 1991: Flora and vegetation of Kira Panagia, N Sporades, Greece. - *Bot. Chron. (Patras)* **10**: 547-566.

SNOGERUP, S. & AL. 2001: The flora of Chios island, Greece. - *Bot. Chron. (Patras)* **14**: 5-197.

SNOGERUP, S. & AL. 2006: Flora and vegetation of Andros, Kikladhes, Greece. - *Ann. Mus. Goulandris* **11**: 85-270.

SNOGERUP, S., BOTHMER, R.V. & GUSTAFSSON, M. 1980: Floristic report from the island of Psathura (Greece). - *Bot. Not.* **133(2)**: 145-148.

SNOGERUP, S., GUSTAFSSON, M. & BOTHMER, R.V. 1990: *Brassica* sect. *Brassica* (Brassicaceae). I. Taxonomy and variation. - *Willdenowia* **19(2)**: 271-365. [Also distributed as *OPTIMA leaflet* no. 190].

SNOGERUP, S. & SNOGERUP, B. 1987: Repeated floristical observations on islets in the Aegean. - *Pl. Syst. Evol.* **155(1-4)**: 143-164.

SNOGERUP, S. & SNOGERUP, B. 1989: Flora and vegetation of the island of Agios Evstratios, Greece. In Phitos, D. (ed.): Abstracts. Sixth OPTIMA meeting, Delphi 10-16 Sept. 1989, p. 138.

SNOGERUP, S. & SNOGERUP, B. 1991: Flora and vegetation of the island of Agios Evstratios, Greece. - *Bot. Chron. (Patras)* **10**: 527-546.

SNOGERUP, S. & SNOGERUP, B. 1993: Additions to the flora of Samos, Greece. - *Fl. Medit.* **3**: 211-222.

SNOGERUP, S. & SNOGERUP, B. 1996: Anatolian and East Aegean elements in the flora of Chios (Greece). In Öztürk, M.A., Seçmen, Ö. & Görk, G. (eds): Plant life in southwest and central Asia, 1: pp. 24-33. - Izmir: Ege University Press.

SNOGERUP, S. & SNOGERUP, B. 2000: New localities for the Greek flora. - *Bot. Chron. (Patras)* **13**: 267-270.

SNOGERUP, S. & SNOGERUP, B. 2001a: Local endemism in European annual *Bupleurum* (Umbelliferae). In Raimondo, F.M. (ed.): Abstracts from the X OPTIMA meeting, Palermo, 13-19 September 2001, p. 152.

SNOGERUP, S. & SNOGERUP, B. 2001b: *Bupleurum* L. (Umbelliferae) in Europe - 1. The annuals, *B.* sect. *Bupleurum* and sect. *Aristata*. - *Willdenowia* **31(2)**: 205-308.

SNOGERUP, S. & SNOGERUP, B. 2003: Local endemism in European annual *Bupleurum* (Umbelliferae). - *Bocconea* **16(2)**: 751-755.

SNOGERUP, S. & SNOGERUP, B. 2004: Changes in the flora of some Aegean islets 1968-2000. - *Pl. Syst. Evol.* **245(3-4)**: 169-213.

SNOGERUP, S. & SNOGERUP, B. 2005: Changes in the flora of some Aegean islets, studied during 30 years. In Karamanos, A.J. & Thanos, C.A. (eds): Biodiversity and natural heritage in the Aegean. Proceedings of the conference "Theophrastus 2000", pp. 235-250. - Athina: Afoi Fragoudi O.E.

SNOGERUP, S. & SNOGERUP, B. 2008: *Chamaesyce serpens* (Kunth) Small. In Greuter, W. & Raus, Th. (eds): Med-Checklist notulae, 27. - *Willdenowia* **38(2)**: 467.

SNOGERUP, S. & SNOGERUP, B. 2009: LBF:s exkursion till Kreta april 2008 [In Swedish]. - *Bot. Not.* **142(1)**: 1-6.

SNOGERUP, S., SNOGERUP, B. & LASSEN, P. 2004: LBF:s exkursion till Zakynthos 7-12 maj 2003 [In Swedish]. - *Bot. Not.* **137(2)**: 12-21.

SNOGERUP, S., SNOGERUP, B. & LASSEN, P. 2005: LBF:s exkursion till Karpathos, 16-23 maj 2004 [In Swedish]. - *Bot. Not.* **138(1)**: 27-38.

SNOGERUP, S., SNOGERUP, B. & STRID, A. 2006: *Acanthus greuterianus* (Acanthaceae), a new species from NW Greece. - *Willdenowia* **36(1)**: 323-327.

SNOGERUP, S. & STRID, A. 2006: *Juncus foliosus*. In Greuter, W. & Raus, Th. (eds): Med-Checklist notulae, 24. - *Willdenowia* **36(2)**: 730.

SNOW, N. & AL. 2018: Monograph of *Diplachne* (Poaceae, Chloridoideae,Cynodonteae). - *PhytoKeys* **93**: 1-102.

SOBEY, D.G. 1981: Biological Flora of the British Isles. *Stellaria media* (L.) Vill. - *J. Ecol.* **69(1)**: 311-335.

SOBIERAJSKA, K. & AL. 2016: Effect of the Aegean Sea barrier between Europe and Asia on differentiation in *Juniperus drupacea* (Cupressaceae) - *Bot. J. Linn. Soc.* **180(3)**: 365-385.

SOBRINO VESPERINAS, E. 1988: Obtainment of some new intergeneric and interspecific hybrids between wild Brassiceae. - *Candollea* **43(2)**: 499-504.

SOBRINO VESPERINAS, E. 1996: Posición taxonómica de *Diplotaxis cretacea* Kotov (Cruciferae). - *Anales Jard. Bot. Madrid* **54(1)**: 182-188.

SOBRINO, E. & AL. 1999: Morfología y taxonomía de la sección *Dendrotelis* del género *Medicago* (Fabaceae). In Correia, A.I.D. & al. (eds): Livro de resumos. Guia de excursões. V jornadas de taxonomia botânica, 16-19/IX 1999, p. 41.

SOBRINO, E. & AL. 2001: Morfología y taxonomía de la sección *Dendrotelis* del género *Medicago* (Fabaceae). - *Portug. Acta Biol.* **19(1-4)**: 225-237.

SOBTI, S.N. 1975: Interspecific hybrids in the genus *Mentha* III. *Mentha arvensis* Linn. × *M. spicata* Linn. - *Cytologia* **40(2)**: 263-267.

SOEST, J.L. VAN 1965: *Taraxacum* sect. *Palustria* Dahlstedt. - *Acta Bot. Neerl.* **14(1)**: 1-53.

SOEST, J.L. VAN 1966: New *Taraxacum* species from Europe. I-III. - *Proc. Kon. Ned. Akad. Wetensch. C* **69**: 432-446, 447-463, 464-479.

SOEST, J.L. VAN 1971: Quelques nouvelles espèces de *Taraxacum*, native d'Europe II. - *Acta Bot. Neerl.* **20(1)**: 141-156.

SOEST, J.L. VAN 1976: New *Taraxacum* species from Europe. V. - *Proc. Kon. Ned. Akad. Wetensch. C* **79(2)**: 171-190.

SOJÁK, J. 1971: Übersicht der tschechoslowakischen *Pilosella*-Arten [In Czech with German summary]. - *Preslia* **43(2)**: 183-186.

SOJÁK, J. 1972: Poznámky k nomenklatuřa *Euphorbia* L. (s.l.). Bemerkungen zur Nomenklatur von *Euphorbia* L. (s.l.). - *Čas. Nár. Mus., Odd. Přír.* **140(3-4)**: 168-178.

SOJÁK, J. 1980a: Fragmenta phytotaxonomica et nomenclatorica (1.). - *Čas. Nár. Mus., Odd. Přír.* **148(2)**: 76-80.

SOJÁK, J. 1980b: Fragmenta phytotaxonomica et nomenclatorica (2.). - *Čas. Nár. Mus., Odd. Přír.* **148(3-4)**: 193-209.

SOJÁK, J. 1981: Fragmenta phytotaxonomica et nomenclatorica (3.). - *Čas. Nár. Mus., Odd. Přír.* **150(3-4)**: 121-146.

SOJÁK, J. 1983: Fragmenta phytotaxonomica et nomenclatorica (4.). - *Čas. Nár. Mus., Odd. Přír.* **152(1)**: 12-25.

SOJÁK, J. 1987: Notes on *Potentilla paradoxa* and *P. supina*. - *Preslia* **59**: 271-272.

SOJÁK, J. 1993: Taxonomische Bemerkungen zu einigen mediterranen *Potentilla*-Sippen. - *Preslia* **65 (2)**: 117-130.

SOJÁK, J. 1996: Notes on *Potentilla* (Rosaceae). XIV. Type specimens in the Lehmann herbarium. - *Preslia* **68(2)**: 97-124.

SOJÁK, J. 2011: Synopsis of *Drymocallis* Fourr. ex Rydb. (Rosaceae - Potentilleae) in the Old World. - *Ann. Naturhist. Mus. Wien* **112B**: 319-328.

SOJÁK, J. & POUZAR, Z. 1984: (751) Proposal to conserve *Medicago* L. with *M. sativa* L. as its type (Fabaceae). - *Taxon* **33(3)**: 519-521. [approved, cf. Taxon 36(4): 737-738, 1987].

SOKOLOVSKAYA, A.P. & PROBATOVA, N.S. 1976: On the karyological characteristics of genera *Milium* L. and *Holcus* L. (Poaceae) [In Russian]. - *Bot. Zhurn. (Moscow & Leningrad)* **61(7)**: 969-973.

SOKOLOVSKAYA, A.P. & PROBATOVA, N.S. 1978: Chromosome numbers of some grasses (Poaceae) of the U.S.S.R. flora. II [In Russian with English summary]. - *Bot. Zhurn. (Moscow & Leningrad)* **63(9)**: 1247-1257.

SOLDANO, A. 1986: Note di aggiornamento nomenclaturale su alcune fanerogame italiane e di altre regioni Europee. - *Atti Soc. Ital. Sci. Nat. Mus. Civico Storia Nat. Milano* **127(3-4)**: 215-220.

SOLDANO, A. 1991a: Le sottospecie di Cesati; altre novità e precisazioni nomenclaturali e tassonomiche su fanerogame d'Italia e dell'area Mediterranea. - *Atti Soc. Ital. Sci. Nat. Mus. Civico Storia Nat. Milano* **131(15)**: 245-256.

SOLDANO, A. 1991b: Paterintà del nome e tipificazione di *Galactites elegans* (Dicotiledoneae, Asteraceae). - *Rivista Piemont. Storia Nat.* **12**: 21-23.

SOLDANO, A. 1992a: Nomenclatural notes on some Compositae from the Mediterranean area. - *Compositae Newslett.* **20/21**: 4-11.

SOLDANO, A. 1992b: Riproposizione di taxa sottspecifici prioritari dovuti a botanici italiani. - *Nat. Bresciana* **27**: 51-56.

SOLDANO, A. 1993a: Il genere *Oenothera* L., subsect. *Oenothera*, in Italia (Onagraceae). - *Nat. Bresciana* **28**: 85-116.

SOLDANO, A. 1993b: Nouvi dati nomenclaturali su piante della flora italiana e mediterrana. - *Atti Soc. Ital. Sci. Nat. Mus. Civico Storia Nat. Milano* **133(10)**: 113-118.

SOLDANO, A. 1993c: Le piante di Allioni ed una negletta opera di F.X. Hartmann. - *Rivista Piemont. Storia Nat.* **14**: 67-75.

SOLDANO, A. 1994a: Sulla corretta datazione e la priorità di taxa apparsi in operi di Desfontaines e Gussone. - *Nat. Bresciana* **29**: 143-152.

SOLDANO, A. 1994b: Le "Rettificazioni ed aggiunte all'Index Kewensis" di Loreto Grande. - *Boll. Mus. Civico Storia Nat. Verona* **18**: 339-347.

SOLDANO, A. 1994c: Neglected name priorities in the European flora. - *Thaiszia* **4(2)**: 119-123.

SOLER, A. 1981: Números cromosómicos para la flora española. 245-256. - *Lagascalia* **10(2)**: 253-256.

SOLER, A. 1983: Revisión de las especies de *Fumaria* de la Península Ibérica e Islas Baleares. - *Lagascalia* **11(2)**: 141-228.

SÖLLNER, R. 1954: Recherches cytotaxinomiques sur le genre *Cerastium*. - *Ber. Schweiz. Bot. Ges.* **64**: 221-354.

SOLOVIEVA, N.M. & AL. 1982: Karyosystematic study of the genus *Ferula* (Umbelliferae) [In Russian]. - *Bot. Zhurn. (Moscow & Leningrad)* **67(9)**: 1228-1237.

SOLOVIEVA, N.M. & AL. 1985: Karyotaxonomic study of some species of *Peucedanum* (Umbelliferae). - *Pl. Syst. Evol.* **151(1-2)**: 89-101.

SOMLYAY, L. 1998: Flowering plant types in the Hungarian Natural History Museum (BP). II. *Achillea* type specimens of József Ujhelyi (1910-1979). - *Ann. Hist.-Nat. Mus. Natl. Hung.* **90**: 133-138.

SOMLYAY, L. 2011: Distribution of *Conringia austriaca* (Brassicaceae) in Hungary and its phytogeographical significance in Central Europe. - *Ann. Hist.-Nat. Mus. Natl. Hung.* **103**: 5-22.

SONCK, C.E. 1984: New *Taraxacum* species from Greece. - *Ann. Bot. Fenn.* **21(2)**: 157-170.

SONCK, C.E. 1985a: New *Taraxacum* species from Greece. II. - *Ann. Bot. Fenn.* **22(2)**: 139-148.

SONCK, C.E. 1985b: New *Taraxacum* species from Greece. III. - *Ann. Bot. Fenn.* **22(3)**: 255-262.

SONCK, C.E. 1986: *Taraxacum subolivaceum* (sect. *Palustria*), species nova, from Greece, and *T. vexatum*, nomen novum. - *Ann. Bot. Fenn.* **23(2)**: 165-168.

SONCK, C.E. 1989: *Taraxacum panhellenicum*, new species of sect. *Erythrocarpa* from Greece, and a new name, *T. praegracilens* - *Ann. Bot. Fenn.* **26(1)**: 51-52. [correction in vol. 26(2) post p. 236].

SONCK, C.E. 1993: New *Taraxacum* species from Greece. IV. - *Ann. Bot. Fenn.* **30(3)**: 205-210.

SONCK, C.E. 1999: New *Taraxacum* species from Greece. V. - *Ann. Bot. Fenn.* **36(3)**: 211-217.

SØNDERGAARD, P. & EGLI, B.R. 2006: *Zelkova abelicea* (Ulmaceae) in Crete: floristics, ecology, propagation and threats. - *Willdenowia* **36(1)**: 317-322.

SØNDERHOUSEN, O. 1977: The fourth tulip from Crete. - *Quart. Bull. Alpine Gard. Soc. Gr. Brit.* **45(2)**: 143-148.

SØNDERHOUSEN, O. 1982: Bulbs of the eastern Mediterranean. - *J. Scott. Rock Gard. Club* **18(1)**: 64-68.

SØNDERHOUSEN, O. 1992: A *Fritillaria* anthology. - *Rock Gard.* **23(1)**: 16-27.

SOÓ, R.V. 1922: Über die mitteleuropäischen Arten und Formen der Gattung *Consolida* (DC.) S.F. Gray. - *Oesterr. Bot. Z.* **71(11-12)**: 233-246.

SOÓ, R.V. 1926-27: Systematische Monographie der Gattung *Melampyrum*. I-II. - *Repert. Spec. Nov. Regni Veg.* **23**: 159-176, 385-397.

SOÓ, R.V. 1927a: Orchideae novae europeae et mediterranae. (Diagnoses, combinationes et nomina nova). - *Repert. Spec. Nov. Regni Veg.* **24**: 25-37.

SOÓ, R.V. 1927b: Systematische Monographie der Gattung *Melampyrum*. III. Systematischer Teil. - *Repert. Spec. Nov. Regni Veg.* **24**: 127-193 + 3 plates.

SOÓ, R.V. 1929a: Revision der Orchideen Südosteuropas und Südwestasiens. - *Bot. Arch.* **23**: 1-196.

SOÓ, R.V. 1929b: Die mittel- und südeuropäischen Arten und Formen der Gattung *Rhinanthus* und ihre Verbreitung in Südosteuropa. - *Repert. Spec. Nov. Regni Veg.* **26**: 179-219.

SOÓ, R.V. 1932: Die Orchideen Europas und des Mediterrangebietes. I. - *Pflanzenareale* **3(7)**: 73-81, Karte 61-70.

SOÓ, R.V. 1956: *Festuca* Studien (mit einem Anhang von A. Horánszky). - *Acta Bot. Acad. Sci. Hung.* **2(1-2)**: 187-220.

SOÓ, R.V. 1958: Über kritische Arten und Formen der bulgarischen Flora. - *Izv. Bot. Inst. (Sofia)* **6**: 363-374.

SOÓ, R.V. 1959: *Ophrys*-Studien. - *Acta Bot. Acad. Sci. Hung.* **5(3-4)**: 437-471.

SOÓ, R.V. 1960a: Über südosteuropäische Salicornien. - *Acta Bot. Acad. Sci. Hung.* **6(3-4)**: 397-403.

SOÓ, R.V. 1960b: Synopsis generis *Dactylorhiza* (*Dactylorchis*) (Pars II. Conspecti orchidearum europaearum et mediterranearum). - *Ann. Univ. Sci. Budapest. Rolando Eötvös, Sect. Biol.* **3**: 335-357.

SOÓ, R.V. 1964: Die *Ranunculus auricomus* L. emend Korsh. Artengruppe in der Flora Ungarns und der Karpaten. I. - *Acta Bot. Acad. Sci. Hung.* **10(1-2)**: 221-237.

SOÓ, R.V. 1965: Die *Ranunculus auricomus* L. emend Korsh. Artengruppe in der Flora Ungarns und der Karpaten. II. - *Acta Bot. Acad. Sci. Hung.* **11(3-4)**: 395-404.

SOÓ, R.V. 1970a: Arten und Unterarten der Gattung *Rhinanthus* in Europa. - *Acta Bot. Acad. Sci. Hung.* **16(1-2)**: 193-206.

SOÓ, R.V. 1970b: Species and subspecies of the genus *Ophrys*. - *Acta Bot. Acad. Sci. Hung.* **16(3-4)**: 373-392.

SOÓ, R.V. 1972a: Systematisch-nomenklatorische Bemerkungen zur Flora Mitteleuropas mit Beziehungen zur südosteuropäischen Flora. - *Feddes Repert.* **83(3)**: 129-212.

SOÓ, R.V. 1972b: Die Arten und Unterarten der Gattung *Orchis* s.str. - *Jahresber. Naturwiss. Vereins Wuppertal* **25**: 37-48.

SOÓ, R.V. 1973a: Zeitgemässe Taxonomie der *Festuca ovina*-Gruppe. - *Acta Bot. Acad. Sci. Hung.* **18(3-4)**: 363-377.

SOÓ, R.V. 1973b: Supplement to species and subspecies of the genus *Ophrys*. - *Acta Bot. Acad. Sci. Hung.* **18(3-4)**: 379-384.

SOÓ, R.V. 1974: Systematisch-nomenklatorische Bemerkungen zur Flora Mitteleuropas mit Beziehungen zur südosteuropäischen Flora. Nachträge und Verbesserungen. - *Feddes Repert.* **85(7-8)**: 433-453.

SOÓ, R.V. 1978: [New combinations in *Ophrys* L.]. In Heywood, V.H. (ed.): Notulae systematicae ad Floram Europaeam spectantes, no. 20. - *Bot. J. Linn. Soc.* **76(4)**: 367-368.

SOÓ, R.V. 1979: Nomina nova in genere *Ophrys*. - *Acta Bot. Acad. Sci. Hung.* **25(3-4)**: 361-363.

SOÓ, R.V. & BORHIDI, A. 1968: Über einige Formenkreise der ungarischen und karpatischen Flora, XII. *Myosotis palustris* und *Lamium Galeobdolon*. - *Acta Bot. Acad. Sci. Hung.* **14(1-2)**: 157-163.

SOÓ, R.V. & BORSOS, O. 1962: Die *Melittis*-Arten und Formen der ungarischen und karpatischen Flora. - *Acta Bot. Acad. Sci. Hung.* **8(1-2)**: 205-212.

SOÓ, R.V. & DRASKOVITS, R. 1968: Über einige Formenkreise der ungarischen und karpatischen Flora. *Angelica silvestris*. - *Ann. Univ. Sci. Budapest. Rolando Eötvös, Sect. Biol.* **9-10**: 383-385.

SOÓ, R.V. & ISÉPY, I. 1968: Über einige Formenkreise der ungarischen und karpatischen Flora, XVI. *Cardamine pratensis*. - *Acta Bot. Acad. Sci. Hung.* **14(3-4)**: 395-401.

SOÓ, R.V. & KOVÁCS-LÁNG, E. 1966: Über einige Formenkreise der ungarischen und karpatischen Flora. II. *Caltha*. - *Ann. Univ. Sci. Budapest. Rolando Eötvös, Sect. Biol.* **8**: 337-344.

SOÓ, R.V. & SIMON, T. 1960: Bemerkungen über südosteuropäische *Fraxinus*- und *Dianthus*-Arten. - *Acta Bot. Acad. Sci. Hung.* **6(1-2)**: 143-153.

SOÓ, R.V. & SZ.-BORSOS, O. 1968: *Erophila* species of the Carpatho-Pannonian flora. - *Acta Bot. Acad. Sci. Hung.* **14(3-4)**: 403-413.

ŠOPOVA, M. 1967: Karyological study of two legumes. - *Godišen Zborn. Prir.-Mat. Fak. Univ. Skopje, Biol.* **19**: 125-130.

ŠOPOVA, M. 1969: Cytological study in genus *Colchicum* from Macedonia. - *Godišen Zborn. Prir.- Mat. Fak. Univ. Skopje, Biol.* **21**: 119-130.

ŠOPOVA, M. 1971a: The karyology of three Macedonian Liliums. - *Godišen Zborn. Prir.-Mat. Fak. Univ. Skopje, Biol.* **23**: 151-161 + 1 plate.

ŠOPOVA, M. 1971b: The cytological study of two *Paeonia* species from Macedonia. - *Fragm. Balcan.* **8(16)**: 137-142.

ŠOPOVA, M. 1971c: The cytology of one more *Crocus* species from Macedonia. - *Fragm. Balcan.* **8 (22)**: 189-195.

ŠOPOVA, M. 1972a: The cytology of ten *Crocus* species from Macedonia. - *Godišen Zborn. Prir.-Mat. Fak. Univ. Skopje, Biol.* **24**: 73-82 + 2 plates and 1 chart.

ŠOPOVA, M. 1972b: Cytological study in the genus *Allium* from Macedonia [In Macedonian with English summary]. - *Godišen Zborn. Prir.-Mat. Fak. Univ. Skopje, Biol.* **24**: 83-102 + 5 plates.

ŠOPOVA, M. & AL. 1990: Chromosome atlas of some Macedonian angiosperms. VI. - *Godišen Zborn. Biol. Fak. Univ. "Kiril i Metodij"* **41-42**: 381-399.

ŠOPOVA, M. & AL. 1992: Chromosome atlas of some Macedonian angiosperms. VII. - *Godišen Zborn. Biol. Fak. Univ. "Kiril i Metodij"* **45**: 239-248 + 16 plates.

ŠOPOVA, M. & AL. 1993: Chromosome atlas of some Macedonian angiosperms. VIII. – *Godišen Zborn. Biol. Fak. Univ. "Kiril i Metodij"* **46**: 191-206.

ŠOPOVA, M. & GELEVSKI, B. 1971: The cytology of two *Sternbergia* species of Macedonia. - *Fragm. Balcan.* **8(14)**: 121-126.

ŠOPOVA, M. & RIZOVA, M. 1975: Chromosome numbers and karyotype study in nine species of *Plantago*. - *Godišen Zborn. Prir.-Mat. Fak. Univ. Skopje, Biol.* **27-28**: 237-251 + 2 plates.

ŠOPOVA, M. & SEKOVSKI, Ž. 1981: Chromosome atlas of some Macedonian angiosperms. II. - *Godišen Zborn. Biol. Fak. Univ. "Kiril i Metodij"* **34**: 65-76 + 13 plates.

ŠOPOVA, M. & SEKOVSKI, Ž. 1982: Chromosome atlas of some Macedonian angiosperms. III. - *Godišen Zborn. Biol. Fak. Univ. "Kiril i Metodij"* **35**: 145-160 + 13 plates.

ŠOPOVA, M. & SEKOVSKI, Ž. 1987: Chromosome atlas of some Macedonian angiosperms. V. - *Godišen Zborn. Biol. Fak. Univ. "Kiril i Metodij"* **39-40**: 353-365 + 7 plates.

ŠOPOVA, M., SEKOVSKI, Ž. & JOVANOVSKA, M. 1983a: Cytotaxonomy of the genus *Muscari* Mill. from S.R. Macedonia. - *Fragm. Balcan.* **11(16)**: 151-161.

ŠOPOVA, M., SEKOVSKI, Ž. & JOVANOVSKA, M. 1983b: Cytological study in the genus *Iris* L. from SR Macedonia. - *Fragm. Balcan.* **11(17)**: 163-173.

ŠOPOVA, M., SEKOVSKI, Ž. & JOVANOVSKA, M. 1983c: Chromosome atlas of some Macedonian angiosperms. IV. - *Godišen Zborn. Biol. Fak. Univ. "Kiril i Metodij"* **36**: 73-86.

ŠOPOVA, M., STAROVA, U. & MATVEJEVA, J. 1984: Study in the genus *Gagea*: 1. Cytotaxonomy of some *Gagea* species from Macedonia. - *Fragm. Balcan.* **12(7)**: 47-61.

SORENG, R.J. & BARRIE, F.R. 1999: (1391) Proposal to conserve the name *Poa pratensis* (Gramineae) with a conserved type. - *Taxon* **48(1)**: 157-159. [approved, cf. Taxon 49(4): 802, 2000].

SORENG, R.J. (ED.) 2003: Catalogue of New World grasses (Poaceae): IV. Subfamily Pooideae. - *Contr. U.S. Natl. Herb.* **48**: 1-730.

SORSA, P. 1980: Spore morphology of the fern genus *Gymnocarpium* and its relations to the taxonomy. - *Ann. Bot. Fenn.* **17(1)**: 86-90.

SORVIG, K. 1983: The genus *Nigella*. - *Plantsman* **4(4)**: 229-235.

SOTIRIS, A. 2012: 100 endemic plants of Greece. - Athens: Wild Greece Editions, 209 pp.

ŠOURKOVÁ, M. 1972: *Pleconax* Rafin. - eine bis heute unbeachtete Silenoideen-Gattung (Caryophyllaceae). - *Oesterr. Bot. Z.* **119(4-5)**: 577-581.

ŠOURKOVÁ, M. 1976: Die Gattung *Bupleurum* L. in der Tschechoslowakei [In Czech with German summary]. - *Zprávy Československ. Bot. Společn. Československ. Akad. Věd* **11(1)**: 1-16.

ŠOURKOVÁ, M. 1977: Taxonomic notes on *Silene* subg. *Otites*. - *Preslia* **49(1)**: 9-12.

ŠOURKOVÁ, M. 1988: *Veratrum flavum*, a forgotten species of Balkan flora. - *Novit. Bot. Univ. Carol.* **4**: 59-66.

ŠOURKOVÁ, M. 1995a: *Silene* L. subgen. *Otites* (Adanson) Reichenb. in the Czech and the Slovak republics. - *Novit. Bot. Univ. Carol.* **8**: 31-45.

ŠOURKOVÁ, M. 1995b: *Silene bazardzica* - a new endemic species of Bulgarian flora. - *Novit. Bot. Univ. Carol.* **8**: 47-49.

SOUTHERN, D.I. 1967: Species relationships in the genus *Tulipa*. - *Chromosoma* **23(1)**: 80-94.

SOWTER, F.A. 1949: Biological Flora of the British Isles. *Arum maculatum* L. - *J. Ecol.* **37(1)**: 207-219.

SOYKAN, A. & MERIÇ, Ç. 2012: Morphological and anatomical studies of *Tulipa orphanidea* (Liliaceae). - *Phytol. Balcan.* **18(1)**: 43-48.

SPACH, E. 1844: Revisio generis *Genista*. - *Ann. Sci. Nat. Bot., Sér. 3* **2(4-5)**: 237-279.

SPACH, E. 1845: Revisio generis *Genista* (suite). - *Ann. Sci. Nat. Bot., Sér. 3* **3(2-3)**: 102-158.

SPACH, E. 1846: Revisio generis *Poterium*. - *Ann. Sci. Nat. Bot., Sér. 3* **5**: 31-44.

SPACH, E. 1849: Conspectus generis *Haplophyllum*. - *Ann. Sci. Nat. Bot., Sér. 3* **11(3)**: 174-192.

SPALIK, K. 1995: Typification of the Boissier names in *Anthriscus* Pers. (Apiaceae). - *Bot. J. Linn. Soc.* **119(1)**: 77-86.

SPALIK, K. 1996: Species boundaries, phylogenetic relationships, and ecological differentiation in *Anthriscus* (Apiaceae). - *Pl. Syst. Evol.* **199(1-2)**: 17-32.

SPALIK, K. 1997: Revision of *Anthriscus* (Apiaceae). - *Polish Bot. Stud.* **13**: 1-69.

SPALIK, K. & AL. 2019: *Laserocarpum*, a new genus of Apiaceae endemic to Greece.- *Acta Soc. Bot. Poloniae* **88(3)**: 9 pp.

SPALIK, K. & JARVIS, C.E. 1989: Typification of Linnaean names now in *Anthriscus* (Apiaceae). - *Taxon* **38(2)**: 288-293.

SPALTON, L.M. 2002: An analysis of the characters of *Bromus racemosus* L., *B. commutatus* Schrad. and *B. secalinus* L. (Poaceae) - *Watsonia* **24(2)**: 193-202.

SPAMPINATO, G. & AL. 2018: Habitat, occurrence and conservation status of the Saharo-Macaronesian and Southern-Mediterranean element *Fagonia cretica* L. (Zygophyllaceae) in Italy. - *J. Arid Land* **10 (1)**: 140-151.

ŠPANIEL, S. 2019: Two new combinations and a new synonym in the genus *Odontarrhena* (Brassicaceae). - *Phytotaxa* **406(4)**: 250-254.

ŠPANIEL, S. & AL. 2015: AlyBase: database of names, chromosome numbers, and ploidylevels of Alysseae (Brassicaceae), with a new generic concept of the tribe. - *Pl. Syst. Evol.* **301(10)**: 2463-2491.

ŠPANIEL, S., MARHOLD, K. & ZOZOMOVÁ-LIHOVÁ, J. 2017: The polyploid *Alyssum montanum - A. repens* complex in the Balkans: a hotspot of species and genetic diversity. - *Pl. Syst. Evol .* **303(10)**: 1443-1465.

ŠPANIEL, S., MARHOLD, K. & ZOZOMOVÁ-LIHOVÁ, J. 2018: *Alyssum montanum - A. repens* group (Brassicaceae) in the Balkans, genetic and morphological patterns and revised taxonomic treatment. - *Bot. Serbica* **42(suppl. 1)**: 42.

ŠPANIEL, S., ZOZOMOVÁ-LIHOVÁ, J. & MARHOLD, K. 2017: Revised taxonomic treatment of the *Alyssum montanum - A. repens* complex in the Balkans: a multivariate morphometric analysis. - *Pl. Syst. Evol.* **303(10)**: 1413-1442.

SPANOS, I.A., DASKALAKOU, E.N. & THANOS, C.A. 2000: Postfire, natural regeneration of *Pinus brutia* forests in Thasos island, Greece. - *Acta Oecol.* **21(1)**: 13-20.

SPANOS, K. & AL. 2012: The importance and conservation values of beech forests in Greece - case study of an ancient beech forest in the NATURA area 'Koula - Haidou' in Xanthi, N. Greece. - *BfN-Skripten* **327**: 141-164.

SPANOS, K.A. & AL. 2021: Forests of Greece, Their multiple functions and uses, Sustainable management and biodiversity conservation in the face of climate change. - *Open J. Ecol.* **11**: 374-406.

SPANOU, S. 2010: Ecological evaluation and creation of an examplar biomonitoring programme for the wider Athens International Airport area. [In Greek with English summary]. - Patras: Ph.D. thesis, Univ. of Patras, 323 pp.

SPANOU, S. & AL. 2006: Notes on flora and vegetation of the sand dunes of western Greece. - *Willdenowia* **36(1)**: 235-246.

SPANOU, S. & AL. 2010: Contribution to the study of the flora of Attiki (Greece), new records from the flora of the wider Athens International Airport area. - *Bot. Chron. (Patras)* **20**: 83-98.

SPARLING, J.H. 1968: Biological Flora of the British Isles. *Schoenus nigricans* L. (*Chaetospora nigricans* Knuth). - *J. Ecol.* **56(3)**: 883-899.

SPASSKAYA, N.A. 1978: On the system of the *Cardamine* L. genus in the European part of the Soviet Union [In Russian with English summary]. - *Vestn. Leningradsk. Univ., Ser. Biol.* **15**: 59-67.

SPASSKAYA, N.A. 1979: Chromosome numbers of species of the genus *Cardamine* L. (Brassicaceae) from the European part of the U.S.S.R. [In Russian with English summary]. - *Bot. Zhurn. (Moscow & Leningrad)* **64(8)**: 1099-1110.

SPEIRS, D.C. 1979: The evolution of *Sempervivum*. - *J. Sempervivum Soc.* **10(1)**: 6-12.

SPENGLER, H. 1919: Die verschiedenen Typen im Korollenbau von *Lithospermum*. - *Oesterr. Bot. Z.* **68(5-7)**: 109-123 + 1 plate.

SPERANZA, M. 1995: Morphology and phenology of *Cynodon dactylon* (L.) Pers. (Gramineae) in Italy. - *Webbia* **49(2)**: 225-237.

SPERANZA, M. & BONOMO, M.C. 1994: Rapporti di affinità su basi morfologiche tra *Avena sterilis* L. subsp. *sterilis*, *Avena sterilis* L. subsp. *ludoviciana* (Durieu) Nyman, *Avena fatua* L. - *Inform. Bot. Ital.* **25(1)**: 96-99.

SPERANZA, M. & CRISTOFOLINI, G. 1986: The genus *Dactylis* L. in Italy. 1. The tetraploid entities. - *Webbia* **39(2)**: 379-396.

SPERANZA, M. & CRISTOFOLINI, G. 1987: The genus *Dactylis* L. in Italy. 2. The diploid entities. - *Webbia* **41(2)**: 213-224.

SPETA, F. 1970: *Veronica hederifolia* agg. in Oberösterreich. - *Mitt. Bot. Arbeitsgem. Oberösterr. Landesmus. Linz* **2**: 19-22.

SPETA, F. 1971a: Chromosomenzahlen einiger Angiospermen. - *Oesterr. Bot. Z.* **119(1-3)**: 1-5.

SPETA, F. 1971b: Beitrag zur Systematik von *Scilla* L. subgen. *Scilla* (inklusive *Chionodoxa* Boiss.). - *Oesterr. Bot. Z.* **119(1-3)**: 6-18.

SPETA, F. 1972: Entwicklungsgeschichte und Karyologie von Elaiosomen an Samen und Früchten. - *Naturk. Jahrb. Stadt Linz* **1972**: 9-65.

SPETA, F. 1974: Cytotaxonomische und arealkundliche Untersuchungen an der *Scilla bifolia*-Gruppe in Oberösterreich, Niederösterreich und Wien. - *Naturk. Jahrb. Stadt Linz* **1973**: 9-54.

SPETA, F. 1975: *Scilla messeniaca* Boiss. (Liliaceae) und ihre verwandtschaftlichen Beziehungen. - *Ann. Mus. Goulandris* **2**: 59-67.

SPETA, F. 1976a: Cytotaxonomischer Beitrag zur Kenntnis der *Scilla nivalis*-Gruppe. - *Linzer Biol. Beitr.* **8(1)**: 293-322. [Also distributed as *OPTIMA leaflet* no. 19].

SPETA, F. 1976b: Über *Chionodoxa* Boiss., ihre Gliederung und Zugehörigkeit zu *Scilla* L. - *Naturk. Jahrb. Stadt Linz* **21**: 9-79. [Also distributed as *OPTIMA leaflet* no. 23].

SPETA, F. 1976c: Cytotaxonomischer Beitrag zur Kenntnis der *Scilla*-Arten Ungarns und Siebenbürgens. - *Naturk. Jahrb. Stadt Linz* **22**: 9-63.

SPETA, F. 1976d: Neue *Scilla*-Arten aus dem östlichen Mittelmeerraum. - *Naturk. Jahrb. Stadt Linz* **22**: 65-72.

SPETA, F. 1976e: Auf den Spuren von *Scilla amoena* L. - *Naturk. Jahrb. Stadt Linz* **22**: 73-102.

SPETA, F. 1979a: Karyological investigations in "Scilla" in regard to their importance for taxonomy. - *Webbia* **34(1)**: 419-431.

SPETA, F. 1979b: Zur Systematik und Karyologie von *Bellevalia hyacinthoides* (Bertol.) K. Persson & Wendelbo (= *Strangweia spicata* Boiss., Liliaceae s.l.). - *Linzer Biol. Beitr.* **11(2)**: 245-266. [Also distributed as *OPTIMA leaflet* no. 103].

SPETA, F. 1980a: Die Gattungen *Chaenorhinum* (DC.) Reichenb. und *Microrrhinum* (Endl.) Fourr. im östlichen Teil ihrer Areale (Balkan bis Indien). - *Stapfia* **7**: 1-72.

SPETA, F. 1980b: Karyosystematik, Kultur und Verwendung der Meerzwiebel (*Urginea* Steinh., Liliaceae s.l.). - *Linzer Biol. Beitr.* **12(1)**: 193-238. [Also distributed as *OPTIMA leaflet* no. 104].

SPETA, F. 1981: Die frühjahrsblühenden *Scilla*-Arten des östlichen Mittelmeerraumes. - *Naturk. Jahrb. Stadt Linz* **25**: 19-198 + 47 plates. [Also distributed as *OPTIMA leaflet* no. 113].

SPETA, F. 1982: Über die Abgrenzung und Gliederung der Gattung *Muscari* und über ihre Beziehungen zu anderen Vertretern der Hyacinthaceae. - *Bot. Jahrb. Syst.* **103(2)**: 247-291.

SPETA, F. 1986a: Heterokarpidie, Dehiszenz, Heterospermie und basifixe Samen bei *Cymbalaria* Hill (Scrophulariaceae) und systematische Schlussfolgerungen. - *Phyton (Horn)* **26(1)**: 23-57.

SPETA, F. 1986b: Über die herbstblühenden Scillen des Mittelmeerraumes. - *Linzer Biol. Beitr.* **18(2)**: 399-416.

SPETA, F. 1987: Über die verwandtschaftlichen Beziehungen von *Brimeura* Salisb.: ein Vergleich mit den Gattungen *Oncostema* Rafin., *Hyacinthoides* Medic. und *Camassia* Lindl. (Hyacinthaceae). - *Phyton (Horn)* **26(2)**: 247-310.

SPETA, F. 1989a: Nomenklatorische, morphologische und karyologische Notizen zu *Allium chamaemoly* L. (Alliaceae) nebst einer Übersicht über die *Allium*-Zwiebeltypen. - *Phyton (Horn)* **29 (2)**: 227-245.

SPETA, F. 1989b: Zur Morphologie und Karyologie von *Cymbalaria ebelii* (Cufod.) Speta (Scrophulariaceae, Antirrhineae). - *Phyton (Horn)* **29(2)**: 247-254.

SPETA, F. 1990a: *Ornithogalum gussonei* Ten., *O. collinum* Guss. und *O. exscapum* Ten., drei häufig verkannte, aus Italien beschriebene Arten (Hyacinthaceae). - *Phyton (Horn)* **30(1)**: 97-171.

SPETA, F. 1990b: *Ornithogalum sibthorpii* Greuter und *O. sigmoideum* Freyn & Sint. sind nicht identisch. - *Linzer Biol. Beitr.* **22(2)**: 787-829.

SPETA, F. 1991a: Zwei neue *Scilla*-Arten (Hyacinthaceae) aus dem östlichen Mittelmeerraum. - *Phyton (Horn)* **31(1)**: 27-33.

SPETA, F. 1991b: *Ornithogalum wiedemannii* Boiss. (Hyacinthaceae) und seine nächste Verwandten. - *Candollea* **46(2)**: 485-501.

SPETA, F. 1991c: *Ornithogalum macrum* Speta (Hyacinthaceae), eine merkwürdige neue Art aus der Türkei. - *Folia Geobot. Phytotax.* **26(3)**: 349-355.

SPETA, F. 1991d: Zwei neue *Scilla*-Arten (Hyacinthaceae) aus der S-Türkei. - *Willdenowia* **21(1-2)**: 157-166.

SPETA, F. 1992: *Ornithogalum improbum* (Hyacinthaceae), eine neue Art vom Bozdağ bei Izmir (Türkei). - *Willdenowia* **22(1-2)**: 119-124.

SPETA, F. 1993: The autumn-flowering squills of the Mediterranean region. In: Demiriz, H. & Özhatay, N. (eds): Proceedings of the V OPTIMA meeting, Istanbul, 8-15.9. 1986, pp. 109-124. - Istanbul Üniversitesi Fen Fakültesi.

SPETA, F. 1994: Hyacinthaceen und Liliaceen s. str. bei Schur. [pp. 193-222 in Speta, F. (ed.): Leben und Werk von Ferdinand Schur]. - *Stapfia* **32**: 1-334.

SPETA, F. 1998a: Systematische Analyse der Gattung *Scilla* L. s.l. (Hyacinthaceae). - *Phyton (Horn)* **38(1)**: 1-141.

SPETA, F. 1998b: Die *Scilla*-Arten (Hyacinthaceae) der griechischen Inseln Kreta und Karpathos. - *Linzer Biol. Beitr.* **30(1)**: 431-437.

SPETA, F. 2000a: *Ornithogalum sphaerolobum* und seine Doppelgänger. - *Preslia* **72(2-4)**: 369-398.

SPETA, F. 2000b: Beitrag zur Kenntnis der Gattung *Prospero* Salisb. (Hyacinthaceae) auf der griechischen Insel Kreta. - *Linzer Biol. Beitr.* **32(2)**: 1323-1326.

SPETA, F. 2001: Die echte und die falsche Meerzwiebel: *Carybdis* Speta und *Stellarioides* Medicus (Hyacinthaceae), mit Neubeschreibungen und Neukombinationen im Anhang. - *Stapfia* **75**: 139-176.

SPETA, F. 2006: Die Gattung *Loncomelos* Raf. (Hyacinthaceae-Ornithogaloideae), vorgestellt anhand dreier neuer Arten. - *Phyton (Horn)* **46(1)**: 1-25.

SPETA, F., SPETA, E. & HOHLA, M. 2019: Vater und Sohn Vierhapper - zwei Leben für die Botanik. Eine Familiengeschiche im Kontext des botanischen Umfeldes. - *Stapfia* **119**: 1-202.

SPILIOPOULOU, K. & AL. 2021: The Natura 2000 network and the ranges of threatened species in Greece. - *Biodivers. Conserv.* **30(4)**: 945-961.

SPITALE, D. & PAPATHEODOULOU, A. 2019: First record of the invasive *Tradescantia fluminensis* Vell. (Commelinaceae) in three rivers in Cyprus. - *Cypricola* **11**: 1-7.

SPRAGUE, T.A. 1909: *Saxifraga scardica*. - *Bot. Mag.* **135**: tab. 8243 + 2 pp. text.

SPRAGUE, T.A. 1914: *Viola gracilis*. - *Bot. Mag.* **140**: tab. 8541 + 2 pp. text.

SPRAGUE, T.A. & HUBBARD, C.E. 1933: *Eragrostis major* and *E. minor*. - *Bull. Misc. Inform. Kew* **1933 (1)**: 15-18.

SPREITZENHOFER, G.C. 1878: Beitrag zur Flora der jonischen Inseln: Corfu, Cephalonia und Ithaca. - *Verh. K.K. Zool.-Bot. Ges. Wien* **27**: 711-734.

SPRENGEL, K. 1818: Species Umbelliferarum minus cognitae. - Hallae: Officina Rengeriana, x + 154 pp. + 7 plates.

SPRENGEL, K. 1822: Pflanzen in Clarke's Reisen. In Sprengel, K.: Neue Entdeckungen im ganzen Umfang der Pflanzenkunde 3: 157-166. - Leipzig: Friedrich Fleischer.

SPRENGEL, K. 1825-28: Caroli Linnaei ... Systema vegetabilium. Edito decima sexta ... – Gottingae [Göttingen]: Librariae Dieterichianae, 5 vols. [cf. TL-2 5: 813, 1985].

SPRENGER, C. 1908: Beitrag zur Kenntnis der *Quercus coccifera* L. - *Mitt. Deutsch. Dendrol. Ges.* **17**: 155-157, 219.

SPRENGER, C. 1912a: Korfus Wälder in unseren Tagen. - *Mitt. Deutsch. Dendrol. Ges.* **21**: 127-132.

SPRENGER, C. 1912b: Wilde Orchideen Korfus. - *Gartenwelt* **16**: 370-374.

SPRENGER, C. 1913a: Korfus Wälder in unseren Tagen (II. Teil). - *Mitt. Deutsch. Dendrol. Ges.* **22**: 208-212.

SPRENGER, C. 1913b: Das spanische Ginster. - *Mitt. Deutsch. Dendrol. Ges.* **22**: 212-218.

SPRENGER, C. 1913c: Im Walde der *Abies cephalonica*. - *Gartenwelt* **17**: 677.

SPRENGER, C. 1914: Dendrologische Mitteilungen aus Korfu. - *Mitt. Deutsch. Dendrol. Ges.* **23**: 221-235.

SPRENGER, C. 1915a: *Cyclamen* auf Korfu. - *Gartenwelt* **19**: 119-120.

SPRENGER, C. 1915b: Dendrologische Mitteilungen aus Leukas. - *Mitt. Deutsch. Dendrol. Ges.* **24**: 1-33.

SPRENGER, C. 1916a: Neue Mitteilungen über den Lorbeer. - *Mitt. Deutsch. Dendrol. Ges.* **25**: 99-103.

SPRENGER, C. 1916b: Ölbaum und Oleaster. - *Mitt. Deutsch. Dendrol. Ges.* **25**: 103-110.

SPRENGER, C. 1916c: Die Gnetaceen (Meerträubelgewächse) Griechenlands. - *Mitt. Deutsch. Dendrol. Ges.* **25**: 110-113.

SPRENGER, C. 1916d: *Erica verticillata* Forsk. - *Gartenwelt* **20**: 42.
SPRENGER, C. 1917: Dendrologische Mitteilungen aus Griechenland. - *Mitt. Deutsch. Dendrol. Ges.* **26**: 144-169.
SPRENGER, C. 1923: Über allerlei Sträucher Griechenlands. - *Mitt. Deutsch. Dendrol. Ges.* **33**: 206-214.
SPRENGER, C. 1925: Über allerlei Sträucher Griechenlands. III. - *Mitt. Deutsch. Dendrol. Ges.* **35**: 116-124.
SPRENGER, C. 1930: Über einige Baumarten Griechenlands. II. - *Mitt. Deutsch. Dendrol. Ges.* **42**: 74-84.
SPRENGER, C. 1931: Vom Efeu in Griechenland. - *Mitt. Deutsch. Dendrol. Ges.* **43**: 303-307 + 2 plates.
SPRENGER, C. 1933: Gehölze mit Nutzfrüchten in Griechenland. I. - *Mitt. Deutsch. Dendrol. Ges.* **45**: 158-168.
SPRENGER, C. 1934: Gehölze mit Nutzfrüchten in Griechenland. II. - *Mitt. Deutsch. Dendrol. Ges.* **46**: 84-97.
SPRINGATE, L. 2000: *Sedum stefco* vs. *Sedum stefscho*. - *Sedum Soc. Newsletter* **55**: 12-13.
SPRUNER, W.V. 1839: [Verzeichnis griechischer Pflanzen]. - *Flora* **22(2)**: 704 + Int.-bl. 17-19, 24-28.
SRAMKO, G. & AL. 2014: Molecular phylogeny and evolutionary history of the Eurasiatic orchid genus *Himantoglossum* s.l. (Orchidaceae). - *Ann. Bot. (Oxford), n.s.* **114(8)**: 1609-1626.
STACE, C.A. 1961: Some studies in *Calystegia*: Compatibility and hybridisation in *C. sepium* and *C. silvatica*. - *Watsonia* **5(2)**: 88-105.
STACE, C.A. 1971a: Notes on the European species of *Convolvulus*. In Heywood, V.H. (ed.): Notulae systematicae ad Floram Europaeam spectantes, no. 8. - *Bot. J. Linn. Soc.* **64(1)**: 56-60.
STACE, C.A. 1971b: Anatomy and taxonomy in *Juncus* subgenus *Genuini*. - *Bot. J. Linn. Soc.* **63, suppl. 1**: 75-81.
STACE, C.A. 1973: Chromosome numbers in the British species of *Calystegia* and *Convolvulus*. - *Watsonia* **9(4)**: 363-367.
STACE, C.A. 1977: The origin of radiate *Senecio vulgaris* L. - *Heredity* **39(3)**: 383-388.
STACE, C.A. 1978a: Changing concepts of the genus *Nardurus* Reichenb. (Gramineae). In Heywood, V.H. (ed.): Notulae systematicae ad Floram Europaeam spectantes, no. 20. - *Bot. J. Linn. Soc.* **76 (4)**: 344-350.
STACE, C.A. 1978b: Notes on *Cutandia* and related genera. In Heywood, V.H. (ed.): Notulae systematicae ad Floram Europaeam spectantes, no. 20. - *Bot. J. Linn. Soc.* **76(4)**: 350-352.
STACE, C.A. 1981: Generic and infrageneric nomenclature of annual Poaceae: Poeae related to *Vulpia* and *Desmazeria*. - *Nordic J. Bot.* **1(1)**: 17-26.
STACE, C.A. & AINSCOUGH, M.M. 1984: Continuing addition to the gene-pool of the *Festuca rubra* aggregate. - *Pl. Syst. Evol.* **147(3-4)**: 227-236.
STACE, C.A. & AUQUIER, P. 1978: Taxonomy and variation of *Vulpia ciliata* Dumort. - *Bot. J. Linn. Soc.* **77(2)**: 107-112.
STACE, C.A. & COTTON, R. 1976: Nomenclature, comparison and distribution of *Vulpia membranacea* (L.) Dumort. and *V. fasciculata* (Forskål) Samp. - *Watsonia* **11(2)**: 117-123.
STACE, C.A. & JARVIS, C.E. 1985: Typification of Linnaean taxa of annual Poaceae: Poeae related to *Vulpia* and *Desmazeria*. - *Bot. J. Linn. Soc.* **91(3)**: 435-444.
STACE, C.A. & TUTIN, T.G. 1978: A new species of *Gaudinia* from Spain. In Heywood, V.H. (ed.): Notulae systematicae ad Floram Europaeam spectantes, no. 20. - *Bot. J. Linn. Soc.* **76(4)**: 353-356.
STACE, C.A. (ED.) 1989: IOPB chromosome data, 1. - *Newslett. Int. Organ. Pl. Biosyst.* **13**: 15-22.
STADLMANN, J. 1910: Ein Beitrag zur Kenntnis der Gattung *Pedicularis* (Sekt. *Comosae* Maxim.). - Wien: Verlag des k.k. Staatsgymnasiums im XIII. Bezirke Wiens, 7 pp.
STAFLEU, F.A. 1970: Boissier's diagnoses. - *Taxon* **19(5)**: 803-805.
STAFLEU, F.A. & COWAN, R.S. 1976: Taxonomic literature. A selective guide to botanical publications and collections with dates, commentaries and types. Second edition. Volume I: A-G. - *Regnum Veg.* **94**: xl + 1136 pp.
STAFLEU, F.A. & COWAN, R.S. 1979: Taxonomic literature. A selective guide to botanical publications and collections with dates, commentaries and types. Second edition. Volume II: H-Le. - *Regnum Veg.* **98**: xviii + 991 pp.
STAFLEU, F.A. & COWAN, R.S. 1981: Taxonomic literature. A selective guide to botanical publications and collections with dates, commentaries and types. Second edition. Volume III: Lh-O. - *Regnum Veg.* **105**: xii + 980 pp.
STAFLEU, F.A. & COWAN, R.S. 1983: Taxonomic literature. A selective guide to botanical publications and collections with dates, commentaries and types. Second edition. Volume IV: P-Sak. - *Regnum Veg.* **110**: ix + 1214 pp.
STAFLEU, F.A. & COWAN, R.S. 1985: Taxonomic literature. A selective guide to botanical publications and collections with dates, commentaries and types. Second edition. Volume V: Sal-Ste. - *Regnum Veg.* **112**: 1066 pp.
STAFLEU, F.A. & COWAN, R.S. 1986: Taxonomic literature. A selective guide to botanical publications and collections with dates, commentaries and types. Second edition. Volume VI: Sti-Vuy. - *Regnum Veg.* **115**: 926 pp.
STAFLEU, F.A. & COWAN, R.S. 1988: Taxonomic literature. A selective guide to botanical publications and collections with dates, commentaries and types. Second edition. Volume VII: W-Z. - *Regnum Veg.* **116**: lvi + 653 pp.
STAFLEU, F.A. & MENNEGA, E.A. 1992: Taxonomic literature. A selective guide to botanical publications and collections with dates, commentaries and types. Supplement I: A-Ba. - *Regnum Veg.* **125**: viii + 453 pp.
STAFLEU, F.A. & MENNEGA, E.A. 1993: Taxonomic literature. A selective guide to botanical publications and collections with dates, commentaries and types. Supplement II: Be-Bo. - *Regnum Veg.* **130**: vi + 464 pp.
STAFLEU, F.A. & MENNEGA, E.A. 1995: Taxonomic literature. A selective guide to botanical publications and collections with dates, commentaries and types. Supplement III: ... - *Regnum Veg.* **132**:
STAFLEU, F.A. & MENNEGA, E.A. 1997: Taxonomic literature. A selective guide to botanical publications and collections with dates, commentaries and types. Supplement IV: Ce-Cz. - *Regnum Veg.* **134**: iv + 613 pp.
STAFLEU, F.A. & MENNEGA, E.A. 1999: Taxonomic literature. A selective guide to botanical publications and collections with dates, commentaries and types. Supplement V: Da-Di. - *Regnum Veg.* **135**: viii + 432 pp.
STAHL-BISKUP, E. & SAEZ, F. (EDS) 2002: Thyme: the genus *Thymus*. - London: Taylor & Francis, xii + 330 pp.
STANEV, S. 1972: On the history of the Bulgarian relict endemics: 2. *Haberlea rhodopensis* Friv. [In Bulgarian with Russian and English summaries]. - *Izv. Prir. Muz. Plovdiv* **2**: 31-42.
STANTON, T.R. 1955: Oat identification and classification. - *Techn. Bull. U.S.D.A.* **1100**: 1-206.
STAPF, O. 1915: The genus *Phelipaea*. - *Bull. Misc. Inform. Kew* **1915(6)**: 285-295 + 1 plate.
STAPF, O. 1916: *Abies cephalonica*. - *Bot. Mag.* **142**: tab. 8691 + 5 pp. text.
STAPF, O. 1918: *Paeonia peregrina*. - *Bot. Mag.* **144**: tab. 8742 + 4 pp. text.
STAPF, O. 1919a: *Iris reichenbachii*. - *Bot. Mag.* **145**: tab. 8812 + 3 pp. text.
STAPF, O. 1919b: *Atraphaxis billardieri*. - *Bot. Mag.* **145**: tab. 8820 + 2 pp. text.
STAPF, O. 1925: *Smilax excelsa*. - *Bot. Mag.* **150**: tab. 9067 + 5 pp. text.
STAPF, O. 1930: *Ephedra procera*. - *Bot. Mag.* **153(4)**: tab. 9204 + 4 pp. text.
STAPLES, G. & JARVIS, C.E. 2006: Typification of Linnaean plant names in Convolvulaceae. - *Taxon* **55 (4)**: 1019-1024.
STAPPERFENNE, H.-J. 1966: *Leucorchis frivaldii* (Hampe) Schltr. - eine Orchidee des südöstlichen Europa. - *Orchidee (Hamburg)* **17(Sonderheft)**: 124-127.
STARKEY, J. 2018: From Ottoman Aleppo to Edinburgh: the botanical legacies of Adam Freer MD and his colleagues. In Aščerić-Todd, I. & al. (eds): Travellers in Ottoman lands. The Botanical Legacy, pp. 123-145. - Oxford: ASTENE and Archaeopress Publishing Ltd, 379 pp.
STARLING, B.N. 1978a: Plant awards 1977-78: *Ranunculus millefoliatus* Vahl. - *Quart. Bull. Alpine Gard. Soc. Gr. Brit.* **46(4)**: 294-295.

STARLING, B.N. 1978b: Plant awards 1977-78: *Achillea ageratifolia* Boiss. - *Quart. Bull. Alpine Gard. Soc. Gr. Brit.* **46(4)**: 307.

STARLING, B.N. 1979: Plant awards 1979-80: *Cystopteris dickieana* Sim. - *Quart. Bull. Alpine Gard. Soc. Gr. Brit.* **47(4)**: 354-355.

STARLING, B.N. 1980a: Plant awards 1979-80: *Origanum* (*Amaracus*) *dictamnus* L. - *Quart. Bull. Alpine Gard. Soc. Gr. Brit.* **48(4)**: 293.

STARLING, B.N. 1980b: Plant awards 1979-80: ×*Jankemonda vandedemii* (Beauverd) Hort. - *Quart. Bull. Alpine Gard. Soc. Gr. Brit.* **48(4)**: 299-300.

STARMÜHLER, W. 1998: (1370) Proposal to conserve the name *Aconitum* (Ranunculaceae) with *A. variegatum* as its type. - *Taxon* **47(3)**: 747-748. [not approved, cf. Taxon 49(2): 274, 2000].

STARÝ, F. 1954: Die Heilpflanzen Albaniens. I. Gemeine oder echte Meereszwiebel - *Urginea maritima* (L.) Baker. - *Preslia* **26**: 365-384.

STATHI, E. & AL. 2020: Population genetic variability and distribution of the endangered Greek endemic *Cicer graecum* under climate change scenarios. - *AoB Plants* **12**: doi:10.1093/aobpla/plaa007.

STAUDT, G. 1953: Die geographische Verbreitung der Gattung *Fragaria* und die Bedeutung für die Phylogenie der Gattung. - *Ber. Deutsch. Bot. Ges.* **66(6)**: 236-238.

STAUDT, G. 1962: Taxonomic studies in the genus *Fragaria*. Typification of *Fragaria* species known at the time of Linnaeus. - *Canad. J. Bot.* **40(6)**: 869-886 + 2 plates.

STAVRIDAKIS, K.G. 2006: Wild edible plants of Crete [In Greek and English]. - Rethimno, Crete: 414 pp.

STEARN, W.T. 1941: *Paeonia rhodia*: the wild paeony of Rhodes. - *Gard. Chron., ser. 3* **110**: 158-160.

STEARN, W.T. 1967: Sibthorp, Smith, the 'Flora Graeca' and the 'Florae Graecae Prodromus'. - *Taxon* **16(3)**: 168-178.

STEARN, W.T. 1973: A synopsis of the genus *Vinca* including its taxonomic and nomenclatural history. In Taylor, W.I. & Farnsworth, N.R. (eds): The *Vinca* alkaloids: botany, chemistry and pharmacology, pp. 19-94. - New York: Dekker.

STEARN, W.T. 1974: Typification of *Cannabis sativa* L. - *Bot. Mus. Leafl.* **23(9)**: 325-336.

STEARN, W.T. 1975a: Multum pro parvo: the nomenclatural history and synonymy of *Neotinea maculata* (Orchidaceae). - *Ann. Mus. Goulandris* **2**: 69-81.

STEARN, W.T. 1975b: The name *Helichrysum virgineum*. In Heywood, V.H. (ed.): Notulae systematicae ad Floram Europaeam spectantes, no. 17. - *Bot. J. Linn. Soc.* **71(1)**: 44-45.

STEARN, W.T. 1976: From Theophrastus and Dioscorides to Sibthorp and Smith: the background and origin of the Flora Graeca. - *Biol. J. Linn. Soc.* **8(4)**: 285-298 + 2 plates.

STEARN, W.T. 1977: *Allium longanum*, a species new for the south-eastern Aegean region. - *Ann. Mus. Goulandris* **3**: 39-44. [Also distributed as *OPTIMA leaflet* no. 55].

STEARN, W.T. 1978a: European species of *Allium* and allied genera of Alliaceae: A synonymic enumeration. - *Ann. Mus. Goulandris* **4**: 83-198.

STEARN, W.T. 1978b: Mediterranean and Indian species of *Drimia* (Liliaceae): A nomenclatural survey with special reference to the medicinal squill, *D. maritima* (syn. *Urginea maritima*). - *Ann. Mus. Goulandris* **4**: 199-210.

STEARN, W.T. 1981: The genus *Allium* in the Balkan Peninsula. - *Bot. Jahrb. Syst.* **102(1-4)**: 201-213.

STEARN, W.T. 1982: The "Flora Europaea" and the Greek flora. - *Ann. Mus. Goulandris* **5**: 123-129.

STEARN, W.T. 1983a: Pauline Haritonidou (1920-1981), botanist. - *Ann. Mus. Goulandris* **6**: 111-115.

STEARN, W.T. 1983b: The Linnaean species of *Ornithogalum* (Liliaceae). - *Ann. Mus. Goulandris* **6**: 139-170.

STEARN, W.T. 1984: Homonyms in the genus *Ornithogalum* (Liliaceae). - *Bot. Helv.* **94(1)**: 189-197.

STEARN, W.T. 1985: Constantinos A. Zahariadi [obituary]. - *Taxon* **34(4)**: 764-765.

STEARN, W.T. 1986: The Greek species of *Symphytum* (Boraginaceae). - *Ann. Mus. Goulandris* **7**: 175-220.

STEARN, W.T. 1990: The Linnaean species of *Hyacinthus* (Liliaceae: Hyacinthaceae). - *Ann. Mus. Goulandris* **8**: 181-222.

STEARN, W.T. 1993: The gender of the generic name *Onosma* (Boraginaceae). - *Taxon* **42(3)**: 679-681.

STEARN, W.T. 1995a: The feminine gender of the generic name *Onosma* (Boraginaceae). - *Ann. Mus. Goulandris* **9**: 235-238.

STEARN, W.T. 1995b: Constantinos Alexandros Zahariadi (1901-1985), botanist. - *Ann. Mus. Goulandris* **9**: 239-243.

STEARN, W.T. 1999: John Lindley's contribution to Sibthorp and Smith's "Flora Graeca". - *Ann. Mus. Goulandris* **10**: 163-170.

STEARN, W.T. & DAVIS, P.H. 1984: Peonies of Greece. A taxonomic and historical survey of the genus *Paeonia* in Greece. - Kifissia: Goulandris Natural History Museum, 133 pp. [With foreword and plates by N. Goulandris].

STEARN, W.T., DAVIS, P.H. & TAN, KIT 2012: Peonies of Greece. A taxonomic and historical survey ofthe genus *Paeonia* in Greece. - Kifissia: Goulandris Nat. Hist. Mus., 133 pp. [With foreword and plates by N. Goulandris]. Second edition of Stearn and Davis (1984) with text revised by Kit Tan.

STEARN, W.T. & ÖZHATAY, N. 1977: *Allium sphaerocephalon* and an allied species of European Turkey, *A. proponticum*. - *Ann. Mus. Goulandris* **3**: 45-50. [Also distributed as *OPTIMA leaflet* no. 56].

STEARN, W.T. & SHAW, J.M.H. 2002: The genus *Epimedium* and other herbaceous Berberidaceae including the genus *Podophyllum*. - Portland, Oregon: Timber Press, xi + 342 pp.

STEBBINS, G.L. 1937: Critical notes on *Lactuca* and related genera. - *J. Bot. (London)* **75**: 12-18.

STEBBINS, G.L. 1981: Chromosomes and evolution in the genus *Bromus* (Gramineae). - *Bot. Jahrb. Syst.* **102(1-4)**: 359-379.

STEBBINS, G.L., JENKINS, J.A. & WALTERS, M.S. 1953: Chromosomes and phylogeny in the Compositae, tribe Cichorieae. - *Univ. Calif. Publ. Bot.* **26(6)**: 401-429.

STEBBINS, G.L. & ZOHARY, D. 1959: Cytogenetic and evolutionary studies in the genus *Dactylis*. I: Morphology, distribution and interrelationships of the diploid subspecies. - *Univ. Calif. Publ. Bot.* **31(1)**: 1-32 + 3 plates.

STECK-BLASER, B. 1992: Karyologische Untersuchungen an *Muscari comosum* (L.) Miller, *M. botryoides* (L.) Miller emend. D.C. und *M. racemosum* (L.) Miller emend. D.C. im Gebiet der Schweiz. - *Bot. Helv.* **102(2)**: 211-227.

STEDJE, B. 2001: The generic delimitation within Hyacinthaceae: a comment on works by F. Speta. - *Bothalia* **31(2)**: 192-195.

STEEN, A, VAN DER 1995: Does endive integrate with wild chicory? In Valdés, B. & Silvestre, S. (eds): Abstracts. VIII OPTIMA meeting, Sevilla, Spain, 25 September - 1 October, 1995, p. 123.

STEFANAKI-NIKIFORAKI, M. 1972: Simvoli is tin meleti tou polimorfismou tou genous *Aegilops* en Elladi. - Athine: Thesis, 109 pp.

STEFANAKI-NIKIFORAKI, M. 1982: Simvoli is tin gnosin tis chloridos periochon tou nomou Messinias entos tis evmesogiakis zonis vlastiseos [In Greek]. - *Geôrgiki Erevna* **6**: 443-454.

STEFANAKI, A. & KOKKINI, S. 2015: Phytogeographical affinities at the crossroads of two continents: Distribution patterns of Lamiaceae in Chios Island (East Aegean Islands, Greece) and Çeşme-Karaburun Peninsula (West Anatolia, Turkey). - *Syst. Biodivers.* **13(4)**: 307-325.

STEFANI, A. & ONNIS, A. 1985: Contributo alla conoscenza della caratteristiche fenologiche di popolazioni naturali di *Dasypyrum villosum* (L.) P. Candargy di diverse provenienze. - *Inform. Bot. Ital.* **15(2-3)**: 213-224.

STEFANI, A. & ONNIS, A. 1988: Significato dell'allogamia in *Dasypyrum villosum* (L.) P. Candargy. - *Inform. Bot. Ital.* **19(3)**: 395-400.

STEFANI, C. DE., FORSYTH MAJOR, C.J. & BARBEY, W. 1892: Samos. Étude géologique, paléontologique et botanique. - Lausanne: Georges Bridel & Cie Éditeurs, 99 pp. + 14 plates.

STEFANI, C. DE., FORSYTH MAJOR, C.J. & BARBEY, W. 1895: Karpathos. Étude géologique, paléontologique et botanique. - Lausanne: Georges Bridel & Cie Éditeurs, 180 pp. + 15 plates.

STEFANIDIS, K. & PAPASTERGIOU, E.S. 2007: Aquatic vegetation and related abiotic environment in a shallow urban lake of Greece. - *Belgian J. Bot.* **140(1)**: 25-38.

STEFANOV, B. 1921: Notizen zur Flora von West-Thrazien [In Bulgarian with German summary]. - *God. Sofiisk. Univ. Fiz.-Mat. Fak.* **15-16**: 1-100 + 1 map.

STEFANOV, B. 1926a: *Colchicum Davidovi* n sp. - *Izv. Bulg. Bot. Druzh.* **1**: 69-70.

STEFANOV, B. 1926b: Monographie der Gattung *Colchicum* L. [In Bulgarian with key in Latin]. - *Sborn. Bălg. Akad. Nauk., Klon. Prir.-Mat.* **22**: 1-100 + 2 maps.

STEFANOV, B. 1927: Herkunft und Entwicklung der Vegetationstypen in den Rhodopen [In Bulgarian with German summary]. - Sofia: Staatsverlag, 204 pp. + 3 tables and 9 maps.

STEFANOV, B. 1928: Über einige recente und fossile Eichenarten in Bulgarien. - *Izv. Bulg. Bot. Druzh.* **2**: 61-86 + 2 plates.

STEFANOV, B. 1929: Fünf neue Arten aus Bulgarien und Thrakien. - *Izv. Bulg. Bot. Druzh.* **3**: 77-85.

STEFANOV, B. 1930: *Poa Borisii* nov. spec. - *Izv. Tsarsk. Prir. Inst. Sofiya* **3**: 130-140.

STEFANOV, B. 1932: Systematische und geographische Studien über die mediterran-orientalischen Arten der Gattung *Hypericum* [In Bulgarian with German summary]. - *God. Sofiisk. Univ. Agron.- Lesov. Fak., 1* **10**: 19-58.

STEFANOV, B. 1933a: Systematische und geographische Studien über die mediterran-orientalischen Arten der Gattung *Hypericum* [cont.; in Bulgarian with German summary]. - *God. Sofiisk. Univ. Agron.-Lesov. Fak., 1* **11**: 139-186.

STEFANOV, B. 1933b: Die mediterran-orientalischen Arten der Gattung *Hypericum* L. – *Pflanzenareale* **4(1)**: 1-10, Karte 1-9.

STEFANOV, B. 1934: Systematische und geographische Studien über die mediterran-orientalischen Arten der Gattung *Hypericum* [end; in Bulgarian with German summary]. - *God. Sofiisk. Univ. Agron.-Lesov. Fak., 1* **12**: 69-100.

STEFANOV, B. 1941: Bemerkungen und Nachträge zur Flora Bulgariens [In Bulgarian]. - *Izv. Tsarsk. Prir. Inst. Sofiya* **14**: 312-320.

STEFANOV, B. & GEORGIEV, T. 1931: Beitrag zur Begrenzung der Arten der Gattung *Centaurea* L. von der Sect. *Cyanus* D.C. [In Bulgarian with German summary]. - *Spis. Bulg. Akad. Nauk* **23**: 133-193.

STEFANOV, B. & JORDANOV, D. 1931a: Materialien zur Kenntnis der Moor-Vegetation in den West- Rhodopen (Dospad-Gebirge). - *Bot. Jahrb. Syst.* **64(4)**: 357-385.

STEFANOV, B. & JORDANOV, D. 1931b: Topographische Flora von Bulgarien. - *Bot. Jahrb. Syst.* **64(4-5)**: 388-536 + 1 map.

STEFANOV, B. & NEDJALKOV, S. 1956: *Quercus thracica* nov. spec. - arbor memorabilis [In Bulgarian with Russian and German summaries]. - *Izv. Bot. Inst. (Sofia)* **5**: 39-53 + 1 plate.

STEFANOVA-GATEVA, B. 1979: Zwei neue Arten der Gattung *Verbascum* in Bulgarien. - *Feddes Repert.* **90(7-8)**: 407-408.

STEFANOVA-GATEVA, B. 1987: Study on the species of *Verbascum* L. in Bulgaria. In: Kuzmanov, B.A. & al. (eds): Proceedings, fourth national conference of botany, pp. 8-13. - Sofia.

STEFANOVA, B. 1974: [Chromosome numbers in *Verbascum* from Bulgaria]. In Löve, Á. (ed.): IOPB chromosome number reports XLIII. - *Taxon* **23(1)**: 195.

STEFANOVIĆ, S. 2006: Distribution, diversity and molecular phylogeny of *Edraianthus* DC. (grass bells, Campanulaceae). In Peev, D.R. (ed.): Book of abstracts, IV Balkan botanical congress, Sofia, Bulgaria, 20-26 June 2006, p. 45.

STEFANOVIĆ, S. & AL. 2008: Molecular phylogeny of *Edraianthus* (Grassy Bells: Campanulaceae) based on non-coding plastid DNA sequences. - *Taxon* **57(2)**: 452-475.

STEFANOVIĆ, S. & LAKUŠIĆ, D. 2009: Molecular reappraisal confirms that the *Campanula trichocalycina-pichleri* complex belongs to *Asyneuma* (Campanulaceae). - *Bot. Serbica* **33(1)**: 21-31.

STEIN, B. 1879: *Haberlea rhodopensis* Frivaldsky. - *Gartenflora* **28**: 323-325 + 1 plate.

STEINBERG, C. 1971: Revisione sistematica e distributiva delle "Adonis" annuali in Italia. - *Webbia* **25 (2)**: 299-351.

STEINBERG, C. 1979: The location of Mediterranean specimens in various European herbaria, seen through specimens of *Adonis microcarpa* DC. - *Webbia* **34(1)**: 543-551.

STEINBERG, C. 1981: Nomenklatorische Typen aus der Gattung *Bromus* in italienischen Herbarien. - *Bot. Jahrb. Syst.* **102(1-4)**: 411-425.

STEINECKE, K. & BENNERT, H.W. 1993: Biosystematic investigations of the *Asplenium obovatum* complex (Aspleniaceae, Pteridophyta). I. Morphology. - *Bot. Jahrb. Syst.* **114(4)**: 481-502.

STEININGER, H. 1886a: Beschreibung der europäischen Arten des Genus *Pedicularis*. - *Bot. Centralbl.* **28**: 215-219, 246-249, 279-282, 313-315.

STEININGER, H. 1886b: Beschreibung der europäischen Arten des Genus *Pedicularis* [cont.]. - *Bot. Centralbl.* **28**: 341-342, 375-377, 388-391.

STEININGER, H. 1887a: Beschreibung der europäischen Arten des Genus *Pedicularis* [cont.]. - *Bot. Centralbl.* **29**: 23-24, 54-58, 85-89, 122-123, 154-157.

STEININGER, H. 1887b: Beschreibung der europäischen Arten des Genus *Pedicularis* [cont.]. - *Bot. Centralbl.* **29**: 185-188, 216-221, 246-250, 278-280.

STEININGER, H. 1887c: Beschreibung der europäischen Arten des Genus *Pedicularis* [cont.]. - *Bot. Centralbl.* **29**: 314-317, 346-349, 375-378.

STEININGER, H. 1887d: Beschreibung der europäischen Arten des Genus *Pedicularis* [cont. and end]. - *Bot. Centralbl.* **30**: 25-28, 56-62, 87-93.

ŠTĚPÁNEK, J. 1982: Die Chromosomenzahlen von tschechoslowakischen Arten der Gattung *Knautia* L. (Dipsacaceae). - *Folia Geobot. Phytotax.* **17(4)**: 359-386.

ŠTĚPÁNEK, J. 1983: (721) Proposal to conserve 2999 *Arabidopsis* Heynh. (Cruciferae) with a conserved type. - *Taxon* **32(4)**: 649-650. [approved, cf. Taxon 36(1): 77, 1987].

ŠTĚPÁNEK, J. & KIRSCHNER, J. 2014: A revision of names in *Taraxacum* sect. *Erythrocarpa* and *T.* sect. *Erythrosperma* (Asteraceae: Cichorieae) published by C. E. Sonck from Greece, with nomenclatural comments. - *Willdenowia* **44(1)**: 137-144.

ŠTĚPÁNEK, J. & KIRSCHNER, J. 2015: *Taraxacum umbrosum* (Asteraceae, Cichorieae), a new species intermediate between sect. *Erythrosperma* and sect. *Erythrocarpa*, widespread in the Balkans. - *Ann. Bot. Fenn.* **52(3-4)**: 160-164.

ŠTĚPÁNEK, J. & KIRSCHNER, J. 2018: Taxonomic revision of selected species in *Taraxacum* sect. *Erythrosperma* (Asteraceae: Cichorieae) from the E Mediterranean region. - *Willdenowia* **48(3)**: 365-369.

ŠTĚPÁNKOVÁ, J. 1993a: *Myosotis margaritae* - a new species from Bulgaria. - *Folia Geobot. Phytotax.* **28(2)**: 279-288.

ŠTĚPÁNKOVÁ, J. 1993b: Reports (179-184). In Kamari, G., Felber, F. & Garbari, F. (eds): Mediterranean chromosome number reports - 3. - *Fl. Medit.* **3**: 344-347.

ŠTĚPÁNKOVÁ, J. 1994a: Remarks on the karyogeography of *Myosotis alpestris* in Europe. - *Preslia* **65(4)**: 325-334.

ŠTĚPÁNKOVÁ, J. 1994b: *Myosotis margaritae* - a new species for Greece, Romania and Serbia. - *Preslia* **66(3)**: 261-264.

ŠTĚPÁNKOVÁ, J. 1994c: *Myosotis michaelae*, a new species of *Myosotis* ser. *Palustres* (Boraginaceae). - *Folia Geobot. Phytotax.* **29(3)**: 375-384.

ŠTĚPÁNKOVÁ, J. 2006: Karyotaxonomy of *Myosotis alpestris* group. - *Preslia* **78(3)**: 345-352.

STEPANOVA, S.I. 1985: The history and taxonomy of Mediterranean species of lupins (*Lupinus* L.) [In Russian]. - *Trudy Prikl. Bot.* **91**: 48-55.

STEPHANIDES, TH. 1940: A synoptic survey of the fresh-water flora of Corfu. - *Actes Inst. Bot. Univ. Athènes* **1**: 211-234.

STEPHANIDES, TH. 1948a: A survey of the freshwater biology of Corfu and of certain other regions of Greece [vascular plants on pp. 130-139, 187-201, 211-212]. - *Prakt. Hellen. Hydroviol. Inst.* **2**: 263 pp.

STEPHANOU, G. & GEORGIADIS, TH. 1982: [Chromosome numbers in *Anthemis* from Greece]. In Löve, Á. (ed.): Chromosome number reports LXXVII. - *Taxon* **31(4)**: 764.

STEPHENSON, R. 1994: *Sedum*: cultivated stonecrops. - Portland, Oregon: Timber Press, 335 pp.

STEPHENSON, R. 2002: *Sedum eriocarpum* from Lefkada (SSS02-23). - *Sedum Soc. Newsletter* **63**: 12-13.
STEPHENSON, R. 2003: *Sedum stefco*: another rare European stonecrop that is a denizen of wet places. - *Cactus & Co.* **7(2)**: 69-75.
STEPHENSON, R. 2007a: Very wet Crete. - *Sedum Soc. Newsletter* **81**: 61-66.
STEPHENSON, R. 2007b: *Sedum creticum* Presl., an evolving species. - *Int. Cactus Adventures* **76**: 24-27.
STEPHENSON, R. & STEPHENSON, J. 1997: Kefalonian stonecrops. - *Sedum Soc. Newsletter* **40**: 11-17.
STEPHENSON, R. & STEPHENSON, J. 2000a: Crete in late summer. - *Sedum Soc. Newsletter* **52**: 37-45.
STEPHENSON, R. & STEPHENSON, J. 2000b: *Sedum tuberiferum* and other Greek Macedonian stonecrops in habitat. - *Sedum Soc. Newsletter* **55**: 15-24.
STEPHENSON, R. & STEPHENSON, J. 2000c: Crete in summer part 2. - *Sedum Soc. Newsletter* **53**: 28-31.
STEPHENSON, R. & STEPHENSON, J. 2002: Bansko and beyond part 1 - the Rila Mountains. - *Sedum Soc. Newsletter* **63**: 15-20.
STEPHENSON, R. & STEPHENSON, J. 2003a: Bansko & beyond: Bulgaria II - Northern Pirin Mountains and the lower Mesta Valley. - *Sedum Soc. Newsletter* **64**: 32-36.
STEPHENSON, R. & STEPHENSON, J. 2003b: Bulgaria 3 - the western Rhodopes. - *Sedum Soc. Newsletter* **65**: 56-63.
STEPHENSON, R. & STEPHENSON, J. 2004: Corfu, towards a checklist. - *Sedum Soc. Newsletter* **73**: 52-57.
STEPHENSON, R. & STEPHENSON, J. 2007: Fireplace of endemism. - *Sedum Soc. Newsletter* **83**: 4-7.
STEPHENSON, R. & STEPHENSON, J. 2008a: Southern Bulgaria revisited (part 2). - *Sedum Soc. Newsletter* **84**: 28-32.
STEPHENSON, R. & STEPHENSON, J. 2008b: Bulgaria part 3. - *Sedum Soc. Newsletter* **85**: 60-64.
STERN, F.C. 1940: *Paeonia Clusii*. - *Bot. Mag.* **162(4)**: tab. 9594 + 3 pp. text.
STERN, F.C. 1946: A study of the genus *Paeonia*. With fifteen illustrations in colour by L. Snelling and S. Ross-Craig. London: Royal Horticultural Society, viii + 155 pp. + 15 plates.
STERN, F.C. 1951: Snowdrops and snowflakes. - *J. Roy. Hort. Soc.* **76(3)**: 73-77 + 2 plates.
STERN, F.C. 1956: Snowdrops and snowflakes. A study of the genera *Galanthus* and *Leucojum*. - London: Royal Horticultural Society, viii + 128 pp. + map and frontispiece.
STERN, W. 1991: *Ophrys sitiaca* auch auf Kos. - *Ber. Arbeitskreis. Heimische Orchid.* **8(1)**: 92.
STERN, W. & DOSTMANN, H. 1989: Kos - der Orchideen wegen? - *Ber. Arbeitskreis. Heimische Orchid.* **6(1)**: 63-84.
STERNBERG, K.M.V. 1807: Systematische Bestimmung derjenigen Pflanzen, welche in Tourneforts Reisen nach dem Oriente abgebildet sind. - *Bot. Zeitung (Regensburg)* **6**: 313-318.
STERNBERG, K.M.V. 1810: Revisio Saxifragarum iconibus illustrata [suppl. 1 in 1822, suppl. 2 in 1831]. - Ratisbonae: H.F.A. Augustin, xiv + 62 pp. + 31 plates [cf. TL-2 5: 901, 1985].
STERNECK, J.V. 1895a: Beitrag zur Kenntnis der Gattung *Alectorolophus* All. - *Oesterr. Bot. Z.* **45**: 7-14, 45-50, 98-103, 126-131, 161-166.
STERNECK, J.V. 1895b: Beitrag zur Kenntnis der Gattung *Alectorolophus* All. [cont.]. - *Oesterr. Bot. Z.* **45**: 272-278, 295-303, 377-382, 415-422.
STERNECK, J.V. 1895c: Beitrag zur Kenntnis der Gattung *Alectorolophus* All. [end]. - *Oesterr. Bot. Z.* **45(12)**: 469-472 + 4 plates.
STERNECK, J.V. 1901: Monographie der Gattung *Alectorolophus*. - *Abh. K.K. Zool.-Bot. Ges. Wien* **1 (2)**: 1-150 + 3 maps and 1 diagram.
STERNECK, J.V. 1911: Botanische Reiseskizzen aus Griechenland. - *Naturwiss. Z. Lotos* **59(2)**: 48-62.
STEVANOSKI, I. & AL. 2020: Disentangling relationships between the amphi-Adriatic *Euphorbia spinosa* and Balkan endemic *E. glabriflora* (Euphorbiaceae). - *Bot. J. Linn. Soc.* **194(3)**: 358-374.
STEVANOVIĆ, B. 1986: Ecophysiological characteristics of the species *Ramonda serbica* Panč. and *Ramonda nathaliae* Panč. et Petrov. - *Ekologija (Belgrade)* **21(2)**: 119-134.
STEVANOVIĆ, V. 1989: On the distribution of the arctic-alpine floristic elements in the Balkans. In Phitos, D. (ed.): Abstracts. Sixth OPTIMA meeting, Delphi 10-16 Sept. 1989, p. 140.
STEVANOVIĆ, V. 1991: Chorology and taxonomic position of Košanin's taxons [In Serbo-Croatian with English summary]. - *Zborn. Rad. Srpska Akad. Nauka* **1991**: 289-306.
STEVANOVIĆ, V. 1996: Analysis of the Central European and Mediterranean orophytic element on the mountains of the W. and Central Balkan Peninsula, with special reference to endemics. - *Bocconea* **5(1)**: 77-97.
STEVANOVIĆ, V. & AL. 1997: Ecological differentiation of the species from *Draba lasiocarpa* complex in the C. Balkans. In: Abstracts from the First Balkan Botanical Congress, Thessaloniki, September 19-22, 1997, p. 47. - Thessaloniki.
STEVANOVIĆ, V. & AL. 2009: Distribution and diversity of arctic-alpine species in the Balkans. - *Pl. Syst. Evol.* **283(3-4)**: 219-235.
STEVANOVIĆ, V. & AL. 2014: Plant species and subspecies discovered by Dr. Josif Pančić 1 - distribution and floristic importance. - *Bot. Serbica* **38(2)**: 251-268.
STEVANOVIĆ, V., IATROÚ, G. & TAN, KIT 1998: The endemic serpentine flora of the Balkan Peninsula. In Moret, J. (ed.): Abstracts of the IX. OPTIMA meeting, Paris, 11-17 May 1998, p. 55. - Paris: Mus. Nat. d'Hist. Naturelle.
STEVANOVIĆ, V., JAKOVLJEVIĆ & MATEVSKI, V. 2014: Chasmophytic communities of endemic and relict species *Ramonda nathaliae* Pančić & Petrović on ophiolithic substrate in Republic of Macedonia. - *Bot. Serbica* **38(1)**: 81-90 + appendix.
STEVANOVIĆ, V. & LAKUŠIĆ, D. 1991: Distribution of the species *Geum reptans* L. (Rosaceae) in Yugoslavia [In Serbo-Croatian with English summary]. - *Glasn. Prir. Mus. Beogradu, Ser. B, Biol. Nauke* **46**: 39-46.
STEVANOVIĆ, V. & LAKUŠIĆ, D. 1995: *Draba lasiocarpa* aggr. (Brassicaceae) in the C. Balkans. In Valdés, B. & Silvestre, S. (eds): Abstracts. VIII OPTIMA meeting, Sevilla, Spain, 25 September – 10 October 1995, p. 90.
STEVANOVIĆ, V. & LAKUŠIĆ, D. 2000: *Draba* (sect. *Aizoides*) *kuemmerlei*, nom. nov. (Brassicaceae) - taxonomic, chorological and ecological characteristics. - *Bot. Chron. (Patras)* **13**: 87-93.
STEVANOVIĆ, V., MATEVSKI, V. & TAN, KIT 2009: *Helianthemum marmoreum* (Cistaceae), a new species from the central Balkans. - *Bot. Serbica* **33(1)**: 13-19.
STEVANOVIĆ, V., MATEVSKI, V. & TAN, KIT 2010: *Jurinea micevskii* (Asteraceae), a new species from the Republic of Macedonia. - *Phytol. Balcan.* **16(2)**: 249-254.
STEVANOVIĆ, V. & NIKETIĆ, M. 1990: *Viola dukadjinica* W. Becker & Košanin, a new species of the Yugoslav flora. - *Razprave IV. Razreda Sazu* **31(20)**: 317-326.
STEVANOVIĆ, V., NIKETIĆ, M. & LAKUŠIĆ, D. 1991: Chorological additions to the flora of eastern Yugoslavia. - *Fl. Medit.* **1**: 121-142.
STEVANOVIĆ, V., NIKETIĆ, M. & LAKUŠIĆ, D. 1993: Distribution of the vascular plants in Yugoslavia (Serbia, Montenegro) and Macedonia. I. - *Bull. Inst. Jard. Bot. Univ. Belgrade, n.s.* **24-25**: 33-54.
STEVANOVIĆ, V., NIKETIĆ, M. & LAKUŠIĆ, D. 2001: *Pedicularis ernesti-mayeri* (*P.* subsect. *Comosae*, Scrophulariaceae), a new species from Mt Prokletije (Yugoslavia). - *Razprave IV. Razreda Sazu* **43 (2)**: 209-228.
STEVANOVIĆ, V., NIKETIĆ, M. & STEVANOVIĆ, B. 1986a: On distribution of endemic and relic species *Ramonda serbica* Panč. (Gesneriaceae) in SR Macedonia (Yugoslavia) [In Serbo-Croatian with English summary]. - *Glasn. Prir. Mus. Beogradu, Ser. B, Biol. Nauke* **41**: 17-25.
STEVANOVIĆ, V., NIKETIĆ, M. & STEVANOVIĆ, B. 1986b: Sympatric area of the sibling and endemo-relict species *Ramonda serbica* Panč. and *R. nathaliae* Panč. et Petrov.

(Gesneriaceae) in southeast Serbia (Yugoslavia). - *Bull. Inst. Jard. Bot. Univ. Belgrade, n.s.* **20**: 45-54.

STEVANOVIĆ, V., NIKETIĆ, M. & STEVANOVIĆ, B. 1987: Phytocoenological characteristics of sympatric habitats of endemo-relic species *Ramonda serbica* Panč. and *R. nathaliae* Panč. et Petrov. [In Serbo-Croatian with English summary]. - *Bull. Inst. Jard. Bot. Univ. Belgrade, n.s.* **21**: 17-26.

STEVANOVIĆ, V., NIKETIĆ, M. & STEVANOVIĆ, B. 1989: Chorological differentiation of the endemic ... species *Ramonda serbica* Panc. and *R. nathaliae* Panc. & Petrov. In Phitos, D. (ed.): Abstracts. Sixth OPTIMA meeting, Delphi 10-16 Sept. 1989, p. 141.

STEVANOVIĆ, V., NIKETIĆ, M. & STEVANOVIĆ, B. 1991: Chorological differentiation of endemo-relic species *Ramonda serbica* Panč. and *R. nathaliae* Panč. & Petrov. (Gesneriaceae) on the Balkan Peninsula. - *Bot. Chron. (Patras)* **10**: 507-515.

STEVANOVIĆ, V. & STEVANOVIĆ, B. 1985: Asplenio cuneifolii - Ramondaetum nathaliae. New chasmophytic community on serpentine rocks in Macedonia [In Serbo-Croatian with English summary]. - *Glasn. Prir. Mus. Beogradu, Ser. B, Biol. Nauke* **40**: 75-87 + 1 table.

STEVANOVIĆ, V. & TAN, KIT 2000: On the distribution of *Viola kosaninii* in the Balkan Peninsula. - *Preslia* **72(2-4)**: 469-474.

STEVANOVIĆ, V., TAN, KIT & IATROÚ, G. 2001: Distribution of obligate serpentine endemics in the Balcans. In Raimondo, F.M. (ed.): Abstracts from the X OPTIMA meeting, Palermo, 13-19 September 2001, p. 153.

STEVANOVIĆ, V., TAN, KIT & IATROÚ, G. 2003: Distribution of the endemic Balkan flora on serpentine I. - obligate serpentine endemics. - *Pl. Syst. Evol.* **242(1-4)**: 149-170.

STEVANOVIĆ, V., TAN, KIT & PETROVA, A.V. 2004: Mapping the endemic flora of the Balkans - a progress report. In Vasić, O. (ed.): Abstracts, XI OPTIMA Meeting, Beograd 5-11.IX.2004, p. 63.

STEVANOVIĆ, V., TAN, KIT & PETROVA, A.V. 2007: Mapping the endemic flora of the Balkans - a progress report. - *Bocconea* **21**: 131-137.

STEVANOVIĆ, V. & VUJNOVIĆ, K. 1990: Morpho-anatomical adaptations of the endemic species *Degenia velebitica* (Deg.) Hay. - *Feddes Repert.* **101(7-8)**: 385-389 + 2 plates.

STEVANOVIĆ, V. (ED.) 1999: The red data book of flora of Serbia. 1. Extinct and critically endangered taxa [In Serbian and English]. - Beograd: Ministry of Environment, 566 pp.

STIEFELHAGEN, H. 1910: Systematische und pflanzengeographische Studien zur Kenntnis der Gattung *Scrophularia*. Vorarbeiten zu einer Monographie. - *Bot. Jahrb. Syst.* **44(4)**: 406-496 + 1 plate.

STINCA, A. & AL. 2021: Italian vascular flora: New findings, updates and explorations of floristic similarities between regions. - *Diversity MDPI* **13, 600**: doi.org/10.3390/d13110600.

STIRTON, C.H. 1981: Tribe 11. Psoraleae (Benth.) Rydb. (1919). In Polhill, R.M. & Raven, P.H. (eds): Advances in legume systematics 1: pp. 337-343. - Kew: Royal Botanic Gardens.

STOEVA, M.P. 1977: [Chromosome numbers in *Poa* from Bulgaria]. In Löve, Á. (ed.): IOPB chromosome number reports LVIII. - *Taxon* **26(5-6)**: 560.

STOEVA, M.P. 1982: [Chromosome numbers in Poaceae from Bulgaria]. In Löve, Á. (ed.): Chromosome number reports LXXVI. - *Taxon* **31(3)**: 579-580.

STOEVA, M.P. 1983: Variation in the populations of *Deschampsia flexuosa* (L.) Trin. [In Bulgarian with English abstract]. In Velčev, V.I. & al. (eds): Third national conference of botany, pp. 341-347. - Sofia: Publishing House of the Bulgarian Academy of Sciences.

STOEVA, M.P. 1984a: Variation of the chromosome number in *Poa bulbosa* L. [In Bulgarian with English abstract]. In: 90th anniversary of Acad. Daki Jordanov, pp. 86-98. - Targovište: "D. Stefanov" Publishing House.

STOEVA, M.P. 1984b: Population variability in *Poa compressa* L. and *P. pratensis* L. [In Bulgarian with English abstract]. In Velčev, V.I. & al. (eds): Săvremenni teoretični i priložni aspecti ... pp. 268-275. - Sofia: Publishing House of the Bulgarian Academy of Sciences.

STOEVA, M.P. 1985: Chromosome numbers of Bulgarian angiosperms. - *Fitologiya* **30**: 78-79.

STOEVA, M.P. 1986: A karyological study of several species from the genus *Poa* L. (blue grass) in Bulgaria [In Bulgarian with Russian and English summaries]. - *Fitologiya* **31**: 21-33.

STOEVA, M.P. 1987a: Chromosome numbers of Bulgarian angiosperms. - *Fitologiya* **33**: 65-66.

STOEVA, M.P. 1987b: A population study of *Carex panicea* L. in Bulgaria [In Bulgarian with English abstract]. In Kuzmanov, B.A. & al. (eds): Proceedings, fourth national conference on botany, pp. 60-67. - Sofia: Publishing House of the Bulgarian Academy of Sciences.

STOEVA, M.P. 1988: Population variability of *Deschampsia cespitosa* (L.) P.B. in Bulgaria [In Bulgarian with Engl. & Russ. summaries]. In Velčev, V.I. & al. (eds): 100th anniversary of Acad. N.A. Stojanov, pp. 107-114. - Sofia: Publishing House of the Bulgarian Academy of Sciences.

STOEVA, M.P. 1991a: Karyological study of Bulgarian representatives of the genus *Papaver* L. sect. *Orthorhoeades* Fedde [In Bulgarian with Russian and English summaries]. - *Fitologiya* **40**: 40-55.

STOEVA, M.P. 1991b: *Botrychium matricariifolium* A. Braun ex Koch - a new species recorded for the Bulgarian flora. - *Fitologiya* **40**: 81-82.

STOEVA, M.P. 1991c: Comparative morphological study of *Nardus stricta* populations in the central Balkan mountains and the western Rhodopes [In Bulgarian with English summary]. - *Fitologiya* **41**: 28-36.

STOEVA, M.P. 1992a: Chromosome numbers of Bulgarian Cyperaceae. - *Fitologiya* **43**: 77-78.

STOEVA, M.P. 1992b: Karyological study of *Bolboschoenus maritimus* (L.) Palla and *Holoschoenus vulgaris* Link (Cyperaceae) in Bulgaria. - *Dokl. Bulg. Akad. Nauk* **45(1)**: 61-63.

STOEVA, M.P. 1992c: Karyological study of the genera *Blysmus* Panzer, *Eriophorum* L. and *Scirpus* L. (Cyperaceae) in Bulgaria. - *Dokl. Bulg. Akad. Nauk* **45(5)**: 113-115.

STOEVA, M.P. 1994: Reports (285-293). In Kamari, G., Felber, F. & Garbari, F. (eds): Mediterranean chromosome number reports - 4. - *Fl. Medit.* **4**: 258-262.

STOEVA, M.P. 2000: Reports (1213-1226) In Kamari, G., Felber, F. & Garbari, F. (eds): Mediterranean chromosome number reports - 10. - *Fl. Medit.* **10**: 423-430.

STOEVA, M.P. & KOŽUHAROV, S.I. 1978: [Chromosome numbers in *Poa* from Bulgaria]. In Löve, Á. (ed.): IOPB chromosome number reports LXII. - *Taxon* **27(5-6)**: 522-523.

STOEVA, M.P. & KOŽUHAROV, S.I. 1983: Caryological study of *Deschampsia cespitosa* (L.) P.B. in Bulgaria [In Bulgarian with English abstract]. In Velčev, V.I. & al. (eds): Third national conference of botany, pp. 223-231. - Sofia: Publishing House of the Bulgarian Academy of Sciences.

STOEVA, M.P. & KOŽUHAROV, S.I. 1984: Population structures and evolutionary trends in section *Bolboforum* Aschers. & Graebn. of genus *Poa* L. I. *Poa badensis* aggr. [In Bulgarian with Russian and English summaries]. - *Fitologiya* **27**: 3-23.

STOEVA, M.P. & KOŽUHAROV, S.I. 1985: Population structures and evolutionary trends in section *Bolboforum* Aschers. & Graebner of genus *Poa* L. II. *P. media* Schur and *P. alpina* L. [In Bulgarian with Russian and English summaries]. - *Fitologiya* **30**: 3-29.

STOEVA, M.P. & POPOVA, E.D. 1990: Population variability of *Carex caryophyllea* Latourr. in Bulgaria [In Bulgarian with Russian and English summaries]. - *Fitologiya* **38**: 41-58.

STOEVA, M.P. & POPOVA, E.D. 1991: Variation patterns within the genus *Carex* L. sect. *Digitatae* (Fries) Christ in Bulgaria. - *Folia Geobot. Phytotax.* **26(4)**: 379-395.

STOEVA, M.P. & POPOVA, E.D. 1992: Taxonomical evaluation of the Bulgarian populations of genus *Carex* L. sect. *Glaucae* (Ascherson) Christ. - *Dokl. Bulg. Akad. Nauk* **45(9)**: 99-101.

STOEVA, M.P. & POPOVA, E.D. 1993a: Taxonomical study of genus *Carex* sect. *Phaestoglochin* and sect. *Stellulatae* in Bulgaria. In Petrova, A.V. (ed.): Abstracts from the VII OPTIMA meeting in Borovetz, Bulgaria, 18-30 July 1993, p. 96.

STOEVA, M.P. & POPOVA, E.D. 1993b: Cytotaxonomic study on *Carex* sect. *Acrocystis* in Bulgaria. - *Fragm. Florist. Geobot.* **38(1)**: 29-43.

STOEVA, M.P., POPOVA, E.D. & UZUNOVA, K.G. 1991: Population variability of *Carex sempervirens* complex and *C. Kitaibeliana* ... [In Bulgarian with Engl. summary]. In Kožuharov, S.I. & Kuzmanov, B.A. (eds): Evolution of flowering ... 2, pp. 140-159. - Sofia: Publishing House of the Bulgarian Academy of Sciences.

STOEVA, M.P. & ŠTĚPÁNKOVÁ, J. 1990: Variation patterns within the *Carex flava* agg. in Bulgaria and Czechoslovakia. - *Preslia* **62(1)**: 1-23.

STÖHR, O. 2010: Die Unterarten und Hybriden von *Asplenium trichomanes* L. im Bundesland Salzburg (Österreich). - *Stapfia* **92**: 29-44.

STOJANOV, N. 1914: Betrachtungen über die Frühlingsflora des Tekir-Gebirges und dessen Meeresstrandes. - *God. Sofiisk. Univ. Fiz.-Mat. Fak.* **8-9**: 1-43 + 1 map.

STOJANOV, N. 1921a: Floristische Materialien von dem Belassiza-Gebirge [In Bulgarian]. - *God. Sofiisk. Univ. Fiz.-Mat. Fak.* **15-16**: 1-133 + 1 map.

STOJANOV, N. 1921b: Über die Vegetation des Ali-Botusch Gebirges [In Bulgarian with German summary]. - *God. Sofiisk. Univ. Fiz.-Mat. Fak.* **17**: 1-35.

STOJANOV, N. 1921c: Die von †Neicev in Macedonien und Adrianopel gesammelten Pflanzen. [In Bulgarian]. - *Trav. Soc. Bulg. Sci. Nat,* **9**: 37-50.

STOJANOV, N. 1922: Phytogeographische und floristische Charakteristik des Pirin-Gebirges [In Bulgarian with German summary]. - *God. Sofiisk. Univ. Fiz.-Mat. Fak.* **18**: 1-27.

STOJANOV, N. 1923: *Convolvulus nitidus* Boiss. in Ost-Mazedonien [In Bulgarian with German summary]. - *God. Sofiisk. Univ. Agron.-Lesov. Fak., 1* **1**: 123-125.

STOJANOV, N. 1926a: Beiträge zur Flora Bulgariens und der angrenzenden Länder. - *Izv. Bulg. Bot. Druzh.* **1**: 71-82.

STOJANOV, N. 1926b: Die Verbreitung der mediterranen Vegetation in Südbulgarien. - *Bot. Jahrb. Syst.* **60(4)**: 375-407 + 1 plate.

STOJANOV, N. 1926c: Über den Formenkreis von *Jasione supina* (Sieb.) DC. - *Notizbl. Bot. Gart. Berlin-Dahlem* **9(87)**: 545-560 + 1 map.

STOJANOV, N. 1927: Über die Verbreitung der orientalischen Buche auf der Balkanhalbinsel. - *Magyar Bot. Lapok* **25**: 131-136.

STOJANOV, N. 1928a: Thracische und macedonische Herbarmaterialien des verstorbenen Prof. Dr. Theodor Nikoloff. - *Spis. Bulg. Akad. Nauk* **18**: 49-209.

STOJANOV, N. 1928b: Über die in Bulgarien vorkommenden Formen von *Verbascum humile* Janka. - *Izv. Tsarsk. Prir. Inst. Sofiya* **1**: 142-144.

STOJANOV, N. 1929: Kritische Studien und kleine Mitteilungen aus dem Herbar des Königlichen Naturhistorischen Museums in Sofia. - *Izv. Tsarsk. Prir. Inst. Sofiya* **2**: 257-272.

STOJANOV, N. 1930a: Kritische Studien und kleine Mitteilungen aus dem Herbar des Königlichen Naturhistorischen Museums in Sofia. II. - *Izv. Tsarsk. Prir. Inst. Sofiya* **3**: 119-137.

STOJANOV, N. 1930b: *Potentila* [sic!] *Regis Borisii*, sp. n. e sect. *Rupestres*, eine neue Pflanzenart aus Bulgarien. - *Izv. Tsarsk. Prir. Inst. Sofiya* **3**: 249-252.

STOJANOV, N. 1930c: Über die Verbreitung und die Formen der in Bulgarien wildwachsenden Kleearten [In Bulgarian with German summary]. - *Sborn. Bălg. Akad. Nauk., Klon. Prir.-Mat.* **26**: 1-159 + 15 plates.

STOJANOV, N. 1930d: Versuch einer Analyse des relikten Elements in der Flora der Balkanhalbinsel. - *Bot. Jahrb. Syst.* **63(5)**: 368-418.

STOJANOV, N. 1931: Beitrag zur Flora des Ali-Botuš-Gebirges (auf Grund der Sammlungen des Herrn A. Drenovsky). - *Izv. Bulg. Bot. Druzh.* **4**: 116-118.

STOJANOV, N. 1932a: The beech woods of the Balkan Peninsula. In Rübel, E. (ed.): Die Buchenwälder Europas. - *Veröff. Geobot. Inst. Rübel Zürich* **8**: 182-218.

STOJANOV, N. 1932b: Floristische Notizen. - *Izv. Bulg. Bot. Druzh.* **5**: 98-110.

STOJANOV, N. 1932c: A new European species of *Fritillaria*. - *Lily Year-Book* **1932**: 94-95.

STOJANOV, N. 1936a: Eine neue *Astragalus*-Art aus Macedonien. - *Izv. Bulg. Bot. Druzh.* **7**: 100-107.

STOJANOV, N. 1936b: Caractère phytogéographique du massif de Rila, des Rhodopes et de Pirine. In: Comptes Rendus du IVe Congrès des Géographes et des Ethnographes Slaves, pp. 143-150. – Sofia.

STOJANOV, N. 1939a: Über den Formenschwarm um *Viscaria atropurpurea* Grsb. - nebst einigen Überlegungen über die Bedeutung der Formenschwärme im Lichte des Artproblems. - *God. Sofiisk. Univ. Fiz.-Mat. Fak., 3 (Estestv. Istorija)* **35**: 311-341.

STOJANOV, N. 1939b: Beitrag zur Kenntnis der Flora Ostthraziens (Die in der Umgebung von Lozengrad von A. Mateeff im J. 1899 gesammelten Pflanzen). - *Trav. Soc. Bulg. Sci. Nat.* **18**: 64-69.

STOJANOV, N. 1939c: Die in den letzten elf Jahren (1928-38) erschienene Literatur über die Flora Bulgariens [In German and Bulgarian]. - *Izv. Tsarsk. Prir. Inst. Sofiya* **12**: 209-230.

STOJANOV, N. 1940: Zur Frage über die Herkunft des arktischen Elements in der Flora der Balkanhalbinsel. - *God. Sofiisk. Univ. Fiz.-Mat. Fak., 3 (Estestv. Istorija)* **36**: 195-250.

STOJANOV, N. 1941: Kritische Studien und kleine Mitteilungen aus dem Herbar des Königlichen Naturhistorischen Museums in Sofia. VI. - *Izv. Tsarsk. Prir. Inst. Sofiya* **14**: 153-167.

STOJANOV, N. 1942: Die Natur der Insel Thasos [In Bulgarian with German summary]. – *Belomorski Pregled* **1(1)**: 73-107.

STOJANOV, N. 1943: Über die Kiefern des Aegäischen Bezirks [In Bulgarian with German summary]. - *God. Sofiisk. Univ. Fiz.-Mat. Fak., 3 (Estestv. Istorija)* **39**: 141-168.

STOJANOV, N. & ACHTAROV, B. 1934: Über den Begriff und die systematische Stellung von *Centaurea affinis* Friv. - *Izv. Tsarsk. Prir. Inst. Sofiya* **7**: 91-105.

STOJANOV, N. & ACHTAROV, B. 1935a: Kritische Studien über die Nelken Bulgariens. - *Sborn. Bălg. Akad. Nauk., Klon. Prir.-Mat.* **29**: 1-101 + 7 plates.

STOJANOV, N. & ACHTAROV, B. 1935b: Studien über die Centaureen Bulgariens. - Sofia: Bulg. Akad. Wiss., Stiftung "Ivan und Janko K. Urumov", 89 pp.

STOJANOV, N. & ACHTAROV, B. 1937: Material zur Kenntnis der Gattung *Anthemis* (vorläufige Mitteilung). - *Notizbl. Bot. Gart. Berlin-Dahlem* **13(119)**: 513-523.

STOJANOV, N. & ACHTAROV, B. 1939: Neues Material zur Kenntnis der Flora des Pirin-Gebirges. - *Izv. Tsarsk. Prir. Inst. Sofiya* **12**: 181-187.

STOJANOV, N. & ACHTAROV, B. 1941: Über einige neue und interessante Arten der Gattung *Anthemis*. - *Sborn. Bălg. Akad. Nauk., Klon. Prir.-Mat.* **61(3)**: 73-86.

STOJANOV, N. & AVRAMOVA, B. 1958: Untersuchung der natürlichen Fundorte von *Atropa belladonna* L. in Bulgarien [In Bulgarian with Russian and German summaries]. - *Izv. Bot. Inst. (Sofia)* **6**: 391-402 + 1 map.

STOJANOV, N. & JORDANOV, D. 1938: Botanische Studien auf dem thessalischen Olymp. I. Floristische Ergebnisse. - *God. Sofiisk. Univ. Fiz.-Mat. Fak., 3 (Estestv. Istorija)* **34**: 147-249 + 6 plates.

STOJANOV, N. & KITANOV, B. 1943a: Beitrag zur Kenntnis der Florula der Insel Samothrake. - *Izv. Bulg. Bot. Druzh.* **9**: 49-51.

STOJANOV, N. & KITANOV, B. 1943b: Einige neue und kritische Pflanzen der Insel Thasos. - *Izv. Bulg. Bot. Druzh.* **9**: 94-105.

STOJANOV, N. & KITANOV, B. 1944: Beitrag zur Kenntnis der Flora und der Vegetationsverhältnisse der Insel Samothrake [In Bulgarian with German summary]. - *God. Sofiisk. Univ. Fiz.-Mat. Fak., 3 (Estestv. Istorija)* **40**: 403-464.

STOJANOV, N. & KITANOV, B. 1945: Flora der Insel Thasos [In Bulgarian with German summary]. - *God. Sofiisk. Univ. Fiz.-Mat. Fak., 3 (Estestv. Istorija)* **41**: 233-320.

STOJANOV, N. & KITANOV, B. 1946: Flora der Insel Thasos [In Bulgarian with German summary]. - *God. Sofiisk. Univ. Fiz.-Mat. Fak., 3 (Estestv. Istorija)* **42**: 89-196.

STOJANOV, N. & KITANOV, B. 1950: Die Pflanzendecke der Insel Thasos [In Bulgarian with Russian and German summaries]. - *Izv. Bot. Inst. (Sofia)* **1**: 214-358.

STOJANOV, N. & KITANOV, B. 1951: Über die systematische Stellung von *Celsia thasia* Stoj. et Kitan. - *Izv. Bot. Inst. (Sofia)* **2**: 181-185.

STOJANOV, N., KITANOV, B. & VĂLEV, S. 1950: Über die Verbreitung von *Carum lumpeanum* Dörfl. et Hay. auf der Balkanhalbinsel [In Bulgarian with Russian and German summaries]. - *Izv. Bot. Inst. (Sofia)* **1**: 410-414.

STOJANOV, N., KITANOV, B. & VELČEV, V.I. 1955: Floristisches Material aus den Ostrhodopen [In Bulgarian with Russian and German abstracts]. - *Izv. Bot. Inst. (Sofia)* **4**: 111-117.

STOJANOV, N. & STEFANOV, B. 1921: Zwei neue Pflanzen aus Bulgarien. - *Oesterr. Bot. Z.* **70(3-5)**: 103-107.

STOJANOV, N. & STEFANOV, B. 1922: Phytogeographische und floristische Charakteristik des Pirin-Gebirges [In Bulgarian with German summary]. - *God. Sofiisk. Univ. Fiz.-Mat. Fak.* **18**: 1-27 + 7 plates.

STOJANOV, N. & STEFANOV, B. 1923a: A new *Veronica* from Bulgaria. - *J. Bot. (London)* **61**: 219-220.

STOJANOV, N. & STEFANOV, B. 1923b: Beitrag zur Flora Bulgariens und Mazedoniens. - *Oesterr. Bot. Z.* **72(1-5)**: 85-92.

STOJANOV, N. & STEFANOV, B. 1924a: *Scabiosa rhodopensis*. - *Bull. Misc. Inform. Kew* **1924(3)**: 98-101.

STOJANOV, N. & STEFANOV, B. 1924b: *Verbascum pseudonobile* spec. nov. - *God. Sofiisk. Univ. Agron.-Lesov. Fak., 1* **2**: 69-73.

STOJANOV, N. & STEFANOV, B. 1924c: Flore de la Bulgarie [several later editions]. - Sofia: iii + 608 pp. [cf. TL-2 4: 14, 1986].

STOJANOV, N. & STEFANOV, B. 1925: Eine neue *Chondrilla* aus Bulgarien. - *Bot. Jahrb. Syst.* **59 Beibl. 134**: 16 + 1 plate.

STOJANOV, N. & STEFANOV, B. 1933: Zwei neue Pflanzenarten aus Bulgarisch-Mazedonien. – *Magyar Bot. Lapok* **32(1-6)**: 1-5.

STOJANOV, N. & STEFANOV, B. 1934: *Sedum tuberiferum* Stoj. et Stef. sp. nova aus Bulgarien. - *Notizbl. Bot. Gart. Berlin-Dahlem* **11(110)**: 1013-1023.

STOJANOV, N., STEFANOV, B. & KITANOV, B. 1966-67: Flora na Bălgarija. Ed. 4, 2 vols., 1326 pp. - Sofija: Nauka i Izkustvo.

STOJANOV, N. & STOITSCHKOV, J. 1931: Die Walnüsse Bulgariens (Vorläufige Mitteilung). - *Izv. Bulg. Bot. Druzh.* **4**: 67-74.

STOKER, F. 1935: A list of lily names and synonyms. - *Lily Year-Book* **1935**: 4-34.

STOKER, F. 1938: The *carniolicum* group of lilies. - *Lily Year-Book* **1938**: 86-107.

STORK, A.L. 1971: Seed characters in European taxa of *Malcolmia* R. Br. (Cruciferae). - *Svensk Bot. Tidskr.* **65(3)**: 283-292.

STORK, A.L. 1972a: *Malcolmia orsiniana* (Ten.) Ten. in Italien und auf dem Balkan. - *Svensk Bot. Tidskr.* **66(3)**: 239-256.

STORK, A.L. 1972b: Studies in the Aegean flora. XX. Biosystematics of the *Malcolmia maritima* complex. - *Opera Bot.* **33**: 1-118 + 1 map.

STORK, A.L. 1972c: Samen und Keimlinge von *Malcolmia* sens. lat. (Cruciferae). - *Svensk Bot. Tidskr.* **66(4)**: 417-436.

STORK, A.L. 1984: Tulipes sauvages et cultivées. - *Sér. Doc. Conserv. Jard. Bot. Genève* **13**: 1-185 + 8 plates.

STOTZ, J., ARX, B.V. & THOMMEN, M. 1986: Etude sur *Ophrys holosericea* (Burm.) Greuter subsp. *elatior* (Gumprecht) Gumprecht. - *Saussurea* **17**: 1-12.

STOYANOV, K.H. 2005: Floristic materials and critical notes on the genus *Orobanche* subgen. *Phelipanche* in Bulgaria. - *Fl. Medit.* **15**: 461-476.

STOYANOV, K.H. 2009a: Chorology and critical notes on genus *Orobanche* (Orobanchaceae) in Bulgaria. In Ivanova, D. (ed.): Proceedings of the IV Balkan botanical congress, pp. 245-254. - Sofia: Institute of Botany, BAS.

STOYANOV, K.H. 2009b: Chorology and critical notes on *Orobanche* subsect. *Minores* in Bulgaria. - *Phytol. Balcan.* **15(3)**: 351-360.

STOYANOV, K.H. 2013: Distribution and environmental characteristics of *Orobanche gracilis* in Bulgaria. - *J. Agric. Forest. Sci.* **12(2)**: 22-30.

STOYANOV, K.H. & RAYCHEVA, T.G. 2022: Current state of *Iris* L., subgenus *Iris* (Iridaceae) in Bulgaria. - *Ecol. Balkan.* **14(1)**: 31-46.

STOYANOV, S. 2010: A new annual *Bupleurum* (Apiaceae) species from Northeastern Bulgaria and Romanian Dobrogea. - *Phytol. Balcan.* **16(1)**: 65-74.

STOYANOV, S. 2019: Contribution to the taxonomy and distribution of *Bupleurum boissieri*, *B. aequiradiatum* and *B. gerardi* (Apiaceae) in the Balkans. - *Phytotaxa* **392(3)**: 197-216.

STOYANOV, S. & AL. 2020: *Seseli besserianum* (Apiaceae), a restored species and a replacement name for *Seseli peucedanifoliium* Besser. - *Phytotaxa* **458(2)**: 120-138.

STOYANOV, S. & BARZOV, Z. 2020: *Papaver nigrotinctum*: first records for the flora ofBulgaria and Republic of North Macedonia. - *Phytol. Balcan.* **26(1)**: 35-42.

STOYANOV, S. & GORANOVA, V.Y 2009: Notes on some critical *Bupleurum* species from sect. *Aristata* in Bulgaria. In Ivanova, D. (ed.): Proceedings of the IV Balkan botanical congress, pp. 177-181. - Sofia: Institute of Botany, BAS.

STOYANOV, S., GORANOVA, V.Y. & STOYKOV, D. 2006a: *Astragalus physocalyx* Fischer - rediscovered in Bulgaria. In Peev, D.R. (ed.): Book of abstracts, IV Balkan botanical congress, Sofia, Bulgaria, 20-26 June 2006, p. 273.

STOYANOV, S., GORANOVA, V.Y. & STOYKOV, D. 2006b: Report 87 [*Astragalus physocalyx*]. In Vladimirov, V. & al. (eds): New floristic records in the Balkans: 2. - *Phytol. Balcan.* **12(2)**: 295.

STOYANOV, S. & MARINOV, Y. 2021: *Thymus aznavourii* Velen. (Lamiaceae): First records for the Bulgarian and Greek flora. - *Dokl. Bulg. Akad. Nauk* **74(3)**: 352-362.

STOYANOVA, K. 2003: Reports (1372-1375). In: Kamari, G., Blanché, C. & Garbari, F. (eds): Mediterranean chromosome number reports - 13. - *Fl. Medit.* **13**: 401-404.

STRANDHEDE, S.-O. 1965: Chromosome studies in *Eleocharis*, subser. *Palustres*. III. Observations on Western European taxa. - *Opera Bot.* **9(2)**: 1-86.

STRANDHEDE, S.-O. 1966: Morphologic variation and taxonomy in European *Eleocharis*, subser. *Palustres*. - *Opera Bot.* **10(2)**: 1-187.

STRANGE, K. & RIX, E.M. 2007: 574. *Linum doerfleri*. - *Curtis's Bot. Mag.* **24(1)**: 12-17.

STRASSER, W. 1981: Vegetations-Studien in der südlichen Ägäis (Kap Sunion, Kos, Rhodos, Kreta). - Steffisburg: Privately printed, 28 pp.

STRASSER, W. 1982: Vegetations-Studien in den Bergen Griechenlands. Bericht über meinen Studienaufenhalt Juni-Juli 1982. - Steffisburg: Privately printed, 61 pp.

STRASSER, W. 1985: Botanische Studien. März/April 1985. Süd-Peloponnes. - Steffisburg: Privately printed, 42 pp.

STRASSER, W. 1986: Ost-Peloponnes. Botanische Studien 1986. - Steffisburg: Privately printed, 48 pp.

STRASSER, W. 1988: West-Kreta. Botanische Studien 1987. - Steffisburg: Privately printed, 46 pp.

STRASSER, W. 1989: West-Kreta. Botanische Exkursionen der Zürcherischen botanischen Gesellschaft 26.3.-6.4. 1989. - Steffisburg: Privately printed, 40 pp.

STRASSER, W. 1990: Korfu. Botanische Streifzüge durch die Insel im Frühling 1990. - Steffisburg: Privately printed, 47 pp.

STRASSER, W. 1992: Botanische Streifzüge durch das nordöstliche Griechenland. Frühjahr 1992. - Steffisburg: Privately printed, 85 pp.

STRASSER, W. 1993: Pflanzen des ostägäischen Raumes (Türkisches Festland und vorgelagerte Inseln). - Ott, Thun: Privately printed, 130 pp.

STRASSER, W. 1994a: Nördl. Peloponnes + Euböa: botanische Streifzüge 1993, mit Bestimmungsschlüsseln für die griechischen *Gagea*- + *Ornithogalum*-Arten. - Steffisburg: Privately printed, 76 pp.

STRASSER, W. 1994b: Westl. Peloponnes + Taygetosgebiet, botanische Streifzüge 1994, mit Bestimmungsschlüsseln für die griechischen *Trifolium*- u. *Bromus*-Arten. - Steffisburg: Privately printed, 51 pp.

STRASSER, W. 1997: Pflanzen des Peloponnes (Süd-Griechenland). - Vaduz/Lichtenstein: A.R.G. Gantner Verlag KG, 321 pp.

STRASSER, W. 1999: Plants of the Peloponnese, southern part of Greece. - Vaduz/Lichtenstein: A.R. G. Gantner Verlag KG, 350 pp.

STRASSER, W. 2002a: Pflanzen des Peloponnes (Süd-Griechenland) mit über 1900 Strichzeichnungen. 2. vollständig überarbeitete und erweiterte Auflage. - Vaduz/Lichtenstein: A.R.G. Gantner Verlag KG, 376 pp. + 24 plates.

STRASSER, W. 2002b: Lesbos / Chios. Botanische Studien. - Steffisburg: Privately published, ii + 64 pp.

STRGAR, V. 1973: Zur Verbreitung von *Sesleria angustifolia* (Hackel & Beck) Deyl in Bosnien und Serbien sowie *Sesleria autumnalis* (Scop.) F.W. Schultz in Mazedonien. - *Biol. Vestn.* **21(2)**: 127-133.

STRGAR, V. 1976: *Sesleria autumnalis* (Scop.) F.W. Schultz in Mazedonien. - *Biol. Vestn.* **24(2)**: 145-150.

STRGAR, V. 1979: Chromosome numbers of three Balkan *Sesleria* species [In Slovenian with English summary]. - *Biol. Vestn.* **27(1)**: 71-74.

STRGAR, V. 1981: Die Sippenstruktur von *Sesleria* auf der Balkanhalbinsel. - *Bot. Jahrb. Syst.* **102(1-4)**: 215-224.

STRGAR, V. 1982: *Sesleria ujhelyii* spec. nova. - *Biol. Vestn.* **30(1)**: 155-170.

STRGULC KRAJŠEK, S., ACCETTO, M. & JOGAN, N. 2015: *Myosotis refracta* (Boraginaceae): a new, unexpected forget-me-not in Slovene flora. In Bogdanović, S. & Jogan, N.: Book of abstracts - 6th Balkan Botanical Congress, p. 111. – Rijeka.

STRID, A. 1965a: Studies in the Aegean flora. VI. Notes on some genera of Labiatae. - *Bot. Not.* **118(1)**: 104-122.

STRID, A. 1965b: Studies in the Aegean flora. VII. Chromosome morphology in the *Nigella arvensis* complex. - *Bot. Not.* **118(2)**: 139-165.

STRID, A. 1968: Stable telocentric chromosomes formed by spontaneous misdivision in *Nigella doerfleri* (Ranunculaceae). - *Bot. Not.* **121(2)**: 153-164.

STRID, A. 1969a: Variation in the satellite chromosomes of *Nigella doerfleri* (Ranunculaceae). - *Bot. Not.* **122(1)**: 9-19.

STRID, A. 1969b: *Garidella unguicularis* Lam. (Ranunculaceae) - new for Europe. - *Bot. Not.* **122(3)**: 330-332.

STRID, A. 1969c: Evolutionary trends in the breeding system of *Nigella* (Ranunculaceae). - *Bot. Not.* **122(3)**: 380-397.

STRID, A. 1970a: Studies in the Aegean flora. XVI. Biosystematics of the *Nigella arvensis* complex with special reference to the problem of non-adaptive radiation. - *Opera Bot.* **28**: 1-169 + 3 plates and 1 map.

STRID, A. 1970b: Biosystematic and cytological studies in the *Nigella arvensis* complex. - Lund: Carl Bloms Boktryckeri A.-B., 9 pp.

STRID, A. 1971a: Biosystematics of *Nigella* in the Aegean area. In Davis, P.H., Harper, P.C. & Hedge, I.C. (eds): Plant life of South-West Asia, pp. 171-180. - Edinburgh: Botanical Society.

STRID, A. 1971b: Past and present distribution of *Nigella arvensis* L. ssp. *arvensis* in Europe. - *Bot. Not.* **124(2)**: 231-236.

STRID, A. 1971c: Chromosome numbers in some Albanian angiosperms. - *Bot. Not.* **124(4)**: 490-496.

STRID, A. 1972: Some evolutionary and phytogeographical problems in the Aegean. In Valentine, D. H. (ed.): Taxonomy, phytogeography and evolution, pp. 289-300. - London & New York: Academic Press.

STRID, A. 1974: Plantegeografi og evolutionsproblemer i det ægæiske øhav [In Danish]. - *Nat. Verden* **8-9**: 280-292.

STRID, A. 1976: Floristic notes from Mt. Olympos and Mt. Falakron (Boz Dagh), northern Greece. - *Bot. Not.* **129(3)**: 251-256.

STRID, A. 1977a: A new species of *Crepis* from Mt Olympus, Greece. - *Bot. Not.* **130(2)**: 143-145.

STRID, A. 1977b: *Sisyrinchium laxum* Otto ex Sims found in N Greece. - *Bot. Not.* **130(4)**: 473-475.

STRID, A. 1978: Contribution to the flora of Mount Kajmakcalan (Voras Oros), northern Greece. - *Ann. Mus. Goulandris* **4**: 211-247.

STRID, A. 1979: The endemic plant species of Mount Olympus. - *Fisis [Journal of the Hellenic Society for the Protection of Nature]* **19-20**: 4-5, 42-54.

STRID, A. 1980a: Wild flowers of Mount Olympus. - Kifissia: Goulandris Nat. Hist. Museum, xxviii + 362 pp. incl. 109 plates.

STRID, A. 1980b: [Miscellaneous chromosome numbers from Greece]. In Löve, Á. (ed.): Chromosome number reports LXIX. - *Taxon* **29(5-6)**: 709-710.

STRID, A. 1981: New species of *Cephalaria* (Dipsacaceae) and *Stipa* (Gramineae) from the Greek mountains. (Materials for the Mountain Flora of Greece, 10). - *Willdenowia* **11(2)**: 301-305.

STRID, A. 1983: Atlas Florae Europaeae notes 2. Changes of taxonomical status in Balkan *Arenaria* and *Minuartia* (Caryophyllaceae). - *Ann. Bot. Fenn.* **20(2)**: 113.

STRID, A. 1984: Distribution patterns and endemism in the Greek mountain flora. - *Norrlinia* **2**: 55-61.

STRID, A. 1986a: The mountain flora of Greece with special reference to the Anatolian element. - *Proc. Roy. Soc. Edinburgh, B, Biol. Sci.* **89**: 59-68.

STRID, A. 1986b: *Adonis cyllenea* (Ranunculaceae) and *Helichrysum taenari* (Asteraceae) rediscovered in Peloponnisos [with an appendix by G. Sfikas]. - *Ann. Mus. Goulandris* **7**: 221-231.

STRID, A. 1986c: [Miscellaneous chromosome numbers from Greece]. In Löve, Á. (ed.): Chromosome number reports XCIII. - *Taxon* **35(4)**: 901-902.

STRID, A. 1987a: Guide to excursion no. 58. The mountain flora and vegetation of northern Greece. - Berlin: XIV International Botanical Congress, 43 pp.

STRID, A. 1987b: Chromosome numbers of Turkish mountain plants. An annotated list of 34 taxa. - *Notes Roy. Bot. Gard. Edinburgh* **44(2)**: 351-356.

STRID, A. 1988: Some autumn-flowering bulbs of Greece. - *Herbertia* **43(2)**: 12-21.

STRID, A. 1989a: The Flora Hellenica project. In Phitos, D. (ed.): Abstracts. Sixth OPTIMA meeting, Delphi, 10-16 Sept. 1989, p. 9.

STRID, A. 1989b: Patterns of distribution and differentiation in Greek mountain plants. In Phitos, D. (ed.): Abstracts. Sixth OPTIMA meeting, Delphi, 10-16 Sept. 1989, p. 19.

STRID, A. 1989c: Endemism and speciation in the Greek flora. In Tan, Kit (ed.): Plant taxonomy, phytogeography and related subjects. The Davis & Hedge festschrift, pp. 27-44. - Edinburgh: Edinburgh Univ. Press.

STRID, A. 1991a: The "Flora Hellenica" project. - *Bot. Chron. (Patras)* **10**: 81-94.

STRID, A. 1991b: Distribution patterns and differentiation in Greek mountain plants. - *Bot. Chron. (Patras)* **10**: 155-161.

STRID, A. 1993a: The central European element in the Greek mountain flora. In Petrova, A.V. (ed.): Abstracts from the VII OPTIMA meeting in Borovetz, Bulgaria, 18-30 July 1993, p. 16.

STRID, A. 1993b: Phytogeographical aspects of the Greek mountain flora. - *Fragm. Florist. Geobot. Suppl.* **2(2)**: 411-433.

STRID, A. 1995: The Flora Hellenica databases. In: Proceedings, 5th scientific conference, Hellenic Botanical Society, Delphi 21-23 October 1994, pp. 61-66. - Thessaloniki: Univ. Studio Press.

STRID, A. 1996a: Flora Hellenica bibliography. A critical survey of floristic, taxonomic and phytogeographical literature relevant to the vascular plants of Greece, 1753-1994. - *Fragm. Florist. Geobot. Suppl.* **4**: i-x, 1-508.

STRID, A. 1996b: The Greek mountain flora, with special reference to the Central European element. - *Bocconea* **5(1)**: 99-112.

STRID, A. 1996c: Phytogeographia Aegaea and the Flora Hellenica Database. - *Ann. Naturhist. Mus. Wien, B* **98B Suppl.**: 279-289.

STRID, A. 1997a: From Flora Graeca to Flora Hellenica - on the botanical exploration of Greece. In: Chorological problems in the European flora. Abstracts, VIII meeting of the Atlas Florae Europaeae committee, p. 44. - Helsinki.

STRID, A. 1997b: Flora Hellenica and its database system. In: Abstracts from the First Balkan Botanical Congress, Thessaloniki, September 19-22, 1997, p. 48. - Thessaloniki.

STRID, A. 1998: Notes on *Lomelosia calocephala* and *Pseudofumaria alba*. In Greuter, W. & Raus, Th. (eds): Med-Checklist notulae, 17. - *Willdenowia* **28(1-2)**: 163-174.

STRID, A. 1999: Flora Hellenica and its database system. In Correia, A.I.D. & al. (eds): Livro de resumos. Guia de excursões. V jornadas de taxonomia botânica, 16-19/IX 1999, p. 17.

STRID, A. 2000a: *Viburnum opulus* L. In Greuter, W. & Raus, Th. (eds): Med-Checklist notulae, 19. - *Willdenowia* **30(2)**: 232.

STRID, A. 2000b: New taxa described in Grisebach's "Spicilegium Florae Rumelicae et Bithynicae" (1843-46). - *Preslia* **72**: 241-321.

STRID, A. 2001: The Flora Hellenica database. - *Portug. Acta Biol.* **19(1-4)**: 49-59.

STRID, A. 2004: *Silene lydia* Boiss., *Lomelosia calocephala* (Boiss.) Greuter & Burdet. In Greuter, W. & Raus, Th. (eds): Med-Checklist notulae, 22. - *Willdenowia* **34(1)**: 72-74.

STRID, A. 2006a: Flora Hellenica bibliography. Second edition. - Kraków: W. Szafer Institute of Botany, xi + 650 pp. + CD.

STRID, A. 2006b: *Carex umbrosa*. In Greuter, W. & Raus, Th. (eds): Med-Checklist notulae, 24. - *Willdenowia* **36(2)**: 727.

STRID, A. 2006c: I ikogenia Gesneriaceae. Ena tropiko stichio stin elliniki chlorida [In Greek]. - *Fisis [Journal of the Hellenic Society for the Protection of Nature]* **115**: 4-7.
STRID, A. 2007a: I ikogenia Berberidaceae stin Ellada [In Greek]. - *Fisis [Journal of the Hellenic Society for the Protection of Nature]* **116**: 10-13.
STRID, A. 2007b: *Hyssopus officinalis* subsp. *aristatus*, *Seseli montanum* subsp. *tommasinii*. In Greuter, W. & Raus, Th. (eds): Med-Checklist notulae, 25. - *Willdenowia* **37(1)**: 210-211.
STRID, A. 2007c: The lilies of Greece. - *Lilies* **75th Anniv. Issue**: 18-26.
STRID, A. 2011a: *Trigonella rotundifolia*. In Greuter, W. & Raus, Th. (eds): Med-Checklist notulae, 30. *Willdenowia* **41(2)**: 318.
STRID, A. 2011b: *Digitalis cariensis* subsp. *ikarica*. In Greuter, W. & Raus, Th. (eds): Med-Checklist notulae, 30. - *Willdenowia* **41(2)**: 323.
STRID, A. 2012: Notes on *Leucanthemum chloroticum*, *Petasites pyrenaicus*, *Drymocallis rupestris* and *Oenenthe pimpinelloides*. In Greuter, W. & Raus, Th. (eds): Med-Checklist notulae, 31. - *Willdenowia* **42(2)**: 289-292.
STRID, A. 2013a: *Klasea radiata* subsp. *cetinjensis* (Rohlena) Greuter & Wagenitz. In Raab-Straube, E. v. & Raus, Th. (eds): Euro+Med-Checklist notulae, 2. - *Willdenowia* **43(2)**: 241.
STRID, A. 2013b: In memoriam: Sven Snogerup. In Gkelis, S. & al. (eds): 13th Panhellenic Scientific Conference, Hellenic Botanical Society, Thessaloniki 3-6 October. Program and book of abstracts, p. 11.
STRID, A. 2014: [Note on *Pyrus pyraster*]. In Raab-Straube, E. v. & Raus, Th. (eds): Euro+Med-Checklist notulae, 3. - *Willdenowia* **44(2)**: 294.
STRID, A. 2015a: *Lilium jankae* G. Don and *Tulipa rhodopaea* (Velen.) Velen. In Raab-Straube, E. v. & Raus, Th. (eds): Euro+Med-Checklist notulae, 4. - *Willdenowia* **45(1)**: 125-126.
STRID, A. 2015b : Atlas of the Aegean Flora. In Bogdanović, S. & Jogan, N.: Book of abstracts - 6th Balkan Botanical Congress, p. 37. - Rijeka.
STRID, A. 2015c: Reliquiae Runemarkianae. Chromosome numbers of angiosperms from the Aegean Islands. - *Phytol. Balcan.* **21(3)**: 245-293.
STRID, A. 2016a: Atlas of the Aegean flora. Part 1: Text & plates. - *Englera* **33(1)**: 1-700.
STRID, A. 2016b: Atlas of the Aegean flora. Part 2: Maps. - *Englera* **33(2)**: 1-878.
STRID, A. 2016c: Rediscovery of *Iris athoa*. - *Species Group of the B.I.S. Newsletter* **Dec. 2016**: 4-7.
STRID, A. 2017a: Reports 158-188. In Vladimirov, V. & al. (eds): New floristic records in the Balkans: 33. - *Phytol. Balcan.* **23(2)**: 304-308.
STRID, A. 2017b: *Allium melanogyne* and *Ornithogalum naxense*. In Raus, Th. & Raab-Straube, E.v. (eds): Euro+Med-Checklist notulae, 8. - *Willdenowia* **47(3)**: 294-296.
STRID, A. 2018a: Mapping the flora of Greece. - *Bot. Serbica* **142 (suppl. 1)**: 73.
STRID, A. 2018b: Reports 124-148. In Vladimirov, V., Aybeke, M. & Tan, Kit (eds): New floristic records in the Balkans: 36. - *Phytol. Balcan.* **24(2)**: 277-280.
STRID, A. 2018c: Reports 206-230. In Vladimirov, V., Aybeke, M. & Tan, Kit (eds): New floristic records in the Balkans: 37. - *Phytol. Balcan.* **24(3)**: 427-432.
STRID, A. 2019a: *Cerastium bulgaricum*, *Rorippa austriaca*, *Euphorbia esula* subsp. *tommasiniana*. In Raab-Straube, E. v. & Raus, Th. (eds): Euro+Med-Checklist notulae, 10. - *Willdenowia* **49(1)**: 97, 100, 102.
STRID, A. 2019b: *Acis orientalis*: a new name for *Leucojum ionicum* (Amaryllidaceae).- *Phytol. Balcan.* **25(1)**: 79-80.
STRID, A. 2019c: Reports 100-103. In Vladimirov, V., Aybeke, M. & Tan, Kit (eds): New floristic records in the Balkans: 39. - *Phytol. Balcan.* **25(2)**: 222-223.
STRID, A. 2019d: *Bidens albus*. In Raab-Straube, E. v. & Raus, Th. (eds): Euro+Med-Checklist notulae, 11. - *Willdenowia* **49(3)**: 424-426.
STRID, A. 2019e: *Centaurea phrygia* subsp. *pseudophrygia*. In Raab-Straube, E. v. & Raus, Th. (eds): Euro+Med-Checklist notulae, 11. - *Willdenowia* **49(3)**: 426-428.
STRID, A. 2019f: Current floristic exploration of Greece. In Daskalakou, E.N. & al. (eds): 16th Conference of the Hellenic Botanical Society – Athens 10-13 October 2019. Program and book of abstracts, p. 9.
STRID, A. 2020a: The botanical exploration of Greece. - *Pl. Syst. Evol.* **306(2)**: 23 pp.
STRID, A. 2020b: Reports 71-73. In Vladimirov, V., Aybeke, M. & Tan, Kit: New floristic records in the Balkans: 43. - *Phytol. Balcan.* **26(3)**: 553-556.
STRID, A. 2021: Reports 109-124. In Vladimirov, V., Aybeke, M. & Tan, Kit (eds): New floristic records in the Balkans: 45. - *Phytol. Balcan.* **27(2)**: 264-267.
STRID, A. & AL. 1989: *Euphorbia sultan-hassei* (Euphorbiaceae), a new species from SW Crete. - *Willdenowia* **19(1)**: 63-67. - [Also distributed as *OPTIMA leaflet* no. 184].
STRID, A. & AL. 2012: Prespa National Park - Devas mountain. In Vrahnakis, M. & al. (eds): Report of the 9th European Dry Grassland Meeting (EDGM), 19-23 May 2012, pp. 98-107. - Prespa: Information Centre of Prespa National Forest Management Body.
STRID, A. & AL. 2017: Additions to the flora of the Prespa National Park, Greece. - *Phytol. Balcan.* **23(2)**: 207-269.
STRID, A. & ANDERSSON, I.A. 1985: Chromosome numbers of Greek mountain plants. An annotated list of 115 species. - *Bot. Jahrb. Syst.* **107(1-4)**: 203-228.
STRID, A., ANDONOVSKI, A. & ANDONOVSKI, V. 2003: The high mountain vegetation of the Balkan Peninsula. In Nagy, L. & al. (eds): Alpine biodiversity in Europe (Ecological studies 167), pp. 113-121. - Berlin & Heidelberg: Springer-Verlag.
STRID, A., BERGMEIER, E. & FOTIADIS, G. 2020: Flora and vegetation of the Prespa National Park. - Agios Germanos: Society for the Protection of Prespa, 522 pp.
STRID, A. & FRANZÉN, R. 1981: [Chromosome numbers in plants from Mt Olympus]. In Löve, Á. (ed.): Chromosome number reports LXXIII. - *Taxon* **30(4)**: 829-842.
STRID, A. & FRANZÉN, R. 1982: New floristic records from the mountains of northern Greece. (Materials for the Mountain Flora of Greece, 12). - *Willdenowia* **12(1)**: 9-28.
STRID, A. & FRANZÉN, R. 1983: [Chromosome numbers in Greek plants]. In Löve, Á. (ed.): IOPB chromosome number reports LXXVIII. - *Taxon* **32(1)**: 138-140.
STRID, A. & FRANZÉN, R. 1984: Chromosome numbers in flowering plants from Greece. (Materials for the Mountain Flora of Greece, 22). - *Willdenowia* **13(2)**: 329-333.
STRID, A. & GUTERMANN, W. 2005: *Euphorbia pinea* L. In Greuter, W. & Raus, Th. (eds): Med- Checklist notulae, 23. - *Willdenowia* **35(1)**: 59.
STRID, A. & HARTVIG, P. 1987: New taxa and new records from the mountains of SW and SC Turkey. - *Bot. Jahrb. Syst.* **108(2-3)**: 301-341.
STRID, A. & KARL, R. 2009: *Bongardia chrysogonum* (Berberidaceae) rediscovered on the East Aegean island of Chios. - *Phytol. Balcan.* **15(3)**: 337-342.
STRID, A. & LASSEN, P. 2000: *Nonea atra, Tripleurospermum oreades, Ajuga salicifolia* subsp. *bessarabica, Onobrychis vicifolia, Sesleria alba*. In Greuter, W. & Raus, Th. (eds): Med-Checklist notulae, 19. - *Willdenowia* **30(2)**: 231-243.
STRID, A. & PAPANICOLAOU, K. 1980: New species of *Aethionema* and *Peucedanum* from the Greek mountains. (Materials for the Mountain Flora of Greece, 5). - *Bot. Not.* **133(4)**: 521-526.
STRID, A. & PAPANICOLAOU, K. 1981: Floristic notes from the mountains of Northern Greece. (Materials for the Mountain Flora of Greece, 7). - *Nordic J. Bot.* **1(1)**: 66-82.
STRID, A. & PAPANICOLAOU, K. 1985: The Greek mountains. In Gómez-Campo, C. (ed.): Plant conservation in the Mediterranean area, pp. 89-111. - Dordrecht, etc.: Dr. W. Junk Publishers.
STRID, A. & RAUS, TH. 2012: *Soldanella chrysosticha* subsp. *chrysosticha*. In Greuter, W. & Raus, Th. (eds): Med-Checklist notulae, 31. - *Willdenowia* **42(2)**: 291.
STRID, A. & RAUS, TH. 2020: *Stellaria ruderalis*. In Raus, Th. & Raab-Straube, E.v. (eds): Euro+Med- Checklist notulae, 12. - *Willdenowia* **50(2)**: 312.
STRID, A. & STRID, B. 2009: Flora Graeca Sibthorpiana. An annotated re-issue. Volumes 1-2 (plates 1-200). - Ruggell: A.R.G. Gantner Verlag, xv + 412 pp.
STRID, A. & STRID, B. 2010: Flora Graeca Sibthorpiana. An annotated re-issue. Volumes 3-4 (plates 201-400). - Ruggell: A.R.G. Gantner Verlag, v + 412 pp.
STRID, A. & STRID, B. 2011: Flora Graeca Sibthorpiana. An annotated re-issue. Volumes 5-6 (plates 401-600). - Ruggell: A.R.G. Gantner Verlag, v + 410 pp.
STRID, A. & STRID, B. 2012: Flora Graeca Sibthorpiana. An annotated re-issue. Volumes 7-8 (plates 601-800). - Ruggell: A.R.G. Gantner Verlag, v + 413 pp.

STRID, A. & STRID, B. 2013: Flora Graeca Sibthorpiana. An annotated re-issue. Volumes 9-10 (plates 801-966). - Koenigstein: Koeltz Scientific Books, viii + 353 pp.

STRID, A. & TAN, KIT 1992: Flora Hellenica and the threatened plants of Greece. - *Opera Bot.* **113**: 55-67.

STRID, A. & TAN, KIT 1998a: Notes on *Galinsoga parviflora* and *Agrimonia procera*. In Greuter, W. & Raus, Th. (eds): Med-Checklist notulae, 17. - *Willdenowia* **28(1-2)**: 163-174.

STRID, A. & TAN, KIT 1999: Notes on *Myosotis speluncicola, Anthemis wiedemanniana, Astragalus gladiatus, Glyceria nemoralis* and *Juncus acutiflorus*. In Greuter, W. & Raus, Th. (eds): Med-Checklist notulae, 18. - *Willdenowia* **29(1-2)**: 51-67.

STRID, A. & TAN, KIT 2000a: The flora and Floras of Turkey and Greece. In Özhatay, N. & al. (eds): Abstracts from the Second Balkan Botanical Congress, Istanbul, Turkey, May 14-18, 2000, p. 4.

STRID, A. & TAN, KIT 2000b: *Viola herzogii* (W. Becker) Bornm. In Greuter, W. & Raus, Th. (eds): Med-Checklist notulae, 19. - *Willdenowia* **30(2)**: 241.

STRID, A. & TAN, KIT 2001: Bulbous and tuberous plants of Greece. In Raimondo, F.M. (ed.): Abstracts from the X OPTIMA meeting, Palermo, 13-19 September 2001, p. 155.

STRID, A. & TAN, KIT 2002: Notes on *Galium paschale* Forssk., *Peucedanum minutifolium* (Janka) Velen., *Agropyron cristatum* (L.) Gaertn., *Piptatherum virescens* (Trin.) Boiss., *Nectaroscordum siculum* (Ucria) Lindl. In Greuter, W. & Raus, Th. (eds): Med-Checklist notulae, 21. - *Willdenowia* **32 (2)**: 203-207.

STRID, A. & TAN, KIT 2003: Bulbous and tuberous plants of Greece. - *Bocconea* **16(2)**: 757-762.

STRID, A. & TAN, KIT 2004a: New species of *Erodium* (Geraniaceae), *Onosma* (Boraginaceae) and *Centaurea* (Asteraceae) from northern Greece. - *Phytol. Balcan.* **9(3)**: 471-478.

STRID, A. & TAN, KIT 2004b: *Galinsoga parviflora* Cav. In Greuter, W. & Raus, Th. (eds): Med-Checklist notulae, 22. - *Willdenowia* **34(1)**: 74.

STRID, A. & TAN, KIT 2005: A new species of *Omphalodes* (Boraginaceae) from southeast Peloponnese, Greece. - *Phytol. Balcan.* **11(1)**: 69-72.

STRID, A. & TAN, KIT 2009: A new species of *Centaurea* (Asteraceae) from the island of Samothraki (NW Greece). - *Phytol. Balcan.* **15(2)**: 185-189.

STRID, A. & TAN, KIT 2017: Recent progress in plant taxonomy and floristic studies in Greece. - *Bot. Serbica* **41(2)**: 123-152.

STRID, A. & TAN, KIT 2019: Report 104. In Vladimirov, V., Aybeke, M. & Tan, Kit (eds): New floristic records in the Balkans: 39. - *Phytol. Balcan.* **25(2)**: 223-225.

STRID, A., TAN, KIT & KOFINAS, G. 2021: Report 77. In Vladimirov, V., Aybeke, M. & Tan, Kit (eds): New floristic records in the Balkans: 44. - *Phytol. Balcan.* **27(1)**: 140-141.

STRID, A. & TAN, KIT (EDS) 1991: Mountain flora of Greece, vol. 2. - Edinburgh: Edinburgh Univ. Press, xxv + 974 pp.

STRID, A. & TAN, KIT (EDS) 1996: Flora and vegetation of the Peloponnese and Kithira. Report of a student excursion from the University of Copenhagen, May 14-28, 1995. - Copenhagen: Botanical Institute, 91 pp. + appendix 24 pp.

STRID, A. & TAN, KIT (EDS) 1997: Flora Hellenica, vol. 1. Gymnospermae to Caryophyllaceae. - Koenigstein: Koeltz Scientific Books, xxxvi + 547 pp., incl. 722 2-colour maps.

STRID, A. & TAN, KIT (EDS) 1998b: Flora and vegetation of North East Greece, including the islands of Thasos and Samothraki. Report of a student excursion from the University of Copenhagen, May 17-31, 1997. - Copenhagen: Botanical Institute, 120 pp. incl. appendix.

STRID, A. & TAN, KIT (EDS) 2000: Flora and phytogeography of NW Greece (Epirus and W Macedonia). Report of a student excursion from the University of Copenhagen, 20 May - 2 June 1999. - Copenhagen: Botanical Institute, 119 pp. incl. tabular summary.

STRID, A. & TAN, KIT (EDS) 2003: Flora Hellenica, vol. 2. Nymphaeaceae to Platanaceae. - Ruggell: A. R.G. Gantner Verlag, xvi + 511 pp., incl. 611 2-colour maps.

STRID, A. & TZANOUDAKIS, D. 1999: Contribution to the study of the Greek insular flora: The flora and the phytogeographical connections of Samothraki. In Elefheriou, E.P. (ed.): Proceedings, 7th Panhellenic scientific conference, Alexandroupolis, pp. 332-335 [In Greek with English abstract]. - Thessaloniki: Univ. Studio Press.

STRID, A. & VASSILIADES, D. 2005: *Ranunculus crenatus* Waldst. & Kit. In Greuter, W. & Raus, Th. (eds): Med-Checklist notulae, 23. - *Willdenowia* **35(1)**: 60.

STRID, A. (ED.) 1972: Evolution in the Aegean. - *Opera Bot.* **30**: 1-83.

STRID, A. (ED.) 1986: Mountain flora of Greece, vol. 1. - Cambridge: Cambridge University Press, xxx + 822 pp.

STROBL, W. & WITTMANN, H. 1985: Beiträge zur Kenntnis von Verbreitung, Soziologie und Karyologie von *Achnatherum calamagrostis* (L.) P.B. im Bundesland Salzburg (Österreich). - *Ber. Bayer. Bot. Ges.* **56**: 95-102.

STROH, G. 1937: Die Gattung *Gagea* Salisb. - *Beih. Bot. Centralbl., Abt. 2* **57(3)**: 485-520.

STROH, G. 1938: Vorläufiges Verzeichnis der altweltlichen Arten der Gattungen *Lithospermum* und *Lithodora*. - *Beih. Bot. Centralbl., Abt. 2* **58(2)**: 203-212.

STROH, G. 1939a: Die Gattung *Onosma* Linn. Versuch einer systematischen Kodifizierung der Arten. - *Beih. Bot. Centralbl., Abt. 2* **59(2-3)**: 430-454.

STROH, G. 1939b: Die Gattung *Gypsophila*. Vorläufiger Katalog. - *Beih. Bot. Centralbl., Abt. 2* **59(2-3)**: 455-477.

STROH, G. 1941: Die Gattung *Myosotis* L. Versuch einer systematischen Übersicht über die Arten. - *Beih. Bot. Centralbl., Abt. 2* **61(1)**: 317-345.

STROH, G. 1942: Die Gattung *Veronica* L. Versuch einer systematischen Kodifizierung der Arten. - *Beih. Bot. Centralbl., Abt. 2* **61(2)**: 384-451.

STROH, G. & STROH, R. 1964: *Ophrys*-Studien auf Rhodos. In Sundermann, H. (ed.): Probleme der Orchideengattung *Ophrys*. - *Jahresber. Naturwiss. Vereins Wuppertal* **19**: 39-40.

STRÖHLE, W. 1999: Bibliographie und Index nomenklatorischer Neuheiten der Jahrgänge 1-30 (1969-1998) des "Mitteilungsblatt Arbeitskreis Heimische Orchideen Baden-Württemberg" und des "Journal Europäischer Orchideen". - *J. Eur. Orch.* **31(3)**: 722-786.

STRÖHLE, W. 2003: Nomenklatorische Liste der europäischen Orchideentaxa - Arten und Unterarten. - *J. Eur. Orch.* **35(4)**: 771-860.

STRÖHLE, W. 2004: *Epipactis baumanniorum* Ströhle nom. nov. ersetzt *Epipactis gracilis* B. & H. Baumann 1988. - *J. Eur. Orch.* **36(4)**: 1043-1044.

STRÖHLEIN, H. & SUNDERMANN, H. 1972: Chromosomenzählungen bei Erdorchideen. - *Jahresber. Naturwiss. Vereins Wuppertal* **25**: 75-80.

STRÖMBERG, R. 1940: Griechische Pflanzennamen. - *Göteborgs Högskolas Årsskrift* **46**: 1-190.

STROTHER, J.L. 1977: Tageteae - systematic review. In Heywood, V.H., Harvorne, J.B. & Turner, B.L. (eds): The biology and chemistry of the Compositae 2: 769-783. - London, etc.: Academic Press.

STUART, D.C. 1965: *Muscari* and allied genera. - *Lily Year-Book* **29**: 125-138.

STUART, D.C. 1966: Cytotaxonomic studies in *Muscari* sensu lato. - Edinburgh: Unpubl. Ph.D. thesis, Univ. of Edinburgh [n.v.].

STUART, D.C. 1970: Chromosome numbers in the genus *Muscari* Mill. - *Notes Roy. Bot. Gard. Edinburgh* **30(1)**: 189-196.

STUBBE, H. 1966: Genetik und Zytologie von *Antirrhinum* L. sect. *Antirrhinum*. - Jena: Gustav Fischer Verlag, 421 pp.

STUCKEY, R.L. 1981: Distributional history of *Lythrum salicaria* L. (purple loose strife) in North America. - *Bartonia* **47**: 3-20.

STUCKEY, R.L. 1985: Distributional history of *Najas marina* (spiny najad) in North America. – *Bartonia* **51**: 22-26.

STUESSY, T.F. 1977: Heliantheae - systematic review. In Heywood, V.H., Harborne, J.B. & Turner, B.L. (eds): The biology and chemistry of the Compositae 2: 621-671. - London, etc.: Academic Press .

STURDY, R. 2001: Crete May 2-9. - *Int. Dendrol. Soc. Year Book* **2000**: 48-58.

STURM, K. 1911: Monographische Studien über *Adoxa Moschatellina* L. - *Vierteljahrsschr. Naturf. Ges. Zürich* **55**: 391-462.

STUTZ, H.C. 1972: On the origin of cultivated rye. - *Amer. J. Bot.* **59(1)**: 59-70.

STYLES, B.T. 1962: The taxonomy of *Polygonum aviculare* and its allies in Britain. - *Watsonia* **5(4)**: 177-214 + 2 plates.

SUÁREZ-CERVERA, M. 1986: Aportación a la cariología del género *Lavandula* L. - *Anales Jard. Bot. Madrid* **42(2)**: 389-394.

SUC, J.-P. 1984: Origin and evolution of the Mediterranean vegetation and climate in Europe. - *Nature* **307(5950)**: 429-432.

SUDRE, H. 1908-13: Rubi Europae vel monographia iconibus illustrata Ruborum Europae. - Paris: Libraire des Sciences Naturelles, 305 pp. + 215 plates.

SUKHORUKOV, A.P. 2007: Einige neue und wenig bekannte Taxa aus der Familie Chenopodiaceae in Europa und im östlichen Mittelmeergebiet. - *Feddes Repert.* **118(3-4)**: 73-83.

SUKHORUKOV, A.P. 2007a: Fruit anatomy and its taxonomic significance in *Corispermum* (Corispermoideae, Chenopodiaceae). - *Willdenowia* **37(1)**: 63-87.

SUKHORUKOV, A.P. & AL. 2021: Evolutionary relationships, biogeography and morphological characters of *Glinus* (Molluginaceae), with special emphasis on the genus composition in Sub- Saharan Africa. - *PhytoKeys* **173**: 1-92.

SUŁEK, E. 1977: Karyological studies in *Polygonatum* Mill. II. Karyotype analysis of *P. verticillatum* (L.) All. - *Acta Biol. Cracov., Ser. Bot.* **20(1-2)**: 25-32.

ŠUMBEROVÁ, K. & AL. 2012: Distribution, habitat ecology, soil seed bank and seed dispersal of threatened *Lindernia procumbens* and alien *Lindernia dubia* (Antirrhinaceae) in the Czech Republic. - *Phyton (Horn)* **52(1)**: 39-72.

ŠUMBEROVÁ, K., TZONEV, R. & VLADIMIROV, V. 2004: *Bidens frondosa* (Asteraceae) - a new alien species for the Bulgarian flora. - *Phytol. Balcan.* **10(2-3)**: 179-181.

SÜNDERMANN, F. 1903: Eine neue *Arabis* aus Macedonien. - *Allg. Bot. Z. Syst.* **9(3)**: 62-63.

SUNDERMANN, H. 1964a: Zum Problem der Artabgrenzung innerhalb der Gattung *Ophrys*. In Sundermann, H. (ed.): Probleme der Orchideengattung *Ophrys*. - *Jahresber. Naturwiss. Vereins Wuppertal* **19**: 9-17.

SUNDERMANN, H. 1964b: Bestimmungsschlüssel und Artenliste der Gattung *Ophrys*. In Sundermann, H. (ed.): Probleme der Orchideengattung *Ophrys*. - *Jahresber. Naturwiss. Vereins Wuppertal* **19**: 63-70.

SUNDERMANN, H. 1969a: Über einige ostmediterrane Orchideen. - *Orchidee (Hamburg)* **20(2)**: 79-83.

SUNDERMANN, H. 1969b: Die Orchideen im südwestlichen Kleinasien. - *Orchidee (Hamburg)* **20(6)**: 309-317.

SUNDERMANN, H. 1972a: Artenproduktion und Konsumbedürfnis. Kritische Bemerkungen zur Benennung und Beschreibung von Splittersippen. - *Orchidee (Hamburg)* **23(4)**: 166-168.

SUNDERMANN, H. 1972b: Die spezifisch ostmediterranen Arten der Gattung *Orchis*. - *Jahresber. Naturwiss. Vereins Wuppertal* **25**: 59-60.

SUNDERMANN, H. 1973: *Himantoglossum* (*Loroglossum*) *hircinum-caprinum-calcaratum-affine*. - *Acta Bot. Acad. Sci. Hung.* **19(1-4)**: 367-374.

SUNDERMANN, H. 1975: Zum Problem der Definition taxonomischer Kategorien (Spezies, Subspezies, Praespezies, Varietät) - dargestellt am Beispiel des Sippenkomplexes *Ophrys fuciflora* (Crantz) Moench - *Ophrys scolopax* Cav. - *Taxon* **24(5-6)**: 615-627.

SUNDERMANN, H. 1977: The genus *Ophrys* - An example of the importance of isolation for speciation. - *Amer. Orchid Soc. Bull.* **46(9)**: 825-831.

SUNDERMANN, H. 1980: Europäische und mediterrane Orchideen. Eine Bestimmungsflora mit Berücksichtigung der Ökologie: 3. Erweiterte und verbesserte Auflage. - Hildesheim: Brücke-Verlag Kurt Schmersow, 279 pp.

SUNDERMANN, H. & BANK, C. VON DER 1977: Neue Beiträge zur Zytotaxonomie der Erdorchideen. II (*Ophrys*, *Orchis*, *Neotinia*, *Steveniella* und *Gennaria*). - *Orchidee (Hamburg)* **28(4)**: 146-147.

SUNDERMANN, H. (ED.) 1964: Probleme der Orchideengattung *Ophrys*. - *Jahresber. Naturwiss. Vereins Wuppertal* **19**: 72 pp. + 6 plates.

SÜNTER, G. 1979: Morphological and cytological studies on the *Dianthus* species of Istanbul area. - *Istanbul Üniv. Fen Fak. Mecm., B* **44**: 31-48 + 2 plates.

SUŠNIK, F. 1967: Zur Chromosomenzahl einiger Pflanzensippen II [In Slovenian with German summary]. - *Biol. Vestn.* **15**: 63-66 + 2 plates.

SUŠNIK, F. & AL. 1972: [Miscellaneous chromosome numbers from Jugoslavia]. In Löve, Á. (ed.): IOPB chromosome number reports XXXVI. - *Taxon* **21(2-3)**: 345.

SUŠNIK, F. & LOVKA, M. 1970: Cytologische und morphologische Untersuchungen der Gattung *Ceterach* Gars. in Slowenien [In Slovenian with German summary]. - *Biol. Vestn.* **18**: 19-25.

SUŠNIK, F. & LOVKA, M. 1973: [Miscellaneous chromosome numbers from Jugoslavia]. In Löve, Á. (ed.): IOPB chromosome number reports XLI. - *Taxon* **22(4)**: 462-463.

SUŠNIK, F., LOVKA, M. & DRUŠKOVIĆ, B. 1974: Chromosome numbers of some plant species from Slovenia I [In Slovenian with English summary]. - *Biol. Vestn.* **22(1)**: 39-56.

ŠUŠTAR, F. 1976: Zur Chorologie und Taxonomie des *Brachypodium pinnatum*-Komplexes in Slowenien [In Slovenian with German summary]. - *Biol. Vestn.* **24(1)**: 1-11.

SUTHERLAND, W.J. 1990: Biological Flora of the British Isles. *Iris pseudacorus* L. - *J. Ecol.* **78(3)**: 833-848.

SUTORÝ, K. 1989: *Cynoglossum pustulatum* Boiss., an overlooked Mediterranean species. - *Čas. Morav. Muz. Brně, Vědy Přír.* **74(1-2)**: 167-173.

SUTORÝ, K. 2000: *Cynoglossum hungaricum* Simonk. In Greuter, W. & Raus, Th. (eds): Med-Checklist notulae, 19. - *Willdenowia* **30(2)**: 230-231.

SUTORÝ, K. 2001: *Cynoglossum montanum* L. and *C. nebrodense* Guss. In Greuter, W. & Raus, Th. (eds): Med-Checklist notulae, 20. - *Willdenowia* **31(2)**: 320.

SUTORÝ, K. 2008: What is *Cynoglossum rotatum* Velenovský? - *Phytol. Balcan.* **14(2)**: 255-256.

SUTORÝ, K. 2012: Typification of Joseph Freyn's names in the genera *Celsia* and *Verbascum* (Scrophulariaceae). - *Willdenowia* **42(1)**: 29-35.

SUTORÝ, K. 2016a: Reports 69-70. In Vladimirov, V. & al. (eds): New floristic records in the Balkans: 29. - *Phytol. Balcan.* **22(1)**: 105.

SUTORÝ, K. 2016b: New names in the "Cynoglossum montanum group" (Boraginaceae) in the Mediterranean area. - *Edinburgh J. Bot.* **73(3)**: 265-275.

SUTORÝ, K. 2017a: Typifications of *Verbascum adrianopolitanum* and *V. banaticum*, and a new hybrid name, *Verbascum* ×*stribrnyi* (Scrophulariaceae). - *Ann. Bot. Fenn.* **54(1-3)**: 125-129.

SUTORÝ, K. 2017b: Report 189. In Vladimirov, V. & al. (eds): New floristic records in the Balkans: 33. - *Phytol. Balcan.* **23(2)**: 308-309.

SUTORÝ, K. 2019: Report 83-86. In Vladimirov, V. & al. (eds): New floristic records in the Balkans: 38. - *Phytol. Balcan.* **25(1)**: 113-114.

SUTORÝ, K. 2019: Reports 105-106. In Vladimirov, V., Aybeke, M. & Tan, Kit (eds): New floristic records in the Balkans: 39. - *Phytol. Balcan.* **25(2)**: 105-106.

SUTORÝ, K. 2021: Taxonomic and nomenclatural notes on *Jacobaea othonnae* (Asteraceae-Senecioneae-Senecioninae). - *Phytotaxa* **510(1)**: 61-68.

SUTORÝ, K. & ŘEPKA, R. 2020: Report 74. In Vladimirov, V., Aybeke, M. & Tan, Kit: New floristic records in the Balkans: 43. - *Phytol. Balcan.* **26(3)**: 556-557.

SUTORÝ, K. & VERLOOVE, P. 2015: *Euphorbia hypericifolia* L. In Raab-Straube, E. v. & Raus, Th. (eds): Euro+Med-Checklist notulae, 4. - *Willdenowia* **45(1)**: 122.

SUTTON, D.A. 1988: A revision of the tribe Antirrhineae. - London & Oxford: Oxford Univ. Press, viii + 575 pp. + 4 microfiches.

SVENSON, H.K. 1929a: Monographic studies in the genus *Eleocharis*. - *Rhodora* **31**: 121-135, 152-163, 167-191, 199-219.

SVENSON, H.K. 1929b: Monographic studies in the genus *Eleocharis* [cont.]. - *Rhodora* **31**: 224-242 + 4 plates.

SVENSON, H.K. 1932: Monographic studies in the genus *Eleocharis*. II. - *Rhodora* **34**: 193-203, 215-227 + 3 plates.

SVENSON, H.K. 1934: Monographic studies in the genus *Eleocharis*. III. - *Rhodora* **36(431)**: 377-389 + 2 plates.

SVENSON, H.K. 1937: Monographic studies in the genus *Eleocharis*. IV. - *Rhodora* **39**: 210-231, 236-273 + 6 plates.

SVENSON, H.K. 1939: Monographic studies in the genus *Eleocharis* - V. - *Rhodora* **41**: 1-19, 43-77, 90-110 + 11 plates.

SVENSSON, H.G. 1928: Zur Entwicklungsgeschichte der Blüten und Samen von *Limosella aquatica* L. - *Svensk Bot. Tidskr.* **22(3)**: 465-476.

SVENSSON, L. 1990: Floral morphology and the effects of crossing distance in *Scleranthus*. - Lund: Ph.D. thesis, 121 pp. [including summary and 6 papers].

SVENSSON, R. & WIGREN, M. 1986: Sminkrotens historia och biologi i Sverige [In Swedish with English summary]. - *Svensk Bot. Tidskr.* **80(2)**: 107-131.

SVENSSON, S. 1984: Chromosome numbers and morphology in the *Capsella bursa-pastoris* complex (Brassicaceae) in Greece. - *Willdenowia* **13(2)**: 267-276.

ŠVEŘEPOVÁ, G. 1966: Zur Unterscheidung der Taxa *Anagallis arvensis* L. und *Anagallis coerulea* Nathh. - *Novit. Bot. Delect. Seminum Horti Bot. Univ. Carol. Prag.* **1966**: 65-75.

ŠVEŘEPOVÁ, G. 1968: Zur Zytotaxonomie der Art *Anagallis arvensis* L. - *Preslia* **40**: 143-146.

ŠVEŘEPOVÁ, G. 1970: Analyse der morphologischen Merkmale einiger Taxa der Gattung *Anagallis* L. (Experimentell taxonomische Studie). - *Preslia* **42**: 350-372 + 1 table.

ŠVEŘEPOVÁ, G. 1972: Zur Zytotaxonomie der Gattung *Anagallis* L. - *Preslia* **44(3)**: 219-226.

ŠVEŘEPOVÁ, G. 1978: Artkreuzung von *Anagallis arvensis* und *Anagallis foemina* und Genese der Sektion *Anagallis*. - *Folia Geobot. Phytotax.* **13(2)**: 175-224.

SVESHNIKOVA, L.I. 1975: On the origin of karyotype of the genus *Galanthus* L. [In Russian]. - *Bot. Zhurn. (Moscow & Leningrad)* **60(12)**: 1760-1768.

SVOMA, E. 1981: Zur systematischen Embryologie der Gattung *Scilla* L. (Liliaceae). - *Stapfia* **9**: 1-124.

SWEET, R. 1823-25: The British flower garden, vol. 1. - London: Simpkin & Marshall, plates 1-100 with text.

SWEET, R. 1825-27: The British flower garden, vol. 2. - London: Simpkin & Marshall, plates 101-200 with text.

SWEET, R. 1827-29: The British flower garden, vol. 3. - London: Simpkin & Marshall, plates 201-300 with text.

SWEET, R. 1829-31: The British flower garden, second series, vol. 1. - London: J. Ridgeway, plates 1-100 with text.

SWEET, R. 1831-33: The British flower garden, second series, vol. 2. - London: J. Ridgeway, plates 101-200 with text.

SWEET, R. 1833-35: The British flower garden, second series, vol. 3. - London: J. Ridgeway, plates 201-300 with text.

SWEET, R. 1835-37: The British flower garden, second series, vol. 4. - London: J. Ridgeway, plates 301-412 with text.

ŚWIETLIŃSKA, Z. 1963: Cytogenetic relationships among *Rumex acetosa*, *Rumex arifolius* and *Rumex thyrsiflorus*. - *Acta Soc. Bot. Poloniae* **32(2)**: 215-279.

SYKORA, K.V., BABALONAS, D. & PAPASTERGIADOU, E. 2003: Strandline and sand-dune vegetation of coasts of Greece and some other Aegean countries. - *Phytocoenologia* **33(2-3)**: 409-446.

SYME, J.T.B. 1863: Remarks on *Gladiolus illyricus*, Koch, and its allies. - *J. Bot. (London)* **1**: 130-134.

SYMEONIDIS, L. & MOUSTAKAS, M. 1986: Biosystematic study of *Hordeum marinum* group in Greece. - *Flora, Morphol. Geobot. Oekophysiol.* **178(3)**: 177-182.

SYMEONIDIS, L., MOUSTAKAS, M. & KARATAGLIS, S. 1987: Biosystematic study of the genus *Taeniatherum* in Greece. - *Flora, Morphol. Geobot. Oekophysiol.* **179(3)**: 171-177.

SYNGE, H. 2005: Knapweed of the Muses refound. - *Plant Talk* **39**: 14-15.

SYNGE, P.M. 1973: *Lilium carniolicum* including var. *albanicum*, var. *bosniacum* and var. *jankae*, and *Lilium rhodopaeum*. - *Lilies* **1973**: 6-9.

SYNGE, P.M. (ED.) 1980: Lilies. A revision of Elwes' Monograph of the genus *Lilium* and its supplements. - London: B.T. Batsford Ltd, 276 pp. + 56 plates.

SYSKA, M. 1995: Die Orchideenflora des westlichen Nestos-Deltas und des angrenzenden Berglandes (Nordostgriechenland). Verbreitung - Ökologie - Gefährdung. - *J. Eur. Orch.* **27(3)**: 339-552.

SYSKA, M. 1996: Die Orchideenflora des westlichen Nestos-Deltas und des angrenzenden Berglandes (Nordost-Griechenland): Verbreitung - Ökologie - Gefährdung. - *Ber. Arbeitsgeb. Entwicklungsforsch. (Münster)* **25**: 42-55.

SZ.-BORSOS, O. 1970: Contributions to the knowledge on the chromosome numbers of phanerogams growing in Hungary and south-eastern Europe. - *Acta Bot. Acad. Sci. Hung.* **16(3-4)**: 255-265.

SZ.-BORSOS, O. 1971: Contributions to the knowledge on the chromosome numbers of phanerogams growing in Hungary and south-eastern Europe. II. - *Acta Bot. Acad. Sci. Hung.* **17(1-2)**: 37-46.

SZABÓ, Z.V. 1905: Monographie der Gattung *Knautia*. - *Bot. Jahrb. Syst.* **36(4)**: 389-442.

SZABÓ, Z.V. 1907: Index criticus specierum atque synosymorum generis *Knautia* (L.) Coult. - *Bot. Jahrb. Syst.* **38 Beibl. 89**: 1-31.

SZABÓ, Z.V. 1911: A *Knautia* génusz monographiája (Monographia gen. "Knautia") [In Hungarian and Latin]. - *Math. Term. Közlem.* **31**: 1-436 + 54 plates and 4 maps.

SZABÓ, Z.V. 1926: Geschichte der Gattung *Cephalaria*. - *Magyar Bot. Lapok* **24(1-12)**: 1-14.

SZABÓ, Z.V. 1934: Nouvelles données à la connaissance des *Knautia* de l'ouest et du sud de l'Europe [In Hungarian and French]. - *Bot. Közlem.* **31(3-4)**: 109-141.

SZABÓ, Z.V. 1940: A *Cephalaria* - génusz monográfiája. Monographia gen. *Cephalaria*. - *Mat. Term. Közlem.* **38(4)**: i-viii, 1-352 + 5 maps.

SZELĄG Z. 2015: Hieracia balcanica X. Typification of the *Hieracium* (Asteraceae) names described by K. H. Zahn from the Balkan Peninsula. - *Polish Bot. J.* **60**: 41-65.

SZELĄG, Z. 2008: Taxonomic and nomenclatural notes on *Pilosella alpicola* agg. (Asteraceae) in the Balkans and Carpathians. - *Ann. Bot. Fenn.* **45(4)**: 301-306.

SZELĄG, Z. 2011: Hieracia balcanica VI. Typification of *Hieracium pavichii* (Asteraceae). - *Polish Bot. J.* **56(1)**: 51-54.

SZELĄG, Z. 2014: Hieracia balcanica IX. Typfication of the *Hieracium* (Asteraceae) names described by J. Freyn from the Balkan Peninsula. - *Polish Bot. J.* **59(2)**: 197-213.

SZELĄG, Z. 2018: Hieracia balcanica XV. Taxonomic and nomenclatural notes on *Hieracium pilosissimum* and *H. divaricatum*, with remarks on the *H. heldreichii* aggregate (Asteraceae). - *Phytotaxa* **356(1)**: 81-90.

SZELĄG, Z. & SOMLYAY, L. 2009: History of discovery and typification of *Haberlea rhodopensis* Friv. (Gesneriaceae). - *Ann. Bot. Fenn.* **46(6)**: 555-558.

SZELĄG, Z. & VLADIMIROV, V. 2019: The species intermediate between *Hieracium petrovae* and *H. olympicum* (Asteraceae): a treatment of *H. kritschimanum* and description of a new species from Greece. - *Phytotaxa* **402(2)**: 107-113.

SZELĄG, Z. & VLADIMIROV, V. 2020: *Pilosella bulgarica* (Asteraceae) a new hybridogenous species from Bulgaria. - *Phytotaxa* **428(3)**: 290-294.

SZELIGA, M., CIURA, J. & TYRKA, M. 2017: Genetic diversity of three European *Veratrum* species revealed by amplified fragment length polymorphism. - *Biodiv. Res. Conserv.* **47(1)**: 1-8.

SZENTPÉTERI, L.J. & MÓNUS, F. 1999: *Epipactis degenii* Szentpéteri & Mónus spec. nov., a new *Epipactis* species from Greece. - *J. Eur. Orch.* **31(3)**: 644-651.

SZIJJ, J. & AL. 1997: Das Nestosdelta in Nordost-Griechenland. Forschungen zur Ökologie und naturschützerischen Wertanalyse. - *Ber. Arbeitsgeb. Entwicklungsforsch. (Münster)* **26**: vi + 146 pp.

SZILÁDY, Z. 1933: Der urzeitliche Nussbaum bei uns und auf der Balkanhalbinsel [In Hungarian and German]. - *Bot. Közlem.* **30(5)**: 121-137.

SZUJKÓ-LACZA, J. 1983: Distribution and diversity of *Gentianella austriaca*, *Gentianella austriaca* ssp. *fatrae* (Borb.) em. SzL. and *Gentianella livonica* (Eschholz apud Grisebach) em. SzL. - *Stud. Bot. Hung.* **16**: 109-116.

SZUJKÓ-LACZA, J., KOVÁTS, D. & RAJCZY, M. 1979: Revision and numerical evaluation of the Hungarian *Spergularia* species (Caryophyllaceae). - *Ann. Hist.-Nat. Mus. Natl. Hung.* **71**: 73-99.

TAARNA, K. 1973: Chromosome counts on *Ziziphora* L. Labiatae. - *Ann. Bot. Fenn.* **10(4)**: 341.

TABORDA DE MORAIS, A. 1936: Estudos na aveias I. As aveias portuguesas de secção *Euavena* Griseb. - *Bol. Soc. Brot., Ser. 2* **11**: 49-86 + 7 plates.

TABORDA DE MORAIS, A. 1939: Estudos na aveias II. As aveias portuguesas de secção *Euavena* Griseb. - *Bol. Soc. Brot., Ser. 2* **13**: 573-709 + 20 plates and 5 tables.

TABUR, Ş., CIVELEK, S. & BAĞCI, E. 2001: Cytotaxonomic studies on some *Vicia* L. species growing in eastern Mediterranean and southern Aegean regions, I. - *Acta Bot. Gallica* **148(2)**: 159-174.

TABUR, Ş., CIVELEK, S. & BAĞCI, E. 2002: Cytotaxonomic studies on some *Vicia* L. species growing in eastern Mediterranean and southern Aegean regions: 2. - *Acta Bot. Hung.* **44(1-2)**: 185-204.

TACIK, T. 1971: Observationes ad *Campanulae* subsectionem *Heterophyllae* (Nyman ex Witasek) Fedorov e Polonia meridionali pertinentes. Pars 1. - *Fragm. Florist. Geobot.* **17(2)**: 221-236.

TÄCKHOLM, G. 1922: Zytologische Studien über die Gattung *Rosa*. - *Acta Horti Berg.* **7(3)**: 97-381.

TAEB, F., EKICI, M. & PODLECH, D. 2012: Taxonomic revision of annual species of *Astragalus* (Fabaceae) in Turkey. - *Feddes Repert.* **123(1)**: 1-26.

TAHIRI, H. & CUBAS, P. 2000: Reports (1201-1207) In Kamari, G., Felber, F. & Garbari, F. (eds): Mediterranean chromosome number reports - 10. - *Fl. Medit.* **10**: 415-419.

TAKÁCS, A. & AL. 2018: Resurrection and typification of *Elatine campylosperma*(Elatinaceae), a long-forgotten waterwort species. - *PeerJ* **6**: e4913; DOI 10.7717/peerj.4913.

TAKÁCS, A. & AL. 2020: Evidence of hybridization between *Galatella villosa* and *G. linosyris*, and a taxonomic reappraisal of the hybrid *G.* ×*subvillosa*. - *Preslia* **92(4)**: 375-390.

TAKHTAJAN, A.L. 1999: Botanical observations in coastal Albania [In Russian with English abstract]. - *Fl. Veget. Pl. Resources Armenia* **12**: 6-10.

TALAVERA, S. 1976: Revisión de las especies españolas del género *Biarum* Schott. - *Lagascalia* **6(2)**: 275-296.

TALAVERA, S. 1977: Números cromosómicos para la flora española. 18-23. - *Lagascalia* **7(2)**: 201-203.

TALAVERA, S. 1979a: Revisión de la sect. *Erectorefractae* Chowdhuri del género *Silene*. – *Lagascalia* **8(2)**: 135-164.

TALAVERA, S. 1979b: Números cromosómicos para la flora española. 99-103. - *Lagascalia* **9(1)**: 122-123.

TALAVERA, S. 1981: Números cromosómicos para la flora española. 201-205. - *Lagascalia* **10(2)**: 232-233.

TALAVERA, S. & AL. 1999: Números cromosomáticos de plantas occidentales, 809-812. - *Anales Jard. Bot. Madrid* **57(1)**: 139.

TALAVERA, S., GARCÍA MURILLO, P. & HERRERA, J. 1993: Chromosome numbers and a new model for karyotype evolution in *Ruppia* L. (Ruppiaceae). - *Aquatic Bot.* **45(1)**: 1-13.

TALAVERA, S., GARCÍA MURILLO, P. & SMIT, H. 1986: Sobre el género *Zannichellia* L. (Zannichelliaceae). - *Lagascalia* **14(2)**: 241-271.

TALAVERA, S. & MUÑOZ GARMENDIA, F. 1989: Sinopsis del género *Silene* L. (Caryophyllaceae) en la Península Ibérica y Baleares. - *Anales Jard. Bot. Madrid* **45(2)**: 407-460.

TALAVERA, S. & VALDÉS, B. 1976: Revisión del género *Cirsium* (Compositae) en la Península Ibérica. - *Lagascalia* **5(2)**: 127-223.

TALAVERA, S. & VELAYOS, M. 1992: Sinopsis del género *Arabis* L. (Brassicaceae) en la Península Ibérica e Islas Baleares. - *Anales Jard. Bot. Madrid* **50(1)**: 146-150.

TALENT, N. & AL. 2008: (1847) Proposal to conserve the name *Crataegus* against *Mespilus* (Rosaceae). - *Taxon* **57(3)**: 1007-1008. [approved, cf. Taxon 60(1): 227, 2011].

TALI, K., FOLEY, M.J.Y. & KULL, T. 2004: Biological Flora of the British Isles. *Orchis ustulata* L. - *J. Ecol.* **92(1)**: 174-184.

TALOUMIS, G. 1971: The plants and flowers of Greece. - *Garden (New York)* **21**: 176-181.

TAMBERG, T.G. 1967: Polyploidy in the genus *Gladiolus* L. - *Genetika* **6**: 123-128.

TAMMARO, F. 1985: Studio biosistematico e biometrico in *Matthiola valesiaca* gr. (Cruciferae) delle Alpi e *M. fruticulosa* (L.) Maire dell'Appennino Centrale. - *Arch. Bot. Biogeogr. Ital.* **61(1-2)**: 1-28.

TAMMARO, F. & PACE, L. 1988: Il genere *Pinguicula* L. (Lentibulariaceae) in Italia centrale e d istituzione di una nuova specie, *P. fiorii* Tamm. & Pace. - *Inform. Bot. Ital.* **19(3)**: 429-436.

TAMMARO, F. & VERI, L. 1974: Il vischio quercino ("Loranthus europaeus" Jacq.) in Abruzzo. - *Inform. Bot. Ital.* **5(3)**: 256-259.

TAMURA, M. & LAUENER, L.A. 1979: A synopsis of *Aconitum* subgenus *Lycoctonum*: II. - *Notes Roy. Bot. Gard. Edinburgh* **37(3)**: 431-466.

TAMURA, M.N. 1993: Biosystematic studies of the genus *Polygonatum* (Liliaceae). III. Morphology of staminal filaments and karyology of eleven Eurasian species. - *Bot. Jahrb. Syst.* **115(1)**: 1-26.

TAN, KIT 1980a: Studies in the Thymelaeaceae I: Germination, seedlings, fruits & seeds. – *Notes Roy. Bot. Gard. Edinburgh* **38(1)**: 149-164.

TAN, KIT 1980b: Studies in the Thymelaeaceae II: A revision of the genus *Thymelaea*. - *Notes Roy. Bot. Gard. Edinburgh* **38(2)**: 189-246.

TAN, KIT 1989: The Flora of Turkey and its supplement - a model for Mediterranean Flora writing? In Phitos, D. (ed.): Abstracts. Sixth OPTIMA meeting, Delphi, 10-16 Sept. 1989.

TAN, KIT 1991: The "Flora of Turkey" and its supplement - a model for Mediterranean Flora writing? - *Bot. Chron. (Patras)* **10**: 41-48.

TAN, KIT 1997a: Biodiversity in the Greek Brassicaceae. In: Chorological problems in the European flora. Abstracts from Atlas Florae Europaeae meeting, Helsinki, August 8-10 1997, p. 46. - Helsinki.

TAN, KIT 1997b: Do you believe in Greek Brassicaceae? In Tsekos, I. & Moustakas, M. (eds): Abstracts from the First Balkan Botanical Congress, Thessaloniki, 19-22 September 1997, p. 25. - Thessaloniki.

TAN, KIT 1999: To Levkoio sti Thaso [In Greek]. - *Fisis [Journal of the Hellenic Society for the Protection of Nature]* **85**: 32.

TAN, KIT 2001: Flora Hellenica and the Endemic Flora of Greece. In: Proceedings of the 10th Panhellenic Pharmaceutical Congress, 17-19 May 2001, 6 pp.

TAN, KIT 2006a: Reports 47-68. In Vladimirov, V., Tan, Kit & Stevanović, V. (eds): New floristic records in the Balkans: 1. - *Phytol. Balcan.* **12(1)**: 115-119.

TAN, KIT 2006b: *Campanula merxmuelleri, Astragalus suberosus*. In Greuter, W. & Raus, Th. (eds): Med-Checklist notulae, 24. - *Willdenowia* **36(2)**: 721, 724.

TAN, KIT & AL. 1999a: Notes on *Genista anatolica*. In Greuter, W. & Raus, Th. (eds): Med-Checklist notulae, 18. - *Willdenowia* **29(1-2)**: 58-59.

TAN, KIT & AL. 1999b: Serpentine endemics in Albania and NW Greece. In Correia, A.I.D. & al. (eds): Livro de resumos. Guia de excursões. V jornadas de taxonomia botânica, 16-19/IX 1999, p. 68.

TAN, KIT & AL. 2004: An autumn-flowering *Leucojum* (Amaryllidaceae) from South Albania and Western Greece. - *Fisis [Journal of the Hellenic Society for the Protection of Nature]* **107**: 39-42.

TAN, KIT & AL. 2009a: *Centaurea wagenitziana* (Asteraceae: Centaureinae), a new species from the Eastern Balkans. - *Phytol. Balcan.* **15(1)**: 51-58.

TAN, KIT & AL. 2009b: Reports 75-89. In Vladimirov, V., Dane, F. & Tan, Kit (eds): New floristic records in the Balkans: 10. - *Phytol. Balcan.* **15(1)**: 132-134.

TAN, KIT & AL. 2009c: Reports 59-69. In Vladimirov, V., Dane, F. & Tan, Kit (eds): New floristic records in the Balkans: 11. - *Phytol. Balcan.* **15(2)**: 284-287.

TAN, KIT & AL. 2010a: *Hypericum boehlingraabei* (Hypericaceae), a new species from the northern Peloponnese (Greece). - *Phytol. Balcan.* **16(2)**: 227-231.

TAN, KIT & AL. 2010b: On a new *Petasites* species from the southern Pindos (Greece). - *Phytol. Balcan.* **16(2)**: 243-248.

TAN, KIT & AL. 2011: The genus *Gymnospermium* (Berberidaceae) in the Balkans. - *Phytotaxon* **25**: 1-17.

TAN, KIT & AL. 2012: Reports 114-116. In Vladimirov, V. & al. (eds): New floristic reports in the Balkans: 18. - *Phytol. Balcan.* **18(1)**: 84-85.

TAN, KIT & AL. 2013: On the identity of *Anchusa leucantha* (Boraginaceae) from northern Greece. - *Phytotaxa* **140(1)**: 35-42.

TAN, KIT & AL. 2015: The rediscovery of *Astragalus autranii* (Fabaceae) on Mt Tomor, south central Albania. - *Phytotaxa* **234(1)**: 83-89.

TAN, KIT & AL. 2015a: Report 116. In Vladimirov, V., Dane, F. & Tan, Kit (eds): New floristic records in the Balkans: 28. - *Phytol. Balcan.* **21(3)**: 382-383.

TAN, KIT & AL. 2016: Reports 92-147. In Vladimirov, V. & al. (eds): New floristic records in the Balkans: 30. - *Phytol. Balcan.* **22(2)**: 276-281.

TAN, KIT & AL. 2020: Reports 62-70. In Vladimirov, V., Aybeke, M. & Tan, Kit (eds): New floristic records in the Balkans: 41. - *Phytol. Balcan.* **26(1)**: 174-178.

TAN, KIT, AUTHIER, P. & VOLD, G. 1998: Notes on *Ranunculus garganicus*. In Greuter, W. & Raus, Th. (eds): Med-Checklist notulae, 17. - *Willdenowia* **28(1-2)**: 163-174.

TAN, KIT & BIEL, B. 2015: The flora of Samothraki with special emphasis on its phytogeographical reletionships. In Bogdanović, S. & Jogan, N.: Book of abstracts - 6th Balkan Botanical Congress, p. 39. - Rijeka.

TAN, KIT & BIEL, B. 2018d: The flora of Amorgos. - *Bot. Serbica* **142 (suppl. 1)**: 73-74.

TAN, KIT, BIEL, B. & SCHULER, A. 2007: The occurrence of *Fimbristylis bisumbellata* and *F. squarrosa* (Cyperaceae) in Greece. - *Phytol. Balcan.* **13(1)**: 81-82.

TAN, KIT, BIEL, B. & SFIKAS, G. 2015: *Legousia snogerupii* (Campanulaceae), a new species from southeastern Kiklades, Greece. - *Phytotaxa* **201(1)**: 63-70.

TAN, KIT, BIEL, B. & ŠILJAK-YAKOVLEV, S. 2014: *Galanthus samothracicus* (Amaryllidaceae) from the island of Samothraki, northeastern Greece. - *Phytol. Balcan.* **20(1)**: 65-70.

TAN, KIT, BIEL, B. & ZIELIŃSKI, J. 2020: *Micromeria zarkosii* (Lamiaceae), an unusual new species from Naxos (central Kiklades, Greece). - *Nordic J. Bot.* **38(1)**: 4 pp.

TAN, KIT, BONETTI, A. & LAFRANCHIS, T. 2010: Reports 74-82. In Vladimirov, V., Dane, F. & Tan, Kit (eds): New floristic records in the Balkans: 13. - *Phytol. Balcan.* **16(1)**: 157-159.

TAN, KIT, ELEFTHERIADOU, E. & THEODOROPOULOS, K. 2004: Additions to Flora Hellenica - *Arabis procurrens*. - *Fisis [Journal of the Hellenic Society for the Protection of Nature]* **105**: 42.

TAN, KIT & ELIAS, TH.S. 1989: Peter Hadland Davis: Biographical sketch; list of publications. In Tan, Kit (ed.): Plant taxonomy, phytogeography and related subjects. The Davis & Hedge festschrift, pp. x-xvi. - Edinburgh: Edinburgh Univ. Press.

TAN, KIT, FREITAG, H. & JAHN, R. 1998: Notes on *Suaeda palaestina*. In Greuter, W. & Raus, Th. (eds): Med-Checklist notulae, 17. - *Willdenowia* **28(1-2)**: 163-174.

TAN, KIT & GEMICI, Y. 1990: A new *Barbarea* from southwest Anatolia. - *Edinburgh J. Bot.* **47(3)**: 287-289.

TAN, KIT, GIANNOPOULOS, K. & VOLD, G. 2013: *Cardamine calliphaea* sp. nov. (Brassicaceae) from southwestern Greece. - *Nordic J. Bot.* **31(3)**: 282-285.

TAN, KIT & IATROÚ, G. 1995a: *Onobrychis aliacmonia* (Leguminosae) extinct? In: Proceedings, 5th sci. conf., Hellenic Bot. Soc., Delphi 21-23 October 1994, pp. 217-219. - Thessaloniki: Univ. Studio Press.

TAN, KIT & IATROÚ, G. 1995b: A new *Colchicum* from the southern Peloponnese. - *Rock Gard.* **24(3)**: 255-257, 269-270.

TAN, KIT & IATROÚ, G. 1996: *Onobrychis aliacmonia* Rech.f. - a new slant on an old story. - *Ann. Naturhist. Mus. Wien, B* **98B Suppl.**: 305-309.

TAN, KIT & IATROÚ, G. 1997: Two rare taxa in the southern Peloponnese. In: Abstracts from the First Balkan Botanical Congress, Thessaloniki, September 19-22, 1997, p. 26. – Thessaloniki.

TAN, KIT & IATROÚ, G. 1998: Two rare taxa in the southern Peloponnese. In Tsekos, I. & Moustakas, M. (eds): Progress in Botanical Reseearch, pp. 153-156. - Dordrecht: Kluwer Press.

TAN, KIT & IATROÚ, G. 1999a: The endemic plants of Greece - volume one. In Elefheriou, E.P. (ed.): Proceedings, 7th Panhellenic scientific conference, Alexandroupolis, pp. 108-112. - Thessaloniki: Univ. Studio Press.

TAN, KIT & IATROÚ, G. 1999b: The endemic plants of Greece. In Correia, A.I.D. & al. (eds): Livro de resumos. Guia de excursões. V jornadas de taxonomia botânica, 16-19/IX 1999, p. 31.

TAN, KIT & IATROÚ, G. 1999c: *Astragalus laconicus* Iatrou & Kit Tan, sp. nova and *Onobrychis peloponnesiaca* (Iatroú & Kit Tan) Iatroú & Kit Tan, comb. & stat. nov. In Greuter, W. & Raus, Th. (eds): Med-Checklist notulae, 18. - *Willdenowia* **29(1-2)**: 56-59.

TAN, KIT, IATROÚ, G. & JOHNSEN, B. 2001: Endemic Plants of Greece - the Peloponnese. - Copenhagen: Gad Publishers, 480 pp. incl. 111 colour plates.

TAN, KIT & ISSIGONI, M. 2007: Reports 82-115. In Vladimirov, V. & al. (eds): New floristic records in the Balkans: 6. - *Phytol. Balcan.* **13(3)**: 446-449.

TAN, KIT & ISSIGONI, M. 2008a: Reports 95-98. In Vladimirov, V., Dane, F. & Tan, Kit (eds): New floristic records in the Balkans: 7. - *Phytol. Balcan.* **14(1)**: 145-146.

TAN, KIT & ISSIGONI, M. 2008b: Reports 53-81. In Vladimirov, V., Dane, F. & Tan, Kit (eds): New floristic records in the Balkans: 8. - *Phytol. Balcan.* **14(2)**: 300-302.

TAN, KIT & ISSIGONI, M. 2015: Reports 82-84. In Vladimirov, V. & al. (eds): New floristic records in the Balkans: 27. - *Phytol. Balcan.* **21(2)**: 205-206.

TAN, KIT, KATSIOTIS, D. & VOLD, G. 2016: Reports 152-158. In Vladimirov, V. & Tan, Kit (eds): New floristic records in the Balkans: 31. - *Phytol. Balcan.* **22(3)**: 453-455.

TAN, KIT & KOFINAS, G. 2020a: Reports 115-116. In Vladimirov, V., Aybeke, M. & Tan, Kit (eds): New floristic records in the Balkans: 42. - *Phytol. Balcan.* **26(2)**: 208-211.

TAN, KIT & KOFINAS, G. 2020b : Reports 75-84. In Vladimirov, V., Aybeke, M. & Tan, Kit: New floristic records in the Balkans: 43. - *Phytol. Balcan.* **26(3)**: 557-559.

TAN, KIT & KOFINAS, G. 2021: Reports 125-126. In Vladimirov, V., Aybeke, M. & Tan, Kit (eds): New floristic records in the Balkans: 45. - *Phytol. Balcan.* **27(2)**: 267-269.

TAN, KIT, KOFINAS, G. & STRID, A. 2020: Report 85-87. In Vladimirov, V., Aybeke, M. & Tan, Kit: New floristic records in the Balkans: 43. - *Phytol. Balcan.* **26(3)**: 560-561.

TAN, KIT, KOFINAS, G. & ZIELIŃSKI, J. 2021: Reports 86-87. In Vladimirov, V., Aybeke, M. & Tan, Kit (eds): New floristic records in the Balkans: 44. - *Phytol. Balcan.* **27(1)**: 145-147.

TAN, KIT, LAFRANCHIS, L. & VOLD, G. 2010: Reports 162-164. In Vladimirov, V. & al. (eds): New floristic records in the Balkans: 14. - *Phytol. Balcan.* **16(3)**: 436.

TAN, KIT & MARCHANT, R. 2021: Report 127. In Vladimirov, V., Aybeke, M. & Tan, Kit (eds): New floristic records in the Balkans: 45. - *Phytol. Balcan.* **27(2)**: 269-271.

TAN, KIT & MULLAJ, A. 2000a: *Anacyclus radiatus* Loisel., *Malcolmia graeca* subsp. *bicolor* (Boiss. & Heldr.) Stork, *Acacia cyanophylla* Lindl., *Lotus cytisoides* L. In Greuter, W. & Raus, Th. (eds): Med-Checklist notulae, 19. - *Willdenowia* **30(2)**: 235-238.

TAN, KIT & MULLAJ, A. 2000b: *Physalis angulata* L., *Isolepis cernua* (Vahl) Roemer & Schultes, *Catapodium marinum* (L.) C.E. Hubb. In Greuter, W. & Raus, Th. (eds): Med-Checklist notulae, 19. - *Willdenowia* **30(2)**: 240-242.

TAN, KIT & MULLAJ, A. 2000c: *Parapholis filiformis* (Roth) C.E. Hubb., *Halophila stipulacea* (Forssk.) Asch., *Juncus subnodulosus* Schrank. In Greuter, W. & Raus, Th. (eds): Med-Checklist notulae 19. - *Willdenowia* **30(2)**: 242-243.

TAN, KIT & MULLAJ, A. 2001a: *Gymnospermium altaicum* subsp. *scipetarum* (E. Mayer & Pulević) Kit Tan & Mullaj, comb. & stat. nov. In Greuter, W. & Raus, Th. (eds): Med-Checklist notulae, 20. - *Willdenowia* **31(2)**: 319-320.

TAN, KIT & MULLAJ, A. 2001b: The flora of Albania. In Özhatay, N. (ed.): Proceedings of the Second Balkan Botanical Congress, 1: 53-56. - Istanbul.

TAN, KIT, MULLAJ, A. & VOLD, G. 1999: Notes on *Galinsoga parviflora*. In Greuter, W. & Raus, Th. (eds): Med-Checklist notulae, 18. - *Willdenowia* **29(1-2)**: 53.

TAN, KIT, NIELSEN, H. & BÖHLING, N. 2004: *Lathyrus hirsutus* L. In Greuter, W. & Raus, Th. (eds): Med-Checklist notulae, 22. - *Willdenowia* **34(1)**: 75.

TAN, KIT, PERDETZOGLOU, D. & RAUS, TH. 2000: *Halimium voldii* Kit Tan, Perdetzoglou & Raus, sp. nova. In Greuter, W. & Raus, Th. (eds): Med-Checklist notulae, 19. - *Willdenowia* **30(2)**: 232-235.

TAN, KIT, PERDETZOGLOU, D. & ROUSSIS, V. 1997: *Biebersteinia orphanidis* (Geraniaceae) from southern Greece. - *Ann. Bot. Fenn.* **34(1)**: 41-45.

TAN, KIT & PETROVA, A. 2009: Nomenclatural notes. - *Phytol. Balcan.* **15(2)**: 291-292.

TAN, KIT, POLYMENAKOS, K. & PANTAKIS, G. 2021: Reports 128-129. In Vladimirov, V., Aybeke, M. & Tan, Kit (eds): New floristic records in the Balkans: 45. - *Phytol. Balcan.* **27(2)**: 271-273.

TAN, KIT, POLYMENAKOS, K. & VOLD, G. 2011: Reports 73-79. In Vladimirov, V. & Tan, Kit (eds): New floristic records in the Balkans: 16. - *Phytol. Balcan.* **17(2)**: 259-260.

TAN, KIT, RAABE, U. & VOLD, G. 2007: Report 63. In Vladimirov, V. & al. (eds): New floristic records in the Balkans: 5. - *Phytol. Balcan.* **13(2)**: 272-273.

TAN, KIT & RUNEMARK, H. 2006: *Teucrium microphyllum*. In Greuter, W. & Raus, Th. (eds): Med- Checklist notulae, 24. - *Willdenowia* **36(2)**: 724.

TAN, KIT, SCHULER, A. & STRID, A. 2007: A pink-flowered variant of *Centaurea prespana* (Asteraceae) in NW Greece. - *Phytol. Balcan.* **13(3)**: 351-352.

TAN, KIT & SFIKAS, G. 2003: A *Consolida* new for Europe at Lemnos. - *Fisis [Journal of the Hellenic Society for the Protection of Nature]* **100**: 38.

TAN, KIT & SFIKAS, G. 2007: Reports 116-122. In Vladimirov, V. & al. (eds): New floristic records in the Balkans: 6. - *Phytol. Balcan.* **13(3)**: 449-450.

TAN, KIT, SFIKAS, G. & STEVANOVIĆ, V. 1998: Notes on *Phelypaea boissieri*. In Greuter, W. & Raus, Th. (eds): Med-Checklist notulae, 17. - *Willdenowia* **28(1-2)**: 163-174.

TAN, KIT, SFIKAS, G. & VOLD, G. 1997: *Viola parnonia* (Violaceae), a new species from southern Greece. - *Ann. Bot. Fenn.* **34(3)**: 149-152.

TAN, KIT, SFIKAS, G. & VOLD, G. 1999: *Juniperus drupacea* (Cupressaceae) in the southern Peloponnese. - *Acta Bot. Fenn.* **162**: 133-135.

TAN, KIT, SFIKAS, G. & VOLD, G. 2002: Notes on *Consolida arenaria* Carlström. In Greuter, W. & Raus, Th. (eds): Med-Checklist notulae, 21. - *Willdenowia* **32(2)**: 203.

TAN, KIT, SFIKAS, G. & VOLD, G. 2008a: Reports 99-104. In Vladimirov, V., Dane, F. & Tan, Kit (eds): New floristic records in the Balkans: 7. - *Phytol. Balcan.* **14(1)**: 146-147.

TAN, KIT, SFIKAS, G. & VOLD, G. 2008b: Reports 82-90. In Vladimirov, V., Dane, F. & Tan, Kit (eds): New floristic records in the Balkans: 8. - *Phytol. Balcan.* **14(2)**: 302-304.

TAN, KIT, ŠILJAK-YAKOVLEV, S. & VOLD, G. 2011: *Geranium kikianum* sp. nov. (Geraniaceae) from the southern Peloponnese, Greece. - *Nordic J. Bot.* **29(1)**: 1-5.

TAN, KIT, STEVANOVIĆ, V. & STRID, A. 2004: Distribution and centres of diversity for endemic geophytic monocots in the Balkans. In Vasić, O. (ed.): Abstracts, XI OPTIMA Meeting, Beograd 5-11.IX.2004, p. 63.

TAN, KIT, STEVANOVIĆ, V. & STRID, A. 2007: Distribution and centres of diversity for endemic geophytic monocots in the Balkans. - *Bocconea* **21**: 139-146.

TAN, KIT, STEVANOVIĆ, V. & VOLD, G. 1998: Notes on *Viola kosaninii*. In Greuter, W. & Raus, Th. (eds): Med-Checklist notulae, 17. - *Willdenowia* **28(1-2)**: 163-174.

TAN, KIT & STRID, A. 1997: A new *Astragalus* (Fabaceae) from southern Peloponnisos. - *Pl. Syst. Evol.* **206**: 47-50.

TAN, KIT & STRID, A. 1998: Notes on *Amsinckia micrantha, Cynoglottis barrelieri* and *Valerianella lasiocarpa*. In Greuter, W. & Raus, Th. (eds): Med-Checklist notulae, 17. - *Willdenowia* **28(1-2)**: 163-174.

TAN, KIT & STRID, A. 1999: Phytogeographical notes on the Parnonas range in the Peloponnese, southern Greece. - *Preslia* **71**: 361-367.

TAN, KIT & STRID, A. 2001: The flora and Floras of Turkey and Greece. In Özhatay, N. (ed.): Proceedings of the Second Balkan Botanical Congress, 1: 63-72. - Istanbul.

TAN, KIT & STRID, A. 2008: *Paronychia manfrediana* (Caryophyllaceae), a new species from northeast Greece. - *Phytol. Balcan.* **14(1)**: 41-44.

TAN, KIT & STRID, A. 2018: Reports 47-48. In Vladimirov, V., Aybeke, M. & Tan, Kit (eds): New floristic records in the Balkans: 35. - *Phytol. Balcan.* **24(1)**: 167-170.

TAN, KIT, STRID, A. & GOULANDRIS, N. 2009: Wild Flowers of Greece. - Kifissia: Goulandris Nat. Hist. Mus., xxxvii + 253 pp. incl. 121 pp. of colour plates.

TAN, KIT, SUDA, J. & RAUS, TH. 2003: (1582) Proposal to conserve the name *Inula verbascifolia* (Willd.) Hausskn. against *I. verbascifolia* Poir. and with a conserved type. - *Taxon* **52(2)**: 358-359. [approved, cf. Taxon 54(2): 530, 2005].

TAN, KIT, TSOUNIS, G. & TSOUNIS, L. 2010: *Micromeria acropolitana* (Lamiaceae) rediscovered in Athens (Greece). - *Phytol. Balcan.* **16(2)**: 237-242.

TAN, KIT & VLADIMIROV, V. 1998: *Swertia punctata* Baumg. in Bulgaria. In: Abstracts from the IX OPTIMA Meeting, Paris, France, 10-18 May 1998, p. 55.

TAN, KIT & VLADIMIROV, V. 2001: *Swertia punctata* Baumg. (Gentianaceae) in Bulgaria. - *Bocconea* **13**: 461-466.

TAN, KIT & VOLD, G. 2000: *Astragalus ponticus* Pallas. In Greuter, W. & Raus, Th. (eds): Med-Checklist notulae, 19. - *Willdenowia* **30(2)**: 237.

TAN, KIT & VOLD, G. 2006a: Reports 31-46. In Vladimirov, V., Tan, Kit & Stevanović, V. (eds): New floristic records in the Balkans: 1. - *Phytol. Balcan.* **12(1)**: 113-115.

TAN, KIT & VOLD, G. 2006b: Reports 132-202. In Vladimirov, V., Dihoru, G. & Tan, Kit (eds): New floristic records in the Balkans: 3. - *Phytol. Balcan.* **12(3)**: 428-433.

TAN, KIT & VOLD, G. 2016: Reports 148-150. In Vladimirov, V. & al. (eds): New floristic records in the Balkans: 30. - *Phytol. Balcan.* **22(2)**: 281-283.

TAN, KIT & VOLD, G. 2017: Reports 190. In Vladimirov, V. & al. (eds): New floristic records in the Balkans: 33. - *Phytol. Balcan.* **23(2)**: 309-311.

TAN, KIT & VOLD, G. 2019: Report 107. In Vladimirov, V., Aybeke, M. & Tan, Kit (eds): New floristic records in the Balkans: 39. - *Phytol. Balcan.* **25(2)**: 227-231.

TAN, KIT, VOLD, G. & SFIKAS, G. 2006a: Reports 88-95. In Vladimirov, V. & al. (eds): New floristic records in the Balkans: 2. - *Phytol. Balcan.* **12(2)**: 296-297.

TAN, KIT, VOLD, G. & SFIKAS, G. 2006b: Reports 203-228. In Vladimirov, V., Dihoru, G. & Tan, Kit (eds): New floristic records in the Balkans: 3. - *Phytol. Balcan.* **12(3)**: 434-435.

TAN, KIT, VOLD, G. & STRID, A. 2005: *Alkanna sfikasiana* and *A. sartoriana* (Boraginaceae) in the Peloponnese, southern Greece. - *Fisis [Journal of the Hellenic Society for the Protection of Nature]* **109**: 43-45.

TAN, KIT, VOLD. G. & SFIKAS, G. 2007: Reports 64-75. In Vladimirov, V. & al. (eds): New floristic records in the Balkans: 5. - *Phytol. Balcan.* **13(2)**: 273-274.

TAN, KIT, VURAL, M. & STRID, A. 2006: Report 69. In Vladimirov, V., Tan, Kit & Stevanović, V. (eds): New floristic records in the Balkans: 1. - *Phytol. Balcan.* **12(1)**: 120.

TAN, KIT & YILDIZ, B. 1988: New *Asyneuma* (Campanulaceae) taxa from Turkey. - *Willdenowia* **18(1)**: 67-80.

TAN, KIT & ZIELIŃSKI, J. 2000a: *Lonicera alpigena* subsp. *hellenica* (Boiss.) Kit Tan & Ziel., comb. & stat. nov. and subsp. *glutinosa* (Vis.) Kit Tan & Ziel., comb. & stat. nov. In Greuter, W. Raus, Th. (eds): Med-Checklist notulae, 19. - *Willdenowia* **30(2)**: 232.

TAN, KIT & ZIELIŃSKI, J. 2000b: *Genista sessilifolia* DC. In Greuter, W. Raus, Th. (eds): Med-Checklist notulae, 19. - *Willdenowia* **30(2)**: 237-238.

TAN, KIT & ZIELIŃSKI, J. 2001: *Micromeria browiczii* (Labiatae), an unusual new species from Zakinthos (Ionian Islands, Greece). - *Polish Bot. J.* **46(1)**: 31-33.

TAN, KIT & ZIELIŃSKI, J. 2017: Reports 191-192. In Vladimirov, V. & al. (eds): New floristic records in the Balkans: 33. - *Phytol. Balcan.* **23(2)**: 311-312.

TAN, KIT & ZOGRAFIDIS, A. 2014: On the identity of *Carduus euboicus* (Asteraceae) from Mt Dirphys, Evvia, Greece. - *Phytotaxa* **161(3)**: 194-200.

TAN, KIT, ZOGRAFIDIS, A. & MERMYGKAS, D. 2014: Report 131. In Vladimirov, V. & Tan, Kit (eds): New floristic records in the Balkans: 24. - *Phytol. Balcan.* **20(1)**: 123-125.

TAN, KIT (ED.) 1989: Plant taxonomy, phytogeography and related subjects. The Davis & Hedge festschrift. - Edinburgh: Edinburgh Univ. Press, xxvi + 351 pp.

TAN, KIT, SFIKAS, G. & VOLD, G. 2008: Reports 76-82. In Vladimirov, V., Dane, F. & Tan, Kit (eds): New floristic records in the Balkans: 9. - *Phytol. Balcan.* **14(3)**: 445-446.

TANDON, S.L. & RAO, G.R. 1966: Interrelationship within the *Solanum nigrum* complex. - *Indian J. Genet. Pl. Breed.* **26**: 130-141.

TANJI, A., BOULET, C. & HAMMOUMI, M. 1985: Etat actuel de l'infestation par *Solanum elaeagnifolium* Cav. pour les différentes cultures du périmètre du Tadla (Maroc). - *Weed Res.* **25(1)**: 1-9.

TARDÍO PATO, F.J., GARCÍA GONZALO, P. & ORTIZ MARCIDE, J.M. 1998: Observaciones cariológical en *Vicia amphicarpa* sensu lato. - *Anales Jard. Bot. Madrid* **56(2)**: 261-268.

TARIMCILAR, G. & KAYNAK, G. 2004: Chromosome counts in some Anatolian species of *Mentha* (Labiatae). - *Fl. Medit.* **14**: 253-262.

TARNAVSCHI, I.T. & SANDA, V. 1985: Taxonomisches Studium der *Dianthus*-Arten aus der Flora Rumäniens. - *Feddes Repert.* **96(1-2)**: 63-72.

TASHEV, A. 2003: *Pinus brutia* Ten.: a new species for the Bulgarian flora. - *Phytol. Balcan.* **9(1)**: 41-52.

TASHEV, A. 2010: Reports 83-84. In Vladimirov, V., Dane, F. & Tan, Kit (eds): New floristic records in the Balkans: 13. - *Phytol. Balcan.* **16(1)**: 159-161.

TASHEV, A. & DIMITROV, D. 2012: *Sesleria rhodopaea* Tashev & Dimitrov (Poaceae) sp. nova - a new graminean from Bulgaria. - *Dokl. Bulg. Akad. Nauk* **65(2)**: 169-172.

TASHEV, A. & PETKOVA, K. 2009: Fruit and seed morphological peculiarities of the critically threatened *Eriolobus trilobatus* (Rosaceae). In Ivanova, D. (ed.): Proceedings of the IV Balkan botanical congress, pp. 55-58. - Sofia: Institute of Botany, BAS.

TASHEV, A. & VITKOVA, A. 2006: New chorological data of *Salix elaeagnos* (Salicaceae) in Bulgaria. - *Nauka za Gorata* **41(1)**: 105-109.

TASHEV, A., VITOVA, A. & RUSSAKOVA, V. 2006: Distribution of *Ophrys apifera* Huds. (Orchidaceae) in Bulgaria. - *Fl. Medit.* **16**: 247-252.

TASKOVA, R. & AL. 1997: A taxonomic study on *Plantago* species in Bulgaria. - *Phytol. Balcan.* **3(2-3)**: 149-154.

TATEOKA, T. 1968: Phytogeographical notes on the genus *Brachypodium* P. Beauv. (Gramineae). - *Bol. Soc. Argent. Bot.* **12**: 44-56.

TAUBENHEIM, G. 1977: Die Verbreitung der Gattung *Epipactis* in der Türkei. - *Orchidee (Hamburg)* **Sonderheft**: 78-88.

TAUBENHEIM, G. 1979: Die Verbreitung der Orchideen in der Türkei IV/1. Ein Beitrag zur "Flora of Turkey". 8.-11. Die Gattungen *Traunsteinera*, *Neotinia*, *Comperia* und *Steveniella*. - *Orchidee (Hamburg)* **30(6)**: 223-228.

TAUBENHEIM, G. 1980: Die Verbreitung der Orchideen in der Türkei IV/2. Ein Beitrag zur "Flora of Turkey". 8.-11. Die Gattungen *Traunsteinera*, *Neotinia*, *Comperia* und *Steveniella* [cont.]. - *Orchidee (Hamburg)* **31(1)**: 5-10.

TAUSCH, F. 1996: *Limodorum abortivum* (L.) Swartz var. *rubrum* auf Rhodos. - *Ber. Arbeitskreis. Heimische Orchid.* **12(2)**: 60.

TAUSCH, J.F. 1828: Bemerkungen über einige Arten der Gattung *Paeonia*. - *Flora* **11(1)**: 81-89.

TAUSCH, J.F. 1829a: Botanische Beobachtungen. - *Flora* **12(1)**: 9-11, 65-73.

TAUSCH, J.F. 1829b: Bemerkungen über *Lavatera*. - *Flora* **12(1)**: 177-183.

TAUSCH, J.F. 1829c: Botanische Beobachtungen. - *Flora* **12(2)**: 641-650.

TAUSCH, J.F. 1831: Botanische Beobachtungen. - *Flora* **14(2)**: 641-646.

TAUSCH, J.F. 1835: Bemerkungen über Galium und einige verwandte Gattungen. - *Flora* **18(1)**: 337-359.

TAYLOR, F.J. 1956: Biological Flora of the British Isles. *Carex flacca* Schreb. (*C. glauca* Scop.; *C. diversicolor* auct., non Crantz). - *J. Ecol.* **44(1)**: 281-290.

TAYLOR, H. 1945: Cyto-taxonomy and phylogeny of the Oleaceae. - *Brittonia* **5(4)**: 337-367.

TAYLOR, K. 1997a: Biological Flora of the British Isles. *Geum urbanum* L. - *J. Ecol.* **85(5)**: 705-720.

TAYLOR, K. 1997b: Biological Flora of the British Isles. *Geum rivale* L. - *J. Ecol.* **85(5)**: 721-731.

TAYLOR, K. 1999: Biological Flora of the British Isles. *Galium aparine* L. - *J. Ecol.* **87(4)**: 713-730.

TAYLOR, K. & MARKHAM, B. 1978: Biological Flora of the British Isles. *Ranunculus ficaria* L. (*Ficaria verna* Huds.; *F. ranunculoides* Moench). - *J. Ecol.* **66(3)**: 1011-1031.

TAYLOR, K., ROWLAND, A.P. & JONES, H.E. 2001: Biological Flora of the British Isles. *Molinia caerulea* (L.) Moench. - *J. Ecol.* **89(1)**: 126-144.

TAYLOR, M. 2005: Illustrated checklist. Orchids of Chios, Inouses & Psara. - Pelineo, Chios: 99 pp. [Revised edition 2012, 139 pp.].

TAYLOR, P. 1989: The genus *Utricularia* - a taxonomic monograph. - *Kew Bull., Addit. Ser.* **14**: i-xi, 1-724.

TAYLOR, R.J. 1967: Interspecific hybridization and its evolutionary significance in the genus *Aquilegia*. - *Brittonia* **19(4)**: 374-390.

TAYLOR, W.C. & HICKEY, R.J. 1992: Habitat, evolution and speciation in *Isoetes*. - *Ann. Missouri Bot. Gard.* **79(3)**: 613-622.

TAYLOR, W.I. & FARNSWORTH, N.R. (EDS) 1973: The *Vinca* alkaloids: botany, chemistry and pharmacology. - New York: Marcel Dekker Inc., xx + 357 pp.

TÉBAR, F.J. & LLORENS, L. 1993: Heterocarpy in *Thymelaea velutina* (Poiret ex Camb.) Endl.: a case of phenotypic adaptation to Mediterranean selective pressures. - *Bot. J. Linn. Soc.* **111(2)**: 295-300.

TEDD, H.G. & TURRILL, W.B. 1933: On the flora of the Nearer East: XIV. The elms of western Thrace. - *Bull. Misc. Inform. Kew* **1933(5)**: 232-240.

TEDD, H.G. & TURRILL, W.B. 1935: *Fraxinus pallisae* and its relationships. - *Bull. Misc. Inform. Kew* **1935(3)**: 132-141.

TEDD, H.G. & TURRILL, W.B. 1951: A study of *Papaver* populations in western Thrace. - *Kew Bull.* **1951(2)**: 192-196.

TEDIN, O. 1925: Vererbung, Variation und Systematik in der Gattung *Camelina*. - *Hereditas (Lund)* **6 (3)**: 275-386.

TEKŞEN, M. & AYTAC, Z. 2011: The revision of the genus *Fritillaria* L. (Liliaceae) in the Mediterranean region (Turkey). - *Turk. J. Bot.* **35(5)**: 447-478.

TEKŞEN, M. & AYTAÇ, Z. 2004: New *Fritillaria* L. taxa from Turkey. - *Israel J. Pl. Sci.* **52(4)**: 347-355 + front cover.

TEKŞEN, M. & ERKUL, S.K. 2015: The synopsis of the genus *Gagea* (Liliaceae) in Turkey. – *Phytotaxa* **230(1)**: 101-129.

TELESCA, L. & AL. 2015: Seagrass meadows (*Posidonia oceanica*) distribution and trajectories of change. - *Sci. Rep.* **5**: DOI: 10.1038/srep12505.

TEMEL, M., TÜMEN, G. & TOKUR, S. 1996: Studies on the taxonomy, morphology and anatomy of *Galanthus gracilis* Čelak. In Öztürk, M.A., Seçmen, Ö. & Görk, G. (eds): Plant life in southwest and central Asia, 1: pp. 295-311. - Izmir: Ege University Press.

TEMESY, E. 1957: Der Formenkreis von *Saxifraga stellaris* L. - *Phyton (Horn)* **7(1-3)**: 40-141 + 2 fold- out maps.

TENORE, M. 1811-15: Flora Napolitana ossia descrizione della piante indigene del regno di Napoli e delle più rare specie di piante esotiche, vol. 1. - Napoli: lxii + 324 pp. + 50 plates [cf. Sabato in *Taxon* 39(3): 409-416, 1990].

TENORE, M. 1831: Sylloge plantarum vascularium florae napolitanae hucusque detectarum. - Napoli: Typographia Fibreni, pp. i-vi, 7-639.

TENORE, M. 1842: Annotazioni alla Flora Greca. - *Rendiconti Reale Accad. Sci. Fis.* **1(2)**: 82-103.

TEOFILOVSKI, A. 2011: Contributions to the flora of the Republic of Macedonia [In Macedonian with English summary]. - Skopje: 141 pp.

TEOFILOVSKI, A. 2017: Contribution to knowledge of the flora of the Republic of Macedonia. - *Bot. Serbica* **41(1)**: 99-103.

TEOFILOVSKI, A. 2018: *Pedicularis hoermanniana* and *Petasites kablikianus*, new species in the flora of the Republic of Macedonia. - *Bot. Serbica* **42(2)**: 259-263.

TEOFILOVSKI, A. 2021: New floristic data in North Macedonia with a first record of *Allium melananthеrum* Pančić. - *Biol. Nyssana* **12(1)**: 1-10.

TEOFILOVSKI, A. & AL. 2012: Chorology and habitats of some plants in the Republic of Macedonia. - *Forest Review (Skopje)* **43**: 24-32.

TEOFILOVSKI, A., NIKOLOV, Z. & MANDŽUKOVSKI, D. 2015: *Juncus filiformis* L. (Juncaceae), a new species in the flora of Macedonia. - *Forest Review (Skopje)* **46**: 71-73.

TEPPNER, H. 1970: Karyotypen europäischer, perennierender Sippen der Gramineen-Gattung *Anthoxanthum*. - *Oesterr. Bot. Z.* **118(3)**: 280-292.

TEPPNER, H. 1971: Cytosystematik, bimodale Chromosomensätze und permanente Anorthoploidie bei *Onosma* (Boraginaceae). - *Oesterr. Bot. Z.* **119(1-3)**: 196-233.

TEPPNER, H. 1972: Cytosystematische Studien an *Onosma* (Boraginaceae). Die Formenkreise von *O. echioides*, *O. helveticum* und *O. arenarium*. - *Ber. Deutsch. Bot. Ges.* **84(11)**: 691-696.

TEPPNER, H. 1980: Karyologie und Systematik einiger Gefässpflanzen der Ostalpen. Botanische Studien im Gebiet der Planneralm (Niedere Tauern, Steiermark), VII. - *Phyton (Horn)* **20(1-2)**: 73-94.

TEPPNER, H. 1981: Karyosystematik von *Onosma stellulatum*, *O. pygmaeum* und *O. leptanthum* (Boraginaceae). - *Bot. Jahrb. Syst.* **102(1-4)**: 297-306.

TEPPNER, H. 1988a: *Onosma epiroticum* spec. nova (Boraginaceae) aus Griechenland. – *Phyton (Horn)* **28(1)**: 51-54.

TEPPNER, H. 1988b: *Onosma kaheirei* spec. nova und *O. erectum* (Boraginaceae) aus Griechenland. - *Phyton (Horn)* **28(1)**: 115-131 + 1 plate.

TEPPNER, H. 1988c: *Onosma stridii* spec. nova (Boraginaceae) aus Griechenland. - *Phyton (Horn)* **28 (2)**: 271-275.

TEPPNER, H. 1988d: *Lathyrus grandiflorus* (Fabaceae-Vicieae): Blüten-Bau, -Funktion und *Xylocopa violacea*. - *Phyton (Horn)* **28(2)**: 321-336 + 1 plate.

TEPPNER, H. 1989: Karyology of some Greek *Onosma* species. In Phitos, D. (ed.): Abstracts. Sixth OPTIMA meeting, Delphi 10-16 Sept. 1989, p. 35.

TEPPNER, H. 1991: Karyology of some Greek *Onosma* species (Boraginaceae). - *Bot. Chron. (Patras)* **10**: 271-292.

TEPPNER, H. 1996: Remarks to the *Onosma* species *O. bourgaei, O. spruneri* and *O. stellulata* (Boraginaceae) offered. In Samenverzeichnis 1996, pp. 33-39. - Graz: Bot. Garten Inst. Bot. Univ. Graz

TEPPNER, H. 2003: The heterodiaspory of *Capsella bursa-pastoris* (Brassicaceae). - *Phyton (Horn)* **43 (2)**: 381-391.

TEPPNER, H. 2004: Notes on *Lagenaria* and *Cucurbita* (Cucurbitaceae) - review and new combinations. - *Phyton (Horn)* **44(2)**: 245-308.

TEPPNER, H. 2008: An asterotrichous, hexaploid *Onosma* from Bulgaria: *O. malkarmayorum* spec. nova (Boraginaceae-Lithospermeae). - *Phyton (Horn)* **48(1)**: 117-132.

TEPPNER, H. & AL. 1994: *Nigritella carpatica* (Orchidaceae-Orchideae) - ein Reliktendemit der Ost- Karpaten. - *Phyton (Horn)* **34(2)**: 169-187 + 1 plate.

TEPPNER, H., EHRENDORFER, F. & PUFF, CH. 1976: Karyosystematic notes on the *Galium palustre*-group (Rubiaceae). - *Taxon* **25(1)**: 95-97.

TEPPNER, H. & IATROÚ, G. 1987: *Onosma sangiasense* spec. nova (Boraginaceae) from Peloponnisos (Greece). - *Phyton (Horn)* **27(2)**: 285-288.

TEPPNER, H. & KARL, R. 2013: *Onosma pseudoeuboica* spec. nova, *O. euboica* and notes on *O. heterophylla* (Boraginaceae-Lithospermeae) from Central Greece. - *Phyton (Horn)* **53(2)**: 185-220.

TEPPNER, H. & KARL, R. 2017: *Onosma kittanae* (Boraginaceae-Lithospermeae). - *Mitt. Naturwiss. Vereines Steiermark* **147**: 77-97.

TEPPNER, H. & KLEIN, E. 1985a: Karyologie und Fortpflanzungsmodus von *Nigritella* (Orchidaceae-Orchideae) inkl. *N. archiducis-joannis* spec. nov. und zweier Neukombinationen. - *Phyton (Horn)* **25(1)**: 147-176 + 2 plates.

TEPPNER, H. & KLEIN, E. 1985b: *Nigritella widderi* spec. nov. (Orchidaceae-Orchideae). – *Phyton (Horn)* **25(2)**: 317-326.

TEPPNER, H. & KLEIN, E. 1990: *Nigritella rhellicani* spec. nova und *N. nigra* (L.) Rchb. f. s.str. (Orchidaceae-Orchideae). - *Phyton (Horn)* **31(1)**: 5-26 + 2 plates.

TEPPNER, H. & WETSCHNIG, W. 1980: Zur Karyologie von *Poa hybrida*, *P. chaixii*, *P. sylvicola* und *P. stiriaca* (Poaceae) unter besonderer Berücksichtigung von B-Chromosomen. - *Phyton (Horn)* **20(1-2)**: 47-63.

TERABAYASHI, S. 1983: Studies in the morphology and systematics of Berberidaceae. V. Floral anatomy of *Caulophyllum* Michx., *Leontice* L., *Gymnospermium* Spach and *Bongardia* Mey. - *Mem. Fac. Sci. Kyoto Univ., Ser. Biol.* **8(2)**: 197-217.

TERLEVIĆ, A. & AL. 2021: *Carex phyllostachys* (Cyperaceae), a new species in Croatia. - *Acta Bot. Croat.* **80(1)**: 106-111.

TERPÓ, A. 1985: Studies on taxonomy and grouping of *Pyrus* species. - *Feddes Repert.* **96(1-2)**: 73-87.

TERRACCIANO, A. 1897: Revisione monografica delle specie del genere *Nigella*. - *Boll. Reale Orto Bot. Palermo* **1(2)**: 122-153.

TERRACCIANO, A. 1898: Revisione monografica delle specie del genere *Nigella* [cont.]. - *Boll. Reale Orto Bot. Palermo* **2(1)**: 19-42.

TERRACCIANO, A. 1905: Gagearum species florae orientalis ad exemplaria imprimis in herbariis Boissier et Barbey servata, comparavit et illustravit. - *Bull. Herb. Boissier, Ser. 2* **5(11-12)**: 1061-1076, 1113-1128.

TERRELL, E.E. 1968: A taxonomic revision of the genus *Lolium*. - *Techn. Bull. U.S.D.A.* **1392**: 1-65.

TERRELL, M. & TERRELL, S. 1974: Orchids and archaeology on the island of Crete. - *Amer. Orchid Soc. Bull.* **43(1)**: 43-46.

TERRISSE, A. 1990: Note historico-nomenclaturale sur le maceron (*Smyrnium olusatrum* L.). - *Bull. Soc. Bot. Centre-Ouest, n.s.* **21**: 71-74.

TERSIJSKI, D. 1983: Scanning electron-microscopic study on the spermoderm of ... *Vicia* sect. *Ervu* [In Bulgarian with English abstract]. In Velčev, V.I. & al. (eds): Third national conference of botany, pp. 199-206. - Sofia: Publishing House of the Bulgarian Academy of Sciences.

TERSIJSKI, D. & DELIPAVLOV, D.D. 1985: Vergleichende taxonomische Untersuchung von *Vicia hirsuta* (L.) S.F. Gray und *Vicia meyeri* Boiss. - *Feddes Repert.* **96(5-6)**: 387-392.

TERSIJSKI, D. & DIMITROV, B. 1984: Karyotype analysis in *Vicia hirsuta* (L.) S.F. Gray and *Vicia meyeri* Boiss. - *Caryologia* **36(4)**: 345-354.

TESCHNER, W. 1975a: Bemerkungen zu einigen Orchideenvorkommen auf Kreta. - *Orchidee (Hamburg)* **26(4)**: 169-170.

TESCHNER, W. 1975b: Eine hybridogene *Orchis*-Sippe auf Kreta? - *Orchidee (Hamburg)* **26(5)**: 217-221.

TĚŠITEL, J. & AL. 2010: Phylogeny, life history evolution and biogeography of the Rhinanthoid Orobanchaceae. - *Folia Geobot.* **45(4)**: 347-367.

THALER, I. 1953: Die Ausbreitung von *Veronica filiformis* Smith. - *Phyton (Horn)* **5(1-2)**: 41-54.

THANOPOULOS, R. 2007: The genus *Medicago* in Greece: 1. A review of species diversity, geographical distribution and ecological adaptation. - *Fl. Medit.* **17**: 217-276.

THANOS, C.A. 2005: Theophrastus on oaks. - *Bot. Chron. (Patras)* **18**: 13-20.

THANOS, C.A. & AL. 1989: Early post-fire regeneration in *Pinus brutia* forest ecosystems of Samos island (Greece). - *Acta Oecol., Oecol. Pl.* **10(1)**: 79-94.

THANOS, C.A. & MARCOU, S. 1991: Post-fire regeneration in *Pinus brutia* forest ecosystems of Samos island (Greece): 6 years after. - *Acta Oecol.* **12**: 633-642.

THELLUNG, A. 1906: Die Gattung *Lepidium* (L.) R. Br. Eine monographische Studie. - *Neue Denkschr. Allg. Schweiz. Ges. Gesammten Naturwiss.* **41**: 1-340.

THELLUNG, A. 1915: Über die in Mitteleuropa vorkommenden *Galinsoga*-Formen. - *Allg. Bot. Z. Syst.* **21(1-4)**: 1-16.

THELLUNG, A. 1926a: Die Linnéschen *Daucus*-Arten im Lichte der Original-Herbarexemplare. - *Repert. Spec. Nov. Regni Veg.* **22**: 305-315.

THELLUNG, A. 1926b: *Daucus*-Studien. I. Zur Gattungsabgrenzung und Sektionseinteilung von *Daucus*. - *Repert. Spec. Nov. Regni Veg.* **23**: 147-159.

THELLUNG, A. 1928: Die Abstammung der Gartenmöhre (*Daucus Carota* subsp. *sativus*) und des Gartenrettichs (*Raphanus Raphanistrum* subsp. *sativus*). - *Repert. Spec. Nov. Regni Veg. Beih.* **46**: 1-7.

THEOCHAROPOULOS, M. & AL. 2001: Geographical distribution and synecology of *Cladium mariscus* L. in Greece. In Kamari, Psaras & Constantinidis (eds): Proceedings of the 8th scientific congress of the Greek Botanical Society, October 5-8, 2000, pp. 161-167 [In Greek with English abstract].

THEOCHAROPOULOS, M. & AL. 2006: Vegetation types with *Cladium mariscus* (Cyperaceae) in Greece. - *Willdenowia* **36(1)**: 247-256.

THEOCHAROPOULOS, M. & AL. 2008: Study of the flora and vegetation and phytogeography of the islets of western Corinth gulf. A forepart of biomonitoring project. In: Praktika 10ou Sinedriou Ellinikis Votanikis Eterias, 5-8 maiou 2005, Ioannina: pp. 223-231 [In Greek with English abstract].

THEOCHAROPOULOS, M. & AL. 2020: Vegetation types with *Quercus ithaburensis* subsp. *macrolepis* in Greece. - *Ecol. Medit.* **46(1)**: 17-40.

THEOCHAROPOULOS, M. & GEORGIADIS, TH. 1984: Contribution à l'étude de la végétation de l'Attique orientale (Nea Makri) en Grèce. (Prise en compte des impacts urbains et touristiques). - *Ecol. Medit.* **10(3-4)**: 133-157.

THEODOROPOULOS, K. 1991: Bestimmung und Klassifizierung der pflanzensoziologischen Vegetationseinheiten im Universitätswald Taxiarchis Chalkidiki [In Greek with German abstract]. - *Sci. Ann. Fac. Agric. Forest. Univ. Thessalonika* **18**: 1-200 + tables.

THEODOROPOULOS, K. & AL. 1999: Vegetation units along the path Skete of the Prodrome - Skete of St. Anna in the Holly [sic!] Mountain. In Eleftheriou, E.P. (ed.): Proceedings, 7th Panhellenic scientific conference, Alexandroupolis, pp. 80-95 [In Greek with English abstract]. - Thessaloniki: Univ. Studio Press.

THEODOROPOULOS, K. & AL. 2002: Notes on *Ononis adenotricha* Boiss. In Greuter, W. & Raus, Th. (eds): Med-Checklist notulae, 21. - *Willdenowia* **32(2)**: 201.

THEODOROPOULOS, K. & ELEFTHERIADOU, E. 1999: Chorology and threats of *Pancratium maritimum* L. in the Strimonikos Gulf. In Elefheriou, E.P. (ed.): Proceedings, 7th Panhellenic scientific conference, Alexandroupolis, pp. 321-326 [In Greek with English abstract]. - Thessaloniki: Univ. Studio Press.

THEODOROPOULOS, K. & ELEFTHERIADOU, E. 2003: I vlastisis thamnonon oxikedris arkevthou (*Juniperus oxycedrus* L. subsp. *oxycedrus*) sto oros Orvilos (A. Makedonia, Nomos Dramas, Ellada). In Platis, P.D. & Papachristou, Th.G. (eds): Proceedings of the 3rd Panhellenic Rangeland Congress in Kapenissi, 4-6 September 2002, pp. 223-231.

THEODOROPOULOS, K. & ELEFTHERIADOU, E. 2005: The vascular flora of a small downy oak (*Quercus pubescens* Willd.) forest in the Thessalian plain (C Greece). - *Bot. Chron. (Patras)* **18(1)**: 211-224.

THEODOROPOULOS, K., REIF, A. & ATHANASIADIS, N. 1995: *Quercus dalechampii* forests in Central Macedonia, Greece. - *Bot. Helv.* **105(1)**: 37-54.

THEODOROVA, T.A. 2015: New nomenclatural combinations in *Nitrosalsola* (Chenopodiaceae). - *Ukrainian Bot. J.* **72(5)**: 442-445.

THERMAN, E. 1953: Chromosomal evolution in the genus *Polygonatum*. - *Hereditas (Lund)* **39(1-2)**: 277-288.

THERMAN, E. 1956: Cytotaxonomy of the tribe Polygonatae. - *Amer. J. Bot.* **43(2)**: 134-142.

THEURILLAT, J.-P. 1979: Étude biosystématique d'*Epilobium dodonaei* Vill. et d'*E. fleischeri* Hochst. (Onagraceae). - *Bull. Soc. Neuchâteloise Sci. Nat.* **102**: 105-128 + 1 plate and 1 table.

THIÉBAUD, M.-A. 1987: Contribution à l'étude des espèces littorales du genre *Elymus* L. – *Candollea* **42(1)**: 327-350.

THIÉBAUD, M.-A. 1996: Complements au Prodrome de la Flore Corse: Valerianaceae. - Geneva, Conservatoire et Jardin Botanique: 116 pp.

THIÉBAUD, M.-A. & DESCHÂTRES, R. 1986: Notes sur le genre *Aegilops* L. en Corse. - *Candollea* **41(1)**: 54-59.

THIEDE, J. & STEPHENSON. R. 2018: The taxonomy and typification of *Sedum creticum*, and the curious protologue of *S.* ×*donatae* (Crassulaceae). - *Willdenowia* **48(1)**: 51-55.

THIELE, G. & THIELE, W. 2001: Ein Beitrag zum Kenntnisstand der Orchideen in Nordwest-Griechenland. - *J. Eur. Orch.* **33(3)**: 891-916.

THIELE, G. & THIELE, W. 2002: Ergänzungen zur Kenntnis der Orchideen in Nordwest-Griechenland. - *J. Eur. Orch.* **34(3)**: 639-654.

THIELE, G. & THIELE, W. 2003: Orchideenkartierung in NW-Griechenland - Teil 3. - *J. Eur. Orch.* **35(2)**: 361-380.

THIELE, G. & THIELE, W. 2007: Wieder einmal Orchideen der Zagoria (Griechenland) - Ein Reisebericht. - *J. Eur. Orch.* **39(2)**: 273-295.

THIELE, G. & THIELE, W. 2010: Unsere Orchideen-Fundorte in Europa. - *J. Eur. Orch.* **42(2)**: 333-411.

THIN, N.N. 1980: New species of genus *Anthemis* in Bulgaria's flora. - *Dokl. Bulg. Akad. Nauk* **33(3)**: 379-382.

THIN, N.N. 1983a: New taxa and combinations of the genus *Anthemis* L. in Bulgaria [In Bulgarian with Russian and English summaries]. - *Fitologiya* **22**: 81-89.

THIN, N.N. 1983b: Taxonomic materials of the genus *Anthemis* L. in Bulgaria [In Bulgarian with Russian and English summaries]. - *Fitologiya* **22**: 90-92.

THIV, M., REYES-BETANCORT, J.A. & FRAGMAN-SAPIR, O. 2019: Lifeforms as criterion for species delimitation: Are *Aristida adscensionis* and *A. caerulescens* (Aristidoideae, Poaceae) two species? - *Phytotaxa* **393(1)**: 67-74.

THOMAS, D.C., WEIGEND, M. & HILGER, H.H. 2008: Phylogeny and systematics of *Lithodora* (Boraginaceae-Lithospermeae) and its affinities to the monotypic genera *Mairetis*, *Halacsya* and *Paramoltkia* based on ITS1 and $trnL_{UAA}$-sequence data and morphology. - *Taxon* **57(1)**: 79-97.

THOMAS, J.L. 1961: *Liquidambar*. - *Arnoldia (Jamaica Plain)* **21(10)**: 59-65.

THOMAS, K. & AL. 2013: Plant genetic resources in a touristic island: the case of Lefkada (Ionian Islands, Greece). - *Genet. Resources Crop Evol.* **60(8)**: 2431-2455.

THOMAS, P.A. & POLWART, A. 2003: Biological Flora of the British Isles. *Taxus baccata* L. - *J. Ecol.* **91 (3)**: 489-524.

THOMAS, R.G. 1958: Sexuality in diploid and hexaploid races of *Mercurialis annua* L. - *Ann. Bot. (Oxford), n.s.* **22(85)**: 55-72.

THOMAS, S.M. 1980: Population studies and delimitation of species in the genus *Petrorhagia* (section *Kohlrauschia*). - London: Ph.D. thesis, Univ. of London, Westfield College, 279 pp.

THOMAS, S.M. 1983: A taxonomic clarification of *Petrorhagia* section *Kohlrauschia* (Caryophyllaceae). - *Bot. J. Linn. Soc.* **87(1)**: 55-75.

THOMAS, S.M. & MURRAY, B.G. 1981: Breeding systems and hybridization in *Petrorhagia* sect. *Kohlrauschia* (Caryophyllaceae). - *Pl. Syst. Evol.* **139(1-2)**: 77-94.

THOMAS, S.M. & MURRAY, B.G. 1983: Chromosome studies in species and hybrids of *Petrorhagia* sect. *Kohlrauschia* (Caryophyllaceae). - *Pl. Syst. Evol.* **141(3-4)**: 243-255.

THOMMEN, E. 1946: Souvenirs d'une herborisation dans le Péloponèse. - *Bull. Soc. Bot. Genève, Sér. 2* **37**: 184-185.

THOMPSON, H.P. 1940: Plant hunting in southern Greece. - *Quart. Bull. Alpine Gard. Soc. Gr. Brit.* **8 (1)**: 44-65.

THOMPSON, J.A. 2004: Towards a taxonomic revision of *Pteridium* (Dennerstaedtiaceae). – *Telopea* **10(4)**: 793-803.

THOMPSON, J.D. 2005: Plant evolution in the Mediterranean. - Oxford: Oxford Univ. Press, 302 pp.

THOMPSON, J.D. & AL. 2005: Ecological differentiation of Mediterranean endemic plants. - *Taxon* **54 (4)**: 967-976.

THOMPSON, J.D., DEBUSSCHE, M. & AFFRE, L. 2001: Ecological and evolutionary aspects of population differentiation in three related *Cyclamen* species in the western Mediterranean. - *Bocconea* **13**: 165-172.

THORNBERG, M. & TAN, KIT 2016: Reports 160-162. In Vladimirov, V. & al. (eds): New floristic records in the Balkans: 30. - *Phytol. Balcan.* **22(2)**: 286.

THORNBERG, M. & TAN, KIT 2017: Reports 71-74. In Vladimirov, V., Aybeke, M. & Tan, Kit (eds): New floristic records in the Balkans: 32. - *Phytol. Balcan.* **23(1)**: 137.

THORNING-LUND, F. 2002: Lefkas 2002. Opsats om fund af orkidéer på Lefkas, Meganissa samt den nordligste del af Ithaka i marts 2002 [In Danish]. - Privately distributed typescript, 11 pp. + colour slides.

THORNING-LUND, F. 2003: Ionien 2003. Artikel om fund af orkidéer på Kefalloniá, den nordlige del af Itháki, den sydlige del af Léfkas samt Meganíssi i april 2003 [In Danish]. - Privately distributed typescript, 12 pp.

THORNING-LUND, F. 2004: Kykladerne m.v. - november 2003. - Privately distributed typescript, 10 pp.

THOROGOOD, CH. 2019: Field guide to the wild flowers of the Eastern Mediterranean. - Royal Botanic Gardens Kew, 656 pp.

THULIN, M. & KOOL, A. 2017: A plant that Linnaeus forgot: taxonomic revision of *Rhodalsine* (Caryophyllaceae). - *Willdenowia* **47(3)**: 317-323.

THULIN, M. & PHILLIPS, S. 2015: The identity of *Pennisetum longistylum* (Poaceae). - *Willdenowia* **45(2)**: 173-175.

THYMAKIS, N. 2003: The Cretan date palm (*Phoenix theophrasti*). - *Medit. Garden* **32**: 32-34.

TICHY, H. 1971: Orchideenbastarde. - *Mitt. Arbeitskreis Heimische Orchid. Baden-Württemberg* **3**: 66-74.

TIGERSCHIÖLD, E. 1981: The *Asplenium trichomanes* complex in East Central Sweden. - *Nordic J. Bot.* **1(1)**: 12-16.

TIKHOMIROV, V.N. & SOKOLOFF, D.D. 1996: On the genera *Hammatolobium* Fenzl and *Tripodion* Medik. (Papilionaceae, Loteae s.l.). - *Feddes Repert.* **107(3-4)**: 209-217.

TIMSON, J. 1963: The taxonomy of *Polygonum lapathifolium* L., *P. nodosum* Pers. and *P. tomentosum* Schrank. - *Watsonia* **5(6)**: 386-395.

TIMSON, J. 1965: A study of hybridization in *Polygonum* section *Persicaria*. - *J. Linn. Soc., Bot.* **59 (377)**: 155-161.

TINEO, V. 1817: Plantarum rariorum Siciliae minus cognitarum pugillus primus ... - Panormi [Palermo]: Typis Regiis, 23 pp.

TINIAKOU, A. 1991a: Cytogeographical study on some species of *Viola* sect. *Viola* (Violaceae) from Greece. - *Willdenowia* **20(1-2)**: 153-158.

TINIAKOU, A. 1991b: *Viola dirphya* (Violaceae), a new species from Evvia island, Greece. – *Candollea* **46(1)**: 119-124.

TINIAKOU, A. 1991c: On some species of *Viola* sect. *Viola* from Greece. - *Bot. Chron. (Patras)* **10**: 661-665.

TINIAKOU, A. 1992: Reports (56-60). In Kamari, G., Felber, F. & Garbari, F. (eds): Mediterranean chromosome number reports - 2. - *Fl. Medit.* **2**: 232-235.

TINIAKOU, A. 1996: The presence of *Viola jordanii* (Violaceae) in Greece. - *Bot. Chron. (Patras)* **12**: 67-70.

TINIAKOU, A. 1997a: Distribution and ecology of the genus *Viola* in Peloponnisos, Greece. In: Abstracts from the First Balkan Botanical Congress, Thessaloniki, September 19-22, 1997, p. 50. - Thessaloniki.

TINIAKOU, A. 1997b: Reports (773-778). In Kamari, G., Felber, F. & Garbari, F. (eds): Mediterranean chromosome number reports - 6. - *Fl. Medit.* **6**: 333-337.

TINIAKOU, A. 2000: Distribution of the genus *Viola* (Violaceae) in Peloponnisos (Greece). - *Bot. Chron. (Patras)* **13**: 357-368.

TINIAKOU, A. & KAMARI, G. 1995: Contribution to the cytogeographical study of the *Iris unguicularis* group. In: Proc., 5th sci. conf., Hellenic Bot. Soc., Delphi 21-23 October 1994, pp. 269-272 [In Greek with English summary]. - Thessaloniki: Univ. Studio Press.

TINIAKOU, A. KAMARI, G. 2004: Taxonomy, distribution and ecology of the *Iris unguicularis* group in Greece. In Vasić, O. (ed.): Abstracts, XI OPTIMA Meeting, Beograd 5-11.IX.2004, p. 108.

TIPPERY, N.P. & AL. 2018: Evidence for allopolyploid speciation in *Nymphoides* (Menyanthaceae). - *Syst. Bot.* **43(1)**: 117-129.

TISON, J.-M. 2004a: Contribution à la connaissance du genre *Gagea* Salisb. (Liliaceae) en Afrique du Nord. - *Lagascalia* **24**: 67-87.

TISON, J.-M. 2004b: Identité et situation taxonomique de *Gagea polmorpha* Boiss. - *Candollea* **59(1)**: 109-117.

TISON, J.-M. 2011: Étude morphologique et anatomique d'un echantillonage d'*Althenia* F. Petit (Potamogetonaceae) des côtes françaises. - *Biocosme Mésogéen* **28(3-4)**: 63-74.

TISON, J.-M. & AL. 2013: Reticulate evolution of the critical Mediterranean *Gagea* sect. *Didymobulbos* (Liliaceae) and its taxonomic implications. - *Pl. Syst. Evol.* **299(2)**: 413-438.

TISON, J.-M., JAUZEIN, PH. & MICHAUD, H. 2014: Flore de la France méditerranéenne continentale. - Turriers: Naturalia Publications, 2078 pp.

TISON, J.-M. & PERRET, P. 2004: Typification d'*Ornithogalum pusillum* F.W. Schmidt et relations taxonomiques entre *Gagea pusilla* (F.W. Schmidt) Sweet, *Ornithogalum clusii* Tausch et *G. clusiana* Schult. & Schult. f. - *Candollea* **59(1)**: 103-108.

TISSOT-DAGUETTE, M. 1979: Nouvelles recherches cytotaxonomiques sur le genre *Onosma* L. - *Compt. Rend. Hebd. Séances Acad. Sci., Sér. D* **288(8)**: 755-758.

TIŢĂ, I. 1990: Contribuţii privind studiul cariotipului la *Sternbergia lutea* (L.) [In Rumanian]. - *Stud. Cercet. Biol. (Bucharest), Ser. Biol. Veg.* **42(2)**: 153-157.

TITZ, E. & TITZ, W. 1981: Die diploiden und tetraploiden Arznei-Baldriane der Schweiz und angrenzender Regionen (Valerianaceae). - *Bot. Helv.* **91**: 169-188.

TITZ, W. 1966: Chromosomenzahlen dreier Angiospermenarten aus Österreich. - *Oesterr. Bot. Z.* **113(2)**: 187-190.

TITZ, W. 1968a: Chromosomenzahlen europäischer *Arabis*-Arten (Cruciferae). - *Ber. Deutsch. Bot. Ges.* **81(6)**: 217-220.

TITZ, W. 1968b: Zur Zytotaxonomie von *Arabis hirsuta* agg. (Cruciferae). I. Allgemeine Grundlagen und die Chromosomenzahlen der in Österreich vorkommenden Sippen. - *Oesterr. Bot. Z.* **115(3)**: 255-290.

TITZ, W. 1969a: Beitrag zur Kenntnis der österreichischen Sippen des *Valeriana officinalis*-Aggregats und ihrer Chromosomenzahlen. - *Oesterr. Bot. Z.* **116(1-5)**: 172-180.

TITZ, W. 1969b: Zur Zytotaxonomie von *Arabis hirsuta* agg. (Cruciferae). II. Morphologische Analyse österreichischer Populationen und die Abgrenzung der Sippen. - *Oesterr. Bot. Z.* **117(1)**: 21-53.

TITZ, W. 1969c: Zur Zytotaxonomie von *Arabis hirsuta* agg. (Cruciferae). IV. Chromosomenzahlen von *A. sagittata* (Bertol.) DC. und *A. hirsuta* (L.) Scop. s.str. aus Europa. - *Oesterr. Bot. Z.* **117(2)**: 195-200.

TITZ, W. 1970: Bestimmungsschlüssel für die in Österreich wild wachsenden Arten der Gattung *Arabis*. - *Oesterr. Bot. Z.* **118(3)**: 301-305.

TITZ, W. 1972a: Zur Evolution der Gattung *Arabis* im Alpenraum. - *Ber. Deutsch. Bot. Ges.* **84(11)**: 697-704.

TITZ, W. 1972b: Evolution of the *Arabis hirsuta* group in Central Europe. - *Taxon* **21(1)**: 121-128.

TITZ, W. 1973: Nomenklatur, Chromosomenzahlen und Evolution von *Arabis auriculata* Lam., *A. nova* Vill. und *A. verna* (L.) R. Br. (Brassicaceae) - *Oesterr. Bot. Z.* **121(1-2)**: 121-131.

TITZ, W. 1976a: Verbreitung und Evolution von *Arabis pumila* Jacq. und *A. soyeri* Reuter et Huet in den Alpen. - *Linzer Biol. Beitr.* **8(2)**: 333-346.

TITZ, W. 1976b: Cytosystematic study on the Iberian taxa of the *Arabis hirsuta* group. - *Feddes Repert.* **87(7-8)**: 493-502.

TITZ, W. 1976c: Über *Arabis ×jaborneggii* A. Neumann ex Titz, hybr. nova, und die Chromosomenzahl von *Arabis vochinensis* Sprengel. - *Carinthia II* **86**: 265-268.

TITZ, W. 1976d: Die ost- und mitteleuropäische Tieflandsart *Arabis nemorensis* (Hoffm.) Koch ist von *A. planisiliqua* (Pers.) Reichenb. abzutrennen. - *Linzer Biol. Beitr.* **8(2)**: 347-356.

TITZ, W. 1977: Notizie critiche sul genere *Arabis* (Brassicaceae) nella Flora d'Italia e dei territori confinanti. - *Giorn. Bot. Ital., n.s.* **111(1-2)**: 1-12.

TITZ, W. 1978a: Experimentelle Systematik und Genetik der kahlen Sippen in der *Arabis hirsuta*- Gruppe (Brassicaceae). - *Bot. Jahrb. Syst.* **100(1)**: 110-139.

TITZ, W. 1978b: Über das Vorkommen von *Arabis sudetica* Tausch (Brassicaceae) in den Ostalpen. - *Carinthia II* **88**: 275-278.

TITZ, W. 1978c: Kritisch-experimentelle Untersuchungen zur Resynthetisierbarkeit von *Arabis hirsuta* s.str. (4x) aus *A. sagittata* (2x) und *A. ciliata* (2x). - *Beitr. Biol. Pflanzen* **54(3)**: 443-466.

TITZ, W. 1979a: Die Interfertilitätsbeziehungen europäischer Sippen der *Arabis hirsuta*-Gruppe (Brassicaceae). - *Pl. Syst. Evol.* **131(3-4)**: 291-310.

TITZ, W. 1979b: Über die Abtrennung des karpatisch-ostsubmediterranen Endemiten *Arabis hornungiana* von *A. hirsuta* s. str. (Brassicaceae). - *Linzer Biol. Beitr.* **11(2)**: 219-226.

TITZ, W. & SCHNATTINGER, R. 1980: Experimentelle und biometrische Untersuchungen über die systematische Relevanz von Samen- und Fruchtmerkmalen an *Arabis glabra* var. *glabra* und var. *pseudoturritis* (Brassicaceae). - *Pl. Syst. Evol.* **134(3-4)**: 269-286.

TITZ, W. & TITZ, E. 1980: "Valeriana versifolia" und andere oktoploide Arznei-Baldriane in den Schweizer Alpen und angrenzenden Regionen. - *Ber. Schweiz. Bot. Ges.* **89(3-4)**: 251-277.

TITZ, W. & TITZ, E. 1982: Analyse der Formenmannigfaltigkeit der *Valeriana officinalis*-Gruppe im zentralen und südlichen Europa. - *Ber. Deutsch. Bot. Ges.* **95(1)**: 155-164.

TOBLER, F. 1912: Die Gattung *Hedera*. Studien über Gestalt und Leben des Efeus, seine Arten und Geschichte. - Jena: Gustav Fischer Verlag, iv + 151 pp.

TOCL, C. & ROHLENA, J. 1902: Additamenta in floram peninsulae Athoae. - *Sitzungsber. Königl. Böhm. Ges. Wiss. Prag, Math.-Naturwiss. Cl.* **1902**: 1-8.

TODD, E.E. 1930: A short survey of the genus *Viola*. Part I. - The *Nomimium* and *Dischidium* sections. - *J. Roy. Hort. Soc.* **55(2)**: 223-243 + 1 plate.

TODOROV, T.D. 1979: About the distribution of Spanish Licorice (*Glycyrrhiza glabra* L.) in Bulgaria [In Bulgarian with Russian and English summaries]. - *Fitologiya* **11**: 68-70.

TODOROVIĆ, B. & STEVANOVIĆ, B. 1994: Adaptive characteristics of the endemic species *Satureja horvatii* Šilić (Lamiaceae) in mountain-mediterranean and mediterranean habitats. - *Bot. J. Linn. Soc.* **114(4)**: 367-376.

TOFTS, R.J. 2004a: Biological Flora of the British Isles. *Geranium robertianum* L. - *J. Ecol.* **92(3)**: 537-555.

TOFTS, R.J. 2004b: Biological Flora of the British Isles. *Geranium purpureum* Vill. - *J. Ecol.* **92(4)**: 720-731.

TOGBY, H.A. 1943: A cytological study of *Crepis fuliginosa*, *C. neglecta*, and their F1 hybrid, and its bearing on the mechanism of phylogenetic reduction in chromosome number. - *J. Genet.* **45(1)**: 67-111.

TOIVONEN, H. 1981: Notes on the nomenclature and taxonomy of *Carex canescens* (Cyperaceae). - *Ann. Bot. Fenn.* **18(2)**: 91-97.

TOKARSKI, M. 1972: Morphological and taxonomical analysis of fruits and seeds of the European and Caucasian species of the genus *Geranium* L. [In Polish with English summary]. - *Monogr. Bot.* **36**: 5-115.

TOMAN, J. 1972: A taxonomic survey of the genera *Petasites* and *Endocellion*. - *Folia Geobot. Phytotax.* **7(4)**: 381-406.

TOMAN, J., HARMATHA, J. & NOVOTNÝ, L. 1968: Verwandschaftsbeziehungen der Gattung *Adenostyles* und die Berechtigung ihrer Einreihung in die Tribus Senecioneae. - *Preslia* **40**: 122-132.

TOMAN, J. & STARÝ, F. 1964: *Digitalis* ×*rhodopaea* hybr. nov. (*Digitalis lanata* Ehrh. × *D. viridiflora* Lindl.). - *Preslia* **36**: 309-310.

TOMAN, J. & STARÝ, F. 1965: Neue Erkenntnisse über die Art *Petasites kablikianus*. - *Preslia* **37(3)**: 264-275.

TOMAN, J. & STARÝ, F. 1966: Über die Verbreitung der Art *Petasites kablikianus*. - *Preslia* **38(2)**: 168-185.

TOMASELLI, R. 1973: *Quercus coccifera* L. ssp. *coccifera* (syn. *Q. calliprinos* Webb) e ssp. *arcuata* (DC.) stat. nov. (syn. *Q. calliprinos* auct.) dell'Anatolia. - *Atti Ist. Bot. Univ. Pavia, Ser. 6* **9**: 131-153.

TOMASELLI, R. 1981: Main physiognomic types and geographic distribution of shrub systems related to Mediterranean climates. In Di Castri, F., Goodall,, D.W. & Specht, R.L. (eds): Mediterranean-type shrublands, pp. 95-106. - Amsterdam, etc.: Elsevier.

TOMB, A.S. 1977: Lactuceae - systematic review. In Heywood, V.H., Harborne, J.B. & Turner, B.L. (eds): The biology and chemistry of the Compositae 2: 1067-1079. - London, etc.: Academic Press.

TOMKOVICH, L.P. 1982: K kariologicheskomu izucheniyu vidov roda *Ferulago* Koch (Umbelliferae) [In Russian]. - *Biol. Nauki (Moscow)* **1(217)**: 66-69.

TOMKOVICH, L.P. & PIMENOV, M.G. 1989: Botanico-geographical analysis of the genus *Ferulago* W.D.J. Koch (Umbelliferae). - *Feddes Repert.* **100(3-4)**: 119-129.

TOMMASINI, M. 1840: Ueber die griechischen Pflanzen des Hrn. Wilhelm v. Spruner. - *Flora* **23(2)**: 728-734.

TOMOVIĆ, G. 2007: Fitogeografska pripadnost, distribucija i centri diverziteta balkanske endemične flore u Srbiji [In Serbian with Endlish abstract].

TOMOVIĆ, G. & AL. 2003: New distribution data on some pontic and submediterranean plant species in Serbia. - *Arch. Biol. Sci., Belgrade* **55(1-2)**: 45-54.

TOMOVIĆ, G. & AL. 2007a: *Fritillaria* (Liliaceae) in Serbia: distribution, habitats and some taxonomic notes. - *Phytol. Balcan.* **13(3)**: 359-370.

TOMOVIĆ, G. & AL. 2007b: New chorological data on some threatened and rare plants in Serbia. - *Arch. Biol. Sci., Belgrade* **59(1)**: 63-73.

TOMOVIĆ, G. & AL. 2014: Balkan endemic plants in Central Serbia and Kosovo regions: distribution patterns, ecological characteristics, and centres of diversity. - *Bot. J. Linn. Soc.* **176(2)**: 173-202.

TOMOVIĆ, G. & NIKETIĆ, M. 2005: *Gagea spathacea* (Hayne) Salisb. - a new species for the flora of Serbia. - *Arch. Biol. Sci., Belgrade* **57(4)**: 291-294.

TOMOVIĆ, G. & NIKETIĆ, M. 2015: Floristic novelties and taxonomic remarks on the genus *Viola* L. (Violaceae) from Serbia. In Bogdanović, S. & Jogan, N.: Book of abstracts - 6th Balkan Botanical Congress, p. 40. - Rijeka.

TOMOVIĆ, G., NIKETIĆ, M. & STEVANOVIĆ, V. *Malosorbus florentina* (Rosaceae-Maloideae) - distribution, synecology and threatened status in Serbia. - *Phyton (Horn)* **43(2)**: 295-306.

TOMOVIĆ, G. & AL. 2009: *Cardamine amara* (Brassicaceae) in Serbia and Republic of Macedonia. - *Biológia (Bratislava)* **64(6)**: 1095-1099.

TOMŠOVIC, P. 1969: Die wichtigsten Ergebnisse der Revision tschechoslowakischer Sumpfkressen (*Rorippa* Scop. em. Reichenb.) [In Czech with German summary]. - *Preslia* **41**: 21-38 + 1 plate and 2 maps.

TOMŠOVIC, P. 1971: Zur Vorkommen seltener *Rorippa*-Arten in Österreich (mit Bemerkungen zu ihrer Arealgrenzen). - *Preslia* **43(4)**: 338-343.

TOMŠOVIC, P. 1974: The diploid *Rorippa islandica* discovered in southern Europe. - *Folia Geobot. Phytotax.* **9(2)**: 209-212.

TONASCIA, N. 1992: Biosystematische Untersuchungen an *Heracleum sphondylium* s.l. in der Schweiz. - *Ber. Geobot. Inst. E.T.H. Stiftung Rübel* **58**: 101-120.

TOORN, J. VAN DER 1980: On the ecology of *Cotula coronopifolia* L. and *Ranunculus sceleratus* L. - *Acta Bot. Neerl.* **29(5-6)**: 385-396.

TOPALI, S. 1941: Nouvelles herborisations en Macédoine grecque. - *Bull. Soc. Bot. Genève, Sér. 2* **32**: 185-186.

TOPALI, S. & BEAUVERD, G. 1934: Excursions botaniques au Liban, à Chypre et en Grèce. - *Bull. Soc. Bot. Genève, Sér. 2* **26**: 155-158.

TOPALI, S. & HUMBERT, H. 1940: Contributions à la flore de la Macédoine grecque. - *Bull. Soc. Bot. Genève, Sér. 2* **31**: 446-448.

TÖREN, J. 1950: Les caractères morphologiques, anatomiques et cytologiques de *Bongardia chrysogonum* Boiss. - *Istanbul Üniv. Fen Fak. Mecm., B* **15(3)**: 239-263.

TÖREN, J. 1954: Recherches sur des écotypes de *Bongardia chrysogonum* (L.) Boiss. – *Istanbul Üniv. Fen Fak. Mecm., B* **19(2)**: 83-123.

TÖREN, J. 1961: Recherches sur les Berberidaceae de la Turquie. I. Caractères morphologiques et anatomiques du *Leontice leontopetalum* L. - *Istanbul Üniv. Fen Fak. Mecm., B* **26(3-4)**: 125-162.

TÖREN, J. 1962: Recherches sur les Berberidaceae de la Turquie. II. Caractères cytologiques du *Leontice leontopetalum* L. - *Istanbul Üniv. Fen Fak. Mecm., B* **27(3-4)**: 229-250 + 3 plates.

TÖREN, J. 1965: Recherches sur les Berberidaceae de la Turquie. IV. Chromosomes somatiques du *Leontice leontopetalum* L. - *Istanbul Üniv. Fen Fak. Mecm., B* **30(3-4)**: 191-203 + 2 plates.

TÖREN, J. 1971a: Investigations on the Berberidaceae of Turkey. VI. Contributions to the cytology of *Bongardia chrysogonum* (L.) Boiss. (ecotype of Urfa). - *Istanbul Üniv. Fen Fak. Mecm., B* **36(1-2)**: 71-79 + 1 plate.

TÖREN, J. 1971b: Investigations on the Berberidaceae of Turkey. VII. *Bongardia chrysogonum* (L.) Boiss.) (Van ecotype). - *Istanbul Üniv. Fen Fak. Mecm., B* **36(1-2)**: 81-88 + 1 plate.

TORGES, E. 1894: Zur Gattung *Calamagrostis* Adans. - *Mitth. Thüring. Bot. Vereins, N.F.* **6**: 14-22.

TORGES, E. 1895: Zur Gattung *Calamagrostis* Adans. [cont.]. - *Mitth. Thüring. Bot. Vereins, N.F.* **8**: 13-16.

TORGES, E. 1898a: Zur Gattung *Calamagrostis* Adans. [cont.]. - *Mitth. Thüring. Bot. Vereins, N.F.* **11**: 78-93.

TORGES, E. 1898b: Zur Gattung *Calamagrostis* Adans. [cont.]. - *Mitth. Thüring. Bot. Vereins, N.F.* **12**: 22-25.

TORNABENE, F. 1887: Flora Sicula viva et exsiccata seu collectio plantarum in Sicilia sponte nascentium hucusque cognitarum juxta methodum naturalem vegetabilium ... - Catania: En typis Francisci Galati, 688 pp. + 4 plates. [Reprinted 1973 by O. Koeltz].

TORNABENE, F. 1889-92: Flora Aetnea seu descriptio plantarum in Monte Aetna sponte nascentium ... Catania: En typis Francisci Galati, 4 vols. [cf. TL-2 6: 399, 1986].

TORNADORE, N. 1982: Contributo alla conoscenza citotassonomica di "Allium moschatum" L. - *Webbia* **35(2)**: 283-293.

TORNADORE, N. 1983: Numeri cromosomici per la Flora Italiana: 806-813. - *Inform. Bot. Ital.* **13(2-3)**: 151-157.

TORNADORE, N. 1986: Il gen. *Ornithogalum* L. (Liliaceae). V. Osservazioni sulla citosistematica di *O. pyrenaicum* L. - *Atti Soc. Tosc. Sci. Nat. Pisa Mem., Ser. B* **92**: 247-257.

TORNADORE, N. 1987: Il genere *Ornithogalum* L. (Liliaceae). VI. *O. narbonense* L. - *Atti Soc. Tosc. Sci. Nat. Pisa Mem., Ser. B* **93**: 111-120.

TORNADORE, N. 1990: Population variability in *Allium sphaerocephalon* L. - *Atti Soc. Tosc. Sci. Nat. Pisa Mem., Ser. B* **96**: 53-62.

TORNADORE, N. & GARBARI, F. 1979: Il genere "Ornithogalum" L. (Liliaceae) in Italia. 3. Contributo alla revisione citotassonomica. - *Webbia* **33(2)**: 379-423.

TORNADORE, N. & MARCUCCI, R. 1988a: Attuali conoscenze cittotassaonomiche sulla flora pugliese. - *Thalassia Salentina* **18**: 77-86.

TORNADORE, N. & MARCUCCI, R. 1988b: Notes on the caryology of some critical taxa of the genus *Ornithogalum* L. (Liliaceae). - *Giorn. Bot. Ital., n.s.* **122(1-2), suppl. 1**: 53.

TORNADORE, N. & ORZA, P. 1988: Il gen. *Ornithogalum* L. (Liliaceae) in Italia. VIII. Il subgen. *Beryllis* (Salisb.) Baker con particolare riguardo ad *O. brevistylum* Wolfner. - *Atti Soc. Tosc. Sci. Nat. Pisa Mem., Ser. B* **94**: 341-356.

TORNADORE, N., POPOVA, M.P. & GARBARI, F. 1974: Numeri cromosomici per la Flora Italiana: 172-181. - *Inform. Bot. Ital.* **6(1)**: 43-54.

TOSHEVA, A. 2003: Reports (1339-1343). In: Kamari, G., Blanché, C. & Garbari, F. (eds): Mediterranean chromosome number reports - 13. - *Fl. Medit.* **13**: 365-369.

TOSHEVA, A. 2005: *Lathyrus filiformis* (Fabaceae) a new species for the Bulgarian flora. - *Fl. Medit.* **15**: 397-402.

TÓTH, L., CSORDÁS, I. & PÁPAY, V. 1978: Chemotaxonomical proofs concerning the classification of *Kickxia elatine* (L.) Dum. and *Kickxia spuria* (L.) Dum. - *Acta Bot. Acad. Sci. Hung.* **24(1-2)**: 179-185.

TOUMI, L. & LUMARET, R. 2010: Genetic variation and evolutionary history of holly oak: a circum-Mediterranean species-complex [*Quercus coccifera* L./*Q. calliprinos* (Webb) Holmboe, Fagaceae]. - *Pl. Syst. Evol.* **290(1)**: 159-171.

TOWNSEND, C.C. 1968: *Parietaria officinalis* and *P. judaica*. - *Watsonia* **6(6)**: 365-370.

TOWNSEND, C.C. 1978: Towards a revision of *Haplophyllum* (Rutaceae): 4. - *Kew Bull.* **33(1)**: 127-131.

TOWNSEND, C.C. 1982: *Valerianella orientalis* (Schlecht.) Boiss. & Bal. In Greuter, W. & Raus, Th. (eds): Med-Checklist notulae, 6. - *Willdenowia* **12(2)**: 199.

TOWNSEND, C.C. 1986: Taxonomic revision of the genus *Haplophyllum* (Rutaceae). - *Hooker's Icon. Pl.* **40(1-3)**: i-iv, 1-336.

TOWNSEND, C.C. & GUEST, E. (EDS) 1966: Flora of Iraq. Vol. 2. - Baghdad: Ministry of Agriculture, viii + 184 pp.

TOWNSEND, C.C. & GUEST, E. (EDS) 1974: Flora of Iraq. Vol. 3. - Baghdad: Ministry of Agriculture, vii + 662 pp.

TOWNSEND, C.C. & GUEST, E. (EDS) 1980a: Flora of Iraq. Vol. 4(1). - Baghdad: Ministry of Agriculture, ix + 627 pp.

TOWNSEND, C.C. & GUEST, E. (EDS) 1980b: Flora of Iraq. Vol. 4(2). - Baghdad: Ministry of Agriculture, pp. 628-1199.

TOWNSEND, C.C. & GUEST, E. (EDS) 1985: Flora of Iraq. Vol. 8. - Baghdad: Ministry of Agriculture, ix + 440 pp.

TOWPASZ, K. 1971: *Ficaria verna* Huds. ssp. *calthifolia* (Rchb.) Vel. in Poland [In Polish with English summary]. - *Fragm. Florist. Geobot.* **17(2)**: 215-219.

TÖZÜN, B. 1961: Morphological and anatomical characteristics of *Digitalis viridiflora* Lindl. - *Istanbul Üniv. Fen Fak. Mecm., B* **26(3-4)**: 111-123.

TRABAUD, L. 1991: Is fire an agent favouring plant invasions? In Groves, R.H. & Di Castri, F. (eds): Biogeography of Mediterranean invasions, pp. 179-190. - Cambridge, etc.: Cambridge Univ. Press.

TRAKOLIS, D., PLATIS, P. & MELIADIS, I. 2000: Biodiversity and conservation actions on Mount Voras, Greece. - *Environmental Management* **26(2)**: 145-151.

TRALAU, H. 1958: Studie über den arktisch-alpinen *Ranunculus platanifolius* L. und den alpinen *Ranunculus aconitifolius* L. - *Beitr. Biol. Pflanzen* **34(3)**: 479-507.

TRÁVNÍČEK, B. 1998: Notes on the taxonomy of *Pseudolysimachion* sect. *Pseudolysimachion* (Scrophulariaceae) in Europe. I. *P. incanum* and *P. spicatum*. - *Preslia* **70(3)**: 193-223.

TREES-FRICK, I. 1993: Karyologische und morphologische Untersuchungen an *Viola reichenbachiana* Jord. ex Boreau und *Viola riviniana* Rchb. - *Mitth. Naturf. Ges. Bern, N.S.* **50**: 81-98.

TREHANE, P. 1995: (1172) Proposal to conserve *Chrysanthemum* L. with a conserved type (Compositae). - *Taxon* **44(3)**: 439-441. [approved, cf. Taxon 47(2): 443, 1998].

TRELA-SAWICKA, Z. 1965: Cytological investigations in the genus *Trapa* L. - *Acta Biol. Cracov., Ser. Bot.* **8(2)**: 237-243.

TRELA-SAWICKA, Z. 1966: Cytological investigations in the genus *Scleranthus* L. - *Acta Biol. Cracov., Ser. Bot.* **9(1)**: 59-63 + 1 plate.

TRELA-SAWICKA, Z. 1968: Cytological investigations in the genus *Thymus* L. - *Acta Biol. Cracov., Ser. Bot.* **11(1)**: 59-69.

TRIANTIS, K.A., GUILHAUMON, F. & WHITTAKER, R.J. 2012: The island species-area relationship: biology and statistics. - *J. Biogeogr.* **39(2)**: 215-231.

TRIEST, L. 1988: A revision of the genus *Najas* L. (Najadaceae) in the Old World. - Bruxelles: Academie Royale des Sciences d'Outre-Mer, 172 pp. + 29 plates.

TRIFONOVA, V.I. 1990: Comparative biomorphological study of the taxonomy and phylogeny of the genera *Consolida* (DC.) S.F. Gray and *Aconitella* Spach. - *Collect. Bot. (Barcelona)* **19**: 97-110.

TRIGAS, P. 2003: Contribution to the study of the endemism of the flora of the island of Evvia (W Aegean, Greece) [In Greek with English summary]. - Patras: Ph.D. thesis, Univ. of Patras, 341 pp.

TRIGAS, P. 2018: A new *Hypericum* (sect. *Drosocarpium*, Hypericaceae) from the Cyclades Islands (Greece). - *Nordic J. Bot.* **36(11)**: 5 pp.

TRIGAS, P. & AL. 2012: Distribution patterns and conservation perspectives of the endemic flora of Peloponnese (Greece). - *Folia Geobot.* **47(4)**: 421-439.

TRIGAS, P. & AL. 2018: Multivariate morphometric analysis of *Petrorhagia* subsect. *Saxifragae*

(Caryophyllaceae) in Greece, with a new species from SE Peloponnisos: *P. laconica*. - *Willdenowia* **48(1)**: 137-146.

TRIGAS, P. & AL. 2021: Some noteworthy new floristic records from Greece. - *Bot. Serbica* **45(2)**: 321-331.

TRIGAS, P. & BAREKA, E.P. 2019: *Allium topalianum* (*A.* sect. *Codonoprasum*, Amaryllidaceae), a new autumn flowering species from Andros and Evvia Islands. In Daskalakou, E.N. & al. (eds): 16th Conference of the Hellenic Botanical Society – Athens 10-13 October 2019. Program and book of abstracts, p. 106.

TRIGAS, P. & BAREKA, P. 2020: *Allium stamatiadae*, a new species of *A.* sect. *Codonoprasum* (Amaryllidaceae) from Andros island (Aegean Archipelago, Greece). - *Phytotaxa* **443(3)**: 287-294.

TRIGAS, P., CONSTANTINIDIS, TH. & TOULOUMENIDOU, T. 2006: A new hexaploid species of *Centaurea* (sect. *Acrolophus*, Asteraceae) from Greece. In Peev, D.R. (ed.): Book of abstracts, IV Balkan botanical congress, Sofia, Bulgaria, 20-26 June 2006, p. 201.

TRIGAS, P., CONSTANTINIDIS, TH. & TOULOUMENIDOU, T. 2008: A new hexaploid species of *Centaurea* section *Acrolophus* (Asteraceae) from Evvia Island, Greece. - *Bot. J. Linn. Soc.* **158(4)**: 762-774.

TRIGAS, P. & IATROU, G. 1997: Contribution to the biodiversity (botanical diversity) and phytogeography of the island of Euboea. In: Abstracts from the First Balkan Botanical Congress, Thessaloniki, September 19-22, 1997, p. 52. - Thessaloniki.

TRIGAS, P. & IATROU, G. 2000: Additions to the flora of Evvia (Greece). - *Bot. Chron. (Patras)* **13**: 273-286.

TRIGAS, P. & IATROU, G. 2003: *Asperula* (sect. *Cynanchicae*) *brachyphylla*, spec. nova (Rubiaceae) from the island of Evvia (Greece). - *Phyton (Horn)* **43(1)**: 29-37.

TRIGAS, P. & IATROÚ, G. 1998: Contribution to the biodiversity (botanical diversity) and phytogeography of the island of Euboea. In Tsekos, I. & Moustakas, M. (eds): Progress in Botanical Reseearch, pp. 161-164. - Dordrecht: Kluwer Press.

TRIGAS, P. & IATROÚ, G. 2005: A new species of *Minuartia* (Caryophyllaceae) from the island of Evvia (Greece). - *Nordic J. Bot.* **23(4)**: 415-425.

TRIGAS, P. & IATROÚ, G. 2006: The local endemic flora of Evvia (W Aegean, Greece). - *Willdenowia* **36(1)**: 257-270.

TRIGAS, P., IATROÚ, G. & KARETSOS, G. 2007: Species diversity, endemism and conservation of the family Caryophyllaceae in Greece. - *Biodivers. Conserv.* **16(2)**: 357-376.

TRIGAS, P., IATROÚ, G. & PANITSA, M. 2008: Vascular plant species diversity, biogeography and vulnerability in the Aegean islands as examplified by Evvia island (W Aegean, Greece). - *Fresenius Environ. Bull.* **17(1)**: 48-57.

TRIGAS, P., IATROÚ, G. & TZANOUDAKIS, D. 2010: *Allium apergii* sp. nov. (Alliaceae, *A.* sect. *Codonoprasum*) from Evvia Island, Greece. - *J. Biol. Res. Thessalon.* **14**: 225-229.

TRIGAS, P., KALPOUTZAKIS, E. & CONSTANTINIDIS, TH. 2017: Two new *Allium* (*A.* sect. *Cupanioscordum*, Amaryllidaceae) species from Greece. - *Phytotaxa* **297(2)**: 179-188.

TRIGAS, P., KARETSOS, G. & TSIFTSIS, S. 2012: Reports 117-131. In Vladimirov, V. & al. (eds): New floristic reports in the Balkans: 18. - *Phytol. Balcan.* **18(1)**: 86-88.

TRIGAS, P., KOKKORIS, I. & KOUGIOUMOUTZIS, K. 2017: The rediscovery of *Silene guicciardii* (Caryophyllaceae) on Mt. Parnassos (Greece) after 160 years: taxonomic re-evaluation and conservation. - *Phytotaxa* **331(2)**: 281-288.

TRIGAS, P., PANITSA, M. & TSIFTSIS, S. 2013: Elevational gradient of vascular plant species richness and endemism in Crete - The effect of post-isolation mountain uplift on acontinental island system. - *PLoS ONE* **8**: e59425, doi:10.1371/journal.pone.0059425.

TRIGAS, P. & TZANOUDAKIS, D. 2000: *Allium runemarkii* (Liliaceae), a new species from the island of Evvia (W Aegean, Greece). - *Nordic J. Bot.* **20(1)**: 89-92.

TRINAJSTIĆ, I. 1968: Über die Formen der *Biserrula pelecinus* L. [In Serbo-Croatian with German summary]. - *Acta Bot. Croat.* **26-27**: 215-219.

TRINAJSTIĆ, I. 1973a: Taxa nova et combinationes novae in flora Jugoslaviae I. In: Suppl. Fl. Anal. Jugosl. 1: 5-11. - Zagreb.

TRINAJSTIĆ, I. 1973b: Conspectus specierum generis *Bulbocapnos* Bernh. (Fumariaceae) in flora Jugoslaviae. In: Suppl. Fl. Anal. Jugosl. 1: 17-20. - Zagreb.

TRINAJSTIĆ, I. 1974a: Beitrag zur Kenntnis der Horologie der neophytischen Art *Tagetes minuta* L. auf dem Gebiete Jugoslawiens [In Serbo-Croatian with German summary]. - *Acta Bot. Croat.* **33**: 231-235.

TRINAJSTIĆ, I. 1974b: Taxa et combinationes novae in Flora Jugoslaviae II. In: Suppl. Fl. Anal. Jugosl. 2: 5-8. - Zagreb.

TRINAJSTIĆ, I. 1975: Taxa et combinationes novae in Flora Jugoslaviae III. In: Suppl. Fl. Anal. Jugosl. 3: 5-8. - Zagreb.

TRINAJSTIĆ, I. 1976a: Beitrag zur Taxonomie des *Iris pallida* Komplexes. - *Biosistematika* **2(1)**: 67-78.

TRINAJSTIĆ, I. 1976b: Taxa et combinationes novae in Flora Jugoslaviae IV. In: Suppl. Fl. Anal. Jugosl. 4: 5-10. - Zagreb.

TRINAJSTIĆ, I. 1978: Taxa et combinationes novae in Flora Jugoslaviae V. In: Suppl. Fl. Anal. Jugosl. 5: 5-8. - Zagreb.

TRINAJSTIĆ, I. 1979: Taxa et combinationes novae in Flora Jugoslaviae VI. In: Suppl. Fl. Anal. Jugosl. 6: 5-9. - Zagreb.

TRINAJSTIĆ, I. 1980a: Taxa et combinationes novae in Flora Jugoslaviae VII. In: Suppl. Fl. Anal. Jugosl. 7: 5-7. - Zagreb.

TRINAJSTIĆ, I. 1980b: Brassicacearum genera nova in Flora Jugoslaviae. In: Suppl. Fl. Anal. Jugosl. 7: 9-13. - Zagreb.

TRINAJSTIĆ, I. 1982: Taxa nova et combinationes novae in Flora Jugoslaviae VIII. In: Suppl. Fl. Anal. Jugosl. 8: 3-11. - Zagreb.

TRINAJSTIĆ, I. 1985: *Viola koritnicensis* Trinajstić spec. nov. - the new species of genus *Viola* from massif of Koritnik at Kosovo (Serbia, Yugoslavia) [In Serbo-Croatian with English summary]. - *Glasn. Prir. Mus. Beogradu, Ser. B, Biol. Nauke* **40**: 27-31.

TRINAJSTIĆ, I. 1986a: Oreomediterranean phytogeographical region [In Serbo-Croatian with English summary]. - *Biosistematika* **11(2)**: 83-89.

TRINAJSTIĆ, I. 1986b: Taxa nova et combinationes novae in Flora Jugoslaviae IX. In: Suppl. Fl. Anal. Jugosl. 9: 3-5. - Zagreb.

TRINAJSTIĆ, I. 1990: *Aristolochia pallida* Willd. (Aristolochiaceae) in the flora of Croatia. - *Acta Bot. Croat.* **49**: 143-146 + 1 plate.

TRINAJSTIĆ, I. & PAVLETIĆ, Z. 1978: *Azolla filiculoides* Lam. in the water-plant vegetation of eastern Croatia [In Serbo-Croatian with English summary]. - *Acta Bot. Croat.* **37**: 159-162.

TRINAJSTIĆ, I. & PAVLETIĆ, Z. 1979: Taxa et combinationes novae generis *Silene* L. in Flora Jugoslaviae. In: Suppl. Fl. Anal. Jugosl. 6: 11-14. - Zagreb.

TRINAJSTIĆ, I. (ED.) 1979: Flora Analytica Jugoslaviae, I/5 [In Serbo-Croatian]. - Zagreb: Tisak Grafičkog Zavoda, pp. 627-728.

TRINIUS, C.B. 1828-36: Species graminum iconibus et descriptionibus illustravit ... - Petropoli [St. Petersburg]: 3 vols. [cf. TL-2 6: 495-496, 1986].

TRIPATHI, R.C., SACHDEVA, S.K. & MALIK, C.P. 1977: Cytogenetic studies and evolutionary patterns in some *Cynodon* species. - *Biol. Zentralbl.* **96(4)**: 423-435.

TRÖHLER, A. 1976: Zytotaxonomische Untersuchungen an *Ranunculus ficaria* L. - *Mitth. Naturf. Ges. Bern, N.S.* **33**: 21-28.

TROÌA, A. 2001: The genus *Isoëtes* L. (Lycophyta, Isoëtaceae): synthesis of karyological data. - *Webbia* **56(1)**: 201-218.

TROÌA, A. & AL. 2012: Morphological, karyological and taxonomic remarks on *Ferulago nodosa* (L.) Boiss. (Apiaceae). - *Pl. Biosyst.* **146 suppl.**: 330-337.

TROÌA, A. & BELLINI, E. 1998: Karyologioal [sic!] investigations on Sicilian populations of *Isoetes duriei* Bory (Lycopodiophyta, Isoetaceae). In Moret, J. (ed.): Abstracts of the IX. OPTIMA meeting, Paris, 11-17 May 1998. - Paris: Mus. Nat. d'Hist. Naturelle.

TROÌA, A. & BELLINI, E. 2001: Karyological observations on *Isoëtes duriei* Bory (Lycophyta, Isoëtaceae) in Sicily. - *Bocconea* **13**: 397-400.

TROÌA, A. & GREUTER, W. 2014: A critical conspectus of Italian *Isoetes* (Isoetaceae). - *Pl. Biosyst.* **148 (1)**: 13-20.
TROÌA, A. & GREUTER, W. 2015: A conspectus of and key to Greek *Isoetes* (Isoetaceae), based on reassessment of Haussknecht's gatherings of 1885. - *Willdenowia* **45(3)**: 391-403.
TROÌA, A. & LANSDOWN, R. 2016: The first confirmed population of the globally endangered *Pilularia minuta* (Marsileaceae) in Sicily. - *Webbia* **71(2)**: 283-286.
TRPIN, D. 1973: Zur Kenntnis und Bewertung von *Herniaria rotundifolia* Visiani. - *Feddes Repert.* **84 (4)**: 295-301.
TRYON, R.M. JR. 1941: A revision of the genus *Pteridium*. - *Rhodora* **43**: 1-31, 37-67 + 4 plates.
TSAKIRI, E., BABALONAS, D. & PAPASTERGIADOU, E. 1995: Growth conditions of the waterfern *Azolla filiculoides* Lam. at Axios delta. In: Proceedings, 5th sci. conf., Hellenic Bot. Soc., Delphi 21-23 October 1994, pp. 117-120 [In Greek with English summary]. - Thessaloniki: Univ. Studio Press.
TSAKIRI, M. & AL. 2017: Contribution to the flora of Chelmos-Vouraikos National Park. In Dousis, M.A. & Thanos, K.A.: 15th conference of the Hellenic Botanical Society, book of abstracts: p. 126.
TSAKIRI, M. & AL. 2020: Contribution to the vascular flora of Chelmos-Vouraikos National Park (N Peloponnese, Greece). - *Phytol. Balcan.* **26(3)**: 523-536.
TSAKIRI, M. & AL. 2021: National Park and UNESCO Global Geopark of Chelmos-Vouraikos (Greece): Floristic diversity, ecosystem services and management implications. - *Land MDPI* **11, 33**: doi.org/10.3390/land11010033.
TSAKIRI, M., KOUGIOUMOUTZIS, K. & IATROÚ, G. 2016: Contribution to the vascular flora of Chalki Island (East Aegean, Greece) and biomonitoring of a local endemic taxon. - *Willdenowia* **46(1)**: 175-190.
TSALIKI, M., BERGMEIER, E. & DIMOPOULOS, P. 2005: Vegetation patterns and plant diversity in mixed oak woodland in the region of Bourazani, Epirus (NW Greece). - *Bot. Chron. (Patras)* **18(1)**: 225-251.
TSCHERMAK-WOESS, E. 1967: Der eigenartige Verlauf der I. meiotischen Prophase von *Rhinanthus*, die Riesenchromosomen und das besondere Verhalten der kurzen Chromosomen in Mitose, Meiose und hochendopolyploiden Kernen. - *Caryologia* **20(2)**: 135-152.
TSCHEULIN, T. & AL. 2007: [Data sheets on *Solanum elaeagnifolium*]. - *OEPP/EPPO Bulletin* **37**: 236-245.
TSIALTAS, I. & AL. 2013: Report 130. In Vladimirov, V., Dane, F. & Tan, Kit (eds): New floristic records in the Balkans: 21. - *Phytol. Balcan.* **19(1)**: 149-151.
TSIALTAS, I. & TAN, KIT 2013: Reports 114-116. In Vladimirov, V. & al. (eds): New floristic records in the Balkans: 22. - *Phytol. Balcan.* **19(2)**: 290.
TSIALTAS, I. & TAN, KIT 2014: Reports 240-241. In Vladimirov, V. & al. (eds): New floristic records in the Balkans: 25. - *Phytol. Balcan.* **20(2-3)**: 217-218.
TSIALTAS, I. & TAN, KIT 2015b: Reports 122-126. In Vladimirov, V., Dane, F. & Tan, Kit (eds): New floristic records in the Balkans: 28. - *Phytol. Balcan.* **21(3)**: 384-385.
TSIALTAS, I., TAN, KIT & VOLD, G. 2014: Reports 132-136. In Vladimirov, V. & Tan, Kit (eds): New floristic records in the Balkans: 24. - *Phytol. Balcan.* **20(1)**: 125-127.
TSIFTIS, S., TSIRIPIDIS, I. & TRIGAS, P. 2011: Identifying important areas for orchid conservation in Crete. - *Eur. J. Envir. Sci.* **1(2)**: 28-37.
TSIFTSIS, S. 2002: I orchidees tou Menikiou Orous [In Greek]. - *Fisis [Journal of the Hellenic Society for the Protection of Nature]* **98**: 27-28.
TSIFTSIS, S. 2004: Nea tis ellinikis chloridas [In Greek]. - *Fisis [Journal of the Hellenic Society for the Protection of Nature]* **105**: 41-42.
TSIFTSIS, S. 2007: Anatoliki Makedonia: katafigio spanion yia tin Ellada idon tis ikogenias Orchidaceae [In Greek]. - *Fisis [Journal of the Hellenic Society for the Protection of Nature]* **118**: 35-38.
TSIFTSIS, S. 2009: The orchids (Orchidaceae) of E. Macedonia: Distribution, ecology and high conservation value areas. [In Greek with English summary]. - Thessaloniki, Aristotle University: 279 pp.
TSIFTSIS, S. 2010: *Epipactis persica* subsp. *persica* (Orchidaceae), a new taxon for the flora of Greece. - *J. Eur. Orch.* **42(3-4)**: 477-486.
TSIFTSIS, S. 2016: Morphological variability of *Himantoglossum* s.s. (Orchidaceae) in Greece. - *Phytotaxa* **245(1)**: 17-30.
TSIFTSIS, S. 2019: Contribution to the orchid flora of mount Simvolo (E Macedonia - NE Greece). - *Bot. Chron. (Patras)* **22**: 117-126.
TSIFTSIS, S. & AL. 2021: *Colchicum tulakii* (Colchicaceae), a new species from central Macedonia, northeastern Greece. - *Nordic J. Bot.* **39(6)**: 6 pp.
TSIFTSIS, S. & ANTONOPOULOS, Z. 2011: *Pseudorchis albida*: an enigmatic orchid of the Greek flora. - *J. Eur. Orch.* **43(4)**: 795-806.
TSIFTSIS, S. & ANTONOPOULOS, Z. 2017: Atlas of the Greek orchids, vol. 1. The genera *Epipactis*, *Cephalanthera*, *Limodorum*, *Neottia*, *Epipogium*, *Corallorhiza, Goodyera, Spiranthes, Platanthera, Gymnadenia, Pseudorchis, Coeloglossum, Dactylorhiza, Serapias, Anacamptis, Neotinea, Orchis, Himantoglossum*. - Rethymno: Mediterraneo Editions, 511 pp.
TSIFTSIS, S., ANTONOPOULOS, Z. & DIMADIS, CH. 2019: *Crocus novicii*, a new species for the flora of Greece. - *Parnassiana Arch.* **7**: 65-68.
TSIFTSIS, S., KARAGIANNAKIDOU, V. & TSIRIPIDIS, I. 2007: The orchid flora of East Macedonia (NE Greece). - *J. Eur. Orch.* **39(3-4)**: 489-526.
TSIFTSIS, S. & TSIRIPIDIS, I. 2016: Threat categories of the Greek orchids (Orchidaceae). - *Bot. Chron. (Patras)* **21**: 43-74.
TSIFTSIS, S., TSIRIPIDIS, I. & KARAGIANNAKIDOU, V. 2006: Study of the orchid flora of Mount Falakro (NE Greece). - *J. Eur. Orch.* **38(1)**: 57-78.
TSIFTSIS, S., TSIRIPIDIS, I. & KARAGIANNAKIDOU, V. 2009: Identifying areas of high importance for orchid conservation in east Macedonia (NE Greece). - *Biodivers. Conserv.* **18(7)**: 1765-1780.
TSIFTSIS, S., TSIRIPIDIS, I. & PAPAIOANNOU, A. 2012: Ecology of the orchid *Goodyera repens* in its southern distribution limits. - *Pl. Biosyst.* **146(4)**: 857-866.
TSIFTSIS, S., TSIRIPIDIS, I. & TRIGAS, P. 2011: Identifying important areas for orchid conservation in Crete. - *Eur. J. Environ. Sci.* **1**: 28-37.
TSIFTSIS, S., TSIRIPIDIS, I. & VIDAKIS, K. 2012: Orchids of Rodopi Mountain-Range National Park. [In Greek and English]. - Paranesti: Management Body of Rodopi Mountain-Range National Park, 197 pp.
TSIFTSIS, S., TSIRIPIDIS, I. & VIDAKIS, K. 2015: Orchids of Northern Pindos National Park [in Greek and English]. - Ioannina: Management Agency of Vikos-Aoos and Pindos National Forests, 204 pp.
TSIMBURLA, E. & YANNITSAROS, A. 1992: A preliminary study of the flora and vegetation of the island Piperi [In Greek with English summary]. In: Proc. 12th Annual Meeting Hellen. Soc. Biol. Sci., Mitilini 27.-29.4. 1990, pp. 172-173. - Athine.
TSINTIDES, T.C. 1995: Ta endemika fita tis Kiprou [In Greek]. - Levkosia: Sigrotima Trapezas Kiprou, 123 pp.
TSINTIDES, T.C. 1998: The endemic plants of Cyprus. - Nicosia: Cyprus Association of Professional Foresters, 123 pp.
TSINTIDES, T.C. & AL. 2007: The Red Data Book of the flora of Cyprus. - Lefkosia: Chr. Nicolaou & Sons Ltd.
TSINTIDES, T.C., HADJIKYRIAKOU, G.N. & CHRISTODOULOU, CH.S. 2002: Trees and shrubs in Cyprus. - Lefkosia: Foundation Anastasios G. Leventis, 442 pp.
TSIOTSIOU, V. & CHRISTODOULAKIS, D. 2004: Contribution to the urban ecology of Greece: The flora of Mesolongi city. - *Fl. Medit.* **14**: 129-151.
TSIOURLIS, G., KONSTANTINIDIS, P. & XOFIS, P. 2007: Taxonomy and ecology of phryganic communities with *Sarcopoterium spinosum* (L.) Spach of the Aegean (Greece). - *Israel J. Pl. Sci.* **55**: 15-34.

TSIRIKA, A. & HARITONIDIS, S. 2005: A survey of the benthic flora in the National Marine Park of Zakynthos (Greece). - *Botanica Marina* **48(1)**: 38-45.

TSIRIPIDIS, I. 2001: I fitokinonies dason oxias tis Rodopis ke ektimisi perivallonton tous gia anadasosi. [In Greek]. - Thessaloniki: Ph.D. thesis, Aristotelian University, 366 pp

TSIRIPIDIS, I. & AL. 2010: Approaching the serpentine factor at a local scale - a study in an ultramafic area in northern Greece. - *Plant Soil* **329**: 35-50.

TSIRIPIDIS, I. & ATHANASIADIS, N. 2002: Ecological differentiation of beech forests in Greek Rhodope mountain range. In Kamari, G. & al. (eds): Ellinini Votaniki Eteria. 9° Panelliniko Epistimoniko Sinedrio. Praktika, pp. 328-336 [In Greek with English abstract].

TSIRIPIDIS, I. & ATHANASIADIS, N. 2003: Contribution to the knowledge of the vascular flora of NE Greece: Floristic composition of the beech (*Fagus sylvatica* L.) forests in the Greek Rodopi. - *Willdenowia* **33(2)**: 273-297.

TSIRIPIDIS, I., BERGMEIER, E. & DIMOPOULOS, P. 2007: Geographical and ecological differentiation in Greek *Fagus* forest vegetation. - *J. Veg. Sci.* **18**: 743-750.

TSIROUKIS, A. & AL. 2009: Conservation status of horse chestnut (*Aesculus hippocastanum* L.) in Greece [In Greek with English summary]. In: 3. Sinedrio E.Oi.E. & E.Z.E., Ioannina 2006, pp. 400-406.

TSITSONI, TH. & KARAGIANNAKIDOU, V. 1992: Phytogeographical analysis of the flora of *Pinus halepensis* forests in the peninsula of Kassandra-Chalkidiki, NE Greece. - *Feddes Repert.* **103(1-2)**: 41-47.

TSITSONI, TH. & KARAGIANNAKIDOU, V. 2000: Site quality and stand structure in *Pinus halepensis* forest of north Greece. - *Forestry?* **73(1)**: 51-64.

TSONEVA, S. & AL. 2012: Atlas of aquatic and wetland plants in Bulgaria. - Sofia: IBER, BAS, GeoSoft Ltd.

TSOUKA, V. & HARITONIDIS, S. 1999: Comparative study of *Posidonia oceanica* (L.) Delile in Thermaikos and in Toronaios Gulf. In Elefheriou, E.P. (ed.): Proceedings, 7th Panhellenic scientific conference, Alexandroupolis, pp. 253-257 [In Greek with English abstract]. - Thessaloniki: Univ. Studio Press.

TSOUNIS, G, TSOUNIS, L. & TAN, KIT 2009: I *Micromeria acropolitana* anakalifthike xana stin Athina [in Greek]. - *Fisis [Journal of the Hellenic Society for the Protection of Nature]* **126**: 22-24.

TSOUNIS, L. , TAN, KIT & KEHAYIAS, G. Report 149. In Vladimirov, V., Aybeke, M. & Tan, Kit (eds): New floristic records in the Balkans: 36. - *Phytol. Balcan.* **24(2)**: 280-281.

TSUNODA, S., HINATA, K. & GÓMEZ-CAMPO, C. (EDS) 1980: *Brassica* crops and wild allies. Biology and breeding. - Tokyo: Japan Scientific Societies Press, xviii + 354 pp.

TUBEUF, K.V., NECKEL, G. & MARZELL, H. 1923: Monographie der Mistel. - München & Berlin: Verlag von R. Oldenbourg, xii + 832 pp. + 6 maps.

TUCCI, G.F. & RICCIARDI, M. 1979: Considerazioni sistematiche e cariologiche sul genere *Centranthus* DC. - *Delpinoa, n.s.* **20**: 51-57 + 2 plates.

TUCKER, A.O., HARLEY, R.M. & FAIRBROTHERS, D.E. 1980: The Linnaean types of *Mentha* (Lamiaceae). - *Taxon* **29(2-3)**: 233-255.

TUISL, G. 1966: Beiträge zur Fruchtanatomie und Pappusstruktur von *Chondrilla juncea* L. und *Chondrilla chondrilloides* (Ard.) Karst. - *Ann. Naturhist. Mus. Wien* **69**: 95-100.

TÜMEN, G. & AL. 2000: Two new records for Turkey: *Satureja icarica* P.H. Davis, *Satureja pilosa* Velen. - *Türk. Bot. Derg.* **24(3)**: 211-214.

TUNTAS, B. 1905: I chloris tis Kithnou [In Greek]. - *Geôgraf. Delt.* **2-3**: 179-185.

TÜRDÜ, P., ZEYBEK, N. & ÖZTÜRK, M.A. 1984: Studies on the morphology, anatomy and ecology of some *Bromus* L. species. - *Ege Üniv. Fen Fak. Derg., B* **7(1)**: 29-38.

TURECEK, G., TITZ, E. & TITZ, W. 1980: Eine neue diploide *Valeriana officinalis*-Sippe im illyrischen Raum. - *Stud. Geobot.* **1(1)**: 157-162.

TURKINGTON, R. & AARSSEN, L.W. 1983: Biological Flora of the British Isles. *Hypochoeris radicata* L. (*Achyrophorus radicatus* (L.) Scop.). - *J. Ecol.* **71(3)**: 999-1022.

TURLAND, N. & BIENVENU, J. 2021: *Sedum dasyphyllum* subsp. *dasyphyllum*. In Raus, Th. & Raab-Straube, E.v. (eds): Euro+Med-Checklist notulae, 14. - *Willdenowia* **51(3)**: 258-359.

TURLAND, N. & CHILTON, L. 2008: Flora of Crete: Supplement II, Additions 1997-2008 [continuously updated]. - Internet Publ.: 66 pp. (http://www.marengowalks.com/).

TURLAND, N.J. 1988: Mountains and gorges in Western Crete. - *Quart. Bull. Alpine Gard. Soc. Gr. Brit.* **56(1)**: 46-52, 57.

TURLAND, N.J. 1990a: Plant portraits: *Arum idaeum*. - *Rock Gard.* **22(1)**: 51-52 + 1 plate.

TURLAND, N.J. 1990b: Cretan bellflowers and St. John's Worts. - *Quart. Bull. Alpine Gard. Soc. Gr. Brit.* **58(4)**: 310-320.

TURLAND, N.J. 1991a: Hardy alpine garden plants from Crete. - *Rock Gard.* **22(3)**: 303-317.

TURLAND, N.J. 1991b: A close look at a Cretan *Daphne*. - *Bot. Mag. (Kew Mag.)* **8(1)**: 37-45.

TURLAND, N.J. 1992a: Studies on the Cretan flora 1. Floristic notes. - *Bull. Brit. Mus. (Nat. Hist.), Bot.* **22(2)**: 159-164.

TURLAND, N.J. 1992b: Studies on the Cretan flora 2. The *Dianthus juniperinus* complex (Caryophyllaceae). - *Bull. Brit. Mus. (Nat. Hist.), Bot.* **22(2)**: 165-169.

TURLAND, N.J. 1992c: Floristic notes from Crete. - *Bot. J. Linn. Soc.* **108(4)**: 345-357.

TURLAND, N.J. 1993: Cretan cliff-dwellers. - *Quart. Bull. Alpine Gard. Soc. Gr. Brit.* **61(4)**: 381-390.

TURLAND, N.J. 1994: Another *Daphne* from Crete. - *Bot. Mag. (Kew Mag.)* **11**: 117-128 + 1 plate.

TURLAND, N.J. 1995a: (1163) Proposal to reject the name *Cenchrus frutescens* L. (Gramineae). - *Taxon* **44(3)**: 419-420. [not approved ; cf. *Taxon* 47(4): 863, 1998].

TURLAND, N.J. 1995b: Neotypification of *Acer orientale* (Aceraceae). - *Taxon* **44(4)**: 597-600.

TURLAND, N.J. 1995c: (1204) Proposal to conserve the name *Centaurea pumilio* L. (Compositae) with a conserved type. - *Taxon* **44(4)**: 643-644. [approved; cf. *Taxon* 47(4): 867, 1998].

TURLAND, N.J. 1995d: Linnaeus's interpretation of Prospero Alpino's *De plantis exoticis*, with special emphasis on the flora of Crete. - *Bull. Nat. Hist. Mus. London, Bot.* **25(2)**: 127-159.

TURLAND, N.J. 1996a: (1224) Proposal to reject the name *Atriplex hastata* L. (Chenopodiaceae). - *Taxon* **45(2)**: 325-326. [approved, cf. *Taxon* 47(4): 870, 1998].

TURLAND, N.J. 1996b: (1227) Proposal to conserve the name *Hedysarum cornutum* L. (*Onobrychis cornuta* Desv.; Leguminosae) against *Hedysarum spinosum* L. - *Taxon* **45(2)**: 331-332. [approved, cf. *Taxon* 47(4): 871, 1998].

TURLAND, N.J. 1996c: (1232) Proposal to reject the name *Statice echinus* L. (Plumbaginaceae). - *Taxon* **45(2)**: 339-340. [approved, cf. *Taxon* 47(4): 872, 1998].

TURLAND, N.J. 1996d: (1251) Proposal to reject the name *Scrophularia aquatica* (Scrophulariaceae). - *Taxon* - **45(3)**: 561-562. [approved, cf. Taxon 48(2): 362, 1999].

TURLAND, N.J. 1996e: (1253) [sphalm. 1553] Proposal to reject the name *Rosa eglanteria* (Rosaceae). - *Taxon* **45(3)**: 565-566. [approved, cf. Taxon 55: 1094, 2005].

TURLAND, N.J. 1996f: (1270) Proposal to reject the name *Malva rotundifolia* (Malvaceae). - *Taxon* **45(4)**: 707-708. [approved, cf. Taxon 48(2): 365, 1999].

TURLAND, N.J. 1997a: (1278) Proposal to reject the name *Quercus aegilops* (Fagaceae). - *Taxon* **46(1)**: 117-118. [approved, cf. Taxon 48(2): 366, 1999].

TURLAND, N.J. 1997b: (1279) Proposal to reject the name *Astragalus capitatus* (Leguminosae). - *Taxon* **46(1)**: 119-120. [approved, cf. Taxon 48(2): 367, 1999].

TURLAND, N.J. 1997c: (1315) Proposal to reject the name *Dianthus arboreus* (Caryophyllaceae). - *Taxon* **46(4)**: 779-783. [approved; cf. *Taxon* 49(2): 262, 2000].

TURLAND, N.J. 2004: (1647) Proposal to conserve the name *Chrysanthemum coronarium* (Compositae) with a conserved type. - *Taxon* **53(4)**: 1072-1074. [approved, cf. Taxon 55(4): 1101-1102, 2005].

TURLAND, N.J. 2005: (1677) Proposal to conserve the name *Galactites tomentosa* against *Centaurea elegans* (Compositae). - *Taxon* **54(1)**: 202-203. [approved, cf. Taxon 56(2): 590, 2007].

TURLAND, N.J. 2006: Lectotypification of *Campanula saxatilis*, *Phyteuma pinnatum* and *Verbascum arcturus*, Linnaean names of three endemic taxa to Crete. - *Willdenowia* **36(1)**: 303-309.

TURLAND, N.J. 2008: *Anthemis samariensis* (Asteraceae, Anthemideae), a new species from the mountains of W Kriti (Greece). - *Willdenowia* **38(1)**: 61-69.

TURLAND, N.J. 2009: *Linum tenuifolium* L. In Greuter, W. & Raus, Th. (eds): Med-Checklist notulae, 28. - *Willdenowia* **39(2)**: 339.

TURLAND, N.J. 2010: *Ranunculus fontanus* C. Presl and *Tamarix hampeana* Boiss. & Heldr. In Greuter, W. & Raus, Th. (eds): Med-Checklist notulae, 29. - *Willdenowia* **40(2)**: 198-199.

TURLAND, N.J. 2013a: *Jacobaea maritima* (L.) Pelser & Meijden subsp. *maritima*. In Raab-Straube, E.v. & Raus, Th. (eds): Euro+Med-Checklist notulae, 2. - *Willdenowia* **43(2)**: 241.

TURLAND, N.J. 2013b: *Fumaria bastardii* Boreau. In Raab-Straube, E. v. & Raus, Th. (eds): Euro+Med-Checklist notulae, 2. - *Willdenowia* **43(2)**: 243.

TURLAND, N.J. & AL. 2004: Weeds of the traditional agriculture of Crete. - *Willdenowia* **34(2)**: 381-406.

TURLAND, N.J. & BERGMEIER, E. 2013: *Lobularia arabica* (Boiss.) Muschl. In Raab-Straube, E. v. & Raus, Th. (eds): Euro+Med-Checklist notulae, 2. - *Willdenowia* **43(2)**: 242.

TURLAND, N.J. & CAFFERTY, S. 1999: (1426) Proposal to conserve the name *Peucedanum nodosum* (Umbelliferae) with a conserved type. - *Taxon* **48(3)**: 595-596. [approved, cf. *Taxon* 50(2): 563-564, 2001].

TURLAND, N.J. & CHILTON, L. 2000: A revision of *Centaurea argentea* (Compositae, Cardueae), an endemic species of Kriti and Kithira (Greece). - *Bot. Chron. (Patras)* **13**: 71-79.

TURLAND, N.J. & CHILTON, L. 2005: *Androcymbium rechingeri* and *Ranunculus cupreus* were reported in error for Karpathos (Greece). - *Bot. Chron. (Patras)* **18(2)**: 5-10.

TURLAND, N.J., CHILTON, L. & PRESS, J.R. 1993: Flora of the Cretan area. Annotated checklist & atlas. - London: H.M.S.O., xii + 439 pp.

TURLAND, N.J. & FOLEY, M.J.Y. 2004: (1650) Proposal to conserve the name *Orobanche sanguinea* (Orobanchaceae) with a conserved type. - *Taxon* **53(4)**: 1077-1078. [withdrawn, cf. *Taxon* 55(4): 1102, 2005].

TURLAND, N.J. & JARVIS, C.E. (EDS) 1997: Typification of Linnaean specific and varietal names in the Leguminosae (Fabaceae). - *Taxon* **46(3)**: 457-485.

TURLAND, N.J., KIRSCHNER, J. & ŠTĚPÁNEK, J. 1996: (1244-1246) Proposals to reject the names *Trifolium agrarium*, *T. filiforme* and *T. procumbens* (Leguminosae). - *Taxon* **45(3)**: 549-551. [all approved, cf. *Taxon* 48(2): 360-361, 1999].

TURLAND, N.J. & KYRIAKOPOULOS, CH.K. 2004: *Silene sartorii* Boiss. & Heldr., *Andryala integrifolia* L. In Greuter, W. & Raus, Th. (eds): Med-Checklist notulae, 22. - *Willdenowia* **34(1)**: 73.

TURLAND, N.J. & RUMSEY, F. 1997: (1318-1319) Proposals to reject the names *Orobanche major* and *O. laevis* (Orobanchaceae). - *Taxon* **46(4)**: 787-791. [both approved, cf. *Taxon* 49(2): 263-264, 2000].

TURLAND, N.J. & WYSE JACKSON, M. 1997: (1313-1314) Proposals to reject the names *Cerastium viscosum* and *C. vulgatum* (Caryophyllaceae). - *Taxon* **46(4)**: 775-778. [both approved, cf. *Taxon* 49(2): 262, 2000].

TURMEL, J.-M. 1948: Répartition géographique des *Eryngium*. I. Ancien monde. - *Bull. Mus. Natl. Hist. Nat., Sér. 2* **20(4)**: 395-401.

TURMEL, J.-M. 1954: Écologie descriptive et expérimentale du genre *Eryngium*. - *Bull. Mus. Natl. Hist. Nat., Sér. 2* **26(1)**: 139-148.

TURNER, B.L. & POWELL, A.M. 1977: Helenieae - systematic review. In Heywood, V.H., Harborne, J.B. & Turner, B.L. (eds): The biology and chemistry of the Compositae 2: 699-737. - London, etc.: Academic Press.

TURNER, I.M. 2013: Notes on the names of two species of *Campanula* (Campanulaceae) validated by R. H. Beddome. - *Edinburgh J. Bot.* **70(1)**: 1-5.

TURNER, R. 1995: Euphorbias. A gardener's guide. - London: B.T. Batsford, Ltd, 192 pp. + 16 pp. of colour plates.

TURRILL, W.B. 1918: Contributions to the flora of Macedonia: I. - *Bull. Misc. Inform. Kew* **1918(8-9)**: 249-341.

TURRILL, W.B. 1919: Contributions to the Flora of Macedonia: II. - *Bull. Misc. Inform. Kew* **1919(3)**: 105-108.

TURRILL, W.B. 1920a: The rev. H.F. Tozer and plants collected by him in the Nearer East. - *Bull. Misc. Inform. Kew* **1920(1)**: 29-31.

TURRILL, W.B. 1920b: The genus *Rosmarinus*. - *Bull. Misc. Inform. Kew* **1920(3)**: 105-108.

TURRILL, W.B. 1920c: Contributions to the flora of Macedonia: III. - *Bull. Misc. Inform. Kew* **1920(6)**: 177-196.

TURRILL, W.B. 1920d: *Arceuthobium oxycedri* and its distribution. - *Bull. Misc. Inform. Kew* **1920(8)**: 264-268.

TURRILL, W.B. 1920e: *Fritillaria pontica*. - *Bot. Mag.* **146**: tab. 8865 + 2 pp. text.

TURRILL, W.B. 1922: A contribution to the flora of the Nearer East. - *Bull. Misc. Inform. Kew* **1922(9)**: 291-298.

TURRILL, W.B. 1923: *Convolvulus nitidus* Boiss. from the Balkan Peninsula. - *New Phytol.* **22(2)**: 95-96.

TURRILL, W.B. 1924: On the flora of the Gallipoli peninsula. - *Bull. Misc. Inform. Kew* **1924(7-10)**: 287-299, 305-331, 337-363, 369-383.

TURRILL, W.B. 1925: Notes on the flora of the Balkan Peninsula. - *Bull. Misc. Inform. Kew* **1925(1)**: 34-35.

TURRILL, W.B. 1926a: On the flora of the Nearer East. - *Bull. Misc. Inform. Kew* **1926(2)**: 101-106.

TURRILL, W.B. 1926b: Sibthorp in the Gallipoli Peninsula. - *Bull. Misc. Inform. Kew* **1926(2)**: 111.

TURRILL, W.B. 1926c: On the flora of the Nearer East: II. - *Bull. Misc. Inform. Kew* **1926(9)**: 375-380.

TURRILL, W.B. 1927a: *Viola delphinantha*, Boiss. - *Hooker's Icon. Pl.* **32(1)**: tab. 3117 + 2 pp. text.

TURRILL, W.B. 1927b: *Veronica rigida*, Turrill. - *Hooker's Icon. Pl.* **32(1)**: tab. 3118 + 1 p. text.

TURRILL, W.B. 1928: A preliminary account of the Bulgarian species of *Veronica*, section *Pseudolysimachium* Koch. - *Izv. Bulg. Bot. Druzh.* **2**: 19-24.

TURRILL, W.B. 1929: The plant-life of the Balkan Peninsula. A phytogeographical study. - Oxford: Clarendon Press, xxiii + 490 pp. + 1 map.

TURRILL, W.B. 1930a: On the flora of the Nearer East: V. - *Bull. Misc. Inform. Kew* **1930(3)**: 122-126.

TURRILL, W.B. 1930b: *Ranunculus psilostachys*. - *Bot. Mag.* **153(3)**: tab. 9188 + 3 pp. text.

TURRILL, W.B. 1930c: Influence of the Suez Canal on the marine phanerogamic flora of the eastern Mediterranean. - *J. Bot. (London)* **68**: 238-240.

TURRILL, W.B. 1930d: *Cerinthe minor*, L., var. *hispida*, Turrill. - *Hooker's Icon. Pl.* **32(2)**: tab. 3133 + 3 pp. text.

TURRILL, W.B. 1930e: *Veronica euxina*, Turrill. - *Hooker's Icon. Pl.* **32(2)**: tab. 3135 + 2 pp. text.

TURRILL, W.B. 1930f: *Veronica orchidea*, Crantz. - *Hooker's Icon. Pl.* **32(2)**: tab. 3136 + 2 pp. text.

TURRILL, W.B. 1930g: *Fagus orientalis*, Lipsky. - *Hooker's Icon. Pl.* **32(2)**: tab. 3137 + 3 pp. text.

TURRILL, W.B. 1931a: *Rorippa pyrenaica* and closely related species. - *Izv. Bulg. Bot. Druzh.* **4**: 48-56.

TURRILL, W.B. 1931b: On the flora of the Nearer East: X. A contribution to the flora of South Macedonia. - *Bull. Misc. Inform. Kew* **1931(9)**: 453-454.

TURRILL, W.B. 1932a: On the flora of the Nearer East: XI. A contribution to the flora of Albania. - *Bull. Misc. Inform. Kew* **1932(4)**: 193-198.

TURRILL, W.B. 1932b: On the flora of the Nearer East: XII. Dr Giuseppi's 1931 collection from Euboea and other parts of Greece. - *Bull. Misc. Inform. Kew* **1932(5)**: 248-250.

TURRILL, W.B. 1932c: On the flora of the Nearer East: XIII. Miscellaneous new records and extensions of known distributions. - *Bull. Misc. Inform. Kew* **1932(9)**: 450-454.

TURRILL, W.B. 1932d: A tentative list of species and synonyms of *Fritillaria*. - *Lily Year-Book* **1932**: 78-83.

TURRILL, W.B. 1932e: *Delphinium macedonicum*, Halácsy et Charrel. - *Hooker's Icon. Pl.* **32(3)**: tab. 3152 + 3 pp. text.

TURRILL, W.B. 1932f: *Tamarix hampeana*, Boiss. et Held., var. *aegaea*, Turrill. - *Hooker's Icon. Pl.* **32(3)**: tab. 3153 + 3 pp. text.

TURRILL, W.B. 1932g: *Cotyledon lassithiensis*, Hayek. - *Hooker's Icon. Pl.* **32(3)**: tab. 3155 + 2 pp. text.

TURRILL, W.B. 1932h: *Onopordon tauricum*, Willd. - *Hooker's Icon. Pl.* **32(3)**: tab. 3156 + 2 pp. text.

TURRILL, W.B. 1932i: *Sideritis scardica*, Griseb. - *Hooker's Icon. Pl.* **32(3)**: tab. 3157 + 2 pp. text.

TURRILL, W.B. 1933a: A study of variation in *Glaucium flavum*. - *Bull. Misc. Inform. Kew* **1933(4)**: 174-184.

TURRILL, W.B. 1933b: *Rhamnus rhodopeus*, Vel. - *Hooker's Icon. Pl.* **32(4)**: tab. 3177 + 2 pp. text.

TURRILL, W.B. 1933c: *Rhamnus rupestris*, Scop., var. *rumeliacus*, Hayek. - *Hooker's Icon. Pl.* **32(4)**: tab. 3178 + 2 pp. text.

TURRILL, W.B. 1933d: *Astragalus sericophyllus* Griseb. - *Hooker's Icon. Pl.* **33(1)**: tab. 3202 + 2 pp. text.

TURRILL, W.B. 1934a: The occurrence of *Fraxinus Pallisae* in Western Thrace. - *Bull. Misc. Inform. Kew* **1934(1)**: 44.

TURRILL, W.B. 1934b: The correlation of morphological variation with distribution in some species of *Ajuga*. - *New Phytol.* **33(3)**: 218-230.

TURRILL, W.B. 1934c: *Silene asterias* Griseb. - *Hooker's Icon. Pl.* **33(2)**: tab. 3226 + 3 pp. text.

TURRILL, W.B. 1934d: *Iris cretensis*. - *Bot. Mag.* **157(3)**: tab. 9369 + 4 pp. text.

TURRILL, W.B. 1934e: *Ranunculus asiaticus* var. *albiflorus*. - *Bot. Mag.* **157(4)**: tab. 9380 + 4 pp. text.

TURRILL, W.B. 1934f: A new species of *Aubrieta*. *Aubrieta glabrescens*, Turrill. - *Gard. Chron., ser. 3* **95**: 384.

TURRILL, W.B. 1934g: *Silene ventricosa* Adamović. - *Hooker's Icon. Pl.* **33(2)**: tab. 3227 + 2 pp. text.

TURRILL, W.B. 1934h: *Lychnis subintegra* (Hayek) Turrill. - *Hooker's Icon. Pl.* **33(2)**: tab. 3228 + 3 pp. text.

TURRILL, W.B. 1934i: *Stachys serbica* Panč. - *Hooker's Icon. Pl.* **33(2)**: tab. 3229 + 2 pp. text.

TURRILL, W.B. 1935a: On the flora of the Nearer East: XVI. Miscellaneous new records. - *Bull. Misc. Inform. Kew* **1935(1)**: 54-57.

TURRILL, W.B. 1935b: *Hypericum thasium* Griseb. - *Hooker's Icon. Pl.* **33(3)**: tab. 3251 + 3 pp. text.

TURRILL, W.B. 1935c: *Astragalus glycyphylloides* DC. - *Hooker's Icon. Pl.* **33(3)**: tab. 3252 + 2 pp. text.

TURRILL, W.B. 1935d: [Description of *Camphorosma nestensis* sp. nova]. - *Bull. Misc. Inform. Kew* **1935(3)**: 142-143.

TURRILL, W.B. 1935e: *Verbascum gracile* Turrill. - *Hooker's Icon. Pl.* **33(3)**: tab. 3253 + 3 pp. text.

TURRILL, W.B. 1935f: *Polygonum icaricum* K. H. Rechinger. - *Hooker's Icon. Pl.* **33(3)**: tab. 3254 + 3 pp. text.

TURRILL, W.B. 1935g: *Geranium aristatum* Freyn & Sint. - *Hooker's Icon. Pl.* **33(4)**: tab. 3276 + 2 pp. text.

TURRILL, W.B. 1935h: *Cephalorrhynchus glandulosus* Boiss. - *Hooker's Icon. Pl.* **33(4)**: tab. 3277 + 3 pp. text.

TURRILL, W.B. 1935i: *Solenanthus scardicus* Bornm. - *Hooker's Icon. Pl.* **33(4)**: tab. 3278 + 4 pp. text.

TURRILL, W.B. 1935j: *Allium bornmülleri* Hayek. - *Hooker's Icon. Pl.* **33(4)**: tab. 3279 + 2 pp. text.

TURRILL, W.B. 1936a: On the flora of the Near East: XVII. New plants and new records. - *Bull. Misc. Inform. Kew* **1936(1)**: 100-103.

TURRILL, W.B. 1936b: Notes on the genus *Sempervivum*. - *Izv. Bulg. Bot. Druzh.* **7**: 124-127.

TURRILL, W.B. 1936c: *Lotus aegaeus*. - *Bot. Mag.* **159(1)**: tab. 9431 + 4 pp. text.

TURRILL, W.B. 1936d: *Delphinium consolida* subsp. *paniculatum*. - *Bot. Mag.* **159(1)**: tab. 9435 + 4 pp. text.

TURRILL, W.B. 1936e: *Scrophularia laciniata* Waldst. & Kit. var. *macedonica* Stoyanoff. - *Hooker's Icon. Pl.* **34(1)**: tab. 3304 + 2 pp. text.

TURRILL, W.B. 1937a: On the flora of the Nearer East: XVIII. New species, new records, and notes. - *Bull. Misc. Inform. Kew* **1937(2)**: 79-86.

TURRILL, W.B. 1937b: Plants used in the Balkan Peninsula in place of tea. - *Proc. Linn. Soc. London* **149(2)**: 67-70.

TURRILL, W.B. 1937c: A contribution to the botany of Athos peninsula. - *Bull. Misc. Inform. Kew* **1937 (4)**: 197-273.

TURRILL, W.B. 1937d: The genus *Fritillaria* in the Balkan Peninsula and Asia Minor. - *J. Roy. Hort. Soc.* **62(8)**: 329-335 + 4 plates.

TURRILL, W.B. 1937e: *Galanthus ikariae*. - *Bot. Mag.* **160(1)**: tab. 9474 + 2 pp. text.

TURRILL, W.B. 1937f: *Viscaria atropurpurea*. - *Bot. Mag.* **160(2)**: tab. 9489 + 4 pp. text.

TURRILL, W.B. 1937g: A new *Sempervivum* from the Balkan Peninsula. - *Gard. Chron., ser. 3* **102**: 302-304.

TURRILL, W.B. 1938a: *Biarum davisii*. - *Gard. Chron., ser. 3* **104**: 437.

TURRILL, W.B. 1938b: *Dianthus leucophoeniceus* Dörfl. et Hayek. - *Hooker's Icon. Pl.* **34(3)**: tab. 3351 + 2 pp. text.

TURRILL, W.B. 1938c: *Cistus ×Skanbergii*. - *Bot. Mag.* **161(1)**: tab. 9514 + 2 pp. text.

TURRILL, W.B. 1938d: *Viola allchariensis* var. *Herzogii*. - *Bot. Mag.* **161(3)**: tab. 9537 + 3 pp. text.

TURRILL, W.B. 1938e: *Silene fabarioides* Hausskn. - *Hooker's Icon. Pl.* **34(3)**: tab. 3353 + 2 pp. text.

TURRILL, W.B. 1938f: *Silene schwarzenbergeri* Halácsy. - *Hooker's Icon. Pl.* **34(3)**: tab. 3354 + 2 pp. text.

TURRILL, W.B. 1940a: *Viola Stojanowii* var. *latifolia*. - *Bot. Mag.* **162(4)**: tab. 9599 + 2 pp. text.

TURRILL, W.B. 1940b: On the flora of the Nearer East: XXII. New records and new species from Greece and the Greek islands. - *Bull. Misc. Inform. Kew* **1940(6)**: 262-266.

TURRILL, W.B. 1941: *Fritillaria Drenovskii*. - *Bot. Mag.* **163(3)**: tab. 9625 + 2 pp. text.

TURRILL, W.B. 1942: Descriptions of some Greek fritillaries. - *Gard. Chron., ser. 3* **111**: 188, 228, 239-240, 248.

TURRILL, W.B. 1943a: *Colchicum catacuzenium*. - *Bot. Mag.* **164(1)**: tab. 9652 + 2 pp. text.

TURRILL, W.B. 1943b: *Styrax officinalis*. - *Bot. Mag.* **164(1)**: tab. 9653 + 4 pp. text.

TURRILL, W.B. 1943c: *Fritillaria tuntasiana* Heldr. ex. Hal. - *Hookers's Icon. Pl.* **35(2)**: tab. 3426 + 1 p. text.

TURRILL, W.B. 1944: *Fritillaria messanensis*. - *Bot. Mag.* **164(2)**: tab. 9659 + 2 pp. text.

TURRILL, W.B. 1948a: *Fritillaria pontica* var. *ionica*. - *Bot. Mag.* **165(1)**: tab. 8 + 3 pp. text.

TURRILL, W.B. 1948b: *Dianthus haematocalyx* var. *alpinus*. - *Bot. Mag.* **165(1)**: tab. 9 + 6 pp. text.

TURRILL, W.B. 1949a: Winter flora on the Athos peninsula. - *Kew Bull.* **1948(3)**: 328.

TURRILL, W.B. 1949b: *Laurentia* in the Balkan Peninsula. - *Kew Bull.* **1949(1)**: 22.

TURRILL, W.B. 1949c: Additional notes to the flora and vegetation of the Athos peninsula, Greece. - *Kew Bull.* **1949(2)**: 155-156.

TURRILL, W.B. 1949d: *Lysimachia atropurpurea*. - *Bot. Mag.* **166(3)**: tab. 77 + 4 pp. text.

TURRILL, W.B. 1950a: *Arum creticum*. - *Bot. Mag.* **167(2)**: tab. 101 + 4 pp. text.

TURRILL, W.B. 1950b: *Prunus webbii*. - *Bot. Mag.* **167(3)**: tab. 118 + 4 pp. text.

TURRILL, W.B. 1950c: *Muscari macrocarpum*. - *Bot. Mag.* **167(4)**: tab. 124 + 4 pp. text.

TURRILL, W.B. 1950d: *Fritillaria sibthorpiana*. - *Bot. Mag.* **167(4)**: tab. 129 + 3 pp. text.

TURRILL, W.B. 1951a: *Saxifraga macedonica*, *Saxifraga sancta*. - *Bot. Mag.* **168(1)**: tab. 137 + 4 pp. text.

TURRILL, W.B. 1951b: *Ramonda nathaliae*. - *Bot. Mag.* **168(2)**: tab. 152 + 4 pp. text.

TURRILL, W.B. 1952a: *Campanula reiseri*. - *Bot. Mag.* **169(1)**: tab. 178 + 3 pp. text.

TURRILL, W.B. 1952b: *Gentiana asclepiadea*. - *Bot. Mag.* **169(2)**: tab. 191 + 5 pp. text.

TURRILL, W.B. 1952c: Wild and cultivated olives. - *Kew Bull.* **1951(3)**: 437-442 + 4 plates.

TURRILL, W.B. 1953: *Geum coccineum*. - *Bot. Mag.* **169(4)**: tab. 212 + 4 pp. text.

TURRILL, W.B. 1954: The lilies of the Balkan Peninsula. - *Lily Year-Book* **1954**: 30-39.

TURRILL, W.B. 1955a: Two new species of monocotyledones from Greece. - *Kew Bull.* **1955(1)**: 59-61.

TURRILL, W.B. 1955b: Some new plants from Greece. - *Kew Bull.* **1955(3)**: 353-357.

TURRILL, W.B. 1956a: *Daphne mezereum*. - *Bot. Mag.* **171(1)**: tab. 272 + 4 pp. text.

TURRILL, W.B. 1956b: *Thymus pastoralis* Turrill. - *Hooker's Icon. Pl.* **36(3)**: tab. 3558 + 2 pp. text.

TURRILL, W.B. 1957: *Genista lydia*. - *Bot. Mag.* **171(3)**: tab. 292 + 3 pp. text.

TURRILL, W.B. 1958a: *Serratula* in Crete. - *Kew Bull.* **1957(3)**: 391-392.

TURRILL, W.B. 1958b: *Silene integripetala*. - *Bot. Mag.* **172(2)**: tab. 328 + 3 pp. text.

TURRILL, W.B. 1959a: *Genista sagittalis*. - *Bot. Mag.* **172(3)**: tab. 332 + 3 pp. text.

TURRILL, W.B. 1959a: *Anthyllis montana*. - *Bot. Mag.* **172(3)**: tab. 333 + 3 pp. text.

TURRILL, W.B. 1959c: *Crocus sieberi*. - *Bot. Mag.* **172(3)**: tab. 340 + 3 pp. text.

TURRILL, W.B. 1960a: A new species of *Aristolochia* from the island of Samos. - *Kew Bull.* **14(1)**: 108-109.

TURRILL, W.B. 1960b: *Crocus goulimyi*. - *Bot. Mag.* **173(1)**: tab. 354 + 2 pp. text.

TURRILL, W.B. 1962a: *Vitex agnus-castus*. - *Bot. Mag.* **174(2)**: tab. 400 + 4 pp. text.

TURRILL, W.B. 1962b: *Bupleurum fruticosum*. - *Bot. Mag.* **174(2)**: tab. 408 + 3 pp. text.

TURRILL, W.B. 1962c: A lily from the Rhodope mountains, Bulgaria. - *Lily Year-Book* **25**: 31-32.

TURRILL, W.B. 1964: *Cyclamen creticum*. - *Bot. Mag.* **175(1)**: tab. 450 + 4 pp. text.

TURRILL, W.B. & SEALY, J.R. 1980: Studies in the genus *Fritillaria* (Liliaceae). Illustrated by Stella Ross-Craig. - *Hooker's Icones Pl.* **39(1-2)**: 1-280 incl. tab. 3801-3850.

TURRISI, R.E. 2001: *Ipomoea imperati* (Vahl) Griseb. (Convolvulaceae), nouvelle entité pour la flore sicilienne. - *Fl. Medit.* **11**: 374-375.

TUTEL, B. 1970: *Asphodelus fistulosus* L. (Çirişotu) [Liliaceae] [In Turkish]. - *Türk Biyol. Derg.* **20**: 125-127.

TUTEL, B. 1980: Studia ad floram turcicam: XIII. The genus *Plantago* in European Turkey. - *Istanbul Üniv. Fen Fak. Mecm., B* **45**: 99-108.

TUTEL, B. 1983: Comparison of the taxonomy and leaf anatomy of the genus *Biebersteinia* with the other genera of Geraniaceae in Turkey. - *Istanbul Üniv. Fen Fak. Mecm., B* **47-48**: 51-87.

TUTEL, B. & DELICE, I. 1978: On the distribution of *Juncus gerardi* Lois. ssp. *libanoticus* (Thiéb.) Snog. in European Turkey. - *Biyol. Derg.* **28(1-4)**: 89-91. [Also distributed as *OPTIMA leaflet* no. 94].

TUTIN, T.G. 1942: Biological Flora of the British Isles. *Zostera* L. - *J. Ecol.* **30(1)**: 217-226.

TUTIN, T.G. 1957a: A contribution to the experimental taxonomy of *Poa annua* L. - *Watsonia* **4(1)**: 1-10.

TUTIN, T.G. 1957b: Biological Flora of the British Isles. *Allium ursinum* L. - *J. Ecol.* **45(3)**: 1003-1010.

TUTIN, T.G. 1963: New taxa and new names in European *Dianthus*. In Heywood, V.H. (ed.): Notulae systematicae ad Floram Europaeam spectantes, no. 2. - *Feddes Repert. Spec. Nov. Regni Veg.* **68 (3)**: 189-193.

TUTIN, T.G. 1964: New taxa and new names in European Ranunculaceae. In Heywood, V.H. (ed.): Notulae systematicae ad Floram Europaeam spectantes, no. 3. - *Feddes Repert. Spec. Nov. Regni Veg.* **69(1)**: 53-55.

TUTIN, T.G. 1967: New combinations and new names in Umbelliferae. In Heywood, V.H. (ed.): Notulae systematicae ad Floram Europaeam spectantes, no. 6. - *Feddes Repert.* **74(1-2)**: 31-34.

TUTIN, T.G. 1968a: Notes on the genus *Athamantha* (Umbelliferae) in Europe. In Heywood, V.H. (ed.): Notulae systematicae ad Floram Europaeam spectantes, no. 7. - *Feddes Repert.* **79(1-2)**: 18-20.

TUTIN, T.G. 1968b: [Notes on *Medicago sativa*]. In Heywood, V.H. (ed.): Notulae systematicae ad Floram Europaeam spectantes, no. 7. - *Feddes Repert.* **79(1-2)**: 53.

TUTIN, T.G. 1968c: [New combinations in *Pimpinella* and *Peucedanum*]. In Heywood, V.H. (ed.): Notulae systematicae ad Floram Europaeam spectantes, no. 7. - *Feddes Repert.* **79(1-2)**: 62.

TUTIN, T.G. 1971: [Notes on *Gentiana*]. In Heywood, V.H. (ed.): Notulae systematicae ad Floram Europaeam spectantes, no. 11. - *Bot. J. Linn. Soc.* **64(4)**: 378.

TUTIN, T.G. 1973a: Notes on *Jasione* L. in Europe. In Heywood, V.H. (ed.): Notulae systematicae ad Floram Europaeam spectantes, no. 14. - *Bot. J. Linn. Soc.* **67(3)**: 276-279.

TUTIN, T.G. 1973b: [Notes on the *Inula candida* group]. In Heywood, V.H. (ed.): Notulae systematicae ad Floram Europaeam spectantes, no. 14. - *Bot. J. Linn. Soc.* **67(3)**: 282-283.

TUTIN, T.G. 1975: [Note on *Jasione*]. In Heywood, V.H. (ed.): Notulae systematicae ad Floram Europaeam spectantes, no. 16. - *Bot. J. Linn. Soc.* **70(1)**: 17-18.

TUTIN, T.G. (ED.) 1993: Flora Europaea. Volume 1: Psilotaceae to Platanaceae. Second edition. - Cambridge: Cambridge Univ. Press, xlvi + 581 pp. + 5 maps.

TUTIN, T.G. & AL. (EDS) 1964: Flora Europaea. Volume 1: Lycopodiaceae to Platanaceae. - Cambridge: Cambridge Univ. Press, xxxv + 464 pp. + 5 maps.

TUTIN, T.G. & AL. (EDS) 1968: Flora Europaea. Volume 2: Rosaceae to Umbelliferae. - Cambridge: Cambridge Univ. Press, xxviii + 455 pp. + 5 maps.

TUTIN, T.G. & AL. (EDS) 1972: Flora Europaea. Volume 3: Diapensiaceae to Myoporaceae. - Cambridge: Cambridge Univ. Press, xxxii + 370 pp. + 5 maps.

TUTIN, T.G. & AL. (EDS) 1976: Flora Europaea. Volume 4: Plantaginaceae to Compositae (and Rubiaceae). - Cambridge: Cambridge Univ. Press, xxxii + 505 pp. + 5 maps.

TUTIN, T.G. & AL. (EDS) 1980: Flora Europaea. Volume 5: Alismataceae to Orchidaceae (Monocotyledons). - Cambridge: Cambridge Univ. Press, xxxviii + 452 pp. + 5 maps.

TUZLACI, E. 1981: The Flora of Marmara Island. - *Istanbul Üniv. Eczac. Fak. Mecm.* **17**: 138-154.

TUZLACI, E. 1983: Eight new taxa of *Asphodeline* from Turkey. - *Notes Roy. Bot. Gard. Edinburgh* **41(1)**: 85-94.

TUZLACI, E. 1984: The distribution of *Helianthemum aegyptiacum* (L.) Miller in Turkey. - *Istanbul Üniv. Eczac. Fak. Mecm.* **20**: 97-104.

TUZLACI, E. 1986: The genus *Anthericum* (Liliaceae) in Turkey. - *Mar. Üniv. Eczac. Derg.* **2(2)**: 119-124.

TUZLACI, E. 1987: Revision of the genus *Asphodeline* (Liliaceae) - A new infrageneric classification. - *Candollea* **42(2)**: 559-576.

TUZLACI, E. 1998: Revision of the genus *Asphodeline* (Liliaceae). II Two new species from Turkey. - *Candollea* **53(2)**: 423-433.

TYREE, L. 2004: Anixi stin periochi tis Knossou [In Greek]. - *Fisis [Journal of the Hellenic Society for the Protection of Nature]* **105**: 7-8.

TZAKOU, O., BAZOS, I. & YANNITSAROS, A. 2009: Volatile constituents of *Phlomis lycia* (Labiatae). In Ivanova, D. (ed.): Proceedings of the IV Balkan botanical congress, pp. 656-658. - Sofia: Institute of Botany, BAS.

TZANOUDAKIS, D. 1977: A cytotaxonomic study of the genus *Paeonia* L. in Greece [In Greek with English summary]. - Patras: Ph.D. thesis, Univ. of Patras, 132 pp.

TZANOUDAKIS, D. 1981: Contribution to the study of Cyclades Flora: The Flora of the island of Yiaros. - *Bot. Chron. (Patras)* **1(2)**: 124-136.

TZANOUDAKIS, D. 1982: A cytotaxonomic study of two Greek mountainous species of *Allium* (sectio *Molium*) [In Greek with English abstract]. - *Bot. Chron. (Patras)* **2(2)**: 143-151.

TZANOUDAKIS, D. 1983a: New taxa of *Allium* from Greece. - *Candollea* **38(1)**: 317-323.

TZANOUDAKIS, D. 1983b: Karyotypes of ten taxa of *Allium* section *Scorodon* from Greece. - *Caryologia* **36(3)**: 259-284.

TZANOUDAKIS, D. 1983c: Karyotypes of four wild *Paeonia* species from Greece. - *Nordic J. Bot.* **3(3)**: 307-318.

TZANOUDAKIS, D. 1985: Kittarologiki meleti ton idon *Allium chamaemoly* L. ke *Allium ursinum* L. In: Elliniki Votaniki Eteria, Praktika 4ou epistimonikou simposiou, 23-24 martiou 1985, pp. 77-83 [In Greek]. - Thessaloniki.

TZANOUDAKIS, D. 1986a: Chromosome studies in the Greek flora. I. Karyotypes of some Aegean angiosperms. - *Bot. Helv.* **96(1)**: 27-36.

TZANOUDAKIS, D. 1986b: Chromosome studies in the Greek Flora. II. Karyotypes of four Aegean endemics of *Allium* sect. *Codonoprasum* (Liliaceae). - *Willdenowia* **16(1)**: 203-211.

TZANOUDAKIS, D. 1986c: Chromosome studies in some species of *Allium* sect. *Allium* in Greece. - *Ann. Mus. Goulandris* **7**: 233-247.

TZANOUDAKIS, D. 1986d: Karyotype variation in *Allium* sect. *Molium* G. Don from Greece. – *Caryologia* **39(1)**: 69-88.

TZANOUDAKIS, D. 1991: Karyotype variation and evolution in the Greek *Allium*. In Hanelt, P., Hammer, K. & Knüpffer, H. (eds): The genus *Allium* - taxonomic problems and genetic resources, pp. 305-320. - Gatersleben.

TZANOUDAKIS, D. 1992: Floristic and phytogeographical studies on the islands of the East Aegean [In Greek with English summary]. - Patras: Erevnitiko Programma G.G.E.T., iii + 120 pp.

TZANOUDAKIS, D. 1999: The genus *Allium* in Cyprus: a preliminary cytotaxonomical study. – *Bocconea* **11**: 105-115.

TZANOUDAKIS, D. 2000: *Allium aegilicum* (Alliaceae), a new autumn-flowering species from the island of Antikithira (Greece). - *Bot. Chron. (Patras)* **13**: 81-86.

TZANOUDAKIS, D. 2001a: The genus *Allium* in Greece: species and identification keys. In Kamari, Psaras & Constantinidis (eds): Proceedings of the 8th scientific congress of the Greek Botanical Society, October 5-8, 2000, pp. 405-418 [In Greek with English abstract].

TZANOUDAKIS, D. 2001b: The genus *Allium* in Greece: A genus representative of the floristic richness and diversity of the area. In Özhatay, N. (ed.): Proceedings of the Second Balkan Botanical Congress, 1: 353-358. - Istanbul.

TZANOUDAKIS, D. 2019: Exploring and registering the Greek Flora: Case studies and experience from the study of the genus *Allium*. In Daskalakou, E.N. & al. (eds): 16th Conference of the Hellenic Botanical Society – Athens 10-13 October 2019. Program and book of abstracts, p. 43.

TZANOUDAKIS, D. & AL. 1989: Cytogeographical studies in some Aegean Liliaceae. In Phitos, D. (ed.): Abstracts. Sixth OPTIMA meeting, Delphi 10-16 Sept. 1989, p. 145.

TZANOUDAKIS, D. & AL. 1991: Cytogeographical studies in some Aegean Liliaceae. - *Bot. Chron. (Patras)* **10**: 761-775.

TZANOUDAKIS, D. & AL. 1997: Contribution to the study of the Greek insular flora: Antikythera and the islets around Kythera. In: Abstracts from the First Balkan Botanical Congress, Thessaloniki, September 19-22, 1997, p. 53. - Thessaloniki.

TZANOUDAKIS, D. & AL. 1998: Contribution to the study of the Greek insular flora: Antikythera and the islets around Kythera. In Tsekos, I. & Moustakas, M. (eds): Progress in Botanical Reseearch, pp. 177-180. - Dordrecht: Kluwer Press.

TZANOUDAKIS, D. & AL. 2006: Floristic and phytosociological investigation of the island Antikythera and nearby islets (SW Aegean, Greece). - *Willdenowia* **36(1)**: 285-301.

TZANOUDAKIS, D. & ARABATZIS, TH. 1997: The genus *Paeonia* in Greece: A subspecies and a cytogeographical distribution pattern new to Greece. In: Abstracts from the First Balkan Botanical Congress, Thessaloniki, September 19-22, 1997, p. 53. - Thessaloniki.

TZANOUDAKIS, D. & IATROÚ, G. 1981: New combinations for two endemic taxa of the Greek flora [In Greek with English abstract]. - *Bot. Chron. (Patras)* **1(1)**: 22-28.

TZANOUDAKIS, D. & IATROÚ, G. 1982: A cytogeographical study of *Allium circinatum* Sieber [In Greek with English abstract]. - *Bot. Chron. (Patras)* **2(2)**: 152-158.

TZANOUDAKIS, D. & KIK, C. 2019: Collecting and regenerating the *Allium ampeloprasum* complex from Greece, its centre of biodiversity. In Daskalakou, E.N. & al. (eds): 16th Conference of the Hellenic Botanical Society – Athens 10-13 October 2019. Program and book of abstracts, p. 44.

TZANOUDAKIS, D. & KOLLMANN, F. 1991: *Allium chalkii* (Liliaceae), a new species from the eastern Aegean island of Chalki (Greece). - *Israel J. Bot.* **40(1)**: 61-64.

TZANOUDAKIS, D. & KYPRIOTAKIS, Z. 1987: Chromosome studies in the Greek flora III. Karyotypes of eight Aegean species. - *Bot. Helv.* **97(2)**: 229-237.

TZANOUDAKIS, D. & KYPRIOTAKIS, Z. 1993: *Allium platakisii*, a new species of the Greek insular flora. - *Fl. Medit.* **3**: 309-314.

TZANOUDAKIS, D. & KYPRIOTAKIS, Z. 1998: A new polyploid *Scilla* (Liliaceae) from the Cretan area (Greece). - *Folia Geobot.* **33**: 103-108.

TZANOUDAKIS, D. & KYPRIOTAKIS, Z. 2008: *Allium brussalisii* (Alliaceae), a new species from Greece. - *Bot. J. Linn. Soc.* **158(1)**: 140-146.

TZANOUDAKIS, D. & PANITSA, M. 1995: The flora of the Greek islands. - *Ecol. Medit.* **2(1-2)**: 195-212.

TZANOUDAKIS, D. & RIX, E.M. 2017: 858. *Paeonia parmassica*. Paeoniaceae. - *Curtis's Bot. Mag.* **34(1)**: 41-50.

TZANOUDAKIS, D. & TAN, KIT 1999: *Allium samothracicum* Tzanoudakis, Strid & Kit Tan, a new species from the North Aegean area, Greece. In Correia, A.I.D. & al. (eds): Livro de resumos. Guia de excursões. V jornadas de taxonomia botânica, 16-19/IX 1999, p. 65.

TZANOUDAKIS, D. & TAN, KIT 2001: *Allium samothracicum* Tzanoud., Strid & Kit Tan, a new species from the North Aegean area, Greece. - *Portug. Acta Biol.* **19(1-4)**: 355-360.

TZANOUDAKIS, D. & TRIGAS, P. 2015: *Allium occultum*, a new species of *A.* sect. *Codonoprasum* (Amaryllidaceae) from Skiros Island (W Aegean, Greece). - *Phytotaxa* **202(2)**: 135-142.

TZANOUDAKIS, D., TSAKIRI, M. & RAUS, TH. 2018: Contribution to the knowledge of the Greek *Allium*: What is *Allium achaium* Boiss. & Orph.? - *Bot. Serbica* **142 (suppl. 1)**: 88.

TZANOUDAKIS, D., TSAKIRI, M. & RAUS, TH. 2019: What is *Allium achaium* Boiss. & Orph.? Disentangling the taxonomy of a Greek mountain species. - *Willdenowia* **49(2)**: 231-239.

TZANOUDAKIS, D. & VOSA, C.G. 1988: The cytogeographical distribution pattern of *Allium* (Alliaceae) in the Greek peninsula and islands. - *Pl. Syst. Evol.* **159(3-4)**: 193-215.

TZONEV, R. 2005: *Sicyos angulatus* (Cucurbitaceae); a new adventive species for the flora of Bulgaria. - *Phytol. Balcan.* **11(1)**: 67-68.

TZONEV, R. 2007: *Eclipta prostrata* (Asteraceae): a new alien species for the Bulgarian flora. - *Phytol. Balcan.* **13(1)**: 79-80.

TZONEV, R. & PANOVSKA, H. 2017: Notes on the distribution of *Hyacinthella leucophaea* subsp. *atchleyi* in Bulgaria. - *Phytol. Balcan.* **23(3)**: 409-412.

TZVELEV, N.N. 1965: De genere *Calamagrostis* Adans. in URSS notulae systematicae [In Russian]. - *Novosti Sist. Vysš. Rast.* **2**: 5-50.

TZVELEV, N.N. 1971a: Notae de gramineis Florae URSS, 6 [In Russian]. - *Novosti Sist. Vysš. Rast.* **7**: 57-83.

TZVELEV, N.N. 1971b: On the taxonomy and phylogeny of genus *Festuca* L. of the U.S.S.R. flora I. The system of the genus and the main trends of evolution [In Russian with English summary]. - *Bot. Zhurn. (Moscow & Leningrad)* **56(9)**: 1252-1262.

TZVELEV, N.N. 1971c: Genus *Alopecurus* L. in USSR [In Russian; English translation of this volume by Amerind Publ. Co., New Delhi, 1978]. - *Novosti Sist. Vysš. Rast.* **8**: 12-22.

TZVELEV, N.N. 1974: De genere *Poa* L. in URSS [In Russian]. - *Novosti Sist. Vysš. Rast.* **11**: 24-41.

TZVELEV, N.N. 1980: Systema specierum generis *Vicia* L. in parte Europaea URSS [In Russian]. - *Novosti Sist. Vysš. Rast.* **17**: 200-208.

TZVELEV, N.N. 1983: Grasses of the Soviet Union. Parts 1 (pp. i-xvi, 1-568) and 2 (pp. i-xvi, 569-1196) [translated from the Russian original, Zlaki SSSR, Leningrad 1976, 788 pp.]. - New Delhi: Oxonian Press Pvt. Ltd.

UBERA, J.L. 1979: Números cromosómicos para la flora española. 104-109. - *Lagascalia* **9(1)**: 123-126.

UBERA, J.L. 1980: Números cromosómicos para la flora española. 132-136. - *Lagascalia* **9(2)**: 257-259.

UBERA, J.L. 1981: Números cromosómicos para la flora española. 214-219. - *Lagascalia* **10(2)**: 237-239.

UBERA, J.L. & RUÍZ DE CLAVIJO, E. 1984: Números cromosómicos para la flora española. 342-348. - *Lagascalia* **12(2)**: 294-296.

UBERA, J.L. & VALDÉS, B. 1983: Revisión del género *Nepeta* (Labiatae) en la Península Ibérica e Islas Baleares. - *Lagascalia* **12(1)**: 3-80.

UBRIZSY, G. & PÉNZES, A. 1960: Beiträge zur Kenntnis der Flora und der Vegetation Albaniens. - *Acta Bot. Acad. Sci. Hung.* **6(1-2)**: 155-167.

UDDLING, Å. 1929: Die Chromosomenzahlen von drei *Circaea*-Arten. - *Hereditas (Lund)* **12**: 294-296.

UHLICH, H. 1994: *Orobanche baumanniorum* var. *longebracteata* (Orobanchaceae) - eine neue Varietät aus Griechenland. - *Willdenowia* **24(1-2)**: 49-51.

UHLICH, H. 2014: Geschichte der Erforschung des Sommerwurzgewächse (Gattungen *Orobanche* und *Phelipanche*). - *Kochia* **8**: 27-61.

UHLICH, H. 2015: Geschichte der Erforschung des Sommerwurzgewächse (Gattungen *Orobanche* und *Phelipanche*). Teil 2: Von Linnés "Species Plantarum" bis Becks Monographie. - *Kochia* **9**: 59-92.

UHLICH, H., GUBLER, E. & RÄTZEL, S. 2020: *Orobanche baumanniorum*. In Raus, Th. & Raab-Straube, E.v. (eds): Euro+Med-Checklist notulae, 12. - *Willdenowia* **50(2)**: 320-322.

UHLICH, H., PUSCH, J. & BARTHEL, K.-J. 1995: Die Sommerwurzarten Europas. - Magdeburg: Westarp Wissenschaften, 235 pp.

UHLICH, H. & RÄTZEL, S. 2021: *Phelipanche gussoneana*. In Raus, Th. & Raab-Straube, E.v. (eds): Euro+Med-Checklist notulae, 14. - *Willdenowia* **51(3)**: 364.

UHRÍKOVÁ, A. 1970: Contribution to the cytology of the taxa *Hieracium aurantiacum* L. and *Hieracium bauhinii* Bess. - *Acta Fac. Rerum Nat. Univ. Comen., Bot.* **18**: 61-67.

UHRÍKOVÁ, A. 1987: Karyological study of the Slovak flora XXII. - *Acta Fac. Rerum Nat. Univ. Comen., Bot.* **34**: 33-37.

UHRÍKOVÁ, A. & AL. 1983: [Chromosome numbers from Czechoslovakia]. In Löve, Á. (ed.): IOPB chromosome number reports LXXX. - *Taxon* **32(3)**: 507.

UHRÍKOVÁ, A. & FERÁKOVÁ, V. 1977: [Chromosome numbers from Czechoslovakia]. In Löve, Á. (ed.): IOPB chromosome number reports LVI. - *Taxon* **26(2-3)**: 263.

UHRÍKOVÁ, A. & FERÁKOVÁ, V. 1978: [Chromosome numbers from Czechoslovakia]. In Löve, Á. (ed.): IOPB chromosome number reports LXI. - *Taxon* **27(4)**: 379.

UHRÍKOVÁ, A. & FERÁKOVÁ, V. 1980: [Chromosome numbers from Czechoslovakia]. In Löve, Á. (ed.): IOPB chromosome number reports LXIX. - *Taxon* **29(5-6)**: 726-727.

UHRÍKOVÁ, A., FERÁKOVÁ, V. & SCHWARZOVÁ, T. 1983: [Chromosome numbers from Czechoslovakia]. In Löve, Á. (ed.): IOPB chromosome number reports LXXX. - *Taxon* **32(3)**: 507-508.

UHRÍKOVÁ, A., FERÁKOVÁ, V. & SCHWARZOVÁ, T. 1985a: Karyologisches Studium der slowakischen Flora - *Acta Fac. Rerum Nat. Univ. Comen., Bot.* **32**: 67-70.

UHRÍKOVÁ, A., FERÁKOVÁ, V. & SCHWARZOVÁ, T. 1985b: Karyologisches Studium der slowakischen Flora. - *Acta Fac. Rerum Nat. Univ. Comen., Bot.* **32**: 71-75.

UHRÍKOVÁ, A. & KRÁLIK, E. 1980: [Chromosome numbers in Orchidaceae from Czechoslovakia]. In Löve, Á. (ed.): IOPB chromosome number reports LXIX. - *Taxon* **29(5-6)**: 729.

UHRÍKOVÁ, A. & MÁJOVSKÝ, J. 1977: [Chromosome numbers from Czechoslovakia]. In Löve, Á. (ed.): IOPB chromosome number reports LVI. - *Taxon* **26(2-3)**: 263.

UHRÍKOVÁ, A. & MÁJOVSKÝ, J. 1978: [Chromosome numbers from Czechoslovakia]. In Löve, Á. (ed.): IOPB chromosome number reports LXI. - *Taxon* **27(4)**: 378-379.

UHRÍKOVÁ, A. & MÁJOVSKÝ, J. 1980: [Chromosome numbers from Czechoslovakia]. In Löve, Á. (ed.): IOPB chromosome number reports LXIX. - *Taxon* **29(5-6)**: 725-726.

UHRÍKOVÁ, A. & MÁJOVSKÝ, J. 1983: [Chromosome numbers from Czechoslovakia]. In Löve, Á. (ed.): IOPB chromosome number reports LXXX. - *Taxon* **32(3)**: 507.

UHRÍKOVÁ, A., MÁJOVSKÝ, J. & SCHWARZOVÁ, T. 1986: Karyological study of the Slovak flora XV. - *Acta Fac. Rerum Nat. Univ. Comen., Bot.* **33**: 73-76.

UHRÍKOVÁ, A. & PACLOVÁ, L. 1986: Karyological study of the Slovak flora XIV. - *Acta Fac. Rerum Nat. Univ. Comen., Bot.* **33**: 69-72.

UHRÍKOVÁ, A., PACLOVÁ, L. & DÚBRAVCOVÁ, Z. 1980: [Chromosome numbers from Czechoslovakia]. In Löve, Á. (ed.): IOPB chromosome number reports LXIX. - *Taxon* **29(5-6)**: 730.

UHRÍKOVÁ, A. & PITONIAK, P. 1980: [Chromosome numbers from Czechoslovakia]. In Löve, Á. (ed.): IOPB chromosome number reports LXIX. - *Taxon* **29(5-6)**: 729.

UHRÍKOVÁ, A. & SCHWARZOVÁ, T. 1976: [Chromosome numbers from Czechoslovakia]. In Löve, Á. (ed.): IOPB chromosome number reports LIII. - *Taxon* **25(4)**: 491.

UHRÍKOVÁ, A. & SCHWARZOVÁ, T. 1977: [Chromosome numbers from Czechoslovakia]. In Löve, Á. (ed.): IOPB chromosome number reports LVI. - *Taxon* **26(2-3)**: 262-263.

UHRÍKOVÁ, A. & SCHWARZOVÁ, T. 1978: [Chromosome numbers from Czechoslovakia]. In Löve, Á. (ed.): IOPB chromosome number reports LXI. - *Taxon* **27(4)**: 379-380.

UHRÍKOVÁ, A. & SCHWARZOVÁ, T. 1980: [Chromosome numbers from Czechoslovakia]. In Löve, Á. (ed.): IOPB chromosome number reports LXIX. - *Taxon* **29(5-6)**: 729-730.

UHRÍKOVÁ, A. & ZÁBORSKÝ, J. 1980: [Chromosome numbers in *Cerastium* from Czechoslovakia]. In Löve, Á. (ed.): IOPB chromosome number reports LXIX. - *Taxon* **29(5-6)**: 729.

UHROVÁ, A. 1935: Revision der Gattung *Coronilla* L. - *Beih. Bot. Centralbl., Abt. 2* **53(1)**: 1-174.

UJHELYI, J. 1959a: Species Sesleriae generis novae. - *Feddes Repert. Spec. Nov. Regni Veg.* **62(1)**: 59-70 + 4 plates.

UJHELYI, J. 1959b: Révision des espèces du genre "Sesleria" en Italie. - *Webbia* **14(2)**: 597-614.

UJHELYI, J. 1960a: Weitere zytotaxonomische Beiträge zur Kenntnis der Gattung *Sesleria*. - *Bot. Közlem.* **48(3-4)**: 278-280.

UJHELYI, J. 1960b: Études taxonomiques sur le groupe *Lotus corniculatus* L. sensu lato. - *Ann. Hist.- Nat. Mus. Natl. Hung.* **52**: 185-200.

UJHELYI, J. 1961: Data to the systematics of the subsectio *Glaucae* of sectio *Bulbosae* of the genus *Koeleria*. - *Ann. Hist.-Nat. Mus. Natl. Hung.* **53**: 207-224.

UJHELYI, J. 1962: Data to the systematics of the subsectio *Bulbosae* of the genus *Koeleria*. II. - *Ann. Hist.-Nat. Mus. Natl. Hung.* **54**: 199-220.

UJHELYI, J. 1963: Data to the systematics of the sectio *Bulbosae* of the genus *Koeleria*. III. - *Ann. Hist.-Nat. Mus. Natl. Hung.* **55**: 187-214.

UJHELYI, J. 1964: Data to the systematics of the sectio *Bulbosae* of the genus *Koeleria*. IV. - *Ann. Hist.-Nat. Mus. Natl. Hung.* **56**: 195-214.

UJHELYI, J. 1965: Data to the sectio *Bulbosae* et sectio *Caespitosae* of the genus *Koeleria*. V. - *Ann. Hist.-Nat. Mus. Natl. Hung.* **57**: 179-202.

UJHELYI, J. 1966: Data to the systematics of the sectio *Bulbosae* and sectio *Caespitosae* of the genus *Koeleria*. VI. - *Ann. Hist.-Nat. Mus. Natl. Hung.* **58**: 177-196.

UJHELYI, J. 1968: Data to the systematics of the sections *Bulbosae* and *Caespitosae* of the genus *Koeleria*. VII. - *Ann. Hist.-Nat. Mus. Natl. Hung.* **60**: 79-107.

UJHELYI, J. 1969: Data to the systematics of the sections *Bulbosae* and *Caespitosae* of the genus *Koeleria*. VIII. - *Ann. Hist.-Nat. Mus. Natl. Hung.* **61**: 165-188.

UJHELYI, J. 1970: Data to the systematics of the sections *Bulbosae* and *Caespitosae* of the genus *Koeleria*. IX. - *Ann. Hist.-Nat. Mus. Natl. Hung.* **62**: 175-195.

UJHELYI, J. 1972: Data to the systematics of the sections *Bulbosae* and *Caespitosae* of the genus *Koeleria*. X. - *Ann. Hist.-Nat. Mus. Natl. Hung.* **64**: 115-128.

UJHELYI, J. 1973a: Data to the systematics of the sections *Bulbosae* and *Caespitosae* of the genus *Koeleria*. XI. - *Ann. Hist.-Nat. Mus. Natl. Hung.* **65**: 83-108.

UJHELYI, J. 1973b: Evolutionary problems of the European Koelerias. - *Symp. Biol. Hung.* **12**: 163-176.

UJHELYI, J. 1974: Data to the systematics of the sections *Bulbosae* and *Caespitosae* of the genus *Koeleria*. XII. - *Ann. Hist.-Nat. Mus. Natl. Hung.* **66**: 101-127.

UJHELYI, J. 1975: New species and new sections of the genus *Achillea* L. (Asteraceae). - *Ann. Hist.- Nat. Mus. Natl. Hung.* **67**: 41-55.

ULBRICH, E. 1905-06: Über die systematische Gliederung und geographische Verbreitung der Gattung *Anemone* L. - *Bot. Jahrb. Syst.* **37(2-3)**: 172-334 + 3 maps.

ULBRICH, E. 1938: Die Arten der Gattung *Helleborus* (Tourn.) L. - *Blätt. Staudenk.* **3**: 18 pp.

ÜNAL, A. 1974: Studies on *Zostera* L. [In Turkish with English summary]. - *Ankara Üniv. Eczac. Fak. Mecm.* **4(1)**: 24-27.

ÜNAL, F., YUZBASIOĞLU, D. & DUMAN, H. 1997: Karyotyping in *Sternbergia clusiana* and *S. colchiciflora* (Amaryllidaceae) from Turkey. - *Hereditas (Lund)* **126(3)**: 277-280.

ÜNAL, M. 1978: Cytological and embryological studies on *Anemone pavonina* Lam. - *Istanbul Üniv. Fen Fak. Mecm., B* **43**: 185-208.

ÜNAL, M. 1979: [Chromosome numbers in *Anemone* from the Istanbul area]. In Löve, Á. (ed.): IOPB chromosome number reports LXV. - *Taxon* **28(5-6)**: 632.

UNGER, F. 1862: Wissenschaftliche Ergebnisse einer Reise in Griechenland und den jonischen Inseln. - Wien: W. Braumüller, xii + 213 pp. + 1 map.

UNGRICHT, S., KNAPP, S. & PRESS, J.R. 1998: A revision of the genus *Mandragora* (Solanaceae). - *Bull. Nat. Hist. Mus. London, Bot.* **28(1)**: 17-40.

UOTILA, P. 1973: Chromosome counts on *Chenopodium* L. from SE Europe and SW Asia. - *Ann. Bot. Fenn.* **10(4)**: 337-340.

UOTILA, P. 1980: Chorological notes on *Chenopodium* in the eastern Mediterranean area. - *Ann. Bot. Fenn.* **17(1)**: 65-68. [Also distributed as *OPTIMA leaflet* no. 109].

UOTILA, P. & AL. 2010: *Typha domingensis* (Typhaceae) new to Serbia. - *Bot. Serbica* **34(2)**: 111-114.

UOTILA, P. & AL. 2021: Phylogeny, biogeography and systematics of Dysphanieae (Amaranthaceae). - *Taxon* **70(3)**: 526-551.

UOTILA, P., BAYTOP, A. & LANDOLT, E. 1984: Duckweeds (Lemnaceae) in Turkey. - *Webbia* **38**: 839-844.

UOTILA, P. & GREUTER, W. 2001: *Potamogeton trichoides* Cham. & Schltdl. In Greuter, W. & Raus, Th. (eds): Med-Checklist notulae, 20. - *Willdenowia* **31(2)**: 327-328.

UOTILA, P., VIERSSEN, W. VAN & WIJK, R.J. VAN 1984: Notes on the morphology and taxonomy of *Zannichellia* in Turkey. - *Ann. Bot. Fenn.* **20(4)**: 351-356. [Also distributed as *OPTIMA leaflet* no. 143].

UPCOTT, M. & LA COUR, L. 1936: The genetic structure of *Tulipa*. I. A chromosome survey. - *J. Genet.* **33(2)**: 237-254.

UPHOF, J.C.TH. 1958: A review of the genus *Gagea* Salisb. - *Pl. Life* **14**: 124-132.

UPHOF, J.C.TH. 1959: A review of the genus *Gagea* Salisb. (cont.). - *Pl. Life* **15**: 151-161.

UPHOF, J.C.TH. 1960: A review of the genus *Gagea* Salisb. (cont.). - *Pl. Life* **16**: 163-176.

UPSON, T. & ANDREWS, S. 2004: The genus *Lavandula*. - Kew: Royal Botanic Gardens, xiv + 442 pp.

URBAN, I. 1873: Prodromus einer Monographie der Gattung *Medicago* L. - *Verh. Bot. Vereins Prov. Brandenburg* **15(1)**: 1-85 + 2 plates.

URBAN, I. 1883: Die *Medicago*-Arten Linné's. - *Ber. Deutsch. Bot. Ges.* **1(6)**: 256-262.

URBANSKA-WORYTKIEWICZ, K. 1975: Cytological variation within *Lemna* L. - *Aquatic Bot.* **1(4)**: 377-394.

URBANSKA-WORYTKIEWICZ, K. 1977: Reproduction in natural triploid hybrids (2n=24) between *Cardamine rivularis* Schur and *C. amara* L. - *Ber. Geobot. Inst. E.T.H. Stiftung Rübel* **44**: 42-85.

URBANSKA-WORYTKIEWICZ, K. 1980: Cytological variation within the family of Lemnaceae. In Landolt, E. (ed.): Biosystematische Untersuchungen in der Familie der Wasserlinsen (Lemnaceae). Vol. 1. - *Veröff. Geobot. Inst. E.T.H. Stiftung Rübel Zürich* **70**: 30-101.

URBANSKA-WORYTKIEWICZ, K. & SCHWANK, O. 1980: Variation within *Lotus corniculatus* L. s.l. from Switzerland. III. Microdifferentiation in *L. alpinus* (DC.) Schleicher above the timberline. - *Ber. Geobot. Inst. E.T.H. Stiftung Rübel* **47**: 29-45.

URUMOV, I. 1907: *Centaurea* novae Bulgaricae. - *Magyar Bot. Lapok* **6(5-7)**: 165-166.

URUMOV, I. 1913: Beiträge zur Flora von Bulgarien. - *Magyar Bot. Lapok* **12(8-9)**: 212-222 + 1 plate.

URUMOV, I. 1914a: Hieracia nova Bulgarica additis Centaureis nonnulis novis, aliisque. - *Magyar Bot. Lapok* **13(6-9)**: 184-189.

URUMOV, I. 1914b: Po florata Čataldža i Bulair (Turska Trakiya) [In Bulgarian]. - *Spis. Bulg. Akad. Nauk* **9**: 133-180.

URUMOV, I. 1920: Neue und seltene Pflanzen Bulgariens. I. - *Magyar Bot. Lapok* **19(1-12)**: 33-40.

URUMOV, I. 1922: Thymi Bulgaricae rariores ac novi. - *Spis. Bulg. Akad. Nauk* **25**: 187-192.

URUMOV, I. 1923a: Contribution à la connaissance de la flore de la Thrace occidentale [In Bulgarian]. - *Spis. Bulg. Akad. Nauk* **28**: 1-107.

URUMOV, I. 1923b: Matériaux pour la flore de Pirin-Planina [In Bulgarian]. - *Spis. Bulg. Akad. Nauk* **28**: 109-178.

URUMOV, I. 1927a: Thymi bulgarici. - *Magyar Bot. Lapok* **25(1-12)**: 227-229.

URUMOV, I. 1927b: Plantae bulgaricae nonnullae novae ac rariores. - *Magyar Bot. Lapok* **26(1-12)**: 100-103.

URUMOV, I. 1929a: Rosae bulgaricae. II. - *Spis. Bulg. Akad. Nauk* **19**: 139-146.

URUMOV, I. 1929b: Rubi bulgaricae. II. – *Izv. Bulg. Bot. Druzh.* **3**: 27-28.

URUMOV, I. 1930: Die Bulgarischen Vertreter der Gattung *Cephalaria* [In Bulgarian with German summary]. - *Spis. Bulg. Akad. Nauk* **22**: 213-222.

URUMOV, I. 1931: Rosae bulgaricae. IV. - *Spis. Bulg. Akad. Nauk* **23**: 195-198.

URUMOV, I. 1934: Rosae bulgaricae. V. - *Spis. Bulg. Akad. Nauk* **24**: 47-61.

UTELLI, A.-B., ROY, B.A. & BALTISBERGER, M. 2000: Molecular and morphological analyses of European *Aconitum* species (Ranunculaceae). - *Pl. Syst. Evol.* **224(3-4)**: 195-212.

UYSAL, I. 1999: Morphological, anatomical and ecological studies on the two Turkish endemic species collected from Kaz Dağı (B1 Balıkesir) "*Allium sibthorpianum* Schultes & Schultes fil. and *Allium reuterianum* Boiss." - *Turk. J. Bot.* **23(2)**: 137-148.

UYSAL, I. 1992: Morphology, anatomy and ecology of endemic species *Dianthus ingoldbyi* Turrill. - *J. Fac. Sci. Ege Unic., Ser. B* **14(1)**: 30-38.

UYSAL, T. & AL. 2007: *Muscari turcicum* (Liliaceae/Hyacinthaceae), a new species from south Anatolia, Turkey. - *Bot. J. Linn. Soc.* **154(2)**: 233-236.

UZEN, E. 1998: Taxonomical and morphological studies on *Asplenium adiantum-nigrum* and *Asplenium onopteris* L. [In Turkish with English summary]. - *OT Sist. Bot. Derg.* **4(1)**: 57-72.

UZUNHISARCIKLI, M.E. & VURAL, M. 2012: The taxonomic revision of *Alcea* and *Althaea* (Malvaceae) in Turkey. - *Turk. J. Bot.* **36(6)**: 603-636.

UZUNOV, D. & AL. 1998: Notes on the distribution of *Serapias vomeracea* (Burm.) Briq. (Orchidaceae) in the Eastern Rhodopes. - *Phytol. Balcan.* **4(3)**: 115-119.

UZUNOV, D. & GUSSEV, CH. 2003: High mountain flora of Bulgaria - Statistics, ecological characteristics and phytogeography. - *Bocconea* **16(2)**: 763-770.

UZUNOV, D. & RAIMONDO, F.M. 2003: Comparative analysis of North Pirin (Mt.) and Madonie Mt. floras. - *Bocconea* **16(2)**: 771-781.

VAARAMA, A. 1951: Chromosome number and cryptic polyploidy in *Lepidium sativum*. - *Hereditas (Lund)* **37(1-2)**: 290-292.

VAARAMA, A. & HIIRSALMI, H. 1967: Chromosome studies on some Old World species of the genus *Scrophularia*. - *Hereditas (Lund)* **58(3)**: 333-358.

VÁCHOVÁ, M. 1980: [Chromosome numbers from Czechoslovakia]. In Löve, Á. (ed.): IOPB chromosome number reports LXIX. - *Taxon* **29(5-6)**: 724.

VÁCHOVÁ, M. 1987: Karyological study of the Slovak flora XXI. - *Acta Fac. Rerum Nat. Univ. Comen., Bot.* **34**: 27-32.

VÁCHOVÁ, M. & CHMELAŘ, J. 1976: [Chromosome numbers in *Salix* from Czechoslovakia]. In Löve, Á. (ed.): IOPB chromosome number reports LIII. - *Taxon* **25(4)**: 490.

VÁCHOVÁ, M. & FERÁKOVÁ, V. 1977: [Chromosome numbers from Czechoslovakia]. In Löve, Á. (ed.): IOPB chromosome number reports LVI. - *Taxon* **26(2-3)**: 264.

VÁCHOVÁ, M. & FERÁKOVÁ, V. 1978: [Chromosome numbers from Czechoslovakia]. In Löve, Á. (ed.): IOPB chromosome number reports LXI. - *Taxon* **27(4)**: 382-383.
VÁCHOVÁ, M. & FERÁKOVÁ, V. 1980: [Chromosome numbers from Czechoslovakia]. In Löve, Á. (ed.): IOPB chromosome number reports LXIX. - *Taxon* **29(5-6)**: 722-723.
VÁCHOVÁ, M. & FERÁKOVÁ, V. 1986: Karyological study of the Slovak flora XII. - *Acta Fac. Rerum Nat. Univ. Comen., Bot.* **33**: 57-62.
VÁCHOVÁ, M. & LHOTSKÁ, M. 1978: [Chromosome numbers from Czechoslovakia]. In Löve, Á. (ed.): IOPB chromosome number reports LXI. - *Taxon* **27(4)**: 383.
VÁCHOVÁ, M. & LHOTSKÁ, M. 1980: [Chromosome numbers from Czechoslovakia]. In Löve, Á. (ed.): IOPB chromosome number reports LXIX. - *Taxon* **29(5-6)**: 723-724.
VÁCHOVÁ, M. & MÁJOVSKÝ, J. 1976: [Chromosome numbers from Czechoslovakia]. In Löve, Á. (ed.): : IOPB chromosome number reports LIII. - *Taxon* **25(4)**: 489.
VÁCHOVÁ, M. & MÁJOVSKÝ, J. 1977: [Chromosome numbers from Czechoslovakia]. In Löve, Á. (ed.): IOPB chromosome number reports LVI. - *Taxon* **26(2-3)**: 263-264.
VÁCHOVÁ, M. & MÁJOVSKÝ, J. 1978: [Chromosome numbers from Czechoslovakia]. In Löve, Á. (ed.): IOPB chromosome number reports LXI. - *Taxon* **27(4)**: 381-382.
VÁCHOVÁ, M. & MÁJOVSKÝ, J. 1980: [Chromosome numbers from Czechoslovakia]. In Löve, Á. (ed.): IOPB chromosome number reports LXIX. - *Taxon* **29(5-6)**: 720-721.
VÁCHOVÁ, M. & PACLOVÁ, L. 1978: [Chromosome numbers from Czechoslovakia]. In Löve, Á. (ed.): IOPB chromosome number reports LXI. - *Taxon* **27(4)**: 384.
VÁCHOVÁ, M. & PACLOVÁ, L. 1980: [Chromosome numbers from Czechoslovakia]. In Löve, Á. (ed.): IOPB chromosome number reports LXIX. - *Taxon* **29(5-6)**: 721.
VÁCHOVÁ, M. & SCHWARZOVÁ, T. 1977: [Chromosome numbers from Czechoslovakia]. In Löve, Á. (ed.): IOPB chromosome number reports LVI. - *Taxon* **26(2-3)**: 264.
VÁCHOVÁ, M. & SCHWARZOVÁ, T. 1978: [Chromosome numbers from Czechoslovakia]. In Löve, Á. (ed.): IOPB chromosome number reports LXI. - *Taxon* **27(4)**: 382.
VÁCHOVÁ, M. & SCHWARZOVÁ, T. 1980: [Chromosome numbers from Czechoslovakia]. In Löve, Á. (ed.): IOPB chromosome number reports LXIX. - *Taxon* **29(5-6)**: 723.
VÁCHOVÁ, M. & ZÁBORSKÝ, J. 1980: [Chromosome numbers from Czechoslovakia]. In Löve, Á. (ed.): IOPB chromosome number reports LXIX. - *Taxon* **29(5-6)**: 724.
VAILLANT, M. LE 1970: Étude comparée en cultures expérimentales de trois espèces de *Crupina* DC. - *Rev. Gén. Bot.* **77(908-910)**: 111-124.
VALANT-VETSCHERA, K.M. 1994: Lectotypification of *Achillea cretica* (Compositae, Anthemideae). - *Taxon* **44(1)**: 91-96.
VALANT-VETSCHERA, K.M. 1995: Lectotypification of Linnaean *Achillea* species: *A. tomentosa*, *A. atrata* and *A. nana*. - *Feddes Repert.* **106(1-2)**: 49-57.
VALANT-VETSCHERA, K.M. 1996: Lectotypification of *Achillea santolina* L. and *Achillea falcata* L. (Compositae-Anthemideae). - *Bot. J. Linn. Soc.* **121(2)**: 159-168.
VALDÉS-BERMEJO, E. & AGUDO MATA, M.P. 1983: Estudios cariológicos en especies ibéricas del género *Centaurea* L. (Compositae). I. - *Anales Jard. Bot. Madrid* **40(1)**: 119-142.
VALDÉS-BERMEJO, E. & SÁNCHEZ-CRESPO, A. 1978: Datos cariológicos y taxonómicos sobre el género *Teucrium* L. (Labiatae) en la Península Ibérica. - *Acta Bot. Malac.* **4**: 27-54.
VALDÉS, B. 1970a: Revisión de las especies europeas de *Linaria* con semillas aladas. - *Publ. Univ. Sevilla (Anal. Univ. Hispal.), Ser. Ci.* **7**: 288 pp.
VALDÉS, B. 1970b: Taxonomía experimental del género *Linaria*. III. Cariología de algunas especies de *Linaria*, *Cymbalaria* y *Chaenorrhinum*. - *Bol. Real Soc. Esp. Hist. Nat., Secc. Biol.* **67(2)**: 243-256.
VALDÉS, B. 1973: Revisión de las especies anuales del género *Anthoxanthum* (Gramineae). - *Lagascalia* **3(1)**: 99-141.
VALDÉS, B. 1978: Números cromosómicos para la flora española. 76-83. - *Lagascalia* **8(1)**: 122-125 + 1 plate.
VALDÉS, B. 2004a: The Euro+Med treatment of Boraginaceae. - *Willdenowia* **34(1)**: 59-61.
VALDÉS, B. 2004b: Some validations in Liliaceae. - *Willdenowia* **34(1)**: 63-64.
VALDÉS, B. & AL. 1995: *Allium ampeloprasum* L. Distribution of polyploids in S. Spain. In Valdés, B. & Silvestre, S. (eds): Abstracts. VIII OPTIMA meeting, Sevilla, Spain, 25 September - 1 October, 1995, p. 76.
VALDÉS, B. & GONZÁLEZ-BERNÁLDEZ, F. 1972: *Trifolium baeticum* Boiss. y *Trifolium pallidum* Waldst. & Kit. - *Lagascalia* **2(2)**: 189-191.
VALDÉS, B., PASTOR, J. & UBERA, J. 1977: Números cromosómicos para la flora española. 1-14. - *Lagascalia* **7(2)**: 192-199.
VALDÉS, B. & SCHOLZ, H. 2006: The Euro+Med treatment of Gramineae - a generic synopsis and some new names. - *Willdenowia* **36(2)**: 657-669.
VALDÉS, B. & SILVESTRE, S. (EDS) 1995: Abstracts. VIII OPTIMA meeting, Sevilla, Spain, 25 September 1 October, 1995. - Sevilla: 140 pp.
VALENTINE, D.H. 1941: Variation in *Viola riviniana* Rchb. - *New Phytol.* **40(3)**: 189-209.
VALENTINE, D.H. 1950: The experimental taxonomy of two species of *Viola*. - *New Phytol.* **49(2)**: 193-212 + 1 plate.
VALENTINE, D.H. 1955: Studies in British *Primulas*. IV. Hybridization between *Primula vulgaris* Huds. and *P. veris* L. - *New Phytol.* **54(1)**: 70-80.
VALENTINE, D.H. 1966: The experimental taxonomy of some *Primula* species. - *Trans. & Proc. Bot. Soc. Edinburgh* **40(2)**: 169-180 + 1 plate.
VALENTINE, D.H. 1980: Ecotypic and polymorphic variation in *Centaurea scabiosa* L. - *Watsonia* **13 (2)**: 103-109.
VALENTINE, D.H. & HARVEY, M.J. 1961: Variation in *Viola rupestris* Schmidt. - *Veröff. Geobot. Inst. E.T.H. Stiftung Rübel Zürich* **36**: 157-163.
VALENTINE, D.H. (ED.) 1972: Taxonomy, phytogeography and evolution. - London & New York: Academic Press, xi + 431 pp.
VALENTINI, G. & MONTECCHI, A. 1990: Una nuova orchidea per l'Italia e l'Europa occidentale, *Pseudorchis frivaldii* (Hampe ex Griseb.) P.F. Hunt. - *Il Fungo?* **8(9)**: 6-7 [n.v.].
VALKOUN, J. 2001: The geographical distribution of wild wheats in their historical setting and current context. In Caligari, P.D.S. & Brandham, P.E. (eds): Wheat taxonomy: the legacy of John Percival. - *The Linnean Special Issue* **3**: 81-94.
VALLÉS, J. 1992: Reports (79-82). In Kamari, G., Felber, F. & Garbari, F. (eds): Mediterranean chromosome number reports - 2. - *Fl. Medit.* **2**: 247-249.
VALLÈS, J. & TORRELL, M. 1996: Reports (552-558). In Kamari, G., Felber, F. & Garbari, F. (eds): Mediterranean chromosome number reports - 5. - *Fl. Medit.* **5**: 357-363.
VALLI, A.-T. 2013: Katagrafi tis chloridas tou Ethnikou Thalassinou Parkou Zakinthou ke vio-parakolouthisi ton apoklistika endimikon fitikon taxa tis Zakinthou. [In Greek with English abstract]. - Patras: M.Sc. thesis, Univ. of Patras, 243 pp. See http://nemertes.lis.upatras.gr/jspui/bitstream/10889/7845/3/Nimertis_Valli%28bio%29.pdf
VALLI, A.-T. & AL. 2019: Determinants of alpha and beta vascular plant diversity in Mediterranean island systems: the Ionian islands, Greece. - *Nordic J. Bot.* **37(1)**: e02156.
VALLI, A.-T. & AL. 2021a: Conservation biology of threatened Mediterranean chasmophytes: The case of *Asperula naufraga* endemic to Zakynthos island (Ionian islands, Greece). - *PLoS ONE* **16(2)**: e0246706.
VALLI, A.-T. & AL. 2021b: Conservation of *Micromeria browiczii* (Lamiaceae), endemic to Zakynthos island (Ionian Islands, Greece). - *Plants MDPI* **10, 778**: 18 pp.
VALLI, A.-T. & ARTELARI, R. 2015a: *Limonium korakonisicum* (Plumbaginaceae), a new species from Zakynthos Island (Ionian Islands, Greece). In Bogdanović, S. & Jogan, N.: Book of abstracts - 6th Balkan Botanical Congress, p. 118. - Rijeka.
VALLI, A.-T. & ARTELARI, R. 2015b: *Limonium korakonisicum* (Plumbaginaceae), a new species from Zakynthos Island (Ionian Islands, Greece). - *Phytotaxa* **217(1)**: 63-72.
VALLIANATOU, I. 2005: Geobotanical study of Salamis, Egina and other islands of the Saronic Gulf (Hellas). [In Greek with English summary]. - University of Athens: Ph.D. thesis, 558 pp.
VALLIANATOU, I. 2011: Egaleo - Pikilo. I krimmeni biopikilotita se ena vouno - ekplixi. [In Greek]. - *Fisis [Journal of the Hellenic Society for the Protection of Nature]* **132**: 19-22.

VALLIANATOU, I., KOUMPLI-SOVANTZI, L. & YANNITSAROS, A. 1994: Flora and vegetation of the island of Psittalia (Saronic Gulf, Greece). - *Candollea* **49(1)**: 209-224.

VALLIANATOU, I. & YANNITSAROS, A. 1992: Adventive flora of the Saronikos Gulf islands. A contribution [In Greek with English summary]. In: Proc. 12th Annual Meeting Hellen. Soc. Biol. Sci., Mitilini, 27-29.4. 1990, pp. 48-50. - Athine.

VALLIANATOU, I. & YANNITSAROS, A. 1993: A contribution to the flora of the island of Aiyina (Saronic Gulf, Greece) [In Greek with English abstract]. In: Proc. 15th Conf. Hellen. Soc. Biol. Sci., Florina-Kastoria, 21-24.4 1993, pp. 406-409. - Florina-Kastoria.

VALLIANATOU, I. & YANNITSAROS, A. 2000: A contribution to the vascular flora of the islands of the Saronic Gulf (Greece). - *Bot. Chron. (Patras)* **13**: 303-323.

VALLIANATOU, I. & YANNITSAROS, A. 2009: Flora and vegetation of the island of Patroklos (Saronic Gulf, Greece). - *Bocconea* **23**: 419-437.

VALLIANATOU, I. & YANNITSAROS, A. 2010: Adventive flora of Salamis, Egina and other islands of the Saronic Gulf (Greece). - *Bot. Chron. (Patras)* **20**: 25-44.

VALLIANATOU, I., YANNITSAROS, A. & ECONOMIDOU, E. 1995: Coastal vegetation of Saronic Gulf islands. In: Praktika 16ou Sinedriou E.E.V.E., Volos, 5.-7.5. 1994, 2, pp. 15-17 [In Greek with English summary].

VALSECCHI, F. 1975: Contributo alla conoscenza sistematica del genere *Genista* in Sardegna: I. *Genista acanthoclada* DC. - *Giorn. Bot. Ital., n.s.* **109(4-5)**: 239-249.

VALSECCHI, F. 1976: Il genere *Anchusa* in Sardegna. - *Webbia* **30(1)**: 43-68. [Also distributed as *OPTIMA leaflet* no. 29].

VALSECCHI, F. 1977: Contributo alla conoscenza del genere *Echium*: I - *Echium* della Sardegna. - *Webbia* **32(1)**: 101-127.

VALSECCHI, F. 1994: Sul complesso *Astragalus tragacantha* L. (Leguminosae) nel Mediterraneo. - *Webbia* **49(1)**: 31-41.

VALSECCHI, F. 1995: Systematic, taxonomic and chorological studies in the *Silene colorata* Poir.- *S. sericea* All. group (Caryophyllaceae). - *Inform. Bot. Ital.* **27(1)**: 59.

VALTUEÑA, F.J. & AL. 2008: Reproductive biology in *Anagyris foetida* L. (Leguminosae), an autumn- winter flowering and ornithophilous Mediterranean shrub. - *Bot. J. Linn. Soc.* **157(3)**: 519-532.

VAMVOUKAKIS, J.A. 1988: *Phoenix theophrasti* on Crete. - *Principes* **32(2)**: 82-83.

VANDAS, C. 1890: Neue Beiträge zur Kenntniss der Flora Bosniens und der Hercegovina. - *Sitzungsber. Königl. Böhm. Ges. Wiss. Prag, Math.-Naturwiss. Cl.* **1890(20)**: 249-285.

VANDAS, C. 1905a: Novae plantae Balcanicae. - *Magyar Bot. Lapok* **4(6-7)**: 109-113.

VANDAS, C. 1905b: Additamenta ad floram Macedoniae et Thessaliae. - *Magyar Bot. Lapok* **4(8-11)**: 262-268.

VANDAS, C. 1909: Reliquiae Formánekianae. Enumeratio critica plantarum vascularum, quam itineribus in Haemo peninsula et Asia Minore (Bithynia) factis collegit Dr. Ed. Formánek. - Brunae [Brno]: Jos. Jelínek, viii + 623 + xxxiii pp.

VANDAS, C. 1911: Plantae novae Formanekianae. - *Repert. Spec. Nov. Regni Veg.* **9**: 524-534.

VANGJELI, J. 1988: [Preliminary biosystematic data about the deciduous species of the genus *Quercus*]. - *Bul. Shkencacet Nat.* **42(2)**: 65-74 [n.v.].

VANGJELI, J. 1989: [Taxonomic evaluation of *Quercus coccifera* according to morphological, anatomical and micrometrical data]. - *Bul. Shkencacet Nat.* **43(2)**: 53-60 [n.v.].

VANGJELI, J. 2015: Excursion Flora of Albania. - Königstein: Koeltz Scientific Books, 661 pp.

VANGJELI, J. 2017: Flora Albanica Atlas. Vol. I: Pteridophyla-Apiaceae. - Schmitten-Obereifenberg: Koeltz Botanical Books, 933 pp.

VANGJELI, J. & AL. (EDS) 2000: Flora e Shqipërisë. Vol. 4. Compositae-Orchidaceae [In Albanian]. - Tiranë: Akad. Shkencave Rep. Shqipërisë, 502 pp.

VANGJELI, J. & RUCI, B. 2000: Flora and vegetation of Albania: an overview. - *Bot. Chron. (Patras)* **13**: 469-478.

VANGJELI, J., RUCI, B. & MULLAJ, A. 1995: Red book: threatened and rare plants species of Albania [In Albanian with English summary]. - Tirana: Inst. of Biological Research, ii + 169 pp.

VARDAR, Y. & AHMET, M. [ÖZTÜRK] 1973: Some ecological aspects of *Myrtus communis* L. - *Bot. Jahrb. Syst.* **93(4)**: 562-567.

VARDAR, Y., SEÇMEN, Ö. & ÖZTÜRK, M.A. 1980: Some distributional problems and biological characteristics of *Ceratonia* in Turkey. - *Portug. Acta Biol., Sér. A, Morfol.* **16(1-4)**: 75-86.

VARDAVAKIS, E., PAVLIDES, G. & LAVRENTIADES, G. 1987: On the vegetation of a typical xerorthent soil of Polygyros area (SE of Thessaloniki). - *Feddes Repert.* **98(3-4)**: 253-264.

VÄRE, H. & AL. 2003: Taxonomic diversity of vascular plants in Europe. In Nagy, L. & al. (eds): Alpine biodiversity in Europe (Ecological studies 167), pp. 131-148. - Berlin & Heidelberg: Springer-Verlag.

VARGA-SIPOS, J. 1976: Chorologische Florenanalyse der alpinen Stufe in den SW-bulgarischen Hochgebirgen Pirin und Rila. - *Acta Biol. Debrecina* **13**: 121-146.

VARSAMIS, G. & AL. 2019: Morphological variability of *Dryas octopetala* in Greece. In Daskalakou, E.N. & al. (eds): 16th Conference of the Hellenic Botanical Society – Athens 10-13 October 2019. Program and book of abstracts, p. 58.

VARSAMIS, G. & AL. 2020: Plant morphological variability at the distribution edges: the case of *Dryas octopetala* (Rosaceae) in northern Greece. - *Willdenowia* **50(2)**: 267-277.

VARSAMIS, G. & AL. 2021: Genetic diversity of alpine *Dryas octopetala* populations at their southern distribution limit in Europe. - *Nordic J. Bot.* **39(4)**: e03150, 8 pp.

VARSAMOPOULOU, C. , DROUZAS, A.D. & HANLIDOU, E. 2013: Investigating hybridization in a mixed population of *Salvia officinalis* L. and *S. ringens* Sm. In Gkelis, S. & al. (eds): 13th Panhellenic Scientific Conference, Hellenic Botanical Society, Thessaloniki 3-6 October. Program and book of abstracts, p. 53.

VASIĆ, O. 1993: On taxonomic interpretation of variability of *Astragalus onobrychis* L. 1753 (Fabaceae). In Petrova, A.V. (ed.): Abstracts from the VII OPTIMA meeting in Borovetz, Bulgaria, 18-30 July 1993, p. 85.

VASIĆ, O. 2001: The flora of Yugoslavia (Serbia and Montenegro) and its investigation. In Özhatay, N. (ed.): Proceedings of the Second Balkan Botanical Congress, 1: 57-62. – Istanbul.

VASIĆ, O. & DIKLIĆ, N. 1996: Flora of Serbia - a review of phytotaxonomic studies published during the period 1981-1993. - *Bocconea* **5(1)**: 45-52.

VASIĆ, O. & NIKETIĆ, M. 1997: On the degree to which taxa *schramii* and *vaillantii* of the genus *Fumaria* are related. In: Abstracts from the First Balkan Botanical Congress, Thessaloniki, September 19-22, 1997, p. 54. - Thessaloniki.

VASIĆ, O. & STEVANOVIĆ, B. 1989: Some taxonomical consideration of *Convolvulus compactus* (Boiss.) Stace complex. In Phitos, D. (ed.): Abstracts. Sixth OPTIMA meeting, Delphi 10-16 Sept. 1989, p. 147.

VASIĆ, O. (ED.) 2004: Abstracts, XI OPTIMA Meeting, Beograd 5-11.IX.2004. - Beograd: 186 pp.

VASIĆ, O. DIKLIĆ, N. 2001: The flora and vegetation of serpentinites in Serbia. - *Bocconea* **13**: 151-164.

VASILEVA, M.G., KLJUYKOV, E.V. & PIMENOV, M.G. 1985: Karyotaxonomic analysis of the genus *Bunium* (Umbelliferae). - *Pl. Syst. Evol.* **149(1-2)**: 71-88.

VASILEVA, M.G. & PIMENOV, M.G. 1991: Karyotaxonomical analysis in the genus *Angelica* (Umbelliferae). - *Pl. Syst. Evol.* **177(3-4)**: 117-138.

VASILIOPOULOS, G., TSIRIPIDIS, I. & KARAGIANNAKIDOU, V. 2007: Do abandoned tree plantations resemble natural riparian forests? A case study from northeast Greece. - *Bot. Helv.* **117**: 125-142.

VASSILCZENKO, I.T. 1953: Obzor vidov roda *Trigonella* L. (A review of species of the genus *Trigonella*) [In Russian]. - *Trudy Bot. Inst. Akad. Nauk S.S.S.R., Ser. 1, Fl. Sist. Vysš. Rast.* **10**: 124-269.

VASSILCZENKO, I.T. 1979: De positione systematica generum *Medicago* L. et *Trigonella* L. [In Russian]. - *Novosti Sist. Vysš. Rast.* **16**: 130-136.

VASSILIADES, D. 1996: *Alkanna sartoriana* (Boraginaceae), an endangered species. - *Bot. Chron. (Patras)* **12**: 71-72.

VASSILIADES, D. 2004: *Astragalus idaeus* (Fabaceae) rediscovered in Crete. - *Bot. Chron. (Patras)* **16**: 13-17.

VASSILIADES, D. & PERSSON, K. 2002: A new winter-flowering species of *Colchicum* from Greece. - *Preslia* **74**: 57-65.

VASSILIADES, D. & YANNITSAROS, A. 2016: *Biebersteinia orphanidis* and its *locus classicus*: A pleasant end to an old story. - *Bot. Chron. (Patras)* **21**: 17-20.

VASSILIADIS, D. & YANNITSAROS, A. 2000: Orphanides's best discovery. - *Bot. Chron. (Patras)* **13**: 241-248.

VASSILIJEV, I.V. 1961: De genere *Betula* L. notulae systematicae et geographicae [In Russian]. - *Bot. Mater. Gerb. Bot. Inst. Komarova Akad. Nauk S.S.S.R.* **21**: 93-103.

VASSILIJEV, I.V. 1971: Conspectus generis *Tilia* L. [In Russian; English translation of this volume by Amerind Publ. Co., New Delhi, 1978]. - *Novosti Sist. Vysš. Rast.* **8**: 202-208.

VÁZQUEZ, F.M. 2000: The genus *Scolymus* Tourn. ex L. (Asteraceae): taxonomy and distribution. - *Anales Jard. Bot. Madrid* **58(1)**: 83-100.

VÁZQUEZ, F.M. & COOMBES, A.J. 2017: (2569) Proposal to conserve the name *Quercus ilex* (Fagaceae) with a conserved type. - *Taxon* **66(6)**: 1473-1475. [approved, cf. Taxon 69(2): 394, 2020].

VÁZQUEZ, F.M., LÓPEZ GONZÁLEZ, G. & DEVESA, J.A. 1995: Tipificación de los taxónes linneanos del género *Stipa* (Gramineae) que viven en la Península Ibérica. - *Anales Jard. Bot. Madrid* **52(2)**: 179-186.

VEDEL, H. 1977: Træer og buske omkring Middelhavet. Tegninger af R. Als og A. Rasmussen [In Danish]. - København: Gyldendal, 128 pp.

VEDEL, H. 1978: Trees and shrubs of the Mediterranean. Illustrated by R. Als and A. Rasmussen. Translated from the Danish by A. Rush, edited and adapted by H. Synge. - Harmondsworth: Penguin Books, 127 pp.

VEDRAN, Š. & AL. 2015: *Vicia loiseleuri* (M. Bieb.) D. Litv. (Fabaceae) in Croatian flora. In Bogdanović, S. & Jogan, N.: Book of abstracts - 6th Balkan Botanical Congress, p. 113. – Rijeka.

VELASCO STEIGRAD, M. 1992: Estudios taxonómicos en las especies de la sect. *Subacaulia* Koch ex Reiche del género *Geranium* L. (Geraniaceae), en la Península Ibérica. - *Bull. Soc. Bot. France, Lett. Bot.* **139(3)**: 295-303.

VELAYOS, M. 1988: Acotaciones a *Ranunculus* subgénero *Batrachium* (DC.) A. Gray: Tratamiento taxonómico general y estudio de la variabilidad de *R. peltatus*. - *Anales Jard. Bot. Madrid* **45(1)**: 103-119.

VELČEV, V.I. 1970: *Convolvulus compactus* Boiss. in Bulgaria [In Bulgarian with Russian and English summaries]. - *Izv. Bot. Inst. (Sofia)* **20**: 115-122.

VELČEV, V.I. 1972: On the taxonomic position of *Convolvulus sündermannii* Bornm. [In Bulgarian with English and Russian summaries]. - *Izv. Bot. Inst. (Sofia)* **22**: 119-126 + 1 plate.

VELČEV, V.I. 2002: New to science species of the genera *Vicia* L. and *Festuca* L. from Bulgaria. - *Phytol. Balcan.* **8(1)**: 3-14.

VELČEV, V.I. & AL. 1962: Neue Materialen zur Flora des südlichen Piringebirges, des Struma-Tals und des Slavjanka-Gebirges [In Bulgarian with Russian and German summaries]. - *Izv. Bot. Inst. (Sofia)* **9**: 177-179.

VELČEV, V.I. & AL. (EDS) 1983: Third national conference of botany. Sofia, 26.-30.X. 1981 [a series of papers, mostly in Bulgarian with English abstracts]. - Sofia: Publishing House of the Bulgarian Academy of Sciences, 1021 pp.

VELČEV, V.I. & AL. (EDS) 1988: 100th anniversary of Academician Nikolaj A. Stojanov [a series of papers, mostly in English or in Bulgarian with English summaries]. - Sofia: Publishing House of the Bulgarian Academy of Sciences, 207 pp.

VELČEV, V.I., BONDEV, I. & KOŽUHAROV, S.I. 1981: Using data of geobotanical mapping of Bulgaria for the purposes of the chorological atlas. In Velčev, V.I. & Kožuharov, S.I. (eds): Mapping the Flora of the Balkan Peninsula, pp. 160-165. - Sofia: Publishing House of the Bulgarian Academy of Sciences.

VELČEV, V.I. & JANKOVIĆ, M.M. 1975: *Pinus heldreichii* Christ in the Balkan Peninsula [In Bulgarian]. In Jordanov, D. & al. (eds): Problems of Balkan flora and vegetation, pp. 286-295 + 1 plate. - Sofia: Publishing House of the Bulgarian Academy of Sciences.

VELČEV, V.I., JORDANOV, D. & GANČEV, S. 1973: Untersuchung an *Ramonda serbica* Panč. in Bulgarien [In Bulgarian with Russian and German summaries]. - *Izv. Bot. Inst. (Sofia)* **24**: 139-167.

VELČEV, V.I., KOŽUHAROV, S.I. & ANČEV, M.E. 1992: Atlas of the endemic plants in Bulgaria [In Bulgarian with English foreword]. - Sofia: Publishing House of the Bulgarian Academy of Sciences, 204 pp.

VELČEV, V.I., KOŽUHAROV, S.I. & KUZMANOV, B.A. 1988: Contribution of Bulgarian botanists to the knowledge of the flora and vegetation of the Balkan Peninsula. In Velčev, V.I. & al. (eds): 100th anniversary of Acad. Nikolaj A. Stojanov, pp. 46-59. - Sofia: Publishing House of the Bulgarian Academy of Sciences.

VELČEV, V.I. & KOŽUHAROV, S.I. (EDS) 1981: Mapping the flora of the Balkan Peninsula. - Sofia: Publishing House of the Bulgarian Academy of Sciences, 247 pp.

VELČEV, V.I., MARKOVA, M.L. & VASILEVA, S.I. 1989: Genus *Arbutus* L. with the species *A. andrachne* L. and *A. unedo* L. - new for Bulgarian dendroflora [In Bulgarian with Russian and English summaries]. - *Fitologiya* **36**: 10-14 + 2 plates.

VELČEV, V.I. & VASSILEV, P.D. 2002: New species of genus *Festuca* L. from Bulgaria. - *Phytol. Balcan.* **8(2)**: 185-190.

VELČEV, V.I., VASSILEV, P.D. & DIMITROVA, M.M. 1983: Biometric investigations on cones of *Pinus heldreichii* Christ in Bulgaria [In Bulgarian with English abstract]. In Velčev, V.I. & al. (eds): Third national conference of botany, pp. 488-498. - Sofia: Publishing House of the Bulgarian Academy of Sciences.

VELČEV, V.I. (ED.) 1984: Red Data Book of the People's Republic of Bulgaria. Volume 1. Plants [In Bulgarian with Russian and English summaries]. - Sofia: Publishing House of the Bulgarian Academy of Sciences, 447 pp.

VELDE, G. VAN DER & HARTOG, C. DEN 1992: Continuing range extension of *Halophila stipulacea* (Forssk.) Aschers. (Hydrocharitaceae) in the Mediterranean - now found at Kefallinia and Ithaki (Ionian Sea). - *Acta Bot. Neerl.* **41(3)**: 345-348.

VELDE, G. VAN DER & HEIJDEN, L.A. VAN DER 1981: The floral biology and seed production of *Nymphoides peltata* (Gmel.) O. Kuntze (Menyanthaceae). - *Aquatic Bot.* **10(3)**: 261-293.

VELDKAMP, J.F. 1976: (413) Proposal for the conservation of *Echinochloa* Beauv. (1812) (Gramineae), against *Tema* Adanson (1763) and *Ornithospermum* Dumoulin (1782) ... - *Taxon* **25(2-3)**: 363. [approved, cf. Taxon 27(2-3): 289, 1978].

VELDKAMP, J.F. 1983: (710) Proposal to conserve *Trisetum* (Gramineae) and its type species. - *Taxon* **32(3)**: 487-488. [not approved, cf. Taxon 36(1): 75, 1987].

VELDKAMP, J.F. 1996: (1221) Proposal to conserve the name *Brachiaria* (Trin.) Griseb. (Gramineae) with a conserved type. - *Taxon* **45(2)**: 319-320. [not approved, cf. Taxon 47(4): 869, 1998].

VELDKAMP, J.F. 2004: Miscellaneous notes on mainly southeast Asian Gramineae. - *Reinwardtia* **12 (2)**: 135-140.

VELDKAMP, J.F. 2015: De nomenclatuur van Speenkruiden (*Ficaria verna* Huds. s.l., Ranunculaceae). - *Gorteria* **37(3-4)**: 84-116.

VELENOVSKÝ, J. 1886a: Beiträge zur Kenntnis der Flora von Ost-Rumelien. - *Oesterr. Bot. Z.* **36(7-8)**: 225-229, 264-268.

VELENOVSKÝ, J. 1886b: Beiträge zur Kenntniss der bulgarischen Flora. - *Abh. Königl. Böhm. Ges. Wiss.* **7(1)**: 1-47.

VELENOVSKÝ, J. 1887: Neue Beiträge zur Kenntniss der Flora von Ost-Rumelien und Bulgarien. - *Sitzungsber. Königl. Böhm. Ges. Wiss. Prag, Math.-Naturwiss. Cl.* **1887(24)**: 436-466 + 1 plate.

VELENOVSKÝ, J. 1890a: Plantae novae bulgaricae. - *Sitzungsber. Königl. Böhm. Ges. Wiss. Prag, Math.-Naturwiss. Cl.* **1889(2)**: 28-39.

VELENOVSKÝ, J. 1890b: Plantae novae bulgaricae. Pars II. - *Sitzungsber. Königl. Böhm. Ges. Wiss. Prag, Math.-Naturwiss. Cl.* **1890(3)**: 39-59.

VELENOVSKÝ, J. 1891a: Flora Bulgarica. Descriptio et enumeratio systematica plantarum vascularium in principatu Bulgariae sponte nascentium ... - Pragae: Fr. Řivnáć, ix + 676 pp. [cf. TL-2 6: 693, 1986].

VELENOVSKÝ, J. 1891b: Ueber zwei verkannte Cruciferen. - *Oesterr. Bot. Z.* **41(4)**: 121-123.

VELENOVSKÝ, J. 1892: Nachträge zur "Flora bulgarica" (Schluss). - *Oesterr. Bot. Z.* **42(1)**: 14-17.

VELENOVSKÝ, J. 1893: Neue Nachträge zur Flora von Bulgarien. - *Sitzungsber. Königl. Böhm. Ges. Wiss. Prag, Math.-Naturwiss. Cl.* **1892(26)**: 369-390.

VELENOVSKÝ, J. 1894: Dritter Nachtrag zur Flora von Bulgarien. - *Sitzungsber. Königl. Böhm. Ges. Wiss. Prag, Math.-Naturwiss. Cl.* **1893(37)**: 1-72.

VELENOVSKÝ, J. 1895: Vierter Nachtrag zur Flora von Bulgarien. - *Sitzungsber. Königl. Böhm. Ges. Wiss. Prag, Math.-Naturwiss. Cl.* **1894(29)**: 1-29.

VELENOVSKÝ, J. 1896: Fünfter Nachtrag zur Flora von Bulgarien. - *Sitzungsber. Königl. Böhm. Ges. Wiss. Prag, Math.-Naturwiss. Cl.* **1895(37)**: 12 pp.

VELENOVSKÝ, J. 1898: Flora Bulgarica. Descriptio et enumeratio systematica plantarum vascularium in principatu Bulgariae sponte nascentium. Supplementum I. - Pragae: Fr. Řivnáć, xx + 404 pp.

VELENOVSKÝ, J. 1899a: Ueber *Micromeria Frivaldskyana* Deg. und *M. balcanica* Vel. - *Oesterr. Bot. Z.* **49(8)**: 291-292.

VELENOVSKÝ, J. 1899b: Sechster Nachtrag zur Flora von Bulgarien. - *Sitzungsber. Königl. Böhm. Ges. Wiss. Prag, Math.-Naturwiss. Cl.* **1898(29)**: 1-8.

VELENOVSKÝ, J. 1900: Siebenter Nachtrag zur Flora von Bulgarien. - *Sitzungsber. Königl. Böhm. Ges. Wiss. Prag, Math.-Naturwiss. Cl.* **1899(40)**: 1-8.

VELENOVSKÝ, J. 1904a: Einige Novitäten aus Bulgarien. - *Allg. Bot. Z. Syst.* **10(3-4)**: 33-35.

VELENOVSKÝ, J. 1904b: Nachträge zur Flora von Bulgarien. - *Sitzungsber. Königl. Böhm. Ges. Wiss. Prag, Math.-Naturwiss. Cl.* **1903(28)**: 1-28 + 1 plate.

VELENOVSKÝ, J. 1905: Beiträge zur Flora des Orients. - *Allg. Bot. Z. Syst.* **11(3)**: 43-45.

VELENOVSKÝ, J. 1906: Vorstudien zu einer Monographie der Gattung *Thymus* L. - *Beih. Bot. Centralbl., Abt. 2* **19**: 271-287.

VELENOVSKÝ, J. 1911: Letzte Nachträge zur Flora der Balkanländer. - *Sitzungsber. Königl. Böhm. Ges. Wiss. Prag, Math.-Naturwiss. Cl.* **1910(8)**: 1-13.

VELENOVSKÝ, J. 1922: Reliquiae Mrkvičkanae. - Pragae: Fr. Řivnáč, 32 pp.

VELKOV, D. 1983: Evolutionary variation of the catkins at the European chestnut (*Castanea sativa* Mill.) [In Bulgarian with English abstract]. In Velčev, V.I. & al. (eds): Third national conference of botany, pp. 348-356. - Sofia: Publishing House of the Bulgarian Academy of Sciences.

VEN, W.T.G. VAN DER & AL. 1996: Genetic variation and systematic relationships in *Vicia*. In Pickersgill, B. & Lock, J.M. (eds): Advances in legume systematics. Part 8: Legumes of economic importance, pp. 31-40. - Kew: Royal Botanic Gardens.

VENT, W. 1962: Monographie der Gattung *Oreoherzogia* W. Vent gen. nov. - *Feddes Repert. Spec. Nov. Regni Veg.* **65(1-2)**: 3-132.

VENT, W. 1972: Beiträge zur Kenntnis der Sippenstruktur einiger *Papaver*-Arten der Sektion *Argemonorhoeades* Fedde (Papaveraceae). - *Feddes Repert.* **83(4)**: 233-243.

VENT, W., MEINHARDT, U. & VOBACH, V. 1973: Beiträge zur Kenntnis der infraspezifischen Struktur von *Frangula alnus* Mill. (Rhamnaceae). - *Gleditschia* **1**: 17-31.

VENTER, H.J.T. 1997: A revision of *Periploca* (Periplocaceae). - *S. African J. Bot.* **63(3)**: 123-128.

VENTURELLA, G. & AL. 2015: Up-to-date report on the distribution of *Helianthemum lippii* (Cistaceae) in Italy. - *Webbia* **70(1)**: 151-154.

VENTURELLA, G., BAUM, B.R. & MANDRACCHIA, G. 2007: The genus *Tamarix* (Tamaricaceae): in Sicily: first contribution. - *Fl. Medit.* **17**: 25-46.

VERDCOURT, B. 1948: Biological Flora of the British Isles. *Cuscuta* L. - *J. Ecol.* **36(2)**: 356-365.

VERDCOURT, B. 1987: On the typification of *Heliotropium supinum* L. (Boraginaceae). - *Kew Bull.* **42 (3)**: 710.

VERDUS, M.C. 1961: Comparaison d'*Erica carnea* et d'*Erica mediterranea*. - *Bull. Soc. Hist. Nat. Toulouse* **96**: 181-187.

VERGOS, S. 1979: Strukturen und Entwicklungsdynamik natürlicher Schwarzkiefernwälder in Nordwest-Griechenland (Revierkomplex Krania-Monahition/Grevena). - Göttingen: Diss., 121 pp.

VERLAQUE, R. 1975: Contribution à l'étude caryologique des Dipsacaceae de la Méditerranée orientale. - *Biol. Gallo-Hellen.* **6(1)**: 75-100.

VERLAQUE, R. 1977a: Rapports entre les Valerianaceae, les Morinaceae et les Dipsacaceae. - *Bull. Soc. Bot. France* **124(7-8)**: 475-482.

VERLAQUE, R. 1977b: Importance du fruit dans la détermination des Dipsacaceae. - *Bull. Soc. Bot. France* **124(9)**: 515-527.

VERLAQUE, R. 1977c: Validité et position systématique du genre *Tremastelma* Rafin. (Dipsacaceae). - *Biol. Écol. Médit.* **4(3)**: 105-115.

VERLAQUE, R. 1977d: [Chromosome numbers in Dipsacaceae]. In Löve, Á. (ed.): IOPB chromosome number reports LVI. - *Taxon* **26(2-3)**: 272-274.

VERLAQUE, R. 1978: Étude cytotaxinomique de quelques Dipsacaceae et Morinaceae du Nord de la Grèce. - *Biol. Écol. Médit.* **5(1)**: 15-30.

VERLAQUE, R. 1980: [Chromosome numbers in Dipsacaceae]. In Löve, Á. (ed.): Chromosome number reports LXVII. - *Taxon* **29(2-3)**: 362-365.

VERLAQUE, R. 1981: Utilisation des caractères du pollen pour la réorganisation taxinomique de la famille des Dipsacaceae. - *Compt. Rend. Hebd. Séances Acad. Sci., Sér. 3, Sci. Vie* **293(7)**: 351-354.

VERLAQUE, R. 1982a: Etude de deux complexes vicariants d'annuelles: *Scabiosa palaestina* L. (s.l.) et *Scabiosa stellata* L. (s.l.) (section *Trochocephalus* Mertens & Koch). - *Bull. Soc. Bot. France, Lett. Bot.* **129(4-5)**: 305-320.

VERLAQUE, R. 1982b: [Chromosome numbers in Dipsacaceae]. In Löve, Á. (ed.): Chromosome number reports LXXVII. - *Taxon* **31(4)**: 770-771.

VERLAQUE, R. 1983a: Etude biosystématique et phylogénétique des Dipsacaceae. - Marseille: Thèse, Univ. Provence. I: Text, 406 pp; II: Plates, maps and tables.

VERLAQUE, R. 1983b: Contribution à l'étude du genre *Morina* L. - *Pollen & Spores* **25(2)**: 143-162.

VERLAQUE, R. 1984a: A biosystematic and phylogenetic study of the Dipsacaceae. In Grant, W.F. (ed.): Plant Biosystematics, pp. 307-320. - Toronto, etc.: Academic Press.

VERLAQUE, R. 1984b: Étude biosystématique et phylogénétique des Dipsacaceae. I. Délimitation des Dipsacaceae à l'intérieur des Dipsacales, rapports avec les autres familles de l'ordre. - *Rev. Gén. Bot.* **91(1079-1081)**: 81-121.

VERLAQUE, R. 1985a: Étude biosystématique et phylogénétique des Dipsacaceae. II. Caractères généraux des Dipsacaceae. - *Rev. Cytol. Biol. Vég. Botaniste* **8(2)**: 117-168.

VERLAQUE, R. 1985b: Étude biosystématique et phylogénétique des Dipsacaceae. III. Tribus des Knautieae et des Dipsaceae. - *Rev. Cytol. Biol. Vég. Botaniste* **8(3-4)**: 171-243.

VERLAQUE, R. 1986a: Étude biosystématique et phylogénétique des Dipsacaceae. IV. Tribu des Scabioseae (phylum no. 1, 2, 3). - *Rev. Cytol. Biol. Vég. Botaniste* **9(1)**: 5-72.

VERLAQUE, R. 1986b: Étude biosystématique et phylogénétique des Dipsacaceae. V. Tribu des Scabioseae (phylum no. 4) et conclusion. - *Rev. Cytol. Biol. Vég. Botaniste* **9(2)**: 97-176.

VERLAQUE, R. 1992: Modalités de la spéciation chez les Genisteae. In: Actes del simposi internacional de botànica Pius Font i Quer, 1988, 2: pp. 49-68.

VERLAQUE, R. & ABOUCAYA, A. 1998: Position des *Delphinium* du ssg. *Staphisagria* dans la tribu des *Delphinieae* Warm. In Moret, J. (ed.): Abstracts of the IX. OPTIMA meeting, Paris, 11-17 May 1998, p. 57. - Paris: Mus. Nat. d'Hist. Naturelle.

VERLAQUE, R. & AL. 1991: Quelques examples de spéciation insulaire en Méditerranée occidentale. - *Bot. Chron. (Patras)* **10**: 137-153.

VERLAQUE, R. & CONTANDRIOPOULOS, J. 1990: Analyse des variations chromosomiques en région méditerranéenne: Polyploïdie, différenciation et adaptation. - *Ecol. Medit.* **16**: 93-112.

VERLAQUE, R. & FILOSA, D. 1992: Rapports (107-117). In Kamari, G., Felber, F. & Garbari, F. (eds): Mediterranean chromosome number reports - 2. - *Fl. Medit.* **2**: 264-272.

VERLAQUE, R. & FILOSA, D. 1993: Rapports (228-233). In Kamari, G., Felber, F. & Garbari, F. (eds): Mediterranean chromosome number reports - 3. - *Fl. Medit.* **3**: 364-367.
VERLAQUE, R., SEIDENBINDER, M. & DONADILLE, P. 1987: Recherches cytotaxonomiques sur la spéciation en région méditerranéenne. I: Espèces à nombre chromosomique stable. - *Biol. Écol. Médit.* **10(4)**: 273-289.
VERLAQUE, R., SEIDENBINDER, M. & REYNAUD, C. 1987: Recherches cytotaxonomiques sur la spéciation en région méditerranéenne. III: Espèces aneuploïdes. - *Biol. Écol. Médit.* **10(4)**: 315-346.
VERLAQUE, R. & VIGNAL, C. 1993: Étude cyto-biogéographique de *Bromus erectus* Hudson (de S.-O. de l'Europe), comparison histologique avec les complexes polyploïdes de *B. riparius* et *B. inermis* (section *Pnigma* Dumort.). - *Rev. Cytol. Biol. Vég. Botaniste* **16**: 47-72.
VERLAQUE, R., VIGNAL, C. & SEIDENBINDER, M. 1987: Recherches cytotaxonomiques sur la spéciation en région méditerranéenne. II: Espèces dysploïdes. - *Biol. Écol. Médit.* **10(4)**: 291-313.
VERLAQUE, R. RAYNAUD, C. & VIGNAL, CH. 1996: Reports (545-551). In Kamari, G., Felber, F. & Garbari, F. (eds): Mediterranean chromosome number reports - 5. - *Fl. Medit.* **5**: 350-357.
VERLOOVE, F. 2012a: New combinations in *Cenchrus* (Paniceae, Poaceae) in Europe and the Mediterranean area. - *Willdenowia* **42(1)**: 77-78.
VERLOOVE, F. 2012b: *Cenchrus echinatus*. In Greuter, W. & Raus, Th. (eds): Med-Checklist notulae, 31. - *Willdenowia* **42(2)**: 293.
VERLOOVE, F. 2014: A conspectus of *Cyperus* s.l. (Cyperaceae) in Europe (incl. Azores, Madeira and Canary Islands), with emphasis on non-native naturalized species. - *Webbia* **69(2)**: 179-223.
VERLOOVE, F. & AL. 2020: Genome size confirms the presence of *Artemisia verlotiorum* (Asteraceae) in Lebanon. - *Turk. J. Bot.* **44(4)**: 585-591.
VERLOOVE, F. & REYNDERS, M. 2007: Studies in the genus *Paspalum* (Paniceae, Poaceae) in Europe - 1. *Paspalum distichum* subsp. *paucispicatum*, an overlooked taxon in France. - *Willdenowia* **37(1)**: 199-204.
VERLOOVE, F. & SÁNCHEZ GULLÓN, E. 2012: A taxonomic revision of non-native *Cenchrus* s.str. (Paniceae, Poaceae) in the Mediterranean area. - *Willdenowia* **42(1)**: 67-75.
VERMA, S.C. 1958: Cytology of *Ophioglossum vulgatum*. - *Acta Bot. Neerl.* **7(4)**: 629-634.
VERMEULEN, P. 1972: Übersicht zur Systematik und Taxonomie der Gattung *Orchis* s.str. - *Jahresber. Naturwiss. Vereins Wuppertal* **25**: 22-36.
VERMEULEN, P. 1976: Was ist *Orchis latifolia* L.? - *Acta Bot. Neerl.* **25(5)**: 371-379.
VERMEULEN, P. 1978: (443) *Orchis latifolia*, nomen rejiciendum propositum. - *Taxon* **27(1)**: 128. [not approved, cf. Taxon 29(4): 493, 1980].
VERROIOS, G. & AL. 1999: A continuous monitoring programme for plants & habitats in the extended area around Athens international airport. In Elefheriou, E.P. (ed.): Proceedings, 7th Panhellenic scientific conference, Alexandroupolis, pp. 224-228 [In Greek with English abstract]. - Thessaloniki: Univ. Studio Press.
VERROIOS, G. & GEORGIADIS, TH. 2011: Aleppo pine forest of northern and western Peloponnisos (southern Greece): Plant communities and diversity. - *Pl. Biosyst.* **145(3)**: 606-619.
VESELÁ, E. 1969: Beitrag zur Taxonomie der Art *Ficaria verna* Huds. - *Preslia* **41(4)**: 313-322.
VESTEK, A. & AL. 2018: Karyological variation in *Prospero autumnale* (L.) Speta (Hyacinthaceae) complex in Pannonian plain and Balkan Peninsula. - *Bot. Serbica* **42 (suppl. 1)**: 54.
VESTEK, A. & AL. 2019: Morpho-anatomical differentiation and genome size variation in three ploidy levels within the B^7 cytotype of *Prospero autumnale* (Hyacinthaceae) complex from the Balkan Peninsula and Pannonian Basin. - *Pl. Syst. Evol.* **305(8)**: 597-609.
VESTERGREN, T. 1939: Systematische Beobachtungen über *Myosotis sylvatica* (Ehrh.) Hoffm. und verwandte Formen. - *Ark. Bot.* **29A(8)**: 1-39.
VEYA, P. & TANDÉ, A. 2018: *Ophrys homeri* × *Ophrys phrygia* - a new hybrid of Chios. - *J. Eur. Orch.* **50(2-4)**: 203-206.
VIANE, R.L.L. & AL. 1993: Observations on some ferns of Poros and the neighbouring Peloponnisos (Greece). In Petrova, A.V. (ed.): Abstracts from the VII OPTIMA meeting in Borovetz, Bulgaria, 18-30 July 1993, p. 61.
VIANE, R.L.L. & AL. 1996: Observations on some ferns of Poros and adjacent parts of the Peloponnesus (Greece). - *Bocconea* **5(1)**: 279-300.
VIANE, R.L.L. & HEEDE, C.J. VAN DEN 2002: Subspecific names for the recently described new Aspleniums from Cyprus. - *GEP News* **10**: 5-6.
VIANO, J. 1971: Contribution à l'étude caryosystématique du genre *Linaria*. - *Caryologia* **24(2)**: 183-201.
VIANO, J. 1982: *Linaria triphylla* (L.) Miller: distribution géographique, typification, taxinomie, variabilité. - *Candollea* **37(1)**: 219-234.
VICENS, J. & MOLERO, J. 1992: Reports (70-78). In Kamari, G., Felber, F. & Garbari, F. (eds): Mediterranean chromosome number reports - 2. - *Fl. Medit.* **2**: 243-247.
VICENS, J., MOLERO, J. & BLANCHÉ, C. 1991: Reports (38-44). In Kamari, G., Felber, F. & Garbari, F. (eds): Mediterranean chromosome number reports - 1. - *Fl. Medit.* **1**: 241-243.
VICENTE, A., ALONSO, M.Á. & CRESPO, M.B. 2015: Epitype designation for the name *Biscutella didyma* L. (Brassicaceae). - *Phytotaxa* **233(3)**: 288-292.
VICHODZEVSKY, N. 1956: Einige für die Flora Bulgariens neue und seltene Arten [In Bulgarian]. - *Izv. Bot. Inst. (Sofia)* **5**: 466-471.
VICIOSO, C. 1952: Tréboles españoles. Revisión del género *Trifolium*. - *Anales Inst. Bot. Cavanilles* **10(2)**: 347-398.
VICIOSO, C. 1953: Tréboles españoles. Revisión del género *Trifolium* [cont.]. - *Anales Inst. Bot. Cavanilles* **11(2)**: 289-383.
VICTOR, D. & CLIFTON, R.T.F. 1997: *Geranium cinereum* complex in Greece. - *Geraniaceae Group News* **65**: 20-22.
VIDA, G. 1963a: A new Asplenium (sectio *Ceterach*) species and the problem of the origin of *Phyllitis hybrida* (Milde) C. Christ. - *Acta Bot. Acad. Sci. Hung.* **9(1-2)**: 197-215.
VIDA, G. 1963b: Systematics of the genus *Dryopteris* sensu lato [In Hungarian with Russian and English summaries]. - *Bot. Közlem.* **50(3)**: 125-133.
VIDA, G. 1971: The nature of polyploidy in *Asplenium ruta-muraria* L. and *Asplenium lepidum* C. Presl. - *Caryologia* **23(4)**: 525-547.
VIDA, G. 1974: Genome analysis of the European *Cystopteris fragilis* complex. I. Tetraploid taxa. - *Acta Bot. Acad. Sci. Hung.* **20(1-2)**: 181-192.
VIDA, G. & AL. 1970: Cytologie der Farngattung *Cheilanthes* in Europa und auf den Kanarischen Inseln. - *Verh. Schweiz. Naturf. Ges.* **150**: 149-150.
VIDA, G. & AL. 1971: Cytologie der Farn-Gattung *Cheilanthes* in Europa und auf den Canarischen Inseln. - *Bauhinia* **4(2)**: 223-253.
VIDA, G., MAJOR, A. & REICHSTEIN, T. 1983: Relations and evolution in *Cheilanthes* (Sinopteridaceae, Pteridophyta) in Macaronesia and Mediterranean area, deduced from genome analysis of their hybrids. - *Acta Bot. Malac.* **8**: 101-126.
VIDA, G. & REICHSTEIN, T. 1971: *Asplenium* ×*lessinense* (*A. fissum* × *A. viride*), ein neuer diploider Farnbastard. - *Candollea* **26(1)**: 191-195 + 4 plates.
VIDA, G. & REICHSTEIN, T. 1975: Taxonomic problems in the fern genus *Polystichum* caused by hybridization. In Walters, S.M. & King, C.J. (eds): European floristic and taxonomic studies, pp. 126-135. - Oxford.
VIEGI, L. 1987: *Erianthus ravennae* (L.) Beauv. (Gramineae) from S. Rossore (Italy). I: Caryological and anatomical features. - *Ann. Bot. (Rome)* **45(2)**: 103-115.
VIEGI, L. & AL. 1972: Numeri cromosomici per la Flora Italiana: 135-144. - *Inform. Bot. Ital.* **4(3)**: 229-236.
VIEGI, L. & CELA RENZONI, G. 1977: Numeri cromosomici per la Flora Italiana: 276-282. - *Inform. Bot. Ital.* **8(3)**: 276-281.
VIEGI, L. & CELA RENZONI, G. 1983: Numeri cromosomici per la Flora Italiana: 831-835. - *Inform. Bot. Ital.* **13(2-3)**: 168-171.

VIEGI, L. & CELA RENZONI, G. 1991: Numeri cromosomici per la Flora Italiana: 1244-1246. - *Inform. Bot. Ital.* **22(3)**: 240-243.

VIERHAPPER, F. 1897: Ueber einen neuen *Dianthus* aus dem Balkan. - *Verh. K.K. Zool.-Bot. Ges. Wien* **47**: 31-35.

VIERHAPPER, F. 1898: Zur Systematik und geographischen Verbreitung einiger alpinen *Dianthus*-Gruppen. - *Sitzungsber. Kaiserl. Akad. Wiss., Math.-Naturwiss. Cl., Abt. 1* **107**: 1057-1170 + 2 plates and 1 map.

VIERHAPPER, F. 1906a: Monographie der alpinen *Erigeron*-Arten Europas und Vorderasiens. - *Beih. Bot. Centralbl., Abt. 2* **19**: 385-560 + 6 plates and 2 maps.

VIERHAPPER, F. 1906b: Aufzählung der von Professor Dr. Oskar Simony im Sommer 1901 in Südbosnien gesammelten Pflanzen. - *Mitt. Naturwiss. Vereins Univ. Wien* **4**: 36-76.

VIERHAPPER, F. 1907a: Die systematische Stellung der Gattung *Scleranthus*. - *Oesterr. Bot. Z.* **57(2-3)**: 41-47, 91-96.

VIERHAPPER, F. 1907b: Versuch einer natürlichen Systematik des *Cirsium arvense* (L.) Scop. - *Oesterr. Bot. Z.* **57(3)**: 106-111.

VIERHAPPER, F. 1909: Eine neue *Soldanella* aus dem Balkan. - *Oesterr. Bot. Z.* **59(4-5)**: 148-150, 202-205.

VIERHAPPER, F. 1914a: Beiträge zur Kenntnis der Flora Kretas. Aufzählung der anlässlich der fünften Wiener Universitätsreise im April 1914 auf Kreta gesammelten Blüten- und Farnpflanzen. - *Oesterr. Bot. Z.* **64(12)**: 465-482.

VIERHAPPER, F. 1914b: Beiträge zur Kenntnis der Flora Griechenlands. Bearbeitung der anlässlich der zweiten Wiener Universitätsreise im April 1911 in Griechenland gesammelten Pflanzen. A. Anthophyta und Pteridophyta. - *Verh. K.K. Zool.-Bot. Ges. Wien* **64**: 239-270 + 1 plate.

VIERHAPPER, F. 1915: Beiträge zur Kenntnis der Flora Kretas. Aufzählung der anlässlich der fünften Wiener Universitätsreise im April 1914 auf Kreta gesammelten Blüten- und Farnpflanzen (Fortsetzung). - *Oesterr. Bot. Z.* **65**: 21-28, 50-75, 119-140, 204-236, 252-265.

VIERHAPPER, F. 1916a: Beiträge zur Kenntnis der Flora Kretas. Aufzählung der anlässlich der fünften Wiener Universitätsreise im April 1914 auf Kreta gesammelten Blüten- und Farnpflanzen (Schluss). - *Oesterr. Bot. Z.* **66(5)**: 150-180.

VIERHAPPER, F. 1916b: Nachtrag zu meinen "Beiträgen zur Kenntnis der Flora Kretas". - *Oesterr. Bot. Z.* **66(7-9)**: 297.

VIERHAPPER, F. 1918: Was ist *Trifolium Pilczii* Adamović? - *Oesterr. Bot. Z.* **67**: 252-264, 328-337.

VIERHAPPER, F. 1919a: Beiträge zur Kenntnis der Flora Griechenlands. Bearbeitung der anlässlich der zweiten Wiener Universitätsreise im April 1911 in Griechenland gesammelten Pflanzen. A. Anthophyta und Pteridophyta. II-IV. - *Verh. Zool.-Bot. Ges. Wien* **69**: 102-156, 157-245, 246-312.

VIERHAPPER, F. 1919b: Über echten und falschen Vikarismus. - *Oesterr. Bot. Z.* **68(1-4)**: 1-22.

VIERHAPPER, F. 1926: [Die Verbreitung der Arten der Gattung *Soldanella* L.] In Hanning, E. & Winkler, H. (eds): Die Pflanzenareale, 1. Reihe, Heft 1: Karte 7-8. - Jena: Gustav Fischer Verlag.

VIERHAPPER, F. 1927: Zwei neue *Nigella*-Arten aus der Verwandtschaft der *N. arvensis*. - *Magyar Bot. Lapok* **25**: 146-150.

VIERHAPPER, F. 1929: August v. Hayek. Ein Nachruf. - *Verh. Zool.-Bot. Ges. Wien* **78**: 136-151.

VIERHAPPER, F. & RECHINGER, K.H. 1935: Bearbeitung der von Ignaz Dörfler im Jahre 1904 auf Kreta gesammelten Blüten- und Farnpflanzen. - *Oesterr. Bot. Z.* **84(2-3)**: 123-157, 161-197.

VIGNAL, C. & REYNAUD, C. 1992: Rapports (118-119). In Kamari, G., Felber, F. & Garbari, F. (eds): Mediterranean chromosome number reports - 2. - *Fl. Medit.* **2**: 272-273.

VIINIKKA, Y. 1976: *Najas marina* L. (Najadaceae). Karyotypes, cultivation and morphological variation. - *Ann. Bot. Fenn.* **13(3)**: 119-131.

VILATERSANA, R. & AL. 2000: Karyology, generic delineation and dysploidy in the genera *Carduncellus, Carthamus* and *Phonus* (Asteraceae). - *Bot. J. Linn. Soc.* **134(3)**: 425-438.

VILATERSANA, R. & BERNAL, M. 1992: Reports (83-91). In Kamari, G., Felber, F. & Garbari, F. (eds): Mediterranean chromosome number reports - 2. - *Fl. Medit.* **2**: 249-255.

VILLA, R. 1978: Numeri cromosomici per la Flora Italiana: 398-401. - *Inform. Bot. Ital.* **10(1)**: 80-84.

VILLA, R. 1979: Numeri cromosomici per la Flora Italiana: 457-463. - *Inform. Bot. Ital.* **10(2)**: 241-248.

VILLA, R. 1981: Numeri cromosomici per la Flora Italiana: 733-741. - *Inform. Bot. Ital.* **12(2)**: 155-160.

VILLA, R. 1990: Numeri cromosomici per la Flora Italiana: 1197-1204. - *Inform. Bot. Ital.* **20(2-3)**: 647-652.

VILLA, R. & SANNA, D. 1984: Numeri cromosomici per la Flora Italiana: 956-961. - *Inform. Bot. Ital.* **15 (1)**: 49-52.

VILLAR, J.E. & AL. 2014: The genus *Tamarix* (Tamaricaceae) in Crete (Greece). - *Willdenowia* **44(3)**: 321-326.

VILLAR, J.L. & AL. 2015: *Tamarix minoa* (Tamaricaceae), a new species from the island of Crete (Greece) based on morphological and plastid molecular sequence data. - *Willdenowia* **45(1)**: 161-172.

VILLARS, D. 1779: Prospectus de l'histoire des plantes de Dauphiné. - Grenoble: Imprimerie royale, 57 pp.

VILLARS, D. 1786-89: Histoire des plantes de Dauphiné. 3 vols. - Grenoble, etc. [cf. TL-2 6: 740, 1986].

VINEY, D.E. 1994: An illustrated Flora of North Cyprus. - Koenigstein: Koeltz Scientific Books, xxix + 697 pp. [reprinted 2011, with corrections].

VINEY, D.E. 1996: An illustrated Flora of North Cyprus. Volume 2: Sedges, grasses and ferns. - Vaduz: A.R.G. Gantner Verlag K.-G., xv + 167 pp.

VISIANI, R. DE 1842: Flora Dalmatica. Vol. 1, pp. i-xii, 1-252, pl. 1-25. - Lipsiae [Leipzig]: F. Hofmeister .

VISIANI, R. DE 1843: Ilustrazione di alcune piante della Grecia e dell'Asia Minore. - *Mem. Reale Ist. Veneto Sci.* **1**: 39-58 + 6 plates.

VISIANI, R. DE 1847: Flora Dalmatica. Vol. 2, pp. i-x, 1-268, pl. 26-51. - Lipsiae [Leipzig]: F. Hofmeister .

VISIANI, R. DE 1850: Flora Dalmatica. Vol. 3(1), pp. i-iv, 1-185. - Lipsiae [Leipzig]: F. Hofmeister .

VISIANI, R. DE 1851: Flora Dalmatica. Vol. 3(2), pp. 186-390, pl. 52-55. - Lipsiae [Leipzig]: F. Hofmeister.

VISIANI, R. DE 1860: Plantarum serbicarum pemptas ossia descrizione di cinque piante Serbiane. - *Mem. Reale Ist. Veneto Sci.* **9**: 165-175 + 5 plates.

VISIANI, R. DE 1872: Flora Dalmatica. Supplementum, pp. 1-189, pl. 1-10. - Lipsiae [Leipzig]: F. Hofmeister.

VISIANI, R. DE 1877: Flora Dalmatica. Supplementum alterum 1, pp. 1-103, i-iii, pl. 1. - Lipsiae [Leipzig]: F. Hofmeister.

VISIANI, R. DE 1881: Flora Dalmatica. Supplementum alterum 2, pp. 1-72, pl. 1-7. - Lipsiae [Leipzig]: F. Hofmeister.

VISIANI, R. DE & PANČIĆ, J. 1861: Plantae serbicae rariores aut novae. - *Mem. Reale Ist. Veneto Sci.* **10**: 427-446 + 7 plates.

VÍT, P. & AL. 2017: Two new polyploid species closely related to *Alnus glutinosa* in Europe and North Africa – An analysis based on morphometry, karyology, flow cytometry and microsatellites. - *Taxon* **66(3)**: 567-583.

VITEK, E. 1989: *Spiranthes spiralis* (L.) Chevall., Neufund für Rhodos. - *Orchidee (Hamburg)* **40(2)**: 61.

VITEK, E. 2000: The evolution within *Carlina* (Compositae). In Özhatay, N. & al. (eds): Abstracts from the Second Balkan Botanical Congress, Istanbul, Turkey, May 14-18, 2000, p. 47.

VITEK, E. & KIEHN, M. 1990: Chromosomenzählungen an *Euphrasia rostkoviana* (Scrophulariaceae) und verwandten Taxa. - *Flora, Morphol. Geobot. Oekophysiol.* **184(1)**: 31-41.

VITEK, E. & KIEHN, M. 1996: *Euphrasia illyrica* und *E. dinarica*, zwei weitere tetraploide Arten in *Euphrasia* subsect. *Angustifoliae* (Scrophulariaceae). - *Ann. Naturhist. Mus. Wien, B* **98B Suppl.**: 311-315.

VLACHOS, A.V. 1996: Chloridiki erevna ke ikologikes monades vlastisis tou orinou sigrotimatos ton Vardousion [In Greek]. - Patras: M.Sc. thesis, Univ. of Patras, 87 pp.

VLACHOS, A.V., GEORGIADIS, T. & TINIAKOU, A. 2009: Floristic research of the mountains of Sterea Ellas (Central Greece) and their affinities with mountains of Peloponissos. - *J. Biol. Res. Thessalon.* **12**: 193-209.

VLACHOS, S., CHRISTODOULAKIS, D. & KAMARI, G. 2003: The flora of Mount Boumistos (NW Sterea Ellas, Greece): Species list and chorological notes. - *Fl. Medit.* **12**: 413-438.

VLADIMIROV, V. 2001a: New chorological data on four alien species in the Bulgarian flora. - *Phytol. Balcan.* **7(1)**: 33-37.

VLADIMIROV, V. 2001b: New records of vascular plants to the Bulgarian flora. - *Phytol. Balcan.* **7(2)**: 195-198.

VLADIMIROV, V. 2003: A new diploid *Hieracium* (Asteraceae: Lactuceae) from Bulgaria. - *Bot. J. Linn. Soc.* **143(2)**: 213-218.

VLADIMIROV, V. 2005: Photoguide to the vascular plants in the Rhodopes [separate editions in English and Bulgarian]. - Defra (U.K.): 96 pp.

VLADIMIROV, V. 2009: *Erigeron sumatrensis* (Asteraceae): a recently recognized alien species in the Bulgarian flora. - *Phytol. Balcan.* **15(3)**: 361-365.

VLADIMIROV, V. 2010: Typification and current taxonomic position of *Senecio arnautorum* Velen. - *Ann. Bot. Fenn.* **47(6)**: 493-495.

VLADIMIROV, V. & AL. 2015: Floristic and habitat diversity of Ali Botush reserve (SW Bulgaria). In Bogdanović, S. & Jogan, N.: Book of abstracts - 6th Balkan Botanical Congress, p. 120. - Rijeka.

VLADIMIROV, V. & AL. 2020: Contribution to the knowledge of habitat and vascular plant diversity of Alibotush Reserve, Slavyanka Mountain (Bulgaria). - *Fl. Medit.* **30**: 19-32 + suppl.

VLADIMIROV, V. & AL. (EDS) 2006: New floristic records in the Balkans: 2. - *Phytol. Balcan.* **12(2)**: 279-301.

VLADIMIROV, V. & AL. (EDS) 2007: New floristic records in the Balkans: 5. - *Phytol. Balcan.* **13(2)**: 261-276.

VLADIMIROV, V. & AL. (EDS) 2012: New floristic records in the Balkans: 18. - *Phytol. Balcan.* **18(1)**: 69-92.

VLADIMIROV, V. & AL. (EDS) 2015: New floristic records in the Balkans: 27. - *Phytol. Balcan.* **21(2)**: 189-219.

VLADIMIROV, V. & AL. (EDS) 2016: New floristic records in the Balkans: 29. - *Phytol. Balcan.* **22(1)**: 93-123.

VLADIMIROV, V. & AL. (EDS) 2017a: New floristic records in the Balkans: 33. - *Phytol. Balcan.* **23(2)**: 281-329.

VLADIMIROV, V. & AL. (EDS) 2017b: New floristic records in the Balkans: 34. - *Phytol. Balcan.* **23(3)**: 413-444.

VLADIMIROV, V. & AL. (EDS) 2019: New floristic records in the Balkans: 38. - *Phytol. Balcan.* **25(1)**: 97-120.

VLADIMIROV, V., AYBEKE, M. & TAN, KIT (EDS) 2017: New floristic records in the Balkans: 32. - *Phytol. Balcan.* **23(1)**: 119-146.

VLADIMIROV, V., AYBEKE, M. & TAN, KIT (EDS) 2018a: New floristic records in the Balkans: 35. - *Phytol. Balcan.* **24(1)**: 155-174.

VLADIMIROV, V., AYBEKE, M. & TAN, KIT (EDS) 2018b: New floristic records in the Balkans: 36 - *Phytol. Balcan.* **24(2)**: 263-292.

VLADIMIROV, V., AYBEKE, M. & TAN, KIT (EDS) 2019: New floristic records in the Balkans: 39. - *Phytol. Balcan.* **25(2)**: 203-238.

VLADIMIROV, V., AYBEKE, M. & TAN, KIT (EDS) 2020a: New floristic records in the Balkans: 41. - *Phytol. Balcan.* **26(1)**: 163-179.

VLADIMIROV, V., AYBEKE, M. & TAN, KIT (EDS) 2020b: New floristic records in the Balkans: 42. - *Phytol. Balcan.* **26(2)**: 187-216.

VLADIMIROV, V., AYBEKE, M. & TAN, KIT (EDS) 2020c: New floristic records in the Balkans: 43. - *Phytol. Balcan.* **26(3)**: 537-572.

VLADIMIROV, V., AYBEKE, M. & TAN, KIT (EDS) 2021a: New floristic records in the Balkans: 44. - *Phytol. Balcan.* **27(1)**: 127-150.

VLADIMIROV, V., AYBEKE, M. & TAN, KIT (EDS) 2021b: New floristic records in the Balkans: 45. - *Phytol. Balcan.* **27(2)**: 239-276.

VLADIMIROV, V., BANCHEVA, S. & DELCHEVA, M. 2015: *Solanum elaeagnifolium* (Solanaceae), a new alien species for the Bulgarian flora. - *Fl. Medit.* **25**: 121-125.

VLADIMIROV, V., COŞKUNÇELEBI, K. & TAN, KIT 2015: A new diploid species of *Pilosella* (Asteraceae) from Turkey. - *Turk. J. Bot.* **39**: 70-75.

VLADIMIROV, V., DANE, F. & TAN, KIT (EDS) 2008a: New floristic records in the Balkans: 7. - *Phytol. Balcan.* **14(1)**: 131-148.

VLADIMIROV, V., DANE, F. & TAN, KIT (EDS) 2008b: New floristic records in the Balkans: 8. - *Phytol. Balcan.* **14(2)**: 291-304.

VLADIMIROV, V., DANE, F. & TAN, KIT (EDS) 2009: New floristic records in the Balkans: 10. - *Phytol. Balcan.* **15(1)**: 115-139.

VLADIMIROV, V., DANE, F. & TAN, KIT (EDS) 2011: New floristic records in the Balkans: 17. - *Phytol. Balcan.* **17(3)**: 361-384.

VLADIMIROV, V., DANE, F. & TAN, KIT (EDS) 2012: New floristic records in the Baklans: 19. - *Phytol. Balcan.* **18(2)**: 205-230.

VLADIMIROV, V., DANE, F. & TAN, KIT (EDS) 2015a: New floristic records in the Balkans: 26. - *Phytol. Balcan.* **21(1)**: 53-91.

VLADIMIROV, V., DANE, F. & TAN, KIT (EDS) 2015b: New floristic records in the Balkans: 28. - *Phytol. Balcan.* **21(3)**: 367-399.

VLADIMIROV, V., DIHORU, G. & TAN, KIT (EDS) 2006: New floristic records in the Balkans: 3. - *Phytol. Balcan.* **12(3)**: 413-440.

VLADIMIROV, V. & DIMITROVA, D. 2006: *Leontodon tuberosus* (Asteraceae: Cichorioideae): a new species to the Bulgarian flora. - *Phytol. Balcan.* **12(1)**: 63-65.

VLADIMIROV, V. & DIMITROVA, D. 2007: Chromosome numbers of selected species from the Bulgarian flora. - *Phytol. Balcan.* **13(3)**: 379-381.

VLADIMIROV, V. & PETROVA, A. 2009: *Senecio inaequidens* (Asteraceae): a new alien species for the Bulgarian flora. - *Phytol. Balcan.* **15(3)**: 373-375.

VLADIMIROV, V., PETROVA, A.S. & ASSYOV, B. 2014: *Euphorbia prostrata* - a new alien species to the Bulgarian flora. - *Dokl. Bulg. Akad. Nauk* **67(4)**: 527-532.

VLADIMIROV, V. & SZELĄG, Z. 2000: Chromosome numbers of eight *Hieracium* species from Bulgaria. In Özhatay, N. & al. (eds): Abstracts from the Second Balkan Botanical Congress, Istanbul, Turkey, May 14-18, 2000, p. 119.

VLADIMIROV, V. & SZELĄG, Z. 2001a: Reports (1271-1277). In Kamari, G., Blanché, C. & Garbari, F. (eds): Mediterranean chromosome number reports. - *Fl. Medit.* **11**: 478-483.

VLADIMIROV, V. & SZELĄG, Z. 2001b: Chromosome numbers in selected species of *Hieracium* subgenus *Pilosella* (Asteraceae) from Bulgaria. - *Polish Bot. J.* **46(2)**: 269-273.

VLADIMIROV, V. & SZELĄG, Z. 2006: A new diploid species of *Hieracium* sect. *Pannosa* (Asteraceae) from Bulgaria. - *Bot. J. Linn. Soc.* **150(2)**: 261-265.

VLADIMIROV, V. & SZELĄG, Z. 2018: Diploid *Hieracium* and *Pilosella* species (Asteraceae) on the Balkan Peninsula. - *Bot. Serbica* **42 (suppl. 1)**: 55.

VLADIMIROV, V., TAN, KIT & STEVANOVIĆ, V. (EDS) 2006: New floristic records in the Balkans: 1. - *Phytol. Balcan.* **12(1)**: 107-128.

VLADIMIROV, V. & TAN, KIT (EDS) 2014: New floristic records in the Balkans: 24. - *Phytol. Balcan.* **20(1)**: 99-137.

VLADIMIROV, V. & TAN, KIT (EDS) 2016: New floristic records in the Balkans: 31. - *Phytol. Balcan.* **22(3)**: 429-467.

VOGEL, A. 1997: Die Verbreitung, Vergesellschaftung und Populationsökologie von *Corrigiola*

litoralis, Illecebrum verticillatum und *Herniaria glabra* (Illecebraceae). - *Diss. Bot.* **289**: x + 282 pp. + tables.
VOGEL, S. 1977: Zur *Ophrys*-Bestäubung auf Kreta. - *Orchidee (Hamburg)* **Sonderheft**: 131-140.
VOGT, R., LACK, H.W. & RAUS, T. 2018: The herbarium of Ignaz Dörfler in Berlin. - *Willdenowia* **48(1)**: 57-92.
VOGT, R. & OBERPRIELER, CH. 1993: Reports (171-176). In Kamari, G., Felber, F. & Garbari, F. (eds): Mediterranean chromosome number reports - 3. - *Fl. Medit.* **3**: 338-340.
VOGT, R. & WILLING, E. 2015: *Otospermum glabrum*. In Raab-Straube, E. & Raus, Th. (eds): Euro+Med-Checklist notulae, 5. - *Willdenowia* **45(3)**: 452.
VOKOU, D., KATRADI, D. & KOKKINI, S. 1996: Ethnobotanical survey of Zagori (Epirus, Greece), a renowned centre of folk medicine in the past. - *J. Ethnopharmacol.* **39**: 187-196.
VOKOU, D., KOKKINI, S. & BESSIÈRE, J.M. 1993: Geographic variation of Greek oregano (*Origanum vulgare* ssp. *hirtum*) essential oils. - *Biochem. Syst. Ecol.* **21(2)**: 287-295.
VOKOU, D., PETANIDOU, T. & BELLOS, D. 1990: Pollination ecology and reproductive potential of *Jankaea heldreichii* (Gesneriaceae); a tertiary relict on Mt Olympus, Greece. - *Biol. Conserv.* **52**: 125-133.
VOLAKAKIS, I. & YANNITSAROS, A. 1995: I chlorida ke i vlastisi tis nisidas Karga (V. Kriti). In: Praktika, 17. Epistimoniko Sinedrio, Elliniki Eteria Viologikon Epistimon, pp. 440-442 [In Greek with English summary]. - Patras.
VOLARIĆ-MRŠIĆ, I. & HORVATIĆ, N. 1977: *Arbutus andrachne* L. on the island of Mljet [In Serbo-Croatian with English abstract]. - *Acta Bot. Croat.* **36**: 177-178 + 1 plate and 1 map.
VOLIOTIS, D. 1967: Untersuchungen über die Vegetation und die Flora des Cholomongebirges unter besonderer Berücksichtigung der aromatischen, Heil- und Bienenzuchtpflanzen [In Greek with German summary]. - *Sci. Ann. Fac. Phys. Math. Aristot. Univ. Thessalonika* **10(4)**: 1-131.
VOLIOTIS, D. 1973: Contribution to the knowledge of the Greek flora: New taxa of the mountainous region Cholomon, Greek Macedonia. - *Feddes Repert.* **84(4)**: 327-328.
VOLIOTIS, D. 1975: Die Vegetationsstufung einiger Gebirge im Nordgriechenland (Voras, Vermion, Pieria). In Jordanov, D. & al. (eds): Problems of Balkan flora and vegetation, pp. 391-399. - Sofia: Publishing House of the Bulgarian Academy of Sciences.
VOLIOTIS, D. 1976a: Über die Vegetation und Flora des Lailias-Gebirges in Nordgriechenland. In Dafis, S. & Landolt, E. (eds): Zur Vegetation und Flora von Griechenland, Band 2. - *Veröff. Geobot. Inst. E.T.H. Stiftung Rübel Zürich* **56**: 21-58.
VOLIOTIS, D. 1976b: Die Gehölzvegetation und die Vegetationszonierung des nordgriechischen Gebirgszuges Voras - Vermion - Pieria - Olymp - Ossa. - *Bot. Jahrb. Syst.* **97(1)**: 120-154.
VOLIOTIS, D. 1976c: Die Gehölzvegetation und die Vegetationszonierung des Gebirgszuges Tymfristos-Oeta-Parnassos (Griechenland). - *Candollea* **31(1)**: 37-51.
VOLIOTIS, D. 1977a: Flore et végétation de la région montagneuse de Lailia (Serres), Grèce [In Greek with French summary]. - *Biol. Gallo-Hellen.* **6, Suppl.**: 1-90.
VOLIOTIS, D. 1977b: O itamos (*Taxus baccata*) ke i simasia tou dia ton anthropon [In Greek].- *Farmak. Deltion, Epist. Ekdosis* **3(1)**: 22-36.
VOLIOTIS, D. 1978: Einige neue und seltene *Ranunculus*-Arten der griechischen Flora. - *Feddes Repert.* **89(2-3)**: 121-132.
VOLIOTIS, D. 1981a: Neue und seltene Taxa für die griechische Flora aus dem Voras-Gebirge. VI. - *Bot. Chron. (Patras)* **1(2)**: 115-123.
VOLIOTIS, D. 1981b: Flora und Vegetation des Voras-Gebirges. - *Sci. Ann. Fac. Phys. Math. Aristot. Univ. Thessalonika* **19**: 189-278. [Dated 1979 but actually published in 1981].
VOLIOTIS, D. 1981c: Über die Orchideen des Voras-Gebirges. - *Mitt. Arbeitskreis Heimische Orchid. Baden-Württemberg* **13(2)**: 205-219.
VOLIOTIS, D. 1981d: Neue und seltene Arten (bzw. Unterarten) für die griechische Flora aus dem Voras-Gebirge. III. - *Acta Bot. Croat.* **40**: 251-260.
VOLIOTIS, D. 1982a: Neue und seltene Arten (bzw. Unterarten) für die griechische Flora aus dem Voras-Gebirge. 4. - *Webbia* **35(2)**: 311-321.
VOLIOTIS, D. 1982b: Über die Orchideen des Tzena-Gebirges. - *Mitt. Arbeitskreis Heimische Orchid. Baden-Württemberg* **14(4)**: 473-482.
VOLIOTIS, D. 1982c: Neue und seltene Arten (bzw. Unterarten) für die griechische Flora aus dem Voras-Gebirge. II. - *Biol. Gallo-Hellen.* **9(2)**: 243-260.
VOLIOTIS, D. 1982d: Neue Funde von neun in Griechenland seltenen Gefässpflanzen aus dem Voras- Gebirge. - *Phyton (Horn)* **22(1)**: 9-22.
VOLIOTIS, D. 1982e: Neue und seltene Arten (bzw. Unterarten) für die griechische Flora aus dem Voras-Gebirge. V. - *Bauhinia* **7(3)**: 155-166.
VOLIOTIS, D. 1983a: Zur Kenntnis einiger Saxifragaceen im griechischen Raum. - *Feddes Repert.* **94 (7-8)**: 563-573.
VOLIOTIS, D. 1983b: Neue und seltene Arten (bzw. Unterarten) für die griechische Flora aus dem Voras-Gebirge. I. - *Feddes Repert.* **94(7-8)**: 575-589.
VOLIOTIS, D. 1983c: Zur Flora des Tzena-Gebirges in Nord-Griechenland. - *Acta Bot. Croat.* **42**: 157-172.
VOLIOTIS, D. 1984a: Die Flora der Nestos-Schlucht in Griechenland. - *Phyton (Horn)* **24(2)**: 153-164 + 1 plate.
VOLIOTIS, D. 1984b: Zur Kenntnis der Flora Griechenlands nebst kritischen Bemerkungen. – *Bauhinia* **8(1)**: 23-30.
VOLIOTIS, D. 1984c: Neue Ergebnisse und kritische Betrachtungen zur Flora Griechenlands. - *Feddes Repert.* **95(5-6)**: 409-424.
VOLIOTIS, D. 1984d: A phenological study of flowering period and flower colours of aromatic plants in Greece. - *Vegetatio* **56(3)**: 129-137.
VOLIOTIS, D. 1985a: Über die Verbreitung, Ökologie und Synökologie von *Ostrya carpinifolia* in Griechenland. - *Acta Bot. Hung.* **31(1-4)**: 339-347.
VOLIOTIS, D. 1985b: Seltene und stenotope Gefässpflanzen aus dem griechischen Nestos-Tal. - *Linzer Biol. Beitr.* **17(1)**: 171-190.
VOLIOTIS, D. 1986: Verbreitung einiger seltenen und subendemischen Arten der griechischen Flora mit taxonomischen und chorologischen Betrachtungen. - *Folia Geobot. Phytotax.* **21(1)**: 91-104.
VOLIOTIS, D. 1987a: Die natürliche Flora von Griechenland. - *Acta Bot. Croat.* **46**: 213-224.
VOLIOTIS, D. 1987b: Über Flora und Vegetation des Pinovon-Gebirges in Nord-Griechenland. - *Bauhinia* **8(4)**: 177-193.
VOLIOTIS, D. 1987c: Historical and environmental significance of the yew (*Taxus baccata* L.). - *Israel J. Bot.* **35(1)**: 47-52.
VOLIOTIS, D. 1988a: The flora of Mt. Pinovon, northern Greece [In Greek with English abstract]. - *Bot. Chron. (Patras)* **8(1-2)**: 5-44.
VOLIOTIS, D. 1988b: Über die griechischen Pulsatillen (Ranunculaceae). Taxonomie, Standorte und Verbreitung. - *Phyton (Horn)* **28(1)**: 81-90.
VOLIOTIS, D. 1989: *Silene bupleuroides* L. subsp. *ganiatsasiana* (Caryophyllaceae). a new campion of Greek flora [In Greek with English summary]. - *Bios (Thessaloniki)* **1989**: 257-272.
VOLIOTIS, D. 1990a: Neue Materialien und kritische Kommentare zu "Floram Hellenicam" (Griechenland). - *Ann. Mus. Goulandris* **8**: 223-236.
VOLIOTIS, D. 1990b: *Silene bupleuroides* L. subsp. *ganiatsasiana* (Caryophyllaceae) a new campion of Greek flora. - *Razprave IV. Razreda Sazu* **31(26)**: 397-413.
VOLIOTIS, D. 1991: *Silene bupleuroides* L. ssp. *ganiatsasiana* (Caryophyllaceae), a new campion in the Greek flora. - *Israel J. Bot.* **40(3)**: 225-236.
VOLIOTIS, D. & ATHANASIADIS, N. 1971: Dendra ke thamni [Trees and shrubs; in Greek]. - Thessaloniki: 294 pp.
VOLIOTIS, D. & DROSSOS, E.G. 1983: A study of an extensive biotope of the aromatic sea daffodil (*Pancratium maritimum*) near Aphytos, Kassandra Peninsula, northern Greece. - *Bauhinia* **7(4)**: 229-242.
VOLIOTIS, D. & KARAGIANNAKIDOU, V. 1984: Verbreitung der aromatischen Orchideen in Griechenland. - *Orchidee (Hamburg)* **35(1)**: 21-27.

VOLKOVA, P. & AL. 2020: Out of Colchis: The colonization of Europe by *Primula vulgaris* Huds. (Primulaceae). - *Acta Soc. Bot. Poloniae* **89(3)**: 15 pp.

VOLLMAR, F. 1940: Einiges über den Formenkreis von *Thalictrum aquilegifolium* L. - *Ber. Bayer. Bot. Ges.* **24**: 62-66.

VONK, D.H. 1979: Biosystematic studies of the *Carex flava* complex I. Flowering. - *Acta Bot. Neerl.* **28(1)**: 1-20.

VOSA, C.G. 1976: Chromosome banding patterns in cultivated and wild barleys (*Hordeum* spp). - *Heredity* **37(3)**: 395-403.

VOSA, C.G. & BARLOW, P.W. 1972: Meiosis and B-chromosomes in *Listera ovata*. - *Caryologia* **25(1)**: 1-8.

VÖTH, W. 1967: *Ophrys bombyliflora* auf Rhodos. - *Orchidee (Hamburg)* **18(2)**: 58-60.

VÖTH, W. 1978: *Himantoglossum affine* auf dem Peloponnes. - *Orchidee (Hamburg)* **29(3)**: 113-114.

VÖTH, W. 1981a: Fundorte griechischer Orchideen. - *Mitt. Arbeitskreis Heimische Orchid. Baden-Württemberg* **13(1)**: 1-89.

VÖTH, W. 1981b: Gattungshybride: *Serapias vomeracea* subsp. *laxiflora* × *Orchis morio* subsp. *picta*. - *Orchidee (Hamburg)* **32(4)**: 138-142.

VÖTH, W. 1982a: Blütenökologische Untersuchungen an *Epipactis atrorubens*, *E. helleborine* und *E. purpurata* in Niederösterreich. - *Mitt. Arbeitskreis Heimische Orchid. Baden-Württemberg* **14(4)**: 393-437.

VÖTH, W. 1982b: Die "ausgeborgten" Bestäuber von *Orchis pallens* L. - *Orchidee (Hamburg)* **33(5)**: 196-203.

VÖTH, W. 1982c: Ergänzung zu Fundorten griechischer Orchideen. - *Mitt. Arbeitskreis Heimische Orchid. Baden-Württemberg* **14(4)**: 483-488.

VÖTH, W. 1984: Bestäubungsbiologische Beobachtungen an griechischen *Ophrys*-Arten. - *Mitt. Arbeitskreis Heimische Orchid. Baden-Württemberg* **16(1)**: 1-20.

VÖTH, W. 1987: Neue bestäubungsbiologische Beobachtungen an griechischen *Ophrys*-Arten. - *Mitt. Arbeitskreis Heimische Orchid. Baden-Württemberg* **19(1)**: 112-118.

VÖTH, W. 1989a: Die Hybride aus den Eltern *Ophrys attica* und *O. heldreichii*. - *Orchidee (Hamburg)* **40(2)**: 56-59.

VÖTH, W. 1989b: Orchideen und Orchideenstandorte bei Galaxidi in Südgriechenland. - *Mitt.-Bl. Wiener-Niederösterr. Arb.-Gem. Heim. Orchid.* **24**: 1-5.

VÖTH, W. & EHRENDORFER, F. 1976: Biometrische Untersuchungen an Populationen von *Ophrys cornuta*, *O. holosericea* und ihren Hybriden (Orchidaceae). - *Pl. Syst. Evol.* **124(4)**: 279-290.

VOTSI, N.-E. & AL. 2013: Distribution pattern of the invasive plant *Solanum elaegnifolium* in Greece. In Gkelis, S. & al. (eds): 13th Panhellenic Scientific Conference, Hellenic Botanical Society, Thessaloniki 3-6 October. Program and book of abstracts, p. 54.

VRAHNAKIS, M.S. & AL. 2014: Floristic diversity of valonia oak silvopastoral woodlands in Greece. - *Agroforest Syst.* **88**: 877-893.

VUILLE, F.-L. 1987: Reproductive biology of the genus *Damasonium* (Alismataceae). - *Pl. Syst. Evol.* **157(1-2)**: 63-71.

VUILLEMIN, F. 1991: Rapports (45-45f). In Kamari, G., Felber, F. & Garbari, F.: Mediterranean chromosome number reports - 1. - *Fl. Medit.* **1**: 243-245.

VUILLEMIN, F. 1992: Rapports (120). In Kamari, G., Felber, F. & Garbari, F. (eds): Mediterranean chromosome number reports - 2. - *Fl. Medit.* **2**: 273-275.

VUILLEUMIER, B.S. 1969: (263) Proposal for the conservation of the generic name (9574) *Leontodon* L. with *L. hispidus* L. as typus conservandus against *Virea* Adanson. - *Taxon* **18(3)**: 343-345. [approved, cf. Taxon 19(5): 815, 1970].

VUKIĆEVIĆ, E., CINCOVIĆ, T. & KOJIĆ, M. 1976: The walnut (*Juglans regia* L.) in west Serbia (I contribution) [In Serbo-Croatian with English summary]. - *Glasn. Prir. Mus. Beogradu, Ser. B, Biol. Nauke* **31**: 55-64.

VUKOJEVIĆ, Đ., MITIĆ, Z.S. & ZLATKOVIĆ, B. 2018: Morphological variability of *Tragopogon pterodes* Pančić ex Petrović achenes: taxonomic evaluation of heterocarpy. - *Pl. Biosyst.* **152(5)**: 937-944.

VUKOJIČIĆ, S. & AL. 2021: Nomenclatural notes and typification of the names of plant taxa described by Josif Pančić from Montenegro. - *Phytotaxa* **490(1)**: 1-17.

VUKOJIČIĆ, S. & AL. 2014: Typification of the plant names described by Nedeljko Košanin from the Balkan Peninsula. - *Phytotaxa* **163(3)**: 104-112.

VUKOVIĆ, N., TOMMASONI, A. & D'ONOFRIO, T. 2013: The orchid *Ophrys speculum* Link (Orchidaceae) in Croatia. - *Acta Bot. Croat.* **72(1)**: 185-191.

VUKSANOVIĆ, S, & AL. 2016: Balkan endemic vascular plants of Montenegro - critical inventory with chorological and life-form analyses. - *Willdenowia* **46(3)**: 387-397.

VUKSANOVIĆ, S. & PETROVIĆ, D. 2004: In spite of prevailing opinion to the contrary *Kickxia cirrhosa* (L.) Fritisch grows on the Balkan Peninsula. In Vasić, O. (ed.): Abstracts, XI OPTIMA Meeting, Beograd 5-11.IX.2004, p. 105.

VUKSANOVIĆ, S. & PETROVIĆ, D. 2007: The occurrence of *Kickxia cirrhosa* (L.) Fritsch in Montenegro supports the earlier records of this species for the Balkan Peninsula. - *Bocconea* **21**: 201-205.

VURAL, M. & TAN, KIT 1983: New taxa and records from Turkey. - *Notes Roy. Bot. Gard. Edinburgh* **41 (1)**: 65-76.

WACQUANT, J.P. 1990: Biogeographical and physiological aspects of the invasion by *Dittrichia* (ex *Inula*) *viscosa* ... In Di Castri, F., Hansen, A.J. & Debussche, M. (eds): Biological invasions ..., pp. 353-364. - Dordrecht, etc.: Kluwer Academic Publishers.

WAGENITZ, G. 1955: Pollenmorphologie und Systematik in der Gattung *Centaurea* L. s.l. - *Flora* **142 (2)**: 213-279.

WAGENITZ, G. 1960: Joseph Bornmüller 1862-1948 [obituary]. - *Willdenowia* **2(3)**: 343-360.

WAGENITZ, G. 1962: Pflanzen von der Orientreise Tourneforts im Herbar Willdenow in Berlin. - *Willdenowia* **3(1)**: 109-136.

WAGENITZ, G. 1965: Zur Systematik und Nomenklatur einiger Arten von *Filago* L. emend. Gaertn. subgen. *Filago* ("Filago germanica"-Gruppe). - *Willdenowia* **4(1)**: 37-59.

WAGENITZ, G. 1968a: *Filago desertorum* Pomel und *F. hurdwarica* (DC.) Wagenitz, zwei verkannte Arten der "Filago germanica"-Gruppe aus Nordafrika, Vorder- und Zentralasien. - *Willdenowia* **4(3)**: 283-298.

WAGENITZ, G. 1968b: Weitere Beiträge zur Kenntnis der Gattung *Filago* (Compositae-Inuleae). - *Willdenowia* **5(1)**: 55-66.

WAGENITZ, G. 1969: Abgrenzung und Gliederung der Gattung *Filago* L. s.l. (Compositae-Inuleae). - *Willdenowia* **5(3)**: 395-444.

WAGENITZ, G. 1970a: Die Gattung *Filago* L. s.l. (Compositae-Inuleae) in der Ägäis. - *Willdenowia* **6 (1)**: 115-138.

WAGENITZ, G. 1970b: Über die Verbreitung einiger *Filago*-Arten. - *Feddes Repert.* **81(1-5)**: 107-117.

WAGENITZ, G. 1971a: *Centaurea pseudocadmea*, eine neue Art der Sektion *Phalolepis* aus Griechenland. - *Ann. Naturhist. Mus. Wien* **75**: 243-247.

WAGENITZ, G. 1971b: Zur taxonomischen Stellung und Nomenklatur von *Micropus longifolius* (Compositae-Inuleae). - *Oesterr. Bot. Z.* **119(4-5)**: 399-403.

WAGENITZ, G. 1972: Beiträge zur Kenntnis der Gattung *Centaurea* L. 1. Zur Taxonomie türkischer Arten der Sektionen *Acrolophus* und *Acrocentron*. - *Willdenowia* **6(3)**: 479-507.

WAGENITZ, G. 1974: Parallele Evolution von Merkmalen in der Gattung *Centaurea*. - *Phyton (Horn)* **16 (1-4)**: 301-312.

WAGENITZ, G. 1975: Floristic connections between the Balkan Peninsula and the Near East as exemplified by the genus *Centaurea*. In Jordanov, D. & al. (eds): Problems of Balkan flora and vegetation, pp. 223-228. - Sofia: Publishing House of the Bulgarian Academy of Sciences.

WAGENITZ, G. 1981: *Centaurea melanocephala* Pančić (Compositae) eine endemische Art Serbiens - kein Sektionsbastard. - *Phyton (Horn)* **21(2)**: 289-294.

WAGENITZ, G. 1983a: Die Gattung *Adenostyles* Cass. (Compositae-Senecioneae). - *Phyton (Horn)* **23 (1)**: 141-159 + 2 plates.

WAGENITZ, G. 1983b: *Centaurea* and the Index Kewensis. - *Taxon* **32(1)**: 107-109.

WAGENITZ, G. 1983c: Die einjährigen *Centaurea*-Arten der Sektion *Cyanus* DC. - *Tuexenia* **3**: 535-544.

WAGENITZ, G. 1986: *Centaurea* in South-West Asia: patterns of distribution and diversity. - *Proc. Roy. Soc. Edinburgh, B, Biol. Sci.* **89**: 11-21.

WAGENITZ, G. 1989: Nahe Verwandtschaft zwischen Arten der *Centaurea*-Sektionen *Acrolophus* und *Phalolepis*. - *Flora, Morphol. Geobot. Oekophysiol.* **182(5-6)**: 341-351.

WAGENITZ, G. & AL. 2018: A synopsis of the *Centaurea soskae* and *triniifolia* group (*Centaurea* sect. *Acrolophus*) in the Prespa area and Northern Pindos. - *Phytotaxa* **348(2)**: 77-89.

WAGENITZ, G. & BEDARFF, U. 1989: Taxonomic notes on some species of the genus *Cichorium* (Compositae-Lactuceae). In Tan, Kit (ed.): Plant taxonomy, phytogeography and related subjects. The Davis & Hedge festschrift, pp. 11-21. - Edinburgh: Edinburgh Univ. Press.

WAGENITZ, G., ERTUGRUL, K. & DURAL, H. 1998: A new species in *Centaurea* sect. *Psephelloideae* (Compositae) from SW Turkey. - *Willdenowia* **28(1-2)**: 157-161.

WAGENITZ, G. & GAMAL-ELDIN, E. 1985: Zur Kenntnis der griechischen *Centaurea*-Arten der Sektion *Acrocentron*. - *Bot. Jahrb. Syst.* **107(1-4)**: 95-127.

WAGENITZ, G. & HELLWIG, P. 1995: Organization of biodiversity within Mediterranean genera: *Centaurea*. In Valdés, B. & Silvestre, S. (eds): Abstracts. VIII OPTIMA meeting, Sevilla, Spain, 25 September - 1 October, 1995, p. 20.

WAGENITZ, G., TAN, KIT & BERGMEIER, E. 2017: Report 88. In Vladimirov, V., Aybeke, M. & Tan, Kit (eds): New floristic records in the Balkans: 32. - *Phytol. Balcan.* **23(1)**: 140-141.

WAGENSOMMER, R.P. & AL. 2017: First record for the flora of Italy and lectotypification of the name *Linum elegans* (Linaceae). - *Phytotaxa* **296(2)**: 161-170.

WAGENSOMMER, R.P., FRÖHLICH, T. & FRÖHLICH, M. 2014: First record of the southeast European species *Cerinthe retorta* Sibth. & Sm. (Boraginaceae) in Italy and considerations on its distribution and conservation status. - *Acta Bot. Gallica* **161(2)**: 111-115.

WAGENSOMMER, R.P., PERRINO, E.V. & SILLETTI, G.N. 2014: *Carex phyllostachys* C. A. Mey. (Cyperaceae) new for Italy and phytogeographical considerations. - *Phyton (Horn)* **54(2)**: 215-222.

WAGENSOMMER, R.P. & VENANZONI, R. 2021: *Geranium lucarinii* sp. nov. and re-evaluation of *G. kikianum* (Geraniaceae). - *Phytotaxa* **489(3)**: 252-262.

WAGNER, I. 1996: Zusammenstellung morphologischer Merkmale und ihrer Ausprägungen zur Unterscheidung von Wild- und Kulturformen des Apfel- (*Malus*) und des Birnbaumes (*Pyrus*). - *Mitt. Deutsch. Dendrol. Ges.* **82**: 87-108.

WAGNER, J. 1932: Lindenstudien. - *Magyar Bot. Lapok* **31(1-12)**: 55-60.

WAGNER, J. 1935: Beiträge zur Flora von Griechenland. - *Repert. Spec. Nov. Regni Veg.* **38**: 281-288 + 3 plates.

WAGNER, M. DE N. 1948: Variações cromosómicas em duas espécies do género *Sideritis* L. - *Agron. Lusit.* **10(3)**: 163-169 + 1 plate.

WAGNER, M. DE N. 1949: Subsídio para o estudo citológico do género *Luzula* DC. (I). - *Genét. Ibér.* **1 (1-2)**: 59-67.

WAGNER, S., DODINET, E. & SELOSSE, M.-A. 2015: A touch of orchids from Samos (Greece). - *Botany Letters* **162(4)**: 251-253.

WAHLSTEEN, E. & TYLER, T. 2019: Morphometric analysis and species delimitation in *Legousia* (Campanulaceaee). - *Willdenowia* **49(1)**: 21-33.

WAINES, J.G. 1975: The biosystematics and domestication of peas (*Pisum* L.). - *Bull. Torrey Bot. Club* **102(6)**: 385-395.

WAINES, J.G. & BARNHART, D. 1992: Biosystematic research in *Aegilops* and *Triticum*. – *Hereditas (Lund)* **116(3)**: 207-212.

WAINES, J.G., HILU, K. & SHARMA, H. 1982: Species formation in *Aegilops* and *Triticum*. In Estes, J.R., Tyrl, R.J. & Brunken, J.N. (eds): Grasses and grasslands. Systematics and ecology, pp. 89-108. - Norman: Univ. of Oklahoma Press.

WAISBECKER, A. 1899: Beiträge zur Kenntnis der Gattung *Odontites*. - *Oesterr. Bot. Z.* **49(12)**: 437-442.

WAISEL, Y. 1959: Ecotypic variation in *Nigella arvensis* L. - *Evolution* **13(4)**: 469-475.

WALAS, Ł. & AL. 2018: The present status and potential distribution of relict populations of *Aesculus hippocastanum* L. in Greece and the diverse infestation by *Cameraria orchidella* Deschka & Dimić. - *Pl. Biosyst.* **152(5)**: 1048-1058.

WALAS, Ł. & AL. 2019: Spatial genetic structure and diversity of natural populations of *Aesculus hippocastanum* L. in Greece. - *PLoS ONE* **14**: doi.org/10.1371/journal.pone.0226225.

WALDMANN, H. 1971: Seltene Orchideen-Bastarde in Griechenland. - *Orchidee (Hamburg)* **22(5)**: 205-207.

WALDSTEIN, F. DE P.A. & KITAIBEL, P. 1799-1812: Descriptiones et icones plantarum rariorum Hungariae. - Viennae: Typis Matthiae Andreae Schmidt, 3 vols. [cf. TL-2 7: 31, 1988].

WALE, R.S. 1938: Some new Sempervivums from the Balkans. - *Quart. Bull. Alpine Gard. Soc. Gr. Brit.* **6(2)**: 97-105.

WALE, R.S. 1940a: Notes on some Graecian Semperviva. - *Bull. Misc. Inform. Kew* **1940(4)**: 141-147 + 1 plate.

WALE, R.S. 1940b: Semperviva of the Balkan Peninsula. - *Quart. Bull. Alpine Gard. Soc. Gr. Brit.* **8 (3)**: 200-218.

WALKER, T.G. 1962: Cytology and evolution in the fern genus *Pteris* L. - *Evolution* **16(1)**: 27-43.

WALLIS, R. 2009: *Galanthus peshmenii* A.P. Davis & C.D. Brickell ´Kastellorhizo´, A.M. - *Alpine Gardener* **77(1)**: 31-32.

WALLISCH, R. 1930: Die Chromosomenverhältnisse bei *Tilia platyphyllos, Tilia cordata* und *Tilia argentea*. - *Oesterr. Bot. Z.* **79(2)**: 97-106.

WALLROTH, F.W. 1842: Monographischer Versuch über die Gewächs-Gattung *Agrimonia* Celsi. In: Beiträge zur Botanik. Eine Sammlung monographischer Abhandlungen über besonders schwierige Gewächs-Gattungen, 1(1): 1-61 + 1 plate. - Leipzig: F. Hofmeister.

WALPOLE, R. 1817: Plants of Greece. Medicinal and economical uses. (From Dr. Sibthorp's papers). In: Memoirs relating to European and Asiatic Turkey, pp. 235-254. - London: Longman, etc.

WALTER, H. 1956: Vegetationsgliederung Anatoliens. - *Flora* **143(2)**: 295-326.

WALTER, H. 1975: Betrachtungen zur Höhenstufenfolge im Mediterrangebiet (insbes. in Griechenland) in Verb. mit dem Wettbewerbsfaktor. In Dafis, S. & Landolt, E. (eds): Zur Vegetation und Flora von Griechenland, Band 1. - *Veröff. Geobot. Inst. E.T.H. Stiftung Rübel Zürich* **55**: 72-83.

WALTER, R. 1973: *Holcus mollis* in Poland: 1. Karyological investigations. - *Acta Biol. Cracov., Ser. Bot.* **16(2)**: 227-233 + 1 plate.

WALTER, R. 1977: *Holcus mollis* L. in Poland. II. Studies in the origin of polyploid types. - *Acta Biol. Cracov., Ser. Bot.* **20(1-2)**: 113-131 + 2 plates.

WALTER, R. & KUTA, E. 1971: Cytological and embryological studies in *Sonchus* L. I. *Sonchus asper* (L.) Hill and *Sonchus oleraceus* L. - *Acta Biol. Cracov., Ser. Bot.* **14(1)**: 103-109 + 2 plates.

WALTERS, J., VEKSLYARSKA, T. & DOBEŠ, CH. 2015: Flow cytometric, chromosomal and morphometric analyses challenge current taxonomic concepts in the *Portulaca oleracea* complex (Portulacaceae, Caryophyllales). - *Bot. J. Linn. Soc.* **179(1)**: 144-156.

WALTERS, S.M. 1949: Biological Flora of the British Isles. *Eleocharis* R. Br. - *J. Ecol.* **37(1)**: 192-206.

WALTERS, S.M. 1953: *Montia fontana* L. - *Watsonia* **3(1)**: 1-6.

WALTERS, S.M. 1962: Generic and specific concepts and the European flora. - *Preslia* **34(2)**: 207-226.

WALTERS, S.M. 1964: [New combinations in *Erophila*]. In Heywood, V.H. (ed.): Notulae systematicae ad Floram Europaeam spectantes, no. 3. - *Feddes Repert. Spec. Nov. Regni Veg.* **69(1)**: 56-57.

WALTERS, S.M. 1967: Chromosome numbers of European species of *Potentilla* L. grex *Aureae* Wolf. - *Acta Fac. Rerum Nat. Univ. Comen., Bot.* **14**: 7-11.

WALTERS, S.M. 1972: Endemism in the genus *Alchemilla* in Europe. In Valentine, D.H. (ed.): Taxonomy, phytogeography and evolution, pp. 301-305. - London & New York: Academic Press.

WALTERS, S.M. 1981: Aims and methods in mapping the endemic vascular plants of the Balkan Peninsula. In Velčev, V.I. & Kožuharov, S.I. (eds): Mapping the flora of the Balkan Peninsula, pp. 83-94. - Sofia: Publishing House of the Bulgarian Academy of Sciences.

WALTERS, S.M. 1992: Notes on *Moehringia* L. (Caryophyllaceae). In Newton, M.E. (ed.): Flora Europaea: Notulae systematicae ad Floram Europaeam spectantes. Series 2. No. 5. - *Bot. J. Linn. Soc.* **109(3)**: 323-324.

WALTERS, S.M. & KING, C.J. (EDS) 1975: European floristic and taxonomic studies. B.S.B.I. conference report no. 15. A conference held in Cambridge 29 June to 2 July 1974. - Oxford: E.W. Classey Ltd, ix + 144 pp.

WALTHER, E. 1949a: Zur Morphologie und Systematik des Arzneibaldrians in Mitteleuropa. - *Mitt. Thüring. Bot. Ges., Beih.* **1**: 1-108.

WALTHER, E. 1949b: *Valeriana*-Studien. I. *Valeriana montana* L. und *Valeriana tripteris* L. - *Mitt. Thüring. Bot. Ges.* **1(1)**: 144-167.

WALTHER, K. 1969: Halmfrucht-Gesellschaften in Griechenland. - *Vegetatio* **18**: 263-272 + 5 tables.

WARBURG, E.F. 1938a: Taxonomy and relationship in the Geraniales in the light of their cytology. Parts I-II. - *New Phytol.* **37(2-3)**: 130-159, 189-210.

WARBURG, O. 1929: Heimat und Geschichte der Lilie (*Lilium candidum*). - *Repert. Spec. Nov. Regni Veg. Beih.* **56**: 167-204.

WARBURG, O. & WARBURG, E.F. 1930: A preliminary study of the genus *Cistus*. - *J. Roy. Hort. Soc.* **55(1)**: 1-52 + 6 plates.

WARD, S. 1981: The cytology of the genus *Cyclamen*. In Bailey, R.H. (ed.): Cyclamen 1980. Proceedings of the second conference of the Cyclamen Society, pp. 17-25. - Ilford, Essex: Cyclamen Society.

WARDLE, P. 1961: Biological Flora of the British Isles. *Fraxinus excelsior* L. - *J. Ecol.* **49**: 739-751.

WARNCKE, M. & MAGNUSSEN, M. 1990: *Anthoxanthum odoratum* L. sensu lato i Europa. – Student report, 25 pp. [photocopied].

WÄRNER, CH. 2011: Biological Flora of Central Europe: *Euphorbia palustris* L. - *Perspect. Pl. Ecol., Evol. Syst.* **13(1)**: 55-69.

WARNOCK, M.J. 1993: (1070) Proposal to conserve 2539 *Delphinium* L. (Ranunculaceae) with a conserved type. - *Taxon* **42(2)**: 453-456. [not approved, cf. Taxon 44(4): 609, 1995].

WARWICK, S.I. 1993a: Guide to the wild germplasm of *Brassica* and allied crops. Part I. Taxonomy and genome status in the tribe Brassiceae (Cruciferae). - *Techn. Bull. Centre Land Biol. Resources Res., Ottawa* **1993-14E**: 33 pp.

WARWICK, S.I. 1993b: Guide to the wild germplasm of *Brassica* and allied crops. Part II. Chromosome numbers in the tribe Brassiceae (Cruciferae). - *Techn. Bull. Centre Land Biol. Resources Res., Ottawa* **1993-15E**: 22 pp.

WARWICK, S.I. 1993c: Guide to the wild germplasm of *Brassica* and allied crops. Part III. Interspecific and intergeneric hybridization in the tribe Brassiceae (Cruciferae). - *Techn. Bull. Centre Land Biol. Resources Res., Ottawa* **1993-16E**: 31 pp.

WARWICK, S.I. 1993d: Guide to the wild germplasm of *Brassica* and allied crops. Part IV. Wild species in the tribe Brassiceae (Cruciferae) as sources of agronomic traits. - *Techn. Bull. Centre Land Biol. Resources Res., Ottawa* **1993-17E**: 19 pp.

WARWICK, S.I. 1994: Guide to the wild germplasm of *Brassica* and allied crops. Part V. Life history and geographical data in the tribe Brassiceae (Cruciferae). - *Techn. Bull. Centre Land Biol. Resources Res., Ottawa* **1994-2E**: 61 pp.

WARWICK, S.I. & AL-SHEHBAZ, I.A. 2003: Nomenclatural notes on *Sisymbrium* (Brassicaceae). – *Novon* **13(2)**: 265-267.

WARWICK, S.I. & AL-SHEHBAZ, I.A. 2006: Brassicaceae: chromosome number index and database on CD-Rom. - *Pl. Syst. Evol.* **259(2)**: 237-248.

WARWICK, S.I. & BLACK, L.D. 1991: Molecular systematics of *Brassica* and allied genera (Subtribe Brassicinae, Brassiceae) - chloroplast genome and cytodeme congruence. - *Theor. Appl. Genet.* **82**: 81-92.

WARWICK, S.I. & BLACK, L.D. 1993: Molecular relationships in the subtribe Brassicinae (Cruciferae, tribe Brassiceae). - *Canad. J. Bot.* **71(7)**: 906-918.

WARWICK, S.I. & BLACK, L.D. 1994: Evaluation of the subtribes Moricandiinae, Savignyinae, Vellinae, and Zillinae (Brassicaceae, tribe Brassiceae) using chloroplast DNA restriction site variation. - *Canad. J. Bot.* **72(11)**: 1692-1701.

WARWICK, S.I., FRANCIS, A. & AL-SHEHBAZ, I.A. 2006: Brassicaceae: Species checklist and database on CD-Rom. - *Pl. Syst. Evol.* **259(2)**: 249-258.

WASILJEW, W.N. 1961: Das System der Gattung *Calamagrostis* Roth. - *Feddes Repert. Spec. Nov. Regni Veg.* **63(3)**: 229-251.

WATERFALL, U.T. 1967: *Physalis* in Mexico, Central America and the West Indies. - *Rhodora* **69**: 82-120, 203-239, 319-329.

WATKINSON, A.R. 1978: Biological Flora of the British Isles. *Vulpia fasciculata* (Forskål) Samp. (*Vulpia membranacea* auct., non (L.) Dumort.; *Festuca uniglumis* Aiton). - *J. Ecol.* **66(3)**: 1033-1049.

WATKINSON, A.R., NEWSHAM, K.K. & FORRESTER, L. 1998: Biological Flora of the British Isles. *Vulpia ciliata* Dumort. ssp. *ambigua* (Le Gall) Stace & Auquier (*Vulpia ambigua* (Le Gall) More, *Festuca ambigua* Le Gall). - *J. Ecol.* **86(4)**: 690-705.

WATSON, L. & DALLWITZ, M.J. 1992: The grass genera of the world. - Wallingford: CAB International, 1038 pp. [revised edition 1994].

WATSON, L., DAMANAKIS, M. & DALLWITZ, M.J. 1988: Ta geni ton agrostodon tis Elladas [The grass genera in Greece; in Greek]. - Iraklion: Typokreta, v + 231 pp.

WATSON, M.F. 1989: Nomenclatural aspects of *Oxalis* section *Corniculatae* in Europe. - *Bot. J. Linn. Soc.* **101(4)**: 347-362.

WATSON, S. & WOODWARD, R. 1974: Erythroniums. - *Quart. Bull. Alpine Gard. Soc. Gr. Brit.* **42(1)**: 36-53.

WATSON, W. 1907: *Ferula communis* var. *brevifolia*. - *Bot. Mag.* **133**: tab. 8157 + 2 pp. text.

WATSON, W. 1908: *Cytisus decumbens*. - *Bot. Mag.* **134**: tab. 8230 + 2 pp. text.

WCISŁO, H. 1967: Karyological studies in Caltha palustris L. s.l. - *Acta Biol. Cracov., Ser. Bot.* **10(1)**: 1-23.

WCISŁO, H. 1972: Karyological studies in *Symphytum* L. - *Acta Biol. Cracov., Ser. Bot.* **15(2)**: 153-163 + 1 plate.

WCISŁO, H. 1978: Chromosome numbers in the genus *Polygonum* L. s.l. in Poland. - *Acta Biol. Cracov., Ser. Bot.* **20(1-2)**: 153-165.

WEARN, J. 2015: Risking their lives to collect plants on the Salonica Front. - *Journal of the Salonica Campaign Society 1915-1918* **31**: 8-14.

WEAVER, R.J. 1960: Grape growing in Greece. - *Econ. Bot.* **14(3)**: 207-224.

WEBB, C.J. 1979: Breeding system and seed set in *Euonymus europaeus* (Celastraceae). - *Pl. Syst. Evol.* **132(4)**: 299-303.

WEBB, D.A. 1961a: What is the type of *Sedum telephium* L.? In Heywood, V.H. (ed.): Notulae systematicae ad Floram Europaeam spectantes, no. 1. - *Feddes Repert. Spec. Nov. Regni Veg.* **64 (1)**: 18-19.

WEBB, D.A. 1961b: [Notes on Crassulaceae]. In Heywood, V.H. (ed.): Notulae systematicae ad Floram Europaeam spectantes, no. 1. - *Feddes Repert. Spec. Nov. Regni Veg.* **64(1)**: 20-24.

WEBB, D.A. 1961c: [Notes on *Parnassia palustris*]. In Heywood, V.H. (ed.): Notulae systematicae ad Floram Europaeam spectantes, no. 1. - *Feddes Repert. Spec. Nov. Regni Veg.* **64(1)**: 24-25.

WEBB, D.A. 1963a: [Note on *Sedum sartorianum*]. In Heywood, V.H. (ed.): Notulae systematicae ad Floram Europaeam spectantes, no. 2. - *Feddes Repert. Spec. Nov. Regni Veg.* **68(3)**: 198.

WEBB, D.A. 1963b: [Note on *Saxifraga stellaris*]. In Heywood, V.H. (ed.): Notulae systematicae ad Floram Europaeam spectantes, no. 2. - *Feddes Repert. Spec. Nov. Regni Veg.* **68(3)**: 198-199.

WEBB, D.A. 1963c: [Notes on *Saxifraga cymosa* and *S. pedemontana*]. In Heywood, V.H. (ed.): Notulae systematicae ad Floram Europaeam spectantes, no. 2. - *Feddes Repert. Spec. Nov. Regni Veg.* **68(3)**: 205-206.

WEBB, D.A. 1963d: [Note on *Saxifraga juniperifolia*]. In Heywood, V.H. (ed.): Notulae systematicae ad Floram Europaeam spectantes, no. 2. - *Feddes Repert. Spec. Nov. Regni Veg.* **68(3)**: 208.

WEBB, D.A. 1966: The flora of European Turkey. - *Proc. Roy. Irish Acad., B* **65B**: 1-100.

WEBB, D.A. 1967a: Generic limits in European Lythraceae. In Heywood, V.H. (ed.): Notulae systematicae ad Floram Europaeam spectantes, no. 6. - *Feddes Repert.* **74(1-2)**: 10-13.

WEBB, D.A. 1967b: *Alcea* L. in Europe. In Heywood, V.H. (ed.): Notulae systematicae ad Floram Europaeam spectantes, no. 6. - *Feddes Repert.* **74(1-2)**: 26-27.

WEBB, D.A. 1967c: [Notes on *Daphne*]. In Heywood, V.H. (ed.): Notulae systematicae ad Floram Europaeam spectantes, no. 6. - *Feddes Repert.* **74(1-2)**: 29-30.

WEBB, D.A. 1971: Taxonomic notes on *Antirrhinum* L. In Heywood, V.H. (ed.): Notulae systematicae ad Floram Europaeam spectantes, no. 10. - *Bot. J. Linn. Soc.* **64(3)**: 271-275.

WEBB, D.A. 1972: [Notes on *Pedicularis brachyodonta*]. In Heywood, V.H. (ed.): Notuale systematicae ad Floram Europaeam spectantes, no. 13. - *Bot. J. Linn. Soc.* **65(4)**: 357.

WEBB, D.A. 1978a: Taxonomic notes on *Narcissus* L. In Heywood, V.H. (ed.): Notulae systematicae ad Floram Europaeam spectantes, no. 20. - *Bot. J. Linn. Soc.* **76(4)**: 298-307.

WEBB, D.A. 1978b: The European species of *Galanthus* L. In Heywood, V.H. (ed.): Notulae systematicae ad Floram Europaeam spectantes, no. 20. - *Bot. J. Linn. Soc.* **76(4)**: 307-313.

WEBB, D.A. 1978c: [*Sternbergia lutea* subsp. *sicula*, comb. et stat. nov.]. In Heywood, V.H. (ed.): Notulae systematicae ad Floram Europaeam spectantes, no. 20. - *Bot. J. Linn. Soc.* **76(4)**: 358.

WEBB, D.A. 1987a: Taxonomic and nomenclatural notes on *Saxifraga*. In Chater, A.O. (ed.): Flora Europaea: Notulae systematicae ad Floram Europaeam spectantes. Series 2. No. 1. - *Bot. J. Linn. Soc.* **95(4)**: 227-251.

WEBB, D.A. 1987b: Typification of the Linnean species of *Saxifraga*. - *Bot. J. Linn. Soc.* **95(4)**: 259-272.

WEBB, D.A. & CHATER, A.O. 1963: [Notes on Polygonaceae]. In Heywood, V.H. (ed.): Notulae systematicae ad Floram Europaeam spectantes, no. 2. - *Feddes Repert. Spec. Nov. Regni Veg.* **68 (3)**: 187-189.

WEBB, D.A. & CHATER, A.O. 1972: [Notes on *Linaria genistifolia*]. In Heywood, V.H. (ed.): Notulae systematicae ad Floram Europaeam spectantes, no. 12. - *Bot. J. Linn. Soc.* **65(2)**: 263-264.

WEBB, D.A. & CHATER, A.O. 1978: Notes on the genus *Iris*. In Heywood, V.H. (ed.): Notulae systematicae ad Floram Europaeam spectantes, no. 20. - *Bot. J. Linn. Soc.* **76(4)**: 315-316.

WEBB, D.A. & FERGUSON, I.K. 1967: [*Thymelaea tartonraira* subsp. *argentea*, stat. nov.]. In Heywood, V.H. (ed.): Notulae systematicae ad Floram Europaeam spectantes, no. 6. - *Feddes Repert.* **74(1-2)**: 28.

WEBB, D.A. & GORNALL, R.J. 1989: Saxifrages of Europe. With notes on African, American and some Asiatic species. - London: Christopher Helm Ltd, viii + 307 pp. + plates.

WEBB, D.A. & WALTERS, S.M. 1972: Taxonomic and nomenclatural notes on *Veronica*. In Heywood, V.H. (ed.): Notulae systematicae ad Floram Europaeam spectantes, no. 12. - *Bot. J. Linn. Soc.* **65 (2)**: 266-269.

WEBER, A. 1976: Die Chromosomenzahlen der in Mitteleuropa vorkommenden Arten von *Orobanche* sect. *Orobanche*. - *Pl. Syst. Evol.* **124(4)**: 303-308.

WEBER, A. 1993: Struktur, Antheseverlauf und Bestäubung der Blüte von *Nigella arvensis* (Ranunculaceae). - *Verh. Zool.-Bot. Ges. Österreich* **130**: 99-125.

WEBER, H.C. 1973: Zur Biologie von *Tozzia alpina* L. (Standort, Wirtspflanzen, Entwicklung und Parasitismus). - *Beitr. Biol. Pflanzen* **49(2)**: 237-249.

WEBER, H.C. 1980: Zur Evolution des Parasitismus bei den Scrophulariaceae und Orobanchaceae. - *Pl. Syst. Evol.* **136(3-4)**: 217-232.

WEBER, H.E. 1986: Zur Nomenklatur und Verbreitung der von K.E.A. Weihe aufgestellten Taxa der Gattung *Rubus* L. (Rosaceae). - *Bot. Jahrb. Syst.* **106(3)**: 289-335.

WEBERLING, F., ENDLICH, B. & ENGEL, K. 1971: Zur systematischen Stellung von *Valeriana pancicii* Halácsy & Baldacci und *V. bertiscea* Pančić. - *Oesterr. Bot. Z.* **119(1-3)**: 94-101.

WEBSTER, R.D. 1993: Nomenclature of *Setaria* (Poaceae: Paniceae). - *Sida* **15(3)**: 447-489.

WEBSTER, S.D. 1990: Three natural hybrids in *Ranunculus* L. subgenus *Batrachium* (DC.) A. Gray. - *Watsonia* **18(2)**: 139-146.

WEBSTER, S.D. & SELL, P.D. 1988: Lectotypification of five names in *Ranunculus* subgenus *Batrachium* (DC.) A. Gray. - *Watsonia* **17(2)**: 179-181.

WEGENER, K.-A. 1967: Chromosomenzahlen aus Wurzelspitzen von *Alchemilla*-Arten der Sektionen *Pentaphyllon* Rothm. und *Brevicaulon* Rothm. - *Biol. Zentralbl.* **86(6)**: 771-792.

WEGMÜLLER, S. 1971: A cytotaxonomic study of *Lamiastrum galeobdolon* (L.) Ehrend. & Polatschek in Britain. - *Watsonia* **8(3)**: 277-288.

WEGMÜLLER, S. 1974: Zytotaxonomische Untersuchungen an Sippen von *Lamiastrum galeobdolon* (L.) Ehrend. et Polatschek s.l. aus dem Gebiet der Schweiz. - *Ber. Schweiz. Bot. Ges.* **83(4)**: 274-294.

WEIDE, H. 1962: Systematische Revision der Arten *Pimpinella saxifraga* L. und *Pimpinella nigra* Willd. in Mitteleuropa. - *Feddes Repert. Spec. Nov. Regni Veg.* **64(2-3)**: 240-268.

WEIGEND, M. 2006: *Urtica dioica* subsp. *cypria*, with a re-evaluation of the *U. dioica* group (Urticaceae) in western Asia. - *Willdenowia* **36(2)**: 811-822.

WEIMARCK, G. 1978: B chromosomes in *Heracleum sphondylium* s.lat. in Europe. In Cauwet-Marc, A. M. & Carbonnier, J.: Les Ombellifères, pp. 293-303. - Perpignan.

WEIN, K. 1925a: Beiträge zur Geschichte der Einführung und Einbürgerung einiger Arten von *Xanthium* in Europa. - *Beih. Bot. Centralbl., Abt. 2* **42(1)**: 151-176.

WEIN, K. 1925b: Was ist *Alyssum paniculatum* Desf.? - *Oesterr. Bot. Z.* **74(7-9)**: 198-201.

WEIN, K. 1930: Ist *Lotus graecus* L. (1767) Synonym von *Dorycnium latifolium* Willd. (1800)? - *Repert. Spec. Nov. Regni Veg.* **28**: 207-208.

WEIN, K. 1932: Die älteste Einführungs- und Einbürgerungsgeschichte des *Erigeron canadensis*. - *Bot. Arch.* **34**: 394-418.

WEIN, K. 1954: Die Geschichte von *Datura stramonium*. - *Kulturpflanze* **2**: 18-71.

WEIN, R.W. 1973: Biological Flora of the British Isles. *Eriophorum vaginatum* L. - *J. Ecol.* **61(2)**: 601-615.

WEINERT, E. 1968: Zur Chorologie der submeridionalen Eichen-Hainbuchenwälder des südöstlichen Europa. - *Feddes Repert.* **78(1-3)**: 131-133 + 1 map.

WEINERT, E. 1970: Zur Nomenklatur und Verbreitung der Gattung *Physospermum* Cuss. - *Hercynia, N.F.* **7(1-3)**: 124-130.

WEINSTEIN, A. 1989: Geographic variation and phenology of *Pinus halepensis*, *P. brutia* and *P. eldarica* in Israel. - *Forest Ecol. Managem.* **27(2)**: 99-108.

WEISS, M.E. 1869: Beiträge zur Flora von Griechenland und Creta. - *Verh. K.K. Zool.-Bot. Ges. Wien* **19**: 37-54, 741-758.

WEISSMANN-KOLLMANN, F. 1965: A taxonomic study in *Dianthus* of Palestine and of the neighbouring countries. - *Israel J. Bot.* **14(3)**: 141-148.

WEISSMANN-KOLLMANN, F. 1966: A taxonomic study in *Dianthus* of Palestine and of the neighbouring countries [cont.]. - *Israel J. Bot.* **14(4)**: 149-170.

WEITZ, S. & FEINBRUN, N. 1972: Cytology and systematics of *Narcissus tazetta* L. in Israel. - *Israel J. Bot.* **21(1)**: 9-20.

WEITZEL, C. & AL. 2014: Resurrecting deadly carrots: towards a revision of *Thapsia*.

(Apiaceae) based on phylogenetic analysis of nrITS sequences and chemical profiles. - *Bot. J. Linn. Soc.* **174 (4)**: 620-636.

WELCH, D. 1967: Notes on *Myosotis scorpioides* agg. - *Watsonia* **6(5)**: 276-279.

WELLIGE, U. & SYSKA, M. 1992: Orchideen am Nestos. In Mattes, H. & Lienau, C. (eds): Colloquium Nordost-Griechenland 1, pp. 50-51. - Münster: Inst. Geographie.

WELLINGHAUSEN, N. & KOCH, H. 1989: Orchideensuche auf Kreta. - *Ber. Arbeitskreis. Heimische Orchid.* **6(1)**: 85-100.

WELLS, A. 1997: Impressions of the Peloponnese. - *Quart. Bull. Alpine Gard. Soc. Gr. Brit.* **65(2)**: 218-222.

WELLS, T.C.E. 1976: Biological Flora of the British Isles. *Hypochoeris maculata* L. (*Achyroprorus maculatus* (L.) Scop.). - *J. Ecol.* **64(2)**: 757-774.

WELTER-SCHULTES, F.W. 1995: La végétation de l'île de Gavdos (Grèce), la plus méridionale de l'Europe. Influence de facteurs historiques et humains. - *Biol. Gallo-Hellen.* **21(2)**: 189-201.

WELTER, A. 1979: Étude d'un complexe polyploïde: *Sempervivum arachnoideum* L. dans la flore orophile d'Europe centrale et méridionale. - *Compt. Rend. Hebd. Séances Acad. Sci., Sér. D* **288(4)**: 387-390.

WENDELBERGER, G. 1963: Die Schwarzföhrenwälder Südosteuropas. - *Mitt. Florist.-Soziol. Arbeitsgem.* **10**: 130-148.

WENDT, N. 1996: *Papaver ramosissimum* Fedde - ein Synonym von *P. rhoeas* L. (Papaveraceae). - *Willdenowia* **25(2)**: 627-631.

WENGER-RAINZE, M. 1970: Contribution à l'étude cytotaxonomique du genre *Gypsophila* L. - *Bull. Soc. Neuchâteloise Sci. Nat.* **93**: 179-186 + 1 plate.

WENZIG, TH. 1883: Die Gattung *Fraxinus* Tourn. - *Bot. Jahrb. Syst.* **4**: 165-188 + 1 plate.

WENZIG, TH. 1886: Die Eichen Europas, Nordafrikas und des Orients. - *Jahrb. Königl. Bot. Gart. Berlin* **4**: 179-213.

WERGER, J. & BURTON, R.E. 1972: Roses. A bibliography of botanical, horticultural and other works related to the genus *Rosa*. - Metuchen, N.J.: Scarecrow Press, vi + 169 pp.

WERNER, K. 1960: Zur Nomenklatur und Taxonomie von *Digitalis* L. - *Bot. Jahrb. Syst.* **79(2)**: 218-254.

WERNER, K. 1964: Die Verbreitung der *Digitalis*-Arten. - *Wiss. Z. Martin-Luther-Univ. Halle-Wittenberg, Math.-Naturwiss. Reihe* **13(6)**: 453-486.

WERNER, K. & EBEL, F. 1994: Zur Lebensgeschichte der Gattung *Helleborus* (Ranunculaceae). - *Flora, Morphol. Geobot. Oekophysiol.* **189(2)**: 97-130.

WESMAEL, A. 1892: Monographie des espèces du genre *Fraxinus*. - *Bull. Soc. Roy. Bot. Belgique* **31 (1)**: 69-117.

WESSELY, I. 1961: Die mitteleuropäischen Sippen der Gattung *Solanum* Sektion *Morella*. - *Feddes Repert. Spec. Nov. Regni Veg.* **63(3)**: 290-321.

WESTBURY, D.B. 2004: Biological Flora of the British Isles. *Rhinanthus minor* L. - *J. Ecol.* **92(5)**: 906-927.

WETSCHNIG, W. 1983: Zur Karyologie von *Dactylis glomerata* L. (Poaceae) am Südost-Rand der Alpen. - *Phyton (Horn)* **23(2)**: 271-305.

WETSCHNIG, W. 1991: Karyotype morphology of some diploid subspecies of *Dactylis glomerata* L. (Poaceae). - *Phyton (Horn)* **31(1)**: 35-55.

WETTSTEIN, R.V. 1886: *Isoetes Heldreichii*. - *Verh. K.K. Zool.-Bot. Ges. Wien* **36**: 239-240 + 1 plate.

WETTSTEIN, R.V. 1887: Monographie der Gattung *Hedraeanthus*. - *Denkschr. Kaiserl. Akad. Wiss., Wien. Math.-Naturwiss. Kl.* **53, 2. Abt.**: 185-218 + 1 plate and 3 maps.

WETTSTEIN, R.V. 1889a: Beitrag zur Flora des Orientes. Bearbeitung der von Dr. A. Heider im Jahre 1885 in Pisidien und Pamphilien gesammelten Pflanzen. - *Sitzungsber. Kaiserl. Akad. Wiss., Math.- Naturwiss. Cl., Abt. 1* **98, 1. Abt.**: 348-398 + 3 plates.

WETTSTEIN, R.V. 1889b: Die Gattungen *Erysimum* und *Cheiranthus*. Ein Beitrag zur Systematik der Cruciferen. - *Oesterr. Bot. Z.* **39(7-8)**: 243-247, 281-284 + 1 plate, 327-330.

WETTSTEIN, R.V. 1892a: Beitrag zur Flora Albaniens. Bearbeitung der von J. Dörfler im Jahre 1890 im Gebiete des Šar-Dagh gesammelten Pflanzen. - *Biblioth. Bot.* **5/26**: 103 pp. + 5 plates.

WETTSTEIN, R.V. 1892b: Die Flora der Balkanhalbinsel und deren Bedeutung für die Geschichte der Pflanzenwelt. - *Monatsbl. Wiss. Club Wien* **3(11)**: 136-140.

WETTSTEIN, R.V. 1895: Globulariaceen-Studien. - *Bull. Herb. Boissier* **3(6)**: 271-290 + 1 plate.

WETTSTEIN, R.V. 1896a: Die europäischen Arten der Gattung *Gentiana* aus der Section *Endotricha* Froel. und ihr entwicklungsgeschichtlicher Zusammenhang. - *Denkschr. Kaiserl. Akad. Wiss., Wien. Math.-Naturwiss. Kl.* **64**: 309-382 + 3 maps and 4 plates.

WETTSTEIN, R.V. 1896b: Monographie der Gattung *Euphrasia*. - Leipzig: Verlag von W. Engelmann, 316 pp. + 14 plates and 4 maps.

WETTSTEIN, R.V. 1917a: Studien über die systematische Gliederung von *Cytinus Hypocistis* L. - *Ber. Deutsch. Bot. Ges.* **35(2)**: 86-99 + 1 plate.

WETTSTEIN, R.V. 1918: *Moltkea Dörfleri* Wettstein und die Abgrenzung der Gattung *Moltkea*. - *Oesterr. Bot. Z.* **67(10-12)**: 361-369.

WEYLAND, H. 2007: *Ophrys episcopalis* Poiret auf der Peloponnes (Griechenland). - *J. Eur. Orch.* **39 (3-4)**: 543-550.

WHALEN, M.D. 1991: Taxonomy of *Saccharum* (Poaceae). - *Baileya* **23(3)**: 109-125.

WHELDON, J.A. & SALMON, C.E. 1925: Notes on the genus *Erythraea*. - *J. Bot. (London)* **63**: 345-352.

WHITAKER, T. & ROLFE, R. 2002: *Fritillaria euboeica* Rix, P.C. - *Alpine Gardener* **70(4)**: 463-464.

WHITEHEAD, F.H. & SINHA, R.P. 1967: Taxonomy and taximetrics of *Stellaria media* (L.) Vill., *S. neglecta* Weihe and *S. pallida* (Dumort.) Piré. - *New Phytol.* **66(4)**: 769-784.

WHITSON, M. 2011: (2016) Proposal to conserve the name *Physalis* (Solanaceae) with a conserved type. - *Taxon* **60(2)**: 608-609. [approved, cf. Taxon 61(5): 1112, 2012].

WHITTEMORE, A.T. 2008: (1824) Proposal to reject the name *Ulmus campestris* (Ulmaceae). - *Taxon* **57(2)**: 656-657. [approved, cf. Taxon 59(4): 1275, 2010].

WICKENS, G.E. 1969: A revision of *Symphytum* L. in Turkey and adjacent areas. - *Notes Roy. Bot. Gard. Edinburgh* **29(2)**: 157-180.

WIDDER, F.J. 1923: Die Arten der Gattung *Xanthium*. Beiträge zu einer Monographie. - *Repert. Spec. Nov. Regni Veg. Beih.* **20**: 1-222 + 4 plates and 4 maps.

WIDDER, F.J. 1925: Übersicht über die bisher in Europa beobachteten *Xanthium*-Arten und Bastarde. - *Repert. Spec. Nov. Regni Veg.* **21**: 273-305.

WIDDER, F.J. 1954: Die Nomenklatur von *Saxifraga aizoides* Linné. - *Phyton (Horn)* **5(3)**: 204-210.

WIDDER, F.J. 1964: Die Veränderlichkeit von *Xanthium spinosum*. - *Phyton (Horn)* **11(1-2)**: 69-82 + 1 plate.

WIDDER, F.J. 1966: Die Nomenklatur von *Minuartia setacea* subsp. *banatica*. - *Phyton (Horn)* **11(3-4)**: 239-244 + 1 plate.

WIDDER, F.J. 1975: Die Gliederung der Gattung *Leontodon*. - *Phyton (Horn)* **17(1-2)**: 23-29.

WIDÉN, C.-J. & AL. 1991: The phloroglucinols of the *Dryopteris villarii* complex and some related ferns (Pteridophyta, Dryopteridaceae). - *Bot. Helv.* **101(1)**: 77-120.

WIDLER-KIEFER, H. & YEO, P.F. 1987: Fertility relationships of *Geranium* (Geraniaceae): sectt. *Ruberta*, *Anemonifolia*, *Lucida* and *Unguiculata*. - *Pl. Syst. Evol.* **155(1-4)**: 283-306.

WIDLER, B.E. & BOCQUET, G. 1979: *Brassica insularis* Moris: Beispiel eines messinischen Verbreitungsmusters. - *Candollea* **34(1)**: 133-151.

WIDMER, E. 1891: Die europäischen Arten der Gattung *Primula*. - München: Verlag von R. Oldenbourg, vii + 154 pp.

WIEDENBEIN, F.W. 1988: Quartärgeologie und Biogeographie der Kykladeninsel Milos. -Diss., Friedrich-Alexander-Univ., Erlangen-Nürnberg, 191 pp.

WIEFELSPÜTZ, W. 1970: Zur Verbreitung der europäischen allogamen *Epipactis*-Arten. In Senghas, K. & Sundermann, H. (eds): Probleme der Orchideengattung *Epipactis*. - *Jahresber. Naturwiss. Vereins Wuppertal* **23**: 38-42.

WIEGERS, J. & MENSCHAAR, M. 1980: Some observations on *Leucojum aestivum* L. [In Dutch with English summary]. - *Gorteria* **10(4)**: 61-69.

WIEGLEB, G. 1989a: Notes on pondweeds - outlines for a monographical treatment of the genus *Potamogeton* L. - *Feddes Repert.* **99(7-8)**: 249-266.

WIEGLEB, G. 1989b: On *Potamogeton coloratus* (Potamogetonaceae) in Turkey. - *Willdenowia* **19(1)**: 121-125.

WIEGLEB, G. 1990: The importance of stem anatomical characters for the systematics of the genus *Potamogeton* L. - *Flora, Morphol. Geobot. Oekophysiol.* **184(3)**: 197-208.

WIEGLEB, G., BOBROV, A.A. & ZALEWSKA-GAŁOSZ, J. 2017: A taxonomic account of *Ranunculus* section *Batrachium* (Ranunculaceae). - *Phytotaxa* **319(1)**: 1-55.

WIEGLEB, G. & KAPLAN, Z. 1998: An account of the species of *Potamogeton* L. (Potamogetonaceae). - *Folia Geobot.* **33(3)**: 241-316.

WIELINGA, A. & WIELINGA, R. 2004: Op zoek naar orchideeen op Lesbos. - *Eurorchis* **16**: 5-11, 33.

WIERSEMA, J.H. & DAHLBERG, J. 2007: The nomenclature of *Sorghum bicolor* (L.) Moench (Gramineae). - *Taxon* **56(3)**: 941-946.

WIGERMO, CH. 2011: Med LBF på Lesbos maj 2010 [In Swedish]. - *Bot. Not.* **144(1)**: 8-12.

WIJMSTRA, T. & SMIT, A. 1976: Palynology of the middle part (30-78 metres) of the 120 m deep section in northern Greece (Macedonia). - *Acta Bot. Neerl.* **25(4)**: 297-312.

WIJNANDS, D.O. 1983: *Petteria ramentacea*. - *Bull. Bot. Tuinen Wageningen* **11(6)**: 12.

WIJNHEIJMER, E.H.M., BRANDENBURG, W.A. & JARVIS, C.E. 1988: Lectotypification of *Daucus carota* L. (Umbelliferae). - *Taxon* **37(1)**: 175-184.

WIKLUND, A. 1985: The genus *Asteriscus* (Asteraceae-Inuleae). - *Nordic J. Bot.* **5(4)**: 299-314.

WIKLUND, A. 1987: The genus *Nauplius* (Asteraceae-Inuleae). - *Nordic J. Bot.* **7(1)**: 1-23.

WIKLUND, A. 1992: The genus *Cynara* L. (Asteraceae-Cardueae). - *Bot. J. Linn. Soc.* **109(1)**: 75-123.

WILBUR, R.L. 1975: A revision of the North American genus *Amorpha* (Legumniosae-Psoraleae). - *Rhodora* **77(811)**: 337-409.

WILD, J.-D., MAYER, E. & GOTTSBERGER, G. 2003: Pollination and reproduction of *Tussilago farfara* (Asteraceae). - *Bot. Jahrb. Syst.* **124(3)**: 273-285.

WILDE-DUYFJES, B.E.E. DE 1973: Typification of 23 *Allium* species described by Linnaeus and possibly occurring in Africa. - *Taxon* **22(1)**: 57-91.

WILDE-DUYFJES, B.E.E. DE 1976: A revision of the genus *Allium* L. (Liliaceae) in Africa. - *Meded. Landbouwhogeschool* **76(11)**: 1-237.

WILDHABER, O.J. 1975: *Orchis patens* ssp. *falcicalcarata* nom. nov. - *Ber. Schweiz. Bot. Ges.* **84(3)**: 252-260.

WILFORD, R. 2000: The cultivation of *Fritillaria* species at Kew. - *Curtis's Bot. Mag., ser. 6* **17(3)**: 149-154.

WILHALM, TH. 2009: *Digitaria ciliaris* in Europe. - *Willdenowia* **39(2)**: 247-259.

WILHELM, M. 1995: Vegetationsuntersuchungen in den Lagunen bei Agiasma, Nordost-Griechenland. In Mattes, H. & Lienau, C. (eds): Colloquium Nordost-Griechenland 3, pp. 2-20. - Münster: Inst. Geographie.

WILLDENOW, C.L. 1797-1825: Caroli a Linné Species Plantarum ... - Berolini: G.C. Nauk, 6 vols. [cf. TL-2 7: 303-304, 1988].

WILLDENOW, C.L. 1803-16: Hortus berolinensis, sive icones et descriptiones, plantarum rariorum vel minus cognitarum ... - Berolini: Imprimis Fr. Schüppel, 2 vols. [cf. TL-2 7: 305, 1988].

WILLIAMS, C.A., DEMISSIE, A. & HARBORNE, J.B. 1983: Flavonoids as taxonomic markers in Old World *Lupinus* species. - *Biochem. Syst. Ecol.* **11(3)**: 221-231.

WILLIAMS, F.N. 1889: Revison of the specific forms of the genus *Gypsophila*. - *J. Bot. (London)* **27**: 321-329.

WILLIAMS, F.N. 1890: Synopsis of the genus *Tunica*. - *J. Bot. (London)* **28**: 193-199.

WILLIAMS, F.N. 1893: A monograph of the genus *Dianthus*, Linn. - *J. Linn. Soc., Bot.* **29(203)**: 346-478.

WILLIAMS, F.N. 1895: On the genus *Arenaria* Linn. - *Bull. Herb. Boissier* **3(11)**: 593-603.

WILLIAMS, F.N. 1896a: A revision of the genus *Silene*, Linn. - *J. Linn. Soc., Bot.* **32**: 1-196.

WILLIAMS, F.N. 1896b: A systematic revision of the genus *Herniaria*. - *Bull. Herb. Boissier* **4(8)**: 556-570.

WILLIAMS, F.N. 1898a: Note monographique sur le genre *Rhodalsine* Gay. - *Bull. Herb. Boissier* **6(1)**: 1-10.

WILLIAMS, F.N. 1898b: Critical notes on some species of *Cerastium*. - *J. Bot. (London)* **36**: 341-344, 382-387.

WILLIAMS, F.N. 1898c: A revision of the genus *Arenaria*, Linn. - *J. Linn. Soc., Bot.* **33(232)**: 326-437.

WILLIAMS, F.N. 1898d: Énumération provisoire des espèces du genre *Cerastium*. - *Bull. Herb. Boissier* **6(11)**: 893-904.

WILLIAMS, F.N. 1899a: An account of *Velezia*. - *J. Bot. (London)* **37**: 25-33.

WILLIAMS, F.N. 1899b: Critical notes on some species of *Cerastium* [cont.]. - *J. Bot. (London)* **37**: 116-124, 209-216, 310-315, 474-477.

WILLIAMS, F.N. 1902: Note synoptique sur le genre *Moenchia*. - *Bull. Herb. Boissier, Ser. 2* **2(7)**: 602-613.

WILLIAMS, F.N. 1906: The genus *Telephium*. - *J. Bot. (London)* **44**: 289-304.

WILLIAMS, J.T. 1963: Biological Flora of the British Isles. *Chenopodium album* L. - *J. Ecol.* **51(3)**: 711-725.

WILLIAMS, J.T. 1969: Biological Flora of the British Isles. *Chenopodium rubrum* L. - *J. Ecol.* **57(3)**: 831-841.

WILLIAMS, J.T., SANCHEZ, A.M.C. & JACKSON, M.T. 1974: Studies on lentils and their variation: 1. The taxonomy of the species. - *J. Soc. Advance Breeding Res. Asia Oceanea (SABRAO)* **6(2)**: 133-145.

WILLIAMS, W., AKHTAR, M.A. & FALUYI, M. 1980: Cross compatibility between European and American *Lupinus* species. - *Bot. J. Linn. Soc.* **81(3)**: 225-232.

WILLING, B. & WILLING, E. 1977: Bibliographie über die Orchideen Europas und der Mittelmeerländer 1744-1976. Spezielle Orchideen-Literatur ohne Berücksichtigung allgemeiner floristischer Werke. - *Willdenowia, Beih.* **11**: 325 pp.

WILLING, B. & WILLING, E. 1980: Ergänzende Beobachtungen zu *Ophrys hebes* aus Süd-Griechenland. - *Orchidee (Hamburg)* **31(4)**: 155-159.

WILLING, B. & WILLING, E. 1981: Orchideen-Notizen aus Arkadien, Süd-Griechenland. – *Orchidee (Hamburg)* **32(5)**: 175-182.

WILLING, B. & WILLING, E. 1983: Beitrag zu Verbreitung der Orchideen Ätoliens und Akarnaniens sowie der Insel Lefkas (NW-Griechenland). - *Mitt. Arbeitskreis Heimische Orchid. Baden-Württemberg* **15(3)**: 351-413.

WILLING, B. & WILLING, E. 1984: Beitrag zur Verbreitung der Orchideen des Epirus (NW-Griechenland). - *Mitt. Arbeitskreis Heimische Orchid. Baden-Württemberg* **16(1)**: 21-104.

WILLING, B. & WILLING, E. 1985a: Beitrag zur Orchideenflora NW-Griechenlands. - Kartierungsergebnisse 1984/85. - *Mitt. Arbeitskreis Heimische Orchid. Baden-Württemberg* **17(4)**: 508-628.

WILLING, B. & WILLING, E. 1985b: Bibliographie über die Orchideen Europas und der Mittelmeerländer: 1. Supplement. - *Englera* **5**: 1-280.

WILLING, B. & WILLING, E. 1986a: Die Gattung *Dactylorhiza* in Griechenland. Teil 1: *Dactylorhiza pindica* B. & E. Willing, spec. nov. - *Ber. Arbeitskreis. Heimische Orchid.* **3(2)**: 200-228.

WILLING, B. & WILLING, E. 1986b: Bemerkungen zu Orchideen-Hybriden aus NW-Griechenland. - *Ber. Arbeitskreis. Heimische Orchid.* **3(2)**: 235-250.

WILLING, B. & WILLING, E. 1986c: Verbreitung der "montanen" Orchideenarten der Peloponnes. - *Jahresber. Naturwiss. Vereins Wuppertal* **39**: 104-120.

WILLING, B. & WILLING, E. 1986d: Quelques nouveaux hybrides d'*Ophrys* du nord-ouest de la Grèce. - *Orchidophile (Ansières)* **17(72)**: 1098-1103.

WILLING, B. & WILLING, E. 1987: Die Gattung *Dactylorhiza* in Griechenland. Teil 2: *Dactylorhiza cordigera* s.l. und *D. baumanniana* s.l. - *Ber. Arbeitskreis. Heimische Orchid.* **4(1)**: 18-85.

WILLING, B. & WILLING, E. 1988a: Die Gattung *Dactylorhiza* in Griechenland. Teil 3: *Dactylorhiza kalopissii* Nelson und Neufunde anderer Arten. - *Mitt. Arbeitskreis Heimische Orchid. Baden-Württemberg* **20(2)**: 391-437.

WILLING, B. & WILLING, E. 1988b: Noch einmal: *Ophrys holoserica* oder *Ophrys holosericea*. - *Ber. Arbeitskreis. Heimische Orchid.* **5(1-2)**: 22-23.

WILLING, B. & WILLING, E. 1989: Die Gattung *Dactylorhiza* in Griechenland. Teil 4: *Dactylorhiza smolikana*, spec. nov., ungefleckte Varianten der *Dactylorhiza baumanniana* und weitere Orchideenfunde aus NW-Griechenland. - *Ber. Arbeitskreis. Heimische Orchid.* **6(1)**: 25-62.

WILLING, B. & WILLING, E. 1991a: Die Gattung *Dactylorhiza* in Griechenland. Teil 5: Untersuchungsergebnisse 1989/1990 sowie weitere Fundmeldungen. - *Ber. Arbeitskreis. Heimische Orchid.* **8(1)**: 4-53.

WILLING, E. 1977a: Bibliographie der Orchideen Europas und der Mittelmeerländer (1. Teil). - *Mitt. Arbeitskreis Heimische Orchid. Baden-Württemberg* **9(1)**: 3-37.

WILLING, E. 1977b: Bibliographie der Orchideen Europas und der Mittelmeerländer (2. Teil). - *Mitt. Arbeitskreis Heimische Orchid. Baden-Württemberg* **9(3)**: 186-199.

WILLING, E. 1977c: Literatur-Schnelldienst "Europäische Orchideen" 1977. - *Mitt. Arbeitskreis Heimische Orchid. Baden-Württemberg* **9(3)**: 211-228.

WILLING, E. 1978: Literatur-Schnelldienst "Europäische Orchideen" 1978. - *Mitt. Arbeitskreis Heimische Orchid. Baden-Württemberg* **10(1-2)**: 99-109.

WILLING, E. 1979a: Literatur-Schnelldienst "Europäische Orchideen" 1979 Nr. 1. - *Mitt. Arbeitskreis Heimische Orchid. Baden-Württemberg* **11(1)**: 66-71.

WILLING, E. 1979b: Literatur-Schnelldienst "Europäische Orchideen" 1979 Nr. 2. - *Mitt. Arbeitskreis Heimische Orchid. Baden-Württemberg* **11(2)**: 163-170.

WILLING, E. 1979c: Literatur-Schnelldienst "Europäische Orchideen" 1979 Nr. 3. - *Mitt. Arbeitskreis Heimische Orchid. Baden-Württemberg* **11(3)**: 235-237.

WILLING, E. 1979d: Fortsetzung des Literatur-Schnelldienstes 1979 Nr. 3. - *Mitt. Arbeitskreis Heimische Orchid. Baden-Württemberg* **11(4)**: 316-320.

WILLING, E. 1984a: Stand der Orchideenkartierung auf den Inseln Kerkira (Korfu) und Paxos (NW- Griechenland). - *Ber. Arbeitskreis. Heimische Orchid.* **1(1)**: 23-40.

WILLING, E. 1984b: Literatur-Dienst "Europäische Orchideen" 1984/1. - *Mitt. Arbeitskreis Heimische Orchid. Baden-Württemberg* **16(1)**: 184-192.

WILLING, E. 1984c: Literatur-Dienst "Europäische Orchideen" 1984/2. - *Mitt. Arbeitskreis Heimische Orchid. Baden-Württemberg* **16(4)**: 664-674.

WILLING, E. 1985a: Literatur-Dienst "Europäische Orchideen" 1985/1. - *Mitt. Arbeitskreis Heimische Orchid. Baden-Württemberg* **17(2)**: 307-320.

WILLING, E. 1985b: Literatur-Dienst "Europäische Orchideen" 1985/2. - *Mitt. Arbeitskreis Heimische Orchid. Baden-Württemberg* **17(4)**: 641-656.

WILLING, E. 1986a: Literatur-Dienst "Europäische Orchideen" 1986/1. - *Ber. Arbeitskreis. Heimische Orchid.* **3(1)**: 169-186.

WILLING, E. 1986b: Literatur-Dienst "Europäische Orchideen" 1986/2. - *Ber. Arbeitskreis. Heimische Orchid.* **3(2)**: 287-308.

WILLING, E. 1987a: Literatur-Dienst "Europäische Orchideen" 1987/1. - *Mitt. Arbeitskreis Heimische Orchid. Baden-Württemberg* **19(1)**: 131-150.

WILLING, E. 1987b: Literatur-Dienst "Europäische Orchideen" 1987/2. - *Mitt. Arbeitskreis Heimische Orchid. Baden-Württemberg* **19(4)**: 939-960.

WILLING, E. 1988a: Literatur-Dienst "Europäische Orchideen" 1988/1. - *Mitt. Arbeitskreis Heimische Orchid. Baden-Württemberg* **20(1)**: 199-216.

WILLING, E. 1988b: Literatur-Dienst "Europäische Orchideen" 1988/2. - *Mitt. Arbeitskreis Heimische Orchid. Baden-Württemberg* **20(4)**: 891-914.

WILLING, E. 1989a: Literatur-Dienst "Europäische Orchideen" 1989/1. - *Ber. Arbeitskreis. Heimische Orchid.* **6(1)**: 163-188.

WILLING, E. 1989b: Literatur-Dienst "Europäische Orchideen" 1989/2. - *Ber. Arbeitskreis. Heimische Orchid.* **6(2)**: 100-124.

WILLING, E. 1989c: *Dactylorhiza* in Griechenland. - *Eurorchis* **1(1)**: 113-128.

WILLING, E. 1990a: Literatur-Dienst "Europäische Orchideen" 1990/1. - *Ber. Arbeitskreis. Heimische Orchid.* **7(1)**: 98-132.

WILLING, E. 1990b: Literatur-Dienst "Europäische Orchideen" 1990/2. - *Ber. Arbeitskreis. Heimische Orchid.* **7(2)**: 83-121.

WILLING, E. 1990c: *Dactylorhiza* in Nordwestgriechenland - neue Erkenntnisse. - *Jahresber. Naturwiss. Vereins Wuppertal* **43**: 58-70.

WILLING, E. 1991a: Literatur-Dienst "Europäische Orchideen" 1991/1. - *Ber. Arbeitskreis. Heimische Orchid.* **8(1)**: 94-119.

WILLING, E. 1991b: Literatur-Dienst "Europäische Orchideen" 1991/2. - *Ber. Arbeitskreis. Heimische Orchid.* **8(2)**: 129-156.

WILLING, E. 1992a: Literatur-Dienst "Europäische Orchideen" 1992/1. - *Ber. Arbeitskreis. Heimische Orchid.* **9(2)**: 149-167.

WILLING, E. 1992b: Literatur-Dienst "Europäische Orchideen" 1992/2. - *Ber. Arbeitskreis. Heimische Orchid.* **9(2)**: 168-188.

WILLING, E. 1993a: Literatur-Dienst "Europäische Orchideen" 1993/1. - *Ber. Arbeitskreis. Heimische Orchid.* **10(1)**: 116-141.

WILLING, E. 1993b: Literatur-Dienst "Europäische Orchideen" 1993/2. - *Ber. Arbeitskreis. Heimische Orchid.* **10(2)**: 71-94.

WILLING, E. 1994: Literatur-Dienst "Europäische Orchideen" 1994/1. - *Ber. Arbeitskreis. Heimische Orchid.* **11(1)**: 139-151.

WILLING, E. 1995a: Literatur-Dienst "Europäische Orchideen" 1994/2. - *Ber. Arbeitskreis. Heimische Orchid.* **11(2)**: 75-90.

WILLING, E. 1995b: Literatur-Dienst "Europäische Orchideen" 1995/1. - *J. Eur. Orch.* **27(4)**: 714-744.

WILLING, E. 1995c: Literatur-Dienst "Europäische Orchideen" 1995/2. - *J. Eur. Orch.* **27(4)**: 745-770.

WILLING, E. 1996a: Literatur-Dienst "Europäische Orchideen" 1995/2. - *Ber. Arbeitskreis. Heimische Orchid.* **12(2)**: 82-96.

WILLING, E. 1996b: Literatur-Dienst "Europäische Orchideen" 1996/1. - *Ber. Arbeitskreis. Heimische Orchid.* **13(1)**: 116-128.

WILLING, E. 1996c: Literatur-Dienst "Europäische Orchideen" 1996/1. - *J. Eur. Orch.* **28(2)**: 416-431.

WILLING, E. 1996d: Literatur-Dienst "Europäische Orchideen" 1996/2. - *Ber. Arbeitskreis. Heimische Orchid.* **13(2)**: 111-128.

WILLING, E. 1996e: Literatur-Dienst "Europäische Orchideen" 1996/2. - *J. Eur. Orch.* **28(4)**: 805-824.

WILLING, E. 1997a: Literatur-Dienst "Europäische Orchideen" 1997/1. - *Ber. Arbeitskreis. Heimische Orchid.* **14(1)**: 141-157.

WILLING, E. 1997b: Literatur-Dienst "Europäische Orchideen" 1997. - *J. Eur. Orch.* **29(4)**: 780-804.

WILLING, E. 1998a: Literatur-Dienst "Europäische Orchideen" 1998/1. - *Ber. Arbeitskreis. Heimische Orchid.* **15(1)**: 108-126.

WILLING, E. 1998b: Literatur-Dienst "Europäische Orchideen" 1998/1. - *J. Eur. Orch.* **30(2)**: 453-474.

WILLING, E. 1998c: Literatur-Dienst "Europäische Orchideen" 1998/2. - *J. Eur. Orch.* **30(4)**: 889-914.

WILLING, E. 1998d: Literatur-Dienst "Europäische Orchideen" 1998/2. - *Ber. Arbeitskreis. Heimische Orchid.* **15(2)**: 76-95.

WILLING, E. 1999a: Literatur-Dienst "Europäische Orchideen" 1999/1. - *Ber. Arbeitskreis. Heimische Orchid.* **16(1)**: 151-172.

WILLING, E. 1999b: Literatur-Dienst "Europäische Orchideen" 1999/2. - *J. Eur. Orch.* **31(3)**: 703-721.

WILLING, E. 2000a: Literatur-Dienst "Europäische Orchideen" 1999/2. - *Ber. Arbeitskreis. Heimische Orchid.* **16(2)**: 110-128.

WILLING, E. 2000b: Literatur-Dienst "Europäische Orchideen" 2000/1. - *J. Eur. Orch.* **32(1)**: 135-155.

WILLING, E. 2000c: Literatur-Dienst "Europäische Orchideen" 2000/1. - *Ber. Arbeitskreis. Heimische Orchid.* **17(1)**: 77-95.
WILLING, E. 2000d: Literatur-Dienst "Europäische Orchideen" 2000/2. - *J. Eur. Orch.* **32(3-4)**: 643-668.
WILLING, E. 2001: Literatur-Dienst "Europäische Orchideen" 2000/2. - *Ber. Arbeitskreis. Heimische Orchid.* **18(1)**: 240-254.
WILLING, E. 2004: Bibliographie über die Orchideen Europas und der Mittelmeerländer. 2. Supplement. - *J. Eur. Orch.* **36(1)**: 3-400.
WILLING, E. & RAUS, TH. 2015: *Senecio inaequidens* DC. In Raab-Straube, E. v. & Raus, Th. (eds): Euro+Med-Checklist notulae, 4. - *Willdenowia* **45(1)**: 121.
WILLING, E. & WILLING, B. 1979: OPTIMA-Projekt "Kartierung der mediterranen Orchideen". 1. Index der Verbreitungskarten für die Orchideen Europas und der Mittelmeerländer. - *Beih. Veröff. Naturschutz Landschaftspflege Baden-Württemberg* **14**: 1-163.
WILLING, E. (sometimes with co-authors) 1978-2020: Lists of Greek collecting localities, herbarium specimens and field records, see www.willing-botanik.de.
WILLIS, K.J. 1992a: The late Quaternary vegetational history of Northwest Greece. I. Lake Gramousti. - *New Phytol.* **121(1)**: 101-117.
WILLIS, K.J. 1992b: The late Quaternary vegetational history of Northwest Greece. II. Rezina marsh. - *New Phytol.* **121(1)**: 119-138.
WILLIS, K.J. 1992c: The late Quaternary vegetational history of Northwest Greece. III. A comparative study of two contrasting sites. - *New Phytol.* **121(1)**: 139-155.
WILLKOMM, M. 1850: Recherches sur l'organographie et la classification des Globulariées. - Leipsick [Leipzig]: Gustave Mayer, 32 pp. + 4 plates.
WILLMOTT, E.A. 1910-14: The genus *Rosa*. 2 vols. (25 fasc.). With drawings by A. Parsons. - London, John Murray: xvi + xxvii + 552 pp. + many plates.
WILLNER, W., DI PIETRO, R. & BERGMEIER, E. 2009: Phytogeographical evidence for post-glacial dispersal limitation of European beech forest species. - *Ecography* **32(6)**: 1011-1018.
WILMANNS, O. 1981: Zur Kenntnis von *Brachypodium rupestre* (Host) Roem. et Schult. - *Göttinger Florist. Rundbr.* **15(4)**: 71-75.
WILMOT-DEAR, M. 1985: *Ceratophyllum* revisited: a study in fruit and leaf variation. - *Kew Bull.* **40(2)**: 243-271.
WILMOTT, A.J. 1918: Two new plants from Macedonia. - *J. Bot. (London)* **56**: 145-146.
WILSKI, P. 1909: Nachtrag zu Band I, Kap. IV: Th. von Heldreich, "Die Flora von Thera". In Hiller v. Gaertringen: Thera. Untersuchungen, Vermessungen und Ausgrabungen in den Jahren 1895-1902, Band IV, 2B, pp. 119-130. - Berlin: Verlag Georg Riemer.
WILSON, F.D. 1982: A cytological basis for the separation of *Geropogon* from *Tragopogon* (Compositae: Lactuceae). - *Brittonia* **34(3)**: 290-293.
WILSON, F.D. 1983: Karyotypes of *Tragopogon* (Compositae: Lactuceae). - *Brittonia* **35(4)**: 341-350.
WILSON, K.L. 1988: *Polygonum* sensu lato (Polygonaceae) in Australia. - *Telopea* **3(2)**: 177-182.
WILSON, K.L. 1989: (931) Proposal to conserve 468 *Scirpus* L. (Cyperaceae) with *S. sylvaticus* L. as type. - *Taxon* **38(2)**: 316-320. [approved, cf. Taxon 42(3): 691, 1993].
WILSON, K.L. 1990: Some widespread species of *Persicaria* (Polygonaceae) and their allies. - *Kew Bull.* **45(4)**: 621-636.
WILTSHIRE, T. 2001: *Cyclamen hederifolium* var. *confusum*. - *Cyclamen* **25(1)**: 26.
WILTSHIRE, T. & DALE, H. 1995: Cyclamen Society expedition to Crete 1995. - *Cyclamen* **19(2)**: 54-55.
WINFIELD, M.O. & AL. 2006: A cytological and molecular study of the genera *Scorzonera* L. and *Podospermum* (L.) DC. (Asteraceae). - *Caryologia* **59(2)**: 153-163.
WINGE, Ø. 1940: Taxonomic and evolutionary studies in *Erophila* based on cytogenetic investigations. - *Compt.-Rend. Trav. Carlsberg Lab., Sér. Physiol.* **23(3)**: 41-78 + 35 plates.
WINTERFELD, G., PERNER, K. & RÖSER, M. 2011: Genome composition and origin of the polyploid Aegean grass *Avenula agropyroides* (Poaceae). - *J. Biogeogr.* **38(4)**: 727-741.
WIRDLECHNER, M.P. 1983: Historical and phenological observations on the spread of *Chaenorrhinum minus* across North America. - *Canad. J. Bot.* **61(1)**: 179-187.
WIRTH, H.W. 1975: Zur Nomenklatur von *Orchis/Dactylorhiza latifolia*. - *Orchidee (Hamburg)* **26(6)**: 263-270.
WIRTH, H.W. 1977: Zur Nomenklatur von *Ophrys fuciflora*. - *Acta Bot. Acad. Sci. Hung.* **23(1-2)**: 285-293.
WIRTH, H.W. 1985: Zur Nomenklatur von *Ophrys speculum*. - *Ber. Arbeitskreis. Heimische Orchid.* **2 (1)**: 89-92.
WIRTH, H.W. 2009: Zur Lectotypisierung von *Ophrys holoserica*. - *Ber. Arbeitskreis. Heimische Orchid.* **26(1)**: 6-18.
WIRTH, H.W. & BLATT, H. 1988: Kritische Anmerkungen zu "Die Gattung *Ophrys* L. eine taxonomische Übersicht". - *Ber. Arbeitskreis. Heimische Orchid.* **5(1-2)**: 4-21.
WIRTH, W. & BLATT, H. 2016: Zur Nomenklatur der Hummel-Ragwurz (*Ophrys fuciflora* (F.W. Schmidt)): Typisierung von *Arachnites fuciflora* F. W. Schmidt. - *Ber. Arbeitskreis. Heimische Orchid.* **33(1)**: 92-102.
WISCHMEYER, U. & HOFFMANN, H. 1995: Bestimmung der chemischen und biologischen Gewässergüte und der Erfassung der Vegetation ... des westlichen Nestos-Deltas. In Mattes, H. & Lienau, C. (eds): Colloquium Nordost-Griechenland 3, pp. 53-69. - Münster: Inst. Geographie.
WISSKIRCHEN, R. 1995: Zur Bestimmung der Unterarten von *Polygonum lapathifolium* L. s.l. - *Florist. Rundbr.* **29(1)**: 1-25.
WISSKIRCHEN, R. & KENT, D.H. 1999: (1436-1437) Proposals to conserve two names in *Persicaria* (Polygonaceae) [both approved; cf. *Taxon* 50(2): 565-566, 2001]. - *Taxon* **48(4)**: 829-830.
WISSKIRCHEN, R., RAABE, U. & KLEINSTEUBER, A. 2013: *Xanthium pungens*. In Raab-Straube, E.v. & Raus, Th.: Notulae ad floram euro-mediterraneam pertinentes No. 30. - *Willdenowia* **43(1)**: 159.
WISSNER, W. & KATING, H. 1974: Botanische und phytochemische Untersuchungen an den europäischen und kleinasiatischen Arten der Gattung *Helleborus*: 1. Zur Verbreitung, Morphologie, Anatomie und Kultur der *Helleborus*-Arten. - *Pl. Med. (Stuttgart)* **26(2)**: 128-143.
WITASEK, J. 1902: Ein Beitrag zur Kenntnis der Gattung *Campanula*. - *Abh. K.K. Zool.-Bot. Ges. Wien* **1(3)**: 1-106.
WITCOMBE, J.R. 1993: A guide to the species of *Aegilops* L.: their taxonomy, morphology and distribution. - Rome: IBPGR, vi + 74 pp.
WITT, J. 1978: *Iris unguicularis*. - *Bull. Amer. Iris Soc.* **228**: 34-38.
WITTMANN, H. 1985: Beitrag zur Systematik der *Ornithogalum*-Arten mit verlängert-traubiger Infloreszenz. - *Stapfia* **13**: 1-117.
WITTMANN, H. 2002: Zur Kenntnis von *Ornithogalum brachystylum*. - *Ber. Bayer. Bot. Ges.* **72**: 51-59.
WITTMANN, H. 2004: Zur Kenntnis von *Ornithogalum prasinantherum* Zahariadi 1977 (Hyacinthaceae). - *Linzer Biol. Beitr.* **36(2)**: 1125-1138.
WITTMANN, H. & STROBL, W. 1987: Untersuchungen am Artenpaar *Agrimonia eupatoria* L. - *A. procera* Wallr. im Bundesland Salzburg (Österreich). - *Linzer Biol. Beitr.* **19(1)**: 91-119.
WOLF, TH. 1908: Monographie der Gattung *Potentilla*. - *Biblioth. Bot.* **16/71**: 715 pp. + 20 plates and 2 maps.
WOLFF, H. 1921: *Biasolettia Bornmülleri* von Thasos. - *Repert. Spec. Nov. Regni Veg.* **17**: 43.
WOLFF, H. 1922: *Athamantha macrosperma* spec. nov. ex Attica. - *Repert. Spec. Nov. Regni Veg.* **18**: 133.
WOLFF, H. 1925a: *Stefanoffia*, eine neue Umbelliferen-Gattung von der Balkanhalbinsel und aus Kleinasien. - *Notizbl. Bot. Gart. Berlin-Dahlem* **9(84)**: 281-282.
WOLFF, H. 1925b: Umbelliferae novae Asiae minoris. II. - *Repert. Spec. Nov. Regni Veg.* **21**: 110-111.
WOLFF, P. 1980: Die Hydrilleae (Hydrocharitaceae) in Europa. - *Göttinger Florist. Rundbr.* **14(2)**: 33-56.

WOLFF, W.J. 1968: The halophilous vegetation of the lagoons of Mesolonghi, Greece. - *Vegetatio* **16 (1-4)**: 95-134.

WOLLIN, HJ. 1972: Über *Orchis boryi* Rchb. f. - *Jahresber. Naturwiss. Vereins Wuppertal* **25**: 95-97.

WOOD, C.E. JR. 1955: Evidence for the hybrid origin of *Drosera anglica*. - *Rhodora* **57(676)**: 105-130.

WOOD, D.J. & BRUMMITT, R.K. 1989: Notes on *Acanthus hirsutus* Boiss. with a new key to Turkish taxa of *Acanthus*. - *Notes Roy. Bot. Gard. Edinburgh* **46(1)**: 7-11.

WOOD, J.J. 1983a: Notes on *Himantoglossum* (Orchidaceae). - *Kew Bull.* **38(1)**: 75-78.

WOOD, J.J. 1983b: Two new combinations in *Ophrys* (Orchidaceae). - *Kew Bull.* **38(1)**: 135-137.

WOOD, J.J. 1985a: *Ophrys spruneri*. - *Bot. Mag. (Kew Mag.)* **2(4)**: tab. 45 + pp. 345-348.

WOOD, J.J. 1985b: Orchids of Europe: *Orchis italica* Poir. & *Orchis spitzelii* Sauter ex W. Koch. - *Orchid Digest* **49(3)**: 113-116.

WOOD, J.J. 1986a: *Ophrys argolica* subsp. *argolica* & *O. oestrifera* subsp. *heldreichii*. - *Bot. Mag. (Kew Mag.)* **3(1)**: tab. 49 + pp. 14-18.

WOOD, J.J. 1986b: *Ophrys lutea* var. *minor* & *O. doerfleri*. - *Bot. Mag. (Kew Mag.)* **3(1)**: tab. 50 + pp. 19-22.

WOOD, J.J. 1986c: *Serapias lingua* & *S. vomeracea* subsp. *orientalis*. - *Bot. Mag. (Kew Mag.)* **3(1)**: tab. 51 + pp. 22-28.

WOOD, J.J. 1986d: *Himantoglossum caprinum*. - *Bot. Mag. (Kew Mag.)* **3(1)**: tab. 52 + pp. 28-31.

WOOD, J.J. 1986e: *Orchis anatolica*. - *Bot. Mag. (Kew Mag.)* **3(1)**: tab. 53 + pp. 32-33.

WOOD, J.J. 1986e: *Orchis simia* & *O. boryi*. - *Bot. Mag. (Kew Mag.)* **3(1)**: tab. 54 + pp. 34-38.

WOOD, J.J. 1993a: *Orchis sancta*. - *Bot. Mag. (Kew Mag.)* **10(1)**: tab. 212 + pp. 9-12.

WOOD, J.J. 1993b: *Limodorum abortivum*. - *Bot. Mag. (Kew Mag.)* **10(4)**: tab. 230-231 + pp. 161-166.

WOOD, J.J. & BELL, S. 1994: *Himantoglossum hircinum*. - *Bot. Mag. (Kew Mag.)* **11(1)**: tab. 240 + pp. 22-28.

WOOD, J.R.I. & AL. 2015: A foundation monograph of *Convolvulus* L. (Convolvulaceae). – *PhytoKeys* **51**: 1-282.

WOODELL, S.R.J. 1965: Natural hybridization between cowslip (*Primula veris* L.) and the primrose (*Primula vulgaris* Huds.) in Britain. - *Watsonia* **6(3)**: 190-202.

WÖRZ, A. 1999a: A taxonomic index of the species of *Eryngium* L. (Apiaceae: Saniculoideae). - *Stuttgarter Beitr. Naturk., A* **596**: 1-48.

WÖRZ, A. 1999b: Systematics and evolution of the genus *Astrantia* L. (Apiaceae-Saniculoideae). - *Bot. Jahrb. Syst.* **121(4)**: 507-536.

WÖRZ, A. 2001: On the intraspecific classification of *Astrantia major* L. (Apiaceae-Saniculoideae). - *Bot. Jahrb. Syst.* **123(3)**: 303-319.

WÖRZ, A. 2002: On the distribution and the relationships of the south-west Asian species of *Eryngium* L. (Apiaceae-Saniculoideae). In Koyuncu, M. (ed.): VIth Plant Life of Southwest Asia Symposium, Abstracts p. 31. - Van: Yüzüncü Yil University.

WÖRZ, A. 2006: Systematics and distribution patterns of the Balkan species of *Eryngium* (Apiaceae, Saniculoideae). - *Phytol. Balcan.* **12(2)**: 221-230.

WÖRZ, A. 2010: The nomenclatural types of some *Eryngium* taxa (Apiaceae-Saniculoideae) from the Balkan Peninsula. - *Phytol. Balcan.* **16(2)**: 233-236.

WÖRZ, A. 2011: Revision of *Eryngium* L. (Apiaceae-Saniculoideae): General part and Palaearctic species. - *Biblioth. Bot.* **159**: 498 pp.

WÖRZ, A. & DIEKMANN, H. 2010: Classification and evolution of the genus *Eryngium* L. (Apiaceae-Saniculoideae): results of fruit anatomical and petal morphological studies. - *Plant Div. Evol.* **128(3-4)**: 387-408.

WRABER, T. 1981: *Vicia loiseleurii* (M.B.) D. Litvinov, a hitherto neglected species from the *V. hirsuta* group. - *Biol. Vestn.* **29(2)**: 181-191.

WRABER, T. 1982: De *Euphorbia pancicii* G. Beck notula. - *Glasn. Republ. Zavoda Zaštitu Prir. Prirodnjačke Zbirke Titogradu* **15**: 77-84.

WRABER, T. 1983: *Androsace mathildae* - neu für die Balkanhalbinsel. - *Maked. Akad. Nauk. Umetn., Oddel. Biol. Med. Nauki* **4(1-2)**: 41-44.

WRABER, T. 1986a: Ein neues *Cynoglossum* (*C. krasniqii*) aus Jugoslawien. - *Candollea* **41(1)**: 145-150.

WRABER, T. 1986b: Die *Solenanthus*-Gruppe der Gattung *Cynoglossum* in der Flora Jugoslawiens. In: Zbornik Radova Simpozijuma "Stogodišnjica Flore Okoline Niša", pp. 67-71. - Niš.

WRABER, T. 1988: Zur Vorkommen von *Centranthus longiflorus* Steven auf der Balkanhalbinsel. - *Biol. Vestn.* **36(3)**: 93-106.

WRABER, T. 1990: Sto znamenitih rastlin na Slovenskem. [Notable plants in Slovenia; in Slovenian with English summary]. - Ljubljana: Prešernova Družba, 239 pp.

WRABER, T. & MIKULETIĆ, V. 1965: *Daphne blagayana* Freyer auf Nordwest-Grenze ihres Areals [In Slovenian with German summary]. - *Biol. Vestn.* **13**: 61-67.

WRABER, T. & SKOBERNE, P. 1989: The red data list of threatened vascular plants in Socialist Republic of Slovenia [In Slovenian with English summary]. - *Varstvo Narave* **14-15**: 429 pp.

WRAY, T. 2004: Stonecrop confusion in Crete. - *Sedum Soc. Newsletter* **74**: 81-84.

WRIGLEY, F. 1986: Taxonomy and chorology of *Silene* section *Otites* (Caryophyllaceae). - *Ann. Bot. Fenn.* **23(1)**: 69-81.

WRÓBEL, A. & AL. 2021: Invasion of *Eragrostis albensis* in Central Europe: distribution patterns, taxonomy and phylogenetic insight into the *Eragrostis pilosa* complex. - *Biol. Invasions* **23**: doi.org/10.1007/s10530-021-02507-6.

WU, Q.-C. & CUTLER, D.F. 1985: Taxonomic, evolutionary and ecological implications of the leaf anatomy of rhizomatous *Iris* species. - *Bot. J. Linn. Soc.* **90(4)**: 253-303.

WUCHERPFENNIG, W. 2009: Was ist *Orchis* (= *Neotinea*) *commutata* Todaro? - *Ber. Arbeitskreis. Heimische Orchid.* **26(1)**: 19-39.

WUCHERPFENNIG, W. 2016: Die westmediterranen Arten der *Ophrys tenthredinifera*-Gruppe - ein Vergleich. - *Ber. Arbeitskreis. Heimische Orchid.* **33(1)**: 110-150.

WUCHERPFENNIG, W. 2019: Über *Limodorum*: Arten, Lebensweise und Taxonomie. - *J. Eur. Orch.* **51(3-4)**: 311-338.

WUCHERPFENNIG, W. 2021: Von *Orchis bifolia* bis *Platanthera fornicata* - die verzwickte Geschichte der Weissen Waldhyazinthen. - *Ber. Arbeitskreis. Heimische Orchid.* **38(1)**: 163-176.

WUNDERLICH, R. 1972: Die systematische Stellung von *Theligonum*. - *Oesterr. Bot. Z.* **119(4-5)**: 329-394.

WYK, B.E. VAN 1991: A synopsis of the genus *Lotononis* (Fabaceae: Crotalarieae). - *Contr. Bolus Herb.* **14**: 1-292.

WYSE JACKSON, M.B. 1992: Taxonomic notes on *Cerastium fontanum* Baumg. emend. Jalas (Caryophyllaceae) in Europe. In Newton, M.E. (ed.): Flora Europaea: Notulae systematicae ad Floram Europaeam spectantes. Series 2. No. 5. - *Bot. J. Linn. Soc.* **109(3)**: 325-328.

XANTHAKIS, M. 2019: Development of a flora database of Kefallinia island, Greece. - *Bot. Chron. (Patras)* **22**: 223-231.

XANTHAKIS, M. & AL. 2015: Study of vertical and horizontal forest structure in Mt. Ainos National Park, Cephalonia Island, Greece. - *Fl. Medit.* **25 (Special issue)**: 245-256.

XENA DE ENRECH, N., CARDONA, A. & MATHEZ, J. 1991: Estudio citotaxonómico del género *Fedia* Gaertn. (Valerianaceae). - *Anales Jard. Bot. Madrid* **48(2)**: 157-169.

XENA DE ENRECH, N. & MATHEZ, J. 1990: Révision du genre *Fedia* Gaertn. emend. Moench (Valerianaceae). - *Naturalia Monspel., Sér. Bot.* **54**: 3-77.

XYSTRAKIS, F. & AL. 2019: Inter-regional beta-diversity patterns of the woody flora of Greece. - *Ann. For. Res.* **62(1)**: 18 pp.

YABUNO, T. 1966: Biosystematic study of the genus *Echinochloa*. - *Jap. J. Bot.* **19(2)**: 277-323.

YAKAR-TAN, N. 1979: [Chromosome numbers in *Digitalis* from the Istanbul area]. In Löve, Á. (ed.): IOPB chromosome number reports LXV. - *Taxon* **28(5-6)**: 631.

YALTIRIK, F. 1963: A study on the floristic analysis of vegetation of the Belgrade forest and composition of the main stand types [In Turkish with English summary]. - *Istanbul Üniv. Orman Fak. Derg., A* **13(1)**: 33-80.

YALTIRIK, F. 1967: Contributions to the taxonomy of woody plants in Turkey. - *Notes Roy. Bot. Gard. Edinburgh* **28(1)**: 9-16.

YALTIRIK, F. 1970: A new subspecies and a review of Turkish *Alnus* [In Turkish]. - *Türk Biyol. Derg.* **20**: 115-121.

YALTIRIK, F. 1978: Taxonomic revision of the indigenous taxa within the family Oleaceae in Turkey. - Istanbul Üniversitesi Orman Fakültesi: 118 pp.

YALTIRIK, F. & BOYDAK, M. 1989: Natural distribution and site characteristics of *Phoenix theophrasti* in Turkey. In Phitos, D. (ed.): Abstracts. Sixth OPTIMA meeting, Delphi 10-16 Sept. 1989, p. 149.

YALTIRIK, F. & BOYDAK, M. 1991: Distribution and ecology of the palm *Phoenix theophrasti* (Palmae) in Turkey. - *Bot. Chron. (Patras)* **10**: 869-872.

YALTIRIK, F. & EFE, A. 2000: *Liquidambar orientalis*. - *Curtis's Bot. Mag., ser. 6* **17(2)**: 66-71 + 1 plate.

YANKOVA, E. & CHERNEVA, Z. 2007: Notes on the species distribution of genus *Angelica* in Bulgaria. - *Phytol. Balcan.* **13(2)**: 189-192.

YANNITSAROS, A. 1969: Contribution to the knowledge of the flora and vegetation of the island of Cythera [In Greek with English summary]. - Athens: Univ. of Athens, 155 pp.

YANNITSAROS, A. 1971: Notes on the flora of Elaphonesos island (Laconia, Greece). - *Biol. Gallo- Hellen.* **3(2)**: 149-162.

YANNITSAROS, A. 1973: Notes on the ecology and distribution of *Capsella grandiflora* (Fauche & Chaub.) Boiss. - *Biol. Gallo-Hellen.* **4(2)**: 163-168.

YANNITSAROS, A. 1974: Studies on the adventive flora of Greece. II. *Coronopus didymus* (L.) Smith (Cruciferae) [In Greek with English summary]. - *Biol. Gallo-Hellen.* **5(2)**: 285-308.

YANNITSAROS, A. 1975a: Contributions to the vernacular names of the Greek plants. I. Some vernacular names of plants from the island of Cythera. - *Biol. Gallo-Hellen.* **6(1)**: 161-167.

YANNITSAROS, A. 1975b: I prostasia tis chloridos ke vlastiseos tis Ellados. [In Greek]. - *O Pan* **393**: 93-99.

YANNITSAROS, A. 1977a: *Linaria hellenica* Turrill, an endemic plant of Lakonia (Greece) that requires protection [In Greek with English summary]. - *Fisis [Journal of the Hellenic Society for the Protection of Nature]* **12**: 13-16 & 34-35.

YANNITSAROS, A. 1977b: Contributions to the vernacular names of the Greek plants. II. Vernacular names of plants from the island of Lesvos. - *Biol. Gallo-Hellen.* **6(2)**: 311-316.

YANNITSAROS, A. 1977c: The genus *Paspalum* L. in Kriti. - *Biol. Gallo-Hellen.* **7(1-2)**: 87-92.

YANNITSAROS, A. 1978: *Tulipa goulimyi* Sealy et Turrill and the dangers which threaten it [In Greek with English summary]. - *Fisis [Journal of the Hellenic Society for the Protection of Nature]* **14**: 5-9 & 36-37.

YANNITSAROS, A. 1979a: A preliminary study for the protection of the flora and vegetation of the island of Lesvos [In Greek with English summary]. - *Fisis [Journal of the Hellenic Society for the Protection of Nature]* **18**: 15-25 & 44-46.

YANNITSAROS, A. 1979b: *Tagetes minuta* L. in Greece. - *Candollea* **34(1)**: 99-107.

YANNITSAROS, A. 1982a: The adventive flora of Greece. A review [In Greek with English abstract]. - *Bot. Chron. (Patras)* **2(2)**: 159-166.

YANNITSAROS, A. 1982b: *Apium leptophyllum* (Pers.) F. Mueller ex Bentham, a new adventive plant for Greece [In Greek with English abstract]. - *Bot. Chron. (Patras)* **2(2)**: 167-172.

YANNITSAROS, A. 1986: New data on the naturalization and distribution of *Coronopus didymus* (Cruciferae) in Greece. - *Willdenowia* **16(1)**: 61-64.

YANNITSAROS, A. 1989: Adventive flora of Crete: History, phytogeography, ecology and agricultural aspects. In Phitos, D. (ed.): Abstracts. Sixth OPTIMA meeting, Delphi 10-16 Sept. 1989, p. 40.

YANNITSAROS, A. 1991: Adventive flora of Crete: history, phytogeography, ecology and agricultural aspects. - *Bot. Chron. (Patras)* **10**: 299-307.

YANNITSAROS, A. 1992: A contribution to the adventive flora of Lesvos (E. Aegean, Greece). In: Proc. 14th Panhellenic Biological Conference, Nicosia-Cyprus, April 28 - May 3, 1992, pp. 157-159.

YANNITSAROS, A. 1995: *Araujia sericifera* Brot. (Asclepiadaceae): ena zizanio ton oporonon esperidoidon tis D. Kritis. In: Perilipsis, 9. Epistimoniko Sinedrio Ellinikis Zizaniologikis Eterias, p. 2 [In Greek]. – Athina.

YANNITSAROS, A. 1998: Additions to the flora of Kithira (Greece). I. - *Willdenowia* **28(1-2)**: 77-94.

YANNITSAROS, A. 1999: Merika spania fita tis periochis tou Mesotopou Lesvou [In Greek]. - *Mesotopos, Politismos Paradosi* **3**: 21-24 [n.v.].

YANNITSAROS, A. 2004a: *Koelreuteria paniculata* Laxmann. In Greuter, W. & Raus, Th. (eds): Med- Checklist notulae, 22. - *Willdenowia* **34(1)**: 76.

YANNITSAROS, A. 2004b: Additions to the flora of Kithira (Greece) II. - *Willdenowia* **34(1)**: 117-128.

YANNITSAROS, A. 2006: *Catharanthus roseus, Asparagus densiflorus*. In Greuter, W. & Raus, Th. (eds): Med-Checklist notulae, 24. - *Willdenowia* **36(2)**: 720, 730.

YANNITSAROS, A. 2009: *Nothoscordum gracile* (Aiton) Stearn. In Greuter, W. & Raus, Th. (eds): Med- Checklist notulae, 28. - *Willdenowia* **39(2)**: 343.

YANNITSAROS, A. & AL. 1995: Flora and vegetation of Strofades Islands (Ionian Sea, Greece). - Athens: Hellenic Society for the Protection of Nature, 26 pp. + appendix.

YANNITSAROS, A. & AL. 2001: New data on the distribution of some plant taxa in the Aegean area. In Kamari, Psaras & Constantinidis (eds): Proceedings, 8th scientific congress, Greek Botanical Society, Patras October 5-8, 2000, pp. 118-121 [In Greek with English abstract].

YANNITSAROS, A. & BAZOS, I. 1993: The flora of Lesvos and the studies of P.C. Candargy [In Greek with English summary]. In: Proc. 15th Conference Hellen. Soc. Biol. Sci, Florina-Kastoria 21-24.4. 1993, pp. 131-134. - Florina-Kastoria.

YANNITSAROS, A. & BAZOS, I. 2001: A genus of the American family Martyniaceae new for Greece. In Kamari, Psaras & Georgiadis (eds): Proceedings, 8th scientific congress, Greek Botanical Society, Patras October 5-8, 2000, pp. 115-117 [In Greek with English abstract].

YANNITSAROS, A. & BAZOS, I. 2006: *Ibicella* (Stapf) van Esteline: A genus of the American family Martyniaceae new for Greece. - *Ann. Mus. Goulandris* **11**: 271-279.

YANNITSAROS, A. & BAZOS, I. 2008: The Gymnospermae of Lesvos (East Aegean Islands, Greece). - *Bot. Chron. (Patras)* **19**: 63-79.

YANNITSAROS, A., CONSTANTINIDIS, TH. & VASSILIADES, D. 1996: The rediscovery of *Biebersteinia orphanidis* Boiss. (Geraniaceae) in Greece. - *Bot. J. Linn. Soc.* **120(3)**: 239-242.

YANNITSAROS, A. & DAMANAKIS, M. 1983: The occurrence of *Galinsoga parviflora* Cav. in Greece [In Greek with English summary]. - *Zizaniology* **1**: 127-131.

YANNITSAROS, A. & ECONOMIDOU, E. 1974: Studies on the adventive flora of Greece. I. General remarks on some recently introduced taxa. - *Candollea* **29(1)**: 111-119.

YANNITSAROS, A. & ECONOMIDOU, E. 1979: I elliniki chlorida pou apilite ke i anangi dimiourgias enos 'Vivliou Erithron Dedomenon' gia tin Ellada. In Antipas, B. (ed.): Praktika sinedriou prostasias panidas-chloridas-viotopon, pp. 121-125. [In Greek]. - Athine.

YANNITSAROS, A. & GREUTER, W. 2004: *Parkinsonia aculeata* L. In Greuter, W. & Raus, Th. (eds): Med- Checklist notulae, 22. - *Willdenowia* **34(1)**: 75-76.

YANNITSAROS, A. & KOUMPLI-SOVANTZI, L. 1989a: A contribution to the hydrophilous flora of Crete. In Phitos, D. (ed.): Abstracts. Sixth OPTIMA meeting, Delphi 10-16 Sept. 1989, p. 150.

YANNITSAROS, A. & KOUMPLI-SOVANTZI, L. 1989b: *Crepis pusilla* (Sommier) Merxm. in Greece. In Phitos, D. (ed.): Abstracts. Sixth OPTIMA meeting, Delphi 10-16 Sept. 1989, p. 151.

YANNITSAROS, A. & KOUMPLI-SOVANTZI, L. 1991: A contribution to the hydrophilous flora of Crete. - *Bot. Chron. (Patras)* **10**: 579-586.

YANNITSAROS, A. & KOUMPLI-SOVANTZI, L. 1992: *Crepis pusilla* (Sommier) Merxm.: A probable case of epizoochorous dispersal. In. Thanos, C.A. (ed.): Plant-animal interactions in mediterranen-type ecosystems (MEDECOS VI), pp. 294-296. - Athens.

YANNITSAROS, A. & PHEREOS, C. 1999: New data on the distribution and the status of populations of *Lilium candidum* L. in Greece. In Elefheriou, E.P. (ed.): Proceedings, 7th Panhellenic scientific conference, Alexandroupolis, pp. 173-178 [In Greek with English abstract]. - Thessaloniki: Univ. Studio Press.

YANNITSAROS, A. & PROTOPAPADAKIS, E. 1989: The summer weedy flora of Citrus orchards in western Crete [In Greek with English abstract]. - *Bios (Thessaloniki)* **1989(1, suppl.)**: 273-283.

YANNITSAROS, A., SARIKA, M. & CHITOS, TH. 2008: The flora of the island of Lake Pamvotis (Epirus, Greece). In: Praktika 10ou Sinedriou Ellinikis Votanikis Eterias, 5-8 maiou 2005, Ioannina: pp. 129-133 [In Greek with English abstract].

YANNITSAROS, A. & VASSILIADES, D. 1994: New records of *Adonis cyllenea* Boiss., Heldr. & Orph. in Boiss. from N. Peloponnisos (Greece). In: Praktika 16ou Sinedriou E.E.V.E. Volos, 2: 4-6 [In Greek with English abstract].

YANNITSAROS, A. & VASSILIADES, D. 1995: The presence of *Coriaria myrtifolia* L. in Greece. In: Proceedings, 5th scientific conference, Hellenic Botanical Society, Delphi 21-23 October 1994, pp. 73-76 [In Greek with English summary]. - Thessaloniki: Univ. Studio Press.

YANNITSAROS, A. & VASSILIADES, D. 1998: Notes on *Typha minima*. In Greuter, W. & Raus, Th. (eds): Med-Checklist notulae, 17. - *Willdenowia* **28(1-2)**: 163-174.

YANNITSAROS, A. & VASSILIADES, D. 2003: *Typha minima* Funck in Greece. - *Phyton (Horn)* **43(2)**: 263-269.

YANNITSAROS, A., VASSILIADES, D. & BAZOS, I. 2006: *Basella rubra*. In Greuter, W. & Raus, Th. (eds): Med-Checklist notulae, 24. - *Willdenowia* **36(2)**: 720-721.

YANNITSAROS, A., ZERVOU, S. & BAZOS, I. 1999: Contribution to the flora of Filipappou, Pnika and Asteroskopio hills (Attiki, Greece). In: Proceedings of 21st Panhellenic Conference of EEBE, Syros, May 28-31, 1999, pp. 49-50 [In Greek and English].

YAPA, K.N. 1994: A phenetic study of species of genus *Cicer*. - Birmingham: M.Sc. thesis, Univ. of Birmingham, 128 pp.

YARCI, C. 1999: Contributions to the flora of the western part of Istranca mountains (Kırklareli/Thrace region). - *Türk. Bot. Derg.* **23(3)**: 211-228.

YARCI, C. 1997: Flora of Demirkoy (Istranca Mountains / Kirklareli European Turkey). - *Fl. Medit.* **7**: 55-99.

YARCI, C. & ÖZCELIK, H. 2002: Wall fora [sic!] of Edirne (Thrace region). - *OT Sist. Bot. Derg.* **9(1)**: 57-66.

YATSKIEVYCH, G. & SMITH, A.R. 2003: Typification of *Notholaena* R.Br. (Pteridaceae). - *Taxon* **52(2)**: 331-336.

YEO, P.F. 1968: A contribution to the taxonomy of the genus *Ruscus*. - *Notes Roy. Bot. Gard. Edinburgh* **28(3)**: 237-264.

YEO, P.F. 1971: Revisional notes on *Euphrasia*. In Heywood, V.H. (ed.): Notulae systematicae ad Floram Europaeam spectantes, no. 11. - *Bot. J. Linn. Soc.* **64(4)**: 353-361.

YEO, P.F. 1974: The biology and systematics of *Geranium*, sections *Anemonifolia* Knuth and *Ruberta* Dum. - *Bot. J. Linn. Soc.* **67(4)**: 285-346.

YEO, P.F. 1976: Materials for a Flora of Turkey XXXIII: *Euphrasia*. - *Notes Roy. Bot. Gard. Edinburgh* **35(1)**: 53-59.

YEO, P.F. 1979: A taxonomic revision of *Euphrasia* in Europe. - *Bot. J. Linn. Soc.* **77(4)**: 223-334.

YEO, P.F. 1984: Fruit-discharge-type in *Geranium* (Geraniaceae): its use in classification and its evolutionary implications. - *Bot. J. Linn. Soc.* **89(1)**: 1-36.

YEO, P.F. 1985: Hardy geraniums. - London: Christopher Helm Ltd, 192 pp. + 16 plates.

YEO, P.F. 1990a: The classification of Geraniaceae. In Vorster, P. (ed.): Proceedings of the international Geraniaceae symposium, pp. 1-22. - Stellenbosch.

YEO, P.F. 1990b: Lectotypification of the name *Erodium* L'Héritier and the correct names of the subdivisions of *Monsonia* Linnaeus. In Vorster, P. (ed.): Proceedings of the international Geraniaceae symposium, pp. 229-233. - Stellenbosch.

YEO, P.F. 2001: Hardy geraniums. New edition. - London: Batsford, 218 pp. + 32 plates.

YEO, P.F. 2004: The morphology and affinities of *Geranium* sections *Lucida* and *Unguiculata*. - *Bot. J. Linn. Soc.* **144(4)**: 409-429.

YEO, P.F. & WIDLER-KIEFER, H. 1990: The chemotaxonomy of *Geranium* (Geraniaceae). - *Pl. Syst. Evol.* **173(1-2)**: 1-15.

YILDIRIM, A. 2000: A new record for the flora of Turkey. - *OT Sist. Bot. Derg.* **5(1)**: 47-48.

YILDIRIM, A. & VURAL, M. 2003: A new record for the flora of Turkey: *Diplachne fusca* (L.) P. Beauv. (Gramineae). - *OT Sist. Bot. Derg.* **10(1)**: 23-26.

YILDIRIM, H. 2016: *Muscari elmasii* sp. nova (Asparagaceae): a new species from western Anatolia, Turkey. - *Turk. J. Bot.* **40(4)**: 380-387.

YILDIRIM, H. & AL. 2016: Two new records and a confirmation for the vascular flora of Turkey. - *Turk. J. Bot.* **40(6)**: 676-681.

YILDIRIM, H. & AL. 2019: A new bellflower, *Campanula phitosiana* sp. nov. (Campanulaceae) from Western Anatolia, Turkey. - *Phytotaxa* **399(1)**: 25-36.

YILDIRIM, H. & AL. 2021: *Scutellaria topcuoglui* (Lamiaceae), a new serpentinomorph species from south-western Anatolia, Turkey. - *Phytotaxa* **328(3)**: 180-190.

YILDIRIMLI, Ş. 1996a: *Lunaria* (Brassicaceae), a new genus record for the flora of Turkey. - *OT Sist. Bot. Derg.* **3(1)**: 65-68.

YILDIRIMLI, Ş. 1996b: Chorology of the Turkish gymnosperms. - *OT Sist. Bot. Derg.* **3(2)**: 113-124.

YILDIRIMLI, Ş. 2002: A preliminary monograph of *Noaea* Moq. (Chenopodiaceae). - *OT Sist. Bot. Derg.* **9(2)**: 1-18.

YILDIRIMLI, Ş. & DINÇ, M. 2003: The first gathering record of *Phragmites frutescens* H. Scholz (Poaceae) from Turkey. - *OT Sist. Bot. Derg.* **10(1)**: 27-30.

YILDIRIMLI, Ş. & KOCA, A. 2003: A new species record for the flora of Turkey. - *OT Sist. Bot. Derg.* **10 (2)**: 53-56.

YILDIZ, B. & AL. 2016: A taxonomic revision of the genus *Cirsium* Mill. sect. *Cirsium* (Asteraceae: Cardueae) in Turkey. - *Turk. J. Bot.* **40(5)**: 514-530 + 67 pp. appendix.

YILDIZ, B., DIRMENCI, T. & ARABACI, T. 2009: A new record for the flora of Turkey: *Cirsium candelabrum* (*Cirsium* sect. *Cirsium*, Asteraceae, Cynareae). - *Turk. J. Bot.* **33(1)**: 47-51.

YILDIZ, K. & ÇIRPICI, A. 1996: Karyological studies on the species of *Silene* L. of Northwest Anatolia [In Turkish with English abstract]. - *Türk. Bot. Derg.* **20 suppl.**: 73-82.

YILDIZ, K. & ÇIRPICI, A.H. 2013: Taxonomic revision of *Silene* (Caryophyllaceae) sections *Siphonomorpha*, *Lasiostemones*, *Sclerocalycinae*, *Chloranthae*, *Tataricae*, and *Otites* in Turkey. - *Turk. J. Bot.* **37(2)**: 191-218.

YILMAZ, G. & DANE, F. 2008: *Verbascum samniticum* Ten, (Scrophulariaceae): A new record for the flora of Turkey. - *Turk. J. Bot.* **32(5)**: 411-414.

YILMAZ, G. & DANE, F. 2011: Studies on *Verbascum ovalifolium* and *V. purpureum* (Scrophulariaceae) from the vicinity of Edirne (European Turkey). - *Phytol. Balcan.* **17(2)**: 205-212.

YILMAZ, G. & DANE, F. 2012: The genus *Verbascum* L. in European Turkey. - *Bot. Serbica* **36(1)**: 9-13.

YILMAZ, G. & KAYNAK, G. 2008: A new species of *Linum* (Linaceae) from west Anatolia, Turkey. - *Bot. J. Linn. Soc.* **156(3)**: 459-462.

YILMAZ, Ö, KAYNAK, G. & VURAL, M. 2003: A new taxon of *Linum* (Linaceae) from NW Anatolia, Turkey. - *Ann. Bot. Fenn.* **40(2)**: 147-150.

YILMAZ, Ö. & KAYNAK, G. 2010: A new taxon of *Linum* (Linaceae) from Southwest Anatolia, Turkey. - *Novon* **20(4)**: 507-511.

YOU, Z.L. & SLEEP, A. 1988: A chromosome count of *Azolla filiculoides* (Azollaceae: Pteridophyta). - *Brit. Fern Gaz.* **13(4)**: 193-198.

YOUNG, A. 1995: Summer in Greece - 1994. - *Saxifr. Mag.* **3**: 52-56.

YOUNG, D.P. 1970a: Bestimmung und Verbreitung der autogamen *Epipactis*-Arten. In Senghas, K. & Sundermann, H. (eds): Probleme der Orchideengattung *Epipactis*. - *Jahresber. Naturwiss. Vereins Wuppertal* **23**: 43-52.

YOUNG, D.P. 1970b: Notiz über einige südwestasiatische *Epipactis*-Arten. In Senghas, K. & Sundermann, H. (eds): Probleme der Orchideengattung *Epipactis*. - *Jahresber. Naturwiss. Vereins Wuppertal* **23**: 106-108.

YOUNG, D.P. 1970c: Bestimmungsschlüssel der europäischen *Epipactis*-Arten. In Senghas, K. & Sundermann, H. (eds): Probleme der Orchideengattung *Epipactis*. - *Jahresber. Naturwiss. Vereins Wuppertal* **23**: 123-124.

YOUNG, I. 2002: Small fritillaries - yellow. - *Rock Gard.* **27(4)**: 340-359.

YOUNG, S.B. 1970: On the taxonomy and distribution of *Vaccinium uliginosum*. - *Rhodora* **72(792)**: 439-459.

YUNCKER, T.G. 1932: The genus *Cuscuta*. - *Mem. Torrey Bot. Club* **18(2)**: 109-331.

YURTSEVA, O.V. & TIKHOMIROV, V.N. 1997: Morphological diversity and taxonomy of the *Pimpinella tragium* Vill. group (Umbelliferae-Apioideae) in the Mediterranean. - *Feddes Repert.* **109(7-8)**: 479-500.

YUZBASIOGLU, S. & VAROL, O. 2004: A new autumn-flowering *Crocus* from SW Turkey. - *Plantsman* **3(2)**: 104-106.

YILDIRIM, H. 2019: A new *Rindera* (Boraginaceae) species from Western Anatolia, Turkey. - *Phytotaxa* **427(3)**: 249-258.

ZABELINA, N.I. & FILIN, V.R. 1974: To the cytology of *Polypodium vulgare* s.l. from the European part of USSR [In Russian with English summary]. - *Vestn. Moskovsk. Univ., Ser. 6, Biol.* **29(2)**: 47-51.

ZÁBORSKÝ, J. & ZAHRADNÍKOVÁ, K. 1976: Some remarks on the variability and distribution of *Cruciata glabra* (L.) Ehrend. [In Slovakian with English and Russian summaries]. - *Biológia, A (Bratislava)* **31 (1)**: 55-61.

ZACHARIADOU, T., TSIFTSIS, S. & ANAGNOSTOPOULOS, A. 2019: Athens International Airport "Eleftherios Venizelos" as an area of interest for important orchid species of Attica. In Daskalakou, E.N. & al. (eds): 16th Conference of the Hellenic Botanical Society – Athens 10-13 October 2019. Program and book of abstracts, p. 67.

ZACHISTAL, H. 2008: *Sedum*-richness of Mt. Timfi and the Vikos Gorge, Greece. - *Sedum Soc. Newsletter* **86**: 77-78.

ZAFFRAN, J. 1966: Une Campanule nouvelle de la flore crétoise: *Campanula aizoides* Zaffran. - *Bull. Soc. Bot. France* **113(1-2)**: 69-73 + 1 plate.

ZAFFRAN, J. 1967: Découverte du genre *Melitella* Somm. en Crète. - *Compt. Rend. Hebd. Séances Acad. Sci., Sér. D* **264(6)**: 805-808.

ZAFFRAN, J. 1968: Une herborisation classique de l'île de Crète: les gorges de Samaria. - *Monde Pl.* **63(361)**: 7-9.

ZAFFRAN, J. 1970: Les pteridophytes de l'île de Crète. - *Israel J. Bot.* **19(2-3)**: 236-244.

ZAFFRAN, J. 1976: Contributions à la flore et à la végétation de la Crète. I. Floristique. - Marseille: Université de Provence, 223 pp. (mimeographed)

ZAFFRAN, J. 1990: Contribution à la flore et à la végétation de Crète. - Marseille: Université de Provence, 615 pp. + 30 tables.

ZAGANIARIS, D.N. 1932: Ta xirofita tis Attikis [In Greek]. - Athine: 47 pp.

ZAGANIARIS, D.N. 1934: I chloris tis Manis [In Greek]. - Athine: 66 pp.

ZAGANIARIS, D.N. 1935a: I chloris tis Lavreotikis [In Greek]. - *Geôrgikon Deltion (Agricultural Bulletin of Greek Agricultural Society)* **28**: 52, 256, 297, 335 [n.v.].

ZAGANIARIS, D.N. 1935b: I chloris tou Porou. Meleti sistimatikis votanikis [In Greek]. - Athine: 24 pp.

ZAGANIARIS, D.N. 1936: I Kouskouta [In Greek]. - *Geôrgikon Deltion (Agricultural Bulletin of Greek Agricultural Society)* **29**: 296-301.

ZAGANIARIS, D.N. 1937a: Les colchicums de la Grèce. - *Geôrgikon Deltion (Agricultural Bulletin of Greek Agricultural Society)* **30**: 28-32, 103-104 [n.v.].

ZAGANIARIS, D.N. 1937b: Peri tou Hysciamou [In Greek]. - *Geôrgikon Deltion (Agricultural Bulletin of Greek Agricultural Society)* **30**: 376-380.

ZAGANIARIS, D.N. 1937c: Peri tis Arrenopteridos [In Greek]. - *Geôrgikon Deltion (Agricultural Bulletin of Greek Agricultural Society)* **30**: 413-416.

ZAGANIARIS, D.N. 1938: Herbarium macedonicum. Primum mille. - *Sci. Ann. Fac. Phys. Math. Aristot. Univ. Thessalonika* **4**: 97-131.

ZAGANIARIS, D.N. 1939a: Herbarium macedonicum. Secundum mille. - *Sci. Ann. Fac. Phys. Math. Aristot. Univ. Thessalonika* **5**: 151-185.

ZAGANIARIS, D.N. 1939b: Ta zizania tis eparchias Thessalonikis [In Greek]. - *Sci. Ann. Fac. Agric. Forest. Univ. Thessalonika* **1**: 327-386.

ZAGANIARIS, D.N. 1940a: Herbarium macedonicum. Tertium et quartum mille. - *Sci. Ann. Fac. Phys. Math. Aristot. Univ. Thessalonika* **6**: 38-141.

ZAGANIARIS, D.N. 1940b: La flore de Poros. - *Actes Inst. Bot. Univ. Athènes* **1**: 237-253.

ZAHARIADI, C. 1973: Quelques taxons rares ou nouvellement découverts de la flore de la Grèce. - *Ann. Mus. Goulandris* **1**: 165-183.

ZAHARIADI, C. 1975a: Le sous-genre *Codonoprasum* (genre *Allium*, fam. Alliaceae) en Grèce et en Roumanie. In Jordanov, D. & al. (eds): Problems of Balkan flora and vegetation, pp. 229-236. - Sofia: Publishing House of the Bulgarian Academy of Sciences.

ZAHARIADI, C. 1975b: Le sous-genre *Codonoprasum* (genre *Allium* L., fam. Alliaceae Agardh, 1858) en Grèce et en Roumaine. II. - *Biol. Gallo-Hellen.* **6(1)**: 27-64.

ZAHARIADI, C. 1977a: Cinq espèces nouvelles du genre *Ornithogalum* (Liliaceae) trouvées en Grèce. - *Ann. Mus. Goulandris* **3**: 51-75.

ZAHARIADI, C. 1977b: Quelques taxons rares ou nouvellement découverts de la flore de la Grèce (II partie). - *Ann. Mus. Goulandris* **3**: 77-104.

ZAHARIADI, C. 1977c: Notes on the intrageneric classification of the genus *Ornithogalum* L. (Liliaceae) [In Russian]. - *Bot. Zhurn. (Moscow & Leningrad)* **62(11)**: 1624-1639.

ZAHARIADI, C. 1978a: Deux taxons nouveaux ou rares d'*Ornithogalum* de la Méditerranée Orientale et quelques considérations sur la notion de l'endémisme. - *Ann. Mus. Goulandris* **4**: 249-261.

ZAHARIADI, C. 1978b: [Notes on *Ornithogalum*]. In Heywood, V.H. (ed.): Notulae systematicae ad Floram Europaeam spectantes, no. 20. - *Bot. J. Linn. Soc.* **76(4)**: 356-357.

ZAHARIADI, C. 1981: Nouvelles espèces du genre *Ornithogalum* du Proche-Orient et de la Grèce. - *Bull. Soc. Bot. France, Lett. Bot.* **128(4-5)**: 303-314.

ZAHARIADI, C. 1983: Quelques taxons nouveaux du genre *Ornithogalum* (Liliaceae) récemment trouvés dans les Balcans et dans le Proche-Orient. - *Ann. Mus. Goulandris* **6**: 171-197.

ZAHARIADI, C., STAMATIADOU, E. & DIMA, A. 1982: Geographical distribution of species of *Ornithogalum* (Liliaceae) in Greece, including two new taxa. - *Ann. Mus. Goulandris* **5**: 131-162.

ZAHARIS, A.S. 1977: Ta dasi tis Kritis apo tin archeotita eos simeron [In Greek]. - Athine: 146 pp.

ZAHAROF-POURPOUTIDI, E. 1987: Viometriki ke kittarologiki meleti tou genous *Fritillaria* L. stin Ellada [In Greek with English summary]. - *Sci. Ann. School Biol. Aristot. Univ. Thessalonika* **4**: i-x, 1-236.

ZAHAROF, E. 1986a: Variation and taxonomy of *Fritillaria graeca* (Liliaceae) in Greece. - *Pl. Syst. Evol.* **154(1-2)**: 41-61.

ZAHAROF, E. 1986b: *Fritillaria rixii* (Liliaceae), a new species from Euboea, Greece. - *Nordic J. Bot.* **6 (6)**: 725-728.

ZAHAROF, E. 1987: *Fritillaria rhodocanakis* subsp. *argolica* (Liliaceae), a new subspecies from Peloponnese, Greece. - *Willdenowia* **16(2)**: 343-348.

ZAHAROF, E. 1988: A phenetic study of *Fritillaria* (Liliaceae) in Greece. - *Pl. Syst. Evol.* **161(1-2)**: 23-34.

ZAHAROF, E. 1989a: Karyological studies of twelve *Fritillaria* species from Greece. - *Caryologia* **42(2)**: 91-102.

ZAHAROF, E. 1989b: Karyotype variation of *Fritillaria graeca* and *F. davisii* from Greece. - *Nordic J. Bot.* **9(4)**: 367-373.

ZAHLBRUCKNER, A. 1897: Eine neue Art der Gattung *Ajuga*. - *Bot. Centralbl.* **69**: 348.

ZAHN, K.H. 1906: Was ist *Hieracium amphibolum* Rehmann? Ein Beitrag zur Kenntnis des *Piloselloidea* sectio *Alpicolina*. - *Allg. Bot. Z. Syst.* **12(3)**: 37-40.

ZAHN, K.H. 1919: Beitrag zur Kenntnis der Hieracien Mazedoniens und der Balkanländer (Sammlungen J. Bornmüller). - *Repert. Spec. Nov. Regni Veg.* **16**: 177-182, 293-300.

ZAHN, K.H. 1928: Hieracia orientalia nova vel minus cognita. - *Repert. Spec. Nov. Regni Veg.* **24**: 378-385.

ZAHN, K.H., HERMANN, F. & GEORGIEV, T. 1931: Neue und weniger bekannte Hieracien (aus dem Pirin-Gebirge, den Rhodopen u.a.). - *Izv. Bulg. Bot. Druzh.* **4**: 75-86.

ZAIKA, M.A. & AL. 2020: *Scorzonera* sensu lato (Asteraceae, Cichorieae) - taxonomic reassessment in the light of new molecular phylogenetic and carpological analyses. - *PhytoKeys* **137**: 1-85.

ZAIMES, G.N. & AL. 2021: Assessment of two adjacent mountainous riparian areas along Nestos river tributaries of Greece. - *Forests MDPI* **12, 128**: https://doi.org/10.3390/f12091284.

ZAJAC, A. 1975: The genus *Cerastium* L. in Poland. Section *Fugacia* and *Caespitosa*. - *Monogr. Bot.* **47**: 1-100 + 8 plates.

ZAJAC, E.U. 1974: Genus *Fumaria* L. in Poland [In Polish with English summary]. - *Zesz. Nauk. Uniw. Jagiellon. Prace Bot.* **2**: 25-112.

ZAKHAROVA, E.A. 2010: Morphological evidence of polyphyletic nature of traditional *Carum* (Apiaceae-Apioideae). - *Plant Div. Evol.* **128(3-4)**: 409-421.

ZAKHAROVA, E.A. & AL. 2016: A taxonomic study of the genus *Hellenocarum* H. Wolff (Umbelliferae - Apioideae) based on morphology, fruit anatomy, and molecular data. - *Turk. J. Bot.* **40(2)**: 176-193.

ZAKHAROVA, E.A., DEGTJAREVA, G. & PIMENOV, M.G. 2012: Redefined generic limits of *Carum* (Umbelliferae, Apioideae) and new systematic placement of some of its taxa. - *Willdenowia* **42(2)**: 149-168.

ZANDEE, M. 1981: Studies in the *Juncus articulatus* L. - *J. acutiflorus* Ehrh. - *J. anceps* Laharpe, - *J. alpinus* Vill. aggregate. I. A cytotaxonomic survey. - *Proc. Kon. Ned. Akad. Wetensch. C* **84(2)**: 243-254.

ZANDEE, M. & GLAS, P.C.G. 1982: Studies in the *Holcus lanatus* - *Holcus mollis* complex (Poaceae [= Gramineae]). II. Comparative morphology of the spikelet. - *Proc. Kon. Ned. Akad. Wetensch. C* **85(3)**: 413-437.

ZANDEE, M. & HAASNOOT, J. 1983: Studies in the *Juncus articulatus* L. - *J. acutiflorus* Ehrh. - *J. anceps* Laharpe - *J. alpinoarticulatus* Chaix aggregate. II. Comparative anatomy of the panicle axis, culm and leaf. - *Proc. Kon. Ned. Akad. Wetensch. C* **86(2)**: 275-304.

ZANDEE, M. & SCHEEPEN, J. VAN 1981: Studies in the *Holcus lanatus* - *Holcus mollis* complex (Poaceae [= Gramineae]). I. A cytotaxonomic survey. - *Proc. Kon. Ned. Akad. Wetensch. C* **84(4)**: 479-491.

ZANIN, A. & LONGHI WAGNER, H.M. 2005: Lectotypifications in the genus *Andropogon* (Poaceae). - *Novon* **15(1)**: 250-252.

ZARE, G. & AL. 2018: Taxonomic notes on the genus *Chaenorhinum* (Plantaginaceae) in Turkey. - *Acta Bot. Croat.* **77(2)**: 209-213.

ZARIKOU, M. & TRIGAS, P. 2012: Reports 215-244. In Vladimirov, V., Dane, F. & Tan, Kit (eds): New floristic records in the Balkans: 20. - *Phytol. Balcan.* **18(3)**: 369-371.

ZARKOS, G. & AL. 2013: Reports 113-115. In Vladimirov, V., Dane, F. & Tan, Kit (eds): New floristic records in the Balkans: 23. - *Phytol. Balcan.* **19(3)**: 394-397.

ZARKOS, G. & AL. 2014a: Reports 142-148. In Vladimirov, V. & Tan, Kit (eds): New floristic records in the Balkans: 24. - *Phytol. Balcan.* **20(1)**: 128-133.

ZARKOS, G. & AL. 2014b: Reports 244-265. In Vladimirov, V. & al. (eds): New floristic records in the Balkans: 25. - *Phytol. Balcan.* **20(2-3)**: 218-223.

ZARKOS, G. & AL. 2015a: Reports 258-263. In Vladimirov., V, Dane, F. & Tan, Kit (eds): New floristic records in the Balkans: 26. - *Phytol. Balcan.* **21(1)**: 87-89.

ZARKOS, G. & AL. 2015b: Reports 121-128. In Vladimirov, V. & al. (eds): New floristic records in the Balkans: 27. - *Phytol. Balcan.* **21(2)**: 215-217.

ZARKOS, G. & AL. 2015c: Reports 137-238. In Vladimirov, V., Dane, F. & Tan, Kit (eds): New floristic records in the Balkans: 28. - *Phytol. Balcan.* **21(3)**: 387-397.

ZARKOS, G. & AL. 2016a: Reports 119-123. In Vladimirov, V. & al. (eds): New floristic records in the Balkans: 29. - *Phytol. Balcan.* **22(1)**: 117-122.

ZARKOS, G. & AL. 2016b: Reports 163-185. In Vladimirov, V. & al. (eds): New floristic records in the Balkans: 30. - *Phytol. Balcan.* **22(2)**: 286-290.

ZARKOS, G. & AL. 2016c: Reports 219-224. In Vladimirov, V. & Tan, Kit (eds): New floristic records in the Balkans: 31. - *Phytol. Balcan.* **22(3)**: 462-464.

ZARKOS, G. & AL. 2017a: Reports 89-114. In Vladimirov, V., Aybeke, M. & Tan, Kit (eds): New floristic records in the Balkans: 32. - *Phytol. Balcan.* **23(1)**: 141-144.

ZARKOS, G. & AL. 2017b: Reports 213-235. In Vladimirov, V. & al. (eds): New floristic records in the Balkans: 33. - *Phytol. Balcan.* **23(2)**: 321-325.

ZARKOS, G. & AL. 2017c: Reports 145-181. In Vladimirov, V. & al. (eds): New floristic records in the Balkans: 34. - *Phytol. Balcan.* **23(3)**: 438-442.

ZARKOS, G. & AL. 2018: Reports 156-180. In Vladimirov, V., Aybeke, M. & Tan, Kit (eds): New floristic records in the Balkans: 36. - *Phytol. Balcan.* **24(2)**: 283-291.

ZARKOS, G. & AL. 2019: Report 90-91. In Vladimirov, V. & al. (eds): New floristic records in the Balkans: 38. - *Phytol. Balcan.* **25(1)**: 116-119.

ZARKOS, G. & AL. 2019: Reports 109-117. In Vladimirov, V., Aybeke, M. & Tan, Kit (eds): New floristic records in the Balkans: 39. - *Phytol. Balcan.* **25(2)**: 233-236.

ZARKOS, G. & AL. 2020: Reports103-108. In Vladimirov, V., Aybeke, M. & Tan, Kit: New floristic records in the Balkans: 43. - *Phytol. Balcan.* **26(3)**: 568-569.

ZARKOS, G., CHRISTODOULOU, V. & TAN, KIT 2011a: Reports 80-88. In Vladimirov, V. & Tan, Kit (eds): New floristic records in the Balkans: 16. - *Phytol. Balcan.* **17(2)**: 260-263.

ZARKOS, G., CHRISTODOULOU, V. & TAN, KIT 2011b: Reports 131-142. In Vladimirov, V, Dane, F. & Tan, Kit (eds): New floristic reports in the Balkans: 17. - *Phytol. Balcan.* **17(3)**: 380-384.

ZARKOS, G., CHRISTODOULOU, V. & TAN, KIT 2012a: Reports 147-149. In Vladimirov, V. & al. (eds): New floristic records in the Balkans: 18. - *Phytol. Balcan.* **18(1)**: 91.

ZARKOS, G., CHRISTODOULOU, V. & TAN, KIT 2012b: Reports 206-214. In Vladimirov, V., Dane, F. & Tan, Kit (eds): New floristic records in the Balkans: 20. - *Phytol. Balcan.* **18(3)**: 367-369.

ZARKOS, G., CHRISTODOULOU, V. & TAN, KIT 2013a: Reports 147-153. In Vladimirov, V., Dane, F. & Tan, Kit (eds): New floristic records in the Balkans: 21. - *Phytol. Balcan.* **19(1)**: 154-156.

ZARKOS, G., CHRISTODOULOU, V. & TAN, KIT 2013b: Reports 147-159. In Vladimirov, V. & al. (eds): New floristic records in the Balkans: 22. - *Phytol. Balcan.* **19(2)**: 294-298.

ZARKOS, G., CHRISTODOULOU, V. & TAN, KIT 2020: Reports 117-124. In Vladimirov, V., Aybeke, M. & Tan, Kit (eds): New floristic records in the Balkans: 42. - *Phytol. Balcan.* **26(2)**: 211-214.

ZARKOS, G., CHRISTODOULOU, V. & TAN, KIT 2021a: Reports 86-87. In Vladimirov, V., Aybeke, M. & Tan, Kit (eds): New floristic records in the Balkans: 44. - *Phytol. Balcan.* **27(1)**: 148-149.

ZARKOS, G., CHRISTODOULOU, V. & TAN, KIT 2021b: Reports 130-134. In Vladimirov, V., Aybeke, M. & Tan, Kit (eds): New floristic records in the Balkans: 45. - *Phytol. Balcan.* **27(2)**: 274-275.

ZARRE, SH. 1998: Systematic revision of *Astragalus* sect. *Adiaspastus*, sect. *Macrophyllium* and sect. *Pterophorus* (Fabaceae). - München: Diss., Ludwig-Maximilians Univ., 201 pp.

ZARRE, SH. & PODLECH, D. 1997: Problems in the taxonomy of the tragacanthic *Astragalus*. - *Sendtnera* **4**: 243-250.

ZARREI, M. & AL. 2007: Systematic revision of the genus *Gagea* Salisb. (Liliaceae) in Iran. - *Bot. J. Linn. Soc.* **154(4)**: 559-588.

ZBILJIĆ, M., LAKUŠIĆ, B. & LAKUŠIĆ, D. 2018: Morpho-anatomical differentiation of *Teucrium montanum* L. (Lamiaceae) in the Balkan Peninsula - *Bot. Serbica* **42 (suppl. 1)**: 50.

ZELENÝ, V. 1972: Der *Leucanthemum vulgare* Lam.-Komplex in der Tschechoslowakei [In Czech with German summary]. - *Zprávy Českoslov. Bot. Společn. Českoslov. Akad. Věd* **7(1)**: 1-19.

ZELTNER, L. 1961: Contribution à l'étude cytologique des genres *Blackstonia* Huds. et *Centaurium* Hill (Gentianacées). - *Ber. Schweiz. Bot. Ges.* **71**: 18-24.

ZELTNER, L. 1962: Deuxième contribution à l'étude cytologique des genres *Blackstonia* Huds. et *Centaurium* Hill (Gentianacées). - *Bull. Soc. Neuchâteloise Sci. Nat.* **85**: 83-95 + 2 plates.

ZELTNER, L. 1963: Recherches sur quelques taxa méditerranéens du genre *Centaurium* Hill (Gentianaceae). - *Bull. Soc. Neuchâteloise Sci. Nat.* **86**: 93-100 + 2 plates.

ZELTNER, L. 1966: Sur quelques taxa nouveaux ou méconnus des genres *Blackstonia* Huds. et *Centaurium* Hill. - *Bull. Soc. Neuchâteloise Sci. Nat.* **89**: 61-73 + 4 plates.

ZELTNER, L. 1970: Recherches de biosystématique sur les genres *Blackstonia* Huds. et *Centaurium* Hill. - *Bull. Soc. Neuchâteloise Sci. Nat.* **93**: 5-56 + 5 plates.

ZELTNER, L. 1978: Notes de cytotaxonomie sur les genres *Blackstonia* Huds. et *Centaurium* Hill en Crète. - *Bull. Soc. Neuchâteloise Sci. Nat.* **101**: 107-117.

ZELTNER, L. 1985: Etude cytotaxonomique et cytogéographique du *Centaurium pulchellum* (Swartz) Druce, sensu lato. - *Bot. Helv.* **95(1)**: 47-57.

ZELTNER, L. 1987: Contribution à l'ètude cytotaxonomique et cytogéographique des genres *Blackstonia* Huds. et *Centaurium* Hill en URSS. - *Bull. Soc. Neuchâteloise Sci. Nat.* **110**: 17-35.

ZELTNER, L. 1991: Contribution à l'étude cytogéographique des genres *Blackstonia* Huds. et *Centaurium* Hill (Gentianaceae) en Turquie, à Rhodes et à Chypre. - *Bull. Soc. Neuchâteloise Sci. Nat.* **114**: 77-103.

ZENNER, G. 1972: Beitrag zur Unterscheidung der Arten von *Polypodium vulgare* L. s.l. in Europa. - *Göttinger Florist. Rundbr.* **6(2)**: 21-63.

ZERBST, K.-J. & BOCQUET, G. 1974: Beiträge zur Kenntnis der Gattung *Digitalis* L. (Scrophulariaceae). I. Taxonomische Stellung von *Digitalis graeca* Ivanina. - *Ber. Deutsch. Bot. Ges.* **87(1)**: 1-12.

ZERBST, K.-J. & BOCQUET, G. 1975: Beiträge zur Kenntnis der Gattung *Digitalis* L. (Scrophulariaceae). III. *Digitalis* x *pelia* Zerbst & Bocquet, hybr. nov. (*Digitalis graeca* Ivanina × *D. grandiflora* Miller). - *Feddes Repert.* **86(4)**: 201-209.

ZERLENTIS, C. 1952: Beitrag zur Pflanzengeographie der Kykladen-Inseln [In Greek with German summary]. - Athine: 155 pp.

ZERLENTIS, C. 1959: Ta fita tou Likavittou [In Greek]. - *Deltion Agrot. Trapezis [Bulletin of the Agricultural Bank]* **109**: 19-28.

ZERLENTIS, C. 1965: Simvole is tin chlorida tou Imittou [In Greek]. - Athine: 32 pp. [Reprinted 1968 in Geôrgikon Deltion (Bulletin Agricole), ser. 3, 19].

ŽERTOVÁ, A. 1961: Studie über die tschechoslowakischen Arten der Gattung *Lotus* L. I. - *Preslia* **33(1)**: 17-35.

ŽERTOVÁ, A. 1963: Beitrag zur Zytologie der Art *Vicia incana* Gouan [In Slovakian with German summary]. - *Biológia, A (Bratislava)* **18(9)**: 697-700.

ŽERTOVÁ, A. 1964: Bemerkungen zu *Lotus alpinus* Schleich. ex Ser. in Österreich. - *Oesterr. Bot. Z.* **111(2-3)**: 337-344.

ZERVOU, S. 2011: The flora and vegetation of Kalimnos island (Dodekanisa, Greece) [In Greek with English summary]. - Ph.D. thesis, National and Kapodistrian University of Athens: 364 pp.

ZERVOU, S., EHRENDORFER, F. & YANNITSAROS, A. 2009: *Galium capitatum* Bory & Chaub. In Greuter, W. & Raus, Th. (eds): Med-Checklist notulae, 28. - *Willdenowia* **39(2)**: 339.

ZERVOU, S., RAUS, TH. & YANNITSAROS, A. 2009: Additions to the flora of the island of Kalimnos (SE Aegean, Greece). - *Willdenowia* **39(1)**: 165-177.

ZERVOU, S. & YANNITSAROS, A. 1999: Flora of the Acropolis of Athens and of the surrounding archaeological area. In Elefheriou, E.P. (ed.): Proceedings, 7th Panhellenic scientific conference, Alexandroupolis, pp. 342-347 [In Greek with English abstract]. - Thessaloniki: Univ. Studio Press.

ZERVOU, S. & YANNITSAROS, A. 2006: Contribution to the adventive flora of Calymnos (East Aegean, Greece). In Peev, D.R. (ed.): Book of abstracts, IV Balkan botanical congress, Sofia, Bulgaria, 20-26 June 2006, p. 206.

ZERVOU, S. & YANNITSAROS, A. 2009: Contribution to the adventive flora of Kalimnos (East Aegean, Greece). In Ivanova, D. (ed.): Proceedings of the IV Balkan botanical congress, pp. 264-267. - Sofia: Institute of Botany, BAS.

ZÉSIGER, F. 1961: Recherches cytotaxinomiques sur les Joubarbes (genres *Sempervivum* L. et *Jovibarba* Opiz). - *Ber. Schweiz. Bot. Ges.* **71**: 113-117.

ZETTERLUND, H. 1992: Bulbs in Gothenburg Botanic Garden. - *Rock Gard.* **22(4)**: 400-419.

ZEVEN. A.C. 1973: *Triticum lorentii* (Hochst.) Zeven, nov. comb. - *Taxon* **22(2-3)**: 321.

ZEYBEK, N. 1969: Die Chromosomenzahlen von *Salicornia*-Arten der Westanatolischen Meeresküsten [In Turkish with German summary]. - *Ege Üniv. Fen Fak. Ilmi Raporlar Ser.* **83**: 1-15 + 2 plates.

ZEYBEK, N. 1988: Taxonomical investigations on Turkish snowdrops (*Galanthus* L.). - *Doğa Türk Bot. Derg.* **12(1)**: 89-103.

ZEYBEK, N. & AL. 1977: Chromosomenzahlen von drei Salzpflanzen aus Westanatolien. - *Ber. Schweiz. Bot. Ges.* **87(1-2)**: 60-62.

ZEYBEK, N., KESERCIOĞLU, T. & MERT, H.H. 1972: Die Chromosomenzahlen von *Salicornia*-Arten aus Anatolien. - *Verh. Schweiz. Naturf. Ges.* **152**: 134-142.

ZEYBEK, N., KESERCIOĞLU, T. & MERT, H.H. 1976: Halophyten von der Türkei. 1. Die Chromosomenzahlen von *Salicornia*-Arten aus Anatolien. In: Kazdaği göknari ve Türkiye florasi uluslararasi simpozyomu bildirileri, pp. 165-175. - Istanbul Üniversitesi Orman Fakültesi Yayinlari.

ZEYBEK, N. & SAUER, E. 1995: Türkiye Kardelenleri (*Galanthus* L.) I. - Beitrag zur Kenntnis der türkischen Schneeglöckchen [sic!] (*Galanthus* L.) I [In Turkish and German]. - Bornova-Izmir: Ege Üniversitesi, iv + 95 pp.

ZHANG, L.-B. & KADEREIT, J.W. 2002: The systematics of *Soldanella* (Primulaceae) based on morphological and molecular (ITS, AFLPs) evidence. - *Nordic J. Bot.* **22(2)**: 129-169.

ZHANG, L.-B. & KADEREIT, J.W. 2004: Nomenclature of *Soldanella* (Primulaceae). - *Taxon* **53(3)**: 741-752.

ZHANG, M.-H. & AL. 2021: Integrative taxonomy of the *Selaginella helvetica* group based on morphological, molecular and ecological data, - *Taxon* **70(6)**: 1163-1187.

ZHUKOVSKY, P.M. 1928: A critical-systematical survey of the species of the genus *Aegilops* L. [In Russian with English summary]. - *Trudy Prikl. Bot.* **18(1)**: 417-609.

ZIBROWIUS, H. 1992: Records of *Halophila stipulacea* from "Calypso" cruises in Greek and Turkish waters, 1955-1977. - *Posidonia Newslett.* **4(2)**: 7-10.

ZIDORN, C. 2012: *Leontodon* and *Scorzoneroides* (Asteraceae, Cichorieae) in Italy. - *Pl. Biosyst.* **146 suppl.**: 41-51.

ZIELIŃSKI, J. 1975: The genus *Cytisus* L. s.l. in Poland [In Polish with English and Russian summaries]. - *Arbor. Kórnickie* **20**: 47-111.

ZIELIŃSKI, J. 1977: *Crataegus curvisepala* Lindman and *C. microphylla* C. Koch in Bulgaria. - *Arbor. Kórnickie* **22**: 29-38.

ZIELIŃSKI, J. 1982: Distribution of *Rosa sempervirens* L. in the eastern Mediterranean region. - *Ann. Mus. Goulandris* **5**: 163-170.

ZIELIŃSKI, J. 1984: *Crataegus Heldreichii* Boiss. in Bulgaria. - *Fragm. Florist. Geobot.* **28(4)**: 625-627.

ZIELIŃSKI, J. 1985: Studies on the genus *Rosa* L. - Systematics of section *Caninae* DC. em. Christ [In Polish with English summary]. - *Arbor. Kórnickie* **30**: 3-109.

ZIELIŃSKI, J. 1986: *Rosa rubiginosa* L. and *R. heckeliana* Tratt. in Bulgaria. - *Fragm. Florist. Geobot.* **30(3)**: 207-211.

ZIELIŃSKI, J. 1987: Floristic notes on the island of Chios. - *Willdenowia* **16(2)**: 373-374.

ZIELIŃSKI, J. 1990: New records for the flora of Evvia. - *Ann. Mus. Goulandris* **8**: 237-240.

ZIELIŃSKI, J. 1991a: *Scorpiurus vermiculatus* (Fabaceae) rediscovered in Greece. - *Willdenowia* **20(1-2)**: 39-41. [Also distributed as OPTIMA leaflet no. 210].

ZIELIŃSKI, J. 1991b: *Cotoneaster tomentosus* (Rosaceae) - a new species for the flora of Turkey. - *Karaca Arbor. Mag.* **1(2)**: 49-52.

ZIELIŃSKI, J. 1991c: The genus *Rosa* L. in Greece. - *Arbor. Kórnickie* **35**: 3-45.

ZIELIŃSKI, J. 1991d: *Polygonum karacae* (Polygonaceae), a new species from SW Turkey. - *Willdenowia* **21(1-2)**: 173-174.

ZIELIŃSKI, J. 1992a: *Vicia parvula* (Fabaceae) - a new species from NW Turkey. - *Karaca Arbor. Mag.* **1(4)**: 121-123.

ZIELIŃSKI, J. 1992b: *Salix xanthicola* (Salicaceae) - a species new to Bulgaria. - *Fragm. Florist. Geobot.* **37(2)**: 499-501.

ZIELIŃSKI, J. 2000: *Amelanchier parciflora* subsp. *chelmea* (Halácsy) Ziel., comb. & stat. nov. In Greuter, W. & Raus, Th. (eds): Med-Checklist notulae, 19. - *Willdenowia* **30(2)**: 239.

ZIELIŃSKI, J., PETROVA, A. & PANCHEVA, Z. 2006: *Salix ×velchevii* and *S. ×ardana* (Salicaceae), two new willow hybrids from the Bulgarian Rhodope Mts. - *Acta Soc. Bot. Poloniae* **75(2)**: 145-148.

ZIELIŃSKI, J., PETROVA, A.V. & TAN, KIT 2004: Taxonomic status of the roses (*Rosa*) described by S.G. Dimitrov from Bulgaria. - *Ann. Bot. Fenn.* **41(6)**: 449-451.

ZIELIŃSKI, J., PETROVA, A.V. & TOMASZEWSKI, D. 2006: *Quercus trojana* subsp. *yaltirikii* (Fagaceae), a new subspecies from southern Turkey. - *Willdenowia* **36(2)**: 845-849.

ZIELIŃSKI, J. & TAN, KIT 2000: *Dichondra repens* (Convolvulaceae), a new alien species in the flora of Turkey. - *Karaca Arbor. Mag.* **5(3)**: 119-124.

ZIELIŃSKI, J., TAN, KIT & TOMASZEWSKI, D. 2004: Notes on the taxonomy of *Genista januensis* and *G. lydia* (Fabaceae). - *Ann. Bot. Fenn.* **41(6)**: 453-457.

ZIJLSTRA, G. 1972: Bemerkungen zur Systematik und Ökologie von *Blackstonia perfoliata* und *B. acuminata* (Gentianaceae). - *Acta Bot. Neerl.* **21(6)**: 587-597.

ZIJLSTRA, G. 1991: (997) Proposal to conserve the spelling of 4618 *Euonymus* (Celastraceae). - *Taxon* **40(1)**: 137-139. [approved, cf. Taxon 43(1): 115, 1994].

ZIJLSTRA, G. 1996: (1249) Proposal to conserve the name *Conopodium* (Umbelliferae) with a conserved type. - *Taxon* **45(3)**: 557-558. [approved, cf. Taxon 48(2): 361, 1999].

ZIKA, P.F., REVEAL, J.L. & JARVIS, C.E. 2008: (1818) Proposal to conserve the name *Impatiens noli-tangere* (Balsaminaceae) with a conserved type. - *Taxon* **57 (2)**: 650-651. [approved, cf. Taxon 59(4): 1274, 2010].

ZIKOS, A. & CONSTANTINIDIS, TH. 2017: The flora of Oligirtos and Farmakas Mountains (NE Peloponnisos, Greece): early results. In Dousis, M.A. & Thanos, K.A.: 15th conference of the Hellenic Botanical Society, book of abstracts: p. 80.

ZIKOS, A. & CONSTANTINIDIS, TH. 2018: Floristic diversity of mountains Oligirtos and Farmakas (north- east Peloponnisos. Greece): Preliminary results. - *Bot. Serbica* **142 (suppl. 1)**: 91-92.

ZIMAN, S., BULAKH, E. & TSARENKO, O. 2011: *Anemone* L. (Ranunculaceae): comparative morphology and taxonomy of the species from the Balkan flora. - *Bot. Serbica* **35(2)**: 87-97.

ZIMMER, B. 1984: *Cystopteris dickieana* R. Sim. In Greuter, W. & Raus, Th. (eds): Med-Checklist notulae, 9. - *Willdenowia* **14(1)**: 38.

ZIMMER, B. 1989: Some features of the Greek fern flora: Remarks on the Greek cheilanthoid species. In Phitos, D. (ed.): Abstracts. Sixth OPTIMA meeting, Delphi, 10-16 Sept. 1989, p. 31.

ZIMMER, B. 1991: Remarks on the Greek cheilanthoid ferns. - *Bot. Chron. (Patras)* **10**: 221-237.

ZIMMER, B. & GREUTER, W. 1994: (1093) Proposal to reject the name *Asplenium ramosum* L. (Pteridophyta). - *Taxon* **43(2)**: 303-304. [approved, cf. Taxon 48(1): 133, 1999].

ZIMMERMANN, W. 1958: Zur Taxonomie der Gattung *Pulsatilla* Miller. - *Feddes Repert. Spec. Nov. Regni Veg.* **61(1)**: 94-100.

ZINDROS, A. & AL. 2020: Tree line shift in the Olympus mountain (Greece) and climate change. - *Forests MDPI* **11, 985** : doi:10.3390/f11090985.

ZIPPEL, E. & RAUS, TH. 2020: *Buglossoides incrassata*. subsp. *splitgerberi*. In Raus, Th. & Raab- Straube, E.v. (eds): Euro+Med-Checklist notulae, 12. - *Willdenowia* **50(2)**: 308-310.

ZIPPEL, E. & WILHALM, TH. 2003: Nachweis und Verbreitung annueller *Buglossoides*-Arten (Lithospermeae, Boraginaceae) in Südtirol (Italien). - *Gredleriana* **3**: 347-360.

ZOGARIS, S., CHATZINIKOLAOU, Y. & DIMOPOULOS, P. 2008: Riparian woodland flora in upland rivers of Western Greece. - *Medit. Marine Sci.* **9(2)**: 87-102.

ZOGOPOULOU, E., KOUGIOUMOUTZIS, K. & PANITSA, M. 2017: Plant diversity patterns and geodiversity: the Methana peninsula. In Dousis, M.A. & Thanos, K.A.: 15th conference of the Hellenic Botanical Society, book of abstracts: p. 82.

ZOGRAFIDIS, A. 2016: Two new infraspecific taxa of *Verbascum delphicum* (Scrophulariaceae, Scrophularieae) from mainland Greece and the island of Evvia. - *PhytoKeys* **74**: 107-122.

ZOGRAFIDIS, A. 2017: New floristic records in Greece. - *Parnassiana Arch.* **5**: 61-66.

ZOGRAFIDIS, A. 2018: Resurrection and typification of *Verbascum auriculatum* (Scrophulariaceae), a long disused name in Flora Graeca Sibthorpiana. - *Phytotaxa* **361(2)**: 233-243.

ZOGRAFIDIS, A. 2019a: Typification of *Verbascum undulatum* (Scrophulariaceae). - *Phytotaxa* **391(3)**: 243-246.

ZOGRAFIDIS, A. 2019b: An introduction to the genus *Verbascum*. In Daskalakou, E.N. & al. (eds): 16th Conference of the Hellenic Botanical Society – Athens 10-13 October 2019. Program and book of abstracts, p. 19.

ZOGRAFIDIS, A. & AL. 2019: Rediscovery of two rare Greek endemic species of *Verbascum* L. In Daskalakou, E.N. & al. (eds): 16th Conference of the Hellenic Botanical Society – Athens 10-13 October 2019. Program and book of abstracts, p. 70.

ZOGRAFIDIS, A. & AL. 2020a: Notes on the generic name *Graecobolanthus* (Caryophylleae, Caryophyllaceae). - *Taxon* **69(5)**: 992-997.

ZOGRAFIDIS, A. & AL. 2020b: Notes on *Verbascum* (Scrophulariaceae) from Greece. - *Phytol. Balcan.* **26(3)**: 485-494.

ZOGRAFIDIS, A., BANCHEVA, S. & TAN, KIT 2014: A new yellow-flowered species of *Centaurea* (Asteraceae) from Mt Imittos, Attica, Greece. - *Phytol. Balcan.* **20(1)**: 57-63.

ZOGRAFIDIS, A. & STRID, A. 2017: A taxonomic revision of the *Verbascum daenzeri* group (Scrophulariaceae). - *Phytol. Balcan.* **23(1)**: 65-71.

ZOGRAFIDIS, A. & TAN, KIT 2013: Reports 116-122. In Vladimirov, V., Dane, F. & Tan, Kit (eds): New floristic records in the Balkans: 23. - *Phytol. Balcan.* **19(3)**: 397.

ZOGRAFIDIS, A. & TAN, KIT 2014a: Reports 149-155. In Vladimirov, V. & Tan, Kit (eds): New floristic records in the Balkans: 24. - *Phytol. Balcan.* **20(1)**: 133-135.

ZOGRAFIDIS, A. & TAN, KIT 2014b: Reports 266-269. In Vladimirov, V. & al. (eds): New floristic records in the Balkans: 25. - *Phytol. Balcan.* **20(2-3)**: 223-224.

ZOGRAFIDIS, A., TAN, KIT & VOLD, G. 2013: Reports 160-168. In Vladimirov, V. & al. (eds): New floristic records in the Balkans: 22. - *Phytol. Balcan.* **19(2)**: 298-300.

ZOHARY, D. 1963: Spontaneous brittle six-row barleys, their nature and origin. In: Proceedings of the First Barley Genetics Symposium, pp. 27-31. - Wageningen: Centre for Agric. Publ. & Docum.

ZOHARY, D. 1969: The progenitors of wheat and barley in relation to domestication ... In Ucko, P.J. & Dimbleby, G.W. (eds): The domestication and exploitation of plants and animals, pp. 47-66. - London: Gerald Duckworth & Co. Ltd.

ZOHARY, D. 1971: Origin of South-West Asiatic cereals: wheats, barley, oats and rye. In Davis, P.H., Harper, P.C. & Hedge, I.C. (eds): Plant life of South-West Asia, pp. 235-263. - Edinburgh: Botanical Society.

ZOHARY, D. 1972: The wild progenitor and the place of origin of the cultivated lentil: *Lens culinaris*. - *Econ. Bot.* **26(4)**: 326-332.

ZOHARY, D. 1986: The origin and early spread of agriculture in the Old World. In Barigozzi, C. (ed.): The origin and domestication of cultivated plants, pp. 3-20. - Amsterdam, etc.: Elsevier.

ZOHARY, D. 1991a: Conservation of wild progenitors of cultivated plants in the Mediterranean basin. - *Bot. Chron. (Patras)* **10**: 467-474.

ZOHARY, D. 1991b: The wild genetic resources of cultivated lettuce (*Lactuca sativa* L.). - *Euphytica* **53**: 31-35.

ZOHARY, D. 1992: Is the European plum, *Prunus domestica* L., a *P. cerasifera* Ehrh. × *P. spinosa* L. allo-polyploid? - *Euphytica* **60**: 75-77.

ZOHARY, D. 1995a: Fig, *Ficus carica* (Moraceae). In Smartt, J. & Simmond, N.W. (eds): Evolution of crop plants (2nd ed.), pp. 366-370. - London: Longman.

ZOHARY, D. 1995b: Olive, *Olea eurpaea* (Oleaceae). In Smartt, J. & Simmond, N.W. (eds): Evolution of crop plants (2nd ed.), pp. 379-382. - London: Longman.

ZOHARY, D. 1995c: The domestication of the grapevine *Vitis vinifera* L. in the Near East. In McGovern, P., Fleming, S. & Katz, S. (eds): The origins and ancient history of wine, pp. 53-61. - New York: Gordon & Breach.

ZOHARY, D. 1996: The genus *Pistacia* L. In Padulosi, S., Caruso, T. & Barone, E. (eds): Taxonomy, distribution, conservation and uses of *Pistacia* genetic resources. Report of a workshop, 29-30 June 1995, Palermo. - Rome: International Plant Genetic Resources Institute.

ZOHARY, D. 1997a: Speciation patterns in predominantly self-pollinated Mediterranean annuals. - *Lagascalia* **19(1-2)**: 283-288.

ZOHARY, D. 1997b: Wild apples and wild pears. - *Bocconea* **7**: 409-416.

ZOHARY, D. 2002: Domestication of the carob (*Ceratonia siliqua* L.). - *Israel J. Pl. Sci.* **50 (suppl. 1)**: 141-145.

ZOHARY, D. & HOPF, M. 1988: Domestication of plants in the Old World. The origin and spread of cultivated plants in West Asia, Europe and the Nile Valley. - Oxford: Clarendon Press, ix + 249 pp. [4th ed, 264 pp, 2012].

ZOHARY, D. & PLITMANN, U. 1979: Chromosome polymorphism, hybridization and colonization in the *Vicia sativa* group (Fabaceae). - *Pl. Syst. Evol.* **131(1-2)**: 143-156.

ZOHARY, M. 1938a: On the vegetative reproduction of some oriental geophytes. - *Palestine J. Bot., Jerusalem Ser.* **1(1)**: 35-41.

ZOHARY, M. 1938b: On the *Plantago* species of the Near East. - *Palestine J. Bot., Jerusalem Ser.* **1 (2)**: 225-232 + 1 plate.

ZOHARY, M. 1950: The segetal plant communities of Palestine. - *Vegetatio* **2(6)**: 387-411 + 2 fold-out tables.

ZOHARY, M. 1952: A monographic study of the genus *Pistacia*. - *Palestine J. Bot., Jerusalem Ser.* **5 (4)**: 187-228.

ZOHARY, M. 1960: The species of *Capparis* in the Mediterranean and the Near Eastern countries. - *Bull. Res. Council Israel, Sect. D, Bot.* **8(1)**: 49-64.

ZOHARY, M. 1961: On the oak species of the Middle East. - *Bull. Res. Council Israel, Sect. D, Bot.* **9 (4)**: 161-186.

ZOHARY, M. 1963: Taxonomical studies in *Alcea* of South-Western Asia. Part II. - *Israel J. Bot.* **12(1)**: 1-26.

ZOHARY, M. 1971a: A revision of the species of *Trifolium* sect. *Trifolium* (Leguminosae). I. Introduction. - *Candollea* **26(2)**: 297-308.

ZOHARY, M. 1971b: The phytogeographical foundations of the Middle East. In Davis, P.H., Harper, P.C. & Hedge, I.C. (eds): Plant life of South-West Asia, pp. 43-52. - Edinburgh: Botanical Society.

ZOHARY, M. 1972a: A revision of the species of *Trifolium* sect. *Trifolium* (Leguminosae). II. Taxonomic treatment. - *Candollea* **27(1)**: 99-158.

ZOHARY, M. 1972b: A revision of the species of *Trifolium* sect. *Trifolium* (Leguminosae). III. Taxonomic treatment (sequel). - *Candollea* **27(2)**: 249-264.

ZOHARY, M. 1972c: Origins and evolution in the genus *Trifolium*. - *Bot. Not.* **125(4)**: 501-511.

ZOHARY, M. 1973: Geobotanical foundations of the Middle East. Vol. 1 (pp. i-xi, 1-340 pp. + 2 maps), vol. 2 (pp. i-xi, 341-739 + 1 map). - Stuttgart: Gustav Fischer Verlag & Amsterdam: Swets & Zeitlinger.

ZOHARY, M. 1975: The phytogeographical delimitation of the mediterranean region towards the east. In Guinochet, M. & al. (eds): La flore du Bassin Méditerranéen, pp. 329-334. - Paris: C.N.R.S.

ZOHARY, M. 1983: The genus *Nigella* (Ranunculaceae) - a taxonomic revision. - *Pl. Syst. Evol.* **142(1-2)**: 71-107.

ZOHARY, M. & HELLER, D. 1970: The *Trifolium* species of sect. *Vesicaria* Crantz. - *Israel J. Bot.* **19(2-3)**: 314-335.

ZOHARY, M. & HELLER, D. 1984: The genus *Trifolium*. - Jerusalem: Israel Academy of Sciences and Humanities, x + 606 pp. + 1 plate.

ZOHARY, M., HEYN, C.C. & HELLER, D. 1980: Conspectus Florae Orientalis. An annotated catalogue of the flora of the Middle East. Fasc. 1. - Jerusalem: The Israel Academy of Sciences and Humanities, xiv + 107 pp. + 1 map.

ZOHARY, M., HEYN, C.C. & HELLER, D. 1983: Conspectus Florae Orientalis. An annotated catalogue of the flora of the Middle East. Fasc. 2. - Jerusalem: The Israel Academy of Sciences and Humanities, xiv + 86 pp. + 1 map.

ZOHARY, M. & ORSHAN, G. 1959: The maquis of *Ceratonia siliqua* in Israel. - *Vegetatio* **8(5-6)**: 285-297.

ZOHARY, M. & ORSHAN, G. 1966: An outline of the geobotany of Crete. - *Israel J. Bot.* **14, suppl.**: 1-49.

ZOLLER, H., GEISSLER, P. & ATHANASIADIS, N. 1977: Beiträge zur Kenntnis der Wälder, Moos- und Flechtenassoziationen in den Gebirgen Nordgriechenlands. - *Bauhinia* **6(1)**: 215-255 + 14 tables and 5 figures.

ZONNEVELD, B.J.M. 1979a: Comments on *Sempervivum ciliosum* Craib. - *J. Sempervivum Soc.* **10(1)**: 20-22.

ZONNEVELD, B.J.M. 1979b: *Sempervivum ciliosum* Craib - further comments. - *J. Sempervivum Soc.* **10(2)**: 38-39.

ZONNEVELD, B.J.M. 1979c: New data on the varieties of *Sempervivum ciliosum* Craib. - *J. Sempervivum Soc.* **10(3)**: 74-78.

ZONNEVELD, B.J.M. 1981: Sempervivums in southern Yugoslavia. - *Quart. Bull. Alpine Gard. Soc. Gr. Brit.* **49(4)**: 322-324, 329.

ZONNEVELD, B.J.M. 1982: Sempervivums in Bulgaria. - *Quart. Bull. Alpine Gard. Soc. Gr. Brit.* **50(2)**: 117-120.

ZONNEVELD, B.J.M. 2009: The systematic value of nuclear genome size for "all" species of *Tulipa* L. (Liliaceae). - *Pl. Syst. Evol.* **282(1-4)**: 217-245.

ZOTOS, A. 2006: Flora, vegetation ecology and management proposals for the wet meadows and reed thickets of the lakes Trichonida and Lysimachia (W. Greece). [In Greek with English summary]. - Thessaloniki: Ph.D. thesis, Aristotelian University, 314 pp.

ZOTOS, A. & AL. 2006: *Ludwigia peploides* subsp. *montevidensis*, a new alien taxon for the flora of Greece and the Balkans. - *J. Biol. Res. Thessalon.* **5**: 71-78.

ZOTOS, A. & AL. 2021: Plant species diversity of the wet meadows under natural and anthropogenic interventions: The case of Lakes Amvrakia and Ozeros (W. Greece). - *Not. Bot. Hort. Agrobot. Cluj-Napoca* **49(3)**: 12435.

ZOTOS, A., RAUS, TH. & DIMOPOULOS, P. 2006: New floristic reports from the lakes Trichonis and Lisimachia (W Greece). - *Willdenowia* **36(2)**: 731-739.

ZOZ, I.G. 1976: De generis *Herniaria* L. (Caryophyllaceae) speciebus in Europa orientali crescentibus [In Russian]. - *Novosti Sist. Vysš. Rast.* **13**: 105-109.

ZUBOV, D. 2017: New *Crocus* species from the south-eastern Aegean, Karpathos island, Greece. - *Int. Rock Gard.* **90**: 13 pp.

ZUBOV, D.A., KONCA, Y. & DAVIS, A.P. 2019: *Galanthus bursanus* (Amaryllidaceae): a new species of snowdrop from the Marmara Sea region, NW Turkey. - *Kew Bull.* **74(2)**: 8 pp.

ŻUK, J. 1963: An investigation on polyploidy and sex determination within the genus *Rumex*. - *Acta Soc. Bot. Poloniae* **32(1)**: 5-67 + 6 plates.

ZÜNDORF, W. 1939: Zytogenetisch-entwicklungsgeschichtliche Untersuchungen in der *Veronica*-Gruppe *Biloba* der Sektion *Alsinebe* Griseb. - *Z. Indukt. Abstammungs-Vererbungsl.* **77(2)**: 195-238.

ZÜNDORF, W. 1940: Geographie und Phylogenie der *Veronica*-Gruppe *Biloba* (Vorläufige Mitteilung). - *Ber. Deutsch. Bot. Ges.* **58(3)**: 125-130.

ZVOLÁNEK, Z. 2010: Ali Botusch at last. - *Alpine Gardener* **78(3)**: 266-277.

ZWIENEN, K.J. VAN & BRINK, M. VAN DEN 2005: In search of autumn-flowering bulbs - ten days in the Peloponnese. - *Rock Gard.* **115**: 2-21.

ZWIENEN, K.J. VAN & LAMBERT, R. 2009: Autumn in Greece - The Peloponnese and eastern Attica. - *Rock Gard.* **122**: 54-71.

COLOUR PLATES 1-268

The following colleagues have kindly provided photos:

Erwin Bergmeier: 36:7, 98:11, 108:4, 207:7, 214:11
Jean Bienvenu: 26:3, 31:11, 32:7, 37:4, 39:4, 44:2, 68:2 (main figure), 123:4, 128:1 (inset), 108:6, 266:5 (inset)
Stylianos Charalambidis: 19:2 (inset of flowers), 25:3 (insets), 69:1 (inset), 88:3 (inset of flowers), 102:12, 104:8 (inset), 123:2, 202:5 (inset), 241:4 (inset)
Ina Dinter: 5:4, 5:9, 6:7, 7:8, 13:12 (inset), 16:3, 20:12 (main figure), 22:8, 26:12, 27:2, 28:7, 36:3, 37:5 (inset), 40:7, 40:12, 44:6, 51:4 (main figure), 54:8, 60:2 (inset), 62:6, 63:3, 65:1 (inset), 65:9 (main figure), 68:8, 68:9 (inset), 72:3 (inset), 75:5 (inset), 78:12, 83:11, 84:12 (inset), 85:9, 86:5, 88:11, 90:7, 96:10, 106:9, 111:7, 113:5, 114:3, 116:2, 116:6, 117:2, 118:2, 128:7, 129:10, 132:4 (inset), 137:3, 139:8 (inset), 140:10, 142:4, 143:11, 145:6, 151:12, 153:2, 153:6, 170:9, 170:10, 178:5, 179:8 (main figure), 192:6, 192:8, 200:9, 203:9, 206:5 (main figure), 218:8 (inset), 225:3 (inset), 226:2 (main figure), 226:3, 250:7, 263:1, 268:8 (inset)
Johannes Flohe: 33:1, 39:11, 57:2, 66:1, 66; 2, 79:11, 86:6, 87:1, 117:7, 165:11 (main figure), 173:5, 185:12, 197:8, 200:1, 200:6, 201:1, 205:1, 223:2 (inset), 223:7, 225: 11, 234:3 (main figure), 237:1 (inset), 237:2, 249:5 (main figure), 258:11, 260:4 (inset right)
Michael Foley: 253:4
Christina Fournaraki: 229:10
Ori Fragman: 89:8
Thomas Giannakis: 126:7, 227:1
Günter Gottschlich: 49:3
Katerina Goula: 6:4, 8:5, 8:10, 8:11, 9:2, 11:1, 24:12, 27:4, 28:5, 30:12, 31:4, 31:5, 31:8, 32:9, 32:11, 33:9, 38:7, 38:9, 38:10, 39:5, 40:11 (inset), 52:12, 60:6, 60:8 (inset), 63:2, 65:1, 65:2, 68:6 (main figure), 70:5, 70:10, 74:1, 74:4, 74:7, 75:5, 77:7, 78:4, 78:8, 79:4, 80:10, 81:11, 89:7, 89:10, 90:2 (inset), 91:4, 91:11, 92:8, 92:11, 93:3, 93:5, 93:8 (inset), 94:12, 98:4 (inset), 99:7, 100:12, 101:7, 101:10, 101:12, 107:4, 112:3, 115:1 (inset of flower), 116:3, 117:4, 124:4, 125:7, 127:1 (main figure), 131:12, 137:2, 138:12, 145:5, 149:10, 152:9, 152:12, 154:2, 154:6, 159:5, 159:8, 167:2, 170:12, 171:6, 172:2, 173:2, 179:4, 180:4, 180:7, 181:2, 181:3 (main figure), 183:3, 190:12, 194:2, 195:7, 196:8, 196:9 (inset), 197:6, 198:8, 200:2, 208:7, 209:7, 211:8, 215:10, 224:12, 234:12 (inset), 237:4, 241:3, 243:4 (inset), 244:11, 247:11 (inset), 248:12, 250:8, 250:9, 253:2, 254:1, 254:2, 257:12, 258:9, 262:3, 264:8, 267:1
Johan Grudemo: 131:11
Doerte Harpke: 165:6
Helmut Hornik: 202:3
Armin Jagel: 2:12, 3:3, 3:6, 7:11, 16:2, 16:5, 17:12, 70:8, 147:7, 168:10, 170:1, 203:7, 204:5, 207:5, 209:4, 226:6, 242:3
Leftherios Kipopoulos: 186:12
Steve Lenton: 11:3, 11:7, 11:12, 12:7, 18:2, 36:11, 170:4, 176:7, 179:4 (main figure), 197:5, 202:7, 265:4
Manolis Manousakis: 20:4, 82:8, 93:6
Themidia Nasopoulou: 197:11
Gisela Nikolopoulou: 16:1, 20:3 (insets), 22:12 (insets), 24:7, 80:11, 103:11, 127:4, 133:2, 242:12 (inset)
Georgios Politis: 67:11
Kostas Polymenakos: 266:9
Uwe Raabe: 4:3, 8:1, 8:2, 12:10, 19:9, 39:8 (inset), 46:11, 62:7, 91:7, 102:4, 102:8, 106:12, 107:2, 115:6, 121:6, 122:5, 131:6, 150:10, 169:3, 178:12, 189:6 (inset), 191:9, 210:6, 211:1 (inset), 211:2, 211:7, 214:9, 216:5, 218:3, 221:11, 225:9, 225:10 (inset), 228:12, 268:4
Stefan Rätzel: 9:12
Janis Rukšāns: 164; 12
Astrid Scharlau: 12:4, 17:6, 22:11, 24:9, 40:11, 43:8, 43:11, 46:5, 46:6, 52:10, 57:6, 59:2, 66:5, 73:4, 81:9, 81:10 (main figure), 92:1, 94:4, 95:2, 95:4, 96:9, 103:8, 105:10, 109:6, 114:5, 124:3 (inset of flowers), 124:7 (main figure), 127:8, 128:4, 128:6, 134:8, 137:4, 138:10, 141:12, 142:2, 142:5, 145:7, 146:7, 147:1, 149:10 (inset), 153:1, 153:7, 155:10, 173:8, 192:10, 195:5, 203:1, 204:10, 205:12, 210:12, 216:4, 218:1, 243:1, 252:10, 261:1
Pascale Servais & Pierre Seba: 24:5, 31:2, 44:9, 114:6 (main figure), 115:1 (inset of fruit), 115:8 (inset), 123:8, 124:3, 124:10, 131:3, 138:7, 146:3, 150:4, 168:5, 183:4 (inset), 212:10, 245:10, 261:11 (inset), 264:10, 264: 11 (insets), 264:12
Lulëzim Shuka: 91:5
Åke Svensson: 220:6
Flemming Thorning-Lund: 66:3, 165:1 (main figure), 228:9
Panayiotis Trigas: 9:11, 10:7, 10:8, 11:4, 11:10, 19:1, 40:3, 42:2, 51:8, 56:5 (inset), 67:2, 70:7 (inset), 71:11, 79:8, 80:7, 82:12, 83:2, 86:2 (inset), 87:11, 89:5, 92:6, 95:8, 97:5, 97:7, 97:10, 97:11, 100:3, 100:10, 102:9, 104:6, 105:7 (inset), 106:10, 111:3, 122:6, 127:1 (inset), 134:2 (inset), 152:1 (inset), 152:5, 171:11, 176:1, 182:7, 184:8, 187:6 (main figure), 190:6, 218:4, 228:11, 246:2, 258:10, 259:3
Spyros Tsiftsis: 152:3, 197:12, 198:3, 202:11, 235:7, 240:7, 241:8, 250:11
Nick Turland: 7:7, 16:10, 18:8 (inset), 19:3, 19:5, 20:2, 21:9, 22:5, 28:4, 42:12, 43:6, 44:10, 49:6, 56:1, 56:7, 64:5, 66:6, 68:2 (inset), 69:8, 70:1, 84:3 (inset), 84:7, 84:9, 85:6, 87:5, 91:8, 93:12, 98:12, 104:3, 109:11, 115:1 (main figure), 115:2, 124:1, 129:6, 134:3, 143:3, 143:10, 143:12, 144:7, 158:3, 162:9, 174:9, 175:1, 219:9, 223:12, 225:2, 231:4, 253:1, 261: 7, 263:6
Peter Wind: 126:10
Aris Zografidis: 34:10, 93:11, 182:2, 240:5, 253:7, 254:10, 257:4, 268:6

All other photos are by the author. About 60 per cent of the Greek species are illustrated. Within each plate figures are numbered from left to right and top to bottom.

And why do you worry about clothes? See how the flowers of the field grow. They do not labor or spin. Yet I tell you that not even Solomon in all his splendor was dressed like one of these.
Matthew 6: 28-29

Plate 1

1. Aspleniaceae: *Asplenium adiantum-nigrum* (Prespa area). - 2. *Asplenium ceterach* (Kriti, Rethimnis). - 3. *Asplenium cuneifolium* (Smolikas). - 4. *Asplenium fissum* (Olimbos). - 5. *Asplenium onopteris* (Athos peninsula). - 6. *Asplenium ruta-muraria* (Falakron). - 7. *Asplenium scolopendrium* (Olimbos). - 8. *Asplenium septentrionale* (Pisoderi area). - 9. *Asplenium trichomanes* (Parnassos). - 10. *Asplenium viride* (Pangeon). - 11. Athyriaceae: *Athyrium filix-femina* (Vrondous, inset from Voras). - 12. Blechnaceae: *Blechnum spicant* (unknown origin).

Plate 2

1. Blechnaceae: *Woodwardia radicans* (W Kriti). - 2. Cystopteridaceae: *Cystopteris fragilis* (Olimbos). - 3. Dennstaedtiaceae: *Pteridium aquilinum* (ESE Karpenisi). - 4. Dryopteridaceae: *Dryopteris filix-mas* (Rodopi, inset from Voras). - 5. *Dryopteris pallida* (Taigetos). - 6. *Polystichum aculeatum* (Varnous). - 7. *Polystichum lonchitis* (Olimbos). - 8. *Polystichum setiferum* (Athos peninsula). - 9. Equisetaceae: *Equisetum arvense* (Pisoderi). - 10. *Equisetum fluviatile* (Voras). - 11. *Equisetum palustre* (Killini). - 12. *Equisetum ramosissimum* (W Peloponnisos).

Plate 3

1. Equisetaceae: *Equisetum telmateia* (NW Peloponnisos). - 2. Isoetaceae: *Isoetes histrix* (Evrou, Dadia). - 3. Marsileaceae: *Marsilea aegyptiaca* (Elafonisos). - 4. *Marsilea quadrifolia* (Lake Kerkini). - 5: Ophioglossaceae: *Botrychium lunaria* (Timfi). - 6. *Ophioglossum lusitanicum* (SE Peloponnisos). - 7. *Ophioglossum vulgatum* (Konitsis, SW of Fourka). - 8. Osmundaceae: *Osmunda regalis* (Ikaria). - 9. Polypodiaceae: *Polypodium cambricum* (Methana peninsula). - 10. Pteridaceae: *Adiantum capillus-veneris* (Samos). - 11: *Anogramma leptophylla* (Rodos). - 12. *Cheilanthes acrostica* (Imittos).

Plate 4

1. Pteridaceae: *Cosentinia vellea* (S Peloponnisos). - 2. *Notholaena marantae* (Evrou, SW of Kotronia). - 3. *Pteris vittata* (Mani peninsula). - 4. Salviniaceae: *Azolla filiculoides* (Lake Megali Prespa). - 5. *Salvinia natans* (Lake Doirani). - 6. Selaginellaceae: *Selaginella denticulata* (Imittos). - 7. Thelypteridaceae: *Thelypteris palustris* (Lake Mikri Prespa). - 8. Cupressaceae: *Cupressus sempervirens* (W Kriti). - 9. *Juniperus communis* subsp. *communis* (Serbia, Daliblatsa). - 10. *Juniperus drupacea* (Parnonas). - 11. *Juniperus excelsa* (Prespa area; inset of twigs, *J. foetidissima* left, *J. excelsa* right). - 12. *Juniperus macrocarpa* (Naxos).

Plate 5

1. Cupressaceae: *Juniperus oxycedrus* subsp. *deltoides* (Prespa area). - 2. *Juniperus phoenicea* subsp. *turbinata* (Imittos). - 3. Ephedraceae: *Ephedra distachya* (E of Asprovalta). - 4. *Ephedra foeminea* (NE Peloponnisos, insets from Thasos). - 5. Pinaceae: *Abies alba* (Varnous). - 6. *Abies borisii-regis* (Mt Athos). - 7. *Abies cephalonica* (Giona, inset of male flowers from Achaia). - 8. *Picea abies* (Rodopi). - 9. *Pinus brutia* (Thasos, inset from Karpathos). - 10. *Pinus halepensis* (Evvia, inset from NE Peloponnisos). - 11. *Pinus heldreichii* (Olimbos). - 12. *Pinus nigra* (Olimbos).

Plate 6

1. Pinaceae: *Pinus peuce* (North Macedonia, Mt Pelister). - 2. *Pinus pinea* (W Peloponnisos). - 3. *Pinus sylvestris* (Pangeon). - 4. Taxaceae: *Taxus baccata* (Kitheronas). - 5. Acanthaceae: *Acanthus greuterianus* (E of Limni Vegoritis). - 6. *Acanthus hungaricus* (Ioanninon, near village Kipi). - 7. *Acanthus spinosus* (Kerkira). - 8. Aceraceae: *Acer campestre* (Prespa area). - 9. *Acer heldreichii* (Varnous). - 10. *Acer hyrcanum* subsp. *reginae-amaliae* (Stix ravine). - 11. *Acer monspessulanum* (Evrou, SW Kotronia; inset from same area, near Lefkimi). - 12. *Acer negundo* (Paranestion).

Plate 7

1. Aceraceae: *Acer opalus* subsp. *obtusatum* (Mt Triklario). - 2. *Acer platanoides* (Killini). - 3. *Acer pseudoplatanus* (Rodopi). - 4. *Acer tataricum* (Florina-Pisoderi). - 5. Adoxaceae: *Adoxa moschatellina* (Above Metsovon). - 6. Agavaceae: *Agave americana* (Naxos). - 7. Aizoaceae: *Aizoanthemopsis hispanica* (Gavdos). - 8. *Aptenia cordifolia* (Lefkas). - 9. *Carpobrotus edulis* (Portugal, Algarve). - 10. *Mesembryanthemum nodiflorum* (Karpathos). - 11. Alismataceae: *Alisma lanceolatum* (W Peloponnisos). - 12. *Alisma plantago-aquatica* (Gramos).

Plate 8

1. Alismataceae: *Baldellia ranunculoides* (Kalogria). - 2. *Damasonium bourgaei* (Kalogria). - 3. Alliaceae: *Allium amethystinum* (Ioanninon, near Kipi). - 4. *Allium ampeloprasum* (Kriti, Anopolis). - 5. *Allium arampatzisii* (Thessaloniki). - 6. *Allium bornmuelleri* (Loutra Arideas). - 7. *Allium bourgeaui* subsp. *creticum* (E Kriti). - 8. *Allium callimischon* (W Peloponnisos). - 9. *Allium carinatum* (Serbia, Mt Tara). - 10. *Allium chalkii* (Rodos). - 11. *Allium chamaemoly* (Akrokorinthos). - 12. *Allium chamaespathum* (Imittos).

Plate 9

1. Alliaceae: *Allium circinatum* (Methoni castle). - 2. *Allium cithaeronis* (Kitheronas). - 3. *Allium cupanii* subsp. *cupanii* (Mt Boutsi). - 4. *Allium cyrilli* (Lake Stimfali, insets from Chios). - 5. *Allium flavum* subsp. *flavum* (Gramos). - 6. *Allium frigidum* (Killini). - 7. *Allium gomphrenoides* (Kithira). - 8. *Allium guttatum* subsp. *tenorei* (Thermopile). - 9. *Allium heldreichii* (Olimbos). - 10. *Allium junceum* (Kastellorizo). - 11. *Allium lagarophyllum* (Iti). - 12. *Allium maniaticum* (Mani peninsula).

Plate 10

1. Alliaceae: *Allium melanantherum* (Falakron). - 2. *Allium melanogyne* (Evrou, Dadia). - 3. *Allium meteoricum* (Meteora monasteries). - 4. *Allium moschatum* (Mt Boutsi). - 5. *Allium neapolitanum* (Rodos). - 6. *Allium nigrum* (Samos). - 7. *Allium occultum* (Skiros). - 8. *Allium orestis* (Evvia, Mt Ochi). - 9. *Allium pallens* (Chios). - 10. *Allium paniculatum* (Evrou, Dadia). - 11. *Allium parnassicum* (Parnassos). -12. *Allium phthioticum* (Killini, inset from Vasilitsa).

Plate 11

1. Alliaceae: *Allium ritsi* (Monemvasia). - 2. *Allium roseum* (NE Peloponnisos, inset from E Kriti). - 3. *Allium rubrovittatum* (Kriti, Katharo plateau). - 4. *Allium runemarkii* (Evvia, Karistos). - 5. *Allium samothracicum* (Samothraki). - 6. *Allium schoenoprasum* (Montenegro, Mt Durmitor). - 7. *Allium scorodoprasum* subsp. *rotundum* (Kriti, Lassithi plateau). - 8. *Allium siculum* subsp. *dioscoridis* (Evrou, Lefkimi). - 9. *Allium sphaerocephalon* subsp. *sphaerocephalon* (Prespa area). - 10. *Allium staticiforme* (Paros). - 11. *Allium subhirsutum* (N Peloponnisos). - 12. *Allium tardans* (Kriti, Kroustas).

Plate 12

1. Alliaceae: *Allium trifoliatum* (Chios). - 2. *Allium ursinum* (Vasilitsa, inset from Vermion). - 3. Amaranthaceae: *Amaranthus albus* (by Lake Kerkini). - 4. *Amaranthus blitioides* (Naxos). - 5. *Amaranthus deflexus* (SE Peloponnisos). - 6. *Amaranthus emarginatus* (by Lake Kerkini). - 7. *Amaranthus graecizans* (Kriti, Katharo plateau). - 8. *Amaranthus hybridus* (Githion-Areopolis). - 9. *Amaranthus hypochondriacus* (Loutra Arideas). - 10. *Amaranthus palmeri* (Sparti-Mistras). - 11. *Amaranthus powellii* (Sparti-Githion). - 12. *Amaranthus retroflexus* (Loutra Arideas).

Plate 13

1. Amaryllidaceae: *Acis orientalis* (Albania, Vlora). - 2. *Galanthus elwesii* (Chios). - 3. *Galanthus ikariae* (Naxos). - 4. *Galanthus peshmenii* (Turkey, Antalya, Gedelma Köyü). - 5. *Galanthus reginae-olgae* (E foothills of Taigetos). - 6. *Galanthus samothracicus* (Athos peninsula). - 7. *Leucojum aestivum* (Evros delta, inset from Litochoron area). - 8. *Narcissus obsoletus* (Navarino, inset from Areopolis). - 9. *Narcissus poeticus* (Prespa area, Mt Sfika; inset from N of Metsovon). - 10. *Narcissus tazetta* (Sicily, Madonie mts). - 11. *Pancratium maritimum* (Elafonisos, insets from Kalogria). - 12. *Sternbergia cluisana* (Turkey, prov. Tunceli; inset from Samos).

Plate 14

1. Amaryllidaceae: *Sternbergia colchiciflora* (Chelmos; inset of flower from Parnassos, of seeds from Panachaiko). - 2. *Sternbergia lutea* subsp. *greuteriana* (Kriti, SE Limnes). - 3. *Sternbergia lutea* subsp. *lutea* (SE Peloponnisos). - 4. Anacardiaceae: *Cotinus coggygria* (Ioanninon, Koukouli; inset from SE Peloponnisos). - 5. *Pistacia atlantica* (Rodos). - 6. *Pistacia lentiscus* (E Kriti). - 7. *Pistacia terebinthus* (Chios). - 8. *Rhus coriaria* (Samos, Karvouni; inset from Fthiotidos, Pavliani). - 9. Anthericaceae: *Anthericum liliago* (Grevenon). - 10. Apiaceae: *Aegopodium podagraria* (Rodopi). - 11. *Ammi majus* (Achaias, Ano Kastritsi). - 12. *Ammi visnaga* (Lesvos).

Plate 15

1. Apiaceae: *Angelica sylvestris* (Pisoderi). - 2. *Apium nodiflorum* (by Lake Megali Prespa). - 3. *Astrantia major* (Fourka pass). - 4. *Athamanta* densa (Olimbos). - 5. *Bifora testiculata* (N Peloponnisos). - 6. *Bubon albanicum* (Kastorias, E of Chrisi). - 7. *Bubon macedonicum* (Aoos river gorge). - 8. *Bunium ferulaceum* (near Lake Stimfalia). - 9. *Bunium microcarpum* (Chios). - 10. *Bupleurum apiculatum* (Konitsis, Agia Paraskevi). - 11. *Bupleurum falcatum* (Olimbos). - 12. *Bupleurum fruticosum* (SE Peloponnisos).

Plate 16

1. Apiaceae: *Bupleurum glumaceum* (Akrokorinthos). - 2. *Bupleurum greuteri* (Elafonisos). - 3. *Bupleurum odontites* (Chalkidiki). - 4. *Bupleurum subovatum* (Samos). - 5. *Bupleurum trichopodum* (Methana peninsula). - 6. *Cachrys cristata* (Evrou, Loutros-Dadia; inset from Samos). - 7. *Carum graecum* (Timfi). - 8. *Carum meoides* (Mt Boutsi). - 9. *Chaerophyllum aromaticum* (Evvia, Dirfi-Xirovouni). - 10. *Chaerophyllum creticum* (Lefka Ori). - 11. *Chaerophyllum heldreichii* (Iti). - 12. *Chaerophyllum nodosum* (Kerkira, inset from Loutra Arideas).

Plate 17

1. Apiaceae: *Conium divaricatum* (Prespa area). - 2. *Crithmum maritimum* (Naxos; inset from Kriti, Lasithiou, Pachia Ammos). - 3. *Daucus carota* (Konitsa). - 4. *Daucus glaber* (Rodos). - 5. *Daucus guttatus* (Litochoron railway station). - 6. *Daucus involucratus* (Naxos). - 7. *Dichoropetalum chryseum* (Rodos). - 8. *Dichoropetalum lavrentiadis* (Voras). - 9. *Dichoropetalum oligophyllum* (Varnous). - 10. *Echinophora spinosa* (Kalogria). - 11. *Echinophora tenuifolia* subsp. *sibthorpiana* (Serres-Drama). - 12. *Elaeoselinum asclepium* (Elafonisos).

Plate 18

1. Apiaceae: *Eryngium amethystinum* (Kristallopigi). - 2. *Eryngium amorginum* (Kriti, Pachia Ammos). - 3. *Eryngium campestre* (Karpathos, inset from Litochoron railway station). - 4. *Eryngium creticum* (Lesvos). - 5. *Eryngium glomeratum* (Kastellorizo). - 6. *Eryngium maritimum* (Litochoron railway station). - 7. *Eryngium palmatum* (North Macedonia, Pelister). - 8. *Eryngium ternatum* (Kriti, Imbros gorge; inset from Rouvas gorge). - 9. *Falcaria vulgaris* (Parnassos). - 10. *Ferula communis* (Kriti, Agia Galini; inset from Attiki, Cape Sounion). - 11. *Ferulago humilis* (Samos). - 12. *Ferulago nodosa* (Kerkira, inset of fruits from Killini).

Plate 19

1. Apiaceae: *Ferulago serpentinica* (N Evvia). - 2. *Ferulago sylvatica* (Evrou, Kotronia; inset of flowers from Athos peninsula, of fruits from Olimbos). - 3. *Ferulago thyrsiflora* (Kriti, N of Preveli monastery; inset from Kotsifos gorge). - 4. *Foeniculum vulgare* (Arachova). - 5. *Geocaryum creticum* (Kriti, Dikti). - 6. *Geocaryum peloponesiacum* (Killini). - 7. *Geocaryum pindicolum* (Giona). - 8. *Hellenocarum multiflorum* (Dodonis, near village Kipi). - 9. *Helosciadium nodiflorum* (Peloponnisos). - 10. *Heptaptera colladonioides* (Mistras). - 11. *Heracleum orphanidis* (Vrondous). - 12. *Heracleum sphondylium* subsp. *pyrenaicum* (Smolikas).

Plate 20

1. Apiaceae: *Heracleum sphondylium* subsp. *ternatum* (Kalavrita). - 2. *Horstrissea dolinicola* (Kriti, Psiloritis). - 3. *Johrenia distans* (Parnassos, insets from Akrokorinthos). - 4. *Kundmannia sicula* (Rodos). - 5. *Lagoecia cuminoides* (Naxos). - 6. *Laserocarpum pseudomeum* (Vardousia, inset from Parnassos). - 7. *Laser trilobum* (Evrou, Lefkimi). - 8. *Laserpitium siler* subsp. *garganicum* (Killini, inset from Giona). - 9. *Lecokia cretica* (Samos). - 10. *Ligusticum mutellina* (Voras). - 11. *Ligusticum olympicum* (Olimbos). - 12. *Malabaila aurea* (Chalkidiki, inset from Argos).

Plate 21

1. Apiaceae: *Malabaila involucrata* (Kerkira). - 2. *Microsciadium minutum* (Samos). - 3. *Oenanthe aquatica* (Lake Kerkini). - 4. *Oenanthe fistulosa* (Kerkira). - 5. *Oenanthe lachenalii* (above Metsovon, inset from Prespa area). - 6. *Oenanthe pimpinelloides* subsp. *incrassans* (Kalavrita). - 7. *Opopanax hispidus* (Kalavrita, inset of fruits from Killini). - 8. *Orlaya daucorlaya* (Prespa area; inset from Dodonis, Aristi-Mesovouni). - 9. *Ormoselenia alpina* (Lefka Ori). - 10. *Pastinaca sativa* subsp. *urens* (Prespa area). - 11. *Peucedanum aegopodioides* (Prespa area). - 12. *Peucedanum austriacum* (Gramos, inset from Olimbos).

Plate 22

1. Apiaceae: *Peucedanum officinale* subsp. *longifolium* (Montenegro, Mt Orjan). - 2. *Physospermum cornubiense* (Athos peninsula; inset from Serbia, Mt Maljen). - 3. *Pimpinella cretica* (Rodos). - 4. *Pimpinella peregrina* (Evvia, Steni Dirfos). - 5. *Pimpinella tragium* subsp. *depressa* (Lefka Ori). - 6. *Pimpinella tragium* subsp. *tragium* (Athos peninsula). - 7. *Prangos ferulacea* (Chelmos, inset from Parnassos). - 8. *Pseudorlaya pumila* (Chalkidiki). - 9. *Ridolfia segetum* (Evvia, Krinthos). - 10. *Sanicula europaea* (ESE of Karpenisi). - 11. *Scaligeria napiformis* (Naxos). - 12. *Scandix australis* subsp. *grandiflora* (Prespa area, insets from Akrokorinthos).

Plate 23

1. Apiaceae: *Scandix macrorhyncha* (E of Katara pass). - 2. *Scandix pecten-veneris* (Argolis, Didima; inset from Mani peninsula). - 3. *Selinum silaifolium* (Olimbos, inset from Athos peninsula). - 4. *Seseli arenarium* (Axios river delta, inset from Arachova). - 5. *Seseli crithmifolium* (Karpathos). - 6. *Seseli montanum* subsp. *tommasinii* (Prespa area). - 7. *Seseli parnassicum* (Parnassos). - 8. *Seseli peucedanoides* (Falakron, inset from Gramos). - 9. *Seseli rhodopeum* (Rodopi). - 10. *Seseli rigidum* (Konitsis, Pades-Distrato). - 11. *Smyrnium creticum* (Athos peninsula). - 12. *Smyrnium olusatrum* (Kithira).

Plate 24

1. Apiaceae: *Smyrnium perfoliatum* subsp. *rotundifolium* (Ano Trikala, inset from Kalavrita). - 2. *Tamamschjanella rhizomatica* (Gramos). - 3. *Thamnosciadium junceum* (Vardousia). - 4. *Thaspia garganica* (Ikaria, inset from Naxos). - 5. *Tordylium aegaeum* (Tilos). - 6. *Tordylium apulum* (Githion-Areopolis, inset of fruits from Kithira). - 7. *Tordylium officinale* (Limni Stimfalia; inset from Limni Doxa, Feneos). - 8. *Torilis leptophylla* (Mt Kitheronas). - 9. *Torilis modosa* (Naxos). - 10. *Trinia glauca* subsp. *glauca* (Falakron). - 11. *Turgenia latifolia* (Killini). - 12. Apocynaceae: *Amsonia orientalis* (Komotini).

Plate 25

1. Apocynaceae: *Nerium oleander* (Karpathos). - 2. *Vinca major* (Kriti, Agia Galini). - 3. Aquifoliaceae: *Ilex aquifolium* (Athos peninsula). - 4. Araceae: *Arisarum vulgare* (Peloponnisos, Vouraikos gorge). - 5. *Arum concinnatum* (Naxos). - 6. *Arum creticum* (Kriti, Kotsifou gorge). - 7. *Arum cylindraceum* (Killini). - 8. *Arum cyrenaicum* (Kriti, Chanion, Kandanos). - 9. *Arum dioscoridis* (Rodos). - 10. *Arum elongatum* (Athos peninsula). - 11. *Arum idaeum* (Lefka Ori). - 12. *Arum italicum* (Ano Kastritsi; inset of inflorescence from Serron, Siderokastro).

Plate 26

1. Araceae: *Arum nigrum* (Athos peninsula). - 2. *Arum purpureospathum* (Kriti, Agia Roumeli). - 3. *Biarum davisii* (Kriti). - 4. *Biarum marmarisense* (Turkey, Marmaris). - 5. *Biarum tenuifolium* (Kerkira, inset of fruits from Prespa area). - 6. *Dracunculus vulgaris* (Rodos, inset from Chios). - 7. Araliaceae: *Hedera helix* (Athos peninsula). - 8. Arecaceae: *Phoenix theophrasti* (E Kriti, Vai). - 9. Aristolochiaceae: *Aristolochia clematitis* (Edessis, Agras-Kerasia). - 10. *Aristolochia cretica* (E Kriti, Zakros). - 11. *Aristolochia elongata/nardiana* [cf. notes] (Pangeon). - 12. *Aristolochia guicciardii* (Rodos).

Plate 27

1. Aristolochiaceae: *Aristolochia hirta* (Chios). - 2. *Aristolochia incisa* (Samos). - 3. *Aristolochia lutea* (Athos peninsula). - 4. *Aristolochia microstoma* (Attiki, Keratea). - 5. *Aristolochia pallida* (Mt Triklario). - 6. *Aristolochia parviflora* (Rodos). - 7. *Aristolochia rotunda* (Pellis, Klissochori). - 8. *Aristolochia sempervirens* (Kithira). - 9. Asclepiadaceae: *Cionura erecta* (Pellis, Skidra; inset from Athos peninsula). - 10. *Cynanchum acutum* (Kavalas, Komotini). - 11. *Gomphocarpus fruticosus* (Naxos, inset of fruit from Kavala area). - 12. *Gomphocarpus physocarpus* (W Kriti, Nea Roumata).

Plate 28

1. Asclepiadaceae: *Periploca angustifolia* (Gavdos). - 2. *Periploca graeca* (Lake Kerkini, inset from Evros delta). - 3. *Vincetoxicum canescens* subsp. *pedunculatum* (Samos). - 4. *Vincetoxicum creticum* (Rethimnis, Mt Kryoneritis). - 5. *Vincetoxicum fuscatum* (Smolikas, Samarina). - 6. *Vincetoxicum hirundinaria* subsp. *nivale* (Pangeon, inset from Falakron). - 7. *Vincetoxicum speciosum* (Pangeon). - 8. Asparagaceae: *Asparagus aphyllus* subsp. *orientalis* (Karpathos, inset from SE Kriti). - 9. *Asparagus officinalis* (Axios river delta). - 10. Asphodelaceae: *Asphodeline liburnica* (Olimbos). - 11. *Asphodeline lutea* (Pangeon, inset from Naxos). - 12. *Asphodeline taurica* (Tzena).

Plate 29

1. Asphodelaceae: *Asphodelus albus* (Varnous, inset of fruits from Gramos). - 2. *Asphodelus fistulosus* (Kithira). - 3. *Asphodelus ramosus* (Athos peninsula). - 4. *Asphodelus tenuifolius* (by Vouliagmeni lagoon). - 5. Asteraceae: *Achillea abrotanoides* (Gramos). - 6. *Achillea ageratifolia* (Olimbos). - 7. *Achillea ambrosiaca* (Olimbos). - 8. *Achillea barbeyana* (Vardousia). - 9. *Achillea chrysocoma* (Voras, inset from Triklario). - 10. *Achillea clusiana* (Voras). - 11. *Achillea clypeolata* (Timfi, inset from Bulgarian Rodopi). - 12. *Achillea coarctata* (Prespa area).

Plate 30

1. Asteraceae: *Achillea cretica* (Rodos). - 2. *Achillea crithmifolia* (Varnous). - 3. *Achillea fraasii* (Timfristos). - 4. *Achillea grandifolia* (Ioanninon, Konitsa-Pades). - 5. *Achillea holosericea* (Parnassos). - 6. *Achillea ligustica* (Evvia, SW of Metochi). - 7. *Achillea millefolium* (above Metsovon). - 8. *Achillea setacea* (Prespa area). - 9. *Achillea taygetea* (Taygetos). - 10. *Achillea umbellata* (Killini). - 11. *Adenostyles alliariae* (Varnous). - 12. *Aetheorhiza bulbosa* (Elafonisos).

Plate 31

1. Asteraceae: *Anacyclus clavatus* (Rodos). - 2. *Andryala integrifolia* (Tilos). - 3. *Antennaria dioica* (Voras). - 4. *Anthemis abrotanifolia* (Psiloritis). - 5. *Anthemis ammanthus* (E Kriti, Cape Sidero). - 6. *Anthemis auriculata* (Athos peninsula). - 7. *Anthemis austriaca* (Prespa area). - 8. *Anthemis brachmannii* (Taigetos, Langada gorge). - 9. *Anthemis chia* (Chios, inset from S Peloponnisos). - 10. *Anthemis cretica* subsp. *columnae* (Timfristos). - 11. *Anthemis glaberrima* (W Kriti, islet of Imeri Gramvousa. - 12. *Anthemis laconica* (Taigetos).

Plate 32

1. Asteraceae: *Anthemis macedonica* (Evrou, Dadia-Pessani). - 2. *Anthemis palaestina* (Samos). - 3. *Anthemis rhodensis* (Rodos). - 4. *Anthemis rigida* subsp. *liguliflora* (Kithira). - 5. *Anthemis rigida* subsp. *rigida* (Rodos). - 6. *Anthemis ruthenica* (Florinis, Mt Devas). - 7. *Anthemis samariensis* (Lefka Ori). - 8. *Anthemis sibthorpii* (Mt Athos). - 9. *Anthemis spruneri* (Parnassos). - 10. *Anthemis tinctoria* subsp. *tinctoria* (Kilkis, Kroussia). - 11. *Anthemis tomentella* (Kriti, Lasithiou, Istro). - 12. *Anthemis tomentosa* (Attiki, Lavrion).

Plate 33

1. Asteraceae: *Anthemis wiedemanniana* (Vourinos). - 2. *Arctium lappa* (Konitsis, NNW of Loutra Amarantou). - 3. *Arctium minus* (Olimbos). - 4. *Artemisia absinthium* (Metsovon). - 5. *Artemisia alba* (Serbia, Mt Maljen; inset from Kastorias, Nestorio). - 6. *Artemisia arborescens* (Naxos). - 7. *Artemisia campestris* (Paranestion). - 8. *Artemisia santonicum* (Axios river delta). - 9. *Artemisia umbelliformis* subsp. *eriantha* (Gramos). - 10. *Aster alpinus* subsp. *cylleneus* (Giona). - 11. *Asteriscus aquaticus* (Rodos). - 12. *Atractylis cancellata* (Chios).

Plate 34

1. Asteraceae: *Bellis annua*. - 2. *Bellis perennis* (SE of Kristallopigi). - 3. *Bellis sylvestris* (E Kriti, 2 km SW of Kellaria). - 4. *Bidens cernuus* (Mikri Prespa, Mikrolimni). - 5. *Bidens frondosus* (NE of Githion). - 6. *Bidens pilosus* (Korinthias, Kiato; inset from Argolidos, Mili). - 7. *Bidens tripartitus* (NNE of Thessaloniki). - 8. *Calendula arvensis* (Athos peninsula). - 9. *Cardopatium corymbosum* (Lesvos). - 10. *Carduus euboicus* (Evvia, Mt Dirfis). - 11. *Carduus kerneri* (Prespa area). - 12. *Carduus nutans* subsp. *taygeteus* (Samos).

Plate 35

1. *Carduus pycnocephalus* (Kithira, inset from Ano Trikala). - 2. *Carduus tmoleus* (Voras). - 3. *Carlina acanthifolia* subsp. *utzka* (Kristallopigi). - 4. *Carlina acaulis* subsp. *caulescens* (Croatia, Velebit mts). - 5. *Carlina barnebyana* (Karpathos). - 6. *Carlina corymbosa* subsp. *graeca* (Killini). - 7. *Carlina gummifera* (Karpathos). - 8. *Carlina lanata* (Lesvos). - 9. *Carlina tragacanthifolia* (Karpathos). - 10. *Carlina vulgaris* (Voras). - 11. *Carthamus boissieri* (Kriti, Rethimnis, Anogia-Axos). - 12. *Carthamus careuleus* (Kriti, Mesa Lakonia).

Plate 36

1. Asteraceae: 1. *Carthamus lanatus* (Killini). - 2. *Carthamus leucocaulos* (Naxos). - 3. *Catananche lutea* (Samos). - 4. *Centaurea achaia* subsp. *achaia* (Attikis, NE Markopoulo). - 5. *Centaurea acicularis* (Ikaria). - 6. *Centaurea adami* (Prespa area). - 7. *Centaurea aegialophila* (E Kriti). - 8. *Centaurea affinis* (Ioanninon, S of Milea; inset from Voras). - 9. *Centaurea alba* subsp. *albanica* (Ioanninon, Vikos village). - 10. *Centaurea amplifolia* (Killini). - 11. *Centaurea argentea* subsp. *chionantha* (Kriti, Monastiraki). - 12. *Centaurea athoa* subsp. *athoa* (Mt Athos).

Plate 37

1. Asteraceae: *Centaurea atropurpurea* (Naxos). - 2. *Centaurea austrobalcanica* (Falakron). - 3. *Centaurea axillaris* (Florinis, Triklario). - 4. *Centaurea baldaccii* (Lefka Ori). - 5. *Centaurea benedicta* (Serron, S of village Kerkini; inset from Chalkidiki). - 6. *Centaurea calcitrapa* (Rodos). - 7. *Centaurea chalcidicaea* (Mt Athos). - 8. *Centaurea chrysocephala* (by Meteora monasteries). - 9. *Centaurea cuneifolia* subsp. *pallida* (Litochoron railway station). - 10. *Centaurea cyanus* (Prespa area). - 11. *Centaurea deustiformis* (Mt Boutsi). - 12. *Centaurea devasiana* (Prespa area, Mt Devas).

Plate 38

1. Asteraceae: *Centaurea diffusa* (Chalkidiki, Nea Triglia). - 2. *Centaurea ebenoides* (Evvia, Neos Pagontas). - 3. *Centaurea epirota* (Timfi). - 4. *Centaurea finazzeri* (Prespa area). - 5. *Centaurea graeca* (Prespa area; inset from Albania, by Lake Ohrid). - 6. *Centaurea grbavacensis* (Tzena). - 7. *Centaurea grisebachii* subsp. *confusa* (Meteora monasteries). - 8. *Centaurea idaea* (Lefka Ori). - 9. *Centaurea immanuelis-loewii* (Thessalonikis, Liti). - 10. *Centaurea incompleta* (foothills of Olimbos). - 11. *Centaurea ipsaria* (Thasos, Ipsario). - 12. *Centaurea jacea* subsp. *angustifolia* (Prespa area).

Plate 39

1. Asteraceae: *Centaurea johnseniana* (Serron, NNW Sidirokastro). - 2. *Centaurea kalambakensis* (Theopetra SE of Kalambaka). - 3. *Centaurea lactucifolia* (Rodos). - 4. *Centaurea lancifolia* (Lefka Ori). - 5. *Centaurea messenicolasiana* (Karditsis, Messinikola village). - 6. *Centaurea musarum* (Parnassos). - 7. *Centaurea nervosa* (Voras). - 8. *Centaurea niederi* (Kalogria). - 9. *Centaurea parilica* (Falakron). - 10. *Centaurea paxorum* (Paxi). - 11. *Centaurea pelia* (Ossa, village Spileo). - 12. *Centaurea pindicola* (Voras).

Plate 40

1. Asteraceae: *Centaurea poculatoris* (S Kriti). - 2. *Centaurea promota* (above Metsovon). - 3. *Centaurea pseudocadmea* (Peloponnisos, Gaidarovouni). - 4. *Centaurea pseudophrygia* (Loutra Arideas). - 5. *Centaurea pumilio* (SW Kriti, Elafonisos). - 6. *Centaurea raphanina* subsp. *mixta* (Attiki, Lavrion). - 7. *Centaurea rechingeri* (Samos). - 8. *Centaurea salonitana* (Vardousia). - 9. *Centaurea samothracica* (Samothraki). - 10. *Centaurea scabiosa* (Montenegro, Scepan polje). - 11. *Centaurea solstitialis* (Naxos, inset from Lesvos). - 12. *Centaurea sonchifolia* (Zakinthos).

Plate 41

1. Asteraceae: *Centaurea soskae* (Prespa area). - 2. *Centaurea spinosa* (Attiki, Lavrion; inset from Samothraki). - 3. *Centaurea spruneri* (Arkadias, Tripoli-Sparti). - 4. *Centaurea stenolepis* (Dramas, Skaloti-Sideronero). - 5. *Centaurea thracica* (Trikalon, Trigono-Orthovounion). - 6. *Centaurea urvillei* (Chios). - 7. *Centaurea vlachorum* (Mt Milea). - 8. *Centaurea vorasiana* (Voras). - 9. *Centaurea xylobasis* (Samos, Kerki). - 10. *Centaurea zuccariniana* (Ioanninon, Aristi). - 11. *Chondrilla juncea* (Prespa area). - 12. *Cichorium intybus* (Prespa area).

Plate 42

1. Asteraceae: *Cichorium pumilum* (Kastellorizo). - 2. *Cichorium spinosum* (Evvia, Karistos). - 3. *Cirsium appendiculatum* (Voras). - 4. *Cirsium arvense* (Bulgaria, Sofia; inset from Prespa area). - 5. *Cirsium candelabrum* (Prespa area, inset from Parnassos). - 6. *Cirsium eriophorum* (Parnassos, inset from Prespa area). - 7. *Cirsium* cf. *heldreichii* (Korinthias, Ano Trikala). - 8. *Cirsium hypopsilum* (Killini). - 9. *Cirsium italicum* (Lesvos). - 10. *Cirsium ligulare* (Pangeon; inset from Bulgaria, Sofia). - 11. *Cirsium mairei* (Giona). - 12. *Cirsium morinifolium* (Lefka Ori).

Plate 43

1. Asteraceae: *Cirsium tymphaeum* (Florinis, Vigla pass). - 2. *Cirsium vulgare* (Prespa area). - 3. *Cotula coronopifolia* (Kalogria). - 4. *Crepis arcuata* (Voras). - 5. *Crepis athoa* (Mt Athos). - 6. *Crepis auriculifolia* (Mt Kedros). - 7. *Crepis baldaccii* (Timfi). - 8. *Crepis foetida* subsp. *commutata* (Naxos). - 9. *Crepis heldreichiana* (Taigetos). - 10. *Crepis incana* (Evvia, Dirfis). - 11. *Crepis multiflora* (Naxos). - 12. *Crepis neglecta* (Athens).

Plate 44

1. Asteraceae: *Crepis rubra* (Korinthias, Lake Stimfalia). - 2. *Crepis sibthorpiana* (Lefka Ori). - 3. *Crepis turcica* (Konitsis, Pades). - 4. *Crepis vesicaria* (Korinthias, Ano Trikala). - 5. *Crepis viscidula* (Varnous). - 6. *Crepis zacintha* (Chalkidiki). - 7. *Crupina crupinastrum* (Athos peninsula). - 8. *Cynara cardunculus* (Lakonias, Githion-Skala). - 9. *Cynara cornigera* (Tilos). - 10. *Cynara cyrenaica* (Kriti, Lasithou, Meseleri). - 11. *Doronicum austriacum* (Florinis, Vigla pass). - 12. *Doronicum caucasicum* (Varnous).

Plate 45

1. Asteraceae: *Doronicum columnae* (Olimbos). - 2. *Echinops graecus* (Naxos). - 3. *Echinops microcephalus* (Thessalonikis, Vertiskos). - 4. *Echinops ritro* (Gramos). - 5. *Echinops sphaerocephalus* (Prespa area). - 6. *Echinops spinosissimus* (Karpathos). - 7. *Erigeron acris* (Olimbos). - 8. *Erigeron alpinus* (Olimbos). - 9. *Erigeron annuus* (Pellis, Megaplatanos). - 10. *Erigeron canadensis* (Serron, Lake Kerkini). - 11. *Erigeron epiroticus* (Olimbos). - 12. *Erigeron glabratus* (Vardousia).

Plate 46

1. Asteraceae: *Erigeron sumatrensis* (Lakonias, NE Githion). - 2. *Eupatorium cannabinum* (Olimbos, Prionia). - 3. *Filago aegaea* subsp. *aegaea* (E Kriti, Sitia). - 4. *Filago arvensis* (Prespa area). - 5. *Filago contracta* (Naxos). - 6. *Filago eriocephala* (Naxos). - 7. *Filago germanica* (Serron, S of village Kerkini). - 8. *Filago pygmaea* (Amorgos). - 9. *Filago pyramidata* (Chios). - 10. *Flaveria bidentis* (W Peloponnisos). - 11. *Galactites tomentosus* (Kerkira). - 12. *Galatella cretica* (Kriti, Sitias, Lastros).

Plate 47

1. Asteraceae: *Galatella linosyris* (Killini). - 2. *Galinsoga parviflora* (W of Florina). - 3. *Geropogon hybridus* (Samos). - 4. *Glebionis coronaria* (Attikis, Cape Sounion). - 5. *Glebionis segetum* (Amorgos). - 6. *Gnaphalium hoppeanum* subsp. *magellense* (Vardousia). - 7. *Gnaphalium roeseri* subsp. *roeseri* (Chelmos, Stix ravine). - 8. *Gnaphalium supinum* (Voras). - 9. *Gnaphalium sylvaticum* (Varnous). - 10. *Gnaphalium uliginosum* (Lake Kerkini). - 11. *Hedypnois rhagadioloides* (Kithira). - 12. *Helianthus laetiflorus* (NNE Thessaloniki).

Plate 48

1. Asteraceae: *Helichrysum amorginum* (Amorgos). - 2. *Helichrysum doerfleri* (E Kriti, Thriptis Ori). - 3. *Helichrysum heldreichii* (W Kriti, Samaria gorge). - 4. *Helichrysum italicum* (Naxos). - 5. *Helichrysum orientale* (Samos). - 6. *Helichrysum sibthorpii* (Mt Athos). - 7. *Helichrysum stoechas* subsp. *barrelieri* (Kriti, Rethimnis, Agia Galini). - 8. *Helichrysum taenari* (Mani peninsula). - 9. *Helminthotheca echioides* (Evvia, Limni). - 10. *Hieracium bosniacum* (Falakron). - 11. *Hieracium chalcidicum* (Parnassos). - 12. *Hieracium pannosum* (Timfi).

Plate 49

1. Asteraceae: *Hieracium sericophyllum* (Falakron). - 2. *Hieracium sparsum* (Voras). - 3. *Hieracium villosum* (Gramos). - 4. *Hymenonema graecum* (Naxos). - 5. *Hymenonema laconicum* (Lakonias, Kallithea). - 6. *Hyoseris lucida* (Kriti, Malia). - 7. *Hyoseris scabra* (Kriti, Rethimnis, Preveli). - 8. *Hypochaeris achyrophorus* (Arkadias, Astros). - 9. *Hypochaeris cretensis* (Prespa area, inset from Kerkira). - 10. *Hypochaeris tenuiflora* (Lefka Ori). - 11. *Inula britannica* (Pellis, Loutra Aridea to Dobro Polje). - 12. *Inula conyzae* (Voras).

Plate 50

1. Asteraceae: *Inula crithmoides* (SW Peloponnisos, Navarino). - 2. *Inula ensifolia* (Loutra Arideas). - 3. *Inula germanica* (Evrou, Valtos-Chandras). - 4. *Inula graveolens* (Evvia, Stropones). - 5. *Inula hirta* (Evrou, Metaxades). - 6. *Inula oculus-christi* (Prespa area, Mt Devas). - 7. *Inula salicina* (Prespa area, Mt Devas). - 8. *Inula verbascifolia* subsp. *aschersoniana* (Thasos). - 9. *Inula viscosa* (WNW Igoumenitsa). - 10. *Jacobaea abrotanifolia* subsp. *carpathica* (Voras). - 11. *Jacobaea gnaphalioides* (Karpathos). - 12. *Jacobaea maritima* subsp. *bicolor* (Korinthias, Egirouses).

Plate 51

1. Asteraceae: *Jacobaea othonnae* (Athos peninsula). - 2. *Jacobaea subalpina* (Voras). - 3. *Jurinea cadmea* (Rodos). - 4. *Jurinea consanguinea* (Samos, inset from Ikaria). - 5. *Jurinea mollis* (Litochoron railway station). - 6. *Jurinea taygetea* (Taigetos). - 7. *Klasea radiata* subsp. *cetinjensis* (Prespa area, Mt Devas). - 8. *Lactuca alpestris* (Lefka Ori). - 9. *Lactuca alpina* (Rodopi). - 10. *Lactuca aurea* (Varnous). - 11. *Lactuca intricata* (Timfi, inset from Olimbos). - 12. *Lactuca perennis* (Prespa area).

Plate 52

1. Asteraceae: *Lactuca serriola* (Bulgaria, Sofia; inset of flowers from Prespa area). - 2. *Lactuca tuberosa* (Rodos). - 3. *Lamyropsis carpini* (Kastorias, Eptachori). - 4. *Lamyropsis cynaroides* (W Kriti, Anopolis). - 5. *Lapsana communis* subsp. *adenophora* (Grevenon, Samaria). - 6. *Leontodon autumnalis* (Smolikas area, Samarina-Fourka). - 7. *Leontodon cichoriaceus* (Timfi). - 8. *Leontodon crispus* subsp. *asper* (Olimbos). - 9. *Leontodon graecus* (Taigetos). - 10. *Leontodon tuberosus* (Naxos). - 11. *Leucanthemum vulgare* subsp. *praecox* (Kastorias, Eptachori). - 12. *Matricaria chamomilla* (Lakonias, Monemvasia).

Plate 53

1. Asteraceae: *Matricaria discoidea* (Voras). - 2. *Notobasis syriaca* (Amorgos). - 3. *Onopordum acanthium* (Prespa area). - 4. *Onopordum argolicum* (Kalavrita, inset from Parnassos). - 5. *Onopordum boissierianum* (Kastellorizo). - 6. *Onopordum bracteatum* subsp. *bracteatum* (Karpathos). - 7. *Onopordum illyricum* (Chios). - 8. *Onopordum laconicum* (S Peloponnisos). - 9. *Onopordum majorii* (Ikaria). - 10. *Onopordum tauricum* (Naxos). - 11. *Otanthus maritimus* (Rodos). - 12. *Pallenis spinosa* (Chios; inset from Kriti, Agia Fotia).

Plate 54

1. Asteraceae: *Petasites albus* (above Metsovon). - 2. *Petasites hybridus* (Rodopi). - 3. *Phagnalon pygmaeum* (Lefka Ori). - 4. *Phagnalon rupestre* (Argos). - 5. *Picnomon acarna* (Kastellorizo). - 6. *Picris hieracioides* (Dramas, N of Prosotsani). - 7. *Picris pauciflora* (Ikaria). - 8. *Picris rhagadioloides* (Thasos). - 9. *Pilosella cymosa* (Prespa area, Mt Devas). - 10. *Pilosella rhodopea* (Voras). - 11. *Pilosella testimonialis* subsp. *pilisquamum* (Taigetos). - 12. *Prenanthes purpurea* (Olimbos).

Plate 55

1. Asteraceae: *Ptilostemon afer* (Falakron). - 2. *Ptilostemon chamaepeuce* (Samos). - 3. *Ptilostemon strictus* (Florinis, Oxia). - 4. *Pulicaria dysenterica* (Korinthias, Ano Trikala). - 5. *Reichardia picroides* (Kriti, Limni Chersonisou). - 6. *Rhagadiolus stellatus* (Argolidos, Didima; inset from Korinthias, Ano Trikala). - 7. *Scolymus grandiflorus* (Rodos). - 8. *Scolymus hispanicus* (Mani peninsula). - 9. *Scolymus maculatus* (Lesvos). - 10. *Scorzonera araneosa* (Amorgos). - 11. *Scorzonera austriaca* (Olimbos). - 12. *Scorzonera cana* (Chelmos).

Plate 56

1. Asteraceae: *Scorzonera cretica* (Kriti, Plakias). - 2. *Scorzonera crocifolia* (Lakonias, S of Monemvasia). - 3. *Scorzonera elata* (Kastellorizo). - 4. *Scorzonera mollis* (Prespa area). - 5. *Scorzonera purpurea* (Voras, inset from Iti). - 6. *Scorzonera sublanata* (Samos). - 7. *Senecio fruticulosus* (Lefka Ori). - 8. *Senecio leucanthemifolius* (Rodos). - 9. *Senecio macedonicus* subsp. *macedonicus* (Falakron). - 10. *Senecio ovatus* (Gramos). - 11. *Senecio rupestris* (Varnous). - 12. *Senecio scopolii* (Grevenon, WNW Kranea).

Plate 57

1. Asteraceae: *Senecio thapsoides* (Chelmos). - 2. *Senecio vernalis* (N Greece, no precise locality available). - 3. *Senecio viscosus* (Katara pass). - 4. *Silybum marianum* (Portugal, Algarve). - 5. *Solidago virgaurea* (Voras, inset from Mt Milea). - 6. *Sonchus asper* (Naxos). - 7. *Sonchus oleraceus* (Argos). - 8. *Staehelina fruticosa* (Karpathos). - 9. *Staehelina petiolata* (Kriti, Kandanos gorge). - 10. *Staehelina uniflosculosa* (Olimbos, inset from Eptachorion). - 11. *Symphyotrichum squamatum* (Lakonias, Paralia Kiparissias). - 12. *Tagetes minuta* (Messinias, Neromilos).

Plate 58

1. Asteraceae: *Tanacetum corymbosum* (Bulgaria, Monastir). - 2. *Tanacetum parthenium* (Voras). - 3. *Tanacetum vulgare* (W of Florina). - 4. *Taraxacum aleppicum* (Samos, Karvouni). - 5. *Taraxacum cylleneum* (Killini). - 6. *Taraxacum epirense* (Parnassos). - 7. *Taraxacum graecofontanum* (Voras). - 8. *Taraxacum megalorhizon* (Karpathos). - 9. *Taraxacum pindicola* (Mt Milea). - 10. *Taraxacum scaturiginosum* (Parnassos). - 11. *Taraxacum serotinum* (Prespa area). - 12. *Telekia speciosa* (North Macedonia, Mt Pelister).

Plate 59

1. Asteraceae: *Tephroseris papposa* (Varnous). - 2. *Tolpis umbellata* (Naxos). - 3. *Tragopogon crocifolius* subsp. *samaritanii* (Timfristos). - 4. *Tragopogon dubius* (Pellis, SE Arnissa). - 5. *Tragopogon longifolius* (Iti). - 6. *Tragopogon porrifolius* (Achaias, Solo; inset of flowers from Lakonias, Areopolis; of fruits from Kriti, Limni Chersonisou). - 7. *Tripleurosperum tempskyanum* (above Metsovon). - 8. *Tripleurospermum tenuifolium* (Prespa area). - 9. *Tripolium pannonicum* (Axios river delta). - 10. *Tussilago farfara* (Taigetos, Langada gorge). - 11. *Urospermum picroides* (Amorgos, inset from Kerkira). - 12. *Xanthium spinosum* (Thessalonikis, Vertiskos).

Plate 60

1. Asteraceae: *Xanthium strumarium* s. lat. (Serron: Serres-Drama). - 2. *Xeranthemum annuum* (Prespa area, inset from Chalkidiki). - 3. *Xeranthemum cylindraceum* (Prespa area). - 4. *Xeranthemum inapertum* (Achaias, Mt Klokos). - 5. Balsaminaceae: *Impatiens balfourii* (Metsovon, inset from Prespa area). - 6. *Impatiens noli-tangere* (Rodopi). - 7. Berberidaceae: *Berberis cretica* (Pilion). - 8. *Berberis vulgaris* (Serbia, Daliblatska; inset from Kastorias, Ambelokipi). - 9. *Bongardia chrysogonum* (Achaias, Selinous-Pteri). - 10. *Gymnospermium peloponnesiacum* (Mt Klokos). - 11. *Leontice leontopetalum* (NE Peloponnisos). - 12. Betulaceae: *Alnus glutinosa* (Prespa area).

Plate 61

1. Betulaceae: *Alnus incana* (Rodopi). - 2. *Betula pendula* (Voras). - 3. *Carpinus betulus* (Ioanninon, Monodendri). - 4. *Carpinus orientalis* (Prespa area). - 5. *Corylus avellana* (Dramas, Sideronero). - 6. *Corylus colurna* (Florinis, Kristallopigi). - 7. *Ostrya carpinifolia* (Athos peninsula, inset from Timfi). - 8. Biebersteiniaceae: *Biebersteinia orphanidis* (Killini, insets from Saitas). - 9. Boraginaceae: *Aegonychon purpurocaeruleum* (Prespa area). - 10. *Alkanna graeca* subsp. *baeotica* (Taigetos). - 11. *Alkanna graeca* subsp. *graeca* (E of Karpenision). - 12. *Alkanna hellenica* (Akrokorinthos).

Plate 62

1. Boraginaceae: *Alkanna methanaea* (Lakonias, Mt Koulouchera). - 2. *Alkanna pindicola* (Prespa area, Mt Devas). - 3. *Alkanna sartoriana* (Argolidos, Viviani). - 4. *Alkanna sieberi* (Kriti, Agia Fotia). - 5. *Alkanna tinctoria* (Rodos). - 6. *Alkanna tubulosa* (Samos). - 7. *Alkanna versicolor* (Parnonas). - 8. *Amsinckia micrantha* (by Lake Kerkini). - 9. *Anchusa aegyptiaca* (Naxos). - 10. *Anchusa azurea* (Samos). - 11. *Anchusa cespitosa* (Lefka Ori). - 12. *Anchusa officinalis* subsp. *leucantha* (Pellis, SE Arnissa).

Plate 63

1. Boraginaceae: *Anchusa officinalis* subsp. *officinalis* (Prespa area). - 2. *Anchusa procera* (Rodopis, Xirolimni lagoon). - 3. *Anchusa thessala* (Chalkidiki). - 4. *Anchusa undulata* subsp. *hybrida* (Lakonias, Mistras). - 5. *Anchusella cretica* (Kalavrita). - 6. *Anchusella variegata* (Kithira). - 7. *Asperugo procumbens* (Prespa area). - 8. *Borago officinalis* (Lakonias, Areopolis). - 9. *Buglossides arvensis* subsp. *arvensis* (Timfristos). - 10. *Buglossoides incrassata* subsp. *incrassata* (NNW Metsovon). - 11. *Cerinthe major* (Chios, inset from Sicily). - 12. *Cerinthe minor* (Gramos, inset from Athos peninsula).

Plate 64

1. Boraginaceae: *Cerinthe retorta* (Lakonias, Kelefa castle). - 2. *Cynoglossum columnae* (Naxos). - 3. *Cynoglossum creticum* (Chios). - 4. *Cynoglossum montanum* (Dramas, Dipotama to Megalo Livadi). - 5. *Cynoglossum sphacioticum* (Lefka Ori). - 6. *Cynoglottis barrelieri* (Florinis, Mt Triklario; inset from Mt Devas). - 7. *Echium angustifolium* (Kriti, Phaistos). - 8. *Echium arenarium* (Kithira). - 9. *Echium italicum* (Messinias, Methoni; inset from Prespa area). - 10. *Echium parviflorum* (SW Kriti, Imbros gorge). - 11. *Echium plantagineum* (Portugal, Algarve; inset from Kithira). - 12. *Echium vulgare* (Kastorias, Nestorio to Nea Kotili).

Plate 65

1. Boraginaceae: *Gastrocotyle macedonica* (Thessalonikis, Platanakia-Thermi; inset from Chalkidiki). - 2. *Halacsya sendtneri* (Smolikas). - 3. *Heliotropium curassavicum* (Methana peninsula). - 4. *Heliotropium dolosum* (Karpathos, inset from Rodos). - 5. *Heliotropium europaeum* (Lakonias, Areopolis). - 6. *Heliotropium halacsyi* (Ioanninon, Metamorfosi). - 7. *Heliotropium hirsutissimum* (Rodos). - 8. *Heliotropium suaveolens* (by Lake Kerkini). - 9. *Hormuzakia aggregata* (Kos, inset from Rodos). - 10. *Lappula squarrosa* (Mt Siniatsikon; inset from Kozanis, Neraida). - 11. *Lithodora hispidula* (Samos). - 12. *Lithodora zahnii* (Mani peninsula).

Plate 66

1. Boraginaceae: *Macrotomia densiflora* (Chelmos, Stix ravine). - 2. *Myosotis alpestris* subsp. *suaveolens* (Falakron). - 3. *Myosotis incrassata* (Karpathos). - 4. *Myosotis nemorosa* (Rodopi, inset from Pisoderi pass). - 5. *Myosotis sicula* (Naxos). - 6. *Myosotis solange* (Lefka Ori). - 7. *Myosotis sylvatica* subsp. *cyanea* (Timfi). - 8. *Neatostema apulum* (Rodos). - 9. *Omphalodes luciliae* subsp. *scopulorum* (Pangeon). - 10. *Omphalodes runemarkii* (SE Peloponnisos, Mt Koulochera). - 11. *Onosma aucheriana* (Chios). - 12. *Onosma elegantissima* (Vourinos).

Plate 67

1. Boraginaceae: *Onosma erecta* (Killini). - 2. *Onosma euboica* (N Evvia). - 3. *Onosma frutescens* (Lakonias, Githion-Areopolis). - 4. *Onosma graeca* (Kriti, Irakliou, Gonie-Kera; inset from Samos). - 5. *Onosma heterophylla* (Prespa area). - 6. *Onosma kaheirei/aucheriana* (Ikaria). - 7. *Onosma kittanae* (Evrou, Dadia-Pessani). - 8. *Onosma leptantha* (Taigetos). - 9. *Onosma paradoxa* (Mt Athos). - 10. *Onosma pygmaea* (Valia Kalda National Park). - 11. *Onosma stridii* (Mt Kallidromon). - 12. *Onosma visianii* (Florinis, Mt Triklarion).

Plate 68

1. Boraginaceae: *Paracaryum aucheri* (Samos). - 2. *Paracaryum lithospermifolium* (Lefka Ori). - 3. *Pulmonaria cesatiana* (Killini; insets from Achaias, above Kamarovrisi). - 4. *Pulmonaria rubra* (Rodopi, inset from Varnous). - 5. *Rindera graeca* (Evvia, Dirfis; inset from Giona). - 6. *Solananthus scardicus* (Gramos). - 7. *Solenanthus stamineus* (Chelmos). - 8. *Symphytum anatolicum* (Samos). - 9. *Symphytum bulbosum* (Killini, inset from Chalkidiki). - 10. *Symphytum circinale* (Rodos). - 11. *Symphytum creticum* (Lakonias, Monemvasia; inset from W Kriti, Imbros gorge). - 12. *Symphytum davisii* subsp. *naxicola* (Naxos; inset subsp. *icaricum* from Ikaria).

Plate 69

1. Boraginaceae: *Symphytum ottomanum* (Evrou, Dadia; inset from Athos peninsula). - 2. *Symphytum tuberosum* subsp. *angustifolium* (Florinis, Pisoderi pass). - 3. Brassicaceae: *Aethionema arabicum* (Rodos). - 4. *Aethionema carlsbergii* (Taigetos). - 5. *Aethionema orbiculatum* (Mt Athos). - 6. *Aethionema saxatile* subsp. *graecum/oreophilum* (Gramos). - 7. *Alliaria petiolata* (Florinis, Mt Triklarion). - 8. *Alyssoides cretica* (Kriti, Mt Giouchtas). - 9. *Alyssoides utriculata* (foothills of Voras, Loutra Arideas). - 10. *Alyssum doerfleri* (Prespa area). - 11. *Alyssum fulvescens* (Chios). - 12. *Alyssum handelii* (Olimbos).

Plate 70

1. Brassicaceae: *Alyssum lassiticum* (Lefka Ori). - 2. *Alyssum minutum* (Timfristos). - 3. *Alyssum pogonocarpum* (Rodos). - 4. *Alyssum simplex* (Parnassos; inset from Kerkira, Pandokrator). - 5. *Alyssum simulans* (Evvia, Xirovouni). - 6. *Alyssum sphacioticum* (Lefka Ori). - 7. *Alyssum spruneri* (Parnassos, inset '*A. densistellatum*' from N Evvia). - 8. *Alyssum strigosum* (Korinthias, Loutraki). - 9. *Alyssum umbellatum* (Evrou, Dadia-Pessani). - 10. *Alyssum xiphocarpum* (Lesvos). - 11. *Arabidopsis thaliana* (Athos peninsula). - 12. *Arabis alpina* subsp. *caucasica* (Vasilitsa, inset from Mt Athos).

Plate 71

1. Brassicaceae: *Arabis bryoides* (Mt Athos). - 2. *Arabis cretica* (Lefka Ori). - 3. *Arabis glabra* (Voras, inset of flowers from Killini). - 4. *Arabis sudetica* (Falakron). - 5. *Arabis turrita* (Florinis, Mt Triklarion; inset from Olimbos). - 6. *Arabis verna* (Achaias, Ano Kastritsi). - 7. *Aubrieta deltoidea* (Taigetos, inset from Giona). - 8. *Aubrieta erubescens* (Mt Athos). - 9. *Aubrieta gracilis* (Timfristos). - 10. *Aubrieta scardica* (Gramos, inset from Olimbos). - 11. *Aubriera scyria* (Evvia, Kimi-Metochi). - 12. *Aubrieta thessala* (Olimbos).

Plate 72

1. Brassicaceae: *Aubrieta corymbosa* (Olimbos). - 2. *Aubrieta moreana* (Achaias, Mt Klokos). - 3. *Aubrieta saxatilis* subsp. *orientalis* (Lakonias, Gerolimin-Vathia; inset from Thessalonikis, Nimfopetra). - 4. *Barbarea longirostris* (Varnous). - 5. *Barbarea sicula* (Parnassos). - 6. *Berteroa incana* (Prespa area). - 7. *Berteroa obliqua* (by Lake Doirani). - 8. *Berteroa orbiculata* (Athos peninsula). - 9. *Biscutella didyma* (Chios). - 10. *Bornmuellera baldaccii* (Valia Kalda National Park). - 11. *Bornmuellera tymphaea* (Valia Kalda National Park, inset from Vasilitsa). - 12. *Brassica cretica* subsp. *aegaea* (Mt Athos).

Plate 73

1. Brassicaceae: *Bunias erucago* (Kerkira). - 2. *Cakile maritima* (Samos). - 3. *Calepina irregularis* (Athos peninsula). - 4. *Capsella bursa-pastoris* (Naxos). - 5. *Capsella grandiflora* (Ioanninon, Mt Douskon; inset from above Metsovon). - 6. *Cardamine acris* subsp. *acris* (Tzena). - 7. *Cardamine amara* subsp. *balcanica* (Rodopi). - 8. *Cardamine bulbifera* (Florinis, Pisoderi pass). - 9. *Cardamine carnosa* (Timfi). - 10. *Cardamine glauca* (Valia Kalda National Park). - 11. *Cardamine graeca* (Kalavrita, inset from Athos peninsula). - 12. *Cardamine impatiens* subsp. *impatiens* (Rodopi).

Plate 74

1. Brassicaceae: *Cardamine matthioli* (Vasilitsa). - 2. *Cardamine plumieri* (Gramos). - 3. *Cardaria draba* (Kristallopigi). - 4. *Carrichtera annua* (Attikis: Asteras Vouliagmenis). - 5. *Clypeola jonthlaspi* (Kalavrita; inset from Prespa area, Mt Devas). - 6. *Coincya nivalis* (Olimbos). - 7. *Coronopus squamatus* (Lakonias, Sikea). - 8. *Crambe hispanica* (Lakonias, WNW of Itilo). - 9. *Didesmus aegyptius* (Kithira). - 10. *Diplotaxis erucoides* (Korinthias, S of Chilomodion). - 11. *Draba cretica* (Lefka Ori). - 12. *Draba heterocoma* subsp. *archipelagi* (Samos, Mt Kerki).

Plate 75

1. Brassicaceae: *Draba lacaitae* (Killini, inset from Giona). - 2. *Draba lasiocarpa* (Falakron; inset left from Valia Kalda National Park, right from Mt Douskon). - 3. *Draba muralis* (Prespa area). - 4. *Draba parnassica* (Evvia, Mt Dirfis; inset from Parnassos). - 5. *Draba strasseri* (Lakonias, Mt Koulochera). - 6. *Enarthrocarpus arcuatus* (Messinias, Methoni castle). - 7. *Erophila verna* subsp. *praecox* (Varnous). - 8. *Eruca vesicaria* (Kithira). - 9. *Erysimum atticum* (Giona). - 10. *Erysimum boryanum* (Parnassos). - 11. *Erysimum bulgaricum* (Evrou, Valtos-Chandras). - 12. *Erysimum calycinum* (Mt Athos).

Plate 76

1. Brassicaceae: *Erysimum cephalonicum* (Achaias, Mt Klokos). - 2. *Erysimum corinthium* (Monemvasia, inset from Vouliagmeni lagoon). - 3. *Erysimum crassistylum* (Prespa area). - 4. *Erysimum creticum* (E Kriti, Zakros). - 5. *Erysimum cuspidatum* (Kastorias, Nestorio; inset of flowers from Vermion, of fruits from Samarina). - 6. *Erysimum graecum* (Argolidos, Tolon). - 7. *Erysimum mutabile* (Lefka Ori). - 8. *Erysimum naxense* (Naxos). - 9. *Erysimum olympicum* (Olimbos). - 10. *Erysimum pseudocuspidatum* (Parnassos). - 11. *Erysimum pusillum* (Taigetos). - 12. *Erysimum senoneri* subsp. *icaricum* (Ikaria).

Plate 77

1. Brassicaceae: *Fibigia clypeata* (Prespa area). - 2. *Hesperis balansae* (Samos). - 3. *Hesperis dinarica* (Bulgaria, Mt Vitosha). - 4. *Hesperis laciniata* subsp. *laciniata* (Serron, Siderokastro; inset from Kastorias, Eptachorion). - 5. *Hesperis theophrasti* subsp. *theophrasti* (Prespa area). - 6. *Hirschfeldia incana* (Korinthias, Ano Trikala). - 7. *Hornungia petraea* (Lakonias, Mt Koulochera). - 8. *Iberis carnosa* (Kerkira). - 9. *Iberis saxatilis* (?Giona). - 10. *Iberis sempervirens* (Florinis, Mt Triklarion; insets from Olimbos). - 11. *Iberis umbellata* (Konitsis, Pades). - 12. *Isatis athoa* (Mt Athos).

Plate 78

1. Brassicaceae: *Isatis tinctoria* (Kilkis, with *Papaver rhoeas*). - 2. *Kernera saxatilis* (Olimbos). - 3. *Lepidium campestre* (Prespa area). - 4. *Lepidium graminifolium* (Attikis, Chalandri). - 5. *Lepidium hirtum* subsp. *nebrodense* (Killini). - 6. *Leptoplax emarginata* (Valia Kalda National Park, inset from Samarina). - 7. *Lunaria annua* subsp. *pachyrhiza* (Mistras, inset from Prespa area). - 8. *Malcolmia africana* (Attikis, Kamariza). - 9. *Malcolmia chia* (Kriti, Kotsifou gorge). - 10. *Malcolmia flexuosa* subsp. *naxensis* (Chios). - 11. *Malcolmia graeca* subsp. *bicolor* (Killini). - 12. *Malcolmia nana* (Rodos).

Plate 79

1. Brassicaceae: *Malcolmia orsiniana* subsp. *angulifolia* (Timfristos). - 2. *Matthiola fruticulosa* subsp. *fruticulosa* (Attikis, Lavrion). - 3. *Matthiola fruticulosa* subsp. *valesiaca* (Pangeon, inset from Siniatsikon). - 4. *Matthiola incana* (Monemvasia). - 5. *Matthiola longipetala* subsp. *pumilio* (Rodos). - 6. *Matthiola tricuspidata* (Naxos). - 7. *Myagrum perfoliatum* (Kastorias, Eptachori). - 8. *Nasturtium officinale* (Andros). - 9. *Neslia apiculata* (by Lake Stimfalia). - 10. *Noccaea brevistyla* subsp. *pseudorivularis* (Parnassos). - 11. *Noccaea bulbosa* (Parnitha). - 12. *Noccaea graeca* (Chelmos).

Plate 80

1. Brassicaceae: *Noccaea microphylla* (Gramos). - 2. *Noccaea ochroleuca* (Falakron; inset from Evrou, Dadia-Pessani). - 3. *Noccaea perfoliata* (Parnassos). - 4. *Noccaea tymphaea* (Voras, inset from Valia Kalda National Park). - 5. *Noccaea zaffranii* (Lefka Ori). - 6. *Odontarrhena chalcidica* (Samarina). - 7. *Odontarrhena euboea* (Evvia, Mantoudi). - 8. *Odontarrhena fragillima* (Lefka Ori). - 9. *Odontarrhena heldreichii* (Katara pass; inset from Konitsis, Pades). - 10. *Odontarrhena lesbiaca* (Lesvos). - 11. *Odontarrhena sibirica* (Evrou, Dadia-Pessani; inset from Korinthos). - 12. *Odontarrhena smolikana* (Smolikas).

Plate 81

1. Brassicaceae: *Phyllolepidum cyclocarpum* (Douskon, inset from Vardousia). - 2. *Raphanus raphanistrum* (Athos peninsula; inset from Ilias, Killini). - 3. *Rapistrum rugosum* (Korinthias, Ano Trikala). - 4. *Ricotia carnosula* (Kastellorizo). - 5. *Ricotia cretica* (Kriti, Kotsifou gorge). - 6. *Rorippa amphibia* (Prespa area). - 7. *Rorippa thracica* (Voras, inset from Vasilitsa). - 8. *Sinapis alba* (Samos). - 9. *Sisymbrium officinale* (Naxos). - 10. *Sisymbrium orientale* (Naxos; inset from Lakonias, Malea peninsula). - 11. *Thlaspi arvense* (Rodopi). - 12. Butomaceae: *Butomus umbellatus* (Prespa area).

Plate 82

1. Buxaceae: *Buxus sempervirens* (N Pindos, inset from Olimbos). - 2. Cactaceae: *Opuntia ficus-indica* (Mani peninsula, inset from Nafplion). - 3. Caesalpiniaceae: *Ceratonia siliqua* (Lakonias: Githion-Skala; insets from Kriti, Lasithiou, Gournia). - 4. *Cercis siliquastrum* (foothills of Mt Athos; inset of fruits from Fokidis, Athanasios Diakos). - 5. Callitrichaceae: *Callitriche cophocarpa* (Voras). - 6. *Callitriche stagnalis* (Evrou, Dadia-Pessani). - 7. Campanulaceae: *Asyneuma canescens* (Falakron). - 8. *Asyneuma giganteum* (Rodos). - 9. *Asyneuma limonifolium* (Olimbos). - 10. *Asyneuma pichleri* (Varnous). - 11. *Asyneuma virgatum* (Chios). - 12. *Campanula aizoon* subsp. *aizoides* (Chelmos).

Plate 83

1. Campanulaceae: *Campanula andrewsii* (Lakonias, NW Daimonia). - 2. *Campanula asperuloides* (Chelmos). - 3. *Campanula calaminthifolia* (Naxos). - 4. *Campanula celsii* (Mt Kitheronas). - 5. *Campanula cervicaria* (Tzena, inset from Vrondous). - 6. *Campanula columnaris* (Vardousia). - 7. *Campanula constantini* (Evvia, Mt Dirfis). - 8. *Campanula cretica* (Kriti, Nea Roumata). - 9. *Campanula delicatula* (Rodos). - 10. *Campanula dimitrii* (Karpathos, Kali Limni). - 11. *Campanula drabifolia* (Lakonias, Githion). - 12. *Campanula erinus* (Rodos).

Plate 84

1. Campanulaceae: *Campanula foliosa* (Varnous). - 2. *Campanula formanekiana* (Florinis, Farangi). - 3. *Campanula garganica* (Kefallinia, Mt Enos). - 4. *Campanula glomerata* (Gramos). - 5. *Campanula hagielia* (Rodos). - 6. *Campanula hawkinsiana* (E of Katara pass). - 7. *Campanula hierapetrae* (Kriti, Thriptis Ori). - 8. *Campanula incurva* (Pierias, Platamonas). - 9. *Campanula jacquinii* subsp. *jacquinii* (Kriti, Mt Kofinas). - 10. *Campanula laciniata* (Karpathos). - 11. *Campanula lavrensis* (Athos peninsula). - 12. *Campanula lingulata* (Athos peninsula, inset from Thasos).

Plate 85

1. Campanulaceae: *Campanula lyrata* (Chios). - 2. *Campanula macrostachya* (Evrou, Metaxades). - 3. *Campanula oreadum* (Olimbos). - 4. *Campanula orphanidea* (Falakron, inset from Mt Athos). - 5. *Campanula pangea* (Pangeon). - 6. *Campanula papillosa* (Taigetos). - 7. *Campanula patula* (Smolikas). - 8. *Campanula pelviformis* (Kriti, Thriptis Ori). - 9. *Campanula phrygia* (Chalkidiki). - 10. *Campanula radicosa* (Vardousia). - 11. *Campanula ramosissima* (Kerkira). - 12. *Campanula rhodensis* (Rodos).

Plate 86

1. Campanulaceae: *Campanula rotundifolia* s. lat. (Olimbos). - 2. *Campanula rupestris* (Parnassos). - 3. *Campanula rupicola* (Parnassos). - 4. *Campanula samothracica* (Samothraki). - 5. *Campanula simulans* (Kos). - 6. *Campanula sparsa* (Katara pass). - 7. *Campanula spatulata* (Parnassos; inset from Lakonias, Kriovrisi). - 8. *Campanula stenosiphon* (Taigetos). - 9. *Campanula tubulosa* (Kriti, Kotsifou gorge). - 10. *Campanula tymphaea* (Katara pass). - 11. *Campanula versicolor* (Olimbos). - 12. *Campanula wanneri* (Xanthis, Mt Koula).

Plate 87

1. Campanulaceae: *Edraianthus graminifolius* (Timfi). - 2. *Jasione heldreichii* (Prespa area; inset from Evrou, SW Mega Derio). - 3. *Jasione laevis* subsp. *orbiculata* (Voras). - 4. *Legousia speculum-veneris* subsp. *speculum-veneris* (by Lake Stimfalia). - 5. *Petromarula pinnata* (Kriti, Chania). - 6. *Phyteuma orbiculare* (Montenegro, Mt Durmitor). - 7. *Solenopsis minuta* subsp. *annua* (W Kriti, Imbros gorge). - 8. Cannabaceae: *Humulus lupulus* (Drama-Nikiforos). - 9. Capparaceae: *Capparis spinosa* subsp. *rupestris* (Amorgos). - 10. Caprifoliaceae: *Lonicera alpigena* subsp. *formanekiana* (Olimbos). - 11. *Lonicera alpigena* subsp. *hellenica* (Stix ravine). - 12. *Lonicera etrusca* (Chios).

Plate 88

1. Caprifoliaceae: *Lonicera implexa* (Kerkira). - 2. *Lonicera nummulariifolia* (Parnassos). - 3. *Sambucus ebulus* (Killini; inset of flowers from Athos peninsula, of fruits from Lake Kerkini). - 4. *Sambucus nigra* (Dramas, Sidironero). - 5. *Sambucus racemosa* (Rodopi). - 6. *Viburnum lantana* (Montenegro, Mr Orjan; inset from Mt Triklarion). - 7. Caryophyllaceae: *Agrostemma githago* (Sparti-Tripoli). - 8. *Arenaria conferta* subsp. *serpentini* (Gramos, Epano Arena). - 9. *Arenaria cretica* (Olimbos). - 10. *Arenaria deflexa* (Rodos). - 11. *Arenaria filicaulis* subsp. *filicaulis* (Thasos). - 12. *Arenaria filicaulis* subsp. *teddii* (Pangeon).

Plate 89

1. Caryophyllaceae: *Arenaria fragillima* (Kriti, Thriptis Ori). - 2. *Arenaria gionae* (Giona). - 3. *Arenaria leucadia* (Lefkas). - 4. *Arenaria luschanii* (Kastellorizo). - 5. *Arenaria rhodia* (Rodos). - 6. *Arenaria rotundifolia* (Voras). - 7. *Bolanthus chelmicus* subsp. *meteoricus* (Kozanis, by Polifitos artificial lake). - 8. *Bolanthus confertifolius* (SW Turkey, Datca peninsula). - 9. *Bolanthus corinthiacus* (Korinthias, Vouliagmeni lagoon). - 10. *Bolanthus creutzburgii* (Psiloritis). - 11. *Bolanthus fruticulosus* (Lakonias, Pounta opposite Elafonisos). - 12. *Bolanthus graecus* (Evvia, Mt Dirfis).

Plate 90

1. Caryophyllaceae: *Bolanthus thymifolius* (Parnassos). - 2. *Bufonia stricta* (Lefka Ori, inset from Mt Gerania). - 3. *Cerastium banaticum* subsp. *speciosum* (Olimbos). - 4. *Cerastium bulgaricum* (Rodopis, NW of Ourania). - 5. *Cerastium candidissimum* (Mt Kitheronas). - 6. *Cerastium cerastoides* (Timfi). - 7. *Cerastium comatum* (Rodos). - 8. *Cerastium decalvans* subsp. *leontopodium* (Timfi). - 9. *Cerastium dichotomum* (Argolidis, Didima). - 10. *Cerastium glomeratum* (Lakonias, Areopolis). - 11. *Cerastium illyricum* subsp. *brachiatum* (Taigetos). - 12. *Cerastium pumilum* subsp. *glutinosum* (Taigetos).

Plate 91

1. Caryophyllaceae: *Cerastium ramosissimum* (Viotias, Ipaton). - 2. *Cerastium rectum* (Varnous). - 3. *Cerastium runemarkii* (Evvia, Mt Ochi). - 4. *Cerastium scaposum* (Lefka Ori). - 5. *Cerastium smolikanum* (Albania, Mt Valamara). - 6. *Cerastium theophrasti* (Olimbos). - 7. *Corrigiola litoralis* (Sparti). - 8. *Dianthus androsaceus* (Taigetos). - 9. *Dianthus aridus* (Evrou, Asimenio). - 10. *Dianthus armeria* (Rodopi). - 11. *Dianthus arpadianus* (Evrou, Nea Makri). - 12. *Dianthus biflorus* (Iti, inset from Parnassos).

Plate 92

1. Caryophyllaceae: *Dianthus cinnamomeus* (Naxos). - 2. *Dianthus corymbosus* (Athos peninsula). - 3. *Dianthus crinitus* (Rodos). - 4. *Dianthus cruentus* (Florinis, Mt Triklario). - 5. *Dianthus deltoides* subsp. *deltoides* (Varnous). - 6. *Dianthus diffusus* (Milos). - 7. *Dianthus elegans* (Ikaria). - 8. *Dianthus formanekii* (Kozanis, Kriovrisi-Vlasti). - 9. *Dianthus fruticosus* subsp. *carpathus* (Karpathos). - 10. *Dianthus giganteus* (Evrou, Lefkimi). - 11. *Dianthus glutinosus* (Lesvos). - 12. *Dianthus gracilis* subsp. *xanthianus* (Thasos).

Plate 93

1. Caryophyllaceae: *Dianthus haematocalyx* subsp. *haematocalyx* (Olimbos). - 2. *Dianthus haematocalyx* subsp. *pindicola* (Katara pass). - 3. *Dianthus ingoldbyi* (Rodopis, Maronia). - 4. *Dianthus integer* subsp. *minutiflorus* (Timfi, inset from Chelmos). - 5. *Dianthus juniperinus* subsp. *bauhiniorum* (Psiloritis). - 6. *Dianthus muglensis* (Rodos). - 7. *Dianthus myrtinervius* subsp. *myrtinervius* (Varnous). - 8. *Dianthus noeanus* (Orvilos). - 9. *Dianthus petraeus* subsp. *orbelicus* (Mt Athos). - 10. *Dianthus pinifolius* subsp. *serbicus* (Prespa area, Mt Devas). - 11. *Dianthus serratifolius* (Imittos). - 12. *Dianthus sphacioticus* (Lefka Ori).

Plate 94

1. Caryophyllaceae: *Dianthus stenopetalus* (Florinis, Mt Triklario). - 2. *Dianthus superbus* (Rodopi). - 3. *Dianthus sylvestris* (Ioanninon, Chrisovitsa; inset from SW of Fourka). - 4. *Dianthus tripunctatus* (Naxos). - 5. *Dianthus tymphresteus* (Chelmos). - 6. *Dianthus viscidus* (Taigetos). - 7. *Dianthus xylorrhizus* (W Kriti, Polirrhinia). - 8. *Dianthus zonatus* (Samos). - 9. *Drypis spinosa* (Timfi). - 10. *Gypsophila muralis* (Evrou, Dadia). - 11. *Gypsophila nana* (Taigetos). - 12. *Gypsophila pallasii* (Orvilos).

Plate 95

1. Caryophyllaceae: *Herniaria glabra* (Varnous). - 2. *Herniaria hirsuta* (Naxos). - 3. *Herniaria parnassica* (Olimbos). - 4. *Illecebrum verticillatum* (Naxos). - 5. *Minuartia attica* (Mt Athos). - 6. *Minuartia baldaccii* (Valia Kalda National Park). - 7. *Minuartia confusa* (Parnassos). - 8. *Minuartia dirphya* (Evvia, Mt Dirfis). - 9. *Minuartia doerfleri* (Falakron). - 10. *Minuartia greuteriana* (Evrou, Dadia). - 11. *Minuartia hamata* (Prespa area, Mt Devas). - 12. *Minuartia hirsuta* (Valia Kalda National Park).

Plate 96

1. Caryophyllaceae: *Minuartia juniperina* (Taigetos). - 2. *Minuartia pseudosaxifraga* (Timfi). - 3. *Minuartia recurva* (Voras). - 4. *Minuartia stellata* (Chelmos). - 5. *Minuartia verna* (E of Katara pass). - 6. *Minuartia wettsteinii* (Thriptis Ori). - 7. *Moehringia pendula* (Rodopi). - 8. *Moehringia trinervia* (Pangeon). - 9. *Moenchia graeca* (Naxos). - 10. *Moenchia mantica* (Chalkidiki). - 11. *Paronychia albanica* subsp. *graeca* (Taigetos). - 12. *Paronychia argentea* (Rodos).

Plate 97

1. Caryophyllaceae: *Paronychia macedonica* (Grevenon, Exarchos). - 2. *Paronychia manfrediana* (Evrou, Dadia-Pessani). - 3. *Paronychia rechingeri* (Pangeon). - 4. *Petrorhagia armerioides* (Amorgos). - 5. *Petrorhagia candica* (SW Kriti). - 6. *Petrorhagia dianthoides* (Kriti, Imbros gorge). - 7. *Petrorhagia fasciculata* (Kriti, Patsianos). - 8. *Petrorhagia glumacea* (Mani peninsula). - 9. *Petrorhagia graminea* (Ioanninon, Likostomo). - 10. *Petrorhagia grandiflora* (Parnonas, Leonidion). - 11. *Petrorhagia laconica* (Malea peninsula, Viglafia). - 12. *Petrorhagia phthiotica* (Iti).

Plate 98

1. Caryophyllaceae: *Petrorhagia saxifraga* (Prespa area). - 2. *Petrorhagia thessala* (Loutra Arideas). - 3. *Polycarpon tetraphyllum* (Athos peninsula). - 4. *Sagina stridii* (Chelmos, inset from Killini). - 5. *Saponaria bellidifolia* (Florinis, Mt Triklario). - 6. *Saponaria calabrica* (Kozanis, Siatista). - 7. *Saponaria glutinosa* (Florinis, Mt Triklario). - 8. *Saponaria intermedia* (Trikalon, Trigono). - 9. *Saponaria officinalis* (Pierias, Litochoron). - 10. *Scleranthus perennis* subsp. *marginatus* (Voras). - 11. *Silene ammophila* subsp. *ammophila* (SE Kriti). - 12. *Silene antri-jovis* (Psiloritis).

Plate 99

1. Caryophyllaceae: *Silene armeria* (Florinis, Vigla pass). - 2. *Silene asterias* (Varnous). - 3. *Silene atropurpurea* (Florinis, Pisoderi). - 4. *Silene auriculata* (Parnassos). - 5. *Silene barbeyana* (Giona). - 6. *Silene behen* (Attiki, Cape Sounion). - 7. *Silene bellidifolia* (Preveza). - 8. *Silene bupleuroides* subsp. *staticifolia* (Fokidos, Stromi). - 9. *Silene caesia* (Parnassos). - 10. *Silene chlorifolia* (Kastorias, NW Eptachori). - 11. *Silene chromodonta* (Olimbos). - 12. *Silene ciliata* subsp. *graefferi* (Olimbos).

Plate 100

1. Caryophyllaceae: *Silene colorata* (E of Nafplion). - 2. *Silene compacta* (Evrou, Dadia-Sapka). - 3. *Silene congesta* (Peloponnisos, S of Egio). - 4. *Silene conglomeratica* (Achaias, Mega Spileo). - 5. *Silene corinthiaca* (Amorgos). - 6. *Silene coronaria* (Athos peninsula). - 7. *Silene cretica* (Mani peninsula). - 8. *Silene cythnia* (Amorgos). - 9. *Silene dichotoma* (Amorgos). - 10. *Silene dirphya* (Evvia, Mt Dirfis). - 11. *Silene discolor* (Rodos). - 12. *Silene echinosperma* (Taigetos, Langada gorge).

Plate 101

1. Caryophyllaceae: *Silene exaltata* (Pierias, Litochoron railway station). - 2. *Silene fabaria* subsp. *fabaria* (Athos peninsula). - 3. *Silene fabarioides* (Kastorias, Eptachorion). - 4. *Silene frivaldskyana* (Paranestion, inset from Prespa area). - 5. *Silene fuscata* (Samos). - 6. *Silene gallica* (Kithira). - 7. *Silene gallinyi* (Prespa area). - 8. *Silene gigantea* subsp. *gigantea* (Samos). - 9. *Silene graeca* (Prespa area). - 10. *Silene grisebachii* (Rodopis, Ptelea lagoon). - 11. *Silene haussknechtii* (Vasilitsa). - 12. *Silene holzmannii* (Attikis, Faliriko delta).

Plate 102

1. Caryophyllaceae: *Silene integripetala* (Lakonias, Githion-Areopolis). - 2. *Silene italica* subsp. *italica* (Prespa area). - 3. *Silene latifolia* (Varnous). - 4. *Silene longipetala* (NE Peloponnisos). - 5. *Silene marschallii* subsp. *guicciardii* (Parnassos). - 6. *Silene melzheimeri* (Smolikas). - 7. *Silene multicaulis* subsp. *sporadum* (Evvia, Steni Dirfios). - 8. *Silene muscipula* (Viotias, Kleidi-Asopia). - 9. *Silene nocturna* (Attikis, Lavrion). - 10. *Silene oligantha* subsp. *oligantha* (Olimbos). - 11. *Silene oligantha* subsp. *pseudoradicosa* (Evvia, Neos Pagontas). - 12. *Silene orphanidis* (Mt Athos).

Plate 103

1. Caryophyllaceae: *Silene paradoxa* (Prespa area). - 2. *Silene parnassica* subsp. *parnassica* (Timfi). - 3. *Silene pentelica* (Ikaria). - 4. *Silene pusilla* subsp. *albanica* (Valia Kalda National Park). - 5. *Silene radicosa* (Parnassos). - 6. *Silene roemeri* (Voras). - 7. *Silene salamandra* (Rodos). - 8. *Silene sartorii* (Naxos). - 9. *Silene saxifraga* (Olimbos). - 10. *Silene schwarzenbergeri* (Ioanninon, S of village Milea). - 11. *Silene sclerocarpa* (Korinthias, Posidonia). - 12. *Silene sedoides* (Athos peninsula).

Plate 104

1. Caryophyllaceae: *Silene subconica* (Prespa area). - 2. *Silene subintegra* (Prespa area). - 3. *Silene succulenta* (Gavdos). - 4. *Silene ungeri* (Kerkira). - 5. *Silene urvillei* (Lesvos). - 6. *Silene variegata* (Lefka Ori). - 7. *Silene viridiflora* (Serron, Mt Beles). - 8. *Silene vulgaris* subsp. *commutata* (Athos peninsula). - 9. *Spergula arvensis* (Athos peninsula). - 10. *Spergularia bocconei* (Korinthias, Vouliagmeni lagoon). - 11. *Spergularia rubra* (Voras). - 12. *Stellaria alsine* (Varnous).

Plate 105

1. Caryophyllaceae: *Stellaria aquatica* (Lake Prespa, inset from Lake Kerkini). - 2. *Stellaria cupaniana* (Ikaria). - 3. *Stellaria glochidisperma* (Varnous). - 4. *Stellaria graminea* (Falakron). - 5. *Stellaria media* (Achaias, Ano Kastritsi). - 6. *Stellaria nemorum* (Rodopi). - 7. *Telephium imperati* (Taigetos, inset from Parnassos). - 8. *Vaccaria hispanica* (Florinis, SE of Perea). - 9. *Velezia quadridentata* (Lesvos). - 10. *Velezia rigida* (Naxos). - 11. Celastraceae: *Euonymus europaeus* (Dodonis, Kostaniani; inset from Serron, Promachonas-Angistro). - 12. *Euonymus latifolius* (Olimbos).

Plate 106

1. Celastraceae: *Euonymus verrucosus* (Prespa area). - 2. Chenopodiaceae: *Arthrocaulon macrostachyum* (Naxos). - 3. *Atriplex halimus* (Korinthias, Xilokastro). - 4. *Atriplex recurva* (Attikis, Charakas beach). - 5. *Atriplex tatarica* (Serron, Amphipolis). - 6. *Beta nana* (Taigetos). - 7. *Beta vulgaris* subsp. *maritima* (Kastellorizo). - 8. *Blitum bonus-henricus* (Rodopi). - 9. *Blitum virgatum* (Kavalas, Akrovouni). - 10. *Caloxylon aegaeum* (Milos). - 11. *Caroxylon carpathum* (Karpathos). - 12. *Chenopodiastrum murale* (Sparti).

Plate 107

1. Chenopodiaceae: *Chenopodium betaceum* (Pellis, Sarakini). - 2. *Chenopodium vulvaria* (Peloponnisos). - 3. *Dysphania ambrosioides* (Pellis, Aridea). - 4. *Dysphania botrys* (Falakron). - 5. *Dysphania pumilio* (Pellis, Aridea). - 6. *Halimione portulacoides* (Achaias, Kalogria). - 7. *Halocnemum strobilaceum* (Axios river delta). - 8. *Noaea mucronata* (Samos). - 9. *Polycnemum arvense* (Prespa area). - 10. *Salicornia europaea* s. lat. (Achaias, Kalogria). - 11. *Salicornia fruticosa* (Achaias, Kalogria). - 12. *Salicornia perennis* (unknown locality).

Plate 108

1. Chenopodiaceae: *Salsola tragus* (Serres-Drama). - 2. *Soda inermis* (Kastellorizo). - 3. *Suaeda maritima* (Axios river delta). - 4. *Suaeda palaestina* (Koufonisi SE Kriti). - 5. *Suaeda vera* (Naxos). - 6. Cistaceae: *Cistus creticus* (Chios). - 7. *Cistus laurifolius* (Evrou, Dadia). - 8. *Cistus monspeliensis* (Mallorca). - 9. *Cistus parviflorus* (Kriti, Agia Fotia). - 10. *Cistus salviifolius* (E Kriti, Vai). - 11. *Cistus sintenisii* (Valia Kalda National Park). - 12. *Fumana aciphylla* (Evrou, Dadia).

Plate 109

1. Cistaceae: *Fumana arabica* (Chios). - 2. *Fumana bonapartei* (E of Katara pass). - 3. *Fumana paphlagonica* subsp. *alpina* (Lefka Ori). - 4. *Fumana procumbens* (Tzena). - 5. *Fumana thymifolia* (E Kriti, Agia Fotia). - 6. *Helianthemum aegyptiacum* (Naxos). - 7. *Helianthemum hymettium* (Lefka Ori). - 8. *Helianthemum nummularium* (Parnassos). - 9. *Helianthemum oelandicum* subsp. *canum* (Prespa area, Mt Devas). - 10. *Helianthemum salicifolium* (Akrokorinthos). - 11. *Helianthemum stipulatum* (Koufonisi SE of Kriti). - 12. *Helianthemum syriacum* (Samos).

Plate 110

1. Cistaceae: *Tuberaria guttata* (Lakonias, Monemvasia). - 2. Colchicaceae: *Androcymbium rechingeri* (SW Kriti, Elafonisos). - 3. *Colchicum asteranthum* (Arkadias, Mt Lirkio). - 4. *Colchicum atticum* (Evvia, Psachna-Kamaritsa). - 5. *Colchicum autumnale* (Falakron). - 6. *Colchicum bivonae* (Mt Athos). - 7. *Colchicum boissieri* (SE Peloponnisos, Geraki-Kosmas; inset from Samos). - 8. *Colchicum burttii* (Chios). - 9. *Colchicum chalcedonicum* (Evvia, Pili-Vlachia). - 10. *Colchicum chimonanthum* (Serron, SE of Nea Zichni). - 11. *Colchicum confusum* (Trikalon, Metsovon-Kalambaka). - 12. *Colchicum cretense* (Lefka Ori).

Plate 111

1. Colchicaceae: *Colchicum cupanii* (Sparti-Tripoli, inset from Kalogria). - 2. *Colchicum doerfleri* (Siatista-Kozani). - 3. *Colchicum euboeum* (Evvia, Mt Dirfis). - 4. *Colchicum graecum* (Panachaikon). - 5. *Colchicum haynaldii* (Prespa area). - 6. *Colchicum lingulatum* (Evvia, Gimno-Kimi). - 7. *Colchicum macrophyllum* (Simi, inset from Rodos). - 8. *Colchicum parlatoris* (SE of Kalamata; inset from Messinias, Kambos). - 9. *Colchicum parnassicum* (Parnassos, inset from Giona). - 10. *Colchicum peloponnesiacum* (Achaias, N of Kalamias). - 11. *Colchicum pulchellum* (Killini). - 12. *Colchicum pusillum* (Kriti, Amnissos).

Plate 112

1. Colchicaceae: *Colchicum rausii* (Karditsis, Argithea-Mouzakion). - 2. *Colchicum sfikasianum* (Lakonias, Monemvasia-Molai). - 3. *Colchicum soboliferum* (Attiki, Oropos). - 4. *Colchicum triphyllum* (Chelmos). - 5. *Colchicum turcicum* (Bulgaria, Kardjalik). - 6. *Colchicum variegatum* (Ikaria). - 7. *Colchicum zahnii* (Lakonias, Tsikalia). - 8. Convallariaceae: *Convallaria majalis* (Florinis, Mt Triklario). - 9. *Polygonatum multiflorum* (Smolikas). - 10. *Polygonatum odoratum* (Florinis, Mt Triklario). - 11. *Polygonatum verticillatum* (Pangeon). - 12. Convolvulaceae: *Calystegia sepium* (Litochoron railway station).

Plate 113

1. Convolvulaceae: *Calystegia soldanella* (E of Asprovalta). - 2. *Convolvulus althaeoides* (Rodos). - 3. *Convolvulus argyrothamnos* (E Kriti, Christos-Metaxachori). - 4. *Convolvulus arvensis* (near Lake Stimfalia, inset from S Italy). - 5. *Convolvulus betonicifolius* (Rodos). - 6. *Convolvulus boissieri* subsp. *thracicus*, ined. (Evrou, Dadia to Mt Sapka). - 7. *Convolvulus cantabrica* (Mt Kitheronas). - 8. *Convolvulus coelesyriacus* (Kastellorizo). - 9. *Convolvulus dorycnium* (Kriti, Skinias). - 10. *Convolvulus holosericeus* (Kozanis, Neraida). - 11. *Convolvulus libanoticus* (Killini). - 12. *Convolvulus lineatus* (Florinis, SW of Perea).

Plate 114

1. Convolvulaceae: *Convolvulus mairei* (Lake Stimfalia). - 2. *Convolvulus oleifolius* (Rodos). - 3. *Convolvulus pentapetaloides* (Rodos). - 4. *Convolvulus scammonia* (Rodos, inset from Samos). - 5. *Convolvulus siculus* (Naxos). - 6. *Cressa cretica* (Tilos, inset from Naxos). - 7. *Cuscuta atrans* (Lefka Ori). - 8. *Cuscuta brevistyla* (Kastellorizo). - 9. *Cuscuta campestris* (Litochoron railway station). - 10. *Cuscuta epithymum* (Ikaria). - 11. *Cuscuta europaea* (Gramos). - 12. *Cuscuta palaestina* (Karpathos).

Plate 115

1. Convolvulaceae: *Ipomoea imperati* (W Kriti, E of Kastelli Kissamou; inset of flower from Rodos, of fruit from Tilos). - 2. *Ipomoea indica* (W Kriti, W of Topolia). - 3. *Ipomoea sagittata* (Attikis, NW of Megara). - 4. Cornaceae: *Cornus mas* (Timfi; inset from Dramas, Skaloti-Sideronero). - 5. *Cornus sanguinea* (Donodis, Miliotades; inset from Ano Trikala). - 6. Crassulaceae: *Crassula tillaea* (Peloponnisos). - 7. *Prometheum tymphaeum* (Timfi). - 8. *Rosularia serrata* (Lefka Ori, inset from Tilos). - 9. *Sedum acre* (Prespa area). - 10. *Sedum album* (Katara pass). - 11. *Sedum alpestre* subsp. *erythraeum* (Voras). - 12. *Sedum amplexicaule* subsp. *tenuifolium* (Mt Kitheronas).

Plate 116

1. Crassulaceae: *Sedum annuum* (Voras). - 2. *Sedum caespitosum* (Thasos). - 3. *Sedum confertiflorum* (Lesvos). - 4. *Sedum dasyphyllum* (Ioanninon, Aristi). - 5. *Sedum eriocarpum* (Ikaria). - 6. *Sedum grisebachii* (Thasos). - 7. *Sedum hispanicum* (Timfi, Vikos balcony). - 8. *Sedum laconicum* (Lefka Ori). - 9. *Sedum litoreum* (Athos peninsula). - 10. *Sedum magellense* subsp. *olympicum* (Olimbos). - 11. *Sedum ochroleucum* (Olimbos). - 12. *Sedum praesidis* (Kriti, Mt Dikti).

Plate 117

1. Crassulaceae: *Sedum rubens* (Prespa area). - 2. *Sedum samium* (Samos). - 3. *Sedum sediforme* (Kriti, Ano Viannos). - 4. *Sedum stefco* (Varnous). - 5. *Sedum tristriatum* (Lefka Ori). - 6. *Sedum urvillei* (Prespa area). - 7. *Sempervivum heuffelii* (Olimbos). - 8. *Sempervivum marmoreum* (Mt Beles). - 9. *Sempervivum ruthenicum* (Voras). - 10. *Umbilicus horizontalis* (Athos peninsula). - 11. *Umbilicus luteus* (Kriti, Psiloritis; inset from Mt Pieria). - 12. Cucurbitaceae: *Bryonia alba* (Prespa area).

Plate 118

1.Cucurbitaceae: *Bryonia cretica* (Naxos; inset from Kriti, Imbros gorge). - 2. *Ecballium elaterium* (Chalkidiki). - 3. Cyperaceae: *Blysmus compressus* (above Metsovon). - 4. *Bolboschoenus maritimus* (Lesvos). - 5. *Carex acuta* (above Metsovon). - 6. *Carex atrata* (Vardousia). - 7. *Carex canescens* (Voras). - 8. *Carex caryophyllea* (Vermion). - 9. *Carex castroviejoi* (Voras). - 10. *Carex davalliana* (Valia Kalda National Park). - 11. *Carex depauperata* (Evrou, NW of Lefkimi). - 12. *Carex distans* (Parnassos).

Plate 119

1. Cyperaceae: *Carex divisa* (Samos). - 2. *Carex echinata* (Varnous). - 3. *Carex elata* (Thesprotias, NE of Parga). - 4. *Carex extensa* (Kavalas, Keramoti). - 5. *Carex ferruginea* (Tzena). - 6. *Carex flacca* subsp. *flacca* (Valia Kalda National Park). - 7. *Carex halleriana* (Florinis, Mt Triklario). - 8. *Carex hirta* (Prespa area). - 9. *Carex illegitima* (Parnassos). - 10. *Carex kitaibeliana* (Olimbos). - 11. *Carex lasiocarpa* (Voras, Dobro Polje). - 12. *Carex lazarei* (Voras).

Plate 120

1. Cyperaceae: *Carex leporina* (Voras). - 2. *Carex limosa* (Voras, Dobro Polje). - 3. *Carex macrolepis* (Vasilitsa). - 4. *Carex nigra* (Voras). - 5. *Carex otrubae* (Vardousia). - 6. *Carex pallescens* (Iti). - 7. *Carex paniculata* (Gramos). - 8. *Carex pendula* (Samos; inset from Achaias, Ano Kastritsi). - 9. *Carex phyllostachys* (Timfi, Vikos balcony). - 10. *Carex pseudocyperus* (Prespa area). - 11. *Carex remota* (Pellis, Loutra Arideas). - 12. *Carex riparia* (Prespa area).

Plate 121

1. Cyperaceae: *Carex rostrata* (Voras, Dobro Polje). - 2. *Carex vesicaria* (Vermion). - 3. *Carex viridula* (Voras). - 4. *Cladium mariscus* (Litochoron railway station). - 5. *Cyperus capitatus* (Rodos). - 6. *Cyperus difformis* (Kalamata). - 7. *Cyperus fuscus* (Pellis, Aridea). - 8. *Cyperus longus* (Pellis, Loutra Arideas). - 9. Cyperus *michelianus* subsp. *pygmaeus* (Lake Kerkini). - 10. *Eleocharis palustris* (Gramos). - 11. *Eleocharis quinqueflora* (Voras). - 12. *Eleocharis uniglumis* (Ioanninon, Samarina-Fourka).

Plate 122

1. Cyperaceae: *Eriophorum angustifolium* (Voras). - 2. *Eriophorum gracile* (Vasilitsa). - 3. *Eriophorum latifolium* (Voras). - 4. *Eriophorum vaginatum* (Voras). - 5. *Fimbristylis bisumbellata* (Sparti). - 6. *Isolepis cernua* (Andros). - 7. *Rhynchospora alba* (Voras, Dobro Polje). - 8. *Schoenoplectus lacustris* (Lake Stimfalia, inset from Lake Prespa). - 9. *Schoenus nigricans* (Vasilitsa). - 10. *Scirpoides holoschoenus* (Kozanis, E of Pentalofos). - 11. *Scirpus sylvaticus* (Prespa area). - 12. Datiscaceae: *Datisca cannabina* (W Kriti, Nea Roumata; inset of female flowers from Samos).

Plate 123

1. Dioscoreaceae: *Tamus communis* (Athos peninsula). - 2. Dipsacaceae: *Cephalaria ambrosioides* (Mt Athos). - 3. *Cephalaria flava* (Mt Athos, inset from Prespa area). - 4. *Cephalaria squamiflora* (Lefka Ori). - 5. *Cephalaria transsylvanica* (Korinthias, Xilokastro; inset from Konitsa). - 6. *Dipsacus fullonum* (Prespa area). - 7. *Dipsacus laciniatus* (Evrou, Asimenio). - 8. *Knautia degenii* (Tilos). - 9. *Knautia integrifolia* (Kilkis, Kroussia; inset from Mani peninsula). - 10. *Knautia magnifica* (Voras). - 11. *Knautia midzorensis* (Rodopi). - 12. *Knautia orientalis* (Evrou, Mikra Doxipara).

Plate 124

1. Dipsacaceae: *Lomelosia albocincta* (Kriti, Mt Kedros). - 2. *Lomelosia crenata* subsp. *crenata* (Kastorias, E of Chrisi). - 3. *Lomelosia divaricata* (Tilos, inset of flowers from Naxos). - 4. *Lomelosia epirota* (Thesprotias, Souli). - 5. *Lomelosia hymettia* (Attikis, Psatha; inset from Lakonias, Charakas). - 6. *Lomelosia minoana* (Kriti, Kalami). - 7. *Lomelosia palaestina* (Naxos; inset from Evvia, Marouli). - 8. *Lomelosia rhodopensis* (Falakron). - 9. *Lomelosia sphaciotica* (Lefka Ori). - 10. *Pterocephalus brevis* (Tilos). - 11. *Pterocephalus perennis* subsp. *perennis* (Evvia, NE of Metochi). - 12. *Pterocephalus pinardii* (Samos).

Plate 125

1. Dipsacaceae: *Pterocephalus plumosus* (Evvia, Marouli). - 2. *Scabiosa atropurpurea* (Chios). - 3. *Scabiosa taygetea* subsp. *portae* (Gramos). - 4. *Scabiosa tenuis* (Konitsis, Agia Paraskevi). - 5. Droseraceae: *Aldrovanda vesiculosa* (Lake Mikri Prespa). - 6. *Drosera anglica* (Voras, Dobro Polje). - 7. *Drosera rotundifolia* (Rodopi, Livaditis). - 8. Elatinaceae: *Elatine alsinastrum* (Kriti, Chanion, Omalos plateau). - 9. Ericaceae: *Arbutus andrachne* (Lakonias, Lambokambos-Kremasti). - 10. *Arbutus unedo* (ditto). - 11. *Arctostaphylos uva-ursi* (Tzena, inset from Voras). - 12. *Bruckenthalia spiculifolia* (Voras).

Plate 126

1. Ericaceae: *Erica arborea* (Athos peninsula). - 2. *Erica carnea* (Valia Kalda National Park). - 3. *Erica manipuliflora* (Evvia, Stropones). - 4. *Moneses uniflora* (Valia Kalda National Park, inset from Smolikas). - 5. *Monotropa hypopitys* (Olimbos). - 6. *Pyrola chlorantha* (Chelmos, Stix ravine). - 7. *Pyrola media* (Tzena). - 8. *Pyrola minor* (Rodopi). - 9. *Rhododendron luteum* (Lesvos). - 10. *Vaccinium myrtillus* (Voras, inset from Tzena). - 11. *Vaccinium uliginosum* (Voras, inset from Tzena). - 12. *Vaccinium vitis-idaea* (Bulgaria, Mt Vitosha).

Plate 127

1. Euphorbiaceae: *Andrachne telephioides* (Rodopis, Maronia; inset from W Kriti). - 2. *Chrozophora tinctoria* (Sithonia, Ormos Panagias). - 3. *Euphorbia acanthothamnos* (Samos; inset of flowers from Amorgos, of fruits from Mani peninsula). - 4. *Euphorbia aleppica* (Akrokorinthos). - 5. *Euphorbia amygdaloides* (Dramas, Megali Livadi). - 6. *Euphorbia apios* (Imittos). - 7. *Euphorbia capitulata* (Olimbos). - 8. *Euphorbia chamaesyce* (Naxos). - 9. *Euphorbia characias* subsp. *wulfenii* (Mistras, inset from Athos peninsula). - 10. *Euphorbia cyparissias* (Florinis, Vigla pass; inset from Dramas, Dipotama). - 11. *Euphorbia deflexa* (Mt Dirfis). - 12. *Euphorbia dendroides* (Mani peninsula).

Plate 128

1. Euphorbiaceae: *Euphorbia dimorphocaulon* (Kriti, Lasithiou, Limnes; inset from Gramvoussa). - 2. *Euphorbia epithymoides* (Falakron). - 3. *Euphorbia esula* subsp. *tommasiniana* (Rodos). - 4. *Euphorbia exigua* (Naxos). - 5. *Euphorbia heldreichii* (Varnous). - 6. *Euphorbia helioscopia* (Naxos). - 7. *Euphorbia herniariifolia* (Samos). - 8. *Euphorbia hypericifolia* (Kastellorizo). - 9. *Euphorbia illirica* (Fthiotidos, Moschchori). - 10. *Euphorbia montenegrina* (Varnous, inset from Metsovon). - 11. *Euphorbia myrsinites* (Florinis, Maniaki). - 12. *Euphorbia niciciana* (Kastorias, Eptachori).

Plate 129

1. Euphorbiaceae: *Euphorbia oblongata* (Athos peninsula). - 2. *Euphorbia paralias* (Kavalas, Keramoti; inset from Achaias, Kalogria). - 3. *Euphorbia peplis* (Naxos). - 4. *Euphorbia pinea* (Messinias, Navarino bay). - 5. *Euphorbia prostrata* (Lakonias, NE of Githion). - 6. *Euphorbia rechingeri* (Lefka Ori). - 7. *Euphorbia rigida* (Chios). - 8. *Euphorbia spinosa* subsp. *glabriflora* (Smolikas). - 9. *Euphorbia sultan-hassei* (SW Kriti, Imbros gorge). - 10. *Euphorbia taurinensis* (Samos). - 11. *Euphorbia thessala* (Prespa area). - 12. *Mercurialis ovata* (Athos peninsula).

Plate 130

1. Euphorbiaceae: *Ricinus communis* (Karpathos). - 2. Fabaceae: *Adenocarpus complicatus* (Chalkidikis, Stratonion). - 3. *Alhagi graecorum* (Samos). - 4. *Amorpha fruticosa* (Evrou, NNE of Orestiada). - 5. *Anagyris foetida* (Kastellorizo, inset from Chios). - 6. *Anthyllis aurea* (Mt Boutsi). - 7. *Anthyllis hermanniae* (Evvia, Kokkinomilea). - 8. *Anthyllis montana* subsp. *jacquinii* (Pangeon). - 9. *Anthyllis vulneraria* subsp. *pindicola* (Vardousia). - 10. *Anthyllis vulneraria* subsp. *rubriflora* (Taigetos). - 11. *Astragalus agraniotii* (Parnonas). - 12. *Astragalus angustiflorus* subsp. *anatolicus* (Rodopis, Drimi to Ano Drosini).

Plate 131

1. Fabaceae: *Astragalus angustifolius* (Mt Athos, inset from Falakron). - 2. *Astragalus austroaegaeus* (Rodos). - 3. *Astragalus boeticus* (Tilos). - 4. *Astragalus creticus* subsp. *rumelicus* (Iti). - 5. *Astragalus depressus* (Vitsi, inset from Giona). - 6. *Astragalus drupaceus* (Lakonias, Krokee). - 7. *Astragalus gladiatus* (Florinis, SW of Perea). - 8. *Astragalus glycyphyllos* subsp. *glycyphyllos* (Mt Milea). - 9. *Astragalus hamosus* (Amorgos). - 10. *Astragalus hellenicus* (Parnitha, inset from Kitheronas). - 11. *Astragalus idaeus* (Psiloritis). - 12. *Astragalus laconicus* (Lakonias, Vlachiotis).

Plate 132

1. Fabaceae: *Astragalus lacteus* (Prespa area). - 2. *Astragalus lesbiacus* (Chios). - 3. *Astragalus mayeri* (Prespa area; inset from Florinis, Perea). - 4. *Astragalus monspessulanus* (Parnassos, inset from Thasos). - 5. *Astragalus nummularius* (Kriti, Mt Dikti). - 6. *Astragalus onobrychis* (Prespa area). - 7. *Astragalus pelecinus* (Rodos). - 8. *Astragalus peregrinus* (SE Kriti, Koufonisi). - 9. *Astragalus ponticus* (Serron, NNE of Siderokastro). - 10. *Astragalus sempervirens* (Parnassos). - 11. *Astragalus sinaicus* (Samos). - 12. *Astragalus sirinicus* (Timfi).

Plate 133

1. Fabaceae: *Astragalus spruneri* (Attikis, Lavrion). - 2. *Astragalus stella* (Akrokorinthos). - 3. *Astragalus suberosus* (Lakonias, NW of Daimonia). - 4. *Astragalus taygeteus* (Taigetos). - 5. *Astragalus thracicus* subsp. *monachorum* (Mt Athos). - 6. *Astragalus vesicarius* (Timfi). - 7. *Bituminaria bituminosa* (Lakonias, Areopolis-Githion; inset from Kerkira). - 8. *Calicotome villosa* (Kithira). - 9. *Chamaecytisus albus* (Xanthis, NE of Dimarion). - 10. *Chaemecytisus austriacus* (above Metsovon). - 11. *Chamaecytisus eriocarpus* (Rodopi). - 12. *Chamaecytisus hirsutus* subsp. *polytrichus* (Vardousia).

Plate 134

1. Fabaceae: *Chamaecytisus spinescens* subsp. *creticus* (Lakonias, Kriovrisi). - 2. *Cicer graecum* (Killini; inset from Chelmos). - 3. *Cicer incisum* (Lefka Ori). - 4. *Colutea arborescens* (Pellis, Sarakini; inset of fruits from Eptachori-Konitsa). - 5. *Colutea insularis* (Rodos). - 6. *Colutea melanocalyx* (Chios). - 7. *Coronilla argentea* (Athos peninsula). - 8. *Coronilla scorpioides* (Naxos). - 9. *Cytisus decumbens* subsp. *pindicolus* (E of Katara pass). - 10. *Cytisus procumbens* (Florinis, Mt Triklario). - 11. *Cytisus villosus* (Athos peninsula). - 12. *Dorycnium graecum* (Athos peninsula).

Plate 135

1. Fabaceae: *Dorycnium herbaceum* (Ioanninon, Konitsa-Elefthero). - 2. *Dorycnium hirsutum* (Ioanninon, SE of Pades). - 3. *Dorycnium rectum* (W Kriti, Nea Roumata). - 4. *Ebenus cretica* (E Kriti, W of Ierapetra). - 5. *Ebenus sibthorpii* (Lakonias, Vlachiotis). - 6. *Erophaca baetica* subsp. *orientalis* (Lakonias, Areopolis). - 7. *Galega officinalis* (Litochoron railway station). - 8. *Genista acanthoclada* (Taigetos). - 9. *Genista anatolica* (Evrou, Dadia). - 10. *Genista carinalis* (Evrou, Dadia-Sapka). - *Genista depressa* (Voras). - 12. *Genista fasselata* (Karpathos).

Plate 136

1. Fabaceae: *Genista halacsyi* (Taigetos). - 2. *Genista hassertiana* (Grevenon, Exarchos). - 3. *Genista januensis* subsp. *lydia* (Prespa area). - 4. *Genista millii* (Evritanias, Karpenisi-Krikelo). - 5. *Genista parnassica* (Parnassos). - 6. *Genista radiata* (Olimbos). - 7. *Genista sagittalis* (Rodopi). - 8. *Genista sakellariadis* (Olimbos, foothills). - 9. *Genista sessilifolia* (Kozanis, Neraida). - 10. *Genista sylvestris* (Kerkira). - 11. *Genista tinctoria* (E of Karpenisi). - 12. *Glycyrrhiza echinata* (Litochoron railway station).

Plate 137

1. Fabaceae: *Gonocytisus dirmilensis* (Fokidos, Athanasios Diakos). - 2. *Hedysarum grandiflorum* subsp. *bulgaricum* (Mt Gerania). - 3. *Hippopcrepis biflora* (Samos). - 4. *Hippocrepis ciliata* (Naxos). - 5. *Hippocrepis comosa* (Falakron). - 6. *Hippocrepis emerus* subsp. *emeroides* (Prespa area). - 7. *Hippocrepis multisiliquosa* (Kithira). - 8. *Hymenocarpus circinnatus* (Arkadias, Astros). - 9. *Laburnum alpinum* (Karditsis, N of Petrilo). - 10. *Lathyrus annuus* (Amorgos). - 11. *Lathyrus aphaca* (Argolidos, Didima). - 12. *Lathyrus articulatus* (Kithira).

Plate 138

1. Fabaceae: *Lathyrus cicera* (Athos peninsula). - 2. *Lathyrus digitatus* (Florinis, Mt Triklario). - 3. *Lathyrus grandiflorus* (Athos peninsula, inset from Voras). - 4. *Lathyrus laxiflorus* (Iti). - 5. *Lathyrus niger* (Athos peninsula). - 6. *Lathyrus nissolia* (Parnassos). - 7. *Lathyrus ochrus* (Tilos). - 8. *Lathyrus pannonicus* (Grevenon, NNE of Antharkia). - 9. *Lathyrus pratensis* (Kalavrita). - 10. *Lathyrus saxatilis* (Naxos). - 11. *Lathyrus setifolius* (Prespa area). - 12. *Lathyrus sylvestris* (Serron, Ano Vrondou).

Plate 139

1. Fabaceae: *Lathyrus venetus* (Florinis, Mt Triklario). - 2. *Lembotropis nigricans* (Prespa area). - 3. *Lens nigricans* (Grevena-Samarina). - 4. *Lotus angustissimus* (Taigetos). - 5. *Lotus corniculatus* (Voras). - 6. *Lotus cytisoides* (E Kriti, Vai). - 7. *Lotus edulis* (E Kriti, Ierapetra). - 8. *Lotus gebelia* (Florinis, SW of Perea; inset from Thasos). - 9. *Lotus halophilus* (Kriti, Rethimnis, Plakias beach). - 10. *Lotus longisiliquosus* (Argolidos, Tolon). - 11. *Lotus ornithopodioides* (E Kriti, Vai). - 12. *Lotus pedunculatus* (Prespa area).

Plate 140

1. Fabaceae: *Lotus peregrinus* (Kithira). - 2. *Lupinus albus* subsp. *graecus* (Kriti, Rethimnis, Aktounda). - 3. *Lupinus angustifolius* (Athos peninsula; inset from Argolidos, Epidavros). - 4. *Lupinus luteus* (Lakonias, Monemvassia). - 5. *Lupinus pilosus* (Kithira). - 6. *Medicago arborea* (Kithira). - 7. *Medicago constricta* (Arkadias, Leonidion). - 8. *Medicago coronata* (Argolidos, Argos). - 9. *Medicago disciformis* (Amorgos). - 10. *Medicago intertexta* (Kerkira). - 11. *Medicago marina* (Rodos). - 12. *Medicago minima* (Arkadias, SE of Leonidion).

Plate 141

1. Fabaceae: *Medicago orbicularis* (Amorgos). - 2. *Medicago polymorpha* (Athos peninsula). - 3. *Medicago prostrata* (Grevenon, Exarchos). - 4. *Medicago rugosa* (Kithira). - 5. *Medicago sativa* subsp. *varia* (Loutra Arideas). - 6. *Medicago scutellata* (Samos). - 7. *Medicago truncatula* (Rethimnis, W of Agia Galini). - 8. *Medicago tuberculata* (Samos). - 9. *Melilotus albus* (Prespa area, inset from Loutra Arideas). - 10. *Melilotus creticus* (Samos). - 11. *Melilotus graecus* (Korinthias, Vouliagmeni lagoon). - 12. *Melilotus indicus* (Naxos).

Plate 142

1. Fabaceae: *Melilotus italicus* (Kerkira). - 2. *Melilotus neapolitanus* (Naxos). - 3. *Melilotus officinalis* (Samarina). - 4. *Melilotus siculus* (Samos). - 5. *Onobrychis aequidentata* (Naxos). - 6. *Onobrychis alba* subsp. *pentelica* (Parnonas). - 7. *Onobrychis aliacmonia* (Kozanis, Aliakmon dam). - 8. *Onobrychis arenaria* subsp. *lasiostachya* (Florinis, SW of Perea). - 9. *Onobrychis caput-galli* (Rodos). - 10. *Onobrychis crista-galli* (Samos). - 11. *Onobrychis ebenoides* (Iti). - 12. *Onobrychis gracilis* (Rodos).

Plate 143

1. Fabaceae: *Onobrychis montana* subsp. *scardica* (Pangeon). - 2. *Onobrychis peloponnesiaca* (Lakonias, Vlachiotis). - 3. *Onobrychis sphaciotica* (Lefka Ori). - 4. *Ononis ornithopodioides* (ESE of Navplio). - 5. *Ononis pubescens* (Xilokastro to Kato Trikala). - 6. *Ononis pusilla* (Mt Kitheronas). - 7. *Ononis ramosissima* (Karpathos). - 8. *Ononis reclinata* (Serron, NNW of Siderokastro). - 9. *Ononis spinosa* (Kriti, Thriptis Ori). - 10. *Ononis talaverae* (Gavdos). - 11. *Ononis variegata* (Kerkira). - 12. *Ononis verae* (W Kriti, Aradena gorge).

Plate 144

1. Fabaceae: *Ononis viscosa* subsp. *breviflora* (Amorgos). - 2. *Oxytropis pilosa* (Tzena). - 3. *Oxytropis purpurea* (Prespa area, above Oxia). - 4. *Pisum sativum* subsp. *biflorum* (Lakonias, Areopolis). - 5. *Podocytisus caramanicus* (Ioanninon, NE of Pirgos; inset of flowers from Konitsa, of fruits from Nestorio). - 6. *Scorpiurus muricatus* (E Kriti, W of Ierapetra). - 7. *Securigera globosa* (W Kriti, Samaria gorge). - 8. *Securigera parviflora* (Amorgos). - 9. *Securigera securidaca* (Lefkas, inset from Athos peninsula). - 10. *Securigera varia* (Ioanninon, Konitsa-Pades). - 11. *Spartium junceum* (Athos peninsula). - 12. *Sulla coronaria* (Sicily, Castellomare del Golfo).

Plate 145

1. Fabaceae: *Sulla spinosissima* (Naxos). - 2. *Teline monspessulana* (Ikaria). - 3. *Tetragonolobus purpureus* (Kriti, Rethimnis, N of Agia Galini). - 4. *Trifolium alpestre* (Iti). - 5. *Trifolium angustifolium* (Lesvos). - 6. *Trifolium argutum* (Kos). - 7. *Trifolium arvense* (Naxos). - 8. *Trifolium aurantiacum* (Lakonias, Areopolis). - 9. *Trifolium aureum* (above Metsovon). - 10. *Trifolium badium* (Gramos, Epano Arena). - 11. *Trifolium campestre* (Lefkas). - 12. *Trifolium cherleri* (Ilias, SSW of Killini).

Plate 146

1. Fabaceae: *Trifolium clypeatum* (Chios). - 2. *Trifolium dalmaticum* (Kerkira). - 3. *Trifolium dasyurum* (Tilos). - 4. *Trifolium dubium* (above Metsovon). - 5. *Trifolium fragiferum* (Litochoron railway station). - 6. *Trifolium globosum* (Athos peninsula). - 7. *Trifolium glomeratum* (Naxos). - 8. *Trifolium grandiflorum* (Prespa area). - 9. *Trifolium heldreichianum* (Falakron). - 10. *Trifolium hirtum* (Prespa area). - 11. *Trifolium hybridum* (Gramos). - 12. *Trifolium incarnatum* (Lakonias, Githion-Areopolis).

Plate 147

1. Fabaceae: *Trifolium infamia-ponertii* (Naxos). - 2. *Trifolium lappaceum* (Kerkira). - 3. *Trifolium lucanicum* (Prespa area). - 4. *Trifolium medium* subsp. *balcanicum* (Konitsis, Samarina-Fourka). - 5. *Trifolium noricum* subsp. *ottonis* (Vardousia). - 6. *Trifolium ochroleucon* (above Metsovon). - 7. *Trifolium ornithopodioides* (Lakonias, Viglafia). - 8. *Trifolium pallescens* (Voras). - 9. *Trifolium parnassi* (Timfristos). - 10. *Trifolium patens* (Prespa area). - 11. *Trifolium patulum* (Konitsis, Konitsa-Elefthero). - 12. *Trifolium petrisavii* (Samos).

Plate 148

1. Fabaceae: *Trifolium physodes* (Amorgos). - 2. *Trifolium pignantii* (Tzena). - 3. *Trifolium purpureum* (Samos). - 4. *Trifolium resupinatum* (Achaias, Kalavrita). - 5. *Trifolium spadiceum* (Rodopi). - 6. *Trifolium spumosum* (Amorgos). - 7. *Trifolium squarrosum* (Kerkira). - 8. *Trifolium stellatum* (Amorgos). - 9. *Trifolium subterraneum* (Achaias, Kalogria). - 10. *Trifolium sylvaticum* (Prespa area). - 11. *Trifolium tomentosum* (Chios). - 12. *Trifolium uniflorum* (Amorgos).

Plate 149

1. Fabaceae: *Trifolium vesiculosum* (Evrou, Metaxades). - 2. *Trifolium xanthinum* (Kerkira). - 3. *Trigonella caerulea* subsp. *procumbens* (Evrou, Valtos-Chandras). - 4. *Trigonella corniculata* subsp. *balansae* (Samos). - 5. *Trigonella gladiata* (Samos). - 6. *Trigonella rotundifolia* (Attikis, Lavrion). - 7. *Tripodion graecum* (Taigetos). - 8. *Tripodion tetraphyllum* (Arkadias, SE of Leonidion). - 9. *Vicia bithynica* (Kriti, Rethimnis, Melambes). - 10. *Vicia cuspidata* (Attikis, Parnitha; inset from Naxos). - 11. *Vicia hirsuta* (Prespa area). - 12. *Vicia hybrida* (Attikis, Imittos; inset from Lakonias, WNW of Itilo).

Plate 150

1. Fabaceae: *Vicia lutea* (Serron, village Kerkini; inset from Prespa area). - 2. *Vicia melanops* (Lakonias, Kelefa castle). - 3. *Vicia onobrychioides* (Prespa area, Mt Devas). - 4. *Vicia palaestina* (Tilos). - 5. *Vicia pannonica* subsp. *striata* (Korinthias, Lake Stimfalia). - 6. *Vicia pinetorum* (Naxos). - 7. *Vicia sativa 'angustifolia'* (Prespa area). - 8. *Vicia sepium* (Varnous). - 9. *Vicia serratifolia* (Prespa area). - 10. *Vicia sibthorpii* (NE Peloponnisos). - 11. *Vicia tenuifolia* (Chelmos, Stix ravine). - 12. *Vicia tetrasperma* (Prespa area).

Plate 151

1. Fabaceae: *Vicia villosa* subsp. *maniatissa* (Lakonias, Pirgos Dirou). - 2. Fagaceae: *Castanea sativa* (Florina to Vigla pass). - 3. *Fagus sylvatica* subsp. *sylvatica* (Athos peninsula; inset of unripe fruit from Evrou, Dadia-Sapka, of ripe fruit from Falakron). - 4. *Quercus aucheri* (Simi). - 5. *Quercus cerris* (Kastorias, Pefkofito). - 6. *Quercus coccifera* (NE Peloponnisos, Methana peninsula). - 7. *Quercus frainetto* (Florina to Vigla pass). - 8. *Quercus ilex* (Evvia, NNW of Kimi). - 9. *Quercus ithaburensis* subsp. *macrolepis* (Lakonias, Lambokambos). - 10. *Quercus pubescens* (Achaias, Diakopto-Kalavrita). - 11. *Quercus trojana* (Prespa area). - 12. Frankeniaceae: *Frankenia hirsuta* (Rodos).

Plate 152

1. Fumariaceae: *Corydalis blanda* subsp. *parnassica* (Parnassos). - 2. *Corydalis cava* (Varnous). - 3. *Corydalis integra* (Samos). - 4. *Corydalis solida* subsp. *incisa* (Parnassos). - 5. *Corydalis thasia* (Andros). - 6. *Corydalis uniflora* (Lefka Ori). - 7. *Corydalis wettsteinii* (Mt Cholomondas). - 8. *Fumaria capreolata* (Athos peninsula). - 9. *Fumaria densiflora* (Attikis, Kamariza). - 10. *Fumaria judaica* (Athens). - 11. *Fumaria macrocarpa* (Lakonias, Githion-Areopolis). - 12. *Fumaria parviflora* (Attikis, Kamariza).

Plate 153

1. Fumariaceae: *Fumaria petteri* (Naxos). - 2. *Hypecoum procumbens* (Samos). - 3. *Pseudofumaria alba* subsp. *leiosperma* (by Lake Megali Prespa). - 4. Gentianaceae: *Blackstonia perfoliata* (Evvia, Neos Pagontas to Psachna). - 5. *Centaurium erythraea* (Olimbos). - 6. *Centaurium maritimum* (Sithonia peninsula). - 7. *Centaurium tenuiflorum* (Naxos). - 8. *Gentiana asclepiadea* (unknown locality). - 9. *Gentiana cruciata* (Pangeon). - 10. *Gentiana lutea* (Iti). - 11. *Gentiana punctata* (Voras). - 12. *Gentiana verna* subsp. *balcanica* (Gramos, Epano Arena).

Plate 154

1. Gentianaceae: *Gentianella bulgarica* (Voras, Dobro Polje). - 2. *Gentianella crispata* (Olimbos). - 3. *Schenkia spicata* (Litochoron railway station). - 4. Geraniaceae: *Erodium absinthoides* (Prespa area, Mt Devas). - 5. *Erodium chrysanthum* (Achaias, Klokos). - 6. *Erodium ciconium* (Florinis, Aminteo). - 7. *Erodium cicutarium* (Kithira). - 8. *Erodium gruinum* (Chios). - 9. *Erodium hartvigianum* (Siatista-Galatini). - 10. *Erodium malacoides* (Athens). - 11. *Erodium vetteri* (Samos, Mt Kerki). - 12. *Geranium aristatum* (Varnous).

Plate 155

1. Geraniaceae: *Geranium asphodeloides* (Imathias, Naousa to Kato Vermio). - 2. *Geranium bohemicum* (Florinis, Vigla pass). - 3. *Geranium brutium* (Olympia; insert from Lakonias, Areopolis). - 4. *Geranium columbinum* (Florinis, Mt Triklario). - 5. *Geranium dissectum* (Lakonias, Areopolis). - 6. *Geranium kikianum* (Taigetos). - 7. *Geranium lucidum* (Ikaria). - 8. *Geranium macrorrhizum* (Olimbos). - 9. *Geranium macrostylum* (Timfristos). - 10. *Geranium molle* (Naxos). - 11. *Geranium peloponnesiacum* (Parnassos). - 12. *Geranium purpureum* (Lakonias, Githion-Areopolis).

Plate 156

1. Geraniaceae: *Geranium reflexum* (Varnous). - 2. *Geranium robertianum* (Athos peninsula). - 3. *Geranium sanguineum* (Olimbos). - 4. *Geranium subcaulescens* (Killini). - 5. *Geranium subcaulescens/subacutum* (Kitheronas). - 6. *Geranium sylvaticum* (Falakron). - 7. *Geranium tuberosum* (Parnassos, Livadia above Arachova). - 8. *Geranium versicolor* (above Metsovon). - 9. Gesneriaceae: *Haberlea rhodopensis* (Pangeon). - 10. *Jankaea heldreichii* (Olimbos). - 11. *Ramonda nathaliae* (Loutra Arideas). - 12. *Ramonda serbica* (North Macedonia).

Plate 157

1. Globulariaceae: *Globularia alypum* (Imittos). - 2. *Globularia bisnagarica* (Falakron). - 3. *Globularia cordifolia* (N Albania, Valbona). - 4. *Globularia stygia* (Killini). - 5. Grossulariaceae: *Ribes alpinum* (Rodopi, Megalo Livadi). - 6. *Ribes uva-crispa* subsp. *austro-europaeum* (Taigetos). - 7. Haloragaceae: *Myriophyllum spicatum* (Lake Mikri Prespa). - 8. Hamamelidaceae: *Liquidambar orientalis* (Rodos). - 9. Hippocastanaceae: *Aesculus hippocastanum* (Konstantinopolis, Hagia Sophia; inset from Ioanninon, Milea). - 10. Hippuridaceae: *Hippuris vulgaris* (Timfi). - 11. Hyacinthaceae: *Bellevalia brevipedicellata* (W Kriti). - 12. *Bellevalia ciliata* (Argolidos, Didima).

Plate 158

1. Hyacinthaceae: *Bellevalia dubia* subsp. *boissieri* (Kithira). - 2. *Bellevalia hyacinthoides* (Lakonias, Richea-Metamorfosis; inset from Argolidos, Mt Didimo). - 3. *Bellevalia juliana* (Kriti, Elounda). - 4. *Bellevalia romana* (unknown locality). - 5. *Bellevalia sitiaca* (E Kriti, Fanaromeni). - 6. *Bellevalia trifoliata* (Chios). - 7. *Drimia maritima* (Achaias, Kalogria; inset of flowers from Mani peninsula). - 8. *Hyacinthella leucophaea* subsp. *atchleyi* (Vourinos). - 9. *Muscari anatolicum* (Turkey, Akseki-Beysehir). - 10. *Muscari armeniacum* (Falakron). - 11. *Muscari botryoides/heldreichii* (Giona). - 12. *Muscari commutatum* (Imittos; inset from Achaias, Ano Kastritsi).

Plate 159

1. Hyacinthaceae: *Muscari comosum* (Amorgos). - 2. *Muscari kerkis* (Samos, Mt Kerki). - 3. *Muscari macrocarpum* (Samos). - 4. *Muscari neglectum* (Chalkidiki). - 5. *Muscari pulchellum* (Attikis, Vouliagmeni). - 6. *Muscari tenuiflorum* (Evrou, Valtos-Chandras). - 7. *Muscari weissii* (Rodos). - 8. *Ornithogalum arabicum* (Imittos). - 9. *Ornithogalum armeniacum* (Samos). - 10. *Ornithogalum atticum* (Thasos). - 11. *Ornithogalum collinum* (W foothills of Taigetos). - 12. *Ornithogalum creticum* (Kriti, Kotsifou gorge).

Plate 160

1. Hyacinthaceae: *Ornithogalum exscapum* (Kavalas, SE foot of Mt Pangeon). - 2. *Ornithogalum fimbriatum* (Evrou, NW of Lefkimi). - 3. *Ornithogalum gussonei* (Rodos). - 4. *Ornithogalum montanum* (Thesprotias, Ori Paramithias). - 5. *Ornithogalum narbonense* (Prespa area). - 6. *Ornithogalum naxense* (Naxos). - 7. *Ornithogalum nutans* (Chios). - 8. *Ornithogalum oligophyllum* (Varnous). - 9. *Ornithogalum oreoides* (Evrou, NW of Lefkimi). - 10. *Ornithogalum pyrenaicum* subsp. *sphaerocarpum* (Athos peninsula). - 11. *Ornithogalum refractum* (Prespa area). - 12. *Ornithogalum umbellatum* (Achaias, Pirgaki).

Plate 161

1. Hyacinthaceae: *Scilla autumnalis* (Imittos). - 2. *Scilla bifolia* (Mt Athos). - 3. *Scilla messeniaca* (W foothills of Taigetos). - 4. *Scilla nana* (Lefka Ori). - 5. Hydrocharitaceae: *Elodea canadensis* (Lake Megali Prespa). - 6. *Hydrocharis morsus-ranae* (Evros delta, inset from Lake Mikri Prespa). - 7. Hydrophyllaceae: *Phacelia tanacetifolia* (Chios). - 8. Hypericaceae: *Hypericum aciferum* (SW Kriti). - 9. *Hypericum aegypticum* subsp. *webbii* (unknown locality). - 10. *Hypericum amblycalyx* (E Kriti, Kato Zakro). - 11. *Hypericum athoum* (Thasos, Mt Ipsario). - 12. *Hypericum atomarium/icaricum* (Ikaria).

Plate 162

1. Hypericaceae: *Hypericum aucheri* (Evrou, Dadia). - 2. *Hypericum barbatum* (Ioanninon, Samarina-Fourka). - 3. *Hypericum cerastioides* (Dramas, Dipotama to Megalo Livadi). - 4. *Hypericum delphicum* (Evvia, Mt Ochi; inset from Steni Dirfios). - 5. *Hypericum empetrifolium* (Rodos). - 6. *Hypericum hircinum* (W Kriti, Nea Roumata). - 7. *Hypericum hirsutum* (Ioanninon, Konitsa-Elefthero). - 8. *Hypericum jovis* (unknown locality on Kriti). - 9. *Hypericum kelleri* (Lefka Ori). - 10. *Hypericum linarioides* subsp. *alpestre* (Falakron). - 11. *Hypericum maculatum* subsp. *immaculatum* (Rodopi). - 12. *Hypericum montbretii* (Beles, inset from Falakron).

Plate 163

1. Hypericaceae: *Hypericum olympicum* (Athos peninsula). - 2. *Hypericum perfoliatum* (Naxos). - 3. *Hypericum perforatum* (Kozanis, Neraida). - 4. *Hypericum rumeliacum* subsp. *apollonis* (Ioanninon, Theotokos). - 5. *Hypericum spruneri* (Kerkira). - 6. *Hypericum taygeteum* (Taigetos, Langada gorge). - 7. *Hypericum tetrapterum* (Prespa area, Sfika). - 8. *Hypericum thasium* (Xanthis, Lagos). - 9. *Hypericum trichocaulon* (Lefka Ori). - 10. *Hypericum triquetrifolium* (Lakonias, Githion-Skala). - 11. *Hypericum vesiculosum* (Evrou, Dadia-Pessani; inset from Parnassos). - 12. Iridaceae: *Crocus athous* (Athos peninsula).

Plate 164

1. Iridaceae: *Crocus biflorus* subsp. *alexandri* (Falakron). - 2. *Crocus biflorus* subsp. *melantherus* (Lakonias, Mt Koulochera). - 3. *Crocus biflorus* subsp. *nubigena* (Lesvos). - 4. *Crocus biflorus* subsp. *stridii* (Mt Chortiatis). - 5. *Crocus boryi* (Mt Koulochera, inset from Monemvassia). - 6. *Crocus cancellatus* subsp. *mazziaricus* (Samos, Karvouni; inset from Achaias, Klokos). - 7. *Crocus cartwrightianus* (Chanion, Akrotiri; inset from Lakonias, Cape Maleas). - 8. *Crocus chrysanthus* (Falakron). - 9. *Crocus cvijicii* (Gramos, Epano Arena). - 10. *Crocus flavus* (Grevena-Kalambaka). - 11. *Crocus fleischeri* (Chios). - 12. *Crocus georgei* (Arkadias, Langadia).

Plate 165

1. Iridaceae: *Crocus goulimyi* (Lakonias, Lambokambos; inset from Areopolis). - 2. *Crocus hadriaticus* (Ioanninon, Dodonis; inset from Kefallinia). - 3. *Crocus laevigatus* (NE Peloponnisos, Korfos; inset from Kithira). - 4. *Crocus nivalis* (Menalon). - 5. *Crocus niveus* (Arkadias, Kosmas-Leonidion; inset from Mani peninsula). - 6. *Crocus novicii* (Mt Douskon). - 7. *Crocus olivieri* (Chios). - 8. *Crocus oreocreticus* (Rethimnis, Spili-Gerakari; inset from Psiloritis). - 9. *Crocus orphei* (Falakron). - 10. *Crocus pallasii* (Chios). - 11. *Crocus pelistericus* (Voras). - 12. *Crocus pulchellus* (Falakron; inset from Bulgaria, SW of Kardjalik).

Plate 166

1. Iridaceae: *Crocus robertianus* (E of Ioannina). - 2. *Crocus sieberi* (Lefka Ori). - 3. *Crocus speciosus* subsp. *hellenicus* (Fokidos, Navpaktos-Lidorikon). - 4. *Crocus tournefortii* (unknown locality). - 5. *Crocus veluchensis* (Voras, with *C. pelistericus*; inset from Varnous). - 6. *Freesia leichtlinii* subsp. *alba* (Lasithiou, Mirsini-Tourloti). - 7. *Gladiolus anatolicus* (Rodos). - 8. *Gladiolus illyricus* (Ikaria). - 9. *Gladiolus imbricatus* (above Metsovon). - 10. *Gladiolus italicus* (Samos, inset from Chios). - 11. *Iris attica* (Imittos; inset from Argolidos, Mt Didima). - 12. *Iris germanica* (Prespa area).

Plate 167

1. Iridaceae: *Iris hellenica* (Chelmos). - 2. *Iris orientalis* (Lesvos). - 3. *Iris planifolia* (Sicily). - 4. *Iris pseudacorus* (Prespa area). - 5. *Iris reichenbachii* subsp. *reichenbachii* (Evrou, Dadia). - 6. *Iris reichenbachii subsp. athoa* (Athos peninsula). - 7. *Iris sintenisii* (Florina to Pisoderi pass). - 8. *Iris suaveolens* (Chios). - 9. *Iris tuberosa* (Rethimnis, Spili-Gerakari; inset from Kalavrita). - 10. *Iris unguicularis* subsp. *carica* (Arkadias, Tripoli-Sparti). - 11. *Iris unguicularis* subsp. *cretica* (Rethimnis, Spili-Gerakari). - 12. *Iris variegata* (Serbia, Deliblatska).

Plate 168

1. Iridaceae: *Moraea mediterranea* (Attikis, Lavrion). - 2. *Moraea sisyrinchium* (Kithira). - 3. *Romulea bulbocodium* (Rethimnis, Kotsifou gorge). - 4. *Romulea linaresii* subsp. *graeca* (Achaias, Fteri). - 5. *Romulea tempskyana* (Tilos). - 6. Juncaceae: *Juncus acutiflorus* (Voras, Dobro Polje). - 7. *Juncus acutus* (Achaias, Kalogria). - 8. *Juncus alpino-articulatus* (Voras). - 9. *Juncus articulatus* (Vasilitsa). - 10. *Juncus capitatus* (SE Peloponnisos). - 11. *Juncus compressus* (by Lake Megali Prespa). - 12. *Juncus conglomeratus* (by Lake Mikri Prespa).

Plate 169

1. Juncaceae: *Juncus effusus* (Rodopi). - 2. *Juncus inflexus* (NNW of Metsovon). - 3. *Juncus pygmaeus* (Achaias, Kalogria). - 4. *Juncus tenuis* (Voras). - 5. *Juncus thomasii* (Voras, inset from Metsovon area). - 6. *Juncus trifidus* (Voras). - 7. *Luzula campestris* (Voras). - 8. *Luzula forsteri* (Vardousia, inset from Olimbos). - 9. *Luzula luzuloides* (Voras). - 10. *Luzula multiflora* (Voras). - 11. *Luzula spicata* (Vardousia). - 12. *Luzula sylvatica* (Olimbos).

Plate 170

1. Juncaginaceae: *Triglochin barrelieri* (Messinias, Pilos). - 2. Lamiaceae: *Acinos alpinus* subsp. *meridionalis* (Taigetos). - 3. *Acinos graveolens* (Rodos). - 4. *Acinos nanus* (Kriti, Mesa Lakonia). - 5. *Acinos suaveolens* (Timfristos). - 6. *Ajuga bombycina* (Kastellorizo). - 7. *Ajuga chamaepitys* (Parnassos). - 8. *Ajuga genevensis* (Rodopi; inset from Florinis, Mt Triklario). - 9. *Ajuga iva* (Kerkira). - 10. *Ajuga laxmannii* (Falakron). - 11. *Ajuga orientalis* (Parnassos). - 12. *Ajuga piskoi* (Kozani).

Plate 171

1. Lamiaceae: *Ajuga pyramidalis* (Dramas, Megalo Livadi). - 2. *Ajuga reptans* (Varnous). - 3. *Ajuga salicifolia* subsp. *bassarabica* (Evrou, Chelidona). - 4. *Ballota acetabulosa* (Amorgos). - 5. *Ballota glandulosissima* (Kastellorizo). - 6. *Ballota nigra* (by Meteora monasteries). - 7. *Ballota pseudodictamnus* (W Kriti, Imbros gorge). - 8. *Betonica alopecuros* (Parnassos). - 9. *Betonica officinalis* (Serron, Mt Menikion). - 10. *Betonica scardica* (North Macedonia, Mt Pelister). - 11. *Calamintha cretica* (W Kriti, Sfakiano gorge). - 12. *Calamintha grandiflora* (Gramos, Epano Arena).

Plate 172

1. Lamiaceae: *Calamintha nepeta* (Evvia, Stropones). - 2. *Clinopodium dalmaticum* subsp. *bulgaricum* (Toxotes-Stavropouli). - 3. *Clinopodium vulgare* (Evrou, Loutros-Dadia). - 4. *Galeobdolon luteum* subsp. *montanum* (Varnous). - 5. *Galeopsis bifida* (Ioanninon, Chrisovitsa). - 6. *Galeopsis speciosa* (Rodopi). - 7. *Glechoma hederacea* (Prespa area). - 8. *Hyssopus officinalis* subsp. *aristatus* (Prespa area). - 9. *Lamium amplexicaule* (Kalavrita). - 10. *Lamium bifidum* subsp. *bifidum* (Achaias, Ano Kastritsi). - 11. *Lamium garganicum* subsp. *striatum* (Athos peninsula). - 12. *Lamium maculatum* (Varnous).

Plate 173

1. Lamiaceae: *Lamium moschatum* (Kastellorizo). - 2. *Lamium purpureum* (Gramos, Gramousta). - 3. *Lavandula stoechas* (Lakonias, Kriovrisi). - 4. *Leonurus cardiaca* (Prespa area). - 5. *Marrubium peregrinum* (Kilkis). - 6. *Marrubium thessalum* (Ossa, inset from Olimbos). - 7. *Marrubium velutinum* subsp. *velutinum* (Vardousia). - 8. *Marrubium vulgare* (Falakron, inset from Naxos). - 9. *Melissa officinalis* (Loutra Arideas). - 10. *Melittis melissophyllum* subsp. *albida* (Prespa area). - 11. *Mentha aquatica* (by Lake Mikri Prespa). - 12. *Mentha pulegium* (Litochoron railway station).

Plate 174

1. Lamiaceae: *Mentha spicata* (Litochoron railway station). - 2. *Mentha suaveolens* (Korinthias, Ano Trikala). - 3. *Micromeria hispida* (Kriti, Thriptis Ori). - 4. *Micromeria juliana* (Taigetos). - 5. *Micromeria myrtifolia* (Naxos). - 6. *Moluccella spinosa* (Lakonias, SE of Richea). - 7. *Nepeta argolica* subsp. *argolica* (Mt Kitheronas). - 8. *Nepeta camphorata* (Taigetos). - 9. *Nepeta melissifolia* (Gavdos). - 10. *Nepeta nuda* subsp. *albiflora* (Falakron). - 11. *Nepeta parnassica* (Parnassos). - 12. *Nepeta scordotis* (Kithira).

Plate 175

1. Lamiaceae: *Nepeta sphaciotica* (Lefka Ori). - 2. *Nepeta spruneri* (Gramos). - 3. *Origanum calcaratum* (Amorgos). - 4. *Origanum dictamnus* (unknown locality in Kriti). - 5. *Origanum microphyllum* (Lefka Ori). - 6. *Origanum onites* (Kastellorizo). - 7. *Origanum scabrum* (Evvia, SE Stropones). - 8. *Origanum sipyleum* (Samos). - 9. *Origanum vetteri* (Karpathos, Mt Kali Limni). - 10. *Origanum vulgare* subsp. *vulgare* (Rodopi). - 11. *Phlomis cretica* (Rethimnis, NW of Agia Galini; inset from S Chios). - 12. *Phlomis fruticosa* (Ioanninon, Aristi-Mesovouni).

Plate 176

1. Lamiaceae: *Phlomis herba-venti* subsp. *pungens* (Parnassos). - 2. *Phlomis lanata* (Kriti, Thriptis Ori). - 3. *Phlomis lycia* (Kastellorizo). - 4. *Phlomis samia* (Olimbos, inset from Mt Athos). - 5. *Phlomis tuberosa* (Evrou, Valtos-Chandras). - 6. *Prasium majus* (Sicily, Castellomare del Golfo). - 7. *Prunella cretensis* (unspecified locality on Kriti). - 8. *Prunella laciniata* (above Metsovon). - 9. *Prunella vulgaris* (Valia Kalda National Park). - 10. *Salvia aethiopis* (Kilkis; inset from Florinis, SW of Perea). - 11. *Salvia amplexicaulis* (Prespa area, Mt Devas). - 12. *Salvia argentea* (Prespa area, Mt Devas).

Plate 177

1. Lamiaceae: *Salvia candidissima* (Kastorias, WNW of Eptachori). - 2. *Salvia fruticosa* (Rethimnis, NW of Agia Galini). - 3. *Salvia glutinosa* (Rodopi). - 4. *Salvia officinalis* (Ioanninon, Aristi-Mesovouni). - 5. *Salvia pomifera* subsp. *pomifera/calycina* (Kithira). - 6. *Salvia pratensis* (Ioanninon, Aristi-Mesovouni). - 7. *Salvia ringens* (Kastorias, Eptachori, inset from Olimbos). - 8. *Salvia sclarea* (Voras, S foothills). - 9. *Salvia teddii* (Falakron). - 10. *Salvia tomentosa* (Evrou, Kirki-Alexandroupolis). - 11. *Salvia verbenaca* (Athos peninsula). - 12. *Salvia verticillata* (Rodopi).

Plate 178

1. Lamiaceae: *Salvia virgata* (Kerkira). - 2. *Salvia viridis* (Rethimnis, NW of Agia Galini). - 3. *Satureja athoa* (Mt Athos). - 4. *Satureja montana* (Olimbos). - 5. *Satureja pilosa* (Menikion). - 6. *Satureja spinosa* (Lefka Ori). - 7. *Satureja thymbra* (Rodos, inset from Kastellorizo). - 8. *Scutellaria albida* subsp. *vacillans* (Athos peninsula). - 9. *Scutellaria alpina* (Falakron). - 10. *Scutellaria columnae* (Serron, above Ano Poroia). - 11. *Scutellaria galericulata* (Voras, Dobro Polje). - 12. *Scutellaria hastifolia* (NW Peloponnisos).

Plate 179

1. Lamiaceae: *Scutellaria hirta* (Lasithiou, Thriptis Ori). - 2. *Scutellaria orientalis* (Prespa area, Mt Devas). - 3. *Scutellaria rupestris* subsp. *parnassica* (Killini). - 4. *Scutellaria sieberi* (main figure from unspecified locality in Kriti; inset from Rethimnis, Meronas). - 5. *Sideritis clandestina* subsp. *clandestina* (Taigetos). - 6. *Sideritis curvidens* (Rodos). - 7. *Sideritis euboea* (Evvia, Mt Dirfis). - 8. *Sideritis lanata* (Kos, inset from Samos). - 9. *Sideritis montana* (Prespa area, Mt Devas). - 10. *Sideritis purpurea* (Arkadias, Panagitsa). - 11. *Sideritis raeseri* (Parnassos, inset from Timfi). - 12. *Sideritis scardica* (Falakron).

Plate 180

1. Lamiaceae: *Sideritis sipylea* (Samos). - 2. *Sideritis syriaca* (Lefka Ori). - 3. *Stachys angustifolia* (Evrou, Dadia). - 4. *Stachys annua* (Grevenon, Paleokastro). - 5. *Stachys arvensis* (Athos peninsula). - 6. *Stachys candida* (Taigetos, Langada gorge). - 7. *Stachys canescens* (Messinias, Gialova lagoon). - 8. *Stachys chrysantha* (Malea peninsula). - 9. *Stachys cretica* subsp. *smyrnaea* (Kos). - 10. *Stachys germanica* subsp. *heldreichii* (Ioanninon, Aristi-Mesovouni). - 11. *Stachys ionica* (Lefkas). - 12. *Stachys iva* (Tzena).

Plate 181

1. Lamiaceae: *Stachys leucoglossa* (Athos peninsula). - 2. *Stachys macrotricha* (Grevenon, Felli). - 3. *Stachys menthifolia* (Zagori, Kapesovo-Vradeto; inset from Vikos gorge). - 4. *Stachys mollissima* (Kerkira). - 5. *Stachys mucronata* (Karpathos). - 6. *Stachys obliqua* (Evrou, Valtos-Chandras). - 7. *Stachys ocymastrum* (Sicily, Castellomare del Golfo). - 8. *Stachys palustris* (Evros delta). - 9. *Stachys plumosa* (Prespa area). - 10. *Stachys recta* subsp. *olympica* (Olimbos). - 11. *Stachys serbica* (Prespa area). - 12. *Stachys spinosa* (SW Kriti, Paleochora).

Plate 182

1. Lamiaceae: *Stachys spinulosa* (Lakonias, Githion-Areopolis). - 2. *Stachys spruneri* (Parnitha). - 3. *Stachys swainsonii* subsp. *swainsonii* (S foothills of Parnassos). - 4. *Stachys sylvatica* (Valia Kalda National Park). - 5. *Stachys tetragona* (Evvia, Metochi-Kimi). - 6. *Stachys tymphaea* (Varnous). - 7. *Stachys virgata* (SE Peloponnisos). - 8. *Teucrium alpestre* (SW Kriti opposite Elafonisi). - 9. *Teucrium aroanium* (Lakonias, Charakas). - 10. *Teucrium brevifolium* (S Chios). - 11. *Teucrium chamaedrys* subsp. *olympicum* (Olimbos). - 12. *Teucrium divaricatum* (Amorgos).

Plate 183

1. Lamiaceae: *Teucrium flavum* (Lakonias, Metamorfosi to Mt Koulochera). - 2. *Teucrium francisci-werneri* (Arkadias, Moni Elonis). - 3. *Teucrium halacsyanum* (Etolias-Akarnanias, Astakos-Mitikas). - 4. *Teucrium microphyllum* (Kriti, Xiloskala to Samaria gorge; inset from Tilos). - 5. *Teucrium montanum* (Samos, Karvouni). - 6. *Teucrium polium* subsp. *capitatum* (Vikos balcony). - 7. *Teucrium scordium* subsp. *scordioides* (Litochoron railway station). - 8. *Teucrium siculum* (Valia Kalda National Park). - 9. *Thymbra calostachya* (Kriti, Kapsa gorge). - 10. *Thymbra capitata* (Karpathos). - 11. *Thymbra spicata* (Fthiotidos, Thermopile). - 12. *Thymus atticus* (Evrou, Dadia-Loutros).

Plate 184

1. Lamiaceae: *Thymus boissieri* (Olimbos). - 2. *Thymus cilicicus* (Samos, Mt Karvouni). - 3. *Thymus comptus* (Evrou, W of Lefkimi). - 4. *Thymus hartvigii* (Vardousia). - 5. *Thymus laconicus* (Lakonias, Geraki-Alopochori). - 6. *Thymus leucotrichus* (Lefka Ori). - 7. *Thymus longicaulis* (Parnassos). - 8. *Thymus parnassicus* (Parnassos). - 9. *Thymus praecox* subsp. *jankae* (Voras). - 10. *Thymus sibthorpii* (Olimbos). - 11. *Thymus sipyleus* (Samos, Mt Kerki). - 12. *Thymus teucrioides* subsp. *alpinus* (Gramos).

Plate 185

1. Lamiaceae: *Thymus thracicus* (Voras). - 2. *Thymus zygioides* (Evrou, Dadia). - 3. *Ziziphora capitata* (Florinis, SW of Perea). - 4. *Ziziphora taurica* (Lesvos, Mt Olimbos). - 5. Lauraceae: *Laurus nobilis* (Athos peninsula). - 6. Lemnaceae: *Spirodela polyrhiza*, with *Salvinia natans* (Lake Kerkini). - 7. *Wolffia arrhiza*, with *Potamogeton* cf. *natans* (Kavala airport). - 8. Lentibulariaceae: *Pinguicula balcanica* (Tzena). - 9. *Pinguicula crystallina* subsp. *hirtiflora* (Ioanninon, NE of Milea). - 10. *Utricularia minor* (Lake Mikri Prespa). - 11. *Utricularia vulgaris* (Lake Stimfalia). - 12. Liliaceae: *Erythronium dens-canis* (Falakron).

Plate 186

1. Liliaceae: *Fritillaria bithynica* (Turkey, Muğla; inset from Samos). - 2. *Fritillaria carica* (Turkey, Samsun Dağ; inset from Chios). - 3. *Fritillaria conica* (Lakonias, Velanidia). - 4. *Fritillaria davisii* (Lakonias, Areopolis). - 5. *Fritillaria drenovskii* (Dramas, S of Granitis; inset from Falakron). - 6. *Fritillaria ehrhartii* (Evvia, W of Karistos). - 7. *Fritillaria epirotica* (Smolikas). - 8. *Fritillaria euboeica* (Evvia, Mt Dirfis). - 9. *Fritillaria graeca* subsp. *graeca* (Taigetos; inset from Korinthias, Angelokastro). - 10. *Fritillaria graeca* subsp. *thessala* (Katara pass). - 11. *Fritillaria gussichiae* (North Macedonia, Veles-Prilep). - 12. *Fritillaria macedonica* (Olimbos).

Plate 187

1. Liliaceae: *Fritillaria messanensis* subsp. *messanensis* (Olimbos). - 2. *Fritillaria montana* (Florinis, Mt Boutsi). - 3. *Fritillaria obliqua* subsp. *tuntasiana* (Kithnos). - 4. *Fritillaria pontica* (Athos peninsula). - 5. *Fritillaria rhodia* (Rodos). - 6. *Fritillaria rhodocanakis* (Idra). - 7. *Fritillaria sibthorpiana* (Turkey, Muğla, E of Bozburu). - 8. *Gagea amblyopetala* (Achaias, E of Pirgaki). - 9. *Gagea dubia* (Argolidos, Mt Didimo). - 10. *Gagea fragifera* (Falakron). - 11. *Gagea graeca* (Naxos). - 12. *Gagea heldreichii* (Kriti, Irakliou, Kofinas Oros).

Plate 188

1. Liliaceae: *Gagea lutea* (Voras, Dobro Polje). - 2. *Gagea minima* (Falakron). - 3. *Gagea omalensis* (Kriti, Chanion, Omalos). - 4. *Gagea peduncularis* (Korinthias, Angeolokastro). - 5. *Gagea rigida* (Kithira). - 6. *Gagea villosa* (Korinthias, Angelokastro). - 7. *Lilium albanicum* (Valia Kalda National Park, inset from Gramos). - 8. *Lilium candidum* (Kilkis, Terpillos). - 9. *Lilium chalcedonicum* (Olimbos). - 10. *Lilium martagon* (Voras). - 11. *Lilium rhodopeum* (Rodopi). - 12. *Tulipa agenensis* (Chios).

Plate 189

1. Liliaceae: *Tulipa australis* (Killini, inset from Valia Kalda National Park). - 2. *Tulipa bakeri* (probably from W Kriti, Omalos). - 3. *Tulipa bithynica* (Turkey, Balıkesir-Erdemit). - 4. *Tulipa cretica* (Lasithiou, Ag. Ioannis to Shinokapsala). - 5. *Tulipa doerfleri* (Rethimnis, Spili-Gerakari). - 6. *Tulipa goulimyi* (Lakonias, Kriovrisi; inset from Leonidion). - 7. *Tulipa hageri* (foothills of Taigetos). - 8. *Tulipa orphanidea* (Arkadias, NE of Levidi; inset from Chelmos). - 9. *Tulipa praecox* (Chios). - 10. *Tulipa saxatilis* (Rethimnis, Spili). - 11. *Tulipa undulatifolia* (S Chios). - 12. Linaceae: *Linum arboreum* (Rethimnis, Kotsifou gorge; inset of fruits from Rodos).

Plate 190

1. Linaceae: *Linum austriacum* (Mt Athos). - 2. *Linum bienne* (Ioanninon, Miliotades). - 3. *Linum capitatum* (Rodopi). - 4. *Linum catharticum* (Ioanninon, Samarina-Fourka). - 5. *Linum decumbens* (Idra). - 6. *Linum elegans* (Athos peninsula). - 7. *Linum flavum* subsp. *albanicum* (Timfi). - 8. *Linum gyaricum* subsp. *icaricum* (Ikaria). - 9. *Linum hellenicum* (Lakonias, Lira). - 10. *Linum hirsutum* (Prespa area, Mt Devas). - 11. *Linum hologynum* (above Metsovon). - 12. *Linum leucanthum* (Imittos).

Plate 191

1. Linaceae: *Linum maritimum* (Litochoron railway station). - 2. *Linum nodiflorum* (Kerkira). - 3. *Linum pubescens* subsp. *sibthorpianum* (Kerkira). - 4. *Linum punctatum* subsp. *pycnophyllum* (Gramos, Epano Arena). - 5. *Linum strictum* subsp. *spicatum* (Amorgos). - 6. *Linum tenuifolium* (Prespa area, Mt Devas). - 7. *Linum virgultorum* (Rodos). - 8. Loranthaceae: *Loranthus europaeus* (Sicily, Madonie mts). - 9. Lythraceae: *Ammannia coccinea* (Kalamata). - 10. *Lythrum junceum* (Rodos). - 11. *Lythrum salicaria* (Lakonias, NE of Githion). - 12. *Lythrum virgatum* (Evrou, WSW of Ellinochori; inset from Kavalas, Keramoti).

Plate 192

1. Malvaceae: *Abutilon theophrasti* (Evrou, WSW of Ellinochori). - 2. *Alcea biennis* (Ikaria). - 3. *Alcea heldreichii* (Samos). - 4. *Alcea setosa* (Florinis, Ag. Athanasios). - 5. *Althaea officinalis* (Litochoron railway station). - 6. *Hibiscus trionum* (Rodopis, Stilario). - 7. *Malva arborea* (Attikis, Lavrion). - 8. *Malva cretica* (Samos). - 9. *Malva moschata* (above Metsovon). - 10. *Malva multiflora* (Naxos). - 11. *Malva neglecta* (Prespa area). - 12. *Malva parviflora* (Kriti, Lasithiou, Vai).

Plate 193

1. Malvaceae: *Malva punctata* (Evvia, Limni). - 2. *Malva setigera* (Prespa area). - 3. *Malva sylvestris* (Chios). - 4. *Malva thuringiaca* (Evritanias, SW of Agrafa). - 5. *Malva unguiculata* (Kriti, Chanion, Vathi). - 6. Melanthiaceae: *Veratrum lobelianum* (Bulgaria, Vitosha; inset of flowers from Metsovon, of fruits from Gramos). - 7. Menyanthaceae: *Nymphoides peltata* (Lake Kerkini). - 8. Molluginaceae: *Glinus lotoides* (Timfi area). - 9. *Mollugo cerviana* (E of Asprovalta). - 10. Moraceae: *Broussonetia papyrifera* (Serron, Kerkini). - 11. *Ficus carica* (Karpathos). - 12. Morinaceae: *Morina persica* (Taigetos).

Plate 194

1. Myrtaceae: *Myrtus communis* (W Kriti, Nea Roumata; inset left from Kalogria, right from Thesprotias, Kallithea). - 2. Nartheciaceae: *Narthecium scardicum* (Grevenon, Mavrovouni). - 3. Nymphaeaceae: *Nuphar lutea* (Lake Mikri Prespa). - 4. *Nymphaea alba* (Lake Kerkini). - 5. Oleaceae: *Fraxinus angustifolia* (Evrou, Dadia) . - 6. *Fraxinus ornus* (Athos peninsula, inset from Prespa area). - 7. *Jasminum fruticans* (Farangi by Limni Vegoritis). - 8. *Ligustrum vulgare* (Loutra Arideas, inset from Vertiskos). - 9. *Olea europaea* subsp. *oleaster* (Athos peninsula). - 10. *Phillyrea latifolia* (Imathias, Vergina). - 11. *Syringa vulgaris* (Pangeon). - 12. Onagraceae: *Circaea lutetiana* (Mt Pieria).

Plate 195

1. Onagraceae: *Epilobium angustifolium* (Varnous). - 2. *Epilobium dodonaei* (Voras, inset from Ioanninon, Milea). - 3. *Epilobium hirsutum* (Evvia, Dirfis-Xirovouni). - 4. *Epilobium parviflorum* (Naxos). - 5. *Epilobium vernonicum* (Voras). - 6. *Oenothera indecora* (Litochoron railway station). - 7. Orchidaceae: *Anacamptis boryi* (Parnonas). - 8. *Anacamptis collina* (Rethimnis, NW of Agia Galini). - 9. *Anacamptis coriophora* (Rodos). - 10. *Anacamptis laxiflora* (Achaias, Kalavrita). - 11. *Anacamptis morio* subsp. *caucasica* (above Metsovon). - 12. *Anacamptis papilionacea* subsp. *aegaea* (Messinias, NE of Pilos).

Plate 196

1. Orchidaceae: *Anacamptis pyramidalis* (Chios). - 2. *Anacamptis sancta* (Rodos). - 3. *Cephalanthera damasonium* (Metsovon, inset from Mt Iti). - 4. *Cephalanthera epipactoides* (Evrou, Dadia-Pessani). - 5. *Cephalanthera longifolia* (Killini). - 6. *Cephalanthera rubra* (Dramas, Prosotsani-Volakas). - 7. *Coeloglossum viride* (Falakron). - 8. *Comperia comperiana* (Lesvos, Mt Olimbos). - 9. *Corallorhiza trifida* (Olimbos, inset from Mt Paikon). - 10. *Dactylorhiza baumanniana* (Iti). - 11. *Dactylorhiza cordigera* (Gramos, Epano Arena). - 12. *Dactylorhiza kalopissii* (above Metsovon).

Plate 197

1. Orchidaceae: *Dactylorhiza romana* (Samos). - 2. *Dactylorhiza saccifera* (above Metsovon). - 3. *Dactylorhiza sambucina* (Varnous). - 4. *Epipactis atrorubens* (Olimbos). - 5. *Epipactis cretica* (Kriti, Kato Symi). - 6. *Epipactis halacsyi* (Taigetos). - 7. *Epipactis helleborine* (Olimbos). - 8. *Epipactis microphylla* (Pangeon). - 9. *Epipactis palustris* (Kastorias, Eptachori-Pendalofos). - 10. *Epipactis subclausa* (Olimbos). - 11. *Epipogium aphyllum* (Olimbos). - 12. *Goodyera repens* (Rodopi).

Plate 198

1. Orchidaceae: *Gymnadenia conopsea* (Rodopi). - 2. *Gymnadenia frivaldii*, with *Dactylorhiza cordigera* (Voras). - 3. *Gymnadenia odoratissima* (Falakron). - 4. *Gymnadenia rehllicani* (Montenegro, Mt Durmitor). - 5. *Himantoglossum jankae* (Iti). - 6. *Himantoglossum robertianum* (Kriti, Rethimnis, Melambes). - 7. *Limodorum abortivum* (Rodos). - 8. *Listera cordata* (Rodopi). - 9. *Listera ovata* (Ioanninon, SW of Fourka; inset from Valia Kalda National Park). - 10. *Neotinea lactea* (Kriti, Rethimnis, Spili-Gerakari). - 11. *Neotinea maculata* (Chios, Mt Pelineo). - 12. *Neotinea tridentata* (Prespa area, Mt Devas).

Plate 199

1. Orchidaceae: *Neotinea ustulata* (Prespa area, Mt Boutsi, inset from Mt Devas). - 2. *Neottia nidus-avis* (Olimbos, inset from above Metsovon). - 3. *Ophrys apifera* (Prespa area). - 4. *Ophrys bertolonii* (Sicily, Segesta). - 5. *Ophrys bombyliflora* (Rethimnis, Melambes). - 6. *Ophrys* cretica (Lasithiou, Mirsini-Tourloti). - 7. *Ophrys delphinensis* (Achaias, Agridi). - 8. *Ophrys ferrum-equinum* (Lakonias,WNW Itilo). - 9. *Ophrys fuciflora/holoserica*, cf. notes (Rethimnis, Agia Galini). - 10. *Ophrys fusca* subsp. *cinerophila* (Rethimnis, Agia Galini). - 11. *Ophrys fusca* subsp. *fusca* (Attikis, Mt Imittos). - 12. *Ophrys fusca* subsp. *iricolor* (Chios).

Plate 200

1. Orchidaceae: *Ophrys helenae* (Ioanninon, Klimatia). - 2. *Ophrys insectifera* (Ioanninon, Kerasovo). - 3. *Ophrys lutea* subsp. *galilaea* (Chios). - 4. *Ophrys omegaifera* (Rethimnis, Melambes). - 5. *Ophrys regis-ferdinandii* (Chios). - 6. *Ophrys reinholdii* (Rodos). - 7. *Ophrys scolopax* subsp. *cornuta* (Kastorias, W of Eptachori). - 8. *Ophrys scolopax* subsp. *heldreichii* (Rethimnis, Melambes). - 9. *Ophrys scolopax* subsp. *rhodia* (Rodos). - 10. *Ophrys speculum* (Spain, Andalusia, Alora). - 11. *Ophrys sphegodes* subsp. *aesculapii* (Messinias, Megalopolis-Kalamata). - 12. *Ophrys sphegodes* subsp. *cretensis* (Lasithiou, Mirsini-Tourloti).

Plate 201

1. Orchidaceae: *Ophrys sphegodes* subsp. *epirotica* (N of Konitsa). - 2. *Ophrys sphegodes* subsp. *mammosa* (Athos peninsula). - 3. *Ophrys sphegodes* subsp. *sphegodes* (Prespa area, Mt Devas). - 4. *Ophrys sphegodes* subsp. *spruneri* (Rethimnis, Agia Galini). - 5. *Ophrys tenthredinifera* (Mallorca). - 6. *Ophrys umbilicata* subsp. *bucephala* (Chios). - 7. *Orchis anatolica* (Chios). - 8. *Orchis anthropophora* (Killini). - 9. *Orchis italica* (Chios). - 10. *Orchis mascula* (Florinis, Mt Sfika). - 11. *Orchis militaris* (Falakron). - 12. *Orchis pallens* (Iti).

Plate 202

1. Orchidaceae: *Orchis pauciflora* (Athos peninsula). - 2. *Orchis provincialis* (Chios). - 3. *Orchis punctulata*, not typical, perhaps *O. punctulata* × *O. purpurea* (Evrou, NE of Alexandroupolis). - 4. *Orchis purpurea* (Kozanis, Tsotili). - 5. *Orchis quadripunctata* (Florinis, Mt Devas; inset from Mt Athos). - 6. *Orchis simia* (Florinis, Mt Devas). - 7. *Orchis sitiaca* (E Kriti, Katharo plateau). - 8. *Orchis spitzelii* subsp. *spitzelii* (Chelmos). - 9. *Platanthera bifolia* (Voras). - 10. *Platanthera chlorantha* (Imathias, Naousa to Kato Vermio). - 11. *Pseudorchis albida* (Orvilos). - 12. *Serapias bergonii* (Rethimnis, Melambes).

Plate 203

1. Orchidaceae: *Serapias lingua* (Naxos). - 2. *Serapias orientalis* (Kithira). - 3. *Serapias vomeracea* (Attiki, Lavrion). - 4. *Spiranthes spiralis* (Thessalonikis, Mt Vertiskos). - 5. Orobanchaceae: *Cistanche phelypaea* (Koufonisi SE of Kriti). - 6. *Orobanche alba* (Florinis, Mt Triklario). - 7. *Orobanche amethystea* (Malea peninsula, Velanidia). - 8. *Orobanche baumanniorum* (Mt Athos). - 9. *Orobanche crenata* (Kerkira). - 10. *Orobanche elatior* (Prespa area, Mt Devas). - 11. *Orobanche gracilis* (Kithira). - 12. *Orobanche hederae* (Samos).

Plate 204

1. Orobanchaceae: *Orobanche minor* (Arkadias, Leonidion-Astros). - 2. *Orobanche pubescens* (Rodos). - 3. *Phelipanche lavandulacea* (Athos peninsula). - 4. *Phelipanche mutelii* var. *nana* (Amorgos). - 5. *Phelipanche mutelii* var. *oxyloba* (Korinthias, Loutraki). - 6. *Phelipanche nowackiana* (Katara pass). - 7. *Phelypaea boissieri* (Prespa area, Mt Devas). - 8. Oxalidaceae: *Oxalis acetosella* (Rodopi). - 9. *Oxalis articulata* (Kerkira). - 10. *Oxalis corniculata* (Naxos). - 11. *Oxalis pes-caprae* (E Kriti, Vai). - 12. Paeoniaceae: *Paeonia clusii* subsp. *clusii* (Lefka Ori).

Plate 205

1. Paeoniaceae: *Paeonia clusii* subsp. *rhodia* (Rodos). - 2. *Paeonia corsica* (Sicily, Madonie mts). - 3. *Paeonia mascula* subsp. *hellenica* (Evvia, Mt Ochi). - 4. *Paeonia mascula* subsp. *mascula* (Samos). - 5. *Paeonia parnassica* (Parnassos). - 6. *Paeonia peregrina* (Lefkas). - 7. Papaveraceae: *Chelidonium majus* (Prespa National Park). - 8. *Glaucium corniculatum* (Rethimnis, Agia Galini). - 9. *Glaucium flavum* (Athos peninsula; inset from Attiki, Lavrion). - 10. *Papaver dubium* subsp. *lecoqii* (Prespa area, Mt Sfika; inset from Iti). - 11. *Papaver hybridum* (Lakonias, Itilo). - 12. *Papaver nigrotinctum* (Naxos).

Plate 206

1. Papaveraceae: *Papaver rhoeas* (Kilkis). - 2. *Papaver somniferum* (Lefkas). - 3. Parnassiaceae: *Parnassia palustris* (Varnous). - 4. Phytolaccaceae: *Phytolacca americana* (Serron, Kerkini). - 5. Plantaginaceae: *Plantago afra* (Rodos, inset from Malea peninsula). - 6. *Plantago albicans* (Argolidos, Tolon). - 7. *Plantago amplexicaulis* (Cape Sounion). - 8. *Plantago argentea* (Prespa area, Mt Sfika). - 9. *Plantago atrata* subsp. *graeca* (Timfi). - 10. *Plantago bellardii* (Cape Sounion). - 11. *Plantago coronopus* subsp. *weldenii* (Rodos). - 12. *Plantago cretica* (Rodos).

Plate 207

1. Plantaginaceae: *Plantago indica* (Kavalas, Keramoti). - 2. *Plantago lagopus* (Athens). - 3. *Plantago lanceolata* (Bulgaria, Sofia). - 4. *Plantago major* subsp. *intermedia* (Lakonias, Githion-Areopolis). - 5. *Plantago maritima* subsp. *crassifolia* (Lakonias, Viglafia). - 6. *Plantago media* subsp. *pindica* (Mt Milea, inset from Falakron). - 7. *Plantago phaeostoma* (Koufonisi off SE Kriti). - 8. *Plantago reniformis* (Montenegro, Mt Durmitor). - 9. Platanaceae: *Platanus orientalis* (Thasos). - 10. Plumbaginaceae: *Acantholimon aegaeum* (Samos). - 11. *Acantholimon androsaceum* (Lefka Ori). - 12. *Acantholimon graecum* (Taigetos).

Plate 208

1. Plumbaginaceae: *Armeria canescens* (Evvia, Mt Ochi; inset from Katara pass). - 2. *Armeria sancta* (Athos peninsula). - 3. *Goniolimon tataricum* (Prespa area). - 4. *Limonium ammophilon* (Rodos). - 5. *Limonium carpathum* (Karpathos). - 6. *Limonium frederici* (W Kriti, Gramvousa). - 7. *Limonium narbonense/compactum* (Evvia, Chiliadou beach). - 8. *Limonium roridum* (Naxos). - 9. *Limonium sinuatum* (Samos). - 10. *Limonium virgatum* (Naxos). - 11. *Plumbago europaea* (Kastoria). - 12. Poaceae: *Achnatherum calamagrostis* (Olimbos; inset from Ioanninon, SE of Fourka).

Plate 209

1. Poaceae: *Aegilops biuncialis* (unknown locality). - 2. *Aegilops comosa* (Evvia, Neos Pagontas to Psachna). - 3. *Aegilops cylindrica* (Bulgaria, Sofia). - 4. *Aegilops geniculata* (Korinthias, Heraion). - 5. *Aegilops neglecta* (Kerkira). - 6. *Aegilops speltoides* (Evrou, Valtos-Chandras). - 7. *Aeluropus lagopoides* (E Kriti, Cape Sidero). - 8. *Aeluropus littoralis* (Lesvos, Alikes Kallonis). - 9. *Agrostis rupestris* (Voras). - 10. *Aira elegans* (Ikaria). - 11. *Alopecurus aequalis* (by lake Megali Prespa). - 12. *Alopecurus gerardi* (Parnassos).

Plate 210

1. Poaceae: *Alopecurus pratensis* (N of Metsovon). - 2. *Alopecurus rendlei* (Argolidos, Didima). - 3. *Andropogon distachyos* (Samos). - 4. *Anthoxanthum odoratum* (Vardousia). - 5. *Apera spica-venti* (Evrou, Valtos-Chandras). - 6. *Arrhenatherum palaestinum* (Peloponnisos). - 7. *Arundo donax* (WNW of Igoumenitsa). - 8. *Arundo plinii* (Rodos; inset from Achaias, Diakopto-Kalavrita). - 9. *Avena barbata* (Rethimnis, Agia Galini). - 10. *Avena sterilis* (Kithira). - 11. *Bothriochloa ischaemum* (Konitsa). - 12. *Brachypodium distachyon* (Naxos).

Plate 211

1. Poaceae: *Briza maxima* (Athos peninsula, inset from Peloponnisos). - 2. *Briza minor* (Peloponnisos). - 3. *Bromus diandrus* (Argolidos, Tolon). - 4. *Bromus racemosus* (Prespa National Park). - 5. *Bromus squarrosus* (Prespa National Park). - 6. *Bromus tectorum* (Prespa National Park). - 7. *Castellia tuberculosa* (Peloponnisos). - 8. *Catapodium marinum* (Argolidos, Tolon). - 9. *Cenchrus longispinus* (Kavalas, Loutra Eleftherion). - 10. *Chrysopogon gryllus* (W of Florina). - 11. *Crypsis alopecuroides* (by Lake Kerkini). - 12. *Cynodon dactylon* (Konitsa).

Plate 212

1. Poaceae: *Cynosurus cristatus* (above Metsovon). - 2. *Cynosurus echinatus* (Kastellorizo). - 3. *Cynosurus effusus* (Korinthias, Ano Trikala). - 4. *Dactylis glomerata* subsp. *hispanica* (Athos peninsula). - 5. *Danthoniastrum compactum* (Timfi, inset from Taigetos). - 6. *Dasypyrum hordeaceum* (Taigetos). - 7. *Dasypyrum villosum* (Lakonias, Githion-Areopolis). - 8. *Deschampsia cespitosa* (Iti). - 9. *Digitaria sanguinalis* (Xanthis, Paschalia). - 10. *Echinaria capitata* (Tilos). - 11. *Echinochloa crus-galli* (Lakonias: Githion-Areopolis). - 12. *Eleusine indica* (Xanthis, Paschalia).

Plate 213

1. Poaceae: *Elymus caninus* (Gramos). - 2. *Elytrigia intermedia* (Korinthias, Xilokastro to Kato Trikala). - 3. *Elytrigia repens* (Gramos). - 4. *Eragrostis cilianensis* (Xanthis, Paschalia). - 5. *Festuca drymeja* (Valia Kalda National Park). - 6. *Festuca olympica* (Olimbos). - 7. *Festuca paniculata* (Voras; inset from Bulgaria, Mt Vitosha). - 8. *Festuca valesiaca* (Prespa area). - 9. *Festucopsis sancta* (Pangeon, inset from Mt Athos). - 10. *Glyceria fluitans* (by lake Megali Prespa). - 11. *Helictochloa agropyroides* (Lakonias, Vlachiotis). - 12. *Helictotrichon convolutum* (Kerkira).

Plate 214

1. Poaceae: *Holcus lanatus* (Florinis, Kotas). - 2. *Hordelymus europaeus* (Gramos). - 3. *Hordeum bulbosum* (Metsovon; inset from Korinthias, Ano Trikala). - 4. *Hordeum murinum* (Naxos). - 5. *Hordeum vulgare* subsp. *spontaneum* (Chios; inset from Attiki, Kaki Thalassa). - 6. *Imperata cylindrica* (Lakonias, NE of Githion). - 7. *Lagurus ovatus* (Rodos). - 8. *Lamarckia aurea* (Athens). - 9. *Lolium temulentum* (Peloponnisos). - 10. *Lygeum spartum* (Lasithiou, Agia Fotia). - 11. *Maillea crypsoides* (Kriti, Elafonisos). - 12. *Melica ciliata* (Parnassos).

Plate 215

1. Poaceae: *Melica minuta* (Athos peninsula). - 2. *Melica uniflora* (Evrou, NW of Lefkimi). - 3. *Milium effusum* (Rodopi, inset from Gramos). - 4. *Nardus stricta* (Varnous). - 5. *Parapholis incurva* (Argolidos, Tolon; inset from Attikis, Lavrion). - 6. *Paspalum distichum* (by Lake Kerkini). - 7. *Phacelurus digitatus* (Evros delta). - 8. *Phalaris minor* (Samos). - 9. *Phleum alpinum* (Taigetos). - 10. *Phleum hirsutum* (Prespa area). - 11. *Phleum phleoides* (Parnassos). - 12. *Phragmites australis* (Lakonias, SW of Skala).

Plate 216

1. Poaceae: *Phragmites frutescens* (Lakonias, SE of Skala). - 2. *Piptatherum holciforme* (Prespa National Park). - 3. *Poa bulbosa* (Chios). - 4. *Poa infirma* (Naxos). - 5. *Poa jubata* (W Peloponnisos). - 6. *Rostraria cristata* (Athens). - 7. *Saccharum ravennae* (Kavalas, Keramoti; inset of culm from Loutra Arideas). - 8. *Secale strictum* (Varnous, Vigla pass). - 9. *Sesleria achtarovii* (Falakron). - 10. *Sesleria robusta* (Prespa area, Mt Sfika). - 11. *Sesleria tenerrima* (Parnassos). - 12. *Sesleria vaginalis* (Timfristos; inset from Valia Kalda National Park).

Plate 217

1. Poaceae: *Setaria viridis* (Konitsa). - 2. *Sorghum halepense* (Patras, inset from Konitsa). - 3. *Sporobolus pungens* (Lakonias, Elafonisos). - 4. *Stipa capensis* (Lakonias, Areopolis-Githion). - 5. *Stipa capillata* (Prespa area, Mt Devas; inset from Serbia, Deliblatska). - 6. *Stipa pennata* (Mt Parnonas). - 7. *Taeniatherum caput-medusae* (Evrou, Metaxades). - 8. *Tragus racemosus* (Vertiskos). - 9. *Trisetaria aurea* (Kerkira). - 10. Polygalaceae: *Polygala alpestris* subsp. *croatica* (Gramos, Epano Arena). - 11. *Polygala comosa* (Ioanninon, Miliotades). - 12. *Polygala major* (Timfristos).

Plate 218

1. Polygalaceae: *Polygala monspeliaca* (Naxos). - 2. *Polygala nicaeensis* (Prespa National Park). - 3. *Polygala rausiana* (Korinthias, Evrostina). - 4. *Polygala subuniflora* (Chelmos, Stix ravine). - 5. *Polygala supina* subsp. *rhodopaea* (Serron, NNW of Siderokastro). - 6. *Polygala venulosa* (E Kriti, Sitia). - 7. Polygonaceae: *Aconogonon alpinum* (Varnous). - 8. *Atraphaxis billardierei* (Chios, inset from Samos). - 9. *Emex spinosa* (Messinias, Methoni). - 10. *Oxyria digyna* (Vardousia). - 11. *Persicaria amphibia* (Lake Stimfalia). - 12. *Persicaria bistorta* (Voras).

Plate 219

1. Polygonaceae: *Persicaria decipiens* (Pellis, Aridea). - 2. *Persicaria hydropiper* (by lake Megali Prespa). - 3. *Persicaria lapathifolia* (by lake Kerkini). - 4. *Persicaria mitis* (by lake Kerkini). - 5. *Polygonum arenastrum* (Timfi). - 6. *Polygonum aviculare* (Prespa National Park). - 7. *Polygonum equisetiforme* (Patras). - 8. *Polygonum icaricum* (Samothraki). - 9. *Polygonum idaeum* (Psiloritis). - 10. *Polygonum maritimum* (Lakonias, Githion-Skala). - 11. Rumex alpinus (Bulgaria, Mt Vitosha; inset from North Macedonia, Mt Pelister). - 12. *Rumex balcanicus* (Voras).

Plate 220

1. Polygonaceae: *Rumex bucephalophorus* subsp. *aegaeus* (Kithira). - 2. *Rumex cristatus* (Prespa National Park). - 3. *Rumex hydrolapathum* (by lake Mikri Prespa). - 4. *Rumex obtusifolius* (Mt Athos). - 5. *Rumex palustris* (Prespa area). - 6. *Rumex pulcher* subsp. *raulinii* (Prespa area). - 7. *Rumex scutatus* (Smolikas). - 8. *Rumex tuberosus* subsp. *creticus* (Chios). - 9. *Rumex vesicarius* (Korinthias, by lake Vouliagmeni). - 10. Portulacaceae: *Montia arvensis* (Varnous). - 11. *Portulaca oleracea* (Naousa). - 12. Posidoniaceae: *Posidonia oceanica* (Naxos).

Plate 221

1. Potamogetonaceae: *Potamogeton crispus* (Lake Megali Prespa, with *Azolla filiculoides*). - 2. *Potamogeton lucens* (Lake Megali Prespa). - 3. *Potamogeton natans* (Lake Mikri Prespa). - 4. *Potamogeton perfoliatus* (Lake Doirani). - 5. *Potamogeton pusillus* (Voras, Dobro Polje). - 6. Primulaceae: *Anagallis arvensis* (Kithira; inset from Kriti, Rethimnis, Plakias beach). - 7. *Anagallis foeminea* (Rodos). - 8. *Androsace maxima* (Florinis, Maniakion). - 9. *Androsace villosa* (Falakron). - 10. *Asterolinon linum-stellatum* (Kithira). - 11. *Coris monspeliensis* (Peloponnisos). - 12. *Cyclamen creticum* (Chanion, NE of Kandanos).

Plate 222

1. Primulaceae: *Cyclamen graecum* (Lakonias, Areopolis; inset leaves of *C. hederifolium* left and *C. graecum* right). - 2. *Cyclamen hederifolium* (Grevenon, S of Spileo). - 3. *Cyclamen persicum* (Chios). - 4. *Cyclamen rhodium* subsp. *peloponnesiacum* (W foothills of Taigetos). - 5. *Cyclamen rhodium* subsp. *rhodium* (Rodos). - 6. *Cyclamen rhodium* subsp. *vividum* (Lakonias, Kiparissia-Charakas). - 7. *Lysimachia atropurpurea* (Evrou, Dadia). - 8. *Lysimachia nummularia* (above Metsovon). - 9. *Lysimachia punctata* (Evrou, Dadia-Sapka). - 10. *Lysimachia serpyllifolia* (Taigetos). - 11. *Lysimachia vulgaris* (Vrondous). - 12. *Primula elatior* (Tzena, inset from Rodopi).

Plate 223

1. Primulaceae: *Primula veris* (Varnous). - 2. *Primula vulgaris* (E foothills of Taigetos; inset subsp *rubra* from Mt Pilion). - 3. *Soldanella chrysosticta* (Mt Beles). - 4. *Soldanella pindicola* (Valia Kalda National Park). - 5. *Soldanella rhodopaea* (Rodopi). - 6. Punicaceae: *Punica granatum* (Mani peninsula). - 7. Rafflesiaceae: *Cytinus hypocistis* (Rodos). - 8. *Cytinus ruber* (Rethimnis, N of Plakias). - 9. Ranunculaceae: *Aconitum lycoctonum* (Rodopi). - 10. *Actaea spicata* (Varnous, inset from Iti). - 11. *Adonis annua* (Lakonias, Areopolis). - 12. *Adonis cretica* (Gavdos).

Plate 224

1. Ranunculaceae: *Adonis cyllenea* (Killini, inset from Oligirtos). - 2. *Adonis flammea* (Parnassos). - 3. *Anemone apennina* subsp. *blanda* (Achaias, Pteri). - 4. *Anemone coronaria* (Chanion, Anopolis). - 5. *Anemone hortensis* subsp. *heldreichii* (Rethimnis, Spili-Gerakari). - 6. *Anemone nemorosa* (Varnous). - 7. *Anemone pavonina* (Chios). - 8. *Anemone ranunculoides* (Tzena). - 9. *Aquilegia ottonis* subsp. *amaliae* (Olimbos). - 10. *Aquilegia vulgaris* (Pangeon). - 11. *Caltha palustris* (Voras, inset from Vrondous). - 12. *Ceratocephala falcata* (Attikis, Megara).

Plate 225

1. Ranunculaceae: *Clematis cirrhosa* (Methana peninsula). - 2. *Clematis elisabethae-carolae* (Lefka Ori). - 3. *Clematis flammula* (Xilokastro to Kato Trikala, inset from Thasos). - 4. *Clematis vitalba* (Kastorias, Eptachori-Pendalofos; inset from Ano Trikala). - 5. *Clematis viticella* (Evrou, SW of Kotronia). - 6. *Consolida ajacis* (Sparti-Tripoli, inset from Ano Trikala). - 7. *Consolida arenaria* (Rodos). - 8. *Consolida regalis* subsp. *paniculata* (Prespa National Park). - 9. *Consolida tenuissima* (Attiki). - 10. *Consolida tuntasiana* (Mt Gerania). - 11. *Delphinium balcanicum* (SW of Mt Vourinos). - 12. *Delphinium fissum* (Pierias, Ag. Dimitrios to Katerini; inset from Pangeon).

Plate 226

1. Ranunculaceae: *Delphinium peregrinum* (Prespa area). - 2. *Delphinium staphisagria* (Kerkira, inset of fruits from Karpathos). - 3. *Garidella nigellastrum* (Rodos). - 4. *Garidella unguicularis* (Irakliou, ENE of Mallia). - 5. *Helleborus odorus* subsp. *cyclophyllus* (Parnassos, inset from Vasilitsa). - 6. *Myosurus heldreichii* (Malea peninsula). - 7. *Nigella arvensis* subsp. *arvensis* (Athos peninsula; inset from Ioanninon, Eleffthero-Paleoseli). - 8. *Nigella arvensis* subsp. *brevifolia* (Rodos). - 9. *Nigella arvensis* subsp. *glauca* (Samos). - 10. *Nigella damascena* (Prespa area). - 11. *Nigella degenii* subp. *degenii* (Naxos). - 12. *Nigella doerfleri* (Amorgos).

Plate 227

1. Ranunculaceae: *Nigella orientalis* (Thessaloniki area). - 2. *Nigella stricta* (W Kriti, Panagia Chrisokalatitis). - 3. *Pulsatilla halleri* subsp. *rhodopaea* (Falakron). - 4. *Ranunculus arvensis* (Argolidos, Didima). - 5. *Ranunculus asiaticus* (Rethimnis, Agia Galini; inset above from Lasithiou, Gournia, below from Chanion, Anopolis-Aradena). - 6. *Ranunculus baudotii* (Achaias, Metochi-Kalogria). - 7. *Ranunculus brevifolius* (Chelmos, inset from Vardousia). - 8. *Ranunculus brutius* (Killini). - 9. *Ranunculus bullatus* (Rethimnis, Spili-Gerakari). - 10. *Ranunculus cacuminis* (Voras). - 11. *Ranunculus chius* (Kefallinia). - 12. *Ranunculus crenatus* (Gramos).

Plate 228

1. Ranunculaceae: *Ranunculus creticus* (Irakliou, Choudetsi). - 2. *Ranunculus cupreus* (Lasithiou, Orino-Thriptis). - 3. *Ranunculus ficaria* (Chanion, Anopolis). - 4. *Ranunculus ficarioides* (Taigetos). - 5. *Ranunculus fontanus* (Voras). - 6. *Ranunculus garganicus* (Kerkira, inset from Prespa National Park). - 7. *Ranunculus gracilis* (Kerkira). - 8. *Ranunculus illyricus* (Falakron). - 9. *Ranunculus incomparabilis* (Ioanninon, by Aoos lake). - 10. *Ranunculus isthmicus* (Kithira). - 11. *Ranunculus lateriflorus* (Iti). - 12. *Ranunculus lingua* (Argolidos, Mili to Nea Chios).

Plate 229

1. Ranunculaceae: *Ranunculus millefoliatus* (Florinis, Pisoderi pass). - 2. *Ranunculus millii* (Lakonias, Areopolis-Githion). - 3. *Ranunculus muricatus* (Rodos). - 4. *Ranunculus neapolitanus* (Athos peninsula; inset from Korinthias, Ano Trikala). - 5. *Ranunculus ophioglossifolius* (N of Metsovon). - 6. *Ranunculus paludosus* (Rethimnis, Melambes). - 7. *Ranunculus pindicola* (N of Metsovon). - 8. *Ranunculus platanifolius* (Bulgaria, Mt Vitosha). - 9. *Ranunculus psilostachys* (Killini). - 10. *Ranunculus radinotrichus* (Lefka Ori). - 11. *Ranunculus repens* (by lake Mikri Prespa). - 12. *Ranunculus rumelicus* (Thasos).

Plate 230

1. Ranunculaceae: *Ranunculus sardous* (Florinis, Mt Triklario). - 2. *Ranunculus sartorianus* (Killini). - 3. *Ranunculus sceleratus* (by Lake Megali Prespa). - 4. *Ranunculus serbicus* (Voras). - 5. *Ranunculus sprunerianus* (Falakron). - 6. *Ranunculus subhomophyllus* (Kriti, Thriptis Ori). - 7. *Ranunculus thasius* (Thasos). - 8. *Ranunculus trichophyllus* (N of Metsovon). - 9. *Ranunculus velutinus* (Parnassos; inset from Korinthias, Ano Trikala). - 10. *Thalictrum aquilegifolium* (Evritanias. ESE of Karpenisi). - 11. *Thalictrum lucidum* (Litochoron railway station). - 12. *Thalictrum minus* subsp. *saxatile* (Timfi).

Plate 231

1. Ranunculaceae: *Thalictrum orientale* (Taigetos, W foothills). - 2. *Trollius europaeus* (Varnous). - 3. Resedaceae: *Reseda alba* (Attikis, Imittos). - 4. *Reseda minoica* (Gavdos). - 5. *Reseda tymphaea* (Ioanninon, S of Milea). - 6. Rhamnaceae: *Frangula rupestris* (Florinis, Kristallopigi; inset from Ioanninon, Pades). - 7. *Paliurus spina-christi* (E of lake Doirani, inset from Litochoron railway station). - 8. *Rhamnus alaternus* (Amorgos). - 9. *Rhamnus alpina* subsp. *fallax* (Croatia, Velebit mts). - 10. *Rhamnus cathartica* (Prespa National Park). - 11. *Rhamnus lycioides* subsp. *oleoides* (Naxos, inset from Amorgos). - 12. *Rhamnus orbiculata* (Ioanninon, Vikos gorge).

Plate 232

1. Rhamnaceae: *Rhamnus saxatilis* subsp. *prunifolia* (Killini; inset from Florinis, Kristallopigi). - 2. *Rhamnus sibthorpiana* (Parnassos, inset from Chelmos). - 3. Rosaceae: *Agrimonia eupatoria* (Pellis, Loutra Arideas). - 4. *Alchemilla bulgarica* (Voras). - 5. *Alchemilla fissa* (Gramos). - 6. *Alchemilla flabellata* (Voras). - 7. *Alchemilla gorcensis* (Mt Beles/Kerkini). - 8. *Alchemilla heterophylla* (Varnous). - 9. *Alchemilla indivisa* (Voras). - 10. *Alchemilla plicatula* (Voras). - 11. *Alchemilla serbica* (Voras). - 12. *Alchemilla xanthochlora* (Iti).

Plate 233

1. Rosaceae: *Amelanchier ovalis* (Olimbos). - 2. *Amelanchier parviflora* subsp. *chelmea* (Chelmos). - 3. *Aremonia agrimonoides* (Grevenon, WSW of Kranea). - 4. *Cotoneaster integerrimus* (Olimbos). - 5. *Cotoneaster tomentosus* (Olimbos). - 6. *Crataegus monogyna* (Imathias, Naousa to Kato Vermio; inset from Prespa area). - 7. *Crataegus orientalis* (Mt Athos; inset from Kastorias, Nea Kotili). - 8. *Crataegus pycnoloba* (Chelmos). - 9. *Dryas octopetala* (Albania, Prokletije mts). - 10. *Drymocallis halacsyana* (Samothraki, Mt Fengari). - 11. *Drymocallis longisepala* subsp. *epirotica* (Mt Milea). - 12. *Drymocallis regis-borisii* (Evrou, NW of Lefkimi; inset from Athos peninsula).

Plate 234

1. Rosaceae: *Filipendula ulmaria*, with *Cirsium tymphaeum* (N of Metsovon). - 2. *Filipendula vulgaris* (Prespa area, Mt Devas). - 3. *Fragaria vesca* (Pangeon, inset from Rodopi). - 4. *Fragaria viridis* (Dramas, Sidironero). - 5. *Geum coccineum* (Varnous). - 6. *Geum montanum* (Voras). - 7. *Geum rhodopeum* (Vrondous). - 8. *Geum rivale* (Rodopi). - 9. *Geum urbanum* (Pangeon). - 10. *Malus sylvestris* (Bulgaria, Sofia; inset from Florinis, Vigla pass). - 11. *Malus trilobata* (Evrou, N of Nipsa). - 12. *Potentilla arcadiensis* (Arkadias, Moni Eloni).

Plate 235

1. Rosaceae: *Potentilla astracanica* (Evrou, NW of Lefkimi). - 2. *Potentilla aurea* subsp. *chrysocraspeda* (Voras). - 3. *Potentilla deorum* (Olimbos). - 4. *Potentilla detommasii* (Falakron). - 5. *Potentilla heptaphylla* subsp. *australis* (Gramos, Epano Arena). - 6. *Potentilla inclinata* (Evrou, Dadia; inset from Olimbos). - 7. *Potentilla kionaea* (Giona). - 8. *Potentilla micrantha* (Valia Kalda National Park). - 9. *Potentilla recta* (Prespa area, Mt Devas). - 10. *Potentilla reptans* (Samarina to Fourka pass). - 11. *Potentilla speciosa* (Parnassos). - 12. *Potentilla supina* (by lake Megali Prespa).

Plate 236

1. Rosaceae: *Potentilla tommasiniana* (Falakron). - 2. *Prunus avium* (Rodopi; inset from Florinis, Vigla pass). - 3. *Prunus cerasifera* (Lesvos). - 4. *Prunus cocomilia* (Varnous; inset of flowers from Parnassos, of fruits from Killini). - 5. *Prunus mahaleb* (Taigetos). - 6. *Prunus prostrata* (Lefka Ori; inset from Samos, Karvouni). - 7. *Prunus spinosa* (Vertiskos; inset from Dramas, Kapnofito-Chalepi). - 8. *Prunus webbii* (Florinis, Maniaki; inset from Karpathos). - 9. *Pyracantha coccinea* (Loutra Arideas). - 10. *Pyrus elaegrifolia* (Evrou, NW of Lefkimi). - 11. *Pyrus pyraster* (Evrou, Pessani; inset from Voras). - 12. *Pyrus spinosa* (Rethimnis, N of Agia Galini; inset of fruits from Lakonias, Nomia).

Plate 237

1. Rosaceae: *Rosa agrestis* (Evrou, Lefkimi; inset from Loutra Arideas). - 2. *Rosa arvensis* (Kerasia to Loutra Arideas). - 3. *Rosa canina* (Evrou, Lefkimi; inset from Florinis, Vigla pass). - 4. *Rosa gallica* (Grevenon, Anthrakia). - 5. *Rosa glauca* (Montenegro, Mt Orjan). - 6. *Rosa heckeliana* (Parnassos, inset from Olimbos). - 7. *Rosa pendulina* (Rodopi, inset from Falakron). - 8. *Rosa pulverulenta* (Chios, Mt Pelineo; inset from Turkey, Koca Tepe). - 9. *Rosa sempervirens* (Kerkira: inset from Lakonias, Mistras). - 10. *Rosa turcica* (Falakron). - 11. *Rosa villosa* (Florinis, Mt Triklario; inset from Montenegro, Mt Orjan). - 12. *Rubus caesius* (Rodopi; inset of fruit from Prespa area).

Plate 238

1. Rosaceae: *Rubus canescens* (Valia Kalda National Park, inset from Taigetos). - 2. *Rubus hirtus* (Olimbos). - 3. *Rubus idaeus* (Rodopi). - 4. *Rubus praecox* (Dramas, Sideronero). - 5. *Rubus sanctus* (Serron, Kerkini; inset from Achaias, Ag. Vasileos). - 6. *Sanguisorba cretica* (W Kriti, Imbros gorge). - 7. *Sanguisorba officinalis* (W of Katara pass, inset from Voras). - 8. *Sarcopoterium spinosum* (Amorgos). - 9. *Sorbus aria* (Kastorias, NNW of Pefkofito; inset from Valia Kalda National Park). - 10. *Sorbus aucuparia* (Varnous, inset from Voras). - 11. *Sorbus austriaca* (Rodopi). - 12. *Sorbus domestica* (Evrou, Mikro Derio).

Plate 239

1. Rosaceae: *Sorbus torminalis* (Mt Beles; inset from Dramas, Sideronero-Skaloti). - 2. *Sorbus umbellata* (Florinis, Mt Boutsi). - 3. *Spiraea chamaedryfolia* (Rodopi). - 4. Rubiaceae: *Asperula abbreviata* (Naxos). - 5. *Asperula arcadiensis* (Killini). - 6. *Asperula aristata* subsp. *condensata* (Gramos). - 7. *Asperula arvensis* (Parnassos). - 8. *Asperula boissieri* (Vardousia). - 9. *Asperula boryana* (Taigetos). - 10. *Asperula crassula* (unknown locality in Kriti). - 11. *Asperula elonea* (Lakonias, N of Charakas). - 12. *Asperula idaea* (Lefka Ori).

Plate 240

1. Rubiaceae: *Asperula lilaciflora* (Ikaria). - 2. *Asperula lutea* (Mt Kitheronas). - 3. *Asperula malevonensis* (Parnonas). - 4. *Asperula muscosa* (Olimbos). - 5. *Asperula naufraga* (Zakinthos). - 6. *Asperula nitida* subsp. *mytilinica* (Chios, Mt Pelineo). - 7. *Asperula oetaea* (Chelmos). - 8. *Asperula pulvinaris* (Parnitha). - 9. *Asperula purpurea* subsp. *purpurea* (Evrou, Dadia). - 10. *Asperula rigida* (Chanion, Kandanos gorge). - 11. *Asperula samia* (Samos, Mt Kerkis). - 12. *Asperula suberosa* (Mt Athos).

Plate 241

1. Rubiaceae: *Asperula suffruticosa* (Evvia, Mt Dirfis). - 2. *Asperula taygetea* (Arkadias, E of Leonidion). - 3. *Crucianella graeca* (Fthiotidos, Domokos). - 4. *Crucianella latifolia* (Kerkira, inset from Athos peninsula). - 5. *Crucianella maritima* (Kerkira). - 6. *Cruciata laevipes* (Prespa area; inset from Dramas, Dipotama to Megalo Livadi). - 7. *Cruciata pedemontana* (Iti). - 8. *Cruciata taurica subsp. euboea* (Evvia, Mt Dirfis). - 9. *Cruciata verna* (Falakron). - 10. *Galium amorginum* (Amorgos). - 11. *Galium anisophyllon* subsp. *plebeium* (Olimbos). - 12. *Galium aparine* (Kastellorizo; inset from Lakonias, Githion-Areopolis).

Plate 242

1. Rubiaceae: *Galium brevifolium* (Samos). - 2. *Galium canum* (Karpathos). - 3. *Galium capitatum* (Malea peninsula). - 4. *Galium corinthiacum* (Korinthias, above Loutraki). - 5. *Galium cyllenium* (Killini). - 6. *Galium debile* (by lake Megali Prespa). - 7. *Galium degenii* (Olimbos). - 8. *Galium fruticosum* (Chanion, Kakopetros). - 9. *Galium graecum* (Samos). - 10. *Galium hellenicum* (Olimbos). - 11. *Galium incanum* (Vardousia). - 12. *Galium melanantherum* (Mt Kitheronas, inset from Akrokorinthos).

Plate 243

1. Rubiaceae: *Galium murale* (Naxos). - 2. *Galium odoratum* (Rodopi). - 3. *Galium rhodopeum* (Falakron). - 4. *Galium rotundifolium* (Stix ravine; inset from Fthiotidos, Mt Oxia). - 5. *Galium thymifolium* (Evvia, Mt Dirfis). - 6. *Galium tricornutum* (Kastorias, Eptachori). - 7. *Galium verrucosum* (Lakonias, Areopolis). - 8. *Galium verticillatum* (Prespa National Park). - 9. *Galium verum* (above Metsovon). - 10. *Plocama calabrica* (Ioanninon, Chrisovitsa). - 11. *Rubia peregrina* (Evvia, Steni Dirfios). - 12. *Rubia tenuifolia* (Samos).

Plate 244

1. Rubiaceae: *Rubia tinctorum* (Lesvos). - 2. *Theligonum cynocrambe* (Lakonias, Areopolis). - 3. *Valantia aprica* (Parnassos). - 4. *Valantia hispida* (Rodos). - 5. *Valantia muralis* (Rodos). - 6. Ruscaceae: *Ruscus aculeatus* (Prespa National Park). - 7. *Ruscus hypoglossum* (Athos peninsula). - 8. Rutaceae: *Dictamus albus* (Prespa National Park). - 9. *Haplophyllum balcanicum* (Menikion). - 10. *Haplophyllum coronatum* (Vourinos; inset from Florinis, Perea). - 11. *Haplophyllum megalanthum* (Lesvos). - 12. *Ruta chalepensis* subsp. *chalepensis* (Naxos).

Plate 245

1. Rutaceae: *Ruta graveolens* (Parnassos; inset from Lakonias, Mistras). - 2. Salicaceae: *Populus tremula* (Dramas, Tsisavos; inset from Athos peninsula). - 3. *Salix alba* (Florinis, Vigla pass). - 4. *Salix amplexicaulis* (Florinis, Vigla pass). - 5. *Salix caprea* (Rodopi). - 6. *Salix xanthicola* (Evrou, Mikro Derio). - 7. Santalaceae: *Arceuthobium oxycedri* (Vardousia, inset from Parnassos). - 8. *Comandra umbellata* subsp. *elegans* (Prespa area, Mt Devas). - 9. *Osyris alba* (Athos peninsula). - 10. *Thesium bergeri* (Tilos). - 11. *Thesium divaricatum* (Kozanis, E of Pentalofos). - 12. *Thesium humile* (Navplias, Tolon).

Plate 246

1. Santalaceae: *Thesium linophyllon* (Prespa area, Mt Triklarion). - 2. *Thesium parnassi* (Chelmos). - 3. *Viscum album* subsp. *abietis* (Athos peninsula, inset of subsp. *creticum* from Thriptis Ori). - 4. Saxifragaceae: *Chrysosplenium alternifolium* (Varnous). - 5. *Saxifraga adscendens* (Ioanninon, Mt Dousko). - 6. *Saxifraga aizoides* (Tzena). - 7. *Saxifraga bulbifera* (above Metsovon, inset from Mt Vitsi). - 8. *Saxifraga carpetana* subsp. *graeca* (Prespa National Park). - 9. *Saxifraga corymbosa* (Timfi). - 10. *Saxifraga exarata* (Olimbos). - 11. *Saxifraga federici-augusti* subsp. *grisebachii* (Voras). - 12. *Saxifraga ferdinandi-coburgi* (Falakron).

Plate 247

1. Saxifragaceae: *Saxifraga glabella* (Olimbos). - 2. *Saxifraga hederacea* (Naxos). - 3. *Saxifraga marginata* (Ioanninon, Mt Dousko). - 4. *Saxifraga paniculata* (Montenegro, Mt Durmitor; inset from Vardousia). - 5. *Saxifraga pedemontana* subsp. *cymosa* (Voras). - 6. *Saxifraga rotundifolia* subsp. *chrysospleniifolia* (Taigetos, Artemisia). - 7. *Saxifraga rotundifolia/heucherifolia* (Voras). - 8. *Saxifraga sancta* (Mt Athos). - 9. *Saxifraga scardica* (Voras). - 10. *Saxifraga sempervivum* (Falakron). - 11. *Saxifraga sibirica* (Samothraki, Mt Fengari; inset from Lesvos, Mt Olimbos). - 12. *Saxifraga sibthorpii* (Parnassos).

Plate 248

1. Saxifragaceae: *Saxifraga spruneri* (Olimbos). - 2. *Saxifraga stellaris* subsp. *engleri* (Voras). - 3. *Saxifraga stribrnyi* (Falakron). - 4. *Saxifraga taygetea* (Vardousia). - 5. Scrophulariaceae: *Antirrhinum majus* (Korinthias, Ano Trikala). - 6. *Antirrhinum siculum* (Kefallinia). - 7. *Bellardia latifolia* (E of Katara pass: inset from Rethimnis, N of Agia Galini). - 8. *Bellardia trixago* (Amorgos). - 9. *Bellardia viscosa* (Rodos). - 10. *Cymbalaria longipes* (Rodos). - 11. *Cymbalaria minor* subsp. *dodecanesi* (Karpathos). - 12. *Cymbalaria muralis* (Lefkas).

Plate 249

1. Scrophulariaceae: *Cymbalaria spetae* (Lakonias, Charakas). - 2. *Digitalis cariensis* subsp. *ikarica* (Ikaria). - 3. *Digitalis ferruginea* (Varnous). - 4. *Digitalis grandiflora* (Olimbos). - 5. *Digitalis laevigata* (Taigetos; inset from Ioanninon, Koukouli). - 6. *Digitalis lanata* (Pangeon). - 7. *Digitalis viridiflora* (Voras). - 8. *Euphrasia minima* (Smolikas). - 9. *Euphrasia pectinata* (Varnous). - 10. *Euphrasia salisburgensis* (Falakron). - 11. *Gratiola officinalis* (Prespa National Park). - 12. *Kickxia commutata* subsp. *graeca* (Achaias, Kalogria).

Plate 250

1. Scrophulariaceae: *Kickxia elatine* subsp. *crinita* (Karpathos). - 2. *Lathraea rhodopea* (Rodopi). - 3. *Limosella aquatica* (by Lake Kerkini). - 4. *Linaria alpina* (Olimbos). - 5. *Linaria chalepensis* (Samos). - 6. *Linaria dalmatica* (Prespa National Park; inset from Pellis, Skidra). - 7. *Linaria genistifolia* (Chalkidikis, Amoliani). - 8. *Linaria micrantha* (Attiki, Kamariza). - 9. *Linaria pelisseriana* (Olimbos, S foothills). - 10. *Linaria peloponnesiaca* (both from Prespa area). - 11. *Linaria tenuis* (Lakonias, N of Neapolis). - 12. *Linaria triphylla* (Argolidos, Epidavros).

Plate 251

1. Scrophulariaceae: *Linaria vulgaris* (Vrondous). - 2. *Macrosyringion glutinosum* (Varnous). - 3. *Melampyrum arvense* (Prespa National Park). - 4. *Melampyrum cristatum* (Dramas, SW of Skaloti). - 5. *Melampyrum heracleoticum* (Prespa area, Mt Devas). - 6. *Misopates orontium* (Kithira). - 7. *Odontites linkii* (Lakonias, Charakas). - 8. *Odontites vulgaris* (Vertiskos). - 9. *Pedicularis brachyodonta* subsp. *moesiaca* (Vourinos). - 10. *Pedicularis friderici-augusti* (Falakron). - 11. *Pedicularis graeca* (Timfi). - 12. *Pedicularis leucodon* (Voras).

Plate 252

1. Scrophulariaceae: *Pedicularis olympica* (Voras). - 2. *Pedicularis orthantha* (Voras). - 3. *Rhinanthus pumilus* (Timfi). - 4. *Rhinanthus rumelicus* (Falakron). - 5. *Rhinanthus sintenisii* (Varnous). - 6. *Rhynchocorys elephas* (Gramos). - 7. *Scrophularia aestivalis* (Olimbos). - 8. *Scrophularia canica* subsp. *bicolor* (Florinis, Koula-Pili; inset from Kerkira). - 9. *Scrophularia heterophylla* (Lakonias, Monemvasia). - 10. *Scrophularia lucida* (Rethimnis, N of Agia Galini; inset from Naxos). - 11. *Scrophularia peregrina* (Athos peninsula). - 12. *Scrophularia scopolii* (Varnous).

Plate 253

1. Scrophulariaceae: *Sibthorpia europaea* (Kriti, Agios Dikeos S of Elos). - 2. *Siphonostegia syriaca* (Pilion). - 3. *Tozzia alpina* subsp. *carpathica* (Varnous). - 4. *Verbascum acaule* (Chelmos). - 5. *Verbascum adenanthum* (Vourinos). - 6. *Verbascum adrianopolitanum* (Evrou, Dadia; inset of leaves and flowers from Metaxades). - 7. *Verbascum antinori* (Lesvos). - 8. *Verbascum aphentulium* (Pilion). - 9. *Verbascum arcturus* (Chanion Sfakia). - 10. *Verbascum baldaccii* (Albania, Prokletije mts). - 11. *Verbascum banaticum* (Evrou, Dadia; inset of flowers from Kavalas, Nea Iraklitsa). - 12. *Verbascum blattaria* (Pellis, Skidra; inset of fruits from Ioanninon, Aristi).

Plate 254

1. Scrophulariaceae: *Verbascum boissieri* (Parnitha). - 2. *Verbascum botuliforme* (Ossa). - 3. *Verbascum cylindrocarpum* (Thasos). - 4. *Verbascum cylleneum* (Killini). - 5. *Verbascum daenzeri* (Korinthias, Korfes). - 6. *Verbascum densiflorum* (Mt Athos; inset from Dramas, Paranestion). - 7. *Verbascum dieckianum* (Pellis, above Notia). - 8. *Verbascum epixanthinum* (Timfristos). - 9. *Verbascum eriophorum* (E foothills of Olimbos; inset from Konitsis, Kefalochori). - 10. *Verbascum euboicum* (Evvia, Mt Dirfis). - 11. *Verbascum flavidum* (Prespa area, Mt Devas). - 12. *Verbascum glabratum* (Kastorias, Nestorio to Nea Kotili).

Plate 255

1. Scrophulariaceae: *Verbascum glandulosum* (Prespa National Park; inset of leaves from Florina area). - 2. *Verbascum glomeratum* (Chios. inset of leaves from Samos). - 3. *Verbascum graecum* (Olimbos; inset of leaves from Pellis, Sarakini; of flowers from Kitheronas). - 4. *Verbascum guicciardii* (Ioanninon, Votonosi). - 5. *Verbascum halacsyanum* (Kavala). - 6. *Verbascum haussknechtii* (Evvia, Limni-Mantoudi). - 7. *Verbascum humile* (Dramas, Paranestion-Dipotama). - 8. *Verbascum ikaricum* (Ikaria). - 9. *Verbascum lasianthum* (Lesvos, Mt Olimbos). - 10. *Verbascum levanticum* (Kastellorizo). - 11. *Verbascum limnense* (Evvia, Limni). - 12. *Verbascum longifolium* (Varnous).

Plate 256

1. Scrophulariaceae: *Verbascum macrurum* (Arkadias, Klitoria-Tripoli). - 2. *Verbascum mallophorum* (Killini; inset of leaves from Metsovon, of flowers from Evvia). - 3. *Verbascum mucronatum* (Lesvos). - 4. *Verbascum mykales* (Samos). - 5. *Verbascum nobile* (Dramas, Nikiforos). - 6. *Verbascum ovalifolium* (Evrou, Kanadas). - 7. *Verbascum pangaeum* (Pangeon). - 8. *Verbascum phlomoides* (Viotias, NE of Livadia; inset of leaves from Konitsa, of flowers from Vigla pass). - 9. *Verbascum pinnatifidum* (Samothraki, insets from Thasos). - 10. *Verbascum pulverulentum* (Grevenon, Dhotsiko-Eptachori). - 11. *Verbascum pycnostachyum* (Samos). - 12. *Verbascum reiseri* (Vardousia).

Plate 257

1. Scrophulariaceae: *Verbascum roripifolium* (Kavalas, Nea Peramos; inset from Paranestion). - 2. *Verbascum rupestre* (Evrou, NW of Lefkimi). - 3. *Verbascum samniticum* (Lakonias, Areopolis). - 4. *Verbascum sinuatum* (Ikaria, inset from W Kriti). - 5. *Verbascum spathulisepalum* (Evrou, Dadia-Sapka). - 6. *Verbascum speciosum* (Prespa area, inset of fruits from Rodopi). - 7. *Verbascum spinosum* (S foothills of Lefka Ori). - 8. *Verbascum symes* (Simi). - 9. *Verbascum syriacum* (Rodos). - 10. *Verbascum undulatum* (S foothills of Parnassos; inset from Ioanninon, Kefalovriso). - 11. *Veronica anagallis-aquatica* (NW of Metsovon). - 12. *Veronica aphylla* (Giona).

Plate 258

1. Scrophulariaceae: *Veronica arvensis* (Erimanthos). - 2. *Veronica barrelieri* (Belles). - 3. *Veronica beccabunga* (Varnous). - 4. *Veronica bellidioides* (Voras). - 5. *Veronica bozamanii* (Florinis, Mt Boutsi). - 6. *Veronica chamaedrys* subsp. *chamaedryoides* (Taigetos). - 7. *Veronica contandriopouli* (Killini). - 8. *Veronica cymbalaria* (Amorgos). - 9. *Veronica dillenii* (Varnous). - 10. *Veronica erinoides* (Stix ravine). - 11. *Veronica glauca* subsp. *peloponnesiaca* (Parnassos). - 12. *Veronica grisebachii* (Samos).

Plate 259

1. Scrophulariaceae: *Veronica hederifolia* subsp. *hederifolia* (Arkadias, E of Tripoli). - 2. *Veronica jacquinii* (Mt Athos, inset from Kerkira). - 3. *Veronica oetaea* (Iti). - 4. *Veronica orsiniana* subsp. *teucrioides* (Voras). - 5. *Veronica persica* (Athos peninsula). - 6. *Veronica serpyllifolia* (Timfristos; inset from Ioanninon, Mt Douskon). - 7. *Veronica thessalica* (Olimbos). - 8. *Veronica thymifolia* (Taigetos, inset from Iti). - 9. *Veronica urticifolia* (Olimbos). - 10. *Veronica verna* (Varnous). - 11. Smilacaceae: *Smilax aspera* (Lakonias, Lambokambos; inset from Evvia, Stropones). - 12. Solanaceae: *Atropa belladonna* (Olimbos).

Plate 260

1. Solanaceae: *Datura innoxia* (Arkadias, Ag. Andreas). - 2. *Datura stramonium* (Pellis, Aridea). - 3. *Datura wrightii* (Lakonias, Githion-Areopolis; inset from Mani peninsula, Vathia). - 4. *Hyoscyamus albus* (Argolidos, Nafplion; inset left from Kriti, Sfakia, right from Akrokorinthos). - 5. *Hyoscyamus aureus* (Rodos). - 6. *Hyoscyamus niger* (Loutra Arideas). - 7. *Lycium schweinfurthii* (Koufonisi SE of Kriti). - 8. *Mandragora officinarum* (Amorgos; inset from Messinias, Methoni). - 9. *Nicotiana glauca* (Lesvos). - 10. *Physalis angulata* (Ioanninon, Metamorfosi). - 11. *Solanum dulcamara* (by lake Megali Prespa). - 12. *Solanum elaeagnifolium* (Chalkidiki, Nea Triglia; inset from Imittos).

Plate 261

1. Solanaceae: *Solanum nigrum* (Naxos). - 2. *Solanum rostratum* (Pierias, Plaka Litochorou). - 3. *Solanum villosum* (Lakonias, Paralia Kiparissias). - 4. Sparganiaceae: *Sparganium erectum* (by lake Mikri Prespa). - 5. *Sparganium minimum* (Voras, Dobro Polje). - 6. Styracaceae: *Styrax officinalis* (Rodos). - 7. Tamaricaceae: *Tamarix minoa* (W Kriti, Georgioupoli). - 8. *Tamarix parviflora* (Kastellorizo). - 9. *Tamarix smyrnensis* (Kithira). - 10. Thymelaeaceae: *Daphne blagayana* (Katara pass). - 11. *Daphne gnidioides* (Kastellorizo, inset from Tilos). - 12. *Daphne gnidium* (Evvia, WNW of Kimi).

Plate 262

1. Thymelaeaceae: *Daphne jasminea* (Korinthias, above Loutraki). - 2. *Daphne kosaninii* (Pangeon). - 3. *Daphne laureola* (Olimbos). - 4. *Daphne mezereum* (Voras). - 5. *Daphne oleoides* (Timfristos, inset from Parnassos). - 6. *Daphne sericea* (Rethimnis, Spili-Gerakari). - 7. *Thymelaea hirsuta* (Rethimnis, Kato Rodakino). - 8. *Thymelaea tarton-raira* (Lakonias, Mt Koulouchera). - 9. Tiliaceae: *Tilia tomentosa* (Ioanninon, village of Vikos). - 10. Trapaceae: *Trapa natans* (Evros delta, inset from Megali Prespa). - 11. Trilliaceae: *Paris quadrifolia* (Rodopi). - 12. Typhaceae: *Typha angustifolia* (Evros delta).

Plate 263

1. Typhaceae: *Typha domingensis* (Samos; inset from Chalkidiki, Xerxes canal). - 2. *Typha latifolia* (by lake Mikri Prespa). - 3. *Typha shuttleworthii* (NW of Metsovon). - 4. Ulmaceae: *Celtis australis* (Drama-Nikiforos). - 5. *Ulmus minor* (Dramas, Kapnofito-Chalepi). - 6. *Zelkova abelicea* (Lefka Ori, Xiloskala). - 7. Urticaceae: *Parietaria judaica* (Athos peninsula). - 8. *Urtica dioica* (Rodopi). - 9. *Urtica pilulifera* (E of Nafplion). - 10. Valerianaceae: *Centranthus calcitrapae* (Andalusia: Parque Natural de Doñana). - 11. *Centranthus longiflorus* subsp. *junceus* (Ioanninon, SE of Pades). - 12. *Centranthus ruber* subsp. *sibthorpii* (Parnassos).

Plate 264

1. Valerianaceae: *Centranthus sieberi* (Lefka Ori). - 2. *Fedia graciliflora* (Sicily, N of Scopellio). - 3. *Valeriana asarifolia* (Rethimnis, Kotsifou gorge). - 4. *Valeriana crinii* subsp. *epirotica* (Timfi). - 5. *Valeriana italica* (Chios). - 6. *Valeriana montana/tripteris* (Pangeon). - 7. *Valeriana officinalis* subsp. *officinalis* (above Metsovon). - 8. *Valeriana olenaea* (Killini). - 9. *Valeriana tuberosa* (Killini, inset from N Pindos). - 10. *Valerianella discoidea* (Tilos). - 11. *Valerianella echinata* (Argolidos, Didima; insets from Tilos). - 12. *Valerianella obtusiloba* (Tilos).

Plate 265

1. Valerianaceae: *Valerianella vesicaria* (Rodos). - 2. Verbenaceae: *Lantana camara* (E Kriti, Sitia). - 3. *Phyla nodiflora* (Magnisias, Nees Pegases). - 4. *Verbena supina* (Lasithiou, Dories). - 5. *Vitex agnus-castus* (Lesvos). - 6. Violaceae: *Viola aetolica* (Valia Kalda National Park). - 7. *Viola alba* subsp. *dehnhardtii* (Athos peninsula). - 8. *Viola albanica* (Gramos, Epano Arena). - 9. *Viola arvensis* (Florinis, Sfika). - 10. *Viola athois* (Mt Athos). - 11. *Viola chelmea* (Killini). - 12. *Viola delphinantha* (Mt Athos).

Plate 266

1. Violaceae: *Viola doerfleri* (Voras). - 2. *Viola dukadjinica* (Gramos, Epano Arena). - 3. *Viola epirota* (Ioanninon, Mt Douskon). - 4. *Viola eximia* (Voras). - 5. *Viola fragrans* (Lefka Ori). - 6. *Viola ganiatsasii* (Dramas, Megalo Livadi). - 7. *Viola graeca* (Giona). - 8. *Viola herzogii* (WSW of Loutra Arideas). - 9. *Viola hymettia* (Imittos). - 10. *Viola kitaibeliana* (Achaias, above Kalavrita). - 11. *Viola mercurii* (Killini). - 12. *Viola odorata* (Taigetos, Langada gorge).

Plate 267

1. Violaceae: *Viola oligyrtia* (Mt Oligirtos). - 2. *Viola orphanidis* (North Macedonia, Mt Pelister). - 3. *Viola perinensis* (Falakron). - 4. *Viola poetica* (Vardousia). - 5. *Viola pseudograeca* (Olimbos). - 6. *Viola rausii* (Pilion). - 7. *Viola reichenbachiana* (Valia Kalda National Park). - 8. *Viola riviniana* (Varnous). - 9. *Viola scorpiuroides* (Lasithiou, Vai). - 10. *Viola striis-notata* (Olimbos). - 11. *Viola tricolor* subsp. *macedonica* (Vrondous, insets from Voras). - 12. *Viola velutina* (Varnous).

Plate 268

1. Violaceae: *Viola voliotisii* (Voras, Dobro Polje). - 2. *Viola vourinensis* (Vourinos). - 3. Vitaceae: *Vitis vinifera* subsp. *sylvestris* (Thesprotias, Agia Kiriaki). - 4. Zannichelliaceae: *Althenia filiformis* (Achaias, Kalogria). - 5. Zygophyllaceae: *Fagonia cretica* (Cyprus, south coast). - 6. *Peganum harmala* (Evvia, Chalkida). - 7. *Tetraena alba* (Cyprus, Akrotiri peninsula; inset from Karpathos). - 8. *Tribulus terrestris* (Magnisias, Nees Pegases; inset from Thasos). - 9-12. Late additions from E Kriti. - 9. Asparagaceae: *Asparagus horridus* (Koufonisi). - 10. Asteraceae: *Chlamydophora tridentata* (Koufonisi). - 11. *Reichardia orientalis* (Koufonisi). - 12. Geraniaceae: *Erodium crassifolium* (Ierapetra).

INDEX TO SCIENTIFIC NAMES

Accepted names of genera are in **boldface**, accepted names of species and subspecies in Roman type, synonyms in *italics*. Ordinary numerals refer to map numbers (not page numbers). Numerals such as 137:9 refer to colour plates in vol. 3. Numerals preceeded by "x" (e.g, x1437) refer to page numbers in the chapter "Excluded species and taxonomic comments" in vol. 3.

Only the more important synonyms have been included, especially those appearing as accepted names in recent Floras, checklists, monographs and taxonomic revisions. For more extensive synonymy, see the Greek Checklist (Dimopoulos & al. 2013, 2016) and Euro+Med PlantBase, as well as standard Floras for the area.